THE MASTER SEMICONDUCTOR REPLACEMENT HANDBOOK

Listed By Industry Standard Number

by TAB Editorial Staff

FIRST EDITION

FIRST PRINTING

Printed in the United States of America

Library of Congress Cataloging in Publication Data

Main entry under title:

The Master semiconductor replacement handbook—
listed by industry standard number.

Includes index.
1. Semiconductors—Catalogs. I. TAB Books.
TK7871.85.M365 621.3815'2'0216 81-18350
ISBN 0-8306-1470-2 AACR2

Preface

Over the years it has been a constant challenge to keep up with the never-ending flood of semiconductor devices on the consumer market. Simply knowing the equipment manufacturer's part number has not been enough to identify the device because a universal standardization of semiconductors has yet to be realized.

In the past few years, leading semiconductor suppliers—Sylvania, RCA, General Electric, Radio Shack, and Motorola—have individually introduced a "general replacement" line of semiconductors for wide ranges of applications that meet the needs of electronic servicemen, hobbyists, experimenters, hams, and engineers. It is to these wide-range semiconductor lines that we have cross-referenced thousands upon thousands of specialized Industry Standard numbers.

Indexed by the Industry Standard number, this list covers over 140,000 diodes, transistors, and ICs. This list will shorten the time it takes to find a replacement device in most cases. Especially when used in conjunction with TAB book No. 1471, *The Master Semiconductor Replacement Handbook–Listed by Manufacturer's Number*, this book will soon become one of your most worthwhile references.

Contents

Introduction

KEY TO MASTER CROSS REFERENCE

Original Part Number	ECG	SK	GE	RS276	MOTOR
Number shown in equipment parts list or schematic					
Sylvania replacement					
RCA replacement					
General Electric replacement					
Radio Shack replacement					
Motorola replacement					

Industry Standard No.	ECG	SK	GE	RS 276-	MOTOR.
.25N10	5019A	3785	ZD-10	562	
.25N100	5050A	3816	ZD-100		
.25N11	5020A	3786	ZD-11	563	
.25N110	5051A	3817	ZD-110		
.25N12	5021A	3787	ZD-12	563	
.25N120	5052A	3818	ZD-120		
.25N13	5022A	3788	ZD-12	563	
.25N130	5053A	3819	ZD-130		
.25N14	5023A	3789	ZD-14	564	
.25N140	5054A	3820	ZD-140		
.25N15	145A	3790/5024A	ZD-15	564	
.25N150	5055A	3821	ZD-150		
.25N16	5025A	3791		564	
.25N17	5026A	3792	ZD-17		
.25N18	5027A	3793	ZD-18		
.25N19	5028A	3794	ZD-19		
.25N20	5029A	3795	ZD-20		
.25N200	5060A	3826	ZD-200		
.25N22	5030A	3796	ZD-22		
.25N24	5031A	3797	ZD-24		
.25N25	5032A	3798	ZD-25		
.25N27	5033A	3799	ZD-27		
.25N30	5035A	3801	ZD-30		
.25N33	5036A	3802	ZD-33		
.25N36	5037A	3803	ZD-36		
.25N39	5038A	3804	ZD-39		
.25N43	5039A	3805	ZD-43		
.25N45	5040A	3806	ZD-47		
.25N47	5040A	3806	ZD-47		
.25N50	5041A	3807	ZD-51		
.25N52	5041A	3807	ZD-51		
.25N56	5042A	3808	ZD-56		
.25N6.8	5014A	3780	ZD-6.8	561	
.25N62	5044A	3810	ZD-62		
.25N68	5045A	3811	ZD-68		
.25N7.5	5015A	3781	ZD-7.5		
.25N75	5046A	3812	ZD-75		
.25N8.2	5016A	3782	ZD-8.2	562	
.25N82	5047A	3813	ZD-82		
.25N9.1	5018A	3784	ZD-9.1	562	
.25N91	5049A	3815	ZD-91		
.25NN15	5024A		ZD-15	564	
.25T10	140A	3061	ZD-10	562	
.25T10.5	140A		ZD-10	562	
.25T10.5A	140A		ZD-10	562	
.25T100	5096A	3346	ZD-100		
.25T100A	5096A	3346	ZD-100		
.25T10A	140A		ZD-10	562	
.25T10B	140A	3061	ZD-10	562	
.25T11	5074A		ZD-11.0	563	
.25T11.5	141A	3092	ZD-11.5		
.25T11.5B	141A	3092	ZD-11.5		
.25T110	151A	3099	ZD-110		
.25T110A	5151A				
.25T110B	151A	3099	ZD-110		
.25T11A	5074A		ZD-11.0	563	
.25T12	142A		ZD-12	563	
.25T12.8	143A	3093	ZD-13	563	
.25T12.8A	143A	3093	ZD-13	563	
.25T12.8B	143A	3093	ZD-13	563	
.25T120	5097A	3347	ZD-120		
.25T120A	5097A	3347	ZD-120		
.25T12A	142A	3062	ZD-12	563	
.25T12B	142A	3062	ZD-12	563	
.25T13	143A	3750	ZD-13	563	
.25T130	5098A	3348	ZD-130		
.25T130A	5098A	3348	ZD-130		
.25T13A	143A	3750	ZD-13	563	
.25T14	144A	3094	ZD-14	564	
.25T140	5099A	3349	ZD-140		
.25T14A	144A	3094	ZD-14	564	
.25T14B	144A	3094	ZD-14	564	
.25T15	145A	3063	ZD-15	564	
.25T15.8	5075A	3751	ZD-16	564	
.25T15.8A	5075A	3751	ZD-16	564	
.25T150	5100A	3350	ZD-150		
.25T150A	5100A	3350	ZD-150		
.25T15A	145A	3063	ZD-15	564	
.25T15B	145A	3063	ZD-15	564	
.25T16	5075A	3751	ZD-16	564	
.25T160	5101A	3351	ZD-160		
.25T160A	5101A	3351	ZD-160		
.25T16A	5075A	3751	ZD-16	564	
.25T17	5076A	9022	ZD-17		
.25T175	5102A	3352	ZD-170		
.25T17A	5076A	9022	ZD-17		
.25T18	5077A	3752	ZD-18		
.25T180	5103A	3353	ZD-180		
.25T180A	5103A	3353	ZD-180		
.25T18A	5077A	3752	ZD-18		
.25T19	5078A	9023	ZD-19		
.25T19A	5078A	9023	ZD-19		
.25T20	5079A	3335	ZD-20		
.25T200	5105A	3355	ZD-200		
.25T200A	5105A	3355	ZD-200		
.25T20A	5079A	3335	ZD-20		
.25T22	5080A	3336	ZD-22		
.25T22A	5080A	3336	ZD-22		
.25T23.5	5081A		ZD-24		
.25T23.5A	5081A				
.25T24	5081A		ZD-24		
.25T24A	5081A		ZD-24		
.25T25	5082A	3753	ZD-25		
.25T25A	5082A	3753	ZD-25		
.25T27	146A	3064	ZD-27		
.25T27A	146A	3064	ZD-27		
.25T27B	146A	3064	ZD-27		
.25T28	5083A	3754	ZD-28		
.25T28A	5083A	3754	ZD-28		
.25T30	5084A	3755	ZD-30		
.25T30A	5084A	3755	ZD-30		
.25T33	147A	3095	ZD-33		
.25T33A	147A	3095	ZD-33		
.25T33B	147A	3095	ZD-33		
.25T36	5085A	3337	ZD-36		
.25T36A	5085A	3337	ZD-36		
.25T39	5086A	3338	ZD-39		
.25T39A	5086A	3338	ZD-39		
.25T43	5087A	3339	ZD-43		
.25T43A	5087A	3339	ZD-43		
.25T47	5088A	3340	ZD-47		
.25T47A	5088A	3340	ZD-47		
.25T5.6	136A	3057	ZD-5.6	561	
.25T5.6A	136A		ZD-5.6	561	
.25T5.6B	136A	3057	ZD-5.6	561	
.25T5.8	5070A	9021	ZD-6.2	561	
.25T5.8A	5070A	9021	ZD-6.2	561	
.25T51	5089A	3341	ZD-51		
.25T51A	5089A	3341	ZD-51		
.25T55	148A	3096	ZD-55		
.25T55B	148A	3096	ZD-55		
.25T56	5090A	3342	ZD-56		
.25T56A	5090A	3342	ZD-56		
.25T6.2	137A	3058	ZD-6.2	561	
.25T6.2A	137A		ZD-6.2	561	
.25T6.2B	137A	3058	ZD-6.2	561	
.25T6.8	5071A	3334	ZD-6.8	561	
.25T6.8A	5071A	3334	ZD-6.8	561	
.25T60	5091A		ZD-60		
.25T60A	5091A		ZD-60		
.25T62	149A	3097	ZD-62		
.25T62A	149A	3097	ZD-62		
.25T62B	149A	3097	ZD-62		
.25T68	5092A	3343	ZD-68		
.25T68A	5092A	3343	ZD-68		
.25T7.1	5071A	3334	ZD-6.8	561	
.25T7.1A	5071A	3334	ZD-6.8	561	
.25T7.5	138A	3059	ZD-7.5		
.25T7.5A	138A		ZD-7.5		
.25T7.5B	138A	3059	ZD-7.5		
.25T75	5093A	3444/123A	ZD-75		
.25T75A	5093A	3444/123A	ZD-75		
.25T8.2	5072A		ZD-8.2	562	
.25T8.2A	5072A		ZD-8.2	562	
.25T8.7	5073A	3749	ZD-9.1	562	
.25T8.7A	5073A	3749	ZD-9.1	562	
.25T8.8	5073A	3749	ZD-9.1	562	
.25T8.8A	5073A	3749	ZD-9.1	562	
.25T82	150A	3098	ZD-82		
.25T82A	150A	3098	ZD-82		
.25T82B	150A	3098	ZD-82		
.25T87	5094A	9024	ZD-87		
.25T87A	5094A	9024	ZD-87		
.25T9.1	139A	3060	ZD-9.1	562	
.25T9.1A	139A		ZD-9.1	562	
.25T9.1B	139A	3060	ZD-9.1	562	
.25T91	5095A	3345	ZD-91		
.25T91A	5095A	3345	ZD-91		
.4M100Z5	5096A				
.4M10AZ5	140A				
.4M10Z5	140A				
.4M110Z5	151A				
.4M11Z5	5074A				
.4M120Z5	5097A				
.4M12AZ5	142A				
.4M12Z5	142A				
.4M130Z5	5098A				
.4M13Z5	143A	3750			
.4M150Z5	5100A				
.4M15Z5	145A				
.4M160Z5	5101A				
.4M16Z5	5075A	3751			
.4M180Z5	5103A				
.4M18Z5	5077A	3752			
.4M200Z5	5105A				
.4M20Z5	5079A				
.4M22Z5	5080A				
.4M24Z5	5081A				
.4M3.3AZ5	5066A				
.4M3.6AZ5	134A				
.4M3.9AZ5	5067A				
.4M30Z5	5084A	3755			
.4M33Z5	147A				
.4M36Z5	5085A				
.4M4.3Z5	5068A				
.4M4.7AZ5	5069A				
.4M43Z5	5087A				
.4M47Z5	5088A				
.4M51Z5	5089A				
.4M56Z5	5090A				
.4M6.2AZ5	137A				
.4M6.8AZ5	5071A				
.4M6.8Z5	5071A				
.4M62Z5	149A				
.4M68Z5	5092A				
.4M7.5AZ5	138A				
.4M7.5Z5	138A				
.4M75Z5	5093A				
.4M8.2AZ5	5072A				
.4M8.2Z5	5072A				
.4M82Z5	150A				
.4M9.1AZ5	139A				
.4M9.1Z5	139A				
.4M91Z5	5095A				
.4T10	140A	3061	ZD-10	562	
.4T100	5096A	3346	ZD-100		
.4T100A	5096A	3346	ZD-100		
.4T100B	5096A	3346	ZD-100		
.4T10A	140A	3061	ZD-10	562	
.4T10B	140A		ZD-10	562	
.4T11	5074A		ZD-11.0	563	
.4T11.5	141A	3092	ZD-11.5		
.4T11.5A	141A	3092	ZD-11.5		
.4T110	151A	3099	ZD-110		
.4T110A	151A	3099	ZD-110		
.4T110B	5096A	3346	ZD-100		
.4T11A	5074A		ZD-11.0	563	
.4T11B	5074A		ZD-11.0	563	
.4T12.8	143A	3093	ZD-13	563	
.4T12.8A	143A	3093	ZD-13	563	
.4T120	5097A	3347	ZD-120		
.4T120A	5097A	3347	ZD-120		
.4T120B	5097A	3347	ZD-120		
.4T12A	142A		ZD-12	563	
.4T12B	142A		ZD-12	563	
.4T13	143A	3750	ZD-13	563	
.4T130	5098A	3348	ZD-130		
.4T130A	5098A	3348	ZD-130		
.4T130B	5098A	3348	ZD-130		
.4T13A	143A	3750	ZD-13	563	
.4T13B	143A	3750	ZD-13	563	
.4T14	144A	3094	ZD-14	564	
.4T14A	144A	3094	ZD-14	564	
.4T15	145A	3063	ZD-15	564	
.4T150	5100A	3350	ZD-150		
.4T150A	5100A	3350	ZD-150		
.4T150B	5100A	3350	ZD-150		
.4T15B	145A		ZD-15	564	
.4T16	5075A	3751	ZD-16	564	
.4T160	5101A	3351	ZD-160		
.4T160A	5101A	3351	ZD-160		
.4T160B	5101A	3351	ZD-160		
.4T16A	5075A	3751	ZD-16	564	
.4T16B	5075A	3751	ZD-16	564	
.4T18	5077A	3752	ZD-18		
.4T180	5103A	3353	ZD-180		
.4T180A	5103A	3353	ZD-180		
.4T180B	5103A	3353	ZD-180		
.4T18A	5077A	3752	ZD-18		
.4T18B	5077A	3752	ZD-18		
.4T20	5079A	3335	ZD-20		
.4T200	5105A	3355	ZD-200		
.4T200A	5105A	3355	ZD-200		
.4T200B	5105A	3355	ZD-200		
.4T20A	5079A	3335	ZD-20		
.4T20B	5079A	3335	ZD-20		
.4T22	5080A	3336	ZD-22		
.4T22A	5080A	3336	ZD-22		

Industry Standard No.	ECG	SK	GE	RS 276-	MOTOR.
.4T22B	5080A	3336	ZD-22		
.4T24	5081A		ZD-24		
.4T24A	5081A		ZD-24		
.4T24B	5081A		ZD-24		
.4T27	146A	3064	ZD-27		
.4T27A	146A	3064	ZD-27		
.4T27B	146A	3064	ZD-27		
.4T30A	5084A	3755	ZD-30		
.4T30B	5084A	3755	ZD-30		
.4T33	147A	3095	ZD-33		
.4T33A	147A	3095	ZD-33		
.4T33B	147A	3095	ZD-33		
.4T36	5085A	3337	ZD-36		
.4T36A	5085A	3337	ZD-36		
.4T36B	5085A	3337	ZD-36		
.4T39	5086A	3338	ZD-39		
.4T39A	5086A	3338	ZD-39		
.4T39B	5086A	3338	ZD-39		
.4T43	5087A	3339	ZD-43		
.4T43A	5087A	3339	ZD-43		
.4T43B	5087A	3339	ZD-43		
.4T47	5088A	3340	ZD-47		
.4T47A	5088A	3340	ZD-47		
.4T47B	5088A	3340	ZD-47		
.4T5.6	136A	3057	ZD-5.6	561	
.4T5.6A	136A	3057	ZD-5.6	561	
.4T5.6B	136A	3057	ZD-5.6	561	
.4T51	5089A	3341	ZD-51		
.4T51A	5089A	3341	ZD-51		
.4T51B	5089A	3341	ZD-51		
.4T55		3096	ZD-55		
.4T55A		3096	ZD-55		
.4T56		3342	ZD-56		
.4T56A		3342	ZD-56		
.4T56B		3342	ZD-56		
.4T6.2		3058	ZD-6.2	561	
.4T6.2A		3058	ZD-6.2	561	
.4T6.2B		3058	ZD-6.2	561	
.4T6.8		3334	ZD-6.8	561	
.4T6.8A		3058	ZD-6.2	561	
.4T6.8B		3334	ZD-6.8	561	
.4T62		3097	ZD-62		
.4T62A		3097	ZD-62		
.4T62B		3097	ZD-62		
.4T68		3343	ZD-68		
.4T68A		3343	ZD-68		
.4T68B		3343	ZD-68		
.4T7.5		3059	ZD-7.5		
.4T7.5A		3059	ZD-7.5		
.4T7.5B		3059	ZD-7.5		
.4T75		3344	ZD-75		
.4T75A		3344	ZD-75		
.4T75B		3344	ZD-75		
.4T8.2			ZD-8.2	562	
.4T8.2A			ZD-8.2	562	
.4T8.2B			ZD-8.2	562	
.4T82		3098	ZD-82		
.4T82A		3098	ZD-82		
.4T82B		3098	ZD-82		
.4T9.1		3060	ZD-9.1	562	
.4T9.1A		3060	ZD-9.1	562	
.4T9.1B			ZD-9.1	562	
.4T91		3345	ZD-91		
.4T91A		3345	ZD-91		
.4T91B		3345	ZD-91		
.5E05		3017B	504A	1104	
.5E1		3017B	504A	1104	
.5E10		3033	531	1114	
.5E12		3125		1114	
.5E2		3017B	504A	1104	
.5E3		3017B	504A	1104	
.5E4		3017B	504A	1104	
.5E5		3017B	504A	1104	
.5E6		3017B	504A	1104	
.5E7		3033	531	1114	
.5E8		3033	531	1114	
.5J05		3017B	504A	1104	
.5J1		3017B	504A	1104	
.5J10		3033	531	1114	
.5J12		3125		1114	
.5J2		3017B	504A	1104	
.5J3		3017B	504A	1104	
.5J4		3017B	504A	1104	
.5J5		3017B	504A	1104	
.5J6		3017B	504A	1104	
.5J7		3033	531	1114	
.5J8		3033	531	1114	
.5M68ZS5	5092A				
.5M7.5ZS5	138A				
.5M75ZS5	5093A				
.5M8.2ZS5	5072A				
.5M8.7ZS5	5073A	3749			
.5M82ZS5	150A				
.5M87ZS5	5094A	9024			
.5M9.1ZS5	139A				
.5M91ZS5	5095A				
0.5Z5.1U	5010A	3776			
.75N110	151A	3099	ZD-110		
.75N12	142A	3062	ZD-12	563	
.75N27	146A	3064	ZD-27		
.75N33	147A	3095	ZD-33		
.75N5	145A	3063	ZD-15	564	
.75N5.1	135A	3056	ZD-5.1		
.75N5.6	136A	3057	ZD-5.6	561	
.75N6.2	137A	3058	ZD-6.2	561	
.75N62	149A	3097	ZD-62		
.75N7.5	138A	3059	ZD-7.5		
.75N82	150A	3098	ZD-82		
.7E05	116		504A	1104	
.7E1	116		504A	1104	
.7E10	125	3033	531	1114	
.7E12	506	3125		1114	
.7E2	116		504A	1104	
.7E3	116		504A	1104	
.7E4	116		504A	1104	
.7E5	116		504A	1104	
.7E6	116		504A	1104	
.7E7	125	3033	531	1114	
.7E8	125	3033	531	1114	
.7J05	116	3017B/117	504A	1104	
.7J1	116	3017B/117	504A	1104	
.7J10	125	3033	531	1114	
.7J12	506	3125		1114	
.7J2	116	3017B/117	504A	1104	
.7J3	116	3017B/117	504A	1104	
.7J5	116	3017B/117	504A	1104	
.7J6	116	3017B/117	504A	1104	
.7J7	125	3033	531	1114	
.7J8	125	3033	531	1114	
.7JZ10	140A	3061	ZD-10	562	
.7JZ100	5096A	3346	ZD-100		
.7JZ100B	5096A				
.7JZ11	5074A		ZD-11.0	563	
.7JZ110	5151A				
.7JZ12	142A	3062	ZD-12	563	
.7JZ120	5097A	3347	ZD-120		
.7JZ13	143A	3750	ZD-13	563	
.7JZ130	5098A	3348	ZD-130		
.7JZ14	144A	3094	ZD-14	564	
.7JZ140	5099A	3349	ZD-140		
.7JZ15	145A	3063	ZD-15	564	
.7JZ150	5100A	3350	ZD-150		
.7JZ16	5075A	3142	ZD-16	564	
.7JZ160	5101A	3351	ZD-160		
.7JZ17	5076A	9022	ZD-17		
.7JZ18	5077A	3752	ZD-18		
.7JZ180	5103A	3353	ZD-180		
.7JZ19	5078A	9023	ZD-19		
.7JZ20	5079A	3335	ZD-20		
.7JZ200	5104A	3354	ZD-190		
.7JZ22	5080A	3336	ZD-22		
.7JZ220	151A(2)				
.7JZ24	5081A	3151	ZD-24		
.7JZ25	5082A	3753	ZD-25		
.7JZ27	146A	3064	ZD-27		
.7JZ30	5084A	3150	ZD-30		
.7JZ33	147A	3095	ZD-33		
.7JZ36	5085A	3337	ZD-36		
.7JZ39	5086A	3338	ZD-39		
.7JZ43	5087A	3339	ZD-43		
.7JZ45	5088A	3340	ZD-47		
.7JZ47	5088A	3340	ZD-47		
.7JZ50	5089A	3341	ZD-51		
.7JZ51	5089A	3341	ZD-51		
.7JZ56	148A	3096	ZD-55		
.7JZ6.8	5071A	3334	ZD-6.8	561	
.7JZ62	149A	3097	ZD-62		
.7JZ68	5092A	3343	ZD-68		
.7JZ7.5	138A	3059	ZD-7.5		
.7JZ75	5093A	3344	ZD-75		
.7JZ8.2	5072A	3136	ZD-8.2	562	
.7JZ82	150A	3098	ZD-82		
.7JZ9.1	139A	3060	ZD-9.1	562	
.7JZ91	5095A	3345	ZD-91		
.7JZM100B	5096A				
.7RM220D	151A(2)				
.7Z100A	5096A	3346	ZD-100		
.7Z100B	5096A	3346	ZD-100		
.7Z100C	5096A	3346	ZD-100		
.7Z100D	5096A	3346	ZD-100		
.7Z10A	140A	3061	ZD-10	562	
.7Z10B	140A	3061	ZD-10	562	
.7Z10C	140A	3061	ZD-10	562	
.7Z10D	140A	3061	ZD-10	562	
.7Z110A	151A	3099	ZD-110		
.7Z110B	151A		ZD-110		
.7Z110C	151A	3099	ZD-110		
.7Z110D	151A	3099	ZD-110		
.7Z11A	5074A	3139	ZD-11.0	563	
.7Z11B	5074A	3139	ZD-11.0	563	
.7Z11C	5074A	3139	ZD-11.0	563	
.7Z11D	5074A	3139	ZD-11.0	563	
.7Z120A	5097A	3347	ZD-120		
.7Z120B	5097A	3347	ZD-120		
.7Z120C	5097A	3347	ZD-120		
.7Z120D	5097A	3347	ZD-120		
.7Z12A	142A	3062	ZD-12	563	
.7Z12B	142A	3062	ZD-12	563	
.7Z12C	142A	3062	ZD-12	563	
.7Z12D	142A		ZD-12	563	
.7Z130A	5098A	3348	ZD-130		
.7Z130B	5098A	3348	ZD-130		
.7Z130C	5098A	3348	ZD-130		
.7Z130D	5098A	3348	ZD-130		
.7Z13A	143A	3750	ZD-13	563	
.7Z13B	143A	3750	ZD-13	563	
.7Z13C	143A	3750	ZD-13	563	
.7Z13D	143A	3750	ZD-13	563	
.7Z140A	5099A	3349	ZD-140		
.7Z140B	5099A	3349	ZD-140		
.7Z140C	5099A	3349	ZD-140		
.7Z140D	5099A	3349	ZD-140		
.7Z14A	145A	3063	ZD-15	564	
.7Z14B	144A	3063/145A	ZD-15	564	
.7Z14C	145A	3063	ZD-15	564	
.7Z150A	5100A	3350	ZD-150		
.7Z150B	5100A	3350	ZD-150		
.7Z150C	5100A	3350	ZD-150		
.7Z150D	5100A	3350	ZD-150		
.7Z15B	145A				
.7Z15D	145A	3063	ZD-15	564	
.7Z160A	5101A	3351	ZD-160		
.7Z160B	5101A	3351	ZD-160		
.7Z160C	5101A	3351	ZD-160		
.7Z160D	5101A	3351	ZD-160		
.7Z16A	5075A	3751	ZD-16	564	
.7Z16B	5075A	3751	ZD-16	564	
.7Z16C	5075A	3751	ZD-16	564	
.7Z16D	5075A	3751	ZD-16	564	
.7Z17A	5076A	9022	ZD-17		
.7Z17B	5076A	9022	ZD-17		
.7Z17C	5076A	9022	ZD-17		
.7Z17D	5076A	9022	ZD-17		
.7Z180A	5103A	3353	ZD-180		
.7Z180B	5103A	3353	ZD-180		
.7Z180C	5103A	3353	ZD-180		
.7Z180D	5103A	3353	ZD-180		
.7Z18A	5077A	3752	ZD-18		
.7Z18B	5077A	3752	ZD-18		
.7Z18C	5077A	3752	ZD-18		
.7Z18D	5077A	3752	ZD-18		
.7Z19A	5078A	9023	ZD-19		
.7Z19B	5078A	9023	ZD-19		
.7Z19C	5078A	9023	ZD-19		
.7Z19D	5078A	9023	ZD-19		
.7Z200A	5105A	3355	ZD-200		
.7Z200B	5105A	3355	ZD-200		
.7Z200C	5105A	3355	ZD-200		
.7Z200D	5105A	3355	ZD-200		
.7Z20A	5079A	3335	ZD-20		
.7Z20B	5079A	3335	ZD-20		
.7Z20C	5079A	3335	ZD-20		
.7Z20D	5079A	3335	ZD-20		
.7Z220A	151A(2)				
.7Z220B	151A(2)				
.7Z220C	151A(2)				
.7Z220D	151A(2)				
.7Z22A	5080A	3336	ZD-22		
.7Z22B	5080A	3336	ZD-22		
.7Z22C	5080A	3336	ZD-22		
.7Z22D	5080A	3336	ZD-22		
.7Z24A	5081A	3151	ZD-24		
.7Z24B	5081A	3151	ZD-24		
.7Z24C	5081A		ZD-24		
.7Z24D	5081A	3151	ZD-24		
.7Z25A	5082A	3753	ZD-25		
.7Z25B	5082A	3753	ZD-25		
.7Z25C	5082A	3753	ZD-25		

Industry Standard No.	ECG	SK	GE	RS 276-	MOTOR.
.7Z25D	5082A	3753	ZD-25		
.7Z27A	146A	3064	ZD-27		
.7Z27B	146A		ZD-27		
.7Z27C	146A	3064	ZD-27		
.7Z27D	146A	3064	ZD-27		
.7Z30A	5084A	3755	ZD-30		
.7Z30B	5084A	3755	ZD-30		
.7Z30C	5084A	3755	ZD-30		
.7Z30D	5084A	3755	ZD-30		
.7Z33A	147A	3095	ZD-33		
.7Z33B	147A	3095	ZD-33		
.7Z33C	147A		ZD-33		
.7Z33D	147A	3095	ZD-33		
.7Z36A	5085A	3337	ZD-36		
.7Z36B	5085A	3337	ZD-36		
.7Z36C	5085A	3337	ZD-36		
.7Z36D	5085A	3337	ZD-36		
.7Z39A	5086A	3338	ZD-39		
.7Z39B	5086A	3338	ZD-39		
.7Z39C	5086A	3338	ZD-39		
.7Z39D	5086A	3338	ZD-39		
.7Z43A	5087A	3339	ZD-43		
.7Z43B	5087A	3339	ZD-43		
.7Z43C	5087A	3339	ZD-43		
.7Z43D	5087A	3339	ZD-43		
.7Z45A	5088A	3340	ZD-47		
.7Z45B	5088A	3340	ZD-47		
.7Z45C	5088A	3340	ZD-47		
.7Z45D	5088A	3340	ZD-47		
.7Z47A	5088A	3340	ZD-47		
.7Z47B	5088A	3340	ZD-47		
.7Z47C	5088A	3340	ZD-47		
.7Z47D	5088A	3340	ZD-47		
.7Z50A	5089A	3341	ZD-51		
.7Z50B	5089A	3341	ZD-51		
.7Z50C	5089A	3341	ZD-51		
.7Z50D	5089A	3341	ZD-51		
.7Z51A	5089A	3341	ZD-51		
.7Z51B	5089A	3341	ZD-51		
.7Z51C	5089A	3341	ZD-51		
.7Z51D	5089A	3341	ZD-51		
.7Z52A	5089A	3341	ZD-51		
.7Z52B	5089A	3341	ZD-51		
.7Z52C	5089A	3341	ZD-51		
.7Z52D	5089A	3341	ZD-51		
.7Z56A	148A	3096	ZD-55		
.7Z56B	5090A	3096/148A	ZD-55		
.7Z56C	148A	3096	ZD-55		
.7Z56D	148A	3096	ZD-55		
.7Z6.8A	5071A	3334	ZD-6.8	561	
.7Z6.8B	5071A	3334	ZD-6.8	561	
.7Z6.8C	5071A	3334	ZD-6.8	561	
.7Z6.8D	5071A	3334	ZD-6.8	561	
.7Z62A	149A	3096/148A	ZD-62		
.7Z62B	149A	3097	ZD-62		
.7Z62C	149A	3097	ZD-62		
.7Z62D	149A	3097	ZD-62		
.7Z68A	5092A	3343	ZD-68		
.7Z68B	5092A	3343	ZD-68		
.7Z68C	5092A	3343	ZD-68		
.7Z68D	5092A	3343	ZD-68		
.7Z7.5A	138A	3059	ZD-7.5		
.7Z7.5B	138A	3059	ZD-7.5		
.7Z7.5C	138A	3059	ZD-7.5		
.7Z7.5D	138A	3059	ZD-7.5		
.7Z75A	5093A	3344	ZD-75		
.7Z75B	5093A	3344	ZD-75		
.7Z75C	5093A	3344	ZD-75		
.7Z75D	5093A	3344	ZD-75		
.7Z8.2A	5072A	3136	ZD-8.2	562	
.7Z8.2B	5072A	3136	ZD-8.2	562	
.7Z8.2C	5072A	3136	ZD-8.2	562	
.7Z8.2D	5072A	3136	ZD-8.2	562	
.7Z82A	150A	3098	ZD-82		
.7Z82B	150A	3098	ZD-82		
.7Z82C	150A	3098	ZD-82		
.7Z82D	150A	3098	ZD-82		
.7Z9.1A	139A	3060	ZD-9.1	562	
.7Z9.1B	139A	3060	ZD-9.1	562	
.7Z9.1C	139A	3060	ZD-9.1	562	
.7Z9.1D	139A	3060	ZD-9.1	562	
.7Z91A	5095A	3345	ZD-91		
.7Z91B	5095A	3345	ZD-91		
.7Z91C	5095A	3345	ZD-91		
.7Z91D	5095A	3345	ZD-91		
.7ZM1	5074A	3139	ZD-11.0	563	
.7ZM100A	5096A	3346	ZD-100		
.7ZM100B	5096A	3346	ZD-100		
.7ZM100C	5096A	3346	ZD-100		
.7ZM100D	5096A	3346	ZD-100		
.7ZM10A	140A	3061	ZD-10	562	
.7ZM10B	140A	3061	ZD-10	562	
.7ZM10C	140A	3061	ZD-10	562	
.7ZM10D	140A		ZD-10	562	
.7ZM110A	151A	3099	ZD-110		
.7ZM110B	151A	3099	ZD-110		
.7ZM110C	151A	3099	ZD-110		
.7ZM110D	151A	3099	ZD-110		
.7ZM11A	5074A	3139	ZD-11.0	563	
.7ZM11B	5074A	3139	ZD-11.0	563	
.7ZM11C	5074A	3139	ZD-11.0	563	
.7ZM120A	5097A	3347	ZD-120		
.7ZM120B	5097A	3347	ZD-120		
.7ZM120C	5097A	3347	ZD-120		
.7ZM120D	5097A	3347	ZD-120		
.7ZM12A	142A	3062	ZD-12	563	
.7ZM12B	142A	3062	ZD-12	563	
.7ZM12C	142A	3062	ZD-12	563	
.7ZM12D	142A	3062	ZD-12	563	
.7ZM130A	5098A	3348	ZD-130		
.7ZM130B	5098A	3348	ZD-130		
.7ZM130C	5098A	3348	ZD-130		
.7ZM130D	5098A	3348	ZD-130		
.7ZM13A	143A	3750	ZD-13	563	
.7ZM13B	143A	3750	ZD-13	563	
.7ZM13C	143A	3750	ZD-13	563	
.7ZM13D	143A	3750	ZD-13	563	
.7ZM140A	5099A	3349	ZD-140		
.7ZM140B	5099A	3349	ZD-140		
.7ZM140C	5099A	3349	ZD-140		
.7ZM140D	5099A	3349	ZD-140		
.7ZM14A	144A		ZD-14	564	
.7ZM14B	144A	3094	ZD-14	564	
.7ZM14C	144A	3094	ZD-14	564	
.7ZM14D	144A	3094	ZD-14	564	
.7ZM150A	5100A	3350	ZD-150		
.7ZM150B	5100A	3350	ZD-150		
.7ZM150C	5100A	3350	ZD-150		
.7ZM150D	5100A	3350	ZD-150		
.7ZM15A	145A	3063	ZD-15	564	
.7ZM15B	145A	3063	ZD-15	564	
.7ZM15C	145A	3063	ZD-15	564	
.7ZM15D	145A	3063	ZD-15	564	
.7ZM160A	5101A	3351	ZD-160		

Industry Standard No.	ECG	SK	GE	RS 276-	MOTOR.
.7ZM160B	5101A	3351	ZD-160		
.7ZM160C	5101A	3351	ZD-160		
.7ZM160D	5101A	3351	ZD-160		
.7ZM16A	5075A	3751	ZD-16	564	
.7ZM16B	5075A	3751	ZD-16	564	
.7ZM16C	5075A	3751	ZD-16	564	
.7ZM16D	5075A	3751	ZD-16	564	
.7ZM17A	5076A	9022	ZD-17		
.7ZM17B	5076A	9022	ZD-17		
.7ZM17C	5076A	9022	ZD-17		
.7ZM17D	5076A	9022	ZD-17		
.7ZM180A	5103A	3353	ZD-180		
.7ZM180B	5103A	3353	ZD-180		
.7ZM180C	5103A	3353	ZD-180		
.7ZM180D	5103A	3353	ZD-180		
.7ZM18A	5077A	3752	ZD-18		
.7ZM18B	5077A	3752	ZD-18		
.7ZM18C	5077A	3752	ZD-18		
.7ZM18D	5077A	3752	ZD-18		
.7ZM19A	5078A	9023	ZD-19		
.7ZM19B	5078A	9023	ZD-19		
.7ZM19C	5078A	9023	ZD-19		
.7ZM19D	5078A	9023	ZD-19		
.7ZM200A	5105A	3355	ZD-200		
.7ZM200B	5105A	3355	ZD-200		
.7ZM200C	5105A	3355	ZD-200		
.7ZM200D	5105A	3355	ZD-200		
.7ZM20A	5079A	3335	ZD-20		
.7ZM20B	5079A	3335	ZD-20		
.7ZM20C	5079A	3335	ZD-20		
.7ZM20D	5079A	3335	ZD-20		
.7ZM220A	151A(2)				
.7ZM220B	151A(2)				
.7ZM220C	151A(2)				
.7ZM22A	5080A	3336	ZD-22		
.7ZM22B	5080A	3336	ZD-22		
.7ZM22C	5080A	3336	ZD-22		
.7ZM22D	5080A	3336	ZD-22		
.7ZM24A	5081A	3151	ZD-24		
.7ZM24B	5081A	3151	ZD-24		
.7ZM24C	5081A	3151	ZD-24		
.7ZM24D	5081A	3151	ZD-24		
.7ZM25A	5082A	3753	ZD-25		
.7ZM25B	5082A	3753	ZD-25		
.7ZM25C	5082A	3753	ZD-25		
.7ZM25D	5082A	3753	ZD-25		
.7ZM27A	146A	3064	ZD-27		
.7ZM27B	146A	3064	ZD-27		
.7ZM27C	146A	3064	ZD-27		
.7ZM27D	146A	3064	ZD-27		
.7ZM30A	5084A	3755	ZD-30		
.7ZM30B	5084A	3755	ZD-30		
.7ZM30C	5084A	3755	ZD-30		
.7ZM30D	5084A	3755	ZD-30		
.7ZM33A	147A	3095	ZD-33		
.7ZM33B	147A	3095	ZD-33		
.7ZM33C	147A	3095	ZD-33		
.7ZM33D	147A	3095	ZD-33		
.7ZM36A	5085A	3337	ZD-36		
.7ZM36B	5085A	3337	ZD-36		
.7ZM36C	5085A	3337	ZD-36		
.7ZM36D	5085A	3337	ZD-36		
.7ZM39A	5086A	3338	ZD-39		
.7ZM39B	5086A	3338	ZD-39		
.7ZM39C	5086A	3338	ZD-39		
.7ZM39D	5086A	3338	ZD-39		
.7ZM43A	5087A	3339	ZD-43		
.7ZM43B	5087A	3339	ZD-43		
.7ZM43C	5087A	3339	ZD-43		
.7ZM43D	5087A	3339	ZD-43		
.7ZM47A	5088A	3340	ZD-47		
.7ZM47B	5088A	3340	ZD-47		
.7ZM47C	5088A	3340	ZD-47		
.7ZM47D	5088A	3340	ZD-47		
.7ZM50A	5089A	3341	ZD-51		
.7ZM50B	5089A	3341	ZD-51		
.7ZM50C	5089A	3341	ZD-51		
.7ZM50D	5089A	3341	ZD-51		
.7ZM51A	5089A	3341	ZD-51		
.7ZM51B	5089A	3341	ZD-51		
.7ZM51C	5089A	3341	ZD-51		
.7ZM51D	5089A	3341	ZD-51		
.7ZM52A	5089A	3341	ZD-51		
.7ZM52B	5089A	3341	ZD-51		
.7ZM52C	5089A	3341	ZD-51		
.7ZM52D	5089A	3341	ZD-51		
.7ZM56A	5090A	3342	ZD-56		
.7ZM56B	5090A	3342	ZD-56		
.7ZM56C	5090A	3342	ZD-56		
.7ZM56D	5090A	3342	ZD-56		
.7ZM6.8A	5071A	3334	ZD-6.8	561	
.7ZM6.8B	5071A	3334	ZD-6.8	561	
.7ZM6.8C	5071A	3334	ZD-6.8	561	
.7ZM6.8D	5071A	3334	ZD-6.8	561	
.7ZM62A	149A	3097	ZD-62		
.7ZM62B	149A	3097	ZD-62		
.7ZM62C	149A	3097	ZD-62		
.7ZM62D	149A	3097	ZD-62		
.7ZM68A	5092A	3343	ZD-68		
.7ZM68B	5092A	3343	ZD-68		
.7ZM68C	5092A	3343	ZD-68		
.7ZM68D	5092A	3343	ZD-68		
.7ZM7.5A	138A	3059	ZD-7.5		
.7ZM7.5B	138A	3059	ZD-7.5		
.7ZM7.5C	138A		ZD-7.5		
.7ZM7.5D	138A	3059	ZD-7.5		
.7ZM75A	5093A	3344	ZD-75		
.7ZM75B	5093A	3344	ZD-75		
.7ZM75C	5093A	3344	ZD-75		
.7ZM75D	5093A	3344	ZD-75		
.7ZM8.2A	5072A	3136	ZD-8.2	562	
.7ZM8.2B	5072A	3136	ZD-8.2	562	
.7ZM8.2C	5072A	3136	ZD-8.2	562	
.7ZM8.2D	5072A	3136	ZD-8.2	562	
.7ZM82A	150A	3098	ZD-82		
.7ZM82B	150A	3098	ZD-82		
.7ZM82C	150A	3098	ZD-82		
.7ZM82D	150A	3098	ZD-82		
.7ZM9.1A	139A	3060	ZD-9.1	562	
.7ZM9.1B	139A	3060	ZD-9.1	562	
.7ZM9.1C	139A	3060	ZD-9.1	562	
.7ZM9.1D	139A	3060	ZD-9.1	562	
.7ZM91A	5095A	3345	ZD-91		
.7ZM91B	5095A	3345	ZD-91		
.7ZM91C	5095A	3345	ZD-91		
.7ZM91D	5095A	3345	ZD-91		
A-0205	170	3649	BR-600	1104	
A-025	169	3678			
A-04	116		504A	1104	
A-04049-B	116	3016	504A	1104	
A-04091-A	116	3016	504A	1104	
A-04092	116	3016	504A	1104	
A-04092-B	116	3016	504A	1104	
A-04093	116	3016	504A	1104	

Industry Standard No.	ECG	SK	GE	RS 276-	MOTOR.
A-04093A	116	3016	504A	1104	
A-04212-A	116	3016	504A	1104	
A-04212-B	116	3016	504A	1104	
A-04226	116	3017B/117	504A	1104	
A-042313	116	3311	504A	1104	
A-04242	116	3016	504A	1104	
A-04901A	116		504A	1104	
A-1.5-01	116	3311	504A	1104	
A-100(RECT.)	116		504A	1104	
A-1005-725	159		82	2032	
A-10105	116	3016	504A	1104	
A-10113	116	3016	504A	1104	
A-10118	116	3016	504A	1104	
A-102	126		52	2024	
A-106			10	2051	
A-11095924	123A	3444	20	2051	
A-11166527	130		14	2041	
A-11237336	123A	3444	20	2051	
A-113110	159		82	2032	
A-113335	5942		5048		
A-113367(DIODE)	519			1122	
A-1141 6062	123A	3444		2051	
A-1141-5932	128		243	2030	
A-1141-6062			20	2051	
A-11790169	519			1122	
A-118038	125		531	1114	
A-120018	128		243	2030	
A-120077	5021A	3787	ZD-12	563	
A-120125(DIODE)	177		300	1122	
A-120278	123A	3444	20	2051	
A-120304	175		246	2020	
A-120327	130		14	2041	
A-120407	5188A				
A-120417	159		82	2032	
A-120420	5005A	3771	ZD-3.3		
A-120526	159		82	2032	
A-125278	5013A		ZD-6.2	561	
A-125332	123A	3444	20	2051	
A-128		3124	10	2051	
A-132591	5802	9005		1143	
A-134166-2	142A		ZD-12	563	
A-1379	123A	3124/289	20	2051	
A-1380	123A	3124/289	20	2051	
A-1384	160	3007	245	2004	
A-140605	130		14	2041	
A-144-3078	5547				
A-14F-C	519				
A-156	123A		20	2051	
A-1567	123A	3124/289	20	2051	
A-158B	123A	3124/289	20	2051	
A-158C	123A	3124/289	20	2051	
A-1634 1125	519			1122	
A-1634-1125			514	1122	
A-168	123A	3124/289	20	2051	
A-18			14	2041	
A-1853-0009-1	106	3984	21	2034	
A-1853-0010-1	106	3984	21	2034	
A-1853-0016-1	159		82	2032	
A-1853-0020-1	159		82	2032	
A-1853-0027-1	159		82	2032	
A-1853-0034-1	106	3984	21	2034	
A-1853-0036-1	159		82	2032	
A-1853-0039-1	159		82	2032	
A-1853-0041-1	129		244	2027	
A-1853-0049-1	159		82	2032	
A-1853-0050-1	234	3247	65	2050	
A-1853-0053-1	218		234		
A-1853-0058-1	159		82	2032	
A-1853-0062-1	159		82	2032	
A-1853-0063-1	218		234		
A-1853-0065-1	159		82	2032	
A-1853-0066-1	234	3247	65	2050	
A-1853-0077-1	234	3247	65	2050	
A-1853-0086-1	234	3247	65	2050	
A-1853-0092-1	159		82	2032	
A-1853-0098-1	234	3247	65	2050	
A-1853-0099-1	159		82	2032	
A-1853-0220-1	218		234		
A-1853-0233-1	153		69	2049	
A-1853-0234-1	153		69	2049	
A-1853-0254-1	153		69	2049	
A-1853-0285-1	159		82	2032	
A-1853-0300-1	234		65	2050	
A-1853-0321-1	159		82	2032	
A-1853-0404-1	199		62	2010	
A-18530053-1	218		234		
A-1854-0003-1	123A	3444	20	2051	
A-1854-0019-1	123A	3444	20	2051	
A-1854-0022-1			243	2030	
A-1854-0023-1	199	3245	62	2010	
A-1854-0025-1	123A	3444	20	2051	
A-1854-0027-1	123A	3444	20	2051	
A-1854-0071-1	123A	3444	20	2051	
A-1854-0087-1	128		243	2030	
A-1854-0088-1	199	3245	62	2010	
A-1854-0090-1	128		243	2030	
A-1854-0092-1	107		11	2015	
A-1854-0094-1	123A	3444	20	2051	
A-1854-0099-1	123A	3444	20	2051	
A-1854-0201-1	123A	3444	20	2051	
A-1854-0215-1	123A	3444	20	2051	
A-1854-0229-1	128				
A-1854-0241-1	123A	3444	20	2051	
A-1854-0246-1	123A	3444	20	2051	
A-1854-0251-1	123A	3444	20	2051	
A-1854-0255-1	123A	3444	20	2051	
A-1854-0284-1	199	3245	62	2010	
A-1854-0291-1	130		14	2041	
A-1854-0294-1	130		14	2041	
A-1854-0354-1	123A	3444	20	2051	
A-1854-0358-1	194	3275	220		
A-1854-0365-1	194	3275	220		
A-1854-0408-1	123A	3444	20	2051	
A-1854-0420-1	152		66	2048	
A-1854-0434-1	123A	3444	20	2051	
A-1854-0449-1	175		246	2020	
A-1854-0458	181		75	2041	
A-1854-0464	152	3893	66	2048	
A-1854-0471-1	123A	3444	20	2051	
A-1854-0474-1	194	3275	220		
A-1854-0485	108	3452	86	2038	
A-1854-0492-1	123A	3444	20	2051	
A-1854-0533	194	3275	220		
A-1854-0541-1	123A	3444	20	2051	
A-1854-0554-1	123A	3444	20	2051	
A-1854-JBD1	108	3452	86	2038	
A-1901-0025-1	177		300	1122	
A-1901-0033-1	177		300	1122	
A-1901-0050-1	519		514	1122	
A-1901-0053-1	177		300	1122	
A-1901-0096-1	177		300	1122	
A-1901-0150-1	177		300	1122	
A-1901-0156-1	177		300	1122	
A-1901-0196-1	519		514	1122	
A-1901-044-1	519		514	1122	
A-1901-0461-1	519		514	1122	
A-1901-1067-1	177		300	1122	
A-1946	116	3016	504A	1104	
A-195C	159	3114/290	82	2032	
A-2008-9140	519		514	1122	
A-2010-5409-B	912		IC-172		
A-2010-5409-RB	912		IC-172		
A-225	543				
A-225-1	160				
A-258-1	126				
A-36617	519		514	1122	
A-415			20	2051	
A-41764	525				
A-41766	5986				
A-42125	519				
A-42126	519				
A-42130	5802				
A-42160	109				
A-42180	177				
A-42196	116				
A-42218	177				
A-42242	519				
A-42340	5315				
A-42342	5890				
A-473			61	2038	
A-494			20	2051	
A-514-027662	102A	3004	53	2007	
A-514-027757	116				
A-54806-A	5547				
A-567			8	2002	
A-567A	123A	3444	20	2051	
A-6-67703	130		14	2041	
A-6-67703-A-7	130		14	2041	
OA-95	110MP				
A-95-5281	116	3016	504A	1104	
A-95-5289	116	3016	504A	1104	
A-95-5295	119	3109	CR-2		
A-95-5296	118	3066	CR-1		
A-DHS-LAA-106	5801				
A-DHZ-LAA-057	5128A				
A-PS-1503399-0-0	218		234		
A-PS-1509730-0-0	181		75	2041	
A-PS-1510196-0-0	181		75	2041	
A.184/5	123	3444	20	2051	
A0-54-175	128	3024	243	2030	
A0-54-195	123A	3020/123	20	2051	
A00	7400			1801	
A01	177	3100/519	300	1122	
A01(MOTOROLA)	177		300	1122	
A01(TRANSISTOR)	160		245	2004	
OA01-08R	5072A		ZD-8.2	562	
OA01-11SE	5074A				
A02	177	3311	1N34AS	1123	
A02(I.C.)	7403				
A03	7404	7404	504A	1802	
A03004	5626	3633			
A0311400	1128	3488	IC-105		
A0375			504A	1104	
A0377	116	3017B/117	504A	1104	
A04	116	3031A	504A	1104	
A04(I.C.)	7405				
A04021	358		42		
A04049			504A	1104	
A04049B	116		504A	1104	
A04091A	116		504A	1104	
A04092	116		504A	1104	
A04092A	116		504A	1104	
A04092B	116		504A	1104	
A04093	116		504A	1104	
A04093-A	506	3032A	511	1114	
A04093-X	506	3043	300	1114	
A04093A	506	3125	504A	1114	
A04093A2			511	1114	
A04093X	506	3125		1114	
A04166-2	145A	3094/144A	ZD-15	564	
A04201	312	3112	6GC1	2028	
A04201(JVC)	113A		6GC1		
A04210-A	166	9075		1152	
A04210A	116		504A	1104	
A04212-A	125	3032A	531	1114	
A04212-B	116		504A	1104	
A04212A	116	3051/156	504A	1104	
A04212B	116	3051/156	504A	1104	
A04226	116(3)				
A04230	506	3017B/117		1114	
A04230-A	506	3043	504A	1114	
A04231-A	166	9075		1152	
A042313	116	3017B/117	504A	1104	
A04233	116	3130	504A	1104	
A04234-2	139A	3059/138A	ZD-9.1	562	
A04241-A	506	3043	504A	1114	
A04242	116		504A	1104	
A04284-A	166	9075		1152	
A04294	506	3125		1114	
A04294-1	506	3125		1114	
A04299-201	502	3108/505	CR-4		
A04299-202	503	3068	CR-5		
A04299-251	503	3108/505	CR-5		
A04331-021	116	3031A	504A	1104	
A04331-023	116	3017B/117	504A	1104	
A04331-043	116	3031A	504A	1104	
A04332-007	138A	3059	ZD-7.5		
A04344-007	138A	3059	ZD-7.5		
A04344-026	5082A	3753	ZD-25		
A04350-022	116		504A	1104	
A0450(HEP)	413				
A0455(HEP)	414				
A04710	119	3109	CR-2		
A04716	166	9075		1152	
A04727	120	3110	CR-3		
A04731	116	3016	504A	1104	
A04735-A	118	3066	CR-1		
A04901A	116		504A	1104	
A0491A	116	3017B/117	504A	1104	
A05	116	3017B/117	504A	1104	
A05(I.C.)	7410			1807	
A0500(HEP)	405				
A0502(HEP)	401				
A054-103	109	3088	1N34AS	1123	
A054-105	109	3087	1N34AS	1123	
A054-108	123A	3444	20	2051	
A054-109	129	3025	244	2027	
A054-114	123A	3444	20	2051	
A054-115	123A	3444	20	2051	
A054-119		3126	90		
A054-142	220	3990		2036	
A054-148	107	3018	11	2015	
A054-150	116	3017B/117	504A	1104	
A054-151	142A	3062	ZD-12	563	
A054-154	130	3027	14	2041	
A054-155	123A	3444	20	2051	
A054-156	128	3024	243	2030	
A054-157	108	3452	86	2038	

Industry Standard No.	ECG	SK	GE	RS 276-	MOTOR.
A054-158	108	3452	86	2038	
A054-159	108	3452	86	2038	
A054-160	128	3024	243	2030	
A054-163	108	3452	86	2038	
A054-164	108	3452	86	2038	
A054-165	703A		IC-12		
A054-170	107	3018	11	2015	
A054-173	123A	3444	20	2051	
A054-175	123A	3444	20	2051	
A054-186	128	3024	243	2030	
A054-187	109	3088	1N34AS	1123	
A054-195	123A	3444	20	2051	
A054-206	128	3024	243	2030	
A054-221	123A	3444	20	2051	
A054-222	123A	3444	20	2051	
A054-223	159	3025/129	82	2032	
A054-224	175	3026	246	2020	
A054-225	123A	3444	20	2051	
A054-226	110MP	3709	1N34AS	1123(2)	
A054-228	177	3100/519	300	1122	
A054-229	156	3081/125	512		
A054-230	116	3311	504A	1104	
A054-231	143A	3750	ZD-13	563	
A054-232	5079A	3335	ZD-20		
A054-233	123A	3444	20	2051	
A054-234	123A	3444	20	2051	
A054-470	108	3452	86	2038	
A059-100	160	3007	245	2004	
A059-101	160	3007	245	2004	
A059-102	160	3007	245	2004	
A059-103	160	3007	245	2004	
A059-104	126	3006/160	52	2024	
A059-105	126	3008	52	2024	
A059-106	102A	3004	53	2007	
A059-107	102A	3004	53	2007	
A059-108	121	3009	239	2006	
A059-109	123A	3444	20	2051	
A059-110	128	3024	243	2030	
A059-111	224	3049	46		
A059-114	116	3032A	504A	1104	
A059-115	104	3009	16	2006	
A059-116	102A	3004	53	2007	
A059-117	138A	3059	ZD-7.5		
A059-118	125	3033	531	1114	
A06	116	3017B/117	504A	1104	
A06(I.C.)	7420			1809	
A06-1-12	123A	3444	20	2051	
A060-100	107	3039/316	11	2015	
A061-105	107	3039/316	11	2015	
A061-106	107	3039/316	11	2015	
A061-107	126	3006/160	52	2024	
A061-108	107	3039/316	11	2015	
A061-109	107	3039/316	11	2015	
A061-110	126	3006/160	52	2024	
A061-111	126	3006/160	52	2024	
A061-112	107	3039/316	11	2015	
A061-114	102A	3004	53	2007	
A061-115	102A	3123	53	2007	
A061-116	131	3052	44	2006	
A061-118	116	3017B/117	504A	1104	
A061-119	139A	3060	ZD-9.1	562	
A065-101	116	3311	504A	1104	
A065-102	123A	3444	20	2051	
A065-103	123A	3444	20	2051	
A065-104	123A	3444	20	2051	
A065-105	158	3004/102A	53	2007	
A065-106	102A	3004	53	2007	
A065-108	123A	3444	20	2051	
A065-109	123A	3444	20	2051	
A065-110	123A	3444	20	2051	
A065-111	131	3052	44	2006	
A065-112	158	3004/102A	53	2007	
A065-113	123A	3444	20	2051	
A066-109	123A	3444	20	2051	
A066-110	312	3448	FET-1	2035	
A066-111	229	3018	61	2038	
A066-112	123A	3444	20	2051	
A066-113	159	3025/129	82	2032	
A066-113(2SC7321)			20	2051	
A066-113A	159	3114/290	82	2032	
A066-113AB	159	3114/290	82	2032	
A066-114	152	3054/196	66	2048	
A066-115	289A	3138/193A	268	2038	
A066-116	195A	3048/329	46		
A066-117	237	3299	46		
A066-118	159	3025/129	82	2032	
A066-118A	159	3114/290	82	2032	
A066-119		3100	1N34AS	1123	
A066-119(1S1555)	177		300	1122	
A066-119(GE)			1N34AS	1123	
A066-119(SI)			300	1122	
A066-119(SILICON)	177				
A066-12	177	3100/519	300	1122	
A066-120	109	3087	1N34AS	1123	
A066-121	109	3090	1N34AS	1123	
A066-121	143A				
A066-122		3750	ZD-13	563	
A066-124	116	3017B/117	504A	1104	
A066-125	904				
A066-133	123A	3750/143A	ZD-13	563	
A066-143	199	3122	62	2010	
A066-199	109	3090			
A066-199(1S1555)	177				
A0666-118	290A	3114/290	269	2032	
A068-100	116	3017B/117	504A	1104	
A068-101	109	3087	1N34AS	1123	
A068-102	116	3017B/117	504A	1104	
A068-103	139A	3060	ZD-9.1	562	
A068-104	125	3033	531	1114	
A068-106	312	3448	FET-1	2035	
A068-107	312	3448	FET-1	2035	
A068-108	123A	3444	20	2051	
A068-109	159	3025/129	82	2032	
A068-109A	159	3114/290	82	2032	
A068-111	108	3452	86	2038	
A068-112	108	3452	86	2038	
A068-113	123A	3444	20	2051	
A068-114	175	3026	246	2020	
A069-101	107	3039/316	11	2015	
A069-102	107	3039/316	11	2015	
A069-102/103	123A	3444	20	2051	
A069-103	107	3039/316	11	2015	
A069-104	123A	3444	20	2051	
A069-104/106	123A	3444	20	2051	
A069-105	102A	3004	53	2007	
A069-106	123A	3444	20	2051	
A069-107	158	3004/102A	53	2007	
A069-109	109	3087	1N34AS	1123	
A069-111	110MP		1N60	1123	
A069-112	116	3311	504A	1104	
A069-114	107	3039/316	11	2015	
A069-115	109	3087	1N34AS	1123	
A069-116	107	3039/316	11	2015	
A069-118	110MP	3709	1N60	1123(2)	
A069-119	107	3039/316	11	2015	
A069-120	123A	3444	20	2051	
A069-121	126	3006/160	52	2024	
A069-122	123A	3444	20	2051	
A07	109	3087	1N34AS	1123	
A072133-001	OBS-NLA				
A08	506	3017B/117	300	1114	
A08(I.C.)	7430				
A08-1050115	181	3036	75	2041	
A08-105018	130		14	2041	
A09	7440				
A090	109	3087	1N34AS	1123	
A1-100-263-001	181				
A1-100-263-002	181				
A1-101-420-015	5072A				
A1-3	126	3007	52	2024	
A1-44				2041	
A1-ANA-AOM-003	116				
A1-DHF-R00-148	519				
A1-DHG-EAA-006	109				
A1-DHL-EAA-009	3024				
A1-DHS-LAA-001	116				
A1-DHS-LAA-003	116				
A1-DHS-LAA-008	5802				
A1-DHS-LAA-010	125				
A1-DHS-LAA-050	519				
A1-DHS-LAA-105	116				
A1-DHS-LAA-AAA	519				
A1-DHS-LBA-002	116				
A1-DHS-LBA-006	5804				
A1-DHS-LDA-002	116				
A1-DHS-LDA-025	116				
A1-DHS-LHA-003	156				
A1-DHS-SAA-013	5986				
A1-DHZ-LAA-001	137A				
A1-DHZ-LAA-020	137A				
A1-DHZ-LAA-021	5072A				
A1-DHZ-LAA-022	5066A				
A1-DHZ-LAA-024	142A				
A1-DHZ-LAA-025	5071A				
A1-DHZ-LAA-027	5086A				
A1-DHZ-LAA-028	5131A				
A1-DHZ-LAA-029	5081A				
A1-DHZ-LAA-033	135A				
A1-DHZ-LAA-037	5123A				
A1-DHZ-LAA-038	5096A				
A1-DHZ-LAA-040	145A				
A1-DHZ-LAA-041	140A				
A1-DHZ-LAA-042	145A				
A1-DHZ-LAA-043	5120A				
A1-DHZ-LAA-045	5088A				
A1-DHZ-LAA-049	5080A				
A1-DHZ-LAA-050	5138A				
A1-DHZ-LAA-055	137A				
A1-DHZ-LAA-056	136A				
A1-DHZ-LAA-058	139A				
A1-DHZ-LAA-060	5130A				
A1-DHZ-LAA-061	5150A				
A1-DHZ-LAA-062	137A				
A1-DHZ-LAA-063	5075A				
A1-DHZ-LAA-066	138A				
A1-DHZ-SAA-001	5188A				
A1-ICA-MAW-001	906				
A1-ICA-MDL-002	987				
A1-ICA-MDR-001	778A				
A1-ICA-MHN-001	941M				
A1-ICA-MLE-004	909				
A1-ICA-MLR-008	941				
A1-ICC-AAL-002	834				
A1-ICD-CLM-002	74C14				
A1-ICD-HFL-001	9668				
A1-ICT-MDL-001	978				
A1-ICT-MLR-001	955M				
A1-ICV-RLG-005	960				
A1-ICV-RLG-006	960				
A1-ICV-RLG-024	972				
A1-ICV-RLQ-005	923				
A1-ICV-RLS-007	923D				
A1-ICY-RLG-004	967				
A1-RES-IA0-028	116				
A1-TSC-H16-003	2013				
A1-TSD-A00-018	184				
A1-TSF-EP0-001	326				
A1-TSS-P00-002	284				
A1-TSS-P00-006	181				
A1-TSS-P00-008	284				
A1-TSS-P00-009	128				
A1-TSS-P00-011	280				
A1-TSS-P00-018	254				
A1-TSS-P00-019	123AP				
A1-TSS-P00-021	123AP				
A1-TSS-P00-022	282				
A1-TSS-P00-024	196				
A1-TSS-P00-025	197				
A1-TSS-P00-026	130				
A1-TSS-P00-027	175				
A1-TSS-P00-031	128				
A1-TSS-P00-036	128				
A1-TSS-P00-040	128				
A1-TSS-P00-041	159				
A1-TSS-P00-043	123A				
A1-TSS-P00-044	286				
A1-TSS-P00-046	129				
A1-TSS-P00-052	102				
A1-TSS-PP0-006	281				
A1-TSS-PP0-007	185				
A1-TSS-PP0-015	129				
A1-TSS-PP0-016	159				
A1-TSS-PP0-017	129				
A1-TSS-PP0-018	129				
A1-TSS-PP0-019	197				
A1-TSS-SND-321	195A				
A1-TSS-SND-440	396				
A1-TSS-SNM-222	123A				
A1-TSS-SNM-223	102A				
A1-TSS-SNR-400	123AP				
A1-TSS-SNR-403	123AP				
A1-TSS-SPD-234	129				
A1-TSS-SPD-415	397				
A1-TSS-SPM-907	159				
A1-TSS-SP0-001	159				
A1-TSS-SPR-001	159				
A1-TSS-WU4-990	6404				
A1-TSU-N00-002	6402				
A10	74H40	9003/5800			
A100			504A	1104	1N4002
A100(RECTIFIER)	116		504A	1104	
A100(TRANSISTOR)	126		52	2024	
A1000	125	3080	531	1114	1N4007
A1004	159				
A1004C	159				
A1004D	159				
A1007	88				
A1007A	88				

Industry Standard No.	ECG	SK	GE	RS 276-	MOTOR.
A1008	292				
A1008K	292				
A1008L	292				
A1008M	292				
A100A	126	3004/102A	52	2024	
A100B	126	3004/102A	52	2024	
A100C	126	3004/102A	52	2024	
A101	126	3007	52	2024	
A101-A			504A	1104	
A101-A(RECT.)	116		504A	1104	
A1010	332				
A10105	116		504A	1104	
A1010K	332				
A1010L	332				
A1010M	332				
A1010V	332				
A1011	398				
A10113	116		504A	1104	
A10118	116		504A	1104	
A1011D	398				
A1011E	398				
A1012	153$				
A1012-O	153$				
A10142	116	3017B/117	504A	1104	
A1015	290A				
A1015-O	290A				
A1015-GR	290A				
A1015-O	290A				
A1015-Y	290A				
A1016	129	3025	244	2027	
A10164	116	3031A	504A	1104	
A10165	116	3031A	504A	1104	
A10169	116	3311	504A	1104	
A1016K	234				
A1017	234				
A1018	288				
A1019	91				
A101A(RECTIFIER)	116		504A	1104	
A101A(TRANSISTOR)	126		52	2024	
A101AA	126	3006/160	52	2024	
A101AY	126	3006/160	52	2024	
A101B	126	3006/160	52	2024	
A101BA	126	3006/160	52	2024	
A101BB	126	3006/160	52	2024	
A101BC	126	3006/160	52	2024	
A101BX	126	3006/160	52	2024	
A101C	126	3006/160	52	2024	
A101CA	126	3006/160	52	2024	
A101CV	126	3006/160	52	2024	
A101CX	126	3006/160	52	2024	
A101E	126	3006/160	52	2024	
A101QA	126	3006/160	52	2024	
A101V	126		52	2024	
A101X	126	3007	52	2024	
A101Y	126	3006/160	52	2024	
A101Z	126	3005	52	2024	
A102	126	3007	52	2024	
A102(JAPAN)	126		52	2024	
A1021	374				
A1021-O	374				
A1021-R	374				
A1021-Y	374				
A1024	112		1N82A		
A1028	281$				
A1029	290A				
A1029B	290A				
A1029C	290A				
A1029D	290A				
A102A	126	3007	52	2024	
A102AA	126	3007	52	2024	
A102AB	126	3007	52	2024	
A102B	126		52	2024	
A102BA	126	3007	52	2024	
A102BN	126	3007	52	2024	
A102CA	126	3007	52	2024	
A102TV	126	3007	52	2024	
A103	126	3007	52	2024	
A1030	290A	3114/290	82	2032	
A1030B	290A				
A1030C	290A				
A1031(ECB)	159				
A1031B(ECB)	159				
A1031C(ECB)	159				
A1031D(ECB)	159				
A1032(ECB)	159				
A1032B(ECB)	159				
A1032C(ECB)	159				
A1033	159				
A1033B	159				
A1033C	159				
A1033D	159				
A103A	126	3006/160	52	2024	
A103B	126	3006/160	52	2024	
A103C	126	3006/160	52	2024	
A103CA	126	3007	52	2024	
A103CAK	126		52	2024	
A103CG	126	3006/160	52	2024	
A103DA	126	3006/160	52	2024	
A104	126	3007	52	2024	
A104(JAPAN)	126		52	2024	
A104-8	225	3045	256		
A1040	290A				
A1040-GR	290A				
A1040-O	290A				
A1040-Y	290A				
A1041	1013				
A1047	374				
A1047D	374				
A1047E	374				
A1047F	374				
A1049	290A$				
A1049-GR	290A$				
A104A	126	3124/289	52	2024	
A104B	126	3124/289	52	2024	
A104D	126	3006/160	52	2024	
A104P	126	3006/160	52	2024	
A104Y	126	3124/289	52	2024	
A105	126	3006/160	245	2004	
A1050A	285				
A1050A-O	285				
A1050A-R	285				
A1050A-Y	285				
A1051A	285				
A1051A-O	285				
A1051A-R	285				
A1051A-Y	285				
A106	126	3444/123A	20	2051	
A106(JAPAN)	123A	3444	20	2051	
A1060	391				
A1061	391				
A1062	391$				
A1063	285				
A1064	285				
A1065	285				
A107	126	3024/128	243	2030	
A107(JAPAN)	160		245	2004	
A1072	281				
A1073	285				
A1074	285				
A1075	93				
A1076	93				
A108	123A	3444	20	2051	
A108(JAPAN)	160		245	2004	
A1086			10	2051	
A1087			10	2051	
A108A	123A	3444	20	2051	
A108B	123A	3444	20	2051	
A109	126	3006/160	52	2024	
A1090	159				
A1090-GR	159				
A1090-O	159				
A1090-Y	159				
A1092	234				
A1094	93				
A10940	93				
A1094R	93				
A1094Y	93				
A1095	93				
A10950	93				
A1095R	93				
A1095Y	93				
A1096	185#				
A109C	909D				
A10A	116	3051/156	512		
A10B	116	3051/156	512		
A10C	116	3031A	504A	1104	
A10D	116	3017B/117	504A	1104	
A10E	116	3017B/117	504A	1104	
A10G05-010-A	123A	3444	20	2051	
A10G05-011-A	123A	3444	20	2051	
A10G05-015-D	123A	3444	20	2051	
A10M	116	3051/156	504A	1104	
A10N	125	3032A	504A	1104	
A10P	125		531	1114	
A11	178MP		504A	1104	
A110	199	3156/172A	62	2010	
A110(JAPAN)	199		62	2010	
A1100(BCE)	290A				
A1100D(BCE)	290A				
A1100E(BCE)	290A				
A1100F(BCE)	290A				
A1100L(BCE)	159				
A1100LE(BCE)	159				
A1100LF(BCE)	159				
A1102	391				
A1103	391				
A1109	108	3452	86	2038	
A110C	910D				
A111	123A	3444	20	2051	
A111(JAPAN)	126		52	2024	
A1110	374#				
A1111	1002	3481			
A1111P	1002	3481			
A1115	290A				
A11159761	519			1122	
A1115D	290A				
A1115E	290A				
OA111SE	5074A				
A111T2	154	3045/225	40	2012	
A112	126		52	2024	
A112-000172	159	3114/290	82	2032	
A112-000185	159	3114/290	82	2032	
A112-000187	159	3114/290	82	2032	
A11200482	159		82	2032	
A1123	288				
A112363	130	3027	14	2041	
A1124C	121	3009	239	2006	
A1128	290A				
A113	126	3006/160	52	2024	
A114	126				
A114(TRANSISTOR)	102A		53	2007	
A11414257	123A		20	2051	
A114A			A114A		1N4934
A114B					1N4935
A114C	125	3031A	531	1114	1N4936
A114D					1N4936
A114E					1N4937
A114F					1N4933
A114M					1N4937
A114N					MR817
A114PD2					MR1-1400
A115	126	3124/289	243	2030	
A115(JAPAN)	102A		53	2007	
A1150	290A				
A1150-O	290A				
A1150-Y	290A				
A115A					MR851
A115B					MR852
A115C					MR854
A115D					MR854
A115E					MR856
A115F					MR850
A115M			A115M		MR856
A116	126	3444/123A	20	2051	
A116(JAPAN)	102A		53	2007	
A116078	159	3114/290	82	2032	
A116081	154	3045/225	40	2012	
A116084	129	3025	244	2027	
A116284	129	3025	244	2027	
A117	126	3006/160	245	2004	
A1170	108	3452	86	2038	
A11744	312	3448	FET-1	2035	
A11745	312	3448	FET-1	2035	
A118	126	3006/160	245	2004	
A118284	159	3114/290	82	2032	
A119	129				
A119730	159	3114/290	82	2032	
A119983	129	3025	244	2027	
A12	102A	3003	53	2007	
A12(I.C.)	7451			1825	
A12-1	225	3045	256		
A12-1-70	123A	3444	20	2051	
A12-1-705	123A	3444	20	2051	
A12-1A9G	123A	3444	20	2051	
A12-2	225	3045	256		
A120	129				
A1200	1003	3288	IC-43		
A1201	1003	3288	IC-43		
A1201B	1003	3288	IC-43		
A1201C	1003	3288	IC-43		
A1201C-W		3288	IC-43		
A1201T	1003	3288	IC-43		
A1202	1003				
A120A	129				
A120P1	159	3114/290	82	2032	
A121	126		52	2024	
A121-1	159	3114/290	82	2032	
A121-1410	101	3861	8	2002	

Industry Standard No.	ECG	SK	GE	RS 276-	MOTOR.
A121-15	101	3861	8	2002	
A121-16	101	3861	8	2002	
A121-17	101	3861	8	2002	
A121-1RED	159	3114/290	82	2032	
A121-21	101	3861	8	2002	
A121-361	154	3045/225	40	2012	
A121-444	159	3114/290	82	2032	
A121-446	159	3114/290	82	2032	
A121-480	108	3452	86	2038	
A121-495	159	3114/290	82	2032	
A121-496	159	3114/290	82	2032	
A121-497	159	3114/290	82	2032	
A121-497WHT	159	3114/290	82	2032	
A121-50	101	3861	8	2002	
A121-585	107	3039/316	11	2015	
A121-585B	107	3039/316	11	2015	
A121-602	159	3114/290	82	2032	
A121-603	159	3114/290	82	2032	
A121-679	159	3114/290	82	2032	
A121-687	107	3039/316	11	2015	
A121-699	159	3114/290	82	2032	
A121-746	159	3114/290	82	2032	
A121-762	101	3861	8	2002	
A121-774	159	3114/290	82	2032	
A1210C-W	1003				
A1214	159	3114/290	82	2032	
A121467	159	3114/290	82	2032	
A12163	121	3009	239	2006	
A121659	159	3114/290	82	2032	
A12178	121	3009	239	2006	
A122	160	3007	245	2004	
A122(JAPAN)	126		52	2024	
A122-1962	103A	3010	59	2002	
A1220	160	3006	245	2004	
A122GRN	159	3114/290	82	2032	
A122YEL	159	3114/290	82	2032	
A123	126		52	2024	
A123-7	116		504A	1104	
A1238	199	3245	62	2010	
A124	126	3006/160	52	2024	
A124047	159	3114/290	82	2032	
A1243	100	3005	1	2007	
A124623	108	3452	86	2038	
A124624	108	3452	86	2038	
A124755	159	3114/290	82	2032	
A125	126	3006/160	52	2024	
A125329	108	3452	86	2038	
A12546	154	3045/225	40	2012	
A12594	159	3114/290	82	2032	
A126	126	3006/160	245	2004	
OA12612	142A	3062	ZD-12	563	
A126524	159	3114/290	82	2032	
A126700	159	3114/290	82	2032	
A126705	154	3045/225	40	2012	
A126707	159	3114/290	82	2032	
A126715	159	3114/290	82	2032	
A126718	159	3114/290	82	2032	
A126719	159	3114/290	82	2032	
A126724	129	3025	244	2027	
A127		3006	245	2004	
A127-7	101	3861	8	2002	
A127364	176	3845	80		
A127712	154	3045/225	40	2012	
A128	123A	3444	20	2051	
A128(JAPAN)	158		53	2007	
A12888	159	3114/290	82	2032	
A128A	123A	3444	20	2051	
A129(TRANSISTOR)	158		53	2007	
A129-30	101	3861	8	2002	
A129-34	159	3114/290	82	2032	
A129509	108	3452	86	2038	
A129510	108	3452	86	2038	
A129511	108	3452	86	2038	
A129512	108	3452	86	2038	
A129513	108	3452	86	2038	
A129571	108	3452	86	2038	
A129572	108	3452	86	2038	
A129573	108	3452	86	2038	
A129574	108	3452	86	2038	
A129697	159	3114/290	82	2032	
A129699	159	3114/290	82	2032	
A129E					MR1376
A129M					MR1376
A12A	102A		53	2007	
A12B	102A	3004	53	2007	
A12C	102A		53	2007	
A12D	102A	3004	53	2007	
A12H	102A		53	2007	
A12V	102A		53	2007	
A13	102A		504A	1104	
A13(DIODE)	177				
A13(RECTIFIER)	116		504A	1104	
A13(TRANSISTOR)	160		245	2004	
A13-0032	130	3027	14	2041	
A13-14604-1A	121MP	3718	239(2)	2006(2)	
A13-14604-1B	121MP	3718	239(2)	2006(2)	
A13-14604-1C	121MP	3718	239(2)	2006(2)	
A13-14604-1D	121MP	3718	239(2)	2006(2)	
A13-14604-1E	121MP	3718	239(2)	2006(2)	
A13-14777-1	121MP	3718	239(2)	2006(2)	
A13-14777-1A	121MP	3718	239(2)	2006(2)	
A13-14777-1B	121MP	3718	239(2)	2006(2)	
A13-14777-1C	121MP	3718	239(2)	2006(2)	
A13-14777-1D	121MP	3718	239(2)	2006(2)	
A13-14778-1A	121MP	3718	239(2)	2006(2)	
A13-14778-1B	121MP	3718	239(2)	2006(2)	
A13-14778-1C	121MP	3718	239(2)	2006(2)	
A13-14778-1D	121MP	3718	239(2)	2006(2)	
A13-15809-1	225	3045	256		
A13-17918-1	130	3027	14	2041	
A13-22741-2	121MP	3013	239(2)	2006(2)	
A13-23594-1	130	3027	14	2041	
A13-28432-1	225	3045	256		
A13-33188-2	130	3027	14	2041	
A13-55062-1	225	3045	256		
A13-86416-1	103A	3010	59	2002	
A13-86420-1	101	3861	8	2002	
A13-87433-1	101	3861	8	2002	
A130	154	3039/316	40	2012	
A130(JAPAN)	160		245	2004	
A130-149	159	3114/290	82	2032	
A130-40315	159	3114/290	82	2032	
A130-40429	159	3114/290	82	2032	
A130-ORN	154		40	2012	
A130-V10	154		40	2012	
A130139	159	3114/290	82	2032	
A1300RN	108	3039/316	86	2038	
A130V100	108		86	2038	
A131	126	3006/160	52	2024	
A1310(IC)	1230				
A1314	128	3024	243	2030	
A132	116	3017B/117	504A	1104	
A132(JAPAN)	126		52	2024	
A132(TRANSISTOR)	154		40	2012	
A132-1	116	3016	504A	1104	
A133	194	3275	220		
A133(JAPAN)	126		52	2024	
A134	126	3006/160	52	2024	
A1341	128	3024	243	2030	
A1342	710		IC-89		
A135	160		245	2004	
A1353	1080	3284			
A136	126		52	2024	
A1364	1004				
A1364N	1004	3365	IC-149		
A1365	712		IC-2		
A13658E	712		IC-2		
A13658E3	712				
A1365P	712				
A1367	1159	3290			
A1368	738	3167	IC-29		
A1369(IC)	1178	3480			
A137	123A	3444	20	2051	
A137(JAPAN)	126		52	2024	
A137(NPN)	123A	3444	20	2051	
A137(PNP)	159		82	2032	
A1373	1094		IC-157		
A1374	1121				
A1375	1122		IC-40		
A1376	1122		IC-40		
A1377	160	3006	245	2004	
A1378	160	3123	245	2004	
A1379	123A	3444	20	2051	
A138	199	3245	62	2010	
A138(JAPAN)	102A		53	2007	
A1380	123A	3444	20	2051	
A1383	126	3006/160	52	2024	
A1384	126	3123	52	2024	
A139	199	3245	62	2010	
A139(JAPAN)	126		52	2024	
A1396	101	3861	8	2002	
A139E	5994				MR1386
A139ER	5995				
A139M					MR1386
A13A2	116	3017B/117	504A	1104	
A13AA2	116	3016	504A	1104	
A13B2	116	3016	504A	1104	
A13C2	116	3031A	504A	1104	
A13D2	116	3017B/117	504A	1104	
A13E2	116	3017B/117	504A	1104	
A13F2	116	3016	504A	1104	
A13M2	116	3032A	531	1114	
A13N1	123A	3444	20	2051	
A14	102A				
A14(TRANSISTOR)	160		245	2004	
A14-1001	126	3006/160	52	2024	
A14-1002	126	3006/160	52	2024	
A14-1003	126	3006/160	52	2024	
A14-1004	102A	3123	53	2007	
A14-1005	102A	3123	53	2007	
A14-1006	102A	3123	53	2007	
A14-1007	102A	3123	53	2007	
A14-1008	102A	3123	53	2007	
A14-1009	102A	3123	53	2007	
A14-1010	102A	3123	53	2007	
A14-586-01	121	3009	239	2006	
A14-601-10	130	3027	14	2041	
A14-601-12	130	3027	14	2041	
A14-601-13	130	3027	14	2041	
A14-602-19	225	3045	256		
A14-602-36	225	3045	256		
A14-602-37	225	3045	256		
A14-602-63	108	3452	86	2038	
A14-603-05	108	3452	86	2038	
A14-603-06	108	3452	86	2038	
A1401-0	77				
A1401-2	316				
A1401-3	316				
A1401-4	316				
A14010-0	316				
A1402-0	278				
A1403-0	77				
A1405-0	76				
A1409	154	3044	40	2012	
A141	102A	3444/123A	20	2051	
A141(JAPAN)	126		52	2024	
A141(NPN)	123A				
A141(PNP)	126				
A1412-1	123A	3444	20	2051	
A1414A	121	3009	239	2006	
A1414A9	121	3009	239	2006	
A1418				2038	
A141B	126		52	2024	
A141C	126		52	2024	
A142	102A	3005	20	2051	
A142(JAPAN)	126		52	2024	
A1423MH2AB1					MR1-1200
A1425MH2AB1					MR1-1200
A142A	126		52	2024	
A142B	126		52	2024	
A142C	126		52	2024	
A143	102A	3444/123A	20	2051	
A143(JAPAN)	126		52	2024	
A144	126		52	2024	
A144A-1	121	3009	239	2006	
A144A-19	121	3009	239	2006	
A144C	126		52	2024	
A145	126		52	2024	
A145A	126		52	2024	
A145C	126		52	2024	
A146	126	3721/100	1	2007	
A1460	199	3245	62	2010	
A1462	108	3452	86	2038	
A1462-19				2004	
A1465-1	160	3006	245	2004	
A1465-19	160	3006	245	2004	
A1465-29	103A	3010	59	2002	
A1465-4	101	3861	8	2002	
A1465-49	101	3861	8	2002	
A1465A	160	3007	245	2004	
A1465A9	160	3007	245	2004	
A1465B	160	3008	245	2004	
A1465B9	160	3008	245	2004	
A1465C	121	3009	239	2006	
A1465C9	121	3009	239	2006	
A1466-1	102A	3003	53	2007	
A1466-19	102A	3003	53	2007	
A1466-19-	102A			2007	
A1466-2	102A	3004	53	2007	
A1466-29	102A	3004	53	2007	
A1466-3	102A	3004	53	2007	
A1466-39	102A	3004	53	2007	
A14665-2	103A	3010	59	2002	
A146B-3	121MP	3015	239(2)	2006(2)	
A146B-39	121MP	3015	239(2)	2006(2)	
A147	126	3721/100	1	2007	
A1471114-1	159	3114/290	82	2032	
A1472-19	123		20	2051	

Industry Standard No.	ECG	SK	GE	RS 276-	MOTOR.
A1473549-1	129	3025	244	2027	
A1473563-1	159	3114/290	82	2032	
A1473570-1	159	3114/290	82	2032	
A1473574-1	159	3114/290	82	2032	
A1473590-1	159	3114/290	82	2032	
A1473591-1	159	3114/290	82	2032	
A1473597-1	159	3114/290	82	2032	
A1473616-1	129	3025	244	2027	
A1474-3	100	3005	1	2007	
A1474-39	100	3005	1	2007	
A14743	175	3131A/369	246	2020	
A1477A	105	3012	4		
A1477A9	105	3012	4		
A1477B	121MP	3013	239(2)	2006(2)	
A1477B9	121MP	3013	239(2)	2006(2)	
A1477C	121	3014	239	2006	
A1477C9	121	3014	239	2006	
A148	126	3721/100	1	2007	
A1484A	121	3009	239	2006	
A1484A9	121	3009	239	2006	
A1488B	160	3006	245	2004	
A1488B9	160	3006	245	2004	
A1488C	100	3005	1	2007	
A1488C9	100	3005	1	2007	
A148P-2	102A	3004	53	2007	
A148P-29	102A	3004	53	2007	
A148P2	102A		53	2007	
A148P2-29	102A		53	2007	
A149	126	3721/100	1	2007	
A1492-1	105	3012	4		
A1492-19	105	3012	4		
A149L-4	102A	3003	53	2007	
A149L-49	102A	3003	53	2007	
A14A		3017B	A14A		1N4002
A14A-70	121	3009	239	2006	
A14A-705	121	3009	239	2006	
A14A10G	121	3717	239	2006	
A14A8-1	160	3007	245	2004	
A14A8-19	160	3007	245	2004	
A14A8-196	160		245	2004	
A14B	116		504A	1104	1N4003
A14C	116	3017B/117	504A	1104	1N4004
A14D	116		504A	1104	1N4004
A14E	116	3017B/117	504A	1104	1N4005
A14E2	116		504A	1104	
A14F	116	3051/156	504A	1104	1N4001
A14M	116		504A	1104	1N4005
A14N	125		512		1N4006
A14P	125	3033	512		1N4007
A14PD					MR1-1400
A14PD1	506			1114	
A14PD2	506			1114	
A14PD3	506			1114	
A15	7474			1818	
A15(TRANSISTOR)	102A		53	2007	
A15-1001	126	3006/160	52	2024	
A15-1002	126	3006/160	52	2024	
A15-1003	126	3006/160	52	2024	
A15-1004	102A	3123	53	2007	
A15-1005	102A	3123	53	2007	
A15-1007	109	3087	1N34AS	1123	
A15-1008	116	3017B/117	504A	1104	
A15-12	338				2N5848
A150	126		52	2024	
A151	126	3444/123A	20	2051	
A151(JAPAN)	160		245	2004	
A1518	108	3452	86	2038	
A1519	108	3452	86	2038	
A152	126	3444/123A	20	2051	
A152(JAPAN)	160		245	2004	
A1520	108	3452	86	2038	
A1521	108	3452	86	2038	
A153	123A	3444	20	2051	
A153(JAPAN)	126			2024	
A154			61	2038	
A154(NPN)	108		86	2038	
A154(PNP)	126		52	2024	
A155			61	2038	
A155(NPN)	108		86	2038	
A155(PNP)	126		52	2024	
A1558-17	159	3114/290	82	2032	
A156	123A	3444	20	2051	
A156(JAPAN)	126		52	2024	
A1565	228A				
A1567	123A	3444	20	2051	
A1567-1	123A	3444	20	2051	
A157	123A	3444	20	2051	
A157(JAPAN)	126		52	2024	
A157A	123A	3444	20	2051	
A157B	123A	3444	20	2051	
A157C	123A	3444	20	2051	
A158	123A	3444	20	2051	
A158A	123A	3444	20	2051	
A158B	123A	3444	20	2051	
A158C	123A	3444	20	2051	
A159	123A	3444	20	2051	
A15927	121	3009	239	2006	
A159A	123A	3444	20	2051	
A159B	123A	3444	20	2051	
A159C	123A	3444	20	2051	
A15A	5801	9004	A15A	1142	MR501
A15B	5802	9005		1143	MR502
A15BK	102A		53	2007	
A15BL	102A		53	2007	
A15BLU	102A		53	2007	
A15C	5803	9006		1144	MR504
A15D	5804	9007		1144	MR504
A15E	5805	9008			MR506
A15F	5800	9003	A15F	1142	MR501
A15H	102A		53	2007	
A15K	102A		53	2007	
A15M	5806	3848			MR506
A15N	5808	9009			MR508
A15R	102A		53	2007	
A15U(DIODE)	5800	9003		1142	
A15U(RECTIFIER)				1141	
A15U(TRANSISTOR)	102A		53	2007	
A15V	102A		53	2007	
A15VR	102A		53	2007	
A15Y	102A		53	2007	
A16	102A		53	2007	
A160	126	3118	82	2032	
A160(JAPAN)	100	3721	1	2007	
A161	159	3118	82	2032	
A162	159	3118	82	2032	
A163	160	3006	245	2004	
A164			61	2038	
A164(NPN)	108		86	2038	
A164(PNP)	126		52	2024	
A165			61	2038	
A165(NPN)	108		86	2038	
A165(PNP)	126		52	2024	
A166	126		52	2024	
A167	100		1	2007	
A168	123A	3444	20	2051	
A168(JAPAN)	100	3721	1	2007	
A168A	100	3721	1	2007	
A168P1	121MP	3013	239(2)	2006(2)	
A169	100	3005	1	2007	
A16A1	103A	3010	59	2002	
A16A2	103A	3010	59	2002	
A17	102A		53	2007	
A17(I.C.)	7490			1808	
A170	129	3025	244	2027	
A170(JAPAN)	100	3721	1	2007	
A171	159	3025/129	82	2032	
A171(JAPAN)	100	3721	1	2007	
A172	100	3005	1	2007	
A172A	100	3005	1	2007	
A173	102A	3005	53	2007	
A173B	102A	3005	53	2007	
A174	102	3004	2	2007	
A1746711	177				
A175	126	3024/128	52	2024	
A176	126		52	2024	
A176-025-9-002	108	3452	86	2038	
A177	159		82	2032	
A177(A)	159		82	2032	
A177A	159	3114/290	82	2032	
A177AB	159	3114/290	82	2032	
A178A	159	3114/290	82	2024	
A178AB	159	3114/290	82	2032	
A178B	159	3114/290	82	2024	
A178BA	159	3114/290	82	2032	
A179A	159	3114/290	82	2024	
A179AC	159	3114/290	82	2032	
A179B	159	3114/290	82	2024	
A179BB	159	3114/290	82	2032	
A179Q	159				
A17H	102A		53	2007	
A18		3027	14	2041	
A18(I.C.)	7493A				
A18-4	130	3027	14	2041	
OA180	109	3091	245	2004	
A180	126				
A180A	6354				
A180AR	6355				
A180B	6354				
A180BR	6355				
A180C	6354				
A180CR	6355				
A180D	6354				
A180DR	6355				
A180E	6356				
A180ER	6357				
A180M	6356				
A180MR	6357				
A180N	6358				
A180NR	6359				
A180P	6358				
A180PR	6359				
A180RA	6355				
A180RB	6355				
A180RC	6355				
A180RD	6355				
A180RE	6357				
A180RM	6357				
A180RN	6359				
A180RP	6359				
A180RS	6359				
A180RT	6359				
A180S	6358				
A180SR	6359				
A180T	6358				
A180TR	6359				
A181	126	3004/102A	53	2007	
A182	102A	3005	1	2007	
A1820-0203-1	941		IC-263	010	
A1820-0219-1	941M		IC-265	007	
A1821-0001-1	912		IC-172		
A1826-0007-1	941		IC-263	010	
A183	126	3004/102A	53	2007	
A1844-17	159	3114/290	82	2032	
A1853-0233-1	197		250	2027	
A1854-0533	194	3275			
A1858	101	3861	8	2002	
A1867-17	159	3114/290	82	2032	
A187TV	102A	3004	53	2007	
A188	126	3006/160	52	2024	
A188103	192		63	2030	
A1884-0209	5697	3522			
A1884-0218	5697	3522			
A189	126	3006/160	52	2024	
A18A					1N3890
A18H	102A		53	2007	
A19	126		52	2024	
A19(I.C.)	7495				
A19-020-072	108	3452	86	2038	
A1901-5338	159		82	2032	
A190429	159	3114/290	82	2032	
A190A	6354				
A190AR	6355				
A190B	6354		A190B		
A190BR	6355				
A190C	6354				
A190CR	6355				
A190D	6354				
A190DR	6355				
A190E	6356				
A190ER	6357				
A190M	6356				
A190MR	6357				
A190N	6358				
A190NR	6359				
A190P	6358				
A190PA	6104	6560			
A190PAR	6105	6561			
A190PB	6104	6560			
A190PBR	6105	6561			
A190PC	6106				
A190PCR	6107				
A190PD	6106				
A190PDR	6107				
A190PE	6106				
A190PER	6107				
A190PR	6359				
A190RA	6355				
A190RB	6355				
A190RC	6355				
A190RD	6355				
A190RE	6357				
A190RM	6357				
A190RN	6359				
A190RP	6359				
A190RS	6359				
A190RT	6359				
A190S	6358				
A190SR	6359				

Industry Standard No.	ECG	SK	GE	RS 276-	MOTOR.
A190T	6358				
A190TR	6359				
A192	312	3448	FET-1	2035	2N4416
A194	312	3112		2028	
A1946	116		504A	1104	
A195	312	3112		2028	
A196	312	3112		2028	
A197	126	3004/102A	2	2007	
A198	102A	3721/100	1	2007	
A198794-1	101	3861	8	2002	
A1A	154	3124/289	40	2012	
A1A1	116	3017B/117	504A	1104	
A1A5	116	3017B/117	504A	1104	
A1A9	116	3017B/117	504A	1104	
A1B	123A	3444	20	2051	
A1B1	116	3016	504A	1104	
A1B5	116	3017B/117	504A	1104	
A1B9	116	3016	504A	1104	
A1C	162	3438	35		
A1C1	116	3017B/117	504A	1104	
A1C5	116	3016	504A	1104	
A1C9	116	3016	504A	1104	
A1D	163A	3439	36		
A1D1	116	3031A	504A	1104	
A1D5	116	3017B/117	504A	1104	
A1D9	116	3031A	504A	1104	
A1DI	163A	3439			
A1DJ	163A	3439	36		
A1E	108	3452	86	2038	
A1E1	116	3031A	504A	1104	
A1E5	116	3031A	504A	1104	
A1E9	116	3031A	504A	1104	
A1F	123A	3444	20	2051	
A1F1	116	3017B/117	504A	1104	
A1F5	116	3017B/117	504A	1104	
A1F9	116	3017B/117	504A	1104	
A1G	107	3018	11	2015	
A1G-1	107	3039/316	11	2015	
A1G-1A	108	3452	86	2038	
A1G1	116	3017B/117	504A	1104	
A1G5	116	3017B/117	504A	1104	
A1G9	116	3017B/117	504-A	1104	
A1H	123A	3444	20	2051	
A1H(MOTOROLA)	163A		36		
A1H1	125	3032A	531	1114	
A1H5	125	3032A	531	1114	
A1H9	125	3032A	531	1114	
A1J	123A	3444	20	2051	
A1K	108	3452	86	2038	
A1K1	125	3032A	531	1114	
A1K5	125	3032A	531	1114	
A1K9	125	3032A	531	1114	
A1L	123A	3444	20	2051	
A1M	154	3044	40	2012	
A1M-1	107		11	2015	
A1M1	125	3033	531	1114	
A1M5	125	3033	531	1114	
A1M9	125	3033	531	1114	
A1N	124	3021	12		
A1P	107	3039/316	11	2015	
A1P-/4923	108		86	2038	
A1P-1	107	3039/316	11	2015	
A1P-1A	107	3039/316	11	2015	
A1P-5	108	3452	86	2038	
A1P/4922	107	3018	11	2015	
A1P/4923	107	3039/316	11	2015	
A1P/4923-1	108	3039/316	86	2038	
A1R	107	3039/316	11	2015	
A1R-1	107	3039/316	11	2015	
A1R-1/4925	108	3452	86	2038	
A1R-1A	108	3452	86	2038	
A1R-2	107	3039/316	11	2015	
A1R-2/4926	108	3452	86	2038	
A1R-24926	108		86	2038	
A1R-2A	108	3452	86	2038	
A1R-5	108	3452	86	2038	
A1R/4924	107	3018	11	2015	
A1R/4925	107	3039/316	11	2015	
A1R/4925A	108	3452	86	2038	
A1R/4926	107	3039/316	11	2015	
A1R/4926A	108	3452	86	2038	
A1S	154	3040	40	2012	
A1T	123A	3444	20	2051	
A1T-1	123A	3444	20	2051	
A1U	161	3018	39	2015	
A1U(3RDIF)	233		210	2009	
A1U(LAST IF)	233			2009	
A1U(LASTIF)			210	2051	
A1V	123A	3444	20	2051	
A1VE	123A	3444	20	2051	
A1W	123A	3444	20	2051	
A1Y	196	3026	241	2020	
A1Z	233	3018	210	2009	
A20	126		52	2024	
A200	116	3311	504A	1104	
A200(REACH)	724		IC-86		
A200-052	159	3114/290	82	2032	
A20030703-0702	159	3114/290	82	2032	
A2006681-95	108	3452	86	2038	
A201	126		28	2017	
A201-0	126		52	2024	
A2019ZC	123A	3444	20	2051	
A201A	126		52	2024	
A201B	126		52	2024	
A201E	126		52	2024	
A201N	126				
A201TVO			52	2024	
A201TVO	126			2024	
A202	126	3005	52	2024	
A202(JAPAN)	126		52	2024	
A202A	126	3005	52	2024	
A202B	126		52	2024	
A202C	126	3005	52	2024	
A202D	126		52	2024	
A203	102A	3007	52	2024	
A203(NPN)	186		28	2017	
A203(PNP)	102A		53	2007	
A20371	109	3087	1N34AS	1123	
A20372	123A	3444	20	2051	
A2039-2	101	3861	8	2002	
A203A	126		52	2024	
A203AA	102A		53	2007	
A203AA(PNP)	126		52	2024	
A203B	102A	3007	53	2007	
A203B(PNP)	126		52	2024	
A203P	102A		53	2007	
A203P(PNP)	126		52	2024	
A204	100	3721	1	2007	
A205	100	3721	1	2007	
A2057013-0004		3114	82	2032	
A2057013-004	159				
A2057013-0701	159	3114/290	82	2032	
A2057013-0702	159	3114/290	82	2032	
A2057013-0703	159	3114/290	82	2032	
A2057A2-198	159	3114/290	82	2032	
A2057B104-8	154	3045/225	40	2012	
A2057B106-12	159	3114/290	82	2032	
A2057B108-6	159	3114/290	82	2032	
A2057B110-9	129	3025	244	2027	
A2057B112-9	129	3025	244	2027	
A2057B114-9	129	3025	244	2027	
A2057B115-9	129	3025	244	2027	
A2057B116-9	129	3025	244	2027	
A2057B121-9	129	3025	244	2027	
A2057B122-9	129	3025	244	2027	
A2057B145-12	129	3025	244	2027	
A2057B163-12	129	3025	244	2027	
A2057B2-115	107	3039/316	11	2015	
A206	100	3721	1	2007	
A207	100	3721	1	2007	
A208	176		28	2017	
A208(JAPAN)	100	3721	1	2007	
A208(NPN)	186		28	2017	
A209	176	3721/100	1	2007	
A2090	154	3045/225	40	2012	
A2090056-1	121	3009	239	2006	
A2090056-27	121	3009	239	2006	
A2090056-5	121	3009	239	2006	
A2091859-0025	121	3009	239	2006	
A2091859-0720	121	3009	239	2006	
A2091859-10	121	3009	239	2006	
A2091859-11	121	3009		2006	
A2092418	101	3011A	8	2002	
A2092418-0711	101	3861	8	2002	
A2092693-0724	108	3452	86	2038	
A2092693-0725	108	3452	86	2038	
A20K	159	3114/290	82	2032	
A20KA	159	3114/290	82	2032	
A21	126		52	2024	
A210	176	3039/316	28	2017	MRF517
A210(JAPAN)	100	3721	1	2007	
A211			28	2017	
A211(JAPAN)	100	3721	1	2007	
A212	100	3721	1	2007	
A21268	138A	3059	ZD-7.5		
A213	126		52	2024	
A214	126		52	2024	
A215	126		245	2004	
A216	126		52	2024	
A217	100	3721	1	2007	
A218	160		245	2004	
A218012DS	121	3009	239	2006	
A219	126		245	2004	
A220	126				
A220(TRANSISTOR)	160		245	2004	
A22008	159	3114/290	82	2032	
A221	126	3007	245	2004	
A222	126	3007	245	2004	
A223	126		245	2004	
A224	160		245	2004	
A225	126		245	2004	
A225(OVEN RECT)	543				
A225(TRANSISTOR)	160				
A226	160		245	2004	
A227	160		245	2004	
A228	160		245	2004	
A229	160		245	2004	
A22B	156	3051	512		
A22B1	156	3051	512		
A22D	156	3051	512		
A22D1	156	3051	512		
A22M	156	3051	512		
A23	116	3311	504A	1104	
A230	160		245	2004	
A231	176		53	2007	
A2311	234		65	2050	
A23114050	159	3114/290	82	2032	
A23114051	159	3114/290	82	2032	
A23114130				2012	
A23114550	159	3114/290	82	2032	
A232	176		53	2007	
A233	160		245	2004	
A2332	161	3716	39	2015	
A233A	160		245	2004	
A233B	160		245	2004	
A233C	160		245	2004	
A234	160	3007	245	2004	
A234A	160		245	2004	
A234B	126	3006/160	52	2024	
A234C	126	3006/160	52	2024	
A235	160		245	2004	
A235A	126	3006/160	52	2024	
A235B	160		245	2004	
A235C	126	3006/160	52	2024	
A236	126		52	2024	
A237	126		52	2024	
A2370773	199	3245	62	2010	
A238	126$		52	2024	
A239	160		245	2004	
A23B	5834		5048		
A23B1	5834		5048		
A24	108	3452	86	2038	
A240	160		245	2004	
A240A	160		245	2004	
A240B	160		245	2004	
A240B2	160		245	2004	
A240BL	160		245	2004	
A241	160		245	2004	
A2410	123A	3444	20	2051	
A24100	123A	3444	20	2051	
A2411	123A	3444	20	2051	
A2412	123A	3444	20	2051	
A2413	123A	3444	20	2051	
A2414	158		53	2007	
A2415	175	3131A/369	246	2020	
A2416	195A	3024/128	46		
A2417	163A	3439	36		
A2418	130	3027	14	2041	
A2419	109	3087	1N34AS	1123	
A242	160		245	2004	
A2420	109	3087	1N34AS	1123	
A2421	116		504A	1104	
A2422	116		504A	1104	
A2423	156		512		
A2428	159		82	2032	
A243	160		245	2004	
A2434	123A	3444	20	2051	
A244	161	3117	39	2015	
A244(AMC)	161	3716	39	2015	
A244(JAPAN)	160		245	2004	
A2448	159	3114/290	82	2032	
A245	108	3452	86	2038	
A245(AMC)	108		86	2038	
A245(JAPAN)	160		245	2004	
A2458	506		CR-2	1114	
A2459	118		CR-1		
A246	123A	3444	20	2051	
A246(AMC)	123A	3444	20	2051	

Industry Standard No.	ECG	SK	GE	RS 276-	MOTOR.
A246(JAPAN)	126		52	2024	
A2460	116	3031A	504A	1104	
A2461	116	3032A	504A	1104	
A2462	116	3032A	504A	1104	
A2464	161	3117	39	2015	
A2465	161	3117	39	2015	
A2466	123A	3444	20	2051	
A2468	123A	3444	20	2051	
A2469	123A	3444	20	2051	
A246V	126		52	2024	
A247	154	3044	40	2012	
A247(AMC)	154		40	2012	
A247(JAPAN)	126		52	2024	
A2470	123A	3444	20	2051	
A2471	128	3024	243	2030	
A2473	109	3088	1N34AS	1123	
A2474	142A	3062	ZD-12	563	
A2475	113A		6GC1		
A2476	109	3087	1N34AS	1123	
A2479	161	3018	39	2015	
A248	102A	3117	39	2015	
A248(AMC)	123A	3444	20	2051	
A248(JAPAN)	100	3721	1	2007	
A2480	161	3124/289	39	2015	
A2481	116	3031A	504A	1104	
A2482	129	3025	244	2027	
A2485	116		504A	1104	
A249	123A	3444	20	2051	
A249(AMC)	128		243	2030	
A2498	108	3452	86	2038	
A2498512	159	3114/290	82	2032	
A2499	123A	3444	20	2051	
A24MW594	108	3452	86	2038	
A24MW595	108	3452	86	2038	
A24MW596	108	3452	86	2038	
A24MW597	108	3452	86	2038	
A24T-016-016	108	3452	86	2038	
A24T-016-01L				2038	
A25-1001	126	3008	52	2024	
A25-1002	126	3008	52	2024	
A25-1003	126	3008	52	2024	
A25-1004	102A	3004	53	2007	
A25-1005	102A	3004	53	2007	
A25-1006	102A	3004	53	2007	
A25-1007	109	3087	1N34AS	1123	
A25-1008	116		504A	1104	
A25-12	338				
A25-28					MRF314
A250	126$	3004/102A	52	2024	
A251	126		52	2024	
A25114130	154	3045/225	40	2012	
A252	126		52	2024	
A253			28	2017	
A253(JAPAN)	126		52	2024	
A254	126		52	2024	
A255	126		52	2024	
A256	126		52	2024	
A257	126		52	2024	
A25762-010	154	3045/225	40	2012	
A25762-012	154	3045/225	40	2012	
A258	126		52	2024	
A259	126		52	2024	
A25A305020101	128		243	2030	
A25A509-016-101	123A		20	2051	
A26	100	3721	1	2007	
OA26	601				
A260	160		52	2024	
A261	160		52	2024	
A262	160		52	2024	
A2620	154	3045/225	40	2012	
A263	160		52	2024	
A264	160		52	2024	
A265	160		52	2024	
A2652-919	312			2028	
A266	126		52	2024	
A267	126		52	2024	
A268	126		52	2024	
A269	126		52	2024	
A27(RCA)	175		246	2020	
A270	195A	3765	46		
A270(JAPAN)	126		52	2024	
A271			28	2017	
A271(JAPAN)	126		52	2024	
A272	152	3893	66	2048	
A272(JAPAN)	126		52	2024	
A273	152	3893	66	2048	
A273(JAPAN)	126		52	2024	
A274(NPN)	195A		46		
A274(PNP)	126		52	2024	
A2746	161	3039/316	39	2015	
A275			28	2017	
A275(JAPAN)	126		52	2024	
A276			66	2048	
A276(JAPAN)	160		245	2004	
A277			66	2048	
A277(JAPAN)	100	3721	1	2007	
A278	171	3201	27		
A278(JAPAN)	100	3721	1	2007	
A279	171	3201	27		
A279(JAPAN)	100	3721	1	2007	
A2798	159	3114/290	82	2032	
A28	126		52	2024	
A280	100		52	2024	
A281	100		52	2024	
A282	102A	3721/100	1	2007	
A283	100	3721	1	2007	
A284	100	3721	1	2007	
A285	160		245	2004	
A286	160		245	2004	
A287	160		245	2004	
A288	126		52	2024	
A288A	126		52	2024	
A289	126		52	2024	
A28A	5818$				1N3890
A28B	5818$				1N3891
A28C	5820$				1N3892
A28D	5820$				1N3893
A28F	5819$				1N3889
A28F	5818$				
A29	126		52	2024	
A290	126		52	2024	
A29035-E			ZD-6.8	561	
A291(TRANSISTOR)	126		52	2024	
A292	126	3007	52	2024	
A2920	316				
A293	126		52	2024	
A294	126	3007	52	2024	
A295	126		52	2024	
A296	126		245	2004	
A297	126		245	2004	
A297074C11	129	3025	244	2027	
A297L012C01	159	3114/290	82	2032	
A297V073C01	129	3025	244	2027	
A297V073C02	129	3025	244	2027	

Industry Standard No.	ECG	SK	GE	RS 276-	MOTOR.
A297V073C03	159	3114/290	82	2032	
A297V073C04	159	3114/290	82	2032	
A297V082B03	129	3025	244	2027	
A298	160		245	2004	
A29A	5819$				
A29B	5819$				
A29C	5821$				
A29D	5821$				
A29V082B03	129		244	2027	
A2A	154	3045/225	40	2012	
A2A1	116	3017B/117	504A	1104	
A2A4	116	3017B/117	504A	1104	
A2A5	116	3017B/117	504A	1104	
A2A9	116	3017B/117	504A	1104	
A2B	123A	3444	20	2051	
A2B1	116	3016	504A	1104	
A2B4	116	3017B/117	504A	1104	
A2B5	116	3017B/117	504A	1104	
A2B9	116	3017B/117	504A	1104	
A2BRN	123A	3444	20	2051	
A2C	108	3452	86	2038	
A2C1	116	3017B/117	504A	1104	
A2C4	116	3017B/117	504A	1104	
A2C5	116	3017B/117	504A	1104	
A2C9	116	3016	504A	1104	
A2D	108	3452	86	2038	
A2D1	116	3017B/117	504A	1104	
A2D4	116	3017B/117	504A	1104	
A2D5	116	3031A	504A	1104	
A2D9	116	3031A	504A	1104	
A2E	130	3027	14	2041	
A2E-2	130	3027	14	2041	
A2E1	116	3017B/117	504A	1104	
A2E4	116	3017B/117	504A	1104	
A2E5	116	3016	504A	1104	
A2E9	116	3017B/117	504A	1104	
A2EBLK	130		14	2041	
A2EBRN	130	3027	14	2041	
A2EBRN-1	130	3027	14	2041	
A2F	108	3452	86	2038	
A2F1	116	3017B/117	504A	1104	
A2F4	116	3017B/117	504A	1104	
A2F5	116	3017B/117	504A	1104	
A2F9	116	3017B/117	504A	1104	
A2FGRN	123A	3444	20	2051	
A2G	108	3452	86	2038	
A2G1	116	3017B/117	504A	1104	
A2G4	116	3017B/117	504A	1104	
A2G5	116	3017B/117	504A	1104	
A2G9	116	3017B/117	504A	1104	
A2H	108	3452	86	2038	
A2H1	125	3032A	531	1114	
A2H4	125	3032A	531	1114	
A2H5	125	3032A	531	1114	
A2H9	125	3032A	531	1114	
A2J	123A	3444	20	2051	
A2K	154	3045/225	40	2012	
A2K1	125	3032A	531	1114	
A2K4	125	3032A	531	1114	
A2K5	125	3032A	531	1114	
A2K9	125	3032A	531	1114	
A2L	108	3452	86	2038	
A2M	107	3018	11	2015	
A2M-1	107	3018	11	2015	
A2M1	125	3033	531	1114	
A2M4	125	3033	531	1114	
A2M5	125	3033	531	1114	
A2M9	129	3033	244	2027	
A2MA	156	3033	512		
A2N	107	3018	11	2015	
A2N-1	107	3018	11	2015	
A2N-2	107	3039/316	11	2015	
A2N-2A	108	3452	86	2038	
A2N2156	213		254		
A2P	107	3039/316	11	2015	
A2P-5	108	3452	86	2038	
A2S	130	3027	14	2041	
A2S-3	130	3027	14	2041	
A2SA550P	129		244	2027	
A2SA564F	129		244	2027	
A2SA564FR	159	3025/129	82	2032	
A2SA666PQR	102A		53	2007	
A2SB240A	104	3009	16	2006	
A2SB242A	104	3009	16	2006	
A2SB248A	104	3009	16	2006	
A2SC538PQR	123A		20	2051	
A2SC538R	199	3444/123A	62	2010	
A2SD2260P	152		66	2048	
A2SD2260P			23	2020	
A2SD226PQ	175		246	2020	
A2T	108	3452	86	2038	
A2T682	103A	3010	59	2002	
A2T919	108	3452	86	2038	
A2U	124	3021	12		
A2V	108	3452	86	2038	
A2W	108	3452	86	2038	
A2Y	107	3018	11	2015	
A2Z	154	3045/225	40	2012	
A3-12					MRF403
A3-28					2N5641
A30	102A	3087	1N60	1123	
A30(DIODE)	109			1123	
A30(TRANSISTOR)	126		52	2024	
A300	116	3031A	504A	1104	1N4004
A300043-06	176	3845	80		
A301	123A	3444	20	2051	
A301(JAPAN)	126		52	2024	
A3011112	128	3024	243	2030	
A302	102A		53	2007	
A30270	159	3114/290	82	2032	
A30278	129	3025	244	2027	
A30290	159	3114/290	82	2032	
A303	102A		53	2007	
A30302	121	3009	239	2006	
A304	100		53	2007	
A305	100	3721	1	2007	
A306	123A	3444	20	2051	
A306(JAPAN)	160		245	2004	
A307	123A	3444	20	2051	
A307(JAPAN)	160		245	2004	
A308	126		52	2024	
A309	126		52	2024	
A31	102A	3721/100	1	2007	
A31-0206	108	3452	86	2038	
A310	154	3045/225	40	2012	
A310(JAPAN)	126		52	2024	
A311	102A	3024/128	243	2030	
A311(JAPAN)	100	3721	1	2007	
A312	102A	3721/100	1	2007	
A313	126		52	2024	
A314	126		52	2024	
A3148	727	3071	IC-210		
A315	126		52	2024	
A316	126		52	2024	

Industry Standard No.	ECG	SK	GE	RS 276-	MOTOR.
A3170717	154	3045/225	40	2012	
A3170757	225	3045	256		
A32	102A		53	2007	
A32-2805-50-1	128		243	2030	
A32-2809	108	3452	86	2038	
A3201(IC)	1179				
A321	102A	3452/108	86	2038	
A321(JAPAN)	160		245	2004	
A322	102A	3122	245	2004	
A322(JAPAN)	160		245	2004	
A322805-50-1	128		243	2030	
A323	123A	3444	20	2051	
A323(JAPAN)	160		245	2004	
A324	126	3444/123A	20	2051	
A324(JAPAN)	160		245	2004	
A324F	126				
A325	126		52	2024	
A326	126		52	2024	
A327(TRANSISTOR)	160		245	2004	
A327A					MR2001S
A327B					MR2002S
A327C					MR2004S
A327F					MR2000S
A329	126		52	2024	
A329A	126		52	2024	
A329B	126		52	2024	
A33	102A		53	2007	
A330	126	3721/100	1	2007	
A3300	1005	3723	IC-42		
A3301	1006	3358	IC-38		
A3301A	1006				
A331	126	3007	52	2024	
A3310	1230				
A332	160	3721/100	1	2007	
A335	126		52	2024	
A3350	1217		IC-76		
A335854	912		IC-172		
A337	126		52	2024	
A338	126		52	2024	
A339	127		52	2024	
A339H	127				
A34-6001-1	121	3009	239	2006	
A34-6002-17	121	3009	239	2006	
A340	127		245	2004	
A340H	127				
A341	126	3007	245	2004	
A341-0A	160		245	2004	
A341-0B	160	3007	245	2004	
A341-0B	160		245	2004	
A342	126		245	2004	
A342A	160		245	2004	
A343	160		52	2024	
A344	123A	3444	20	2051	
A344(JAPAN)	126		52	2024	
A345	123A	3444	20	2051	
A345(JAPAN)	160		245	2004	
A346	123A	3444	20	2051	
A346(JAPAN)	160		245	2004	
A347	160		245	2004	
A34715	121	3009	239	2006	
A348	160		245	2004	
A349	160		245	2004	
A35	102A	3040	40	2012	
A35(JAPAN)	154		40	2012	
A350	126	3008	52	2024	
A35084	121	3009	239	2006	
A350A	100	3008	1	2007	
A350C	126		52	2024	
A350H	126		52	2024	
A350R	126		52	2024	
A350T	126		52	2024	
A350TY	126		52	2024	
A350Y	126		52	2024	
A351	126		52	2024	
A3513	159	3114/290	82	2032	
A351A	126		52	2024	
A351B	126		52	2024	
A352	126	3008	52	2024	
A35201	121	3009	239	2006	
A3523	129	3025	244	2027	
A35260	121	3009	239	2006	
A352A	126		52	2024	
A352B	126		52	2024	
A353	126	3008	52	2024	
A353-9008-001	129	3025	244	2027	
A3533	129	3025	244	2027	
A3533-1	129	3025	244	2027	
A353A	126		52	2024	
A353C	126		52	2024	
A354	126	3007	52	2024	
A3540	159	3114/290	82	2032	
A3545	225	3045	256		
A3547	225	3045	256		
A3549	159	3114/290	82	2032	
A354A	126	3007	52	2024	
A354B	126	3007	52	2024	
A355	126		52	2024	
A3552	225	3045	256		
A3553	225	3045	256		
A3559	159	3114/290	82	2032	
A355A	126		52	2024	
A356	126		52	2024	
A3562	159	3114/290	82	2032	
A3563	159	3114/290	82	2032	
A357	126		52	2024	
A3574	159	3114/290	82	2032	
A358	126$	3007	52	2024	
A3581	159	3114/290	82	2032	
A3582	225	3045	256		
A359	126		52	2024	
A36	102A	3005	1	2007	
A360	160	3007	245	2004	
A3607	101	3861	8	2002	
A3609	101	3861	8	2002	
A361	160		245	2004	
A3616-1	129	3025	244	2027	
A362	160$		245	2004	
A363	160$		245	2004	
A364	126		52	2024	
A365	126		52	2024	
A36508	109	3091	1N34AS	1123	
A36539	142A	3062	ZD-12	563	
A36577	129	3025	244	2027	
A366	126		52	2024	
A367	126	3007	52	2024	
A368	126	3007	52	2024	
A36896	121	3009	239	2006	
A369	126		52	2024	
A37	126	3861/101	52	2024	
A372	160$		245	2004	
A373	102A$		245	2004	
A374	102A		53	2007	
A375	126		245	2004	
A376	160	3006	245	2004	
A376(JAPAN)	126		52	2024	
A377	160		245	2004	
A3772.01			11	2015	
A378	160	3006	245	2004	
A379	160	3006	245	2004	
A38	126	3006/160	52	2024	
A380	126	3008	52	2024	
A381	126		52	2024	
A382	126		52	2024	
A383	126		52	2024	
A384	126		52	2024	
A385	126	3008	52	2024	
A385A	126		52	2024	
A385D	126		52	2024	
A39	126	3006/160	52	2024	
A3902441	130		14	2041	
A390M	6110				
A391	100		52	2024	
A391593	181		75	2041	
A392	100		52	2024	
A393	100		52	2024	
A393A	126		52	2024	
A394	100		52	2024	
A395	100		52	2024	
A396(TRANSISTOR)	102		2	2007	
A397	102	3722	2	2007	
A398	102		52	2024	
A399	102		52	2024	
A3A	108	3452	86	2038	
A3A1	116	3311	504A	1104	
A3A3	116	3311	504A	1104	
A3A5	116	3311	504A	1104	
A3A9	116	3311	504A	1104	
A3B1	116	3016	504A	1104	
A3B3	116	3016	504A	1104	
A3B5	116	3016	504A	1104	
A3B9	116	3016	504A	1104	
A3C	108	3984/106	86	2038	
A3C1	116	3017B/117	504A	1104	
A3C3	116	3016	504A	1104	
A3C5	116	3017B/117	504A	1104	
A3C9	116	3017B/117	504A	1104	
A3D	108	3452	86	2038	
A3D1	116	3017B/117	504A	1104	
A3D3	116	3016	504A	1104	
A3D5	116	3031A	504A	1104	
A3D9	116	3031A	504A	1104	
A3E	123A	3444	20	2051	
A3E1	116	3031A	504A	1104	
A3E3	116	3031A	504A	1104	
A3E5	116	3031A	504A	1104	
A3E9	116	3031A	504A	1104	
A3F	123A	3444	20	2051	
A3F1	116	3017B/117	504A	1104	
A3F3	116	3017B/117	504A	1104	
A3F5	116	3017B/117	504A	1104	
A3F9	116	3017B/117	504A	1104	
A3G	123A	3444	20	2051	
A3G1	116	3017B/117	504A	1104	
A3G3	116	3017B/117	504A	1104	
A3G5	116	3017B/117	504A	1104	
A3G9	116	3017B/117	504A	1104	
A3H	108	3452	36	2038	
A3H-1	163A	3439	36		
A3H1	125	3032A	531	1114	
A3H3	125	3032A	531	1114	
A3H5	125	3032A	531	1114	
A3H9	125	3032A	531	1114	
A3J	199	3245	62	2010	
A3K	199	3024/128	62	2010	
A3K1	125	3032A	531	1114	
A3K3	125	3032A	531	1114	
A3K5	125	3032A	531	1114	
A3K9	125	3032A	531	1114	
A3L	157	3747	232		
A3L4-6001	130		14	2041	
A3L4-6001-01	130	3027	14	2041	
A3M	154	3045/225	40	2012	
A3M1	125	3033	531	1114	
A3M3	125	3033	531	1114	
A3M5	125	3033	531	1114	
A3M9	125	3033	531	1114	
A3MA	154		40	2012	
A3N	123A	3444	20	2051	
A3N71	108	3452	86	2038	
A3N72	108	3452	86	2038	
A3N73	108	3452	86	2038	
A3P	108	3452	86	2038	
A3R	108	3452	86	2038	
A3S	123A	3444	20	2051	
A3SB	108	3452	86	2038	
A3T	123A	3444	20	2051	
A3T201	101	3861	8	2002	
A3T202	101	3861	8	2002	
A3T203	101	3861	8	2002	
A3T2221	123A	3444	20	2051	
A3T2221A	123A	3444	20	2051	
A3T2222	123A	3444	20	2051	
A3T2222A	123A	3444	20	2051	
A3T2484	192	3137	63	2030	
A3T2894	159	3118	82	2032	
A3T2906	159	3118	82	2032	
A3T2906A	159	3118	82	2032	
A3T2907	159	3118	82	2032	
A3T2907A	193	3138	67	2023	
A3T3011	123A	3444	20	2051	
A3T918	161	3716	39	2015	
A3T929	123A	3444	20	2051	
A3T930	123A	3444	20	2051	
A3TE120	130	3027	14	2041	
A3TE230	130	3027	14	2041	
A3TE240	130	3027	14	2041	
A3TX003	130	3027	14	2041	
A3TX004	130	3027	14	2041	
A3U	130	3027	14	2041	
A3U-4	130	3027	14	2041	
A3W	123A	3444	20	2051	
A3Y	124	3104A	12		
A3Z	123A	3444	20	2051	
A40	102A	3721/100	1	2007	
A40-6704	113A	3119/113	6GC1		
A40-6722	113A	3119/113	6GC1		
A400		3031A	504A	1104	MRF904
A400(RECTIFIER)	116		504A	1104	
A400(TRANSISTOR)	126		52	2024	
A401	126		245	2004	MRF914
A402	159		82	2032	
A403	160		245	2004	
A4030	1008		IC-271		
A4030P	1009	3499	IC-77		
A4031	1010				
A4031P	1010				
A4032	1011				
A4032P	1011		IC-79		

Industry Standard No.	ECG	SK	GE	RS 276-	MOTOR.
A4037764-2	129	3025	244	2027	
A404	160		245	2004	
A40410	129	3025	244	2027	
A405	160		245	2004	
A4050	1150				
A4050P	1150		IC-286		
A4051	1150				
A4051P	1150		IC-286		
A406	102A		53	2007	MRF965
A407	102A		53	2007	
A408	126		245	2004	
A4086	159	3114/290	82	2032	
A4087	159	3114/290	82	2032	
A409	126		245	2004	
A40A	5942		5048		1N3209
A40B	5944		5048		1N3210
A40C	5946		MR-2		1N3211
A40D	948		MR-2		1N3212
A40E	5946				1N3213
A40F	5940		5048		1N3208
A40M	5952	3585	A40M		1N3214
A41	160		245	2004	
A410			245	2004	
A4100	1180				
A4101	1180				
A4101(IC)	1180				
A411		3006	245	2004	
A412	126		52	2024	
A4126	159	3114/290	82	2032	
A413	126		245	2004	
A4131P	1010				
A414	100	3721	1	2007	
A41440	159	3114/290	82	2032	
A415	123A	3444	20	2051	
A415(JAPAN)	100	3721	1	2007	
A416	121	3717	239	2006	
A417	102$	3452/108	86	2038	
A417(JAPAN)	160		245	2004	
A417-115	154	3045/225	40	2012	
A417-116	159	3114/290	82	2032	
A417-132	159	3114/290	82	2032	
A417-138	129	3025	244	2027	
A417-153	159	3114/290	82	2032	
A417-154	107	3039/316	11	2015	
A417-170	129	3025	244	2027	
A417-176	159	3114/290	82	2032	
A417-182	159	3114/290	82	2032	
A417-184	159	3114/290	82	2032	
A417-19	161	3039/316	39	2015	
A417-190	107	3039/316	11	2015	
A417-196	159	3114/290	82	2032	
A417-200	159	3114/290	82	2032	
A417-201	159	3114/290	82	2032	
A417-205	107	3039/316	11	2015	
A417-234	129	3025	244	2027	
A417-235	159	3114/290	82	2032	
A417-43	129	3025	244	2027	
A417-62	121	3009	239	2006	
A417014	181		75	2041	
A417032	152	3893	66	2048	
A417033	130		14	2041	
A417034	192		63	2030	
A417756	197		250	2027	
A418	108	3452	86	2038	
A419	108	3452	86	2038	
A419(JAPAN)	160		245	2004	
A41A	5945		5096		
A41B	5945	3609/5986	5096		
A41C	5989		5101		
A41D	5949		5100		
A41E	5953	3501/5994	5104		
A41F	5941		5100		
A41M	5953	3501/5994	5104		
A42	160	3006	245	2004	
A420	108	3452	86	2038	
A420(JAPAN)	160		245	2004	
A4201	113A	3119/113	6GC1		
A421	160		245	2004	
A4212-A	116	3031A	504A	1104	
A4212A	125		531	1114	
A422	160		245	2004	
A422-A	116		504A	1104	
A424	316				
A4247	121	3009	239	2006	
A425	160		245	2004	
A426	160		245	2004	
A427	108	3452	86	2038	
A427(JAPAN)	108		86	2038	
A428	126		52	2024	
A429		3434	223		
A429(ECB)	288				
A429-0981-12	161	3039/316	39	2015	
A429-0(ECB)	288				
A42946	113A	3119/113	6GC1		
A42946B	113A	3119/113	6GC1		
A429R(ECB)	288				
A429Y	288		223		
A42X00022-01	5834		5004		
A42X00041-01	5019A	3785	ZD-10	562	
A42X00269-01	116		504A	1104	
A42X00286-01	102	3722	2	2007	
A42X00290-01	156		512		
A42X00340-01	109	3090	1N34AS	1123	
A42X00374-01	116		504A	1104	
A42X00390-01	177		300	1122	
A42X00434-01	123A		20	2051	
A42X00454-01	5192A				
A42X00460-01	135A		ZD-5.1		
A42X00480-01	5074A		ZD-11.0	563	
A42X210	103A	3010	59	2002	
A43	126	3006/160	52	2024	
A430	316	3039	39	2015	
A430(JAPAN)	160		245	2004	
A43021415	123A	3444	20	2051	
A43023843	130	3027	14	2041	
A43023845	159	3114/290	82	2032	
A431	160		245	2004	
A4310	159	3114/290	82	2032	
A431A	160		245	2004	
A432	160		245	2004	
A432A	160		245	2004	
A433	160		245	2004	
A434	160		245	2004	
A4347	121	3009	239	2006	
A435	160	3006	245	2004	
A435A	160		245	2004	
A435B	160		245	2004	
A436	126		52	2024	
A437	126		52	2024	
A438	126		52	2024	
A44	126	3005	1	2007	
A440	160		245	2004	MM4049
A4400	1193				
A4400(IC)	1193				
A4400Y	1211				
A440A	160	3007	245	2004	
A4430	1211	3739			
A4442	159	3114/290	82	2032	
A446	102$		52	2024	
A447	126		52	2024	
A4478	129	3025	244	2027	
A448			245	2004	
A4496-R	374				
A44A					1N3492
A44B					1N3493
A44C			A44C		1N3494
A44D			A44D		1N3495
A44E					MR327
A44F					1N3491
A44M			A44M		MR328
A45	126	3006/160	52	2024	
A45-1	126	3006/160	52	2024	
A45-2	126	3006/160	52	2024	
A45-3	126	3006/160	52	2024	
A450	160$		245	2004	
A450H	160$		245	2004	
A451	160$	3039/316	39	2015	
A451(JAPAN)	160		245	2004	
A451H	160$		245	2004	
A452	160$		245	2004	
A452H	160$		245	2004	
A453	126		52	2024	
A454	123A	3444	20	2051	
A454(JAPAN)	126		52	2024	
A455	709	3039/316	17	2051	
A455(IC)	709		IC-11		
A455(TRANSISTOR)	126		52	2024	
A456	126		52	2024	
A457	126		52	2024	
A46-86101-3	107	3039/316	11	2015	
A46-86109-3	107	3039/316	11	2015	
A46-86110-3	107	3039/316	11	2015	
A46-86133-3	107	3039/316	11	2015	
A46-8614-3	101	3861	8	2002	
A46-86301-3	107	3039/316	11	2015	
A46-86302-3	107	3039/316	11	2015	
A46-86303-3				2015	
A46-867-3	154	3045/225	40	2012	
A460	160		245	2004	
A46051-1	124		12		
A461	160		245	2004	
A462	160		245	2004	
A463	160		245	2004	
A464	160		245	2004	
A4648	154	3045/225	40	2012	
A465-181-19	161	3039/316	39	2015	
A466	128	3024	243	2030	
A466(JAPAN)	126		52	2024	
A466-2	126		52	2024	
A466-3	126		52	2024	
A466A			39	2015	
A466BLK	126		52	2024	
A466BLU	126		52	2024	
A466YEL	126		52	2024	
A467	290A	3039/316	61	2038	
A467(JAPAN)			82	2032	
A467-0	290A		82	2032	
A467-0(JAPAN)			82	2032	
A467-Y	290A		82	2032	
A467-Y(JAPAN)			82	2032	
A467G	290A		82	2032	
A467G(JAPAN)			82	2032	
A467G-0	290A		82	2032	
A467G-0(JAPAN)			82	2032	
A467G-R	290A		82	2032	
A467G-R(JAPAN)			82	2032	
A467G-Y	290A		82	2032	
A467G-Y(JAPAN)			82	2032	
A468	126	3008	52	2024	
A469	126	3007	52	2024	
A470	126	3007	52	2024	
A4700	101	3861	8	2002	
A471	126	3007	52	2024	
A471-1	126		52	2024	
A471-2	126		52	2024	
A471-3	126		52	2024	
A472		3007	20	2051	
A472(JAPAN)	126		52	2024	
A472(NPN)	123A				
A472(PNP)	126				
A472A	126		52	2024	
A472B	126		52	2024	
A472C	126		52	2024	
A472D	126		52	2024	
A472E	126		52	2024	
A473			86	2038	
A473(JAPAN)	153		69	2049	
A473(NPN)	108				
A473(PNP)	153				
A473-GR	153		69	2049	
A473-O	153		69	2049	
A473-R	153		69	2049	
A473-Y			69	2049	
A47392R-O	128		243	2030	
A473Y			69	2049	
A474	126	3005	52	2024	
A4745	159	3114/290	82	2032	
A476	126		52	2024	
A477	126		52	2024	
A478	102A$		52	2024	
A4789	108	3452	86	2038	
A479	102A$		52	2024	
A48-10075A01	121MP	3013	239(2)	2006(2)	
A48-10075A02	121MP	3013	239(2)	2006(2)	
A48-10075A03	121MP	3013	239(2)	2006(2)	
A48-10075A04	121MP	3013	239(2)	2006(2)	
A48-10075A05	121MP	3013	239(2)	2006(2)	
A48-10075A06	121MP	3013	239(2)	2006(2)	
A48-10075A07	121MP	3013	239(2)	2006(2)	
A48-10075A08	121MP	3013	239(2)	2006(2)	
A48-10103A01	121MP	3013	239(2)	2006(2)	
A48-10103A02	121MP	3013	239(2)	2006(2)	
A48-10103A03	121MP	3013	239(2)	2006(2)	
A48-10103A04	121MP	3013	239(2)	2006(2)	
A48-10103A05	121MP	3013	239(2)	2006(2)	
A48-10103A06	121MP	3013	239(2)	2006(2)	
A48-10103A07	121MP	3013	239(2)	2006(2)	
A48-10103A08	121MP	3013	239(2)	2006(2)	
A48-10103A09	121MP	3013	239(2)	2006(2)	
A48-10103A10	121MP	3013	239(2)	2006(2)	
A48-10103A11	121MP	3013	239(2)	2006(2)	
A48-124216	101	3861	8	2002	
A48-124217	101	3861	8	2002	
A48-124218	101	3861	8	2002	
A48-124220	101	3011	8	2002	
A48-124221	101	3861	8	2002	
A48-125233	101	3861	8	2002	
A48-125234	101	3861	8	2002	

Industry Standard No.	ECG	SK	GE	RS 276-	MOTOR.
A48-125235	101	3861	8	2002	
A48-125236	101	3011	8	2002	
A48-128239	101	3861	8	2002	
A48-134520	101	3861	8	2002	
A48-134648	225	3045	256		
A48-134700	101	3861	8	2002	
A48-134727	121	3009	239	2006	
A48-134731	121	3009	239	2006	
A48-134789	108	3452	86	2038	
A48-134819	154	3045/225	40	2012	
A48-134837	108	3452	86	2038	
A48-134843	154	3045/225	40	2012	
A48-134845	108	3452	86	2038	
A48-134853	154	3045/225	40	2012	
A48-134898	154	3045/225	40	2012	
A48-134902	108	3452	86	2038	
A48-134904	108	3452	86	2038	
A48-134907	121	3009	239	2006	
A48-134919	154	3045/225	40	2012	
A48-134922	108	3452	86	2038	
A48-134923	108	3452	86	2038	
A48-134924	108	3452	86	2038	
A48-134925	108	3452	86	2038	
A48-134926	108	3452	86	2038	
A48-134927	154	3045/225	40	2012	
A48-134931	101	3861	8	2002	
A48-134945	108	3452	86	2038	
A48-134961	108	3452	86	2038	
A48-134962	108	3452	86	2038	
A48-134963	108	3452	86	2038	
A48-134964	108	3452	86	2038	
A48-134965	108	3452	86	2038	
A48-134966	108	3452	86	2038	
A48-134981	108	3452	86	2038	
A48-137002	154	3045/225	40	2012	
A48-137035	154	3045/225	40	2012	
A48-137071	108	3452	86	2038	
A48-137075	108	3452	86	2038	
A48-137076	108	3452	86	2038	
A48-137077	108	3452	86	2038	
A48-137102	121	3009	239	2006	
A48-137197	108	3452	86	2038	
A48-137213	121	3009	239		
A48-137214	121	3009	239	2006	
A48-137215	121	3009	239	2006	
A48-137216	121	3009	239	2006	
A48-137217	121	3009	239	2006	
A48-137218	121	3009	239	2006	
A48-137219	121	3009	239	2006	
A48-137220	121	3009	239	2006	
A48-40247G01	108	3452	86	2038	
A48-43351A02	108	3452	86	2038	
A48-63076A81	121	3039/316	239	2006	
A48-63076A82	108	3452	86	2038	
A48-63078A52	109	3090	1N34AS	1123	
A48-64978A10	121MP	3013	239(2)	2006(2)	
A48-64978A11	121MP	3013	239(2)	2006(2)	
A48-64978A24	121MP	3013	239(2)	2006(2)	
A48-869254	103A	3010	59	2002	
A48-869283	103A	3010	59	2002	
A48-869476	103A	3010	59	2002	
A48-869476A	103A	3010	59	2002	
A48-97046A05	108	3452	86	2038	
A48-97046A06	108	3452	86	2038	
A48-97046A07	108	3452	86	2038	
A48-97127A06	108	3452	86	2038	
A48-97127A12	108	3452	86	2038	
A48-97127A18	108	3452	86	2038	
A480	108	3452	86	2038	
A480(JAPAN)	159		82	2032	
A4802-00004	159	3114/290	82	2032	
A481	161	3039/316	39	2015	
A4815	159	3114/290	82	2032	
A4819	154	3045/225	40	2012	
A481A0028	123A	3444	20	2051	
A481A0030	159		82	2032	
A481A0031	123A	3444	20	2051	
A482	161	3117	39	2015	
A482(JAPAN)	159		82	2032	
A4822-130-40348	159	3114/290	82	2032	
A483	218$	3039/316	39	2015	
A4838	154	3045/225	40	2012	
A484	161	3513	39	2015	
A484(ADMIRAL)			39	2015	
A484(ZENITH)			11	2015	
A4843	154	3045/225	40	2012	
A4844	159	3114/290	82	2032	
A484BL	323				
A484R	323				
A484Y	323				
A485	316	3024/128	86	2038	2N2857
A4851	108	3452	86	2038	
A4853	154	3045/225	40	2012	
A485BL	323				
A485R	323				
A485Y	323				
A486	161	3039/316	39	2015	BFW92A
A486(JAPAN)	323		218	2026	
A486BL	323				
A486E8JAPAN)				2026	
A486R	323				
A488-A0001	177		300	1122	
A488-A0060	519			1122	
A489	161	3039/316	39	2015	
A489(JAPAN)	153		69	2049	
A489-0	153		69	2049	
A489-R	153		69	2049	
A489-Y	153		69	2049	
A489751-028	159	3114/290	82	2032	
A489751-031	159	3114/290	82	2032	
A489Y	153				
A49	102A	3004	52	2024	
A490	161	3024/128	243	2030	BFX89
A490(JAPAN)	153		82	2032	
A490(POWER)	153		69	2049	
A490(RF)	153				
A490-0	153				
A490-LBGL1	153				
A490-0	292				
A490Y			82	2032	
A492	161	3039/316	39	2015	
A493	290A		82	2032	
A493-0	159		82	2032	
A493BL	234				
A493GR	234		82	2032	
A493Y	290A		82	2032	
A494	234	3444/123A	20	2051	
A494(JAPAN)	123A	3444	20	2051	
A494-GR	234		82	2032	
A494-0	234		82	2032	
A494-Y	234		82	2032	
A49463	358		42		
A495		3025	11	2015	
A495(JAPAN)	290A		82	2032	
A495(PNP)	290A				
A495-G	290A		82	2032	
A495-0	290A	3025/129	82	2032	
A495-R	159		82	2032	
A495-Y	290A		82	2032	
A4950		3025	82	2032	
A495A	290A		82	2032	
A495D	290A		82	2032	
A495FA-1	290A				
A495G	290A$				
A495G-0	290A$				
A495G-GR	290A$		82	2032	
A495G-0	290A		82	2032	
A495G-R	290A$		82	2032	
A495G-Y	290A$		82	2032	
A495GR	290A				
A495H	290A				
A495J	290A				
A4950	290A				
A495Q	290A				
A495TM-O	290A				
A495W	290A		82	2032	
A495X	290A				
A495Y	290A	3025/129	82	2032	
A496	290A		11	2015	
A496(JAPAN)	185		58	2025	
A496-	374				
A496-0	374		58	2025	
A496-R			58	2025	
A496-Y			58	2025	
A496Y	374		58	2025	
A497	108	3025/129	86	2038	
A497(JAPAN)	129		244	2027	
A498(F.E.T.)	222			2036	
A498(JAPAN)	129		244	2027	
A498Y	129		244	2027	
A499	106		82	2032	
A499-0	106		82	2032	
A499-R	106		82	2032	
A499-Y	106		82	2032	
A4A	123A	3444	20	2051	
A4A-1-70	121	3009	239	2006	
A4A-1-705	121	3009	239	2006	
A4A-1A9G	121	3009	239	2006	
A4A1	116	3017B/117	504A	1104	
A4A5	116	3311	504A	1104	
A4A9	116	3311	504A	1104	
A4B	199	3038	62	2010	
A4B.5-100A					MDA920A3
A4B.5-200A					MDA920A4
A4B.5-400A					MDA920A6
A4B.5-50A					MDA920A2
A4B.5-600A					MDA920A7
A4B.5-800A					MDA920A8
A4B1	116	3016	504A	1104	
A4B1-100B					MDA920A3
A4B1-200B					MDA920A4
A4B1-400B					MDA920A6
A4B1-50B					MDA920A2
A4B1-600B					MDA920A7
A4B1-800B					MDA920A8
A4B2-100C					MDA920A3
A4B2-100D					MDA920A3
A4B2-200C					MDA920A4
A4B2-200D					MDA920A4
A4B2-400C					MDA920A6
A4B2-400D					MDA920A6
A4B2-50C					MDA920A2
A4B2-50D					MDA920A2
A4B2-600C					MDA920A7
A4B2-600D					MDA920A7
A4B2-800C					MDA920A8
A4B2-800D					MDA920A8
A4B3-100D					MDA970A2
A4B3-100F					MDA970A2
A4B3-200D					MDA970A3
A4B3-200F					MDA970A3
A4B3-50D					MDA970A1
A4B3-50F					MDA970A1
A4B5	116	3017B/117	504A	1104	
A4B6-100E					MDA2501
A4B6-100F					MDA2501
A4B6-200E					MDA2502
A4B6-200F					MDA2502
A4B6-400E					MDA2504
A4B6-400F					MDA2504
A4B6-50E					MDA2500
A4B6-50F					MDA2500
A4B6-600E					MDA3506
A4B6-600F					MDA3506
A4B9	116	3017B/117	504A	1104	
A4C1	116	3017B/117	504A	1104	
A4C5	116	3017B/117	504A	1104	
A4C9	116	3017B/117	504A	1104	
A4D1	116	3031A	504A	1104	
A4D5	116	3031A	504A	1104	
A4D9	116	3017B/117	504A	1104	
A4E	107	3039/316	11	2015	
A4E-5	108	3452	86	2038	
A4E1	116	3031A	504A	1104	
A4E5	116	3017B/117	504A	1104	
A4E9	116	3017B/117	504A	1104	
A4F	199	3122	62	2010	
A4F1	116	3017B/117	504A	1104	
A4F5	116	3017B/117	504A	1104	
A4F9	116	3017B/117	504A	1104	
A4G	108	3452	86	2038	
A4G1	116	3017B/117	504A	1104	
A4G5	116	3017B/117	504A	1104	
A4G9	116	3017B/117	504A	1104	
A4H	154		40	2012	
A4H1	125	3032A	531	1114	
A4H5	125	3032A	531	1114	
A4H9	125	3032A	531	1114	
A4I7014	181		75	2041	
A4J	130	3027	14	2041	
A4J BLK	130				
A4J RED	130				
A4J(RED)			14	2041	
A4JBLK			14	2041	
A4JBRN	130		14	2041	
A4JD3B1	101	3861	8	2002	
A4JRED	130	3027			
A4JRED-1	130	3027	14	2041	
A4JX2A822	103A	3010	59	2002	
A4K	184		57	2017	
A4K1	125	3032A	531	1114	
A4K5	125	3032A	531	1114	
A4K9	125	3032A	531	1114	
A4L	103A	3835	59	2002	
A4M	123A	3444	20	2051	
A4M1	125	3033	531	1114	
A4M5	125	3033	531	1114	
A4M9	125	3033	531	1114	

Industry Standard No.	ECG	SK	GE	RS 276-	MOTOR.
A4N	123A	3444	20	2051	
A4P	123A	3444	20	2051	
A4R	123A	3444	20	2051	
A4S	130	3027	14	2041	
A4S-1	130	3027	14	2041	
A4T	108	3452	86	2038	
A4U	123A	3444	20	2051	
A4V	123A	3444	20	2051	
A4Y-1	107	3039/316	11	2015	
A4Y-1A	108	3452	86	2038	
A4Y-2	123A	3444	20	2051	
A4Z	130	3027	14	2041	
A50		3311	504A	1104	1N4001
A50(RECTIFIER)	116		504A	1104	
A50(TRANSISTOR)	102A		53	2007	
A500	106	3017B/117	504A	1104	2N6603
A500(JAPAN)	159		82	2032	
A500-0	106		82	2032	
A500-R	106		82	2032	
A500-Y	106		82	2032	
A501	129		244	2027	2N6603
A5010005	519			1122	
A502	290A		82	2032	
A503	129		244	2027	
A503-254915-535	5191AK				
A503-254915-69	5257A				
A503-254915-85	5127A				
A503-0	129		244	2027	
A503-R	129		244	2027	
A503-Y	129		244	2027	
A503GR	129		244	2027	
A504	129		244	2027	
A504-R	129		244	2027	
A504-Y	129		244	2027	
A504GR	129	3203/211	244	2027	
A505	374		58	2025	
A505-0	374		58	2025	
A505-R	374		58	2025	
A505-Y	374		58	2025	
A506	160		245	2004	
A507	160		245	2004	
A508	160		245	2004	
A509			269	2032	
A509(ECB)	159				
A509-0(ECB)	159				
A509-D(ECB)	159				
A509-0			269	2032	
A509-0(ECB)	159				
A509-X(ECB)	159				
A509-Y			269	2032	
A509-Y(ECB)	159				
A509G(ECB)	159				
A509G-0(ECB)	159				
A509G-Y(ECB)	159				
A509GR			269	2032	
A509GR(ECB)	159				
A509Q(ECB)	159				
A509R			269	2032	
A509R(ECB)	159				
A509Y			269	2032	
A509Y(ECB)	159				
A51	126	3006/160	245	2004	
A510	323		82	2032	2N6604
A510-0	323		82	2032	
A510-R	323		82	2032	
A511	323		82	2032	2N6604
A511-0	323		82	2032	
A511-R	323		82	2032	
A511BDC	4511B				
A512	323		82	2032	
A512-0	323		82	2032	
A512-R	323		82	2032	
A513	323		82	2032	
A513-0	323		82	2032	
A513-R	323		82	2032	
A514-0033903	177				
A514-019136	116				
A514-022057		3087	1N34AS	1123	
A514-023553	101	3010	8	2002	
A514-023626	116	3016	504A	1104	
A514-025607	116	3016	504A	1104	
A514-026814	6027				
A514-027051	5002A				
A514-027662	100	3004/102A	53	2007	
A514-027757	116		504A	1104	
A514-028072	116	3016	504A	1104	
A514-028073	116	3016	504A	1104	
A514-030963	5962				
A514-031162	5962	3016			
A514-032815	100	3862/103	8	2002	
A514-033338	289A	3444/123A	20	2051	
A514-033377	121	3009			
A514-033596	125				
A514-033903	116	3016	504A	1104	
A514-0339903	116		504A	1104	
A514-035596		3051	504A	1104	
A514-038984		3060	512		
A514-040296	312	3118	FET-1	2028	
A514-042791	177	3087	1N34AS	1123	
A514-043826	5514				
A514-044910	159	3114/290	82	2032	
A514-047801	2040				
A514-047828	210				
A514-047829	211				
A514-047830	196	3192/186	28	2017	
A514-047837	211				
A514-047970	5072A				
A514-048551	754				
A514-052689	5963				
A514-053752	754				
A514-053981	975				
A514-054214	130				
A514-054489	74107				
A514-057814	5673				
A514-058337	5414				
A514-058621	88				
A514-058969	3020				
A514-058992	5014A				
A514-059342	5650				
A514-059463	7400				
A514-059465	7402				
A514-059466	7420				
A514-059468	74150				
A514-059469	74154				
A514-059471	5543				
A514-060021	7405				
A514-060260	2042				
A514-061662	912				
A514-061735	2040				
A514-061987	7404				
A514-061988	7486				
A514-062023	992				
A514-062141	975				
A514-062142	4016B				
A514-062143	4051B				
A514-062144	4071B				
A514-062145	4081B				
A514-062212	975				
A514-063293	778A				
A514-063882	7412				
A514-063883	74195				
A514-064-32	2040				
A514-100243	519				
A514-101082	74LS00				
A514-101084	74LS03				
A514-101085	74LS04				
A514-101086	74LS05				
A514-101087	74LS10				
A514-101088	74LS12				
A514-101089	74LS20				
A514-101090	74LS30				
A514-101092	74LS75				
A514-101093	74LS86				
A514-101094	74LS107				
A514-101095	74LS151				
A514-101096	74LS164				
A514-101097	74LS195A				
A514-101099	4013B				
A514-101391	4017B				
A514-101392	4019B				
A514-101393	4081B				
A514-101395	4001B				
A514-101396	4555B				
A514-101397	4512B				
A514-101550	4069				
A514-101844	4012B				
A514-101845	4023B				
A514-102343	4002B				
A514-102579	74LS42				
A514-102916	4024B				
A515			14	2041	
A516	323		244	2027	
A516A	323		244	2027	
A517	126	3007	52	2024	
A518	126	3007	52	2024	
A52	102A	3004	52	2024	
A522	130	3027	14	2041	
A522(JAPAN)	159		82	2032	
A522-3	130	3027	14	2041	
A5226-1	159	3114/290	82	2032	
A522A	159		82	2032	
A523			14	2041	
A523(NPN)	130				
A523(PNP)	129				
A523A	129				
A524	106	3984			
A525	160	3006	245	2004	
A5253	121	3009	239	2006	
A525A	160	3006	245	2004	
A525B	160	3006	245	2004	
A527	129	3025	244	2027	
A528	129	3025	244	2027	
A53	102A	3004	52	2024	
A530	159	3025/129	82	2032	
A530H	159	3025/129	82	2032	
A532	129	3025	244	2027	
A5320111	159	3114/290	82	2032	
A532A	129	3025	244	2027	
A532B	129	3025	244	2027	
A532C	129	3025	244	2027	
A532D	129	3025	244	2027	
A532E	129	3025	244	2027	
A532F	129	3025	244	2027	
A537	129	3025	244	2027	
A537A	129	3025	244	2027	
A537AA	129	3025	244	2027	
A537AB	129	3025	244	2027	
A537AC	129	3025	244	2027	
A537AH	129	3025	244	2027	
A537B	129	3025	244	2027	
A537C	129	3025	244	2027	
A537H	129	3025	244	2027	
A538	102A	3004	53	2007	
A539	290A	3025/129	82	2032	
A539K	290A				
A539L	290A	3025/129	82	2032	
A539M	290A				
A539S	290A	3025/129	82	2032	
A54-3	152	3041	66	2048	
A54-96-001	123A		20	2051	
A54-96-002	123A		20	2051	
A54-96-003	159		82	2032	
A542			82	2032	
A543		3137	63	2030	
A544	159	3025/129	82	2032	
A545	193A	3138	67	2023	
A545GRN	193A	3138	67	2023	
A545K	193A		67	2023	
A545KLM	193A	3138	67	2023	
A545L	193A	3138	67	2023	
A545LM	193A		67	2023	
A545M	193				
A546	129	3025	244	2027	
A546A	129	3025	244	2027	
A546AA	129				
A546AB	129				
A546AD	129				
A546AE	129				
A546AF	129				
A546AH	129				
A546AI	129				
A546AJ	129				
A546AK	129				
A546AL	129				
A546B	129	3025	244	2027	
A546D	129				
A546E	129	3025	244	2027	
A546F	129				
A546H	129	3025	244	2027	
A546I	129				
A546J	129				
A547	129/427	3138/193A	67	2023	
A547A		3138	67	2023	
A547AA	129/427				
A547AB	129/427				
A547AD	129/427				
A547AE	129/427				
A547AF	129/427				
A547AH	129/427				
A547AI	129/427				
A547AJ	129/427				
A547B	129/427				
A547D	129/427				
A547E	129/427				
A547F	129/427				
A547H	129/427				
A547I	129/427				
A547J	129/427				

Industry Standard No.	ECC	SK	GE	RS 276-	MOTOR.
A548	159	3025/129	244	2027	
A548A	159				
A548B	159				
A548C	159				
A549A	288	3434	223		
A549AH	288	3434	223		
A55	126	3005	1	2007	
A550	159	3025/129	82	2032	
A550A	159	3025/129	82	2032	
A550AO	159				
A550AP	159				
A550AQ	159	3025/129	82	2032	
A550AR	159				
A550AS	159				
A5500	159				
A550Q	159		82	2032	
A550R	159	3025/129	82	2032	
A550S	159		82	2032	
A551	129	3025	244	2027	
A551-008	941D				
A551C	129	3025	244	2027	
A551D	129	3025	244	2027	
A551E	129	3025	244	2027	
A552	129	3025	244	2027	
A553	129				
A554	159				
A554A	129				
A555	290A				
A556	290A				
A556-142	109	3087	1N34AS	1123	
A558	129				
A558C	290A				
A559	129				
A559A	129				
A56	126	3006/160	245	2004	
A56(GERMANIUM)	160				
A56(SILICON)	234				
A560	129	3025	244	2027	
A561	290A	3025/129	82	2032	MRF962
A561-O	290A	3025/129	82	2032	
A561-R	290A	3025/129	82	2032	
A561-Y	290A	3025/129	82	2032	
A561GR	290A$	3025/129	82	2032	
A561R	290A				
A561Y	290A				
A562	290A	3114/290	269	2032	
A562-O	290A			2032	
A562-O	290A	3114/290	269	2032	
A562-R	290A	3025/129	269	2032	
A562-Y	290A	3025/129	269	2032	
A562G	290A		269	2032	
A562GR	290A$		269	2032	
A562Q	290A				
A562R	290A		269	2032	
A562Y	290A	3114/290	269	2032	
A562Y-TM	290A				
A564	290A		65	2050	
A564-O	234	3247	65	2050	
A564-O	159	3247/234	82	2050	
A564-Q	234				
A5640R				2050	
A564A	290A	3247/234	65	2050	
A564ABQ	290A	3247/234	65	2050	
A564AO	290A	3247/234	65	2050	
A564AP	290A	3247/234	65	2050	
A564AQ	290A	3247/234	65	2050	
A564AQGD	290A				
A564AR	290A				
A564AS	234	3247	65	2050	
A564AT	234	3247	65	2050	
A564D	290A				
A564F	290A	3247/234	65	2050	
A564FQ	290A	3247/234	65	2050	
A564FR	290A	3247/234	65	2050	
A564G	290A				
A564H	290A				
A564J	290A	3247/234	65	2050	
A564P	290A	3247/234	65	2050	
A564POR	290A	3247/234	65	2050	
A564PQ	290A				
A564PQR	290A				
A564Q	290A	3247/234	65	2050	
A564QGD	290A				
A564QHD	290A	3247/234	65	2050	
A564QR	290A	3247/234	65	2050	
A564R	290A	3247/234	65	2050	
A564S	234	3247	65	2050	
A564T	234	3247	65	2050	
A564Y	290A				
A565	159	3025/129	82	2032	
A565A	159	3025/129	82	2032	
A565B	159	3025/129	82	2032	
A565C	159	3025/129	82	2032	
A565D	159		82	2032	
A565K	159		82	2032	
A565Q	234	3247	65	2050	
A566	218	3083/197	234		
A566A	218	3083/197	234		
A566B	218	3083/197	234		
A566C	218$	3083/197	234		
A566H	218				
A566HA	218				
A566HB	218				
A567	123A	3444	20	2051	
A567(JAPAN)	159		82	2032	
A567A	159	3444/123A	20	2051	
A567B	159				
A567C	159				
A567D(EBC)	234				
A568	290A	3025/129	82	2032	
A568-OR	290A				
A568B	290A				
A568D	290A				
A568E	290A				
A568F	290A				
A568G	290A				
A568GN	290A				
A568H	290A				
A568J	290A				
A568K	290A				
A568L	290A				
A568M	290A				
A568R	290A				
A568X	290A				
A568Y	290A				
A569	290A	3025/129	82	2032	
A569J	159	3025/129	82	2032	
A57	160	3006	52	2024	
A570	290A	3025/129			
A570(TRANSISTOR)	159		82	2032	
A570A	6116				
A570B	6116				
A570C	6116				
A570D	6116				

Industry Standard No.	ECC	SK	GE	RS 276-	MOTOR.
A570E	6116				
A570M	6116				
A571	129	3025	244	2027	
A572	130	3247/234	14	2041	
A572-1	130	3247/234	14	2041	
A573	290A				
A573501	154	3045/225	40	2012	
A573D	290A				
A575	290A				
A575K	290A		67	2023	
A575L	290A		67	2023	
A576-0001-002	159	3114/290	82	2032	
A576-0001-013	159	3114/290	82	2032	
A578		3247	65	2050	
A578(ECB)	234				
A578A		3247	65	2050	
A578A(ECB)	234				
A578B		3247	65	2050	
A578B(ECB)	234				
A578C		3247	65	2050	
A578C(ECB)	234				
A579		3247	65	2050	
A579(ECB)	234				
A579A		3247	65	2050	
A579A(ECB)	234				
A579B		3247	65	2050	
A579B(ECB)	234				
A579C		3247	65	2050	
A579C(ECB)	234				
A57A144-12	108	3452	86	2038	
A57A145-12	108	3452	86	2038	
A57B124-10	121	3009	239	2006	
A57C12-1	154	3045/225	40	2012	
A57C12-2	154	3045/225	40	2012	
A57C5	103A	3010	59	2002	
A57D1-122	154	3045/225	40	2012	
A57L5-1	121	3009	239	2006	
A57M2-16	154	3045/225	40	2012	
A57M2-17	154	3045/225	40	2012	
A57M3-7	121	3009	239	2006	
A57M3-8	121	3009	239	2006	
A58	160	3007	52	2024	
A580	129				
A580-040215	181	3535	75	2041	
A580-040315	181	3535	75	2041	
A580-040515	181		75	2041	
A580-080215	181		75	2041	
A580-080315	181	3535	75	2041	
A580-080515	181	3535	75	2041	
A580B	129				
A580G	129				
A580Y	129				
A581	129				
A581B	129				
A581G	129				
A581Y	129				
A59	160	3006	52	2024	
A592D2(BCE)	159				
A592Y	159	3025/129	82	2032	
A593	123A	3444	20	2051	
A594	129	3025	244	2027	
A594-O	129	3025	244	2027	
A594-R	129	3025	244	2027	
A594-Y	129	3025	244	2027	
A595	102A		53	2007	
A595C	129	3025	244	2027	
A59625-1	159	3114/290	82	2032	
A59625-10	159	3114/290	82	2032	
A59625-11	159	3114/290	82	2032	
A59625-12	159	3114/290	82	2032	
A59625-2	159	3114/290	82	2032	
A59625-3	159	3114/290	82	2032	
A59625-4	159	3114/290	82	2032	
A59625-5	159	3114/290	82	2032	
A59625-6	159	3114/290	82	2032	
A59625-7	159	3114/290	82	2032	
A59625-8	159	3114/290	82	2032	
A59625-9	159	3114/290	82	2032	
A597	129$	3025/129		2027	
A5A	196	3041	241	2020	
A5A-1	184	3041	57	2017	
A5A-1B	152	3893	66	2048	
A5A-2	184	3041	57	2017	
A5A-3	184	3041	57	2017	
A5A-4	184	3190	57	2017	
A5A-5	184	3190	57	2017	
A5A-IB	152			2048	
A5A1	116		504A	1104	
A5A2	116		504A	1104	
A5A5	116	3311	504A	1104	
A5A9	116	3017B/117	504A	1104	
A5B	128	3024	243	2030	
A5B1	116	3017B/117	504A	1104	
A5B2	116	3016	504A	1104	
A5B5	116	3016	504A	1104	
A5B9	116	3016	504A	1104	
A5C	123A	3444	20	2051	
A5C1	116	3016	504A	1104	
A5C2	116	3017B/117	504A	1104	
A5C5	116	3016	504A	1104	
A5C9	116	3017B/117	504A	1104	
A5D1	116	3017B/117	504A	1104	
A5D2	116	3016	504A	1104	
A5D5	116	3016	504A	1104	
A5D9	116	3017B/117	504A	1104	
A5E	188		226	2018	
A5E1	116	3017B/117	504A	1104	
A5E2	116	3017B/117	504A	1104	
A5E5	116	3017B/117	504A	1104	
A5E9	116	3017B/117	504A	1104	
A5F	190	3104A	217		
A5F1	116	3017B/117	504A	1104	
A5F2	116	3017B/117	504A	1104	
A5F5	116	3017B/117	504A	1104	
A5F9	116	3017B/117	504A	1104	
A5FF	190		217		
A5G	184	3190	57	2017	
A5G1	116	3017B/117	504A	1104	
A5G2	116	3017B/117	504A	1104	
A5G5	116	3017B/117	504A	1104	
A5G9	116	3017B/117	504A	1104	
A5H	123A	3444	20	2051	
A5H1	125	3032A	531	1114	
A5H2	125	3032A	531	1114	
A5H5	125	3032A	531	1114	
A5H9	125	3032A	531	1114	
A5J	108	3452	86	2038	
A5K	123A	3444	20	2051	
A5K1	125	3032A	531	1114	
A5K2	125	3032A	531	1114	
A5K5	125	3032A	531	1114	
A5K9	125	3032A	531	1114	
A5L	123A	3444	20	2051	
A5M	123A	3444	20	2051	

Industry Standard No.	ECG	SK	GE	RS 276-	MOTOR.
A5M1	125	3033	531	1114	
A5M2	125	3033	531	1114	
A5M5	125	3033	531	1114	
A5M9	125	3033	531	1114	
A5N	123A	3444	20	2051	
A5P	123A	3444	20	2051	
A5R	123A	3444	20	2051	
A5S	123A	3444	20	2051	
A5T	191	3044/154	249		
A5T-1	191	3104A	249		
A5T2192	128		47	2030	MPS6531
A5T2193	128				MPS6531
A5T2222	123A	3444	20	2051	2N4401
A5T2222A	128				
A5T2243	282				MPS6530
A5T2484	128				
A5T2604	159		82	2032	
A5T2605	159		82	2032	2N5086
A5T2609					2N5086
A5T2907	159	3025/129	82	2032	2N4403
A5T2907A	159				
A5T3391	123AP		47	2030	2N5210
A5T3391A	123AP		47	2030	2N5210
A5T3392	123AP		81	2051	MPS3392
A5T3496	159				
A5T3497	397				2N5400
A5T3504	159		48		2N3906
A5T3505	159		67	2023	2N3906
A5T3565	123AP		81	2051	MPS6514
A5T3571	278				MPS6595
A5T3572	278				MPS6595
A5T3638	159		82	2032	MPS3638
A5T3638A	159		82	2032	MPS3638A
A5T3644	159	3025/129	82	2032	MPS6516
A5T3645	159	3025/129	82	2032	MPSA55
A5T3707	123AP		62	2010	MPS3707
A5T3708	123AP		62	2010	MPS3708
A5T3709	123AP		62	2010	MPS3709
A5T3710	123AP		62	2010	MPS3710
A5T3711	123AP		62	2010	
A5T3798	159				
A5T3821	457	3448	FET-1	2035	
A5T3903	123A	3444	20	2051	2N3903
A5T3904	123A	3444	20	2051	2N3904
A5T3905	159	3025/129	82	2032	2N3905
A5T3906	159	3025/129	82	2032	2N3906
A5T4026	383		67	2023	MPSA55
A5T4027	159				MPSA56
A5T4028	383		67	2023	MPSA55
A5T4029	159				MPSA56
A5T404			82	2032	MPS404
A5T404A			82	2032	MPS404A
A5T4058			65	2050	MPS6522
A5T4059	288		65	2050	MPS6516
A5T4060	288		65	2050	MPS6516
A5T4061			65	2050	MPS6517
A5T4062			65	2050	MPS6518
A5T4123	123AP		210	2051	2N4123
A5T4124	123AP		20	2051	2N4124
A5T4125	159	3025/129	82	2032	2N4125
A5T4126	159	3025/129	82	2032	2N4126
A5T4248	159		82	2032	2N5086
A5T4249	159		82	2032	2N5086
A5T4250	159		48		2N5087
A5T4260					MPS3638
A5T4261					MPS3640
A5T4402	159		21	2034	2N4402
A5T4403	159		21	2034	2N4403
A5T4409	128		81	2051	2N4409
A5T4410	282		222		2N4410
A5T5058	287		235	2012	
A5T5059	287	3433	222		
A5T5086	159		82	2032	2N5086
A5T5087	159		48		2N5087
A5T5172	123AP		81	2051	MPS5172
A5T5209	128		81	2051	2N5209
A5T5210	128		47	2030	2N5210
A5T5219	123AP		81	2051	2N5219
A5T5220	123AP		81	2051	2N5220
A5T5221	159		82	2032	2N5221
A5T5223	123AP		81	2051	2N5223
A5T5225	123AP		81	2051	2N5225
A5T5226	159		82	2032	2N5226
A5T5227			82	2032	2N5227
A5T5400	397				2N5400
A5T5401	397				2N5401
A5T5550	287				2N5550
A5T5551	287				2N5551
A5T6222			47	2030	
A5T6461	287				
A5T6462	287				
A5T6463	287				
A5T6464	287				
A5T6538	123AP				
A5T6539	123AP				
A5T6540	128				
A5T6541	128				
A5T718A	128				
A5T918	278				
A5T929	123AP				
A5T930	123AP				
A5TA	191	3232	249		
A5U	123A	3444	20	2051	
A5V	181		75	2041	
A5W	123A	3444	20	2051	
A5Y	184	3190	57	2017	
A60	126	3007	52	2024	
A600	116	3017B/117	504A	1104	
A600(RECTIFIER)	116		504A	1104	
A603	159	3025/129	82	2032	
A604	129	3025	244	2027	
A6046940	312				
A606	129	3025	244	2027	
A606S	129	3025	244	2027	
A607	129/427				
A607A	129/427				
A607B	129/427				
A607C	129/427				
A607D	129/427				
A607K	129/427				
A607L	129/427				
A607M	129/427				
A607N	129/427				
A607S	129/427				
A607SA	129/427				
A607SB	129/427				
A607SC	129/427				
A607SD	129/427				
A608	290A				
A608-C		3025	82	2032	
A608-C(ECB)	159				
A608-D			82	2032	
A608-E		3025	82	2032	
A608-E(ECB)	159				
A608-F	159	3025/129	82	2032	
A608-F(ECB)	159				
A608A		3025	82	2032	
A608A(ECB)	159				
A608B		3025	82	2032	
A608B(ECB)	159				
A608C		3025	82	2032	
A608C(ECB)	159				
A608D		3025	82	2032	
A608D(ECB)	159				
A608E		3025	82	2032	
A608E(ECB)	159				
A608E-NP(ECB)	159				
A608F		3025	82	2032	
A608F(ECB)	159				
A608F-NP(ECB)	159				
A608F-NP(ECB)	159				
A608G		3025	82	2032	
A608G(ECB)	159				
A608K(ECB)	159				
A608KD(ECB)	159				
A608KE(ECB)	159				
A608KF(ECB)	159				
A608KG(ECB)	159				
A608KNP(ECB)	159				
A608KNPD(ECB)	159				
A608KNPE	159				
A608KNPF	159				
A608KNPG	159				
A608NP(ECB)	159				
A609			82	2032	
A609(ECB)	159				
A609A			82	2032	
A609A(ECB)	159				
A609B			82	2032	
A609B(ECB)	159				
A609C			82	2032	
A609C(ECB)	159				
A609D			82	2032	
A609D(ECB)	159				
A609E			82	2032	
A609E(ECB)	159				
A609F			82	2032	
A609F(ECB)	159				
A609G			82	2032	
A609G(ECB)	159				
A61	160	3007	245	2004	
A610	290A	3025/129	82	2032	
A610074-1	159	3114/290	82	2032	
A610075-1	154	3045/225	40	2012	
A610083	159	3114/290	82	2032	
A610083-1	159	3114/290	82	2032	
A610083-2	159	3114/290	82	2032	
A610083-3	159	3114/290	82	2032	
A610110-1	159	3114/290	82	2032	
A610120-1	159	3114/290	82	2032	
A610B	290A	3025/129	82	2032	
A610S					MPF256
A611	290A	3025/129	82	2032	
A611-4E	290A	3025/129	82	2032	
A613	218	3083/197	234		
A614	218	3083/197	234		
A615-1008	160	3006	245	2004	
A615-1009	160	3006	245	2004	
A615-1010	102A	3004	53	2007	
A615-1011	102A	3004	53	2007	
A615-1012	109	3087	1N34AS	1123	
A616	218	3083/197	234		
A617K	159		82	2032	
A6181-1	154	3045/225	40	2012	
A618K	159		82	2032	
A62-18427	121	3009	239	2006	
A62-19581	108	3452	86	2038	
A620	290A				
A620WL	290A				
A623	307	3203/211	274		
A623-O	307	3203/211	274		
A623A	307	3203/211	274		
A623B	307	3203/211	274		
A623C	307	3203/211	274		
A623D	307	3203/211	274		
A623G	307	3203/211	274		
A623R	307	3203/211	274		
A623Y	307	3203/211	274		
A624	307	3203/211	274		
A624A	307	3203/211	274		
A624B	307	3203/211	274		
A624C	307	3203/211	274		
A624D	307	3203/211	274		
A624E	307		274		
A624G	307		274		
A624GN	307	3203/211	274		
A624L	307	3203/211	274		
A624LG	307	3203/211	274		
A624R	307	3203/211	274		
A624Y	307	3203/211	274		
A625	129				
A625B	129				
A625G	129				
A625Y	129				
A626	88	3359/281	74	2043	
A626K	88				
A626L	88	3359/281			
A626M	88	3359/281			
A626N	88	3359/281			
A627	88	3359/281	74	2043	
A627K	88				
A627L	88	3359/281			
A627M	88	3359/281			
A627N	88	3359/281			
A628		3025	82	2032	
A628(BCE)	290A				
A628A		3025	82	2032	
A628A(BCE)	290A				
A628AC(BCE)	290A				
A628AD(BCE)	290A				
A628AE(BCE)	290A				
A628AF(BCE)	290A				
A628AG	234				
A628C(BCE)	290A				
A628D		3025	82	2032	
A628D(BCE)	290A				
A628E		3025	82	2032	
A628E(BCE)	290A				
A628EF(BCE)	290A				
A628F		3025	82	2032	
A628F(BCE)	290A				
A628G	234				
A629	234	3114/290	82	2032	
A63-18426	154	3045/225	40	2012	
A63-18427	121	3009	239	2006	
A6314340	198				
A6315611	235				
A6317800	165	3115	38		

Industry Standard No.	ECG	SK	GE	RS 276-	MOTOR.
A6317801	165				
A6317802	165				
A6317900	238	3115/165	38		
A6317901	238		259		
A6319400	171	3220/198	251		
A6319403	376	3220/198	251		
A6319500	375	3929			
A631961	235				
A6319611	235				
A6319640	322	3251/306			
A6321241	297				
A6324922	287				
A6324924	287				
A6324942	287				
A6325060	287				
A6325070	287				
A634	187A		248		
A634A	187A		248		
A634B	187A		248		
A634C	187A		248		
A634D	187A		248		
A634K	187A		248		
A634L	187A		248		
A634M	187A		248		
A634N	187A				
A636	187		248		
A636(4)K	187				
A636(4)L	187				
A636-4K	187				
A636-4L	187				
A6364K	187A				
A6364L	187A				
A636A	187		248		
A636B	187		248		
A636C	187		248		
A636D	187		248		
A636K	187		248		
A636L	187		248		
A636M	187		248		
A637	288	3025/129	82	2032	
A638	288	3434	223		
A639		3434	223		
A639(ECB)	288				
A639S(ECB)	288				
A64	102A	3004	53	2007	
A640	234	3039/316	82	2032	
A640(JAPAN)	159		82	2032	
A640A	234	3025/129	82	2032	
A640B	234	3025/129	82	2032	
A640C	234	3025/129	82	2032	
A640D	234	3025/129	82	2032	
A640E	234	3025/129	82	2032	
A640F	234				
A640L	234	3039/316	82	2033	
A640M	234	3025/129	82	2032	
A640S	234	3132	82	2033	
A641	234	3841/294	65	2050	
A641(JAPAN)	234		65	2050	
A641(NPN)	123A	3444	20	2051	
A641(PBP)				2050	
A641(PNP)	234	3841/294	65	2050	
A641A	234	3841/294	65	2050	
A641B	234	3841/294	65	2050	
A641C	234	3841/294	65	2050	
A641D	234	3841/294	65	2050	
A641L	234	3841/294	65	2050	
A641M	234	3841/294	65	2050	
A641S	234	3841/294	65	2050	
A642	290A	3114/290	82	2032	
A642(JAPAN)	290A		82	2032	
A642-254	108	3452	86	2038	
A642-260	108	3452	86	2038	
A642-268	108	3452	86	2038	
A642-271	121MP	3013	239(2)	2006(2)	
A642A	290A	3114/290	82	2032	
A642B	290A	3114/290	82	2032	
A642C	290A	3114/290	82	2032	
A642D	290A	3114/290	82	2032	
A642E	290A	3114/290	82	2032	
A642F	290A	3114/290	82	2032	
A642L	290A	3114/290	62	2010	
A642L(NPN)	199			2010	
A642R	290A				
A642S		3114	212	2010	
A642S(NPN)	199		62	2010	
A642V	290A				
A642W	290A		82	2032	
A642Y	290A				
A643	193A	3138	67	2023	
A643A	193A	3138	67	2023	
A643B	193A	3138	67	2023	
A643C	193A	3138	67	2023	
A643D	193A	3138	67	2023	
A643E	193A	3138	67	2023	
A643F	193A	3138	67	2023	
A643L	193A	3039/316	67	2023	
A643R	193A	3138	67	2023	
A643S	193A	3039/316	67	2023	
A643V	193A	3138	67	2023	
A643W	193A	3025/129	67	2023	
A644L	199	3245	62	2010	
A644S	199	3245	62	2010	
A645	211		253	2026	
A645L	199		62	2010	
A645S	199		62	2010	
A646	211		248		
A648	281	3173/219	74	2043	
A648A	281	3173/219	74	2043	
A648B	281	3173/219	74	2043	
A648C	281	3173/219	74	2043	
A649	281				
A649A	281				
A649B	281				
A649C	281				
A649L	123A	3444	20	2051	
A649S	123A	3444	20	2051	
A65	102A	3004	53	2007	
A65-09-220	290A	3114/290	269	2032	
A65-1-1A9G	160		245	2004	
A65-1-70	160	3006	245	2004	
A65-1-705	160	3006	245	2004	
A65-1A9G	160	3006	245	2004	
A65-2-70	103A	3010	59	2002	
A65-2-705	103A	3010	59	2002	
A65-2A9G	103A	3010	59	2002	
A65-4-70	101	3861	8	2002	
A65-4-705	101	3011	8	2002	
A65-4A9G	101	3861	8	2002	
A65-P11305-0001A	519			1122	
A65-P11311-0001	177		300	1122	
A65-P11324-0001	519			1122	
A6501760	153				
A65023-63	159				
A650232E	159	3114/290	221		
A650235G	290A	3114/290	221		
A6502360	294				
A650236A	159				
A650237F	290A				
A650238A	290A	3114/290	221	2032	
A650372D	290A	3138/193A	269	2032	
A650923F	290A	3114/290	269	2032	
A6509240	290A				
A650925C	290A				
A650925H	290A	3114/290	269	2032	
A6531540	234				
A6532300	292	3930			
A6532841	162	3434/288	221		
A6532921	129	3841/294	244	2027	
A6532941		3841	244	2027	
A656	88		248		
A656A	88				
A656L	88		248		
A656M	88		248		
A657	88				
A657A	88				
A658	88		74	2043	
A658A	88				
A659	290A		82	2032	
A659A	290A				
A659B	290A				
A659C	290A		82	2032	
A659D	290A		82	2032	
A659E	290A		82	2032	
A659F	290A		82	2032	
A659G	290A				
A659L	290A				
A659P	290A				
A659R	290A				
A65A-70	160	3007	245	2004	
A65A-705	160	3007	245	2004	
A65A19G	160	3007	245	2004	
A65B-70	126	3008	52	2024	
A65B-705	126	3008	52	2024	
A65B19G	126	3008	52	2024	
A65C-19G	121	3717	239	2006	
A65C-70	121	3009	239	2006	
A65C-705	121	3009	239	2006	
A65C19G	121	3009	239	2006	
A66	102A	3004	53	2007	
A66-1-70	102A	3003	53	2007	
A66-1-705	102A	3003	53	2007	
A66-1A9G	102A	3003	53	2007	
A66-2-70	102A	3004	53	2007	
A66-2-705	102A	3004	53	2007	
A66-2A9G	102A	3004	53	2007	
A66-3-3A9G	102A		53	2007	
A66-3-70	102A	3004	53	2007	
A66-3-705	102A	3004	53	2007	
A66-3A9G	102A	3004	53	2007	
A66-P11138-0001	128		243	2030	
A660031	121MP	3013	239(2)	2006(2)	
A660097	121	3009	239	2006	
A661	290A	3450/298	272		
A661-0	290A		272		
A6610140	153				
A6610160	153				
A661GR	290A$	3450/298	272		
A661R	290A	3450/298	272		
A661Y	290A	3450/298	272		
A663	219	3173	74	2043	
A663-BL	219				
A663-R	219				
A663-Y	219				
A666	290A	3841/294	65	2050	
A6661QRS			82	2032	
A666A	290A	3841/294	65	2050	
A666AP	290A				
A666AQ	290A				
A666AR	290A				
A666AS	234				
A666AT	234				
A666H	290A	3841/294	65	2050	
A666HR	290A	3841/294	65	2050	
A666P	290A				
A666Q	290A	3841/294			
A666QRS	290A		65	2050	
A666R	290A	3841/294	65	2050	
A666S	234	3841/294	65	2050	
A666T	234				
A667-GRN	199	3245	62	2010	
A667-ORG	199	3245	62	2010	
A667-RED	199	3245	62	2010	
A667-YEL	199	3245	62	2010	
A667RED	108	3039/316	86	2038	
A668-GRN	199	3245	62	2010	
A668-ORG	199	3245	62	2010	
A668-YEL	199	3245	62	2010	
A669	159	3114/290	82	2032	
A669-GRN	199	3245	62	2010	
A669-YEL	199	3245	62	2010	
A66X0043-001	177		300	1122	
A67	102A	3006/160	52	2024	
A67-07-244	199	3245	62	2010	
A67-08-760	319	3039/316	280		
A67-15-280	195A	3048/329	46		
A67-33-340	199	3245	62	2010	
A67-33-540	289A	3122	268	2038	
A67-37-940	229	3018	61	2038	
A67-70-960	297		271	2030	
A67-76-200	236	3239	216	2053	
A670	153	3274	69	2049	
A670720K	123A	3245/199	61	2038	
A670722D	123A	3122	61	2051	
A6707244	123AP				
A670729B	123AP	3245/199	61	2051	
A6708760	316				
A6708850	233	3018	61	2038	
A6708861	233				
A670A	153	3274	69	2049	
A670B	153	3274	69	2049	
A670C	153	3274	69	2049	
A671	378	3441/292			
A671-0	153				
A6716560	295				
A671656K	295	3253	270		
A671A	378	3441/292			
A671B	378	3441/292			
A671C	378	3441/292			
A671K	378	3441/292			
A671KA	378	3441/292			
A671KB	378	3441/292			
A671KC	378	3441/292			
A672	159	3114/290		2032	
A672A	159	3114/290	82	2032	
A672B	159	3114/290	82	2032	
A672C	159	3114/290	82	2032	
A673	290A	3114/290	48		
A673(B)	290A				

Industry Standard No.	ECG	SK	GE	RS 276-	MOTOR.
A673-0	290A				
A673-0	290A				
A6733280	199				
A6733340	199				
A673351K	289A	3122	210	2038	
A6733540	289A				
A673354K	289A	3122	210	2038	
A673355H	289A	3122	210	2038	
A673355K	289A	3122	210	2038	
A6733560	289A				
A67335H	289A				
A6737940	229				
A6738040	229				
A6738701	162	3115/165	35		
A673A	290A	3114/290	269	2032	
A673AA	290A	3114/290	269	2032	
A673AB	290A	3114/290	269	2032	
A673AC	290A	3114/290	269	2032	
A673AD	290A	3114/290	269	2032	
A673AE	290A	3114/290	269	2032	
A673AK	290A				
A673AS	290A		269	2032	
A673ASC	290A		269	2032	
A673B	290A	3114/290	269	2032	
A673C	290A	3114/290	269	2032	
A673C2	290A$		269	2032	
A673D	290A	3114/290	269	2032	
A673WT	290A$		269	2032	
A675	290A	3466/159	82	2032	
A675315F	171	3044/154	27		
A675318A	287	3044/154	27		
A6754170	199				
A6754194H			62	2010	
A675419H	199	3122	62		
A675A	290A		82	2032	
A675B	290A		82	2032	
A675C	290A		82	2032	
A677	290A	3114/290	244	2027	
A677-5	290A				
A677-6	290A				
A6770960	297				
A6771373	238	3115/165	38		
A6773802	94	3836/284	35		
A677HL	290A				
A678	290A	3114/290	82	2032	
A678(C)	290A				
A678(SONY)			82	2032	
A678-5	290A				
A678-6	290A				
A678760A	198	3219	251		
A678760C	198		32		
A678970A	198				
A678970H	198	3219			
A678971C	198	3219	32		
A678971D	198	3220	251		
A678E	290A				
A678HL	290A				
A679	281	3359	263	2043	
A679B	281				
A679R	281		263	2043	
A679Y	281		263	2043	
A68-23-560	152	3054/196	66	2048	
A680	281		263	2043	
A680R	281		263	2043	
A680Y	281		263	2043	
A682	374				
A682-0	374				
A682-Y	374				
A6823540	152				
A6823560	152				
A683	294	3025/129	48		
A683401	175				
A683Q	294		48		
A683R	294		48		
A683RS	294				
A683S	294		48		
A684	294	3841	48		
A6841440	196				
A6841460	152				
A6843401	175		246	2020	
A684NC	294				
A684P	294				
A684Q	294		48		
A684R	294		48		
A684S	294				
A685	290A				
A69	126	3006/160	52	2024	
A690D	159		82	2032	
A690V081H97	121	3009	239	2006	
A691M4		3100	300	1122	
A691M5	177	3100/519	300	1122	
A691M5-2	177	3100/519	300	1122	
A692T5	116	3311	504A	1104	
A692T5-0	116	3311	504A	1104	
A692X13-4	109	3088	1N34AS	1123	
A692X16-0	116	3311	504A	1104	
A694X1	5022A	3788	ZD-12	563	
A694X1-0A	5022A	3788	ZD-12	563	
A695(BCE)	159				
A695C		3450	272		
A695C(BCE)	159				
A695D		3450	272		
A695D(BCE)	159				
A695D2	298				
A695E(BCE)	159				
A695Y	290A				
A696		3434	82		
A696(BCE)	290A				
A696A(BCE)	290A				
A696C		3434	223		
A696C(BCE)	290A				
A696D(BCE)	290A				
A696E		3434	223		
A696E(BCE)	290A				
A697(BCE)	290A				
A697C(BCE)	290A				
A697D(BCE)	290A				
A697E(BCE)	290A				
A699	187A		248		
A699-0	187A		248		
A699-OR	187A				
A699-0	187A		248		
A699A	187A		248		
A699A-Q	187A		248		
A699A0	187A		248		
A699AP	187A		248		
A699AQ	187A		248		
A699AR	187A		248		
A699CF	187A		248		
A699P	187A		248		
A699Q	187A		248		
A699R	187A		248		
A69IM4				1122	
A6A	163A	3439	36		
A6A1	156	3051	512		
A6A5	156	3051	512		
A6A9	156	3051	512		
A6B	108	3452	86	2038	
A6B-3-70	121MP	3015	239(2)	2006(2)	
A6B-3-705	121MP	3015	239(2)	2006(2)	
A6B-3A9G	121MP	3015	239(2)	2006(2)	
A6B1	156	3051	512		
A6B5	156	3051	512		
A6B9	156	3051	512		
A6C	184	3041	57	2017	
A6C GRN	184	3190			
A6C(GRN)			57	2017	
A6C-1	184	3041	57	2017	
A6C-1 RED	184	3054/196			
A6C-1(RED)			57	2017	
A6C-1-RED			57	2017	
A6C-2	184	3041	57	2017	
A6C-2 BLK	184	3190			
A6C-2(BLK)			57	2017	
A6C-2-BLACK		3190	57	2017	
A6C-3	184	3041	57	2017	
A6C-3 WHT	184	3190			
A6C-3(WHT)			57	2017	
A6C-3-WHITE		3190	57	2017	
A6C-4	184		57	2017	
A6C-GREEN		3190	57	2017	
A6C1	156		512		
A6C5	156	3051	512		
A6C9	156	3051	512		
A6D	188	3041	226	2018	
A6D-1	188	3054/196	226	2018	
A6D-2	188	3054/196	226	2018	
A6D-3	188	3054/196	226	2018	
A6D1	156	3051	512		
A6D5	156	3051	512		
A6D9	156	3051	512		
A6E	108	3452	86	2038	
A6E1	156	3051	512		
A6E5	156	3051	512		
A6E9	156	3051	512		
A6F	108	3452	86	2038	
A6F1	156	3051	512		
A6F5	156	3051	512		
A6F9	156	3051	512		
A6G	188	3199	226	2018	
A6G1	156	3051	512		
A6G5	156	3051	512		
A6G9	156	3051	512		
A6H	123AP	3854	20	2051	
A6H1	156	3051	512		
A6H5	156	3051	512		
A6H9	156	3051	512		
A6HD	123A	3444	20	2051	
A6J	123A	3444	20	2051	
A6K	123A	3444	20	2051	
A6K1	156	3051	512		
A6K5	156	3051	512		
A6K9	156	3051	512		
A6L	130	3133/164	14	2041	
A6LBLK	130	3027	14	2041	
A6LBLK-1	130	3027	14	2041	
A6LBRN	130	3027	14	2041	
A6LBRN-1	130	3027	14	2041	
A6LRED	130	3027	14	2041	
A6LRED-1	130	3027	14	2041	
A6M	163A	3439	36		
A6M1	156	3051	512		
A6M5	156	3051	512		
A6M9	156	3051	512		
A6N	130	3027	14	2041	
A6N-6	130	3027	14	2041	
A6N1	506			1114	
A6N5	506			1114	
A6N9	506			1114	
A6P	108	3452	86	2038	
A6R	123A	3444	20	2051	
A6S	123A	3444	20	2051	
A6T	108	3452	86	2038	
A6T5222	123AP				
A6U	108	3452	86	2038	
A6V	107	3039/316	11	2015	
A6V-5	108	3452	86	2038	
A6W	184	3190	57	2017	
A6Y	190		217		
A6Z	163A	3439	36		
A6ZH	163A	3439	36		
OA7	109	3091		1123	
A7-12	130	3027	14	2041	
A7-13	130	3027	14	2041	
A70	126	3008	245	2004	
OA70	109				
A70-12	339				
A700	153		248		
A7000900	109				
A7001800	109	3087	1N34AS	1123	
A7001957	109				
A700B	153		248		
A700Y	153		248		
A701			82	2032	
A701(ECB)	159				
A701584-00	159	3114/290	82	2032	
A701589-00	159	3114/290	82	2032	
A701E(ECB)	159				
A701F			82	2032	
A701F(ECB)	159				
A701FJ			82	2032	
A701FJ(ECB)	159				
A701FO			82	2032	
A701FO(ECB)	159				
A701G(ECB)	159				
A702		3438	35		
A702(ECB)	159				
A702(TO3 NPN)	162				
A702(TO92 PNP)	234				
A702E(ECB)	159				
A702F(ECB)	159				
A702G(ECB)	159				
A703	307		274		
A703C	307		274		
A703D	307		274		
A703E	307		274		
A704	290A				
A705	290A	3841/294	36		
A706	189				
A706-2	189				
A706-3	189				
A706-4	189				
A706H	189				
A706J	211				
A707	298		67	2023	
A707V	298	3450	67	2023	
A708	129	3203/211	244	2027	
A708A	129	3203/211	244	2027	

Industry Standard No.	ECG	SK	GE	RS 276-	MOTOR.
A708B	129	3203/211	244	2027	
A708C	129	3203/211	244	2027	
A709	288				
A709(CBE)	159				
A70A	6154				
A70B	6154				
A70C	6154				
A70D	6154				
A70E	6156				
A70F	160	3006	245	2004	
A70L	160	3006	245	2004	
A70M	6156				
A70MA	160	3006	245	2004	
A70N	6158				
A70P	6158				
A70S	6158				
A70T	6158				
A71	160	3008	52	2024	
A7102001	116	3311	504A	1104	
A7110021	5010A				
A7110102	5014A				
A7110503	5024A				
A712	397				
A713AS	232				
A713S	232				
A714	219				
A714C	281				
A714D	281				
A714E	281				
A714L	281				
A714LC	219				
A714LD	219				
A714LE	281				
A715	185#		58	2025	
A715A	185#		58	2025	
A715B	185#		58	2025	
A715C	185#		58	2025	
A715D	185#		58	2025	
A715FB	108		86	2038	
A715WB	185#				
A715WBP	185#	9076/187A	248		
A715WT	185#				
A715WT-B	185#	9076/187A			
A715WT-C	185#	9076/187A			
A715WT-D	185#	9076/187A			
A715WTA	185#	9076/187A	58	2025	
A715WTB	185	9076/187A	58	2025	
A715WTC	185	9076/187A	58	2025	
A715WTD	185#	9076/187A	58	2025	
A71687-1	159	3114/290	82	2032	
A717	129	3203/211	253	2026	
A718	159	3025/129	82	2032	
A719	290A	3025/129	269	2032	
A719-O	290A				
A719A	290A				
A719B	290A				
A719K	290A				
A719P	290A		269	2032	
A719PQ	290A				
A719PQR	290A				
A719Q	290A	3025/129	82	2032	
A719QR	290A				
A719R	290A		269	2032	
A719RS	290A		269	2032	
A719S	290A	3025/129	269	2032	
A719W	290A				
A71A	160				
A71A(RECTIFIER)	6155				
A71A(TRANSISTOR)	126		52	2024	
A71AB	160	3008	52	2024	
A71AC	160	3008	52	2024	
A71B	160				
A71B(RECTIFIER)	6155				
A71B(TRANSISTOR)	126		52	2024	
A71BS	160	3008	52	2024	
A71C	6155				
A71D	160				
A71D(RECTIFIER)	6155				
A71D(TRANSISTOR)	126		52	2024	
A71E	6157				
A71M	6157				
A71N	6159				
A71P	6159				
A71S	6159				
A71T	6159				
A71Y	160		52	2024	
A72	126	3006/160	52	2024	
A72-49-600	519	3100	300	1122	
A72-83-300	6408	9084A		1050	
A72-86-700	5072A	3136	ZD-8.2	562	
A720	290A	3450/298	272		
A720657	5547				
A720A	290A				
A720AP	290A				
A720AQ	290A				
A720AR	290A				
A720AS	290A				
A720NC	290A				
A720NCP	290A				
A720NCQ	290A				
A720NCR	290A				
A720NCS	290A				
A720P	290A	3450/298	272		
A720Q	290A	3450/298	272		
A720R	290A	3450/298	272		
A720S	290A	3450/298	272		
A721	234		65	2050	
A721S	234	3247	65	2050	
A721T	234	3247	65	2050	
A721U	234	3247	65	2050	
A722	234	3247	65	2050	
A722S	234		65	2050	
A722T	234	3247	65	2050	
A722U	234	3247	65	2050	
A723	290A	3025/129	82	2032	
A723A	290A	3025/129	82	2032	
A723B	290A	3025/129	82	2032	
A723C	290A	3025/129	82	2032	
A723D	290A	3025/129	82	2032	
A723E	290A	3025/129	82	2032	
A723F	290A	3025/129	82	2032	
A723R	290A				
A7246602	177	3175	300	1122	
A7246653	177				
A7246711	177	3100/519	300	1122	
A7246727	177		300	1122	
A7246771	177				
A7246950	5007A				
A7249601	519	3100	514	1122	
A725			65	2050	
A725(BCE)	234				
A7252802	614	3327			
A7252805	614	3327			
A7252824	614				
A7253	154	3045/225	40	2012	
A725EH2AB1	125	3033	531	1114	
A725F		3247	65	2050	
A725F(BCE)	234				
A725G		3247	65	2050	
A725G(BCE)	234				
A725H		3247	65	2050	
A725H(BCE)	234				
A725Y(BCE)	234				
A725YF(BCE)	234				
A725YG(BCE)	234				
A725YH(BCE)	234				
A726		3247	65	2050	
A726(BCE)	234				
A726F		3247	65	2050	
A726F(BCE)	234				
A726G		3247	65	2050	
A726G(BCE)	234				
A726H		3247	65	2050	
A726H(BCE)	234				
A726Y(BCE)	234				
A726YF(BCE)	234				
A726YG(BCE)	234				
A7270300	138A				
A7272130	5093A	3344	ZD-75		
A7275400		9091	300	1122	
A7279039	121	3009	239	2006	
A7279049	121	3009	239	2006	
A728	290A				
A7285774	121	3009	239	2006	
A7285778	121	3009	239	2006	
A7285900	136A		ZD-5.6	561	
A7286100	137A				
A7286107	5013A				
A7286120	5013A	3779			
A7286201	5071A	3334	ZD-6.8	561	
A7286300	5071A				
A7287100	5019A				
A7287500	142A	3062	ZD-12	563	
A7287513	142A	3062	ZD-12	563	
A7288504	5081A	3151	ZD-24		
A7289047	121	3009	239	2006	
A728A	290A				
A728AD	290A				
A728AE	290A				
A728AF	234				
A728C	290A				
A728D	290A				
A728E	290A				
A728F	234				
A728G	234				
A7290594	121	3009	239	2006	
A7291252	121	3009	239	2006	
A7297043	121	3009	239	2006	
A7297092	121	3009	239	2006	
A7297093	121	3009	239	2006	
A72BLU	126	3006/160	52	2024	
A72BRN	126	3006/160	52	2024	
A72ORN	126	3006/160	52	2024	
A72WHT	126	3006/160	52	2024	
A73	160	3091		1123	
OA73	109				
A73-16-179	177		300	1122	
A730	193A	3138	82	2032	
A730P	193A				
A730Q	193A				
A730R	193A				
A730S	193A				
A731	193A	3025/129	82	2032	
A731P	193A				
A731Q	193A				
A731R	193A				
A731S	193A				
A732	129				
A733	290A	3025/129	48		
A733(P)	290A				
A73301	294				
A73302	294				
A733A	290A	3025/129	48		
A733AP	290A				
A733AQ	290A				
A733B	290A	3025/129	48		
A733C	290A	3025/129	48		
A733D	290A	3025/129	48		
A733E	290A	3025/129	48		
A733F	290A	3025/129	48		
A733GR	294				
A733H	290A	3025/129	82		
A733I	290A	3025/129	48		
A733IQ	290A		48		
A733K	290A				
A733P	290A	3025/129	82		
A733P1	290A				
A733P2	290A				
A733PQ	290A				
A733Q	290A	3025/129	48		
A733QP	290A		48		
A733R	290A	3025/129	82		
A736	129		244	2027	
A738B	185#				
A738C	185#		58	2025	
A738D	185#		58	2025	
A74	126	3007	245	2004	
A74-3-3A9G	100	3721	1	2007	
A74-3-70	100	3005	1	2007	
A74-3-705	100	3005	1	2007	
A74-3A9G	100	3005	1	2007	
A740	398	3930			
A740A	292	3930			
A740AB	292				
A741H	106$				
A742			244	2027	
A743	374#		58	2025	
A743A	374#		58	2025	
A743AA	374#		58	2025	
A743AB	374#		58	2025	
A743AC	374#		58	2025	
A743AD	374#		58	2025	
A743B	374#		58	2025	
A743C	374#		58	2025	
A743D	374#		58	2025	
A744	219		74	2043	
A745	281				
A745A	281				
A746	281	3846/285	74	2043	
A747	281	3846/285	20	2051	
A747A	281	3846/285	20	2051	
A747B	199		62	2010	
A748	153	3444/123A	20	2051	
A748B	123A	3444	20	2051	
A748C	199	3245	62	2010	
A748Q	153		69	2049	
A749		3124	62	2010	
A749B	123A	3444	20	2051	

Industry Standard No.	ECG	SK	GE	RS 276-	MOTOR.
A749C	199	3245	62	2010	
A75	129	3006/160	52	2024	
A75-68-500	116	3311	504A	1104	
A7509201	116	3313	504A	1104	
A751	383	3138/193A	67	2023	
A751Q	383		67	2023	
A751QR	383		67	2023	
A751R	193A		67	2023	
A751S	193A		67	2023	
A752	383	3138/193A	67	2023	
A752P	383		67	2023	
A752Q	383		67	2023	
A752R	383		67	2023	
A752S	383		67	2023	
A753	281		263	2043	
A753-4004-248	159	3114/290	82	2032	
A753A	281		263	2043	
A753B	281		263	2043	
A753C	281		263	2043	
A754	153		69	2049	
A7547053		3313	504A	1104	
A754A	153		69	2049	
A754B	153		69	2049	
A754C	153		69	2049	
A754D	153		69	2049	
A755	153		69	2049	
A755A	153		69	2049	
A755B	153		69	2049	
A755C	153		69	2049	
A755D	153		69	2049	
A756	88		263	2043	
A7568250	506	3125	504A	1104	
A7568300	552	3125	511	1114	
A7568500	116	3311	504A	1104	
A7568615	116	3311			
A7568700	125	3313/116	504A	1104	
A756A	88		263	2043	
A756B	88		263	2043	
A756C	88		263	2043	
A757	88	3122	263	2043	
A7570013-01	159	3114/290	82	2032	
A7572100	116	3311	504A	1104	
A7572200	116	3311	504A	1104	
A7576004-01	159	3114/290	82	2032	
A75779500				1114	
A7579500	525	3081/125	510	1114	
A757A	88		263	2043	
A757B	88		263	2043	
A757C	88		263	2043	
A758	281	3025/129	263	2043	
A758011			510	1114	
A7580111	125	3081	510	1114	
A7580910	506	3125	511	1114	
A7582000	506	3130	511	1114	
A758A	281		263	2043	
A758B	281	3025/129	263	2043	
A758C	281		263	2043	
A759			221	2024	
A759500	125	3032A	509	1114	
A759A	159	3025/129	82	2032	
A759B	159	3025/129	82	2032	
A75B	126	3004/102A	52	2024	
A76	126	3006/160	52	2024	
A76-11770	121	3009	239	2006	
A76228	199		62	2010	
A764	197	3083	234		
A765(TRANSISTOR)	218		234		
A77	126	3007	52	2024	
A772.01	107	3841/294			
A772738	108	3452	86	2038	
A772739	108	3452	86	2038	
A772B1	107	3841/294	11	2015	
A772D1	108	3841/294	86	2038	
A772EH	107	3841/294	11	2015	
A772FE	107	3841/294	11	2015	
A775	292	3441			
A775A	292	3441			
A775AA	292	3441			
A775AB	292	3441			
A775AC	292	3441			
A775B	292	3441			
A775C	292	3441			
A777	298	3450	272		
A777R	298	3450	272		
A778	288	3450/298	223		
A778A	288$	3434/288	223		
A778AK	288	3434	223		
A778K	288	3434	223		
A779	187$		248		
A779K	187$		248		
A779KB	187$		248		
A779KC	187$		248		
A779KD	187$		248		
A77A	126	3006/160	52	2024	
A77A-70	105	3012	4		
A77A-705	105	3012	4		
A77A19G	105	3012	4		
A77B	126	3006/160	52	2024	
A77B-70	121MP	3718	239(2)	2006(2)	
A77B-705	121MP	3718	239(2)	2006(2)	
A77B19G	121MP	3718	239(2)	2006(2)	
A77C	126	3006/160	52	2024	
A77C-70	121	3014	239	2006	
A77C-705	121	3014	239	2006	
A77C19G	121	3014	239	2006	
A77D	126	3006/160	52	2024	
A78	102A	3006/160	52	2024	
A780AK	211$	9076/187A	248		
A780AKA	211$	9076/187A	248		
A780AKB	211$	9076/187A	248		
A780AKC	211$	9076/187A	248		
A780AKD		9076	248		
A781	106		IC-6		
A78331	159	3114/290	82	2032	
A78B	102A		52	2024	
A78C	102A		52	2024	
A78D	102A		52	2024	
A79	102A	3006/160	245	2004	
A7978850	552	3051/156	511	1114	
A7978855	552	3125	511	1114	
A7A	123A	3444	20	2051	
A7A1	156	3081/125	512		
A7A30	108	3452	86	2038	
A7A31	108	3452	86	2038	
A7A32	108	3452	86	2038	
A7A5	156	3081/125	512		
A7A9	156	3081/125	512		
A7B	124	3021	12		
A7B5	156		512		
A7B9	156		512		
A7C	116	3016	504A	1104	
A7C(MOTOROLA)			57	2017	
A7C1	156		57		
A7C5	156		57		
A7C7	156		57		
A7C9	156		57		
A7D	116	3016	504A	1104	
A7D1	156		57		
A7D5	156		57		
A7D9	156		57		
A7E	116	3017B/117	504A	1104	
A7E(TRANSISTOR)	123A	3444	20	2051	
A7E1	156		57		
A7E5	156		57		
A7E9	156		57		
A7F	190		217		
A7F1	156		57		
A7F5	156		57		
A7F9	156		57		
A7G	116	3016	504A	1104	
A7G1	156	3051	512		
A7G5	156	3051	512		
A7G9	156	3051	512		
A7H1	156	3051	512		
A7H5	156	3051	512		
A7H9	156	3051	512		
A7K1	156	3051	512		
A7K5	156	3051	512		
A7K9	156	3051	512		
A7M	130	3027	14	2041	
A7M-5(TRANSISTOR)	130		14	2041	
A7M1	156	3051	512		
A7M5	156	3051	512		
A7M9	156	3051	512		
A7N	108	3452	86	2038	
A7N1	506			1114	
A7N5	506			1114	
A7N9	506			1114	
A7P	108	3452	86	2038	
A7R	123A	3444	20	2051	
A7S	123A	3444	20	2051	
A7S4511H	610	3323			
A7S4511T	610	3323			
A7S4513H	611	3324			
A7S4513T	611	3324			
A7S4514H	612	3325			
A7S4514T	612	3325			
A7S6011H	610	3323			
A7S6011T	610	3323			
A7S6013H	611	3324			
A7S6013T	611	3324			
A7S6014H	612	3325			
A7S6017T	613	3326			
A7T	123A	3444	20	2051	
A7T3391	289A		47	2030	2N5210
A7T3391A	289A		47	2030	2N5210
A7T3392	289A		81	2051	MPS3392
A7T5172	289A		81	2051	MPS5172
A7U	108	3452	86	2038	
A7V	108	3452	86	2038	
A7W	108	3452	86	2038	
A7Y	123A	3444	20	2051	
A7Z	184		57	2017	
OA8-1	160	3007	245	2004	
OA8-1-12	160	3007	245	2004	
OA8-1-12-7	160	3007	245	2004	
A8-1-70	160	3007	245	2004	
A8-1-70-1	160		245	2004	
A8-1-70-12	160	3007	245	2004	
A8-1-70-12-7	160	3007	245	2004	
A8-1-A-4-7B	160		245	2004	
A8-10	160	3007	245	2004	
A8-11	160	3007	245	2004	
A8-12	160	3007	245	2004	
A8-13	160	3007	245	2004	
A8-14	160	3007	245	2004	
A8-15	160	3007	245	2004	
A8-16	160	3007	245	2004	
A8-17	160	3007	245	2004	
A8-18	160	3007	245	2004	
A8-19	160	3007	245	2004	
A8-1A	160	3007	245	2004	
A8-1A0	160	3007	245	2004	
A8-1A0R	160	3007	245	2004	
A8-1A1	160	3007	245	2004	
A8-1A19	160	3007	245	2004	
A8-1A2	160	3007	245	2004	
A8-1A21	160	3007	245	2004	
A8-1A3	160	3007	245	2004	
A8-1A3P	160	3007	245	2004	
A8-1A4	160	3007	245	2004	
A8-1A4-7	160	3007	245	2004	
A8-1A4-7B	160		245	2004	
A8-1A5	160	3007	245	2004	
A8-1A5L	160	3007	245	2004	
A8-1A6	160	3007	245	2004	
A8-1A6-4	160	3007	245	2004	
A8-1A7	160	3007	245	2004	
A8-1A7-1	160	3007	245	2004	
A8-1A8	160	3007	245	2004	
A8-1A82	160	3007	245	2004	
A8-1A9	160	3007	245	2004	
A8-1A9G	160	3007	245	2004	
A80	126	3006/160	52	2024	
A80-12G					MRF492
A800	125	3032A	531	1114	1N4006
A800-511-00	129	3025	244	2027	
A800-516-00	129	3025	244	2027	
A800-523-01	159	3114/290	82	2032	
A800-523-02	159	3114/290	82	2032	
A800-527-00	159	3114/290	82	2032	
A80052402	130	3027	14	2041	
A8015613	129	3025	244	2027	
A80414120	130	3027	14	2041	
A80414130	130	3027	14	2041	
OA81	109	3091	245	1123	
A815185	129	3025	244	2027	
A815185E	129	3025	244	2027	
A815199	159	3114/290	82	2032	
A815199-6	159	3114/290	82	2032	
A815203-5	121MP	3013	239(2)	2006(2)	
A815211	159	3114/290	82	2032	
A815213	159	3114/290	82	2032	
A815229	159	3114/290	82	2032	
A815247	159	3114/290	82	2032	
A82	126	3006/160	52	2024	
A825	290A		65	2050	
A826	290A	3114/290			
A826P	290A	3114/290			
A826Q	290A	3114/290			
A826RY	290A	3114/290			
A829	159	3114/290	82	2032	
A829A	159	3114/290	82	2032	
A829B	159	3114/290	82	2032	
A829C	159	3114/290	82	2032	
A829D	159	3114/290	82	2032	
A829E	159	3114/290	82	2032	
A829F		3114	82	2032	

Industry Standard No.	ECG	SK	GE	RS 276-	MOTOR.
A83	126	3006/160	52	2024	
A833	159	3114/290	82	2032	
A836	234		244	2027	
A836D	234		244	2027	
A836E	234		244	2027	
A837	88		263	2043	
A83P2B	129	3025	244	2027	
A84	160	3006	245	2004	
A840	288$	3434/288	223		
A8405	159	3114/290	82	2032	
A844	234	3114/290	244	2027	
A844D	234		244	2027	
A844E	234		244	2027	
A84A-70	121	3009	239	2006	
A84A-705	121	3009	239	2006	
A84A19G	121	3009	239	2006	
A85	126	3007	52	2024	
A8540	159	3114/290	82	2032	
A86		3006	245	2004	
A86-10-2	101	3861	8	2002	
A86-213-2	154	3045/225	40	2012	
A86-214-2	154	3045/225	40	2012	
A86-215-2	154	3045/225	40	2012	
A86-3-1	115		6GX1		
A86-316-2	154	3045/225	40	2012	
A86-4-1	114	3120	6GD1	1104	
A86-44-2	101	3861	8	2002	
A86-565-2	194	3275	220		
A86-9-1	113A	3119/113	6GC1		
A86X0030-100	121MP	3013	239(2)	2006(2)	
A87	126	3006/160	52	2024	
A88	123A	3444	20	2051	
A88(JAPAN)	123A	3444	20	2051	
A88-70	160		245	2004	
A88-705	160		245	2004	
A880-250-107	129	3025	244	2027	
A8867	159	3114/290	82	2032	
A88B-70	160	3006	245	2004	
A88B-705	160	3006	245	2004	
A88B19G	160	3006	245	2004	
A88C-70	100	3005	1	2007	
A88C-705	100	3005	1	2007	
A88C19G	100	3005	1	2007	
A89	160	3006	245	2004	
A8A	194		220		
A8B	123A	3444	20	2051	
A8C	185	3191	58	2025	
A8D	128		243	2030	
A8E	196	3054	241	2020	
A8F	241	3188A/182	57	2020	
A8G	123A	3444	20	2051	
A8J	188	3199	226	2018	
A8K	196	3054	241	2020	
A8L	123A	3444	20	2051	
A8M	162	3438	35		
A8N	194	3275	220		
A8P	130		14	2041	
A8P-2-70	102A	3004	53	2007	
A8P-2-705	102A	3004	53	2007	
A8P-2A9G	102A	3004	53	2007	
A8P-404-ORN	121	3717	239	2006	
A8P404-ORN	121	3717	239	2006	
A8P404F	121	3009	239	2006	
A8R	199	3245	62	2010	
A8S	108	3452	86	2038	
A8T	163A	3439	36		
A8T3391	123AP		47	2030	2N5210
A8T3391A	123AP		47	2030	2N5210
A8T3392	123AP		81	2051	MPS3392
A8T3702	159		82	2032	MPS3702
A8T3703	159		82	2032	MPS3703
A8T3704			47	2030	MPS3704
A8T3705			47	2030	MPS3705
A8T3706	123AP		47	2030	MPS3706
A8T3707	123AP		81	2051	
A8T3708	123AP		81	2051	MPS3708
A8T3709	123AP		81	2051	MPS3709
A8T3710	123AP		81	2051	MPS3710
A8T3711	123AP		81	2051	MPS3711
A8T4026	294		67	2023	MPSA55
A8T4027	159				MPSA56
A8T4028	294		67	2023	MPSA55
A8T4029	159				MPSA56
A8T404			82	2032	MPS404
A8T404A	159		82	2032	MPS404A
A8T4058	159		82	2032	MPS6522
A8T4059	159		82	2032	MPS6516
A8T4060	159		82	2032	MPS6516
A8T4061	159		82	2032	MPS6517
A8T4062	159		82	2032	MPS6518
A8T5172	123AP		81	2051	MPS5172
A8U	130	3133/164	14	2041	
A8V	154	3044	40	2012	
A8VA	154		40	2012	
A8W	130		14	2041	
A8Y	241	3188A/182	57	2020	
A8Z	194	3045/225	220		
OA9	109	3091		1123	
A9-175	199	3245	62	2010	
OA90	109	3006/160		1123	
A90(TRANSISTOR)	126		52	2024	
A909-1008	160	3006	245	2004	
A909-1009	126	3006/160	52	2024	
A909-1010	126	3006/160	52	2024	
A909-1011	102A	3004	20	2007	
A909-1012	102A	3004	20	2007	
A909-1013	102A	3004	20	2007	
A909-1015	109	3087	1N34AS	1123	
A909-1016	110MP	3709		1123(2)	
A909-1017	109	3087	1N34AS	1123	
A909-1018	116	3311	504A	1104	
A909-1019	116	3031A	504A	1104	
A909-27125-160	108	3452	86	2038	
OA90G	109	3088			
OA90LF	109	3088	1N60		
A90M	6356			1123	
OA90MLF	109	3088			
A90T2	108	3452	86	2038	
OA90Z	109	3088		1123	
OA90ZA			1N60	1123	
OA91	109	3087		1123	
A916-31025-58	108		86	2038	
A916-31025-5B	108	3452	86	2038	
A92	126	3007	52	2024	
A92-1-70	105	3012	4		
A92-1-705	105	3012	4		
A92-1A9G	105	3012	4		
A921-59B	108	3452	86	2038	
A921-62B	108	3452	86	2038	
A921-63B	108	3452	86	2038	
A921-64B	108	3452	86	2038	
A921-70B	159	3114/290	82	2032	
A9218	519			1122	
A9228-3	519			1122	
A93	126	3007	52	2024	
A937	123A	3444	20	2051	
A937-1	123A	3444	20	2051	
A937-3	123A	3444	20	2051	
A94	126	3006/160	52	2024	
A94004	121	3009	239	2006	
A94037	159	3114/290	82	2032	
A94063	129	3025	244	2027	
A945-0	159		82	2032	
OA95	110MP	3709	1N34AS	1123(2)	
A95-5280	113A	3119/113	6GC1		
A95-5281	116		504A	1104	
A95-5286	358		42		
A95-5289	116		504A	1104	
A95-5295	119		CR-2		
A95-5296	118	3066	CR-1		
A95-5297	113A	3119/113	6GC1		
A95-5314	500A	3304	527		
A95115	101	3861	8	2002	
A952	159	3114/290			
A95211	101	3861	8	2002	
A95227	159	3114/290	82	2032	
A95232	159	3114/290	82	2032	
A964-17887	121	3009	239	2006	
A970246	159	3114/290	82	2032	
A970248	159	3114/290	82	2032	
A970251	159	3114/290	82	2032	
A970254	159	3114/290	82	2032	
A97A83	121	3009	239	2006	
A984193	159	3114/290	82	2032	
OA99	109	3088			
A9903	6408	9084A			
A991-01-0098	159	3114/290	82	2032	
A991-01-1225	159	3114/290	82	2032	
A991-01-1316	108	3452	86	2038	
A991-01-1319	159	3114/290	82	2032	
A991-01-3058	159	3114/290	82	2032	
A992-00-1192	121	3009	239	2006	
A99S07	101	3861	8	2002	
A99SK5		3861	8	2002	
A99SK7	101	3861	8	2002	
A9A	108	3452	86	2038	
A9B	123A	3444	20	2051	
A9C	172A	3156	64		
A9D	108	3452	86	2038	
A9E	123A	3444	20	2051	
A9F	123A	3444	20	2051	
A9G	123A	3444	20	2051	
A9H	123A	3444	20	2051	
A9J	123A	3444	20	2051	
A9K	162	3438	35		
A9L-4-70	102A	3003	53	2007	
A9L-4-705	102A	3003	53	2007	
A9L-4A9G	102A	3003	53	2007	
A9M	191	3054/196	249	1123	
A9N	181		75	2041	
A9P	162	3438	35		
A9R	163A	3439	36		
A9S	123A	3444	20	2051	
A9T	123A	3444	20	2051	
A9U	123A	3444	20	2051	
A9V	152	3893	66	2048	
A9W	123A	3444	20	2051	
A9Y	123A	3444	20	2051	
AA015	112	3089	1N82A		
AA1	100	3005	1	2007	
AA10	147A	3095	ZD-33		
AA10-1	147A	3095	ZD-33		
AA100	116	3016	504A	1104	1N4002
AA1000	125	3033	531	1114	1N4007
AA101	5402	3638			
AA107	5401	3638/5402			
AA108	5402	3638			
AA109	5402	3638			
AA111	109	3087	1N34AS	1123	
AA112	109	3087	1N34AS	1123	
AA112P	109	3087	1N34AS	1123	
AA113	109	3087	1N34AS	1123	
AA113852-2	909		IC-249		
AA114	109	3090	1N34AS	1123	
AA115	5402	3638			
AA116	109		1N34AS	1123	
AA117	109	3087	1N34AS	1123	
AA118	109	3087	1N34AS	1123	
AA119	109	3709/110MP	1N34AS	1123	
AA120	109	3091	1N34AS	1123	
AA121	109	3090	1N34AS	1123	
AA123	109	3090	1N34AS	1123	
AA130	109	3091	1N34AS	1123	
AA131	109	3087	1N34AS	1123	
AA132	109	3087	1N34AS	1123	
AA133	225	3045	256		
AA133(DIODE)	109	3090			
AA134	109	3087	1N34AS	1123	
AA135	109	3087	1N34AS	1123	
AA136	109	3087	1N34AS	1123	
AA137	109	3087	1N34AS	1123	
AA138	109	3091	1N34AS	1123	
AA139	109	3090	1N34AS	1123	
AA140	109	3091	1N34AS	1123	
AA142	109	3091	1N34AS	1123	
AA143	109	3087	1N34AS	1123	
AA143S	109	3090	1N34AS	1123	
AA144	109	3090	1N34AS	1123	
AA1M	225	3045	256		
AA2	101	3861	8	2002	
AA200	116	3016	504A	1104	1N4003
AA218	109	3087	1N34AS	1123	
AA2A	225	3045	256		
AA2K	225	3045	256		
AA2SB240A	121	3009	239	2006	
AA2Z	225	3045	256		
AA3	160		245	2004	
AA300	116	3031A	504A	1104	1N4004
AA310	225	3045	256		
AA3M	225	3045	256		
AA4	121	3717	239	2005	
AA400	116	3017B/117	504A	1104	1N4004
AA5	105	3012	4	2005	
AA50	116	3017B/117	504A	1104	1N4001
AA500	116	3017B/117	504A	1104	1N4005
AA600	116	3017B/117	504A	1104	1N4005
AA779	109	3088	1N34AS	1123	
AA8-1-70	160	3007	245	2004	
AA8-1-705	160	3007	245	2004	
AA8-1A9G	160	3007	245	2004	
AA800	125	3032A	531	1114	1N4006
AAY-22	116	3016	504A	1104	
AAY10-120	5872		5032		
AAY139	109	3087	1N34AS	1123	
AAY15	109	3087	1N34AS	1123	
AAY18	109	3087	1N34AS	1123	
AAY22	109	3091	1N34AS	1123	
AAY27	109	3091	1N34AS	1123	

Industry Standard No.	ECG	SK	GE	RS 276-	MOTOR.
AAY30	109	3087	1N34AS	1123	
AAY33	109	3087	1N34AS	1123	
AAY46	109	3087	1N34AS	1123	
AAZ10	109	3090	1N34AS	1123	
AAZ18	109	3087	1N34AS	1123	
AAZ18D	116	3311	504A	1104	
AB100	156		512		MR501
AB1000	156	3051	512		MR510
AB200	156		512		MR502
AB2027	146A	3064	ZD-27		
AB300	156		512		MR504
AB400	156		512		MR504
AB50	156	3081/125	512		MR501
AB500	156		512		MR506
AB600	156	3051	512		MR506
AB700	156		512		
AB800	156	3051	512		MR508
ABF			57	2017	
AC-107	102A	3004	53	2007	
AC-113	102A	3004	53	2007	
AC-113A	102A	3004	53	2007	
AC-114	102A	3004	53	2007	
AC-116	102A	3004	53	2007	
AC-117	102A	3004	53	2007	
AC-117A	102A	3004	53	2007	
AC-117P	102A	3004	53	2007	
AC-121IV	102A		53	2007	
AC-122	102A	3004	53	2007	
AC-123	102A	3004	53	2007	
AC-125	102A	3004	53	2007	
AC-126	102A	3004	53	2007	
AC-127	103A	3010	59	2002	
AC-128	102A	3004	53	2007	
AC-132	102A	3004	53	2007	
AC-150	102A	3004	53	2007	
AC-151	102A	3004	53	2007	
AC-152	102A	3004	53	2007	
AC-154	102A	3004	53	2007	
AC-155	102A	3004	53	2007	
AC-156	102A	3004	53	2007	
AC-157	103A	3862/103	59	2002	
AC-161	102A	3004	53	2007	
AC-162	102A	3004	53	2007	
AC-165	102A	3004	53	2007	
AC-166	102A	3004	53	2007	
AC-167	102A	3004	53	2007	
AC-168	102A	3004	53	2007	
AC-169	102A	3004	53	2007	
AC-172	103A	3010	59	2002	
AC-175A	103A	3444/123A	20	2051	
AC-175B	103A	3444/123A	20	2051	
AC-175P	103A	3444/123A	20	2051	
AC-N7B	102A		53	2007	
AC06BT	56004	3658			
AC06DT	56006	3659			
AC100	156		512		MR501
AC100(PHILCO)	283		36		
AC1000	5809	9010			MR510
AC105	102A	3004	53	2007	
AC106	102A	3004	53	2007	
AC107	160		245	2004	
AC107M	160		245	2004	
AC107N			50	2004	
AC108	102A		53	2007	
AC109	102A		53	2007	
AC10BT	56004	3658			
AC10DT	56006	3659			
AC110	102A		53	2007	
AC113	102A		53	2007	
AC114	102A		53	2007	
AC115	102A		53	2007	
AC116	102A		53	2007	
AC117	102A		53	2007	
AC117A	102A		53	2007	
AC117B	102A		53	2007	
AC117P	102A		53	2007	
AC118	102A		53	2007	
AC119	102A		53	2007	
AC120	102A	3424/5158A	53	2007	
AC121	102A		53	2007	
AC121-IV	102A		53	2007	
AC121-V	102A		53	2007	
AC121-VI	102A		53	2007	
AC121-VII	102A		53	2007	
AC121IV			53	2007	
AC121V	102A		53	2007	
AC121VI	102A		53	2007	
AC121VII	102A		53	2007	
AC122	102A		53	2007	
AC122-30	102A		53	2007	
AC122-GRN		3123	53	2007	
AC122-RED		3123	53	2007	
AC122-VIO			53	2007	
AC122-YEL		3123	53	2007	
AC122GRN	102A		53	2007	
AC122RED	102A		53	2007	
AC122YEL	102A		53	2007	
AC123	102A		53	2007	
AC124	102A		53	2007	
AC125	102A		53	2007	
AC126	158		53	2007	
AC127	103A	3010	59	2002	
AC127-01	103A	3835	59	2002	
AC127-132	103A	3010	59	2002	
AC128	158	3004/102A	53	2007	
AC128-01	158		53	2007	
AC128/01	158	3004/102A	53	2007	
AC12801	158		53	2007	
AC128K	158		53	2007	
AC129	160		245	2004	
AC130	101	3861	8	2002	
AC131	102A		53	2007	
AC131-30	158		53	2007	
AC132	102A		53	2007	
AC132-01	102A		53	2007	
AC133A	102A		53	2007	
AC134	102A		53	2007	
AC135	102A	3123	53	2007	
AC136	102A		53	2007	
AC137	102A		53	2007	
AC138	158		53	2007	
AC139	158		53	2007	
AC141	103A	3835	59	2002	
AC141B	103A	3835	59	2002	
AC141K	103A	3835	59	2002	
AC142	158		53	2007	
AC142K	158		53	2007	
AC148	121	3717	239	2006	
AC150	102A	3311	53	2007	
AC150-GRN			51	2004	
AC150-YEL			51	2004	
AC150GRN	102A		53	2007	
AC150YEL	102A		53	2007	
AC151	102A	3004	53	2007	
AC151-IV	102A		53	2007	
AC151-RIV	102A		53	2007	
AC151-RV	102A		53	2007	
AC151-RVI	102A		53	2007	
AC151-V	102A		53	2007	
AC151-VI	102A		53	2007	
AC151-VII	102A		53	2007	
AC151IV	102A		53	2007	
AC151R	102A	3004	53	2007	
AC151RIV	102A		53	2007	
AC151RV	102A		53	2007	
AC151RVI	102A		53	2007	
AC151V	102A		53	2007	
AC151VI	102A		53	2007	
AC151VII	102A		53	2007	
AC152	102A		53	2007	
AC152-IV	102A			2007	
AC152-V	102A			2007	
AC152-VI	102A			2007	
AC152IV	102A		53	2007	
AC152V	102A		53	2007	
AC152VI	102A		53	2007	
AC153	158	3004/102A	53	2007	
AC153K	158	3004/102A	53	2007	
AC154	158		53	2007	
AC155	102A		53	2007	
AC156	102A		53	2007	
AC157	103		8	2002	
AC160	102A		53	2007	
AC160-GRN			51	2004	
AC160-RED			51	2004	
AC160-VIO			51	2004	
AC160-YEL			51	2004	
AC160A	102A		53	2007	
AC160B	102A		53	2007	
AC160GRN	102A		53	2007	
AC160RED	102A		53	2007	
AC160YEL	102A		53	2007	
AC161	102A	3123	53	2007	
AC162	102A		53	2007	
AC163	102A		53	2007	
AC164	126	3006/160	52	2024	
AC165	102A		53	2007	
AC166	158		53	2007	
AC167	158		53	2007	
AC168	102A		53	2007	
AC169	126		52	2024	
AC170	102A	3123	53	2007	
AC171	102A		53	2007	
AC172	103A		59	2002	
AC173	102A		53	2007	
AC175	103A	3124/289	59	2002	
AC175A			59	2002	
AC175B			59	2002	
AC175P			59	2002	
AC176	103A	3010	59	2002	
AC176K	158		53	2007	
AC178	158		53	2007	
AC179	103A	3835	59	2002	
AC180	158		53	2007	
AC180K	158		53	2007	
AC181	103A	3011	59	2002	
AC181K	103A	3835	59	2002	
AC182	102A	3123	53	2007	
AC183	103A		59	2002	
AC184	102A		53	2007	
AC185	103A	3835	59	2002	
AC186	103A	3835	59	2002	
AC187	103A	3010	59	2002	
AC187/01	101	3861	8	2002	
AC18701	103A		59		
AC187K	103A	3835	59	2002	
AC187R	103A	3835	59	2002	
AC188	158	3004/102A	53	2007	
AC188/01	158	3004/102A	53	2007	
AC18801	158		53	2007	
AC188K	158		53	2007	
AC18M/U				2002	
AC191	102A	3004	53	2007	
AC192	102A	3004	53	2007	
AC193	158	3004/102A	53	2007	
AC193K	158	3004/102A	53	2007	
AC194	103A	3010	59	2002	
AC194K	103A	3010	59	2002	
AC200	156	3051	512		MR502
AC30	156	3311	512		
AC300	5804	9007		1144	MR504
AC400	5804	9007		1144	MR504
AC50	156	9003/5800	512		MR501
AC50(DIODE)	177		300	1122	
AC50(RECTIFIER)	5800	3848		1142	
AC500	5806	3848			MR506
AC600	5806	3848			MR506
AC800	5808	9009			MR508
AC880					MR508
AC9082	159	3114/290	82	2032	
AC9083	159	3114/290	82	2032	
AC9084	159	3114/290	82	2032	
AC9085	159	3114/290	82	2032	
ACDC9000-1	159	3114/290	82	2032	
ACR810-101	126	3008	52	2024	
ACR810-102	126	3008	52	2024	
ACR810-103	126	3008	52	2024	
ACR810-104	100	3004/102A	1	2007	
ACR810-105	100	3004/102A	1	2007	
ACR810-106	100	3004/102A	1	2007	
ACR810-107	109	3087	1N34AS	1123	
ACR810-108	116	3031A	504A	1104	
ACR83-1001	126	3008	52	2024	
ACR83-1002	126	3008	52	2024	
ACR83-1003	126	3008	52	2024	
ACR83-1004	100	3004/102A	1	2007	
ACR83-1005	100	3004/102A	1	2007	
ACR83-1006	100	3004/102A	1	2007	
ACR83-1007	109	3087		1123	
ACR83-1008	116	3311	504A	1104	
ACY-17	102A$	3004/102A	2	2007	
ACY-18	102A$	3004/102A	2	2007	
ACY-19	102A$	3004/102A	2	2007	
ACY-20	102A$	3004/102A	2	2007	
ACY-21	102A$	3004/102A	53	2007	
ACY-22	102A$	3003	53	2007	
ACY-23	102A$	3004/102A	53	2007	
ACY-32	102A	3004	53	2007	
ACY16	102A	3009	239	2006	
ACY17	102A$	3722/102	2	2007	
ACY17-1	102A$	3004/102A	53	2007	
ACY18	102A$	3722/102	2	2007	
ACY18-1	102A$	3004/102A	53	2007	
ACY19	102A$	3722/102	2	2007	
ACY19-1	102A$	3004/102A	53	2007	
ACY20	102A$		53	2007	
ACY20-1	102A$	3004/102A	53	2007	
ACY21	102A$		53	2007	

Industry Standard No.	ECG	SK	GE	RS 276-	MOTOR.
ACY21-1	102A$	3004/102A	53	2007	
ACY22	102A$		53	2007	
ACY22-1	102A$	3004/102A	53	2007	
ACY23	102A$	3123	53	2007	
ACY23-V	102A$		53	2007	
ACY23-VI	102A$		53	2007	
ACY23V	102A$		53	2007	
ACY23VI	102A		53	2007	
ACY24	160		245	2004	
ACY27	102A	3004	53	2007	
ACY28	102A	3004	53	2007	
ACY29	102A	3004	53	2007	
ACY30	102A	3004	53	2007	
ACY31	102A	3004	53	2007	
ACY32	102A	3123	53	2007	
ACY32-V	102A		53	2007	
ACY32-VI	102A		53	2007	
ACY32V	102A		53	2007	
ACY32VI	102A		53	2007	
ACY33	158		53	2007	
ACY33-VI	158		53	2007	
ACY33-VII	158		53	2007	
ACY33-VIII	158		53	2007	
ACY33VI	158		53	2007	
ACY33VII	158		53	2007	
ACY34	102A	3004	53	2005	
ACY35	102A	3004	53	2005	
ACY36	102A	3004	53	2007	
ACY38	102A	3123	53	2007	
ACY39	176	3845	80		
ACY40	102	3005	1	2007	
ACY41	102	3004	53	2007	
ACY41-1	102A	3004	53	2007	
ACY44	102A	3004	53	2007	
ACY44-1	102A	3004	53	2007	
ACZ	160		245	2004	
ACZ21			51	2004	
AD-140	121	3009	239	2006	
AD-148	121	3009	239	2006	
AD-149	121	3009	239	2006	
AD-150	121	3009	239	2006	
AD-152	131	3009	239	2006	
AD-156	131	3009	239	2006	
AD-157	155	3009	239	2006	
AD-159	121	3009	239	2006	
AD-162			30	2006	
AD-1UF	116		504A	1104	
AD-29B	506	3080	511	1114	
AD-29S	506	3130	511	1114	
AD10	116	3311	504A	1104	
AD100	116	3311	504A	1104	
AD100(DIODE)	177		300	1122	
AD103	179	3014	239	2006	
AD104	179	3009	239	2006	
AD105	179	3009	239	2006	
AD130	121	3009	239	2006	
AD130-III	121		239	2006	
AD130-IV	121		239	2006	
AD130-V	121		239	2006	
AD131	121	3009	239	2006	
AD131-III	121	3717	239	2006	
AD131-IV	121	3717	239	2006	
AD131-V	121	3717	239	2006	
AD132	121	3014	239	2006	
AD133	179	3642	76		
AD138	121	3009	239	2006	
AD138/50	121	3642/179	239	2006	
AD13850	121	3717	239	2006	
AD139	104	3009	16	2006	
AD140	121	3014	239	2006	
AD1403AN					MC1403U
AD1403N					MC1403U
AD1408-7D					MC1408-L7
AD1408-8D					MC1408L8
AD142	179	3642	76		
AD143	121	3642/179	76		
AD143B	121	3642/179	76		
AD143R	121	3717	239	2006	
AD145	121	3717	239	2006	
AD148			3	2006	
AD149	104	3013	16	2006	
AD149-01	121	3009	239	2006	
AD149-02	121	3009	239	2006	
AD149-IV	121	3717	239	2006	
AD149-V	121	3717	239	2006	
AD149B	121	3009	239	2006	
AD149C	121	3009	239	2006	
AD150	121	3311	239	2006	
AD150(DIODE)	177		300	1122	
AD150-IV	121	3717	239	2006	
AD150-V	121	3717	239	2006	
AD1508-8D					MC1508L8
AD152	131	3198	44	2006	
AD153	121	3717	239	2006	
AD155	131		44	2006	
AD156	131	3198	44	2006	
AD157	155	3839	43		
AD159	121	3717	239	2006	
AD160	175		246	2020	
AD161	155	3839	43		
AD162	131	3198	44	2006	
AD164	131	3198	44	2006	
AD165	155	3839	43		
AD166	127	3764	25		
AD167	127	3764	25		
AD169	131	3198	44	2006	
AD200	116	3311	504A	1104	
AD200(DIODE)	177		300	1122	
AD262	131		44	2006	
AD263	131		44	2006	
AD29	156		512		
AD29A-4	159	3114/290	82		
AD29A-5	159	3114/290	82		
AD29A-6	159	3114/290	82		
AD29A-9	159	3114/290	82		
AD29E-1	159	3114/290	82	2032	
AD29E-2	159	3114/290	82	2032	
AD29E10	159	3114/290	82	2032	
AD29E4	159	3114/290	82	2032	
AD29E5	159	3114/290	82	2032	
AD29E6	159	3114/290	82	2032	
AD29E7	159	3114/290	82	2032	
AD29E8	159	3114/290	82	2032	
AD29E9	159	3114/290	82	2032	
AD29S	506	3130	511	1114	
AD30	177		300	1122	
AD301AL					LM301AH
AD301AN	975	3596/976			
AD30A1	159	3114/290	82	2032	
AD30A2	159	3114/290	82	2032	
AD30A3	159	3114/290	82	2032	
AD30A4	159	3114/290	82	2032	
AD30A5	159	3114/290	82	2032	
AD4001	116	3017B/117	504A	1104	

Industry Standard No.	ECG	SK	GE	RS 276-	MOTOR.
AD4002	125		531	1114	
AD4003	125		531	1114	
AD4004	125		531	1114	
AD4005	125		531	1114	
AD4006	125		531	1114	
AD4007	125	3033	531	1114	
AD50	116	3311	504A	1104	
AD505J					MC1776CG
AD505K					MC1776CG
AD505S					MC1776G
AD509J					LM301AH
AD509K					LM301AH
AD509S					LM101AH
AD518J					LM301AH
AD518K					LM301AH
AD518S					LM101AH
AD530					MC1595L
AD531					MC1595L
AD532J					MC1595G
AD559JD					MC1408L8
AD559K					MC1408L8
AD559KD					MC1408L8
AD559S					MC1508L8
AD559SD					MC1508L8
AD580J					MC1403U
AD580K					MC1403P1
AD580M					MC1403AP1
AD580S					MC1503U
AD580T					MC1503AU
AD710CH	910		IC-251		
AD710CN	910D		IC-252		
AD711CH	911		IC-253		
AD711CN	911D		IC-254		
AD741		3514		010	
AD741C	941	3514		010	
AD741CH	941	3514	IC-263	010	
AD741CJ					MC1741CG
AD741CN	941D	3552/941M	IC-264		
AD741J					MC1741G
AD741K					MC1741G
AD741L					MC1741G
AD741S					MC1741SG
AD7520D					MC3410L
AD7520F					MC3410L
AD7520N					MC3410L
AD8	506		511	1114	
ADDAC-08AD					DAC-08AQ
ADDAC-08CD					DAC-08CQ
ADDAC-08D					DAC-08Q
ADDAC-08ED					DAC-08EQ
ADDAC-08HD					DAC-08HQ
ADY-26	105		4		
ADY-27	121	3004/102A	53	2007	
ADY22	121	3009	239	2006	
ADY23	121	3014	239	2006	
ADY24	121	3009	239	2006	
ADY26	213	3012/105	254		
ADY27	121	3717	239	2006	
ADY28	121	3009	239	2006	
ADZ11	105	3012	4		
ADZ12	105	3012	4		
AE-50	102A	3004	53	2007	
AE-904-03	743		IC-214		
AE-907	737	3375	IC-16		
AE10	177	3311	300	1122	
AE100	177	3311	300	1122	
AE150	177	3311	300	1122	
AE1A	156	3311	504A	1104	
AE1B	156	3051	512		
AE1C	156	3051	512		
AE1D	156	3051	512		
AE1E	156	3051	512		
AE1F	156	3051	512		
AE1G	156	3051	512		
AE200	177	3311	300	1122	
AE30	177	3311	300	1122	
AE3A	5800	9003		1142	
AE3B	5801	9004		1142	
AE3C	5802	9005		1143	
AE3D	5804	9007		1144	
AE3E	5806	3051/156			
AE3F	5808	3848/5806			
AE3G	5809	9010			
AE50	177	3311	300	1122	
AE900	786	3140	IC-227		
AE902	786	3140	IC-227		
AE904	743	3172	IC-214		
AE904-02	743		IC-214		
AE907	737	3375	IC-16		
AE907-51	737	3375	IC-16		
AE920	1232	3287/1132		703	
AEX-82308		3027	255	2041	
AEX-85715	129	3004/102A	244	2027	
AEX79846	130		14	2041	
AEX9846	130	3027	14	2041	
AF-101	100	3005	1	2007	
AF-105A	160	3006	245	2004	
AF-106		3006	50	2004	
AF-109	160	3006	245	2004	
AF-121	160	3006	245	2004	
AF-137			50	2004	
AF-166	160	3006	245	2004	
AF-182	160	3006	245	2004	
AF1(DIODE)	506			1114	
AF101	160		245	2004	
AF102	160	3006	245	2004	
AF105	126	3008	52	2024	
AF105A	160		245	2004	
AF106	160	3006	245	2004	
AF106A	160		245	2004	
AF107	160		245	2004	
AF108	126	3006/160	52	2024	
AF109	160		245	2004	
AF109R	160		245	2004	
AF110	160	3006	245	2004	
AF111	160	3006	245	2004	
AF112	160	3006	245	2004	
AF113	160	3006	245	2004	
AF114	126	3006/160	245	2004	
AF114N	160	3006	245	2004	
AF115	126	3006/160	245	2004	
AF115N	160	3006	245	2004	
AF116	126	3006/160	245	2004	
AF116N	160	3006	245	2004	
AF117	126	3006/160	245	2004	
AF117C	160		245	2004	
AF117N	160	3006	245	2004	
AF118	160	3006	245	2004	
AF119	160	3006	245	2004	
AF120	126	3006/160	52	2024	
AF121	160		245	2004	
AF121S	160		245	2004	
AF122	160		245	2004	
AF124	126	3006/160	245	2004	

Industry Standard No.	ECG	SK	GE	RS 276-	MOTOR.
AF125	126	3006/160	245	2004	
AF126	126	3006/160	245	2004	
AF127	126	3005	245	2004	
AF127/01	160		245	2004	
AF128	160	3005	245	2004	
AF129	126	3006/160	52	2024	
AF130	126	3006/160	52	2024	
AF131	126	3006/160	52	2024	
AF132	126	3006/160	52	2024	
AF133	126	3006/160	52	2024	
AF134	126	3008	52	2024	
AF135	126	3008	52	2024	
AF136	126	3008	52	2024	
AF137	126	3008	52	2024	
AF137A	160		245	2004	
AF138	126	3006/160	52	2024	
AF138/20		3008	51	2004	
AF138/290	100	3721	1	2007	
AF139	160		245	2004	
AF142	160		245	2004	
AF143	160		245	2004	
AF144	126	3006/160	52	2024	
AF146	160	3006	245	2004	
AF147	160	3006	245	2004	
AF148	160	3006	245	2004	
AF149	160	3006	245	2004	
AF150	160	3006	245	2004	
AF164	160	3006	245	2004	
AF165	160	3006	245	2004	
AF166	160	3008	245	2004	
AF167	160	3006	245	2004	
AF168	160	3006	245	2004	
AF169	160	3006	245	2004	
AF170	160	3006	245	2004	
AF171	160	3006	245	2004	
AF172	160	3006	245	2004	
AF178	160	3006	245	2004	
AF179	160	3006	245	2004	
AF180	160	3006	245	2004	
AF181	160	3006	245	2004	
AF182	160		245	2004	
AF185	160	3006	245	2004	
AF186	160	3006	245	2004	
AF186G	160	3006	245	2004	
AF186W	160	3006	245	2004	
AF187	102A		53	2007	
AF188	102A	3123	53	2005	
AF192	101	3861	8	2002	
AF193	160	3006	245	2004	
AF200	160	3006	245	2004	
AF200U	160	3722/102	2	2007	
AF201	160	3006	245	2004	
AF201C	160	3006	245	2004	
AF201U	160	3722/102	2	2007	
AF202	160	3006	245	2004	
AF202L	160	3006	245	2004	
AF202S	160	3006	245	2004	
AF21490	159	3114/290	82	2032	
AF239	160	3006	245	2004	
AF239S	160	3006	245	2004	
AF240	160	3006	245	2004	
AF251	160		245	2004	
AF252	160		245	2004	
AF253	160		245	2004	
AF256	160	3006	245	2004	
AF267	160		245	2004	
AF279	160	3006	245	2004	
AF280	121	3717	239	2006	
AF306	160		50	2004	
AF3570	159	3114/290	82	2032	
AF3590	159	3114/290	82	2032	
AF699	159	3114/290	82	2032	
AFC3527			20	2051	
AFCS1170F	159	3114/290	82	2032	
AFS-160-1017	109	3087	1N34AS	1123	
AFS-160-1020	110MP	3709		1123(2)	
AFS-160-1021	116(4)	3017B/117	504A(4)	1104	
AFS24226	159	3114/290	82	2032	
AFT0019M	159	3114/290	82	2032	
AFT052	159	3114/290	82	2032	
AFT1341	159	3114/290	82	2032	
AFT1746	159	3114/290	82	2032	
AFY10	126	3006/160	52	2024	
AFY11	126	3006/160	52	2024	
AFY12	160	3006	245	2004	
AFY14	160	3006	245	2004	
AFY15	160	3006	245	2004	
AFY16	160	3006	245	2004	
AFY18	126	3006/160	52	2024	
AFY18C	160		245	2004	
AFY18D	160		245	2004	
AFY18E	160		245	2004	
AFY19	126		245	2004	
AFY34	160	3006	245	2004	
AFY37	160	3006	245	2004	
AFY39	160	3006	245	2004	
AFY40	160	3006	245	2004	
AFY40K	160	3006	245	2004	
AFY40R	160	3006	245	2004	
AFY41	160	3006	245	2004	
AFY42	160	3006	245	2004	
AFZ11	160	3006	245	2004	
AFZ12	160	3006	245	2004	
AFZ23	100	3005	1	2007	
AG100D	116	3016	504A	1104	
AG100G	116	3031A	504A	1104	
AG100J	116	3017B/117	504A	1104	
AG134	126	3006/160	52	2024	
AG30	177	3100/519	300	1122	
AG6	188	3054/196	226	2018	
AGH-001	1212		725		
AH1005	125	3033	531	1114	
AH1010	125	3033	531	1114	
AH1015	125	3033	531	1114	
AH14	125		531	1114	
AH805	125	3033	531	1114	
AH810	125	3033	531	1114	
AH814			509	1114	
AH815	125	3033	531	1114	
AIE	161	3716	39	2015	
AJ-30	116	3016	504A	1104	
AJ10	116	3016	504A	1104	
AJ103	179	3642	76		
AJ15	116	3016	504A	1104	
AJ20	116	3017B/117	504A	1104	
AJ25	116	3016	504A	1104	
AJ30	116		504A	1104	
AJ35	116	3016	504A	1104	
AJ40	116	3016	504A	1104	
AJ5	116	3017B/117	504A	1104	
AJ50	116	3017B/117	504A	1104	
AJ60	116	3017B/117	504A	1104	
AL100	127	3035	25		
AL101	121	3717	239	2006	

Industry Standard No.	ECG	SK	GE	RS 276-	MOTOR.
AL102	127	3035	25		
AL103	127	3035	25		
AL113	179	3642	76		
AL210	160	3035	245	2004	
ALB6494612	196		241	2020	
ALC-1A	178MP	3100/519	300(2)	1122(2)	
ALC1	113A	3119/113	6GC1		
ALC1A	113A	3119/113	6GC1		
ALCI	113A		6GC1		
ALD-3141	123A		20	2051	
ALD-35	123A	3444	20	2051	
ALS-8922	123A	3122	20	2051	
ALZ10	100	3005	1	2007	
AM			18	2030	
AM-010	116	3016	504A	1104	
AM-020	116	3016	504A	1104	
AM-025	116	3016	504A	1104	
AM-030	116	3016	504A	1104	
AM-035	116	3016	504A	1104	
AM-060	116	3017B/117	504A	1104	
AM-22	116	3016	504A	1104	
AM-33	116	3016	504A	1104	
AM-6-5	116	3031A	504A	1104	
AM-G-10	116	3311	504A	1104	
AM-G-11	170	3649	BR-600	1152	
AM-G-22	116	3031A	504A	1104	
AM-G-5	116	3031A	504A	1104	
AM-G-5A	125	3031A		1114	
AM-G-5C	125	3031A	531	1114	
AM-G22	116		504A	1104	
AM005	116	3017B/117	504A	1104	
AM010	116	3016	504A	1104	
AM020	116	3016	504A	1104	
AM025	116		504A	1104	
AM030	116		504A	1104	
AM035	116		504A	1104	
AM040	116	3031A	504A	1104	
AM050	116	3017B/117	504A	1104	
AM060	116		504A	1104	
AM1	5830	3051/156	5004		
AM11	5832	3051/156	5004		
AM12	5832	3051/156	5004		
AM13	116	3016	5004	1104	
AM166039F					LM301AH
AM166039T					LM301AH
AM2	5830	3051/156	5004		
AM21	5834	3016	5004	1104	
AM22	5834		5004	1104	
AM23	116	3016	5004	1104	
AM24	5834	3016	5004	1104	
AM25LS07DC					SN74LS378J
AM25LS07DM					SN54LS378J
AM25LS07FM					SN54LS378W
AM25LS07PC					SN74LS378N
AM25LS08DC					SN74LS379J
AM25LS08DM					SN54LS379J
AM25LS08FM					SN54LS379W
AM25LS08PC					SN74LS379N
AM25LS09DC					SN74LS399J
AM25LS09DM					SN54LS399J
AM25LS09FM					SN54LS399W
AM25LS09PC					SN74LS399N
AM25LS14DC					SN74LS384J
AM25LS14DM					SN54LS384J
AM25LS14FM					SN54LS384W
AM25LS14PC					SN74LS384N
AM25LS15DC					SN74LS385J
AM25LS15DM					SN54LS385J
AM25LS15FM					SN54LS385W
AM25LS15PC					SN74LS385N
AM25LS22DC					SN74LS322J
AM25LS22DM					SN54LS322J
AM25LS22FM					SN54LS322W
AM25LS22PC					SN74LS322N
AM25LS23DC					SN74LS323J
AM25LS23DM					SN54LS323J
AM25LS23FM					SN54LS323W
AM25LS23PC					SN74LS323N
AM25LS2518DC					SN74LS388J
AM25LS2518DM					SN54LS388J
AM25LS2518FM					SN54LS388W
AM25LS2518PC					SN74LS388N
AM26LS30DC					MC3487L
AM26LS30PC					MC3487P
AM26LS31DC					AM26LS31DC
AM26LS31PC					AM26LS31PC
AM26LS32DC					MC3486L
AM26LS32PC					MC3486P
AM26LS33DC					MC3486L
AM26LS33PC					MC3486P
AM26S10DC					MC26S10L
AM26S10PC					MC26S10P
AM26S11DC					MC26S11L
AM26S11PC					MC26S11P
AM2708					MCM2708
AM2716					MCM2716
AM27S31					MCM7641
AM27S33					MCM7643
AM3	116	3017B/117	5004	1104	
AM300	177	3311	300	1122	
AM300A	177	3311	300	1122	
AM301	177	3311	300	1122	
AM301A	177	3311	300	1122	
AM302	177		300	1122	
AM302A	177	3311	300	1122	
AM303	177	3311	300	1122	
AM303A	177	3311	300	1122	
AM304	177	3311	300	1122	
AM304A	177	3311	300	1122	
AM305	177		300	1122	
AM305A	177	3311	300	1122	
AM306	177	3311	300	1122	
AM306A	177	3311	300	1122	
AM307	177	3311	300	1122	
AM307A	177	3311	300	1122	
AM308	177		300	1122	
AM308A	177	3311	300	1122	
AM31	5836	3051/156	5004		
AM32	5836	3016	5004	1104	
AM3235	130	3027	14	2041	
AM33	116		5004	1104	
AM34	5836	3051/156	5004		
AM4	5830	3051/156	5004		
AM4044					MCM66L41
AM405	116	3017B/117	504A	1104	
AM41	5838	3051/156	5004		
AM410	116	3016	504A	1104	
AM415	116	3017B/117	504A	1104	
AM42	5838	3016	5004	1104	
AM420	116	3016	504A	1104	
AM425	116	3016	504A	1104	
AM43	116	3016	5004	1104	
AM430	116	3016	504A	1104	
AM435	116	3016	504A	1104	

Industry Standard No.	ECG	SK	GE	RS 276-	MOTOR.
AM44	5838	3051/156	5004		
AM440	116	3016	504A	1104	
AM445	116	3017B/117	504A	1104	
AM450	116	3017B/117	504A	1104	
AM460	116	3017B/117	504A	1104	
AM5	5830	3051/156	5004		
AM51	5840	3500/5882	5008		
AM52	5840	3584/5862	5008		
AM53	116	3500/5882	5008	1104	
AM54	5840	3500/5882	5008		
AM54LS138J					SN54LS138J
AM54LS138W					SN54LS138W
AM54LS139J					SN54LS139J
AM54LS139W					SN54LS139W
AM54LS151J					SN54LS151J
AM54LS151W					SN54LS151W
AM54LS153J					SN54LS153J
AM54LS153W					SN54LS153W
AM54LS157J					SN54LS157J
AM54LS157W					SN54LS157W
AM54LS158J					SN54LS158J
AM54LS158W					SN54LS158W
AM54LS164J					SN54LS164J
AM54LS164W					SN54LS164W
AM54LS174J					SN54LS174J
AM54LS174W					SN54LS174W
AM54LS175J					SN54LS175J
AM54LS175W					SN54LS175W
AM54LS181J					SN54LS181J
AM54LS181W					SN54LS181W
AM54LS190J					SN54LS190J
AM54LS190W					SN54LS190W
AM54LS191J					SN54LS191J
AM54LS191W					SN54LS191W
AM54LS192J					SN54LS192J
AM54LS192W					SN54LS192W
AM54LS193J					SN54LS193J
AM54LS193W					SN54LS193W
AM54LS251J					SN54LS251J
AM54LS251W					SN54LS251W
AM54LS253J					SN54LS253J
AM54LS253W					SN54LS253W
AM54LS273J					SN54LS273J
AM54LS273W					SN54LS273W
AM54LS299J					SN54LS299J
AM54LS299W					SN54LS299W
AM54LS374J					SN54LS374J
AM54LS374W					SN54LS374W
AM54LS377J					SN54LS377J
AM54LS377W					SN54LS377W
AM62		3500	504A	1104	
AM620	177	3175	300	1122	
AM620A	177	3175	300	1122	
AM626	177	3175	300	1122	
AM626A	177		300	1122	
AM63	116	7042/5842	5008	1104	
AM632	177	3175	300	1122	
AM632A	177	3175	300	1122	
AM64	5842	7042	5008		
AM65	116	3017B/117	504A	1104	
AM66	116	3017B/117	504A	1104	
AM725A31T					MC1556G
AM741CT	941	3514			
AM741HC		3514		010	
AM748CT	1171	3565			
AM74LS138J					SN74LS138J
AM74LS138N					SN74LS138N
AM74LS139J					SN74LS139J
AM74LS139N					SN74LS139N
AM74LS151J					SN74LS151J
AM74LS151N					SN74LS151N
AM74LS153J					SN74LS153J
AM74LS153N					SN74LS153N
AM74LS157J					SN74LS157J
AM74LS157N					SN74LS157N
AM74LS158J					SN74LS158J
AM74LS158N					SN74LS158N
AM74LS164J					SN74LS164J
AM74LS164N					SN74LS164N
AM74LS174J					SN74LS174J
AM74LS174N					SN74LS174N
AM74LS175J					SN74LS175J
AM74LS175N					SN74LS175N
AM74LS181J					SN74LS181J
AM74LS181N					SN74LS181N
AM74LS190J					SN74LS190J
AM74LS190N					SN74LS190N
AM74LS191J					SN74LS191J
AM74LS191N					SN74LS191N
AM74LS192J					SN74LS192J
AM74LS192N					SN74LS192N
AM74LS193J					SN74LS193J
AM74LS193N					SN74LS193N
AM74LS251J					SN74LS251J
AM74LS251N					SN74LS251N
AM74LS253J					SN74LS253J
AM74LS253N					SN74LS253N
AM74LS273J					SN74LS273J
AM74LS273N					SN74LS273N
AM74LS299J					SN74LS299J
AM74LS299N					SN74LS299N
AM74LS374J					SN74LS374J
AM74LS374N					SN74LS374N
AM74LS377J					SN74LS377J
AM74LS377N					SN74LS377N
AM9			40	2012	
AM9016					MCM4116
AM9114					MCM2114
AM9147					MCM2147
AM91L14					MCM21L14
AM9208					MCM68308
AM9218					MCM68A316E
AM9232					MCM68A332
AMD746	790	3077	IC-5		
AMD780	714	3075	IC-4		
AMD781	715	3076	IC-6		
AMF-121	130		14	2041	
AMF104	130	3027	14	2041	
AMF105	130	3027	14	2041	
AMF106	163A		36		
AMF115	130	3027	14	2041	
AMF116	130	3027	14	2041	
AMF117	130	3027	14	2041	
AMF117A	130	3027	14	2041	
AMF118	130	3027	14	2041	
AMF118A	130	3027	14	2041	
AMF119	130	3027	14	2041	
AMF119A	130	3027	14	2041	
AMF120	130	3027	14	2041	
AMF120A	130	3027	14	2041	
AMF201	130		14	2041	
AMF201B	181		75	2041	
AMF201C	181		75	2041	
AMF210	130	3027	14	2041	

Industry Standard No.	ECG	SK	GE	RS 276-	MOTOR.
AMF210A	130	3027	14	2041	
AMF210B	181		75	2041	
AMLM101					LM101AH
AMLM101A					LM101AH
AMLM101AD					L
AMLM101AF					L
AMLM101D					LM101AH
AMLM101F					LM101AH
AMLM105					LM105H
AMLM105F					LM105H
AMLM105H					LM105H
AMLM107					LM107H
AMLM107D		3596			LM107H
AMLM107F					LM107H
AMLM111D		3668			LM111J
AMLM111H					LM111H
AMLM201	1171	3565			LM201AH
AMLM201A	1171				LM201AH
AMLM201AD					LM201AN
AMLM201AF					LM201AH
AMLM201D					LM201AN
AMLM201F					LM201AH
AMLM205					LM205H
AMLM205F					LM205H
AMLM205H					LM205H
AMLM207					LM207H
AMLM207D		3596			LM207H
AMLM207F					LM207H
AMLM211D					LM211J
AMLM211H					LM211H
AMLM301	1171	3565			LM301AH
AMLM301A	1171	3565			LM301AH
AMLM301AD					LM301AJ
AMLM301D					LM301AJ
AMLM305					LM305H
AMLM305A					LM305H
AMLM305F					LM305H
AMLM305H					LM305H
AMLM311D		3668			LM311J-8
AMLM311H					LM311H
AMP-2971-4	128		243	2030	
AMP2919-2	130		14	2041	
AMP2970-2	129		244	2027	
AMU3F7733312					MC1733L
AMU3F7733393					MC1733CL
AMU3F7748312					MC1748G
AMU3I7741312					MC1741F
AMU3I7741393					MC1741CL
AMU5B7733312					MC1733G
AMU5B7733393					MC1733CG
AMU5B7741312					MC1741G
AMU5B7741393	941		IC-263	010	MC1741CG
AMU5B7747312					MC1747G
AMU5B7747393	947		IC-268		MC1747CG
AMU5B7748312					MC1748G
AMU5B7748393					MC1748CG
AMU5R7723312					MC1723G
AMU5R7723393	923		IC-259		MC1723CG
AMU6A7723312					MC1723L
AMU6A7723393	923D		IC-260	1740	MC1723CL
AMU6A7733312					MC1733L
AMU6A7733393					MC1733CL
AMU6A7741312					MC1741L
AMU6A7741393	941D		IC-264		MC1741CL
AMU6A7748312					MC1748G
AMU6A7748393					MC1748CP1
AMU6W7747312					MC1747L
AMU6W7747393					MC1747CL
AN	123A	3444	20	2051	
AN-1	116	3016	504A	1104	
AN-206	1057		IC-47		
AN-214P	1058		IC-49		
AN-240	712		IC-2		
AN-G-5B	116	3031A	504A	1104	
AN-G5B	125		531	1114	
AN103	1192	3445			
AN103-0	1192	3445			
AN136	1053		IC-44		
AN155	147A	3095	ZD-33		
AN203	1054	3457	IC-45		
AN203AA	1054	3457	IC-45		
AN203BA	1054	3457	IC-45		
AN203BB	1054	3457	IC-45		
AN203C	1054	3457	IC-45		
AN206	1057		IC-46		
AN206AB	1057		IC-47		
AN206B	1057		IC-47		
AN206S	1057		IC-47		
AN208	1136		IC-284		
AN208LL	1136		IC-284		
AN210	1055	3494	IC-47		
AN210A	1055	3494	IC-47		
AN210B	1055	3494	IC-47		
AN210C	1055	3494	IC-47		
AN210D	1055	3494	IC-47		
AN211	1056	3458	IC-48		
AN211-AB	1056	3458			
AN211A	1056	3458	IC-48		
AN211AB	1056	3458	IC-48		
AN211B	1056	3458	IC-48		
AN214	1058	3459	IC-49		
AN214P	1058	3459	IC-49		
AN214PQR	1058	3459	IC-49		
AN214Q	1058	3459	IC-49		
AN214QR	1058	3459			
AN214R	1058	3459	IC-49		
AN216	1125		IC-282		
AN217	1060	3460	IC-50		
AN217AA	1060	3460	IC-50		
AN217AB	1060	3460	IC-50		
AN217BA	1060	3460	IC-50		
AN217BB	1060	3460	IC-50		
AN217CA	1060	3460	IC-50		
AN217CB	1060	3460	IC-50		
AN217P	1060	3460	IC-50		
AN217PBB	1060	3460	IC-50		
AN220	1061	3228	IC-51		
AN221	712	3072	IC-2		
AN225	1062		IC-52		
AN227	1063		IC-53		
AN229	1064		IC-54		
AN230	1065		IC-55		
AN231	1066		IC-56		
AN234	1067		IC-57		
AN236	1123	3743	IC-281		
AN238	749	3168	IC-97		
AN238B	749	3171/744	IC-97		
AN239	1161	3968	IC-300		
AN2390A	1161	3968			
AN239B	1161	3968			
AN239Q	1161	3968	IC-300		
AN239QA	1161	3968	IC-300		
AN239QB	1161	3968	IC-300		
AN240	712	3072	IC-148		

Industry Standard No.	ECG	SK	GE	RS 276-	MOTOR.
AN240D	712	3072	IC-148		
AN240P	712	3072	IC-148		
AN240PN	712	3072	IC-148		
AN240S	712		IC-148		
AN241	1162	3072/712	IC-2		
AN241A		3072	IC-2		
AN241B		3072	IC-2		
AN241C		3072	IC-2		
AN241D	1162	3072/712	IC-2		
AN241P	1162	3072/712	IC-320		
AN241PD	1162	3072/712	IC-148		
AN242	1068	3226/1071	IC-58		
AN245	1164	3727	IC-301		
AN246	1164	3727	IC-246		
AN247	1173	3729	IC-302		
AN247P	1173	3729	IC-302		
AN253	1072	3295	IC-59		
AN253AB	1072		IC-59		
AN253B	1072		IC-59		
AN253BB	1072		IC-59		
AN260	1074	3495	IC-60		
AN262	1263	3920	IC-293		
AN271	1181	3706	IC-61		
AN271A	1181	3706			
AN271B	1181	3706	IC-61		
AN274	1137		IC-62		
AN277	1073	3496	IC-63		
AN277-AB	1073		IC-63		
AN277A	1073		IC-63		
AN277AB	1073	3496	IC-63		
AN277B	1073	3496	IC-63		
AN277BA	1073	3496	IC-63		
AN288	1069	3227	IC-64		
AN289	1182	3227/1069	IC-303		
AN318	1268	3921			
AN328	1070		IC-65		
AN331	1168	3728	IC-66		
AN342	1071	3226	IC-67		
AN343	1163	3226/1071	IC-68		
AN360	1223	3493	IC-295		
AN362	1248	3497	IC-292		
AN366	1242	3483	IC-291		
AN374	1224		IC-296		
AN377	788	3147			
AN380	1191	3725/1196			
AN612	1249	3963			
ANJ101	159		82	2032	
A01	126	3008	1	2024	
A04091A			52	2024	
A04092			504A	1104	
A04092A			504A	1104	
A04092B			504A	1104	
A04093			504A	1104	
A04093-A			504A	1104	
A04093-X	506			1114	
A04093A			504A	1104	
A04166-2			ZD-15	564	
A04210A			504A	1104	
A04212-A			504A	1104	
A04212-B			504A	1104	
A04230-A	506			1114	
A04233			504A	1104	
A04234-2			ZD-9.1	562	
A04241-A	506			1114	
A04716			504A	1104	
A04901A			504A	1104	
A07	101	3011	8	2002	
APB-11A-1008	102A	3004	53	2007	
APB-11H-1001	160	3006	245	2004	
APB-11H-1004	160	3006	245	2004	
APB-11H-1007	126	3006/160	52	2024	
APB-11H-1008	102	3004	2	2007	
APB-11H-1010	102A	3004	53	2007	
APS-AH1	501B	3069/501A	520		
AQ2	128	3024	243	2030	
AQ2(PHILCO)	116		504A	1104	
AQ3	128	3024	243	2030	
AQ3(PHILCO)	116		504A	1104	
AQ4	123	3124/289	20	2051	
AQ5	128	3024	243	2030	
AQ6	123	3124/289	20	2051	
AR-10	121	3009	239	2006	
AR-102	102A	3004	53	2007	
AR-103	102A	3004	53	2007	
AR-104	102A	3004	53	2007	
AR-105	102A	3123	53	2007	
AR-107	123A	3444	20	2051	
AR-108	123A	3444	20	2051	
AR-11	121	3009	239	2006	
AR-12	121	3009	239	2006	
AR-13	121	3009	239	2006	
AR-14	121	3009	239	2006	
AR-15	124	3021	12		
AR-17	152	3893	66	2048	
AR-18	124	3021	12		
AR-200	123A	3444	20	2051	
AR-201	123A	3444	20	2051	
AR-202	123A	3444	20	2051	
AR-203(R)			63	2030	
AR-22(DIO)			504A	1104	
AR-22(XSTR)			66	2048	
AR-23(DIO)			509	1114	
AR-23(XSTR)			69	2049	
AR-24			509	1114	
AR-25	153		69	2049	
AR-29			29	2025	
AR-30	153	3084	241	2020	
AR-4	121	3009	239	2006	
AR-5	121	3009	239	2006	
AR-6	121	3009	239	2006	
AR-7	121	3009	239	2006	
AR-8	121	3009	239	2006	
AR-9	121	3009	239	2006	
AR10	121	3717	239	2006	
AR102	158	3004/102A	53	2007	
AR103	160	3006	245	2004	
AR104	160	3006	245	2004	
AR105	160	3008	245	2004	
AR107	123A	3444	20	2051	
AR108	123A	3444	20	2051	
AR11	121	3717	239	2006	
AR111	161	3716	39	2015	
AR12	121	3717	239	2006	
AR13	121	3717	239	2006	
AR14	121	3717	239	2006	
AR15	130		14	2041	
AR15-L8-0026	130		14	2041	
AR16	116	3017B/117	504A	1104	1N4001
AR17	116		504A	2020	1N4002
AR17(GREY)	152		66	2048	
AR17(PHILCO)			246	2020	
AR17A	175		246	2020	
AR17B	175		246	2020	
AR17GREY			66	2048	

Industry Standard No.	ECG	SK	GE	RS 276-	MOTOR.
AR18	116	3026	504A	1104	1N4003
AR18(PHILCO)			246	2020	
AR19	116	3031A	504A	1104	1N4004
AR20	116	3031A	504A	1104	1N4004
AR200	108	3018	86	2038	
AR200(GREEN)	107		11	2015	
AR200(W)			20	2051	
AR200W	108	3452	86	2038	
AR200WHITE			20	2051	
AR201	108	3452	86	2038	
AR201(Y)			20	2051	
AR201(YELLOW)	108		86	2038	
AR201Y	108	3452	86	2038	
AR201YELLOW			20	2051	
AR202	108	3452	86	2038	
AR202(GREEN)	108		86	2038	
AR202G	108	3452	86	2038	
AR202GREEN			20	2051	
AR203	128	3024	243	2030	
AR203(R)			63	2030	
AR203(RED)	128		243	2030	
AR203R	128	3024	243	2030	
AR203RED			63	2030	
AR204	123A	3444	20	2051	
AR205	123A	3444	20	2051	
AR206	123A	3444	20	2051	
AR207	192		63	2030	
AR208	123A	3444	20	2051	
AR209	161	3018	39	2015	
AR21	116	3017B/117	504A	1104	1N4005
AR210	161	3018	39	2015	
AR211	161	3018	39	2015	
AR212	108	3452	86	2038	
AR213	161	3018	39	2015	
AR213(V)			62	2010	
AR213(VIOLET)	161	3716	39	2015	
AR213V	161	3018	39	2015	
AR213VIOLET			62	2010	
AR218	229	3018	39	2015	
AR218(ORANGE)	161	3716	39	2015	
AR218(RED)	161	3716	39	2015	
AR218(RO)			62	2010	
AR218ORANGE			63	2030	
AR218RED			63	2030	
AR218RO	161	3018	39	2015	
AR219	161	3018	39	2015	
AR219(YY)			60	2015	
AR219YY	107	3018	11	2015	
AR22	116	3017B/117	504A	1104	1N4005
AR22(PHILCO)	182	3188A	55		
AR22(RECTIFIER)	116		504A	1104	
AR22(TRANSISTOR)	184		57	2017	
AR220	161	3018	39	2015	
AR220(GY)			60	2015	
AR220(YELLOW)	107		11	2015	
AR220GREEN			60	2015	
AR220GY	107	3018	11	2015	
AR220YELLOW			60	2015	
AR221	108	3452	86	2038	
AR222	161	3018	39	2015	
AR222(BLUE)	107		11	2015	
AR222(BY)			60	2015	
AR222(YELLOW)	107		11	2015	
AR222BLUE			60	2015	
AR222BY	107	3018	11	2015	
AR222YELLOW			60	2015	
AR224	107	3018	60	2015	
AR224(WHITE)	107		11	2015	
AR224(YELLOW)	107		11	2015	
AR224WHITE			60	2015	
AR224YELLOW			60	2015	
AR23	125		531	1114	1N4006
AR23(PHILCO)	183	3189A	56	2027	
AR23(TRANSISTOR)	185		58	2025	
AR24	125		531	1114	1N4007
AR24(PHILCO)	186		28	2017	
AR24(RED)	196		241	2020	
AR24(TRANSISTOR)	184		57	2017	
AR24RED			241	2020	
AR25	187	3193	29	2025	
AR25(G)			69	2049	
AR25(GREEN)	153		69	2049	
AR25(ORANGE)	108		86	2038	
AR25(WHITE)	108		86	2038	
AR25A					MR2500
AR25B					MR2501
AR25D					MR2502
AR25F					MR2504
AR25G	187	3193	29	2025	MR2504
AR25H					MR2506
AR25J					MR2506
AR25K					MR2508
AR25M					MR2510
AR26	211	3084	253	2026	
AR27	197	3084	250	2027	
AR27(GREEN)	153		69	2049	
AR27GREEN			69	2049	
AR28	196	3041	241	2020	
AR28(RED)	196		241	2020	
AR28RED			241	2020	
AR29	211	3053	253	2026	
AR30	196	3084	241	2020	
AR303		3122	61	2051	
AR304	129	3025	67	2027	
AR304(GREEN)	159		82	2032	
AR304(RED)	129		244	2027	
AR304GREEN			67	2023	
AR304RED			67	2023	
AR306	123A	3444	20	2051	
AR306(BLUE)	123A	3444	20	2051	
AR306(ORANGE)	123A	3444	20	2051	
AR306BLUE			18	2030	
AR306ORANGE			18	2030	
AR308	129	3025	67	2023	
AR308(VIOLET)	159		82	2032	
AR308VIOLET				2027	
AR313	108	3452	86	2038	
AR35	196	3087	241	2020	
AR37	197	3084	250	2027	
AR37(GREEN)	153		69	2049	
AR37GREEN			69	2049	
AR38	196	3041	241	2020	
AR38(RED)	196		241	2020	
AR38RED			241	2020	
AR4	104		16	2006	
AR410	6402	3628			
AR44	153	3084	69	2049	MDA920A6
AR5	104	3719	16	2006	
AR501	222	3050/221	FET-4	2036	
AR502	222	3050/221	FET-4	2036	
AR6	104	3719	16	2006	
AR7	104	3719	16	2006	
AR7C	116	3017B/117	504A	1104	
AR8	121	3717	239	2006	

Industry Standard No.	ECG	SK	GE	RS 276-	MOTOR.
AR882			504A	1104	
AR8P404R	121	3009	239	2006	
AR9	121	3717	239	2006	
AS-14	116		504A	1104	
AS-15	116		504A	1104	
AS-2	116		504A	1104	
AS-3	116		504A	1104	
AS-4	116		504A	1104	
AS-5	116		504A	1104	
AS11	116	3311	504A	1104	
AS14	116	3016	504A	1104	
AS15	116	3016	504A	1104	
AS2	116	3016	504A	1104	
AS215	121	3009	239	2006	
AS3	116	3016	504A	1104	
AS33867	102A		53	2007	
AS33868	102A	3004	53	2007	
AS3428	101	3861	8	2002	
AS34280	160		245	2004	
AS3A	5800	9003		1142	
AS3B	5801	9004		1142	
AS3C	5802	9005		1143	
AS3D	5804	9007		1144	
AS3E	5806	3848			
AS3F	5808	9009	53		
AS4	116	3016	504A	1104	
AS477	176	3123	80		
AS5	116	3016	504A	1104	
AS6	116	3017B/117	504A	1104	
ASL802001-1	912		IC-172		
ASY-24		3008	245	2004	
ASY-26		3005	1	2007	
ASY-27		3005	1	2007	
ASY-62		3861	8	2002	
ASY-72			8	2002	
ASY12-1	158	3004/102A	53	2007	
ASY12-2	158	3004/102A	53	2007	
ASY13-1	158	3004/102A	53	2007	
ASY13-2	158		53	2007	
ASY14	158		53	2007	
ASY14-1	158	3123	53	2007	
ASY14-2		3123	53	2007	
ASY14-3		3123	53	2007	
ASY24			53	2007	
ASY26		3123	1	2007	
ASY26-RT				2005	
ASY27		3721	1	2007	
ASY28		3011	8	2002	
ASY29		3011	8	2002	
ASY30		3006	52	2024	
ASY31		3123	53	2007	
ASY32		3123	53	2005	
ASY48		3004	53	2007	
ASY48-IV		3004	53	2007	
ASY48-V		3004	53	2007	
ASY48-VI		3004	53	2007	
ASY49		3004	53	2007	
ASY50		3004	53	2007	
ASY51		3123	53	2007	
ASY52		3004	53	2007	
ASY53		3011	8	2002	
ASY54		3004	53	2007	
ASY55		3004	53	2007	
ASY56		3004	53	2007	
ASY56N			1	2007	
ASY57		3006	52	2024	
ASY57N			1	2007	
ASY58		3006	52	2024	
ASY58N			1	2007	
ASY59		3006	52	2024	
ASY61		3010	59	2002	
ASY62			59	2002	
ASY63		3006	245	2004	
ASY63N			50	2004	
ASY67		3123	245	2004	
ASY70		3004	53	2007	
ASY70-IV		3004	53	2007	
ASY70-VI		3004	53	2007	
ASY70IV			53	2007	
ASY70V			53	2007	
ASY70VI			53	2007	
ASY71			53	2007	
ASY72		3011A	8	2002	
ASY73		3861	8	2002	
ASY74	101	3861	8	2002	
ASY75	101	3861	8	2002	
ASY76	100	3005	1	2007	
ASY77	100	3005	1	2007	
ASY80	100	3005	1	2007	
ASY81	176	3123	80		
ASY86	101	3861	8	2002	
ASY87	101	3861	8	2002	
ASY88	101	3861	8	2002	
ASY89	101	3861	8	2002	
ASY90	102A	3004	53	2007	
ASY91	102A	3004	53	2007	
ASZ10	126	3006/160	52	2024	
ASZ11	126	3006/160	52	2024	
ASZ15	121	3009	239	2006	
ASZ16	121	3009	239	2006	
ASZ17	121	3009	239	2006	
ASZ18	121	3009	239	2006	
ASZ20	160	3006	245	2004	
ASZ20N	160	3006	245	2004	
ASZ21	126	3006/160	245	2004	
ASZ30	126	3006/160	52	2024	
AT-1	160	3006	245	2004	
AT-10	130	3027	14	2041	
AT-11	159	3114/290	82	2032	
AT-12	128	3024	243	2030	
AT-12(PHILCO)	128		243	2030	
AT-14	160	3006	245	2004	
AT-15	100	3005	1	2007	
AT-1856	130	3027	14	2041	
AT-2	160	3006	245	2004	
AT-3	160	3006	245	2004	
AT-4	160	3005	245	2004	
AT-5	100	3008	1	2007	
AT-50	102A	3004	53	2007	
AT-6	160	3004/102A	245	2004	
AT-6A	102A	3004	53	2007	
AT-7	128	3024	243	2030	
AT-8	160	3006	245	2004	
AT-9	160	3006	245	2004	
AT/RF1	160	3006	245	2004	
AT/RF2	160	3006	245	2004	
AT/S13	160	3006	245	2004	
AT0017					MRF904
AT0017A					MRF904
AT004					MRF904
AT0045					MRF904
AT100H	102A	3123	53	2007	
AT100M	102A	3123	53	2007	
AT100N	102A	3123	53	2007	
AT10H	102A	3004	53	2007	
AT10M	102A	3004	53	2007	
AT10N	102A	3004	53	2007	
AT11		3114	21	2034	
AT1138	179	3642	76		
AT1138A	179	3642	76		
AT1138B	179	3642	76		
AT12	128		243	2030	
AT13	160		245	2004	
AT14	160		245	2004	
AT1425					BFR90
AT15	160		245	2004	
AT16	160	3245/199	245	2004	
AT17	160	3245/199	245	2004	
AT1825					2N6604
AT1833	179	3642	76		
AT1834	179	3642	76		
AT1845					2N6603
AT1845A					2N6603
AT1856	130		14	2041	
AT200	127	3035	25		
AT20H	102A	3004	53	2007	
AT20M	102A	3004	53	2007	
AT20N	102A	3004	53	2007	
AT25					MRF901
AT25A					MRF901
AT25B					MRF901
AT2625					MRF902
AT2645					2N6603
AT2645A					2N6603
AT2715					MRF962
AT2848	129	3025	244	2027	
AT30H	102A	3004	53	2007	
AT30M	102A	3004	53	2007	
AT30N	102A	3004	53	2007	
AT310	107	3039/316	11	2015	
AT311	107	3039/316	11	2015	
AT312	107	3039/316	11	2015	
AT313	107	3039/316	11	2015	
AT314	107	3039/316	11	2015	
AT315	107	3039/316	11	2015	
AT316	107	3039/316	11	2015	
AT318	107	3039/316	11	2015	
AT319	107	3039/316	11	2015	
AT321	107	3039/316	11	2015	
AT322	107	3039/316	11	2015	
AT323	107	3039/316	11	2015	
AT324	107	3039/316	11	2015	
AT325	107	3039/316	11	2015	
AT326	107	3039/316	11	2015	
AT3260	130	3027	14	2041	
AT327	107	3039/316	11	2015	
AT328	107	3039/316	11	2015	
AT329	123A	3444	20	2051	
AT330	107	3039/316	11	2015	
AT331	159	3114/290	82	2032	
AT331A	159	3114/290	82	2032	
AT332	159	3114/290	82	2032	
AT332A	159	3114/290	82	2032	
AT333	159	3025/129	82	2032	
AT335	123A	3444	20	2051	
AT335A	159	3114/290	82	2032	
AT336	123A	3444	20	2051	
AT337	123A	3444	20	2051	
AT338	107	3039/316	11	2015	
AT339	128	3024	243	2030	
AT340	161	3117	39	2015	
AT341	161	3117	39	2015	
AT342	161	3117	39	2015	
AT343	161	3117	39	2015	
AT344	161	3117	39	2015	
AT345	161	3117	39	2015	
AT346	161	3117	39	2015	
AT347	123A	3444	20	2051	
AT348	123A	3444	20	2051	
AT349	123A	3444	20	2051	
AT350	154	3044	40	2012	
AT351	154	3044	40	2012	
AT370	123A	3444	20	2051	
AT380	128	3024	243	2030	
AT381	128	3024	243	2030	
AT382	128	3024	243	2030	
AT383	128	3024	243	2030	
AT384	128	3024	243	2030	
AT385	128	3024	243	2030	
AT386	128	3024	243	2030	
AT387	128	3024	243	2030	
AT388	128	3024	243	2030	
AT391	128	3025/129	244	2027	
AT392	128	3025/129	244	2027	
AT393	128	3025/129	244	2027	
AT394	129	3025	244	2027	
AT395	129	3025	244	2027	
AT396	129	3025	244	2027	
AT397	129	3025	244	2027	
AT398	129	3025	244	2027	
AT4	160		245	2004	
AT400	123A	3444	20	2051	
AT401	123A	3444	20	2051	
AT402	123A	3444	20	2051	
AT403	123A	3444	20	2051	
AT404	123A	3444	20	2051	
AT405	123A	3444	20	2051	
AT406	123A	3444	20	2051	
AT407	123A	3444	20	2051	
AT410	159	3114/290	82	2032	
AT410-1	159	3114/290	82	2032	
AT412	159	3114/290	82	2032	
AT412-1	159	3114/290	82	2032	
AT413	159	3114/290	82	2032	
AT413-1	159	3114/290	82	2032	
AT414	159	3114/290	82	2032	
AT414-1	159	3114/290	82	2032	
AT415	159	3114/290	82	2032	
AT415-1	159	3114/290	82	2032	
AT416	159	3114/290	82	2032	
AT416-1	159	3114/290	82	2032	
AT417	159	3114/290	82	2032	
AT417-1	159	3114/290	82	2032	
AT418	159	3114/290	82	2032	
AT418-1	159	3114/290	82	2032	
AT419	159	3114/290	82	2032	
AT419-1	159	3114/290	82	2032	
AT420	123A	3444	20	2051	
AT421	123A	3444	20	2051	
AT422	123A	3444	20	2051	
AT423	123A	3444	20	2051	
AT424	123A	3444	20	2051	
AT425	123A	3444	20	2051	
AT426	123A	3444	20	2051	
AT427	123A	3444	20	2051	
AT430	159	3114/290	82	2032	
AT430-1	159	3114/290	82	2032	
AT431	159	3114/290	82	2032	

Industry Standard No.	ECG	SK	GE	RS 276-	MOTOR.
AT431-1	159	3114/290	82	2032	
AT432	159	3114/290	82	2032	
AT432-1	159	3114/290	82	2032	
AT433	159	3114/290	82	2032	
AT433-1	159	3114/290	82	2032	
AT434	159	3114/290	82	2032	
AT434-1	159	3114/290	82	2032	
AT435	159	3114/290	82	2032	
AT435-1	159	3114/290	82	2032	
AT436	159	3114/290	82	2032	
AT436-1	159	3114/290	82	2032	
AT437	159	3114/290	82	2032	
AT437-1	159	3114/290	82	2032	
AT438	159	3114/290	82	2032	
AT438-1	159	3114/290	82	2032	
AT440	128	3024	243	2030	
AT441	128	3024	243	2030	
AT442	128	3024	243	2030	
AT443	128	3024	243	2030	
AT444	128	3024	243	2030	
AT445	128	3024	243	2030	
AT446	128	3024	243	2030	
AT451	159	3114/290	82	2032	
AT451-1	159	3114/290	82	2032	
AT452	159	3114/290	82	2032	
AT452-1	159	3114/290	82	2032	
AT453	159	3114/290	82	2032	
AT453-1	159	3114/290	82	2032	
AT454	159	3114/290	82	2032	
AT454-1	159	3114/290	82	2032	
AT455	159	3114/290	82	2032	
AT455-1	159	3114/290	82	2032	
AT460	129	3025	244	2027	
AT461	129	3025	244	2027	
AT462	129	3025	244	2027	
AT463	129	3025	244	2027	
AT464	129	3025	244	2027	
AT465	129	3025	244	2027	
AT466	129	3025	244	2027	
AT467	129	3025	244	2027	
AT468	129	3025	244	2027	
AT470	128	3024	243	2030	
AT471	128	3024	243	2030	
AT472	128	3024	243	2030	
AT473	128	3024	243	2030	
AT474	128	3024	243	2030	
AT475	128	3024	243	2030	
AT476	128	3024	243	2030	
AT477	128	3024	243	2030	
AT478	128	3024	243	2030	
AT479	128	3024	243	2030	
AT480	129	3025	244	2027	
AT481	129	3025	244	2027	
AT482	129	3025	244	2027	
AT483	129	3025	244	2027	
AT484	129	3025	244	2027	
AT485	129	3025	244	2027	
AT490	123A	3444	20	2051	
AT491	123A	3444	20	2051	
AT492	123A	3444	20	2051	
AT493	123A	3444	20	2051	
AT494	123A	3444	20	2051	
AT495	123A	3444	20	2051	
AT5	160		245	2004	
AT50	102A		53	2007	BFR90
AT51					BFR90
AT5156	129		244	2027	
AT52	101	3861	8	2002	BFR90
AT520	108	3452	86	2038	
AT521	101	3861	8	2002	
AT53	101	3861	8	2002	
AT551	101	3861	8	2002	
AT6	102A	3123	53	2007	
AT6A	102A	3123	53	2007	
AT7	192		63	2030	
AT71	101	3861	8	2002	
AT72	101	3861	8	2002	
AT73R	101	3861	8	2002	
AT74	102A		53	2007	
AT74S	102A		53	2007	
AT75R	101	3861	8	2002	
AT76R	101	3861	8	2002	
AT77	101	3861	8	2002	
AT874	102A	3004	53	2007	
ATAF1	102A	3004	53	2007	
ATAF2	102A	3004	53	2007	
ATC-SR-3	156	3032A	512		
ATC-TR-13	128	3024	243	2030	
ATC-TR-14	121	3009	239	2006	
ATC-TR-15	130	3027	14	2041	
ATC-TR-19	152	3041	66	2048	
ATC-TR-4	128	3024	243	2030	
ATC-TR-5	121	3009	239	2006	
ATC-TR-6	121	3009	239	2006	
ATC-TR-7	128	3024	243	2030	
ATGP	102A	3004	53	2007	
ATRF1	126	3006/160	52	2024	
ATRF2	126	3006/160	52	2024	
ATS13	126	3006/160	52	2024	
AU100N	158	3004/102A	53	2007	
AU101	127	3014	25		
AU102	121	3009	239	2006	
AU103	127	3035	25		
AU104	127	3035	25		
AU105	127	3035	25		
AU106	127	3035	25		
AU107	127	3035	25		
AU108	127	3035	25		
AU110	127	3035	25		
AU111	127	3035	25		
AU112	127	3035	25		
AU113	127	3035	25		
AU2012	142A	3062	ZD-12	563	
AUY-21	121	3009	239	2006	
AUY10	121	3009	239	2006	
AUY19	121	3014	239	2006	
AUY20	121	3014	239	2006	
AUY21	104	3719	16	2006	
AUY21A	121	3009	239	2006	
AUY22	104	3719	16	2006	
AUY22A	121	3009	239	2006	
AUY28	127	3035	25		
AUY29	179	3642	76		
AUY31	121	3717	239	2006	
AUY33	121	3717	239	2006	
AUY37	179	3642	76		
AUY38	127	3035	25		
AV-2015	145A	3063	ZD-15	564	
AV0000105-0	177	3100/519	300	1122	
AV01-07	141A	3092	ZD-11.5		
AV10	147A	3095	ZD-33		
AV105	121	3009	239	2006	
AV2012	142A	3062	ZD-12	563	
AV2015	145A	3063	ZD-15	564	
AV2027	146A	3064	ZD-27		
AV2027A	146A	3064	ZD-27		
AV2033	147A	3095	ZD-33		
AV2055	148A	3096	ZD-55		
AV2062	149A	3097	ZD-62		
AV2082	150A	3098	ZD-82		
AV2110	151A	3099	ZD-110		
AV5	142A	3062	ZD-12	563	
AW-01-07	5071A		ZD-9.1	562	
AW-01-08	5072A	3059/138A	ZD-8.2	562	
AW-01-09	139A	3060	ZD-9.1	562	
AW-01-10	140A	3061	ZD-10	562	
AW-01-12	142A	3031A	ZD-12	1104	
AW-01-12C	142A	3062	ZD-12	563	
AW-01-22		3336	ZD-22		
AW-01-33	147A	3095	ZD-27		
AW01			ZD-12	563	
AW01(RCA)	143A	3750	ZD-13	563	
AW01-0	5070A	9021			
AW01-06	5070A	9021	ZD-6.1	561	
AW01-07		3334	ZD-6.8	561	
AW01-08	5072A		ZD-8.2	562	
AW01-08J	5072A		ZD-8.2	562	
AW01-09	139A	3060	ZD-9.1	562	
AW01-10	140A		ZD-10	562	
AW01-11	5074A	3139	ZD-11	563	
AW01-12			ZD-12	563	
AW01-12C	142A		ZD-12	563	
AW01-12V	142A	3062	ZD-12	563	
AW01-13	5022A	3788	ZD-12	563	
AW01-15	145A	3063	ZD-15	564	
AW01-16	5075A	3751	ZD-16	564	
AW01-18	5077A	3752	ZD-18		
AW01-20	5079A	3335	ZD-20		
AW01-22	5080A	3336	ZD-22		
AW01-24	5081A	3151	ZD-24		
AW01-27	146A	3064	ZD-27		
AW01-30	5084A	3755	ZD-30		
AW01-33	147A	3095	ZD-33		
AW01-7	5071A	3334	ZD-6.8	561	
AW01-9	139A	3060	ZD-9.1	562	
AW01-9/CP3112030	139A		ZD-9.1	562	
AW0122	5080A	3336	ZD-22		
AW09	139A	3060	ZD-9.1	562	
AWH-24	123A	3444	20	2051	
AWOL-13	5022A	3093	ZD-12	563	
AX-7		3059	ZD-7.5		
AX12	142A		ZD-12	563	
AX91770	108	3452	86	2038	
AYY10-120	5852	3500/5882	5016		
AZ-050	135A	3056	ZD-5.1		
AZ-052	136A	3056/135A	ZD-5.6	561	
AZ-054	136A		ZD-5.6	561	
AZ-056	136A	3057	ZD-5.6	561	
AZ-058	5070A	9021	ZD-6.2	561	
AZ-061	137A	3058	ZD-6.2	561	
AZ-063	137A	3058	ZD-6.2	561	
AZ-065	5071A	3334	ZD-6.8	561	
AZ-067	5071A	3334	ZD-6.8	561	
AZ-069	5071A	3334	ZD-6.8	561	
AZ-071	138A		ZD-7.5		
AZ-073	138A	3059	ZD-7.5		
AZ-075	138A	3059	ZD-7.5		
AZ-077	5072A	3059/138A	ZD-8.2	562	
AZ-079	5072A		ZD-8.2	562	
AZ-081	5072A	3136	ZD-8.2	562	
AZ-083	5072A	3136	ZD-8.2	562	
AZ-085	5073A	3749	ZD-9.1	562	
AZ-088	5073A	3749	ZD-9.1	562	
AZ-090	139A	3060	ZD-9.1	562	
AZ-092	139A	3060	ZD-9.1	562	
AZ-094	140A		ZD-10	562	
AZ-096	140A		ZD-10	562	
AZ-098	140A	3061	ZD-10	562	
AZ-100	140A	3061	ZD-10	562	
AZ-105	5074A		ZD-11.0	563	
AZ-110	5074A	3139		563	
AZ-115	141A	3092	ZD-11.5		
AZ-120	142A	3062	ZD-12	563	
AZ-125	143A	3750	ZD-13	563	
AZ-130	143A	3750	ZD-13	563	
AZ-135	144A		ZD-14	564	
AZ-140	144A	3094	ZD-14	564	
AZ-145	145A		ZD-15	564	
AZ-15	145A	3063	ZD-15	564	
AZ-150	145A	3063	ZD-15	564	
AZ-157	5075A		ZD-16	564	
AZ-162	5075A	3142	ZD-16	564	
AZ-167	5076A	9022	ZD-17		
AZ-172	5076A	9022	ZD-17		
AZ-177	5077A	3145	ZD-18		
AZ-182	5077A	3752	ZD-18		
AZ-187	5078A	9023	ZD-19		
AZ-192	5078A	9023	ZD-19		
AZ-197	5079A	3335	ZD-20		
AZ10	140A	3061	ZD-10	562	
AZ11	5074A	3139	ZD-11	563	
AZ110			ZD-11.0	563	
AZ12	142A	3062	ZD-12	563	
AZ13	143A	3750	ZD-12	563	
AZ15	145A	3063	ZD-15	564	
AZ18	5027A	3793	ZD-18		
OAZ213	142A	3062	ZD-12	563	
AZ22	5030A	3796	ZD-22		
AZ27	146A	3064	ZD-27		
AZ27A	146A	3064	ZD-27		
AZ3.3	5066A	3330	ZD-3.3		
AZ3.6	134A	3055	ZD-3.6		
AZ3.9	5067A	3331	ZD-3.9		
AZ4.3	5068A	3332	ZD-4.3		
AZ4.7	5069A	3056/135A	ZD-5.1		
AZ5.1	135A	3056	ZD-5.1		
AZ5.6	136A	3057	ZD-5.6	561	
AZ6.2	137A	3058	ZD-6.2	561	
AZ6.8	5071A	3334	ZD-6.8	561	
AZ7	144A	3094	ZD-14	564	
AZ7.5	138A	3059	ZD-7.5		
AZ746	5005A	3771	ZD-3.3		
AZ746A	5066A	3771/5005A	ZD-3.3		
AZ747	5006A	3772	ZD-3.6		
AZ747A	134A	3055	ZD-3.6		
AZ748	5007A	3773	ZD-3.9		
AZ748A	5067A	3331	ZD-3.9		
AZ749	5008A	3774	ZD-4.3		
AZ749A	5068A	3774/5008A	ZD-4.3		
AZ750	5009A		ZD-4.7		
AZ750A	5069A		ZD-4.7		
AZ751	5010A	3776	ZD-5.1		
AZ751A	5010A	3776	ZD-5.1		
AZ752	5011A	3777	ZD-5.6	561	
AZ752A	136A	3777/5011A	ZD-5.6	561	
AZ753	5013A	3779	ZD-6.2	561	
AZ753A	137A	3779/5013A	ZD-6.2	561	
AZ754	5014A	3780	ZD-6.8	561	

Industry Standard No.	ECG	SK	GE	RS 276-	MOTOR.
AZ754A	5071A	3780/5014A	ZD-6.8	561	
AZ755	5015A	3781	ZD-7.5		
AZ755A	138A	3781/5015A	ZD-7.5		
AZ756	5016A	3782	ZD-8.2	562	
AZ756A	5072A	3782/5016A	ZD-8.2	562	
AZ757	5018A	3784	ZD-9.1	562	
AZ757A	139A	3784/5018A	ZD-9.1	562	
AZ758	5019A	3785	ZD-10	562	
AZ758A	140A	3785/5019A	ZD-10	562	
AZ759	5021A	3787	ZD-12	563	
AZ759A	142A	3787/5021A	ZD-12	563	
AZ8.2	5072A	3136	ZD-8.2	562	
AZ9.1	139A	3060	ZD-9.1	562	
AZ957	5014A	3780	ZD-6.8	561	
AZ957A	5014A	3780	ZD-6.8	561	
AZ957B	5071A		ZD-6.8	561	
AZ958	5015A	3781	ZD-7.5		
AZ958A	5015A	3781	ZD-7.5		
AZ958B	138A	3059	ZD-7.5		
AZ959	5016A	3782	ZD-8.2	562	
AZ959A	5016A	3782	ZD-8.2	562	
AZ959B	5072A	3136	ZD-8.2	562	
AZ960	5018A	3784	ZD-9.1	562	
AZ960A	5018A	3784	ZD-9.1	562	
AZ960B	139A	3784/5018A	ZD-9.1	562	
AZ961	5019A	3785	ZD-10	562	
AZ961A	5019A	3785	ZD-10	562	
AZ961B	140A	3785/5019A	ZD-10	562	
AZ962	5020A	3786	ZD-11	563	
AZ962A	5020A	3786	ZD-11	563	
AZ962B	5074A	3786/5020A	ZD-11	563	
AZ963	5021A	3787	ZD-12	563	
AZ963A	5021A	3787	ZD-12	563	
AZ963B	142A	3787/5021A	ZD-12	563	
AZ964	5022A	3788	ZD-12	563	
AZ964A	5022A	3788	ZD-12	563	
AZ964B	143A	3788/5022A	ZD-12	563	
AZ965	5024A	3790	ZD-15	564	
AZ965A	5024A	3790	ZD-15	564	
AZ965B	145A	3790/5024A	ZD-15	564	
AZ966	5025A	3791		564	
AZ966A	5025A	3791		564	
AZ966B	5075A	3791/5025A		564	
AZ967	5027A	3752/5077A	ZD-18		
AZ967A	5027A	3752/5077A	ZD-18		
AZ967B	5077A	3752	ZD-18		
AZ968	5029A	3795	ZD-20		
AZ968A	5029A	3795	ZD-20		
AZ968B	5079A	3795/5029A	ZD-20		
AZ969	5030A	3796	ZD-22		
AZ969A	5030A	3796	ZD-22		
AZ969B	5080A	3796/5030A	ZD-22		
AZ970	5031A	3797	ZD-24		
AZ970A	5031A	3797	ZD-24		
AZ970B	5081A	3151	ZD-24		
AZ971	5033A	3799	ZD-27		
AZ971A	5033A	3799	ZD-27		
AZ971B	146A	3064	ZD-27		
AZ972	5035A	3755/5084A	ZD-30		
AZ972A	5035A	3755/5084A	ZD-30		
AZ972B	5084A	3755	ZD-30		
AZ973	5036A	3802	ZD-33		
AZ973A	5036A	3802	ZD-33		
AZ973B	147A	3802/5036A	ZD-33		
AZ974	5037A	3803	ZD-36		
AZ974A	5037A	3803	ZD-36		
AZ974B	5085A	3803/5037A	ZD-36		
AZ975	5038A	3804	ZD-39		
AZ975A	5038A	3804	ZD-39		
AZ975B	5086A	3804/5038A	ZD-39		
AZ976	5039A	3805	ZD-43		
AZ976A	5039A	3805	ZD-43		
AZ976B	5087A	3805/5039A	ZD-43		
AZG			11	2015	
AZQQ-001GEA	1165	3827/1166			
AZY	107	3039/316	11	2015	
B 722246-2	123A			2051	
B-1058	102A	3004	53	2007	
B-12822-2	130		14	2041	
B-12822-4	181		75	2041	
B-1338	123A	3124/289	20	2051	
B-1421	123A	3444	20	2051	
B-1426	159	3114/290	82	2032	
B-1433	123A	3444	20	2051	
B-1501U	116	3311	504A	1104	
B-1511	121	3014	239	2006	
B-1599	177	3100/519	300	1122	
B-1666	123A	3444	20	2051	
B-169	123A	3444	20	2051	
B-1695	185	3083/197	58	2025	
B-1702	177	3100/519	300	1122	
B-1790	184	3054/196	57	2017	
B-1808	5072A	3136	ZD-8.2	562	
B-1823	152	3054/196	66	2048	
B-1842	123A	3444	20	2051	
B-1872	123A	3444	20	2051	
B-1881U	116(2)	3311	504A(2)	1104	
B-1882U	116(2)	3311	504A(2)	1104	
B-1910	199	3124/289	62	2010	
B-1914	121	3009	239	2006	
B-22-3	102A	3004	53	2007	
B-22-4	102A	3004	53	2007	
B-23	102A	3004	53	2007	
B-23-1	102A	3004	53	2007	
B-23-2	102A	3004	53	2007	
B-24-1	102A	3004	53	2007	
B-26	102A	3004	53	2007	
B-26-1	102A	3123	53	2007	
B-28	109			1123	
B-30	109	3091		1123	
B-30P	109	3088		1123	
B-31	116	3016	504A	1104	
B-315-1	158	3004/102A	53	2007	
B-324	158	3004/102A	53	2007	
B-3P	109	3088		1123	
B-46-110	178MP	3119/113	6GC1	1122(2)	
B-5	176	3123	80		
B-6001	312	3112	FET-1	2028	
B-6002	145A	3063	ZD-15	564	
B-6288			82	2032	
B-6340			243	2030	
B-66		3124	17	2051	
B-722246-2			20	2051	
B-75561-31	159		82	2032	
B-75568-2	107		11	2015	
B-75583-1	123A	3444	20	2051	
B-75583-2	123A	3444	20	2051	
B-75583-202			20	2051	
B-75583-I02	123A			2051	
B-75589-13	123A	3444	20	2051	
B-75589-3	123A	3444	20	2051	
B-75608-3	123A		20	2051	
B-F1A	102A	3004	53	2007	
B-T-1000-139			11	2015	

Industry Standard No.	ECG	SK	GE	RS 276-	MOTOR.
B-T1000-139	107		11	2015	
B00	7442	7442			
B004B				1152	
B004C				1172	
B004E				1173	
B0075660	7473		7473	1803	
B01	7475	7475		1806	
B01-02	116	3311	504A	1104	
B0102	116		504A	1104	
B02	7476	7476		1813	
B02022	5455	3597			
B02023	5455	3597			
B02610	5452	6752			
B02611	5453	6753			
B02612	5454	6754			
B02644	5452	6752			
B02645	5453	6753			
B02646	5454	6754			
B03002	5455	3597			
B03003	5455	3597			
B0301-049	181	3036	75	2041	
B0305401	1235	3637			
B0306000	1004	3365	IC-149		
B0306004	1004	3365	IC-149		
B0311400	1128	3488	IC-105		
B0311402	1128	3488	IC-105		
B0311405	1128	3488			
B0313300	1134	3489	IC-106		
B0313400	1133	3490	IC-107		
B0313600	1131	3286	IC-109		
B0313700	1132	3287	IC-110		
B0313800	1130	3478	IC-111		
B0315500	1200	3714			
B0316403	1236	3072/712	IC-148		
B0319200	1155	3231	IC-179	705	
B03604	5452	6752			
B03605	5453	6753			
B03606	5454	6754			
B03641	5452	6752			
B03642	5453	6753			
B03643	5454	6754			
B048010	1208	3712			
B0480800	1207	3713			
B0480810	1208	3712			
B0480820	1197	3733			
B071	138A	3059	ZD-7.5		
B090	139A	3060	ZD-9.1	562	
B094	140A	3061	ZD-10	562	
B1-12					2N4427
B1-8.2	5072A	3136	ZD-8.2	562	
B100	102	3061/140A	ZD-2	2007	1N4002
B100(ZENER)	140A		ZD-10	562	
B1000	177				1N4007
B10064	104	3009	16	2006	
B10069	104	3009	16	2006	
B101	102	3004	2	2007	
B10142	127	3764	25		
B10142A	127	3764	25		
B10142B	127	3764	25		
B10143	127	3764	25		
B10143A	127	3764	25		
B10143B	127	3764	25		
B10162	121	3014	239	2006	
B10163	121	3014	239	2006	
B1017	121	3009	239	2006	
B102		3004	2	2007	
B102000	127	3035	25		
B102001	127	3035	25		
B102002	127	3035	25		
B102003	127	3035	25		
B1022	102	3004	2	2007	
B1022-1	102A	3004	53	2007	
B103	102	3005	1	2007	
B103000	127	3035	25		
B103001	127	3035	25		
B103002	127	3035	25		
B103003		3035	25		
B103004	127	3035	25		
B104		3005	1	2007	
B1045494P1	114		6GD1	1104	
B10474	121	3014	239	2006	
B10475	121	3014	239	2006	
B105	158	3004/102A	53	2007	
B1058	102	3722	2	2007	
B1058-1	102A	3004	53	2007	
B106		3004	53	2007	
B107	121	3004/102A	239	2006	
B107A	121	3009	239	2006	
B108	158	3004/102A	53	2007	
B1085	121	3014	239	2006	
B108A		3004	53	2007	
B108B		3004	53	2007	
B109		3004	53	2007	
B10912	121	3009	239	2006	
B10913	121	3009	239	2006	
B10A	5322	3680			
B10B	5322	3680			
B10D	5324	3681			
B10F	5326	3682			
B10Z	5322	3680			
B110	177	3004/102A			
B110(TRANSISTOR)	102A		53	2007	
B111	177	3004/102A	53	2007	
B111K	102A	3004	53	2007	
B112	102A	3004	53	2007	
B113	102A	3004	53	2007	
B113000	179	3642	76		
B114	102A	3004	53	2007	
B115	102A	3004	53	2007	
B1151	121	3009	239	2006	
B1151A	121	3009	239	2006	
B1151B	121	3009	239	2006	
B1152	121	3009	239	2006	
B1152A	121	3009	239	2006	
B1152B	121	3009	239	2006	
B1154	102	3004	2	2007	
B1154-1	102A	3004	53	2007	
B116	102A	3004	53	2007	
B117	102A	3004	53	2007	
B1178	127	3764	25		
B117K	102A	3004	53	2007	
B1181	121	3009	239	2006	
B119	121	3009	239	2006	
B119A	121	3009	239	2006	
B12-02	116		504A	1104	
B12-1-A-21	123A	3444	20	2051	
B12-12	350				2N6081
B12-1A21	123A	3444	20	2051	
B12-28	357				MRF314
B120	102A	3004	53	2007	
B1215			16	2006	
B122	121	3009	239	2006	
B123	104	3009	16	2006	
B123A	104	3009	16	2006	

Industry Standard No.	ECG	SK	GE	RS 276-	MOTOR.
B124		3009	239	2006	
B126	121	3009	239	2006	
B126A	121	3009	239	2006	
B126F	121	3035	239	2006	
B126V	121	3717	239	2006	
B127	121	3009	239	2006	
B1274	121	3009	239	2006	
B1274A	121	3009	239	2006	
B1274B	121	3009	239	2006	
B127A	121	3009	239	2006	
B128	127	3035	239	2006	
B128A	127	3009	239	2006	
B128V	121	3035	239	2006	
B129	127	3009	239	2006	
B130	131		44	2006	
B130A	131		44	2006	
B131	152	3893	66	2048	
B131(JAPAN)	121		239	2006	
B131A	121	3009	239	2006	
B132	153	3009	69	2049	
B132(JAPAN)	121		239	2006	
B132A	121		239	2006	
B133550	130		14	2041	
B133577	130		14	2041	
B133578	123A	3444	20	2051	
B133684	130		14	2041	
B133685	130		14	2041	
B133823	175		246	2020	
B134	121	3009	239	2006	
B134(JAPAN)	102A		53	2007	
B134-D	102A		53	2007	
B134-E	102A		53	2007	
B134A	121	3014	239	2006	
B134C	121	3014	239	2006	
B135	102A	3004	53	2007	
B13517-2	912		IC-172		
B135B	102A	3004	53	2007	
B135C	102A	3004	53	2007	
B135E	102A	3004	53	2007	
B136	102A	3004	53	2007	
B136-2	102A		53	2007	
B136-3	102A		53	2007	
B1368	121	3014	239	2006	
B1368A	121	3009	239	2006	
B1368B	121	3009	239	2006	
B1368C	121	3009	239	2006	
B1368D	121	3009	239	2006	
B1368E	121	3014	239	2006	
B1368F	121	3014	239	2006	
B136A	102A	3004	53	2007	
B136B	102A	3004	53	2007	
B136C	102A	3004	53	2007	
B136U	102A	3004	53	2007	
B137	121	3009	239	2006	
B138	121	3009	239	2006	
B140	144A	3009	16	2006	
B141	104	3009	16	2006	
B142	104	3009	16	2006	
B142B	104	3005	16	2006	
B142C	104	3005	16	2006	
B143	104	3009	16	2006	
B143000	186	3192	28	2017	
B143001	186	3192	28	2017	
B143003	186	3192	28	2017	
B143004	152	3054/196	66	2048	
B143009	186	3192	28	2017	
B143010	186	3192	28	2017	
B143011	152	3054/196	66	2048	
B143012	152	3054/196	66	2048	
B143015	186	3192	28	2017	
B143016	186	3192	28	2017	
B143018	152	3054/196	66	2048	
B143019	152	3054/196	66	2048	
B143024	186	3192	28	2017	
B143025	186	3192	28	2017	
B143026	152	3054/196	66	2048	
B143027	152	3054/196	66	2048	
B143P	104	3009	16	2006	
B144	104	3009	16	2006	
B144P	104	3009	16	2006	
B145	104	3009	16	2006	
B146	104	3009	16	2006	
B147	104	3009	16	2006	
B149	104	3009	16	2006	
B14A-1-21	121	3717	239	2006	
B15-12	350				2N6081
B151(TRANSISTOR)	104		16	2006	
B152	127	3009			
B152(TRANSISTOR)	104		16	2006	
B153	102A	3004	2	2007	
B154	102A	3004	2	2007	
B155	102A	3004	53	2007	
B155A	102A	3004	53	2007	
B155B	102A	3004	53	2007	
B156	102A	3003	53	2007	
B156A	102A	3003	53	2007	
B156AA	102A	3003	53	2007	
B156AB	102A	3003	53	2007	
B156AC	102A	3003	53	2007	
B156B	102A	3003	53	2007	
B156C	102A	3003	53	2007	
B156D	102A	3003	53	2007	
B156P	102A	3003	53	2007	
B157	126	3004/102A	53	2007	
B158	126	3004/102A	53	2007	
B159	126	3004/102A	53	2007	
B160	126	3004/102A	53	2007	
B161	102	3004	2	2007	
B162		3004	2	2007	
B163	102	3004	2	2007	
B164		3004	2	2007	
B165	102	3004	2	2007	
B166		3004	2	2007	
B167	158	3004/102A	53	2007	
B168	102A	3004	53	2007	
B169	123A	3444	53	2051	
B169(JAPAN)	123A	3444	53	2051	
B170	102A	3004	53	2007	
B170000	130	3027	14	2041	
B170000-ORG			14	2041	
B170000-ORN				2041	
B170000-RED	130		14	2041	
B170000-YEL			75	2041	
B170000BLK	130		14	2041	
B170000BRN	130		14	2041	
B170001	130	3027	14	2041	
B170001-BLK	130		14	2041	
B170001-BRN	130		14	2041	
B170001-ORG	130		14	2041	
B170001-RED	130		14	2041	
B170001-YEL			75	2041	
B170001BLK	130		14	2041	
B170001BRN	130		14	2041	
B170002	130	3027	14	2041	
B170002-ORG	130		14	2041	
B170002-RED	130		14	2041	
B170002-YEL			75	2041	
B170003	130	3027	14	2041	
B170003-BLK	181		75	2041	
B170003-BRN	181		75	2041	
B170003-ORG			75	2041	
B170003-RED	181		75	2041	
B170003-YEL			75	2041	
B170004	130	3027	14	2041	
B170004-BLK	181		75	2041	
B170004-BRN	181		75	2041	
B170004-ORG			75	2041	
B170004-RED	181		75	2041	
B170004-YEL			75	2041	
B170005	130	3027	14	2041	
B170005-BLK	181		75	2041	
B170005-BRN	181		75	2041	
B170005-ORG			75	2041	
B170005-RED	181		75	2041	
B170005-YEL			75	2041	
B170006	130	3027	14	2041	
B170006-BLK	181		75	2041	
B170006-BRN	181		75	2041	
B170006-ORG			75	2041	
B170006-RED	181		75	2041	
B170006-YEL			75	2041	
B170007	130	3027	14	2041	
B170007-BLK	181		75	2041	
B170007-BRN	181		75	2041	
B170007-ORG			75	2041	
B170007-RED	181		75	2041	
B170007-YEL			75	2041	
B170008	162	3027/130	35		
B170008-BLK	181		75	2041	
B170008-BRN	181		75	2041	
B170008-ORG			75	2041	
B170008-RED	181		75	2041	
B170008-YEL			75	2041	
B170009	130	3027	14	2041	
B170010	130	3027	14	2041	
B170011	130	3027	14	2041	
B170012	130	3027	14	2041	
B170013	130	3027	14	2041	
B170014	130	3027	14	2041	
B170015	130	3027	14	2041	
B170016	130	3027	14	2041	
B170017	181	3027/130	75	2041	
B170018	130	3027	14	2041	
B170019	130	3027	14	2041	
B170020	130	3027	14	2041	
B170021	130	3027	14	2041	
B170022	130	3027	14	2041	
B170023	181	3027/130	75	2041	
B170024	130	3027	14	2041	
B170025	130	3027	14	2041	
B170026	181	3027/130	75	2041	
B171	102A	3004	53	2007	
B171(JAPAN)	102A		53	2007	
B171A	102A	3004	53	2007	
B171B	102A	3004	53	2007	
B172	102A	3004	53	2007	
B172-10	156	3051	512		
B172-100	156	3051	512		
B172-20	156	3051	512		
B172-30	156		512		
B172-40	156		512		
B172-5	156	3051	512		
B172-50	156		512		
B172-60	156	3051	512		
B172-70	156	3051	512		
B172-80	156	3051	512		
B172-90	156	3051	512		
B172A	102A	3004	53	2007	
B172AF	102A	3004	53	2007	
B172B	102A	3004	53	2007	
B172C	102A	3004	53	2007	
B172D	102A	3004	53	2007	
B172E	102A	3004	53	2007	
B172F	102A	3004	53	2007	
B172H	102A	3004	53	2007	
B172P	102A	3004	53	2007	
B172R	102A	3004	53	2007	
B173	102A	3004	53	2007	
B17307	130		14	2041	
B173A	102A	3004	53	2007	
B173B	102A	3004	53	2007	
B173C	102A	3004	53	2007	
B173L	102A	3004	53	2007	
B174	102A	3004	53	2007	
B175	102A	3004	53	2007	
B175A	102A	3004	53	2007	
B175B	102A	3004	53	2007	
B175C	102A	3004	53	2007	
B175E	102A	3004	53	2007	
B176	102A	3004	53	2007	
B176-0	102A	3004	53	2007	
B176-P	102A	3004	53	2007	
B176-PR	102A	3004	53	2007	
B176000	163A	3439	36		
B176001	162	3438	35		
B176002	162	3438	35		
B176003	162	3438	35		
B176004	163A	3439	36		
B176005	163A	3439	36		
B176006	163A	3439	36		
B176007	163A	3439	36		
B176009	163A	3439	36		
B176010	163A	3439	36		
B176011	163A	3439	36		
B176013	163A	3439	36		
B176014	163A	3439	36		
B176015	163A	3439	36		
B176024	163A	3439	36		
B176025	163A	3439	36		
B176026	163A	3439	36		
B176027	163A	3439	36		
B176028	163A	3439	36		
B176029	163A	3439	36		
B176B	102A		53	2007	
B176M	102A	3004	53	2007	
B176P	102A		53	2007	
B176PRC	102A	3004	53	2007	
B176R	102A	3004	53	2007	
B177	102A	3009	239	2006	
B177(JAPAN)	102A		53	2007	
B177000	181		75	2041	
B178	102A	3004	239	2006	
B178(JAPAN)	102A		53	2007	
B178-0	102A	3004	53	2007	
B178-S	102A	3004	53	2007	
B178A	102A	3004	53	2007	
B178C	102A	3004	53	2007	
B178D	102A	3004	53	2007	

Industry Standard No.	ECG	SK	GE	RS 276-	MOTOR.
B178M	102A	3004	53	2007	
B178N	102A	3004	53	2007	
B178T	102A	3004	53	2007	
B178U	102A	3004	53	2007	
B178V	102A	3004	53	2007	
B178X	102A	3004	53	2007	
B178Y	102A	3004	53	2007	
B179	121	3014	239	2006	
B180		3004	53	2007	
B180A		3004	53	2007	
B181		3004	53	2007	
B181A		3004	53	2007	
B183	102A	3004	53	2007	
B184	102A	3004	53	2007	
B185	102A		53	2007	
B185(O)	102A		53	2007	
B185(O)	102A		53	2007	
B185AA	102A		53	2007	
B185F	102A		53	2007	
B185P	102A		53	2007	
B186	102A	3004	53	2007	
B186(O)	102A		53	2007	
B186(SANYO)			53	2007	
B186-1	102A	3004	53	2007	
B186-K	102A	3004	53	2007	
B186A	102A	3004	53	2007	
B186AG	102A	3004	53	2007	
B186B	102A	3004	53	2007	
B186BY	102A	3004	53	2007	
B186G	102A	3004	53	2007	
B186H	102A		53	2007	
B186L	102A	3004	53	2007	
B186Y	102A	3004	53	2007	
B187	102A	3004	53	2007	
B187(1)			53	2007	
B187(SANYO)			53	2007	
B187AA	102A	3004	53	2007	
B187B	102A	3004	53	2007	
B187C	102A	3004	53	2007	
B187D	102A	3004	53	2007	
B187G	102A	3004	53	2007	
B187K	102A		53	2007	
B187R	102A	3004	53	2007	
B187RED	102A		53	2007	
B187S	102A	3004	53	2007	
B187Y	102A		53	2007	
B187YEL	102A		53	2007	
B188	102A	3004	53	2007	
B189	102A/410	3004/102A	53	2007	
B19		3719	16	2006	
B1904	121	3009	239	2006	
B191-5	5940		5048		
B196	5635	3533			
B199	102	3722	2	2007	
B1A	182	3188A	55		
B1A1	116	3311	504A	1104	
B1A5	116	3311	504A	1104	
B1A9	116	3311	504A	1104	
B1B	116	3016	504A	1104	
B1B1	116	3017B/117	504A	1104	
B1B5	116	3017B/117	504A	1104	
B1B9	116	3017B/117	504A	1104	
B1C	184	3178A	57	2017	
B1C-1	184		57	2017	
B1C-2	184		57	2017	
B1C1	116	3016	504A	1104	
B1C5	116	3017B/117	504A	1104	
B1C9	116	3017B/117	504A	1104	
B1D	152		66	2048	
B1D-1	184		57	2017	
B1D1	116	3031A	504A	1104	
B1D5	116	3031A	504A	1104	
B1D9	116	3017B/117	504A	1104	
B1E	191	3103A/396	249		
B1E(QUASAR)	191		249		
B1E-1	191	3232	249		
B1E1	116	3031A	504A	1104	
B1E1(QUASAR)	191		249		
B1E5	116	3017B/117	504A	1104	
B1E9	116	3017B/117	504A	1104	
B1F	184		57	2017	
B1F1	116	3017B/117	504A	1104	
B1F5	116	3017B/117	504A	1104	
B1F9	116	3017B/117		1104	
B1G	191	3232	249		
B1G1	116	3017B/117	504A	1104	
B1G5	116	3017B/117	504A	1104	
B1G9	116	3017B/117	504A	1104	
B1H	161	3018	39	2015	
B1H(DIODE)			504A	1104	
B1H(XSTR)			60	2015	
B1H1	125	3032A	510,531	1114	
B1H5	125	3032A	510,531	1114	
B1H9	125	3032A	510,531	1114	
B1J	159	3114/290	82	2032	
B1K	123A	3444	20	2051	
B1K1	125	3032A	510,531	1114	
B1K5	125	3032A	510,531	1114	
B1K9	125	3032A	510,531	1114	
B1M	188		226	2018	
B1M1	125	3033	510,531	1114	
B1M5	125	3033	510,531	1114	
B1M9	125	3033	510,531	1114	
B1N	123A	3444	20	2051	
B1N-1	159	3114/290	82	2032	
B1N-2	159	3114/290	82	2032	
B1P	163A	3439	36		
B1P-1	163A	3439	36		
B1P7201	123A	3444	20	2051	
B1P9			504A	1104	
B1R	247	3444/123A	20	2051	
B1T	283	3439/163A	36		
B1U	241	3188A/182	57	2020	
B1U148	184	3054/196	57	2017	
B1V	286	3131A/369	267		
B1W	123A	3444	20	2051	
B1Y	286	3194	267		
B1Z	175	3538	246	2020	
B2			214	2038	
B2-8Z					2N6080
B20		3719	16	2006	
B20-001			3	2006	
B200	102A	3004	2	2007	1N4003
B200A	102A	3004	2	2007	
B200C40	116	3016	504A	1104	
B201	102A	3004	53	2007	
B202	102	3004	2	2007	
B203AA	179	3004/102A	2	2007	
B204	179	3642	76		
B205	179	3642	76		
B206	179	3642	76		
B2090	139A	3060	ZD-9.1	562	
B21		3719	16	2006	
B211032	179	3642			
B215	121	3009	239	2006	
B216	121	3009	239	2006	
B216A	121	3009	239	2006	
B217	121	3009	239	2006	
B217A	121	3009	239	2006	
B217G	121	3009	239	2006	
B217U	121	3009	239	2006	
B218	102	3004	2	2007	
B219	102	3004	2	2007	
B22	102A		53	2007	
B22-3	102A		53	2007	
B22-4	102A		53	2007	
B220	102	3004	2	2007	
B220A	102	3004	2	2007	
B221	102	3004	2	2007	
B221A	102	3004	2	2007	
B222	102	3004	2	2007	
B223	102	3004	2	2007	
B223-10	156		512		
B223-20	525		512		
B223-30	156		512		
B223-40	156		512		
B223-50	156		512		
B223-60	156		512		
B224	102A	3009	239	2006	
B224(JAPAN)	102	3722	2	2007	
B225	102	3004	2	2007	
B226	102A	3004	2	2007	
B227	102A	3004	2	2007	
B228	127	3009	16	2006	
B229	127	3009	239	2006	
B22A	102A		53	2007	
B22B	102A		53	2007	
B22I	102A		53	2007	
B22R	102A		53	2007	
B22Y	102A		53	2007	
B23	102A		53	2007	
B23-1	102A		53	2007	
B23-2	102A		53	2007	
B23-79	197	3084	250	2027	
B23-82	196	3054	241	2020	
B230	127	3009	16	2006	
B231	127	3035	25		
B232	127	3035	25		
B232008	124		12		
B233	127	3764	16	2006	
B234	127	3035	25		
B234N	127	3035	25		
B235	213	3012/105	4		
B235A	213	3006/160	4		
B236	105	3012	4		
B237	105	3012	4		
B237-12A	105	3012	4		
B237-12B	105	3012	4		
B238	158	3004/102A	53	2007	
B238-12A	158	3012/105	53	2007	
B238-12B	158	3012/105	53	2007	
B238-12C	158	3012/105	53	2007	
B239		3009	239	2006	
B239A		3009	239	2006	
B24	102A		53	2007	
B24-06B	125		509	1114	
B24-06C	506		509	1114	
B24-1	102A		53	2007	
B240		3006	245	2004	
B240A		3006	245	2004	
B241		3004	2	2007	
B246	104	3009	16	2006	
B247	104	3009	239	2006	
B248	104	3009	239	2006	
B248A	104	3009	239	2006	
B249	179	3009	16	2006	
B249A	179	3009	16	2006	
B25	121	3717	239	2006	
B25-12	351				2N6082
B25-28	359				MRF314
B250	104	3719	16	2006	
B250A	104	3009	16	2006	
B250C100	116	3016	504A	1104	
B250C100TD	116	3016	504A	1104	
B250C125	116	3016	504A	1104	
B250C125K4	116	3016	504A	1104	
B250C125N2	116	3016	504A	1104	
B250C125X4	116	3016	504A	1104	
B250C150	116	3016	504A	1104	
B250C150K4	116	3016	504A	1104	
B250C75	116	3016	504A	1104	
B250C75K4	116	3016	504A	1104	
B250C75K41	116	3017B/117	504A	1104	
B250C75K45	116	3016	504A	1104	
B250C75K5	116		504A	1104	
B250C7K4S	116	3016	504A	1104	
B251	127	3035	25		
B251A	127	3035	25		
B252	127	3035	25		
B252A	127	3035	25		
B253	127	3035	25		
B253A	127	3035	25		
B254	226	3009	16	2006	
B255	226	3009	16	2006	
B256	226	3009	16	2006	
B257	102A	3004	2	2007	
B258	213S	3012/105	4		
B259	213	3012/105	4		
B25B	121	3717	239	2006	
B26	104	3009	53	2007	
B26(JAPAN)	104	3719	16	2006	
B260	213	3012/105	4		
B261	102A	3004	53	2007	
B262	102A	3004	53	2007	
B2620	506	3130	511	1114	
B263	102A	3004	53	2007	
B264	102A	3004	53	2007	
B265	102A	3004	53	2007	
B266	102A	3004	53	2007	
B266A-1	159	3025/129	82	2032	
B266B-1	159	3025/129	82	2032	
B266P	102A	3004	53	2007	
B266Q	102A	3004	53	2007	
B267	102A	3004	53	2007	
B268	102A	3004	53	2007	
B269	102A	3004	53	2007	
B269-3345	519		514	1122	
B26A	104	3719	16	2006	
B27	104	3719	16	2006	
B270	102A	3004	53	2007	
B270A	102A	3004	53	2007	
B270B	102A	3004	53	2007	
B270C	102A	3004	53	2007	
B270D	102A	3004	53	2007	
B270E	102A	3004	53	2007	
B271	102A	3004	53	2007	
B272	102A	3004	53	2007	
B273	102A	3004	53	2007	

Industry Standard No.	ECG	SK	GE	RS 276-	MOTOR.
B274(JAPAN)	127	3764	25		
B274(SYLVANIA)	128		243	2030	
B275	127	3035	25		
B276	127	3035	25		
B28	109	3087		1123	
B28(JAPAN)	104	3719	16	2006	
B282	127	3009	239	2006	
B283	121	3009	239	2006	
B284	121	3009	239	2006	
B285	127	3009	239	2006	
B29	104	3719	16	2006	
B290	100	3005	1	2007	
B291	100	3005	1	2007	
B292	100	3005	1	2007	
B292A	100	3005	1	2007	
B293	102A	3004	53	2007	
B294		3017B	504A	1104	
B294(RECTIFIER)	116		504A	1104	
B294(TRANSISTOR)	102A		53	2007	
B295	121$	3009	239	2006	
B296	127	3035	25		
B299	102A	3004	53	2007	
B2A	159	3114/290	82	2032	
B2A1	116	3311	504A	1104	
B2A5	116	3311	504A	1104	
B2A9	116	3017B/117	504A	1104	
B2B	175	3111	246	2020	
B2B1	116	3017B/117	504A	1104	
B2B5	116	3017B/117	504A	1104	
B2B9	116	3017B/117	504A	1104	
B2C1	116	3017B/117	504A	1104	
B2C5	116	3016	504A	1104	
B2C9	116	3017B/117	504A	1104	
B2D	123A	3444	20	2051	
B2D1	116	3017B/117	504A	1104	
B2D5	116	3017B/117	504A	1104	
B2D9	116	3017B/117	504A	1104	
B2E	159	3114/290	82	2032	
B2E1	116	3017B/117	504A	1104	
B2E5	116	3017B/117	504A	1104	
B2E9	116	3017B/117	504A	1104	
B2F1	116	3017B/117	504A	1104	
B2F5	116	3017B/117	504A	1104	
B2F9	116	3017B/117	504A	1104	
B2G	159	3114/290	82	2032	
B2G1	116	3017B/117	504A	1104	
B2G5	116	3017B/117	504A	1104	
B2G9	116	3017B/117	504A	1104	
B2H	162	3559	35		
B2H1	125	3081	510,531	1114	
B2H5	125	3081	510,531	1114	
B2H9	125	3081	510,531	1114	
B2J	184	3054/196	57	2017	
B2K	186	3192			
B2K1	125	3032A	510,531	1114	
B2K5	125	3032A	510,531	1114	
B2K9	125	3032A	510,531	1114	
B2L	283	3467	36		
B2M	188	3199	83	2018	
B2M-1	159	3114/290	82	2032	
B2M-2	159	3114/290	82	2032	
B2M-3	159	3114/290	82	2032	
B2M1	125	3033	510,531	1114	
B2M5	125	3033	510,531	1114	
B2M9	125	3033	510,531	1114	
B2P	286	3194	267		
B2S	159	3114/290	82	2032	
B2SB241	104	3009	16	2006	
B2SB244	104	3009	16	2006	
B2V			57	2017	
B2W	121	3114/290	239	2006	
B2Y	159	3114/290	82	2032	
B2Z	107	3018	17	2051	
B3			214	2038	
B3-12					2N6080
B3-28	346				2N5641
B30	109	3087		1123	
B30(JAPAN)	104	3719	16	2006	
B30-12	351				2N6083
B300	127	3035	25		1N4004
B301	127	3035	25		
B302	102A	3004	53	2007	
B303(TRANSISTOR)	102A		53	2007	
B303-0	102A		53	2007	
B3030		3004	53	2007	
B303A	102A	3004	53	2007	
B303B	102A	3004	53	2007	
B303C	102A	3004	53	2007	
B303H	102A		53	2007	
B303K	102A	3004	53	2007	
B304	158	3004/102A	53	2007	
B304A	158	3004/102A	53	2007	
B309	127	3035	25		
B30C1000	166	9075	504A	1152	
B30C250	116	3016	504A	1104	
B30C250-1	116	3017B/117	504A	1104	
B30C250KP	167	3647	510	1172	
B30C350-1	116	3016	504A	1104	
B30C500	116		504A	1104	
B30C50KP	167	3647	510	1172	
B30C600	116	3016	504A	1104	
B30C600CB	116	3016	504A	1104	
B31	116		504A	1104	
B31(RECTIFIER)	116		504A	1104	
B31(TRANSISTOR)	104	3719	16	2006	
B310	127	3035	25		
B310(ZENER)	5085A	3337	ZD-36		
B311	127	3035	25		
B312	127	3035	25		
B313	127	3035	25		
B314		3005	1	2007	
B315	102A	3004	53	2007	
B316	102A	3004	53	2007	
B317	102A	3004	53	2007	
B318	127	3035	25		
B319	127	3035	25		
B32	102A	3004	53	2007	
B32-0	102A	3004	53	2007	
B32-1	102A	3004	53	2007	
B32-2	102A	3004	53	2007	
B32-4	102A	3004	53	2007	
B320	127	3035	25		
B321	102A	3004	53	2007	
B322	102A	3004	53	2007	
B323	102A	3004	53	2007	
B324	158	3004/102A	53	2007	
B324A	158	3004/102A	53	2007	
B324B	158	3004/102A	53	2007	
B324D	158	3004/102A	53	2007	
B324E	158	3004/102A	53	2007	
B324E-1	158	3004/102A	53	2007	
B324F	158		53	2007	
B324G	158	3004/102A	53	2007	
B324H	158	3004/102A	53	2007	
B324I	158	3004/102A	53	2007	
B324J	158		53	2007	
B324K	158	3004/102A	53	2007	
B324L	158	3004/102A	53	2007	
B324N	158	3004/102A	53	2007	
B324P	158	3004/102A	53	2007	
B324S	158		53	2007	
B324V	158	3004/102A	53	2007	
B326	102A	3004	2	2007	
B327	102A	3004	2	2007	
B328	102	3004	2	2007	
B329	102A	3004	53	2007	
B329K	102A	3004	53	2007	
B32N	102A	3004	53	2007	
B33	102A	3095/147A	53	2007	
B33-4	102A	3004	53	2007	
B331	213	3012/105	254		
B331H	213		254		
B331HA	213		254		
B331HB	213		254		
B331HC	213		254		
B332	213	3012/105	254		
B332H	213		254		
B332HA	213		254		
B332HB	213		254		
B332HC	213		254		
B333	213	3012/105	254		
B335	102	3722	2	2007	
B336	102A	3004	53	2007	
B337	121	3009	239	2006	
B337A	121	3004/102A	239	2006	
B337B	121	3013	16(2)	2006	
B337BK	121	3717	239	2006	
B337H	121	3004/102A	239	2006	
B337HA	121	3717	239	2006	
B337HB	121	3717	239	2006	
B338	121	3009	239	2006	
B338H	121	3717	239	2006	
B338HA	121	3004/102A	239	2006	
B338HB	121	3004/102A	239	2006	
B339	179	3642	76		
B339H	179	3642	76		
B33C	102A	3004	53	2007	
B33D	102A	3004	53	2007	
B33E	102A	3004	53	2007	
B33F	102A	3004	53	2007	
B34	102	3004	2	2007	
B340	179	3642	76		
B340H	179	3642	76		
B341	127	3764	25		
B341H	127	3764	25		
B341V	127	3764	25		
B342	127	3035	25		
B343	127	3035	25		
B345	102A	3004	53	2007	
B346	102A	3004	53	2007	
B3465	195A	3048/329	46		
B3468	195A	3765	46		
B346K	102A	3004	53	2007	
B346Q	102A		53	2007	
B347	102A	3004	2	2007	
B348	102A	3004	53	2007	
B348Q	102A		53	2007	
B348R	102A	3004	53	2007	
B349	102A	3004	53	2007	
B34N	102	3035	2	2007	
B350	102	3004	2	2007	
B35016	5636	3939			
B351	105	3012	4		
B352	213	3012/105	48		
B352D	213		48		
B353	213	3012/105	4		
B3531	186	3192	28	2017	
B3533	186	3192	28	2017	
B3537	186	3192	28	2017	
B3538	186	3192	28	2017	
B354		3012	4		
B3540	186	3192	28	2017	
B3541	186	3192	28	2017	
B3542	186	3192	28	2017	
B3547	152	3054/196	66	2048	
B3548	152	3054/196	66	2048	
B355	104	3009	16	2006	
B3550	152	3054/196	66	2048	
B3551	152	3054/196	66	2048	
B356	104	3009	16	2006	
B357	127	3035	25		
B3570	186	3192	28	2017	
B3576	186	3192	28	2017	
B3577	152	3054/196	66	2048	
B3578	152	3054/196	66	2048	
B358	127	3035	25		
B3580	152	3054/196	66	2048	
B3584	152	3054/196	66	2048	
B3585	152	3054/196	66	2048	
B3586	152	3054/196	66	2048	
B3588	152	3054/196	66	2048	
B3589	152	3054/196	66	2048	
B359	127	3035	25		
B35C600	116	3016	504A	1104	
B360	127	3035	25		
B3606	186	3192	28	2017	
B3607	186	3192	28	2017	
B3608	186	3192	28	2017	
B3609	186	3192	28	2017	
B361	127	3035	25		
B3610	186	3192	28	2017	
B3611	186	3192	28	2017	
B3612	186	3192	28	2017	
B3613	186	3192	28	2017	
B3614	186	3192	28	2017	
B362	127	3035	25		
B364	158	3004/102A	53	2007	
B365	158	3004/102A	53	2007	
B36564	116	3311	504A	1104	
B365B	158		53	2007	
B366	127	3004/102A	53	2007	
B367	131	3198	44	2006	
B367(A)	131		44	2006	
B367A	131	3198	44	2006	
B367B	131	3198	44	2006	
B367C	131	3198	44	2006	
B367H	131	3198	44	2006	
B368	131$	3198/131	44	2006	
B368A	131$	3198/131	44	2006	
B368B	131$	3198/131	44	2006	
B368H	131$	3198/131	44	2006	
B37	102A	3004	53	2007	
B370	102A	3004	53	2007	
B370A	102A	3004	53	2007	
B370AA	102A	3004	53	2007	
B370AB	102A	3004	53	2007	
B370AC	102A	3004	53	2007	
B370AHA	102A	3004	53	2007	

Industry Standard No.	ECG	SK	GE	RS 276-	MOTOR.
B370AHB	102A	3004	53	2007	
B370B	102A	3004	53	2007	
B370C	102A	3004	53	2007	
B370D	102A	3004	53	2007	
B370P	102A	3004	53	2007	
B370PB	102A		53	2007	
B370V	102A	3004	53	2007	
B371	102A	3004	53	2007	
B371D	102A	3004	53	2007	
B372	176	3004/102A	53	2007	
B373	176	3004/102A	53	2007	
B3746	128	3024	243	2030	
B3747	186	3192	28	2017	
B3748	186	3192	28	2017	
B375	127		25		
B375-2B	127	3035	25		
B375-5B	127	3035	25		
B3750	186	3192	28	2017	
B375A	127		25		
B375A-2B	127	3035	25		
B375A-5B	127	3035	25		
B375A-NB	127	3035	25		
B375TV	127	3035	25		
B376	102A	3004	53	2007	
B376G	102A	3004	53	2007	
B377	102A	3004	53	2007	
B377B	102		53	2007	
B378	102A	3005	53	2007	
B378A	102A	3004	2	2007	
B379	102	3004	2	2007	
B379-2	102	3004	2	2007	
B379A	102	3004	2	2007	
B379B	102	3004	2	2007	
B37A	102A	3004	53	2007	
B37B	102A	3004	53	2007	
B37C	102A	3004	53	2007	
B37E	102A	3004	53	2007	
B37F	102A	3004	53	2007	
B38	102	3004	2	2007	
B380	102	3004	2	2007	
B380A	102	3004	2	2007	
B381	102A	3004	53	2007	
B382	102A	3004	53	2007	
B383	102A	3004	53	2007	
B383-1	102A	3004	53	2007	
B383-2	102A	3004	53	2007	
B384	102A	3006/160	52	2024	
B385	102A	3006/160	52	2024	
B386	102A	3004	53	2007	
B387	102A	3005	1	2007	
B389	102A	3004	53	2007	
B39	102A	3004	53	2007	
B390	127	3035	25		
B391	127	3009	239	2006	
B392	100	3005	1	2007	
B393	100	3005	1	2007	
B394	100	3005	1	2007	
B395	100	3005	1	2007	
B396	102	3005	1	2007	
B3A1	116	3311	504A	1104	
B3A5	116	3017B/117	504A	1104	
B3A9	116	3017B/117	504A	1104	
B3B1	116	3017B/117	504A	1104	
B3B5	116	3017B/117	504A	1104	
B3B9	116	3017B/117	504A	1104	
B3C1	116	3017B/117	504A	1104	
B3C5	116	3017B/117	504A	1104	
B3C9	116	3017B/117	504A	1104	
B3D1	116	3017B/117	504A	1104	
B3D5	116	3031A	504A	1104	
B3D9	116	3031A	504A	1104	
B3E1	116	3031A	504A	1104	
B3E5	116	3031A	504A	1104	
B3E9	116	3017B/117	504A	1104	
B3F1	116	3017B/117	504A	1104	
B3F5	116		504A	1104	
B3F9	116	3017B/117	504A	1104	
B3G1	116	3017B/117	504A	1104	
B3G5	116	3017B/117	504A	1104	
B3G9	116	3017B/117	504A	1104	
B3H1	125	3032A	510,531	1114	
B3H5	125	3032A	510,531	1114	
B3H9	125	3032A	510,531	1114	
B3K1		3032A	509	1114	
B3K5	125	3032A	510,531	1114	
B3K9	125	3032A	510,531	1114	
B3M1	125	3033	510,531	1114	
B3M5	125	3033	510,531	1114	
B3M9	125	3033	510,531	1114	
B3N1	506		511	1114	
B3N5	506		511	1114	
B3N9	506		511	1114	
B4			214	2038	
B40	102A	3004	53	2007	
B40-12	320				2N6084
B40-28	360				MRF315
B400	102A	3004	53	2007	1N4004
B4004B	102A		53	2007	
B400A	102A	3004	53	2007	
B400B	102A	3004	53	2007	
B400K	102A	3004	53	2007	
B401	100	3005	1	2007	
B402	100	3005	1	2007	
B403	100	3005	1	2007	
B405	158	3004/102A	53	2007	
B405-2C	158	3004/102A	53	2007	
B405-3C	158	3004/102A	53	2007	
B405-4C	158	3004/102A	53	2007	
B405A	158	3004/102A	53	2007	
B405B	158	3004/102A	53	2007	
B405C	158	3004/102A	53	2007	
B405D	158	3004/102A	53	2007	
B405E	158	3004/102A	53	2007	
B405G	158		53	2007	
B405H	158	3004/102A	53	2007	
B405K	158		53	2007	
B405R	158	3004/102A	53	2007	
B405RE	158	3004/102A	53	2007	
B407	121	3009	239	2006	
B407-O			239	2006	
B407-O	121	3009		2006	
B407TV	121	3009	239	2006	
B408	102A	3005	1	2007	
B41	104		16	2006	
B410	127	3014	25		
B411	153	3014	25		
B413	131	3009	16	2006	
B414	226	3009	16	2006	
B415	158	3004/102A	53	2007	
B415A	158	3004/102A	53	2007	
B415B	158		53	2007	
B416	100	3005	1	2007	
B417	100	3005	1	2007	
B42	121		16	2006	

Industry Standard No.	ECG	SK	GE	RS 276-	MOTOR.
B421		3004	53	2007	
B422	102	3004	2	2007	
B423	102	3004	2	2007	
B424	121	3009	239	2006	
B425	121	3009	25		
B425Y	127	3035	25		
B426	121	3009	239	2006	
B426BL	121	3009	239	2006	
B426R	121	3009	239	2006	
B426Y	121	3009	239	2006	
B427	102A	3004	53	2007	
B428	102A	3004	53	2007	
B43	102A	3004	53	2007	
B430	213	3012/105	4		
B431	158	3004/102A	53	2007	
B432	127	3035	25		
B433	213	3012/105	4		
B434	153	3274	69	2049	
B434-O			69	2049	
B434-O	153	3274		2049	
B434-R	153	3274	69	2049	
B434-Y	153	3274	69	2049	
B435	153	3274	69	2049	
B435-O			69	2049	
B435-O	153	3274		2049	
B435-R	153	3274	69	2049	
B435-Y	153	3274	69	2049	
B439	102A	3004	53	2007	
B439A	102A	3004	53	2007	
B43A	102A	3004	53	2007	
B44	102A	3004	53	2007	
B440	102A	3004	53	2007	
B443	102A	3004	53	2007	
B443A	102A	3004	53	2007	
B443B	102A	3004	53	2007	
B444	126	3006/160	52	2024	
B444A	126	3006/160	52	2024	
B444B	126	3006/160	52	2024	
B445	226$	3009	239	2006	
B446	226$	3009	239	2006	
B447	127	3035	25		
B448	131	3198	44	2006	
B449	121	3009	239	2006	
B449F	121	3009	239	2006	
B449P	121	3009	239	2006	
B45-12					2N6084
B450	158	3004/102A	53	2007	
B450A	158	3004/102A	53	2007	
B451	158	3004/102A	53	2007	
B452	158	3004/102A	53	2007	
B452A	158	3004/102A	53	2007	
B453	158	3004/102A	53	2007	
B454		3004	53	2007	
B455		3004	53	2007	
B457	158	3004/102A	53	2007	
B457-C	158		53	2007	
B457A	158	3004/102A	53	2007	
B458	131		44	2006	
B458A	131		44	2006	
B459	102A	3004	53	2007	
B459-O			53	2007	
B459-O	102A	3004		2007	
B459A	102A	3004	53	2007	
B459B	102A	3004	53	2007	
B459C	102A	3004	53	2007	
B459D	102A	3004	53	2007	
B46	102A	3004	53	2007	
B46-110	113A		6GC1		
B460	102A	3004	53	2007	
B460A	102A	3004	53	2007	
B460B	102A	3004	53	2007	
B461	176	3123	80		
B462	131$	3198/131	44	2006	
B463	131	3052	44	2006	
B463BL	131	3052	44	2006	
B463E	131	3052	44	2006	
B463R	131	3052	44	2006	
B463Y	131	3052	44	2006	
B464	127	3035	25		
B465	127	3035	25		
B466	226$	3198/131	44	2006	
B467	226$	3198/131	44	2006	
B468	127	3035	25		
B468A	127	3035	25		
B468B	127	3035	25		
B468C	127	3035	25		
B468D	127	3035	25		
B47	102A	3004	53	2007	
B470	102A	3004	2	2007	
B471	121	3009	239	2006	
B471-2	121	3009	239	2006	
B471A	121	3009	239	2006	
B471B	121	3009	239	2006	
B472	121$	3009	239	2006	
B472A	121$	3009	239	2006	
B472B	121$	3009	239	2006	
B473	131	3198	44	2006	
B473D	131	3198	44	2006	
B473F	131	3198	44	2006	
B473H	131		44	2006	
B474	226	3082	49	2025	
B474-2	226		49	2025	
B474-3	226	3082	49	2025	
B474-4	226	3082	49	2025	
B474-6D	226	3082	49	2025	
B474MP	226	3082	49	2025	
B474S	226	3082	49	2025	
B474V10	226	3082	49	2025	
B474V4	226	3082	49	2025	
B474Y	226	3082	49	2025	
B475	102A	3004	53	2007	
B475A	102A	3004	53	2007	
B475B	102A	3004	53	2007	
B475D	102A	3004	53	2007	
B475E	102A	3004	53	2007	
B475F	102A	3004	53	2007	
B475G	102A	3004	53	2007	
B475P	102A	3004	53	2007	
B475Q	102A	3004	53	2007	
B476	176		53	2007	
B48	102	3004	2	2007	
B481	226	3198/131	44	2006	
B481D	131	3198	44	2006	
B481E	131	3198	44	2006	
B482	102A	3004	53	2007	
B483	179	3642	76		
B484	179	3642	76		
B485	179$	3642/179	76		
B486	102	3004	2	2007	
B49	102A	3004	53	2007	
B492	176		80		
B492B	176		80		
B494	158	3004/102A	53	2007	
B495	158	3004/102A	53	2007	

Industry Standard No.	ECG	SK	GE	RS 276-	MOTOR.
B495A	158	3004/102A	53	2007	
B495C	158	3004/102A	53	2007	
B495D	158	3004/102A	53	2007	
B495T	158	3004/102A	53	2007	
B496	102A	3004	53	2007	
B497	102A	3004	2	2007	
B498	102A	3004	53	2007	
B4A-1-A-21	121	3717	239	2006	
B4A1	116	3311	504A	1104	
B4A5	116	3311	504A	1104	
B4A9	116	3017B/117	504A	1104	
B4B1	116	3017B/117	504A	1104	
B4B5	116	3017B/117	504A	1104	
B4B9	116	3016	504A	1104	
B4C1	116	3017B/117	504A	1104	
B4C5	116	3016	504A	1104	
B4C9	116	3016	504A	1104	
B4D1	116	3017B/117	504A	1104	
B4D5	116	3031A	504A	1104	
B4D9	116	3031A	504A	1104	
B4E1	116	3031A	504A	1104	
B4E5	116	3017B/117	504A	1104	
B4E9	116	3017B/117	504A	1104	
B4F1	116	3017B/117	504A	1104	
B4F5	116	3017B/117	504A	1104	
B4F9	116	3017B/117	504A	1104	
B4G1	116	3017B/117	504A	1104	
B4G5	116	3017B/117	504A	1104	
B4G9	116	3017B/117	504A	1104	
B4H1	125	3032A	531	1114	
B4H5	125	3032A	531	1114	
B4H9	125	3032A	531	1114	
B4K1	125	3032A	531	1114	
B4K5	125	3032A	531	1114	
B4K9	125	3032A	531	1114	
B4M1	125	3033	531	1114	
B4M5	125	3033	531	1114	
B4M9	125	3033	531	1114	
B5	102A	3004	53	2007	
B5-8Z					2N6081
B50	116	3004/102A	2	2007	1N4001
B500					1N4005
B5000	152	3893	66	2048	
B5001	186	3192	28	2017	
B5002	152	3054/196	66	2048	
B502	197	3085	250	2027	
B5020	130	3027	14	2041	
B5021	152	3054/196	66	2048	
B5022	152	3054/196	66	2048	
B503	197	3085	250	2027	
B5031	152	3054/196	66	2048	
B5032	152	3054/196	66	2048	
B506	88		263	2043	
B506A	88		263	2043	
B506C	285		263	2043	
B506D	285		263	2043	
B507	153	3083/197	69	2049	
B508	153	3083/197	69	2049	
B509	153	3083/197	69	2049	
B51	116	3123	245	2004	
B51(JAPAN)		3722	2	2007	
B510	323	3029/130MP	244	2027	
B510S	323	3203/211	244	2027	
B511	153	3274	69	2049	
B511C	153	3274	69	2049	
B511D	153	3274	69	2049	
B511E	187A	3274/153			
B512	153	3083/197	69	2049	
B512A	197	3083	69	2049	
B513	153	3083/197	69	2049	
B513A	197	3083	69	2049	
B514	153	3083/197	69	2049	
B515	153	3083/197	69	2049	
B516C	102A		53	2007	
B516CD	102A		53	2007	
B516D	102A		53	2007	
B516P	102A		53	2007	
B52	102	3004	2	2007	
B522-893	113A	3119/113	6GC1		
B525D		3450	272		
B525E		3450	272		
B527-062	113A	3119/113	6GC1		
B53	102	3004	2	2007	
B532	281	3846/285	266		
B532P	281	3846/285	266		
B532Q	281	3846/285	266		
B532R	281	3846/285	266		
B532S	281	3846/285	266		
B534	102A		53	2007	
B534A	102A		53	2007	
B535	158		53	2007	
B536	292	3441			
B537	292	3441			
B537L	292	3441			
B537LM	292		69	2049	
B539	285	3846	266		
B539R	285	3846	20	2051	
B54	102A	3004	53	2007	
B541	88		263	2043	
B542	159	3114/290			
B544	294	3841	48		
B544D	294		48		
B544E	294		48		
B544F	294		48		
B544P1	193A		48		
B544P1D	193A		48		
B544P1E	193A		48		
B544P1F	193A		48		
B546EX	398	3930			
B546K	398	3930			
B54731-30	159		82	2032	
B5493957-4	129	3025	244	2027	
B5493957-5	129	3025	244	2027	
B5493957-6	129	3025	244	2027	
B54B	102A	3004	53	2007	
B54E	102A		53	2007	
B54F	102A	3004	53	2007	
B54Y	102A	3004	53	2007	
B55	116	3004/102A			
B55(TRANSISTOR)	102A		53	2007	
B550	88		250	2027	
B552	285	3846			
B554	285	3846	266		
B554-0	285		266		
B554-R	285	3846	266		
B555	281	3359	263	2043	
B555-0			263	2043	
B555-0	281	3359		2043	
B556	281	3359			
B556R	281	3359			
B56	102A	3004	53	2007	
B56-15	5802	9005		1143	
B56-33	5801	9004		1142	
B560	298	3004/102A	53	2007	

Industry Standard No.	ECG	SK	GE	RS 276-	MOTOR.
B561			82	2032	
B561B			82	2032	
B561C			82	2032	
B562	294		223		
B562B	294		223		
B562C	294		223		
B564	294	3841			
B56A	102A	3004	53	2007	
B56B	102A		53	2007	
B56C	102A	3004	53	2007	
B57	102A		53	2007	
B58	158		53	2007	
B588-20	525	3925			
B59	116	3004/102A	53	2007	
B5A	102A		53	2007	
B5A(MOTOROLA)	165		38		
B5A1	116	3017B/117	504A	1104	
B5A5	116	3017B/117	504A	1104	
B5A9	116	3017B/117	504A	1104	
B5B1	116	3017B/117	504A	1104	
B5B5	116	3017B/117	504A	1104	
B5B9	116	3017B/117	504A	1104	
B5C	175	3538	246	2020	
B5C1	116	3016	504A	1104	
B5C5	116	3017B/117	504A	1104	
B5C9	116	3017B/117	504A	1104	
B5D	154	3045/225	40	2012	
B5D1	116	3031A	504A	1104	
B5D5	116	3017B/117	504A	1104	
B5D9	116	3017B/117	504A	1104	
B5E	196	3054	66	2048	
B5E1	116	3017B/117	504A	1104	
B5E5	116	3017B/117	504A	1104	
B5E9	116	3017B/117	504A	1104	
B5F1	116	3017B/117	504A	1104	
B5F5	116	3017B/117	504A	1104	
B5F9	116	3017B/117	504A	1104	
B5G1	116	3017B/117	504A	1104	
B5G5	116	3017B/117	504A	1104	
B5G9	116	3017B/117	504A	1104	
B5H1	125	3032A	510,531	1114	
B5H5	125	3032A	510,531	1114	
B5H9	125	3032A	510,531	1114	
B5K1	125	3032A	510,531	1114	
B5K5	125	3032A	510,531	1114	
B5K9	125	3032A	510,531	1114	
B5M1	125	3033	510,531	1114	
B5M5	125	3033	510,531	1114	
B5M9	125	3033	510,531	1114	
B60	102A	3004	53	2007	
B600	88				1N4005
B601-1006	160	3006	245	2004	
B601-1007	160	3006	245	2004	
B601-1008	160	3006	245	2004	
B601-1009	102A	3004	53	2007	
B601-1010	158	3004/102A	53	2007	
B601-1011	109	3087		1123	
B601-1012	116	3311	504A	1104	
B60A	102A	3004	53	2007	
B60C300	116(4)	3031A	504A(4)	1104	
B61	102A	3004	53	2007	
B614 007 0	177			1122	
B614-007-0			300	1122	
B62	226	3009	16	2006	
B63	131		44	2006	
B64	127	3009	239	2006	
B645	285$	3846/285			
B65	102A	3004	53	2007	
B65-1-A-21	160		245	2004	
B65-1A21	160		245	2004	
B65-2-A-21	103A		59	2002	
B65-2A21	103A		59	2002	
B65-4-A-21	101		8	2002	
B65-4A21	101	3861	8	2002	
B65A-1-21	160		245	2004	
B65B-1-21	160		245	2004	
B65C-1-21	121	3717	239	2006	
B66	102A	3004	53	2007	
B66-1-A-21	102A		53	2007	
B66-1A21	102A		53	2007	
B66-2-A-21	102A		53	2007	
B66-2A21	102A		53	2007	
B66-3-A-21	102A		53	2007	
B66-3A21	102A		53	2007	
B66H	102A	3004	53	2007	
B66X0035-001	118	3066	CR-1		
B66X0036-001	119	3109	CR-2		
B66X0040-001	142A		ZD-12	563	
B66X0040-002	5079A	3335	ZD-20		
B66X0040-003	139A		ZD-9.1	562	
B66X0040-004	5141A		5ZD-30		
B66X0040-005	140A		ZD-10	562	
B66X0040-006	130		14	2041	
B66X0040-007	5081A		ZD-24		
B66X0040-008	5100A		ZD-150		
B66X0041-001	120	3110	CR-3		
B67		3004	53	2007	
B678760A	198		251		
B67A		3004	53	2007	
B69	121	3009	239	2006	
B692X13	109	3087		1123	
B6A1	156	3051	512		
B6A5	156	3051	512		
B6A9	156	3051	512		
B6B-3-A-21	121MP	3718	239(2)	2006(2)	
B6B-3A21	121MP	3718	239(2)	2006(2)	
B6B1	156	3051	512		
B6B5	156	3051	512		
B6B9	156	3051	512		
B6C1	156		512		
B6C5	156	3051	512		
B6C9	156	3051	512		
B6D1	156		512		
B6D5	156		512		
B6D9	156		512		
B6E1	156		512		
B6E5	156		512		
B6E9	156		512		
B6F1	156		512		
B6F5	156		512		
B6F9	156		512		
B6G1	156	3051	512		
B6G5	156	3051	512		
B6G9	156	3051	512		
B6H1	156	3051	512		
B6H5	156	3051	512		
B6H9	156	3051	512		
B6K1	156	3051	512		
B6K5	156	3051	512		
B6K9	156	3051	512		
B6M1	156	3051	512		
B6M5	156	3051	512		
B6M9	156	3051	512		
B6N1	506		511	1114	

Industry Standard No.	ECG	SK	GE	RS 276-	MOTOR.
B6N5	506		511	1114	
B6N9	506		511	1114	
B6P	123AP	3444/123A	20	2051	
B71	102A	3004	2	2007	
B72	102	3004	2	2007	
B73	102A	3004	53	2007	
B73A	102A		53	2007	
B73B	102A	3004	53	2007	
B73C	102A	3004	53	2007	
B73GR	102A	3004	53	2007	
B74	158	3006/160	245	2004	
B74-3-A-21	100	3721	1	2007	
B74-3A21	100	3721	1	2007	
B75	102A	3004	53	2007	
B7579500	125		510	1114	
B75A	100	3004/102A	1	2007	
B75AH	102A	3004	53	2007	
B75B	102A	3004	53	2007	
B75C	102A	3004	53	2007	
B75F	102A	3004	53	2007	
B75H	102A	3004	53	2007	
B75LB	102A	3004	53	2007	
B76	102A	3004	53	2007	
B77	102A	3004	53	2007	
B77(B)	102A	3004	53	2007	
B77A	102A	3004	53	2007	
B77A-1-21	105		4		
B77AA	102A	3004	53	2007	
B77AB	102A	3004	53	2007	
B77AC	102A	3004	53	2007	
B77AD	102A	3004	53	2007	
B77AH	102A	3004	53	2007	
B77AP	102A	3004	53	2007	
B77B	102A	3004	53	2007	
B77B-1-21	121	3717	239	2006	
B77B-11	102A	3004	53	2007	
B77C	102A	3004	53	2007	
B77C-1-21	121MP	3718	239(2)	2006(2)	
B77D	102A	3004	53	2007	
B77H	102A	3004	53	2007	
B77V	102A	3004	53	2007	
B77VRED	102A	3004	53	2007	
B78	102A	3004	53	2007	
B79	102A	3004	53	2007	
B7978850	506		511	1114	
B7A1	156	3081/125	512		
B7A5	156	3081/125	512		
B7A9	156	3081/125	512		
B7B1	156		512		
B7B5	156		512		
B7B9	156		512		
B7C1	156		512		
B7C5	156		512		
B7C9	156		512		
B7D1	156		512		
B7D5	156		512		
B7D9	156		512		
B7E1	156		512		
B7E5	156		512		
B7E9	156		512		
B7F1	156		512		
B7F5	156		512		
B7F9	156		512		
B7G1	156	3051	512		
B7G5	156	3051	512		
B7G9	156	3051	512		
B7H1	156	3051	512		
B7H5	156	3051	512		
B7H9	156	3051	512		
B7K1	156	3051	512		
B7K5	156	3051	512		
B7K9	156	3051	512		
B7M1	156	3051	512		
B7M5	156	3051	512		
B7M9	156	3051	512		
B7N1	506		511	1114	
B7N5	506		511	1114	
B7N9	506		511	1114	
B8-12					MRF212
B80	131	3009	239	2006	
B800					1N4006
B81		3009	239	2006	
B82		3009	239	2006	
B83	104	3009	16	2006	
B83-7	5697	3522			
B84	121	3009	239	2006	
B84A-1-21	121	3717	239	2006	
B85	127	3004/102A	2	2007	
B87	127	3004/102A	2	2007	
B8780010	123		20	2051	
B87J0007	128		243	2030	
B88B-1-21	160		245	2004	
B88C-1-21	100	3721	1	2007	
B89	102A	3004	53	2007	
B89A	102A	3004	53	2007	
B89AH	102A	3004	53	2007	
B89H	102A	3004	53	2007	
B8P-2-A-21	102A		53	2007	
B8P-2A21	102A		53	2007	
B90	102A	3004	53	2007	
B91	116	3004/102A	53	2007	
B92	102A	3004	2	2007	
B92-1-A-21	101	3861	8	2002	
B92-1A21	105	3012	4		
B94	102A	3004	53	2007	
B9426	108		86	2038	
B95	102A	3004	2	2007	
B97	102A	3004	53	2007	
B98	102	3004	2	2007	
B9L-4-A-21	102A		53	2007	
B9L-4A21	102A		53	2007	
B9TUI			FET-2	2035	
BA-100	116	3016	504A	1104	
BA-104	116	3016	504A	1104	
BA-142-01	116		504A	1104	
BA100	116		504A	1104	1N4002
BA1000					1N4007
BA1003	172A	3156	64		
BA102			90		MV1636
BA104	116	3100/519	504A	1104	
BA105	116	3016	504A	1104	
BA108	116	3311	504A	1104	
BA111			90		MV1644
BA114	177	3311	300	1122	
BA119	116	3311	504A	1104	
BA120	611		504A	1104	
BA121	611	3324			1N5443A
BA124		3126	90		
BA125					1N5449A
BA127	116	3016	504A	1104	
BA128	116	3311	504A	1104	
BA129	116	3311	504A	1104	
BA130	116	3311	504A	1104	
BA1310	1226	3763			
BA1310P	1226	3763			
BA133	506		509	1114	
BA136A	5800	9003		1142	
BA138		3126			1N5441A
BA142-01			504A	1104	
BA145	506	3032A		1114	
BA147	177	3100/519	300	1122	
BA148	506	3311	512		
BA149					1N5139
BA150					1N5453A
BA153	116	3311	504A	1104	
BA164				1011	
BA167	177	3100/519	300	1122	
BA168	177	3100/519	300	1122	
BA170	177	3100/519	300	1122	
BA174	177	3100/519	300	1122	
BA178	5800	9003		1142	
BA182	555	3100/519	300	1122	
BA187	177		300	1122	
BA200					1N4003
BA202	519		514	1122	
BA216	177	3100/519	514	1122	
BA217				1011	
BA218				1011	
BA219	177	3100/519	514	1122	
BA223-20	525	3925			
BA243	610		300	1122	
BA243A			300	1122	
BA244	177	3322	300	1122	
BA300					1N4004
BA301	1135	3876	IC-318		
BA301B	1135	3876			
BA316	177	3100/519	300	1122	
BA317	519	3100	514	1122	
BA318	1241	3828			
BA400					1N4004
BA401	1104	3225	IC-91		
BA402	1100	3223	IC-92		
BA50					1N4001
BA500					1N4005
BA511		3827	IC-315		
BA511A	1165	3827/1166	IC-315		
BA521	1166	3827	IC-316	704	
BA521A	1166	3827	IC-315		
BA6	158	3005	53	2007	
BA600					1N4005
BA67	123A	3444	20	2051	
BA6A	158	3005	53	2007	
BA71	123A	3444	20	2051	
BA8-1A-21	160		245	2004	
BA800					1N4006
BAC SHIMI	519			1122	
BACSH2M1	123A	3444	20	2051	
BACSH2M2	123A	3444	20	2051	
BACSH2M3	123A	3444	20	2051	
BACSHIMI	519			1122	
BACSIM1	519			1122	
BACT2F	123A	3444	20	2051	
BAM120					MRF317
BAM20					MRF314
BAM40					MRF315
BAM80					MRF316
BAV205	5838	3051/156	5004		
BAV206	5840	3500/5882	5008		
BAV207	5842	7042	5008		
BAV208	5846	3516	5012		
BAV209	5846	3516	5012		
BAV215	5838		5004		
BAV216	5840	3500/5882	5008		
BAV217	5842	7042	5008		
BAV218	5846	3516	5012		
BAV219	5846	3516	5012		
BAV225	5838		5004		
BAV226	5840	3500/5882	5008		
BAV227	5842	7042	5008		
BAV228	5846	3516	5012		
BAV229	5846	3516	5012		
BAV305	5838		5004		
BAV306	5840	3584/5862	5008		
BAV307	5842	7042	5008		
BAV308	5846	3516	5012		
BAV309	5846	3516	5012		
BAV315	5838		5004		
BAV316	5840	3584/5862	5008		
BAV317	5842	7042	5008		
BAV318	5846	3516	5012		
BAV319	5846	3516	5012		
BAV325	5838		5004		
BAV326	5840	3584/5862	5008		
BAV327	5842	7042	5008		
BAV328	5846	3516	5012		
BAV329	5846	3516	5012		
BAV405	5858		5020		
BAV406	5860		5024		
BAV407	5862		5024		
BAV408	5866		5028		
BAV409	5866		5028		
BAV415	5858		5020		
BAV416	5860		5024		
BAV417	5862		5024		
BAV418	5866		5028		
BAV419	5866		5028		
BAV425	5858		5020		
BAV426	5860		5024		
BAV427	5862		5024		
BAV428	5866		5028		
BAV429	5866		5028		
BAV505	5878		5036		
BAV506	5880		5040		
BAV507	5882		5040		
BAV508	5886		5044		
BAV509	5886		5044		
BAV515	5878		5036		
BAV516	5880		5040		
BAV517	5882		5040		
BAV518	5886		5044		
BAV519	5886		5044		
BAV525	5878		5036		
BAV526	5880		5040		
BAV527	5882		5040		
BAV528	5886		5044		
BAV529	5886		5044		
BAV706	5860	3500/5882	5024		
BAV707	5862	3500/5882	5024		
BAV708	5858		5028		
BAV709	5866		5028		
BAV715	5858		5020		
BAV716	5860	3500/5882	5024		
BAV717	5862	3500/5882	5024		
BAV718	5866		5028		
BAV719	5866		5028		
BAV725	5858		5020		
BAV726	5860	3500/5882	5024		
BAV727	5862	3500/5882	5024		

Industry Standard No.	ECG	SK	GE	RS 276-	MOTOR.
BAV728	5866		5028		
BAV729	5866		5028		
BAV805	5878		5036		
BAV806	5880	3500/5882	5040		
BAV807	5882	3500	5040		
BAV808	5886		5044		
BAV809	5886		5044		
BAV815	5878		5036		
BAV816	5880	3500/5882	5040		
BAV817	5882	3500	5040		
BAV818	5886		5044		
BAV819	5886		5044		
BAV825	5878		5036		
BAV826	5880	3500/5882	5040		
BAV827	5882	3500	5040		
BAV828	5886		5044		
BAV829	5886		5044		
BAW10TF20	177		300	1122	
BAW11TF21	177		300	1122	
BAW12TF22	177	3175	300	1122	
BAW13TF23	177	3175	300	1122	
BAW16	177	3175	300	1122	
BAW17	177	3175	300	1122	
BAW18	177	3175	300	1122	
BAW24	177	3100/519	300	1122	
BAW24A	177	3100/519	300	1122	
BAW24B	177	3100/519	300	1122	
BAW25	177	3100/519	300	1122	
BAW25A	177	3100/519	300	1122	
BAW25B	177	3100/519	300	1122	
BAW45	177	3100/519	300	1122	
BAW53	177	3100/519	300	1122	
BAW59	177	3100/519	300	1122	
BAW59A	177	3100/519	300	1122	
BAW59B	177	3100/519	300	1122	
BAW62	519	3100		1122	
BAW63A	177	3100/519	300	1122	
BAW63B	177	3100/519	300	1122	
BAW99	177	3100/519	300	1122	
BAX-13	519	3100	514	1122	
BAX-16	177	3100/519	300	1122	
BAX12	116	3017B/117	504A	1104	
BAX13	519	3100	514	1122	
BAX16	177		300	1122	
BAX28	177	3100/519	300	1122	
BAX30	177	3100/519	300	1122	
BAX74	177	3100/519	300	1122	
BAX79C16				564	
BAX87	177	3100/519	300	1122	
BAX87A	177	3100/519	300	1122	
BAX88TP11	177	3100/519	300	1122	
BAX89A	177	3100/519	300	1122	
BAX91C/TF102	519		514	1122	
BAX95TP600	519		514	1122	
BAY15	125	3032A	531	1114	
BAY16	125	3032A	531	1114	
BAY17	177	3016	300	1122	
BAY18	177	3175	300	1122	
BAY19	177	3175	300	1122	
BAY20	177	3175	300	1122	
BAY23	125	3033	531	1114	
BAY31	177	3100/519	300	1122	
BAY36	177	3100/519	300	1122	
BAY41	177	3100/519	300	1122	
BAY41A	177	3100/519	300	1122	
BAY41B	177	3100/519	300	1122	
BAY44	116	3311	504A	1104	
BAY52	177	3100/519	300	1122	
BAY64	116	3016	504A	1104	
BAY68	177	3100/519	300	1122	
BAY72				1122	
BAY73	177		300	1122	
BAY86	116	3311	504A	1104	
BAY87	116	3311	504A	1104	
BAY90	125	3033	531	1114	
BB-10	506		511	1114	
BB-109		3017B	504A	1104	
BB-10S	506	3130	511	1114	
BB-2	116		509	1114	
BB-4	506	3843	511	1114	
BB-6	552	3016	511	1114	
BB-6S	116		504A	1104	
BB10	552	9000	531	1114	
BB104	614	3126	90		
BB104B					MV104
BB104G					MV104G
BB105	611	3327/614			
BB105A					BB105A
BB105B	616	3319	90		BB105B
BB105G					BB105G
BB107	116	3017B/117	504A	1104	
BB109	616	3327/614	90		
BB109G	614	3327	90		
BB113					MVAM115
BB117	116	3017B/117	504A	1104	
BB121A	612	3325			
BB121B	616	3320			
BB122	612	3325	90		
BB127	116	3017B/117	504A	1104	
BB142	612	3325	B		
BB1A	116	3017B/117	504A	1104	
BB2	506	3032A	511	1114	
BB205B					BB205B
BB205G					BB205G
BB2A	552	3032A			
BB2A184	116	3311	504A	1104	
BB4	506		511	1114	
BB6	552		511	1114	
BB6S	506	3130	511	1114	
BB71	123A	3444	20	2051	
BBGS	506		511	1114	
BBIA		3311	504A	1104	
BBY17	610	3323			
BC-107	123A	3444	20	2051	
BC-1072	123A	3122	20	2051	
BC-107A	123A	3444	20	2051	
BC-108	123A	3124/289	20	2051	
BC-1082	123A	3122	20	2051	
BC-1086	123A	3122	20	2051	
BC-108B	123A		20	2051	
BC-1096	123A	3122	20	2051	
BC-109B	123A		20	2051	
BC-114	123A		20	2051	
BC-119	128		243	2030	
BC-121	123A	3444	20	2051	
BC-122	123A	3444	20	2051	
BC-123	123A	3444	20	2051	
BC-138	128		243	2030	
BC-140	128		243	2030	
BC-140A	128		243	2030	
BC-140B	128		243	2030	
BC-140C	128		243	2030	
BC-140D	128		243	2030	
BC-141	128		243	2030	

Industry Standard No.	ECG	SK	GE	RS 276-	MOTOR.
BC-142	128		243	2030	
BC-148A	123A	3444	20	2051	
BC-148B	123A	3444	20	2051	
BC-148C	123A	3444	20	2051	
BC-167-B	123A	3444	20	2051	
BC-169-C			10	2051	
BC-1690	123A	3122	20	2051	
BC-169B	123A		20	2051	
BC-169C	123A		20	2051	
BC-207	116		504A	1104	
BC-261	159		82	2032	
BC-307	116		504A	1104	
BC-71	123A	3444	20	2051	
BC100	154	3044	40	2012	
BC1029LB	917		IC-258		
BC103	128	3024	243	2030	
BC103C	128	3024	243	2030	
BC107	123A	3444	20	2051	
BC1073	104	3009	16	2006	
BC1073A	104	3009	16	2006	
BC107A	123A	3444	20	2051	
BC107B	123A	3444	20	2051	
BC108	123A	3444	20	2051	
BC108A	123A	3444	20	2051	
BC108B	123A	3444	20	2051	
BC108C	199	3444/123A	20	2051	
BC109	123A	3444	20	2051	
BC1096	123A	3444	20	2051	
BC109B	123A	3444	62	2010	
BC109BP	123AP	3444/123A	20	2051	
BC109C	199	3444/123A	20	2051	
BC110	123A	3444	20	2051	
BC111		3452	86	2038	
BC112	108	3452	86	2038	
BC113	123A	3444	20	2051	
BC114	123A	3444	20	2051	
BC114TR	123A	3444	20	2051	
BC115	123A	3444	20	2051	
BC116	159	3025/129	82	2032	
BC116A	159	3025/129	82	2032	
BC117	128	3024	40	2012	
BC118	123A	3444	20	2051	
BC119	128	3444/123A	243	2030	
BC120	128	3024	243	2030	
BC121	107	3245/199	11	2015	
BC122	107	3245/199	11	2015	
BC123	107	3245/199	11	2015	
BC125	123A	3444	20	2051	
BC125A	123A	3244	220		
BC125B	123A	3444	20	2051	
BC126	159	3114/290	82	2032	
BC126-1	159	3114/290	82	2032	
BC126A			21	2034	
BC127	199	3039/316	62	2010	
BC1274	104	3009	16	2006	
BC1274A	104	3009	16	2006	
BC1274B	104	3009	16	2006	
BC128	199	3039/316	62	2010	
BC129	123A	3444	20	2051	
BC130	123A	3444	20	2051	
BC130A				2014	
BC131	123A	3444	20	2051	
BC132	123A	3444	20	2051	
BC134	123A	3444	20	2051	
BC135	123A	3444	20	2051	
BC136	123A	3444	20	2051	
BC137	159	3114/290	82	2032	
BC137-1	159	3114/290	82	2032	
BC138	128	3024	243	2030	
BC139	129	3025	244	2027	
BC140	128	3024	243	2030	
BC140-10	128	3024	243	2030	
BC140-16	128	3024	243	2030	
BC140-6	128	3024	243	2030	
BC140C	128	3024	243	2030	
BC140D	128	3024	243	2030	
BC141	128	3024	243	2030	
BC141-10	128	3024	243	2030	
BC141-16	128	3024	243	2030	
BC141-6	128	3024	243	2030	
BC142	128	3024	243	2030	
BC143	129	3025	244	2027	
BC144	128	3024	243	2030	
BC145	324	3044/154	40	2012	
BC146	199	3444/123A	20	2051	
BC147	123A	3444	20	2051	
BC1478	123A	3444	20	2051	
BC147A	123A	3444	20	2051	
BC147B	123A	3444	20	2051	
BC148	123A	3444	20	2051	
BC148A	123A	3444	20	2051	
BC148B	123A	3124/289	20	2051	
BC148C	123A	3444	20	2051	
BC149	123A	3444	20	2051	
BC149A	123A	3444	20	2051	
BC149B	123A	3444	20	2051	
BC149C	123A	3444	20	2051	
BC149G	123A	3444	20	2051	
BC150	123A	3444	20	2051	
BC151	123A	3444	20	2051	
BC152	123A	3444	20	2051	
BC153	234	3114/290	65	2050	
BC153-1	159	3114/290	82	2032	
BC154	234	3114/290	65	2050	
BC154-1	159	3114/290	82	2032	
BC155	107	3039/316	11	2015	
BC155A	108	3452	86	2038	
BC155B	123A	3444	20	2051	
BC155C	199	3245	62	2010	
BC156	107	3039/316	11	2015	
BC156A	108	3452	86	2038	
BC156B	123A	3444	20	2051	
BC156C	199	3245	62	2010	
BC157	159	3114/290	82	2032	
BC157-1	159	3114/290	82	2032	
BC157A	159		82	2032	
BC157B	159	3025/129	82	2032	
BC158	159	3118	82	2032	
BC158-1	159	3114/290	82	2032	
BC158A	159	3118	82	2032	
BC158A-1	159	3114/290	82	2032	
BC158B	159	3114/290	82	2032	
BC158B-1	159	3114/290	82	2032	
BC158C		3247	65	2050	
BC158VI			221	2024	
BC159	159	3118	82	2024	
BC159-1	159	3114/290	82	2032	
BC159A	159	3114/290	82	2024	
BC159A-1	159	3114/290	82	2032	
BC159B	159	3114/290	82	2024	
BC159B-1	159	3114/290	82	2032	
BC159C		3247	65	2050	
BC160	129	3025	244	2027	
BC160-10	129	3025	244	2027	

Industry Standard No.	ECG	SK	GE	RS 276-	MOTOR.
BC160-16	129	3025	244	2027	
BC160-6	129	3025	244	2027	
BC161	129	3025	244	2027	
BC161-06			67	2023	
BC161-10	129	3025	244	2027	
BC161-16	129	3025	244	2027	
BC161-6	129	3025	244	2027	
BC167	123AP	3444/123A	20	2051	BC237
BC167A	289A		210	2051	
BC167B	289A	3444/123A	20	2051	
BC168	123AP	3444/123A	20	2051	BC238
BC168A	289A	3444/123A	20	2051	
BC168B	123A	3444	20	2051	
BC168C	199	3444/123A	20	2051	
BC169	123A	3444	20	2051	BC239
BC169A	123A	3444	20	2051	
BC169B	123A	3444	20	2051	
BC169C	199	3444/123A	20	2051	
BC169CL	123A	3444	20	2051	
BC170	123A	3122	62	2010	BC238
BC170A	123A	3122	62	2010	
BC170B	123A	3122	62	2010	
BC170C	123A	3122	62	2010	
BC171	123A	3444	20	2051	BC237
BC171A	123A	3444	20	2051	
BC171B	123A	3444	20	2051	
BC172	123A	3444	20	2051	BC238
BC172A	123A	3444	20	2051	
BC172B	123A	3444	20	2051	
BC172C	123A	3444	20	2051	
BC173	123A	3124/289	62	2010	BC239
BC173,B			18	2030	
BC173A	123A	3124/289	62	2010	
BC173B	123A	3245/199	62	2010	
BC173C	123A	3122	243	2010	
BC174	128	3024	243	2030	BC174
BC174A	159	3024/128	243	2030	
BC174B	159	3024/128	243	2030	
BC175	123A	3444	20	2051	
BC177	234	3004/102A	65	2050	
BC177-1	159	3114/290	82	2032	
BC177A	234	3114/290	65	2050	
BC177A-1	159	3114/290	82	2032	
BC177B	234	3114/290	65	2050	
BC177B-1	159	3114/290	82	2032	
BC177V	159	3114/290	82	2032	
BC177V-1	159	3114/290	82	2032	
BC177V1	159	3114/290	82	2032	
BC177V1-1	159	3114/290	82	2032	
BC177VI			21	2034	
BC178	234	3114/290	65	2050	
BC178-1	159	3114/290	82	2032	
BC178A	234	3114/290	65	2050	
BC178A-1	159	3114/290	82	2032	
BC178B	234	3053	65	2050	
BC178B-1	159	3114/290	82	2032	
BC178C	234	3247	65	2050	
BC178D	159	3114/290	82	2032	
BC178D-1	159	3114/290	82	2032	
BC178V	159	3114/290	82	2032	
BC178V-1	159	3114/290	82	2032	
BC178V1	159	3114/290	82	2032	
BC178V1-1	159	3114/290	82	2032	
BC178VI			21	2034	
BC179	234	3114/290	65	2050	
BC179-1	159	3114/290	82	2032	
BC179A	234	3114/290	65	2050	
BC179A-1	159	3114/290	82	2032	
BC179B	234	3114/290	65	2050	
BC179B-1	159	3114/290	82	2032	
BC179C	234	3247	65	2050	
BC17LB-1				2032	
BC180	123A	3444	20	2051	
BC180B	123A	3444	20	2051	
BC181	159		82	2032	BC308
BC181A	159	3114/290	82	2032	
BC181A-1	159	3114/290	82	2032	
BC182	123A	3444	20	2051	BC182
BC182A	289A	3444/123A	20	2051	
BC182K	199	3245	212	2010	
BC182KA	123A	3245/199	212	2010	
BC182KB	123A	3245/199	212	2010	
BC182L	123A	3444	20	2051	
BC182LA	289A	3244	220		
BC183	123A	3444	20	2051	BC183
BC183A	123AP	3444/123A	20	2051	
BC183B	123AP	3444/123A	20	2051	
BC183K		3245	212	2010	
BC183KA		3245	212	2010	
BC183KB	123A	3245/199	212	2010	
BC183KC	123A	3245/199	212	2010	
BC183L	199	3444/123A	20	2051	
BC183LA	289A	3244	220		
BC184	123A	3444	20	2051	BC184
BC184B	123A	3444	20	2051	
BC184K	123A	3245/199	212	2010	
BC184KB	123A	3245/199	212	2010	
BC184KC		3245	212	2010	
BC184L	199	3444/123A	20	2051	
BC185	123A	3444	20	2051	
BC186	159	3444/123A	20	2051	
BC187	159	3114/290	82	2032	
BC187-1	159	3114/290	82	2032	
BC188	107	3039/316	11	2015	
BC189	107	3039/316	11	2015	
BC190A			63	2030	
BC190B	123A		82	2030	
BC192	159	3114/290	82	2032	
BC192-1	159	3114/290	82	2032	
BC194	108	3452	86	2038	
BC194B	108	3452	86	2038	
BC195	108	3452	86	2038	
BC195CD	108	3452	86	2038	
BC196	159	3114/290	82	2032	
BC196-1	159	3114/290	82	2032	
BC196A	159	3114/290	82	2032	
BC196A-1	159	3114/290	82	2032	
BC196B	159	3118	82	2032	
BC196B-1	159	3114/290	82	2032	
BC196V1	159	3114/290	82	2032	
BC196V1-1	159	3114/290	82	2032	
BC197	123A	3444	20	2051	
BC197A	123A	3444	20	2051	
BC197B	123A	3444	20	2051	
BC198	123A	3444	20	2051	
BC198A			17	2051	
BC199	123A	3444	20	2051	
BC200	159	3118	82	2032	
BC200-1	159	3114/290	82	2032	
BC201	159	3114/290	82	2024	
BC201-1	159	3114/290	82	2032	
BC202	159	3114/290	82	2032	
BC202-1	159	3114/290	82	2032	
BC203	159	3114/290	82	2032	
BC203-1	159	3114/290	82	2032	
BC204	234	3114/290	65	2050	
BC204-1	159	3114/290	82	2032	
BC204A	234	3114/290	65	2050	
BC204A-1	159	3114/290	82	2032	
BC204B	234	3114/290	65	2050	
BC204B-1	159	3114/290	82	2032	
BC204V	159	3114/290	82	2032	
BC204V-1	159	3114/290	82	2032	
BC204V1	159	3114/290	82	2032	
BC204V1-1	159	3114/290	82	2032	
BC204VI			21	2034	
BC205	159	3114/290	82	2024	
BC205-1	159	3114/290	82	2032	
BC205A	159	3114/290	82	2024	
BC205A-1	159	3114/290	82	2032	
BC205B	234	3114/290	82	2050	
BC205B-1	159	3114/290	82	2032	
BC205C	234	3984/106	21	2050	
BC205L	106	3984	21	2034	
BC205V	159	3114/290	82	2024	
BC205V-1	159	3114/290	82	2032	
BC205V1	159	3114/290	82	2032	
BC205V1-1	159	3114/290	82	2032	
BC205VI			22	2024	
BC206	234	3114/290	65	2050	
BC206-1	159	3114/290	82	2032	
BC206A	159	3247/234	65	2050	
BC206B	234	3114/290	65	2050	
BC206B-1	159	3114/290	82	2032	
BC206C	234	3247	65	2050	
BC207	123A	3444	20	2051	
BC207A	123A	3444	20	2051	
BC207B	123A	3444	20	2051	
BC207BL	123A	3444	20	2051	
BC208	123A	3444	20	2051	
BC208A	123A	3444	20	2051	
BC208AL	123A	3444	20	2051	
BC208B	123A	3444	20	2051	
BC208BL	123A	3444	20	2051	
BC208C	123A	3444	20	2051	
BC208CL	123A	3444	20	2051	
BC209	123A	3444	20	2051	
BC209/7825B	199	3122	62	2010	
BC209A	123A	3444	20	2051	
BC209B	123A	3444	20	2051	
BC209BL	123A	3444	20	2051	
BC209C	123A	3444	20	2051	
BC209CL	123A	3444	20	2051	
BC20C	199	3245	62	2010	
BC210	123A	3444	20	2051	
BC211	128	3024	243	2030	
BC212	159	3114/290	82	2032	BC212
BC212-1	159	3114/290	82	2032	
BC212A	159		67	2023	
BC212B	159		67	2023	
BC212K	159	3114/290	82	2032	
BC212K-1	159	3114/290	82	2032	
BC212KA	159	3114/290	82	2032	
BC212KA-1	159	3114/290	82	2032	
BC212KB	159	3114/290	82	2032	
BC212KB-1	159	3114/290	82	2032	
BC212L	159	3114/290	82	2032	
BC212L-1	159	3114/290	82	2032	
BC212LA	159	3114/290	82	2032	
BC212LA-1	159	3114/290	82	2032	
BC212LB	159	3114/290	82	2032	
BC212LB-1	159	3114/290	82	2032	
BC212V1			82	2032	
BC212VI	159		21	2032	
BC213	159	3114/290	82	2032	BC213
BC213-1	159	3114/290	82	2032	
BC213A	159		82	2032	
BC213B	159		67	2023	
BC213K	159	3114/290	82	2032	
BC213K-1	159	3114/290	82	2032	
BC213KA	159	3114/290	82	2032	
BC213KA-1	159	3114/290	82	2032	
BC213KB	159	3114/290	82	2032	
BC213KB-1	159	3114/290	82	2032	
BC213KC	159	3114/290	82	2032	
BC213KC-1	159	3114/290	82	2032	
BC213L	159	3114/290	82	2032	
BC213L-1	159	3114/290	82	2032	
BC213LA	159	3114/290	82	2032	
BC213LA-1	159	3114/290	82	2032	
BC213LB	159	3114/290	82	2032	
BC213LB-1	159	3114/290	82	2032	
BC213LC	159	3114/290	82	2032	
BC213LC-1	159	3114/290	82	2032	
BC214	159	3114/290	82	2032	BC214
BC214-1	159	3114/290	82	2032	
BC214A	159		82	2032	
BC214B	159		67	2023	
BC214K	159	3114/290	82	2032	
BC214K-1	159	3114/290	82	2032	
BC214KA	159	3114/290	82	2032	
BC214KA-1	159	3114/290	82	2032	
BC214KB	159	3114/290	82	2032	
BC214KB-1	159	3114/290	82	2032	
BC214KC	159	3114/290	82	2032	
BC214KC-1	159	3114/290	82	2032	
BC214L	159	3114/290	82	2032	
BC214L-1	159	3114/290	82	2032	
BC214LA	159	3114/290	82	2032	
BC214LA-1	159	3114/290	82	2032	
BC214LB	159	3114/290	82	2032	
BC214LB-1	159	3114/290	82	2032	
BC214LC	159	3114/290	82	2032	
BC214LC-1	159	3114/290	82	2032	
BC216	128	3024	243	2030	
BC216A	128	3024	243	2030	
BC216B	128	3024	243	2030	
BC220	123A	3444	20	2051	
BC221	159	3114/290	82	2032	
BC221-1	159	3114/290	82	2032	
BC222	123A	3444	20	2051	
BC223A	123A	3444	20	2051	
BC223B	123A	3444	20	2051	
BC224	234	3138/193A	67	2023	BC351
BC225	234	3118	65	2050	
BC225-1	159	3114/290	82	2032	
BC226	192	3137	63	2030	
BC231					2N4403
BC231A	290A	3025/129	82	2032	
BC231B	290A	3138/193A	67	2023	
BC231M	129		48		
BC232					2N4401
BC232A	289A	3024/128	243	2030	
BC232B	289A	3024/128	243	2030	
BC233A	123A	3444	20	2051	
BC236		3244	220		
BC237	159	3444/123A	20	2051	BC237
BC237A	123AP	3444/123A	20	2051	

Industry Standard No.	ECG	SK	GE	RS 276-	MOTOR.
BC237B	123AP	3444/123A	20	2051	
BC238	229	3444/123A	20	2051	BC238
BC238-16		3450	82	2032	
BC238A	123A	3444	20	2051	
BC238B	123A	3444	20	2051	
BC238C	123A	3444	62	2010	
BC239	123AP	3444/123A	20	2051	BC239
BC239A	123AP	3245/199	212	2010	
BC239B	123AP	3122	62	2010	
BC239C	123A	3245/199	62	2010	
BC250	159	3114/290	82	2032	BC308
BC250-1	159	3114/290	82	2032	
BC250A	159	3114/290	82	2024	
BC250A-1	159	3114/290	82	2032	
BC250B	159	3114/290	82	2024	
BC250B-1	159	3114/290	82	2032	
BC250C	159	3114/290	82	2024	
BC250C-1	159	3114/290	82	2032	
BC251	159	3114/290	82	2032	BC309
BC251A	159	3114/290	82	2032	
BC251A-1	159	3114/290	82	2032	
BC251B	159	3114/290	82	2032	
BC251B-1	159	3114/290	82	2032	
BC251C	159	3114/290	82	2032	
BC251C-1	159	3114/290	82	2032	
BC252	159	3114/290	82	2024	BC309
BC252-1	159	3114/290	82	2032	
BC252A	159	3114/290	82	2024	
BC252A-1	159	3114/290	82	2032	
BC252B	159	3114/290	82	2024	
BC252B-1	159	3114/290	82	2032	
BC252C	159	3114/290	82	2032	
BC252C-1	159	3114/290	82	2032	
BC253	159	3114/290	82	2024	BC415
BC253-1	159	3114/290	82	2032	
BC253A	159	3114/290	82	2024	
BC253A-1	159	3114/290	82	2032	
BC253B	159	3114/290	82	2024	
BC253B-1	159	3114/290	82	2032	
BC253C	159	3114/290	82	2024	
BC253C-1	159	3114/290	82	2032	
BC254	128	3024	243	2030	MPS8098
BC255	128	3024	243	2030	MPSA05
BC256	159	3114/290	82	2032	MPS8598
BC256-1	159	3114/290	82	2032	
BC256A	159	3114/290	82	2032	
BC256A-1	159	3114/290	82	2032	
BC256B	159	3114/290	82	2032	
BC256B-1	159	3114/290	82	2032	
BC256C-1	159		82	2032	
BC257	159	3118	82	2032	BC350
BC257-1	159	3114/290	82	2032	
BC257A	193A	3138	67	2023	
BC257B	193A	3138	67	2023	
BC257VI	159		82	2032	
BC258	159	3114/290	82	2032	BC351
BC258-1	159	3114/290	82	2032	
BC258A	193A	3138	67	2023	
BC258B	193A	3138	67	2023	
BC258C		3247	65	2050	
BC258VI	159		82	2024	
BC259	159	3114/290	82	2024	2N5087
BC259-1	159	3114/290	82	2032	
BC259A	193A	3138	67	2023	
BC259B	193A	3138	67	2023	
BC25BB	159	3114/290	82	2032	
BC260	159	3114/290	82	2032	
BC260-1	159	3114/290	82	2032	
BC260A	159	3114/290	82	2024	
BC260A-1	159	3114/290	82	2032	
BC260B	159	3114/290	82	2024	
BC260B-1	159	3114/290	82	2032	
BC260C	159	3114/290	82	2024	
BC260C-1	159	3114/290	82	2032	
BC261	159	3025/129	82	2032	
BC261A	159	3114/290	82	2032	
BC261A-1	159	3114/290	82	2032	
BC261B	159	3025/129	82	2032	
BC261B-1	159	3114/290	82	2032	
BC261C	159	3114/290	82	2032	
BC261C-1	159	3114/290	82	2032	
BC262	159	3114/290	82	2032	
BC262-1	159	3114/290	82	2032	
BC262A	159	3114/290	82	2024	
BC262A-1	159	3114/290	82	2032	
BC262B	159	3114/290	82	2024	
BC262B-1	159	3114/290	82	2032	
BC262C	159	3114/290	82	2032	
BC262C-1	159	3114/290	82	2032	
BC263	159	3114/290	82	2032	
BC263A	159	3114/290	82	2024	
BC263A-1	159	3114/290	82	2032	
BC263B	159	3114/290	82	2024	
BC263B-1	159	3114/290	82	2032	
BC263C	159	3114/290	82	2024	
BC263C-1	159	3114/290	82	2032	
BC264C	312			2028	
BC266	159	3114/290	82	2032	
BC266-1	159	3114/290	82	2032	
BC266A	159	3114/290	82	2032	
BC266B	159	3114/290	82	2032	
BC267	123A	3444	20	2051	
BC268	123A	3444	20	2051	
BC269	123A	3444	20	2051	
BC270	123A	3444	20	2051	
BC280	123A	3444	20	2051	
BC280A	123A	3444	20	2051	
BC280B	123A	3444	20	2051	
BC280C	123A	3444	20	2051	
BC281	159	3114/290	82	2032	
BC281-1	159	3114/290	82	2032	
BC281A	159	3114/290	82	2032	
BC281A-1	159	3114/290	82	2032	
BC281B	159	3114/290	82	2032	
BC281B-1	159	3114/290	82	2032	
BC281C	159	3114/290	82	2032	
BC281C-1	159	3114/290	82	2032	
BC282	123A	3444	20	2051	
BC283	159	3114/290	82	2032	
BC283-1	159	3114/290	82	2032	
BC284	123A	3444	20	2051	
BC284A	123A	3444	20	2051	
BC284B	123A	3444	20	2051	
BC285	154	3444/123A	20	2051	
BC286	128	3024	243	2030	
BC287	129	3025	244	2027	
BC288	128	3024	243	2030	
BC289	123A	3444	20	2051	
BC289A	123A	3444	20	2051	
BC289B	123A	3444	20	2051	
BC290	123A	3444	20	2051	
BC290B	123A	3444	20	2051	
BC290C	123A	3444	20	2051	
BC291	159	3114/290	82	2032	
BC291-1	159	3114/290	82	2032	
BC291A	159	3114/290	82	2032	
BC291A-1	159	3114/290	82	2032	
BC291D	159	3114/290	82	2032	
BC291D-1	159	3114/290	82	2032	
BC292	159	3114/290	82	2032	
BC292A	159	3114/290	82	2032	
BC292A-1	159	3114/290	82	2032	
BC292D	159	3114/290	82	2032	
BC292D-1	159	3114/290	82	2032	
BC295	108	3039/316	86	2038	
BC297	159	3138/193A	67	2023	
BC298	159	3138/193A	67	2023	
BC300-4	154		230		
BC301	128	3024	243	2030	
BC301-4	128		230		
BC302-4	128		63	2030	
BC302-5	128		63	2030	
BC302-6	128		63	2030	
BC303	129	3025	244	2027	
BC304-4			67	2023	
BC304-5			67	2023	
BC304-6			67	2023	
BC307	159	3017B/117	82	2032	BC307
BC307A	159	3114/290	82	2032	
BC307B	159	3138/193A	67	2023	
BC307C	159	3247/234	65	2050	
BC307V1	159		82	2032	
BC307VI			221	2032	
BC308	159	3114/290	82	2024	BC308
BC308A	159	3114/290	82	2032	
BC308B	159	3247/234	82	2032	
BC308C	159	3247/234	65	2050	
BC308V1	159		82	2032	
BC308VI			221	2024	
BC309	159	3114/290	82	2024	BC309
BC309(ALGG)				2024	
BC309(SIEG)				2024	
BC309A	159	3114/290	82	2024	
BC309B	159		82	2024	
BC309C	159	3138/193A	67	2023	
BC310	128		243	2030	
BC311	129		244	2027	
BC312	128		251		
BC313	129	3024/128	243	2030	
BC315	159	3025/129	82	2032	2N5087
BC317	123AP	3444/123A	20	2051	BC317
BC317A	123AP	3444/123A	20	2051	
BC317B	123AP	3444/123A	20	2051	
BC318	123AP	3444/123A	20	2051	BC318
BC318A	123AP	3444/123A	20	2051	
BC318B	123AP	3444/123A	20	2051	
BC318C	123AP	3444/123A	20	2051	
BC319	123AP	3444/123A	20	2051	BC319
BC319B	123AP	3444/123A	20	2051	
BC319C	123AP	3444/123A	20	2051	
BC320	159	3247/234	65	2022	BC320
BC320A	234	3247	65	2050	
BC320C	234	3247	65	2050	
BC321	234	3247	65	2050	BC321
BC321A	234	3247	65	2050	
BC321B	234	3247	65	2050	
BC321C	234	3247	65	2050	
BC322	234	3247	65	2050	BC322
BC322B	234	3247	65	2050	
BC322C	234	3247	65	2050	
BC325	159	3114/290	82	2032	
BC325A	159	3114/290	82	2032	
BC326	159	3114/290	82	2032	
BC326A	159	3114/290	82	2032	
BC327	159	3200/189	218	2026	BC327
BC327-16	298	3138/193A	67	2023	
BC327-25	298	3450			
BC328	159	3200/189	218	2026	BC328
BC328-16	298	3138/193A	82	2032	
BC329					2N5089
BC330					BC330
BC331					BC347
BC332					BC237
BC333					BC238
BC3337			47	2030	
BC334					BC308
BC335					BC413
BC336					BC415
BC337	159	3124/289	226	2018	BC337
BC337-16	123AP		47	2030	
BC338	123AP	3124/289	226	2018	BC338
BC338-16	123AP	3137/192A	63	2030	
BC340-06			47	2030	
BC340-10	128	3024	243	2030	
BC340-16	128	3024	243	2030	
BC340-6	128	3024	243	2030	
BC341-06			18	2030	
BC341-10	128	3024	243	2030	
BC341-6	128	3024	243	2030	
BC347					BC347
BC348					BC348
BC349					BC349
BC350					BC350
BC351					BC351
BC352					BC352
BC354					BC351
BC355					BC351
BC357					BC351
BC358					BC348
BC360-10	129	3025	244	2027	
BC360-16	129	3025	244	2027	
BC360-6	129	3025	244	2027	
BC361-06			67	2023	
BC361-10	129	3025	244	2027	
BC361-6	129	3025	244	2027	
BC370	193	3138	67	2023	
BC377	123A	3137/192A	63	2030	
BC378	123A	3137/192A	63	2030	
BC381	159	3025/129	82	2032	BC328
BC382	123A		220		BC237
BC382B	123AP	3245/199	62	2010	
BC382C	123AP	3245/199	62	2010	
BC383	123A	3244	220		BC238
BC383B	123AP	3245/199	62	2010	
BC383C	123AP	3245/199	62	2010	
BC384	123A	3244	220		BC414
BC384B	123AP	3245/199	212	2010	
BC384C	199	3245	62	2010	
BC385					BC237A
BC385A		3245	212	2010	
BC385B	199	3245	62	2010	
BC386					BC238
BC386A	199	3245	62	2010	
BC386B	199	3245	62	2010	
BC387					2N4401
BC388					2N4403
BC393	288	3434			
BC394	154		235	2012	

Industry Standard No.	ECG	SK	GE	RS 276-	MOTOR.
BC395		3244	220		
BC396	129		244	2027	
BC400	159	3114/290	82	2032	
BC404			82	2032	
BC404A	129	3025	244	2027	
BC404V1	129		244	2027	
BC404VI		3025	82	2032	
BC405	159	3025/129	82	2032	
BC405A	159	3025/129	82	2032	
BC405B	159	3025/129	82	2032	
BC406	193	3138	67	2023	
BC406B	159	3025/129	82	2032	
BC408	123A	3444	20	2051	
BC408A	199	3245			
BC408B	199	3245	62	2010	
BC408BR	199	3245			
BC408C	199	3245	62	2010	
BC409	123A	3444	20	2051	
BC412	195A	3048/329	46		
BC413	123AP				BC413
BC413B	123AP	3245/199	62	2010	
BC413C	123AP	3245/199	62	2010	
BC414	123A				BC414
BC414B	123AP	3124/289	62	2010	
BC414C	123AP	3245/199	62	2010	
BC415			21	2034	BC415
BC415A	159	3025/129	82	2032	
BC415B	159	3025/129	82	2032	
BC416			21	2034	BC416
BC416A	159	3025/129	82	2032	
BC416B	159	3025/129	82	2032	
BC417	159	3025/129	82	2032	
BC418	159	3025/129	82	2032	
BC418A			221	2032	
BC418B			48	2024	
BC419	159	3025/129	82	2024	
BC419A			221	2024	
BC419B			48	2024	
BC424					MPSA06
BC425					MPSA05
BC426					MPSA56
BC427					MPSA55
BC429	128	3024	243	2030	
BC430	188	3054/196	226	2018	
BC431	128		81	2051	MPSA05
BC432	159		82	2032	MPSA55
BC442	107	3039/316	11	2015	
BC445	159				BC445
BC446	159				BC446
BC447	159				BC447
BC448					BC448
BC449					BC449
BC450	383				BC450
BC456	123A	3018	61	2038	
BC460	323	3083/197	218	2026	
BC461	323	3083/197	218	2026	
BC461-4	129		67	2023	
BC461-5	129		67	2023	
BC461-6	129		67	2023	
BC477	129	3025	244	2027	
BC477A	129	3025	244	2027	
BC477V1	129		244	2027	
BC478	159	3025/129	82	2032	
BC478A	159	3025/129	82	2032	
BC478B	159	3247/234	82	2032	
BC479	159	3025/129	82	2032	
BC479B	159	3025/129	82	2032	
BC485	159				BC485
BC486	159				BC486
BC487	159				BC487
BC488	159				BC488
BC489	159				BC489
BC490	159				BC490
BC507A	123A	3444	20	2051	
BC507B	123A	3444	20	2051	
BC508			63	2030	
BC508A	123A	3444	20	2051	
BC508B	123A	3444	20	2051	
BC508C	123A	3444	20	2051	
BC509			63	2030	
BC509B	123A	3444	20	2051	
BC509C	123A	3444	20	2051	
BC510	192	3137	63	2030	
BC510B	123A	3444	20	2051	
BC510C	107	3039/316	11	2015	
BC512	159	3025/129	82	2032	BC307
BC512A	159	3025/129	82	2032	
BC512B	193	3138	67	2023	
BC513	159	3025/129	82	2032	BC308
BC513A	159	3025/129	82	2032	
BC513B	193	3138	67	2023	
BC514	193	3138	67	2023	BC309
BC514A	159	3025/129	82	2032	
BC514B	193	3138	67	2023	
BC516					MPSA63
BC517					MPSA13
BC520	199	3245	62	2010	
BC520B	199	3245	62	2010	
BC520C	199	3245	62	2010	
BC521	199	3245	62	2010	
BC521C	199	3245	62	2010	
BC521D	199	3245	62	2010	
BC522	199	3245	62	2010	
BC522C	199	3245	62	2010	
BC522D	199	3245	62	2010	
BC522E	199	3245	62	2010	
BC523	199	3245	62	2010	
BC523B	199	3245	62	2010	
BC523C	199	3245	62	2010	
BC524					BC414
BC526A	234	3247	65	2050	
BC526B	234	3247	65	2050	
BC526C	234	3247	65	2050	
BC527	129		244	2027	MPSA55
BC528	129		244	2027	2N5400
BC530					2N5401
BC532	194	3275	220		2N5550
BC533	194	3275	220		2N5551
BC534	129		244	2027	MPSA56
BC535	128		243	2030	MPSA55
BC537	128		243	2030	
BC538	128		243	2030	
BC546	199	3444/123A	20	2010	BC546
BC546A	199	3444/123A	20	2051	
BC546B	199	3444/123A	20	2051	
BC547	123AP				BC547
BC547B	123AP		20	2051	
BC547C	123AP	3124/289	20		
BC548	123AP	3122	17	2051	BC548
BC548B	123AP	3245/199			
BC548C	199	3245			
BC548VI			17	2051	
BC549	123AP				BC549
BC550	123AP				BC550

Industry Standard No.	ECG	SK	GE	RS 276-	MOTOR.
BC556	159		82	2032	BC556
BC556A	159		82	2032	
BC556VI			82	2032	
BC557	159		82	2032	BC557
BC557A	159		82	2032	
BC557B	159		48		
BC557VI			82	2032	
BC558	159	3466	82	2032	BC558
BC558A	159		48		
BC558B	159	3466	48		
BC558VI			82	2032	
BC559	159	3118	82	2032	BC559
BC559A	159		48		
BC559B	159		48		
BC560	123AP				BC560
BC560A	159		48		
BC582	123AP	3137/192A	63	2030	BC237
BC582A	123AP	3137/192A	63	2030	
BC582B	123AP	3137/192A	63	2030	
BC583	123A	3444	20	2051	BC238
BC583A	123AP	3137/192A	63	2030	
BC583B	123AP	3137/192A	63	2030	
BC584		3137	63	2030	BC239
BC635	315				MPSA05
BC636	383				MPSA55
BC637	315				MPSA05
BC638	383				MPSA55
BC639	128				MPSA06
BC640	383				MPSA56
BC6500	108	3452	86	2038	
BC682	128	3244	220		
BC71	123A	3444	20	2051	
BC714					BC414
BC727	129		244	2027	2N4403
BC728	129		244	2027	2N4403
BC737	199		243	2030	2N4401
BC738	199		243	2030	2N4401
BCF5	117	3017B	504A	1104	
BCM1002-1	160	3006	245	2004	
BCM1002-18	158	3004/102A	53	2007	
BCM1002-2	107	3039/316	11	2015	
BCM1002-3	158	3004/102A	53	2007	
BCM1002-4	102A	3004	53	2007	
BCM1002-5	102A	3004	53	2007	
BCM1002-6	131		44	2006	
BCR25A-10	5686	3521			
BCR25A-8	5685	3521/5686			
BCW29	106	3118	21	2034	
BCW29R	106	3118	21	2034	
BCW30	106	3118	21	2034	
BCW30R	106	3118	21	2034	
BCW31	108	3452	86	2038	
BCW31R	108	3452	86	2038	
BCW32	108	3452	86	2038	
BCW32R	108	3452	86	2038	
BCW34	123A	3444	20	2051	
BCW35	159	3114/290	82	2032	
BCW36	123A	3444	20	2051	
BCW37	159	3114/290	82	2032	
BCW37A	159	3114/290	82	2032	
BCW44	128	3137/192A	63	2030	
BCW45	159	3138/193A	67	2023	
BCW46	128	3024	243	2030	
BCW47	128	3024	243	2030	
BCW48	128	3024	243	2030	
BCW48A	123AP		17	2051	
BCW49	128	3024	243	2030	
BCW50	171	3103A/396	27		
BCW51	192	3137	63	2030	
BCW52	193	3138	67	2023	
BCW56	159	3114/290	82	2032	
BCW56A	159	3114/290	82	2032	
BCW57	159	3118	82	2032	
BCW57A	159	3114/290	82	2032	
BCW58	159	3118	82	2032	
BCW58A	159	3114/290	82	2032	
BCW58B	159	3247/234	65	2050	
BCW59	159	3118	82	2032	
BCW59A	159	3114/290	82	2032	
BCW59B		3247	65	2050	
BCW60A	123A	3444	20	2051	
BCW60AA	123A	3444	20	2051	
BCW60AB			210	2051	
BCW60AC			210	2051	
BCW60B			210	2051	
BCW60C			210	2051	
BCW61	159	3118	82	2032	
BCW61A	193	3118	67	2023	
BCW61B	193	3118	67	2023	
BCW61BA		3138	67	2023	
BCW61BB		3138	67	2023	
BCW61BC		3247	65	2050	
BCW61BD		3247	65	2050	
BCW61C	159	3118	82	2032	
BCW61D	159	3118	82	2032	
BCW62	159	3025/129	82	2032	
BCW62A	159	3025/129	82	2032	
BCW62B	193	3138	67	2023	
BCW63	159	3025/129	82	2032	
BCW63A	159	3025/129	82	2032	
BCW63B	193	3138	67	2023	
BCW64	193	3138	67	2023	
BCW64A	159	3025/129	82	2032	
BCW64B	193	3138	67	2023	
BCW65EA			20	2051	
BCW65EB			20	2051	
BCW65EC			47	2030	
BCW66EF			47	2030	
BCW66EG			47	2030	
BCW66EH			47	2030	
BCW66EW			47	2030	
BCW69	159	3025/129	82	2032	
BCW69R	159	3025/129	82	2032	
BCW70		3138	67	2023	
BCW70R		3138	67	2023	
BCW71	108	3452	86	2038	
BCW71R	108	3452	86	2038	
BCW72	108	3452	86	2038	
BCW72R	108	3452	86	2038	
BCW73-16	192	3137	63	2030	
BCW74-16	192	3137	63	2030	
BCW75-10	193	3138	67	2023	
BCW75-16	193	3138	67	2023	
BCW76-10	193	3138	67	2023	
BCW76-16	193	3138	67	2023	
BCW77-16	192	3137	63	2030	
BCW78-16	192	3137	63	2030	
BCW79-10	159	3025/129	82	2032	
BCW79-16	159	3025/129	82	2032	
BCW79-25	129	3203/211	253	2026	
BCW80-10	159	3025/129	82	2032	
BCW80-16	159	3025/129	82	2032	
BCW80-25	129	3203/211	253	2026	
BCW82	192	3137	63	2030	

Industry Standard No.	ECG	SK	GE	RS 276-	MOTOR.
BCW82A	192	3137	63	2030	
BCW82B	192	3137	63	2030	
BCW83	123A	3444	20	2051	
BCW83A	192A	3137	63	2030	
BCW83B	192	3137	63	2030	
BCW84	192	3137	63	2030	
BCW86	159	3138/193A	67	2023	212B
BCW87		3245	212	2010	BC239
BCW88			65	2050	BC239
BCW90	128	3137/192A	63	2030	
BCW90A	128		47	2030	
BCW90B	128		47	2030	
BCW90C	128		47	2030	
BCW90K	192	3137	63	2030	
BCW90KA	128		47	2030	
BCW90KB	128		47	2030	
BCW90KC	128		47	2030	
BCW91	128	3137/192A	63	2030	
BCW91A	128		63	2030	
BCW91B	128		63	2030	
BCW91K	192	3137	63	2030	
BCW91KA	192		63	2030	
BCW91KB	192		63	2030	
BCW92	193	3138	67	2023	
BCW92A	159		48		
BCW92K	193	3138	67	2023	
BCW92KA	159		48		
BCW92KB	159		48		
BCW93	193	3138	67	2023	
BCW93A	159		67	2023	
BCW93B	159		67	2023	
BCW93K	193	3138	67	2023	
BCW93KA	159		67	2023	
BCW93KB	159		67	2023	
BCW94	123AP	3444/123A	20	2051	
BCW94A	123AP		20	2051	
BCW94B	123AP		20	2051	
BCW94C	123AP		20	2051	
BCW94K	192	3024/128	243	2030	
BCW94KA	123AP		20	2051	
BCW94KB	123AP		20	2051	
BCW94KC	123AP		20	2051	
BCW95	128	3024	243	2030	
BCW95A	128		81	2051	
BCW95B	128		81	2051	
BCW95K	192	3024/128	243	2030	
BCW95KA	128		81	2051	
BCW95KB	128		81	2051	
BCW96	159	3025/129	82	2032	
BCW96A	159		82	2032	
BCW96B	159		82	2032	
BCW96K	193	3025/129	82	2032	
BCW96KA	159		82	2032	
BCW96KB	159		82	2032	
BCW97	159	3025/129	82	2032	
BCW97A	159		82	2032	
BCW97B	159		82	2032	
BCW97K	193	3025/129	82	2032	
BCW97KA	159		82	2032	
BCW97KB	159		82	2032	
BCW98A		3245	212	2010	
BCW98B		3245	212	2010	
BCW98C		3245	212	2010	
BCW98D		3245	212	2010	
BCX10	129	3203/211	253	2026	
BCX17			221	2032	
BCX17R			221	2032	
BCX19			210	2051	
BCX19R			210	2051	
BCX20			210	2051	
BCX20R			210	2051	
BCX25					BCX25
BCX26	159				BCX26
BCX27					BCX27
BCX28					BCX28
BCX29					BCX29
BCX30	383				BCX30
BCX40					MPSA06
BCX45	159				BCX45
BCX46	159				BCX46
BCX47	159				BCX47
BCX48					BCX48
BCX49	159				BCX49
BCX50	159				BCX50
BCX58	123AP				BCX58
BCX58IX			47	2030	
BCX58VII			17	2051	
BCX58VIII			17	2051	
BCX58X			47	2030	
BCX59	123AP				BCX59
BCX59IX			47	2030	
BCX59VII			47	2030	
BCX59VIII			47	2030	
BCX59X			47	2030	
BCX70AG			210	2051	
BCX70AH			210	2051	
BCX70AJ			210	2051	
BCX71BG		3138	67	2023	
BCX71BH		3138	67	2023	
BCX71BJ		3247	65	2050	
BCX71BK		3247	65	2050	
BCX73					2N4401
BCX73-16			47	2030	
BCX73-25			47	2030	
BCX73-40			47	2030	
BCX74					2N4401
BCX74-16			47	2030	
BCX74-25			47	2030	
BCX74-40			47	2030	
BCX75					2N4403
BCX76					2N4403
BCX78					2N3906
BCX78VII			21	2034	
BCX79					2N3906
BCX79VII			21	2034	
BCY-50	123A		20	2051	
BCY-58	123A		20	2051	
BCY10	159	3114/290	82	2032	
BCY10A	159	3114/290	82	2032	
BCY11	159	3114/290	82	2032	
BCY11A	159	3114/290	82	2032	
BCY12	159	3114/290	82	2032	
BCY12A	159	3114/290	82	2032	
BCY13	123A	3444	20	2051	
BCY15	123A	3444	20	2051	
BCY16	123A	3444	20	2051	
BCY17	129	3025	244	2027	
BCY18	129	3025	244	2027	
BCY19	129	3025	244	2027	
BCY21	129	3025	244	2027	
BCY22	129	3025	244	2027	
BCY23	129	3025	244	2027	
BCY24	129	3025	244	2027	
BCY25	129	3025	244	2027	

Industry Standard No.	ECG	SK	GE	RS 276-	MOTOR.
BCY26	129	3025	244	2027	
BCY27	129	3025	244	2027	
BCY28	129	3025	244	2027	
BCY29	129	3114/290	244	2027	
BCY30	129	3025	244	2027	
BCY31	129	3025	244	2027	
BCY32	129	3025	244	2027	
BCY33	129	3025	244	2027	
BCY34	129	3025	244	2027	
BCY35	159	3114/290	82	2032	
BCY35A	159	3114/290	82	2032	
BCY36	123A	3444	20	2051	
BCY37	159	3114/290	82	2032	
BCY37A	159	3114/290	82	2032	
BCY38	159	3114/290	82	2032	
BCY38A	159	3114/290	82	2032	
BCY39	159	3114/290	82	2032	
BCY39A	159	3114/290	82	2032	
BCY40	159	3114/290	82	2032	
BCY40A	159	3114/290	82	2032	
BCY42	123A	3444	20	2051	
BCY43	123A	3444	20	2051	
BCY443	128	3024	243	2030	
BCY46	128	3024	243	2030	
BCY47	128	3024	243	2030	
BCY48	128	3024	243	2030	
BCY49	128	3024	243	2030	
BCY50	123A	3444	20	2051	
BCY501	123A	3444	245	2004	
BCY50I	123A	3122			
BCY51	123A	3444	20	2051	
BCY511	123A	3444	245	2051	
BCY51I	123A	3122		2004	
BCY54	159	3114/290	82	2032	
BCY54A	159	3114/290	82	2032	
BCY55			212	2010	
BCY56	123A	3444	20	2051	
BCY57	123A	3444	20	2051	
BCY58	159	3444/123A	20	2009	
BCY58A	123A	3444			
BCY58B	123A	3444	20	2051	
BCY58C	123A	3444	20	2051	
BCY58D	123A	3444	20	2051	
BCY58VII	123A		88	2030	
BCY58VIII	123A		88	2030	
BCY59	123A	3444	20	2051	
BCY59A	123A	3444	20	2051	
BCY59B	123A	3444	20	2051	
BCY59C	123A	3444	20	2051	
BCY59D	123A	3444	20	2051	
BCY59VII	123A		88	2030	
BCY59VIII	123A		88	2030	
BCY65	128	3024	243	2030	
BCY66	128	3024	243	2030	
BCY67	129	3025	244	2027	
BCY69	123A	3444	20	2051	
BCY70	159	3114/290	82	2032	
BCY70A	159	3114/290	82	2032	
BCY71	159	3114/290	82	2032	
BCY71A	159	3114/290	82	2032	
BCY72	159	3114/290	82	2024	
BCY72A	159	3114/290	82	2032	
BCY771X	159		82	2032	
BCY77VII	159		82	2032	
BCY77VIII	159		82	2032	
BCY78	234	3114/290	65	2050	
BCY78A	129	3114/290	67	2023	
BCY78B	129		67	2023	
BCY78V11	159	3138/193A			
BCY78V111	159	3138/193A			
BCY78VII	193		67	2023	
BCY78VIII	193		67	2023	
BCY79	159	3114/290	65	2050	
BCY79A		3114	67	2023	
BCY79B			67	2023	
BCY79VII	159		67	2023	
BCY79VIII	193		67	2023	
BCY84A	123A	3444	20	2051	
BCY85	128	3275/194	220		
BCY86	128	3275/194	220		
BCY87	108	3452	86	2038	
BCY88	108	3452	86	2038	
BCY89	108	3452	86	2038	
BCY90	159	3114/290	82	2032	
BCY90A	159	3114/290	82	2032	
BCY90B	159	3114/290	82	2032	
BCY90B-1	159	3114/290	82	2032	
BCY91	159	3114/290	82	2032	
BCY91A	159	3114/290	82	2032	
BCY91B	159	3114/290	53	2007	
BCY91B-1	159	3114/290	82	2032	
BCY92	159	3114/290	82	2032	
BCY92A	159	3114/290	82	2032	
BCY92B	159	3114/290	82	2032	
BCY92B-1	159	3114/290	82	2032	
BCY93	159	3114/290	82	2032	
BCY93A	159	3114/290	82	2032	
BCY93B	159	3114/290	82	2032	
BCY93B-1	159	3114/290	82	2032	
BCY94	159	3114/290	82	2032	
BCY94A	159	3114/290	82	2032	
BCY94B	159	3114/290	82	2032	
BCY94B-1	159	3114/290	82	2032	
BCY95	159	3114/290	82	2032	
BCY95A	159	3114/290	82	2032	
BCY95B	159	3114/290	82	2032	
BCY95B-1	159	3114/290	82	2032	
BCY96	129	3025	244	2027	
BCY96B	129	3025	244	2027	
BCY97	129	3025	244	2027	
BCY97B	129	3025	244	2027	
BCY98	159	3114/290	82	2032	
BCY98A	159	3114/290	82	2032	
BCY98B	159	3114/290	82	2032	
BCY98B-1	159	3114/290	82	2032	
BCZ10	159	3114/290	82	2032	
BCZ10A	159	3114/290	82	2032	
BCZ10B	159	3114/290	82	2032	
BCZ10C	159	3114/290	82	2032	
BCZ11	159	3114/290	82	2032	
BCZ11A	159	3114/290	82	2032	
BCZ12	159	3114/290	82	2032	
BCZ12A	159	3114/290	82	2032	
BCZ12B	159	3114/290	82	2032	
BCZ12C	159	3114/290	82	2032	
BCZ13	159	3114/290	82	2032	
BCZ13A	159	3114/290	82	2032	
BCZ13B	159	3118	82	2032	
BCZ13C	159	3118	82	2032	
BCZ13D	159	3118	82	2032	
BCZ13E	159	3118	82	2032	
BCZ13F	159	3118	82	2032	
BCZ13G	159	3118	82	2032	
BCZ13H	159	3118	82	2032	

Industry Standard No.	ECG	SK	GE	RS 276-	MOTOR.
BCZ14	159	3114/290	82	2032	
BCZ14A	159	3114/290	82	2032	
BCZ14B	159	3118	82	2032	
BCZ14C	159	3118	82	2032	
BCZ14D	159	3118	82	2032	
BCZ14E	159	3118	82	2032	
BCZ14F	159	3118	82	2032	
BD-00072	103A	3835	59	2002	
BD-107	116		504A	1104	
BD-107(RECT.)	116		504A	1104	
BD-127			504A	1104	
BD-131	241		57	2020	
BD-132	185		58	2027	
BD-1A	116(2)		504A	1104	
BD-3A184	166	9075	BR-600	1152	
BDOA		3647	504A	1104	
BD1			BD1	1123	
BD106	152	3054/196	66	2048	
BD106A	152	3054/196	66	2048	
BD106B	152	3054/196	66	2048	
BD107		3054	28	2017	
BD107A	152	3054/196	66	2048	
BD107B	152	3054/196	66	2048	
BD109	152	3054/196	66	2048	
BD109-6			66	2048	
BD111	130	3027	14	2041	
BD111A	328		75	2041	
BD112	328	3027/130	14	2041	
BD113	229	3027/130	14	2041	
BD115		3045	40	2012	
BD116	328	3027/130	14	2041	
BD117	284	3017B/117	504A	1104	
BD118	328	3027/130	14	2041	
BD119	286		12		
BD120	286		12		
BD121	328	3895	14	2041	
BD123	328	3895	262	2041	
BD124	152	3054/196	66	2048	
BD127	157	3017B/117	12		
BD130	130	3027	14	2041	
BD131	184	3190	57	2017	
BD132	185	3191	58	2027	
BD135	184	3054/196	57	2017	
BD136	185	3083/197	58	2025	
BD137	184	3054/196	57	2017	
BD138	185	3083/197	58	2025	
BD139	373	3054/196	57	2017	
BD140	185	3083/197	58	2025	
BD141	280		262	2041	
BD142	130	3027	14	2041	
BD144	162	3438	35		
BD145	328	3027/130	14	2041	
BD148	182	3188A	55		
BD149	184	3190	57	2017	
BD151	185		58	2025	
BD152	185		58	2025	
BD153	184	3122	57	2017	
BD154	184		57	2017	
BD155	184	3190	57	2017	
BD156	185		58	2025	
BD157	157	3747	232		
BD158	157	3747	232		
BD160	389	3438	35		
BD162	152	3054/196	66	2048	
BD163	152	3054/196	66	2048	
BD165	184		57	2017	
BD166	185		58	2025	
BD167	184		57	2017	
BD168	185		58	2025	
BD169	184	3188A/182	55		
BD170	374	3189A/183	56	2027	
BD175	184		57	2017	
BD176	185		58	2025	
BD177	184		57	2017	
BD178	185		58	2025	
BD179	184	3188A/182	55		
BD180	185	3189A/183	56	2027	
BD181	130	3535/181	14	2041	
BD182	130	3535/181	14	2041	
BD183	185	3535/181	265	2047	
BD183ELK	130		14	2041	
BD184	284		262	2041	
BD185	184		57	2017	
BD186	185		58	2025	
BD187	184		57	2017	
BD188	185		58	2025	
BD189	184		57	2017	
BD190	185		58	2025	
BD195	182		57	2017	
BD196	183		58	2025	
BD197	182		57	2017	
BD199	182		55		
BD2				1123	
BD200	183	3189A	56	2027	
BD201	331	3188A/182			
BD202	332	3189A/183			
BD203	331	3188A/182			
BD204	332	3189A/183			
BD205	182		57	2017	
BD206	183		58	2025	
BD207	182		57	2017	
BD208	183		58	2025	
BD215	157	3103A/396	232		
BD216	157	3103A/396	232		
BD220	152	3893	66	2048	
BD221	152	3893	66	2048	
BD222	152	3893	66	2048	
BD223	153		69	2049	
BD224	153		69	2049	
BD225	153		69	2049	
BD231A	152	3893	66	2048	
BD232	157	3747			
BD233	184	3190			
BD237	184	3188A/182			
BD238	185	3189A/183			
BD239	152	3893	66	2048	
BD239A	152	3893	66	2048	
BD240	153		69	2049	
BD240A	153		69	2049	
BD241	152	3893	66	2048	
BD242	153		69	2049	
BD242A	331		69	2049	
BD242B	153		69	2049	
BD243	152	3893	66	2048	
BD243B	152		66	2048	
BD244	153	3189A/183	69	2049	
BD244A	153	3189A/183	69	2049	
BD244B	153	3189A/183	69	2049	
BD245	390		14	2041	
BD245A	390		14	2041	
BD245B	390	3836/284	265	2047	
BD245C	390	3836/284	265	2047	
BD246	393	3437/180	74	2043	
BD246A	393	3437/180	74	2043	
BD246B	393	3437/180	74	2043	
BD246C	393	3437/180	74	2043	
BD249	181		75	2041	
BD249A	181		75	2041	
BD249B	181		75	2041	
BD249C	181		75	2041	
BD250	180		74	2043	
BD250A	180		74	2043	
BD250B	180		74	2043	
BD250C	180	3437	74	2043	
BD262	254	3997			
BD262A	254	3997	69	2049	
BD263	253	3996			
BD263A	259	3978			
BD265	263	3180			
BD267	263	3180			
BD267A	263	3180			
BD271	152	3893	66	2048	
BD272	153		69	2049	
BD273	196		241	2020	
BD274	292		250	2027	
BD275	196		241	2020	
BD276	197		250	2027	
BD278	196	3534	241	2020	
BD3				1123	
BD3A-184	117		504A	1104	
BD3A-1B4	166	9075	BR-600	1152	
BD4				1123	
BD433	152	3054/196	66	2048	
BD434	153	3083/197	69	2049	
BD435	152	3893	66	2048	
BD436	153		69	2049	
BD437	152	3893	66	2048	
BD438	153		69	2049	
BD439	152	3893	66	2048	
BD440	153		69	2049	
BD441	196		241	2020	
BD442	197		250	2027	
BD461	182	3188A	55		
BD462	185		56	2027	
BD463	182	3188A	55		
BD464	183	3189A	56	2027	
BD5				1123	
BD533	152	3893	66	2048	
BD534	153		69	2049	
BD535	152	3893	66	2048	
BD536	153		69	2049	
BD537	196		241	2020	
BD538	197		250	2027	
BD561	184		57	2017	
BD562	185		58	2025	
BD575	241	3188A/182	57	2020	
BD576	242	3189A/183	58	2027	
BD577	241	3188A/182	57	2020	
BD578	242	3189A/183	58	2027	
BD580	242		58	2027	
BD585	241	3188A/182	57	2020	
BD586	242	3189A/183	58	2027	
BD587	241	3188A/182	57	2020	
BD588	242	3189A/183	58	2027	
BD589	241	3188A/182	57	2020	
BD590	242	3189A/183	58	2027	
BD6				1123	
BD643	261	3180/263			
BD645	261	3180/263			
BD647	263	3180			
BD680	254	3189A/183			
BD695A	259	3978			
BD696A	260	3979			
BD7				1123	
BD71	108	3452	86	2038	
BDOA	166			1152	
BDX10	130	3027	262	2041	
BDX11	280		262	2041	
BDX12	162	3438	35		
BDX13	181	3535	75	2041	
BDX18	219	3173	74	2043	
BDX18N	219		74	2043	
BDX20	285		74	2043	
BDX23	284	3467/283	36		
BDX24	280		262	2041	
BDX27	153		69	2049	
BDX27-10	153		69	2049	
BDX27-6	153		69	2049	
BDX28	153		69	2049	
BDX28-10	153		69	2049	
BDX28-6	153		69	2049	
BDX32	389	3710/238	259		
BDX33	263	3180			
BDX33A	263	3180			
BDX33B	263	3180			
BDX34	264	3181A			
BDX34A	264	3181A			
BDX34B	264	3181A			
BDX40	284		14	2041	
BDX41	284	3036	75	2041	
BDX50	284	3621	265	2047	
BDX51	284	3621	265	2047	
BDX53	263	3180			
BDX53A	263	3180			
BDX53B	263	3180			
BDX54	264	3181A			
BDX54A	264	3181A			
BDX54B	264	3181A			
BDX60	284	3621	265	2047	
BDX61	181	3621	265	2047	
BDX62	244	3859/252			
BDX62A	244	3859/252			
BDX63	243	3182			
BDX63A	243	3948/247			
BDX64	248	3949			
BDX64A	248	3949			
BDX65	247	3948			
BDX65A	247	3948			
BDX70	196		241	2020	
BDX71	196		241	2020	
BDX72	196		241	2020	
BDX73	196		241	2020	
BDX74	152	3893	66	2048	
BDX75	152	3893	66	2048	
BDX87	247	3948			
BDX87A	247	3948			
BDX87B	247	3948			
BDX87C	247	3948			
BDX88	248	3949			
BDX88A	248	3949			
BDX88B	248	3949			
BDX88C	248	3949			
BDY-10	130	3036	75	2041	
BDY10	130		75	2041	
BDY11	130	3036	75	2041	
BDY12	152	3054/196	66	2048	
BDY13	152	3054/196	66	2048	
BDY15	155	3839	43		

Industry Standard No.	ECG	SK	GE	RS 276-	MOTOR.
BDY15A	155	3839	43		
BDY15B	155	3839	43		
BDY15C	155	3839	43		
BDY16	175	3839/155	246	2020	
BDY16A	155	3839	43		
BDY16B	155	3839	43		
BDY17	130	3027	14	2041	
BDY18	284	3621	265	2047	
BDY19	284	3621	265	2047	
BDY20	130	3621	265	2047	
BDY23	280	3027/130	14	2041	
BDY24	283	3467	36		
BDY25	283	3467	36		
BDY26	283	3467	36		
BDY27	283	3467	36		
BDY28	283	3467	36		
BDY34	152	3054/196	66	2048	
BDY37	284	3260			
BDY38	130	3027	14	2041	
BDY39	130	3027	14	2041	
BDY53	130	3027	14	2041	
BDY54	280	3438	35		
BDY55	284	3621	265	2047	
BDY56	280	3621	265	2047	
BDY57	327	3561	75	2041	
BDY58	327	3561	75	2041	
BDY60	282	3748	76		
BDY61	328	3642/179	76		
BDY62	121	3009	239	2006	
BDY63	280		262	2041	
BDY65	190	3104A	217		
BDY66		3104A	217		
BDY69	281		263	2043	
BDY70	129	3025	244	2027	
BDY71	175	3538	246	2020	
BDY72	175	3538	246	2020	
BDY73	130	3621	265	2047	
BDY74	284		262	2041	
BDY76	181	3836/284	75	2041	
BDY77	284	3621	265	2047	
BDY78	384	3538	246	2020	
BDY79	384	3538	246	2020	
BDY82	153		69	2049	
BDY83	153		69	2049	
BDY90	328	3895			
BDY90(AUDIO)			262	2041	
BDY91	328	3895			
BDY91(AUDIO)			262	2041	
BDY92	328	3895			
BDY92(AUDIO)			262	2041	
BE-55	614	3327	90		
BE-66	123A	3124/289	17	2051	
BE107	116	3017B/117	504A	1104	
BE117	116	3017B/117	504A	1104	
BE127	116	3017B/117	504A	1104	
BE173	107		11	2015	
BE6	100	3005	1	2007	
BE6A	100	3005	1	2007	
BE71	128	3024	243	2030	
BF-115	108	3019	86	2038	
BF-180			17	2051	
BF-200(PENNCREST)			39	2015	
BF-214	123A	3444	20	2051	
BF-215	123A	3444	20	2051	
BF-226	123A	3444	20	2051	
BF-255			20	2051	
BF-832	159	3025/129	82	2032	
BF1	102A	3004	53	2007	
BF108	154	3045/225	40	2012	
BF109	154	3045/225	40	2012	
BF110	154	3045/225	40	2012	
BF111	154	3045/225	40	2012	
BF114	154	3045/225	40	2012	
BF115		3444	20	2051	
BF117	154	3045/225	40	2012	
BF118	154	3045/225	40	2012	
BF119	154	3045/225	40	2012	
BF120	154		222		
BF121	107	3039/316	11	2015	
BF123	107	3039/316	11	2015	
BF125	107	3039/316	11	2015	
BF127	107	3039/316	11	2015	
BF140	154	3045/225	40	2012	
BF140A	154	3045/225	40	2012	
BF140R	154	3045/225	40	2012	
BF140S	154	3045/225	40	2012	
BF152	311	3452/108	86	2038	
BF153	311	3452/108	86	2038	
BF154	311	3452/108	86	2038	
BF155	161	3117	39	2015	
BF155R	154	3045/225	40	2012	
BF155S	154	3045/225	40	2012	
BF156	154	3045/225	40	2012	
BF157	154	3045/225	40	2012	
BF157B	154	3045/225	40	2012	
BF158	311	3452/108	86	2038	
BF159	311	3452/108	86	2038	
BF160	311	3452/108	86	2038	
BF161	161	3117	39	2015	
BF162	108	3452	86	2038	
BF163	108	3452	86	2038	
BF164	108	3452	86	2038	
BF165	108	3452	86	2038	
BF166	161	3117	39	2015	
BF167	161	3117	39	2015	
BF168	161	3117	39	2015	
BF169	161	3117	39	2015	
BF173	161	3018	39	2015	
BF173A	108	3452	86	2038	
BF174	154	3045/225	40	2012	
BF175	161	3117	39	2015	
BF176	108	3452	86	2038	
BF177	128	3045/225	40	2012	
BF178	154	3045/225	40	2012	
BF179	154	3045/225	40	2012	
BF179A	154	3045/225	40	2012	
BF179B	154	3045/225	40	2012	
BF179C	154	3045/225	40	2012	
BF180	161	3117	39	2015	
BF181	161	3117	39	2015	
BF182	161	3117	39	2015	
BF183	161	3117	39	2015	
BF183A	123A	3444	20	2051	
BF184	161	3117	39	2015	
BF185	161	3117	39	2015	
BF186	154	3045/225	40	2012	
BF187	108	3452	86	2038	
BF188	108	3452	86	2038	
BF189	123A	3444	20	2051	
BF194	229	3018	11	2015	
BF194A	108	3452	86	2038	
BF194B	107	3018	11	2015	
BF195	229	3018	11	2015	
BF195C	107	3039/316	11	2015	
BF195D	107	3018	11	2015	
BF196	229	3039/316	11	2015	
BF197	107	3039/316	11	2015	
BF198	229	3039/316	11	2015	BF198
BF199	229	3039/316	39	2015	BF199
BF1A	102A	3004	53	2007	
BF2	102A	3004	53	2007	
BF200	161	3018	39	2015	
BF200(ZENITH)			11	2015	
BF206	161	3117	39	2015	
BF207	161	3117	39	2015	
BF208	161	3117	39	2015	
BF209	161	3117	39	2015	
BF212	161	3117	39	2015	
BF213	161	3117	39	2015	
BF214	161	3117	39	2015	
BF215	161	3117	39	2015	
BF216	107	3039/316	11	2015	
BF217	107	3039/316	11	2015	
BF218	107	3039/316	11	2015	
BF219	107	3039/316	11	2015	
BF220	107	3039/316	11	2015	
BF222	161	3117	39	2015	
BF223	108	3452	86	2038	
BF224	229	3452/108	86	2038	
BF224J	123A	3444	20	2051	
BF225	229	3452/108	86	2038	
BF225J	123A	3444	20	2051	
BF226	161	3117	39	2015	
BF227			39	2015	
BF228		3244	220		
BF229	107	3039/316	11	2015	
BF230	107	3039/316	11	2015	
BF232	161	3117	39	2015	
BF233-2	229	3452/108	86	2038	
BF233-3	229	3039/316	86	2038	
BF233-4	108	3452	86	2038	
BF233-5	161	3452/108	86	2038	
BF234	229	3452/108	86	2038	
BF235	108	3452	86	2038	
BF236	108	3452	86	2038	
BF237	123AP	3452/108	86	2038	
BF238	123AP	3452/108	86	2038	
BF240	229	3039/316	11	2015	
BF240B			61	2038	
BF241	229	3039/316	11	2015	
BF241C			61	2038	
BF241D			61	2038	
BF243	234		65	2050	MPSH54
BF244	312	3112		2028	
BF244A	312	3448	FET-1	2035	
BF244B	312	3448	FET-1	2035	
BF244C	312	3448	FET-1	2035	
BF245	312	3112		2028	
BF245A	312	3116	FET-2	2028	
BF245B	312	3116		2028	
BF245C	312	3116		2028	
BF246	312	3112		2028	
BF247	312	3112		2028	
BF248	123A	3444	20	2051	
BF249	159	3444/123A	20	2009	
BF250	123A	3444	20	2051	
BF251	161	3117	39	2015	
BF252	161	3117	39	2015	
BF253	107		11	2015	
BF254	229	3018	86	2038	BF254
BF254(SIEG)				2013	
BF254B			39	2015	
BF255	229	3122	20	2038	BF255
BF255(SIEG)				2033	
BF255C			39	2015	
BF255D			39	2015	
BF256	312	3448	FET-1	2035	
BF256A	312	3116		2028	
BF256B	312	3116		2028	
BF256C	312	3116		2028	
BF257	396	3045/225	40	2012	
BF258	396	3045/225	40	2012	
BF259	154	3045/225	40	2012	
BF260	161	3117	39	2015	
BF261	161	3117	39	2015	
BF262	108	3452	86	2038	
BF26235-5	117	3017B	504A	1104	
BF263	108	3452	86	2038	
BF264	108	3452	86	2038	
BF270	161	3117	39	2015	
BF271	161	3117	39	2015	
BF273	161	3117	39	2015	
BF273C	161	3117	39	2015	
BF273D	161	3117	39	2015	
BF274	161	3117	39	2015	
BF274B	161	3117	39	2015	
BF274C	161	3117	39	2015	
BF279			39	2015	
BF287	161	3117	39	2015	
BF288	161	3117	39	2015	
BF290	161	3117	39	2015	
BF291	123A	3444	20	2051	
BF291A	123A	3444	20	2051	
BF291B	123A	3444	20	2051	
BF292	154	3201/171	27		
BF292A	154	3045/225	40	2012	
BF292B	154	3045/225	40	2012	
BF292C	154	3045/225	40	2012	
BF293	123A	3444	20	2051	
BF293A	123A	3444	20	2051	
BF293D	123A	3444	20	2051	
BF294	154	3045/225	40	2012	
BF297	287	3201/171	220		BF391
BF298	287	3434/288	223		BF392
BF299	288	3434	223		BF393
BF302	161	3117	39	2015	
BF303	311	3117	277		
BF304	161	3117	39	2015	
BF305	154	3045/225	40	2012	
BF306	161	3117	261		
BF308	229		277		
BF309	278		277		
BF310	229	3039/316	11	2015	MPSH04
BF311	229	3039/316	11	2015	MPSH24
BF314	229	3039/316	277		BF366
BF315	106	3984	21	2034	
BF316	106	3984	21	2034	
BF321A	123		20	2051	
BF321B	123A	3444	20	2051	
BF321C	123A	3444	20	2051	
BF321D	123A	3444	20	2051	
BF321E	123A	3444	20	2051	
BF321F	123A	3444	20	2051	
BF322	128		261		
BF323	129		48		
BF324	395				BF506
BF325	278		261		
BF329	107		11	2015	

Industry Standard No.	ECG	SK	GE	RS 276-	MOTOR.
BF330	278		261		
BF332	107	3039/316	11	2015	
BF333	107	3039/316	11	2015	
BF333C	107		11	2015	
BF333D	107		11	2015	
BF334	107	3039/316	11	2015	
BF335	107	3039/316	11	2015	
BF336	154		40	2012	
BF337	154		40	2012	
BF338	154		40	2012	
BF339	106	3984	21	2034	
BF340	159	3118	62	2010	MPSH54
BF340A	159	3118	82	2032	
BF340B	159	3118	82	2032	
BF340C	159	3118	82	2032	
BF340D	159	3118	82	2032	
BF341	341	3118	62	2010	MPSH54
BF341A	159	3118	82	2032	
BF341B	159	3118	82	2032	
BF341C	159	3118	82	2032	
BF341D	159	3118	82	2032	
BF342	159	3118	62	2010	
BF342A	159	3118	82	2032	
BF342B	159	3118	82	2032	
BF342C	159	3118	82	2032	
BF342D	159	3118	82	2032	
BF343	159		62	2010	
BF344	161	3716	39	2015	
BF345	161	3716	39	2015	
BF348	312	3112		2028	
BF355	154		40	2012	
BF357	108	3452	86	2038	MPSH33
BF362			39	2015	BF362
BF363			39	2015	BF363
BF364			61	2038	
BF365	229		61	2038	
BF366					BF366
BF367	229				BF367
BF368	123AP				BF368
BF369	123AP				BF369
BF371					BF371
BF373					BF373
BF374					BF374
BF375					BF375
BF384					BF374
BF385					BF375
BF387	128		243	2030	
BF391	287				BF391
BF392	287				BF392
BF393	287				BF393
BF394	293				BF394
BF395	293				BF395
BF397			223		BC490
BF398			223		MPSA93
BF4-05L					1N4001
BF4-100L					1N4007
BF4-10L					1N4002
BF4-120L					MR1-1200
BF4-20L					1N4003
BF4-40L					1N4004
BF4-60L					1N4005
BF4-80L					1N4006
BF411		3244	220		
BF412		3244	220		2N5550
BF413					2N5551
BF414					BF506
BF422	373	3433/287			
BF423	383	3434/288			
BF440	159	3444/123A	20	2051	
BF441	159	3444/123A	20	2051	
BF450	159				BF506
BF451	159				BF506
BF456	373	3201/171	27		
BF457	373	3201/171	27		
BF458	157	3201/171	27		
BF459	157	3747	27		
BF479					BF479
BF494	229		61	2038	BF254
BF495	229		61	2038	BF255
BF5	121	3009	239	2006	
BF5-05L					MR501
BF5-100L					MR510
BF5-10L					MR501
BF5-20L					MR502
BF5-40L					MR504
BF5-60L					MR506
BF5-80L					MR508
BF500					BF509
BF506					BF506
BF509					BF509
BF516	161	3118	39	2015	
BF523	395				MPSH24
BF540	159		221		MPSH54
BF541	159		221		MPSH54
BF542	159		221		
BF594	229	3452/108	86	2038	
BF595	229	3452/108	86	2038	
BF596			20	2051	
BF597			62	2010	
BF6-05L					MR501
BF6-100L					MR510
BF6-10L					MR501
BF6-20L					MR502
BF6-40L					MR504
BF6-60L					MR506
BF6-80L					MR508
BF65	234		65	2050	
BF679					BF679
BF680					BF680
BF689	161	3716			
BF706	159				BF506
BF709	395				BF509
BF71	123A	3444	20	2051	
BF906					BF506
BFR11	123A	3444	20	2051	
BFR16	123A	3137/192A	63	2030	
BFR18	128	3201/171	27		
BFR19	128	3201/171	27		
BFR20	128	3201/171	27		
BFR21	154		230		
BFR22	128	3104A	217		
BFR24	129	3748/282			
BFR25	199	3245	62	2010	
BFR26	123A	3444	20	2051	
BFR28			86	2038	
BFR36	128	3024	243	2030	
BFR39					MPSA06
BFR40	297		63	2030	MPSA05
BFR40T05	128		63	2030	
BFR41	297		47	2030	MPSA05
BFR41T05	128		47	2030	
BFR49					2N6603
BFR50	128				2N4401
BFR51	128		63	2030	2N4401
BFR52	128		47	2030	2N4401
BFR53					MMBR920
BFR57	154	3201/171	27		
BFR58	154	3201/171	27		
BFR59	154	3201/171	27		
BFR61			67	2023	
BFR62	383		48		
BFR63	76				MRF511
BFR64	76				MRF511
BFR65	76				MRF511
BFR77	128	3748/282	264		
BFR78	128	3748/282	264		
BFR80			67	2023	MPSA55
BFR80T05	383		67	2023	
BFR81			48		MPSA55
BFR81T05	383		48		
BFR90					BFR90
BFR91					BFR91
BFR92					MMBR920
BFR93					MMBR930
BFR94	76				MRF511
BFR95					MRF517
BFR96					BFR96
BFR99	395				2N4959
BFS11			39	2015	
BFS13E	107	3039/316	11	2015	
BFS13F	107	3039/316	11	2015	
BFS13G	107	3039/316	11	2015	
BFS14A	159	3118	82	2032	
BFS14B	159	3118	82	2032	
BFS14C	159	3118	82	2032	
BFS14D	159	3118	82	2032	
BFS14E	107	3039/316	11	2015	
BFS14F	107	3039/316	11	2015	
BFS14G	107	3039/316	11	2015	
BFS15E	107	3039/316	11	2015	
BFS15F	107	3039/316	11	2015	
BFS15G	107	3039/316	11	2015	
BFS16	159	3118	82	2032	
BFS16A	159	3118	82	2032	
BFS16B	159	3118	82	2032	
BFS16C	159	3118	82	2032	
BFS16D	159	3118	82	2032	
BFS16E	107	3039/316	11	2015	
BFS16F	107	3118	11	2015	
BFS16G	107	3039/316	11	2015	
BFS17	108	3452	86	2038	
BFS17R	108	3452	86	2038	
BFS18	107	3039/316	11	2015	
BFS18CA	161	3132	39	2015	
BFS18R	107	3039/316	11	2015	
BFS19	107	3039/316	11	2015	
BFS19CB			210	2051	
BFS19R	107	3039/316	11	2015	
BFS20	107	3039/316	11	2015	
BFS20R	107	3039/316	11	2015	
BFS21	312	3112		2028	
BFS21A	312	3112		2028	
BFS22	195A	3748/282	46		
BFS22A	282	3748			2N3924
BFS23	195A	3748/282	46		
BFS23A	282	3748			
BFS26	159	3118	82	2032	
BFS26A	159	3118	82	2032	
BFS26B	159	3118	82	2032	
BFS26C	159	3118	82	2032	
BFS26D	159	3118	82	2032	
BFS26E	159	3118	82	2032	
BFS26F	159	3118	82	2032	
BFS26G	159	3118	82	2032	
BFS27E	161	3132	39	2015	
BFS27F	161	3132	39	2015	
BFS27G	161	3132	39	2015	
BFS28	221	3050		2036	
BFS28R	221	3050		2036	
BFS29	128	3024	243	2030	
BFS29P	192	3137	63	2030	
BFS30	192	3137	63	2030	
BFS30P	192	3137	63	2030	
BFS31	159	3118	82	2032	
BFS31P	123A	3244	20	2051	
BFS32	159	3118	82	2032	
BFS32P	159	3118	82	2032	
BFS33	159		82	2032	
BFS33P	159	3118	82	2032	
BFS34	193		67	2023	
BFS34P	159	3118	82	2032	
BFS36	128	3122	243	2030	
BFS36A	123A	3122	20	2051	
BFS36B	123A	3122	20	2051	
BFS36C	123A	3122	20	2051	
BFS37	159	3025/129	82	2032	
BFS37A	159	3025/129	82	2032	
BFS38	123A	3122	20	2051	
BFS38A	123A	3122	20	2051	
BFS39	123AP	3137/192A	63	2030	
BFS40	159	3025/129	82	2032	
BFS40A	159	3025/129	82	2032	
BFS41	159	3025/129	82	2032	
BFS42	123A	3122	20	2051	
BFS42A	123A	3122	20	2051	
BFS42B	123A	3122	20	2051	
BFS42C	123A	3122	20	2051	
BFS43	123A	3122	20	2051	
BFS43A	123A	3122	20	2051	
BFS43B	123A	3122	20	2051	
BFS43C	123A	3122	20	2051	
BFS44		3138	67	2023	
BFS45		3138	67	2023	
BFS50	311	3192/186	28	2017	
BFS51	186	3192	28	2017	
BFS55A			212	2010	
BFS58			86	2038	
BFS59		3192	63	2030	BC338
BFS60		3137	63	2030	BC485A
BFS61					BCX47
BFS62	161	3452/108	86	2038	
BFS68	312	3448	FET-1	2035	
BFS68P	312	3448	FET-1	2035	
BFS69	159	3025/129	82	2032	
BFS80					2N4416A
BFS96	129	3138/193A	67	2023	BC328
BFS97	129	3138/193A	67	2023	BC486
BFS98	129				BCX48
BFS99	171	3201	27		BCX29
BFT21	159		48		
BFT22	159		48		
BFT24					MRF931
BFT30	128		63	2030	
BFT31	128		47	2030	
BFT33	282	3748	264		
BFT34	282	3748	264		
BFT39	128	3748/282	264		
BFT40	128	3748/282	264		

Industry Standard No.	ECG	SK	GE	RS 276-	MOTOR.
BFT41	128	3748/282	264		
BFT42	282	3748	264		
BFT43	282	3748	264		
BFT54	128		47	2030	
BFT55	123A		62	2010	
BFT60	129		244	2027	
BFT61	129		244	2027	
BFT62	129		244	2027	
BFT70	159		67	2023	
BFT71	159		48		
BFT79	129		244	2027	
BFT80	129		244	2027	
BFT81	129		244	2027	
BFV10			20	2051	
BFV11			20	2051	
BFV12			20	2051	
BFV14			47	2030	
BFV20	159	3118	82	2032	
BFV21	159	3118	82	2032	
BFV22	159	3118	82	2032	
BFV25	159	3114/290	82	2032	
BFV26	159	3114/290	82	2032	
BFV27			39	2015	
BFV28			39	2015	
BFV29	159	3118	82	2032	
BFV30	159	3118	82	2032	
BFV33	159	3118	82	2032	
BFV34			82	2032	
BFV37		3245	212	2010	
BFV38		3245	212	2010	
BFV40			210	2051	
BFV41			210	2051	
BFV42			20	2051	
BFV43			20	2051	
BFV44			20	2051	
BFV45			210	2051	
BFV46			20	2051	
BFV47			20	2051	
BFV49			210	2051	
BFV50			20	2051	
BFV51			20	2051	
BFV52			47	2030	
BFV53			20	2051	
BFV54			20	2051	
BFV55			20	2051	
BFV59			39	2015	
BFV60			39	2015	
BFV61			39	2015	
BFV62		3245	212	2010	
BFV80			39	2015	
BFV82	159	3025/129	82	2024	
BFV82A	159	3025/129	82	2024	
BFV82B	159	3025/129	82	2024	
BFV82C	159	3025/129	82	2024	
BFV83	108	3452	86	2038	
BFV83A	108	3452	86	2038	
BFV83B	123A	3444	20	2051	
BFV83C	123A	3444	20	2051	
BFV85	123A	3444	20	2051	
BFV85A	123A	3444	20	2051	
BFV85B	123A	3444	20	2051	
BFV85C	123A	3444	20	2051	
BFV85D	108	3452	86	2038	
BFV85E	108	3452	86	2038	
BFV85F	108	3452	86	2038	
BFV85G	108	3452	86	2038	
BFV86	159	3025/129	82	2032	
BFV86A	159	3025/129	82	2032	
BFV86B	159	3025/129	82	2032	
BFV86C	159	3025/129	82	2032	
BFV87	123A	3444	20	2051	
BFV88	123A	3444	20	2051	
BFV88A	123A	3444	20	2051	
BFV88B	123A	3444	20	2051	
BFV88C	123A	3444	20	2051	
BFV89	199	3245	62	2010	
BFV89A	199	3245	62	2010	
BFW10	312	3448	FET-1	2035	
BFW11	312	3448	FET-1	2035	
BFW12	312	3448	FET-1	2035	
BFW13	312	3112		2028	
BFW16A	278		261		MRF517
BFW17A	278		261		MRF517
BFW19	278		261		
BFW20	159	3138/193A	67	2023	
BFW22	159		65	2050	
BFW24	282		264		
BFW25	282	3748	264		
BFW26	282	3748	264		
BFW29	123A	3137/192A	63	2030	
BFW30	278	3132	261		
BFW31	159		67	2023	
BFW32	123A	3444	20	2051	
BFW33	154	3024/128	243	2030	
BFW37	154	3045/225	40	2012	
BFW41	161		39	2015	
BFW43	288		223		
BFW44	129		244	2027	
BFW45	154	3045/225	40	2012	
BFW46	123A	3444	20	2051	2N3924
BFW47	278		261		2N3553
BFW54	312	3448	FET-1	2035	
BFW55	312	3448	FET-1	2035	
BFW56	312	3448	FET-1	2035	
BFW57		3244	220		
BFW58		3244	220		
BFW59	123A	3444	20	2051	
BFW60	123A	3444	20	2051	
BFW61	5547	3448	FET-1	2035	
BFW63	161	3132	39	2015	
BFW64	108	3452	86	2038	
BFW68	123A	3444	20	2051	
BFW87	159	3083/197	218	2026	
BFW88	189	3083/197	218	2026	
BFW89	159	3059/138A	82	2032	
BFW90	159	3025/129	82	2032	
BFW91	129	3025	244	2027	
BFW92					BFW92A
BFW93					BFW92A
BFX12	106	3984	21	2034	
BFX13	106	3984	21	2034	
BFX17	128	3024	243	2030	
BFX18	108	3452	86	2038	
BFX19	108	3452	86	2038	
BFX20	108	3452	86	2038	
BFX21	108	3452	86	2038	
BFX29	159	3025/129	82	2032	
BFX30	159	3025/129	82	2032	
BFX31	161	3117	39	2015	
BFX32	107	3039/316	11	2015	
BFX33	311		261		
BFX34	282	3045/225	40	2012	
BFX35	159		82	2032	
BFX37	234		65	2050	
BFX38	129	3025	244	2027	
BFX39	129	3025	244	2027	
BFX40	129	3025	244	2027	
BFX41	129	3025	244	2027	
BFX43	123A	3452/108	86	2038	
BFX44	123A	3452/108	86	2038	
BFX45	108	3452	86	2038	
BFX47			39	2015	
BFX48	106	3984	21	2034	
BFX50	128	3024	243	2030	
BFX51	128	3024	243	2030	
BFX52	128	3024	243	2030	
BFX53	192	3137	63	2030	
BFX55	278		28	2017	
BFX59			283	2016	
BFX59F	123A	3137/192A	63	2030	
BFX60	161	3117	39	2015	
BFX61	128	3024	243	2030	
BFX62	161	3117	39	2015	
BFX65	159	3025/129	82	2032	
BFX68	128	3024	243	2030	
BFX68A	128	3024	243	2030	
BFX69	128	3024	243	2030	
BFX69A	128	3024	243	2030	
BFX73	161	3117	39	2015	
BFX74	128	3024	243	2030	
BFX74A	129	3025	244	2027	
BFX77	161	3117	39	2015	
BFX84	128	3024	243	2030	
BFX85	128	3024	243	2030	
BFX86	128	3024	243	2030	
BFX87	129	3025	244	2027	
BFX88	129	3025	244	2027	
BFX89	161	3716	39	2015	BFX89
BFX92	123A	3444	20	2051	
BFX92A	123A		63	2030	
BFX93	123A	3444	20	2051	
BFX94	123A	3024/128	243	2030	
BFX95	123A	3024/128	243	2030	
BFX95A	123A	3444	20	2051	
BFX96	123	3024/128	243	2030	
BFX96A	128	3192/186	28	2017	
BFX97	128	3024	243	2030	
BFX97A	128		47	2030	
BFX98	154		40	2012	
BFY	123A	3444	20	2051	
BFY-22	123A	3124/289	20	2051	
BFY-23	123A	3124/289	20	2051	
BFY-23A	123A	3124/289	20	2051	
BFY-24	123A	3124/289	20	2051	
BFY-29	123A	3124/289	20	2051	
BFY-30	123A	3124/289	20	2051	
BFY-37	108	3019	86	2038	
BFY-39	123A	3124/289	20	2051	
BFY-47	108	3018	86	2038	
BFY-48	108	3018	86	2038	
BFY10	123A	3018	243	2030	
BFY11	123A	3018	243	2030	
BFY12	123A	3024/128	243	2030	
BFY13	128	3024	243	2030	
BFY14	128	3024	243	2030	
BFY15	128	3024	243	2030	
BFY16	192	3024/128	63	2030	
BFY167				2011	
BFY17	128	3024	243	2030	
BFY18	123A	3025/129	244	2027	
BFY19	123A	3024/128	243	2030	
BFY22	123A	3444	20	2051	
BFY23	123A	3444	20	2051	
BFY23A	123A	3444	20	2051	
BFY24	123A	3444	20	2051	
BFY25	123A	3444	20	2051	
BFY26	123A	3444	20	2051	
BFY27	128	3024	243	2030	
BFY28	123A	3444	20	2051	
BFY29	123A	3444	20	2051	
BFY30	123A	3444	20	2051	
BFY33	123A	3444	20	2051	
BFY34	128	3024	243	2030	
BFY37	123A	3452/108	20	2051	
BFY371	123A	3444	20	2051	
BFY37I	160	3122	245	2051	
BFY39	123A	3444	20	2051	
BFY39-1	199	3122	62	2010	
BFY39-2	199	3245	62	2010	
BFY39-3	199	3245	62	2010	
BFY39/1	123A	3444	20	2051	
BFY39/2	123A	3444	20	2051	
BFY39/3	123A	3444	20	2051	
BFY391	123A	3444	20	2051	
BFY39I	123A	3122		2051	
BFY40	128	3024	243	2030	
BFY41	154	3245/199	251		
BFY43	154	3045/225	40	2012	
BFY44	128	3024	243	2030	
BFY45	154	3045/225	40	2012	
BFY46	128	3024	243	2030	
BFY47	107	3024/128	11	2015	
BFY48	107		11	2015	
BFY49	107	3018	11	2015	
BFY50	128	3104A	243	2030	
BFY51	128	3024	243	2030	
BFY52	128	3024	243	2030	
BFY53	128	3024	243	2030	
BFY55	128	3024	243	2030	
BFY56	128	3024	243	2030	
BFY56A	128	3024	243	2030	
BFY57	154	3045/225	40	2012	
BFY63	311	3444/123A	20	2051	
BFY64	129		244	2027	
BFY65	154	3045/225	40	2012	
BFY66	128	3024	243	2030	
BFY67	128	3024	243	2030	
BFY67A	128	3024	243	2030	
BFY67C	128	3024	243	2030	
BFY68	128	3024	243	2030	
BFY68A	128	3024	243	2030	
BFY69	107	3039/316	11	2015	
BFY69A	107	3039/316	11	2015	
BFY69B	161	3039/316	39	2015	
BFY70	128	3024	243	2030	
BFY72	123A	3024/128	243	2030	
BFY73	123A	3444	20	2051	
BFY74	123A	3444	20	2051	
BFY75	123A	3444	20	2051	
BFY76	123A	3444	20	2051	
BFY77	123A	3444	20	2051	
BFY78	108	3452	86	2038	
BFY79	161	3117	39	2015	
BFY80	128	3045/225	40	2012	
BFY87	161	3716	39	2015	
BFY87A	161	3716	504A	1104	
BFY90	311	3452/108	86	2038	BFY90
BFY90B	108	3452	86	2038	

Industry Standard No.	ECG	SK	GE	RS 276-	MOTOR.
BFY94	159		82	2032	
BFY99	128	3024	243	2030	
BG-66	123A	3124/289	17	2051	
BG-94	123A	3124/289	210	2051	
BG71	123A	3444	20	2051	
BGY22					MHW401
BGY22A					MHW401
BGY23					MHW709
BGY23A					MHW709
BGY36					MHW612A
BGY50					MHW1121
BGY51					MHW1122
BGY52					MHW1171
BGY53					MHW1172
BGY54					MHW1171
BGY55					MHW1172
BGY56					MHW1221
BGY57					MHW1222
BGY59					MHW1342
BH4R1	116		504A	1104	
BH4R6	156	3051	512		
BH71	123A	3444	20	2051	
BI-82	106	3118	22	2032	
BI71	123A	3444	20	2051	
BIP7201	123		20	2051	
BJ-155	614	3327	90		
BJ10	159	3118	82	2032	
BJ11	159	3118	82	2032	
BJ11A	159	3118	82	2032	
BJ11B	159	3118	82	2032	
BJ12	159	3118	82	2032	
BJ12B	159	3118	82	2032	
BJ12C	159	3118	82	2032	
BJ13	159	3118	82	2032	
BJ13A	159	3118	82	2032	
BJ13B	159	3118	82	2032	
BJ14	159	3118	82	2032	
BJ14A	159	3118	82	2032	
BJ15	159	3118	82	2032	
BJ160	159	3118	82	2032	
BJ161	159	3118	82	2032	
BJ161A	159	3118	82	2032	
BJ161B	159	3118	82	2032	
BJ161C	159	3118	82	2032	
BJ1A	159	3114/290	82	2032	
BJ1B	159	3114/290	82	2032	
BJ1C	159	3114/290	82	2032	
BJ2	159	3114/290	82	2032	
BJ2A	159	3114/290	82	2032	
BJ2B	159	3114/290	82	2032	
BJ2C	159	3114/290	82	2032	
BJ2D	159	3114/290	82	2032	
BJ3	159	3114/290	82	2032	
BJ3A	159	3114/290	82	2032	
BJ3B	159	3114/290	82	2032	
BJ4	159	3114/290	82	2032	
BJ4A	159	3114/290	82	2032	
BJ4B	159	3114/290	82	2032	
BJ4C	159	3114/290	82	2032	
BJ4D	159	3114/290	82	2032	
BJ5	159	3114/290	82	2032	
BJ56	159	3114/290	82	2032	
BJ5A	159	3114/290	82	2032	
BJ5B	159	3114/290	82	2032	
BJ6	159	3118	82	2032	
BJ6A	159	3118	82	2032	
BJ6B	159	3118	82	2032	
BJ6C	159	3118	82	2032	
BJ6D	159	3118	82	2032	
BJ7	159	3118	82	2032	
BJ7A	159	3118	82	2032	
BJ7B	159	3118	82	2032	
BJ7C	159	3118	82	2032	
BJ7D	159	3118	82	2032	
BJ8	159	3118	82	2032	
BJ8A	159	3118	82	2032	
BJ8B	159	3118	82	2032	
BJ8C	159	3118	82	2032	
BJ8D	159	3118	82	2032	
BJ9	159	3118	82	2032	
BJ9A	159	3118	82	2032	
BJ9B	159	3118	82	2032	
BJ9C	159	3118	82	2032	
BJ9D	159	3118	82	2032	
BLW60	325				MRF450A
BLW64					MRF208
BLW75					MRF226
BLX13					MRF426A
BLX14					MRF464A
BLX15					MRF428A
BLX65					MRF629
BLX66					MRF616
BLX67	362				2N5944
BLX68	363				2N5945
BLX69A					2N6136
BLX88	311		261		
BLX89	278		261		
BLX91					MRF313A
BLX92					MRF5174
BLX93					MRF321
BLX94A					MRF5177A
BLX95					MRF326
BLX96					MRF321
BLX97					MRF321
BLX98					MRF323
BLY10	130	3027	14	2041	
BLY11	130	3027	14	2041	
BLY12	328	3027/130	14	2041	
BLY15	236	3027/130	14	2041	
BLY15A	175		246	2020	
BLY20	186	3192	28	2017	
BLY21	152	3893	66	2048	
BLY27	123AP		63	2030	
BLY28	192		63	2030	
BLY3			28	2017	
BLY33	278		261		
BLY34	278		261		
BLY35	154	3893/152	66	2048	
BLY36	152	3893	66	2048	
BLY37	186	3192	28	2017	
BLY38	186	3192	28	2017	
BLY47	280	3027/130	14	2041	
BLY47A	175	3131A/369	246	2020	
BLY48	280	3027/130	14	2041	
BLY48A	175	3131A/369	246	2020	
BLY49	162	3438	35		
BLY49A	175	3131A/369	246	2020	
BLY50	162	3438	35		
BLY53	186	3192	28	2017	
BLY53A					2N5946
BLY57	475				2N3926
BLY58	476				2N3927
BLY59					2N3375
BLY60					2N3632
BLY61		3024	243	2030	
BLY62	186	3192	28	2017	
BLY63	152	3893	66	2048	
BLY78	186	3192	28	2017	
BLY79			66	2048	
BLY87A					MRF212
BLY88	152	3893	66	2048	
BLY88A	350				2N6081
BLY89	152	3893	66	2048	
BLY89A	351				2N6082
BLY90					MRF243
BLY91	186	3192	28	2017	
BLY91A	357				2N5641
BLY92			66	2048	
BLY92A	360				2N5643
BLY93	359		63	2030	
BLY93A	360				2N5643
BLY99	278		261		
BM100-28					MRF317
BM15-12					MRF215
BM30-12	477				MRF216
BM45-12					MRF243
BM70-12	352				MRF245
BM80-12	352				MRF245
BM80-28					MRF316
BMT1991	159	3114/290	82	2032	
BMT2303	159	3114/290	82	2032	
BMT2411	159	3114/290	82	2032	
BMT2412	159	3114/290	82	2032	
BN-66	123A	3124/289	20	2051	
BN7133	130	3027	14	2041	
BN7168	175	3131A/369	246	2020	
BN7214	130	3027	14	2041	
BN7253	154	3045/225	40	2012	
BN7517	123A	3444	20	2051	
BN7518	123A	3444	20	2051	
BN7551	142A		ZD-12	563	
BO-71	107		11	2015	
BO8875/2	159		82	2032	
BO94	140A		ZD-10	562	
BP101					MRD3050
BP102					MRD3050
BP5263	160		245	2004	
BP67	123A	3444	20	2051	
BPS-S-50	519		514	1122	
BPW14					MRD300
BPW16					MRD160
BPW17					MRD160
BPW24					L14H1
BPW30	3036				MRD360
BPX25					MRD300
BPX25A	3036				MRD370
BPX29					MRD310
BPX29A					MRD370
BPX37					MRD300
BPX38	3031				MRD3055
BPX43	3032				MRD300
BPX58					MRD300
BPX59	3036				MRD360
BPX70,C,D,E					MRD450
BPX72,C,D,E					MRD450
BPX81					MRD160
BPY62	3032				MRD3050
BQ-94	123A	3124/289	210	2051	
BQ67	123A	3444	20	2051	
BR-1	166	9075	BR-600	1152	
BR-108	506		511	1114	
BR-55	614	3327	90		
BR-66	123A	3124/289	20	2051	
BR-67	123A	3122			
BR-82	159	3118	82	2032	
BR-832	159	3025/129	82	2032	
BR100B	186	3192	28	2017	
BR101B	152	3054/196	66	2048	
BR103	5400	3950			
BR22					MDA920A4
BR24					MDA920A6
BR26					MDA920A7
BR410					MDA920A9
BR42	116	3016	504A	1104	MDA920A4
BR44	116	3016	504A	1104	
BR46	116	3017B/117	504A	1104	MDA920A7
BR47	116	3017B/117	504A	1104	
BR48	116	3017B/117	504A	1104	MDA920A8
BR505					MDA920A2
BR51					MDA920A3
BR510					MDA920A9
BR51400-1	116	3311	504A	1104	
BR51401-2	116	3311	504A	1104	
BR52	116	3016	504A	1104	MDA920A4
BR54					MDA920A6
BR56					MDA920A7
BR58					MDA920A8
BR67	123A	3122	20	2051	
BR81D	166	9075			
BR82D	167	3647			
BR84D	168	3648			
BR86D	169	3678			
BR88D	170	3649			
BRC-116	130		14	2041	
BRC-5496	196		241	2020	
BRC5296	152	3893	66	2048	
BRC6109	197		250	2027	
BRN-SPEC-24-12	519		514	1122	
BRX44	5400	3950			
BRX45	5401	3638/5402			
BRX46	5402	3638			
BRX47	5404	3627			
BRX48	5405	3951			
BRX49	5405	3951			
BRY55-100	5402	3638			
BRY55-200	5404	3627			
BRY55-30	5400	3950			
BRY55-300	5405	3951			
BRY55-60	5401	3638/5402			
BS-1	167	3647	504A	1172	
BS-66	123A		20	2051	
BS-94	123A	3124/289	210	2051	
BS-B1				1152	
BS-B2				1172	
BS-B4				1173	
BS1	116		504A	1104	
BS10-01A	5673		SC245B		
BS10-02A	5673		SC245B		
BS10-03A	5675		SC245D		
BS10-04A	5675		SC245D		
BS10-05A	5676		SC245E		
BS10-06A	5677		SC245M		
BS2	116		504A	1104	
BS475	123A	3444	20	2051	
BS6-01A	5673		SC240B		
BS6-02A	5673		SC240B		
BS6-03A	5675		SC240D		
BS6-04A	5675		SC240D		

Industry Standard No.	ECG	SK	GE	RS 276-	MOTOR.
BS6-05A	5676		SC240E		
BS6-06A	5677		SC240M		
BS67	123A	3444	20	2051	
BS7-02A	56004	3658	SC141B		
BS7-04A	56006	3659	SC141D		
BS7-05A	56008	3660	SC141E		
BS7-06A	56008	3660			
BS8-01A	5673		SC245B		
BS8-02A	5673		SC245B		
BS8-03A	5675		SC245B		
BS8-04A	5675		SC245D		
BS8-05A	5676		SC245E		
BS8-06A	5677		SC245M		
BS9-02A	56004	3658	SC143B		
BS9-04A	56006		SC143D		
BS9-05A	56008	3660	SC143E		
BS9-06A	56008	3660			
BS9011G	123AP	3444/123A	20	2051	
BSA01	177	3100/519	300	1122	
BSA02	177	3100/519	300	1122	
BSA11	177	3100/519	300	1122	
BSS10	123A	3444	20	2051	
BSS13	278		261		
BSS14	282	3748	264		
BSS15	282	3748	264		
BSS16	282	3748	264		
BSS17	129		244	2027	
BSS18	129		244	2027	
BSS19	171	3201	27		
BSS20	171	3201	27		
BSS21	123A	3444	20	2051	
BSS23	192	3137	63	2030	
BSS26	123A	3452/108	86	2038	
BSS28	128		261		
BSS29	128		261		
BSS33	154		32		
BSS34					MPSA06
BSS35					2N5550
BSS38		3244	220		BCX27
BSS40	128		47	2030	
BSS41	128		47	2030	
BSS42	282	3748	264		
BSS43	282	3748	264		
BSS44	129	3748/282	264		
BSTC0113	5483	3942			
BSTC0126	5485	3943			
BSTC0140	5487	3944			
BSTC0313	5524	6624			
BSTC0326	5527	6627			
BSTC0353	5531	6631			
BSTC062686T	231	3857			
BSTC0626T	230	3042			
BSTC063386T	231	3042/230			
BSTC063389T	231	3857			
BSTC0640T	230	3042			
BSTCC012686T	310	3856			
BSTCC0126T	310	3856			
BSTCC013386T	310	3856			
BSTCC0133T	310	3856			
BSTCC014086T	310	3856			
BSTCC0140T	310	3856			
BSTCC014686T	310	3856			
BSTCC0146T	310	3856			
BSTD0353	5531	6631			
BSV15	129	3025	244	2027	
BSV16	129	3025	244	2027	
BSV21	159	3114/290	82	2032	
BSV21A	159	3114/290	82	2032	
BSV22					2N4416
BSV35	108	3452	86	2038	
BSV35A	123A	3444	20	2051	
BSV35B	108	3452	86	2038	
BSV35C	108	3452	86	2038	
BSV35D	108	3452	86	2038	
BSV40	123A	3444	20	2051	
BSV41	123A	3444	20	2051	
BSV43A	159	3025/129	82	2032	
BSV43B	193	3138	67	2023	
BSV44A	159	3025/129	82	2032	
BSV44B	193	3138	67	2023	
BSV45A	159	3025/129	82	2032	
BSV45B	193	3138	67	2023	
BSV47A	159	3025/129	82	2032	
BSV47B	193	3138	67	2023	
BSV48A	159	3025/129	82	2032	
BSV48B	193	3138	67	2023	
BSV49A	159	3025/129	82	2032	
BSV49B	193	3138	67	2023	
BSV51	128	3024	243	2030	
BSV52	108	3452	86	2038	
BSV52R	108	3452	86	2038	
BSV53	123A	3444	20	2051	
BSV53P	107	3039/316	11	2015	
BSV54	123A	3444	20	2051	
BSV54P	107	3039/316	11	2015	
BSV55A	159	3118	82	2032	
BSV55AP	159	3118	82	2032	
BSV55P	159	3118	82	2032	
BSV59	123A	3444	20	2051	
BSV60	282	3054/196	28	2017	
BSV65FA			20	2051	
BSV65FB			20	2051	
BSV68	288		221		
BSV69	128	3137/192A	63	2030	
BSV82	129		244	2027	
BSV83	129		244	2027	
BSV84	128	3444/123A	20	2051	
BSV85			47	2030	
BSV86	123AP	3444/123A	20	2051	
BSV87	123AP	3444/123A	20	2051	
BSV88	123A	3444	20	2051	
BSV89	123A	3444	20	2051	
BSV90	123A	3444	20	2051	
BSV91	123A	3444	20	2051	
BSV95	278		261		
BSV96	159	3025/129	82	2032	
BSV97	159	3025/129	82	2032	
BSV98	159	3025/129	82	2032	
BSW-21A	159	3114/290	82	2032	
BSW-22	159	3114/290	82	2032	
BSW-22A	159	3114/290	82	2032	
BSW-24	159	3114/290	82	2032	
BSW-44	159	3114/290	82	2032	
BSW-44A	159	3114/290	82	2032	
BSW-45	159	3114/290	82	2032	
BSW-45A	159	3114/290	82	2032	
BSW-68	396	3245/199	251		
BSW-72	159	3114/290	82	2032	
BSW-73	159	3114/290	82	2032	
BSW-74	159	3114/290	82	2032	
BSW-75	159	3114/290	82	2032	
BSW10	128	3024	243	2030	
BSW11	123A	3444	20	2051	
BSW12	123A	3444	20	2051	
BSW19	123A	3444	20	2051	
BSW21	159	3114/290	82	2032	
BSW21A	159	3114/290	82	2032	
BSW22	159	3114/290	82	2032	
BSW22A	159	3114/290	82	2032	
BSW23	129	3025	244	2027	
BSW24	159	3114/290	82	2032	
BSW26	128	3024	243	2030	
BSW27	128	3024	243	2030	
BSW28	128	3024	243	2030	
BSW29	128	3024	243	2030	
BSW32	289A	3045/225	40	2012	MPS8099
BSW33	123A	3444	20	2051	
BSW34	123A	3444	20	2051	
BSW35	128	3024	243	2030	
BSW39	123A	3444	20	2051	
BSW40	129		244	2027	
BSW41	123A	3444	20	2051	
BSW42	123A	3444	20	2051	BC238B
BSW42A	123A	3444	20	2051	
BSW43	123A	3444	20	2051	
BSW43A	123A	3444	20	2051	
BSW44	159	3114/290	82	2032	BC308A
BSW44A	159	3114/290	82	2032	
BSW45	159	3114/290	82	2032	
BSW45A	159	3114/290	82	2032	
BSW49	128		47	2030	
BSW51	123A	3444	20	2051	
BSW52	123A	3444	20	2051	
BSW53	123A	3444	20	2051	
BSW58	123AP	3444/123A	20	2051	
BSW65	128	3024	243	2030	
BSW66	128	3024	243	2030	
BSW67	282	3104A	217		
BSW68	282	3104A	217		
BSW69	396	3044/154	40	2012	
BSW70	154	3045/225	82	2012	
BSW72	159	3114/290	82	2032	
BSW73	159	3114/290	82	2032	
BSW74	159	3114/290	82	2032	
BSW75	159	3114/290	82	2032	
BSW82	123A	3444	20	2051	
BSW83	123A	3444	20	2051	
BSW84	123A	3444	20	2051	
BSW85	123A	3444	20	2051	
BSW88	123A	3444	20	2051	
BSW88A			210	2051	
BSW88B			210	2051	
BSW89	123A	3444	20	2051	
BSW89A			210	2051	
BSW89B			210	2051	
BSW92	123A	3444	20	2051	BC238
BSX12	108	3452	86	2038	
BSX12A	128	3137/192A	63	2030	
BSX19	123A	3444	20	2051	
BSX20	311	3444/123A	20	2051	
BSX21	194	3045/225	40	2012	
BSX22	195A	3024/128	243	2030	
BSX23	128	3024	243	2030	
BSX24	123A	3444	20	2051	
BSX25	123A	3444	20	2051	
BSX26	108	3452	86	2038	
BSX27	108	3452	86	2038	
BSX28	311	3452/108	86	2038	
BSX29	106	3984	21	2034	
BSX30	123A	3444	20	2051	
BSX32	311	3048/329	46		
BSX33	128	3024	243	2030	
BSX35	108	3452	86	2038	
BSX36	159	3114/290	82	2032	
BSX38	123A	3444	20	2051	
BSX38A	123A		210	2051	
BSX38B	123A		210	2051	
BSX39	311	3444/123A	20	2051	
BSX40	129	3025	244	2027	
BSX41	129	3025	244	2027	
BSX44	123A	3444	20	2051	
BSX45	128	3024	243	2030	
BSX45-10	128	3137/192A	63	2030	
BSX45-16	128	3137/192A	63	2030	
BSX45-6	128	3137/192A	63	2030	
BSX46	128	3024	243	2030	
BSX46-10	128	3137/192A	63	2030	
BSX46-16	128	3137/192A	63	2030	
BSX46-6	128	3137/192A	63	2030	
BSX48	123A	3444	20	2051	
BSX49	123A	3444	20	2051	
BSX51	123A	3444	20	2051	
BSX51A	123A	3444	20	2051	
BSX52	123A	3444	20	2051	
BSX52A	123A	3444	20	2051	
BSX53	123A	3444	20	2051	
BSX54	123A	3444	20	2051	
BSX59	128	3048/329	46		
BSX60	128	3024	243	2030	
BSX61	128	3024	243	2030	
BSX62	128	3024	243	2030	
BSX62-16	128	3748/282			
BSX62B	128	3024	243	2030	
BSX62C	128	3024	243	2030	
BSX62D	128	3024	243	2030	
BSX63	282	3024/128	243	2030	
BSX63-10	282	3748			
BSX63-16	282	3748			
BSX63B	128	3024	243	2030	
BSX63C	128	3024	243	2030	
BSX66	123A	3444	20	2051	
BSX67	123A	3444	20	2051	
BSX68	123A	3444	20	2051	
BSX69	123A	3444	20	2051	
BSX70	128	3024	243	2030	
BSX71	128	3024	243	2030	
BSX72	128	3024	243	2030	
BSX75	123A	3444	20	2051	
BSX76	123A	3444	20	2051	
BSX77	123A	3444	20	2051	
BSX78	123A	3444	20	2051	
BSX79	123A	3444	20	2051	
BSX79A			210	2051	
BSX79B			210	2051	
BSX80	123A	3444	20	2051	
BSX81	123A	3444	20	2051	
BSX81A			210	2051	
BSX81B			210	2051	
BSX87	123A	3444	20	2051	
BSX87A	311	3452/108	86	2038	
BSX88	123A	3452/108	86	2038	
BSX88A	311	3452/108	86	2038	
BSX89	123A	3444	20	2051	
BSX90	123A	3444	20	2051	
BSX91	123A	3444	20	2051	
BSX92	108	3452	86	2038	
BSX93	108	3452	86	2038	
BSX94A	123A	3444	20	2051	

Industry Standard No.	ECG	SK	GE	RS 276-	MOTOR.
BSX95	128	3024	243	2030	
BSX96	128	3024	243	2030	
BSX97	123A	3444	20	2051	
BSY-40	159	3114/290	82	2032	
BSY-41	159	3114/290	82	2032	
BSY-62	108	3019	86	2038	
BSY-72	108	3019	86	2038	
BSY-73	108	3019	86	2038	
BSY-74	108	3019	86	2038	
BSY-80	108	3019	86	2038	
BSY-95	108	3019	86	2038	
BSY10	123A	3124/289	20	2051	
BSY11	123A	3124/289	20	2051	
BSY165	123A	3444	20	2051	
BSY168	123A	3444	20	2051	
BSY17	123A	3444	20	2051	
BSY18	123A	3444	20	2051	
BSY19	123A	3444	20	2051	
BSY20	123A	3444	20	2051	
BSY21	123A	3444	20	2051	
BSY22	108	3452	86	2038	
BSY23	108	3452	86	2038	
BSY24	123A		20	2051	
BSY25	123A	3024/128	243	2030	
BSY26	123A	3444	20	2051	
BSY27	123A	3444	20	2051	
BSY28	123A	3444	20	2051	
BSY29	123A	3444	20	2051	
BSY34	123A	3444	20	2051	
BSY38	123A	3444	20	2051	
BSY39	123A	3444	20	2051	
BSY40	159	3114/290	82	2024	
BSY41	159	3114/290	82	2024	
BSY44	128	3024	243	2030	
BSY45	128	3024	243	2030	
BSY46	128	3024	243	2030	
BSY48	123A	3444	20	2051	
BSY49	123A	3444	20	2051	
BSY51	128	3024	243	2030	
BSY52	128	3024	243	2030	
BSY53	128	3024	243	2030	
BSY54	128	3024	243	2030	
BSY55	128	3024	243	2030	
BSY56	128	3024	243	2030	
BSY58	123A	3444	20	2051	
BSY59	123A	3444	20	2051	
BSY61	123A	3444	20	2051	
BSY62	123A	3444	20	2051	
BSY62A	123A	3444	20	2051	
BSY62B	192	3137	63	2030	
BSY63	123A	3444	20	2051	
BSY68		3244	220		
BSY70	123A	3452/108	86	2038	
BSY71	128	3024	243	2030	
BSY72	123A	3452/108	86	2038	
BSY73	123A	3444	20	2051	
BSY74	123A	3444	20	2051	
BSY75	123A	3444	20	2051	
BSY76	123A	3444	20	2051	
BSY77	128	3024	243	2030	
BSY78	128	3024	243	2030	
BSY79	194	3044/154	243	2030	
BSY80	123A	3444	20	2051	
BSY81	128	3024	243	2030	
BSY82	128	3024	243	2030	
BSY83	128	3024	243	2030	
BSY84	128	3024	243	2030	
BSY85	128	3024	243	2030	
BSY86	128	3024	243	2030	
BSY87	128	3024	243	2030	
BSY88	128	3024	243	2030	
BSY89	123A	3444	20	2051	
BSY90	128	3024	243	2030	
BSY91	123A	3024/128	243	2030	
BSY92	128	3024	243	2030	
BSY93	123A	3444	20	2051	
BSY95	123A	3444	20	2051	
BSY95A	123A	3444	20	2051	
BT-82	159	3114/290			
BT-94	123A	3124/289	214	2051	
BT100A-300R	5456			1020	
BT101-300R	5484	3943/5485			
BT101-500R	5486	3944/5487		1020	
BT102-300R	5484	3943/5485			
BT102-500R	5486	3944/5487			
BT121	230	3042			
BT32	6408	9084A			
BT67	123A	3444	20	2051	
BT71	123A	3444	20	2051	
BT82			82	2032	
BT832	159	3025/129	82	2032	
BT929			39	2015	
BT930		3452	86	2038	
BTB1000	156	3033	512		
BTD0105	5640		SC136A		
BTD0110	5640		SC136A		
BTD0120	5641		SC136B		
BTD0140	5642		SC136D		
BTM-50	5800	9003		1142	
BTM1000	156	3033	512		
BTM50	116		504A	1104	
BTU0340	5677	3520			
BTU0510	5673	3520/5677			
BTU0520	5673	3520/5677			
BTU0530	5675	3520/5677			
BTU0535	5677	3520			
BTU0540	5675	3508			
BTU0550	5676	3520/5677			
BTU0560	5677	3520			
BTU0610	5682	3521/5686			
BTU0620	5683	3521/5686			
BTU0630	5684	3521/5686			
BTU0640	5685	3521/5686			
BTU0650	5686	3521			
BTU0660	5697	3522	SC240M		
BTU3605	5697	3522			
BTU650	5686	3521			
BTW20-400	5685	3521/5686			
BTW20-500	5686	3521			
BTW20-600	5697	3522			
BTW20400	5685	3521/5686			
BTW20500	5686	3521			
BTW20600	5697	3522			
BTX-068	199	3245	62	2010	
BTX-070	123A		20	2051	
BTX-071	129		244	2027	
BTX-084	234 5	3247	65	2050	
BTX-094	123A	3444	20	2051	
BTX-095	123A	3444	20	2051	
BTX-096	123A	3444	20	2051	
BTX-097	129	3025	244	2027	
BTX-2367B	123A	3444	20	2051	
BTX0605	56022		SC240B2		
BTX0610	56022		SC240B2		

Industry Standard No.	ECG	SK	GE	RS 276-	MOTOR.
BTX0620	56022		SC240B2		
BTX0640	56024		SC240D2		
BTX0660	56026		SC240M2		
BTX068	123A	3444	20	2051	
BTX070	102A	3444/123A	53	2007	
BTX071	103A	3835	59	2002	
BTX2367B	123A		10	2051	
BTX2505	56022		SC260B2		
BTX2510	56022		SC260B2		
BTX2520	56022		SC260B2		
BTX2540	56024		SC260D2		
BTX2560	56026		SC260M2		
BTX31-100R	5491	6791			
BTX31-200R	5492	6792			
BTX31-300R	5494	6794			
BTX31-400R	5494	6794			
BTX31-500R	5496	6796			
BTX31-50R	5491	6791			
BTX31-600R	5496	6796			
BTX33-100R	5542	6642			
BTX33-50R	5541	6642/5542			
BTX68-500R	5496	6796			
BTX68-600R	5496	6796			
BTX81-100R	5542	6642			
BTX94-200	5693	3652			
BTX94-400	5695	3509			
BTX94-600	5697	3522			
BTY87-600R	5529	6629			
BTY87-700R	5530	6631/5531			
BTY87-800R	5531	6631			
BTY91-100R	5522	6622			
BTY91-200R	5524	6624			
BTY91-300R	5526	6627/5527			
BTY91-400R	5527	6627			
BTY91-500R	5528	6629/5529			
BTY91-600R	5529	6629			
BTY91-700R	5530	6631/5531			
BTY91-800R	5531	6631			
BU029	117(4)		504A	1104	
BU105	165	3115	38		BU205
BU106	283	3438	35		
BU107	162	3438	35		
BU108	165	3115	38		BU208
BU109	163A	3439	36		
BU110	283	3438	35		
BU111	283	3439/163A	36		
BU115	165	3115	38		
BU120	283	3438	35		
BU126	283				MJ3030
BU180					MJE5741
BU180A					MJE5742
BU204		3115	237		BU204
BU205	165	3710/238	237		BU205
BU207	165	3111			BU207
BU207A	163A	3111			
BU208	283	3115/165	38		BU208
BU310	283	3945/327			
BU311	283	3945/327			
BU312	283	3945/327			
BU326	283	3710/238			
BU326A	283	3710/238			
BU67	123A	3444	20	2051	
BU71	123A	3444	20	2051	
BUC 97704-2	123A	3444		2051	
BUC97704-2			20	2051	
BUX10	327	3945			
BUX11	327	3945			
BUX12	327	3945			
BUX39	327	3621			
BUX40	327	3621			
BUX80					2N6547
BUX81					MJ13335
BUX82					2N6545
BUX83					2N6545
BUX84					MJE13005
BUX85					MJE13005
BUX86					MJE13003
BUX87					MJE13003
BUY10	130	3027	14	2041	
BUY11	130	3027	14	2041	
BUY20	162	3438	35		
BUY21	162	3438	35		
BUY21A	162	3438	35		
BUY22	162	3438	35		
BUY24	152	3054/196	66	2048	
BUY35	283	3438	35		
BUY38	384	3538	246	2020	
BUY43	181	3535	75	2041	
BUY44	280	3438	35		
BUY46	181	3535	75	2041	
BUY51A	327	3945			
BUY52A	327	3945			
BUY53A	327	3945			
BUY54A.	327	3945			
BUY69B	283	3467	36		
BUY69C	283	3467	36		
BV25	116	3311	504A	1104	
BV67	123A	3444	20	2051	
BV71	123A	3444	20	2051	
BW67	123A	3444	20	2051	
BW71	123A	3444	20	2051	
BWS66	198	3220	251		
BWS67	198	3220	251		
BX-324	158	3004/102A	53	2007	
BX-495	158(2)	3004/102A	53		
BX090	139A	3060	ZD-9.1	562	
BX67	123A	3444	20	2051	
BX71	123A	3444	20	2051	
BX909	139A	3060	ZD-9.1	562	
BXY10	125	3033	510,531	1114	
BXY63	139A	3060	ZD-9.1	562	
BY100	125	3032A	510,531	1114	
BY1001	125	3033	510,531	1114	
BY1002	125	3033	510,531	1114	
BY1008	125	3032A	510,531	1114	
BY101	116	3016	504A	1104	
BY102	116	3017B/117	504A	1104	1N4003
BY103	506	3032A	511	1114	
BY104	125	3033	510,531	1114	
BY105	125	3051/156	510,531	1114	
BY106	116	3017B/117	504A	1104	
BY107	116	3017B/117	504A	1104	
BY108	125	3032A	510,531	1114	
BY109	125	3032A	510,531	1114	
BY1101	125	3033	510,531	1114	
BY1102	125	3033	510,531	1114	
BY111	116	3017B/117	504A	1104	1N4001
BY112	116	3016	504A	1104	1N4004
BY113	116	3016	504A	1104	1N4003
BY114	116		504A	1104	
BY115	116	3017B/117	504A	1104	
BY116	116	3017B/117	504A	1104	1N4004
BY117	116	3017B/117	504A	1104	
BY118	125	3032A	510,531	1114	

Industry Standard No.	ECG	SK	GE	RS 276-	MOTOR.
BY119	125	3032A	510,531	1114	
BY12 00	125			1114	
BY1200			510,531	1114	
BY1201	125	3033	510,531	1114	
BY1202	125	3033	510,531	1114	
BY121	116	3017B/117	504A	1104	1N4001
BY122	166	3016	504A	1152	
BY123	167	3016	504A	1172	1N4003
BY124	116	3017B/117	504A	1104	1N4004
BY125	116	3017B/117	504A	1104	1N4004
BY126	116	3311	504A	1104	1N4006
BY126/100	156		512		
BY126/200	156		512		
BY126/300	156		512		
BY126/400	156		512		
BY126/50	156		512		
BY127	506	3017B/117	504A	1114	MR1-1400
BY127/500	125		510,531	1114	
BY127/600	125		510,531	1114	
BY127/700	125		510,531	1114	
BY128	125	3033	510,531	1114	1N4007
BY129	125	3032A	510,531	1114	
BY130	116	3017B/117	504A	1104	
BY133	506		511	1114	
BY134	116		504A	1104	
BY135	116		504A	1104	
BY141	116	3311	504A	1104	1N4001
BY153	116	3016	504A	1104	
BY156	156		512		
BY157	506		511	1114	
BY158	506		511	1114	
BY164	166	9075	504A	1152	
BY172	156	3051	512		
BY173	156	3051	512		
BY179	168	3648		1173	
BY18					1N3882
BY184					MR1-1600
BY2001	5848	7048	5012		
BY2002	5848	7048	5012		
BY201	5804		5004		
BY201-2	5834		5004		
BY202	5834		5004		
BY203	5834		5004		
BY204	5806		5004		
BY205	5838		5004		
BY206	5804	3500/5882	5008		
BY207	5806	7042/5842	5008		
BY208	5808	3516/5846	5012		
BY209	5846	3516	5012		
BY2101	5848	7048	5012		
BY2102	5848	7048	5012		
BY211	5830		5004		
BY212	5832		5004		
BY213	5834		5004		
BY214	5812		5004		
BY215	5838		5004		
BY216	5840	3500/5882	5012		
BY217	5806	7042/5842	5008		
BY218	5846	3516			
BY219	5846	3516	5012		
BY2201	5848	7048	5012		
BY2202	5848	7048	5012		
BY221	5830		5004		
BY222	5832		5004		
BY223	5834		5004		
BY224	5817		5004		
BY225	5838		5004		
BY226	5805	3500/5882	5008		
BY227	5842	7042	5008		
BY228	5846	3516	5012		
BY229	5846	3516	5012		
BY3001	5848	7048	5012		
BY3002	5848	7048	5012		
BY301	5830		5004		
BY302	5832		5004		
BY303	5834		5004		
BY304	5856		5004		
BY305	5838		5004		
BY306	5840	3584/5862	5008		
BY307	5842	7042	5008		
BY308	5846	3516	5012		
BY309	5846	3516	5012		
BY3101	5848	7048	5012		
BY3102	5848	7048	5012		
BY311	5830		5004		
BY312	5832		5004		
BY313	5834		5004		
BY314	5836		5004		
BY315	5838		5004		
BY316	5840	3584/5862	5008		
BY317	5842	7042	5008		
BY318	5846	3516	5012		
BY319	5846	3516	5012		
BY3201	5848	7048	5012		
BY3202	5848	3516/5846	5012		
BY321	5830		5004		
BY322	5832		5004		
BY323	5834		5004		
BY324	5836		5004		
BY325	5838		5004		
BY326	5840	3584/5862	5008		
BY327	5842	7042	5008		
BY328	5846	3516	5012		
BY329	5846	3516	5012		
BY4001	5868	7068	5028		
BY4002	5868	7068	5028		
BY401	116		5016		
BY402	116		5016		
BY403	116		5016		
BY404	116		5020		
BY405	116		5020		
BY406	552		5024		
BY407	552		5024		
BY408	5866		5028		
BY409	5866		5028		
BY4101	5868	7068	5028		
BY4102	5868	7068	5028		
BY411	5850		5016		
BY412	5852		5016		
BY413	5854		5016		
BY414	5856		5020		
BY415	5858		5020		
BY416	5860		5024		
BY417	5862		5024		
BY418	5866		5028		
BY419	5866		5028		
BY4201	5868	7068	5028		
BY4202	5868	7068	5028		
BY421	5850		5016		
BY422	5852		5016		
BY423	5854		5016		
BY424	5856		5020		
BY425	5858		5020		
BY426	5860		5024		
BY427	5862		5024		
BY428	5866		5028		
BY429	5866		5028		
BY5001	5890	7090	5044		
BY5002	5890	7090	5044		
BY501	5870		5028		
BY502	5872		5032		
BY503	5874		5032		
BY504	5876		5036		
BY505	5878		5036		
BY506	5880		5040		
BY507	5882		5040		
BY508	5886		5044		
BY509	5886		5044		
BY5101	5890	7090	5044		
BY5102	5890	7090	5044		
BY511	5870		5032		
BY512	5872		5032		
BY513	5874		5032		
BY514	5876		5036		
BY515	5878		5036		
BY516	5880		5040		
BY517	5882		5040		
BY518	5886		5044		
BY519	5886		5044		
BY5201	5890	7090	5044		
BY5202	5890	7090	5044		
BY521	5870		5032		
BY522	5872		5032		
BY523	5874		5032		
BY524	5876		5036		
BY525	5878		5036		
BY526	5880		5040		
BY527	5882		5040		
BY528	5886		5044		
BY529	5886		5044		
BY67	123A	3444	20	2051	
BY7001	5868	7068	5028		
BY7002	5868	7068	5028		
BY701	5850		5016		
BY702	5852		5016		
BY703	5854		5016		
BY704	5856		5020		
BY705	5858		5020		
BY706	5860	3500/5882	5024		
BY707	5862	3500/5882	5024		
BY708	5866		5028		
BY709	5866		5028		
BY71	123A	3444	20	2051	
BY7101	5868	7068	5028		
BY7102	5868	7068	5028		
BY711	5850		5016		
BY712	5852		5016		
BY713	5854		5016		
BY714	5856		5020		
BY715	5858		5020		
BY716	5860	3500/5882	5024		
BY717	5862	3500/5882	5024		
BY718	5866		5028		
BY719	5866		5028		
BY7201	5868	7068	5028		
BY7202	5868	7068	5028		
BY721	5850		5016		
BY722	5852		5016		
BY723	5854		5016		
BY724	5856		5020		
BY725	5858		5020		
BY726	5860	3500/5882	5024		
BY727	5862	3500/5882	5024		
BY728	5866		5028		
BY729	5866		5028		
BY8001	5890	7090	5044		
BY8002	5890	7090	5044		
BY801	5870		5032		
BY802	5872		5032		
BY803	5874		5032		
BY805	5878		5036		
BY806	5880	3500/5882	5040		
BY807	5882	3500	5040		
BY808	5886		5044		
BY809	5886		5044		
BY8101	5890	7090	5044		
BY8102	5890	7090	5044		
BY811	5870		5032		
BY812	5872		5032		
BY813	5874		5032		
BY814	5876		5036		
BY815	5878		5036		
BY816	5880	3500/5882	5040		
BY817	5882	3500	5040		
BY818	5886		5044		
BY819	5886		5044		
BY8201	5890	7090	5044		
BY8202	5890	7090	5044		
BY821	5870		5032		
BY822	5872		5032		
BY823	5874		5032		
BY824	5876		5036		
BY825	5878		5036		
BY826	5880	3500/5882	5040		
BY827	5882	3500	5040		
BY828	5886		5044		
BY829	5886		5044		
BYW29-100					MR2401F
BYW29-150					MR2402F
BYW29-50					MR2400F
BYW30-100					SPECIAL
BYW30-150					SPECIAL
BYW30-50					SPECIAL
BYW31-100					SPECIAL
BYW31-150					SPECIAL
BYW31-50					SPECIAL
BYX10	125	3032A	510,531	1114	MR1-1600
BYX12/400	125		510,531	1114	
BYX13/600	125		510,531	1114	
BYX15	5999		5109		
BYX16	5998		5108		
BYX20200R					1N3493R
BYX21100					1N3492
BYX21200					1N3493
BYX21200R					1N3493R
BYX216400					1N3495
BYX21L100					1N3492
BYX21L200					1N3493
BYX21L400R					1N3495R
BYX22-1200	506		511	1114	
BYX22/200	116	3017B/117	504A	1104	
BYX22/400	116	3031A	504A	1104	
BYX22/600	116	3017B/117	504A	1104	
BYX22/800	125	3033	510,531	1114	
BYX26-60	156	3016	512		
BYX30-200,R					1N3901
BYX30-300,R					1N3902
BYX30-400,R					1N3903

Industry Standard No.	ECG	SK	GE	RS 276-	MOTOR.
BYX30-500,R					MR1386
BYX30-600,R					MR1386
BYX30/150	5802	9005		1143	
BYX36-150	156		512		
BYX36-300	116		504A	1104	
BYX36-600	116		504A	1104	
BYX36/150	116	3016	504A	1104	
BYX36/300	116	3016	504A	1104	
BYX36/600	116	3031A	504A	1104	
BYX36150					1N4003
BYX36300					1N4003
BYX36600					1N4004
BYX38	156	3051	512		
BYX38-1200,R					MR1130
BYX38-300	5898		5068		
BYX38-300,R					MR1122
BYX38-300R	5899		5069		
BYX38-600	5924	7104/5904	5072		
BYX38-600,R					MR1126
BYX38-600R	5925		5073		
BYX38-900	5910	7110	5076		
BYX38-900,R					MR1130
BYX38-900R	5911	7111	5077		
BYX38/300	5836	3500/5882	5004		
BYX38/300R	5835		5005		
BYX38/600	5842		5008		
BYX38/600R	5839		5005		
BYX38/900	5848	7048	5012		
BYX38/900R	5843	3517/5883	5009		
BYX38900	5882	3500			
BYX39-1000	5890	7068/5868	5028		
BYX39-1000R	5891	7069/5869	5029		
BYX39-600	5882		5024		
BYX39-600R	5883		5025		
BYX39-800	5886		5028		
BYX39-800R	5887		5029		
BYX39/1000	5890		5044		
BYX39/600	5882	3500	5040		
BYX39/800	5886		5044		
BYX39600	5882	3500			
BYX40-1000	5890	7090	5044		
BYX40-1000R	5891	7091	5045		
BYX40-600	5882		5040		
BYX40-600R	5883		5041		
BYX40-800	5886		5044		
BYX40-800R	5887		5045		
BYX40600	5882	3500			
BYX42-1200,R					MR1128
BYX42-300	5876		5036		
BYX42-300,R					MR1122
BYX42-300R	5879		5037		
BYX42-600	5882		5040		
BYX42-600,R					MR1124
BYX42-600R	5883		5041		
BYX42-900	5890	7090	5044		
BYX42-900,R					MR1126
BYX42-900R	5891	7091	5045		
BYX42/300	5878	3500/5882	5036		
BYX42/600	5882	3500	5040		
BYX42/900	5890	7090	5044		
BYX42600	5882	3500			
BYX45-1000R	156		512		
BYX45-600R	156		512		
BYX45-800R	156		512		
BYX46-200	5916	7096/5896	5064		
BYX46-200R	5917		5065		
BYX46-300	5920		5068		
BYX46-300R	5921		5068		
BYX46-400	5920	7100/5900	5068		
BYX46-400R	5921		5069		
BYX46-500	5924		5072		
BYX46-500R	5925		5073		
BYX46-600	5924	7104/5904	5072		
BYX46-600R	5925		5073		
BYX48-300	5876		5036		
BYX48-300R	5859		5021		
BYX48-600	5882		5040		
BYX48-600R	5863		5025		
BYX48-900	5890	7090	5044		
BYX48-900R	5869	7069	5029		
BYX48/300	5878	3500/5882	5036		MR1124
BYX48/600	5882	3500	5040		MR1126
BYX48/900	5890		5044		MR1130
BYX48600	5882	3500			
BYX50-200	5818$		5016		
BYX50-200R	5819$		5017		
BYX50-300	5820$		5020		
BYX50-300R	5821$		5021		
BYX50-400	5858		5020		
BYX50-400R	5859		5021		
BYX50-500	5862		5024		
BYX50-500R	5863		5025		
BYX50-600	5862		5024		
BYX50-600R	5863		5025		
BYX52-300	5988	3608/5990	5100		
BYX52-300R	5989	3518/5995	5101		
BYX52-600	5994	3501	5104		
BYX52-600R	5995	3518	5105		
BYX52-900	6002	7202	5108		
BYX52-900R	6003	7203	5109		
BYX55009		3032A	509	1114	
BYX56-1000	6002	7202	5108		
BYX56-1000R	6003	7203	5109		
BYX56-600	5994	3501	5104		
BYX56-600R	5995	3518	5105		
BYX56-800	5998		5108		
BYX56-800R	5999		5109		
BYX57-500	506		511	1114	
BYX57-600	506		511	1114	
BYX58-100	177		300	1122	
BYX58-200	177		300	1122	
BYX58-50	177		300	1122	
BYX60-100	116			1104	
BYX60-200	116		504A	1104	
BYX60-300	116		504A	1104	
BYX60-400	116		504A	1104	
BYX60-50	116		504A	1104	
BYX60-500	116		504A	1104	
BYX60-600	116		504A	1104	
BYX60-700	125		510,531	1114	
BYX61-100	5818$		5064		
BYX61-300	5820$		5068		
BYX61-400	5820$	7100/5900	5068		
BYX61-50	5818$		5064		
BYX62-600	5822	7104/5904	5072		
BYX66-1000	5932	7110/5910	5076		
BYX66-400	5920	7100/5900	5068		
BYX66-600	5822	7104/5904	5072		
BYX66-800	5908		5076		
BYY-31	116	3016	504A	1104	
BYY-32	116	3016	504A	1104	
BYY-35	116	3016	504A	1104	
BYY20					1N3493R
BYY20/200					1N3493R
BYY21/200					1N3493R
BYY31	116	3051/156	504A	1104	1N4003
BYY32	116		504A	1104	1N4003
BYY33	116	3016	504A	1104	1N4004
BYY34	116	3016	504A	1104	1N4004
BYY35	116		504A	1104	
BYY36	116	3033	504A	1104	
BYY37	125	3033	510,531	1114	
BYY60	156	3051	512		
BYY61	156	3051	512		
BYY88	5874	3500/5882			
BYY89	116	3016	504A	1104	
BYY91	125	3032A	510,531	1114	
BYZ13	5944	3060/139A	5048		
BZ-050	135A		ZD-5.1		
BZ-052	136A	3057	ZD-5.1	561	
BZ-054	136A		ZD-5.6	561	
BZ-056	136A		ZD-5.6	561	
BZ-058	5070A		ZD-6.2	561	
BZ-061	137A		ZD-6.2	561	
BZ-063	137A		ZD-6.2	561	
BZ-065	5071A	3334	ZD-6.8	561	
BZ-067	5071A	3334	ZD-6.8	561	
BZ-069	5071A	3334	ZD-6.8	561	
BZ-071	138A	3059	ZD-7.5		
BZ-073	138A		ZD-7.5		
BZ-077	5072A		ZD-8.2	562	
BZ-079	5072A		ZD-8.2	562	
BZ-080	5072A	3136	ZD-9.1	562	
BZ-0800	5072A	3059/138A	ZD-8.2	562	
BZ-081	5072A	3059/138A	ZD-8.2	562	
BZ-083	5072A		ZD-8.2	562	
BZ-085	5073A	3749	ZD-9.1	562	
BZ-088	5073A	3749	ZD-9.1	562	
BZ-090	139A	3060	ZD-9.1	562	
BZ-090(DIODE)				1123	
BZ-090(ZENER)	139A	3060	ZD-9.1	562	
BZ-0900	139A	3060	ZD-9.1	562	
BZ-0901	139A	3060	ZD-9.1	562	
BZ-092	139A		ZD-9.1	562	
BZ-094	139A	3060	ZD-9.1	562	
BZ-096	140A	3060/139A	ZD-10	562	
BZ-098	140A		ZD-10	562	
BZ-100		3061	ZD-10	562	
BZ-105	5074A		ZD-11.0	563	
BZ-110	5074A		ZD-11.0	563	
BZ-115	141A		ZD-11.5		
BZ-12	142A	3062	ZD-12	563	
BZ-120	142A		ZD-12	563	
BZ-125	143A	3093	ZD-13	563	
BZ-130	143A	3750	ZD-13	563	
BZ-135	144A		ZD-14	564	
BZ-140	144A		ZD-14	564	
BZ-145	145A		ZD-15	564	
BZ-150	145A	3063	ZD-15	564	
BZ-157	5075A		ZD-16	564	
BZ-162	5015A		ZD-16	564	
BZ-167	5076A	9022	ZD-17		
BZ-172	5076A	9022	ZD-17		
BZ-177	5077A	3145	ZD-18		
BZ-182	5077A	3752	ZD-18		
BZ-187	5078A	9023	ZD-19		
BZ-192	5078A	9023	ZD-19		
BZ-197	5079A	3335	ZD-20		
BZ-210	5079A	3335	ZD-20		
BZ-220	5080A	3336	ZD-22		
BZ-230	5081A	3151	ZD-24		
BZ-240	5081A	3151	ZD-24		
BZ-250	5082A	3753	ZD-25		
BZ-260	146A	3148	ZD-27		
BZ-270	146A		ZD-27		
BZ-280	5083A	3754	ZD-28		
BZ-290	5084A	3755	ZD-30		
BZ-300	5084A	3755	ZD-30		
BZ-310	147A	3755/5084A	ZD-33		
BZ-320	147A		ZD-33		
BZ-330	147A		ZD-33		
BZ-340	5085A	3337	ZD-36		
BZ-350	5085A	3337	ZD-36		
BZ-X19	145A	3063	ZD-15	564	
BZ/88/C4V7	135A		ZD-5.1		
BZ052	136A	3056/135A	ZD-5.6	561	
BZ058		9021	ZD-5.8		
BZ066			ZD-6.6	561	
BZ071	138A	3334/5071A	ZD-7.5		
BZ075	138A		ZD-7.5		
BZ080	5072A	3059/138A	ZD-8.2	562	
BZ081	5072A	3136	ZD-8.2	562	
BZ085	5072A		ZD-8.2	562	
BZ090	139A	3060	ZD-9.1	562	
BZ090.1Z9	139A	3060	ZD-9.1	562	
BZ094	139A		ZD-9.1	562	
BZ1-9	139A	3060	ZD-9.1	562	
BZ1-90			ZD-9.1	562	
BZ100	140A	3056/135A	ZD-5.1		
BZ102-2V8	5003A	3769			
BZ102-3V4	5005A	3771	ZD-3.3		
BZ1021V4	177	3100/519	300	1122	
BZ110	5074A	3139	ZD-11	563	
BZ120	142A	3062	ZD-12	563	
BZ130	143A	3750	ZD-13	563	
BZ140	144A	3094		564	
BZ150	145A	3063	ZD-15	564	
BZ162	5075A	3751	ZD-16	564	
BZ177	5077A		ZD-18		
BZ192	5078A		ZD-19		
BZ210	5079A	3335	ZD-20		
BZ230	5080A	3336	ZD-22		
BZ240	5081A		ZD-24		
BZ260	146A		ZD-27		
BZ270	146A		ZD-27		
BZ290	5084A		ZD-30		
BZ3.6	134A	3055	ZD-3.6		
BZ3.9	5067A	3055/134A	ZD-3.6		
BZ310	5084A		ZD-30		
BZ360		3337	ZD-36		
BZ4.3	5068A	3332	ZD-4.3		
BZ4.7	5069A	3056/135A	ZD-5.1		
BZ5.1	135A	3056	ZD-5.1		
BZ5.6	136A	3057	ZD-5.6	561	
BZ6.2	137A	3057/136A	ZD-5.6	561	
BZ6.8	5071A	3334	ZD-6.8	561	
BZ67	123A	3444	20	2051	
BZ7.5	138A		ZD-7.5		
BZ71	123A	3444	20	2051	
BZ79C10	140A		ZD-10	562	
BZ8.2	5072A	3136	ZD-8.2	562	
BZ9.1	139A	3060	ZD-9.1	562	
BZS2.7	5002A	3768			
BZV17C13	143A	3750			
BZV17C16	5075A	3751			
BZV17C18	5077A	3752			
BZX-46C15		3063	ZD-15	564	
BZX-71C15	145A		ZD-15	564	

Industry Standard No.	ECG	SK	GE	RS 276-	MOTOR.
BZX-79C15		3063	ZD-15	564	
BZX10	137A	3058	ZD-6.2	561	
BZX14	139A	3066/118	ZD-9.1	562	
BZX17	142A	3062	ZD-12	563	
BZX19	145A	3063	ZD-15	564	
BZX25	146A	3064	ZD-27		
BZX25A	146A	3064	ZD-27		
BZX27	147A	3095	ZD-33		
BZX29C10	140A		ZD-10	562	
BZX29C11	5074A		ZD-11.0	563	
BZX29C12	142A		ZD-12	563	
BZX29C13	143A	3750	ZD-13	563	
BZX29C15	145A		ZD-15	564	
BZX29C16	5075A	3751	ZD-16	564	
BZX29C18	5077A	3752	ZD-18		
BZX29C20	5079A	3335	ZD-20		
BZX29C22	5080A	3336	ZD-22		
BZX29C24	5081A		ZD-24		
BZX29C27	146A		ZD-27		
BZX29C30	5084A	3755	ZD-30		
BZX29C33	147A		ZD-33		
BZX29C36	5085A	3337	ZD-36		
BZX29C39	5086A	3338	ZD-39		
BZX29C43	5087A	3339	ZD-43		
BZX29C47	5088A	3340	ZD-47		
BZX29C4V7	5069A		ZD-4.7		
BZX29C51	5089A	3341	ZD-51		
BZX29C56	5090A	3342	ZD-56		
BZX29C5V1	135A		ZD-5.1		
BZX29C5V6	136A		ZD-5.6	561	
BZX29C62	149A	3097	ZD-62		
BZX29C6V2	137A		ZD-6.2	561	
BZX29C6V8	5071A	3334	ZD-6.8	561	
BZX29C7V5	138A		ZD-7.5		
BZX29C82	150A	3098	ZD-82		
BZX29C8V2	5072A	3136	ZD-8.2	562	
BZX29C9V1	139A		ZD-9.1	562	
BZX46C10	140A	3061	ZD-10	562	
BZX46C11	5074A		ZD-11	563	
BZX46C12	142A	3062	ZD-12	563	
BZX46C13	143A	3750	ZD-12	563	
BZX46C15	145A	3063		564	
BZX46C16	5075A	3751		564	
BZX46C18	5077A		ZD-18		
BZX46C20	5079A	3335	ZD-20		
BZX46C22	5080A	3336	ZD-22		
BZX46C24	5081A		ZD-24		
BZX46C27	146A		ZD-27		
BZX46C2V7	5063A	3837			
BZX46C30	5084A	3755	ZD-30		
BZX46C33	147A		ZD-33		
BZX46C36	5085A	3337	ZD-36		
BZX46C39	5038A	3804	ZD-39		
BZX46C3V0	5065A	3838			
BZX46C3V3	5066A		ZD-3.3		
BZX46C3V6	134A		ZD-3.6		
BZX46C3V9	5067A		ZD-3.9		
BZX46C43	5087A	3339	ZD-43		
BZX46C4V3	5068A		ZD-4.3		
BZX46C4V7	5069A		ZD-4.M		
BZX46C5V1	5010A	3776	ZD-5.1		
BZX46C5V6	136A		ZD-5.6	561	
BZX46C6V2	137A		ZD-6.2	561	
BZX46C6V8	5071A		ZD-6.8	561	
BZX46C7V5	138A		ZD-7.5		
BZX46C8V2	5072A	3136	ZD-8.2	562	
BZX46C9V1	139A	3060	ZD-9.1	562	
BZX55C10	140A		ZD-10	562	
BZX55C11	5074A		ZD-11	563	
BZX55C12	142A		ZD-12	563	
BZX55C13	143A	3750	ZD-12	563	
BZX55C15	145A		ZD-14	564	
BZX55C16	5075A	3751		564	
BZX55C18	5077A	3752	ZD-18		
BZX55C20	5079A	3335	ZD-20		
BZX55C22	5079A	3336/5080A	ZD-22		
BZX55C24	5079A		ZD-24		
BZX55C27	5079A		ZD-27		
BZX55C2V7	5063A	3837			
BZX55C30	5079A		ZD-30		
BZX55C33	5079A		ZD-33		
BZX55C36	5085A	3337	ZD-36		
BZX55C39	5086A	3339/5087A	ZD-43		
BZX55C3V0	5065A	3838			
BZX55C3V3	5066A		ZD-3.3		
BZX55C3V6	5006A	3772	ZD-3.6		
BZX55C3V9	5067A		ZD-3.9		
BZX55C43	5087A		ZD-60		
BZX55C4V3	5068A		ZD-4.3		
BZX55C4V7	5069A		ZD-4.7		
BZX55C5V6	136A	3057	ZD-5.6	561	
BZX55C6V2	137A	3058	ZD-6.2	561	
BZX55C6V8	5071A		ZD-6.8	561	
BZX55C7V5	138A		ZD-7.5		
BZX55C8V2	5072A	3136	ZD-8.2	562	
BZX55C9V1	139A	3060	ZD-9.1	562	
BZX55D10	5125A	3391	5ZD-10		
BZX55D5V6	5117A	3383			
BZX55D6V8	5120A	3386	5ZD-6.8		
BZX55D8V2	5122A	3388	5ZD-8.2		
BZX58C8V2	5072A	3136			
BZX59C13	143A	3750			
BZX59C16	5075A	3751			
BZX59C18	5077A	3752			
BZX60C30	5084A	3755			
BZX60C36	5085A	3337	ZD-36		
BZX61-C75					1M75ZS5
BZX61-C7V5					1M7.5ZS5
BZX61C8V2	5122A	3136/5072A			
BZX63C8V2	5072A	3136			
BZX64C13	143A	3750			
BZX64C16	5075A	3751			
BZX64C18	5077A	3752			
BZX65C30	5084A	3755			
BZX69C8V2	5072A	3136			
BZX70-C10					5M10ZS5
BZX70-C75					5M75ZS5
BZX70C10	5125A	3391			
BZX70C11	5126A	3392	5ZD-11		
BZX70C12	5127A	3393	5ZD-12		
BZX70C18	5133A	3399	5ZD-18		
BZX70C22	5136A	3402	5ZD-22		
BZX70C24	5137A	3403	5ZD-24		
BZX70C27	5139A	3405	5ZD-27		
BZX70C30	5141A	3407	5ZD-30		
BZX70C33	5142A	3408	5ZD-33		
BZX70C36	5143A	3409			
BZX70C39	5144A	3410			
BZX70C43	5145A	3411			
BZX70C47	5146A	3412			
BZX70C51	5147A	3413			
BZX70C56	5148A	3414			
BZX70C62	5150A	3416			
BZX70C68	5151A	3417			
BZX70C75	5152A	3418			
BZX70C7V5	5121A	3387			
BZX70C8V2	5122A	3388			
BZX70C9V1	5124A	3390			
BZX71C10	5019A	3785	ZD-10	562	
BZX71C11	5020A	3786	ZD-11	563	
BZX71C12	5021A	3787	ZD-12	563	
BZX71C13	5022A	3788	ZD-12	563	
BZX71C15	5024A	3790	ZD-15	564	
BZX71C16	5025A	3791		564	
BZX71C18	5027A	3793	ZD-18		
BZX71C20	5029A	3795	ZD-20		
BZX71C22	5030A	3796	ZD-22		
BZX71C24	5031A	3797	ZD-24		
BZX71C5V1	5010A	3776	ZD-5.1		
BZX71C5V6	5011A	3777	ZD-5.6	561	
BZX71C6V2	5013A	3779	ZD-6.2	561	
BZX71C6V8	5014A	3780	ZD-6.8	561	
BZX71C7V5	5015A	3781	ZD-7.5		
BZX71C8V2	5016A	3782	ZD-8.2	562	
BZX71C9V1	139A	3060	ZD-9.1	562	
BZX71C9V2	5018A	3784	ZD-9.1	562	
BZX72	5018A	3784	ZD-9.1	562	
BZX72A	5018A	3784	ZD-9.1	562	
BZX72B	5018A	3784	ZD-9.1	562	
BZX72C	5018A	3784	ZD-9.1	562	
BZX76	143A	3750			
BZX79-C10					BZX79-C10
BZX79-C4V7					BZX79-C4V7
BZX79-C75					BZX79-C75
BZX79-C7V5		3059	ZD-7.5		
BZX79-C9V1					BZX79-C9V1
BZX79C10	140A	3061	ZD-10	562	
BZX79C11	5074A		ZD-11	563	
BZX79C12	142A	3062	ZD-12	563	
BZX79C13	143A	3750	ZD-12	563	
BZX79C15	5024A	3790	ZD-15	564	
BZX79C16	5075A	3751			
BZX79C18	5077A		ZD-18		
BZX79C20	5079A	3335	ZD-20		
BZX79C22	5080A	3336	ZD-22		
BZX79C24	5081A		ZD-24		
BZX79C27	146A	3064	ZD-27		
BZX79C27A	146A	3064	ZD-27		
BZX79C30	5084A	3755	ZD-30		
BZX79C33	147A		ZD-33		
BZX79C36	5085A	3337	ZD-36		
BZX79C39	5086A	3338	ZD-39		
BZX79C43	5087A	3339	ZD-43		
BZX79C47	5088A	3340	ZD-47		
BZX79C4V7	5069A		ZD-4.7		
BZX79C51	5089A	3341	ZD-51		
BZX79C56	5090A	3342	ZD-56		
BZX79C5V1	5010A	3776	ZD-5.1		
BZX79C5V6	136A	3057	ZD-5.6	561	
BZX79C62	149A		ZD-62		
BZX79C68	5092A	3343	ZD-68		
BZX79C6V2	137A	3058	ZD-6.2	561	
BZX79C6V8	5071A	3334	ZD-6.8	561	
BZX79C75	5046A	3812	ZD-75		
BZX79C7V5	138A		ZD-7.5		
BZX79C8V2	5072A		ZD-8.2	562	
BZX79C9V1	139A	3060	ZD-9.1	562	
BZX80C8V2	5072A	3136			
BZX81C13	143A	3750			
BZX81C16	5075A	3751			
BZX81C18	5077A	3752			
BZX82C30	5084A	3755			
BZX83C10	140A		ZD-10	562	
BZX83C11	5074A		ZD-11	563	
BZX83C12	142A		ZD-12	563	
BZX83C13	143A		ZD-12	563	
BZX83C15	145A		ZD-15	564	
BZX83C16	5075A	3751		564	
BZX83C18	5077A	3752	ZD-18		
BZX83C20	5079A	3335	ZD-20		
BZX83C22	5080A	3336	ZD-22		
BZX83C24	5081A		ZD-24		
BZX83C27	146A		ZD-27		
BZX83C2V7	5063A	3837			
BZX83C30	5084A	3755	ZD-30		
BZX83C33	147A		ZD-33		
BZX83C36	5085A	3337	ZD-36		
BZX83C39	5086A	3338	ZD-39		
BZX83C3V0	5065A	3838			
BZX83C3V3	5066A		ZD-3.3		
BZX83C3V6	5006A	3772	ZD-3.6		
BZX83C3V9	5007A		ZD-3.9		
BZX83C43	5087A	3339	ZD-43		
BZX83C4V3	5008A	3774	ZD-4.3		
BZX83C4V7	5069A		ZD-4.7		
BZX83C5V1	5010A	3776	ZD-5.1		
BZX83C5V6	136A		ZD-5.6	561	
BZX83C6V2	137A		ZD-6.0	561	
BZX83C6V8	5071A	3334	ZD-6.8	561	
BZX83C7V5	138A		ZD-7.5		
BZX83C8V2	5072A	3136	ZD-8.2	562	
BZX83C9V1	5018A	3784	ZD-9.1	562	
BZX85C13	143A	3750			
BZX85C16	5075A	3751			
BZX85C18	5077A	3752			
BZX85C30	5084A	3755			
BZX85C43	5087A	3339	ZD-43		
BZX85C4V3	5068A		ZD-4.3		
BZX85C8V2	5072A	3136			
BZX94C62	149A	3097	ZD-62		
BZX96C13	143A	3750			
BZX96C16	5075A	3751			
BZX96C18	5077A	3752			
BZX96C2V7	5063A	3837			
BZX96C3	5065A	3838			
BZX96C30	5084A	3755			
BZX96C3V3	5066A	3330	ZD-3.3		
BZX96C8V2	5072A	3136			
BZY-19	145A	3063	ZD-15	563	
BZY-83C15		3063	ZD-15	564	
BZY-83D15	145A		ZD-15	564	
BZY18	142A	3062	ZD-12	563	
BZY19	145A	3063	ZD-15	563	
BZY29C33	147A	3095	ZD-33		
BZY56	5069A	3056/135A	ZD-4.7		
BZY57	5010A	3776	ZD-5.1		
BZY58	136A	3057	ZD-5.6	561	
BZY59	137A	3058	ZD-6.2	561	
BZY60	5071A	3334	ZD-6.8	561	
BZY61	138A		ZD-7.5		
BZY62	5072A	3136	ZD-8.2	562	
BZY63	139A		ZD-9.1	562	
BZY64	5008A	3774	ZD-4.3		
BZY65	5010A	3776	ZD-5.1		
BZY66	5013A	3779	ZD-6.2	561	
BZY67	5015A	3781	ZD-7.5		
BZY68	5018A	3784	ZD-9.1	562	
BZY69	5021A	3787	ZD-12	563	

Industry Standard No.	ECG	SK	GE	RS 276-	MOTOR.
BZY78	5011A	3777	ZD-5.6	561	
BZY78P	5011A	3777	ZD-5.6	561	
BZY83/C10	140A		ZD-10	562	
BZY83/C15	145A	3063	ZD-15	564	
BZY83/C5V6	136A		ZD-5.6	561	
BZY83/D10	140A		ZD-10	562	
BZY83/D15	145A	3063	ZD-15	564	
BZY83/D5V6	136A		ZD-5.6	561	
BZY83C10	140A	3061	ZD-10	562	
BZY83C11	5074A		ZD-11	563	
BZY83C12	142A	3062	ZD-12	563	
BZY83C13V5	5022A	3788	ZD-12	563	
BZY83C15	145A	3063	ZD-15	564	
BZY83C16	145A		ZD-15	564	
BZY83C16V5	5026A	3792	ZD-17		
BZY83C18	5077A	3752	ZD-18		
BZY83C20	5079A	3335	ZD-20		
BZY83C22	5080A	3336	ZD-22		
BZY83C24V5	5032A	3798	ZD-25		
BZY83C4V7	5069A	3056/135A	ZD-4.7		
BZY83C5V1	5010A	3776	ZD-5.1		
BZY83C5V6	136A	3057	ZD-5.6	561	
BZY83C6V2	137A	3058	ZD-6.2	561	
BZY83C6V8	5071A	3334	ZD-6.8	561	
BZY83C7V5	138A		ZD-7.5		
BZY83C8V2	5072A	3136	ZD-8.2	562	
BZY83C9V1	5018A	3784	ZD-9.1	562	
BZY83D10	5019A	3785	ZD-10	562	
BZY83D12	5021A	3787	ZD-12	563	
BZY83D15	5024A	3790	ZD-15	564	
BZY83D18	5027A	3793	ZD-18		
BZY83D22	5030A	3796	ZD-22		
BZY83D4V7	5009A	3056/135A	ZD-4.7		
BZY83D5V6	5011A	3777	ZD-5.6	561	
BZY83D6V8	5014A	3780	ZD-6.8	561	
BZY83D8V2	5016A	3782	ZD-8.2	562	
BZY84D5V6	5178A	3057/136A			
BZY84D8V2	5183A	3136/5072A			
BZY85/C15		3063	ZD-15	564	
BZY85/C27		3064	ZD-27		
BZY85/C3V6	134A		ZD-3.6		
BZY85/C5V6			ZD-5.6	561	
BZY85/D15		3063	ZD-15	564	
BZY85/D5V6			ZD-5.6	561	
BZY85B2V7	5002A	3768			
BZY85B3V3	5066A	3330	ZD-3.3		
BZY85B8V2	5072A	3136			
BZY85C10	140A	3061	ZD-10	562	
BZY85C11	5074A		ZD-11	563	
BZY85C12	142A	3062	ZD-12	563	
BZY85C13	5022A	3788	ZD-12	563	
BZY85C13V5	143A	3750	ZD-14	564	
BZY85C15	145A	3063	ZD-15	564	
BZY85C16	5025A	3751/5075A		564	
BZY85C16V5	5026A	3792	ZD-17		
BZY85C18	5077A	3752	ZD-18		
BZY85C20	5079A	3335	ZD-20		
BZY85C22	5080A	3336	ZD-22		
BZY85C24	5081A		ZD-24		
BZY85C24V5	5081A		ZD-25		
BZY85C27	146A	3064	ZD-27		
BZY85C2V7	5063A	3837			
BZY85C3	5065A	3838			
BZY85C30	5084A	3755	ZD-30		
BZY85C33	5036A	3802	ZD-33		
BZY85C3V3	5066A	3055/134A	ZD-3.3		
BZY85C3V6	134A	3055	ZD-3.6		
BZY85C3V9	5067A	3055/134A	ZD-5.9		
BZY85C4V3	5068A		ZD-4.3		
BZY85C4V7	5069A	3056/135A	ZD-4.7		
BZY85C5V1	5010A	3776	ZD-5.1		
BZY85C5V6	136A	3057	ZD-5.6	561	
BZY85C6V2	137A	3058	ZD-6.2	561	
BZY85C6V8	5071A	3334	ZD-6.8	561	
BZY85C7V5	138A		ZD-7.5		
BZY85C8V2	5016A	3782	ZD-8.2	562	
BZY85C9V1	5018A	3784	ZD-9.1	562	
BZY85D10	5019A	3785	ZD-10	562	
BZY85D12	5021A	3787	ZD-12	563	
BZY85D15	5024A	3790	ZD-15	564	
BZY85D18	5027A	3793	ZD-18		
BZY85D22	5030A	3796	ZD-22		
BZY85D4V7	5009A	3775	ZD-4.7		
BZY85D5V6	5011A	3777	ZD-5.6	561	
BZY85D6V8	5014A	3780	ZD-6.8	561	
BZY85D8V2	5016A	3782	ZD-8.2	562	
BZY88-C30					.5M30Z5
BZY88-C3V3					.5M3.3AZ5
BZY88/6V2			ZD-6.2	561	
BZY88/C5V6			ZD-5.6	561	
BZY88/C6V2		3058	ZD-6.2	561	
BZY88/C9V1			ZD-9.1	562	
BZY88/C9V5			ZD-9.1	562	
BZY88C10	5019A	3785	ZD-10	562	
BZY88C11	5020A	3786	ZD-11	563	
BZY88C12	5021A	3787	ZD-12	563	
BZY88C13	5022A	3788	ZD-12	563	
BZY88C15	5024A	3790	ZD-15	564	
BZY88C16	5025A	3791		564	
BZY88C18	5027A	3793	ZD-18		
BZY88C20	5029A	3795	ZD-20		
BZY88C22	5030A	3796	ZD-22		
BZY88C24	5031A	3797	ZD-24		
BZY88C2V7	5063A	3837			
BZY88C30	5035A	3801	ZD-30		
BZY88C3V0	5065A	3838			
BZY88C3V3	5063A	3330/5066A	ZD-3.3		
BZY88C3V6	5065A		ZD-3.6		
BZY88C3V9	5066A	3055/134A	ZD-3.9		
BZY88C4V3	134A	3332/5068A	ZD-4.3		
BZY88C4V7	5009A	3775	ZD-4.7		
BZY88C5V1	5010A	3776	ZD-5.1		
BZY88C5V6	5011A	3777	ZD-5.6	561	
BZY88C6V2	5012A	3778	ZD-6.0	561	
BZY88C6V8	5014A	3780	ZD-6.8	561	
BZY88C7V5	5015A	3781	ZD-7.5		
BZY88C8V2	5016A	3782	ZD-8.2	562	
BZY88C9V1	5018A	3784	ZD-9.1	562	
BZY91-C75					50M75ZS5
BZY91-C7V5					50M7.5ZS5
BZY92/C15		3063	ZD-15	564	
BZY92/C27		3064	ZD-27		
BZY92/C5V6			ZD-5.6	561	
BZY92C10	140A		ZD-10	562	
BZY92C11	5074A		ZD-11.0	563	
BZY92C12	142A		ZD-12	563	
BZY92C13	143A		ZD-13	563	
BZY92C15	145A	3063	ZD-15	564	
BZY92C16	5075A	3751	ZD-16	564	
BZY92C18	5077A	3752	ZD-18		
BZY92C20	5079A	3335	ZD-20		
BZY92C22	5080A	3336	ZD-22		
BZY92C24	5081A		ZD-24		
BZY92C27	146A	3064	ZD-27		
BZY92C30	5084A		ZD-30		
BZY92C33	147A	3095	ZD-33		
BZY92C36	5085A	3337	ZD-36		
BZY92C3V9	5067A	3331	ZD-3.9		
BZY92C4V3	5068A	3332	ZD-4.3		
BZY92C4V7	5069A		ZD-4.7		
BZY92C5V1	135A	3056	ZD-5.1		
BZY92C5V6	136A	3057	ZD-5.6	561	
BZY92C6V2			ZD-6.2	561	
BZY92C6V8	5071A	3334	ZD-6.8	561	
BZY92C7V5	138A		ZD-7.5		
BZY92C8V2	5072A	3136	ZD-8.2	562	
BZY92C9V1	139A		ZD-9.1	562	
BZY93-C75					10M75Z5
BZY93-C7V5					10M7.5Z5
BZY94/C12		3062	ZD-12	563	
BZY94/C15		3063	ZD-15	564	
BZY94C10	5019A	3785	ZD-10	562	
BZY94C11	5020A	3786	ZD-11	563	
BZY94C12	5021A	3787	ZD-12	563	
BZY94C13	5022A	3788	ZD-12	563	
BZY94C15	5024A	3790	ZD-15	564	
BZY94C16	5025A	3791		564	
BZY94C18	5027A	3793	ZD-18		
BZY94C20	5029A	3795	ZD-20		
BZY94C22	5030A	3796	ZD-22		
BZY94C24	5031A	3797	ZD-24		
BZY94C27	5033A	3799	ZD-27		
BZY94C30	5035A	3801	ZD-30		
BZY94C33	5036A	3802	ZD-33		
BZY94C36	5037A	3803	ZD-36		
BZY94C39	5038A	3804	ZD-39		
BZY94C43	5039A	3805	ZD-43		
BZY94C47	5040A	3806	ZD-47		
BZY94C51	5041A	3807	ZD-51		
BZY94C56	5042A	3808	ZD-56		
BZY94C62	5044A	3810	ZD-62		
BZY94C68	5045A	3811	ZD-68		
BZY94C75	5046A	3812	ZD-75		
BZY95C10	5125A	3391			
BZY95C11	5126A	3392			
BZY95C12	5127A	3393			
BZY95C13	5128A	3394	5ZD-13		
BZY95C15	5130A	3396	5ZD-15		
BZY95C16	5131A	3397	5ZD-16		
BZY95C18	5133A	3399			
BZY95C20	5135A	3401	5ZD-20		
BZY95C22	5136A	3402			
BZY95C24	5137A	3403			
BZY95C27	5139A	3405			
BZY95C30	5141A	3407			
BZY95C33	5142A	3408			
BZY95C36	5143A	3409			
BZY95C39	5144A	3410			
BZY95C43	5145A	3411			
BZY95C47	5146A	3412			
BZY95C51	5147A	3413			
BZY95C56	5148A	3414			
BZY95C62	5150A	3416			
BZY95C68	5151A	3417			
BZY95C75	5152A	3418			
BZY96-C4V7					1M4.7AZ5
BZY96-C75					1M75Z5
BZY96C10	5125A	3391			
BZY96C4V7	5115A	3381			
BZY96C5V1	5116A	3382			
BZY96C5V6	5117A	3383			
BZY96C6V2	5119A	3385			
BZY96C6V8	5120A	3386			
BZY96C7V5	5121A	3387			
BZY96C8V2	5122A	3388			
BZY96C9V1	5124A	3390			
BZZ10	5012A	3778	ZD-6.0	561	
BZZ11	5014A	3780	ZD-6.8	561	
BZZ12	5015A	3781	ZD-7.5		
BZZ13	5016A	3782	ZD-8.2	562	
C-10-20A	178MP	3100/519	300(2)	1122(2)	
C-2C	506		511	1114	
C-36582	312			2028	
C-4401	177		300	1122	
C-505	177		300	1122	
C-6BX212	6401			2029	
C-F8	237	3299	46		
C00 68602300	123A	3444		2051	
C00-686-0241			82	2032	
C00-68602300			20	2051	
C00686-0258-0	159		82	2032	
C0068602720	159		82	2032	
C02	5072A	3136	ZD-8.2	562	
C0410	125	3032A	509	1114	
C05-03C	113A	3119/113	6GC1		
C06C	506	3843			
C08P1	113A	3119/113	6GC1		
C08P1R	115	3121	6GX1		
C0C13000-1C			63	2030	
C1-1002	724		IC-86		
C1-1004	718	3159	IC-8		
C1-12	362				MRF616
C1-12Z	362				2N5944
C1-28	311				MRF313
C1.0E02	125	3033	510,531	1114	
C1/2-12					MRF626
C10	113A	3119/113	6GC1		
C10-02A	113A	3119/113	6GC1		
C10-12A	364				2N5946
C10-13B	113A	3119/113	6GC1		
C10-15B	113A		6GC1		
C10-15C	113A		6GC1		
C10-16A	113A		6GC1		
C10-18B	113A	3119/113	6GC1		
C10-1B	113A	3119/113	6GC1		
C10-20A	178MP		300(2)	1122(2)	
C10-22C	178MP	3100/519	300(2)	1122(2)	
C10-2NA		3119	6GC1		
C10-31A	113A	3119/113	6GC1		
C10-38C	113A	3119/113	6GC1		
C10-47B	113A	3119/113	6GC1		
C100		3122	20	2051	
C100-OY		3122	20	2051	
C1000	199	3122	62	2010	
C1000-BL	199	3122	62	2010	
C1000-GR	199	3122	62	2010	
C1000-Y	199	3122	20	2051	
C1000Y	199	3122	62	2010	
C1001	278		261		
C1002			212	2010	
C1003			81	2051	
C1004		3133	37		
C1004A		3710	37		
C1006		3245	62	2010	
C1006A		3245	62	2010	
C1006B		3245	62	2010	
C1006C		3245	62	2010	
C1007	123A	3444	20	2051	

Industry Standard No.	ECG	SK	GE	RS 276-	MOTOR.
C1008	128	3024	243	2030	
C101		3025	12		
C101(JAPAN)	124		12		
C1010		3245	62	2010	
C1010A		3245	62	2010	
C1010B		3245	62	2010	
C1010C		3245	62	2010	
C10110	116	3016	504A	1104	
C1012	154	3044	40	2012	
C1012A	154	3044	40	2012	
C1013	300	3464	247		
C1013C	300	3464	273		
C1013D	300	3464	273		
C1014	300	3464	273		
C1014B	300	3464	273		
C1014C	300	3464	273		
C1014CD	300	3464	273		
C1014D	300	3464	273		
C1014D1	300	3464	273		
C10159	116	3311	504A	1104	
C1017	299	3298	236		
C10176	116	3017B/117	504A	1104	
C1018	299	3298	236		
C1018B	210		252	2018	
C101A		3021	12		
C101B		3298	236		
C101X		3006	12		
C102		3006	244	2027	
C102(JAPAN)	105		4		
C10215-2	160	3006	245	2004	
C10227	102A	3004	53	2007	
C1023	108	3452	86	2038	
C1023(JAPAN)	107		11	2015	
C1023-0	107	3039/316	11	2015	
C1023-Y	107	3039/316	11	2015	
C10230-3	102A	3004	53	2007	
C1023G	107		11	2015	
C1024	175	3538	246	2020	
C1024-D2	175	3538	246	2020	
C1024B	175	3538	246	2020	
C1024C	175	3538	246	2020	
C1024D	175	3538	246	2020	
C1024E	175	3538	246	2020	
C1024F	175		246	2020	
C1025	175	3626	246	2020	
C1025B	160		245	2004	
C1025CTV	175		246	2020	
C1026	107	3039/316	11	2015	
C10260	160		245	2004	
C10261	160		245	2004	
C10262	160		245	2004	
C1026G	107	3039/316	11	2015	
C1026Y	107	3039/316	11	2015	
C1027	163A		36		
C10279-1	123A	3444	20	2051	
C10279-3	123A	3444	20	2051	
C10291	126	3008	52	2024	
C103	123A	3122	20	2051	
C103(JAPAN)	123A	3444	20	2051	
C1030	87	3619	262	2041	
C1030A	87	3619	262	2041	
C1030B	87	3619	262	2041	
C1030C	87	3619	262	2041	
C1030D	87	3619	262	2041	
C1030P	280		262	2041	
C1032	107	3039/316	11	2015	
C1032G	107		11	2015	
C1032Y	107	3039/316	11	2015	
C1033	287	3122	20	2051	
C1033A	287	3122	20	2051	
C1034	277	3133/164	260		
C1035	161	3132	39	2015	
C1035C	161	3132	39	2015	
C1035D	161	3132	39	2015	
C1035E	161	3132	39	2015	
C1036	161	3132	39	2015	
C103A		3122			2N5062
C103A(SCR)	5402	3638			
C103A(TRANSISTOR)	123A		20	2051	
C103B	5404		C103B		2N5064
C103Q	5400	3950			
C103Y	5400	3950	C103Y		2N5060
C103YY	5401	3638/5402	C103YY		2N5061
C104	123A	3122	20	2051	
C1044	161		39	2015	
C1045	164	3115/165	37		
C1045B	164	3115/165	37		
C1045C	164	3115/165	37		
C1045D	164	3115/165	37		
C1045E	164	3115/165	37		
C1045R	164	3115/165	37		
C1046	165	3115	38		
C1047	107	3039/316	61	2038	
C1047B	107		61	2038	
C1047C	107		61	2038	
C1047D	107		61	2038	
C1047E	107		61	2038	
C1048	154	3040	40	2012	
C1048B	154	3040	40	2012	
C1048C	154	3040	40	2012	
C1048D	154	3040	40	2012	
C1048E	154	3040	40	2012	
C1048F	154	3040	40	2012	
C104A	123A	3122	20	2051	
C105	123A	3122	20	2051	
C1050D	94		35		
C1050E	94		35		
C1050F	94		35		
C1051	280		262	2041	
C1051C	280		262	2041	
C1051D	280		262	2041	
C1051E	280		262	2041	
C1051F	280		262	2041	
C1051LC	280		262	2041	
C1051LD	280		262	2041	
C1051LE	280		262	2041	
C1051LF	280		262	2041	
C1055H	384		234		
C1056	154	3044	40	2012	
C1059	124	3021	12		
C106	282$		264		
C106(PNP)	129		244	2027	
C1060	152	3893	66	2048	
C1060A	152	3893	66	2048	
C1060B	152	3893	66	2048	
C1060BM	152	3893	66	2048	
C1060C	152	3893	66	2048	
C1060D	152	3893	66	2048	
C1061	152	3893	66	2048	
C1061A	152	3893	66	2048	
C1061B	152	3893	66	2048	
C1061C	152	3893	66	2048	
C1061D	152	3893	66	2048	
C1061T	152	3893	66	2048	
C1061T-B	152		66	2048	
C1061TB	152	3893	66	2048	
C106A	5454		MR-5	1067	
C106A1	5454	6754		1067	C106A1
C106A2	5454	6754		1067	
C106A3	5454	6754		1067	
C106A4	5454	6754		1067	
C106B	5455	3597	MR-5	1067	
C106B1	5455	3597		1067	C106B1
C106B2	5455	3597	C106B2	1067	
C106B3	5455	3597		1067	
C106B4	5455	3597		1067	
C106C	5456	3598/5457	C106C	1020	
C106C1	5456	3598/5457		1020	C106C1
C106C2	5456	3598/5457		1020	
C106C3	5456	3598/5457		1020	
C106C4	5456	3598/5457		1020	
C106D	5457	3598	C106D		
C106D1	5457	3598			C106D1
C106D2	5457	3598			
C106D3	5457	3598			
C106D4	5457	3598			
C106E1					C106E1
C106F	5453		MR-5	1067	
C106F1	5453	6753		1067	C106F1
C106F2	5453	6753		1067	
C106F3	5453	6753		1067	
C106F4	5453	6753		1067	
C106M1					C106M1
C106Q	5452	6752		1067	
C106Q1	5452	6752		1067	
C106Q2	5452	6752		1067	
C106Q3	5452	6752		1067	
C106Q4	5452	6752		1067	
C106Y	5452	6752	MR-5	1067	
C106Y1	5452	6752		1067	C106Y1
C106Y2	5452	6752		1067	
C106Y3	5452	6752		1067	
C106Y4	5452	6752		1067	
C1070		3716	39	2015	
C1071	123A	3444	20	2051	
C1072		3024	243	2030	
C1072A		3024	243	2030	
C1079	280	3535/181	262	2041	
C1079R	280		262	2041	
C1079Y	280		262	2041	
C107A	5454	6754		1067	C106A
C107A1	5454	6754		1067	
C107A2	5454	6754		1067	
C107A3	5454	6754		1067	
C107B	5455	3597		1067	C106B
C107B1	5455	3597		1067	
C107B2	5455	3597		1067	
C107B3	5455	3597		1067	
C107B4	5455	3597		1067	
C107C	5456			1020	C106C
C107C1	5456	3598/5457		1020	
C107C2	5456	3598/5457		1020	
C107C3	5456	3598/5457		1020	
C107C4	5456	3598/5457		1020	
C107D	5457	3598			C106D
C107D1	5457	3598			
C107D2	5457	3598			
C107D3	5457	3598			
C107D4	5457	3598			
C107E					C106E
C107F	5453			1067	C106F
C107F1	5453	6753		1067	
C107F2	5453	6753		1067	
C107F3	5453	6753		1067	
C107F4	5453	6753		1067	
C107M					C106M
C107Q	5452	6752		1067	C106Q
C107Q1	5452	6752		1067	
C107Q2	5452	6752		1067	
C107Q3	5452	6752		1067	
C107Q4	5452	6752		1067	
C107Y	5452	6752		1067	C106Y
C107Y1	5452	6752		1067	
C107Y2	5452	6752		1067	
C107Y3	5452	6752		1067	
C107Y4	5452	6752		1067	
C108	128	3024	243	2030	
C1080	280	3535/181	262	2041	
C1080R	280		262	2041	
C1080Y	280		262	2041	
C1086	389	3710/238	38		
C1086M	389		38		
C109	128	3024	243	2030	
C1095	186A		247	2052	
C1095(6)	186A	3357	247	2052	
C1095L	186A		247	2052	
C1095M	186A		247	2052	
C1096	186A		247	2052	
C1096(M)	186A		247	2052	
C1096-3ZM	186A		247	2052	
C10963ZM	186A	3202/210	247	2052	
C10964ZL	186A	3202/210	247	2052	
C1096A	186A	3248	247	2052	
C1096B	186A	3248	247	2052	
C1096C	186A	3248	247	2052	
C1096D	186A	3248	247	2052	
C1096K	186A	3248	247	2052	
C1096L	186A	3248	247	2052	
C1096M	186A	3248	247	2052	
C1096N	186A		247	2052	
C1096W	186A	3248	247	2052	
C1098	186		247	2052	
C1098A	186		247	2052	
C1098B	186		247	2052	
C1098C	186A		247	2052	
C1098D	186		247	2052	
C1098L	186		247	2052	
C1098M	186		247	2052	
C109A	128	3024	243	2030	
C10A	5482				2N4169
C10B	5483				2N4170
C10F	5481				2N4168
C10G					2N4170
C10V					2N4167
C110	5480	3122	20	2051	
C1100	389		38		
C1101	164	3133	37		
C1101A	164	3133	37		
C1101B	164	3133	37		
C1101C	164	3133	37		
C1101D	164	3133	37		
C1101E	164	3133	37		
C1101F	164		37		
C1101L	164		37		
C1102	124	3021	12		
C1102(M)	124		12		
C11021	102A	3004	53	2007	
C1102A	124	3021	12		

Industry Standard No.	ECG	SK	GE	RS 276-	MOTOR.
C1102B	124	3021	12		
C1102C	124	3021	12		
C1102K	124		12		
C1102L	124		12		
C1102M	124		12		
C1103	154	3044	40	2012	
C1103(A)	154		40	2012	
C1103A	154	3044	40	2012	
C1103B	154	3044	40	2012	
C1103C	154	3044	40	2012	
C1103L	154	3044	40	2012	
C1104	286	3021/124	12		
C1104A	286	3021/124	12		
C1104B	286	3021/124	12		
C1104C	286	3021/124	12		
C1105	124	3021	12		
C1105A	124	3021	12		
C1105B	124	3021	12		
C1105C	124	3021	12		
C1105K	124		12		
C1105L	124		12		
C1105M	124		12		
C1106	94		35		
C1106A	94		35		
C1106B	94		35		
C1106C	94		35		
C1106K	94		35		
C1106L	94		35		
C1106M	94		35		
C1107Q	291		247	2052	
C111	123A	3122	20	2051	
C1111	87		262	2041	
C1112	87	3836/284			
C1114	283		36		
C1115	280	3836/284	262	2041	
C1116	284	3836	40	2012	
C1116-O	284		40	2012	
C1116A	284	3836			
C1117	161	3039/316			
C111B	108	3452	86	2038	
C111E	123A	3122	20	2051	
C112(JAPAN)	128		243	2030	
C112(PNP)	129		244	2027	
C1123	123AP		11	2015	
C1124	190		251		
C1126	229		11	2015	
C1127	171		217		
C1128	161		39	2015	
C1128(3RD IF)	233			2009	
C1128(3RD-IF)			210	2051	
C1128(S)	229		39	2015	
C1128D	229		20	2051	
C1129	229		39	2015	
C1129(R)	229		39	2015	
C113	128	3024	243	2030	
C114	128	3024	243	2030	
C115	128	3024	243	2030	
C1151	164	3133	37		
C1151A	164		37		
C1152	162		35		
C1152F	162		35		
C1152G	162		35		
C1153	389		38		
C1154	389	3439/163A	36		
C1155		3104A	217		
C1156		3104A	217		
C1156(IC)	1194	3484			
C1156H	1194	3484			
C1156H(IC)	1194	3484			
C1156H2(IC)	1194	3484			
C1157	190	3104A	217		
C1158	313$	3452/108	86	2038	
C1159	313$		11	2015	
C116	123		46		
C1160	175	3538	246	2020	
C1160K	175	3021/124	246	2020	
C1160L	175		246	2020	
C1161	175	3538	246	2020	
C1162	184		57	2017	
C1162A	184#		57	2017	
C1162B	184#		57	2017	
C1162C	184#		270	2017	
C1162CP	184#		57	2017	
C1162D	184#		57	2017	
C1162MP	184#		57	2017	
C1162WB	184#		57	2017	
C1162WBP	184#		247	2052	
C1162WTB	184#		57	2017	
C1162WTC	184#		57	2017	
C1162WTD	184		57	2017	
C1166	289A		271	2030	
C1166-O	289A		271	2030	
C11660			271	2030	
C1166D	289A		271	2030	
C1166GR	289A		271	2030	
C11660	289A		271	2030	
C1166R	289A		271	2030	
C1166Y	289A		271	2030	
C1167	389		38		
C1168	124	3271	12		
C1168X	124		12		
C116T	195A		46		
C117		3024	243	2030	
C1170	389	3115/165	38		
C1170A	238	3710	38		
C1170B	389		38		
C1171	164	3115/165	38		
C1172	238		259		
C1172A	238		259		
C1172B	238		259		
C1173	152		66	2048	
C1173(RF-PWR)			216	2053	
C1173-O	152		66	2048	
C1173-GR	152	3197/235	66	2048	
C1173-O	152		66	2048	
C1173-R	152	3197/235	66	2048	
C1173-Y	152	3197/235	66	2048	
C1173C	152	3197/235	66	2048	
C1173R	152	3197/235	66	2048	
C1173X	152		66	2048	
C1173XO	152		66	2048	
C1173Y	152	3197/235	66	2048	
C1174	389		38		
C1175	289A	3122	20	2051	
C1175C	289A	3122	20	2051	
C1175D	289A	3122	20	2051	
C1175E	289A	3122	243	2051	
C1175F	289A	3122	20	2051	
C118		3025	244	2027	
C118(JAPAN)	128		243	2030	
C1182B	161	3132	39	2015	
C1182C	161	3132	39	2015	
C1182D	161	3132	39	2015	
C1184	164	3133	37		
C1184A	164	3133	37		
C1184B	164	3133	37		
C1184C	164	3133	37		
C1184D	164	3133	37		
C1184E	164	3133	37		
C1185	162	3438	35		
C1185A	162	3438	35		
C1185B	162	3438	35		
C1185C	162	3438	35		
C1185K	162	3438	35		
C1185L	162	3438	35		
C1185M	162	3438	35		
C1187	229	3246	61	2038	
C118M					SPECIAL
C119		3025	244	2027	
C119(JAPAN)	128		243	2030	
C1195	94		35		
C11A	5482				2N4169
C11B	5483				2N4170
C11C	5484				2N4171
C11D	5485				2N4172
C11E	5486				2N4173
C11F	5481				2N4168
C11G	5483				2N4170
C11H	5484				2N4171
C11M	5487				2N4174
C11S1C1E1C	116	3016	504A	1104	
C11U	5481				2N4167
C12(TRANSISTOR)	128		243	2030	
C12-12	364				2N5646
C12-28					MRF321
C120	123A	3024/128	243	2030	
C1204	289A	3245/199	62	2010	
C1204B	289A	3245/199	62	2010	
C1204C	289A	3245/199	62	2010	
C1204D	199	3245	62	2010	
C1205	289A	3452/108	86	2038	
C1205A	289A	3452/108	86	2038	
C1205B	289A	3452/108	86	2038	
C1205C	289A	3452/108	86	2038	
C1209			271	2030	
C1209C			271	2030	
C121	128	3024	243	2030	
C1210		3842	222		
C1210C		3842	222		
C1210D		3842	222		
C1210E		3842	222		
C1211		3122	47	2030	
C1211C			47	2030	
C1211D			47	2030	
C1211E			47	2030	
C1212	373		57	2017	
C1212A	373		57	2017	
C1212AA	373		57	2017	
C1212AB	373		57	2017	
C1212AC	373		57	2017	
C1212AD	373		57	2017	
C1212AWT	373		57	2017	
C1212AWTA	373		57	2017	
C1212AWTB	373		57	2017	
C1212AWTC	373		57	2017	
C1212AWTD	373		57	2017	
C1212B	373		57	2017	
C1212C	373		57	2017	
C1212D	373		57	2017	
C1212WT	373#		57	2017	
C1212WTA	373#		57	2017	
C1212WTB	373#		57	2017	
C1212WTC	373#		57	2017	
C1212WTD	373#		57	2017	
C1213		3122	268	2038	
C1213A	289A	3122	268	2038	
C1213AA	289A	3122	268	2038	
C1213AB	289A	3122	268	2038	
C1213AC	289A	3122	268	2038	
C1213AD	289A	3122	268	2038	
C1213AK	289A		268	2038	
C1213AKA	289A		268	2038	
C1213AKB	289A		268	2038	
C1213AKC	289A		268	2038	
C1213AKD	289A		268	2038	
C1213B	289A	3122	268	2038	
C1213BC	289A		268	2038	
C1213C	289A	3122	268	2038	
C1213CD	289A		268	2038	
C1213D	289A	3122	268	2038	
C1214	289A	3124/289	268	2038	
C1214A	289A	3124/289	268	2038	
C1214B	289A	3124/289	268	2038	
C1214C	289A	3124/289	268	2038	
C1214D	289A		268	2038	
C1215	199		62	2010	
C1217		3201	27		
C1218		3024	243	2030	
C122		3018	20	2051	
C122(TRANSISTOR)	128		243	2030	
C1220E	289A	3024/128	243	2030	
C1222	199	3117	62	2010	
C1222A	199	3245	62	2010	
C1222B	199	3245	62	2010	
C1222C	199	3245	62	2010	
C1222D	199	3245	62	2010	
C1226	186A		247	2052	
C1226-O	186A		247	2052	
C1226A	186A	3357	247	2052	
C1226AC	186A		247	2052	
C1226AO	186A		247	2052	
C1226AP	186A	3357	247	2052	
C1226AQ	186A	3357	247	2052	
C1226AR	186A		247	2052	
C1226C	186A	3357	247	2052	
C1226CF	186A		247	2052	
C1226F	186A	3357	247	2052	
C1226P	186A	3357	247	2052	
C1226Q	186A	3357	247	2052	
C1226QR	186A		247	2052	
C1226R	186A		247	2052	
C122A	5462	3686			
C122A1	5462	3686			C122A1
C122A2	5462	3686			
C122A3	5462	3686			
C122A4	5462	3686			
C122A5	5462	3686			
C122A6	5462	3686			
C122B	5463	3572	C122B		
C122B1	5463	3572			C122B1
C122B2	5463	3572			
C122B3	5463	3572			
C122B4	5463	3572			
C122B5	5463	3572			
C122B6	5463	3572			
C122C	5465	3558			
C122C1	5465	3558			C122C1
C122C2	5465	3558			

Industry Standard No.	ECG	SK	GE	RS 276-	MOTOR.
C122C3	5465	3558			
C122C4	5465	3558			
C122C5	5465	3558			
C122C6	5465	3558			
C122D	5465	3558	C122D		
C122D1	5465	3558			C122D1
C122D2	5465	3558			
C122D3	5465	3558			
C122D4	5465	3558			
C122D5	5465	3558			
C122D6	5465	3558			
C122E1			C122E1		C122E1
C122F	5461	3685	C122F		
C122F1	5461	3685			C122F1
C122F2	5461	3685			
C122F3	5461	3685			
C122F4	5461	3685			
C122F5	5461	3685			
C122F6	5461	3685			
C122M1					C122M1
C122N1	5468				C122N1
C122S1	5468				C122S1
C123(TRANSISTOR)	128		243	2030	
C1235	124		12		
C1237	235	3197	215		
C1237E	235	3197	216	2053	
C1239	282/427		46		
C124	128	3024	243	2030	
C1243	300	3464	273		
C1243-24	300	3464	252	2018	
C1243C	300	3464	273		
C1243C1	300	3464	273		
C1243C2	300	3464	273		
C1243D	300	3464	273		
C1243D1	300	3464	273		
C1243D2	300	3464	273		
C1243E	300	3464	273		
C1244	123A	3444	20	2051	
C1247A	289A	3842			
C1247AF	289A	3842			
C125		3006	245	2004	
C126A					2N6395
C126B					2N6396
C126C					MCR220-5
C126D					2N6397
C126E					MCR220-7
C126F					2N6394
C126M					2N6398
C127	123A	3452/108	86	2038	
C12711		3444	20	2051	
C1278		3866	220		
C1278S		3866	220		
C1279	287	3244	220		
C1279S	287	3244	220		
C127A					2N6401
C127B					2N6402
C127C					MCR221-5
C127D					2N6403
C127E					MCR221-7
C127F					2N6400
C127M					2N6404
C128	101		8	2002	
C1280	172A	3156	64		
C1280A	172A	3156	64		
C1280AS	172A	3156	64		
C1280S	172A	3156	64		
C129	101		8	2002	
C1293A(LAST IF)	233			2009	
C1293A(LAST-IF)			210	2051	
C1293B	107		210	2051	
C1293B(3RD IF)				2009	
C1293B(3RD-IF)			210	2009	
C1293B(LAST IF)	233			2009	
C1293B(LAST-IF)			210	2009	
C1295	165		38		
C1295-0	165		38		
C1295A	165		38		
C13(TRANSISTOR)	101		8	2002	
C130	128	3024	86	2030	
C1303	473		261		
C1304	286	3021/124	12		
C1304BK	286		12		
C1306	235		215		
C1306I	199		215		
C1307	236	3197/235	216	2053	
C1307-1	236	3239			
C1307K	236	3239			
C1308	238		259		
C1308K	238		259		
C1309	165		38		
C131	123A	3122	20	2051	
C1312			62	2010	
C1312F		3899	62	2010	
C1312G		3899	62	2010	
C1312H		3899	62	2010	
C1312Y		3899	62	2010	
C1312YF		3899	62	2010	
C1312YG		3899	62	2010	
C1312YH		3899	62	2010	
C1313	199		62	2010	
C1313F	199	3250/315	62	2010	
C1313G	199	3250/315	62	2010	
C1313H	199	3250/315	62	2010	
C1313Y	199	3245	62	2010	
C1313YF	199	3250/315	62	2010	
C1313YG	199	3250/315	62	2010	
C1313YH	199	3250/315	62	2010	
C1316	277		260		
C1317	289A		268	2038	
C1317B	289A		268	2038	
C1317P	289A		268	2038	
C1317Q	289A		268	2038	
C1317R	289A	3138/193A	268	2038	
C1317S	289A		268	2038	
C1317T	289A		268	2038	
C1318	289A	3124/289	271	2030	
C1318C	297		271	2030	
C1318Q	289A	3124/289	271	2030	
C1318R	289A		271	2030	
C1318S	289A		271	2030	
C132	123A	3122	20	2051	
C1325	238	3710	38		
C1325A	238		38		
C1327	199		62	2010	
C1327FS	199		62	2010	
C1327TV	199	3899			
C1327U	199		62	2010	
C1327XAN	199	3899			
C1327Y	199	3899			
C1328	199		62	2010	
C1328T	199		62	2010	
C1328U	199		62	2010	
C133	123A	3122	20	2051	
C1330	192A	3137	63	2030	
C1330A	192A	3137	63	2030	
C1330B	192A	3137	63	2030	
C1330C	192A	3137	63	2030	
C1330D	192A	3137	63	2030	
C1330L	192A	3137	63	2030	
C1330R	192A		63	2030	
C1335	289A		62	2010	
C1335A	199	3899	62	2010	
C1335B	199	3899	62	2010	
C1335C	199	3899	62	2010	
C1335D	199	3899	62	2010	
C1335E	199	3899	62	2010	
C1335F	199	3899	62	2010	
C134	123A	3122	20	2051	
C1342	107	3124/289	268	2038	
C1342A	107	3124/289	268	2038	
C1342B	107	3124/289	268	2038	
C1342C	107	3124/289	268	2038	
C1343	280		262	2041	
C1343A	280		262	2041	
C1343B	280		262	2041	
C1343C	280		262	2041	
C1343H	280		262	2041	
C1343HA	280		262	2041	
C1343HB	280		262	2041	
C1344	199		62	2010	
C1344C	199		62	2010	
C1344D	199		62	2010	
C1344E	199		62	2010	
C1344F	199		62	2010	
C1345	199		62	2010	
C1345C	199		62	2010	
C1345F	199		62	2010	
C1346	192A	3137	63	2030	
C1346R	192A	3137	63	2030	
C1346S	192A		63	2030	
C1347	192A	3137	63	2030	
C1347Q	192		63	2030	
C1347R	192		63	2030	
C1347S	192		63	2030	
C1348	389		259		
C134B	123A	3122	20	2051	
C135	123A	3122	20	2051	
C1358	165		38		
C1358K	165		38		
C1358P	165		38		
C1358Q	165		38		
C1358R	165		38		
C1359	85	3122	61	2038	
C1359A	85		61	2038	
C1359B	85	3122	61	2038	
C1359C	85	3122	61	2038	
C135A	5542				2N3896
C135B					2N3897
C135C	5544				MCR3935-5
C135D	5545				2N3898
C135E	5546				MCR3935-7
C135F	5541				MCR3935-2
C135M	5547				2N3899
C136	123A	3122	20	2051	
C1360		3039	11	2015	
C1361	289A		20	2051	
C1362	123AP		20	2051	
C1363	289A		20	2051	
C1364	289A		20	2051	
C1364A	289A		20	2051	
C1367	164		37		
C1367A	164		37		
C1368	184#		57	2017	
C1368C	184#		57	2017	
C1368D	184#		247	2052	
C137(TRANSISTOR)	123A		20	2051	
C1372Y	289A	3122	20	2051	
C1377	236	3197/235	216	2053	
C137E	5546		C137E		MCR3935-7
C137M	5547		C137M		2N3899
C137N	5548	6648	C137N		MCR3935-10
C137P		6649	C137P		
C137S	5548	6648			MCR3935-9
C138(TRANSISTOR)	123A		20	2051	
C1382	373	3747/157	232		
C1382-0	373	3747/157	232		
C1382Y	373	3747/157	232		
C1383	293	3849	47	2030	
C1383P	293	3849	47	2030	
C1383Q	293	3849	47	2030	
C1383R	293	3849	47	2030	
C1383R/494		3849	63	2030	
C1383S	293		47	2030	
C1383X	293		47	2030	
C1384	293	3849	47	2030	
C1384P	293	3849			
C1384Q	293	3849	47	2030	
C1384R	293	3849	47	2030	
C1384S	293	3849	47	2030	
C1386C	184		57	2017	
C138A		3122	20	2051	
C139(TRANSISTOR)	123A		20	2051	
C13901		3444	20	2051	
C1390A	107	3018	11	2015	
C1390I	123A	3444	20	2051	
C1390J	123A	3444	20	2051	
C1390K	123A	3444	20	2051	
C1390V	123A	3444	20	2051	
C1390W	123A	3444	20	2051	
C1390WH	123A	3444	20	2051	
C1390WI	123A	3444	20	2051	
C1390WX	123A	3444	20	2051	
C1390WY	123A	3444	20	2051	
C1390X	123A	3444	20	2051	
C1390XJ	123A	3444	20	2051	
C1390XK	123A	3444	20	2051	
C1390YM	123A	3444	20	2051	
C1391	124		12		
C1391VL	124		12		
C1393	229		20	2051	
C1394	229		11	2015	
C1398	152		66	2048	
C1398Q	152		66	2048	
C14	101		8	2002	
C140(TRANSISTOR)	128		243	2030	
C1402	280		262	2041	
C1407		3137	63	2030	
C1407P			63	2030	
C1407R			63	2030	
C1407S			63	2030	
C1409	375	3440/291			
C1409A	375	3440/291			
C1409AA	375	3440/291			
C1409AB	375	3440/291			
C1409AC	375	3440/291			
C1409B	375	3440/291			
C1409C	375	3440/291			
C1410	375	3440/291			

Industry Standard No.	ECG	SK	GE	RS 276-	MOTOR.
C1410A	375	3440/291			
C1410AA	375	3440/291			
C1410AB	375	3440/291			
C1410AC	375	3440/291			
C1410B	375	3440/291			
C1410C	375	3440/291			
C1413	165		38		
C1413A	165		38		
C1416	123A		62	2051	
C1416BL	123A		62	2010	
C1417	107		11	2015	
C1417C	107	3293	11	2015	
C1417D	107		11	2015	
C1417D(1)			11	2015	
C1417D(U)	107			2015	
C1417DU	107		11	2015	
C1417F	107		11	2015	
C1417G	107		11	2015	
C1417H	107		11	2015	
C1417U	107		11	2015	
C1417V	107		11	2015	
C1417VW	107		11	2015	
C1417W	107		11	2015	
C1418	152		66	2048	
C1418A	152		66	2048	
C1418B	152		66	2048	
C1418C	152		66	2048	
C1418D	152		66	2048	
C1419	152		66	2048	
C1419A	152		66	2048	
C1419B	152		66	2048	
C1419C	152		66	2048	
C1419D	152		66	2048	
C1424	316		280		
C1429	152		243	2018	
C1429-1	152		243	2030	
C1429-2	152		243	2030	
C1437	100	3004/102A	1	2007	
C1438		3004	1	2007	
C1446	376		217		
C1446B	376		251		
C1446C	376		251		
C1446LB	376		251		
C1446PQ	376		251		
C1446Q	376		251		
C1447-O	396	3219	251		
C1447LB	396	3219	251		
C1447R	396	3219	251		
C1448	375		251		
C1448A	375		241	2020	
C1448P	375	3929	251		
C1448Q	375	3929	251		
C1448R	375	3929	251		
C1448S	375	3929	251		
C1449	184	3253/295	270		
C1449L	184	3253/295			
C1449M	184	3253/295			
C1450S	286		66	2048	
C147	128	3048/329	243	2030	
C1472K	172A	3156	64		
C1475	315	3250	279		
C1475-1	315	3250	279		
C1475-2	315		279		
C1475-3	315		279		
C1475-4	315		279		
C1475A	315	3250			
C1475D	315	3250			
C1475E	315	3250			
C1477	283	3467	36		
C148(TRANSISTOR)	108		86	2038	
C15(TRANSISTOR)	123A		20	2051	
C15-1	123A	3444	20	2051	
C15-2	123A	3444	20	2051	
C15-3	123A	3444	20	2051	
C150	5480		243	2030	
C1501	288		223		
C1501P	157	3747			
C1501Q	157	3747	232		
C1501R	157	3747	232		
C1505	198	3219	251		
C1505I	198		251		
C1505K	198	3219	251		
C1505L	198	3219	251		
C1505LA	198	3219	251		
C1505M	198	3219	251		
C1506	198	3219	251		
C1507	198		251		
C1507A	198	3219	251		
C1507K	198	3219	251		
C1507L	198	3219	251		
C1507LM	198	3219	251		
C1507M	198	3219	251		
C1509	297		271	2030	
C1509P	297		271	2030	
C1509Q	297		271	2030	
C1509R	297		271	2030	
C1509S	297		271	2030	
C150T	128		243	2030	
C151	123	3265	20	2051	
C1514	376		251		
C1518	293	3849			
C1518R	293	3849			
C1518S	293	3849			
C152(TRANSISTOR)	123		243	2030	
C1520	198	3220	251		
C1520-1A	198	3220	251		
C1520-3A	198	3220	251		
C1520I	198		251		
C1520K	198	3220	251		
C1520KL	198	3220	251		
C1520L	198	3220	251		
C1520M	198	3220	251		
C1521K	198$	3220/198	251		
C1521L	198	3220	251		
C1521LM	198	3220	251		
C1537	199	3245	62	2010	
C1537-3			62	2010	
C1537-O	199			2010	
C1537B	199	3245	62	2010	
C1537S	199	3245	62	2010	
C1538	199	3245	62	2010	
C1538A	199	3245	62	2010	
C1538S	199	3245	62	2010	
C1538SA	199	3245	62	2010	
C154	152	3040	66	2048	
C1542	199	3444/123A	20	2051	
C1547	161		280		
C154B	152	3040	66	2048	
C154C	154	3040	66	2048	
C154H	5484	3040	66	2048	
C155	107	3452/108	86	2038	
C156	107	3452/108	86	2038	
C1566	157#		232		
C1569	376		251		
C1569-O	198	3219	251		
C1569-R	376	3219			
C1569LBQ	376	3219	251		
C1569LBR	376	3219	251		
C1569LBY	376	3219	251		
C1569R	376	3219	251		
C157	123	3122	20	2051	
C157(IC)	1171	3565			
C1573	399	3433/287	222		
C1573P		3433	222		
C157A(IC)	1171	3565			
C158	123	3122	20	2051	
C1585	388	3836/284			
C1585F	388	3836/284			
C1585H	388	3836/284			
C1588	278		261		
C159	123A	3122	20	2051	
C15A	5482				2N4169
C15B	5483		C15B		2N4170
C15C	5484				2N4171
C15D	5485		C15D		2N4172
C15F	5481				2N4168
C15G	5483				2N4170
C15H	5485				2N4171
C15M	5487				2N4174
C16	123A	3719/104	20	2051	
C160	123A	3122	20	2051	
C1624	291	3219			
C1625	291	3219			
C1629A	86	3563			
C1629M	86	3563			
C163	311	3048/329	46		
C1633	85	3245/199	62	2010	
C1639	123A	3444	20	2051	
C1648	199	3245	62	2010	
C1658Q		3275	220		
C166	123A	3122	20	2051	
C1667	130	3836/284	265	2047	
C1667Q	130	3836/284			
C1667R	130	3836/284			
C167	123A	3122	20	2051	
C1670	382$		217		
C1674	107		61	2038	
C1674K	107		61	2038	
C1674L	107	3018	61	2038	
C1674M	107		61	2038	
C1675	85	3124/289	61	2038	
C1675K	85		61	2038	
C1675L	85	3124/289	61	2038	
C1675M	85	3124/289	61	2038	
C1678	236	3197/235	215		
C1679	236	3197/235	215		
C1681BL	199	3245	62	2010	
C1683LA	375		251		
C1683P	375		251		
C1683R	375		251		
C1684	289A		62	2010	
C1685	85	3124/289	220		
C1685P	85	3124/289	220		
C1685Q	85	3124/289	62	2010	
C1685S	199	3124/289	220		
C1686	229		61	2038	
C1687	229	3246	20	2051	
C1688	229$		20	2051	
C16A	123A	3122	20	2051	
C17	123A	3452/108	86	2038	
C170	123A	3444	86	2038	
C171	123A	3039/316	11	2015	
C172	123A	3452/108	53	2038	
C1723	376$	3219			
C1728	302	3252	275		
C1728-3	302	3252	275		
C1728D	302	3252	275		
C172A	123A	3452/108	86	2038	
C173	101		8	2002	
C1739	123A	3444	20	2051	
C174	123A	3039/316	11	2015	
C1740	85	3122			
C1740P	85	3122			
C1740Q	85	3122			
C1740R	85	3122			
C1740S	85	3122			
C174A	161	3039/316	11	2015	
C175	101		8	2002	
C1755	376		251		
C1755A	376		251		
C1755C	376		251		
C1756	376		251		
C1756A	376	3219	251		
C1756C	376	3219	251		
C1756D	376	3219	251		
C1756K	376	3219	251		
C1757	376		251		
C175B	102A		53	2007	
C176	101		8	2002	
C1760	306	3251	276		
C1760-2	306	3251			
C1760-3	306	3251	276		
C1760E	306	3251			
C177	101		8	2002	
C178	101		8	2002	
C179	103A	3011A	59	2002	
C17A	123A	3452/108	86	2038	
C18	123A	3122	20	2051	
C180(TRANSISTOR)	103A		59	2002	
C181	103A	3835	59	2002	
C1815	85	3245/199			
C1816	236$	3197/235	216	2053	
C1816E	236$	3197/235			
C1816H	236$	3197/235			
C1817	236$	3239/236			
C182	313	3122	278	2016	
C1828	369	3131A	246	2020	
C182Q	313	3039/316	278	2016	
C183	313		278	2016	
C183E	313		278	2016	
C183J	313		278	2016	
C183K	313		278	2016	
C183L	313		278	2016	
C183M	313		278	2016	
C183P	313		278	2016	
C183Q	313		278	2016	
C183R	313		278	2016	
C183W	313		278	2016	
C184	313		278	2016	
C1847VQ	373#	9041/373			
C184H	313		278	2016	
C184J	313		278	2016	
C184L	313		278	2016	
C185	313	3039/316	278	2016	
C185(TRANSISTOR)	313		278	2016	
C185A	313	3039/316	278	2016	
C185J	313		278	2016	
C185M	313	3039/316	278	2016	

Industry Standard No.	ECG	SK	GE	RS 276-	MOTOR.
C185Q	313	3039/316	278	2016	
C185R	313	3039/316	278	2016	
C185V	108	3452	86	2038	
C186	107	3039/316	11	2015	
C1869	328	3836/284			
C187	107	3039/316	11	2015	
C188	128	3024	243	2030	
C188A	128	3024	243	2030	
C188AB	128	3024	243	2030	
C189	128	3024	243	2030	
C1894	238	3111			
C19	128	3024	243	2030	
C190	128	3024	243	2030	
C1908E		3124	11	2015	
C1909	235	3197	215		
C191	123A	3122	20	2051	
C192	123A		20	2051	
C193	123A		20	2051	
C194	123A		20	2051	
C1944	236	3239			
C1945295DY1	123A	3444	20	2051	
C195	123A		20	2051	
C1957	295		270		
C196	123A		20	2051	
C1964	236		216	2053	
C1969	236	3239			
C197	123A		20	2051	
C1973	85	3849/293	279		
C1974	236		215		
C1975	235	3197	215		
C199	123A	3452/108	86	2038	
C1A	178MP	3100/519	300(2)	1122(2)	
C1B	116	3016	504A	1104	
C1H	116	3016	504A	1104	
C2-8Z					2N5945
C20	128	3717/121	243	2030	
C200(TRANSISTOR)	123A		20	2051	
C201	129	3025	244	2027	
C201(JAPAN)	123A		20	2051	
C202	102A	3025/129	244	2027	
C202(JAPAN)	123A		20	2051	
C203	123A	3122	20	2051	
C204	123A	3122	20	2051	
C2043	236	3239			
C205	123A	3122	20	2051	
C205D					MCR22-6
C206	161	3018	86	2038	
C208	161		243	2030	
C2098	236	3239			
C20A	5491	6791	MR-3		2N4169
C20B	5492	6792	MR-3		2N4170
C20C			MR-3		2N4171
C20D	5496		MR-3		2N4172
C20F	5491	6791	MR-3		2N4168
C20U	5491	6791	MR-3		2N4167
C21	130	3027	14	2041	
C210	128	3024	243	2030	
C211	128	3024	243	2030	
C212	128	3104A	220		
C2122	283	3467			
C213	282	3024/128	243	2030	
C21382	116		504A	1104	
C21383			ZD-9.1	562	
C214	282	3024/128	243	2030	
C21480	109	3087		1123	
C215	282	3024/128	243	2030	
C216	128	3024	243	2030	
C217	128	3024	243	2030	
C218	128	3024	20	2051	
C218A	128	3122	20	2051	
C22		3024	243	2030	
C220	128	3024	243	2030	
C220A	5491	3589			
C220B	5492	3589	C220B		
C220C	5494	3580/5507	C220C		
C220D	5494	3580/5507	C220D		
C220E	5496		C220E		
C220F	5491	3589			
C220U	5491	3589			
C221	128	3024	243	2030	
C222	128	3024	243	2030	
C2228	399	3244			
C2229	399	3244			
C2229Y	399	3244			
C222A	5514	3613			
C222B	5514	3613	C222B		
C222D	5515		C222D		
C222F	5514	3613			
C222U	5514	3613			
C223	282	3024/128			
C223(TRANSISTOR)	128		243	2030	
C224	282	3024/128	243	2030	
C225	282	3024/128	243	2030	
C2256	388	3836/284			
C226	128	3024	243	2030	
C227	128	3024	243	2030	
C228	128	3024	243	2030	
C228A					C228A
C228A3	5562				C228A3
C228B					C228B
C228B3	5562				C228B3
C228C					C228C
C228C3	5564				C228C3
C228D			C228D		C228D
C228D3	5564				C228D3
C228E					C228E
C228E3	5566				C228E3
C228F	5541				C228F
C228F3	5562				C228F3
C228M					C228M
C228M3	5566				C228M3
C229	128	3024	243	2030	
C229A	5517				C229A
C229A3					C229A3
C229B	5517				C229B
C229B3					C229B3
C229C	5518				C229C
C229C3					C229C3
C229D	5518				C229D
C229D3					C229D3
C229E	5519				C229E
C229E3					C229E3
C229F	5517				C229F
C229F3					C229F3
C229M	5519				C229M
C229M3					C229M3
C22B	5514	3589			
C23	128/400	3719/104	243	2030	
C230	123A	3122	20	2051	
C2300.037-096			20	2051	
C23018	519		514	1122	
C230A					C230A
C230A3					C230A3
C230B			C230B		C230B
C230B3					C230B3
C230C			C230C		C230C
C230C3					C230C3
C230D	5527		C230D		C230D
C230D3					C230D3
C230E					C230E
C230E3					C230E3
C230F					C230F
C230F3					C230F3
C230M					C230M
C230M3					C230M3
C231	123A	3024/128	243	2030	
C231A					C231A
C231A3					C231A3
C231B					C231B
C231B3					C231B3
C231C					C231C
C231C3					C231C3
C231D			C231D		C231D
C231D3					C231D3
C231E					C231E
C231E3					C231E3
C231F					C231F
C231F3					C231F3
C231M					C231M
C231M3					C231M3
C232	123A	3122	243	2030	
C232A					C232A
C232B					C232B
C232C					C232C
C232D			C232D		C232D
C232E					C232E
C232F					C232F
C232M					C232M
C233	123A	3122	243	2030	
C2337A	87	3836/284			
C233A					C233A
C233B					C233B
C233C					C233C
C233D					C233D
C233E					C233E
C233F					C233F
C233M					C233M
C234	282	3122	243	2030	
C235	282	3512	47	2030	
C235-0		3512	47	2030	
C235-0	282	3104A		2030	
C236	128	3024	243	2030	
C237	123A	3122	20	2051	
C238	123A	3444	20	2030	
C239	123A	3122	20	2051	
C23H12B			504A	1104	
C24	128/400	3719/104	243	2030	
C240	130	3438	35		
C241	130	3438	35		
C242	130	3438	35		
C243	280	3438	35		
C244	385	3027/130	14	2041	
C245	385	3438	35		
C246	385	3438	35		
C247	128	3024	243	2030	
C2475078-3	108	3452	86	2038	
C248	123A	3024/128	243	2030	
C248507	116	3031A	504A	1104	
C2485076-3	128	3024	243	2030	
C2485077-2	128	3024	243	2030	
C2485078-1	108	3018	86	2038	
C2485079-1	108	3018	86	2038	
C249	128	3024	243	2030	
C25-12					2N6136
C25-28					MRF323
C250	123A		39	2015	
C251	161		39	2015	
C251A	161		39	2015	
C252	161		39	2015	
C253	161		39	2015	
C2538-11	123A	3444	20	2051	
C254	128	3538	246	2020	
C255110-011	177		300	1122	
C256125-011	519		514	1122	
C26	123A	3719/104	20	2051	
C261	195A	3137/192A	63	2030	
C263		3452	86	2038	
C266	313	3039/316	11	2015	
C267		3122	20	2051	
C267A	123A	3122	20	2051	
C268		3024	243	2030	
C268A		3024	243	2030	
C269		3452	86	2038	
C27	123A	3024/128	243	2030	
C270	165	3438	35		
C271	108	3452	86	2038	
C272	108	3452	86	2038	
C273	154	3044	40	2012	
C277C	103A	3010	59	2002	
C28	123	3122	20	2051	
C281	123A	3122	20	2051	
C281A	123A	3444	20	2051	
C281B	123A	3444	20	2051	
C281C	123A	3444	20	2051	
C281C-EP	123A	3444	20	2051	
C281D	123A	3444	20	2051	
C281EP	123A	3444	20	2051	
C281H	123A	3444	20	2051	
C281HA	123A	3444	20	2051	
C281HB	123A	3444	20	2051	
C281HC	123A	3444	20	2051	
C282	123A	3039/316	20	2051	
C282(TRANSISTOR)	108		86	2038	
C282H	123A	3444	20	2051	
C282HA	123A	3444	20	2051	
C282HB	123A	3444	20	2051	
C282HC	123A	3444	20	2051	
C283	123A	3122	20	2051	
C283(TRANSISTOR)	123A		20	2051	
C284	123A	3444	20	2051	
C284H	123A	3444	20	2051	
C284HA	123A	3444	20	2051	
C284HB	123A	3444	20	2051	
C285	311	3444/123A	20	2051	
C285A	311	3444/123A	20	2051	
C286(TRANSISTOR)	108		86	2038	
C287(TRANSISTOR)	313		278	2016	
C287A		3039	278	2016	
C288		3039	278	2016	
C288A		3039	278	2016	
C289	108	3452	86	2038	
C29	123	3122	20	2051	
C290		3048	46		
C290(TRANSISTOR)	195A		46		
C291(TRANSISTOR)	128		243	2030	
C292	282	3024/128	243	2030	
C293	282	3048/329	46		
C296	161	3452/108	86	2038	

Industry Standard No.	ECG	SK	GE	RS 276-	MOTOR.
C299	282/427	3122	256		
C2AJ102	116		504A	1104	
C2M100-28					MRF327
C2M50-28R					2N6439
C2M60-28R					2N6439
C2M70-28R					MRF327
C3-12	363				2N5645
C3-28					MRF5175
C30	128	3024	243	2030	
C300	123A	3444	20	2051	
C3000P	7400			1801	
C3002P	7402			1811	
C3004P	7404			1802	
C301	123A	3444	20	2051	
C3010P	7410			1807	
C301A(PNP)	129		244	2027	
C302		3122	21	2034	
C302(JAPAN)	123A	3444	20	2051	
C302(PNP)	129		244	2027	
C3020P	7420			1809	
C3041P	7441	7441		1804	
C306	128	3024	243	2030	
C307	128	3024	243	2030	
C3073P	7473			1803	
C3075P	7475			1806	
C308	128	3024	243	2030	
C309	128	3024	243	2030	
C30A	5522	6622	MR-3		2N683
C30B	5524	6624	MR-3		2N685
C30C	5526	6627/5527	MR-3		2N687
C30D	5527	6627	MR-3		2N688
C30F	5521	6621	MR-3		2N682
C30U	5520	6621/5521	MR-3		2N681
C31	123A	3024/128	243	2030	
C310	282	3024/128	243	2030	
C311	195A	3048/329	46		
C312	224	3049	46		
C3123	108	3452	86	2038	
C313	161	3452/108	86	2038	
C313(JAPAN)	161	3716	39	2015	
C313C	161	3132	39	2015	
C313H	161	3132	39	2015	
C315	123A	3444	20	2051	
C316	123A	3452/108	86	2038	
C317	123A	3444	20	2051	
C317-O		3444	20	2051	
C317C	123A	3444	20	2051	
C318	123A	3452/108	86	2038	
C318(JAPAN)	123A	3444	20	2051	
C318A	123A	3452/108	86	2038	
C318A(JAPAN)	123A	3444	20	2051	
C319	311	3444/123A	20	2051	
C31A		3581			2N683
C31B		3581			2N685
C31C	1048	3582/5545			2N687
C31D		3582			2N688
C31F		3581			2N682
C31U		3581			2N681
C32	123A	3722/102	243	2030	
C320	472	3444/123A	20	2051	
C321	123A	3444	20	2051	
C321H	123A	3444	20	2051	
C321HA	123A	3444	20	2051	
C321HB	123A	3444	20	2051	
C321HC	123A	3444	20	2051	
C323	123A	3444	20	2051	
C324	123A	3444	20	2051	
C324A	123A	3444	20	2051	
C324H	123A	3444	20	2051	
C324HA	123A	3444	20	2051	
C325	316	3054/196	66	2048	
C325A	316		66	2048	
C325E	152	3054/196	66	2048	
C328A	316	3444/123A	20	2051	
C329	316		11	2015	
C329B	316		11	2015	
C329C	316		11	2015	
C32A	128	3024	243	2030	2N683
C32B					2N685
C32C					2N687
C32D					2N688
C32F					2N682
C32U					2N681
C33	161		20	2051	
C333H	123A$		261		
C333HA	123A$		261		
C333HB	123A$		261		
C333HC	123A$		261		
C335	123A	3245/199	62	2010	
C337B	123A	3009	239	2006	
C33A					MCR3818-3
C33B					2N5165
C33C					MCR3818-5
C33D					2N5166
C33F					2N5164
C33U					MCR3818-1
C34(TRANSISTOR)	103A		59	2002	
C341V		3035	25		
C348	123A	3132	39	2015	
C35(TRANSISTOR)	103A		59	2002	
C350(TRANSISTOR)	123A		20	2051	
C350H	123A	3444	20	2051	
C351	107	3293	11	2015	
C351(FA)	107		11	2015	
C352	123A	3444	20	2051	
C352(JAPAN)	123A	3444	20	2051	
C352A	123A	3444	20	2051	
C352A(JAPAN)	128		243	2030	
C353	128	3024	243	2030	
C353A	128	3024	243	2030	
C356		3444	20	2051	
C35A	5542		MR-3		C35A
C35B	5543		MR-3		C35B
C35C	5544		MR-3		C35C
C35D	5545		MR-3		C35D
C35E	5546				C35E
C35F	5541		MR-3		C35F
C35G	5543		MR-3		C35G
C35H	5544				C35H
C35M	5547		C35M		C35M
C35N					C35N
C35S					C35S
C35U	5540		MR-3		C35U
C36	103	3719/104			
C36(TRANSISTOR)	101		8	2002	
C360	123A	3444	20	2051	
C360D	123A	3444	20	2051	
C361	289A	3039/316	11	2015	
C362	289A	3039/316	11	2015	
C363	85	3444/123A	20	2051	
C36566			255	2041	
C36577		3025	244	2027	
C36578		3039	39	2015	
C36579		3024	243	2030	
C36580		3444	20	2051	
C36583	152	3893	66	2048	
C366	85	3444/123A	20	2051	
C367	85	3444/123A	20	2051	
C368	199	3245	62	2010	
C368BL	199	3245	62	2010	
C368GR	199	3245	62	2010	
C368V	199	3245	62	2010	
C369	199		20	2051	
C369BL	199	3250/315	20	2051	
C369G	199	3250/315	20	2051	
C369G-BL	199	3250/315	20	2051	
C369G-GR	199	3250/315	20	2051	
C369G-V	199		20	2051	
C369GBL	199		20	2051	
C369GGR	199		20	2051	
C369GR	199	3250/315	20	2051	
C369V	199		20	2051	
C36A	5502		MR-3		2N1844
C36B	5504		MR-3		2N1846
C36C	5506		MR-3		2N1848
C36D	5507		MR-3		2N1849
C36E	5508				2N1850
C36F	5501		MR-3		2N1843
C36G	5503		MR-3		2N1845
C36H	5505				2N1847
C36U	5500		MR-3		2N1842
C37	123A	3122	20	2051	
C37(TRANSISTOR)	123A		20	2051	
C370	289A	3452/108	86	2038	
C370F	289A	3452/108	86	2038	
C370G	289A	3452/108	86	2038	
C370H	289A	3452/108	86	2038	
C370J	289A	3452/108	86	2038	
C370K	289A	3452/108	86	2038	
C371	289A	3122	20	2051	
C371(O)	289A	3122	20	2051	
C371-O	289A	3122	20	2051	
C371-0	289A	3122	20	2051	
C371-R	289A	3122	20	2051	
C371-R-1	289A	3122	62	2010	
C3710		3122	20	2051	
C371B	289A	3122	20	2051	
C371G	289A	3122	20	2051	
C371O	289A	3122			
C371R	289A	3122	20	2051	
C372	85	3245/199	20	2051	
C372(O)		3245	20	2051	
C372-O		3245	20	2051	
C372-1	85	3245/199	20	2051	
C372-2	85	3245/199	20	2051	
C372-0	85	3245/199	20	2051	
C372-R	85	3245/199	20	2051	
C372-Y	85	3245/199	20	2051	
C372-Z	85	3245/199	20	2051	
C372GR	85	3245/199	20	2051	
C372H	85	3245/199	20	2051	
C372Y	85	3245/199	20	2051	
C373	85	3245/199	62	2010	
C373BL	85	3245/199	62	2010	
C373G	85	3245/199	62	2010	
C373GR	85	3245/199	62	2010	
C373W	85	3245/199	62	2010	
C374	199	3444/123A	62	2010	
C374-BL	199	3444/123A	62	2010	
C374-V	199	3444/123A	62	2010	
C374JA	199	3444/123A	62	2010	
C375	107	3132	11		
C375-O	107		277		
C375-0	311	3132	11	2015	
C375-Y	107	3132	277		
C376	85	3024/128	243	2030	
C377	85		20	2051	
C378	289A	3444/123A	20	2051	
C379	85	3444/123A	20	2051	
C37A	5522	6622	MR-3		MCR3918-3
C37B	5524	6624	MR-3		2N5169
C37C	5526	6627/5527	MR-3		MCR3918-5
C37D	5527	6627	MR-3		2N5170
C37E	5528	6629/5529			MCR3918-7
C37F	5521	6621	MR-3		2N5168
C37G		3581	MR-3		2N5169
C37H		3582			MCR3918-5
C37M	5529	6629			2N5171
C37U	5520	6621/5521	MR-3		MCR3918-1
C38			20	2051	
C38(TRANSISTOR)	123A		20	2051	
C380	85	3245/199	61	2038	
C380(TRANSISTOR)	229		61	2038	
C380-O	85	3245/199	61	2038	
C380-0	85		61	2038	
C380-O/4454C			61	2038	
C380-Y	85	3245/199			
C3800P	7490			1808	
C3801P	7492	7492		1819	
C380A	85	3245/199	61	2038	
C380A(O)	85		61	2038	
C380A-O(TV)	85		61	2038	
C380A-R	85	3245/199	61	2038	
C380A-R(TV)	85		61	2038	
C380AO	85	3245/199	61	2038	
C380ATV	85		61	2038	
C380AY	85	3245/199	61	2038	
C380D	85		61	2038	
C380R	85	3245/199	61	2038	
C380R/4454C			61	2038	
C380Y	85	3245/199	61	2038	
C381	107		11	2015	
C381-O	107		11	2015	
C381-0	107		11	2015	
C381-R	107		11	2015	
C381BN	107		11	2015	
C381R	107		11	2015	
C382		3132	283	2016	
C382-BK(1)			283	2016	
C382-BK(2)			283	2016	
C382-GR		3132	283	2016	
C382-GY		3132	283	2016	
C382BK		3132	283	2016	
C382BL		3132	283	2016	
C382BN		3132	283	2016	
C382BR		3132	283	2016	
C382G		3132	283	2016	
C382R		3132	283	2016	
C382V		3132	283	2016	
C383	85	3124/289	283	2016	
C383G	85	3124/289	283	2016	
C383T	85		283	2016	
C383W	85		283	2016	
C383Y	85	3124/289	283	2016	
C384	107	3132	86	2038	
C384-O	107		86	2038	
C384Y	107		86	2038	
C385(TRANSISTOR)	107		11	2015	

Industry Standard No.	ECG	SK	GE	RS 276-	MOTOR.
C385A	107	3293	11	2015	
C386	107	3452/108	86	2038	
C386A-O(TV)	107		86	2038	
C387	311		86	2038	
C387(TRANSISTOR)	107		280		
C387A	311		86	2038	
C387A(FA-3)	108	3293/107	86	2038	
C387A(FA-3)TRAN	316		280		
C387A(TRANSISTOR)	316		280		
C387G	108		86	2038	
C387G(TRANSISTOR)	107		280		
C388	85		210	2051	
C388(TRANSISTOR)	233		210	2009	
C388A	85		210	2009	
C388ATV	85		210	2009	
C389	161	3019	39	2015	
C389-O	161	3019	39	2015	
C389R	161	3019	39	2015	
C38A	5541				2N3896
C38B	5543				2N3897
C38C	5544				MCR3935-5
C38D	5545				2N3898
C38E	5546		C38E		MCR3935-7
C38F	5541				MCR3935-2
C38G	5543				2N3897
C38H	5544				MCR3935-5
C38U	5540	6642/5542			MCR3935-1
C39(TRANSISTOR)	108		86	2038	
C39-207	123A	3444	20	2051	
C390	316	3132			
C390(TRANSISTOR)			39	2015	
C392	316	3452/108	86	2038	
C394	289A	3124/289	61	2015	
C394(TRANSISTOR)			61	2038	
C394-O	289A	3124/289	61	2038	
C394-O	289A	3124/289	61	2038	
C394GR	289A	3124/289	61	2038	
C394R	289A	3124/289	61	2038	
C394W	289A	3124/289	61	2038	
C394Y	289A	3124/289	61	2038	
C395	123A	3444	20	2051	
C395A	123A	3444	20	2051	
C395R	123A	3444	20	2051	
C396	311	3444/123A	20	2051	
C397(TRANSISTOR)	108		86	2038	
C398	319P	3018			
C398(FA-1)	319P	3018	39	2015	
C398(TRANSISTOR)			39	2015	
C399	319P	3018	39	2015	
C39A	108	3452	86	2038	
C40	108	3452	86	2038	
C40-28					MRF326
C400	123A	3444	20	2051	
C400-O	123A	3444	20	2051	
C400-GR	123A		62	2010	
C400-R	123A	3444	20	2051	
C400-Y	123A	3444	20	2051	
C401	289A		21	2034	
C401(JAPAN)	123A	3124/289	20	2051	
C4011	137A		ZD-6.2	561	
C4015	139A		ZD-9.1	562	
C4016	140A	3061	ZD-10	562	
C4018	142A	3062	ZD-12	563	
C402	289A	3025/129	244	2027	
C402(JAPAN)	123A	3444	20		
C4020	145A	3063	ZD-15	564	
C4026	146A	3064	ZD-27		
C402A	289A	3444/123A	20		
C403	289A		20		
C403(C)	289A		20	2051	
C403A	289A		20		
C403B(SONY)	289A		243	2030	
C403C	289A		20	2051	
C403C(SONY)			243	2030	
C404	289A	3444/123A	20	2051	
C405	123A	3452/108	86	2038	
C406	123A	3011A	86	2038	
C407	385	3044/154	40	2012	
C407(JAPAN)	162		35		
C408	385	3439/163A	36		
C409	385	3439/163A	36		
C41	163A	3722/102	36		
C410	385	3439/163A	36		
C41001	129		244	2027	
C411	385	3439/163A	36		
C412	278	3439/163A	36		
C413E					MPF4391
C413N					2N4091
C41TV	163A	3439	36		
C42	163A	3438	35		
C420	128	3024	243	2030	
C423	311	3444/123A	20	2051	
C423B	123A	3444	20	2051	
C423C	123A	3444	20	2051	
C423D	123A	3444	20	2051	
C423E	123A	3444	20	2051	
C423F	123A	3444	20	2051	
C424	123A	3452/108	86	2038	
C424(JAPAN)	123A	3444	20	2051	
C424D	123A	3444	20	2051	
C425	128	3444/123A	20	2051	
C425B	123A	3444	20	2051	
C425C	123A	3444	20	2051	
C425D	123A	3444	20	2051	
C425E	123A	3444	20	2051	
C425F	123A	3444	20	2051	
C426	128	3024	243	2030	
C428	311		47	2030	
C429	313	3452/108	86	2038	
C429J	313	3452/108	86	2038	
C429X	313	3452/108	86	2038	
C42A	163A	3438	35		
C43	163A	3438	35		
C430	313	3452/108	86	2038	
C430H	313	3452/108	86	2038	
C430W	313	3452/108	86	2038	
C44	163A	3438	35		
C440	311	3122	20	2051	
C441	311	3122	20	2051	
C44189	910		IC-251		
C442	311	3122	20	2051	
C443	128	3024	243	2030	
C444	123A	3444	20	2051	
C45	123A	3122	20	2051	
C450	123A	3444	20	2051	
C454	85	3124/289	86	2038	
C454(A)	85	3124/289	20	2051	
C454A	85	3124/289	20	2051	
C454B	85	3124/289	268	2051	
C454C	85	3124/289	20	2051	
C454D	85	3124/289	20	2051	
C454L	85	3124/289	20	2051	
C454LA	85	3124/289	20	2051	
C455	289A	3138/193A	268	2038	
C456	123A	3048/329	46		
C456-O	195A	3048/329	46		
C456A	195A	3048/329	46		
C456D	195A	3048/329	46		
C458	85	3124/289	268	2038	
C458(C)	85	3124/289	268	2038	
C458A	85	3124/289	268	2038	
C458AD	85		268	2038	
C458B	85	3124/289	268	2038	
C458B-D		3124	268	2038	
C458BC	85	3124/289	268	2038	
C458BK			268	2038	
C458BL	85	3124/289	268	2038	
C458BLG		3124	268	2038	
C458C	85	3124/289	268	2038	
C458CLG	85	3124/289	268	2038	
C458CM	85	3124/289	268	2038	
C458D	85	3124/289	268	2038	
C458G	85	3124/289	268	2038	
C458GLB	85	3124/289	268	2038	
C458K	85		268	2038	
C458KA	85		268	2038	
C458KB	85		268	2038	
C458KC	85		268	2038	
C458KD	85		268	2038	
C458L	85	3124/289	268	2038	
C458LB	85	3124/289	268	2038	
C458LG	85	3124/289	268	2038	
C458LG(B)	85	3124/289	20	2051	
C458LGA	85	3124/289	268	2038	
C458LGB	85	3124/289	268	2038	
C458LGBM	85	3124/289	268	2038	
C458LGC	85	3124/289	268	2038	
C458LGD	85	3124/289	268	2038	
C458LGO	85	3124/289	268	2038	
C458LGS	85	3124/289	268	2038	
C458M	85		268	2038	
C458P	85	3124/289	268	2038	
C458RGS	85	3124/289	268	2038	
C458TOK	85		268	2038	
C458VC	85		268	2038	
C459	85	3122	243	2030	
C459B	85	3122	243	2030	
C459D	85	3122	268	2038	
C46	128	3024	243	2030	
C460	85	3122	210	2038	
C460(A)	85		210	2051	
C460(B)	85		210	2051	
C460A	85	3122	210	2009	
C460B	85	3122	210	2009	
C460C	85	3122	210	2009	
C460D	85	3122	210	2009	
C460G	85		210	2009	
C460GB	85	3122	210	2009	
C460H	85	3122	210	2009	
C460K	85	3122	210	2009	
C460L	85	3122	210	2009	
C461	85	3122	61	2038	
C461A	85	3122	61	2038	
C461B	85	3122	61	2038	
C461C	85	3122	61	2038	
C461E	85	3122	61	2038	
C461L	85	3122	61	2038	
C463	161	3039/316	283	2016	
C463(Y)	131		44	2006	
C463H	161	3132	283	2016	
C464	316	3132	39	2015	
C464C	316	3138/193A	268	2015	
C465	316	3132	39	2015	
C466	316	3132	39	2015	
C466H	316	3132	39	2015	
C468	123A	3444	20	2051	
C468(LGR)	123A	3138/193A	268	2038	
C468A	123A	3444	20	2051	
C469	313	3039/316	278	2016	
C469A	313	3039/316	278	2016	
C469F	313	3039/316	278	2016	
C469K	313		278	2016	
C469Q	313	3039/316	278	2016	
C469R	313	3039/316	278	2016	
C47	128	3122	20	2051	
C470	154	3044	40	2012	
C472Y	123A	3044/154	40	2012	
C475	313	3444/123A	20	2051	
C475K	313	3444/123A	20	2051	
C476	313	3122	20	2051	
C477	161	3452/108	86	2038	
C478	123A	3048/329	46		
C478(D)	123A	3048/329	46		
C478-4	123A	3048/329	46		
C478D	123A	3048/329	46		
C479	128	3024	243	2030	
C48	128	3122	20	2051	
C481	128	3765/195A	46		
C481X	195A	3765	46		
C482	128	3512	46		
C482-GR	195A	3512	46		
C482-O	195A	3512	46		
C482-Y	195A	3512	46		
C482GR	195A	3512	46		
C482X	195A	3512	46		
C482Y	195A	3512	46		
C484	324		264		
C484BL	324		264		
C484R	324	3024/128	264		
C484Y	324	3024/128	264		
C485	324	3748/282	264		
C485BL	324	3748/282	264		
C485C	324	3748/282	264		
C485Y	324	3748/282	264		
C486	324	3748/282	243		
C486BL	324	3748/282	264		
C486Y	324	3748/282	264		
C487	175	3538	246	2020	
C488	175	3538	246	2020	
C489	175	3538	246	2020	
C48C	128		20	2051	
C49	154	3024/128	243	2030	
C490	175	3271	246	2020	
C491	175	3271	246	2020	
C491BL	175	3271	246	2020	
C491R	175	3271	246	2020	
C491Y	175	3271	246	2020	
C492	280		35		
C493	87	3027/130	255	2041	
C493-BL	87		255	2041	
C493-R	87		255	2041	
C493-Y	87		255	2041	
C494		3027	255	2041	
C494-BL			255	2041	
C494-R			255	2041	
C494-Y			255	2041	
C494BL			255	2041	
C495	295		57	2017	

Industry Standard No.	ECG	SK	GE	RS 276-	MOTOR.
C495-0	184		57	2017	
C495-R	295		57	2017	
C495-Y	295		270		
C495T	295		270		
C495T(CB)	295		270		
C495Y	295		270		
C496	295		255		
C496(O)	295		270		
C496-0	295		270		
C496-0	295		270		
C496-R	295		270		
C496-Y	295		270		
C496Y	184		57	2017	
C497	128	3024	243	2030	
C497-0	128	3024	243	2030	
C497-R	128	3024	243	2030	
C497-Y	128	3024	243	2030	
C498	128	3024	243	2030	
C498-0	128	3024	243	2030	
C498-R	128	3024	243	2030	
C498-Y	128	3024	243	2030	
C499	382	3244	220		
C499-R(FA-1)	194		220		
C499-RY	194		220		
C499-Y(FA-1)	194		220		
C499R	194	3244	220		
C499Y	194	3244	220		
C49Y	128	3027/130	14	2041	
C5-12	363				2N5945
C5-8Z					2N5946
C50(TRANSISTOR)	101		8	2002	
C50-28					2N6439
C500	154	3040	40	2012	
C5005	109	3087		1123	
C500R	154	3040	40	2012	
C500Y	154	3040	40	2012	
C501(TRANSISTOR)	128		243	2030	
C502	195A	3048/329	46	2027	
C502(JAPAN)	195A		46		
C503	128	3024	243	2030	
C503-0	128	3024	243	2030	
C503-Y	128	3024	243	2030	
C503GR	128		243	2030	
C504	128	3024	243	2030	
C504-0	128	3024	243	2030	
C504-Y	128	3024	243	2030	
C504GR	128	3024	243	2030	
C505	154	3044	40	2012	
C505-0	154	3044	40	2012	
C505-R	154	3044	40	2012	
C506	154	3044	40	2012	
C506-0	154	3044	40	2012	
C506-R	154	3044	40	2012	
C507	154	3044	40	2012	
C507-0	154	3044	40	2012	
C507-R	154	3044	40	2012	
C507-Y	154	3044	40	2012	
C508	384		267		
C509	289A		268	2038	
C509(O)	289A			2030	
C509(O)			243	2030	
C509(TRANSISTOR)	289A		268	2038	
C509G	289A		243	2030	
C509Y	289A		268	2038	
C50A	101	3011	8	2002	
C50BA042	121	3009	239	2006	
C51	128	3024	243	2030	
C511A				1067	
C511B				1067	
C511C				1020	
C511F				1067	
C511G				1067	
C511H				1020	
C511U				1067	
C512	324	3024/128	243	2030	
C512-0	324	3024/128	243	2030	
C512-R	324	3024/128	243	2030	
C513	324	3024/128	243	2030	
C513-0	324	3024/128	243	2030	
C513R	324	3024/128	243	2030	
C514	124	3021	12		
C515	124	3021	12		
C515A	124	3021	12		
C515A(BK)	124		12		
C515AX	124		12		
C516	128	3024	243	2030	
C517	237	3049/224	46		
C517C	237	3049/224	46		
C519	87	3438	35		
C51909B			82	2032	
C519A	87	3438	35		
C52(TRANSISTOR)	123A		20	2051	
C520	87	3027/130			
C520(TRANSISTOR)			14	2041	
C520A	87	3027/130	14	2041	
C521	87	3027/130	14	2041	
C521A	87	3027/130	14	2041	
C522	427/282	3103A/396	256		
C522-0		3103A	256		
C522-R	225	3103A/396	256		
C523	427/282	3103A/396	256		
C523-0		3103A	256		
C523-R	225	3103A/396	256		
C523383			17	2051	
C524	427/282	3103A/396	256		
C524-0		3103A	256		
C524-R	225	3103A/396	256		
C525	237	3103A/396	256		
C525-0	225		256		
C525-R	237	3103A/396	256		
C526	154	3040	40	2012	
C529	85	3122	20	2051	
C529A	85	3444/123A	20	2051	
C53	123A	3122	20	2051	
C532000585	234		65	2050	
C535	107		61	2038	
C5359		3452	86	2038	
C535A	107		61	2038	
C535B	107		61	2038	
C535C	107		61	2038	
C535G	107		61	2038	
C536	199	3122	62	2010	
C536A		3122	62	2010	
C536AG		3122	62	2010	
C536B		3122	62	2010	
C536C		3122	62	2010	
C536D	289A	3122	62	2010	
C536DK		3122	62	2010	
C536E	289A	3122	62	2010	
C536ED		3122	62	2010	
C536EH		3122	62	2010	
C536EJ		3122	62	2010	
C536EN		3122	62	2010	
C536ER		3122	62	2010	

Industry Standard No.	ECG	SK	GE	RS 276-	MOTOR.
C536ET		3122	62	2010	
C536EZ		3122	62	2010	
C536F		3122	62	2010	
C536F1		3122	62	2010	
C536F2		3122	62	2010	
C536FC		3122		2010	
C536FP			62	2010	
C536FS	199	3122	62	2010	
C536FS6		3122	62	2010	
C536FZ		3122	62	2010	
C536G	199	3122	62	2010	
C536GF		3122	62	2010	
C536GK		3122	62	2010	
C536GT		3122	62	2010	
C536GV		3122	62	2010	
C536GY		3122	62	2010	
C536H	199	3122	62	2010	
C536W		3122	62	2010	
C537	289A		20	2009	
C537(F)	199		20	2051	
C537(G)	199		20	2051	
C537-01	289A	3122	20	2051	
C537A	199	3122	20	2051	
C537B	199	3122	20	2051	
C537C	289A	3122	20	2051	
C537D	289A	3122	20	2051	
C537D2	289A	3122	20	2051	
C537E	289A	3122	20	2051	
C537EF	289A	3122	20	2051	
C537EH	289A	3122	20	2051	
C537EJ	289A	3122	20	2051	
C537EK	289A	3122	20	2051	
C537F	289A	3122	20	2051	
C537F1	289A	3122	20	2051	
C537F2	289A	3122	20	2051	
C537FC	289A	3122	20	2051	
C537FV	199	3122	20	2051	
C537G	199	3122	20	2051	
C537G1			20	2051	
C537GF	123A	3122	20	2051	
C537GI	199	3122	20	2051	
C537H	199	3122	20	2051	
C537HT	199	3122	20	2051	
C537W	199	3122	20	2051	
C538	123A	3444	20	2051	
C538A	123A	3444	20	2051	
C538AQ	123A	3444	20	2051	
C538P	123A	3444	20	2051	
C538Q	123A	3444	20	2051	
C538R	123A	3444	20	2051	
C538S	123A	3444	20	2051	
C538T	123A	3444	20	2051	
C539	123A	3444	20	2051	
C539K	123A	3444	20	2051	
C539L	123A	3444	20	2051	
C539R	123A	3444	20	2051	
C539S	123A	3444	20	2051	
C54	123A	3122	20	2051	
C540	313	3444/123A	20	2051	
C544	107	3452/108	86	2038	
C544C	107	3452/108	86	2038	
C544D	107	3452/108	86	2038	
C544E	107	3452/108	86	2038	
C545	107	3039/316	11	2015	
C545A	107	3039/316	11	2015	
C545B	107	3039/316	11	2015	
C545C	107	3039/316	11	2015	
C545D	107	3039/316	11	2015	
C545E	107	3039/316	11	2015	
C55	123A	3122	20	2051	
C555A	703A	3157	IC-12		
C555H(IC)	1092	3472			
C558	389	3439/163A	36		
C56	107	3018	11	2015	
C560	128	3024	243	2030	
C561	108	3452	86	2038	
C562	161		39	2015	
C5620			244	2027	
C5620	161			2027	
C562Y	161		39	2015	
C563	161	3132	39	2015	
C563A	161	3132	39	2015	
C563A(3RDIP)			210	2009	
C564	123A		243	2030	
C564A	123A		243	2030	
C564P	123A	3025/129	243	2030	
C564Q	123A	3025/129	243	2030	
C564R	123A		243	2030	
C564S	123A		243	2030	
C564T	123A		243	2030	
C566	316$		20	2051	
C566(IC)	1052	3249			
C566H(IC)	1052	3249			
C566H-L	1052	3249			
C566H-M	1052	3249			
C566H-N	1052	3249			
C566H2(IC)	1052	3249			
C567	316	3132	39	2015	
C568	316		86	2038	
C571	346	3048/329	46		
C573(IC)	1184	3486			
C573C(IC)	1184	3486			
C576(IC)	1185	3468			
C576H(IC)	1185	3468			
C577(IC)	1082	3461			
C577H	1082	3461			
C577H(IC)	1082	3461			
C058	102A	3044/154	40	2012	
C580	128	3048/329	46		
C582	124	3021	12		
C582A	124	3021	12		
C582B	124		12		
C582BC	124		12		
C582BX	124		12		
C582BY	124		12		
C582C	124	3021	12		
C583	316		261		
C583C	316		261		
C586	280	3438	35		
C587	123A	3444	20	2051	
C587A	123A	3444	20	2051	
C588	123A	3444	20	2051	
C589	154	3044	40	2012	
C58A	154	3044	40	2012	
C59	154	3024/128	243	2030	
C590	154	3024/128	268	2038	
C590Y	289A		268	2038	
C592A2	1195	3469			
C592H2	1195	3469			
C593	123A	3444	20	2051	
C594	123A	3024/128	243	2030	
C595	123A		20	2051	
C596		3444	20	2051	
C596(IC)	1187	3742			

Industry Standard No.	ECG	SK	GE	RS 276-	MOTOR.
C5A					2N2324
C5B					2N2326
C5F					2N2323
C5G					2N2325
C5U					2N2322
C60	103A	3011	1N60	1123	
C60(DIODE)	109	3090		1123	
C60(TRANSISTOR)	101		8	2002	
C6001	760	3157/703A	IC-12		
C601	311		277		
C6012	762				MZC2.7A10
C602E	123A	3122	20	2051	
C6032					MZC47A10
C605	313	3039/316	278	2016	
C605(NEC)	313		278	2016	
C605(Q)	313		278	2016	
C6052P	941M	3553	IC-265	007	
C6055L	725	3162	IC-19		
C6056P	722	3161	IC-9		
C6057P	790	3454	IC-230		
C6059P	746	3234	IC-217		
C605K	313		278	2016	
C605L	313		278	2016	
C605M	313		278	2016	
C605Q	313	3018	11	2016	
C605TW	313	3039/316	278	2016	
C606	313	3039/316	278	2016	
C606(NEC)	313		278	2016	
C6060P	748	3236			
C6061P	750	3280	IC-219		
C6062P	708	3135/709	IC-10		
C6063P	712	3072	IC-2		
C6065P	719		IC-28		
C6066P	798	3216	IC-234		
C6068P	720		IC-7		
C6069G	780	3141	IC-222		
C6070P	714	3075	IC-4		
C6071P	715	3076	IC-6		
C6072P	739	3235	IC-30		
C6074P	718	3159	IC-8		
C6075P	738	3167	IC-29		
C6076P	749	3168	IC-97		
C6077P	782		IC-224		
C6079P	747	3279	IC-218		
C608	237	3765/195A	46		
C6080P	721		IC-14		
C6081P	779A	3240/779-1	IC-221		
C6082P	708	3135/709	IC-10		
C6083P	712	3072	IC-2		
C6085P	731	3170	IC-13		
C6089	705A	3134	IC-3		
C608E	237	3765/195A	46		
C608T	237	3765/195A	46		
C609	237	3765/195A	46		
C609(TRANSISTOR)	237		46		
C6090	804	3455	IC-27		
C6091	816	3242			
C6096P	801	3160	IC-35		
C6099P	795	3237	IC-232		
C609F	237		46		
C609F(TRANSISTOR)	237		46		
C609T	237	3765/195A	46		
C609T(TRANSISTOR)	237		46		
C61	128	3024	243	2030	
C610		3025	82	2032	2N4092
C6100P	783	3215	IC-225		
C6101P	723	3144	IC-15		
C6102P	778A	3465	IC-220		
C6103P	909		IC-249		
C611	316		86	2038	MFE2095
C611(TRANSISTOR)	108		86	2038	
C611A				1067	
C611B			C611B	1067	
C611F				1067	
C611G				1067	
C611U				1067	
C612	316	3452/108	86	2038	MFE2095
C613		3452	86	2038	MFE2095
C614	195A	3048/329	46		MFE2093
C6141990	177		300	1122	
C614D	195A	3048/329	46		
C614E	195A	3048/329	46		
C614F	195A	3048/329	46		
C614G	195A	3048/329	46		
C615	195A	3048/329	46		MFE2095
C615A	195A	3048/329	46		
C615B	195A	3048/329	46		
C615C	195A	3048/329	46		
C615D	195A	3048/329	46		
C615E	195A	3048/329	46		
C615F	195A	3048/329			
C615G	195A	3048/329	46		
C619		3444	20	2051	
C619B		3444	20	2051	
C619C		3444	20	2051	
C619D		3444	20	2051	
C62(TRANSISTOR)	123A		20	2051	
C620			20	2051	MFE2093
C620C			20	2051	
C620CD			20	2051	
C620D			20	2051	
C620DE			20	2051	
C620E			20	2051	
C621	85	3444/123A	20	2051	MFE2093
C622	123A	3444	20	2051	MFE2093
C623					MFE2093
C624					MFE2093
C625					MFE2093
C627	154	3044	40	2012	
C628	311	3024/128	243	2030	
C629	107	3039/316	11	2015	
C63	123A	3122	20	2051	
C63(JAPAN)	108		86	2038	
C631	85		20	2051	
C631A	85		20	2051	
C632	85		20	2051	
C632A	85		20	2051	
C633	85	3124/289	20	2051	
C633-7	85	3124/289	20	2051	
C633A	85	3124/289	20	2051	
C633G	85	3124/289	20	2051	
C633H	85	3124/289	20	2051	
C634	85	3124/289	20	2051	
C634A	85	3124/289	20	2051	
C635A	124	3021	12		
C636	187		29	2025	
C64	128	3122	20	2051	
C64(JAPAN)	154		40	2012	
C640	313	3444/123A	20	2051	
C640B	313	3444/123A	20	2051	
C641	85	3765/195A	86	2038	
C641B	85	3452/108	86	2038	
C642	164	3133	37		
C642A	164	3710/238	37		

Industry Standard No.	ECG	SK	GE	RS 276-	MOTOR.
C643	389		259		
C643A	238		259		
C644	199	3245	62	2010	
C644C	199	3245	62	2010	
C644F	199		62	2010	
C644F/494	199	3245	62	2010	
C644FR	199	3245	62	2010	
C644FS	199	3245	62	2010	
C644H	199	3245	62	2010	
C644HR	199	3245	62	2010	
C644HS	199		47	2030	
C644P	199	3245	62	2010	
C644PJ	199	3245	62	2010	
C644Q	199	3245	62	2010	
C644R	199	3245	62	2010	
C644RST	199	3245	62	2010	
C644S	199	3245	62	2010	
C644S/494	199	3245	62		
C644ST				2010	
C644T	199	3245	62	2010	
C645		3018	11	2015	
C645A		3039	11	2015	
C645B		3039	11	2015	
C645C		3039	11	2015	
C645G			11	2015	
C645N			11	2015	
C646	328	3563	255	2041	
C647	328		255	2041	
C647Q	328		14	2041	
C647R	328		14	2041	
C648	123A	3245/199	62	2010	
C648H	123A	3245/199	62	2010	
C649	123A	3039/316	86	2038	
C65	154	3044	40	2012	
C650	123A	3122	62	2010	MFE2093
C650B	123A	3122	20	2010	
C651	278	3122	261		MFE2093
C652	311	3039/316	277		MFE2093
C653	316				MFE2093
C654	311	3444/123A	20	2051	
C655		3444	20	2051	
C656		3039	11	2015	
C657	107	3039/316	11	2015	
C658	107	3452/108	86	2038	
C658A	107	3452/108	86	2038	
C659	107	3452/108	86	2038	
C65B	154	3044	40	2012	
C65N	154	3044	40	2012	
C65Y	154	3044	40	2012	
C65YA	154	3044	40	2012	
C65YB	154	3044	40	2012	
C65YTV	154	3044	40	2012	
C66	154	3044	40	2012	
C66-P11111-0001	123A	3444	20	2051	
C66-P11150-00001	123A		20	2051	
C662	108	3452	86	2038	
C663	161	3132	39	2015	
C664	130		14	2041	
C6644H	199		62	2010	
C664B	130	3027	14	2041	
C664C	130	3027	14	2041	
C665	87		262	2041	
C665H	87		262	2041	
C665HA	87		262	2041	
C665HB	87		262	2041	
C668	229		11	2038	
C668-O	107		61	2038	
C668A	107		61	2038	
C668B			61	2038	
C668B1	107		61	2038	
C668BC2	107		61	2038	
C668C	107		61	2038	
C668C1	107		61	2038	
C668CD	107		61	2038	
C668D	107		61	2038	
C668DO	107		61	2038	
C668D1	107		61	2038	
C668DE	107		61	2038	
C668DO	107		61	2038	
C668DV	107		61	2038	
C668DX	107		61	2038	
C668DZ	107		61	2038	
C668E	107		61	2038	
C668E1	107		61	2038	
C668E2	107		61	2038	
C668EP	107		61	2038	
C668EV	107		61	2038	
C668EX	107		61	2038	
C668F	107		61	2038	
C6690		3112			2N4220
C6691		3112			2N4220
C6692		3112			MFE2095
C67		3122	20	2051	
C670	124		12		
C673	234		65	2050	2N5361
C673C2	159		82	2032	
C673D	159		82	2032	
C674	107	3112		2028	2N5361
C674(JAPAN)	107			1172	
C674B	107	3132	39	2015	
C674C	107	3132	39	2015	
C674CV	107	3716/161	39	2015	
C674D	107	3132	39	2015	
C674E	107	3132	39	2015	
C674F	107	3132	39	2015	
C674G	107	3132	39	2015	
C675	87	3467/283	36		
C68		3122	20	2051	
C680	286		246	2020	MFE2093
C680A	286		246	2020	MFE2093
C680R	286		246	2020	
C681	283	3439/163A	36		MFE2093
C681A	283	3439/163A	36		MFE2093
C681B	283	3439/163A	36		
C681YL	283	3439/163A	36		
C682	161	3117	39	2015	MFE2094
C682A	161	3117	39	2015	MFE2094
C682B	161	3117	39	2015	
C683	161	3018	39	2015	MFE2094
C683A	161	3018	39	2015	MFE2094
C683B	161	3018	39	2015	
C683V	161	3018	39	2015	
C684	107		86	2038	MFE2095
C684(JAPAN)	108		86	2038	
C684A	107		86	2038	MFE2095
C684A(JAPAN)	108		86	2038	
C684B	107		86	2038	
C684BK	107		86	2038	
C684F	107		86	2038	
C685	124	3021	12		MFE2095
C685(O)	124		12		
C685A	124	3021	12		MFE2095
C685B	124	3021	12		
C685GU	124	3021	12		

Industry Standard No.	ECG	SK	GE	RS 276-	MOTOR.
C685P	124	3021	12		
C685S	124		12		
C685Y	124	3021	12		
C686	154	3044	40	2012	
C686-248-0	159		82	2032	
C6862400	123A	3444	20	2051	
C687	280	3079	35		
C688		3452	86	2038	
C689		3444	20	2051	
C689H		3444	20	2051	
C69	154	3024/128	243	2030	
C693	199	3124/289	62	2010	
C693(JAPAN)	199	3124/289	62	2010	
C693A	199	3124/289	62	2010	
C693B	199	3124/289	62	2010	
C693C	199	3124/289	62	2010	
C693D	199	3124/289	62	2010	
C693E	199	3124/289	62	2010	
C693E(JAPAN)	199		62	2010	
C693EB	199	3124/289	62	2010	
C693ET	199	3124/289	62	2010	
C693F	199	3124/289	62	2010	
C693FC	199	3124/289	62	2010	
C693FL	199	3124/289	62	2010	
C693FU	199	3124/289	62	2010	
C693G	199	3124/289	62	2010	
C693G(JAPAN)	199		62	2010	
C693GL	199	3124/289	62	2010	
C693GS	199	3124/289	62	2010	
C693GU	199	3124/289	62	2010	
C693GZ	199	3124/289	62	2010	
C693H	199	3124/289	62	2010	
C694	199	3245	62	2010	
C694D	123A	3444	20	2051	
C694E	199	3245	62	2010	
C694F	199	3245	62	2010	
C694G	199	3245	62	2010	
C694Z	199	3245	62	2010	
C695	313	3039/316	11	2015	
C696	128	3765/195A	46		
C696A	128	3765/195A	46		
C696B	128	3765/195A	46		
C696D	128	3765/195A	46		
C696E	128	3765/195A	46		
C696F	128	3765/195A	46		
C696G	128	3765/195A	46		
C696H	128	3765/195A	46		
C696I	128	3765/195A	46		
C697	282/427		46		
C697A	282/427	3299/237	46		
C697D		3299	46		
C697F	282/427	3299/237	46		
C6A			C6A		MCR1906-3
C6B			C6B		MCR1906-4
C6B3	5400	3950			
C6F			C6F		MCR1906-2
C6G					MCR1906-4
C6U			C6U		MCR1906-1
C7		3247	65	2050	
C70	154	3044	40	2012	
C701		3122	277		
C701(TRANSISTOR)	123		20	2051	
C702	123	3122	20	2051	
C705	107	3039/316	11	2015	
C705B	107	3039/316	11	2015	
C705C	107	3039/316	11	2015	
C705D	107	3039/316	11	2015	
C705E	107	3039/316	11	2015	
C705F	107	3039/316	11	2015	
C705TV	107		11	2015	
C707	316	3039	11	2015	
C7076		3245	212	2010	
C707H	316	3039	11	2015	
C708	128	3024	243	2030	
C708A	128	3024	243	2030	
C708AA	128	3024	243	2030	
C708AB	128	3024	243	2030	
C708AC	128	3024	243	2030	
C708AH	128		243	2030	
C708AHA	128		243	2030	
C708AHB	128		243	2030	
C708AHC	128		243	2030	
C708B	128	3024	243	2030	
C708C	128	3024	243	2030	
C709	85	3444/123A	20	2051	
C709B	85	3444/123A	20	2051	
C709C	85	3444/123A	20	2051	
C709CD	85	3444/123A	20	2051	
C709D	85	3444/123A	20	2051	
C71	101		8	2002	
C710		3356	20	2051	
C710(B)		3356	20	2051	
C710(C)				2031	
C710(D)		3356	20	2051	
C710-1		3356	20	2051	
C710-2		3356	20	2051	
C710-4		3356	20	2051	
C710B		3356	20	2051	
C710B2				2031	
C710BC		3356	20	2051	
C710C		3356	20	2051	
C710D		3356	20	2051	
C710DE				2031	
C710E		3356	20	2051	
C711	85	3899	62	2010	
C711(E)	85	3899	62	2010	
C711A	85	3899	62	2010	
C711AE	85	3899	62	2010	
C711AF	85	3899			
C711AG	199	3899			
C711D	85	3899	62	2010	
C711E	85	3899	62	2010	
C711F	85	3899	62	2010	
C711FG		3899	62	2010	
C711G		3899	62	2010	
C712		3124	20	2051	
C712A		3124	20	2051	
C712C		3124	20	2051	
C712D		3124	20	2051	
C712E		3124	20	2051	
C712W		3124	20	2051	
C713		3444	20	2051	
C714		3444	20	2051	
C715	85	3444/123A	62	2009	
C715(JAPAN)	199	3444/123A	20	2051	
C715A	85	3444/123A	20	2051	
C715B	85	3444/123A	20	2051	
C715C	85	3444/123A	20	2051	
C715D	85	3444/123A	20	2051	
C715E	85	3444/123A	20	2051	
C715EJ	85	3444/123A	20	2051	
C715EV	85	3444/123A	11	2051	
C715F	85	3444/123A	20	2051	
C715XL	85	3444/123A	20	2051	

Industry Standard No.	ECG	SK	GE	RS 276-	MOTOR.
C716	289A	3444/123A	20	2051	
C716B	289A	3444/123A	20	2051	
C716C	289A	3444/123A	20	2051	
C716D	289A	3444/123A	20	2051	
C716E	289A	3444/123A	20	2051	
C716F	289A	3444/123A	20	2051	
C716G		3444	20	2051	
C717	107	3132	39	2015	
C717(FINAL IF)	107			2009	
C717(FINAL-IF)			210	2051	
C717B	107	3132	39	2015	
C717BK	107	3132	39	2015	
C717BLK	107	3132	39	2015	
C717C	107	3132	39	2015	
C717E	107	3132	39	2015	
C72	101		8	2002	
C720	161		20	2051	
C722	107		61	2038	
C725	289A	3444/123A	20	2051	
C725-0	289A		20	2051	
C727	287	3044/154	40	2012	
C728	287	3044/154	40	2012	
C73	100		1	2007	
C73(JAPAN)	101		8	2002	
C731R			243	2030	
C732	199	3245	62	2010	
C732BL	199	3245	62	2010	
C732GR	199	3245	62	2010	
C732S	199	3245	62	2010	
C732V	199	3245	62	2010	
C732Y	199	3245	62	2010	
C733	199	3245	62	2010	
C733-0	199	3245	62	2010	
C733-O	199	3245	62	2010	
C7335-BL	199	3444/123A	20	2051	
C733BL	199	3245	62	2010	
C733GR	199	3245	62	2010	
C733R	199	3245	62	2010	
C733V	199	3245	62	2010	
C733Y	199	3245	62	2010	
C734	289A	3245/199	243	2030	
C734-0	289A	3245/199	243	2030	
C734-O	289A	3245/199	243	2030	
C734-R	289A	3245/199	243	2030	
C734-Y	289A	3245/199	243	2030	
C734GR	289A	3245/199	243	2030	
C7340	289A		243	2030	
C734Y	289A	3245/199	243	2030	
C735	289A	3122	268	2038	
C735(FA-3)	108	3122	86	2038	
C735(0)	289A	3122	268	2038	
C735-0	289A	3122	268	2038	
C735-O	289A	3122	20	2051	
C735-R	123A	3122	20	2051	
C735ORN	123A	3444	20	2051	
C735B	289A		268	2038	
C735F	289A		268	2038	
C735FA3	289A		268	2038	
C735GR	289A		268	2038	
C735H	289A		268	2038	
C735J	289A		268	2038	
C735K	289A		268	2038	
C735L	289A		268	2038	
C735Y	289A	3122	268	2038	
C736	87	3027/130	14	2041	
C738		3018	11	2015	
C738C		3039	11	2015	
C738D		3018	11	2015	
C739		3018	11	2015	
C739C		3018	11	2015	
C74	123	3452/108	86	2038	
C740	108	3452	86	2038	
C741	486	3444/123A	20	2051	
C742			20	2051	
C743A		3044	40	2012	
C744	128		63	2030	
C7446AP(HEP)	7446	7446			
C746A		3044	40	2012	
C748		3452	86	2038	
C75	100		1	2007	
C75(JAPAN)	101		8	2002	
C752	85	3444/123A	20	2051	
C752G	85	3444/123A	20	2051	
C756	282	3748	264		
C756-1	282	3748	264		
C756-1-1	282	3748	264		
C756-1-2	282	3748	264		
C756-1-3	282	3748	264		
C756-1-4	282	3748	264		
C756-2	282	3748	264		
C756-2-1	282	3748	264		
C756-2-2	282	3748	264		
C756-2-3	282	3748	264		
C756-2-4	282	3748	264		
C756-2-5	282	3748	264		
C756-3	282	3748	264		
C756-3-1	282	3748	264		
C756-3-2	282	3748	264		
C756-3-3	282	3748	264		
C756-3-4	282	3748	264		
C756-4	282		264		
C756-4-1	282		264		
C756-4-2	282		264		
C756-4-3	282		264		
C756-4-4	282		264		
C756-5	282	3748			
C756A	282	3765/195A	264		
C756A-1	282	3765/195A	264		
C756A-2	282	3765/195A	264		
C756A-3	282	3765/195A	264		
C756A-4	282		264		
C756C	282	3765/195A	264		
C756D	282	3765/195A	264		
C756E	282	3765/195A			
C76	100		1	2007	
C76(JAPAN)	101		8	2002	
C760			17	2051	
C761	161	3039/316	39	2015	
C761Y	161	3039/316	39	2015	
C761Z	161	3039/316	39	2015	
C762	161	3039/316	39	2015	
C763		3122	11	2038	
C763(C)		3122	11	2015	
C763B		3122	61	2038	
C763C		3122	61	2038	
C763CD			61	2038	
C763D		3122	61	2038	
C764	312	3112	27	2028	
C765		3027	14	2041	
C768	328	3027/130	14	2041	
C77	101		8	2002	
C772			61	2038	
C772B		3132	11	2038	
C772BG		3132	61	2038	

Industry Standard No.	ECG	SK	GE	RS 276-	MOTOR.
C772BH		3132	61	2038	
C772BV		3132	61	2038	
C772BX		3132	61	2038	
C772BY		3132	61	2038	
C772C		3132	61	2038	
C772C1		3132	61	2038	
C772C2		3132	61	2038	
C772CK		3132	61	2038	
C772CL		3132	61	2038	
C772CS	229	3132	61	2038	
C772CU		3132	11	2038	
C772CV		3132	61	2038	
C772CX		3132	11	2038	
C772D		3132	61	2038	
C772DJ		3132	61	2038	
C772DU		3132	61	2038	
C772DV		3132	61	2038	
C772DX		3132	61	2038	
C772DY		3132	61	2038	
C772E		3132	61	2038	
C772F		3132	61	2038	
C772K		3132	61	2038	
C772KB		3132	61	2038	
C772KC		3132	61	2038	
C772KD		3132	61	2038	
C772KD1		3132	61	2038	
C772KD2		3132	61	2038	
C772R		3132	61	2038	
C772R(JAPAN)	229		61	2038	
C772RB-D		3132	61	2038	
C772RD		3132	61	2038	
C773		3444	20	2051	
C773C		3444	20	2051	
C773D		3444	20	2051	
C773E		3122	20	2051	
C774	128	3048/329	46		
C775	128	3048/329	46		
C776	195A	3024/128	46		
C776(Y)	195A	3024/128	46		
C776Y	195A	3024/128	46		
C777	237	3299	46		
C778	237	3299	46		
C778B	237	3299	46		
C779	286		267		
C779-O	286		267		
C779-R	286	3194	267		
C779-Y	286	3194	267		
C77C	101		8	2002	
C78	101		8	2002	
C780	194	3433/287	222		
C780AG	194	3433/287	222		
C780AG-O	194	3104A	222		
C780AG-R	194	3104A	222		
C780AG-Y	194		222		
C780G	194	3433/287	222		
C781	128	3765/195A	46		
C782	286	3131A/369	12		
C783	286	3194	12		
C784		3246	61	2038	
C784-O		3246	61	2038	
C784-O		3246	61	2038	
C7840	229		61	2038	
C784A			61	2038	
C784BN		3246	61	2038	
C784R		3246	61	2038	
C784Y		3246	61	2038	
C785	199	3246/229	11	2015	
C785(O)	229		61	2038	
C785BN	229	3246	61	2038	
C785D	229	3246	61	2038	
C785E	229	3246	61	2038	
C785R	229	3246	61	2038	
C786	161		61	2038	
C786R	161		39	2015	
C787	161	3132	39	2015	
C788	154	3044	40	2012	
C789	152	3054/196	66	2048	
C789-O	152	3054/196	66	2048	
C789-O	152	3054/196	66	2048	
C789-R	152	3054/196	66	2048	
C789-Y	152	3054/196	66	2048	
C79	108	3452	86	2038	
C791	175	3562	246	2020	
C793	130	3027	14	2041	
C793BL	130	3027	14		
C793R	130	3027	14	2041	
C793Y	130	3027	14	2041	
C794R	130	3027	14	2041	
C795	124	3021	12		
C795A	124	3021	12		
C796	123A	3444	20	2051	
C797	128	3024	243	2030	
C798	195A	3024/128	243	2030	
C799	237	3299	216	2053	
C799K	237	3299	46		
C79BL				2041	
C7A					MCR1906-3
C7A4	5454	6754		1067	
C7B					MCR1906-4
C7G					MCR1906-4
C7U					MCR1906-1
C8		3247	65	2050	
C80	161	3452/108	86	2038	
C800		3039	11	2015	
C803	195A	3024/128	243	2030	
C805	154	3044	40	2012	
C806	283	3439/163A	36		
C806A	283	3439/163A	36		
C807	385	3467/283	36		
C807A	385	3467/283	36		
C814	192A	3137	63	2030	
C815	289A	3124/289	20	2030	
C815(M)	289A	3124/289	20	2051	
C815A	289A	3124/289	20	2051	
C815B	289A	3124/289	20	2051	
C815C	289A	3124/289	20	2051	
C815F	289A	3124/289	20	2051	
C815K	289A	3124/289	20	2051	
C815L	289A	3124/289	20	2051	
C815M	289A	3124/289	20	2051	
C815S	289A	3124/289	20	2051	
C815SA	289A	3124/289	20	2051	
C815SC	289A	3124/289	20	2051	
C816	128	3024	243	2030	
C816K	128	3024	243	2030	
C818	154	3044	40	2012	
C821	346		261		
C822	346		261		
C825	286		20	2051	
C826	128	3024	243	2030	
C827	128	3024	243	2030	
C828	85	3122	62	2010	
C828-O	85		62	2010	
C828-OP	85		62	2010	
C828A	85		62	2010	
C828AP	85	3866	62	2010	
C828AQ	85	3866	62	2010	
C828AR	85	3866	62	2010	
C828AS	85	3866	20	2051	
C828E	85		62	2010	
C828F	85		62	2010	
C828FR	85	3866	62	2010	
C828H	85		62	2010	
C828K	85		62	2010	
C828LR	85	3866	62	2010	
C828LS	85	3866	62	2010	
C828N	85		62	2010	
C828P	85	3866	62	2010	
C828Q	85	3866	62	2010	
C828QRS	85		62	2010	
C828R	85	3866	62	2010	
C828R/494	199	3866	62	2010	
C828S	85	3866	62	2010	
C828T	85	3866	62	2010	
C828W	85		62	2010	
C828Y	85	3122	62	2010	
C829	85	3122	20	2051	
C829A	85	3122	20	2051	
C829B	85	3122	20	2051	
C829BC	85	3122	20	2051	
C829C	85	3122	20	2051	
C829D	85	3122	20	2051	
C829R	85	3122	20	2051	
C829X	85	3122	20	2051	
C829Y	85	3122	20	2051	
C82S	108	3452	86	2038	
C83-829	116	3017B/117	504A	1104	
C83-880	116	3017B/117	504A	1104	
C830	175	3562	246	2020	
C830A	175	3562	246	2020	
C830B	175	3562	246	2020	
C830C	175	3562	246	2020	
C833BL	384	3245/199	62	2010	
C835	107		11	2015	
C836M	107	3444/123A	20	2051	
C837	107		11	2015	
C837F	107		11	2015	
C837H	107		11	2015	
C837K	107	3132	11	2015	
C837L	107	3132	11	2015	
C837WF	107		11	2015	
C838	289A	3122	62	2051	
C838(H)	289A	3122	62	2010	
C838(J)	289A	3122	62	2010	
C838(K)	289A	3122	62	2010	
C838(M)	289A	3122	62	2010	
C838A	289A	3122	20	2051	
C838B	289A	3122	20	2051	
C838C	289A	3122	20	2051	
C838D	289A	3122	20	2051	
C838E	289A	3122	20	2051	
C838F	289A	3122	20	2051	
C838H	289A	3122	20	2051	
C838J	289A	3122	20	2051	
C838K	289A	3122	20	2051	
C838L	289A	3122	62	2051	
C838M	289A	3122	20	2051	
C838R	289A	3122	20	2051	
C839	85		20	2030	
C839(H)	85		20	2051	
C839(J)	85		20	2051	
C839(M)	85		20	2051	
C839A	85		20	2051	
C839B	85		20	2051	
C839C	85		20	2051	
C839D	85		20	2051	
C839E	85		20	2051	
C839F	85		20	2051	
C839H	85		20	2051	
C839J	85		20	2051	
C839L	85		20	2051	
C839M	85		20	2051	
C839N	85		20	2051	
C839S	85		20	2051	
C840	175	3538	246	2020	
C840A	175	3538	246	2020	
C840AC	175	3538	246	2020	
C840H	175		246	2020	
C840PQ	175		246	2020	
C844	311	3444/123A	20	2051	
C847	123A	3444	20	2051	
C848	123A	3444	20	2051	
C849	123A	3444	20	2051	
C850	123A	3444	20	2051	
C851			14	2041	
C853	192A	3137	63	2030	
C853A	192A	3137	63	2030	
C853B	192A	3137	63	2030	
C853C	192A	3137	63	2030	
C853KLN	192A	3122	63	2030	
C853L	192A	3122	63	2030	
C856	154		40	2012	
C856-02	154		40	2012	
C856C	154		40	2012	
C857	154	3044	40	2012	
C857H	154	3044	40	2012	
C858	199		62	2010	
C858E	199		62	2010	
C858F	199		62	2010	
C858FG	199		62	2010	
C858G	199		62	2010	
C859	199	3245	62	2010	
C859E	199	3245	62	2010	
C859F	199	3245	62	2010	
C859FG	199	3245	62	2010	
C859G	199	3245	62	2010	
C859GK	199	3122	62	2010	
C860	161	3132	39	2015	
C860C	161	3132	39	2015	
C860D	161	3132	39	2015	
C860E	161	3132	39	2015	
C863	161	3132	39	2015	
C864	161	3132	39	2015	
C867	277		12		
C868		3044	40	2012	
C869		3044	40	2012	
C87	123A	3122	20	2051	
C870	85	3444/123A	20	2051	
C870BL	85	3444/123A	20	2051	
C870E	85	3444/123A	20	2051	
C870F	85	3444/123A	20	2051	
C871	85	3124/289	20	2051	
C871BL	85	3124/289	20	2051	
C871D	85	3124/289	20	2051	
C871E	85	3124/289	20	2051	
C871F	85	3124/289	20	2051	
C871G	85	3124/289	20	2051	
C875	128	3024	243	2030	

Industry Standard No.	ECG	SK	GE	RS 276-	MOTOR.
C875-1	128	3024	243	2030	
C875-1C	128	3024	243	2030	
C875-1D	128	3024	243	2030	
C875-1E	128	3024	243	2030	
C875-1F	128	3024	243	2030	
C875-2	128	3024	243	2030	
C875-2C	128	3024	243	2030	
C875-2D	128	3024	243	2030	
C875-2E	128	3024	243	2030	
C875-2F	128	3024	243	2030	
C875-3	128	3024	243	2030	
C875-3C	128	3024	243	2030	
C875-3D	128	3024	243	2030	
C875-3E	128	3024	243	2030	
C875-3F	128	3024	243	2030	
C875BR	128		243	2030	
C875C	128		243	2030	
C875D	128	3024	243	2030	
C875E	128	3024	243	2030	
C875F	128	3024	243	2030	
C876	128	3024	243	2030	
C876C	128	3024	243	2030	
C876D	128	3024	243	2030	
C876E	128	3024	243	2030	
C876F	128		243	2030	
C876TV	128	3024	243	2030	
C876TVD	128		243	2030	
C876TVE	128	3024	243	2030	
C876TVEF	128	3024	243	2030	
C88	128	3044/154	40	2012	
C881	192A	3137	63	2030	
C881A	192A		63	2030	
C881B	192A		63	2030	
C881C	192A		63	2030	
C881D	192A		63	2030	
C881K	192A	3137	63	2030	
C881L	192A	3137	63	2030	
C881M	192A	3137			
C889	130	3438	35		
C88A	128	3044/154	40	2012	
C89	101		8	2002	
C894	289A	3444/123A	20	2051	
C895	162		246	2020	
C896	123A	3444	20	2051	
C897	280		262	2041	
C897A	280		262	2041	
C897B	280		262	2041	
C897C	280		262	2041	
C898	280		262	2041	
C898A	280		262	2041	
C898B	280		262	2041	
C898C	280		262	2041	
C899	123A	3444	20	2051	
C899K	123A	3444	20	2051	
C90	101		8	2002	
C900	199	3899	62	2010	
C900(L)	199	3899	62	2010	
C900A	199	3899	62	2010	
C900AF	199	3899			
C900B	199	3899	62	2010	
C900C	199	3899	62	2010	
C900D	199	3899	62	2010	
C900E	199	3899	62	2010	
C900EF	199	3899			
C900EU	199	3899			
C900F	199	3899	62	2010	
C900L	199	3899	62	2010	
C900M	199	3899	62	2010	
C900U	199	3899	62	2010	
C900U/E	199	3899			
C900UE	199	3899			
C900V	199	3899			
C900VE	199	3899			
C901	163A		36		
C901A	163A		36		
C903D	229	3018	61	2038	
C907	123A	3245/199	62	2010	
C907A	123A	3245/199	62	2010	
C907AC	123A	3245/199	62	2010	
C907AD	123A	3245/199	62	2010	
C907AH	123A	3245/199	62	2010	
C907C	123A	3245/199	62	2010	
C907D	123A	3245/199	62	2010	
C907H	123A	3122	62	2010	
C907HA	123A	3122	62	2010	
C9080	129		244	2027	
C9081	129		244	2027	
C9082	159	3114/290	82	2032	
C9083	159	3114/290	82	2032	
C9084	159	3114/290	82	2032	
C9085	159	3114/290	82	2032	
C91			8	2002	
C912	85	3452/108	86	2038	
C913	123A$	3444/123A	20	2051	
C917	383		61	2038	
C917K	161		61	2038	
C918	161	3132	39	2015	
C92-0025-RO	724	3524/784			
C92-0025-RO	784		IC-86		
C920	229	3018	11	2038	
C920E		3018	11	2015	
C920Q		3039	61	2038	
C920R		3039	61	2038	
C921		3039	11	2015	
C921C1		3039	11	2015	
C921K			11	2015	
C921L		3039	11	2015	
C921M		3039	11	2015	
C922	107	3018	11	2015	
C922A	107	3039/316	11	2015	
C922B	107	3039/316	11	2015	
C922C	107	3039/316	11	2015	
C922K	107		11	2015	
C922L	107		11	2015	
C922M	107		11	2015	
C923	199	3124/289	62	2010	
C923A	199	3124/289	62	2010	
C923B	199	3124/289	62	2010	
C923C	199	3124/289	62	2010	
C923D	199	3124/289	62	2010	
C923E	123A	3124/289	20	2051	
C923F	199	3124/289	62	2010	
C924	107	3452/108	86	2038	
C924E	107	3452/108	86	2038	
C924F	107	3452/108	86	2038	
C924M	107	3452/108	86	2038	
C925		3444	20	2051	
C926			40	2012	
C926A		3244	40	2012	
C927	161	3246/229	283	2016	
C927A	161		283	2016	
C927B	161	3246/229	283	2016	
C927C	161	3246/229	283	2016	
C927C1				2016	

Industry Standard No.	ECG	SK	GE	RS 276-	MOTOR.
C927CJ	161	3246/229	39	2015	
C927CU	161	3246/229	283		
C927CW	161	3246/229	283	2016	
C927D	161	3246/229	283	2016	
C927DD	161	3246/229			
C927E	161	3246/229	283	2016	
C928	161	3132	39	2015	
C928B	161	3132	39	2015	
C928C	161	3132	39	2015	
C928D	161	3132	39	2015	
C928E	161	3132	39	2015	
C929	107		61	2038	
C929-O	107	3132	61	2038	
C929B	107		61	2038	
C929C	107		61	2038	
C929C1	107		61	2038	
C929D	107		61	2038	
C929D1	107		61	2038	
C929DE	107	3132	61	2038	
C929DP	107		61	2038	
C929DU	107	3132	61	2038	
C929DV	107		61	2038	
C929E	107		61	2038	
C929ED	107		61	2038	
C929F	107		61	2038	
C929FK	107		61	2038	
C930	107	3356	61	2038	
C930B	107	3356	61	2038	
C930BK	107		61	2038	
C930BV	107	3356	61	2038	
C930C	107		61	2038	
C930CK	107	3356	62	2010	
C930CL	107		61	2038	
C930CS	107		61	2038	
C930D	107	3356	61	2038	
C930DE	107		11	2015	
C930DH	107	3356	62	2010	
C930DS	107		61	2038	
C930DT	107	3356	61	2038	
C930DT-2	107		61	2038	
C930DX	107		61	2038	
C930DZ	107	3356	62	2010	
C930E	107	3356	61	2038	
C930EP	107		61	2038	
C930ET	107		11	2015	
C930EV	107	3356	61	2038	
C930EX	107		61	2038	
C930F	107		61	2038	
C931			247	2052	
C931C			247	2052	
C931D			247	2052	
C931E			247	2052	
C932		3054	66	2048	
C932E		3054	66	2048	
C933	289A		20	2051	
C933BB	289A	3122	20	2051	
C933C	289A	3122	20	2051	
C933D	289A	3122	20	2051	
C933E	289A	3122	20	2051	
C933F	289A	3122	20	2051	
C933FP	289A	3122	20	2051	
C933FPC	289A	3122	20	2051	
C933FPD	289A	3122	20	2051	
C933FPE	289A	3122	20	2051	
C933FPF	289A	3122	20	2051	
C933FPG		3122	20	2051	
C933G		3122	20	2051	
C934	128	3024	243	2030	
C934(RECTIFIER)	117		504A	1104	
C934(TRANSISTOR)	123A		20	2051	
C934C	123A	3444	20	2051	
C934D	123A	3444	20	2051	
C934E	123A	3444	20	2051	
C934F	123A	3444	20	2051	
C934G	123A	3444	20	2051	
C934P	123A	3444	20	2051	
C935	162		35		
C936	164	3133	37		
C937	389		259		
C937(BK)	389		259		
C937(YL)	389		259		
C937-01	389		259		
C937A	389		259		
C937B	389		259		
C937YL	389		259		
C938	289A	3444/123A	20	2051	
C938A	289A	3444/123A	20	2051	
C938B	289A	3444/123A	20	2051	
C938C	289A	3444/123A	20	2051	
C939	328	3619	36		
C939D	328	3619	268	2038	
C939L	328	3619	36		
C94	282	3112			2N5457
C940	283	3439/163A	36		
C940L	283	3439/163A	36		
C940M	283	3439/163A	36		
C941	289A	3124/289	20	2051	
C941-0	289A		20	2051	
C941-O	289A	3124/289	20	2051	
C941-R	289A	3124/289	20	2051	
C941-Y	289A	3124/289	20	2051	
C941R	289A		20	2051	
C9426	229	3018	61	2038	
C942H2	1195	3469			
C943	123A	3444	20	2051	
C943A	123A	3444	20	2051	
C943B	123A	3444	20	2051	
C943C	123A	3444	20	2051	
C944	85		20	2051	
C944A	85		20	2051	
C944B	85		20	2051	
C944C	85		20	2051	
C944D	85		20	2051	
C944K	85		20	2051	
C945	85	3124/289	62	2010	
C945(R)	85	3124/289	62	2010	
C945-O	85		62	2010	
C945A	85	3124/289	62	2010	
C945AP	85	3124/289	62	2010	
C945AQ	85	3124/289	62	2010	
C945B	85	3124/289	62	2010	
C945C	85	3124/289	62	2010	
C945D	85	3124/289	62	2010	
C945E	85	3124/289	62	2010	
C945F	85	3124/289	62	2010	
C945G	85	3124/289	62	2010	
C945H	85	3124/289	62	2010	
C945K	85	3124/289	62	2010	
C945L	85	3124/289	62	2010	
C945M	85	3124/289	62	2010	
C945O	85		62	2010	
C945P	85	3124/289	62	2010	
C945Q	85	3124/289	62	2010	
C945QL	85	3124/289	62	2010	

Industry Standard No.	ECG	SK	GE	RS 276-	MOTOR.
C945QP	85	3124/289	62	2010	
C945R	85	3124/289	62	2010	
C945S	85	3124/289	62	2010	
C945T	85	3124/289	62	2010	
C945TQ	85	3124/289	62	2010	
C945TR	85	3124/289	62	2010	
C945X	85	3124/289	62	2010	
C947	161	3132	39	2015	
C948	161	3132	39	2015	
C94A		3112			2N5457
C94E	457	3112			2N5457
C95	128	3112	40		2N5457
C957	319P	3132	39	2015	
C959	396	3024/128	243	2030	
C959A	128	3024	243	2030	
C959B	128	3024	243	2030	
C959C	128	3024	243	2030	
C959D	128	3024	243	2030	
C959M	128	3024	243	2030	
C959S	128	3024	243	2030	
C959SA	128	3024	243	2030	
C959SB	128	3024	243	2030	
C959SC	128	3024	243	2030	
C959SD	128	3024	243	2030	
C95A		3112			2N5457
C95E		3112			2N5459
C960	427/128	3444/123A	20	2051	
C960(JAPAN)	128/400		256		
C9604	123A	3444	20	2051	
C960A	427/128		256		
C960B	427/128	3103A/396	256		
C960C	427/128	3103A/396	256		
C960D	427/128	3103A/396	256		
C960S	427/128	3103A/396	256		
C960SA	427/128	3103A/396	256		
C960SB	427/128	3103A/396	256		
C960SC	427/128	3103A/396	256		
C960SD	427/128	3103A/396	256		
C9634	199		86	2038	
C966	123A	3444	20	2051	
C967	123A	3444	20	2051	
C968	123A		20	2051	
C968P	123A		20	2051	
C96E		3112			2N5459
C97		3024	243	2030	
C971		3137	63	2030	
C972	128	3024	243	2030	
C972C	128	3024	243	2030	
C972D	128	3024	243	2030	
C972E	128	3024	243	2030	
C97A		3024	243	2030	
C98	123A	3452/108	86	2038	
C982	172A	3156	64		
C983		3244	251		
C983-O		3244	251		
C983-Y		3244	251		
C983A		3244	251		
C983C		3244	251		
C983R		3244	251		
C983S		3244	251		
C983Y		3244	251		
C984	128	3444/123A	20	2051	
C984A	128	3444/123A	20	2051	
C984B	128	3444/123A	20	2051	
C984C	128	3444/123A	20	2051	
C99	123A	3452/108	86	2038	
C991	311	3444/123A	20	2051	
C992	311	3444/123A	20	2051	
C993	123A		243	2030	
C995	154	3044	40	2012	
C996	154/427	3103A/396	256		
C997	161	3132	39	2015	
C999	389	3115/165	38		
C999A	389	3115/165	38		
C99CD	128		243	2030	
C9FF-10A580	128		243	2030	
CA-092	139A	3060	ZD-9.1	562	
CA-3053	724		IC-86		
CA-90			1N60	1123	
CA-9011H	123A	3444	20	2051	
CA10	116	3311	504A	1104	
CA100	116	3311	504A	1104	MHW1171
CA101AT		3565			LM101AH
CA101T		3565			LM101AH
CA1020A	116	3017B/117	504A	1104	
CA102BA	116	3017B/117	504A	1104	
CA102DA	116	3016	504A	1104	
CA102FA	116	3031A	504A	1104	
CA102HA	116	3017B/117	504A	1104	
CA102MA	116	3017B/117	504A	1104	
CA102PA	125	3032A	510,531	1114	
CA102RA	125	3032A	510,531	1114	
CA102VA	125	3033	510,531	1114	
CA107T		3690			LM107H
CA108AS					LM108AJ-8
CA108AT					LM108AH
CA108S					LM108J-8
CA108T					LM108H
CA1190GQ	1231	3832			
CA1310E	801	3160	IC-35		MC1310P
CA1352E	749	3168	IC-97		MC1352P
CA1365	712	3072			
CA1391E	815	3255			MC1391P
CA1394E	836	3206			MC1394P
CA1398E	738	3167	IC-29		MC1398P
CA139AG		3569			LM139AJ
CA139G		3569			LM139J
CA1458	778A	3555			
CA1458E	778A	3465			
CA1458G	778A	3465			
CA1458S	778A	3551	IC-220		MC1458CP1
CA1458T		3555			MC1458G
CA150	116	3311	504A	1104	
CA152VA	125	3033	510,531	1114	
CA1558S					MC1558U
CA1558T					MC1558G
CA20	116		504A	1104	
CA200	116	3311	504A	1104	MHW1172
CA2002	1232	3852		703	
CA2004	1232	3853			
CA201AT	1171	3565			LM201AH
CA201T	1171	3565			LM201AH
CA2065	712	3072			
CA207T		3690			LM207H
CA208AT					LM208AH
CA208S					LM208J-8
CA208T					LM208H
CA2100					MHW1171
CA2100R					MHW1171R
CA2111		3135	IC-10		
CA2111A		3135	IC-10		
CA2111AE	708	3135/709	IC-10		MC1357P
CA2111AQ	708	3135/709	IC-10		MC1357PQ
CA2111E	708	3135/709	IC-10		
CA2200					MHW1172
CA2200R					MHW1172R
CA224E	987	3643			
CA224G	987	3643			
CA2300					MHW1221
CA2301					MHW1222
CA239AE	834	3569			LM239AN
CA239AG	834	3569			LM239AJ
CA239E	834	3569			LM239N
CA239G	834	3569			LM239J
CA2418					MHW1182
CA250	116	3031A	504A	1104	
CA2600					MHW1342
CA2601BU					MHW1343
CA2603					MHW1344
CA2700					MHW1392
CA2800					MHW1172
CA2810					MHW1342
CA2812					MHW593
CA2818					MHW1182
CA2820					MHW590
CA2830					MHW592
CA2840					MHW1222
CA2850					MHW1182
CA2870					MHW1342
CA2875					MHW1182
CA2876					MHW1221
CA2D2	102A	3123	53	2007	
CA3000	900	3547	IC-245		MC1550G
CA3000T	900	3547			
CA3001	901	3549	IC-180		MC1550G
CA3001T	901	3549			
CA3002					MC1550G
CA3004					MC1550G
CA3005					MC1550G
CA3006					MC1550G
CA3007					MC1550G
CA3008					MC1709F
CA3008A					MC1709F
CA3010	903	3540	IC-288		MC1709G
CA3010A	903	3540			MC1709G
CA3010AT	903	3540			
CA3010T	903	3540			
CA3011	726	3129			MC1590G
CA3011T	726	3129			
CA3012	726	3129			MC1590G
CA3012T	726	3129			
CA3013	704	3022/1188	IC-205		MC1357P
CA3013T	704	3022/1188			
CA3014	704	3022/1188	IC-205		MC1357P
CA3014T	704	3022/1188			
CA3015	903	3540	IC-288		MC1709G
CA3015A	903	3540			MC1709G
CA3015AT	903	3540			
CA3015T	903	3540			
CA3016					MC1709F
CA3016A					MC1709F
CA3018	904	3542	IC-289		
CA3018A	904	3542	IC-289		
CA3018T	904	3542			
CA3019	905	3546	IC-290		
CA3019T	905	3546			
CA301AE	975	3641			
CA301AT	1171	3565			LM301AH
CA301E	975	3641			
CA3020	784	3524	IC-236		MC1554G
CA3020A		3524	IC-236		MC1454G
CA3020B		3524	IC-236		
CA3020T	784	3524			
CA3021					MC1590G
CA3022					MC1590G
CA3023					MC1590G
CA3026	906	3548	IC-246		CA3054
CA3026T	906	3548			
CA3028	724	3525	IC-86		
CA3028A	724	3525	IC-86		MC1550G
CA3028AF	724	3525	IC-86		MC1550G
CA3028AS	724	3525	IC-86		MC1550G
CA3028AT	724	3525			
CA3028B	724	3525	IC-86		MC1550G
CA3028BF	724	3525	IC-86		MC1550G
CA3028BS	724	3525	IC-86		MC1550G
CA3028T	724	3525			
CA3029	908	3539	IC-248		MC1709P2
CA3029A		3539			MC1709P2
CA3029E	908	3539			
CA3030	908	3539	IC-248		MC1709P2
CA3030A	908	3539	IC-248		MC1709P2
CA3030AE	908	3539			
CA3030E	908	3539			
CA3031					MC1712G
CA3032					MC1712CG
CA3033					MC1533L
CA3033A					MC1533L
CA3035	785	3254	IC-226		MC1352P
CA3035T	785	3254			
CA3035V	785	3254	IC-226		
CA3035V1	785	3254	IC-226		MC1352P
CA3037	908	3539	IC-248		MC1709L
CA3037A	908	3539	IC-248		MC1709L
CA3037AE	908	3539			
CA3037E	908	3539			
CA3038	908	3539	IC-248		MC1709L
CA3038A	908	3539	IC-248		MC1709L
CA3038AE	908	3539			
CA3038E	908	3539			
CA3039	907	3545	IC-247		
CA3039T	907	3545			
CA3040					MC1510G
CA3041	706	3101	IC-43		MC1351P
CA3041T	706	3101			
CA3042	710	3102	IC-89		MC1357P
CA3042E	710	3102			
CA3043	786	3140	IC-227		MC1357P
CA3043T	786	3140			
CA3044	711	3070	IC-207		MC1364P
CA3044T	711	3070			
CA3044V1	711	3070	IC-207		MC1364P
CA3045	912	3543	IC-172		MC3346P
CA3045E	912	3543			
CA3045F	912	3543	IC-172		MC3346P
CA3046	912	3543	IC-172		MC3346P
CA3046E	912	3543			
CA3047	913		IC-255		MC1433L
CA3047A	913		IC-255		MC1433L
CA3048	727	3071	IC-210		MC3301P
CA3049	906	3548	IC-246		
CA3049T	906	3548	IC-246		
CA3052	727	3071	IC-210		MC3301P
CA3052E	727	3071	IC-210		
CA3053	724	3525	IC-86		MC1550G
CA3053F					MC1550G
CA3053S					MC1550G
CA3053T	724	3525			

Industry Standard No.	ECG	SK	GE	RS 276-	MOTOR.
CA3054	917	3544	IC-258		CA3054
CA3054E	917	3544			
CA3056		3541			MC1741CG
CA3056A		3541	IC-255		MC1741G
CA3058	914	3541	IC-256		CA3059
CA3058E	914	3541			
CA3059	914	3541	IC-256		CA3059
CA3059E	914	3541			
CA3064		3141	IC-222		MC1364P
CA3064/5A	780	3141	IC-222		
CA3064E	783	3215			MC1364P
CA3064T	780	3141	IC-222		
CA3065	712	3072	IC-226		MC1358P
CA3065/7F	712	3072	IC-148		
CA3065E	712	3072	IC-148		
CA3065F	712	3072			
CA3065F.C.	712	3072			
CA3065FC	712	3072	IC-148		
CA3065RCA	712	3072	IC-148		
CA3066	728	3073	IC-22		MC1399P
CA3066AE	728	3073			
CA3066E	728	3073	IC-22		
CA3067	729	3074	IC-23		MC1323P
CA3067AE	729	3074			
CA3067E	729	3074	IC-23		
CA3068	730	3143			MC1352P
CA3070	714	3075	IC-4		MC1399P
CA3070E	714	3075			
CA3070G	714	3075	IC-4		
CA3071	715	3076	IC-6		MC1399P
CA3071E	715	3076			
CA3072	713	3077/790	IC-5		MC1323P
CA3072E	713	3077/790			
CA3075	723	3144	IC-15		
CA3075D	723	3144	IC-15		
CA3075E	723	3144	IC-15		
CA3076	781	3169	IC-223		MC1590G
CA3076T	781	3169			
CA3078AS		3566			MC1776G
CA3078AT		3566			MC1776G
CA3078S		3566			MC1776CG
CA3078T		3566			MC1776CG
CA3079	914	3541	IC-256		CA3059
CA3079E	914	3541			
CA307E	976	3596			
CA307T		3690			LM307H
CA3081	916	3550	IC-257		
CA3081E	916	3550			
CA3085		9036			MC1723G
CA3085A		9036			MC1723G
CA3085AF					MC1723L
CA3085AS					MC1723G
CA3085B					MC1723G
CA3085BF					MC1723L
CA3085BS					MC1723G
CA3085F					MC1723L
CA3085S					MC1723G
CA3086	912	3543	IC-172		MC3386P
CA3086E	912	3543			
CA3086F					MC3346P
CA3088E	787	3146	IC-228		
CA3089E	788	3147	IC-229		
CA3089F	788	3147			
CA308AS					LM308N
CA308AT	938				LM308AH
CA308S					LM308H
CA3090		3078	IC-229		
CA30900	789	3078			
CA3090AQ	789	3078			MC1310P
CA3090E	789	3078			
CA3090Q	789	3078			
CA3091D					MC1594L
CA311E	922M	3668			
CA3120E	731	3170	IC-13		MC1344P
CA3121		3149	IC-33		
CA3121E	791	3149	IC-231		
CA3121E-G	791	3149			
CA3121G	791	3149			
CA3123E	744	3171	IC-215	2022	
CA3125E	798	3216	IC-234		MC1323P
CA3126		3158	IC-233		
CA3126EM	797	3158	IC-233		
CA3126EM1	797	3158	IC-233		
CA3126EMI	797		IC-233		
CA3126Q	797	3158	IC-233		
CA3130AE	930	3696			
CA3130E	930	3696			
CA3130S	930	3568			
CA3134	1175	3212			
CA3134E					TDA1190Z
CA3134EM	1175	3212			TDA1190Z
CA3134GQM	1175	3212			
CA3134Q	1175	3212			
CA3134QM	1175	3212			TDA1190Z
CA3135G	818	3207			
CA3136A					MC3346P
CA3136E	843	3208			
CA3136G	843	3208			
CA3137	1176	3210			
CA3137E	1176	3210			MC1323P
CA3137EM	1176	3210			
CA3137EM1	1176	3210			
CA3139					CA3139
CA3139E	1174	3186			
CA3139Q	1174	3186			
CA3143	1177	3213			
CA3143E	1177	3213			
CA3144G	985	3214			
CA3146					MC3346P
CA3151G	986	3918			
CA3156	822	3919			
CA3159	820	3927			
CA3168E	2024	3667			
CA3170E	982	3205			
CA3172	821	3882			
CA3172G	821	3882			
CA3189E	788	3829			
CA3221	819	3928			
CA3221G	819	3928			
CA324E	987	3643			
CA324G	987	3643			
CA3302E					MC3302P
CA339	834	3569			
CA339AE	834	3569			LM339AN
CA339AG	834	3569			LM339AJ
CA339E	834	3569			LM339N
CA339G	834	3569			LM339J
CA3401E	992	3688			MC3401P
CA344V1	711	3070	IC-207		
CA358AS	928	3691			
CA358AT	928	3691			
CA358S	928	3691			
CA358T	928	3691			
CA3741	941	3514	IC-255	010	

Industry Standard No.	ECG	SK	GE	RS 276-	MOTOR.
CA3741CS	941	3553	IC-255	010	
CA3741CT	941	3514	IC-263	010	
CA3741S	941	3553	IC-255	010	
CA3741T	941	3514	IC-255	010	
CA3747CT	941	3526/947	IC-268		
CA3748CT	1171	3565			
CA401B					MHW1182
CA416					MHW1182
CA418					MHW1182
CA50	116	3311	504A	1104	
CA555CE	955M	3564		1723	
CA555CG	955M	3564			
CA601B/U					MHW1342
CA6078AS					MC1776G
CA6078AT					MC1776G
CA636					MHW1342
CA6664			504A	1104	
CA6665			504A	1104	
CA6741S					MC1776G
CA6741T					MC1776G
CA723	923	3164			
CA723CE	923D				MC1723CP
CA723E	923D	3165			
CA723T	923	3164			
CA7248			504A	1104	
CA741CE	941M	3552			
CA741CG	941M	3552			
CA741CS	941M	3553	IC-265	007	MC1741CP1
CA741CT	941	3514	IC-263	010	MC1741CG
CA741S		3553			MC1741U
CA741T					MC1741G
CA747CE	947D	3556	IC-268		MC1747CL
CA747CF	947D	3556			MC1747CL
CA747CG	947D	3556			
CA747CT	947	3526	IC-267		MC1747CG
CA747E		3556			MC1747L
CA747F		3556			MC1747L
CA747T		3526	IC-268		MC1747G
CA748CS					MC1748CP1
CA748CT	1171	3645			MC1748CG
CA748S					MC1748U
CA748T		3645			MC1748G
CA758E	743	3172	IC-214		MC1310P
CA801					MHW590
CA804					MHW590
CA810M	1115	3184	IC-278		
CA810Q	1115A	3917	IC-278		
CA8314			504A	1104	
CA860					MHW590
CA870					MHW590
CA90	177	3088	300	1122	
CAC5028A			11	2015	
CAM-12	123A	3444	20	2051	
CB-103			1	2007	
CB10	116	3311	504A	1104	
CB100	156	3033	512		
CB103	126		52	2024	
CB106	109	3088		1123	
CB150	116	3311	504A	1104	
CB156	160	3007	245	2004	
CB157	126	3008	52	2024	
CB158	126	3008	52	2024	
CB161	102A	3004	53	2007	
CB163	109	3087		1123	
CB1P4	105	3012	4		
CB20	116		504A	1104	
CB200	116	3311	504A	1104	
CB244	160	3007	245	2004	
CB246	123A	3444	20	2051	
CB248	102A	3004	53	2007	
CB249	102A	3004	53	2007	
CB250	116	3031A	504A	1104	
CB254	160	3007	245	2004	
CB393	109	3090		1123	
CB5	116	3017B/117	504A	1104	
CB50	116	3311	504A	1104	
CC102BA	116	3016	504A	1104	
CC102DA	116	3016	504A	1104	
CC102FA	116	3017B/117	504A	1104	
CC102HA	116	3031A	504A	1104	
CC102KA	116	3017B/117	504A	1104	
CC102MA	116	3017B/117	504A	1104	
CC102PA	125	3032A	510,531	1114	
CC102RA	125	3032A	510,531	1114	
CC102VA	125	3033	510,531	1114	
CC1168F	123A	3444	20	2051	
CC152VA	125	3033	510,531	1114	
CC59018F			17	2051	
CCS-2006D		3452	17	2051	
CCS1235G	123A	3444	20	2051	
CCS2001H	123AP	3444/123A	20	2051	
CCS2004	123A	3444	20	2051	
CCS2004B	199		62	2010	
CCS2004D303	199	3124/289	62	2010	
CCS2005B	159	3114/290	82	2032	
CCS2006D	108		86	2038	
CCS2008F015	108	3018	86	2038	
CCS4004	123A	3444	20	2051	
CCS6168	123A	3444	20	2051	
CCS6168F	123A	3444	20	2051	
CCS6168G	128	3024	243	2030	
CCS6225F	161	3117	39	2015	
CCS6226G	161	3117	39	2015	
CCS6227F	161	3117	39	2015	
CCS6228F	159		82	2032	
CCS6229H	128	3020/123	243	2030	
CCS9015	159	3114/290	82	2032	
CCS9016D	108	3452	86	2038	
CCS9016E	108	3452	86	2038	
CCS9017	161	3117	39	2015	
CCS9017G925	161	3117	39	2015	
CCS9018E	123A	3444	20	2051	
CCS9018F	108	3452	86	2038	
CCS9018H924	161	3117	39	2015	
CD-00-9	177		300	1122	
CD-0000	109	3088		1123	
CD-0000N	109	3090		1123	
CD-0014N	109	3100/519		1123	
CD-0021	177	3100/519	300	1122	
CD-0033			ZD-6.2	561	
CD-0071		3126	90		
CD-2		3311		1104	
CD-20	177	3100/519	300	1122	
CD-2N	116	3311	504A	1104	
CD-37	177	3100/519	300	1122	
CD-37A	177		300	1122	
CD-4	116	3017B/117	504A	1104	
CD-5038	519		514	1122	
CD-6439	911		IC-253		
CD-8457			504A	1104	
CD-84857	177	3100/519	300	1122	
CD-860037	116		504A	1104	
CD000	109	3087		1123	
CD0000		3122	1N34AS	1123	

Industry Standard No.	ECG	SK	GE	RS 276-	MOTOR.
CD0000/7825B	109			1123	
CD0000NC	177	3100/519	300	1122	
CD0014		3087	300	1122	
CD0014(MORSE)	177		300	1122	
CD0014NA	123A	3444	20	2051	
CD0014NG	123A	3444	20	2051	
CD0015N	123A	3444	20	2051	
CD0021	123A	3444	20	2051	
CD0033	137A		ZD-6.2	561	
CD0120	136A		ZD-5.6	561	
CD05	116		504A	1104	
CD10000-1B	159		82	2032	
CD101	109	3088	1N60	1123	
CD1111	116	3311	504A	1104	
CD1112	116	3311	504A	1104	
CD1113	116	3311	504A	1104	
CD1114	116	3311	504A	1104	
CD1115	116	3031A	504A	1104	
CD1116	116	3031A	504A	1104	
CD1117	416	3311		1104	
CD1121	116	3311	504A	1104	
CD1122	116	3016	504A	1104	
CD1123	116	3016	504A	1104	
CD1124	116	3016	504A	1104	
CD1125	116	3016	504A	1104	
CD1126	116	3016	504A	1104	
CD1127	116	3016	504A	1104	
CD1141	116	3311	504A	1104	
CD1142	116	3311	504A	1104	
CD1143	116	3311	504A	1104	
CD1147	116	3311	504A	1104	
CD1148	116	3311	504A	1104	
CD1149	116	3311	504A	1104	
CD1151		3031A	504A	1104	
CD12000	123A	3444	20	2051	
CD1224	177		300	1122	
CD13332	116	3311	504A	1104	
CD13333	116	3311	504A	1104	
CD13334		3311	504A	1104	
CD13335	116	3311	504A	1104	
CD13336	116	3311	504A	1104	
CD13337	116	3311	504A	1104	
CD13338	116	3311	504A	1104	
CD13339	116	3311	504A	1104	
CD1752					MRF317
CD1802					MRF226
CD1803					MRF209
CD1979					MRF321
CD2	116(2)		504A(2)	1104	
CD2035					MRF5175
CD2087					MRF5175
CD2088					MRF321
CD2089					MRF323
CD22100					MC142100
CD2514					2N6081
CD2545	317				MRF450
CD2810					MRF321
CD31-00002	134A	3055	ZD-3.6		
CD31-00006	135A		ZD-5.1		
CD31-00007	136A		ZD-5.6	561	
CD31-00008	137A		ZD-6.2	561	
CD31-00012	139A	3060	ZD-9.1	562	
CD31-00015	142A	3062	ZD-12	563	
CD31-00017	145A	3063	ZD-15	564	
CD31-00023	146A	3064	ZD-27		
CD31-00025	147A	3095	ZD-33		
CD31-10361	146A	3064	ZD-27		
CD31-10365	145A	3063	ZD-15	564	
CD31-12019	139A		ZD-9.1	562	
CD31-12022	142A	3062	ZD-12	563	
CD31-12024	145A	3063	ZD-15	564	
CD31-12030	146A	3064	ZD-27		
CD31-12032	147A	3095	ZD-33		
CD31-12039	139A	3060	ZD-9.1	562	
CD3100001					1N4728
CD3100025					1N4753
CD3112016					1N4736
CD3112018			Z11	562	
CD3112032					1N4752
CD3112039	139A	3060	ZD-9.1	562	
CD3122	135A	3056	ZD-5.1		
CD3122055	139A		ZD-9.1	562	
CD3168					1N5262
CD3171	149A	3097	ZD-62		
CD3174	150A	3098	ZD-82		1N5268
CD3212048					1M8.2ZS
CD3212055	5078A	9023	ZD-9.1	562	
CD3212055(ZENER)	139A		ZD-9.1	562	
CD3212062					1M33ZS
CD3214738					1M8.2ZS
CD3214752					1M33ZS
CD3400					MRF315
CD3401					MRF316
CD3403					MRF317
CD3550					MRF315
CD37	177	3100/519	300	1122	
CD37A	177	3100/519	300	1122	
CD37A2	116	3311	504A	1104	
CD38	123A	3444	20	2051	
CD3907562					.4M8.2Z
CD3909732					.4M33Z
CD4	116		504A	1104	
CD4000AE	4000	4000			
CD4000BE	4000	4000	4000		
CD4000UB					MC14000UB
CD4001	4001B		4001	2401	
CD4001AD	4001B	4001			
CD4001AE	4001B	4001	4001		
CD4001B					MC14001B
CD4001BE	4001B	4001	4001	2401	
CD4001CN	4001B		4001		
CD4001UB					MC14001UB
CD4002	4002B		4002		
CD4002AD	4002B	4002			
CD4002AE	4002B	4002			
CD4002B					MC14002B
CD4002BD	4002B	4002			
CD4002BE		4002	4002		
CD4002UB					MC14002UB
CD4006A	4006B	4006			
CD4006AE	4006B	4006			
CD4006B	4006B	4006			MC14006B
CD4006BE	4006B	4006			
CD4007UB					MC14007UB
CD4007UBE	4007	4007	4007		
CD4008	4008B	4008			
CD4008B	4008B				MC14008B
CD4008BCN	4008B	4008			
CD4008BE	4008B	4008			
CD4009UB					MC14049UB
CD4009UBE	4049	4009	4009		
CD40101B					MC14531B
CD40106B	40106B				MC14584B
CD40108B					MC14580B

Industry Standard No.	ECG	SK	GE	RS 276-	MOTOR.
CD40109B					MC14504B
CD4010AE	4050B	4010			
CD4010B	4050B				MC14050B
CD4010BE	4050B	4010			
CD4011	4011B		4011	2411	
CD4011AE	4011B	4011			
CD4011B	4011B				MC14011B
CD4011BE	4011B	4011	4011	2411	
CD4011UB					MC14011UB
CD4012	4012B		4012		
CD4012AE	4012B	4012			
CD4012BE	4012B	4012	4012	2412	
CD4012UB					MC14012B
CD4013	4013B		4013	2413	
CD4013AE	4013B	4013	4013	2413	
CD4013B					MC14013B
CD4013BE	4013B	4013	4013	2413	
CD4014AE	4014B	4014			
CD4014B		4014			MC14014B
CD4014BE	4014B	4014			
CD4015AE	4015B	4015			
CD4015B					MC14015B
CD4015BE	4015B	4015	4015		
CD4016	4016B		4016		
CD40160B					MC14160B
CD40161B		40161			MC14161B
CD40161BE	40161B	40161			
CD40162B					MC14162B
CD40163B					MC14163B
CD4016AE	4016B	4016	4016		
CD4016B	4016B				MC14016B
CD4016BE	4016B	4016	4016		
CD40174B					MC14174B
CD40175B					MC14175B
CD4017AD	4017B	4017			
CD4017AE	4017B	4017			
CD4017B					MC14017B
CD4017BE	4017B	4017	4017	2417	
CD4018	4018B	4022			
CD40181B					MC14581B
CD40182B	40182B				MC14582B
CD4018AE	4018B	4018			
CD4018B		4018			MC14018B
CD4018BCN	4018B	4018			
CD4018BE	4018B	4018			
CD4019	4019B		4019		
CD40192	40192B	40192			
CD40192B		40192			MC14510B
CD40192BC	40192B	40192			
CD40192BCN	40192B	40192			
CD40192BE	40192B	40192			
CD40193B	4516B				MC14516B
CD40194B					MC14194B
CD4019AE	4019B	4019			
CD4019B					MC14519B
CD4019BE	4019B	4019	4019		
CD4020AE	4020B	4020			
CD4020B		4020			MC14020B
CD4020BE	4020B	4020	4020	2420	
CD4021AE	4021B	4021			
CD4021B		4021			MC14021B
CD4021BE	4021B	4021	4021	2421	
CD4022	4022B	4022			
CD4022AE	4022B	4022			
CD4022B	4022B	4022			MC14022B
CD4022BCN	4022B	4022			
CD4022BE	4022B	4022			
CD4023	4023B		4023		
CD4023AE	4023B	4023			
CD4023B					MC14023B
CD4023BE	4023B	4023	4023	2423	
CD4023C	4023B	4023	4023		
CD4023UB					MC14023UB
CD4023UBE	4023B		4023		
CD4024AE	4024B	4024			
CD4024B	4024B				MC14024B
CD4024BE	4024B	4024	4024		
CD4025	4025B		4025		
CD4025AE	4025B	4025			
CD4025BE	4025B	4025	4025		
CD4025UB					MC14025UB
CD4027AE	4027B	4027			
CD4027B	4027B				MC14027B
CD4027BE	4027B	4027	4027	2427	
CD4028	4028B	4028			
CD4028AE	4028B	4028			
CD4028B		4028			MC14028B
CD4028BC	4028B	4028			
CD4028BCN	4028B	4028			
CD4028BE	4028B	4028			
CD4029AE	4029B	4029			
CD4029B		4029			MC14029B
CD4029BE	4029B	4029			
CD4030AE	4030B	4030			
CD4030B					MC14070B
CD4030BD	4030B	4030			
CD4030BE		4030	4030		
CD4032B					MC14032B
CD4034B	4034B				MC14034B
CD4035AE	4035B	4035			
CD4035B	4035B				MC14035B
CD4038B					MC14038B
CD4040	4040B		4040		
CD4040AE	4040B	4040			
CD4040B		4040			MC14040B
CD4040BE	4040B	4040	4040		
CD4042	4042B		4042		
CD4042AE	4042B	4042			
CD4042B		4042			MC14042B
CD4042BE	4042B	4042	4042		
CD4043	4043B	4043			
CD4043AE	4043B	4043			
CD4043B	4043B	4043			MC14043B
CD4043BE	4043B	4043			
CD4043CN	4043B	4043			
CD4044	4044B	4044			
CD4044AE	4044B	4044			
CD4044B	4044B	4044			MC14044B
CD4044BE	4044B	4044			
CD4044CN	4044B	4044			
CD4046A	980	4046			
CD4046AE	980	4046			
CD4046B		4046			MC14046B
CD4047AE	4047B	4047			
CD4047BE	4047B	4047			
CD4049UB					MC14049UB
CD4049UBE	4049	4049	4049	2449	
CD4050	4050B		4050		
CD4050AE	4050B	4050			
CD4050B		4050			MC14050B
CD4050BE	4050B	4050	4050	2450	
CD4051AE	4051B	4051			
CD4051B		4051			MC14051B
CD4051BE	4051B	4051	4051	2451	

Industry Standard No.	ECG	SK	GE	RS 276-	MOTOR.
CD4052	4052B		4052		
CD4052AE	4052B	4052			
CD4052B		4052			MC14052B
CD4052BE	4052B	4052	4052		
CD4053	4053B	4053			
CD4053B	4053B	4053			MC14053B
CD4053BCN	4053B	4053			
CD4053BE	4053B	4053			
CD4055AE	4055B	4055			
CD4055B		4055			MC14543B
CD4055BE	4055B	4055	4055		
CD4056B		4056			MC14543B
CD4060AE	4060B	4060			
CD4060B	4060B	4060			MC14060B
CD4060BE	4060B	4060			
CD4062A					MC14562B
CD4063B					MC14585B
CD4066AE	4066B	4066			
CD4066B	4066B	4066			MC14066B
CD4066BE	4066B	4066			
CD4067B					MC14529B
CD4068B					MC14068B
CD4069UB	4069	4069			MC14069UB
CD4069UBE	4069	4069			
CD4070B		4070			MC14070B
CD4070BE	4070B	4070			
CD4071B		4071			MC14071B
CD4071BE	4071B	4071			
CD4072B		4072			MC14072B
CD4072BE	4072B	4072			
CD4073B					MC14073B
CD4075B	4075B	4075			MC14075B
CD4075BE	4075B	4075			
CD4076B	4076B				MC14076B
CD4077B					MC14077B
CD4078B	4078B				MC14078B
CD4081B	4081B	4081			MC14081B
CD4081BE	4081B	4081	4081		
CD4082B					MC14082B
CD4085B	4085B				MC14506B
CD4086B					MC14506B
CD4093B		4093			MC14093B
CD4093BE	4093B	4093			
CD4094B		4094			MC14094B
CD4098B	4098B	4098			
CD4098BE	4098B	4098			
CD4099B					MC14099B
CD4112					1N3154
CD4115					1N3157
CD4116	142A	3062	ZD-12	563	
CD4117	142A	3062	ZD-12	563	
CD4118	142A	3062	ZD-12	563	
CD4121	142A	3062	ZD-12	563	
CD4122	142A	3062	ZD-12	563	
CD437	159		82	2032	
CD4404					MC14404
CD4406					MC14406
CD4407					MC14407
CD441	199		62	2010	
CD4413					MC14413
CD4414					MC14414
CD4416					MC14416
CD4418					MC14418
CD445	159		82	2032	
CD446	123A	3444	20	2051	
CD4502B					MC14502B
CD4503B					MC14503B
CD4508B					MC14508B
CD4510B	4510B				MC14510B
CD4511	4511B	4511			
CD4511B		4511			MC14511B
CD4511BE	4511B	4511			
CD4511E	4511B	4511			
CD4512B					MC14512B
CD4514B	4514B				MC14514B
CD4515B	4515B				MC14515B
CD4516B	4516B				MC14516B
CD4517B					MC14517B
CD4518B	4518B	4518			MC14518B
CD4518BE	4518B	4518	4518	2490	
CD4520B	4520B	4520			MC14520B
CD4520BE	4520B	4520			
CD4527B					MC14527B
CD4532B					MC14532B
CD4555	4555B	4555			
CD4555B		4555			MC14555B
CD4555BE	4555B	4555			
CD4556B		4556			MC14556B
CD4556BE	4556B	4556			
CD461	130		14	2041	
CD461-014-614	130		14	2041	
CD48	136A	3057	ZD-5.6	561	
CD500	234	3016	65	2050	
CD5002	519		514	1122	
CD5003	177		300	1122	
CD522	218		234		
CD541	909	3590	IC-249		
CD562	199		62	2010	
CD5918					MRF321
CD5919A					MRF323
CD5944					2N5944
CD5945	363				2N5945
CD5946	364				2N5946
CD600	177	3017B/117			
CD6016	519		514	1122	
CD6016-013-689	519		514	1122	
CD6019	123A	3444	20	2051	
CD6105					MRF5177A
CD6105A					MRF5177A
CD6150	123A	3444	20	2051	
CD6153	128		243	2030	
CD6153-2	128		243	2030	
CD6157	123A	3444	20	2051	
CD6161P1N013-75	519		514	1122	
CD6375	123A	3444	20	2051	
CD7012	335				MRF454
CD8000	123A	3444	20	2051	
CD8000-1	123A	3122	20	2051	
CD8457	177	3100/519	300	1122	
CD8547			300	1122	
CD860011	178MP	3100/519	300(2)	1122(2)	
CD86003	177	3100/519	300	1122	
CD860037	177	3100/519	300	1122	
CD9525	123A	3444	20	2051	
CDC-13000-1	123A	3122	20	2051	
CDC-13000-1D	123A	3122	20	2051	
CDC-8000-1			17	2051	
CDC-8000-1D	192	3137	63	2030	
CDC-8001	123A	3018	20	2051	
CDC-9000			52	2024	
CDC-9000-1B	129	3025	244	2027	
CDC-9000-1D	159	3025/129	82	2032	
CDC-9002-IC			21	2034	
CDC10000-1B	159	3114/290	82	2032	

Industry Standard No.	ECG	SK	GE	RS 276-	MOTOR.
CDC12000			20	2051	
CDC12000-1C	123A	3444	20	2051	
CDC12018C	123A	3444	20	2051	
CDC1201BC	123A	3444	20	2051	
CDC12030B	107	3018	11	2015	
CDC120700	128		243	2030	
CDC12077F	123A	3444	20	2051	
CDC12108		3020	20	2051	
CDC12112C	108	3452	86	2038	
CDC12112D	108	3452	86	2038	
CDC12112E	108		86	2038	
CDC12112F	108	3452	86	2038	
CDC13000			20	2051	
CDC13000-1	123A	3444	20	2051	
CDC13000-1B	123A	3444	20	2051	
CDC13000-1B	123A	3444	20	2051	
CDC13000-1C	123A	3444	20	2051	
CDC13000-1D	123A	3444	20	2051	
CDC13000C			20	2051	
CDC13016A	123A	3444	20	2051	
CDC13019B			20	2051	
CDC13500-1	123A	3444	20	2051	
CDC15018			20	2051	
CDC2010	123A	3444	20	2051	
CDC2010C	123A	3444	20	2051	
CDC2010D	123A	3444	20	2051	
CDC25100-6	123A	3444	20	2051	
CDC2510C	123A	3444	20	2051	
CDC2510C-G	123A	3444	20	2051	
CDC2510G	123A	3444	20	2051	
CDC4023A130	123A	3444	20	2051	
CDC430	123A	3444	20	2051	
CDC4306813	123A	3444	20	2051	
CDC496	159	3025/129	82	2032	
CDC5000	107	3018	11	2015	
CDC5000-1B	108	3018	86	2038	
CDC5008	128	3024	243	2030	
CDC5028A	128	3024	243	2030	
CDC5038A	108	3452	86	2038	
CDC5071A	108	3452	86	2038	
CDC5075B	108	3452	86	2038	
CDC587			18	2030	
CDC60132			20	2051	
CDC731	312	3448	FET-1	2035	
CDC744	154	3045/225	40	2012	
CDC745			18	2030	
CDC745(ZENITH)	123A	3444	20	2051	
CDC746	123A	3444	20	2051	
CDC746(ZENITH)	159		82	2032	
CDC8000	123A	3444	20	2051	
CDC8000-1	128	3020/123	243	2030	
CDC8000-1B	123A	3444	20	2051	
CDC8000-1C	128		243	2030	
CDC8000-1D	192		63	2030	
CDC8000-CM	128		243	2030	
CDC8001	123A	3444	20	2051	
CDC8002	128	3020/123	243	2030	
CDC8002-1	128	3020/123	243	2030	
CDC8011B	123A	3122	20	2051	
CDC8021	123A	3444	20	2051	
CDC8054	123A	3444	20	2051	
CDC8201			10	2051	
CDC8457			300	1122	
CDC86X7-5	123A	3444	20	2051	
CDC9000-1	159	3025/129	82	2032	
CDC9000-1B	192	3025/129	244	2027	
CDC9000-1C			67	2023	
CDC9000-1D	159		82	2032	
CDC90001B			67	2023	
CDC9002	129	3025	244	2027	
CDC9002-1B	128	3020/123	243	2030	
CDC9002-1C	128	3025/129	243	2030	
CDG-00	177	3100/519	300	1122	
CDG-20	177		300	1122	
CDG-20/494	177		300	1122	
CDG-21	177	3100/519	300	1122	
CDG-22	177	3100/519	300	1122	
CDG-24	177	3130	300	1122	
CDG00	177	3100/519	300	1122	
CDG005	116	3311	504A	1104	
CDG025	177	3100/519	300	1122	
CDG22	177		300	1122	
CDG24	177	3100/519	300	1122	
CDG24/3490	177		300	1122	
CDG27	177	3100/519	300	1122	
CDJ-00	177	3100/519	300	1122	
CDK-4			504A	1104	
CDQ10001	123A	3444	20	2051	
CDQ10002	123A	3444	20	2051	
CDQ10003	123A	3444	20	2051	
CDQ10004	123A	3444	20	2051	
CDQ10005	123A	3444	20	2051	
CDQ10006	123A	3444	20	2051	
CDQ10007	123A	3444	20	2051	
CDQ10008	123A	3444	20	2051	
CDQ10009	123A	3444	20	2051	
CDQ10010	123A	3444	20	2051	
CDQ10011	128	3024	243	2030	
CDQ10012	128	3024	243	2030	
CDQ10013	154	3045/225	40	2012	
CDQ10014	128	3024	243	2030	
CDQ10015	154	3045/225	40	2012	
CDQ10016	123A	3444	20	2051	
CDQ10017	123A	3444	20	2051	
CDQ10018	123A	3444	20	2051	
CDQ10019	123A	3444	20	2051	
CDQ10020	123A	3444	20	2051	
CDQ10021	123A	3444	20	2051	
CDQ10022	123A	3444	20	2051	
CDQ10023	123A	3444	20	2051	
CDQ10024	123A	3444	20	2051	
CDQ10025	123A	3444	20	2051	
CDQ10026	123A	3444	20	2051	
CDQ10027	123A	3444	20	2051	
CDQ10028	123A	3444	20	2051	
CDQ10032	123A	3444	20	2051	
CDQ10033	128	3024	243	2030	
CDQ10034	154	3045/225	40	2012	
CDQ10035	123	3020	20	2051	
CDQ10036	123	3020	20	2051	
CDQ10037	154	3045/225	40	2012	
CDQ1004	154		40	2012	
CDQ10044		3024	63	2030	
CDQ10045	154	3024/128	40	2012	
CDQ10046	154	3024/128	40	2012	
CDQ10047	154	3045/225	40	2012	
CDQ10048	128	3024	243	2030	
CDQ10049	154	3045/225	40	2012	
CDQ10051	128	3024	243	2030	
CDQ10052	128	3024	243	2030	
CDQ10053	128	3024	243	2030	
CDQ10057	128	3024	243	2030	
CDQ10058		3024	243	2030	
CDQ1018			20	2051	

Industry Standard No.	ECG	SK	GE	RS 276-	MOTOR.
CDQ1021			20	2051	
CDQ1024			20	2051	
CDR-2	116	3031A	504A	1104	
CDR-4	116	3017B/117	504A	1104	
CDR69/121/207	911		IC-253		
CDS-16B			504A	1104	
CDT-1322	179		76		
CDT1309	121	3009	239	2006	
CDT1310	121	3009	239	2006	
CDT1311	121	3009	239	2006	
CDT1319	121	3009	239	2006	
CDT1320	121	3014	239	2006	
CDT1321	121	3014	239	2006	
CDT1322		3642	25		
CDT1349	121	3009	239	2006	
CDT1349A	121	3009	239	2006	
CDT1350	121	3009	239	2006	
CDT1350A	121	3009	239	2006	
CDZ-318-75	138A	3059	ZD-7.5		
CDZ-9V	139A		ZD-9.1	562	
CDZ-C9V	139A		ZD-9.1	562	
CDZ15000	128	3024	243	2030	
CDZ318-75			ZD-13	563	
CE0360/7839	102A	3004	20	2007	
CE03607839			2	2007	
CE0361/7839	102A		20	2007	
CE0362/7839	102A	3004	20	2007	
CE0363/7839	158	3004/102A	53	2007	
CE0378/7839		3126	90		
CE0398/7839	116	3031A	504A	1104	
CE0495/7839	109	3088		1123	
CE213811	102A		53	2007	
CE37	177	3100/519	300	1122	
CE4001B	123AP	3444/123A	20	2051	
CE4001C	123AP	3452/108	86	2038	
CE4001E	123A	3444	20	2051	
CE4002D	129	3025	244	2027	
CE4003D	128	3024	243	2030	
CE4003E	123A	3444	20	2051	
CE4004C	123AP	3444/123A	20	2051	
CE4005C	159	3114/290	82	2032	
CE4008B	229	3854/123AP	61	2038	
CE4008C	229	3854/123AP	61	2038	
CE4008D	229	3854/123AP	61		
CE4010D	108	3452	86	2038	
CE4010E	108	3452	86	2038	
CE4012D	159	3466	82		
CE4013E	123AP	3854	20	2051	
CE502	116	3016	504A	1104	
CE504	116	3017B/117	504A	1104	
CE506	116	3017B/117	504A	1104	
CE508	125	3032A	510,531	1114	
CE510	125	3033	510,531	1114	
CEC6050	116	3017B/117	504A	1104	
CER-69			504A	1104	
CER-71			504A	1104	
CER-710A			504A	1104	
CER500	116	3017B/117	504A	1104	
CER500,A,B,C					1N4005
CER500A	116	3017B/117	504A	1104	
CER500B	116	3017B/117	504A	1104	
CER500C	116	3017B/117	504A	1104	
CER67	116	3311	504A	1104	
CER67,A,B,C					1N4001
CER670	116	3017B/117	504A	1104	
CER670A	116	3311	504A	1104	
CER670B	116	3017B/117	504A	1104	
CER670C	116	3017B/117	504A	1104	
CER67A	116		504A	1104	
CER67B	116		504A	1104	
CER67C	116	3311	504A	1104	
CER68	116	3016	504A	1104	
CER68,A,B,C					1N4002
CER680	116	3016	504A	1104	
CER680A	116	3016	504A	1104	
CER680B	116	3016	504A	1104	
CER680C	116	3017B/117	504A	1104	
CER68A	116	3016	504A	1104	
CER68B	116	3016	504A	1104	
CER68C	116	3017B/117	504A	1104	
CER69	116	3016	504A	1104	
CER69,A,B,C					1N4003
CER690	116	3016	504A	1104	
CER690A	116	3017B/117	504A	1104	
CER690B	116	3016	504A	1104	
CER690C	116	3017B/117	504A	1104	
CER69A	116		504A	1104	
CER69B	116	3017B/117	504A	1104	
CER69C	116	3017B/117	504A	1104	
CER6B	116	3017B/117	504A	1104	
CER70	116	3016	504A	1104	
CER70,A,B,C					1N4004
CER700	116	3016	504A	1104	
CER700A	116	3016	504A	1104	
CER700B	116	3031A	504A	1104	
CER700C	116	3016	504A	1104	
CER70A	116	3016	504A	1104	
CER70B	116	3031A	504A	1104	
CER70C	116	3031A	504A	1104	
CER71	116	3017B/117	504A	1104	
CER71,A,B,C					1N4005
CER710	116	3017B/117	504A	1104	
CER710A	116	3017B/117	504A	1104	
CER710B	116	3017B/117	504A	1104	
CER710C	116	3017B/117	504A	1104	
CER71A	116	3017B/117	504A	1104	
CER71B	116	3017B/117	504A	1104	
CER71C	116	3017B/117	504A	1104	
CER72	125	3032A	510,531	1114	
CER72,A,B,C,D					1N4006
CER720	125	3032A	510,531	1114	
CER720A	125	3032A	510,531	1114	
CER720B	125	3032A	510,531	1114	
CER720C	125	3032A	510,531	1114	
CER72A	125	3032A	510,531	1114	
CER72B	125	3032A	510,531	1114	
CER72C	125	3032A	510,531	1114	
CER72D	125	3032A	510,531	1114	
CER72F	125	3032A	510,531	1114	
CER73	125	3033	510,531	1114	
CER73,A,B,C,D					1N4007
CER730	125	3033	510,531	1114	
CER730A	125	3033	510,531	1114	
CER730B	125	3033	510,531	1114	
CER730C	125	3033	510,531	1114	
CER73A	125	3033	510,531	1114	
CER73B	125	3033	510,531	1114	
CER73C	125	3033	510,531	1114	
CER73D	125	3033	510,531	1114	
CER73F	125	3033	510,531	1114	
CF-092	139A	3060	ZD-9.1	562	
CF-1295H			18	2030	
CF-2	123A		20	2051	
CF1	108	3452	86	2038	

Industry Standard No.	ECG	SK	GE	RS 276-	MOTOR.
CF102DA	116	3016	504A	1104	
CF102PA	125	3032A	510,531	1114	
CF102RA	125	3032A	510,531	1114	
CF102VA	156	3033	512		
CF11	313		278	2016	
CF14	233		210	2009	
CF152VA	156	3051	512		
CF2	123A	3444	20	2051	
CF2386		3112			2N5459
CF3	610	3126	90		
CF5	123A	3122	20	2051	
CF6	195A	3048/329	46		
CF8	237	3299	46		
CG1	123A	3444	20	2051	
CG12-E			1N60	1123	
CG12E	109	3087		1123	
CG23018	519		514	1122	
CG64H	109	3087		1123	
CG65H	109	3087		1123	
CG66H	109	3087		1123	
CG74H	109	3087		1123	
CG86H	109	3091		1123	
CGD1029	109	3091		1123	
CGD462	109	3087		1123	
CGD591	109	3091		1123	
CGD685	109	3087		1123	
CGE-50	160		245	2004	
CGE-500	177		300	1122	
CGE-51	160		245	2004	
CGE-52	102A		53	2007	
CGE-53	158		53	2007	
CGE-60	319P		283	2016	
CGE-61	233		210	2009	
CGE-62	199	3245	62	2010	
CGE-63	192	3137	63	2030	
CGE-64	172A		64		
CGE-66	152		66	2048	
CGE-67	193	3138	67	2023	
CGE-68	124		12		
CGE-69	153		69	2049	
CGE-75	283	3467	36		
CGE-78MP	283		36		
CGE-79	238	3710	259		
CGJ-1	506	3843	511	1114	
CH119D	116	3016	504A	1104	
CH4728		3330	ZD-3.3		
CH4730		3331	ZD-3.9		
CH4731		3332	ZD-4.3		
CH4761		3344	ZD-75		
CH4763		3345	ZD-91		
CH4764		3346	ZD-100		
CH5275		3349	ZD-140		
CH5278		3352	ZD-170		
CHA4Z3.6	134A		ZD-3.6		
CHA4Z5.1	135A	3056	ZD-5.1		
CHA4Z5.6	136A	3057	ZD-5.6	561	
CHAZ6.2	137A	3058	ZD-6.2	561	
CHAZ6.2A	137A	3058	ZD-6.2	561	
CHAZ62.A			ZD-6.2	561	
CHE					MRF626
CHMZ3.6	134A	3055	ZD-3.6		
CHMZ5.1	135A	3056	ZD-5.1		
CHMZ5.6	136A	3057	ZD-5.6	561	
CHZ10	140A	3061	ZD-10	562	
CHZ10A	140A	3061	ZD-10	562	
CHZ210A			ZD-10	562	
CI-1003		3168	IC-15		
CI-8.2	5072A		ZD-8.2	562	
CI2711	123A	3122		2051	
CI2712	123A	3122	20	2051	
CI2713	123A	3122	20	2051	
CI2714	123A	3122	20	2051	
CI2923	123A	3122	20	2051	
CI2924	123A	3122	20	2051	
CI2925	123A	3122	20	2051	
CI2926	123A	3122	20	2051	
CI3390	123A	3122	20	2051	
CI3391	123A	3122	20	2051	
CI3391A	123A	3122	20	2051	
CI3392	123A	3122	20	2051	
CI3393	123A	3122	20	2051	
CI3394	123A	3122	20	2051	
CI3395	123A	3122	20	2051	
CI3396	123A	3122	20	2051	
CI3397	123A	3122	20	2051	
CI3398	123A	3122	20	2051	
CI3402	123A	3122	20	2051	
CI3403	123A	3122	20	2051	
CI3404	123A	3122	20	2051	
CI3405	123A	3122	20	2051	
CI3414	123A	3122	20	2051	
CI3415	123A	3122	20	2051	
CI3416	123A	3122	20	2051	
CI3417	123A	3122	20	2051	
CI3704	128	3024	243	2030	
CI3705	128	3024	243	2030	
CI3706	128	3024	243	2030	
CI3900	123A	3122	20	2051	
CI3900A	123A	3122	20	2051	
CI3901	123A	3122	20	2051	
CI4256	123A	3122	20	2051	
CI4424	123A	3122	20	2051	
CI4425	123A	3122	20	2051	
CIC-300	124		12		
CII-225-Q	130		14	2041	
CII73Y		3893	66	2048	
CIJ70645	110MP			1123(2)	
CIL-531		3452	20	2051	
CIL-532		3452	20	2051	
CIL511	108		86	2038	
CIL512	108		86	2038	
CIL513	108		86	2038	
CIL521	108		86	2038	
CIL522	108		86	2038	
CIL523	108		86	2038	
CIL531	108		86	2038	
CIL532	108		86	2038	
CIL533	108		86	2038	
CJ-5206	123A	3122	20	2051	
CJ-5207	123A	3122	20	2051	
CJ-5208	123A	3020/123	20	2051	
CJ-5209				2027	
CJ-5210			18	2030	
CJ-5211			20	2051	
CJ-5212			20	2051	
CJ5201	123A	3444	20	2051	
CJ5202	123A	3444	20	2051	
CJ5203	123A	3444	20	2051	
CJ5204	102A		53	2007	
CJ5206		3444	20	2051	
CJ5206A	192		63	2030	
CJ5207		3444	20	2051	
CJ5209	129	3025	244	2027	
CJ5210	128	3024	243	2030	

Industry Standard No.	ECG	SK	GE	RS 276-	MOTOR.
CJ5211	123A	3444	20	2051	
CJ5212	123A	3444	20	2051	
CJ5213	128	3024	243	2030	
CJ5214	128	3024	243	2030	
CJ5215	128	3024	243	2030	
CK-706			1N60	1123	
CK13	102A	3005	53	2007	
CK13A	102A	3005	53	2007	
CK14	100	3005	1	2007	
CK14A	100	3005	1	2007	
CK16	100	3005	1	2007	
CK16A	100	3005	1	2007	
CK17	100	3005	1	2007	
CK17A	100	3005	1	2007	
CK22	102A	3004	53	2007	
CK22A	102A	3004	53	2007	
CK22B	102A	3004	53	2007	
CK22C	102A	3004	53	2007	
CK25	100	3005	1	2007	
CK25A	100	3005	1	2007	
CK26	100	3005	1	2007	
CK261	101	3861	8	2002	
CK262	101	3861	8	2002	
CK26A	100	3005	1	2007	
CK27	100	3005	1	2007	
CK27A	100	3005	1	2007	
CK28	126	3008	52	2024	
CK28A	126	3008	52	2024	
CK4	126	3006/160	52	2024	
CK419	123A	3024/128	243	2030	
CK420	123A	3245/199	212	2010	
CK421	123A	3156/172A	212	2010	
CK422	123A	3024/128	243	2030	
CK474	123A	3024/128	243	2030	
CK475	123A	3024/128	243	2030	
CK476	123A	3452/108	86	2038	
CK477	123A	3452/108	86	2038	
CK4A	126	3006/160	52	2024	
CK64	102A	3004	53	2007	
CK64A	102A	3004	53	2007	
CK64B	102A	3004	53	2007	
CK64C	102A	3004	53	2007	
CK65	102A	3004	53	2007	
CK65A	102A	3004	53	2007	
CK65B	102A	3004	53	2007	
CK65C	102A	3004	53	2007	
CK66	102A	3004	53	2007	
CK661	100	3005	1	2007	
CK662	100	3005	1	2007	
CK66A	102A	3004	53	2007	
CK66B	102A	3004	53	2007	
CK66C	102A	3004	53	2007	
CK67	102A	3004	53	2007	
CK67A	102A	3004	53	2007	
CK67B	102A	3004	53	2007	
CK67C	102A	3004	53	2007	
CK705	109	3087		1123	
CK706	109	3087		1123	
CK706A	109	3087		1123	
CK706P	109	3087		1123	
CK710	112		1N82A		
CK715	109	3087		1123	
CK721	102	3004	2	2007	
CK722	102	3004	2	2007	
CK725	102	3004	2	2007	
CK727	102	3004	2	2007	
CK731	112		1N82A		
CK751	102	3004	2	2007	
CK754	102	3004	2	2007	
CK759	100	3005	1	2007	
CK759A	100	3004/102A	1	2007	
CK760	100	3005	1	2007	
CK760A	100	3004/102A	1	2007	
CK761	100	3005	1	2007	
CK762	126	3006/160	52	2024	
CK766	126	3005	52	2024	
CK766A	126	3006/160	52	2024	
CK768	100	3005	1	2007	
CK776	100	3005	1	2007	
CK776A	100	3005	1	2007	
CK790	102	3004	2	2007	
CK791	102	3004	2	2007	
CK793	102	3004	2	2007	
CK794	102	3004	2	2007	
CK83				2005	
CK870		3004	2	2007	
CK871	102	3004	2	2007	
CK872	102	3004	2	2007	
CK875	102	3004	2	2007	
CK878	102	3004	2	2007	
CK879	102	3004	2	2007	
CK882	102	3004	2	2007	
CK888	102	3004	2	2007	
CK891	102A		53	2007	
CK892	102A		53	2007	
CK942	159	3114/290	82	2032	
CL010	116	3311	504A	1104	
CL025	116	3311	504A	1104	
CL05	116	3017B/117	504A	1104	
CL1	116	3016	504A	1104	
CL1.5	116	3017B/117	504A	1104	
CL100			LED56		MLED930
CL1020					1N5297
CL110	3028		H11B1		MLED930
CL110A	3028				MLED930
CL110B					MLED930
CL1520					1N5302
CL2	116	3016	504A	1104	
CL2210					1N5283
CL2220					1N5306
CL3	116	3016	504A	1104	
CL3310					1N5287
CL3320					1N5310
CL4	116	3031A	504A	1104	
CL4710					1N5290
CL4720					1N5314
CL5	116	3017B/117	504A	1104	
CL6	116	3017B/117	504A	1104	
CL6810					1N5293
CL7	116	3032A	504A	1104	
CL8	116	3032A	504A	1104	
CLI-10					4N33
CLI-2					4N38
CLI-3					4N35
CLI-5	3040				4N26
CLM05	116	3017B/117	504A	1104	
CLM1	116	3017B/117	504A	1104	
CLR2050	3032				MRD370
CLR2060	3036				MRD360
CLR2110					MRD310
CLR2140					MRD310
CLR2150					MRD300
CLR2160					MRD300
CLR2170					MRD370
CLR2180	3036				MRD360
CM	113A	3119/113	6GC1		
CM-9	736		IC-17		
CM0770	128	3024	243	2030	
CM10-12A	365				MRF641
CM20-12A	366				MRF644
CM25-28					MRF325
CM2550	121	3014	239	2006	
CM30-12A	367				MRF646
CM4000		4000	4000		
CM4001		4001	4001	2401	
CM4002		4002	4002		
CM4006A	4006B	4006			
CM4006AE	4006B	4006			
CM4007		4007	4007		
CM4009		4009	4009		
CM4011		4011	4011	2411	
CM4012		4012	4012	2412	
CM4013		4013	4013	2413	
CM4014AE	4014B	4014			
CM4015		4015	4015		
CM4016		4016	4016		
CM4017		4017	4017	2417	
CM4018AE	4018B	4018			
CM4019		4019	4019		
CM4020			4020	2420	
CM4020AE	4020B	4020			
CM4021			4021	2421	
CM4022AE	4022B	4022			
CM4023		4023	4023	2423	
CM4024		4024	4024		
CM4025		4025	4025		
CM4027		4027	4027	2427	
CM4028AE	4028B	4028			
CM4030		4030	4030		
CM4035	4035B	4035			
CM4042		4042	4042		
CM4046	980	4046			
CM4049		4049	4049	2449	
CM4050		4050	4050	2450	
CM4051		4051	4051	2451	
CM4081		4081	4081		
CM45-12A	367				MRF646
CM45-28					MRF326
CM4511	4511B	4511			
CM4518	4518B	4518	4518		
CM50-12A	368				MRF648
CM60-12A	368				MRF648
CM600					2N4392
CM601					2N4091
CM602					2N4091
CM603	466				2N4091
CM640					2N3824
CM641					2N4393
CM642					2N4092
CM643					2N4391
CM644					2N4092
CM645					2N4092
CM646					2N4091
CM647	466				2N4391
CM650					MFE2012
CM651					MFE2012
CM697					MPF4392
CM7163			10	2051	
CM80-28					MRF327
CM80-28R					MRF327
CM8470			504A	1104	
CM8640E			2	2007	
CMC334-423	123A		20	2051	
CM0770			20	2051	
CMP-01CJ					MC1556G
CMP-01CP					MC1556P
CMX740					MPF4391
CN2484	128	3024	243	2030	
CNY17,18					4N25
CNY21					4N25
C02	5072A		ZD-8.2	562	
C049	116	3031A	504A	1104	
COC13000-1C			63	2030	
COD1-6045			504A	1104	
COD1-6046			504A	1104	
COD1-6047			504A	1104	
COD1-6048			504A	1104	
COD11556		3017B	504A	1104	
COD1531	116	3017B/117	504A	1104	
COD1532	116	3016	504A	1104	
COD1533	116	3016	504A	1104	
COD1534	116	3017B/117	504A	1104	
COD1535	116	3017B/117	504A	1104	
COD1536	116	3017B/117	504A	1104	
COD1537	125		510,531	1114	
COD1538	125		510,531	1114	
COD1551	116	3016	504A	1104	
COD1552	116	3017B/117	504A	1104	
COD15524		3016	504A	1104	
COD1553	116	3016	504A	1104	
COD15534		3016	504A	1104	
COD1554	116	3017B/117	504A	1104	
COD15544		3016	504A	1104	
COD1555	116	3017B/117	504A	1104	
COD1556	116	3017B/117	504A	1104	
COD15564		3017B	504A	1104	
COD1575			504A	1104	
COD16047	116		504A	1104	
COD1611	116	3017B/117	504A	1104	
COD1612	116	3016	504A	1104	
COD1613	116	3016	504A	1104	
COD1614	116	3017B/117	504A	1104	
COD1615	116	3017B/117	504A	1104	
COD1616	116	3017B/117	504A	1104	
COD1617	125	3033	510,531	1114	
COD1618	125	3033	510,531	1114	
CODI115524			504A	1104	
CODI11556	116	3017B/117	504A	1104	
CODI1531	116	3017B/117	504A	1104	
CODI15524	116	3017B/117		1104	
CODI15534	116	3017B/117	504A	1104	
CODI15544	116	3017B/117	504A	1104	
CODI15564	116	3017B/117	504A	1104	
CODI537	125	3032A		1114	
CODI538		3032A		1114	
CODI6045	116		504A	1104	
CODI6047	116		504A	1104	
CODI617	125	3032A		1114	
CP102	125	3032A	510,531	1114	
CP102BA	116	3016	504A	1104	
CP102DA	116	3016	504A	1104	
CP102FA	116	3031A	504A	1104	
CP102HA	116	3031A	504A	1104	
CP102KA	116	3017B/117	504A	1104	
CP102MA	116	3017B/117	504A	1104	
CP102PA	125	3032A	510,531	1114	
CP102RA	125	3032A	510,531	1114	
CP102VA	125	3033	510,531	1114	

Industry Standard No.	ECG	SK	GE	RS 276-	MOTOR.
CP103	125	3033	510,531	1114	
CP152VA	125	3033	510,531	1114	
CP2357	128		243	2030	
CP3212055	139A		ZD-9.1	562	
CP400	130		14	2041	
CP401	130		14	2041	
CP404	130		14	2041	
CP405	130		14	2041	
CP406	130		14	2041	
CP407	130		14	2041	
CP408	130	3024/128	14	2041	
CP409	130	3024/128	243	2030	
CP650					MPF4391
CP651					MPF4391
CP652					MPF4392
CP653					MPF4392
CP800	158		53	2007	
CP801	158		53	2007	
CP802	158		53	2007	
CP803	158		53	2007	
CQ1	102A		53	2007	
CQT1075	121	3009	239	2006	
CQT1076	121	3009	239	2006	
CQT1077	121	3009	239	2006	
CQT1110	121	3014	239	2006	
CQT1110A	121	3009	239	2006	
CQT1111	121	3009	239	2006	
CQT1111A	121	3009	239	2006	
CQT1112	121	3014	239	2006	
CQT1129	121	3717	239	2006	
CQT940A	121	3009	239	2006	
CQT940B	121	3009	239	2006	
CQT940BA	121	3009	239	2006	
CQY10					MLED930
CQY11,B,C					MLED930
CQY12,B					MLED930
CQY13					4N26
CQY14					4N26
CQY15					4N26
CQY31					MLED930
CQY32					MLED930
CQY40,41					4N26
CQY80					MOC1005
CR/E	116		504A	1104	
CR0000	109	3090		1123	
CR01C	5524	6624			
CR01D	5526	6627/5527			
CR01E	5527	6627			
CR02A-1(GAK)	5401	3638/5402			
CR02A-2(GAK)	5402	3638			
CR02A-4(GAK)	5404	3627			
CR02A-8(GAK)	5405	3951			
CR02AM-1(GAK)	5401	3638/5402			
CR02AM-2(GAK)	5402	3638			
CR02AM-4(GAK)	5404	3627			
CR02AM-6(GAK)	5405	3951			
CR02AM-8(GAK)	5405	3951			
CR02C	5517	3615			
CR02D	5518	3653			
CR02E	5518	3653			
CR05A-1	5401	3638/5402			
CR05A-10	5406	3952			
CR05A-12	5406	3952			
CR05A-2	5402	3638			
CR05A-4	5404	3627			
CR05A-6	5405	3951			
CR05A-8	5405	3951			
CR1-051C	5408		C6F		
CR1-101C	5408		C6A		
CR1-201C	5408		C6B		
CR1-301C	5409		C6C		
CR101/6515	109	3090		1123	
CR102/6515	109	3090		1123	
CR1034	116		504A	1104	
CR1035	116	3016	504A	1104	
CR10C-10	5528	6629/5529			
CR10C-12	5529	6629			
CR10C-16	5531	6631			
CR10C-2	5522	6622			
CR10C-4	5524	6624			
CR10C-6	5526	6627/5527			
CR10C-8	5527	6627			
CR12M01FY	5522	6622			
CR12M02FY	5524	6624			
CR12M03FY	5526	6627/5527			
CR12M04FY	5527	6627			
CR12M06FY	5529	6629			
CR12M08FY	5531	6631			
CR12MX5FY	5521	6621			
CR12U01FY	5522	6622			
CR12U02FY	5524	6624			
CR12U03FY	5526	6627/5527			
CR12U04FY	5527	6627			
CR12U06FY	5529	6629			
CR12U08FY	5531	6631			
CR12UX5FY	5521	6621			
CR20F-2	5542	6642			
CR20M01FY	5542	6642			
CR20M01FY-1	5542	6642			
CR20U01FY	5542	6642			
CR20U01FY-1	5542	6642			
CR20UX5FY	5541	6642/5542			
CR20UX5FY-1	5541	6642/5542			
CR24M01FY	5542	6642			
CR24M01FY-1	5542	6642			
CR24MX5FY	5541	6642/5542			
CR24MX5FY-1	5541	6642/5542			
CR24U01FY	5542	6642			
CR24U01FY-1	5542	6642			
CR24UX5FY	5541	6642/5542			
CR24UX5FY-1	5541	6642/5542			
CR2AM-1	5453	6753			
CR2AM-2	5454	6754			
CR2AM-4	5455	3597			
CR32077-1	358		42		
CR32077R	358		42		
CR353			300	1122	
CR3AMZ-8	5457	3598			
CR4U02FY	5483	3942			
CR4U03FY	5484	3943/5485			
CR4U04FY	5485	3943			
CR4U05FY	5486	3944/5487			
CR4U06FY	5487	3944			
CR501/6515	116(2)		504A(2)	1104	
CR801			504A	1104	
CR951		3051	512		
CR952		3051	512		
CR953		3051	512		
CR954		3051	512		
CR955		3051	512		
CR956		3062	ZD-12	563	
CRT1544	121	3009	239	2006	
CRT1545	121	3009	239	2006	
CRT1552	121	3009	239	2006	
CRT1553	121	3009	239	2006	
CRT1602	121	3014	239	2006	
CRT3602A	121	3009	239	2006	
CS-1120C1			20	2051	
CS-1120C2			20	2051	
CS-1120D1			20	2051	
CS-1120H			20	2051	
CS-1120I			10	2051	
CS-1235F	123A	3122	20	2051	
CS-1238F			20	2051	
CS-1238P			20	2051	
CS-1244X		3452	17	2051	
CS-1258			17	2051	
CS-1259			10	2051	
CS-12941			10	2051	
CS-1294F			21	2034	
CS-1305			17	2051	
CS-1330		3018	20	2051	
CS-1352			63	2030	
CS-1359		3452	17	2051	
CS-1361E		3452	20	2051	
CS-1361F			20	2051	
CS-1361G		3124	10	2051	
CS-1372			18	2030	
CS-1386E	123A	3018	20	2051	
CS-1386H		3452	20	2051	
CS-16E			504A	1104	
CS-2004C	108	3020/123	86	2038	
CS-2005B	159	3114/290	82	2032	
CS-2005C	159	3114/290	82	2032	
CS-2007G	108	3452	86	2038	
CS-2007H	108	3122	86	2038	
CS-2008F	108	3452	86	2038	
CS-2142	193		67	2023	
CS-2143	192	3024/128	63	2030	
CS-3001B	108	3452	86	2038	
CS-3024	172A	3156	64		
CS-460B	123A	3452/108	20	2051	
CS-461B	108	3452	86	2038	
CS-6168F			10	2051	
CS-6168G	123A		20	2051	
CS-6168H	123A	3122	20	2051	
CS-6225E	107		11	2015	
CS-6225F			20	2051	
CS-6225G	123A	3018	20	2051	
CS-6227E			20	2051	
CS-6227F			20	2051	
CS-6227G	108	3452	86	2038	
CS-6228G	129	3025	244	2027	
CS-9011	123A	3122	20	2051	
CS-90111			20	2051	
CS-9011F			20	2051	
CS-9011G	123A		20	2051	
CS-9011L	123A		20	2051	
CS-9012HF			22	2032	
CS-9012HH				2027	
CS-9013		3124	20	2051	
CS-9013HE			10	2051	
CS-9013HG			18	2030	
CS-9013HH			21	2034	
CS-9014	123A	3444	39	2015	
CS-9014B	123A	3444	20	2051	
CS-9014D	123A	3444	20	2051	
CS-9015D	159		82	2032	
CS-9016	229	3452/108	86	2038	
CS-9016D	229	3018	11	2015	
CS-9016F	229		86	2038	
CS-9018	108	3018	86	2038	
CS-9018D			17	2051	
CS-9018E	108		86	2038	
CS-9018F	108		86	2038	
CS-9018G	108		86	2038	
CS-9018H	108		86	2038	
CS-901ZHF			22	2032	
CS-9104	123A		20	2051	
CS-9125B	123A	3124/289	20	2051	
CS10-02M		3613	C222U		
CS10-02N	5491	6791			
CS10-05M		3613	C222F		
CS10-05N	5491	6791			
CS10-1M		3613	C222A		
CS10-1N	5491	6791			
CS10-2M		3613	C222B		
CS10-2N	5492	6792			
CS10-4N	5494	6794			
CS10-6N	5496	6796			
CS1014	108	3452	86	2038	
CS1014D	108	3452	86	2038	
CS1014E	108	3452	86	2038	
CS1014F	108	3452	86	2038	
CS1014G	161	3132	39	2015	
CS1014H	161	3117	39	2015	
CS1018	108	3452	86	2038	
CS1068			20	2051	
CS1120C	161	3132	39	2015	
CS1120D	161	3132	39	2015	
CS1120E	161	3132	39	2015	
CS1120F	161	3117	39	2015	
CS1120H	161	3132	39	2015	
CS1120I			17	2051	
CS1121G	159		82	2032	
CS1124G	106	3984	21	2034	
CS1129E	128	3024	243	2030	
CS1166	123A	3444	20	2051	
CS1166D	123A	3444	20	2051	
CS1166D-G		3124	10	2051	
CS1166E	123A	3444	20	2051	
CS1166F	123A	3444	20	2051	
CS1166G	123A	3444	20	2051	
CS1166H	123A	3444	20	2051	
CS1166H/F	123A	3444	20	2051	
CS1168E	108	3452	86	2038	
CS1168F	123A	3444	20	2051	
CS1168G	123A	3444	20	2051	
CS1168H	123A	3444	20	2051	
CS1170F	159	3025/129	82	2032	
CS1221F	159		82	2032	
CS1225D	108	3452	86	2038	
CS1225E	108	3452	86	2038	
CS1225F	108	3452	86	2038	
CS1225H	128	3024	243	2030	
CS1225HF	128	3024	243	2030	
CS1226	108	3452	86	2038	
CS1226E	108	3452	86	2038	
CS1226F	108	3452	86	2038	
CS1226G	108	3452	86	2038	
CS1226H	108	3452	86	2038	
CS1226N			20	2051	
CS1227	108	3452	86	2038	
CS1227D	108	3452	86	2038	
CS1227E	108	3452	86	2038	
CS1227F	108	3452	86	2038	
CS1227G	108	3452	86	2038	
CS1228	159	3025/129	82	2032	

Industry Standard No.	ECG	SK	GE	RS 276-	MOTOR.
CS1228E	159	3114/290	82	2032	
CS1229	123A	3444	20	2051	
CS1229A	123A	3444	20	2051	
CS1229B	123A	3444	20	2051	
CS1229C	123A	3444	20	2051	
CS1229D	123A	3444	20	2051	
CS1229E	123A	3444	20	2051	
CS1229F	123A	3444	20	2051	
CS1229G	123A	3444	20	2051	
CS1229H	123A	3444	20	2051	
CS1229K	128	3024	243	2030	
CS1229N			18	2030	
CS122A					CS122A
CS122B					CS122B
CS122C					CS122C
CS122D					CS122D
CS122E					CS122E
CS122M					CS122M
CS122N					CS122N
CS122S					CS122S
CS1235C	123A	3444	20	2051	
CS1235E	123A	3444	20	2051	
CS1235G	123A	3444	20	2051	
CS1236D	123A	3444	20	2051	
CS1236H	123A	3444	20	2051	
CS1237	129	3025	244	2027	
CS1238	108	3452	86	2038	
CS1238F	123A	3122	20	2051	
CS1238G	108	3452	86	2038	
CS1238H	108	3452	86	2038	
CS1238I	108	3452	86	2038	
CS1238P	123A	3444	20	2051	
CS1243E	108	3452	86	2038	
CS1243H	108		86	2038	
CS1244H	108		86	2038	
CS1244J	108		86	2038	
CS1244X	108	3018	86	2038	
CS1245F	123A	3444	20	2051	
CS1245G	123A	3444	20	2051	
CS1245H	123A	3444	20	2051	
CS1245I	123A	3444	20	2051	
CS1245T	123A	3444	20	2051	
CS1248	128	3024	243	2030	
CS1248I	128	3024	243	2030	
CS1248T	128	3024	243	2030	
CS1250E	123A	3444	20	2051	
CS1250F	128		243	2030	
CS1251E	159	3025/129	82	2032	
CS1251F	159	3114/290	82	2032	
CS1252B	108	3452	86	2038	
CS1252C	108	3452	86	2038	
CS1255H	128	3024	243	2030	
CS1255HF	193	3024/128	67	2023	
CS1255M			63	2030	
CS1256H	129	3195/311	244	2027	
CS1256HG	192	3195/311	63	2030	
CS1257	123A	3444	20	2051	
CS1258	123A	3444	20	2051	
CS1259	123A	3444	20	2051	
CS1283A	123A	3444	20	2051	
CS1284B	161	3716	39	2015	
CS1284F	161	3716	39	2015	
CS1284G	161	3117	39	2015	
CS1284H	161	3117	39	2015	
CS1286			20	2051	
CS1288	123A	3444	20	2051	
CS1289	123A	3444	20	2051	
CS1293	108	3452	86	2038	
CS1294E	159	3114/290	82	2032	
CS1294H	159	3114/290	82	2032	
CS1295E	123A	3444	20	2051	
CS1295G	123A	3444	20	2051	
CS1295H	128	3024	243	2030	
CS1298	159		82	2032	
CS1303	159	3114/290	82	2032	
CS1305	128	3024	243	2030	
CS1308	159	3114/290	82	2032	
CS1312G	129	3114/290	244	2027	
CS131A	6354	6554			
CS131AZ	6354	6554			
CS131B	6354	6554			
CS131C	6354	6554			
CS131D(AXIAL)	116		504A	1104	
CS131E	6356	6556			
CS131F	6356	6556			
CS131Z	6354	6554			
CS1330	161	3018	39	2015	
CS1330A	108	3018	86	2038	
CS1330B	108		86	2038	
CS1330C	108		86	2038	
CS1330D			17	2051	
CS13401	123A	3444	20	2051	
CS1340D	108	3452	86	2038	
CS1340E	108	3452	86	2038	
CS1340F	108	3452	86	2038	
CS1340G	108	3452	86	2038	
CS1340H	108	3452	86	2038	
CS1340I	123A	3444	20	2051	
CS1344	123A	3444	20	2051	
CS1345	123A	3444	20	2051	
CS1347	154		40	2012	
CS1348	123A	3444	20	2051	
CS1349	123A	3444	20	2051	
CS1350	108	3452	86	2038	
CS1351	108	3452	86	2038	
CS1352	128		243	2030	
CS1353	123A	3444	20	2051	
CS1354	159		82	2032	
CS1359	108		86	2038	
CS1360	108		86	2038	
CS1361E	108	3018	86	2038	
CS1361F	108	3018	86	2038	
CS1361G	123A	3444	20	2051	
CS1362	123A	3444	20	2051	
CS1363	123A	3444	20	2051	
CS1368	123A	3444	20	2051	
CS1368A	123A	3444	20	2051	
CS1368B	123A	3444	20	2051	
CS1368C	123A	3444	20	2051	
CS1368D	123A	3444	20	2051	
CS1369	129	3025	244	2027	
CS1369D			21	2034	
CS1370	123A	3444	20	2051	
CS1371	123A	3444	20	2051	
CS1372	123A	3444	20	2051	
CS1383	123A	3444	20	2051	
CS1386H	108		86	2038	
CS1420	123A	3444	20	2051	
CS1453E	123A	3444	20	2051	
CS1453F	128	3124/289	243	2030	
CS1453G	128	3024	243	2030	
CS1460E	161	3132	39	2015	
CS1460H	161	3132	39	2015	
CS1461J	161		39	2015	

Industry Standard No.	ECG	SK	GE	RS 276-	MOTOR.
CS1461X	161	3132	39	2015	
CS1462F	161	3132	39	2015	
CS1462I	128	3024	243	2030	
CS1463A	123A		20	2051	
CS1464H	128	3024	243	2030	
CS1465H	129		244	2027	
CS1508G	108	3452	86	2038	
CS1509E	108	3452	86	2038	
CS1509F	108	3452	86	2038	
CS1518E	108	3452	86	2038	
CS1555	108	3452	86	2038	
CS1585	123A	3444	20	2051	
CS1585E/F	123A	3444	20	2051	
CS1585G	123A	3444	20	2051	
CS1585H	107	3018	11	2015	
CS1589E	161	3132	39	2015	
CS1589F	161	3132	39	2015	
CS1589S	161	3132	39	2015	
CS1591LE	128	3024	243	2030	
CS1594E	161	3132	39	2015	
CS1596E	161	3132	39	2015	
CS16-08		6648	C137N		
CS16-10		6649	C137P		
CS1609F	128	3024	243	2030	
CS1613	128	3024	243	2030	
CS1625	123A	3444	20	2051	
CS1627	159		82	2032	
CS1661	108	3452	86	2038	
CS1664	128		243	2030	
CS1665	123A	3444	20	2051	
CS16E	116	3311	504A	1104	
CS1711	128	3024	243	2030	
CS1758	158		53	2007	
CS1759	103A	3835	59	2002	
CS1834	108	3452	86	2038	
CS183E	123		20	2051	
CS184E	123		20	2051	
CS184J	108	3452	86	2038	
CS1893	128	3024	243	2030	
CS1909B	293		47	2030	
CS1910B	294		48		
CS1990	128	3024	243	2030	
CS20-02R	5520	6621/5521			
CS20-05M		3613	C232P		
CS20-05N	5521	6621			
CS20-05R	5521	6621			
CS20-1.5R	5523	6624/5524			
CS20-1M		3613	C232A		
CS20-1N	5522	6622			
CS20-1R	5522	6622			
CS20-2.5R	5525	6627/5527			
CS20-2M		3613	C232B		
CS20-2N	5524	6624			
CS20-2R	5524	6624			
CS20-3R	5526	6627/5527			
CS20-4N	5527	6627			
CS20-4R	5527	6627			
CS20-5R	5528	6629/5529			
CS20-6N	5529	6629			
CS20-6R	5529	6629			
CS2001	123A	3124/289	20	2051	
CS2001H	123AP	3444/123A	20	2051	
CS2004	123A	3124/289	20	2051	
CS2004C	123AP	3444/123A	20	2051	
CS2004D	123A	3444	20	2051	
CS2005		3114	21	2034	
CS2005B			21	2034	
CS2006	123A	3444	20	2051	
CS2006F	229	3018	61	2038	
CS2006G	108	3452	86	2038	
CS2007G			20	2051	
CS2007H			20	2051	
CS2008	108	3018	20	2038	
CS2008G	108	3452	20	2038	
CS2008H	108	3452	86	2038	
CS2008H552	108		86	2038	
CS2023	192	3137	63	2030	
CS2082		3138	67	2023	
CS2142		3025	67	2023	
CS220-5					CS220-5
CS220-7					CS220-7
CS220-9					CS220-9
CS221-5					CS221-5
CS221-7					CS221-7
CS221-9					CS221-9
CS2218	123A	3444	20	2051	
CS2219	123A	3444	20	2051	
CS2221	123A	3444	20	2051	
CS2222	123A	3444	20	2051	
CS2369	123A	3444	20	2051	
CS2481	123A	3444	20	2051	
CS2484	128	3024	243	2030	
CS25-02R	5520	6621/5521			
CS25-05N	5521	6621			
CS25-05R	5521	6621			
CS25-1N	5522	6622			
CS25-1R	5522	6622			
CS25-2N	5524	6624			
CS25-2R	5524	6624			
CS25-3R	5526	6627/5527			
CS25-4N	5527	6627			
CS25-4R	5527	6627			
CS25-6N	5529	6629			
CS25-6R	5529	6629			
CS25-8N	5531	6631			
CS25-8R	5531	6631			
CS2711	123A	3444	20	2051	
CS2712	123A	3444	20	2051	
CS2713	123A	3444	20	2051	
CS2714	123A	3444	20	2051	
CS2715	161	3117	39	2015	
CS2716	161	3117	39	2015	
CS2922	123A	3444	20	2051	
CS2923	123A	3444	20	2051	
CS2924	123A	3444	20	2051	
CS2925	123A	3444	20	2051	
CS2941			22	2032	
CS3001B			20	2051	
CS3082	2023	3694			
CS3390	123A	3444	20	2051	
CS3391	123A	3444	20	2051	
CS3391A	123A	3444	20	2051	
CS3392	123A	3444	20	2051	
CS3393	123A	3444	20	2051	
CS3394	123A	3444	20	2051	
CS3395	123A	3444	20	2051	
CS3396	123A	3444	20	2051	
CS3397	123A	3444	20	2051	
CS3398	123A	3444	20	2051	
CS3402	123A	3444	20	2051	
CS3403	123A	3444	20	2051	
CS3404	123A	3444	20	2051	
CS3405	123A	3444	20	2051	
CS3414	123A	3444	20	2051	

Industry Standard No.	ECG	SK	GE	RS 276-	MOTOR.
CS3415	123A	3444	20	2051	
CS3416	123A	3444	20	2051	
CS3417	123A	3444	20	2051	
CS35-02M		3615	C229U		
CS35-02N	5540	6642/5542			
CS35-02R	5540	6642/5542			
CS35-05M		3615	C229F		
CS35-05N	5541	6642/5542			
CS35-05R	5541	6642/5542			
CS35-1.5R	5543	3581			
CS35-1M		3615	C229A		
CS35-1N	5542	6642			
CS35-1R	5542	6642			
CS35-2.5R	5544	3581/5543			
CS35-2M		3615	C229B		
CS35-2N	5543	3581			
CS35-4N	5545	3582			
CS35-4R	5545	3582			
CS35-6N	5547	3505			
CS360	123A	3444	20	2051	
CS3605	123A	3444	20	2051	
CS3606	123A	3444	20	2051	
CS3607	123A	3444	20	2051	
CS3662	161	3117	39	2015	
CS3663	161	3117	39	2015	
CS3702	159	3114/290	82	2032	
CS3703	159	3114/290	82	2032	
CS3704	128	3024	243	2030	
CS3705	128	3024	243	2030	
CS3706	128	3024	243	2030	
CS3707	161	3117	39	2015	
CS3708	161	3117	39	2015	
CS3709	161	3117	39	2015	
CS3710	161	3117	39	2015	
CS3711	161	3117	39	2015	
CS3843	123A	3444	20	2051	
CS3844	123A	3444	20	2051	
CS3845	123A	3444	20	2051	
CS3854	123A	3444	20	2051	
CS3854A	123A	3444	20	2051	
CS3855	123A	3444	20	2051	
CS3855A	123A	3444	20	2051	
CS3859	123A	3444	20	2051	
CS3859A	123A	3444	20	2051	
CS3860	123A	3444	20	2051	
CS3900	123A	3444	20	2051	
CS3900A	123A	3444	20	2051	
CS3901	123A	3444	20	2051	
CS3903	123A	3444	20	2051	
CS3904	123A	3444	20	2051	
CS3906	159	3114/290	82	2032	
CS4001	311		11	2015	
CS4003	123A	3245/199	212	2010	
CS4005	128		213		
CS4006	128	3244	220		
CS4007	123A	3244	220		
CS4012	159		221		
CS4013	159		221		
CS4021	123A		60	2015	
CS4060	123A		60	2015	
CS4061	123A	3245/199	212	2010	
CS4193	123A		11	2015	
CS4194	123A		11	2015	
CS429J	108	3452	86	2038	
CS430H	108	3452	86	2038	
CS4424	123A	3444	20	2051	
CS4425	123A	3444	20	2051	
CS461F	108		86	2038	
CS469F	108		86	2038	
CS5088	123A	3444	20	2051	
CS5369	123A	3444	20	2051	
CS5447	159	3114/290	82	2032	
CS5448	159	3114/290	82	2032	
CS5449	128	3024	243	2030	
CS5450	128	3024	243	2030	
CS5451	128	3024	243	2030	
CS5995	703A		IC-12		
CS6168F	123A	3444	20	2051	
CS6168G	128	3444/123A	243	2030	
CS6225E		3018	17	2051	
CS6225F	108	3452	86	2038	
CS6226F	108	3452	86	2038	
CS6227E	108	3452	86	2038	
CS6227F	108	3452	86	2038	
CS6228F	159		82	2032	
CS6229F	123A	3444	20	2051	
CS6229G	123A	3444	20	2051	
CS6395					CS6395
CS6396					CS6396
CS6397					CS6397
CS6398					CS6398
CS6399					CS6399
CS6401					CS6401
CS6402					CS6402
CS6403					CS6403
CS6404					CS6404
CS6405					CS6405
CS6505					CS6505
CS6506					CS6506
CS6507					CS6507
CS6508					CS6508
CS6509					CS6509
CS696	123A	3444	20	2051	
CS697	123A	3444	20	2051	
CS706	123A	3444	20	2051	
CS718	123A	3444	20	2051	
CS718A	123A	3444	20	2051	
CS72-3					CS72-3
CS72-4					CS72-4
CS72-5					CS72-5
CS72-6					CS72-6
CS72-7					CS72-7
CS72-8					CS72-8
CS720A	123A	3444	20	2051	
CS7228G	129	3114/290	244	2027	
CS7229F	128		243	2030	
CS7229G	123A	3444	20	2051	
CS8-02N	5491	6791			
CS8-05N	5491	6791			
CS8-1N	5491	6791			
CS8-2N	5492	6792			
CS8-3N	5494	6794			
CS8-4N	5494	6794			
CS8-6N	5496	6796			
CS8050	297	3122	271	2030	
CS9001	108	3452	86	2038	
CS9010					MPSH32
CS9011	123A	3444	20	2051	2N4124
CS9011(E)(F)			20	2051	
CS9011(EF)			20	2051	
CS9011(GH)			20	2051	
CS9011/3490	123A	3444	20	2051	
CS90111	123A	3444	20	2051	
CS9011D	123A	3444	20	2051	
CS9011E	123A	3444	20	2051	
CS9011F	123A	3444	20	2051	
CS9011G	123A	3444	20	2051	
CS9011G/3490	123A	3444	20	2051	
CS9011H	123A	3444	20	2051	
CS9011I	123A	3444	20	2051	
CS9011N			20	2051	
CS9012	159	3114/290	82	2032	
CS9012/3490	159		244	2027	
CS90121			244	2027	
CS9012E	159	3114/290	82	2032	
CS9012E-F	159	3114/290	82	2032	
CS9012F	159		244	2027	
CS9012FC	159		244	2027	
CS9012FG	159		82	2032	
CS9012H	159	3114/290	244	2027	
CS9012HE	159	3114/290	244	2027	
CS9012HF	159	3114/290	244	2027	
CS9012HP(LAST IF)	233			2009	
CS9012HP(LAST-IF)			210	2051	
CS9012HG	159	3114/290	82	2032	
CS9012HG/3490	159		244	2027	
CS9012HH	159	3114/290	82	2032	
CS9012HH/3490	159		82	2032	
CS9012I	159			2027	
CS9013	123A	3124/289	20	2051	
CS9013(HG)			57	2017	
CS9013/3490	123A		243	2030	
CS9013A	123A	3124/289	20	2051	
CS9013B	123A	3124/289	20	2051	
CS9013C	123A	3124/289	20	2051	
CS9013D	123A	3124/289	20	2051	
CS9013E	123A	3124/289	243	2030	
CS9013E-F		3124	18	2030	
CS9013F	123A	3124/289	243	2030	
CS9013G	123A	3124/289	243	2030	
CS9013H	123A	3124/289	243	2030	
CS9013HE	123A	3124/289	20	2051	
CS9013HF	123A	3124/289	20	2051	
CS9013HG	123A	3124/289	20	2051	
CS9013HG/3490	123A		243	2030	
CS9013HH	123A	3124/289	20	2051	
CS9014	123A	3444	20	2051	
CS9014(C)			20	2051	
CS9014/3490	123A	3444	20	2051	
CS9014A	123A	3444	20	2051	
CS9014B	123A	3444	20	2051	
CS9014B-C		3124	20	2051	
CS9014C	123A	3444	20	2051	
CS9014C/3490	123A	3444	20	2051	
CS9014D	123A	3444	20	2051	
CS9014G	123A	3444	20	2051	
CS9015	159	3444/123A	20	2051	
CS9015B	159	3444/123A	20	2051	
CS9015C	159	3114/290	82	2032	
CS9015C2	159	3114/290	82	2032	
CS9015D	159	3114/290	82	2032	
CS9016	229	3018	39	2015	MPSH30
CS9016(G)			20	2051	
CS9016/3490	229	3452/108	86	2038	
CS9016D	229	3444/123A	20	2051	
CS9016E	229	3039/316	61	2038	
CS9016EF	229	3444/123A	20	2051	
CS9016F	229	3452/108	61	2038	
CS9016F(TRUETONE)			17	2051	
CS9016F(WESTGHSE)			20	2051	
CS9016FG	229	3444/123A	20	2051	
CS9016G	229	3122	86	2038	
CS9016G/3490	229		86	2038	
CS9016H	229		86	2038	
CS9017					2N4124
CS9017F	161	3122	11	2015	
CS9017G	161	3122	39	2015	
CS9017H	161	3132	39	2015	
CS9018	108	3452	86	2038	MPS918
CS9018/3490	108	3452	86	2038	
CS9018D	108	3018	86	2038	
CS9018E	108	3452	86	2038	
CS9018EF	108	3444/123A	20	2051	
CS9018F	108	3452	86	2038	
CS9018F/3490	108	3452	86	2038	
CS9018FG	108	3444/123A	20	2051	
CS9018G	108	3452	86	2038	
CS9019					MPSH30
CS901ZHF	159		82	2032	
CS9020E	159		82	2032	
CS9020F	159		82	2032	
CS9021G-I	108	3018	86	2038	
CS9021HF	129		244	2027	
CS9021HP(LAST IF)	233			2009	
CS9021HP(LAST-IF)			210	2051	
CS9021HG(LAST IF)	233			2009	
CS9021HG(LAST-IF)			210	2051	
CS9022LE	128	3020/123	243	2030	
CS9101B	123A	3444	20	2051	
CS9102	159	3114/290	21	2034	
CS9102B	129	3025	244	2027	
CS9103B	128	3024	243	2030	
CS9103C	128	3024	243	2030	
CS9104		3444	62	2010	
CS9123C1			21	2034	
CS9124-C2	108	3452	86	2038	
CS9124B1	108		86	2038	
CS9125-B1	108	3452	86	2038	
CS9125B	123A	3444	20	2051	
CS9126	123A	3124/289	20	2051	
CS9128		3025	21	2034	
CS9128-B2	159	3114/290	82	2032	
CS9128C1	159		82	2032	
CS9129	159	3025/129	82	2032	
CS9129(B)			21	2034	
CS9129-B1		3114	21	2034	
CS9129B	159	3025/129	82	2032	
CS9129B1	159		82	2032	
CS9129B2	159	3114/290	82	2032	
CS918	108	3452	86	2038	
CS925M	123A	3444	20	2051	
CS929	161	3117	39	2015	
CS93-1007	175		246	2020	
CS930	161	3117	39	2015	
CS9417	192	3122	63	2030	
CS956	128	3024	243	2030	
CS9600-4	123A	3444	20	2051	
CS9600-5	123A	3444	20	2051	
CST1739	104	3719	16	2006	
CST1740	104	3719	16	2006	
CST1741	104	3719	16	2006	
CST1742	104	3719	16	2006	
CST1743	121	3717	239	2006	
CST1744	121	3717	239	2006	
CST1745	121	3717	239	2006	
CST1746	121	3717	239	2006	
CT-2002		3088	1N60	1123	
CT-2005			1N60	1123	

Industry Standard No.	ECG	SK	GE	RS 276-	MOTOR.
CT-2017		3126	90		
CT-3003			504A	1104	
CT-3005			504A	1104	
CT02C	5492	6792			
CT02D	5494	6794			
CT02E	5494	6794			
CT03C	5514	3613			
CT03D	5515	6615			
CT03E	5515	6615			
CT04C	5483	3942			
CT04D	5484	3943/5485			
CT04E	5485	3943			
CT04F	5486	3944/5487			
CT06B	5462	3686			
CT06C	5463	3572			
CT06D	5465	3687			
CT06E	5465	3687			
CT100	116		504A	1104	
CT1009	102A	3004	53	2007	
CT1012	107	3018	11	2015	
CT1013	107	3018	11	2015	
CT1017	158	3004/102A	53	2007	
CT11-1A	5640	3583/5641			
CT11-2A	5641	3583			
CT1122	121	3009	239	2006	
CT1124	121	3009	239	2006	
CT1124A	121	3009	239	2006	
CT1124B	121	3009	239	2006	
CT12C	5463	3572			
CT12E	5465	3687			
CT1300	108		86	2038	
CT15-10					CT15-10
CT15-3					CT15-3
CT15-4					CT15-4
CT15-5					CT15-5
CT15-6					CT15-6
CT15-7					CT15-7
CT15-8					CT15-8
CT15-9					CT15-9
CT1500	108		86	2038	
CT15A10					CT15A10
CT15A3					CT15A3
CT15A4					CT15A4
CT15A5					CT15A5
CT15A6					CT15A6
CT15A7					CT15A7
CT15A8					CT15A8
CT15A9					CT15A9
CT16XT	113A		6GC1		
CT200	116	3016	504A	1104	
CT2002	109	3088		1123	
CT2005	110MP	3088		1123(2)	
CT2006		3126	90		
CT2007	109	3088		1123	
CT2008	110MP	3088	1N60	1123(2)	
CT220-3					CT220-3
CT220-5					CT220-5
CT220-7					CT220-7
CT220-9					CT220-9
CT221-3					CT221-3
CT221-5					CT221-5
CT221-7					CT221-7
CT221-9					CT221-9
CT223-10					CT223-10
CT223-3					CT223-3
CT223-4					CT223-4
CT223-5					CT223-5
CT223-6					CT223-6
CT223-7					CT223-7
CT223-8					CT223-8
CT223-9					CT223-9
CT223A10					CT223A10
CT223A3					CT223A3
CT223A4					CT223A4
CT223A5					CT223A5
CT223A6					CT223A6
CT223A7					CT223A7
CT223A8					CT223A8
CT223A9					CT223A9
CT300	116	3016	504A	1104	
CT3003	116	3031A	504A	1104	
CT3005	116		504A	1104	
CT302-2N	5693	3652			
CT302-4N	5695	3509			
CT302-6N	5697	3522			
CT31-1A	5640	3583/5641			
CT31-2A	5641	3583			
CT33-1A	5650	3583/5641			
CT33-2A	5651	3583/5641			
CT401-2N	5693	3652			
CT401-4N	5695	3509			
CT401-6N	5697	3522			
CT402-2N	5693	3652			
CT461	109	3087		1123	
CT600	116	3017B/117	504A	1104	
CT6342					CT6342
CT6342A					CT6342A
CT6343					CT6343
CT6343A					CT6343A
CT6344					CT6344
CT6344A					CT6344A
CT6345					CT6345
CT6345A					CT6345A
CT6346					CT6346
CT6346A					CT6346A
CT6347					CT6347
CT6347A					CT6347A
CT6348					CT6348
CT6348A					CT6348A
CT6349					CT6349
CT6349A					CT6349A
CT6776	154		40	2012	
CTC-TX100	317		281		
CTC-TX50	318		282		
CTC14					MRF464A
CTC15					MRF428A
CTC2001					MRF2001
CTC2002					MRF2003
CTC2003					MRF2003
CTC2005					MRF2005
CTC2010					MRF2010
CTN100			510	1114	
CTN1000			510	1114	
CTN200	116		504A	1104	
CTN300			510	1114	
CTN400			510	1114	
CTN50			510	1114	
CTN500			510	1114	
CTN600			510	1114	
CTN800			510	1114	
CTP-2001-1001	102A	3004	53	2007	
CTP-2001-1002	102A	3004	53	2007	
CTP-2001-1003	102A	3004	53	2007	
CTP-2001-1004	102A	3004	53	2007	
CTP-2001-1007	123A	3444	20	2051	
CTP-2001-1008	123A	3444	20	2051	
CTP-2001-1009	102A	3004	53	2007	
CTP-2001-1010	109	3087		1123	
CTP-2001-1011	116	3017B/117	504A	1104	
CTP-2001-1012	116	3017B/117	504A	1104	
CTP-2006-1001	102A	3004	53	2007	
CTP-2006-1002	102A	3004	53	2007	
CTP-2006-1003	102A	3004	53	2007	
CTP-2006-1004	109	3087		1123	
CTP-461			1N295	1123	
CTP100			510	1114	
CTP1000			510	1114	
CTP1032	102A	3004	53	2007	
CTP1033	102A	3004	53	2007	
CTP1034	102A	3004	53	2007	
CTP1035	102A	3004	53	2007	
CTP1036	102A	3004	53	2007	
CTP1104	104	3719	16	2006	
CTP1106	104	3009	16	2006	
CTP1108	104	3719	16	2006	
CTP1109	104	3719	16	2006	
CTP1111	121	3717	239	2006	
CTP1117	104	3719	16	2006	
CTP1119	104	3009	16	2006	
CTP1124	121	3717	239	2006	
CTP1133	121	3014	239	2006	
CTP1135	121	3014	239	2006	
CTP1136	121	3717	239	2006	
CTP1137	121	3717	239	2006	
CTP1265	121	3009	239	2006	
CTP1266	121	3009	239	2006	
CTP1306	121	3009	239	2006	
CTP1307	121	3009	239	2006	
CTP1500	121	3717	239	2006	
CTP1503	121	3717	239	2006	
CTP1504	121	3014	239	2006	
CTP1508	121	3009	239	2006	
CTP1509	105	3012	4		
CTP1511	121	3009	239	2006	
CTP1512	105	3009	4		
CTP1513	121	3009	239	2006	
CTP1514	121	3717	239	2006	
CTP1544	179		76		
CTP1545	179		76		
CTP1550	121	3717	239	2006	
CTP1551	121	3717	239	2006	
CTP1552	179		76		
CTP1553	179		76		
CTP1728	121	3717	239	2006	
CTP1729	121	3717	239	2006	
CTP1730	121	3009	239	2006	
CTP1731	121	3717	239	2006	
CTP1732	121	3717	239	2006	
CTP1733	121	3717	239	2006	
CTP1735	121	3717	239	2006	
CTP1736	121	3717	239	2006	
CTP1739	121	3717	239	2006	
CTP1740			16	2006	
CTP200			510	1114	
CTP2076-1001	102A	3004	53	2007	
CTP2076-1002	102A	3004	53	2007	
CTP2076-1003	102A	3004	53	2007	
CTP2076-1004	102A	3004	53	2007	
CTP2076-1005	102A	3004	53	2007	
CTP2076-1006	102A	3004	53	2007	
CTP2076-1007	102A	3004	53	2007	
CTP2076-1008	102A	3004	53	2007	
CTP2076-1009	102A	3004	53	2007	
CTP2076-1010	102A	3004	53	2007	
CTP2076-1011	102A	3004	53	2007	
CTP2076-1012	102A	3004	53	2007	
CTP300			510	1114	
CTP3500	121	3717	239	2006	
CTP3503	121	3717	239	2006	
CTP3504	121	3717	239	2006	
CTP3508	121	3717	239	2006	
CTP3544	179		76		
CTP3545	179		76		
CTP3552	179		76		
CTP3553	179		76		
CTP400			510	1114	
CTP461	109	3087		1123	
CTP50			510	1114	
CTP500			510	1114	
CTP573	109	3091		1123	
CTP600			510	1114	
CTP800			510	1114	
CU-12E	5455	3597		1067	
CU01C	5483	3942			
CU01D	5484	3943/5485			
CU01E	5485	3943			
CU01F	5486	3944/5487			
CU06VC	230	3042			
CU06VD	230	3042			
CU06VE	230	3042			
CU06VF	230	3042			
CU06VG	230	3042			
CU12E	5455	3597			
CV03B(KAG)	5402	3638			
CV03C(KAG)	5404	3627			
CV03D(KAG)	5405	3951			
CV03E(KAG)	5405	3951			
CV04C	5404	3627			
CV04D	5405	3951			
CV04E	5405	3951			
CV12B	5454	6754			
CV12C	5455	3597			
CV12E	5457	3598			
CV1620	610	3323			
CV1626	612	3325			
CV1634	613	3326			
CV425	109	3087		1123	
CV442	109	3087		1123	
CV5007	610	3323			
CV5022	613	3326			
CV832	613	3326			
CW01B(GAK)	5402	3638			
CW01C(GAK)	5404	3627			
CX-0031			1N60	1123	
CX-0033			1N34AS	1123	
CX-0035			504A	1104	
CX-0036		3091	1N60	1123	
CX-0037		3016	504A	1104	
CX-0039		3016	504A	1104	
CX-0040		3016	504A	1104	
CX-0041		3091	90		
CX-0042		3091	1N60	1123	
CX-0045		3088	1N60	1123	
CX-0047		3311	1N34AS	1123	
CX-0048		3016	504A	1104	
CX-0049		3016	ZD-20		
CX-0052		3062	ZD-12	563	
CX-0054		3032A	504A	1104	

Industry Standard No.	ECG	SK	GE	RS 276-	MOTOR.
CX-0055	177		300	1122	
CX-093	712		IC-2		
CX-9001			504A	1104	
CX0036	109	3087		1123	
CX0037	116	3016	504A	1104	
CX0039	116		504A	1104	
CX0040	116		504A	1104	
CX0041	109	3087		1123	
CX0042	110MP			1123(2)	
CX0047	116	3017B/117	504A	1104	
CX0048	116	3017B/117	504A	1104	
CX0049	116	3017B/117	504A	1104	
CX0050	135A		ZD-5.1		
CX0051	140A	3061	ZD-10	562	
CX089D	1096	3709/110MP			
CX093D	712		IC-148		
CX095	1308	3833			
CX095A	1308	3833			
CX095C	1308	3833			
CX9000	358		42		
CX9001	116	3016	504A	1104	
CXL100		3721	1	2007	
CXL101			8	2002	
CXL102		3722	2	2007	
CXL1024		3253	270		
CXL1025		3155	722		
CXL1027		3153	721		
CXL102A		3004	52	2024	
CXL103			8	2002	
CXL104		3719	16	2006	
CXL104MP		3720	13	2006(2)	
CXL105		3012	4		
CXL106		3118	21	2034	
CXL107		3293	11	2015	
CXL108		3452	86	2038	
CXL109		3090	1N34AS	1123	
CXL1090		3291	723		
CXL110		3709	1N34AS	1123	
CXL112		3089	1N82A		
CXL113		3119	6GC1		
CXL114		3120	6GD1	1104	
CXL115		3121	6GX1		
CXL116		3313	504A	1104	
CXL117		3313	504A	1104	
CXL118		3066	CR-1		
CXL119		3109	CR-2		
CXL120		3110	CR-3		
CXL121		3717	239	2006	
CXL121MP		3718	239	2006	
CXL123		3580	MR-4		
CXL123A		3444	20	2051	
CXL124		3021	12		
CXL125		3051	512		
CXL126			52	2024	
CXL127		3764	25		
CXL128		3024	18	2030	
CXL129		3025	244	2027	
CXL130		3027	14	2041	
CXL130MP		3029	15MP	2041(2)	
CXL131		3198	44	2006	
CXL131MP			44	2006	
CXL132			FET-1	2028	
CXL133			FET-1	2028	
CXL134		3055	ZD-3.6		
CXL135		3056	ZD-5.1		
CXL137		3058	ZD-6.2	561	
CXL138		3059	ZD-7.5		
CXL139		3060	ZD-9.1	562	
CXL140		3061	ZD-10	562	
CXL141		3092	ZD-11	563	
CXL142		3062	ZD-12	563	
CXL143		3093	ZD-13	563	
CXL144		3094		564	
CXL145		3063	ZD-15	564	
CXL146		3064	ZD-27		
CXL147		3095	ZD-33		
CXL148		3096	ZD-55		
CXL149		3097	ZD-62		
CXL150		3098	ZD-82		
CXL151		3099	ZD-110		
CXL152		3054	66	2048	
CXL153		3083	69	2049	
CXL154		3044	40	2012	
CXL156		3051	512		
CXL157		3747	232		
CXL158		3004	53	2007	
CXL159		3466	81	2051	
CXL160		3006	245	2004	
CXL161		3716	39	2015	
CXL162		3438	35		
CXL163		3439	36		
CXL164		3111	37		
CXL165		3115	38		
CXL166		3105	BR-206	1152	
CXL167		3105	BR-206	1172	
CXL168		3106		1173	
CXL171		3201	27		
CXL175		3261	246	2020	
CXL177		3175	300	1122	
CXL178MP			300	1122	
CXL180		3437	74	2043	
CXL181		3535	75	2041	
CXL182		3188A	55		
CXL183		3189A	56	2027	
CXL184		3190	57	2017	
CXL185		3191	58	2025	
CXL186		3192	28	2017	
CXL187		3193	29	2025	
CXL188		3199	217		
CXL189		3200	218	2026	
CXL190		3232	217		
CXL191		3232	249		
CXL192		3137	63	2030	
CXL193		3138	67	2023	
CXL194		3275	220		
CXL195		3765	219		
CXL196		3054	241	2020	
CXL197		3083	250	2027	
CXL198		3220	251		
CXL199		3245	62	2010	
CXL210		3298	236		
CXL211		3203	253	2026	
CXL218		3625	234		
CXL219		3173	74	2043	
CXL220			FET-3	2036	
CXL221			FET-3	2036	
CXL222		3065	FET-4	2036	
CXL223			255	2041	
CXL224		3049	46		
CXL226		3082	49	2025	
CXL226MP		3086	49	2025	
CXL228		3103A	257		
CXL238		3710	259		
CXL5005		3330	ZD-3.3		

Industry Standard No.	ECG	SK	GE	RS 276-	MOTOR.
CXL5006		3055	ZD-3.6		
CXL5007		3331	ZD-3.9		
CXL5008		3332	ZD-4.3		
CXL5009		3333	ZD-4.7		
CXL5010		3056	ZD-5.1		
CXL5011		3057	ZD-5.6	561	
CXL5012		3778		561	
CXL5013		3058	ZD-6.2	561	
CXL5014		3334	ZD-6.8	561	
CXL5015		3059	ZD-7.5		
CXL5016			ZD-8.2	562	
CXL5017		3783		562	
CXL5018		3060	ZD-9.1	562	
CXL5019		3061	ZD-10	562	
CXL502		3067	CR-4		
CXL5020			ZD-11	563	
CXL5021		3062	ZD-12	563	
CXL5022		3788		563	
CXL5023		3789		564	
CXL5024		3094	ZD-15	564	
CXL5025		3791		564	
CXL5027		3752	ZD-18		
CXL5029		3335	ZD-20		
CXL503		3068	CR-5		
CXL5030		3336	ZD-22		
CXL5033		3064	ZD-27		
CXL5035		3755	ZD-30		
CXL5036		3095	ZD-33		
CXL5038		3338	ZD-39		
CXL5039		3339	ZD-43		
CXL504		3108	CR-6		
CXL5040		3340	ZD-47		
CXL5041		3341	ZD-51		
CXL5042		3342	ZD-56		
CXL5044		3097	ZD-62		
CXL5045		3343	ZD-68		
CXL5046		3344	ZD-75		
CXL5047		3098	ZD-82		
CXL5049		3345	ZD-91		
CXL505		3108	CR-7		
CXL5050		3346	ZD-100		
CXL5051		3099	ZD-110		
CXL5052		3347	ZD-120		
CXL5053		3348	ZD-130		
CXL5054		3349	ZD-140		
CXL5055		3350	ZD-150		
CXL5056		3351	ZD-160		
CXL5057		3352	ZD-170		
CXL5058		3353	ZD-180		
CXL5059		3354	ZD-190		
CXL506		3125	511	1114	
CXL5060		3355	ZD-200		
CXL5066		3330	ZD-3.3		
CXL5067		3331	ZD-3.9		
CXL5068		3333	ZD-4.7		
CXL5069		3333	ZD-4.7		
CXL507		3315	511	1114	
CXL5070				561	
CXL5071		3334	ZD-6.8	561	
CXL5072			ZD-8.2	562	
CXL5073		3749	ZD-8.7	562	
CXL5074			ZD-11	563	
CXL5075		3751		564	
CXL5077		3752	ZD-18		
CXL5079		3335	ZD-20		
CXL508		3756	SS-3A3		
CXL5080		3336	ZD-22		
CXL5084		3755	ZD-30		
CXL5085		3338	ZD-39		
CXL5086		3338	ZD-39		
CXL5087		3339	ZD-43		
CXL5088		3340	ZD-47		
CXL5089		3341	ZD-51		
CXL509		3757	SS-3AT2		
CXL5090		3342	ZD-56		
CXL5092		3343	ZD-68		
CXL5093		3344	ZD-75		
CXL5095		3345	ZD-91		
CXL5096		3346	ZD-100		
CXL5097		3347	ZD-120		
CXL5098		3350	ZD-150		
CXL5099		3349	ZD-140		
CXL510		3758	SS-3DB3		
CXL5100		3350	ZD-150		
CXL5101		3351	ZD-160		
CXL5102		3352	ZD-170		
CXL5103		3353	ZD-180		
CXL5104		3354	ZD-190		
CXL5105		3355	ZD-200		
CXL511		3759	SS-2AV2		
CXL5111		3377	5ZD-3.3		
CXL5112		3378	5ZD-3.6		
CXL5113		3379	5ZD-3.9		
CXL5114		3380	5ZD-4.3		
CXL5115		3381	5ZD-4.7		
CXL5116		3382	5ZD-5.1		
CXL5117		3383	5ZD-5.6		
CXL5118		3384	5ZD-6.0		
CXL5119		3385	5ZD-6.2		
CXL512		3760	SS-6DW4		
CXL5120		3386	5ZD-6.8		
CXL5121		3387	5ZD-7.5		
CXL5122		3388	5ZD-8.2		
CXL5123		3389	5ZD-8.7		
CXL5124		3390	5ZD-9.1		
CXL5125		3391	5ZD-10		
CXL5126		3392	5ZD-11		
CXL5127		3393	5ZD-12		
CXL5128		3394	5ZD-13		
CXL5129		3395	5ZD-14		
CXL513		3443	513		
CXL5130		3396	5ZD-15		
CXL5131		3397	5ZD-16		
CXL5132		3398	5ZD-17		
CXL5133		3399	5ZD-18		
CXL5134		3400	5ZD-19		
CXL5135		3401	5ZD-20		
CXL5136		3402	5ZD-22		
CXL5137		3403	5ZD-24		
CXL5138		3404	5ZD-25		
CXL5139		3405	5ZD-27		
CXL5140		3406	5ZD-28		
CXL5141		3407	5ZD-30		
CXL5142		3408	5ZD-33		
CXL519		3100	514	1122	
CXL5452		3597	MR-5	1067	
CXL5453		3597	MR-5	1067	
CXL5454		3597	MR-5	1067	
CXL5455		3597	MR-5	1067	
CXL5456		3598		1020	
CXL5470		3580	MR-4		
CXL5471		3580	MR-4		
CXL5472		3580	MR-4		
CXL5473		3580	MR-4		

Industry Standard No.	ECG	SK	GE	RS 276-	MOTOR.
CXL5474		3580	MR-4		
CXL5491		3582	MR-3		
CXL5492		3582	MR-3		
CXL5494		3582	MR-3		
CXL5500		3582	MR-3		
CXL5501		3582	MR-3		
CXL5502		3582	MR-3		
CXL5503		3582	MR-3		
CXL5505		3582	MR-3		
CXL5506		3582	MR-3		
CXL5507		3582	MR-3		
CXL5514		3582	MR-3		
CXL5520		3582	MR-3		
CXL5521		3582	MR-3		
CXL5522		3582	MR-3		
CXL5523		3582	MR-3		
CXL5525		3582	MR-3		
CXL5526		3582	MR-3		
CXL5527		3582	MR-3		
CXL5540		3582	MR-3		
CXL5541		3582	MR-3		
CXL5542		3582	MR-3		
CXL5544		3582	MR-3		
CXL5545		3582	MR-3		
CXL5800		3848	512		
CXL5801		3848	512		
CXL5802		3848	512		
CXL5803		3848	512		
CXL5804		3848	512		
CXL5805		3848	512		
CXL5806		3848	512		
CXL5808		3051	512		
CXL5809		3051	512		
CXL5830		3603	MR-1		
CXL5832		3603	MR-1		
CXL5834		3603	MR-1		
CXL5836		3608	MR-2		
CXL5850		3603	MR-1		
CXL5852		3603	MR-1		
CXL5854		3603	MR-1		
CXL5856		3603	MR-1		
CXL5858		3608	MR-2		
CXL5870		3603	MR-1		
CXL5872		3603	MR-1		
CXL5874		3603	MR-1		
CXL5876		3608	MR-2		
CXL5878		3608	MR-2		
CXL5944		3608	MR-2		
CXL5946		3608	MR-2		
CXL5948		3608	MR-2		
CXL6400				2029	
CXL6401				2029	
CXL6407				1050	
CXL6408		3523		1050	
CXL6409				2029	
CXL703A		3157	IC-12		
CXL704		3023	IC-205		
CXL705A		3134	IC-3		
CXL706		3288	IC-43		
CXL708		3135	IC-10		
CXL709		3135	IC-11		
CXL710		3102	IC-89		
CXL711		3070	IC-207		
CXL713		3077	IC-5		
CXL714		3075	IC-4		
CXL715		3076	IC-6		
CXL718		3159	IC-8		
CXL722		3161	IC-9		
CXL725		3162	IC-19		
CXL726		3129	IC-81		
CXL727		3071	IC-210		
CXL728		3073	IC-22		
CXL729		3074	IC-23		
CXL731		3170	IC-13		
CXL734		3163	IC-211		
CXL738		3167	IC-29		
CXL747		3279	IC-218		
CXL749		3168	IC-97		
CXL780		3141	IC-20		
CXL781		3169	IC-223		
CXL783		3215	IC-21		
CXL786		3140	IC-227		
CXL787		3146	IC-228		
CXL788		3147	IC-229		
CXL790		3454	IC-18		
CXL949		3166	IC-25		
CY100	125	3033	510,531	1114	
CY40	116	3031A	504A	1104	
CY50	116	3017B/117	504A	1104	
CY80	156	3032A	512		
CZ-092	139A	3060	ZD-9.1	562	
CZ-094	139A	3060			
CZ-92	139A	3060	ZD-9.1	562	
CZ092	139A	3060	ZD-9.1	562	
CZ094	139A	3060	ZD-9.1	562	
CZ13B	143A	3750			
CZ16B	5075A	3751			
CZ17B	5076A	9022			
CZ18B	5077A	3752			
CZ19B	5078A	9023			
CZ25B	5082A	3753			
CZ8.2B	5072A	3136			
CZD010	140A	3061	ZD-10	562	
CZD010-5	140A	3061	ZD-10	562	
CZD012-5	142A	3062	ZD-12	563	
CZD014	144A	3094	ZD-14	564	
CZD014-5	144A	3094	ZD-14	564	
CZD015	145A	3063	ZD-15	564	
CZD015-5	145A	3063	ZD-15	564	
D-00169C	109	3087		1123	
D-00184R	177	3175	300	1122	
D-00204R	109	3087		1123	
D-00269C	109	3087		1123	
D-00284R		3088	1N60	1123	
D-00369C	138A	3059	ZD-7.5		
D-00384R	116	3311	504A	1104	
D-00469C	137A	3058	ZD-6.2	561	
D-00484R	142A	3062	ZD-12	563	
D-00569C	139A		ZD-9.1	562	
D-00669C	109	3087		1123	
D-05	116	3031A	504A	1123(2)	
D-12	177		300	1122	
D-18	139A	3060	ZD-9.1	562	
D-2	109			1123	
D-215-1	519		514	1122	
D-258	124		12		
D-50492-01	159		82	2032	
D-F1	101		8	2002	
D-F1A	102A		53	2007	
D-GM-2			1N34AS	1123	
D004	116	3016	504A	1104	
D006	107	3018	11	2015	
D008	102A	3004	53	2007	
D009	229	3018	61	2038	
D00D	192		ZD-63	2030	
D01-100	116	3017B/117	504A	1104	
D010	110MP	3709	1N60	1123(2)	
D018	102A	3004	53	2007	
D019	100	3005	1	2007	
D020	126	3008	52	2024	
D021	102A	3004	53	2007	
D0250	175		246	2020	
D026	126	3007	52	2024	
D028	116	3031A	504A	1104	
D030	102A	3004	53	2007	
D031	102A	3004	53	2007	
D031(CHAN.MASTER)			20	2051	
D032	109	3090			
D038	102A	3004	53	2007	
D043	102A	3004	53	2007	
D048	123A	3444	20	2051	
D05		3311	504A	1104	
D053	123A	3444	20	2051	
D057	199		62	2010	
D058	108	3452	86	2038	
D059	102A		53	2007	
D063	160	3006	245	2004	
D069	108	3452	86	2038	
D072	229	3018	61	2038	
D073	160	3007	245	2004	
D078	100	3005	1	2007	
D079	160	3006	245	2004	
D080	127	3034	25		
D081	127	3009	25		
D083	103A	3010	59	2002	
D085	103A	3010	59	2002	
D086	160	3006	245	2004	
D087	108	3452	86	2038	
D088	108	3452	86	2038	
D093	109	3087		1123	
D1-12B					MRF817
D1-12E					MRF838A
D1-13	147A		ZD-33		
D1-2	109	3087		1123	
D1-20	177	3100/519	300	1122	
D1-26			504A	1104	
D1-52S	116	3017B/117	504A	1104	
D1-7	116	3017B/117	504A	1104	
D1-8	139A	3060	ZD-9.1	562	
D1/2-12					MRF816
D100	116		504A	1104	1N4002
D1000	125	3033	510,531	1114	1N4007
D100A	103A		59	2002	
D101	101		8	2002	
D101167	116	3031A	504A	1104	
D10167	116	3017B/117	504A	1104	
D10168	116	3031A	504A	1104	
D101B	102A		53	2007	
D102	175		246	2020	
D102-C	175			2020	
D102-O			246	2020	
D102-R	175		246	2020	
D102-Y	175		246	2020	
D103	175		246	2020	
D103-O	175			2020	
D103-O			246	2020	
D103-R	175		246	2020	
D103-Y	175		246	2020	
D104	103A		59	2002	
D105	103A		59	2002	
D105B	102A		53	2007	
D105C	125		510,531	1114	
D108	125	3033	510,531	1114	
D10B1051	107	3039/316	11	2015	
D10B1055	107	3039/316	11	2015	
D10G1051	107	3039/316	11	2015	
D10G1052	107	3039/316	11	2015	
D11	103		8	2002	
D110	280	3438	35		
D110-O		3438	35		
D110-R	280	3438	35		
D110-Y	280	3438	35		
D1101	312	3112	FET-1	2028	2N4220A
D1102	312	3112	FET-1	2028	MFE2094
D1103					MFE2093
D111	280	3438	35		
D111-O		3438	35		
D111-R	280	3438	35		
D111-Y	280	3438	35		
D113	181		75	2041	
D113-O	181			2041	
D113-O			75	2041	
D113-R	181		75	2041	
D113-Y	181		75	2041	
D114	181		75	2041	
D114-O	181			2041	
D114-O			75	2041	
D114-R	181		75	2041	
D114-Y	181		75	2041	
D116	130	3086/226MP	49(2)	2025(2)	
D117	328		53	2007	
D1172S		3017B	504A	1104	
D1177			FET-1	2028	MFE2095
D1178			FET-1	2028	MFE2094
D1179					MFE2093
D118	280		262	2041	
D1180	456		FET-1	2028	2N4222A
D1181			FET-1	2028	2N4220A
D1182					2N4220A
D1183					2N4223
D1184					2N4220A
D1185					2N4220A
D118BL	280		262	2041	
D118R	280		262	2041	
D118Y	280		262	2041	
D119	280		262	2041	
D119BL	280		262	2041	
D119R	280		262	2041	
D119Y	280		262	2041	
D11C10B1	128	3024	243	2030	
D11C11B1	128	3024	243	2030	
D11C1B1	128	3024	243	2030	
D11C1F1			17	2051	
D11C201B20	124	3021	12		
D11C203B20	124	3021	12		
D11C205B20	124	3021	12		
D11C207B20	124	3021	12		
D11C210B20	124	3021	12		
D11C211B20	124	3021	12		
D11C2D1B20			17	2051	
D11C3B1	128	3024	243	2030	
D11C3F1			63	2030	
D11C5B1	128	3024	243	2030	
D11C5F1			17	2051	
D11C7B1	128	3024	243	2030	
D12	130		14	2041	
D120	324	3748/282	264		
D1201		3051			2N4224

Industry Standard No.	ECG	SK	GE	RS 276-	MOTOR.
D1201A	125	3311	510,531	1114	1N4002
D1201B	125	3311	510,531	1114	1N4003
D1201D	125	3312	510,531	1114	1N4004
D1201F	116	3051/156	504A	1104	1N4001
D1201M	125	3313/116	510,531	1114	1N4005
D1201N	125	3051/156	510,531	1114	1N4006
D1201P	125	3051/156	510,531	1114	1N4007
D1202					2N3821
D1203					MFE2093
D120A	324	3748/282	264		
D120B	324	3748/282	264		
D120C	324	3748/282	264		
D120H	324	3202/210	264		
D120HA	324	3202/210	264		
D120HB	324	3202/210	264		
D120HC	324	3202/210	264		
D121	282	3748	264		
D121A	282	3748	264		
D121B	282	3748	264		
D121H	282	3202/210	264		
D121HA	282	3202/210	264		
D121HB	282	3202/210	264		
D124(TRANSISTOR)	280		262	2041	
D124A	87		262	2041	
D124AH	87		262	2041	
D124AHA	87		262	2041	
D124B	280		262	2041	
D125	87		262	2041	
D125A	87		262	2041	
D125AH	87		262	2041	
D125AHA	87		262	2041	
D125AHB	87		262	2041	
D126	87	3004/102A	53	2041	
D126A	87		262	2041	
D126AH	87		262	2041	
D126AHA	87		262	2041	
D126AHB	87		262	2041	
D126H	87		262	2041	
D126HA	87		262	2041	
D126HB	87		262	2041	
D127	103A		59	2002	
D127A	103A		59	2002	
D128	103A		59	2002	
D128A	103A		59	2002	
D129	175		246	2020	
D129-BL	175		246	2020	
D129-R	175		246	2020	
D129-Y	175		246	2020	
D13	105		4		
D130	175		246	2020	
D130-R	175		246	2020	
D130-Y	175		246	2020	
D1301	312	3112		2028	2N4222A
D1302	312	3112		2028	2N4220A
D1303	312	3112		2028	2N4220A
D130BL	175		246	2020	
D132			75	2041	
D133			17	2051	
D133(CHAN.MASTER)			86	2038	
D134	126	3004/102A	52	2024	
D135	102A	3004	53	2007	
D13K1	6402	3628			
D13Q500					MBS4991
D13T1	6402	3628	X17		MPU6027
D13T2					MPU6028
D14	106		21	2034	
D141	152	3893	66	2048	
D141(CHAN.MASTER)	108		86	2038	
D141H01	152	3893	66	2048	
D141H9Z	152	3893	66	2048	
D142	175		246	2020	
D1420			FET-1	2028	
D1421			FET-1	2028	
D142M	175		246	2020	
D143	175		246	2020	
D144	175		246	2020	
D1445			504A	1104	
D144S	116	3016	504A	1104	
D145	175		246	2020	
D146	226		14	2041	
D146UK	130		246	2041	
D147			14	2041	
D149	128		245	2004	
D15	87		14	2041	
D150	175		246	2020	
D151	328		14	2041	
D152	375	3048/329	46		
D152(CHAN.MASTER)	195A		46		
D154	152		66	2048	
D155	175		246	2020	
D155H	175		246	2020	
D155K	175		246	2020	
D155L	175		246	2020	
D156	158	3004/102A	53	2007	
D157	124		12		
D157B	124		12		
D158	286		12		
D159	286	3018	12		
D15A	116	3017B/117	504A	1104	
D15C	116	3016	504A	1104	
D16	87		14	2041	
D160	107	3018	11	2015	
D161	328		8	2002	
D162	103A		8	2002	
D163	130		14	2041	
D164	130		14	2041	
D165	385	3438	35		
D166	385	3438	35		
D1666	108	3452	86	2038	
D167	103A		8	2002	
D16E7	123A	3444	20	2051	
D16E9	123A	3444	20	2051	
D16EC18	123A	3444	20	2051	
D16G6	107	3124/289	11	2015	
D16K1	107	3039/316	11	2015	
D16K2	107	3039/316	11	2015	
D16K3	107	3039/316	11	2015	
D16K4	107	3039/316	11	2015	
D16P1	172A		64		
D16P2	172A	3156	64		
D16U3	113A	3311	6GC1		
D16U4	113A	3311	6GC1		
D17	87		ZD-14	564	
D170	103A		59	2002	
D170A	103A		59	2002	
D170AA	103A		59	2002	
D170AB	103A		59	2002	
D170AC	103A		59	2002	
D170B	103A		59	2002	
D170BC	103A		59	2002	
D170C	103A		59	2002	
D170PB	103A		59	2002	
D171	283		53	2007	
D172	130		245	2004	

Industry Standard No.	ECG	SK	GE	RS 276-	MOTOR.
D172-F	506	3125	511	1114	
D173	160		245	2004	
D174	160	3006	245	2004	
D175	160		14	2041	
D176	139A	3060	ZD-9.1	562	
D177	328	3438	35		
D178	103A		59	2002	
D178A	103A		59	2002	
D178Q	103A		59	2002	
D178T	103A		59	2002	
D18	87	3016	504A	1104	
D180	87		53	2007	
D180A	87		255	2041	
D180B	87		255	2041	
D180C	87		255	2041	
D180D	87		255	2041	
D180M	87		14	2041	
D181	385	3438	35		
D182			243	2030	
D183			243	2030	
D184			66	2048	
D186	103A		59	2002	
D186A	103A		59	2002	
D186B	103A		59	2002	
D187	103A		59	2002	
D187A	103A		59	2002	
D187R	103A		59	2002	
D187Y	103A		59	2002	
D188	87		14	2041	
D188A	87		14	2041	
D188B	87		14	2041	
D188C	87		14	2041	
D189	280	3438	35		
D189A	280	3438	35		
D18A12	184	3054/196	57	2017	
D19	101		8	2002	
D190	124		12		
D191	103A		59	1104	
D192	103A		59	2002	
D193	103A	3004/102A	53	2002	
D194	103A		59	2002	
D195	103A	3011A	8	2002	
D195A	103A		8	2002	
D198	94		35		
D198H	94		36		
D198HQ	94		35		
D198HR	94		35		
D198Q	94		35		
D198R	94		35		
D198S	94		35		
D198V	94		35		
D199	164		35		
D1A	123A	3444	20	2051	
D1A(ZENER)	5072A		ZD-8.2	562	
D1B	113A	3119/113	6GC1		
D1C	113A	3119/113	6GC1		
D1D	125	3130	510,531	1114	
D1E	116	3017B/117	504A	1104	
D1F	506	3011	511	1114	
D1F(ZENER)	149A		ZD-62		
D1G	139A	3066/118	ZD-9.1	562	
D1H	116	3016	504A	1104	
D1J	116	3017B/117	504A	1104	
D1J70542		3088	1N60	1123	
D1J70543		3088	1N60	1123	
D1J70544		3031A	504A	1104	
D1J70545		3100	300	1122	
D1K	125	3033	510,531	1114	
D1L	116	3017B/117	504A	1104	
D1M	136A		ZD-5.6	561	
D1R35	110MP			1123	
D1R38	123A	3444	20	2051	
D1R39	110MP			1123	
D1S	143A	3062/142A	ZD-13	563	
D1T	144A	3094	ZD-14	564	
D1U	5072A		ZD-8.2	562	
D1W	146A	3064	ZD-27		
D1Y	107	3039/316	11	2015	
D1Z			ZD-15	564	
D1ZBLU	145A	3063	ZD-15	564	
D1ZRED	144A	3094	ZD-14	564	
D1ZVIO	143A	3093	ZD-13	563	
D1ZYEL	145A	3063	ZD-15	564	
D2-1	116		504A	1104	
D2-12B					MRF817
D2-12E					MRF870A
D2-77-1	177		300	1122	
D20	101		8	2002	
D200	389	3017B/117	504A	1104	
D200(IR)				1122	
D200A	389		38		
D200MP			300	1122	
D200MP(IR)				1122	
D201	87		90	2041	
D201(0)	87			2041	
D201(0)			14	2041	
D201M	87		14	2041	
D201Y	87		14	2041	
D202	87	3438	35		
D203	87	3438	35		
D204	128		243	2030	
D204L	128	3024	243	2030	
D205	128/427		256		
D21	101		8	2002	
D2101S	515	9098			
D2102S	515	9098			
D2103S	515	9098			
D2103SF	515	9098			
D211	130		14	2041	
D212	130		14	2041	
D213	280		35		
D21489			504A	1104	
D215	128		8	2002	
D217	87		35		
D218	87		35		
D219	128		243	2030	
D22	101		8	2002	
D220	116	3311	504A	1104	
D220-1M	515	9098			
D220-1N	515	9098			
D2201A	515	9098			1N4934
D2201B	515	9098			1N4935
D2201D	515	3317			1N4936
D2201F	515	9098			1N4933
D2201M	515	3318	504A	1104	1N4937
D2201N	515	9098	504A	1104	MR816
D220M	116		504A	1104	
D221	128		243	2030	
D223	427/324	3049/224	46		
D226	175		246	2020	
D226-0	175		246	2020	
D226A	175		246	2020	
D226AP	175		246	2020	
D226B	175		246	2020	

Industry Standard No.	ECG	SK	GE	RS 276-	MOTOR.
D226BP	175		246	2020	
D226P	175		246	2020	
D226Q	175		246	2020	
D227	85		20	2030	
D227(DIODE)	116		504A	1104	
D227A	85		20	2051	
D227B	85		20	2051	
D227C	85		20	2051	
D227D	85		20	2051	
D227E	85		20	2051	
D227F	85		20	2051	
D227L	85		20	2051	
D227R	85		20	2051	
D227S	85		20	2051	
D227W	85		20	2051	
D228	192A		63	2030	
D23	101		8	2002	
D232	181	3020/123	62	2010	
D233	128	3137/192A	243	2030	
D234	152	3054/196	66	2048	
D234-0	152	3054/196	66	2048	
D234-O	152	3054/196	66	2048	
D234-R	152		66	2048	
D234-Y	152		66	2048	
D235	152	3054/196	66	2048	
D235-0	152	3054/196	66	2048	
D235-O	152	3054/196	66	2048	
D235-R	152		66	2048	
D235-Y	152		66	2048	
D235D	152	3054/196	66	2048	
D235G	152		66	2048	
D235GR	152		66	2048	
D235R	152	3054/196	66	2048	
D235Y	235	3054/196	66	2048	
D236(RECT.)	166			1152	
D236(TRANSISTOR)	175		246	2020	
D238	175		246	2020	
D238F	175		246	2020	
D24	124		12		
D2406A	5818				1N3880
D2406B	5818				1N3881
D2406C	5820				1N3882
D2406D	5820				1N3883
D2406F	5818				1N3879
D2406M	5822				MR1366
D2412A	5818	3586			1N3890
D2412B	5818	3586			1N3891
D2412C	5820	3587			1N3892
D2412D	5820	3587			1N3893
D2412F	5818	3586			1N3889
D2412M	5822	3587			MR1376
D241H			14	2041	
D246	389		38		
D24A3391			212	2010	
D24A3391A			212	2010	
D24A3392			212	2010	
D24A3393			212	2010	
D24A3394	108	3452	86	2038	
D24A3900			212	2010	
D24A3900A			212	2010	
D24B	124		12		
D24C	124		12		
D24CK	124		12		
D24D	124		12		
D24E	124		12		
D24F	124		12		
D24K	124		12		
D24KC	124		12		
D24KD	124		12		
D24KE	124		12		
D24Y	124		12		
D24YB	124		12		
D24YD	124		12		
D24YE	124		12		
D24YF	124		12		
D24YK	124		12		
D24YLC	124		12		
D24YLD	124		12		
D25	116	3016	504A	1104	
D2520A	6006				1N3900
D2520B	6006				1N3901
D2520C	6008				1N3902
D2520D	6008				1N3903
D2520F	6006				1N3899
D2520M	6010				MR1386
D254	175		246	2020	
D2540A	6006	3588			1N3910
D2540B	6006	3588			1N3911
D2540C					1N3912
D2540D	6008	3589			1N3913
D2540F	6006	3588			1N3909
D2540M	6010	3589			MR1396
D255	175		246	2020	
D257	196		246	2020	
D25A	116	3016	504A	1104	
D25B	116	3016	504A	1104	
D25C	116	3016	504A	1104	
D260015	614	3327			
D260028	614	3327			
D2600EF	116	9098/515	504A	1104	
D2600F	515	9098			
D2600M	515	9098			
D2601A					MR811
D2601B					MR812
D2601D					MR814
D2601E	515	9098			
D2601EF	515	9098			
D2601F					MR810
D2601M	515	9098			MR816
D2601N		9098			MR818
D261	192A		63	2030	
D261A	192A		63	2030	
D261B	192A		63	2030	
D261C	192A		63	2030	
D261D	192A		63	2030	
D261E	192A		63	2030	
D261F	192A		63	2030	
D261L	192A		63	2030	
D261P	192A		63	2030	
D261R	192A		63	2030	
D261V	192A		63	2030	
D261W	192A		63	2030	
D26A	87		14	2041	
D26B	87		14	2041	
D26B1	107	3039/316	11	2015	
D26B2	107	3039/316	11	2015	
D26C	87		14	2041	
D26C1	107	3039/316	11	2015	
D26C2	107	3039/316	11	2015	
D26C3	107	3039/316	11	2015	
D26C4		3156	212	2010	
D26C5		3156	212	2010	
D26E-1			212	2010	
D26E-2			214	2038	
D26E-3			214	2038	
D26E-4			39	2015	
D26E-5		3245	212	2010	
D26E-6			214	2038	
D26E-7		3245	212	2010	
D26E2	107	3039/316	11	2015	
D26G-1			39	2015	
D26G1	107	3039/316	11	2015	
D27C1	152	3893	66	2048	
D27C2	152	3893	66	2048	
D27C3	152	3893	66	2048	
D27C4	152	3893	66	2048	
D27D1			29	2025	
D27D2			29	2025	
D27D3			29	2025	
D27D4			29	2025	
D28	116	3016	504A	1104	
D283	162		35		
D284	384	3438	35		
D285	162	3438	35		
D286	109	3088		1123	
D286D	198		251		
D287	284	3836	265	2047	
D287A	284	3836			
D287AQ	284	3836			
D287AR	284	3836			
D287AS	284	3836			
D287B	284	3836			
D287BQ	284	3836			
D287BR	284	3836			
D287BS	284	3836			
D287C	284	3836			
D287CQ	284	3836			
D287CR	284	3836			
D287CS	284	3836			
D288	196		66	2048	
D288A	196		66	2048	
D288B	196		66	2048	
D288C	196		66	2048	
D288L	196		66	2048	
D289	196		66	2048	
D289A	196		66	2048	
D289B	196		66	2048	
D289C	196		66	2048	
D28A1	152	3893	66	2048	
D28A10	152	3893	66	2048	
D28A12	152	3054/196	66	2048	
D28A13	152	3893	66	2048	
D28A2	152	3893	66	2048	
D28A3	152	3893	66	2048	
D28A4	152	3893	66	2048	
D28A5	152	3893	66	2048	
D28A6	152	3041	66	2048	
D28A7	152	3893	66	2048	
D28A9	152	3893	66	2048	
D28B	171		27		
D28D07			63	2030	
D28D1	152	3893	66	2048	
D28D10	152	3893	66	2048	
D28D2	152	3893	66	2048	
D28D3	152	3893	66	2048	
D28D4	152	3893	66	2048	
D28D5	152	3893	66	2048	
D28D7	152	3893	66	2048	
D28D8			28	2017	
D29	175		246	2020	
D290	175		246	2020	
D290L	175		246	2020	
D291	175		246	2020	
D292	175		246	2020	
D292(CHAN.MASTER)	107		11	2015	
D294	123A	3444	20	2051	
D297	175		246	2020	
D299	389		38		
D299SL	389		38		
D29A10		3244	220		
D29A11		3244	220		
D29A4	159	3025/129	82	2032	
D29A5	159	3114/290	82	2032	
D29A6	159	3118	82	2032	
D29A7		3118	220		
D29A8		3118	220		
D29A9	159	3114/290	82	2032	
D29E08			67	2023	
D29E08J1			67	2023	
D29E09			67	2023	
D29E09J1			67	2023	
D29E1	159	3114/290	82	2032	
D29E10	193	3114/290	67	2023	
D29E10J1	193		67	2023	
D29E1J1	193		67	2023	
D29E2	159	3114/290	82	2032	
D29E2J1	193		67	2023	
D29E4	159	3114/290	82	2032	
D29E4J1	193		67	2023	
D29E5	159	3114/290	82	2032	
D29E5J1	193		67	2023	
D29E6	159	3114/290	82	2032	
D29E6J1	193		67	2023	
D29E7	159	3114/290	82	2032	
D29E7J1	193		67	2023	
D29E8	159	3114/290	82	2032	
D29E8J1			67	2023	
D29E9	193	3114/290	67	2023	
D29E9J1	193		67	2023	
D29F1		3247	65	2050	
D29F2		3247	65	2050	
D29F3		3247	65	2050	
D29F4		3247	65	2050	
D29F5			82	2032	
D29F6			82	2032	
D29F7			82	2032	
D2E	147A	3095	ZD-33		
D2G	141A	3092	ZD-11.5		
D2G-1	141A	3092	ZD-11.5		
D2G-2	141A	3092	ZD-11.5		
D2G-3	142A	3062	ZD-12	563	
D2G-4	142A	3062	ZD-12	563	
D2H	183	3016	56	2027	
D2H(DIODE)	116		504A	1104	
D2J	116	3031A	504A	1104	
D2M	502	3067	CR-4		
D2N	503	3068	CR-5		
D2R31	109	3088		1123	
D2R38	123A	3444	20	2051	
D2T2218					MD2218
D2T2218A					MD2218A
D2T2219					MD2219
D2T2219A					MD2219A
D2T2904					MD2904
D2T2904A					MD2904A
D2T2905					MD2905
D2T2905A					MD2905A
D2T918					MD918

Industry Standard No.	ECG	SK	GE	RS 276-	MOTOR.
D2U			1N34AS	1123	
D2X4	116		504A	1104	
D2Y	505	3108	CR-7		
D2Z	139A	3066/118	ZD-9.1	562	
D2Z-2	146A	3064	ZD-27		
D30	6408		59	1104	1N5762
D30-O	103A		59	2002	
D30-N	103A		59	2002	
D300	389	3100/519	504A	1104	1N4004
D3005VN	130		14	2041	
D300B	389		38		
D301	99	3311			
D308	199	3039/316	62	2010	
D30A1	159	3114/290	82	2032	
D30A2	159	3118	82	2032	
D30A3	159	3118	82	2032	
D30A4	159	3114/290	82	2032	
D30A5	159	3114/290	82	2032	
D31	103A		59	2002	
D312	164		37		
D313C	152		241	2020	
D313D	152		241	2020	
D313F			241	2020	
D315	152		246	2020	
D316	280	3836/284			
D317	152		66	2048	
D317A	396		66	2048	
D317F	152		66	2048	
D317P	152		66	2048	
D318	152		66	2048	
D318A	396		66	2048	
D319	284		75	2041	
D31D	103A		59	2002	
D32			59	2002	
D320	162		35		
D3202U		3523			1N5761A
D3202Y	6408	3523/6407			1N5761
D320F	173BP	3999			
D321	283	3467	36		
D322	280		262	2041	
D322A	280		262	2041	
D322B	280		262	2041	
D322C	280		262	2041	
D323	280		262	2041	
D3232					MC3232AP
D323A	280		262	2041	
D323B	280		262	2041	
D323C	280		262	2041	
D324	124		12		
D3242					MC3242AP
D325	152		66	2048	
D325C	152		66	2048	
D325D	152		66	2048	
D325E	152		66	2048	
D325F			66	2048	
D327	85	3444/123A	20	2030	
D327A	85	3444/123A	20	2051	
D327B	85	3444/123A	20	2051	
D327C	85	3444/123A	20	2051	
D327D	85	3444/123A	20	2051	
D327E	85	3444/123A	20	2051	
D327F	.85	3444/123A	20	2051	
D328	123A	3444	20	2051	
D328S	128	3202/210	252	2018	
D32K1			17	2051	
D32K2			17	2051	
D32P1			62	2010	
D32P2			62	2010	
D32P3			62	2010	
D32P4			62	2010	
D33	103A		59	2002	
D330D	152		66	2048	
D334	280	3438	35		
D334A	280	3438	35		
D334R	280	3438	35		
D3356	177		300	1122	
D336	297		63	2030	
D336R	297		63	2030	
D336Y	297		63	2030	
D33C	103A		59	2002	
D33D21	297	3444/123A	20	2051	
D33D21J1	192A		63	2030	
D33D22	297	3444/123A	20	2051	
D33D22J1	192A		63	2030	
D33D22J2			47	2030	
D33D24	297	3122	20	2051	
D33D24J1	192		63	2030	
D33D25	297	3444/123A	20	2051	
D33D25J1	192A		63	2030	
D33D26	297	3444/123A	20	2051	
D33D26J1	192A		63	2030	
D33D27	297	3444/123A	20	2051	
D33D27J1	192A		63	2030	
D33D28	123A	3444	20	2051	
D33D28J1			63	2030	
D33D29	297	3122	63	2030	
D33D29J1	192A		63	2030	
D33D30	297	3122	63	2030	
D33D30J1	192A		63	2030	
D34	103A		59	2002	
D34003220-001	177		300	1122	
D34013890-002	74H04	74H04			
D34014094	175		246	2020	
D341	130		74	2043	
D341H	130		74	2043	
D342	123A	3444	20	2051	
D343	152		66	2048	
D3436			504A	1104	
D35	103A		59	2002	
D3500	238		259		
D350Q	165		259		
D351	283	3467	36		
D352	158		53	2007	
D352D	103A		59	2002	
D352E	103A		59	2002	
D352F	103A		59	2002	
D353	162	3100/519	300	1122	
D3530	112		1N82A		
D355		3849	300	1122	
D355C			271	2030	
D355E		3849	271	2030	
D356C	291	3188A/182			
D356D	291	3188A/182			
D357	291	3440			
D357C	291	3440			
D357D	291	3440			
D358C	291	3440			
D358D	291	3440			
D359	152		66	2048	
D359C	152		66	2048	
D359C2	152		66	2048	
D359D	152		66	2048	
D359D1	152		66	2048	
D359D2	152		66	2048	

Industry Standard No.	ECG	SK	GE	RS 276-	MOTOR.
D359E			66	2048	
D36	103A		8	2002	
D360	152		215	2048	
D3601		3197	215		
D360C	152		66	2048	
D360D	152		215		
D360E			66	2048	
D365H	152		66	2048	
D366-0	152		66	2048	
D36667	1115	3184/115			
D366P	152		66	2048	
D366Q	152		66	2048	
D367	103A		59	2002	
D367A	103A		59	2002	
D367B	103A		59	2002	
D367C	103A		59	2002	
D367D	103A		59	2002	
D367E	103A		59	2002	
D367F	103A		59	2002	
D367P	103A		59	2002	
D368	238		IC-116		
D37	103A		59	2002	
D372BL	123A	3444	20	2051	
D379	280	3836/284	265	2047	
D379P	280	3836/284	265	2047	
D379Q	280	3836/284	265	2047	
D379R	280	3836/284	265	2047	
D379S	280	3836/284	265	2047	
D37A	103A		59	2002	
D37B	103A		59	2002	
D37C	103A		59	2002	
D38	103A		59	2002	
D380	165		259		
D382	152		66	2048	
D382LM	152		66	2048	
D383	283	3467	36		
D386A	375		251		
D386D	375		251		
D386Y	375		251		
D387AP	375		251		
D389	152		66	2048	
D389-0	152		66	2048	
D389-0P	152		66	2048	
D389A	196		66	2048	
D389AFP	152		66	2048	
D389B	152		66	2048	
D389BL	152		66	2048	
D389LB	152		66	2048	
D3A	507	3031A	300	1122	
D3F	139A	3066/118	ZD-9.1	562	
D3G	178MP	3121/115	300(2)	1122(2)	
D3H	137A	3058	ZD-6.2	561	
D3N	141A	3092	ZD-11.5		
D3R	116	3017B/117	504A	1104	
D3R38	116	3017B/117	504A	1104	
D3R39	116	3017B/117	504A	1104	
D3U	116	3016	504A	1104	
D3V	116	3017B/117	504A	1104	
D3W	142A	3093	ZD-12	563	
D3Y	137A		ZD-6.2	561	
D3Z	116	3311	504A	1104	
D4	113A	3119/113	6GC1		
D400	293		504A	1104	
D400(RECT.)	116		504A	1104	
D400(TRANSISTOR)	293		47	2030	
D4000411	909		IC-249		
D400D	382		47	2030	
D400E	382		47	2030	
D400F	382		47	2030	
D400P1		3849	47	2030	
D400P1D		3849	47	2030	
D400P1E		3849	47	2030	
D400P1F		3849	47	2030	
D401EK	375	3929			
D401K	375	3929			
D401L	375	3929			
D401LA	291	3929			
D401M	375	3929			
D40C1	265	3860	D40C1		D40C1
D40C2	266				D40C2
D40C3	267		23		
D40C4	265	3860	D40C4		D40C4
D40C5	266		D40C5		D40C5
D40C7	265	3860			
D40D1	210	3054/196	252	2018	D40D1
D40D10	210	3202	252	2018	D40D10
D40D11	210	3202	252	2018	D40D11
D40D13					D40D13
D40D14			252	2018	D40D14
D40D2	210	3192/186	252	2018	D40D2
D40D3	210	3192/186	252	2018	D40D3
D40D4	210	3192/186	252	2018	D40D4
D40D5	210	3192/186	252	2018	D40D5
D40D7	210	3192/186	252	2018	D40D7
D40D8	210	3192/186	252	2018	D40D8
D40E1					D40E1
D40E5					D40E5
D40E7			D40E7		D40E7
D40K1	268				D40K1
D40K2	268		D40K2		D40K2
D40K3					D40K3
D40K4					D40K4
D40N1	171	3201	27		D40N1
D40N2			27		D40N2
D40N3	171	3201	27		D40N3
D40N4			27		D40N4
D40N5	171	3201	27		
D40P1					D40P1
D40P3	171				D40P3
D40P5					D40P5
D41	130		14	2041	
D418	165		38		
D41D1	211	3203	253	2026	D41D1
D41D10	211	3203	253	2026	D41D10
D41D11	211	3203	253	2026	D41D11
D41D13					D41D13
D41D14			253	2026	D41D14
D41D2	211	3083/197	253	2026	D41D2
D41D4	211	3203	253	2026	D41D4
D41D5	211	3203	253	2026	D41D5
D41D7	211	3203	253	2026	D41D7
D41D8	211	3203	253	2026	D41D8
D41E1					D41E1
D41E5					D41E5
D41E7					D41E7
D41K1	269				D41K1
D41K2	269				D41K2
D41K3					D41K3
D41K4					D41K4
D424	284		265	2047	
D424-0	375		265	2047	
D424-R	284		265	2047	
D425	280		262	2041	
D4250			262	2041	

Industry Standard No.	ECG	SK	GE	RS 276-	MOTOR.
D4250	280			2041	
D42C1	186	3192	28	2017	MDS26
D42C2	186	3192	28	2017	MDS26
D42C3	186	3192	28	2017	MDS26
D42C4	186	3192	28	2017	MDS27
D42C5	186	3192	28	2017	MDS27
D42C6	186	3192	28	2017	MDS27
D42C7	186	3192	28	2017	MDS27
D42C8	186	3192	28	2017	MDS27
D42C9					MDS27
D43	103A		8	2002	
D431-F	506	3125	511	1114	
D43A	103A		8	2002	
D43C1	187	3193	29	2025	MDS76
D43C2	187	3193	29	2025	MDS76
D43C3	187	3193	29	2025	MDS76
D43C4	187	3193	29	2025	MDS77
D43C5	187	3193	29	2025	MDS77
D43C6	187	3193	29	2025	MDS77
D43C7	153		69	2049	MDS77
D43C8	187	3193	29	2025	MDS77
D43C9					MDS77
D44	101		8	2002	
D44C1	377	3273	247	2052	MJE180
D44C10	377				MJE182
D44C11	377				MJE182
D44C12	377				MJE182
D44C2	377	3893/152	66	2048	MJE180
D44C3	377	3893/152	66	2048	MJE180
D44C4	377	3893/152	66	2048	MJE181
D44C5	377	3893/152	66	2048	MJE181
D44C6	377	3893/152	66	2048	MJE181
D44C7	377	3893/152	66	2048	MJE181
D44C8	377	3893/152	66	2048	MJE181
D44C88	186	3192	28	2017	
D44C8B	377	3893/152	66	2048	
D44C9	377	3893/152	66	2048	MJE181
D44CBB	377	3054/196	66	2048	
D44D1					2N6386
D44D2					2N6386
D44D3					2N6043
D44D4					2N6043
D44D5					2N6044
D44D6					2N6044
D44E1	263	3180			2N6386
D44E2	263	3180			2N6387
D44E3	263	3180			2N6388
D44H1			55		D44H10
D44H10	377	3188A/182	55		D44H10
D44H11	377	3188A/182	D44H11		D44H11
D44H2			55		D44H11
D44H4		3188A	55		D44H10
D44H5		3188A	55		D44H11
D44H7	377	3188A/182	55		D44H10
D44H8		3188A	55		D44H11
D44R1	198	3219	251		TIP47
D44R2	198	3219	251		TIP47
D44R3	198		251		TIP48
D44R4	198	3220	251		TIP48
D44R5	198		32		TIP47
D44R6	198		32		TIP48
D45	162	3438	35		
D458	283	3439/163A	36		
D45C	116	3016	504A	1104	
D45C1	378	3083/197	69	2049	MJE170
D45C10	378				MJE172
D45C11	378				MJE172
D45C12	378				MJE172
D45C2	378	3083/197	69	2049	MJE170
D45C3	378	3083/197	69	2049	MJE170
D45C4	378	3274/153	69	2049	MJE171
D45C5	378	3274/153	69	2049	MJE171
D45C6	378	3274/153	69	2049	MJE171
D45C7	378	3083/197	69	2049	MJE171
D45C8	378	3083/197	D45C8	2049	MJE171
D45C9	378	3083/197	69	2049	MJE171
D45CZ	116	3017B/117	504A	1104	
D45E1	264	3181A			TIP125
D45E2	264	3181A			TIP125
D45E3	264	3181A			TIP126
D45H1			56	2027	D45H10
D45H10		3189A	56	2027	D45H10
D45H11	378				D45H11
D45H12					D45H11
D45H2		3189A	56	2027	D45H11
D45H4		3274	56	2027	D45H10
D45H5		3274	56	2027	D45H11
D45H7	378	3189A/183	56	2027	D45H10
D45H8	332	3189A/183	56	2027	D45H11
D45H9					D45H11
D46	162	3438	35		
D468B	287		222		
D468C	287		222		
D47	328	3438	35		
D471	293		268	2038	
D471L	293		268	2038	
D48	116	3017B/117	504A	1104	
D49	291		246	2020	
D4C28	128	3117	39	2030	
D4C29	128	3117	39	2030	
D4C30	128	3117	39	2030	
D4C31	128	3117	39	2030	
D4D20	128	3117	39	2030	
D4D21	128	3117	39	2030	
D4D22	161	3117	39	2015	
D4D24	123	3122	17	2051	
D4D25	123	3444	20	2051	
D4D26	123	3444	20	2051	
D4G	502	3067	CR-4		
D4H	147A	3095	ZD-33		
D4J	147A	3095	ZD-33		
D4L	125	3033	510,531	1114	
D4M	116	3016	504A	1104	
D4N	141A	3092	ZD-11.5		
D4P	144A	3094	ZD-14	564	
D4R	109	3087		1123	
D4R26	116	3311	504A	1104	
D4R39	116	3017B/117	504A	1104	
D5	114	3120	6GD1	1104	
D50	87	3017B/117	504A	1104	1N4001
D500	116	3017B/117	504A	1104	1N4005
D51	87	3119/113	6GC1		
D513	152		66	2048	
D53	87		14	2041	
D5403	HIDIV-1	3868/DIV-1	FR-8		
D5404	HIDIV-2	3869/DIV-2	FR-9		
D55	181		75	2041	
D555CJ					MC1555G
D55A	181		75	2041	
D56	277		246	2020	
D562	108	3452	86	2038	
D56W1	165		38		BU208
D56W2	165		38		BU208
D56W3					BU207

Industry Standard No.	ECG	SK	GE	RS 276-	MOTOR.
D56W4					BU207
D57	175		246	2020	
D58	175		246	2020	
D582	284	3182/243			
D582A	284	3182/243			
D59	162	3438	35		
D5B	145A	3063	ZD-15	564	
D5E-37	6401			2029	
D5E-43	6409			2029	
D5E-44	6401			2029	
D5E-45	6409			2029	
D5E37	6401			2029	
D5E43	6409			2029	
D5E44	6401			2029	
D5E45	6409			2029	
D5G	552	3175	504A	1104	
D5H	552	3175	511	1114	
D5K1			D5K1		2N6114
D5K2			D5K2		2N6115
D5R35	116		504A	1104	
D5R39	116	3017B/117	504A	1104	
D5V	177	3100/519	300	1122	
D5W	137A	3058	ZD-6.2	561	
D6	115	3121	6GX1		
D6.2	137A	3058	ZD-6.2	561	
D60	162	3438	35		
D600	373	3017B/117	504A	1104	
D61	103A		59	2002	
D62	103A		59	2002	
D625	228A	3103A/396			
D63	103A		59	2002	
D64	103A		59	2002	
D6462	177	3031A	300	1122	
D65	126		52	2024	
D65-1	103A		59	2002	
D65C	116	3017B/117	504A	1104	
D66	126		52	2024	
D6623	116	3017B/117	504A	1104	
D6623A	116	3016	504A	1104	
D6624	116	3017B/117	504A	1104	
D6624A	116	3016	504A	1104	
D6625	116	3017B/117	504A	1104	
D6625A	116	3016	504A	1104	
D665	388	3836/284			
D67	280	3438	35		
D6726	177	3087	300	1122	
D67B	280	3438	35		
D67C	280	3438	35		
D67D	280	3438	35		
D67E	280	3438	35		
D68	130	3017B/117	504A	1104	
D68B			255	2041	
D68C			255	2041	
D68D			255	2041	
D68E			255	2041	
D69	130		14	2041	
D6C	513	3443	513		
D6HZ	116	3017B/117	504A	1104	
D6M	5072A	3136	ZD-8.2	562	
D6U	5080A	3336	ZD-22		
D7	115	3121	6GX1		
D70	175		246	2020	
D71	175		246	2020	
D71L	175		268	2038	
D72	103A		59	1122	
D72-2C	103A		59	2002	
D72-3C	103A		59	2002	
D72-4C	103A		59	2002	
D720D			59	2002	
D720E			59	2002	
D72A	103A		59	2002	
D72B	103A		59	2002	
D72C	103A		59	2002	
D72RE	103A		59	2002	
D73	87		262	2041	
D73A	87		262	2041	
D73B	87		262	2041	
D73C	87		262	2041	
D73D	280		262	2041	
D73E	87		262	2041	
D74	87		262	2041	
D74A	87		262	2041	
D74B	87		262	2041	
D74C	87		262	2041	
D74D	87		262	2041	
D74E	87		262	2041	
D75	103A		59	2002	
D75A	103A		59	2002	
D75AH	103A		59	2002	
D75B	103A		59	2002	
D75C	103A		59	2002	
D75H	103A		59	2002	
D77			300	1122	
D77(DIODE)	177		300	1122	
D77(TRANSISTOR)	103A		59	2002	
D77A	103A		59	2002	
D77AH	103A		59	2002	
D77B	103A		59	2002	
D77C	103A		59	2002	
D77D	103A		59	2002	
D77H	103A		59	2002	
D77P	103A		59	2002	
D78	282	3748	264		
D78A	282	3748	264		
D78B	282	3748	264		
D78C	282	3748	264		
D78D	282	3748	264		
D79	282/427		256		
D79A	282/427		256		
D79B	282/427		256		
D79C	282/427		256		
D79D	282/427		256		
D7A30	128	3024	243	2030	
D7A31	128	3024	243	2030	
D7A32	128	3024	243	2030	
D7E	177		300	1122	
D7L	5076A	9022	ZD-17		
D7M		3126	90		
D7N	614	3327	90		
D7Z	177	3100/519	300	1122	
D80	87		14	2041	
D800	125	3032A	510,531	1114	1N4006
D81	87	3027/130	14	2041	
D818	389	3999/173BP	305		
D81K			504A	1104	
D82	87		14	2041	
D8216					MC8T26AL
D8226					MC8T28L
D82A			14	2041	
D83	87	3438	35		
D84	87	3438	35		
D8410	109	3087		1123	
D85C	125	3032A	510,531	1114	
D88	87	3032A	510,531	1114	

Industry Standard No.	ECG	SK	GE	RS 276-	MOTOR.
D88A	87	3438	35		
D8G	506	3130	511	1114	
D8HZ	125	3032A	510,531	1114	
D8L	506	3130	511	1114	
D8M	506	3130	511	1114	
D8P	143A	3750			
D8U	5073A	3749	ZD-9.1	562	
D90	152		66	2048	
D91	152		66	2048	
D911138-1	128		243	2030	
D911138-2	128		243	2030	
D911138-3	128		243	2030	
D911138-4	128		243	2030	
D911138-5	128		243	2030	
D911138-6	128		243	2030	
D911138-7	128		243	2030	
D912	123A	3444	20	2051	
D917254-2	123A	3444	20	2051	
D919039-2	519		514	1122	
D91F	152		66	2048	
D92	291		246	2020	
D921881-1	123A	3444	20	2051	
D926640-1	123		20	2051	
D928121	123A	3444	20	2051	
D92D	291	3562	246	2020	
D94	375		12		
D96	103A		59	2002	
D9634	199	3124/289	86	2038	
D975	112		1N82A		
D9M	5073A	3749			
DA000	116	3016	504A	1104	
DA001	116	3017B/117	504A	1104	
DA002	116	3017B/117	504A	1104	
DA058	125	3032A	510,531	1114	
DA101				1122	
DA101A				1122	
DA101B				1122	
DA102				1122	
DA102A				1122	
DA102B				1122	
DA205			300	1122	
DA2068	125	3032A	510,531	1114	
DA90	109	3090		1123	
DAAY001002	109	3088	1N60	1123	
DAAY002001	116	3017B/117	504A	1104	
DAAY003002	177	3100/519	300	1122	
DAAY004001	177	3100/519	300	1122	
DAAY010092	139A	3060	ZD-9.1	562	
DAC-01					MC1506L
DAC-08AQ					DAC-08AQ
DAC-08BC					DAC-08EP
DAC-08BM					DAC-08Q
DAC-08CP					DAC-08CP
DAC-08CQ					DAC-08CQ
DAC-08EP					DAC-08EP
DAC-08EQ					DAC-08EQ
DAC-08HP					DAC-08HP
DAC-08HQ					DAC-08HQ
DAC-08Q					DAC-08Q
DAC-IC10BC					MC3410L
DANZ00600				1123	
DANZ006000	109	3088		1123	
DANZ0060000	109			1123	
DASZ158800	519	3100	300	1122	
DAT1A	126	3006/160	52	2024	
DAT2	126	3006/160	52	2024	
DB202				1102	
DB204				1103	
DB206				1104	
DB210				1114	
DBAZ073304	294	3114/290	32		
DBBY001003	282	3748	264		
DBBY003001	186A	3248	247	2052	
DBBY003002	186A	3357	247	2052	
DBBY005001	235	3197	215		
DBCZ037300	199	3122	62	2010	
DBCZ0373000	119	3122	CR-2		
DBCZ039404	229	3018	61	2038	
DBCZ073304	199	3124/289	20	2051	
DBCZ073503	289A	3138/193A	268	2038	
DBCZ073504	289A	3122	210	2051	
DBCZ083905	123A	3018	61	2038	
DBCZ083906	123A	3444	20	2051	
DBCZ094504	199	3122	62	2010	
DBCZ101800	299	3298	236		
DBCZ136406	123A	3444	20	2051	
DBCZ176003	306	3251	276		
DBCZ373000	199		62	2010	
DC-10	102A	3123	53	2007	
DC-12	158		53	2007	
DC-13	109	3087		1123	
DC-9	102A		53	2007	
DC6(QUASAR)	513		513		
DC8457	177	3100/519	300	1122	
DCT310	614	3327			
DD-000	116	3017B/117	504A	1104	
DD-003	116	3016	504A	1104	
DD-006	116	3016	504A	1104	
DD-007	116	3017B/117	504A	1104	
DD-79D107-1	130		14	2041	
DD003			504A	1104	
DD006			504A	1104	
DD007			504A	1104	
DD04	113A	3119/113	6GC1		
DD05	114	3120	6GD1		
DD056	116		504A	1104	
DD058	125	3032A	510,531	1114	
DD06	115	3121	6GX1		
DD07		3121	6GX1		
DD175C	116	3017B/117	504A	1104	
DD176C	116	3016	504A	1104	
DD177C	116	3031A	504A	1104	
DD2066	116	3031A	504A	1104	
DD2068	125	3032A	510,531	1114	
DD2320	116	3311	504A	1104	
DD2321	116	3016	504A	1104	
DD236	116	3031A	504A	1104	
DD266	116	3017B/117	504A	1104	
DD268	125	3032A	510,531	1114	
DDAY-048001			514	1122	
DDAY001001	109	3088	1N60	1123	
DDAY001002	109	3088	1N60	1123	
DDAY001003	109	3088			
DDAY001004	109	3088	1N60	1123	
DDAY001010	109	3088	1N60	1123	
DDAY001022	109	3088	1N60	1123	
DDAY002001	116	3311	504A	1104	
DDAY002002	116	3017B/117	504A	1104	
DDAY004001	177	3100/519	300	1122	
DDAY006002		3126	90		
DDAY006003	613	3126			
DDAY008001	137A	3058	ZD-6.2	561	
DDAY008003	5072A		ZD-8.2	562	
DDAY009001	136A	3056/135A	ZD-5.1		

Industry Standard No.	ECG	SK	GE	RS 276-	MOTOR.
DDAY009003	5075A	3751	ZD-16	564	
DDAY009007	5075A		ZD-16	564	
DDAY010002	139A	3060	ZD-9.1	562	
DDAY010005	139A		ZD-9.1	562	
DDAY010007	5015A		ZD7.5		
DDAY030002	612	3327/614			
DDAY042001	116(4)	3311	504A	1104	
DDAY047001	519	3100	514	1122	
DDAY047005	519	3100	300	1122	
DDAY048001	177	3100/519	300	1122	
DDAY048007	519	3100	514	1122	
DDAY048008	177	3100/519	300	1122	
DDAY048012	519	3100	514	1122	
DDAY048013	519		300	1122	
DDAY048014	519	3100	514	1122	
DDAY049001	116	3311			
DDAY063001	519	3175			
DDAY067001	612	3325			
DDAY069001	519	3100	514	1122	
DDAY090001	553	3175	300	1122	
DDAY101001			1N60	1123	
DDAY103001	116	3311	504A	1104	
DDAY108001		3059	ZD-7.5		
DDAY126001	139A	3060	ZD-9.1	562	
DDBY002001	116	3311	504A	1104	
DDBY003001	290A	3114/290	48	2032	
DDBY003002	294	3138/193A	48		
DDBY003003	294	3138/193A	65	2050	
DDBY004001	290A	3114/290	269	2032	
DDBY008001	129	3114/290	244	2027	
DDBY104002	294	3434/288	48		
DDBY209003	123A	3018	211	2016	
DDBY216002	229	3122	61	2038	
DDBY219001	229	3122	61	2038	
DDBY222002	123A	3444	20	2051	
DDBY224001	199	3122	62	2010	
DDBY224002	199	3124/289			
DDBY224003	199	3124/289	212	2009	
DDBY224004	199	3124/289	212	2010	
DDBY224006	199	3124/289	212	2010	
DDBY227001	186A	3248	247	2052	
DDBY227002	233	3248			
DDBY227004	186A	3357	28	2052	
DDBY228001	152	3197/235	215	2053	
DDBY230001	235	3197	215		
DDBY231002	236	3197/235	216	2053	
DDBY233001	123A	3124/289	210	2051	
DDBY246001	295		57		
DDBY256001	295	3253	270		
DDBY256002	295	3253			
DDBY257001	235	3197	333		
DDBY259001	229	3122	213	2038	
DDBY259002	229	3124/289	213	2038	
DDBY261002	229	3039/316	60	2038	
DDBY262001	199	3122	61	2010	
DDBY269001	107	3293	17		
DDBY270001	123AP	3122	210		
DDBY27001				2051	
DDBY272001	293		285	2030	
DDBY273001	289A	3124/289	210	2009	
DDBY276001	293	3849			
DDBY277002	233		61	2015	
DDBY278001	152		60	2048	
DDBY278002	152	3197/235	66	2048	
DDBY283001	289A	3124/289	210	2051	
DDBY287001	229	3124/289	61	2038	
DDBY288001	295		270		
DDBY289001	235	3197	337		
DDBY295002	107	3132	61		
DDBY299001		3245	62	2010	
DDBY301001		3122	210	2051	
DDBY307001	236	3239			
DDBY307003			216	2053	
DDBY403001	130	3027			
DDBY407001	196	3054	66	2048	
DDBY407004	196	3054	241	2020	
DDBY410001	293		243	2030	
DDBY410002	128	3020/123	243		
DDBY41002				2030	
DDBY4233001			62	2010	
DDCY001001	312	3116			
DDCY001002	453	3834/132	FET-2	2035	
DDCY002002	312	3448	FET-1	2035	
DDCY006001	312	3116	FET-2	2035	
DDCY007002	290A	3114/290	269	2032	
DDCY103001	220	3990	FET-4	2036	
DDCY104001	222	3050/221	FET-4	2036	
DDCY104003	222	3050/221	FET-4	2036	
DDE-201	116	3311	504A	1104	
DDEY001001	973	3233			
DDEY002001	724	3525			
DDEY004001	1100	3223	IC-92		
DDEY019001	974	3965			
DDEY020001	973	3233			
DDEY026001	975	3641			
DDEY029001	7490	7490		1808	
DDEY030001	7400	7400		1801	
DDEY046001	977	3462			
DDEY061001	1194	3484			
DDEY064001	1082	3461	IC-140		
DDEY088001	977	3462			
DDEY089001	4011B		4011	2411	
DDEY091001	1194	3484			
DDEY093001	977	3462			
DDEY097001	1195	3469			
DDEY123001	1170	3745	IC-65		
DDEY146001	1278	3726			
DDFY004002	177		300	1122	
DDFY017001	6402	3628			
DDMV-1	109	3087		1123	
DDMV-2	109	3087		1123	
DE-201		3032A	504A	1104	
DE-3181	102	3722	2	2007	
DE1004					MPE3003
DE104				1122	
DE110				1122	
DE111				1122	
DE112				1122	
DE113				1122	
DE114				1122	
DE115				1122	
DE14	116	3031A	504A	1104	
DE14A	116	3017B/117	504A	1104	
DE16	116	3017B/117	504A	1104	
DE16A	116	3017B/117	504A	1104	
DE201	116		504A	1104	
DED4191	159		82	2032	
DEEY004001	1100	3223			
DER1	125	3033	510,531	1114	
DF-2	130	3029	14	2041	
DF1	101		8	2002	
DFFY007001	177	3100/519	300	1122	
DG-1N	506	3125	511	1114	
DG-1NR	506	3125	511	1114	

Industry Standard No.	ECG	SK	GE	RS 276-	MOTOR.
DG13	127	3113/516			
DG1K				1114	
DG1M				1104	
DG1MR				1104	
DG1N	506	3125	511	1104	
DG1N60	109	3088	1N60	1123	
DG1NR	506	3043	511	1114	
DG1PR	116	3031A	504A	1104	
DG1S34	109	3087	1N34AS	1123	
DGIN			511	1114	
DGM-2	109	3087		1123	
DGM-3	109	3087		1123	
DGM3			1N34AS	1123	
DH-001	116	3031A	504A	1104	
DH-002		3311	504A	1104	
DH14	116	3017B/117	504A	1104	
DH14A	116	3017B/117	504A	1104	
DH16	116	3017B/117	504A	1104	
DH16A	116	3017B/117	504A	1104	
DH4R2	116	3016	504A	1104	
DHD800	177	3091	300	1122	
DHD8001	177	3100/519	300	1122	
DHD805	177	3311	300	1122	
DHD805(ZENER)	145A		ZD-15	564	
DHD806	177	3311	300	1122	
DHE					MRF838A
DI-1	109	3100/519		1123	
DI-1649	116	3017B/117	504A	1104	
DI-172S	116		504A	1104	
DI-2	109	3087		1123	
DI-20		3100	300	1122	
DI-3	177	3100/519	300	1122	
DI-410					1N4007
DI-42					1N4003
DI-42S	116	3016	504A	1104	
DI-44					1N4004
DI-46	116	3017B/117	504A	1104	1N4005
DI-48					1N4006
DI-5	5002A	3768			
DI-510					1N4007
DI-52					1N4003
DI-52S	116	3016	504A	1104	
DI-54					1N4004
DI-55	116	3017B/117	504A	1104	
DI-56	116		504A	1104	1N4005
DI-58					1N4006
DI-645	116		504A	1104	
DI-646	116		504A	1104	
DI-647	116	3017B/117	504A	1104	
DI-648	116	3017B/117	504A	1104	
DI-649	116	3017B/117	504A	1104	
DI-650	125	3033	510,531	1114	
DI-7	116	3016	504A	1104	
DI-705	116	3017B/117	504A	1104	
DI-71	116	3017B/117	504A	1104	
DI-710					1N4007
DI-72					1N4003
DI-72S	116	3017B/117	504A	1104	
DI-74					1N4004
DI-746					1N746
DI-759					1N759
DI-76					1N4005
DI-78					1N4006
DI-8		3060	ZD-9.1	562	
DI-8(COURIER)			ZD-9.1	562	
DI-8(DIODE)	109	3090		1123	
DI-957					1N957A
DI-976					1N976A
DI1649			504A	1104	
DI42S			504A	1104	
DI46			504A	1104	
DI52S			504A	1104	
DI56		3017B	504A	1104	
DI645		3016	504A	1104	
DI646		3016	504A	1104	
DI650		3032A	509	1114	
DI7			504A	1104	
DI72S			504A	1104	
DIA	5072A		ZD-8.2	562	
DIB	113A		6GC1		
DIC	113A	3100/519	6GC1		
DICR1	125	3033	510,531	1114	
DID	125	3033	510,531	1114	
DIE	116	3017B/117	504A	11044	
DIJ	116	3017B/117	504A	1104	
DIJ61224	109	3088		1123	
DIJ70485		3126	90		
DIJ70486	177	3100/519	300	1122	
DIJ70488	116	3311	504A	1104	
DIJ70542	109	3087		1123	
DIJ70543	110MP	3709		1123(2)	
DIJ70544	116		504A	1104	
DIJ70545	177		300	1122	
DIJ70644	110MP	3088	1N34AS	1123(2)	
DIJ70645	109	3088		1123	
DIJ70646	109	3088		1123	
DIJ70695			504A	1104	
DIJ71273	177		300	1122	
DIJ71711	177	3100/519	300	1122	
DIJ71776	109	3087		1123	
DIJ71777		3126	90		
DIJ71778	110MP	3087	1N34AS	1123(2)	
DIJ71895-1	178MP	3100/519	300(2)	1122(2)	
DIJ71958	116	3311	504A	1104	
DIJ71959	116	3031A	504A	1104	
DIJ71960	177	3100/519	300	1122	
DIJ71961	502	3067	CR-4		
DIJ72163	506	3100/519	511	1114	
DIJ72164	177	3100/519	300	1122	
DIJ72165	139A	3060	ZD-9.1	562	
DIJ72166	109	3088		1123	
DIJ72167	506	3130	511	1114	
DIJ72168	116	3016	504A	1104	
DIJ72169	5096A		ZD-100		
DIJ72170	506	3130	511	1114	
DIJ72171	506	3130	511	1114	
DIJ72172	5073A	3749	ZD-9.1	562	
DIJ72174	506	3130	511	1114	
DIJ72290	506	3130	511	1114	
DIJ72291	506	3109/119	511	1114	
DIJ72292	156	3051	512		
DIJ72293	137A	3058	ZD-6.2	561	
DIJ72294	109	3088		1123	
DIJ72349	109	3087		1123	
DIK	125	3033	510,531	1114	
DIL	116	3017B/117	504A	1104	
DIM	136A		ZD-5.6	561	
DIS	143A	3750	ZD-13	563	
DIS-1S	116		504A	1104	
DIS1S		3311	504A	1104	
DIT	144A		ZD-14	564	
DK19	109	3091		1123	
DK20	109	3087		1123	
DK21	109	3087		1123	

Industry Standard No.	ECG	SK	GE	RS 276-	MOTOR.
DM-104		3462	VR-100		
DM-106	981	3724			
DM-11	709		IC-11		
DM-11A	709		IC-16		
DM-14	718	3159	IC-8		
DM-19	805	3163/736	IC-238		
DM-20	744		IC-215	2022	
DM-24	719		IC-28		
DM-26	721		IC-14		
DM-30	721		IC-14		
DM-31	709	3135	IC-11		
DM-32	744	3171	IC-24	2022	
DM-35	806	3734	IC-239		
DM-41	737	3375	IC-16		
DM-44	743	3172	IC-214		
DM-50	983	3887			
DM-51	788	3147	IC-229		
DM-54	719		IC-28		
DM-87	834	3569			
DM-9	736		IC-17		
DM-92		3630	VR-108		
DM10P					MRF1015MB
DM11	709	3135	IC-11		
DM11A	709	3375/737	IC-16		
DM14	718		IC-8		
DM1800N					MC1800P
DM1801N					MC1801P
DM19		3163	IC-211		
DM20	744		IC-215	2022	
DM24	719		IC-28		
DM26	721		IC-14		
DM30	721		IC-14		
DM30-12BA					MRF844
DM30P					MRF1035MB
DM31	709	3135	IC-11		
DM32	744	3171	IC-215	2022	
DM35	806	3734	IC-239		
DM41	737	3375	IC-16		
DM44	743		IC-214		
DM50P					MRF1090MB
DM54	719		IC-28		
DM54LS00J					SN54LS00J
DM54LS00W					SN54LS00W
DM54LS01J					SN54LS01J
DM54LS01W					SN54LS01W
DM54LS02J					SN54LS02J
DM54LS02W					SN54LS02W
DM54LS03J					SN54LS03J
DM54LS03W					SN54LS03W
DM54LS04J					SN54LS04J
DM54LS04W					SN54LS04W
DM54LS05J					SN54LS05J
DM54LS05W					SN54LS05W
DM54LS08J					SN54LS08J
DM54LS08W					SN54LS08W
DM54LS09J					SN54LS09J
DM54LS09W					SN54LS09W
DM54LS10J					SN54LS10J
DM54LS10W					SN54LS10W
DM54LS11J					SN54LS11J
DM54LS11W					SN54LS11W
DM54LS12J					SN54LS12J
DM54LS12W					SN54LS12W
DM54LS132J					SN54LS132J
DM54LS132W					SN54LS132W
DM54LS136J					SN54LS136J
DM54LS136W					SN54LS136W
DM54LS138J					SN54LS138J
DM54LS138W					SN54LS138W
DM54LS139J					SN54LS139J
DM54LS139W					SN54LS139W
DM54LS13J					SN54LS13J
DM54LS13W					SN54LS13W
DM54LS14J					SN54LS14J
DM54LS14W					SN54LS14W
DM54LS151J					SN54LS151J
DM54LS151W					SN54LS151W
DM54LS153J					SN54LS153J
DM54LS153W					SN54LS153W
DM54LS155J					SN54LS155J
DM54LS155W					SN54LS155W
DM54LS156J					SN54LS156J
DM54LS156W					SN54LS156W
DM54LS157J					SN54LS157J
DM54LS157W					SN54LS157W
DM54LS158J					SN54LS158J
DM54LS158W					SN54LS158W
DM54LS15J					SN54LS15J
DM54LS15W					SN54LS15W
DM54LS164J					SN54LS164J
DM54LS164W					SN54LS164W
DM54LS168J					SN54LS168J
DM54LS168W					SN54LS168W
DM54LS169J					SN54LS169J
DM54LS169W					SN54LS169W
DM54LS170J					SN54LS170J
DM54LS170W					SN54LS170W
DM54LS173J					SN54LS173J
DM54LS173W					SN54LS173W
DM54LS174J					SN54LS174J
DM54LS174W					SN54LS174W
DM54LS175J					SN54LS175J
DM54LS175W					SN54LS175W
DM54LS189J					SN54LS189J
DM54LS189W					SN54LS189W
DM54LS190J					SN54LS190J
DM54LS190W					SN54LS190W
DM54LS191J					SN54LS191J
DM54LS191W					SN54LS191W
DM54LS192J					SN54LS192J
DM54LS192W					SN54LS192W
DM54LS193J					SN54LS193J
DM54LS193W					SN54LS193W
DM54LS196J					SN54LS196J
DM54LS196W					SN54LS196W
DM54LS197J					SN54LS197J
DM54LS197W					SN54LS197W
DM54LS20J					SN54LS20J
DM54LS20W					SN54LS20W
DM54LS21J					SN54LS21J
DM54LS21W					SN54LS21W
DM54LS22J					SN54LS22J
DM54LS22W					SN54LS22W
DM54LS248J					SN54LS248J
DM54LS248W					SN54LS248W
DM54LS249J					SN54LS249J
DM54LS249W					SN54LS249W
DM54LS253J					SN54LS253J
DM54LS253W					SN54LS253W
DM54LS266J					SN54LS266J
DM54LS266W					SN54LS266W
DM54LS26J					SN54LS26J
DM54LS26W					SN54LS26W
DM54LS279J					SN54LS279J

Industry Standard No.	ECG	SK	GE	RS 276-	MOTOR.
DM54LS279W					SN54LS279W
DM54LS27J					SN54LS27J
DM54LS27W					SN54LS27W
DM54LS283J					SN54LS283J
DM54LS283W					SN54LS283W
DM54LS30J					SN54LS30J
DM54LS30W					SN54LS30W
DM54LS32J					SN54LS32J
DM54LS32W					SN54LS32W
DM54LS353J					SN54LS353J
DM54LS353W					SN54LS353W
DM54LS37J					SN54LS37J
DM54LS37W					SN54LS37W
DM54LS386J					SN54LS386J
DM54LS386W					SN54LS386W
DM54LS38J					SN54LS38J
DM54LS38W					SN54LS38W
DM54LS40J					SN54LS40J
DM54LS40W					SN54LS40W
DM54LS42J					SN54LS42J
DM54LS42W					SN54LS42W
DM54LS47J					SN54LS47J
DM54LS47W					SN54LS47W
DM54LS48J					SN54LS48J
DM54LS48W					SN54LS48W
DM54LS49J					SN54LS49J
DM54LS49W					SN54LS49W
DM54LS51J					SN54LS51J
DM54LS51W					SN54LS51W
DM54LS54J					SN54LS54J
DM54LS54W					SN54LS54W
DM54LS55J					SN54LS55J
DM54LS55W					SN54LS55W
DM54LS670J					SN54LS670J
DM54LS670W					SN54LS670W
DM54LS73J					SN54LS73AJ
DM54LS73W					SN54LS73AW
DM54LS74J					SN54LS74AJ
DM54LS74W					SN54LS74AW
DM54LS75J					SN54LS75J
DM54LS75W					SN54LS75W
DM54LS76J					SN54LS76AJ
DM54LS76W					SN54LS76AW
DM54LS77J					SN54LS77J
DM54LS77W					SN54LS77W
DM54LS78J					SN54LS78AJ
DM54LS78W					SN54LS78AW
DM54LS83AJ					SN54LS83AJ
DM54LS83AW					SN54LS83AW
DM54LS85J					SN54LS85J
DM54LS85W					SN54LS85W
DM54LS86J					SN54LS86J
DM54LS86W					SN54LS86W
DM71LS95J					SN54LS795J
DM71LS95W					SN54LS795W
DM71LS96J					SN54LS796J
DM71LS96W					SN54LS796W
DM71LS97J					SN54LS797J
DM71LS97W					SN54LS797W
DM71LS98J					SN54LS798J
DM71LS98W					SN54LS798W
DM7400	7400	7400		1801	
DM7400N	7400	7400	7400	1801	
DM7401N	7401	7401			
DM7402N	7402	7402	7402	1811	
DM7403N	7403	7403			
DM7404N	7404	7404	7404	1802	
DM7405N	7405	7405			
DM7406N	7406	7406	7406	1821	
DM7407N	7407	7407			
DM7408N	7408	7408	7408	1822	
DM7409N	7409	7409			
DM74107N	74107	74107			
DM74109N	74109	74109			
DM7410J				1807	
DM7410N	7410	7410	7410	1807	
DM7411N	7411	7411			
DM74121N	74121	74121			
DM74123N			74123	1817	
DM74125	74125	74125			
DM74125N	74125	74125			
DM74126	74126	74126			
DM7413	7413	7413			
DM74132	74132	74132			
DM74132N	74132	74132			
DM7413N	7413	7413	7413	1815	
DM74145N	74145	74145	74145	1828	
DM7414N	7414	7414			
DM74150N	74150	74150	74150	1829	
DM74151N	74151	74151			
DM74153N	74153	74153			
DM74154	74154	74154			
DM74154N	74154		74154	1834	
DM74155N	74155	74155			
DM74156N	74156	74156			
DM74164N	74164	74164			
DM7416N	7416	7416			
DM74174N	74174	74174			
DM74175N	74175	74175			
DM7417N	7417	7417			
DM74180N	74180	74180			
DM74181N	74181	74181			
DM74190N	74190	74190			
DM74191N	74191	74191			
DM74192N		74192	74192	1831	
DM74193N		74193	74193	1820	
DM74196N	74196	74196	74196	1833	
DM74198N	74198	74198			
DM74199N	74199	74199			
DM7420J				1809	
DM7420N	7420	7420	7420	1809	
DM7423N	7423	7423			
DM7425N	7425	7425			
DM7426N	7426	7426			
DM7427J				1823	
DM7427N	7427	7427	7427	1823	
DM7427W				1823	
DM7430N	7430	7430			
DM7432J				1824	
DM7432N	7432	7432	7432	1824	
DM7437N	7437	7437			
DM7438N	7438	7438			
DM7440N	7440	7440			
DM7441AN	7441	7441			
DM7441N			7441	1804	
DM7442N	7442	7442			
DM7445N	7445	7445			
DM7447AN	7447	7447			
DM7447N			7447	1805	
DM7448N			7448	1816	
DM7450N	7450	7450			
DM7451J				1825	
DM7451N	7451	7451	7451	1825	
DM7453N	7453	7453			

Industry Standard No.	ECG	SK	GE	RS 276-	MOTOR.
DM7454N	7454	7454			
DM7460N	7460	7460			
DM7470N	7470	7470			
DM7472N	7472	7472			
DM7473N	7473	7473	7473	1803	
DM7474N	7474	7474	7474	1818	
DM7475N	7475	7475	7475	1806	
DM7476N	7476	7476	7476	1813	
DM7485N		7485	7485	1826	
DM7486J				1827	
DM7486N	7486	7486	7486	1827	
DM7489N	7489	7489			
DM7490	7490	7490	7490	1808	
DM7490N	7490	7490	7490	1808	
DM7492N	7492	7492	7492		
DM7493N	7493A	7493			
DM7495N	7495	7495			
DM7496N	7495	7496			
DM74H00J	74H00	74H00			
DM74H00N	74H00	74H00			
DM74H04J	74H04	74H04			
DM74H04N	74H04	74H04			
DM74LS00J					SN74LS00J
DM74LS00N	74LS00			1900	SN74LS00N
DM74LS01J	74LS01				SN74LS01J
DM74LS01N	74LS01				SN74LS01N
DM74LS02J					SN74LS02J
DM74LS02N	74LS02				SN74LS02N
DM74LS03J					SN74LS03J
DM74LS03N	74LS03				SN74LS03N
DM74LS04J					SN74LS04J
DM74LS04N					SN74LS04N
DM74LS05J					SN74LS05J
DM74LS05N	74LS05				SN74LS05N
DM74LS08J					SN74LS08J
DM74LS08N	74LS08				SN74LS08N
DM74LS09J					SN74LS09J
DM74LS09N					SN74LS09N
DM74LS10J					SN74LS10J
DM74LS10N	74LS10				SN74LS10N
DM74LS11J					SN74LS11J
DM74LS11N	74LS11				SN74LS11N
DM74LS12J					SN74LS12J
DM74LS12N					SN74LS12N
DM74LS132J	74LS132				SN74LS132J
DM74LS132N	74LS132				SN74LS132N
DM74LS136J					SN74LS136J
DM74LS136N					SN74LS136N
DM74LS138J					SN74LS138J
DM74LS138N	74LS138				SN74LS138N
DM74LS139J	74LS139				SN74LS139J
DM74LS139N	74LS139				SN74LS139N
DM74LS13J	74LS13				SN74LS13J
DM74LS13N	74LS13				SN74LS13N
DM74LS14J					SN74LS14J
DM74LS14N	74LS14				SN74LS14N
DM74LS151J					SN74LS151J
DM74LS151N	74LS151				SN74LS151N
DM74LS153J					SN74LS153J
DM74LS153N	74LS153				SN74LS153N
DM74LS155J	74LS155				SN74LS155J
DM74LS155N	74LS155				SN74LS155N
DM74LS156J					SN74LS156J
DM74LS156N					SN74LS156N
DM74LS157J					SN74LS157J
DM74LS157N	74LS157				SN74LS157N
DM74LS158J					SN74LS158J
DM74LS158N					SN74LS158N
DM74LS15J	74LS15				SN74LS15J
DM74LS15N	74LS15				SN74LS15N
DM74LS163N	74LS163A	74LS163			
DM74LS164J					SN74LS164J
DM74LS164N	74LS164				SN74LS164N
DM74LS168J					SN74LS168J
DM74LS168N					SN74LS168N
DM74LS169J					SN74LS169J
DM74LS169N					SN74LS169N
DM74LS170J	74LS170				SN74LS170J
DM74LS170N	74LS170				SN74LS170N
DM74LS173J					SN74LS173J
DM74LS173N					SN74LS173N
DM74LS174J					SN74LS174J
DM74LS174N	74LS174				SN74LS174N
DM74LS175J					SN74LS175J
DM74LS175N	74LS175				SN74LS175N
DM74LS189J					SN74LS189J
DM74LS189N					SN74LS189N
DM74LS190J	74LS190				SN74LS190J
DM74LS190N	74LS190				SN74LS190N
DM74LS191J					SN74LS191J
DM74LS191N	74LS191				SN74LS191N
DM74LS192J	74LS192				SN74LS192J
DM74LS192N	74LS192				SN74LS192N
DM74LS193J					SN74LS193J
DM74LS193N	74LS193				SN74LS193N
DM74LS196J					SN74LS196J
DM74LS196N		74LS196			SN74LS196N
DM74LS197J					SN74LS197J
DM74LS197N					SN74LS197N
DM74LS20J					SN74LS20J
DM74LS20N	74LS20				SN74LS20N
DM74LS21J	74LS21				SN74LS21J
DM74LS21N	74LS21				SN74LS21N
DM74LS22J	74LS22				SN74LS22J
DM74LS22N	74LS22				SN74LS22N
DM74LS248J					SN74LS248J
DM74LS248N		74LS248			SN74LS248N
DM74LS249J					SN74LS249J
DM74LS249N		74LS249			SN74LS249N
DM74LS253J					SN74LS253J
DM74LS253N	74LS253				SN74LS253N
DM74LS266J					SN74LS266J
DM74LS266N	74LS266				SN74LS266N
DM74LS26J	74LS26				SN74LS26J
DM74LS26N	74LS26				SN74LS26N
DM74LS279J	74LS279				SN74LS279J
DM74LS279N	74LS279				SN74LS279N
DM74LS27J					SN74LS27J
DM74LS27N	74LS27				SN74LS27N
DM74LS283J					SN74LS283J
DM74LS283N					SN74LS283N
DM74LS289J					SN74LS289J
DM74LS289N					SN74LS289N
DM74LS30J					SN74LS30J
DM74LS30N	74LS30				SN74LS30N
DM74LS32J					SN74LS32J
DM74LS32N	74LS32				SN74LS32N
DM74LS353J					SN74LS353J
DM74LS353N					SN74LS353N
DM74LS37J	74LS37				SN74LS37J
DM74LS37N	74LS37				SN74LS37N
DM74LS386J					SN74LS386J
DM74LS386N					SN74LS386N
DM74LS38J					SN74LS38J

Industry Standard No.	ECG	SK	GE	RS 276-	MOTOR.
DM74LS38N	74LS38				SN74LS38N
DM74LS40J	74LS40				SN74LS40J
DM74LS40N	74LS40				SN74LS40N
DM74LS42J					SN74LS42J
DM74LS42N	74LS42				SN74LS42N
DM74LS47J					SN74LS47J
DM74LS47N					SN74LS47N
DM74LS48J					SN74LS48J
DM74LS48N					SN74LS48N
DM74LS49J					SN74LS49J
DM74LS49N		74LS49			SN74LS49N
DM74LS51J					SN74LS51J
DM74LS51N	74LS51				SN74LS51N
DM74LS54J	74LS54				SN74LS54J
DM74LS54N	74LS54				SN74LS54N
DM74LS55J	74LS55				SN74LS55J
DM74LS55N	74LS55				SN74LS55N
DM74LS670J	74LS670				SN74LS670J
DM74LS670N	74LS670				SN74LS670N
DM74LS73J					SN74LS73AJ
DM74LS73N					SN74LS73AN
DM74LS74J					SN74LS74AJ
DM74LS74N	74LS74A				SN74LS74AN
DM74LS75J					SN74LS75J
DM74LS75N	74LS75				SN74LS75N
DM74LS76J	74LS76A				SN74LS76AJ
DM74LS76N	74LS76A				SN74LS76AN
DM74LS77J					SN74LS77J
DM74LS77N					SN74LS77N
DM74LS78J					SN74LS78AJ
DM74LS78N					SN74LS78AN
DM74LS83AJ					SN74LS83AJ
DM74LS83AN					SN74LS83AN
DM74LS85J					SN74LS85J
DM74LS85N	74LS85				SN74LS85N
DM74LS86J					SN74LS86J
DM74LS86N					SN74LS86N
DM74LS90N	74LS90	74LS90			
DM74S00J	74S00	74S00			
DM74S00N	74S00	74S00			
DM74S04J	74S04	74S04			
DM74S04N	74S04	74S04			
DM75491N					MC75491P
DM75492N					MC75492P
DM7822J					MC1489AL
DM7837J					MC3437L
DM7838J					MC3438L
DM7887J					MC3490P
DM7887N					MC3490P
DM7889J					MC3491P
DM7889N					MC3491P
DM7897J					MC3494P
DM7897N					MC3494P
DM81LS95J					SN74LS795J
DM81LS95N					SN74LS795N
DM81LS96J					SN74LS796J
DM81LS96N					SN74LS796N
DM81LS97J					SN74LS797J
DM81LS97N					SN74LS797N
DM81LS98J					SN74LS798J
DM81LS98N					SN74LS798N
DM8822J					MC1489AL
DM8822N					MC1489AP
DM8837N					MC3437P
DM8838N					MC3438P
DM8861N					MC75491P
DM8863N					MC75492P
DM8887J					MC3490P
DM8889J					MC3491P
DM8897J					MC3494P
DM9	736		IC-17		
DM9093N	9093				MC853P
DM9094N	9094				MC856P
DM9097N	9097				MC855P
DM9099N	9099				MC852P
DM930N	9930				MC830P
DM932N	9932				MC832P
DM933N	9933				MC833P
DM93415					MCM93415
DM93425					MCM93425
DM935N	9935				MC840P
DM936N	9936				MC836P
DM937N	9937				MC837P
DM944N	9944				MC844P
DM945N	9945				MC845P
DM946N	9946				MC846P
DM948N	9948				MC848P
DM949N	9949				MC849P
DM957N	9157				MC857P
DM958N					MC858P
DM961N	9961				MC861P
DM962N	9962				MC862P
DM963N	9963				MC863P
DMB10-12BA					MRF840
DMB20-12BA					MRF842
DMB45-12BA					MRF846
DMB5-12BA					MRF840
DME10					MRF1015MB
DME150					MRF1150MB
DME25					MRF1035MB
DME250					MRF1250M
DME375					MRF1325M
DME50					MRF1090MB
DME7					MRF1008MB
DN	123A	3444	20	2051	
DN20-00453	128		243	2030	
DN3066A					MFE2095
DN3067A					MFE2094
DN3068A					MFE2093
DN3069A					2N4221A
DN3070A					2N4220A
DN3071A					2N4220A
DN3356A					2N4091
DN3366A					2N4091
DN3367A					2N4091
DN3368A					2N4221A
DN3369A					2N422/A
DN3370A					MFE2093
DN3436A					2N3436
DN3437A					2N3437
DN3438A					2N3438
DN3458A					2N4222A
DN3459A					2N4220A
DN3460A					2N4220A
DND800	177	3100/519	300	1122	
DNX1					2N3822
DNX1A					2N3822
DNX2					MFE2094
DNX2A					MFE2094
DNX3					MFE2093
DNX3A					MFE2093
DNX4	459				2N3822
DNX4A	459				2N3822
DNX5					2N3821

Industry Standard No.	ECG	SK	GE	RS 276-	MOTOR.
DNX5A					2N3821
DNX6					MFE2093
DNX6A					MFE2093
DNX7					2N3824
DNX7A					2N3824
DNX8					2N3821
DNX8A					2N3821
DNX9					2N3821
DNX9A					2N3821
D019			2	2007	
D031			20	2051	
D038			2	2007	
D043		3004	2	2007	
D063			51	2004	
D078			2	2007	
D079			51	2004	
D080			16	2006	
D081			3	2006	
D083			8	2002	
D085			8	2002	
D086			51	2004	
D087			11	2015	
D088			11	2015	
DP100	125	3033	510,531	1114	
DP13B00-2				1151	
DP13B01-2				1152	
DP13B02-2				1172	
DP13B04-2				1173	
DR1	116	3311	504A	1104	
DR100	116	3017B/117	504A	1104	
DR1000	125	3033	510,531	1114	
DR1100	116	3311	504A	1104	
DR1575	177		300	1122	
DR1PR	116	3031A	504A	1104	
DR2	116	3311	504A	1104	
DR200	116	3016	504A	1104	
DR202				1102	
DR204				1103	
DR206				1104	
DR210				1114	
DR291	109	3087		1123	
DR3	116	3311	504A	1104	
DR300	116	3016	504A	1104	
DR351	109	3087		1123	
DR352	109	3087		1123	
DR365	109	3091		1123	
DR385	109	3087		1123	
DR4	116	3031A	504A	1104	
DR400	116	3016	504A	1104	
DR426	109	3087		1123	
DR427	116	3091	504A	1104	
DR434	109	3087		1123	
DR435	116	3016	504A	1104	
DR449	109	3087		1123	
DR464	109	3091		1123	
DR5	116	3031A	504A	1104	
DR500	116	3017B/117	504A	1104	
DR5101	116	3016	504A	1104	
DR5102	116	3017B/117	504A	1104	
DR600	116		504A	1104	
DR668	116	3311	504A	1104	
DR669	116	3311	504A	1104	
DR670	116	3031A	504A	1104	
DR671	116	3031A	504A	1104	
DR695	116	3031A	504A	1104	
DR698	116	3031A	504A	1104	
DR699	116	3031A	504A	1104	
DR700	125	3032A	510,531	1114	
DR800	125	3032A	510,531	1114	
DR826	116	3031A	504A	1104	
DR848	116	3311	504A	1104	
DR863	116	3311	504A	1104	
DR900	125	3033	510,531	1114	
DRC-81252	102	3722	2	2007	
DRC-87540	123		20	2051	
DRS1	116	3031A	504A	1104	
DRS101			504A	1104	
DRS102	116	3016	504A	1104	
DRS104	116	3017B/117	504A	1104	
DRS106	116	3051/156	504A	1104	
DRS107	125	3032A	510,531	1114	
DRS108	125	3051/156	510,531	1114	
DRS2	156	3033	512		
DS-0065	116	3016	504A	1104	
DS-1		3011	504A	1104	
DS-1.5-2	156	3051	512		
DS-10			504A	1104	
DS-102	222	3065	FET-4	2036	
DS-105	222	3050/221	FET-4	2036	
DS-106	222	3065	FET-4	2036	
DS-107	135A		ZD-5.1		
DS-108	142A	3062	ZD-24	563	
DS-110	144A		ZD-14	564	
DS-111	78	3054/196			
DS-112	79	3239/236			
DS-113	519		514	1122	
DS-113A	506	3130	511	1114	
DS-113B	506	3130	511	1114	
DS-117	177	3100/519	300	1122	
DS-13	116	3017B/117	504A	1104	
DS-13(COURIER)	116		504A	1104	
DS-130	116	3311	504A	1104	
DS-130B	116	3311	504A	1104	
DS-130C	116	3311	504A	1104	
DS-130E	116	3311	504A	1104	
DS-130Y	116	3311			
DS-130YB	116	3311	504A	1104	
DS-130YE	116		504A	1104	
DS-131	116		504A	1104	
DS-131A	116		504A	1104	
DS-131B	116		504A	1104	
DS-132	116		504A	1104	
DS-132A	116		504A	1104	
DS-132B	116		504A	1104	
DS-13A(SANYO)	116		504A	1104	
DS-13B(SANYO)	116		504A	1104	
DS-14	102A	3004	53	2007	
DS-14(DIODE)	116		504A	1104	
DS-149	5072A	3136	ZD-8.2	562	
DS-159	5071A	3058/137A	ZD-6.8	561	
DS-15A(SANYO)	506		511	1114	
DS-15B(SANYO)	506		511	1114	
DS-16A	116	3031A	504A	1104	
DS-16A(SANYO)	125		510,531	1114	
DS-16B		3032A	504A	1104	
DS-16B(SANYO)	116		504A	1104	
DS-16B(SYLVANIA)	116		504A	1104	
DS-16C(SANYO)	116		504A	1104	
DS-16D	116	3017B/117	504A	1104	
DS-16D(SANYO)	116		504A	1104	
DS-16E		3311	504A	1104	
DS-16E(SANYO)	116		504A	1104	
DS-16ND			504A	1104	
DS-16NE		3311	504A	1104	

Industry Standard No.	ECG	SK	GE	RS 276-	MOTOR.
DS-16NY	116	3017B/117	504A	1104	
DS-16YA	116	3017B/117	504A	1104	
DS-17			504A	1104	
DS-17-6A	116		504A	1104	
DS-18	116		504A	1104	
DS-18(DELCO)	109	3090		1123	
DS-189	5074A	3095/147A	ZD-33	563	
DS-19	102A	3031A	53	2007	
DS-19(RECTIFIER)	116		504A	1104	
DS-1K	125		510,531	1114	
DS-1M	116	3031A	504A	1104	
DS-1N		3311	504A	1104	
DS-1P	116	3017B/117	504A	1104	
DS-1U	507	3311	504A	1104	
DS-22		3007	1	2007	
DS-24		3006	52	2024	
DS-25		3008	52	2024	
DS-26		3004	53	2007	
DS-27	109	3087			
DS-28(DELCO)	100	3721	1	2007	
DS-2M		3311	511	1114	
DS-2N	156	3031A	512		
DS-31	5071A	3100/519	300	1122	
DS-31(DELCO)	177		300	1122	
DS-32			1N34AS	1123	
DS-33			1N34AS	1123	
DS-34		3006	245	2004	
DS-35		3006	245	2004	
DS-36		3006	245	2004	
DS-37			245	2004	
DS-38		3311	504A	1104	
DS-39		3087	1N60	1123	
DS-41		3006	245	2004	
DS-410	177	3100/519	300	1122	
DS-410(AMPEX)	177		300	1122	
DS-410(MOTOROLA)	123A		20	2051	
DS-42		3006	245	2004	
DS-43			ZD-9.1	562	
DS-430			504A	1104	
DS-44		3010	20	2051	
DS-442	177	3175	300	1122	
DS-442(SEARS)	177		300	1122	
DS-45			20	2051	
DS-46		3124	20	2051	
DS-47	123A	3124/289	20	2051	
DS-49			ZD-8.2	562	
DS-501	105	3012	4		
DS-503	104	3009	4		
DS-509	130	3027	14	2041	
DS-51		3006	50	2004	
DS-512	128	3048/329	243	2030	
DS-513	152	3893	66	2048	
DS-514	130	3027	14	2041	
DS-515	121	3009	239	2006	
DS-519		3036	14	2041	
DS-52		3006	50	2004	
DS-520		3009	239	2006	
DS-53		3006	2	2007	
DS-55	614	3327	90		
DS-56		3006	245	2004	
DS-60	172A	3241/232	64		
DS-62		3006	245	2004	
DS-63		3006	245	2004	
DS-64		3006	245	2004	
DS-65		3006	50	2004	
DS-66	123A	3122	20	2051	
DS-66L	123A	3041	20	2051	
DS-66W			20	2051	
DS-67	123A	3444	20	2051	
DS-67W		3020	20	2051	
DS-68	106	3118	21	2034	
DS-71	107	3018	11	2015	
DS-72	107	3018	11	2015	
DS-73		3018	11	2015	
DS-74	107	3854/123AP	61	2015	
DS-75		3124	17	2051	
DS-76	123A	3122	20	2051	
DS-77			20	2051	
DS-78	161	3018	39	2015	
DS-781	107	3293	11	2015	
DS-79	116	3311	504A	1104	
DS-79(DELCO)	116		504A	1104	
DS-8	102A	3861/101	53	2007	
DS-81		3018	61	2038	
DS-82	159	3114/290	82	2032	
DS-83	159	3114/290	82	2032	
DS-85	161	3854/123AP	39	2015	
DS-86	129	3114/290	244	2027	
DS-88	312	3116	FET-1	2035	
DS-94	289A	3124/289	210	2009	
DS-97	177	3175	300	1122	
DS-IM	116	3311	504A	1104	
DS0026CG					MMH0026CG
DS0026CH					MMH0026CG
DS0026CJ					MMH0026CL
DS0026CN					DS0026CP1
DS0026G					MMH0026G
DS0026H					DS0026G
DS0026J					DS0026L
DS0056CG					MMH0026CG
DS0056CH					MMH0026CG
DS0056CJ					MMH0026CL
DS0056CN					MMH0026CP1
DS0056G					MMH0026G
DS0056H					MMH0026G
DS0056J					MMH0026L
DS0065			504A	1104	
DS1	116		504A	1104	
DS1-002-0	177		300	1122	
DS1005-1X8628	519		514	1122	
DS104	177	3100/519	ZD-6.2	561	
DS11	101		8	2002	
DS113B	506	3130	511	1114	
DS117			300	1122	
DS118A				1143	
DS118B				1144	
DS12	101		8	2002	
DS13	102A	3245/199	62	2007	
DS130	116		504A	1104	
DS130B	116		504A	1104	
DS130C				1103	
DS130D	116			1102	
DS130E	116		504A	1104	
DS130NB				1104	
DS130NC				1103	
DS130ND	116	3311	504A	1104	
DS130NE				1102	
DS130Y	116		504A	1104	
DS130YC	116		504A	1104	
DS130YE	116	3311	504A	1104	
00000DS131	116	3313	504A	1104	
DS131A	116	9001/113A	6GC1	1102	
DS131B	116	9001/113A	504A	1104	
DS132A	116	9002	504A	1104	
DS132B		9002	504A	1104	
DS133B	506	3130	511	1114	
DS13B				1144	
DS14		3123	52	2024	
DS1488J	75188				MC1488L
DS1488N	75188				MC1488P
DS1489AJ					MC1489AL
DS1489AN					MC1489AP
DS1489J	75189				MC1489L
DS1489N	75189				MC1489P
DS149	5072A		ZD-8.2	562	
DS159			ZD-6.2	561	
DS15A			509	1114	
DS15C				1102	
DS15E				1104	
DS16	102A	3123	53	2007	
DS160(G.E.)	116		504A	1104	
DS16A		3017B	504A	1104	
DS16C			504A	1104	
DS16E	116		504A	1104	
DS16N	116		504A	1104	
DS16NA	125		510,531	1114	
DS16NB	116	3017B/117	504A	1104	
DS16NC	116	3017B/117	504A	1104	
DS16ND	116	3017B/117	504A	1104	
DS16NE	116	3125	504A	1104	
DS16NE(SONY)	506		511	1114	
DS17	116	9001/113A	504A(2)	1104	
DS17(ADMIRAL)	116		504A	1104	
DS17N	116		504A	1104	
DS18	116	9002	504A	1104	
DS189	5074A	3092/141A	ZD-11.0	563	
DS18N	116		504A	1104	
DS19	116(2)		2		
DS1B	123A	3444	20	2051	
DS1K	116	3081/125	504A	1104	
DS1K7	116		504A	1104	
DS1M	156	3051	512	1114	
DS1N	116		504A	1104	
DS1P	116		504A	1104	
DS2	101		8	2002	
DS2-27				1123	
DS21	100	3721	1	2007	
DS22	100	3123	1	2007	
DS23	100	3721	1	2007	
DS24	126		52	2024	
DS25	126		52	2024	
DS26	102A		53	2007	
DS27	109	3087		1123	
DS29	102A		53	2007	
DS2K	116	3311	504A	1104	
DS2M	125	3081	510,531	1114	
DS2N	116	3031A	504A	1104	
DS2N22			504A	1104	
DS2P				1144	
DS3	102A		53	2007	
DS31	177		1N34AS	1123	
DS31(DELCO)				1123	
DS32			1N34AS	1123	
DS33			1N295	1123	
DS33(DELCO)	109			1123	
DS34	160		245	2004	
DS3486J					MC3486L
DS3486N					MC3486P
DS3487J					MC3487L
DS3487N					MC3487P
DS35	160		245	2004	
DS36	126		52	2024	
DS3612H					MC1472U
DS3612N					MC1472P1
DS3632H					MC1472U
DS3632J					MC1472U
DS3632N					MC1472P1
DS3650J					MC3450L
DS3650N					MC3450P
DS3651J					MC3430L
DS3651N					MC3430P
DS3652J					MC3452L
DS3652N					MC3452P
DS3653J					MC3432L
DS3653N					MC3432P
DS3674J					MC3460L
DS3674N					MC3460P
DS37	160		245	2004	
DS38	126	3311	504A	1104	
DS38(CRAIG)	116		504A	1104	
DS38(DELCO)	160		245	2004	
DS38(SANYO)	116		504A	1104	
DS39	110MP			1123(2)	
DS4	101		8	2002	
DS41	160		245	2004	
DS410	177	3088	300	1122	
DS410(COURIER)	177		300	1122	
DS410(EMERSON)	177		300	1122	
DS410(FANON)	177		300	1122	
DS410(G.E.)	109	3090		1123	
DS410(OLYMPIC)	177		300	1122	
DS410R	177	3175	300	1122	
DS410R(G.E.)	109	3090		1123	
DS42	160		245	2004	
DS43			ZD-9.1	562	
DS430	116		504A	1104	
DS44	123A	3444	20	2051	
DS441	177		300	1122	
DS442	177		300	1122	
DS442FM	177		300	1122	
DS443	177	3100/519	300	1122	
DS448	519	3100	300	1122	
DS45	123A	3444	20	2051	
DS46	123A	3444	20	2051	
DS47	123A	3444	20	2051	
DS48			ZD-9.1	562	
DS5	101		8	2002	
DS50	136A	3057	ZD-5.6	561	
DS501	105		4		
DS502	105		4		
DS503	104	3719	16	2006	
DS504	105		4		
DS505	105		4		
DS506	105		4		
DS509	130		14	2041	
DS51	126		52	2024	
DS512	128		243	2030	
DS513	152		66	2048	
DS514		3027	19	2041	
DS515	121	3717	239	2006	
DS519	130		14	2041	
DS52	126		52	2024	
DS520	121	3717	239	2006	
DS525	105		4		
DS53	100	3721	1	2007	
DS55	614	3087	90		
DS55107J					MC55107L
DS55107W					MC75107L

Industry Standard No.	ECG	SK	GE	RS 276-	MOTOR.
DS55108J					MC55108L
DS55108W					MC55108L
DS55110J					MC75S110L
DS55121J					MC8T13L
DS55121W					MC8T13L
DS55122J					MC8T14L
DS55122W					MC8T14L
DS55325J					MC55325L
DS56	160		245	2004	
DS570	105	3012	4		
DS58(SANYO)	116		504A	1104	
DS5BN-6				1104	
DS6	101		8	2002	
DS62	160		245	2004	
DS63	160		245	2004	
DS64	160		245	2004	
DS65	160		245	2004	
DS66	123A	3122	20	2051	
DS67	123A	3444	20	2051	
DS67W	123A	3444	20	2051	
DS68	106	3118	21	2034	
DS7	101		8	2002	
DS71	107		11	2015	
DS72	107		11	2015	
DS73	107		11	2015	
DS74	107	3854/123AP	11	2015	
DS75	107		11	2015	
DS75107J					MC75107L
DS75107N					MC75107P
DS75108J					MC75108L
DS75108N					MC75108P
DS75110J					MC75S110L
DS75110N					MC75S110P
DS75121J					MC8T13L
DS75121N					MC8T13P
DS75122J					MC8T14L
DS75122N					MC8T14P
DS75123J					MC8T23L
DS75123N					MC8T23P
DS75124J					MC8T24L
DS75124N					MC8T24P
DS75125/27					MC75125/27
DS75128/29					MC75128/29
DS75207J					MC75107L
DS75207N					MC75107P
DS75208J					MC75108L
DS75208N					MC75108P
DS75325J					MC75325L
DS75325N					MC75325P
DS75491J					MC75491P
DS75491N					MC75491P
DS75492J					MC75492P
DS75492N					MC75492P
DS76	123A	3122	20	2051	
DS77	123A	3444	20	2051	
DS78	161	3018	39	2015	
DS7837J					MC3437L
DS7837W					MC3437L
DS7838J					MC3438L
DS7838W					MC3438L
DS7887J					MC3490P
DS7889J					MC3491P
DS7897J					MC3494P
DS79	116		504A	1104	
DS8	101		8	2002	
DS81	161	3018	39	2015	
DS82	159		82	2032	
DS83	159		82	2032	
DS85	161	3854/123AP	39	2015	
DS86			21	2034	
DS8641N/J					DS8641N/J
DS88	312	3448	FET-1	2035	
DS8833J					MC8T28L
DS8833N					MC8T28P
DS8834J					MC8T26AL
DS8834N					MC8T26AP
DS8835J					MC8T26AL
DS8835N					MC8T26AP
DS8837J					MC3437L
DS8837N					MC3437P
DS8838J					MC3438L
DS8838N					MC3438P
DS8839J					MC8T28L
DS8839N					MC8T28P
DS8887J					MC3490P
DS8887N					MC3490P
DS8889J					MC3491P
DS8889N					MC3491P
DS8897J					MC3494P
DS8897N					MC3494P
DS89	117		504A	1104	
DS8T26AJ/N					MC8T26AL/P
DS8T26AMJ/N					MC8T26AL/P
DS8T28J/N					MC8T28J/P
DS8T28MJ/N					MC8T28LP
DS9	101		8	2002	
DS94	123A	3124/289			
DS96	106	3984	21	2034	
DS97	177	3175		1123	
DS97(DELCO)	177		300	1122	
DS99			ZD-9.1	562	
DSA150	177	3100/519	300	1122	
DSI-104-2	519		514	1122	
DSR1201					MR501
DSR1203					MR504
DSR1205					MR506
DSS16685	109	3087		1123	
DSS953	156	3100/519	300	1122	
DSZ3006					5M6.0ZS5
DSZ3100					5M100ZS5
DT100	105		4		
DT1003	154	3045/225	40	2012	
DT1040	121	3014	239	2006	
DT1110	128	3024	243	2030	
DT1111	128	3024	243	2030	
DT1112	128	3024	243	2030	
DT1120	128	3024	243	2030	
DT1121	128	3024	243	2030	
DT1122	128	3024	243	2030	
DT1311	128	3024	243	2030	
DT1321	128	3024	243	2030	
DT1510	128	3024	243	2030	
DT1511	128	3024	243	2030	
DT1512	128	3024	243	2030	
DT1520	128	3024	243	2030	
DT1521	128	3024	243	2030	
DT1522	128	3024	243	2030	
DT1602	154	3045/225	40	2012	
DT1603	154	3045/225	40	2012	
DT161	123A	3444	20	2051	
DT1610	123	3027/130	20	2051	
DT1612	154	3045/225	40	2012	
DT1613	154	3045/225	40	2012	
DT1621	128	3024	243	2030	

Industry Standard No.	ECG	SK	GE	RS 276-	MOTOR.
DT18(CHAN.MASTER)	116		504A	1104	
DT230A			300	1122	1N4002
DT230B	506		511	1114	
DT230F			300	1122	1N4001
DT230G			300	1122	1N4003
DT230H					1N4004
DT3301	175	3131A/369	246	2020	
DT3302	175	3131A/369	246	2020	
DT401	104	3014	16	2006	
DT4011	130	3027	14	2041	
DT41	121	3717	239	2006	
DT4110	130	3027	14	2041	
DT4111	130	3027	14	2041	
DT4120	130	3027	14	2041	
DT4121	130	3027	14	2041	
DT4303	283	3438	35		
DT4304	283	3438	35		
DT4305	283	3438	35		
DT4306	283	3438	35		
DT6110	121	3717	239	2006	
DT80	105		4		
DTA1011			3	2006	
DTG-110			16	2006	
DTG1010	127	3764	25		
DTG1011	121	3009	239	2006	
DTG1040	121	3009	239	2006	
DTG110	121	3642/179	239	2006	
DTG110(SEARS)			3	2006	
DTG110(WARDS)			16	2006	
DTG110A	179	3642	76		
DTG110B	179	3642	76		
DTG1110	127	3764	25		
DTG1200	179	3642	76		
DTG2000	179	3642	76		
DTG2000A	179	3642	76		
DTG2100	179	3642	76		
DTG2100A	179	3642	76		
DTG2200	179	3642	76		
DTG2400	179	3642			
DTG400M	179	3642	76		
DTG600	179	3642	76		
DTG601	179	3642	76		
DTG602	179	3642	76		
DTG603	179	3642	76		
DTG603M	179	3642	76		
DTN206	128		243	2030	
DTS013	164	3115/165	37		
DTS0710	162	3438	35		
DTS0713	164		37		
DTS1010					2N6056
DTS1020					MJ3001
DTS103	284	3438	35		
DTS104	284	3438	35		
DTS105	284	3836			
DTS106	284	3836			
DTS107	284	3836			
DTS310	283				2N6306
DTS311	283				2N6306
DTS3705	162	3438	35		
DTS3705A	162	3438	35		
DTS3705B	162	3438	35		
DTS401	162	3438	35		2N3902
DTS4010					MJ3041
DTS402	283	3439/163A	35		2N3902
DTS4025					MJ3041
DTS4026					MJ10012
DTS403	283	3559/162	73		2N6308
DTS4039					MJ10000
DTS4040					MJ10000
DTS4041					MJ10000
DTS4045					MJ10000
DTS4059					MJ10000
DTS4060					MJ10001
DTS4061					MJ10000
DTS4065					MJ10001
DTS4066	98				MJ10000
DTS4067	98				MJ10000
DTS4074	98				MJ10004
DTS4075	98				MJ10004
DTS409	388	3559/162	73		2N6308
DTS410	385	3438	35		MJ410
DTS411	94	3438	35		MJ411
DTS413	283	3438	35		MJ413
DTS423	385	3438	35		MJ423
DTS423M	283	3438	35		
DTS424	283	3439/163A	36		2N6308
DTS425	283	3439/163A	36		2N6545
DTS430	385	3438	35		2N6307
DTS431	283	3438	35		MJ431
DTS431M	283	3438	35		
DTS515					2N6306
DTS516					2N6306
DTS517					2N6306
DTS518					2N6307
DTS519					2N6308
DTS660					2N6233
DTS663					2N6235
DTS665					2N6235
DTS701	164	3115/165	37		BU204
DTS702	165	3115	38		BU205
DTS704	165	3115	38		
DTS712					BU207
DTS714					BU208
DTS801	389	3115/165	37		BU205
DTS802	165	3115	38		BU207
DTS804	165	3115	38		BU208
DTS812					BU207
DTS814					BU208
DU1	160		245	2004	
DU1000	125	3033	510,531	1114	
DU12	160		245	2004	
DU2	160		245	2004	
DU3	102A		20	2007	
DU4	158		53	2007	
DU400	116	3017B/117	504A	1104	
DU47	105	3012	4		
DU5	102A		53	2007	
DU6	104	3009	16	2006	
DU600	116	3017B/117	504A	1104	
DU7	105	3012	4		
DU800	125	3033	510,531	1114	
DV2209	614		90		
DVMV1Y	601	3463			
DVR-8	964	3630			
DVV004	614	3327	90		
DW 6505	123A	3444			
DW 7375	123A	3444			
DW-6505			20	2051	
DW-6982			243	2030	
DW-7375			20	2051	
DW-7655			65	2050	
DW-7655-LV00223			65	2050	
DW6034/M	123A	3444	20	2051	
DW6195	128		243	2030	

Industry Standard No.	ECG	SK	GE	RS 276-	MOTOR.
DW7655		3247		2050	
DX-0061	136A	3057	ZD5.6	561	
DX-0087		3060	ZD-9.1	562	
DX-0099	116		504A	1104	
DX-0150	601	3463	514	1122	
DX-0161	109	3088	1N60	1123	
DX-0162	109	3088	1N60	1123	
DX-0241	109		1N34AS	1123	
DX-0255		3100	300	1122	
DX-0270	177	3175	514	1122	
DX-0273	519	3100	514	1122	
DX-0299	519	3100	514	1122	
DX-0307	612	3325			
DX-0445	116	3311	504A	1104	
DX-0475	116	3311	504A	1104	
DX-0529	134A		ZD3.6		
DX-0530	5013A		ZD-6.2	561	
DX-0543		3175	300	1122	
DX-0697	605	3864			
DX-0721		3311	504A	1104	
DX-0722	614	3325/612			
DX-0723	610	3323			
DX-0725	109	3088	1N60	1123	
DX-0727			ZD-5.6	561	
DX-0728		3060	ZD-9.1	562	
DX-0729	5072A		ZD-8.2	562	
DX-0749	612	3325			
DX-0750	601	3463			
DX-0752	601	3463			
DX-081			ZD-8.2	562	
DX-0987	135A	3056	ZD5.1		
DX-1007	614	3327			
DX-1034		3061	ZD-10		
DX-1039		3311	504A	1104	
DX-1080		3327	90		
DX-1131			504A	1104	
DX-1132		3057	ZD-5.6	561	
DX-1195		3095	ZD-33		
DX-1229		3051	512		
DX1018	108		86	2038	
DX1194			ZD-6.2	561	
DX520	116		504A	1104	
DX6873	109	3090		1123	
DX7429	519		514	1122	
DZ-081	5072A	3136	X11	562	
DZ-12			ZD-12	563	
DZ0820	5072A	3136	ZD-8.2	562	
DZ10	139A	3060	ZD-9.1	562	
DZ10A	140A	3061	ZD-10	562	
DZ12	143A	3093	ZD-13	563	
DZ12A	142A	3062	ZD-12	563	
DZ13	143A	3750	ZD-13	563	
DZ15A	145A	3063	ZD-15	564	
DZ27A	146A	3064	ZD-27		
DZ33A	147A	3095	ZD-33		
DZ82A	150A	3098	ZD-82		
E-01381	128		243	2030	
E-044A	102A		53	2007	
E-065	126	3008	52	2024	
E-066	126	3008	52	2024	
E-067	126	3008	52	2024	
E-068	126	3008	52	2024	
E-070	102A	3004	53	2007	
E-0704W	116		504A	1104	
E-075L	116	3016	504A	1104	
E-158		3004	2	2007	
E-167-228	128		243	2030	
E-185B121712			18	2030	
E-2462		3006	50	2004	
E-2465	102A	3004	53	2007	
E-2466	103A	3010	59	2002	
E-2491B	128		243	2030	
E-41			504A	1104	
E.S57X9			511	1114	
E0011	614	3126	90		
E0018	109	3087		1123	
E0105			2	2007	
E03090-002	167	3647	504A	1172	
E03155-001	116	3311	504A	1104	
E03155-002	116	3311	504A	1104	
E044A			2	2007	
E066			2	2007	
E067			2	2007	
E068			2	2007	
E070			51	2004	
E0704-W			504A	1104	
E075L			504A	1104	
E0771-3		3060	ZD-9.1	562	
E0771-6	144A	3094	ZD-14	564	
E0771-7	142A	3062	ZD-12	563	
E0788-C		3031A	504A	1104	
E0788C	116		504A	1104	
E09-306112			300	1122	
E1	116	3016	504A	1104	1N4002
E10	125	3033	510,531	1114	1N4007
E100		3112			MPF102
E101	457	3112			2N5457
E1010			504A	1104	
E1011	116	3017B/117	504A	1104	
E10116	116		504A	1104	
E10157	116	3017B/117	504A	1104	
E10171	116	3031A	504A	1104	
E10172	116	3311	504A	1104	
E1018N	116		504A	1104	
E102	457	3016	52	2024	2N5457
E102(ELCOM)	116		504A	1104	
E103	312	3112	FET-1	2028	2N5459
E1031RXT			1N60	1123	
E1042		3105	512(4)		
E105	126		52	2024	MPF4391
E106		3017B			MPF4391
E106(ELCOM)	116		504A	1104	
E107					MPF4391
E108		3032A			MPF4391
E108(ELCOM)	125		510,531	1114	
E109					MPF4391
E110					MPF4391
E111	466				MPF4391
E112					MPF4392
E1121R	177	9091	300	1122	
E1124	116	3017B/117	504A	1104	
E113	466				MPF4393
E1138R			300	1122	
E114					2N5486
E1145R			511	1114	
E1145RED			511	1114	
E1145RJH			511	1114	
E1146J			504A	1104	
E1146R			504A	1104	
E1153RB			504A	1104	
E1156RD			504A	1104	
E1157RNA			504A	1104	
E1158ZA				561	
E1176ALC1A	113A	3119/113	6GC1		

Industry Standard No.	ECG	SK	GE	RS 276-	MOTOR.
E1176R	178MP	3100/519	300(2)	1122(2)	
E12					MR1-1200
E1229R	519	3100			
E1237RF	116	3313			
E125C200	116	3017B/117	504A	1104	
E1286R	519	3100			
E13-000-03	123A	3444	20	2051	
E13-000-04	123A	3444	20	2051	
E13-001-02	159	3114/290	82	2032	
E13-001-03	159	3114/290	82	2032	
E13-001-04	159	3114/290	82	2032	
E13-002-03	123A	3444	20	2051	
E13-003-00	123A	3444	20	2051	
E13-003-01	123A	3444	20	2051	
E13-004-00	233	3018	210	2009	
E13-005-02	123A	3444	20	2051	
E13-006-02	159	3114/290	82	2032	
E13-007-00	171	3104A	27		
E13-008-00	162	3438	35		
E13-009-00	165	3111	237		
E13-010-00	191	3104A	249		
E13-011-00	191	3104A	249		
E13-012-00	182	3104A	55		
E13-013-03	120	3110	CR-3		
E13-013-04	120	3110	CR-3		
E13-017-01	177	3100/519	300	1122	
E13-020-00	116		504A	1104	
E13-021-01	505	3108	CR-7		
E13-021-03	505	3108	CR-7		
E13-20-00	116	3311	504A	1104	
E1303Z	5012A	3778			
E1304R	125	3032A			
E1315R	156	3051			
E132	102A		53	2007	
E1322Z	5012A	3778			
E1324R	519	3100			
E1339R	552	3843			
E135	116	3017B/117	504A	1104	
E14					MR1-1400
E1410	116	3016	504A	1104	
E1411	116		504A	1104	
E1412	116		504A	1104	
E1412R	552	3998/506			
E1413	116	3017B/117	504A	1104	
E1415	116	3031A	504A	1104	
E143	116		504A	1104	
E1440	116	3017B/117	504A	1104	
E146	116	3016	504A	1104	
E1466Z	5036A	3095/147A			
E14C350	116	3016	504A	1104	
E150L	116		504A	1104	
E158	102A		53	2007	
E1704R	125	3032A			
E1717R	125	3032A			
E174					MPF970
E175					MPF970
E176					MPF971
E181	102A		53	2007	
E181A	102A		53	2007	
E181B	102A		53	2007	
E181C	102A		53	2007	
E181D	102A		53	2007	
E1852	142A	3062	ZD-12	563	
E185B121712	157	3747	232		
E1A	107	3039/316	11	2015	
E1M3	125	3032A	510,531	1114	
E1N3	125	3032A	510,531	1114	
E2	116	3017B/117	504A	1104	1N4003
E201		3112			2N5457
E202		3112			2N5458
E202(ELCOM)	156		512		
E203	466	3112			2N5459
E20C30			1N34AS	1123	
E21	116	3311	504A	1104	
E210		3444	20	2051	2N5484
E211	451		82	2032	2N5485
E21135	110MP	3709	1N60	1123(2)	
E212		3444	20	2051	2N5486
E213	159		82	2032	
E21430	109	3090		1123	
E21431	137A	3058	ZD-6.2	561	
E214B			2	2007	
E230		3112			2N5457
E231		3112			2N5458
E232	466S	3112			2N5459
E241	102A	3004	53	2007	
E24100	116	3081/125	504A	1104	
E24101	156	3081/125	512		
E24103	123A	3444	20	2051	
E24104	102A		53	2007	
E24105	103A	3835	59	2002	
E24106	102A	3004	53	2007	
E24107	131	3052	44	2006	
E2412	100	3005	1	2007	
E241A	102A		53	2007	
E241B	102A		53	2007	
E2427	101		8	2002	
E2428	101		8	2002	
E2429	101		8	2002	
E2430	123A	3444	20	2051	
E2431	123A	3444	20	2051	
E2434	108		86	2038	
E2435	108		86	2038	
E2436	123A	3444	20	2051	
E2438	160		245	2004	
E2439	160		245	2004	
E2440	160		245	2004	
E2441	128	3024	243	2030	
E2444	123A	3444	20	2051	
E2445	102A		53	2007	
E2447	103A		59	2002	
E2448	102A		53	2007	
E2449	128	3024	243	2030	
E2450	160	3006	245	2004	
E2451	126	3005	52	2024	
E2452	123A	3444	20	2051	
E2453	102A	3004	53	2007	
E2454	123A	3444	20	2051	
E2455	123A	3444	20	2051	
E2459	123A	3444	20	2051	
E2460	124	3021	12		
E2461	123A	3444	20	2051	
E2462	160		245	2004	
E2465	102A	3004	53	2007	
E2466	103A	3010	59	2002	
E2467	102A	3004	53	2007	
E2474	160		245	2004	
E2475	160		245	2004	
E2476	102A		53	2007	
E2477	160		245	2004	
E2478	160		245	2004	
E2479	160		245	2004	
E2480	102A		53	2007	
E2481	102A		53	2007	

Industry Standard No.	ECG	SK	GE	RS 276-	MOTOR.
E2482	102A	3123	53	2007	
E2484	110MP	3709		1123(2)	
E2486	137A	3058	ZD-6.2	561	
E2495	727	3071	IC-210		
E2496	152	3893	66	2048	
E2497	123A	3444	20	2051	
E2498	129	3025	244	2027	
E2499	123A	3444	20	2051	
E25C5	116		504A	1104	
E262	142A	3062	ZD-12	563	
E270					MPF970
E271					MPF970
E295ZZ01			1N34AS	1123	
E3	116	3017B/117	504A	1104	1N4004
E300	312	3116	FET-2	2035	2N5486
E3006	116		504A	1104	
E300L	116		504A	1104	
E302(ELCOM)	5802	9005		1143	
E304		3116			2N5486
E304(ELCOM)	5804			1144	
E305	451	3116			2N5485
E306(ELCOM)	5806	3848			
E308					U308
E309					U309
E30C60			1N34AS	1123	
E310					U310
E318-1	128		243	2030	
E4	158	3016	53	2007	1N4004
E4002	101	3861	8	2002	
E41	116	3017B/117	504A	1104	
E4676B	116	3017B/117	504A	1104	
E5	116	3017B/117	504A	1104	
E500L			504A	1104	
E570022-01	175	3131A/369	246	2020	
E6	116	3017B/117	504A	1104	1N4005
E629	108	3452	86	2038	
E650L	116		504A	1104	
E660	113A		6GC1		
E7441			504A	1104	
E750(ELCOM)	116		504A	1104	
E752(ELCOM)	116		504A	1104	
E756(ELCOM)	116		504A	1104	
E758(ELCOM)	125		510,531	1114	
E760(ELCOM)	125		510,531	1114	
E8	125	3033	510,531	1114	1N4006
E84	140A	3061	ZD-10	562	
E842			86	2038	
E843			86	2038	
E844			86	2038	
E9	125	3033	510,531	1114	
E9625	229	3018	61	2038	
EA-15X8517			18	2030	
EA0002	160	3006	245	2004	
EA0007	160	3006	245	2004	
EA0009	102A	3004	53	2007	
EA0013	108	3452	86	2038	
EA0015	116		504A	1104	
EA0016	116	3017B/117	504A	1104	
EA0031	116		504A	1104	
EA005	116	3311	504A	1104	
EA0053	160	3006		2004	
EA0081	128	3024	18	2030	
EA0086	129		244	2027	
EA0087	129	3025	244	2027	
EA0088			21	2034	
EA0090	128	3024	243	2030	
EA0091	108	3452	86	2038	
EA0092	123A	3444	20	2051	
EA0093	108	3452	86	2038	
EA0094	108	3452	86	2038	
EA0095	108	3452	86	2038	
EA010	116		504A	1104	
EA020	116		504A	1104	
EA030	116		504A	1104	
EA040	116	3031A	504A	1104	
EA050	116	3017B/117		1104	
EA060	116	3017B/117	504A	1104	
EA080	125	3032A	510,531	1114	
EA1-380			57	2017	
EA100	125		510,531	1114	
EA1072	116	3017B/117	504A	1104	
EA1080	123A	3444	20	2051	
EA1081	102A	3123	53	2007	
EA1082	121	3717	239	2006	
EA1123	109	3018		1123	
EA1128	123A	3444	20	2051	
EA1129	123A	3444	20	2051	
EA1135	123A	3444	20	2051	
EA1145	123A	3444	20	2051	
EA1318	142A	3062	ZD-12	563	
EA1337	160	3006	245	2004	
EA1338	160	3006	245	2004	
EA1339	160	3006	245	2004	
EA1340	160	3006	245	2004	
EA1341	121	3009	239	2006	
EA1342	160	3006	245	2004	
EA1343	108	3452	86	2038	
EA1344	123A	3444	20	2051	
EA1345	123A	3444	20	2051	
EA1346	102A	3004	53	2007	
EA1385		3126	90		
EA1405	177	3100/519	300	1122	
EA1406	123A	3444	20	2051	
EA1407	123A	3444	20	2051	
EA1408	123A	3444	20	2051	
EA1448	116	3031A	504A	1104	
EA1451	123A	3444	20	2051	
EA1452	123A	3444	20	2051	
EA1499	123A	3444	20	2051	
EA1549	128	3124/289	243	2030	
EA1562	108	3452	86	2038	
EA1563	108	3452	86	2038	
EA1564	123A	3444	20	2051	
EA1578	123A	3444	20	2051	
EA1581	123A	3444	20	2051	
EA15X1	123A	3444	20	2051	
EA15X10	121	3009	239	2006	
EA15X100	130	3027	14	2041	
EA15X101	123A	3444	20	2051	
EA15X102	128	3024	243	2030	
EA15X103	123A	3444	20	2051	
EA15X11	160	3006	245	2004	
EA15X111	123A	3444	20	2051	
EA15X112	123A	3444	20	2051	
EA15X113	108	3452	86	2038	
EA15X12	121	3009	239	2006	
EA15X121	175		246	2020	
EA15X123	181		75	2041	
EA15X124	180		74	2043	
EA15X13	160	3006	245	2004	
EA15X130	229	3018	61	2038	
EA15X131	108	3452	86	2038	
EA15X132	108	3452	86	2038	
EA15X133	126	3006/160	52	2024	
EA15X134	107	3018	11	2015	
EA15X135	107	3018	11	2015	
EA15X136	123A	3444	20	2051	
EA15X137	123A	3444	20	2051	
EA15X139	131	3052	44	2006	
EA15X14	116	3311	504A	1104	
EA15X140	126	3006/160	52	2024	
EA15X141	126	3006/160	52	2024	
EA15X142	123A	3444	20	2051	
EA15X143	123A	3444	20	2051	
EA15X144	128	3024	243	2030	
EA15X15	121	3009	239	2006	
EA15X152	199	3020/123	62	2010	
EA15X153	123A	3444	20	2051	
EA15X154	131	3052	44	2006	
EA15X157	123A	3444	20	2051	
EA15X160	186	3192	28	2017	
EA15X161	199	3122	62	2010	
EA15X162	123A	3444	20	2051	
EA15X163	123A	3444	20	2051	
EA15X164	102A	3004	53	2007	
EA15X165	312	3448	FET-1	2035	
EA15X167	123A	3444	20	2051	
EA15X168	123A	3444	20	2051	
EA15X169	312	3112	FET-1	2028	
EA15X173	121	3009	239	2006	
EA15X18	123A	3444	20	2051	
EA15X180	123A		61	2038	
EA15X185	294	3138/193A	82	2032	
EA15X189	123A	3018	20	2051	
EA15X19	102A	3004	53	2007	
EA15X190	229	3122	20	2051	
EA15X192	312	3448	FET-1	2035	
EA15X193	312	3448	FET-1	2035	
EA15X194	129	3114/290	244	2027	
EA15X2	102A	3004	53	2007	
EA15X20	123A	3444	20	2051	
EA15X203	102A	3004	53	2007	
EA15X207	102A	3004	53	2007	
EA15X212	102A	3004	53	2007	
EA15X213	199		20	2051	
EA15X22	123A	3444	20	2051	
EA15X23	102A	3004	53	2007	
EA15X233	159		82	2032	
EA15X237	123A	3124/289	20	2051	
EA15X238			FET-3	2036	
EA15X239	107	3018	20	2051	
EA15X24	123A	3444	20	2051	
EA15X240	123A	3018	20	2051	
EA15X241	199	3020/123	210	2051	
EA15X242	290A	3114/290	67	2023	
EA15X243	185	3191	58	2025	
EA15X244	184	3190	57	2017	
EA15X245	199	3020/123	62	2010	
EA15X246	324	3444/123A	20	2051	
EA15X247	172A	3156	64		
EA15X248	186	3192	28	2017	
EA15X249	289A	3024/128	243	2030	
EA15X25	102A	3004	53	2007	
EA15X250	293	3512	47	2030	
EA15X251	229		212	2010	
EA15X2522			47	2030	
EA15X256	289A	3122	268	2038	
EA15X257	158	3004/102A	53	2007	
EA15X258	199	3122	62	2010	
EA15X259	199	3122	62	2010	
EA15X26	121	3009	239	2006	
EA15X264	199	3124/289	62	2010	
EA15X266	172A	3156	64		
EA15X267	289A	3122	268	2038	
EA15X268	294	3114/290	48		
EA15X269	184		57	2017	
EA15X27	160		245	2004	
EA15X270	185		58	2025	
EA15X272	123A	3444	20	2051	
EA15X273	193		67	2023	
EA15X274	192		63	2030	
EA15X28	102A	3004	53	2007	
EA15X288	199	3245	62	2010	
EA15X29	160		245	2004	
EA15X3	102A	3004	53	2007	
EA15X30	160		245	2004	
EA15X31	123A	3444	20	2051	
EA15X3118			69	2049	
EA15X325	199	3122	61	2010	
EA15X326	158	3004/102A	53	2007	
EA15X327	152	3054/196	66	2048	
EA15X328	153	3083/197	69	2049	
EA15X33	121	3009	239	2006	
EA15X330	123A	3444	20	2051	
EA15X331	123A	3444	20	2051	
EA15X332	265	3860	64		
EA15X333	152	3054/196	66	2048	
EA15X334	153	3083/197	69	2049	
EA15X335	199	3124/289			
EA15X336	199	3122	211	2016	
EA15X338	229	3018			
EA15X339	229	3018			
EA15X340	199	3124/289			
EA15X341	159	3114/290			
EA15X343	289A	3124/289			
EA15X345	298	3450			
EA15X346	199	3124/289			
EA15X347	299	3137/192A			
EA15X349	192	3020/123	63	2030	
EA15X35	121	3009	239	2006	
EA15X350	229	3018	213	2038	
EA15X351	229	3018	60	2015	
EA15X352	199		62	2010	
EA15X353	199	3124/289	62	2010	
EA15X354	199	3124/289	62	2010	
EA15X355	199	3137/192A	20	2051	
EA15X356	278	3020/123	81	2051	
EA15X36	102A	3004	53	2007	
EA15X360	184	3024/128	268	2038	
EA15X361	199	3122	20	2051	
EA15X362	295		270		
EA15X363	235	3197	215		
EA15X364	189	3356	211	2051	
EA15X365	123AP		211	2016	
EA15X367	123A		211	2016	
EA15X37	123A	3444	20	2051	
EA15X370	123A	3018	20	2051	
EA15X371	123A	3020/123	20	2051	
EA15X372	235	3197	270		
EA15X373	123A	3038	210	2051	
EA15X374	229	3124/289	213	2038	
EA15X376	229		61	2038	
EA15X378	297	3124/289	210	2030	
EA15X379	123A		20	2051	
EA15X38	121	3009	239	2006	
EA15X380	306	3251	276		
EA15X381	235	3239/236	215		
EA15X383	306		336		

Industry Standard No.	ECG	SK	GE	RS 276-	MOTOR.
EA15X385	294	3114/290	48		
EA15X386	199	3899	85	2010	
EA15X393	229		61	2038	
EA15X394	312	3116	FET-2	2035	
EA15X395	290A	3114/290	269	2032	
EA15X396	294	3019	61	2038	
EA15X397	293		47	2030	
EA15X4	102A	3004	53	2007	
EA15X40	160	3006	245	2004	
EA15X400	312	3116	FET-2	2035	
EA15X401	312	3116	FET-2	2035	
EA15X402	222	3050/221	FET-4	2036	
EA15X404	199	3124/289	212	2010	
EA15X405	222	3050/221	FET-4	2036	
EA15X4064	107	3018	11	2015	
EA15X408	297	3122	81	2051	
EA15X41	160	3006	245	2004	
EA15X412	123A	3047	210		
EA15X413	293	3849	47	2030	
EA15X414	235	3197	215		
EA15X415			214	2038	
EA15X43	160	3006	245	2004	
EA15X437	229		62	2051	
EA15X44	123A	3444	20	2051	
EA15X441			211	2051	
EA15X446	312		FET-1	2028	
EA15X45	123A	3444	20	2051	
EA15X450	229	3132			
EA15X4531	192	3137	63	2030	
EA15X4631		3138	223		
EA15X48	108	3452	86	2038	
EA15X49	108	3452	86	2038	
EA15X5	160	3124/289	245	2004	
EA15X50	108	3452	86	2038	
EA15X51	108	3452	86	2038	
EA15X52	123A	3444	20	2051	
EA15X53	121	3009	239	2006	
EA15X54	108	3452	86	2038	
EA15X55	108	3452	86	2038	
EA15X56	123A	3444	20	2051	
EA15X57	128		243	2030	
EA15X58	123A	3444	20	2051	
EA15X59	123A	3444	20	2051	
EA15X6	105	3012	4		
EA15X60			20	2051	
EA15X63	123A	3444	20	2051	
EA15X66	160		245	2004	
EA15X67	102A	3004	53	2007	
EA15X68	123A	3444	20	2051	
EA15X6840	102A	3004	53	2007	
EA15X69	106	3118	21	2034	
EA15X7	102A	3004	53	2007	
EA15X70	106	3118	21	2034	
EA15X71	106	3118	21	2034	
EA15X7112	123A	3444	20	2051	
EA15X7113	108		20	2051	
EA15X7114	108	3018			
EA15X7115	123A	3444	20	2051	
EA15X7117	107		17	2051	
EA15X7118	123A	3122	20	2051	
EA15X7119	123A	3444	20	2051	
EA15X7120	123A	3018	20	2051	
EA15X7121		3054	66	2048	
EA15X7125	107		11	2015	
EA15X7140	108	3018	11	2015	
EA15X7141	107	3452/108	86	2038	
EA15X7173	319P	3018	283	2016	
EA15X7174	319P	3018	283	2016	
EA15X7175	123A	3444	20	2051	
EA15X7176	123A	3444	20	2051	
EA15X7177	108	3452	86	2038	
EA15X7178	319P	3018	283	2016	
EA15X7179	319P	3018	283	2016	
EA15X72	123A	3444	20	2051	
EA15X7215	107	3018	11	2015	
EA15X7228	108	3452	86	2038	
EA15X7231	107	3293	11	2015	
EA15X7232	123A	3444	20	2051	
EA15X7233	229	3018	61	2038	
EA15X7234	229	3018	61	2038	
EA15X7235	229	3018	61	2038	
EA15X7236	229	3018	61	2038	
EA15X7243	108	3452	86	2038	
EA15X7244	229	3018	61	2038	
EA15X7245	199	3124/289	62	2010	
EA15X7262	123AP		20	2051	
EA15X7263	108	3452	86	2038	
EA15X7264	107	3018	11	2015	
EA15X73	123A	3444	20	2051	
EA15X75	123A	3444	20	2051	
EA15X7514	123A	3444	20	2051	
EA15X7517	123A	3444	20	2051	
EA15X7519	192	3137	63	2030	
EA15X7583	199	3124/289	62	2010	
EA15X7585	108	3122			
EA15X7586	123A	3444	20	2051	
EA15X7587	108	3452	86	2038	
EA15X7588	123AP	3444/123A	20	2051	
EA15X7589	123A	3444	20	2051	
EA15X7590	123A	3444			
EA15X7592	294	3024/128	48		
EA15X76	123A	3444	20	2051	
EA15X7635	128	3024	212	2051	
EA15X7638	123A	3444	20	2051	
EA15X7639	294	3024/128	212	2010	
EA15X7643	123A	3122	20	2051	
EA15X77	123A	3444	20	2051	
EA15X7722	108	3452	86	2038	
EA15X8	102A	3004	53	2007	
EA15X8118	153	3083/197	69	2049	
EA15X8119	152	3054/196	66	2048	
EA15X8122	189		20	2051	
EA15X8130	153	3083/197	69	2049	
EA15X83	123A	3444	20	2051	
EA15X84	123A	3444	20	2051	
EA15X8442	102A	3004	53	2007	
EA15X8443	103A	3835	59	2002	
EA15X8444	102A		53	2007	
EA15X85	123A	3444	20	2051	
EA15X8502	123A	3020/123	20	2051	
EA15X8511	123A	3444	20	2051	
EA15X8517	293	3854/123AP	47	2030	
EA15X8518	123AP	3444/123A	20	2051	
EA15X8521	297	3137/192A	47	2030	
EA15X8522	298	3138/193A	48		
EA15X8524	298	3715	48		
EA15X8529	123A	3122	10	2051	
EA15X8589	108	3018	86	2038	
EA15X86	123A	3444	20	2051	
EA15X8601	123AP	3083/197	61	2038	
EA15X8602	123AP	3054/196	66	2048	
EA15X8605	185	3083/197	58	2025	
EA15X8608	108	3452	86	2038	
EA15X8609	108		86	2038	
EA15X8610	108	3452	86	2038	
EA15X88	121	3009	239	2006	
EA15X89	123A	3444	20	2051	
EA15X9	123A	3444	20	2051	
EA15X90			20	2051	
EA15X91	123A	3444	20	2051	
EA15X94	108	3452	86	2038	
EA15X96	123A	3444	20	2051	
EA15X98	123A	3444	20	2051	
EA15X99	152	3893	66	2048	
EA1628	123A	3444	20	2051	
EA1629	123A	3444	20	2051	
EA1630	123A	3444	20	2051	
EA1638	123A	3444	20	2051	
EA1661	177	3100/519	300	1122	
EA1672	116	3311	504A	1104	
EA1684	128	3024	243	2030	
EA1695	123A	3444	20	2051	
EA1696	123A	3444	20	2051	
EA1697	123A	3444	20	2051	
EA1698	128	3124/289	243	2030	
EA16X1	109	3087		1123	
EA16X101	177	3100/519	300	1122	
EA16X105	177	3100/519			
EA16X106	612	3325			
EA16X109	140A	3061			
EA16X11	109	3087		1123	
EA16X110	177	3100/519	300	1122	
EA16X117	613	3126	90		
EA16X118	136A		ZD-5.6	561	
EA16X122	519		300	1122	
EA16X123	5071A	3334	ZD-6.8	561	
EA16X124	145A	3063	ZD-15	564	
EA16X125	138A		ZD-7.5		
EA16X126	5069A		ZD-4.7		
EA16X127	614	3126	90		
EA16X13			504A	1104	
EA16X134	177	3100/519	300	1122	
EA16X135	519	3100	514	1122	
EA16X136	135A	3056	ZD-5.0		
EA16X140	109	3087	1N34AS	1123	
EA16X146	177	3100/519	300	1122	
EA16X149	156	3311	504A	1104	
EA16X150	140A	3061	ZD-10	562	
EA16X152	519		514	1122	
EA16X157	140A	3061	ZD-10	562	
EA16X162	142A	3062	ZD-12	563	
EA16X163	5012A	3778			
EA16X166	610	3126	90		
EA16X171	177		300	1122	
EA16X177	614	3327	90		
EA16X19		3126	90		
EA16X192	611	3324			
EA16X2	116	3031A	504A	1104	
EA16X20	177		300	1122	
EA16X21	116	3311	504A	1104	
EA16X22	109	3087		1123	
EA16X27	109	3088		1123	
EA16X28		3126	90		
EA16X29	142A	3062	ZD-12	563	
EA16X30	116	3311	504A	1104	
EA16X31		3062	90		
EA16X33	116	3031A	504A	1104	
EA16X34	116	3031A	504A	1104	
EA16X35	1003	3072/712	IC-43		
EA16X38	5080A	3336	ZD-22		
EA16X39	177	3100/519	300	1122	
EA16X4	141A	3092	ZD-11.5		
EA16X402	5073A		ZD-8.7	562	
EA16X406	135A	3056			
EA16X415	139A	3060			
EA16X429	519	3175			
EA16X430	116	3311			
EA16X48	109	3088	1N60	1123	
EA16X49	177	3100/519	300	1122	
EA16X5	109	3088		1123	
EA16X55	116	3311	504A	1104	
EA16X6	141A	3062/142A	ZD-11.5		
EA16X60	177	3100/519	300	1122	
EA16X61	177	3100/519	300	1122	
EA16X62	139A	3060	ZD-9.1	562	
EA16X68	601	3100/519	300	1122	
EA16X69	601	3100/519	300	1122	
EA16X71	116	3311	504A	1104	
EA16X73	601	3463	300	1122	
EA16X74	139A		ZD-9.1	562	
EA16X75	177		300	1122	
EA16X77	142A		ZD-12	563	
EA16X8	116	3016	504A	1104	
EA16X80	137A		ZD-6.2	561	
EA16X81	143A	3750	ZD-13	563	
EA16X82		3060	ZD-9.1	562	
EA16X84	177	3100/519	300	1122	
EA16X88	138A	3059	ZD-7.5		
EA16X89	5067A		ZD-3.9		
EA16X9	109	3087		1123	
EA16X90	614	3126			
EA16X91	519	3100			
EA16X92	116	3100/519	504A	1104	
EA16X95	177	3175			
EA16X97	109	3088		1123	
EA1700	121	3009	239	2006	
EA1703	123A	3444	20	2051	
EA1716	123A	3444	20	2051	
EA1718	123A	3444	20	2051	
EA1733	108	3452	86	2038	
EA1735	123A	3444	20	2051	
EA1740	130	3027	14	2041	
EA1760	717		IC-209		
EA1778			20	2051	
EA1793	108	3452	86	2038	
EA1872	123A	3444	20	2051	
EA1873	128	3020/123	243	2030	
EA1SY11			50	2004	
EA2131	107	3018	11	2015	
EA2132	107	3018	11	2015	
EA2133	126	3006/160	52	2024	
EA2134	102A	3003	53	2007	
EA2135	102A	3004	53	2007	
EA2136	158	3004/102A	53	2007	
EA2137	110MP	3088	1N60	1123(2)	
EA2138		3126	90		
EA2140	116	3031A	504A	1104	
EA2176	102A	3004	53	2007	
EA2271	123A	3444	20	2051	
EA2308/8308					MCM68A308
EA2316E/8316E					MCM68316E
EA2429	199	3018	62	2010	
EA2488	186	3192	28	2017	
EA2489	123A	3444	20	2051	
EA2490	123A	3444	20	2051	
EA2491	126	3006/160	52	2024	
EA2493	108	3452	86	2038	
EA2494	107	3018	11	2015	

Industry Standard No.	ECG	SK	GE	RS 276-	MOTOR.
EA2495	108	3452	86	2038	
EA2496	107	3018	11	2015	
EA2497	126	3006/160	52	2024	
EA2498	126	3006/160	52	2024	
EA2499	116	3031A	504A	1104	
EA2500	142A		ZD-12	563	
EA2501	116	3031A	504A	1104	
EA2502	109	3088		1123	
EA2503		3126	90		
EA2600	108	3452	86	2038	
EA2601	108	3452	86	2038	
EA2602	108	3452	86	2038	
EA2603	108	3452	86	2038	
EA2604	108	3452	86	2038	
EA2605	108	3452	86	2038	
EA2606	110MP	3088	1N60	1123(2)	
EA2607	177	3100/519	300	1122	
EA2608	136A	3100/519	ZD-5.6	561	
EA2708					MCM2708
EA2716					MCM2716
EA2738	199	3122	62	2010	
EA2739	123A	3444	20	2051	
EA2740	123A	3444	212	2010	
EA2741	116	3311	504A	1104	
EA2770	123A	3444	20	2051	
EA2770(N)			10	2051	
EA2771	199	3124/289	62	2010	
EA2812	107	3018	11	2015	
EA3127	109	3088		1123	
EA3149	123A	3444	20	2051	
EA3211		3122	62	2010	
EA3278	312	3112	FET-1	2028	
EA3282	1003	3288	IC-43		
EA33X0367	1003		IC-43		
EA33X8351	1027		721		
EA33X8356	1005	3723	IC-42		
EA33X8363	1180	3888	IC-41		
EA33X8364	799	3238	IC-34		
EA33X8367	1003	3288	IC-43		
EA33X8368	1036		IC-168		
EA33X8372	1142	3485	IC-128		
EA33X8378	722	3161			
EA33X8380	1055	3494			
EA33X8381	1181	3706			
EA33X8385	7473			1803	
EA33X8388	1167	3732			
EA33X8389	1155	3231	IC-179	705	
EA33X8394	1192	3445			
EA33X8396	1155	3231	IC-179	705	
EA33X8398	1193	3701			
EA33X8500	1100	3223	IC-92		
EA33X8508	1249	3963			
EA3406	108	3452	86	2038	
EA3447	177	3100/519	300	1122	
EA3674	186	3192	28	2017	
EA3713	229	3018	60	2015	
EA3714	159	3114/290	82	2032	
EA3715	153		69	2049	
EA3716	152	3893	66	2048	
EA3717		3126	90		
EA3718	109	3088		1123	
EA3719	142A	3062	ZD-12	563	
EA3763	199	3018	62	2010	
EA3827	116	3311	504A	1104	
EA3866	139A	3060	ZD-9.1	562	
EA3989	116	3311	504A	1104	
EA3990	192	3122	63	2030	
EA4025	199	3124/289	62	2010	
EA4055	152		28	2017	
EA4085	152	3893	28	2017	
EA4112	123A	3444	20	2051	
EA5711	116		504A	1104	
EA57X1	116	3016	504A	1104	
EA57X10	116	3016	504A	1104	
EA57X11	116	3016	504A	1104	
EA57X14	116	3311	504A	1104	
EA57X3	116	3016	504A	1104	
EA57X8	116	3016	504A	1104	
EA75X1	116	3017B/117	504A	1104	
EA8332					MCM68A332
EAI-380	184		57	2017	
EB-134144	194		220		
EB0001	102A	3004	53	2007	
EB0003	102A	3004	53	2007	
EB1(ELCOM)	166			1152	
EB10(ELCOM)	167			1172	
EB11(ELCOM)	168			1173	
EB2(ELCOM)	168			1173	
EB3(ELCOM)	166			1152	
EB4(ELCOM)	167			1172	
EB6(ELCOM)	168			1173	
EB9(ELCOM)	166			1152	
EC100	125	3051/156	510,531	1114	
EC103A	5402	3638	C103A		MCR104
EC103B	5404	3627	C103B		MCR120
EC103C					MCR100-5
EC103D	5405	3951	C203D		MCR100-6
EC103M	5406	3952			
EC103Y	5400	3950	C103Y		MCR102
EC106A1	5454	6754			
EC106A2	5454	6754			
EC106B1	5455	3597			
EC106B2	5455	3597			
EC106F1	5453	6753			
EC106F2	5453	6753			
EC106Y1	5452	6752			
EC106Y2	5452	6752			
EC107A1	5413	3954/5414			
EC107A2	5413	3954/5414			
EC107B1	5414	3954			
EC107B2	5414	3954			
EC107F1	5412	3953			
EC107F2	5412	3953			
EC107Y1	5411	3953/5412			
EC107Y2	5411	3953/5412			
EC401	116	3311	504A	1104	
EC402	116	3311	504A	1104	
EC961	130	3027	14	2041	
ECG100		3721	1	2007	
ECG1003		3288	IC-43		
ECG1004		3365	IC-36		
ECG1005		3723	IC-42		
ECG1006		3358	IC-38		
ECG1009		3499	IC-77		
ECG101		3861	8	2002	
ECG1010		3376	IC-37		
ECG1011		3492	IC-79		
ECG101L				2002	
ECG102		3722	2	2007	
ECG1024		3152	720		
ECG1025		3155	722		
ECG1027		3153	721		
ECG1028		3436	724		
ECG1029		3368	IC-162		

Industry Standard No.	ECG	SK	GE	RS 276-	MOTOR.
ECG102A		3004	52	2007	
ECG103		3862	8	2002	
ECG1035		3369	IC-167		
ECG1036		3370	IC-168		
ECG1037		3371	IC-170		
ECG1039		3366	IC-159		
ECG103A		3835	59	2002	
ECG104		3719	16	2006	
ECG1040		3362	IC-161		
ECG1041		3361	IC-160		
ECG1043		3363	IC-169		
ECG1046		3471	IC-118		
ECG1049		3470	IC-121		
ECG104MP		3720	16(TWO)	2006(2)	
ECG105		3012	4		
ECG1050		3475	IC-133		
ECG1052		3249	IC-135		
ECG1054		3457	IC-45		
ECG1055		3494	IC-47		
ECG1056		3458	IC-48		
ECG1058		3459	IC-49		
ECG106		3984	21	2034	
ECG1060		3460	IC-50		
ECG1061		3228	IC-51		
ECG1062		3235	IC-52		
ECG1069		3227	IC-64		
ECG107		3293	11	2015	
ECG1071		3226	IC-67		
ECG1072		3295	IC-59		
ECG1073		3496	IC-63		
ECG1074		3495	IC-60		
ECG1078		3498	IC-136		
ECG108		3452	11	2038	
ECG1080		3284	IC-98		
ECG1081A		3474	IC-197		
ECG1082		3461	IC-140		
ECG1085		3476	IC-104		
ECG1087		3477	IC-103		
ECG109		3090	1N34AS	1123	
ECG1090		3291	723		
ECG1092		3472	IC-130		
ECG1096		3703	IC-312		
ECG110		3709		1123(2)	
ECG1100		3223	IC-92		
ECG1101		3283	IC-95		
ECG1102		3224	IC-93		
ECG1103		3281	IC-94		
ECG1104		3225	IC-91		
ECG1105		3285	IC-101		
ECG1109		3711	IC-99		
ECG1115		3184	IC-278		
ECG1115A		3917	IC-278		
ECG1116		3969	IC-279		
ECG112		3089	1N82A		
ECG1123		3743	IC-281		
ECG1128		3488	IC-105		
ECG113		3119	6GC1		
ECG1130		3478	IC-111		
ECG1131		3286	IC-109		
ECG1132		3287	IC-110		
ECG1133		3490	IC-107		
ECG1134		3489	IC-106		
ECG1135		3876	IC-314		
ECG113A		9001	6GC1		
ECG114		3120	6GD1	1104	
ECG1140		3473	IC-138		
ECG1142		3485	IC-128		
ECG115		3121	6GX1		
ECG1150		3890	IC-286		
ECG1153		3282	IC-182		
ECG1154		3230	IC-287		
ECG1155		3231	IC-179	705	
ECG1158		3289	IC-71		
ECG1159		3290	IC-72		
ECG116		3313	504A	1104	
ECG1160		3243	IC-196		
ECG1161		3968	IC-300		
ECG1162		3072	IC-2		
ECG1164		3727	IC-301		
ECG1165		3827	IC-315		
ECG1166		3827	IC-316	704	
ECG1167		3732	IC-317		
ECG1168		3728	IC-66		
ECG117		3017B	504A	1104	
ECG1170		3874	IC-65		
ECG1171		3565	IC-173		
ECG1173		3729	IC-302		
ECG1179		3737	IC-75		
ECG118		3066	CR-1		
ECG1180		3888	IC-80		
ECG1181		3706	IC-61		
ECG1184		3486	IC-137		
ECG1185		3468	IC-139		
ECG1186		3168	IC-175		
ECG1187		3742	IC-176		
ECG119		3109	CR-2		
ECG120		3110	CR-3		
ECG1205		3482	IC-178		
ECG121		3717	76		
ECG1213		3704	IC-150		
ECG1215		3738	IC-74		
ECG1217		3700	IC-76		
ECG121MP		3718	76(2)	2006(2)	
ECG122		3942	MR-4		
ECG1223		3493	IC-295		
ECG1227		3762	IC-70		
ECG123		3020	20	2051	
ECG1232		3852		703	
ECG1234		3487	IC-181		
ECG1236		3072	IC-114		
ECG1237		3707	IC-154		
ECG123A		3444	20	2051	
ECG123AP		3854	20	2051	
ECG124		3021	12		
ECG1242		3483	IC-291		
ECG1243		3731	IC-113		
ECG1248		3497	IC-292		
ECG125		3081	510	1114	
ECG126			52	2024	
ECG127		3764	25		
ECG128		3024	18	2030	
ECG129		3025	244	2027	
ECG130		3027	14	2041	
ECG130MP		3029	15MP	2041(2)	
ECG131		3198	44	2006	
ECG131MP		3840	44(TWO)	2006(2)	
ECG132		3834	FET-1	2035	
ECG133			FET-2	2028	
ECG134		3055	ZD-3.6		
ECG134A		3055	ZD-3.6		
ECG135		3056	ZD5.1		
ECG135A		3056	ZD-5.0		
ECG136		3057	ZD-5.6	561	
ECG136A		3057	ZD-5.6	561	

Industry Standard No.	ECG	SK	GE	RS 276-	MOTOR.
ECG137		3058	ZD-6.2	561	
ECG137A		3058	ZD-6.2	561	
ECG138		3059	ZD-7.5		
ECG138A		3059	ZD-7.5		
ECG139		3060	ZD-9.1	562	
ECG139A		3060	ZD-9.1	562	
ECG140		3061	ZD-10	562	
ECG140A		3061	ZD-10	562	
ECG141		3092	ZD-11.5		
ECG141A		3092	ZD-11.5		
ECG142		3062	ZD-12	563	
ECG142A		3062	ZD-12	563	
ECG143		3750	ZD-13	563	
ECG143A		3750	ZD-13	563	
ECG144		3094	ZD-14	564	
ECG144A		3094	ZD-14	564	
ECG145		3063	ZD-15	564	
ECG145A		3063	ZD-15	564	
ECG146		3064	ZD-27		
ECG146A		3064	ZD-27		
ECG147		3095	ZD-33		
ECG147A		3095	ZD-33		
ECG148		3096	ZD-55		
ECG148A		3096	ZD-55		
ECG149		3097	ZD-62		
ECG149A		3097	ZD-62		
ECG150		3098	ZD-82		
ECG150A		3098	ZD-82		
ECG151		3099	ZD-110		
ECG151A		3099	ZD-110		
ECG152		3893	66	2048	
ECG153		3274	69	2049	
ECG154		3044	40	2012	
ECG155		3839	43		
ECG156		3051	512		
ECG157		3747	232		
ECG158		3004	53	2007	
ECG159		3466	82	2032	
ECG160		3006	52	2004	
ECG161		3716	39	2015	
ECG162		3559	35		
ECG163		3439	36		
ECG163A		3439	36		
ECG164		3133	37		
ECG165		3115	38		
ECG166		9075	BR-206	1152	
ECG167		3647	BR-206	1172	
ECG168		3648	BR-1000	1173	
ECG169		3678	BR-1000		
ECG170		3649	BR-1000		
ECG171		3201	27		
ECG172		3156	64		
ECG173A		3999	305		
ECG175		3261	246	2020	
ECG176		3845	80		
ECG177		9091	300	1122	
ECG178MP				1122(2)	
ECG179		3642	76		
ECG180		3437	74	2043	
ECG181		3535	75	2041	
ECG182		3188A	55		
ECG183		3189A	56	2027	
ECG184		3190	57	2017	
ECG185		3191	58	2025	
ECG186		3192	28	2017	
ECG186A		3357	247	2052	
ECG187		3193	29	2025	
ECG187A		9076	248		
ECG188		3199	217	2018	
ECG189		3200	218	2026	
ECG190		3232	217		
ECG191		3232	249		
ECG192			63	2030	
ECG192-1				2030	
ECG193			67	2023	
ECG193-1				2023	
ECG194		3275	220		
ECG195		3048	46		
ECG195A		3765	46		
ECG196		3054	241	2020	
ECG197		3083	250	2027	
ECG198		3220	251		
ECG199		3245	62	2010	
ECG209		KH3403	A-122		
ECG210		3202	252	2018	
ECG211		3203	253	2026	
ECG218		3625	234		
ECG219		3173	74	2043	
ECG220		3990	FET-3	2036	
ECG221		3050	FET-4		
ECG222		3065	FET-4	2036	
ECG223			255	2041	
ECG224		3049	46		
ECG225		3045	256		
ECG226		3082	49	2025	
ECG226MP		3086	49(TWO)	2025(2)	
ECG229		3246	61	2038	
ECG230		3042	700		
ECG231		3857	700		
ECG232		3241	258		
ECG233		3854	210	2009	
ECG234		3247	65	2050	
ECG235		3197	215		
ECG236		3239	216	2053	
ECG237		3299	46		
ECG238		3710	259		
ECG241		3188A		2020	
ECG242		3189A	58	2027	
ECG265		3860	D4004		
ECG278		3218	261		
ECG280		3297	262	2041	
ECG280MP		3360	262MP	2041(2)	
ECG281		3359	263	2043	
ECG282		3748	264		
ECG283		3467	36		
ECG284		3836	265	2047	
ECG285		3846	266		
ECG286		3194	267		
ECG287		3433	222		
ECG288		3434	221		
ECG289		3124	89	2038	
ECG289MP			268MP	2038(2)	
ECG290		3114	82	2032	
ECG293		3849	47	2030	
ECG293MP			47MP	2030(2)	
ECG294		3841	48		
ECG295		3253	270	2030	
ECG297		3449	222,271	2030	
ECG297MP			271MP	2030	
ECG298		3450	223		
ECG299		3298	236		
ECG300		3464	273		
ECG302		3252	275		
ECG306		3251	276		

Industry Standard No.	ECG	SK	GE	RS 276-	MOTOR.
ECG3060				067	
ECG307		3203	274		
ECG311		3195	277		
ECG312			FET-2	2035	
ECG313			39	2016	
ECG315		3250	279		
ECG316		3039	280		
ECG319			283	2016	
ECG321		3844	277		
ECG322		3252	337		
ECG326		3746		2037	
ECG328		3895	265		
ECG329		3048	46		
ECG390		3958		2020	
ECG392		3960		2029	
ECG4000		4000	4000		
ECG4001			4001	2401	
ECG4001B		4001	4001	2401	
ECG4002B		4002	4002		
ECG4007B		4007	4007		
ECG4011			4011	2411	
ECG4011B		4011	4011	2411	
ECG4012			4012	2412	
ECG4012B		4012	4012	2412	
ECG4013			4013	2413	
ECG4013B		4013	4013	2413	
ECG4015B		4015	4015		
ECG4016B		4016	4016		
ECG4017			4017	2417	
ECG4017B		4017	4017	2417	
ECG4019B		4019	4019		
ECG4020			4020	2420	
ECG4020B		4020	4020	2420	
ECG4021			4021	2421	
ECG4021B		4021	4021	2421	
ECG4023			4023	2423	
ECG4023B		4023	4023	2423	
ECG4024B		4024	4024		
ECG4025B		4025	4025		
ECG4027			4027	2427	
ECG4027B		4027	4027	2427	
ECG4028				2428	
ECG4028B		4028		2428	
ECG4030B		4030	4030		
ECG4040B		4040	4040		
ECG4042B		4042	4042		
ECG4049			4049	2449	
ECG4049B		4049	4049	2449	
ECG4050			4050	2450	
ECG4050B		4050	4050	2450	
ECG4051			4051	2451	
ECG4051B		4051	4051	2451	
ECG4052B		4052	4052		
ECG4055B		4055	4055		
ECG4066B		4066		2466	
ECG4081B		4081	4081		
ECG4511B		4511		2447	
ECG4518			4518	2490	
ECG4518B		4518	4518	2490	
ECG500		3304	517		
ECG5005A		3771	ZD-3.3		
ECG5006A		3772	ZD-3.6		
ECG5007A		3773	ZD-3.9		
ECG5008A		3774	ZD-4.3		
ECG5009A		3775	ZD-4.7		
ECG500A		3304	527		
ECG501		3069	520		
ECG5010A		3776	ZD-5.1		
ECG5011			ZD-5.6	561	
ECG5011A		3777	ZD-5.6	561	
ECG5012			ZD-6.0	561	
ECG5012A		3778	ZD-6.0	561	
ECG5013			ZD-6.2	561	
ECG5013A		3779	ZD-6.2	561	
ECG5014			ZD-6.8	561	
ECG5014A		3780	ZD-6.8	561	
ECG5015A		3781	ZD-7.5		
ECG5016			ZD-8.2	562	
ECG5016A		3782	ZD-8.2	562	
ECG5017			ZD-9.1	562	
ECG5017A		3783		562	
ECG5018			ZD-9.1	562	
ECG5018A		3784	ZD-9.1	562	
ECG5019			ZD-10.0	562	
ECG5019A		3785	ZD-10	562	
ECG501A		3069	520		
ECG502		3067	CR-4		
ECG5020			ZD-11.0	563	
ECG5020A		3786	ZD-11	563	
ECG5021			ZD-12.0	563	
ECG5021A		3787	ZD-12.0	563	
ECG5022			ZD-13	563	
ECG5022A		3788	ZD-13	563	
ECG5023			ZD-14	564	
ECG5023A		3789	ZD-14	564	
ECG5024			ZD-15.0	564	
ECG5024A		3790	ZD-15	564	
ECG5025			ZD-16	564	
ECG5025A		3791	ZD-16	564	
ECG5026A		3792	ZD-17		
ECG5027A		3793	ZD-18		
ECG5028A		3794	ZD-19		
ECG5029A		3795	ZD-20		
ECG503		3068	CR-5		
ECG5030A		3796	ZD-22		
ECG5031A		3797	ZD-24		
ECG5032A		3798	IC-25		
ECG5033A		3799	ZD-27		
ECG5034A		3800	ZD-28		
ECG5035A		3801	ZD-30		
ECG5036A		3802	ZD-33		
ECG5037A		3803	ZD-36		
ECG5038A		3804	ZD-39		
ECG5039A		3805	ZD-43		
ECG504		3108	CR-6		
ECG5040A		3806	ZD-47		
ECG5041A		3807	ZD-51		
ECG5042A		3808	ZD-56		
ECG5043A		3809	ZD-60		
ECG5044A		3810	ZD-62		
ECG5045A		3811	ZD-68		
ECG5046A		3812	ZD-75		
ECG5048A		3814	ZD-87		
ECG5049A		3815	ZD-91		
ECG505		3108	CR-7		
ECG5050A		3816	ZD-100		
ECG5051A		3817	ZD-110		
ECG5052A		3818	ZD-120		
ECG5053A		3819	ZD-130		
ECG5054A		3820	ZD-140		
ECG5055A		3821	ZD-150		
ECG5056A		3822	ZD-160		
ECG5057A		3823	ZD-170		
ECG5058A		3824	ZD-180		

Industry Standard No.	ECG	SK	GE	RS 276-	MOTOR.
ECG5059A		3825	ZD-190		
ECG506		3998	511	1114	
ECG5060A		3826	ZD-200		
ECG5066		3330	ZD-3.3		
ECG5066A		3330	ZD-3.3		
ECG5067		333105067A	ZD-3.9		
ECG5067A		3331	ZD-3.9		
ECG5068		3332	ZD-4.3		
ECG5068A		3332	ZD-4.3		
ECG5069		3333	ZD-4.7		
ECG5069A		3333	ZD-4.7		
ECG507		3315	A14F		
ECG5070		9021	ZD-6.0	561	
ECG5070A		9021	ZD-6.0	561	
ECG5071		3334	ZD-6.8	561	
ECG5071A		3334	ZD-6.8	561	
ECG5072		3136	ZD-8.2	562	
ECG5072A		3136	ZD-8.2	562	
ECG5073		3749	ZD-8.7	562	
ECG5073A		3749	ZD-8.7	562	
ECG5074		3139	ZD-11	563	
ECG5074A		3139	ZD-11	563	
ECG5075		3751	ZD-16	564	
ECG5075A		3751	ZD-16	564	
ECG5076A		9022	ZD-17		
ECG5077		3752	ZD-18		
ECG5077A		3752	ZD-18		
ECG5078A		9023	ZD-19		
ECG5079		3335	ZD-20		
ECG5079A		3335	ZD-20		
ECG508		3756	SS-3A3		
ECG508/R-3A3		3756	SS-3A3		
ECG5080		3336	ZD-22		
ECG5080A		3336	ZD-22		
ECG5081		3151	ZD-24		
ECG5081A		3151	ZD-24		
ECG5082		3753	ZD-25		
ECG5082A		3753	ZD-25		
ECG5083		3754	ZD-28		
ECG5083A		3754	ZD-28		
ECG5084		3755	ZD-30		
ECG5084A		3755	ZD-30		
ECG5085		3337	ZD-36		
ECG5085A		3337	ZD-36		
ECG5086		3338	ZD-39		
ECG5086A		3338	ZD-39		
ECG5087		3339	ZD-43		
ECG5087A		3339	ZD-43		
ECG5088		3340	ZD-47		
ECG5088A		3340	ZD-47		
ECG5089		3341	ZD-51		
ECG5089A		3341	ZD-51		
ECG509		3757	SS-3AT2		
ECG509/R-3AT2		3757	SS-3AT2		
ECG5090		3342	ZD-56		
ECG5090A		3342	ZD-56		
ECG5092		3343	ZD-68		
ECG5092A		3343	ZD-68		
ECG5093		3344	ZD-75		
ECG5093A		3344	ZD-75		
ECG5094A		9024	ZD-87		
ECG5095		3345	ZD-91		
ECG5095A		3345	ZD-91		
ECG5096		3346	ZD-100		
ECG5096A		3346	ZD-100		
ECG5097A		3347	ZD-120		
ECG5098		3348	ZD-130		
ECG5098A		3348	ZD-130		
ECG5099		3349	ZD-140		
ECG5099A		3349	ZD-140		
ECG510		3758	SS-3DB3		
ECG510/R-3DB3		3758	SS-3DB3		
ECG5100A		3350	ZD-150		
ECG5101		3351	ZD-160		
ECG5101A		3351	ZD-160		
ECG5102		3352	ZD-170		
ECG5102A		3352	ZD-170		
ECG5103		3353	ZD-180		
ECG5103A		3353	ZD-180		
ECG5104		3354	ZD-190		
ECG5104A		3354	ZD-190		
ECG5105		3355	ZD-200		
ECG5105A		3355	ZD-200		
ECG511		3759	SS-2AV2		
ECG511/R-2AV2		3759	SS-2AV2		
ECG5111		3377	5ZD-3.3		
ECG5111A		3377	5ZD-3.3		
ECG5112		3378	5ZD-3.6		
ECG5112A		3378	5ZD-3.6		
ECG5113		3379	5ZD-3.9		
ECG5113A		3379	5ZD-3.9		
ECG5114		3380	5ZD-4.3		
ECG5114A		3380	5ZD-4.3		
ECG5115		3381	5ZD-4.7		
ECG5115A		3381	5ZD-4.7		
ECG5116		3382	5ZD-5.1		
ECG5116A		3382	5ZD-5.1		
ECG5117		3383	5ZD-5.6		
ECG5117A		3383	5ZD-5.6		
ECG5118		3384	5ZD-6.0		
ECG5118A		3384	5ZD-6.0		
ECG5119		3385	5ZD-6.2		
ECG5119A		3385	5ZD-6.2		
ECG512		3760	SS-6DW4		
ECG512/R-6DW4		3760	SS-DDW4		
ECG5120		3386	5ZD-6.8		
ECG5120A		3386	5ZD-6.8		
ECG5121		3387	5ZD-7.5		
ECG5121A		3387	5ZD-7.5		
ECG5122		3388	5ZD-8.2		
ECG5122A		3388	5ZD-8.2		
ECG5123		3389	5ZD-8.7		
ECG5123A		3389	5ZD-8.7		
ECG5124		3390	5ZD-9.1		
ECG5124A		3390	5ZD-9.1		
ECG5125		3391	5ZD-10		
ECG5125A		3391	5ZD-10		
ECG5126		3392	5ZD-11		
ECG5126A		3392	5ZD-11		
ECG5127		3393	5ZD-12		
ECG5127A		3393	5ZD-12		
ECG5128		3394	5ZD-13		
ECG5128A		3394	5ZD-13		
ECG5129		3395	5ZD-14		
ECG5129A		3395	5ZD-14		
ECG513		3443	513		
ECG5130		3396	5ZD-15		
ECG5130A		3396	5ZD-15		
ECG5131		3397	5ZD-16		
ECG5131A		3397	5ZD-16		
ECG5132		3398	5ZD-17		
ECG5132A		3398	5ZD-17		
ECG5133		3399	5ZD-18		
ECG5133A		3399	5ZD-18		
ECG5134		3400	5ZD-19		
ECG5134A		3400	5ZD-19		
ECG5135		3401	5ZD-20		
ECG5135A		3401	5ZD-20		
ECG5136		3402	5ZD-22		
ECG5136A		3402	5ZD-22		
ECG5137		3403	5ZD-24		
ECG5137A		3403	5ZD-24		
ECG5138		3404	5ZD-25		
ECG5138A		3404	5ZD-25		
ECG5139		3405	5ZD-27		
ECG5139A		3405	5ZD-27		
ECG5140		3406	5ZD-28		
ECG5140A		3406	5ZD-28		
ECG5141		3407	5ZD-30		
ECG5141A		3407	5ZD-30		
ECG5142		3408	5ZD-33		
ECG5142A		3408	5ZD-33		
ECG517		3301	519		
ECG519		3100	514	1122	
ECG521		3304	539		
ECG522		3303	523		
ECG523		3306	521		
ECG524		3329	750		
ECG525		3925	525		
ECG526A		3306	528		
ECG529		3307	529		
ECG530		3308	540		
ECG5304		3106	BR-425		
ECG5305		3676	BR-600		
ECG5306		3677	BR-1000		
ECG5307		3107	BR-1000		
ECG531		3301	524		
ECG532		3303	525		
ECG533		3302	526		
ECG534		3305	534		
ECG535		3307	535		
ECG536		3900	522		
ECG538		3310	537		
ECG539		3309	538		
ECG5400		3950	2N5060		
ECG5401		3638	C103YY		
ECG5402		3638	2N2324		
ECG5403		3627	C6B		
ECG5404		3627	C6B		
ECG5411		3953	C106Y	1067	
ECG5412		3953	C106A	1067	
ECG5413		3954	C106A	1067	
ECG5414		3954	C106B	1067	
ECG5415		3955	C106D	1020	
ECG5421			C106F	1067	
ECG5422			C106A	1067	
ECG5423			C106A	1067	
ECG5431			C106F	1067	
ECG5432			C106A	1067	
ECG5433			C106A	1067	
ECG5442		3634	C122F		
ECG5444		3634	C122B		
ECG5446		3635	C122D		
ECG5452		6752	MR-5	1067	
ECG5453		6753	MR-5	1067	
ECG5454		6754	MR-5	1067	
ECG5455		3597	MR-5	1067	
ECG5456		3598	C106C	1020	
ECG5457		3598	C106D		
ECG5461		3685	C122F		
ECG5462		3686	C122B		
ECG5463		3572	C122B		
ECG5465		3687	C122D		
ECG5491		6791	MR-3		
ECG5492		6792	MR-3		
ECG5494		6794	MR-3		
ECG5500		3579	MR-3		
ECG5501		3579	MR-3		
ECG5502		3579	MR-3		
ECG5503		3579	MR-3		
ECG5504		3579	MR-3		
ECG5505		3580	MR-3		
ECG5506		3580	MR-3		
ECG5507		3580	MR-3		
ECG5514		3613	C230B		
ECG5515		6615	C230D		
ECG5516		6616	C234M		
ECG5520		6621	MR-3		
ECG5521		6621	MR-3		
ECG5522		6622	MR-3		
ECG5523		6624	MR-3		
ECG5524		6624	MR-3		
ECG5525		6627	MR-3		
ECG5526		6627	MR-3		
ECG5527		6627	MR-3		
ECG5528		6629	C147M		
ECG5529		6629	C147M		
ECG5530		6631	C147N		
ECG5531		6631	C147N		
ECG5540		6642	MR-3		
ECG5541		6642	MR-3		
ECG5542		6642	MR-3		
ECG5543		3581	MR-3		
ECG5544		3582	MR-3		
ECG5545		3582	MR-3		
ECG5546		3505	C147N		
ECG5547		3505	C147N		
ECG5600		3664	SC141B	1001	
ECG5601		3664	SC141B	1001	
ECG5602		3664	SC141B	1001	
ECG5603		3665	SC141B	1001	
ECG5604		3666	SC141D	1000	
ECG5605		3666	SC141D	1000	
ECG5611		3631	SC146B		
ECG5612		3631	SC146B		
ECG5613		3631	SC146B		
ECG5614		3632	SC146B		
ECG5615		3633	SC146D		
ECG5616		3633	SC146D		
ECG5621		3631	SC146B		
ECG5622		3631	SC146B		
ECG5623		3631	SC146B		
ECG5624		3632	SC146B		
ECG5625		3633	SC146D		
ECG5626		3633	SC146D		
ECG5631		3937	SC146B		
ECG5632		3937	SC146B		
ECG5633		3938	SC146B		
ECG5634		3533	SC146D		
ECG5635		3533	SC146D		
ECG5636		3939	SC151M		
ECG5800		9003	512	1142	
ECG5801		9004	512	1142	
ECG5802		9005	512	1143	
ECG5803		9006	512	1144	
ECG5804		9007	512	1144	
ECG5805		9008	512		
ECG5806		3848	512		

Industry Standard No.	ECG	SK	GE	RS 276-	MOTOR.
ECG5808		9009	512		
ECG5809		9010	512		
ECG5830		7042	5004		
ECG5831		7049	5005		
ECG5832		7042	5004		
ECG5833		7049	5005		
ECG5834		7042	5004		
ECG5835		7049	5005		
ECG5836		7042	5004		
ECG5837		7049	5005		
ECG5838		7042	5004		
ECG5839		7049	5005		
ECG5840		7042	5008		
ECG5841		7049	5009		
ECG5842		7042	5008		
ECG5843		7049	5009		
ECG5846		7048	5012		
ECG5847		7049	5013		
ECG5848		7048	5012		
ECG5849		7049	5013		
ECG5850		3601	5016		
ECG5851		7069	5017		
ECG5852		3600	5016		
ECG5853		7069	5017		
ECG5854		3600	5016		
ECG5855		7069	5017		
ECG5856		3599	5020		
ECG5857		7069	5021		
ECG5858		3599	5020		
ECG5859		7069	5021		
ECG5860		3584	5024		
ECG5861		3517	5025		
ECG5862		3584	1N1348A		
ECG5863		7069	5025		
ECG5866		7068	5028		
ECG5867		7069	5029		
ECG5868		7068	5028		
ECG5869		7069	5029		
ECG5870		3604	5032		
ECG5871		3517	5033		
ECG5872		3603	5032		
ECG5873		3517	5033		
ECG5874		3603	5032		
ECG5875		3517	5033		
ECG5876		3602	5036		
ECG5877		3517	5037		
ECG5878		3602	5036		
ECG5879		3517	5037		
ECG5880		3500	5040		
ECG5881		3517	5041		
ECG5882		3500	5040		
ECG5883		3517	5041		
ECG5886		7090	5044		
ECG5887		7091	5045		
ECG5890		7090	5044		
ECG5891		7091	5045		
ECG5892		7096	5064		
ECG5893		7111	5065		
ECG5894		7096	5064		
ECG5895		7111	5065		
ECG5896		7096	5064		
ECG5897		7111	5065		
ECG5898		7100	5068		
ECG5899		7111	5069		
ECG5900		7100	5068		
ECG5901		7111	5069		
ECG5902		7104	5072		
ECG5903		7111	5073		
ECG5904		7104	5072		
ECG5905		7111	5073		
ECG5908		7110	5076		
ECG5909		7111	5077		
ECG5910		7110	5076		
ECG5911		7111	5077		
ECG5940		3607	5048		
ECG5941		3518	5097		
ECG5942		3606	5048		
ECG5943		7153	5097		
ECG5944		3606	5048		
ECG5945		7153	5097		
ECG5946		3605	5100		
ECG5947		7153	5101		
ECG5948		3605	5100		
ECG5949		7153	5101		
ECG5950		3585	5104		
ECG5951		7153	5105		
ECG5952		3585	5104		
ECG5953		7153	5105		
ECG5980		3610	5096		
ECG5981		3698	5097		
ECG5982		3609	5096		
ECG5983		3698	5097		
ECG5986		3609	5096		
ECG5987		3698	5097		
ECG5988		3608	5100		
ECG5989		3518	5101		
ECG5990		3608	5100		
ECG5991		3518	5101		
ECG5992		3501	5104		
ECG5993		3518	5105		
ECG5994		3501	5104		
ECG5995		3518	5105		
ECG5998		7202	5108		
ECG5999		7203	5109		
ECG6002		7202	5108		
ECG6003		7203	5109		
ECG6020		7220	5128		
ECG6021		7227	5129		
ECG6022		7226	5128		
ECG6023		7227	5129		
ECG6026		7226	5128		
ECG6027		7227	5129		
ECG6030		7234	5132		
ECG6031		7245	5133		
ECG6034		7234	5132		
ECG6035		7245	5133		
ECG6038		7240	5136		
ECG6039		7245	5137		
ECG6040		7240	5136		
ECG6042		7244	5140		
ECG6043		7245	5141		
ECG6044		7244	5140		
ECG6045		7245	5141		
ECG6050		7254	5128		
ECG6051		7261	5129		
ECG6054		7254	5128		
ECG6055		7261	5129		
ECG6058		7260	5132		
ECG6059		7261	5133		
ECG6060		7260	5132		
ECG6061		7261	5133		
ECG6064		7264	5136		
ECG6065		7273	5137		
ECG6068		7272	5140		
ECG6069		7273	5141		
ECG6072		7272	5140		
ECG6073		7273	5141		
ECG614		3327	90		
ECG6400				2029	
ECG6401				2029	
ECG6407				1050	
ECG6408				1050	
ECG6409				2029	
ECG703		3157	IC-12		
ECG703A		3157	IC-12		
ECG704		3023	IC-205		
ECG705A		3134	IC-3		
ECG708		3135	IC-10		
ECG709		3135	IC-11		
ECG710		3102	IC-89		
ECG711		3070	IC-207		
ECG712		3072	IC-148		
ECG713		3077	IC-5		
ECG714		3075	IC-4		
ECG715		3076	IC-6		
ECG718		3159	IC-8		
ECG720		9014	IC-7		
ECG722		3161	IC-9		
ECG723		3144	IC-15		
ECG724		3525	IC-86		
ECG725		3162	IC-19		
ECG726		3129	IC-81		
ECG727		3071	IC-210		
ECG728		3073	IC-22		
ECG729		3074	IC-23		
ECG731		3170	IC-13		
ECG732		3455	IC-27		
ECG733	804	3455			
ECG734		3163	IC-17		
ECG736		3163	IC-17		
ECG737		3375	IC-16		
ECG738		3167	IC-29		
ECG739		3235	IC-30		
ECG7400		7400	7400	1801	
ECG7401		7401	IC-194		
ECG7402		7402	7402	1811	
ECG7404		7404	7404	1802	
ECG7406		7406	7406	1821	
ECG7408		7408	7408	1822	
ECG740A		3328	IC-31		
ECG7410		7410	7410	1807	
ECG74123		74123	74123	1817	
ECG7413		7413	7413	1815	
ECG74145		74145	74145	1828	
ECG74150		74150	74150	1829	
ECG74154		74154	74154	1834	
ECG74192		74192	74192	1831	
ECG74193		74193	74193	1820	
ECG74196		74196	74196	1833	
ECG742		3453	IC-213		
ECG7420		7420	7420	1809	
ECG7427		7427	7427	1823	
ECG743		3172	IC-32		
ECG7432		7432	7432	1824	
ECG744		3171	IC-24	2022	
ECG7441		7441	7441	1804	
ECG7447		7447	7447	1805	
ECG7448		7448	7448	1816	
ECG745		3276	IC-216		
ECG7451		7451	7451	1825	
ECG746		3234	IC-217		
ECG747		3279	IC-218		
ECG7473		7473	7473	1803	
ECG7474		7474	7474	1818	
ECG7475		7475	7475	1806	
ECG7476		7476	7476	1813	
ECG748		3236	IC-183		
ECG7485		7485	7485	1826	
ECG7486		7486	7486	1827	
ECG749		3168	IC-97		
ECG7490		7490	7490	1808	
ECG7492		7492	7492	1819	
ECG74C00				2301	
ECG74C02				2302	
ECG74C04				2303	
ECG74C08				2305	
ECG74C192				2321	
ECG74C193				2322	
ECG74C74				2310	
ECG74C76				2312	
ECG74LS00		74LS00		1900	
ECG74LS02		74LS02		1902	
ECG74LS04		74LS04		1904	
ECG74LS08		74LS08		1908	
ECG74LS10		74LS10		1910	
ECG74LS123		74LS123		1926	
ECG74LS151		74LS151		1929	
ECG74LS157		74LS157		1930	
ECG74LS161A		74LS161		1931	
ECG74LS174		74LS174		1932	
ECG74LS175		74LS175		1934	
ECG74LS193		74LS193		1936	
ECG74LS20		74LS20		1912	
ECG74LS27		74LS27		1913	
ECG74LS30		74LS30		1914	
ECG74LS32		74LS32		1915	
ECG74LS367		74LS367		1835	
ECG74LS51		74LS51		1917	
ECG74LS73		74LS73		1918	
ECG74LS74A		74LS74		1919	
ECG74LS75		74LS75		1920	
ECG74LS85		74LS85		1922	
ECG74LS93		74LS93		1925	
ECG750		3280	IC-219		
ECG75491B				1701	
ECG75492B				1702	
ECG778		3465	IC-220		
ECG779-1		3240	IC-221		
ECG780		3141	IC-20		
ECG781		3169	IC-223		
ECG783		3215	IC-21		
ECG784		3524	IC-236		
ECG785		3254	IC-226		
ECG786		3140	IC-227		
ECG787		3146	IC-228		
ECG788		3829	IC-229		
ECG790		3077	IC-18		
ECG791		3149	IC-33		
ECG795		3237	IC-232		
ECG797		3158	IC-233		
ECG798		3216	IC-234		
ECG799		3238	IC-34		
ECG801		3160	IC-35		
ECG802		3277	IC-235		
ECG803		3278	IC-237		
ECG804		3455	IC-27		
ECG806		3734	IC-239		
ECG807		3451	IC-240		

Industry Standard No.	ECG	SK	GE	RS 276-	MOTOR.
ECG815		3255	IC-244		
ECG823				1731	
ECG900		3547	IC-245		
ECG901		3549	IC-180		
ECG903		3540	IC-288		
ECG904		3542	IC-289		
ECG905		3546	IC-290		
ECG906		3548	IC-246		
ECG907		3545	IC-247		
ECG908		3539	IC-248		
ECG909D		3590	IC-250		
ECG912		3543	IC-172		
ECG914		3541	IC-256		
ECG915				017	
ECG916		3550	IC-257		
ECG917		3544	IC-258		
ECG923		3164	VR-115		
ECG923D		3165	IC-260	1740	
ECG941		3514	IC-263	010	
ECG941D		3552	IC-264		
ECG941M		3552	IC-265	007	
ECG947		3526	IC-268		
ECG949		3166	IC-25		
ECG955M		3564	IC-269	1723	
ECG960		3591	VR-102		
ECG961		3671	VR-105		
ECG964		3630	VR-108		
ECG966		3592	VR-111		
ECG967		3673	VR-114		
ECG977		3462	VR-100		
ECG978		3689		1728	
ECG980		4046		2446	
ECG981		3724	VR-106		
ECG987		3643		1711	
ECG990				702	
ECG992		3688		1713	
ECR-600-2	116		504-A	1104	
ECR600-2		3017B	504A	1104	
ED-1402	123A	3444	20	2051	
ED-1502C	229	3018	61	2038	
ED-4	116	3016	504A	1104	
ED-46	109	3088		1123	
ED-5	116		504A	1104	
ED-6	116	3016	504A	1104	
ED-60	109	3088		1123	
ED12(ELCOM)	109	3090		1123	
ED1402	123A	3444			
ED1402A	123A	3444	20	2051	
ED1402A/09-305066	123A		20	2051	
ED1402B	123A	3444	20	2051	
ED1402C	123AP	3444/123A	20	2051	
ED1402C/30	123A	3444			
ED1402C/30V	123A	3444			
ED1402D	123AP	3444/123A	20	2051	
ED1402D/30V	123A	3444			
ED1402E	123AP		20	2051	
ED1502	229	3039/316	20	2038	
ED1502A	229	3018	61	2038	
ED1502B	107	3018	39	2038	
ED1502C	107		61	2038	
ED1502D	107	3246/229	61	2038	
ED1502E	107	3311	60	2038	
ED1502P			20	2038	
ED1502R	233		20	2051	
ED150Z	123A	3444	20	2051	
ED1602C		3466	82	2032	
ED1602D			82	2032	
ED1702L	123A	3444	20	2051	
ED1702L/09-305068	123A		20	2051	
ED1702M	123AP		47	2030	
ED1702N	293		62	2010	
ED1704L			20	2051	
ED1802-0			21	2034	
ED1802M	159	3247/234	48		
ED1802N	294		21	2034	
ED1802N,M			21	2034	
ED1804	116	3016	504A	1104	
ED1892	116	3016	504A	1104	
ED2(ELCOM)	112		1N82A		
ED21(ELCOM)	177		300	1122	
ED2106	116	3016	504A	1104	
ED2107	116		504A	1104	
ED2108	116	3016	504A	1104	
ED2109	116	3016	504A	1104	
ED2110	116	3016	504A	1104	
ED219464	109	3087		1123	
ED224548	116		504A	1104	
ED224550	116	3016	504A	1104	
ED23(ELCOM)	506		511	1114	
ED2842	116	3016	504A	1104	
ED2843	116	3016	504A	1104	
ED2844	116	3016	504A	1104	
ED2845	116	3017B/117	504A	1104	
ED2846	116	3017B/117	504A	1104	
ED2847	125	3033	510,531	1114	
ED2848	125	3033	510,531	1114	
ED2849	125	3033	510,531	1114	
ED29(ELCOM)	156		512		
ED2910	125	3033	510,531	1114	
ED2911	125	3032A	510,531	1114	
ED2912	125	3033	510,531	1114	
ED2913	125	3033	510,531	1114	
ED2914	116	3311	504A	1104	
ED2915	116	3016	504A	1104	
ED2916	116	3016	504A	1104	
ED2917	116	3016	504A	1104	
ED2918	116	3016	504A	1104	
ED2919	116	3016	504A	1104	
ED2920	116	3017B/117	504A	1104	
ED2921	116	3017B/117	504A	1104	
ED2922	125	3033	510,531	1114	
ED2923	125	3032A	510,531	1114	
ED2924	125	3033	510,531	1114	
ED3(ELCOM)	112		1N82A		
ED3000	116	3311	504A	1104	
ED3000A	116	3311	504A	1104	
ED3000B	116	3017B/117	504A	1104	
ED3001	116	3017B/117	504A	1104	
ED3001A	116	3016	504A	1104	
ED3001B	116	3017B/117	504A	1104	
ED3002	116	3017B/117	504A	1104	
ED3002A	116	3016	504A	1104	
ED3002B	116	3017B/117	504A	1104	
ED3003	116	3017B/117	504A	1104	
ED3003A	116	3031A	504A	1104	
ED3003B	116	3031A	504A	1104	
ED3003S	116	3016	504A	1104	
ED3004	116	3017B/117	504A	1104	
ED3004A	116	3031A	504A	1104	
ED3004B	116	3031A	504A	1104	
ED3005	116	3017B/117	504A	1104	
ED3005A	116	3017B/117	504A	1104	
ED3005B	116	3017B/117	504A	1104	
ED3006	116	3017B/117	504A	1104	

Industry Standard No.	ECG	SK	GE	RS 276-	MOTOR.
ED3006A	116	3017B/117	504A	1104	
ED3006B	116	3017B/117	504A	1104	
ED3007	125	3032A	510,531	1114	
ED3007A	125	3032A	510,531	1114	
ED3007B	125	3032A	510,531	1114	
ED3008	125	3033	510,531	1114	
ED3008A	125	3033	510,531	1114	
ED3008B	125	3033	510,531	1114	
ED3009	125	3033	510,531	1114	
ED3009A	125	3033	510,531	1114	
ED3009B	125	3033	510,531	1114	
ED3010	125	3033	510,531	1114	
ED3010A	125	3033	510,531	1114	
ED3010B	125	3033	510,531	1114	
ED31(ELCOM)	177		300	1122	
ED3100					1N4001
ED3101					1N4002
ED3102					1N4003
ED3104					1N4004
ED3106					1N4005
ED3108					1N4006
ED3110					1N4007
ED32(ELCOM)	177		300	1122	
ED329128	116		504A	1104	
ED329130	116		504A	1104	
ED4		3087	504A	1104	
ED4(ELCOM)	109	3090		1123	
ED46	109	3088		1123	
ED491130	5073A	3749	ZD-9.1	562	
ED494583	116	3311	504A	1104	
ED498150	5078A	9023	ZD-19		
ED5		3016	504A	1104	
ED51	160		245	2004	
ED511097	116	3311	504A	1104	
ED511918	5081A	3151	ZD-24		
ED514721	177	3100/519	300	1122	
ED515790	177	3100/519	300	1122	
ED516420	177	3100/519	300	1122	
ED52	100	3005	1	2007	
ED53	100	3005	1	2007	
ED536062	177	3100/519	300	1122	
ED54B	100	3005	1	2007	
ED55	102A	3004	53	2007	
ED56	102A	3004	53	2007	
ED560913	177	3100/519	300	1122	
ED57	102A	3004	53	2007	
ED592K	229	3018	61	2038	
ED592M	108	3018	86	2038	
ED6		3091	504A	1104	
ED6(ELCOM)	109	3090		1123	
ED6.2EB	137A	3058	ZD-6.2	561	
ED60	109	3087		1123	
ED7	116	3089/112	504A	1104	
ED7(ELCOM)	112		1N82A		
ED8(ELCOM)	156		512		
ED8307					MR1366
ED8310					MR1376
ED9(ELCOM)	109	3090		1123	
EDC TR11-4	128			2030	
EDC-Q10-1	123A	3444	20	2051	
EDC-TR-11-1			62	2010	
EDC-TR11-4			243	2030	
EDC-TR11-5			243	2030	
EDG-0001	109	3090			
EDG-0003	109	3088	1N60	1123	
EDG-0006	109	3088	1N60	1123	
EDG-1	109	3090		1123	
EDG-3	109	3088		1123	
EDG-6	109	3087		1123	
EDH6023	519		514	1122	
EDJ-363	116	3032A	504A	1104	
EDMF-15B	503	3068	CR-5		
EDMF-25B	504		CR-6		
EDMF25B	504	3108/505	CR-6		
EDO 219	123A			2051	
EDO-219			20	2051	
EDS-0001	177	3100/519	300	1122	
EDS-0002	116	3311	504A	1104	
EDS-0004	116	3016	509	1114	
EDS-0014	177	3100/519	300	1122	
EDS-0017	116	3311	504A	1104	
EDS-0024	116	3100/519	300	1104	
EDS-0042	614	3126	90		
EDS-1	507	3100/519	300	1122	
EDS-100	123A	3444	20	2051	
EDS-11			504A	1104	
EDS-17	116	3311	504A	1104	
EDS-24	116	3017B/117	504A	1104	
EDS-25	177	3100/519	300	1122	
EDS-31		3126	90		
EDS-4	116	3017B/117	504A	1104	
EDZ-0045	139A		ZD-9.1	562	
EDZ-11		3142	ZD-16	564	
EDZ-14			ZD-8.2	562	
EDZ-19	136A	3057	ZD-5.6	561	
EDZ-2	138A	3059	ZD-7.5		
EDZ-20	5071A		ZD-6.8	561	
EDZ-23	139A	3060	ZD-9.1	562	
EDZ-24	5072A	3136	DZ-8.2	562	
EE100	102	3722	2	2007	MLED60
EE60					MLED60
EER600-2	117	3017B	504A	1104	
EF1(ELCOM)	312			2028	
EF100	125	3033	510,531	1114	
EF2(ELCOM)	312		FET-1	2035	
EF200	525	3925	39	2015	
EF3	312		FET-1	2035	
EF4(ELCOM)	222			2036	
EF5(ELCOM)	222			2036	
EG100	116		504A	1104	
EG100H	116	3017B/117	504A	1104	
EGA01-11	5074A	3139	ZD-11.0	563	
EGF211B1					MDA2501
EGF212B1					MDA2502
EGF213B1					MDA2504
EGF214B1					MDA2504
EGF215B1					MDA3506
EGF216B1					MDA3506
EGL211B1					MDA2501
EGL212B1					MDA2502
EGL213B1					MDA2504
EGL214B1					MDA2504
EGL215B1					MDA3506
EGL216B1					MDA3506
EGT211B1					MDA2501
EGT212B1					MDA2502
EGT213B1					MDA2504
EGT214B1					MDA2504
EGT215B1					MDA3506
EGT216B1					MDA3506
EH16X20	177		300	1122	
EH2C11/7+12/1			504A	1104	
EI3-006-02	159		82	2032	
EI517342	1011		IC-79		

Industry Standard No.	ECG	SK	GE	RS 276-	MOTOR.
EICM-0037	1052	3249			
EICM-0060	1155	3231	IC-179	705	
EICM-14	1103	3281	IC-94		
EJL1B1					MDA970A1
EJL2B1					MDA970A3
EJT1B1					MDA970A1
EJT2B1					MDA970A3
EK136	102A		53	2007	
EK159	100	3005	1	2007	
EL131	312	3746/326	FET-1	2028	
EL214	128	3024	243	2030	
EL231			11	2015	
EL232	123A	3444	20	2051	
EL238			20	2051	
EL264	129	3025	244	2027	
EL401	172A	3156	64		
EL403	289A	3124/289	210	2038	
EL434	107	3018	11	2015	
EL642			20	2051	
EL75	229	3246	61	2038	
EM-1171			504A	1104	
EM1021	116	3031A	504A	1104	
EM1J2	116	3017B/117	504A	1104	
EM401	116	3311	504A	1104	
EM402	116	3017B/117	504A	1104	
EM403	116	3017B/117	504A	1104	
EM404	116	3017B/117	504A	1104	
EM405	116	3017B/117	504A	1104	
EM406	116	3017B/117	504A	1104	
EM407	116	3017B/117	504A	1104	
EM408	116	3032A	504A	1104	
EM410	116	3033	504A	1104	
EM501	116	3017B/117	504A	1104	1N4002
EM502	116	3016	504A	1104	1N4003
EM503	116	3031A	504A	1104	1N4004
EM504	116	3031A	504A	1104	1N4004
EM505	116	3017B/117	504A	1104	1N4005
EM506	116	3017B/117	504A	1104	1N4005
EM507	125	3032A	510,531	1114	
EM508	125	3032A	510,531	1114	1N4006
EM510	125	3033	510,531	1114	1N4007
EM9-8	914		IC-256		
EMS 72272	519			1122	
EMS-73500	123A	3444	20	2051	
EMS72258	519		514	1122	
EMS72272			514	1122	
EMS73278	128		243	2030	
EMS73279	128		243	2030	
EN10			17	2051	
EN1132	159		82	2032	2N3905
EN1364	1231	3832			
EN1613	128	3245/199	243	2030	MPS6530
EN1711	128	3245/199	243	2030	MPS6530
EN2219	123A	3122	20	2051	2N4401
EN2222	123A	3444	20	2051	MPS2222
EN2369A	123A				MPS2369
EN2484	199	3444/123A	20	2051	MPSA05
EN2894	159		82	2032	
EN2894A	159	3114/290	82	2032	2N5226
EN2905	129		244	2027	2N4403
EN2907	159	3118	82	2032	MPS2907
EN30			10	2051	
EN3009	123A	3444	20	2051	MPS2369
EN3011	123A	3452/108	86	2038	MPS6516
EN3012					MPS3646
EN3013	123A	3444	20	2051	MPS3646
EN3014	123A	3444	20	2051	MPS2369
EN3250	159		82	2032	MPS6516
EN3502	129		244	2027	2N3906
EN3504	159	3118	82	2032	2N3906
EN3903	123A		210	2051	2N3903
EN3904	123A		243	2030	2N3904
EN3905	159		82	2032	2N3905
EN3906	159		82	2032	2N3906
EN3962	159		82	2032	2N5086
EN40			20	2051	
EN4123					2N4123
EN4124					2N4124
EN4125					2N4125
EN4126					2N4126
EN5172	107		11	2015	MPS5172
EN697	123A	3444	20	2051	MPS6530
EN706	123A	3444	20	2051	MPS706
EN708	123A	3444	20	2051	MPS834
EN718A	108	3452	86	2038	MPSH04
EN722	159	3118	82	2032	MPS3703
EN744	123A	3452/108	86	2038	2N4265
EN870	128	3244	220		2N4410
EN871		3244	220		2N4410
EN914	123A	3452/108	86	2038	MPSH32
EN915					MPS3826
EN916	108	3452	86	2038	MPS6512
EN918	108	3452	86	2038	MPS918
EN930	123A	3444	20	2051	MPS6514
EN956	123A	3245/199	212	2010	MPS6531
E0-44A		3004	2	2007	
E0105	100	3005	1	2007	
E044A	102A		53	2007	
E065	100	3005	1	2007	
E066	100	3005	1	2007	
E067	100	3005	1	2007	
E068	100	3005	1	2007	
E070	126		52	2024	
E0704	116	3017B/117	504A	1104	
E0771-3	139A		ZD-9.1	562	
EP-100			66	2048	
EP-1428-2H			504A	1104	
EP-276			241	2020	
EP-2798	113A	3100/519	6GC1		
EP-3572	113A		6GC1		
EP-3641-1	503		CR-5		
EP-421			69	2049	
EP-422			66	2048	
EP-5219-1	115		6GX1		
EP-5619-2/7628			504A	1104	
EP-5641H-2	504	3108/505	CR-6		
EP-797			66	2048	
EP-801			66	2048	
EP-802			69	2049	
EP-943			69	2049	
EP-944			66	2048	
EP-976			218	2026	
EP100	186	3192	28	2017	
EP1000	125	3033	510,531	1114	
EP101A	187	3193	29	2025	
EP1259			504A	1104	
EP1259-2	116		504A	1104	
EP1428-2H	116	3017B/117	504A	1104	
EP15X1	123AP	3444/123A	20	2051	
EP15X10	5404	3232/191	SP-1507		
EP15X105	375	3440/291			
EP15X106	5404	3627	60	2015	
EP15X107			40	2012	
EP15X108	162	3836/284			
EP15X11	196	3054	241	2020	
EP15X12	238	3710	259		
EP15X123			60	2015	
EP15X126	165	3710/238	36		
EP15X13	193	3138	67	2023	
EP15X14	196	3054	241	2020	
EP15X15	153	3083/197	69	2049	
EP15X16	124	3021	SP-1511		
EP15X17	159	3114/290	82	2032	
EP15X18	154	3044	40	2012	
EP15X19	291	3054/196	241		
EP15X2	123AP	3849/293	20	2051	
EP15X20	233	3122	210	2009	
EP15X21	193	3114/290	67	2023	
EP15X22	152	3054/196	66	2048	
EP15X23	153	3083/197	69	2049	
EP15X24	187	3193	253	2025	
EP15X25	186	3192	252	2017	
EP15X26	159	3114/290	82	2032	
EP15X27	171	3201	27		
EP15X28	165	3115	SP-1518		
EP15X29	81	3036	SP-1519	2041	
EP15X3	199	3245	62	2010	
EP15X30	196	3054	SP-1520	2020	
EP15X32	373	3192/186	28	2017	
EP15X33	128	3024	243	2030	
EP15X34	286	3192/186	28	2017	
EP15X35	124	3021	12		
EP15X36			FET-4	2036	
EP15X37			11	2015	
EP15X38			11	2015	
EP15X39			20	2051	
EP15X4	129	3025	244	2027	
EP15X40			FET-4	2036	
EP15X41			61	2038	
EP15X42			61	2038	
EP15X43	196	3054	241	2020	
EP15X44	197	3083	250	2027	
EP15X45	165	3710/238	SP-1526		
EP15X47	123A	3444	20	2051	
EP15X48	290A		82	2032	
EP15X48(NPN)	123A		20	2051	
EP15X48(PNP)	159		82	2032	
EP15X49	123A	3444	20	2051	
EP15X5	128	3250/315	243	2030	
EP15X50	198	3220	251		
EP15X51	193	3114/290	67	2023	
EP15X53	159	3466	82	2032	
EP15X54	233	3246/229	210	2009	
EP15X55	108	3452	86	2038	
EP15X6	229	3854/123AP	61	2038	
EP15X60	159	3466	21	2032	
EP15X61	171	3201	27		
EP15X64	222	3065		2036	
EP15X68	152	3054/196	66	2048	
EP15X7	123A	3444	20	2051	
EP15X8	123AP	3444/123A	20	2051	
EP15X86	123AP	3854		2051	
EP15X87	123AP	3854			
EP15X88	123AP	3444/123A		2051	
EP15X89	123AP	3854			
EP15X9	123AP		20	2051	
EP15X90	159	3114/290	21	2034	
EP15X92			FET-2	2035	
EP16X1	178MP	3100/519	300(2)	1122(2)	
EP16X10	506	3175	511	1114	
EP16X11	506	3998	511	1114	
EP16X12	138A	3059	ZD-7.5		
EP16X13	116	3126	504A	1104	
EP16X15		3843	300	1122	
EP16X16	5081A	3151	ZD-24		
EP16X17	5100A		ZD-150		
EP16X2	5136A	3402	SP-1901		
EP16X20	177	3100/519	300	1122	
EP16X21	110MP	3087	1N34AS	1123(2)	
EP16X22	177	3175	300	1122	
EP16X23	177	3100/519	300	1122	
EP16X24	506	3998	511	1114	
EP16X25	135A	3056	ZD-5.0		
EP16X27	177	3100/519	300	1122	
EP16X3	109	3088	1N60	1123	
EP16X30			300	1122	
EP16X36	142A	3062	ZD-12	563	
EP16X4	177	3031A	300	1122	
EP16X5	145A	3063	ZD-15	564	
EP16X58	5024A	3790			
EP16X59	519	3100			
EP16X6	113A	3119/113	6GC1		
EP16X68	145A	3063			
EP16X7	128	3124/289	243	2030	
EP2					4N26
EP20				2024	
EP200	116	3017B/117	504A	1104	
EP25			22	2032	
EP2798			300	1122	
EP3053	152	3893	66	2048	
EP3149	116	3017B/117	504A	1104	
EP35			67	2023	
EP36X12	138A	3059	ZD-7.5		
EP400	116	3017B/117	504A	1104	
EP431	292	3441			
EP57X1	116	3017B/117	504A	1104	
EP57X10	552	3998/506			
EP57X12	116	3311	504A	1104	
EP57X2	505	3108	CR-7		
EP57X3	513	3443	513		
EP57X4	552	3312	504A	1104	
EP57X5	552	3125	511	1114	
EP5X5	116		504A	1104	
EP600	116	3017B/117	504A	1104	
EP62X41	522	3303	517		
EP62X61	522		517		
EP62X84	522	3303			
EP6X10	177	3100/519	300	1122	
EP6X11	506	3130	511	1114	
EP800	125	3033	510,531	1114	
EP84X1	1186		IC-97		
EP84X10	1186	3168/749	IC-97		
EP84X11	7420		7420	1809	
EP84X12	738	3167	IC-29		
EP84X19	74123		74123	1817	
EP84X2	712	3072	IC-148		
EP84X3	713	3077/790	IC-5		
EP84X35	1196	3725			
EP84X4	783	3215	IC-225		
EP84X5	783	3215	IC-225		
EP84X6	712	3072	IC-2		
EP84X7	728	3073	IC-22		
EP84X8		3168	IC-225		
EP84X9	713	3077/790	IC-5		
EPB22-15	506	3130	511	1114	
EPS57X5	116	3311			
EPX15X17	159		82	2032	
EPX2			20	2051	

Industry Standard No.	ECG	SK	GE	RS 276-	MOTOR.
EPX54	161	3132			
EPY62-1	3032				MRD3055
EPY62-2	3032				MRD3056
EPY62-3					MRD310
EQ-09R	139A	3060	ZD-9.1	562	
EQA 9	139A	3060			
EQA-01-07S	5014A		ZD-7.5		
EQA-01-09P	139A	3060			
EQA-0109S	139A	3060	ZD-9.1	562	
EQA-9			ZD-9.1	562	
EQA.9			ZD-9.1	562	
EQA01-01			ZD-9.1	562	
EQA01-05	135A		ZD-5.0		
EQA01-05S	135A	3056	ZD-5.1		
EQA01-05T	135A	3057/136A	ZD-5.6	561	
EQA01-06	5013A	9021/5070A	ZD-6.8	561	
EQA01-067			ZD-6.2	561	
EQA01-06R	5070A	9021			
EQA01-06S	5012A	9021/5070A	ZD-6.2	561	
EQA01-06SB	5012A	9021/5070A		561	
EQA01-06T	137A	9021/5070A	ZD-6.2	561	
EQA01-07	5071A	3334	ZD-6.8	561	
EQA01-072	138A	3059	ZD-7.5		
EQA01-07R	5071A		ZD-6.8	561	
EQA01-07RE	5071A	3334	ZD-6.8	561	
EQA01-07S	5014A	3059/138A	ZD-6.8	561	
EQA01-08	5072A	3136	ZD-8.2	562	
EQA01-08R	5072A	3059/138A	ZD-8.2	562	
EQA01-08S	5072A	3136			
EQA01-09	139A	3060	ZD-9.1	562	
EQA01-09(R)			ZD-9.1	562	
EQA01-09R	139A	3060	ZD-9.1	562	
EQA01-09S	139A	3060			
EQA01-10	5019A			562	
EQA01-10S	140A		ZD-10	562	
EQA01-11	5074A	3092/141A	ZD-11	563	
EQA01-115			ZD-11	563	
EQA01-11S	5020A	3139/5074A	ZD-11.0	563	
EQA01-11SE	5074A		ZD-11	563	
EQA01-11Z	5074A	3092/141A	ZD-11	563	
EQA01-12	142A	3062	ZD-13	563	
EQA01-12R	142A	3062	ZD-12	563	
EQA01-12S	142A	3062	ZD-12	563	
EQA01-12Z	142A		ZD-12	563	
EQA01-13	143A	3750	ZD-13	563	
EQA01-13R	143A		ZD-13	563	
EQA01-14				564	
EQA01-14R	144A	3094		564	
EQA01-14RD	144A	3094		564	
EQA01-15			ZD-15	564	
EQA01-15R	145A	3063	ZD-15	564	
EQA01-16	5075A	3142	ZD-16	564	
EQA01-17R	5076A	9022	ZD-17		
EQA01-18	5077A	3752	ZD-18		
EQA01-19R	5078A	9023			
EQA01-20RB	5079A		ZD-20		
EQA01-21R	5079A		ZD-20		
EQA01-24	5081A		ZD-24		
EQA01-24RR	5081A	3151	ZD-24		
EQA01-30	5084A	3755	ZD-30		
EQA01-32R	147A	3095	ZD-33		
EQA01-33RF	147A	3095			
EQA01-35	5085A	3337	ZD-36		
EQA01-35RC	5143A	3409			
EQA01-Z4RA	5084A	3755	ZD-30		
EQA0105T	136A	3057	ZD-5.6	561	
EQA09R	139A		ZD-9.1	562	
EQA107RE			ZD-6.8	561	
EQB-0108	5072A	3136	ZD-8.2	562	
EQB01	142A	3062	ZD-12	563	
EQB01-011Z			ZD-11	563	
EQB01-02R	142A		ZD-12	563	
EQB01-06	5070A	9021	ZD-6.2	561	
EQB01-06A	5070A	9021			
EQB01-06V	5070A	9021			
EQB01-07	138A	3059	ZD-7.5		
EQB01-08	5072A	3136	ZD-8.2	562	
EQB01-09	139A	3060	ZD-9.1	562	
EQB01-09S	139A	3060			
EQB01-10	140A	3061	ZD-10	562	
EQB01-11	5074A	3092/141A	ZD-11.0	563	
EQB01-11V	5074A		ZD-11.0	563	
EQB01-11Z	5074A	3092/141A	ZD-11.0	563	
EQB01-12	142A	3062	ZD-12	563	
EQB01-12A	142A	3062	ZD-12	563	
EQB01-12B	142A	3062	ZD-12	563	
EQB01-12BV	142A	3062	ZD-12	563	
EQB01-12R	142A	3062	ZD-12	563	
EQB01-12Z	142A	3062	ZD-12	563	
EQB01-13	143A		ZD-13	563	
EQB01-14	144A		ZD-14	564	
EQB01-15	145A	3063	ZD-15	564	
EQB01-15Z	145A	3063	ZD-15	564	
EQB01-15ZB	145A		ZD-15	564	
EQB01-16	5075A	3751	ZD-16	564	
EQB01-18	5077A	3752	ZD-18		
EQB01-19	5078A	9023	ZD-19		
EQB01-24	5081A		ZD-24		
EQB01-33	147A	3095	ZD-33		
EQB01-90S	139A		ZD-9.1	562	
EQB01-10	140A			562	
EQF-0009	312	3018	FET-2	2035	
EQF-3		3112	FET-1	2028	
EQF-4	312	3448	FET-1	2035	
EQG-12A			ZD-12	563	
EQG-15	102A	3004	53	2007	
EQG-6	121	3009	239	2006	
EQG-8	104	3009	16	2006	
EQG-9	102A	3004	53	2007	
EQG01-12A	142A	3062	ZD-12		
EQH-1	103A	3010	59	2002	
EQR-0016	294	3025/129	65		
EQR-0038	290A	3114/290	21	2034	
EQR-1		3114	21	2034	
EQS-0018	233	3018	61	2038	
EQS-0061	199	3124/289	212	2010	
EQS-0100	123A	3018	61	2038	
EQS-0140	152	3893			
EQS-0159	236	3197/235	216	2053	
EQS-0160	235	3197	215		
EQS-0165	297	3122	210	2051	
EQS-0184	306	3251	276		
EQS-0192	289A	3122	210	2051	
EQS-0196	123A	3018	211	2016	
EQS-0198	123A	3018	61	2038	
EQS-10		3124	62	2010	
EQS-100	123A	3444	20	2051	
EQS-11		3124	62	2010	
EQS-13	123A	3444	20	2051	
EQS-131	199	3024/128	62	2010	
EQS-139	229	3018	61	2038	
EQS-140	152		66	2048	
EQS-141	236	3197/235	216	2053	
EQS-18	107	3018	11	2015	

Industry Standard No.	ECG	SK	GE	RS 276-	MOTOR.
EQS-19	313	3018	278	2016	
EQS-20	313	3124/289	278	2016	
EQS-21	108	3452	86	2038	
EQS-22	123A	3444	20	2051	
EQS-5	123A	3444	20	2051	
EQS-56	195A	3049/224	46		
EQS-57	195A	3048/329	46		
EQS-60	237	3299	46		
EQS-61	123A	3444	20	2051	
EQS-62		3124	17	2051	
EQS-64		3020	20	2051	
EQS-66		3041	28	2017	
EQS-67		3041	28	2017	
EQS-78	199	3024/128	62	2010	
EQS-86	237	3299	46		
EQS-89	186A	3357	247	2017	
EQS-9	123A	3444	20	2051	
EQS131	128	3024	243	2030	
EQS140	152	3893	66	2048	
ER1	116	3017B/117	504A	1104	1N4001
ER1(ELCOM)	118		CR-1		
ER10(ELCOM)	5892		5064		
ER1001	125	3033	510,531	1114	
ER101	116		504A	1104	
ER102	117	3051/156	504A	1104	
ER102D	116	3016	504A	1104	
ER103D	116	3031A	504A	1104	
ER103E	116	3016	504A	1104	
ER104D	116	3017B/117	504A	1104	
ER105D	116	3017B/117	504A	1104	
ER106D	116		504A	1104	
ER107D	125	3033	510,531	1114	
ER108D	125	3033	510,531	1114	
ER10R(ELCOM)	5893		5065		
ER11	116	3500/5882	504A	1104	1N4002
ER11(ELCOM)	5894		5064		
ER11R(ELCOM)	5895		5065		
ER12	116	3017B/117	504A	1104	
ER12(ELCOM)	5896		5064		
ER12R(ELCOM)	5897		5065		
ER13(ELCOM)	5898		5068		
ER13R(ELCOM)	5899		5069		
ER14(ELCOM)	5900		5068		
ER14R(ELCOM)	5901		5069		
ER15(ELCOM)	5902		5072		
ER15R(ELCOM)	5903		5073		
ER15X10	121	3009	239	2006	
ER15X11	160	3006	245	2004	
ER15X12	160	3006	245	2004	
ER15X13	160	3006	245	2004	
ER15X14	160	3008	245	2004	
ER15X15	160	3008	245	2004	
ER15X16	160	3008	245	2004	
ER15X17	102A	3004	53	2007	
ER15X18	102A	3004	53	2007	
ER15X19	160		245	2004	
ER15X20	160		245	2004	
ER15X21	160		245	2004	
ER15X22	102A	3123	53	2007	
ER15X23	102A	3123	53	2007	
ER15X24	160	3006	245	2004	
ER15X25	160	3006	245	2004	
ER15X26	160	3006	245	2004	
ER15X4	160		245	2004	
ER15X5	160		245	2004	
ER15X6	160		245	2004	
ER15X7	102A		53	2007	
ER15X9	102A		53	2007	
ER16(ELCOM)	5904		5072		
ER17(ELCOM)	5908		5076		
ER17R(ELCOM)	5909		5077		
ER18(ELCOM)	5910		5076		
ER181	116	3017B/117	504A	1104	1N4001
ER182	116	3017B/117	504A	1104	1N4002
ER183	116	3017B/117	504A	1104	1N4003
ER184	116	3017B/117	504A	1104	1N4004
ER185	116	3017B/117	504A	1104	1N4005
ER186	125	3032A	510,531	1114	1N4006
ER187	125	3033	510,531	1114	1N4007
ER18R(ELCOM)	5911		5077		
ER1B22-15			504A	1104	
ER2	116	3311	504A	1104	1N4935
ER2(ELCOM)	119		CR-2		
ER20(ELCOM)	5980		5096		
ER2000					MR501
ER2001					MR501
ER2002					MR502
ER2003					MR504
ER2004					MR504
ER2005					MR506
ER2006					MR506
ER201	116		504A	1104	
ER20R(ELCOM)	5981		5097		
ER21	116	3016	504A	1104	1N4003
ER21(ELCOM)	5986		5096		
ER21R(ELCOM)	5983		5097		
ER22	116	3016	504A	1104	
ER22(ELCOM)	5986		5096		
ER22R(ELCOM)	5987		5097		
ER23(ELCOM)	5988		5100		
ER23R(ELCOM)	5989		5101		
ER24(ELCOM)	5990		5100		
ER24R(ELCOM)	5991		5101		
ER25(ELCOM)	5992		5104		
ER25R(ELCOM)	5995		5105		
ER26(ELCOM)	5994		5104		
ER26R(ELCOM)	5995		5105		
ER27(ELCOM)	5998		5108		
ER28(ELCOM)	6002		5108		
ER28R(ELCOM)	6003		5109		
ER301	116		504A	1104	
ER308	125	3032A	510,531	1114	
ER31	116	3016	504A	1104	1N4004
ER310	125	3033	510,531	1114	
ER33(ELCOM)	504		CR-6		
ER34(ELCOM)	505		CR-7		
ER381	116	3017B/117	504A	1104	
ER4					1N4936
ER401	116		504A	1104	
ER41	116	3016	504A	1104	1N4004
ER410			504A	1104	
ER42	116	3017B/117	504A	1104	
ER5(ELCOM)	120		CR-3		
ER501	116		504A	1104	
ER51	116	3017B/117	504A	1104	1N4005
ER510			504A	1104	
ER57X2	116	3016	504A	1104	
ER57X3	116	3016	504A	1104	
ER57X4	116	3016	504A	1104	
ER6					1N4937
ER601	116	3017B/117	504A	1104	
ER61	116	3017B/117	504A	1104	1N4005
ER62	116	3017B/117	504A	1104	
ER801	125	3032A	510,531	1114	

Industry Standard No.	ECG	SK	GE	RS 276-	MOTOR.
ER81	125	3032A	510,531	1114	1N4006
ER824-06	506		511	1114	
ER9(ELCOM)	513		513		
ERA0106R		3057	ZD-5.6	561	
ERB-07RE	5071A		ZD-6.8	561	
ERB-24-06A	552		509	1104	
ERB01-05	135A		ZD-5.0		
ERB01-07			ZD-6.8	561	
ERB11-01	116	3311	504A	1104	
ERB12-01	125	3311			
ERB12-02	125	3311	504A	1104	
ERB12-11			504A	1104	
ERB22-15	116	3031A	504A	1104	
ERB24	552		504A	1104	
ERB24(GE)	116		504A	1104	
ERB24-02B	506	3316	504A	1104	
ERB24-04A	506	3317	511	1114	
ERB24-04B	506	3317	511	1114	
ERB24-04C	552	3317	511	1114	
ERB24-04D	552	3317	511	1114	
ERB24-06	552	3318	504A	1104	
ERB24-06A		3318	509	1104	
ERB24-06B	552	3318	509	1114	
ERB24-06C	552	3318	509	1114	
ERB24-06D	506	3318	509	1114	
ERB2406A	552	3016	504A	1104	
ERB26-20	525	3925	530		
ERB26-20L	525	3925	530		
ERB26-20M	525	3925	530		
ERB26-20MV	525	3925	530		
ERB28-04	552	9000	504A	1104	
ERB28-04D		9000	504A	1104	
ERB29-04	552	3843			
ERC01-06	125	3032A	509	1114	
ERC04-06	552	3032A			
ERC04-10	125	3080	509	1114	
ERC24-04	552	9000	504A	1104	
ERC26-13	506	3843			
ERC26-131			511	1114	
ERC26-13L	525	3843	511	1114	
ERC27-13	506	3130	511	1114	
ERCC4-06	125	3081	510	1114	
ERD1000	125	3033	510,531	1114	
ERD300	116	3016	504A	1104	
ERD400	116	3016	504A	1104	
ERD700	125	3032A	510,531	1114	
ERD800	125	3032A	510,531	1114	
ERD900	125	3033	510,531	1114	
ERS100			18	2030	
ERS120	396		18	2030	
ERS140	396		235	2012	
ERS160	396		235	2012	
ERS180	396		235	2012	
ERS200	396		235	2012	
ERS225	396		235	2012	
ERS250	396		235	2012	
ERS275	396		235	2012	
ERS301	396		235	2012	
ERS325	396		235	2012	
ERS350	396		32		
ERS375	396		32		
ERS401	396		32		
ERS425	396		32		
ERS450	396		32		
ERV-02F2150	116	3031A	504A	1104	
ES-16X16			511	1114	
ES-48	124		12		
ES-88	195A		46		
ES-923933-1	149A		ZD-62		
ES1(ELCOM)	160		245	2004	
ES10	105	3012	4		
ES10(ELCOM)	105		4		
ES10110	131	3052	44	2006	
ES10186	107	3018	11	2015	
ES10187	108	3452	86	2038	
ES10188	107	3018	11	2015	
ES10189	109	3088		1123	
ES10190		3126	90		
ES10222	123A	3444	20	2051	
ES10223	123A	3444	20	2051	
ES10224	109	3088		1123	
ES10225	109	3088		1123	
ES10231	109	3124/289	62	2010	
ES10232	123A	3444	20	2051	
ES10233	116	3311	504A	1104	
ES10234	142A	3062	ZD-12	563	
ES103	196		241	2020	
ES104	197		250	2027	
ES105	253	3996			
ES106	254	3997			
ES107	232	3241	258		
ES108	243	3182			
ES109	244	3859/252			
ES11(ELCOM)	160		245	2004	
ES112	234		65	2050	
ES113	241	3188A/182	57	2020	
ES114	242	3189A/183	58	2027	
ES13	104	3009	16	2006	
ES13(ELCOM)	104	3719	16	2006	
ES14	126	3006/160	52	2024	
ES14(ELCOM)	160		245	2004	
ES15(ELCOM)	160		245	2004	
ES15046	107	3018	11	2015	
ES15047	107	3018	11	2015	
ES15048		3124	20	2051	
ES15049	199	3124/289	62	2010	
ES15050	123A	3444	20	2051	
ES15051	187	3193	29	2025	
ES15052	199	3122	62	2010	
ES15054	109	3088		1123	
ES15055		3126	90		
ES15056	116	3311	504A	1104	
ES15057	177		300	1122	
ES15102				2015	
ES15226	128	3122	243	2030	
ES15227	186	3192	28	2017	
ES15X1	123A	3444	20	2051	
ES15X10	108	3452	86	2038	
ES15X100	102A		53	2007	
ES15X101	159		82	2032	
ES15X102	107	3018	11	2015	
ES15X104	161	3117	39	2015	
ES15X105	161	3117	39	2015	
ES15X106	233	3117	210	2009	
ES15X107	159	3044/154	40	2012	
ES15X11	123A	3444	20	2051	
ES15X119			39	2015	
ES15X12	152	3444/123A	20	2051	
ES15X120			11	2015	
ES15X121			11	2015	
ES15X122	161	3132	39	2015	
ES15X123	161	3117	39	2015	
ES15X125	241	3440/291	241		
ES15X126	163A	3439	36		
ES15X127	233	3018	210	2009	
ES15X128	159	3025/129	82	2032	
ES15X14	123A	3444	20	2051	
ES15X16	123A	3444	20	2051	
ES15X17	121	3009	239	2006	
ES15X18	108	3452	86	2038	
ES15X19	108	3452	86	2038	
ES15X2	108	3452	86	2038	
ES15X20	123A	3444	20	2051	
ES15X22	233		210	2009	
ES15X23	123A	3444	20	2051	
ES15X24	123A	3444	20	2051	
ES15X3	108	3452	86	2038	
ES15X30	108	3452	86	2038	
ES15X31	102A	3004	53	2007	
ES15X32	102A	3004	53	2007	
ES15X37	123A	3444	20	2051	
ES15X4	102A	3004	53	2007	
ES15X42	123A	3444	20	2051	
ES15X43	121	3009	239	2006	
ES15X45	104	3009	16	2006	
ES15X48	103A	3010	59	2002	
ES15X49	102A	3004	53	2007	
ES15X50	102A	3004	53	2007	
ES15X51	104	3009	16	2006	
ES15X52	127	3034	25		
ES15X53	102A	3004	53	2007	
ES15X54	127	3035	25		
ES15X55	102A	3004	53	2007	
ES15X56	161	3018	39	2015	
ES15X57	161	3018	39	2015	
ES15X58	123A	3444	20	2051	
ES15X59	154	3044	40	2012	
ES15X6	108	3452	86	2038	
ES15X60	108	3452	86	2038	
ES15X61	126	3008	52	2024	
ES15X62	123A	3444	20	2051	
ES15X63	102A	3008	53	2007	
ES15X64	123A	3444	20	2051	
ES15X65	161	3117	39	2015	
ES15X66	107	3117	11	2015	
ES15X67	107	3117	11	2015	
ES15X68	123A	3444	20	2051	
ES15X69	108	3452	86	2038	
ES15X7	123A	3444	20	2051	
ES15X70	123A	3444	20	2051	
ES15X71	103A	3010	59	2002	
ES15X72	102A	3010	53	2007	
ES15X73	126	3004/102A	52	2024	
ES15X74	103A	3010	59	2002	
ES15X75	158	3004/102A	53	2007	
ES15X76	123A	3444	20	2051	
ES15X77	127	3035	25		
ES15X78	104	3009	16	2006	
ES15X79	161	3117	39	2015	
ES15X8	102A	3004	53	2007	
ES15X80	107	3018	11	2015	
ES15X81	107	3018	11	2015	
ES15X82	107	3044/154	11	2015	
ES15X83	123A	3444	20	2051	
ES15X84	123A	3444	20	2051	
ES15X85	123A	3444	20	2051	
ES15X86	152	3893	66	2048	
ES15X87	161	3039/316	39	2015	
ES15X88	161	3039/316	39	2015	
ES15X89	154		40	2012	
ES15X9	159	3114/290	82	2032	
ES15X90	290A	3114/290	82	2032	
ES15X91	228A		257		
ES15X92	312	3834/132	FET-1	2035	
ES15X93	128	3024	243	2030	
ES15X94	165	3115	38		
ES15X95	124	3021	12		
ES15X96	161	3018	39	2015	
ES15X97	107	3018	11	2015	
ES15X98	175	3026	246	2020	
ES15X99	158		53	2007	
ES16(ELCOM)	130		14	2041	
ES1627	177	3100/519	300	1122	
ES16X10			504A	1104	
ES16X103	109	3090		1123	
ES16X12	109	3087		1123	
ES16X13	116	3311	504A	1104	
ES16X14	109	3091		1123	
ES16X16	506	3043	511	1114	
ES16X2	109	3087		1123	
ES16X20	506		511	1114	
ES16X21	110MP	3709	1N60	1123(2)	
ES16X23	177	3100/519	300	1122	
ES16X24	177	3059/138A	300	1122	
ES16X25	116	3032A	504A	1104	
ES16X27	177	3175	300	1122	
ES16X28	506	3043	511	1114	
ES16X29	5080A	3336	ZD-22		
ES16X3	109	3088		1123	
ES16X30	177	3100/519	300	1122	
ES16X31	506	3043	511	1114	
ES16X32	177	3100/519	300	1122	
ES16X33	156	3051	512		
ES16X38			504A	1104	
ES16X4		3091	300	1122	
ES16X40	177	3100/519	300	1122	
ES16X41	5079A	3754/5083A	ZD-28		
ES16X5	109	3087		1123	
ES16X6	109	3088		1123	
ES16X7	109	3087		1123	
ES16X70	110MP	3709		1123(2)	
ES16X9			504A	1104	
ES17(ELCOM)	102A		53	2007	
ES18	121	3014	239	2006	
ES18(ELCOM)	104	3719	16	2006	
ES19	102A	3123	53	2007	
ES19(ELCOM)	160		245	2004	
ES1KX122			61	2038	
ES2(ELCOM)	102A		53	2007	
ES20(ELCOM)	123A		20	2051	
ES21	105	3012	4		
ES21(ELCOM)	121	3717	239	2006	
ES22(ELCOM)	128		243	2030	
ES23	102A	3123	53	2007	
ES23(ELCOM)	160		245	2004	
ES24(ELCOM)	124		12		
ES25	100	3005	1	2007	
ES25(ELCOM)	160		245	2004	
ES26	100	3005	1	2007	
ES26(ELCOM)	102A		53	2007	
ES27(ELCOM)	179		76		
ES28(ELCOM)	155		43		
ES29(ELCOM)	131		44	2006	
ES3	126	3008	52	2024	
ES3(ELCOM)	102A		53	2007	
ES31(ELCOM)	130		14	2041	
ES3110	126	3006/160	52	2024	
ES3111	126	3006/160	52	2024	

Industry Standard No.	ECG	SK	GE	RS 276-	MOTOR.
ES3112	126	3006/160	52	2024	
ES3113	126	3006/160	52	2024	
ES3114	126	3006/160	52	2024	
ES3115	126	3006/160	52	2024	
ES3116	126	3006/160	52	2024	
ES3120	102A	3004	53	2005	
ES3121	102A	3123	53	2005	
ES3122	102A	3004	53	2005	
ES3123	102A	3004	53	2005	
ES3124	102A	3004	53	2005	
ES3125	102A	3004	53	2007	
ES3126	102A	3004	53	2007	
ES32(ELCOM)	154		40	2012	
ES3266	108	3452	86	2038	
ES33(ELCOM)	162		35		
ES34(ELCOM)	159		82	2032	
ES35(ELCOM)	130MP		15	2041(2)	
ES36(ELCOM)	175		246	2020	
ES36X103		3088	1N60	1123	
ES37(ELCOM)	158		53	2007	
ES38(ELCOM)	103A		59	2002	
ES39(ELCOM)	157	3747	232		
ES4(ELCOM)	158		53	2007	
ES40(ELCOM)	194		220		
ES4053	722	3161	IC-9		
ES41	126	3006/160	52	2024	
ES41(ELCOM)	160		245	2004	
ES42(ELCOM)	127	3764	25		
ES43(ELCOM)	181		75	2041	
ES44(ELCOM)	175		246	2020	
ES45(ELCOM)	175		246	2020	
ES46	123	3122	20	2051	
ES46(ELCOM)	123A		20	2051	
ES47(ELCOM)	6400A			2029	
ES47X1	116		504A	1104	
ES49(ELCOM)	218		234		
ES5	101	3861	8	2002	
ES5(ELCOM)	103A		59	2002	
ES50(ELCOM)	131		44	2006	
ES501	105	3012	4		
ES501(ELCOM)	105		4		
ES503	121	3009	239	2006	
ES503(ELCOM)	104	3719	16	2006	
ES51		3025	504A	1104	
ES51(ELCOM)	129		244	2027	
ES51X65	128		243	2030	
ES53(ELCOM)	123A		20	2051	
ES54(ELCOM)	161	3716	39	2015	
ES55(ELCOM)	162		35		
ES56(ELCOM)	128		243	2030	
ES57(ELCOM)	128			2030	
ES57X1	116	3017B/117	504A	1104	
ES57X11	505	3108	SP-2002		
ES57X12	177	3032A	300	1122	
ES57X13	503	3068	CR-5		
ES57X14		3081	512(4)		
ES57X2	116	3017B/117	504A	1104	
ES57X21	503	3068	CR-5		
ES57X4	116	3017B/117	504A	1104	
ES57X5	116	3031A	504A	1104	
ES57X6	113A	3119/113	6GC1		
ES57X7	138A	3059	ZD-7.5		
ES57X8	506	3031A	511	1114	
ES57X9	506	3031A	511	1114	
ES58		3054	57	2017	
ES58(ELCOM)	184		57	2017	
ES59(ELCOM)	179		76		
ES6(ELCOM)	103A		59	2002	
ES60(ELCOM)	185		58	2025	
ES61(ELCOM)	226	3082	49	2025	
ES62(ELCOM)	128		243	2030	
ES63(ELCOM)	176		80		
ES64(ELCOM)	219		74	2043	
ES65(ELCOM)	159		82	2032	
ES66(ELCOM)	182	3188A	55		
ES67(ELCOM)	183	3189A	56	2027	
ES68(ELCOM)	185		58	2025	
ES69(ELCOM)	390		255	2041	
ES7	104	3009	16	2006	
ES7(ELCOM)	131		44	2006	
ES71(ELCOM)	162		35		
ES73(ELCOM)	161	3716	39	2015	
ES74(ELCOM)	180		74	2043	
ES75(ELCOM)	172A		64		
ES75X1	712		IC-2		
ES76(ELCOM)	171		27		
ES8(ELCOM)	160		245	2004	
ES80(ELCOM)	152		66	2048	
ES81	189	3054/196	218	2026	
ES81(ELCOM)	153		69	2049	
ES82(ELCOM)	188		226	2018	
ES83(ELCOM)	189		218	2026	
ES84	163A	3439	36		
ES84(ELCOM)	163A		36		
ES84X1	1008		IC-271		
ES84X3	712		IC-2		
ES85(ELCOM)	123A		20	2051	
ES86(ELCOM)	161	3716	39	2015	
ES89	171	3201	27		
ES89(ELCOM)	171		27		
ES9	104	3009	16	2006	
ES9(ELCOM)	104		16	2006	
ES90	219		74	2043	
ES90(ELCOM)	219		74	2043	
ES91(ELCOM)	176		80		
ES94	225		256		
ES94(ELCOM)	225		256		
ES95	165		38		
ES95(ELCOM)	165		38		
ES96	191	3232	249		
ES96(ELCOM)	191		249		
ES97	194	3275	220		
ES97(ELCOM)	194		220		
ESA-06	168	3312	504A	1104	
ESA-10C	116	3312	504A	1104	
ESA-10N	116	3312	504A	1104	
ESA06	168		504A	1104	
ESA213	126	3006/160	52	2024	
ESA233	126	3006/160	52	2024	
ESD918964P	128		243	2030	
ESI-UPC1020H		3243	IC-196		
ESI-UPC30C		3470	IC-121		
ESK-1		3051	504A	1104	
ESK1			504A	1104	
ESK1/02	116	3017B/117			
ESK1/06	116	3017B/117	504A	1104	
ESKE125C500	125	3032A	510,531	1114	
ESKE40C500	116	3017B/117	504A	1104	
ESPA05-30	527	9035			
ESQ-141	236	3239	216	2053	
ET1	160	3005	245	2004	
ET10	103	3862	8	2002	
ET11	103	3862	8	2002	
ET12	126	3006/160	52	2024	

Industry Standard No.	ECG	SK	GE	RS 276-	MOTOR.
ET1511	102A		53	2007	
ET15X1	102A		53	2007	
ET15X10	123A	3444	20	2051	
ET15X11	123A	3444	20	2051	
ET15X12	123A	3444	20	2051	
ET15X13	123A	3444	20	2051	
ET15X14	123A	3444	20	2051	
ET15X15	123A	3444	20	2051	
ET15X16	123A	3444	20	2051	
ET15X17	121	3009	239	2006	
ET15X18	108	3452	86	2038	
ET15X19	108	3452	86	2038	
ET15X2	107	3019	11	2015	
ET15X20	123A	3444	20	2051	
ET15X21	108		86	2038	
ET15X23	108		86	2038	
ET15X24	123A	3444	20	2051	
ET15X25	102A		53	2007	
ET15X26	127	3035	25		
ET15X27	123A	3444	20	2051	
ET15X29	160		245	2004	
ET15X3	107		11	2015	
ET15X30	108	3452	86	2038	
ET15X31	102A		53	2007	
ET15X32	102A		53	2007	
ET15X33	129	3025	244	2027	
ET15X34	154	3044	40	2012	
ET15X36	128	3024	243	2030	
ET15X37	123A	3444	20	2051	
ET15X38	129		244	2027	
ET15X39	129		244	2027	
ET15X4	121	3009	239	2006	
ET15X40	127	3035	25		
ET15X41	123A	3444	20	2051	
ET15X42	123A	3444	20	2051	
ET15X43	121	3009	239	2006	
ET15X45	123A	3444	20	2051	
ET15X5	121	3009	239	2006	
ET15X54			10	2051	
ET15X7	108		86	2038	
ET15X8	128	3024	243	2030	
ET15X9	108		86	2038	
ET16X1	109	3087		1123	
ET16X10	113A		6GC1		
ET16X11	113A		6GC1		
ET16X13	112		1N82A		
ET16X14	112	3089	1N82A		
ET16X15	142A	3062	ZD-12	563	
ET16X16	117		504A	1104	
ET16X16X			504A	1104	
ET16X17	136A		ZD-5.6	561	
ET16X19	109	3087		1123	
ET16X20	109	3087		1123	
ET16X21	109	3087		1123	
ET16X6	112		1N82A		
ET16X7	113A		6GC1		
ET2	160	3005	245	2004	
ET200	116	3016	504A	1104	
ET234843	123A	3444	20	2051	
ET234854	289A	3124/289	268	2038	
ET238894	123A	3444	20	2051	
ET3	102	3004	2	2007	
ET329218	289A	3124/289	268	2038	
ET350335	234	3114/290	65	2050	
ET35035				2050	
ET352146	289A		268	2038	
ET368021	123A	3444	20	2051	
ET379262	199	3245	62	2010	
ET379462	199	3122	62	2010	
ET380834	199	3245	62	2010	
ET392927	210		252	2018	
ET398711	199	3124/289	62	2010	
ET398777	199	3245	62	2010	
ET4	102	3004	2	2007	
ET400	116	3031A	504A	1104	
ET412626	123A	3444	20	2051	
ET41X27	358A		41		
ET41X37	109	3087		1123	
ET41X47	358A		41		
ET453611	186A	3357	247	2052	
ET491051	312	3112	FET-1	2028	
ET495371	186	3192	28	2017	
ET5	102	3004	2	2007	
ET517263	199	3122	62	2010	
ET517375	235	3141/780	215		
ET517994	199	3245	62	2010	
ET51X25	116	3017B/117	504A	1104	
ET52X25	116	3017B/117	504A	1104	
ET539122	294	3114/290	48		
ET55-25	116	3016	504A	1104	
ET55X25	116	3017B/117	504A	1104	
ET55X29			504A	1104	
ET57X25	116	3016	504A	1104	
ET57X26	120	3110	CR-3		
ET57X29	116	3016	504A	1104	
ET57X30	116	3016	504A	1104	
ET57X31	119	3109	CR-2		
ET57X32	118	3066	CR-1		
ET57X33	116	3110/120	504A	1104	
ET57X35	116	3016	504A	1104	
ET57X38			504A	1104	
ET57X39	506	3125	511	1114	
ET57X40	506	3125	511	1114	
ET58X32	118	3066	CR-1		
ET6	104	3009	504A	2006	
ET600	116	3017B/117		1104	
ET7	105	3012	4		
ET8	101	3011	8	2002	
ET9	101	3011	8	2002	
ETD-10D1	116		504A	1104	
ETD-10D2	116		504A	1104	
ETD-1N60	109	3087		1123	
ETD-HZ7C	5071A	3334	ZD-6.8	561	
ETD-RD75	138A	3059	ZD-7.5		
ETD-RD9.1FB	139A		ZD-9.1	562	
ETD-SD46	109	3090		1123	
ETD-SG9150	177	3100/519	300	1122	
ETD-V06C	116	3032A	504A	1104	
ETD1S2788	177	3100/519	300	1122	
ETDCDG21	177	3100/519	300	1122	
ETI-21	1002		IC-69		
ETI-22	801	3160	IC-35		
ETI-23	1155	3231	IC-179	705	
ETP2008	154		235	2012	
ETP3114	154		231		
ETP3923	154		235	2012	
ETP5092	396		32		
ETS-003	130	3027	14	2041	
ETS-005	390		255	2041	
ETS-017	175		246	2020	
ETS-068	123A	3444	20	2051	
ETS-069	129	3025	244	2027	
ETS-070	128	3024	243	2030	
ETS-071	129	3025	244	2027	

Industry Standard No.	ECG	SK	GE	RS 276-	MOTOR.
ETT-CDC-12000	123A	3122	20	2051	
ETTB-2SB176	102A	3004	53	2007	
ETTB-2SB176A	102A	3004	53	2007	
ETTB-2SB176B	102A	3004	53	2007	
ETTB-2SB176R	102A	3004	53	2007	
ETTB-367B	131	3198	44	2006	
ETTB-75LB	102A		53	2007	
ETTC-2SC490	175		246	2020	
ETTC-458LG	123A	3444	20	2051	
ETTC-930D	229	3122	210	2038	
ETTC-945	199	3124/289	212	2010	
ETTC-CD12000	123A	3444	20	2051	
ETTC-CD13000	123A	3444	20	2051	
ETTC-CD8000	123A	3444	20	2051	
ETTD-235	152	3054/196	66	2048	
ETX18	123A	3444	20	2051	
EU15X1	108	3452	86	2038	
EU15X2	108	3452	86	2038	
EU15X27	128	3024	243	2030	
EU15X3	108	3452	86	2038	
EU15X34	128	3024	243	2030	
EU15X4		3089	1N82A		
EU15X6	108	3452	86	2038	
EU16X1	109	3091		1123	
EU16X11	177	3175	300	1122	
EU16X14	112	3089	1N82A		
EU16X19	110MP	3087	1N34AS	1123(2)	
EU16X2	109	3088		1123	
EU16X20	116	3130	300	1114	
EU16X4	112	3089	1N82A		
EU16X7			504A	1104	
EU16X8	114	3120	6GD1	1104	
EU30X87		3126	90		
EU57X30			504A	1104	
EU57X31	119	3109	CR-2		
EU57X32	118	3066	CR-1		
EU57X38	120	3017B/117	CR-3		
EU57X40	116	3017B/117	504A	1104	
EV16X20		3130	511	1114	
EVD-3		3100	300	1122	
EVM511			504A	1104	
EVR11	141A	3092	ZD-11.5		
EVR110	151A	3099	ZD-110		
EVR110A	151A	3099	ZD-110		
EVR110B	151A	3099	ZD-110		
EVR11A	141A	3092	ZD-11.5		
EVR11B	141A	3092	ZD-11.5		
EVR12	142A	3062	ZD-12	563	
EVR12A	142A	3062	ZD-12	563	
EVR12B	142A	3062	ZD-12	563	
EVR15	145A	3063	ZD-15	564	
EVR15A	145A	3063	ZD-15	564	
EVR15B	145A	3063	ZD-15	564	
EVR27	146A	3064	ZD-27		
EVR27A	146A	3064	ZD-27		
EVR27B	146A	3064	ZD-27		
EVR4	136A	3057	ZD-5.6	561	
EVR4A	136A	3057	ZD-5.6	561	
EVR4B	136A	3057	ZD-5.6	561	
EVR5	137A	3058	ZD-6.2	561	
EVR56	148A	3096	ZD-55		
EVR56A	148A	3096	ZD-55		
EVR56B	148A	3096	ZD-55		
EVR5A	137A	3058	ZD-6.2	561	
EVR5B	137A	3058	ZD-6.2	561	
EVR82	150A	3098	ZD-82		
EVR82A	150A	3098	ZD-82		
EVR82B	150A	3098	ZD-82		
EVR9	139A	3060	ZD-9.1	562	
EVR9A	139A	3066/118	ZD-9.1	562	
EVR9B	139A	3066/118	ZD-9.1	562	
EVS-K112C	113A		6GC1		
EW162	108	3452	86	2038	
EW163	107	3018	11	2015	
EW164	107	3018	11	2015	
EW165	107	3124/289	11	2015	
EW165Y	123A	3444	20	2051	
EW166	109	3087		1123	
EW167	109	3087		1123	
EW168	110MP	3091	1N60	1123(2)	
EW169	156	3032A	512		
EW169B	116(2)		504A(2)	1104	
EW181	123A	3444	20	2051	
EW182	123A	3444	20	2051	
EW183	153	3025/129	69	2049	
EW183B	175	3026	246	2020	
EW212	161	3716	39	2015	
EWQ202	159	3114/290	82	2032	
EWS-78	199	3124/289	62	2010	
EX-141216	128		243	2030	
EX142-X			300	1122	
EX15X25	102A	3004	53	2007	
EX16X10		3100	300	1122	
EX16X27			300	1122	
EX215-X	513		513		
EX39-X	705A		IC-3		
EX4035	722	3161	IC-9		
EX4053	722	3161	IC-9		
EX42-X	714		IC-4		
EX46-X	713		IC-5		
EX48-X	715		IC-6		
EX499-X			20	2051	
EX500-X			20	2051	
EX524-X			255	2041	
EX62-X	790	3454	IC-230		
EX695-X			20	2051	
EX699-X			21	2034	
EX743-X			40	2012	
EX744-X			18	2030	
EX746-X			22	2032	
EX748-X			20	2051	
EX76-X			504A	1104	
EX888-X			20	2051	
EXV420DIR5JB			1N60	1123	
EYV-320D1R2J	177		300	1122	
EYV-420D1R5JA	116	3017B/117	504A	1104	
EYV420D1R5JB	109	3088		1123	
EYV420DIR5JB			1N60	1123	
EYZP-307-900	5529	6629			
EYZP-384	177		300	1122	
EYZP-546	159		82	2032	
EYZP-623	159		82	2032	
EYZP-632	123A	3444	20	2051	
EYZP-791	123A	3444	20	2051	
EYZP-808	234		65	2050	
EYZP623				2032	
EYZP632				2051	
EYZP791				2051	
EYZP808				2050	
EZ10(ELCOM)	140A		ZD-10	562	
EZ11(ELCOM)	5074A		ZD-11.0	563	
EZ110(ELCOM)	151A		ZD-110		
EZ12(ELCOM)	142A		ZD-12	563	
EZ13(ELCOM)	143A	3750	ZD-13	563	
EZ14(ELCOM)	144A		ZD-14	564	
EZ15(ELCOM)	145A		ZD-15	564	
EZ150	145A	3063	ZD-15	564	
EZ16(ELCOM)	5075A	3751	ZD-16	564	
EZ18(ELCOM)	5077A	3752	ZD-18		
EZ20(ELCOM)	5079A	3335	ZD-20		
EZ22(ELCOM)	5080A	3336	ZD-22		
EZ24(ELCOM)	5081A		ZD-24		
EZ27(ELCOM)	141A		ZD-11.5		
EZ3(ELCOM)	5066A	3330	ZD-3.3		
EZ3-12(ELCOM)	5127A	3393	5ZD-12		
EZ3-14(ELCOM)	5129A	3395	5ZD-14		
EZ3-15(ELCOM)	5130A	3396	5ZD-15		
EZ3-16(ELCOM)	5131A	3397	5ZD-16		
EZ3-20(ELCOM)	5135A	3401			
EZ3-30(ELCOM)	5141A	3407	5ZD-30		
EZ3-7(ELCOM)	5120A	3386	5ZD-6.8		
EZ30(ELCOM)	5084A	3755	ZD-30		
EZ33(ELCOM)	147A		ZD-33		
EZ36(ELCOM)	5085A	3337	ZD-36		
EZ39(ELCOM)	5086A	3338	ZD-39		
EZ4(ELCOM)	5067A		ZD-3.9		
EZ47(ELCOM)	5088A	3340	ZD-47		
EZ5(ELCOM)	5069A		ZD-4.7		
EZ5-30(ELCOM)	5141A	3407	5ZD-30		
EZ56(ELCOM)	5090A	3342	ZD-56		
EZ5R6(ELCOM)	136A		ZD-5.6	561	
EZ6(ELCOM)	5071A		ZD-6.8	561	
EZ62(ELCOM)	149A		ZD-62		
EZ68(ELCOM)	5092A	3343	ZD-68		
EZ7(ELCOM)	138A		ZD-7.5		
EZ8(ELCOM)	5072A		ZD-8.2	562	
EZ82(ELCOM)	150A		ZD-82		
EZ9(ELCOM)	139A		ZD-9.1	562	
F-05	116		504A	1104	
F-14A	125	3311	504A	1104	
F-14C	116		504A	1104	
F-302-1			20	2051	
F-302-1532			20	2051	
F-302-2532			20	2051	
F-67-E			16	2006	
F-677				564	
F-7404PC		7404	7404		
F-74174PC	74174	74174			
F-74175PC	74175	74175			
F-7417PC	7417	7417			
F-9311	74154	74154			
F0010	452	3112			
F0015	312	3116			
F0021	312	3116			
F0082M	725		IC-19		
F05		3016	504A	1104	
F1	116	3016	504A	1104	1N4934
F10	125	3033	510,531	1114	1N4007
F101(SCR)	5401	3638/5402			
F10100					MC10500
F10101					MC10501
F10102					MC10502
F10103					MC10503
F10104					MC10504
F10105					MC10505
F10106					MC10506
F10107					MC10507
F10109					MC10509
F10110					MC10110
F10111					MC10111
F10113					MC10513
F10114					MC10514
F10115					MC10515
F10116					MC10516
F10117					MC10517
F10118					MC10518
F10119					MC10519
F10121					MC10521
F10123					MC10123
F10124	116	3311	504A	1104	MC10524
F10125					MC10525
F10130					MC10530
F10131					MC10531
F10132					MC10132
F10133					MC10533
F10134					MC10134
F10135					MC10535
F10136					MC10536
F10137					MC10537
F10141					MC10541
F10145A					MCM10545
F10148	116	3016	504A	1104	
F10153					MC10553
F10158					MC10558
F10159					MC10559
F10160					MC10560
F10161					MC10561
F10162					MC10562
F10164					MC10564
F10165					MC10565
F10166					MC10166
F10168					MC10568
F10170					MC10570
F10171					MC10571
F10172					MC10572
F10173					MC10173
F10174					MC10574
F10175					MC10575
F10176					MC10576
F10177					MC10177
F10179					MC10579
F10180	116	3311	504A	1104	MC10580
F10181					MC10581
F10186					MC10586
F102(SCR)	5402	3638			
F10210					MC10610
F10211					MC10611
F10212					MC10612
F10231					MC10631
F102A(SCR)	5402	3638			
F103(SCR)	5403	3627/5404			
F103P		3118	21	2034	
F104(SCR)	5404	3627			
F10405					MCM10147
F10410					MCM10544
F10415					MCM10546
F11034	125		510,531	1114	
F1110(GAK)	5401	3638/5402			
F1110(SCR)	5401	3638/5402			
F1111(GAK)	5402	3638			
F1112(GAK)	5404	3627			
F1112(SCR)	5404	3627			
F1113(GAK)	5405	3951			
F1114(GAK)	5405	3951			
F114E	506	3843	511	1114	
F1168	230	3042			
F1168A	230	3042			
F1169	5402	3638			

Industry Standard No.	ECG	SK	GE	RS 276-	MOTOR.
F1170	5404	3627			
F1177(SCR)	5402	3638			
F1178(SCR)	5404	3627			
F1180	5531	6631			
F12					MR1-1200
F121-433804	123A	3124/289	20	2051	
F121-546			20	2051	
F121-60216			21	2034	
F121-603			21	2034	
F12100B					MR1130
F1271	5400	3950			
F136	109	3087		1123	
F136555	725		IC-19		
F14			504A	1104	MR1-1400
F14-C			504A	1104	
F1422	230	3042			
F1462	312	3448	FET-1	2035	
F1463	312	3448	FET-1	2035	
F14A	116	3311	504A	1104	
F14AP				1102	
F14B	125	3031A	504A	1104	
F14BP				1102	
F14C	125	3311	510,531	1114	
F14CP				1103	
F14D	125	3031A	504A	1114	
F14E	125	3017B/117	510,531	1114	
F14F	125	3017B/117	510,531	1114	
F14H	125	3033A	510,531	1114	
F14J	125	3033A	510,531	1114	
F15810	108		86	2038	
F15835	108		86	2038	
F15840	123A	3444	20	2051	
F15840-1	123A	3444	20	2051	
F15841	108		86	2038	
OF160			504A	1104	
F161-249-1001	949		IC-25		
OF162			300	1122	
OF164	116	3311	504A	1104	
F16H1	137A		ZD-6.2	561	
F16K					MCM4116
OF173			1N34AS	1123	
F1826-0044	725		IC-19		
F2	116	3017B/117	504A	1104	1N4935
F20			3	2006	
F20-1001	108	3452	86	2038	
F20-1002	108	3452	86	2038	
F20-1003	108	3452	86	2038	
F20-1004	108	3452	86	2038	
F20-1005	108	3452	86	2038	
F20-1006	102A	3004	53	2007	
F20-1007	102A	3004	53	2007	
F20-1008	102A	3004	53	2007	
F20-1009	102A	3004	53	2007	
F20-1010	109	3087		1123	
F20-1012	109	3087		1123	
F20-1013	109	3087		1123	
F20-1014	109	3087		1123	
F20-1015	116	3017B/117	504A	1104	
F20-1016	116	3017B/117	504A	1104	
F2004	454	3991			
F2005	220	3531			
F20303	109	3087		1123	
F2041		3025	67	2023	
F209	159	3119/113	82	2032	
F21490	159	3114/290	82	2032	
F215-1001	108	3452	86	2038	
F215-1002	108	3452	86	2038	
F215-1003		3452	86	2038	
F215-1004		3452	86	2038	
F215-1005		3452	86	2038	
F215-1006	102A	3004	53	2007	
F215-1007	102A	3004	53	2007	
F215-1008	102A	3004	53	2007	
F215-1009	102A	3004	53	2007	
F215-1010	109	3087		1123	
F215-1012	109	3087		1123	
F215-1013	109	3087		1123	
F215-1014	109	3087		1123	
F215-1016	116	3017B/117	504A	1104	
F215-1017	116	3017B/117	504A	1104	
F222	123A		20	2051	
F229(SCR)	5403	3627/5404			
F2427	108	3452	86	2038	
F2443	123A	3444	20	2051	
F2448	123A	3444	20	2051	
F2450	108	3452	86	2038	
F248	314	3898			
F2480			2	2007	
F248A	314	3898			
F24T-011-013			11	2015	
F24T-011-015	229		11	2015	
F24T-016-024	108		11	2015	
F2560-1	729		IC-23		
F2584	123A	3444	20	2051	
F2633	108	3452	86	2038	
F2634	108	3452	86	2038	
F2636	108	3452	86	2038	
F2708					MCM2708
F2708-1					MCM27A08
F3	116	3017B/117	504A	1104	1N4004
F302-1	123A	3444	20	2051	
F302-1532	123A	3444	20	2051	
F302-2	123A	3444	20	2051	
F302-2532	123A	3444	20	2051	
F306-001	123A	3444	20	2051	
F306-022	123A	3444	20	2051	
F318-1	128		243	2030	
F3516E					MCM68A316E
F3519	123A	3444	20	2051	
F3530	108	3452	86	2038	
F3532	123A	3444	20	2051	
F3535	161	3117	39	2015	
F3549	129		244	2027	
F3559	159		82	2032	
F3560	128	3024	243	2030	
F3561	128	3024	18	2030	
F3565	128		243	2030	
F3569	123A	3444	20	2051	
F3570	159	3114/290	82	2032	
F3571	123A	3444	20	2051	
F3574	161	3025/129	39	2015	
F3589	128	3024	243	2030	
F3590	159	3114/290	82	2032	
F3597	159		82	2032	
F361			243	2030	
F366	123A	3444	20	2051	
F4	116	3016	504A	1104	1N4936
F4-005	725		IC-19		
F4001		4001	4001	2401	
F4001B	4001B				MC14001B
F4002		4002	4002		
F4002B					MC14002B
F4006	4006B	4006			
F4006B	4006B				MC14006B
F4006BC	4006B	4006			
F4007		4007	4007		
F4007UB					MC14007B
F4008B	4008B				MC14008B
F4011	4011B	4011	4011	2411	
F4011B	4011B				MC14011B
F4011PC	4011B		4011	2411	
F4012		4012	4012	2412	
F4012B	4012B				MC14012B
F4013	4013B	4013	4013	2413	
F4013B	4013B				MC14013B
F4014	4014B	4014			
F4014B	4014B				MC14014B
F4015	4015B	4015	4015		
F4015B	4015B				MC14015B
F4016	4016B	4016	4016		
F40160B	40160B				MC14160B
F40161B	40161B				MC14161B
F40162B	40162B				MC14162B
F40163B	40163B				MC14163B
F4016B	4016B				MC14016B
F4017	4017B	4017	4017	2417	
F40174B	40174B				MC14174B
F40175B	40175B				MC14175B
F4017B	4017B				MC14017B
F4018	4018B	4022			
F4018B	4018B	4018			MC14018B
F4019	4019B	4019	4019		
F40192B	40192B				MC14510B
F40193B	40193B				MC14516B
F40194B	40194B				MC14194B
F4019B	4019B				MC14519B
F4020	4020B	4020	4020	2420	
F4020B	4020B				MC14020B
F4020PC	4020B	4020			
F4021	4021B	4021	4021	2421	
F4021B	4021B				MC14021B
F4021PC	4021B	4021			
F4022	4022B	4022			
F4022B	4022B				MC14022B
F4023		4023	4023	2423	
F4023B	4023B				MC14023B
F4024		4024	4024		
F4024B	4024B				MC14024B
F4025		4025	4025		
F4025B	4025B				MC14025B
F4027		4027	4027	2427	
F4027B	4027B				MC14027B
F4028B	4028B				MC14028B
F4029	4029B	4029			
F4029B	4029B				MC14029B
F4030		4030	4030		
F4030B	4030B				MC14070B
F4031B					MC14557B
F4035	4035B	4035			
F4035B	4035B				MC14035B
F4040		4040	4040		
F4040B	4040B				MC14040B
F4042		4042	4042		
F4042B	4042B				MC14042B
F4043	4043B	4043			
F4043B	4043B				MC14043B
F4044B	4044B				MC14044B
F4046	980	4046			
F4046B	980				MC14046B
F4049		4049	4049	2449	
F4049B	4049				MC14049UB
F4049PC	4049	4049			
F4050		4050	4050	2450	
F4050B	4050B				MC14050B
F4051		4051	4051	2451	
F4051B	4051B				MC14051B
F4052		4052	4052		
F4052B					MC14052B
F4053B	4053B				MC14053B
F4055PC	4055B	4055			
F4066	4066B	4066			
F4066B	4066B				MC14066B
F4067B					MC14529B
F4068B	4068B				MC14068B
F4069UB					MC14069UB
F4070B					MC14070B
F4071B	4071B				MC14071B
F4072	4072B	4072			
F4073B	4073B				MC14073B
F4075	4075B	4075			
F4075B	4075B				MC14075B
F4076B	4076B				MC14076B
F4077B	4077B				MC14077B
F4078B	4078B				MC14078B
F4081		4081	4081		
F4081B	4081B				MC14081B
F4085B	4085B				MC14506B
F4086B	4086B				MC14506B
F4093B					MC14093B
F4117-1	949		IC-97		
F442-41	725		IC-19		
F4510B	4510B				MC14510B
F4511	4511B	4511			
F4511B					MC14511B
F4511BPC	4511B	4511			
F4512B	4512B				MC14512B
F4514B	4514B				MC14514B
F4515B	4515B				MC14515B
F4516B	4516B				MC14516B
F4518		4518	4518	2490	
F4518B	4518B				MC14518B
F4518PC	4518B	4518			
F4520B	4520B				MC14520B
F4522B	4522B				MC14522B
F4526B					MC14526B
F4528B					MC14528B
F4531B					MC14531B
F4532B					MC14532B
F4539B	4539B				MC14539B
F4555B	4555B				MC14555B
F4556B	4556B				MC14556B
F4581B					MC14581B
F4582B					MC14582B
F4583B	40100B				MC14583B
F4706	108	3039/316	86	2038	
F4709	128	3024	243	2030	
F4846-1	949		IC-97		
F5	116	3017B/117	504A	1104	1N4937
F501	161	3716	39	2015	
F501(ZENITH)	161	3716	39	2015	
F501-16	161	3716	39	2015	
F50116			39	2015	
F502	161	3716	39	2015	
F502(ZENITH)	161	3716	39	2015	
F521(SCR)	5401	3638/5402			
F522(SCR)	5402	3638			
F523	161	3716	39	2015	
F523(SCR)	5404	3627			

Industry Standard No.	ECG	SK	GE	RS 276-	MOTOR.
F5404DM	7404	7404			
F549-1	159		82	2032	
F572(SCR)	5404	3627			
F572-1	123A	3444	20	2051	
F587	123A	3444	20	2051	
F6	116	3017B/117	504A	1104	1N4005
F610		3126	90		
F623	5517	3615			
F625-1	128		243	2030	
F656(GAK)	5401	3638/5402			
F656(SCR)	5402	3638			
F657(GAK)	5402	3638			
F657(SCR)	5402	3638			
F658(GAK)	5404	3627			
F658(SCR)	5404	3627			
F659(GAK)	5405	3951			
F664S	5402	3638			
F67E	121	3009	239	2006	
F68708					MCM68708
F68B10					MCM68B10
F68B308					MCM68B308
F699	159	3114/290	82	2032	
F69916			18	2030	
F7316			17	2051	
F73216	160	3018	245	2004	
F7474PC	7474		7474	1818	
F75116	123A	3444	20	2051	
F757PC	722		IC-9		
F767PC	722	3161	IC-9		
F78L05AC	977	3462			
F78L05AV	977	3591/960			
F78L62AC	988	3973			
F78L62WV	988	3973			
F8	125	3032A	510,531	1114	1N4006
F8(TRANSISTOR)	282	3748			
F88-9779F	736		IC-17		
F889779F	736		IC-17		
F903B-10	5312	3985			
F903B-11	5312	3105			
F903B-12	5313	3106/5304			
F903B-14	5314	3106/5304			
F903B-16	5315	3107/5307			
F903B-18	5316	3107/5307			
F903B-20	5312	3105			
F903B-21	5312	3105			
F903B-22	5313	3106/5304			
F903B-24	5314	3106/5304			
F903B-26	5315	3107/5307			
F903B-28	5316	3107/5307			
F9316PC	74161	74161			
F947(KAG)	5404	3627			
F948(KAG)	5405	3951			
F949(KAG)	5405	3951			
F9600	123A	3444	20	2051	
F9623	123A	3444	20	2051	
F9623(G.E.)	108		86	2038	
F9623F			20	2051	
F9625	108	3452	86	2038	
F96N	108	3452	86	2038	
FA-1			1N60	1123	
FA-1(DIODE)	109	3090		1123	
FA-1(SEARS)	123A		20	2051	
FA-1(TRANSISTOR)	175		246	2020	
FA-2	614	3327			
FA-4			510	1114	
FA111	177		300	1122	
FA2310E				1122	
FA2310U				1122	
FA2320E				1122	
FA2320U				1122	
FA2330E				1122	
FA2330U				1122	
FA4	116	3031A	504A	1104	
FA6	116	3017B/117	504A	1104	
FA8	125	3032A	510,531	1114	
FA8005	141A	3092	ZD-11.5		
FA8006	141A	3092	ZD-11.5		
FA8007	141A	3092	ZD-11.5		
FA8008	141A	3092	ZD-11.5		
FA8009	145A	3063	ZD-15	564	
FA8010	145A	3063	ZD-15	564	
FA8011	145A	3063	ZD-15	564	
FA8012	145A	3063	ZD-15	564	
FB-200	168	3648	BR-600	1173	
FB05	166	9075			
FB100	166	9075			
FB1000	170	3649			
FB1043	109	3126		1123	
FB200	168	3648	BR-600	1173	
FB400	168	3648			
FB401	160	3006	245	2004	
FB402	160	3006	245	2004	
FB403	160	3006	245	2004	
FB420	102A	3004	53	2007	
FB421	102A	3004	53	2007	
FB440	160	3006	245	2004	
FB600	169	3678			
FB6853	123A	3444	20	2051	
FB800	170	3649			
FBC237	123A	3444	20	2051	
FBN-2N1183	176	3123	80		
FBN-35469	175	3026	246	2020	
FBN-35903	175		246	2020	
FBN-36220	130	3510	14	2041	
FBN-36485	130	3510	14	2041	
FBN-36486	175	3026	246	2020	
FBN-36488	175	3538	246	2020	
FBN-36603	130	3510	14	2041	
FBN-36972	181	3510	75	2041	
FBN-36973	181	3510	75	2041	
FBN-37605	124	3021	12		
FBN-38021	124		12		
FBN-38022	181	3511	75	2041	
FBN-38982	124	3021	12		
FBN-CP2293	102	3722	53	2007	
FBN-CP34634	128	3512	243	2030	
FBN-CP34759	176	3123	80		
FBN-L108	124	3044/154	12		
FBN-L109	128	3512	243	2030	
FBN-L113	128	3512	243	2030	
FBN-L115	128		243	2030	
FBN-L148	128	3512	243	2030	
FBT-00-015	912		IC-172		
FBTX070	128		243	2030	
FC-52	614	3325/612			
FC-52M	614	3325/612			
FC5006	108	3452	86	2038	
FCD,810,A					4N27
FCD0003PC	177	3100/519	300	1122	
FCD0014NCS	177	3100/519	300	1122	
FCD0070ANC	614	3126	90		
FCD0070PC		3126	90		
FCD0074		3126	90		
FCD810	3040		H11A5		
FCD810,B,C,D					MOC1006
FCD820	3040		H11A2		
FCD820,C,D					MOC1005
FCD825,B					4N25
FCD825C,D					4N35
FCD830,C,D					MOC1005
FCD831,C,D					MOC1006
FCD836	3040				4N27
FCD836C,D					MOC1006
FCD850C,D					4N29
FCD855C,D					4N29
FCD860C,D					4N32
FCD865C,D					4N32
FCS-9012-HH	159		82	2032	
FCS-9012F	159	3025/129	82	2032	
FCS-9012G	159	3025/129	82	2032	
FCS-9013F	123AP	3024/128	81	2051	
FCS-9013G	123AP	3024/128	81	2051	
FCS-9015C			21	2034	
FCS-9016F			211	2016	
FCS-9016G	123AP		211	2016	
FCS-9018F			20	2051	
FCS1168E	108	3452	86	2038	
FCS1168E641	108	3452	86	2038	
FCS1168F813	123A	3444	20	2051	
FCS1168G	123A	3444	20	2051	
FCS1168G704	123A	3444	20	2051	
FCS1170F	159	3025/129	82	2032	
FCS1225E	108	3452	86	2038	
FCS1227E	108	3452	86	2038	
FCS1227E814	108	3452	86	2038	
FCS1227F	108	3452	86	2038	
FCS1227F743	108	3452	86	2038	
FCS1227G	108	3452	86	2038	
FCS1227G810	108	3452	86	2038	
FCS1229	128	3024	243	2030	
FCS1229F	123A	3444	20	2051	
FCS1229G	123A	3444	20	2051	
FCS1795D	129	3025	244	2027	
FCS8050C	289A	3122	210	2051	
FCS8550	290A	3114/290	269	2032	
FCS8550C	290A		269	2032	
FCS9011E	123AP	3452/108	86	2038	
FCS9011F	123AP		20	2051	
FCS9011G	123AP	3444/123A	20	2051	
FCS9011H	123AP	3452/108	86	2038	
FCS9012	159	3025/129	244	2027	
FCS9012H	159	3114/290	82	2032	
FCS9012HE	159	3114/290	82	2032	
FCS9012HG	159	3025/129	244	2027	
FCS9012HH			21	2034	
FCS9013	123AP	3024/128	81	2051	
FCS9013F	123AP	3024/128	81	2051	
FCS9013G	123AP	3024/128	81	2051	
FCS9013H	123AP	3024/128	81	2051	
FCS9013HG	123AP	3020/123	243	2030	
FCS9013HH	123AP	3024/128	243	2030	
FCS9014	123AP	3452/108	243	2038	
FCS9014(B)			20	2051	
FCS9014B	123AP	3452/108	86	2038	
FCS9014C	123AP	3024/128	243	2030	
FCS9014D	123AP	3452/108	86	2038	
FCS9015B	159	3114/290	82	2032	
FCS9015C	159	3118	21	2034	
FCS9015D	159		21	2034	
FCS9016	123AP	3246/229	20	2051	
FCS9016D	123AP	3246/229	86	2038	
FCS9016E	123AP	3246/229	20	2051	
FCS9016F	123AP	3246/229	61	2016	
FCS9016G	123AP	3246/229	211	2016	
FCS9018	108	3452			
FCS9018D	108	3452	86	2038	
FCS9018E	108	3452	86	2038	
FCS9018F	108	3452	86	2038	
FCS9018G	108	3452	86	2038	
FCS9018H	108	3452	86	2038	
FCS9066	108	3452	86	2038	
FCS916F				2038	
FD-1029-4&				2032	
FD-1029-BG	102	3722	2	2007	
FD-1029-DF	116		504A	1104	
FD-1029-DG	116		504A	1104	
FD-1029-EE	102	3004	2	2007	
FD-1029-ET	121	3717	239	2006	
FD-1029-FY	128		243	2030	
FD-1029-GE	128		243	2030	
FD-1029-GM	128		243	2030	
FD-1029-GP	519		514	1122	
FD-1029-JA	123A		20	2051	
FD-1029-JB	298	3450	272		
FD-1029-JE	5882	3500	5040		
FD-1029-JN	128		243	2030	
FD-1029-JP	129		82	2032	
FD-1029-LL	123A	3444	20	2051	
FD-1029-LU	284		262	2041	
FD-1029-LW	281		263	2043	
FD-1029-MB	159		244	2027	
FD-1029-MC	177		300	1122	
FD-1029-ML	129		244	2027	
FD-1029-MM	128	3024	243	2030	
FD-1029-NG	123A	3444	20	2051	
FD-1029-NS	192		63	2030	
FD-1029-PA	128		243	2030	
FD-1029-PB	286	3194	267		
FD-1029-PP	123A	3444	20	2051	
FD-1029-PT	123A		20	2051	
FD-1029-RB	129		244	2027	
FD-1029ET	121	3717	239	2006	
FD-1073-AP	716		IC-208		
FD-1073-BF	7400	7400		1801	
FD-1073-BG	7402	7402		1811	
FD-1073-BH	7403	7403			
FD-1073-BJ	7404	7404		1802	
FD-1073-BL	7407	7407			
FD-1073-BM	7408	7408		1822	
FD-1073-BN	7410	7410		1807	
FD-1073-BR	7420	7420		1809	
FD-1073-BS	7430	7430			
FD-1073-BU	7440	7440			
FD-1073-BW	7451	7451		1825	
FD-1073-CA	7486	7486		1827	
FD-1073-CH	7496	7496			
FD-1073-CW	7489	7489			
FD-1073-DA	7425	7425			
FD-1073-DF	74155	74155			
FD-1073-DH	74160	74160			
FD-1073-DK	74163	74163			
FD-1073-DL	74175	74175			
FD-1073-DM	74177	74177			
FD-1073-DP	74180	74180			
FD-1073-DR	74191	74191			
FD-1073-EB	7426	7426			
FD-1073-EE	74126	74126			
FD-1073-FB	74174	74174			

Industry Standard No.	ECG	SK	GE	RS 276-	MOTOR.
FD-4500-11	124		12		
FD01880	519		514	1122	
FD06193	519		514	1122	
FD100	519		300	1122	
FD1029EE	102	3722	2	2007	
FD111	519	3100	514	1122	
FD1599	116		504A	1104	
FD1708	177		300	1122	
FD1843	177		300	1122	
FD1980	109	3087		1123	
FD200	177		300	1122	
FD222	177	3087	300	1122	
FD3	116	3016	504A	1104	
FD300			504A	1104	
FD333	519$	3311	504A	1104	
FD3389	116	3311	504A	1104	
FD444	177	3311			
FD4500AL	181		75	2041	
FD6	116		504A	1104	
FD600	519		514	1122	
FD6451	177		300	1122	
FD6489	177		300	1122	
FD777	519	3100	514	1122	
FDH-9	177		300	1122	
FDH400	116	3100/519	504A	1104	
FDH444	519	3100	300	1122	
FDH600	506		511	1114	
FDH6229	177	3100/519	300	1122	
FDH666		3100	300	1122	
FDH694	177		300	1122	
FDH900	177	3100/519	300	1122	
FDH999	519			1122	
FDM1006			300	1122	
FDM1007			300	1122	
FDN600			300	1122	
FDN666		3100	300	1122	
FE-100		3112	FET-2	2035	
FE0654A					2N5486
FE0654B	451	3112			2N5485
FE0654C		3112			2N5484
FE0655A					2N5638
FE0655B					2N5639
FE0655C					2N5640
FE100	312	3112	FET-2	2028	MFE2095
FE100A	312	3112		2028	MFE2095
FE102	312	3112		2028	MFE2094
FE102A	312	3112		2028	MFE2094
FE104					MFE2093
FE104A	312	3112		2028	MFE2093
FE202					MFE2093
FE204					MFE2093
FE300					2N4220A/2A
FE302					2N4220A/2A
FE304					2N4220A
FE3819	457	3116			2N5458
FE400	312	3112		2028	2N3436
FE402	312	3112		2028	2N3437
FE402A	312	3112		2028	
FE404					2N3436
FE404A	312	3112		2028	
FE4302	457	3116			
FE5245			FET-2	2035	2N5486
FE5246		3116			2N5486
FE5247		3116			2N5486
FE5457		3116			2N5457
FE5458		3116			2N5458
FE5459		3116			2N5459
FE5484		3116			2N5484
FEL2113	709	3375/737	IC-11		
FF274	715	3076	IC-6		
FF400	312	3112		2028	
FG-12377	177		300	1122	
FG-2N	506	3311	511	1114	
FG-2N/10D-4			511	1114	
FG-2N2			511	1114	
FG-2NA	116	3031A	504A	1104	
FG2N		3130	511	1114	
FG2PC	506	3130	511	1114	
FG79	199		62	2010	
FGM03-4	500A	3304			
FH100	112		1N82A		
FH70401E	748	3236			
FI 1023	107			2015	
FI-1007	159		82	2032	
FI-1008	159		82	2032	
FI-1019	106		21	2034	
FI-1021	234		65	2050	
FI-1023			11	2015	
FI0049					MFE3021
FI100	464				2N4352
FJ1033			86	2038	
FJH101	7430	7430			
FJH111	7420	7420		1809	
FJH121	7410	7410		1807	
FJH131	7400	7400		1801	
FJH141	7440	7440			
FJH161	7451	7451		1825	
FJH171	7453	7453			
FJH181	7454	7454			
FJH191	7480	7480			
FJH211	7483	7483			
FJH221	7402	7402		1811	
FJH231	7401	7401			
FJH241	7404	7404		1802	
FJH251	7405	7405			
FJH261	7442	7442			
FJH311	7401	7401			
FJH321	7405	7405			
FJJ101	7470	7470			
FJJ111	7472	7472			
FJJ121	7473	7473		1803	
FJJ131	7474	7474		1818	
FJJ141	7490	7490		1808	
FJJ152	7492	7492		1819	
FJJ181	7475	7475		1806	
FJJ191	7476	7476		1813	
FJK101	74121	74121			
FJL101	7441	7441		1804	
FK2369A	108	3452	86	2038	
FK2484	108	3452	86	2038	
FK2894	106	3984	21	2034	
FK3014	108	3452	86	2038	
FK3299	108	3452	86	2038	
FK3300	108	3452	86	2038	
FK3484			17	2051	
FK3494			17	2051	
FK914	108	3452	86	2038	
FK918	108	3452	86	2038	
FL274	712		IC-2		
FLH101	7400	7400		1801	
FLH121	7420	7420		1809	
FLH131	7430	7430			
FLH141	7440	7440			
FLH161	7451	7451		1825	

Industry Standard No.	ECG	SK	GE	RS 276-	MOTOR.
FLH171	7453	7453			
FLH181	7454	7454			
FLH191	7402	7402		1811	
FLH201	7401	7401			
FLH211	7404	7404		1802	
FLH221	7480	7480	IC-115		
FLH241	7483	7483			
FLH271	7405	7405			
FLH281	7442	7442			
FLH291U	7426	7426			
FLH341	7486	7486		1827	
FLH351	7413	7413		1815	
FLH361	7443	7443			
FLH371	7444	7444			
FLH381	7408	7408		1822	
FLH401	74181	74181			
FLH421	74180	74180			
FLH431	7485	7485		1826	
FLH481	7406	7406		1821	
FLH481T	7416	7416			
FLH491T	7417	7417			
FLH501	7412	7412			
FLH511	7423	7423			
FLH521	7425	7425			
FLH531	7437	7437			
FLH541	7438	7438			
FLH551	7448	7448		1816	
FLH581	7411	7411			
FLH601	74132	74132			
FLH611	7422	7422			
FLH621	7427	7427		1823	
FLH631		7432		1824	
FLH661	7428	7428			
FLJ101	7470	7470			
FLJ111	7472	7472			
FLJ121	7473	7473		1803	
FLJ131	7476	7476		1813	
FLJ141	7474	7474		1818	
FLJ151	7475	7475		1806	
FLJ161	7490	7490		1808	
FLJ171	7492	7492		1819	
FLJ191	7495	7495			
FLJ201	74190	74190			
FLJ211	74191	74191			
FLJ231	7494	7494			
FLJ241	74192	74192		1831	
FLJ251	74193	74193		1820	
FLJ261	7496	7496			
FLJ271	74107	74107			
FLJ311	74198	74198			
FLJ321	74199	74199			
FLJ331	7497	7497			
FLJ341	74110	74110			
FLJ351	74111	74111			
FLJ381	74196	74196		1833	
FLJ401	74160	74160			
FLJ411	74161	74161			
FLJ421	74162	74162			
FLJ431	74163	74163			
FLJ441	74164	74164			
FLJ451	74165	74165			
FLJ461	74166	74175			
FLJ531	74174	74174			
FLJ541	74175	74175			
FLJ561	74195	74195			
FLK101	74121	74121			
FLK111	74122	74122			
FLK121	74123	74123		1817	
FLL101	74141	74141			
FLL111	7445	7445			
FLL111T	74145	74145		1828	
FLL121	7446	7446			
FLL121T	7447	7447		1805	
FLL121U	7446	7446			
FLL121V	7447	7447		1805	
FLQ101	7489	7489			
FLQ111	7481	7481			
FLQ131	74170	74170			
FLY101	7460	7460			
FLY111	74150	74150		1829	
FLY121	74151	74151			
FLY131	74153	74153			
FLY141	74154	74154		1834	
FLY151	74155	74155			
FLY161	74156	74156			
FM1613	108	3452	86	2038	
FM1711	108	3452	86	2038	
FM1893	154		40	2012	
FM1J2	116	3017B/117	504A	1104	
FM211N		3311	504A	1104	
FM2368	108	3452	86	2038	
FM2369	108	3452	86	2038	
FM2846	108	3452	86	2038	
FM2894	106	3984	21	2034	
FM3014	108	3452	86	2038	
FM4027					MCM4027A
FM708	108	3452	86	2038	
FM709	108	3452	86	2038	
FM720A	108	3452	86	2038	
FM870	108	3452	86	2038	
FM871	108	3452	86	2038	
FM910	108	3452	86	2038	
FM911	108	3452	86	2038	
FM914	108	3452	86	2038	
FMPS-A20			20	2051	
FMPSA20	123A	3444	20	2051	
FN-51-1A			20	2051	
FO5	116	3017B/117	504A	1104	
FOS100			212	2010	
FOS101			212	2010	
FOS104			17	2051	
FP-2	HIDIV-1	3868/DIV-1	FR-8		
FP-3	HIDIV-2	3869/DIV-2	FR-9		
FP4339					2N4220A'S
FP4340					2N4220A'S
FPA103,4,5					MRS100
FPE100	3028				MLED930
FPE510					MLED930
FPQ3467					MPQ3467
FPQ3468					MPQ3467
FPQ3724					MPQ3724
FPQ3725					MPQ3725
FPR40-1001	123A	3444	20	2051	
FPR40-1003	108	3452	86	2038	
FPR40-1004	102A	3123	53	2007	
FPR40-1005	102A	3123	53	2007	
FPR40-1006	109	3087		1123	
FPR50-1001	123A	3444	20	2051	
FPR50-1002	123A	3444	20	2051	
FPR50-1003	108	3452	86	2038	
FPR50-1004	108	3452	86	2038	
FPR50-1005	102A	3004	53	2007	
FPR50-1006	102A	3004	53	2007	
FPR50-1011	116	3311	504A	1104	

Industry Standard No.	ECG	SK	GE	RS 276-	MOTOR.
FPT100,B					MRD160
FPT120,C					MRD300
FPT131					MRD160
FPT132					MRD160
FPT220					MRD160
FPT400	3036				MRD360
FPT450A					MRD300
FPT500,A					MRD300
FPT510,A					MRD3055
FPT520					MRD300
FPT520A					MRD300
FPT530A					MRD300
FPT550A					MRD300
FPT560					MRD300
FPT570	3036				MRD360
FQ01E	5685	3521/5686			
FQ3467					MQ3467
FQ3468					MQ3467
FQ3724					MQ3725
FQ3725					MQ3725
FR-1	116	3017B/117	504A	1104	
FR-10	125		510,531	1114	
FR-1033	118	3066	CR-1		
FR-1H	116	3017B/117	504A	1104	
FR-1H(M)	116	3017B/117	504A	1104	
FR-1HM			504A	1104	
FR-1M	116	3017B/117	504A	1104	
FR-1MD	116	3311	504A	1104	
FR-1N	116		504A	1104	
FR-1P	116	3017B/117	504A	1104	
FR-1U	519	3017B/117	504A	1104	
FR-2	116	3017B/117	504A	1104	
FR-202	116		504A	1104	
FR-2M			504A	1104	
FR-2P	116	3017B/117	504A	1104	
FR-U			504A	1104	
FR01E	5685	3521/5686			
FR1		3311	504A	1104	1N4934
FR10		3031A	504A	1104	
FR1D			300	1122	
FR1M	116	3017B/117	504A	1104	
FR1MB	116		504A	1104	
FR1MD		3031A	504A	1104	
FR1P		3016	504A	1104	
FR1U		3311	504A	1104	
FR2	116	3313	504A	1104	1N4935
FR2(S1B1)	116		504A	1104	
FR2-005			511	1114	
FR2-02	116	3081/125	504A	1104	
FR2-02(DIO)			504A	1104	
FR2-02(RECT)			511	1114	
FR2-020			504A	1104	
FR2-02C	116	3032A	511	1114	
FR2-04			511	1114	
FR2-06	125	3032A	504A	1104	
FR2-0Z	116			1104	
FR2-10			511	1114	
FR2-0Z			504A	1104	
0000FR202	116	3130	504A	1104	
FR206		3017B	512		
FR2M			504A	1104	
FR2P	116	3311	504A	1104	
FR3					1N4936
FR4					1N4936
FR6					1N4937
FRB-564	175		246	2020	
FRH-101	116	3311	504A	1104	
FRH101	116	3311	504A	1104	
0000000FRI	116	3017B/117	504A	1104	
FRS3693	229	3444/123A	20	2051	
FS-1133			17	2051	
FS-2299			21	2034	
FS-7812	966	3592			
FS06B	5632	3937			
FS06C	5633	3938			
FS06D	5634	3533/5635			
FS06E	5635	3533			
FS102	5402	3638			
FS1168E641			20	2051	
FS1168F813			20	2051	
FS1221	123A	3444	20	2051	
FS12E	5635	3533			
FS1308	107	3039/316	11	2015	
FS1331	128	3024	243	2030	
FS152	5403	3627/5404			
FS1682	161	3716	39	2015	
FS19	109	3087		1123	
FS1974	123A	3444	20	2051	
FS1978	195A	3765	46		
FS1990	234		65	2050	
FS2003-1	130	3027	14	2041	
FS202	5404	3627			
FS2042		3024	63	2030	
FS2043		3122	20	2051	
FS2299	160	3007	245	2004	
FS24226	159	3114/290	82		
FS24954	159	3118	82	2032	
FS26382			21	2034	
FS27233	128	3024	243	2030	
FS27604		3124	20	2051	
FS302	5405	3951			
FS3266	108	3452	86	2038	
FS32669			11	2015	
FS326690	108	3452	86	2038	
FS35529	108	3452	86	2038	
FS3683	108	3452	86	2038	
FS36999	123A	3444	20	2051	
FS402	5405	3951			
FS503PA	6158	7358			
FS503RA	6158	7358			
FS503VA	6158	7358			
FS703PA	6158	7358			
FS703RA	6158	7358			
FS703VA	6158	7358			
FSA1169	115	3121	6GX1		
FSA1177	113A	3119/113	6GC1		
FSA1178	113A	3119/113	6GC1		
FSA1202	115	3121	6GX1		
FSE1001	108	3452	86	2038	
FSE3001	108	3452	86	2038	
FSE4002	199		62	2010	
FSE5002	108	3452	86	2038	
FSIJIIL				2032	
FSP-1	108	3452	86	2038	
FSP-164	108	3452	86	2038	
FSP-165	108	3452	86	2038	
FSP-166	108	3452	86	2038	
FSP-166-1	108	3452	86	2038	
FSP-215	108	3452	86	2038	
FSP-242-1	108	3452	86	2038	
FSP-270-1	108	3452	86	2038	
FSP-288-1	116		504A	1104	
FSP-289-1	108	3452	86	2038	
FSP-42	108	3452	86	2038	
FSP-42-1	108	3452	86	2038	
FSP1			11	2015	
FST2	116	3016	504A	1104	
FST3	116	3016	504A	1104	
FT-1	116	3017B/117	504A	1104	
FT-10			504A	1104	
FT-1M	506	3043	511	1114	
FT-1N			504A	1104	
FT-1P	506	3017B/117	511	1114	
FT001	128	3024	243	2030	
FT0019H	128	3024	243	2030	
FT0019M	159	3114/290	82	2032	
FT002	128	3024	243	2030	
FT003	128	3024	243	2030	
FT004	128	3024	243	2030	
FT004A	128	3024	243	2030	
FT005	123A	3444	20	2051	
FT006	123A	3444	20	2051	
FT008	123A	3122	20	2051	
FT008A	123A	3122	20	2051	
FT023	123A	3122	20	2051	
FT024	123A	3122	20	2051	
FT025	123A	3122	20	2051	
FT026	123A	3122	20	2051	
FT027	128	3024	243	2030	
FT052	159	3114/290	82	2032	
FT053	123A	3122	20	2051	
FT0601	222			2036	
FT0650	452				2N4416
FT0654A		3116			2N3824
FT0654B		3116			2N3824
FT0654C		3116			2N4221
FT0654D		3116			2N4221
FT0654E					2N4220
FT0655A					2N4091
FT0655B					2N4092
FT0655C					2N4093
FT06B	5632	3937			
FT06C	5633	3938			
FT06D	5634	3533/5635			
FT06E	5635	3533			
FT1	177		504A	1122	
FT1(SHARP)	177		504A	1122	
FT10	125	3032A	510,531	1114	
FT107B			85	2010	
FT118	161	3117	39	2015	
FT12C	5633	3938			
FT12E	5635	3533			
FT1315	108	3452	86	2038	
FT1324B	108	3452	86	2038	
FT1324C	108	3452	86	2038	
FT1341	159	3114/290	82	2032	
FT14A	116		504A	1104	
FT1702	106		21	2034	
FT1718					MD3251
FT1718A					MD3251
FT1718B					MD3250
FT1718C					MD3251
FT1718D					MD3250
FT1718E					MD3251
FT1746	159	3114/290	82	2032	
FT19H			82	2032	
FT19M			82	2032	
FT1M	506		511	1114	
FT1N	116	3016	504A	1104	
FT1P			511	1114	
FT2955					FT2955
FT2974					2N2974
FT2975					2N2975
FT2978					2N2978
FT2979					2N2979
FT300	124	3021	12		
FT3055	331				FT3055
FT317					FT317
FT317A					FT317A
FT317B					FT317B
FT34C	154	3045/225	40	2012	
FT34D	154	3045/225	40	2012	
FT34G			FET-2	2035	
FT34Y	312	3448	FET-1	2035	
FT3567			20	2051	
FT3568			81	2051	
FT3569			20	2051	
FT359					FT359
FT3638	159		82	2032	
FT3641	154		40	2012	
FT3642			47	2030	
FT3643	123A		20	2051	
FT3644			67	2023	
FT3838					2N3838
FT40			17	2051	
FT401					FT401
FT4017					2N3806
FT4018					2N3806
FT4019					2N3807
FT402					FT402
FT4020					2N3809
FT4021					2N3808
FT4022					2N3809
FT4023					2N3811
FT4024					2N3810
FT4025					2N3811
FT410	283				MJ410
FT411	94				MJ411
FT413					MJ413
FT417					FT417
FT417A					FT417A
FT417B					FT417B
FT423					MJ423
FT430	97				2N6307
FT431			36		MJ431
FT45	161	3716	39	2015	
FT47					TIP47
FT48					TIP48
FT49					TIP49
FT50					TIP50
FT5040			22	2032	
FT5041			82	2032	
FT527	235	3197	215		
FT57					MFE3006
FT701					MFE3021
FT703					MFE3003
FT704	464				2N4352
FT709	108	3452	86	2038	
FTIM			511	1114	
FTR118	229				MPSH32
FTR129	229				MPSH24
FTR129A					MPSH11
FTR158	229				MPSH32
FTR168	108				MPSH08
FU-1M	552	3043	511	1114	
FU-1MA	552	3016	504A	1104	
FU-1N	552	3032A	504A	1104	
FU-1NA	552	3016	504A	1104	

Industry Standard No.	ECG	SK	GE	RS 276-	MOTOR.
FU-IM	552	3017B/117	504A	1104	
FU10	506	3125	511	1114	
FU1K	506	3125	511	1114	
FU1M	116	3017B/117	504A	1104	
FU1N	552		504A	1104	
FU1NA	552		511	1114	
FU1U	116	3031A	504A	1104	
FU5B7709393	909	3590	IC-249		
FU5B7710393	910		IC-251		
FU5B7740393	940		IC-262		
FU5B7741393	941	3514	IC-263	010	
FU5B7749393	949		IC-25		
FU5D770331X	703A	3157	IC-12		
FU5D770339	703A	3157	IC-12		
FU5E7064393	780	3141	IC-222		
FU5F7711393	911		IC-253		
FU5F7715393	915			017	
FU5F7747393	947		IC-268		
FU5T7725393	925		IC-261		
FU6A7709393	909D	3590	IC-250		
FU6A7710393	910D		IC-252		
FU6A7711393	911D		IC-254		
FU6A7723393	923D		IC-260	1740	
FU6A7741393	941D	3552/941M	IC-264		
FU6A7742393	979	3886			
FU8B770339X	703A		IC-12		
FU9A7723393	923D	3165	IC-260	1740	
FU9A7741393	941D	3552/941M	IC-264		
FU9T7741393	941M	3553	IC-265	007	
FUIU			504A	1104	
FUN14LH026	720	9014	IC-7		
FV-1043	610		90		
FV-11	601	3463			
FV-21	138A		ZD-7.5		
FV-22	137A	3058	ZD-6.2	561	
FV-23	109	3087		1123	
FV-24	5070A	9021	ZD-6.2	561	
FV1043	610	3323	90		
FV21	138A	3059	ZD-7.5		
FV214	605	3864			
FV22	137A	3059/138A	ZD-6.2	561	
FV23	109	3087		1123	
FV2369A	108	3452	86	2038	
FV2484	108	3452	86	2038	
FV2747C	100	3452/108	1	2007	
FV2894	106	3984	21	2034	
FV3014	108	3452	86	2038	
FV3299	108	3452	86	2038	
FV3300	108	3452	86	2038	
FV914	108	3452	86	2038	
FV918	108	3452	86	2038	
FW100	116	9075/166	504A	1104	
FW200	167	3647	504A	1104	
FW300			BR-600	1173	
FW400	116	3648/168	504A	1104	
FW50	166	9075	BR-600	1152	
FW500	116	3016	504A	1104	
FW600	116	3017B/117	504A	1104	
FW600A	116	3017B/117	504A	1104	
FWB3001	167	3647		1172	
FWB3001A					MDA101A
FWB3002	167	3647		1172	
FWB3002A					MDA102A
FWB3003	168	3648		1173	
FWB3004	168	3648		1173	
FWB3004A		3106			MDA104A
FWB3005	169	3107/5307			
FWB3006	169	3107/5307			
FWB3006A					MDA106A
FWB3008A					MDA108A
FWB3010A					MDA110A
FWL100	116		504A	1104	MDA101A
FWL1000					MDA110A
FWL200	116		504A	1104	MDA102A
FWL300	116	3016	504A	1104	MDA104A
FWL400					MDA104A
FWL50					MDA100A
FWL500					MDA106A
FWL600					MDA106A
FWL700					MDA108A
FWL800					MDA108A
FWLA100					MDA201
FWLA1000					MDA210
FWLA200					MDA202
FWLA300					MDA204
FWLA400					MDA204
FWLA50					MDA200
FWLA500					MDA206
FWLA600					MDA206
FWLA700					MDA208
FWLA800					MDA208
FWLC100				1172	MDA970A2
FWLC1000					SPECIAL
FWLC200				1172	MDA970A3
FWLC300				1173	MDA970A4
FWLC400				1173	MDA970A5
FWLC50				1172	MDA970A1
FWLC500					SPECIAL
FWLC600					SPECIAL
FWLC800					SPECIAL
FWLD100	5312	3985			MDA970A2
FWLD1000					SPECIAL
FWLD200	5313	3986			MDA970A3
FWLD300					MDA970A4
FWLD400	5314	3987			MDA970A5
FWLD50	5312	3985			MDA970A1
FWLD500					SPECIAL
FWLD600	5315	3988			SPECIAL
FWLD800	5316	3989			SPECIAL
FX2368			20	2051	
FX274	714		IC-4		
FX3013			61	2038	
FX3014			61	2038	
FX3300			210	2051	
FX3724			47	2030	
FX3962			67	2023	
FX3964		3247	65	2050	
FX4046			210	2051	
FX4960			63	2030	
FX709			11	2015	
FX914			61	2038	
FX918			61	2038	
FZ101			20	2051	
FZ1215			ZD-12	563	
FZ12A	142A	3062	ZD-12	563	
FZ12T10	142A	3062	ZD-12	563	
FZ12T5	142A	3062	ZD-12	563	
FZ14T10	144A	3094	ZD-14	564	
FZ14T5	144A	3094	ZD-14	564	
FZ15A	145A	3063	ZD-15	564	
FZ15T10	145A	3063	ZD-15	564	
FZ15T5	145A	3063	ZD-15	564	
FZ27A	146A	3064	ZD-27		
FZ27T10	146A	3064	ZD-27		

Industry Standard No.	ECG	SK	GE	RS 276-	MOTOR.
FZ27T5	146A	3064	ZD-27		
FZ3.6T10	134A	3055	ZD-3.6		
FZ3.6T5	134A	3055	ZD-3.6		
FZ33A	147A	3095	ZD-33		
FZ5.1T10	135A	3056	ZD-5.1		
FZ5.1T5	135A	3056	ZD-5.1		
FZ5.6T10	136A	3057	ZD-5.6	561	
FZ5.6T5	136A	3057	ZD-5.6	561	
FZ6.2T10	137A	3058	ZD-6.2	561	
FZ82A	150A	3098	ZD-82		
FZ901	135A	3057/136A	ZD-5.1		
FZH111					MC672L
FZH115					MC672TL
FZH131					MC660L
FZH135					MC660TL
FZH141					MC662L
FZH145					MC662TL
FZH191					MC671L
FZH195					MC671TL
FZH201					MC677L
FZH205					MC677TL
FZJ121					MC663L
FZJ125					MC663TL
FZJ131					MC682L
FZJ135					MC682TL
FZJ141					MC684L
FZJ145					MC684TL
FZJ151					MC685L
FZJ155					MC685TL
FZK101					MC667L
FZK105					MC667TL
G-1010			1N60	1123	
G00-003-A	109	3088		1123	
G00-004-A	109	3087		1123	
G00-008-A	109	3088		1123	
G00-009-A	109	3088		1123	
G00-012-A	177		300	1122	
G00-012A	177		300	1122	
G00-013-8	109	3087		1123	
G00-014-A	177	3100/519	300	1122	
G00-502-A			504A	1104	
G00-502A	116	3031A	52	1104	
G00-534-A	116	3311	504A	1104	
G00-535-B	116	3311	504A	1104	
G00-535A	116	3311	504A	1104	
G00-536-A	116		504A	1104	
G00-536A	116		504A	1104	
G00-543-A	116	3311	504A	1104	
G00-551-A	116	3311	504A	1104	
G00-803-A	177	3100/519	300	1122	
G00003A	109	3090		1123	
G00004A	110MP	3709	1N34AS	1123(2)	
G00009A	109	3090		1123	
G0002	160		245	2004	
G0005	158	3004/102A	53	2007	
G0006	102	3004	2	2007	
G0007	102A	3004	53	2007	
G0008	126	3006/160	52	2024	
G0010	126		52	2024	
G004	102A		53	2007	
G005-036C	123A	3444	20	2051	
G005-036E	123A	3444	20	2051	
G00502A			504A	1104	
G00535	116	3311	504A	1104	
G00535A	116		504A	1104	
G01	116	3311	504A	1104	
G01-012-F	140A	3061	ZD-10	562	
G01-036-G	139A		ZD-9.1	562	
G01-036-H	139A		ZD-9.1	562	
G01-036A	137A	3058	ZD-6.2	561	
G01-037-A	142A	3062	ZD-12	563	
G01-083-A	177	3100/519	300	1122	
G01-209-B	177	3100/519	300	1122	
G01-209B	177		300	1122	
G01-217-A	177		300	1122	
G01-401-A		3126	90		
G01-406-A	613	3126	90		
G01-407-A		3126	90		
G01-803-A	177	3088	300(2)	1122(2)	
G01-803A	177	3100/519	300	1122	
G01036	139A	3060	ZD-9.1	562	
G01036A	137A		ZD-6.2	561	
G01036G	139A		ZD-9.1	562	
G01209	177	3100/519	300	1122	
G01209B	177		300	1122	
G01211	116	3100/519	504A	1104	
G01803	177	3100/519	300	1122	
G01803A	177		300	1122	
G01A	112	3089	1N82A		
G02	116	3311	5036	1104	
G03-007C	184	3054/196	57	2017	
G03-017-B	294	3025/129	65		
G03-404-B	129	3513	244	2027	
G03-404-C	129	3513	244	2027	
G03-406-C	185		58	2025	
G03-407-Y	159	3118	82	2032	
G03007	159	3114/290	82	2032	
G03007C	184		57	2017	
G03014	294	3114/290	48		
G03703C	187	3193	29	2025	
G04-041B	107	3018	11	2015	
G04-701-A	121	3717	239	2006	
G04-704-A	131	3198	44	2006	
G04-711-E	158	3004/102A	53	2007	
G04-711-F	158	3004/102A	53	2007	
G04-711-G	158	3004/102A	53	2007	
G04-711-H	158	3004/102A	53	2007	
G04041B	107		11	2015	
G05-003-A	233	3018	11	2015	
G05-003-B	233	3018	11	2015	
G05-004A	108	3452	86	2038	
G05-010-A	123A	3444	20	2051	
G05-011-A	123A	3444	20	2051	
G05-012-G	199	3245	62	2010	
G05-015-D	123A	3444	20	2051	
G05-015C	123A	3047	20	2051	
G05-034-D	123A	3444	20	2051	
G05-035-D	123A	3444	20	2051	
G05-035-D,E			62	2010	
G05-035-E	199	3124/289	62	2010	
G05-035D		3124	62	2010	
G05-035E	123A	3124/289	20	2051	
G05-036-B	199	3444/123A	20	2051	
G05-036-C	123A	3444	20	2051	
G05-036-C,D,E			20	2051	
G05-036-D	199	3444/123A	20	2051	
G05-036-E	123A	3444	20	2051	
G05-036B	85		20	2051	
G05-036C	85		20	2051	
G05-036D	123A	3124/289	20	2051	
G05-036E	123A		20	2051	
G05-037-A	229	3122	20	2051	
G05-037-B	229	3444/123A	20	2051	
G05-037-D	229	3132	61	2038	

Industry Standard No.	ECG	SK	GE	RS 276-	MOTOR.
G05-037B	123A	3132	20	2051	
G05-050-C	107	3018	11	2015	
G05-055-C	289A	3122	243	2030	
G05-055-D	289A	3024/128	268	2038	
G05-055-E	289A	3122	243	2030	
G05-063-R	229	3018	61	2038	
G05-064-A	123A	3444	20	2051	
G05-065-A	229	3018	61	2038	
G05-066A	229	3124/289	213	2038	
G05-406-C	187A	9076	248		
G05-413A	289A	3024/128	243	2030	
G05-413B	85	3020/123	20	2051	
G05-413C	85	3020/123	18	2030	
G05-413D	85	3020/123	18	2030	
G05-415-A	293	3849			
G05-415-B	293	3024/128	47	2030	
G05-416-C	186A	3357	247	2052	
G05-706-D	199	3124/289	212	2010	
G05-706-E	199	3124/289	212	2010	
G05004A	108		86	2038	
G05015C	123A	3444	20	2051	
G05035D	199	3245	62	2010	
G05035E	123A	3444	20	2051	
G05036	123A	3444	20	2051	
G05036B	123A	3444	20	2051	
G05036C	123A	3444	20	2051	
G05036D	123A	3444	20	2051	
G05036E	123A	3444	20	2051	
G05037B	123A	3444	20	2051	
G05059	123A	3444	20	2051	
G05413A	128		243	2030	
G05413B	128		243	2030	
G05415	293	3122	47	2030	
G05705	152	3893	66	2048	
G06-711-B			28	2017	
G06-714C	186	3192	28	2017	
G06-717-B	184	3041	28	2017	
G06-717-C	184	3041	57	2017	
G06-717-D	184		57	2017	
G06-717B			57	2017	
G06-723D	103A	3835			
G06714C	186	3192	28	2017	
G08-005L	312	3112	FET-1	2028	
G08-007-B	312		FET-2	2035	
G08005L	312		FET-1	2028	
G09-003-C	1103	3281			
G09-006-A	1029		IC-162		
G09-006-B	1029		IC-162		
G09-006-C	1029		IC-162		
G09-007-A	1103	3281	IC-94		
G09-007-B	1103	3281	IC-94		
G09-007-C	1103	3281			
G09-007-D	1103	3281			
G09-008-B	1087	3477	IC-103		
G09-008-C	1087	3477	IC-103		
G09-008-D	1087		IC-103		
G09-008-E	1087	3477	IC-103		
G09-009-A	1006	3358	IC-38		
G09-009-B	1103	3281			
G09-009-D	1103	3281			
G09-010-A	1029		IC-162		
G09-011-A	1038		IC-274		
G09-012-A	1038		IC-274		
G09-012-B	1038		IC-274		
G09-012-C	1038		IC-274		
G09-013-A	1058	3459	IC-49		
G09-015-B	1003	3288	IC-43		
G09-017-A	1087		IC-103		
G09-017-B	1087	3477	IC-103		
G09-017-C	1087	3477	IC-103		
G09-017-D	1087	3477	IC-103		
G09-018-A	788	3829			
G09-028-A	801	3829/788	IC-35		
G09-029-B	1052	3281/1103	IC-94		
G09-029-C	1103	3281	IC-94		
G09-029-D	1103	3281	IC-94		
G09-036-G	1029		IC-162		
G09007	1103	3281	IC-94		
G09007B	1103	3281	IC-94		
G09009	1006		IC-38		
G09009A	1006		IC-38		
G09015	1003	3328/740A	IC-43		
G0Q-535-B	116		504A	1104	
G1	116	3017B/117	504A	1104	1N4002
G10	125	3033	510,531	1114	1N4007
G100	109	3090		1123	
G100A			504A	1104	1N4001
G100B	116		504A	1104	1N4002
G100D	116	3016	504A	1104	1N4003
G100F					1N4004
G100G	116	3016	504A	1104	1N4004
G100H					1N4005
G100J	116	3017B/117	504A	1104	1N4005
G100K	125	3032A	510,531	1114	1N4006
G100M	125	3033	510,531	1114	1N4007
G1010	109	3087		1123	
G101079	101	3861	8	2002	
G1010A	177	3100/519	300	1122	
G10119	116		504A	1104	
G11	102A		53	2007	
G110		3012	4		
G12	102A		53	2007	MR1-1200
G1242	116	3311	504A	1104	
G1288	109	3087		1123	
G12T10	142A	3062	ZD-12	563	
G12T20	142A	3062	ZD-12	563	
G12T5	142A	3062	ZD-12	563	
G13	160		245	2004	
G130			504A	1104	
G13638	159	3114/290	82	2032	
G14	102A		53	2007	MR1-1400
G156	109	3087		1123	
G157	109	3087		1123	
G158	109	3091		1123	
G159	109	3087		1123	
G15T10	145A	3063	ZD-15	564	
G15T20	145A	3063	ZD-15	564	
G16			8	2002	
G16506	101	3861	8	2002	
G17			8	2002	
G18		3010	8	2002	
G181-725-001	181	3036	75	2041	
G19	121	3009	239	2006	
G198	109	3091		1123	
G199	109	3087		1123	
G1A					1N4001
G1B					1N4002
G1D					1N4003
G1F					1N4004
G1G					1N4004
G1H					1N4005
G1HA	109	3087		1123	
G1J					1N4005
G1K					1N4006

Industry Standard No.	ECG	SK	GE	RS 276-	MOTOR.
G1M					1N4007
G2	116	3016	504A	1104	MR250-2
G200	109	3091		1123	
G212	199		62	2010	
G222	613	3311	504A	1104	
G23-45	130	3027	14	2041	
G23-46	128		243	2030	
G23-67	130		14	2041	
G23-76	390		255	2041	
G27T10	146A	3064	ZD-27		
G27T20	146A	3064	ZD-27		
G27T5	146A	3064	ZD-27		
G296	116	3311	504A	1104	
G297	109	3087		1123	
G2A	116	3311	504A	1104	1N5391
G2B					1N5392
G2D					1N5393
G2G					1N5395
G2J					1N5397
G2K					1N5398
G2M					1N5399
G3	116	3017B/117	504A	1104	
G3115	311		277		
G390076S1	909	3590	IC-249		
G390077S1	910		IC-251		
G390503S1	915			017	
G390506-S1	940		IC-262		
G390507S2	923D		IC-260	1740	
G395967	123A	3444	20	2051	
G395967-2	123A	3444	20	2051	
G3A					MR500
G3A3	6154	7354			
G3B					MR501
G3B3	6154	7354			
G3C3	6154	7354			
G3D					MR502
G3D3	6154	7354			
G3E3	6154	7354			
G3F					MR504
G3F3	6154	7354			
G3G					MR504
G3G3	6154	7354			
G3H					MR506
G3H3	6156	7356			
G3J					MR506
G3K					MR508
G3K3	6156	7356			
G3M					MR510
G3M3	6158	7358			
G3N3	6158	7358			
G4	116	3031A	504A	1104	
G409	109	3087		1123	
G4090	230	3042			
G498	109	3087		1123	
G4A3	6154	7354			
G4C3	6154	7354			
G4D3	6154	7354			
G4E3	6154	7354			
G4F3	6154	7354			
G4G3	6154	7354			
G4K3	6156	7356			
G4M3	6158	7358			
G4N1	162	3438	35		
G4N3	6158	7358			
G4Z110					.5M110Z10
G4Z7.5					.5M7.5Z10
G5	116	3017B/117	504A	1104	
G5.1T10	135A	3056	ZD-5.1		
G5.1T20	135A	3056	ZD-5.1		
G5.1T5	135A	3056	ZD-5.1		
G5.6T10	136A	3057	ZD-5.6	561	
G5.6T20	136A	3057	ZD-5.6	561	
G5.6T5	136A	3057	ZD-5.6	561	
G5006	5862	3500/5882			
G5010	5882	3500			
G5019	113A	3119/113	6GC1		
G506	610		5016		
G580	109	3087		1123	
G59122	113A		6GC1		
G5A3	6154	7354			
G5A7A66-2	128		243	2030	
G5B3	6154	7354			
G5C	109	3087		1123	
G5C3	6154	7354			
G5D3	6154	7354			
G5E3	6154	7354			
G5F	109	3087		1123	
G5F3	6154	7354			
G5H3	6156	7356			
G5K	109	3087		1123	
G5K3	6156	7356			
G5M3	6158	7358			
G6	116	3017B/117	504A	1104	1N4005
G6004	5862	3012/105			
G6006	5862	3500/5882			
G6010	5882	3500			
G6013	121	3014	239	2006	
G6016	131	3198	44	2006	
G657	116	3016	504A	1104	
G657061	519		514	1122	
G657123	519		514	1122	
G659	116	3016	504A	1104	
G7	112		1N82A		
G700	116	3091	504A	1104	
G701	116	3016	504A	1104	
G702	116	3091	504A	1104	
G766	109	3087		1123	
G769	109	3090		1123	
G770	109	3090		1123	
G788	109	3087		1123	
G789	109	3091		1123	
G790	109	3091		1123	
G7A	112		1N82A		
G7D	109	3087		1123	
G7E	109	3087		1123	
G7F	109	3087		1123	
G7G	109	3087		1123	
G8	125	3032A	510,531	1114	1N4006
G814	109	3087		1123	
G815	109	3087		1123	
G816	109	3087		1123	
G820	109	3087		1123	
G821	109	3091		1123	
G822	109	3091		1123	
G823	109	3091		1123	
G824	109	3091		1123	
G825	109	3087		1123	
G844	109	3087		1123	
G845	109	3091		1123	
G846	109	3087		1123	
G847	109	3087		1123	
G868	109	3091		1123	
G869	109	3091		1123	
G9.1T10	139A	3060	ZD-9.1	562	

Industry Standard No.	ECG	SK	GE	RS 276-	MOTOR.
G9.1T20	139A	3060	ZD-9.1	562	
G9.1T5	139A	3060	ZD-9.1	562	
G9423	229	3018	61	2038	
G9600	123A	3444	20	2051	
G9600(G.E.)	123A		20	2051	
G9623	123A	3444	20	2051	
G9625	229	3018	61	2038	
G9696	123A	3444	20	2051	
GA-10	100	3004/102A			
GA-52829			52	2024	
GA-53149			52	2024	
GA-53242			52	2024	
GA4Z12.0					.5M12AZ
GA4Z2.4					.5M2.4AZ
GA52829	100	3005	1	2007	
GA53149	100	3005	1	2007	
GA53242	100	3005	1	2007	
GA53270	101	3861	8	2002	
GARE	137A		ZD-6.2	561	
GB-1	109	3087		1123	
GBS3115F	5673	3520/5677			
GBS3215F	5673	3520/5677			
GBS3315F	5675	3520/5677			
GBS3415F	5675	3520/5677			
GBS3515F	5676	3520/5677			
GC1003	160		245	2004	
GC1004	160		245	2004	
GC1005	160		245	2004	
GC1006	160		245	2004	
GC1007	160		245	2004	
GC1034	101	3861	8	2002	
GC1035	101	3861	8	2002	
GC1036	101	3011	8	2002	
GC1081	126		52	2024	
GC1092	160	3006	245	2004	
GC1093	160		245	2004	
GC1093X3	160	3006	245	2004	
GC1097	102A		53	2007	
GC1101	102A(2)		53(2)		
GC1134	102A		53	2007	
GC1136	102A		53	2007	
GC1137	103A	3835	59	2002	
GC1142	160		245	2004	
GC1143	102A		53	2007	
GC1144	123A	3444	20	2051	
GC1145	102A	3025/129	53	2007	
GC1146	160		245	2004	
GC1148	160	3008	245	2004	
GC1149	160		245	2004	
GC1150	102A		53	2007	
GC1155	160		245	2004	
GC1159	100	3721	1	2007	
GC1182	160		245	2004	
GC1183	102A		53	2007	
GC1184	102A	3004	53	2007	
GC1185	103A	3835	59	2002	
GC1186	102A	3123	53	2007	
GC1187	102A	3123	53	2007	
GC1257	102A	3123	53	2007	
GC1302	100	3005	1	2007	
GC1422	102A		53	2007	
GC1423	103A	3835	59	2002	
GC148	103A	3835	59	2002	
GC1573	160		245	2004	
GC1615-1	128		243	2030	
GC181	100	3005	1	2007	
GC182	100	3005	1	2007	
GC250	102A	3004	53	2007	
GC282	160	3008	245	2004	
GC283	160	3008	245	2004	
GC284	160	3008	245	2004	
GC285	103A		59	2002	
GC286	103A		59	2002	
GC31	100	3721	1	2007	
GC32	100	3721	1	2007	
GC33	100	3721	1	2007	
GC34	100	3721	1	2007	
GC343	102A	3004	53	2007	
GC35	100	3721	1	2007	
GC360	100	3005	1	2007	
GC387	160		245	2004	
GC388	160		245	2004	
GC389	160		245	2004	
GC4022	100	3004/102A	1	2007	
GC4045	121	3717	239	2006	
GC4057			3	2006	
GC4062	121	3717	239	2006	
GC408	102A	3004	53	2007	
GC4087	121	3009	239	2006	
GC4094	131	3035	44	2006	
GC4097	121	3717	239	2006	
GC4111	121	3717	239	2006	
GC4125	121	3009	239	2006	
GC4144	102A	3123	53	2007	
GC4156	121	3009	239	2006	
GC4251	121	3009	239	2006	
GC4267-2	121	3009	239	2006	
GC452	101	3861	8	2002	
GC453	101	3861	8	2002	
GC454	101	3861	8	2002	
GC460	160	3005	245	2004	
GC461	160	3005	245	2004	
GC462	160	3005	245	2004	
GC463	103A	3835	59	2002	
GC464	102A	3123	53	2007	
GC465	103A	3835	59	2002	
GC466	102A	3004	53	2007	
GC467	103A	3835	59	2002	
GC5000	102A	3123	53	2007	
GC5010			2	2007	
GC5012	109	3087		1123	
GC520	102A	3004	53	2007	
GC521	102A	3005	53	2007	
GC532	100	3005	1	2007	
GC551	102A	3004	53	2007	
GC552	102A	3004	53	2007	
GC578	102A	3004	53	2007	
GC579	102A	3004	53	2007	
GC580	102A	3004	53	2007	
GC581	102A	3004	53	2007	
GC588	102A		53	2007	
GC60	100	3005	1	2007	
GC608	103A	3835	59	2002	
GC609	103A	3835	59	2002	
GC61	100	3005	1	2007	
GC630	160		245	2004	
GC630A	160	3123	245	2004	
GC631	160		245	2004	
GC639	102A	3004	53	2007	
GC640	102A	3123	53	2007	
GC641	121	3009	239	2006	
GC680	102A	3004	53	2007	
GC681	102A		53	2007	
GC682	102A	3004	53	2007	

Industry Standard No.	ECG	SK	GE	RS 276-	MOTOR.
GC691	121	3717	239	2006	
GC692	121	3717	239	2006	
GC733B			2	2007	
GC783	123		20	2051	
GC784	123		20	2051	
GC856	102A	3004	53	2007	
GC864	102A	3004	53	2007	
GD-25	109	3087		1123	
GD-26	109	3087		1123	
GD-29	109	3088		1123	
GD-30	109	3087		1123	
GD-510			300	1122	
GD10				1123	
GD1001	109	3087		1123	
GD101	519		514	1122	
GD102	519		514	1122	
GD11E	109	3091		1123	
GD12	116	3087	504A	1104	
GD12E	109	3016		1123	
GD13E	109	3087		1123	
GD1E	109	3087		1123	
GD1O	109	3090			
GD1P	109	3090		1123	
GD3638	177	3087	300	1122	
GD3638-00	109	3090		1123	
GD400	109	3016		1123	
GD401	109	3087		1123	
GD402	109	3087		1123	
GD403	109	3087		1123	
GD404	109	3087		1123	
GD405	109	3091		1123	
GD406	109	3087		1123	
GD409	109	3087		1123	
GD4E	109	3087		1123	
GD5004	109	3088		1123	
GD510			1N34AS	1123	
GD556	109	3090		1123	
GD5E	109	3087		1123	
GD663	109	3087		1123	
GD6E	109	3087		1123	
GD72E/3	109	3091		1123	
GD72E/4	109	3091		1123	
GD72E/5	109	3016		1123	
GD73E/4	109	3087		1123	
GD73E/5	109	3087		1123	
GD74E/3	109	3087		1123	
GD74E/4	109	3087		1123	
GD74E/5	109	3087		1123	
GD8E	109	3087		1123	
GE-1	100	3722/102	1	2007	
GE-10	123A	3122	10	2051	
GE-10A	199			2010	
GE-11	107	3293	11	2015	
GE-12	124	3021	12		
GE-130	116		1N60		
GE-13MP	104MP	3013	13MP	2006(2)	
GE-14	130	3027	14	2041	
GE-15MP	130MP	3029	15MP	2041(2)	
GE-16	121	3717	16	2006	
GE-17	123A	3854	17	2051	
GE-18	128	3847	18	2030	
GE-19	130	3510	19	2041	
GE-1N506			504A	1104	
GE-1N5061	116	3313	504A	1104	
GE-2	102	3721/100	2	2007	
GE-20	123A	3444	20	2051	
GE-208	160			2004	
GE-21	159	3025/129	21	2032	
GE-210	123A	3124/289	210	2051	
GE-211	123A	3849/293	211	2016	
GE-212	199	3245	212	2010	
GE-213	316	3444/123A	213		
GE-214	316	3039	214	2038	
GE-215		3197	215		
GE-216		3197	216	2053	
GE-217	188	3232/191	217	2018	
GE-218	189		218	2026	
GE-219	195A	3049/224	219		
GE-21A	129			2027	
GE-22	159	3118	22	2032	
GE-220	287	3275/194	220		
GE-221	288	3715	221		
GE-222	287	3433	222		
GE-223	288	3434	223		
GE-224	191	3232	224		
GE-226			226	2018	
GE-228		3528	228		
GE-229		3103A	229		
GE-23	175	3026	23	2020	
GE-231		3104A	231		
GE-232	157	3747	232		
GE-235	287	3044/154	235	2012	
GE-236	299	3202/210	236		
GE-237	165	3115	237		
GE-238	165	3957	238		
GE-239	179	3642	239	2006	
GE-240	213	3012/105	240		
GE-241	196	3054	241	2020	
GE-243	128	3024	243	2030	
GE-244	129	3025	244	2027	
GE-245	160	3006	245	2004	
GE-246	175	3261	246	2020	
GE-247	186A	3192/186	247	2052	
GE-248	187A	3193/187	248		
GE-249	191	3232	249		
GE-24MP	175(2)	3028	24MP		
GE-25	127	3764	25		
GE-250	197	3083	250	2027	
GE-251	198		251		
GE-252	210	3199/188	252	2018	
GE-253	211	3200/189	253	2026	
GE-255	390		255	2041	
GE-256	225	3045	256		
GE-257	228A		257		
GE-258	232	3241	258		
GE-259	238	3710	259		
GE-26	153	3085	26	2049	
GE-260	277	3844/321	260		
GE-261	278		261		
GE-262	280	3297	262	2041	
GE-262MP	280MP	3360	262MP	2041(2)	
GE-263	281	3359	263	2043	
GE-264	282		264		
GE-265	284	3836	265	2047	
GE-265MP	284MP		265MP		
GE-266	285	3846	266		
GE-267	286	3194	267		
GE-268	289A	3122	268	2038	
GE-268MP	289MP		268MP	2038(2)	
GE-269	290A	3114/290	269	2032	
GE-27	171	3201	27		
GE-270	295	3253	270		
GE-271	297	3449	271	2030	
GE-271MP	297MP		271MP		

Industry Standard No.	ECG	SK	GE	RS 276-	MOTOR.
GE-272	298	3450	272		
GE-273	300	3464	273		
GE-273MP	300MP		273MP		
GE-274	307		274		
GE-275	302	3252	275		
GE-276	306	3251	276		
GE-277	311	3195	277		
GE-278	313		278	2016	
GE-279	315	3250	279		
GE-28	186	3192	28	2017	
GE-280	316		280		
GE-281	317		281		
GE-282	317		282		
GE-283			283	2016	
GE-284	320		284		
GE-285		3049	285		
GE-287		3176	287		
GE-288	320	3177/351	288		
GE-29	187	3193	29	2025	
GE-296	320		296		
GE-297		9038	297		
GE-298		3177	298		
GE-299		3218	299		
GE-3	121	3009	3	2006	
GE-30	131	3052	30	2006	
GE-300	177	9091	300	1122	
GE-305	173BP	3999	305		
GE-31MP	131MP	3840	31MP	2006(2)	
GE-32	198	3220	32		
GE-320	195A	3195/311	320		
GE-322	235	3197	322		
GE-3229			66	2048	
GE-325	198	3219	325		
GE-3265	123A	3444		2051	
GE-327		3132	327	2038	
GE-329	235	3239/236	329		
GE-331		3253	331		
GE-332	236	3239	332	2053	
GE-333		3197	333		
GE-334	294	3841	334	2032	
GE-336		3253	336		
GE-337	236		337		
GE-339		3847	339		
GE-34	183			2027	
GE-35	162	3995	35		
GE-36	283	3439/163A	36		
GE-37	164	3710/238	37		
GE-38	165	3115	38		
GE-39	161	3716	39	2015	
GE-4	105	3012	4		
GE-40	154	3044	40	2012	
GE-4000	4000	4000	4000		
GE-4001	4001B	4001	4001	2401	
GE-4002	4002B	4002	4002		
GE-4007		4007	4007		
GE-4009		4009	4009		
GE-4011	4011B	4011	4011	2411	
GE-4012	4012B	4012	4012	2412	
GE-4013	4013B	4013	4013	2413	
GE-4015	4015B	4015	4015		
GE-4016	4016B	4016	4016		
GE-4017	4017B	4017	4017	2417	
GE-4019	4019B	4019	4019		
GE-4020	4020B	4020	4020	2420	
GE-4021	4021B	4021	4021	2421	
GE-4023	4023B	4023	4023	2423	
GE-4024	4024B	4024	4024		
GE-4025	4025B	4025	4025		
GE-4027	4027B	4027	4027	2427	
GE-4030	4030B	4030	4030		
GE-4040	4040B	4040	4040		
GE-4042	4042B	4042	4042		
GE-4049	4049	4049	4049	2449	
GE-4050	4050B	4050	4050	2450	
GE-4051	4051B	4051	4051	2451	
GE-4052	4052B	4052	4052		
GE-4055	4055B	4055	4055		
GE-4081		4081	4081		
GE-41	358A		41		
GE-42	358		42		
GE-43	155	3839	43		
GE-44	131	3198	44	2006	
GE-45	195A	3024/128	45		
GE-4518	4518B	4518	4518	2490	
GE-46	282		46		
GE-47	293	3124/289	47	2030	
GE-47MP	293MP		47MP	2030(2)	
GE-48	294	3114/290	48		
GE-49	226	3082	49	2025	
GE-5	101	3011	5	2002	
GE-50	160		50	2004	
GE-5004	5838	7042/5842	5004		
GE-5005	5839	7049/5849	5005		
GE-5008	5842	7042	5008		
GE-5009	5843	7049/5849	5009		
GE-5012		7048	5012		
GE-5013	5849	7049	5013		
GE-5016	5854	3600	5016		
GE-5017	5855	7069/5869	5017		
GE-5020	5858	3599	5020		
GE-5021	5859	7069/5869	5021		
GE-5024	5862	3584	5024		
GE-5025	5863	7069/5869	5025		
GE-5028	5868	7068	5028		
GE-5029	5869	7069	5029		
GE-5032	5874	3603	5032		
GE-5033	5875	3517/5883	5033		
GE-5036	5878	3602	5036		
GE-5037	5879	3517/5883	5037		
GE-504	116		504A	1104	
GE-5040	5882	3500	5040		
GE-5041	5883	3517	5041		
GE-5044	5890	7090	5044		
GE-5045	5891	7091	5045		
GE-5048	5944	3606	5048		
GE-504A	116	3017B/117	504A	1104	
GE-505				1114	
GE-5064	5916	7096/5896	5064		
GE-5065	5917	7111/5911	5065		
GE-5068	5920	7100/5900	5068		
GE-5069		7111	5069		
GE-5072		7104	5072		
GE-5073	5925	7111/5911	5073		
GE-5076	5932	7110/5910	5076		
GE-5077	5933	7111/5911	5077		
GE-509	125	3033A	509	1114	
GE-5096	5986	3609	5096		
GE-5097	5987	3698	5097		
GE-509A	125			1114	
GE-51	160		51	2004	
GE-510	156	3081/125	510	1114	
GE-5100	5990	3608	5100		
GE-5101	5991	3518/5995	5101		
GE-5104	5994	3501	5104		

Industry Standard No.	ECG	SK	GE	RS 276-	MOTOR.
GE-5105	5995	3518	5105		
GE-5108	6002	7202	5108		
GE-5109	6003	7203	5109		
GE-511	506	3998	511	1114	
GE-512	156	3051	512		
GE-5128	6054	7254	5128		
GE-5129	6055	7261/6061	5129		
GE-513	513	3443	513		
GE-5132	6060	7260	5132		
GE-5133	6061	7261	5133		
GE-5136	6064	7264	5136		
GE-5137	6065	7273/6073	5137		
GE-514	519	3100	514	1122	
GE-5140	6072	7272	5140		
GE-5141	6073	7273	5141		
GE-516	522	3303	516		
GE-517	522	3303	517		
GE-518	500A	3304	518		
GE-519	532	3301/531	519		
GE-52	102A	3004	52	2007	
GE-520	501B	3069/501A	520		
GE-522	536A	3900			
GE-523	522	3303			
GE-524	531	3300/517			
GE-525	532	3301/531			
GE-526	533	3302			
GE-527	500A	3304			
GE-53	158	3004/102A	53	2007	
GE-530		3925	530		
GE-531	125	3081	531	1114	
GE-534	534	3305			
GE-535	535	3307/529			
GE-536	537	3302/533			
GE-537	538	3310			
GE-538	539	3309			
GE-539	521	3304/500A			
GE-540	530	3308	540		
GE-55		3188A	55		
GE-56		3189A	56	2027	
GE-57		3190	57	2017	
GE-58		3191	58	2025	
GE-59		3835	59	2002	
GE-6	101	3011	6	2002	
GE-60	161	3854/123AP	60	2015	
GE-61	108	3246/229	61	2038	
GE-62	199	3245	62	2010	
GE-63	192		63	2030	
GE-64	172A	3156	64		
GE-65	234	3114/290	65	2050	
GE-66	152		66	2048	
GE-663				2030	
GE-67	193		67	2023	
GE-69		3083	69	2049	
GE-6GC1		3119	6GC1		
GE-6GD1	113A	3120/114	6GD1		
GE-6GX1		3121	6GX1		
GE-7	101	3011	7	2002	
GE-700	5513	3042/230	700		
GE-701	310	3855/308			
GE-702	308	3856/310			
GE-72	162	3268	72		
GE-720	1024	3152	720		
GE-721	1027	3153	721		
GE-722	1025	3155	722		
GE-723	1090	3291	723		
GE-724	1028	3436	724		
GE-725	1212		725		
GE-73	163A	3559/162	73		
GE-74	180	3437	74	2043	
GE-7400	7400	7400	7400	1801	
GE-7402	7402	7402	7402	1811	
GE-7404	7404	7404	7404	1802	
GE-7406	7406	7406	7406	1821	
GE-7408	7408	7408	7408	1822	
GE-7410	7410	7410	7410	1807	
GE-74123	74123	74123	74123	1817	
GE-7413	7413	7413	7413	1815	
GE-74145	74145	74145	74145	1828	
GE-74150	74150	74150	74150	1829	
GE-74154	74154	74154	74154	1834	
GE-74192	74192	74192	74192	1831	
GE-74193	74193	74193	74193	1820	
GE-74196	74196	74196	74196	1833	
GE-7420	7420	7420	7420	1809	
GE-7427	7427	7427	7427	1823	
GE-7432	7432	7432	7432	1824	
GE-7441	7441	7441	7441	1804	
GE-7447	7447	7447	7447	1805	
GE-7448	7448	7448	7448	1816	
GE-7451	7451	7451	7451	1825	
GE-7473	7473	7473	7473	1803	
GE-7474	7474	7474	7474	1818	
GE-7475	7475	7475	7475	1806	
GE-7476	7476	7476	7476	1813	
GE-7485	7485	7485	7485	1826	
GE-7486	7486	7486	7486	1827	
GE-7490	7490	7490	7490	1808	
GE-7492	7492	7492	7492	1819	
GE-75	181	3535	75	2041	
GE-750		3329	750		
GE-76	179	3642	76		
GE-8	103	3862	8	2002	
GE-81	123A	3479	81	2051	
GE-82	159	3466	82	2032	
GE-83	188	3199	83	2017	
GE-84	189	3200	84	2025	
GE-85	199		85	2010	
GE-86	316	3293/107	86	2038	
GE-88	192A	3137	88	2030	
GE-89	193	3138	89	2023	
GE-9	160	3006	51	2004	
GE-90	614	3126	90		
GE-9A	160			2004	
GE-FET-1	312	3112	FET-1	2028	
GE-FET-2	312		FET-2	2035	
GE-FET-3		3116	FET-3	2036	
GE-FET-4	222	3065	FET-4	2036	
GE-FR8	HIDIV-1	3868/DIV-1	FR-8		
GE-FR9	HIDIV-2	3869/DIV-2	FR-9		
GE-M100	160			2004	
GE-X1	5483	3942	X1		
GE-X10	6401		X10	2029	
GE-X11	5072A	3136	ZD-8.2	562	
GE-X12	5673	3508/5675	X12		
GE-X13	6408	3523/6407	X13	1050	
GE-X16	5485		X16		
GE-X16A1938	123A	3444		2051	
GE-X17	6402	3628	X17		
GE-X18	186	3893/152	X18	2017	
GE-X3	5521	6621			
GE-X36	116			1104	
GE-X4	5543	3606/5944	X4		
GE-X5	5453	6753	X5	1067	
GE-X66	109	3087		1123	

Industry Standard No.	ECG	SK	GE	RS 276-	MOTOR.
GE-X8	103	3011A	X8	2002	
GE-X9	102A	3004	X9	2007	
GE129	107		1N60	2015	
GE296X	172A	3156			
GE3265			10	2051	
GE3638	109	3087		1123	
GE414	177		300	1122	
GE42-7	116	3031A		1104	
GE505	125			1114	
GE5060	98				MJ10000
GE5061	98				MJ10000
GE5062	98				MJ10001
GE56	183			2027	
GE5ZD-10	5125A	3391	5ZD-10		
GE5ZD-11	5126A	3392	5ZD-11		
GE5ZD-12	5127A	3393	5ZD-12		
GE5ZD-13	5128A	3394	5ZD-13		
GE5ZD-14	5129A	3395	5ZD-14		
GE5ZD-15	5130A	3396	5ZD-15		
GE5ZD-16	5131A	3397	5ZD-16		
GE5ZD-17	5132A	3398	5ZD-17		
GE5ZD-18	5133A	3399	5ZD-18		
GE5ZD-19	5134A	3400	5ZD-19		
GE5ZD-20	5135A	3401	5ZD-20		
GE5ZD-22	5136A	3402/5136	5ZD-22		
GE5ZD-24	5137A	3403/5137	5ZD-24		
GE5ZD-25	5138A	3404	5ZD-25		
GE5ZD-27	5139A	3405	5ZD-27		
GE5ZD-28	5140A	3406	5ZD-28		
GE5ZD-3.3	5111A	3377	5ZD-3.3		
GE5ZD-3.6	5112A	3378	5ZD-3.6		
GE5ZD-3.9	5113A	3379	5ZD-3.9		
GE5ZD-30	5141A	3407	5ZD-30		
GE5ZD-33	5142A	3408	5ZD-33		
GE5ZD-4.3	5114A	3380	5ZD-4.3		
GE5ZD-4.7	5115A	3381	5ZD-4.7		
GE5ZD-5.0		3382	5ZD-5.0		
GE5ZD-5.1	5116A	3382	5ZD-5.1		
GE5ZD-5.6	5117A	3383	5ZD-5.6		
GE5ZD-5.8		3383	5ZD-5.8		
GE5ZD-6.0	5118A	3384	5ZD-6.0		
GE5ZD-6.2	5119A	3385	5ZD-6.2		
GE5ZD-6.6		3386	5ZD-6.6		
GE5ZD-6.8	5120A	3386	5ZD-6.8		
GE5ZD-7.5	5121A	3387	5ZD-7.5		
GE5ZD-8.0		3388	5ZD-8.0		
GE5ZD-8.1		3388	5ZD-8.1		
GE5ZD-8.2	5122A	3388	5ZD-8.2		
GE5ZD-8.7		3389	5ZD-8.7		
GE5ZD-9.1	5124A	3390	5ZD-9.1		
GE6.2			ZD-6.2	561	
GE6.2A			ZD-6.2	561	
GE6.2B			ZD-6.2	561	
GE6060					MJ10004
GE6061					MJ10004
GE6062					MJ10005
GE6063	177	3100/519	300	1122	
GE6366	116		504A	1104	
GEA-121				2025	
GEB100	5312	3985			
GEB101	5312	3985			
GEB102	5313	3986			
GEB104	5314	3987			
GEB106	5315	3988			
GEB108	5316	3989			
GEBR-1000		3649	BR-1000	1114	
GEBR-206		3986	BR-206		
GEBR-425		3681	BR-425		
GEBR-600	169	3676/5305	BR-600		
GECR-1	118	3066	CR-1		
GECR-2	119	3109	CR-2		
GECR-3	120	3110	CR-3		
GECR-4	502	3067	CR-4		
GECR-5	503	3068	CR-5		
GECR-6	504	3108/505	CR-6		
GECR-7	505	3108	CR-7		
GED05B850	109	3087		1123	
GEFR-10	HIDIV-3	3870/DIV-3	FR-10		
GEFR-8		3868	FR-8		
GEFR-9		3869	FR-9		
GEIC-10	708	3135/709	IC-10		
GEIC-101	1105	3285	IC-101		
GEIC-102	1022		IC-102		
GEIC-103	1087	3477	IC-103		
GEIC-104	1085	3476	IC-104		
GEIC-105	1128	3488	IC-105		
GEIC-106	1134	3489	IC-106		
GEIC-107	1133	3490	IC-107		
GEIC-109	1131	3286	IC-109		
GEIC-11	709	3375/737	IC-11		
GEIC-110	1132	3287	IC-110		
GEIC-111	1130	3478	IC-111		
GEIC-112		3294	IC-112		
GEIC-113		3731	IC-113		
GEIC-114		3072	IC-114		
GEIC-115	748	3236	IC-115		
GEIC-116	1047		IC-116		
GEIC-118	1046	3471	IC-118		
GEIC-119	1124		IC-119		
GEIC-12		3157	IC-12		
GEIC-120	1089*		IC-120		
GEIC-121	1049	3470	IC-121		
GEIC-122	1048		IC-122		
GEIC-123	1050	3475	IC-123		
GEIC-124	1077		IC-124		
GEIC-125	1076		IC-125		
GEIC-126	1091*		IC-126		
GEIC-127	1084		IC-127		
GEIC-128	1142	3485	IC-128		
GEIC-13	731	3170	IC-13		
GEIC-130	1092	3472	IC-130		
GEIC-131	1051		IC-131		
GEIC-132	1079		IC-132		
GEIC-133	1050	3475	IC-133		
GEIC-135	1052	3249	IC-135		
GEIC-136	1078	3498	IC-136		
GEIC-137	1184	3486	IC-137		
GEIC-138	1140	3473	IC-138		
GEIC-139	1185	3468	IC-139		
GEIC-14	721		IC-14		
GEIC-140	1082	3461	IC-140		
GEIC-142	1086*		IC-142		
GEIC-143	1061	3228	IC-143		
GEIC-147	712	3072	IC-147		
GEIC-148	712	3072	IC-148		
GEIC-149	1004	3365	IC-149		
GEIC-15	723	3144	IC-15		
GEIC-150	1213	3704	IC-150		
GEIC-153	1229		IC-153		
GEIC-154		3707	IC-154		
GEIC-155	801	3160	IC-155		
GEIC-156	1122		IC-156		
GEIC-157	1094		IC-157		
GEIC-158	1121		IC-158		

Industry Standard No.	ECG	SK	GE	RS 276-	MOTOR.
GEIC-159	1039	3366	IC-159		
GEIC-16	737	3375	IC-16		
GEIC-160	1041	3361	IC-160		
GEIC-161		3362	IC-161		
GEIC-162	1029	3368	IC-162		
GEIC-163	1030		IC-163		
GEIC-164	1032		IC-164		
GEIC-165	1033		IC-165		
GEIC-166	OBS-NLA		IC-166		
GEIC-167	1035	3369	IC-167		
GEIC-168	1036	3370	IC-168		
GEIC-169	1043	3363	IC-169		
GEIC-17	736	3163	IC-17		
GEIC-170	1037	3371	IC-170		
GEIC-171	1087	3477	IC-171		
GEIC-172	912	3543	IC-172		
GEIC-173	975	3565/1171	IC-173		
GEIC-175	1186	3168/749	IC-175		
GEIC-176	1187	3742	IC-176		
GEIC-178	1205	3482	IC-178		
GEIC-179	1155	3231	IC-179	705	
GEIC-18	790	3077	IC-18		
GEIC-180	901	3549	IC-180		
GEIC-181		3487	IC-181		
GEIC-182	1153	3282	IC-182		
GEIC-183	748	3236	IC-183		
GEIC-184	1062	3235/739			
GEIC-185	909		IC-185		
GEIC-19	725	3162	IC-19		
GEIC-190	960	3591	IC-190		
GEIC-191	964	3630	IC-191		
GEIC-193			IC-193	1818	
GEIC-194	7401	7401	IC-194		
GEIC-196	1160	3243	IC-196		
GEIC-197	1160	3474/1081A	IC-197		
GEIC-199	1172		IC-199		
GEIC-2	712	3072	IC-2		
GEIC-20	780	3141	IC-20		
GEIC-201	1093		IC-201		
GEIC-203	1230		IC-203		
GEIC-204	1225		IC-204		
GEIC-205	704	3022/1188	IC-205		
GEIC-207	711	3070	IC-207		
GEIC-208	716		IC-208		
GEIC-209	717		IC-209		
GEIC-21	783	3215	IC-21		
GEIC-210	727	3071	IC-210		
GEIC-211		3163	IC-211		
GEIC-212	735		IC-212		
GEIC-213	742	3453	IC-213		
GEIC-214	743	3172	IC-214		
GEIC-215	744	3171	IC-215	2022	
GEIC-216	745	3276	IC-216		
GEIC-217	746	3234	IC-217		
GEIC-218	747	3279	IC-218		
GEIC-219	750	3280	IC-219		
GEIC-22	728	3073	IC-22		
GEIC-220	778A	3465	IC-220		
GEIC-221	779A	3240/779-1	IC-221		
GEIC-222	780	3141	IC-222		
GEIC-223	781	3169	IC-223		
GEIC-224	782		IC-224		
GEIC-225	783	3215	IC-225		
GEIC-226	785	3254	IC-226		
GEIC-227	786	3140	IC-227		
GEIC-228	787	3146	IC-228		
GEIC-229	788	3147	IC-229		
GEIC-23	729	3074	IC-23		
GEIC-230	790	3077	IC-230		
GEIC-231	791	3149	IC-231		
GEIC-232	795	3237	IC-232		
GEIC-233	797	3158	IC-233		
GEIC-234	798	3216	IC-234		
GEIC-235	802	3277	IC-235		
GEIC-236	784	3524	IC-236		
GEIC-237	803	3278	IC-237		
GEIC-238	805		IC-238		
GEIC-239	806		IC-239		
GEIC-24	744	3171	IC-24	2022	
GEIC-240	807	3451	IC-240		
GEIC-241	810		IC-241		
GEIC-243	814		IC-243		
GEIC-244	815	3255	IC-244		
GEIC-245	900	3547	IC-245		
GEIC-246	906	3548	IC-246		
GEIC-247	907	3545	IC-247		
GEIC-248	908	3539	IC-248		
GEIC-249	909		IC-249		
GEIC-25	949	3166	IC-25		
GEIC-250	909D	3590	IC-250		
GEIC-251	910		IC-251		
GEIC-252	910D		IC-252		
GEIC-253	911		IC-253		
GEIC-254	911D		IC-254		
GEIC-255	913		IC-255		
GEIC-256	914	3541	IC-256		
GEIC-257	916	3550	IC-257		
GEIC-258	917	3544	IC-258		
GEIC-259		3164	IC-259		
GEIC-26	748	3236	IC-26		
GEIC-260	923D	3165	IC-260	1740	
GEIC-261	925		IC-261		
GEIC-262	940		IC-262		
GEIC-263	941	3514	IC-263	010	
GEIC-264	941D	3552/941M	IC-264		
GEIC-265	941M	3552	IC-265	007	
GEIC-266	946		IC-266		
GEIC-267	947		IC-267		
GEIC-268	947	3526	IC-268		
GEIC-269	955M	3564	IC-269	1723	
GEIC-27	804	3455	IC-27		
GEIC-273	1031		IC-273		
GEIC-276	1107		IC-276		
GEIC-277	OBS-NLA		IC-277		
GEIC-278	1115	3184	IC-278		
GEIC-279	1116	3969	IC-279		
GEIC-28	719		IC-28		
GEIC-280	1117		IC-280		
GEIC-281	1123	3743	IC-281		
GEIC-284	1136		IC-284		
GEIC-285	1149		IC-285		
GEIC-286	1150	3890	IC-286		
GEIC-287	1154	3230	IC-287		
GEIC-288	903	3540	IC-288		
GEIC-289	904	3542	IC-289		
GEIC-29	738	3167	IC-29		
GEIC-290	905	3546	IC-270		
GEIC-291		3483	IC-291		
GEIC-292		3497	IC-292		
GEIC-293		3920	IC-293		
GEIC-295	1223	3493	IC-295		
GEIC-296	1224		IC-296		
GEIC-3	705A	3134	IC-3		
GEIC-30	739	3235	IC-30		

Industry Standard No.	ECG	SK	GE	RS 276-	MOTOR.
GEIC-300	1161	3968	IC-300		
GEIC-301	1164	3727	IC-301		
GEIC-302	1173	3729	IC-302		
GEIC-303	1182		IC-303		
GEIC-31	740A	3328	IC-31		
GEIC-311	1098		IC-311		
GEIC-312	1096		IC-312		
GEIC-313	1126		IC-313		
GEIC-314		3876	IC-314		
GEIC-315		3827	IC-315		
GEIC-316		3827	IC-316	704	
GEIC-317		3732	IC-317		
GEIC-32	743	3172	IC-32		
GEIC-33	791	3149	IC-33		
GEIC-34	799	3238	IC-34		
GEIC-35	801	3160	IC-35		
GEIC-36	1004	3365	IC-36		
GEIC-37	1010	3376	IC-37		
GEIC-38	1006	3358	IC-38		
GEIC-39	1121		IC-39		
GEIC-4	714	3075	IC-4		
GEIC-40	1122		IC-40		
GEIC-41	1080	3284	IC-41		
GEIC-42	1005	3723	IC-42		
GEIC-43	1003	3288	IC-43		
GEIC-44	1053		IC-44		
GEIC-45	1054	3457	IC-45		
GEIC-46	1057		IC-46		
GEIC-47	1055	3494	IC-47		
GEIC-48	1056	3458	IC-48		
GEIC-49	1058	3459	IC-49		
GEIC-5	713	3077/790	IC-5		
GEIC-50	1060	3460	IC-50		
GEIC-51	1061	3228	IC-51		
GEIC-52	1062	3235/739	IC-52		
GEIC-53	1063		IC-53		
GEIC-54	1064		IC-54		
GEIC-55	1065		IC-55		
GEIC-56	1066		IC-56		
GEIC-57	1067		IC-57		
GEIC-58	1068		IC-58		
GEIC-59	1072	3295	IC-59		
GEIC-6	715	3076	IC-6		
GEIC-60	1074	3495	IC-60		
GEIC-61	1181	3706	IC-61		
GEIC-62	1137		IC-62		
GEIC-63	1073	3496	IC-63		
GEIC-64	1069	3227	IC-64		
GEIC-65	1070		IC-65		
GEIC-66	1168	3728	IC-66		
GEIC-67	1071	3226	IC-67		
GEIC-68	1163		IC-68		
GEIC-69	1002	3481	IC-69		
GEIC-7	720	9014	IC-7		
GEIC-70	1227	3762	IC-70		
GEIC-71	1158	3289	IC-71		
GEIC-72	1159	3290	IC-72		
GEIC-73	1094		IC-73		
GEIC-74	1215	3738	IC-74		
GEIC-75	1179	3737	IC-75		
GEIC-76	1217	3700	IC-76		
GEIC-77	1009	3499	IC-77		
GEIC-79	1011	3492	IC-79		
GEIC-8	718	3159	IC-8		
GEIC-80	1180	3888	IC-80		
GEIC-81	726	3129	IC-81		
GEIC-82	1188	3022	IC-82		
GEIC-83	726	3129	IC-83		
GEIC-84	704	3023	IC-84		
GEIC-86	724	3525	IC-86		
GEIC-87	1108		IC-87		
GEIC-88	1061	3228	IC-88		
GEIC-89	710	3102	IC-89		
GEIC-9	722	3161	IC-9		
GEIC-90	1106		IC-90		
GEIC-91	1104	3225	IC-91		
GEIC-92	1100	3223	IC-92		
GEIC-93	1102	3224	IC-93		
GEIC-94	1103	3281	IC-94		
GEIC-95	1101	3283	IC-95		
GEIC-96	1004	3365	IC-96		
GEIC-97	749	3168	IC-97		
GEIC-98	1080	3284	IC-98		
GEIC-99	1109	3711	IC-99		
GEL2111	708	3135/709	IC-10		
GEL2111AL1	708	3135/709	IC-10		
GEL2111F1	708	3135/709	IC-10		
GEL2113	709	3375/737	IC-11		
GEL2113AL1	709	3375/737	IC-11		
GEL2113F1	709	3375/737	IC-11		
GEL2114	713	3077/790	IC-5		
GEL234	717		IC-209		
GEL234F1	717		IC-209		
GEL234F2	717		IC-209		
GEL239F2	721		IC-14		
GEL277	804	3455	IC-27		
GEL3072F1	790	3077	IC-5		
GEMR-1	5874	3603	5032		
GEMR-2	5990	3608	5100		
GEMR-3	5545	3582	MR-3		
GEMR-4	5494	3582/5545	MR-4		
GEMR-5	5455	3597	MR-5	1067	
GEMR-6	186	3192	MR-6	2017	
GER-A	102A	3004	53	2007	
GER-A-D	160	3006	245	2004	
GER4001		3311		1114	1N4001
GER4002		3311		1114	1N4002
GER4003		3311		1114	1N4003
GER4004		3031A		1114	1N4004
GER4005		3017B		1114	1N4005
GER4006		3032A		1114	1N4006
GER4007		3033A		1114	1N4007
GES2222	123AP		S2222		
GES2222A	123AP		S2222A		
GESS-2AV2	511	3759	SS-2AV2		
GESS-3A3	508	3756	SS-3A3		
GESS-3AT2	509	3757	SS-3AT2		
GESS-3DB3	510	3758	SS-3DB3		
GESS-6DW4	512	3760	SS-6DW4		
GET-103			2	2007	
GET-113		3004	2	2007	
GET-113A		3004	2	2007	
GET-114		3004	2	2007	
GET-572		3009	3	2006	
GET-672		3006	51	2004	
GET-672A		3006	51	2004	
GET-673		3006	51	2004	
GET-692		3008	51	2004	
GET-873A		3008	51	2004	
GET-883		3008	51	2004	
GETO-50P			53	2007	
GET103	102A		53	2007	
GET113	102A		53	2007	
GET113A	102A		53	2007	

Industry Standard No.	ECG	SK	GE	RS 276-	MOTOR.
GET114	102A		53	2007	
GET2221	123A	3444	20	2051	
GET2221A			20	2051	
GET2222	123A	3444	20	2051	
GET2222A			20	2051	
GET2369	123A	3444	20	2051	
GET2483		3122	62	2010	
GET2904		3025	21	2034	
GET2905		3025	21	2034	
GET2906		3114	48		
GET2907		3114	48		
GET3013	123A	3444	20	2051	
GET3014	123A	3444	20	2051	
GET3416	123A	3124/289			
GET3562			212	2010	
GET3563			61	2038	
GET3638	159	3114/290	82	2032	
GET3638A	159	3118	82	2032	
GET3646	123A	3444	20	2051	
GET3903			210	2051	
GET3904			20	2051	
GET3905			21	2034	
GET3906			21	2034	
GET4870	6401			2029	
GET4871	6409			2029	
GET5116	160	3123	245	2004	
GET5117	160	3123	245	2004	
GET5172	123A	3124/289			
GET5305	172A	3156	64		
GET5306	172A	3156	64		
GET5307	172A	3156	64		
GET5308	172A	3156	64		
GET5308A	172A	3156	64		
GET572	121	3717	239	2006	
GET671	160		245	2004	
GET672	160		245	2004	
GET672A	160		245	2004	
GET673	160		245	2004	
GET691	160		245	2004	
GET692	160		245	2004	
GET693	160		245	2004	
GET706	123A	3444	20	2051	
GET708	123A	3444	20	2051	
GET871	126	3006/160	52	2024	
GET872	126	3006/160	52	2024	
GET873	126		52	2024	
GET873A	126		52	2024	
GET874	126		52	2024	
GET875	126	3006/160	52	2024	
GET880	100	3005	1	2007	
GET881	100	3005	1	2007	
GET882	100	3005	1	2007	
GET883	126		52	2024	
GET885	126	3006/160	52	2024	
GET887	100	3005	1	2007	
GET888	100	3005	1	2007	
GET889	100	3005	1	2007	
GET890	100	3005	1	2007	
GET891	100	3005	1	2007	
GET892	100	3005	1	2007	
GET895	100	3005	1	2007	
GET896	100	3005	1	2007	
GET897	100	3005	1	2007	
GET898	102	3004	2	2007	
GET914	123A	3444	20	2051	
GET929	128	3122	243	2030	
GET930	128	3122	243	2030	
GETO-50P	158	3004/102A	53	2007	
GEVR-100		3462	VR-100		
GEVR-102		3591	VR-102		
GEVR-105		3671	VR-105		
GEVR-106		3724	VR-106		
GEVR-108		3630	VR-108		
GEVR-111		3592	VR-111		
GEVR-114		3673	VR-114		
GEX36			504A	1104	
GEX66			504A	1104	
GEX8			8	2002	
GEZD-10	140A	3061	ZD-10	562	
GEZD-10-4	140A			562	
GEZD-10.4			ZD-10.4	562	
GEZD-100	5096A	3346	ZD-100		
GEZD-11	5074A	3139	ZD-11	563	
GEZD-11.5	141A	3092	ZD-11.5		
GEZD-110	151A	3099			
GEZD-12	142A	3062	ZD-12	563	
GEZD-120	5097A	3347	ZD-120		
GEZD-13	143A	3750	ZD-13	563	
GEZD-130	5098A	3348	ZD-130		
GEZD-14	144A	3094	ZD-14	564	
GEZD-140	5099A	3349	ZD-140		
GEZD-15	145A	3063	ZD-15	564	
GEZD-150	5100A	3350	ZD-140		
GEZD-16	5075A	3751	ZD-16	564	
GEZD-160	5101A	3351	ZD-160		
GEZD-17	5076A	9022	ZD-17		
GEZD-170	5102A	3352	ZD-170		
GEZD-18	5077A	3752	ZD-18		
GEZD-180	5103A	3353	ZD-180		
GEZD-19	5078A	9023	ZD-19		
GEZD-190	5104A	3354	ZD-190		
GEZD-20	5079A	3335	ZD-20		
GEZD-200	5105A	3355	ZD-200		
GEZD-22	5080A	3336	ZD-22		
GEZD-24	5081A	3151	ZD-24		
GEZD-25	5082A	3753			
GEZD-27	146A	3064	ZD-27		
GEZD-28	5083A	3754	ZD-28		
GEZD-3.3	5066A	3330	ZD-3.3		
GEZD-3.6	134A	3055	ZD-3.6		
GEZD-3.9	5067A	3331	ZD-3.9		
GEZD-30	5084A	3755	ZD-30		
GEZD-33	147A	3095	ZD-33		
GEZD-36	5085A	3337	ZD-36		
GEZD-39	5086A	3338	ZD-39		
GEZD-4.3	5068A	3332	ZD-4.3		
GEZD-4.7	5069A	3333	ZD-4.7		
GEZD-43	5087A	3339	ZD-43		
GEZD-47	5088A	3340	ZD-47		
GEZD-5.0	135A	3056	ZD-5.0		
GEZD-5.1	135A	3056	5ZD-5.1		
GEZD-5.6	136A	3057	ZD-5.6	561	
GEZD-51	5089A	3341	ZD-51		
GEZD-55	148A	3096	ZD-55		
GEZD-56	5090A	3342	ZD-56		
GEZD-6.0	5070A	9021	ZD-5.0	561	
GEZD-6.2	137A	3058	ZD-6.2	561	
GEZD-6.6			ZD-6.6	561	
GEZD-6.8	5071A	3334	ZD-6.8	561	
GEZD-60	5091A		ZD-60		
GEZD-62	149A	3097	ZD-62		
GEZD-68	5092A	3343	ZD-68		
GEZD-7.5	138A	3059	ZD-7.5		
GEZD-75	5093A	3344	ZD-75		

Industry Standard No.	ECG	SK	GE	RS 276-	MOTOR.
GEZD-8.1		3136	ZD-8.1		
GEZD-8.2	5072A	3136	ZD-8.2	562	
GEZD-8.7		3749		562	
GEZD-82	150A	3098	ZD-82		
GEZD-87	5094A	9024	ZD-87		
GEZD-9.1	139A	3060	ZD-9.1	562	
GEZD-91	5095A	3345	ZD-91		
GEZJ252A			504A	1104	
GEZJ252B			504A	1104	
GF20	102	3123	2	2007	
GF21	102	3123	2	2007	
GF32	102	3123	2	2007	
GFT3008/40	102	3009	2	2007	
GFT44	160		245	2004	
GFT45	160		245	2004	
GG686					MRD300
GGE-51	160		245	2004	
GH1F	525	3925	530		
GH3F		3998	511	1114	
GI-100-1			504A	1104	
GI-1N4385	116	3311	504A	1104	
GI-2711			10	2051	
GI-2712			17	2051	
GI-2714			210	2051	
GI-2715			60	2015	
GI-2716			60	2015	
GI-2921			60	2015	
GI-2922			60	2015	
GI-2923			60	2015	
GI-2924			212	2010	
GI-2925			212	2010	
GI-2926			211	2016	
GI-3008	116	3017B/117		1104	
GI-300B	116	3311			
GI-300D	116	3311	504A	1104	
GI-3391			212	2010	
GI-3391A			212	2010	
GI-3392			212	2010	
GI-3393			212	2010	
GI-3394			212	2010	
GI-3395			212	2010	
GI-3396			212	2010	
GI-3397			212	2010	
GI-3398			212	2010	
GI-3403			210	2051	
GI-3405			210	2051	
GI-3415			210	2051	
GI-3416			210	2051	
GI-3417			210	2051	
GI-3605			60	2015	
GI-3606			60	2015	
GI-3607			60	2015	
GI-3638			221	2024	
GI-3638A			221	2024	
GI-3641			210	2051	
GI-3642			210	2051	
GI-3643			210	2051	
GI-3704			20	2051	
GI-3705			20	2051	
GI-3706			20	2051	
GI-3707			212	2010	
GI-3708			212	2010	
GI-3709			212	2010	
GI-3710			212	2010	
GI-3711			212	2010	
GI-3721			212	2010	
GI-3900			212	2010	
GI-3900A			212	2010	
GI-P100-D	116	3031A	504A	1104	
GI-TVC3	120	3110	CR-3		
GI08B	116	3311	504A	1104	
GI1	100	3005	1	2007	
GI1-1000					MR1-1000
GI1-1200					MR1-1200
GI1-1400					MR1-1400
GI1-1600					MR1-1600
GI10	123A	3444	20	2051	
GI2	102	3004	2	2007	
GI237	125		510,531	1114	
GI2500					MR2500
GI2501					MR2501
GI2502					MR2502
GI2504					MR2504
GI2506					MR2506
GI2508					MR2508
GI2510					MR2510
GI2711	123A	3444	20	2051	
GI2712	123A	3444	20	2051	
GI2713	123A	3444	20	2051	
GI2714	123A	3444	20	2051	
GI2715	123A	3444	20	2051	
GI2716	123A	3444	20	2051	
GI2921	123A	3444	20	2051	
GI2922	123A	3444	20	2051	
GI2923	123A	3444	20	2051	
GI2924	123A	3444	20	2051	
GI3	126		52	2024	
GI3002	116	3312	504A	1104	
GI3638	159	3114/290	82	2032	
GI3638A	159	3114/290	82	2032	
GI3641	123A	3444	20	2051	
GI3643	123A	3444	20	2051	
GI3644	159	3114/290	82	2032	
GI3702	159	3114/290	82	2032	
GI3703	159	3444/123A	82	2032	
GI3704	123A	3444	20	2051	
GI3705	123A	3444	20	2051	
GI3706	123A	3122	20	2051	
GI3707	123A	3444	20	2051	
GI3708	123A	3444	20	2051	
GI3709	123A	3444	20	2051	
GI3710	123A	3444	20	2051	
GI3711	123A	3444	20	2051	
GI3793	161	3716	39	2015	
GI3992-17	116	3311	504A	1104	
GI4	102	3004	2	2007	
GI411	116	3311	504A	1104	
GI420	116	3311	504A	1104	
GI5	101	3861	8	2002	
GI500					MR500
GI501					MR501
GI502					MR502
GI504					MR504
GI506					MR506
GI508					MR508
GI510					MR510
GI6	101	3861	8	2002	
GI6506	101	3861	8	2002	
GI7		3861	8	2002	
GI750					MR750
GI751					MR751
GI752					MR752
GI754					MR754
GI756					MR756

Industry Standard No.	ECG	SK	GE	RS 276-	MOTOR.
GI758					MR758
GI8	103	3862	8	2002	
GI810					MR810
GI811					MR811
GI812					MR812
GI814					MR814
GI816					MR816
GI817					MR817
GI818					MR818
GI820					MR820
GI821					MR821
GI822					MR822
GI824					MR824
GI826					MR826
GI850					MR850
GI851					MR851
GI852					MR852
GI854					MR854
GI856					MR856
GI9	105		4		
GI910					MR910
GI911					MR911
GI912					MR912
GI914					MR914
GI916					MR916
GI917					MR917
GI918					MR918
GIP100D			504A	1104	
GJ4M	116	3016	504A	1104	
GLA100	140A	3061	ZD-10	562	
GLA100A	140A	3061	ZD-10	562	
GLA100B	140A	3061	ZD-10	562	
GLA39	134A	3055	ZD-3.6		
GLA39A	134A	3055	ZD-3.6		
GLA39B	5067A	3055/134A	ZD-3.6		
GLA47	135A	3056	ZD-5.1		
GLA47A	135A	3056	ZD-5.1		
GLA47B	5069A	3056/135A	ZD-5.1		
GLA51	135A		ZD-5.1		
GLA51A	135A		ZD-5.1		
GLA51B	135A		ZD-5.1		
GLA56	136A	3057	ZD-5.6	561	
GLA56A	136A	3057	ZD-5.6	561	
GLA56B	136A	3057	ZD-5.6	561	
GLA62	137A	3058	ZD-6.2	561	
GLA62A	137A	3058	ZD-6.2	561	
GLA62B	137A	3058	ZD-6.2	561	
GLA82B	5072A	3136			
GLA91	139A	3060	ZD-9.1	562	
GLA91A	139A	3060	ZD-9.1	562	
GLA91B	139A	3060	ZD-9.1	562	
GLAZ2.6A					1N702A
GLAZ6.8A					1N710A
GLZ100A					1N738A
GLZ24A					1N769A
GLZ7.0A					1N763A
GLZ7.5A					1N711A
GM-1	156			1103	
GM-1A	156			1104	
GM-1C				1114	
GM-1Z	156			1102	
GM-3	156		512	1144	
GM-3B	156		512		
GM-3Y	156		512	1143	
GM-3Z	156		512	1143	
GM-770			11	2015	
GM0290	160		245	2004	
GM0375	160		245	2004	
GM0376	160		245	2004	
GM0377	160		245	2004	
GM0378	168	3648	52	1173	
GM0380	108		86	2038	
GM1J2	125		510,531	1114	
GM290	160		245	2004	
GM290A	160		245	2004	
GM308	161	3019	39	2015	
GM378	160		245	2004	
GM378A	160		245	2004	
GM378RED	160		245	2004	
GM380	126		52	2024	
GM3Y	156	3051	512		
GM3Z	156		512		
GM428	121	3717	239	2006	
GM656A	160		245	2004	
GM760	159	3114/290	82	2032	
GM770	107	3039/316	11	2015	
GM875	160		245	2004	
GM876	160		245	2004	
GM877	160		245	2004	
GM878	160		245	2004	
GM878A	160		245	2004	
GM878B	160		245	2004	
GME.022				2038	
GME040-1	159	3114/290	82	2032	
GME0404	106	3118	21	2034	
GME0404-1		3118	21	2034	
GME0404-2	106	3118	21	2034	
GME1001	123A	3444	20	2051	
GME1002	123A	3444	20	2051	
GME2001	123A	3444	20	2051	
GME2002	123A	3444	20	2051	
GME3001	108	3452	86	2038	
GME3002	108	3452	86	2038	
GME4001	123A	3444	20	2051	
GME4002	123A	3444	20	2051	
GME4003	123A	3444	20	2051	
GME404-1	106		21	2034	
GME6001			11	2015	
GME6002			11	2015	
GME6003	123A	3444	20	2051	
GME9001	108	3452	86	2038	
GME9002	108	3452	86	2038	
GME9021	108	3452	86	2038	
GME9022	108	3452	86		
GM0-380			17	2051	
GM0290	160		245	2004	
GM0375	160		245	2004	
GM0376	160		245	2004	
GM0377	160		245	2004	
GM0378	160		245	2004	
GM0380	161	3117	39	2015	
G00535	116		504A	1104	
G01036	139A		ZD-9.1	562	
G01209	177	3100/519	300	1122	
G01211	177	3100/519	300	1122	
G01803	177	3100/519	300	1122	
G03-007C			28	2017	
G03-404-B				2027	
G03007	159		82	2032	
G03014	159		82	2032	
G04-041B			20	2051	
G04-701-A			16	2006	
G04-703-A	176		80		
G04-704-A			30	2006	

Industry Standard No.	ECG	SK	GE	RS 276-	MOTOR.
G04-711-E			53	2007	
G04-711-F			53	2007	
G04-711-G			53	2007	
G04-711-H			53	2007	
G05-003-A			20	2051	
G05-003-B			20	2051	
G05-004A			11	2015	
G05-010-A	123A		20	2051	
G05-011-A	123A		20	2051	
G05-015-C			20	2051	
G05-015-D	128		243	2030	
G05-034-D			18	2030	
G05-035-D			62	2010	
G05-035-D,E			62	2010	
G05-035-E			62	2010	
G05-036-C,D,E			20	2051	
G05-036B			20	2051	
G05-036C			20	2051	
G05-036D			18	2030	
G05-036E			20	2051	
G05-037B			20	2051	
G05-050-C			17	2051	
G05-055-D			18	2030	
G05-413A			18	2030	
G05036	123A	3444	20	2051	
G05059	123A		20	2051	
G05415	293	3122	47	2030	
G05705	152	3893	66	2048	
G06-714C			28	2017	
G06-717-B			28	2017	
G08-005L			FET-1	2028	
G09007	1103		IC-94		
G09009	1006		IC-38		
G09015	1003		IC-43		
G00-003-A	109	3087		1123	
G00-502-A	116	3017B/117	504A	1104	
GP-1	116	3017B/117	504A	1104	
GP-25B			504A	1104	
GP-354	109	3090		1123	
GP05A	116		504A	1104	
GP08B	116	3311	504A	1104	
GP08D	116		504A	1104	
GP1			504A	1104	
GP10A					1N4001
GP10B					1N4002
GP10D					1N4003
GP10G					1N4004
GP10J					1N4005
GP10K					1N4006
GP10M					1N4007
GP139	102A	3004	53	2007	
GP139A	102A	3004	53	2007	
GP139B	102A	3004	53	2007	
GP1432	121	3717	239	2006	
GP1493	121	3009	239	2006	
GP1494	121	3009	239	2006	
GP15A		3081			1N5391
GP15B		3081			1N5392
GP15D		3081			1N5393
GP15G		3081			1N5395
GP15J		3081			1N5397
GP15K		3081			1N5398
GP15M		3081			1N5399
GP1622	105	3009	4		
GP1882	121	3009	239	2006	
GP2-345	177	3100/519	300	1122	
GP2-345/MA161	177		300	1122	
GP20A		3081			1N5391
GP20B		3081			1N5392
GP20D		3081			1N5393
GP20G		3081			1N5395
GP20J		3081			1N5397
GP20K		3081			1N5398
GP20M		3081			1N5399
GP230	116	3311	504A	1104	
GP2354	109	3087		1123	
GP2364				1123	
GP250	116	3311	504A	1104	
GP25A		3848			MR500
GP25B		3848			MR501
GP25D		3848			MR502
GP25G	156	3848/5806	504A	1104	MR504
GP25J		3848			MR506
GP25K		3051			MR508
GP25M		3051			MR510
GP290	176	3845	80		
GP30A		3848			MR500
GP30B		3848			MR501
GP30D		3848			MR502
GP30G		3848			MR504
GP30J		3848			MR506
GP30K		3051			MR508
GP30M		3051			MR510
GP420	121	3717	239	2006	
GPM1NA	109	3087		1123	
GPM1NB	109	3087		1123	
GPM1NC				1123	
GPM2NA	109	3091		1123	
GPT-16	121	3009	239	2006	
GPT16			16	2006	
GR1	5818				1N4934
GR102	195A	3765	46		
GR2	5820				1N4935
GR4	5820				1N4936
GR6					1N4937
GRASS-R2982	130		14	2041	
GRF-3	312	3448	FET-1	2035	
GRU-2A	552	3998/506			
GRU2A	506	3998	511	1114	
GS600,3,6,9,10					MRD300
GS612	3032				MRD3050
GS670	3032				MRD3050
GS680					MRD300
GS683					MRD300
GS686					MRD300
GS73B/3	109	3090		1123	
GS9014	123AP	3122	10	2051	
GS90141				2051	
GS9014I	123AP	3122	20		
GS9014J	123AP	3122	20	2051	
GS9014K	123AP	3122	20	2051	
GS9015H	159	3138/193A	82	2032	
GS9015I	159	3138/193A	82	2032	
GS9015J	159	3138/193A	82	2032	
GS9018F	123A	3122	10	2051	
GS9019H	287	3244	222		
GS9019I	287		222		
GS9022F	159	3434/288	223		
GS9022G	159	3434/288	223		
GS9022H	288	3434	223		
GS9022I	288	3434	223		
GS9023H	123AP	3122	20	2051	
GS9023I	123AP	3122	10	2051	
GS9023J	123AP	3122	20	2051	
GS9023K	123A	3122	20	2051	
GSM482	116	3311	504A	1104	
GSM483	116	3311	504A	1104	
GSM51	116	3311	504A	1104	
GSM52	116	3311	504A	1104	
GSM53	116	3311	504A	1104	
GSM54	116	3311	504A	1104	
GSR1	125	3033	510,531	1114	
GT-109		3004	2	2007	
GT-1200		3011	5	2002	
GT-269		3123	2	2007	
GT-32	6407	3004/102A			
GT-34HV		3123	52	2024	
GT-34S		3123	2	2007	
GT-35	6408		8	2002	
GT-66			52	2024	
GT-759R		3005	2	2007	
GT-760R		3005	2	2007	
GT-761R		3005	2	2007	
GT-762R		3123	2	2007	
GT-903			8	2002	
GT-905			8	2002	
GT-947			8	2002	
GT1			50	2004	
GT100			1	2007	
GT109	102A		53	2007	
GT109R	102A	3123	53	2007	
GT11	100	3123	1	2007	
GT12	100	3123	1	2007	
GT1200	101		8	2002	
GT1201			5	2002	
GT1202	101	3861	8	2002	
GT122	102		53	2007	
GT1223	102A	3123	53	2007	
GT123	102		53	2007	
GT13	100	3123	1	2007	
GT132	102A	3123	53	2007	
GT14	102A		53	2007	
GT14H	102A		53	2005	
GT153	100	3005	1	2007	
GT15T5	145A	3063	ZD-15	564	
GT1604	100	3005	1	2007	
GT1605	100	3005	1	2007	
GT1606	100	3005	1	2007	
GT1607	100	3005	1	2007	
GT1608	101	3861	8	2002	
GT1609	101	3861	8	2002	
GT1644	159	3114/290	82	2032	
GT1658	103A		59	2002	
GT1665	102A		53	2007	
GT167	101		59	2002	
GT18	102A		53	2007	
GT2			50	2004	
GT20	102		53	2004	
GT20H	102A		53	2005	
GT20R	102A	3123	53	2007	
GT210H			1	2007	
GT222	102		53	2007	
GT229	101		8	2002	
GT24H			50	2004	
GT269	100	3005	1	2007	
GT2693	102	3721/100	1	2007	
GT2694	100	3005	1	2007	
GT2695	102	3721/100	1	2007	
GT2696	102		53	2005	
GT2765	101	3861	8	2002	
GT2766	101	3861	8	2002	
GT2767	101	3861	8	2002	
GT2768	103A		59	2002	
GT2883	102		53	2007	
GT2884	101	3011	8	2002	
GT2885	102		53	2007	
GT2886	101	3011	8	2002	
GT2887	102		53	2007	
GT2888	101	3011	8	2002	
GT2906	101	3861	8	2002	
GT3			50	2004	
GT31	102A		53	2007	
GT3150	101	3861	8	2002	
GT32	102A		53	2007	1N5761
GT33	102A		53	2005	
GT336	103A	3010	59	2002	
GT34	102		53	2004	
GT34HV	102A		53	2007	
GT34S	100	3005	1	2007	
GT35	103A	3835	59	2002	1N5762
GT364	103	3835	59	2002	
GT365	103	3835	59	2002	
GT366	103	3835	59	2002	
GT40	102A		245	2004	
GT41	102A		53	2007	
GT42	102A		53	2007	
GT43	102A		52	2024	
GT44	102A		53	2007	
GT45	102A		53	2007	
GT46	102A	3006/160	52	2024	
GT47	102A	3006/160	52	2024	
GT5116	126	3123	52	2024	
GT5117	126	3006/160	52	2024	
GT5148	126	3006/160	52	2024	
GT5149	126	3006/160	52	2024	
GT5151	126	3006/160	52	2024	
GT5153	100	3006/160	52	2024	
GT66	126	3006/160	52	2024	
GT74	102		53	2007	
GT75	102		53	2007	
GT751	102A	3123	53	2007	
GT758	102		53	2007	
GT759	102		53	2007	
GT759R	102A		53	2007	
GT760	100		53	2004	
GT760R	102A		53	2007	
GT761	100	3721	1	2007	
GT761R	100	3721	1	2007	
GT762	100	3721	1	2007	
GT762R	100	3005	1	2007	
GT763	102A		53	2007	
GT764	100	3005	1	2007	
GT766	100	3123	1	2007	
GT766A	100	3123	1	2007	
GT792	101		8	2002	
GT81	102		53	2007	
GT81H			1	2007	
GT81HS	102A		53	2007	
GT81R	102A	3123	53	2007	
GT82	102		53	2007	
GT83	100	3005	1	2007	
GT832	100	3005	1	2007	
GT87	100	3005	1	2007	
GT88	100	3005	1	2007	
GT903	103A	3835	59	2002	
GT904	101		8	2002	
GT905	101	3861	8	2002	
GT905R	101		8	2002	

Industry Standard No.	ECG	SK	GE	RS 276-	MOTOR.
GT947	101	3861	8	2002	
GT948	101		8	2002	
GT948R			8	2002	
GT949	103		8	2002	
GT949R	101		8	2002	
GTE-2		3004	2	2007	
GTE02-01	5542	6642			
GTE1	100	3123	1	2007	
GTE2	100	3721	1	2007	
GTJ33141	126	3006/160	52	2024	
GTJ33229	126	3006/160	52	2024	
GTJ33230	126	3006/160	52	2024	
GTJ33231	102A		53	2007	
GTJ33232	102A		53	2007	
GTV	100	3004/102A	1	2007	
GTX-2001		3009	3	2006	
GTX2001	121	3717	239	2006	
GU-3SZ				1143	
GV5760	177	3124/289	300	1122	
GV6063	123A	3124/289	20	2051	
GVL 20077	123A			2051	
GVL20077		3024	20	2051	
GVL20226	519		514	1122	
GVL20265	519		514	1122	
GVL20327-1	519		514	1122	
GZ27	146A	3064	ZD-27		
GZ4.3	5175A		ZD-4.3		
GZ5.1	5177A	3056/135A	ZD-5.1		
GZ6.2	137A	3058	ZD-6.2	561	
GZ8.2	5122A	3136/5072A			
GZ9.1	5185A	3060/139A	ZD-9.1	562	
H)-00005				563	
H)-00008				1104	
H)-00012				562	
H)-00018				1104	
H-1567		3124	20	2051	
H-881	125	3016	510,531	1114	
H087		3091	1N60	1123	
H089	142A	3062		563	
H091	109	3091		1123	
H10	102A		53	2007	
H100	116	3017B/117	504A	1104	
H1000	125	3033	510,531	1114	1N4007
H10174			504A	1104	
H102	123A	3444	20	2051	
H102D2					MC672TL
H102D6					MC672L
H103A	121	3717	239	2006	
H103D1					MC671L
H103D2					MC671TL
H103D6					MC671L
H103SH	5640		SC136A		
H104			10	2051	
H104D1					MC660L
H104D2					MC660TL
H104D6					MC660L
H105D1					MC673L
H105D2					MC673TL
H105D6					MC673L
H109	156	3032A	512		
H110D1					MC663L
H110D2					MC663TL
H110D6					MC663L
H111D1					MC688L
H111D2					MC688TL
H111D6					MC688L
H112D1					MC681L
H112D2					MC681TL
H112D6					MC681L
H113D1					MC689L
H113D2					MC689TL
H113D6					MC689L
H113SH	5640		SC136A		
H114D1					MC691L
H114D2					MC691TL
H114D6					MC691L
H115D1					MC678L
H115D2					MC678TL
H115D6					MC678L
H117D1					MC667
H117D6					MC667
H118D1					MC680L
H118D2					MC680TL
H118D6					MC680L
H119D1					MC677L
H119D2					MC677TL
H119D6					MC677L
H11A1	3041		H11A1		H11A1
H11A10					4N26
H11A2	3042		H11A2		H11A2
H11A3	3041		H11A3		H11A3
H11A4			H11A4		H11A4
H11A5			H11A5		H11A5
H11A5100	3041				H11A5100
H11A520					H11A520
H11A550					H11A550
H11B1	3045		H11B1		H11B1
H11B2	3041		H11B2		H11B2
H11B3			H11B3		H11B3
H11C1	3046		H11C1		H11C1
H11C2	3046		H11C2		H11C2
H11C3	3046		H11C3		H11C3
H11C4	3046		H11C4		H11C4
H11C5	3046		H11C5		H11C5
H11C6	3046		H11C6		H11C6
H12	105		4		
H122D1					MC668L
H122D2					MC668TL
H122D6					MC668L
H123SH	5641		SC136B		
H124D2					MC661TL
H124D6					MC661L
H12A	105		4		
H133SH	5642		SC136D		
H143SH	5642		SC136D		
H1567	123A		20	2051	
H156D1					MC685L
H156D2					MC685TL
H156D6					MC685L
H157D1					MC684L
H157D2					MC684TL
H157D6					MC684L
H158D1					MC676L
H158D2					MC676TL
H158D6					MC676L
H15B1,2					MRS100
H1P1	125	3311			
H1V	108	3452	86	2038	
H200	116	3017B/117	504A	1104	
H20052			504A	1104	
H221(GE)	783		IC-225		
H300	116	3017B/117	504A	1104	
H316	109	3091		1123	
H386-9	505	3108	CR-7		

Industry Standard No.	ECG	SK	GE	RS 276-	MOTOR.
H386C1D2Y	505	3108	CR-7		
H400	116	3031A	504A	1104	
H442	108	3452	86	2038	
H445-2	513	3443	513		
H475			504A	1104	
H484	513	3443	513		
H50	116	3311	504A	1104	
H500	116	3017B/117	504A	1104	
H585	116		504A	1104	
H598	503	3068	CR-5		
H598-10	503	3068	CR-5		
H5A3	6354	6554			
H5C3	6354	6554			
H5D3	6354	6554			
H5E3	6354	6554			
H5F3	6354	6554			
H5G3	6354	6554			
H5H3	6356	6556			
H5K3	6356	6556			
H5M3	6358	6558			
H5N3	6358	6558			
H600	116	3017B/117	504A	1104	
H614		3091	1N60	1123	
H615	113A	3119/113	6GC1		
H616	116	3031A	504A	1104	
H617	116	3031A	504A	1104	
H618	116	3311	504A	1104	
H619	116	3311	504A	1104	
H620	116	3311	504A	1104	
H621	125	3033	510,531	1114	
H623	177		300	1122	
H624			1N34AS	1123	
H625	116	3032A	504A	1104	
H626	116	3033	504A	1104	
H7126-3	116	3017B/117	504A	1104	
H74A1					4N26
H74C1					H74C1
H74C2					H74C2
H7618	188		217	2018	
H781	116	3017B/117	504A	1104	
H783	116		504A	1104	
H800	125	3033	510,531	1114	1N4006
H815	506	3125	511	1114	
H816	504		CR-6		
H8287	110MP			1123	
H8287-4	110MP			1123	
H8513	177		300	1122	
H881			504A	1104	
H889	177	3311	300	1122	
H890	507		504A	1104	
H891	116(3)		504A	1104	
H931	123A		20	2051	
H932	154		40	2012	
H933	123A		20	2051	
H934	123A		20	2051	
H9423	123A	3018	20	2051	
H9618	123A	3018	20	2051	
H9623	123A	3018	86	2038	
H9625	108	3452	86	2038	
H9696	123A	3124/289	20	2051	
HA-00102	100	3008	1	2007	
HA-00354			51	2004	
HA-00495	159	3114/290	82	2032	
HA-00496	185	3083/197	58	2025	
HA-0054			21	2034	
HA-00564	234		65	2050	
HA-00610	159		82	2032	
HA-00634	187	3193	29	2025	
HA-00636	187	3193	29	2025	
HA-00643	193	3138	67	2023	
HA-00699	153	3083/197	69	2049	
HA-00733	294	3138/193A	65	2050	
HA-1			511	1114	
HA-101	126		52	2024	
HA-12	100	3721	1	2007	
HA-1202	1041	3009	IC-160		
HA-15	126	3721/100	52	2024	
HA-201		3123	2	2007	
HA-234	160		245	2004	
HA-234B			51	2004	
HA-235A			51	2004	
HA-235C			51	2004	
HA-268	160	3004/102A	245	2004	
HA-269	126	3006/160	52	2024	
HA-330		3123	2	2007	
HA-350	126	3006/160	52	2024	
HA-350A	126	3006/160	52	2024	
HA-353	126	3006/160	52	2024	
HA-353C	126	3007	52	2024	
HA-354	126	3006/160	52	2024	
HA-354B	126	3008	52	2024	
HA-49	126	3008	52	2024	
HA-505			29	2025	
HA-52		3008	1	2007	
HA-53		3008	1	2007	
HA00052	100	3721	1	2007	
HA00053	100	3721	1	2007	
HA00562	290A	3114/290	269	2032	
HA100	116	3017B/117	504A	1104	
HA1000	125	3033	510,531	1114	
HA102	126		52	2024	
HA103	126		52	2024	
HA104	126		52	2024	
HA1040	160		245	2004	
HA1101	726	3129			
HA1102	726	3129			
HA1103	704	3023	IC-205		
HA1104	704	3023	IC-205		
HA1108	1061	3228	IC-51		
HA1110	901	3549	IC-180		
HA11107	712	3072	IC-148		
HA11112	1189	3705			
HA11115W	720	9014	IC-7		
HA1115	720	9014	IC-7		
HA1115W	720	9014	IC-7		
HA1117	1122		IC-40		
HA1118	1094		IC-157		
HA11228	1213	3704	IC-150		
HA1124	712	3072	IC-147		
HA1124D	712	3072	IC-2		
HA1124DS	712		IC-148		
HA1124S	712	3077/790	IC-2		
HA1125	712	3072	IC-148		
HA1125A	712	3072	IC-2		
HA1126	1004	3365			
HA1126D	1004	3365	IC-149		
HA1126DW	1004		IC-149		
HA1127	912	3543	IC-172		
HA1128	712	3072	IC-148		
HA1133	730	3143			
HA1137	788	3829	IC-229		
HA1137P	788	3829	IC-229		
HA1137W	788	3829	IC-229		
HA1139	749	3171/744	IC-97		

Industry Standard No.	ECG	SK	GE	RS 276-	MOTOR.
HA1139A	1169	3708	IC-24	2022	
HA1140	1080	3284	IC-98		
HA1141	712	3072	IC-2		
HA1144	1213	3704	IC-150		
HA1148	1229		IC-153		
HA1150		3160	IC-35		
HA1151	1237	3707	IC-154		
HA1152	1080	3284			
HA1154	712	3072	IC-2		
HA1156	801	3160	IC-155		
HA1156-6C	801	3160	IC-155		
HA1156W	801	3160	IC-35		
HA1157	1122		IC-156		
HA1158	1094		IC-157		
HA1158-O	1094		IC-157		
HA11580	1196	3725			
HA1159	121		IC-158		
HA1177	1196	3725	IC-157		
HA1197	1214	3736			
HA1199	1290				HA1199
HA12	160	3008	245	2004	
HA1201	1039	3366	IC-159		
HA1202	1041	3361	IC-160		
HA1203	1042	3367			
HA1211	1040	3362	IC-161		
HA1301	903	3540	IC-288		
HA1302	784	3524	IC-236		
HA1306	1029	3368	IC-162		
HA1306P	1029		IC-162		
HA1306P-U	1029		IC-162		
000HA1306U	1029	3368	IC-162		
HA1306W	1029	3368	IC-162		
HA1306WU	1029		IC-162		
HA1308	1030		IC-163		
HA1309	1030		IC-163		
HA1310	1031		IC-273		
HA1311	1032		IC-164		
HA1311W	1032		IC-164		
HA1312	1033		IC-165		
HA1312W	1033		IC-165		
HA1313	OBS-NLA		IC-166		
HA1314	1035	3369	IC-167		
HA1316	1036	3370	IC-168		
HA1318	1038		IC-274		
HA1318P	1038		IC-274		
HA1318P-U	1038		IC-274		
HA1318PU	1038		IC-274		
HA1318PU-1	1038		IC-274		
HA1318PU-2	1038		IC-274		
HA1318PU-3	1038		IC-274		
HA1319	1043	3363	IC-169		
HA1322	1037	3371	IC-170		
HA1322A	1037	3371			
HA1322C	1037	3371	IC-170		
HA1325	OBS-NLA	3373/1238			
HA1339	1169	3708	IC-64		
HA1339A	1169	3708	IC-322		
HA1342	1239	3708/1169			
HA1342A	1239	3708/1169	IC-335		
HA1350			53	2007	
HA1360	102A		53	2007	
HA1364	1231	3832			
HA1366	1260	3744	IC-336		
HA1366R	1261	3872			
HA1366W	1260	3744			
HA1366WR	1261	3872			
HA1406	1087	3364	IC-103		
HA1406-2	1087	3364	IC-103		
HA1406-3	1087	3364	IC-103		
HA1406-4	1087	3364	IC-103		
HA15	100	3005	1	2007	
HA17711	911D		IC-254		
HA17723	923D		IC-260	1740	
HA17741M	941	3514	IC-263	010	
HA17805	960	3591			
HA17806	962	3669			
HA17808	964	3630			
HA17812	966	3592			
HA17815	968	3593			
HA17824	972	3670			
HA200	116	3016	504A	1104	
HA2000					3N156A
HA2001	312	3112		2028	3N155
HA201	126	3008	52	2024	
HA2010	312	3112		2028	3N156A
HA202	100	3721	1	2007	
HA2190	160		245	2004	
HA235	126		52	2024	
HA2356	160		245	2004	
HA235A	160	3006	245	2004	
HA240	160	3006	245	2004	
HA266	160		245	2004	
HA267	160		245	2004	
HA269			51	2004	
HA30	160		245	2004	
HA300	116	3017B/117	504A	1104	
HA3210	160		245	2004	
HA330	126	3008	52	2024	
HA342	126		52	2024	
HA3480	160		245	2004	
HA350			51	2004	
HA353		3006	51	2004	
HA354			51	2004	
HA3670	160		245	2004	
HA400	116	3031A	504A	1104	
HA4400	160		245	2004	
HA471	126	3006/160	52	2024	
HA49	100	3721	1	2007	
HA50	116	3311	504A	1104	
HA500	116	3017B/117	504A	1104	
HA5001	101	3861	8	2002	
HA5002	101	3861	8	2002	
HA5003	101	3861	8	2002	
HA5005	101	3861	8	2002	
HA5009	101	3861	8	2002	
HA5010	103A	3835	59	2002	
HA5011	101	3861	8	2002	
HA5012	101	3861	8	2002	
HA5014	101	3861	8	2002	
HA5016	103	3861/101	8	2002	
HA5020	101	3861	8	2002	
HA5021	101	3861	8	2002	
HA5022	101	3861	8	2002	
HA5023	101	3861	8	2002	
HA5024	101	3861	8	2002	
HA5025	101	3861	8	2002	
HA5026	101	3861	8	2002	
HA505	153		69	2049	
HA52	126		52	2024	
HA525	160	3006	245	2004	
HA53	126		52	2024	
HA54	102A		53	2007	
HA56	102A		53	2007	
HA600	116	3017B/117	504A	1104	

Industry Standard No.	ECG	SK	GE	RS 276-	MOTOR.
HA70	160		245	2004	
HA7501			82	2032	
HA7502			82	2032	
HA7506			82	2032	
HA7507			22	2032	
HA7510			82	2032	
HA7520			67	2023	
HA7521			67	2023	
HA7522			67	2023	
HA7523			67	2023	
HA7524			67	2023	
HA7526			67	2023	
HA7527			67	2023	
HA7528			67	2023	
HA7530	129	3025	244	2027	
HA7531	129	3025	244	2027	
HA7533			82	2032	
HA7534			82	2032	
HA7536			22	2032	
HA7537			82	2032	
HA7538			82	2032	
HA7543			82	2032	
HA7597			82	2032	
HA7598			82	2032	
HA7599			82	2032	
HA7630	129	3025	244	2027	
HA7631	129	3025	244	2027	
HA7632	129	3025	244	2027	
HA7633			82	2032	
HA7723			67	2023	
HA7730			67	2023	
HA7732			67	2023	
HA7734			67	2023	
HA7735			67	2023	
HA7736			67	2023	
HA7737			67	2023	
HA7804			22	2032	
HA7806			22	2032	
HA7808			22	2032	
HA7810			22	2032	
HA7815			82	2032	
HA800	125	3032A	510,531	1114	
HA9048	159		221	2024	
HA9049	159		221	2024	
HA9054			221	2024	
HA9055			221	2024	
HA9079			221	2024	
HA9500	129		21	2034	
HA9501	129		21	2034	
HA9502	129		21	2034	
HAM-1	102A		53	2007	
HAR10	116	3017B/117	504A	1104	
HAR15	116	3017B/117	504A	1104	
HAR20	116	3016	504A	1104	
HAW1183B	176		80		
HAW1183C	176		80		
HB-00054	102A	3004	53	2007	
HB-00056	102A	3004	53	2007	
HB-00156	102A	3004	53	2007	
HB-00171	102A	3004	53	2007	
HB-00172	102A	3004	53	2007	
HB-00173	102A	3004	53	2007	
HB-00175	102A	3004	53	2007	
HB-00176	102A	3004	53	2007	
HB-00178	102A	3004	2	2007	
HB-00186	102A	3004	53	2007	
HB-00187	102A	3004	53	2007	
HB-00303			2	2007	
HB-00324	158	3004/102A	53	2007	
HB-00370			2	2007	
HB-00405	158	3004/102A	53	2007	
HB-00564	187	3193	29	2025	
HB-12B			ZD-13	563	
HB-156		3004	2	2007	
HB-172	102A	3004	53	2007	
HB-173	102A	3004	53	2007	
HB-175	102A	3004	53	2007	
HB-186		3123	52	2024	
HB-187		3123	52	2024	
HB-32		3004	52	2024	
HB-33		3004	52	2024	
HB-475	158	3004/102A	53	2007	
HB-54		3004	2	2007	
HB-56		3004	2	2007	
HB-75B			2	2007	
HB-77C		3004	2	2007	
HB-7B			ZD-6.8	561	
HB-85	102A	3004	53	2007	
HB100					MR501
HB1000		3051			MR510
HB156	102A		53	2007	
HB156C	102A		53	2007	
HB171	102A		53	2007	
HB172	102A		53	2007	
HB175	102A		53	2007	
HB176	102A		53	2007	
HB178	102A		53	2007	
HB186	102A	3004	53	2007	
HB187	102A	3004	53	2007	
HB2	116	3016	504A	1104	
HB200					MR502
HB263	102A		53	2007	
HB270	102A		53	2007	
HB3	116	3311	504A	1104	
HB300					MR504
HB32	102A		53	2007	
HB324	158		53	2007	
HB33	102A		53	2007	
HB365	158	3004/102A	53	2007	
HB367	131	3052	44	2006	
HB400					MR504
HB415	158		53	2007	
HB422	126	3004/102A	52	2024	
HB459	102A	3004	53	2007	
HB461	176		80		
HB475	158		53	2007	
HB50		3081			MR501
HB500					MR506
HB54	102A		53	2007	
HB55	102A		53	2007	
HB56	102A		53	2007	
HB600		3051			MR506
HB75	102A	3004	53	2007	
HB75C	102A		53	2007	
HB77	102A	3004	53	2007	
HB77B	102A	3004	53	2007	
HB77C	102A		53	2007	
HB800		3051			MR508
HBF4001		4001	4001	2401	
HBF4002		4002	4002		
HBF4007		4007	4007		
HBF4009		4009	4009		
HBF4011		4011	4011	2411	
HBF4012		4012	4012	2412	

Industry Standard No.	ECG	SK	GE	RS 276-	MOTOR.
HBF4013		4013	4013	2413	
HBF4016		4016	4016		
HBF4017		4017	4017	2417	
HBF4018AE	4018B	4018			
HBF4020			4020	2420	
HBF4020A	4020B	4020			
HBF4021			4021	2421	
HBF4021A	4021B	4021			
HBF4023		4023	4023	2423	
HBF4024		4024	4024		
HBF4025		4025	4025		
HBF4027		4027	4027	2427	
HBF4035	4035B	4035			
HBF4052A	4052B	4052			
HC-00176	102A		53	2007	
HC-00184	313	3018			
HC-00268	128	3024	220		
HC-00372	123A		61	2038	
HC-00373	199	3122	62	2010	
HC-00380	107	3018	11	2015	
HC-00381	107	3038			
HC-00394	229	3018	61	2038	
HC-00458			10	2051	
HC-00460			20	2051	
HC-00461	229		20	2051	
HC-00481		3014	31		
HC-00496	184	3054/196	57	2017	
HC-00509	128		81	2051	
HC-00535	229	3018	17	2051	
HC-00536	199	3124/289	62	2010	
HC-00537	123A	3444	20	2051	
HC-00644		3020	18	2030	
HC-00668	108	3452	86	2038	
HC-00693	123A	3444	20	2051	
HC-00711	199	3018	62	2010	
HC-00730	101	3011	8	2002	
HC-00732	199	3122	62	2010	
HC-00735	123A	3122	210	2051	
HC-00772	108	3452	86	2038	
HC-00784	107	3018	11	2015	
HC-00828	123A	3444	20	2051	
HC-00829	107	3018	11	2015	
HC-00838		3444	20	2051	
HC-00839	108	3452	86	2038	
HC-00871	123A	3444	20	2051	
HC-00900	199	3124/289	62	2010	
HC-00920	107		11	2015	
HC-00921	123A	3444	20	2051	
HC-00923	199	3124/289	62	2010	
HC-00924	123A	3444	20	2051	
HC-00929	229	3018	62	2010	
HC-00930	107	3018	62	2010	
HC-00945	123A	3124/289	20	2051	
HC-01000	199	3245	62	2010	
HC-01047	107	3018	11	2015	
HC-01060			28	2017	
HC-01096	186	3192	28	2017	
HC-01098	186	3192	28	2017	
HC-01209	128	3024	243	2030	
HC-01226	186	3192	28	2017	
HC-01317	128	3122	243	2030	
HC-01318	128	3122	243	2030	
HC-01335	199	3122	62	2010	
HC-01359	107	3018	11	2015	
HC-01390	289A	3444/123A	20	2051	
HC-01417	107	3444/123A	20	2051	
HC-01820	123A	3444	20	2051	
HC-01830		3452	86	2038	
HC-2	525	3925	530		
HC-30	116	3087	504A	1104	
HC-373			17	2051	
HC-39		3311	504A	1104	
HC-495			28	2017	
HC-50102	5402	3638			
HC-535			17	2051	
HC-537		3018	62	2010	
HC-56	123A	3124/289	20	2051	
HC-561		3018	17	2051	
HC-668			60	2015	
HC-68		3311	504A	1104	
HC-772		3018	17	2051	
HC00730			5	2002	
HC00838	123A		20	2051	
HC00930	229	3039/316	60	2038	
HC01820			20	2051	
HC01830			17	2051	
HC1000109	703A	3157	IC-12		
HC1000109-0	703A	3157	IC-12		
HC1000111-0	703A	3157	IC-12		
HC1000114-0	726	3129			
HC1000117-0	718	3159	IC-8		
HC10001200	1170	3745	IC-65		
HC10002050	1100	3223	IC-92		
HC1000217		3514	IC-263	010	
HC1000217-0	941D	3552/941M	IC-264		
HC10004010	801	3160	IC-35		
HC1000403	1005	3723	IC-42		
HC10004110		7400	7400	1801	
HC1000417	799	3238	IC-34		
HC1000505	1085	3476	IC-104		
HC10006170	801		IC-35		
HC1000703	1027		721		
HC10008060	1254	3880			
HC1001001	1043		IC-169		
HC10034050	1192		IC-181		
HC2	525		530		
HC206	108		86	2038	
HC30	156		512		
HC300					1N4722
HC371	123A	3444	20	2051	
HC372	123A	3444	20	2051	
HC373	123A	3444	20	2051	
HC380	107		11	2015	
HC381	107	3018	11	2015	
HC39	156		512		
HC394	107		11	2015	
HC454	107		11	2015	
HC458	123A	3444	20	2051	
HC460	107	3018	11	2015	
HC461	107	3018	11	2015	
HC495	152	3893	66	2048	
HC50	156	3081/125	512		
HC500	116	3017B/117	504A	1104	1N4723
HC515	124	3021	12		
HC535	107		11	2015	
HC535A	107		11	2015	
HC535B	107		11	2015	
HC537	107	3039/316	11	2015	
HC539	123A	3444	20	2051	
HC545	107	3039/316	11	2015	
HC561	123A	3444	20	2051	
HC645	108		86	2038	
HC668	108	3452	86	2038	
HC67	116		504A	1104	1N4001
HC670	116	3017B/117	504A	1104	
HC68	116		504A	1104	1N4002
HC680	116	3017B/117	504A	1104	
HC69	116		504A		1N4003
HC70					1N4004
HC700	116	3017B/117	504A	1104	
HC71	116	3051/156	504A	1104	1N4005
HC710	116	3051/156	504A	1104	
HC72	125	3051/156	510,531	1114	1N4006
HC720	125	3051/156	510,531	1114	
HC73	125	3051/156	510,531	1114	1N4007
HC730	125	3051/156	510,531	1114	
HC772	108	3452	86	2038	
HC784	107	3018	11	2015	
HC80	116	3016	504A	1104	
HC829	108		86	2038	
HCF4006B	4006B	4006			
HCL-29	123A	3444	20	2051	
HCL-6066	123A	3444	20	2051	
HCL9				1104	
HCV	116	3311	504A	1104	
HD-00072	103A	3010	59	2002	
HD-00227	123A	3444	20	2051	
HD-00261	192	3122	63	2030	
HD-00471	289A	3024/128	47	2017	
HD-1	116	3017B/117	504A	1104	
HD-1000101	109			1123	
HD-187	101	3011	8	2002	
HD-2000106	177		300	1122	
HD-2000108			511	1114	
HD-2000308	116		504A	1104	
HD-20008	506		511	1114	
HD-3000301	116		504A	1104	
HD-3000409	138A		ZD-7.5		
HD1-74C00	74C00			2301	
HD1-74C02	74C02			2302	
HD1-74C08	74C08			2305	
HD1-74C192				2321	
HD1-74C193				2322	
HD1-74C74	74C74			2310	
HD1-74C76	74C76			2312	
HD1-74C90	74C90			2315	
HD10-001-01	109	3087		1123	
HD10000101	109	3090		1123	
HD10000302	109	3090		1123	
HD1000101		3087	1N34AS	1123	
HD1000101-0	109	3090		1123	
HD10001010	109	3090	1N34AS	1123	
HD1000105	110MP	3088	1N60	1123(2)	
HD1000105-0	110MP		1N60	1123	
HD10001050	109	3088		1123	
HD1000301	109	3088		1123	
HD1000302	109	3088		1123	
HD1000303	110MP	3059/138A		1123	
HD1000303-0				1123	
HD1000309	138A		ZD-7.5		
HD1468	109	3090		1123	
HD20-003-01	116	3031A	504A	1104	
HD2000	116	3031A	504A	1104	
HD2000-301	116	3031A	504A	1104	
HD20000703	116		504A	1104	
HD20000903	116		504A	1104	
HD2000106			300	1122	
HD2000108		3051	512		
HD2000110	116	3311	504A	1104	
HD2000110-0	116		504A	1104	
HD20001100	116	3311	504A	1104	
HD20001210	519		514	1122	
HD2000206	177	3100/519	300	1122	
HD2000301	116	3311	504A	1104	
HD2000301-0	116	3031A	504A	1104	
HD20003010	116	3311	504A	1104	
HD2000305			504A	1104	
HD2000307	116	3031A	504A	1104	
HD2000308		3031A	504A	1104	
HD2000413	116	3311	504A	1104	
HD2000501	116		504A	1104	
HD2000510	116		504A	1104	
HD20005100	156	3051	512		
HD2000610	116	3121/115		1104	
HD2000703	116	3311	504A	1104	
HD20007030	116		504A	1104	
HD2000710	113A	3119/113	6GC1		
HD2000803	116(2)	3311			
HD2000903	116	3311	504A	1104	
HD2001105	177	3100/519	300	1122	
HD20011050	177	3100/519	300	1122	
HD2001310	116	3311	504A	1104	
HD200207	116		504A	1104	
HD200301	116	3017B/117	504A	1104	
HD2149	109	3090		1123	
HD2155	109	3090		1123	
HD2501	7440	7440			
HD2501P	7440	7440			
HD2502	7460	7460			
HD2502P	7460	7460			
HD2503	74100	7400		1801	
HD2503P	74100	7400		1801	
HD2504	7420	7420		1809	
HD2504P	7420	7420		1809	
HD2505	7451	7451		1825	
HD2505P	7451	7451		1825	
HD2506	7450	7450			
HD2506P	7450	7450			
HD2507	7410	7410		1807	
HD2507P	7410	7410		1807	
HD2508	7430	7430			
HD2508P	7430	7430			
HD2509	7401	7401			
HD2509P	7401	7401			
HD2510	7474	7474		1818	
HD2510P	7474	7474		1818	
HD2511	7402	7402		1811	
HD2511P	7402	7402		1811	
HD2512	7453	7453			
HD2512P	7453	7453			
HD2514	7454	7454			
HD2514P	7454	7454			
HD2515	7473	7473		1803	
HD2515P	7473	7473		1803	
HD2516	7476	7476		1813	
HD2516P	7476	7476		1813	
HD2517	7475	7475		1806	
HD2517P	7475	7475		1806	
HD2518	7441	7441		1804	
HD2518P	7441	7441		1804	
HD2519	7490	7490		1808	
HD2519P	7490	7490		1808	
HD2520P	7493A	7493			
HD2521	7492	7492		1819	
HD2521P	7492	7492		1819	
HD2522	7404	7404		1802	
HD2522P	7404	7404		1802	
HD2523	7405	7405			

Industry Standard No.	ECG	SK	GE	RS 276-	MOTOR.
HD2523P	7405	7405			
HD2526	7486	7486		1827	
HD2526P	7486			1827	
HD2529	7472	7472			
HD2529P	7472	7472			
HD2530	74107	74107			
HD2531	7445	7445			
HD2531P	7445	7445			
HD2532	7447	7447		1805	
HD2532P	7447	7447		1805	
HD2533	7494	7494			
HD2533P	7494	7494			
HD2534	7495	7495			
HD2534P	7495	7495			
HD2535	7483	7483			
HD2535P	7483	7483			
HD2536	7442	7442			
HD2536P	7442	7442			
HD2537	7443	7443			
HD2537P	7443	7443			
HD2538	7444	7444			
HD2538P	7444	7444			
HD2539	7470	7470			
HD2539P	7470	7470			
HD2541	74192	74192		1831	
HD2541P	74192	74192		1831	
HD2542	74193	74193		1820	
HD2542P	74193	74193		1820	
HD2543	74121	74121			
HD2543P	74121	74121			
HD2544	7438	7438			
HD2544P	7438	7438			
HD2545	7413	7413		1815	
HD2545P	7413	7413		1815	
HD2546	7496	7496			
HD2546P	7496	7496			
HD2547	74181	74181			
HD2547P	74181	74181			
HD2548	74150	74150		1829	
HD2548P	74150	74150		1829	
HD2549	74151	74151			
HD2549P	74151	74151			
HD2550	7408	7408		1822	
HD2550P	7408	7408		1822	
HD2552	7437	7437			
HD2552P	7437	7437			
HD2555	74145	74145		1828	
HD2555P	74145	74145		1828	
HD2558	74141	74141			
HD2558P	74141	74141			
HD2560	7426	7426			
HD2560P	7426	7426			
HD2561	74123	74123		1817	
HD2561P	74123	74123		1817	
HD2564	74153	74153			
HD2564P	74153	74153			
HD2572	74196	74196		1833	
HD2572P	74196	74196		1833	
HD2580	74154	74154		1834	
HD2580P	74154	74154		1834	
HD30000109-0	5074A		ZD-11.0	563	
HD3000101-0	142A	3062	ZD-12	563	
HD3000109-0		3092	ZD-11	563	
HD30001090	5074A	3092/141A	ZD-11.0	563	
HD3000113	5074A	3092/141A	ZD-11.0	563	
HD30001200	140A	3061	ZD-10	562	
HD3000201-0	5084A	3755	ZD-30		
HD3000309	138A	3059	ZD-7.5		
HD3000309-0	138A		ZD-7.5		
HD3000401	139A	3060	ZD-9.1	562	
HD3000409		3059	ZD-8.2	562	
HD3001009	5074A	3092/141A	ZD-11.0	563	
HD30010090	5074A	3092/141A	ZD-11.0	563	
HD3001109		3094	ZD-15	564	
HD3001109-0	144A		ZD-14	564	
HD30011090	144A	3094	ZD-14	564	
HD30017090	139A	3060	ZD-9.1	562	
HD3001809	5074A	3061/140A	ZD-11.0	563	
HD30019090	138A		ZD-7.5		
HD3002109	144A		ZD-14	564	
HD3002409	142A		ZD-12	563	
HD3003109	5072A	3059/138A	ZD-8.2	562	
HD3003309	5010A	3776	ZD-5.1		
HD30042090	136A	3057	ZD-5.1		
HD3033090	5011A	3777			
HD4000		4000	4000		
HD40001060	614	3327	90		
HD4000109	109	3126		1123	
HD40001330	612	3325			
HD4000909	614	3126	90		
HD4001			4001	2401	
HD4002		4002	4002		
HD4009		4009	4009		
HD4011		4011	4011	2411	
HD4012		4012	4012	2412	
HD4013		4013	4013	2413	
HD4015		4015	4015		
HD4016	4016B	4016	4016		
HD4017		4017	4017	2417	
HD4019		4019	4019		
HD4020			4020	2420	
HD4021			4021	2421	
HD4023		4023	4023	2423	
HD4024		4024	4024		
HD4025		4025	4025		
HD4027		4027	4027	2427	
HD4030		4030	4030		
HD4049		4049	4049	2449	
HD4050		4050	4050	2450	
HD4518			4518	2490	
HD6001	177		300	1122	
HD6125	177		300	1122	
HD6147	116	3017B/117	504A	1104	
HD6865	116	3017B/117	504A	1104	
HD7400	7400	7400	7400	1801	
HD7400P	7400	7400	7400	1801	
HD7402	7402	7402		1811	
HD7402P	7402	7402		1811	
HD7403	7403	7403			
HD7403P	7403	7403			
HD7404	7404			1802	
HD7404P	7404			1802	
HD7405	7405	7405			
HD7405P	7405	7405			
HD7406	7406	7406		1821	
HD7406P	7406	7406		1821	
HD7407	7407	7407			
HD7407P	7407	7407			
HD7410	7410	7410		1807	
HD74107	74107	74107			
HD74107P	74107	74107			
HD7410P	7410	7410		1807	
HD7412	7412	7412			
HD74121	74121	74121			

Industry Standard No.	ECG	SK	GE	RS 276-	MOTOR.
HD74121P	74121	74121			
HD74126	74126	74126			
HD74126P	74126	74126			
HD7412P	7412	7412			
HD74132	74132	74132			
HD74132P	74132	74132			
HD7414	7414	7414			
HD7414P	7414	7414			
HD74150	74150	74150		1829	
HD74150P	74150	74150		1829	
HD74151A	74151	74151			
HD74151AP	74151	74151			
HD74155	74155	74155			
HD74155P	74155	74155			
HD74156	74156	74156			
HD74156P	74156	74156			
HD7416	7416	7416			
HD74160	74160	74160			
HD74160P	74160	74160			
HD74161	74161	74161			
HD74161P	74161	74161			
HD74162	74162	74162			
HD74162P	74162	74162			
HD74163	74163	74163			
HD74163P	74163	74163			
HD74164	74164	74164			
HD74164P	74164	74164			
HD74166P	74166	74175			
HD7416P	7416	7416			
HD74174	74174	74174			
HD74174P	74174	74174			
HD74175	74175	74175			
HD74175P	74175	74175			
HD74180	74180	74180			
HD74180P	74180	74180			
HD74190	74190	74190			
HD74190P	74190	74190			
HD74191	74191	74191			
HD74191P	74191	74191			
HD74199	74199	74199			
HD74199P	74199	74199			
HD7420	7420	7420		1809	
HD7420P	7420	7420		1809	
HD7422	7422	7422			
HD7427	7427	7427		1823	
HD7427P	7427	7427		1823	
HD7430	7430	7430			
HD7430P	7430	7430			
HD7432	7432	7432		1824	
HD7432P	7432	7432		1824	
HD7440	7440	7440			
HD7440P	7440	7440			
HD7442A	7442	7442			
HD7442AP	7442	7442			
HD7443A	7443	7443			
HD7443AP	7443	7443			
HD7444A	7444	7444			
HD7444AP	7444	7444			
HD7451	7451	7451		1825	
HD7451P	7451	7451		1825	
HD7454	7454	7454			
HD7454P	7454	7454			
HD7460	7460	7460			
HD7460P	7460	7460			
HD7472	7472	7472			
HD7472P	7472	7472			
HD7473AP	7473		7473	1803	
HD7473P	7473		7473	1803	
HD7474	7474	7474	7474	1818	
HD7474P	7474	7474	7474	1818	
HD7475	7475	7475		1806	
HD7475P	7475	7475		1806	
HD7485	7485	7485		1826	
HD7485P	7485	7485		1826	
HD7486	7486	7486		1827	
HD7486P	7486	7486		1827	
HD7490A	7490			1808	
HD7490AP	7490			1808	
HD7492A	7492	7492		1819	
HD7492AP	7492	7492		1819	
HD7493AP	7493A	7493			
HD7496	7496	7496			
HD7496P	7496	7496			
HD74S00	74S00	74S00			
HD74S00P	74S00	74S00			
HD74S04	74S04	74S04			
HD74S04P	74S04	74S04			
HD9-74C00				2301	
HD9-74C02				2302	
HD9-74C08				2305	
HD9-74C192				2321	
HD9-74C193				2322	
HD9-74C74				2310	
HD9-74C76				2312	
HD9-74C90				2315	
HE-00829	107		300	2015	
HE-00930	199	3245	62	2010	
HE-0A90		3087	1N60	1123	
HE-10001	109	3087		1123	
HE-10002	110MP	3087		1123	
HE-10003	109	3087		1123	
HE-10024	109	3088		1123	
HE-10025	109	3088		1123	
HE-10027	109	3090		1123	
HE-10030	177		300	1122	
HE-10040	109	3087		1123	
HE-10041	612	3126			
HE-10044	110MP	3088	1N60	1123(2)	
HE-1024	109	3088		1123	
HE-1N34	109	3087		1123	
HE-1N34A	109	3087		1123	
HE-1N60	109	3087		1123	
HE-1N60P	109	3087		1123	
HE-1S188	109	3087		1123	
HE-1S426	109	3087		1123	
HE-1S446	109	3087		1123	
HE-1S555		3126	90		
HE-20011	116	3017B/117	504A	1104	
HE-20049	138A	3059	ZD-7.5		
HE-7	138A	3092/141A	ZD-7.5		
HE-7A	138A	3092/141A	ZD-7.5		
HE-CD0000		3088	1N60	1123	
HE-M8489	177		300	1122	
HE-0A90	109			1123	
HE-SD1	116	3017B/117	504A	1104	
HEP-0700	112		1N82A		
HEP-2	160	3006	245		
HEP-237	213		254		
HEP-246		3191	58		
HEP-248	219	3173	74	2043	
HEP-594	718	3159	IC-8		
HEP-595	720	9014	IC-7		
HEP-6005				2006	
HEP-625	179		76		

Industry Standard No.	ECG	SK	GE	RS 276-	MOTOR.
HEP-626	179		76		
HEP-705	219	3173	74	2043	
HEP-707		3438	35		
HEP-714		3232	D44R1		
HEP-740		3710	36		
HEP-75	311	3466/159	277		
HEP-C3000P	7400	7400		1801	
HEP-C3001P	7401	7401			
HEP-C3002P	7402	7402		1811	
HEP-C3004P	7404	7404		1802	
HEP-C3010P	7410	7410		1807	
HEP-C3020P	7420	7420		1809	
HEP-C3030P	7430	7430			
HEP-C3040P	7440	7440			
HEP-C3041L	7441	7441			
HEP-C3041P	7441	7441		1804	
HEP-C3050L	7450	7450			
HEP-C3050P	7450	7450			
HEP-C3073P	7473	7473		1803	
HEP-C3075P	7475	7475		1806	
HEP-C3800P	7490	7490		1808	
HEP-C3801L	7492	7492			
HEP-C3801P	7492	7492		1819	
HEP-C3806P	974	3965			
HEP-C4000P	4001B	4000			
HEP-C4001P	4011B	4001			
HEP-C4002P	4023B	4002			
HEP-C4020P	4013B	4020			
HEP-C4021P	4027B	4021			
HEP-C4030P	4017B	4030			
HEP-C4040P	4042B	4040			
HEP-C4050P	4016B	4050			
HEP-C6001	760	3157/703A			
HEP-C6050G	973	3233			
HEP-C6052P	941M	3552	IC-265	007	
HEP-C6055L	725	3162	IC-19		
HEP-C6056P	722	3161	IC-9		
HEP-C6057P	790	3454	IC-230		
HEP-C6059P	746	3234	IC-217		
HEP-C6060P	748	3236			
HEP-C6061P	750	3280	IC-219		
HEP-C6062P	708	3135/709	IC-10		
HEP-C6063P	712	3072	IC-2		
HEP-C6065P	719		IC-28		
HEP-C6066P	798	3216	IC-234		
HEP-C6068P	720	9014	IC-7		
HEP-C6069G	780	3141	IC-222		
HEP-C6069P	780	3141	IC-222		
HEP-C6070P	714	3075	IC-4		
HEP-C6071P	715	3076	IC-6		
HEP-C6072P	739	3235	IC-30		
HEP-C6074P	718	3159	IC-8		
HEP-C6075P	738	3167	IC-29		
HEP-C6076P	749	3168	IC-97		
HEP-C6077P	782		IC-224		
HEP-C6079P	747	3279	IC-218		
HEP-C6080P	721		IC-14		
HEP-C6081P	779A	3240/779-1			
HEP-C6082P	708	3135/709	IC-10		
HEP-C6083P	712	3072	IC-2		
HEP-C6085P	731	3172/743	IC-13		
HEP-C6089	705A	3134	IC-3		
HEP-C6090	804	3455	IC-27		
HEP-C6091G	816	3242			
HEP-C6094P	718	3159	IC-8		
HEP-C6095P	720		IC-7		
HEP-C6096P	801	3160	IC-35		
HEP-C6099P	795	3237	IC-232		
HEP-C6100P	783	3215	IC-225		
HEP-C6101P	723	3144	IC-15		
HEP-C6102P	778A	3465	IC-220		
HEP-C6103P	909	3552/941M	IC-249		
HEP-C6105P	972	3670			
HEP-C6107P	975	3641			
HEP-C6110P	960	3591			
HEP-C6111P	962	3669			
HEP-C6113P	966	3592			
HEP-C6114P	968	3593			
HEP-C6116P	972	3670			
HEP-C6118P	961	3671			
HEP-C6120P	963	3672			
HEP-C6122P	967	3673			
HEP-C6123P	969	3674			
HEP-C6131P	978	3689			
HEP-C6132	977	3462			
HEP-C7400P	7400	7400			
HEP-C7401P	7401	7401			
HEP-C7402P	7402	7402			
HEP-C7403P	7403	7403			
HEP-C7404P	7404	7404			
HEP-C7405P	7405	7405			
HEP-C7406P	7406	7406			
HEP-C7407P	7407	7407			
HEP-C7408P	7408	7408			
HEP-C7409P	7409	7409			
HEP-C74107P	74107	74107			
HEP-C7410P	7410	7410			
HEP-C7411P	7411	7411			
HEP-C74121P	74121	74121			
HEP-C74123P	74123	74123			
HEP-C74125P	74125	74125			
HEP-C7413P	7413	7413			
HEP-C74141P	74141	74141			
HEP-C74145P	74145	74145			
HEP-C7414P	7414	7414			
HEP-C74150P	74150	74150			
HEP-C74151AP	74151	74151			
HEP-C74153P	74153	74153			
HEP-C74154P	74154	74154			
HEP-C74155P	74155	74155			
HEP-C74156P	74156	74156			
HEP-C74160AP	74160	74160			
HEP-C74161AP	74161	74161			
HEP-C74162AP	74162	74162			
HEP-C74163AP	74163	74163			
HEP-C74164P	74164	74164			
HEP-C74165P	74165	74165			
HEP-C74166P	74166	74175			
HEP-C7416P	7416	7416			
HEP-C74170P	74170	74170			
HEP-C74174P	74174	74174			
HEP-C74175P	74175	74175			
HEP-C74176P	74176	74176			
HEP-C74177P	74177	74177			
HEP-C7417P	7417	7417			
HEP-C74180P	74180	74180			
HEP-C74181P	74181	74181			
HEP-C74190P	74190	74190			
HEP-C74191P	74191	74191			
HEP-C74192P	74192	74192			
HEP-C74193P	74193	74193			
HEP-C74195P	74195	74195			
HEP-C74196P	74196	74196			
HEP-C74198P	74198	74198			

Industry Standard No.	ECG	SR	GE	RS 276-	MOTOR.
HEP-C74199P	74199	74199			
HEP-C7420P	7420	7420			
HEP-C7423P	7423	7423			
HEP-C7425P	7425	7425			
HEP-C7426P	7426	7426			
HEP-C7427P	7427	7427			
HEP-C7430P	7430	7430			
HEP-C7432P	7432	7432			
HEP-C7437P	7437	7437			
HEP-C7438P	7438	7438			
HEP-C7440P	7440	7440			
HEP-C7445P	7445	7445			
HEP-C7446AP	7446	7446			
HEP-C7447AP	7447	7447			
HEP-C7448P	7448	7448			
HEP-C7454P	7454	7454			
HEP-C7460P	7460	7460			
HEP-C7470P	7470	7470			
HEP-C7472P	7472	7472			
HEP-C7473P	7473	7473			
HEP-C7474P	7474	7474			
HEP-C7475P	7475	7475			
HEP-C7483P	7483	7483			
HEP-C7490AP	7490	7490			
HEP-C7492AP	7492	7492			
HEP-C7493AP	7493A	7493			
HEP-C7495P	7495	7495			
HEP-C7496P	7496	7496			
HEP-F0010	312	3112			
HEP-F0015	312	3116	FET-1		
HEP-F0021	312	3116	FET-2		
HEP-F1035	326	3746			
HEP-F2004	221	3050			
HEP-F2005	220	3990		2036	
HEP-F2007	222	3065			
HEP-G0001	126	3006/160	52		
HEP-G0002	160	3006	245		
HEP-G0003	160	3006	245	2005	
HEP-G0004	102A	3004	53		
HEP-G0005	100	3004/102A	53		
HEP-G0006	100	3004/102A	53		
HEP-G0007	100	3004/102A	53		
HEP-G0008	126	3006/160	52		
HEP-G0009	126	3007	52		
HEP-G0010	126	3006/160	52		
HEP-G0011	101	3835/103A	8		
HEP-G003	160		245		
HEP-G004	102A		53	2007	
HEP-G008	127		25		
HEP-G6000	104	3014	16		
HEP-G6001	179	3642	76		
HEP-G6003	104	3014	16		
HEP-G6004	105	3012	4		
HEP-G6005	121	3717	239		
HEP-G6006	213		254		
HEP-G6007	127	3034	25		
HEP-G6008	127	3035	25		
HEP-G6009	179	3642	76		
HEP-G6010	213		254		
HEP-G6011	176	3123	80		
HEP-G6013	121	3014	239		
HEP-G6014	179	3642	76		
HEP-G6015	179	3717/121	76		
HEP-G6016	131	3198	44		
HEP-G6018	179	3642	76		
HEP-R0050	125	3311			
HEP-R0051	125	3311			
HEP-R0052	125	3311			
HEP-R0053	125	3031A			
HEP-R0054	125	3017B/117			
HEP-R0055	125	3032A			
HEP-R0056	125	3033A			
HEP-R0070	5801	9004			
HEP-R0071	5802	9005			
HEP-R0072	5804	9007			
HEP-R0074	5808	9009			
HEP-R0076	5808	9009			
HEP-R0078	5809	9010			
HEP-R0090	5800	9003		1142	
HEP-R0091	5801	9004		1142	
HEP-R0092	5802	9005		1143	
HEP-R0094	5804	9007		1144	
HEP-R0096	5806	3848			
HEP-R0097	5808	9009			
HEP-R0098	5809	9010			
HEP-R0130	5870	3604	5032		
HEP-R0131	5874	3603	5032		
HEP-R0132	5874	3603	5032		
HEP-R0134	5878	3602	5036		
HEP-R0136	5882	3500	5040		
HEP-R0137	5886		5044		
HEP-R0138	5890	7090	5044		
HEP-R0160	5940	3585/5952			
HEP-R0161	5944	3585/5952	5048		
HEP-R0162	5944	3585/5952	5048		
HEP-R0164	5948	3585/5952			
HEP-R0166	5952	3585			
HEP-R0250	5980	3610	5096		
HEP-R0251	5986	3609	5096		
HEP-R0253	5986	3609	5096		
HEP-R0254	5990	3608	5100		
HEP-R0255	5990	3608	5100		
HEP-R0256	5994	3501	5104		
HEP-R0257	5994	3501	5104		
HEP-R0600	156	3175			
HEP-R0602	156	3175			
HEP-R0604	156	3175	512		
HEP-R0606	506	3175	511		
HEP-R0700	112	3089	1N82A		
HEP-R0801	166	9075			
HEP-R0803	167	3105			
HEP-R0804	168	3106/5304			
HEP-R0805	169	3107/5307			
HEP-R0851	166	9075			
HEP-R0852	166	9075			
HEP-R0853	167	3647			
HEP-R0855	168	3648			
HEP-R0856	169	3678			
HEP-R1001	5400	3950			
HEP-R1002	5401	3638/5402			
HEP-R1003	5402	3638			
HEP-R1004	5403	3627/5404			
HEP-R1005	5404	3627			
HEP-R1202	5515	6615			
HEP-R1205	5515	6615			
HEP-R1218	5414	3954		1067	
HEP-R1220	5442	3634/5444			
HEP-R1221	5444	3634			
HEP-R1222	5446	3635			
HEP-R1223	5448	3636			
HEP-R1243	5483	3942			
HEP-R1244	5484	3943/5485			
HEP-R1245	5485	3943			
HEP-R1246	5486	3944/5487			

Industry Standard No.	ECG	SK	GE	RS 276-	MOTOR.
HEP-R1247	5487	3944			
HEP-R1300	5500	3579/5504			
HEP-R1301	5501	3579/5504			
HEP-R1302	5502	3579/5504			
HEP-R1304	5504	3579			
HEP-R1306	5506	3580/5507			
HEP-R1307	5507	3580			
HEP-R1471	5541	6642/5542			
HEP-R1472	5542	6642			
HEP-R1473	5543	3581			
HEP-R1475	5545	3582			
HEP-R1721	5622	3631/5623			
HEP-R1722	5623	3631			
HEP-R1723	5624	3632			
HEP-R1725	5626	3633			
HEP-R1781	5681	3521/5686			
HEP-R1782	5682	3521/5686			
HEP-R1783	5683	3521/5686			
HEP-R1785	5685	3521/5686			
HEP-R2002	6407	9083A			
HEP-R2500	610	3323			
HEP-R2501	611	3324			
HEP-R2503	614	3327			
HEP-R3201	522	3303			
HEP-R3203	500A	3304			
HEP-R9001	178MP		300(2)		
HEP-R9002	178MP		300(2)		
HEP-R9003	178MP		300(2)		
HEP-R9134	109	3087		1123	
HEP-R9135	109	3087		1123	
HEP-S0001	194	3275	220		
HEP-S0002	192	3137	63		
HEP-S0003	192	3137	63		
HEP-S0004	123A	3444	20		
HEP-S0005	194	3275	220		
HEP-S0006	234	3247	65	2024	
HEP-S0008	229	3132	61		
HEP-S0010	229	3039/316	61		
HEP-S0011	123A	3444	20		
HEP-S0012	129	3118			
HEP-S0013	159	3118			
HEP-S0014	123A	3444			
HEP-S0015	123A	3275/194			
HEP-S0016	108	3452	86		
HEP-S0017	161	3018			
HEP-S0019	159	3715			
HEP-S0020	229	3246	61		
HEP-S0021	108	3452	86		
HEP-S0022	123A	3444	20		
HEP-S0023	199	3245	62		
HEP-S0024	199	3245	62		
HEP-S0025	123A	3444			
HEP-S0026	159	3466	82		
HEP-S0030	123A	3444	20		
HEP-S0031	159	3247/234	82		
HEP-S0032	159	3114/290	82		
HEP-S0033	229	3246	61		
HEP-S0034	123A	3044/154			
HEP-S3001	195A	3048/329	46		
HEP-S3002	282	3024/128			
HEP-S3006	320	3176/350			
HEP-S3010	282	3512			
HEP-S3011	282	3024/128			
HEP-S3012	129	3025	244		
HEP-S3013	311	3048/329	277		
HEP-S3019	190	3232/191	217		
HEP-S3020	188	3199			
HEP-S3021	191	3232	249		
HEP-S3022	191	3232	249		
HEP-S3023	210	3512	252		
HEP-S3024	210	3199/188	252		
HEP-S3025	210	3202	252		
HEP-S3026	188	3199	226		
HEP-S3027	211	3513	253	2026	
HEP-S3028	211	3200/189	253	2026	
HEP-S3029	211	3203	253	2026	
HEP-S3030	189	3200	218	2026	
HEP-S3031	189	3200	218	2026	
HEP-S3032	189	3200	218	2026	
HEP-S3033	154	3044	40		
HEP-S3034	154	3044	40		
HEP-S3035	154	3044	40		
HEP-S5000	184	3190	57		
HEP-S5001	182	3188A	55		
HEP-S5002	183	3189A	56	2027	
HEP-S5003	184	3054/196	57		
HEP-S5004	182	3188A	55		
HEP-S5005	183	3189A	56	2027	
HEP-S5006	185	3191	58		
HEP-S5007	185	3083/197	58		
HEP-S5008	183	3189A	56	2027	
HEP-S5009	183	3189A	56	2027	
HEP-S5010	183	3189A	56	2027	
HEP-S5011	124	3021	12		
HEP-S5012	175	3538	246		
HEP-S5013	129	3513	244		
HEP-S5014	128	3512	243		
HEP-S5015	157	3103A/396	232		
HEP-S5018	218	3083/197	234		
HEP-S5019	175	3026	246		
HEP-S5020	162	3438	35		
HEP-S5022	159	3466			
HEP-S5023	129	3025	244		
HEP-S5024	154	3103A/396	40		
HEP-S5025	154	3044	40		
HEP-S5026	154	3024/128	40		
HEP-S7000	181	3535	75		
HEP-S7001	180	3437	74	2043	
HEP-S7002	130	3036	14		
HEP-S7003	219	3173	74	2043	
HEP-S7004	130	3027	14		
HEP-S9001	6402	3628			
HEP-S9002	6409			2029	
HEP-S9100	172A	3156	64		
HEP-S9101	253	3996			
HEP-S9102	257	3180/263			
HEP-S9120	232	3241	258		
HEP-S9121	254	3997			
HEP-S9140	243	3182			
HEP-S9141	244	3183A			
HEP-Z0206	5005A	3771	ZD-3.3		
HEP-Z0206A	5005A	3771			
HEP-Z0208	5007A	3773	ZD-3.9		
HEP-Z0208A	5007A	3773			
HEP-Z0210	5009A	3775	ZD-4.7		
HEP-Z0211	5010A	3776	ZD-5.1		
HEP-Z0211A	5010A	3776			
HEP-Z0212	5011A	3777	ZD-5.6	561	
HEP-Z0212A	5011A	3777			
HEP-Z0214	5013A	3779	ZD-6.2	561	
HEP-Z0214A	5013A	3779			
HEP-Z0215	5014A	3780	ZD-6.8	561	
HEP-Z0215A	5014A	3780			
HEP-Z0216	138A	3781/5015A	ZD-7.5		

Industry Standard No.	ECG	SK	GE	RS 276-	MOTOR.
HEP-Z0216A	5015A	3781			
HEP-Z0217	5016A	3782	ZD-8.2	562	
HEP-Z0217A	5016A	3782			
HEP-Z0219	5018A	3784	ZD-9.1	562	
HEP-Z0219A	5018A	3784			
HEP-Z0220	5019A	3785	ZD-10	562	
HEP-Z0220A	5019A	3785			
HEP-Z0222	5021A	3787	ZD-12	563	
HEP-Z0225	5024A	3790	ZD-15	564	
HEP-Z0228	5027A	3793	ZD-18		
HEP-Z0230	5029A	3795	ZD-20		
HEP-Z0231	5030A	3796	ZD-22		
HEP-Z0234	5033A	3799	ZD-27		
HEP-Z0254	5098A	3819/5053A	ZD-130		
HEP-Z0255	5099A	3820/5054A	ZD-140		
HEP-Z0401	5066A	3330	ZD-3.3		
HEP-Z0402	134A	3055	ZD-3.6		
HEP-Z0403	5067A	3331	ZD-3.9		
HEP-Z0405	5069A	3333	ZD-4.7		
HEP-Z0406	135A	3056	ZD-5.1		
HEP-Z0407	136A	3057	ZD-5.6	561	
HEP-Z0408	137A	3058	ZD-6.2	561	
HEP-Z0409	5071A	3334	ZD-6.8	561	
HEP-Z0410	138A	3059	ZD-7.5		
HEP-Z0411	5072A	3136	ZD-8.2	562	
HEP-Z0412	139A	3060	ZD-9.1	562	
HEP-Z0413	140A	3061	ZD-10	562	
HEP-Z0414	5074A	3139	ZD-11.0	563	
HEP-Z0415	142A	3062	ZD-12	563	
HEP-Z0416	143A	3750	ZD-13	563	
HEP-Z0417	144A	3094	ZD-14		
HEP-Z0418	145A	3063	ZD-15	564	
HEP-Z0419	5075A	3751	ZD-16	564	
HEP-Z0420	5077A	3752	ZD-18		
HEP-Z0421	5079A	3335	ZD-20		
HEP-Z0422	5080A	3336	ZD-22		
HEP-Z0423	5081A	3151	ZD-24		
HEP-Z0424	146A	3064	ZD-27		
HEP-Z0425	5083A	3754	ZD-28		
HEP-Z0426	147A	3095	ZD-33		
HEP-Z0427	5085A	3337	ZD-36		
HEP-Z0428	5086A	3338	ZD-39		
HEP-Z0430	5088A	3340	ZD-47		
HEP-Z0432	5090A	3342	ZD-56		
HEP-Z0433	149A	3097	ZD-62		
HEP-Z0436	150A	3098	ZD-82		
HEP-Z0438	5096A	3346	ZD-100		
HEP-Z0439	151A	3099	ZD-110		
HEP-Z0440	5097A	3347	ZD-120		
HEP-Z0442	5100A	3350	ZD-150		
HEP-Z0444	5103A	3353	ZD-180		
HEP-Z0445	5105A	3355	ZD-200		
HEP-Z212				561	
HEP-Z2500	5111A	3377			
HEP-Z2502	5113A	3379			
HEP-Z2504	5115A	3381			
HEP-Z2506	5117A	3383			
HEP-Z2508	5119A	3385			
HEP-Z2510	5121A	3387	5ZD-7.5		
HEP-Z2513	5124A	3390	5ZD-9.1		
HEP-Z2514	5125A	3391	5ZD-10		
HEP-Z2516	5127A	3393	5ZD-12		
HEP-Z2519	5130A	3396	5ZD-15		
HEP-Z2522	5133A	3399	5ZD-18		
HEP-Z2525	5136A	3402	5ZD-22		
HEP-Z2526	5137A	3403	5ZD-24		
HEP-Z2528	5139A	3405	5ZD-27		
HEP-Z2530	5141A	3407	5ZD-30		
HEP-Z2531	5142A	3408	5ZD-33		
HEP-Z2537	5148A	3414			
HEP-Z2542	5153A	3419			
HEP-Z2545	5156A	3422			
HEP-Z2547	5158A	3424			
HEP-Z2548	5159A	3425			
HEP-Z2551	5162A	3428			
HEP1	126		52	2024	
HEP101	140A		ZD-10	562	
HEP102	134A		ZD-3.6		
HEP103	137A		ZD-6.2	561	
HEP104	139A		ZD-9.1	562	
HEP105	142A		ZD-12	563	
HEP134	109			1123	
HEP135	109			1123	
HEP151	5940		5048		
HEP153	5944		5048		
HEP154	116		504A	1104	
HEP155				1102	
HEP156	116		504A	1104	
HEP157	116		504A	1104	
HEP158	116		504A	1104	
HEP159	125		510,531	1114	
HEP160	125		510,531	1114	
HEP161	5800		504A	1142	
HEP162	5802		504A	1143	
HEP165	113A		6GC1		
HEP166	115		6GX1		
HEP170	125		510,531	1114	
HEP175	166		BR-600	1152	
HEP176	167		BR-600	1172	
HEP177	168		BR-600	1173	
HEP178	169		BR-600		
HEP2			1	2007	
HEP200	104	3719	16	2006	
HEP230	104	3719	16	2006	
HEP231	105		4		
HEP232	121	3717	239	2006	
HEP233	105		4		
HEP234	127	3764	25		
HEP235	127	3035	25		
HEP236	179		76		
HEP238	176		80		
HEP240	124		12		
HEP241	175		246	2020	
HEP242	129		244	2027	
HEP243	128		243	2030	
HEP244	157	3747	232		
HEP245	184		57	2017	
HEP246	185		58	2025	
HEP247	130		14	2041	
HEP248				2043	
HEP250	102A		53	2007	
HEP251	102A		53	2007	
HEP252	102	3722	2	2007	
HEP253	158		53	2007	
HEP254	158		53	2007	
HEP280	102A		53	2007	
HEP281	102A		53	2007	
HEP3	160		245	2004	
HEP300	5470		MR-4		
HEP302	5472		MR-4		
HEP304	5501		MR-3		
HEP306	5504		MR-3		
HEP310	6400A		2N2160	2029	
HEP311	6407	9083A		1050	

Industry Standard No.	ECG	SK	GE	RS 276-	MOTOR.
HEP320	5400	3950			
HEP50	123A		20	2051	
HEP51	129		244	2032	
HEP52	159		82	2032	
HEP53	123A		20	2051	
HEP54	123A		20	2051	
HEP55	123A		20	2051	
HEP56	108		86	2038	
HEP57	159		82	2024	
HEP591	704		IC-205		
HEP594	718		IC-8		
HEP595	720		IC-7		
HEP602	135A		ZD-5.1		
HEP603	136A		ZD-5.6	561	
HEP604	141A		ZD-11.5	563	
HEP605	143A		ZD-13	563	
HEP606	144A		ZD-14	564	
HEP607	145A		ZD-15	564	
HEP608	146A		ZD-27		
HEP609	147A		ZD-33		
HEP610	148A		ZD-55		
HEP611	149		ZD-62		
HEP612	150A		ZD-82		
HEP613	151A		ZD-110		
HEP620	5515	6615			
HEP621	5515	6615			
HEP622	5446	3635			
HEP623	121	3717	239	2006	
HEP624	121	3717	239	2006	
HEP626		3642	16	2006	
HEP627	179		76		
HEP628	121	3717	239	2006	
HEP629	102	3722	2	2007	
HEP630	102	3722	2	2007	
HEP631	102	3722	2	2007	
HEP632	102	3722	2	2007	
HEP633	102	3722	2	2007	
HEP634	102A		53	2007	
HEP635	126		52	2024	
HEP636	126		52	2024	
HEP637	160		245	2004	
HEP638	126		52	2024	
HEP639	126		52	2024	
HEP640	126		52	2024	
HEP641	101		8	2002	
HEP642	131		44	2006	
HEP643	131		44	2006	
HEP700	185		58	2025	
HEP701	184		57	2017	
HEP702	218		234		
HEP703	175		246	2020	
HEP703X2			24MP	2020	
HEP704	130		14	2041	
HEP704X2			15MP	2041(2)	
HEP705			19	2041	
HEP706	154		40	2012	
HEP707	162		35		
HEP708	159		82	2032	
HEP709	161	3716	39	2015	
HEP710	129		244	2027	
HEP711			18	2030	
HEP712	154		40	2012	
HEP713	154		40	2012	
HEP714	190		217		
HEP715	159		82	2032	
HEP716	159		82	2032	
HEP717	159		82	2032	
HEP718	108		86	2038	
HEP719	161	3716	39	2015	
HEP720	108		86	2038	
HEP721	107		11	2015	
HEP722	108		86	2038	
HEP723	108		86	2038	
HEP724	123A		20	2051	
HEP725	123A		20	2051	
HEP726	199		62	2010	
HEP727	108		86	2038	
HEP728	123A		20	2051	
HEP729	123A		20	2051	
HEP730	199		62	2010	
HEP731	107		11	2015	
HEP732	107		11	2015	
HEP733	108		86	2015	
HEP734	107		11	2015	
HEP735	123A		20	2051	
HEP736	128		243	2030	
HEP737	199		62	2010	
HEP738	123A		20	2051	
HEP739	159		82	2032	
HEP740	163A		36		
HEP75			88	2030	
HEP76			89	2023	
HEP801	312		FET-1	2028	
HEP802	312		FET-1	2035	
HEPC3000L				1801	
HEPC3000P				1801	
HEPC3002P				1811	
HEPC3004L				1802	
HEPC3004P				1802	
HEPC3010L				1807	
HEPC3010P				1807	
HEPC3020L				1809	
HEPC3020P				1809	
HEPC3041L				1804	
HEPC3041P				1804	
HEPC3073L				1803	
HEPC3073P				1803	
HEPC3075L				1806	
HEPC3075P				1806	
HEPC3800L				1808	
HEPC3800P				1808	
HEPC3801L				1819	
HEPC3801P				1819	
HEPC4001P			4001	2401	
HEPC4020P			4020	2420	
HEPC4021P			4021	2421	
HEPC4050P			4050	2450	
HEPC4051P			4051	2451	
HEPC6052P			IC-265	007	
HEPC6065P	719		IC-20		
HEPC6094P	1006		IC-8		
HEPC6096P	801		IC-35		
HEPC7400P			7400	1801	
HEPC7402P			7402	1811	
HEPC7404P			7404	1802	
HEPC7406P			7406	1821	
HEPC7408P			7408	1822	
HEPC7410P			7410	1807	
HEPC74123P			74123	1817	
HEPC7413P			7413	1815	
HEPC74145P			74145	1828	
HEPC74150P			74150	1829	
HEPC74154P			74154	1834	
HEPC74192P			74192	1831	

Industry Standard No.	ECG	SK	GE	RS 276-	MOTOR.
HEPC74193P			74193	1820	
HEPC74196P			74196	1833	
HEPC7420P			7420	1809	
HEPC7427P			7427	1823	
HEPC7432P			7432	1824	
HEPC7441AP			7441	1804	
HEPC7447AP			7447	1805	
HEPC7448P			7448	1816	
HEPC7451P			7451	1825	
HEPC7473P			7473	1803	
HEPC7474P	7474		7474	1818	
HEPC7475P			7475	1806	
HEPC7476P			7476	1813	
HEPC7485P			7485	1826	
HEPC7486P			7486	1827	
HEPC7490AP			7490	1808	
HEPC7492AP			7492	1819	
HEPF0010			FET-2	2035	
HEPF0015			FET-1	2028	
HEPF0021			FET-2	2035	
HEPF2004			FET-4	2036	
HEPF2005			FET-3	2036	
HEPF2007			FET-4	2036	
HEPF2007A			FET-4	2036	
HEPG0001			52	2024	
HEPG0002			245	2004	
HEPG0003			245	2004	
HEPG0004			1	2007	
HEPG0005			1	2007	
HEPG0005P/G			1	2007	
HEPG0006			1	2007	
HEPG0006P/G			1	2007	
HEPG0007			1	2007	
HEPG0007P/G			1	2007	
HEPG0008			52	2024	
HEPG0009				2024	
HEPG0010				2024	
HEPG0011				2002	
HEPG6000			16	2006	
HEPG6000P/G			16	2006	
HEPG6003			16	2006	
HEPG6003P/G			16	2006	
HEPG6005				2006	
HEPG6005P/G			16	2006	
HEPG6013			16	2006	
HEPG6013P/G			16	2006	
HEPG6016	131		44	2006	
HEPR0050			504A	1104	
HEPR0051			504A	1104	
HEPR0052			504A	1104	
HEPR0053			504A	1104	
HEPR0054			504A	1104	
HEPR0055			509	1114	
HEPR0056			509	1114	
HEPR0070				1143	
HEPR0071				1143	
HEPR0072				1144	
HEPR0080			512	1141	
HEPR0081			512	1143	
HEPR0082			512	1143	
HEPR0084			512	1144	
HEPR0090			512	1141	
HEPR0091			512	1143	
HEPR0092			512	1143	
HEPR0094			512	1144	
HEPR0170				1114	
HEPR0600			511	1114	
HEPR0602			511	1114	
HEPR0604			511	1114	
HEPR0606			511	1114	
HEPR0801			BR-600	1151	
HEPR0802			BR-600	1152	
HEPR0803			BR-600	1172	
HEPR0804			BR-600	1173	
HEPR0841				1151	
HEPR0842				1152	
HEPR0843				1172	
HEPR0845				1173	
HEPR0851				1151	
HEPR0852				1152	
HEPR0853				1172	
HEPR0855				1173	
HEPR1215				1067	
HEPR1216				1067	
HEPR1217				1067	
HEPR1218				1067	
HEPR2002				1050	
HEPR3010			511	1114	
HEPR9003			6GD1	1104	
HEPR9134			1N34AS	1123	
HEPR9134A			1N34AS	1123	
HEPR9135			1N60	1123	
HEPR9137				1122	
HEPS0002			210	2051	
HEPS0003				2030	
HEPS0004			20	2051	
HEPS0005	287		220		
HEPS0006			65	2050	
HEPS0008	229		61	2038	
HEPS0010	229		61	2038	
HEPS0011			210	2051	
HEPS0012	129		21	2034	
HEPS0013			21	2034	
HEPS0014			47	2030	
HEPS0015	123AP		20	2051	
HEPS0016			86	2038	
HEPS0017			61	2038	
HEPS0019			21	2034	
HEPS0020	108		86	2038	
HEPS0021			224	2038	
HEPS0022				2051	
HEPS0023				2010	
HEPS0024			212	2010	
HEPS0025	123AP		210	2051	
HEPS0026			82	2032	
HEPS0027	287		222		
HEPS0028	288		223		
HEPS0029	288		221		
HEPS0030	123AP		85	2010	
HEPS0031	234		65	2050	
HEPS0032	159		221		
HEPS0033	229		61	2038	
HEPS3005	347		286		
HEPS3006	349		287		
HEPS3007	351		288		
HEPS3008	311		277		
HEPS3009	320		289		
HEPS3012			244	2027	
HEPS3013	311		277		
HEPS3014	311		291		
HEPS3023			87	2018	
HEPS3024			83	2018	
HEPS3025				2018	
HEPS3026				2018	

Industry Standard No.	ECG	SK	GE	RS 276-	MOTOR.
HEPS3027			84	2026	
HEPS3028			84	2026	
HEPS3029				2026	
HEPS3030				2026	
HEPS3031			227	2026	
HEPS3032			227	2026	
HEPS3033			230	2012	
HEPS3034			230	2012	
HEPS3035			230	2012	
HEPS5000				2017	
HEPS5002			56	2027	
HEPS5003				2017	
HEPS5005				2027	
HEPS5006			58	2025	
HEPS5007			58	2025	
HEPS5008				2027	
HEPS5009				2027	
HEPS5010				2027	
HEPS5012				2020	
HEPS5013				2027	
HEPS5019	175		246	2020	
HEPS5023				2027	
HEPS5024	154		40	2012	
HEPS5025			230	2012	
HEPS5026			18	2030	
HEPS7000			75	2041	
HEPS7001			74	2043	
HEPS7002			14	2041	
HEPS7003				2043	
HEPS7004			75	2041	
HEPS9002			2N2160	2029	
HEPS9120	232		258		
HEPZ0212			ZD-5.6	561	
HEPZ0212A				561	
HEPZ0214			ZD-6.2	561	
HEPZ0214A				561	
HEPZ0215			ZD-6.8	561	
HEPZ0215A				561	
HEPZ0217			ZD-8.2	562	
HEPZ0217A				562	
HEPZ0219			ZD-9.1	562	
HEPZ0219A				562	
HEPZ0220			ZD-10.0	562	
HEPZ0220A				562	
HEPZ0222			ZD-12.0	563	
HEPZ0222A				563	
HEPZ0225			ZD-15.0	564	
HEPZ0225A				564	
HEPZ0254	5053A		ZD-130		
HEPZ0407			ZD-5.6	561	
HEPZ0408			ZD-6.2	561	
HEPZ0409			ZD-6.8	561	
HEPZ0411			ZD-8.2	562	
HEPZ0412			ZD-9.1	562	
HEPZ0413			ZD-10	562	
HEPZ0414			ZD-11	563	
HEPZ0415			ZD-12.0	563	
HEPZ0416			ZD-12.0	563	
HEPZ0417			ZD-15.0	564	
HEPZ0418			ZD-15.0	564	
HEPZ0419			ZD-15.0	564	
HEPZ2504	5115A		5ZD-4.7		
HEPZ2551	5100A		ZD-150		
HF-OSW05	116	3016	504A	1104	
HF-1	558	3843	511	1114	
HF-10024	109	3088		1123	
HF-17	506		511	1114	
HF-19D	179	3642	76		
HF-1A	558	3843	504A	1114	
HF-1B	506	3016	511	1114	
HF-1C	558	3515	504A	1114	
HF-18334	139A		ZD-9.1	562	
HF-18339	139A		ZD-9.1	562	
HF-18553		3126	90		
HF-1Z	558	3016	511	1114	
HF-20004	139A	3060	ZD-9.1	562	
HF-20005	612	3126	90		
HF-20007	614	3126	90		
HF-20008	109	3090		1123	
HF-20011	139A	3060	ZD-9.1	562	
HF-20014	177		300	1122	
HF-20032			504A	1104	
HF-20033	5071A		ZD-6.8	561	
HF-20034	177		300	1122	
HF-20035		3126	90		
HF-20041	142A		ZD-12	563	
HF-20042	116		504A	1104	
HF-20046		3126	90		
HF-20047	116	3311	504A	1104	
HF-20048	177	3100/519	300	1122	
HF-20050	116		504A	1104	
HF-20052	116	3031A	504A	1104	
HF-20060	177	3100/519	300	1122	
HF-20061		3100	300	1122	
HF-20062	610	3323	90		
HF-20063	519	3100	514	1122	
HF-20064	177	3100/519	300	1122	
HF-20065	139A	3060	ZD-9.1	562	
HF-20067	116	3311	504A	1104	
HF-20071	125	3081	510,531	1114	
HF-20080	612		90		
HF-20083	116	3311	504A	1104	
HF-20084	116	3311	504A	1104	
HF-20088	110MP	3709	1N34AS	1123(2)	
HF-20094		3126	90		
HF-20095	177	3100/519	300	1122	
HF-20105	601		1N60	1104	
HF-20116		3126	90		
HF-20124	519	3100	300	1122	
HF-35			51	2004	
HF-40	123A		20	2051	
HF-47	106	3114/290	21	2034	
HF-DS410	177	3100/519	300	1122	
HF-MV2	177	3100/519	300	1122	
HF-SD-1A		3017B	504A	1104	
HF-SD-1C			504A	1104	
HF-SD-1Z	506		504A	1104	
HF-SD1			504A	1104	
HF-SD1/BB-4			511	1114	
HFOW05			504A	1104	
HF1	558		511	1114	
HF10	129		244	2027	
HF11			28	2017	
HF11(PHILCO)	128		243	2030	
HF12			29	2025	
HF12(PHILCO)	129		244	2027	
HF12H	160	3007	245	2004	
HF12M	160	3007	245	2004	
HF12N	160	3008	245	2004	
HF15			28	2017	
HF16			29	2025	
HF17			62	2010	
HF19	121	3009	239	2006	
HF1A	558		511	1114	
HF1B	558		511	1114	
HF1C	558		511	1114	
HF1Z	558		511	1114	
HF2	123A	3444	20	2051	
HF20	121	3009	239	2006	
HF2000411B	312		FET-1	2035	
HF200191A	312	3448	FET-1	2035	
HF200191A-0	312	3448	FET-1	2035	
HF200191A/			FET-1	2028	
HF200191AO	312		FET-2	2035	
HF200191AO		3116	FET-2	2035	
HF200191B-0	312	3448	FET-1	2035	
HF200191BO	312		FET-1	2035	
HF200191BO		3116	FET-1	2028	
HF200301B		3112	FET-1	2028	
HF200301BO	312		FET-1	2028	
HF200301BO		3112	FET-1	2028	
HF200301C-0	312		FET-2	2028	
HF200301C-0			FET-1	2028	
HF200301E	222	3112	FET-1	2036	
HF20032	177		300	1122	
HF200411B	312	3448	FET-1	2035	
HF200411CO	312	3448	FET-2	2035	
HF20060			300	1122	
HF20066	116	3311	504A	1104	
HF20H	160	3007	245	2004	
HF20M	160	3007	245	2004	
HF22	130		14	2041	
HF23	180	3437	74	2043	
HF24	181		75	2041	
HF25	180		74	2043	
HF3	123A	3444	20	2051	
HF309301E	199	3018	62	2010	
HF3H	160	3005	245	2004	
HF3M	160	3005	245	2004	
HF4	123A	3444	20	2051	
HF40	194	3137/192A	220		
HF47	234	3114/290	21	2050	
HF5	123A	3444	20	2051	
HF50	183	3123	20	2051	
HF50H	160		245	2004	
HF50M	160	3006	245	2004	
HF51	187	3083/197	29	2025	
HF57	152	3054/196	66	2048	
HF58	153	3083/197	69	2049	
HF6	123A	3444	20	2051	
HF6H	160	3005	245	2004	
HF6M	160	3005	245	2004	
HF7	123A	3444	20	2051	
HF8	123A	3444	20	2051	
HF9	128		243	2030	
HFOW05	116	3017B/117	504A	1104	
HFR-10					1N4934
HFR-150					1N4935
HFR-200					1N4935
HFR-5					1N4933
HFSD-1	506	3109/119	511	1114	
HFSD-14	506	3130	511	1114	
HFSD-1A	506	3043	511	1114	
HFSD-1B	506	3125	511	1114	
HFSD-1C	506	3130	511	1114	
HFSD-1C(SEARS)	506		511	1114	
HFSD-1Z	506	3043	511	1114	
HFSD1	506	3016	511	1114	
HFSD12	156	3031A	512		
HFSD1A			504A	1104	
HFSD1B		3080	504A	1104	
HFSD1Z		3016	511	1114	
HFSG005	116	3017B/117	504A	1104	
HG-2					MR250-2
HG-3					MR250-3
HG-4					MR250-4
HG-5					MR250-5
HG-R10			504A	1104	
HG1090	109	3091		1123	
HG5002	109	3087		1123	
HG5004	109	3087		1123	
HG5006	109	3087		1123	
HG5007	109	3087		1123	
HG5008	109	3087		1123	
HG5009	109	3087		1123	
HG5078	109	3087		1123	
HG5079	109	3087		1123	
HG5085	109	3087		1123	
HG5088	109	3087		1123	
HG5808	109	3091		1123	
HGR-10	116	3017B/117	504A	1104	1N4002
HGR-20	116		504A	1104	1N4003
HGR-30	116		504A	1104	1N4004
HGR-40	116		504A	1104	1N4004
HGR-5	116		504A	1104	1N4001
HGR-60	116		504A	1104	1N4005
HGR1	116	3311	504A	1104	
HGR2	116	3016	504A	1104	
HGR3	116	3017B/117	504A	1104	
HGR4	116	3031A	504A	1104	
HIFI	116		504A	1104	
HJ15	102A	3123	53	2007	
HJ15D	160		245	2004	
HJ17	102A	3123	53	2007	
HJ17D	102A	3123	53	2007	
HJ22	102A	3123	53	2007	
HJ226	158	3123	53	2007	
HJ228	158	3005	53	2007	
HJ22D	102A		53	2005	
HJ23	102A	3123	53	2007	
HJ230	158	3123	53	2007	
HJ23D	102A		53	2005	
HJ315	158	3123	53	2007	
HJ32	160		245	2004	
HJ34	160		245	2004	
HJ34A	160		245	2004	
HJ35	104	3719	16	2006	
HJ37	160	3123	245	2004	
HJ41	100	3005	1	2007	
HJ43	102A	3004	53	2007	
HJ50	102A		53	2007	
HJ51	102A		53	2007	
HJ54	102A	3123	53	2007	
HJ55	160		245	2004	
HJ56	160		245	2004	
HJ57	160		245	2004	
HJ60	102A		53	2007	
HJ606	102A	3005	53	2007	
HJ60A	160	3123	245	2004	
HJ60C	160	3005	245	2004	
HJ62	102A	3123	53	2007	
HJ70	160		245	2004	
HJ71	158		53	2005	
HJ72	158		53	2005	
HJ73	158		53	2007	
HJ74	158		53	2007	
HJ75			50	2004	
HJX2	102A	3123	53	2007	

Industry Standard No.	ECG	SK	GE	RS 276-	MOTOR.
HK-00049	312	3116	FET-2	2035	
HK-00330	312		FET-1	2028	
HKT-158	123A	3444	20	2051	
HKT-161	123A	3444	20	2051	
HL18998	7400	7400		1801	
HL18999	7474	7474		1818	
HL19000	7404	7404		1802	
HL19001	7410	7410		1807	
HL19002	7473	7473		1803	
HL19003	7420	7420		1809	
HL19004	7402	7402		1811	
HL19006	7493A	7493			
HL19008	74121	74121			
HL19009	7442	7442			
HL19010	7476	7476		1813	
HL19011	7440	7440			
HL19012	7475	7475		1806	
HL19013	7430	7430			
HL19014	7486	7486		1827	
HL19015	7490	7490		1808	
HL24510	941		IC-263	010	
HL24592	925		IC-261		
HL24593	941		IC-263	010	
HL24630	909	3590	IC-249		
HL55763	7408			1822	
HL55861	74150			1829	
HL55862	7404			1802	
HL56420	7400			1801	
HL56421	7404			1802	
HL56422	7420			1809	
HL56425	7474			1818	
HL56426	7485			1826	
HL56429	74192			1831	
HL56430	74193			1820	
HL56899	7410			1807	
HM-00049	100	3005	1	2007	
HM-08014	102A	3004	53	2007	
HM10	5019A	3785	ZD-10	562	
HM100	5050A	3816	ZD-100		
HM100A	5050A	3816	ZD-100		
HM10A	5019A	3785	ZD-10	562	
HM10B	140A	3785/5019A	ZD-10	562	
HM11	5020A	3786	ZD-11	563	
HM110	5051A	3817	ZD-110		
HM110A	5051A	3817	ZD-110		
HM11A	5020A	3786	ZD-11	563	
HM12	5021A	3787	ZD-12	563	
HM120	5052A	3818	ZD-120		
HM120A	5052A	3818	ZD-120		
HM12A	5021A	3787	ZD-12	563	
HM12B	142A	3787/5021A	ZD-12	563	
HM13	5022A	3788	ZD-12	563	
HM130	5053A	3819	ZD-130		
HM130A	5053A	3819	ZD-130		
HM13A	5022A	3788	ZD-12	563	
HM15	5024A	3790	ZD-15	564	
HM150	5055A	3821	ZD-150		
HM150A	5055A	3821	ZD-150		
HM15A	5024A	3790	ZD-15	564	
HM15B	145A	3790/5024A	ZD-15	564	
HM16	5025A	3791		564	
HM160	5056A	3822	ZD-160		
HM160A	5056A	3822	ZD-160		
HM16A	5025A	3791		564	
HM18	5027A	3793	ZD-18		
HM180	5058A	3824	ZD-180		
HM180A	5058A	3824	ZD-180		
HM18A	5027A	3793	ZD-18		
HM20	5029A	3795	ZD-20		
HM200	5060A	3826	ZD-200		
HM200A	5060A	3826	ZD-200		
HM20A	5029A	3795	ZD-20		
HM22	5030A	3796	ZD-22		
HM22A	5030A	3796	ZD-22		
HM24	5031A	3797	ZD-24		
HM24A	5031A	3797	ZD-24		
HM27	5033A	3799	ZD-27		
HM27A	5033A	3799	ZD-27		
HM27B	146A	3064	ZD-27		
HM30	5035A	3801	ZD-30		
HM30A	5035A	3801	ZD-30		
HM33	5036A	3802	ZD-33		
HM33A	5036A	3802	ZD-33		
HM36	5037A	3803	ZD-36		
HM36A	5037A	3803	ZD-36		
HM39	5038A	3804	ZD-39		
HM39A	5038A	3804	ZD-39		
HM43	5039A	3805	ZD-43		
HM435101	65101S				MCM5101
HM43A	5039A	3805	ZD-43		
HM468A10					MCM68A10
HM47	5040A	3806	ZD-47		
HM4716	2117				MCM4116
HM472114					MCM2114
HM47A	5040A	3806	ZD-47		
HM4816					MCM4517
HM4847	2147				MCM2147
HM4864					MCM6665
HM51	5041A	3807	ZD-51		
HM51A	5041A	3807	ZD-51		
HM56	5042A	3808	ZD-56		
HM56A	5042A	3808	ZD-56		
HM6.8	5014A	3780	ZD-6.8	561	
HM6.8A	5014A	3780	ZD-6.8	561	
HM6116P					MCM65116
HM6148P					MCM65148
HM62	5044A	3810	ZD-62		
HM62A	5044A	3810	ZD-62		
HM68	5045A	3811	ZD-68		
HM68A	5045A	3811	ZD-68		
HM7.5	5015A	3781	ZD-7.5		
HM7.5A	5015A	3781	ZD-7.5		
HM75	5046A	3812	ZD-75		
HM75A	5046A	3812	ZD-75		
HM7641					MCM7641
HM7643					MCM7643
HM7681					MCM7681
HM7685					MCM7685
HM8.2	5016A	3782	ZD-8.2	562	
HM8.2A	5016A	3782	ZD-8.2	562	
HM82	5047A	3813	ZD-82		
HM82A	5047A	3813	ZD-82		
HM9.1	5018A	3784	ZD-9.1	562	
HM9.1A	5018A	3784	ZD-9.1	562	
HM9.1B	139A	3784/5018A	ZD-9.1	562	
HM91	5049A	3815	ZD-91		
HM91A	5049A	3815	ZD-91		
HN-00002	177	3100/519	300	1122	
HN-00003	109	3088		1123	
HN-00005	142A		ZD-12	563	
HN-00008	116	3311	504A	1104	
HN-00012	139A		ZD-9.1	562	
HN-00018	116	3311	504A	1104	
HN-00024	139A		ZD-9.1	562	

Industry Standard No.	ECG	SK	GE	RS 276-	MOTOR.
HN-00029	116		504A	1104	
HN-00032	116	3311	504A	1104	
HN-00033		3126	90		
HN-00045	614		90		
HN-00052	610		90		
HN-00061	140A	3061	ZD-10	562	
HN-0047	177		300	1122	
HN462316EP					MCM68A316E
HN462532					MCM2532
HN462716					MCM2716
HN46332					MCM68A332
HN46364					MCM68365
HN48016					MCM2816
HNIL301AJ	9301				MC662P
HNIL301AL	9301				MC662L
HNIL301CJ	9301				MC662P
HNIL301CL	9301				MC662L
HNIL302AJ	9302				MC693P
HNIL302AL	9302				MC693L
HNIL302CJ	9302				MC693P
HNIL302CL	9302				MC693L
HNIL303AJ	9303				MC668P
HNIL303AL	9303				MC668L
HNIL303CJ	9303				MC668P
HNIL303CL	9303				MC668L
HNIL311AJ	9311				MC664P
HNIL311AL	9311				MC664L
HNIL311CJ	9311				MC664P
HNIL311CL	9311				MC664L
HNIL312AJ	9312				MC688P
HNIL312AL	9312				MC688L
HNIL312CJ	9312				MC688P
HNIL312CL	9312				MC688L
HNIL321AJ	9321				MC672P
HNIL321AL	9321				MC672L
HNIL321CJ	9321				MC672P
HNIL321CL	9321				MC672L
HNIL322AJ	9322				MC660P
HNIL322AL	9322				MC660L
HNIL322CJ	9322				MC660P
HNIL322CL	9322				MC660L
HNIL324AJ	9324				MC668P
HNIL324AL	9324				MC668L
HNIL324CJ	9324				MC668P
HNIL324CL	9324				MC668L
HNIL325AJ	9325				MC671P
HNIL325AL	9325				MC671L
HNIL325CJ	9325				MC671P
HNIL325CL	9325				MC671L
HNIL326AJ	9326				MC670P
HNIL326AL	9326				MC670L
HNIL326CJ	9326				MC670P
HNIL326CL	9326				MC670L
HNIL331AJ	9331				MC669P
HNIL331AL	9331				MC669L
HNIL331CJ	9331				MC669P
HNIL331CL	9331				MC669L
HNIL332AJ	9332				MC681P
HNIL332AL	9332				MC681L
HNIL332CJ	9332				MC681P
HNIL332CL	9332				MC681L
HNIL333AJ	9333				MC697P
HNIL333AL	9333				MC697L
HNIL333CJ	9333				MC697P
HNIL333CL	9333				MC697L
HNIL334AJ	9334				MC678P
HNIL334AL	9334				MC678L
HNIL334CJ	9334				MC678P
HNIL334CL	9334				MC678L
HNIL335AJ	9335				MC677P
HNIL335AL	9335				MC677L
HNIL335CJ	9335				MC677P
HNIL335CL	9335				MC677L
HNIL336AJ					MC681P
HNIL336AL					MC681L
HNIL336CJ					MC681P
HNIL336CL					MC681L
HNIL341AJ					MC673P
HNIL341AL					MC673L
HNIL341CJ					MC673P
HNIL341CL					MC673L
HNIL342AJ	9342				MC667P
HNIL342AL	9342				MC667L
HNIL342CJ	9342				MC667P
HNIL342CL	9342				MC667L
HNIL347AJ					MC675P
HNIL347AL					MC675L
HNIL347CJ	9347				MC675P
HNIL347CL	9347				MC675L
HNIL361AJ	9361				MC665P
HNIL361AL					MC665L
HNIL361CJ					MC665P
HNIL361CL	9361				MC665L
HNIL362AJ					MC666P
HNIL362AL					MC666L
HNIL362CJ					MC666P
HNIL362CL					MC666L
HNIL363AJ	9363				MC691P
HNIL363AL	9363				MC691L
HNIL363CJ	9363				MC691P
HNIL363CL	9363				MC691L
HNIL368AJ	9368				MC694P
HNIL368AL	9368				MC694L
HNIL368CJ	9368				MC694P
HNIL368CL	9368				MC694L
HNIL370AJ	9370				MC682P
HNIL370AL	9370				MC682L
HNIL370CJ	9370				MC682P
HNIL370CL	9370				MC682L
HNIL371AJ	9371				MC684P
HNIL371AL	9371				MC684L
HNIL371CJ	9371				MC684P
HNIL371CL	9371				MC684L
HNIL372AJ	9372				MC685P
HNIL372AL	9372				MC685L
HNIL372CJ	9372				MC685P
HNIL372CL	9372				MC685L
HNIL375AJ	9375				MC686P
HNIL375AL	9375				MC686L
HNIL375CJ	9375				MC686P
HNIL375CL	9375				MC686L
HNIL380AJ	9380				MC676P
HNIL380AL	9380				MC676L
HNIL380CJ	9380				MC676P
HNIL380CL	9380				MC676L
HNIL381AJ					MC676P
HNIL381AL					MC676L
HNIL381CJ					MC676P
HNIL381CL					MC676L
HNIL382AJ	9382				MC676P
HNIL382AL	9382				MC676L
HNIL382CJ	9382				MC676P
HNIL382CL	9382				MC676L
H0300	127	3764	25		

Industry Standard No.	ECG	SK	GE	RS 276-	MOTOR.
H070		3066	CR-1		
H089			ZD-12	563	
H091		3087		1123	
H099		3110	CR-3		
HP-5A	116	3311	504A	1104	
HP205	116	3017B/117	504A	1104	
HP5082-2800	519		514	1122	
HPF1025	723		IC-15		
HR-05A	116	3031A	504A	1104	
HR-1	102A		53	2007	
HR-101A	121	3717	239	2006	
HR-107(PHILCO)	175		246	2020	
HR-11	123A	3444	20	2051	
HR-11A	123A	3444	20	2051	
HR-11B	123A	3444	20	2051	
HR-13	123A	3444	20	2051	
HR-13A	123A	3444	20	2051	
HR-14	199	3245	62	2010	
HR-14A	123A	3444	20	2051	
HR-15	199	3124/289	62	2010	
HR-15A	123A	3444	20	2051	
HR-16	123A	3444	20	2051	
HR-16A	123A	3444	20	2051	
HR-17	123A	3444	20	2051	
HR-17A	123A	3444	20	2051	
HR-18	123A	3444	20	2051	
HR-18A		3444	20	2051	
HR-19	123A	3444	20	2051	
HR-19A	123A	3444	20	2051	
HR-19E	179	3642	76		
HR-1S85		3126	90		
HR-2	102A	3004	53	2007	
HR-20	160		245	2004	
HR-200	5802	9005	512	1143	
HR-20A	160		245	2004	
HR-21	160		245	2004	
HR-21A	160		245	2004	
HR-22	160		245	2004	
HR-22A	160		245	2004	
HR-22B	160		245	2004	
HR-24	160		245	2004	
HR-24A			245	2004	
HR-25	160		245	2004	
HR-25A	160		245	2004	
HR-26	160		245	2004	
HR-26A	160		245	2004	
HR-27	160		245	2004	
HR-27A	160		245	2004	
HR-2A		3017B	504A	1104	
HR-2A(PENNCREST)	116		504A	1104	
HR-3	102A	3004	53	2007	
HR-30		3004	52	2024	
HR-32	123A	3444	20	2051	
HR-36	123A	3444	20	2051	
HR-37	123A	3444	20	2051	
HR-38	123A	3444	20	2051	
HR-39	102A	3004	53	2007	
HR-4	102A	3005	53	2007	
HR-40		3006	50	2004	
HR-43		3006	50	2004	
HR-45		3006	50	2004	
HR-47	199	3124/289	62	2010	
HR-48	123A	3444	20	2051	
HR-4A	102A		53	2007	
HR-5	102A	3005	53	2007	
HR-58			11	2015	
HR-59		3018	11	2015	
HR-5A	116	3017B/117	504A	1104	
HR-5AX2	116		504A	1104	
HR-5B	116	3311	504A	1104	
HR-6	102A		53	2007	
HR-60		3018	20	2051	
HR-61			2	2007	
HR-67	192		63	2030	
HR-68	196	3054	241	2020	
HR-69	186	3192	28	2017	
HR-7	102A	3004	53	2007	
HR-70	187	3193	29	2025	
HR-71	159	3114/290	82	2032	
HR-72	193		67	2023	
HR-73	210	3202	252	2018	
HR-74	211	3203	253	2026	
HR-75	199	3245	62	2010	
HR-7A	102A		53	2007	
HR-8	102A	3005	53	2007	
HR-84			21	2034	
HR-8A	102A		53	2007	
HR-9	102A	3005	53	2007	
HR-9A	102A		53	2007	
HR1			51	2004	
HR10	116		504A	1104	
HR100					1N5401
HR101	104	3719	16	2006	
HR101A		3009	3	2006	
HR102	121	3009	239	2006	
HR102C	121	3009	239	2006	
HR103	121	3009	239	2006	
HR103A		3009	3	2006	
HR104961C	187	3193	29	2025	
HR105	121	3717	239	2006	
HR105611C	193		67	2023	
HR105A	121	3717	239	2006	
HR105B	121	3717	239	2006	
HR106	124	3021	12		
HR107	124	3021	12		
HR107H	175	3131A/369	246	2020	
HR11	116	3124/289	504A	1104	
HR11A		3124	20	2051	
HR11B		3124	20	2051	
HR12A	128		243	2030	
HR12B	128		243	2030	
HR12C	128		243	2030	
HR12D	128		243	2030	
HR12E	128		243	2030	
HR12F	128		243	2030	
HR13	116	3024/128	504A	1104	
HR13A		3024	20	2051	
HR14		3124	20	2051	
HR14A		3124	20	2051	
HR15		3124	52	2024	
HR15A		3124	20	2051	
HR16		3124	20	2051	
HR16A		3124	20	2051	
HR17		3124	20	2051	
HR17A		3124	20	2051	
HR18		3124	20	2051	
HR18A		3124	20	2051	
HR19		3124	20	2051	
HR19A		3124	20	2051	
HR2			51	2004	
HR20		3006	51	2004	
HR200	5082A		ZD-25		1N5402
HR20A		3006	51	2004	
HR21		3006	51	2004	
HR21A		3006	51	2004	
HR22		3006	51	2004	
HR22A		3006	51	2004	
HR22B		3006	51	2004	
HR24		3006	51	2004	
HR24A	160	3006	51	2004	
HR25		3006	51	2004	
HR25A		3006	51	2004	
HR26		3006	51	2004	
HR26A		3006	51	2004	
HR27		3006	51	2004	
HR27A		3006	51	2004	
HR28	128	3024	243	2030	
HR29	128	3024	243	2030	
HR2A		3004	2	2007	
HR3			2	2007	
HR30	158	3004/102A	53	2007	
HR32			20	2051	
HR36	123A	3124/289	20	2051	
HR37		3124	20	2051	
HR38			63	2030	
HR39			11	2015	
HR4		3123	2	2007	
HR40	160	3006	245	2004	
HR400					1N5404
HR40836	160		245	2004	
HR40837	160		245	2004	
HR41	160		245	2004	
HR42	160	3006	245	2004	
HR43	160	3006	245	2004	
HR43835	160		245	2004	
HR44	160	3006	245	2004	
HR448636	160		245	2004	
HR45	160	3006	245	2004	
HR45838	160		245	2004	
HR45910	160		245	2004	
HR45913	160		245	2004	
HR46	160	3006	245	2004	
HR47	199	3124/289	62	2010	
HR48		3020	20	2051	
HR4A		3005	1	2007	
HR5		3031A	1	2007	
HR50	160	3008	245	2004	
HR51	160	3008	245	2004	
HR52	160	3008	245	2004	
HR53	102A	3004	53	2007	
HR58	108	3452	86	2038	
HR59	108	3452	86	2038	
HR5A	116	3311	504A	1104	
HR5A8E	116	3311	504A	1104	
HR5B	116		504A	1104	
HR6		3005	2	2007	
HR60	108	3452	86	2038	
HR600		3051			1N5406
HR61	102A		53	2007	
HR62	123A		20	2051	
HR63	123A	3018	20	2051	
HR64	123A		20	2051	
HR65	123A	3122	20	2051	
HR66	123A	3122	20	2051	
HR67		3024	63	2030	
HR69			66	2048	
HR7			51	2004	
HR70			69	2049	
HR71	159	3114/290	82	2032	
HR72	193		67	2023	
HR76	229	3018	11	2015	
HR77	229	3018	61	2038	
HR78	229	3018	61	2038	
HR79	107	3018	11	2015	
HR7A		3004	2	2007	
HR8		3123	2	2007	
HR80	107	3018	11	2015	
HR81	128	3018	243	2030	
HR82	192		63	2030	
HR83	192		63	2030	
HR84		3018	21	2034	
HR84(NPN)	123A		20	2051	
HR84(PNP)	159		82	2032	
HR85	210	3202	252	2018	
HR86	211	3203	253	2026	
HR87	229	3018	61	2038	
HR8A		3005	1	2007	
HR9		3123	2	2007	
HR9A		3005	1	2007	
HRF100					MR851
HRF200					MR852
HRF400					MR854
HRF600					MR856
HRN1020					MFE3003
HRN1030					MFE3003
HRN8318D					MFE3003
HRN8338D					3N155
HRN8346D					MFE3003
HRN8350					MFE3003
HRN8353					MFE3003
HRN8360					2N4352
HRN8363					3N155
HS-1168	123A	3444	20	2051	
HS-1225	108	3452	86	2038	
HS-1226	108	3452	86	2038	
HS-1227	108	3452	86	2038	
HS-1229	123A	3444	20	2051	
HS-15	102	3004	2	2007	
HS-15/12		3067	CR-4		
HS-20	505	3067/502	CR-5		
HS-22D	102	3004	2	2007	
HS-40014	123AP	3018	86	2038	
HS-40016	297	3024/128	81	2051	
HS-40017	108	3452	20	2051	
HS-40019	229	3452/108	86	2038	
HS-40020	108	3018	20	2051	
HS-40027	298	3114/290	82	2032	
HS-40030	123AP	3444/123A	20	2051	
HS-40031	159	3025/129	82	2032	
HS-40035	159	3114/290	21	2034	
HS-40037	123A	3444	20	2051	
HS-40039	128	3444/123A	243	2051	
HS-40040	159	3114/290	82	2032	
HS-40044	123A	3444	20	2051	
HS-40045	108	3039/316	86	2038	
HS-40046	123A	3444	20	2051	
HS-40047	108	3452	86	2038	
HS-40050	159	3025/129	82	2032	
HS-40053	106	3114/290	21	2034	
HS-40054	229	3018	211	2016	
HS-40055	108	3018	20	2051	
HS-40057	290A	3114/290	269	2032	
HS-7/1	118	3066	CR-1		
HS-8/1	118	3066	CR-1		
HS-90028	159		82	2032	
HS07		3577	C116F21		
HS1/7	118	3066	CR-1		
HS1/9	118	3066	CR-1		

Industry Standard No.	ECG	SK	GE	RS 276-	MOTOR.
HS10/1	503	3066/118	CR-5		
HS1001	116	3311	504A	1104	
HS1002	116	3311	504A	1104	
HS1003	116	3311	504A	1104	
HS1007	116	3311	504A	1104	
HS1008	116	3311	504A	1104	
HS1009	116	3311	504A	1104	
HS1010	116	3311	504A	1104	
HS1011		3067	504A	1104	
HS1012	116	3311	504A	1104	
HS102	102A	3004	53	2007	
HS1020	116	3311	504A	1104	
HS133	112		1N82A		
HS15			2	2007	
HS15(SANYO)	503		CR-5		
HS15-1C	503		CR-5		
HS15/1	505	3067/502	CR-7		
HS15/1(SONY)	502		CR-4		
HS15/16	503	3068	CR-5		
HS15/1A(SONY)	502		CR-4		
HS15/1B	502		CR-4		
HS15/1C	503	3068	CR-5		
HS151A(SONY)	502		CR-4		
HS17		3577	C116A21		
HS170	102	3123	2	2007	
HS17D	102A	3123	53	2007	
HS20	502	3067	CR-7		
HS20(SONY)	505		CR-5		
HS20-1	505	3068/503	CR-5		
HS20-1A	505	3068/503	CR-5		
HS20-1BS	505		CR-5		
HS20/1	505	3067/502	CR-6		
HS20/16S	505	3108	CR-7		
HS20/1AS	505	3108	CR-6		
HS20/1B	505	3068/503	CR-5		
HS20/1C	505	3067/502	CR-4		
HS2027	5002A	3768			
HS2033A	5005A	3771	ZD-3.3		
HS2036	134A	3055	ZD-3.6		
HS2039	134A	3055	ZD-3.6		
HS2047	135A		ZD-5.1		
HS2051	135A	3056	ZD-5.1		
HS2056	136A		ZD-5.6	561	
HS2062	137A	3058	ZD-6.2	561	
HS2091	139A	3060		562	
HS2100	140A	3061	ZD-10	562	
HS2120	142A	3062	ZD-12	563	
HS2150	145A	3063	ZD-15	564	
HS2165			ZD-16	564	
HS2270	146A	3064	ZD-27		
HS22D		3123	52	2024	
HS2330	147A	3095	ZD-33		
HS23D	102	3722	2	2007	
HS25-1	505	3108	CR-6		
HS25-1C	505		CR-6		
HS25/1		3108	CR-6		
HS25/16	505	3108	CR-7		
HS25/1AS	505		CR-5		
HS25/1B	505	3108	CR-7		
HS25/1BS	505	3108	CR-7		
HS25/1C	505	3108	CR-6		
HS2516	505		CR-7		
HS251C	505		CR-7		
HS27		3577	C116B21		
HS290	102	3722	2	2007	
HS29D		3123	52	2024	
HS30	504		CR-6		
HS30-16	504	3108/505	CR-6		
HS30-1B	504	3108/505	CR-6		
HS30-1C	505	3108	CR-7		
HS30/1C	505	3108	CR-7		
HS3103	116	3031A	504A	1104	
HS3104	116	3031A	504A	1104	
HS3108	125	3032A	510,531	1114	
HS3110	125	3033	510,531	1114	
HS37		3578	C116C21		
HS40021	161	3132	39	2015	
HS40022	161	3132	39	2015	
HS40023	161	3132	39	2015	
HS40024	161	3132	39	2015	
HS40025	161	3132	39	2015	
HS40026	128		243	2030	
HS40032	129	3025	244	2027	
HS40046	123A		20	2051	
HS40049	108		86	2038	
HS47		3578	C116D21		
HS5	102A		53	2007	
HS5810	192	3137	63	2030	
HS5811	193	3138	67	2023	
HS5812	192	3137	63	2030	
HS5813	193	3138	67	2023	
HS5814	192	3137	63	2030	
HS5815	193	3138	67	2023	
HS5816	192		63	2030	
HS5817	193	3138	67	2023	
HS5818	192	3137	63	2030	
HS5819	193	3138	67	2023	
HS5820	192	3137	63	2030	
HS5821	193	3138	67	2023	
HS5822	192	3137	63	2030	
HS5823	193	3138	67	2023	
HS6/1	502	3068/503	CR-4		
HS6010	192		63	2030	
HS6011	193	3138	67	2023	
HS6012	192		63	2030	
HS6013	193	3138	67	2023	
HS6014	192	3137	63	2030	
HS6015	193	3138	67	2023	
HS6016	192	3137	63	2030	
HS6017	193	3138	67	2023	
HS7/1		3066	CR-1		
HS7/17N	118		CR-1		
HS7027	5002A	3768			
HS7033	5005A	3771	ZD-3.3		
HS7036	134A	3055	ZD-3.6		
HS7039	134A	3055	ZD-3.6		
HS7051	135A	3056	ZD-5.1		
HS7056	136A	3057	ZD-5.6	561	
HS7062	137A		ZD-6.2	561	
HS7091	139A	3060	ZD-9.1	562	
HS7100	140A	3061	ZD-10	562	
HS7120	142A	3062	ZD-12	563	
HS7150	145A	3063	ZD-15	564	
HS7270	146A	3064	ZD-27		
HS7330	147A	3095	ZD-33		
HS8/1		3066	CR-1		
HS811		3066	CR-1		
HS9/1	118	3066	CR-1		
HS9/17N	505	3108	CR-7		
HS902A	112		1N82A		
HSC-1	156		512		
HSC1		3033	509	1114	
HSFD-1A	116	3032A	504A	1104	
HSFD-1A(SONY)	506		511	1114	
HSFD-1A(SONY-58				1114	
HSFD-1Z		3100	504A	1104	
HSFD1	506		511	1114	
HSS1	502		CR-4		
HST-9201			14	2041	
HST-9205			14	2041	
HST-9206			14	2041	
HST-9210			14	2041	
HST5001			212	2010	
HST5906			216	2053	
HT-23G	177		300	1122	
HT040519C			8	2002	
HT040519C(2SB405)			53	2007	
HT040519C(2SD72)			59	2002	
HT100	159	3114/290	82	2024	
HT1001510	102A	3005	53	2007	
HT101	159	3114/290	82	2024	
HT101011X	126	3006/160	52	2024	
HT101021A	126	3008	52	2024	
HT102341	160	3006	245	2004	
HT102341A	160	3006	245	2004	
HT102341B	160	3006	245	2004	
HT102341C	160	3006	245	2004	
HT10234UB				2004	
HT102351	160	3006	245	2004	
HT102351A	160	3006	245	2004	
HT103501	160	3006	245	2004	
HT103501A	126	3006/160	52	2024	
HT103531C	126	3006/160	52	2024	
HT103541B	126	3006/160	52	2024	
HT104861	128	3024	243	2030	
HT104861A	128	3024	243	2030	
HT104861B	128	3024	243	2030	
HT104941B-0	159		82	2032	
HT104941B-0			22	2032	
HT104941C-0	159		82	2032	
HT104941C-0			21	2034	
HT104941CO	159		82	2032	
HT104941CO		3114	21	2034	
HT104942A	159		82	2032	
HT104951A-0	159		82	2032	
HT104951A-0			21	2034	
HT104951B	159		82	2032	
HT104951B-0	159	3025/129	82	2032	
HT104951C	159	3025/129	82	2032	
HT104961B	185	3191	58	2025	
HT104961C	184	3190	57	2017	
HT104971A				2027	
HT104971A-0	129		244	2027	
HT104971A-0				2027	
HT104971AO	129		244	2027	
HT104971AO		3025		2027	
HT105611	159	3025/129	82	2032	
HT105611A	159	3025/129	82	2032	
HT105611B	159	3025/129	82	2032	
HT105611BO	159	3114/290	82	2032	
HT105611BO			22	2032	
HT105611C	159	3114/290	82	2032	
HT105612B	159		82	2032	
HT105621B-0	159		82	2032	
HT105621B-0			21	2034	
HT105621BO	159		82	2032	
HT105621BO		3114	22	2032	
HT105641C	234	3025/129	65	2050	
HT105641D	234		65	2050	
HT105641H	234		65	2050	
HT105642B	234	3025/129	65	2050	
HT106731B	159	3025/129	82	2032	
HT106731BO	290A	3114/290	269	2032	
HT107211T	234	3114/290	65	2050	
HT107331Q0	294	3138/193A	48		
HT2000771C				2007	
HT200540	102A	3004	53	2007	
HT200540A	102A	3004	53	2007	
HT200541	102A	3004	53	2007	
HT200541A	102A	3004	53	2007	
HT200541B	102A	3006/160	53	2007	
HT200541B-0	102A	3004	53	2007	
HT200541C	160	3006	245	2004	
HT200561	102A	3004	53	2007	
HT200561A	102A	3004	53	2007	
HT200561B	102A	3004	53	2007	
HT200561C	102A	3004	53	2007	
HT200561C-O	102A	3004	53	2007	
HT200751B	102A	3004	53	2007	
HT200770B	102A	3004	53	2007	
HT200771	102A	3004	53	2007	
HT200771A	102A	3004	53	2007	
HT200771B	102A	3004	53	2007	
HT200771C	102A	3004	53	2007	
HT201721A	102A	3004	53	2007	
HT201721D	102A	3004	53	2007	
HT201725A	102A	3004	53	2007	
HT201782A	102A	3004	53	2007	
HT201861A	102A		53	2007	
HT201871L	102A		53	2007	
HT203243A	158	3004/102A	53	2007	
HT203701	102A	3004	53	2007	
HT203701A	102A	3004	53	2007	
HT203701B	102A	3004	53	2007	
HT204051	158	3004/102A	53	2007	
HT204051B	158	3004/102A	53	2007	
HT204051C	158	3004/102A	53	2007	
HT204051D	158	3004/102A	53	2007	
HT204051E	158	3004/102A	53	2007	
HT204053	158	3004/102A	53	2007	
HT204053A	158	3004/102A	53	2007	
HT204053B	158	3004/102A	53	2007	
HT2040710 A	121	3717			
HT2040710A			239	2006	
HT204071D			3	2006	
HT20436A	131		44	2006	
HT20451A	158		53	2007	
HT2046710	131	3198	44	2006	
HT204736	131	3052	44	2006	
HT3036201			20	2051	
HT303620B	123A	3444	20	2051	
HT3036210	123A	3444	20	2051	
HT3037010	108	3452	86	2038	
HT303711A	107	3018	11	2015	
HT303711A-O	108		86	2038	
HT303711A-O			17	2051	
HT303711AO	123A		20	2051	
HT303711AO		3452	20	2051	
HT303711B	107	3018	11	2015	
HT303711B-O	123A	3018	20	2051	
HT303711B-O			17	2051	
HT303711BO	123A		20	2051	
HT303711BO		3444	20	2051	
HT303711C	107	3018	11	2015	
HT3037201	108	3452	86	2038	
HT3037201A	108	3452	86	2038	
HT3037201B	108	3452	86	2038	
HT303720A	108	3452	86	2038	

Industry Standard No.	ECG	SK	GE	RS 276-	MOTOR.
HT303721-0	123A	3444	20	2051	
HT3037210	123A	3444	20	2051	
HT3037210-0	123A	3444	20	2051	
HT3037210-0			10	2051	
HT303721D	123A	3444	20	2051	
HT303730	123A	3444	20	2051	
HT303730A	123A	3444	20	2051	
HT3037310	107	3039/316	11	2015	
HT30373100	123A	3444	20	2051	
HT303801	107	3018	11	2015	
HT3038013-B	107	3039/316	11	2015	
HT303801A	107	3018	11	2015	
HT303801A0	107		11	2015	
HT303801A0		3039	20	2051	
HT303801B	107	3018	11	2015	
HT303801B-0	107	3018	11	2015	
HT303801B-0			20	2051	
HT303801B0	107		11	2015	
HT303801B0			20	2051	
HT303801C	107	3018	11	2015	
HT303801C0	107		11	2015	
HT303801C0		3039	20	2051	
HT30387100	316	3039			
HT303941	108	3452	86	2038	
HT303941A	108	3452	86	2038	
HT303941B	108	3452	86	2038	
HT304508K			10	2051	
HT304531	123A	3444	20	2051	
HT304531A	123A	3444	20	2051	
HT304531B	123A	3444	20	2051	
HT304531C	123A	3444	20	2051	
HT304540A0	108	3452	86	2038	
HT304540A0			20	2051	
HT304540B0	123A	3444	20	2051	
HT304540B0			20	2051	
HT304580	123A	3444	20	2051	
HT304580A	123A	3444	20	2051	
HT304580B	123A	3444	20	2051	
HT304580C0	123A	3444	20	2051	
HT304580C0			10	2051	
HT304580K	123A	3444	20	2051	
HT304580Y0	123A		20	2051	
HT304580Y0		3444	10	2051	
HT304580Z	123A	3444	20	2051	
HT304581	123A	3444	20	2051	
HT304581A	123A	3444	20	2051	
HT304581B	123A	3444	20	2051	
HT304581B-0	123A	3444	20	2051	
HT304581B-0			10	2051	
HT304581C	123A	3444	20	2051	
HT304601B0	233		61	2015	
HT304601B0		3122	20	2051	
HT304601C0	108		86	2038	
HT304601C0		3452	20	2051	
HT304611B	108	3452	86	2038	
HT304861	128	3024	243	2030	
HT304861A	128	3024	243	2030	
HT304861B	123A	3444	20	2051	
HT304911	175	3026	246	2020	
HT304911A	175	3026	246	2020	
HT304911B	175	3026	246	2020	
HT30491C	128	3024	243	2030	
HT30494	130	3027	14	2041	
HT304941X	130		14	2041	
HT30494X	130	3027	14	2041	
HT304961B	184	3190	57	2017	
HT304961C	184	3190	57	2017	
HT304961C-0	186	3192	28	2017	
HT304961C-0			28	2017	
HT304971A	128	3024	243	2030	
HT304971A0	128		243	2030	
HT304971A0		3024	18	2030	
HT304971B	128	3020/123	243	2030	
HT305351B0	233		86	2038	
HT305351C0	108		86	2038	
HT305351C0		3452	17	2051	
HT305361E	123A	3444	20	2051	
HT305361G	123A	3444	20	2051	
HT305371E	123A	3444	20	2051	
HT305642B	234	3114/290	65	2050	
HT306441	123A	3444	20	2051	
HT306441A	123A	3444	20	2051	
HT306441A0	128		243	2030	
HT306441A0			18	2030	
HT306441B	123A	3444	20	2051	
HT306441B-0	123A	3444	20	2051	
HT306441B-0			18	2030	
HT306441B0	123A		20	2051	
HT306441B0		3124	18	2030	
HT306441C	123A	3444	20	2051	
HT306441C-0	123A		20	2051	
HT306441C-0			18	2030	
HT306442A	199	3124/289	62	2010	
HT306442B	199	3124/289	62	2010	
HT306451	108	3452	86	2038	
HT306451A	107	3018	11	2015	
HT306451B	108	3452	86	2038	
HT306451H			17	2051	
HT306681C	108		86	2038	
HT306962A-0	108	3452	86	2038	
HT306962A-0			17	2051	
HT307321A	123A	3444	20	2051	
HT307321B-0	123A	3444	20	2051	
HT307321B-0			20	2051	
HT307322A	123A	3444	20	2051	
HT3073310		3124	20	2051	
HT30733100	199	3124/289	62	2010	
HT307331B	123A	3444	20	2051	
HT307331C	123A	3444	20	2051	
HT307331C0	123A		20	2051	
HT307341	128	3024	243	2030	
HT307341A	128	3024	243	2030	
HT307341B	123A	3444	20	2051	
HT307341C-0	123A	3444	20	2051	
HT307341C-0			18	2030	
HT307342B	128		243	2030	
HT307342C	128		243	2030	
HT307720B	107	3018	11	2015	
HT307721C	107	3018	11	2015	
HT307721D	107	3018	11	2015	
HT307902B	184		57	2017	
HT3082810			18	2030	
HT308281B	123A	3444	20	2051	
HT308281C	123A	3444	20	2051	
HT308281D	199	3122	62	2010	
HT308281G	123A	3444	20	2051	
HT308281H	199	3245	62	2010	
HT3082810	123A	3444	20	2051	
HT308282A	123A	3444	20	2051	
HT308282A-0	123A	3444	20	2051	
HT308282A-0			20	2051	
HT308282B	199	3444/123A	62	2010	
HT308291A	123A	3018	11	2015	
HT308291A-0		3018		2015	
HT308291A-0	123A		11	2015	
HT308291B	123A	3018	11	2015	
HT308291B-0	123A	3018	11	2015	
HT308291B-0			20	2051	
HT308291B0	107	3444/123A	20	2038	
HT308291B0	123A	3018	20	2051	
HT308291C	123A	3122	11	2015	
HT308291D0	123A		20	2038	
HT308301B0	152	3893	66	2048	
HT308301B0			23	2020	
HT309002A0	229	3124/289	62	2010	
HT309291C	108		86	2038	
HT309291E	107	3018	11	2015	
HT309301C	199	3018	62	2010	
HT309301D	107	3018	11	2015	
HT309301E	199	3018	62	2010	
HT309301F	199	3018	62	2010	
HT309451L0	199		212	2010	
HT309451R0	199	3124/289	212		
HT309680B	128	3024	243	2030	
HT309714A-0	128	3048/329	243	2030	
HT309841B0	128		243	2030	
HT309841B0			18	2030	
HT309842A-0	123A	3444	20	2051	
HT309842A-0			18	2030	
HT310001F			212	2010	
HT310002A	199	3245	62	2010	
HT311621B	152	3054/196	66	2048	
HT312131C0	289A	3024/128	268	2038	
HT313171R	289A	3122	268	2038	
HT313172A	289A	3122	268	2038	
HT313181C	128		243	2030	
HT313271T	199	3122	62	2010	
HT313272B	199	3245	62	2010	
HT313592B	229	3038	86	2038	
HT313681B0	184	3190	57	2017	
HT313831X	293	3849	47	2030	
HT313831X0	293	3849			
HT313832B	293	3849			
HT313832C	293	3849	47	2030	
HT313841R	293	3849	47	2030	
HT314071Q	192		63	2030	
HT31518210	293	3849			
HT315182C0	293	3849			
HT316751M0	229	3124/289	213	2038	
HT318291D0		3122	20		
HT31909100	235	3197			
HT31909100			63	2030	
HT32					1N5761
HT340519C			8	2002	
HT35					1N5762
HT36441B	199	3124/289	62	2010	
HT38281D			20	2051	
HT400	123A	3444	20	2051	
HT400721	103A	3010	59	2002	
HT400721A	103A	3010	59	2002	
HT400721B	103A	3010	59	2002	
HT400721C	103A	3010	59	2002	
HT400721D	103A	3010	59	2002	
HT400721E	103A	3010	59	2002	
HT400723	103A	3835	59	2002	
HT400723A	103A	3835	59	2002	
HT400723B	103A	3835	59	2002	
HT400770B	103A	3010	59	2002	
HT401	123A	3444	20	2051	
HT401191	130	3027	14	2041	
HT401191A	130	3027	14	2041	
HT401191B	130	3027	14	2041	
HT401193A0	280		262	2041	
HT401301B0	175		246	2020	
HT402352B	152	3054/196	66	2048	
HT403131E0			241	2020	
HT403151E		3538	246	2020	
HT403152A	175		246	2020	
HT403152B	175		246	2020	
HT403523A	103A	3835	59	2002	
HT404001E	192	3137	63	2030	
HT600011F	159	3025/129	82	2032	
HT600011H	159		82	2032	
HT6000210	159	3025/129	82	2032	
HT70011100	242	3191/185			
HT70011100			58	2027	
HT8000101	154	3040	40	2012	
HT800011F	123A	3444	20	2051	
HT800011G	123A	3444	20	2051	
HT800011H	123A	3444	20		
HT800011K	123A	3444	20		
HT800012F			18	2030	
HT8000710	116	3031A	504A	1104	
HT80011H				2051	
HT80011K				2051	
HT8001210	128	3024	243	2030	
HT8001310	128	3024	243	2030	
HT8001610	161	3132	39	2015	
HT8001710	161	3132	39	2015	
HT8001810	123A	3444	20	2051	
HT800191H		3020	18	2030	
HT9000410-0			19	2041	
HT9000410-0			19	2041	
HV-100	116		504A	1104	
HV-15	604			1123	
HV-23	600	3863	300	1122	
HV-23(ELGIN)	600		504A	1104	
HV-23BL		3863	1N34AS	1123	
HV-23G(BL)			300	1122	
HV-23GYL	600	3863			
HV-25	177	3100/519	300	1122	
HV-25(DIODE)	177		300	1122	
HV-25(RCA)	177		300	1122	
HV-26	116		504A	1104	
HV-26G	116		504A	1104	
HV-27	177	3100/519	300	1122	
HV-46	605	3864	504A	1104	
HV-46GR		3864	504A	1104	
HV-80	601	3463	514	1122	
HV-801			1N34AS	1123	
HV0000102	102A		53	2007	
HV0000105	116	3017B/117	504A	1104	
HV0000105-0	116	3017B/117	504A	1104	
HV00001050	177	3100/519	300	1122	
HV00001060	601	3463			
HV0000202	102A	3017B/117	1N60	1123	
HV0000206	601	3463	300	1122	
HV0000302	107	3039/316	11	2015	
HV0000405	158	3100/519	53	2007	
HV0000405-0			52	2024	
HV0000406	116		504A	1104	
HV0000502	601	3463	300	1122	
HV00005020	601	3463			
HV0000705	116	3311	504A	1104	
HV12	102A	3123	53	2007	
HV15	604	3087		1123	
HV15(TRANSISTOR)	102A		53	2007	
HV15PD	525	3925			

Industry Standard No.	ECG	SK	GE	RS 276-	MOTOR.
HV16	102A	3123	53	2007	
HV17	102A	3123	53	2007	
HV17B	102A		53	2007	
HV19	102A		53	2007	
HV20	525	3925			
HV20C	525	3925			
HV23			300	1122	
HV23(DIODE)		3100	300	1122	
HV23(TRANSISTOR)	600		20	2051	
HV23-G	600	3863			
HV23D	600	3863			
HV23E	600	3863			
HV23F	600		20		
HV23GBL	600	3863			
HV25	123A	3444	20	2051	
HV25(HITACHI)	177		300	1122	
HV26	600		504A	1104	
HV46	605	3864	504A	1104	
HV46#OR#(DIODE)			504A	1104	
HV46(DIODE)			504A	1104	
HV46OR	177		300	1122	
HV46GR	605	3031A	504A	1104	
HV46GR(DIODE)			504A	1104	
HV46RD	605	3864			
HV80	601	3463	300	1122	
HV9601		3126	90		
HVE15	525	3925			
HVT-22DA	502	3067	CR-4		
HVT400	523	3306			
HW10	140A		ZD-10	562	
HW100	5096A	3346	ZD-100		
HW100A	5096A	3346	ZD-100		
HW10A	140A		ZD-10	562	
HW11	5074A	3092/141A	ZD-11.0	563	
HW110	151A	3099	ZD-110		
HW110A	151A	3099	ZD-110		
HW110B	151A	3099			
HW11A	5074A	3092/141A	ZD-11.0	563	
HW11B	141A	3092	ZD-11.5		
HW12	142A		ZD-12	563	
HW120	5097A	3347	ZD-120		
HW120A	5097A	3347	ZD-120		
HW12A	142A	3062	ZD-12	563	
HW12B	142A	3062	ZD-12	563	
HW13	143A	3750	ZD-13	563	
HW130	5098A	3348			
HW130A	5098A	3348			
HW13A	143A	3750	ZD-13	563	
HW13B	143A	3750	ZD-13	563	
HW15	145A	3063	ZD-15	564	
HW150	5100A	3350	ZD-150		
HW150A	5100A	3350	ZD-150		
HW15A	145A	3063	ZD-15	564	
HW15B	145A	3063	ZD-15	564	
HW16	5075A	3751	ZD-16	564	
HW160	5101A	3351	ZD-160		
HW160A	5101A	3351	ZD-160		
HW16A	5075A	3751	ZD-16	564	
HW18	5077A	3752	ZD-18		
HW180	5103A		ZD-180		
HW180A	5103A		ZD-180		
HW18A	5077A	3752	ZD-18		
HW20	5079A	3335	ZD-20		
HW200	5105A	3355	ZD-200		
HW200A	5105A	3355	ZD-200		
HW20A	5079A	3335	ZD-20		
HW22	5080A	3336	ZD-22		
HW22A	5080A	3336	ZD-22		
HW24	5081A	3151	ZD-24		
HW24A	5081A	3151	ZD-24		
HW27	146A	3064	ZD-27		
HW27A	146A	3064	ZD-27		
HW27B	146A	3064	ZD-27		
HW30	5084A	3755	ZD-30		
HW30A	5084A	3755	ZD-30		
HW33	147A	3095	ZD-33		
HW33A	147A	3095	ZD-33		
HW33B	147A	3095	ZD-33		
HW36	5085A	3337	ZD-36		
HW36A	5085A	3337	ZD-36		
HW39	5086A	3338	ZD-39		
HW39A	5086A	3338	ZD-39		
HW43	5087A	3339	ZD-43		
HW43A	5087A	3339	ZD-43		
HW47	5088A	3340	ZD-47		
HW47A	5088A	3340	ZD-47		
HW51	5089A	3341	ZD-51		
HW51A	5089A	3341	ZD-51		
HW56	5090A	3342	ZD-56		
HW56A	5090A	3342	ZD-56		
HW56B	148A	3342/5090A	ZD-55		
HW6.8	5071A	3334	ZD-6.8	561	
HW6.8A	5071A	3334	ZD-6.8	561	
HW62	149A	3097	ZD-62		
HW62A	149A	3097	ZD-62		
HW62B	149A	3097	ZD-62		
HW68	5092A	3343	ZD-68		
HW68A	5092A	3343	ZD-68		
HW7.5	138A		ZD-7.5		
HW7.5A	138A		ZD-7.5		
HW75	5093A	3344	ZD-75		
HW75A	5093A	3344	ZD-75		
HW8.2	5072A		ZD-8.2	562	
HW8.2A	5072A	3060/139A	ZD-8.2	562	
HW82	150A	3098	ZD-82		
HW82A	150A	3098	ZD-82		
HW82B	150A	3098	ZD-82		
HW9.1	139A	3060	ZD-9.1	562	
HW9.1A	139A	3060	ZD-9.1	562	
HW9.1B	139A	3060	ZD-9.1	562	
HW9.81B				562	
HW91	5095A	3345	ZD-91		
HW91A	5095A	3345	ZD-91		
HX-50063	123A	3124/289	20	2051	
HX-50072	123A	3124/289	20	2051	
HX-50091	229	3246	61		
HX-50092	123AP	3854	20	2051	
HX-50094	159	3466	82		
HX-50097	123AP	3854	20	2051	
HX-50102	5402	3638			
HX-50103	192		63	2030	
HX-50104	193		67	2023	
HX-50105	159	3124/289	82	2032	
HX-50106	299	3137/192A	236		
HX-50107	199	3124/289	62	2010	
HX-50108	287	3433	222		
HX-50109	288	3434	223		
HX-50110	229	3018	20	2051	
HX-50112	159	3114/290	82	2032	
HX-50113	123A	3018	20	2051	
HX-50126	229	3018			
HX-50127	123A	3018			
HX-50128	159	3466			
HX-50129	293	3137/192A			

Industry Standard No.	ECG	SK	GE	RS 276-	MOTOR.
HX-50130	294	3138/193A			
HX-50159	5400	3950			
HX-50161	123AP	3854	20	2051	
HX-50176	159	3466	82		
HX50001	108		86	2038	
HX50002	123A	3444	20	2051	
HX50003	107		11	2015	
HXTR2102					2N6604
HXTR6104					2N6603
HXTR6105					2N6603
HY3045801C	123A	3444	20	2051	
HZ-11	5019A	3785	ZD-11.0	563	
HZ-11A	5019A	3785	ZD-10	563	
HZ-11B	5020A	3785/5019A			
HZ-11C	5020A	3785/5019A		563	
HZ-12	5021A	3787	ZD-12	563	
HZ-12A	5021A	3787	ZD-12	563	
HZ-12A3	5021A	3787			
HZ-12B	5022A	3787/5021A	ZD-12	563	
HZ-12C	5022A	3787/5021A		563	
HZ-15	5024A			564	
HZ-20VC	5079A		ZD-20		
HZ-212	142A	3062	ZD-12	563	
HZ-6B	5070A	9021	ZD-6.2	561	
HZ-6C	137A	9021/5070A		561	
HZ-7	5014A	3780	ZD-7.5	561	
HZ-7(B)			ZD-6.8	561	
HZ-7A	5014A	3780	ZD-7.5	561	
HZ-7B	5014A	3780	ZD-6.8	561	
HZ-7C	5014A	3780		561	
HZ-7GR	5015A	3058/137A	ZD-7.5		
HZ-7SV	5015A	3059/138A	ZD-7.5		
HZ-7SVA	5015A		ZD-7.5		
HZ-7SVB	5015A		ZD-7.5		
HZ-9	5073A	3784/5018A	ZD-9.1	562	
HZ-9B	5073A	3784/5018A	ZD-9.1	562	
HZ-9C	5018A	3784		562	
HZ11			ZD-11	563	
HZ11(H)				563	
HZ11A	5019A		ZD-10	562	
HZ11B	5020A		ZD-10	562	
HZ11C	5020A		ZD-11.0	563	
HZ11H			ZD-11	563	
HZ11L				563	
HZ11Y	5020A		ZD-11.0	563	
HZ12			ZD-13	563	
HZ12(H)				563	
HZ12H			ZD-13	563	
HZ12L				563	
HZ15			ZD-15	564	
HZ15(H)				564	
HZ15H			ZD-15	564	
HZ15L				564	
HZ16			ZD-16	564	
HZ16(H)				564	
HZ16H	5075A	3751	ZD-16	564	
HZ16L				564	
HZ35	5037A	3409/5143A			
HZ5	5069A	3333	ZD-4.7		
HZ5223	5063A	3837			
HZ5225	5065A	3838			
HZ5233	5070A	9021			
HZ5238	5073A	3749			
HZ5243	143A	3750			
HZ5246	5075A	3751			
HZ5247	5076A	9022			
HZ5248	5077A	3752			
HZ5249	5078A	9023			
HZ5253	5082A	3753			
HZ5255	5083A	3754			
HZ5256	5084A	3755			
HZ6			ZD-5.6	561	
HZ6(H)				561	
HZ6B	5012A		ZD-5.6	561	
HZ6B2	5012A	3057/136A	ZD-5.6	561	
HZ6C	5012A	9021/5070A	ZB-6.2	561	
HZ6L				561	
HZ7	5014A		ZD-6.8	561	
HZ7(H)				561	
HZ7A	5014A		ZD-7.5	561	
HZ7B	5014A		ZD-6.8	561	
HZ7C	5014A		ZD-7.5		
HZ7L				561	
HZ9	5018A		ZD-9.1	562	
HZ9(H)				562	
HZ9B	5073A		ZD-8.7	562	
HZ9C2	5018A		ZD-9.1	562	
HZ9H			ZD-9.1	562	
HZ9L				562	
HZT33	615	3802/5036A			
I-B198			58	2025	
I//L				2006	
I/O,				2038	
I12032	160		245	2004	
I2114					MCM2114
I2708					MCM2708
I2716					MCM2716
I2816					MCM2816
I3-33187-13	5097A		ZD-120		
I472446-I	121	3717	239	2006	
I473608-2	128		243	2030	
I473679-1	128		243	2030	
I473688-1	5404	3627			
I50865	106	3984	21	2034	
I51-0141-00	175		246	2020	
I538		3311	504	1104	
I6114	175		246	2020	
I6191	128		243	2030	
I6289	5635	3533			
I6342	197		250	2027	
I81030	159		82	2032	
I85436			300	1122	
I9623				2038	
I9631	128	3018	243	2030	
I964(6	177		300	1122	
I9680	159	3114/290	82	2032	
I9A115180-2	126		52	2024	
I9A115728-2	123A	3444	20	2051	
IB	503		CR-5		
IB198				2025	
IC-06-1	722		IC-9		
IC-06-1(MAGNAVOX)	722		IC-9		
IC-06-2	722		IC-9		
IC-06-2(MAGNAVOX)	722		IC-9		
IC-1(DYNALCO)	912		IC-172		
IC-1(PHILCO)	703A		IC-12		
IC-1(RCA)	711		IC-207		
IC-1-3304	912		IC-172		
IC-10(ELCOM)	710		IC-09		
IC-10(PHILCO)	706		IC-43		
IC-100(ELCOM)	7492			1819	
IC-101(ELCOM)	7447			1805	
IC-102(ELCOM)	7408			1822	
IC-103(ELCOM)	7413	7413		1815	

Industry Standard No.	ECG	SK	GE	RS 276-	MOTOR.
IC-104(ELCOM)	7406			1821	
IC-107(ELCOM)	910D		IC-252		
IC-108(ELCOM)	911D		IC-254		
IC-109(ELCOM)	941D		IC-264		
IC-11(PHILCO)	790		IC-230		
IC-12		3075	IC-10		
IC-12(ELCOM)	708		IC-10		
IC-12(PHILCO)	714		IC-4		
IC-126	1105		IC-101		
IC-13		3076	IC-6		
IC-13(PHILCO)	715		IC-6		
IC-132	1079		IC-132		
IC-132(SHARP)	1079		IC-132		
IC-136	1051		IC-131		
IC-14		3077	IC-30		
IC-14(PHILCO)	739		IC-30		
IC-142	1050	3475	IC-123		
IC-15		3072	IC-2		
IC-15(PHILCO)	712		IC-2		
IC-16(PHILCO)	780		IC-222		
IC-17		3143	IC-8		
IC-17(PHILCO)	746		IC-217		
IC-18(PHILCO)	747	3279	IC-218		
IC-19		3134	IC-221		
IC-2(DYNALCO)	912		IC-172		
IC-2(PHILCO)	703A		IC-12		
IC-2(RCA)	710		IC-89		
IC-2-274(DYNALCO)	912		IC-172		
IC-20		3165	IC-260	1740	
IC-20(PHILCO)	923D		IC-260	1740	
IC-200(ELCOM)	745	3276	IC-216		
IC-201(ELCOM)	722		IC-9		
IC-202(ELCOM)	747	3279	IC-218		
IC-203(ELCOM)	749		IC-97		
IC-204(ELCOM)	750	3280	IC-219		
IC-205(ELCOM)	708		IC-10		
IC-21(PHILCO)	782		IC-224		
IC-210(ELCOM)	909		IC-249		
IC-212(ELCOM)	911		IC-253		
IC-218(ELCOM)	713		IC-5		
IC-23(ELCOM)	784		IC-236		
IC-24(ELCOM)	706	3288/1003	IC-43		
IC-246(ELCOM)	946		IC-266		
IC-25		3134	IC-3		
IC-26(ELCOM)	748	3236			
IC-26(PHILCO)	712		IC-2		
IC-27(ELCOM)	909D		IC-250		
IC-28	749		IC-97		
IC-28(ELCOM)	785		IC-226		
IC-287(ELCOM)	814		IC-243		
IC-289(ELCOM)	798		IC-234		
IC-293(ELCOM)	739		IC-30		
IC-294(ELCOM)	746		IC-217		
IC-295(ELCOM)	941M		IC-265	007	
IC-296(ELCOM)	312		IC-220		
IC-297(ELCOM)	738	3167	IC-29		
IC-3(PHILCO)	717		IC-209		
IC-30	799		IC-34		
IC-301(ELCOM)	724		IC-86		
IC-304(ELCOM)	912		IC-172		
IC-305(ELCOM)	727		IC-210		
IC-306(ELCOM)	914		IC-256		
IC-307(ELCOM)	780		IC-222		
IC-31	1062	3075/714			
IC-317(ELCOM)	913		IC-255		
IC-318(ELCOM)	913		IC-255		
IC-319(ELCOM)	786		IC-227		
IC-320(ELCOM)	729		IC-23		
IC-322(ELCOM)	731		IC-13		
IC-323(ELCOM)	787		IC-228		
IC-324(ELCOM)	788	3829	IC-229		
IC-325(ELCOM)	791	3149	IC-231		
IC-33	731	3170	IC-13		
IC-33(PHILCO)	731		IC-13		
IC-4(PHILCO)	703A		IC-12		
IC-40(ELCOM)	941		IC-263	010	
IC-401	706	3288/1003	IC-43		
IC-403(ELCOM)	1053		IC-44		
IC-405(ELCOM)	721		IC-14		
IC-408(ELCOM)	724		IC-86		
IC-42(ELCOM)	911		IC-253		
IC-43(ELCOM)	923		IC-259		
IC-457(ELCOM)	783		IC-225		
IC-46(ELCOM)	947		IC-268		
IC-48(ELCOM)	725		IC-19		
IC-49(ELCOM)	716		IC-208		
IC-5(ELCOM)	703A		IC-12		
IC-5(PHILCO)	782		IC-224		
IC-502(ELCOM)	719		IC-28		
IC-505(ELCOM)	1061	3228			
IC-509(ELCOM)	1083		IC-275		
IC-512(ELCOM)	1039		IC-159		
IC-51A(TELE-SIG)	909D		IC-250		
IC-520(ELCOM)	955M			1723	
IC-521(ELCOM)	799	3238	IC-34		
IC-523(ELCOM)	782		IC-224		
IC-53(ELCOM)	923D		IC-260	1740	
IC-531(ELCOM)	923D		IC-260	1740	
IC-533(ELCOM)	1046		IC-118		
IC-534(ELCOM)	1136		IC-284		
IC-535(ELCOM)	1123	3743	IC-281		
IC-537(ELCOM)	1071		IC-67		
IC-540(ELCOM)	1137		IC-62		
IC-542(ELCOM)	1003		IC-43		
IC-543(ELCOM)	1006		IC-38		
IC-545(ELCOM)	1109	3711	IC-99		
IC-546(ELCOM)	712		IC-2		
IC-552(ELCOM)	1056	3458	IC-48		
IC-553(ELCOM)	1073		IC-63		
IC-554(ELCOM)	1054	3457	IC-45		
IC-6	711	3070	IC-207		
IC-6(ELCOM)	703A		IC-12		
IC-6(PHILCO)	711		IC-207		
IC-601(ELCOM)	735		IC-212		
IC-602(ELCOM)	804		IC-27		
IC-604(ELCOM)	736		IC-17		
IC-606(ELCOM)	743		IC-214		
IC-607(ELCOM)	744		IC-215	2022	
IC-7(PHILCO)	703A		IC-12		
IC-71(ELCOM)	7430	7430			
IC-7400	7400	7400	7400		
IC-7402	7402	7402	7402		
IC-7404	7404	7404	7474		
IC-7408	7408	7408	7408		
IC-7410	7410	7410	7410	1807	
IC-74123	74123	74123	74123	1817	
IC-7420	7420	7420			
IC-7430	7430	7430			
IC-7474	7474	7474	7474		
IC-7475	7475	7475			
IC-7476	7476	7476	7476		
IC-7492		7492	7492		
IC-74LS04	74LS04	74LS04			
IC-74LS138	74LS138	74LS138			
IC-74LS163	74LS163A	74LS163			
IC-8	748	3101/706			
IC-8(ELCOM)	704		IC-205		
IC-80(ELCOM)	7400			1801	
IC-82(ELCOM)	7402			1811	
IC-84(ELCOM)	7404			1802	
IC-86(ELCOM)	7410			1807	
IC-87(ELCOM)	7420			1809	
IC-89(ELCOM)	7441			1804	
IC-8B	704		IC-205		
IC-9(ELCOM)	711		IC-207		
IC-9(PHILCO)	718	3159	IC-8		
IC-91		3157	IC-12		
IC-91(ELCOM)	7451			1825	
IC-91(PHILCO)	703A		IC-12		
IC-95(ELCOM)	7473			1803	
IC-96(ELCOM)	7475			1806	
IC-97(ELCOM)	7474			1818	
IC-98(ELCOM)	7490			1808	
IC-99(ELCOM)	7476			1813	
IC1(RCA)	711		IC-207		
IC101(COLUMBIA)	172A		64		
IC101-109	703A	3157	IC-12		
IC1303P	725	3162	IC-19		
IC13A(ELCOM)	717		IC-209		
IC2(ELCOM)	704		IC-205		
IC2(RCA)	710		IC-89		
IC20(ELCOM)	784		IC-236		
IC201(ELCOM)	722		IC-9		
IC202(ELCOM)	747		IC-218		
IC203(ELCOM)	749		IC-97		
IC204(ELCOM)	750		IC-219		
IC21	782		IC-224		
IC210(ELCOM)	909		IC-249		
IC218(ELCOM)	713		IC-5		
IC23(ELCOM)	784		IC-236		
IC25A(ELCOM)	705A		IC-3		
IC27(ELCOM)	909D		IC-250		
IC28(ELCOM)	785		IC-226		
IC293(ELCOM)	739		IC-30		
IC294(ELCOM)	746		IC-217		
IC297(ELCOM)	738		IC-29		
IC3(PHILCO)	717		IC-209		
IC301(ELCOM)	724		IC-86		
IC3014	704		IC-205		
IC302(ELCOM)	711		IC-207		
IC305(ELCOM)	727		IC-210		
IC307(ELCOM)	783		IC-222		
IC308(ELCOM)	712		IC-2		
IC309(ELCOM)	723		IC-15		
IC310(ELCOM)	781		IC-223		
IC312(ELCOM)	714		IC-4		
IC313(ELCOM)	715		IC-6		
IC314(ELCOM)	728		IC-22		
IC315(ELCOM)	729		IC-23		
IC317(ELCOM)	941		IC-263	010	
IC319(ELCOM)	786		IC-227		
IC35(ELCOM)	915			017	
IC36(ELCOM)	925		IC-261		
IC37(ELCOM)	940		IC-262		
IC38(ELCOM)	949		IC-25		
IC40	703A	3514/941	IC-12		
IC40(ELCOM)	941		IC-263	010	
IC401	706	3288/1003	IC-43		
IC405(ELCOM)	721		IC-14		
IC411(ELCOM)	910		IC-251		
IC42(ELCOM)	911		IC-253		
IC43(ELCOM)	923		IC-259		
IC46(ELCOM)	947		IC-268		
IC48(ELCOM)	725		IC-19		
IC49(ELCOM)	716		IC-208		
IC500		3157	IC-12		
IC500(IR)	703A		IC-12		
IC501		3023	IC-205		
IC501(IR)	704		IC-205		
IC502	705A	3134	IC-3		
IC502(ELCOM)	719		IC-28		
IC502(IR)	705A		IC-3		
IC504	708		IC-10		
IC504(IR)	708		IC-10		
IC505	709	3135	IC-11		
IC505(IR)	709		IC-11		
IC506(IR)	710		IC-89		
IC507	712	3072	IC-2		
IC507(IR)	712		IC-2		
IC508	713	3077/790	IC-5		
IC508(IR)	713		IC-5		
IC509	714	3075	IC-4		
IC509(IR)	714		IC-4		
IC510	715	3076	IC-6		
IC510(IR)	715		IC-6		
IC511	718	3159	IC-8		
IC511(IR)	718		IC-8		
IC512(IR)	720		IC-7		
IC513	722	3161	IC-9		
IC513(IR)	722		IC-9		
IC53(ELCOM)	923D		IC-260	1740	
IC700	1024		720		
IC7420			7420	1809	
IC743038	195A	3765	46		
IC743039	195A	3765	46		
IC743040	229	3018	61	2038	
IC743041	229	3018	61	2038	
IC743042	123A	3444	20	2051	
IC743043	290A	3114/290	269	2032	
IC743044	289A	3138/193A	268	2038	
IC743045	235	3197	215		
IC743046	312	3448	FET-1	2035	
IC743047	139A	3060	ZD-9.1	562	
IC743048	116	3017B/117	504A	1104	
IC743049	519	3100	514	1122	
IC743050	109	3087		1123	
IC743051	177	3100/519	300	1122	
IC8B	748	3101/706			
ICB8000C					LM111J
ICB8001C					LM111J
ICB8741C					MC1741CG
ICF-1	703A	3157	IC-12		
ICF-1-6826	703A	3157	IC-12		
ICH8500ATV					MC1776CG
ICH8500TV					MC1776CG
ICL101ALNDP					LM101AH
ICL101ALNFB					LM101AH
ICL101ALNTY					LM101AH
ICL301ALNPA					LM301AH
ICL301ALNTY					LM301AH
ICL741CLNPA					MC1741CP1
ICL741CLNTY					MC1741CP1
ICL741LNDP					MC1741L
ICL741LNFB					MC1741L
ICL741LNTY					MC1741L
ICL748CTY	1171	3565			
ICL748TY	1171	3565			
ICL8001CTZ					LM111J

Industry Standard No.	ECG	SK	GE	RS 276-	MOTOR.
ICL8001MTZ					LM111J
ICL8007CTA					MC1709CG
ICL8007MTA					MC1709CG
ICL8008CPA					LM301AN
ICL8008CTY					LM301AN
ICL8013A					MC1594G
ICL8013B					MC1594G
ICL8013C					MC1594G
ICL8017CTW					LM301AN
ICL8017MTW					LM301AN
ICL8021C					MC1776G
ICL8021M					MC1776G
ICL8022C					MC1776G
ICL8022M					MC1776G
ICL8043CDE					MC1776G
ICL8043CPE					MC1776G
ICL8043MDE					MC1776G
ICL8048CDE					MC1776G
ICL8048DPE					MC1776G
ICS	504		CR-6		
ICT-10					ICTE-10
ICT-12					ICTE-12
ICT-15					ICTE-15
ICT-18					ICTE-18
ICT-22					ICTE-22
ICT-36					ICTE-36
ICT-45					ICTE-45
ICT-5					ICTE-5
ICT-8					ICTE-8
ID100	5400	3950	C103Y		
ID101	5401	3638/5402	C103YY		
ID102	5402	3638	C103A		
ID103	5403	3627/5404	C103B		
ID104	5404	3627	C103B		
ID105	5405	3951	C203C		
ID106	5405	3951	C203D		
IE-850	103A	3010	59	2002	
IE12Z5			ZD-12	563	
IE460B	229	3018	20	2051	
IE703E	703A	3157	IC-12		
IE850			20	2051	
IF-65	102A	3004	53	2007	
IF65			2	2007	
IG-100	105	3012	4		
IH5101IIE					MC1545G
IH5101MIE					MC1545G
IL1	3041		H11A3		4N25
IL12	3042		H11A5		4N27
IL15			H11A5		4N27
IL16			H11A5		4N27
IL5	3042		H11A1		4N25
IL74			H11A5		4N27
ILA30			H11B3		4N33
ILA55			H11B255		4N33
ILCA2-30			H11B3		4N33
ILCA2-55			H13B1		4N33
IM2114					MCM2114
IM2114L					MCM21L14
IM2147	2147				MCM2147
IM4004			504A	1104	
IM4116					MCM4116
IM55S08					MCM93415
IM55S18					MCM93425
IM5625					MCM7641
IM5626					MCM7643
IM56S18					MCM7681
IM56S26					MCM7643
IM56S45					MCM7641
IM6316					MCM68A316E
IM6364					MCM68365
IM6508	6508				MCM6508
IM6518					MCM6518
IM6551					MCM5101
IM7027					MCM4027A
IM7114					MCM2114
IM7141					MCM6641
IM7141L					MCM66L41
IN60-1	109	3087		1123	
IN962B			ZD-11	563	
INJ33349			1N60	1123	
INJ60034			1N60	1123	
INJ60284	109	3087		1123	
INJ61224	110MP			1123	
INJ61225	137A		ZD-6.2	561	
INJ61227	116(2)	3016	504A(2)	1104	
INJ61433	177	3100/519	300	1122	
INJ61434			504A	1104	
INJ61675	109	3088		1123	
INJ61677	177	3100/519	300	1122	
INJ61725	177	3100/519	300	1122	
INJ61726	116	3017B/117	504A	1104	
INJ70973			1N60	1123	
INJ70980			1N34AS	1123	
INTRON-108	128		243	2030	
INTRON-127	175		246	2020	
IP100	5400	3638/5402			
IP101	5401	3638/5402			
IP102	5402	3638			
IP103	5403	3627/5404			
IP104	5404	3627			
IP105	5405	3951	C203C		
IP106	5405	3951	C203D		
IP20-0001	123A	3444	20	2051	
IP20-0002	123AP	3356	20	2051	
IP20-0003	123A	3444	20	2051	
IP20-0004	195A	3765	46		
IP20-0004(DRIVER)	195A		46		
IP20-0005	237	3299	46		
IP20-0006	123A	3444	20	2051	
IP20-0007	152	3054/196	66	2048	
IP20-0008	103A	3010	59	2002	
IP20-0009	159	3114/290	82	2032	
IP20-0010	312	3112	FET-2	2028	
IP20-0011	312	3448	FET-1	2035	
IP20-0012	312	3448	FET-1	2035	
IP20-0014	1102	3224	IC-93		
IP20-0015	109	3087		1123	
IP20-0016	109	3088		1123	
IP20-0017		3126	90		
IP20-0018	177	3100/519	ZD-15	564	
IP20-0019	139A	3060	ZD-9.1	562	
IP20-0020	109	3100/519		1123	
IP20-0021	519	3100	300	1122	
IP20-0022	116		504A	1104	
IP20-0023	177		300	1122	
IP20-0024	116	3017B/117	504A	1104	
IP20-0025	116	3016	504A	1104	
IP20-0026	125	3017B/117	510,531	1114	
IP20-0027	138A	3059	ZD-7.5		
IP20-0028	390	3027/130	255	2041	
IP20-0029	123A	3018	211	2016	
IP20-0032	123A	3124/289	44	2006	
IP20-0033	300	3464			
IP20-0034	199	3018	62	2010	
IP20-0035	312	3116	FET-2	2035	
IP20-0036	152	3054/196	66	2048	
IP20-0037	229	3018	60	2015	
IP20-0038	229	3018	60	2015	
IP20-0039	123A	3018	210	2051	
IP20-0040	123A	3124/289	61	2038	
IP20-0041	289A	3122	20	2051	
IP20-0046	290A	3114/290	21	2034	
IP20-0048	195A	3765	219		
IP20-0054	116	3311	504A	1104	
IP20-006			61	2038	
IP20-0060	109	3088	1N60	1123	
IP20-0061	519	3100	300	1122	
IP20-0062	5070A	9021			
IP20-0076	103A	3010	59	2002	
IP20-0078	312	3116	FET-2	2035	
IP20-0083	235	3041	66	2048	
IP20-0086	135A	3056	ZD-5.1		
IP20-0093	1104	3225			
IP20-0095	1097	3446			
IP20-0103	295		270		
IP20-0110	229	3018	61	2038	
IP20-0112	298	3450			
IP20-0118	109	3087			
IP20-0120	601	3311	504A	1104	
IP20-0122	123A	3122	20	2051	
IP20-0123	299		236		
IP20-0131	299	3298	236		
IP20-0135	235	3197	215		
IP20-0139	5010A	3776			
IP20-0141	1100	3223			
IP20-0142	1103	3281			
IP20-0144	220	3990			
IP20-0145	177	3100/519	300	1122	
IP20-0148	152	3893			
IP20-0151	614	3327	90		
IP20-0154	236	3239	216	2053	
IP20-0155	235	3197	322		
IP20-0156	295	3054/196			
IP20-0157	222	3050/221	FET-4	2036	
IP20-0159	159	3114/290	221		
IP20-0160	103A	3010	59	2002	
IP20-0161	1155	3231	IC-179	705	
IP20-0164	156	3051	512		
IP20-0165	289A	3024/128	81	2051	
IP20-0166		3126	90		
IP20-0167	116	3311	504A	1104	
IP20-0172	300	3464	273		
IP20-0174	703A	3157	IC-12		
IP20-0179	289A	3122	61	1123	
IP20-0184	177	3100/519	300	1122	
IP20-0185	612	3325	90		
IP20-0186	5071A	3334	ZD-6.8	561	
IP20-0191	289A	3124/289	210	2038	
IP20-0192	290A	3114/290	269	2032	
IP20-0203	5070A	9021	ZD-6.2	561	
IP20-0204	611	3324	90		
IP20-0205	7400			1801	
IP20-0206	7474			1818	
IP20-0210	974	3965			
IP20-0211	298	3024/128	89	2023	
IP20-0212	175	3026	246	2020	
IP20-0213	290A	3114/290	221		
IP20-0214	293	3020/123	81	2051	
IP20-0216	519	3100	300	1122	
IP20-0217	129	3114/290	244	2027	
IP20-0218	222	3065	FET-4	2036	
IP20-0230	184		247	2030	
IP20-0231	289A		268	2038	
IP20-0233	140A	3061	ZD-10	562	
IP20-0245	199	3124/289			
IP20-0253	988	3831			
IP20-0274	519	3175			
IP20-0282	519	3100	300	1122	
IP20-0283	109	3088	1N60	1123	
IP20-0284	137A	3056/135A			
IP20-0305	312		FET-3	2036	
IP20-0316	7474		7474	1818	
IP20-0321	235	3197			
IP20-0323	152		66	2048	
IP20-0331	229		215		
IP20-0333	1248	3497			
IP20-0342	611	3324			
IP20-0422	519	3100			
IP20-0429	1192	3445			
IP20-0432	312	3834/132			
IP20-0434	199	3124/289			
IP20-0435	152	3197/235			
IP20-0436	123AP	3132			
IP20-0437	123AP	3449/297			
IP20-0438	236	3197/235			
IP20-0439	156	3051			
IP20-0440	116	3311			
IP20-0441	519	3100			
IP20-0442	519	3100			
IP20-0443	612	3325			
IP20-0444	116	3463/601			
IP20-0445	605	3864			
IP20-0446	5015A	3781			
IR-28-R/508	910D		252		
IR-RE50	131		44	2006	
IR-TR53	128	3024		2030	
IR-TR56	187		29	2025	
IR-TR57	175		246	2020	
IR-TR59	130	3027	14	2041	
IR-TR60	157	3103A/396	232		
IR-TR61	162	3995	35		
IR1000	243				MJ1000
IR1001	243				MJ1001
IR1010					2N6056
IR1020					MJ3001
IR106A1	5454	6754		1067	
IR106A1-C	5454	6754		1067	
IR106A2	5454	6754		1067	
IR106A3	5454	6754		1067	
IR106A4	5454	6754		1067	
IR106A41	5454	6754		1067	
IR106B1	5455	3597		1067	
IR106B1-C	5455	3597		1067	
IR106B2	5455	3597		1067	
IR106B3	5455	3597		1067	
IR106B4	5455	3597		1067	
IR106B41	5455	3597		1067	
IR106C1	5456	3635/5446		1020	
IR106C1-C	5456			1020	
IR106C2	5456			1020	
IR106C3	5456			1020	
IR106C4	5456			1020	
IR106C41	5456			1020	
IR106F1	5453	6753		1067	
IR106F2	5453	6753		1067	
IR106F3	5453	6753		1067	
IR106F4	5453	6753		1067	
IR106F41	5453	6753		1067	

Industry Standard No.	ECG	SK	GE	RS 276-	MOTOR.
IR106Q1	5452	6752		1067	
IR106Q2	5452	6752		1067	
IR106Q3	5452	6752		1067	
IR106Q4	5452	6752		1067	
IR106Q41	5452			1067	
IR106Y1	5452	6752		1067	
IR106Y1-C	5452	6752		1067	
IR106Y2	5452	6752		1067	
IR10D3L	506		511	1114	
IR10E6J	116		504A	1104	
IR10E6X	519	3100	514	1122	
IR122A	5462	3686			
IR122A-C	5462	3686			
IR122B	5463	3572			
IR122B-C	5463	3572			
IR122C	5465	3687			
IR122C-C	5465	3687			
IR122D	5465	3687			
IR122D-C	5465	3687			
IR122F	5461	3685			
IR122F-C	5461	3685			
IR1601	5471	3942/5483			
IR1602	5472	3942/5483			
IR1774	5483	3942			
IR1776	5484	3943/5485			
IR1777	5485	3943			
IR1778	5486	3944/5487			
IR1D	116		504A	1104	
IR1F	506		511	1114	
IR20	116	3031A	504A	1104	
IR2160	6400A		2N2160	2029	
IR2500	246				MJ2500
IR2501	246				MJ2501
IR2A	116	3311	504A	1104	
IR2E	116	3032A	504A	1104	
IR3000	245				MJ3000
IR3001	245				MJ3001
IR30A		3581	C230A		
IR30B		3581	C230B		
IR30C	5544	3582/5545	C230C		
IR30D	5545	3582	C230D		
IR30F		3581	C230F		
IR30U		3581	C230U		
IR31A		3581	C231A		
IR31B		3581	C231B		
IR31C		3582	C231C		
IR31D		3582	C231D		
IR31F		3581	C231F		
IR31U		3581	C231U		
IR3771					2N3771
IR3772					2N3772
IR3773					2N3773
IR3D	506		511	1114	
IR401					2N3902
IR402	283				2N3902
IR403	283				2N6308
IR4039	98				MJ10000
IR4040	98				MJ10000
IR4041	98				MJ10000
IR4045	98				MJ10000
IR4050	98				MJ10000
IR4055	98				MJ10000
IR4059	98				MJ10000
IR4060	98				MJ10001
IR4061	98				MJ10000
IR4065	98				MJ10001
IR409	162				2N6308
IR410	94				MJ410
IR411	94				MJ411
IR413					MJ413
IR423	162				MJ423
IR424					2N6308
IR425					2N6545
IR430					2N6307
IR431					MJ431
IR4502					MJ4502
IR5000	98				MJ10000
IR5001	246				MJ10000
IR5002					MJ10001
IR5060					MJ10000
IR5061					MJ10000
IR5062					MJ10001
IR515	283				2N6250
IR516	283				2N6250
IR517	283				2N6251
IR518	283				2N6546
IR519	283				2N6547
IR5252					MJ10003
IR5261					MJ10002
IR5JA	109	3087		1123	
IR6000	98				MJ10004
IR6001	98				MJ10004
IR6002	98				MJ10005
IR6060	98				MJ10004
IR6061	98				MJ10004
IR6062	98				MJ10005
IR6251	97				MJ10006
IR6252	97				MJ10007
IR6302					2N5630
IR640	245				MJ3000
IR641	245				MJ3001
IR642	270				2N6578
IR645	246				MJ2500
IR646	246				MJ2501
IR647	271				2N6052
IR660					MJ410
IR663					MJ423
IR665					MJ12003
IR682	5521	6621			
IR683	5522	6622			
IR684	5523	6624/5524			
IR685	5524	6624			
IR686	5525	6627/5527			
IR687	5526	6627/5527			
IR688	5527	6627			
IR689	5528	6629/5529			
IR690	5529	6629			
IR692	5531	6631			
IR701					BU204
IR801					BU205
IR802					MJ802
IR900	244				MJ900
IR901	244				MJ901
IR914			1N91	1122	
IRC10	5454	6754			
IRC20	5455	3597	MR-5	1067	
IRC5	5453	6753		1067	
IRD54	6408	9084A			
IRD54-C	6408	9084A			
IRIF			511	1114	
IRL40					MLED930
IRL60					MLED60
IRTR-50		3052	30	2006	
IRTR-51				2010	

Industry Standard No.	ECG	SK	GE	RS 276-	MOTOR.
IRTR-52			67	2023	
IRTR-53			63	2030	
IRTR-54			21	2034	
IRTR-55		3054	28	2017	
IRTR-56			29	2025	
IRTR-57		3026	23	2020	
IRTR-59				2041	
IRTR-60		3104A	27		
IRTR-61			14	2041	
IRTR-62			20	2051	
IRTR-63		3047	215		
IRTR-65		3048	45		
IRTR-70			61	2038	
IRTR-71			39	2015	
IRTR01			16	2006	
IRTR04			2	2007	
IRTR05			1	2007	
IRTR06			51	2004	
IRTR07			51	2004	
IRTR08			6	2002	
IRTR09			8	2002	
IRTR10			7	2002	
IRTR11			1	2007	
IRTR12			50	2004	
IRTR14			53	2007	
IRTR16			3	2006	
IRTR17			51	2004	
IRTR18			51	2004	
IRTR19			21	2034	
IRTR20			22	2032	
IRTR21			47	2030	
IRTR22			17	2051	
IRTR24			17	2051	
IRTR26			19	2041	
IRTR28			21	2034	
IRTR30			21	2034	
IRTR31			48	2032	
IRTR33			47	2030	
IRTR36			75	2041	
IRTR50		3052	30		
IRTR51	199	3122	17	2010	
IRTR52		3118	82	2032	
IRTR53			18	2030	
IRTR55		3054	28	2017	
IRTR56		3083	29	2025	
IRTR57	196	3026	241	2020	
IRTR57X2			24MP	2020	
IRTR58	218		234		
IRTR59		3036		2041	
IRTR61			14	2041	
IRTR61X2			15MP	2041(2)	
IRTR62	123A	3444	20	2051	
IRTR63	123A	3444	20	2051	
IRTR64	195A	3048/329	46		
IRTR65	195A	3048/329	46		
IRTR67	163A	3439	36		
IRTR68	164	3133	38		
IRTR69	172A	3156	64		
IRTR70	161	3132		2015	
IRTR71	161	3018	39	2015	
IRTR72	188	3199	226	2018	
IRTR73	189	3200	218	2026	
IRTR74	190	3104A	217		
IRTR75	191	3103A/396	249		
IRTR76	152	3054/196	57	2048	
IRTR77	153	3083/197	58	2049	
IRTR78	154	3045/225	40	2012	
IRTR79	171	3104A	27		
IRTR80		3039	86	2038	
IRTR82	176	3123	80		
IRTR83		3019	86	2038	
IRTR84	158		53	2007	
IRTR85	102A	3004	53	2007	
IRTR86	123	3124/289	20	2051	
IRTR87	128	3024	18	2030	
IRTR88	129	3025	244	2027	
IRTR91	155	3839	43		
IRTR92	196	3054	241	2020	
IRTR93	165	3115	38		
IRTR94	226MP		49(2)	2025(2)	
IRTR94MP	226MP	3086	49(2)	2025(2)	
IRTR95	108		39	2038	
IRTRFE100		3116	FET-1	2028	
IRW106A1				1067	
IRW106C1				1020	
IRW106Y1				1067	
IS-446D	178MP	3100/519	300(2)	1122(2)	
IS010	5550		C126F		
IS020	5550		ZJ436F		
IS110	5552		C126A		
IS120	5552		ZJ436A		
IS210	5552		C126B		
IS220	5552		ZJ436B		
IS310	5554		C126C		
IS320	5554		ZJ436C		
IS410	5554		C126D		
IS420	5554		ZJ436D		
IS510	5556		C126E		
IS520	5556		ZJ436E		
IS610	5556		C126M		
IS620	5556		ZJ436M		
ISBF1	102A		53	2007	
ISD-162	103	3862	8	2002	
ISD162			8	2002	
ISPT030	56022		SC265B2		
ISPT040	56022		SC265B2		
ISPT130	56022		SC265B2		
ISPT140	56022		SC265B2		
ISPT230	56022		SC265B2		
ISPT240	56022		SC265B2		
ISPT330	56024		SC265D2		
ISPT340	56024		SC265D2		
ISPT430	56024		SC265D2		
ISPT440	56024		SC265D2		
ISPT530	56026		SC265E2		
ISPT540	56026		SC265E2		
ISPT630	56026		SC265M2		
ISPT640	56026		SC265M2		
IT010	5645	3658/56004	SC147B		
IT015	56004#	3658/56004			
IT06	5645	3658/56004	SC140B		
IT08	5645	3658/56004	SC142B		
IT108	312	3116	FET-2	2035	2N5486
IT109					2N5486
IT10D4K	116		504A	1104	
IT110	5645	3658/56004	SC147B		
IT115	56004#	3658/56004			
IT120			210	2051	
IT121			210	2051	
IT122			212	2010	
IT16	5645	3658/56004	SC140B		
IT1700					MFE3003
IT1701					MFE3003
IT1702					MFE3003

Industry Standard No.	ECG	SK	GE	RS 276-	MOTOR.
IT1750					3N155
IT18	5645		SC142B		
IT205A			50	2004	
IT210	5645	3112	SC147B		
IT215	56004#	3658/56004			
IT22	109	3087		1123	
IT2218			20	2051	
IT2219			20	2051	
IT2221			20	2051	
IT2222			20	2051	
IT23	109	3087		1123	
IT23G	109	3087		1123	
IT26	5645	3658/56004			
IT2604		3118	221		
IT2605		3118	221		
IT261	109	3090			
IT2700					3N170
IT2701					3N170
IT28	56004	3658	SC142B		
IT310	5645	3659/56006	SC147D		
IT315	56006	3659			
IT36	5645	3659/56006	SC142B		
IT38	5645	3659/56006	SC142D		
IT40	177	3100/519			
IT410	5645	3659/56006	SC147D		
IT415	56006#	3659/56006			
IT46	5645	3659/56006	SC140D		
IT48	5645	3659/56006	SC142D		
IT510	5645	3660/56008	SC147E		
IT515	56008#	3660/56008			
IT56	5645	3660/56008	SC140E		
IT58	5645	3660/56008	SC142E		
IT6	612	3325			
IT610	5645	3660/56008	SC147M		
IT615	56008#	3660/56008			
IT66	5645	3660/56008	SC140M		
IT68	5645	3660/56008	SC142M		
IT918			39	2015	
IT918A			212	2010	
IT929			39	2015	
IT930			39	2015	
ITC918A			212	2010	
ITE3066	457	3112			2N5639
ITE3067	457	3112			2N5639
ITE3068		3112			2N5639
ITE4117		3112			2N5640
ITE4118		3112			2N5640
ITE4119		3112			2N5640
ITE4338		3112			2N5486
ITE4339	457	3112			2N5486
ITE434					2N5486
ITE4340	457	3112			2N5486
ITE4341		3112			2N5486
ITE4391	467				MPF4391
ITE4392					MPF4392
ITE4393					MPF4393
ITE4416		3116	FET-2	2035	2N4416
ITE4867		3112			2N5457
ITE4868		3112			2N5458
ITE4869		3112			2N5459
ITS5817					1N5817
ITS5818					1N5818
ITS5819					1N5819
ITS5823					1N5823
ITS5824					1N5824
ITS5825					1N5825
ITT-310	613		90		
ITT-310S	613		90		
ITT-401		3126	90		
ITT-410	612	3325	90		
ITT102	109	3088	1N60	1123	
ITT1330		3279	IC-218		MC1330P
ITT1352			IC-97		MC1352P
ITT1800-1D					MC1900L
ITT1800-5D					MC1800L
ITT1800-5N					MC1800P
ITT1806-1D					MC1906L
ITT1806-5D					MC1806L
ITT1806-5N					MC1806P
ITT1807-1D					MC1907L
ITT1807-5D					MC1807L
ITT1807-5N					MC1807P
ITT1808-1D					MC1908L
ITT1808-5D					MC1808L
ITT1808-5N					MC1808P
ITT200	519		514	1122	
ITT210	611	3126	90		
ITT301	109	3088	1N60	1123	
ITT3064		3141	IC-225		MC1364P
ITT3065			IC-2		MC1358P
ITT3066					MC1399P
ITT310	613	3327/614	90		
ITT310G	613	3327/614	90		
ITT310S	613	3327/614			
ITT350	116	3311	504A	1104	
ITT3701					TDA1190Z
ITT3707					MC1399P
ITT3710					MC1391P
ITT3714					MC1394P
ITT402	116		504A	1104	
ITT410	614	3325/612			
ITT413	177		300	1122	
ITT552	2011	3975			
ITT554	2012	9092			
ITT556	2013	9093			
ITT558	2014	9094			
ITT641					MC1385P
ITT652	2011	3975			MC1411P
ITT654	2012	9092			MC1412P
ITT656	2013	9093			MC1413P
ITT658	2014	9094			
ITT709(D.I.P.)	909D		IC-250		
ITT710(D.I.P.)	910D		IC-252		
ITT711(D.I.P.)	911D		IC-254		
ITT711-5(D.I.P.)	911D		IC-254		
ITT718	109	3088	1N60	1123	
ITT7215	177	3100/519	300	1122	
ITT723(D.I.P.)	923D		IC-260	1740	
ITT723-5(D.I.P.)			IC-260	1740	
ITT73	117	3100/519	300	1104	
ITT73C	519	3100			
ITT73N	519	3100	514	1122	
ITT7400N	7400	7400	7400	1801	
ITT7401N	7401	7401	IC-194		
ITT7402N	7402	7402	7402	1811	
ITT7404N	7404	7404	7404	1802	
ITT7405N	7405	7405			
ITT7406N	7406	7406	7406	1821	
ITT7408N	7408	7408	7408	1822	
ITT741(D.I.P.)	941D		IC-264		
ITT741(METAL CAN)	941			010	
ITT741(METAL-CAN)			IC-263	010	
ITT741-5(D.I.P.)	941D		IC-264		
ITT74107N	74107	74107			
ITT7410N	7410	7410	7410	1807	
ITT74121N	74121	74121			
ITT74122N	74122	74122			
ITT74123N	74123	74123	74123	1817	
ITT7412N	7412	7412			
ITT7413N	7413	7413	7413	1815	
ITT74141N	74141	74141			
ITT74145N	74145	74145	74145	1828	
ITT74150N	74150	74150	74150	1829	
ITT74151N	74151	74151			
ITT74153N	74153	74153			
ITT74154N	74154		74154	1834	
ITT74161N	74161	74161			
ITT74164N	74164	74164			
ITT7416N	7416	7416			
ITT74192N	74192	74192	74192	1831	
ITT74193N	74193	74193	74193	1820	
ITT74195N	74195	74195			
ITT7420N	7420	7420	7420	1809	
ITT7428	7428	7428			
ITT7428N	7428	7428			
ITT7430N	7430	7430			
ITT7432N	7432	7432	7432	1824	
ITT7433	7433	7433			
ITT7433N	7433	7433			
ITT7437N	7437	7437			
ITT7438N	7438	7438			
ITT7440N	7440	7440			
ITT7442N	7442	7442			
ITT7443N	7443	7443			
ITT7444N	7444	7444			
ITT7445N	7445	7445			
ITT7446AN	7446	7446			
ITT7447AN	7447	7447	7447	1805	
ITT7448N	7448	7448	7448	1816	
ITT7451N	7451	7451	7451	1825	
ITT7454N	7454	7454			
ITT7473N	7473	7473	7473	1803	
ITT7475N	7475	7475	7475	1806	
ITT7476N	7476		7476	1813	
ITT7480N	7480	7480			
ITT7486N	7486	7486	7486	1827	
ITT7490N	7490	7490	7490	1808	
ITT7492N	7492	7492	7492	1819	
ITT7493N	7493A	7493			
ITT7494N	7494	7494			
ITT7495AN	7495	7495			
ITT74H00N	74H00	74H00			
ITT903-1D					MC930L
ITT903-5D					MC830L
ITT903-5N					MC830P
ITT921	177	3100/519	300	1122	
ITT932-1D					MC932L
ITT932-5D					MC832L
ITT932-5N					MC832P
ITT933-1D					MC933L
ITT933-5D					MC833L
ITT933-5N					MC833P
ITT935-1D					MC940L
ITT935-5D					MC840L
ITT935-5N					MC840P
ITT936-1D					MC936L
ITT936-5D	9936				MC836L
ITT936-5N					MC836P
ITT937-1D					MC937L
ITT937-5D					MC837L
ITT937-5N					MC837P
ITT938-1D					MC935L
ITT938-5D					MC835L
ITT938-5N					MC835P
ITT944-1D					MC944L
ITT944-5D					MC844L
ITT944-5N					MC844P
ITT945-1D					MC945L
ITT945-5D					MC845L
ITT945-5N					MC845P
ITT946-1D					MC946L
ITT946-5D	9946				MC846L
ITT946-5N					MC846P
ITT948-1D					MC948L
ITT948-5D					MC848L
ITT948-5N					MC848P
ITT949-1D					MC949L
ITT949-5D					MC849L
ITT949-5N					MC849P
ITT951-1D					MC951L
ITT951-5D					MC851L
ITT951-5N					MC851P
ITT961-1D					MC961L
ITT961-5D					MC861L
ITT961-5N					MC861P
ITT962-1D					MC962L
ITT962-5D					MC862L
ITT962-5N					MC862P
ITT963-1D					MC963L
ITT963-5D					MC863L
ITT963-5N					MC863P
ITT992	116	3311	504A	1104	
IW01-08J			ZD-8.2	562	
IW01-09J	5074A		ZD-11.0	563	
IX8055-379005N	519		514	1122	
J-05			504A	1104	1N4001
J-1			504A	1104	1N4002
J-10			509	1114	1N4007
J-2			504A	1104	1N4003
J-4			504A	1104	1N4004
J-6			504A	1104	1N4005
J-8			509	1114	1N4006
J10	156	3033	512		
J100	116	3017B/117	504A	1104	
J1000-7400	7400	7400		1801	
J1000-7402	7402	7402		1811	
J1000-7404	7404			1802	
J1000-7410	7410	7410		1807	
J1000-74121	74121	74121			
J1000-7447	7447	7447		1805	
J1000-7476	7476	7476		1813	
J1000-7490	7490	7490		1808	
J1000-7492	7492	7492		1819	
J1000-NE555	955M			1723	
J101183	116	3311	504A	1104	
J107	107	3018	11	2015	
J108	108	3452	86	2038	
J139A	123A	3444	20	2051	
J187	108	3452	86	2038	
J2-4570			504A	1104	
J20437	116	3311	504A	1104	
J20438	142A		ZD-12	563	
J241	110MP	3709	1N34AS	1123(2)	
J241015	129	3025	244	2027	
J241015(2SA695C)			67	2023	
J241015(2SC1209C)			63	2030	
J241054	123A	3444	20	2051	
J241099	123A	3444	20	2051	

Industry Standard No.	ECG	SK	GE	RS 276-	MOTOR.
J241100	116	3311	504A	1104	
J241102	116	3311	504A	1104	
J241105		3126	90		
J241111(2SB370)			53	2007	
J241111(2SD170)			54	2002	
J241142	116	3311	504A	1104	
J241164	102A	3004	53	2007	
J241177	107	3018	11	2015	
J241178	158	3004/102A	53	2007	
J241179	137A	3058	ZD-6.2	561	
J241181		3126	90		
J241182	177	3100/519	300	1122	
J241183	506	3311	511	1114	
J241184		3126	90		
J241185	103A	3010	59	2002	
J241186	136A		ZD-5.6	561	
J241188	107	3018	11	2015	
J241189	107	3018	11	2015	
J241190	102A	3004	53	2007	
J241209	116		504A	1104	
J241210	116		504A	1104	
J241211	116		504A	1104	
J241212	117		504A	1104	
J241213	177		300	1122	
J241214	116		504A	1104	
J241215	506		511	1114	
J241216	5082A	3753	ZD-25		
J241217	141A		ZD-11.5		
J241218	138A		ZD-7.5		
J241219	749		IC-97		
J241220	1050		IC-123		
J241222	712		IC-2		
J241224	218		234		
J241225	159		82	2032	
J241226	159		82	2032	
J241227	238	3710	259		
J241228	124		12		
J241229	124		12		
J241230	123A	3444	20	2051	
J241231	175		246	2020	
J241232	116		504A	1104	
J241233	198	3220	251		
J241234	177		300	1122	
J241235	177		300	1122	
J241236	231	3857			
J241239		3126	90		
J241241	152	3893	66	2048	
J241242	177	3100/519	300	1122	
J241245	109	3090		1123	
J241250	184	3104A	57	2017	
J241251	123A	3444	20	2051	
J241252	196	3054	241	2020	
J241253	159	3114/290	82	2032	
J241255	128	3122	243	2030	
J241256	128	3122	243	2030	
J241258	218	3085	234		
J241259	159	3114/290	82	2032	
J241260	116	3311	504A	1104	
J24127	116		504A	1104	
J241271	116	3031A	504A	1104	
J24186	136A	3057	ZD-5.6	561	
J242	109	3087		1123	
J24262	5070A	9021	ZD-6.2	561	
J243	109	3087		1123	
J24366	131MP	3840	44(2)	2006(2)	
J2441	109	3087		1123	
J24458	123A	3444	20	2051	
J24561	107	3018	11	2015	
J24562	107	3018	11	2015	
J24563	107	3018	11	2015	
J24564	123A	3444	20	2051	
J24565	123A	3444	20	2051	
J24566	124	3021	12		
J24567	110MP	3088	1N60	1123(2)	
J24569		3126	90		
J24570	116	3031A	504A	1104	
J24596	108	3452	86	2038	
J24620	160		245	2004	
J24621	160		245	2004	
J24622	160		245	2004	
J24623	160		245	2004	
J24624	123A	3444	20	2051	
J24625	123A	3444	20	2051	
J24626	102A		53	2007	
J24628	110MP		1N60	1123	
J24630	116		504A	1104	
J24631	137A		ZD-6.2	561	
J24632	139A		ZD-9.1	562	
J24635	108	3452	86	2038	
J24636	108	3452	86	2038	
J24637	108	3452	86	2038	
J24639	102A		53	2007	
J24640	159	3114/290	82	2032	
J24641	123A	3444	20	2051	
J24642	175		246	2020	
J24643	110MP			1123	
J24645	116		504A	1104	
J24647	116		504A	1104	
J24658	123A	3444	20	2051	
J24701	107	3018	11	2015	
J24752	123A	3444	20	2051	
J24753	123A	3444	20	2051	
J24755	177	3100/519	300	1122	
J24756	116		504A	1104	
J24812	199	3018	62	2010	
J24813	108	3452	86	2038	
J24814	108	3452	86	2038	
J24817	123A	3444	20	2051	
J24820	110MP	3088	1N60	1123	
J24832			21	2034	
J24833			1	2007	
J24834			2	2007	
J24838			1N60	1123	
J24842			17	2051	
J24843			20	2051	
J24844			61	2038	
J24845			20	2051	
J24846			10	2051	
J24852	108	3452	86	2038	
J24855	123A	3444	20	2051	
J24858		3126	90		
J24863	108	3452	86	2038	
J24868	103A	3010	59	2002	
J24869	102A	3004	53	2007	
J24870	102A	3004	53	2007	
J24871	116	3311	504A	1104	
J24872	140A	3061	ZD-10	562	
J24874	123A	3444	20	2051	
J24875	199	3122	62	2010	
J24877	116	3311	504A	1104	
J24878	123A	3444	20	2051	
J24903	107		11	2015	
J24904	108	3452	86	2038	
J24905	108	3452	86	2038	
J24906	123A	3444	20	2051	
J24907	123A	3444	20	2051	
J24908	129		244	2027	
J24909	123A	3444	20	2051	
J24911	109	3090		1123	
J24912	177		300	1122	
J24913	109	3090		1123	
J24914	109	3090		1123	
J24915	108		86	2038	
J24916	123A	3444	20	2051	
J24919	117		504A	1104	
J24920	116		504A	1104	
J24921	108	3452	86	2038	
J24923	108	3452	86	2038	
J24932	199	3122	62	2010	
J24933	108	3452	86	2038	
J24934	102A	3004	53	2007	
J24935	116	3311	504A	1104	
J24939	116	3311	504A	1104	
J24940	116	3311	504A	1104	
J24950		3126	90		
J308	312	3116	FET-2	2035	J308
J309					J309
J310					J310
J310159	102A		53	2007	
J310224	102A		53	2007	
J310249	123A	3444	20	2051	
J310250	123A	3444	20	2051	
J310251	160		245	2004	
J310252	102A		53	2007	
J320020	116	3017B/117	504A	1104	
J320041	109	3087		1123	
J39C		3126	90		
J4-1000	7400	7400	7400	1801	
J4-1002	7402	7402	IC-194	1811	
J4-1004	7404	7404	7404	1802	
J4-1010	7410	7410	7410		
J4-1047	7447	7447	7447	1805	
J4-1075	7475	7475	7475	1806	
J4-1076	7476	7476	7476	1813	
J4-1090	7490	7490	7490	1808	
J4-1092	7492	7492	7492	1819	
J4-1121	74121	74121			
J4-1215			IC-265	007	
J4-1555	955M		IC-269	1723	
J4-1600				1011	
J4-1601				1103	
J4-1602				1144	
J4-1605				1172	
J4-1610				1122	
J4-1619				561	
J4-1620				562	
J4-1626				2014	
J4-1725				1067	
J4-1730(DIAC)				1050	
J4A3	6354	6554			
J4B3	6354	6554			
J4C3	6354	6554			
J4D3	6354	6554			
J4E3	6354	6554			
J4F3	6354	6554			
J4G3	6354	6554			
J4H3	6356	6556			
J4K3	6356	6556			
J4M3	6358	6558			
J4N3	6358	6558			
J5062	160		245	2004	
J5063	102A		53	2007	
J5064	102A		53	2007	
J5A3	6354	6554			
J5B3	6354	6554			
J5C3	6354	6554			
J5D3	6354	6554			
J5E3	6354	6554			
J5F3	6354	6554			
J5G3	6354	6554			
J5H3	6356	6556			
J5K3	6356	6556			
J5M3	6358	6558			
J5N3	6358	6558			
J685	109	3087		1123	
J775-0	5322	3680			
J775-1	5322	3680			
J775-2	5322	3680			
J775-4	5324	3681			
J775-6	5326	3682			
J961B(G.E.)	123A		20	2051	
J9680	159	3114/290	82	2032	
J9697	159	3114/290	82	2032	
JA-1010				1807	
JA-H	123A	3444	20	2051	
JA-KCDP	156	3081/125	512		
JA-L	123A	3444	20	2051	
JA1050	159	3114/290	82	2032	
JA1050B	159	3114/290			
JA1050G	159	3114/290	82	2032	
JA1050GL	159	3114/290	82	2032	
JA1200	123A	3444	20	2051	
JA1300	194	3024/128			
JA1350	123AP	3122	62	2010	
JA1350B	199	3122	62	2010	
JA1350W	199	3122	62	2010	
JA7010	199	3122	62		
JA7072	184	3104A			
JAM702C	116		504A	1104	
JB-00030	116		504A	1104	
JB-00036	117		504A	1104	
JB-BB1A	116	3031A	504A	1104	
JB00036	166	9075		1152	
JB16C4	116(4)	3016	504A(4)	1104	
JC-00012	116	3311	504A	1104	
JC-00014	116	3311	504A	1104	
JC-00017	506	3311	511	1114	
JC-00025			504A	1104	
JC-00028	116		504A	1104	
JC-00032	116		504A	1104	
JC-00033	116	3311	504A	1104	
JC-00035	116	3311	504A	1104	
JC-00037	116		504A	1104	
JC-00044	116	3311	504A	1104	
JC-00045	116	3311	504A	1104	
JC-00047	116	3311	504A	1104	
JC-00049	116		504A	1104	
JC-00051	116	3311	504A	1104	
JC-00053	116	3311			
JC-00055	116	3311	504A	1104	
JC-00059	156	3311	504A	1104	
JC-0025	117		504A	1104	
JC-10D1	116	3017B/117	504A	1104	
JC-8	748	3236			
JC-DS16E	116	3311	504A	1104	
JC-KS05	116	3311	504A	1104	
JC-SD-1X	116	3017B/117	504A	1104	
JC-SD-1Z	116	3017B/117	504A	1104	

Industry Standard No.	ECG	SK	GE	RS 276-	MOTOR.
JC-SD1Z			504A	1104	
JC-SG005	116	3031A	504A	1104	
JC-V03C	116(2)		504A(2)	1104	
JC00049	116	3311	504A		
JC100	105	3012	4		
JCN1	116	3016	504A	1104	
JCN2	116	3017B/117	504A	1104	
JCN3	116		504A	1104	
JCN4	116	3016	504A	1104	
JCN5	116	3017B/117	504A	1104	
JCN6	116	3017B/117	504A	1104	
JCN7	125	3032A	510,531	1114	
JCV-2	116	3016	504A	1104	
JCV-3	116	3016	504A	1104	
JCV2			504A	1104	
JCV3			504A	1104	
JCV7	125	3032A	510,531	1114	
JD-00040	116	3311	504A	1104	
JD-BB1A	116	3017B/117	504A	1104	
JD-SD1D		3311	504A	1104	
JD-SD1Z			504A	1104	
JDSD1D	116		504A	1104	
JE1033B	312	3448	FET-1	2035	
JE9011	123A	3444	210		
JE9011G	123AP	3444/123A	20	2051	
JE9011H	123AP	3444/123A	20	2051	
JE9014	199	3444/123A			
JE9014B	123AP	3444/123A			
JE9014BE	199	3444/123A			
JE9014C	199	3433/287	222		
JE9014D	199	3444/123A			
JE9014E	199	3444/123A			
JE9015B	294	3434/288	223		
JEM1	105	3012	4		
JEM2	105	3012	4		
JEM3	105	3012	4		
JEM4	105	3012	4		
JEM5	105	3012	4		
JF-1033	312		FET-2	2035	
JF1033	312	3448	FET-2	2035	
JF1033-RED	312	3448			
JF1033B	312	3448	FET-2	2035	
JF1033G	312	3448	FET-2	2035	
JF1033J	312	3448			
JF1033S	312	3834/132	FET-1	2035	
JF1033Y	312	3448			
JF1034	312	3112			
JJ1100		3067	CR-4		
JJ650		3066	CR-1		
JK1033B	312	3448			
JL-40A			300	1122	
JL40A	177	3100/519	300	1122	
JLM-20	123A	3444	20	2051	
JM-40	177	3100/519	300	1122	
JM40	177	3100/519	300	1122	
JM401	177		300	1122	
JN271	123A	3444	20	2051	
JNJ61673	312	3112		2028	
J01006					MRF315
J02000					MRF5177A
J02005					MRF5177A
J02007A					2N6439
J02009					MRF325
J02014					MRF326
J02015A					MRF327
J02016					MRF327
J02401					MRF326
J03020					MRF644
J03025	366				MRF644
J03030	367				MRF646
J03037					MRF646
J03040	367				MRF646
J03055	368				MRF648
J04020					MRF215
J04030					MRF216
J04036					MRF216
J04040					MRF216
J04045					MRF216
J04070	352				MRF247
J04075	352				MRF247
J04080					MRF245
JP40	104	3009	16	2006	
JP5062	160	3006	245	2004	
JP5063	102A	3004	53	2007	
JP5064	102A	3004	53	2007	
JP575005	110MP	3709	1N60	1123(2)	
JP575995	110MP	3709		1123(2)	
JR05		3123	52	2024	
JR10	160	3007	245	2004	
JR100	160	3006	245	2004	
JR15	102A	3005	53	2007	
JR200	160	3006	245	2004	
JR30	160	3006	245	2004	
JR30X	160	3007	245	2004	
JR40	104	3009	8	2006	
JR5	102A	3004	53	2007	
JSP6009	242	3191/185	58	2027	
JSP7001	123A	3246/229	20	2038	
JSP7001B	123A	3246/229	20	2051	
JSP7005	108	3018	86	2038	
JSP7006	108	3018	86	2038	
JT-1601-40	123A	3444	20	2051	
JT-1601-41	112	3089	1N82A		
JT-E1014	109	3087		1123	
JT-E1024D	116	3017B/117	504A	1104	
JT-E1031	109	3087		1123	
JT-E1064	116	3017B/117	504A	1104	
JT-E1095	118	3066	CR-1		
JZ13B	143A	3750			
JZ16B	5075A	3751			
JZ17B	5076A	9022			
JZ18B	5077A	3752			
JZ19B	5078A	9023			
JZ25B	5082A	3753			
JZ30B	5084A	3755			
JZ8.2B	5072A	3136			
K0120SA	177		300	1122	
K04774	104	3009	16	2006	
K071687	106	3984	21	2034	
K071818-001	129		244	2027	
K071961-001	128		243	2030	
K071962-001	129		244	2027	
K071964-001	130		14	2041	
K1.3G22-1A	166			1152	
K1.3G22A	116		504A	1104	
K10	312	3116	FET-1	2028	
K1001					MFE3005
K1003					MFE3005
K1004					MFE3001
K11	312	3448			
K11(1-GATE)			FET-1	2028	
K11(1GATE)				2035	
K11-0(1-GATE)			FET-1	2028	
K11-0	312	3448			
K11-0(1 GATE)				2035	
K11-R	312	3448			
K11-R(1 GATE)				2035	
K11-R(1-GATE)			FET-1	2028	
K11-Y	456	3448			
K11-Y(1 GATE)				2035	
K11-Y(1-GATE)			FET-1	2028	
K112	113A	3119/113	6GC1		
K112C	113A	3119/113	6GC1		
K112D	114		6GD1	1104	
K115J510-1	113A	3119/113	6GC1		
K115J510-2	113A	3119/113	6GC1		
K115J511-2	109	3087		1123	
K117J460-1	113A	3119/113	6GC1		
K117J460-2	113A	3119/113	6GC1		
K1181		3025	244	2027	
K118I	129			2027	
K118J966-1	114	3120	6GD1	1104	
K118J966-2	115	3121	6GX1		
K118J966-3		3121	6GX1		
K118J966-4	115	3121	6GX1		
K118J9663	115	3121	6GX1		
K119J804-5	177		300	1122	
K12	312	3448			
K12(1 GATE)				2035	
K12(1-GATE)			FET-1	2028	
K12-3(1-GATE)			FET-1	2028	
K12-GR(1 GATE)				2035	
K12-GR(1-GATE)			FET-1	2028	
K12-0	312	3448			
K12-0(1 GATE)				2035	
K12-R(1 GATE)				2035	
K12-R(1-GATE)			FET-1	2028	
K1201					MFE3005
K1202					MFE3004
K120Y	312	3448			
K120Y(1 GATE)				2035	
K120Y(1-GATE)			FET-1	2028	
K121J688-1	107		11	2015	
K122	113A	3119/113	6GC1		
K122-J176-1		9086	CR-1		
K122-J176-2		3066	CR-1		
K122-J177P1		3109	CR-2		
K122C	113A	3119/113	6GC1		
K122D	114	3120	6GD1	1104	
K122J176-1	118	9086/518	CR-1		
K122J176-2	118	3066	CR-1		
K122J177-P1		3109	CR-2		
K122J177P1	119		CR-2		
K13	312	3448			
K13(1 GATE)				2035	
K13(1-GATE)			FET-1	2028	
K14-0066-1			53	2007	
K14-0066-12			10	2051	
K14-0066-13			18	2030	
K14-0066-4			1	2007	
K14-0066-6			61	2038	
K15-0(1-GATE)			FET-1	2028	
K15-GR(1 GATE)				2035	
K15-GR-(1-GATE)			FET-1	2028	
K15-0	312	3448			
K15-0(1 GATE)				2035	
K15-R	312	3448			
K15-R(1 GATE)				2035	
K15-R(1-GATE)			FET-1	2028	
K15-Y	312	3448			
K15-Y(1 GATE)				2035	
K15-Y(1-GATE)			FET-1	2028	
K1501					3N157A
K1502					3N157A
K1504	464				2N4352
K16	312		277		
K1615	113A	3119/113	6GC1		
K1616	114	3120	6GD1		
K1617	115	3121	6GX1		
K16H	312		277		
K16HB	312		277		
K16HC	312		277		
K17	312	3448	FET-1	2028	
K17(1 GATE)				2035	
K17(1-GATE)			FET-1	2028	
K17-0(1 GATE)				2035	
K170(1 GATE)				2035	
K170(1-GATE)			FET-1	2028	
K170R(1 GATE)				2035	
K170R(1-GATE)			FET-1	2028	
K17A	312	3448			
K17A(1 GATE)				2035	
K17A(1-GATE)			FET-1	2028	
K17B	312	3448			
K17B(1 GATE)				2035	
K17B(1-GATE)			FET-1	2028	
K17BL	312	3448			
K17BL(1 GATE)				2035	
K17BL(1-GATE)			FET-1	2028	
K17GR	312	3977/456			
K17GR(1 GATE)				2035	
K17GR(1-GATE)			FET-1	2028	
K17R	312	3448			
K17R(1 GATE)				2035	
K17R(1-GATE)			FET-1	2028	
K17Y	312	3448			
K17Y(1 GATE)				2035	
K17Y(1-GATE)			FET-1	2028	
K19	459	3448	FET-1	2028	
K19(1 GATE)	459			2035	
K19(1-GATE)			FET-1	2028	
K19(GR)	459	3116	FET-1	2028	
K198L			FET-1	2028	
K19BL	459	3448			
K19BL(1 GATE)	459			2035	
K19BL(1-GATE)			FET-1	2028	
K19GC	459	3448	FET-1	2028	
K19GC(1 GATE)	459			2035	
K19GC(1-GATE)			FET-1	2028	
K19GE	459	3448			
K19GE(1 GATE)				2035	
K19GE(1-GATE)			FET-1	2028	
K19GR	459	3448	FET-1	2028	
K19GR(1 GATE)	459			2035	
K19GR(1-GATE)	.		FET-1	2028	
K19Y	459	3448	FET-1	2028	
K19Y(1 GATE)	459			2035	
K19Y(1-GATE)			FET-1	2028	
K1A5	116	3311	504A	1104	
K1B5	116	3016	504A	1104	
K1C5	116	3016	504A	1104	
K1D5	116	3017B/117	504A	1104	
K1E5	116	3017B/117	504A	1104	
K1F5	116	3017B/117	504A	1104	
K1G5	116	3017B/117	504A	1104	
K1H5	116	3032A	504A	1104	
K1K5	125	3032A	510,531	1114	
K1M5	125	3033	510,531	1114	

Industry Standard No.	ECG	SK	GE	RS 276-	MOTOR.
K200	116	3017B/117	504A	1104	
K2001	107	3039/316	11	2015	
K2109			86	2038	
K2110			86	2038	
K2111			86	2038	
K2112			86	2038	
K2113			86	2038	
K2114			86	2038	
K2115	161		86	2038	
K2116			86	2038	
K2117			86	2038	
K2118			86	2038	
K2119			11	2015	
K2120			11	2015	
K2121			11	2015	
K2122			11	2015	
K2123			11	2015	
K2124			11	2015	
K2125			11	2015	
K2126			11	2015	
K2127			11	2015	
K22(1 GATE)				2035	
K22(1-GATE)			FET-1	2028	
K22-Y	222	3116			
K22-Y(1 GATE)				2035	
K22-Y(1-GATE)			FET-1	2028	
K22Y(1 GATE)	312			2035	
K22Y(1-GATE)			FET-1	2028	
K23	312	3448			
K23(1 GATE)	312			2035	
K23(1-GATE)			FET-1	2028	
K23A	312	3116	FET-1	2028	
K24	312	3112			
K24(1 GATE)	312			2028	
K24154	1142	3485	IC-128		
K24C(1 GATE)				2028	
K24D(1 GATE)				2028	
K24DR(1 GATE)				2028	
K24E(1 GATE)				2028	
K24F(1 GATE)				2028	
K24G(1 GATE)				2028	
K25	312	3448			
K25(1 GATE)	312			2035	
K25(1-GATE)			FET-1	2028	
K2501	107	3039/316	11	2015	
K2502	107	3039/316	11	2015	
K2503	107	3039/316	11	2015	
K2509	107	3039/316	11	2015	
K25C	312	3448			
K25C(1 GATE)				2035	
K25C(1-GATE)			FET-1	2028	
K25D	312	3448			
K25D(1 GATE)				2035	
K25D(1-GATE)			FET-1	2028	
K25E	312	3448			
K25E(1 GATE)				2035	
K25E(1-GATE)			FET-1	2028	
K25ET	312	3448			
K25ET(1 GATE)				2035	
K25ET(1-GATE)			FET-1	2028	
K25F	312	3448			
K25F(1 GATE)	312			2035	
K25F(1-GATE)			FET-1	2028	
K25G	312	3448			
K25G(1 GATE)				2035	
K25G(1-GATE)			FET-1	2028	
K2601C			11	2015	
K2602C			11	2015	
K2603C			11	2015	
K2604			11	2015	
K2604C			11	2015	
K2615			11	2015	
K2616			11	2015	
K2857C			39	2015	
K2857P			39	2015	
K2A5	116	3311	504A	1104	
K2B5	116	3017B/117	504A	1104	
K2C5	116	3016	504A	1104	
K2CDP22/1B		3107	BR-600		
K2CDP221B	166			1152	
K2D5	116	3031A	504A	1104	
K2E5	116	3031A	504A	1104	
K2F5	116	3017B/117	504A	1104	
K2G	116		504A	1104	
K2G5	116	3017B/117	504A	1104	
K2H5	125	3032A	510,531	1114	
K2K5	125	3032A	510,531	1114	
K2M5		3033	509	1114	
K3	109	3087		1123	
K30(Y)	312	3112			
K30-O(1 GATE)				2035	
K30-O(1-GATE)			FET-1	2028	
K30-O	312	3112			
K30-O(1 GATE)				2028	
K300					J310
K30A(1 GATE)				2035	
K30A(1-GATE)			FET-1	2028	
K30AD(1 GATE)				2035	
K30AD(1-GATE)			FET-1	2028	
K30AGR(1 GATE)				2035	
K30AGR(1-GATE)			FET-1	2028	
K30B	312	3448			
K30B(1 GATE)	312			2035	
K30B(1-GATE)			FET-1	2028	
K30C	312	3448			
K30C(1 GATE)				2035	
K30C(1-GATE)			FET-1	2028	
K30D(1 GATE)				2035	
K30D(1-GATE)			FET-1	2028	
K30GR(1 GATE)	312			2035	
K30GR(1-GATE)			FET-1	2028	
K30R	312	3448			
K30R(1 GATE)				2035	
K30R(1-GATE)			FET-1	2028	
K30Y		3448	FET-1	2028	
K30Y(1 GATE)				2035	
K30Y(1-GATE)			FET-1	2028	
K31	312	3448			
K31(1 GATE)				2035	
K31(1-GATE)			FET-1	2028	
K31C	312	3448	FET-1	2028	
K31C(1 GATE)	312			2035	
K31C(1-GATE)			FET-1	2028	
K32A(2 GATE)				2036	
K32B(2 GATE)				2036	
K32C(2 GATE)				2036	
K32D(2 GATE)				2036	
K33	312	3448			
K33(1 GATE)	312			2035	
K33(1-GATE)			FET-1	2028	
K33(E)		3834	FET-1	2028	
K33E	312	3448	FET-1	2028	
K33E(1 GATE)	312			2035	
K33E(1-GATE)			FET-1	2028	
K33F	312	3448	FET-1	2028	
K33F(1 GATE)	312			2035	
K33F(1-GATE)			FET-1	2028	
K33GR	312	3448	FET-1	2028	
K33GR(1 GATE)				2035	
K33GR(1-GATE)			FET-1	2028	
K34	312	3448	FET-1	2028	
K34(1 GATE)	312			2035	
K34(1-GATE)			FET-1	2028	
K34(E)	312	3116	FET-1	2028	
K34B	312	3448			
K34B(1-GATE)			FET-1	2028	
K34B(1GATE)				2035	
K34C	312	3448	FET-1	2028	
K34C(1 GATE)	312			2035	
K34C(1-GATE)			FET-1	2028	
K34D	312	3448	FET-1	2028	
K34D(1 GATE)				2035	
K34D(1-GATE)			FET-1	2028	
K34E	312	3116	FET-1	2028	
K34GR	312	3112			
K34S747	112	3089			
K35(1 GATE)	312			2028	
K35-0(1 GATE)				2028	
K35-1(1 GATE)				2028	
K35-2(1 GATE)				2028	
K35-BL	221	3050			
K35A(1 GATE)				2028	
K35BL(1 GATE)				2028	
K35C(1 GATE)				2028	
K35G(1 GATE)				2036	
K35GN(1 GATE)				2028	
K35R(1 GATE)				2028	
K35Y(1 GATE)				2028	
K3683C			39	2015	
K3683P			39	2015	
K37	312	3050/221			
K37(1 GATE)				2035	
K37(1-GATE)			FET-1	2028	
K37H	312	3050/221			
K37H(1 GATE)				2035	
K37H(1-GATE)			FET-1	2028	
K37L	312	3050/221			
K37L(1 GATE)				2035	
K37L(1-GATE)			FET-1	2028	
K3880C			39	2015	
K3880P			86	2038	
K390	221	3065/222			
K39Q(2 GATE)	222			2036	
K39R(2 GATE)	222			2036	
K3B5	116	3017B/117	504A	1104	
K3C5	116	3017B/117	504A	1104	
K3D5	116	3017B/117	504A	1104	
K3E	112	3089	1N82A		
K3F5	116	3017B/117	504A	1104	
K3G5	116	3017B/117	504A	1104	
K3H5	125	3032A	510,531	1114	
K3K5	125	3032A	510,531	1114	
K3M5	125	3033	510,531	1114	
K4-500	158		53	2007	
K4-501	103A	3835	59	2002	
K4-505	159	3114/290	82	2032	
K4-506	123A	3444	20	2051	
K4-510	108	3452	86	2038	
K4-520	104	3009	16	2006	
K4-521	104	3009	16	2006	
K4-525	130	3027	14	2041	
K4-550	109	3087		1123	
K4-555	116	3017B/117	504A	1104	
K4-557	116	3031A	504A	1104	
K4-584				1067	
K4-586(DIAC)				1050	
K40(2 GATE)	222			2036	
K4002	107	3039/316	11	2015	
K40A(2 GATE)				2036	
K40B(2 GATE)				2036	
K40C(2 GATE)				2036	
K40D(2 GATE)				2036	
K40M(2 GATE)				2036	
K417-68	126		52	2024	
K42		3116	FET-1	2028	
K45	222	3065			
K45(2 GATE)	222			2036	
K45B(2 GATE)				2036	
K47	457	3116	FET-1	2028	
K47M	457	3448	FET-1	2028	
K47M(1 GATE)				2035	
K47M(1-GATE)			FET-1	2028	
K49		3116	FET-1	2028	
K49F		3448	FET-1	2028	
K49F(1 GATE)				2035	
K49F(1-GATE)			FET-1	2028	
K49H		3448	FET-1	2028	
K49H(1 GATE)	312			2035	
K49H(1-GATE)			FET-1	2028	
K49HK(1 GATE)	312			2035	
K49HK(1-GATE)			FET-1	2028	
K49I		3448	FET-1	2028	
K49I(1 GATE)	312			2035	
K49I(1-GATE)			FET-1	2028	
K49M(1 GATE)	312			2035	
K49M(1-GATE)			FET-1	2028	
K4A5	116	3311	504A	1104	
K4B5	116	3016	504A	1104	
K4C5	116	3017B/117	504A	1104	
K4D5	116	3031A	504A	1104	
K4E5	116	3031A	504A	1104	
K4F5	116	3017B/117	504A	1104	
K4G5	116	3017B/117	504A	1104	
K4H5	125	3032A	510,531	1114	
K4K5	125	3032A	510,531	1114	
K4M5	125	3033	510,531	1114	
K52	109	3090		1123	
K55(1 GATE)	312			2035	
K55(1-GATE)			FET-2	2035	
K55D(1 GATE)	312			2035	
K55D(1-GATE)			FET-2	2035	
K55E(1 GATE)	312			2035	
K55E(1-GATE)			FET-2	2035	
K5A5	116	3017B/117	504A	1104	
K5B5	116	3017B/117	504A	1104	
K5C5	116	3016	504A	1104	
K5D5	116	3031A	504A	1104	
K5E5	116	3031A	504A	1104	
K5F5	116	3017B/117	504A	1104	
K5G5	116	3017B/117	504A	1104	
K5H5	125	3032A	510,531	1114	
K5K5	125	3032A	510,531	1114	
K5M5	125	3033	510,531	1114	
K6	109	3087		1123	
K60	109	3087		1123	
K75508-1	160		245	2004	
K8532799	116(2)		504A	1104	
K8533058-1	116	3017B/117	504A	1104	

Industry Standard No.	ECG	SK	GE	RS 276-	MOTOR.
KB533137	113A	3119/113	6GC1		
KB82	109	3087		1123	
KB9682	159	3114/290	82	2032	
KA1Z25			18	2030	
KA4559	312	3448	FET-1	2035	
KB-162	601		300	1122	
KB-162C5	601	3463	300	1122	
KB-162C5A	601	3463	300	1122	
KB-165	601		300	1122	
KB-182	116	3017B/117	504A	1104	
KB-262	605		300	1122	
KB-265	601		300	1122	
KB-269	601		300	1122	
KB102	177	3100/519	300	1122	
KB162	601	3463			
KB162N	601	3463	300	1122	
KB162W	601	3463			
KB165	601	3463	300	1122	
KB169	601	3463	300	1122	
KB262	605	3864	504A	1104	
KB265	601	3463	300	1122	
KB265A	601	3463	300	1122	
KB265A(RECT)	116		504A	1104	
KB265E	605	3100/519			
KB265F	601	3100/519			
KB269	601	3463	300	1122	
KB462F	607	3864/605			
KB8339	123A	3444	20	2051	
KB8416			17	2051	
KBD06	5305	3676			
KBD08	5306	3677			
KBD10	5307	3107			
KBF-02	125(4)	3081/125	510	1114	
KBH08	5344				MDA 3508
KBH10	5344				MDA 3510
KBH25-05	5322	3679			
KBH2502	5322	3680			
KBH2504	5324	3681			
KBH2506	5326	3682			
KBH2508	5327				MDA 3508
KBH2510	5328				MDA 3510
KBH508					MDA 3508
KBH510					MDA 3510
KBHG08					MDA 3508
KBHG10					MDA 3510
KBHG2508					MDA 3508
KBHG2510					MDA 3510
KBHG508					MDA 3508
KBHG510					MDA 3510
KBL005	5312	3985			MDA970A1
KBL02	5313	3986	BR-1000	1172	MDA970A3
KBL04	5314	3987			MDA970A5
KBL06	5315	3988			
KBL08	5316	3989			
KBP005	166	9075			MDA100A
KBP02	167	3105			MDA102A
KBP04	168	3648			MDA104A
KBP06	169	3678			MDA106A
KBP08	170				MDA108A
KBP10	170				MDA110A
KBPC005	5312				MDA970A1
KBPC02	5313				MDA2502
KBPC04	5314				MDA2504
KBPC06	5315				MDA3506
KBPC08	5316				MDA3508
KBPC10					MDA3510
KBPC10-005					MDA800
KBPC10-02					MDA802
KBPC10-04					MDA804
KBPC10-06					MDA806
KBPC10005					MDA2500
KBPC1002					MDA2502
KBPC1004					MDA2504
KBPC1005	5312	3985			MDA2500
KBPC1006					MDA3506
KBPC102	5313	3986			MDA2502
KBPC104	5314	3987			MDA2504
KBPC106	5315	3988			MDA3506
KBPC108	5316	3989			MDA3508
KBPC110					MDA3510
KBPC12-005					MDA1200
KBPC12-02					MDA1202
KBPC12-04					MDA1204
KBPC12-06					MDA1206
KBPC2005					MDA970A1
KBPC202					MDA970A3
KBPC25-005	5322	3679			MDA2500
KBPC25-02	5322				MDA2502
KBPC25-04	5324	3681			MDA2504
KBPC25-06	5326	3682			MDA2506
KBPC25005		3679			MDA2500
KBPC2502		3680			MDA2502
KBPC2504		3681			MDA2504
KBPC2506		3682			MDA3506
KBPC2508	5327				MDA3508
KBPC2510	5328				MDA3510
KBPC35-005					MDA3500
KBPC35-02					MDA3502
KBPC35-04					MDA3504
KBPC35-06					MDA3506
KBPC35-08					MDA3508
KBPC35-10					MDA3510
KBPC6005	5312	3985			MDA2500
KBPC602	5313	3986			MDA2502
KBPC604	5314	3987			MDA2504
KBPC606	5315	3988			MDA3506
KBPC608	5316	3989			
KBPC8005	5312	3985			MDA2500
KBPC802	5313	3986			MDA2502
KBPC804	5314	3987			MDA2504
KBPC806	5315	3988			MDA3506
KBPC808	5316	3989			
KBPS005					MDA920A2
KBPS02					MDA920A3
KBPS04					MDA920A6
KBPS06					MDA920A7
KBPS08					MDA920A8
KBPS10					MDA920A9
KBS005					MDA970A1
KBS02					MDA970A3
KC-0691L/8	116	3017B/117	504A	1104	
KC-08C221	116(4)		504A(4)	1104	
KC-1.3C3X11/1	116(3)		504A(3)	1104	
KCO-8CP			504A	1104	
KCO-8CP11/1&12/1			504A	1104	
KCO-8CP11/1+12/1			504A	1104	
KCO.8	116		504A	1104	
KCO.8C22/1		3107	BR-600	1152	
KCO.8CP			504A	1104	
KCO.8CP11			504A	1104	
KCO.8CP11/1			504A	1104	
KCO.8CP11/1&121Y			504A	1104	
KCO.8CP11/1+121Y			504A	1104	
KCO.8CP11/H12/1	116(3)		504A(3)	1104	
KCO.8CP12/1			504A	1104	
KC06911	116		504A	1104	
KC06E11/8	116		504A	1104	
KC08C11/10		3016	504A	1104	
KC08C11/8		3016	504A	1104	
KC08C1110	116		504A	1104	
KC08C21/5		3016	504A	1104	
KC08C215	116		504A	1104	
KC08C22/19			504A	1104	
KC08C221	116		504A	1104	
KC08C2219	166	9075		1152	
KC08CP111	116(3)		504A(3)	1104	
KC08CP121	116(3)		504A(3)	1104	
KC0911/8			504A	1104	
KC1.3C	116	3311	504A	1104	
KC1.3C22/1			504A	1104	
KC1.3C3X11/1			504A	1104	
KC1.3G	116	3311		1104	
KC1.3G12/1X2	116	3311	504A	1104	
KC1.3G22/12	116		504A	1104	
KC1.3G22/1A		3311	504A	1104	
KC112A	113A		6GC1		
KC1322/1			504A	1104	
KC13C221	116		504A	1104	
KC2AP22/1B			BR-600	1152	
KC2AP221B	166	9075		1152	
KC2BP22/1B	116		504A	1104	
KC2D22/1	166	9075	504A	1152	
KC2D22/1A		3107	BR-600		
KC2D221	116		504A	1104	
KC2DP	116		504A	1104	
KC2DP11/1			504A	1104	
KC2DP11/1412/1	116(3)	3017B/117	504A(3)	1104	
KC2DP12/1	116(2)	3031A	504A(2)	1104	
KC2DP12/1N	116	3311	504A	1104	
KC2DP12/2			504A	1104	
KC2DP121N	116		504A	1104	
KC2DP122	116		504A	1104	
KC2DP22/1	116(4)	3311	504A(4)	1104	
KC2DP22/1B	166	3107/5307	BR-600	1152	
KC2DP22/1C		3107	BR-600		
KC2DP221	116		504A	1104	
KC2DP221B	116		504A	1104	
KC2DP221C	166	3647/167		1152	
KC2G	177(3)	3110/120	300(3)		
KC2G11/1			504A	1104	
KC2G11/1&12/1			504A	1104	
KC2G11/1+12/1			504A	1104	
KC2G12/1			504A	1104	
KCD-80P11/1+12/1				1114	
KCD-80P11/1912/1				1114	
KCDP12-1	156	3081/125	512		
KCO-8CP	116		504A	1104	
KCO-8CP11/1	116	3017B/117			
KCO-8CP11/1 12/1				1104	
KCO-8CP11/1+12/1			504A	1104	
KD2101	102A	3004	53	2007	
KD2102	123A	3124/289	20	2051	
KD2103	116	3016	504A	1104	
KD2104	116	3016	504A	1104	
KD2114	904	3542			
KD2115	784	3524	IC-236		
KD2118	128	3024	243	2030	
KD2119	108	3452	86	2038	
KD2120	129	3025	244	2027	
KD2121	176	3123	80		
KD2122(I.C.)	912		IC-172		
KD2124	103	3862	8	2002	
KD2130	222	3065		2036	
KD2501	140A		ZD-10	562	
KD2503	137A	3058	ZD-6.2	561	
KD2504	139A		ZD-9.1	562	
KD2505	142A		ZD-12	563	
KD27	109	3090		1123	
KD300A	177	3100/519	300	1122	
KD5000			214	2038	
KD6311	1046	3471	IC-118		
KDC-80P11/1912/1	506		511		
KDC.80P11/1&12/1			511	1114	
KDC.80P11/1+12/1			511	1114	
KDD-0013	109	3087	1N34AS		
KDD0015	5018A	3784	ZD-9.1		
KDD0032	116	3311	504A		
KDD0041	519	3100			
KDT410					MJ410
KDT411	94				MJ411
KDT413					MJ413
KDT423					MJ423
KDT430					2N6307
KDT431					MJ431
KDT515					2N6306
KDT516					2N6306
KDT517					2N6306
KDT518					2N6307
KDT519					2N6308
KE-1007	130		14	2041	
KE-262	116	3031A	504A	1104	
KE1007-0004-00	129		244	2027	
KE3684	457	3116			2N5457
KE3685	457				2N5458
KE3687	457				2N5457
KE3823					2N5668
KE3970	467				2N5638
KE3971					2N5639
KE3972					2N5640
KE4091	467				2N5638
KE4092					2N5639
KE4093					2N5640
KE4220	457				2N5457
KE4221					2N5458
KE4222					2N5459
KE4223		3116			MPF102
KE4224		3116			MPF102
KE4391	467				MPF4391
KE4392					MPF4392
KE4393					MPF4393
KE4416	312	3116	FET-2	2035	2N5486
KE4856	467				MPF4391
KE4857					MPF4392
KE4858					MPF4393
KE4859	467				2N4859
KE4860					2N4860
KE4861					2N4861
KE510					2N5640
KE5103	457	3116			2N3823
KE5104		3116			2N3823
KE5105	312	3116	FET-2	2035	2N3823
KE511					2N5640
KGE41007	116	3311	504A	1104.	
KGE41054	128	3122	243	2030	
KGE41055	123	3122	20	2051	
KGE41061	186	3192	28	2017	
KGE41414	293	3849	47	2030	
KGE41959	109	3087	1N34AS	1123	

Industry Standard No.	ECG	SK	GE	RS 276-	MOTOR.
KGE46109	177	3100/519	300	1122	
KGE46146	199	3124/289	212	2010	
KGE46338	123AP	3122	213		
KGE46441		3477	IC-103		
KGE46465	177	9091	300	1122	
KGS1000	102A		53	2007	
KIA7205AP	1155	3231	IC-179	705	
KIA7205P	1155	3231	IC-179	705	
KIA7310P	229	3445/1192			
KL08	5316	3989			
KLH1422	123A	3444	20	2051	
KLH4567	116	3016	504A	1104	
KLH4577			504A	1104	
KLH4745	196		241	2020	
KLH4746	159	3114/290	82	2032	
KLH4763	177	3100/519	300	1122	
KLH4781	197		250	2027	
KLH4792	107	3018	11	2015	
KLH4793	722	3161	IC-9		
KLH4794	781	3169	IC-223		
KLH4795	726	3129			
KLH5353	197		250	2027	
KLH5489	718	3159	IC-8		
KLH5807	128		243	2030	
KLH5808	129		244	2027	
KLH704	123A	3444	20	2051	
KM917E	123AP	3854			
KM917F	123AP	3854		2051	
KM917G	123AP	3854		2051	
KM918E	108	3293/107			
KM918F	108	3293/107			
K04774	104	3009	16	2006	
K0S25201-1			1N34AS	1123	
K0S25201-2			1N34AS	1123	
K0S25211-2			ZD-6.8	561	
K0S25642-10			51	2004	
K0S25642-20			53	2007	
K0S25642-40			53	2007	
K0S25643-10			53	2007	
K0S25643-15			53	2007	
K0S25651-20			53	2007	
K0S25657-53			53	2007	
K0S25671-20			61	2038	
K0S25671-21			61	2038	
K0S25671-23			61	2038	
KP3946					2N6274
KP3948					2N6274
KPG6682	123A	3444	20	2051	
KR-162		3100		1122	
KR-Q0001	102A		53	2007	
KR-Q0002	102A		53	2007	
KR-Q0004	102A		53	2007	
KR-Q0005	109	3090		1123	
KR-Q1010	102A	3004	53	2007	
KR-Q1011	102A	3004	53	2007	
KR-Q1012	102A	3004	53	2007	
KR-Q1013	123A	3444	20	2051	
KR8417			17	2051	
KS-05	116	3311	504A	1104	
KS-05X	116	3031A	504A	1104	
KS-19938	130		14	2041	
KS-20033L2	219		74	2043	
KS-20971-L1	909D	3590	IC-250		
KS-21177	941M		IC-265	007	
KS030A	5065A	3838			
KS033A	5066A	3771/5005A	ZD-3.3		
KS033B	5005A	3771	ZD-3.3		
KS036A	134A	3055			
KS05			504A	1104	
KS056A	136A		ZD-5.6	561	
KS056B			ZD-5.6	561	
KS062A	137A		ZD-6.2	561	
KS082A	5072A	3136			
KS100A	140A	3061	ZD-10	562	
KS100B	140A	3061	ZD-10	562	
KS10969-L5	7416	7416			
KS120A	142A	3062	ZD-12	563	
KS120B	142A	3062	ZD-12	563	
KS130A	143A	3750			
KS150A	145A	3063	ZD-15	564	
KS150B	145A	3063	ZD-15	564	
KS160A	5075A	3751			
KS180A	5077A	3752			
KS20180-L1	128		243	2030	
KS2030A	5065A	3838			
KS2033A	5066A	3771/5005A	ZD-3.6		
KS2033B	5005A	3771	ZD-3.3		
KS2039A	5067A	3055/134A	ZD-3.6		
KS2039B	134A	3055	ZD-3.6		
KS2047A	5069A	3056/135A	ZD-5.1		
KS2047B	135A	3056	ZD-5.1		
KS2051A	135A	3056	ZD-5.1		
KS2051B	135A	3056	ZD-5.1		
KS2056A	136A	3057	ZD-5.6	561	
KS2056B	136A	3057	ZD-5.6	561	
KS2062A	137A	3058	ZD-6.2	561	
KS2062B	137A	3058	ZD-6.2	561	
KS2082A	5072A	3136			
KS2091A	139A	3060	ZD-9.1	562	
KS2091B	139A	3060	ZD-9.1	562	
KS20967-L1	7400	7400		1801	
KS20967-L2	7404	7404		1802	
KS20967-L3	7486	7486		1827	
KS20969-L3	7492			1819	
KS20969-L4	7495	7495			
KS2100A	140A	3061	ZD-10	562	
KS2100B	140A	3061	ZD-10	562	
KS2120A	142A	3062	ZD-12	563	
KS2120B	142A	3062	ZD-12	563	
KS21282-L1	7408	7408		1822	
KS21282-L3	7432	7432		1824	
KS2130A	143A	3750			
KS2150A	145A	3063	ZD-15	564	
KS2150B	145A	3063	ZD-15	564	
KS2160A	5075A	3751			
KS30A	134A	3055	ZD-3.6		
KS30AF	134A	3055	ZD-3.6		
KS30B	134A	3055	ZD-3.6		
KS30BF	134A	3055	ZD-3.6		
KS31A	134A	3055	ZD-3.6		
KS32A	134A	3055	ZD-3.6		
KS32AF	134A	3055	ZD-3.6		
KS32B	134A	3055	ZD-3.6		
KS32BF	134A	3055	ZD-3.6		
KS34A	135A	3056	ZD-5.1		
KS34AF	135A	3056	ZD-5.1		
KS34B	135A	3056	ZD-5.1		
KS34BF	135A	3056	ZD-5.1		
KS35A	135A	3056	ZD-5.1		
KS35AF	135A	3056	ZD-5.1		
KS36A	136A	3057	ZD-5.6	561	
KS36AF	136A	3057	ZD-5.6	561	
KS36B	136A	3057	ZD-5.6	561	
KS36BF	136A	3057	ZD-5.6	561	
KS37A	137A	3058	ZD-6.2	561	
KS37AF	137A	3058	ZD-6.2	561	
KS41A	139A	3060	ZD-9.1	562	
KS41AF	139A	3060	ZD-9.1	562	
KS42A	140A	3061	ZD-10	562	
KS42AF	140A	3061	ZD-10	562	
KS42B	140A	3061	ZD-10	562	
KS42BF	140A	3061	ZD-10	562	
KS44A	142A	3062	ZD-12	563	
KS44AF	142A	3062	ZD-12	563	
KS44B	142A	3062	ZD-12	563	
KS44BF	142A	3062	ZD-12	563	
KS46	145A	3063	ZD-15	564	
KS46AF	145A	3063	ZD-15	564	
KS46BF	145A	3063	ZD-15	564	
KSA495Y	159	3114/290	221	2032	
KSA542Y	159	3466			
KSA614		3085	234		
KSA614Y	292	3085			
KSA634-0		3193	29		
KSA634-Y	187A	3193/187	29		
KSA733-0	290A		48		
KSB564-Y	298	3841/294			
KSC1096-0	186A	3192/186	28	2052	
KSC1096-Y	186A	3192/186	28	2052	
KSC11870	229*		61	2038	
KSC1187R	229	3246	61	2038	
KSC1507	198	3219	251		
KSC1520	198	3220	251		
KSC16740	229*		61	2038	
KSC1674R	229	3132	61	2038	
KSC815-0	289A		210	2051	
KSC815-0		3124	210	2051	
KSC945Y	289A	3124/289	212	2010	
KSD1051	130		14	2041	
KSD1052	280		75	2041	
KSD1055	130		14	2041	
KSD1056	130		14	2041	
KSD1057	280		75	2041	
KSD2203	181		75	2041	
KSD261Y	192	3137			
KSD3055	181		14	2041	
KSD3771			75	2041	
KSD3772			75	2041	
KSD4150	287		27		
KSD415R	287	3178A	27		
KSD471-Y	293	3849			
KSD9701	284	3836	75	2041	
KSD9701A	284	3836	75	2041	
KSD9702	284	3836			
KSD9702A	284	3836			
KSD9703	284	3836			
KSD9703A	284	3836			
KSD9704	284	3836	75	2041	
KSD9705	284	3836			
KSD9706	284	3836			
KSD9707	284	3836	14	2041	
KSKE125C200	125	3051/156	510,531	1114	
KSKE12C500	125	3032A	510,531	1114	
KSKE40C200	116	3016	504A	1104	
KSKE40C500	116	3016	504A	1104	
KS033A	134A	3055	ZD-3.6		
KS033B	134A	3055	ZD-3.6		
KS036A	134A	3055	ZD-3.6		
KS039A	134A	3055	ZD-3.6		
KS039B	134A	3055	ZD-3.6		
KS047A	135A	3056	ZD-5.1		
KS047B	135A	3056	ZD-5.1		
KS051A	135A	3056	ZD-5.1		
KS056A	136A	3057	ZD-5.6	561	
KS056B	136A	3057	ZD-5.6	561	
KS062A	137A	3058	ZD-6.2	561	
KS091A	139A		ZD-9.1	562	
KSP1171	328	3895			
KSP1172	328	3895			
KSP1173	328	3895			
KSP1174	328	3895			
KSP1175	328	3895			
KSP1176	328	3895			
KT1017	121	3009	239	2006	
KT205	501B	3069/501A			
KT218			20	2051	
KT600			47	2030	
KT600F			47	2030	
KT600T			47	2030	
KTR0710C	123A	3018	211		
KTR0815C	123A	3038	210		
KTR0839C	123A	3018	61		
KTR0945C	199	3124/289			
KTR1017	299	3298	236		
KTR1096C	186A	3248	28		
KTR1687C	123A	3018	20		
KV-1	102A	3004	53	2007	
KV-2			2	2007	
KV-4	102A	3004	53	2007	
KV-4D			2	2007	
KV1	102A		53	2007	
KV109G	612	3325			
KV110E	613	3326			
KV110V	613	3326			
KV112	611	3324			
KV112A	611	3324			
KV151	611	3324			
KV2	102A		53	2007	
KV4	102A	3123	53	2007	
KV4801	610	3323			
KV4801A	610	3323			
KV620	610	3323			
KV624	611	3324			
KV626	612	3325			
KV634	613	3326			
KV638	614	3327			
KVR10	140A		ZD-10	562	
KX-1	177	3100/519	300	1122	
KY5DPF		3329	750		
KZ-8A	5072A	3136	ZD-8.2	562	
KZ6	5071A	3334	ZD-6.8	561	
KZ6A	5071A	3334	ZD-6.8	561	
KZ6ZA	5071A	3334	ZD-6.8	561	
L-417-29BLK	121	3009	239	2006	
L-417-29GRN	121	3009	239	2006	
L-417-29WHT	121	3009	239	2006	
L-417-60	121	3009	239	2006	
L-612099	7408	7408		1822	
L-612107	7432	7432		1824	
L-612150	74153	74153			
L-612158	7416	7416			
L-612161	74164	74164			
L103T1	703A	3157	IC-12		
L123-T1	923		IC-259		
L129	960	3591			
L130	966	3592			
L131	968	3593			
L1444					L1444

Industry Standard No.	ECG	SK	GE	RS 276-	MOTOR.
L144AP					LM324N
L148T1	1171	3565			
L148T2	1171	3565			
L14F1	3036		L14F1		MRD360
L14F2	3036				MRD370
L14G1					MRD300
L14G2			L14G2		MRD310
L14G3					MRD310
L14H1	3034		L14H1		L14H1
L14H2	3034				L14H2
L14H3	3034				L14H3
L2001F31					T2300PB
L2001F51					T2301PB
L2001F71					T2302PB
L2001F91					SC136B
L2001M3	5651	3506/5642			T2300PB
L2001M4	5651	3583/5641			
L2001M5	5651	3583/5641			T2301PB
L2001M7	5651	3506/5642			T2302PB
L2001M9	5641	3506/5642			SC136B
L2003M3					T2300PB
L2003M5					T2301PB
L2003M7					T2302PB
L2003M9					SC136B
L2004F31					2N6071B
L2004F51					2N6071A
L2004F71					2N6071
L2004F91					2N6071
L200E3					MAC96A4
L200E5	5655				MAC95A4
L200E7	5655				MAC94A4
L200E9	5655				MAC94-4
L201	2011	3975			MC1411P
L202	2012	9092			MC1412P
L203	2013	9093			MC1413P
L204	2014	9094			
L2091241-2	160	3007	245	2004	
L2091241-3	160	3008	245	2004	
L3/2H		3311	504A	1104	
L32H	116		504A	1104	
L4			212	2010	
L4001F31					T2300PD
L4001F51					T2301PD
L4001F71					T2302PD
L4001F91					SC136D
L4001M3	5652	3506/5642			T2300PD
L4001M5	5652				T2301PD
L4001M7	5652	3506/5642			T2302PD
L4001M9	5642	3506			SC136D
L4003M3					T2300PD
L4003M5					T2301PD
L4003M7					T2302PD
L4003M9					SC136D
L4004F31					2N6073B
L4004F51					2N6073A
L4004F79					2N6073
L4004F91					2N6073
L400E3					MAC96A6
L400E5	5656				MAC95A6
L400E7	5656				MAC94A6
L400E9	5656				MAC94-6
L5			212	2010	
L5021	102A	3003	53	2007	
L5022	102A	3012/105	53	2007	
L5022A	102A	3004	53	2007	
L5025	102A	3004	53	2007	
L5025A	102A	3004	53	2007	
L5108	160	3123	245	2004	
L5121	160	3123	245	2004	
L5122	160	3123	245	2004	
L5181	160		245	2004	
L532 008 012	128			2030	
L532-008-012			243	2030	
L532000162	128		243	2030	
L6			212	2010	
L6/2H		3311	504A	1104	
L62H	116		504A	1104	
L7			212	2010	
L8/2H		3311	504A	1104	
L82H	125		510,531	1114	
L842			86	2038	
LA-120	1003		IC-43		
LA-1201	1003		IC-2		
LA-3301	1006		IC-38		
LA1111	1002	3481	IC-69		
LA1111P	1002	3481	IC-69		
LA1200	1003	3288	IC-43		
LA1201	1003	3288	IC-43		
LA1201(C)	1003	3288			
LA1201-B2	1003	3288			
000LA1201B	1003	3288	IC-43		
LA1201C	1003	3288	IC-43		
LA1201C-W	1003	3288	IC-43		
LA1201T	1003	3288	IC-43		
LA1201W	1003	3288	IC-43		
LA1202	1003	3288	IC-43		
LA1222	1227	3762	IC-70		
LA1230	788	3829			
LA1240	1214	3736			
LA1342	710	3102	IC-89		
LA1353	1080	3284	IC-98		
LA1355	1215	3738			
LA1363	712	3072	IC-148		
LA1364	1004	3365			
LA1364N	1004	3365	IC-149		
LA1365	712	3072	IC-148		
LA1366N	1158	3289	IC-71		
LA1367	1159	3290	IC-72		
LA1368	738	3167	IC-29		
LA1369	1178	3480	IC-323		
LA1373	1094		IC-157		
LA1374	1121		IC-39		
LA1375	1122		IC-40		
LA1376	1122		IC-40		
LA1390	1216	3735			
LA1390B	1216	3735			
LA1390C	1216	3735			
LA300	116	3016	504A	1104	
LA3101	1221		IC-202		
LA3115	721	3702			
LA3148	727	3071	IC-210		
LA3155	1215	3738	IC-74		
LA3201	1179	3737	IC-324		
LA3300	1005	3723	IC-42		
LA3301	1006	3358	IC-38		
LA3310	1230		IC-203		
LA3311	1225		IC-204		
LA3350	1217	3700	IC-76		
LA4030	1008		IC-271		
LA4030P	1009	3499	IC-77		
LA4031P	1010	3376	IC-37		
LA4032P	1011	3492	IC-79		
LA4050P	1150	3890	IC-286		
LA4051P	1150	3890	IC-286		

Industry Standard No.	ECG	SK	GE	RS 276-	MOTOR.
LA4100	1180	3888	IC-80		
LA4101	1180	3888	IC-41		
LA4102	1228	3889	IC-332		
LA4400	1193	3701			
LA44000Y	1211	3739			
LA4400FR	1193	3701			
LA4400Y	1211	3739			
LA4420	1211	3739	IC-329		
LA4422	1155	3231			
LA4430	1211	3739			
LA600	116	3031A	504A	1104	
LA703E	703A	3157	IC-12		
LA800	116	3017B/117	504A	1104	
LAA300	116	3017B/117	504A	1104	
LAA600	116	3017B/117	504A	1104	
LAA800	116	3017B/117	504A	1104	
LAS1505	309K	3629			
LB3000	7400	7400		1801	
LB3001	7410	7410		1807	
LB3002	7420	7420		1809	
LB3003	7430	7430			
LB3006	7404	7404		1802	
LB3008	7402	7402		1811	
LB3009	7440	7440			
LB3150	7490			1808	
LC.09M			504A	1104	
LC0.09M1113	506		511	1114	
LD111CJ					MC1405L
LD128(RCA)	112		1N82A		
LD261					MLED60
LD34R		3100	514	1122	
LDA400			210	2051	
LDA400MP			210	2051	
LDA401			210	2051	
LDA401MP			210	2051	
LDA402			210	2051	
LDA404	128	3024	243	2030	
LDA405	128	3024	243	2030	
LDA406	128	3024	243	2030	
LDA408	128	3024	243	2030	
LDA410			62	2010	
LDA450	129	3025	244	2027	
LDA452			21	2034	
LDA454			65	2050	
LDA455			65	2050	
LDF603	456	3116			2N4221A
LDF604	456	3116			2N4221A
LDF605		3112			2N4222A
LDF691					2N4391
LDF692					2N4392
LDF693					2N4393
LDS200	128	3024	243	2030	
LDS201	128	3024	243	2030	
LDS202			21	2034	
LDS203			21	2034	
LDS206			10	2051	
LDS207			62	2010	
LDS210			210	2051	
LDS257			65	2050	
LED55B	3028		LED55B		
LED55C	3028		LED55C		
LED56,F					MLED930
LF152D					LF155J
LF155AH					LF155AH
LF155AJG					LF155AJ
LF155AL					LF155AH
LF155H					LF155H
LF155JG					LF155J
LF155L					LF155H
LF156AH					LF156AH
LF156AJG					LF156AJ
LF156AL					LF156AH
LF156H					LF156H
LF156JG					LF156J
LF156L					LF156H
LF157AH					LF157AH
LF157AJG					LF157AJ
LF157AL					LF157AH
LF157H					LF157H
LF157JG					LF157J
LF157L					LF157H
LF252D					LF255J
LF255H					LF255H
LF255JG					LF255J
LF255L					LF255H
LF255P					LF255J
LF256H					LF256H
LF256JG					LF256J
LF256L					LF256H
LF256P					LF256J
LF257H					LF257H
LF257JG					LF257J
LF257L					LF257H
LF257P					LF257J
LF347AN					MC34004AP
LF347BN					MC34004BP
LF347N					MC34004P
LF351AH					MC34001AG
LF351AN					MC34001AP
LF351BH					MC34001BG
LF351BN					MC34001BP
LF351H					MC34001G
LF351N					MC34001P
LF352D					LF355J
LF353AH					MC34002AG
LF353AN					MC34002AP
LF353BH					MC34002BG
LF353BN					MC34002BP
LF353H					MC34002G
LF353N					MC34002P
LF355AH	937$				LF355AH
LF355AJG					LF355AJ
LF355AL	937$				LF355AH
LF355AP	937M$				LF355AN
LF355BH					LF355BH
LF355BJ					LF355BJ
LF355BN	937M$				LF355BN
LF355H	937				LF355H
LF355JG	937M$				LF355J
LF355L	937				LF355H
LF355N	937M				LF355N
LF355P	937M				LF355N
LF356AH	937$				LF356AH
LF356AJG					LF356AJ
LF356AL	937$				LF356AH
LF356AP					LF356AN
LF356BH					LF356BH
LF356BJ					LF356BJ
LF356BN	937M$				LF356BN
LF356H	937				LF356H
LF356JG	937M$				LF356J
LF356L	937				LF356H
LF356N	937M				LF356N
LF356P	937M				LF356N

Industry Standard No.	ECG	SK	GE	RS 276-	MOTOR.
LF357AH	937$				LF357AH
LF357BH					LF357BH
LF357BJ					LF357BJ
LF357BN	937MS				LF357BN
LF357H	937				LF357H
LF357JG	937MS				LF357J
LF357L	937				LF357H
LF357N	937M				LF357N
LF357P	937M				LF357N
LH0001ACD					MC1776CG
LH0001ACF					MC1776CG
LH0001ACH					MC1776CG
LH0001AD					MC1776G
LH0001AF					MC1776G
LH0001AH					MC1776G
LH0002CH					MC1538R
LH0002H					MC1538R
LH0004CH					MC1436G
LH0004H					MC1536G
LH0042CH					MC1776G
LH101F					MC1741F
LH101H		3514			MC1741G
LH201F					MC1741F
LH201H		3514			MC1741G
LH2101AD					MC1537L
LH2101AF					MC1537L
LH2201AD					MC1537L
LH2201AF					MC1537L
LH2301AD					MC1437L
LH2301AF					MC1437L
LH740A	940		IC-262		
LH740AC	940		IC-262		
LH740ACH					LF355H
LH740AH					LF155H
LID929			212	2010	
LID930			212	2010	
LJ-152	159		82	2032	
LJ152	159	3114/290	82	2032	
LJ152(0)	159		82	2032	
LJ152(0)			21	2034	
LJ152-0	159		82	2032	
LJ152B	159	3114/290	82	2032	
LJ152G	159	3114/290	21	2034	
LL-2	116	3016	504A	1104	
LL2			504A	1104	
LLB-23	123A		20	2051	
LM-1129	123A	3018	20	2051	
LM-1130	123A	3018	20	2051	
LM-1132	123A	3018	20	2051	
LM-1133	123A	3018	20	2051	
LM-1147	123A	3018	20	2051	
LM-1148	123A	3018	20	2051	
LM-1149	159	3114/290	82	2032	
LM-1150	159	3114/290	82	2032	
LM-1151	159	3114/290	82	2032	
LM-1153	159	3114/290	82	2032	
LM-1154	171	3104A	27		
LM-1155	123A	3018	20	2051	
LM-1156	191	3104A	249		
LM-1157	182	3104A	55		
LM-1158	116	3311	504A	1104	
LM-1159	177	3100/519	300	1122	
LM-1160	116	3311	504A	1104	
LM-1862	116	3311	504A	1104	
LM-2589	159		82	2032	
LM100F					LM105H
LM100H					LM105H
LM101AD					LM101AH
LM101AF					LM101AH
LM101AH		3565			LM101AH
LM101AJ					LM101AJ
LM101AJ-14					LM101AJ
LM101AJG					LM101AJ
LM101AL					LM101AH
LM101D					LM101AJ
LM101F					LM101AH
LM101H		3565			LM101AH
LM101J-14					LM101AJ
LM104F					LM104H
LM104H					LM104H
LM104J					LM104H
LM104L					LM104H
LM105F					LM105H
LM105H					LM105H
LM105JG					LM105H
LM105L					LM105H
LM106H					MC1710G
LM107F		3596			LM107H
LM107H		3690			LM107H
LM107L					LM107H
LM108AD					LM108AJ
LM108AF					LM108AF
LM108AH					LM108AH
LM108AJ					LM108J-8
LM108D					LM108J
LM108F					LM108F
LM108H					LM108H
LM1090E	123A	3444	20	2051	
LM1090F	123A	3444	20	2051	
LM1090G	123A	3444	20	2051	
LM109H					LM109H
LM109K					LM109K
LM109LA					LM109K
LM1110A	108	3452	86	2038	
LM1110B	107	3018	11	2015	
LM1117	199	3245	62	2010	
LM1117C	199	3245	62	2010	
LM1117D	123A	3444	20	2051	
LM111D		3668			LM111J
LM111H		3567			LM111H
LM1120B	107	3018	11	2015	
LM1120C	107	3018	11	2015	
LM1123H	108	3452	86	2038	
LM112D					MC1556L
LM112F					MC1556L
LM112H					MC1556G
LM1133	233	3444/123A	210	2009	
LM1138	229	3122	61	2038	
LM1138E	229		61	2038	
LM1138E/F	229	3122	61	2038	
LM1138F	229		61	2038	
LM1138G	229		61	2038	
LM1138G/F	229	3122	61	2038	
LM1138G/H	108	3452	86	2038	
LM1138H	229		61	2038	
LM1138H/I	229	3122	61	2038	
LM1138I	229		61	2038	
LM1153	159		82	2032	
LM1157	182	3188A	55		
LM1159			300	1122	
LM1160			504A	1104	
LM117H					LM117H
LM117K					LM117K
LM118D					MC1741SL
LM118F					MC1741SL
LM118H					MC1741SG
LM120H-12					LM120H-12
LM120H-15					LM120H-15
LM120H-18					LM120H-18
LM120H-24					LM120H-24
LM120H-5.0					LM120H-5.0
LM120H-5.2					MC7905.2CK
LM120H-6.0					LM120H-6.0
LM120H-8.0					LM120H-8.0
LM120K-12					LM120K-12
LM120K-15					LM120K-15
LM120K-18					LM120K-18
LM120K-24					LM120K-24
LM120K-5.0					LM120K-5.0
LM120K-5.2					MC7905.2CK
LM120K-6.0					LM120K-6.0
LM120K-8.0					LM120K-8.0
LM122F					MC1555G
LM122H					MC1555G
LM124AD		3643			LM124J
LM124AF					LM124J
LM124AJ					LM124J
LM124D		3643			LM124J
LM124F					LM124J
LM124J					LM124J
LM125H					MC1568G
LM126H					MC1568G
LM128H					MC1568G
LM1303	725	3162	IC-19		
LM1303N	725	3162	IC-19		
LM1303P		3162	IC-19		
LM1304	718	3159	IC-8		
LM1304N	718	3159	IC-8		
LM1305	720	9014	IC-7		
LM1305N	720	9014	IC-7		
LM1307	722	3161	IC-9		
LM1307N	722	3161	IC-9		
LM1310	801	3160	IC-35		
LM1310N	801	3160	IC-35		MC1310P
LM1351N	748	3236	IC-115		MC1351P
LM1391N	815				MC1391P
LM1394N	836				MC1394P
LM139AD		3569			LM139AJ
LM139AJ					LM139AJ
LM139D		3569			LM139J
LM139J					LM139J
LM1403	123A	3444	20	2051	
LM1404	159	3114/290	82	2032	
LM1408J6					MC1408L6
LM1408J7					MC1408L7
LM1408J8					MC1408L8
LM1408N6					MC1408P6
LM1408N7					MC1408P7
LM1408N8					MC1408P8
LM140K-12					LM140K-12
LM140K-15					LM140K-15
LM140K-18					LM140K-18
LM140K-24					LM140K-24
LM140K-5.0					LM140K-5.0
LM140K-6.0					LM140K-6.0
LM140K-8.0					LM140K-8.0
LM140LAH-12					MC78L12ACG
LM140LAH-15					MC78L15ACG
LM140LAH-18					MC78L18ACG
LM140LAH-24					MC78L24ACG
LM140LAH-5.0					MC78L05ACG
LM140LAH-6.0					MC78L06ACG
LM140LAH-8.0					MC78L08ACG
LM1414J					MC1414L
LM1414N					MC1414P
LM1415-6	123A	3444	10	2051	
LM1415-7	123A	3444	10	2051	
LM143D					MC1536G
LM143F					MC1536G
LM143H					MC1536G
LM1458H		3555			MC1458G
LM1458J					MC1458U
LM1458N	778A	3465	IC-220		MC1458P1
LM1458N-14					MC1458P2
LM145K					MC7905CK
LM1488J					MC1488L
LM1488N					MC1488P
LM1489AJ					MC1489AL
LM1489AN					MC1489AP
LM1489J					MC1489L
LM1489N					MC1489P
LM148D					LM148J
LM148F					MC4741L
LM148J					LM148J
LM1496	973	3233			
LM1496H	973	3233			MC1496G
LM1496J	973D	3892			MC1496L
LM1496N	973D	3892			MC1496P
LM1496N(IC)	973D	3892			
LM149D					MC4741L
LM149F					MC4741L
LM1501H	128(2)	3024/128	63		
LM1502H	129(2)	3025/129	67		
LM1514J					MC1514L
LM1540	123A	3444	20	2051	
LM1540C	199	3124/289	62	2010	
LM1558H					MC1558G
LM1558J					MC1558U
LM1566F	123A	3444	20	2051	
LM158AH		3691			LM158H
LM158H					LM158H
LM158JG		3692			LM158J
LM158L		3691			LM158H
LM1596H					MC1596G
LM1596J					MC1596L
LM1614D	123A		20	2051	
LM1614M	123A		20	2051	
LM163J					MC3450L
LM171H					MC1590G
LM1795	159	3138/193A	82	2032	
LM1800	743		IC-214		
LM1800AN					MC1310P
LM1800N	743	3172	IC-214		MC1310P
LM1805					MC1385P
LM1808N	826				TDA1190Z
LM1818	123A	3444	20	2051	
LM1820	744		IC-24	2022	
LM1820A			IC-24	2022	
LM1820N	744		IC-24	2022	
LM1828N					MC1323P
LM1829	797	3158			
LM1834		3126	90		
LM1841		3375	IC-16		
LM1841N	737	3375	IC-16		MC1356P
LM1845	731	3149/791	IC-13		
LM1845N					MC1344P
LM1848N					MC1323P
LM1850N					MC3426L

Industry Standard No.	ECG	SK	GE	RS 276-	MOTOR.
LM1862	116		504A	1104	
LM1900D					MC3301L
LM1932	506	3125	511	1114	
LM193AH					LM193AH
LM193H					LM193H
LM193JG					LM193H
LM193U					LM193H
LM200F					LM205H
LM200H					LM205H
LM201AD					LM201AJ
LM201AF					LM201AH
LM201AH	1171	3565			LM201AH
LM201AJ					LM201AJ
LM201AJ-14					LM201AJ
LM201AJG					LM201AJ
LM201AL	1171				LM201AH
LM201AN					LM201AN
LM201AP					LM201AN
LM201D					LM201AJ
LM201F					LM201AH
LM201H	1171	3565			LM201AH
LM201J					LM201AJ
LM201J-14					LM201AJ
LM204F					LM204H
LM204H					LM204H
LM205F					LM205H
LM205H					LM205H
LM206H					MC1710CG
LM207F		3596			LM207H
LM207H		3690			LM207H
LM208AD					LM208AJ
LM208AF					LM208AF
LM208AH					LM208AH
LM208AJ					LM208AJ-8
LM208D					LM208J-8
LM208F					LM208F
LM208H					LM208H
LM209H					LM209H
LM209K					LM209K
LM2111	708	3135/709	IC-10		
LM2111N	708	3135/709	IC-10		MC1357P
LM2113N			IC-11		MC1357P
LM211D		3668			LM211J
LM211H		3567			LM211H
LM212D					MC1556L
LM212F					MC1556L
LM212H					MC1456G
LM2152	123AP	3444/123A	20	2051	
LM217H					LM217H
LM217K					LM217K
LM218D					MC1741SL
LM218F					MC1741SL
LM218H					MC1741SG
LM220H-12					MC7912CK
LM220H-15					MC7915CK
LM220H-18					MC7918CK
LM220H-24					MC7924CK
LM220H-5.0					MC7905CK
LM220H-5.2					MC7905.2CK
LM220H-6.0					MC7906CK
LM220H-8.0					MC7908CK
LM220K-12					MC7912CK
LM220K-15					MC7915CK
LM220K-18					MC7918CK
LM220K-24					MC7924CK
LM220K-5.0					MC7905CK
LM220K-5.2					MC7905.2CK
LM220K-6.0					MC7906CK
LM220K-8.0					MC7908CK
LM222H					MC1555G
LM224AD	987	3643			LM224J
LM224AF	987				LM224J
LM224AJ	987				LM224J
LM224AN	987	3643			
LM224D	987	3643			LM224J
LM224F					LM224L
LM224J	987				LM224J
LM224N	987	3643			
LM225H					MC1568G
LM226H					MC1568G
LM228H					MC1568G
LM239AD	834	3569			LM239AJ
LM239AJ	834				LM239AJ
LM239AN	834	3569			
LM239D		3569			LM239J
LM239J					LM239J
LM240LAH-12					MC78L12ACG
LM240LAH-15					MC78L15ACG
LM240LAH-18					MC78L18ACG
LM240LAH-24					MC78L24ACG
LM240LAH-5.0					MC78L05ACG
LM240LAH-6.0					MC78L06CG
LM240LAH-8.0					MC78L08ACG
LM240LAZ-12					MC78L12ACP
LM240LAZ-15					MC78L15ACP
LM240LAZ-18					MC78L18ACP
LM240LAZ-24					MC78L24ACP
LM240LAZ-5.0		3462			MC78L05ACP
LM240LAZ-6.0					MC78L06ACP
LM240LAZ-8.0		3724			MC78L08ACP
LM243H					MC1536G
LM245K					MC7905CK
LM248D					LM248J
LM248J					LM248J
LM249D					MC4741L
LM249J					MC4741L
LM258AH		3691			LM258H
LM258H		3691			LM258H
LM2682	128	3854/123AP	243		
LM2701	190	3219	217		
LM271H					MC1590G
LM2900J					MC3301L
LM2900N	992	3688			MC3301P
LM2901N	834	3569			LM2901N
LM2902J	987				LM2902J
LM2902N	987	3643			LM2902N
LM2903		9011			LM2903N
LM2903JG					LM2903N
LM2903P					LM2903N
LM2903U					LM2903N
LM2904N	928M	3692			LM2904N
LM2905N					MC1455P1
LM293AH					LM293AH
LM293H					LM293H
LM293P					LM293H
LM293U					LM293H
LM300F					LM305H
LM300H					LM305H
LM3011H	726				MC1550G
LM301AD		3565			LM301AJ
LM301AF					LM301AH
LM301AG	1171	3565			
LM301AH	1171	3565			LM301AH
LM301AJ					LM301AJ
LM301AJG					LM301AJ
LM301AL	1171	3565			LM301AH
LM301AN	975	3565/1171			LM301AN
LM301AN-8	975	3641			
LM301AP	975	3641			LM301AN
LM301AP1	975	3641			
LM301AT	975	3565/1171			
LM301AV	975	3641			
LM3026					CA3054
LM3028	724	3525	IC-86		
LM3028A	724	3525	IC-86		
LM3028B	724	3525	IC-86		
LM3028BH	724	3525			
LM302H					LM310H
LM3045					MC3346P
LM3045D	912	3543	IC-172		
LM3046N	912	3543			MC3346P
LM304F					LM304H
LM304H					LM304H
LM304J					LM304H
LM304L					LM304H
LM304N					LM304H
LM3053	724	3163/736	IC-86		
LM3054	917				CA3054
LM3054N	917	3544			
LM305AH					LM305H
LM305AJG					LM305H
LM305AL					LM305H
LM305AP					LM305H
LM305F					LM305H
LM305H					LM305H
LM305JG					LM305H
LM305L					LM305H
LM305P					LM305H
LM3064	780	3141	IC-222		
LM3064H	780	3141	I-222		
LM3064N	783		IC-21		MC1364P
LM3065	712	3072	IC-2		
LM3065N	712	3072	IC-148		MC1358P
LM3066	728		IC-22		
LM3066N		3073			MC1399P
LM3067	729		IC-23		
LM3067N	729	3074			MC1323P
LM306H					MC1710CG
LM3070	714		IC-4		
LM3070N	714	3075	IC-4		MC1399P
LM3071	715		IC-6		
LM3071N	715	3076	IC-6		MC1399P
LM3072	790	3077			
LM3075	723		IC-15		
LM3075N	723	3144	IC-15		MC1375P
LM307AN-8	976	3596			
LM307F		3596			LM307H
LM307H		3690			LM307H
LM307J	976	3596			
LM307L					LM307H
LM307N	976	3596	IC-6		LM307N
LM307P	976	3596			LM307N
LM3086N	912	3543	IC-172		MC3386P
LM308AD					LM308AJ
LM308AF					LM308AJ
LM308AH	938				LM308AH
LM308AH-1					LM308AH
LM308AH-2					LM308AH
LM308AJ					LM308AJ-8
LM308D	938M				LM308J
LM308H	938				LM308H
LM308N	938M				LM308N
LM309H					LM309H
LM309K	309K				LM309K
LM309KC	309K	3629			LM309K
LM309LA					LM309K
LM311D		3668			LM311J
LM311H	922	3567			LM311H
LM311N	922M	3668			LM311N
LM311N-14		3668			LM311J
LM311P	922M	3668			
LM3126					MC1399P
LM312D					MC1456L
LM312F					MC1456L
LM312H					MC1456G
LM3146					MC3346P
LM3146A					MC3346P
LM317H					LM317H
LM317K			VR-116		LM317K
LM317P					LM317T
LM317T	956				LM317T
LM318D					MC1741SCL
LM318F					MC1741SCL
LM318H	918				MC1741SCG
LM318N	918M				MC1741SCP1
LM320H-12					LM320H-12
LM320H-15					LM320H-15
LM320H-18					LM320H-18
LM320H-24					LM320H-24
LM320H-5.0					LM320H-5.0
LM320H-5.2					MC7905.2CK
LM320H-6.0					LM320H-6.0
LM320H-8.0					LM520H-8.0
LM320K-12					LM320K-12
LM320K-15					LM320K-15
LM320K-18					LM320K-18
LM320K-24					LM320K-24
LM320K-5.0					LM320K-5.0
LM320K-6.0					LM320K-6.0
LM320K-8.0					LM320K-8.0
LM320MP-12			VR-113		MC7912CT
LM320MP-15					MC7915CT
LM320MP-18					MC7918CT
LM320MP-24	971				MC7924CT
LM320MP-5.0					MC7905CT
LM320MP-5.2			VR-104		MC7905.2CT
LM320MP-6.0					MC7906CT
LM320MP-8.0					MC7908CT
LM320T-12	967	3673	VR-114		LM320T-12
LM320T-15	969	3674			LM320T-15
LM320T-18					LM320T-18
LM320T-24	971				LM320T-24
LM320T-5.0	961	3671			LM320T-5.0
LM320T-5.2			VR-105		MC7905.2CT
LM320T-6.0	963	3672			LM320T-6.0
LM320T-8.0					LM320T-8.0
LM320T12	967	3673			
LM320T15	969	3674			
LM320T6.0	963	3672			
LM320T8.0	964	3630			
LM322H					MC1455G
LM322N					MC1455P1
LM324	987	3643			
LM324AD	987	3643			
LM324AJ	987				LM324J
LM324AN	987	3643			LM324N
LM324D	987	3643			
LM324J	987	3643			LM324J

Industry Standard No.	ECG	SK	GE	RS 276-	MOTOR.
LM324N	987	3643			LM324N
LM325AN					MC1468L
LM325H					MC1468G
LM325N					MC1468L
LM326H					MC1468G
LM326N					MC1468L
LM328AN					MC1468L
LM328H					MC1468G
LM328N					MC1468L
LM3301N	992	3688			MC3301P
LM3302J					MC3302L
LM3302N		3569			MC3302P
LM335	309K	3629			
LM339	834	3569			
LM339AD	834	3569			LM339AJ
LM339AN	834	3569			LM339AN
LM339N	834	3569			LM339N
LM3401N	992	3688			MC3401P
LM340K-12					LM340K-12
LM340K-15					LM340K-15
LM340K-18					LM340K-18
LM340K-24					LM340K-24
LM340K-5.0					LM340K-5.0
LM340K-6.0					LM340K-6.0
LM340K-8.0					LM340K-8.0
LM340KC-12					MC7812CK
LM340KC-15					MC7815CK
LM340KC-18					MC7818CK
LM340KC-24					MC7824CK
LM340KC-5.0					MC7805CK
LM340KC-6.0					MC7806CK
LM340KC-8.0					MC7808CK
LM340LAH-12					MC78L12ACG
LM340LAH-15					MC78L15ACG
LM340LAH-18					MC78L18ACG
LM340LAH-24					MC78L24ACG
LM340LAH-5.0					MC78L05ACG
LM340LAH-6.0					MC78L06ACG
LM340LAH-8.0					MC78L08ACG
LM340LAZ-12					MC78L12ACP
LM340LAZ-15					MC78L15ACP
LM340LAZ-18					MC78L18ACP
LM340LAZ-24					MC78L24ACP
LM340LAZ-5.0					MC78L05ACP
LM340LAZ-6.0					MC78L06ACP
LM340LAZ-8.0					MC78L08ACP
LM340T-12	966	3592	VR-111		MC7812CT
LM340T-12R	966	3592			
LM340T-15	968	3593			MC7815CT
LM340T-15R	968	3593			
LM340T-18	958				MC7818CT
LM340T-24	972	3670			MC7824CT
LM340T-24R	972	3670			
LM340T-5	960	3591			
LM340T-5.0	960	3591	VR-102		MC7805CT
LM340T-5.0R	960	3591			
LM340T-6	962	3669			
LM340T-6.0	962	3669			MC7806CT
LM340T-6.0R	962	3669			
LM340T-8.0	964	3630	VR-108		MC7808CT
LM340T12	966	3592			
LM340T15	968	3593			
LM340T24	972	3670			
LM340T5.0	960	3591			
LM340T6.0	962	3669			
LM340T8.0	964	3630			
LM340U15	968	3593			
LM340U24	972	3670			
LM340U5	960	3591			
LM340U6	962	3669			
LM340U8	964	3630			
LM341-12	966	3592			
LM341-15	968	3593			
LM341-24	972	3670			
LM341-5	960	3591			
LM341-6	962	3669			
LM341P-12	966	3592	VR-110		MC78M12CT
LM341P-15	968	3593			MC78M15CT
LM341P-18					MC78M18CT
LM341P-24	972	3670			MC78M24CT
LM341P-5.0	960	3591	VR-101		MC78M05CT
LM341P-6.0	962	3669			MC78M06CT
LM341P-8.0	964	3630	VR-107		MC78M08CT
LM341P12	966	3592			
LM341P15	968	3593			
LM341P5.0	960	3591			
LM341P8.0	964	3630			
LM342P-12	966	3592			MC78M12CT
LM342P-15					MC78M15CT
LM342P-18					MC78M18CT
LM342P-24					MC78M24CT
LM342P-5.0					MC78M05CT
LM342P-6.0					MC78M06CT
LM342P-8.0					MC78M08CT
LM342P12	966	3592			
LM342P15	968	3593			
LM342P5.0	960	3462/977			
LM342P8.0	964	3724/981			
LM343D					MC1436G
LM343H					MC1436G
LM345K					MC7905CK
LM348D					LM348J
LM348J	948				LM348J
LM348N	948				LM348N
LM349D					MC4741CL
LM349J					MC4741CL
LM349N					MC4741CL
LM358AH	928	3691			LM358H
LM358AN	928M	3692			LM358N
LM358AT	928M	3691			
LM358H	928	3691			LM358H
LM358JG	928M	3692			LM358J
LM358L	928	3691			LM358H
LM358N	928M	3692			LM358N
LM358P	928M	3692			LM358N
LM358T	928	3691			
LM363AJ					MC3450L
LM363AN					MC3450P
LM363J					MC3450L
LM363N					MC3450P
LM371H					MC1590G
LM376JG					LM305H
LM376L					LM305H
LM376N					LM305H
LM376P					LM305H
LM377	990	3455/804	IC-27		
LM377N	990	3455/804	IC-27		
LM378	804	9068			
LM380	740A	3328	IC-31		
LM380N	740A	3328			
LM383	1232			703	
LM383T	1232	3852		703	
LM386N	823				MC1306P
LM387	824	9013			
LM387N	824	9013			
LM3900	992	3688			
LM3900N	992	3688			MC3401P
LM3905N					MC1455P1
LM393AH					LM393AH
LM393AN					LM393AN
LM393H					LM393H
LM393JG					LM393N
LM393N					LM393N
LM393P					LM393N
LM393U					LM393N
LM4250CH					MC1776CG
LM4250CN					MC1776CP1
LM4250H					MC1776G
LM511160	181		75	2041	
LM55107AJ					MC55107L
LM55108AJ					MC55108L
LM55109J					MC75S110L
LM55110J					MC75S110L
LM55121J					MC8T13L
LM55122J					MC8T14L
LM55123J					MC8T23L
LM55124J					MC8T24L
LM5525J					MC5525L
LM5528J					MC5528L
LM5529J					MC5539L
LM55325N					MC55325L
LM5534J					MC5534L
LM5535J					MC5535L
LM5538J					MC5538L
LM555CH		3693			MC1455G
LM555CN		3564			MC1455P1
LM555H		3693			MC1555G
LM556	978	3689			
LM556CD					MC3456L
LM556CJ					MC3456L
LM556CN	978	3689			MC3456P
LM556D					MC3556L
LM556J					MC3556L
LM565CH					NE565N
LM565CN	989	3595			NE565N
LM565H					NE565N
LM567	832	9089			
LM567CN	832	9089			
LM567GN	832	9089			
LM703L	703A	3157	IC-12		
LM703LN	703A	3157	IC-12		MC1350P
LM709AH					MC1709AG
LM709AJ					MC1709AL
LM709C	909	3590	IC-249		
LM709CH	909	3590	IC-249		MC1709CG
LM709CJ					MC1709CL
LM709CN	909D	3590	IC-250		MC1709CP2
LM709CN-8					MC1709CP1
LM709H					MC1709G
LM709J					MC1709L
LM710C	910		IC-251		
LM710CH	910	3553	IC-251		MC1710CG
LM710CN	910D		IC-252		MC1710CP
LM710H					MC1710G
LM711C	911		IC-253		
LM711CH	911		IC-253		MC1711CG
LM711CN	911D		IC-254		MC1711CP
LM711H					MC1711G
LM723	923	3164	IC-259		
LM723C	923D	3164/923	IC-260	1740	
LM723CD	923D	3165	IC-260	1740	LM723CJ
LM723CH	923	3165	VR-115		LM723CH
LM723CJ					LM723CJ
LM723CN	923D	3164/923			LM723CN
LM723D					LM723J
LM723H		3164			LM723H
LM723J					LM723J
LM725CH	925		IC-261		
LM733CD					MC1733CL
LM733CH					MC1733CG
LM733CJ					MC1733CL
LM733CN					MC1733CP
LM733D					MC1733L
LM733H					MC1733G
LM733J					MC1733L
LM741		3514	IC-263	010	
LM741AD					MC1741L
LM741AF					MC1741F
LM741AH					MC1741G
LM741AJ-14					MC1741L
LM741C	941	3514	IC-263	010	
LM741CD					LM1741CJ
LM741CF					LM741CF
LM741CH	941	3514	IC-263	010	LM741CH
LM741CJ					LM741CJ
LM741CJ-14					LM741CJ-14
LM741CN	941M		IC-264		LM741CN
LM741CN-14					LM741CN-14
LM741D					LM741J-14
LM741ED					MC1741CL
LM741EH					MC1741CG
LM741EJ					MC1741CU
LM741EJ-14					MC1741CL
LM741EN					MC1741CP1
LM741F					LM741F
LM741H		3514	IC-263	010	LM741H
LM741J-14					LM741J-14
LM746N	790	3454	IC-18		MC1323P
LM747C	947	3526	IC-267		
LM747CD	947D				LM747CJ
LM747CF					LM747CF
LM747CH	947	3526	IC-267		LM747CH
LM747CJ	947D				LM747CJ
LM747CN	947D				LM747CN
LM747D					LM747J
LM747F					LM747F
LM747H					LM747H
LM747J					LM747J
LM748CH	1171	3565			MC1748CG
LM748CJ					MC1748CU
LM748CN	975	3641			MC1748CP1
LM748H		3645			MC1748G
LM748J					MC1748U
LM75107AJ					MC75107L
LM75107AN					MC75107P
LM75108AJ					MC75108L
LM75108AN					MC75108P
LM75110J					MC75S110L
LM75110N					MC75S110P
LM75121J					MC8T13L
LM75121N					MC8T13P
LM75122J					MC8T14L
LM75122N					MC8T14P
LM75123J					MC8T23L
LM75123N					MC8T23P
LM75124J					MC8T24L
LM75124N					MC8T24P
LM75207L					MC75107L

Industry Standard No.	ECG	SK	GE	RS 276-	MOTOR.
LM75207N					MC75107P
LM75208J					MC75108L
LM75208N					MC75108P
LM7524J					MC7524L
LM7524N					MC7524P
LM7525J					MC7525L
LM75324J					MC75325L
LM75324N					MC75325P
LM75325J					MC75325P
LM75325N					MC75325L
LM7805KC					MC7805CK
LM7806KC					MC7806CK
LM7808KC					MC7808CK
LM7812KC					MC7812CK
LM7815KC					MC7815CK
LM7818KC					MC7818CK
LM7824KC					MC7824CK
LM78L05	977	3462	VR-100		
LM78L05ACH	977	3831			MC78L05ACG
LM78L05ACZ	977	3462	VR-100		MC78L05ACP
LM78L05CH	977	3831			MC78L05CG
LM78L05CZ	977	3462			MC78L05CP
LM78L08ACH	981				MC78L08ACG
LM78L08ACZ	981	3724	VR-106		MC78L08ACP
LM78L08CH	981	3724			MC78L08CG
LM78L08CZ	981	3724			MC78L08CP
LM78L12ACH	950				MC78L12ACG
LM78L12ACZ	950		VR-109		MC78L12ACP
LM78L12CH	950				MC78L12CG
LM78L12CZ	950				MC78L12CP
LM78L15ACH	951				MC78L15ACG
LM78L15ACZ	951				MC78L15ACP
LM78L15CH	951				MC78L15CG
LM78L15CZ	951				MC78L15CP
LM78L18ACH					MC78L18ACG
LM78L18ACZ					MC78L18ACP
LM78L18CH					MC78L18CG
LM78L18CZ					MC78L18CP
LM78L24ACH					MC78L24ACG
LM78L24ACZ					MC78L24ACP
LM78L24CH					MC78L24CG
LM78L24CZ					MC78L24CP
LM7905CT	961	3671			
LM7906CT	963	3672			
LM7912CT	967	3673			
LM7915CT	969	3674			
LM7924CT	971	3675			
LM8000	192	3854/123AP	270		
LMT540C	199		62	2010	
LMZX-10	140A		ZD-10	562	
LN75116	181	3511	75	2041	
LN75497	130		14	2041	
LN76963	129		244	2027	
LN78533	390		255	2041	
LNA382A	5072A	3136			
LP1H	116	3311	504A	1104	
LP2H	116	3311	504A	1104	
LP3H	116	3311	504A	1104	
LP4H	116	3311	504A	1104	
LPM11	141A	3092	ZD-11.5		
LPM110	151A	3099	ZD-110		
LPM110A	151A	3099	ZD-110		
LPM11A	5074A	3092/141A	ZD-11.5		
LPM12	142A	3062	ZD-12	563	
LPM12A	142A	3062	ZD-12	563	
LPM13	143A	3750	ZD-13	563	
LPM13A	143A	3750	ZD-13	563	
LPM15	145A	3063	ZD-15	564	
LPM15A	145A	3063	ZD-15	564	
LPM18A	5077A	3752			
LPM27	146A	3064	ZD-27		
LPM27A	146A	3064	ZD-27		
LPM3.6	134A	3055	ZD-3.6		
LPM3.6A	134A	3055	ZD-3.6		
LPM30A	5084A	3755			
LPM33	147A	3095	ZD-33		
LPM33A	147A	3095	ZD-33		
LPM4.7	135A	3056	ZD-5.1		
LPM4.7A	135A	3056	ZD-5.1		
LPM5.6	136A	3057	ZD-5.6	561	
LPM5.6A	136A	3057	ZD-5.6	561	
LPM56	148A	3096	ZD-55		
LPM56A	5090A	3096/148A	ZD-55		
LPM62	149A	3097	ZD-62		
LPM62A	149A	3097			
LPM8.2A	5072A	3136			
LPM82	150A	3098	ZD-82		
LPM82A	150A	3098	ZD-82		
LPM9.1	139A	3060	ZD-9.1	562	
LPM9.1A	139A	3060	ZD-9.1	562	
LPT					MRD450
LPT100A					MRD450
LPT100B					MRD450
LPZ200,A					1M200Z10,5
LPZT12	142A	3062	ZD-12	563	
LPZT15	145A	3063	ZD-15	564	
LPZT27	146A	3064	ZD-27		
LPZT33	147A	3095	ZD-33		1M33ZZ10
LPZT8.2					1M8.2ZZ10
LR100CH	140A	3061	ZD-10	562	
LR120CH	142A	3062	ZD-12	563	
LR130C	143A	3750			
LR150CH	145A	3063	ZD-15	564	
LR160C	5075A	3751			
LR180C	5077A	3752			
LR300C	5084A	3755			
LR33H	5005A	3771	ZD-3.3		
LR39CH	134A	3055	ZD-3.6		
LR47CH	135A	3056	ZD-5.1		
LR51CH	135A	3056	ZD-5.1		
LR56CH	135A	3056	ZD-5.1		
LR62CH	137A	3058	ZD-6.2	561	
LR82C	5072A	3136			
LR91CH	139A	3060	ZD-9.1	562	
LRQ849	123A	3444	20	2051	
LRR-100	116	3017B/117	504A	1104	
LRR-200	116	3016	504A	1104	
LRR-300	116	3016	504A	1104	
LRR-400	116	3016	504A	1104	
LRR-50	116	3016	504A	1104	
LRR-500	116	3016	504A	1104	
LRR100			504A	1104	
LRR200			504A	1104	
LRR300			504A	1104	
LRR400			504A	1104	
LRR50			504A	1104	
LRR500			504A	1104	
LS-0031-AR-218	199	3245	62	2010	
LS-0066	152	3893	66	2048	
LS-0067	153		69	2049	
LS-0079-01	159		82	2032	
LS-0079-02	159		82	2032	
LS-0085-01	123A	3444	20	2051	
LS-0095-AR-213	199	3245		2010	

Industry Standard No.	ECG	SK	GE	RS 276-	MOTOR.
LS-0095-AR-213	199		62	2010	
LS1496	973	3233			
LS3705	123A	3444	20	2051	
LS52	104	3719	16	2006	
LS5484	312	3448	FET-1	2035	
LS5485	312	3116	FET-1	2035	
LS555-2	978	3689			
LT1001					MRF517
LT1016	108	3452	86	2038	
LT1016(E)			20	2051	
LT1016D	108	3452	86	2038	
LT1016E	108	3452	86	2038	
LT1016H	229	3018	61	2038	
LT1016I	108	3018	86	2038	
LT1016I,H			20	2051	
LT1016T,H			20	2051	
LT2001					MRF511
LT3072					MRF904
LTE1016	229	3018	61	2038	
LTE1016(G.E.)	108		86	2038	
LTH1016	229		61	2038	
LTH1016(G.E.)	108		86	2038	
LTI1016	107	3018	11	2015	
LU2N544	126		52	2024	
LVA382A	5072A	3136			
LVA82A	5072A	3136			
LZ10	140A	3061	ZD-10	562	
LZ4.7		3056	ZD-5.1		
LZ5.6	136A	3057	ZD-5.6	561	
M,KIM,I-/U				1123	
M-0027	116	3016	504A	1104	
M-1002-17 NC	123A	3444			
M-1002-17-NC			20	2051	
M-1002-17NC				2051	
M-1002-2	123A	3444	20	2051	
M-128J509-1	161	3716	39	2015	
M-128J510-1	107		11	2015	
M-128J511-3	161	3716	39	2015	
M-161					MRD160
M-162					MRD160
M-163					MRD450
M-164					MRD450
M-165					MRD450
M-204B			504A	1104	
M-31	116	3311	504A	1104	
M-4721			10	2051	
M-75517-1			2	2007	
M-75536-1	128		243	2030	
M-75536-2	128		243	2030	
M-75543-1			23	2020	
M-75557-1	123A		20	2051	
M-75557-2	123A		20	2051	
M-75557-3	123A		20	2051	
M-75557-4	123A		20	2051	
M-75557-5	123A		20	2051	
M-75557-6	123A		20	2051	
M-8489A			300	1122	
M-8513-A-01	600	3863			
M-8513A			300	1122	
M-8513R			1N60	1123	
M-8641	123A		20	2051	
M-8641A	158		53	2007	
M-P3D	102	3722	2	2007	
MO					1N4007
M0027			504A	1104	
M012	108		86	2038	
M024	108	3452	86	2038	
M094-585-46	159		82	2032	
MOTMJE371			58	2025	
MOTMJE521			57	2017	
M1-301	177	3100/519	300	1122	
M100		3112	51	2028	2N3796
M100449	358		42		
M100A					1N4001
M100B					1N4002
M100D					1N4003
M100F					1N4004
M100G					1N4004
M100H					1N4005
M100J					1N4005
M100K					1N4006
M100M					1N4007
M101	462	3112		2028	2N3997
M102	125	3033	510,531	1114	
M103					MFE3003
M104					3N158
M105064	120	3110	CR-3		
M105330	358		42		
M108	102A	3123	53	2007	
M109474	113A	3119/113	6GC1		
M10Z	140A	3061	ZD-10	562	
M113					3N155
M113HA	282	3748	264		
M114					2N4352
M116	465				2N4351
M117	465				2N4351
M119					MFE3003
M12	116	3016	504A	1104	
M1200					MR1-1200
M124J779-1	116	3016	504A	1104	
M128-J422PT.1		3110	CR-3		
M128-J753		3110	CR-3		
M128J422-1	120		CR-3		
M128J753	120		CR-3		
M12Z	142A		ZD-12	563	
M135	138A		ZD-7.5		
M1391P	815		IC-244		
M14	116	3017B/117	504A	1104	
M140-1	107	3018	11	2015	
M140-3	123A	3444	20	2051	
M1400-1	107	3018	11	2015	
M150	116		504A	1104	
M150-1	177	3100/519	300	1122	
M15Z	145A	3063	ZD-15	564	
M172A	116	3311	504A	1104	
M1H	116	3016	504A	1104	
M1S-12795B	130		14	2041	
M1X	128		243	2030	
M2					1N4003
M2.5A		3051	512		
M2032	916	3550			
M2032-330	916	3550	IC-257		
M2032-330-BA	916		IC-257		
M21C/Q	5404	3627			
M21C/R	5404	3627			
M21C/Y	5404	3627			
M21C1R	5404	3627			
M21Y	5404	3627			
M22	116	3017B/117	504A	1104	
M2207	5071A		ZD-6.8	561	
M23C		3504		1067	
M24	123A	3444	20	2051	
M2497	116		504A	1104	
M24A	123A	3444	20	2051	

Industry Standard No.	ECG	SK	GE	RS 276-	MOTOR.
M24B	123A	3444	20	2051	
M25	123A	3444	20	2051	
M25A	123A	3444	20	2051	
M25A2			20	2051	
M25B	123A	3444	20	2051	
M25B2	123A	3444	20	2051	
M26	177		300	1122	
M27Z	146A	3064	ZD-27		
M2N168A	101	3861	8	2002	
M300-1300A	128		243	2030	
M3016	116	3311	504A	1104	
M31001	123A	3444	20	2051	
M34A	109			1123	
M351	160		245	2004	
M3519	123A	3444	20	2051	
M3567-2	184		57	2017	
M4			221		1N4004
M401	161	3018	39	2015	
M41032A			510	1114	
M41223-2	116		504A	1104	
M42	116	3017B/117	504A	1104	
M4206	121	3717			
M4246	121	3717			
M4313	102A	3123	53	2007	
M4315	102A	3004	53	2007	
M4327	102	3722	2	2007	
M433	172A	3156	64		
M4331	121	3009	239	2006	
M4363	126	3006/160	52	2024	
M4363BLU	126		52	2024	
M4363GRN	126		52	2024	
M4363ORN	126		52	2024	
M4363WHT	126		52	2024	
M4364	126	3005	52	2024	
M4365	126	3005	52	2024	
M4366	126	3006/160	52	2024	
M4367	126	3006/160	52	2024	
M4368	126	3006/160	52	2024	
M4388	126		52	2024	
M4389	100		1	2007	
M4398	102A	3123	53	2007	
M4439	160		245	2004	
M4442	159	3114/290	82	2032	
M4450	158		53	2007	
M4454	126	3006/160	52	2024	
M4456	126	3006/160	52	2024	
M4457	126	3006/160	52	2024	
M4459	127	3764	25		
M446	159	3114/290	82	2032	
M4462	102A		53	2007	
M4463	104	3009	16	2006	
M4464	123A	3444	20	2051	
M4465	123A	3444	20	2051	
M4466	102A		53	2007	
M4466ORN	158	3123	53	2007	
M4468	102A		53	2007	
M4468BRN	158	3123	53	2007	
M4469	102A		53	2007	
M4469RED	158	3123	53	2007	
M447	123A	3444	20	2051	
M4470	102A		53	2007	
M4470ORN	158	3123	53	2007	
M4471	102A		53	2007	
M4471YEL	158	3123	53	2007	
M4472	102A		53	2007	
M4472GRN	158	3123	53	2007	
M4473	158		53	2007	
M4474	102A		53	2007	
M4474YEL	158	3123	53	2007	
M4475	102A		53	2007	
M4475GRN	158	3123	53	2007	
M4476	102A		53	2007	
M4476BLU	158	3123	53	2007	
M4477	102A		53	2007	
M4477PUR	158		53	2007	
M4478	129	3025	244	2027	
M4482	102	3722	2	2007	
M4483	102	3722	2	2007	
M4484	160		245	2004	
M4485	160		245	2004	
M4486	160		245	2004	
M4501	126	3006/160	52	2024	
M4504	160		245	2004	
M4506	160		245	2004	
M4507	160		245	2004	
M4509	126	3006/160	52	2024	
M4510	158	3123	53	2007	
M4524	160		245	2004	
M4525	159		82	2032	
M4526	160		245	2004	
M4545	126	3007	52	2024	
M4545BLU	126		52	2024	
M4545WHT	126		52	2024	
M4552	144A		ZD-14	564	
M4553	102A		53	2007	
M4553BLU	102A	3123	53	2007	
M4553BRN	102A	3123	53	2007	
M4553GRN	102A	3123	53	2007	
M4553ORN	102A	3123	53	2007	
M4553PUR	105	3012	4		
M4553RED	102A	3123	53	2007	
M4553YEL	102A	3123	53	2007	
M4562	158		53	2007	
M4563	102A	3123	53	2007	
M4564	102A	3123	53	2007	
M4565	102A	3123	53	2007	
M4567	102	3004	2	2007	
M4570	104	3009	16	2006	
M4573	102A	3123	53	2007	
M4582	121	3009	239	2006	
M4582BRN	121	3009	239	2006	
M4583	121	3009	239	2006	
M4583RED	121	3009	239	2006	
M4584	121	3009	239	2006	
M4584GRN	121	3009	239	2006	
M4586	126	3006/160	52	2024	
M4589	126	3006/160	52	2024	
M4590	159	3114/290	82	2032	
M4594	123A	3444	20	2051	
M4595	102A		53	2007	
M4596	102	3004	2	2007	
M4597	102	3004	2	2007	
M4597GRN	102	3004	2	2007	
M4597RED	102	3004	2	2007	
M4603	126	3006/160	52	2024	
M4604	126	3006/160	52	2024	
M4605	126	3008	52	2024	
M4605RED	126		52	2024	
M4606	121	3717	239	2006	
M4607	102	3004	2	2007	
M4608	121	3009	239	2006	
M4619	104	3009	16	2006	
M4619RED	121	3009	239	2006	
M4620	104	3009	16	2006	
M4620GRN	121	3009	239	2006	
M4621	126	3008	52	2024	
M4622	105	3012	4		
M4623	127	3764	25		
M4624	123A	3444	20	2051	
M4627	102	3004	2	2007	
M4630	123A	3444	20	2051	
M4632	126	3006/160	52	2024	
M4640	179		76		
M4640P	179		76		
M4648	154		40	2012	
M4649	104	3009	16	2006	
M4652	127	3764	25		
M4653	139A	3066/118	ZD-9.1	562	
M4659	144A	3094	ZD-14	564	
M4661	121	3717			
M4663	147A	3095	ZD-33		
M4689	128	3024	243	2030	
M4697	160		245	2004	
M4699	150A	3098	ZD-82		
M4700	101	3861	8	2002	
M4701	179	3642	76		
M4702	179	3642	76		
M4704	150A	3098	ZD-82		
M4705	123A	3444	20	2051	
M4706			20	2051	
M4709	108		86	2038	
M4714	123A	3444	20	2051	
M4715	130	3027	14	2041	
M4722	121	3717	239	2006	
M4722BLU	121	3009	239	2006	
M4722GRN	121	3009	239	2006	
M4722ORN	5483	3942			
M4722PUR	121	3009	239	2006	
M4722RED	121	3009	239	2006	
M4722YEL	121	3009	239	2006	
M4727	104	3719	16	2006	
M4728	151A	3099	ZD-110		
M4730	121	3009	239	2006	
M4732	123A	3444	20	2051	
M4733	108	3452	86	2038	
M4734	123A	3444	20	2051	
M4736		3016	504A	1104	
M4737	123A	3444	20	2051	
M4739	123A	3444	20	2051	
M4745	159	3114/290	82	2032	
M4746			11	2015	
M4756	161	3019	39	2015	
M4757	107	3018	11	2015	
M4765	123A	3444	20	2051	
M4766	121	3009	239	2006	
M4767	121	3009	239	2006	
M4768	123A	3444	20	2051	
M4789	107		11	2015	
M4815	159	3025/129	82	2032	
M4815D	159		82	2032	
M4816			ZD-9.1	562	
M4819	154	3044	40	2012	
M4820	108		86	2038	
M4821	123A	3444	20	2051	
M4825	107		11	2015	
M4826	108		86	2038	
M4834	123A	3444	20	2051	
M4837	108	3452	86	2038	
M4838	154	3124/289	40	2012	
M4839	154	3045/225	40	2012	
M484			20	2051	
M4840	123A	3444	20	2051	
M4840A	107		11	2015	
M4841	123A	3444	20	2051	
M4842	123A	3444	20	2051	
M4842A	123A	3444	20	2051	
M4842C	123A	3444	20	2051	
M4843	154	3044	40	2012	
M4844	123A	3444	20	2051	
M4845	108	3452	86	2038	
M4850	141A	3092	ZD-11.5		
M4851	141A	3092	ZD-11.5		
M4852	123A	3444	20	2051	
M4853	154	3020/123	40	2012	
M4854	123A	3444	20	2051	
M4855	108	3452	86	2038	
M4857	108	3452	86	2038	
M4858	147A	3095	ZD-33		
M4860	160		245	2004	
M4872	124	3021	12		
M4882	130	3027	14	2041	
M4885	124	3021	12		
M4887A			16	2006	
M4888	121	3009	239	2006	
M4888A	121	3014	239	2006	
M4888B	121	3009	239	2006	
M4898	123A	3444	20	2051	
M4900	162	3438	35		
M4901	163A	3439	36		
M4904	107	3018	11	2015	
M4906	123A	3444	20	2051	
M4910			21	2034	
M4918			18	2030	
M4919			18	2030	
M4926	123A	3444	20	2051	
M4927	154	3045/225	40	2012	
M4933	123A	3444	20	2051	
M4935	123A	3444	20	2051	
M4936			23	2020	
M4937	123A	3444	20	2051	
M4937(3RD IF)	233			2009	
M4937(3RD-IF)			210	2051	
M4941			20	2051	
M4943	159	3025/129	82	2032	
M4952			20	2051	
M4953			20	2051	
M4970			17	2051	
M4974	104	3719	16	2006	
M4974/P1R	104	3719	16	2006	
M4989	159	3025/129	82	2032	
M4995	163A	3439	36		
M4998	157	3747	232		
M4B-31-22	5312	3311	BR-206		
M4B31-13		3313		1104	
M4HZ	116	3017B/117	504A	1104	
M4Z10	140A	3061	ZD-10	562	
M4Z10-20	140A	3061	ZD-10	562	
M4Z10A	140A	3061	ZD-10	562	
M4Z110	151A	3099	ZD-110		
M4Z12	142A	3062	ZD-12	563	
M4Z12-20	142A	3062	ZD-12	563	
M4Z12A	142A	3062	ZD-12	563	
M4Z15	145A	3063	ZD-15	564	
M4Z15-20	145A	3063	ZD-15	564	
M4Z15A	145A	3063	ZD-15	564	
M4Z27	146A	3064	ZD-27		
M4Z27-20	146A	3064	ZD-27		
M4Z27A	146A	3064	ZD-27		

Industry Standard No.	ECG	SK	GE	RS 276-	MOTOR.
M4Z3.3	134A	3055	ZD-3.6		
M4Z3.3A	134A	3055	ZD-3.6		
M4Z3.9	134A	3055	ZD-3.6		
M4Z3.9A	134A	3055	ZD-3.6		
M4Z33	147A	3095	ZD-33		
M4Z4.7	135A	3056	ZD-5.1		
M4Z4.7-20	135A	3056	ZD-5.1		
M4Z4.7A	135A	3056	ZD-5.1		
M4Z5.1	135A	3056	ZD-5.1		
M4Z5.1-20	135A	3056	ZD-5.1		
M4Z5.1A	135A	3056	ZD-5.1		
M4Z5.6	136A	3057	ZD-5.6	561	
M4Z5.6-20	136A	3057	ZD-5.6	561	
M4Z5.6A	136A	3057	ZD-5.6	561	
M4Z6.2	137A	3058	ZD-6.2	561	
M4Z6.2-20	137A	3058	ZD-6.2	561	
M4Z6.2A	137A	3058	ZD-6.2	561	
M4Z62	149A	3097	ZD-62		
M4Z62A	149A	3097	ZD-62		
M4Z82	150A	3098	ZD-82		
M4Z9.1	139A	3060	ZD-9.1	562	
M4Z9.1-20	139A	3060	ZD-9.1	562	
M4Z9.1A	139A	3060	ZD-9.1	562	
M500	116	3017B/117	504A	1104	
M500,A,B,C					1N4005
M500A	116	3017B/117	504A	1104	
M500B	116	3017B/117	504A	1104	
M500C	116	3017B/117	504A	1104	
M501	121	3717	239	2006	
M51	109	3087		1123	
M5106	1097	3446			
M5106P	1097	3446			
M511					MFE3003
M511-035-0001	923		IC-259		
M5111A					MFE3003
M5113	704	3023	IC-205		
M5113T	704		IC-205		
M5115	1110	3229			
M5115P	1110	3229	IC-319		
M5115P-9085	1110	3229			
M5115PA	1110	3229			
M5115PR	1110	3229			
M5115PRA	1110	3229			
M51247	1173	3729	IC-302		
M51247P	1173		IC-302		
M5131P	1096	3703			
M5133P	913		IC-255		
M5134	1096	3703			
M5134-8266	1096	3703			
M5134P	1096	3703	IC-312		
M5135P	1004	3102/710			
M5141T	941	3514	IC-263	010	
M5143	712		IC-2		
M5143P	712	3072	IC-148		
M5146P	1308	3833			
M51513L	1256	3873			
M51521	1170	3874			
M51521L	1170	3874	IC-65		
M5152L	1170		IC-65		
M5153P	801	3160			
M5169	747		IC-218		
M5169P	747	3279	IC-218		
M51709T	909		IC-249		
M5176P	747	3279			
M5183	749	3168	IC-97		
M5183-8098	749	3168	IC-97		
M5183P	749	3168	IC-97		
M51841P	955M		IC-269	1723	
M5190	738	3167	IC-29		
M5190P	738	3167	IC-29		
M5191P	1050	3475	IC-133		
M5192P	1183	3480/1178	IC-325		
M5285	102A		53	2007	
M530	128		243	2030	
M5310	7430	7430			
M5310P	7430	7430			
M53200	7400	7400		1801	
M53200P	7400	7400		1801	
M53201	7401	7401			
M53201P	7401	7401			
M53202	7402	7402		1811	
M53202P	7402	7402		1811	
M53203	7403	7403			
M53203P	7403	7403			
M53204	7404	7404		1802	
M53204P	7404	7404		1802	
M53205	7405	7405			
M53205P	7405	7405			
M53207P	7407	7407			
M53209P	7409	7409			
M53210	7410	7410	7410	1807	
M53210P	7410	7410	7410	1807	
M53217P	7417	7417			
M53220	7420	7420		1809	
M53220P	7420	7420		1809	
M53225P	7425	7425			
M53230	7430	7430			
M53230P	7430	7430			
M53240	7440	7440			
M53240P	7440	7440			
M53241	7441	7441		1804	
M53241P	7441	7441		1804	
M53242	7442	7442			
M53242P	7442	7442			
M53243	7443	7443			
M53243P	7443	7443			
M53244	7444	7444			
M53244P	7444	7444			
M53247	7447	7447		1805	
M53247P	7447	7447		1805	
M53248	7448	7448		1816	
M53248A	7448	7448		1816	
M53248P	7448	7448		1816	
M53250	7450	7450			
M53250P	7450	7450			
M53253	7453	7453			
M53260	7460	7460			
M53260P	7460	7460			
M53270	7470	7470			
M53270P	7470	7470			
M53272	7472	7472			
M53272P	7472	7472			
M53273	7473	7473		1803	
M53273P	7473	7473	7473	1803	
M53274	7474	7474		1818	
M53274P	7474		IC-193	1818	
M53275	7475	7475		1806	
M53275P	7475	7475		1806	
M53276	7476	7476		1813	
M53276P	7476	7476		1813	
M53280	7480	7480			
M53280P	7480	7480			
M53286	7486	7486		1827	
M53286P	7486	7486		1827	

Industry Standard No.	ECG	SK	GE	RS 276-	MOTOR.
M53289P	7489	7489			
M53290	7490	7490		1808	
M53290P	7490	7490	7490	1808	
M53292	7492			1819	
M53292P	7492	7492		1819	
M53293	7493A	7493			
M53293P	7493A	7493			
M53295	7495	7495			
M53295P	7495	7495			
M53296	7496	7496			
M53296P	7496	7496			
M53307	74107	74107			
M53307P	74107	74107			
M53321	74121	74121			
M53321P	74121	74121			
M53322P	74122	74122			
M53325P	74125	74125			
M53326P	74126	74126			
M53345P	74145	74145			
M53351	74151	74151			
M53351P	74151	74151			
M53355P	74155	74155			
M53356P	74156	74156			
M53360P	74160	74160			
M53362P	74162	74162			
M53363P	74163	74163			
M53365P	74165	74165			
M53366P	74166	74175			
M53370P	74170	74170			
M53374P	74174	74174			
M53375P	74175	74175			
M53380	74180	74180			
M53380P	74180	74180			
M53381P	74181	74181			
M53390P	74190	74190			
M53391P	74191	74191			
M53392	74192			1831	
M53392P	74192			1831	
M53393	74193	74193		1820	
M53393P	74193	74193		1820	
M53398P	74198	74198			
M53399P	74199	74199			
M5362	7442	7442			
M5362P	7442	7442			
M5374	7474	7474		1818	
M5374P	7474	7474		1818	
M5395	7495	7495			
M5395P	7495	7495			
M54	123A	3444	20	2051	
M546	161	3018	39	2015	
M54A	123A	3444	20	2051	
M54B	123A	3444	20	2051	
M54BLK	123A	3444	20	2051	
M54BLU	123A	3444	20	2051	
M54BRN	123A	3444	20	2051	
M54C	123A	3444	20	2051	
M54D	123A	3444	20	2051	
M54E	123A	3444	20	2051	
M54GRN	123A	3444	20	2051	
M54ORN	123A	3444	20	2051	
M54RED	123A	3444	20	2051	
M54WHT	123A	3444	20	2051	
M54YEL	123A	3444	20	2051	
M57704H					MHW710-2
M57704L					MHW710-1
M57704M					MHW710-1
M57706					MHW612
M57710					MHW613
M57712					MHW613
M57715					MHW612
M58472P				1803	
M6			221		1N4005
M6.8		3126	90		
M60	109	3087		1123	
M604	166	9075	504A	1152	
M604HT	116	3031A	504A	1104	
M612	161	3018	39	2015	
M613	161	3018	39	2015	
M614	161	3018	39	2015	
M62	116	3017B/117	504A	1104	
M644	159	3114/290	82	2032	
M652/PIC			21	2034	
M652PIC	129			2027	
M652PIC			244	2027	
M65A	159	3114/290	82	2032	
M65B	159	3114/290	82	2032	
M65C	159	3114/290	82	2032	
M65D	159	3114/290	82	2032	
M65E	159	3114/290	82	2032	
M65F	159	3114/290	82	2032	
M67	116	3311	504A	1104	
M67,A,B,C					1N4001
M670	116	3311	504A	1104	
M670A	116	3311	504A	1104	
M670B	116	3017B/117	504A	1104	
M670C	116	3017B/117	504A	1104	
M671	123A	3444	20	2051	
M67A	116	3311	504A	1104	
M67B	116	3311	504A	1104	
M67C	116	3311	504A	1104	
M68	116		504A	1104	
M68,A,B,C					1N4002
M680	116	3017B/117	504A	1104	
M680A	116	3016	504A	1104	
M680B	116	3016	504A	1104	
M680C	116	3016	504A	1104	
M68A	116	3016	504A	1104	
M68B	116	3016	504A	1104	
M68C	116	3017B/117	504A	1104	
M69	116	3016	504A	1104	
M69,A,B,C					1N4005
M690	116	3016	504A	1104	
M690A	116	3017B/117	504A	1104	
M690B	116	3016	504A	1104	
M690C	116	3016	504A	1104	
M6931			82	2032	
M69A	116		504A	1104	
M69B	116		504A	1104	
M69C	116		504A	1104	
M6HZ	116		504A	1104	
M7		3247	65	2050	
M70	116	3017B/117	504A	1104	
M70,A,B,C					1N4004
M700	116	3017B/117	504A	1104	
M7002	154	3045/225	40	2012	
M7003	123A	3444	20	2051	
M7006		3122	20	2051	
M700A	116	3031A	504A	1104	
M700B	116	3031A	504A	1104	
M700C	116	3016	504A	1104	
M7014	123	3122	20	2051	
M7015	123A	3444	20	2051	
M701B	116	3017B/117	504A	1104	
M702	125	3032A	510,531	1114	

Industry Standard No.	ECG	SK	GE	RS 276-	MOTOR.
M702B	125	3033	510,531	1114	
M702C	116(2)	3016	504A(2)	1104	
M7031	121	3009	239	2006	
M7033	123A	3444	20	2051	
M70A	116	3031A	504A	1104	
M70B	116	3031A	504A	1104	
M70C	116	3016	504A	1104	
M71	116	3017B/117	504A	1104	
M71,A,B,C					1N4005
M710	116	3017B/117	504A	1104	
M7108	123A	3444	20	2051	
M7108/A5N	123A	3444	20	2051	
M7109	123A	3444	20	2051	
M7109/A5P	123A	3444	20	2051	
M710A	116	3017B/117	504A	1104	
M710B	116	3017B/117	504A	1104	
M710C	116	3017B/117	504A	1104	
M7127	159	3114/290	82	2032	
M7127/P2S	159		82	2032	
M7171	123A	3444	20	2051	
M71A	116	3017B/117	504A	1104	
M71B	116	3017B/117	504A	1104	
M71C	116	3017B/117	504A	1104	
M72	125	3032A	510,531	1114	
M72,A,B,C					1N4006
M720	125	3032A	510,531	1114	
M720A	125	3032A	510,531	1114	
M720B	125	3032A	510,531	1114	
M720C	125	3032A	510,531	1114	
M72A	125	3032A	510,531	1114	
M72B	125	3032A	510,531	1114	
M72C	125	3032A	510,531	1114	
M72D	156	3032A	512		
M73	125	3033	510,531	1114	
M73,A,B,C					1N4007
M730	125	3033	510,531	1114	
M730A	125	3033	510,531	1114	
M730B	125	3033	510,531	1114	
M730C	125	3033	510,531	1114	
M7310	197	3083	250	2027	
M7342	127	3764	25		
M7342/P4F	127	3764	25		
M73A	125	3033	510,531	1114	
M73B	125	3033	510,531	1114	
M73C	125	3033	510,531	1114	
M7476	191	3044/154	249		
M7511	163A	3439	36		
M75205-1	5870		5032		
M75205-2	5871		5033		
M7543-1	130		14	2041	
M75516-2	160	3007	245	2004	
M75516-2P	160	3007	245	2004	
M75516-2R	160	3007	245	2004	
M755162-P	160	3007	245	2004	
M755162-R	160	3007	245	2004	
M75517-1	102A	3004	53	2007	
M75517-2	102A		53	2007	
M75537-2	176	3845	80		
M75543-1	152	3893	66	2048	
M75545-1	161	3716	39	2015	
M75547-1	108	3452	86	2038	
M75547-2	108	3452	86	2038	
M75549-2	130		14	2041	
M75561-10RK	220	3990		2036	
M75561-17	129		244	2027	
M75561-23	221			2036	
M75561-23RN	221			2036	
M75561-7	102A		53	2007	
M75561-8	129		244	2027	
M75565-1	123		20	2051	
M758	136A	3057	ZD-5.6	561	
M76	160		245	2004	
M7641	7473		7473	1803	
M77	160		245	2004	
M773	123A	3444	20	2051	
M773RED	123A	3444	20	2051	
M774	123A	3122	20	2051	
M7740RN	123A	3122	20	2051	
M775	123A	3444	20	2051	
M775BRN	123A	3444	20	2051	
M776	123A	3444	20	2051	
M776GRN	123A	3444	20	2051	
M779	123A(2)	3122	20		
M779BLU	123A	3444	20	2051	
M78	160		245	2004	
M780	123A(2)	3122	20		
M780WHT	123A	3444	20	2051	
M783	123A	3444	20	2051	
M783RED	123A	3444	20	2051	
M784	123A	3444	20	2051	
M7840RN	123A	3444	20	2051	
M785	123A	3444	20	2051	
M785YEL	123A	3444	20	2051	
M786	123A(2)	3122	20		
M787	123A(2)	3122	20		
M787BLU	123A	3444	20	2051	
M78A	160		245	2004	
M78B	160		245	2004	
M78BLK	160		245	2004	
M78C	160		245	2004	
M78D	160		245	2004	
M78GRN	160		245	2004	
M78RED	160		245	2004	
M78YEL	160		245	2004	
M791	123A	3444	20	2051	
M8					1N4006
M8014		3123	90		
M8014(TRANSISTOR)	100		1	2007	
M8062A	102A		53	2007	
M8062B	102A		53	2007	
M8062C	102A		53	2007	
M8105	123A	3444	20	2051	
M8116	126		52	2024	
M8120	101	3861	8	2002	
M8124	160		245	2004	
M818	123A	3444	20	2051	
M818WHT	123A	3444	20	2051	
M819	154		40	2012	
M82	125	3032A	510,531	1114	
M822	123A	3444	20	2051	
M8221	123A	3444	20	2051	
M8222	116	504A	1104		
M822A	123A	3444	20	2051	
M822A-BLU	123A	3444	20	2051	
M822B	123A	3444	20	2051	
M823	123A	3444	20	2051	
M823B	123A	3444	20	2051	
M823WHT	123A	3444	20	2051	
M827	123A	3444	20	2051	
M827BRN	123A	3444	20	2051	
M828	129	3025	244	2027	
M828GRN	123A	3444	20	2051	
M829A	159	3114/290	82	2032	
M829B	159	3114/290	82	2032	
M829C	159	3114/290	82	2032	
M829D	159	3114/290	82	2032	
M829E	159	3114/290	82	2032	
M829F	159	3114/290	82	2032	
M833	159	3114/290	82	2032	
M8399	116	3311	504A	1104	
M84	104	3009	16	2006	
M844	184	3190	57	2017	
M847			20	2051	
M847BLK	123A	3444	20	2051	
M8482	112	3089	1N82A		
M8482C	112		1N82A		
M8489	519	3087	300	1122	
M8489-A	109	3090		1123	
M8489A	519		300	1122	
M84B	121	3009	239	2006	
M8513	600	3863	300	1122	
M8513(LAFAYETTE)	600		300	1122	
M8513-0	519	3463/601	514	1122	
M8513-R	601	3463			
M8513A	601	3863/600	300	1122	
M8513A-0	601	3863/600			
M8513A-R	600	3863	300	1122	
M8513A0	600		300	1122	
M8513AO	600	3863			
M8513R	600	3863	300	1122	
M852	184	3190	57	2017	
M8534992	113A	3119/113	6GC1		
M8569	178MP		300(2)	1122(2)	
M8604	102A		53	2007	
M8604A	102A		53	2007	
M8640	158	3100/519	53	2007	
M8640A	158	3123	53	2007	
M8640E		3100	52	2024	
M8640E(C-M)	177		300	1122	
M8640E(DIO)			300	1122	
M8640E(XSTR)			2	2007	
M8D03	727	3071	IC-210		
M8D03(MOTOROLA)	727		IC-210		
M8HZ	125	3032A	510,531	1114	
M9002	158	3004/102A	53	2007	
M9010	108		86	2038	
M9032	107	3018	11	2015	
M9052	487	3765/195A			
M9083	121	3717			
M9090	127	3764	25		
M9091	121	3717			
M9092	103	3722/102	8	2002	
M9093	103	3862	8	2002	
M9095	123A	3444	20	2051	
M91	107	3039/316	11	2015	
M9104	121	3717			
M912	157	3103A/396	232		
M9125	195A	3765			
M9134			28	2017	
M9138	128	3024	243	2030	
M9141	121	3717	239	2006	
M9142	121	3717	239	2006	
M9145	129		244	2027	
M9147	195A	3765			
M9148	102	3004	2	2007	
M9159	123A	3444	20	2051	
M9170	123		243	2030	
M9174	195A	3765			
M9177	176	3845	80		
M9181	179	3642	76		
M9184	128		243	2030	
M9186	195A	3765			
M9187	195A	3765			
M9197	199	3245	62	2010	
M9198	102A	3004	53	2007	
M91A	123A	3444	20	2051	
M91A01	116	3017B/117	504A	1104	
M91A02	116	3017B/117	504A	1104	
M91A03	116	3017B/117	504A	1104	
M91A06	125	3032A	510,531	1114	
M91B	123A	3444	20	2051	
M91BGRN	123A	3444	20	2051	
M91C	123A	3444	20	2051	
M91CM624	123A	3444	20	2051	
M91D	123A	3444	20	2051	
M91E	123A	3444	20	2051	
M91F	123A	3444	20	2051	
M91FM624	123A	3444	20	2051	
M9202	121	3717	239	2006	
M9206	116	3016	504A	1104	
M9209	128		243	2030	
M9221	128		243	2030	
M9225	175		246	2020	
M9226	123A	3444	20	2051	
M9228	128	3024	243	2030	
M9235	116(4)	3017B/117	504A	1104	
M9237	121	3717	239	2006	
M924		3018	20	2051	
M9241	121	3717	239	2006	
M9244	130		14	2041	
M9248		3444	20	2051	
M9249			53	2007	
M9250		3722	2	2007	
M9251	105		4		
M9255	121	3717	239	2006	
M9256	6401			2029	
M9257	129		244	2027	
M9259	130		14	2041	
M9260	195A	3765			
M9263	121	3717	239	2006	
M9264	6400A			2029	
M9265	198		251		
M9266	161	3716	39	2015	
M9269	199	3245	62	2010	
M9271	195A	3765			
M9274	175		246	2020	
M9278	130		14	2041	
M9282	123		20	2051	
M9285	195A	3765			
M9286	198		251		
M9287	124		12		
M9293	199	3245	62	2010	
M9294	121	3717			
M9301	175		246	2020	
M9302	130		14	2041	
M9306	314	3898			
M9308	129		244	2027	
M9309	175		246	2020	
M9312	116	3311	504A	1104	
M9314	116	3311	504A	1104	
M9316	175		246	2020	
M9317	116	3311	504A	1104	
M9319	116	3311	504A	1104	
M9320	198	3220	251		
M9321	130		14	2041	
M9329	199	3245	62	2010	
M9334	159		82	2032	
M9338	199	3245	62	2010	

Industry Standard No.	ECG	SK	GE	RS 276-	MOTOR.
M9342	121	3717	239	2006	
M9344	180		74	2043	
M9348	153	3274	69	2049	
M9355	195A	3765			
M9359	180		74	2043	
M9380	128		243	2030	
M9384	199	3245	62	2010	
M9389	289A		268	2038	
M9393	175		246	2020	
M9400	129		244	2027	
M9408	162	3438	35		
M9409	199		62	2010	
M9412	234	3247	65	2050	
M9416	199	3245	62	2010	
M9420	121	3717			
M9426	129		244	2027	
M9430	195A	3765			
M9431	195A	3765			
M9432	129		244	2027	
M9435	129		244	2027	
M9436	121	3717	239	2006	
M9447	199	3245	62	2010	
M9450	161	3716	39	2015	
M9458	121	3717			
M9465	198	3220	251		
M9467	234		65	2050	
M9474	199	3245	62	2010	
M9475	123		20	2051	
M9480	181		75	2041	
M9481	161	3039/316	39	2015	
M9482	108	3039/316	86	2038	
M9486	199	3245	62	2010	
M9491	123		243	2030	
M95	109	3087		1123	
M9514	159	3114/290	82	2032	
M9515	130		14	2041	
M9517	121	3717			
M9519	128		243	2030	
M9520	129		244	2027	
M9521	192		63	2030	
M9525	123A	3444	20	2051	
M9526	159		82	2032	
M9527	159		82	2032	
M9531	159	3114/290	82	2032	
M9532	123A	3444	20	2051	
M9536	225		256		
M9547	199	3245	62	2010	
M9550	121	3717	239	2006	
M9556	184	3041	57	2017	
M9562	128		243	2030	
M9563	123A	3444	20	2051	
M9568	123A	3444	20	2051	
M9570	123A	3444	20	2051	
M9571	159		82	2032	
M9575	108	3452	86	2038	
M9576	152	3893	66	2048	
M9582	185	3041	57	2017	
M9591	346		243	2030	
M9594	195	3245	62	2010	
M9599	311		277		
M9610	476		246	2020	
M9618	184		57	2017	
M9628	181		75	2041	
M9631	346		243	2030	
M9639	181		75	2041	
M9640	188		226	2018	
M9641	189		218	2026	
M9649	159		82	2032	
M9661	152	3893	66	2048	
M9666	390		255	2041	
M9676	196		241	2020	
M9676/NPN	196		241	2020	
M9677	197		250	2027	
M9677/PNP	197		250	2027	
M9701	197		250	2027	
M9703	346		243	2030	
M9715	181		75	2041	
M9Z	139A	3060	ZD-9.1	562	
MA-110		3016	BR-600	1104	
MA-161	177	3100/519			
MA-23B	604	3004/102A		1123	
MA-25A	604	3004/102A	300	1122	
MA-26	601		300	1122	
MA-26-1	601		504A	1104	
MA-900			1N60	1123	
MA0401	129		21	2034	
MA0402	129		21	2034	
MA0404			22	2032	
MA0404-1			21	2034	
MA0404-2			21	2034	
MA0411			21	2034	
MA0413			21	2034	
MA0414			22	2024	
MA1	160		245	2004	
MA10	172A	3156	64		
MA100	100	3051/156	1	2007	
MA101	116		504A	1104	
MA102	116		504A	1104	
MA1047	5069A		ZD-4.7		
MA1051	5010A	3776			
MA1056	136A		ZD-5.6	561	
MA1062	137A		ZD-6.2	561	
MA1062LF	5013A	3779			
MA1068	5071A		ZD-6.8	561	
MA1075	138A		ZD-7.5		
MA1082	5072A	3136	ZD-8.2	562	
MA1091	5018A	3784	ZD-9.1	562	
MA110	116		504A	1104	
MA1100	140A	3061	ZD-10	562	
MA112	102	3123	53	2007	
MA1120	142A	3787/5021A	ZD-12	563	
MA113	102	3025/129	53	2007	
MA1130	143A	3750	ZD-13	563	
MA1130LF	5022A	3788			
MA114	102		53	2007	
MA115	102	3123	53	2007	
MA1150	145A		ZD-15	564	
MA116	102	3123	53	2007	
MA1160			ZD-16	564	
MA117	102	3123	53	2007	
MA1180	5077A	3752			
MA1220	5080A		ZD-22		
MA1240	5081A		ZD-24		
MA1318	102A		53	2007	
MA150	519	3100	300	1122	
MA150DD	519	3100			
MA150TA	519	3100	300	1122	
MA161	519	3100	514	1122	
MA161C	519	3100			
MA162	519	3100	514	1122	
MA1700	102	3004	2	2007	
MA1702	158		53	2007	
MA1703	102		53	2007	
MA1704	102		53	2007	

Industry Standard No.	ECG	SK	GE	RS 276-	MOTOR.
MA1705	102		52	2024	
MA1706	102		53	2007	
MA1707	102		53	2007	
MA1708	102		53	2007	
MA2	116	3016	504A	1104	
MA203	116	3016	504A	1104	
MA206	102A	3100/519	53	2007	
MA21				1123	
MA211	116		504A	1104	
MA215	116	3016	504A	1104	
MA23	604	3123	53	2007	
MA23(B)			1N60	1123	
MA23A	604	3123	245	2004	
MA23B	604		53	2007	
MA240	102A		53	2007	
MA242	116	3016	504A	1104	
MA242C			504A	1104	
MA242RC	156	3051	512		
MA25	604	3004/102A	53	2007	
MA25A	604			1123	
MA26	601	3463	300	1122	
MA26A	601	3463	300	1122	
MA26G	601	3864/605			
MA26W	605	3864	300		
MA26WO	601	3864/605			
MA26WA	605	3864	300	1122	
MA26WB	605	3864			
MA26WD	605	3864			
MA286	100	3005	1	2007	
MA287	100	3005	1	2007	
MA288	100	3005		2007	
MA3	109	3090		1123	
MA301	612	3325			
MA302	613	3326			
MA303	614	3327			
MA3065	712	3072	IC-148		
MA311		3126	90		
MA320	612	3325			
MA320B					BB105B
MA330	611	3324			
MA332	613	3326			
MA350	116	3311	504A	1104	
MA351	116	3311	504A	1104	
MA393	102A	3008	53	2007	
MA393A	102A	3008	53	2007	
MA393B	102A	3008	53	2007	
MA393C	102A	3008	53	2007	
MA393E	102A		53	2007	
MA393G	102A	3008	53	2007	
MA393R	102A	3008	53	2007	
MA4101	123A		61	2038	
MA4103	123A		61	2038	
MA4104	123A		61	2038	
MA4388	612	3325			
MA45071	612	3325			
MA45111	610	3323			
MA45112	611	3324			
MA45114	613	3326			
MA45115	614	3327			
MA45131	610	3323			
MA45132	611	3324			
MA45134	613	3326			
MA45151	612	3325			
MA45158	610	3323			
MA45160	611	3324			
MA45161	612	3325			
MA451670	611	3324			
MA45168	610	3323			
MA45171	612	3325			
MA4670	121	3009	239	2006	
MA47				1123	
MA50		3500	504A	1104	
MA51A	109	3087		1123	
MA55	109	3087		1123	
MA6001	123A		210	2051	
MA6002	123A		210	2051	
MA6003	123A		210	2051	
MA6101	128	3197/235	215		
MA6102	123A	3197/235	215		
MA62	6408	3523/6407			
MA729	720	9014			
MA732	718	3159			
MA767	722	3161	IC-9		
MA767PC	722		IC-9		
MA7805	960	3591			
MA790			1N295	1123	
MA8	109	3087		1123	
MA8001	128		47	2030	
MA815	102A		53	2007	
MA881	102A		53	2007	
MA882	102A		53	2007	
MA883	102A		53	2007	
MA884	102A		53	2007	
MA885	102A		53	2007	
MA886	102A		53	2007	
MA887	102A		53	2007	
MA888	102A		53	2007	
MA889	102A		53	2007	
MA890	102A	3123	53	2007	
MA891	102A	3123	53	2007	
MA892	102A	3123	53	2007	
MA893	102A	3123	53	2007	
MA894	102	3123	53	2007	
MA895	102	3123	53	2007	
MA896	102	3123	53	2007	
MA897	102	3123	53	2007	
MA898	102	3123	53	2007	
MA899	102	3123	53	2007	
MA90	109	3087		1123	
MA900	102	3087	53	2007	
MA901	100	3123	53	2007	
MA902	102A	3123	53	2007	
MA903	102A	3123	53	2007	
MA904	102A	3123	53	2007	
MA909	102A		53	2007	
MA910	102A		53	2007	
MA9426	123A	3444	20	2051	
MAC10-1	5611	3631/5623			
MAC10-2	5612	3631/5623			
MAC10-3	5613	3631/5623			
MAC10-4	5614	3632/5624			2N6151
MAC10-5	5615	3633/5626			
MAC10-6	5616	3633/5626			2N6152
MAC10-8	5618				2N6153
MAC11-1	5621	3631/5623			
MAC11-2	5622	3631/5623			
MAC11-3	5623	3631			
MAC11-4	5624	3632			2N6154
MAC11-5	5625	3633/5626			
MAC11-6	5626	3633			2N6155
MAC11-8	5628				2N6156
MAC15-10	56010				MAC15-10
MAC15-4	56004	3658			MAC15-4
MAC15-5	56006	3659			MAC15-5

Industry Standard No.	ECG	SK	GE	RS 276-	MOTOR.
MAC15-6	56006	3659			MAC15-6
MAC15-7	56008	3660			MAC15-7
MAC15-8	56008	3660			MAC15-8
MAC15-9	56010				MAC15-9
MAC15A-4	56004	3658			
MAC15A-5	56006	3659			
MAC15A-6	56006	3659			
MAC15A-7	56008	3660			
MAC15A-8	56008	3660			
MAC15A10					MAC15A10
MAC15A4					MAC15A4
MAC15A5					MAC15A5
MAC15A6					MAC15A6
MAC15A7					MAC15A7
MAC15A8					MAC15A8
MAC15A9					MAC15A9
MAC20-10					MAC20-10
MAC20-4					MAC20-4
MAC20-5					MAC20-5
MAC20-6					MAC20-6
MAC20-7					MAC20-7
MAC20-8					MAC20-8
MAC20-9					MAC20-9
MAC20A10					MAC20A10
MAC20A4					MAC20A4
MAC20A5					MAC20A5
MAC20A6					MAC20A6
MAC20A7					MAC20A7
MAC20A8					MAC20A8
MAC20A9					MAC20A9
MAC220-2					MAC220-2
MAC220-3					MAC220-3
MAC220-5					MAC220-5
MAC220-7					MAC220-7
MAC220-9					MAC220-9
MAC221-2					MAC221-2
MAC221-3					MAC221-3
MAC221-5					MAC221-5
MAC221-7					MAC221-7
MAC221-9					MAC221-9
MAC222-1					MAC222-1
MAC222-10					MAC222-10
MAC222-2					MAC222-2
MAC222-3					MAC222-3
MAC222-4					MAC222-4
MAC222-5					MAC222-5
MAC222-6					MAC222-6
MAC222-7					MAC222-7
MAC222-8					MAC222-8
MAC222-9					MAC222-9
MAC222A1					MAC222A1
MAC222A10					MAC222A10
MAC222A2					MAC222A2
MAC222A3					MAC222A3
MAC222A4					MAC222A4
MAC222A5					MAC222A5
MAC222A6					MAC222A6
MAC222A7					MAC222A7
MAC222A8					MAC222A8
MAC222A9					MAC222A9
MAC223-1					MAC223-1
MAC223-10					MAC223-10
MAC223-2					MAC223-2
MAC223-3					MAC223-3
MAC223-4					MAC223-4
MAC223-5					MAC223-5
MAC223-6					MAC223-6
MAC223-7					MAC223-7
MAC223-8					MAC223-8
MAC223-9					MAC223-9
MAC223A1					MAC223A1
MAC223A10					MAC223A10
MAC223A2					MAC223A2
MAC223A3					MAC223A3
MAC223A4					MAC223A4
MAC223A5					MAC223A5
MAC223A6					MAC223A6
MAC223A7					MAC223A7
MAC223A8					MAC223A8
MAC223A9					MAC223A9
MAC228-1					MAC228-1
MAC228-10					MAC228-10
MAC228-2					MAC228-2
MAC228-3					MAC228-3
MAC228-4					MAC228-4
MAC228-5					MAC228-5
MAC228-6					MAC228-6
MAC228-7					MAC228-7
MAC228-8					MAC228-8
MAC228-9					MAC228-9
MAC228A1					MAC228A1
MAC228A10					MAC228A10
MAC228A2					MAC228A2
MAC228A3					MAC228A3
MAC228A4					MAC228A4
MAC228A5					MAC228A5
MAC228A6					MAC228A6
MAC228A7					MAC228A7
MAC228A8					MAC228A8
MAC228A9					MAC228A9
MAC25-10					MAC25-10
MAC25-4	56014				MAC25-4
MAC25-5	56014				MAC25-5
MAC25-6	56014				MAC25-6
MAC25-7					MAC25-7
MAC25-8					MAC25-8
MAC25-9					MAC25-9
MAC25A10					MAC25A10
MAC25A4	56014				MAC25A4
MAC25A5	56014				MAC25A5
MAC25A6	56014				MAC25A6
MAC25A7					MAC25A7
MAC25A8					MAC25A8
MAC25A9					MAC25A9
MAC3010-15					MAC3010-15
MAC3010-25					MAC3010-25
MAC3010-4					MAC3010-4
MAC3010-40					MAC3010-40
MAC3010-8					MAC3010-8
MAC3020-15					MAC3020-15
MAC3020-25					MAC3020-25
MAC3020-4					MAC3020-4
MAC3020-40					MAC3020-40
MAC3020-8					MAC3020-8
MAC3030-15					MAC3030-15
MAC3030-25					MAC3030-25
MAC3030-4					MAC3030-4
MAC3030-40					MAC3030-40
MAC3030-8					MAC3030-8
MAC3040-15					MAC3040-15
MAC3040-25					MAC3040-25
MAC3040-4					MAC3040-4
MAC3040-40					MAC3040-40
MAC3040-8					MAC3040-8
MAC35-1					2N6157
MAC35-2					2N6157
MAC35-3					2N6157
MAC35-4					2N6157
MAC35-5					2N6158
MAC35-6					2N6158
MAC35-7					2N6159
MAC35-8					2N6159
MAC36-1	5680	3521/5686			2N6160
MAC36-2	5681	3521/5686			2N6160
MAC36-3	5682	3521/5686			2N6160
MAC36-4	5683	3521/5686			2N6160
MAC36-5	5684	3521/5686			2N6161
MAC36-6	5685	3521/5686			2N6161
MAC36-7	5686	3521			2N6162
MAC36-8			SC260M		2N6162
MAC37-1			SC261B		2N6157
MAC37-2			SC261B		2N6157
MAC37-3			SC261B		2N6157
MAC37-4			SC261B		2N6157
MAC37-5			SC261D		2N6158
MAC37-6			SC261D		2N6158
MAC37-7			SC261E		2N6159
MAC37-8					2N6159
MAC38-1					2N6160
MAC38-2			SC260B		2N6160
MAC38-3			SC260B		2N6160
MAC38-4			SC260B		2N6160
MAC38-5			SC260D		2N6161
MAC38-6			SC260D		2N6161
MAC38-7			SC260E		2N6162
MAC38-8					2N6162
MAC40688	56022		SC265B2		T6420B
MAC40689	56024		SC265D2		T6420D
MAC40690	56026		SC265M2		T6420M
MAC40795			SC246M		T4101M
MAC40796			SC245M		T4111M
MAC40798	5677	3520			
MAC40799			SC245B2		T4121B
MAC40800			SC245D2		T4121D
MAC40801			SC245M2		T4121M
MAC40802					2N6145
MAC40803					2N6146
MAC40804					2N6147
MAC4688	56022		SC265B2		
MAC4689	56024		SC265D2		
MAC4690	56026		SC265M2		
MAC5-2	5661	3631/5623			
MAC5-3	5662	3631/5623			
MAC5-4	5667A	3632/5624			
MAC5-5	5667A	3632/5624			
MAC5-6	5667A	3633/5626			
MAC50-10					MAC50-10
MAC50-4					MAC50-4
MAC50-5					MAC50-5
MAC50-6					MAC50-6
MAC50-7					MAC50-7
MAC50-8					MAC50-8
MAC50-9					MAC50-9
MAC50A10					MAC50A10
MAC50A4					MAC50A4
MAC50A5					MAC50A5
MAC50A6					MAC50A6
MAC50A7					MAC50A7
MAC50A8					MAC50A8
MAC50A9					MAC50A9
MAC5569	5673	3520/5677			
MAC5570	5675	3520/5677			
MAC5571					2N5571
MAC5572					2N5572
MAC5573	5673	3520/5677			2N5573
MAC5574	5675	3520/5677			2N5574
MAC77-1	5600	3664/5602		1001	2N6068
MAC77-2	5601	3664/5602		1001	2N6069
MAC77-3	5602	3664		1001	2N6070
MAC77-4	5603	3665		1001	2N6071
MAC77-5	5604	3666/5605		1000	2N6072
MAC77-6	5605	3666		1000	2N6073
MAC77-7	5606				2N6074
MAC77-8	5607				2N6075
MAC91-1					MAC91-1
MAC91-2					MAC91-2
MAC91-3					MAC91-3
MAC91-4					MAC91-4
MAC91-5					MAC91-5
MAC91-6					MAC91-6
MAC91-7					MAC91-7
MAC91-8					MAC91-8
MAC91A1					MAC91A1
MAC91A2					MAC91A2
MAC91A3					MAC91A3
MAC91A4					MAC91A4
MAC91A5					MAC91A5
MAC91A6					MAC91A6
MAC91A7					MAC91A7
MAC91A8					MAC91A8
MAC92-1	5655				MAC92-1
MAC92-2	5655				MAC92-2
MAC92-3	5655				MAC92-3
MAC92-4	5655				MAC92-4
MAC92-5	5656				MAC92-5
MAC92-6	5656				MAC92-6
MAC92-7	5657				MAC92-7
MAC92-8	5657				MAC92-8
MAC92A1					MAC92A1
MAC92A2					MAC92A2
MAC92A3					MAC92A3
MAC92A4					MAC92A4
MAC92A5					MAC92A5
MAC92A6					MAC92A6
MAC92A7					MAC92A7
MAC92A8					MAC92A8
MAC93-1					MAC93-1
MAC93-2					MAC93-2
MAC93-3					MAC93-3
MAC93-4					MAC93-4
MAC93-5					MAC93-5
MAC93-6					MAC93-6
MAC93-7					MAC93-7
MAC93-8					MAC93-8
MAC93A1					MAC93A1
MAC93A2					MAC93A2
MAC93A3					MAC93A3
MAC93A4					MAC93A4
MAC93A5					MAC93A5
MAC93A6					MAC93A6
MAC93A7					MAC93A7
MAC93A8					MAC93A8
MAC94-1	5655				MAC94-1
MAC94-2	5655				MAC94-2
MAC94-3	5655				MAC94-3
MAC94-4	5655				MAC94-4
MAC94-5	5656				MAC94-5
MAC94-6	5656				MAC94-6

Industry Standard No.	ECG	SK	GE	RS 276-	MOTOR.
MAC94-7	5657				MAC94-7
MAC94-8	5657				MAC94-8
MAC94A1					MAC94A1
MAC94A2					MAC94A2
MAC94A3					MAC94A3
MAC94A4					MAC94A4
MAC94A5					MAC94A5
MAC94A6					MAC94A6
MAC94A7					MAC94A7
MAC94A8					MAC94A8
MAC95-1	5655				MAC95-1
MAC95-2	5655				MAC95-2
MAC95-3	5655				MAC95-3
MAC95-4	5655				MAC95-4
MAC95-5	5656				MAC95-5
MAC95-6	5656				MAC95-6
MAC95-7	5657				MAC95-7
MAC95-8	5657				MAC95-8
MAC95A1					MAC95A1
MAC95A2					MAC95A2
MAC95A3					MAC95A3
MAC95A4					MAC95A4
MAC95A5					MAC95A5
MAC95A6					MAC95A6
MAC95A7					MAC95A7
MAC95A8					MAC95A8
MAC96-1					MAC96-1
MAC96-2					MAC96-2
MAC96-3					MAC96-3
MAC96-4					MAC96-4
MAC96-5					MAC96-5
MAC96-6					MAC96-6
MAC96-7					MAC96-7
MAC96-8					MAC96-8
MAC96A1					MAC96A1
MAC96A2					MAC96A2
MAC96A3					MAC96A3
MAC96A4					MAC96A4
MAC96A5					MAC96A5
MAC96A6					MAC96A6
MAC96A7					MAC96A7
MAC96A8					MAC96A8
MAQ7786	123A	3444	20	2051	
MAS20	126	3006/160	52	2024	
MAS21	126	3006/160	52	2024	
MAS22	126	3008	52	2024	
MAS23	126	3006/160	52	2024	
MB-01		3031A		1152	
MB-02				1172	
MB-04				1173	
MB-1D	525	3125	511	1114	
MB-1F	525	3843	511	1114	
MB-4-01				1152	
MB-4-02				1172	
MB-4-04				1173	
MB01	116		504A	1104	
MB12A10V05	5340	3679			
MB12A10V10	5340	3679			
MB12A10V20	5340	3680/5322			
MB12A10V30	5342	3681/5324			
MB12A10V40	5342	3681/5324			
MB12A10V60	5342	3682/5326			
MB12A25V05	5322	3679			
MB12A25V10	5322	3679			
MB12A25V20	5322	3680			
MB12A25V30	5324	3681			
MB12A25V40	5324	3681			
MB214					1N4934
MB2147					MCM2147
MB215					1N4935
MB217					1N4936
MB218					1N4937
MB219					1N4937
MB220					MR817
MB221					1N4934
MB222					1N4935
MB224					1N4936
MB225					1N4937
MB226					1N4937
MB228					MR501
MB229					MR502
MB230					MR504
MB231					MR504
MB232					MR506
MB233		3051			MR506
MB234		3051			MR508
MB235		3051			MR510
MB236					1N4002
MB237					1N4003
MB238					1N4004
MB239					1N4004
MB240					1N4005
MB241		3051			1N4005
MB242		3051			1N4006
MB243		3051			1N4007
MB244	116	3016	504A	1104	1N4002
MB245					1N4003
MB246					1N4004
MB247					1N4004
MB248					1N4005
MB249					1N4005
MB250					1N4006
MB251					1N4007
MB257	116	3017B/117	504A	1104	
MB258	116	3017B/117	504A	1104	
MB269	116	3017B/117	504A	1104	
MB270	116	3017B/117	504A	1104	
MB3104	1195	3469			
MB3202	1082	3461	IC-140		
MB3710	1194	3484			
MB400	7400	7400		1801	
MB401	7410	7410		1807	
MB402	7420	7420		1809	
MB403	7430	7430			
MB4044					MCM6641
MB405	7450	7450			
MB406	7460	7460			
MB407	7472	7472			
MB408	7480	7480			
MB410	74107	74107			
MB411	7453	7453			
MB416	7401	7401			
MB417	7402	7402		1811	
MB418	7404	7404		1802	
MB420	7474	7474		1818	
MB433	7438	7438			
MB434	7438	7438			
MB435	7437	7437			
MB442	7442	7442			
MB443	74145	74145		1828	
MB445	74151	74151			
MB447	74180	74180			
MB448	7485	7485		1826	
MB449	7486	7486		1827	

Industry Standard No.	ECG	SK	GE	RS 276-	MOTOR.
MB451	74160	74160			
MB452	7496	7496			
MB453	7495	7495			
MB455	74198	74198			
MB456	74191	74191			
MB457	74190	74190			
MB460	74170	74170			
MB513AR	177		300	1122	
MB601	7400	7400		1801	
MB602	7410	7410		1807	
MB603	7420	7420		1809	
MB604	7430	7430			
MB605	7440	7440			
MB609	7472	7472			
MB6356	5073A	3749	ZD-8.2	562	
MB8114					MCM2114
MB8116					MCM4116
MB8216					MCM4116
MB8227					MCM4027A
MB8264					MCM6665
MB84011	4011B		4011	2411	
MB84011-U	4011B			2411	
MB84011M	4011B		4011		
MB84011U	4011B		4011	2411	
MB84011V	4011B		4011	2401	
MB84055	4055B	4055			
MB8518H					MCM2708
MBD101	112	3089	1N82A		MBD101
MBD102					MBD102
MBD201					MBD201
MBD301					MBD301
MBD501					MBD501
MBD502					MBD502
MBD701					MBD701
MBD702					MBD702
MBM2716					MCM2716
MBS4991	6403		504A		MBS4991
MBS4992					MBS4992
MC-1310P	801		IC-35		
MC-1312P	799		IC-34		
MC-14011CP			4011	2411	
MC-301			300	1122	
MC-7402				1811	
MC010	116	3016	504A	1104	
MC015	116	3017B/117	504A	1104	
MC020	116	3311	504A	1104	
MC020A	116	3311	504A	1104	
MC021	116	3311	504A	1104	
MC021A	116	3311	504A	1104	
MC022	116	3311	504A	1104	
MC022A	116	3311	504A	1104	
MC023	116	3311	504A	1104	
MC023A	116	3311	504A	1104	
MC025	116		504A	1104	
MC030	116	3031A	504A	1104	
MC030A	116	3031A	504A	1104	
MC030B	116	3031A	504A	1104	
MC035	116		504A	1104	
MC040	116	3031A	504A	1104	
MC040A	116	3031A	504A	1104	
MC070	125	3032A	510,531	1114	
MC070A	125	3032A	510,531	1114	
MC080	125	3032A	510,531	1114	
MC080A	125	3032A	510,531	1114	
MC090	125	3033	510,531	1114	
MC090A	125	3033	510,531	1114	
MC100	610	3033	510,531	1114	
MC100A	610	3033	512		
MC101	160	3007	245	2004	
MC103	160		245	2004	
MC1303	725	3162	IC-19		
MC1303L	725	3162	IC-19		
MC1303P	725	3162	IC-19		
MC1304P	718	3159	IC-8		
MC1304PQ	718	3159	IC-8		
MC1305		9014	IC-7		
MC1305P	720	9014	IC-7		
MC1305P-C	720	9014	IC-7		
MC1305PQ	720		IC-7		
MC1306P	745	3276	IC-216		
MC1307	722		IC-9		
MC1307P	722	3161	IC-9		
MC1307PQ	722	9014/720	IC-9		
MC1310	801	3160	IC-35		
MC1310A	801	3160	IC-35		MC1310P
MC1310P	801	3160	IC-35		
MC1311P	743	3172	IC-214		
MC1312P	799	3238	IC-34		
MC1314G	704		IC-205		
MC1314P	802	3277	IC-235		
MC1315P	803	3278	IC-237		
MC1316	814		IC-243		
MC1316P	814		IC-243		
MC1316PC	814		IC-243		
MC1324		3235	IC-30		
MC1324P	739	3235	IC-30		
MC1324PQ	739	3235	IC-30		
MC1326P	739	3235	IC-30		
MC1326PQ	739	3235	IC-30		
MC1328P	790	3454	IC-230		
MC1328PQ	790	3454	IC-230		
MC1329P	713	3077/790	IC-5		
MC1329PQ	713		IC-5		
MC1330A1P	747	3279			
MC1330P	747	3279	IC-218		
MC1339P	721		IC-14		
MC1339PQ	721		IC-14		
MC1344	779A	3240/779-1			
MC1344P	779A	3240/779-1	IC-221		
MC1345P	779A	3240/779-1			
MC1345PQ	779A	3240/779-1			
MC1349P	795	3237	IC-232		
MC1350	746	3234	IC-217		
MC1350P	746	3234	IC-217		
MC1350PQ	746		IC-217		
MC1351		3236	IC-183		
MC1351P	748	3236	IC-183		
MC1351PQ	748	3236	IC-183		
MC1352P	749	3168	IC-97		
MC1352PQ	749	3168	IC-97		
MC1353P		3284	IC-98		
MC1355P	750	3280	IC-219		
MC1355PQ	750	3280	IC-219		
MC1356P	737	3375			
MC1357	708	3135/709	IC-10		
MC1357A		3135	IC-10		
MC1357P	708	3135/709	IC-10		
MC1357PQ	708	3135/709	IC-11		
MC1358	712	3072	IC-148		
MC1358P	712	3072	IC-148		
MC1358PQ	712	3072	IC-148		
MC1364	783	3215	IC-225		
MC1364G	780	3215/783	IC-222		
MC1364P	783	3215	IC-225		

Industry Standard No.	ECG	SK	GE	RS 276-	MOTOR.
MC1364PQ	783	3215	IC-225		
MC1370P	714	3075	IC-4		
MC1370PQ	714	3075	IC-4		
MC1371P	715	3076	IC-6		
MC1371PQ	715	3076	IC-6		
MC1375P	723	3168/749	IC-15		
MC1375PQ	723		IC-15		
MC1389P	788	3147	IC-229		
MC1391P	815	3255	IC-244		
MC1398	738	3167	IC-29		
MC1398P	738	3167	IC-29		
MC1398PQ	738	3167	IC-29		
MC1399P	835	3829/788			
MC14000		4000	4000		
MC14001		4001	4001	2401	
MC14001B	4001B	4001			
MC14002		4002	4002		
MC14002B	4002B	4002			
MC14006	4006B	4006			
MC14006B	4006B	4006			
MC14006BCL	4006B	4006			
MC14006BCP	4006B	4006			
MC14007	4007	4007	4007		
MC14007B	4007	4007			
MC14008	4008B	4008			
MC14011		4011	4011	2411	
MC14011B	4011B	4011	4011	2411	
MC14011BCL	4011B	4011			
MC14011BCP	4011B	4011	4011		
MC14011CP	4011B		4011	2411	
MC14012		4012	4012	2412	
MC14013		4013	4013	2413	
MC14013B	4013B	4013			
MC14014B	4014B	4014			
MC14014BCL	4014B	4014			
MC14014BCP	4014B	4014			
MC14015		4015	4015		
MC14016		4016	4016		
MC14016B	4016B	4016			
MC14017		4017	4017	2417	
MC14017B	4017B	4017			
MC14018	4018B	4022			
MC14018BCL	4018B	4018			
MC14018BCP	4018B	4018			
MC14020			4020	2420	
MC14020B	4020B	4020			
MC14020CP	4020B	4020			
MC14021			4021	2421	
MC14021B	4021B	4021			
MC14021CP	4021B	4021			
MC14022	4022B	4022			
MC14023		4023	4023	2423	
MC14023B	4023B	4023			
MC14024		4024	4024		
MC14024B	4024B	4024			
MC14025		4025	4025		
MC14025B	4025B	4025	4025		
MC14025CP	4025B		4025		
MC14027		4027	4027	2427	
MC14027B	4027B	4027			
MC14028BCL	4028B	4028			
MC14028BCP	4028B	4028			
MC14035	4035B	4035			
MC14035B	4035B	4035			
MC14040B	4040B	4040			
MC14042		4042	4042		
MC14043	4043B	4043			
MC14043BCL	4043B	4043			
MC14043BCP	4043B	4043			
MC14044	4044B	4044			
MC14044BCL	4044B	4044			
MC14044BCP	4044B	4044			
MC14046	980	4046			
MC14049		4049	4049	2449	
MC14049B			4049	2449	
MC14049CP	4049		4049	2449	
MC14050		4050	4050	2450	
MC14051	4051B	4051	4051	2451	
MC14051B	4051B	4051			
MC14052B	4052B	4052			
MC14053	4053B	4053			
MC14053B	4053B	4053			
MC14053BCL	4053B	4053			
MC14053BCP	4053B	4053			
MC14066	4066B	4066			
MC14066B	4066B	4066			
MC14071B	4071B	4071			
MC14072	4072B	4072			
MC14072B	4072B	4072			
MC14075B	4075B	4075			
MC14081		4081	4081		
MC14081B	4081B	4081			
MC1408B					MC1408P8
MC1408F					MC1408L8
MC14093B	4093B	4093			
MC1411P	2011	3975			
MC1412P	2012	9092			
MC1413P	2013	9093			
MC1416P	2014	9094			
MC145109	1167	3732	IC-317		
MC14511	4511B	4511			
MC14511L	4511B	4511			
MC14518			4518	2490	
MC14518B	4518B	4518			
MC14520B	4520B	4520			
MC14555BCL	4555B	4555			
MC14555BCP	4555B	4555			
MC14556BCL	4556B	4556			
MC14556BCP	4556B	4556			
MC1455P			IC-269	1723	
MC1455P1	955M	3564	IC-269	1723	
MC1458	778A	3465	IC-220		
MC14584	40106B	4016			
MC1458CP	778A	3465			
MC1458CP1	778A	3465	IC-220		
MC1458G	778A	3555			
MC1458JG					MC1458U
MC1458L		3555			MC1458G
MC1458P	778A	3465			MC1458P1
MC1458P1	778A	3465	IC-220		
MC1458V	778A		IC-220		
MC1461R	946		IC-266		
MC1496	973	3233			
MC1496A	973D	3892			
MC1496G	973	3233			
MC1496N	973D	3892			
MC1496P	973D	3892			
MC1520		3311	504A	1104	
MC1521	116	3311	504A	1104	
MC1522		3311	504A	1104	
MC1523		3031A	504A	1104	
MC1524		3031A	504A	1104	
MC1527		3032A	510,531	1114	
MC1528		3032A	510,531	1114	

Industry Standard No.	ECG	SK	GE	RS 276-	MOTOR.
MC1529		3033	510,531	1114	
MC1550G	816	3242			
MC1558JG					MC1558U
MC1558L					MC1558G
MC170	116	3311	504A	1104	
MC1709CG	909	3551	IC-249		
MC1709CP1	909	3590	IC-249		
MC1709CP2	909D	3590	IC-250		
MC1709P2	909D	3590	IC-250		
MC1710CG	910	3553	IC-251		
MC1710CL	910D		IC-252		
MC1710CP2	910D		IC-252		
MC1710P2	910D		IC-252		
MC1711CG	911		IC-253		
MC1711CL	911D		IC-254		
MC1723CG	923	3164	IC-259		
MC1723CL	923D	3165	IC-260	1740	
MC1723G	923	3164	IC-259		
MC1741CG	941	3514	IC-263	010	
MC1741CL	941D		IC-264		
MC1741CP1	941M	3553	IC-265	007	
MC1741CP2	941D	3514/941	IC-264		
MC1747CL	947D	3556			
MC1747CP	947D	3556			
MC1747CP1	947D	3556			
MC1747CP2	947D	3556			
MC1748P1	975	3641			
MC19	116	3311	504A	1104	
MC1938	738	3167	IC-29		
MC1Y	601	3463			
MC2	177		300	1122	
MC201	612		90		
MC2326	177	3100/519	300	1122	
MC2526	109	3087		1123	
MC3000L	74H00	74H00			
MC3000P	74H00	74H00			
MC3001	1192	3445			
MC3008L	74H04	74H04			
MC3008P	74H04	74H04			
MC301	553	3323/610	300	1122	
MC308	109	3090		1123	
MC3301L	992	3688			
MC3301P	992	3688			
MC3302	834	3569			
MC3302P	834	3569			
MC3340P	829	3688/992			
MC3340PA	829	3891			
MC3346P	912	3543	IC-172		
MC3456P	978	3689			
MC4040P	974	3965			
MC4044P	974	3965			
MC4080	1083		IC-275		
MC4080-1	1083		IC-275		
MC4080-2	1083		IC-275		
MC4080-3	1083		IC-275		
MC40803	1083		IC-275		
MC456	116	3311	504A	1104	
MC5321	177	3100/519	300	1122	
MC6007	136A	3057	ZD-5.6	561	
MC6007A	136A	3057	ZD-5.6	561	
MC6008	137A	3058	ZD-6.2	561	
MC6008A	137A	3058	ZD-6.2	561	
MC6012A	5072A	3136			
MC6014	140A	3061	ZD-10	562	
MC6014A	140A	3061	ZD-10	562	
MC6016	142A	3062	ZD-12	563	
MC6016A	142A	3062	ZD-12	563	
MC6017A	143A	3750			
MC6018	145A	3063	ZD-15	564	
MC6018A	145A	3063	ZD-15	564	
MC6019A	5075A	3751			
MC6020A	5077A	3752			
MC6024	146A	3064	ZD-27		
MC6024A	146A	3064	ZD-27		
MC6025A	5084A	3755			
MC6026	147A	3095	ZD-33		
MC6026A	147A	3095	ZD-33		
MC6107	136A	3057	ZD-5.6	561	
MC6107A	136A	3057	ZD-5.6	561	
MC6108	137A	3058	ZD-6.2	561	
MC6108A	137A	3058	ZD-6.2	561	
MC6112A	5072A	3136			
MC6116	142A	3062	ZD-12	563	
MC6116A	142A	3062	ZD-12	563	
MC6118	145A	3063	ZD-15	564	
MC6118A	145A	3063	ZD-15	564	
MC6124	146A	3064	ZD-27		
MC6124A	146A	3064	ZD-27		
MC6126	147A	3095	ZD-33		
MC6126A	147A	3095	ZD-33		
MC6130,A					1M47ZS10,5
MC6400,MC6401					1N821
MC6402,MC6403					1N823
MC6404,MC6405					1N825
MC6406,MC6407					1N827
MC6416					1N935
MC6417					1N935A
MC6418					1N936
MC6419					1N936A
MC6420					1N937
MC6421					1N937A
MC6422					1N938
MC6423					1N939A
MC6424,MC6425					1N829
MC6428					1N937
MC6429					1N939A
MC7400				1801	
MC7400L	7400	7400		1801	
MC7400N				1801	
MC7400P	7400	7400		1801	
MC7401L	7401	7401			
MC7401P	7401	7401			
MC7402L	7402	7402		1811	
MC7402P	7402	7402		1811	
MC7403L	7403	7403			
MC7403P	7403	7403			
MC7404L	7404	7404		1802	
MC7404P	7404	7404		1802	
MC7405L	7405	7405			
MC7405P	7405	7405			
MC7406L	7406	7406		1821	
MC7406P	7406	7406		1821	
MC7407L	7407	7407			
MC7407P	7407	7407			
MC7408L	7408	7408		1822	
MC7408P	7408	7408		1822	
MC7409L	7409	7409			
MC7409P	7409	7409			
MC74107P	74107	74107			
MC7410F				1807	
MC7410L	7410	7410		1807	
MC7410P	7410	7410		1807	
MC74121P	74121	74121			
MC74145P	74145	74145		1828	

Industry Standard No.	ECG	SK	GE	RS 276-	MOTOR.
MC74150P	74150	74150		1829	
MC74151P	74151	74151			
MC74153P	74153	74153			
MC74155P	74155	74155			
MC74156P	74156	74156			
MC74164AP	74164	74164			
MC74165P	74165	74165			
MC7416L	7416	7416			
MC7416P	7416	7416			
MC74176P	74176	74176			
MC74177P	74177	74177			
MC7417L	7417	7417			
MC7417P	7417	7417			
MC74180P	74180	74180			
MC74181P	74181	74181			
MC74192P	74192	74192		1831	
MC74193P	74193	74193		1820	
MC74195P	74195	74195			
MC7420F				1809	
MC7420L	7420	7420		1809	
MC7420P	7420	7420		1809	
MC7426L	7426	7426			
MC7426P	7426	7426			
MC7430L	7430	7430			
MC7430P	7430	7430			
MC7437P	7437	7437			
MC7438L	7438	7438			
MC7438P	7438	7438			
MC7440L	7440	7440			
MC7440P	7440	7440			
MC7441AL	7441	7441		1804	
MC7441AP	7441	7441			
MC7442L	7442	7442			
MC7442P	7442	7442			
MC7443L	7443	7443			
MC7443P	7443	7443			
MC7444L	7444	7444			
MC7444P	7444	7444			
MC7445L	7445	7445			
MC7445P	7445	7445			
MC7446L	7446	7446			
MC7446P	7446	7446			
MC7447L	7447	7447		1805	
MC7447P	7447	7447		1805	
MC7448L	7448	7448		1816	
MC7448P	7448	7448		1816	
MC7450L	7450	7450			
MC7450P	7450	7450			
MC7451F				1825	
MC7451L	7451	7451		1825	
MC7451P	7451	7451		1825	
MC7453L	7453	7453			
MC7453P	7453	7453			
MC7454L	7454	7454			
MC7454P	7454	7454			
MC7460L	7460	7460			
MC7460P	7460	7460			
MC7470L	7470	7470			
MC7470P	7470	7470			
MC7472L	7472	7472			
MC7472P	7472	7472			
MC7473L	7473	7473		1803	
MC7473P	7473	7473		1803	
MC7475L	7475	7475		1806	
MC7475P	7475	7475		1806	
MC7476L	7476	7476		1813	
MC7476P	7476	7476		1813	
MC7480L	7480	7480			
MC7480P	7480	7480	IC-115		
MC7483L	7483	7483			
MC7483P	7483	7483			
MC7486F				1827	
MC7486L	7486	7486		1827	
MC7486P	7486	7486		1827	
MC7490L	7490	7490		1808	
MC7490P	7490	7490		1808	
MC7492L	7492	7492		1819	
MC7492P	7492	7492		1819	
MC7493L	7493A	7493			
MC7493P	7493A	7493			
MC7494L	7494	7494			
MC7494P	7494	7494			
MC7495L	7495	7495			
MC7495P	7495	7495			
MC7496L	7496	7496			
MC7496P	7496	7496			
MC7705CP	960	3591			
MC7706CP	962	3669			
MC7712CP	966	3592			
MC7715CP	968	3593			
MC7724CP	972	3670			
MC7805	960	3591			
MC7805CP	960	3591			
MC7805P	960	3591			
MC7806CP	962	3669			
MC7808CP	964	3630			
MC7812CP	966	3592			
MC7815CP	968	3592/966			
MC7824CP	972	3670			
MC78L05	977	3462	VR-100		
MC78L05ACP	977	3462	VR-100		
MC78L05C	977	3462	VR-100		
MC78L05CP	977	3462	VR-100		
MC78L08ACP	981	3724	VR-106		
MC78L08CP	981	3724			
MC78L12ACP	950		VR-109		
MC78M06CP	962	3669			
MC78M08CP	964	3630			
MC78M12CP	966	3592			
MC78M12CT	966		VR-110		
MC78M15CP	968	3593			
MC78M24CP	972	3670			
MC7905CP	961	3671			
MC7906CP	963	3672			
MC7912CP	967	3673			
MC7912CT	967	3673	VR-114		
MC7915CP	969	3674			
MC7924CP	971	3675			
MC7924CT	971	3675			
MC830F					MC930F
MC832F					MC932F
MC833F					MC933F
MC834F					MC934F
MC835F					MC935F
MC836F					MC936F
MC837F					MC937F
MC838F					MC938F
MC839F					MC939F
MC840F					MC940F
MC841F					MC941F
MC844F					MC944F
MC845F					MC945F
MC846F					MC946F
MC847F					MC947F

Industry Standard No.	ECG	SK	GE	RS 276-	MOTOR.
MC848F					MC948F
MC849F					MC949F
MC850F					MC950F
MC851F					MC951F
MC852F					MC952F
MC853F					MC953F
MC855F					MC955F
MC856F					MC956F
MC857F					MC957F
MC858F					MC958F
MC861F					MC961F
MC862F					MC962F
MC863F					MC963F
MC9427	123A	3018	20	2051	
MCA230			H11B3		MCA230
MCA231			H11B2		MCA231
MCA255			H11B255		MCA255
MCR100-1					MCR100-1
MCR100-2					MCR100-2
MCR100-3		3638			MCR100-3
MCR100-4		3627			MCR100-4
MCR100-5		3951			MCR100-5
MCR100-6	5405	3951			MCR100-6
MCR100-7		3952			MCR100-7
MCR100-8	5406	3952			MCR100-8
MCR101	5400	3950			MCR101
MCR102	5400	3950			MCR102
MCR103	5401	3638/5402			MCR103
MCR104	5402	3638			MCR104
MCR106-1	5411	3953/5412		1067	MCR106-1
MCR106-2	5412	3953		1067	MCR106-2
MCR106-3	5413	3954/5414		1067	MCR106-3
MCR106-4	5414	3954		1067	MCR106-4
MCR106-5					MCR106-5
MCR106-6		3955	C106D1		MCR106-6
MCR106-7					MCR106-7
MCR106-8		3956	C106M1		MCR106-8
MCR106-9					MCR106-9
MCR107-1	5411	3953/5412		1067	MCR106-1
MCR107-2	5412	3953		1067	MCR106-2
MCR107-3	5413	3954/5414		1067	MCR106-3
MCR107-4	5414	3954		1067	MCR106-4
MCR107-5					MCR106-5
MCR107-6					MCR106-6
MCR107-7		3636			MCR106-7
MCR107-8		3636	C107M1		MCR106-8
MCR115	5403	3627/5404			MCR115
MCR120	5404	3627			MCR120
MCR1305R2					2N5168
MCR1305R4					2N5169
MCR1305R6					2N5170
MCR1305R8					2N5171
MCR1308-1					MCR3918-1
MCR1308-2					2N5168
MCR1308-3					MCR3918-3
MCR1308-4					2N5169
MCR1308-5					MCR3918-5
MCR1308-6					2N5170
MCR1308-7					MCR3918-7
MCR1308-8					2N5170
MCR1308-9					MCR3918-9
MCR1308R2					2N5168
MCR1308R4					2N5169
MCR1308R6					2N5170
MCR1308R8					2N5171
MCR1330					MPU131
MCR1350					MPU132
MCR1604-1					2N4183
MCR1604-2					2N4184
MCR1604-3					2N4185
MCR1604-4					2N4186
MCR1604-5					2N4187
MCR1604-6					2N4188
MCR1604-7					2N4189
MCR1604-8					2N4190
MCR1718-5			2N3652		MCR1718-5
MCR1718-6			2N3653		MCR1718-6
MCR1718-7					MCR1718-7
MCR1718-8					MCR1718-8
MCR1906-1					MCR1906-1
MCR1906-2					MCR1906-2
MCR1906-3					MCR1906-3
MCR1906-4					MCR1906-4
MCR1906-5					MCR1906-5
MCR1906-6					MCR1906-6
MCR1906-7					MCR1906-7
MCR1906-8					MCR1906-8
MCR201					MCR201
MCR202					MCR202
MCR203					MCR203
MCR204					MCR204
MCR205					MCR205
MCR206					MCR206
MCR2064-6		3582	MR-4		
MCR220-5			C126C		MCR220-5
MCR220-7			C126E		MCR220-7
MCR220-9			C126S		MCR220-9
MCR221-5			ZJ436C		MCR221-5
MCR221-7			ZJ436E		MCR221-7
MCR221-9			ZJ436S		MCR221-9
MCR2305-1					2N4167
MCR2305-2	5483	3942			2N4168
MCR2305-3					2N4169
MCR2305-4	5483	3942			2N4170
MCR2305-5					2N4171
MCR2305-6					2N4172
MCR2305-7					2N4173
MCR2305-8					2N4174
MCR2315-1	5480				2N4167
MCR2315-2	5481				2N4168
MCR2315-3					2N4169
MCR2315-4	5483	3942			2N4170
MCR2315-5	5484	3943/5485			2N4171
MCR2315-6	5485	3943			2N4172
MCR2315-7					2N4173
MCR2315-8					2N4174
MCR2604-1		3582	MR-4		2N4183
MCR2604-2		3582	MR-4		2N4184
MCR2604-3		3582	MR-4		2N4185
MCR2604-4	5515	6615	MR-4		
MCR2604-5	5515	6615	MR-4		2N4187
MCR2604-6	5515	6615			2N4188
MCR2604-7					2N4189
MCR2604-8					2N4190
MCR2604L1					2N4183
MCR2604L2					2N4184
MCR2604L3					2N4185
MCR2604L4					2N4186
MCR2604L5					2N4187
MCR2604L6					2N4188
MCR2604L7					2N4189
MCR2604L8					2N4190
MCR2605-1		3582	MR-4		
MCR2605-2		3582	MR-4		

Industry Standard No.	ECG	SK	GE	RS 276-	MOTOR.
MCR2605-3		3582	MR-4		
MCR2605-4		3582	MR-4		
MCR2605-5		3582	MR-4		
MCR2605-6		3582	MR-4		
MCR2614L-4	5515	6615			
MCR2614L-5	5515	6615			
MCR2614L-6	5515	6615			
MCR2818-1					MCR3818-1
MCR2818-2					2N5164
MCR2818-3					MCR3818-3
MCR2818-4					2N5165
MCR2818-5					MCR3818-5
MCR2818-6					2N5166
MCR2818-7					MCR3818-7
MCR2818-8					2N5167
MCR2818R2					2N5164
MCR2818R4					2N5165
MCR2818R6					2N5166
MCR2818R8					2N5167
MCR2835-1	5517	3615			MCR3835-1
MCR2835-2	5517	3615			MCR3835-2
MCR2835-3	5517	3615			2N3870
MCR2835-4	5517	3615			2N3871
MCR2835-5	5518				MCR3835-5
MCR2835-6	5518				2N3872
MCR2835-7	5519				MCR3835-7
MCR2835-8	5519				2N3873
MCR2918-1			MR-3		MCR3918-1
MCR2918-2					2N5168
MCR2918-3			MR-3		MCR3918-3
MCR2918-4					2N5169
MCR2918-5			MR-3		MCR3918-5
MCR2918-6					2N5170
MCR2918-7					MCR3918-7
MCR2918-8					2N5171
MCR2918R2					2N5168
MCR2918R4					2N5169
MCR2918R6					2N5170
MCR2918R8					2N5171
MCR2935-1	5540	6642/5542			MCR3935-1
MCR2935-2	5541	6642/5542			MCR3935-2
MCR2935-3	5542	6642			2N3896
MCR2935-4	5543				2N3897
MCR2935-5	5544				MCR3935-5
MCR2935-6	5545				2N3898
MCR2935-7	5546				MCR3935-7
MCR2935-8	5547				2N3899
MCR3000-1	5442	3572/5463			
MCR3000-10C					MCR3000-10
MCR3000-10					MCR3000-1
MCR3000-2	5444	3572/5463			2N4441
MCR3000-3	5444	3572/5463			
MCR3000-3C					MCR3000-3
MCR3000-4	5444	3572/5463			2N4442
MCR3000-5	5446	3635			
MCR3000-5C					MCR3000-5
MCR3000-6	5446	3635			2N4443
MCR3000-7	5448	3636			
MCR3000-7C					MCR3000-7
MCR3000-8	5448	3636			2N4444
MCR3000-9C					MCR3000-9
MCR306-1					2N6236
MCR306-2					2N6237
MCR306-3					2N6238
MCR306-4					2N6239
MCR306-5					2N6240
MCR306-6					2N6240
MCR3818-1	5514	3613			MCR3818-1
MCR3818-10					MCR3818-10
MCR3818-2	5514	3613			MCR3818-2
MCR3818-3	5514	3613			MCR3818-3
MCR3818-4	5514	3613			MCR3818-4
MCR3818-5	5515	6615			MCR3818-5
MCR3818-6	5515	6615			MCR3818-6
MCR3818-7	5516	6616			MCR3818-7
MCR3818-8	5516	6616			MCR3818-8
MCR3818-9					MCR3818-9
MCR3835-1	5517	3615			MCR3835-1
MCR3835-10					MCR3835-10
MCR3835-2	5517	3615			MCR3835-2
MCR3835-3	5517	3615			2N3870
MCR3835-4	5517	3615			2N3871
MCR3835-5	5518				MCR3835-5
MCR3835-6	5518				2N3872
MCR3835-7	5519				MCR3835-7
MCR3835-8	5519				2N3873
MCR3835-9					MCR3835-9
MCR3918-1	5520	6621/5521			MCR3918-1
MCR3918-10					MCR3918-10
MCR3918-2	5521				MCR3918-2
MCR3918-3	5522	6621/5521			MCR3918-3
MCR3918-4	5524	6624			MCR3918-4
MCR3918-5	5526	6629/5529			MCR3918-5
MCR3918-6	5527				MCR3918-6
MCR3918-7	5528				MCR3918-7
MCR3918-8	5529	6629			MCR3918-8
MCR3918-9					MCR3918-9
MCR3935-1	5540	6642/5542			MCR3935-1
MCR3935-10	5548				MCR3935-10
MCR3935-2	5541	6642/5542			MCR3935-2
MCR3935-3	5542	6642			2N3896
MCR3935-4	5543	3581			2N3897
MCR3935-5	5544	3582/5545			MCR3935-5
MCR3935-6	5545	3582			2N3898
MCR3935-7	5546	3582/5545			MCR3935-7
MCR3935-8	5547	3505			2N3899
MCR3935-9	5548				MCR3935-9
MCR4018-3					2N6167
MCR4018-4					2N6168
MCR4018-5					2N6169
MCR4018-6					2N6169
MCR4018-7					2N6170
MCR4018-8					2N6170
MCR4035-3					2N6171
MCR4035-4					2N6172
MCR4035-5					2N6173
MCR4035-6					2N6173
MCR4035-7					2N6174
MCR4035-8					2N6174
MCR406-1	5421	3570		1067	
MCR406-2	5422	3570		1067	
MCR406-3	5423	3570		1067	
MCR406-4	5414	3570		1067	
MCR407-1	5431	3954/5414		1067	
MCR407-2	5432			1067	
MCR407-3	5433			1067	
MCR407-4	5414			1067	
MCR600-1					MCR102
MCR600-2					MCR103
MCR600-3					MCR104
MCR600-3.5					MCR115
MCR600-4					MCR120
MCR63-1					MCR63-1
MCR63-10					MCR63-10

Industry Standard No.	ECG	SK	GE	RS 276-	MOTOR.
MCR63-2					MCR63-2
MCR63-3					MCR63-3
MCR63-4					MCR63-4
MCR63-5					MCR63-5
MCR63-6					MCR63-6
MCR63-7					MCR63-7
MCR63-8					MCR63-8
MCR63-9					MCR63-9
MCR64-1					MCR64-1
MCR64-10					MCR64-10
MCR64-2					MCR64-2
MCR64-3					MCR64-3
MCR64-4					MCR64-4
MCR64-5					MCR64-5
MCR64-6					MCR64-6
MCR64-7					MCR64-7
MCR64-8					MCR64-8
MCR64-9					MCR64-9
MCR649AP1					MCR649AP1
MCR649AP10					MCR649AP10
MCR649AP2					MCR649AP2
MCR649AP3					MCR649AP3
MCR649AP4					MCR649AP4
MCR649AP5					MCR649AP5
MCR649AP6					MCR649AP6
MCR649AP7					MCR649AP7
MCR649AP8					MCR649AP8
MCR649AP9					MCR649AP9
MCR649P1					MCR649P1
MCR649P10					MCR649P10
MCR649P2					MCR649P2
MCR649P3					MCR649P3
MCR649P4					MCR649P4
MCR649P5					MCR649P5
MCR649P6					MCR649P6
MCR649P7					MCR649P7
MCR649P8					MCR649P8
MCR649P9					MCR649P9
MCR65-1					MCR65-1
MCR65-10					MCR65-10
MCR65-2					MCR65-2
MCR65-3					MCR65-3
MCR65-4					MCR65-4
MCR65-5					MCR65-5
MCR65-6					MCR65-6
MCR65-7					MCR65-7
MCR65-8					MCR65-8
MCR65-9					MCR65-9
MCR67-1					MCR67-1
MCR67-2					MCR67-2
MCR67-3					MCR67-3
MCR67-6					MCR67-6
MCR68-1					MCR68-1
MCR68-2					MCR68-2
MCR68-3					MCR68-3
MCR68-6					MCR68-6
MCR681	5520	6621/5521			
MCR682	5521	6621			
MCR683	5522	6622			
MCR684	5523	6624/5524			
MCR685	5524	6624			
MCR686	5525	6627/5527			
MCR687	5526	6627/5527			
MCR688	5527	6627			
MCR689	5528	6629/5529			
MCR69-1					MCR69-1
MCR69-2					MCR69-2
MCR69-3					MCR69-3
MCR69-6					MCR69-6
MCR70-1					MCR70-1
MCR70-2					MCR70-2
MCR70-3					MCR70-3
MCR70-6					MCR70-6
MCR71-1					MCR71-1
MCR71-2					MCR71-2
MCR71-3					MCR71-3
MCR71-6					MCR71-6
MCR72-1					MCR72-1
MCR72-2					MCR72-2
MCR72-3					MCR72-3
MCR72-4					MCR72-4
MCR72-5					MCR72-5
MCR72-6					MCR72-6
MCR72-7					MCR72-7
MCR72-8					MCR72-8
MCR729-10					MCR729-10
MCR729-5					MCR729-5
MCR729-6					MCR729-6
MCR729-7					MCR729-7
MCR729-8					MCR729-8
MCR729-9					MCR729-9
MCR808-1					MCR3818-1
MCR808-10					2N2647-10
MCR808-2					2N5164
MCR808-3					MCR3818-3
MCR808-4					2N5165
MCR808-5					MCR3818-5
MCR808-6					2N5166
MCR808-7					MCR3818-7
MCR808-8					2N5167
MCR808-9					MCR3818-9
MCR808R2					2N5164
MCR808R4					2N5165
MCR808R6					2N5166
MCR808R8					2N5167
MCR914-1					2N1595
MCR914-2					2N1595
MCR914-3					2N1596
MCR914-4					2N1597
MCR914-5					2N1598
MCR914-6					2N1599
MCS2	3046		H11C3		MCS2
MCS2400	3046		H11C6		MCS2400
MCT2	3042		H11A2		MCT2
MCT26	3040		H11A5		MCT26
MCT271					MCT271
MCT272					MCT272
MCT273					MCT273
MCT274					MCT274
MCT275	3043				MCT275
MCT277					MCT276
MCT2E	3042		H11A3		MCT2E
MCT492P	7492	7492		1819	
MCV	116	3311	504A	1104	
MD-34	109			1123	
MD-60A	110MP	3087	1N34AS	1123	
MD04	116	3311	504A	1104	
MD1120					MD1120
MD1120F					MD1120F
MD1121					MD1121
MD1121F					MD1121F
MD1122	81				MD1122
MD1122F					MD1122F
MD1123					MD1123

Industry Standard No.	ECG	SK	GE	RS 276-	MOTOR.
MD1123F					MD3250AF
MD1124					MD3250A
MD1124F					MD3250AF
MD1125					MD3250A
MD1125F					MD3250AF
MD1126	81				MD3250A
MD1126F					MD3250AF
MD1127	81				MD2369
MD1127F					MD2369F
MD1128					MD2369
MD1128F					MD2369F
MD1129	81				MD1129
MD1129F					MD1129F
MD1130	82				MD1130
MD1130F					MD1130F
MD1131	81				MD918
MD1131F					MD918F
MD1132	81				MD1132
MD1132F					MD918AF
MD1133					MD3725
MD1133F					MD3725F
MD1134					MD2369
MD134	116	3311	504A	1104	
MD135	116	3311	504A	1104	
MD136	116	3311	504A	1104	
MD137	116	3031A	504A	1104	
MD138	116	3031A	504A	1104	
MD139				1104	
MD2060F					MD2060F
MD2218	81				MD2218
MD2218A	81				MD2218A
MD2218AF					MD2218AF
MD2218F					MD2218F
MD2219	81				MD2219
MD2219A	81				MD2219A
MD2219AF					MD2219AF
MD2219F					MD2219F
MD234				1011	
MD235				1011	
MD236				1102	
MD2369					MD2369
MD2369A	81				MD2369A
MD2369AF					MD2369AF
MD2369B	81				MD2369B
MD2369BF					MD2369BF
MD2369F					MD2369F
MD2904	82				MD2904
MD2904A	82				MD2904A
MD2904AF					MD2904AF
MD2904F					MD2904F
MD2905	82				MD2905
MD2905A					MD2905A
MD2905AF					MD2905AF
MD2905F					MD2905F
MD2906					MD2904
MD2956	330	3012/105			
MD3001					MD6001
MD3002					MD6002
MD3133	82				MD2904
MD3133F			21	2034	MD2904F
MD3134	82				MD2905A
MD3134F			21	2034	MD2905AF
MD3250	82				MD3250
MD3250A	82				MD3250A
MD3250AF					MD3250AF
MD3250F					MD3250F
MD3251	82				MD3251
MD3251A	82				MD3251A
MD3251AF					MD3251AF
MD3251F					MD3251F
MD34	109	3087		1123	
MD3409	81				MD3409
MD3410	81				MD3410
MD3467					MD3467
MD3467F					MD3467F
MD34A	109	3087		1123	
MD35				1123	
MD3725					MD3725
MD3725F					MD3725F
MD3762					MD3762
MD3762F					MD3762F
MD4049					MD4049
MD420	160		245	2004	
MD4368BB	8368$	4368			
MD46	109	3087		1123	
MD4957					MD4957
MD5000					MD5000
MD5000A					MD5000A
MD5000B					MD5000B
MD501	102A		53	2007	
MD501B	102A		53	2007	
MD54				1123	
MD548					MD5000B
MD56				1123	
MD60	109	3087		1123	
MD6001					MD6001
MD6001F					MD6001F
MD6002					MD6002
MD6002F					MD6002F
MD6003					MD6003
MD6003F					MD6003F
MD604A	109	3090		1123	
MD60A	110MP		1N60	1123	
MD6100					MD6100
MD6100F					MD6100F
MD7000	81				MD7000
MD7001	82				MD7001
MD7001F					MD7001F
MD7002	81				MD7002
MD7002A					MD7002A
MD7002B					MD7002B
MD7003	82				MD7003
MD7003A					MD7003A
MD7003AF					MD7003AF
MD7003B					MD7003B
MD7003F					MD7003F
MD7004					MD7004
MD7005					MD7005
MD7005F					MD7005F
MD7006					MD7003
MD7006A					MD7003A
MD7006B					MD7003B
MD7007					MD7007
MD7007A					MD7007A
MD7007B					MD7007B
MD7007BF					MD7007BF
MD7007F					MD7007F
MD7008					MD8001
MD7011					MD6003
MD7021					MD7021
MD7021F					MD7021F
MD708					MD708
MD708A					MD708A
MD708AF					MD708AF
MD708B					MD708B
MD708BF					MD708BF
MD708F					MD708F
MD7091					MD8003
MD752	136A	3057	ZD-5.6	561	
MD752A	136A	3057	ZD-5.6	561	
MD753	137A	3058	ZD-6.2	561	
MD753A	137A	3058	ZD-6.2	561	
MD757	139A	3060	ZD-9.1	562	
MD757A	139A	3060	ZD-9.1	562	
MD759	142A	3062	ZD-12	563	
MD759A	142A	3062	ZD-12	563	
MD8001	81				MD8001
MD8002					MD8002
MD8003					MD8003
MD918					MD918
MD918A	81				MD918A
MD918AF					MD918AF
MD918B	81				MD918B
MD918BF					MD918BF
MD918F					MD918F
MD981					MD8003
MD981F					MD8003
MD982					MD982
MD982F					MD982F
MD983					MD8003
MD984	82				MD984
MD984F					MD3250F
MD985					MD985
MD985F					MD985F
MD986					MD986
MD986F					MD986F
MD990					MD2904
MDA100	166	9075		1151	
MDA101	166	9075		1152	
MDA102	167	3647		1172	
MDA104	168	3106/5304	504A	1104	
MDA106	169	3678			
MDA108	170	3677/5306			
MDA110	170	3649			
MDA200	166	3647/167			
MDA201	166	3647/167			
MDA202	167	3647			
MDA204	168	3648			
MDA206	169	3678			
MDA920-1	166	9075		1152	
MDA920-2		3105	504A	1104	
MDA920-4	167	3647	504A	1172	
MDA920-6	168	3648	504A	1173	
MDA920-7	169	3107/5307	504A		
MDA942-1				1151	
MDA942-2				1152	
MDA942-3				1172	
MDA942-4				1173	
MDA942-5				1173	
MDA942A-1	166			1151	
MDA942A-2	166			1152	
MDA942A-3	167			1172	
MDA942A-4				1173	
MDA942A-5	168			1173	
MDP173	177	3100/519	300	1122	
MDS1678	235	3197	215		
MDS20	171				MDS20
MDS21	171				MDS21
MDS26					MDS26
MDS27					MDS27
MDS31	160		245	2004	
MDS32	160		245	2004	
MDS33	160		245	2004	
MDS33A	160		245	2004	
MDS33C	160		245	2004	
MDS33D	160		245	2004	
MDS34	160		245	2004	
MDS35	126	3008	52	2024	
MDS36	160	3123	245	2004	
MDS37	160		245	2004	
MDS38	160		245	2004	
MDS39	160		245	2004	
MDS40	160		245	2004	
MDS60					MDS60
MDS6518			67	2023	
MDS76					MDS76
MDS77					MDS77
ME-1	123A	3444	20	1122	
ME-2	123A	3444	20	2051	
ME-3	123A	3444	20	2051	
ME-4	177	3100/519	300	1122	
ME-5	210	3041	252	2018	
ME0401			21	2034	
ME0402			21	2034	
ME0404	159	3025/129	82	2032	
ME0404-1	159	3114/290	82	2032	
ME0404-2	159	3114/290	82	2032	
ME0411		3114	221		
ME0412		3114	221		
ME0413		3114	221		
ME0414		3118	221		
ME0463			21	2034	
ME0475		3118	221		
ME1001	123A	3444	20	2051	
ME1002	123A	3444	20	2051	
ME1075	128	3024	243	2030	
ME1100		3244	220		
ME1110	154	3045/225	40	2012	
ME1120	154	3045/225	40	2012	
ME1138	229	3018	61	2038	
ME2001	123A	3444	20	2051	
ME2002	123A	3444	20	2051	
ME213	123A	3444	20	2051	
ME213A	123A	3444	20	2051	
ME216	123A	3444	20	2051	
ME217	123A	3444	20	2051	
ME3001		3452	86	2038	
ME3002	107	3452/108	86	2038	
ME3011	107	3452/108	86	2038	
ME4001	123A	3444	20	2051	
ME4002	123A	3444	20	2051	
ME4003	123A	3444	20	2051	
ME4003C	123A	3444	20	2051	
ME408-02C	5072A		ZD-8.2	562	
ME409-02B	139A		ZD-9.1	562	
ME4101	123A	3444	20	2051	
ME4102	123A	3444	20	2051	
ME4103	123A	3444	20	2051	
ME4104	123A	3444	20	2051	
ME5001	108	3452	86	2038	
ME501	159	3114/290	82	2032	
ME502			22	2032	
ME503			21	2034	
ME511			21	2034	
ME512			21	2034	
ME513			67	2023	
ME60	3005				MLED60
ME6001	123A	3444	20	2051	

Industry Standard No.	ECG	SK	GE	RS 276-	MOTOR.
ME6002	123A	3444	20	2051	
ME6003	123A	3444	20	2051	
ME61					MLED60
ME702					MLED900
ME8001	128	3024	243	2030	
ME8002	128	3024	243	2030	
ME8003	128	3024	243	2030	
ME8201	108	3452	86	2038	
ME900	123A	3444	20	2051	
ME9001	123AP	3018	61	2051	
ME9002	107	3444/123A	20	2009	
ME9003	107		61	2038	
ME900A	123A	3444	20	2051	
ME901	123A	3444	20	2051	
ME901A	123A	3444	20	2051	
ME9021	108	3452	86	2038	
ME9022	107	3452/108	86	2038	
MEF-25	123A	3444	20	2051	
MEM4001		4001	4001	2401	
MEM4007		4007	4007		
MEM4011		4011	4011	2411	
MEM4013			4013	2413	
MEM4016		4016	4016		
MEM4049		4049	4049	2449	
MEM4050		4050	4050	2450	
MEM4051		4051	4051	2451	
MEM511					3N156
MEM511C					3N156
MEM520					3N156
MEM520C					3N158
MEM554		3065			3N128
MEM554C					3N128
MEM556					3N158
MEM556C					3N158
MEM557					3N128
MEM557C					3N128
MEM560					3N155
MEM560C					3N155
MEM561C					MFE3003
MEM562	465				2N4351
MEM562C	465				2N4351
MEM563	465				MFE3002
MEM563C	465				2N4351
MEM564C	222		FET-4	2036	3N201
MEM571C					3N128
MEM614					3N201
MEM616	222				3N201
MEM617					3N202
MEM618					3N203
MEM620					MFE131
MEM621					MFE131
MEM622					MFE131
MEM630	222	3065	FET-4	2036	MFE131
MEM631					MFE131
MEM632					MFE131
MEM633					MFE131
MEM640					3N201
MEM641					3N201
MEM642					3N201
MEM643					MFE131
MEM644					MFE131
MEM645					MFE131
MEM655					MFE3004
MEM660					3N128
MEM667					2N3797
MEM668					MFE3001
MEM670					MFE3001
MEM680	222				3N211
MEM680Y	222		FET-4	2036	
MEM681					3N212
MEM682					3N213
MEM711					2N4351
MEM803					2N4352
MEM804					3N155
MEM806					2N4352
MEM806A					2N4352
MEM807					2N4352
MEM807A					2N4352
MEM814					3N155
MEM817					2N4352
MEM857					MFE5000
MER-65-L11324	519		514	1122	
MEZ12T10	142A	3062	ZD-12	563	
MEZ12T5	142A	3062	ZD-12	563	
MEZ15T10	145A	3063	ZD-15	564	
MEZ15T5	145A	3063	ZD-15	564	
MEZ27T10	146A	3064	ZD-27		
MEZ27T5	146A	3064	ZD-27		
MEZ5.6T10	136A	3057	ZD-5.6	561	
MEZ5.6T5	136A	3057	ZD-5.6	561	
MF-15/1B	502		CR-5		
MF-30/16	504		CR-6		
MF-55-62	121			2006	
MF1161	107	3039/316	11	2015	
MF1162	107	3039/316	11	2015	
MF1163	107	3039/316	11	2015	
MF1164	107	3039/316	11	2015	
MF12/16	502	3067	CR-4		
MF12/1B		3067	CR-4		
MF20/1B	503	3068	CR-5		
MF25/16	504	3108/505			
MF25/1B	505	3108	CR-6		
MF30/1B	504	3108/505	CR-6		
MF3304	159	3114/290	82	2032	
MFC6010	760		IC-12		
MFE-3008			FET-4	2036	
MFE120					MFE120
MFE121	222	3050/221	FET-4	2036	MFE121
MFE122					MFE122
MFE130	222	3065	FET-4	2036	MFE130
MFE130-712	222	3050/221	FET-4	2036	
MFE131	222	3050/221	FET-4	2036	MFE131
MFE132					MFE132
MFE140					MFE140
MFE2000		3116			MFE2000
MFE2001		3116			MFE2001
MFE2004	466				MFE2004
MFE2005	466				MFE2005
MFE2006	466				MFE2006
MFE2010					MFE2010
MFE2011					MFE2011
MFE2012					MFE2012
MFE2013					MFE2013
MFE2093/5					MFE2093/5
MFE2097					2N4092/3
MFE2098	466				2N4091
MFE2133					2N4392
MFE3001		3112			MFE3001
MFE3002					MFE3002
MFE3003					MFE3003
MFE3004	220			2036	MFE3004
MFE3005	221			2036	MFE3005
MFE3006	222	3065		2036	
MFE3007	222			2036	
MFE3008	222			2036	
MFE4007		3112			MFE4007
MFE4008		3112			MFE4008
MFE4009		3116			MFE4009
MFE4010		3116			MFE4010
MFE4011		3116			MFE4011
MFE4012		3116			MFE4012
MFE521					MFE521
MFP121	222	3050/221	FET-4	2036	
MG9623	123A	3444	20	2051	
MGLA100	140A	3061	ZD-10	562	
MGLA100A	140A	3061	ZD-10	562	
MGLA100B	140A	3061	ZD-10	562	
MGLA39	134A	3055	ZD-3.6		
MGLA39A	5067A	3055/134A	ZD-3.6		
MGLA39B	134A	3055	ZD-3.6		
MGLA47	135A	3056	ZD-5.1		
MGLA47A	5069A	3056/135A	ZD-5.1		
MGLA47B	135A	3056	ZD-5.1		
MGLA51	135A	3056	ZD-5.1		
MGLA51A	135A	3056	ZD-5.1		
MGLA51B	135A	3056	ZD-5.1		
MGLA56	136A	3057	ZD-5.6	561	
MGLA56A	136A	3057	ZD-5.6	561	
MGLA56B	136A	3057	ZD-5.6	561	
MGLA62	137A	3058	ZD-6.2	561	
MGLA62A	137A	3058	ZD-6.2	561	
MGLA62B	137A	3058	ZD-6.2	561	
MGLA82B	5072A	3136			
MGLA91	139A	3060	ZD-9.1	562	
MGLA91A	139A	3060	ZD-9.1	562	
MGLA91B	139A	3060	ZD-9.1	562	
MH0026CF					MMH0026C1
MH0026CG					MMH0026CG
MH0026CH					MMH0026CG
MH0026CN					MMH0026CP1
MH0026F					MMH0026CL
MH0026G					MMH0026CG
MH0026H					MMH0026CG
MH1002	522	3303			
MH1002A01	522	3303			
MH1030A01	538	3310	537		
MH1030A02	539	3309	538		
MH1201	529	3307	529,535		
MH1201A01	529	3307			
MH1201A08	549	3901/554			
MH1201A10	549	3901/554			
MH1201A11	549	3901/554			
MH1201A12	557	3307/529			
MH1201C01	529	3307			
MH1201F02	554	3901			
MH1203	523	3306	521,528		
MH1203A01	523	3306			
MH1203B01	523	3306			
MH1203B02	523	3306			
MH1203B03	523	3306			
MH1203B10	523	3306			
MH1203B13	523	3306			
MH1204	523	3306	521,528		
MH1204A01	523	3306			
MH1205	523	3306	521,528		
MH1205A01	523	3306			
MH1206	530	3308	540		
MH1206A01	530	3308			
MH1207	530	3308	540		
MH1207A01	530	3308			
MH1209	530	3308	540		
MH1209A01	530	3308			
MH1220	530	3308			
MH1220A01	530	3308	540		
MH1221	523	3306			
MH1221A01	523	3306	528		
MH1221A02		3306	521		
MH1222	529	3307			
MH1222A01	529	3307	529		
MH1222A02	529	3307	529		
MH1222A03	535	3307/529	535		
MH1222A04	549	3901/554			
MH1222E04	549	3901/554			
MH1501	192	3024/128	63	2030	
MH1502	193	3025/129	67	2023	
MH353	536A	3900	522		
MH353A01	536A	3900			
MH354	500A	3304			
MH383-B01	500A	3304			
MH401	525	3925			
MH500	116	3017B/117	504A	1104	
MH67	116	3311	504A	1104	
MH670	116	3017B/117	504A	1104	
MH68	116	3017B/117	504A	1104	
MH680	116	3016	504A	1104	
MH70	116	3031A	504A	1104	
MH700	116	3017B/117	504A	1104	
MH71	116	3051/156	504A	1104	
MH710	116	3051/156	504A	1104	
MH72	125	3051/156	510,531	1114	
MH720	125	3051/156	510,531	1114	
MH730	125	3051/156	510,531	1114	
MH913	522	3303	523		
MH913A01	522	3303			
MH914	500A	3304	527,539		
MH914A01	500A	3304			
MH915A09	532	3301/531			
MH915A12	531	3300/517			
MH915A13	531	3300/517			
MH915C01	532	3300/517	525		
MH915C02	531	3300/517	524		
MH919	522	3303	523		
MH919A01	522	3303			
MH919D01	522	3303	517		
MH920	500A	3304	527,539		
MH920A01	500A	3304			
MH920A06	500A	3304			
MH920A07	500A	3304			
MH920A10	521	3304/500A			
MH931	522	3303	523		
MH931A01	522	3303			
MH931A04	522	3303			
MH931A07	522	3303			
MH932	500A	3304	527,539		
MH932A01	500A	3304			
MH9410A	128	3024	243	2030	
MH943	534	3305	534		
MH943A01	534	3305			
MH9460A	129	3025	244	2027	
MH9623	123A	3444	20	2051	
MH9630	123A	3444	20	2051	
MH970A02	533	3302			
MH970C01	537		536		
MH970C02	533	3302	526		
MH983A01	522		523		
MH983A02	522	3303	523		
MH983A03	522	3303	523		
MH983A04	522	3303	523		

Industry Standard No.	ECG	SK	GE	RS 276-	MOTOR.
MH985A01	534	3305	534		
MH987	500A	3304			
MH987A01	500A	3304	527		
MH987A02	500A	3304	527		
MH987A03	500A	3304	527		
MH987A04	521	3304/500A	539		
MH988A03	536A	3900	522		
MHM1001	108	3452	86	2038	
MHM1101	108	3452	86	2038	
MHQ2221					MHQ2221
MHQ2222					MHQ2222
MHQ2369					MHQ2369
MHQ2483					MHQ2483
MHQ2484					MHQ2483
MHQ2906					MHQ2906
MHQ2907					MHQ2907
MHQ3467					MHQ3467
MHQ3546					MHQ3546
MHQ3798					MHQ3798
MHQ3799					MHQ3799
MHQ4001A					MHQ4001A
MHQ4002A					MHQ4002A
MHQ4013					MHQ4013
MHQ4014		4014			MHQ4014
MHQ6001					MHQ6001
MHQ6002					MHQ6002
MHQ6100					MHQ6100
MHQ918					MHQ918
MHT180	105	3012	4		
MHT18010	105	3012	4		
MHT1802	105		4		
MHT1803	105		4		
MHT1804	105		4		
MHT1807	105	3012	4		
MHT1808	105	3012	4		
MHT1809	105	3012	4		
MHT181	105	3012	4		
MHT2002	101	3861	8	2002	
MHT2003	101	3861	8	2002	
MHT2004	101	3861	8	2002	
MHT2008	101	3011	8	2002	
MHT2009	101	3861	8	2002	
MHT2010	101	3861	8	2002	
MHT230	105	3012	4		
MHT2305	105	3012	4		
MHT2414	128	3024	243	2030	
MHT2418	128	3024	243	2030	
MHT4401	128	3024	243	2030	
MHT4402			235	2012	
MHT4411	128	3024	243	2030	
MHT4412	128	3024	243	2030	
MHT4413	128	3024	243	2030	
MHT4451	128	3024	243	2030	
MHT4483	128	3024	243	2030	
MHT4511	128	3024	243	2030	
MHT4512	128	3024	243	2030	
MHT4513	128	3024	243	2030	
MHT5901			23	2020	
MHT5906	152	3893	66	2048	
MHT5911			23	2020	
MHT7401	128	3024	243	2030	
MHT7411	128	3024	243	2030	
MHT7412	128	3024	243	2030	
MHT7414	128	3024	243	2030	
MHT7417	128	3024	243	2030	
MHT7601	130	3027	14	2041	
MHT7602	130	3027	14	2041	
MHT7603	130	3027	14	2041	
MHT7607	130	3027	14	2041	
MHT7608	130	3027	14	2041	
MHT7609	130	3027	14	2041	
MHT9001	128	3024	243	2030	
MHT9002	128	3024	243	2030	
MHT9004	128	3024	243	2030	
MHT9005	128	3024	243	2030	
MHW401					MHW401
MHW580					MHW1342
MHW594					MHW1171
MHW595					MHW1172
MI-152	125(2)	9001/113A			
MI-15R	116		504A	1104	
MI-15S	116		504A	1104	
MI-301	177	3100/519	300	1122	
MI152R	116	9002			
MI1546	108		86	2038	
MI301	177	3100/519	300	1122	
MI9623	123A	3444	20	2051	
MI9630	123A	3444	20	2051	
MIC709-1					MC1709G
MIC709-5					MC1709CG
MIC709-5(D.I.P.)	909D		IC-250		
MIC710-1C					MC1710G
MIC710-5C	910				MC1710CG
MIC711-1C					MC1711G
MIC711-5(D.I.P.)	911D		IC-254		
MIC711-5C					MC1711CG
MIC712-1B					MC1712F
MIC712-1C					MC1712G
MIC712-1D					MC1712L
MIC712-5B					MC1712CF
MIC712-5C					MC1712CG
MIC712-5D					MC1712CL
MIC723-1					MC1723G
MIC723-5		3164			MC1723CG
MIC723-5(D.I.P.)			IC-260	1740	
MIC7400J	7400	7400		1801	
MIC7400N	7400	7400		1801	
MIC7401J	7401	7401			
MIC7401N	7401	7401			
MIC7402J	7402	7402		1811	
MIC7402N	7402	7402		1811	
MIC7404J	7404	7404		1802	
MIC7404N	7404	7404		1802	
MIC7405J	7405	7405			
MIC741-1C					MC1741G
MIC741-1D					MC1741L
MIC741-5(D.I.P.)	941D		IC-264		
MIC741-5C					MC1741CG
MIC741-5D					MC1741CL
MIC74107J	74107	74107			
MIC74107N	74107	74107			
MIC7410J	7410	7410		1807	
MIC7410N	7410	7410		1807	
MIC74121J	74121	74121			
MIC74121N	74121	74121			
MIC7413J	7413	7413			
MIC7413N	7413	7413		1815	
MIC74145J	74145	74145		1828	
MIC74145N	74145	74145		1828	
MIC74150J	74150	74150		1829	
MIC74150N	74150	74150		1829	
MIC74151J	74151	74151			
MIC74151N	74151	74151			
MIC74154J	74154	74154		1834	

Industry Standard No.	ECG	SK	GE	RS 276-	MOTOR.
MIC74154N	74154	74154		1834	
MIC74155J	74155	74155			
MIC74156J	74156	74156			
MIC74156N	74156	74156			
MIC74180J	74180	74180			
MIC74180N	74180	74180			
MIC7420J	7420	7420		1809	
MIC7420N	7420	7420		1809	
MIC7428	7428	7428			
MIC7428J	7428	7428			
MIC7430J	7430	7430			
MIC7430N	7430	7430			
MIC7433A	7433	7433			
MIC7440J	7440	7440			
MIC7440N	7440	7440			
MIC7441AJ	7441	7441		1804	
MIC7441AN	7441	7441		1804	
MIC7442J	7442	7442			
MIC7442N	7442	7442			
MIC7443J	7443	7443			
MIC7443N	7443	7443			
MIC7444J	7444	7444			
MIC7444N	7444	7444			
MIC7445J	7445	7445			
MIC7445N	7445	7445			
MIC7451J	7451	7451		1825	
MIC7451N	7451	7451		1825	
MIC7454J	7454	7454			
MIC7454N	7454	7454			
MIC7473J	7473	7473		1803	
MIC7473N	7473	7473		1803	
MIC7474J	7474	7474		1818	
MIC7474N	7474	7474		1818	
MIC7475J	7475	7475		1806	
MIC7475N	7475	7475		1806	
MIC7476J	7476	7476		1813	
MIC7476N	7476	7476		1813	
MIC7481J	7481	7481			
MIC7481N	7481	7481			
MIC7486J	7486	7486		1827	
MIC7486N	7486	7486		1827	
MIC7490J	7490	7490		1808	
MIC7490N	7490	7490		1808	
MIC7492J	7492	7492		1819	
MIC7492N	7492	7492		1819	
MIC7493J	7493A	7493			
MIC7493N	7493A	7493			
MIC7494J	7494	7494			
MIC7494N	7494	7494			
MIC7495J	7495	7495			
MIC7495N	7495	7495			
MIC77413J				1815	
MIS-14150-18A	101		8	2002	
MIS13674/47	234		65	2050	
MIS14150-18	101	3861	8	2002	
MIS14150/37	154		40	2012	
MJ1000	243	3182			MJ1000
MJ10000	98				MJ10000
MJ10001	98				MJ10001
MJ10002	97				MJ10002
MJ10003	97				MJ10003
MJ10004	98				MJ10004
MJ10005	98				MJ10005
MJ10006	97				MJ10006
MJ10007	97				MJ10007
MJ10008	98				MJ10008
MJ10009	98				MJ10009
MJ1001	243	3182			MJ1001
MJ10011					MJ10011
MJ10012	98				MJ10012
MJ10013					MJ10013
MJ10014					MJ10014
MJ10015	99				MJ10015
MJ10016					MJ10016
MJ10020					MJ10020
MJ10021					MJ10021
MJ10022	99				MJ10022
MJ10023	99				MJ10023
MJ10024					MJ10024
MJ10025					MJ10025
MJ105	165		38		BU205
MJ11011					MJ11011
MJ11012					MJ11012
MJ11013					MJ11013
MJ11014					MJ11014
MJ11015					MJ11015
MJ11016					MJ11016
MJ11028					MJ11028
MJ11029					MJ11029
MJ11030					MJ11030
MJ11031					MJ11031
MJ11032					MJ11032
MJ11033					MJ11033
MJ1200					(2) 2N6300
MJ12002		3710			MJ12002
MJ12003	389	3710/238			MJ12003
MJ12004		3710			MJ12004
MJ12005		3710			MJ12005
MJ1201					(2) 2N6301
MJ12010					MJ12010
MJ13010					2N6547
MJ13014	385	3946			MJ13014
MJ13015					MJ13015
MJ13330					MJ13330
MJ13331					MJ13331
MJ13332		9039			MJ13332
MJ13333		9039			MJ13333
MJ13334		9039			MJ13334
MJ13335	386	9039			MJ13335
MJ14000					MJ14000
MJ14001					MJ14001
MJ14002					MJ14002
MJ14003					MJ14003
MJ15001		9033			MJ15001
MJ15002		9034			MJ15002
MJ15003		9033			MJ15003
MJ15004		9034			MJ15004
MJ15011					MJ15011
MJ15012					MJ15012
MJ15015					MJ15015
MJ15016					MJ15016
MJ15022		3947			MJ15022
MJ15023					MJ15023
MJ15024	388	3947			MJ15024
MJ15025					MJ15025
MJ16010					MJ16010
MJ16012					MJ16012
MJ1800	94	3438	35		MJ1800
MJ205					BU205
MJ2249	384	3131A/369	246	2020	2N3766
MJ2250	384	3131A/369	246	2020	2N3767
MJ2251	124	3021	12		2N3738
MJ2252	124	3261/175	12		MJ2252
MJ2253	218		234		2N3740

Industry Standard No.	ECG	SK	GE	RS 276-	MOTOR.
MJ2254	218		234		2N3741
MJ2267	285		74	2043	2N6594
MJ2268	285		74	2043	MJ2955
MJ2300					MJ2300
MJ2305					MJ2305
MJ2500	246	3183A			MJ2500
MJ2501	246	3859/252			MJ2501
MJ2800			19	2041	
MJ2801	130	3027	14	2041	2N6569
MJ2802	130	3027	14	2041	2N5881
MJ2840	284	3027/130	14	2041	2N5877
MJ2841	284	3027/130	14	2041	2N5878
MJ2901	219		74	2043	2N6594
MJ2940			74	2043	2N5875
MJ2941			74	2043	2N5876
MJ2955	219	3173	74	2043	MJ2955
MJ2955A					MJ2955A
MJ3000	245	3182/243			MJ3000
MJ3001	245	3182/243			MJ3001
MJ3010	162	3438	35		2N6542
MJ3011	163A	3439	36		2N6542
MJ3012					2N6542
MJ3026	162	3438	35		MJ3029
MJ3027	162	3438	35		MJ3029
MJ3028	94		36		MJ3029
MJ3029	97	3438	35		MJ3029
MJ3030	283	3439/163A	36		MJ3030
MJ3040	98				MJ3040
MJ3041					MJ3041
MJ3042	98				MJ3042
MJ3055					2N3055
MJ3101	384	3131A/369	246	2020	2N3766
MJ3201	124	3021	12		2N3738
MJ3202	124	3261/175	12		2N3739
MJ3237					MJ3237
MJ3238					MJ3238
MJ3247					MJ3247
MJ3248					MJ3248
MJ3260	283				2N5838
MJ34000	243	3182			
MJ3430	385	3559/162	73		2N6307
MJ3480	165				BU208
MJ3520	245	3858/251			MJ3000
MJ3521	245	3858/251			MJ3001
MJ3583					2N6420
MJ3584					2N6421
MJ3585					2N6422
MJ3701	218	3257	234		2N4898
MJ3702					2N4898
MJ3703					2N4899
MJ3704					2N4900
MJ3738					2N6424
MJ3739					2N6425
MJ3760					MJ3030
MJ3761					MJ3030
MJ3771	181		75	2041	2N3771
MJ3772	181		75	2041	MC5193F
MJ3773		3836			2N3773
MJ3801					MJ3001
MJ3802					MJ3001
MJ400	124	3261/175	12		2N3739
MJ4000	243	3182			2N6055
MJ4001	243	3182			
MJ4010	244	3183A			2N6053
MJ4011	244	3183A			2N6054
MJ4030	250	3859/252			MJ4030
MJ4031	250	3859/252			MJ4031
MJ4032	250	3859/252			MJ4032
MJ4033	249	3858/251			MJ4033
MJ4034	249	3858/251			MJ4034
MJ4035	249	3858/251			MJ4035
MJ410	94	3269			MJ410
MJ4101	175	3131A/369	246	2020	
MJ4102	175	3131A/369	246	2020	
MJ4104					2N4231A
MJ411	94	3560			MJ411
MJ413	283	3467	36		MJ413
MJ420	154	3045/225	40	2012	
MJ4200					(2) 2N6294
MJ4201					(2) 2N6295
MJ421	396	3045/225	40	2012	
MJ4210					(2) 2N6296
MJ4211					(2) 2N6297
MJ423	385	3467/283	36		MJ423
MJ4237					MJ4237
MJ4238					MJ4238
MJ424	283				2N6308
MJ4240					2N6423
MJ4247					MJ4247
MJ4248					MJ4248
MJ425	283				2N6545
MJ431	283	3467	36		MJ431
MJ4360					MJ4360
MJ4361					MJ4361
MJ4380					MJ4380
MJ4381					MJ4381
MJ4400					MJ4400
MJ4401					MJ4401
MJ450	180		74	2043	2N4398
MJ4502	180	3437	74	2043	MJ4502
MJ4645	129	3053			MJ4645
MJ4646		3053			MJ4646
MJ4647					MJ4647
MJ4648		3053			MJ4647
MJ480	130	3027	14	2041	2N3713
MJ481	130	3027	14	2041	2N3713
MJ490	281		74	2043	2N3789
MJ491	281		74	2043	2N3789
MJ5038					2N5038
MJ5039					2N5039
MJ5202	175	3131A/369	246	2020	
MJ5203	175	3131A/369	246	2020	
MJ5204	175	3131A/369	246	2020	
MJ5257			75	2041	
MJ6257			75	2041	2N6257
MJ6302	280		265	2047	2N5630
MJ6502					MJ6502
MJ6503					MJ6503
MJ6700					MJ6700
MJ6701					2N6186
MJ7000					2N6338
MJ701					MJ12002
MJ702					MJ12002
MJ703N	703A		IC-12		
MJ704					MJ12002
MJ7160			35		MJ13014
MJ7161			35		MJ13015
MJ721					MJ12002
MJ723					MJ12002
MJ7260					2N6546
MJ7261					2N6547
MJ802	181	3535	75	2041	MJ802
MJ8020					MJ12004
MJ804					MJ804

Industry Standard No.	ECG	SK	GE	RS 276-	MOTOR.
MJ8100			29	2025	MJ8100
MJ8101					2N6190
MJ8400	165	3115	38		MJ12004
MJ8500					MJ8500
MJ8501					MJ8501
MJ8502					MJ8502
MJ8503					MJ8503
MJ8504					MJ8504
MJ8505					MJ8505
MJ900	244	3859/252			MJ900
MJ9000	283	3467	35		MJ9000
MJ901	244	3859/252			MJ901
MJ920					(2) 2N6298
MJ921					(2) 2N6299
MJE-200	184	3104A	57		
MJE-200E		3197	28	2052	
MJE-2020			56	2027	
MJE-220			57	2017	
MJE-32B			250	2027	
MJE-371			29	2025	
MJE-42			250	2027	
MJE-521	184		58	2027	
MJE1000					MJE1000
MJE1001					MJE1001
MJE101			69	2049	2N5974
MJE102			69	2049	2N5975
MJE103			69	2049	2N5976
MJE104	183	3189A	56	2027	2N5976
MJE105	183	3189A	56	2027	MJE105
MJE105K	242		58	2027	TIP42A
MJE1090	258	3181A/264			MJE1090
MJE1091	258				MJE1091
MJE1092	258				MJE1092
MJE1093	258				MJE1093
MJE1100	257	3180/263			MJE1100
MJE1101	257	3180/263			MJE1101
MJE1102	257	3180/263			MJE1102
MJE1103	257	3180/263			MJE1103
MJE12007					MJE12007
MJE1290		3274	56	2027	MJE1290
MJE1291			56	2027	MJE1291
MJE13002					MJE13002
MJE13003					MJE13003
MJE13004					MJE13004
MJE13005					MJE13005
MJE13006	379				MJE13006
MJE13007	379				MJE13007
MJE13008					MJE13008
MJE13009	379				MJE13009
MJE15028					MJE15028
MJE15029					MJE15029
MJE15030					MJE15030
MJE15031					MJE15031
MJE1660			55		MJE1660
MJE1661			55		MJE1661
MJE170		3193	250	2027	MJE170
MJE171		3193	248		MJE171
MJE172					MJE172
MJE180	184	3192/186	57	2017	MJE180
MJE181	182	3192/186	57		MJE181
MJE182	182	3188A	55		MJE182
MJE200	184	3197/235	57	2017	MJE200
MJE200E	186A		57	2017	
MJE201			66	2048	2N5977
MJE2010			250	2027	TIP42
MJE2011			55		TIP42A
MJE202			66	2048	2N5978
MJE2020			56	2027	TIP41
MJE2021			241	2020	TIP41A
MJE203			55		2N5978
MJE204	182		55		2N5979
MJE205	182	3188A	55		MJE205
MJE2050					MJE200
MJE2055					MJE3055
MJE205K	241	3188A/182	57	2020	TIP41A
MJE2090	262	3181A/264			TIP125
MJE2091	262				TIP125
MJE2092	262				TIP126
MJE2093	262				TIP126
MJE210			250	2027	MJE210
MJE2100	261	3180/263			TIP120
MJE2101	261	3180/263			TIP120
MJE2102		3180			TIP121
MJE2103	261	3180/263			TIP121
MJE2150					MJE210
MJE2160					TIP48
MJE220			57	2017	MJE220
MJE221			57	2017	MJE221
MJE222			57	2017	MJE222
MJE223			57	2017	MJE223
MJE224			57	2017	MJE224
MJE225			57	2017	MJE225
MJE230			58	2025	MJE230
MJE231			58	2025	MJE231
MJE232			58	2025	MJE232
MJE233			58	2025	MJE233
MJE234			58	2025	MJE234
MJE235			58	2025	MJE235
MJE2360	198	3220	251		MJE2360T
MJE2360T	198				MJE2360T
MJE2361	198		251		MJE2361T
MJE2361T	198				MJE2361T
MJE2370	242	3189A/183	58	2027	TIP32
MJE2371	242	3189A/183	58	2027	TIP32A
MJE2380			241	2020	
MJE2381			241	2020	
MJE2382			241	2020	
MJE2383			241	2020	
MJE240			D44C12		MJE240
MJE241			D44C11		MJE241
MJE242			D44C10		MJE242
MJE243					MJE243
MJE244					MJE244
MJE2480	241	3188A/182	57	2020	TIP31
MJE2481	241	3188A/182	57	2020	TIP31A
MJE2482	241	3188A/182	57	2020	2N6121
MJE2483	241	3188A/182	57	2020	2N6122
MJE2490	242	3274/153	58	2027	TIP32
MJE2491	242	3189A/183	58	2027	TIP32A
MJE250			D45C12		MJE250
MJE251			D45C11		MJE251
MJE252			D45C10		MJE252
MJE2520	241	3188A/182	57	2020	TIP31
MJE2521	241	3188A/182	57	2020	
MJE2522	241	3188A/182	57	2020	
MJE2523	241	3188A/182	57	2020	
MJE253					MJE253
MJE254					MJE254
MJE270					MJE270
MJE271					MJE271
MJE2801	182	3188A	55		MJE2801
MJE2801K					MJE2801T
MJE2801T					MJE2801T
MJE29			241	2020	TIP29

Industry Standard No.	ECG	SK	GE	RS 276-	MOTOR.
MJE2901	183	3189A	56	2027	MJE2901
MJE2901K					MJE2901T
MJE2901T					MJE2901T
MJE2940	150	3189A/183	14	2041	
MJE2955	183		56	2027	MJE2955
MJE2955K					MJE2955T
MJE2955T					MJE2955T
MJE29A			241	2020	TIP29A
MJE29B			241	2020	TIP29B
MJE29C					TIP29C
MJE30			250	2027	TIP30
MJE3054	241	3188A/182	57	2020	
MJE3055	182	3188A	55		MJE3055
MJE3055K					MJE3055T
MJE3055T					MJE3055T
MJE30A			250	2027	TIP30A
MJE30B			250	2027	TIP30B
MJE30C					TIP30C
MJE31			241	2020	MJE31
MJE31A			241	2020	MJE31A
MJE31B			241	2020	MJE31B
MJE31C					MJE31C
MJE32			250	2027	MJE32
MJE32A			250	2027	MJE32A
MJE32B					MJE32B
MJE32C					MJE32C
MJE33					TIP41
MJE3300					MJE3300
MJE3301					MJE3301
MJE3302					MJE3302
MJE3310					MJE3310
MJE3311					MJE3311
MJE3312					MJE3312
MJE3370		3191	58	2025	MJE370
MJE3371		3191	58	2025	2N5193
MJE33A					TIP41A
MJE33B					TIP41B
MJE33C					TIP41C
MJE34					TIP42
MJE340	157	3747	232		MJE340
MJE340K			82		TIP48
MJE341	157	3747	232		MJE341
MJE341K			232		TIP47
MJE3439			80		MJE3439
MJE344	157	3747	232		MJE344
MJE3440	157	3747	232		MJE3440
MJE344K			232		TIP47
MJE345			27		MJE3439
MJE34A					TIP42A
MJE34B					TIP42B
MJE34C					TIP42C
MJE350		3747			MJE350
MJE3520	249	3190/184	57	2017	MJE520
MJE3521		3190	57	2017	2N5190
MJE370	185	3191	58	2025	MJE370
MJE370K	242	3189A/183	58	2027	TIP32
MJE371	185	3189A/183	58	2020	MJE371
MJE371K	242	3189A/183	58	2027	TIP32
MJE3738					TIP47
MJE3739					TIP48
MJE3740	242	3189A/183	58	2027	
MJE3741	242	3189A/183	58	2027	
MJE41			241	2020	TIP41
MJE41A			241	2020	TIP41A
MJE41B			241	2020	TIP41B
MJE41C					TIP41C
MJE42					TIP42
MJE423	162	3438	35		
MJE42A			250	2027	TIP42A
MJE42B			250	2027	TIP42B
MJE42C					TIP42C
MJE4340					MJE4340
MJE4341					MJE4341
MJE4342					MJE4342
MJE4350					MJE4350
MJE4351					MJE4351
MJE4352					MJE4352
MJE47					TIP47
MJE48					TIP48
MJE482	184	3190	57	2017	2N5190
MJE483	184	3190	57	2017	2N5191
MJE484					2N5192
MJE488	184	3190	57	2017	
MJE49					TIP49
MJE4918					TIP30
MJE4919					TIP30A
MJE492	185	3191	58	2025	2N5193
MJE4920					TIP30B
MJE4921					TIP29
MJE4922					TIP29A
MJE4923					TIP29B
MJE493	185	3191	58	2025	2N5194
MJE494					2N5195
MJE51					MJE51T
MJE5170					MJE5170
MJE5171					MJE5171
MJE5172					MJE5172
MJE5180					MJE5180
MJE5181					MJE5181
MJE5182					MJE5182
MJE5190					2N6121
MJE5191					2N6122
MJE5192					2N6123
MJE5193					2N6124
MJE5194					2N6125
MJE5195					2N6126
MJE51T					MJE51T
MJE52					MJE52T
MJE520	184	3190	57	2017	MJE520
MJE520K	241	3188A/182	57	2020	TIP31
MJE521	184	3190	57	2017	MJE521
MJE521K	241	3188A/182	57	2020	TIP31
MJE52T					MJE52T
MJE53					MJE53T
MJE53T					MJE53T
MJE5655					TIP47
MJE5656					TIP48
MJE5657					TIP49
MJE5730					MJE5730
MJE5731					MJE5731
MJE5732					MJE5732
MJE5740					MJE5740
MJE5741					MJE5741
MJE5742					MJE5742
MJE5780					MJE5780
MJE5781					MJE5781
MJE5782					MJE5782
MJE5850					MJE5850
MJE5851					MJE5851
MJE5852					MJE5852
MJE5960					2N6489
MJE5974					TIP42
MJE5975					TIP42A
MJE5976					TIP42B

Industry Standard No.	ECG	SK	GE	RS 276-	MOTOR.
MJE5977					TIP41
MJE5978					TIP41A
MJE5979					TIP41B
MJE5980					2N6489
MJE5981					2N6490
MJE5982					2N6491
MJE5983					2N6486
MJE5984					2N6487
MJE5985					2N6488
MJE6040	260	3979			MJE6040
MJE6041	260	3979			MJE6041
MJE6042	260	3979			MJE6042
MJE6043	259	3978			MJE6043
MJE6044	259	3978			MJE6044
MJE6045	259	3978			MJE6045
MJE700	254	3997			MJE700
MJE700T					MJE700T
MJE701	254	3997			MJE701
MJE701T					MJE701T
MJE702	254	3997			MJE702
MJE702T					MJE702T
MJE703	254	3997			MJE703
MJE703T					MJE703T
MJE710	185		250	2027	MJE710
MJE711	185	3191	58	2025	MJE711
MJE712			56	2027	MJE712
MJE720		3190	57	2017	MJE720
MJE721		3190	57	2017	MJE721
MJE722			55		MJE722
MJE800	253	3996			MJE800
MJE800T					MJE800T
MJE801	253	3996			MJE801
MJE801T					MJE801T
MJE802	253	3996			MJE802
MJE802T					MJE802T
MJE803	253	3996			MJE803
MJE803T					MJE803T
MJE8500					MJE8500
MJE8501					MJE8501
MJE8502					MJE8502
MJE8503					MJE8503
MJE9400	184	3188A/182	57	2017	
MJE9411T	108	3020/123	20	2051	
MJE9450	185		58	2025	
MJE9742	157	3747	232		
MJES191	241	3188A/182	57	2020	
MJF10335	312	3116	FET-1	2035	
MJF1033G	312	3116	FET-2	2035	
MJG5194	242		58	2027	
MJR1C	116		504A	1104	
MK-10	312		FET-1	2028	
MK-10-2			FET-1	2028	
MK-10-E	312	3116	FET-1	2035	
MK10	312	3448	FET-1	2035	
MK10-2	312	3448	FET-1	2035	
MK102		3116	FET-2	2035	
MK2147	2147				MCM2147
MK2716					MCM2716
MK34000					MCM68A316E
MK36000-4					MCM68B364
MK3800	1136		IC-284		
MK4027					MCM4027A
MK4104					MCM65147
MK4116					MCM4116
MK4164					MCM6664
MK4332					MCM4132
MK4516					MCM4516
MK4802					MCM4016
MK5485	312	3112	FET-1	2035	
ML101AF					LM101AH
ML101AM					LM101AH
ML101AT					LM101AH
ML101F					LM101AH
ML101M					LM101AH
ML101T					LM101AH
ML107F					LM107H
ML107M					LM107H
ML107T					LM107H
ML108AF					MC1556G
ML108AM					LM108AJ
ML108AT					LM108AH
ML108M					LM108J
ML108T					LM108H
ML111M					LM111J
ML111S					LM111J
ML111T					LM111H
ML118F					MC1741SG
ML118M					MC1741SG
ML118T					MC1741SG
ML1436T					MC1436G
ML1437P					MC1437P
ML1458P					MC1458P2
ML1458S	778A				MC1458P1
ML1458T					MC1458G
ML1488M					MC1488L
ML1489AM					MC1489AL
ML1489M					MC1489L
ML1536T					MC1536G
ML1537M					MC1537L
ML1558M					MC1558L
ML1558T					MC1558G
ML201AF					LM201AH
ML201AM					LM201AH
ML201AT	1171	3565			LM201AH
ML201F					LM201AH
ML201M					LM201AH
ML201T	1171				LM201AH
ML207F					LM207H
ML207M					LM207H
ML207T					LM207H
ML208AF					MC1556G
ML208AM					LM208AJ
ML208AT					LM208AH
ML208M					LM208J
ML208T					LM208H
ML211M					LM211J
ML211S					LM211N
ML211T					LM211H
ML218F					MC1741SG
ML218M					MC1741SG
ML218T					MC1741SG
ML2812	177		300	1122	
ML301AP					LM301AN
ML301AS	975				LM301AN
ML301AT	1171				LM301AH
ML301P					LM301AN
ML301S					LM301AN
ML301T	1171	3565			LM301AN
ML3046P					MC3346P
ML307P					LM307H
ML307S	976	3596			LM307N
ML307T					LM307H
ML308AM					LM308AJ
ML308AT					LM308AH

Industry Standard No.	ECG	SK	GE	RS 276-	MOTOR.
ML308M					LM308J
ML308T					LM308H
ML311M					LM311J
ML311P					LM311J
ML311S					LM311N
ML311T					LM311H
ML318M					MC1741SCP1
ML318T					MC1741SCG
ML4250CS					MC1776CG
ML4250CT					MC1776CG
ML4250T					MC1776G
ML4251CS					MC1776CG
ML4251CT					MC1776CG
ML4251T					MC1776G
ML6503M					MC1537L
ML709AF					MC1709AF
ML709AM					MC1709AL
ML709AT					MC1709AG
ML709CP					MC1709CP2
ML709CT	909				MC1709CG
ML709F					MC1709F
ML709M					MC1709L
ML709T					MC1709G
ML723CF					MC1723CL
ML723CM					MC1723CL
ML723CP					MC1723CL
ML723CT					MC1723CG
ML723F					MC1723L
ML723M					MC1723L
ML723T					MC1723G
ML741AF					MC1556G
ML741AM					MC1556G
ML741AT					MC1556G
ML741CP					MC1741CP2
ML741CS	941M				MC1741CP1
ML741CT	941				MC1741CG
ML741F					MC1741F
ML741M					MC1741L
ML741T					MC1741G
ML747CP	947D	3556			MC1747CL
ML747CT	947				MC1747CG
ML747F					MC1747F
ML747M					MC1747L
ML747T					MC1747G
ML748CP					LM301AN
ML748CS	975	3641			LM301AN
ML748CT					MC1748CG
ML748F					MC1748G
ML748M					MC1748G
ML748T					MC1748G
ML7503M					MC1437L
MLM201AG	1171	3641/975			
MLM224L	987	3643			
MLM224P	987	3643			
MLM239AL	834	3569			
MLM301AG	1171	3641/975			
MLM301AU	975	3641			
MLM307P1	976	3596			
MLM307U	976	3596			
MLM309K	309K	3629			
MLM324L	987	3643			
MLM324P	987	3643			
MLM324P1	987	3643			
MLM339AL	834	3569			
MLM358G	928	3691			
MLM358P1	928M	3692			
MLM358U	928M	3692			
MLNA339A	5067A		ZD-3.9		
MLV4370A		3980			1N4370A
MLV4371A	5063A	3837			
MLV4372A	5065A	3838			1N4372A
MLV746A	5066A		ZD-3.3		1N746A
MLV756A	5072A	3136			
MLV759A					1N759A
MMO	116	3311	504A	1104	
MM10	125	3033	510,531	1114	
MM1139	160		245	2004	
MM1151	102A		53	2007	
MM1152	102A		53	2007	
MM1153	102A		53	2007	
MM1154	102A		53	2007	
MM1199	160		245	2004	
MM1367/2SC684	108	3452	86	2038	
MM1382	108	3452	86	2038	
MM1387	108		86	2038	
MM1500					MRF905
MM1500A					MRF905
MM1501					MRF905
MM1501A					MRF905
MM1510					2N5851
MM1511					2N5852
MM1549					2N5635
MM1550					2N5636
MM1551					2N5637
MM1557	357				2M5641
MM1558	359				2N5642
MM1559	360				2N5643
MM1561					2N6166
MM1601	347				2N5589
MM1602	349				2N5590
MM1603	351				2N5591
MM1605					2N5841
MM1606					2N5842
MM1607					2N5842
MM1608					2N5846
MM1612					2N6255
MM1618	337				2N5847
MM1619			66	2048	
MM1620	337				2N5849
MM1622	339				2N5849
MM1632					2N5941
MM1633					2N5942
MM1646	339				2N5849
MM1660					2N5644
MM1661					2N5645
MM1662					2N5646
MM1665					2N6136
MM1666	351				2N6082
MM1667					2N6083
MM1668	320				2N6084
MM1669					2N6084
MM1680	348				2N6080
MM1681	350	3176			2N6081
MM1713					2N4072
MM1742	102A		53	2007	
MM1755	123A	3444	20	2051	
MM1756	123A	3444	20	2051	
MM1757	123A	3444	20	2051	
MM1758	123A	3444	20	2051	
MM1803	311	3452/108	86	2038	
MM1809	128	3047	243	2030	
MM1809A	128	3047	243	2030	
MM1810	195A	3048/329	46		
MM1810A	128	3024	243	2030	
MM1941	311	3452/108	86	2038	
MM1943	128	3024	243	2030	2N4072
MM1945	108	3452	86	2038	2N4072
MM2	116	3016	504A	1104	
MM2090					3N124
MM2091					3N125
MM2092					3N126
MM2102					2N4351
MM2114	2114				MCM2114
MM2147	2147				MCM2147
MM2148					MCM2148
MM2193A			63	2030	
MM2260	154		40	2012	
MM2261			63	2030	
MM2266	128	3024	243	2030	
MM2270			63	2030	
MM2503	160		245	2004	
MM2532					MCM2532
MM2550	160	3123	245	2004	
MM2552	160	3123	245	2004	
MM2554	160	3123	245	2004	
MM2708					MCM2708
MM2716					MCM2716
MM2894	160		245	2004	
MM3	116	3017B/117	504A	1104	
MM3000	154	3045/225	40	2012	
MM3002	154	3045/225	40	2012	
MM3004			28	2017	
MM3005	282	3045/225	40	2012	
MM3006	282	3748			
MM3007	282	3748			
MM3009	154	3045/225	40	2012	
MM3014	190		217		
MM306			243	2030	
MM3100	154	3045/225	40	2012	
MM3101	154	3045/225	40	2012	
MM3726	159	3114/290	82	2032	
MM380	160		245	2004	
MM3903	123A		210	2051	
MM3904	123A		20	2051	
MM3905	159	3114/290	82	2032	
MM3906	159	3114/290	82	2032	
MM4	116	3017B/117	504A	1104	
MM4000	397	4000	223		
MM4001	397	4001	4001	2401	
MM4002	397	4002	4002		
MM4005	129	3025	244	2027	
MM4006	129	3025	244	2027	
MM4007		4007	4007		
MM4008	129		67	2023	
MM4009	129	4009	4009		
MM4011		4011	4011	2411	
MM4012		4012	4012	2412	
MM4013		4013	4013	2413	
MM4015		4015	4015		
MM4016A		4016	4016		
MM4017		4017	4017	2417	
MM4019		3025	244	2027	
MM4020	353		69	2049	2N6094
MM4020(IC)			4020	2420	
MM4021	354		4021	2421	2N6095
MM4022	355				2N6096
MM4023	356	4023	4023	2423	2N6097
MM4024		4024	4024		
MM4025		4025	4025		
MM4027		4027	4027	2427	
MM4030		4030	4030		
MM4042		4042	4042		
MM4048	159	3114/290	82	2032	
MM4049	395	4049	4049	2449	
MM4050		4050	4050	2450	
MM4051		4051	4051	2451	
MM4052			82	2032	
MM4164					MCM6665
MM439	395				2N4959
MM4429			28	2017	
MM4430			28	2017	
MM4500					2N5583
MM4518			4518	2490	
MM486	128	3024	243	2030	
MM487	128	3024	243	2030	
MM488	128	3024	243	2030	
MM5	116	3017B/117	504A	1104	
MM5000	160		245	2004	
MM5001	160		245	2004	
MM5002	160		245	2004	
MM511	128	3024	243	2030	
MM512	128	3024	243	2030	
MM513	128	3024	243	2030	
MM5177					MRF5177
MM52132					MCM68A332
MM52164					MCM68A364
MM5257					MCM6641
MM5290					MCM4116
MM5295					MCM4517
MM54C00					MC14011B
MM54C02					MC14001B
MM54C04					MC14069UB
MM54C08					MC14081B
MM54C10					MC14023B
MM54C107					MC14027B
MM54C14					MC14584B
MM54C151					MC14512B
MM54C154					MC14514B
MM54C157					MC14519B
MM54C160					MC14160B
MM54C161					MC14161B
MM54C162					MC14162B
MM54C163					MC14163B
MM54C164					MC14015B
MM54C165					MC14021B
MM54C173					MC14076B
MM54C174					MC14174B
MM54C175					MC14175B
MM54C192					MC14510B
MM54C193					MC14516B
MM54C195					MC14035B
MM54C20					MC14012B
MM54C200					MCM14537
MM54C221					MC14538B
MM54C30					MC14068B
MM54C32					MC14071B
MM54C42					MC14028B
MM54C48					MC14558B
MM54C73					MC14027B
MM54C74					MC14013B
MM54C76					MC14027B
MM54C83					MC14008B
MM54C85					MC14585B
MM54C86					MC14070B
MM54C89					MCM14505
MM54C90					MC14518B
MM54C901					MC14049UB
MM54C902					MC14050B

Industry Standard No.	ECG	SK	GE	RS 276-	MOTOR.
MM54C905					MC14559B
MM54C93					MC14520B
MM54C95					MC14035B
MM55104	1255				MC145104
MM55106					MC145106
MM55107					MC145107
MM55114					MC145104
MM55116					MC145106
MM55126					MC145106
MM6	116	3017B/117	504A	1104	
MM6508					MCM6508
MM6518					MCM6518
MM7	125	3032A	510,531	1114	
MM7087	154	3045/225	40	2012	
MM7088	154	3045/225	40	2012	
MM709	108	3039/316	86	2038	
MM74C00	74C00				MC14011B
MM74C00N	74C00			2301	
MM74C02	74C02				MC14001B
MM74C02N				2302	
MM74C04	74C04				MC14069UB
MM74C08	74C08				MC14081B
MM74C08N				2305	
MM74C10	74C10				MC14023B
MM74C107					MC14027B
MM74C14	74C14				MC14584B
MM74C151	74C151				MC14512B
MM74C154	74C154				MC14515B
MM74C157	74C157				MC14519B
MM74C160	40160B				MC14160B
MM74C161	74C161				MC14161B
MM74C162	40162B				MC14162B
MM74C163	40163B				MC14163B
MM74C164	74C164				MC14015B
MM74C165					MC14021B
MM74C173	74C173				MC14076B
MM74C174	74C174				MC14174B
MM74C175	74C175				MC14175B
MM74C192	74C192	40192			MC14510B
MM74C192N				2321	
MM74C193	74C193				MC14516B
MM74C193N				2322	
MM74C195	40195B				MC14035B
MM74C20	74C20				MC14012B
MM74C200					MCM14537
MM74C221	74C221				MC14538B
MM74C30					MC14068B
MM74C32					MC14071B
MM74C42	74C42				MC14028B
MM74C48	74C48				MC14558B
MM74C73	74C73				MC14027B
MM74C74	74C74				MC14013B
MM74C74N				2310	
MM74C76	74C76				MC14027B
MM74C76N				2312	
MM74C83					MC14008B
MM74C85	74C85				MC14585B
MM74C86	4070B				MC14070B
MM74C89					MCM14505
MM74C90	74C90				MC14518B
MM74C901					MC14049UB
MM74C902					MC14050B
MM74C905					MC14549B
MM74C90N				2315	
MM74C93	74C93				MC14520B
MM74C95					MC14035B
MM8	125	3032A	510,531	1114	
MM8002	77		261		2N5943
MM8003	76				MRF511
MM8004	195A	3048/329	46		MRF8004
MM8006	161		11	2015	2N5031
MM8007	161		11	2015	2N5032
MM8008					MRF905
MM8010					MRF905
MM8011					MRF905
MM8012	76				2N5947
MM8020					2N5836
MM8021					2N5837
MM8023					2N5943
MM80C97	80C97				MC14503B
MM9	125	3033	510,531	1114	
MM999	159		82	2032	
MMCM2907		3118	21	2034	
MMCM918			86	2038	
MMCM930			212	2010	
MMT2222			210	2051	
MMT3014			20	2051	
MMT3798		3114	67	2023	
MMT3903			210	2051	
MMT3904			20	2051	
MMT3905		3118	21	2034	
MMT3906		3118	48		
MMT70		3245	212	2010	
MMT71		3118	221	2024	
MMT72			20	2051	
MMT75		3118	221		
MMT8015			11	2015	
MMT918			61	2038	
MMZ12(06)	143A		ZD-13	563	
MN-53		3004	53	2007	
MN194	121	3009	239	2006	
MN22	121	3009	239	2006	
MN23	121	3009	239	2006	
MN24	121	3717	239	2006	
MN25	121	3717	239	2006	
MN26	121	3717	239	2006	
MN29	121	3717	239	2006	
MN29BLK	121	3009	239	2006	
MN29GRN	121	3009	239	2006	
MN29PUR	121	3009	239	2006	
MN29WHT	121	3009	239	2006	
MN32	121	3717	239	2006	
MN34A	109	3087		1123	
MN46	121	3009	239	2006	
MN48	121	3717	239	2006	
MN49	121	3717	239	2006	
MN51	109	3090		1123	
MN52	102A		53	2007	
MN53	102A		53	2007	
MN53BLU	102A	3123	53	2007	
MN53GRN	102A	3123	53	2007	
MN53RED	102A	3123	53	2007	
MN54			20	2051	
MN60	102A	3087	53	2007	
MN60(DIODE)	109			1123	
MN61	127	3764	25		
MN63	127	3764	25		
MN64	127	3764	25		
MN73	121	3717	239	2006	
MN73BLK	121	3009	239	2006	
MN73WHT	121	3009	239	2006	
MN76	104	3009	16	2006	
MOS3635	221	3050	FET-4	2036	
MOTMJE371	185		58	2025	

Industry Standard No.	ECG	SK	GE	RS 276-	MOTOR.
MOTMJE521	184	3190	57	2017	
MP-01	116	3016	504A	1104	
MP-02				1172	
MP-04				1173	
MP-5115	116	3311	504A	1104	
MP01			504A	1104	
MP010ABO					MDA920A4
MP010ABB					MDA920A3
MP010ABH					MDA920A6
MP010ABM					MDA920A7
MP010ABS					MDA920A8
MP010ABZ					MDA920A9
MP011BBB					MDA960A2
MP011BBD					MDA960A3
MP012FBB					MDA970A2
MP012HBB					MDA801
MP012HBD					MDA802
MP012HBH					MDA804
MP012HBM					MDA806
MP012JBB					MDA2501
MP012JBD					MDA2502
MP012JBH					MDA2504
MP012JBM					MDA3506
MP013MBB					MDA2501
MP013MBD					MDA2502
MP013MBF					MDA2504
MP013MBH					MDA2504
MP013MBK					MDA3506
MP013MBM					MDA3506
MP013RBB					MDA2501
MP013RBD					MDA2502
MP013RBF					MDA2504
MP013RBH					MDA2504
MP013RBK					MDA3506
MP013RBM					MDA3506
MP01FBD					MDA970A3
MP100	116	3311	504A	1104	
MP1003-1	116	3017B/117	504A	1104	
MP1003-2	116	3017B/117	504A	1104	
MP1003-4	116	3017B/117	504A	1104	
MP1014	121	3009	239	2006	
MP1014-1	102A		53	2007	
MP1014-2	123A	3122	20	2051	
MP1014-4	102A		53	2007	
MP1014-5	102A		53	2007	
MP1014-6	102A		53	2007	
MP110B	179	3642	76		
MP1509-1	121	3014	239	2006	
MP1509-2	121	3009	239	2006	
MP1509-3	121	3009	239	2006	
MP1612	127	3764	25		
MP1612A	127	3764	25		
MP1612B	127	3764	25		
MP1613	127	3764	25		
MP2060	121	3009	239	2006	
MP2060-1	121	3009	239	2006	
MP2061	121	3719/104	239	2006	
MP2062	121	3717	239	2006	
MP2137A	121	3009	239	2006	
MP2138A	121	3009	239	2006	
MP2139A	121	3009	239	2006	
MP2142A	121	3009	239	2006	
MP2143A	121	3009	239	2006	
MP2144A	121	3009	239	2006	
MP2200A	179		76		
MP225	116	3031A	504A	1104	
MP2300A	179	3642	76		
MP300	116	3031A	504A	1104	
MP3611	121	3009	239	2006	
MP3612	121	3009	239	2006	
MP3613	121	3009	239	2006	
MP3614	121	3009	239	2006	
MP3615	121	3009	239	2006	
MP3617	121	3009	239	2006	
MP3730	127	3764	25		
MP3730A	127	3764			
MP3730B	127	3764			
MP3731	127	3764	25		
MP400	116	3031A	504A	1104	
MP4300158	519		514	1122	
MP4906063	128		243	2030	
MP500	213		254		
MP500(RECT.)	116		504A	1104	
MP500A	105		4		
MP501	105		4		
MP501A	105		4		
MP502	105		4		
MP502A	105		4		
MP503	105	3012	4		
MP503A	105	3012	4		
MP504	105		4		
MP504A	105		4		
MP505	105		4		
MP505A	105		4		
MP506	105		4		
MP506A	105		4		
MP507	105	3012	4		
MP507A	105	3012	4		
MP5113	116	3311	504A	1104	
MP525	104	3012/105	4		
MP549	5802	3051/156	510	1114	
MP600	179	3642	76		
MP601	179	3642	76		
MP602	179	3642	76		
MP603	179	3642	76		
MP651	116	3031A	504A	1104	
MP8111			66	2048	
MP8112			66	2048	
MP8213		3197	215		
MP8221		3197	215		
MP8222		3197	215		
MP8223		3197	215		
MP8231		3197	215		
MP8232		3197	215		
MP9602	116	3311	504A	1104	
MPC1001H2	1127		IC-197		
MPC14305	960		VR-102		
MPC17C	1047		IC-116		
MPC20C	1075A	3877			
MPC23C	1046		IC-118		
MPC30C	1049	3470			
MPC31C	1048		IC-122		
MPC32C	1050	3072/712	IC-123		
MPC3500	177	3100/519	300	1122	
MPC41C	1093		IC-201		
MPC46C	1077		IC-124		
MPC47C	1076		IC-125		
MPC48C	1091*		IC-126		
MPC48C-1	1091*		IC-126		
MPC48C1	1091*		IC-126		
MPC554	1142	3485	IC-128		
MPC554C	1142	3485	IC-128		
MPC558	1051		IC-131		
MPC561	1079		IC-132		

Industry Standard No.	ECG	SK	GE	RS 276-	MOTOR.
MPC561C	1079		IC-132		
MPC562C	1050	3475	IC-133		
MPC566H	1052	3249			
MPC566HB	1103	3281	IC-94		
MPC566HC	1103	3281	IC-94		
MPC566HD	1103	3281	IC-94		
MPC570C	1086*		IC-142		
MPC571C	1078	3498	IC-136		
MPC575C2	1140	3473	IC-138		
MPC577H	1082	3461	IC-140		
MPC595C	1186		IC-175		
MPC596C	1187	3742			
MPC596C2	1187		IC-176		
MPC5MH	1058	3459			
MPF-102	312	3116	FET-1	2028	
MPF-106			FET-2	2035	
MPF-121	222			2036	
MPF101	312	3112	FET-1	2028	
MPF102	451	3448	FET-1	2035	
MPF103	312	3112	FET-2	2028	2N5457
MPF104	312	3112	FET-2	2028	2N5458
MPF105	312		FET-1	2028	2N5459
MPF106	451	3448	FET-1	2035	2N5485
MPF107	312	3448	FET-1	2035	2N5486
MPF108		3116	FET-1	2028	MPF108
MPF109	459	3112	FET-1	2028	MPF109
MPF110					MPF110
MPF111	457	3448	FET-1	2035	MPF111
MPF112	312		FET-1	2028	MPF112
MPF121	222	3065	FET-4	2036	
MPF151					2N5460
MPF152					2N5461
MPF153					2N5462
MPF154					2N5463
MPF155					2N5464
MPF156					2N5465
MPF157					MFE3004
MPF158					MFE3005
MPF159	465				2N4351
MPF160	464				2N4352
MPF161	326				MPF161
MPF256					MPF256
MPI40100	5344	9105			
MPL1000			82	2032	
MPM5006			20	2051	
MPN-3401	177	3100/519	300	1122	
MPN3401	555		300	1122	MPN3401
MPN3402					MPN3402
MPN3403					MPN3403
MPN3404					MPN3404
MPN3500					MPN3500
MPN3503					MPN3503
MPN3504					MPN3504
MPQ1000					MPQ1000
MPQ2001					MPQ2001
MPQ2221					MPQ2221
MPQ2222					MPQ2222
MPQ2369					MPQ2369
MPQ2483					MPQ2483
MPQ2484					MPQ2484
MPQ2906					MPQ2906
MPQ2907					MPQ2907
MPQ3303					MPQ3303
MPQ3467					MPQ3467
MPQ3546					MPQ3546
MPQ3725					MPQ3725
MPQ3725A					MPQ3725A
MPQ3762					MPQ3762
MPQ3798					MPQ3798
MPQ3799					MPQ3799
MPQ3904					MPQ3904
MPQ3906					MPQ3906
MPQ6001					MPQ6001
MPQ6002					MPQ6002
MPQ6100					MPQ6100
MPQ6100A					MPQ6100A
MPQ6426					MPQ6426
MPQ6427					MPQ6427
MPQ6501					MPQ6501
MPQ6502					MPQ6502
MPQ6600					MPQ6600
MPQ6600A					MPQ6600A
MPQ6700					MPQ6700
MPQ7041					MPQ7041
MPQ7042					MPQ7042
MPQ7043					MPQ7043
MPQ7051					MPQ7051
MPQ7052					MPQ7052
MPQ7053					MPQ7053
MPQ7091					MPQ7091
MPQ7092					MPQ7092
MPQ7093					MPQ7093
MPQ918					MPQ918
MPR10					1N4007
MPR12					MR1-1200
MPR15					MR1-1600
MPS 9623G	123A			2051	
MPS-2716		3124	20	2051	
MPS-3563			20	2051	
MPS-3638A			82	2032	
MPS-3640			21	2034	
MPS-3702			82	2032	
MPS-3705			20	2051	
MPS-5172			212	2010	
MPS-6571			20	2051	
MPS-706		3122	20	2051	
MPS-9630I			20	2051	
MPS-A05	287	3479	62	2010	
MPS-A06	287	3479	63	2030	
MPS-A09			62	2010	
MPS-A10	233		20	2051	
MPS-A12	172A	3156	64		
MPS-A13	172A	3156			
MPS-A14	172A	3156			
MPS-A20	123AP	3444/123A			
MPS-A43	287	3433			
MPS-A55	159	3466	21	2034	
MPS-A56	159	3466	67	2023	
MPS-A65	232	3241	258		
MPS-A66	232	3241	258		
MPS-A92	288	3434			
MPS-A93	288	3434			
MPS-H10	229	3246			
MPS-H17	229		61	2038	
MPS-H32			17	2051	
MPS-U01	188	3024/128	83	2018	
MPS-U03	190		217		
MPS-U04	190		217		
MPS-U06	188	3199			
MPS-U07			226	2018	
MPS-U10	191	3232	224		
MPS-U31	322	3252/302	275		
MPS-U51	189	3025/129	84	2026	
MPS-U55	189		218	2026	

Industry Standard No.	ECG	SK	GE	RS 276-	MOTOR.
MPS-U56	189	3200	218	2026	
MPS1097	160	3006	245	2004	
MPS1572	159	3114/290	82	2032	
MPS1893	104	3719			
MPS2369	123AP	3039/316	11	2015	
MPS25	121	3009	239	2006	
MPS2711	123AP	3444/123A	20	2051	
MPS2712	123AP	3444/123A	20	2051	
MPS2713	123AP	3444/123A	20	2051	
MPS2714	123AP	3444/123A	20	2051	
MPS2715	123AP	3444/123A	20	2051	
MPS2716	123AP	3444/123A	20	2051	
MPS2823	107	3039/316	11	2015	
MPS2894	107	3039/316	11	2015	
MPS2923	123AP	3444/123A	20	2051	
MPS2924	123AP	3444/123A	20	2051	
MPS2925	123AP	3444/123A	20	2051	
MPS2926	123AP	3444/123A	20	2051	
MPS2926-BRN			17	2051	
MPS2926-ORG			17	2051	
MPS2926-RED			17	2051	
MPS2926-YEL			17	2051	
MPS2926BRN	123A	3444	20	2051	
MPS2926GRN	123A	3444	20	2051	
MPS2926ORN	123A	3444	20	2051	
MPS2926RED	123A	3444	20	2051	
MPS2926YEL	123A	3444	20	2051	
MPS3392	123AP	3444/123A	20	2051	
MPS3393	123AP	3444/123A	20	2051	
MPS3394	123AP	3444/123A	20	2051	
MPS3395	123AP	3444/123A	20	2051	
MPS3396	123AP	3444/123A	20	2051	
MPS3397	199	3444/123A	20	2051	
MPS3398	123AP	3444/123A	20	2051	
MPS3534			21	2034	
MPS3536	108	3452	86	2038	
MPS3563	108	3018	61	2038	
MPS3569	194	3433/287			
MPS3638	159	3025/129	82	2032	
MPS3638A	159	3025/129	82	2032	
MPS3639	234	3114/290	82	2050	
MPS3640	159	3114/290	82	2032	
MPS3642	123AP	3024/128	243	2030	
MPS3643			20	2051	
MPS3644	159		21	2034	
MPS3645	159		21	2034	
MPS3646	311	3444/123A	20	2051	
MPS3693	123AP	3018	61	2038	
MPS3694	123AP	3124/289	61	2038	
MPS3702	159	3466	82	2032	
MPS3703	159	3114/290	82	2032	
MPS3704	123AP	3444/123A	20	2051	
MPS3705	123AP	3444/123A	20	2051	
MPS3706	123AP	3444/123A	20	2051	
MPS3707	123AP	3444/123A	20	2051	
MPS3708	123AP	3444/123A	20	2051	
MPS3709	123AP	3444/123A	20	2051	
MPS3710	123AP	3122	62	2010	
MPS3711	123AP	3444/123A	20	2051	
MPS3721	123AP	3444/123A	20	2051	
MPS3826	123AP	3444/123A	20	2051	
MPS3827	123AP	3444/123A	20	2051	
MPS393			20	2051	
MPS3992	123A	3444	20	2051	
MPS404	159	3118	82	2024	
MPS404A	159	3118	82	2032	
MPS4145			11	2015	
MPS4354	159		67	2023	
MPS4355	159	3025/129	67	2023	
MPS5086	159	3114/290	82	2032	
MPS5172	123AP	3444/123A	20	2051	
MPS5668	312	3126	FET-1	2028	
MPS6076	159			2024	
MPS6134	159	3118	82	2032	
MPS6351	123A	3444	20	2051	
MPS6413	123A	3444	20	2051	
MPS6434			21	2034	
MPS6507	108	3122	61	2038	
MPS6511	108	3452	86	2038	
MPS6511-S	108	3122	61	2038	
MPS6512	123AP	3444/123A	61	2038	
MPS6513	123AP	3444/123A	61	2038	
MPS6514	123AP	3444/123A	61	2051	
MPS6515	123AP	3444/123A	210	2038	
MPS6516	159	3715	82	2032	
MPS6517	159	3715	82	2032	
MPS6518	159	3466	82	2032	
MPS6519	159	3466	82	2032	
MPS6520	123AP	3444/123A	20	2051	
MPS6521	123AP	3444/123A	20	2051	
MPS6522	159	3466	82	2032	
MPS6523	159	3466	82	2032	
MPS6524	159	3025/129	82	2032	
MPS6528	107	3039/316	11	2015	
MPS6529	107	3039/316	11	2015	
MPS6530	123AP	3444/123A	20	2051	
MPS6531	123AP	3444/123A	11	2015	
MPS6532	123AP	3444/123A	11	2015	
MPS6533	159	3466	82	2032	
MPS6533M	159		21	2034	
MPS6534	159	3466	82	2032	
MPS6534M	159	3118	82	2032	
MPS6535	159	3466	82	2032	
MPS6535M	159	3138/193A	21	2023	
MPS6539	229	3452/108	86	2038	
MPS654			11	2015	
MPS6540	229	3039/316	11	2015	
MPS6541	229	3452/108	86	2038	
MPS6542	229	3452/108	86	2038	
MPS6543	229	3452/108	86	2038	
MPS6544	123AP	3444/123A	20	2051	
MPS6545	123AP	3444/123A	20	2051	
MPS6546	229	3452/108	86	2038	
MPS6547	229	3452/108	86	2038	
MPS6548	229	3452/108	86	2038	
MPS6552	123A	3444	20	2051	
MPS6553	123A	3444	20	2051	
MPS6554	123A	3444	20	2051	
MPS6555	123AP	3444/123A	20	2051	
MPS6556	123A	3444	20	2051	
MPS6560	123AP	3275/194	222		
MPS6561	123AP	3444/123A	20	2051	
MPS65611	123A	3444	20	2051	
MPS6562	159	3114/290	82	2032	
MPS6563	159	3114/290	82	2032	
MPS6564	123AP	3245/199	212	2010	
MPS6565	123AP	3444/123A	20	2051	
MPS6566	123AP	3444/123A	20	2051	
MPS6567	229	3444/123A	20	2051	
MPS6568	108	3444/123A	20	2051	
MPS6568A	229		61	2038	
MPS6569	229	3039/316	11	2015	
MPS6570	229	3039/316	11	2015	
MPS6571	123AP	3444/123A	20	2051	

Industry Standard No.	ECG	SK	GE	RS 276-	MOTOR.
MPS6572		3124	62	2010	
MPS6573	123AP	3444/123A	20	2051	
MPS6574	123AP	3444/123A	20	2051	
MPS6575	123AP	3444/123A	20	2051	
MPS6576	123AP	3444/123A	20	2051	
MPS6580	159	3118	21	2034	
MPS6590	123A	3444	20	2051	
MPS6591	159		220		
MPS706	123AP		86	2038	
MPS706A	123AP	3452/108	86	2038	
MPS7513	229	3018	61	2038	
MPS8000	159	3854/123AP	81		
MPS8001	123AP	3854	210	2051	
MPS805			86	2038	
MPS8097	123AP		210	2051	
MPS8098	123AP	3275/194	81	2051	
MPS8099	123AP	3275/194	81	2051	
MPS834	123AP	3039/316	11	2015	
MPS8598	159	3715	82	2032	
MPS8599	159	3715	82	2032	
MPS918	108	3039/316	11	2015	
MPS9185	123A	3444	20	2051	
MPS9410A	128	3024	243	2030	
MPS9410AJ	128	3854/123AP	243	2030	
MPS9410AK		3854	243	2030	
MPS9410H	128	3137/192A	243	2030	
MPS9411	299	3298	236		
MPS9411A	299		236		
MPS9411A1	299	3434/288	236		
MPS9411G	299	3137/192A			
MPS9411H	299	3137/192A			
MPS9411T	299	3137/192A	236		
MPS9416	297	3122	271	2030	
MPS9416A	297	3122	271	2030	
MPS9416AT	297		271	2030	
MPS9416ST	297		271	2030	
MPS9417A	297	3137/192A	81	2030	
MPS9417A-T			81	2051	
MPS9417AT	297		271	2030	
MPS9417T	297		271	2030	
MPS9418	123AP	3854	268	2038	
MPS9418S	297	3854/123AP	63	2030	
MPS9418T	123AP	3854	63	2030	
MPS9423	123A	3444	20	2051	
MPS94231			86	2038	
MPS9423F	108		86	2038	
MPS9423G	108	3018	86	2038	
MPS9423H	108	3018	86	2038	
MPS9423I	108	3018	20	2038	
MPS9426	229	3246	61	2038	
MPS9426A	229	3246	20	2051	
MPS9426A.B		3018	20	2051	
MPS9426B	229	3246	61	2038	
MPS9426BC	229	3246	61	2038	
MPS9426C	229	3246	61	2038	
MPS9427	229	3018	61	2038	
MPS9427B	229		61	2038	
MPS9427B.C		3018	20	2051	
MPS9427C	229	3444/123A	20	2051	
MPS9433	123A	3444	62	2051	
MPS9433J	123A	3124/289	62	2010	
MPS9433K	199	3137/192A	20	2051	
MPS9433T	123AP	3444/123A			
MPS9434J	123A	3444	62	2051	
MPS9434K	123A	3444	62	2051	
MPS943J	123A	3444			
MPS943K	123A	3444			
MPS9444	177	3100/519	300	1122	
MPS9460A	129	3025	244	2027	
MPS9460H	129	3138/193A	244	2027	
MPS9461A1	287		222		
MPS9466A	293	3849	272	2030	
MPS9466AT	298	3450	272		
MPS9467A	298	3138/193A	82		
MPS9467A-T			82	2032	
MPS9467T	298	3450	272		
MPS9468	298	3450	223		
MPS9468S	298	3450	67		
MPS9468T	298	3450	67		
MPS9476AT	298	3450	272		
MPS9600	123A	3444	20	2051	
MPS9600(G)			20	2051	
MPS9600-5	123A	3444	245	2004	
MPS9600F	123A	3444	20	2051	
MPS9600G	123A	3444	20	2051	
MPS9600G/H	123A	3444	86	2038	
MPS9600H	123A	3444	86	2038	
MPS9600I	123A	3444			
MPS9601	108	3311	86	2038	
MPS9602	116	3311	504A		
MPS9604	229	3018	61	2038	
MPS96041			20	2051	
MPS9604D	123A	3444	20	2051	
MPS9604E	123A	3444	20	2051	
MPS9604F		3018	20	2051	
MPS9604FG	108		20	2051	
MPS9604I	123A	3444	20	2051	
MPS9604R	123A	3444	20	2051	
MPS9606	177	3100/519	300	1122	
MPS9606(H,I)			300	1122	
MPS96061			300	1122	
MPS9606H	177	3100/519	300	1122	
MPS9606I	177	3100/519		1122	
MPS9606T			20	2051	
MPS9611-5	123A	3444	20	2051	
MPS9616	123A	3444	20	2051	
MPS9616A	123A	3137/192A	47	2030	
MPS9616J	123A	3124/289	243	2030	
MPS9618	123A	3444	20	2051	
MPS9618(J)			20	2051	
MPS9618H	123A	3444	20	2051	
MPS9618I	123A	3444	20	2051	
MPS9618J	123A	3444	20	2051	
MPS9623	123AP	3444/123A	20	2051	
MPS9623C	123A	3444	11	2015	
MPS9623C(P)			20	2051	
MPS9623E	123A	3444	20	2051	
MPS9623E.G		3018	20	2051	
MPS9623F	123A	3444	20	2051	
MPS9623G	123A	3444	20	2051	
MPS9623G/H	123A	3444	20	2051	
MPS9623H	123AP	3444/123A	20	2051	
MPS9623H/I	123A	3444	20	2051	
MPS9623I	123AP	3444/123A	20	2051	
MPS9623I/J	123A	3444	20	2051	
MPS9623J			20	2051	
MPS9625	229	3018	61	2038	
MPS9625C	229	3018	61	2038	
MPS9625D	229	3018	61	2038	
MPS9625E	229	3018	61	2038	
MPS9625F	229	3018	61	2038	
MPS9625G	229	3018	20	2038	
MPS9625H	229	3122	61	2038	
MPS9626	123A	3444	20	2051	
MPS9626G	123AP		20	2051	
MPS9626H	123AP		20	2051	
MPS9626I	123AP		20	2051	
MPS9630	123A	3124/289	20	2051	
MPS9630H	123A	3124/289	20	2051	
MPS9630H.I		3124	20	2051	
MPS9630I	123A	3124/289	20	2051	
MPS9630J	123A	3124/289	62	2010	
MPS9630K	123AP	3124/289	62	2010	
MPS9630T	123AP	3124/289	20	2051	
MPS9631	123A	3444	20	2051	
MPS9631(I)			20	2051	
MPS9631I	123A	3444	20	2051	
MPS9631J	123A	3444	20	2051	
MPS9631K	123A	3444	20	2051	
MPS9631S	123AP		20	2051	
MPS9631T	123A	3444	20	2051	
MPS9632	123A	3444	20	2051	
MPS9632(I)			20	2051	
MPS9632(K)			20	2051	
MPS96321			20	2051	
MPS9632H	123A	3444	20	2051	
MPS9632I	123A	3444	20	2051	
MPS9632J	123A	3444	20	2051	
MPS9632K	123A	3444	20	2051	
MPS9632T	123A	3018	20	2051	
MPS9633	199	3124/289	18	2030	
MPS9633C	199	3444/123A	62	2010	
MPS9633D	199	3124/289	62	2010	
MPS9633G	199	3444/123A	20	2051	
MPS9634	123AP	3854	62	2010	
MPS9634B	199	3854/123AP	60	2010	
MPS9634C	123AP	3854	62	2010	
MPS9634D	123AP	3854	20	2051	
MPS9644	177	3100/519	300	1122	
MPS9646	177	3100/519	300	1122	
MPS964611	177		300	1122	
MPS9646G	177	3100/519	300	1122	
MPS9646H	177	3100/519	300	1122	
MPS9646I	177		300	1122	
MPS9646J	177	3100/519	300	1122	
MPS9646M	177		300	1122	
MPS9666	159	3114/290	82	2032	
MPS9680	159	3466	82	2032	
MPS96800			22	2032	
MPS9680H	159		82	2032	
MPS9680H/E			65	2050	
MPS9680H/I	159	3466	82	2032	
MPS9680I	159	3466	82	2032	
MPS9680I/J	159	3114/290	82	2032	
MPS9680J	159	3466	82	2032	
MPS9680T	159	3466	82	2032	
MPS9681	159	3466	82	2032	
MPS9681I	159	3466	82	2032	
MPS9681J	159	3466	21	2032	
MPS9681K	159	3466	82	2032	
MPS9681T	159	3466	82	2032	
MPS9682	159	3114/290	82	2032	
MPS9682(I)			22	2032	
MPS9682-I	159		82	2032	
MPS9682I	159	3114/290	82	2032	
MPS9682J	159	3114/290	82	2032	
MPS9682K	159	3114/290	82	2032	
MPS9682T	159	3114/290	21	2034	
MPS9696	289A	3449/297	271	2030	
MPS9696F	289A	3124/289	20	2051	
MPS9696G	289A	3449/297	271	2038	
MPS9696H	123A	3124/289	20	2051	
MPS9696I	289A	3444/123A	20	2051	
MPS9700D	123A	3444	61	2051	
MPS9700E	123A	3444	61	2051	
MPS9700F	123A	3444	20	2051	
MPS9750D	159	3466	82	2032	
MPS9750F	159	3466	82	2032	
MPS9750G	159	3466	82	2032	
MPS97500F			82	2032	
MPSA05	194	3122	20	2051	
MPSA06	287	3479	243	2030	
MPSA09	123AP	3245/199	212	2010	
MPSA10	123AP	3245/199	212	2010	
MPSA10-BLU		3245	212	2010	
MPSA10-GRN		3245	212	2010	
MPSA10-RED		3245	212	2010	
MPSA10-WHT		3245	212	2010	
MPSA10-YEL		3245	212	2010	
MPSA12	172A		64		
MPSA13	172A		64		
MPSA14	172A		64		
MPSA16		3122	62	2010	
MPSA17		3122	62	2010	
MPSA18			85	2010	
MPSA20	123AP	3479	20	2051	
MPSA20-BLU		3245	212	2010	
MPSA20-GRN		3245	212	2010	
MPSA20-RED		3245	212	2010	
MPSA20-WHT		3245	212	2010	
MPSA20-YEL		3245	212	2010	
MPSA42	287	3044/154	222		
MPSA43	287		222		
MPSA55	159	3466	82	2032	
MPSA56	159	3025/129	82	2032	
MPSA66	232		258		
MPSA70	159	3466	82	2032	
MPSA70-YEL			65	2050	
MPSA92	288		223		
MPSA93	288		223		
MPSD01	287		222		
MPSD03	194		222		
MPSD05	123AP		210	2051	
MPSD06	123AP		62	2010	
MPSD52	288		223		
MPSD53	383		223		
MPSD55	159		22	2032	
MPSD56	159		65	2050	
MPSEL239			20	2051	
MPSH02	229		17	2051	
MPSH04	194		220		
MPSH05	194		220		
MPSH07	108		20	2051	
MPSH08	108		11	2015	
MPSH09			11	2015	
MPSH10	229		61	2038	
MPSH11	229		61	2038	
MPSH17	229		81	2051	
MPSH19	229		61	2038	
MPSH20			20	2051	
MPSH24			11	2015	
MPSH30	229		61	2038	
MPSH31	229		61	2038	
MPSH32	229		17	2051	
MPSH34			39	2015	
MPSH37	123AP		61	2038	
MPSH54	159		221		
MPSH55	159		221		

Industry Standard No.	ECG	SK	GE	RS 276-	MOTOR.
MPSH81	159	3118	21	2034	
MPSK20	123AP	3245/199	212	2010	
MPSK21	123AP	3245/199	212	2010	
MPSK22	123AP	3245/199	212	2010	
MPSK70	159		65	2050	
MPSK71	159		65	2050	
MPSK72	159	3247/234	65	2050	
MPSL01	287	3275/194	220		
MPSL51		3025	221		
MPSU01	188	3199	226	2018	MPSU01
MPSU01A	188	3024/128	226	2018	MPSU01A
MPSU02	188	3199	226	2018	MPSU02
MPSU03	190		217		MPSU03
MPSU04	190	3219	217		
MPSU05	188	3199	226	2018	MPSU05
MPSU06	188		226	2018	MPSU06
MPSU07	188				MPSU07
MPSU10	191	3232	249		MPSU10
MPSU11	191	3232	249		MPSU10
MPSU12					MPSU45
MPSU31	322		275		MPSU31
MPSU45	272		D40K2		MPSU45
MPSU47					MPSU31
MPSU51	189	3200	218	2026	MPSU51
MPSU51A	189	3200	218	2026	MPSU51A
MPSU52	189	3025/129	29	2026	MPSU52
MPSU55	189	3025/129	218	2026	MPSU55
MPSU56	189		218	2026	MPSU56
MPSU57			227		MPSU57
MPSU60					MPSU60
MPSU95	273		D41K2		MPSU95
MPT-10					MPTE-10
MPT-12					MPTE-12
MPT-15					MPTE-15
MPT-18					MPTE-18
MPT-22					MPTE-22
MPT-36					MPTE-36
MPT-45					MPTE-45
MPT-5					MPTE-5
MPT-8					MPTE-8
MPT20	6406	9082A			1N5758
MPT24					1N5759
MPT28	6407	9083A		1050	1N5760
MPT32	6408			1050	1N5761
MPT36					1N5762
MPU131	6402	3628			MPU131
MPU132					MPU132
MPU133					MPU133
MPU231					2N6116
MPU232					2N6117
MPU233					2N6118
MPU6027					MPU6027
MPU6028					MPU6028
MPX-25	116		504A	1104	
MPX215	116	3016	504A	1104	
MPX25			504A	1104	
MPX9410H	128		243	2030	
MPX9623		3122	20	2051	
MPX9623H			20	2051	
MPX9623H/I	123A	3444	20	2051	
MPX9623I			20	2051	
MPX9630I	123A	3444	20	2051	
MPX9681J	159		82	2032	
MQ1	123A	3444	20	2051	
MQ1120					MQ1120
MQ2	123A	3444	20	2051	
MQ2218					MQ2218
MQ2219A					MQ2219A
MQ2221					MQ2218
MQ2222A					MQ2219A
MQ2369					MQ2369
MQ2484					MQ2484
MQ2904					MQ2904
MQ2905A					MQ2905A
MQ2906					MQ2904
MQ2907A					MQ2905A
MQ3/2		3311	504A	1104	
MQ32	116		504A	1104	
MQ3251					MQ3251
MQ3467					MQ3467
MQ3725					MQ3725
MQ3762					MQ3725
MQ3798					MQ3798
MQ3799					MQ3799
MQ3799A		3118			MQ3799A
MQ6/2		3311	504A	1104	
MQ6001					MQ6001
MQ6002					MQ6002
MQ6100					MQ6100
MQ62	116		504A	1104	
MQ7001					MQ7001
MQ7003					MQ7003
MQ7004					MQ7004
MQ7005					MQ7005
MQ7007					MQ7007
MQ7021					MQ7021
MQ8/2		3311	504A	1104	
MQ82	116		504A	1104	
MQ918					MQ918
MQ930					MQ930
MQ982					MQ982
MR-1	156	3016	504A	1104	
MR-150-01	116	3311	504A	1104	
MR-1C	156	3051	510	1114	
MR-1M	125		509	1114	
MR-852	506	3125	511	1114	
MR100					1N5392
MR1000					1N5399
MR100C-H	140A	3061	ZD-10	562	
MR100H	140A	3061	ZD-10	562	
MR1030A	156	3051	512		
MR1030B		3051	512		
MR1031		3051	510	1114	
MR1031A	156	3051	512		
MR1031D	156	3051	512		
MR1032A	156	3051	512		
MR1032B	156	3051	512		
MR1033A		3051	512		
MR1033B		3051	512		
MR1034A		3051	512		
MR1034B		3051	512		
MR1035A		3051	512		
MR1035B		3051	512		
MR1036A		3051	512		
MR1036B		3051	512		
MR1038A		3051	512		
MR1038B		3051	512		
MR1040A		3051	512		
MR1040B		3051	512		
MR1120	5870	3500/5882	5032		
MR1120R	5871	3517/5883	5033		
MR1121	5872	3500/5882	5032		
MR1121R	5873	3517/5883	5033		
MR1122	5874	3500/5882	5032		
MR1122R	5875	3517/5883	5033		
MR1123	5876	3500/5882	5036		
MR1123R	5880	3517/5883	5037		
MR1124	5878	3500/5882	5036		
MR1124R	5879	3517/5883	5037		
MR1125	5880	3500/5882	5040		
MR1125R	5881	3517/5883	5041		
MR1126	5882	3500	5040		
MR1126R	5883	3517	5041		
MR1128	5886		5044		
MR1128R	5887		5045		
MR1130	5890	7090	5044		
MR1130R	5891	7091	5045		
MR1200					MR1-1200
MR120C-H	142A		ZD-12	563	
MR1237FB		3016	504A	1104	
MR1237FL		3016	504A	1104	
MR1237SB	6354	3016	504A	1104	
MR1237SL	116	3016	504A	1104	
MR1247FB		3016	504A	1104	
MR1247FL		3016	504A	1104	
MR1247SB		3016	504A	1104	
MR1247SL		3016	504A	1104	
MR1267		3016	504A	1104	
MR1337-1	552	3017B/117	504A		
MR1337-2	552	3017B/117	504A		
MR1337-3	552	3017B/117	504A		
MR1337-4	552	3031A	504A		
MR1337-5	552	3017B/117	504A		
MR1500					MR1-1600
MR150C-H	145A	3063	ZD-15	564	
MR1C	156	3311	504A	1104	
MR1M	116	3080	509	1114	
MR200					1N5393
MR2064	116		504A	1104	
MR2065	116		504A	1104	
MR2069	156		512		
MR2070	156		512		
MR21CR	5404	3627			
MR2261	116	3311	504A	1104	
MR2262	117		504A	1104	
MR2266	551	3032A	511	1114	
MR2271			504A	1104	
MR2272	515	3031A	511	1114	
MR2273	551	3016	511	1114	
MR2369	156	3051	512		
MR33C-H	5005A	3771	ZD-3.3		
MR36H	134A	3055	ZD-3.6		
MR3932	123A	3444	20	2051	
MR3933	128	3024	243	2030	
MR3934	129	3025	244	2027	
MR39C-H	134A	3055	ZD-3.6		
MR400					1N5395
MR47C-H	135A	3056	ZD-5.1		
MR501	5802	9005		1143	
MR502				1143	
MR504				1144	
MR51C-H	135A	3056	ZD-5.1		
MR51E-H	135A	3056	ZD-5.1		
MR56C-H	136A	3057	ZD-5.6	561	
MR600					1N5397
MR62C-H	137A	3058	ZD-6.2	561	
MR62E-H	137A	3058	ZD-6.2	561	
MR62H	137A	3058	ZD-6.2	561	
MR750	5812	3639			
MR751	5812	3639			
MR752	5814	3639/5812			
MR754	5814	9096			
MR756	5815	3640			
MR760	5817	9097			
MR800					1N5398
MR801	506	3032A	511	1114	
MR811	506	3125	511	1114	
MR812	506	3125	511	1114	
MR814	506	3125	511	1114	
MR816	506	3125	511	1114	
MR91C-H	139A	3060	ZD-9.1	562	
MR91E-H	139A	3060	ZD-9.1	562	
MR9600	116	3311	504A	1104	
MR9601	116	3311	504A	1104	
MR9602	116	3311	504A	1104	
MR9604	123A	3444	20	2051	
MR990	125	3033	510,531	1114	
MRA1720-2					MRF2003M
MRA1720-20					MRF2016M
MRA1720-5					MRF2005M
MRA1720-9					MRF2010M
MRAL1720-2					MRF2003M
MRAL1720-20					MRF2016M
MRAL1720-5					MRF2005M
MRAL1720-9					MRF2010M
MRAL2023-1.5					MRF2003M
MRAL2023-12					MRF2016M
MRAL2023-3					MRF2003M
MRAL2023-6					MRF2010M
MRB-20C	116(4)	3016	504A	1104	
MRD150			39	2015	
MRD3050	3032		L14G2		
MRD3051	3032		L14G2		
MRD3054	3032		L14G2		
MRD3055	3032		L14G2		
MRD3056	3032		L14G1		
MRD450			39	2015	
MRF100					1N4934
MRF200					1N4934
MRF201					2N6255
MRF203					MRF245
MRF207	346	9038	231		
MRF209	350	3176			
MRF221	350F	3176/350			
MRF233	350	3176			
MRF238	345		298		
MRF305					MRF325
MRF306					2N6439
MRF400					1N4936
MRF402	473		320		
MRF406	338F		292		
MRF415					2N6366
MRF416					2N6367
MRF417					2N6368
MRF418					MRF460
MRF419					2N6370
MRF420	335				MRF454
MRF451	333				MRF453
MRF452	333				MRF453
MRF453	333		282		
MRF454	335		281		
MRF472	295		57		
MRF501	161		86	2038	
MRF502	161	3132	39	2015	
MRF504					MRF511
MRF515	486		299		
MRF5175					MRF321
MRF5176					MRF323

Industry Standard No.	ECG	SK	GE	RS 276-	MOTOR.
MRF5178					2N6439
MRF519					MRF517
MRF600					1N4937
MRF601					2N6256
MRF602					2N6136
MRF605					2N6439
MRF607	472		294		
MRF618	365				MRF641
MRF619	366				MRF644
MRF620	366				MRF644
MRF621	367				MRF646
MRF8004	195A	3765	46		
MRF901					2N6603
MRF902					2N6603
MRF911					2N6604
MRF912					2N6604
MRS6548			11	2015	
MRV-20C		3016	504A	1104	
MS-1				1102	
MS-2				1102	
MS-4				1103	
MS1010			63	2030	
MS11H	116	3311	504A	1104	
MS12H	116	3311	504A	1104	
MS13H	116	3311	504A	1104	
MS14H	116	3311	504A	1104	
MS1H	116	3311	504A	1104	
MS22B	123A	3444	20	2051	
MS2991	128	3024	243	2030	
MS2H	116	3311	504A	1104	
MS35H	116	3031A	504A	1104	
MS3694		3018	17	2051	
MS36H	116	3031A	504A	1104	
MS3H	116	3311	504A	1104	
MS4H	116	3311	504A	1104	
MS5	5870		5032		
MS50	116	3311	504A	1104	
MS510			63	2030	
MS5C			504A	1104	
MS5H	116	3031A	504A	1104	
MS701T	108	3452	86	2038	
MS7500		3126	90		
MS7501S	108	3452	86	2038	
MS7501T	108	3452	86	2038	
MS7502R	123A	3444	20	2051	
MS7502S	108	3452	86	2038	
MS7502T	108	3452	86	2038	
MS7503R	123A	3444	20	2051	
MS7504	172A	3156	64		
MS7505	159	3114/290	82	2032	
MS7506G	129		244	2027	
MS7506H	128	3124/289	243	2030	
MS7506J	128		243	2030	
MS91X3	505		CR-7		
MS9667	159	3114/290	82	2032	
MS9681	159	3114/290	82	2032	
MSA7505			66	2048	
MSA8505			66	2048	
MSA8508			28	2017	
MSC1250M					MRF1250M
MSC1325M					MRF1325M
MSC2001					MRF2001
MSC2003					MRF2003
MSC2005					MRF2005
MSC2010					MRF2010
MSC2302					MRF2003
MSC2304					MRF2005
MSC2307					MRF2010
MSC82001					MRF2001
MSC82003					MRF2003
MSC82005					MRF2005
MSC82005M					MRF2005M
MSC82010					MRF2010
MSC82012M					MRF2010M
MSC82020M					MRF2016M
MSC82201					MRF2001
MSC82203					MRF2003
MSC82304M					MRF2005M
MSC82310M					MRF2010M
MSC82313M					MRF2016M
MSD6101	276	3296			
MSD7000	276	3296			
MSG7506	128	3124/289	243	2030	
MSJ7505	129	3025	244	2027	
MSK5405	128	3124/289	243	2030	
MSM4001	4001B	4001			
MSM4002	4002B	4002			
MSM4007	4007	4007			
MSM4011	4011B	4011			
MSM4011RS	4011B	4011	4011		
MSM4012	4012B	4012			
MSM4013	4013B	4013			
MSM4015	4015B	4015			
MSM4016	4016B	4016	4016		
MSM4017	4017B	4017			
MSM4019	4019B	4019			
MSM40192	40192B	40192			
MSM4020	4020B	4020			
MSM4023	4023B	4023	4023		
MSM4025	4025B	4025			
MSM4027	4027B	4027			
MSM4028	4028B	4028			
MSM4030	4030B	4030			
MSM4040	4040B	4040			
MSM4042	4042B	4042			
MSM4043	4043B	4043			
MSM4044	4044B	4044			
MSM4049	4049	4049			
MSM4050	4050B	4050	4050		
MSM4051	4051B	4051			
MSM4052	4052B	4052			
MSM4053	4053B	4053			
MSM4066	4066B	4066			
MSM4069	4069	4069			
MSM4069RS	4069	4069			
MSM4071	4071B	4071			
MSM4072	4072B	4072			
MSM4075	4075B	4075			
MSM4081	4081B	4081			
MSM4518	4518B	4518			
MSM4520	4520B	4520			
MSM4556	4556B	4556			
MSM63			82	2032	
MSM631			82	2032	
MSP10A	225		256		
MSP1161	213		254		
MSP15A	225		256		
MSP20A	225		256		
MSP25A	225		256		
MSP999058-1	128		243	2030	
MSR-500	116	3311	504A	1104	
MSR-V5	116	3031A	504A	1104	
MSR500	116	3311	504A	1104	
MSR7502	108	3452	86	2038	

Industry Standard No.	ECG	SK	GE	RS 276-	MOTOR.
MSR7503	123A	3444	20	2051	
MSS-1000	116		504A	1104	
MSS1000	177	3100/519	300	1122	
MSS7501	108	3452	86	2038	
MSS7502	108	3452	86	2038	
MST-10	128	3024	243	2030	
MST10S	154		231		
MST20S	154		32		
MST30S	154		32		
MST40S	396		32		
MST7501	108	3452	86	2038	
MT021	116	3311	504A	1104	
MT021A	116	3311	504A	1104	
MT022	116	3017B/117	504A	1104	
MT022A	116	3017B/117	504A	1104	
MT0404	159	3114/290	82	2024	
MT0404-1	159	3114/290	82	2032	
MT0404-2	159	3114/290	82	2032	
MT0411	159	3114/290	82	2032	
MT0412	159	3118	82	2032	
MT0413	159	3114/290	82	2032	
MT0414		3118	221		
MT0463		3118	21	2034	
MT100	108		86	2038	
MT101	108	3452	86	2038	
MT102	108		86	2038	
MT102351A	160		245	2004	
MT104	123A	3444	20	2051	
MT106	108		86	2038	
MT107	108	3452	86	2038	
MT1070			28	2017	
MT1075		3244	220		
MT1100		3244	220		
MT1131	159	3114/290	82	2032	
MT1131A	159	3114/290	82	2032	
MT1132	159	3114/290	82	2032	
MT1132A	159	3114/290	82	2032	
MT1132B	159	3114/290	82	2032	
MT1254	159	3114/290	82	2032	
MT1255	159	3114/290	82	2032	
MT1256	159	3114/290	82	2032	
MT1257	159	3114/290	82	2032	
MT1258	159	3114/290	82	2032	
MT1259	159	3114/290	82	2032	
MT14	116	3017B/117	504A	1104	
MT1420	159	3114/290	82	2032	
MT1613	128	3024	243	2030	
MT1711	128	3024	243	2030	
MT1889			1N60	1123	
MT1893	154	3045/225	40	2012	
MT1991	159	3114/290	82	2032	
MT2303	159	3114/290	82	2032	
MT24	116	3017B/117	504A	1104	
MT2411	159	3114/290	82	2032	
MT2412	159	3114/290	82	2032	
MT3001			39	2015	
MT3002			39	2015	
MT3011			214	2038	
MT3202	124	3021	12		
MT4101	123A	3444	20	2051	
MT4102	123A	3444	20	2051	
MT4102A	123A	3444	20	2051	
MT4103	123A	3444	20	2051	
MT4104			39	2015	
MT44	116	3031A	504A	1104	
MT5	5892		5064		
MT6001	123A	3444	20	2051	
MT6002	123A	3444	20	2051	
MT6003	123A	3444	20	2051	
MT64	116	3017B/117	504A	1104	
MT696	123A	3444	20	2051	
MT697	123A	3444	20	2051	
MT698	154	3045/225	40	2012	
MT699	154	3045/225	40	2012	
MT706	123A	3444	20	2051	
MT706A	123A	3444	20	2051	
MT706B	123A	3444	20	2051	
MT707	123A	3444	20	2051	
MT708	123A	3444	20	2051	
MT726	159	3114/290	82	2032	
MT743	108	3452	86	2038	
MT744	108	3452	86	2038	
MT753	108	3452	86	2038	
MT84	125	3032A	510,531	1114	
MT869	159	3114/290	82	2032	
MT870	154	3045/225	40	2012	
MT871	154	3045/225	40	2012	
MT9001	123A	3444	20	2051	
MT9002	123A	3444	20	2051	
MT9003			214	2038	
MT910	154	3045/225	40	2012	
MT911	154	3045/225	40	2012	
MT912	154	3045/225	40	2012	
MTE200E	199	3124/289			
MTM1224					MTM1224
MTM1225					MTM1225
MTM474					MTM474
MTM475					MTM475
MTM564					MTM564
MTM565					MTM565
MTP1224					MTP1224
MTP1225					MTP1225
MTP474					MTP474
MTP475					MTP475
MTP564					MTP564
MTP565					MTP565
MTZ607	136A	3057	ZD-5.6	561	
MTZ607A	136A	3057	ZD-5.6	561	
MTZ608	137A	3058	ZD-6.2	561	
MTZ608A	137A	3058	ZD-6.2	561	
MTZ612A	5072A	3136			
MTZ613	139A	3060	ZD-9.1	562	
MTZ613A	139A	3060	ZD-9.1	562	
MTZ614	140A	3061	ZD-10	562	
MTZ614A	140A	3061	ZD-10	562	
MTZ616	142A	3062	ZD-12	563	
MTZ616A	142A	3062	ZD-12	563	
MTZ617A	143A	3750			
MTZ618	145A	3063	ZD-15	564	
MTZ618A	145A	3063	ZD-15	564	
MTZ619A	5075A	3751			
MTZ620A	5077A	3752			
MTZ624	146A	3750/143A	ZD-27		
MTZ624A	146A	3750/143A	ZD-27		
MTZ625A	5084A	3755			
MU-26-1C	390		255	2041	
MU10					MU10
MU20					MU20
MU2646					MU2646
MU4891					MU4891
MU4892					MU4892
MU4893					MU4893
MU4894					MU4894
MU9610	188		226	2018	

Industry Standard No.	ECG	SK	GE	RS 276-	MOTOR.
MU9610F	188		226	2018	
MU9610T	188		226	2018	
MU9611	188	3054/196	226	2018	
MU9611G	188		226	2018	
MU9611T	188	3054/196	226	2018	
MU9660	189		218	2026	
MU9660S	189		218	2026	
MU9660T	189		218	2026	
MU9661	189	3083/197	218	2026	
MU9661T	189	3083/197	218	2026	
MU970					2N2646
MU971					2N2647
MUS4987	6404				MBS4991
MUS4988					MBS4992
MV-11	601	3463			
MV-12	612	3100/519	300	1122	
MV-12A,B,C					MR1-1200
MV-12H	601	3463			
MV-13	601	3864/605	504A	1104	
MV-13(BIAS)			504A	1104	
MV-13(DIO)			300	1122	
MV-13YH	605	3864			
MV-16A,B,C					MR1-1600
MV-2	601		300	1122	
MV-201	611	3324	90		
MV-3				1122	
MV-5	116		504A	1104	
MV1	601	3463	504A	1104	
MV104					MV104
MV104G					MV104G
MV109					MV109
MV11	116	3463/601	504A	1104	
MV11T	605	3463/601			
MV12	612	3325			
MV121LP	601	3463			
MV12A	612	3325			
MV13YH	605	3864			
MV1401					MV1401
MV1403					MV1403
MV1403H					MV1403H
MV1404					MV1404
MV1404H					MV1404H
MV1405	616				MV1405
MV1405H					MV1405H
MV1410	613	3326			
MV1410A	613	3326			
MV1620	610	3323			MV1620
MV1622					MV1622
MV1624	611				MV1624
MV1626	612	3325			MV1626
MV1628					MV1628
MV1630					MV1630
MV1632					MV1632
MV1634	613	3326			MV1634
MV1636					MV1636
MV1638	614	3327			MV1638
MV1640					MV1640
MV1642					MV1642
MV1644					MV1644
MV1646					MV1646
MV1648					MV1648
MV1650					MV1650
MV1652					MV1652
MV1654					MV1654
MV1666					MV1666
MV1720A	610	3323			
MV1726	612	3325			
MV1726A	612	3325			
MV1734	613	3326			
MV1734A	613	3326			
MV1738	614	3327			
MV1738A	614	3327			
MV1866					MV1866
MV1868					MV1868
MV1870					MV1870
MV1871					MV1871
MV1872					MV1872
MV1874					MV1874
MV1876					MV1876
MV1877					MV1877
MV1878					MV1878
MV1Y	601	3463			
MV20					MHW612A
MV201	611	3324			
MV209	614				MV209
MV20C	525	3925			
MV2101	610	3323			MV2101
MV2102					MV2102
MV2103	611				MV2103
MV2104	612	3325			MV2104
MV2105	612	3325			MV2105
MV2106					MV2106
MV2107	613	3326			MV2107
MV2108					MV2108
MV2109	614	3327	90		MV2109
MV2110					MV2110
MV2111	614	3327			MV2111
MV2112					MV2112
MV2113					MV2113
MV2114					MV2114
MV2115					MV2115
MV2201	610	3323			MV2201
MV2203	611	3126	90		MV2203
MV2205		3126	90		MV2205
MV2209	614	3327	90		MV2209
MV2301					MV2301
MV2302					MV2302
MV2303					MV2303
MV2304					MV2304
MV2305					MV2305
MV2306					MV2306
MV2307					MV2307
MV2308					MV2308
MV3	601	3463	300	1122	
MV3(DIODE)	177		300	1122	
MV3(RECTIFIER)	116		504A	1104	
MV30					MHW613A
MV3007	610	3323			
MV3007E	610	3323			
MV3012	612	3325			
MV3012E	612	3325			
MV3015	612	3325			
MV3015E	612	3325			
MV3020	613	3326			
MV3020E	613	3326			
MV3027	613	3326			
MV3027E	613	3326			
MV3039	614	3327			
MV3039E	614	3327			
MV309	616				MV309
MV310					MV310
MV3102					MV3102
MV3103					MV3103
MV3140					MV3140
MV3141					MV3141
MV3142					MV3142
MV33	614	3327			
MV33A	614	3327			
MV3501	610	3323			
MV3504	612	3325			
MV3507	613	3326			
MV4	177	3090/109	504A	1122	
MV401	612	3325			
MV6113	613	3326			
MV830					MV830
MV831					MV831
MV832	613	3326			MV832
MV833					MV833
MV834	614	3327			MV834
MV835					MV835
MV836					MV836
MV837					MV837
MV838					MV838
MV839					MV839
MV840					MV840
MV9052	610	3323			
MV9053	610	3323			
MV9054	610	3323			
MV9151	612	3325			
MV9152	612	3325			
MV9153	612	3325			
MV9154	612	3325			
MV9251	613	3326			
MV9252	613	3326			
MV9253	613	3326			
MV9254	613	3326			
MV9351	614	3327			
MV9352	614	3327			
MV9353	614	3327			
MV9354	614	3327			
MV9600	614	3327	90		
MV9601	614	3327	90		
MV9602		3126	90		
MVA-05A	116	3311	504A	1104	
MVA-05A(DIO)			504A	1104	
MVA-05A(RECT)			504A	1104	
MVAM108					MVAM108
MVAM109					MVAM109
MVAM115					MVAM115
MVAM125					MVAM125
MVB6113	613	3326			
MVB6116	613	3326			
MVB6124	612	3325			
MVB6125	612	3325			
MVB6126	612	3325			
MVB6128	614	3327			
MVB6129	614	3327			
MVB6130	614	3327			
MVB6136	613	3326			
MVE6124	612	3325			
MVE6125	612	3325			
MVE6139	610	3323			
MVE6141	610	3323			
MW4104					MCM4027
MX-3198	977	3462			
MX-3256	1242	3445/1192			
MX-3309	74145	74145			
MX-3364		3184	IC-278		
MX-3369	973D	3233/973			
MX-3370	1195	3469			
MX-3372	1194	3484			
MX-3389	1082	3461	IC-140		
MX-3452	964	3630			
MX1.5					MHW401
MX12					MHW710
MX15					MHW710
MX3336	746		IC-217		
MX7.5					MHW709
MXC-1312	799	3238	IC-34		
MXC1312A	799	3238	IC-34		
MXC1312P	799	3238	IC-34		
MY-1	116	3311	504A	1104	
MY201	612	3325			
MZ-00	177	3100/519	300	1122	
MZ-08	5072A	3136	ZD-8.2	562	
MZ-10	5019A	3061/140A	ZD-10	562	
MZ-1000-15			ZD-12	563	
MZ-11	142A	3062	ZD-11	563	
MZ-110MA	141A	3092	ZD-11.5		
MZ-12	142A		ZD-12	563	
MZ-12B			ZD-12	563	
MZ-204B	5067A		ZD-3.9		
MZ-206		9021	ZD-5.6	561	
MZ-207	5071A	3334	ZD-6.8	561	
MZ-208	5072A	3136	ZD-8.2	562	
MZ-209	139A	3060	ZD-9.1	562	
MZ-210	140A		ZD-10	562	
MZ-210B	140A	3061	ZD-10	562	
MZ-212	142A	3062	ZD-12	563	
MZ-306C	5012A	3778			
MZ-5	5010A	3057/136A	ZD-5.6	561	
MZ-6	5013A	3334/5071A	ZD-6.8	561	
MZ-7	5071A		ZD-6.8	561	
MZ-8	5072A		ZD-8.2	562	
MZ-9			ZD-9.1	562	
MZ-92-9.1B			ZD-9.1	562	
MZ090	139A		ZD-9.1	562	
MZ10	140A	3061	ZD-10	562	
MZ1000-1					1N4728
MZ1000-10					1N4737
MZ1000-11				562	1N4738
MZ1000-12				562	1N4739
MZ1000-13	140A	3061	ZD-10	562	1N4740
MZ1000-14				563	1N4740
MZ1000-15	142A	3062	ZD-12	563	1N4742
MZ1000-16				563	1N4743
MZ1000-17	145A	3063	ZD-15	564	1N4744
MZ1000-18				564	1N4745
MZ1000-19					1N4746
MZ1000-2					1N4729
MZ1000-20	5079A		ZD-20		1N4747
MZ1000-21					1N4748
MZ1000-22					1N4749
MZ1000-23			ZD-27		1N4750
MZ1000-24			ZD-30		1N4751
MZ1000-25					1N4752
MZ1000-26	5143A	3409	ZD-36		1N4753
MZ1000-27					1N4754
MZ1000-28					1N4755
MZ1000-29					1N4756
MZ1000-3					1N4730
MZ1000-30					1N4757
MZ1000-31					1N4758
MZ1000-32					1N4759
MZ1000-33					1N4760
MZ1000-34					1N4761
MZ1000-35					1M86Z010
MZ1000-36					1N4763

Industry Standard No.	ECG	SK	GE	RS 276-	MOTOR.
MZ1000-37					1N4764
MZ1000-4					1N4731
MZ1000-5					1N4732
MZ1000-6					1N4733
MZ1000-7				561	1N4734
MZ1000-8				561	1N4735
MZ1000-9				561	1N4736
MZ1004	5007A	3773	ZD-3.9		
MZ1005	135A	3056	ZD-5.1		
MZ1006			ZD-6.0	561	
MZ1007	138A	3059	ZD-7.5		
MZ1008			ZD-8.2	562	
MZ1009				562	
MZ1010	140A	3061	ZD-10	562	
MZ1012	142A	3062	ZD-12	563	
MZ1014	144A	3094	ZD-14	564	
MZ1016			ZD-16	564	
MZ10A	140A	3061	ZD-10	562	
MZ11		3092	ZD-11	563	
MZ110MA	141A		ZD-11.5		
MZ12	142A	3062	ZD-12	563	
MZ120	5166A				5M200ZS5
MZ122					5M110ZSB5
MZ12A	142A	3062	ZD-12	563	
MZ12B	142A	3062	ZD-12	563	
MZ14	144A			564	
MZ15A	145A	3063	ZD-15	564	
MZ15T20	145A	3063	ZD-15	564	
MZ204B	5067A	3331			
MZ205	135A	3056	ZD-5.0		
MZ206	5070A	3056/135A	ZD-5.6	561	
MZ206B	5070A	9021			
MZ207	5071A	3058/137A	300	1122	
MZ207(ZENER)	5071A	3334	ZD-6.8	561	
MZ207-02A	5071A	3334	ZD-6.8	561	
MZ207A			ZD-6.8	561	
MZ207B		3059	ZD-6.8	561	
MZ207C			ZD-6.8	561	
MZ208	5072A		ZD-8.2	562	
MZ209	139A		ZD-9.1	562	
MZ209A			ZD-9.1	562	
MZ209B			ZD-9.1	562	
MZ209C			ZD-9.1	562	
MZ212	142A	3062	ZD-12	563	
MZ214				564	
MZ216				564	
MZ216B	5075A	3751			
MZ218B	5077A	3752			
MZ220					5M200ZS10
MZ222					5M110ZSB10
MZ224A	5031A	3797	ZD-24		
MZ2360	177	3100/519	300	1122	
MZ2361	177	3100/519	300	1122	
MZ240					5M200ZSB10
MZ250	5018A		ZD-9.1	562	
MZ27A	146A	3064	ZD-27		
MZ306	5070A	9021	ZD-6.0	561	
MZ306B	5012A	9021/5070A			
MZ306C	5070A	9021	ZD-6.0	561	
MZ307			ZD-6.8	561	
MZ308			ZD-8.0	562	
MZ309	139A		ZD-9.1	562	
MZ309B	139A		ZD-9.1	562	
MZ310	5156A	3060/139A	ZD-9.1	562	
MZ310A	140A	3061	ZD-10	562	
MZ312				563	
MZ314				564	
MZ316				564	
MZ316B	5075A	3751			
MZ318B	5077A	3752			
MZ320					5M200ZS20
MZ322					5M110ZSB20
MZ33A	147A	3095	ZD-33		
MZ340					5M200ZSB20
MZ406B	5070A	9021			
MZ408-02C	5072A		ZD-8.2	562	
MZ409-02B	139A		214	562	
MZ416B	5075A	3751			
MZ418B	5077A	3752			
MZ4618	5001A	3767			
MZ4620	134A	3055	ZD-3.6		
MZ4622	134A		ZD-3.6		
MZ4624	135A	3056	ZD-5.1		
MZ4625	135A	3056	ZD-5.1		
MZ4626	136A	3057	ZD-5.6	561	
MZ4627	137A	3058	ZD-6.2	561	
MZ5.1T5	135A		ZD-5.1		
MZ500-1	5000A	3766			1N5221A
MZ500-10	5011A	3057/136A	ZD-5.6	561	1N5232A
MZ500-11	5013A	3058/137A	ZD-6.2	561	1N5234A
MZ500-12	5014A			561	1N5235A
MZ500-13	5015A				1N5236A
MZ500-14	5016A	3136/5072A		562	1N5237A
MZ500-15	5018A	3060/139A	ZD-9.1	562	1N5239A
MZ500-16	5019A	3061/140A	ZD-10	562	1N5240A
MZ500-17	5020A	3092/141A	ZD-11.5	563	1N5241A
MZ500-18	5021A	3062/142A	ZD-12	563	1N5242A
MZ500-19	5022A			563	1N5243A
MZ500-2	5002A	3768			1N5223A
MZ500-20	5024A	3063/145A	ZD-15	564	1N5245A
MZ500-21	5025A		ZD-16	564	1N5246A
MZ500-22	5027A				1N5248A
MZ500-23	5029A				1N5250A
MZ500-24	5030A				1N5251A
MZ500-25	5031A				1N5252A
MZ500-26	5033A	3064/146A	ZD-27		1N5254A
MZ500-27	5035A				1N5256A
MZ500-28	5036A	3095/147A	ZD-33		1N5257A
MZ500-29	5037A				1N5258A
MZ500-3	5004A				1N5225A
MZ500-30	5038A				1N5259A
MZ500-31	5039A				1N5260A
MZ500-32	5040A				1N5261A
MZ500-33	5041A				1N5262A
MZ500-34	5042A				1N5263A
MZ500-35	5044A		ZD-62		1N5265A
MZ500-36	5045A				1N5266A
MZ500-37	5046A				1N5267A
MZ500-38	5047A	3098/150A	ZD-82		1N5268A
MZ500-39	5049A	3345/5095A	ZD-91		1N5270A
MZ500-4	5005A	3771	ZD-3.3		1N5226A
MZ500-40	5050A				1N5271A
MZ500-5	134A	3055	ZD-3.6		1N5227A
MZ500-6	5007A	3055/134A	ZD-3.6		1N5228A
MZ500-7	5008A				1N5229A
MZ500-8	5009A	3056/135A	ZD-5.1		1N5230A
MZ500-9	5010A	3056/135A	ZD-5.1		1N5231A
MZ500.10			ZD-5.6	561	
MZ500.15			ZD-9.1	562	
MZ500.18		3062	ZD-12	563	
MZ500.20		3335	ZD-20		
MZ500.26		3064	ZD-27		
MZ500.5		3055	ZD-3.6		
MZ5210					5M100ZS10
MZ5220					5M200ZS10
MZ5222					5M110ZSB10
MZ5240					5M200ZSB10
MZ5555					1N6283A
MZ5556					1N6287A
MZ5557					1N6289A
MZ5558					1N6303A
MZ5806					5M6.8ZS10
MZ5890					5M90ZS10
MZ5915	5130A	3396	5ZD-15		
MZ5918	5133A	3399	5ZD-18		
MZ5922	5136A	3402	5ZD-22		
MZ5927	5139A	3405	5ZD-27		
MZ5933	5142A	3408	5ZD-33		
MZ5956	5148A	3414			
MZ5A	136A	3057	ZD-5.6	561	
MZ6	137A	3334/5071A	ZD-6.8	561	
MZ6.2	137A	3058	ZD-6.2	561	
MZ6.2A	137A	3058	ZD-6.2	561	
MZ6.2B	137A	3058	ZD-6.2	561	
MZ6.2T5	137A	3058	ZD-6.2	561	
MZ605	137A	3058	ZD-6.2	561	
MZ610	137A	3058	ZD-6.2	561	
MZ620	137A	3058	ZD-6.2	561	
MZ623-12					1N4745A
MZ623-12A					1N4745A
MZ623-12B					1N4745A
MZ623-14					1N4746A
MZ623-14A					1N4746A
MZ623-14B					1N4746A
MZ623-18					1N4749A
MZ623-18A					1N4749A
MZ623-18B					1N4749A
MZ623-25					1N4755A
MZ623-25A					1N4755A
MZ623-25B					1N4755A
MZ623-9					1N4743A
MZ623-9A					1N4743A
MZ623-9B					1N4743A
MZ640	137A	3058	ZD-6.2	561	
MZ70-2.4B	5000A	3766			
MZ70-2.5A	5001A	3767			
MZ70-2.5B	5001A	3767			
MZ70-2.7B	5002A	3768			
MZ70-3.3A	5005A	3770/5004A	ZD-3.6		
MZ70-3.3B	5005A	3770/5004A	ZD-3.6		
MZ706	5120A				5M6.8ZS5
MZ70MA	5070A	9021	ZD-6.2	561	
MZ8	5072A	3136	ZD-8.2	562	
MZ806					5M6.8ZS10
MZ82A	150A	3098	ZD-82		
MZ9	131	3060/139A	ZD-44	2006	
MZ9.1B			ZD-9.1	562	
MZ906					5M6.8ZS20
MZ92-10					1N758
MZ92-100					1N985A
MZ92-10A	140A	3061	ZD-10	562	
MZ92-11					.5M11AZ10
MZ92-110					1N986A
MZ92-12					1N759
MZ92-120					1N987A
MZ92-12A	142A	3062	ZD-12	563	
MZ92-13					1N964A
MZ92-130					1N988A
MZ92-14					.5M14Z10
MZ92-140					.4M140Z10
MZ92-15					1N965A
MZ92-150					1N989A
MZ92-15A	145A	3063	ZD-15	564	
MZ92-16					1N966A
MZ92-160					1N990A
MZ92-17					.5M17Z10
MZ92-170					.4M170Z10
MZ92-18					1N967A
MZ92-180					1N991A
MZ92-19					.5M19Z10
MZ92-190					.4M190Z10
MZ92-2.4	5000A	3766			1N4370
MZ92-2.5	5001A	3767			.5M2.5AZ10
MZ92-2.7	5002A				1N4371
MZ92-2.8					.5M2.8AZ10
MZ92-20					1N968A
MZ92-200					1N992A
MZ92-22					1N969A
MZ92-24					1N970A
MZ92-24B	5080A	3336	ZD-22		
MZ92-25					.5M25Z10
MZ92-27					1N971A
MZ92-27A	146A	3064	ZD-27		
MZ92-28					.5M28Z10
MZ92-3.0					1N4372
MZ92-3.3	5005A	3770/5004A	ZD-3.3		1N746
MZ92-3.6					1N747
MZ92-3.9					1N748
MZ92-3.9A	134A	3055	ZD-3.6		
MZ92-30					1N972A
MZ92-33					1N973A
MZ92-36					1N974A
MZ92-39					1N975A
MZ92-4.3					1N749
MZ92-4.7					1N750
MZ92-4.7A	135A	3056	ZD-5.1		
MZ92-43					1N976A
MZ92-47					1N977A
MZ92-5.1					1N751
MZ92-5.1A	135A	3056	ZD-5.1		
MZ92-5.6					1N752
MZ92-5.6A	136A	3057	ZD-5.6	561	
MZ92-51					1N978A
MZ92-56					1N979A
MZ92-6,0B			ZD-6.2	561	
MZ92-6.0					.5M6.0AZ10
MZ92-6.0B	5070A	9021	ZD-6.2	561	
MZ92-6.2					1N753
MZ92-6.2A	137A	3058	ZD-6.2	561	
MZ92-6.8					1N754
MZ92-60					.5M60Z10
MZ92-62					1N980A
MZ92-68					1N981A
MZ92-7.5					1N755
MZ92-75					1N982A
MZ92-8.2		3136			1N756
MZ92-8.7					.5M8.7AZ10
MZ92-82					1N983A
MZ92-87					.5M87Z10
MZ92-9.1		3060			1N757
MZ92-9.1A	139A	3060	ZD-9.1	562	
MZ92-9.1B	139A	3060	ZD-9.1	562	
MZ92-91					1N984A
MZP5221,A,B					IN5221,A,B
MZP5270,A,B					IN5270,A,B
MZX9.1	139A	3060	ZD-9.1	562	
N-02	116	3311	504A	1104	
N-020	160		245	2004	

Industry Standard No.	ECG	SK	GE	RS 276-	MOTOR.
N-121122	130		14	2041	
N-41	116	3017B/117	504A	1104	
N-52329	130		14	2041	
N-7400A				1801	
N-7404A			7404	1802	
N-7408A			7408	1822	
N-7473A			7473A	1803	
N-756A	5017A	3783	ZD-9.1	562	
N-EA15X130	107	3018	11	2015	
N-EA15X131	108	3452	86	2038	
N-EA15X132	108	3984/106	86	2038	
N-EA15X133	126	3006/160	52	2024	
N-EA15X134	107	3018	11	2015	
N-EA15X135	107	3018	11	2015	
N-EA15X136	123A	3444	20	2051	
N-EA15X137	123A	3444	20	2051	
N-EA15X138	123A	3444	20	2051	
N-EA15X139	131MP	3840	44(2)	2006(2)	
N-EA16X27	110MP	3709		1123(2)	
N-EA16X29	142A	3062	ZD-12	563	
N-EA16X30	116	3311	504A	1104	
N-EA2136	158	3004/102A	53	2007	
N020	126		52	2024	
ON020540	128		243	2030	
N0282CT	181		75	2041	
N0400	162	3438	35		
ON047204-2	123A		20	2051	
ON120623	519		514	1122	
N121122	130		14	2041	
N13T1	6402	3628			
ON143285	519		514	1122	
ON187840	941		IC-263	010	
N1X	128	3045/225	40	2012	
N201AY	123A	3444	20	2051	
ON206068	519		514	1122	
ON271			17	2051	
N2A		3307	529	1123	
N2A(DIODE)	109			1123	
N2A-1	529	3307	529		
N2A-2	529	3307	529		
N2XA	154	3045/225	40	2012	
N3563	123A	3124/289	20	2051	
N4000		4000	4000		
N4001		4001	4001	2401	
N4002		4002	4002		
N4007		4007	4007		
N4009		4009	4009		
N4011		4011	4011	2411	
N4012		4012	4012	2412	
N4013		4013	4013	2413	
N4015		4015	4015		
N4016		4016	4016		
N4019		4019	4019		
N4021			4021	2421	
N4023		4023	4023	2423	
N4025		4025	4025		
N4027	2104	4027	4027	2427	
N4030		4030	4030		
N4049		4049	4049	2449	
N4050		4050	4050	2450	
N4081		4081	4081		
ON47204-1	123A	3444	20	2051	
N48	109	3087		1123	
N4967	176	3845	80		
N4T	123A	3018	20	2051	
N5065A					MC1358P
N5070	714	3075	IC-4		
N5070B	714		IC-4		MC1399P
N5071	715	3076	IC-6		
N5071A	715		IC-6		MC1399P
N5072	713		IC-5		
N5072A	713		IC-5		MC1323P
N5111	708	3135/709	IC-10		
N5111A	708		IC-10		
N52A1T	1171	3565			
N53A1T	1171	3565			
N5406		3175	300	1122	
N5406(RCA)	177		300	1122	
N5556T					MC1456G
N5556V					MC1456P1
N5558F					MC1458L
N5558T		3551			MC1458G
N5558V	778A	3465	IC-220		MC1458P1
N5595A					MC1495L
N5595F					MC1495L
N5596A		3892			MC1496L
N5596K	973	3233			MC1496G
N5709A	909D	3590	IC-250		MC1709CP2
N5709G					MC1709CF
N5709T	909	3590	IC-249		MC1709CG
N5709V					MC1709CP1
N5710A	910D		IC-252		MC1710CP
N5710T	910		IC-251		MC1710CG
N5711A	911D		IC-254		MC1711CP
N5711K	911		IC-253		MC1711CG
N5723A	923D	3165	IC-260	1740	MC1723CP
N5723L	923	3164	IC-259		
N5723T	923				MC1723CG
N5733K					MC1733CG
N5740T	940		IC-262		
N5741A	941D		IC-264		MC1741CP2
N5741T	941	3514	IC-263	010	MC1741CG
N5741V	941M		IC-265	007	MC1741CP1
N5747A	947D				MC1747CL
N5747F	947D				MC1747CL
N5747K	947		IC-267		
N5748A					MC1747CG
N5748T					MC1748CG
N57B2-11	160	3006	245	2004	
N57B2-13	160	3006	245	2004	
N57B2-14	160	3006	245	2004	
N57B2-15	102A	3004	53	2007	
N57B2-17	126	3008	52	2024	
N57B2-18	126	3008	52	2024	
N57B2-19	126	3008	52	2024	
N57B2-22	160	3006	245	2004	
N57B2-23	126	3008	52	2024	
N57B2-25	102A	3004	53	2007	
N57B2-3	102A	3004	53	2007	
N57B2-6	102A	3004	53	2007	
N57B2-7	102A	3004	53	2007	
N57B2-8	103A	3010	59	2002	
N57B4-2	121	3009	239	2006	
N57B4-4	121	3009	239	2006	
N7400A	7400	7400	7400	1801	
N7400F	7400	7400		1801	
N7400N			7400	1801	
N7401A	7400	7401		1801	
N7401F	7400	7401		1801	
N7402A	7402	7402	7402	1811	
N7402F	7402	7402		1811	
N7403A	7403	7403			
N7403F	7403	7403			
N7404A	7404	7404	7404	1802	
N7404F	7404	7404		1802	
N7405A	7405	7405			
N7405F	7405	7405			
N7406A	7406	7406	7406	1821	
N7406F	7406	7406		1821	
N7407A	7407	7407			
N7407F	7407	7407			
N7408A	7408	7408	7408	1822	
N7408F	7408	7408		1822	
N7409A	7409	7409			
N7409F	7409	7409			
N74107A	74107	74107			
N74107F	74107	74107			
N74109	74109	74109			
N74109B	74109	74109			
N74109F	74109	74109			
N7410A	7410	7410	7410	1807	
N7410F	7410	7410		1807	
N7410N				1807	
N7411A	7411	7411			
N7411F	7411	7411			
N74121A	74121	74121			
N74121F	74121	74121			
N74122A	74123	74122		1817	
N74122F	74122	74122			
N74123B	74123	74123	74123	1817	
N74123F	74123	74123		1817	
N74132B	74132	74132			
N74132F	74132	74132			
N7413A	7413	7413	7413	1815	
N7413F	7413	7413		1815	
N74141B	74141	74141			
N74145			74145	1828	
N74145B	74145	74145	74145	1828	
N7414B	7414	7414			
N7414F	7414	7414			
N74150B				1829	
N74150F	74150	74150		1829	
N74150N	74150	74150	74150	1829	
N74151A			7451	1825	
N74151B	74151	74151			
N74151F	74151	74151			
N74153B	74153	74153			
N74153F	74153	74153			
N74154F	74154	74154		1834	
N74154N	74154		74154	1834	
N74155B	74155	74155			
N74155F	74155	74155			
N74156B	74156	74156			
N74156F	74156	74156			
N74160B	74160	74160			
N74160F	74160	74160			
N74161B	74161	74161			
N74161F	74161	74161			
N74162	74162	74162			
N74162B	74162	74162			
N74162F	74162	74162			
N74163B	74163	74163			
N74163F	74163	74163			
N74164A	74164	74164			
N74164F	74164	74164			
N74165B	74165	74165			
N74165F	74165	74165			
N74166B	74166	74175			
N74166F	74166	74175			
N7416A	7416	7416			
N7416F	7416	7416			
N74170B	74170	74170			
N74170F	74170	74170			
N74174	74174	74174			
N74175B	74175	74175			
N74175F	74175	74175			
N7417A	7417	7417			
N7417F	7417	7417			
N74180A	74180	74180			
N74180F	74180	74180			
N74181F	74181	74181			
N74181N	74181	74181			
N74190B	74190	74190			
N74190F	74190	74190			
N74191B	74191	74191			
N74191F	74191	74191			
N74192B	74192	74192	74192	1831	
N74192F	74192	74192		1831	
N74193B	74193	74193	74193	1820	
N74193F	74193	74193		1820	
N74195B	74195	74195			
N74195F	74195	74195			
N74196A			74196	1833	
N74198F	74198	74196			
N74198N	74198	74198			
N74199F	74199	74199			
N74199N	74199	74199			
N7420A	7420	7420	7420	1809	
N7420F	7420	7420		1809	
N7420N				1809	
N7426A	7426	7426			
N7426F	7426	7426			
N7427A			7427	1823	
N7427F				1823	
N7427N				1823	
N7430A	7430	7430			
N7430F	7430	7430			
N7432A	7432	7432	7432	1824	
N7432F	7432	7432		1824	
N7432N				1824	
N7437A	7437	7437			
N7437F	7437	7437			
N7438A	7438	7438			
N7438F	7438	7438			
N7440A	7440	7440			
N7440F	7440	7440			
N7441B	7441	7441		1804	
N7441F	7441	7441		1804	
N7442A	7442	7442			
N7442BA	7442	7442			
N7442BF	7442	7442			
N7442F	7442	7442			
N7443A	7443	7443			
N7443F	7443	7443			
N7444B	7444	7444			
N7444F	7444	7444			
N7445B	7445	7445			
N7446B	7446	7446			
N7447B	7447	7447	7447	1805	
N7447F	7447	7447	7447	1805	
N7448B	7448	7448	7448	1816	
N7450A	7450	7450			
N7450F	7450	7450			
N7451A	7451	7451	7451	1825	
N7451F	7451	7451		1825	
N7451N				1825	
N7453A	7453	7453			
N7453F	7453	7453			

Industry Standard No.	ECG	SK	GE	RS 276-	MOTOR.
N7454A	7454	7454			
N7454F	7454	7454			
N7460A	7460	7460			
N7460F	7460	7460			
N7470A	7470	7470			
N7470F	7470	7470			
N7472A	7472	7472			
N7472F	7472	7472			
N7473	7473		7473	1803	
N7473A	7473	7473	7473	1803	
N7473F	7473	7473		1803	
N7474A	7474	7474	7474	1818	
N7474F	7474	7474		1818	
N7475B	7475	7475	7475	1806	
N7476B	7476	7476	7476	1813	
N7476F	7476			1813	
N7480A	7480	7480			
N7480F	7480	7480			
N7483B	7483	7483			
N7483F	7483	7483			
N7485A	7485	7485	7485	1826	
N7485B			7485	1826	
N7485F	7485	7485		1826	
N7485N				1826	
N7486A	7486	7486	7486	1827	
N7486F	7486	7486		1827	
N7486N				1827	
N7490A	7490	7490	7490	1808	
N7490F	7490	7490		1808	
N7492A	7492	7492	7492	1819	
N7492F	7492	7492		1819	
N7493A	7493A	7493			
N7493F	7493A	7493			
N7494B	7494	7494			
N7494F	7494	7494			
N7495A	7495	7495			
N7495F	7495	7495			
N7496B	7496	7496			
N7496F	7496	7496			
N74H00A	74H00	74H00			
N74H00F	74H00	74H00			
N74H04A	74H04	74H04			
N74H04F	74H04	74H04			
N74LS00	74LS00			1900	
N74LS00F					SN74LS00J
N74LS00N					SN74LS00N
N74LS01F	74LS01				SN74LS01J
N74LS01N	74LS01				SN74LS01N
N74LS02F					SN74LS02J
N74LS02N					SN74LS02N
N74LS03F					SN74LS03J
N74LS03N					SN74LS03N
N74LS04F					SN74LS04J
N74LS04N					SN74LS04N
N74LS05F					SN74LS05J
N74LS05N					SN74LS05N
N74LS08F					SN74LS08J
N74LS08N					SN74LS08N
N74LS09F					SN74LS09J
N74LS09N					SN74LS09N
N74LS10F					SN74LS10J
N74LS10N					SN74LS10N
N74LS11F					SN74LS11J
N74LS11N					SN74LS11N
N74LS12F					SN74LS12J
N74LS12N					SN74LS12N
N74LS132F	74LS132				SN74LS132J
N74LS132N	74LS132				SN74LS132N
N74LS136F					SN74LS136J
N74LS136N					SN74LS136N
N74LS138F					SN74LS138J
N74LS138N					SN74LS138N
N74LS139F	74LS139				SN74LS139J
N74LS139N	74LS139				SN74LS139N
N74LS13F	74LS13				SN74LS13J
N74LS13N	74LS13				SN74LS13N
N74LS145F					SN74LS145J
N74LS145N					SN74LS145N
N74LS14F					SN74LS14J
N74LS14N					SN74LS14N
N74LS151F					SN74LS151J
N74LS151N					SN74LS151N
N74LS153F					SN74LS153J
N74LS153N					SN74LS153N
N74LS157F					SN74LS157J
N74LS157N					SN74LS157N
N74LS158F					SN74LS158J
N74LS158N					SN74LS158N
N74LS15F					SN74LS15J
N74LS15N					SN74LS15N
N74LS164F					SN74LS164J
N74LS164N					SN74LS164N
N74LS170F	74LS170				SN74LS170J
N74LS170N	74LS170				SN74LS170N
N74LS174F					SN74LS174J
N74LS174N					SN74LS174N
N74LS175F					SN74LS175J
N74LS175N					SN74LS175N
N74LS181F					SN74LS181J
N74LS181N					SN74LS181N
N74LS190F					SN74LS190J
N74LS190N					SN74LS190N
N74LS191F					SN74LS191J
N74LS191N					SN74LS191N
N74LS192F	74LS192				SN74LS192J
N74LS192N	74LS192				SN74LS192N
N74LS193F					SN74LS193J
N74LS193N					SN74LS193N
N74LS196F					SN74LS196J
N74LS196N					SN74LS196N
N74LS197F					SN74LS197J
N74LS197N					SN74LS197N
N74LS20F					SN74LS20J
N74LS20N					SN74LS20N
N74LS21F	74LS21				SN74LS21J
N74LS21N	74LS21				SN74LS21N
N74LS221F					SN74LS221J
N74LS221N					SN74LS221N
N74LS22F					SN74LS22J
N74LS22N					SN74LS22N
N74LS251F					SN74LS251J
N74LS251N					SN74LS251N
N74LS253F					SN74LS253J
N74LS253N					SN74LS253N
N74LS260F					SN74LS260J
N74LS260N					SN74LS260N
N74LS266F					SN74LS266J
N74LS266N					SN74LS266N
N74LS26F	74LS26				SN74LS26J
N74LS26N	74LS26				SN74LS26N
N74LS27F					SN74LS27J
N74LS27N					SN74LS27N
N74LS283F					SN74LS283J
N74LS283N					SN74LS283N
N74LS28F					SN74LS28J
N74LS28N					SN74LS28N
N74LS293F					SN74LS293J
N74LS293N					SN74LS293N
N74LS30F					SN74LS30J
N74LS30N					SN74LS30N
N74LS32F					SN74LS32J
N74LS32N					SN74LS32N
N74LS33F					SN74LS33J
N74LS33N					SN74LS33N
N74LS37F	74LS37				SN74LS37J
N74LS37N	74LS37				SN74LS37N
N74LS386F					SN74LS386J
N74LS386N					SN74LS386N
N74LS38F					SN74LS38J
N74LS38N					SN74LS38N
N74LS40F	74LS40				SN74LS40J
N74LS40N	74LS40				SN74LS40N
N74LS42F					SN74LS42J
N74LS42N					SN74LS42N
N74LS51F					SN74LS51J
N74LS51N					SN74LS51N
N74LS54F	74LS54				SN74LS54J
N74LS54N	74LS54				SN74LS54N
N74LS55F					SN74LS55J
N74LS55N					SN74LS55N
N74LS670F	74LS670				SN74LS670J
N74LS670N	74LS670				SN74LS670N
N74LS73F					SN74LS73AJ
N74LS73N					SN74LS73AN
N74LS74F					SN74LS74AJ
N74LS74N					SN74LS74AN
N74LS75F					SN74LS75J
N74LS75N					SN74LS75N
N74LS76F					SN74LS76AJ
N74LS76N	74LS76A				SN74LS76AN
N74LS78F					SN74LS78AJ
N74LS78N					SN74LS78AN
N74LS85F					SN74LS85J
N74LS85N					SN74LS85N
N74LS86F					SN74LS86J
N74LS86N					SN74LS86N
N74LS90F	74LS90				SN74LS90J
N74LS90N	74LS90				SN74LS90N
N74LS92F					SN74LS92J
N74LS92N					SN74LS92N
N74LS93F					SN74LS93J
N74LS93N					SN74LS93N
N74LS95BF					SN74LS95BJ
N74LS95BN					SN74LS95BN
N74S00A	74S00	74S00			
N74S00F	74S00	74S00			
N74S04A	74S04	74S04			
N74S04F	74S04	74S04			
N8T13B					MC8T13P
N8T13F					MC8T13L
N8T14B					MC8T14P
N8T14E					MC8T14L
N8T15A					MC1488L
N8T15F					MC1488L
N8T16A					MC1489L
N8T23B					MC8T23P
N8T23E					MC8T23L
N8T24B					MC8T24P
N8T24E					MC8T24L
N8T26AB	6880				MC8T26AP
N8T26AE	6880				MC8T26AL
N8T26B	6880				MC8T26AP
N8T28B	6889				MC8T28P
N8T37A					MC3437P
N8T38A					MC3438P
N8T95B					MC8T95P
N8T95F					MC8T95L
N8T96B					MC8T96P
N8T96F					MC8T96L
N8T97B					MC8T97P
N8T97F					MC8T97L
N8T98B					MC8T98P
N8T98F					MC8T98L
NA-1114-1001	160		245	2004	
NA-1114-1002	160	3008	245	2004	
NA-1114-1004	102A		53	2007	
NA-1114-1005	102A		53	2007	
NA-1114-1006	102A		53	2007	
NA-1114-1007	102A		53	2007	
NA-1114-1008	102A		53	2007	
NA-1114-1009	102A		53	2007	
NA-1114-1010	102A		53	2007	
NA-1114-1011	102A		53	2007	
NA-22	116	3016	504A	1104	
NA-25	116	3016	504A	1104	
NA-33	116	3016	504A	1104	
NA-36	116	3016	504A	1104	
NA-46	116	3016	504A	1104	
NA0305	5940		5048		
NA0505	5940		5048		
NA1	5940		5048		
NA1005	5944		5048		
NA1022-1001	160	3008	245	2004	
NA1022-1007	102A		53	2007	
NA104	125	3033	510,531	1114	
NA105	125	3033	510,531	1114	
NA11	5834		5048		
NA13	116	3016	504A	1104	
NA1505	5944		5048		
NA156H	1194	3484			
NA16221435	909		IC-249		
NA20	101	3861	8	2002	
NA2002	1232			703	
NA2005	5944		5048		
NA21	5834	3016	5048	1104	
NA22	116		504A	1104	
NA25	116		504A	1104	
NA30	103A	3010	59	2002	
NA32	116	3016	504A	1104	
NA33	116		504A	1104	
NA35	116	3016	504A	1104	
NA36	116		504A	1104	
NA42	116	3016	504A	1104	
NA45	116	3016	504A	1104	
NA46	116		504A	1104	
NA5015-1012				2024	
NA5018-1001	160		245	2004	
NA5018-1002	126		52	2024	
NA5018-1003	126		52	2024	
NA5018-1004	126		52	2024	
NA5018-1005	126		52	2024	
NA5018-1006	126		52	2024	
NA5018-1007	126		52	2024	
NA5018-1008	126		52	2024	
NA5018-1009	126		52	2024	
NA5018-1010	126		52	2024	
NA5018-1011	126		52	2024	
NA5018-1012	126		52		

Industry Standard No.	ECG	SK	GE	RS 276-	MOTOR.
NA5018-1013	102A		53	2007	
NA5018-1014	102A		53	2007	
NA5018-1015	102A		53	2007	
NA5018-1016	102A		53	2007	
NA5018-1022	126		52	2024	
NA5018-1219	126		52	2024	
NA5018-1220	126		52	2024	
NA603	5850		5048		
NA615	5854		5048		
NA62	117	3017B	504A	1104	
NA63	117	3017B	504A	1104	
NA65	117	3017B	504A	1104	
NA66	117	3017B	504A	1104	
NA74	125	3032A	510,531	1114	
NA75	125	3032A	510,531	1114	
NA76	125	3032A	510,531	1114	
NA84	125	3032A	510,531	1114	
NA85	125	3032A	510,531	1114	
NA86	125	3032A	510,531	1114	
NA92H2	1195	3469			
NAM383	1232			703	
NAP-T-Z-10	102A		53	2007	
NAP-TZ-10	102A	3004	53	2007	
NAP-TZ-8	121	3009	239	2006	
NB011		3124	269		
NB011(NPN)			20	2051	
NB013	199	3124/289	62	2010	
NB021	294	3138/193A	48		
NB121	159	3114/290	82	2032	
NB211	297	3122	271	2030	
NB211E1			271	2030	
NB211EI	297	3122	271		
NBS12-100					MDA2501
NBS12-200					MDA2502
NBS12-300					MDA2504
NBS12-400					MDA2504
NBS12-50					MDA2500
NBS12-600					MDA2506
NBS20-100					MDA2501
NBS25-200					MDA2502
NBS25-300					MDA2504
NBS25-400					MDA2504
NBS25-50					MDA2500
NBS25-600					MDA2506
NBS30-100					MDA3501
NBS30-1000					MDA3510
NBS30-200					MDA3502
NBS30-300					MDA3504
NBS30-400					MDA3504
NBS30-50					MDA3500
NBS30-600					MDA3506
NBS30-800					MDA3508
NC1200					MR1-1200
NC1709CP1	909		IC-249		
NC207AL	123A		20	2051	
NC29	109	3087		1123	
NC30	102A		53	2007	
NC32	158		53	2007	
NC33	103A	3835	59	2002	
NC34	176	3845	80		
NCB14	295	3253	270		
NCB35	235	3197	215		
NCBV14	186A	3253/295	270	2052	
NCBW35	235	3197	215		
NCR046	128		243	2030	
NCR047	128		243	2030	
NCS9018D	107	3018	11	2015	
ND5700					2N6442
ND5701					2N6442
ND5702					2N6442
ND62WV	988	3973			
NE-446AQ	177		300	1122	
NE02103					2N6604
NE02107					2N6603
NE02108					2N6603
NE02112					MRF904
NE02133					MMBR901
NE02135					MRF901
NE0801					MRF838
NE0803					MRF840
NE0810					MRF840
NE22120	76				MRF511
NE22154					MRF511
NE269	102A	3004	53	2007	
NE32707					2N6603
NE32708					2N6603
NE32712					MRF904
NE32730					MRF901
NE41603					MRF962
NE41607					MRF962
NE41610					MRF965
NE41703					2N6603
NE41707					2N6603
NE41708					2N6603
NE41712					MRF904
NE41735					MRF901
NE4304	312	3112		2028	
NE501A					MC1733CL
NE501K					MC1733CG
NE515A					MC1420G
NE515G					MC1520F
NE515K					MC1420G
NE516A					MC1420G
NE516G					MC1520F
NE516K					MC1420G
NE531G					MC1439G
NE531T					MC1439G
NE531V					MC1439P
NE533G					MC1776CG
NE533T					MC1776CG
NE533V					MC1776CG
NE537G					MC1456G
NE537T					MC1456G
NE540L					MC1554G
NE550A	923D				MC1723CP
NE550L					MC1723CG
NE555	955M	3564			
NE555JG	955M	3564			MC1455U
NE555L		3693			MC1455G
NE555P		3564			MC1455P1
NE555T		3693			MC1455G
NE555V	955M	3564		1723	MC1455P1
NE556A	978	3689			MC3456P
NE556I					MC3456L
NE556V	978	3689			
NE565A	989	3595			NE565N
NE565K					NE565N
NE567	832	9089			
NE567C	832	9089			
NE567N	832	9089			
NE567V	832	9089			
NE57510					MRF905
NE592A					NE592A
NE592K					NE592K

Industry Standard No.	ECG	SK	GE	RS 276-	MOTOR.
NE59312					2N4957
NE64310					MRF905
NE66912					2N6304
NE71112					2N5583
NE73412					2N5031
NE73435					BFR90
NE74014	278				2N5109
NE74020					MRF511
NE74054					MRF511
NE74113					MRF517
NE74114					MRF517
NE7805	960	3591			
NE7806	962	3669			
NE7812	966	3592			
NE7815	968	3593			
NE7824	972	3670			
NE78L05	977	3462			
NE78L08	981	3724			
NE78M05	960	3591			
NE78M24	972	3670			
NF4302	457	3112		2028	2N5458
NF4303	457	3112		2028	2N5458
NF4304	457				2N5458
NF4445					MFE2012
NF500	116	3116	504A	1104	MPF102
NF501	116	3116	504A	1104	MPF102
NF506	116	3116	504A	1104	MPF102
NF510					2N4093
NF511					2N4093
NF52/					2N5639
NF520	312	3112		2028	
NF521					2N3970
NF522	312	3112		2028	2N5640
NF523	312	3112		2028	2N3971
NF530					2N5638
NF531	312	3112		2028	2N5639
NF532					2N5640
NF533	312	3112		2028	2N5640
NF5457					2N5457
NF5458					2N5458
NF5459					2N5459
NF5485	312	3112		2028	
NF5486	312	3112		2028	
NF550	116	3116	504A	1104	
NF580					MFE2012
NF581					MFE2012
NF582					MFE2012
NF583					MFE2011
NF584					MFE2012
NF585					MFE2011
NGP3002	109	3091		1123	
NGP3003	5802	9005		1143	
NGP5002	137A		ZD-6.2	561	
NGP5007	139A	3066/118	ZD-9.1	562	
NGP5010	142A	3062	ZD-12	563	
NJ100A	107	3018	11	2015	
NJ100B	123A	3444	20	2051	
NJ101B	159	3114/290	82	2032	
NJ102C	123A	3124/289	20	2051	
NJ107	128		243	2030	
NJ181B	102A		53	2007	
NJ202B	107	3293	11	2015	
NJ703N	703A	3157	IC-12		
NJM-703N	703A	3087	IC-12		
NJM2016	1194	3484			
NJM2201	1082	3461	IC-140		
NJM703	703A	3157	IC-12		
NJM78L05	977	3462			
NJM78L05A	977	3462	VR-100		
NJM78M18A	958	3699			
NK12-1A19	123A		20	2051	
NK1302	160		245	2004	
NK1404	160		245	2004	
NK14A119	121	3717	239	2006	
NK4A-1A19	121	3717	239	2006	
NK65-1A19	160		245	2004	
NK65-2A19	103A		59	2002	
NK65-4A19	101	3861	8	2002	
NK65A119	160		245	2004	
NK65B119	126		52	2024	
NK65C119	121	3717	239	2006	
NK66-1A19	102A		53	2007	
NK66-2A19	102A		53	2007	
NK66-3A19	102A		53	2007	
NK6B-3A19	121MP	3718	239(2)	2006(2)	
NK74-3A19	100	3721	1	2007	
NK77A119	105		4		
NK77B119	121MP	3718	239(2)	2006(2)	
NK77C119	121	3717	239	2006	
NK84AA19	121	3717	239	2006	
NK88B119	160		245	2004	
NK88C119	100	3721	1	2007	
NK8P-2A19	102A		53	2007	
NK92-1A19	105		4		
NK9L-4A19	102A		53	2007	
NKA8-1A19	160		245	2004	
NKT-401	121	3009	239	2006	
NKT-402	121	3009	239	2006	
NKT-403	121	3009	239	2006	
NKT-404	121	3009	239	2006	
NKT-405	121	3009	239	2006	
NKT-415	121	3009	239	2006	
NKT-416	121	3009	239	2006	
NKT-451	121	3009	239	2006	
NKT-452	121	3009	239	2006	
NKT-453	121	3009	239	2006	
NKT-454	121	3009	239	2006	
NKT-501	121	3009	239	2006	
NKT-503	121	3009	239	2006	
NKT-504	121	3009	239	2006	
NKT102	158		53	2007	
NKT103	126	3006/160	52	2024	
NKT10339	123A	3122	20	2051	
NKT104	158		53	2007	
NKT10419	123A	3122	20	2051	
NKT10439	123A	3122	20	2051	
NKT105	158	3005	53	2007	
NKT10519	123A	3122	20	2051	
NKT106	158	3005	53	2007	
NKT107	158	3004/102A	53	2007	
NKT108	158	3004/102A	53	2007	
NKT109	158	3004/102A	53	2007	
NKT11	102A		53	2007	
NKT12	102A	3721/100	1	2007	
NKT121	160	3006	245	2004	
NKT122	160	3006	245	2004	
NKT123	102A	3004	53	2007	
NKT12329	123A	3122	20	2051	
NKT124	160	3006	245	2004	
NKT12429	123A	3122	20	2051	
NKT125	160	3006	245	2004	
NKT126	102A	3004	53	2007	
NKT127	126	3006/160	52	2024	
NKT128	100	3005	1	2007	

Industry Standard No.	ECG	SK	GE	RS 276-	MOTOR.
NKT129	100	3005	1	2007	
NKT131	126	3006/160	52	2024	
NKT132	126	3005	52	2024	
NKT133	102	3004	53	2007	
NKT13329	123A	3122	20	2051	
NKT13429	123A	3122	20	2051	
NKT141	100	3005	1	2007	
NKT142	100	3005	1	2007	
NKT143	100	3005	1	2007	
NKT144	100	3005	1	2007	
NKT151	126	3006/160	52	2024	
NKT152	126	3006/160	52	2024	
NKT153/25	158	3006/160	53	2007	
NKT15325	126	3006/160	52	2024	
NKT154/25	158	3006/160	53	2007	
NKT15425	126		52	2024	
NKT162	100	3005	1	2007	
NKT16229	161	3132	39	2015	
NKT163	100	3005	1	2007	
NKT163/25	158	3005	53	2007	
NKT16325	100	3005	1	2007	
NKT164	100	3005	1	2007	
NKT164/25	158	3005	53	2007	
NKT16425	100	3005	1	2007	
NKT202	100	3005	1	2007	
NKT203	100	3004/102A	1	2007	
NKT20329	159	3114/290	82	2032	
NKT20329A	159	3114/290	82	2032	
NKT20339	159	3114/290	82	2032	
NKT204	100	3004/102A	1	2007	
NKT205	100	3004/102A	1	2007	
NKT206	100	3005	1	2007	
NKT207	100	3005	1	2007	
NKT208	102A		53	2007	
NKT211	158	3004/102A	53	2007	
NKT212	158	3123	53	2007	
NKT213	158	3004/102A	53	2007	
NKT214	158	3004/102A	53	2007	
NKT215	158	3004/102A	53	2007	
NKT216	158	3004/102A	53	2007	
NKT217	158	3004/102A	53	2007	
NKT218	158	3004/102A	53	2007	
NKT219	158		53	2007	
NKT221	102	3005	1	2007	
NKT222	102	3004	53	2007	
NKT222S1	100	3005	1	2007	
NKT222S2	100	3005	1	2007	
NKT223	102	3004	2	2007	
NKT223A	102A	3004	53	2007	
NKT224	102	3004	2	2007	
NKT224A	102A	3004	53	2007	
NKT224J	176	3123	80		
NKT225	102	3004	2	2007	
NKT225A	102A	3004	53	2007	
NKT225J	176	3123	80		
NKT226	102	3004	2	2007	
NKT226A	102A	3004	53	2007	
NKT226J	176	3123	80		
NKT227	102	3004	2	2007	
NKT227A	102A	3004	53	2007	
NKT228	102	3004	2	2007	
NKT228A	102A	3004	53	2007	
NKT229			53	2007	
NKT231	102	3004	2	2007	
NKT231A	102A	3004	53	2007	
NKT232	102	3004	2	2007	
NKT232A	102A	3004	53	2007	
NKT24	102A	3123	53	2007	
NKT242	102	3123	52	2024	
NKT243	100	3005	1	2007	
NKT244	102A	3004	53	2007	
NKT245	102		53	2007	
NKT246	102A	3004	53	2007	
NKT247	102	3004	2	2007	
NKT247A	102A	3004	53	2007	
NKT247J	176	3123	80		
NKT249	126	3006/160	52	2024	
NKT25	102A		53	2007	
NKT251	160	3006	245	2004	
NKT252	126	3006/160	52	2024	
NKT253	126	3006/160	52	2024	
NKT254	126	3006/160	52	2024	
NKT255	126	3006/160	52	2024	
NKT261	100	3005	1	2007	
NKT262	100	3005	1	2007	
NKT263	100	3005	1	2007	
NKT264	100	3005	1	2007	
NKT265	126	3006/160	52	2024	
NKT270	126	3006/160	52	2024	
NKT271	158	3004/102A	53	2007	
NKT272	158		53	2007	
NKT273	102A	3004	53	2007	
NKT274	158	3123	53	2007	
NKT275	158	3004/102A	53	2007	
NKT275A	158	3004/102A	53	2007	
NKT275E	158	3004/102A	53	2007	
NKT275J	158	3004/102A	53	2007	
NKT278	102A	3004	53	2007	
NKT281	158		53	2007	
NKT303	102A		53	2007	
NKT308	102A	3004	53	2007	
NKT32	102A	3123	53	2007	
NKT33	102A		53	2007	
NKT35219	161	3117	39	2015	
NKT4	102A	3123	53	2007	
NKT401	121	3717	239	2006	
NKT402	121	3717	239	2006	
NKT403	121	3717	239	2006	
NKT404	121	3717	239	2006	
NKT405	121	3717	239	2006	
NKT4055	219	3173	74	2043	
NKT406	121	3717	239	2006	
NKT415	121	3717	239	2006	
NKT416	121	3717	239	2006	
NKT42	100	3005	1	2007	
NKT43	100	3005	1	2007	
NKT450	121	3009	239	2006	
NKT451	131		44	2006	
NKT452	131		44	2006	
NKT452-S1	131	3014	44	2006	
NKT453	131		44	2006	
NKT454	121	3717	239	2006	
NKT5	102A	3123	53	2007	
NKT501	121	3717	239	2006	
NKT503	121	3717	239	2006	
NKT504	121	3717	239	2006	
NKT52	102A		53	2007	
NKT53	102A		53	2007	
NKT54	102A		53	2007	
NKT618	126		52	2024	
NKT618J	176	3123	80		
NKT62	100	3005	1	2007	
NKT63	100	3005	1	2007	
NKT64	100	3005	1	2007	

Industry Standard No.	ECG	SK	GE	RS 276-	MOTOR.
NKT674F	126		245	2004	
NKT675	126	3006/160	52	2024	
NKT676	126	3006/160	52	2024	
NKT677	126	3006/160	52	2024	
NKT677F	126		245	2004	
NKT701	103A	3010	59	2002	
NKT703	103A	3010	59	2002	
NKT713	103A	3010	59	2002	
NKT717	103A	3010	59	2002	
NKT72	100	3005	1	2007	
NKT73	100	3005	1	2007	
NKT732			8	2002	
NKT734	101	3861	8	2002	
NKT736	101	3861	8	2002	
NKT74	100	3005	1	2007	
NKT751	103A	3010	59	2002	
NKT752	103A	3010	59	2002	
NKT753	101	3861	8	2002	
NKT773	103A	3835	59	2002	
NKT781	103A	3010	59	2002	
NKT800112	312	3112		2028	
NKT800113	312	3112		2028	
NKT80111	457	3112		2028	2N5457
NKT80112	457				2N5457
NKT80113	457				2N5457
NKT80211	312	3112		2028	2N5459
NKT80212	312	3112		2028	2N5459
NKT80213	312	3112		2028	2N5459
NKT80214	312	3112		2028	2N5459
NKT80215	457	3112		2028	2N5458
NKT80216	457	3112		2028	2N5458
NL-10	116	3016	504A	1104	
NL-102	123A		20	2051	
NL-2N683	5522	6622			
NL-2N684	5523	6624/5524			
NL-2N685	5524	6624			
NL-2N686	5525	6627/5527			
NL-2N687	5526	6627/5527			
NL-2N688	5527	6627			
NL-2N689	5528	6629/5529			
NL-2N690	5529	6629			
NL-2N692	5531	6631			
NL-C35A	5542	6642	MR-3		
NL-C35B	5543		MR-3		
NL-C35C	5546		MR-3		
NL-C35D	5545		MR-3		
NL-C35F	5541	6642/5542	MR-3		
NL-C35G	5543		MR-3		
NL-C35U	5540	6642/5542	MR-3		
NL-C36M	5529	6629			
NL-C36N	5531	6631			
NL-C37A	5522	6622	MR-3		
NL-C37B	5524	6624	MR-3		
NL-C37C	5526	3582/5545	MR-3		
NL-C37D	5527	6627	MR-3		
NL-C37E	5528	6629/5529			
NL-C37M	5529	6629			
NL-C37N	5531	6631			
NL-C38A	5541	6642/5542			
NL-C38U	5540	6642/5542			
NL10	116		504A	1104	
NL100B	107	3018	11	2015	
NL15	116	3016	504A	1104	
NL20	116	3016	504A	1104	
NL25	116	3016	504A	1104	
NL30	116	3016	504A	1104	
NL40	116	3017B/117	504A	1104	
NL5	116	3311	504A	1104	
NL50	116	3017B/117	504A	1104	
NL60	116	3017B/117	504A	1104	
NLC36S		6631	C36S		
NN50	116		504A	1104	
NN650			82	2032	
NN7000	192		63	2030	
NN7001	192		63	2030	
NN7002	192		63	2030	
NN7003	192		63	2030	
NN7004	192		63	2030	
NN7005	192		63	2030	
NN7500	193		67	2023	
NN7501	193		67	2023	
NN7502	193		67	2023	
NN7503	193		67	2023	
NN7504	193		67	2023	
NN7505	193		67	2023	
NN7511	193		67	2023	
NN9017			20	2051	
NP50A	116	3017B/117	504A	1104	
NP60A	116	3017B/117	504A	1104	
NPC0010	116	3311	504A	1104	
NPC004	109	3090			
NPC0050	116	3017B/117	504A	1104	
NPC0100	116	3017B/117	504A	1104	
NPC069	123AP	3245/199	212	2010	
NPC069-98	123AP	3245/199	212	2010	
NPC079	159		221		
NPC1075	233	3039/316	210	2009	
NPC108	451	3116	504A	1104	2N5485
NPC108A	451	3116	504A	1104	2N5485
NPC115	128	3024	243	2030	
NPC167	161	3716	39	2015	
NPC173	108	3452	86	2038	
NPC187	128	3024	243	2030	
NPC188	108	3452	86	2038	
NPC189	128	3024	243	2030	
NPC211N	457	3112		2028	2N5457
NPC212N	457	3112		2028	2N5457
NPC213N	457				2N5458
NPC214N	457	3112		2028	2N5457
NPC215N	457	3112		2028	2N5457
NPC216N	457	3112		2028	2N5458
NPC312N	312	3112		2028	
NPC737	123A	3444	20	2051	
NPS404			22	2032	
NPS404A			82	2032	
NPS6512			210	2051	
NPS6513			210	2051	
NPS6514			20	2051	
NPS6516			21	2034	
NPS6517			21	2034	
NPS6518	153	3083/197	69	2049	
NPS6520			20	2051	
NPSA20			62	2010	
NR-071AU	123A	3124/289	20	2051	
NR-10	101		8	2002	
NR-141ES	192	3137	63	2030	
NR-141ET	192		63	2030	
NR-261AS			20	2051	
NR-421AS	199	3245	62	2010	
NR-431AG	123A	3018	20	2051	
NR-431AS	123A	3018	20	2051	
NR-461AS	123A		20	2051	
NR-601AT	159		82	2032	
NR-621AU	234	3114/290	65	2050	

Industry Standard No.	ECG	SK	GE	RS 276-	MOTOR.
NR-671ET	193		67	2023	
NR041	123A	3444	20		
NR041E	123A	3444	20	2051	
NR041I				2051	
NR05	101	3861	8	2002	
NR071AU	123A	3444	20	2051	
NR091ET	123A	3444	20	2051	
NR10	101	3861	8	2002	
NR20	103	3862	8	2002	
NR201AY	123A	3444	20	2051	
NR261AS	123A	3444	20	2051	
NR271AY	123A	3444	20	2051	
NR30	101	3861	8	2002	
NR421	107	3018	11	2015	
NR421DG	107	3018	11	2015	
NR461	123A	3444	20	2051	
NR461AA	161	3132	39	2015	
NR461AF	107	3018	11	2015	
NR461EH	123A	3444	20	2051	
NR5	101	3861	8	2002	
NR577H	1082	3461			
NR601BT	159	3114/290	82	2032	
NR621AT	159	3114/290	82	2032	
NR621EU	159	3114/290	82	2032	
NR631AY	159	3114/290	82	2032	
NR700	101	3861	8	2002	
NR7916	312	3448	FET-1	2035	
NS-3065	712	3072	IC-2		
NS08	5461	3685			
NS1000	159	3114/290	82	2032	1N4933
NS1000A	159	3114/290	82	2032	
NS1001	159	3114/290	82	2032	1N4934
NS1001A	159	3114/290	82	2032	
NS1002					1N4935
NS1004					1N4936
NS1005					1N4937
NS1006					1N4937
NS1110	124		12		
NS12006	5822	3500/5882	5040		MR1376
NS121	102	3722	2	2007	
NS1355	192		63	2030	
NS1356	108	3452	86	2038	
NS1500	123A	3444	20	2051	
NS1510	107	3039/316	11	2015	
NS15835L1	506		511	1114	
NS1672	159	3114/290	82	2032	
NS1672A	159	3114/290	82	2032	
NS1673	159	3114/290	82	2032	
NS1674	159	3114/290	82	2032	
NS1674A	159	3114/290	82	2032	
NS1675	159	3114/290	82	2032	
NS1675A	159	3114/290	82	2032	
NS18	5462	3686			
NS1861	159	3114/290	82	2032	
NS1861A	159	3114/290	82	2032	
NS1862	159	3114/290	82	2032	
NS1863	159	3114/290	82	2032	
NS1863A	159	3114/290	82	2032	
NS1864	159	3114/290	82	2032	
NS1864A	159	3114/290	82	2032	
NS1900	192		63	2030	
NS1960	192		63	2030	
NS1972	123A	3444	20	2051	
NS1973	123A	3444	20	2051	
NS1974	123A	3444	20	2051	
NS1975	123A	3444	20	2051	
NS20-42			40	2012	
NS2000					MR850
NS2001					MR851
NS2002					MR852
NS2003					MR854
NS2004					MR854
NS2005					MR856
NS2006	156	3051	512		MR856
NS2007	156	3051	512		
NS2008	156	3051	512		
NS2100	128	3024	243	2030	
NS2101	128	3024	243	2030	
NS3000					MR850
NS30000	6006				1N3909
NS30001	6006				1N3910
NS30002	6006				1N3911
NS30003	6008				1N3912
NS30004	6010				1N3913
NS3001					MR851
NS3002					MR852
NS3003					MR854
NS3004					MR854
NS3005					MR856
NS3006	156	3051	512		MR856
NS3007	156	3051	512		
NS3008	156	3051	512		
NS3039	107	3039/316	11	2015	
NS3040	107	3039/316	11	2015	
NS3041	107	3039/316	11	2015	
NS316	312	3448	FET-1	2035	
NS32	102	3722	2	2007	
NS3300	108	3452	86	2038	
NS345	161	3716	39		
NS3455				2015	
NS3700					2N2643
NS38	5465	3687			
NS381	108	3452	86	2038	
NS382	108	3452	86	2038	
NS3903	123A		210	2051	
NS3904	123A		20	2051	
NS3905	159		21	2034	
NS3906	159		21	2034	
NS404	159	3114/290	82	2032	
NS406	161	3716	39	2015	
NS45006			283	2016	
NS475	123A	3444	20	2051	
NS476	123A	3444	20	2051	
NS477	123A	3444	20	2051	
NS478	123A	3444	20	2051	
NS479	123A	3444	20	2051	
NS48	5465	3687			
NS480	123A	3444	20	2051	
NS48004			40	2012	
NS500					1N4933
NS501					1N4934
NS502					1N4935
NS504					1N4936
NS505					1N4937
NS506					1N4937
NS6000	5818				1N3879
NS6001	159	3114/290	82	2032	1N3880
NS6002	5818				1N3881
NS6003	5820				1N3882
NS6004	5820				1N3883
NS6005					MR1366
NS6006					MR1366
NS6062	159	3114/290	82	2032	
NS6062A	159	3114/290	82	2032	
NS6063	159	3114/290	82	2032	
NS6063A	159	3114/290	82	2032	
NS6064	159	3114/290	82	2032	
NS6064A	159	3114/290	82	2032	
NS6065	159	3114/290	82	2032	
NS6065A	159	3114/290	82	2032	
NS6112	108	3452	86	2038	
NS6113	108	3452	86	2038	
NS6114	123A	3444	20	2051	
NS6115	123A	3444	20	2051	
NS6207	123A	3444	20	2051	
NS6210	123A	3444	20	2051	
NS6211	159	3114/290	82	2032	
NS6211A	159	3114/290	82	2032	
NS6212	154	3045/225	40	2012	
NS6241	159	3114/290	82	2032	
NS661	159	3114/290	82	2032	
NS662	159	3114/290	82	2032	
NS663	159	3114/290	82	2032	
NS664	159	3114/290	82	2032	
NS665	159	3114/290	82	2032	
NS666	159	3114/290	82	2032	
NS667	159	3114/290	82	2032	
NS668	159	3114/290	82	2032	
NS7261	108	3452	86	2038	
NS7262	123A	3444	20	2051	
NS7267	108	3452	86	2038	
NS7301					2N2643
NS7302					2N2643
NS7303					2N2976
NS7304					2N2976
NS7305					2N2976
NS731	123A	3444	20	2051	
NS731A	123A	3444	20	2051	
NS732	159	3114/290	82	2032	
NS732A	159	3114/290	82	2032	
NS733	123A	3444	20	2051	
NS733A	123A	3444	20	2051	
NS734	123A	3444	20	2051	
NS734A	123A	3444	20	2051	
NS9400	128	3024	243	2030	
NS9420	128	3024	243	2030	
NS949	123A	3444	20	2051	
NS950	192		63	2030	
NS9500	128	3024	243	2030	
NS9540	128	3024	243	2030	
NS9710	108	3039/316	86	2038	
NS9728	128	3024	243	2030	
NS9729	128	3024	243	2030	
NS9730	128	3024	243	2030	
NS9731	128	3024	243	2030	
NSD102					MDS27
NSD103					MDS27
NSD104					2N6552
NSD105					2N6552
NSD106					2N6553
NSD123					2N6591
NSD127					2N6591
NSD128					2N6592
NSD129					2N6593
NSD130					2N6557
NSD132					2N6557
NSD133					2N6558
NSD134					2N6558
NSD135					2N6559
NSD151	268				2N6549
NSD152	268				2N6548
NSD202					MDS77
NSD203					MDS77
NSD204					2N6555
NSD205					2N6555
NSD206					2N6556
NSD457					2N6591
NSD458					2N6593
NSD459					2N6558
NSE-181			247	2052	
NSP105					TIP42A
NSP2010					TIP42
NSP2011					TIP42A
NSP2021					TIP41A
NSP205					TIP41A
NSP2090	262				TIP125
NSP2091	262				TIP125
NSP2092					TIP126
NSP2093					TIP126
NSP2100	261				TIP120
NSP2101	261				TIP120
NSP2102	261				TIP121
NSP2103	261				TIP121
NSP2370					TIP32
NSP2480					TIP31
NSP2481					TIP31A
NSP2490					TIP32
NSP2491					TIP32A
NSP2520					TIP31
NSP2955					MJE2955T
NSP3054					TIP31A
NSP3055					MJE3055T
NSP370					TIP32
NSP371					TIP32
NSP41					TIP41
NSP41A					TIP41A
NSP41B					TIP41B
NSP41C					TIP41C
NSP42					TIP42
NSP42A					TIP42A
NSP42B					TIP42B
NSP42C					TIP42C
NSP4918					TIP30
NSP4919					TIP30A
NSP4920					TIP30B
NSP4921					TIP29
NSP4922					TIP29A
NSP4923					TIP29B
NSP5190					2N6121
NSP5191					2N6122
NSP5192					2N6123
NSP5193					2N6124
NSP5194					2N6125
NSP5195					2N6126
NSP520					TIP31
NSP521					TIP31
NSP575					TIP29A
NSP576					TIP30A
NSP577					TIP29A
NSP578					TIP30A
NSP579					TIP29B
NSP580					TIP30B
NSP581					TIP29C
NSP582					TIP30C
NSP585					TIP29A
NSP586					TIP30A
NSP587					TIP29A
NSP588					TIP30A

Industry Standard No.	ECG	SK	GE	RS 276-	MOTOR.
NSP589					TIP29B
NSP590					TIP30B
NSP595					TIP31A
NSP596					TIP32A
NSP597					TIP31A
NSP5974					TIP42
NSP5975					TIP42A
NSP5976					TIP42B
NSP5977					TIP41
NSP5978					TIP41A
NSP5979					TIP41B
NSP598					TIP32A
NSP5980					2N6489
NSP5981					2N6490
NSP5982					2N6491
NSP5983					2N6486
NSP5984					2N6487
NSP5985					2N6488
NSP599					TIP31B
NSP600					TIP32B
NSP695	261				TIP120
NSP695A	261				TIP100
NSP696	262				TIP125
NSP696A	262				TIP105
NSP697	261				TIP120
NSP697A	261				TIP100
NSP698	262				TIP125
NSP698A	262				TIP105
NSP699	261				TIP121
NSP699A	261				TIP101
NSP700	262				TIP126
NSP700A	262				TIP106
NSP701					TIP122
NSP702					TIP127
NSQ4110L1	5626	3633			
NSQ8050L2	716		IC-208		
NSQ8050L2-W2	716		IC-208		
NSR7140					SPECIAL
NSR7141					SPECIAL
NSR7142					SPECIAL
NSR7143					SPECIAL
NSR8140					SPECIAL
NSR8141					SPECIAL
NSR8142					SPECIAL
NSR8143					SPECIAL
NSS1021	125	3033	510,531	1114	
NSS3058A					MDA100A
NSS3059A					MDA101A
NSS3060A					MDA102A
NSS3061A					MDA104A
NSS3062A					MDA104A
NSS3063A					MDA106A
NSS3064A					MDA106A
NSS3065A					MDA108A
NSS3066A					MDA110A
NT010	56004	3658			
NT015	56004	3658			
NT06	56004	3658			
NT08	56004	3658			
NT101	5401	3638/5402			
NT102	6402	3628			
NT110	56004	3658			
NT1101C	5063A	3837			
NT1113C	143A	3750			
NT1116C	5075A	3751			
NT1118C	5077A	3752			
NT1130C	5084A	3755			
NT115	56004	3658			
NT1301C	5063A	3837			
NT16	56004	3658			
NT18	56004	3658			
NT210	56004	3658			
NT215	56004	3658			
NT26	56004	3658			
NT28	56004	3658			
NT310	56006	3659			
NT315	56006	3659			
NT36	56006	3659			
NT38	56006	3659			
NT410	56006	3659			
NT415	56006	3659			
NT46	56006	3659			
NT48	56006	3659			
NT50C13	143A	3750			
NT50C16	5075A	3751			
NT50C18	5077A	3752			
NT50C30	5084A	3755			
NT510	56008	3660			
NT515	56008	3660			
NT5506	5063A	3837			
NT5507	5065A	3838			
NT5523	143A	3750			
NT5525	5075A	3751			
NT5526	5077A	3752			
NT5531	5084A	3755			
NT5551	5105A		ZD-200		
NT55C13	143A	3750			
NT55C16	5075A	3751			
NT55C18	5077A	3752			
NT55C2V7	5063A	3837			
NT55C3	5065A	3838			
NT55C30	5084A	3755			
NT56	56008	3660			
NT58	56008	3660			
NT610	56008	3660			
NT615	56008	3660			
NT66	56008	3660			
NT68	56008	3660			
NT7706	5063A	3837			
NT77C2V7	5063A	3837			
NT9970	5065A	3838			
NTC-10		3114	221	2032	
NTC-11		3122	212	2030	
NTC-12	165	3710/238	38		
NTC-13	109	3088	1N60	1123	
NTC-14	109	3087	1N34AS	1123	
NTC-15		3100	514	1122	
NTC-16	503	3068	CR-5		
NTC-17	506	3125	511	1114	
NTC-18		3080	509	1114	
NTC-19	116	3313	504A	1104	
NTC-20	5074A	3139	ZD-11	563	
NTC-21	712	3072	IC-148		
NTC-4		3018	61	2038	
NTC-5	108	3452	86	2038	
NTC-6		3114	221	2032	
NTC-7	199	3124/289	212	2010	
NTC-8		3244	27		
NTC-9	198	3220	251		
NU34	109	3087		1123	
NU398B	116	3017B/117	504A	1104	
NV004	614	3327	90		
NV009	614	3327	90		
NZ-206	5070A		ZD-6.2	561	

Industry Standard No.	ECG	SK	GE	RS 276-	MOTOR.
00415	159		82	2032	
0101	116	3311	504A	1104	
01F-SL8020	703A		IC-12		
0234	116		504A	1104	
02Z-10A	140A	3061	ZD-10	562	
02Z12A	142A		ZD-12	563	
02Z12GR	5021A		ZD-12	563	
02Z5.6A	136A		ZD5.6	561	
02Z8.2A	5072A		ZD-8.2	562	
04-8054-3			504A	1104	
04-8054-4			504A	1104	
04-8054-7			504A	1104	
05Z8.2U			ZD-8.2	562	
07B27	708		IC-10		
07B2B	708		IC-10		
09-306113	177	3100/519	300	1122	
09-306195	177		300	1122	
09-309060	123A		20	2051	
OA-10	128	3024	243	2030	
OA-90	109	3088		1123	
OA-90(G)	109			1123	
OA-90G	109	3088		1123	
OA-91	109	3087		1123	
OA-95	110MP		300	1122	
OA10	5800			1142	
OA126-12	142A	3062	ZD-12	563	
OA126/10	5020A	3061/140A	ZD-11	563	
OA126/12	142A	3062	ZD-12	563	
OA126/14	144A	3094	ZD-14	564	
OA126/5	5069A	3056/135A	ZD-4.7		
OA12610	140A		ZD-10	562	
OA12612	142A		ZD-12	563	
OA12614	144A		ZD-14	564	
OA1265	135A		ZD-5.1		
OA127	116	3311	504A	1104	
OA128	116	3311	504A	1104	
OA129	116	3311	504A	1104	
OA130	116	3311	504A	1104	
OA131	116	3031A	504A	1104	
OA132	116	3031A	504A	1104	
OA134Q	109	3087		1123	
OA150	109	3087		1123	
OA159	109	3087		1123	
OA160	109	3087		1123	
OA161	109	3087		1123	
OA172	109	3087		1123	
OA174	109	3087		1123	
OA180	116	3016	504A	1104	
OA200	177	3016	300	1122	
OA2002				1122	
OA2005				1122	
OA202	177	3016	300	1122	
OA205	177	3100/519	300	1122	
OA210	116	3016	504A	1104	
OA211	116	3032A	504A	1104	
OA214	116	3016	504A	1104	
OA47	109	3087		1123	
OA50	109	3087		1123	
OA541	109	3087		1123	
OA6	109	3087		1123	
OA7	109	3087		1123	
OA70	109	3087		1123	
OA71	109	3087		1123	
OA71C	109	3087		1123	
OA72	109	3087		1123	
OA73	109	3087		1123	
OA73C	109	3087		1123	
OA74	109	3087		1123	
OA74A	109	3087		1123	
OA79	109	3091		1123	
OA8	116		504A	1104	
OA81	109	3031A		1123	
OA81C	109	3087		1123	
OA85	109	3087		1123	
OA85C	109	3087		1123	
OA9	109	3087		1123	
OA90	109	3088	1N60	1123	
OA90-FM			1N60	1123	
OA909	109	3087		1123	
OA90FM	110MP		1N60	1123(2)	
OA90G	109		1N60	1123	
OA90GA	109	3088	1N60	1123	
OA90LF	109		1N60	1123	
OA90M	109	3088	1N60	1123	
OA90MLF	109		1N60	1123	
OA90Z	109	3088		1123	
OA90ZA	109			1123	
OA91	109	3087	1N34AS	1123	
OA92	109			1123	
OA95	110MP	3087	1N34AS	1123	
OA99	109	3088	1N60	1123	
OA9D	109			1123	
OA9Z	109			1123	
OAZ200	5069A	3056/135A	ZD-5.1		
OAZ201	135A	3056	ZD-5.1		
OAZ202	136A	3057	ZD-5.6	561	
OAZ203	137A	3058	ZD-6.2	561	
OAZ204	5071A		ZD-6.8	561	
OAZ206	5072A	3136	ZD-8.2	562	
OAZ207	139A	3060	ZD-9.1	562	
OAZ209	135A	3056	ZD-5.1		
OAZ210	137A		ZD-6.2	561	
OAZ212	139A	3060	ZD-9.1	562	
OAZ213	142A	3062	ZD-12	563	
OAZ230	5188A	3062/142A	ZD-12	563	
OAZ240	135A	3056	ZD-5.1		
OAZ241	135A	3056	ZD-5.1		
OAZ242	136A	3057	ZD-5.6	561	
OAZ243	137A		ZD-6.2	561	
OAZ244	5014A		ZD-6.8	561	
OAZ246	5016A	3136/5072A	ZD-8.2	562	
OAZ247	139A	3060	ZD-9.1	562	
OAZ269	135A		ZD-5.1		
OAZ270	137A	3058	ZD-6.2	561	
OAZ272	139A	3060	ZD-9.1	562	
OAZ273	142A	3062	ZD-12	563	
OC-130	100	3005	1	2007	
OC-140	100	3005	1	2007	
OC-16	121	3009	239	2006	
OC-169	160	3006	245	2004	
OC-170	126	3006/160	245	2004	
OC-171	126	3006/160	245	2004	
OC-22	121	3009	239	2006	
OC-23	121	3009	239	2006	
OC-24	121	3009	239	2006	
OC-25	121	3009	239	2006	
OC-26	121	3009	239	2006	
OC-28	121	3009	239	2006	
OC-29	121	3009	239	2006	
OC-304	176	3004/102A	80		
OC-304/1	158		53	2007	
OC-304/2	158		53	2007	
OC-304/3	158		53	2007	
OC-305/1	158		53	2007	

Industry Standard No.	ECG	SK	GE	RS 276-	MOTOR.
OC-305/2	158		53	2007	
OC-306/1	158		53	2007	
OC-306/2	158		53	2007	
OC-306/3	158		53	2007	
OC-307	102A	3004	20	2007	
OC-308	102A	3004	20	2007	
OC-30A	131	3009	239	2006	
OC-318	102A	3003	20	2007	
OC-330	102A	3004	20	2007	
OC-34	102A	3004	20	2007	
OC-340	102A	3004	20	2007	
OC-341	102A	3004	20	2007	
OC-342	102A	3004	20	2007	
OC-343	102A	3004	20	2007	
OC-35	121	3009	239	2006	
OC-350	102A	3004	53	2007	
OC-351	102A	3004	53	2007	
OC-36	121	3009	239	2006	
OC-360	102A	3004	53	2007	
OC-362	102A	3004	53	2007	
OC-363	102A	3004	53	2007	
OC-364	102A	3004	53	2007	
OC-38	102A	3004	53	2007	
OC-390	126	3008	52	2024	
OC-410	100	3005	1	2007	
OC-44	100	3005	1	2007	
OC-45	100	3005	1	2007	
OC-46	100	3005	1	2007	
OC-47	100	3005	1	2007	
OC-602	102A	3004	53	2007	
OC-604	102A	3004	53	2007	
OC-612	126	3008	52	2024	
OC-613	126	3008	52	2024	
OC-614	126	3008	52	2024	
OC-615	160	3006	245	2004	
OC-65	102A	3004	53	2007	
OC-66	102A	3004	53	2007	
OC-70	102A	3003	53	2007	
OC-71	102A	3004	53	2007	
OC-71A	102A	3004	53	2007	
OC-71N	102A	3004	53	2007	
OC-72	102A	3003	53	2007	
OC-73	102A	3003	53	2007	
OC-74	102A	3003	53	2007	
OC-75	102A	3003	53	2007	
OC-75N	102A	3004	53	2007	
OC-77	102A	3004	53	2007	
OC-79	102A	3004	53	2007	
OC-81DD	102A	3004	53	2007	
OC-975	126		52	2024	
OC110	102A		53	2007	
OC120	102A		53	2007	
OC122			53	2007	
OC123	102A	3123	53	2007	
OC130	126	3006/160	52	2024	
OC139	103A	3011	8	2002	
OC140	103A		8	2002	
OC141	103A	3011	8	2002	
OC16	104		16	2006	
OC169	160		245	2004	
OC169R	160	3006	245	2004	
OC170	160		245	2004	
OC170N	160		245	2004	
OC170R	160	3006	245	2004	
OC170V	160	3006	245	2004	
OC171	160		245	2004	
OC171N	160			2004	
OC171R	160	3006	245	2004	
OC171V	160	3006	245	2004	
OC19	121	3009	239	2006	
OC20	121	3009	239	2006	
OC200	159	3114/290	82	2032	
OC201	159	3114/290	82	2024	
OC202	159	3114/290	82	2024	
OC203	159	3114/290	82	2032	
OC204	159	3114/290	82	2032	
OC205	159	3114/290	82	2032	
OC206	159	3114/290	82	2032	
OC207	159	3114/290	82	2032	
OC22	104		16	2006	
OC23	104		16	2006	
OC24	104		16	2006	
OC25	104		16	2006	
OC26	104		16	2006	
OC27	104	3009	16	2006	
OC28	104		16	2006	
OC29	104		16	2006	
OC30	131	3009	44	2006	
OC302	102A		53	2007	
OC303	102A		53	2007	
OC304	102A	3123	53	2007	
OC304-1	102A	3004	53	2007	
OC304-2	102A		53	2007	
OC304-3	102A		53	2007	
OC304N	102A		53	2007	
OC305	102A	3004	53	2007	
OC305-1	102A	3004	53	2007	
OC305-2	102A	3004	53	2007	
OC306	102A		53	2007	
OC306-1	102A		53	2007	
OC306-2	102A		53	2007	
OC306-3	102A		53	2007	
OC307	102A	3123	53	2007	
OC307-1	102A	3123	53	2007	
OC307-2	102A	3123	53	2007	
OC307-3	102A	3123	53	2007	
OC308	102A	3123	53	2007	
OC309	102A	3123	53	2007	
OC309-1	102A	3123	53	2007	
OC309-2	102A	3123	53	2007	
OC309-3	102A	3123	53	2007	
OC30A	131	3198	44	2006	
OC30B	131		44	2006	
OC318	102A		53	2007	
OC32	102		2	2007	
OC320	160		245	2004	
OC33	102A		53	2007	
OC330	102A	3123	53	2007	
OC331	126	3006/160	52	2024	
OC34	102A	3123	53	2007	
OC340	102A	3123	53	2007	
OC341	126		52	2024	
OC342	126		52	2024	
OC343	126		52	2024	
OC35	104		16	2006	
OC350	102A	3123	53	2007	
OC351	126		52	2024	
OC36	104		16	2006	
OC360	102A		53	2007	
OC361	126		52	2024	
OC362	126		52	2024	
OC363	126		52	2024	
OC364	102A		53	2007	
OC38	102A	3123	53	2007	
OC390	126		52	2024	
OC40	126		52	2024	
OC400	160		245	2004	
OC41	102	3004	2	2007	
OC410	160		245	2004	
OC41A	102A	3004	53	2007	
OC41N	102A		52	2024	
OC42	126	3006/160	52	2024	
OC42N	102A		52	2024	
OC43	126	3006/160	52	2024	
OC430	159	3114/290	82	2032	
OC430K	159	3114/290	82	2032	
OC43N	126		52	2024	
OC44	126		52	2024	
OC440	159	3114/290	82	2032	
OC440K	159	3114/290	82	2032	
OC443	159	3114/290	82	2032	
OC443K	159	3114/290	82	2032	
OC445	159	3114/290	82	2032	
OC445K	159	3114/290	82	2032	
OC449	159	3114/290	82	2032	
OC449K	159	3114/290		2032	
OC44N	126		52	2024	
OC45	102A	3123	53	2007	
OC450	159	3114/290	82	2032	
OC45N	126		52	2024	
OC46	102A	3123	53	2007	
OC460	159	3114/290	82	2032	
OC460K	159	3114/290	82	2032	
OC463	159	3114/290	82	2032	
OC463K	159	3114/290	82	2032	
OC465	159	3114/290	82	2032	
OC465K	159	3114/290	82	2032	
OC466	159	3114/290	82	2032	
OC466K	159	3114/290	82	2032	
OC467	159	3114/290	82	2032	
OC467K	159	3114/290	82	2032	
OC468	159	3114/290	82	2032	
OC468K	159	3114/290	82	2032	
OC469	159	3114/290	82	2032	
OC469K	159	3114/290	82	2032	
OC46N	126		52	2024	
OC47	102A		53	2007	
OC470	159	3114/290	82	2032	
OC470K	159	3114/290	82	2032	
OC47N	126		52	2024	
OC50	126		52	2024	
OC53	160	3123	245	2004	
OC54	160	3123	245	2004	
OC55	160	3123	245	2004	
OC56	102A	3123	53	2007	
OC57			53	2007	
OC58		3003	53	2007	
OC59		3123	53	2007	
OC60		3003	53	2007	
OC601	102A		53	2007	
OC602	102A		53	2007	
OC602SP	102A		53	2007	
OC602SQ	102A		53	2007	
OC603	102A		53	2007	
OC604	102A		53	2007	
OC604SP	102A		53	2007	
OC612	126		52	2024	
OC613	126		52	2024	
OC614	126		52	2024	
OC615	126		52	2024	
OC615N	160		245	2004	
OC65	102A	3123	53	2007	
OC66	102A	3123	53	2007	
OC70	102A	3123	53	2007	
OC700	159	3114/290	82	2032	
OC700A	159	3114/290	82	2032	
OC700B	159	3114/290	82	2032	
OC702	159	3114/290	82	2032	
OC702A	159	3114/290	82	2032	
OC702B	159	3114/290	82	2032	
OC704	159	3114/290	82	2032	
OC70N	102A		53	2007	
OC71	102A	3123	53	2007	
OC711	102A	3004	53	2007	
OC71A	102A		53	2007	
OC71N	102A	3123	53	2007	
OC72	102A	3003	53	2007	
OC73	102A		53	2007	
OC74	102A	3123	53	2007	
OC740	159	3114/290	82	2032	
OC7400			82	2032	
OC740G	159	3114/290	82	2032	
OC740M	159	3114/290	82	2032	
OC7400	159	3114/290	21	2032	
OC742	159	3114/290	82	2032	
OC7420			82	2032	
OC74200				2032	
OC742G	159	3114/290	82	2032	
OC742M	159	3114/290	82	2032	
OC7420	159	3114/290	21		
OC74N	102A		53	2007	
OC75	102A		53	2007	
OC75N	102A		53	2007	
OC76	102A	3004	53	2007	
OC77	102A		53	2007	
OC77M	102A	3123	53	2007	
OC78	102A		2	2007	
OC79	102A		53	2007	
OC80	102A		53	2007	
OC81	102A	3004	53	2007	
OC810	102A		53	2007	
OC81D	102A	3004	53	2007	
OC81DD	102A		53	2007	
OC81DN	102A		53	2007	
OC81N			53	2007	
OC83	158		53	2007	
OC83N			53	2007	
OC84	158		53	2007	
OC84N			53	2007	
OF-129	102A	3004	53	2007	
OF129	102A		53	2007	
OF156	177	3100/519	300	1122	
OF160	116		504A	1104	
OF162	177	3100/519	300	1122	
OF164	116		504A	1104	
OF173	109	3087		1123	
OF66	506	3100/519	511	1114	
OG-30L125	116	3031A	504A	1104	
ON047204-2	123A		20	2051	
ON174	160	3006	245	2004	
ON271	123A	3124/289	20	2051	
ON274	123A	3124/289	20	2051	
ON47204-1	123A		20	2051	
ON67A	109			1123	
OOV60529	128	3024	243	2030	
OP-01C					MC1536
OP-01G					MC1536
OP-01H					MC1536

Industry Standard No.	ECG	SK	GE	RS 276-	MOTOR.
OP-01J					MC1536G
OP-01L					MC1536G
OP-01P					MC1536P
OP-08					MC1776
OP-08A					MC1776
OP-08B					MC1776
OP-08C					MC1776
OP-08E					MC1776
OP130	3028		LED56		MLED930
OP131	3028		LED55B		MLED930
OP160					MLED900
OP500					MRD450
OP800	3032				MRD3055
OP801	3032				MRD3050
OP802					MRD310
OP803					MRD300
OP804					MRD300
OP805					MRD300
OP830					MRD300
OPI110					MOC1005
OPI2150			H11A4		MOC1006
OPI2151	3040		H11A4		4N27
OPI2152	3040		H11A2		4N26
OPI2153			H11A1		4N26
OPI2250			H11A3		MOC1006
OPI2251			H11A3		MOC1006
OPI2252	3040		H11A3		4N25
OPI2253			H11A1		4N25
OPI4201					MOC3003
OPI4202					MOC3002
OPI4401					MOC3001
OPI4402					MOC3000
OPI4501					OPI4501
OPI4502					OPI4502
ORF-2	128	3024	243	2030	
OS-16308	116	3311	504A	1104	
OS-1D				1102	
OS-4D				1103	
OS-6D				1104	
OS16308	116		504A	1104	
OS492	176		80		
OS536G			10	2051	
OS70	109			1123	
OSD-0033	138A		ZD-6.8	561	
OSD0033		3059	ZD-7.5		
OSS-16308	116	3311	504A	1104	
OSS-16685	109			1123	
OSS-36885	116		504A	1104	
OSS16308	109	3087		1123	
OSS16685	109	3087		1123	
OSS36503	116	3017B/117	504A	1104	
OSS36685	116	3017B/117	504A	1104	
OSS36885	116	3311	504A	1104	
OV02	506		511	1114	
OY-5061	116	3016	504A	1104	
OY-5062	116	3016	504A	1104	
OY101	116	3016	504A	1104	
OY5061	116		504A	1104	
OY5062	116		504A	1104	
OY5063	116	3016	504A	1104	
OY5064	116	3016	504A	1104	
OY5065	116		504A	1104	
OY5066	116		504A	1104	
OY5067	125	3032A	510,531	1114	
OZ10T10	140A		ZD-10	562	
OZ10T5	140A		ZD-10	562	
OZ12T10	142A		ZD-12	563	
OZ12T5	142A		ZD-12	563	
OZ15T10	145A	3063	ZD-15	564	
OZ15T5	145A	3063	ZD-15	564	
OZ27T10	146A		ZD-27		
OZ27T5	146A		ZD-27		
OZ3.6T10	134A		ZD-3.6		
OZ3.6T5	134A		ZD-3.6		
OZ5.1T10	135A		ZD-5.1		
OZ5.1T5	135A		ZD-5.1		
OZ5.6T10			ZD-5.6	561	
OZ5.6T5	136A		ZD-5.6	561	
OZ6.2T10	137A		ZD-6.2	561	
OZ6.2T5	137A		ZD-6.2	561	
P-10115	116	3031A	504A	1104	
P-10954-1	130		14	2041	
P-10954-2	130		14	2041	
P-11748-1	128		243	2030	
P-11810-1	181		75	2041	
P-11901-1	130		14	2041	
P-11901-3	181		75	2041	
P-11903-1	128		243	2030	
P-31898	121	3009	239	2006	
P-6006	177	3100/519	300	1122	
P-8393	123A		20	2051	
P-T-30	121	3009	239	2006	
P/FTV/117	152	3893	66	2048	
P/N10000020	123A	3444	20	2051	
P/N14-603-02	161	3716	39	2015	
P/N297L010C01	234		65	2050	
P00347100	159		82	2032	
P00347101	159		82	2032	
P04-41-0025-001	519		514	1122	
P04-42-0011	519		514	1122	
P04-44-0028	123A	3444	20	2051	
P04-45-0014-P2	123A	3444	20	2051	
P04-45-0014-P5	123A	3444	20	2051	
P04-45-0015-P1	159		82	2032	
P04-45-0016-P1	159		82	2032	
P04-45-0026-P5	128		243	2030	
P04-450016-002	159		82	2032	
P04410025-003	519		514	1122	
P04410042-001	519		514	1122	
P04410042-002	519		514	1122	
P04440028-001	123A	3444	20	2051	
P04440028-009	123A	3444	20	2051	
P04440028-014	123A	3444	20	2051	
P04440028-8	123A	3444	20	2051	
P04440032-001	123A	3444	20	2051	
P0445-0034-1	181		75	2041	
P0445-0034-2	181		75	2041	
P04450016-004	159		82	2032	
P04450026P5	128		243	2030	
P04450032-1	218		234		
P04450034-1	181		75	2041	
P04450034-2	181		75	2041	
P04450037	181		75	2041	
P04450040-002		3510	255	2041	
P100	116	3017B/117	504A	1104	
P1000	125	3033	510,531	1114	
P1000A	159		82	2032	
P1003					2N5266
P1004	5942		5048		2N5267
P1005					2N5270
P1006	5942		5048		
P100A	116	3081/125	504A	1104	1N5391
P100B	116	3017B/117	504A	1104	1N5392
P100D	116		504A	1104	1N5393

Industry Standard No.	ECG	SK	GE	RS 276-	MOTOR.
P100G	116		504A	1104	1N5395
P100J	116	3051/156	504A	1104	1N5397
P100K					1N5398
P100M					1N5399
P10115	116		504A	1104	
P10115A	125	3032A	510,531	1114	
P10155	109	3087		1123	
P10156	116	3311	504A	1104	
P10156A	125	3032A	510,531	1114	
P1027					2N5266
P1028					2N5268
P1029					2N5270
P10619-1	130		14	2041	
P1069E					MPF970
P1086E					MPF970
P1087		3112	FET-1	2028	
P1087E					MPF970
P1117E					MPF970
P1118E					MPF971
P1119E					MPF971
P1172	177	3100/519	300	1122	
P1172-1	177	3100/519	300	1122	
P12407-1	234		65	2050	
P15	113A	3119/113	6GC1		
P150A	116	3081/125	504A	1104	
P150B	116		504A	1104	
P150D	116		504A	1104	
P150G	116		504A	1104	
P150J	116	3051/156	504A	1104	
P15153	123A	3444	20	2051	
P16	114	3120	6GD1		
P17	115	3121	6GX1		
P1901-48	199	3245	62	2010	
P1901-50	123A	3444	20	2051	
P1901-70	159		82	2032	
P1A	121	3009	239	2006	
P1A5	116	3311	504A	1104	
P1B	159	3004/102A	82	2032	
P1B5	116	3311	504A	1104	
P1C	159	3114/290	82	2032	
P1C5	116	3016	504A	1104	
P1CG	159		82	2032	
P1D	159	3114/290	82	2032	
P1D5	116	3016	504A	1104	
P1E	219		74	2043	
P1E-1	121	3009	239	2006	
P1E-1BLK	219		74	2043	
P1E-1BLU	219		74	2043	
P1E-1GRN	219		74	2043	
P1E-1RED	219		74	2043	
P1E-1V10			74	2043	
P1E-1VIO	219			2043	
P1E-2BLK	219		74	2043	
P1E-2BLU	219		74	2043	
P1E-2GRN	219		74	2043	
P1E-2RED	219		74	2043	
P1E-2V10			74	2043	
P1E-2VIO	219			2043	
P1E-3BLK	219		74	2043	
P1E-3BLU	219		74	2043	
P1E-3GRN	219		74	2043	
P1E-3RED	219		74	2043	
P1E-3V10			74	2043	
P1E-3VIO	219			2043	
P1F	127	3035	25		
P1G	121	3034	239	2006	
P1H	159	3025/129	82	2032	
P1J	159	3025/129	82	2032	
P1K	121	3009	239	2006	
P1KBLK	121	3009	239	2006	
P1KBLU	121	3009	239	2006	
P1KBRN	121	3009	239	2006	
P1KGRN	121	3009	239	2006	
P1KORN	121	3009	239	2006	
P1KRED	121	3009	239	2006	
P1KYEL	121	3009	239	2006	
P1L	102A	3006/160	53	2007	
P1L4956	102A		53	2007	
P1M	129		244	2027	
P1N	159	3114/290	82	2032	
P1N-1	159	3114/290	82	2032	
P1N-2	159	3114/290	82	2032	
P1N-3	159	3114/290	82	2032	
P1P	159	3114/290	82	2032	
P1P-1	159	3114/290	82	2032	
P1R	104	3009	16	2006	
P1T	121	3009	239	2006	
P1V	185	3041	58	2025	
P1V-1	185		58	2025	
P1V-2	185	3041	58	2025	
P1V-3	185	3041	58	2025	
P1V-4	153	3041	69	2049	
P1W	159	3025/129	82	2032	
P1Y	127	3035	25		
P20	116	3311	504A	1104	
P200	116	3017B/117	504A	1104	
P2004	5942		5048		
P2006	5942		5048		
P2010	5942		5048		
P2015	727		IC-210		
P21309	112		1N82A		
P21316	116	3031A	504A	1104	
P21317	116	3031A	504A	1104	
P21344	139A	3060	ZD-9.1	562	
P21443	116	3017B/117	504A	1104	
P218-1	188	3199	226	2018	
P218-2	189	3200	218	2026	
P2271	130	3027	14	2041	
P2440	156		512		
P2A	159	3114/290	82	2032	
P2A5	116	3311	504A	1104	
P2B	153		69	2049	
P2B5	116	3311	504A	1104	
P2C	104	3719	16	2006	
P2C5	116	3016	504A	1104	
P2D	121	3009	239	2006	
P2D5	116	3016	504A	1104	
P2DBLU	121	3009	239	2006	
P2DBRN	121	3009	239	2006	
P2DGRN	121	3009	239	2006	
P2DORN	121	3009	239	2006	
P2DRED	121	3009	239	2006	
P2DYEL	121	3009	239	2006	
P2E	159	3025/129	82	2032	
P2F	176	3845	80		
P2G	159	3114/290	82	2032	
P2GE	159		82	2032	
P2H	159		82	2032	
P2J	180		74	2043	
P2K	153	3083/197	69	2049	
P2L	159		82	2032	
P2M-1	159	3114/290	82	2032	
P2M-2	159	3114/290	82	2032	
P2M-3	159	3114/290	82	2032	

Industry Standard No.	ECG	SK	GE	RS 276-	MOTOR.
P2P	159	3114/290	82	2032	
P2R	104	3009	16	2006	
P2S	159	3466	82	2032	
P2T	185	3041	58	2025	
P2T-1	185	3041	58	2025	
P2T-2	185	3083/197	58	2025	
P2T-3	185	3083/197	58	2025	
P2T-4	185	3083/197	58	2025	
P2U	189	3084	218	2026	
P2U-1	189	3203/211	218	2026	
P2U-2	211	3084	253	2026	
P2V	189	3200	218	2026	
P2W	159	3025/129	82	2032	
P2Y	159	3114/I90	82	2032	
P2Z	179	3642	76		
P3/2H	116	3311	504A	1104	
P300A					MR500
P300B					MR501
P300D					MR502
P300F					MR504
P300G					MR504
P300H					MR506
P300J					MR506
P300K					MR508
P300M					MR510
P3139	130	3027	14	2041	
P3172	175	3131A/369	246	2020	
P31898	121	3717	239	2006	
P32H	116		504A	1104	
P3309	5079A	3335	ZD-20		
P346	108	3452	86	2038	
P38103/507-10	145A		ZD-15	564	
P3A	197		250	2027	
P3A5	116	3311	504A	1104	
P3B	158		53	2007	
P3B5	116	3311	504A	1104	
P3C	159	3114/290	82	2032	
P3C5	116	3017B/117	504A	1104	
P3CA	159	3114/290	82	2032	
P3D	102A		53	2007	
P3D5	116	3016	504A	1104	
P3E	131		44	2006	
P3EBLK	104	3009	16	2006	
P3EBLU	104	3719	16	2006	
P3EGRN	104	3719	16	2006	
P3ERED	104	3719	16	2006	
P3H	127	3764	25		
P3J	127	3764	25		
P3K	189		218	2026	
P3M	185	3083/197	58	2025	
P3N	185	3083/197	58	2025	
P3N-1	185	3083/197	58	2025	
P3N-2	185	3083/197	58	2025	
P3N-3	185	3083/197	58	2025	
P3N-4	185	3083/197	58	2025	
P3N-5	183	3189A	56	2027	
P3P	185	3083/197	58	2025	
P3P-1	185	3083/197	58	2025	
P3P-2	185	3083/197	58	2025	
P3P-3	185	3083/197	58	2025	
P3P-4	185	3083/197	58	2025	
P3P-5	185	3083/197	58	2025	
P3R	131		44	2006	
P3R-1	131		44	2006	
P3R-2	131		44	2006	
P3R-3	131	3052	44	2006	
P3R-4	131		44	2006	
P3S	185	3191	58	2025	
P3T	131	3052	44	2006	
P3T-1	131	3198	44	2006	
P3T-2	131	3198	44	2006	
P3U	197	3083	250	2027	
P3UA	183	3189A	56	2027	
P3V	242	3189A/183	58	2027	
P3W	219		74	2043	
P3Y	189	3118	218	2026	
P3Z	159		82	2032	
P400	116	3017B/117	504A	1104	
P4069	128		243	2030	
P4326	177	3100/519	300	1122	
P480A0018	128		243	2030	
P480A0022	159		82	2032	
P480A0023	159		82	2032	
P480A0027	159		82	2032	
P480A0028	123A	3444	20	2051	
P480A0029	123A	3444	20	2051	
P4A5	116	3311	504A	1104	
P4B	193	3118	67	2023	
P4B5	116	3311	504A	1104	
P4C	159		82	2032	
P4C5	116	3016	504A	1104	
P4D	104	3719	16	2006	
P4D5	116	3017B/117	504A	1104	
P4E	185		58	2025	
P4E-1	183		56	2027	
P4E-2	183		56	2027	
P4E-3	242		58	2027	
P4E-4	242		58	2027	
P4F	127	3035	25		
P4G	159		82	2032	
P4H	127	3764	25		
P4J	242	3189A/183	58	2027	
P4J148	152	3893	66	2048	
P4K	159		82	2032	
P4L	104	3719	16	2006	
P4M	121	3717	239	2006	
P4N	104	3719	16	2006	
P4P	159		82	2032	
P4R	159		82	2032	
P4S	185	3191	58	2025	
P4T	183	3189A	56	2027	
P4U	185	3191	58	2025	
P4V	185	3179A	58	2025	
P4V-1	185	3189A/183	58	2025	
P4V-2	183	3189A	56	2027	
P4W	153		69	2049	
P4W-1	185	3191	58	2025	
P4W-2	185	3191	58	2025	
P4Y	159		82	2032	
P4Z	128		243	2030	
P50200-11	130		14	2041	
P5034	130	3510	14	2041	
P504	5940		5048		
P506	5940		5048		
P5100	125	3033	510,531	1114	
P5148	175		246	2020	
P5149	130	3510	14	2041	
P5152	123A	3444	20	2051	
P5153	123A	3444	20	2051	
P580	125	3032A	510,531	1114	
P5A5	116	3311	504A	1104	
P5B	159	3025/129	82	2032	
P5B5	116	3017B/117	504A	1104	
P5C	159	3114/290	82	2032	
P5C5	116	3016	504A	1104	
P5D	159	3118	82	2032	
P5D5	116	3017B/117	504A	1104	
P5F	242	3189A/183	58	2027	
P5H	153	3083/197	69	2049	
P5L	185	3083/197	58	2025	
P5M	187	3193			
P5R	185	3189A/183	58	2025	
P5S	242	3189A/183	58	2027	
P5U	153		69	2049	
P6/2H	116	3311	504A	1104	
P600	116	3017B/117	504A	1104	
P6009	242	3191/185	58	2027	
P6022A	175		246	2020	
P6128	175		246	2020	
P62H	116		504A	1104	
P633024	128	3024	243	2030	
P633024G	128	3024	243	2030	
P633567	123A	3444	20	2051	
P64447	123A	3444	20	2051	
P6450026	128		243	2030	
P6460006	102	3722	2	2007	
P6460037	102	3722	2	2007	
P6480001	121	3717	239	2006	
P6500A	181	3036	75	2041	
P67	159	3138/193A	21	2034	
P6786	128		243	2030	
P6804	175		246	2020	
P69941	199		62	2010	
P6A5	116	3311	504A	1104	
P6B5	116	3311	504A	1104	
P6C5	116	3017B/117	504A	1104	
P6D5	116	3017B/117	504A	1104	
P6RP10	125	3033	510,531	1114	
P6RP8	125	3032A	510,531	1114	
P7109	156		512		
P7394	177	3100/519	300	1122	
P75534	121	3009	239	2006	
P75534-1	121	3009	239	2006	
P75534-2	104	3719	16	2006	
P75534-3	104	3719	16	2006	
P75534-4	121	3717	239	2006	
P75534-5	121	3717	239	2006	
P7776	117		504A	1104	
P7A5	116	3311	504A	1104	
P7B5	116	3017B/117	504A	1104	
P7C5	116	3031A	504A	1104	
P7D5	116	3017B/117	504A	1104	
P8/2H	116	3311	504A	1104	
P800	125	3051/156	510,531	1114	
P82H	116		504A	1104	
P8393	123A	3444	20	2051	
P8394	123A	3444	20	2051	
P8870	121	3009	239	2006	
P8890	121	3014	239	2006	
P8890A	121	3014	239	2006	
P8890L	121	3717	239	2006	
P8B	197		69	2049	
P8H	153	3083/197	69	2049	
P9459	116		504A	1104	
P9623	123A	3444	20	2051	
P9962-1	128		243	2030	
P9962-2	128		243	2030	
P9962-4	128		243	2030	
P9962-5	128		243	2030	
PA-069	116	3016	504A	1104	
PA-10556	108	3016	86	2038	
PA-10889-1	121	3009	239	2006	
PA-10889-2	121	3009	239	2006	
PA-10890	121	3009	239	2006	
PA-10890-1	121	3009	239	2006	
PA-3	116	3016	504A	1104	
PA-320	116	3016	504A	1104	
PA-320A	116	3016	504A	1104	
PA-320B	116	3016	504A	1104	
PA-400	1090		723		
PA-500	1090		723		
PA069	116		504A	1104	
PA070	116	3031A	504A	1104	
PA071	116	3016	504A	1104	
PA1000	159	3114/290	82	2032	
PA1001	159	3114/290	82	2032	
PA1001A	159	3114/290	82	2032	
PA10556	116		504A	1104	
PA10880	160		245	2004	
PA10887	116		504A	1104	
PA10889-1	158	3123	53	2007	
PA10889-2	158		53	2007	
PA10890	104	3719	16	2006	
PA10890-1	121	3717	239	2006	
PA200	116	3017B/117	504A	1104	
PA234	717		IC-209		
PA239	721		IC-14		
PA239A					MC1303P
PA243	717		IC-209		
PA277	804	3455	IC-27		
PA3	116		504A	1104	
PA300	116	3033	504A	1104	
PA305	116	3016	504A	1104	1N4001
PA305A	116	3311	504A	1104	
PA310	116	3016	504A	1104	1N4002
PA310A	116	3017B/117	504A	1104	
PA315	116	3016	504A	1104	1N4003
PA315A	116	3016	504A	1104	
PA320	116		504A	1104	1N4003
PA320A	116		504A	1104	
PA320B	116		504A	1104	
PA325	116	3017B/117	504A	1104	1N4004
PA325A	116	3016	504A	1104	
PA325B	116	3016	504A	1104	
PA330	116	3031A	504A	1104	1N4004
PA330A	116	3016	504A	1104	
PA330B	116	3016	504A	1104	
PA340	116	3031A	504A	1104	1N4004
PA340A	116	3016	504A	1104	
PA340D	116	3016	504A	1104	
PA350	116	3017B/117	504A	1104	1N4005
PA350A	116	3017B/117	504A	1104	
PA360	116	3017B/117	504A	1104	1N4005
PA360A	116	3017B/117	504A	1104	
PA380	125	3032A	510,531	1114	
PA400	116	3017B/117	504A	1104	
PA401	1090		723		
PA501	1148		723		
PA501X	1090		723		
PA600	116	3017B/117	504A	1104	
PA7000/591	909D	3590	IC-250		
PA7001/0001	123A	3444	20	2051	
PA7001/501	909	3590	IC-249		
PA7001/502	909D	3590	IC-250		
PA7001/503	941		IC-263	010	
PA7001/505	923		IC-259		
PA7001/518	7430	7430			
PA7001/519	7420	7420		1809	

Industry Standard No.	ECG	SK	GE	RS 276-	MOTOR.
PA7001/520	7410	7410		1807	
PA7001/521	7400	7400		1801	
PA7001/522	7440	7440			
PA7001/523	7451	7451		1825	
PA7001/525	7402	7402		1811	
PA7001/526	7401	7401			
PA7001/527	7404	7404		1802	
PA7001/528	7405	7405			
PA7001/529	7474	7474		1818	
PA7001/531	7473	7473		1803	
PA7001/539	7454	7454			
PA7001/591	909D		IC-250		
PA7001/593	74145	74145		1828	
PA7615	116		504A	1104	
PA7703	703A	3157	IC-12		
PA7703E	703A	3157	IC-12		
PA7703X	703A	3157	IC-12		
PA7709	909	3590	IC-249		
PA7709C	909	3590	IC-249		
PA7710-31	910		IC-251		
PA7710-39	910		IC-251		
PA7711-31	911		IC-253		
PA7741	941	3514	IC-263	010	
PA7741C	941	3514	IC-263	010	
PA8260	199	3245	62	2010	
PA8261	145A		ZD-15	564	
PA8543	199	3245	62	2010	
PA8645	116		504A	1104	
PA8900	186	3192	28	2017	
PA9004	199	3245	62	2010	
PA9005	199	3245	62	2010	
PA9006	123A	3444	20	2051	
PA9154	160		245	2004	
PA9155	160		245	2004	
PA9156	102A		53	2007	
PA9157	102A		53	2007	
PA9158	158		53	2007	
PA9160	116		504A	1104	
PA9267	137A		ZD-6.2	561	
PA9483	192		63	2030	
PA9D522/1			BR-600	1152	
PADT20	160	3006	245	2004	
PADT21	160	3006	245	2004	
PADT22	160	3006	245	2004	
PADT23	160	3005	245	2004	
PADT24	160	3006	245	2004	
PADT25	160	3006	245	2004	
PADT26	160	3006	245	2004	
PADT27	160	3006	245	2004	
PADT28	160	3006	245	2004	
PADT30	160	3006	245	2004	
PADT31	160	3006	245	2004	
PADT35	160	3123	245	2004	
PADT40	160		245	2004	
PADT50	104	3009	16	2006	
PADT51	160	3006	245	2004	
PAR-12	121	3009	239	2006	
PAR12	121	3717	239	2006	
PB-998005			52	2024	
PB05	5322	3679			
PB10	5322	3679			
PB110	121	3009	239	2006	
PB40	5324	3681			
PB60	5326	3682			
PBC107	199	3245	62	2010	
PBC107A	199	3245	62	2010	
PBC107B	199	3245	62	2010	
PBC108	199	3245	62	2010	
PBC108A	199	3245	62	2010	
PBC108B	199	3245	62	2010	
PBC108C	199	3245	62	2010	
PBC109	199	3245	62	2010	
PBC109B	199	3245	62	2010	
PBC109C	199	3245	62	2010	
PBC182	108	3452	86	2038	
PBC183	289A	3444/123A	20	2051	
PBC184	192	3244	63	2030	
PBD352301	2011	3975			
PBD352302	2012	9092			
PBD352303	2013	9093			
PBE3014-1	158	3004/102A	53	2007	
PBE3014-2	158	3004/102A	53	2007	
PBE3020-1	158		53	2007	
PBE3020-2	103A	3835	59	2002	
PBE3162	158	3004/102A	53	2007	
PBE3162-1	102A		53	2007	
PBE3162-2	102A		53	2007	
PBE3322	109	3087		1123	
PBX103	102	3722	2	2007	
PBX113	102	3722	2	2007	
PBZO	5322	3680			
PC-20003	1003		IC-43		
PC-20005	1027	3153	721		
PC-20006	1024	3152	720		
PC-20007	1006	3358	IC-38		
PC-20008	720	9014	IC-7		
PC-20012	1030		IC-163		
PC-20015	812		IC-242		
PC-20018	720		IC-7		
PC-20024	722	3161	IC-9		
PC-20030	722	3161	IC-9		
PC-20045	801	3160			
PC-20051	1082	3461			
PC-20066			214	2038	
PC-20069	1060	3460	IC-50		
PC-20070	1226	3763			
PC-20071	1135	3876			
PCO.2P11/2			504A	1104	
PC0211/2			504A	1104	
PC02P1/2			504A	1104	
PC0600A	610	3323			
PC0601A	611	3324			
PC0603A	613	3326			
PC0604A	614	3327			
PC0610A	610	3323			
PC0611A	611	3324			
PC0620A	610	3323			
PC1026C	1226	3763			
PC1066T	158		53	2007	
PC1067T	158		53	2007	
PC1068T	158		53	2007	
PC107	610	3323			
PC107A	610	3323			
PC112					1N5140
PC113					1N5144
PC114					1N5148
PC115					1N5140
PC116					1N5144
PC117					1N5148
PC122					1N5148
PC124					1N5142
PC125					1N5142
PC126					1N5142
PC127	614	3327			

Industry Standard No.	ECG	SK	GE	RS 276-	MOTOR.
PC127A	614	3327			
PC128	614	3327			1N5146
PC128A	614	3327			
PC129					1N5146
PC130					1N5146
PC132	611	3324			
PC132A	611	3324			
PC133	613	3326			
PC133A	613	3326			
PC135	611	3324			1N5140
PC135A	611	3324			
PC136	613	3326			1N5144
PC136A	613	3326			
PC137					1N5148
PC138	610	3323			
PC138A	610	3323			
PC139	610	3323			1N5139
PC139A	610	3323			
PC140	610	3323			1N5739
PC1400A	610	3323			
PC1401A	611	3324			
PC1403A	613	3326			
PC1404A	614	3327			
PC140A	610	3323			
PC141	610	3323			1N5139
PC1410A	610	3323			
PC1411A	611	3324			
PC1414A	614	3327			
PC141A	610	3323			
PC1879-004	140A	3061	ZD-10	562	
PC20018	720	9014	IC-7		
PC202	610	3323			
PC203	610	3323			
PC204	610	3323			
PC205	610	3323			
PC207	610	3323			
PC208	610	3323			
PC209	610	3323			
PC210	610	3323			
PC211	610	3323			
PC212	610	3323			
PC213	610	3323			
PC214	610	3323			
PC215	610	3323			
PC216	610	3323			
PC217	610	3323			
PC218	610	3323			
PC219	610	3323			
PC220	610	3323			
PC221	610	3323			
PC222	610	3323			
PC223	610	3323			
PC224	610	3323			
PC225	610	3323			
PC226	610	3323			
PC227	610	3323			
PC228	610	3323			
PC229	610	3323			
PC230	610	3323			
PC231	610	3323			
PC232	610	3323			
PC233	610	3323			
PC234	610	3323			
PC235	610	3323			
PC2600A	610	3323			
PC2601A	611	3324			
PC2603A	613	3326			
PC2604A	614	3327			
PC2610A	610	3323			
PC2611A	611	3324			
PC2613A	613	3326			
PC2614A	614	3327			
PC3002	158		53	2007	
PC3003	158		53	2007	
PC3004	121	3717	239	2006	
PC3005	158		53	2007	
PC3006	158		53	2007	
PC3007	158		53	2007	
PC3009	158		53	2007	
PC3010	131		44	2006	
PC303	611	3324			
PC304	611	3324			
PC305	611	3324			
PC309	611	3324			
PC310	611	3324			
PC315	611	3324			
PC316	611	3324			
PC317	611	3324			
PC321	611	3324			
PC322	611	3324			
PC323	611	3324			
PC327	611	3324			
PC328	611	3324			
PC329	611	3324			
PC333	611	3324			
PC334	611	3324			
PC335	611	3324			
PC4004	116		504A	1104	
PC4900-1	128		53	2030	
PC503	3040				4N26
PC504	613	3326			
PC505	613	3326			
PC510	613	3326			
PC511	613	3326			
PC516	613	3326			
PC517	613	3326			
PC523	613	3326			
PC529	613	3326			
PC534	613	3326			
PC535	613	3326			
PC554	1142	3485	IC-128		
PC604	614	3327			
PC605	614	3327			
PC610	614	3327			
PC611	614	3327			
PC616	614	3327			
PC617	614	3327			
PC622	614	3327			
PC623	614	3327			
PC628	614	3327			
PC629	614	3327			
PC634	614	3327			
PC635	614	3327			
PD-6018	145A	3063	ZD-15	564	
PD-6018A	145A	3063	ZD-15	564	
PD-6059	145A	3063	ZD-15	564	
PD05	166	9075			
PD10	166	9075			
PD100	170	3051/156	512		
PD101	116	3311	504A	1104	
PD1011	166	9075	BR-600	1152	
PD102	116	3311	504A	1104	
PD1020	169	3017B/117	BR-600		
PD103	116	3311	504A	1104	

Industry Standard No.	ECG	SK	GE	RS 276-	MOTOR.
PD104	116	3311	504A	1104	
PD105	116	3311	504A	1104	
PD106	116	3311	504A	1104	
PD107	116	3311	504A	1104	
PD107A	116	3311	504A	1104	
PD108	116	3311	504A	1104	
PD110	116	3031A	504A	1104	
PD111	116	3031A	504A	1104	
PD114	125	3032A	510,531	1114	
PD115	125	3032A	510,531	1114	
PD116	125	3033	510,531	1114	
PD122	116	3311	504A	1104	
PD125	116	3311	504A	1104	
PD129	116	3311	504A	1104	
PD130	116	3311	504A	1104	
PD131	116	3311	504A	1104	
PD132	116	3311	504A	1104	
PD133	116	3311	504A	1104	
PD134	116	3311	504A	1104	
PD135	116	3031A	504A	1104	
PD137	177	3100/519	300	1122	
PD154	116	3017B/117	504A	1104	
PD155	116	3017B/117	504A	1104	
PD20	167	3647			
PD40	168	3648			
PD60	169	3051/156	512		
PD6000	5002A	3768			
PD6000A	5002A	3768			
PD6002	5005A	3771	ZD-3.3		
PD6002A	134A	3771/5005A	ZD-3.6		
PD6003	134A	3055	ZD-3.6		
PD6003A	134A	3055	ZD-3.6		
PD6004	134A	3055	ZD-3.6		
PD6004A	5067A	3055/134A	ZD-3.6		
PD6006	135A	3056	ZD-5.1		
PD6006A	5069A	3056/135A	ZD-5.1		
PD6007	135A	3056	ZD-5.1		
PD6007A	135A	3056	ZD-5.1		
PD6008	136A	3057	ZD-5.6	561	
PD6008A	136A	3057	ZD-5.6	561	
PD6009	137A	3058	ZD-6.2	561	
PD6009A	137A	3058	ZD-6.2	561	
PD6010	5014A	3780	ZD-6.8	561	
PD6010A	5071A	3780/5014A	ZD-6.8	561	
PD6012	5016A	3782	ZD-8.2	562	
PD6012A	5072A	3782/5016A	ZD-8.2	562	
PD6013	139A	3060	ZD-9.1	562	
PD6013A	139A	3060	ZD-9.1	562	
PD6014	140A	3061	ZD-10	562	
PD6014A	140A	3061	ZD-10	562	
PD6016	142A	3062	ZD-12	563	
PD6016A	142A	3062	ZD-12	563	
PD6017A	143A	3750			
PD6018	145A	3063	ZD-15	564	
PD6018A	145A	3063	ZD-15	564	
PD6019A	5075A	3751			
PD6020	5027A	3793	ZD-18		
PD6020A	5077A	3752	ZD-18		
PD6041	5063A	3837			
PD6042	5065A	3838			
PD6043	5066A	3055/134A	ZD-3.3		
PD6044	134A	3055	ZD-3.6		
PD6045	5067A	3055/134A	ZD-3.6		
PD6047	5069A	3056/135A	ZD-5.1		
PD6048	135A	3056	ZD-5.1		
PD6049	136A	3057	ZD-5.6	561	
PD6049C	136A	3057	ZD-5.6	561	
PD6050	137A	3058	ZD-6.2	561	
PD6051	5071A	3334	ZD-6.8	561	
PD6053	5072A	3136	ZD-8.2	562	
PD6054	139A	3060	ZD-9.1	562	
PD6055	140A	3061	ZD-10	562	
PD6057	142A	3062	ZD-12	563	
PD6058	143A	3750			
PD6059	145A	3063	ZD-15	564	
PD6060	5075A	3751			
PD6061	5077A	3752	ZD-18		
PD80	170	3051/156	512		
PD910	116	3311	504A	1104	
PD913	125	3032A	510,531	1114	
PD914	125	3032A	510,531	1114	
PD915	125	3033	510,531	1114	
PD916	125	3033	510,531	1114	
PE-401	116	3016	504A	1104	
PE05	5312	3985			
PE10	5312	3985			
PE20	5313	3986			
PE3001	123AP	3444/123A	20	2051	
PE3015			17	2051	
PE3100					MPSH10
PE40	5314	3987			
PE401	116		504A	1104	
PE401N	116	3016	504A	1104	
PE402	116	3016	504A	1104	
PE403	116	3016	504A	1104	
PE404	116	3017B/117	504A	1104	
PE405	116	3017B/117	504A	1104	
PE406	116	3017B/117	504A	1104	
PE408	125	3032A	510,531	1114	
PE410	125	3033	510,531	1114	
PE5010			17	2051	MPSH07
PE5013			17	2051	MPSH07
PE5015					MPSH07
PE502	116	3016	504A	1104	
PE5025	233		210	2009	MPSH37
PE5029					MPSH08
PE5030B					MPSH37
PE5031			11	2015	MPSH34
PE504	116	3031A	504A	1104	
PE506	116	3017B/117	504A	1104	
PE508	125	3032A	510,531	1114	
PE510	125	3033	510,531	1114	
PE60	5315	3988			
PE6020					MPSA05
PE6021					MPSA06
PE6022					MPSA05
PE6023					MPSA06
PE80	5316	3989			
PE9001	123AP				2N3904
PEP1001	108	3452	86	2038	
PEP2	123A	3444	20	2051	
PEP2001	128	3024	243	2030	
PEP5	123A	3444	20	2051	
PEP6	123A	3444	20	2051	
PEP7	123A	3444	20	2051	
PEP8	123A	3444	20	2051	
PEP9	123A	3444	20	2051	
PEP95	172A	3156	64		
PET-101-1	108	3452	86	2038	
PET1001	192	3137	63	2030	
PET1002	123A	3444	20	2051	
PET1075	108	3452	86	2038	
PET1075A	154		40	2012	
PET2001	123A	3444	20	2051	
PET2002	123A	3444	20	2051	
PET3001	108	3452	86	2038	
PET3704	123A	3444	20	2051	
PET3705	123A	3444	20	2051	
PET3706	123A	3444	20	2051	
PET4001	123A	3024/128	243	2051	
PET4002	123A	3444	20	2051	
PET4003	199	3156/172A	62	2010	
PET6001	123A	3444	20	2051	
PET6002	123A	3444	20	2051	
PET8000	123A	3444	20	2051	
PET8001	123A	3444	20	2051	
PET8002	123A	3444	20	2051	
PET8003	123A	3444	20	2051	
PET8004	123A	3444	20	2051	
PET8201	108	3452	86	2038	
PET8250	108	3452	86	2038	
PET8251	108	3452	86	2038	
PET8300	108	3452	86	2038	
PET9002	123A	3444	20	2051	
PET9021	128	3024	243	2030	
PET9022	128	3024	243	2030	
PF-AR15	130		14	2041	
PF-AR18	175		246	2020	
PP05	5304	3105			
PP10	5304	3105			
PP100	5307	3107			
PP20	5304	3105			
PP40	5304	3106			
PP60	5305	3676			
PP80	5306	3677			
PG207	610				1N5139
PG210					MV1866
PG215					MV1868
PG222					MV1872
PG233					MV1876
PG239					MV1877
PG247					MV1878
PG307	610				1N5139
PG310	611				MV1866
PG315					MV1868
PG322	613				MV1872
PG330240	128	3024	243	2030	
PG333	614	3327			MV1876
PG333A	614	3327			
PG339					MV1877
PG347					MV1878
PGR-24	5802	9005		1143	
PH-108	116	3016	504A	1104	
PH0401H					MRF5174
PH0403H					MRF5175
PH0406H					MRF5175
PH0412H					MRF321
PH0425H					MRF325
PH0450D					2N6439
PH0450H					2N6439
PH0501H					MRF5175
PH0503H					MRF5175
PH0506H					MRF321
PH0512H					MRF321
PH0525H					MRF325
PH0550H					2N6439
PH1021	116	3016	504A	1104	
PH108	506		511	1114	
PH109	125	3033	510,531	1114	
PH204	116	3311	504A	1104	
PH208	116	3017B/117	504A	1104	
PH25C22	116		504A	1104	
PH25C22/1	116	3016	504A	1104	
PH25C22/21			504A	1104	
PH25C221	166	9075		1152	
PH404	116	3017B/117	504A	1104	
PH8193					MRF905
PH9-221	166	9075		1152	
PH9D5			510	1114	
PH9D522/1	169	3032A	BR-600	1104	
PH9D5221	166	9075		1152	
PH9D522M	166	9075		1152	
PH9DS22	116		504A	1104	
PH9DS221	166	9075		1152	
PI-10,131	159		82	2032	
PIC	159	3004/102A	82	2032	
PIK	121	3717	239	2006	
PIL/4956	160		245	2004	
PIT-37	199		62	2010	
PIT-50	159		82	2032	
PIT-74	128		243	2030	
PIT-79	159		82	2032	
PIT-81	159		82	2032	
PIV	197	3083	250	2027	
PIV-1	197	3083	250	2027	
PIV-2	197	3083	250	2027	
PIV-3	197	3083	250	2027	
PL-150-001-9-005	109			1123	
PL-150-006-9-001	109			1123	
PL-150-040-9-002	177		300	1122	
PL-151-030-9-001	177		300	1122	
PL-151-030-9-005	116		504A	1104	
PL-151-032-9-004	177		300	1122	
PL-151-035-9-001	177		300	1122	
PL-151-040-9-001	177		300	1122	
PL-151-040-9-002	177		300	1122	
PL-151-040-9-003	116		504A	1104	
PL-151-045-9-001	116		504A	1104	
PL-151-045-9-002	177		300	1122	
PL-151-045-9-003	506		511	1114	
PL-151-045-9-004	116		504A	1104	
PL-152-042-9-001	5072A		ZD-8.2	562	
PL-152-044-9-001	139A		ZD-9.1	562	
PL-152-047-9-001	137A		ZD-6.2	561	
PL-152-051-9-001	139A		ZD-9.1	562	
PL-152-052-9-002	5070A		ZD-6.2	561	
PL-152-054-9-001	5072A		ZD-8.2	562	
PL-172-010-9-001	152		66	2048	
PL-172-013-9001	236	3239	216	2053	
PL-172-014-9-001	152		66	2048	
PL-172-014-9-002	282		264		
PL-172-024-9-001	186A		247	2017	
PL-172-024-9-002	235		215		
PL-172-024-9-003	236		216	2053	
PL-172-024-9-004			247	2052	
PL-176-025-9-001	199		62	2010	
PL-176-026-9-001	107		11	2015	
PL-176-029-9-001	123A		20	2051	
PL-176-029-9-002	195A		46		
PL-176-031-9-001	289A		268	2038	
PL-176-031-9-002	199		62	2010	
PL-176-037-9-001	107		11	2015	
PL-176-042-9-001	229		61	2038	
PL-176-042-9-002	123A		20	2051	
PL-176-042-9-003	199		62	2015	
PL-176-042-9-004	123A		20	2051	
PL-176-042-9-005	186A		28	2017	
PL-176-042-9-006	123A		20	2051	

Industry Standard No.	ECG	SK	GE	RS 276-	MOTOR.
PL-176-047-9-001	289A		268	2038	
PL-176-047-9-002			211	2016	
PL-176-049-9-002	123A		11	2015	
PL-177-006-9-001	159		82	2032	
PL-177-006-9-002	290A		269	2032	
PL-182-009-9-001	312		FET-1	2035	
PL-182-014-9-002	123A		20	2051	
PL-307-007-9-002	1102		IC-93		
PL-307-047-9-002	724		IC-86		
PL100Z		3346	ZD-100		
PL1021	161	3132	39	2015	
PL1022	161	3132	39	2015	
PL1023	161	3132	39	2015	
PL1024	107	3039/316	11	2015	
PL1025	107	3039/316	11	2015	
PL1026	107	3039/316	11	2015	
PL1031	159	3025/129	82	2032	
PL1033	159	3025/129	82	2032	
PL1034	159	3025/129	82	2032	
PL1051	108	3039/316	86	2038	
PL1052	123A	3444	20	2051	
PL1053	108	3452	86	2038	
PL1054	123A	3444	20	2051	
PL1055	108	3452	86	2038	
PL1061	108	3452	86	2038	
PL1062	108	3452	86	2038	
PL1063	108	3452	86	2038	
PL1064	108	3452	86	2038	
PL1065	108	3452	86	2038	
PL1066	161	3132	39	2015	
PL1067	161	3132	39	2015	
PL1081	108	3452	86	2038	
PL1082	108	3452	86	2038	
PL1083	128	3024	243	2030	
PL1084	128	3024	243	2030	
PL1085	171	3103A/396	27		
PL1091	312	3448	FET-1	2035	
PL1092	312	3448	FET-1	2035	
PL1093	312	3448	FET-1	2035	
PL1094	312	3448	FET-1	2035	
PL1101	159	3025/129	82	2032	
PL1102	159	3025/129	82	2032	
PL1103	159	3025/129	82	2032	
PL1104	159	3025/129	82	2032	
PL110Z		3099	ZD-110		
PL1111	161	3132	39	2015	
PL1112	161	3132	39	2015	
PL1113	107	3039/316	11	2015	
PL130Z		3348	ZD-130		
PL160Z		3351	ZD-160		
PL180Z		3353	ZD-180		
PL4021			20	2051	
PL4031			21	2034	
PL4032			21	2034	
PL4033			21	2034	
PL4034			21	2034	
PL4051			20	2051	
PL4052			20	2051	
PL4053			20	2051	
PL4054			20	2051	
PL4055			20	2051	
PL4061			212	2010	
PL4062			212	2010	
PL4V3Z		3332	ZD-4.3		
PL51Z		3341	ZD-51		
PL56Z		3342	ZD-56		
PL91Z		3345	ZD-91		
PLE-48	152	3054/196	66	2048	
PLE37	185		58	2025	
PLE52	184		57	2017	
PLL02	1167	3732	IC-317		
PLL02A	1167	3732	IC-317		
PLL02A-F	1167		IC-317		
PLL02A-G	1233		IC-317		
PLL02AG	1233		IC-317		
PLZ52	184		57	2017	
PM1120	234	3247	65	2050	
PM1121	123A	3444	20	2051	
PM194	107	3039/316	11	2015	
PM195	107	3018	11	2015	
PM195A	108	3018	86	2038	
PM26K380					MJ13015
PM27K380					2N6543
PMC-QP0010	130		14	2041	
PMC-QP0012	130		14	2041	
PMC-QP0040	181		75	2041	
PMC-QS-0280	129		244	2027	
PMC-QS-0320	128		243	2030	
PMC-QS-0400	128		243	2030	
PMD10K-100					2N6059
PMD10K-40					2N6057
PMD10K-60					2N6057
PMD10K-80					2N6058
PMD11K-100					2N6052
PMD11K-40					2N6050
PMD11K-60					2N6050
PMD11K-80					2N6051
PMD12K-100					2N6059
PMD12K-40					MJ1000
PMD12K-60					MJ1000
PMD12K-80					MJ1001
PMD13K-100					2N6052
PMD13K-40					MJ900
PMD13K-60					MJ900
PMD13K-80					MJ901
PMD1600K					2N6282
PMD1601K					2N6282
PMD1602K					2N6283
PMD1603K					2N6284
PMD16K-100					2N6284
PMD16K-40					2N6282
PMD16K-60					2N6282
PMD16K-80					2N6283
PMD1700K					2N6285
PMD1701K					2N6285
PMD1702K					2N6286
PMD1703K					2N6287
PMD17K-100					2N6287
PMD17K-40					2N6285
PMD17K-60					2N6285
PMD17K-80					2N6286
PMD20K-120					2N6578
PMD25K-120					2N6578
PMT1767	108	3452	86	2038	
PN	159		82	2032	
PN107	123A		210	2051	
PN108			210	2051	
PN109			210	2051	
PN1613	128				MPS6530
PN1711	128				MPS6530
PN1893					2N4410
PN204	116		504A	1104	
PN2218					2N4401
PN2218A					2N4401

Industry Standard No.	ECG	SK	GE	RS 276-	MOTOR.
PN2219					2N4401
PN2219A					2N4401
PN2221			47	2030	2N4401
PN2221A					2N4401
PN2222	123AP	3854	47	2030	MPS2222
PN2222A		3854			MPS2222A
PN2369	123AP		81	2051	
PN2369A	123AP		81	2051	
PN2484	123AP		81	2051	MPSA05
PN26	175	3538	246	2020	
PN2904	159		21	2034	2N4403
PN2904A					2N4403
PN2905			21	2034	
PN2906	159		21	2034	2N4403
PN2906A	159		21	2034	2N4403
PN2907	159		21	2034	MPS2907
PN2907A	159		21	2034	MPS2907A
PN3250					MPS6516
PN3250A					2N3905
PN3251					2N4125
PN3251A					2N3906
PN350	130	3027	14	2041	
PN3565	123AP				MPS6514
PN3566	123AP				2N4401
PN3567	194				MPS6530
PN3568	194				MPSA06
PN3569	194				MPSA05
PN3641	123AP				MPS6530
PN3642	123AP				MPS6530
PN3643	123AP				MPS6531
PN3644	159				MPS6516
PN3645	159				MPSA55
PN3688					MPSH24
PN3689					MPSH24
PN3690					MPSH20
PN3691	123AP				MPS6512
PN3692	123AP				MPS6513
PN4248					2N5086
PN4249					2N5086
PN4250					2N5087
PN4250A					2N5087
PN4257					MPS4257
PN4257A					MPS4257
PN4258					MPS4258
PN4258A					MPS4258
PN4357					2N5401
PN4416	312				MPF256
PN4888					2N5401
PN4889					2N5401
PN4916	159				2N5208
PN4917	159				2N5208
PN5126					MPSH10
PN5128	123AP				MPS6514
PN5129	123AP				MPS6514
PN5130	108				MPS3563
PN5131	123AP				MPS3646
PN5132	123AP				MPS6539
PN5133	123AP				MPS2714
PN5134	123AP				2N5224
PN5135	123AP				MPS2711
PN5136	123AP				MPS3706
PN5137	123AP				MPS6560
PN5138	159				MPS6516
PN5139	108				MPS6516
PN5142	159				MPS3638
PN5143	159				MPS3638
PN5855					MPS6566
PN5856					MPSA06
PN5857					MPSA56
PN5858					MPSA56
PN5910					MPS3640
PN5964					MPSA42
PN5965					2N5551
PN66	152	3054/196	66	2048	
PN70	159		21	2034	
PN71	159		21	2034	
PN72	159		22	2024	
PN929			212	2010	
PN930	123AP		212	2010	
P093	109	3087		1123	
POWER-12			3	2006	
POWER-25			3	2006	
POWER-299			3	2006	
POWER-99			3	2006	
POWER12	121	3009	239	2006	
POWER25	121	3009	239	2006	
POWER299	121	3014	239	2006	
POWER40	105	3012	4		
POWER500	105	3012	4		
POWER60	105	3012	4		
POWER80	105	3012	4		
POWER99	121	3009	239	2006	
PP3000	130	3027	14	2041	
PP3001	181	3535	75	2041	
PP3003	130	3027	14	2041	
PP3004	181	3535	75	2041	
PP3006	130		14	2041	
PP3007	181	3535	75	2041	
PP3250	152	3054/196	66	2048	
PP3310	152	3054/196	66	2048	
PP3312	152	3054/196	66	2048	
PPR1006	186	3192	28	2017	
PPR1008	186	3192	28	2017	
PQ27	160	3006	245	2004	
PQ28	158	3004/102A	53	2007	
PQ29	158	3004/102A	53	2007	
PQ30	160	3006	245	2004	
PQ31	104	3009	16	2006	
PR-620	145A	3063	ZD-15	564	
PR3-3			30	2006	
PR515	135A	3056	ZD-5.1		
PR605	135A	3056	ZD-5.1		
PR6105-PR6450					1N825
PR6105A-PR6450A					1N827
PR617	142A	3062	ZD-12	563	
PR620	145A	3063	ZD-15	564	
PR804	135A	3056	ZD-5.1		
PR9000	5846	3516	5012		
PR9000R	5847		5013		
PR9001	5846	3516	5012		
PR9001R	5847		5013		
PR9002	5848	7048	5012		
PR9002R	5849	7049	5013		
PR9003	5848	7048	5012		
PR9003R	5849	7049	5013		
PR9004	5866		5028		
PR9004R	5867		5029		
PR9005	5866		5028		
PR9005R	5867		5029		
PR9006	5868	7068	5028		
PR9006R	5869	7069	5029		
PR9007	5868	7068	5028		
PR9007R	5869	7069	5029		
PR9008	5886		5044		

Industry Standard No.	ECG	SK	GE	RS 276-	MOTOR.
PR9008R	5887		5045		
PR9009	5886		5044		
PR9009R	5887		5045		
PR9010R	5891	7091	5045		
PR9011	5890	7090	5044		
PR9011R	5891	7091	5045		
PR9012	5998		5108		
PR9012R	5999		5109		
PR9013	5998		5108		
PR9013R	5999		5109		
PR9014	6002	7202	5108		
PR9014R	6003	7203	5109		
PR9015	6002	7202	5108		
PR9015R	6003	7203	5109		
PR9023	5998		5108		
PR9024	5998		5108		
PR9025	6002	7202	5108		
PR9026	6002	7202	5108		
PR9034	6020	7220	5128		
PR9034R	6021		5129		
PR9035	6026		5128		
PR9035R	6027		5129		
PR9036	6026	7226	5128		
PR9036R	6027	7227	5129		
PR9037	6034		5132		
PR9037R	6035		5133		
PR9038	6034	7234	5132		
PR9038R	6035		5133		
PR9039	6040		5136		
PR9039R	6041		5137		
PR9040	6040	7240	5136		
PR9040R	6041		5137		
PR9041	6042		5140		
PR9041R	6043		5142		
PR9042	6042		5140		
PR9042R	6043		5142		
PR9043	6044	7244	5140		
PR9043R	6045	7245	5141		
PR9044	6044	7244	5140		
PR9044R	6045	7245	5141		
PR9110-PR9450					1N937
PR9110A-PR9450A					1N938
PRC10A	5483	3942			
PRC15A	5483	3942			
PRC20A	5483	3942			
PRC2A	5483	3942			
PRC5A	5483	3942			
PRD105	137A				MZ605
PRD110	137A				MZ610
PRD120	137A				MZ620
PRD140	137A				MZ640
PRD160	137A				MZ640
PRS3017	135A	3056	ZD-5.1		
PRT-101	123A	3444	20	2051	
PRT-104	123A	3444	20	2051	
PRT-104-1	123A	3444	20	2051	
PRT-104-2	123A	3444	20	2051	
PRT-104-3	123A	3444	20	2051	
PRT-104-4	128	3024	243	2030	
PRT101	154		40	2012	
PS-025	116	3016	504A	1104	
PS-035	116	3016	504A	1104	
PS-040	116	3016	504A	1104	
PS-060	116	3017B/117	504A	1104	
PS-1	121	3009	239	2006	
PS-10068	145A	3063	ZD-15	564	
PS-120	116	3016	504A	1104	
PS-801	923D		IC-260	1740	
PS-8917	145A	3063	ZD-15	564	
PS005	116	3311	504A	1104	
PS010		3016	504A	1104	
PS015	116	3016	504A	1104	
PS020		3613	504A	1104	
PS025	116		504A	1104	
PS030		3017B	504A	1104	
PS030(SCR)	5517	3615			
PS035		3615	504A	1104	
PS035(SCR)	5517	3615			
PS040	116		504A	1104	
PS050	116	3017B/117	504A	1104	
PS060	116		504A	1104	
PS08	5514	3613	C222F		
PS1	121	3717	239	2006	
PS10018B	5072A	3136			
PS10019B	139A	3066/118	ZD-9.1	562	
PS10022B	142A	3062	ZD-12	563	
PS10023B	143A	3750			
PS10024B	145A	3063	ZD-15	564	
PS10026B	5077A	3752			
PS10062	5072A	3136			
PS10063	139A	3066/118	ZD-9.1	562	
PS10066	142A	3062	ZD-12	563	
PS10067	143A	3750			
PS10068	145A	3063	ZD-15	564	
PS10069	5075A	3751			
PS10070	5077A	3752			
PS105	116	3311	504A	1104	
PS110		3017B	504A	1104	
PS11327	124		12		
PS1140	125	3033	510,531	1114	
PS120			504A	1104	
PS125	116	3016	504A	1104	
PS130		3016	504A	1104	
PS130(SCR)	5517	3615			
PS1325	137A	3058	ZD-6.2	561	
PS135		3615	504A	1104	
PS135(SCR)	5517	3615			
PS140	116	3016	504A	1104	
PS150	116	3017B/117	504A	1104	
PS1511	140A	3061	ZD-10	562	
PS1512	140A	3061	ZD-10	562	
PS1513	140A	3061	ZD-10	562	
PS1514	140A	3061	ZD-10	562	
PS1515	140A	3061	ZD-10	562	
PS1516	140A	3061	ZD-10	562	
PS1517	140A	3061	ZD-10	562	
PS160	116	3017B/117	504A	1104	
PS18	5514	3613	C222A		
PS209800	123A	3444	20	2051	
PS220	5514	3613	C232B		
PS2207	116	3311	504A	1104	
PS2208	116	3311	504A	1104	
PS2209	116	3311	504A	1104	
PS2247	116	3016	504A	1104	
PS2249	116	3017B/117	504A	1104	
PS230	5517	3615			
PS2346	156	3051	512		
PS2347	156	3051	512		
PS235	5517	3615			
PS2411	116	3311	504A	1104	
PS2412	116	3016	504A	1104	
PS2413	116	3016	504A	1104	
PS2415	116	3017B/117	504A	1104	
PS2416	125	3032A	510,531	1114	
PS2417	125	3033	510,531	1114	
PS28	5514	3613	C222B		
PS30802	519		514	1122	
PS310	5515	6615			
PS320	5515	6615	C232C		
PS3534	5001A	3767			
PS3534A	5001A	3767			
PS3535					1N4570A
PS3539					1N4573A
PS3546					1N4565A
PS3549					1N4568A
PS38	5515	6615	C222C		
PS405	116	3017B/117	504A	1104	1N4001
PS410		3017B	504A	1104	1N4002
PS415	116	3016	504A	1104	1N4003
PS420		3016	504A	1104	1N4003
PS425	116	3016	504A	1104	1N4004
PS430		3016	504A	1104	1N4004
PS435		3016	504A	1104	1N4004
PS440	116	3016	504A	1104	1N4004
PS450	116	3017B/117	504A	1104	1N4005
PS4559	116	3311	504A	1104	
PS4560	116	3031A	504A	1104	
PS460	116	3017B/117	504A	1104	1N4005
PS4725	116	3311	504A	1104	
PS48	5515	6615	C222D		
PS510	5516	6616			
PS520	5516	6616	C232E		
PS5300	116	3031A	504A	1104	
PS5301	116	3031A	504A	1104	
PS5302	116	3031A	504A	1104	
PS58	5516	6616	C222E		
PS6010-1	199	3245	62	2010	
PS603	116	3311	504A	1104	
PS604	116	3311	504A	&50	
PS605	116	3311	504A	&-150	
PS609	116	3311	504A		
PS610		3311	504A	1104	
PS611	116	3311	504A	1104	
PS615	116	3311	504A	1104	
PS616	116	3311	504A	1104	
PS617	116	3311	504A	1104	
PS620	5516		C232M		
PS621	116	3311	504A	1104	
PS622	116	3311	504A	1104	
PS623	116	3311	504A	1104	
PS627	116	3031A	504A	1104	
PS628	116	3031A	504A	1104	
PS629	116	3031A	504A	1104	
PS632	116	3031A	504A		
PS6325	150A	3098	ZD-82		
PS633	116	3031A	504A	1104	
PS636	116	3031A	504A		
PS637	116	3031A	504A	1104	
PS63A205	519		514	1122	
PS6465	5002A	3768			
PS6468	135A	3056	ZD-5.1		
PS68	5516	6616	C222M		
PS700	177	3100/519	300	1122	
PS720	177	3100/519	300	1122	
PS8900	5065A	3838			
PS8901	5066A	3330	ZD-3.3		
PS8903	5067A	3055/134A	ZD-3.6		
PS8905	5069A	3056/135A	ZD-5.1		
PS8906	5116A	3056/135A	ZD-5.1		
PS8907	136A	3057	ZD-5.6	561	
PS8908	137A	3058	ZD-6.2	561	
PS8909	5071A	3334	ZD-6.8	561	
PS8911	5072A	3136	ZD-8.2	562	
PS8912	139A	3060	ZD-9.1	562	
PS8913	140A	3061	ZD-10	562	
PS8915	142A	3062	ZD-12	563	
PS8916	143A	3750			
PS8917	145A	3063	ZD-15	564	
PS8918	5075A	3751			
PS8919	5077A	3752	ZD-18		
PSV111-08	611	3324			
PSV111-14	611	3324			
PSV111-16	611	3324			
PSV114-06	613	3326			
PSV114-14	613	3326			
PSV123-14	611	3324			
PSV123-16	611	3324			
PSV126-06	613	3326			
PSV126-14	613	3326			
PT-029	186A	3357	247	2052	
PT-12	121	3009	239	2006	
PT-150	121	3009	239		
PT-155	121	3009	239	2006	
PT-176	121	3009	239	2006	
PT-234	121	3009	239	2006	
PT-235	121	3009	239	2006	
PT-235A	121	3009	239	2006	
PT-236	121	3009	239	2006	
PT-236A	121	3009	239	2006	
PT-236B	121	3009	239	2006	
PT-242	121	3009	239	2006	
PT-25	121	3009	239	2006	
PT-255	121	3009	239	2006	
PT-256	121	3009	239	2006	
PT-285	121	3009	239	2006	
PT-285A	121	3009	239	2006	
PT-3	116	3009	504A	1104	
PT-301	121	3009	239	2006	
PT-301A	121	3009	239	2006	
PT-307	121	3009	239	2006	
PT-307A	121	3009	239	2006	
PT-3A	121	3009	239	2006	
PT-40	121	3009	239	2006	
PT-50	121	3009	239	2006	
PT-501	105	3012	4		
PT-505			504A	1104	
PT-510	116	3016	504A	1104	
PT-515		3016	504A	1104	
PT-520		9005	504A	1104	
PT-525		9007	504A	1104	
PT-530	116	3005	504A	1104	
PT-530A	100	3005	1	2007	
PT-540		9007	504A	1104	
PT-550	116	3016	504A	1104	
PT-554	121	3009	239	2006	
PT-555	121	3009	239	2006	
PT-560	116		504A	1104	
PT-5B	116		504A	1104	
PT-6	121	3009	239	2006	
PT-72130-1	5804	9007		1144	
PT0-139	102A		53	2007	
PT0139	158		53	2007	
PT06	121	3717	239	2006	
PT12	121	3717	239	2006	
PT150	104	3719	239		
PT1537	195A	3048/329	46		
PT1544	128		243	2030	
PT1545	128		243	2030	

Industry Standard No.	ECG	SK	GE	RS 276-	MOTOR.
PT155	121	3717	239	2006	
PT1558	123A	3444	20	2051	
PT1559	123A	3444	20	2051	
PT1610	123A	3444	20	2051	
PT176	121	3717	239	2006	
PT1835	123A	3444	20	2051	
PT1836	123A	3444	20	2051	
PT1837	123A	3444	20	2051	
PT1941	130		14	2041	
PT200					MFE3004
PT201	105		4		MFE3004
PT2040A	195A	3122	46		
PT234	121	3717	239	2006	
PT235	121	3717	239	2006	
PT235A	104	3719	16	2006	
PT236	104	3719	16	2006	
PT236A	104	3719	16	2006	
PT236B	121	3717	239	2006	
PT236C	121	3009	239	2006	
PT242	121	3717	239	2006	
PT25	121	3717	239	2006	
PT250	105		4		
PT2523	171	3201	27		
PT2524	171	3201	27		
PT2525	171	3201	27		
PT2525A	198		251		
PT2540	192		63	2030	
PT255	121	3717	239	2006	
PT256	121	3717	239	2006	
PT2620	186	3192	28	2017	
PT2635	152	3893	66	2048	
PT2640	186	3192	28	2017	
PT2660	186	3192	28	2017	
PT2677C	195A	3048/329	46		
PT2760	123A	3444	20	2051	
PT285	121	3717	239	2006	
PT285A	121	3717	239	2006	
PT2896	123A	3444	20	2051	
PT2A	160	3006	245	2004	
PT2S	160	3006	245	2004	
PT3	116	3717/121	239	2006	
PT30	121	3009	239	2006	
PT301	121	3717	239	2006	
PT301A	121	3717	239	2006	
PT307	121	3717	239	2006	
PT307A	121	3717	239	2006	
PT3141	123A	3444	20	2051	
PT3141A	123A	3444	20	2051	
PT3141B	123A	3444	20	2051	
PT3141C	195A	3048/329	46		
PT3151A	123A	3444	20	2051	
PT3151B	123A	3444	20	2051	
PT3151C	123A	3444	20	2051	
PT31961	195A	3765	46		
PT32	131		44	2006	
PT320					MFE3003
PT336B			16	2006	
PT3500	123A	3122	20	2051	
PT3501	473	3444/123A	20	2051	MRF230
PT3502	128	3024	243	2030	
PT3503	186	3192	28	2017	MRF232
PT3537					2N5944
PT3570	76				2N5947
PT3571	77				2N5943
PT3571A	77				2N5943
PT366B	121	3717	239	2006	
PT3A	121	3717	239	2006	
PT4	155	3839	43		
PT4-2268-011	519		514	1122	
PT4-2268-01B	519		514	1122	
PT4-2287-01	177		300	1122	
PT4-2311-011	519		514	1122	
PT4-7158	123A	3444	20	2051	
PT4-7158-012	123A	3444	20	2051	
PT4-7158-013	123A	3444	20	2051	
PT4-7158-01A	123A	3444	20	2051	
PT4-7158-021	123A	3444	20	2051	
PT4-7158-022	123A	3444	20	2051	
PT4-7158-023	123A	3444	20	2051	
PT4-7158-02A	123A	3444	20	2051	
PT40	121	3717	239	2006	
PT40063	519		514	1122	
PT4537					2N6080
PT4544					MRF212
PT4555					MRF234
PT4556					MRF450A
PT4570					2N5947
PT4572A					2N5947
PT4574					MRF511
PT4578					MRF517
PT4579	278				2N5943
PT4690	186	3192	28	2017	
PT4800	123A	3444	20	2051	
PT4816	107	3039/316	11	2015	
PT4830	107	3039/316	11	2015	
PT5	116	3017B/117	504A	1104	
PT50	121	3717	239	2006	
PT501	121	3717	239	2006	
PT505	116	9003/5800	504A	1142	1N4001
PT510	116		504A	1104	1N4002
PT515			4		1N4003
PT515(SEMITRON)	213		254		
PT520	116		504A	1143	1N4003
PT525	116		504A	1144	1N4004
PT530	116		53	2007	1N4004
PT530A	102A		53	2007	
PT540	116		504A	1144	1N4004
PT550	116		504A	1104	1N4005
PT554	104	3719	16	2006	
PT555	121	3717	239	2006	
PT560	116	3017B/117	504A	1104	1N4005
PT5693	152	3893	66	2048	
PT5695					MRF233
PT5701	473				MRF402
PT580	125	3032A	510,531	1114	1N4006
PT5B	116	3017B/117	504A	1104	
PT6	121	3717	239	2006	
PT600	186	3192	28	2017	
PT601	186	3192	28	2017	
PT6022/1	169	3678			
PT612	128	3024	243	2030	
PT627	123A	3444	20	2051	
PT6618	186	3192	28	2017	
PT665	152	3893	66	2048	
PT6669	186	3192	28	2017	
PT6696	186	3192	28	2017	
PT6942	328	3895			
PT6994	328	3895			
PT6995	328	3895			
PT6996	327	3945			
PT6D22-1	166	9075		1152	
PT6D22/1		3017B	BR-600	1152	
PT703	123A	3444	20	2051	
PT720	123A	3444	20	2051	
PT72130	116	3017B/117	504A	1104	
PT7903	328	3895			
PT7904	328	3895			
PT7905	328	3895			
PT7906	328	3895			
PT7907	328	3895			
PT7908	328	3895			
PT7909	327	3945			
PT7910	327	3945			
PT7930			75	2041	
PT7931			75	2041	
PT850	128	3024	243	2030	
PT850A	128	3024	243	2030	
PT851	123A	3444	20	2051	
PT8549					2N6081
PT855	160	3046	245	2004	
PT8551					2N3553
PT856	160	3047	245	2004	
PT857	195A	3765	46		
PT8717					MRF231
PT8740					MRF629
PT8740A	472				MRF607
PT8769	350	3176			MRF233
PT8809	362				2N5944
PT8810	363				2N5945
PT8811	364				2N5946
PT8825					2N6136
PT8828					MRF212
PT8837	350	3176			2N6081
PT8838	320				2N6084
PT8851	337				2N5847
PT8851A	338$				2N5847
PT8852	338F$				2N5848
PT8852A	338$				2N5848
PT8854	333				MRF492
PT8854A	334				MRF492A
PT886	123A	3444	20	2051	
PT8861					MRF232
PT8861A					MRF232
PT8863					MRF234
PT8863A	345				MRF234
PT8864					2N6083
PT8864A	345				2N6083
PT8866	341				MRF237
PT887	123A	3444	20	2051	
PT8871					2N6080
PT8871A					2N6080
PT8873F	350F				MRF221
PT8874F	320F				MRF224
PT8877	341				MRF237
PT888	128	3047	243	2030	
PT8889					MRF526
PT896	128	3024	243	2030	
PT897	123A	3444	20	2051	
PT898	123A	3444	20	2051	
PT9700					MRF5174
PT9701					MRF5175
PT9702					MRF323
PT9703					MRF321
PT9704					MRF5177A
PT9704A					MRF5177A
PT9730					2N5641
PT9731	359				MRF314
PT9732	357				2N5641
PT9733	360				MRF315
PT9734					MRF314
PT9776	333				MRF455
PT9776A	334				MRF455A
PT9780	471				MRF464
PT9780A					MRF464A
PT9782					MRF317
PT9782A					MRF317
PT9783					MRF466
PT9784					MRF455
PT9784A					MRF455A
PT9785	470				MRF421
PT9787					2N6370
PT9787A					2N6370
PT9788					MRF401
PT9788A					MRF401
PT9790					MRF428
PT9790A					MRF428A
PT9795					MRF433
PT9795A					2N6081
PT9796					MRF449
PT9796A					MRF449A
PT9797	335				MRF450
PT9797A	336				MRF450A
PT9847					MRF412
PT9C22/1		3081	510	1114	
PTC-703	709	3135	IC-11		
PTC-715	714		IC-4		
PTC101	128	3122	20	2030	
PTC102	100	3005	1	2007	
PTC103	159	3118	82	2032	
PTC104	124	3131A/369	12		
PTC105	121	3009	16	2006	
PTC106	105	3012	4		
PTC107	160	3006	245	2004	
PTC108	101	3011	8	2002	
PTC109	102A	3004	2	2007	
PTC110	184	3054/196	28	2017	
PTC111	185	3083/197	29	2025	
PTC112	175	3026	246	2020	
PTC113	218		234		
PTC114	104	3719	3	2006	
PTC115	123A	3444	335	2051	
PTC116	181	3036	75	2041	
PTC117	154	3044	40	2012	
PTC118	162	3438	35		
PTC119	130	3027	14	2041	
PTC120	131	3052	44	2006	
PTC121	123AP	3444/123A	62	2051	
PTC122	127	3764	25		
PTC123	324	3124/289	243		
PTC124	228A	3103A/396	257		
PTC125	194	3024/128	18		
PTC126	161	3018	39	2015	
PTC127	194	3053	220		
PTC129	163A	3439	36		
PTC130	164	3133	37		
PTC131	159	3118	67	2032	
PTC132	161	3716	39	2015	
PTC134		3010	59	2002	
PTC135		3004	53	2007	
PTC136	123A	3444	20	2051	
PTC137		3054	55		
PTC138		3014	76		
PTC139		3124	210	2051	
PTC140		3027	14	2041	
PTC141		3025	218	2026	
PTC142		3841	29	2025	
PTC143			28	2017	
PTC144		3849	46		

Industry Standard No.	ECG	SK	GE	RS 276-	MOTOR.
PTC151	312	3448	FET-1	2035	
PTC152	312	3448	FET-2	2035	
PTC153	172A	3156	64		
PTC156				2007	
PTC160				2004	
PTC161		3448		2035	
PTC170	287	3433			
PTC172			74	2043	
PTC173			75	2041	
PTC177			67	2023	
PTC178			63	2030	
PTC181				2036	
PTC186		3197	216	2053	
PTC193		3464		2017	
PTC201	5804	9007	504A	1144	
PTC202	116	3313	504A	1104	
PTC203	125	3032A	509	1114	
PTC204	125	3051/156	512	1114	
PTC205	125	3081	510,531	1114	
PTC206	109	3088	1N91	1123	
PTC206M		3088	1N60	1123	
PTC207	109	3087	1N91	1123	
PTC207M		3087	1N34AS	1123	
PTC208	118	3066	CR-1		
PTC209	119	3109	CR-2		
PTC210	502	3067	CR-4		
PTC211	503	3068	CR-5		
PTC212	505	3108	CR-6		
PTC213	505	3108	CR-7		
PTC214	135A	3175	ZD-5.1	1122	
PTC214M			300	1122	
PTC215	135A		300(2)	1122	
PTC216		3998	511	1114	
PTC217		3089	1N82A		
PTC2677C	195A	3765	46		
PTC401	168	3648		1173	
PTC402	169	3107/5307	BR-600		
PTC403			504A	1104	
PTC404	120	3110	CR-3		
PTC406	178MP	3120/114	300(2)	1122(2)	
PTC407	178MP	3119/113	6GC1		
PTC501	134A	3055	ZD-3.6		
PTC502	136A	3057	ZD-5.6	561	
PTC503	137A	3058	ZD-6.2	561	
PTC504	138A	3059	ZD-7.5		
PTC505	139A	3060	ZD-9.1	562	
PTC506	140A	3061	ZD-10	562	
PTC507	142A	3062	ZD-12	563	
PTC508	143A	3750	ZD-13	563	
PTC509	145A	3063	ZD-15	564	
PTC510	5077A	3752	ZD-18		
PTC511	144A	3064/146A	ZD-14	564	
PTC512	147A	3095	ZD-33		
PTC513	5090A	3096/148A	ZD-56		
PTC514	149A	3097	ZD-62		
PTC515	150A	3098	ZD-82		
PTC516	151A	3099	ZD-110		
PTC517	5086A		ZD-39		
PTC518	5088A		ZD-47		
PTC601	536A	3900			
PTC602	501B	3069/501A	520		
PTC603	500A		517		
PTC604	358		42		
PTC605	358		42		
PTC606	358A		41		
PTC701	708	3135/709	IC-10		
PTC703	709		IC-11		
PTC705	713	3077/790			
PTC708	705A	3134	IC-3		
PTC709	718	3159	IC-8		
PTC711	719		IC-28		
PTC713	720	9014	IC-7		
PTC715	714	3075	IC-4		
PTC717	721		IC-14		
PTC719	715	3076	IC-6		
PTC721	722	3358/1006	IC-9		
PTC723	723	3144			
PTC726	712	3072	IC-2		
PTC732	804	3455			
PTC733	791	3149			
PTC734		3171	IC-24	2022	
PTC735	801	3160			
PTC736	736		IC-220		
PTC739		3480	IC-30		
PTC740	740A	3328			
PTC741	738	3167			
PTC754	1004	3365	IC-312		
PTC780	1155	3231			
PTG132	161	3132	39	2015	
PTO-6	121	3717	239	2006	
PTO139	102A	3123	53	2007	
PU6C22	116		504A	1104	
PV-8	116	3016	504A	1104	
PV8	116		504A	1104	
PXB-103	102A	3004	43	2007	
PXB-113	102A	3004	43	2007	
PXB103	102	3722	2	2007	
PXB113	102	3722	2	2007	
PXC-101	102A	3004	53	2007	
PXC-101AB	102A	3004	53	2007	
PXC101	102	3722	2	2007	
PXC101A	102	3722	2	2007	
PXC101AB	102	3722	2	2007	
PY-5	116	3017B/117	504A	1104	
PZ-140B					1N3493
PZ-140D					1N3495
PZ-140F					MR328
PZW	5070A	9021	ZD-62	561	
Q-0-172	128		243	2030	
Q-00169	312	3448	FET-1	2035	
Q-00169A	312	3448	FET-1	2035	
Q-00169B	312	3448	FET-1	2035	
Q-00169C	312	3448	FET-1	2035	
Q-00184R	312	3448	FET-1	2035	
Q-00269	123A	3444	20	2051	
Q-00269A	123A	3444	20	2051	
Q-00269B	123A	3444	20	2051	
Q-00269C	123A	3444	20	2051	
Q-00284R	229	3018	61	2038	
Q-00284R-3	108	3018	86	2038	
Q-00369	123A	3444	20	2051	
Q-00369A	123A	3444	20	2051	
Q-00369B	123A	3444	20	2051	
Q-00369C	123A	3444	20	2051	
Q-00384R	233		210	2009	
Q-00384R-3	108	3018	86	2038	
Q-00469	199	3020/123	62	2010	
Q-00469A	199	3124/289	62	2010	
Q-00469B	199	3124/289	62	2010	
Q-00469C	199	3122	62	2010	
Q-00484R	123A	3444	20	2051	
Q-00484R-1	108	3452	86	2038	
Q-00569	123A	3444	20	2051	
Q-00569A	123A	3444	20	2051	

Industry Standard No.	ECG	SK	GE	RS 276-	MOTOR.
Q-00569B	123A	3444	20	2051	
Q-00569C	199	3122	62	2010	
Q-00584			11	2015	
Q-00584R	233		210	2009	
Q-00584R-3	108	3452	86	2038	
Q-00669	123A	3444	20	2051	
Q-00669A	123A	3444	20	2051	
Q-00669B	123A	3444	20	2051	
Q-00669C	123A	3444	20	2051	
Q-00684R	123A	3444	20	2051	
Q-00769	175	3026	246	2020	
Q-00769A	175	3026	246	2020	
Q-00769B	175	3026	246	2020	
Q-00769C	175	3026	246	2020	
Q-00784R	289A	3138/193A	268	2038	
Q-00869	128	3047	243	2030	
Q-00869A	128	3047	243	2030	
Q-00869B	128	3047	243	2030	
Q-00869C	195A	3048/329	46		
Q-00969	128	3047	243	2030	
Q-00969A	128	3047	243	2030	
Q-00969B	128	3047	243	2030	
Q-00969C	195A	3048/329	46	2030	
Q-00984R	159	3025/129	82	2032	
Q-01069	195A	3048/329	46		
Q-01069A	195A	3048/329	46		
Q-01069B	195A	3048/329	46		
Q-01069C	237	3049/224	46		
Q-01084R	131		44	2006	
Q-01115C	107	3444/123A	11	2015	
Q-0115C	123A	3122	20	2051	
Q-01169	128	3024	243	2030	
Q-01169A	128	3024	243	2030	
Q-01169B	128	3024	243	2030	
Q-01169C	128	3024	243	2030	
Q-01184R	195A	3048/329	46		
Q-01269	175	3026	246	2020	
Q-01269B	175	3026	246	2020	
Q-01269C	175	3026	246	2020	
Q-01284R	195A	3048/329	46		
Q-01369C	1104	3225	IC-91		
Q-01384R	237	3299	46		
Q-01484R	724		IC-86		
Q-0172	128		243	2030	
Q-02115C	123A	3444	20	2051	
Q-03115C	123A	3444	20	2051	
Q-04115C	123A	3444	20	2051	
Q-05115C	123A	3444	20	2051	
Q-06115C	123A	3444	20	2051	
Q-07115C	123A	3444	20	2051	
Q-08115C	199	3122	62	2010	
Q-09115C	199	3122	62	2010	
Q-1	102	3005	2	2007	
Q-10115C	123A	3444	20	2051	
Q-10484R	724		IC-86		
Q-11115C	152	3054/196	66	2048	
Q-12115C	152	3054/196	66	2048	
Q-13115C	199	3122	62	2010	
Q-14115C	123A	3444	20	2051	
Q-15115C	123A	3444	20	2051	
Q-16	102A	3004	53	2007	
Q-16115C	123A	3444	20	2051	
Q-17115C	299	3047	236		
Q-18115C	282	3049/224	264		
Q-1A	100	3005	1	2007	
Q-1N4001	961	3311			
Q-1N914	519	3100	514	1122	
Q-2	101	3861	8	2002	
Q-20115C	116	3017B/117	504A	1104	
Q-21115C	177	3100/519	300	1122	
Q-22115C	109	3088		1123	
Q-23115C	519	3100	514	1122	
Q-24115C	177	3100/519	300	1122	
Q-25115C	139A	3060	ZD-9.1	562	
Q-26115C	116	3017B/117	504A	1104	
Q-2N5225	123AP	3444/123A	20	2051	
Q-2N5226	159	3466	82		
Q-3	101	3861	8	2002	
Q-35	123A	3444	20	2051	
Q-36	159	3114/290	82	2032	
Q-36A	159	3114/290	82	2032	
Q-4	102	3005	2	2007	
Q-5	101	3861	8	2002	
Q-6	102A	3004	53	2007	
Q-7	102A	3004	53	2007	
Q-8	102A	3004	53	2007	
Q-9	101	3861	8	2002	
Q-RF-2	123A	3444	20	2051	
Q-SE1001	123A	3444	20	2051	
Q0-419	159			2032	
Q0415	159		82	2032	
QOV60526	102A	3004	53	2007	
QOV60527	103A	3010	59	2002	
QOV60528	102A	3004	53	2007	
QOV60529	123A	3020/123	20	2051	
QOV60530	123A	3020/123	20	2051	
QOV60537	103A	3010	20	2002	
QOV60538	123A	3004/102A	20	2051	
Q1	102A	3124/289	53	2007	
Q1-7C	102A	3004	53	2007	
Q1/6515	107		11	2015	
Q11/6515	102A	3722/102	53	2007	
Q12/6515	102A		53	2007	
Q16	5636		53	2007	
Q1B	116	3016	504A	1104	
Q1H	116	3016	504A	1104	
Q2		3124	39	2015	
Q2-7C	102A	3004	53	2007	
Q2/6515	107		11	2015	
Q2001M	5641		SC136B		
Q2001M3	5651				T2302PB
Q2001M4					SC136B
Q2001MS2	5642	3506			
Q2001P	5641		SC136B		
Q2003P3			SC136B		T2302PB
Q2003P4					SC136B
Q2004F31					T2302PB
Q2004F41					SC136B
Q2006F31					SC136B
Q2006F41					SC136B
Q2006H	5673		SC240B		
Q2006L4	5645	3507			
Q2008F41					T2800B
Q2008H	5673		SC245B		
Q200E3					MAC94A4
Q200E4					MAC94-4
Q2010F41					SC146B
Q2010H	5673	3508/5675	SC245B		
Q2015G			SC251B		SC251B
Q2015H	5673	3520/5677	SC250B		SC250B
Q2015L5	56004	3658			
Q2015R5	56004	3658	SC251B		
Q2025C	56014		SC260B4		
Q2025D	56022		SC260B2		

Industry Standard No.	ECG	SK	GE	RS 276-	MOTOR.
Q2025G			SC261B		SC261B
Q2025H	5693	3652	SC260B		SC260B
Q2025P					MAC525-4
Q2040D	56022		SC265B2		T6420B
Q205(DYNACO)	722		IC-9		
Q2N1526	126	3008	52	2024	
Q2N2428	102A	3004	53	2007	
Q2N2613	102A	3004	53	2007	
Q2N406	102A	3004	53	2007	
Q2N4105	103A	3010	59	2002	
Q2N4106	158	3004/102A	53	2007	
Q2T2222					MPQ2222
Q2T2905					MPQ2907
Q2T3244					MPQ2907
Q2T3725					MPQ3725
Q3			39	2015	
Q3/2	116	3311	504A	1104	
Q3/6515	123A	3444	20	2051	
Q301	107		11	2015	
Q32	116		504A	1104	
Q34450	124	3021	12		
Q35218	102A	3004	53	2007	
Q35242	123A	3444	20	2051	
Q35259	161	3018	39	2015	
Q4	102A		53	2011	
Q4/6515	107		11	2015	
Q4001M	5642		SC136D		
Q4001M3	5652				T2302PD
Q4001M4					SC136D
Q4003P3			SC136D		T2302PD
Q4003P4					SC136D
Q4004F31					T2302PD
Q4004F41					SC136D
Q4006F41					SC136D
Q4006H	5675		SC240D		
Q4006L4	5645	3507			
Q4008F41					T2800D
Q4008H	5675		SC245D		
Q400E3					MAC94A6
Q400E4					MAC94-6
Q4010F41					SC146D
Q4010H	5675	3508	SC245D		
Q4015G			SC251D		SC251D
Q4015H	5675	3520/5677	SC250D		SC250D
Q4015L5	56006	3659			
Q4015R5	56006	3659	SC151D		
Q4025C	56014		SC260D4		
Q4025D	56024		SC260D2		
Q4025G			SC261D		SC261D
Q4025H	5695	3509	SC260D		SC260D
Q4025P					MAC525-6
Q40263	102A	3004	53	2007	
Q40359	160	3008	245	2004	
Q4040D	56024		SC265D2		
Q49	109	3087		1123	
Q4B	116	3016	504A	1104	
Q5			39	2015	
Q50	109	3087		1123	
Q5004F41					SC136E
Q5006H	5677		SC240E		
Q5008H	5677		SC245E		
Q5010H	5677		SC245E		
Q5015G			SC251E		SC251E
Q5015H			SC250E		SC250E
Q5015L5	56008	3660			
Q5015R5	56008	3660	SC151E		
Q5025D	56026		SC260E2		
Q5025G			SC261E		SC261E
Q5025H	5697	3522	SC260E		SC260E
Q5025P					MAC525-7
Q5030	127	3035	25		
Q5039	103A	3010	59	2002	
Q5040D	56026		SC265E2		
Q5044	160	3006	245	2004	
Q5050	103A	3010	59	2002	
Q5053	123A	3444	20	2010	
Q5053A	199	3122			
Q5053C			20	2051	
Q5053D	199	3124/289	62	2010	
Q5053E	199	3124/289	62	2010	
Q5053F	199	3124/289	62	2010	
Q5053G	199	3124/289	62	2010	
Q5073D			20	2051	
Q5073E			20	2051	
Q5073F			20	2051	
Q5075CLY	124	3021	12		
Q5075DLY	124	3021	12		
Q5075DXY	124	3021	12		
Q5075ELY	124	3021	12		
Q5075EXY	124	3021	12		
Q5075FXY	124	3021	12		
Q5075ZMY	124	3021	12		
Q5077A	159	3114/290			
Q50787	233			2009	
Q50787Z	233			2009	
Q5078Z	199	3132	213	2010	
Q5083B			19	2041	
Q5083C	162	3452/108	35		
Q5087Z	159	3114/290	82	2032	
Q5099E			18	2030	
Q5099F			18	2030	
Q51	109	3087		1123	
Q5100A	129	3114/290	244	2027	
Q5101D			23	2020	
Q5102	234	3114/290	65	2050	
Q5102P	234	3025/129	65	2050	
Q5102Q	234	3025/129	65	2050	
Q5102R	234	3025/129	65	2050	
Q5104Z	124	3021	12		
Q510ZQ			21	2034	
Q5110Z			19	2041	
Q5111	238	3710			
Q5111K	238	3710			
Q5111ZK	238	3710	38		
Q5113ZLM	124	3021	12		
Q5113ZMM	124	3021	12		
Q5113ZMY	124		12		
Q5116C	290A	3114/290	269	2032	
Q5116CA	290A	3114/290	21	2034	
Q5119	162	3438	35		
Q5119D	162	3115/165	36		
Q5120P	162	3452/108	35		
Q5120Q	162	3452/108	35		
Q5120R	162	3452/108	35		
Q5121	199	3245	62	2010	
Q51210	199	3245	62	2010	
Q5121O	199	3124/289		2010	
Q5121Q	199	3025/129	62	2010	
Q5121R	199	3124/289		2010	
Q5123E	123A	3444	20	2051	
Q5123F	123A	3444	20	2051	
Q5124	297	3122	210	2030	
Q5120R			62	2010	
Q51327	165		38		

Industry Standard No.	ECG	SK	GE	RS 276-	MOTOR.
Q5134Z	175	3131A/369	246	2020	
Q5135	159	3114/290	221	2032	
Q5135-OY	159	3114/290			
Q5137BA	291		27		
Q5138	198	3220	251		
Q5138K	198	3220	251		
Q5138L	198	3220	251		
Q5138M	198	3220	251		
Q5140XP	165	3115	38		
Q5140Z	165	3115	38		
Q5140ZXP	165	3115	38		
Q5140ZXQ	165	3111	38		
Q5140ZXR	165	3111	38		
Q514Z	165	3115	38		
Q5160	198	3220	271	2030	
Q5160-O		3220	251		
Q5160-O	198		32		
Q5160R	198		251		
Q5160Y	198	3220	251		
Q5161Z	238	3710	259		
Q5163	375	3220/198			
Q5163D	375	3104A	32		
Q5175	198		251		
Q5175TM	198		32		
Q5178	198		32		
Q5180	199	3124/289	62	2010	
Q5182		3444	81	2051	
Q5183	199	3124/289	212	2010	
Q5183C	289A	3124/289			
Q5183P	199	3124/289	212	2010	
Q5199	162		73		
Q52	116	3311	504A	1104	
Q5202	198	3220			
Q5205	290A	3114/290	269	2032	
Q5206	162		73		
Q5207	238	3710	38		
Q5207Z	238	3710	38		
Q5209	294	3513	48		
Q5210	199	3244	27		
Q5217		3045	222	2012	
Q53	116	3311	504A	1104	
Q54	116	3311	504A	1104	
Q55	116	3311	504A	1104	
Q56	116	3311	504A	1104	
Q57	116	3311	504A	1104	
Q58	116	3311	504A	1104	
Q59	116	3031A	504A	1104	
Q6	158		53	2007	
Q6/2	116	3311	504A	1104	
Q6/6515	107		11	2015	
Q60	116	3031A	504A	1104	
Q6004F41					SC136M
Q6006F51					SC136M
Q6006H	5677		SC240M		
Q6008F51					T2802M
Q6008H	5677		SC245M		
Q6010F51					SC146M
Q6010H	5677	3520	SC245M		
Q6015G			SC251M		SC251M
Q6015H		3520	SC250M		SC250M
Q6015L5	56008	3660			
Q6015R5	56008	3660	SC151M		
Q6025D	56026		SC260M2		
Q6025G			SC261M		SC261M
Q6025H	5697	3522	SC260M		SC260M
Q6025P					MAC525-8
Q6040D	56026		SC265M2		
Q62	116		504A	1104	
Q6521	288	3114/290	221		
Q6522	288		221		
Q7	158	3045/225	53	2007	
Q7-C1318XDN				2030	
Q7/6515	102A		53	2007	
Q8	158			2007	
Q8/2	116	3311	504A	1104	
Q8/6515	102A		53	2007	
Q82	116		504A	1104	
Q9/6515	126		52	2024	
QA-1	160		245	2004	
QA-10	128	3024	243	2030	
QA-11	129	3025	244	2027	
QA-12	123A	3444	20	2051	
QA-13	123A	3444	20	2051	
QA-14	123A	3444	20	2051	
QA-15	123A	3444	20	2051	
QA-16	123A	3444	20	2051	
QA-17	129	3025	244	2027	
QA-18	312	3116	FET-1	2028	
QA-19	123A	3444	20	2051	
QA-20	312	3112	FET-1	2028	
QA-21	159	3114/290	82	2032	
QA-21A	159	3114/290	82	2032	
QA-8	128	3024	243	2030	
QA-9	129	3025	244	2027	
QA01-07R			ZD-15	564	
QA01-07RE	5071A		ZD-6.8	561	
QA01-082	5072A	3136	ZD-8.2	562	
QA01-08R	5072A	3059/138A	ZD-8.2	562	
QA01-11.5E			ZD-11	563	
QA01-115E	5074A	3139			
QA01-11SE	5074A	3139	ZD-11	563	
QA01-12S			ZD-12	563	
QA01-14RD				564	
QA01-25RA	5081A	3064/146A	ZD-24		
QA1-11M	5074A		ZD-11	563	
QA107RE	5071A		ZD-6.8	561	
QA11	124		12		
QA111M	5074A		ZD-11	563	
QA111SE	5074A		ZD-11	563	
QA703E	703A	3157	IC-12		
QA8	128		243	2030	
QB01-11ZB			ZD-11	563	
QB01-15ZB	145A		ZD-15	564	
QB01-18	5077A	3752	ZD-18		
QB106P	5070A	9021	ZD-6.0		
QB111			ZD-11	563	
QB111Z	5074A		ZD-11	563	
QB118		3752	ZD-18		
QC-1	358		42		
QD-CS2688DJ	614	3327	90		
QD-CTT310XQ	614	3327			
QD-CTT410KQ	612	3325			
QD-CTT410XQ	612	3325			
QD-G1N60PXT	109	3088	1N60	1123	
QD-G1N60XXT	109	3088	1N60	1123	
QD-G1S32XXT	109	3088	1N60	1123	
QD-S1S953XA	156		300	1122	
QD-SMA150XN	177	3100/519	300	1122	
QD-SS1555XT	177	3100/519	300	1122	
QD-SS1885XT	116	3311	504A	1104	
QD-SSR1KX4P	116	3311	504A	1104	
QD-SS53XXA	177		300	1122	
QD-SV06CXXB	116	3017B/117	504A	1104	
QD-ZBZ162XJ	5075A	3751	ZD-16	564	

Industry Standard No.	ECG	SK	GE	RS 276-	MOTOR.
QD-ZMZ205XE	135A	3056	ZD-5.0		
QD-ZMZ306CE	5070A		ZD-6.0	561	
QD-ZMZ310XE	614	3327			
QD-ZMZ408CE	5072A		ZD-8.2	562	
QD-ZMZ409BE	139A		ZD-9.1	562	
QD-ZRD56EAA	136A	9021/5070A	ZD-6.0	561	
QD-ZRD56FAA	136A	3057	ZD-5.6	561	
QD-ZRD9EXAA	139A	3060	ZD-9.1	562	
QDCTT310XQ	614	3327			
QDCTT410XQ	612	3126	90		
QDSSR3AMBE	125	3051/156	510	1142	
QDZMZ205XE	135A		ZD-5.0		
QG-0074			3	2006	
QG-0076			2	2007	
QG0074		3009	3	2006	
QG0076	102A		53	2007	
QG0254	123A	3444	20	2051	
QKT-0033XBE			FET-2	2035	
QKT0033XBE	312	3116	FET-2	2035	
QN-BD1126A	601	3463			
QN2613	102	3722	2	2007	
QO-419			82	2032	
QOV60526	102A		53	2007	
QOV60527	103A		59	2002	
QOV60528	102A		53	2007	
QOV60529	123A	3124/289	20	2051	
QOV60530	123A	3444	20	2051	
QOV60537	103A		59	2002	
QOV60538	102A	3444/123A	53	2007	
QOV60539			2	2007	
QP-1	179	3642	76		
QP-10	179	3642	76		
QP-11	130	3027	14	2041	
QP-12	130	3027	14	2041	
QP-13	185	3191	58	2025	
QP-14	184	3190	57	2017	
QP-1A	179	3642	76		
QP-2	179	3642	239		
QP-3	179	3642	76		
QP-31	197	3083	250	2027	
QP-4	179	3642	76		
QP-5	179	3642	76		
QP-6	179	3642	76		
QP-7	179	3642	76		
QP-8	130	3027	14	2041	
QP-8-P		3027	255	2041	
QP001200A	130		14	2041	
QP1	179	3717/121	239	2006	
QP1A	179	3717/121	239	2006	
QP2	179	3717/121	239	2006	
QP6	179	3717/121	239	2006	
QP7	179	3717/121	239	2006	
QP8	130		14	2041	
QP8-6623N	105	3012	4		
QQ-OPLL02AQ	1167	3732			
QQ-M00566AA	1052	3249			
QQ-M07205AT	1155	3231	IC-179		
QQ-MB521AX	1166	3827			
QQ-MBA511AX	1165	3827/1166			
QQ-MBA511AY	1165	3827/1166			
QQ-MBA511BX	1165	3827/1166			
QQ-MBA521AX	1166		IC-282	704	
QQ-MC3001DT	192	3137			
QQ-MM566AA	1052	3249			
QQ-M07205AT			IC-179	705	
QQ-OPLL02AO	1167	3732			
QQ-OPLL02AN	1167	3732			
QQC61209	158		53	2007	
QQC61210	102A	3124/289	53	2007	
QQC61689	159	3114/290	82	2032	
QQC61689A	159	3114/290	82	2032	
QQM07205AT	1155	3231	IC-179	705	
QQMAN612AN	1249	3963			
QQV60526	158		53	2007	
QQV60527	103A	3835	59	2002	
QQV60528	102A		53	2007	
QQV60529	128	3024	243	2030	
QQV60537	103A	3835	59	2002	
QQV60538	158		53	2007	
QQV60539	102A		53	2007	
QQV61772	103A	3835	59	2002	
QR2378	102A	3004	53	2007	
QRF-2	128		243	2030	
QRF-3	312	3448	FET-1	2035	
QRF3	312		FET-1	2035	
QRG-3	312	3448	FET-1	2035	
QRT-101		3262		2032	
QRT-104		3245		2033	
QRT-107		3854		2013	
QRT-119				2002	
QRT-122				2002	
QRT-187				2014	
QRT-200		3087		1123	
QRT-210		3033A		1114	
QRT-212		3311		1011	
QRT-213		3311		1102	
QRT-214		3312		1103	
QRT-215		3313		1104	
QRT-218		3316		1122	
QRT-230		3647		1172	
QRT-236		3057		561	
QRT-237		3058		561	
QRT-238		3334		561	
QRT-240		3060		562	
QRT-241		3061		562	
QRT-242		3139		563	
QRT-243		3062		563	
QRT-244		3093		563	
QRT-245		3063		564	
QRT-246		3751		564	
QRT-257		3136		562	
QRT105	199		62	2010	
QRT106	234		65	2050	
QS-0254			10	2051	
QS0254			10	2051	
QS054	123A	3122	20	2051	
QS1306	235	3197	215		
QS316	159		82	2032	
QSC380	123A	3444	20	2051	
QSC509	297	3124/289	271	2030	
QSC756	282	3748			
QSE1001	123A	3018	20	2051	
QSE3001	108	3452	86	2038	
QSE5020	161	3018	39	2015	
QT-A0683XBN	294	3841			
QT-A0719AXN			269	2032	
QT-A0719XAN	290A	3114/290	269		
QT-A0719XCN	290A	3114/290	21	2034	
QT-A0719XHN	290A	3114/290	269	2032	
QT-A0733XAA	294	3114/290	48		
QT-A0733XBA	294	3114/290			
QT-A0733XON		3138	48	2027	
QT-C0372XAT	123A	3122	61	2051	
QT-C0460CBB	233	3018	61		
QT-C0710XAE	123AP	3356	211	2016	
QT-C0710XBE	123A	3356	211	2051	
QT-C0710XEE	123AP		211	2051	
QT-C0735XBT	289A	3122	210	2051	
QT-C0828XAN	199	3122	61	2010	
QT-C0828XDN	199	3122	61	2010	
QT-C0829XAN	123A	3122	20	2051	
QT-C0829XBN	123A	3122	20	2038	
QT-C0829XON		3122	20		
QT-C0839XDA	123A	3018	61	2038	
QT-C0900XBA	199	3124/289	62	2010	
QT-C0900XBD	199		62	2010	
QT-C0900XCA	199	3124/289	62	2010	
QT-C0945AAA	199	3124/289			
QT-C0945ACA	199	3124/289	212	2010	
QT-C0945AEA	199	3124/289			
QT-C0945AGA	199	3124/289	212	2010	
QT-C0945LAA	199	3124/289			
QT-C1047XAN	229		60	2015	
QT-C1047XBN	229		60	2015	
QT-C1306XZA	235	3197	215		
QT-C1307XZA	236	3239	216	2053	
QT-C1318XAN	297	3124/289	210		
QT-C1318XDN	297	3122	210	2030	
QT-C131BXDN	128	3122	210	2051	
QT-C1359XAN	229	3038	212	2010	
QT-C1687XAN	123A	3038	20	2051	
QT-C1760XAS	306		276		
QT-C1760XCS	306		276		
QT-C1846XAN	295		336		
QT-C2074XAT	322	3252/302			
QT-CBC546AA	199		20	2051	
QT-C0372XAT	123A		61		
QT-C0710XBE	123AP		211		
QT-C0828XDN	199		61		
QT-C0829XBN	123A		20		
QT-C0900XCA	199		62		
QT-C0945ACA	199		212		
QT-CQ4600CBB			61	2038	
QT-D0313XAC	196	3620	66	2048	
QT-D0325XAC	152	3893	28	2017	
QT-K0023AAS	312		FET-2	2035	
QT-K0033XBE	312		FET-2	2035	
QTC1969XBE	236	3239			
QV-BD11225A	601	3463			
QV-BD11246A	601	3463			
QVD1KP114	109	3087	1N34AS	1123	
QVDM8513ART	601	3463			
QVM800B	604	3088	1N60	1123	
QZ-15T10	145A	3063	ZD-15	564	
QZ-15T5	145A	3063	ZD-15	564	
QZ10T10	140A	3061	ZD-10	562	
QZ10T5	140A	3061	ZD-10	562	
QZ12T10	142A	3062	ZD-12	563	
QZ12T5	142A	3062	ZD-12	563	
QZ14T10	144A	3094	ZD-14	564	
QZ14T5	144A	3094	ZD-14	564	
QZ1575	145A		ZD-15	564	
QZ15T10	145A	3063	ZD-15	564	
QZ15T5	145A	3063	ZD-15	564	
QZ27T10	146A	3064	ZD-27		
QZ27T10A	146A	3064	ZD-27		
QZ27T5	146A	3064	ZD-27		
QZ27T5A	146A	3064	ZD-27		
QZ3.6T10	134A	3055	ZD-3.6		
QZ3.6T5	134A	3055	ZD-3.6		
QZ5.1T10	135A	3056	ZD-5.1		
QZ5.1T5	135A	3056	ZD-5.1		
QZ5.6T10	136A	3057	ZD-5.6	561	
QZ5.6T5	136A	3057	ZD-5.6	561	
QZ6.2T10	137A	3058	ZD-6.2	561	
QZ6.2T5	137A	3058	ZD-6.2	561	
R-1	116	3031A		1144	
R-106379	116	3016	504A	1104	
R-113321	116	3016	504A	1104	
R-113392	116	3016	504A	1104	
R-119	100	3005	1	2007	
R-120	102A	3004	53	2007	
R-125	101	3861	8	2002	
R-12C	513	3443	513		
R-1348	142A		ZD-12	563	
R-135	101	3861	8	2002	
R-136	101	3861	8	2002	
R-137	101	3861	8	2002	
R-152	102A	3004	53	2007	
R-1533	101	3861	8	2002	
R-154B	116		504A	1104	
R-16	102A	3004	53	2007	
R-163	100	3005	1	2007	
R-164	102A	3004	53	2007	
R-186	100	3005	1	2007	
R-1A	116	3016		1143	
R-1B	5802	9005		1143	
R-1S188	109	3090		1123	
R-1S333Y	5072A		ZD-8.2	562	
R-2-02	116		504A	1104	
R-2001	127	3764	25		
R-202	101	3861	8	2002	
R-203	101	3861	8	2002	
R-227	100	3005	1	2007	
R-23-1003	102A	3004	53	2007	
R-23-1004	102A	3004	53	2007	
R-24-1001	102A	3004	53	2007	
R-24-1002	102A	3004	53	2007	
R-242	102A	3004	53	2007	
R-244	100	3005	1	2007	
R-245	102A	3004	53	2007	
R-291	102A	3004	53	2007	
R-2A	5944		5048		
R-2B	5944		5048		
R-2SA222	160	3008	245	2004	
R-2SB186	102A	3004	53	2007	
R-2SB187	102A	3004	53	2007	
R-2SB303	102A	3004	53	2007	
R-2SB405	158	3004/102A	53	2007	
R-2SB474	226MP	3086	49(2)	2025(2)	
R-2SB492	176	3845	80		
R-2SC535	107	3018	11	2015	
R-2SC537	123A	3444	20	2051	
R-2SC545	108	3452	86	2038	
R-2SC668	108	3452	86	2038	
R-2SC772	107	3018	11	2015	
R-2SC858	107	3124/289	11	2015	
R-2SD187	103A	3010	59	2002	
R-3	120	3110	CR-3		
R-33	101	3861	8	2002	
R-34	101	3861	8	2002	
R-3530-1	108		86	2038	
R-3552-1	128		243	2030	
R-3553-1	128		243	2030	
R-3555	128		243	2030	
R-3580-1	128		243	2030	
R-3A	5944		5048		
R-424	100	3005	1	2007	

Industry Standard No.	ECG	SK	GE	RS 276-	MOTOR.
R-425	100	3005	1	2007	
R-488	100	3005	1	2007	
R-4A	5804	9007		1144	
R-5	5808	9009			
R-506	100	3005	1	2007	
R-530	102A	3004	53	2007	
R-539	126	3008	52	2024	
R-56	158	3004/102A	53	2007	
R-593	102A	3004	53	2007	
R-5B	5802	9005		1143	
R-5C	5802	9005		1143	
R-608A	102A	3004	53	2007	
R-62	101		8	2002	
R-63	101	3011	8	2002	
R-63HZ	5994	3501	5104		
R-64	102A	3004	53	2007	
R-66	102A	3004	53	2007	
R-67	102A	3004	53	2007	
R-6A	5944		5048		
R-6B	5944		5048		
R-7026	177	3311	300	1122	
R-7027	177	3100/519	300	1122	
R-7029			1N60	1123	
R-7051	109	3090		1123	
R-7092	177		300	1122	
R-7093	137A		ZD-6.2	561	
R-7094	138A		ZD-7.5		
R-7096	117		504A	1104	
R-7097	5019A		ZD-9.1	562	
R-7103	142A	3062	ZD-12	563	
R-83	102A	3004	53	2007	
R-S-1720	116		504A	1104	
R-S1264	116		504A	1104	
R-S1347	116	3311	504A	1104	
R-S1805	5804	9007		1144	
R0080	156	3081/125	512		
R0081	156	3081/125	512		
R0082	156	3081/125	512		
R0086	156	3081/125	512		
R008A	156	3081/125	512		
R0090	156	3081/125	512		
R0091	156	3051	512		
R0092	131	3051/156	44	2006	
R0094	156	3051	512		
R0096	156	3051	512		
R0097	156	3051	512		
R0098	156	3051	512		
R0130	5882	3500	5040		
R0131	5882	3500	5040		
R0132	5882	3500	5040		
R0134	5882	3500	5040		
R0136	5882	3500	5040		
R0160	5994	3501	5104		
R0161	5994	3501	5104		
R0162	5994	3501	5104		
R0164	5994	3501	5104		
R0250	5994	3501	5104		
R0251	5994	3501	5104		
R0253	5994	3501	5104		
R0254	5994	3501	5104		
R0255	5994	3501	5104		
R0256	5994	3501	5104		
R0257	5994	3501	5104		
R02A	125	3081			
R02Z				1143	
R06-1001	108	3452	86	2038	
R06-1002	108	3452	86	2038	
R06-1003	108	3452	86	2038	
R06-1004	108	3452	86	2038	
R06-1005	108	3452	86	2038	
R06-1006	108	3452	86	2038	
R06-1007	102A	3004	53	2007	
R06-1008	102A	3004	53	2007	
R06-1009	102A	3004	53	2007	
R06-1010	102A	3004	53	2007	
R0606	506	3515	511	1114	
R07105	5074A		ZD-11.5		
R07A	5071A	3059/138A	ZD-6.8	561	
R080	125	3032A	510,531	1114	
R1		3066	CR-1	1104	
R100-1	102A		53	2007	
R100-8	102A		53	2007	
R100-9	102A		53	2007	
R1000	156	3033	512		1N4007
R101-2	102A		53	2007	
R101-3	102A		53	2007	
R101-4	102A		53	2007	
R102-41	121	3717	239	2006	
R1035	116	3017B/117	504A	1104	
R104-5	126		52	2024	
R104-6	126		52	2024	
R104-7	126		52	2024	
R104-8	126		52	2024	
R105064	118	3110/120	CR-1		
R105330	358		42		
R106379	116		504A	1104	
R109328	113A	3119/113	6GC1		
R109474	113A	3119/113	6GC1		
R10D1	116		504A	1104	
R10DC	116	3031A	504A	1104	
R1106	109	3087		1123	
R1107	109	3087		1123	
R1109	109	3087		1123	
R1117-1	124		12		
R112524			1N60	1123	
R113321	116		504A	1104	
R113392	116		504A	1104	
R117	101	3861	8	2002	
R118	107	3039/316	11	2015	
R119	100	3721	1	2007	
R12	101		8	2002	
R120	102A	3123	53	2007	
R1215(HEP)				1067	
R1217(HEP)				1067	
R1221	5444	3634			
R1222	5446	3635			
R1223	5448	3636			
R122C	116	3311	504A	1104	
R123	128		243	2030	
R123-2	128		243	2030	
R123-3	128		243	2030	
R123-4	128		243	2030	
R123-5	128		243	2030	
R1241	5529	6629			
R1242	5529	6629			
R1243	5529	6629			
R1244	5529	6629			
R1245	5529	6629			
R1246	5529	6629			
R1247	5529	6629			
R125	101		8	2002	
R1273	102A		53	2007	
R1274	102A		53	2007	
R1300	5529	3504/5508			
R1301	5529	3504/5508			
R1302	5529	3504/5508			
R1303	5529	3504/5508			
R1305	5981		5097		
R1306	5529	3504/5508			
R1307	5529	3504/5508			
R1329	116	3017B/117	504A	1104	
R135	101		8	2002	
R135-1	130		14	2041	
R136	101		8	2002	
R137	101		8	2002	
R14	101		8	2002	
R14010658			514	1122	
R1420010					1N4933
R1420110					1N4934
R1420210					1N4935
R1420410					1N4936
R1420610					1N4937
R145B	169	3107/5307	BR-600		
R1471	5547	3505			
R1473	5547	3505			
R1475	5547	3505			
R15	124		12		
R15003	128		243	2030	
R15003P1	128		243	2030	
R152	102A	3123	53	2007	
R1530	103A		59	2002	
R1531	103A	3835	59	2002	
R1532	103A	3835	59	2002	
R1533	101		8	2002	
R1534	103A	3835	59	2002	
R1537	103A	3835	59	2002	
R1538	103A		59	2002	
R1539	160		245	2004	
R1540	102A		53	2007	
R1541	102A		53	2007	
R1542	102A		53	2007	
R1543	102A		53	2007	
R1544	102A		53	2007	
R1545	103A	3835	59	2002	
R1546	102A	3835/103A	53	2007	
R1547	103A	3835	59	2002	
R1548	102A		53	2007	
R1549	103A	3835	59	2002	
R154B	166	3311	504A	1152	
R1550	160		245	2004	
R1553	103A	3835	59	2002	
R1554	160		245	2004	
R1555	102A		53	2007	
R16	102A		53	2007	
R163	100	3721	1	2007	
R164	102A	3123	53	2007	
R1667	109	3087		1123	
R170			510	1114	
R1711	5601	3664/5602			
R1712	5602	3664			
R1715	5605	3666			
R1750	5675	3508			
R1751	5675	3508			
R1752	5675	3508			
R177	103A	3835	59	2002	
R1781	5685	3509/5695			
R1782	5685	3521/5686			
R1783	5685	3521/5686			
R1785	5685	3521/5686			
R186	100	3721	1	2007	
R1889	109	3087		1123	
R1A		3067	504A	1104	
R1B	116	3017B/117	504A	1104	
R1K	116		504A	1104	
R1N1102	722		IC-9		
R1Z			511	1114	
R2	119	3109	CR-2		
R200					1N4003
R2001	127	3764	25		
R2003	127	3764	25		
R2005	5913		5065		
R2010	5917		5065		
R20100	5933		5077		
R2015	5917		5065		
R202	101		8	2002	
R2020	5917		5065		
R2025	5921		5069		
R203	101		8	2002	
R2030	5921		5069		
R2035	5921		5069		
R2040	5921		5069		
R2045	5925		5073		
R204B	166	9075	504A	1152	
R2050	5925		5073		
R2060	5925		5073		
R2070	5929		5077		
R2080	5929		5077		
R2090	5933		5077		
R2096	175		246	2020	
R210	156	3081/125	512	1114	
R2130	5929		5077		
R2159	116	3017B/117	504A	1104	
R2164	109	3087		1123	
R2252	116	3017B/117	504A	1104	
R227	100	3721	1	2007	
R2270-60106	128		243	2030	
R2270-75116	181		75	2041	
R2270-75497	130		14	2041	
R2270-7693				2027	
R2270-76963	129	3025	244	2027	
R2270-77873D	128		243	2030	
R2270-78399	130		14	2041	
R22707-8399	130		14	2041	
R227075497	130	3027	14	2041	
R227077499	130		14	2041	
R227077873D	128		243	2030	
R227078533			255	2041	
R23-1003	102A		53	2007	
R23-1004	102A		53	2007	
R2334	109	3087		1123	
R2350	102A		53	2007	
R2351	102A		53	2007	
R2352	102A		53	2007	
R2353	102A		53	2007	
R2355	102A		53	2007	
R2356	103A	3835	59	2002	
R2359	103A	3835	59	2002	
R2360	103A	3835	59	2002	
R2364	103A	3835	59	2002	
R2365	103A		59	2002	
R2366	102A		53	2007	
R2367	102A		53	2007	
R2373	102A		53	2007	
R2374	103A		59	2002	
R2375	103A	3835	59	2002	
R24-1001	102A		53	2007	
R24-1002	102A		53	2007	
R24-1003	102A	3004	53	2007	

Industry Standard No.	ECG	SK	GE	RS 276-	MOTOR.
R24-1004	102A	3004	53	2007	
R242	102A	3123	53	2007	
R2432-1	710	3102	IC-89		
R2434-1	706	3102/710	IC-43		
R2434-1(RCA)	710		IC-89		
R244	100	3721	1	2007	
R2442	116	3017B/117	504A	1104	
R2444	124	3021	12		
R2445-1		3070	IC-207		
R2445-1(RCA)	711		IC-207		
R24451	711	3070			
R2446	121	3717	239	2006	
R2446-1	121	3717	239	2006	
R245	102A		53	2007	
R2460-1	117		504A	1104	
R2460-4	117		504A	1104	
R2460-9	127	3764	25		
R2473	108	3452	86	2038	
R2474-2	154		40	2012	
R2476	108	3452	86	2038	
R2477	108	3452	86	2038	
R2482-1	102A		53	2007	
R2494-1	127	3764	25		
R250	156	3081/125	512	1114	
R2500-1	127	3764	25		
R250F	156		512	1114	
R2516		3072	IC-2		
R2516-1	712	3072	IC-2		
R2516-1(RCA)	712		IC-2		
R25161	712	3072	IC-2		
R255	102A	3123	53	2007	
R2559-1	728		IC-22		
R2560-1	729		IC-23		
R258	100	3005	1	2007	
R264-1	197		250	2027	
R265A	121	3717	239	2006	
R2675	102A		53	2007	
R2677	102A		53	2007	
R2683	160		245	2004	
R2684	160		245	2004	
R2685	160		245	2004	
R2686	160		245	2004	
R2687	160		245	2004	
R2688	160		245	2004	
R2689	102A		53	2007	
R2694	160		245	2004	
R2695	160		245	2004	
R2696	160		245	2004	
R2697	160		245	2004	
R2749	102A		53	2007	
R2749M	102A		53	2007	
R289	102A		53	2007	
R290	102A	3123	53	2007	
R291	102A	3123	53	2007	
R2964	127	3764	25		
R2982	130	3027	14	2041	
R2D	142A	3062	ZD-12	563	
R2E	145A	3063	ZD-15	564	
R2SB492	102		2	2007	
R3	120	3110	CR-3		
R3/2H	116	3311	504A	1104	
R3012	525	3925			
R3020606	5822				MR1366
R3020612	5822				MR1376
R302506					MR1366
R302512					MR1376
R3057	113A	3119/113	6GC1		
R3110	5987	3698	5097		
R31100	6003		5109		
R3115	5987	3698	5097		
R3120	5987	3698	5097		
R3125	5991	3518/5995	5101		
R3130	5991	3518/5995	5101		
R3135	5991	3518/5995	5101		
R3140	5991	3518/5995	5101		
R3145	5995	3518	5105		
R3150	5995	3518	5105		
R3160	5995	3518	5105		
R3170	5999		5109		
R3180	5999		5109		
R3190	6003		5109		
R3205	5981		5097		
R3210	5987	3698	5097		
R32100	6003		5109		
R3215	5987	3698	5097		
R3220	5987	3698	5097		
R3225	5991	3518/5995	5101		
R3230	5991	3518/5995	5101		
R3235	5991	3518/5995	5101		
R324	102A	3123	53	2007	
R3240	5991	3518/5995	5101		
R3245	5995	3518	5105		
R3250	5995	3518	5105		
R3260	5995	3518	5105		
R3270	5999		5109		
R3273-P1	123A		20	2051	
R3273-P2	123A		20	2051	
R3275	102A		53	2007	
R3276	102A		53	2007	
R3277	160		245	2004	
R3278	160		245	2004	
R3279	160		245	2004	
R3280	5999		5109		
R3280(RCA)	102A		53	2007	
R3282	102A		53	2007	
R3283	123A	3444	20	2051	
R3284	102A		53	2007	
R3285	116	3017B/117	504A	1104	
R3286	102A		53	2007	
R3287	160		245	2004	
R3288	160		245	2004	
R3290	6003		5109		
R3293	103A	3835	59	2002	
R3293(GE)	123A		20	2051	
R3299	102A		53	2007	
R33	103A		59	2002	
R3301	102A		53	2007	
R3309	160		245	2004	
R3314	113A	3119/113	6GC1		
R336	126		52	2024	
R337	126		52	2024	
R338	160		245	2004	
R339	160		245	2004	
R34	103A		59	2002	
R34-6016-58	123A		20	2051	
R340	123A	3444	20	2051	
R3400006					MR750
R3400106					MR751
R3400206					MR752
R3400306					MR754
R3400406					MR754
R3400506					MR754
R3400606					MR756
R3400706					MR756

Industry Standard No.	ECG	SK	GE	RS 276-	MOTOR.
R3400806					MR758
R3400906					MR760
R3401006					MR760
R3405	6021		5129		
R341	160		245	2004	
R3410	6027		5129		
R3410-P1	519		514	1122	
R34100	6045		5141		
R3415	6027	7227	5129		
R3420	6027	7227	5129		
R3425	6035		5133		
R3430	6035		5133		
R3435	6035		5133		
R3440	6035		5133		
R3445	6041		5137		
R3450	6041		5137		
R3460	6041		5137		
R3470	6043		5142		
R3480	6043		5142		
R3490	6045		5141		
R35		3123	53	2007	
R350	156	3051	512		
R3502	704	3023	IC-205		
R3502(RCA)	704		IC-205		
R3502-1	704	3023	IC-205		
R3502-2	704	3023	IC-205		
R3503	124		12		
R3508	128		243	2030	
R350F	156		512		
R3512-1	121	3717	239	2006	
R3514-1	127	3764	25		
R3515	104	3009	16	2006	
R3515(RCA)	121	3717	239	2006	
R3520-1	124		12		
R3528-1	785	3254	IC-226		
R3528-1(RCA)	785	3717/121	IC-226		
R3533	113A	3119/113	6GC1		
R3555-3	128		243	2030	
R3573-1	103A	3835	59	2002	
R3576-1	233		210	2009	
R3578-1	102A		53	2007	
R3583-7	230	3042			
R3583-8	230	3042			
R3585-5	230	3042			
R3585-6	231	3857			
R3585-7	230	3042			
R3585-8	231	3857			
R3593	128		243	2030	
R3598-2	102A		53	2007	
R36	176	3123	80		
R3605	6021		5129		
R3608	128		243	2030	
R3608-1	128	3024	243	2030	
R3608-2	128	3024	243	2030	
R3610	6055		5129		
R36100	6073		5141		
R3611-1	152	3893	66	2048	
R3613-3(RCA)	191	3232	249		
R3615	6055		5129		
R3620	6055		5129		
R3625	6061		5133		
R3630	6061		5133		
R3635	6061		5133		
R364	102A		53	2007	
R3640	6061		5133		
R3645	6065		5137		
R3647	124		12		
R3650	6065		5137		
R3651-1	222			2036	
R3660	6065		5137		
R3670	6069		5140		
R3676-1	233		210	2009	
R3677-1	780	3141	IC-222		
R3677-1(RCA)	780		IC-222		
R3677-2	780	3141	IC-222		
R3677-2(RCA)	780		IC-222		
R3677-3	780	3141	IC-222		
R3677-3(RCA)	780		IC-222		
R3679	128		243	2030	
R3680	6069		5141		
R3681-1	152	3893	66	2048	
R3690	6073		5141		
R400					1N4004
R4020530	6010				MR1396
R4020620	6010				MR1386
R4020630	6010				MR1396
R4057	123A	3122	20	2051	
R41	101		8	2002	
R4100070	6054		5128		
R4100170	6054		5128		
R4100822	5998		5108		
R4100840	5998		5108		
R4100860	6042		5140		
R4101022	6002	7202	5108		
R4101040	6002	7202	5108		
R4101060	6044	7244	5140		
R4101070	6072	7272	5140		
R4101822	5999		5109		
R4108070	6068		5140		
R4110070	6055		5129		
R4110170	6055		5129		
R4110840	5999		5109		
R4110860	6043		5142		
R4111022	6003	7203	5109		
R4111040	6003	7203	5109		
R4111060	6045	7245	5141		
R4111070	6073	7273	5141		
R4118070	6069		5141		
R4192	124		12		
R4193	124		12		
R4194	124		12		
R4195	124		12		
R4196	124		12		
R424	160		245	2004	
R424-1	160	3123	245	2004	
R425	160	3007	245	2004	
R428	102A	3123	53	2007	
R4348	102A		53	2007	
R4349	102A		53	2007	
R4369	130	3027	14	2041	
R43HZ	5990	3608	5100		
R4409	113A	3119/113	6GC1		
R4437-1	787	3146	IC-228		
R4437-1(RCA)	787		IC-228		
R4437-2	787	3146	IC-228		
R4437-2(RCA)	787		IC-228		
R4438-1	788	3147	IC-229		
R4438-1(RCA)	788	3829	IC-229		
R4438-2	788	3147	IC-229		
R4438-2(RCA)	788	3829	IC-229		
R46	102	3722	2	2007	
R4666	113A	3119/113	6GC1		
R47M10					MHW709
R47M13					MHW710

Industry Standard No.	ECG	SK	GE	RS 276-	MOTOR.
R47M15					MHW710
R488	100	3721	1	2007	
R497	160	3123	245	2004	
R4A	116	3017B/117	504A	1104	
R4HZ	5990	3608	5100		
R5	125	3033	510,531	1114	
R5048	128	3024	243	2030	
R5050	103A	3835	59	2002	
R5051	102A		53	2007	
R5052	102A		53	2007	
R5053	102A		53	2007	
R5054	103A	3835	59	2002	
R5055	102A		53	2007	
R5056	103A		59	2002	
R506	100	3123	1	2007	
R5096	109	3087		1123	
R5097	102A		53	2007	
R5098	102A		53	2007	
R5099	102A		53	2007	
R5100	102A		53	2007	
R5101	158		53	2007	
R5102	160		245	2004	
R5103	160		245	2004	
R515	160	3123	245	2004	
R5158-1	797	3158	IC-233		
R515A	160		245	2004	
R516	121	3009	239	2006	
R516(T.I.)	160		245	2004	
R516A	160		245	2004	
R5179	103A		59	2002	
R5180	103A	3835	59	2002	
R5181	102A		53	2007	
R5182	158		53	2007	
R52	102A	3123	53	2007	
R530	102A		53	2007	
R537	102A		53	2007	
R539	160		245	2004	
R5522	109	3087		1123	
R5523	102A		53	2007	
R5524	102A		53	2007	
R5525	102A		53	2007	
R558	160	3004/102A	245	2004	
R558(T.I.)	102A		53	2007	
R56	102A		53	2007	
R563	160		245	2004	
R563(T.I.)	102A		53	2007	
R564	160		245	2004	
R565	160		245	2004	
R5708	102A		53	2007	
R579	160		245	2004	
R579(T.I.)	102A		53	2007	
R581	126	3006/160	52	2024	
R582	123A	3444	20	2051	
R592	101	3861	8	2002	
R593	160	3123	245	2004	
R593A	160		245	2004	
R5970	116	3017B/117	504A	1104	
R5971	116	3017B/117	504A	1104	
R5B	125	3033	510,531	1114	
R5C	125	3033	510,531	1114	
R6	125	3033	510,531	1114	
R6-0A	176		80		
R6/2H	116	3311	504A	1104	
R60-1001	126	3008	52	2024	
R60-1002	160	3008	245	2004	
R60-1003	160	3008	245	2004	
R60-1004	102A	3004	53	2007	
R60-1005	102A	3004	53	2007	
R60-1006	102A	3004	53	2007	
R60-1007	109	3087		1123	
R600					1N4005
R6000124	6354	6554			
R6000124R	6355	6555			
R6000128	6354	6554			
R6000128R	6355	6555			
R6000224	6354	6554			
R6000224R	6355	6555			
R6000324	6354	6554			
R6000324R	6355	6555			
R6000328	6354	6554			
R6000328R	6355	6555			
R6000424	6354	6554			
R6000424R	6355	6555			
R6000428	6354	6554			
R6000524	6356	6556			
R6000528	6356	6556			
R6000624	6356	6556			
R6000628	6356	6556			
R6000724	6358	6558			
R6000724R	6359	6559			
R6000728	6358	6558			
R6000728R	6359	6559			
R6000824	6358	6558			
R6000824R	6359	6559			
R6000828	6358	6558			
R6000828R	6359	6559			
R6000924	6358	6558			
R6000924R	6359	6559			
R6000928	6358	6558			
R6000928R	6359	6559			
R6001024	6358	6558			
R6001024R	6359	6559			
R6001028	6358	6558			
R6001028R	6359	6559			
R6048	116	3017B/117	504A	1104	
R608	102A		53	2007	
R608A	102A		53	2007	
R61	103A	3835	59	2002	
R6110	116	3017B/117	504A	1104	
R612-1	152	3893	66	2048	
R6171-11	515	9098			
R6171-12	515	9098			
R6171-19	515	9098			
R6171-20	515	9098			
R62	103A	3835	59	2002	
R621-1	152	3893	66	2048	
R62194	108	3452	86	2038	
R623-1	196		241	2020	
R63	103A		59	2002	
R632-1	152		66	2048	
R632-2	152	3893	66	2048	
R64	102A	3123	53	2007	
R640-1	196		241	2020	
R6422	116	3017B/117	504A	1104	
R65	102A	3004	53	2007	
R6553	102A		53	2007	
R66	102A	3123	53	2007	
R66-8504		3311	504A	1104	
R67	102A	3123	53	2007	
R684	126		52	2024	
R6922	102A		53	2007	
R6HZ	5994	3501	5104		
R7028	109	3088		1123	
R7029	109	3088		1123	

Industry Standard No.	ECG	SK	GE	RS 276-	MOTOR.
R7030	138A	3059	ZD-7.5		
R7048	102A		53	2007	
R711X					R711X
R7124	102A		53	2007	
R7127	102A		53	2007	
R712X					R712X
R714	126	3008	52	2024	
R714X					R714X
R715	126	3008	52	2024	
R7162	116	3017B/117	504A	1104	
R7163	123A	3444	20	2051	
R7164	102A		53	2007	
R7165	123A	3444	20	2051	
R7166	102A		53	2007	
R7167	104	3009	16	2006	
R7248	116	3017B/117	504A	1104	
R7249	123A	3444	20	2051	
R7253	121	3009	239	2006	
R7271	116	3017B/117	504A	1104	
R7343	123A	3444	20	2051	
R7359	123A	3444	20	2051	
R7360	123A	3444	20	2051	
R7361	123A	3444	20	2051	
R7362	103A	3835	59	2002	
R7363	102A		53	2007	
R7489	102A		53	2007	
R7490	102A		53	2007	
R7491	102A		53	2007	
R7582	123A	3444	20	2051	
R7612	102A		53	2007	
R7613	128	3024	243	2030	
R7615	156		512		
R7620	104	3009	16	2006	
R7682	116	3017B/117	504A	1104	
R7743	109	3087		1123	
R7885	160		245	2004	
R7886	160		245	2004	
R7887	123A	3444	20	2051	
R7888	102A		53	2007	
R7889	102A		53	2007	
R7890	105		4		
R7891	160		245	2004	
R7892	109	3087		1123	
R7893	109	3087		1123	
R7894	142A		ZD-12	563	
R79	103A	3835	59	2002	
R7953	123A	3444	20	2051	
R7954	116	3017B/117	504A	1104	
R7962	160		245	2004	
R8	125	3032A	510,531	1114	
R8/2H	125	3311	510,531	1114	
R80	103A	3835	59	2002	
R800					1N4006
R8022	177		300	1122	
R8023	177		300	1122	
R8024	116	3017B/117	504A	1104	
R8060	109	3087		1123	
R8061	109	3087		1123	
R8066	123A	3444	20	2051	
R8067	123A	3444	20	2051	
R8068	123A	3444	20	2051	
R8069	123A	3444	20	2051	
R8070	123A	3444	20	2051	
R8115	123A	3444	20	2051	
R8116	123A	3444	20	2051	
R8117	123A	3444	20	2051	
R8118	123A	3444	20	2051	
R8119	123A	3444	20	2051	
R8120	123A	3444	20	2051	
R8121	102A		53	2007	
R8158	124		12		
R8219	109	3087		1123	
R8223	123A	3444	20	2051	
R8224	123A	3444	20	2051	
R8225	123A	3444	20	2051	
R8240	160		245	2004	
R8241	160		245	2004	
R8242	160		245	2004	
R8243	123A	3444	20	2051	
R8244	123A	3444	20	2051	
R8257	109	3087		1123	
R8259	123A	3444	20	2051	
R8260	123A	3444	20	2051	
R8261	123A	3444	20	2051	
R83	102A	3123	53	2007	
R8305	123A	3444	20	2051	
R8310	102A		53	2007	
R8311	102A		53	2007	
R8312	123A	3444	20	2051	
R8313	104	3009	16	2006	
R8314		3100	300	1122	
R8364	142A		ZD-12	563	
R8470	116	3017B/117	504A	1104	
R8471	118		CR-1		
R8472	119		CR-2		
R8473	116	3017B/117	504A	1104	
R8474	113A	3119/113	6GC1		
R8475	109	3087		1123	
R8477	120		CR-3		
R8528	123A	3444	20	2051	
R8529	107	3039/316	11	2015	
R8530	123A	3444	20	2051	
R8543	123A	3444	20	2051	
R855-2	116	3016	504A	1104	
R8551	123A	3444	20	2051	
R8552	123A	3444	20	2051	
R8553	123A	3444	20	2051	
R8554	123A	3444	20	2051	
R8555	123A	3444	20	2051	
R8556	123A	3444	20	2051	
R8557	123A	3444	20	2051	
R8559	160		245	2004	
R8560	120	3110	CR-3		
R8620	123A	3444	20	2051	
R8645	156		512		
R8646	123A	3444	20	2051	
R8647	123A	3444	20	2051	
R8648	123A	3444	20	2051	
R8649	128(2)		243(2)		
R8658	123A	3444	20	2051	
R8659	121	3009	239	2006	
R868	102A		53	2007	
R8685	160		245	2004	
R8686	160		245	2004	
R8687	102A		53	2007	
R8688	102A		53	2007	
R8692	160		245	2004	
R8693	160		245	2004	
R8694	160		245	2004	
R8695	102A		53	2007	
R8697	102A		53	2007	
R87	102A		53	2007	
R8703	160		245	2004	
R8704	160		245	2004	

Industry Standard No.	ECG	SK	GE	RS 276-	MOTOR.
R8705	160		245	2004	
R8706	102A		53	2007	
R8707	102A		53	2007	
R8721	117		504A	1104	
R8881	160		245	2004	
R8882	160		245	2004	
R8883	102A		53	2007	
R8884	102A		53	2007	
R8885	102A		53	2007	
R8886	102A		53	2007	
R8887	109	3087		1123	
R8889	123A	3122	20	2051	
R8900	123A(2)	3122	20(2)		
R8914	123A	3444	20	2051	
R8915	128	3024	243	2030	
R8916	123A	3444	20	2051	
R8963	123A	3444	20	2051	
R8964	123A	3444	20	2051	
R8965	123A	3444	20	2051	
R8966	123A	3444	20	2051	
R8967	159	3114/290	82	2032	
R8967A	159	3114/290	82	2032	
R8968	123A	3444	20	2051	
R8969	159	3114/290	82	2032	
R8969A	159	3114/290	82	2032	
R8970	109	3087		1123	
R8971	102A		53	2007	
R8989	124		12		
R9004	123A	3444	20	2051	
R9005	123A	3444	20	2051	
R9006	123A	3444	20	2051	
R9025	123A	3444	20	2051	
R9071	123A	3444	20	2051	
R9381	5079A	3335	ZD-20		
R9382	171	3103A/396	27		
R9383	171	3103A/396	27		
R9384	123A	3444	20	2051	
R9385	123A	3444	20	2051	
R9470	116	3017B/117	504A	1104	
R9483	123A	3444	20	2051	
R9531	160	3006	245	2004	
R9532	160	3006	245	2004	
R9533	158	3004/102A	53	2007	
R9534	158	3004/102A	53	2007	
R9590	109	3087		1123	
R9597	116	3017B/117	504A	1104	
R9600	229	3006/160	61	2038	
R9601	160	3006	245	2004	
R9602	160	3006	245	2004	
R9603	158	3004/102A	53	2007	
R9604	158	3004/102A	53	2007	
R98	102A	3123	53	2007	
R9A	116		504A	1104	
RA-1	116	3312	504A	1104	
RA-1A	116	3032A			
RA-1B	125	3311	504A	1104	
RA-1C	125			1114	
RA-1Z	552	3311	504A	1114	
RA-1ZC	116		504A	1114	
RA-2	125	3032A	509	1114	
RA-26		3061	ZD-10	562	
RA-2C	125			1114	
RA-Z	506	3130	511	1114	
RA1	506	3100/519	511	1114	
RA132AA	156	3051	512		
RA132BA	116	3016	504A	1104	
RA132DA	156	3051	512		
RA132MA	156	3051	512		
RA132PA	156	3051	512		
RA132RA	156	3051	512		
RA132VA	156	3051	512		
RA1A			300	1122	
RA1B	116	3311	504A	1104	
RA1Y	116		504A	1104	
RA1Z	116	3311	504A	1114	
RA1ZC	116		504A	1104	
RA2	125		504A	1104	
RA26		3311	ZD-51		
RA26(ZENER)			ZD-10	562	
RA2C	125	3081	510	1114	
RB1000	5307	3107			
RB6-PS-801	923D		IC-260	1740	
RB600	5305	3676			
RB800	5306	3677			
RC-1700	130		14	2041	
RC-2	525	3925	530		
RC-2V	525	3925	530		
RC080	125	3032A	510,531	1114	
RC1414DC					MC1414L
RC1414DP					MC1414P
RC1437D					MC1437L
RC1437DP					MC1437P
RC1458DN	778A				MC1458P1
RC1458T		3555			MC1458G
RC1488DC	75188				MC1488L
RC1489ADC					MC1489AL
RC1489DC	75189				MC1489L
RC1556T					MC1456CG
RC1558T					MC1558G
RC1700	130	3027	14	2041	
RC2	525	3925	530		
RC2270	128	3024	243	2030	
RC3302DB					MC3302P
RC4131DP					MC1471SCP1
RC4131T					MC1741SG
RC4136D					MC3403L
RC4136DP	997				MC3403P
RC4136J					MC3403L
RC4136N					MC3403P
RC4195T					MC1468G
RC4195TK					MC1468R
RC4444R					MC3416L
RC4558DN	778A				MC4558CP1
RC4558JG	778A				MC4558CU
RC4558L					MC4558CG
RC4558P	778A				MC4558CP1
RC4558T					MC4558CG
RC4739DB	725	3162			
RC555DE	955M	3564			
RC555NB	955M	3564			
RC555T	955M	3693			
RC702T					MC1712CG
RC709D					MC1709CL
RC709DC	909D	3590			
RC709DN					MC1709CP1
RC709DP	909D	3590			MC1709CP2
RC709T	909	3590	IC-249		MC1709CG
RC710DC	910D				MC1710CL
RC710DP	910D				MC1710CP
RC710T	910		IC-251		MC1710CG
RC711DC	911D				MC1711CL
RC711DP	911D				MC1711CP
RC711T	911		IC-253		MC1711CG
RC723D	923D				MC1723CL

Industry Standard No.	ECG	SK	GE	RS 276-	MOTOR.
RC723T	923				MC1723CG
RC733D					MC1733CL
RC733T					MC1733CG
RC741D					MC1741CL
RC741DN	941M	3552			MC1741CP1
RC741DP	941M	3552			MC1741CP2
RC741Q					MC1741CF
RC741T	941	3514	IC-263	010	MC1741CG
RC747D					MC1747CL
RC747DB	947D	3556			
RC747DC	947D	3556			
RC747DP	947D	3556			
RC747T					MC1747CG
RC748ND	975	3641			
RC748T	1171	3565			MC1748CG
RC75107AD					MC75107L
RC75107ADP					MC75107P
RC75108AD					MC75108L
RC75108ADP					MC75108P
RC75109D					MC75S110L
RC75109DP					MC75S110P
RC75110D					MC75S110L
RC75110DP					MC75S110P
RC75325DD					MC75325L
RC7805	960	3591			
RC7806	962	3669			
RC7812	966	3592			
RC7815	968	3593			
RC7824	972	3670			
RC8T13DD					MC8T13L
RC8T13MP					MC8T13P
RC8T14DD					MC8T14L
RC8T14MP					MC8T14P
RC8T23DD					MC8T23L
RC8T23MP					MC8T23P
RC8T24DD					MC8T24L
RC8T24MP					MC8T24P
RCA1000	243	3182			MJ1000
RCA1001	243	3182			MJ1001
RCA106A	5454	6754			
RCA106B	5455	3597			
RCA106C	5456	3571			
RCA106D	5457	3571			
RCA106F	5453	6753			
RCA106Q	5452	3570			
RCA106Y	5452	3570			
RCA107A	5454	6754			
RCA107B	5455	3597			
RCA107F	5453	6753			
RCA107Q	5452	6752			
RCA107Y	5452	6752			
RCA120	261	3896			TIP120
RCA121	261	3896			TIP121
RCA122					TIP122
RCA125	262	3897			TIP125
RCA126	262	3897			TIP126
RCA1A01	128	3024	243	2030	
RCA1A02	129	3025	244	2027	
RCA1A03	282	3512	264		
RCA1A05	129	3025	244	2027	
RCA1A06	128	3024	243	2030	
RCA1A07	128	3024	243	2030	
RCA1A08	129	3025	244	2027	
RCA1A09	396	3103A			
RCA1A10	397	3053			
RCA1A11	396	3103A			
RCA1A15	396	3103A	243	2030	
RCA1A16	397	3053	244	2027	
RCA1A17	128	3024	243	2030	
RCA1A18	128	3024	243	2030	
RCA1A19	129	3025	244	2027	
RCA1B01	284	3621	265	2047	2N5878
RCA1B04	283	3217	36		MJ15022
RCA1B05	283	3217	36		MJ15024
RCA1B06	280	3297	262	2041	MJ15003
RCA1B07	245	3182/243			
RCA1B08	246	3183A			
RCA1B09	283	3467			MJ15024
RCA1C03	291	3440			FT317
RCA1C04	292	3083/197			FT417
RCA1C05	196	3054	241		2N6130
RCA1C06	197	3083	250	2027	2N6133
RCA1C07		3188A			MJE3055T
RCA1C08		3189A			MJE2955T
RCA1C09		3620			MJE3055T
RCA1C10	196	3054	241	2020	2N6292
RCA1C11	197	3083	250	2027	2N6107
RCA1C12	291	3440			FT317A
RCA1C13	292	3083/197			FT417A
RCA1C14	196	3054	241	2020	2N6290
RCA1C15		3180			2N6388
RCA1C16		3181A			2N6668
RCA1E02		3261			2N3583
RCA1E03		3623			2N6420
RCA29	152	3054/196	66	2048	TIP29
RCA29/SDH	152		66	2048	
RCA29A	152	3054/196	66	2048	TIP29A
RCA29A/SDH	152		66	2048	
RCA29B	291	3440			TIP29B
RCA29BSDH	291	3440			
RCA29C	291	3440			TIP29C
RCA29CSDH	291	3440			
RCA30	153	3083/197	69	2049	TIP30
RCA3054	196				2N6122
RCA3055		3188A			2N6487
RCA30A	153	3083/197	69	2049	TIP30A
RCA30B	292	3441			TIP30B
RCA30C	292				TIP30C
RCA31		3054			TIP31
RCA31A		3054			TIP31A
RCA31B		3054			TIP31B
RCA31C					TIP31C
RCA32		3083			TIP32
RCA32A		3083			TIP32A
RCA32B		3083			TIP32B
RCA32C		3083			TIP32C
RCA34098	126		52	2024	
RCA34099	126		52	2024	
RCA34100	126		52	2024	
RCA34101	102A		53	2007	
RCA34106	102A		53	2007	
RCA3441		3929			FT317B
RCA3517	102A		53	2007	
RCA35953	102A		53	2007	
RCA35954	102A		53	2007	
RCA370		3083	D45C2		
RCA371		3083	D45C6		
RCA3858	102A		53	2007	
RCA40231	103A	3854/123AP	59	2002	
RCA40245	161	3716	39	2015	
RCA40246	161	3716	39	2015	
RCA40250	152	3026	23	2020	
RCA40395	102A		53	2007	
RCA40396N	103A		59	2002	

Industry Standard No.	ECG	SK	GE	RS 276-	MOTOR.
RCA40396P	102A		53	2007	
RCA41		3188A			TIP41
RCA410		3560			MJ410
RCA411	94	3560			MJ411
RCA413		3560			MJ413
RCA41A		3188A			TIP41A
RCA41B		3188A			TIP41B
RCA41C					TIP41C
RCA42		3189A			TIP42
RCA423		3560			MJ423
RCA42A		3189A			TIP42A
RCA42B		3189A			TIP42B
RCA42C					TIP42C
RCA431		3560			MJ431
RCA44098	126		52	2024	
RCA520		3054	D44C2		
RCA521		3054	D44C6		
RCA6263		3929			FT317B
RCA8203	262	3181A/264			2N6666
RCA8203A	264	3221			2N6667
RCA8203B	264				2N6668
RCA8350	246	3183A			2N6648
RCA8350A	246	3183A			2N6649
RCA8350B	246	3183A			2N6650
RCA8766					MJ10002
RCA8766A					MJ10002
RCA8766B	97				MJ10003
RCA8766C					MJ10003
RCA8766D					MJ10003
RCA8766E					MJ10003
RCA8767					2N6546
RCA8767A					2N6547
RCA8767B					2N6547
RCA900	246	3183A			
RCA901	246	3183A			
RCA9113					2N6546
RCA9113A					2N6547
RCA9113B					2N6547
RCC-7022	116		504A	1104	
RCC7022	120	3110	CR-3		
RCC7225	116(4)		504A(4)	1104	
RCP111A	171	3201	27		2N6557
RCP111B	171	3201	27		2N6557
RCP111C	171	3201	27		2N6558
RCP111D	171	3201	27		2N6559
RCP113A	171	3201	27		2N6557
RCP113B	171	3201	27		2N6557
RCP113C	171	3201	27		2N6558
RCP113D	171	3201	27		2N6559
RCP115	171	3201	27		2N6591
RCP115A	171	3201	27		
RCP115B	171	3201	27		2N6557
RCP117	171	3201	27		2N6591
RCP117B	171	3201	27		2N6557
RCP131A					2N6592
RCP131B					2N6593
RCP131C					2N6558
RCP131D					2N6559
RCP133A					2N6592
RCP133B					2N6593
RCP133C					2N6558
RCP133D					2N6559
RCP135					2N6553
RCP135B					2N6557
RCP137					2N6553
RCP137B					2N6557
RCP700A		3179A			2N6554
RCP700B		3179A			2N6554
RCP700C		3179A			2N6555
RCP700D		3179A			2N6556
RCP701A		3178A			2N6551
RCP701B		3178A			2N6551
RCP701C		3178A			2N6552
RCP701D		3178A			2N6553
RCP702A		3179A			2N6554
RCP702B		3179A			2N6554
RCP702C		3179A			2N6555
RCP702D		3179A			2N6556
RCP703A		3178A			2N6551
RCP703B		3178A			2N6551
RCP703C		3178A			2N6552
RCP703D		3178A			2N6553
RCP704		3179A			2N6554
RCP704B		3179A			2N6554
RCP705		3178A			2N6551
RCP705B		3178A			2N6551
RCP706		3179A			2N6554
RCP706B		3179A			2N6554
RCP707		3178A			2N6551
RCP707B		3178A			2N6551
RCS242	130	3027	14	2041	2N3055A
RCS257					2N3772
RCS258		3036			2N5885
RCS29		3626			2N4231A
RCS29A		3626			2N4232A
RCS29B		3626			2N4233A
RCS30		3625			2N6312
RCS30A		3625			2N6313
RCS30B		3625			2N6314
RCS30C		3625			2N6420
RCS31		3626			2N4231A
RCS31A		3626			2N4232A
RCS31B		3626			2N4233A
RCS32		3625			2N6312
RCS32A		3625			2N6313
RCS32B		3625			2N6314
RCS32C		3625			2N6421
RCS559					2N6211
RCS560					2N6211
RCS564		3559			2N6249
RCS579		3439			2N6306
RCS617					2N5882
RCS618		3437			2N5880
RCS880	397	3053			
RCS881	397	3053			
RCS882	397	3053			
RD-1003	116	3311			
RD-10EB	140A	3061	ZD-10	562	
RD-11B	140A		ZD-10	562	
RD-11E			ZD-11.0	563	
RD-13A	143A		ZD-13	563	
RD-13AD			ZD-13	563	
RD-13AK	143A		ZD-13	563	
RD-13AK-P	143A		ZD-13	563	
RD-13AKP	143A		ZD-13	563	
RD-13AL	5072A		ZD-8.2	562	
RD-13AM			ZD-13	563	
RD-13AN	5022A		ZD-12	563	
RD-13E	143A		ZD-13	563	
RD-13M	143A		ZD-13	563	
RD-15E	145A		ZD-15	564	
RD-16H	145A		ZD-15	564	
RD-16HA	145A		ZD-15	564	
RD-19A	5079A		ZD-20		
RD-19AM	5079A		ZD-20		
RD-2.2E	5072A	3136	ZD-20	562	
RD-20E	5079A		ZD-20		
RD-24A	5081A	3151	ZD-24		
RD-26235-1	116	3016	504A	1104	
RD-29799P	116	3016	504A	1104	
RD-3	116		504A	1104	
RD-31903P	116	3016	504A	1104	
RD-3472	116	3016	504A	1104	
RD-35A	147A	3095	ZD-33		
RD-5.1EB	139A		ZD-5.1		
RD-5A	135A	3056	ZD-5.1		
RD-6.8EB	5071A		ZD-6.8	561	
RD-6A	136A		ZD-5.6	561	
RD-6AM		3057	ZD-5.6	561	
RD-6L	137A		ZD-6.2	561	
RD-7.5EB	138A	3059	ZD-7.5		
RD-7A	138A	3058/137A	ZD-9.1	562	
RD-7AM			ZD-6.8	561	
RD-7E			ZD-6.8	561	
RD-8.2#E			X11	562	
RD-8.2A			ZD-8.2	562	
RD-8.2E	5072A		ZD-8.2	562	
RD-8.2E(C)			ZD-8.2	562	
RD-8.2EB			ZD-8.2	562	
RD-8.2EC	5072A		ZD-8.7	562	
RD-8.2FB			ZD-8.2	562	
RD-9.1E	139A		ZD-9.1	562	
RD-91	139A	3060	ZD-9.1	562	
RD-91E	139A		ZD-9.1	562	
RD-96	139A		ZD-9.1	562	
RD-9A	139A		ZD-9.1	562	
RD-9A(L)	5074A		ZD-11.5		
RD-9AL			ZD-9.1	562	
RD-9E	140A		ZD-10	562	
RD-9L	139A		ZD-9.1	562	
RD10E	5019A	3061/140A	ZD-10	562	
RD10EA	140A		ZD-10	562	
RD10EB	5019A	3061/140A	ZD-10	562	
RD10EB-2	140A	3061	ZD-10	562	
RD10EB1	5019A	3785			
RD10F	140A	3061	ZD-10	562	
RD10FA	140A		ZD-10	562	
RD10FB	140A		ZD-10	562	
RD10H			ZD-10	562	
RD10M			ZD-10	562	
RD11A	140A	3786/5020A	ZD-10	562	
RD11AL	5019A	3785	ZD-10	562	
RD11AM	5019A	3785	ZD-10	562	
RD11AN	5020A	3786	ZD-11	563	
RD11E	5020A	3786	ZD-11	563	
RD11EA	5074A		ZD-11	563	
RD11EB	5020A		ZD-11	563	
RD11EC	142A	3786/5020A	ZD-12	563	
RD11EM	5074A	3786/5020A	ZD-11	563	
RD11F	5074A	3139	ZD-11.0	563	
RD11FA	5074A		ZD-11.0	563	
RD11FB	5074A	3139	ZD-11	563	
RD11M			ZD-11	563	
RD12	142A	3787/5021A	ZD-12	563	
RD12E	5021A	3787	ZD-12	563	
RD12EA	5021A	3787	ZD-12	563	
RD12EB	5021A	3787	ZD-12	563	
RD12EB1Z	5021A	3787		563	
RD12EC	142A	3062	ZD-12	563	
RD12F	142A		ZD-12	563	
RD12FA	142A		ZD-12	563	
RD12FB	142A	3062	ZD-12	563	
RD12M			ZD-12	563	
RD13	143A	3788/5022A	ZD-13	563	
RD1343	177		300	1122	
RD13A	143A	3788/5022A	ZD-13	563	
RD13AK	143A	3788/5022A	ZD-13	563	
RD13AKP	143A	3788/5022A	ZD-13	563	
RD13AL	5021A	3788/5022A	ZD-12	563	
RD13AM	5022A	3788	ZD-12	563	
RD13AN	143A	3788/5022A	ZD-13	563	
RD13AN-P	143A		ZD-13	563	
RD13ANP		3788	ZD-13	563	
RD13B	143A	3788/5022A	ZD-13	563	
RD13B-Z	143A	3788/5022A	ZD-13	563	
RD13C	5128A	3788/5022A			
RD13D	5128A	3788/5022A			
RD13E	5022A	3788	ZD-13	563	
RD13EA	5022A	3788	ZD-12	563	
RD13EB	5022A	3788	ZD-12	563	
RD13EB2Z	5022A	3788			
RD13F	143A	3750	ZD-13	563	
RD13FA	143A	3750	ZD-13	563	
RD13FB	143A	3750	ZD-13	563	
RD13K	142A		ZD-12	563	
RD13M		3750	ZD-13	563	
RD15E	145A	3790/5024A	ZD-15	564	
RD15EA	5024A	3790	ZD-15	564	
RD15EB	145A	3063	ZD-15	564	
RD15EC	145A	3063			
RD15F	145A	3063	ZD-15	564	
RD15FA	145A	3063	ZD-15	564	
RD15FB	145A	3063	ZD-15	564	
RD15M			ZD-15	564	
RD15S	145A	3063	ZD-15	564	
RD16A	5025A	3791	ZD-15	564	
RD16A-N	5075A		ZD-16	564	
RD16AL	5023A	3791/5025A	ZD-14	564	
RD16AM	5024A	3791/5025A	ZD-15	564	
RD16AN	5026A	3792	ZD-17		
RD16B		3791	ZD-16	564	
RD16C-Y	5075A		ZD-16	564	
RD16E	5076A	3142	ZD-16	564	
RD16E-M	5075A	3751	ZD-16	564	
RD16E-N	5075A		ZD-16	564	
RD16EB	5075A	3751	ZD-16	564	
RD16EC	5076A	9022	ZD-17		
RD16F	5075A	3751	ZD-16	564	
RD16FA	5075A	3751	ZD-16	564	
RD16FB	5075A	3751	ZD-16	564	
RD16H	145A	3063	ZD-15	564	
RD16M			ZD-16	564	
RD18EB	5077A	3752			
RD18F	5077A	3752	ZD-18		
RD18FA	5077A	3752	ZD-18		
RD18FB	5077A	3752	ZD-18		
RD19		3794	ZD-19		
RD19A	5028A	3794	ZD-19		
RD19AL	5028A	3794	ZD-19		
RD19AM	5028A	3794	ZD-19		
RD2.7EB	5063A	3837			
RD2.7FB	5063A	3982/5064A			
RD20E	5079A	3335	ZD-20		
RD20F	5079A	3335	ZD-20		
RD20FA	5079A	3335	ZD-20		
RD20FB	5079A	3335	ZD-20		
RD22	5080A	3336	ZD-22		
RD22E	5080A	3336	ZD-22		

Industry Standard No.	ECG	SK	GE	RS 276-	MOTOR.
RD22F	5080A	3336	ZD-22		
RD22FA	5080A	3336	ZD-22		
RD22FB	5080A	3336	ZD-22		
RD24A	5081A	3151	ZD-24		
RD24AL	5081A	3151	ZD-24		
RD24AM	5081A	3151	ZD-24		
RD24AN	5081A	3151	ZD-25		
RD24EB	5031A	3797			
RD24F	5081A	3151	ZD-24		
RD24FA	5081A	3151	ZD-24		
RD24FB	5081A	3151	ZD-24		
RD24FC	5082A		ZD-25		
RD250	116	3017B/117	504A	1104	
RD26235-1	116		504A	1104	
RD27E	5033A		ZD-28		
RD27EB	5033A		ZD-27		
RD27F	146A		ZD-27		
RD27FA	146A		ZD-27		
RD27FB	146A		ZD-27		
RD29799P	116		504A	1104	
RD29A	5034A	3800	ZD-28		
RD29AL	5033A	3799	ZD-27		
RD29AM	5034A	3800	ZD-28		
RD29AN	5035A	3801	ZD-30		
RD3.0E		3330	ZD-3.3		
RD3.0EB	5065A	3838			
RD3.0F		3330	ZD-3.3		
RD3.3E		3330	ZD-3.3		
RD3.3F		3330	ZD-3.3		
RD3.9E		3331	ZD-3.9		
RD30F	5084A	3755	ZD-30		
RD30FA	5084A	3755	ZD-30		
RD30FB	5084A		ZD-30		
RD31903F	116		504A	1104	
RD33F	147A		ZD-33		
RD33FA	147A		ZD-33		
RD33FB	147A		ZD-33		
RD3472	116		504A	1104	
RD35A		3803	ZD-36		
RD35AE	5038A	3803/5037A	ZD-39		
RD35AL	5036A	3802	ZD-33		
RD35AM	5037A	3802/5036A	ZD-36		
RD36E		3337	ZD-36		
RD36EB	5037A	3803	ZD-36		
RD36F	5085A	3337	ZD-36		
RD36FA	5085A	3337	ZD-36		
RD36FB	5085A	3337	ZD-36		
RD39E		3338	ZD-39		
RD39F	5086A	3338	ZD-39		
RD39FA	5086A	3338	ZD-39		
RD39FB	5086A	3338	ZD-39		
RD3A-1B4			504A	1104	
RD4.3E	5008A	3332/5068A			
RD4.3EB	5008A		ZD-4.3		
RD4.3EC	5069A		ZD-4.7		
RD4.7E		3333	ZD-4.7		
RD4.7EB	5009A		ZD-4.7		
RD4.7F		3333	ZD-4.7		
RD4A	5007A	3773	ZD-3.9		
RD4AL	5006A	3772	ZD-3.6		
RD4AM	5007A	3773	ZD-3.9		
RD4AN	5008A	3774	ZD-4.3		
RD4R7EB	5009A	3775	ZD-4.7		
RD5.1E	5010A	3056/135A	ZD-5.1		
RD5.1EB	135A	3776/5010A	ZD-5.1		
RD5.1EC	5011A	3776/5010A	ZD-5.1		
RD5.6	136A	3777/5011A			
RD5.68	135A	3056			
RD5.6B			ZD-5.6	561	
RD5.6E	136A	3057	ZD5.6	561	
RD5.6E-B	136A			561	
RD5.6EB	5011A	3777	ZD-5.6	561	
RD5.6EBZ7S	5011A	3777			
RD5.6EC	136A	9021/5070A	ZD-6.0		
RD5.6ED			ZD-5.6	561	
RD5.6EK	136A		ZD-5.6	561	
RD5.6ES	5011A	9021/5070A			
RD5.6F	136A		ZD-5.6	561	
RD5.6FA	136A		ZD-5.6	561	
RD5.6FB	136A		ZD-5.6	561	
RD5.6M			ZD-5.6	561	
RD50DC					MC515L
RD53DC					MC465L
RD5A		3333	ZD-4.7		
RD5AL	5009A	3775	ZD-4.7		
RD5AM	5009A	3775	ZD-4.7		
RD5AN	5010A	3776	ZD-5.1		
RD5B	135A	3056	ZD-5.1		
RD5B-C	5011A	3056/135A			
RD5R6EB	5011A	9021/5070A	ZD-6.0	561	
RD6	136A	3057	ZD-5.6	561	
RD6.2E	5013A	3779	ZD-6.2	561	
RD6.2EB	5013A	3779	ZD-6.2	561	
RD6.2EB2	5013A	3779			
RD6.2EC	5014A		ZD-6.6	561	
RD6.2F	137A		ZD-6.2	561	
RD6.2FA	137A		ZD-6.2	561	
RD6.2FB	137A		ZD-6.2	561	
RD6.8E	5014A	3780	ZD-6.8	561	
RD6.8E-B1			ZD-6.8	561	
RD6.8EB	5014A	3780	ZD-6.8	561	
RD6.8F	5071A	3780/5014A	ZD-6.8	561	
RD6.8FA	5071A	3780/5014A	ZD-6.8	561	
RD6.8FB	5071A	3780/5014A	ZD-6.8	561	
RD6A	136A	9021/5070A	ZD-6.0	561	
RD6A(M)			ZD-5.6	561	
RD6AL	5011A	3777	ZD-5.6	561	
RD6AM	136A		ZD-5.6	561	
RD6AN	5011A	9021/5070A	ZD-6.0	561	
RD6B			ZD-5.8	561	
RD7.5E	138A		ZD-7.5		
RD7.5EB	5015A		ZD-7.5		
RD7.5ED	5015A		ZD-7.5		
RD7.5F	138A		ZD-7.5		
RD7.5FA	138A		ZD-7.5		
RD7.5FB	138A		ZD-7.5		
RD7A	5014A	3059/138A	ZD-9.1	562	
RD7AM	5014A	3780	ZD-6.8	561	
RD7AN	5014A	3780	ZD-6.8	561	
RD7B			ZD-6.8	561	
RD7H				561	
RD7R5EB	138A	3059	ZD-7.5		
RD8	5072A	3136	ZD-8.2	562	
RD8.2	5072A		ZD-8.2	562	
RD8.2C	5072A	3136	ZD-8.2	562	
RD8.2E	5015A	3136/5072A	ZD-8.2	562	
RD8.2EA	5072A		ZD-8.2	562	
RD8.2EB	5072A	3136	ZD-8.2	562	
RD8.2EB2Z	5016A	3782			
RD8.2EC	5072A	3136	ZD-8.7	562	
RD8.2EK	5072A		ZD-8.2	562	
RD8.2F	5072A	3136	ZD-8.2	562	
RD8.2FA	5072A		ZD-8.2	562	
RD8.2FB	5072A		ZD-8.2	562	

Industry Standard No.	ECG	SK	GE	RS 276-	MOTOR.
RD8.2M			ZD-8.2	562	
RD8H				562	
RD9	139A	3784/5018A	ZD-9.1	562	
RD9.1	139A	3060			
RD9.1E	139A	3784/5018A	ZD-9.1	562	
RD9.1EA	139A		ZD-9.1	562	
RD9.1EB	5018A	3784	ZD-9.1	562	
RD9.1EB2	5018A			562	
RD9.1EC	139A	3784/5018A		562	
RD9.1ED	139A		ZD-9.1	562	
RD9.1EK	139A		ZD-9.1	562	
RD9.1F	139A		ZD-9.1	562	
RD9.1FA	139A		ZD-9.1	562	
RD9.1FB	139A		ZD-9.1	562	
RD9.1M			ZD-9.1	562	
RD9037	116		504A	1104	
RD91E	139A	3345/5095A			
RD9A	139A	3784/5018A	ZD-9.1	562	
RD9A(10)	139A		ZD-9.1	562	
RD9A-N	139A		ZD-9.1	562	
RD9AL	5073A	3784/5018A	ZD-9.1	562	
RD9AM	5018A	3784	ZD-9.1	562	
RD9AN	5018A	3784	ZD-9.1	562	
RD9B	5073A	3749	ZD-8.7	562	
RD9E	5073A	3784/5018A			
RE-107			ZD-5.6	561	
RE-109			ZD-6.2	561	
RE-110			ZD-6.8	561	
RE-112			ZD-8.2	562	
RE-114			ZD-9.1	562	
RE-115			ZD-10	562	
RE-116			ZD-11	563	
RE-118			ZD-12	563	
RE-119			ZD-13	563	
RE-121			ZD-15	564	
RE-122			ZD-16	564	
RE-168			MR-5	1067	
RE-171			MR-5	1067	
RE-2	116		8		
RE-27			245	2005	
RE-3	116		1		
RE-49			504A	1104	
RE-5			8	2002	
RE-50			504A	1104	
RE-51			510	1114	
RE-52			300	1102	
RE-61			263	2043	
RE-82			74	2043	
RE-90			509	1114	
RE-94			514	1122	
RE1	100	3005	1	2007	
RE10	108	3122	17	2051	
RE100	5066A	3330	ZD-3.3		
RE1001	108	3452	86	2038	
RE1002	108	3452	86	2038	
RE101	134A	3055	ZD-3.6		
RE102	5067A	3331	ZD-3.9		
RE103	5068A	3332	ZD-4.3		
RE104	5069A	3333	ZD-4.7		
RE105	135A	3056			
RE106			ZD-5.6	561	
RE107	136A	3057	ZD-5.6	561	
RE108	5070A	9021		561	
RE109	137A	3058	ZD-6.2	561	
RE11	121	3717	16	2006	
RE110	5071A	3334	ZD-6.8	561	
RE111	138A	3059	ZD-7.5		
RE112	5072A	3136	ZD-8.2	562	
RE113	5073A	3749	ZD.8.7	562	
RE114	139A	3060	ZD-9.1	562	
RE115	140A	3061	ZD-10	562	
RE116	5074A	3139	ZD-11	563	
RE117	5074A	3092/141A	ZD-11.5		
RE118	142A	3062	ZD-12	563	
RE119	143A	3750	ZD-13	563	
RE11MP	121	3717	16	2006	
RE12	123	3020	20	2051	
RE120	144A	3094		564	
RE121	145A	3063	ZD-15	564	
RE122	5075A	3751	ZD-16	564	
RE123	5076A	9022	ZD-17		
RE124	5077A	3752	ZD-18		
RE125	5078A	9023	ZD-19		
RE126	5079A	3335	ZD-20		
RE127	5080A	3336	ZD-22		
RE128	5081A	3151	ZD-24		
RE129	5082A	3753	ZD-25		
RE13	123A	3124/289	20	2051	
RE130	147A	3095	ZD-33		
RE131	5086A	3338			
RE132	149A	3097	ZD-62		
RE133	5096A	3346	ZD-100		
RE14	124	3261/175	12		
RE15	126	3006/160	52	2024	
RE16	127	3764	25		
RE167	118	3066	CR-1		
RE168	5454	6754	MR-5	1067	
RE17	128	3024	18	2030	
RE170	5542	6642	MR-3		
RE171	5455	3597	MR-5	1067	
RE172	5483	3942			
RE173	5543	3581	MR-3		
RE174	5457	3598			
RE175	5485	3505/5547	MR-3		
RE176	5545	3582	MR-3		
RE177	5487	3944			
RE178	5547	3505			
RE179	5640	3583/5641			
RE18	129	3025	21	2034	
RE181	5682	3509/5695			
RE182	5641	3583			
RE184	5683	3509/5695			
RE185	5642	3506			
RE187	5685	3521/5686			
RE188	5643	3519			
RE19	130	3027	14	2041	
RE190	6408	3523/6407	X13	1050	
RE191	198	3220	32		
RE192	199	3245	62	2010	
RE193	234	3247	65	2050	
RE195	613	3126	90		
RE196	192	3137	63	2030	
RE197	193	3138	67	2023	
RE198	237	3299	46		
RE199	222	3065	FET-4	2036	
RE2	552	3011	504A	1104	
RE20	131	3198	30	2006	
RE200	503	3068	CR-5		
RE2001	161	3716	39	2015	
RE2002	161	3716	39	2015	
RE201	236	3239	216	2053	
RE202	224	3049	46		
RE203	235	3197	256		
RE204	302	3252	275		

Industry Standard No.	ECG	SK	GE	RS 276-	MOTOR.
RE205	196	3054			
RE206	282	3748			
RE207	287	3433			
RE208	288	3434			
RE209	295	3253			
RE20MP	131	3840	44	2006	
RE21	196	3054	66	2048	
RE210	297	3449			
RE211	298	3450			
RE212	299	3252/302			
RE213	300	3464			
RE214	500A	3304			
RE215	523	3306			
RE216	502	3067			
RE217	504	3108/505			
RE218	5404	3627			
RE219	294	3841			
RE22	197	3083	69	2049	
RE220	293	3849			
RE225	513	3443			
RE23	154	3045/225	40	2012	
RE24	157	3747	232		
RE243	521	3304/500A			
RE25	158	3004/102A	53	2007	
RE255		3647		1152	
RE26	159	3025/129	22	2032	
RE27	160	3006	245	2004	
RE28	161	3716	39	2015	
RE29	162	3559	36		
RE3		3004	504A	1104	
RE3(TRANSISTOR)	102	3722			
RE30	163A	3439	36		
RE300-IC	703A	3157	IC-12		
RE3001	107		11	2015	
RE3002	107		11	2015	
RE301-IC	709	3135	IC-11		
RE302-IC	705A	3134	IC-3		
RE303-IC	783	3215	IC-225		
RE304-IC	749	3168	IC-97		
RE305-IC	712	3072	IC-148		
RE306-IC	713	3077/790	IC-5		
RE307-IC	714	3075	IC-4		
RE308-IC	715	3076	IC-6		
RE309-IC	722	3161	IC-9		
RE31	164	3133	37		
RE310-IC	747	3279	IC-218		
RE311-IC	708	3135/709	IC-10		
RE312-IC	739	3235	IC-30		
RE313-IC	738	3167	IC-225	2025	
RE314-IC	742		IC-213		
RE315-IC	737		IC-16		
RE316-IC	736		IC-17		
RE317-IC	743	3172	IC-214		
RE318-IC	721		IC-14		
RE319-IC	718	3159	IC-8		
RE32	165	3115	38		
RE320-IC	720	9014	IC-7		
RE321-IC	740A	3328	IC-31		
RE322-IC	804		IC-27		
RE323-IC	790		IC-230		
RE324-IC	791	3149	IC-231		
RE325-IC	1004	3365	IC-149		
RE326-IC	1006	3358	IC-38		
RE327-IC	1052	3249	IC-135		
RE328-IC	1080	3284	IC-98		
RE329-IC	1103	3281	IC-94		
RE33	172A	3156	64		
RE330-IC	1142		IC-128		
RE331-IC	748	3236	IC-115		
RE332-IC	788	3829	IC-229		
RE333-IC	797	3158	IC-233		
RE334-IC	801	3160	IC-35		
RE335-IC	1003	3288	IC-43		
RE336-IC	1005	3723	IC-42		
RE337-IC	1046	3471	IC-118		
RE338-IC	1058	3459	IC-49		
RE339-IC	1081A*	3474/1081A			
RE34	175	3538	246	2020	
RE340-M	1087	3477			
RE341-M	1082	3461	IC-140		
RE342-M	1092	3472	IC-130		
RE343-M	1100	3223			
RE344-M		3223	IC-92		
RE345-IC		3144	IC-15		
RE346-IC		3074	IC-23		
RE348-M		3155	722		
RE349-M		3153	721		
RE35	176	3845	80		
RE36	179	3642	76		
RE360-IC	1233	3732/1167			
RE362-IC		3171		2022	
RE37	181	3535	75	2041	
RE38		3188A	55		
RE382-IC		7400	7400		
RE39	183	3189A	56	2027	
RE4	102A	3004	52	2024	
RE40	184	3190	57	2017	
RE4001	199	3245	62	2010	
RE4002	199	3245	62	2010	
RE4010	199	3245	62	2010	
RE41	185	3191	58	2025	
RE42	186	3192	28	2017	
RE43	187	3193	29	2025	
RE44		3232	217		
RE45	312	3116	FET-2	2035	
RE46	312	3112	FET-1	2028	
RE47	109	3090	1N34AS	1123	
RE48	112	3089	1N82A		
RE49	116	3080	504A	1104	
RE5	103	3010	8	2002	
RE50	117	3017B	504A	1104	
RE5001	161	3716	39	2015	
RE5002	161	3716	39	2015	
RE504	117	3031A	504A	1104	
RE51	125	3081	510	1114	
RE52	177	3175	300	1122	
RE53	106	3118	21	2034	
RE54	156	3839/155	79		
RE55		3843	511	1114	
RE58	323	3259			
RE59	324	3748/282			
RE6	103A	3010	59	2002	
RE60	287	3244			
RE61	281	3173/219	263	2043	
RE62	234	3114/290	221		
RE63	290A	3118	82	2032	
RE64	199	3245	212	2010	
RE66	194	3479	210	2051	
RE67	199	3122	210	2051	
RE68	218	3052	30		
RE69	160	3007		2005	
RE7	104	3719	16	2006	
RE70	128	3265	63	2030	
RE73	171	3201	27		

Industry Standard No.	ECG	SK *	GE	RS 276-	MOTOR.
RE74	180	3437	74	2043	
RE75	188	3199	217		
RE76	189	3200	218	2026	
RE77	191	3232	224		
RE78	194	3275	220		
RE79	195A	3765	219		
RE79A	195A	3765	219		
RE7MP	104MP	3720	13MP		
RE8	105	3012	4		
RE80	210	3202	236		
RE81	211	3203		2026	
RE82	219	3173	74	2043	
RE83	226	3082	49	2025	
RE83MP	226	3086	49	2025	
RE85	326	3746			
RE86	110MP	3709	1N34AS	1123	
RE87	113A	3119/113	6GC1		
RE88	114	3120	6GD1	1104	
RE89	115	3121	6GX1		
RE9	107	3018	11	2015	
RE90	125	3080	509	1114	
RE91	120	3110	CR-3		
RE92	156	3051	512		
RE93P	173BP	3999	305		
RE94	519	3100	514	1122	
RE97	5063A	3837			
RE99	5065A	3838			
RECT-SI-1019	525	3925			
RECT-SI-1020	552	3998/506			
RECT-SI-154	552	3998/506			
RECT-U-1012	116	3647/167			
REF-01AJ					MC1500AU10
REF-01CJ					MC1404U10
REF-01CP					MC1404U10
REF-01DJ					MC1404U10
REF-01EJ					MC1400AU10
REF-01HJ					MC1400U10
REF-01HP					MC1400U10
REF-01J					MC1500AU10
REF-01PP					MC1404U10
REF-02AJ					MC1500AU5
REF-02CJ					MC1404U5
REF-02CP					MC1404U5
REF-02DJ					MC1404U5
REF-02DP					MC1404U5
REF-02EJ					MC1400AU5
REF-02HJ					MC1400U5
REF-02HP					MC1400U5
REF-02J					MC1500AU5
REJ701148-1	113A	3119/113	6GC1		
REJ70148	113A	3119/113	6GC1		
REJ70148A	113A	3119/113	6GC1		
REJ70643	116		504A	1104	
REJ70931	116		504A	1104	
REJ71253	177	3100/519	300	1122	
REN100		3005	1	2007	
REN1003		3288	IC-43		
REN1004	1004	3365	IC-149		
REN1005	1005	3723	IC-42		
REN1006	1006	3358	IC-38		
REN101		3011	8	2002	
REN102		3004	2	2007	
REN1024		3152	720		
REN1025		3155	722		
REN1027		3153	721		
REN1028		3436	724		
REN102A		3004	52	2007	
REN103		3862	8	2002	
REN1037		3371	IC-170		
REN103A		3835	59	2002	
REN104		3719	16	2006	
REN1046		3471	IC-118		
REN105		3012	4		
REN1050		3475	IC-133		
REN1052		3249	IC-135		
REN1056	1056	3458	IC-48		
REN1058	1058	3459	IC-49		
REN106		3118	21	2034	
REN107		3293	11	2015	
REN1073		3496	IC-63		
REN108		3122	11	2038	
REN1080	1080	3284	IC-98		
REN1082	1082	3461	IC-140		
REN1087		3477	IC-103		
REN109		3090	1N34AS	1123	
REN1090		3291	723		
REN1092		3472	IC-130		
REN1096		3703	IC-312		
REN110		3709		1123(2)	
REN1100		3223	IC-92		
REN1102		3224	IC-93		
REN1103		3281	IC-94		
REN1127		3243	IC-197		
REN1128	1128	3488	IC-105		
REN113		3119	6GC1		
REN1130	1130	3478	IC-111		
REN1131		3286	IC-109		
REN1132		3287	IC-110		
REN1133	1133	3490	IC-107		
REN1134	1134	3489	IC-106		
REN1135	1135	3876	IC-314		
REN114		3120	6GD1	1104	
REN1142		3485	IC-128		
REN115		3121	6GX1		
REN1153		3282	IC-182		
REN1155	1155	3231	IC-179	705	
REN116		3313	504A	1104	
REN1161		3968	IC-300		
REN1162		3072	IC-2		
REN1164		3727	IC-301		
REN1165		3827	IC-315		
REN1166		3827	IC-316	704	
REN1167	1167	3732	IC-317		
REN117		3017B	504A	1104	
REN118		3066	CR-1		
REN1192	1192	3445			
REN120		3110	CR-3		
REN121		3717	76	2006	
REN121MP		3718	76	2006(2)	
REN123		3020	20	2051	
REN123A		3444	20	2051	
REN124		3261	12		
REN125		3081	510	1114	
REN126		3006	52	2024	
REN127		3764	25		
REN128		3024	18	2030	
REN129		3025	244	2027	
REN130		3027	14	2041	
REN131		3198	44	2006	
REN131MP		3840	44	2006(2)	
REN132		3834	FET-1	2035	
REN133		3112		2028	
REN134		3055	ZD-3.6		
REN135		3056	ZD-5.1		

Industry Standard No.	ECG	SK	GE	RS 276-	MOTOR.
REN136		3057	ZD-5.6	561	
REN137		3058	ZD-6.2	561	
REN138		3059	ZD-7.5		
REN139		3060	ZD-9.1	562	
REN140		3061	ZD-10	562	
REN141		3092	ZD-11.5		
REN142		3062	ZD-12	563	
REN143		3750	ZD-13	563	
REN144		3094	ZD-14	564	
REN145		3063	ZD-15	564	
REN146		3064	ZD-27		
REN147		3095	ZD-33		
REN148		3096	ZD-55		
REN149		3097	ZD-62		
REN150		3098	ZD-82		
REN151		3099	ZD-110		
REN152		3893	66	2048	
REN153		3274	69	2049	
REN154		3045	40	2012	
REN155		3839	43		
REN156		3051	512		
REN157		3747	232		
REN158		3004	53	2007	
REN159		3025	82	2032	
REN160		3006	51	2004	
REN161		3716	39	2015	
REN162		3559	35		
REN163		3439	36		
REN164		3133	37		
REN165		3115	38		
REN166		3647		1152	
REN171		3201	27		
REN172		3156	64		
REN175		3261	246	2020	
REN176		3845	80		
REN177		3175	400	1122	
REN179		3642	76		
REN180		3437	74	2043	
REN181		3535	75	2041	
REN182		3188A	55		
REN183		3189A	56	2027	
REN184		3190	57	2017	
REN185		3191	58	2025	
REN186		3192	28	2017	
REN186A		3192	247	2052	
REN187		3193	29	2025	
REN188		3199	217	2018	
REN189		3200	218	2026	
REN190		3232	217		
REN191		3232	249		
REN192		3137	63	2030	
REN193		3138	67	2023	
REN194		3275	220		
REN195A		3765	46		
REN196		3054	241	2020	
REN197		3189A	250	2027	
REN198		3220	251		
REN199		3245	62	2010	
REN210		3202	252	2018	
REN211		3203	253	2026	
REN218		3625	234		
REN219		3173	74	2043	
REN220				2036	
REN221		3050		2036	
REN222		3065		2036	
REN223		3563		2041	
REN224		3049	46		
REN225		3045	256		
REN226		3082	49	2025	
REN226MP		3086	49	2025(2)	
REN229		3246	61	2038	
REN233		3854	210	2009	
REN234		3247	65	2050	
REN235		3197	215		
REN236		3239	216	2053	
REN237		3299	46		
REN238		3710	259		
REN280		3297	266	2041	
REN282		3748	264		
REN287		3433	222		
REN288		3434	221		
REN289		3124	81	2038	
REN290		3114	269	2032	
REN293		3849	47	2030	
REN294		3841	48		
REN295		3253	270		
REN297		3449	222	2030	
REN298		3450	223		
REN299		3252	236		
REN300		3464	273		
REN302		3252	275		
REN306		3251	276		
REN311		3194	277		
REN315		3250	279		
REN500A	500A	3304			
REN502		3067	CR-4		
REN503		3068	CR-5		
REN504		3108	CR-6		
REN506		3843	511	1114	
REN5066		3330	ZD-3.3		
REN5067		3331	ZD-3.9		
REN5068		3332	ZD-4.3		
REN5069		3333	ZD-4.7		
REN5070		9021	ZD-6.2	561	
REN5071		3334	ZD-6.8	561	
REN5072		3136	ZD-8.2	562	
REN5073		3749	ZD-9.1	562	
REN5074		3139	ZD-11.0	563	
REN5075		3751	ZD-16	564	
REN5077		3752	ZD-18		
REN5078		9023	ZD-19		
REN5079		3335	ZD-20		
REN5080		3336	ZD-22		
REN5081		3151	ZD-24		
REN5082		3753	ZD-25		
REN5086		3338	ZD-39		
REN5096		3346	ZD-100		
REN513	513	3443	513		
REN521	521	3304/500A			
REN523	523	3306	521		
REN5400	5400	3950			
REN5454		3597		1067	
REN5455		3597		1067	
REN61		3173	263	2043	
REN62		3114	221	2050	
REN63		3118	82	2032	
REN64		3245	212	2010	
REN6401				2029	
REN6408		3523		1050	
REN66		3479	210		
REN67		3122	210	2010	
REN68		3052	30		
REN69		3007		2004	
REN70		3265	63	2030	

Industry Standard No.	ECG	SK	GE	RS 276-	MOTOR.
REN703A		3157	IC-12		
REN705A		3134	IC-3		
REN708		3135	IC-10		
REN709		3135	IC-11		
REN712	712	3072	IC-148		
REN713		3077	IC-5		
REN714	714	3075	IC-4		
REN715		3076	IC-6		
REN718		3159	IC-8		
REN722		3161	IC-9		
REN723	723	3144	IC-15		
REN729		3074	IC-23		
REN731		3170	IC-13		
REN736	736	3163	IC-17		
REN737		3375	IC-16		
REN738	738	3167	IC-29		
REN739		3235	IC-30		
REN740		3328	IC-213		
REN7400	7400	7400	7400	1801	
REN74123		74123	74123	1817	
REN742		3453	IC-213		
REN743		3172	IC-214		
REN744		3171	IC-24	2022	
REN747		3279	IC-218		
REN7473		7473	7473	1803	
REN7474	7474	7474	7474	1818	
REN748		3236	IC-115		
REN749		3168	IC-97		
REN7493A	7493A	7493			
REN780		3141	IC-222		
REN783		3215	IC-225		
REN788	788	3147	IC-229		
REN791	791	3149	IC-231		
REN797		3158	IC-233		
REN801	801	3160	IC-35		
REN804		3455	IC-27		
REN85		3746		2037	
REN90		3080	509	1114	
REN94		3100	514	1122	
REN977	977	3462	VR-100		
RER-023	178MP		6GC1	1122	
RER023	113A	3119/113	6GC1		
RET20	116	3017B/117	504A	1104	
RF-1A	552		511	1114	
RF-3160	116	3016	504A	1104	
RF-32101-8	116	3016	504A	1104	
RF-32101R	116	3016	504A	1104	
RF-6235-1	116	3016	504A	1104	
RF-8.2E	5072A	3059/138A	ZD-8.2	562	
RF03C06	503	3068	CR-5		
RF1	552	3843			
RF1003	350F				MRF221
RF1004					MRF223
RF100DC					MC528L
RF100K					MC528F
RF101DC					MC573L
RF101K					MC573F
RF102DC					MC423L
RF103DC					MC174L
RF10K35		3068	511	1114	
RF110DC					MC524L
RF110K					MC524F
RF111DC					MC574L
RF111K					MC574F
RF112DC					MC424L
RF113DC					MC474L
RF120DC					MC2123L
RF120K					MC2123F
RF121DC					MC2173L
RF121K					MC2173F
RF122DC					MC2023L
RF123DC					MC2073L
RF130DC					MC2124L
RF130K					MC2124F
RF131DC					MC2174L
RF131K					MC2174F
RF132DC					MC2024L
RF1811	109	3087		1123	
RF200	161	3716	39	2015	
RF2081	477				MRF216
RF2092					MRF460
RF2123	345				MRF238
RF2125					MRF450
RF2127	352				MRF245
RF2135					MRF223
RF2142					2N6367
RF2143					MRF454
RF2144	320F				MRF224
RF2146					MRF476
RF2147					MRF475
RF231K					MC2152F
RF26231-1	116	3017B/117	504A	1104	
RF26234-1	116	3017B/117	504A	1104	
RF26235-1	116		504A	1104	
RF26235-2	116	3017B/117	504A	1104	
RF26235-5	116		504A	1104	
RF29799P	116	3016	504A	1104	
RF30DC					MC521L
RF30K					MC521F
RF3160	116		504A	1104	
RF31903P	116	3017B/117	504A	1104	
RF31DC					MC571L
RF31K					MC571F
RF32101-8	116		504A	1104	
RF32101-9	116	3017B/117	504A	1104	
RF32101R	116		504A	1104	
RF32102R	119	9086/518	CR-2		
RF32103-1	118	3066	CR-1		
RF32103R	118	3066	CR-1		
RF32412-3	116(3)	3016	504A(3)	1104	
RF32426-7	178MP		300(2)	1122(2)	
RF32645	116		504A	1104	
RF32DC					MC421L
RF32K					MC421F
RF33426-7	113A	3119/113	6GC1		
RF33550-1	109	3091		1123	
RF33976	116	3017B/117	504A	1104	
RF33DC					MC471L
RF34383	116	3016	504A	1104	
RF34661	177		300	1122	
RF3472	116	3017B/117	504A	1104	
RF34720			504A	1104	
RF35123			1N34AS	1123	
RF400					BB105G
RF50K					MC515F
RF51DC					MC565L
RF51K					MC565F
RF52DC					MC415L
RF5464	113A	3119/113	6GC1		
RF5464-1P	113A	3119/113	6GC1		
RF5465	113A	3119/113	6GC1		
RF5465-1P	113A	3119/113	6GC1		
RF5794	114	3120	6GD1		
RF60034	109	3087		1123	

Industry Standard No.	ECG	SK	GE	RS 276-	MOTOR.
RF60DC					MC516L
RF60K					MC516F
RF61DC					MC566L
RF62DC					MC416L
RF63DC					MC466L
RF70K					MC520F
RF71DC					MC570L
RF7313		3021	12		
RF8.2E	5072A	3136	ZD-8.2	562	
RF90K					MC503F
RFA70597	116	3032A	504A	1104	
RFA70600	116	3032A	504A	1104	
RFC61197	116	3031A	504A	1104	
RFJ-30704	116	3016	504A	1104	
RFJ-31218	116	3016	504A	1104	
RFJ-31362	116	3016	504A	1104	
RFJ-31363	116	3016	504A	1104	
RFJ-33292	116	3016	504A	1104	
RFJ-60366	116	3016	504A	1104	
RFJ30704	116		504A	1104	
RFJ31218	116		504A	1104	
RFJ31362	116		504A	1104	
RFJ31363	116		504A	1104	
RFJ33292	116		504A	1104	
RFJ60172			1N34AS	1123	
RFJ60173			504A	1104	
RFJ60174			504A	1104	
RFJ60286	116		504A	1104	
RFJ60313	113A	3119/113	6GC1		
RFJ60366	116		504A	1104	
RFJ60614	109	3087		1123	
RFJ60869	116	3017B/117	504A	1104	
RFJ6134	116		504A	1104	
RFJ70147	5804	9007	504A	1144	
RFJ70148	113A	3119/113	6GC1		
RFJ70149	506	3017B/117	511	1114	
RFJ70431	113A		6CG1		
RFJ70432	116	3031A	504A	1104	
RFJ70487	116	3311	504A	1104	
RFJ70643	115	3311	6GX1		
RFJ70703	116	3031A	504A	1104	
RFJ70931	116	3031A	504A	1104	
RFJ70970	116	3017B/117	504A	1104	
RFJ70971	506	3016	511	1114	
RFJ70974	116	3016	504A	1104	
RFJ70976	506	3017B/117	511	1114	
RFJ70977	116	3017B/117	504A	1104	
RFJ711122	116		504A	1104	
RFJ71122	116	3017B/117	504A	1104	
RFJ71123	156	3051	512		
RFJ71480	139A	3060	ZD-9.1	562	
RFJ72360	116	3311	504A	1104	
RFJ72787	116	3311	504A	1104	
RFJZ0432	116		504A	1104	
RFL-30596	116	3016	504A	1104	
RFL30596	116		504A	1104	
RFM-33160	116	3016	504A	1104	
RFM33160	116		504A	1104	
RFP-33118	116	3016	504A	1104	
RFP33118	116		504A	1104	
RFS61436	117	3017B	504A	1104	
RFV60500	116	3016	504A	1104	
RG1-D					1N4935
RG1-K					MR817
RG1-M					MR818
RG1004	116	3016	504A	1104	
RG100B	116	3311	504A	1104	
RG100D	116	3311	504A	1104	
RG100DC					MC504L
RG100G	116	3031A	504A	1104	
RG100J	116	3017B/117	504A	1104	
RG100K					MC504F
RG101DC					MC554L
RG101K					MC554F
RG102DC					MC404L
RG103DC					MC454L
RG110DC					MC505L
RG110K					MC505F
RG111DC					MC555L
RG111K					MC555F
RG1122					1N4001
RG1123					1N4002
RG1127	116		504A	1104	
RG112DC					MC405L
RG113DC					MC455L
RG120DC					MC506L
RG120K					MC506F
RG121DC					MC556L
RG122DC					MC406L
RG123DC					MC456L
RG130DC					MC507L
RG130K					MC507F
RG131DC					MC557L
RG131K					MC557F
RG132DC					MC407L
RG133DC					MC457L
RG140DC					MC508L
RG140K					MC508F
RG141DC					MC558L
RG141K					MC558F
RG142DC					MC408L
RG143DC					MC458L
RG150DC					MC509L
RG150K					MC509F
RG151DC					MC559L
RG151K					MC559F
RG152DC					MC409L
RG153DC					MC459L
RG160DC					MC519L
RG160K					MC519F
RG161DC					MC569L
RG161K					MC569F
RG162DC					MC419L
RG163DC					MC469L
RG170DC					MC510L
RG170K					MC510F
RG171DC					MC560L
RG171K					MC560F
RG172DC					MC410L
RG172K					MC410F
RG173DC					MC460L
RG180DC					MC511L
RG180K					MC511F
RG181DC					MC561L
RG181K					MC561F
RG182DC					MC411L
RG183DC					MC461L
RG190DC					MC512L
RG190K					MC512F
RG191DC					MC562L
RG191K					MC562F
RG192DC					MC412L
RG193DC					MC462L
RG1A					1N4933
RG1B					1N4934
RG1D					1N4935
RG1F					1N4936
RG1G					1N4936
RG1H					1N4937
RG1J					1N4937
RG1K					MR817
RG1M					MR818
RG200DC					MC2111L
RG200K					MC2111F
RG201DC					MC2161L
RG201K					MC2161F
RG202DC					MC2011L
RG203DC					MC2061L
RG210DC					MC2100L
RG210K					MC2100F
RG211DC					MC2150L
RG211K					MC2150F
RG212DC					MC2000L
RG213DC					MC2050L
RG220DC					MC2101L
RG220K					MC2101F
RG221DC					MC2151L
RG221K					MC2151F
RG222DC					MC2001L
RG223DC					MC2051L
RG230DC					MC2107L
RG230K					MC2102F
RG231DC					MC2152L
RG232DC					MC2002L
RG233DC					MC2052L
RG240DC					MC2103L
RG240K					MC2103F
RG241DC					MC2153L
RG241K					MC2153F
RG242DC					MC2003L
RG243DC					MC2053L
RG250DC					MC2104L
RG250K					MC2104F
RG251DC					MC2154L
RG251L					MC2154F
RG252DC					MC2004L
RG253DC					MC2054L
RG260DC					MC2105L
RG260K					MC2105F
RG261DC					MC2155L
RG261K					MC2155F
RG262DC					MC2006L
RG263DC					MC2055L
RG263K					MC2055F
RG270DC					MC2106L
RG270K					MC2106F
RG271DC					MC2156L
RG271K					MC2156F
RG273DC					MC2056L
RG280DC					MC527L
RG280K					MC527F
RG281DC					MC577L
RG281K					MC577F
RG282DC					MC427L
RG283DC					MC477L
RG290DC					MC528L
RG290K					MC528F
RG291DC					MC578L
RG291K					MC578F
RG292DC					MC428L
RG293DC					MC478L
RG3-A					MR850
RG300DC					MC2112L
RG300K					MC2112F
RG301DC					MC2162L
RG301K					MC2162F
RG302DC					MC2012L
RG303DC					MC2062L
RG310DC					MC2113L
RG310K					MC2113F
RG311DC					MC2163L
RG311K					MC2163F
RG312DC					MC2013L
RG313DC					MC2063L
RG320K					MC2107F
RG321DC					MC2157L
RG321K					MC2157F
RG322DC					MC2007L
RG323DC					MC2057L
RG370DC					MC529L
RG370K					MC529F
RG371DC					MC579L
RG371K					MC579F
RG372DC					MC429L
RG373DC					MC479L
RG380DC					MC2116L
RG380K					MC2116F
RG381DC					MC2166L
RG381K					MC2166F
RG382DC					MC2016L
RG383DC					MC2066L
RG3A					MR850
RG3B					MR851
RG3D					MR852
RG3F					MR854
RG3G					MR854
RG3H					MR856
RG3J					MR856
RG3K					MR917
RG3M					MR918
RG40DC					MC500L
RG40K					MC500F
RG41DC					MC550L
RG41K					MC550F
RG42DC					MC400L
RG43DC					MC450L
RG50DC					MC501L
RG50K					MC501F
RG51DC					MC551L
RG51K					MC551F
RG52DC					MC401L
RG53DC					MC451L
RG60DC					MC502L
RG60K					MC502F
RG61DC					MC552L
RG61K					MC552F
RG62DC					MC402L
RG63DC					MC452L
RG70DC					MC520L
RG71K					MC570F
RG72DC					MC420L
RG73DC					MC470L
RG80DC					MC526L
RG80K					MC526F
RG81DC					MC576L
RG82DC					MC426L
RG83DC					MC476L
RG90DC					MC503L

Industry Standard No.	ECG	SK	GE	RS 276-	MOTOR.
RG91DC					MC953L
RG91K					MC553F
RG92DC					MC403L
RG93DC					MC453L
RGP-10D	116	3311	504A	1104	
RGP10A					1N4933
RGP10B					1N4934
RGP10D					1N4935
RGP10F					1N4936
RGP10G	552				1N4936
RGP10H					MR818
RGP10J	552				1N4937
RGP10K					MR817
RGP15A		3081			1N4933
RGP15B		3081			1N4934
RGP15D		3081			1N4935
RGP15F		3081			1N4936
RGP15G		3081			1N4936
RGP15H		3081			1N4937
RGP15J		3081			1N4937
RGP15K		3081			MR817
RGP15M		3081			MR818
RGP20A		3081			1N4933
RGP20B		3081			1N4934
RGP20D		3081			1N4935
RGP20F		3081			1N4936
RGP20G		3081			1N4936
RGP20H		3081			1N4937
RGP20J		3081			1N4937
RGP20K		3081			MR817
RGP20M		3081			MR818
RGP25A		3848			MR850
RGP25B		3848			MR851
RGP25D		3848			MR852
RGP25F		3848			MR854
RGP25G		3848			MR854
RGP25H		3848			MR856
RGP25J		3848			MR856
RGP25K		3051			MR917
RGP25M		3051			MR918
RGP30A		3848			MR850
RGP30B		3848			MR851
RGP30D		3848			MR852
RGP30F		3848			MR854
RGP30G		3848			MR854
RGP30H		3848			MR856
RJP30J		3848			MR856
RGP30K		3051			MR917
RGP30M		3051			MR918
RH-1	552		511	1114	
RH-1B	506		511	1114	
RH-1C			511	1114	
RH-1V	552	3843	511	1114	
RH-1X0001TAZZ	748	3236	IC-115		
RH-1X0004CEZZ	749	3168	IC-97		
RH-1X0015TAZZ	1197	3878/1245			
RH-1X0020CEZZ	1004	3365	IC-149		
RH-1X0021CEZZ	1080	3284			
RH-1X0022CEZZ	1051		IC-131		
RH-1X0023CEZZ	1079	3495/1074	IC-60		
RH-1X0024CEZZ	1105	3285	IC-101		
RH-1X0025CEZZ	1050	3475	IC-133		
RH-1X0032CEZZ	1089*		IC-120		
RH-1X0038CEZZ	712		IC-2		
RH-1X0043CEZZ	712	3072	IC-148		
RH-1X1020AFZZ	1115A	3184/1115	IC-278		
RH-1X1038AFZZ	1208	3712			
RH-DX0003BEZZ	116		504A	1104	
RH-DX0003TAZZ	116	3032A	504A	1104	
RH-DX0004TAZZ	506		511	1114	
RH-DX0008CEZZ	116	3017B/117	504A	1104	
RH-DX0017CEZZ	506	3130	511	1114	
RH-DX0025CEZZ	116	3100/519	504A	1104	
RH-DX0026AGZZ	116		504A	1104	
RH-DX0028CEZZ	506	3843	511	1114	
RH-DX0029CEZZ	506	3130	511	1114	
RH-DX0033TAZZ	177	3100/519	514	1122	
RH-DX0038CEZZ	116	3312	504A	1104	
RH-DX0039TAZZ	506	3311	504A	1104	
RH-DX0041CEZZ	116	3081/125	510	1114	
RH-DX0042CEZZ	116	3081/125	510	1114	
RH-DX0043TAZZ	552	3313/116	504A	1104	
RH-DX0045CEZZ	506	3130	511	1114	
RH-DX0046CEZZ	177	3100/519	300	1122	
RH-DX0048CEZZ	519	3100	514	1122	
RH-DX0051CEZZ	506	3130	511	1114	
RH-DX0054CEZZ	177	3100/519	300	1122	
RH-DX0055TAZZ	116	3311	504A	1104	
RH-DX0056CEZZ	116	3017B/117	504A	1104	
RH-DX0056TAZZ	506	3130	511	1114	
RH-DX0059TAZZ	116	3311	504A	1104	
RH-DX0062CEZZ	506	3843	511	1114	
RH-DX0063CEZZ	506	3312	504A	1104	
RH-DX0064CEZZ	116	3017B/117	504A	1104	
RH-DX0065CEZZ	506	3130	511	1114	
RH-DX0065CZZ	506		510	1114	
RH-DX0066TAZZ	116(4)	3311	504A	1104	
RH-DX0067TAZZ	552		504A	1104	
RH-DX0068TAZZ	116	3016	504A	1104	
RH-DX0069TAZZ	116		504A	1104	
RH-DX0072CEZZ	116		504A	1104	
RH-DX0073CEZZ	552	3843	511	1114	
RH-DX0073TAZZ	116(4)	3311	504A	1104	
RH-DX0077CEZZ	525	3925	530		
RH-DX0079TAZZ			504A	1104	
RH-DX0081CEZZ	506	3016	511	1114	
RH-DX0081TAZZ	116	3312	504A	1104	
RH-DX0083CEZZ	177	3311	504A	1104	
RH-DX0085TAZZ	552	3843	511		
RH-DX0086TAZZ	552	3843	511		
RH-DX0090CEZZ	506		511	1114	
RH-DX0091CEZZ	156	3051	512		
RH-DX0092CEZZ	506	3017B/117	504A	1104	
RH-DX0096CEZZ	506	3130	511	1114	
RH-DX0100CEZZ	552	3925/525	530		
RH-DX0101CEZZ	552	3843	504A		
RH-DX0104CEZZ	506	3843	511		
RH-DX0105TAZZ	552	9098/515			
RH-DX0106CEZZ	552	9098/515	512		
RH-DX0114CEZZ	552	3313/116	504A		
RH-DX0123CEZZ	552	9000			
RH-DX1005AFZZ	116		504A	1104	
RH-DY0003SEZZ	116	3311	504A	1104	
RH-EX0011CEZZ	143A	3062/142A	ZD-13	563	
RH-EX0012CEZZ	5082A	3064/146A	ZD-25		
RH-EX0013CEZZ	5078A	9023	ZD-19		
RH-EX0015CEZZ	5074A	3063/145A	ZD-12	563	
RH-EX0017CEZZ	142A	3093	ZD-13	563	
RH-EX0019CEZZ	5070A	9021	ZD-7.5		
RH-EX0019TAZZ	142A	3750/143A	ZD-13	563	
RH-EX0021TAZZ	5020A	3061/140A	ZD-10	562	
RH-EX0022CEZZ	5077A	3752	ZD-19		
RH-EX0024CEZZ	137A		ZD-6.2	561	
RH-EX0024TAZZ	135A	3056	ZD-5.1		
RH-EX0027CEZZ	5078A	9023			
RH-EX0033CEZZ	147A	3095	ZD-33		
RH-EX0034CEZZ	5081A	3155/1025	ZD-24		
RH-EX0037CEZZ	5079A	3335	ZD-20		
RH-EX0038CEZZ	142A	3062	ZD-12	563	
RH-EX0047CEZZ	142A	3062	ZD-12	563	
RH-EX0048CEZZ	5013A	3058/137A	ZD-6.2	561	
RH-EX0053CEZZ	5037A	3803	ZD-36		
RH-EX0054CEZZ	5097A	3347	ZD-120		
RH-EX0057CEZZ	5090A	3096/148A	ZD-55		
RH-EX0062CEZZ	506		504A	1104	
RH-EX0065CEZZ	5025A	3142	ZD-16	564	
RH-EX0072CEZZ	5079A		ZD-20		
RH-EX013CEZZ	5078A	9023	ZD-19		
RH-EX022CEZZ	5078A		ZD-19		
RH-IX0001PAZZ	1187	3742	IC-176		
RH-IX0001TAZZ	748	3236	IC-183		
RH-IX0004CE	749		IC-97		
RH-IX0004CEZZ	749	3168	IC-97		
RH-IX0005PAZZ	7476	7476			
RH-IX0012PAZZ		7404	7404		
RH-IX0014PAZZ		3591	VR-102		
RH-IX0017TAZZ	712		IC-148		
RH-IX0020CE	1004		IC-149		
RH-IX0020CEZZ	1004	3365	IC-149		
RH-IX0021CEZZ	1080		IC-98		
RH-IX0024CEZZ	1105	3285			
RH-IX0025CEZZ	1050		IC-133		
RH-IX0032CEZZ	1089*		IC-120		
RH-IX0037CEZZ	147A	3095	ZD-33		
RH-IX0038PAZZ	7406			1821	
RH-IX0039PAZZ	74107	74107			
RH-IX0040PAZZ	74121	74121			
RH-IX0041PAZZ	74123	74123		1817	
RH-IX0043CEZZ	712	3072	IC-148		
RH-IX0047CEZZ	1189	3705			
RH-IX0111CEZZ	977	3591/960			
RH-IX1020AFZZ	1115A		IC-279		
RH-IX1039AFZZ	1197	3733			
RH-IX1061AFZZ	1207	3713			
RH-VX0004TAZZ	601			1122	
RH-VX0009TAZZ	601	3463			
RH1		3843	511	1114	
RH120	123A	3122	20	2051	
RH1B		3312	511	1114	
RH1M	552	3998/506	504A	1104	
RH1Z	552			1104	
RHDX0033TAZZ	177		300	1122	
RHDX0043TAZZ	116		504A	1104	
RI20DC					MC4328L
RI4010658	519			1122	
RIN1102	722		IC-9		
RIV020					MR852
RIV040					MR854
RIV060					MR856
RJP60313	113A		6GC1		
RKZ12003	519		514	1122	
RKZ120101	519		514	1122	
RL005	125		510,531	1114	1N4933
RL010	125		510,531	1114	1N4934
RL020	125		510,531	1114	1N4935
RL040	125		510,531	1114	1N4936
RL060	125		510,531	1114	1N4937
RL080					MR817
RL100					MR818
RL10DC					MC4326L
RL10K					MC4326F
RL111DC					MC5111L
RL111K					MC5111F
RL113DC					MC5113L
RL11DC					MC4327L
RL11K					MC4327F
RL121DC					MC5121L
RL121K					MC5121F
RL123DC					MC5123L
RL12DC					MC4026L
RL13DC					MC4027L
RL20K					MC4328F
RL21DC					MC4329L
RL21K					MC4329F
RL22DC					MC4028L
RL232G	109	3087		1123	
RL23DC					MC4029L
RL246	109	3087		1123	
RL252	109	3091		1123	
RL30DC					MC4330L
RL30K					MC4330F
RL31	109	3087		1123	
RL31DC					MC4331L
RL31K					MC4331F
RL32	109	3087		1123	
RL32DC					MC4030L
RL32G	109	3087		1123	
RL33DC					MC4031L
RL34	109	3087		1123	
RL34G	109	3087		1123	
RL41	109	3016		1123	
RL41DC					MC4332L
RL41G	109	3087		1123	
RL41K					MC4332F
RL42	109	3091		1123	
RL43DC					MC4032L
RL52	109	3087		1123	
RL61DC					MC4335L
RL61K					MC4335F
RL63DC					MC4035L
RL709T	909		IC-249		
RL711T	911		IC-253		
RL71DC					MC4337L
RL71K					MC4337F
RL73DC					MC4037L
RL80DC					MC4304L
RL80K					MC4304F
RL81DC					MC4305L
RL81K					MC4305F
RL82DC					MC4004L
RL83DC					MC4005L
RLB-17	199	3245	62	2010	
RLF1G	116	3016	504A	1104	
RLH111	7410			1807	
RM-1	506		511		
RM-1V	116	3312	504A	1104	
RM-25	5089A	3341	ZD-51		
RM-26	5092A		504A	1104	
RM-2AV	125	3032A	510	1114	
RM-2C	125	3080	509	1114	
RM1	552	3843		1103	
RM1514DC					MC1514L
RM1537D					MC1537L
RM1A	116			1104	
RM1C				1114	
RM1Z	552	3311	504A	1104	
RM1ZM	116	3311	504A	1104	
RM1ZV	116		504A	1104	

Industry Standard No.	ECG	SK	GE	RS 276-	MOTOR.
RM2				1103	
RM25	5089A	3341	ZD-51		
RM257	5090A	3096/148A	ZD-55		
RM26	116	3343/5092A	ZD-51		
RM26(ZENER)	5089A		ZD-68		
RM2A	125	3032A		1104	
RM2C	125		509	1114	
RM2Z				1102	
RM4136D					MC3503L
RM4136J					MC3503L
RM4195T					MC1568G
RM4195TK					MC1568R
RM4558D					MC4558U
RM4558JG					MC4558U
RM4558L					MC4558G
RM4558T					MC4558G
RM55107AD					MC55107L
RM55325DD					MC55325L
RM702Q					MC1712F
RM702T					MC1721G
RM709D					MC1709L
RM709Q					MC1709F
RM709T					MC1709G
RM710D					MC1710L
RM710T					MC1710G
RM711DC					MC1711L
RM711T					MC1711G
RM723D					MC1723L
RM723T		3164			MC1723G
RM733D					MC1733L
RM733T					MC1733G
RM741D					MC1741L
RM741DP					MC1741P
RM741Q					MC1741F
RM741T					MC1741G
RM747D					MC1747L
RM747T					MC1747G
RM748T					MC1748G
RMC005					1N4933
RMC010					1N4934
RMC020					1N4935
RMC040					1N4936
RMC060					1N4937
RMC080					MR817
RMC100					MR818
RMVTC00921-3	114		6GD1	1104	
RN1015	5944		5048		
RN1120	5986	3609	5096		
RN2015	5944		5048		
RN5015	5952	3585			
RN515	5940		5048		
RO-3-9316A					MCM68A316E
RO-3-9316B					MCM68316E
RO-3-9316C					MCM68A316E
RO-3-9332C					MCM68A332
RO2A			510	1114	
RP-9A	139A	3060	ZD-9.1	562	
RQ-409S	116(2)	3311	504A	1104	
RQ-444S	229	3038	212	2010	
RR12766	717		IC-209		
RR7504	123A	3444	20	2051	
RR8068	123A	3444	20	2051	
RR8070	108	3452	86	2038	
RR8116	108	3452	86	2038	
RR8118	108	3452	86	2038	
RR8119	108	3452	86	2038	
RR8914	123A	3444	20	2051	
RR8989	108	3452	86	2038	
RR8999				2038	
RRB24-06	116	3017B/117	504A	1104	
RS-05	5800	9003		1142	
RS-101	160	3006	245	2004	
RS-102	158	3004/102A	53	2007	
RS-103	160	3006	245	2004	
RS-104	101	3861	8	2002	
RS-105	104	3009		2006	
RS-1055			16	2006	
RS-106	105	3012	4		
RS-107	123A	3444	20	2051	
RS-108	123A	3444	20	2051	
RS-109	108	3452	86	2038	
RS-110	159	3025/129	82	2032	
RS-1192	102A	3004	53	2007	
RS-1264	116	3016	504A	1104	
RS-1347	116	3017B/117	504A	1104	
RS-1524	103A	3010	59	2002	
RS-1536	101	3861	8	2002	
RS-1537	101	3861	8	2002	
RS-1538	101	3861	8	2002	
RS-1539	100	3005	1	2007	
RS-1540	102A	3004	53	2007	
RS-1541	102A	3004	53	2007	
RS-1542	102A	3004	53	2007	
RS-1543	102A	3004	53	2007	
RS-1544	102A	3004	53	2007	
RS-1545	101	3861	8	2002	
RS-1546	102A	3004	53	2007	
RS-1547	101	3861	8	2002	
RS-1548	102A	3004	53	2007	
RS-1550	100	3005	1	2007	
RS-1553	101	3861	8	2002	
RS-1554	126	3008	52	2024	
RS-1555	102A	3004	53	2007	
RS-2001	101	3861	8	2002	
RS-2002	160	3006	245	2004	
RS-2003	126	3006/160	52	2024	
RS-2004	102	3004	2	2007	
RS-2005	102	3004	2	2007	
RS-2006	104	3009	16	2006	
RS-2007	158	3004/102A	53	2007	
RS-2008	154	3044	40	2012	
RS-2009	123A	3444	20	2051	
RS-2011	161	3132	39	2015	
RS-2013	123A	3444	20	2051	
RS-2014	128	3024	243	2030	
RS-2015	108	3452	86	2038	
RS-2016	123A	3444	20	2051	
RS-2017	184	3054/196	57	2017	
RS-2018	210		252	2018	
RS-2019	182	3188A	55		
RS-2020	184	3054/196	57	2017	
RS-2021	159	3025/129	82	2032	
RS-2022	159	3025/129	82	2032	
RS-2023	159	3025/129	82	2032	
RS-2024	159	3025/129	82	2032	
RS-2025	185	3083/197	58	2025	
RS-2026	211		253	2026	
RS-2027	183	3189A	56	2027	
RS-2028	312	3448	FET-1	2035	
RS-2029	6409			2029	
RS-2350	102A	3004	53	2007	
RS-2351	102A	3004	53	2007	
RS-2352	102A	3004	53	2007	
RS-2353	102A	3004	53	2007	

Industry Standard No.	ECG	SK	GE	RS 276-	MOTOR.
RS-2354	102A	3004	53	2007	
RS-2355	102A	3004	53	2007	
RS-2359	101	3861	8	2002	
RS-2360	101	3861	8	2002	
RS-2364	101	3861	8	2002	
RS-2365	101	3861	8	2002	
RS-2366	101	3861	8	2002	
RS-2367	102A	3004	53	2007	
RS-2373	101	3861	8	2002	
RS-2374	101	3861	8	2002	
RS-2375	101	3861	8	2002	
RS-253C			300	1122	
RS-253S		3100	300	1122	
RS-2675	102A	3004	53	2007	
RS-2677	102A	3004	53	2007	
RS-267S		3100	300	1122	
RS-2683	100	3005	1	2007	
RS-2684	100	3005	1	2007	
RS-2685	100	3005	1	2007	
RS-2686	100	3005	1	2007	
RS-2687	100	3005	1	2007	
RS-2688	100	3005	1	2007	
RS-2689	102A	3004	53	2007	
RS-2690	100	3005	1	2007	
RS-2691	100	3005	1	2007	
RS-2692	100	3005	1	2007	
RS-2694	100	3005	1	2007	
RS-2695	100	3005	1	2007	
RS-2696	100	3005	1	2007	
RS-2697	102A	3004	53	2007	
RS-272US		3100	300	1122	
RS-279US	199	3124/289	62	2010	
RS-280S		3100	300	1122	
RS-3275	102A	3004	53	2007	
RS-3276	102A	3004	53	2007	
RS-3277	100	3005	1	2007	
RS-3278	100	3005	1	2007	
RS-3279	100	3005	1	2007	
RS-3282	102A	3004	53	2007	
RS-3283	102A	3004	53	2007	
RS-3284	102A	3004	53	2007	
RS-3285	102A	3004	53	2007	
RS-3286	102A	3004	53	2007	
RS-3288	100	3005	1	2007	
RS-3289	102A	3004	53	2007	
RS-3299	102A	3004	53	2007	
RS-3301	102A	3004	53	2007	
RS-3308	102A	3004	53	2007	
RS-3309	100	3005	1	2007	
RS-3310	102A	3004	53	2007	
RS-3316	102A	3004	53	2007	
RS-3316-1	102A	3004	53	2007	
RS-3316-2	102A	3004	53	2007	
RS-3318	102A	3004	53	2007	
RS-3322	126	3008	52	2024	
RS-3323	126	3008	52	2024	
RS-3324	126	3008	52	2024	
RS-35	147A	3095	ZD-33		
RS-35/AW01-33	147A		ZD-33		
RS-3570	116	3016	504A	1104	
RS-3727	116	3016	504A	1104	
RS-3858-1	121	3009	239	2006	
RS-3862	126	3008	52	2024	
RS-3863	126	3008	52	2024	
RS-3866	126	3008	52	2024	
RS-3867	100	3005	1	2007	
RS-3868	100	3005	1	2007	
RS-3892	160	3006	245	2004	
RS-3898	160	3006	245	2004	
RS-3900	160	3006	245	2004	
RS-3901			50	2004	
RS-3902	160	3006	245	2004	
RS-3903	160	3006	245	2004	
RS-3904	102A	3004	53	2007	
RS-3907	100	3005	1	2007	
RS-3911	160	3006	245	2004	
RS-3913	100	3005	1	2007	
RS-3914			1	2007	
RS-3915			1	2007	
RS-3925	102A	3004	53	2007	
RS-3926			50	2004	
RS-3929	100	3005	1	2007	
RS-406	102A	3004	53	2007	
RS-5008	102A	3004	53	2007	
RS-5104	100	3005	1	2007	
RS-5105	100	3005	1	2007	
RS-5106	100	3005	1	2007	
RS-5107	126	3008	52	2024	
RS-5108	126	3008	52	2024	
RS-5109	126	3008	52	2024	
RS-5201	126	3008	52	2024	
RS-5205	126	3008	52	2024	
RS-5206	102A	3004	53	2007	
RS-5207	126	3008	52	2024	
RS-5208	160	3006	245	2004	
RS-5209	126	3008	52	2024	
RS-5301	126	3008	52	2024	
RS-5305	126	3008	52	2024	
RS-5306	126	3008	52	2024	
RS-5311	102A	3004	53	2007	
RS-5312	126	3008	52	2024	
RS-5313	126	3008	52	2024	
RS-5401	100	3005	1	2007	
RS-5406	102A	3004	53	2007	
RS-5502	102A	3004	53	2007	
RS-5504	100	3005	1	2007	
RS-5505	102A	3004	53	2007	
RS-5506	102A	3004	53	2007	
RS-5511	100	3005	1	2007	
RS-5530	102A	3004	53	2007	
RS-5531	102A	3004	53	2007	
RS-5532	102A	3004	53	2007	
RS-5533	102A	3004	53	2007	
RS-5534	102A	3004	53	2007	
RS-5535	102A	3004	53	2007	
RS-5536	102A	3004	53	2007	
RS-5540	100	3005	1	2007	
RS-5541	102A	3004	53	2007	
RS-5542	102A	3004	53	2007	
RS-5544	102A	3004	53	2007	
RS-5551	102A	3004	53	2007	
RS-5552	102A	3004	53	2007	
RS-5553	102A	3004	53	2007	
RS-5554	102A	3004	53	2007	
RS-5555	102A	3004	53	2007	
RS-5556	102A	3004	53	2007	
RS-5557	102A	3004	53	2007	
RS-5558	102A	3004	53	2007	
RS-5602	102A	3004	53	2007	
RS-5613	121	3009	239	2006	
RS-5704	102A	3004	53	2007	
RS-57042	102A	3004	53	2007	
RS-57062	102A	3004	53	2007	
RS-5708	102A	3004	53	2007	

Industry Standard No.	ECG	SK	GE	RS 276-	MOTOR.
RS-5708-2	102A	3004	53	2007	
RS-5709	102A	3004	53	2007	
RS-5711	102A	3004	53	2007	
RS-5717	102A	3004	53	2007	
RS-5717-3	102A	3004	53	2007	
RS-5717-6	102A	3004	53	2007	
RS-5720	102A	3004	53	2007	
RS-5731	102A	3004	53	2007	
RS-5733	102A	3004	53	2007	
RS-5734	102A	3004	53	2007	
RS-5735	176	3845	80		
RS-5736	102A	3004	53	2007	
RS-5737	102A	3004	53	2007	
RS-5742	102A	3004	53	2007	
RS-5743	102A	3004	53	2007	
RS-5743-1	102A		53	2007	
RS-5743-2	102A	3004	53	2007	
RS-5743-3	102A	3004	53	2007	
RS-5744	102A	3004	53	2007	
RS-5744-3	102A	3004	53	2007	
RS-5749	102A	3004	53	2007	
RS-5752	126	3008	52	2024	
RS-5753	126	3008	52	2024	
RS-5753-2			2	2007	
RS-5754	126	3008	52	2024	
RS-5755	126	3008	52	2024	
RS-5756	126	3008	52	2024	
RS-5757	126	3008	52	2024	
RS-5758	126	3008	52	2024	
RS-5759	126	3008	52	2024	
RS-5760	126	3008	52	2024	
RS-5761	126	3008	52	2024	
RS-5762	126	3008	52	2024	
RS-5802	126	3008	52	2024	
RS-5818	126	3008	52	2024	
RS-5835	121	3009	239	2006	
RS-5851	123A	3124/289	20	2051	
RS-5852	102A	3004	53	2007	
RS-5853	123A	3124/289	20	2051	
RS-5854	102A	3004	53	2007	
RS-5855	121	3009	239	2006	
RS-5856	123A	3124/289	20	2051	
RS-5857	123A	3124/289	20	2051	
RS-6344	116	3016	504A	1104	
RS-6461	116	3016	504A	1104	
RS-6471	116	3016	504A	1104	
RS-684	126	3008	52	2024	
RS-685	126	3008	52	2024	
RS-686	102A	3004	53	2007	
RS-687	102A	3004	53	2007	
RS-7102	108	3018	86	2038	
RS-7103	123A	3124/289	20	2051	
RS-7104	108	3018	86	2038	
RS-7105	123A	3124/289	20	2051	
RS-7106	108	3018	86	2038	
RS-7107	108	3018	86	2038	
RS-7108	108	3018	86	2038	
RS-7109	108	3018	86	2038	
RS-7110	108	3018	86	2038	
RS-7113	108	3452	86	2038	
RS-7114	108	3452	86	2038	
RS-7115	108	3452	86	2038	
RS-7124			20	2051	
RS-7127			17	2051	
RS-7129			20	2051	
RS-7201			11	2015	
RS-7202	108	3452	86	2038	
RS-7212			11	2015	
RS-7315	124	3021	12		
RS-7316	124	3021	12		
RS-7317	124	3021	12		
RS-7318	124	3021	12		
RS-7409	123A	3020/123	20	2051	
RS-7411	123A	3020/123	20	2051	
RS-7412	123A	3020/123	20	2051	
RS-7413	123A	3020/123	20	2051	
RS-7504	123A	3020/123	20	2051	
RS-7511	123A	3452/108	20	2051	
RS-7512	108	3452	86	2038	
RS-7606	123A	3020/123	20	2051	
RS-7607	123A	3020/123	20	2051	
RS-7609	123A	3020/123	20	2051	
RS-7610	123A	3020/123	20	2051	
RS-7611	123A	3020/123	20	2051	
RS-7612	123A	3020/123	20	2051	
RS-7613	123A	3020/123	20	2051	
RS-7614	123A	3020/123	20	2051	
RS-7622	123A	3020/123	20	2051	
RS-7623	123A	3020/123	20	2051	
RS-7665A	159		82	2032	
RS-805US	199	3122	61	2038	
RS-8628	1074		IC-60		
RS10	116	3017B/117	504A	1104	
RS1049	123A	3444	20	2051	
RS1059	123A	3444	20	2051	
RS1177A	525	3925			
RS1192	102A		53	2007	
RS1234	117		504A	1104	
RS1264	116		504A	1104	
RS128	123A	3122	20	2051	
RS1290	139A	3058/137A	ZD-9.1	562	
RS1296(SEARS)	506		511	1114	
RS132	128	3035	243	2030	
RS1348	142A	3062	ZD-12	563	
RS136	123A	3122	20	2051	
RS1428	177	3100/519	300	1122	
RS15048	123A	3444	20	2051	
RS1513	101	3011	8	2002	
RS1524	103A		59	2002	
RS1530	101	3011	8	2002	
RS1531	101		8	2002	
RS1532	101	3011	8	2002	
RS1533	103A	3835	59	2002	
RS1534	101	3011	8	2002	
RS1536	101		8	2002	
RS1537	101		8	2002	
RS1538	101		8	2002	
RS1539	160	3123	245	2004	
RS1540	102A	3123	53	2007	
RS1541	102A		53	2007	
RS1542	102A	3017B/117	53	1104	
RS1543	102A	3123	53	2007	
RS1544	102A	3123	53	2007	
RS1545	102A		53	2007	
RS1546	102A	3123	53	2007	
RS1547	101		8	2002	
RS1548	102A	3123	53	2007	
RS1549	103A	3835	59	2002	
RS1550	160	3123	245	2004	
RS1553	101		8	2002	
RS1554	160	3008	245	2004	
RS1555	102A	3123	53	2007	
RS1720	116		504A	1104	
RS1726	108	3018	86	2038	

Industry Standard No.	ECG	SK	GE	RS 276-	MOTOR.
RS1749	116		504A	1104	
RS1805	116	3017B/117	504A	1104	
RS1811	109	3091		1123	
RS1823	117		504A	1104	
RS1832	116	3017B/117	504A	1104	
RS1FM-12			504A	1104	
RS2003	102A			2005	
RS2018	152	3202/210			
RS2026	153	3203/211			
RS220AF	116	3016	504A	1104	
RS230AF	116	3016	504A	1104	
RS2350	102A	3123	52	2007	
RS2351	102A	3123	53	2007	
RS2352	102A	3123	53	2007	
RS2353	102A	3123	53	2007	
RS2354	102A	3123	53	2007	
RS2355	102A	3123	53	2007	
RS2356	101	3011	8	2002	
RS2359	101		8	2002	
RS2360	101		8	2002	
RS2364	101		8	2002	
RS2365	101		8	2002	
RS2366	101		8	2002	
RS2367	102A	3123	53	2007	
RS2373	102A		53	2007	
RS2374	102A		53	2007	
RS2375	101		8	2002	
RS2675	102A	3123	53	2007	
RS2677	102A	3123	53	2007	
RS2679	160	3123	245	2004	
RS2680	160	3123	245	2004	
RS2683	160	3123	245	2004	
RS2684	160	3123	245	2004	
RS2685	160	3123	245	2004	
RS2686	160	3123	245	2004	
RS2687	160	3123	245	2004	
RS2688	160	3123	245	2004	
RS2689	102A	3123	53	2007	
RS2690	100	3721	1	2007	
RS2691	100	3721	1	2007	
RS2692	100	3721	1	2007	
RS2694	160	3123	245	2004	
RS2695	160	3123	245	2004	
RS2696	100	3123	1	2007	
RS2697	102A	3123	53	2007	
RS276-1804	7441	7441			
RS276-1830	309K	3629			
RS2801	109	3087		1123	
RS2867	102A		53	2007	
RS2914	123A		20	2051	
RS3211	102A	3123	53	2007	
RS322				2007	
RS3275	102A	3123	53	2007	
RS3276	102	3123	2	2007	
RS3277	160	3123	245	2004	
RS3278	160	3123	245	2004	
RS3279	160	3123	245	2004	
RS3280	102A	3123	53	2007	
RS3281	100	3005	1	2007	
RS3282	102A	3123	53	2007	
RS3283	102A	3123	53	2007	
RS3284	102A		53	2007	
RS3285	102A	3123	53	2007	
RS3286	102A	3123	53	2007	
RS3287	100	3005	1	2007	
RS3288	160	3123	245	2004	
RS3289	102A		53	2007	
RS3293	102A		53	2007	
RS3299	102A	3123	53	2007	
RS3301	102A	3123	53	2007	
RS3306	101	3011	8	2002	
RS3308	102A		53	2007	
RS3309	160	3123	245	2004	
RS3310	102A	3003	53	2007	
RS3316	102A		53	2007	
RS3316-1	102A		53	2007	
RS3316-2	102A		53	2007	
RS3318	102A		53	2007	
RS3322	160		245	2004	
RS3323	160		245	2004	
RS3324	160		245	2004	
RS3358-1	121	3009	239	2006	
RS3359-1	121	3009	239	2006	
RS3570	116		504A	1104	
RS3668	160		245	2004	
RS3717	102A	3004	53	2007	
RS3726	102A		53	2007	
RS3727	116		504A	1104	
RS3857	102A		53	2007	
RS3858	121	3009	239	2006	
RS3858-1	121	3009	239	2006	
RS3862	160		245	2004	
RS3863	160		245	2004	
RS3864	160		245	2004	
RS3866	102A		53	2007	
RS3867	158		53	2007	
RS3868	160		245	2004	
RS3880	102A	3004	53	2007	
RS3892	100	3721	1	2007	
RS3897	102A	3123	53	2007	
RS3898	160	3123	245	2004	
RS3900	160		245	2004	
RS3901	160		245	2004	
RS3902	160		245	2004	
RS3903	160		245	2004	
RS3904	102A		53	2007	
RS3905	160	3006	245	2004	
RS3906	160	3123	245	2004	
RS3907	160		245	2004	
RS3911	160		245	2004	
RS3912	160	3006	245	2004	
RS3913	102A		53	2007	
RS3914	100	3005	1	2007	
RS3915	100	3005	1	2007	
RS3925	102A		53	2007	
RS3926	102A		53	2007	
RS3929	160		245	2004	
RS3931	103A	3835	59	2002	
RS3959	121	3009	239	2006	
RS3959-1	121	3014	239	2006	
RS3986	160	3006	245	2004	
RS3995	160	3006	245	2004	
RS4001		4001	4001	2401	
RS4011		4011	4011	2411	
RS4013		4013	4013	2413	
RS4017		4017	4017	2417	
RS4020		4020	4020	2420	
RS4027		4027	4027	2427	
RS4049		4049	4049	2449	
RS4050		4050	4050	2450	
RS406	102A	3123	53	2007	
RS4518			4518	2490	
RS5008	102A		53	2007	
RS5101	160		245	2004	

Industry Standard No.	ECG	SK	GE	RS 276-	MOTOR.
RS5102	102A		53	2007	
RS5103	102A		53	2007	
RS5104	100	3123	1	2007	
RS5105	100	3721	1	2007	
RS5106	160	3123	245	2004	
RS5107	160		245	2004	
RS5108	160	3123	245	2004	
RS5109	160	3008	245	2004	
RS5201	160		245	2004	
RS5202	102A		53	2007	
RS5203	102A		53	2007	
RS5204	160		245	2004	
RS5205	160		245	2004	
RS5206	160	3004/102A	245	2004	
RS5207	160	3008	245	2004	
RS5208	160	3123	245	2004	
RS5209	160		245	2005	
RS5243-2	102A		53	2007	
RS5301	160		245	2004	
RS5302	100	3005	1	2007	
RS5303	100	3005	1	2007	
RS5305	160		245	2004	
RS5306	160		245	2004	
RS5311	160	3004/102A	245	2004	
RS5312	160	3123	245	2004	
RS5313	160	3123	245	2004	
RS5314	160	3008	245	2004	
RS5317			51	2004	
RS5401	102A	3123	53	2007	
RS5402	100	3005	1	2007	
RS5403	100	3005	1	2007	
RS5406	102A	3003	53	2007	
RS5502	102A	3123	53	2007	
RS5503	102A	3123	53	2007	
RS5504	100	3123	1	2007	
RS5505	102A	3123	53	2007	
RS5506	102A	3123	53	2007	
RS5507	102A	3123	53	2007	
RS5511	100	3721	1	2007	
RS5530	102A	3123	53	2007	
RS5531	102A		53	2007	
RS5532	102A	3123	53	2007	
RS5533	102A	3004	53	2007	
RS5534	102A	3004	53	2007	
RS5535	102A	3123	53	2007	
RS5536	102A	3004	53	2007	
RS5540	100	3721	1	2007	
RS5541	102A	3004	53	2007	
RS5542	102A	3123	53	2007	
RS5543	102A	3004	53	2007	
RS5544	102A	3123	53	2007	
RS5545	102A	3004	53	2007	
RS5551	102A		53	2007	
RS5552	102A		53	2007	
RS5553	102A		53	2007	
RS5554	102A		53	2007	
RS5555	102A		53	2007	
RS5556	102A		53	2007	
RS5557	102A	3722/102	53	2007	
RS5558	102A		53	2007	
RS5563	102A	3123	53	2007	
RS5564	102A	3123	53	2007	
RS5565	102A	3123	53	2007	
RS5566	102A	3123	53	2007	
RS5567	102A	3123	53	2007	
RS5568	102A	3123	53	2007	
RS5602	102A		53	2007	
RS5603	102A		53	2007	
RS5605	102A		53	2007	
RS5607	102A	3123	53	2007	
RS5608	102A	3123	53	2007	
RS5610	102A	3123	53	2007	
RS5612	121	3009	239	2006	
RS5613	121	3009	239	2006	
RS5614	121	3009	239	2006	
RS5616	121	3009	239	2006	
RS5704	102A	3123	53	2007	
RS5704-2	102A		53	2007	
RS57042	102A		53	2007	
RS57062	102A		53	2007	
RS5708	102A		53	2007	
RS5708-2	102A		53	2007	
RS5709	102A		53	2007	
RS5711	102A		53	2007	
RS5715	176	3123	80		
RS5715-1	176	3123	80		
RS5717	102A	3004	53	2007	
RS5717-1	102A		53	2007	
RS5717-3	102A		53	2007	
RS5717-6	102A		53	2007	
RS5720	102A		53	2007	
RS5731	102A	3004	53	2007	
RS5732	102A	3004	53	2007	
RS5733	102A	3123	53	2007	
RS5734	102A	3123	53	2007	
RS5735	102A	3845/176	53	2007	
RS5736	102A	3123	53	2007	
RS5737	102A	3004	53	2007	
RS5738	102A		53	2007	
RS5740	102A		53	2007	
RS5740-1	102A	3123	53	2007	
RS5742	102A		53	2007	
RS5743	102A	3123	53	2007	
RS5743-1	102A	3004	53	2007	
RS5743-2	102A		53	2007	
RS5743-3	102A		53	2007	
RS5743.3	100	3721	1	2007	
RS57433	102A		53	2007	
RS5744	102A		53	2007	
RS5744-3	102A		53	2007	
RS5745	102A	3004	53	2007	
RS5746	102A	3004	53	2007	
RS5747	102A	3004	53	2007	
RS5748	102A	3004	53	2007	
RS5749	102A	3123	53	2007	
RS5750	102A	3004	53	2007	
RS5751	102A	3004	53	2007	
RS5752	102A		53	2007	
RS5753	126		52	2024	
RS5753-2	126	3006/160	52	2024	
RS5754	126		52	2024	
RS5755	126		52	2024	
RS5756	126		52	2024	
RS5757	126		52	2024	
RS5758	126		52	2024	
RS5759	126		52	2024	
RS5760	126		52	2024	
RS5761	126		52	2024	
RS5762	126		52	2024	
RS5765	102A	3123	53	2007	
RS5766	102A	3004	53	2007	
RS5767	102A	3004	53	2007	
RS5768	102A	3123	53	2007	
RS5788	176	3845	80		

Industry Standard No.	ECG	SK	GE	RS 276-	MOTOR.
RS5802	126		52	2024	
RS5818	160		245	2004	
RS5825	102	3722	2	2007	
RS5835	121	3717	239	2006	
RS5851	123A	3444	20	2051	
RS5852	102A	3123	53	2007	
RS5853	123A	3444	20	2051	
RS5854	102A	3123	53	2007	
RS5855	121MP	3718	239(2)	2006(2)	
RS5856	123A	3444	20	2051	
RS5857	123A	3444	20	2051	
RS593	160	3123	245	2004	
RS6344	116		504A	1104	
RS6461	116		504A	1104	
RS6471	116	3031A	504A	1104	
RS6523	108		86	2038	
RS6705	116	3016	504A	1104	
RS6821	126	3006/160	52	2024	
RS6822	126	3006/160	52	2024	
RS6824	100	3005	1	2007	
RS684	160	3123	245	2004	
RS6840	102A	3004	53	2007	
RS6843	102	3004	2	2007	
RS6843A	102A	3004	53	2007	
RS6846	102	3004	2	2007	
RS6846A	102A	3004	53	2007	
RS685	160	3123	245	2004	
RS686	160	3123	245	2004	
RS687	160	3123	245	2004	
RS7101	108	3452	86	2038	
RS7102	108		86	2038	
RS7103	123A	3444	20	2051	
RS7104	108		86	2038	
RS7105	123A	3444	20	2051	
RS7106	108		86	2038	
RS7107	108		86	2038	
RS7108	123A	3452/108	20	2051	
RS7109	108		86	2038	
RS7110	108		86	2038	
RS7111	123A	3444	20	2051	
RS7112	108	3018	86	2038	
RS7113	108		86	2038	
RS7114	229	3018	61	2038	
RS7115	108		86	2038	
RS7116	108	3452	86	2038	
RS7117	108	3452	86	2038	
RS7118	108	3452	86	2038	
RS7119	108	3452	86	2038	
RS7120	108	3452	86	2038	
RS7121	123A	3444	20	2051	
RS7122	108	3452	86	2038	
RS7123	108	3452	86	2038	
RS7124	108	3452	86	2038	
RS7125	108	3452	86	2038	
RS7126	108	3452	86	2038	
RS7127	123A	3444	20	2051	
RS7128	108	3452	86	2038	
RS7129	123A	3444	20	2051	
RS7132	123A	3444	20	2051	
RS7133	123A	3444	20	2051	
RS7135	108	3452	86	2038	
RS7136	123A	3444	20	2051	
RS7138	108	3452	86	2038	
RS7139	108	3452	86	2038	
RS7140	108	3452	86	2038	
RS7141	108	3452	86	2038	
RS7142	108	3452	86	2038	
RS7143	107	3018	11	2015	
RS7144	108	3452	86	2038	
RS7145	108	3452	86	2038	
RS7160	123A	3444	20	2051	
RS7161	108	3452	86	2038	
RS7162	108	3452	86	2038	
RS7163	108	3452	86	2038	
RS7164	108	3452	86	2038	
RS7165	108	3452	86		
RS7166	108	3452	86	2038	
RS7167	108	3452	86	2038	
RS7168	108	3452	53	2038	
RS7169	108	3452	86	2038	
RS7170	108	3452	86	2038	
RS7173	229	3018	61	2038	
RS7174	108	3452	86	2038	
RS7175	108	3452	86	2038	
RS7176	108	3452	86	2038	
RS7177	108	3452	86	2038	
RS7201	108	3452	86	2038	
RS7202	108		86	2038	
RS7209	108	3452	86	2038	
RS7210	108	3452	86	2038	
RS7211	108	3452	86	2038	
RS7212	108	3452	86	2038	
RS7214	108	3452	86	2038	
RS7215	108	3452	86	2038	
RS7216	108	3452	86	2038	
RS7217	108	3452	86	2038	
RS7218	108	3452	86	2038	
RS7219	108	3452		2038	
RS7220	108	3452	86	2038	
RS7221	108	3452	86	2038	
RS7222	107	3018	11	2015	
RS7223	123A	3444	20	2051	
RS7224	123A	3444	20	2051	
RS7225	108	3452	86	2038	
RS7226	123A	3444	20	2051	
RS7227	108	3452	86	2038	
RS7228	108	3452	86	2038	
RS7229	108	3452	86	2038	
RS7230	108	3452	86	2038	
RS7231	108	3452	86	2038	
RS7232	123A	3444	20	2051	
RS7233	229	3018	61	2038	
RS7234	123A	3444	20	2051	
RS7235	123A	3444	20	2051	
RS7236	123A	3444	20	2051	
RS7237	108	3452	86	2038	
RS7238		3018	20	2051	
RS7241	123A	3444	20	2051	
RS7242	123A	3444	20	2051	
RS7310	124	3021	12		
RS7311	124	3021	12		
RS7312	124	3021	12		
RS7313	124	3021	12		
RS7315	124	3021	12		
RS7316	124	3021	12		
RS7317	124	3021	12		
RS7318	124	3021	12		
RS7320	124	3021	12		
RS7321	124	3021	12		
RS7327	124	3021	12		
RS7328	124	3021	12		
RS7329	124	3021	12		
RS7330	124	3021	12		
RS7333	108	3452	86	2038	

Industry Standard No.	ECG	SK	GE	RS 276-	MOTOR.
RS7334	108	3452	86	2038	
RS7365	124	3021	12		
RS7366	124	3021	12		
RS7367	124	3021	12		
RS7368	124	3021	12		
RS7400			7400	1801	
RS7402			7402	1811	
RS7404			7404	1802	
RS7405	123A	3444	20	2051	
RS7406	123A	3444	20	2051	
RS7406(IC)			7406	1821	
RS7407	123A	3444	20	2051	
RS7408	123A	3444	20	2051	
RS7408(IC)			7408	1822	
RS7409	123A	3444	20	2051	
RS7410	123A	3444	20	2051	
RS7410(IC)			7410	1807	
RS7411	123A	3444	20	2051	
RS7412	123A	3444	20	2051	
RS74123			74123	1817	
RS7413	123A	3444	20	2051	
RS7413(IC)		7413	7413	1815	
RS74145		74145	74145	1828	
RS7415	123A	3444	20	2051	
RS74150			74150	1829	
RS74154		74154	74154	1834	
RS74192			74192	1831	
RS74193			74193	1820	
RS74196		74196	74196	1833	
RS7420			7420	1809	
RS7421	123A	3444	20	2051	
RS7427			7427	1823	
RS7432			7432	1824	
RS7441			7441	1804	
RS7447			7447	1805	
RS7448			7448	1816	
RS7451			7451	1825	
RS7473			7473	1803	
RS7474			7474	1818	
RS7475			7475	1806	
RS7476			7476	1813	
RS7485			7485	1826	
RS7486			7486	1827	
RS7490			7490	1808	
RS7492			7492	1819	
RS7504	123A	3444	20	2051	
RS7510	123A	3444	20	2051	
RS7511	108	3444/123A	86	2038	
RS7512	108		86	2038	
RS7513	123A	3444	20	2051	
RS7513-15	123A	3444	20	2051	
RS7514	123A	3444	20	2051	
RS7515	123A	3444	20	2051	
RS7516	123A	3444	20	2051	
RS7517	123A	3444	20	2051	
RS7517-19	123A	3444	20	2051	
RS7518	123A	3020/123	20	2051	
RS7519	123A	3020/123	20	2051	
RS7520	108	3452	86	2038	
RS7521	123A	3444	20	2051	
RS7522	108	3452	86	2038	
RS7523	107	3018	11	2015	
RS7524	108	3452	86	2038	
RS7525	123A	3444	20	2051	
RS7526	123A	3444	20	2051	
RS7527	123A	3444	20	2051	
RS7528	123A	3444	20	2051	
RS7529	123A	3444	20	2051	
RS7530	123A	3444	20	2051	
RS7532	108	3452	86	2038	
RS7533	108	3452	86	2038	
RS7542	123A	3444	20	2051	
RS7543	123A	3444	20	2051	
RS7544	123A	3444	20	2051	
RS7555	123A	3444	20	2051	
RS7568	102A	3004	53	2007	
RS7606	123A	3444	20	2051	
RS7607	123A	3444	20	2051	
RS7609	123A	3444	20	2051	
RS7610	123A	3444	20	2051	
RS7611	123A	3444	20	2051	
RS7612	123A	3444	20	2051	
RS7613	123A	3444	20	2051	
RS7614	123A	3444	20	2051	
RS7620	123A	3444	20	2051	
RS7621	123A	3444	20	2051	
RS7622	123A	3444	20	2051	
RS7623	123A	3444	20	2051	
RS7624	123A	3444	20	2051	
RS7625	123A	3444	20	2051	
RS7626	123A	3444	20	2051	
RS7627	123A	3444	20	2051	
RS7628	123A	3444	20	2051	
RS7634	123A	3444	20	2051	
RS7635	123A	3444	20	2051	
RS7636	123A	3444	20	2051	
RS7637	123A	3444	20	2051	
RS7638	123A	3444	20	2051	
RS7639	123A	3444	20	2051	
RS7640	123A	3444	20	2051	
RS7641	123A	3444	20	2051	
RS7642	123A	3444	20	2051	
RS7643	123A	3444	20	2051	
RS7665	159	3114/290	82	2032	
RS7672	128		243	2030	
RS7678	128	3024	243	2030	
RS7814	123A	3444	20	2051	
RS7916	312	3448	FET-1	2035	
RS8100	129	3025	244	2027	
RS8101	128	3024	243	2030	
RS8102	129	3025	244	2027	
RS8103	128	3024	243	2030	
RS8104	129	3025	244	2027	
RS8105	128	3024	243	2030	
RS8106	129	3025	244	2027	
RS8107	128	3024	243	2030	
RS8108	129	3025	244	2027	
RS8109	128	3024	243	2030	
RS8110	129	3025	244	2027	
RS8111	128	3024	243	2030	
RS8112	129	3025	244	2027	
RS8113	128	3024	243	2030	
RS8406	158	3004/102A	53	2007	
RS8407	103A	3010	59	2002	
RS8420	103A	3835	59	2002	
RS8421	102A		53	2007	
RS8424	158		53	2007	
RS8430	116		504A	1104	
RS8441	103A	3010	59	2002	
RS8442	123A	3444	20	2051	
RS8443	103A	3835	59	2002	
RS8444	102A		53	2007	
RS8445	103A	3010	59	2002	
RS8446	158	3004/102A	53	2007	

Industry Standard No.	ECG	SK	GE	RS 276-	MOTOR.
RS8503	123A	3444	20	2051	
RS86057332	123A	3444	20	2051	
RS9510	107		11	2015	
RS9511	107		11	2015	
RS9512	107		11	2015	
RSI350			53	2007	
RSKK36	176	3123	80		
RSLNA0004CEZZ	120	3110	CR-3		
RSLNB0001CEZZ	113A	3119/113	6GC1		
RSLND0003CEZZ	506	3130	511	1114	
RSWY/16L7Z			1N295	1123	
RT-100	123A		20	2051	
RT-101			48	2032	
RT-102			20	2051	
RT-103			82	2032	
RT-104		3245	212	2010	
RT-105		3245	212	2010	
RT-107		3245	212	2010	
RT-108		3245	212	2010	
RT-109		3244	220		
RT-110		3044	235	2012	
RT-112		3245	212	2010	
RT-113			39	2015	
RT-114		3024	18	2030	
RT-115		3118	21	2034	
RT-118		3275	221		
RT-119		3011A	8	2002	
RT-120		3275	221		
RT-121		3114	67	2023	
RT-122				2002	
RT-124			16	2006	
RT-126			21	2034	
RT-127			16	2006	
RT-131			14	2041	
RT-133				2048	
RT-135		3220	32		
RT-141	128		243	2030	
RT-148			74	2043	
RT-149			75	2041	
RT-150			66	2048	
RT-151			69	2049	
RT-152			28	2017	
RT-154	130		14	2041	
RT-155			69	2049	
RT-157				2026	
RT-1689			1N60	1123	
RT-175			FET-2	2035	
RT-176			FET-1	2028	
RT-180				2036	
RT-185	102A		53	2007	
RT-187				2014	
RT-188	128	3024	243	2030	
RT-200				1123	
RT-2016	109			1123	
RT-210			510	1114	
RT-212			504A	1104	
RT-213			504A	1104	
RT-214			504A	1104	
RT-215		3017B	504A	1104	
RT-218			300	1122	
RT-230		3647		1172	
RT-236				561	
RT-237				561	
RT-238				561	
RT-240		3060	ZD-9.1	562	
RT-241		3061	ZD-10	562	
RT-242		3139	ZD-11	563	
RT-243		3062	ZD-12	563	
RT-244			ZD-13	563	
RT-245		3063	ZD-15	564	
RT-246		3146	ZD-16	564	
RT-250		3095	ZD-33		
RT-257				562	
RT-2669	116	3031A	504A	1104	
RT-3858	116	3017B/117	504A	1104	
RT-4232	116	3311	504A	1104	
RT-4625	102A	3004	53	2007	
RT-4764		3031A	512		
RT-61012	190		217		
RT-61014	126	3008	52	2024	
RT-61015	158	3004/102A	53	2007	
RT-61016	158	3004/102A	53	2007	
RT-6922		3056	ZD-5.1		
RT-8667			FET-2	2035	
RT-929-H	123A		20	2051	
RT-929H	123A	3124/289	20	2051	
RT-930H	123A	3452/108	20	2051	
RT05					1N3889
RT10					1N3890
RT100	123A	3444	20	2051	
RT1008	109	3087		1123	
RT1106	109	3087		1123	
RT1108	109	3087		1123	
RT1116	171	3201	27		
RT112	107		11	2015	
RT114	123A	3444	20	2051	
RT115		3118	21	2034	
RT1184	109	3087		1123	
RT121	158	3004/102A	53	2007	
RT1306	136A		ZD-5.6	561	
RT1306(G.E.)	136A		ZD-5.6	561	
RT141	128	3024	243	2030	
RT150		3054	66	2048	
RT151		3083	69	2049	
RT152		3054	28	2017	
RT154	128	3024	243	2030	
RT155		3083	69	2049	
RT1595	116		504A	1104	
RT1667	110MP	3088	1N60	1123	
RT1669	177		300	1122	
RT1686	506		511	1114	
RT1689	177	3100/519	504A	1104	
RT175	312	3112	FET-1	2035	
RT176		3112	FET-1	2028	
RT180	222			2036	
RT1840	116	3016	504A	1104	
RT185	102A	3004	53	2007	
RT188	128	3024	243	2030	
RT1893	171	3201	27		
RT20					1N3891
RT2016	123A	3444	20	2051	
RT2061(G.E.)	177		300	1122	
RT210		3051	510	1114	
RT213	116	3017B/117	504A	1104	
RT214		3017B	504A	1104	
RT215	116		504A	1104	
RT218	177	3100/519	300	1122	
RT2230	102A		53	2007	
RT2309	199	3124/289	62	2010	
RT2329	102A	3004	53	2007	
RT2330	102A	3004	53	2007	
RT2331	102A	3004	53	2007	
RT2332	123A	3444	20	2051	
RT2334	109	3087		1123	

Industry Standard No.	ECG	SK	GE	RS 276-	MOTOR.
RT235		3056	ZD-5.0		
RT2451	110MP	3088	1N60	1123(2)	
RT2452	109	3087		1123	
RT248		3336	ZD-22		
RT2694	109	3087		1123	
RT2709	102A		53	2007	
RT2914	123A	3444	20	2051	
RT2915	108	3018	86	2038	
RT30					1N3892
RT304				2051	
RT3063	123A	3444	20	2051	
RT3064	123A	3444	20	2051	
RT3065	159	3114/290	82	2032	
RT3065A	159	3114/290	82	2032	
RT3069	108	3452	86	2038	
RT3070	108	3452	86	2038	
RT3071	159	3114/290	82	2032	
RT3071A	159	3114/290	82	2032	
RT3072	109	3087		1123	
RT3095	108	3452	86	2038	
RT3096	103A	3010	59	2002	
RT3097	102A	3004	53	2007	
RT3098	102A	3004	53	2007	
RT3099	109	3087		1123	
RT3225	108	3452	86	2038	
RT3226	108	3452	86	2038	
RT3227	108	3452	86	2038	
RT3228	123A	3444	20	2051	
RT3229	102A	3003	53	2007	
RT3230	102A	3004	53	2007	
RT3231	102A	3004	53	2007	
RT3232	108	3452	86	2038	
RT3233	109	3087		1123	
RT3336	109	3087		1123	
RT3361	126	3008	52	2024	
RT3362	126	3008	52	2024	
RT3363	102A	3004	53	2007	
RT3364	102A	3004	53	2007	
RT3365	102A	3004	53	2007	
RT3443	116	3031A	504A	1104	
RT3449	102A	3004	53	2007	
RT3466	160	3008	245	2004	
RT3467	102A	3008	53	2007	
RT3468	102A	3004	53	2007	
RT3469	109	3087		1123	
RT3564	102A		53	2007	
RT3565	123A	3444	20	2051	
RT3566	102A	3004	53	2007	
RT3567	123A	3444	20	2051	
RT3568	102A	3004	53	2007	
RT3574	138A		ZD-7.5		
RT3585			504A	1104	
RT3671	145A	3063	ZD-15	564	
RT3671A	145A	3063	ZD-15	564	
RT3858	116		504A	1104	
RT3981	116	3017B/117	504A	1104	
RT40					1N3893
RT4050	116	3017B/117	504A	1104	
RT4069		3031A	504A	1104	
RT4232	116		504A	1104	
RT4293	109	3087		1123	
RT4525	160	3004/102A	245	2004	
RT4624	102A	3123	53	2007	
RT4625	158		53	2007	
RT4644	109	3087		1123	
RT476			20	2051	
RT4760	123A	3124/289	20	2051	
RT4761	199	3444/123A	20	2051	
RT4762	131	3444/123A	44	2006	
RT4762MHF25			16	2006	
RT4764	116		504A	1104	
RT482	128	3024	243	2030	
RT483	128	3024	243	2030	
RT484	128	3024	243	2030	
RT4880	110MP	3088	1N60	1123(2)	
RT5061	108	3452	86	2038	
RT5063	126	3006/160	52	2024	
RT5070	116	3031A	504A	1104	
RT5151	128	3024	243	2030	
RT5152	128	3024	243	2030	
RT5200	108	3452	86	2038	
RT5201	108	3452	86	2038	
RT5202	123A	3444	20	2051	
RT5203	128		243	2030	
RT5204	128	3039/316	243	2030	
RT5205	108	3452	86	2038	
RT5206	123A	3444	20	2051	
RT5207	123A	3444	20	2051	
RT5208	199	3020/123	62	2010	
RT5212	110MP	3088	81	1123(2)	
RT5213	109	3088		1123	
RT5214	109	3090		1123	
RT5215	5071A		ZD-6.8	561	
RT5216	117		504A	1104	
RT5217	177	3100/519	300	1104	
RT5230	129	3025	244	2027	
RT5379	110MP	3088	1N60	1123	
RT5385			504A	1104	
RT5401	128	3024	243	2030	
RT5402	128	3024	243	2030	
RT5403	128	3024	243	2030	
RT5404	128	3024	243	2030	
RT5435	199	3124/289	62	2010	
RT5464	107	3018	11	2015	
RT5465	107	3018	11	2015	
RT5466	126	3006/160	52	2024	
RT5467	126	3008	52	2024	
RT5468	102A	3004	53	2007	
RT5470	110MP	3088	1N60	1123(2)	
RT5471	5071A	3334	ZD-6.8	561	
RT5472	116	3031A	504A	1104	
RT5473		3126	90		
RT5520	126		52	2024	
RT5521	102A		53	2007	
RT5522	102A		53	2007	
RT5551	123A	3444	20	2051	
RT5554	177	3100/519	300	1122	
RT5637	102A	3004	53	2007	
RT5738	110MP	3709	1N60	1123(2)	
RT5793	136A	3058/137A	ZD-6.2	561	
RT5900	108	3452	86	2038	
RT5901	123A	3452/108	86	2038	
RT5902	108	3452	86	2038	
RT5903	108	3452	86	2038	
RT5904	108	3452	86	2038	
RT5905	199	3124/289	243	2030	
RT5906	128	3124/289	243	2030	
RT5907	128	3124/289	243	2030	
RT5908	109	3090		1123	
RT5909	177	3031A	300	1122	
RT5911	116	3031A	504A	1104	
RT5912	110MP	3709	1N60	1123(2)	
RT5939	109	3090		1123	
RT60					MR1376
RT61012	109	3087		1123	
RT61014	126		52	2024	
RT61015	102A		53	2007	
RT61016	102A		53	2007	
RT6105	5072A		ZD-8.2	562	
RT6105(G.E.)	5072A		ZD-8.2	562	
RT6119	109	3090		1123	
RT6157	108	3452	86	2038	
RT6158	108	3452	86	2038	
RT6159	108	3452	86	2038	
RT6160	108	3452	86	2038	
RT6178		3126	90		
RT6179	109	3090		1123	
RT6180	109	3090		1123	
RT6181	109	3090		1123	
RT6182	109	3090		1123	
RT6183	109	3090		1123	
RT6184	109	3090		1123	
RT6189	109	3090		1123	
RT6201	229	3018	86	2038	
RT6202	229	3452/108	86	2038	
RT6203	108	3452	86	2038	
RT6204	233	3018	11	2015	
RT6205	102A	3004	53	2007	
RT6322	116	3311	504A	1104	
RT6332	116	3311	504A	1104	
RT6600	108	3452	86	2038	
RT6600MHF25			20	2051	
RT6601	107	3018	11	2015	
RT6602	107	3018	11	2015	
RT6604	102A	3004	53	2007	
RT6605	116		504A	1104	
RT6619	109	3090		1123	
RT6728	110MP	3088	IN60	1123(2)	
RT6729	116(2)	3031A	504A	1104	
RT6731	110MP	3126	90		
RT6732	123A	3018	11	2015	
RT6733	123A	3018	20	2038	
RT6734	102A	3004	53	2007	
RT6735		3004	53	2007	
RT6736	102A	3004	53	2007	
RT6737	199	3122	61	2038	
RT6787	229	3039/316	60	2015	
RT6789		3126	90		
RT6790	1003		IC-43		
RT6791	116	3031A	504A	1104	
RT6921	123	3124/289	20	2051	
RT6921MHF25			20	2051	
RT6922	5010A	3776	ZD-5.1		
RT69221	123A	3444	20	2051	
RT6923	5072A	3059/138A	ZD-8.2	562	
RT697M	123A	3444	20	2051	
RT6988	160	3006	245	2004	
RT6989	123A	3122	20	2051	
RT6990	102A	3004	53	2007	
RT6991	108	3018	86	2038	
RT699M	128		243	2030	
RT7311	124	3021	12		
RT7320	108	3452	86	2038	
RT7321	108	3452	86	2038	
RT7322	123A	3444	20	2051	
RT7323	108	3452	86	2038	
RT7324	108	3452	86	2038	
RT7325	123A	3444	20	2051	
RT7326	123A	3444	212	2051	
RT7327	123A	3444	20	2051	
RT7329	519		90		
RT7330	109	3090		1123	
RT7399	1003	3288	IC-43		
RT7400	103A	3010	59	2002	
RT7401	102A	3004	53	2007	
RT7402	166	3648/168		1152	
RT7511	123A	3444	20	2051	
RT7514	123A	3444	20	2051	
RT7515	123A	3444	20	2051	
RT7517	123A	3444	20	2051	
RT7518	123A	3444	20	2051	
RT7528	123A	3444	20	2051	
RT7538	110MP	3709	1N34AS	1123(2)	
RT7539	137A		ZD-6.2	561	
RT7557	123A	3444	212	2010	
RT7558	158	3004/102A	53	2007	
RT7559	199	3124/289	62	2010	
RT7634	116	3031A	504A	1104	
RT7636	109	3087		1123	
RT7638	123A	3444	20	2051	
RT7689	109	3087		1123	
RT7703	229	3039/316	11	2015	
RT7704	107	3018	11	2015	
RT7845	123A	3444	20	2051	
RT7846	158	3004/102A	53	2007	
RT7848	5800	9003		1142	
RT7849	116	3031A	504A	1104	
RT7850	116	3031A	504A	1104	
RT7851		3088		1123	
RT7943	123A	3444	20	2051	
RT7944	103A	3010	59	2002	
RT7945	128	3024	243	2030	
RT7946	177	3100/519	300	1122	
RT8047	199	3124/289	62	2010	
RT8193	123A	3452/108	86	2038	
RT8195	123A	3444	20	2051	
RT8197	123A	3444	20	2051	
RT8198	123A	3444	20	2051	
RT8199	117		504A	1104	
RT8200	5022A	3788	ZD-12	563	
RT8201	123A	3444	20	2051	
RT8231	116	3311	504A	1104	
RT8330		3018	17	2051	
RT8331	312	3116	FET-1	2035	
RT8332	123A	3444	20	2051	
RT8333	229	3039/316	60	2038	
RT8335		3189A		2048	
RT8336		3188A		2049	
RT8337		3122	62	2010	
RT8338				2049	
RT8339	139A	3060	ZD-9.1	562	
RT8340	116		504A	1104	
RT8442	102A		53	2007	
RT8527	229	3452/108	86	2038	
RT8602	102A	3004	53	2007	
RT8665	125	3081	510,531	1114	
RT8666	199	3122	62	2010	
RT8667	312	3448	FET-1	2035	
RT8668	107	3018	11	2015	
RT8669	107	3018	11	2015	
RT8670	290A	3114/290	269	2032	
RT8671	109	3087		1123	
RT8779		3126	90		
RT8839	117	3031A	504A	1104	
RT8840	116	3311	504A	1104	
RT8841	116	3311	504A	1104	
RT8842	158	3004/102A	53	2007	
RT8863	199	3122	62	2010	

Industry Standard No.	ECG	SK	GE	RS 276-	MOTOR.
RT8895	159	3114/290	82	2032	
RT929H	123A	3444	20	2051	
RT930H	108	3444/123A	86	2038	
RTC0120	5404	3627	C103B		
RTC0125	5405	3951			
RTC0130	5405	3951	C103C		
RTC0201	5400	3950			
RTC0203	5400	3950			
RTC0206	5401	3638/5402			
RTC0210	5402	3638			
RTC0215	5404	3627			
RTC0220	5404	3627			
RTC0225	5405	3951			
RTC0230	5405	3951			
RTC0301	5400	3950			
RTC0303	5400	3950			
RTC0306	5401	3638/5402			
RTC0310	5402	3638			
RTC0315	5404	3627			
RTC0320	5404	3627			
RTC0325	5405	3951			
RTC0330	5405	3951			
RTC0401	5400	3950			
RTC0403	5400	3950			
RTC0406	5401	3638/5402			
RTC0410	5402	3638			
RTC0415	5404	3627			
RTC0420	5404	3627			
RTC0425	5405	3951			
RTC0430	5405	3951			
RTD2103	5408	3577			
RTD2106	5408	3578			
RTD2110	5408	3578			
RTJ0203	5400	3950			
RTJ0206	5401	3638/5402			
RTJ0210	5402	3638			
RTJ0215	5404	3627			
RTJ0220	5404	3627			
RTJ0225	5405	3951	C203C		
RTJ0230	5405	3951	C203C		
RTL0605	5461	3685			
RTL0610	5462	3686			
RTL0620	5463	3572			
RTL0640	5465	3687			
RTL0805	5461	3685			
RTL0810	5462	3686			
RTL0820	5463	3572			
RTL0840	5465	3687			
RTN0102	5470		C15U		
RTN0105	5470		C15F		
RTN0110	5471		C15A		
RTN0120	5472		C15B		
RTN0130	5474		C15C		
RTN0140	5474		C15D		
RTN0150	5476		C15E		
RTN0160	5476		C15M		
RTN0202	5481		C220U		
RTN0205	5481		C220F		
RTN0210	5482		C220A		
RTN0220	5483	3942	C220B		
RTN0230	5485	3943	C220C		
RTN0240	5485	3943	C220D		
RTN0250	5487	3944	C220E		
RTN0260	5487	3944	C220M		
RTN1001	804	3455	IC-27		
RTN1102	722		IC-9		
RTR0302		3572	C122U		
RTR0305		3572	C122F		
RTR0310		3572	C122A		
RTR0320		3572	C122B		
RTR0330		3573	C122C		
RTR0340		3573	C122D		
RTS0202	5514	3613	C222U		
RTS0205	5514	3613	C222F		
RTS0210	5514	3613	C222A		
RTS0220	5514	3613	C222B		
RTS0230	5515	6615	C222C		
RTS0240	5515	6615	C222D		
RTS0250	5516	6616	C222E		
RTS0260	5516	6616	C222M		
RTS0502	5514	3613			
RTS0505	5514	3613			
RTS0510	5514	3613			
RTS0520	5514	3613			
RTS0530	5515	6615			
RTS0540	5515	6615			
RTS0550	5516	6616			
RTS0560	5516	6616			
RTS0702	5517	3615	C229U		
RTS0705	5517	3615	C229F		
RTS0710	5517	3615	C229A		
RTS0720	5517	3615	C229B		
RTS0730	5518		C229C		
RTS0740	5518		C229D		
RTS0750	5519		C229E		
RTS0760	5519		C229M		
RTU0102	5500	3589	C230U		
RTU0105	5501		C230F		
RTU0110	5502	3589	C230A		
RTU0120	5504	3589	C230B		
RTU0130	5507	3580	C230C		
RTU0140	5507	3580	C230D		
RTU0150	5509		C230E		
RTU0160	5529	6629	C230M		
RTU0202	5520	6621/5521			
RTU0205	5521	3581/5543			
RTU0210	5522	6622			
RTU0220	5524	6624			
RTU0230	5527	6627			
RTU0240	5527	6627			
RTU0250	5529	6629			
RTU0260	5529	6629			
RTU0705		6642	C228F		
RTU0710	5542	6642			
RTU0720	5543	3581			
RTU0730	5545	3582			
RTU0740	5545	3582			
RU-1	552	3313/116			
RU-2V	506	9000/552			
RU-IN	506		511	1114	
RU1	552	9000	504A	1104	
RU1A				1104	
RU1C				1114	
RU2	552	9000	511	1114	
RU2V			511	1114	
RV-2289	116	3311	504A	1104	
RV06	116	3311	504A	1104	
RV06/7825B	116	3311	504A	1104	
RV1017	177	3126	300	1122	
RV1059	159	3114/290	221		
RV1068	229	3132	61	2038	
RV1180	102A		53	2007	
RV1181	136A	3057	ZD-5.6	561	
RV1189	116	3311	504A	1104	
RV1226	177	3100/519	300	1122	
RV1424	116	3311	504A	1104	
RV1467	108	3452	86	2038	
RV1468	108	3452	86	2038	
RV1469	108	3452	86	2038	
RV1470	108	3452	86	2038	
RV1471	123A	3444	20	2051	
RV1472	129	3025	244	2027	
RV1473	128	3018	243	2030	
RV1474	123A	3444	20	2051	
RV1475	102A	3004	53	2007	
RV1476	5804	9007	504A	1144	
RV1477		3126	90		
RV1478	116	3088	504A	1104	
RV1479	109	3090		1123	
RV2068	293	3024/128	243	2030	
RV2069	129	3025	244	2027	
RV2070	199	3124/289	62	2010	
RV2071	177	3100/519	300	1122	
RV2072	116	3311	504A	1104	
RV2213	139A		ZD-9.1	562	
RV2220	116		504A	1104	
RV2248	199	3124/289	62	2010	
RV2249	123A	3444	20	2051	
RV2250	116	3311	504A	1104	
RV2260	159	3114/290	89	2023	
RV2289	116	3311	504A	1104	
RV2327	116		504A	1104	
RV2351	159	3114/290	82	2032	
RV2353	234	3025/129	65	2050	
RV2354	199	3018	61	2010	
RV2355	290A	3114/290	269	2032	
RV2356	187	3084	29	2025	
RV3301DB					MC3301P
RV6.2	137A		ZD-6.2	561	
RVD0.8C22/1A				1104	
RVD08C22/1A	116	3031A	504A	1104	
RVD10D1	116	3311	504A	1104	
RVD10DC1	116	3031A	504A	1104	
RVD10DC1R	116		504A	1104	
RVD10E1	116	3017B/117	504A	1104	
RVD10E11F			504A	1104	
RVD10E1LF	116	3311	504A	1104	
RVD12B-1	166	9075	BR-600	1152	
RVD14740				562	
RVD1E1LF				1102	
RVD1K110	109	3087		1123	
RVD1N34A	109			1123	
RVD1N4738	5072A	3059/138A	ZD-8.2	562	
RVD1N4739	139A	3060	ZD-9.1	562	
RVD1N4740	140A	3061	ZD-10	562	
RVD1S854	116	3311	504A	1104	
RVD2-1K110	109	3087		1123	
RVD2DP	116	3031A	504A	1104	
RVD2DP22/18			504A	1104	
RVD2DP22/1B		9075		1104	
RVD2DP22/1C	166	9075	BR-600	1152	
RVD2DP221B	166			1152	
RVD2DP221P	116		504A	1104	
RVD2P22/1B	116		504A	1104	
RVD4B265J2	116	3311	504A	1104	
RVD8MB4	166	3105	BR-600	1152	
RVDC08P1	116(2)	3119/113	6GC1		
RVDC08P1R	113A		6GC1		
RVDCD0033	5070A	3058/137A	ZD-6.2	561	
RVDD1245	166	9075		1152	
RVDD124B	166		BR-600	1152	
RVDDS-410	116	3311	504A	1104	
RVDEQA0106S	5070A		ZD-6.2	561	
RVDEQA0107S	138A	3059	ZD-7.5		
RVDFV211	601	3463			
RVDGP05A	116		504A	1104	
RVDKB162C5	116	3311	504A	1104	
RVDKB167	600	3864/605			
RVDKB262	605	3864			
RVDKB265J2	177	3100/519	300	1122	
RVDMZ-206	5070A		ZD-6.2	561	
RVDMZ206				561	
RVDMZ208	138A		ZD-7.5		
RVDMZ209	139A		ZD-9.1	562	
RVDR154B	166	9075	BR-600	1152	
RVDR5R6EB	5070A	9021	ZD-6.2	561	
RVDRD11AN	139A		ZD-9.1	562	
RVDRD5A1E	5010A	3776	ZD-5.1		
RVDRD5R6EB	5070A	9021	ZD-6.0	561	
RVDRD7AN	5071A	3334	ZD-6.8	561	
RVDRD7R5E	158		ZD-7.5		
RVDRD7R5EB	138A	3059	ZD-7.5		
RVDRD7R5FB	138A		ZD-7.5		
RVDRD9R1E	139A	3060			
RVDSC-15		3327	90		
RVDSC20		3327	90		
RVDSD-1	116		504A	1104	
RVDSD-1U	116		504A	1104	
RVDSD-1Y	116	3031A	504A	1104	
RVDSD-1Z	156	3051	512		
RVDSD113	614	3327	90		
RVDSG-5N	156	3051	512		
RVDSG-5P	156	3051	512		
RVDSR3AM2N	116	3311	504A	1104	
RVDVD1150L	601	3463	300	1122	
RVDVD1150M	601	3463			
RVDVD1210L	605	3864	300	1122	
RVDVD1210M	605	3864	300	1122	
RVDVD1211L	605	3864	504A	1104	
RVDVD1212L	605	3311	300	1122	
RVDVD1213	605	3864	300	1122	
RVDVD1250M	177	3100/519	300	1122	
RVDVD1251L	601	3463			
RVDVD1252L	605	3864			
RVILA3301	1006	3358	IC-38		
RVIMC4080	1083		IC-275		
RVIUPC20C2	1075A		IC-117		
RVIUPC22C	1078	3498	IC-136		
RVIUPC575	1140		IC138		
RVIUPD610	1254	3880			
RVS2SC645	108	3452	86	2038	
RVTCS1381	123A	3444	20	2051	
RVTCS1382	159		82	2032	
RVTCS1383	123A	3444	20	2051	
RVTCS1384	108	3452	86	2038	
RVTCS1473	199		62	2010	
RVTMC00921-3			6GD1	1104	
RVTMK10-2	312	3448	FET-1	2035	
RVTMK10-E	312	3448	FET-1	2035	
RVTS22410	123A	3444	20	2051	
RVTS22411	159	3114/290	82	2032	
RX090	139A		ZD-9.1	562	
RYN12104	128		243	2030	
RYN121105	123A	3444	20	2051	
RYN121105-3	123A	3444	20	2051	
RYN121105-4	123A	3444	20	2051	
RZ-15AB	145A	3063	ZD-15	564	
RZ-3	156	3051	512		

Industry Standard No.	ECG	SK	GE	RS 276-	MOTOR.
RZ10	140A	3061	ZD-10	562	
RZ11			ZD-11	563	
RZ12	142A	3062	ZD-12	563	
RZ13	143A	3750	ZD-12	563	
RZ14	5023A	3789	ZD-14	564	
RZ15	145A	3063	ZD-15	564	
RZ15A	145A	3063	ZD-15	564	
RZ16	5025A	3791		564	
RZ18	5077A	3752	ZD-18		
RZ20	5029A	3795	ZD-20		
RZ22	5080A	3336	ZD-22		
RZ22A	5030A	3336/5080A	ZD-22		
RZ25	5032A	3798	ZD-25		
RZ27	146A	3064	ZD-27		
RZ27A	146A	3064	ZD-27		
RZ27AC	146A	3064	ZD-27		
RZ3.3	5066A	3055/134A	ZD-3.3		
RZ3.6	134A	3055	ZD-3.6		
RZ3.9	5067A	3055/134A	ZD-3.9		
RZ30	5035A	3801	ZD-30		
RZ33A	5036A	3802	ZD-33		
RZ39A	5038A	3804	ZD-39		
RZ4.7	5069A	3056/135A	ZD-4.7		
RZ47A	5040A	3806	ZD-47		
RZ5.1	5010A	3056/135A	ZD-5.1		
RZ5.6	136A	3057	ZD-5.6	561	
RZ50	5041A	3807	ZD-51		
RZ56A	5042A	3808	ZD-56		
RZ6.2	137A	3058	ZD-6.2	561	
RZ6.8	5071A	3334	ZD-6.8	561	
RZ68A	5045A	3811	ZD-68		
RZ7.5	138A		ZD-7.5		
RZ8.2	5072A	3136	ZD-8.2	562	
RZ82A	5047A	3813	ZD-82		
RZ9.1	139A	3060	ZD-9.1	562	
RZZ10	140A	3061	ZD-10	562	
RZZ11	5074A		ZD-11	563	
RZZ12	142A	3062	ZD-12	563	
RZZ13	143A		ZD-12	563	
RZZ15	5023A	3063/145A	ZD-14	564	
RZZ156	145A		ZD-15	564	
RZZ18	5027A	3793	ZD-18		
RZZ22	5030A	3796	ZD-22		
RZZ27	5033A	3799	ZD-27		
RZZ27-12	146A	3799/5033A	ZD-27		
RZZ3.3	5066A	3330	ZD-3.3		
RZZ3.6	134A	3055	ZD-3.6		
RZZ3.9	5067A	3331	ZD-3.9		
RZZ4.3	5068A		ZD-4.3		
RZZ4.7	5069A	3056/135A	ZD-4.7		
RZZ5.1	5010A	3056/135A	ZD-5.1		
RZZ5.6	136A	3057	ZD-5.6	561	
RZZ6.2	137A	3058	ZD-6.2	561	
RZZ6.8	5071A	3334	ZD-6.8	561	
RZZ7.5	138A		ZD-7.5		
RZZ8.2	5072A	3136	ZD-8.2	562	
RZZ9.1	139A	3060	ZD-9.1	562	
S	139A		ZD-9.1	562	
S-05	116	3017B/117	504A	1104	
S-05-005	116	3031A	504A	1104	
S-05-01	116	3031A	504A	1104	
S-05-02				1102	
S-05-04				1103	
S-05-06				1104	
S-05-10				1114	
S-05/01	125	3032A	510,531	1114	
S-050	116	3031A	504A	1104	
S-0501	116	3031A	504A	1104	
S-1.5	116	3017B/117	504A	1104	
S-1.5-0	116	3311	504A	1104	
S-1.5-02				1102	
S-1.5-04				1103	
S-1.5-06				1104	
S-1.5-10				1114	
S-10	116	3016	504A	1104	
S-1019		3019	11	2015	
S-1019(UHF)			11	2015	
S-102V			504A	1104	
S-1037	108	3452	86	2038	
S-1041	108	3019	86	2038	
S-1058	108	3452	86	2038	
S-1059	108	3452	86	2038	
S-1060	108	3848/5806	86	2038	
S-1061	123A	3124/289	20	2051	
S-1062	108	3452	86	2038	
S-1065	123A	3124/289	20	2051	
S-1066	123A	3124/289	20	2051	
S-1068	123A	3124/289	20	2051	
S-1078	108	3452	86	2038	
S-1079	108	3018	86	2038	
S-1128	123A	3124/289	20	2051	
S-1143	123A	3124/289	20	2051	
S-1153	108	3452	86	2038	
S-12	5802			1143	
S-1221	123A	3124/289	20	2051	
S-1221A	123A	3020/123	20	2051	
S-1227	108	3452	86	2038	
S-1245	123A	3124/289	20	2051	
S-1276	108	3452	86	2038	
S-1286	108	3018	86	2038	
S-1296	108	3452	86	2038	
S-12A	5802			1143	
S-1313			11	2015	
S-1316	108	3452	86	2038	
S-1317	108	3452	86	2038	
S-1318	108	3452	86	2038	
S-1331W	123A	3020/123	20	2051	
S-1348	102A	3004	53	2007	
S-1349	102A	3004	53	2007	
S-1360	108	3452	86	2038	
S-1361	108	3452	86	2038	
S-1362	108	3452	86	2038	
S-1363	123A	3124/289	20	2051	
S-1364	123A	3124/289	20	2051	
S-1367A	159		82	2032	
S-1403	123A	3124/289	20	2051	
S-1408	108		86	2038	
S-1409	108		86	2038	
S-1453			8	2002	
S-15	552	3515	504A	1104	
S-15-10			504A	1104	
S-1512	123A		20	2051	
S-1533	123A		20	2051	
S-1556-2	121	3717	239	2006	
S-1559	123A	3124/289	20	2051	
S-15C	558	3515			
S-15H	506	3130	511	1114	
S-1639	102A	3004	53	2007	
S-1640	160	3006	245	2004	
S-17	116	3016	504A	1104	
S-17A	116	3016	504A	1104	
S-2	116	3311	504A	1104	
S-20446	195A		46		
S-2064-G	519		514	1122	
S-21271	109	3090		1123	
S-2200-1135	74H04	74H04			
S-2617		3019	11	2015	
S-262	116	3016	504A	1104	
S-3016R	177	3100/519	300	1122	
S-305	130	3027	14	2041	
S-305-PD	130		14	2041	
S-305A	130	3027	14	2041	
S-310E	152		66	2048	
S-320F	129			1123	
S-34	506	3100/519	511	1114	
S-356	130	3027	14	2041	
S-371	160		245	2004	
S-39T	121	3009	239	2006	
S-3A1					MR501
S-3A10					MR510
S-3A2					MR502
S-3A3		3756			MR504
S-3A4					MR504
S-3A5					MR506
S-3A6					MR506
S-3A8					MR508
S-3MX	116	3016	504A	1104	
S-40T	121	3009	239	2006	
S-40TB	121	3009	239	2006	
S-41T	121	3009	239	2006	
S-42T	121	3009	239	2006	
S-437	129	3025	244	2027	
S-437F	129		244	2027	
S-43T	121	3009	239	2006	
S-46T	121	3009	239	2006	
S-48T	121	3009	239	2006	
S-49T	121	3009	239	2006	
S-4C	558	3017B/117	511	1114	
S-500	125		510,531	1114	
S-500B	116	3017B/117	504A	1104	
S-500C	116	3017B/117	504A	1104	
S-522	192		63	2030	
S-5277B	116	3311	504A	1104	
S-5328E	108	3019	86	2038	
S-55TB	126	3006/160	52	2024	
S-5670-E	108	3019	86	2038	
S-5745					MDA2501
S-5745-1					MDA2502
S-5745-2					MDA2504
S-5745-3					MDA2504
S-5745-4					MDA3506
S-5745-5					MDA3506
S-58TB	121	3009	239	2006	
S-5959					MDA960A3
S-5A1					MR751
S-5A2		3639			MR752
S-5A3		3640			MR754
S-5A4		3640			MR754
S-5A5					MR756
S-5A6					MR756
S-5S	116	3311	504A	1104	
S-5SR	116		504A	1104	
S-6211					MDA920A4
S-6211-1					MDA920A6
S-6211-2					MDA920A7
S-6211-3					MDA920A8
S-6211-4					MDA920A9
S-70T	126	3008	52	2024	
S-798			504A	1104	
S-80T	126	3007	52	2024	
S-810			504A	1104	
S-85	184	3190	57	2017	
S-86		3190	57	2017	
S-873TB	160		245	2004	
S-874TB	160	3007	245	2004	
S-87TB	126	3008	52	2024	
S-88576		3089	1N82A		
S-88TB	126	3008	52	2024	
OS-90	109			1123	
S-95101	126	3008	52	2024	
S-95102	126	3008	52	2024	
S-95103	126	3008	52	2024	
S-95125	108	3018	86	2038	
S-95125A	108	3018	86	2038	
S-95126	108	3018	86	2038	
S-95126A	108	3018	86	2038	
S-95201	102A	3004	53	2007	
S-95202	103A	3010	59	2002	
S-95204	102A	3004	53	2007	
S-95253	121	3009	239	2006	
S-95253-1	121	3009	239	2006	
S-05	116		504A	1104	
S0002	233	3122	210	2009	
S001465	199	3245	62	2010	
S001466	123A	3444	20	2051	
S0015	123A	3444	20	2051	
S0016	108	3452	86	2038	
S001683	128		243	2030	
S0020	108	3246/229	86	2038	
S0021	108	3452	86	2038	
S0022	123A	3444	20	2051	
S0023	199		62	2010	
S0024	233	3245/199	210	2009	
S0025	123A	3444	20	2051	
S0026	159	3114/290	82	2032	
S006793	199	3245	62	2010	
S006927	199	3245	62	2010	
S007220	128		243	2030	
S007764	199	3245	62	2010	
S01	160	3006	245	2004	
S010G	125	3033	510,531	1114	
S011-0	5074A	3092/141A	ZD-11.5		
S017446	159		82	2032	
S019843	159		82	2032	
S02	160	3006	245	2004	
S022010	123A	3444	20	2051	
S022011	123A	3444	20	2051	
S022012	159		82	2032	
S023735	159		82	2032	
S024428	123A	3444	20	2051	
S024987	123A	3444	20	2051	
S025	102A	3123	53	2007	
S025232	123A	3444	20	2051	
S025289	123A	3444	20	2051	
S026094	234	3247	65	2050	
S028	101	3861	8	2002	
S03	160	3006	245	2004	
S0301JS2	5452	6752			
S0301JS3	5452	6752			
S0301LS3	5411	3953/5412			
S0301M		3577	C6F		2N1595
S0301MS1					2N4212
S0301MS2			C5F		2N2322
S0301MS3			C6F		MCR1906-1
S0301RS3	5411	3953/5412			
S0303LS2					MCR106-1
S0303LS3	5411	3953/5412			MCR106-1
S0303M	5408				SPECIAL

Industry Standard No.	ECG	SK	GE	RS 276-	MOTOR.
S0303MS2			C106Y1		2N6236
S0303MS3			C107Y1		2N6236
S0303RS2	5452	6752	C106Y2		
S0303RS3	5411	3953/5412	C107Y2		
S0304F1					SPECIAL
S0306F1					SPECIAL
S0306FS21					MCR72-1
S0306FS31					MCR72-1
S0306G	5514	3613	C222U		
S0306H	5491	6791	C220U		
S0306L	5461	3685	C122U		
S0306R	5461	3685			
S0308F1					MCR3000-1
S0308F3					MCR3000-1
S0308FS31					MCR72-1
S0308FS4					MCR72-1
S0308G	5514	3613	C222U		
S0308H	5491	6791	C220U		
S0308L	5461	3685	C122U		
S0308R	5461	3685			
S0310F1					SPECIAL
S0310F3					SPECIAL
S0310G	5514	3613	C222U		
S0310H	5491	6791	C220U		
S0310L	5550		C123F		
S0315G	5514	3613			SPECIAL
S0315H	5501				SPECIAL
S0316H		3505	C230U		
S031A	123A	3444	20	2051	
S0320G	5514	3613			
S0320H	5521	6621			
S0325C					MCR649-1
S0325G	5517	3615	C232U		SPECIAL
S0325H	5521	6621	C230U		2N681
S0335G	5517	3615	C229U		MCR3835-1
S0335H	5541	6615/5515	C228U		MCR3935-1
S037	123A	3444	20	2051	
S04	177	3100/519	300	1122	
S04-1	177	3100/519	300	1122	
S04A-1	177	3100/519	300	1122	
S04B-1	177	3100/519	300	1122	
S05	116		504A	1104	
S0501	116	3311	504A	1104	
S0501JS2	5453	6753			
S0501JS3	5453	6753			
S0501LS3	5412	3953			
S0501M	5408	3577	C6F		2N1595
S0501MS1					2N4213
S0501MS2			C5F		2N2323
S0501MS3			C6F		MCR1906-2
S0503LS2					MCR106-2
S0503LS3	5412	3953			MCR106-2
S0503M	5408	3577			SPECIAL
S0503MS2			C106F1		2N6237
S0503MS3			C107F1		2N6237
S0503RS3	5412	3953			
S0504F1					SPECIAL
S0505FS21					MCR72-2
S0505FS31					MCR72-2
S0506F1					SPECIAL
S0506FS21					MCR72-2
S0506FS31					MCR72-2
S0506G	5514	3613	C222F		
S0506H	5491	6791	C220F		
S0506L	5461	3685	C122F		
S0506R	5461	3685			
S0508F1					2N4441
S0508F3					2N4441
S0508G	5514	3613	C222F		
S0508H	5491	6791	C220F		
S0508L	5461	3685	C122F		
S0510F1					SPECIAL
S0510F3					SPECIAL
S0510G	5514	3613	C222F		
S0510H	5491	6791	C220F		
S0510L	5550		C123F		
S0515G	5514	3613			SPECIAL
S0515H	5501				SPECIAL
S0516H		3505	C230F		
S0520G	5514	3613			
S0520H	5521	6621			
S0525C					MCR649-2
S0525G	5517	3615	C232F		SPECIAL
S0525H	5521	6621	C230F		2N682
S0535G			C229F		MCR3835-2
S0535H	5541	6615/5515	C228F		MCR3935-2
S065	158		53	2007	
S065A	102A		53	2007	
S0702	139A		ZD-9.1	562	
S0704	123A	3444	20	2051	
S074-005-0001	519		514	1122	
S074-007-0001	177		300	1122	
S074-007-001				1122	
S082A	112		1N82A		
S088	102A		53	2007	
S0F					MR818
S0M					1N4007
S1-1	116	3016	504A	1104	
S1-11	335	3139/5074A			
S1-B01-02	116	3031A	504A	1104	
S1-RECT-102	116	3311	504A	1104	
S1-RECT-154	116	3311	504A		
S1-RECT-155	116	3312	504A		
S1-RECT-158	506	3125	511		
S1-RECT-35	177	3100/519	300	1122	
S1.5	116	3311	504A	1104	
S1.5-01	116		504A	1104	
S10	116		504A	1104	
S10-12					MRF433
S10-28					2N6370
S100	125	3032A	510,531	1114	
S100-12	470				MRF421
S100-28	471				MRF422
S100-50					MRF428
S1000	5848	7048	5012		
S1001JS2	5454	6754			
S1001JS3	5454	6754			
S1001LS3	5413	3954/5414			
S1001M	5408	3577	C6A		2N1596
S1001MS1					2N4214
S1001MS2			C5A		2N2324
S1001MS3			C6A		MCR1906-3
S1003LS2					MCR106-3
S1003LS3	5413				MCR106-3
S1003M	5408	3503			SPECIAL
S1003MS2			C106A1		2N6238
S1003MS3			C107A1		2N6238
S1003RS2	5454	3570	C106A1		
S1003RS3	5413		C107A1		
S1004F1					SPECIAL
S1006B		3614	C220A2		
S1006F1					SPECIAL
S1006FS21					MCR72-3
S1006FS31					MCR72-3
S1006G	5514	3613	C222A		
S1006H	5491	6791	C220A		
S1006L	5462	3686	C122A		
S1006R	5462	3686			
S1008B		3614	C220A2		
S1008F1					MCR3000-3
S1008F3					MCR3000-3
S1008FS21					MCR72-3
S1008FS31					MCR72-3
S1008G	5514	3613	C222A		
S1008H	5491	3589	C220A		
S1008L	5462	3686	C122A		MCR3000-3
S1008R	5462	3686			
S1009	108	3452	86	2038	
S101	116	3016	504A	1104	
S1010	5801	9004	11	1142	1N4002
S10100	5809	9010			1N4007
S1010B		3614	C220A2		
S1010F1					SPECIAL
S1010F3					SPECIAL
S1010G	5514	3613	C222A		
S1010H	5491	6791	C220A		
S1010L	5552		C123A		
S10110	506		511	1114	MR1-1200
S10120	506		511	1114	MR1-1200
S10149	116(4)	3031A	504A(4)	1104	
S10153	184		57	2017	
S1015G	5514	3613			SPECIAL
S1015H	5502				SPECIAL
S1016	123A	3122	20	2051	
S1016B		3614	C230A2		
S1016G		3613	C232A		
S1016H		3505	C230A		
S1019	108	3452	86	2038	
S102	116	3016	504A	1104	
S1020	5802	9005		1143	1N4003
S1020G	5514	3613			
S1020H	5522	6622			
S1025C					MCR649-3
S1025G	5517	3615	C232A		SPECIAL
S1025H	5522	6622	C230A		2N683
S103	116	3016	504A	1104	
S1030	5803	9006		1144	1N4004
S1035G		3615	C229A		2N3870
S1035H	5542	6615/5515	C228A		2N3896
S1037	108		86	2038	
S104	116	3016	504A	1104	
S1040	5804	9007		1144	1N4004
S1041	107		11	2015	
S1041-160N	107	3019	11	2015	
S1044	108	3452	86	2038	
S1047	159		82	2032	
S105	116		504A	1104	
S1050	5805	9008			1N4005
S1058	108		86	2038	
S1059	108		86	2038	
S105A	5800	9003		1142	
S106	116		504A	1104	
S1060	5806		17		1N4005
S1061	123A	3444	20	2051	
S1062	108	3444	86	2038	
S1065	123A	3444	20	2051	
S1066	123A	3444	20	2051	
S1068	123A	3444	20	2051	
S1069	123A	3444	20	2051	
S106A	5454		C106A1		C106A1
S106A1	5454	6754	C106A1		
S106B	5455	3597	C106B1		C106B1
S106B1	5455	3597	C106B1		
S106C	5456	3598/5457	C106C1		C106C1
S106C1	5456		C106C1		
S106D	5457	3598	C106D1		C106D1
S106D1	5457		C106D1		
S106E			C106E1		C106E1
S106F	5453	6753	C106F1		C106F1
S106F1	5453	6753	C106F1		
S106M			C106M1		C106M1
S106Q	5452	3597/5455	C106Q1		
S106Y	5452	6752	C106Y1		
S106Y1	5452	6752	C106Y1		
S107	116		504A	1104	
S1070	5808	9009			1N4006
S1074	123A	3444	20	2051	
S1074(R)			20	2051	
S1074R	123A	3444	20	2051	
S1076	108	3452	86	2038	
S1078	108	3452	86	2038	
S1079	108	3452	86	2038	
S107A	5454	6754			C106A1
S107A1	5454	6754	C107A1		
S107B	5455	3597			C106B1
S107B1	5455	3597	C107B1		
S107C	5456	3598/5457			C106C1
S107C1	5456		C107C1		
S107D	5457	3598			C106D1
S107D1	5457		C107D1		
S107E					C106E1
S107F	5453	6753			C106F1
S107F1	5453	6753	C107F1		
S107M					C106M1
S107Q	5452	6752			
S107Y	5452	6752			
S107Y1	5452	6752	C107Y1		
S108	116		504A	1104	
S1080			20		1N4006
S1080(TRANSISTOR)	123A		20	2051	
S1090	5809	9010	20		1N4007
S10A	116	3017B/117	504A	1104	
S10AN12	5890	7090	5044		
S10AN6	5890	7090	5044		
S10AR1	125	3033	510,531	1114	
S10AR2	125	3051/156	510,531	1114	
S10B01-02	116	3311	504A	1104	
S10BR2		3051	512		
S10CN1	5808	9009			
S10GR2	5809	9009/5808			
S10M1	5808	9009			
S11	177		300	1122	
S110A	5809	9010			
S1122	107	3039/316	11	2015	
S1126	107	3039/316	11	2015	
S1128	123A	3444	20	2051	
S1142	108	3452	86	2038	
S1143	123A	3444	20	2051	
S115	116	3017B/117	503A	1104	
S1153	108		86	2038	
S12-1-A-3P	123A	3444	20	2051	
S1201F	125		510,531	1114	
S1202	157		232		
S12020-04	152	3054/196	66	2048	
S1210	5801	9004		1142	
S12100	5809	9010			
S12110	506		511	1114	
S1211N	312	3112		2028	

Industry Standard No.	ECG	SK	GE	RS 276-	MOTOR.
S12120	506		511	1114	
S1212N	312	3112		2028	
S12130	506		511	1114	
S1213N	312	3112		2028	
S12140	506		511	1114	
S1214N	312	3112		2028	
S1215N	312	3112		2028	
S1216N	312	3112		2028	
S1220	5802	9005		1143	
S1221	123A	3444	20	2051	
S1221A	123A	3444	20	2051	
S1221N	312	3112		2028	
S1222N	312	3112		2028	
S1223N	312	3112		2028	
S1224N	312	3112		2028	
S1225N	312	3112		2028	
S1226	123A	3444	20	2051	
S1226N	312	3112		2028	
S1227	108		86	2038	
S122A	5462	3686			
S122B	5463	3572			
S122C	5465	3558			
S122D	5465	3558			
S122F	5461	3685			
S1230	5803	9006		1144	
S1231N	312	3112		2028	
S1232N	312	3112		2028	
S1233N	312	3112		2028	
S1234N	312	3112		2028	
S1235N	312	3112		2028	
S1236N	312	3112		2028	
S1240	5804	3122	20	1144	
S1241	123A	3122	20	2051	
S1241N	312	3448	FET-1	2035	
S1242	123A	3444	20	2051	
S1242N	312	3448	FET-1	2035	
S1243	123A	3444	20	2051	
S1243N	116	3116	504A	1104	
S1245	123A	3444	20	2051	
S124AHB	280		262	2041	
S1250	5805	9008			
S1260	5806	3848			
S1270	5808	9009			
S1272	123A	3444	20	2051	
S1276	108		86	2038	
S1280	5808	9009			
S1286	161	3452/108	39	2015	
S129	116	3311	504A	1104	
S1290	5809	9010			
S1296	108		86	2038	
S12CN1	5808	9009			
S12M1	5809	9010			
S13	116	3016	504A	1104	
S130-138	161	3716	39	2015	
S130-251	161	3716	39	2015	
S1307	123A	3444	20	2051	
S1308	107	3039/316	11	2015	
S1309	123A	3444	20	2051	
S1313	108	3452	86	2038	
S1316	108		86	2038	
S1317	108		86	2038	
S1318	108		86	2038	
S133-1	123A	3122	20	2051	
OS133.0008T		3466	82	2032	
OS133.1003T		3715	82	2032	
S1331	123A	3444	20	2051	
S1331N	123A	3444	20	2051	
S1331W	123A	3444	20	2051	
S1332	126		52	2024	
S1348	102A		53	2007	
S1349	102A		53	2007	
S1350	159	3114/290	82	2032	
S1350A	159	3114/290	82	2032	
S13551M	309K	3629			
S1360	108		86	2038	
S1361	108		86	2038	
S1362	108		86	2038	
S1363	123A	3444	20	2051	
S1364	123A	3444	20	2051	
S1366	154	3044	40	2012	
S1367	159	3114/290	82	2032	
S1368	128	3024	243	2030	
S1369	123A	3444	20	2051	
S1373	123A	3444	20	2051	
S1374	123A	3444	20	2051	
S1384	110MP	3709		1123(2)	
S13A	5804	9007		1144	
S14	116	3311	504A	1104	
S1403	123A	3444	20	2051	
S1405	123A	3444	20	2051	
S1407	154		40	2012	
S1408	108	3452	86	2038	
S1409	108	3452	86	2038	
S1419	123A	3444	20	2051	
S1420	123A	3444	20	2051	
S1421		3024	18	2030	
S1428	177	3100/519	300	1122	
S1429-3	123A	3444	20	2051	
S1430	129	3025	244	2027	
S1431	129	3025	244	2027	
S1432	123A	3444	20	2051	
S1443	123A	3444	20	2051	
S1453	123A	3444	20	2051	
S1475	123A	3444	20	2051	
S1476	123A	3444	20	2051	
S1477		3025	22	2032	
S1487	123A	3444	20	2051	
S14A	5804	9007		1144	
S14A-1-3P	121		239	2006	
0000000S15	116	3313	504A	1104	
S15-50					MRF427
S1502	123A	3444	20	2051	
S1510	123A	3444	20	2051	
S1512		3444	20	2051	
S1514	128	3024	243	2030	
S1516	128	3024	243	2030	
S1517	128	3024	243	2030	
S1520	129	3025	244	2027	
S1523	128	3024	243	2030	
S1525	128	3024	243	2030	
S1526	123A	3444	20	2051	
S1527	123A	3444	20	2051	
S1529	123A	3444	20	2051	
S1530	123A	3444	20	2051	
S1533		3444	20	2051	
S154			511	1114	
S1556-2	121	3717	239	2006	
S1559	123A	3444	20	2051	
S15649	123A	3444	20	2051	
S15650	161	3117	39	2015	
S15657	161	3117	39	2015	
S15658	161	3117	39	2015	
S15659	161	3117	39	2015	
S15660	128	3024	243	2030	
S1568	123A	3444	20	2051	
S1570	123A	3444	20	2051	
S16	116	3051/156	504A	1104	
S160			300	1122	
S1600	116		504A	1104	
S1619	123A	3444	20	2051	
S1620	123A	3444	20	2051	
S1629	123A	3444	20	2051	
S1636	108	3452	86	2038	
S1639	102A		53	2007	
S1640	160		245	2004	
S1642	128	3024	243	2030	
S1644	128		243	2030	
S1671	128		243	2030	
S1672	158		53	2007	
S1674	108	3452	86	2038	
S1674A	108	3452	86	2038	
S1682	108	3452	86	2038	
S1683		3024	18	2030	
S1685			14	2041	
S1689	128	3024	243	2030	
S1691	130	3027	14	2041	
S1692	130	3027	14	2041	
S1697	123A	3444	20	2051	
S1698	129	3025	244	2027	
S169N	123A	3122	20	2051	
S16A	116	3051/156	504A	1104	
S16B	116	3016	504A	1104	
S17	116		504A	1104	
S17074	113A		6GC1		
S175-28					MRF422
S175-50					MRF428
S1761	123A	3444	20	2051	
S1761A	123A	3444	20	2051	
S1761B	123A	3444	20	2051	
S1761C	123A	3444	20	2051	
S1762	128	3024	243	2030	
S1764	123A	3444	20	2051	
S1765	123A	3444	20	2051	
S1766	123A	3444	20	2051	
S1768	123A	3444	20	2051	
S1769	154		40	2012	
S1770	123A	3444	20	2051	
S1772	123A	3444	20	2051	
S1773	128	3024	243	2030	
S1777	128	3024	243	2030	
S1784	123A	3444	20	2051	
S1785	123A	3444	20	2051	
S17862	154	3045/225	40	2012	
S1788			20	2051	
S17900	128	3024	243	2030	
S17A	116		504A	1104	
S18	116	3051/156	504A	1104	
S180	177	3100/519	300	1122	
S18000	128	3024	243	2030	
S1801-02	116	3311	504A	1104	
S18100	159		82	2032	
S18200	128		243	2030	
S1835	123A	3444	20	2051	
S1863	129	3025	244	2027	
S1864	128	3024	243	2030	
S1865	130	3027	14	2041	
S1871	123A	3444	20	2051	
S1874	128	3024	243	2030	
S1878			10	2051	
S1889	159		82	2032	
S1891	123A	3444	20	2051	
S1891A	123A	3444	20	2051	
S1891B	123A	3444	20	2051	
S1897	107	3039/316	11	2015	
S18A	116	3017B/117	504A	1104	
S18B	116	3017B/117	504A	1104	
S19	116	3016	504A	1104	
S1905	130	3027	14	2041	
S1905A	130	3027	14	2041	
S1907	130	3027	14	2041	
S191G	116		504A	1104	
S19386	128	3024	243	2030	
S1955	123A	3444	20	2051	
S1977634	130	3027	14	2041	
S1983	129	3025	244	2027	
S1990	234		65	2050	
S1993	123A	3444	20	2051	
S19A	116	3017B/117	504A	1104	
S1A	116	3031A	504A	1104	
S1A06	116	3016	504A	1104	
S1A060	116	3017B/117	504A	1104	
S1A100				1114	
S1A10F					MR818
S1A12F					SPECIAL
S1A1F					1N4934
S1A2F					1N4935
S1A3	137A	3058	ZD-6.2	561	
S1A3F					1N4936
S1A4F					1N4936
S1A5F					1N4937
S1A60	116	3016	504A	1104	
S1ABF					MR817
S1AGF					1N4937
S1AN12	5874		5032		
S1AN15	5916		5064		
S1AN15R	5917		5065		
S1AN31	5986	3609	5096		
S1AN31R	5987	3698	5097		
S1AN40	5986	3609	5096		
S1AN40R	5987	3698	5097		
S1AN55	6026		5128		
S1AN55R	6027		5129		
S1AN6	5854		5016		
S1AR1	116	3017B/117	504A	1104	
S1AR2	116	3016	504A	1104	
S1B	116	3017B/117	504A	1104	
S1B-01-02	116	3311	504A	1104	
S1B-0306	125	3081	510,531	1114	
00000S1B01	116	3311	504A	1104	
S1B01-01	116	3311	504A	1104	
S1B01-02	116	3017B/117	504A	1104	
S1B01-0226		3031A	504A	1104	
S1B01-04	506		511	1114	
S1B01-06	116	3017B/117	504A	1104	
S1B01-06B	116	3017B/117			
S1B0101CR	116		504A	1104	
S1B0102	116		504A	1104	
S1B02	116		504A	1104	
S1B02-0	116(2)	3106/5304	504A(2)	1104	
S1B02-03C	156	3311	512		
S1B02-06CE	116	3311		1104	
S1B02-06CRE	116	3311		1104	
S1B02-C	116	3311	504A.	1104	
S1B02-CR	116(2)	3311	504A(2)	1104	
S1B02-CR1	178MP	3031A	300(2)	1122(2)	
S1B0201B	166	9075.	BR-600	1152	
S1B0201CR	116	3311	504A	1104	
S1B02C			53	2007	

Industry Standard No.	ECG	SK	GE	RS 276-	MOTOR.
S1B1		3311	504A	1104	
S1BD1-02	116	3031A	504A	1104	
S1BN15	5916		5064		
S1BN15R	5917		5065		
S1BN31	5986	3609	5096		
S1BN31R	5987	3518/5995	5097		
S1BN40	5986	3609	5096		
S1BN40R	5987	3698	5097		
S1BN55	6026		5128		
S1BN55R	6027		5129		
00000S1B01				1104	
S1BR2	5854		5016		
S1BR5	5944		5048		
S1C	125	3032A	510,531	1114	
S1CN1	116	3017B/117	504A	1104	
S1CR5	5944		5048		
S1D	125	3033	510,531	1114	
S1D153	152	3054/196	66	2048	
S1D23-13	506	3130	511	1114	
S1D23-15	506		511	1114	
S1D26	506		511	1114	
S1D30-13	506	3130	511	1114	
S1D30-15	506	3130	511	1114	
S1D50B851-A	177	3100/519	300	1122	
S1D51C052-19	116	3311	504A	1104	
S1D51C169-1	116	3311	504A	1104	
S1DR5	5944		5048		
S1ER5	5944		5048		
S1G10				1102	
S1G20				1102	
S1G40				1103	
S1G40Z				1103	
S1G60Z				1104	
S1GR2	5801	9004		1142	
S1L200			504A	1104	
S1M1	5800	9003		1142	
S1M2	5800	9003		1142	
S1P	506	3130	511	1114	
S1PB15				1172	
S1Q20Z				1172	
S1QB10				1152	
S1QB20				1172	
S1QB20Z				1172	
S1QB40				1173	
S1QB40Z				1173	
S1R-80	558		504A	1104	
S1R12B	142A	3062	ZD-12	563	
S1R13B	142A	3062	ZD-12	563	
S1R20	177	3032A	300	1122	
S1R80	552	9000	511	1114	
S1RB	166	9075	BR-600	1152	
S1RB-10				1152	
S1RB10	166	9075	512	1152	
S1RB20				1172	
S1RB20Z				1172	
S1RB40				1173	
S1RB40Z				1173	
S1RBA-10	116(4)	3311			
S1RBA10				1152	
S1RBA20				1172	
S1RBA40				1173	
S1RBA40Z				1173	
S1RC20	116	3017B/117	504A(2)	1104	
S1RC20R	116	3017B/117	504A	1104	
S1S11			1N34AS	1123	
S1S2			300	1122	
S1S20	109			1123	
S1SD-1			504A	1104	
S1SD-1HF			504A	1104	
S1SD-1X			504A	1104	
S1SM-150-01	116	3017B/117	504A	1104	
S1SM-150-02	116	3016	504A	1104	
S1SW-05-02	116	3016	504A	1104	
S2			21	2034	
S20	125	3032A	510,531	1114	
S200	116	3036	504A	1104	
S2001JS2	5455	3597			
S2001JS3	5455	3597			
S2001LS3	5414	3954			
S2001M	5408	3577	C6B		2N1597
S2001MS1					2N4216
S2001MS2			C5B		2N2326
S2001MS3			C6B		MCR1906-4
S2001RS3	5414	3954			
S2002	161	3124/289	39	2015	
S2003-1	130	3027	14	2041	
S2003LS2					MCR106-4
S2003LS3	5414	3954			MCR106-4
S2003M	5408	3503			SPECIAL
S2003MS2			C106B1		2N6239
S2003MS3			C107B1		2N6239
S2003RS2	5455	3570	C106B1		
S2003RS3	5414	3954	C107B1		
S2004F1					SPECIAL
S2005	5912		5064		
S2006B		3614	C220B2		
S2006F1					SPECIAL
S2006FS21					MCR72-4
S2006FS31					MCR72-4
S2006G	5514	3613	C222B		
S2006H	5492	6792	C220B		
S2006L	5463	3502/5513	C122B		
S2006R	5463	3572			
S2008B		3614	C220B2		
S2008F1					2N4442
S2008F3					2N4442
S2008FS21					MCR72-4
S2008FS31					MCR72-4
S2008G	5514	3613	C222B		
S2008H	5492	6792	C220B		
S2008L	5463	3572	C122B		
S2008R	5463	3572			
S201	116	3016	504A	1104	
S2010	5916		5064		
S20100	5910		5076		
S2010B		3614	C220B2		
S2010F1					SPECIAL
S2010F3					SPECIAL
S2010G	5514	3613	C222B		
S2010H	5492	6792	C220B		
S2010L	5552		C123B		
S2015	5916		5064		
S2015G	5514	3613			SPECIAL
S2015H	5504				SPECIAL
S2016B		3614	C230B2		
S2016H		3505	C230B		
S202	116	3016	504A	1104	
S2020	5916	3019	5064		
S2020G	5514	3613			
S2025	5920		5068		
S2025C					MCR649-4
S2025G	5517	3615	C232B		SPECIAL
S2025H	5524	3505/5547	C230B		2N685
S203	116	3016	504A	1104	

Industry Standard No.	ECG	SK	GE	RS 276-	MOTOR.
S2030	5920		5068		
S2034	123A	3122	20	2051	
S2035	5920		5068		
S2035G	5517	3615			2N3871
S2035H	5543	3505/5547			2N3897
S2038		3019	62	2010	
S204	116	3016	504A	1104	
S2040	5920		5068		
S2041	153	3083/197	69	2049	
S2041635	102A		53	2007	
S2042	152	3893	69	2048	
S2042634	103A	3835	59	2002	
S2043	123A	3444	20	2051	
S2044	123A	3444	20	2051	
S20445	195A	3047			
S20446	237	3299	46		
S2045	5924	3122	5072		
S205	116	3017B/117	504A	1104	
S2050	5924		5072		
S2059	124	3021	12		
S206	116	3017B/117	504A	1104	
S2060	5924		5072		
S2060A	5454	3570		1067	2N6238
S2060B	5455	3570		1067	2N6239
S2060C	5456	3557		1020	MCR106-5
S2060D		3557	C106D1		2N6240
S2060E		3571	C106E1		MCR106-7
S2060F	5461	3685			2N6237
S2060M		3571	C106M1		2N6241
S2060Q	5461	3685			2N6236
S2060Y	5461	3685			2N6237
S2061A	5454	6754		1067	2N6238
S2061B	5455	3597		1067	2N6234
S2061C	5456			1020	MCR106-5
S2061D	5457				2N6240
S2061E			C107E1		MCR106-7
S2061F	5461	3685			2N6237
S2061M			C107M1		2N6241
S2061Q	5461	3685			2N6236
S2061Y	5461	3685			2N6237
S2062A	5454	6754		1067	2N6238
S2062B	5455	3597		1067	2N6239
S2062C	5457				MCR106-5
S2062D		3571	C107D1		2N6240
S2062E			C107E1		MCR106-7
S2062F	5453	6753		1067	2N6237
S2062M			C107M1		2N6241
S2062Q	5452	6752		1067	2N6236
S2062Y	5452	6752		1067	2N6237
S2070	5928		5076		
S208	125	3032A	510,531	1114	
S2080	5928		5076		
S2085			1N34AS	1123	
S2087G	177		300	1122	
S2090	5932	3122	5076		
S2091	159	3114/290	82	2032	
S20ND400	116	3016	504A	1104	
S20NH400	116	3016	504A	1104	
S21	116	3124/289	504A	1104	
S210	125	3033	510,531	1114	
S2104		3024	63	2030	
S2105	5912		MR-1		
S2110	5916		MR-1		
S2114	2114				MCM2114
S2114L					MCM21L14
S2115	5916		MR-1		
S2117	129	3025	244	2027	
S2118	128	3024	243	2030	
S2120	5916		MR-1		
S2121	123A	3444	20	2051	
S2122	123A	3444	20	2051	
S2123	123A	3444	20	2051	
S2124	123A	3444	20	2051	
S2125	5920	3122	MR-1		
S21271	109	3090		1123	
S2128	159	3114/290	82	2032	
S2129	159	3114/290	82	2032	
S2130	5920	3114/290	MR-1		
S2131	107	3039/316	11	2015	
S2132	107	3039/316	11	2015	
S2133	107	3039/316	11	2015	
S2134	107	3039/316	11	2015	
S2135	5920	3039/316	MR-1		
S2140	5920	3122	MR-1		
S2147					MCM2147
S21520	175	3026	246	2020	
S21549	128	3024	243	2030	
S2159	108	3452	86	2038	
S21648	107	3039/316	11	2015	
S217	116	3016	504A	1104	
S2171	123A	3444	20	2051	
S2172	123A	3444	20	2051	
S218	116	3017B/117	504A	1104	
S219	116	3017B/117	504A	1104	
S22	116	3311	504A	1104	
0S22.3504-040		3004	2	2007	
0S22.3504-060		3004	2	2007	
0S22.3505-910		3004	2	2007	
0S22.3511-770		3007	245	2004	
0S22.3511-780		3007	245	2004	
0S22.3511-790		3007	245	2004	
0S22.3516-380		3007	52	2024	
0S22.3640-050		3009	239	2006	
0S22.3640-080		3018	86	2038	
0S22.3640-081		3048	46		
0S22.3640-082		3048	46		
0S22.3901-001		3087		1123	
0S22.3902-001		3087		1123	
0S22.3905-001		3017B	504A	1104	
S220	116	3311	504A	1104	
S2209	128	3024	243	2030	
S221	116	3016	504A	1104	
S222	116	3016	504A	1104	
S2224	108		86	2038	
S2225	123A	3444	20	2051	
S223	116	3016	504A	1104	
S224	116	3016	504A	1104	
S2241	130	3027	14	2041	
S22543	123A	3444	20	2051	
S2274	129		244	2027	
S229	5802	9005		1143	
S22A	116	3311	504A	1104	
S23	116	3017B/117	504A	1104	
S230	116	3017B/117	504A	1104	
S23130			82	2032	
S232	116	3017B/117	504A	1104	
S2321	175		246	2020	
S233	116	3033	504A	1104	
S234	116	3017B/117	504A	1104	
S235	116	3016	504A	1104	
S23579			20	2051	
S236	5802	9005		1143	
S2368	129	3025	244	2027	
S2369	128	3024	243	2030	

Industry Standard No.	ECG	SK	GE	RS 276-	MOTOR.
S2370	129	3025	244	2027	
S2371	128	3024	243	2030	
S238	116	3016	504A	1104	
S239	116	3017B/117	504A	1104	
S2392	130	3027	14	2041	
S2397	123A	3444	20	2051	
S2398C	129	3025	244	2027	
S23A	116	3017B/117	504A	1104	
S24	125	3033	510,531	1114	
S240	116	3016	504A	1104	
S2400	128	3024	243	2030	
S2400A	128	3024	243	2030	
S2400B	128	3024	243	2030	
S2401	128	3024	243	2030	
S2401A	128	3024	243	2030	
S2401B	128	3024	243	2030	
S2401C	128	3024	243	2030	
S2402	128	3024	243	2030	
S2402A	128	3024	243	2030	
S2402B	128	3024	243	2030	
S2402C	128	3024	243	2030	
S2403B	130	3027	14	2041	
S2403C	130	3027	14	2041	
S241	116	3017B/117	504A	1104	
S24226	159		82	2032	
S2427	128	3024	243	2030	
S243	116	3017B/117	504A	1104	
S2438			61	2038	
S24591	123A	3444	20	2051	
S24592	199		62	2010	
S24594	129		244	2027	
S24596	123A	3444	20	2051	
S24597	129	3025	244	2027	
S24598	128	3024	243	2030	
S24612	129	3025	244	2027	
S24612A	129	3025	244	2027	
S24614	128	3024	243	2030	
S24615	129	3025	244	2027	
S24616	128	3024	243	2030	
S2471			14	2041	
S2486	175		246	2020	
S2487	128	3024	243	2030	
S250	116	3017B/117	504A	1104	
S251	116	3017B/117	504A	1104	
S252	116	3017B/117	504A	1104	
S2525	159	3114/290	82	2032	
S2526	128	3024	243	2030	
S25261			39	2015	
S2527	175		246	2020	
S253	116	3016	504A	1104	
S2530	283		36		
S254	116	3016	504A	1104	
S2544	236	3239			
S255	116	3017B/117	504A	1104	
S256	116	3017B/117	504A	1104	
S257	125	3032A	510,531	1114	
S258	125	3032A	510,531	1114	
S25805	108	3018	86	2038	
S2581	123A	3444	20	2051	
S2582	123A	3444	20	2051	
S2590	123A	3444	20	2051	
S2593	123A	3444	20	2051	
S25941			11	2015	
S25A05	5980				1N1183
S25A1	5986				1N1184
S25A10	6002				1N3768
S25A3	5990				1N1187
S25A4	5990				1N1188
S25A6	5994				1N1190
S25A8	5998				1N3766
S26	116	3016	504A	1104	
S260	125	3033	510,531	1114	
S2600B		3577	C122B		C122B1
S2600D		3578	C122D		C122D1
S2600M		3503	C122M		C122M1
S2610B		3577			S2800B
S2610D	5507	3578			S2800D
S2610M	5529	3503			S2800M
S2617	108	3019	86	2038	
S2617(UHF)			11	2015	
S262	116	3017B/117	504A	1104	
S2620B		3577	C122B		S2800B
S2620D		3578	C122D		S2800D
S2620M		3503	C122M		S2800M
S2635	123A	3444	20	2051	
S2636	123A	3444	20	2051	
S2645	159	3114/290	82	2032	
S2645A	159	3114/290	82	2032	
S2648	128	3024	243	2030	
S26822			20	2051	
S2710B	5513	3502			S2800B
S2710D	5513	3502			S2800D
S2710M	5513	3502			S2800M
S2716			61	2038	
S2718			61	2038	
S2719			86	2038	
S27233	128	3024	243	2030	
S2741	130	3027	14	2041	
S27604	161	3124/289	39	2015	
S2771	129	3025	244	2027	
S27893			61	2038	
S2794	128	3024	243	2030	
S28	125	3032A	510,531	1114	
S2800A	5462	3572/5463	C122A		
S2800A1					S2800A1
S2800B	5463	3572	C122B		
S2800B1					S2800B1
S2800C	5465	3558			
S2800C1					S2800C1
S2800D	5465	3558	C122D		
S2800D1					S2800D1
S2800E1					S2800E1
S2800F	5461	3572/5463			
S2800F1					S2800F1
S2800M1					S2800M1
S2800N1					S2800N1
S2800S1					S2800S1
S2935	123A	3444	20	2051	
S2944	123A	3444	20	2051	
S29445	123A	3444	20	2051	
S2984	123A	3444	20	2051	
S2985	123A	3444	20	2051	
S2986	154	3045/225	40	2012	
S2988	159	3114/290	82	2032	
S2988A	159	3114/290	82	2032	
S2989	123A	3444	20	2051	
S2991	129	3025	244	2027	
S2992	128	3024	243	2030	
S2993	129	3025	244	2027	
S2994	129	3025	244	2027	
S2995	129	3025	244	2027	
S29956			62	2010	
S2996	123A	3444	20	2051	
S2997	123A	3444	20	2051	

Industry Standard No.	ECG	SK	GE	RS 276-	MOTOR.
S2998	123A	3444	20	2051	
S2999	123A	3444	20	2051	
S2A06	116	9003/5800	504A	1104	
S2A10	116	3016	504A	1104	
S2A100	5809	9010			
S2A20	5802	9005		1143	
S2A30	5804	9007	504A	1144	
S2A40	5804	9007		1144	
S2A50	5806	3848			
S2A60	5806	3848			
S2A80	5808	9009			
S2AN12	5874		5032		
S2AN6	5874		5032		
S2AO6	5800			1142	
S2AR1	116	3017B/117	504A	1104	
S2AR2	116	3017B/117	504A	1104	
S2C30	116	3016	504A	1104	
S2C40	116	3016	504A	1104	
S2C40A	116	3016	504A	1104	
S2CN1	5802	9005		1143	
S2D153	196	3054	241	2020	
S2E100	125	3033	510,531	1114	
S2E20	116	3016	504A	1104	
S2E60	116	3032A	504A	1104	
S2E60-1	116	3016	504A	1104	
S2F					1N4935
S2F10				1102	
S2F20				1102	
S2GR2	5802	9005		1143	
S2M					1N4003
S2M1	5801	9004		1142	
S2M2	5800	9003		1142	
S2PB	166	9075		1152	
S2PB10				1152	
S2PB20				1172	
S2PB40				1173	
S2Q100				1114,	
S2Q20				1102	
S2Q40				1103	
S2Q60				1104	
S2RB40				1173	
S2TB10				1152	
S2TB20				1172	
S2TB40				1173	
S2V	125	3081	510	1114	
S2VB	116(4)	3311	504A	1104	
S2VB10	5313	3051/156	BR-206		
S2VC	116(2)	3081/125	510	1114	
S2VC10		3311	510	1114	
S2VC10R	125		510	1114	
S2Z20				1143	
S2Z40				1144	
S3			21	2034	
S30	116	3017B/117	504A	1104	
S3001	195A	3048/329	46		
S3002	154	3040	40	2012	
S3004	159	3048/329	82	2032	
S3004-1715	137A	3058	ZD-6.2	561	
S3004-1716	177	3100/519	300	1122	
S3004-1718	137A	3060/139A	ZD-6.2	561	
S3007	351	3177			
S3008		3048	277		
S3008(MOT)	311		277		
S3011A	605	3327/614			
S3012	129	3025	244	2027	
S3013	311	3048/329	277		
S3013(MOT)	311		277		
S3016	601	3017B/117	300	1122	
S3016R	601	3017B/117	300	1122	
S3019	108	3019	86	2038	
S3020	108	3039/316	86	2038	
S3021		3104A	224		
S3030G	109	3090		1123	
S3032		3200	227		
S3033	154	3044	40	2012	
S3034	154	3044	40	2012	
S3035	154	3044	40	2012	
S305	130		14	2041	
S305A	130	3027	14	2041	
S305D	130		14	2041	
S306A	175		246	2020	
S3072C	177	3100/519	300	1122	
S3073	610		90		
S31	116	3016	504A	1104	
S3105	5980	3610	5096		
S3110	5982	3609/5986	5096		
S31100	6002		5108		
S3115	5986	3609	5096		
S3120	5986	3609	5096		
S3125	5988	3608/5990	5100		
S3130	5988	3608/5990	5100		
S3135	5990	3608	5100		
S3140	5990	3608	5100		
S3145	5992	3501/5994	5104		
S3150	5992	3501/5994	5104		
S31551			82	2032	
S3160	5994	3501	5104		
S3170	5998		5108		
S3180	5998		5108		
S31866			210	2051	
S32		3016	504A	1104	
S3205	5980	3610	5096		
S3210	5982	3609/5986	5096		
S3215	5986	3609	5096		
S3220	5986	3609	5096		
S3225	5988	3608/5990	5100		
S3230	5988	3608/5990	5100		
S3235	5990	3608	5100		
S3240	5990	3608	5100		
S32417			86	2038	
S3245	5992	3501/5994	5104		
S3250	5994	3501	5104		
S32550			20	2051	
S3260	5992	3501/5994	5104		
S32669	108	3019	86	2038	
S326690	108	3018	86	2038	
S3270	5998		5108		
S3280	5998		5108		
S3290	6002		5108		
S32903			57	2017	
S33	116	3016	504A	1104	
S33291	124		12		
S33529			69	2049	
S33530			66	2048	
S33755			210	2051	
S3386	129		244	2027	
S33886	129		244	2027	
S33886A	129	3025	244	2027	
S33990			86	2038	
S34	506	3016	504A	1104	
S3405	6020		5128		
S3410	6022		5128		
S34100	6044		5140		
S3415	6026		5128		

Industry Standard No.	ECG	SK	GE	RS 276-	MOTOR.
S3420	6026		5128		
S3425	6030		5132		
S3430	6030		5132		
S3435	6034		5132		
S3440	6034		5132		
S3445	6040		5136		
S3450	6038		5136		
S34540	123A	3444	20	2051	
S3460	6040		5136		
S3470	6042		5140		
S3480	6042		5140		
S3490	6044		5140		
S35	116	3311	504A	1104	
S35232			39	2015	
S35233			86	2038	
S353	130	3510	14	2041	
S354	175		246	2020	
S35486	180		74	2043	
S35487	181		75	2041	
S356	130		14	2041	
S3577G	109	3090		1123	
S36	116	3031A	504A	1104	
S3603G	109	3090		1123	
S3605	6048		5128		
S3610	6050		5128		
S36100	6072	7272	5140		
S3615	6054		5128		
S3620	6054		5128		
S3625	6058		5132		
S3630	6058		5132		
S3635	6060		5132		
S3639	159	3114/290	82	2032	
S3640	6060	3114/290	5132		
S3640A	159	3114/290	82	2032	
S3645	6064		5136		
S3650	6064		5136		
S3655	159	3025/129	82	2032	
S3655A	159	3114/290	82	2032	
S3660	6064		5136		
S3670	6068		5140		
S3680	6068		5140		
S3690	6072	7272	5140		
S36951			268	2038	
S36999	123A	3444	20	2051	
S37	5870		5032		
S3700B	230	3042			
S3700D	230	3042			
S3700M	230	3042			
S3702S	231	3857			
S3702SF	231	3857			
S3703SF	230	3042			
S3704A	231	3857			
S3704B	231	3857			
S3704D	231	3857			
S3704M	231	3857			
S3704S	231	3857			
S3705M	230	3042			
S3706E	231	3857			
S37162			241	2020	
S37165			69	2049	
S37166			66	2048	
S37182			20	2051	
S37214			20	2051	
S37423			20	2051	
S3771	130		14	2041	
S3776B	109	3090		1123	
S38	5870		5032		
S3800S	308	3855			
S3800SF	310	3856			
S3838GA	109	3090	1N60	1123	
S38763			20	2051	
S38787			20	2051	
S38854			20	2051	
S3885G	109	3090		1123	
S39	5874		5032		
S39094			82	2032	
S39261			69	2049	
S39262			66	2048	
S39509			218	2026	
S39560			40	2012	
S3A025					1N5400
S3A05	5800	9003			
S3A06	116	3016	504A	1104	
S3A1	5801			1142	1N5401
S3A10	5809	9010			1N5409
S3A10F					MR918
S3A12F					SPECIAL
S3A1F					MR851
S3A2	5802	9005		1143	1N5402
S3A2F					MR852
S3A3	5804			1144	1N5403
S3A3F					MR854
S3A4	5804			1144	1N5404
S3A4F					MR854
S3A5	5806	3848			1N5405
S3A5F					MR856
S3A6	5806	3051/156			1N5406
S3A6F					MR856
S3A7					1N5408
S3A8	5808	9009			1N5408
S3A8F					MR917
S3A9					1N5409
S3AN12	5878		5036		
S3AN6	5878		5036		
S3AR1	116	3016	504A	1104	
S3BA05				1142	
S3CN1	5804	9007		1144	
S3G10				1143	
S3G20				1143	
S3G4	125	3081	510,531	1114	
S3G40				1144	
S3G40Z				1144	
S3GR2	5804	9007		1144	
S3M1	5802	9005		1143	
S3MX	116		504A	1104	
S3V-10	156	3051			
S3V20	156	3051	512		
S4			21	2034	
0S4.000654			300	1122	
S40	116	3017B/117	504A	1104	
S400	5878		5036		
S4001	116	3311	504A	1104	
S4001M	5409	3502/5513	C6D		2N1599
S4001MS1					2N4219
S4001MS2			C5D		2N2329
S4001MS3			C6D		MCR1906-6
S4002	161	3124/289	39	2015	
S4003M	5409	3503			SPECIAL
S4003MS2			C106D1		2N6240
S4003MS3			C107D1		2N6240
S4003RS2	5457		C106D1		
S4003RS3	5457		C107D1		
S4004F1					SPECIAL
S4006B			C220D2		MCR729-6

Industry Standard No.	ECG	SK	GE	RS 276-	MOTOR.
S4006F1					SPECIAL
S4006FS21					MCR72-6
S4006FS31					MCR72-6
S4006G	5515	6615	C222D		
S4006H	5494	6794	C220D		
S4006L	5465	3687	C122D		
S4006R	5465	3687			
S4008F1					2N4443
S4008F3					2N4443
S4008FS21					MCR72-6
S4008FS31					MCR72-6
S4008G	5515	6615	C222D		
S4008H	5494	6794	C220D		
S4008L	5465	3687	C122D		MCR3000-6
S4008R	5465	3687			
S4010F1					SPECIAL
S4010F3					SPECIAL
S4010G	5515	6615	C222D		
S4010H	5494	6794	C220D		
S4010L	5554		C123D		
S4015					MCM2115
S4015G	5515	6615			SPECIAL
S4015H	5507				SPECIAL
S4016H		3505	C230D		
S40204	161	3018	39	2015	
S40205	154	3044	40	2012	
S4020G	5515	6615			
S4020H	5527	6627			
S4025					MCM2125
S4025C					MCR649-6
S4025G	5518		C232D		SPECIAL
S4025H	5527	6627	C230D		2N688
S4035G	5518		C229D		2N3872
S4035H	5545	3505/5547	C228D		2N3898
S40545	107		11	2015	
S409F	128	3024	243	2030	
S40A	116	3311	504A	1104	
S40A1	5986				1N1184A
S40A10	6002				1N3768
S40A2	5986				1N1186A
S40A3	5990				1N1187A
S40A4	5990				1N1188A
S40A5	5994				1N1189A
S40A6	5994				1N1190
S40A8	5998				1N3766
S410	5874		5032		
S411	5874		5032		
S413	5878		5036		
S413796	102	3722	2	2007	
S415	5882	3051/156	5040		
S417	5886	3051/156	5044		
S420	5874		5032		
S421	5874		5032		
S423	5878		5036		
S4248	102A	3004	53	2007	
S4249	159		82	2032	
S425	5882		5040		
S4264					MCM68365
S427	5886		5044		
S428	116	3016	504A	1104	
S43	116	3311	504A	1104	
S431	116	3311	504A	1104	
S44	116	3016	504A	1104	
S46	116	3016	504A	1104	
S47	116	3311	504A	1104	
S48	116	3311	504A	1104	
S49	116	3311	504A	1104	
0S492	176	3845	80		
S4A-1-A-3P	121		239	2006	
S4A06	116	3016	504A	1104	
S4AN12	5878		5036		
S4AN6	5878		5036		
S4A06			504A	1104	
S4AR1	116	3016	504A	1104	
S4AR2	116	3017B/117	504A	1104	
S4AR30	116	3017B/117	504A	1104	
S4C			504A	1104	
S4CN1	5802			1143	
S4F					1N4936
S4FN300	116	3016	504A	1104	
S4GR2	5802	9005		1143	
S4M					1N4004
S4M1	5804	9007		1144	
S4VB10	5312	3051/156	BR-206	1152	
S4VB20				1172	
S4VB40				1173	
S50	5886		5044		
S50-12	335				MRF450
S50-28	317				MRF464
S500	106	3500/5882	21	2034	
S5000	173BP	3190/184	305		
S5006B					MCR729-7
S5008L					MCR3000-7
S500B	5804	9007		1144	
S500C	5804	9007		1144	
S501	106		21	2034	
S5015		3103A	232		
S5018		3083	234		
S5019		3026	233		
S502	177	3100/519	300	1122	
S5020		3111	39	2015	
S5021	161	3018	39	2015	
S5024		3103A	229		
S5025		3044	230		
S5025C					MCR649-7
S5025G					MCR3835-7
S5025H					MCR3935-7
S502A	177	3100/519	300	1122	
S502B	177	3100/519	300	1122	
S5035G					MCR3835-7
S5035H					MCR3935-7
S504-0	129		244	2027	
S506	177	3100/519	300	1122	
S506A	177	3100/519	300	1122	
S506B	177	3100/519	300	1122	
S5089-A	116	3311	504A	1104	
S509	177	3100/519	300	1122	
S509A	177	3100/519	300	1122	
S509B	177	3100/519	300	1122	
S51	5886		5044		
S5101	65101				MCM5101
S5105	6354	6554			
S5105R	6355	6555			
S5110	6354	6554			
S51100	6358	6558			
S51100R	6359	6559			
S5110R	6355	6555			
S5115R	6355	6555			
S5120	6354	6554			
S5120R	6355	6555			
S5125	6354	6554			
S5125R	6355	6555			
S5135	6354	6554			
S5135R	6355	6555			

Industry Standard No.	ECG	SK	GE	RS 276-	MOTOR.
S5140	6354	6554			
S5140R	6355	6555			
S5145	6356	6556			
S5150	6356	6556			
S5160	6356	6556			
S5170	6358	6558			
S5170R	6359	6559			
S5180	6358	6558			
S5180R	6359	6559			
S5190	6358	6558			
S5190R	6359	6559			
S52	5874		5032		
S520	159		82	2032	
S5233B	5070A	9021			
S5237B	5072A	3136			
S5238B	5073A	3749			
S5243B	143A	3750			
S5246B	5075A	3751			
S5247B	5076A	9022			
S5248B	5077A	3752			
S5249B	5078A	9023			
S5253B	5082A	3753			
S5255B	5083A	3754			
S5256B	5084A	3755			
S5269B	5094A	9024			
S5295C	506	3843			
S5295G	552	3843	511	1114	
S5295J	552	3843	511	1114	
S53	5878		5036		
S5305	6354	6554			
S5305R	6355	6555			
S5310	6354	6554			
S53100	6358	6558			
S53100R	6359	6559			
S5310R	6355	6555			
S5315	6354	6554			
S5315R	6355	6555			
S5320	6354	6554			
S5325	6354	6554			
S5325R	6355	6555			
S5327E	108	3039/316	86	2038	
S5328E	108		86	2038	
S5330	6354	6554			
S5330R	6355	6555			
S5335	6354	6554			
S5335R	6355	6555			
S5340	6354	6554			
S5340R	6355	6555			
S5345	6356	6556			
S5350	6356	6556			
S5360	6356	6556			
OS536G	123A	3444	20	2051	
S5370	6358	6558			
S5370R	6359	6559			
S5380	6358	6558			
S5380R	6359	6559			
S5390	6358	6558			
S5390R	6359	6559			
S5472	157	3054/196	232		
S54LS00F					SN54LS00J
S54LS00W					SN54LS00W
S54LS01F					SN54LS01J
S54LS01W					SN54LS01W
S54LS02F					SN54LS02J
S54LS02W					SN54LS02W
S54LS03F					SN54LS03J
S54LS03W					SN54LS03W
S54LS04F					SN54LS04J
S54LS04W					SN54LS04W
S54LS05F					SN54LS05J
S54LS05W					SN54LS05W
S54LS08F					SN54LS08J
S54LS08W					SN54LS08W
S54LS09F					SN54LS09J
S54LS09W					SN54LS09W
S54LS10F					SN54LS10J
S54LS10W					SN54LS10W
S54LS11F					SN54LS11J
S54LS11W					SN54LS11W
S54LS12F					SN54LS12J
S54LS12W					SN54LS12W
S54LS132F					SN54LS132J
S54LS132W					SN54LS132W
S54LS136F					SN54LS136J
S54LS136W					SN54LS136W
S54LS138F					SN54LS138J
S54LS138W					SN54LS138W
S54LS139F					SN54LS139J
S54LS139W					SN54LS139W
S54LS13F					SN54LS13J
S54LS13W					SN54LS13W
S54LS145F					SN54LS145J
S54LS145W					SN54LS145W
S54LS14F					SN54LS14J
S54LS14W					SN54LS14W
S54LS151F					SN54LS151J
S54LS151W					SN54LS151W
S54LS153F					SN54LS153J
S54LS153W					SN54LS153W
S54LS157F					SN54LS157J
S54LS157W					SN54LS157W
S54LS158F					SN54LS158J
S54LS158W					SN54LS158W
S54LS15F					SN54LS15J
S54LS15W					SN54LS15W
S54LS164F					SN54LS164J
S54LS164W					SN54LS164W
S54LS170F					SN54LS170J
S54LS170W					SN54LS170W
S54LS174F					SN54LS174J
S54LS174W					SN54LS174W
S54LS175F					SN54LS175J
S54LS175W					SN54LS175W
S54LS181F					SN54LS181J
S54LS181W					SN54LS181W
S54LS190F					SN54LS190J
S54LS190W					SN54LS190W
S54LS191F					SN54LS191J
S54LS191W					SN54LS191W
S54LS192F					SN54LS192J
S54LS192W					SN54LS192W
S54LS193F					SN54LS193J
S54LS193W					SN54LS193W
S54LS196F					SN54LS196J
S54LS196W					SN54LS196W
S54LS197F					SN54LS197J
S54LS197W					SN54LS197W
S54LS20F					SN54LS20J
S54LS20W					SN54LS20W
S54LS21F					SN54LS21J
S54LS21W					SN54LS21W
S54LS221F					SN54LS221J
S54LS221W					SN54LS221W
S54LS22F					SN54LS22J

Industry Standard No.	ECG	SK	GE	RS 276-	MOTOR.
S54LS22W					SN54LS22W
S54LS251F					SN54LS251J
S54LS251W					SN54LS251W
S54LS253F					SN54LS253J
S54LS253W					SN54LS253W
S54LS260F					SN54LS260J
S54LS260W					SN54LS260W
S54LS266F					SN54LS266J
S54LS266W					SN54LS266W
S54LS26F					SN54LS26J
S54LS26W					SN54LS26W
S54LS27F					SN54LS27J
S54LS27W					SN54LS27W
S54LS283F					SN54LS283J
S54LS283W					SN54LS283W
S54LS28F					SN54LS28J
S54LS28W					SN54LS28W
S54LS293F					SN54LS293J
S54LS293W					SN54LS293W
S54LS30F					SN54LS30J
S54LS30W					SN54LS30W
S54LS32F					SN54LS32J
S54LS32W					SN54LS32W
S54LS33F					SN54LS33J
S54LS33W					SN54LS33W
S54LS37F					SN54LS37J
S54LS37W					SN54LS37W
S54LS386F					SN54LS386J
S54LS386W					SN54LS386W
S54LS38F					SN54LS38J
S54LS38W					SN54LS38W
S54LS40F					SN54LS40J
S54LS40W					SN54LS40W
S54LS42F					SN54LS42J
S54LS42W					SN54LS42W
S54LS51F					SN54LS51J
S54LS51W					SN54LS51W
S54LS54F					SN54LS54J
S54LS54W					SN54LS54W
S54LS55F					SN54LS55J
S54LS55W					SN54LS55W
S54LS670F					SN54LS670J
S54LS670W					SN54LS670W
S54LS73F					SN54LS73AJ
S54LS73W					SN54LS73AW
S54LS74F					SN54LS74AJ
S54LS74W					SN54LS74AW
S54LS75F					SN54LS75J
S54LS75W					SN54LS75W
S54LS76F					SN54LS76AJ
S54LS76W					SN54LS76AW
S54LS78F					SN54LS78AJ
S54LS78W					SN54LS78AW
S54LS85F					SN54LS85J
S54LS85W					SN54LS85W
S54LS86F					SN54LS86J
S54LS86W					SN54LS86W
S54LS90F					SN54LS90J
S54LS90W					SN54LS90W
S54LS92F					SN54LS92J
S54LS92W					SN54LS92W
S54LS93F					SN54LS93J
S54LS93W					SN54LS93W
S54LS95BF					SN54LS95BJ
S54LS95BW					SN54LS95BW
S55	5882		5040		
S5556T					MC1556G
S5558E					MC1558L
S5558T					MC1558G
S5596F					MC1596L
S5596K					MC1596G
S56	5882	3051/156	5040		
S5670E	108		86	2038	
S57	5886	3051/156	5044		
S5709G					MC1709F
S5709T					MC1709G
S5710T					MC1710G
S5711K					MC1711G
S5723T					MC1723G
S5733K					MC1733G
S5741T		3514			MC1741G
S58	5886	3051/156	5044		
S59	5890	3051/156	5044		
S5A025					MR500
S5A1					MR501
S5A10					MR510
S5A10F					MR918
S5A12F					SPECIAL
S5A2					MR502
S5A3					MR504
S5A4					MR504
S5A5					MR506
S5A6					MR506
S5A8					MR508
S5A8F					MR917
S5AN12	5882	3500	5040		
S5AN6	5882	3500	5040		
S5AR1	116	3017B/117	504A	1104	
S5AR2	116	3017B/117	504A	1104	
S5CN1	5804	9007		1144	
S5GR2	5806	3848			
S5M1	5804	9007		1144	
S5S	116(2)	3311	504A(2)	1104	
S5SR	116	3311	504A	1104	
S5VB10	5313		BR-206		
S6-10	140A	3061	ZD-10	562	
S6-10A	140A	3061	ZD-10	562	
S6-10B	140A	3061	ZD-10	562	
S6-10C	140A	3061	ZD-10	562	
S6-3	166			1152	
S600	5882	3051/156	5040		
S6000C		3575	C126C		MCR220-5
S6000E		3576	C126E		MCR220-7
S6000S			C126S		MCR220-9
S6001M	5410				SPECIAL
S6003M	5410				SPECIAL
S6003RS2		3571	C106M1		
S6004F1					SPECIAL
S6005	116	3017B/117	504A	1104	
S6006F1					SPECIAL
S6006G	5516	6616	C222M		
S6006H	5496	6796	C220M		
S6006L		3503	C122M		
S6008F					2N4444
S6008F3					2N4444
S6008G	5516	6616	C222M		
S6008H	5496	6796	C220M		
S6010F1					SPECIAL
S6010F3					SPECIAL
S6010G	5516	6616	C222M		
S6010H	5496	6796	C220M		
S6010L	5556		C123M		
S6015G	5516	6616			SPECIAL
S6015H	5529	6629			SPECIAL

Industry Standard No.	ECG	SK	GE	RS 276-	MOTOR.
S6016H		3505	C230M		
S6020G	5516	6616			
S6020H	5529	6629			
S6025G	5519		C232M		SPECIAL
S6025H	5529	6629	C230M		2N690
S6035G	5519		C229M		2N3873
S6035H	5547	3505	C228M		2N3899
S608	159		82	2032	
S6080A	231	3857			
S6080B	231	3042/230			
S6089	230	3042			
S61	125	3033	510,531	1114	
S6100C		3575	ZJ436C		MCR221-5
S6100E		3576	ZJ436E		MCR221-7
S6100S			ZJ436S		MCR221-9
S6142G	5457	3598			
S62	125	3033	510,531	1114	
S6200A	5514	3613	C232A		S6200A
S6200B	5514	3613			S6200B
S6200D	5515	6615			S6200D
S6200M	5516	6616			S6200M
S6210A		3579			S6210A
S6210B		3579			S6210B
S6210D		3580			S6210D
S6210M		3504			S6210M
S6220A			C230A2		S6220A
S6220B		3614	C230B2		S6220B
S6220D			C230D2		S6220D
S6220M			C230M2		S6220M
S6230A			C230A8		SPECIAL
S6230B			C230B8		SPECIAL
S6230D			C230D8		SPECIAL
S6230M			C230M8		SPECIAL
S6240A			C230A4		SPECIAL
S6240B			C230B4		SPECIAL
S6240D			C230D4		SPECIAL
S6240M			C230M4		SPECIAL
S6250A			C230A6		SPECIAL
S6250B			C230B6		SPECIAL
S6250D			C230D6		SPECIAL
S6250M			C230M6		SPECIAL
S63	125		510,531	1114	
S6420A	5562	3655			2N6171
S6420B	5562	3655			
S6420D		3656			2N6172
S6420M	5566	3657			2N6174
S6430A			C228A8		SPECIAL
S6430B			C228B8		SPECIAL
S6430D			C228D8		SPECIAL
S6430M			C228M8		SPECIAL
S6493M					SPECIAL
S65-1-A-3P	160		245	2004	
S65-2-A-3P	103A		59	2002	
S65-4-A-3P	101		8	2002	
S6508	6508				MCM6508
S6518					MCM6518
S65A-1-3P	160		245	2004	
S65B-1-3P	126		52	2024	
S65C-1-3P	121	3717	239	2006	
S66-1-A-3P	102A		53	2007	
S66-2-A-3P	102A		53	2007	
S66-3-A-3P	102A		53	2007	
S67794	124		12		
S67809	121	3717	239	2006	
S6801	123A	3444	20	2051	
S6804	128	3024	243	2030	
S6810	6810				MCM6810
S6830					MCM68A30A
S68316B					MCM68316E
S68332					MCM68A332
S684	160		245	2004	
S685	102A		53	2007	
S686	102A		53	2007	
S687	102A		53	2007	
S6A1					MR751
S6A10					MR760
S6A2					MR752
S6A3					MR754
S6A4					MR754
S6A5					MR756
S6A6					MR756
S6A8					MR758
S6AN12	5882	3500	5040		
S6AN6	5882	3500	5040		
S6AR1	116	3017B/117	504A	1104	
S6AR2	116	3051/156	504A	1104	
S6B-3-A-3P	121MP		239(2)	2006(2)	
S6BR2	156	3051	512		
S6F					1N4937
S6GR2	5806	3848			
S6M					1N4005
S6M1	5806	3848			
S7-8			300	1122	
S70.00.730			52	2024	
S70.01.704			20	2051	
S70T	160		245	2004	
S715	186		28	2017	
S72	116	3311	504A	1104	
S73	116	3311	504A	1104	
S74-3-A-3P	100		1	2007	
S75	116	3311	504A	1104	
S750	125	3031A	510,531	1114	
S7502	5405	3951			
S7503	5405	3951			
S750C	125	3031A	510,531	1114	
S77	116	3017B/117	504A	1104	
S77A-1-3P	105		4		
S77B-1-3P	121MP	3718	239(2)	2006(2)	
S77C-1-3P	121	3717	239	2006	
S79	116	3017B/117	504A	1104	
S7AN12	5886		5044		
S7AN6	5886		5044		
S7AR1	125	3032A	510,531	1114	
S7BR2	156	3051	512		
S8-01	5944		5048		
S8-06	156	3051	512		
S8-10	156	3051	512		
S80-12	335				MRF454
S800	5886	3051/156	5044		
S801	186	3192	28	2017	
S8015H	5531	6631			
S8020H	5531	6631			
S8025H	5531	3505/5547	C37N		
S808	156	3051	512		
S81	116	3016	504A	1104	
S82	116	3016	504A	1104	
S83	116	3016	504A	1104	
S84	116	3017B/117	504A	1104	
S84A-1-3P	121	3717	239	2006	
S85	116	3017B/117	504A	1104	
S855	120		CR-3		
S86	116	3017B/117	504A	1104	
S8660	184		57	2017	
S8660121-808	184	3190	57	2017	
S87TB	160		245	2004	
S88569	501B	3069/501A	520		
S88570	500A	3304			
S88571	500A	3304			
S88576	112		1N82A		
S88B-1-3P	160		245	2004	
S88C-1-3P	100	3721	1	2007	
S88TB	160		245	2004	
S8AN12	5886		5044		
S8AN6	5886		5044		
S8AR1	125	3032A	510,531	1114	
S8AR2	125	3051/156	510,531	1114	
S8BR2	156	3051	512		
S8CN1	5808	9009			
S8F					MR817
S8GR2	5808	9009			
S8M					1N4006
S8M1	5806	3848			
S8P-2-A-3P	102A		53	2007	
S8T13E					MC8T13L
S8T14E					MC8T14L
S91	116	3016	504A	1104	
S91-A	116	3016	504A	1104	
S91-H			504A	1104	
S9100	172A	3156	64		
S9101	253	3996			
S9101(IC)	705A		IC-3		
S9121	254	3997			
S913	118	3066	CR-1		
S9140	243	3182			
S9140(IC)	705A		IC-3		
S9141	244	3183A			
S9141(IC)	705A		IC-3		
S91A	5800			1142	
S91B	5801			1142	
S91H	116	3016		1142	
S92	116	3017B/117	504A	1104	
S92-1-A-3P	105		4		
S92-A	116		504A	1104	
S92-H			504A	1104	
S926	118	3066	CR-1		
S92A	5800	9003		1142	
S92H	116	9003/5800		1142	
S93	116	3017B/117	504A	1104	
S93.20.709		3647		1172	
S93.20.714		3647		1172	
S93.24.401		3090		1123	
S93.24.601		3090		1123	
S93.24.604		3090		1123	
S93A	116	3016	504A	1104	
S93H	116	3031A	504A	1104	
S93SE133	130	3027	14	2041	
S93SE140	130MP		15	2041(2)	
S93SE165	130	3027	14	2041	
S94	116	3016	504A	1104	
S95	116	3017B/117	504A	1104	
S95101	160		245	2004	
S95102	160		245	2004	
S95103	160		245	2004	
S95104	160		245	2004	
S95106	160		245	2004	
S95125	108		86	2038	
S95125A	108		86	2038	
S95126	108		86	2038	
S95126A	108		86	2038	
S9516				2004	
S95201	102A		53	2007	
S95202	123A		20	2051	
S95203	102A	3123	53	2007	
S95204	102A		53	2007	
S95206	102A		53	2007	
S95207	102A		53	2007	
S95214	102A	3123	53	2007	
S95218	102A		53	2007	
S9524				2007	
S95252	157	3103A/396	232		
S95253	121	3717	239	2006	
S95253-1	121	3717	239	2006	
S9631	123A	3444	20	2051	
S99101	160	3008	245	2004	
S99102	160	3008	245	2004	
S99103	160	3008	245	2004	
S99104	160	3008	245	2004	
S99201	102A	3004	53	2007	
S99203	102A	3004	53	2007	
S99218	102A		53	2007	
S9923				2007	
S99252	157	3103A/396	232		
S9AN12	5890	7090	5044		
S9AN6	5890	7090	5044		
S9AR1	125	3033	510,531	1114	
S9BR2	156	3051	512		
S9L-4-A-3P	102A		53	2007	
SA-2	156	3017B/117	510,531	1114	
SA-2A	156	3081/125			
SA-2B	156	3032A	504A	1104	
SA-2C	156	3081/125	510	1114	
SA-2Z	156			1143	
SA-7	101	3861	8	2002	
SA-93792	177		300	1122	
SA-93794	139A		ZD-9.1	562	
SA102	126		52	2024	
SA10M1	5809	9010			
SA128	102A		53	2007	
SA128-1	102A	3123	53	2007	
SA15V	102A		53	2007	
SA197	102A	3123	53	2007	
SA197-1	102A	3123	53	2007	
SA197-2	102A	3123	53	2007	
SA197-3	102A	3123	53	2007	
SA1M1	5801	9004		1142	
SA2	116	3017B/117	504A	1104	
SA204	102A	3123	53	2007	
SA205BLU	102A	3123	53	2007	
SA205BRN	102A	3123	53	2007	
SA205GRN	102A	3123	53	2007	
SA205ORN	102A	3123	53	2007	
SA205RED	102A	3123	53	2007	
SA205VIO	102A		53	2007	
SA205WHT	102A	3123	53	2007	
SA205YEL	102A	3123	53	2007	
SA240	102A	3123	53	2007	
SA29	158		53	2007	
SA2B	116	3017B/117	504A	1104	
SA2H			510	1114	
SA2M1	5802	9005		1143	
SA2Z	125	3051/156	510,531	1114	
SA310	159	3114/290	82	2032	
SA310A	159	3114/290	82	2032	
SA311	159	3114/290	82	2032	
SA311A	159	3114/290	82	2032	
SA312	159	3114/290	82	2032	
SA312A	159	3114/290	82	2032	
SA313	159	3114/290	82	2032	

Industry Standard No.	ECG	SK	GE	RS 276-	MOTOR.
SA314	159	3114/290	82	2032	
SA314A	159	3114/290	82	2032	
SA315	159	3114/290	82	2032	
SA315A	159	3114/290	82	2032	
SA316	159	3114/290	82	2032	
SA316A	159	3114/290	82	2032	
SA318-2	158		53	2007	
SA318-3	158		53	2007	
SA33	102A		53	2007	
SA33BRN	102A		53	2007	
SA33RED	102A		53	2007	
SA354B	103	3862	8	2002	
SA3B	116		504A	1104	
SA3M1	5804	9007		1144	
SA410	159		82	2032	
SA410A	159		82	2032	
SA411	159		82	2032	
SA411A	159		82	2032	
SA412	159	3114/290	82	2032	
SA412A	159	3114/290	82	2032	
SA413	159	3114/290	82	2032	
SA413A	159	3114/290	82	2032	
SA414	159	3114/290	82	2032	
SA414A	159	3114/290	82	2032	
SA415	159	3114/290	82	2032	
SA415A	159	3114/290	82	2032	
SA416	159		82	2032	
SA416A	159		82	2032	
SA495	106	3984	21	2034	
SA495A	106	3984	21	2034	
SA496	106	3984	21	2034	
SA496A	106	3984	21	2034	
SA496B	106	3984	21	2034	
SA50	159	3114/290	82	2032	
SA50A	159	3114/290	82	2032	
SA51	159	3114/290	82	2032	
SA52	159	3114/290	82	2032	
SA529	102A		53	2007	
SA52A	159	3114/290	82	2032	
SA52AC	159	3114/290	82	2032	
SA52B	159	3114/290	82	2032	
SA52BC	159	3114/290	82	2032	
SA53	159	3114/290	82	2032	
SA532T	928	3025/129			
SA537	106	3984	21	2034	
SA538	106	3984	21	2034	
SA539	106	3984	21	2034	
SA53A	159	3114/290	82	2032	
SA54	159	3114/290	82	2032	
SA540	106	3984	21	2034	
SA55	159	3114/290	82	2032	
SA55A	159	3114/290	82	2032	
SA56	159	3114/290	82	2032	
SA565	102A		53	2007	
SA56A	159	3114/290	82	2032	
SA5M1	5806	3848			
SA646	102A		53	2007	
SA681	158		53	2007	
SA6M1	5806	3848			
SA7	101		8	2002	
SA70	159	3114/290	82	2032	
SA70A	159	3114/290	82	2032	
SA8-1-A-3P	160		245	2004	
SA821	137A	3058	ZD-6.2	561	
SA821A	137A	3058	ZD-6.2	561	
SA823	137A	3058	ZD-6.2	561	
SA823A	137A	3058	ZD-6.2	561	
SA825	137A	3058	ZD-6.2	561	
SA825A	137A	3058	ZD-6.2	561	
SA827	137A	3058	ZD-6.2	561	
SA827A	137A	3058	ZD-6.2	561	
SA829	137A	3058	ZD-6.2	561	
SA829A	137A	3058	ZD-6.2	561	
SA853	102A		53	2007	
SA8M1	5808	9009			
SAB1044	161	3716	39	2015	
SAB3469	161	3716	39	2015	
SAC-1843			11	2015	
SAC203	56006	3659			
SAC40	106	3984	21	2034	
SAC40A	106	3984	21	2034	
SAC40B	106	3984	21	2034	
SAC42	106	3984	21	2034	
SAC42A	106	3984	21	2034	
SAC42B	106	3984	21	2034	
SAC44	106		21	2034	
SAE-1	280		262	2041	
SAE-2	281		263	2043	
SAJ72155	1076		IC-125		
SAJ72156	1091*		IC-126		
SAJ72157	1046	3471	IC-118		
SAJ72158	1047		IC-116		
SAJ72159	1077		IC-124		
SAJ72160	1086*		IC-142		
SAJ72161	1089*		IC-120		
SAJ72162	1048		IC-122		
SATCHZ339	108	3039/316	86	2038	
SAW-1S1941			504A	1104	
SAW-1S1944			504A	1104	
SAW-2SB56			52	2024	
SAW-2SC372GR			20	2051	
SAW-2SC372Y			20	2051	
SAW-2SC945R			20	2051	
SB-01				1152	
SB-02				1172	
SB-03			504A	1104	
SB-04				1173	
SB-1	506	3130	511	1114	
SB-100	126	3008	52	2024	
SB-1000	125		510,531	1114	
SB-1B			511	1114	
SB-1Z	116		504A	1104	
SB-2	506	3017B/117	511	1114	
SB-2(CENTERING)	507		511	1114	
SB-2B	506	3017B/117	511	1114	
SB-2C	525	3843	511	1114	
SB-2CGL	525	3100/519	511	1114	
SB-2CH	506	3130	511	1114	
SB-2T	506	3031A	511	1114	
SB-3	116	3311	504A	1104	
SB-3-02	116		504A	1104	
SB-309A	116		504A	1104	
SB-309C	116		504A	1104	
SB-3F	156(4)		512(4)		
SB-3F01	116	3031A	504A	1104	
SB-3N	116	3017B/117	504A	1104	
SB01	116	3016	504A	1104	
SB01-02	177	3100/519	300	1122	
SB0319	152	3893	66	2048	
SB0419	218		234		
SB1					MDA920A3
SB1-01-04	116	3017B/117	504A	1104	
SB10					MDA920A9
SB100	126		52	2024	

Industry Standard No.	ECG	SK	GE	RS 276-	MOTOR.
SB1000			504A	1104	
SB101	160		245	2004	
SB102	160		245	2004	
SB103	160		245	2004	
SB168	102A		53	2007	
SB169	102A		53	2007	
SB1Z	5800	9003		1142	
SB2					MDA920A4
SB200	126		52	2024	
SB2C	506	3843	510,531	1114	
SB2CH			510	1114	
SB2G			511	1114	
SB302	116	3016	504A	1104	
SB309A	166			1152	
SB309C	166			1152	
SB315	116	3016	504A	1104	
SB332	116	3016	504A	1104	
SB333	116	3016	504A	1104	
SB393	116	3016	504A	1104	
SB4					MDA920A6
SB5	130		14	2041	
SB5122	126		52	2024	
SB6	130		14	2041	MDA920A7
SB7	130		14	2041	
SB8					MDA920A8
SB821	137A	3058	ZD-6.2	561	
SB821A	137A	3058	ZD-6.2	561	
SB823	137A	3058	ZD-6.2	561	
SB823A	137A	3058	ZD-6.2	561	
SB825	137A	3058	ZD-6.2	561	
SB825A	137A	3058	ZD-6.2	561	
SB827	137A	3058	ZD-6.2	561	
SB827A	137A	3058	ZD-6.2	561	
SB829	137A	3058	ZD-6.2	561	
SB829A	137A	3058	ZD-6.2	561	
SBR-260	116		504A	1104	
SBR05	166	9075			
SBR102	170	3649			
SBR10A1	5340				MDA2501
SBR10A10	5344				SPECIAL
SBR10A2	5340				MDA2502
SBR10A3					MDA2504
SBR10A4	5342				MDA2504
SBR10A5					MDA2506
SBR10A6	5342				MDA2506
SBR10A8	5344				SPECIAL
SBR12	166	9075			
SBR22	167	3647			
SBR42	168	3648			
SBR62	169	3678			
SBR6A1	5340				MDA2501
SBR6A10	5344				SPECIAL
SBR6A2	5340				MDA2502
SBR6A3					MDA2504
SBR6A4	5342				MDA2504
SBR6A5					MDA2506
SBR6A6	5342				MDA2506
SBR6A8	5344				SPECIAL
SBR82	170	3649			
SC-1016	519		514	1122	
SC-110	116	3017B/117	504A	1104	
SC-12	102A	3004	53	2007	
SC-15	613	3327/614	90		
SC-16	116		504A	1104	
SC-20	613	3327/614	90		
SC-29	113A		6GC1		
SC-4			504A	1104	
SC-4004	124	3021	12		
SC-4044	123A	3124/289	20	2051	
SC-4131	124	3021	12		
SC-4131-1	124	3021	12		
SC-4167	124	3021	12		
SC-4244	123A	3021/124	20	2051	
SC-56	103A	3010	53	2002	
SC-6	109			1123	
SC-63	102A		53	2007	
SC-65	123A	3124/289	20	2051	
SC-66	102A	3004	53	2007	
SC-68	102A	3004	53	2007	
SC-69	102A	3004	53	2007	
SC-7	5457	3598			
SC-70	121	3009	239	2006	
SC-71	160	3006	245	2004	
SC-72	160	3006	245	2004	
SC-727	124	3021	12		
SC-73	102A	3004	53	2007	
SC-74	160	3006	245	2004	
SC-78	160	3006	245	2004	
SC-79	160	3006	245	2004	
SC-80	160	3006	245	2004	
SC-832	123A	3124/289	20	2051	
SC0321			75	2041	
SC0328			75	2041	
SC0421			74	2043	
SC0428			74	2043	
SC05	116	3311	504A	1104	
SC0515	5912		5048		
SC05E	116	3017B/117	504A	1104	
SC1	116	3017B/117	504A	1104	
SC10	125	3033	510,531	1114	
SC1001	123A	3444	20	2051	
SC1007	160		245	2004	
SC1010	123A	3444	20	2051	
SC1042	295		270		
SC108			17	2051	
SC108A			17	2051	
SC108B			20	2051	
SC109A			20	2051	
SC10A	125	3033	510,531	1114	
SC110	116		504A	1104	
SC115	5916		5048		
SC1168G	123A	3444	20	2051	
SC1168H	123A	3444	20	2051	
SC116P1	725		IC-19		
SC1182	914		IC-256		
SC12	102A		53	2007	
SC12(DIODE)	177		300	1122	
SC1227F	108	3018	86	2038	
SC1227G	108	3018	86	2038	
SC1229E	128	3024	243	2030	
SC1229G	123A	3444	20	2051	
SC1294H	129		244	2027	
SC136A					SC136A
SC136B			SC136B		SC136B
SC136C					SC136C
SC136D			SC136D		SC136D
SC136E					SC136E
SC136M					SC136M
SC1414	116		504A	1104	
SC141A	56004	3658			SC141A
SC141A1	56004	3658			SC141A1
SC141A2	56004	3658			
SC141A3	56004	3658			
SC141A4	56004	3658			

Industry Standard No.	ECG	SK	GE	RS 276-	MOTOR.
SC141A5	56004	3658			
SC141A6	56004	3658			
SC141B	56004	3658	SC141B		SC141B1
SC141B1	56004	3658			
SC141B2	56004	3658			SC141B2
SC141B3	56004	3658			SC141B3
SC141B4	56004	3658			SC141B4
SC141B5	56004	3658			SC141B5
SC141B6	56004	3658			SC141B6
SC141C	56006	3659			SC141C1
SC141C1	56006	3659			
SC141C2	56006	3659			SC141C2
SC141C3	56006	3659			SC141C3
SC141C4	56006	3659			SC141C4
SC141C5	56006	3659			SC141C5
SC141C6	56006	3659			SC141C6
SC141D	5635	3659/56006	SC141D		SC141D1
SC141D1	56006	3659			
SC141D2	56006	3659			SC141D2
SC141D3	56006	3659			SC141D3
SC141D4	56006	3659			SC141D4
SC141D5	56006				SC141D5
SC141D6	56006	3659			SC141D6
SC141E	56008	3660			SC141E1
SC141E1	56008	3660			
SC141E2	56008	3660			SC141E2
SC141E3	56008	3660			SC141E3
SC141E4	56008	3660			SC141E4
SC141E5	56008	3660			SC141E5
SC141E6	56008	3660			SC141E6
SC141F	56004	3658			
SC141F1	56004	3658			
SC141F2	56004	3658			
SC141F3	56004	3658			
SC141F4	56004	3658			
SC141F5	56004	3658			
SC141F6	56004	3658			
SC141M	5637				SC141M1
SC141M2					SC141M2
SC141M3					SC141M3
SC141M4					SC141M4
SC141M5					SC141M5
SC141M6					SC141M6
SC142B	5645		SC142B		
SC142D	5645		SC142D		
SC142E	5645		SC142E		
SC1431	116		504A	1104	
SC1431(GE)	177		300	1122	
SC143B					MAC222-4
SC143D					MAC222-6
SC143E					MAC222-7
SC143M					MAC222-8
SC146A	5632	3937			SC146A1
SC146A1	5632	3937			
SC146A2	5632	3937			SC146A2
SC146A3	5632	3937			SC146A3
SC146A4	5632	3937			SC146A4
SC146A5	5632	3937			SC146A5
SC146A6	5632	3937			SC146A6
SC146B	5633	3938	SC146B		SC146B1
SC146B2	5633	3938			SC146B2
SC146B3	5633	3938			SC146B3
SC146B4	5633	3938			SC146B4
SC146B5	5633	3938			SC146B5
SC146B6	5633	3938			SC146B6
SC146C	5634	3533/5635			SC146C1
SC146C1	5634	3533/5635			
SC146C2	5634	3533/5635			SC146C2
SC146C3	5634	3533/5635			SC146C3
SC146C4	5634	3533/5635			SC146C4
SC146C5	5634	3533/5635			SC146C5
SC146C6					SC146C6
SC146D	5635	3533	SC146D		SC146D1
SC146D1	5635	3533			
SC146D2	5635	3533			SC146D2
SC146D3	5635	3533			SC146D3
SC146D4	5635	3533			SC146D4
SC146D5	5635	3533			SC146D5
SC146D6	5635	3533			SC146D6
SC146E	5636	3939			SC146E1
SC146E1	5636	3939			
SC146E2					SC146E2
SC146E3	5636	3939			SC146E3
SC146E4	5636	3939			SC146E4
SC146E5	5636	3939			SC146E5
SC146E6	5636	3939			SC146E6
SC146F	5631	3937/5632			
SC146F1	5631	3937/5632			
SC146F2	5631	3937/5632			
SC146F3	5631	3937/5632			
SC146F4	5631	3937/5632			
SC146F5	5631	3937/5632			
SC146F6	5631	3937/5632			
SC146M	5637		SC146M		SC146M1
SC146M2					SC146M2
SC146M3					SC146M3
SC146M4					SC146M4
SC146M5					SC146M5
SC146M6					SC146M6
SC146N					SC146N1
SC146N2					SC146N2
SC146N3					SC146N3
SC146N4					SC146N4
SC146N5					SC146N5
SC146N6					SC146N6
SC146S					SC146S1
SC146S2					SC146S2
SC146S3					SC146S3
SC146S4					SC146S4
SC146S5					SC146S5
SC146S6					SC146S6
SC147A	123AP		212	2010	
SC147B	123AP		212	2010	
SC148	123AP		212	2010	
SC148A	123AP		212	2010	
SC148B	123AP		212	2010	
SC148C	123AP		212	2010	
SC149	123AP		212	2010	
SC149B	56004	3938/5633	212	2010	2N6342A
SC149D	56006	3659			2N6343A
SC149E	56008	3939/5636			SPECIAL
SC149M	56008	3660			2N6344A
SC15		3087	90		
SC151B	56004	3938/5633			MAC15-4
SC151D	56006	3659	SC151D		MAC15-6
SC151E	56008	3939/5636			MAC15-7
SC151M	56008	3660	SC151M		MAC15-8
SC158B			65	2050	
SC159			221	2024	
SC159A			221	2024	
SC159B			65	2050	
SC16			504A	1104	
SC160B	56014	3992			MAC525-4
SC160D	56014	3993			MAC525-6
SC160E					MAC525-7
SC160M		3994			MAC525-8
SC1611	464				2N4352
SC1612	464				2N4352
SC1613	464				2N4352
SC1614	464				2N4352
SC1631	116		504A	1104	
SC1631(GE)	116		504A	1104	
SC17063L	725		IC-19		
SC174	123AP		220		
SC174A	123AP		220		
SC174B	123AP		220		
SC1940G	704		IC-205		
SC2	116	3017B/117	504A	1104	
SC20	614		1N34AS	1123	
SC20575	703A		IC-12		
SC215	5916		5048		
SC23-3	116		504A	1104	
SC23-9	116		504A	1104	
SC240B	5673	3520/5677	SC240B		T4121B
SC240B13	5673	3520/5677			
SC240D	5675	3520/5677	SC240D		T4121D
SC240D13	5675	3520/5677			
SC240E	5676	3520/5677			T4121E
SC240E13	5676	3520/5677			
SC240M	5677				T4121M
SC241B			SC241B		T4101B
SC241D			SC241D		T4101D
SC241E			SC241E		T4101E
SC241M					T4101M
SC245A					SC245A
SC245B	5673	3520/5677	SC245B		SC245B
SC245B13	5673	3520/5677			
SC245C					SC245C
SC245D	5675	3520/5677	SC245D		SC245D
SC245D13	5677	3520			
SC245D2	56024		SC245D2		
SC245E	5676	3520/5677			SC245E
SC245E13	5676	3520/5677			
SC245M	5677				SC245M
SC245N					SC245N
SC245S					SC245S
SC246A					SC246A
SC246B			SC246B		SC246B
SC246C					SC246C
SC246D					SC246D
SC246E					SC246E
SC246M					SC246M
SC246N					SC246N
SC246S					SC246S
SC250A					SC250A
SC250B	5673	3520/5677	SC250B		SC250B
SC250B13	5673	3520/5677			
SC250B2	56022		SC250B2		
SC250B3	56022		SC250B3		
SC250C					SC250C
SC250D	5675	3520/5677	SC250D		SC250D
SC250D13	5675	3520/5677			
SC250D2	56024		SC250D2		
SC250E	5676	3520/5677			SC250E
SC250E13	5676	3520/5677			
SC250M					SC250M
SC250N					SC250N
SC250S					SC250S
SC251A					SC251A
SC251B					SC251B
SC251C					SC251C
SC251D			SC251D		SC251D
SC251E			SC251E		SC251E
SC251M					SC251M
SC251N					SC251N
SC251S					SC251S
SC256B			67	2023	
SC258B			65	2050	
SC259			221	2024	
SC259A			221	2024	
SC259B			65	2050	
SC260A					SC260A
SC260B	5693		SC260B		SC260B
SC260B2	56022		SC260B2		
SC260C					SC260C
SC260D	5695		SC260D		SC260D
SC260D2	56024		SC260D2		
SC260E	5697				SC260E
SC260M	5697				SC260M
SC260M2	56026		SC260M2		
SC260M3	56026		SC260M3		
SC261A					SC261A
SC261B					SC261B
SC261C					SC261C
SC261D					SC261D
SC261E					SC261E
SC261M					SC261M
SC265B	5693				2N5444
SC265D	5695				2N5445
SC265D2	56024	3509/5695	SC265D2		
SC265E	5697				T6410E
SC265M	5697		SC265M		2N5446
SC266B					2N5441
SC266D					2N5442
SC266E					T6400E
SC266M					2N5443
SC2914	725		IC-19		
SC2914P	725	3162	IC-19		
SC2C			504A	1104	
SC305	116	3016	504A	1104	
SC350	123A	3444	20	2051	
SC35A	5673	3508/5675			
SC35B	5673	3508/5675			
SC35D	5675	3508			
SC35F	5673	3508/5675			
SC365	129		244	2027	
SC4	116	3017B/117	504A	1104	
SC4004	124	3021	12		
SC400Y	124		12		
SC4010	123A	3444	20	2051	
SC4044	123A	3444	20	2051	
SC40A	5667A	3508/5675			
SC40B	5673	3508/5675			
SC40D	5675	3508			
SC40F	5661	3508/5675			
SC4116	116	3016	504A	1104	
SC4131	411/128		12		
SC4131-1	411/128		20		
SC4133	184		57	2017	
SC4244	411/128	3444/123A	12		
SC4274	131		44	2006	
SC43	100	3005	1	2007	
SC4303	152	3893	66	2048	
SC4303-1	152	3893	66	2048	
SC4303-2	152	3893	66	2048	
SC4308	152	3893	66	2048	
SC44	100	3005	1	2007	
SC441	198	3220	251		

Industry Standard No.	ECG	SK	GE	RS 276-	MOTOR.
SC45	102A		53	2007	
SC45A	5675	3508			
SC45B	5673	3508/5675			
SC45D	5675	3508			
SC45F	5675	3508			
SC46	100	3005	1	2007	
SC50A	5675	3508			
SC50B	5673	3508/5675			
SC50D	5675	3508			
SC50F	5675	3508			
SC5116L	725	3162	IC-19		
SC5116P	725		IC-19		
SC5117P		3159	IC-8		
SC5118P	720	9014	IC-7		
SC5150P	735		IC-212		
SC5150R	735		IC-212		
SC5172P	719		IC-28		
SC5175G	941	3514	IC-263	010	
SC5177P	718	3159	IC-8		
SC5182P	720	9014	IC-7		
SC5199P	718	3159	IC-8		
SC5204P	798	3216	IC-234		
SC5245P	748	3236	IC-115		
SC54	109	3087		1123	
SC5425P	748	3236			
SC56	103A	3010	53	2002	
SC5740PQ	722	3161	IC-9		
SC5741P	720	9014	IC-7		
SC5743P	722	3161	IC-9		
SC5747	799	3238	IC-34		
SC6	117	3087	504A	1104	
SC609	237	3299	46		
SC60B	5683	3509/5695			
SC60D	5685	3521/5686			
SC60D13	5685	3521/5686			
SC60E	5685	3521/5686			
SC60E13	5685	3521/5686			
SC63	102A	3004	53	2007	
SC65	123A		20	2051	
SC66	102A	3004	53	2007	
SC68	102A	3004	53	2007	
SC69	102A	3004	53	2007	
SC70	121	3009	239	2006	
SC71	160		245	2004	
SC72	160		245	2004	
SC727	124	3021	12		
SC73	102A	3004	53	2007	
SC74	160		245	2004	
SC741T	941		IC-263	010	
SC777	124	3021	12		
SC78	160		245	2004	
SC785	123		20	2051	
SC79	160		245	2004	
SC8	125	3032A	510,531	1114	
SC80	160		245	2004	
SC821	137A	3058	ZD-6.2	561	
SC821A	137A	3058	ZD-6.2	561	
SC823	137A	3058	ZD-6.2	561	
SC823A	137A	3058	ZD-6.2	561	
SC825	137A	3058	ZD-6.2	561	
SC825A	137A	3058	ZD-6.2	561	
SC827	137A	3058	ZD-6.2	561	
SC827A	137A	3058	ZD-6.2	561	
SC829	137A	3058	ZD-6.2	561	
SC829A	137A	3058	ZD-6.2	561	
SC832	123A	3444	20	2051	
SC842	123A	3444	20	2051	
SC843	129	3045/225	244	2027	
SC843(T.I.)			40	2012	
SC8705P	722		IC-9		
SC8707	782		IC-224		
SC8707P	782		IC-224		
SC8718	748	3236			
SC8A	125	3032A	510,531	1114	
SC91	139A	3060	ZD-9.1	562	
SC92A					MAC92-2
SC92B					MAC92-4
SC92D					MAC92-6
SC92F					MAC92-1
SC9314P	718	3159	IC-8		
SC9426P	722	3161	IC-9		
SC9430P	747	3279	IC-218		
SC9431P	749	3168	IC-97		
SC9436P	712	3072	IC-2		
SC9578P	814		IC-243		
SCA-45A	513		513		
SCA05	116	3311	504A	1104	
SCA1	116	3016	504A	1104	
SCA10	125	3033	510,531	1114	
SCA1103	116		504A	1104	
SCA2	116	3017B/117	504A	1104	
SCA3	116	3031A	504A	1104	
SCA3021	316		86	2038	
SCA3244	161		212	2010	
SCA4	116	3031A	504A	1104	
SCA45A	513		513		
SCA5	116	3017B/117	504A	1104	
SCA6	116	3017B/117	504A	1104	
SCA8	125	3032A	510,531	1114	
SCAJ05					MDA2500
SCAJ1					MDA2501
SCAJ2					MDA2502
SCAJ3					MDA2504
SCAJ4					MDA2504
SCAJ6					MDA3506
SCBA-1	5322	3680			
SCBA-2	5322	3680			
SCBA-4	5324	6624/5524			
SCBA-6	5326	3682			
SCBA05	5322	3680			MDA2500
SCBA1	5322				MDA2501
SCBA2	5322				MDA2502
SCBA3					MDA2504
SCBA4	5324				MDA2504
SCBA6	5326				MDA3506
SCBR05					MDA920A2
SCBR05F	125		510,531	1114	
SCBR1					MDA920A3
SCBR10	156	3051	512		MDA920A9
SCBR1F	125		510,531	1114	
SCBR2					MDA920A4
SCBR35F				1114	
SCBR4					MDA920A6
SCBR6	156	3051	512		MDA920A7
SCBR6F	125		510,531	1114	
SCC321			75	2041	
SCC421			74	2043	
SCD T322	123A	3444			
SCD T330	152	3893			
SCD T334	153			2049	
SCD-T320			14	2041	
SCD-T322			20	2051	
SCD-T326			82	2032	
SCD-T330			66	2048	

Industry Standard No.	ECG	SK	GE	RS 276-	MOTOR.
SCD-T334			69	2049	
SCD321			75	2041	
SCD421			74	2043	
SCDT323	123A		20	2051	
SCE1	116	3017B/117	504A	1104	
SCE10	125	3033	510,531	1114	
SCE2	116	3017B/117	504A	1104	
SCE3			504A	1104	
SCE321			75	2041	
SCE4	116	3031A	504A	1104	
SCE421			74	2043	
SCE6	116	3017B/117	504A	1104	
SCE8	125	3032A	510,531	1114	
SCI44191005	128		243	2030	
SCI444103053	128		243	2030	
SCI444204037	129		244	2027	
SCI444291004	129		244	2027	
SCL4000	4000	4000	4000		
SCL4000B					MC14000UB
SCL4001	4001B	4001	4001	2401	
SCL4001B					MC14001B
SCL4001UB					MC14001UB
SCL4002		4002	4002		
SCL4002B					MC14002B
SCL4006ABE	4006B	4006			
SCL4006B		4006			MC14006B
SCL4007		4007	4007		
SCL4007UB					MC14007UB
SCL4008B					MC14008B
SCL4008BE	4008B	4008			
SCL4009		4009	4009		
SCL4009UB					MC14049UB
SCL4010B					MC14050B
SCL4011		4011	4011	2411	
SCL4011B					MC14011B
SCL4011UB					MC14011UB
SCL4012		4012	4012	2412	
SCL4012B					MC14012B
SCL4013		4013	4013	2413	
SCL4013B					MC14013B
SCL4014B		4014			MC14014B
SCL4014BE	4014B	4014			
SCL4015		4015	4015		
SCL4015B					MC14015B
SCL4016		4016	4016		
SCL4016B					MC14016B
SCL4017		4017	4017	2417	
SCL4017B					MC14017B
SCL4018B		4018			MC14018B
SCL4018BE	4018B	4018			
SCL4019		4019	4019		
SCL4019B					MC14019B
SCL4020			4020	2420	
SCL4020AE	4020B	4020			
SCL4020B					MC14020B
SCL4021			4021	2421	
SCL4021B					MC14021B
SCL4022ABE	4022B	4022			
SCL4022B		4022			MC14022B
SCL4023		4023	4023	2423	
SCL4023B					MC14023B
SCL4024		4024	4024		
SCL4024B					MC14024B
SCL4025		4025	4025		
SCL4025B					MC14025B
SCL4027		4027	4027	2427	
SCL4027B					MC14027B
SCL4028B		4028			MC14028B
SCL4028BE	4028B	4028			
SCL4029B					MC14029B
SCL4030		4030	4030		
SCL4030B					MC14070B
SCL4034AB					MC14034B
SCL4035	4035B	4035			
SCL4035B					MC14035B
SCL4040B					MC14040B
SCL4042		4042	4042		
SCL4042B					MC14042B
SCL4043ABE	4043B	4043			
SCL4043B		4043			MC14043B
SCL4044ABE	4044B	4044			
SCL4044B		4044			MC14044B
SCL4046	980	4046			
SCL4046B					MC14046B
SCL4049		4049	4049	2449	
SCL4049UB					MC14049UB
SCL4050		4050	4050	2450	
SCL4050B					MC14050B
SCL4051		4051	4051	2451	
SCL4051B					MC14051B
SCL4052B					MC14052B
SCL4053B		4053			MC14053B
SCL4053BE	4053B	4053			
SCL4060B					MC14060B
SCL4066B					MC14066B
SCL4068B					MC14068B
SCL4069UB					MC14069UB
SCL4070B					MC14070B
SCL4071B					MC14071B
SCL4072	4072B	4072			
SCL4072B					MC14072B
SCL4073B					MC14073B
SCL4075B					MC14075B
SCL4076B					MC14076B
SCL4077B					MC14077B
SCL4078B					MC14078B
SCL4081		4081	4081		
SCL4081B					MC14081B
SCL4082B					MC14082B
SCL4085B					MC14085B
SCL4086B					MC14086B
SCL4093B					MC14093B
SCL4094B					MC14094B
SCL4099B	4099B				MC14099B
SCL4160B	40160B				MC14160B
SCL4161B	40161B				MC14161B
SCL4162B	40162B				MC14162B
SCL4163B	40163B				MC14163B
SCL4174B					MC14174B
SCL4192B					MC14192B
SCL4193B					MC14193B
SCL4502B					MC14502B
SCL4508B					MC14508B
SCL4510B					MC14510B
SCL4511B					MC14511B
SCL4512B					MC14512B
SCL4514B					MC14514B
SCL4515B					MC14515B
SCL4516B					MC14516B
SCL4517B					MC14517B
SCL4518			4518	2490	
SCL4518AE	4518B	4518			
SCL4518B					MC14518B
SCL4520B					MC14520B

Industry Standard No.	ECG	SK	GE	RS 276-	MOTOR.
SCL4522B					MC14522B
SCL4526B					MC14526B
SCL4527B					MC14527B
SCL4528B					MC14528B
SCL4531B					MC14531B
SCL4532B					MC14532B
SCL4543B					MC14543B
SCL4555B					MC14555B
SCL4555BE	4555B	4555			
SCL4556B		4556			MC14556B
SCL4581B					MC14581B
SCL4582B					MC14582B
SCL4584B					MC14584B
SCL4585B					MC14585B
SC05	116	3017B/117	504A	1104	
SC05E	116	3017B/117	504A	1104	
SCP5E	116	3311	504A	1104	
SCR-02	5483	3942			
SCR-02-C	5483	3942			
SCR-03-C	5521	6621			
SCR-04-C	5524	6624			
SCR104	5444	3634			
SCR104(ELCOM)	5423			1067	
SCR108(ELCOM)	5402	3638			
SCR15	5401	3638/5402			
SCR200	5404	3627			
SCR204	5444	3634			
SCR218	5404	3627			
SCR218(ELCOM)	5404	3627			
SCR255	5541	6615/5515			
SCR502	5501	3505/5547			
SCR5020	5402	3638			
SCR520	5483	3942			
SCR538	5444	3634			
SCR604(ELCOM)	5448	3636			
SCR613	5444	3634			
SCS11C1					MOC3003
SCS11C3					MOC3002
SCS11C4					MOC3001
SCS11C6					MOC3000
SCT1	125		510,531	1114	
SCT2	125		510,531	1114	
SCT3	125		510,531	1114	
SCT4	125		510,531	1114	
SCT5	125		510,531	1114	
OSD-0033	138A		ZD-7.5	561	
SD-02	116	3311	504A	1104	
SD-05					1N4001
SD-1	506	3017B/117	511	1114	1N4002
SD-1(BOOST)	506		511	1114	
SD-1-211B	116		504A	1104	
SD-1-211C	116		504A	1104	
SD-1-30DA	506		511	1114	
SD-101	116	3017B/117	504A	1104	
SD-109	123A		20	2051	
SD-11	5077A	3752	ZD-18		
SD-110	177	3100/519	300	1122	
SD-12	109	3087		1123	
SD-13	116	3311	504A	1104	
SD-13(PHILCO)	109			1123	
SD-14	109	3130		1123	
SD-15	139A		ZD-9.1	562	
SD-150	109	3090		1123	
SD-16	109			1123	
SD-16A	116	3017B/117	504A	1104	
SD-16D	116	3311	504A	1104	
SD-18	116		504A	1104	
SD-19	166			1152	
SD-1A	116	3017B/117	504A	1104	
SD-1A4F	506	3031A	511	1114	
SD-1AHF	507	3043	504A	1104	
SD-1AUF	177	3017B/117	300	1122	
SD-1B		3033	504A	1104	
SD-1B(HF)			511	1114	
SD-1BHF	506	3032A	511	1114	
SD-1C	506	3080	511	1114	
SD-1C-4F	116		504A	1104	
SD-1C-UF	116		504A	1104	
SD-1CUF	116		504A	1104	
SD-1HF	116		504A	1104	
SD-1L	116	3016	504A	1104	
SD-1LA	116	3311	504A	1104	
SD-1N4148	519	3175			
SD-1N60	110MP	3088	1N60	1123	
SD-1N60S	109	3088	1N60	1123	
SD-1S1555	177	3175	300	1122	
SD-1U			504A	1104	
SD-1UF	116	3017B/117	504A	1104	
SD-1VHF	506	3125	511	1114	
SD-1X		3311	504A	1104	
SD-1Y		3051	504A	1104	
SD-1Z	116	3081/125	511	1114	
SD-2	116	3016	504	1104	1N4003
SD-201	116	3016	504A	1104	
SD-22	506		511	1114	
SD-24		3126	90		
SD-27	139A	3060	ZD-9.1	562	
SD-2N5062	5402	3638	MR-5		
SD-31	5079A	3335	ZD-20		
SD-32	136A		ZD-5.6	561	
SD-33	143A	3750	ZD-12	563	
SD-34	177		300	1122	
SD-39		3126	90		
SD-4					1N4004
SD-404	112	3089	1N82A		
SD-41					SD41
SD-43	177	3100/519	300	1122	
SD-46	109			1123	
SD-46-2	109	3087		1123	
SD-470		3031A	504A	1104	
SD-49	5071A	3059/138A	ZD-6.8	561	
SD-5	177	3100/519	300	1122	
SD-5(PHILCO)			300	1122	
SD-51	112	3089	1N82A		SD51
SD-5171					SD5171
SD-56	109			1123	
SD-6	135A		ZD-5.1		1N4005
SD-60	109	3090		1123	
SD-60P				1011	
SD-61P				1103	
SD-630	177	3087	300	1122	
SD-632	139A		ZD-9.1	562	
SD-6AUF	125	3032A	510,531	1114	
SD-7	5808	9009	300	1122(2)	
SD-7(PHILCO)			300(2)	1122	
SD-71					MBR7545
SD-72					MBR7545
SD-75					MBR7545
SD-8			X11	562	1N4006
SD-8(PHILCO)			ZD-9.1	562	
SD-8(ZENER)	5072A		ZD-8.2	562	
SD-80	116		504A	1104	
SD-82A	112		1N82A		
SD-91	116		504A	1104	
SD-91A	116		504A	1104	
SD-91S	116		504A	1104	
SD-92	116		504A	1104	
SD-92A	116		504A	1104	
SD-92S	116		504A	1104	
SD-93	116		504A	1104	
SD-93A	116		504A	1104	
SD-94	116		504A	1104	
SD-94A	116		504A	1104	
SD-94AB	116		504A	1104	
SD-95	116		504A	1104	
SD-95A	116		504A	1104	
SD-ERB24-04D	552	3317	511	1114	
SD-ERB26-20		3925	530		
SD-ERC26-13	506	3125	511	1114	
SD-W04	167	3647	BR-206	1172	
SD-Y	116		504A	1104	
SD020	110MP	3088	1N60	1123(2)	
SD040	116	3311	504A	1104	
SD05	116	3311	504A	1104	
SD07	116	3016	504A	1104	
SD1	116	3017B/117		1104	
SD1-1	116		504A	1104	
SD1-11	506	3130	511	1114	
SD1-211B	116		504A	1104	
SD1-211C	116		504A	1104	
SD1-Z			504A	1104	
SD10			504A	1104	
SD100	177	3100/519	300	1122	
SD1005	76				MRF511
SD1006	77				2N5943
SD101	5804	9007	504A	1144	
SD1012					2N5590
SD1013					MRF340
SD1014	350	3176			MRF233
SD1015					MRF315
SD1018-4					MRF224
SD1019					MRF317
SD1019-5					2N6166
SD102	116	3311	504A	1104	
SD1020	311				MRF402
SD1020-6					MRF313A
SD1020-7					MRF313
SD1023	186	3192	28	2017	
SD103	116		504A	1104	
SD104	116	3031A	504A	1104	
SD105	139A	3031A	ZD-9.1	562	
SD1068					MRF230
SD1069	350	3176			2N5847
SD1074	318		280		MRF453
SD1076	335		281		MRF454
SD1077	329				MRF8004
SD1078					MRF454
SD1080	346	9038			MRF207
SD1080-2					MRF628
SD1080-4					MRF604
SD1080-6					MRF627
SD1080-7					MRF626
SD1087	365				MRF641
SD1088	366				MRF644
SD1089	367				MRF646
SD109	123A	3444	20	2051	
SD1095					MRF840
SD1096					MRF842
SD1098	484				MRF844
SD1099					MRF846
SD110	177		300	1122	
SD1101		3017B	504A	1104	
SD1102		3017B	504A	1104	
SD1103		3017B	504A	1104	
SD1104		3017B	504A	1104	
SD111	612		90		
SD1115-4					MRF607
SD1124	352				MRF245
SD1127	341				MRF237
SD113	614	3327	90		
SD1131	361				MRF629
SD1133	350		296		MRF212
SD1133-1					MRF212
SD1134	362				2N5944
SD1134-1	472				MRF225
SD1135	363				2N5945
SD1136	364				2N5946
SD1143	350		296		MRF212
SD1147					MRF5175
SD1148					MRF321
SD1149					MRF323
SD115	612	3325			
SD1166					MRF403
SD1167	337				2N5847
SD1168	338				2N5848
SD1169	339				2N5849
SD1174					2N6255
SD1177					2N5589
SD12	109	3016		1123	
SD12(PHILCO)	177		300	1122	
SD1200	311				2N3866
SD1212		3197	215		
SD1212-4	78				MRF476
SD1214-4	79				MRF475
SD1216	351				2N5591
SD1218	351	3176/350			MRF209
SD1219					MRF316
SD1220-1					2N5641
SD1222-5					2N5642
SD1224-2					2N5643
SD1224-4					MRF466
SD1229	351				2N6083
SD1229-1					MRF222
SD1232					MRF517
SD1242-5					2N5641
SD1244-6	359				2N5642
SD1245					MRF321
SD1256					2N5589
SD1262					MRF226
SD1272	345				MRF239
SD1285	338F				MRF406
SD1288	336$				MRF453A
SD1289	333				MRF453
SD1290	339				2N5849
SD1295					MRF421
SD1299					MRF326
SD12B	109	3087		1123	
SD12E	109	3087		1123	
SD12M	109	3087		1123	
SD13	116	3087	504A	1104	
SD1300	316				BFY90
SD1303					2N3839
SD1308					MRF905
SD1315					MRF511
SD1345		3893	66	2048	
SD1347-7	329				MRF402
SD14	109	3087		1123	
SD1402					MRF559

Industry Standard No.	ECG	SK	GE	RS 276-	MOTOR.
SD1403					MRF428
SD1404					MRF427
SD1405	335				MRF458
SD1407	471				MRF422
SD1409					MRF338A
SD141		3026	23	2020	
SD1410	481				MRF840
SD1412	483				MRF841
SD1415					MRF216
SD1416	352				MRF247
SD1421	485				MRF842
SD1424					MRF449A
SD1425					MRF8004
SD1427					MRF243
SD1428	477				MRF216
SD1429	365				MRF641
SD1433	364				2N5946
SD1434	367				MRF646
SD1438					MRF317
SD1444	361				MRF629
SD1445		3274	69	2049	
SD1449					MRF421
SD1450					MRF422
SD1451	318				MRF453
SD1452					MRF458
SD1460					MRF648
SD1461					MRF313A
SD1462					MRF313A
SD1464					MRF325
SD1465					MRF5177A
SD1466					MRF326
SD1467					MRF326
SD1468					MRF327
SD1469					MRF328
SD1480					MRF317
SD1484	346	9038			
SD1487	470				MRF316
SD15	109	3087		1123	
SD16	109	3087		1123	
SD165	177	3017B/117	300	1122	
SD18	116	3311	504A	1104	
000000SD1A	506	3017B/117	511	1114	
000000SD1AB	116	3017B/117	504A	1104	
SD1BHF			511	1114	
SD1C	125	3033	510,531	1114	
SD1CUF	116	3016	504A	1104	
SD1DM-4	116		504A	1104	
SD1HF	116	3016	504A	1104	
SD1L	116		504A	1104	
SD1LA	116		504A	1104	
SD1X	116	3016	504A	1142	
000000SD1Y	116	3311	504A	1142	
SD1Z	116	3016	504A	1143	
SD1ZHF	116		504A	1104	
SD2	125	3033	510,531	1114	
SD20				1011	
SD201	116		504A	1104	
SD202	116	3031A	504A	1104	
SD21A	109	3087		1123	
SD22		3130	511	1114	
SD23	116		504A	1104	
SD24		3126	90		
SD241					SD241
SD27	139A		ZD-10	562	
SD282A	112		1N82A		
SD2A	116	3017B/117	504A	1104	
SD2B	116	3032A	504A	1104	
SD2C	125		510,531	1114	
SD31	5080A	3336	ZD-22		MBR3545
SD32	5011A	3777	ZD-5.6	561	MBR3545
SD33	142A		ZD-12	563	
SD34	109	3087		1123	
SD3420-2					MRD510
SD345	152	3054/196	66	2048	
SD39	614	3126	90		
SD4	116	3119/113	504A	1104	
SD4(DUAL)	113A		6GC1		
SD42	5071A	3058/137A	ZD-6.8	561	
SD43			300	1122	
SD445	153		69	2049	
SD45	116	3016	504A	1104	
SD46	109	3090		1123	
SD46(4)	110MP	3709	1N34AS	1123(2)	
SD46-2	109			1123	
SD461			1N60	1123	
SD46R	110MP		1N34AS	1123(2)	
SD470	116	3031A	504A	1104	
SD49	5071A	3058/137A	ZD-6.8	561	
SD5	116	3120/114	504A	1104	
SD5(DUAL)	114		6GD1	1104	
SD500	177	3031A	300	1122	
SD500C	116	3031A	504A	1104	
SD501			504A	1104	
SD51	112	3016	1N82A		
SD53	142A	3062	ZD-12	563	
SD5400-1	3036				MRD370
SD5400-2	3036				MRD36/
SD5400-3	3036				MRD36/
SD5420-2					MRD500
SD5440-1	3032		L14G2		
SD5440-2	3032		L14G2		
SD5443-1					MRD310
SD5443-2					MRD300
SD5443-3					MRD300
SD5443-4					MRD300
SD56	109	3087		1123	
SD6	116	3121/115	504A	1104	
SD6(DUAL)	5802			1143	
SD60	109			1123	
SD600	177	3100/519	300	1122	
SD600C	116		504A	1104	
SD630	177	3100/519	300	1122	
SD632	139A	3060	ZD-9.1	562	
SD632(10)	139A		ZD-9.1	562	
SD7	125	3033	300(2)	1122(2)	
SD7(RECT)			510,531	1114	
SD701-02	177	3087	300	1122	
SD8	125	3032A	510,531	1114	
SD80	116	3016	504A	1104	
SD800	125	3033	510,531	1114	
SD82	112	3089	1N82A		
SD82A	112	3089	1N82A		
SD82AG	112	3089	1N82A		
SD91	5800	9003	504A	1142	
SD910	125	3033	510,531	1114	
SD910A	125	3033	510,531	1114	
SD910S	125	3051/156	510,531	1114	
SD91A	5802	9005	504A	1143	
SD91S	5802	9005	504A	1143	
SD92	5802	9005	504A	1143	
SD92A	5802	9005	504A	1143	
SD92S	5802	9005		1143	
SD93	5804	9007	504A	1144	
SD93A	5804	9007	504A	1144	

Industry Standard No.	ECG	SK	GE	RS 276-	MOTOR.
SD93S	116	3017B/117	504A	1104	
SD94	5804	9007	504A	1144	
SD94A	5804	9007	504A	1144	
SD94AB	116	3016	504A	1104	
SD94B	116	9007/5804	504A	1104	
SD94S	116	3017B/117	504A	1104	
SD95	116	3016	504A	1104	
SD950	116		504A	1104	
SD95A	116	3016	504A	1104	
SD96	116	3017B/117	504A	1104	
SD96A	116	3017B/117	504A	1104	
SD96S	116	3051/156	504A	1104	
SD974	177	3100/519	300	1122	
SD98	125	3032A	510,531	1114	
SD98A	125	3032A	510,531	1114	
SD98S	125	3051/156	510,531	1114	
SDA129A					MDA2500
SDA129B					MDA2501
SDA129C					MDA2502
SDA129D					MDA2504
SDA129E					MDA2506
SDA129F					MDA2508
SDA129G					MDA2510
SDA130A					MDA2500
SDA130B					MDA2501
SDA130C					MDA2502
SDA130D					MDA2504
SDA130E					MDA2506
SDA130F					MDA2508
SDA130G					MDA2510
SDA345	152	3054/196	66	2048	
SDA445	153	3083/197	69	2049	
SDA980-1,F					MDA2500,F
SDA980-10					SPECIAL
SDA980-2,F					MDA2501,F
SDA980-3,F					MDA2502,F
SDA980-4,F					MDA2504,F
SDA980-5,F					MDA2504,F
SDA980-6,F					MDA2506,F
SDA980-8					SPECIAL
SDA985-1					MDA2500
SDA985-2					MDA2501
SDA985-3					MDA2502
SDA985-4					MDA2504
SDA985-5					MDA2504
SDA990-1					MDA3500
SDA990-10					MDA3510
SDA990-2					MDA3501
SDA990-3					MDA3502
SDA990-4					MDA3504
SDA990-5					MDA3504
SDA990-6					MDA3506
SDA990-8					MDA3508
SDB345	152	3054/196	66	2048	
SDB445	153	3083/197	69	2049	
SDD-C10	113A	3119/113	CGC1		
SDD1220	128	3024	243	2030	
SDD3000	123A	3444	20	2051	
SDD4	113A	3119/113	6GC1		
SDD420	128	3024	243	2030	
SDD421	123A	3444	20	2051	
SDD5	114	3120	6GD1		
SDD6	115	3121	6GX1		
SDD820			11	2015	
SDD821	123A	3444	20	2051	
SDH-2	109	3090		1123	
SDI345			66	2048	
SDI445	153		69	2049	
SDJ345	152	3893	66	2048	
SDJ445	153	3274	69	2049	
SDK345	152	3054/196	66	2048	
SDK445	153	3274	69	2049	
SDL345	152	3893	66	2048	
SDL445	153	3083/197	69	2049	
SDM20301		3182			MJ4033
SDM20302		3182			MJ4033
SDM20303		3182			MJ4034
SDM20304		3182			MJ4035
SDM20311					MJ4033
SDM20312					MJ4033
SDM20313					MJ4034
SDM20314					MJ4035
SDM20321					MJ4033
SDM20322					MJ4033
SDM20323					MJ4034
SDM20324					MJ4035
SDM21301	271				MJ4030
SDM21302	244				MJ4030
SDM21303	271				MJ4031
SDM21304					MJ4032
SDM21311	246				MJ4030
SDM21312	246				MJ4030
SDM21313					MJ4031
SDM21314					MJ4032
SDM345	152	3054/196	66	2048	
SDM445	153	3083/197	69	2049	
SDM6000					MJ10012
SDM6001					MJ10012
SDM6002					MJ10012
SDM6003					MJ10012
SDN1010					2N6056
SDN1020					MJ3001
SDN345	152	3054/196	66	2048	
SDN4040					MJ10000
SDN4045					MJ10000
SDN445	153	3083/197	69	2049	
SDN6000	98				MJ10000
SDN6001					MJ10000
SDN6002					MJ10001
SDN6060	98				MJ10000
SDN6061	98				MJ10000
SDN6062	98				MJ10000
SDN6251	97				MJ10002
SDN6252	98				MJ10002
SDN6253	98				MJ10003
SDR-25			504A	1104	
SDR25	116	3311	504A	1104	
SDS-113	116	3031A	504A	1104	
SDS113	116		504A	1104	
SDS240	161	3019	39	2015	
SDS420	161	3132	39	2015	
SDT-3048	121	3717	239	2006	
SDT-410	162		35		
SDT-411	385	3559/162	36		
SDT-413	283	3559/162	36		
SDT-423	163A	3559/162	36		
SDT-430	385		36		
SDT-445	153		69	2049	
SDT1001		3559	73		
SDT1002		3559	73		
SDT1003		3559	73		
SDT1004		3559	73		
SDT1005		3559	73		
SDT1011		3559	73		

Industry Standard No.	ECG	SK	GE	RS 276-	MOTOR.
SDT1012		3559	73		
SDT1013		3559	73		
SDT1014		3559	73		
SDT1015		3559	73		
SDT1050					2N5838
SDT1051	283				2N5840
SDT1052					2N6543
SDT1053					2N6543
SDT1054					2N6543
SDT1055	283				2N5838
SDT1056	283				2N3902
SDT1057					2N6545
SDT1058					2N6545
SDT1059					2N6545
SDT1060	283				2N5838
SDT1061	283				2N3902
SDT1062					2N6545
SDT1063					2N6545
SDT1064					2N6545
SDT12301					2N5039
SDT12302					2N5347
SDT12303					2N5347
SDT12305					2N5347
SDT12306					2N5347
SDT12307					2N5347
SDT1301					2N6235
SDT1302					2N6235
SDT1303					2N6235
SDT1304					2N6235
SDT13301					2N6546
SDT13302					2N6547
SDT13303					2N6547
SDT13304					SDT13304
SDT13305					SDT13305
SDT1611	385	3559/162	36		
SDT1612	385	3559/162	36		
SDT1613	385	3559/162	36		
SDT1614	385	3559/162	36		
SDT1615	385		36		
SDT1616	385	3559/162	36		
SDT1617	385	3559/162	36		
SDT1618	385	3559/162	36		
SDT1621			75	2041	
SDT1622			75	2041	
SDT1623			75	2041	
SDT1631			75	2041	
SDT1632			75	2041	
SDT1633			75	2041	
SDT3125					2N6186
SDT3126					2N6186
SDT3321	129	3025	244	2027	MJ8100
SDT3322	129	3025	244	2027	MJ8100
SDT3323					2N6190
SDT3324					2N6192
SDT3325	187	3193	29	2025	MJ8100
SDT3326	186	3192	28	2017	MJ8100
SDT3327					2N6190
SDT3328					2N6192
SDT3401					2N5347
SDT3402					2N5347
SDT3403					2N5347
SDT3404					2N5349
SDT3405					2N5347
SDT3406					2N5347
SDT3407					2N5347
SDT3408					2N5349
SDT3421	282	3192/186	28	2017	2N4877
SDT3422	282	3748	66	2048	2N4877
SDT3423	282	3748			2N5336
SDT3424	282	3748			2N5338
SDT3425	282	3192/186	28	2017	2N4877
SDT3426	282	3192/186	28	2017	2N4877
SDT3427	282	3748			2N5336
SDT3428	282	3748			2N5338
SDT3429	282	3748			
SDT3442	282	3748			
SDT3501	129	3025	244	2027	2N3719
SDT3502	129	3025	244	2027	2N3720
SDT3503	129	3025	244	2027	2N6303
SDT3504		3513			2N6192
SDT3505	189	3193/187	29	2025	2N3867
SDT3506	189	3193/187	29	2025	2N3868
SDT3507		3513			2N6303
SDT3508		3513			2N6193
SDT3509	153	3085	69	2049	
SDT3510	153		69	2049	
SDT3513	153	3085	69	2049	
SDT3514	153		69	2049	
SDT3550	189	3193/187	29	2025	
SDT3552	189	3193/187	29	2025	
SDT3553	189	3193/187	29	2025	
SDT3575	218	3257	234		
SDT3576	218		234		
SDT3577	218		234		
SDT3578	218		234		
SDT3579	218		234		
SDT3701	153	3257	69	2049	
SDT3702	153		69	2049	
SDT3703	153	3257	69	2049	
SDT3704	153		69	2049	
SDT3706	153	3257	69	2049	
SDT3707	153	3085	69	2049	
SDT3708	153		69	2049	
SDT3709	153	3257	69	2049	
SDT3710	153	3085	69	2049	
SDT3711	153		69	2049	
SDT3712	153	3257	69	2049	
SDT3713	153		69	2049	
SDT3715	153	3257	69	2049	
SDT3716	153	3257	69	2049	
SDT3717	153		69	2049	
SDT3720	153	3257	69	2049	
SDT3721	153	3257	69	2049	
SDT3722	153		69	2049	
SDT3725	153	3257	69	2049	
SDT3726	153	3257	69	2049	
SDT3727	153		69	2049	
SDT3729	153	3257	69	2049	
SDT3730	153		69	2049	
SDT3733	153	3257	69	2049	
SDT3760	180		74	2043	
SDT3764	180		74	2043	
SDT3765	180		74	2043	
SDT3766	180		74	2043	
SDT3775	187	3193	29	2025	2N3867
SDT3776	187	3193	29	2025	2N3868
SDT3777					2N6303
SDT3778	187	3193	29	2025	2N3867
SDT3826	180		74	2043	
SDT3827	180		74	2043	
SDT3875	180		74	2043	
SDT3876	180		74	2043	
SDT3877	180		74	2043	
SDT401	162				2N6543
SDT402	389	3559/162	73		2N6543
SDT410	94	3560			MJ410
SDT411	94	3560			MJ411
SDT413	162	3560			MJ413
SDT423	385	3560			MJ423
SDT424	389				2N6308
SDT425					2N6545
SDT430		3559	73		2N6307
SDT4301	282	3192/186	28	2017	
SDT4302	282	3192/186	28	2017	
SDT4304	282	3192/186	28	2017	
SDT4305	282	3192/186	28	2017	
SDT4306	282	3748			
SDT4307	282	3192/186	28	2017	
SDT4308	282	3192/186	28	2017	
SDT4309	282	3748			
SDT431	385	3560	73		MJ431
SDT4310	282	3192/186	28	2017	
SDT4311	282	3192/186	28	2017	
SDT4312	282	3748			
SDT445			69	2049	
SDT4451					2N4877
SDT4452					2N5336
SDT4453					2N4877
SDT4454					2N5336
SDT4455	186	3192	28	2017	2N5337
SDT4456					2N5337
SDT4483	186	3192	28	2017	2N4877
SDT4551	186	3192	28	2017	
SDT4553	186	3192	28	2017	
SDT4583	186	3192	28	2017	
SDT4611	186	3192	28	2017	
SDT4612	186	3192	28	2017	
SDT4614	186	3192	28	2017	
SDT4615	186	3192	28	2017	
SDT4901					2N3583
SDT4902					2N6233
SDT4903					2N6234
SDT4904					2N3585
SDT4905					2N3585
SDT5001	282	3192/186	28	2017	
SDT5002		3512	D40E7		
SDT5006	282	3192/186	28	2017	
SDT5007		3512	D40E7		
SDT5011	186	3192	28	2017	
SDT5012		3512	D40E7		
SDT5101					TIP41A
SDT5102	152	3893	66	2048	TIP41A
SDT5103					TIP41A
SDT5111					TIP42A
SDT5112	153		69	2049	TIP42A
SDT5113					TIP42A
SDT520					2N6306
SDT521					2N6306
SDT522	283				2N6306
SDT525					2N6306
SDT527	283				2N6306
SDT530					2N6306
SDT531					2N6306
SDT532	283				2N6306
SDT535					2N6306
SDT536					2N6307
SDT537	283				2N6307
SDT540					2N6307
SDT541					2N6307
SDT542	283				2N6307
SDT545					2N6308
SDT546					2N6308
SDT547	283				2N6308
SDT550					2N6308
SDT5501	282	3192/186	28	2017	2N5337
SDT5502		3512	D40E7		2N5337
SDT5503	282	3512			2N5337
SDT5504	282				2N5539
SDT5506	282	3192/186	28	2017	2N4877
SDT5507	282	3512	D40E7		2N4877
SDT5508	282	3512			2N5336
SDT5509	282				2N5338
SDT551					2N6308
SDT5511	.86	3192	28	2017	2N5337
SDT5512		3512	D40E7		2N5337
SDT5513		3512			2N5337
SDT5514					2N5339
SDT552					2N6308
SDT5901	186	3192	28	2017	2N3766
SDT5902	186	3192	28	2017	2N3766
SDT5903					2N3767
SDT5904					2N5050
SDT5905					2N5050
SDT5906	186	3192	28	2017	2N3766
SDT5907	152	3893	66	2048	2N3766
SDT5908					2N3767
SDT5909					2N5050
SDT5910					2N5050
SDT5911		3026			2N5427
SDT5912		3026			2N5427
SDT5913					2N5427
SDT5914					2N5429
SDT5951					2N5051
SDT5952					2N3583
SDT5953					2N5052
SDT5954					2N5051
SDT5955					2N3583
SDT5956					2N5052
SDT6001	152	3893	66	2048	
SDT6011	152	3893	66	2048	
SDT6013	152	3893	66	2048	
SDT6031	152		66	2048	
SDT6101	186	3192	28	2017	
SDT6102	186	3192	28	2017	
SDT6103	152	3893	66	2048	
SDT6104	186	3192	28	2017	
SDT6105	186	3192	28	2017	
SDT6106	186	3192	28	2017	
SDT6308					2N5347
SDT6309					2N5347
SDT6310					2N5347
SDT6311					2N5347
SDT6312					2N5347
SDT6313					2N5347
SDT6314					2N5347
SDT6315					2N5347
SDT6316					2N5347
SDT6408					2N5347
SDT6409					2N5347
SDT6410					2N5347
SDT6411					2N5347
SDT6412					2N5347
SDT6413					2N5347
SDT6414					2N5347
SDT6415					2N5347
SDT6416					2N5347
SDT6901	384	3131A/369	246	2020	2N5050
SDT6902					2N5051

Industry Standard No.	ECG	SK	GE	RS 276-	MOTOR.
SDT6903					2N5052
SDT6904					2N5052
SDT6905	384	3131A/369	246	2020	
SDT7201					2N6306
SDT7202					2N6306
SDT7203					2N6306
SDT7204					2N6307
SDT7205					2N6308
SDT7206		3079			2N6341
SDT7207					2N6306
SDT7208					2N6306
SDT7209					2N6307
SDT7401	282	3192/186	28	2017	
SDT7402	282	3748	66	2048	
SDT7403	282	3748			
SDT7411	282	3192/186	28	2017	
SDT7412	282	3748	66	2048	
SDT7413	282	3748			
SDT7414	282	3748	28	2017	
SDT7415	282	3748	66	2048	
SDT7416	282	3748			
SDT7417	282	3748			
SDT7418	282	3748			
SDT7419	282	3748			
SDT7511	152	3893	66	2048	
SDT7512	152	3893	66	2048	
SDT7514	152	3893	66	2048	
SDT7515	152	3893	66	2048	
SDT7603		3027			2N6338
SDT7604		3079			2N6339
SDT7605					2N6341
SDT7609		3027			2N6338
SDT7610		3269			2N6339
SDT7611		3269			2N6341
SDT7612		3269			2N6249
SDT7731		3027			2N5881
SDT7732		3027			2N5881
SDT7733		3027			2N5882
SDT7734		3260			2N5629
SDT7735		3260			2N5630
SDT7736					2N5631
SDT7A01					2N5428
SDT7A02					2N5428
SDT7A03					2N5428
SDT7A07					2N5427
SDT7A08					2N5427
SDT7A09					2N5427
SDT9001	186	3192	28	2017	
SDT9002	186	3192	28	2017	
SDT9003	186	3192	28	2017	
SDT9004	187	3193	29	2025	
SDT9005	186	3192	28	2017	
SDT9006	186	3192	28	2017	
SDT9007	186	3192	28	2017	
SDT9008	186	3192	28	2017	
SDT9009	152	3893	66	2048	
SDT9201	130	3027	14	2041	2N6569
SDT9202	284	3437/180	74	2043	2N5878
SDT9203	284	3079			2N5632
SDT9204	284	3079			2N5633
SDT9205	130	3027	14	2041	2N6569
SDT9206	130	3027	14	2041	2N3055
SDT9207	284	3027/130	74	2043	2N5878
SDT9208	284	3079			2N5632
SDT9209	284	3079			2N5633
SDT9210	130	3027	14	2041	2N6569
SDT9261	130		14	2041	
SDT9301	130		77		2N4231A
SDT9302	130		77		2N4232A
SDT9303	130		75	2041	2N4233A
SDT9304	130		77		2N4231A
SDT9305	130		77		2N4232A
SDT9306	130		75	2041	2N4233A
SDT9307	130		77		2N3713
SDT9308	130	3027	77		2N3715
SDT9309	130		75	2041	2N3716
SDT9701	181		74	2043	2N5303
SDT9702		3260			2N5629
SDT9703		3260			2N5630
SDT9704	181		74	2043	2N5882
SDT9705		3260			2N5629
SDT9706		3260			2N5630
SDT9707	181		74	2043	2N3055
SDT9901	284		35		
SDT9902	284		35		
SDT9903	284		35		
SDT9904	284		35		
SDTB01					2N5346
SDTB02					2N5346
SDTB03					2N5348
SDTB05					2N5346
SDTB06					2N5346
SDTB07					2N5346
SE-0.5A	5804			1144	
SE-0.5B	116		504A	1104	
SE-05	116	3016	504A	1104	
SE-05-01	116		504A	1104	
SE-05-02			504A	1104	
SE-05-2	117		504A	1104	
SE-05-A				1103	
SE-05-B				1104	
SE-05-D				1114	
SE-0566	123A	3444	20	2051	
SE-05A	116	3017B/117	504A	1104	
SE-05C	156		512		
SE-05X	116	3016	504A	1104	
SE-1.5A	156		512		
SE-1001	229	3122	61	2038	
SE-1002	123A		20	2051	
SE-1010		3019	11	2015	
SE-1019	108	3019	86	2038	
SE-1044	108	3019	86	2038	
SE-1331	123A	3124/289	20	2051	
SE-1419	108	3018	86	2038	
SE-2	116		504A	1104	
SE-2001	123A	3124/289	20	2051	
SE-3001	108	3018	86	2038	
SE-3002	108	3019	86	2038	
SE-3005	108	3018	86	2038	
SE-3019	108	3019	86	2038	
SE-3033	130	3027	14	2041	
SE-3034	224	3049	46		
SE-3646	123A		20	2051	
SE-4001	123A	3124/289	20	2051	
SE-4002	123A	3124/289	20	2051	
SE-40022	121	3009	239	2006	
SE-4010	123A		20	2051	
SE-4020			18	2030	
SE-5	116		504A	1104	
SE-5-0399	127	3764	25		
SE-5-0819	102	3004	2	2007	
SE-5001	108	3019	86	2038	
SE-5002	108	3018	86	2038	
SE-5003	108	3018	86	2038	

Industry Standard No.	ECG	SK	GE	RS 276-	MOTOR.
SE-5006	123A	3018	20	2051	
SE-5006-14	108		86	2038	
SE-5020	108	3018	86	2038	
SE-5021	108	3018	86	2038	
SE-5023	161	3039/316	39	2015	
SE-5024			39	2015	
SE-5025	161		39	2015	
SE-5050	108	3018	86	2038	
SE-6001	123A	3124/289	20	2051	
SE-6002	123A	3020/123	20	2051	
SE-7001	103A	3010	59	2002	
SE-8001	128	3024	243	2030	
SE-8010	128	3047	243	2030	
SE0	125		510,531	1114	
SE0001			61	2038	
SE002(1)			11	2015	
SE05	116	3031A	504A	1104	
SE05A	5804	9007		1144	
SE05B	116	3017B/117	504A	1104	
SE05C	125	3032A	510,531	1114	
SE05D	116	3033	504A	1104	
SE05S	116	3016	504A	1104	
SE05SS	116	3017B/117	504A	1104	
SE0S-01	116		504A	1104	
SE1001	161	3018	61	2038	MPSA18
SE1001-1	229	3122	61	2038	
SE1001-2	229	3122	61	2038	
SE1002	161	3444/123A	39	2015	MPSA18
SE1002-1	161	3117	39	2015	
SE1002-2	161	3117	39	2015	
SE1010	161	3019	39	2015	
SE1012	192		63	2030	
SE1019	108	3019	86	2038	
SE1044	108		86	2038	
SE1331	123A	3444	20	2051	
SE1419	108		86	2038	
SE1450 SERIES					MLED930
SE15B	156	3051	512		
SE15C	156	3051	512		
SE15D	156	3051	512		
SE1730	125	3033	510,531	1114	
SE2001	161	3444/123A	39	2015	2N4124
SE2002	161	3039/316	39	2015	
SE2020	107		11	2015	
SE2397	107	3039/316	11	2015	
SE2400	160		245	2004	
SE2401	123A	3444	20	2051	
SE2402	123A	3444	20	2051	
SE3001	161	3018	39	2015	MPS918
SE3001R	108		86	2038	
SE3001Y	108	3019	86	2038	
SE3002	161	3018	86	2038	MP5918
SE3003	108	3039/316	86	2038	
SE3005	161	3117	39	2015	
SE3019	161	3132	39	2015	
SE3030	162		35		
SE3031	162		35		
SE3032	280	3027/130	19	2041	
SE3033	130	3027	14	2041	
SE3034	224	3049	46		
SE3035	130	3027	14	2041	
SE3036	130	3027	14	2041	
SE3040	384		246	2020	
SE3041	384		246	2020	
SE30B26A	116	3016	504A	1104	
SE3646		3444	60	2015	
SE3819	312	3112		2028	
SE4001	123A	3444	61	2051	MPS8097
SE4002	123A	3444	20	2051	MPS8097
SE40022	121	3717	239	2006	
SE4010	123A	3250/315	20	2051	MPS8097
SE4020	123A	3117	39	2015	
SE4020/6-04			39	2015	
SE4021	161	3117	39	2015	
SE4022	161	3117	39	2015	
SE4172	123A	3444	20	2051	
SE46	116	3017B/117	504A	1104	
SE5-0127	199		62	2010	
SE5-0128	123A	3444	20	2051	
SE5-0247-C	519		514	1122	
SE5-0249	161	3716	39	2015	
SE5-0250	161		39	2011	
SE5-0253	123A	3444	20	2051	
SE5-0274	123A	3444	20	2051	
SE5-0366	194		220		
SE5-0367	123A	3444	20		
SE5-0370	159		82	2032	
SE5-0399	127	3035	25		
SE5-0452	128		243	2030	
SE5-0456	519		514	1122	
SE5-0565	199		62	2010	
SE5-0567	123A	3444	20	2051	
SE5-0569	199		62	2010	
SE5-0608	123A	3444	20	2051	
SE5-0745	194	3275	220		
SE5-0798	159		82	2032	
SE5-0831	159		82	2032	
SE5-0848	123A	3444	20	2051	
SE5-0854	123A	3444	20	2051	
SE5-0855	123A	3444	20	2051	
SE5-0887	123A	3444	20	2051	
SE5-0888	123A	3444	20	2051	
SE5-0909	234	3247	65	2050	
SE5-0938	199	3245	62	2010	
SE5-0938-54	123A	3444	20	2051	
SE5-0938-55	199	3245	62	2010	
SE5-0938-56	199	3245	62	2010	
SE5-0938-57	199	3245	62	2010	
SE5-0949	159		82	2032	
SE5-0958	128		243	2030	
SE5-0963	152	3893	66	2048	
SE5-0964	153		69	2049	
SE5-0966	177		300	1122	
SE5-0996	312			2028	
SE5-1057	159		82	2032	
SE5-1223	159		82	2032	
SE50	229		61	2038	
SE5001	161	3018	39	2015	
SE5002	161	3018	39	2015	
SE5003	161	3132	39	2015	
SE5004	161	3018	39	2015	
SE5006	108	3444/123A	86	2038	
SE500G	108		86	2038	
SE5010	108		86	2038	
SE5015	108		86	2038	
SE501K					MC1733G
SE5020	161	3018	86	2038	MPSH30
SE5021	161	3018	39	2015	MPSH30
SE5022	161	3117	39	2015	MPSH30
SE5023	161	3018	39	2015	MPSH30
SE5024	161	3018	39	2015	MPSH30
SE5025	229	3124/289	39	2015	
SE5029	108		86	2038	
SE5030	108	3018	86	2038	

Industry Standard No.	ECG	SK	GE	RS 276-	MOTOR.
SE5030A	123A	3444	20	2051	
SE5030B	123A	3444	20	2051	
SE5031	108		86	2038	
SE5032	161	3716	39	2015	
SE5035	161	3039/316	86	2038	MPSH02
SE5036	108		86	2038	
SE50399	127	3764	25		
SE504	108		86	2038	
SE5040	108		86	2038	
SE5050	161	3018	39	2015	MPSH30
SE5051	161	3117	39	2015	MPSH30
SE5052	161	3117	39	2015	MPSH30
SE5055	319P	3018	283	2016	MPSH32
SE5056	108	3018	86	2038	
SE5151	123A	3444	20	2051	
SE515G					MC1520F
SE515K					MC1520G
SE516A					MC1520G
SE516G					MC1520F
SE516K					MC1520G
SE521	108		86	2038	
SE528E					MC1544L
SE528R					MC1544L
SE5315	222			2036	
SE531G					MC1539G
SE531T					MC1539G
SE533G					MC1776G
SE533T					MC1776G
SE537G					MC1556G
SE537T					MC1556G
SE550L					MC1723G
SE555JG		3564			MC1555U
SE555L		3693			MC1555G
SE555T		3693			MC1555G
SE556A					MC3556L
SE565A					MLM565CP
SE565K					MLM565CP
SE592A					SE592L
SE592K					SE592G
SE6	116		504A	1104	
SE6(DUAL)	115		6GX1		
SE6001	128	3444/123A	243	2030	2N4401
SE6002	161	3444/123A	39	2015	2N4401
SE6006	128	3018	243	2030	
SE6010	123A	3444	20	2051	
SE6020	123A	3024/128	243	2030	
SE6020A	128	3024	243	2030	
SE6021	128	3024	243	2030	
SE6021A	128	3024	243	2030	
SE6022	128	3024	243	2030	
SE6023	128	3024	243	2030	
SE7001	154		40	2012	
SE7002	154	3045/225	40	2012	
SE7005	128	3024	243	2030	
SE700	124	3021	12		
SE7010	154	3045/225	40	2012	
SE7015	128	3024	243	2030	
SE7016	154	3045/225	40	2012	
SE7017	154	3045/225	40	2012	
SE7020	124	3021	12		
SE7030	124	3021	12		
SE7050	154		40	2012	
SE7055	154	3103A/396	40	2012	2N4124
SE7056	154	3103A/396	40	2012	2N4124
SE8001	128	3024	40	2030	
SE8002	128	3024	40	2030	
SE8010	128	3024	40	2030	
SE8012	128	3024	243	2030	
SE8040	123A	3444	20	2051	
SE8041	128	3024	243	2030	
SE8042	128	3024	243	2030	
SE8510	128	3024	243	2030	
SE8520	128	3024	243	2030	
SE8521	128	3024	243	2030	
SE8540	129	3025	244	2027	
SE8541	129	3025	244	2027	
SE8542	129	3025	244	2027	
SE9002	130	3027	14	2041	
SE9020	162		35		
SE9060	175		246	2020	
SE9061	175		246	2020	
SE9062	175		246	2020	
SE9063	175		246	2020	
SE9080	130	3027	14	2041	
SE9300	263				SE9300
SE9301	263				SE9301
SE9302	263				SE9302
SE9303	245				SE9303
SE9304	245				SE9304
SE9305	97				SE9305
SE9306					MJ4033
SE9307					MJ4034
SE9308					MJ4035
SE9331					2N3739
SE9400	264				SE9400
SE9401	264				SE9401
SE9402					SE9402
SE9403					SE9403
SE9404					SE9404
SE9405					SE9405
SE9406					MJ4030
SE9407					MJ4031
SE9408					MJ4032
SE9560	218		234		
SE9561	218		234		
SE9562	218		234		
SE9563	218		234		
SE9570	185	3191	58	2025	
SE9571	185	3191	58	2025	
SE9572	185	3191	58	2025	
SE9573	185	3191	58	2025	
SEC1078	192		63	2030	
SEC1079	192		63	2030	
SEC1477	192		63	2030	
SEC1479	192		63	2030	
SEK-0367				2051	
SELEN-26	113A	3119/113	6GC1		
SELEN-30	502		CR-4		
SELEN-38	113A	3119/113	6GC1		
SELEN-40	505	3108	CR-7		
SELEN-42	120	3110	CR-3		
SELEN-44	166	9075	510	1152	
SELEN-48	119	3109	CR-2		
SELEN-52	503	3068	CR-5		
SELEN-58	502	3067	CR-4		
SELEN-70	116	3311	504A	1104	
SELEN-701	116	3311	504A	1104	
SELEN-92	506	3130	511	1114	
SELEN30		3068	CR-4		
SEN105					1N4001
SEN105FR					1N4933
SEN110					1N4002
SEN1100					1N4007
SEN110FR					1N4934

Industry Standard No.	ECG	SK	GE	RS 276-	MOTOR.
SEN120					1N4003
SEN120FR					1N4936
SEN130					1N4004
SEN140					IN4004
SEN140FR					1N4936
SEN150					1N4005
SEN150FR					1N4937
SEN160					1N4005
SEN160FR					1N4937
SEN180					1N4006
SEN205					MR501
SEN205FR					MR850
SEN210					MR501
SEN2100					MR510
SEN210FR					MR851
SEN220					MR502
SEN220FR					MR852
SEN230FR					MR854
SEN240					MR504
SEN240FR					MR854
SEN250FR					MR856
SEN260					MR506
SEN260FR					MR856
SEN280					MR508
SEN2A1	166	9075		1152	
SEN2A10	170	3649			
SEN2A1FR					SPECIAL
SEN2A2	167	3647		1172	
SEN2A2FR					SPECIAL
SEN2A4	168	3648		1173	
SEN2A4FR					SPECIAL
SEN2A6	169	3678			
SEN2A6FR					SPECIAL
SEN2A8	170	3649			
SEN2A8FR					SPECIAL
SEN300					MR504
SEN305					MR501
SEN305FR					MR850
SEN310					MR501
SEN3100					MR510
SEN310FR					MR851
SEN320					MR502
SEN320FR					MR852
SEN330FR					MR854
SEN340					MR504
SEN340FR					MR854
SEN350					MR506
SEN350FR					MR856
SEN360					MR506
SEN360FR					MR856
SEN380					MR508
SEN3A1					MDA201
SEN3A1FR					SPECIAL
SEN3A2					MDA202
SEN3A2FR					SPECIAL
SEN3A4					MDA204
SEN3A4FR					SPECIAL
SEN3A6					MDA206
SEN3A6FR					SPECIAL
SEN3A8					MDA208
SEN3A8FR					SPECIAL
SES3819	312	3448	FET-1	2035	SPECIAL
SES5001					SPECIAL
SES5002					SPECIAL
SES5003					SPECIAL
SES5301					SPECIAL
SES5302					SPECIAL
SES5303					SPECIAL
SES5401					SPECIAL
SES5401C					SPECIAL
SES5402					SPECIAL
SES5402C					SPECIAL
SES5403					SPECIAL
SES5403C					SPECIAL
SES5601C					SPECIAL
SES5602C					SPECIAL
SES5603C					SPECIAL
SES5701					SPECIAL
SES5702					SPECIAL
SES5703					SPECIAL
SES5801					SPECIAL
SES5802					SPECIAL
SES5803					SPECIAL
SES632	130		14	2041	
SES881	130		14	2041	
SF-1	525	3125	511	1114	
SF.T124	158		53	2007	
SF.T130	158		53	2007	
SF.T131P	158		53	2007	
SF.T163	160		245	2004	
SF.T171	100		1	2007	
SF.T172	100	3721	1	2007	
SF.T173	100	3721	1	2007	
SF.T174	100	3721	1	2007	
SF.T184	101	3123	8	2002	
SF.T187	171	3201	27		
SF.T191			16	2006	
SF.T212	127	3764	25		
SF.T213	127	3764	25		
SF.T214	127	3764	25		
SF.T221	158		53	2007	
SF.T222	102	3722	2	2007	
SF.T223	158		53	2007	
SF.T227	102	3722	2	2007	
SF.T237	102A	3123	53	2007	
SF.T238	127	3764	25		
SF.T239	127	3764	25		
SF.T240	127	3764	25		
SF.T250	127	3764	25		
SF.T251	102A	3123	53	2007	
SF.T252	102A	3123	53	2007	
SF.T253	102A	3123	53	2007	
SF.T264	105		4		
SF.T306	102A	3123	53	2007	
SF.T316	160		245	2004	
SF.T317	160		245	2004	
SF.T318	102	3123	2	2007	
SF.T319	160		245	2004	
SF.T320	160		245	2004	
SF.T321	158	3123	53	2007	
SF.T322	158		53	2007	
SF.T337	102A	3123	53	2007	
SF.T351	158	3123	53	2005	
SF.T352	158	3123	53	2007	
SF.T353	176	3123	80		
SF.T354	160		245	2004	
SF.T357	160		245	2004	
SF.T358	160		245	2004	
SF.T377	103A		59	2002	
SF.T440	192		63	2030	
SF.T443	192		63	2030	
SF.T443A	192		63	2030	
SF.T445	192		63	2030	
SF.T714	192		63	2030	

Industry Standard No.	ECG	SK	GE	RS 276-	MOTOR.
SF1	506		510,531	1114	
SF10					SPECIAL
SF10-01					MC513L
SF10-02					MC513F
SF100-01					MC528L
SF100-02					MC528F
SF1001	123A	3444	20	2051	
SF101-01					MC573L
SF101-02					MC573F
SF102-01					MC423L
SF102-02					MC423F
SF103-01					MC174L
SF103-02					MC473F
SF10B12	5522	6622			
SF10D12	5524	6624			
SF10F12	5526	6627/5527			
SF10G12	5527	6627			
SF10J11	5529	6629			
SF10J12	5529	6629			
SF10L11	5531	6631			
SF10L12	5531	6631			
SF11					SPECIAL
SF11-01					MC653L
SF11-02					MC563F
SF110-01					MC524L
SF110-02					MC524F
SF111-01					MC574L
SF111-02					MC574F
SF112-01					MC424L
SF112-02					MC424F
SF113-01					MC474L
SF113-02					MC474F
SF115			61	2038	
SF115A			61	2038	
SF115B			61	2038	
SF115C			61	2038	
SF115D			61	2038	
SF115E			61	2038	
SF12					SPECIAL
SF12-01					MC413L
SF12-02					MC413F
SF120-01					MC2123L
SF120-02					MC2123F
SF121-01					MC2173L
SF121-02					MC2173F
SF122-01					MC2023L
SF122-02					MC2023F
SF123-01					MC2073L
SF123-02					MC2073F
SF13-01					MC463L
SF13-02					MC463F
SF130-01					MC2124L
SF130-02					MC2124F
SF131-01					MC2174L
SF131-02					MC2174F
SF132-01					MC2024L
SF132-02					MC2024F
SF167			39	2015	
SF16A11	5521	6621			
SF16B11	5522	6622			
SF16B12	5522	6622			
SF16D11	5524	6624			
SF16D12	5524	6624			
SF16E11	5525	6627/5527			
SF16F11	5526	6627/5527			
SF16F12	5526	6627/5527			
SF16G11	5527	6627			
SF16G12	5527	6627			
SF16H11	5528	6629/5529			
SF16J11	5529	6629			
SF16J12	5529	6629			
SF16L11	5531	6631			
SF16L12	5531	6631			
SF16Z11	5520	6621/5521			
SF1713	123A	3444	20	2051	
SF1714	123A	3122	20	2051	
SF1726	123A	3444	20	2051	
SF173			39	2015	
SF1730	123A	3444	20	2051	
SF194			39	2015	
SF194B			39	2015	
SF195			39	2015	
SF195C			39	2015	
SF195D			39	2015	
SF196			39	2015	
SF197			39	2015	
SF1A11	5412	3953		1067	
SF1A11A	5412	3953		1067	
SF1B11	5413	3954/5414		1067	
SF1B11A	5413	3954/5414		1067	
SF1CN1	125		510,531	1114	
SF1D11	5414	3954		1067	
SF1D11A	5414	3954		1067	
SF1R3B41	5454	3557			
SF1R3D41	5455	3557		1067	
SF1R3G41	5457	3557			
SF20-01					MC514L
SF20-02					MC514F
SF21-01					MC564L
SF21-02					MC564F
SF22-01					MC414L
SF22-02					MC414F
SF23-01					MC464L
SF23-02					MC464F
SF294			39	2015	
SF294B			39	2015	
SF295			39	2015	
SF295C			39	2015	
SF295D			39	2015	
SF30-01					MC521L
SF30-02					MC521F
SF31-01					MC571L
SF31-02					MC571F
SF310			39	2015	
SF314			39	2015	
SF315					SPECIAL
SF3151					SPECIAL
SF3152					SPECIAL
SF32-01					MC421L
SF32-02					MC421F
SF33-01					MC471L
SF33-02					MC471F
SF330					SPECIAL
SF3301					SPECIAL
SF3302					SPECIAL
SF334			60	2015	
SF334B			60	2015	
SF335			60	2015	
SF335C			60	2015	
SF335D			60	2015	
SF3B14	5511	3683			
SF3CN1	125		510,531	1114	
SF3D12	5483	3942			
SF3D14	5511	3683			
SF3G12	5486	3944/5487			
SF3G14	5512	3684			
SF3G41	5465	3687			
SF4	125		510,531	1114	
SF415					SPECIAL
SF4151					SPECIAL
SF4152					SPECIAL
SF46					SPECIAL
SF461					SPECIAL
SF462					SPECIAL
SF4CN1	125		510,531	1114	
SF5	125		510,531	1114	
SF50-01					MC515L
SF50-02					MC515F
SF51-01					MC565L
SF51-02					MC565F
SF515					SPECIAL
SF5151					SPECIAL
SF5152					SPECIAL
SF52-01					MC415L
SF52-02					MC415F
SF53-01					MC465L
SF53-02					MC465F
SF530					SPECIAL
SF5301					SPECIAL
SF5302					SPECIAL
SF5D12	5483	3942			
SF5F12	5484	3943/5485			
SF5G12	5485	3943			
SF5J12	5487	3944			
SF6	125		510,531	1114	
SF60-01					MC516L
SF60-02					MC516F
SF61-01					MC566L
SF61-02					MC566F
SF62-01					MC416L
SF62-02					MC416F
SF63-01					MC466L
SF63-02					MC466F
SF703SF	230	3042			
SF80-01					MC522L
SF80-02					MC522F
SF8014	159	3114/290	82	2032	
SF81-01					MC572L
SF81-02					MC572F
SF82-01					MC422L
SF82-02					MC422F
SF83-01					MC472L
SF83-02					MC472F
SFB1087	451				2N5485
SFB1091					3N126
SFB1092					2N5484
SFB6183	116	3116	504A	1104	
SFB8970	222			2036	
SFC-1616			FET-4	2036	
SFC-1617			FET-4	2036	
SFC2209R	309K	3629			
SFC2301A	1171	3565			
SFC2309R	309K	3629			
SFC2748C	1171	3565			
SFC2748M	1171	3565			
SFC2805EC	960	3591			
SFC2805RC	309K	3629			
SFC2806EC	962	3669			
SFC2812EC	966	3592			
SFC2824EC	972	3670			
SFC6032	723	3144	IC-15		
SFC8999	222	3050/221	FET-4	2036	
SFD-23	129		244	2027	
SFD107	109	3091		1123	
SFD111			1N34AS	1123	
SFD112	109	3090		1123	
SFD2285	222	3065	FET-4	2036	
SFD43	177	3100/519	300	1122	
SFD83	177	3100/519	300	1122	
SFE145	312	3112	FET-1	2028	
SFE253			FET-4	2036	
SFE303	221	3050	FET-4	2036	
SFE303424	222	3065	FET-4	2036	
SFE425	222			2036	
SFE427			FET-4	2036	
SFIR3B41	5454	6754		1067	
SFIR3D41	5455	3597		1067	
SFOR2B41	5402	3638			
SFOR2D41	5404	3627			
SFOR2G41	5405	3951			
SFOR2G41(GAK)	5405	3951			
SFR135	116	3016	504A	1104	
SFR151	116	3016	504A	1104	
SFR152	116	3016	504A	1104	
SFR153	116	3016	504A	1104	
SFR154	116	3031A	504A	1104	
SFR155	116	3017B/117	504A	1104	
SFR156	116	3017B/117	504A	1104	
SFR164	116	3031A	504A	1104	
SFR251	116	3016	504A	1104	
SFR252	116	3016	504A	1104	
SFR253	116	3031A	504A	1104	
SFR254	116	3017B/117	504A	1104	
SFR255	116	3017B/117	504A	1104	
SFR256	116	3017B/117	504A	1104	
SFR258	125	3032A	510,531	1114	
SFR264	116	3017B/117	504A	1104	
SFR266	116	3017B/117	504A	1104	
SFR268	125	3032A	510,531	1114	
SFT-163	160	3006	245	2004	
SFT-184	103A	3861/101	59	2002	
SFT-265	105	3012	4		
SFT-266	105	3012	4		
SFT-267	105	3012	4		
SFT-298	101	3861	8	2002	
SFT-306	102A	3004	53	2007	
SFT-307	100	3005	1	2007	
SFT-315	126	3008	52	2024	
SFT-316	126	3008	52	2024	
SFT-317	126	3008	52	2024	
SFT-319	100	3005	1	2007	
SFT-320	126	3008	52	2024	
SFT-322	102A	3004	53	2007	
SFT-323	102A	3004	53	2007	
SFT-327	102A	3004	53	2007	
SFT-337	102A	3004	53	2007	
SFT-352	102A	3004	53	2007	
SFT-353	102A	3845/176	53	2007	
SFT-357	126	3008	52	2024	
SFT-358	160	3006	245	2004	
SFT104	109	3087		1123	
SFT108	109	3087		1123	
SFT120	160		245	2004	
SFT121	158		53	2007	
SFT122	158		53	2007	
SFT123	158		53	2007	
SFT124	158		53	2007	
SFT125	158		53	2007	

Industry Standard No.	ECG	SK	GE	RS 276-	MOTOR.
SFT125P	158		53	2007	
SFT130	158		53	2007	
SFT131	158		53	2007	
SFT131P	158		53	2007	
SFT143	158		53	2007	
SFT144	158		53	2007	
SFT145	158		53	2007	
SFT146	158		53	2007	
SFT151	102A		53	2007	
SFT152	102A		53	2007	
SFT162	160		245	2004	
SFT163	160		245	2004	
SFT171	160		245	2004	
SFT172	160		245	2004	
SFT173	160		245	2004	
SFT174	160		245	2004	
SFT184			8	2002	
SFT186	154	3045/225	40	2012	
SFT187	154	3045/225	40	2012	
SFT190	121	3009	239	2006	
SFT191	121	3009	239	2006	
SFT192	121	3009	239	2006	
SFT212	121	3009	239	2006	
SFT213	121	3014	239	2006	
SFT214	121	3009	239	2006	
SFT221	102	3004	2	2007	
SFT221A	102A	3004	53	2007	
SFT222	102	3004	2	2007	
SFT222A	102A	3004	53	2007	
SFT223	100	3005	1	2007	
SFT226	100	3005	1	2007	
SFT227	100	3005	1	2007	
SFT228	100	3005	1	2007	
SFT229	100	3005	1	2007	
SFT232	102A		53	2007	
SFT237	100	3721	1	2007	
SFT238	121	3009	239	2006	
SFT239	121	3014	239	2006	
SFT240	121	3009	239	2006	
SFT241	158		53	2007	
SFT242	158		53	2007	
SFT243	158		53	2007	
SFT250	121	3009	239	2006	
SFT251	100	3005	1	2007	
SFT252	100	3005	1	2007	
SFT253	100	3005	1	2007	
SFT259	101	3861	8	2002	
SFT260	101	3861	8	2002	
SFT261	101	3861	8	2002	
SFT264	105	3012	4		
SFT265	105		4		
SFT266	105		4		
SFT267	105		4		
SFT268	160		245	2004	
SFT288	100	3721	1	2007	
SFT298	101		8	2002	
SFT306	102A		53	2007	
SFT307	126		52	2024	
SFT308	126	3008	52	2024	
SFT315	160		245	2004	
SFT316	160		245	2004	
SFT317	126		52	2024	
SFT318	126		52	2024	
SFT319	126	3721/100	52	2024	
SFT320	126		52	2024	
SFT321	102A		53	2007	
SFT322	102A		53	2007	
SFT323	102A		53	2007	
SFT325	158/410		53		
SFT327	102A		53	2007	
SFT337	102A	3123	53	2007	
SFT337B	102A		53	2007	
SFT337V	102A		53	2007	
SFT351	102A	3004	53	2007	
SFT352	102A		53	2007	
SFT353	102A	3845/176	53	2007	
SFT354	126		52	2024	
SFT357	126		52	2024	
SFT357P	126	3006/160	52	2024	
SFT358	160		245	2004	
SFT367	158		53	2007	
SFT377	103A	3835	59	2002	
SFT443	128	3024	243	2030	
SFT443A	128	3024	243	2030	
SFT445	128	3024	243	2030	
SFT523	158		53	2007	
SFT526	102A		53	2007	
SFT713	123A	3444	20	2051	
SFT714	123A	3444	20	2051	
SFZ708	136A		ZD-5.6	561	
SFZ716	142A	3062	ZD-12	563	
SG-005	116	9007/5804	504A	1104	
SG-105	116		504A	1104	
SG-1198	116	3016	504A	1104	
SG-205	116		504A	1104	
SG-264A	276	3296			
SG-305	116	3017B/117	504A	1104	
SG-5N	156	3051	512		
SG-5P	156	3051	512		
SG-5T				1102	
SG-613	276	3296			
SG-805	116	9009/5808	504A	1104	
SG-9150	177		300	1122	
SG005	5804		504A	1144	
SG100-01					MC504L
SG100-02					MC504F
SG100T					MC1723G
SG101-01					MC554L
SG101-02					MC554F
SG101AD					LM101AH
SG101AT		3565			LM101AH
SG101J					LM101AH
SG101T		3565			LM101AH
SG102-01					MC404L
SG102-02					MC404F
SG103-01					MC454L
SG103-02					MC454F
SG104T					LM104H
SG105	116	3016	504A	1104	
SG105N					LM105H
SG105T					LM105H
SG107J					LM107H
SG107T					LM107H
SG108AJ					LM108AJ
SG108AT					LM108AH
SG108J					LM108J
SG108T					LM108H
SG109K					LM109K
SG109T					LM109H
SG110-01					MC505L
SG110-02					MC505F
SG111-01					MC555L
SG111-02					MC555F
SG1118AJ					LM108AJ
SG1118AT					LM108AH
SG1118J					LM108J
SG1118T					LM108H
SG111D					LM111J
SG111T		3567			LM111H
SG112-01					MC405L
SG112-02					MC405F
SG113-01					MC455L
SG113-02					MC455F
SG118J					MC1741SL
SG118T					MC1741SG
SG1198	116		504A	1104	
SG120-01					MC506L
SG120-02					MC506F
SG120K-05					LM120K-05
SG120K-12					LM120K-12
SG120K-15					LM120K-15
SG120K-5.2					MC7905.2CK
SG120T-05					LM120T-05
SG120T-12					LM120T-12
SG120T-15					LM120T-15
SG120T-5.2					MC7905.2CK
SG121-01					MC556L
SG121-02					MC406F
SG1217					MC1741G
SG1217J					MC1741SL
SG1217T					MC1741SG
SG122-01					MC406L
SG122-02					MC406F
SG123-01					MC456L
SG123-02					MC456F
SG124J					LM124J
SG1250T					MC1776G
SG130-01					MC507L
SG130-02					MC507F
SG131-01					MC557L
SG131-02					MC557F
SG132-01					MC407L
SG132-02					MC407F
SG133-01					MC457L
SG133-02					MC457F
SG140-01					MC508L
SG140-02					MC508F
SG1401N					MC1533G
SG1401T					MC1533G
SG1402N					MC1594L
SG1402T					MC1594L
SG140K-05					LM140K-5.0
SG140K-06					LM140K-6.0
SG140K-08					LM140K-8.0
SG140K-12					LM140K-12
SG140K-15					LM140K-15
SG140K-18					LM140K-18
SG140K-24					LM140K-24
SG141-01					MC558L
SG141-02					MC558F
SG142-01					MC408L
SG142-02					MC408F
SG143-01					MC458L
SG143-02					MC458F
SG1436CT					MC1436CG
SG1436M					MC1436U
SG1436T					MC1436G
SG1456CT					MC1456CG
SG1456T					MC1456G
SG1458M					MC1458P1
SG1458T		3555			MC1458G
SG1468J					MC1468L
SG1468N					MC1468L
SG1468T					MC1468G
SG1495D					MC1495L
SG1495N					MC1495L
SG1496	973	3233			
SG1496D					MC1496L
SG1496N					MC1496L
SG1496T	973	3233			MC1496G
SG150-01					MC509L
SG150-02					MC509F
SG1501AD					MC1568L
SG1501AT					MC1568G
SG1501D					MC1568L
SG1501T					MC1568G
SG1502D					MC1568L
SG1502N					MC1568L
SG1503					MC1503U
SG151-01					MC559L
SG151-02					MC559F
SG152-01					MC409L
SG152-02					MC409F
SG1524J					MC3520L
SG153-01					MC459L
SG153-02					MC459F
SG1536T					MC1536G
SG1556T					MC1556G
SG1558T					MC1558G
SG1595D					MC1595L
SG1596D					MC1596L
SG1596T					MC1596G
SG160-01					MC519L
SG160-02					MC519F
SG161-01					MC569L
SG161-02					MC569F
SG162-01					MC419L
SG162-02					MC419F
SG163-01					MC469L
SG163-02					MC469F
SG1660D					LM301AH
SG1660J					LM308J
SG1660M					LM308N
SG1660T					LM308H
SG170-01					MC510L
SG170-02					MC510F
SG171-01					MC560L
SG171-02					MC560F
SG172-01					MC410L
SG172-02					MC410F
SG173-01					MC460L
SG173-02					MC460F
SG1760D					LM307H
SG1760F					LM307H
SG1760J					LM308J
SG1760M					LM&08N
SG1760T					LM308H
SG180-01					MC511L
SG180-02					MC511F
SG181-01					MC561L
SG181-02					MC561F
SG182-01					MC411L
SG182-02					MC411F
SG183-01					MC461L
SG183-02					MC461F
SG190-01					MC512L
SG190-02					MC512F
SG191-01					MC562L

Industry Standard No.	ECG	SK	GE	RS 278-	MOTOR.
SG191-02					MC562F
SG1910					MZ2360
SG1912					MZ2360
SG192-01					MC412L
SG192-02					MC412F
SG1920					MZ2361
SG1922					MZ2361
SG193-01					MC462L
SG193-02					MC462F
SG200-01					MC2111L
SG200-02					MC2111F
SG2001N	2011	3975			
SG2002N	2012	9092			
SG2003N	2013	9093			
SG2004N	2014	9094			
SG200T					MC1723G
SG201-01					MC2161L
SG201-02					MC2161F
SG201AD					LM201AH
SG201AM					LM201AN
SG201AN					LM201AN
SG201AT	1171	3565			LM201AH
SG201J					LM201AH
SG201M					LM201AN
SG201N					LM201AN
SG201T	1171	3565			LM201AH
SG202-01					MC2011L
SG202-02					MC2011F
SG203-01					MC2061L
SG203-02					MC2061F
SG204T					LM204H
SG205	5802	9005	504A	1143	
SG205N					LM205H
SG205T					LM205H
SG207J					LM207H
SG207M					LM207H
SG207N		3596			LM207H
SG207T		3690			LM207H
SG208AJ					LM208AJ
SG208AM					LM208AJ-8
SG208AT					LM208AH
SG208J					LM208J
SG208M					LM208J-8
SG208T					LM208H
SG209K					LM209K
SG209T					LM209H
SG210-01					MC2100L
SG210-02					MC2100F
SG211-01					MC2150L
SG211-02					MC2150F
SG2118AJ					LM208AJ
SG2118AM					LM208AJ-8
SG2118AT					LM208AH
SG2118J					LM208J
SG2118M					LM208J-8
SG2118T					LM208H
SG211D		3668			LM211J
SG211M		3668			LM211N
SG211T		3567			LM211H
SG212-01					MC2000L
SG212-02					MC2000F
SG213-01					MC2050L
SG213-02					MC2050F
SG218J					MC1741SL
SG218M					MC1741SL
SG218T					MC1741SG
SG220-01					MC2101L
SG220-02					MC2101F
SG221-01					MC2151L
SG221-02					MC2151F
SG222-01					MC2001L
SG222-02					MC2001F
SG223-01					MC2051L
SG223-02					MC2051F
SG224J	987				LM224J
SG224N	987				LM224N
SG2250T					MC1776G
SG230-01					MC2102L
SG230-02					MC2102F
SG231-02					MC2152F
SG2310-1					MC2152L
SG232-01					MC2002L
SG232-02					MC2002F
SG233-01					MC2052L
SG233-02					MC2052F
SG240-01					MC2103L
SG240-02					MC2103F
SG2401N					MC1433G
SG2402N					MC1494L
SG2402T					MC1494L
SG241-01					MC2153L
SG241-02					MC2153F
SG242-01					MC2003L
SG242-02					MC2003F
SG243-01					MC2053L
SG243-02					MC2053F
SG250-01					MC2104L
SG250-02					MC2104F
SG2501AD					MC1468L
SG2501AT					MC1468G
SG2501D					MC1468L
SG2501N					MC1468L
SG2501T					MC1468G
SG2502D					MC1468L
SG2502N					MC1468L
SG2502T					MC1468G
SG2503					MC1403AU
SG251-01					MC2154L
SG251-02					MC2004F
SG252-01					MC2004L
SG2524J					MC3520L
SG253-01					MC2054L
SG253-02					MC2054F
SG260-01					MC2105L
SG260-02					MC2105F
SG261-01					MC2155L
SG261-02					MC2155F
SG262-01					MC2005L
SG262-02					MC2005F
SG263-01					MC2055L
SG263-02					MC2055F
SG270-01					MC2106L
SG270-02					MC2106F
SG271-01					MC2156L
SG271-02					MC2156F
SG272-01					MC2006L
SG272-02					MC2006F
SG273-01					MC2056L
SG273-02					MC2076F
SG280-01					MC527L
SG280-02					MC572F
SG281-01					MC577L
SG281-02					MC577F
SG282-01					MC427L

Industry Standard No.	ECG	SK	GE	RS 276-	MOTOR.
SG282-02					MC427F
SG283-01					MC477L
SG283-02					MC477F
SG290-01					MC528L
SG290-02					MC528F
SG291-01					MC578L
SG291-02					MC578F
SG292-01					MC428L
SG292-02					MC428F
SG293-01					MC478L
SG293-02					MC478F
SG300-01					MC2112L
SG300-02					MC2112F
SG300N					MC1723CP
SG300T					MC1723CG
SG301-01					MC2162L
SG301-02					MC2162F
SG301AD					LM301AH
SG301AM	975	3641			LM301AN
SG301AN					LM301AN
SG301AT	1171	3565			LM301AH
SG302-01					MC2012L
SG302-02					MC2012F
SG303-01					MC2062L
SG303-02					MC2062F
SG304T					LM304H
SG305	5804	9007	504A	1144	
SG3058	914	3541			
SG3059	914	3541			
SG305AT					LM305H
SG305N					LM305H
SG305T					LM305H
SG3079	914	3541			
SG307J					LM307N
SG307M	976	3596			LM307N
SG307N	976	3596			6
SG307T		3690			LM307H
SG3081J	916	3550			
SG3081N	916	3550			
SG3082N	2023	3694			
SG308AJ					LM308AJ
SG308AM	938M				LM308AN
SG308AT	938				LM308AH
SG308J					LM308J
SG308M	938M				LM308N
SG308T	938				LM308H
SG309K	309K	3629			LM309K
SG309T					LM309H
SG310-01					MC2113L
SG310-02					MC2113F
SG311-01					MC2163L
SG311-02					MC2163F
SG3118AJ					MLM308AL
SG3118AM					MLM308AP1
SG3118AT					MLM308AG
SG3118J					MLM308L
SG3118M					MLM308P1
SG3118T					MLM308G
SG311D		3668			LM311J
SG311M		3668			LM311N
SG311T		3567			LM311H
SG312-01					MC2013L
SG312-02					MC2013F
SG313-01					MC2063L
SG313-02					MC2063F
SG3182	177		300	1122	
SG3183	177		300	1122	
SG318J					MC1741SCL
SG318M					MC1741CP1
SG318T					MC1741CG
SG3198	177		300	1122	
SG320-01					MC2107L
SG320-02					MC2107F
SG320K-05					LM320K-5.0
SG320K-12					LM320K-12
SG320K-15					LM320K-15
SG320K-5.2					MC7905.2CK
SG320T-05					LM320T-5.0
SG320T-12					LM320T-12
SG320T-15					LM320T-15
SG320T-5.2					MC7905.2CT
SG321-01					MC2157L
SG321-02					MC2157F
SG322-01					MC2007L
SG322-02					MC2007F
SG323	116	9067	504A	1104	
SG323-01					MC2057L
SG323-02					MC2057F
SG324J					LM324J
SG324N	987				LM324N
SG3250T					MC1776G
SG340-05K	309K	3629			
SG3400	116		504A	1104	
SG3401N		3688			MC1433G
SG3401T					MC1433G
SG3402N					MC1494L
SG3402T					MC1494L
SG340K-05					MC7805CK
SG340K-06					MC7806CK
SG340K-08					MC7808CK
SG340K-12					MC7812CK
SG340K-15					MC7815CK
SG340K-18					MC7818CK
SG340K-24					MC7824CK
SG3432	177		300	1122	
SG3501AD					MC1468L
SG3501AT					MC1468G
SG3501D					MC1468L
SG3501N					MC1468L
SG3501T					MC1468G
SG3502D					MC1468L
SG3502G					MC1468G
SG3502N					MC1468L
SG3503					MC1403U
SG351-01					MC2165L
SG351-02					MC2165F
SG3516	177		300	1122	
SG3524J					MC3420L
SG353-01					MC2065L
SG353-02					MC2065F
SG3583	177		300	1122	
SG370-01					MC529L
SG370-02					MC529F
SG371-01					MC579L
SG371-02					MC579F
SG372-01					MC429L
SG372-02					MC429F
SG373-01					MC479L
SG373-02					MC479F
SG380-01					MC2116L
SG380-02					MC2116F
SG381-01					MC2166L
SG381-02					MC2166F
SG382-01					MC2016L

Industry Standard No.	ECG	SK	GE	RS 276-	MOTOR.
SG382-02					MC2016F
SG383-01					MC2066L
SG383-02					MC2066F
SG40-01					MC500L
SG40-02					MC500F
SG41-01					MC550L
SG41-02					MC550F
SG42-01					MC400L
SG42-02					MC400F
SG4250CM					MC1776CP1
SG4250CT					MC1776CG
SG4250T					MC1776G
SG43-01					MC450L
SG43-02					MC450F
SG4501D					MC1468L
SG4501N					MC1468L
SG4501T					MC1468G
SG50-01					MC501L
SG50-02					MC501F
SG5013	128	3024	243	2030	
SG5028	177		300	1122	
SG505	116	3017B/117	504A	1104	
SG51-01					MC551L
SG51-02					MC551F
SG52-01					MC401L
SG52-02					MC401F
SG53-01					MC451L
SG53-02					MC451F
SG5392	177		300	1122	
SG5400	177		300	1122	
SG555CM	955M				MC1455P1
SG555CT					MC1455G
SG555T					MC1555G
SG556CJ					MC3456L
SG556CN	978	3689			MC3456P
SG556J					MC3556L
SG556N					MC3556L
SG60-01					MC502L
SG60-02					MC502F
SG608	276	3296			
SG608AA	276	3296			
SG61-01					MC552L
SG61-02					MC552F
SG613	276	3296			
SG613-1	276	3296			
SG613-2	276	3296			
SG613-3	276	3296			
SG62-01					MC402L
SG62-02					MC402F
SG63-01					MC452L
SG63-02					MC452F
SG70-01					MC520L
SG70-02					MC520F
SG709CN	909D		IC-250		
SG709CT	909	3590	IC-249		
SG71-01					MC570L
SG71-02					MC570F
SG710CD	910D		IC-252		MC1710CL
SG710CN	910D		IC-252		MC1710CP
SG710CT	910		IC-251		MC1710CG
SG710D					MC1710L
SG710N					MC1710CP
SG710T					MC1710G
SG711CD	911D		IC-254		MC1711CL
SG711CN	911D		IC-254		MC1711CP
SG711CT	911		IC-253		MC1711CG
SG711D					MC1711L
SG711N					MC1711CP
SG711T					MC1711G
SG72-01					MC420L
SG72-02					MC420F
SG723CD	923D	3165	IC-260	1740	MC1723CL
SG723CN	923D		IC-260	1740	MC1723CP
SG723CT	923	3164	IC-259		MC1723CG
SG723D					MC1723L
SG723T		3164			MC1723G
SG73-01					MC470L
SG73-02					MC470F
SG733CD					MC1733CL
SG733CN					MC1733CP
SG733CT					MC1733CG
SG733D					MC1733L
SG733N					MC1733L
SG733T					MC1733G
SG741CD	941D	3552/941M	IC-264		MC1741CL
SG741CF					MC1741CF
SG741CM	941M	3514/941	IC-265	007	MC1741CP1
SG741CN	941D		IC-264		MC1741CP2
SG741CT	941	3514	IC-263	010	MC1741CG
SG741D					MC1741L
SG741F					MC1741F
SG741SCM					MC1741SCP1
SG741SCT					MC1741SCG
SG741ST					MC1741SG
SG741T		3514			MC1741G
SG747CD	947D	3556			
SG747CJ					MC1747CL
SG747CN	947D	3556			MC1747CP2
SG747CT	947	3526	IC-268		MC1747CG
SG747J					MC1747L
SG747T					MC1747G
SG748CD					MC1748CP1
SG748CM		3644			MC1748CP1
SG748CN		3644			MC1748CP1
SG748CT		3645			MC1748CG
SG748D					MC1748G
SG748T		3645			MC1748G
SG777CJ					LM308AJ
SG777CM					LM308AN
SG777CN					LM308AN
SG777CT					LM308AH
SG777J					LM108AJ
SG777T					LM108AH
SG7805CK	309K	3629			MC7805CK
SG7805K					MC7805CK
SG7806CK					MC7806CK
SG7806K					MC7806CK
SG7808CK					MC7808CK
SG7808K					MC7808CK
SG7812CK					MC7812CK
SG7812K					MC7812CK
SG7815CK					MC7815CK
SG7815K					MC7815CK
SG7818CK					MC7818CK
SG7818K					MC7818CK
SG7824CK					MC7824CK
SG7824K					MC7824CK
SG80-01					MC526L
SG80-02					MC526F
SG805	5808		504A	1104	
SG81-01					MC576L
SG81-02					MC576F
SG82-01					MC426L
SG82-02					MC426F

Industry Standard No.	ECG	SK	GE	RS 276-	MOTOR.
SG83-01					MC476L
SG83-02					MC476F
SG90-01					MC503L
SG90-02					MC503F
SG91-01					MC553L
SG91-02					MC553F
SG9150	177	3100/519	300	1122	
SG92-01					MC403L
SG92-02					MC403F
SG93-01					MC453L
SG93-02					MC453F
SGB-9742			86	2038	
SGC-7202			20	2051	
SGC7202			243	2030	
SGR100	116	3016	504A	1104	1N4002
SGR1000A					1N4007
SGR200A					1N4003
SGR400A					1N4004
SGR600A					1N4005
SGR800A					1N4006
SGS87231	161	3039/316	39	2015	
SH-1	116	3017B/117	504A	1104	
SH-1A	116	3031A	504A	1104	
SH-1B			504A	1104	
SH-1DE	116		504A	1104	
SH-1S			504A	1104	
SH-1Z				1102	
SH0013HC					MMH0026CG
SH0013HM					MMH0026G
SH1	116		504A	1104	
SH1064	123A	3444	20	2051	
SH15	116		504A	1104	
SH16G11	5685	3521/5686			
SH16H11	5686	3521			
SH16J11	5697	3522			
SH1A	116		504A	1104	
SH1B	125	3032A	510,531	1114	
SH1C	125	3033	510,531	1114	
SH1S	5802	9005	504A	1143	
SH4D05	116	3311	504A	1104	
SH4D1	116	3016	504A	1104	
SH4D2	116	3016	504A	1104	
SH4D3	116	3031A	504A	1104	
SH4D4	116	3031A	504A	1104	
SH4D6	116	3017B/117	504A	1104	
SH4D8	125	3032A	510,531	1114	
SH8090FM					MC1508L8
SHA7520	193		67	2023	
SHA7521	193		67	2023	
SHA7522	193		67	2023	
SHA7523	193		67	2023	
SHA7524	193		67	2023	
SHA7526	193		67	2023	
SHA7527	193		67	2023	
SHA7528	193		67	2023	
SHA7530	159	3114/290	82	2032	
SHA7531	159	3114/290	82	2032	
SHA7532	159	3114/290	82	2032	
SHA7533	159	3114/290	82	2032	
SHA7534	159	3114/290	82	2032	
SHA7536	159	3114/290	82	2032	
SHA7537	159	3114/290	82	2032	
SHA7538	159	3114/290	82	2032	
SHA7597	193		67	2023	
SHA7598	193		67	2023	
SHA7599	193		67	2023	
SHAD-1	116		504A	1104	
SHAD1	116	3016	504A	1104	
SI-05A			510	1114	
SI-1000E					1N4007
SI-100E					1N4002
SI-10A		3081	510	1114	MR508
SI-1A		3081	510	1114	MR501
SI-200E					1N4003
SI-2A		3081	510	1114	MR502
SI-300E					1N4004
SI-3A		3081	510	1114	MR504
SI-400E					1N4004
SI-4A		3081	510	1114	MR504
SI-500E					1N4005
SI-50E					1N4001
SI-5A		3081	510	1114	MR506
SI-600	169	3017B/117			
SI-600E					1N4005
SI-6A		3081	510	1114	MR506
SI-7A		3081	510	1114	
SI-800E					1N4006
SI-8A		3081	510	1114	MR508
SI-CA3064E	783	3215			
SI-CA3065E	712	3072			
SI-HA11580	1196	3725			
SI-MC1352P	749	3168	IC-97		
SI-MC1358P	712	3072	IC-148		
SI-MC1364P	783	3215	IC-225		
SI-REC-73			504A	1104	
SI-RECT-044	116	3032A	504A	1104	
SI-RECT-100	116		504A	1104	
SI-RECT-100-102			504A	1104	
SI-RECT-102	116	3032A	504A	1104	
SI-RECT-102A	125	3032A	510,531	1114	
SI-RECT-110	156(4)	3051/156	512		
SI-RECT-110/SB-3F	116		504A	1104	
SI-RECT-112	156(4)	3051/156	504A	1104	
SI-RECT-112/SB-3	156		512		
SI-RECT-114	506	3130	511	1114	
SI-RECT-122	116		504A	1104	
SI-RECT-124	506		511	1114	
SI-RECT-126			504A	1104	
SI-RECT-136	125	3080	510,531	1114	
SI-RECT-140			ZD-6.2	561	
SI-RECT-140/TR-6	137A		ZD-6.2	561	
SI-RECT-140/TR6S	5070A		ZD-6.2	561	
SI-RECT-144	116	3031A	504A	1104	
SI-RECT-152	177	3100/519	300	1122	
SI-RECT-154	116		504A	1104	
SI-RECT-155	116		504A	1104	
SI-RECT-156	116		504A	1104	
SI-RECT-158			511	1114	
SI-RECT-162	506	3130	511	1114	
SI-RECT-168	5010A		ZD-5.1		
SI-RECT-170	506		511	1114	
SI-RECT-174	113A	3031A	6GC1		
SI-RECT-178	115	3311	6GX1		
SI-RECT-178B	115		6GX1		
SI-RECT-178BF	113A		6GC1		
SI-RECT-180	506		511	1114	
SI-RECT-182			511	1114	
SI-RECT-2	116	3017B/117	504A	1104	
SI-RECT-20			504A	1104	
SI-RECT-204	125	3081	510,531	1114	
SI-RECT-206	506	3130	511	1114	
SI-RECT-208	156	3051	512		
SI-RECT-218	116		504A	1104	
SI-RECT-220			504A	1104	

Industry Standard No.	ECG	SK	GE	RS 276-	MOTOR.
SI-RECT-222	116	3031A	504A	1104	
SI-RECT-224	506	3130	511	1114	
SI-RECT-226	116	3031A	504A	1104	
SI-RECT-228	142A	3062	ZD-12	563	
SI-RECT-23		3126	90		
SI-RECT-230	142A	3062	ZD-12	563	
SI-RECT-25	116	3311	504A	1104	
SI-RECT-27	116	3017B/117	504A	1104	
SI-RECT-33	116	3017B/117	504A	1104	
SI-RECT-34	116		504A	1104	
SI-RECT-35	177	3100/519	300	1122	
SI-RECT-36	506	3125	511	1114	
SI-RECT-37	116	3311	504A	1104	
SI-RECT-39	116	3017B/117	504A	1104	
SI-RECT-44	506	3125	511	1114	
SI-RECT-48	116	3017B/117	504A	1104	
SI-RECT-49	116	3017B/117	504A	1104	
SI-RECT-53	116		504A	1104	
SI-RECT-59	116	3031A	504A	1104	
SI-RECT-69	116	3311	504A	1104	
SI-RECT-73	116		504A	1104	
SI-RECT-74	116	3311	504A	1104	
SI-RECT-75	116	3311	504A	1104	
SI-RECT-77	116	3311	504A	1104	
SI-RECT-84	116		504A	1104	
SI-RECT-92	116	3017B/117	504A	1104	
SI-RECT-94	116		504A	1104	
SI-RECT102			504A	1104	
SI-TA7074P	749	3168			
SI-TA7075P		3284	IC-98		
SI-UPC580C	1196	3725			
SI05	5800	9003		1142	
SI05A	5800	9003		1142	
SI1	5801	9004		1142	1N5392
SI10	5809	9010			1N5399
SI1000E	125	3033	510,531	1114	
SI100E	116	3016	504A	1104	
SI10A	156	9010/5809	512	1114	
SI12					MR1-1200
SI15					MR1-1600
SI1A	5802	9005		1143	
SI2					1N5393
SI3					1N5394
SI341P			21	2034	
SI342P			21	2034	
SI343P			21	2034	
SI351P			21	2034	
SI352P			21	2034	
SI353P			21	2034	
SI4					1N5395
SI5	5806	3848			1N5396
SI50E	116	3017B/117	504A	1104	
SI6					1N5397
SI7					1N5398
SI8					1N5398
SI8A	156	3051	512	1114	
SI9					1N5399
SI91G	116	3016	504A	1104	
SIB-01-02	116		504A	1104	
SIB-01-022	116	3311	504A	1104	
SIB-0306	125	3081	510,531	1114	
SIB-C1-02	177		300	1122	
SIBO-1	116		511	1114	
SIB01	116	3311	511	1114	
SIB01-01	552	3311	504A	1104	
SIB01-02	552	3311	504A	1104	
SIB01-04	552	3130	504A	1114	
SIB01-06	116	3017B/117	504A	1104	
SIB01-06B	116		504A	1104	
SIB0102	116	3017B/117	504A	1104	
SIB02-03C			504A	1104	
SIB02-03CR	116		504A	1104	
SIB02-CR	116	3311	504A	1104	
SIB02-CR1	178MP		300(2)	1122(2)	
SIB020-1B	166			1152	
SIB0201CR	116	3031A	504A	1104	
SIB03-10				1114	
SIBOL	116		504A	1104	
SIBOL-02	116		504A	1104	
SIB1			504A	1104	
SIB508794-1				1152	
SIB50B794-1	166	9075	BR-600		
SID01E	116	3031A	504A	1104	
SID01K	125	3033	510,531	1114	
SID01L	116	3017B/117	504A	1104	
SID02E	116	3031A	504A	1104	
SID02K	125	3033	510,531	1114	
SID02L	116	3017B/117	504A	1104	
SID23-13			511	1114	
SID23-15			511	1114	
SID2A-1	113A		6GC1		
SID2A-3	113A		6GC1		
SID30-13	525	3843	511	1114	
SID30-132	525	9000/552	511	1114	
SID30-15	551	3125	511	1114	
SID3B-3	115		6GX1		
SID50894	177		300	1122	
SID50B851	177		300	1122	
SID51C052-19	156		512		
SID51C169	116(4)		504A	1104	
SIDA0.5K	5806	3848			
SIDA0.5N	5808	9009			
SIDA0.5P	5809	9010			
SIDA05K	156	3051	512		
SIDA05N	156	3051	512		
SIDA05P	156	3051	512		
SIG1/100	5802	9005		1143	
SIG1/200	116	9005/5802	504A	1143	
SIG1/400	116	9007/5804		1104	
SIG1/600	116	3017B/117	504A	1104	
SIG1/800	5808	9009			
SIL				1114	
SIL-200	116	3016	504A	1104	
SIL200	116		504A	1104	
SIL4046	980	4046			
SIPT040	56022		SC265B2		
SIPT140	56022		SC265B2		
SIPT230	56022		SC265B2		
SIPT240	56022		SC265B2		
SIPT330	56024		SC265D2		
SIPT340	56024		SC265D2		
SIPT430	56024		SC265D2		
SIPT440	56024		SC265D2		
SIPT530	56026		SC265E2		
SIPT540	56026		SC265E2		
SIPT630	56026		SC265M2		
SIPT640	56026		SC265M2		
SIR-80	116	3016	504A	1104	
SIR-RECT-44	116		504A	1104	
SIR20			504A	1104	
SIR60	552	3313/116	504A	1104	
SIRB-10	166	3647/167		1152	
SIRB10		3105	512		
SIRECT-102	116		504A	1104	

Industry Standard No.	ECG	SK	GE	RS 276-	MOTOR.
SIRECT-2	116		504A	1104	
SIRECT-36	116		504A	1104	
SIRECT-48	116		504A	1104	
SIRECT-59	116		504A	1104	
SIRECT-92	116		504A	1104	
SIS11			1N34AS	1123	
SIS2			1N34AS	1123	
SIS20	109	3087		1123	
SISD-1	506	3125	511	1114	
SISD-1HF	125	3125	510,531	1114	
SISD-1X	116		504A	1104	
SISD-K	116	3017B/117	504A	1104	
SISM-150-01	116	3017B/117	504A	1104	
SISW-05-02	116	3017B/117	504A	1104	
SISW-0502	116		504A	1104	
SIT20/100	5542	6615/5515			
SIT20/50	5541	6615/5515			
SIT4/200	5483	3942			
SIT4/300	5484	3943/5485			
SIT4/400	5485	3943			
SJ-570	116	3020/123	20	1104	
SJ051A	5940		5048		
SJ051E	116	3311	504A	1104	
SJ051F	116	3017B/117	504A	1104	
SJ052A	5940		5048		
SJ052E	116	3311	504A	1104	
SJ052F	116	3017B/117	504A	1104	
SJ053E	5850		5048		
SJ053EK	5870		5032		
SJ053F	5800	9003		1142	
SJ053K	5850		5048		
SJ054E	5870		5048		
SJ054EK	5870		5032		
SJ054K	5870		5048		
SJ10003EK	5890	7090	5044		
SJ1000F	5809	9010			
SJ1003F	156	3051	512		
SJ101A	5834		5048		
SJ101F	116	3016	512	1104	
SJ102A	5834		5048		
SJ102F	116	3016	504A	1104	
SJ103E	5854		5048		
SJ103EK	5872		5032		
SJ103F	5801	9004		1142	
SJ103K	5854		5048		
SJ104E	5870		5048		
SJ104EK	5872		5032		
SJ104K	5870		5048		
SJ1106	130	3027	14	2041	
SJ1152	153		69	2049	
SJ1165	124	3021	12		
SJ1171	153		69	2049	
SJ1172	175	3261	246	2020	
SJ1201	124	3021	12		
SJ1272	180		74	2043	
SJ1284	153	3274	69	2049	
SJ1286	124	3021	12		
SJ130	124		12		
SJ1470	130	3027	14	2041	
SJ1902	247	3948			
SJ1903	248	3949			
SJ1925			FET-1	2028	
SJ2000	130	3027	14	2041	
SJ2001	219	3173	74	2043	
SJ2008	130	3027	14	2041	
SJ2009	218	3625	234		
SJ201A	5834		5048		
SJ201F	116	3016	504A	1104	
SJ2023	180		74	2043	
SJ2024	180		74	2043	
SJ202A	5834		5048		
SJ202F	116	3016	504A	1104	
SJ2031	129		244	2027	
SJ2032	128		243	2030	
SJ203E	5854		5048		
SJ203EK	5874		5032		
SJ203F	5802	9005		1143	
SJ203K	5854		5048		
SJ2047	181		75	2041	
SJ204E	5870		5048		
SJ204EK	5874		5032		
SJ204K	5870		5048		
SJ2064	181		75	2041	
SJ2095	175	3261	246	2020	
SJ2519	284	3836	265	2047	
SJ2520	285	3846	266		
SJ285	185	3191	58	2025	
SJ301F	116	3016	504A	1104	
SJ302F	116	3016	504A	1104	
SJ304EK	5878		5036		
SJ3408	175	3131A/369	246	2020	
SJ3423	162		35		
SJ3447	175	3131A/369	246	2020	
SJ3464	130	3027	14	2041	
SJ3477	162		35		
SJ3478	162		35		
SJ3507	180		74	2043	
SJ3519	181		75	2041	
SJ3520	219	3173	74	2043	
SJ3604	130	3027	14	2041	
SJ3629			210	2051	
SJ3636	180		74	2043	
SJ3637	180	3437	74	2043	
SJ3648	175	3131A/369	246	2020	
SJ3678	130	3027	14	2041	
SJ3679	219	3173	74	2043	
SJ3680	175	3261	246	2020	
SJ401F	116	3017B/117	504A	1104	
SJ402F	116	3017B/117	504A	1104	
SJ403F	5804	9007		1144	
SJ404EK	5878		5036		
SJ501F	116	3017B/117	504A	1104	
SJ5196	162		35		
SJ5525	164	3115/165	37		
SJ5526	165	3111	38		
SJ570	123A	3444	20	2051	
SJ601F	116	3017B/117	504A	1104	
SJ603	5882	3500	5040		
SJ603E	5882	3500	5040		
SJ603EK	5882		5040		
SJ603F	5806	3848			
SJ603K	5882	3500	5040		
SJ604	5882	3500	5040		
SJ604E	5882	3500	5040		
SJ604EK	5882		5040		
SJ604K	5882	3500	5040		
SJ60F	116	3017B/117	504A	1104	
SJ619	130	3027	14	2041	
SJ619-1	130	3027	14	2041	
SJ652	218	3625	234		
SJ803F	5808	9009			
SJ805	124	3021	12		
SJ806	124	3021	12		
SJ811	175	3261	246	2020	

Industry Standard No.	ECG	SK	GE	RS 276-	MOTOR.
SJ8165K	130	3027			
SJ820	130	3027	14	2041	
SJ821	219	3173	74	2043	
SJ822	218	3625	234		
SJ8701	130	3027	14	2041	
SJ9110	130	3027	14	2041	
SJE-5038	186		28	2017	
SJE-513	152		66	2048	
SJE-515	152		66	2048	
SJE-5402	186		28	2017	
SJE-649	184		57	2017	
SJE100	184	3054/196	57	2017	
SJE103	190	3104A	217		
SJE1032	190	3232/191	217		
SJE106	184	3054/196	57	2017	
SJE108	185	3083/197	58	2025	
SJE111	185	3083/197	58	2025	
SJE112	185	3083/197	58	2025	
SJE113	184	3054/196	57	2017	
SJE114	185	3083/197	58	2025	
SJE133	184	3190	57	2017	
SJE1518	185	3083/197	58	2025	
SJE1519	184	3054/196	57	2017	
SJE1520	184	3054/196	57	2017	
SJE202	185	3083/197	58	2025	
SJE203	184	3054/196	57	2017	
SJE205	157	3103A/396	232		
SJE210	185	3083/197	58	2025	
SJE211	184	3054/196	57	2017	
SJE218	157	3103A/396	232		
SJE220	182	3188A	55		
SJE221	185	3083/197	58	2025	
SJE222	184	3054/196	57	2017	
SJE227	185	3083/197	58	2025	
SJE228	184	3054/196	57	2017	
SJE229	184	3054/196	57	2017	
SJE231	185	3083/197	58	2025	
SJE232	157	3103A/396	232		
SJE237	184	3054/196	57	2017	
SJE241	185	3083/197	58	2025	
SJE242	184	3054/196	57	2017	
SJE243	185	3083/197	58	2025	
SJE244	184	3054/196	57	2017	
SJE245	185	3083/197	58	2025	
SJE246	184	3054/196	57	2017	
SJE248	184	3054/196	57	2017	
SJE253	184	3054/196	57	2017	
SJE254	184	3054/196	57	2017	
SJE255	184	3041	57	2017	
SJE256	185	3041	58	2025	
SJE257	185	3083/197	58	2025	
SJE261	184	3054/196	57	2017	
SJE262	184	3054/196	57	2017	
SJE264	180	3437	74	2043	
SJE265	185	3083/197	58	2025	
SJE267	185	3083/197	58	2025	
SJE271	184	3054/196	57	2017	
SJE272	184	3054/196	57	2017	
SJE273	185	3083/197	58	2025	
SJE274	184	3054/196	57	2017	
SJE275	185	3083/197	58	2025	
SJE276	185	3083/197	58	2025	
SJE277	185	3083/197	58	2025	
SJE278	184	3054/196	57	2017	
SJE279	185	3083/197	58	2025	
SJE280	184	3054/196	57	2017	
SJE283	185	3083/197	58	2025	
SJE284	184	3041	57	2017	
SJE288	185	3041	58	2025	
SJE289	184	3041	57	2017	
SJE290	157	3747	232		
SJE305	184	3054/196	57	2017	
SJE320	184	3041	57	2017	
SJE340	184	3054/196	57	2017	
SJE3754	157	3747	232		
SJE400	157	3103A/396	232		
SJE401	184	3041	57	2017	
SJE402	184	3041	57	2017	
SJE403	185	3083/197	58	2025	
SJE404	184	3054/196	57	2017	
SJE405	185	3054/196	57	2017	
SJE407	184	3054/196	57	2017	
SJE408	185	3083/197	58	2025	
SJE42	152	3054/196	66	2048	
SJE5018	182	3188A	55		
SJE5019	182	3188A	55		
SJE5020	182	3188A	55		
SJE513	152	3054/196	66	2048	
SJE514	153	3025/129	69	2049	
SJE515	152	3054/196	66	2048	
SJE516	182	3188A	55		
SJE517	183	3189A	56	2027	
SJE527	184	3190	57	2017	
SJE5402	184	3192/186	57	2017	
SJE5439	241	3188A/182	57	2020	
SJE5441	241	3188A/182	57	2020	
SJE5442	242	3189A/183	58	2027	
SJE583	184	3054/196	57	2017	
SJE584	185	3083/197	58	2025	
SJE633	185	3191	58	2025	
SJE634	184	3190	57	2017	
SJE649			57	2017	
SJE667	184	3190			
SJE669	184	3190	57	2017	
SJE677			69	2049	
SJE678			66	2048	
SJE687			218	2026	
SJE694			66	2048	
SJE695			69	2049	
SJE721	184	3054/196	57	2017	
SJE723	185	3083/197	58	2025	
SJE724	184	3041	57	2017	
SJE736	185	3083/197	58	2025	
SJE737	184	3054/196	57	2017	
SJE743	185	3041	58	2025	
SJE764	180	3437	74	2043	
SJE768	185	3083/197	58	2025	
SJE769	184	3054/196	57	2017	
SJE781	184	3041	57	2017	
SJE783	152	3893	66	2048	
SJE784	184	3054/196	57	2017	
SJE785	184	3041	57	2017	
SJE797	185	3083/197	58	2025	
SJE799	185	3083/197	58	2025	
SJE802	157	3747			
SK-1B	116		504A	1104	
SK-1W50	5089A	3341	ZD-51		
SK-1W55	5090A	3342	ZD-56		
SK-1W80	109			1123	
SK-218	116		510,531	1114	
SK-31024-3	108	3018	86	2038	
SK-3960	123A		20		
SK-7	101	3861	8	2002	
SK1320	108		86	2038	

Industry Standard No.	ECG	SK	GE	RS 276-	MOTOR.
SK1639	159	3114/290	82	2032	
SK1639D	159	3114/290	82	2032	
SK1640	159	3114/290	82	2032	
SK1640A	123A	3444	20	2051	
SK1641	123A	3444	20	2051	
SK16510006-2	129		244	2027	
SK16510006-4	129		244	2027	
SK1700	716		IC-208		
SK1856	159	3114/290	82	2032	
SK1856A	159	3114/290	82	2032	
SK19	312	3448	FET-1	2035	
SK1FM	116	3017B/117	504A	1104	
SK1K-2	116	3311	504A	1104	
SK1W55	148A		ZD-55		
SK2604	159	3114/290	82	2032	
SK2604A	159	3114/290	82	2032	
SK3003	102A		52	2007	
SK3004	102A		53	2007	
SK3005	100		1	2007	
SK3006	160		245	2004	
SK3006/160				2005	
SK3007	160		52	2004	
SK3008	126		52	2024	
SK3009	121		239	2006	
SK3010	103A		59	2002	
SK3011	101		8	2002	
SK3012	105		240		
SK3013	121MP		239(2)	2006(2)	
SK3014	121		239	2006	
SK3015	121MP		239(2)	2006(2)	
SK3016	116		504A	1104	
SK3017	117		504A	1104	
SK3017A	116		504A	1104	
SK3017B				1104	
SK3017B/117				1104	
SK3018	161		39	2038	
SK3019	108		214	2038	
SK3020	123A		20	2051	
SK3021	124		12		
SK3022	704		IC-82		
SK3024	128		243	2030	
SK3025	129		244	2027	
SK3026	175		246	2020	
SK3027	130		14	2041	
SK3028	175(2)		246(2)		
SK3029	130MP		265MP	2041(2)	
SK3030	116		504A	1104	
SK3031	116		504A	1104	
SK3031A			504A	1103	
SK3032	125		510	1114	
SK3033	125		509	1114	
SK3033A			509	1114	
SK3034	127		25		
SK3035	127		25		
SK3036	181		75	2041	
SK3037	181(2)		75(2)		
SK3038	123A		20	2051	
SK3039	316		214	2038	
SK3040	154		235	2012	
SK3041	152		241	2048	
SK3043	515			1114	
SK3044	396		40	2012	
SK3045	225		256		
SK3046	123A		18	2051	
SK3047	128		219	2030	
SK3048	195A		219		
SK3049	224		46		
SK3050	222		FET-4	2036	
SK3050/221			FET-4	2036	
SK3051	156		512		
SK3052	131		30	2006	
SK3054	196		241	2020	
SK3055	134A		ZD-3.6		
SK3056	135A		ZD-5.1		
SK3057	136A		ZD-5.6	561	
SK3057/136A			ZD-5.6	561	
SK3058	137A		ZD-6.2	561	
SK3058/137A			ZD-6.2	561	
SK3059	138A		ZD-7.5		
SK3060	139A		ZD-9.1	562	
SK3060/139A			ZD-9.1	562	
SK3061	140A		ZD-10	562	
SK3061/140A			ZD-10	562	
SK3062	142A		ZD-12	563	
SK3062/142A			ZD-12	563	
SK3063	145A		ZD-15	564	
SK3063/145A			ZD-15	564	
SK3064	146A		ZD-27		
SK3065	222		FET-4	2036	
SK3065/222			FET-4	2036	
SK3066	118		CR-1		
SK3067	502		CR-4		
SK3068	503		CR-5		
SK3069	501B		520		
SK3070	711		IC-207		
SK3071	727		IC-210		
SK3072	712		IC-2		
SK3073	728		IC-22		
SK3074	729		IC-23		
SK3075	714		IC-4		
SK3076	715		IC-6		
SK3077	713		IC-5		
SK3079	162		265		
SK3080	125		509	1114	
SK3081	156		510	1114	
SK3081/125				1114	
SK3082	226		49	2025	
SK3083	197		250	2027	
SK3084	197		250	2027	
SK3085	197		234	2027	
SK3086	226MP		49(2)	2025(2)	
SK3087	109			1123	
SK3088	109			1123	
SK3089	112		1N82A		
SK3090	109			1123	
SK3091	109			1123	
SK3092	141A		ZD-11.5		
SK3093	143A		ZD-13	563	
SK3094	144A		ZD-14	564	
SK3094/144A				564	
SK3095	147A		ZD-33		
SK3096	148A		ZD-55		
SK3097	149A		ZD-62		
SK3098	150A		ZD-82		
SK3099	151A		ZD-110		
SK3100	519		300	1122	
SK3100/519				1122	
SK3101	706		IC-43		
SK3102	710		IC-89		
SK3103	228A		257		
SK3104	228A		257		
SK3105	166		BR-600	1152	
SK3106	168		BR-600	1173	
SK3106/5304				1173	

Industry Standard No.	ECG	SK	GE	RS 276-	MOTOR.
SK3107	170		BR-1000		
SK3108	505		CR-7		
SK3109	119		CR-2		
SK3110	120		CR-3		
SK3111	165		38		
SK3112	312		FET-1	2028	
SK3114	159		334	2032	
SK3115	165		259		
SK3116	312		FET-3	2035	
SK3117	161		39	2038	
SK3118	159		82	2032	
SK3119	113A		6GC1		
SK3120	114		6GD1		
SK3121	115		6GX1		
SK3122	289A		268	2051	
SK3123	176		80		
SK3124	289A		210	2051	
SK3125	525		511		
SK3128	515		511		
SK3129	726		IC-81		
SK3130	506		511	1114	
SK3131	175		260		
SK3132	233		39	2009	
SK3133	164		38		
SK3134	705A		IC-3		
SK3135	708		IC-10		
SK3136	5072A		ZD-8.2	562	
SK3136/5072A				562	
SK3137	192A		88	2030	
SK3138	193		89	2023	
SK3139	5074A		ZD-11	563	
SK3139/5074A				563	
SK3140	786		IC-227		
SK3141	780		IC-222		
SK3142	5075A		ZD-16	564	
SK3144	723		IC-15		
SK3145	5077A		ZD-18		
SK3146	787		IC-228		
SK3147	788		IC-229		
SK3148	146A		ZD-27		
SK3149	791		IC-231		
SK3150	5084A		ZD-30		
SK3151	5081A		ZD-24		
SK3152	1024		720		
SK3153	1027		721		
SK3155	1025		722		
SK3156	172A		64		
SK3157	703A		IC-12		
SK3159	718		IC-8		
SK3160	720		IC-7		
SK3161	722		IC-9		
SK3162	725		IC-19		
SK3163	736		IC-211		
SK3164	923		IC-259		
SK3165	923D		IC-260	1740	
SK3165/923D			IC-260	1740	
SK3166	949		IC-25		
SK3167	738		IC-29		
SK3169	781		IC-223		
SK3170	731		IC-29		
SK3171	744		IC-215	2022	
SK3171/744			IC-24	2022	
SK3172	743		IC-214		
SK3173	219		266	2043	
SK3174	116		504A	1104	
SK3175	506		511	1114	
SK3183	246		263	2043	
SK3184	1115		IC-278		
SK3190	184		57	2017	
SK3191	185		58	2025	
SK3194	286		267		
SK3195	311		45		
SK3204	731		IC-13		
SK3211	797		IC-233		
SK3215	780		IC-225		
SK3216	798		IC-234		
SK3223	1100		IC-92		
SK3224	1102		IC-93		
SK3225	1104		IC-91		
SK3226	1071		IC-67		
SK3227	1069		IC-64		
SK3228	1061		IC-51		
SK3230	1154		IC-287		
SK3231	1155		IC-179	705	
SK3232	191		224		
SK3234	746		IC-217		
SK3235	739		IC-30		
SK3236	748		IC-115		
SK3237	795		IC-232		
SK3238	799		IC-34		
SK3240	779A		IC-221		
SK3243	1160		IC-196		
SK3244	287		222		
SK3249	1052		IC-135		
SK3250	315		279		
SK3252	306		275		
SK3253	295		270		
SK3254	785		IC-226		
SK3255	815		IC-244		
SK3260	284		265		
SK3263	123A		285		
SK3264	123A		285		
SK3270	284		75		
SK3276	745		IC-216		
SK3277	802		IC-235		
SK3278	803		IC-237		
SK3279	747		IC-218		
SK3280	750		IC-219		
SK3281	1103		IC-94		
SK3282	1153		IC-182		
SK3283	1101		IC-95		
SK3284	1080		IC-98		
SK3285	1105		IC-101		
SK3286	1131		IC-109		
SK3287	1132		IC-110		
SK3288	1003		IC-43		
SK3289	1158		IC-71		
SK3291	1090		723		
SK3295	1072		IC-59		
SK3300	531		524		
SK3301	532		525		
SK3302	533		526		
SK3303	522		523		
SK3304	500A		527		
SK3305	534		534		
SK3306	523		528		
SK3307	529		529		
SK3308	530		540		
SK3311	116			1102	
SK3312	116			1103	
SK3313	116			1104	
SK3313/116				1104	
SK3330	5066A		ZD-3.3		
SK3331	5067A		ZD-3.9		
SK3332	5068A		ZD-4.3		
SK3333	5069A		ZD-4.7		
SK3334	5071A		ZD-6.8	561	
SK3334/5071A				561	
SK3335	5079A		ZD-20		
SK3336	5080A		ZD-22		
SK3337	5085A		ZD-36		
SK3338	5086A		ZD-39		
SK3339	5087A		ZD-43		
SK3340	5088A		ZD-47		
SK3341	5089A		ZD-51		
SK3342	5090A		ZD-5.6	561	
SK3343	5092A		ZD-68		
SK3345	5095A		ZD-91		
SK3347	5097A		ZD-120		
SK3348	5098A		ZD-130		
SK3349	5099A		ZD-140		
SK3350	5100A		ZD-150		
SK3351	5101A		ZD-160		
SK3352	5102A		ZD-170		
SK3353	5103A		ZD-180		
SK3354	5104A		ZD-190		
SK3355	5105A		ZD-200		
SK3356	107		211	2031	
SK3358	1006		IC-38		
SK3361	1041		IC-160		
SK3362	1040		IC-161		
SK3363	1043		IC-169		
SK3365	1004		IC-149		
SK3366	1039		IC-159		
SK3368	1029		IC-162		
SK3369	1035		IC-167		
SK3370	1036		IC-168		
SK3371	1037		IC-170		
SK3376	1010		IC-37		
SK3377	5111A		5ZD-3.3		
SK3378	5112A		5ZD-3.6		
SK3379	5113A		5ZD-3.9		
SK3380	5114A		5ZD-4.3		
SK3381	5115A		5ZD-4.7		
SK3382	5116A		5ZD-5.1		
SK3383	5117A		5ZD-5.6		
SK3384	5118A		5ZD-6.0		
SK3385	5119A		5ZD-6.2		
SK3386	5120A		5ZD-6.8		
SK3387	5121A		5ZD-7.5		
SK3388	5122A		5ZD-8.2		
SK3389	5123A		5ZD-8.7		
SK3390	5124A		5ZD-9.1		
SK3391	5125A		5ZD-10		
SK3392	5126A		5ZD-11		
SK3393	5127A		5ZD-12		
SK3394	5128A		5ZD-13		
SK3395	5129A		5ZD-14		
SK3396	5130A		5ZD-15		
SK3397	5131A		5ZD-16		
SK3398	5132A		5ZD-17		
SK3399	5133A		5ZD-18		
SK3400	5134A		5ZD-19		
SK3401	5135A		5ZD-20		
SK3402	5136A		5ZD-22		
SK3403	5137A		5ZD-24		
SK3404	5138A		5ZD-25		
SK3405	5139A		5ZD-27		
SK3406	5140A		5ZD-28		
SK3407	5141A		5ZD-30		
SK3408	5142A		5ZD-33		
SK3433	287		222		
SK3434	288		223		
SK3434A	123A	3122	20	2051	
SK3435	1028		724		
SK3436	1028		723		
SK3440	291		241		
SK3441	292		250		
SK3444/123A				2051	
SK3449	297		271		
SK3450	298		272		
SK3457	1054		IC-45		
SK3458	1056		IC-48		
SK3459	1058		IC-49		
SK3460	1060		IC-50		
SK3461	1082		IC-140		
SK3462	977		VR-100		
SK3464	300		273		
SK3467	283		36		
SK3468	1185		IC-139		
SK3470	1049		IC-121		
SK3471	1046		IC-118		
SK3473	1140		IC-138		
SK3474	1160		IC-197		
SK3475	1050		IC-123		
SK3476	1085		IC-104		
SK3477	1087		IC-103		
SK3478	1130		IC-111		
SK3480	1178		IC-323		
SK3485	1142		IC-128		
SK3486	1184		IC-137		
SK3488	1128		IC-105		
SK3489	1134		IC-106		
SK3490	1133		IC-107		
SK3492	1011		IC-79		
SK3494	1055		IC-47		
SK3495	1074		IC-60		
SK3496	1073		IC-63		
SK3498	1078		IC-136		
SK3499	1009		IC-77		
SK3500	5882		MR-1		
SK3501	5994		5136		
SK3502	5513		700		
SK3504	5529		C137M		
SK3505	5547		C137M		
SK3510	130		14	2041	
SK3511	181		75	2041	
SK3512	282		46	2030	
SK3514	941		IC-263	010	
SK3514/941			IC-263	010	
SK3515	5805		511		
SK3516	5847		5013		
SK3517	5883		5041		
SK3518	5995		5137		
SK3523	'6408			1050	
SK3523/6407				1050	
SK3524	784		IC-236		
SK3525	724		IC-86		
SK3526	947		IC-268		
SK3531			FET-2	2036	
SK3535	181			2041	
SK3535/181				2041	
SK3538	286		246	2020	
SK3539	908		IC-248		
SK3540	903		IC-288		
SK3541	914		IC-256		
SK3542	904		IC-289		
SK3543	912		IC-172		

Industry Standard No.	ECG	SK	GE	RS 276-	MOTOR.
SK3544	917		IC-258		
SK3545	907		IC-247		
SK3546	905		IC-290		
SK3547	900		IC-245		
SK3548	906		IC-246		
SK3549	901		IC-180		
SK3550	916		IC-257		
SK3551			IC-257	2009	
SK3552	909D		IC-249	007	
SK3552/941M			IC-265	007	
SK3553	909		IC-250		
SK3554	917		IC-258		
SK3555	911		IC-253		
SK3556	911D		IC-254		
SK3561			75	2041	
SK3562	175		246		
SK3563	284		75	2041	
SK3564	955M		IC-269	1723	
SK3564/955M			IC-269	1723	
SK3582	5545		MR-3&4		
SK3584	5862		5024		
SK3590	909D		IC-250		
SK3591	960		VR-102		
SK3592	966		VR-112		
SK3597	5455		MR-5	1067	
SK3597/5455				1067	
SK3599	5858		5024		
SK3600	5854		5016		
SK3601	5850		5016		
SK3602	5878		5036		
SK3603	5874		5036		
SK3604	5870		5032		
SK3608	6034		5132		
SK3609	6026		5128		
SK3610	6020		5128		
SK3702	721		IC-14		
SK3716				2011	
SK3716/161				2011	
SK3728	1168		IC-66		
SK3730	1222		245	2004	
SK3749			ZD-8.7	562	
SK3749/5073A			ZD-8.7	562	
SK3750			ZD-13	563	
SK3750/143A			ZD-13	563	
SK3751			ZD-16	564	
SK3751/5075A			ZD-16	564	
SK3770	5004A		245	2004	
SK3777				561	
SK3777/5011				561	
SK3778				561	
SK3778/5012				561	
SK3779				561	
SK3779/5013				561	
SK3780				561	
SK3780/5014				561	
SK3780/5014A				561	
SK3782				562	
SK3782/5016				562	
SK3783				562	
SK3783/5017				562	
SK3784				562	
SK3784/5018				562	
SK3784/5018A				562	
SK3785				562	
SK3785/5019				562	
SK3786				563	
SK3786/5020				563	
SK3787				563	
SK3787/5021				563	
SK3787/5021A				563	
SK3788				563	
SK3788/5022				563	
SK3789				564	
SK3789/5023				564	
SK3790				564	
SK3790/5024				564	
SK3790/5024A				564	
SK3791				564	
SK3791/5025				564	
SK3803	159		ZD-36		
SK3960	392		20		
SK4000	4000		4000		
SK4001	4001B		4001		
SK4002	4002B		4002		
SK4007	4007		4007		
SK4009	4049		4009		
SK4011	4011B		4011		
SK4012	4012B		4012		
SK4013	4013B		4013		
SK4015	4015B		4015		
SK4016	4016B		4016		
SK4017	4017B		4017		
SK4019	4019B		4019		
SK4020			4020	2420	
SK4021			4021	2421	
SK4023	4023B		4023		
SK4024	4024B		4024		
SK4025	4025B		4025		
SK4027	4027B		4027		
SK4030	4030B		4030		
SK4040	4040B		4040		
SK4042	4042B		4042		
SK4046/980				2446	
SK4049	4049		4049	2449	
SK4050	4050B		4050	2450	
SK4051	4051B		4051	2451	
SK4081	4081B		4081		
SK4518			4518	2490	
SK5184A	124	3021	12		
SK5797	159		82	2032	
SK5798	159		82	2032	
SK5801	123A	3444	20	2051	
SK5915	123A	3444	20	2051	
SK6345	159	3114/290	82	2032	
SK6346	159	3114/290	82	2032	
SK6347	159	3114/290	82	2032	
SK6347A	159	3114/290	82	2032	
SK7	101		8	2002	
SK707380	5633	3938			
SK7181	108	3039/316	86	2038	
SK7400	7400		7400	1801	
SK7401	7401		IC-194		
SK7402	7402		7402	1811	
SK7404	7404		7404	1802	
SK7406	7406		7406	1821	
SK7408			7408	1822	
SK7410	7410		7410	1807	
SK74123	74123		74123	1817	
SK7413			7413	1815	
SK74145			74145	1828	
SK74150	74150		74150	1829	
SK74154			74154	1834	
SK74192	74192		74192	1831	
SK74193	74193		74193	1820	
SK74196			74196	1833	
SK7420	7420		7420	1809	
SK7427	7427		7427	1823	
SK7432	7432		7432	1824	
SK7441			7441	1804	
SK7447	7447		7447	1805	
SK7448			7448	1816	
SK7451	7451		7451	1825	
SK7473	7473		7473	1803	
SK7474	7474		7474	1818	
SK7475	7475		7475	1806	
SK7476	7476		7476	1813	
SK7485	7485		7485	1826	
SK7486	7486		7486	1827	
SK7490	7490		7490	1808	
SK7492			7492	1819	
SK74LS00	74LS00			1900	
SK74LS02	74LS02			1902	
SK74LS04	74LS04			1904	
SK74LS08	74LS08			1908	
SK74LS10	74LS10			1910	
SK74LS123	74LS123			1926	
SK74LS151	74LS151			1929	
SK74LS157	74LS157			1930	
SK74LS174	74LS174			1932	
SK74LS175	74LS175			1934	
SK74LS193				1936	
SK74LS20	74LS20			1912	
SK74LS27	74LS27			1913	
SK74LS30	74LS30			1914	
SK74LS32	74LS32			1915	
SK74LS367	74LS367			1835	
SK74LS51	74LS51			1917	
SK74LS73	74LS73			1918	
SK74LS74	74LS74A			1919	
SK74LS75	74LS75			1920	
SK74LS85	74LS85			1922	
SK74LS93	74LS93			1925	
SK7664	159		82	2032	
SK8215	123A	3444	20	2051	
SK8251	123A	3444	20	2051	
SK8261	154		40	2012	
SK8937			39	2015	
SKA-4061			82	2032	
SKA-4074			86	2038	
SKA-4075			11	2015	
SKA-4076			61	2038	
SKA-4590			61	2038	
SKA-4802			62	2010	
SKA-5248			86	2038	
SKA-5541			39	2015	
SKA-5886			86	2038	
SKA-6256			20	2051	
SKA-6437			20	2051	
SKA-8105			20	2051	
SKA0030			20	2051	
SKA1079	129		244	2027	
SKA1080	123A	3444	20	2051	
SKA1117	123A	3444	20	2051	
SKA1279	159		82	2032	
SKA1395	123A	3444	20	2051	
SKA1416	107		11	2015	
SKA4074	108	3018	86	2038	
SKA4075	108		86	2038	
SKA4076	108		86	2038	
SKA4129	159	3114/290	82	2032	
SKA4141	123A	3444	20	2051	
SKA4410	192		63	2030	
SKA4525	108		86	2038	
SKA4616	128		243	2030	
SKA4621	129		244	2027	
SKA4768	161		39	2015	
SKA5248			86	2038	
SKA5886			86	2038	
SKA6250			82	2032	
SKA9013	108	3018	86	2038	
SKA9096			86	2038	
SKB8339	123A	3444	20	2051	
SKN170/04	6354	6554			
SKN170/04R	6355	6555			
SKN170/08	6358	6558			
SKN170/08R	6359	6559			
SKU25/06UNF	5697	3522			
SKU30/04	5695	3509			
SKU30/05	5697	3522			
SKU30/06	5697	3522			
SKVT7504	815	3255			
SKWHG7006	159		82	2032	
SL-030	116		504A	1104	
SL-030T	116	3031A	504A	1104	
SL-100	108	3019	86	2038	
SL-2	116	3016	504A	1104	
SL-3	116	3016	504A	1104	
SL-4	116	3017B/117	504A	1104	
SL-5	5802	9005	504A	1143	
SL-833	116		504A	1104	
SL-833A	116		504A	1104	
SL030/3490	116		504A	1104	
SL0305,T			504A	1104	
SL030S	116		504A	1104	
SL030T	116		504A	1104	
SL051A	5940		5048		
SL07040	909	3590	IC-249		
SL07055	909	3590	IC-249		
SL07059	703A		IC-12		
SL07682	716		IC-208		
SL07684	910		IC-251		
SL08058	703A		IC-12		
SL08066	909	3590	IC-249		
SL08797	941D		IC-264		
SL10	5890	7090	5044		MR1130
SL100	108		86	2038	MR1121
SL1000	5890	7090	5044		MR1130
SL1000X	5890	7090	5044		MR1130
SL103	116		5032	1122	
SL119			82	2032	
SL14971	7408	7408		1822	
SL14972	7438	7438			
SL16793	7400	7400		1801	
SL16794	7401	7401			
SL16795	7402	7402		1811	
SL16796	7404	7404		1802	
SL16797	7405	7405			
SL16798	7408	7408		1822	
SL16799	7411	7411			
SL16800	7420	7420		1809	
SL16801	7410	7410		1807	
SL16802	7430	7430			
SL16803	7440	7440			
SL16804	7450	7450			
SL16805	7453	7453			
SL16806	7473	7473		1803	
SL16807	7474	7474		1818	
SL16808	7476	7476		1813	

Industry Standard No.	ECG	SK	GE	RS 276-	MOTOR.
SL16809	7492	7492		1819	
SL17242	7473	7473		1803	
SL17869	7408	7408		1822	
SL18387	74107	74107			
SL2	116		504A	1104	
SL200	106	3984	21	2034	MR1122
SL201	106	3984	21	2034	
SL20575	703A		IC-12		
SL20721	705A	3134	IC-3		
SL20755	713		IC-5		
SL20783	923		IC-259		
SL20927	909	3590	IC-249		
SL20929	941		IC-263	010	
SL21017	715		IC-6		
SL21122	714	3075	IC-4		
SL21384	915			017	
SL21385	923D		IC-260	1740	
SL21436	910		IC-251		
SL21441	707	3134/705A	IC-3		
SL21577	915			017	
SL21584	949		IC-25		
SL21619	923		IC-259		
SL21654	712	3072	IC-2		
SL21673	941		IC-263	010	
SL21823	910		IC-251		
SL21829	911		IC-253		
SL21864	722	3161	IC-9		
SL21885	923		IC-259		
SL21895	703A		IC-12		
SL21923	909	3590	IC-249		
SL22044	911		IC-253		
SL22108	725	3162	IC-19		
SL22211	725		IC-19		
SL22273	912		IC-172		
SL22310	923D		IC-260	1740	
SL22348	925		IC-261		
SL22623	912		IC-172		
SL22745	941M		IC-265	007	
SL22756	718	3159	IC-8		
SL22757	722		IC-9		
SL22819	703A		IC-12		
SL22935	923D		IC-260	1740	
SL23059	941M		IC-265	007	
SL23145	722		IC-9		
SL23147	712		IC-2		
SL23252	941M		IC-265	007	
SL23256	912		IC-172		
SL23296	923		IC-259		
SL23297	911		IC-253		
SL23299	736		IC-17		
SL23324	909	3590	IC-249		
SL23325	923D		IC-260	1740	
SL23326	941D		IC-264		
SL23418	705A		IC-3		
SL23421	703A		IC-12		
SL23422	910		IC-251		
SL23423	911		IC-253		
SL23424	725		IC-19		
SL23425	949		IC-25		
SL23426	722		IC-9		
SL23482	909	3590	IC-249		
SL23485	923		IC-259		
SL23486	941		IC-263	010	
SL23496	941M		IC-265	007	
SL23546	923		IC-259		
SL23560	705A		IC-3		
SL23648	723		IC-15		
SL23649	781	3169	IC-223		
SL23829	912		IC-172		
SL23903	790		IC-5		
SL23904	714		IC-4		
SL23905	715		IC-6		
SL23971			IC-24	2022	
SL23985	725		IC-19		
SL24618	OBS-NLA		IC-277		
SL25529	309K	3629			
SL3	5878		5036	1104	MR1123
SL300	123A	3444	20	2051	MR1123
SL301C	128		243	2030	
SL301CE	128		243	2030	
SL3081	916	3550			
SL3081DG	916	3550			
SL3081DP	916	3550			
SL3082	2023	3694			
SL3101	129	3025	244	2027	
SL3111	129		244	2027	
SL400	5878		5036		MR1124
SL403			268	2038	
SL5	116	3500/5882	504A	1104	MR1125
SL50	5870		5032		MR1120
SL500	5882	3500	5040		MR1125
SL53431	7401	7401			
SL57308	7407	7407			
SL57313	7412	7412			
SL57316	7417	7417			
SL57318	7423	7423			
SL57319	7425	7425			
SL57320	7426	7426			
SL57330	7443	7443			
SL57331	7444	7444			
SL57332	7445	7445			
SL57333	7446	7446			
SL57334	7447	7447			
SL57336	7450	7450			
SL57338	7453	7453			
SL57340	7460	7460			
SL57341	7470	7470			
SL57342	7472	7472			
SL57347	7480	7480			
SL57356	7494	7494			
SL57358	7496	7496			
SL57361	74122	74122			
SL57368	74155	74155			
SL57369	74156	74156			
SL57371	74165	74165			
SL57372	74166	74175			
SL57374	74175	74175			
SL57375	74176	74176			
SL57376	74177	74177			
SL57377	74178	74178			
SL57378	74179	74179			
SL57379	74180	74180			
SL57380	74181	74181			
SL57382	74190	74190			
SL57383	74191	74191			
SL57388	74198	74198			
SL57389	74199	74199			
SL58236	74160	74160			
SL58238	74162	74162			
SL58911	4020B	4020			
SL58912	4021B	4021			
SL58925	4052B	4052			
SL58935	4518B	4518			
SL600	5882	3500	5040		MR1126

Industry Standard No.	ECG	SK	GE	RS 276-	MOTOR.
SL608	125	3033	510,531	1114	1N4006
SL610	125	3033	510,531	1114	1N4007
SL7059	703A	3157	IC-12		
SL708	125	3033	510,531	1114	1N4006
SL710	125	3033	510,531	1114	1N4007
SL7283	703A	3157	IC-12		
SL7308	703A	3157	IC-12		
SL7531	703A	3157	IC-12		
SL7593	703A	3157	IC-12		
SL7790	703A		IC-12		
SL7990	123A	3444	20	2051	
SL8					MR1128
SL800	5886		5044		MR1128
SL800X	5886		5044		MR1128
SL8020	703A	3157	IC-12		
SL833	116	3016	504A	1104	
SL833A	116	3016	504A	1104	
SL91	116	3016	504A	1104	1N4002
SL92	116	3017B/117	504A	1104	1N4003
SL93	116	3031A	504A	1104	1N4004
SLA-11					1N4001
SLA-12					1N4002
SLA-13					1N4003
SLA-14					1N4004
SLA-15					1N4004
SLA-16					1N4005
SLA-17					1N4005
SLA-18					1N4006
SLA-19					1N4007
SLA-20					MR1-1200
SLA-21					MR501
SLA-22					MR501
SLA-23					MR502
SLA-24					MR504
SLA-25					MR504
SLA-26					MR506
SLA-27					MR506
SLA-28					MR508
SLA-29					MR510
SLA-445	116	3017B/117	504A	1104	
SLA01	125	3033	510,531	1114	
SLA1000					MDA920A9
SLA1095	116	3017B/117	504A	1104	
SLA1096	116	3017B/117	504A	1104	
SLA1100	116	3016	504A	1104	
SLA1101	116	3016	504A	1104	
SLA1102	116	3016	504A	1104	
SLA1103	5804	9007		1144	
SLA1104	116	3017B/117	504A	1104	
SLA1105	116	3017B/117	504A	1104	
SLA11AB	116	3311	504A	1104	
SLA11C	116	3017B/117	504A	1104	
SLA12AB	116	3017B/117	504A	1104	
SLA12C	116	3017B/117	504A	1104	
SLA13AB	116	3017B/117	504A	1104	
SLA13C	116	3017B/117	504A	1104	
SLA1487	116	3016	504A	1104	
SLA1488	116	3016	504A	1104	
SLA1489	116	3016	504A	1104	
SLA1490	116	3031A	504A	1104	
SLA1491	116	3017B/117	504A	1104	
SLA1492	116	3017B/117	504A	1104	
SLA14AB	116	3031A	504A	1104	
SLA14C	116	3031A	504A	1104	
SLA15AB	116	3031A	504A	1104	
SLA15C	116	3017B/117	504A	1104	
SLA1692	116	3016	504A	1104	
SLA1693	116	3016	504A	1104	
SLA1694	116	3016	504A	1104	
SLA1695	116	3016	504A	1104	
SLA1696	116	3017B/117	504A	1104	
SLA1697	116	3017B/117	504A	1104	
SLA16AB	116	3017B/117	504A	1104	
SLA16C	116	3017B/117	504A	1104	
SLA17AB	116	3017B/117	504A	1104	
SLA17C	116	3017B/117	504A	1104	
SLA18AB	125	3032A	510,531	1114	
SLA18C	125	3032A	510,531	1114	
SLA19AB	125	3033	510,531	1114	
SLA19C	125	3033	510,531	1114	
SLA200					MDA920A4
SLA21A	5800	9003		1142	
SLA21B	5800	9003		1142	
SLA21C	5800	9003		1142	
SLA22A	5801	9004		1142	
SLA22B	5801	9004		1142	
SLA22C	5801	9004		1142	
SLA23B	5802	9005		1143	
SLA23C	5802	9005		1143	
SLA24A	5804	9007		1144	
SLA24B	5804	9007		1144	
SLA24C	5804	9007		1144	
SLA25A	5804	9007		1144	
SLA25B	5804	9007		1144	
SLA25C	5804	9007		1144	
SLA2610	116	3016	504A	1104	
SLA2611	116	3016	504A	1104	
SLA2612	116	3016	504A	1104	
SLA2613	116	3017B/117	504A	1104	
SLA2614	116	3017B/117	504A	1104	
SLA2615	116	3017B/117	504A	1104	
SLA2616	125	3032A	510,531	1114	
SLA2617	125	3033	510,531	1114	
SLA26A	5806	3848			
SLA26B	5806	3848			
SLA26C	5806	3848			
SLA27A	5806	3848			
SLA27B	5806	3848			
SLA27C	5806	3848			
SLA28A	5808	9009			
SLA28B	5808	9009			
SLA28C	5808	9009			
SLA29A	5809	9010			
SLA29B	5809	9010			
SLA29C	5809	9010			
SLA300					MDA920A5
SLA3193	116	3016	504A	1104	
SLA3194	116	3017B/117	504A	1104	
SLA3195	116	3017B/117	504A	1104	
SLA3196	125	3032A	510,531	1114	
SLA400					MDA920A6
SLA440	116	3017B/117	504A	1104	
SLA440B	116	3016	504A	1104	
SLA441	116	3016	504A	1104	
SLA441B	116	3016	504A	1104	
SLA442	116	3016	504A	1104	
SLA442B	116	3016	504A	1104	
SLA443	116		504A	1104	
SLA443B	116	3017B/117	504A	1104	
SLA444	116	3017B/117	504A	1104	
SLA444A			504A	1104	
SLA444B	116	3017B/117	504A	1104	
SLA445	116		504A	1104	
SLA445B	116	3017B/117	504A	1104	

Industry Standard No.	ECG	SK	GE	RS 276-	MOTOR.
SLA500					MDA920A7
SLA5191					MR501
SLA5197	5800	9003		1142	
SLA5198	5801	9004		1142	MR501
SLA5199	5802	9005		1143	MR502
SLA5200	5804	9007		1144	MR504
SLA5201	5806	3848			MR506
SLA5202	5808	9009			
SLA5203	5809	9010			
SLA536	116	3311	504A	1104	
SLA537	116	3016	504A	1104	
SLA538	116	3016	504A	1104	
SLA539	116	3017B/117	504A	1104	
SLA540	116	3017B/117	504A	1104	
SLA547	116	3017B/117	504A	1104	
SLA560	125	3032A	510,531	1114	
SLA561	125	3033	510,531	1114	
SLA599	116	3017B/117	504A	1104	
SLA599A	116	3311	504A	1104	
SLA600	116	3017B/117	504A	1104	MDA920A7
SLA600A	116	3017B/117	504A	1104	
SLA601	116	3016	504A	1104	
SLA601A	116	3017B/117	504A	1104	
SLA602	116	3017B/117	504A	1104	
SLA602A	116	3016	504A	1104	
SLA603	116	3016	504A	1104	
SLA603A	116	3016	504A	1104	
SLA604	116	3017B/117	504A	1104	
SLA604A	116	3017B/117	504A	1104	
SLA605	116	3017B/117	504A	1104	
SLA605A	116	3017B/117	504A	1104	
SLA606	116	3017B/117	504A	1104	
SLA606A	116	3017B/117	504A	1104	
SLA700					MDA920A8
SLA800					MDA920A8
SLA900					MDA920A9
SLA01	125			1114	
SLBA05					MDA2500
SLBA1					MDA2501
SLBA2					MDA2502
SLBA3					MDA2504
SLBA4					MDA2504
SLBA6					MDA3506
SLEN-26	113A	3119/113	6GC1		
SM-1	116		504A	1104	
SM-1-005	116	3311	504A	1104	
SM-1-02				1102	
SM-1-47	116	3311	504A	1104	
SM-10	116	3016	504A	1104	
SM-150	125		510,531	1114	
SM-150-005	125		510,531	1114	
SM-150-02	125	3017B/117	510,531	1114	
SM-150-02(BOOST)			511	1114	
SM-150-04				1103	
SM-150-06				1104	
SM-150-10				1114	
SM-150-6(FOCUS)	506		511	1114	
SM-150-A				1103	
SM-150-B				1104	
SM-150-D				1114	
SM-150A	125		510,531	1114	
SM-150B	125	3016	504A	1104	
SM-1K	116	3311	504A	1104	
SM-217	100	3005	1	2007	
SM-4304-S	161	3018	39	2015	
SM-4508-B	123A	3020/123	20	2051	
SM-5564	123A	3020/123	20	2051	
SM-5643	123A	3020/123	20	2051	
SM-716	123A		20	2051	
SM-7815	123A	3020/123	20	2051	
SM-7836	123A	3020/123	20	2051	
SM-7991	195A		46		
SM-A-595819-1	704	3022/1188	IC-205		
SM-A-595830-12	107		11	2015	
SM-A-618687-1	175		246	2020	
SM-A-726655	123A	3444	20	2051	
SM-A-726658	159		82	2032	
SM-A-726664	123A	3444	20	2051	
SM-B-523974	159		82	2032	
SM-B-574495	159		82	2032	
SM-B-610342	123A	3444	20	2051	
SM-B-686767	123A	3444	20	2051	
SM-C-583256	123A	3444	20	2051	
SM-C-706156	519		514	1122	
SM07275			20	2051	
SM07286			20	2051	
SM0843	102A		53	2007	
SM1-02	116	9000/552	504A	1104	
SM10	116		504A	1104	
SM10-01					MC4326L
SM10-02					MC4326F
SM100	125	3033	510,531	1114	
SM101	125	3033	510,531	1114	
SM103	125	3033	510,531	1114	
SM105	116	3017B/117	504A	1104	
SM11	116	3016	504A	1104	
SM11-01					MC4327L
SM11-02					MC4327F
SM110	116	3017B/117	504A	1104	
SM111-01					MC5111L
SM111-02					MC5111F
SM113-01					MC5113L
SM113-02					MC5113F
SM12-01					MC4026L
SM12-02					MC4026F
SM120	116	3016	504A	1104	
SM121-01					MC5121L
SM121-02					MC5121F
SM123-01					MC5123L
SM123-02					MC5123F
SM1297	160		245	2004	
SM13-01					MC4027L
SM13-02					MC4027F
SM130	116	3016	504A	1104	
SM131-01					MC5131L
SM131-02					MC5131F
SM133-01					MC5133L
SM133-02					MC5133F
SM140	116	3017B/117	504A	1104	
SM141-01					MC5141L
SM141-02					MC5141F
SM143-01					MC5143L
SM143-02					MC5143F
SM150	125	3017B/117	504A	1114	
SM150-01	125	3311	510,531	1114	
SM150-02	125		510,531	1114	
SM150-11	125	3311	510,531	1114	
SM150-6	125		510,531	1114	
SM1507	159		82	2032	
SM150A	125	3017B/117	510,531	1114	
SM150B	125		510,531	1114	
SM150C	125	3032A	510,531	1114	
SM150D	125	3033	510,531	1114	
SM150S	125	3017B/117	510,531	1114	

Industry Standard No.	ECG	SK	GE	RS 276-	MOTOR.
SM150SS	125	3017B/117	510,531	1114	
SM151-01					MC5151L
SM151-02					MC5151F
SM153-01					MC5153L
SM153-02					MC5153F
SM160	116	3017B/117	504A	1104	
SM1600	160		245	2004	
SM163-01					MC5163L
SM163-02					MC5163F
SM170	125	3032A	510,531	1114	
SM173-01					MC5173L
SM173-02					MC5173F
SM180	125	3032A	510,531	1114	
SM181-01					MC5181L
SM181-02					MC5181F
SM183-01					MC5183L
SM183-02					MC5183F
SM191-01					MC5195L
SM191-02					MC5191F
SM193-01					MC5193L
SM193-02					MC5193F
SM20	116	3925/525	504A	1104	
SM20-01					MC4328L
SM20-02					MC4328F
SM200	125	3033	510,531	1114	
SM205	116	3311	504A	1104	
SM21-01					MC4329L
SM21-02					MC4329F
SM210	116	3017B/117	504A	1104	
SM217	160	3123	245	2004	
SM22-01					MC4028F
SM220	116	3016	504A	1104	
SM23-01					MC4029L
SM23-02					MC4029F
SM230	116	3016	504A	1104	
SM240	116	3016	504A	1104	
SM2491	160		245	2004	
SM2492	160			2004	
SM249I			245	2004	
SM250	116	3017B/117	504A	1104	
SM260	116	3017B/117	504A	1104	
SM270	125	3032A	510,531	1114	
SM2700	123A	3444	20	2051	
SM2701	123A	3444	20	2051	
SM2716	128		243	2030	
SM2718	129		244	2027	
SM280	125	3032A	510,531	1114	
SM30	116	3016	504A	1104	
SM30-01					MC4330L
SM30-02					MC4330F
SM300	125	3033	510,531	1114	
SM3014	160		245	2004	
SM30D11	5693	3652			
SM30G11	5695	3509			
SM31	116	3017B/117	504A	1104	
SM31-01					MC4331L
SM31-02					MC4331F
SM3104	123A	3444	20	2051	
SM3117A	123A	3444	20	2051	
SM32-01					MC4030L
SM32-02					MC4030F
SM33-01					MC4031L
SM33-02					MC4031F
SM3505	123A	3444	20	2051	
SM3978	128	3024	243	2030	
SM3986	123A	3444	20	2051	
SM3987	129	3025	244	2027	
SM4	116	3311	504A	1104	
SM40	116	3031A	504A	1104	
SM41-01					MC4332L
SM41-02					MC4332F
SM43-01					MC4032L
SM43-02					MC4032F
SM4304-S			61	2038	
SM4508-B	123A	3444	20	2051	
SM4547	159	3114/290	82	2032	
SM4574A	159		82	2032	
SM4719	159	3114/290	82	2032	
SM483	116	3016	504A	1104	
SM486	116	3016	504A	1104	
SM487	116	3016	504A	1104	
SM488	116	3016	504A	1104	
SM5	116	3311	504A	1104	
SM50	116	3017B/117	504A	1104	
SM505	116	3311	504A	1104	
SM51	116	3017B/117	504A	1104	
SM510	116	3016	504A	1104	
SM5109	1167	3732			
SM512	116	3017B/117	504A	1104	
SM513	116	3031A	504A	1104	
SM514	116	3017B/117	504A	1104	
SM515	116	3017B/117	504A	1104	
SM516	116	3017B/117	504A	1104	
SM517	125	3032A	510,531	1114	
SM518	125	3032A	510,531	1114	
SM520	125	3033	510,531	1114	
SM5379	123A	3444	20	2051	
SM5564	123A	3444	20	2051	
SM5643	123A	3444	20	2051	
SM576-1	123A	3444	20	2051	
SM576-2	123A	3444	20	2051	
SM5796	108	3039/316	86	2038	
SM5981	123A	3444	20	2051	
SM60	116	3017B/117	504A	1104	
SM61-01					MC4335L
SM61-02					MC4335F
SM62186			21	2034	
SM6251	128	3024	243	2030	
SM63	7475	7475		1806	
SM63-01					MC4035L
SM63-02					MC4035F
SM645	116	3016	504A	1104	
SM646	116	3016	504A	1104	
SM6727	154		40	2012	
SM6728	129		244	2027	
SM6762			10	2051	
SM6773	123A	3444	20	2051	
SM6814	198	3220	251		
SM70	125	3032A	510,531	1114	
SM705	116	3311	504A	1104	
SM71	125	3032A	510.531	1114	
SM71-01					MC4337L
SM71-02					MC4337F
SM710	116	3016	504A	1104	
SM716	123A	3444	20	2051	
SM720	116	3016	504A	1104	
SM73		3032A	509	1114	
SM73(DIODE)	125		510.531	1114	
SM73(I.C.)	7475			1806	
SM73-01					MC4037L
SM73-02					MC4037F
SM730	116	3016	504A	1104	
SM740	116	3031A	504A	1104	
SM750	116	3017B/117	504A	1104	

Industry Standard No.	ECG	SK	GE	RS 276-	MOTOR.
SM7545	123A	3444	20	2051	
SM760	116	3017B/117	504A	1104	
SM770	125	3032A	510,531	1114	
SM780	125	3032A	510,531	1114	
SM7815	123A	3444	20	2051	
SM7836	123A	3444	20	2051	
SM7989	224	3049	46		
SM7991	128	3047	243	2030	
SM80	125	3032A	510,531	1114	
SM80-01					MC4304L
SM80-02					MC4304F
SM800	125	3033	510,531	1114	
SM81	125	3032A	510,531	1114	
SM81-01					MC4305L
SM81-02					MC4305F
SM8112	123A	3444	20	2051	
SM8113	123A	3444	20	2051	
SM82-01					MC4004L
SM82-02					MC4004F
SM83	125	3032A	510,531	1114	
SM83-01					MC4005L
SM83-02					MC4005F
SM8341	102A		53	2007	
SM843	102A		53	2007	
SM862	126		52	2024	
SM8978	123A	3444	20	2051	
SM90-01					MC5090L
SM90-02					MC5090F
SM9008	123A	3444	20	2051	
SM9135	123A	3444	20	2051	
SM92-01					MC5092L
SM92-02					MC5092F
SM9253	123A	3444	20	2051	
SMB-541191	177		300	1122	
SMB-706009D	128		243	2030	
SMB447610	102	3004	2	2007	
SMB447610A	102	3722	2	2007	
SMB454549	102	3004	2	2007	
SMB454760	160		245	2004	
SMB620782-1	176	3845	80		
SMB621960	102	3722	2	2007	
SMC-583259	128	3044/154	243	2030	
SMC-620774-1	128		243	2030	
SMC449077	129		244	2027	
SMC7410N				1807	
SMC7420N				1809	
SMC7451N				1825	
SMC750123-1	941		IC-263	010	
SMT100					2N3808
SMT101					2N3808
SMT102					2N3808
SMT103					2N3809
SMT104					2N3810
SMT105					2N3810
SMV1172	610	3126	90		
SN-1	116	3017B/117	504A	1104	
SN-1Z	116	3017B/117	504A	1104	
SN-400-319-P1	159		82	2032	
SN-7474			7474	1818	
SN-76666N	712		IC-148		
SN0303	116		504A	1104	
SN1	116		504A	1104	
SN151800J	9800				MC1800L
SN151800N	9800				MC1800P
SN151800U					MC1800F
SN151801J	9801				MC1801L
SN151801N	9801				MC1801P
SN151801U					MC1801F
SN151802J	9802				MC1802L
SN151802N	9802				MC1802P
SN151802U					MC1802F
SN151803J	9803				MC1903L
SN151803N	9803				MC1803P
SN151803U					MC1803F
SN151804J	9804				MC1804L
SN151804N	9804				MC1804P
SN151804U					MC1804F
SN151805J	9805				MC1805L
SN151805N	9805				MC1805P
SN151805U					MC1805F
SN151806J	9806				MC1806L
SN151806N	9806				MC1806P
SN151806U					MC1806F
SN151807J	9807				MC1807L
SN151807N	9807				MC1807P
SN151807U					MC1807F
SN151808J	9808				MC1808L
SN151808N	9808				MC1808P
SN151808U					MC1808F
SN151809J	9809				MC1809L
SN151809N	9809				MC1809P
SN151809U					MC1809F
SN151810J	9810				MC1810L
SN151810N	9810				MC1810P
SN151810U					MC1810F
SN151811J	9811				MC1811L
SN151811N	9811				MC1811P
SN151811U					MC1811F
SN151812J	9812				MC1812L
SN151812N	9812				MC1812P
SN151812U					MC1812F
SN151820J					MC1820L
SN151820N					MC1820P
SN151820U					MC1820F
SN151900J					MC1900L
SN151900U					MC1900F
SN151901J					MC1901L
SN151901U					MC1901F
SN151902J					MC1902L
SN151902U					MC1902F
SN151904J					MC1904L
SN151904U					MC1904F
SN151905J					MC1905L
SN151905U					MC1905F
SN151906J					MC1906L
SN151906U					MC1906F
SN151907J					MC1907L
SN151907U					MC1907F
SN151908J					MC1908L
SN151908U					MC1908F
SN151909J					MC1909L
SN151909U					MC1909F
SN151910J					MC1910L
SN151910U					MC1910F
SN151911J					MC1911L
SN151911U					MC1911F
SN151912J					MC1912L
SN151912U					MC1912F
SN151920J					MC1920L
SN151920U					MC1920F
SN158093J	9093				MC853L
SN158093N	9093				MC853P
SN158093U					MC953F
SN158094J	9094				MC856L

Industry Standard No.	ECG	SK	GE	RS 276-	MOTOR.
SN158094N	9094				MC856P
SN158094U					MC956F
SN158097J	9097				MC855L
SN158097N	9097				MC855P
SN158097U					MC955F
SN158099J	9099				MC852L
SN158099N	9099				MC852P
SN158099U					MC952F
SN15830J	9930				MC830L
SN15830N	9930				MC830P
SN15830U					MC930F
SN15832J	9932				MC832L
SN15832N	9932				MC832P
SN15832U					MC932F
SN15833J	9933				MC833L
SN15833N	9933				MC833P
SN15833U					MC933F
SN15834J					MC834L
SN15834N					MC834P
SN15834U					MC934F
SN15835J	9935				MC840L
SN15835N	9935				MC840P
SN15835U					MC940F
SN15836J	9936				MC836L
SN15836N	9936				MC836P
SN15836U					MC936F
SN15837J	9937				MC837L
SN15837N	9937				MC837P
SN15837U					MC937F
SN15838J	9135				MC835L
SN15838N	9135				MC835P
SN15838U					MC935F
SN15844J	9944				MC834L
SN15844N	9944				MC844P
SN15845J	9945				MC848L
SN15845N	9945				MC845P
SN15845U					MC945F
SN15846J	9946				MC846L
SN15846N	9946				MC846P
SN15846U					MC946F
SN15848N	9948				MC848P
SN15848U					MC948F
SN15849J	9949				MC849L
SN15849N	9949				MC849P
SN15849U					MC849F
SN15850J	9950				MC850L
SN15850N	9950				MC850P
SN15850U					MC950F
SN15851J	9951				MC851L
SN15851N	9951				MC851P
SN15851U					MC951F
SN15857J	9157				MC857L
SN15857N	9157				MC857P
SN15857U					MC957F
SN15858J	9158				MC858L
SN15858N	9158				MC858P
SN15858U					MC958F
SN15861J	9961				MC861L
SN15861N	9961				MC861P
SN15861U					MC961F
SN15862J	9962				MC862L
SN15862N	9962				MC862P
SN15862U					MC962F
SN15863J	9963				MC863L
SN15863N	9963				MC863P
SN15863U					MC863F
SN159093J					MC953L
SN159093U					MC953F
SN159094J					MC956L
SN159094U					MC956F
SN159097J					MC955L
SN159097U					MC955F
SN159099J					MC952L
SN159099U					MC952F
SN15930J					MC930L
SN15930U					MC930F
SN15932J					MC932L
SN15932U					MC932F
SN15933J					MC933L
SN15933U					MC933F
SN15934J					MC934L
SN15934U					MC934F
SN15935J					MC940L
SN15935U					MC940F
SN15936J					MC936L
SN15936U					MC936F
SN15937J					MC937L
SN15937U					MC937F
SN15938J					MC935L
SN15938U					MC935F
SN15944J					MC944L
SN15944U					MC944F
SN15945J					MC945L
SN15945U					MC945F
SN15946J					MC946L
SN15946U					MC946F
SN15948J					MC948L
SN15948U					MC948F
SN15949J					MC949L
SN15949U					MC949F
SN15950J					MC950L
SN15950U					MC950F
SN15951J					MC951L
SN15951U					MC951F
SN15957J					MC957L
SN15957U					MC957F
SN15958J					MC958L
SN15958U					MC958F
SN15961J					MC961L
SN15961U					MC961F
SN15962J					MC962L
SN15962U					MC962F
SN15963J					MC963L
SN166	128	3024	243	2030	
SN167	128	3024	243	2030	
SN16963U					MC963F
SN2388	749		IC-97		
SN4448	519	3100	514	1122	
SN52101AL		3565			LM101AH
SN52104L					LM101H
SN52105L					LM105H
SN52106J					MC1710L
SN52106L					MC1710G
SN52107L		3690			LM107H
SN52108AL					LM108AH
SN52108L					LM108H
SN52109L					LM109H
SN52301	975	3641			
SN52510J					MC1710L
SN52510L					MC1710G
SN52514J					MC1514L
SN52555L		3693			MC1555G
SN52558L					MC1558G
SN52702AFA					MC1712F

Industry Standard No.	ECG	SK	GE	RS 276-	MOTOR.
SN52702AJ					MC1712L
SN52702AL					MC1712G
SN52702FA					MC1712F
SN52702J					MC1712L
SN52702L					MC1712G
SN52709AFA					MC1709AF
SN52709AJ					MC1709AL
SN52709AL					MC1709AG
SN52709FA					MC1709F
SN52709J					MC1709L
SN52709L					MC1709G
SN52710FA					MC1710F
SN52710J					MC1710L
SN52710L					MC1710G
SN52711FA					MC1711F
SN52711J					MC1711L
SN52711L					MC1711G
SN52723FA					MC1723F
SN52723J					MC1723L
SN52723L					MC1723G
SN52733J					MC1733L
SN52733L					MC1733G
SN52741FA					MC1741F
SN52741J					MC1741L
SN52741L		3514			MC1741G
SN52747FA					MC1747F
SN52747J					MC1747L
SN52747L		3526			MC1747G
SN52748L		3645			MC1748G
SN52770L					MC1556G
SN52771L					MC1556G
SN52810FA					MC1710F
SN52810J					MC1710L
SN52810L					MC1710G
SN52811FA					MC1711F
SN52811J					MC1711L
SN52811L					MC1711G
SN54LS00 SERIES					NOTE 1
SN55107AJ					MC55107L
SN55107BJ					MC55107L
SN55108AJ					MC55108L
SN55108BJ					MC75108L
SN55109J					MC75S110L
SN5510FA					MC1510F
SN5510L					MC1510G
SN55110J					MC75S110L
SN55244J					MC1544L
SN55325J					MC55325L
SN60	101	3861	8	2002	
SN72301	975	3641			
SN72301AL	1171	3565			LM301AH
SN72301AP	975	3641			LM301AN
SN72304L					LM304H
SN72305AL					LM305H
SN72305L					LM305H
SN72306J					MC1710CL
SN72306L					MC1710CG
SN72306N					MC1710CP
SN72307	976	3596			
SN72307L		3690			LM307H
SN72307P	976	3596			
SN72308AL					LM308AH
SN72308L					LM308H
SN72309L					LM309H
SN72311L					LM311H
SN72311P		3668			LM311N
SN72376L					LM305H
SN72440J					MC3370P
SN72440N					MC3370P
SN72471P	941M		IC-265	007	
SN72510J					MC1710CL
SN72510L					MC1710CG
SN72510N					MC1710CP
SN72514J					MC1414L
SN72514N					MC1414P
SN72555JP	955M			1723	
SN72555L		3693			MC1455G
SN72555P	955M	3564		1723	MC1455P1
SN72558	778A	3465	IC-220		
SN72558JG	778A	3465			
SN72558L		3555			MC1458G
SN72558P	778A	3465	IC-220		MC1458P1
SN72702J					MC1712CL
SN72702L					MC1712CG
SN72709J	909D				MC1709CL
SN72709L	909	3551	IC-249		MC1709CG
SN72709N	909D	3552/941M	IC-250		MC1709CP2
SN72709P					MC1709CP1
SN72710J	910D				MC1710CL
SN72710L	910	3553	IC-251		MC1710CG
SN72710N	910D		IC-252		MC1710CP
SN72711J	911D				MC1711CL
SN72711L	911		IC-253		MC1711CG
SN72711N	911D		IC-254		MC1711CP
SN72720J					MC1710CL
SN72720L					MC1710CG
SN72720N					MC1710CP
SN72723J	923D				MC1723CL
SN72723L	923				MC1723CG
SN72723N	923D	3164/923			
SN72733J					MC1733CL
SN72733L					MC1733CG
SN72741FA					MC1741CF
SN72741J	941D				MC1741CL
SN72741L	941	3514	IC-263	010	
SN72741N	941D	3552/941M	IC-264		MC1741CP2
SN72741P	941M	3552	IC-265	007	MC1741CP1
SN72747FA					MC1747CF
SN72747J	947D	3556			MC1747CL
SN72747JA	947D	3556			
SN72747L	947				MC1747CG
SN72747N	947D	3556			MC1747CP2
SN72748JG	975	3641			
SN72748L	1171	3645			MC1748CG
SN72748P	975	3644			MC1748CP1
SN72770L					MC1456G
SN72771L					MC1456G
SN72810J					MC1710CL
SN72810L					MC1710CG
SN72810N					MC1710CP
SN72811J					MC1711CL
SN72811L					MC1711CG
SN72811N					MC1711CP
SN72905	961	3591/960			MC7905CT
SN72906	963	3669/962			MC7906CT
SN72908		3630			MC7908CT
SN72912	967	3592/966			MC7912CT
SN72915	969	3593/968			MC7915CT
SN72924	972	3670			
SN72L022P					LM358N
SN72L044JA					LM324N
SN72L044N					LM324N
SN7400	7400	7400	7400	1801	
SN7400A	7400	7400	7400	1801	
SN7400N	7400	7400	7400	1801	
SN7400N-10			7400	1801	
SN7401N	7401	7401			
SN7402	7402		7402	1811	
SN7402N	7402	7402	7402	1811	
SN7402N-10			7402	1811	
SN7403	7403	7403			
SN7403N	7403	7403			
SN7404N	7404	7404	7404	1802	
SN7404N-10			7404	1802	
SN7405N	7405	7405			
SN7406N	7406	7406		1821	
SN7406N-10			7406	1821	
SN7407N	7407	7407			
SN7408N	7408	7408	7408	1822	
SN7408N-10			7408	1822	
SN7409N	7409	7409			
SN74107N	74107	74107			
SN74109	74109	74109			
SN74109N	74109	74109			
SN7410J				1807	
SN7410N	7410	7410	7410	1807	
SN7410N-10			7410	1807	
SN74110	74110	74110			
SN74110N	74110	74110			
SN74111	74111	74111			
SN74111N	74111	74111			
SN74121N	74121	74121			
SN74122N	74122	74122			
SN74123N	74123	74123	74123	1817	
SN74123N-10			74123	1817	
SN74125	74125	74125			
SN74125N	74125	74125			
SN74126	74126	74126			
SN74126N	74126	74126			
SN7412N	7412	7412			
SN74132	74132	74132			
SN74132N	74132	74132			
SN7413N	7413	7413		1815	
SN7413N-10			7413	1815	
SN74141N	74141	74141			
SN74145	74145	74145			
SN74145N	74145	74145		1828	
SN74145N-10			74145	1828	
SN7414N	7414	7414			
SN74150N	74150	74150		1829	
SN74150N-10			74150	1829	
SN74151N	74151	74151			
SN74153N	74153	74153			
SN74154N	74154	74154		1834	
SN74154N-10			74154	1834	
SN74155N	74155	74155			
SN74156N	74156	74156			
SN74160N	74160	74160			
SN74161N	74161	74161			
SN74162N	74162	74162			
SN74163N	74163	74163			
SN74164N	74164	74164			
SN74165N	74165	74165			
SN74166N	74166	74175			
SN7416N	7416	7416			
SN74170N	74170	74170			
SN74174N	74174	74174			
SN74175N	74175	74175			
SN74176N	74176	74197			
SN74177N	74177	74177			
SN74178N	74178	74178			
SN74179N	74179	74179			
SN7417N	7417	7417			
SN74180N	74180	74180			
SN74181N	74181	74181			
SN74190N	74190	74190			
SN74191N	74191	74191			
SN74192			74192	1831	
SN74192N	74192	74192	74192	1831	
SN74192N-10			74192	1831	
SN74193N	74193	74193		1820	
SN74193N-10			74193	1820	
SN74196J	74196	74196		1833	
SN74196N	74196	74196		1833	
SN74196N-10			74196	1833	
SN74198N	74198	74198			
SN74199N	74199	74199			
SN7420J	7420			1809	
SN7420N	7420	7420	7420	1809	
SN7420N-10			7420	1809	
SN7422	7422	7422			
SN7422N	7422	7422			
SN7423N	7423	7423			
SN7425N	7425	7425			
SN7426	7426	7426			
SN7426N	7426	7426			
SN7427			7427	1823	
SN7427J				1823	
SN7427N	7427	7427	7427	1823	
SN7427N-10			7427	1823	
SN7428	7428	7428			
SN7428N	7428	7428			
SN7430N	7430	7430			
SN7432J				1824	
SN7432N	7432	7432		1824	
SN7432N-10			7432	1824	
SN7433	7433	7433			
SN7433N	7433	7433			
SN7437N	7437	7437			
SN7438N	7438	7438			
SN7440N	7440	7440			
SN7441N-10			7441	1804	
SN7442N	7442	7442			
SN7445N	7445	7445			
SN7446AN	7446	7446			
SN7447AN	7447	7447		1805	
SN7447N-10			7447	1805	
SN7448N	7448	7448		1816	
SN7448N-10			7448	1816	
SN7450N	7450	7450			
SN7451				1825	
SN7451J				1825	
SN7451N	7451	7451		1825	
SN7451N-10			7451	1825	
SN7453N	7453	7453			
SN7454N	7454	7454			
SN7460N	7460	7460			
SN7470N	7470	7470			
SN7472N	7472	7472			
SN7473	7473		7473	1803	
SN7473N	7473	7473	7473	1803	
SN7473N-10			7473	1803	
SN7474	7474		7474	1818	
SN74741L					MC1741CG
SN7474M				1818	
SN7474N	7474	7474	7474		
SN7474N-10			7474	1818	
SN7475N	7475	7475		1806	

Industry Standard No.	ECG	SK	GE	RS 276-	MOTOR.
SN7475N-10			7475	1806	
SN7476B				1813	
SN7476N	7476	7476	7476	1813	
SN7476N-10			7476	1813	
SN7480N	7480	7480			
SN7481	7481	7481			
SN7481A	7481	7481			
SN7481AN	7481	7481			
SN7483AN	7483	7483			
SN7485J				1826	
SN7485N	7485	7485		1826	
SN7485N-10			7485	1826	
SN7486J				1827	
SN7486N	7486	7486		1827	
SN7486N-10			7486	1827	
SN7489N	7489	7489			
SN7490AN	7490	7490		1808	
SN7490N	7490		7490	1808	
SN7490N-10			7490	1808	
SN7492AN	7492	7492	7492	1819	
SN7492N-10			7492	1819	
SN7493AN	7493A	7493			
SN7494N	7494	7494			
SN7495AN	7495	7495			
SN7496N	7496	7496			
SN7497N	7497	7497			
SN74H00N	74H00	74H00			
SN74H04N	74H04	74H04			
SN74LS00 SERIES					NOTE 1
SN74LS00N	74LS00	74LS00		1900	
SN74LS02N	74LS02	74LS02		1902	
SN74LS03N	74LS03	74LS03			
SN74LS04N	74LS04	74LS04		1904	
SN74LS05N	74LS05	74LS05			
SN74LS08N	74LS08	74LS08		1908	
SN74LS107N	74LS107	74LS107			
SN74LS109AN	74LS109A	74LS109			
SN74LS109N	74LS109A	74LS109			
SN74LS10N	74LS10	74LS10		1910	
SN74LS11N	74LS11	74LS11			
SN74LS123N		74LS123		1926	
SN74LS138N	74LS138	74LS138			
SN74LS14N	74LS14	74LS14			
SN74LS151N	74LS151	74LS151		1929	
SN74LS153N	74LS153	74LS153			
SN74LS157N	74LS157	74LS157		1930	
SN74LS161AN	74LS161A	74LS161			
SN74LS161N	74LS161A	74LS161			
SN74LS163AN	74LS163A	74LS163			
SN74LS163N	74LS163A	74LS163			
SN74LS164N	74LS164	74LS164			
SN74LS174N	74LS174	74LS174		1932	
SN74LS175N	74LS175	74LS175		1934	
SN74LS191N	74LS191	74LS191			
SN74LS193N	74LS193	74LS193		1936	
SN74LS195AN	74LS195A	74LS195			
SN74LS195N	74LS195A	74LS195			
SN74LS20N	74LS20	74LS20		1912	
SN74LS221N	74LS221	74LS221			
SN74LS240N	74LS240	74LS240			
SN74LS248N	74LS248	74LS248			
SN74LS249N	74LS249	74LS249			
SN74LS253N	74LS253	74LS253			
SN74LS257N	74LS257	74LS257			
SN74LS258N	74LS258	74LS258			
SN74LS259N	74LS259	74LS259			
SN74LS266N	74LS266	74LS266			
SN74LS273N	74LS273	74LS273			
SN74LS27N	74LS27	74LS27		1913	
SN74LS280N	74LS280	74LS280			
SN74LS298N	74LS298	74LS298			
SN74LS30N	74LS30	74LS30		1914	
SN74LS32N	74LS32	74LS32		1915	
SN74LS367N	74LS367	74LS367		1835	
SN74LS377N	74LS377	74LS377			
SN74LS38N	74LS38	74LS38			
SN74LS42N	74LS42	74LS42			
SN74LS49N	74LS49	74LS49			
SN74LS51N	74LS51	74LS51		1917	
SN74LS73N	74LS73	74LS73		1918	
SN74LS74AN	74LS74A	74LS74			
SN74LS74N	74LS74A	74LS74			
SN74LS75N	74LS75			1920	
SN74LS83AN	74LS83A	74LS83			
SN74LS83N	74LS83A	74LS83			
SN74LS85N	74LS85	74LS85		1922	
SN74LS86N	74LS86	74LS86			
SN74LS90	74LS90	74LS90			
SN74LS90N	74LS90	74LS90			
SN74LS93	74LS93			1925	
SN74LS93N	74LS93	74LS93		1925	
SN74S00N	74S00	74S00			
SN74S04N	74S04	74S04			
SN74S214					MCM93425
SN74S314					MCM93415
SN74S474					MCM7641
SN74S476					MCM7643
SN74S478					MCM7681
SN75107AJ					MC75107L
SN75107AN					MC75107P
SN75107BJ					MC75107L
SN75107BN					MC75107P
SN75108AJ					MC75108L
SN75108AN					MC75108P
SN75108BJ					MC75108L
SN75108BN					MC75108P
SN75110N	722	3161	IC-9		
SN75121J					MC8T13L
SN75121N					MC8T13P
SN75122J					MC8T14L
SN75122N					MC8T14P
SN75123J					MC8T23L
SN75123N					MC8T23P
SN75124J					MC8T24L
SN75124N					MC8T24P
SN75125J					MC75125L
SN75125N					MC75125P
SN75126J					MC3481/5L
SN75126N					MC3481/5P
SN75127J					MC75127L
SN75127N					MC75127P
SN75128J					MC75128L
SN75128N					MC75128P
SN75129J					MC75129L
SN75129N					MC75129P
SN75138J					MC3443P
SN75138N					MC3443P
SN75140P					MC75140P1
SN75150J					MC1488L
SN75150N					MC1488P
SN75154J					MC1489L
SN75154N					MC1489P
SN75188J	75188				MC1488L
SN75188N	75188				MC1488P
SN75189AJ					MC1489AL
SN75189AN					MC1489AP
SN75189J					MC1489L
SN75189N	75189				MC1489P
SN75207J					MC75107L
SN75207N					MC75107P
SN75208J					MC75108L
SN75208N					MC75108P
SN75362P					MMH0026CP
SN75369P					MMH0026CP1
SN75461N					MC75491P
SN75466J					MC1411L
SN75466N					MC1411P
SN75467J					MC1412L
SN75467N					MC1412P
SN75468J					MC1413L
SN75468N					MC1413P
SN75475JG					MC1472U
SN75475P					MC1472P1
SN75476N	2011	3975			
SN75477N	2012	9092			
SN75478N	2013	9093			
SN75491N	75491B				MC75491P
SN75492N	75492B				MC75492P
SN76000P					MC1306P
SN76021N	810		IC-241		
SN76021ND	810		IC-241		
SN76104	718	3159	IC-8		
SN76104N	718	3159	IC-8		MC1310P
SN76105	720	9014	IC-7		
SN76105N	720	9014	IC-7		MC1310P
SN76107N	722		IC-9		
SN7610N-07	722	3161	IC-9		
SN76110	722	3161	IC-9		
SN76110N	722	3161	IC-9		
SN76111	719		IC-28		
SN76111N					MC1310P
SN76113N	722				MC1310P
SN76115	801	3160	IC-35		
SN76115N	801	3160	IC-35		MC1310P
SN76116N	743	3172			MC1310P
SN76117N					MC1310P
SN76130N	721				MC1303P
SN76131	725	3162	IC-19		
SN76131N	719	3162/725	IC-28		MC1303P
SN76149N					MC1303P
SN76177	804	3455	IC-27		
SN76177ND	804	3455	IC-27		
SN76242	714	3075	IC-4		
SN76242N	714	3075	IC-4		MC1399P
SN76242N-07	714		IC-4		
SN76243	715	3076	IC-6		
SN76243N	715	3076	IC-6		MC1399P
SN76246	790	3077	IC-5		
SN76246N	790	3077	IC-5		MC1323P
SN76266	728	3073	IC-22		
SN76266N	728		IC-22		
SN76267	729	3074	IC-23		
SN76267N	729		IC-23		
SN76298N	738	3167			MC1398P
SN76514L					MC1496G
SN76514N					MC1496P
SN76530P	747	3279	IC-218		
SN76547N	830	9030			
SN76564		3141	IC-21		
SN76564N	783	3215	IC-225		MC1364P
SN76565N	783				MC1364P
SN76591P	815	3255	IC-244		MC1391P
SN76594P					MC1394P
SN76600	746	3234	IC-217		
SN766008	746	3234			
SN76600N	746		IC-217		
SN76600P	746	3234	IC-217		MC1350P
SN766110	722		IC-9		
SN76635N	744			2022	
SN76642	709		IC-11		
SN76642N	709	3135	IC-11		MC1357P
SN76642P	709	3135			
SN76643A		3135	IC-10		
SN76643N	708	3135/709	IC-10		
SN76644N					MC1352P
SN76650	749	3171/744	IC-97		
SN76650N	749	3171/744	IC-97		MC1352P
SN76651N	748	3236			MC1351P
SN76653N	708		IC-10		MC1352P
SN76660N					MC7357P
SN76664N	712	3072	IC-148		
SN76665	712	3072	IC-2		
SN76665N	712	3072	IC-2		MC1364P
SN76666	712	3072	IC-148		
SN76666N	712	3072	IC-148		MC1358P
SN76669N	737	3375			MC1356P
SN76675N	723		IC-15		MC1375P
SN76676L	781	3169	IC-223		
SN76678P	736				MC1355P
SN76689	788	3829			
SN76689N	788	3829			
SN80		3861	8	2002	
SN0303	116(2)		504A(2)	1104	
SNT204	159	3114/290	82	2032	
SNT204A	159	3114/290	82	2032	
SNW-Q-1			2	2007	
SNW-Q-2			8	2002	
SNW-Q-3			8	2002	
SNW-Q-4			2	2007	
SNW-Q-5			8	2002	
SNW-Q-6			2	2007	
SO-1	160		245	2004	
SO-2	160		245	2004	
SO-25	102A	3004	53	2007	
SO-3	160		245	2004	
SO-632	139A	3060	ZD-9.1	562	
SO-65A	160	3007	245	2004	
SO-88	102A	3004	53	2007	
SO1	160		245	2004	
SO10G	125		510,531	1114	
SO19806	199		62	2010	
SO2	16C		245	2004	
SO25	102A		53	2007	
SO25094	199		62	2010	
SO3	160		245	2004	
SO46	110MP	3709		1123(2)	
SO5	116		504A	1104	
SO501	116		504A	1104	
SO65	102A	3007	53	2007	
SO65A	126		52	2024	
SO88	102A		53	2007	
SOA	125		510,531	1114	
SOD-200D	156		512		
SOD100AL	5802	9005		1143	
SOD100AS	5834		5048		
SOD100BL	5802	9005		1143	
SOD100BS	5834		5048		
SOD100CL	5802	9005		1143	

Industry Standard No.	ECG	SK	GE	RS 276-	MOTOR.
SOD100CS	5834		5048		
SOD100DS	5802	9005		1143	
SOD200AL	5802	9005		1143	
SOD200BL	5802	9005		1143	
SOD200BS	5834		5048		
SOD200CL	5802	9005		1143	
SOD200CS	5834		5048		
SOD200D	116		504A	1104	
SOD200DL	5802	9005		1143	
SOD200DS	5834		5048		
SOD20AS	5830		5048		
SOD30AL	5800	9003		1142	
SOD30AS	5830		5048		
SOD30BL	5800	9003		1142	
SOD30BS	5830		5048		
SOD30CS	5830		5048		
SOD30DL	5800	9003		1142	
SOD30DS	5830		5048		
SOD50AL	5800	9003		1142	
SOD50AS	5830		5048		
SOD50BL	5800	9003		1142	
SOD50BS	5830		5048		
SOD50CL	5800	9003		1142	
SOD50DL	5800	9003		1142	
SOD50DS	5830		5048		
SONY 095	1308	3833			
SONY 095A	1308	3833			
SONY 095C	1308	3833			
SOS0121			244	2027	
SOS0121	129	3025		2027	
SP-1	116	3016	504A	1104	
SP-1.5A	5800	9003		1142	
SP-1108	121	3009	239	2006	
SP-148-3	121	3009	239	2006	
SP-1482-5	121	3009	239	2006	
SP-1483	121	3009	239	2006	
SP-1484	121	3642/179	239	2006	
SP-1556-2	121	3009	239	2006	
SP-1603	121	3009	239	2006	
SP-1603-1	121	3009	239	2006	
SP-1603-2	121	3009	239	2006	
SP-2158	124	3835/103A	12		
SP-404T	121	3009	239	2006	
SP-441	121	3009	239	2006	
SP-485	121	3009	239	2006	
SP-486	121	3009	239	2006	
SP-486W	121	3009	239	2006	
SP-634	121	3009	239	2006	
SP-649	121	3009	239	2006	
SP-649-1	121	3009	239	2006	
SP-70	198		251		
SP-834	121	3009	239	2006	
SP-8660			57	2017	
SP-880	121	3009	239	2006	
SP-880-1	121	3009	239	2006	
SP-880-3	121	3009	239	2006	
SP-891	121	3009	239	2006	
SP-891B	121	3009	239	2006	
SP-891W	121	3009	239	2006	
SP1	116		504A	1104	
SP101	177	3100/519		1122	
SP10100					MC10500
SP10101					MC10501
SP10102					MC10502
SP10103					MC10503
SP10104					MC10504
SP10105					MC10505
SP10106					MC10506
SP10107					MC10507
SP10109					MC10509
SP10110					MC10110
SP10111					MC10111
SP10113					MC10513
SP10114					MC10514
SP10115					MC10515
SP10116					MC10516
SP10117					MC10517
SP10118					MC10518
SP10119					MC10519
SP10121					MC10521
SP10124					MC10524
SP10125					MC10525
SP10128					MC10128
SP10129					MC10129
SP10130					MC10530
SP10131					MC10531
SP10134					MC10134
SP10135					MC10535
SP10136					MC10536
SP10137					MC10537
SP10138					MC10538
SP1013A	121	3717	239	2006	
SP1013B	104	3009	16	2006	
SP10141					MC10541
SP10144					MCM10544
SP10145					MCM10545
SP10160					MC10560
SP10161					MC10561
SP10162					MC10562
SP10164					MC10564
SP10165					MC10565
SP10170					MC10570
SP10171					MC10571
SP10172					MC10572
SP10173					MC10173
SP10174					MC10574
SP10175					MC10575
SP10176					MC10576
SP10178					MC10578
SP10179					MC10579
SP10180					MC10580
SP10181					MC10581
SP10211					MC10611
SP10212					MC10612
SP10216					MC10616
SP10231					MC10631
SP1029	179	3642	76		
SP10800					2N3817
SP10801					2N3817
SP1081	213		254		
SP10810					2N3817
SP1105	213		254		
SP1108	104	3719	16	2006	
SP1118	121	3717	239	2006	
SP1137	104	3719	16	2006	
SP12271	106	3984	21	2034	
SP1271	121	3014	239	2006	
SP1323	104	3009	16	2006	
SP1403	104	3009	16	2006	
SP148-3	121	3717	239	2006	
SP1481	121	3009	239	2006	
SP1481-1	121	3009	239	2006	
SP1481-2	121	3009	239	2006	
SP1481-3	121	3009	239	2006	
SP1481-4	121	3009	239	2006	
SP1481-5	121	3009	239	2006	
SP1482	121	3009	239	2006	
SP1482-2	121	3009	239	2006	
SP1482-3	121	3009	239	2006	
SP1482-4	121	3009	239	2006	
SP1482-5	121	3717	239	2006	
SP1482-6	121	3009	239	2006	
SP1482-7	121	3009	239	2006	
SP1483	121	3717	239	2006	
SP1483-1	121	3009	239	2006	
SP1483-2	121	3009	239	2006	
SP1483-3	121	3009	239	2006	
SP1484			3	2006	
SP1484-1	179		76		
SP1484-2	179		76		
SP1484-3	179		76		
SP1484-4	179		76		
SP1484-5	179		76		
SP1550-3	121	3009	239	2006	
SP1556	121	3009	239	2006	
SP1556-1	121	3009	239	2006	
SP1556-2	121	3717	239	2006	
SP1556-3	121	3009	239	2006	
SP1556-4	121	3717	239	2006	
SP1563-2	121	3009	239	2006	
SP1595BLK	121	3009	239	2006	
SP1595BLU	121	3009	239	2006	
SP1595GRN	121	3009	239	2006	
SP1595RED	121	3009	239	2006	
SP1596BLK	121	3009	239	2006	
SP1596BLU	121	3009	239	2006	
SP1596GRN	121	3009	239	2006	
SP1596RED	121	3009	239	2006	
SP1600	104	3719	16	2006	
SP1603	121	3717	239	2006	
SP1603-1	121	3717	239	2006	
SP1603-2	121	3717	239	2006	
SP1603-3	121	3009	239	2006	
SP1619	104	3719	16	2006	
SP1650	179	3642	76		
SP1651	104	3009	16	2006	
SP1657	121	3717	239	2006	
SP1742	127	3764	25		
SP176	104	3009	16	2006	
SP1801	121	3717	239	2006	
SP1817	179	3642	76		
SP1844	121	3717	239	2006	
SP1927	104	3719	16	2006	
SP1938	121	3009	239	2006	
SP1950	121	3009	239	2006	
SP1994	213		254		
SP1K-1	116	3311	504A	1104	
SP1K-2	116	3017B/117	504A	1104	
SP1U-2	116	3031A	300	1104	
SP2-01				1152	
SP2-02				1172	
SP2-04				1173	
SP2045	121	3009	239	2006	
SP2046	121	3009	239	2006	
SP2048	121	3009	239	2006	
SP2072	121	3009	239	2006	
SP2076	121	3009	239	2006	
SP2077	179	3642	76		
SP2094	121	3717	239	2006	
SP2155	121	3009	239	2006	
SP2158	103A		59	2002	
SP2188	103A	3835	59	2002	
SP2221QD					MHQ2221
SP2221QF					MQ2218
SP2222AF					MD2219AF
SP2222AQF					MQ2219A
SP2222QD					MHQ2222
SP2222QF					MQ2219A
SP2223AF					MD2219AF
SP2234	104	3009	16	2006	
SP2247	104	3719	16	2006	
SP230	104	3009	16	2006	
SP2341	121	3717	239	2006	
SP2358	104	3009	16	2006	
SP2361	121	3009	239	2006	
SP2361BLU	121	3009	239	2006	
SP2361BRN	121	3009	239	2006	
SP2361GRN	121	3009	239	2006	
SP2361ORN	121	3009	239	2006	
SP2361RED	121	3009	239	2006	
SP2361YEL	121	3009	239	2006	
SP2369QD					MHQ2369
SP2395	121	3717	239	2006	
SP2411	127	3764			
SP2426	105		4		
SP2431	104	3719	16	2006	
SP2431P	105		4		
SP2483QF					MQ2484
SP2484F					2N3045
SP2484QD					MHQ2484
SP2484QF					MQ2484
SP2493	121	3014	239	2006	
SP2512	127	3764			
SP2541	121	3014	239	2006	
SP2551	131	3052	44	2006	
SP26	104	3719	16	2006	
SP2610	179	3642			
SP2708	179	3642	76		
SP2905					MD2905
SP2906AQF					MQ2905A
SP2906QD					MHQ2906
SP2906QF					MQ2904
SP2907AF					MD2905AF
SP2907AQF					MQ2905A
SP2907QD					MHQ2907
SP2946F					MD1120F
SP30QF					MD1120F
SP3136F					MD7001F
SP3251AQD					MHQ3251
SP3253QD					MHQ2907
SP328QF					MD930
SP329QF					MD930F
SP334	121	3009	239	2006	
SP3467F					MD3467F
SP3467QD					MHQ3467
SP3724QD					MHQ4013
SP3725F					MD3725F
SP3725QD					MHQ4014
SP3725QF					MQ3725
SP3762QD					MHQ3467
SP3762QF					MQ3467
SP3763QD					2N5146
SP3763QF					MQ3467
SP39	127	3035	25		
SP404	121	3717	239	2006	
SP404T	121	3717	239	2006	
SP4168	108	3018	86	2038	
SP4231	130		14	2041	

Industry Standard No.	ECG	SK	GE	RS 276-	MOTOR.
SP441	121	3717	239	2006	
SP441D	104	3009	16	2006	
SP441G	121	3014	239	2006	
SP441S	104	3009	16	2006	
SP4436	161	3716	39	2015	
SP47	104	3009	16	2006	
SP485	121	3717	239	2006	
SP485B	121	3009	239	2006	
SP485BLK	121	3009	239	2006	
SP485BLU	121	3009	239	2006	
SP485BRN	121	3009	239	2006	
SP485W	121	3009	239	2006	
SP485WHT	121	3009	239	2006	
SP486	104	3719	16	2006	
SP486W	121	3717	239	2006	
SP486WHT	121	3009	239	2006	
SP62	127	3035	25		
SP634	121	3717	239	2006	
SP649	121	3717	239	2006	
SP649-1	121	3717	239	2006	
SP70	159		82	2032	
SP744	121	3014	239	2006	
SP819R	121	3009	239	2006	
SP82A	112	3089	1N82A		
SP834	121	3717	239	2006	
SP838	179	3642	76		
SP838-1	179	3642	76		
SP8400	154	3045/225	40	2012	
SP8401	154	3045/225	40	2012	
SP8402	192		63	2030	
SP8416	152	3893	66	2048	
SP8660	152	3054/196	66	2048	
SP875	121	3009	239	2006	
SP880	104	3719	16	2006	
SP880-1	121	3717	239	2006	
SP880-3	121	3717	239	2006	
SP891	104	3014	16	2006	
SP891-B			16	2006	
SP8918	184		57	2017	
SP891B	121	3717	239	2006	
SP891BLU	121	3009	239	2006	
SP891G	121	3009	239	2006	
SP891GRN	121	3009	239	2006	
SP891R	121	3009	239	2006	
SP891W	121	3717	239	2006	
SP891WHT	121	3009	239	2006	
SP90	159		82	2032	
SP918AF					MD918AF
SP918BF					MD918BF
SPC40	123A	3444	20	2051	
SPC40411			75	2041	
SPC411	94	3560			
SPC413	283	3560			
SPC42	123A	3444	20	2051	
SPC430	94	3559/162	73		
SPC431	283	3560	73		
SPC50	123A	3444	20	2051	
SPC51	123A	3444	20	2051	
SPC52	123A	3444	20	2051	
SPD-80059	181		75	2041	
SPD-80060	181		75	2041	
SPD-80061	181		75	2041	
SPD-80062	130		14	2041	
SPD-80123	128		243	2030	
SPF024	312	3112		2028	
SPF215				2036	
SPF274	222	3050/221	FET-4	2036	
SPF512	222	3050/221	FET-4	2036	
SPF530			61	2038	
SPF609			FET-4	2036	
SPN-01	116	3016	504A	1104	
SPN01			504A	1104	
SPS-0121	129		244	2027	
SPS-0122	128	3024	243	2030	
SPS-1351	108	3018	86	2038	
SPS-1352	108		86	2038	
SPS-1353	108		86	2038	
SPS-1473	108	3018	86	2038	
SPS-1473RT			11	2015	
SPS-1475	123A		20	2051	
SPS-1475(YT)			10	2051	
SPS-1475YT	123A		20	2051	
SPS-1476	123A	3018	62	2010	
SPS-1539	159	3138/193A	82	2032	
SPS-1539(WT)			62	2010	
SPS-1539WT			22	2032	
SPS-2111	108		86	2038	
SPS-2265			11	2015	
SPS-2266			11	2015	
SPS-2320	108		86	2038	
SPS-29	129	3025	244	2027	
SPS-4075	123A	3020/123	20	2051	
SPS-4076	129	3025	244	2027	
SPS-4077	128	3024	243	2030	
SPS-4078	129	3025	244	2027	
SPS-41	123A	3122	20	2051	
SPS-4145	108		86	2038	
SPS-4396			20	2051	
SPS-4423			86	2038	
SPS-856	107		11	2015	
SPS-860	107		11	2015	
SPS-917			86	2038	
SPS-934			20	2051	
SPS-952	123A	3025/129	82	2032	
SPS-952-2	123A	3018	211	2016	
SPS010	5491	6791			
SPS0121	129	3025	244	2027	
SPS0122	128	3024	243	2030	
SPS020	5521	6621	C230F		
SPS030	5541	6615/5515			
SPS035	5541	6615/5515			
SPS08	5491	6791	C220F		
SPS1045	123A	3444	20	2051	
SPS1082			20	2051	
SPS1097	159	3114/290	82	2032	
SPS1097A	159	3114/290	82	2032	
SPS110	5491	6791			
SPS1107	188		226	2018	
SPS12	159		82	2032	
SPS120	5522	6622	C230A		
SPS130	5542	6615/5515			
SPS135	5542	6615/5515			
SPS1351	108	3018	86	2038	
SPS1352	108	3018	86	2038	
SPS1353	108	3018	86	2038	
SPS1436	152	3893	66	2048	
SPS1437	153	3083/197	69	2049	
SPS1475	123A	3444	20	2051	
SPS1523	159	3114/290	82	2032	
SPS1523A	159	3114/290	82	2032	
SPS1593WT			22	2032	
SPS18	5491	6791	C220A		
SPS1802			20	2051	
SPS1817			20	2051	

Industry Standard No.	ECG	SK	GE	RS 276-	MOTOR.
SPS1846			61	2038	
SPS1977			20	2051	
SPS20	108	3039/316	86	2038	
SPS210	5492	6792			
SPS2104			20	2051	
SPS2110	108	3018	86	2038	
SPS2111	108	3018	86	2038	
SPS2129			20	2051	
SPS2130			243	2030	
SPS2131			244	2027	
SPS2135			86	2038	
SPS2142			20	2051	
SPS2164			20	2051	
SPS2167			11	2015	
SPS2194			20	2051	
SPS22	159			2032	
SPS220		3018	86	2038	
SPS2216			62	2010	
SPS2217			62	2010	
SPS2224	108	3018	11	2038	
SPS2225	123A	3444	20	2051	
SPS2226	129	3025	244	2027	
SPS2265	108	3018	86	2038	
SPS2265-2	108	3018	86	2038	
SPS2266	108		86	2038	
SPS2269	159	3114/290	82	2032	
SPS2270	123A	3444	20	2051	
SPS2271	199	3122	62	2010	
SPS2272	159	3114/290	82	2032	
SPS2274	159	3466	82	2032	
SPS2279	159		82	2032	
SPS2320		3018	86	2038	
SPS2415			20	2051	
SPS2425			11	2015	
SPS2526			21	2034	
SPS2664			86	2038	
SPS28	5492	6792	C220B		
SPS3003	108	3039/316	86	2038	
SPS3015	123A	3444	20	2051	
SPS310	5494	6794			
SPS320	5526	6627/5527	C230C		
SPS3329	159		82	2032	
SPS3370	108	3018	86	2038	
SPS3724	159	3114/290	82	2032	
SPS3724A	159	3114/290	82	2032	
SPS3735	123A	3444	20	2051	
SPS3751	123A	3444	20	2051	
SPS3786	159	3114/290	82	2032	
SPS3786A	159	3114/290	82	2032	
SPS3787	107	3039/316	11	2015	
SPS38		3039	86	2038	
SPS3900	123A	3444	20	2051	
SPS3907	123A	3444	20	2051	
SPS3908	123A	3444	20	2051	
SPS3909	123A	3444	20	2051	
SPS3912	128		243	2030	
SPS3914	128		243	2030	
SPS3915	123A		20	2051	
SPS3923	123A	3444	20	2051	
SPS3924	159	3114/290	82	2032	
SPS3924A	159	3114/290	82	2032	
SPS3925	123A	3444	20	2051	
SPS3926	123A	3444	20	2051	
SPS3927	159	3114/290	82	2032	
SPS3927A	159	3114/290	82	2032	
SPS3929	108	3039/316	86	2038	
SPS3930	123A	3444	20	2051	
SPS3931	159	3114/290	82	2032	
SPS3931A	159	3114/290	82	2032	
SPS3936	123A	3444	20	2051	
SPS3937	108	3122	86	2038	
SPS3938	123A	3444	20	2051	
SPS3940	123A	3122	20	2051	
SPS3948	108	3039/316	86	2038	
SPS3951	123A	3444	20	2051	
SPS3952	108	3039/316	86	2038	
SPS3957C	123A	3444	20	2051	
SPS3967	123A	3444	20	2051	
SPS3968	108	3039/316	86	2038	
SPS3971	108	3039/316	86	2038	
SPS3972	123A	3444	20	2051	
SPS3973	123A	3444	20	2051	
SPS3987	159	3114/290	82	2032	
SPS3988	159	3114/290	82	2032	
SPS3988A	159	3114/290	82	2032	
SPS3990	159	3114/290	82	2032	
SPS3990A	159	3114/290	82	2032	
SPS3999	123A	3444	20	2051	
SPS4	108	3039/316	86	2038	
SPS40	108	3018	86	2038	
SPS4000			82	2032	
SPS4002	108	3039/316	86	2038	
SPS4003	123A	3444	20	2051	
SPS4004	123A	3444	20	2051	
SPS4005	108	3039/316	86	2038	
SPS4006	123A	3444	20	2051	
SPS4007	159	3114/290	82	2032	
SPS4007A	159	3114/290	82	2032	
SPS4008	108	3039/316	86	2038	
SPS4009	123A	3444	20	2051	
SPS4010	129	3025	244	2027	
SPS4013	159	3114/290	82	2032	
SPS4013A	159	3114/290	82	2032	
SPS4014	159	3114/290	82	2032	
SPS4014A	159	3114/290	82	2032	
SPS4016	108	3039/316	86	2038	
SPS4017	123A	3444	20	2051	
SPS4018	159	3114/290	82	2032	
SPS4018A	159	3114/290	82	2032	
SPS4019	159	3114/290	82	2032	
SPS4019A	159	3114/290	82	2032	
SPS401K	159		82	2032	
SPS4020	123A	3444	20	2051	
SPS4025	159	3114/290	82	2032	
SPS4025A	159	3114/290	82	2032	
SPS4026	159	3114/290	82	2032	
SPS4026A	159	3114/290	82	2032	
SPS4027	159	3114/290	82	2032	
SPS4027A	159	3114/290	82	2032	
SPS4028	159	3114/290	82	2032	
SPS4028A	159	3114/290	82	2032	
SPS4029	123A	3444	20	2051	
SPS4030	108	3039/316	86	2038	
SPS4031	159	3114/290	82	2032	
SPS4031A	159	3114/290	82	2032	
SPS4032	123A	3444	20	2051	
SPS4034	123A	3444	20	2051	
SPS4037	123A	3444	20	2051	
SPS4038	128	3024	243	2030	
SPS4039	123A	3444	20	2051	
SPS4040	123A	3444	20	2051	
SPS4041	123A	3444	20	2051	
SPS4042	123A	3444	20	2051	
SPS4043	108	3039/316	86	2038	

Industry Standard No.	ECG	SK	GE	RS 276-	MOTOR.
SPS4044	123A	3444	20	2051	
SPS4045	123A	3444	20	2051	
SPS4049	123A	3444	20	2051	
SPS4050	108	3039/316	86	2038	
SPS4051	108	3039/316	86	2038	
SPS4052	123A	3444	20	2051	
SPS4053	123A	3444	20	2051	
SPS4054	159	3114/290	82	2032	
SPS4054A	159	3114/290	82	2032	
SPS4055	123A	3444	20	2051	
SPS4056	159	3114/290	82	2032	
SPS4056A	159	3114/290	82	2032	
SPS4059	123A	3444	20	2051	
SPS4060	123A	3444	20	2051	
SPS4061	123A	3444	20	2051	
SPS4062	123A	3444	20	2051	
SPS4063	123A	3444	20	2051	
SPS4064	159	3114/290	82	2032	
SPS4064A	159	3114/290	82	2032	
SPS4066	123A	3444	20	2051	
SPS4067	123A	3444	20	2051	
SPS4068	108	3039/316	86	2038	
SPS4069	123A	3444	20	2051	
SPS4072	159	3114/290	82	2032	
SPS4072A	159	3114/290	82	2032	
SPS4073	159	3114/290	82	2032	
SPS4073A	159	3114/290	82	2032	
SPS4074	123A	3444	20	2051	
SPS4075	123A	3444	20	2051	
SPS4076	159	3114/290	82	2032	
SPS4076A	159	3114/290	82	2032	
SPS4077	123A	3444	20	2051	
SPS4078	159	3114/290	82	2032	
SPS4078A	159	3114/290	82	2032	
SPS4079	108		86	2038	
SPS4080	108	3039/316	86	2038	
SPS4081	123A	3444	20	2051	
SPS4082	159	3114/290	82	2032	
SPS4082A	159	3114/290	82	2032	
SPS4083	123A	3444	20	2051	
SPS4084	123A	3444	20	2051	
SPS4085	123A	3444	20	2051	
SPS4086	159	3114/290	82	2032	
SPS4086A	159	3114/290	82	2032	
SPS4087	159	3114/290	82	2032	
SPS4087A	159	3114/290	82	2032	
SPS4088	123A	3444	20	2051	
SPS4089	123A	3444	20	2051	
SPS4090	159	3114/290	82	2032	
SPS4090A	159	3114/290	82	2032	
SPS4091	108	3039/316	86	2038	
SPS4092	211	3203	253	2026	
SPS4095	123A	3444	20	2051	
SPS4099	189		218	2026	
SPS41		3444	20	2051	
SPS410	5494	6794			
SPS4143	107	3018	11	2015	
SPS4145	108	3039/316	86	2038	
SPS4167	108	3018	86	2038	
SPS4168	107	3018	11	2015	
SPS4169	123A	3444	20	2051	
SPS4199	123A	3444	20	2051	
SPS42	159	3114/290	82	2032	
SPS420	5527	6627	C230D		
SPS4236			20	2051	
SPS4237	185		58	2025	
SPS4272	199		62	2010	
SPS428	108	3019	86	2038	
SPS429			11	2015	
SPS42A	159	3114/290	82	2032	
SPS43-1	107	3039/316	11	2015	
SPS4300	128	3024	243	2030	
SPS4301	129	3025	244	2027	
SPS4302	159	3114/290	82	2032	
SPS4303	123A	3444	20	2051	
SPS4309	128	3024	243	2030	
SPS4310	129	3025	244	2027	
SPS4311	128		243	2030	
SPS4312	129	3025	244	2027	
SPS4313	123A	3444	20	2051	
SPS4314	159	3114/290	82	2032	
SPS4314A	159	3114/290	82	2032	
SPS4330			82	2032	
SPS4331			61	2038	
SPS4338			82	2032	
SPS4343	161		39	2015	
SPS4344			20	2051	
SPS4345	123A	3444	20	2051	
SPS4347	123A	3444	20	2051	
SPS4348	159	3114/290	82	2032	
SPS4348A	159	3114/290	82	2032	
SPS4354	159	3114/290	82	2032	
SPS4354A	159	3114/290	82	2032	
SPS4355	159	3114/290	82	2032	
SPS4356	123A	3444	20	2051	
SPS4359	123A	3444	20	2051	
SPS4360	123A	3444	20	2051	
SPS4361	128		243	2030	
SPS4363	123A	3444	20	2051	
SPS4365	159	3114/290	82	2032	
SPS4365A	159	3114/290	82	2032	
SPS4367	123A	3444	20	2051	
SPS4368	123A	3444	20	2051	
SPS4382	123A	3444	20	2051	
SPS4390			20	2051	
SPS4391			243	2030	
SPS4392			20	2051	
SPS4397			82	2032	
SPS4399	108		86	2038	
SPS4401			21	2034	
SPS4423	108		86	2038	
SPS4436			61	2038	
SPS4443			20	2051	
SPS4446	123A	3444	20	2051	
SPS4450	123A	3444	20	2051	
SPS4451	123A	3444	20	2051	
SPS4452	159	3114/290	82	2032	
SPS4452A	159	3114/290	82	2032	
SPS4453	123A	3444	20	2051	
SPS4455	123A	3444	20	2051	
SPS4456	123A	3444	20	2051	
SPS4457	123A	3444	20	2051	
SPS4458	159	3114/290	82	2032	
SPS4458A	159	3114/290	82	2032	
SPS4459	123A	3444	20	2051	
SPS4460	159	3114/290	82	2032	
SPS4460A	159	3114/290	82	2032	
SPS4461	128		243	2030	
SPS4462	129	3025	244	2027	
SPS4472	123A	3444	20	2051	
SPS4473	159	3114/290	82	2032	
SPS4473A	159	3114/290	82	2032	
SPS4476	123A	3444	20	2051	
SPS4477	129		244	2027	
SPS4478	123A	3444	20	2051	
SPS4480	159	3114/290	82	2032	
SPS4480A	159	3114/290	82	2032	
SPS4489	159	3114/290	82	2032	
SPS4489A	159	3114/290	82	2032	
SPS4490	128		243	2030	
SPS4491	123A	3444	20	2051	
SPS4492	129	3025	244	2027	
SPS4493	123A	3444	20	2051	
SPS4494	123A	3444	20	2051	
SPS4495	128		243	2030	
SPS4497	129	3025	244	2027	
SPS4498			20	2051	
SPS4610	108	3039/316	86	2038	
SPS47	159		82	2032	
SPS477				2051	
SPS48	5494	3580/5507	C220D		
SPS4813	159		82	2032	
SPS4814	199	3245	62	2010	
SPS4815	159		82	2032	
SPS49	128		243	2030	
SPS4920	123A	3444	20	2051	
SPS4942	123A	3444	20	2051	
SPS5000	123A	3190/184	20	2051	
SPS5006	123A	3191/185	20	2051	
SPS5006-1	123A	3444	20	2051	
SPS5006-2	123A	3444	20	2051	
SPS5007	159	3114/290	82	2032	
SPS5007-1	159	3114/290	82	2032	
SPS5007-1A	159	3114/290	82	2032	
SPS5007-2	159	3114/290	82	2032	
SPS5007-2A	159	3114/290	82	2032	
SPS5007A	159	3114/290	82	2032	
SPS5008	159	3466	82	2032	
SPS510	5496	6796			
SPS514	159	3114/290	82	2032	
SPS514A	159	3114/290	82	2032	
SPS520	5528	6629/5529	C230E		
SPS5328			245	2004	
SPS5450	194	3024/128	220		
SPS5451	292	3114/290	21		
SPS5457	123A	3444	20	2051	
SPS5458	159		82	2032	
SPS5569			11	2015	
SPS58	5496		C220E		
SPS5809	128		243	2030	
SPS610	5496	6796			
SPS6109	159	3114/290	82	2032	
SPS6109A	159	3114/290	82	2032	
SPS6111	123A	3444	20	2051	
SPS6112	123A	3444	20	2051	
SPS6113	123A	3444	20	2051	
SPS6124	128	3024	243	2030	
SPS6125	129		244	2027	
SPS6155			86	2038	
SPS620	5529	6629	C230M		
SPS627			20	2051	
SPS6571	123A	3444	20	2051	
SPS668			82	2032	
SPS6682			11	2015	
SPS68	5496		C220M		
SPS6953			82	2032	
SPS699			20	2051	
SPS7359	123AP	3854			
SPS7652	123A	3444	20	2051	
SPS816			11	2015	
SPS817	123A	3444	20	2051	
SPS817N	123A	3444	20	2051	
SPS820	5531	6631			
SPS837	189		218	2026	
SPS856	107	3039/316	11	2015	
SPS860	107	3039/316	11	2015	
SPS868	123A	3444	20	2051	
SPS871	126	3003	52	2024	
SPS906			39	2015	
SPS907			20	2051	
SPS91	121	3717	239	2006	
SPS915			86	2038	
SPS917			86	2038	
SPS918			210	2051	
SPS919			39	2015	
SPS920			61	2038	
SPS952			212	2010	
SPT010	5673		SC245B		
SPT015	5673		SC250B		
SPT025	5681		SC260B		
SPT030	5693	3652			
SPT040	5693	3652	SC265B		
SPT06	5673		SC240B		
SPT08	5673		SC245B		
SPT110	5673		SC245B		
SPT115	5673		SC250B		
SPT125	5682		SC260B		
SPT130	5693	3652			
SPT140	5693	3652	SC265B		
SPT16	5673		SC240B		
SPT18	5673		SC245B		
SPT210	5673		SC245B		
SPT215	5673		SC250B		
SPT225	5683		SC260B		
SPT230	5693	3652	SC265B		
SPT240	5693	3652	SC265B		
SPT26	5673		SC240B		
SPT28	5673		SC245B		
SPT310	5675		SC245D		
SPT315	5675		SC250D		
SPT325	5684		SC260D		
SPT330	5695	3509	SC265D		
SPT340	5695	3509			
SPT3440	198		251		
SPT36	5675		SC240D		
SPT3713	130		14	2041	
SPT38	5675		SC245D		
SPT410	5675		SC245D		
SPT415	5675		SC250D		
SPT425	5685		SC260D		
SPT430	5695	3509	SC265D		
SPT440	5695	3509			
SPT46	5675		SC240D		
SPT48	5675		SC245D		
SPT510	5676		SC245E		
SPT515	5676		SC250E		
SPT525	5686	3521	SC260E		
SPT530	5697	3522	SC265E		
SPT540	5697	3522	SC265E		
SPT56	5676		SC240E		
SPT58	5676		SC245E		
SPT610	5677		SC245M		
SPT615	5677		SC250M		
SPT625	5697		SC260M		
SPT630	5697	3522	SC265M		
SPT640	5697	3522	SC265M		
SPT66	5677		SC240M		
SPT68	5677		SC245M		

Industry Standard No.	ECG	SK	GE	RS 276-	MOTOR.
SPX2			H11A550		4N35
SPX26			H11A520		4N27
SPX28			H11A520		4N27
SPX2E			H11A550		4N35
SPX35					4N35
SPX36					4N35
SPX37					4N35
SPX4			H11A550		4N35
SPX5			H11A550		4N35
SPX6					4N35
SQ-7	103A	3861/101	59	2002	
SQ46	109	3087		1123	
SQ7	101		8	2002	
SQD-2170	123		20	2051	
SR-0004	113A	3119/113	6GC1		
SR-0005			504A	1104	
SR-0007			504A	1104	
SR-0008			504A	1104	
SR-05K-2	116	3311	504A	1104	
SR-1	116	3311	504A	1104	
SR-100	5802	9005		1143	
SR-101-1	116	3016	504A	1104	
SR-101-2	116	3016	504A	1104	
SR-112	116	3016	504A	1104	
SR-114	5802	9005		1143	
SR-120	116	3016	504A	1104	
SR-120-1	116	3016	504A	1104	
SR-13	178MP	3119/113	300(2)	1122(2)	
SR-130-1	116		504A	1104	
SR-131-1	116	3311	504A	1104	
SR-132-1	116		504A	1104	
SR-136	116		504A	1104	
SR-13H	116	3031A	504A	1104	
SR-14	116	3016	504A	1104	
SR-15	506	3120/114	6GD1		
SR-150-01	116	3017B/117	504A	1104	
SR-150-1	116		504A	1104	
SR-1668	116	3311	504A	1104	
SR-17	116		504A	1104	
SR-18	116	9009/5808	504A	1104	
SR-1849-1	116	3031A	504A	1104	
SR-1HM-2			504A	1104	
SR-1K	116		504A	1104	
SR-1K-2	116	3311	504A	1104	
SR-1K-2A	125	3032A	510,531	1114	
SR-1K-2B	125	3032A	510,531	1114	
SR-1K-2C	125	3032A	510,531	1114	
SR-1K-2D	125	3032A	510,531	1114	
SR-1K2	116		504A	1104	
SR-1Z	116		504A	1104	
SR-2	125	3031A	510,531	1114	
SR-20	178MP		6GC1	1122(2)	
SR-22	116	3016	504A	1104	
SR-23	116	3016	504A	1104	
SR-24	116	3016	504A	1104	
SR-27	116	3016	504A	1104	
SR-28	116	3016	504A	1104	
SR-3	116	3016	504A	1104	
SR-30	116	3016	504A	1104	
SR-31	119	3109	CR-2		
SR-32	118	3066	CR-1		
SR-34	177		300	1122	
SR-35	125	3033	510,531	1114	
SR-37	120	3110	CR-3		
SR-390	116		504A	1104	
SR-4		3031A	504A	1104	
SR-401	116	3016	504A	1104	
SR-499	156	3051	512		
SR-5	116	3031A	504A	1104	
SR-76			504A	1104	
SR-846-2	116	3016	504A	1104	
SR-889	116	3016	504A	1104	
SR-9001	125	3032A	510,531	1114	
SR-9002	113A		6GC1		
SR-9005	120	3016	CR-3	1104	
SR-9007	125	3033	510,531	1114	
SR-IK-2	116		504A	1104	
SR0004	178MP		6GC1	1122	
SR05K-2	116	3017B/117	504A	1104	
SR1	102A	3017B/117	53	2007	
SR1-2			504A	1104	
SR1-K2	116	3017B/117	504A	1104	
SR10	114	3120	6GD1		
SR100	116		504A	1104	
SR101-1	116		504A	1104	
SR101-2	116		504A	1104	
SR1024	116	3016	504A	1104	
SR105	116	3017B/117	504A	1104	
SR10D2	5874		5048		
SR10D4	5942		5048		
SR1104	116	3311	504A	1104	
SR112	116		504A	1104	
SR114	116		504A	1104	
SR1156	506		511	1114	
SR120	116		504A	1104	
SR1266	116	3017B/117	504A	1104	
SR13	113A		6GC1		
SR130-1	108		86	2038	
SR131-1	116	3311	504A	1104	
SR132-1	5800	9003	504A	1142	
SR135-1	116	3017B/117	504A	1104	
SR136	5800	9003			
SR1378-1	116		504A	1104	
SR1378-2	5804	9007		1144	
SR1378-3	116		504A	1104	
SR13H	116		504A	1104	
SR14	114		6GD1	1104	
SR1422	5800	9003		1142	
SR144	116	3017B/117	504A	1104	
SR145	116	3016	504A	1104	
SR1493	116	3017B/117	504A	1104	
SR15	114	3120	6GD1	1104	
SR150	125	3033	510,531	1114	
SR151	116	3016	504A	1104	
SR152	116	3311	504A	1104	
SR153	5940		5048		
SR154	156	3051	512		
SR1549	116		504A	1104	
SR1598	5800	9003		1142	
SR16	117		504A	1104	
SR1643A	5800	9003		1142	
SR1650	5808	9009			
SR1668	116	3016	504A	1104	
SR1692	116	3017B/117	504A	1104	
SR1693	116	3016	504A	1104	
SR1694	116	3031A	504A	1104	
SR1695	116	3031A	504A	1104	
SR17	116		504A	1104	
SR1731-1	116	3017B/117	504A	1104	
SR1731-2	116	3017B/117	504A	1104	
SR1731-3	116	3311	504A	1104	
SR1731-4	116	3017B/117	504A	1104	
SR1731-5	116	3016	504A	1104	
SR1742	116	3017B/117	504A	1104	

Industry Standard No.	ECG	SK	GE	RS 276-	MOTOR.
SR1762	5802	9005		1143	
SR1766	116	3016	504A	1104	
SR18	5808		504A	1104	
SR1984	116	3016	504A	1104	
SR1A-1	116	3017B/117	504A	1104	
SR1A-12	117	3017B	504A	1104	
SR1A-2	116	3017B/117	504A	1104	
SR1A-4	116	3017B/117	504A	1104	
SR1A-8	116	3017B/117	504A	1104	
SR1A1	116		504A	1104	
SR1A12	116		504A	1104	
SR1A2	116		504A	1104	
SR1A4	116		504A	1104	
SR1A8	116		504A	1104	
SR1B	166	3105		1152	
SR1C-12				1104	
SR1C-20				1114	
SR1D-1				1011	
SR1D-2				1102	
SR1D-4				1102	
SR1D-6				1103	
SR1D-8				1103	
SR1D1M	116	3311	504A	1104	
SR1DM	116	3311	504A	1104	
SR1DM-1	116	3311	504A	1104	
SR1DM-10				1104	
SR1DM-2	5800	9003	504A	1142	
SR1DM-4	116	9003/5800	504A	1104	
SR1DM-6				1103	
SR1DM-8				1103	
SR1DM1	116	3017B/117	504A	1104	
SR1DMX	116	9003/5800	504A	1104	
SR1E	116	3017B/117	504A	1104	
SR1EM	116	9003/5800	504A	1104	
SR1EM-1	116	9003/5800	504A	1104	
SR1EM-10				1104	
SR1EM-12				1104	
SR1EM-2	116	9003/5800	504A	1104	
SR1EM-20				1114	
SR1EM-4	156	3017B/117	512	1102	
SR1EM-6				1103	
SR1EM-8				1103	
SR1EM-X	116		504A	1104	
SR1EM1	116		504A	1104	
SR1EM2	116		504A	1104	
SR1FM	116		504A	1104	
SR1FM-1	116	3311	504A	1104	
SR1FM-12	125	3130	511	1114	
SR1FM-2	506	3043	511	1114	
SR1FM-20				1114	
SR1FM-4	116	3311	511	1114	
SR1FM-8		3032A	504A	1104	
SR1FM10	116		504A	1104	
SR1FM12	116		504A	1104	
SR1FM2	116			1142	
SR1FM20	125		510,531	1114	
SR1FM4	116	3017B/117	504A	1104	
SR1FM6	116			1144	
SR1FM8	116			1144	
SR1FMA	116	3017B/117	504A	1104	
SR1HM-12	506		504A	1104	
SR1HM-16	506		504A	1104	
SR1HM-2	506	3311	504A	1104	
SR1HM-4	506	3311	504A	1104	
SR1HM-8	506	3130	511	1114	
SR1K	116	3175	504A	1104	
SR1K-1	116	3311	504A	1104	
SR1K-12				1104	
SR1K-1K	116	3311	504A	1104	
SR1K-2	116		504A	1104	
SR1K-2/494	116		504A	1104	
SR1K-20				1114	
SR1K-4	116	3311	504A	1104	
SR1K-8	116	3016	504A	1104	
SR1K-Z	116	3017B/117	504A	1104	
SR1K/494	116		504A	1104	
SR1K08	116	3017B/117	504A	1104	
SR1K1	116	3311	504	1104	
SR1K2	116	3311	504A	1104	
SR1K8	116		504A	1104	
SR1T	116	3016	504A	1104	
SR1Z	116	3017B/117	504A	1104	
SR20	113A	3119/113	6GC1		
SR200	116	3017B/117	504A	1104	
SR200B	116	3016	504A	1104	
SR20226			61	2038	
SR20234			20	2051	
SR205	116	3017B/117	504A	1104	
SR21	113A		6GC1		
SR210	5802	9005		1143	
SR2121	116	3016	504A	1104	
SR22	116		504A	1104	
SR220	5802	9005		1143	
SR23	116	3100/519	504A	1104	
SR2301	116		504A	1104	
SR2301A	116	3017B/117	504A	1104	
SR24	116	3130	504A	1104	
SR2462	552				1N4004
SR25	506	3130	511	1114	
SR27	116	3016	504A	1104	
SR28	116		504A	1104	
SR29	113A	3119/113	6GC1		
SR2956	330	3012/105			
SR2A-1	116	3016	504A	1104	
SR2A-12	125	3017B/117	510,531	1114	
SR2A-2	116	3016	504A	1104	
SR2A-4	116	3016	504A	1104	
SR2A-8	125	3016	510,531	1114	
SR2A1	116		504A	1104	
SR2A12	116		504A	1104	
SR2A2	116		504A	1104	
SR2A4	116		504A	1104	
SR2A8	116		504A	1104	
SR2B12	156	3051	512		
SR2B16	156	3051	512		
SR2B20	156	3051	512		
SR3	116	3017B/117	504A	1104	
SR30	116		504A	1104	
SR3010	116	3017B/117	504A	1104	
SR31	119		CR-2		
SR32	118		CR-1		
SR35	116		504A	1104	
SR3502					1N4002
SR3512					1N4001
SR3582	116	3017B/117	504A	1104	
SR37	120		CR-3		
SR390	116		504A	1104	
SR390-2	117	3016	504A	1104	
SR3943	116	3016	504A	1104	
SR3946					1N4005
SR3AM	125	3051/156	510	1114	
SR3AM-2	156		510	1114	
SR3AM-3	5801	9004	510	1142	
SR3AM-4		3051	510	1114	

Industry Standard No.	ECG	SK	GE	RS 276-	MOTOR.
SR3AM-6			512	1144	
SR3AM-8	117	9007/5804	504A	1104	
SR3AM1	5800	9003		1142	
SR3AM10	5806	3848			
SR3AM2	125	3051/156	504A	1142	
SR3AM4	125			1143	
SR3AM6	5804	9007		1144	
SR3AM8	5804			1144	
SR3BM-6	116	3017B/117	504A	1104	
SR4	116	3017B/117	504A	1104	
SR40	116	3031A	504A	1104	
SR401	116		504A	1104	
SR405	116	3017B/117	504A	1104	
SR475	5800	9003		1142	
SR499	156		512		
SR5	116		504A	1104	
SR50	116	3017B/117	504A	1104	
SR500	116	3016	504A	1104	
SR5005					MR5005
SR500B	116	3016	504A	1104	
SR5010					MR5010
SR5020					MR5020
SR50253-2	167	3647	504A	1172	
SR5030					MR5030
SR5040					MR5040
SR50411-1	116	3017B/117	504A	1104	
SR50517	117		504A	1104	
SR507	5802	9005		1143	
SR6	113A	3119/113	6GC1		
SR60	116	3017B/117	504A	1104	
SR605	116	3017B/117	504A	1104	
SR6134	116	3017B/117	504A	1104	1N4003
SR6323					1N4001
SR6324	116	3016	504A	1104	
SR6325	116	3017B/117	504A	1104	
SR6385	116	3016	504A	1104	1N4003
SR6404					1N4006
SR6415	116	3017B/117	504A	1104	
SR6560	116	3016	504A	1104	1N4002
SR6567	116	3017B/117	504A	1104	
SR6569					1N4004
SR6592					1N4006
SR6593					1N4007
SR6617	116	3016	504A	1104	
SR6723	116		504A	1104	
SR6724	116		504A	1104	
SR710					SPECIAL
SR710F					SPECIAL
SR711					SPECIAL
SR711F					SPECIAL
SR712					SPECIAL
SR712F					SPECIAL
SR713					SPECIAL
SR713F					SPECIAL
SR714					SPECIAL
SR714F					SPECIAL
SR716					SPECIAL
SR716F					SPECIAL
SR75844	123A	3444	20	2051	
SR76	116	3016	504A	1104	
SR806-126	116	3017B/117	504A	1104	
SR846-2	116	3017B/117	504A	1104	
SR846-3	116	3017B/117	504A	1104	
SR851	116	3017B/117	504A	1104	
SR851-121	116	3017B/117	504A	1104	
SR889	116	3017B/117	504A	1104	
SR9000	118	3066	CR-1		
SR9001	119	3109	CR-2		
SR9002	113A	3119/113	6GC1		
SR9005	116	3017B/117	504A	1104	
SRIDM			504A	1104	
SRIDM-1		3031A	504A	1104	
SRIDM-4			504A	1104	
SRIEM-1			504A	1104	
SRIEM-2			504A	1104	
SRIEM-4			504A	1104	
SRIFM-12			511	1114	
SRIFM-2			511	1114	
SRIFM-4			511	1114	
SRIHM-8			511	1114	
SRIK-2	116		504A	1104	
SRIK-8			504A	1104	
SRK-2	116	3031A	504A	1104	
SRK1	116	3311	504A	1104	
SRLPM-1	116	3031A	504A	1104	
SRR13	113A	3119/113	6GC1		
SRS105					1N4001
SRS110					1N4002
SRS1100					1N4007
SRS120					1N4003
SRS140					1N4004
SRS160					1N4005
SRS180					1N4006
SRS205					MR501
SRS210					MR501
SRS2100					MR510
SRS220					MR502
SRS240					MR504
SRS260					MR506
SRS280					MR508
SRS305					MR501
SRS310					MR501
SRS3100					MR510
SRS320					MR502
SRS360					MR506
SRS380					MR508
SRSFR105					1N4933
SRSFR110					1N4934
SRSFR1100					MR818
SRSFR120					1N4935
SRSFR140					1N4936
SRSFR150					1N4937
SRSFR160					1N4937
SRSFR180					MR817
SRSFR205					MR850
SRSFR210					MR851
SRSFR220					MR852
SRSFR230					MR854
SRSFR240					MR854
SRSFR250					MR856
SRSFR260					MR856
SRSFR305					MR850
SRSFR310					MR851
SRSFR320					MR852
SRSFR330					MR854
SRSFR340					MR854
SRSFR350					MR856
SRSFR360					MR856
SS-1	116		504A	1104	
SS-2N5062	5402	3638			
SS-3704	312		FET-1	2028	
SSQ001	102A	3004	53	2007	
SS00010	116	3031A	504A	1104	
SS0001A	102A	3004	53	2007	

Industry Standard No.	ECG	SK	GE	RS 276-	MOTOR.
SS00010	116		504A	1104	
SS0002	102A	3004	53	2007	
SS0002A	102A	3004	53	2007	
SS0003	158	3004/102A	53	2007	
SS0003A	102A	3004	53	2007	
SS0004	102A	3004	53	2007	
SS0004A	102A	3004	53	2007	
SS0005	102A		53	2007	
SS0005A	102A	3004	53	2007	
SS0007	109	3088		1123	
SS0008	109	3088		1123	
SS0009	116	3031A	504A	1104	
SS000I				2007	
SS1					MZ2360
SS1-145128	123A	3444	20	2051	
SS1-2					MZ2361
SS1123	191		249		
SS155	160		245	2004	
SS16	112	3089			
SS1606	159	3114/290	82	2032	
SS1606A	121	3114/290	239	2006	
SS1906	159	3114/290	82	2032	
SS1906A	159	3114/290	82	2032	
SS1912	154		40	2012	
SS2308	123A	3444	20	2051	
SS2409	213		254		
SS2503	159	3114/290	82	2032	
SS2503A	159	3114/290	82	2032	
SS2504	123A	3444	20	2051	
SS29A4			65	2050	
SS29A5			65	2050	
SS321	116	3311	504A	1104	
SS322	116	3311	504A	1104	
SS324	116	3311	504A	1104	
SS334	116	3311	504A	1104	
SS337	116	3311	504A	1104	
SS3534-4	312	3448	FET-1	2035	
SS3586	312	3112		2028	
SS3672	312			2028	
SS3694	123A	3444	20	2051	
SS3704	312	3448	FET-1	2035	
SS3735	312		FET-1	2028	
SS3935	195A	3765	46		
SS4042	161	3039/316	39	2015	
SS4312	278		261		
SS455	116	3016	504A	1104	
SS524	154	3045/225	40	2012	
SS6111	154		40	2012	
SS6724			82	2032	
SS9327	190		217		
SS9328			20	2051	
SSA43	159	3114/290	82	2032	
SSA43A	159	3114/290	82	2032	
SSA43A-1	159	3114/290	82	2032	
SSA46	159	3114/290	82	2032	
SSA46A	159	3114/290	82	2032	
SSA48	159	3114/290	82	2032	
SSA48A	159	3114/290	82	2032	
SSD974	116	3017B/117	504A	1104	
SSI1120	5804	9007		1144	
SSIB0140	156	3051	512		
SSIB0180	156	3051	512		
SSIB0640	156	3051	512		
SSIB0680	156	3051	512		
SSIC0810	5802	9005		1143	
SSIC0820	5804	9007		1144	
SSIC0840	156	3848/5806	512		
SSIC0860	5809	9010			
SSIC0880	5809	9010	512		
SSIC1110	5802	9005		1143	
SSIC1140	156	3848/5806	512		
SSIC1160	5809	9010			
SSIC1180	156	9010/5809	512		
SSIC1210	5802	9005		1143	
SSIC1220	5804	9007		1144	
SSIC1240	5806	3848	512		
SSIC1260	156	9010/5809			
SSIC1280	5809	9010	512		
SSIC1660A	5809	9010			
SSIC1740	156	3051	512		
SSIC1780	156	3051	512		
SSIC1960A	5809	9010			
SSL34,54					MLED930
SSL4,F					MLED930
SSS101AJ					LM101AH
SSS101AL					LM101AH
SSS107J		3690			LM107H
SSS107P		3596			LM107H
SSS1408A-6Z					MC1408L6
SSS1408A-7Z					MC1408L7
SSS1408A-8Z					MC1408L8
SSS1458J		3555			MC1458G
SSS1508A-8Z					MC1508L8
SSS1558J					MC1558G
SSS201AJ	1171	3565			LM201AH
SSS201AL					LM201AH
SSS201AP					LM201AN
SSS207J		3690			LM207H
SSS207P					LM207H
SSS301AJ	1171	3565			LM301AH
SSS301AL					LM301AH
SSS301AP	975	3641			LM301AN
SSS741BJ					MC1741G
SSS741BL					MC1741F
SSS741BP					MC1741P2
SSS741CJ	941	3514			MC1741CG
SSS741CL					MC1741CF
SSS741CP					MC1741CP2
SSS741GJ					MC1741SG
SSS741GP					MC1741SG
SSS741J					MC1741G
SSS741L					MC1741F
SSS741P					MC1741P2
SSS747B2					MC1747F
SSS747BP					MC1747L
SSS747CK	947				MC1747CG
SSS747CM					MC1747CF
SSS747CP	947D	3556			MC1747CL
SSS747CY	947D	3556			
SSS747GK					MC1747G
SSS747GM					MC1747F
SSS747GP					MC1747L
SSS747L					MC1747F
SSS747P					MC1747L
ST 254 Q	128			2030	
ST-021660	129		244	2027	
ST-103	126	3007	52	2024	
ST-106	105	3012	4		
ST-107	105	3012	4		
ST-108	105	3012			
ST-109	105	3012	4		
ST-110	105	3012	4		
ST-111	105	3012	4		
ST-112	105	3012	4		

Industry Standard No.	ECG	SK	GE	RS 276-	MOTOR.
ST-12	116		504A	1104	
ST-122	102A	3004	53	2007	
ST-123	102A	3004	53	2007	
ST-1242	123A	3020/123	20	2051	
ST-1243	123A	3020/123	20	2051	
ST-1244	123A	3020/123	20	2051	
ST-125	160		245	2004	
ST-1290	123A	3020/123	20	2051	
ST-14	116	3017B/117	504A	1104	
ST-172	101	3861	8	2002	
ST-2	6407			1050	
ST-201	128		243	2030	
ST-2040P	116	3016	504A	1104	
ST-213	128		243	2030	
ST-235	121	3009	239	2006	
ST-254-Q			243	2030	
ST-28A	176	3845	80		
ST-28B	100	3005	1	2007	
ST-28C	100	3005	1	2007	
ST-2N6558	191	3232	27		
ST-2SA673C	290A	3114/290			
ST-2SA844C	234	3114/290			
ST-2SB568C	292	3441			
ST-2SC1106	162	3559	73		
ST-2SC1213D-24	289A	3122			
ST-2SC1514	198	3219	251		
ST-2SC1514-05	376	3219			
ST-2SC383W	199	3124/289	213	2010	
ST-2SD467B	297	3449			
ST-2SD467C	297	3449			
ST-2SD478C	291	3440			
ST-301	102A	3004	53	2007	
ST-302	102A	3004	53	2007	
ST-304	102A	3004	53	2007	
ST-332	102A	3004	53	2007	
ST-370	100	3845/176	1	2007	
ST-37C	100	3005	1	2007	
ST-37D	100	3005	1	2007	
ST-41	5005A	3771	ZD-3.3		
ST-BF459	157	3747			
ST-BU208	165	3115	38		
ST-LM2152	123AP	3444/123A	20	2051	
ST-LM2682	128	3024	243	2030	
ST-MJE9742	157	3747	232		
ST-MPS9433	123A	3444	62	2051	
ST-MPS9433T	123AP	3444/123A			
ST-MPS9682J	159	3466	82	2032	
ST-MPS9700D	123A	3444	61	2051	
ST-MPS9700E	123A	3444	61	2051	
ST-MPS9700F	123A	3444	61	2051	
ST-MPS9750D	159	3466	82	2032	
ST.082.112.005	123A	3444	20	2051	
ST.082.114.015	123A	3444	20	2051	
ST.082.114.016	199	3245	62	2010	
ST.082.115.015	159		82	2032	
ST/123/CR	519		514	1122	
ST/146/CR	519		514	1122	
ST/217/Q	123A	3444	20	2051	
ST01	123A	3444	20	2051	
ST02	123A	3444	20	2051	
ST03	123A	3444	20	2051	
ST04	123A	3444	20	2051	
ST05	123A	3444	20	2051	
ST06	123A	3444	20	2051	
ST07279			245	2004	
ST10	108	3039/316	86	2038	
ST101	181	3036	75	2041	
ST1026	108	3039/316	86	2038	
ST1050	108	3039/316	86	2038	
ST1051	108	3039/316	86	2038	
ST106	105		4		
ST107	105		4		
ST108	105		4		
ST109	105		4		
ST11	107		11	2015	
ST110	105		4		
ST111	105		4		
ST112	105		4		
ST12	108	3016	86	2038	
ST122	102A	3123	53	2007	
ST123	102A	3123	53	2007	
ST1242	123A	3444	20	2051	
ST1243	123A	3444	20	2051	
ST1244	123A	3444	20	2051	
ST129-1	159		82	2032	
ST1290	123A	3444	20	2051	
ST13	108	3039/316	86	2038	
ST1336	108	3039/316	86	2038	
ST14	108	3031A	86	2038	
ST1402D	85	3444/123A	20	2051	
ST1402E	123AP	3444/123A	20	2051	
ST15	108	3039/316	86	2038	
ST150	123A	3444	20	2051	
ST1502C	107	3854/123AP	61		
ST1502D	107	3854/123AP	61		
ST1504	192		63	2030	
ST1505	192		63	2030	
ST1506	123A	3444	20	2051	
ST151	123A	3444	20	2051	
ST152	123A	3444	20	2051	
ST153	123A	3444	20	2051	
ST154	123A	3444	20	2051	
ST155	123A	3444	20	2051	
ST156	123A	3444	20	2051	
ST157	123A	3444	20	2051	
ST16	116	3017B/117	504A	1104	
ST160	123A	3444	20	2051	
ST1602D	159	3466	82		
ST1607	123A	3444	20	2051	
ST161	123A	3444	20	2051	
ST162	123A	3444	20	2051	
ST163	123A	3444	20	2051	
ST1694	123A	3039/316	86	2038	
ST1702M	85	3854/123AP	20	2051	
ST1702N	85	3854/123AP	20	2051	
ST172	101		8	2002	
ST175	123A	3444	20	2051	
ST176	123A	3444	20	2051	
ST177	123A	3444	20	2051	
ST178	123A	3444	20	2051	
ST18	125	3032A	510,531	1114	
ST180	123A	3444	20	2051	
ST181	123A	3444	20	2051	
ST182	123A	3444	20	2051	
ST2	6408	3523/6407	ST2		
ST2(GE)	6407	3523			
ST2-10	5874		5032		
ST2-20	5874		5032		
ST2-2517	106	3984	21	2034	
ST2-30	5878		5032		
ST2-40	5878		5032		
ST2040P	116		504A	1104	
ST2100P					MR1130
ST210E					1N3209
ST210P		3500			MR1121
ST213	128		243	2030	
ST220E					1N3210
ST220P		3500			MR1122
ST22546-1	519		514	1122	
ST230E					1N3211
ST230P		3500			MR1123
ST235	121	3717	239	2006	
ST240E					1N3212
ST240P		3500			MR1124
ST250	123A	3444	20	2051	
ST250E					1N3213
ST250P		3500			MR1125
ST251	123A	3444	20	2051	
ST25A	123A	3444	20	2051	
ST25B	123A	3444	20	2051	
ST25C	123A	3444	20	2051	
ST260E					1N3214
ST260P		3500			MR1126
ST27020	153		69	2049	
ST280P					MR1128
ST28A	102A	3845/176	53	2007	
ST28B	102A	3123	53	2007	
ST28C	102A	3123	53	2007	
ST29	108	3039/316	86	2038	
ST29045	180		74	2043	
ST29046	180		74	2043	
ST29047	180		74	2043	
ST2FR10P	5818				1N3890
ST2FR20P	5818				1N3891
ST2FR30P	5820				1N3892
ST2FR40P	5820				1N3893
ST2FR60P	5822				MR1376
ST2T003	199	3245	62	2010	
ST3-10	5990	3608	5100		
ST3-20	5990	3608	5100		
ST3-30	5990	3608	5100		
ST3-40	5990	3608	5100		
ST30	108	3039/316	86	2038	
ST301	102A	3123	53	2007	
ST302	102A	3123	53	2007	
ST303	102A	3123	53	2007	
ST3030	123	3039/316	86	2038	
ST3031	123	3039/316	86	2038	
ST304	102A	3123	53	2007	
ST31	108	3039/316	86	2038	
ST32	108	3039/316	86	2038	
ST32012-0037	519		514	1122	
ST33	108	3039/316	86	2038	
ST33026	199	3245	62	2010	
ST332	102A	3123	53	2007	
ST34	108	3039/316	86	2038	
ST35	108	3039/316	86	2038	
ST370	102A	3123	53	2007	
ST37C	102A	3123	53	2007	
ST37D	102A	3123	53	2007	
ST37E	102A		53	2007	
ST370	176		80		
ST382	102A		53	2007	
ST39	5007A	3773			
ST4-10	5986	3608/5990	5100		
ST4-20	5986	3608/5990	5100		
ST4-30	5990	3608	5100		
ST4-40	5990	3608	5100		
ST40	108	3039/316	86	2038	
ST402	192		63	2030	
ST403	123A	3444	20	2051	
ST41	108	3039/316	86	2038	
ST410P		3016			1N1184A
ST415	108	3039/316	86	2038	
ST4150	128		243	2030	
ST42	108	3039/316	86	2038	
ST4201	128		243	2030	
ST4202	128		243	2030	
ST4203	128		243	2030	
ST4204	128		243	2030	
ST420P		3016			1N1186A
ST43	108	3039/316	86	2038	
ST430P		3017B			1N1187A
ST4341	128		63	2030	
ST44	108	3039/316	86	2038	
ST440P		3032A			1N1188A
ST45	108	3039/316	86	2038	
ST450P		3032A			1N1189A
ST460P		3033			1N1190A
ST47	112		1N82A		
ST47025			86	2038	
ST4FR10P	6006$				MR861
ST4FR20P	6006$				MR862
ST4FR30P	6008$				MR864
ST4FR40P	6008$				MR864
ST4FR60P	6010$				MR866
ST50	123A	3444	20	2051	
ST501	123A	3444	20	2051	
ST502	123A	3444	20	2051	
ST503	123A	3444	20	2051	
ST504	123A	3444	20	2051	
ST5060	123A	3444	20	2051	
ST5061	192		63	2030	
ST51	123A	3444	20	2051	
ST53	123A	3444	20	2051	
ST54	123A	3444	20	2051	
ST55	123A	3444	20	2051	
ST56	123A	3444	20	2051	
ST5641	161	3716	39	2015	
ST57	123A	3444	20	2051	
ST58	123A	3444	20	2051	
ST59	123A	3444	20	2051	
ST60	108	3039/316	86	2038	
ST61	108	3039/316	86	2038	
ST61000	193		67	2023	
ST6110	108		82	2032	
ST6120	108	3039/316	86	2038	
ST62	108	3039/316	86	2038	
ST62180			82	2032	
ST63	123A	3444	20	2051	
ST64	123A	3444	20	2051	
ST6510	107		11	2015	
ST6511	123A	3444	20	2051	
ST6512	123A	3444	20	2051	
ST6573	128	3024	243	2030	
ST6574	128	3024	243	2030	
ST70	108	3039/316	86	2038	
ST71	108	3039/316	86	2038	
ST7100	199	3245	62	2010	
ST72	108	3039/316	86	2038	
ST72039	129	3025	244	2027	
ST72040	129	3025	244	2027	
ST80	108	3039/316	86	2038	
ST8014	159	3025/129	82	2032	
ST8033	159	3114/290	82	2032	
ST8033A	159	3114/290	82	2032	
ST8034	159	3114/290	82	2032	
ST8034A	159	3114/290	82	2032	

Industry Standard No.	ECG	SK	GE	RS 276-	MOTOR.
ST8035	159	3114/290	82	2032	
ST8035A	159	3114/290	82	2032	
ST8036	159	3114/290	82	2032	
ST8036A	159	3114/290	82	2032	
ST8065	159	3114/290	82	2032	
ST8065A	159	3114/290	82	2032	
ST8190	159		82	2032	
ST82	108	3039/316	86	2038	
ST84027	198	3220	251		
ST84028	198	3220	251		
ST84029	198	3220	251		
ST8500	159	3114/290	82	2032	
ST8500A	159	3114/290	82	2032	
ST8509	159	3114/290	82	2032	
ST8509A	159	3114/290	82	2032	
ST9	108	3039/316	86	2038	
ST903	108	3039/316	86	2038	
ST904	108	3039/316	86	2038	
ST904A	108	3039/316	86	2038	
ST905	108	3039/316	86	2038	
ST910	108	3039/316	86	2038	
STA7604	327	3945			
STA9364	162	3559	73		
STB01-02	177	3100/519	300	1122	
STB4	5002A	3767/5001A			
STB567					MZ2361
STB576	177	3031A	300	1122	
STBOL-02	116		504A	1104	
STC-1035	130	3027	14	2041	
STC-1035A	130	3027	14	2041	
STC-1036	130	3027	14	2041	
STC-1036A	130	3027	14	2041	
STC-1085	130		14	2041	
STC-4401	175	3131A/369	246	2020	
STC1035	130		14	2041	
STC1035A	130		14	2041	
STC1036	130		14	2041	
STC1036A	130		14	2041	
STC1080	130	3027	14	2041	
STC1081	130	3027	14	2041	
STC1082	130	3027	14	2041	
STC1083	130	3027	14	2041	
STC1084	130	3027	14	2041	
STC1085		3027	35		
STC1094	181		75	2041	
STC1300	152	3893	66	2048	
STC1336	192		63	2030	
STC1800	186	3192	28	2017	
STC1850	152	3893	66	2048	
STC1860	152	3893	66	2048	
STC1862	186	3192	28	2017	
STC2220	181		75	2041	
STC2221	181		75	2041	
STC2224	181		75	2041	
STC2225	181		75	2041	
STC2228	181		75	2041	
STC2229	181		75	2041	
STC4252	130	3036	14	2041	
STC4253	130	3027	14	2041	
STC4254	130	3036	14	2041	
STC4255	130	3036	14	2041	
STC4401	175		246	2020	
STC5202	153		69	2049	
STC5203	153		69	2049	
STC5205	153		69	2049	
STC5206	153		69	2049	
STC5303	153		69	2049	
STC5610	129	3025	244	2027	
STC5611	129	3025	244	2027	
STC5612	129	3025	244	2027	
STC5802			69	2049	
STC5803			69	2049	
STC5805	153		69	2049	
STC5806	153		69	2049	
STD9007		3192	28	2017	
STE400	108		86	2038	
STFCN10			504A	1104	
STH7251	197		250	2027	
STK-011	1024	3152	720		
STK-015	1027	3153	721		
STK-016	1323		721		
STK-018	1139	3155/1025			
STK-020	1025	3155	722		
STK-020B	1025	3155	722		
STK-020D	1025	3155	722		
STK-020E	1025	3155	722		
STK-020F	1025	3155	722		
STK-020K	1025	3155	722		
STK-020L	1025	3155	722		
STK-025	1090	3291	724		
STK-025G	1090	3291	724		
STK-027	1274	3155/1025			
STK-032	1090	3291	723		
STK-036	1148	3292			
STK-056	1028	3436	724		
STK-056A	1028		724		
STK054	1028		724		
STK435	1218		IC-330		
STK439	1219		IC-331		
STM101	500A	3304			
STM73Q	159		82	2032	
STPT10					MRD160
STPT15					MRD160
STPT260	3036				MRD360
STPT300					MRD300
STPT310					MRD360
STPT45					MRD450
STPT51	3032				MRD3050
STPT53	3032				MRD3056
STPT80	3032				MRD3056
STPT83	3032				MRD3054
STPT84	3032				MRD3056
STR106	5667A	3508/5675			
STR117	5682	3508/5675			
STR130	5667A	3632/5624			
STR206	5667A	3508/5675			
STR208	5667A	3632/5624			
STR217	5683	3508/5675			
STR225	5683	3521/5686			
STR230	5667A	3508/5675			
STR406	5667A	3508/5675			
STR408	5667A	3633/5626			
STR417	5685	3521/5686			
STR425	5685	3521/5686			
STR430	5667A	3508/5675			
STS402	283	3559/162	73		
STS403	283	3559/162	73		
STS409	283	3559/162	73		
STS410	94	3560			
STS411	94	3560			
STS413	283	3560			
STS430	385	3559/162	73		
STS431	385	3560	73		
STT2300			75	2041	
STT2405		3239	216	2053	
STT2406		3239	216	2053	
STT3500			75	2041	
STT4451	186	3192	28	2017	
STT4483		3239	216	2053	
STT9001		3239	216	2053	
STT9002		3239	216	2053	
STT9004		3239	216	2053	
STT9005		3239	216	2053	
STV-3		3311	504A	1104	
STV-3H	605	3864			
STX0010	175	3131A/369	246	2020	
STX0011	129	3025	244	2027	
STX0013	184	3054/196	57	2017	
STX0014	130	3027	14	2041	
STX0015	124	3021	12		
STX0016	175	3131A/369	246	2020	
STX0020	185	3083/197	58	2025	
STX0026	184	3041	57	2017	
STX0027	130	3027	14	2041	
STX0028	154		40	2012	
STX0029	153		69	2049	
STX0030	153		69	2049	
STX0032	130	3027	14	2041	
STX0033	126		52	2024	
STX0034	126		52	2024	
STX0036	160		245	2004	
STX0085	160		245	2004	
STX0087	160		245	2004	
STX0089	160		245	2004	
STX0090	160		245	2004	
STX0096	102A		53	2007	
STX0099	102A		53	2007	
STX0104	102A		53	2007	
STX0105	102A		53	2007	
STX0110	102A		53	2007	
STX0114	102A		53	2007	
STX0121	158		53	2007	
STX0123	102A		53	2007	
STX0224	158		53	2007	
STX0260	102A		53	2007	
STX0263	102A		53	2007	
STX0264	102A		53	2007	
STX0265	102A		53	2007	
STX0268	102A		53	2007	
STX0269	102A		53	2007	
STX3326	196	3054	241	2020	
STX49007	941M		IC-265	007	
SU-31	109	3088		1123	
SU110	6401			2029	
SU2076	312	3112		2028	
SU2077	312	3112		2028	
SU2080	312			2028	
SU2081	312	3112		2028	
SU44	6409			2029	
SU5	145A	3063	ZD-15	564	
SUC650	137A	3058	ZD-6.2	561	
SUM6010	137A	3058	ZD-6.2	561	
SUM6011	137A	3058	ZD-6.2	561	
SUM6020	137A	3058	ZD-6.2	561	
SUM6021	137A	3058	ZD-6.2	561	
SV-01A	116		504A	1104	
SV-01B	116		504A	1104	
SV-02	605	3864	ZD-6.2	561	
SV-02(RECTIFIER)	116		504A	1104	
SV-03	605	3864	504A	1104	
SV-04	605	3864	300	1122	
SV-05	116		504A	1104	
SV-1020A	145A	3063	ZD-15	564	
SV-12388	116	3136/5072A	504A	1104	
SV-1238E	116	3016	504A	1104	
SV-138A	145A	3063	ZD-15	564	
SV-1R				1102	
SV-3	177	3100/519	300	1122	
SV-31	109			1123	
SV-31(DIODE)	109	3090		1123	
SV-3A	605	3864	300	1122	
SV-3B	177	3100/519	300	1122	
SV-3C	177	3100/519	300	1122	
SV-4015A	145A	3063	ZD-15	564	
SV-4A	605	3864			
SV-4R				1103	
SV-6R				1104	
SV-7	138A		ZD-7.5		
SV-8	177	3100/519	300	1122	
SV-9	601	3100/519	ZD-9.1	562	
SV-9(ZENER)	139A		ZD-9.1	562	
SV-KB262	605	3864			
SV01A	116	3311	504A	1104	
SV02A	116	3058/137A	504A	1104	
SV03	606	3864/605		1143	
SV04	607		504A	1104	
SV05			504A	1104	
SV1000					MR331
SV1017	142A	3062	ZD-12	563	
SV1019	144A	3094	ZD-14	564	
SV1020	145A	3063	ZD-15	564	
SV11021	141A	3092	ZD-11.5		
SV121	135A		ZD-5.1		
SV122	135A	3056	ZD-5.1		
SV123	136A	3057	ZD-5.6	561	
SV1238	116		504A	1104	
SV12388	116	3017B/117	504A	1104	
SV12388E	116	3017B/117	504A	1104	
SV1238E	116		504A	1104	
SV124	5070A	9021	ZD-6.0		
SV133	140A	3061	ZD-10	562	
SV135	142A	3062	ZD-12	563	
SV136	143A	3750			
SV137	144A	3094	ZD-14	564	
SV138	145A	3063	ZD-15	564	
SV139	5075A	3751	ZD-16	564	
SV141	5076A	9022	ZD-17		
SV142	5077A	3752	ZD-18		
SV143	5078A	9023	ZD-19		
SV144	5079A		ZD-20		
SV14B				1122	
SV14C				1122	
SV168	5080A		ZD-22		
SV172	5083A	3754	ZD-28		
SV28	5063A	3837			
SV29	5065A	3838			
SV30	109	3087		1123	
000000SV31	158	3004/102A	53	2007	
SV4012	142A	3062	ZD-12	563	
SV4012A	142A	3062	ZD-12	563	
SV4014	144A	3094	ZD-14	564	
SV4014A	144A	3094	ZD-14	564	
SV4015	145A	3063	ZD-15	564	
SV4015A	145A	3063	ZD-15	564	
SV4018	5027A	3793	ZD-18		
SV4018A	5027A	3793	ZD-18		
SV4027	146A	3064	ZD-27		
SV4027-6	146A	3064	ZD-27		

Industry Standard No.	ECG	SK	GE	RS 276-	MOTOR.
SV4027A	146A	3064	ZD-27		
SV4027A-6	146A	3064	ZD-27		
SV4033	147A	3095	ZD-33		
SV4033A	147A	3095	ZD-33		
SV4055	148A	3096	ZD-55		
SV4055A	148A	3096	ZD-55		
SV4062	149A	3097	ZD-62		
SV4062A	149A	3097	ZD-62		
SV4082	150A	3098	ZD-82		
SV4082A	150A	3098	ZD-82		
SV45	143A	3750			
SV47	5075A	3751			
SV48	5077A	3752			
SV53	5084A	3755			
SV7056					2N6558
SV7401					MZ605
SV800					MR330
SV9		3060	300	1122	
SVC-101	610	3327/614			
SVC-201	614	3327	901		
SVC-251	614	3327			
SVC0053	614	3327	90		
SVC101	614		90		
SVC1125	141A	3092	ZD-11.5		
SVC1150	141A	3092	ZD-11.5		
SVC201	614		90		
SVC251	614		90		
SVC625	137A	3058	ZD-6.2	561	
SVC650	137A	3058	ZD-6.2	561	
SVD-1S1717	139A		ZD-9.1	562	
SVD0278.2A				562	
SVD0279.5A				562	
SVD02Z8		3059	ZD-8.2	562	
SVD02Z8.2			ZD-8.2	562	
SVD02Z8.2A	5072A		ZD-8.2	562	
SVD02Z9.5A	139A	3060	ZD-9.1	562	
SVD0A70		3088	1N60	1123	
SVD0A79	109	3087		1123	
SVD0A90			1N60	1123	
SVD10D-1	116	3311	504A	1104	
SVD12B2B1P-M	166	9075	BR-600	1152	
SVD1S1717	139A		ZD-9.1	562	
SVD1S1850	116	3119/113	504A	1104	
SVD1S1850R		3121	6GX1		
SVD20A70	109	3090		1123	
SVD20A79	109	3087		1123	
SVDAS2HB10	156	3051	512		
SVDMA26	117	3463/601	504A	1104	
SVDMA26-1	109	3463/601		1123	
SVDMA26-2			300#(2)	1122	
SVDOA70	109	3090		1123	
SVDOA90	109	3090		1123	
SVDS1RB10	166	9075	BR-600	1152	
SVDSC20	109	3327/614		1123	
SVDVD1121	177	3100/519	300	1122	
SVDVD1223	116	3311	504A	1104	
SVI-LA3300	1005	3723	IC-42		
SVI-ST020F	1025		722		
SVI-STK020F	1025		722		
SVM11020	141A	3092	ZD-11.5		
SVM1105	141A	3092	ZD-11.5		
SVM111	141A	3092	ZD-11.5		
SVM601	137A	3058	ZD-6.2	561	
SVM6010	137A	3058	ZD-6.2	561	
SVM6011	137A	3058	ZD-6.2	561	
SVM602	137A	3058	ZD-6.2	561	
SVM6020	137A	3058	ZD-6.2	561	
SVM6021	137A	3058	ZD-6.2	561	
SVM605	137A	3058	ZD-6.2	561	
SVM61	137A	3058	ZD-6.2	561	
SVT200-10					2N6306
SVT250-10					2N6306
SVT250-5					2N5838
SVT300-10					2N6307
SVT300-5					2N6542
SVT350-3					2N6545
SVT350-5					2N5840
SVT400-3					2N6545
SVT400-5					2N6543
SVT450-3					2N6545
SVT450-5					2N6543
SVT6000	98				MJ10004
SVT6001	98				MJ10004
SVT6002	98				MJ10005
SVT6060	98				MJ10004
SVT6061	98				MJ10004
SVT6062	98				MJ10005
SVT6251	97				MJ10006
SVT6252	97				MJ10006
SVT6253					MJ10007
SW-05	5800	9003	504A	1142	
SW-05-005	116	3031A	504A	1104	
SW-05-01		3016		1102	
SW-05-02	116	3311	504A	1104	
SW-05-A				1103	
SW-05-B				1104	
SW-05-D				1114	
SW-05V	116	3016	504A	1104	
SW-1	156	3051	512		
SW-1-01	156	3051	512	1143	
SW-1-02	156		512	1143	
SW-1-03				1103	
SW-1-04				1103	
SW-1-06				1104	
SW-1-10				1114	
SW-1A	116	3017B/117	504A	1104	
SW0.5A	116		504A	1104	
SW01	142A	3062	ZD-12	563	
SW05	116	3848	504A	1104	
SW05-01	116		504A	1104	
SW05-02	116		504A	1104	
SW0501		3031A	504A	1104	
SW05A	116	3016	504A	1104	
SW05B	116	3017B/117	504A	1104	
SW05C	125	3032A	510,531	1114	
SW05D	125	3033	510,531	1114	
SW05S	116	3016	504A	1104	
SW05SS	116	3016	504A	1104	
SW05V	116		504A	1104	
SW1	156		512		
SW1800-2F					MC1800F
SW1800-2M					MC1800P
SW1800-2P					MC1800L
SW1801-2F					MC1801F
SW1801-2M					MC1801P
SW1801-2P					MC1801L
SW1802-2F					MC1802F
SW1802-2M					MC1802P
SW1802-2P					MC1802L
SW1805-2F					MC1805F
SW1805-2M					MC1805P
SW1805-2P					MC1805L
SW1806-2F					MC1806F
SW1806-2M					MC1806P

Industry Standard No.	ECG	SK	GE	RS 276-	MOTOR.
SW1806-2P					MC1806L
SW1807-2F					MC1807F
SW1807-2M					MC1807P
SW1807-2P					MC1807L
SW1808-2F					MC1808F
SW1808-2M	9808				MC1808P
SW1808-2P	9808				MC1807L
SW1810-2F					MC1810F
SW1810-2M					MC1810P
SW1810-2P					MC1810L
SW1812-2F					MC1812F
SW1812-2M					MC1812P
SW1812-2P					MC1812L
SW1814-2F					MC1814F
SW1814-2M					MC1814P
SW1814-2P					MC1814L
SW1900-1F					MC1900F
SW1900-1P					MC1900L
SW1901-1F					MC1901F
SW1901-1P					MC1901L
SW1902-1F					MC1902F
SW1902-1P					MC1902L
SW1905-1F					MC1905F
SW1905-1P					MC1905L
SW1906-1F					MC1906F
SW1906-1P					MC1906L
SW1907-1F					MC1907F
SW1907-1P					MC1907L
SW1908-1F					MC1908F
SW1908-1P					MC1908L
SW1910-1F					MC1910F
SW1910-1P					MC1910L
SW1912-1F					MC1912F
SW1912-1P					MC1912L
SW1914-1F					MC1914F
SW1914-1P					MC1914L
SW1C	125	3032A	510,531	1114	
SW1D	125	3033	510,531	1114	
SW4001		4001	4001	2401	
SW4002		4002	4002		
SW4011		4011	4011	2411	
SW4012		4012	4012	2412	
SW4013		4013	4013	2413	
SW4015		4015	4015		
SW4016		4016	4016		
SW4017		4017	4017	2417	
SW4019		4019	4019		
SW4020			4020	2420	
SW4021			4021	2421	
SW4023		4023	4023	2423	
SW4024		4024	4024		
SW4025		4025	4025		
SW4027		4027	4027	2427	
SW4030		4030	4030		
SW4049		4049	4049	2449	
SW4050		4050	4050	2450	
SW705-1F					MC953F
SW705-1P					MC953L
SW705-2F					MC953F
SW705-2M	9093				MC853P
SW705-2P	9093				MC853L
SW706-1F					MC952F
SW706-1P					MC952L
SW706-2F					MC952F
SW706-2M	9099				MC852P
SW706-2P	9099				MC852L
SW708-1F					MC956F
SW708-1P					MC956L
SW708-2F					MC956F
SW708-2M	9094				MC856P
SW708-2P	9094				MC856L
SW709-1F					MC955F
SW709-1P					MC955L
SW709-2F					MC944F
SW709-2M	9097				MC855P
SW709-2P	9097				MC855L
SW930-1F					MC930F
SW930-1P					MC930L
SW930-2F					MC930F
SW930-2M	9930				MC830P
SW930-2P	9930				MC830L
SW932-1F					MC932F
SW932-1P					MC932L
SW932-2F					MC932F
SW932-2M	9932				MC832P
SW932-2P	9932				MC832L
SW933-1F					MC933F
SW933-1P					MC933L
SW933-2F					MC933F
SW933-2M	9933				MC833P
SW933-2P	9933				MC833L
SW935-1F					MC940F
SW935-1P					MC940L
SW935-2F					MC940F
SW935-2M	9935				MC840P
SW935-2P	9935				MC840L
SW936-1F					MC936F
SW936-1P					MC936L
SW936-2F					MC936F
SW936-2M	9936				MC836P
SW936-2P	9936				MC836L
SW937-1F					MC937F
SW937-1P					MC937L
SW937-2F					MC837L
SW937-2M	9937				MC837P
SW937-2P	9937				MC837L
SW944-1F					MC944F
SW944-1P					MC944L
SW944-2F					MC944F
SW944-2M	9944				MC844P
SW944-2P	9944				MC844L
SW945-1F					MC945F
SW945-1P					MC945L
SW945-2F					MC945F
SW945-2M	9945				MC845P
SW945-2P	9945				MC845L
SW946-1F					MC946F
SW946-1P					MC946L
SW946-2F					MC946F
SW946-2M	9946				MC846P
SW946-2P	9946				MC846L
SW948-1F					MC948F
SW948-1P					MC948L
SW948-2F					MC948F
SW948-2M	9948				MC848P
SW948-2P	9948				MC848L
SW949-1F					MC949F
SW949-1P					MC949L
SW949-2F					MC949F
SW949-2M	9949				MC849P
SW949-2P	9949				MC849L
SW950-1F					MC950F
SW950-1P					MC950L
SW950-2F					MC950F

Industry Standard No.	ECG	SK	GE	RS 276-	MOTOR.
SW950-2M	9950				MC850P
SW950-2P	9950				MC850L
SW951-1F					MC951F
SW951-1P					MC951L
SW951-2F					MC951F
SW951-2M	9951				MC851P
SW951-2P	9951				MC851L
SW957-1F					MC957F
SW957-1P					MC957L
SW957-2F					MC957F
SW957-2M	9157				MC857P
SW957-2P	9157				MC857L
SW958-1F					MC958F
SW958-1P					MC958L
SW958-2F					MC958F
SW958-2M	9158				MC858P
SW958-2P	9158				MC858L
SW961-1F					MC961F
SW961-1P					MC961L
SW961-2F					MC961F
SW961-2M	9961				MC861P
SW961-2P	9961				MC861L
SW962-1F					MC962F
SW962-1P					MC962L
SW962-2F					MC962F
SW962-2M					MC862P
SW962-2P	9962				MC862L
SW963-1F					MC963F
SW963-1P					MC963L
SW963-2F					MC963F
SW963-2M	9963				MC863P
SW963-2P	9963				MC863L
SWC	125		510,531	1114	
SWD	125		510,531	1114	
SW0.5A			504A	1104	
SWT1728	126		50	2004	
SWT3588	160		50	2004	
SX-1.5-01				1143	
SX-1.5-02				1143	
SX-1.5-04				1144	
SX-3825	108	3018	86	2038	
SX-58Y	312			2028	
SX-642	116	3016	504A	1104	
SX-990	109	3091		1123	
SX100	5156A	3422			
SX110	5157A	3423			
SX120	5158A	3424			1M120ZS5
SX30	5141A				1M30ZS5
SX3001	108	3039/316	86	2038	
SX33	5142A	3408	5ZD-33		
SX36	5143A	3409			
SX3702	159	3025/129	82	2032	
SX3702A	159	3114/290	82	2032	
SX3709	123A	3122	20	2051	
SX3711	123A	3122	20	2051	
SX3819	312	3448	FET-1	2035	
SX3825	108	3018	86	2038	
SX3826	229	3018	61	2038	
SX3827	108		86	2038	
SX39	5144A	3410			
SX408	229	3018	61	2038	
SX43	5145A	3411			
SX47	5146A	3412			
SX51	5147A	3413			
SX55	123A	3122	20	2051	
SX56	5148A	3414			
SX60M	154	3045/225	40	2012	
SX61	129	3025	244	2027	
SX61M	159	3114/290	82	2032	
SX61MA	159	3114/290	82	2032	
SX62	5150A	3416			
SX623	116	3016	504A	1104	
SX631	116	3016	504A	1104	
SX633	116		504A	1104	
SX641	116	3311	504A	1104	
SX642	116		504A	1104	
SX643	116	3311	504A	1104	
SX644	116	3031A	504A	1104	
SX645	116	3031A	504A	1104	
SX68	5151A	3417			
SX75	5152A	3418			
SX751	5872		5032		
SX752	5874		5032		
SX753	5878		5036		
SX754	5878		5036		
SX780	177	3100/519	300	1122	
SX82	5153A	3419			
SX91	5155A	3421			
SY-23				1103	
SY-24				1103	
SY101			6	2002	
SY2114	2114				MCM2114
SY2147	2147				MCM2147
SY2316B					MCM68A316E
SY2332					MCM68A332
SY2364					MCM68365
SY2716					MCM2716
SY5101	65101				MCM5101
SY61583	176	3123	80		
SYL-101	101	3861	8	2002	
SYL-102	101	3861	8	2002	
SYL-103	103A		59	2002	
SYL-104	103A		59	2002	
SYL-105	100	3005	1	2007	
SYL-106	100	3005	1	2007	
SYL-107	102A	3004	53	2007	
SYL-108	102A	3004	53	2007	
SYL-1182	123A	3020/123	20	2051	
SYL-1297	101	3861	8	2002	
SYL-1310	101	3861	8	2002	
SYL-1311	101	3861	8	2002	
SYL-1329	103A		59	2002	
SYL-1396	103A		59	2002	
SYL-152	103A	3010	59	2002	
SYL-1524	103A		59	2002	
SYL-1583	102A	3004	53	2007	
SYL-160	100	3005	1	2007	
SYL-1608	100	3005	1	2007	
SYL-1668	102A	3004	53	2007	
SYL-1987	101	3861	8	2002	
SYL-2130	101	3861	8	2002	
SYL-2131	101	3861	8	2002	
SYL-2132	101	3861	8	2002	
SYL-2134	103A		59	2002	
SYL-2135	103A		59	2002	
SYL-2136	103A		59	2002	
SYL-2245	101	3861	8	2002	
SYL-2246	101	3861	8	2002	
SYL-2248	100	3005	1	2007	
SYL-2249	100	3005	1	2007	
SYL-2250	100	3005	1	2007	
SYL-2300	108	3019	86	2038	
SYL-4131	108	3019	86	2038	
SYL-4315	103A		59	2002	

Industry Standard No.	ECG	SK	GE	RS 276-	MOTOR.
SYL-4339		3861	8	2002	
SYL101	101		8	2002	
SYL102	101		8	2002	
SYL103	103	3862	8	2002	
SYL104	103	3862	8	2002	
SYL105	100	3123	1	2007	
SYL106	100	3123	1	2007	
SYL107	102	3123	2	2007	
SYL107A	102A	3004	53	2007	
SYL108	102	3123	2	2007	
SYL108A	102A	3004	53	2007	
SYL109	121	3009	239	2006	
SYL1182	123A	3444	20	2051	
SYL1279	101	3011	8	2002	
SYL128	110MP	3709		1123(2)	
SYL1297	103	3862	8	2002	
SYL1310	101		8	2002	
SYL1311	101		8	2002	
SYL1312	101	3011	8	2002	
SYL1313	101	3011	8	2002	
SYL1326	101	3011	8	2002	
SYL1327	101	3011	8	2002	
SYL1329	103	3862	8	2002	
SYL1380	101	3011	8	2002	
SYL1396	103	3862	8	2002	
SYL1408	101	3011	8	2002	
SYL1430	102	3722	2	2007	
SYL1454	101	3011	8	2002	
SYL1468	103A		59	2002	
SYL152	123		20	2051	
SYL1524	103	3862	8	2002	
SYL1535	102	3722	2	2007	
SYL1536	103	3862	8	2002	
SYL1537	101	3011	8	2002	
SYL1538	103	3862	8	2002	
SYL1539	103	3862	8	2002	
SYL1547	103	3862	8	2002	
SYL1583	102	3722	2	2007	
SYL1583A	102A	3004	53	2007	
SYL1588	100	3005	1	2007	
SYL1591	101	3011	8	2002	
SYL160	100	3123	1	2007	
SYL1608	100	3123	1	2007	
SYL1617	101	3011	8	2002	
SYL1655	102	3004	2	2007	
SYL1655A	102A	3004	53	2007	
SYL1665	102	3004	2	2007	
SYL1665A	102A	3004	53	2007	
SYL1668	102	3123	2	2007	
SYL1668A	102A	3004	53	2007	
SYL1690	100	3005	1	2007	
SYL1697	100	3005	1	2007	
SYL1717	100	3005	1	2007	
SYL1750	101	3011	8	2002	
SYL1941	101		8	2002	
SYL1987	101		8	2002	
SYL2120	100	3005	1	2007	
SYL2130	101		8	2002	
SYL2131	101		8	2002	
SYL2132	101		8	2002	
SYL2134	103	3862	8	2002	
SYL2135	103	3862	8	2002	
SYL2136	103	3862	8	2002	
SYL2189	160		245	2004	
SYL2245	101		8	2002	
SYL2246	101		8	2002	
SYL2247	100	3005	1	2007	
SYL2248	102	3722	2	2007	
SYL2248A	102A	3004	53	2007	
SYL2249	102	3722	2	2007	
SYL2249A	102A	3004	53	2007	
SYL2250	100	3721	1	2007	
SYL2300	102	3722	2	2007	
SYL2300A	102A	3004	53	2007	
SYL2650	103	3862	8	2002	
SYL3460	123A	3444	20	2051	
SYL3613	102	3004	2	2007	
SYL3613A	102A	3004	53	2007	
SYL4131	108		86	2038	
SYL4275	159		82	2032	
SYL4280	128	3024	243	2030	
SYL4315	103	3862	8	2002	
SYL4339	101		8	2002	
SYL792		3861	8	2002	
SZ-11			ZD-11	563	
SZ-150	145A	3063	ZD-15	564	
SZ-15B	145A	3063	ZD-15	564	
SZ-200-10			ZD-10	562	
SZ-200-11			ZD-11	563	
SZ-200-12			ZD-12	563	
SZ-200-13	143A	3750	ZD-13	563	
SZ-200-14			ZD-14	564	
SZ-200-15	145A	3063	ZD-15	564	
SZ-200-15A	145A	3063	ZD-15	564	
SZ-200-16			ZD-16	564	
SZ-200-8	139A		ZD-9.1	562	
SZ-200-9			ZD-9.1	562	
SZ-200-9V	139A	3060	ZD-9.1	562	
SZ-7	138A	3059	ZD-7.5		
SZ-9	139A	3060	ZD-9.1	562	
SZ-AW01-11	5074A	3139	ZD-11	563	
SZ-BZ-120	142A	3062			
SZ-EQB01-12A	142A	3062	ZD-12	563	
SZ-RD11FB	5074A	3139	ZD-11	563	
SZ-RD12EB	5021A	3787	ZD-12	563	
SZ-RD12FB	142A	3062	ZD-12	563	
SZ-RD6.2E	5013A	3779			
SZ-RD6.2EB	137A	3058	ZD-6.8	561	
SZ-UZ-12B	5021A	3787	ZD-12	563	
SZ-UZ6.2B	5013A	3787/5021A			
SZ-UZP-11B	5074A	3139			
SZ-WZ-063	5013A	3779			
SZ-WZ-120	5021A	3787			
SZ-YZ063	137A	3779/5013A	ZD-6.2	561	
SZ10	140A	3061	ZD-10	562	
SZ10C	5125A	3391			
SZ11	5074A	3092/141A	ZD-11.5	563	
SZ11.0	141A	3092	ZD-11.5		
SZ11763	151A		ZD-110		
SZ119	5081A		ZD-24		
SZ12	142A	3062	ZD-12	563	
SZ12.0	142A	3062	ZD-12	563	
SZ1200	148A	3096	ZD-55		
SZ12C	5127A	3393	5ZD-12		
SZ13	143A	3750	ZD-13	563	
SZ13.0	143A	3750	ZD-13	563	
SZ13447	5009A	3775			
SZ13448	5014A	3780			
SZ13B	143A	3750			
SZ14	144A	3094	ZD-14	564	
SZ14026	5077A	3752			
SZ15	145A	3063	ZD-15	564	
SZ15.0	145A	3063	ZD-15	564	
SZ150	145A	3063	ZD-15	564	

Industry Standard No.	ECG	SK	GE	RS 276-	MOTOR.
SZ15C	5130A	3396	5ZD-15		
SZ16			ZD-16	564	
SZ16.0,A					1M16ZS10,5
SZ16B	5075A	3751			
SZ18	5027A	3793	ZD-18		
SZ18B	5077A	3752			
SZ18C	5133A	3793/5027A	5ZD-18		
SZ2.7	5063A	3837			
SZ20	5029A	3795	ZD-20		
SZ200	135A	3056	ZD-5.1		
SZ200-5	135A	3056	ZD-5.1		
SZ200-8.2	5072A	3136	ZD-8.2	562	
SZ22C	5136A	3402	5ZD-22		
SZ27C	5139A	3405	5ZD-27		
SZ3.6	134A	3055	ZD-3.6		
SZ33C	5142A	3408			
SZ4.7	135A	3056	ZD-5.1		
SZ5	135A	3056	ZD-5.1		
SZ5.1	135A	3056	ZD-5.1		
SZ5.6	136A	3057	ZD-5.6	561	
SZ56A	5117A	3383			
SZ6.2	137A	3058	ZD-6.2	561	
SZ6.2A	137A	3058	ZD-6.2	561	
SZ62A	5119A	3385			
SZ671-B	142A	3062	ZD-12	563	
SZ671-G	142A	3062	ZD-12	563	
SZ671-O	5074A		ZD-11.0	563	
SZ671-W	5074A		ZD-11.0	563	
SZ7	5014A	3780	ZD-6.8	561	
SZ75A	5121A	3387	5ZD-7.5		
SZ9	139A	3060	ZD-9.1	562	
SZ9.1	139A	3060	ZD-9.1	562	
SZ91A	5124A	3390	5ZD-9.1		
SZ961-B	145A		ZD-15	564	
SZ961-R	144A	3094	ZD-14	564	
SZ961-V	143A	3750	ZD-13	563	
SZ961-Y	144A		ZD-14	564	
SZA-13			ZD-13	563	
SZA-6			ZD-6.0	561	
SZA100	156	3051	512		
SZA13	143A	3750			
SZC9	139A	3060	ZD-9.1	562	
SZL9	140A	3061	ZD-10	562	
SZP-9			ZD-9.1	562	
SZT-9	139A	3060	ZD-9.1	562	
SZT8	142A	3059/138A	ZD-7.5	563	
SZT9	139A	3060	ZD-9.1	562	
T 112	159			2032	
T-00014	102A	3003	53	2007	
T-0150	116	3016	504A	1104	
T-04689	128		243	2030	
T-065	116	3016	504A	1104	
T-075	116	3016	504A	1104	
T-100	116	3016	504A	1104	
T-1000	125		510,531	1114	
T-1000X	125		510,531	1114	
T-10010	519		514	1122	
T-101	121	3009	239	2006	
T-109	100	3005	1	2007	
T-116	100	3005	1	2007	
T-126	102A	3004	53	2007	
T-127	121	3009	239	2006	
T-129	102A	3004	53	2007	
T-13	116	3016	504A	1104	
T-130	102A	3004	53	2007	
T-131	158		53	2007	
T-1363	126	3008	52	2024	
T-1364	126	3008	52	2024	
T-13G	116	3016	504A	1104	
T-14	116	3016	504A	1104	
T-1416	123A	3020/123	20	2051	
T-1460	126	3008	52	2024	
T-14G	116	3016	504A	1104	
T-152148	100	3005	1	2007	
T-163	160		245	2004	
T-1877	100	3721	1	2007	
T-200	116	3016	504A	1104	
T-2028	160		245	2004	
T-2029	160		245	2004	
T-203	107		11	2015	
T-2030	160		245	2004	
T-2038	100	3721	1	2007	
T-2039	100	3721	1	2007	
T-2040	100	3721	1	2007	
T-2062		3102	511	1114	
T-2091	100	3721	1	2007	
T-2122	102	3722	2	2007	
T-22G	116	3016	504A	1104	
T-23	102A	3004	53	2007	
T-23-71	390		255	2041	
T-235	121	3717	239	2006	
T-2357			21	2034	
T-2439	100	3005	1	2007	
T-2440	100	3005	1	2007	
T-2441	100	3005	1	2007	
T-246	159		82	2032	
T-251	159		82	2032	
T-255	123A		20	2051	
T-256	123A	3020/123	20	2051	
T-26G	116	3016	504A	1104	
T-278	160		245	2004	
T-279	160		245	2004	
T-291	128		243	2030	
T-300	116	3016	504A	1104	
T-3321	102A	3004	53	2007	
T-3322	102A	3004	53	2007	
T-339	128	3024	243	2030	
T-3321	102A		53	2007	
T-3323	102A		53	2007	
T-340	129		244	2027	
T-342	184	3024/128	57	2017	
T-344	184	3190	57	2017	
T-345	185		58	2025	
T-348	126		52	2024	
T-3568			60	2015	
T-39	102A	3004	53	2007	
T-396	185	3191	58	2025	
T-399	123A		20	2051	
T-400	116	3016	504A	1104	
T-450	116	3016	504A	1104	
T-4590	116		504A	1104	
T-46	100	3005	1	2007	
T-47	100	3005	1	2007	
T-48	100	3005	1	2007	
T-481	154	3040	40	2012	
T-482	129	3025	244	2027	
T-483	108	3018	86	2038	
T-484	108	3018	86	2038	
T-486	108	3018	86	2038	
T-50	116	3016	504A	1104	
T-500	116	3016	504A	1104	
T-521482	100	3005	1	2007	
T-52149	100	3005	1	2007	
T-521492	100	3005	1	2007	
T-52150	102A	3004	53	2007	
T-52150Z	102A	3004	53	2007	
T-52151	102A	3004	53	2007	
T-52151Z	102A	3004	53	2007	
T-550	116	3016	504A	1104	
T-600	116	3017B/117	504A	1104	
T-6028	126	3008	52	2024	
T-6029	126	3008	52	2024	
T-6030	126	3008	52	2024	
T-6031	126	3008	52	2024	
T-6032	126	3008	52	2024	
T-650	116	3016	504A	1104	
T-700-709			1N295	1123	
T-750-713			1N34AS	1123	
T-750-714			ZD-9.1	562	
T-78	100	3005	1	2007	
T-81	103A	3010	59	2002	
T-82			1	2007	
T-95	102A	3004	53	2007	
T-99	160	3006	245	2004	
T-E01029D	116		504A	1104	
T-E0137	358		42		
T-E0317	358		42		
T-E1011	116	3017B/117	504A	1104	
T-E1014	109	3090		1123	
T-E1024	116	3017B/117	504A	1104	
T-E1024C	116	3017B/117	504A	1104	
T-E1024D	116	3017B/117	504A	1104	
T-E1029	507	3016	504A	1104	
T-E1031	109	3088		1123	
T-E1042	166	9075	504A	1152	
T-E1050	116	3017B/117	504A	1104	
T-E1061	113A		6GC1		
T-E1061A	113A		6GC1		
T-E1064	116	3016	504A	1104	
T-E1068	142A	3062	ZD-12	563	
T-E1077	142A	3062	ZD-12	563	
T-E1078	116	3016	504A	1104	
T-E1078A	116	3016	504A	1104	
T-E1080	116	3017B/117	504A	1104	
T-E1086	113A		6GC1		
T-E1086A	113A	3119/113	6GC1		
T-E1088	116(3)		504A	1104	
T-E1089	116	3016	504A	1104	
T-E1090	116	3017B/117	504A	1104	
T-E1095	118	3066	CR-1		
T-E1097	116	3017B/117	504A	1104	
T-E1098	177		300	1122	
T-E1102	116		504A	1104	
T-E1102A	116	3031A	504A	1104	
T-E1105	110MP	3709		1123(2)	
T-E1106	142A	3093	ZD-12	563	
T-E1107	119	3109	CR-2		
T-E1108	507	3031A	504A	1104	
T-E1116			1N34AS	1123	
T-E1118	177	3100/519	300	1122	
T-E1119	177	3100/519	300	1122	
T-E1121	177	3100/519	300	1122	
T-E1124	116	3031A	504A	1104	
T-E1133	116	3311	504A	1104	
T-E1138	116	3100/519	504A	1104	
T-E1140	142A	3062	ZD-12	563	
T-E1144	116	3043	504A	1104	
T-E1145	506	3120/114	511	1114	
T-E1146	506		511	1114	
T-E1148	116		504A	1104	
T-E1153	506	3043	511	1114	
T-E1155	116	3311	504A	1104	
T-E1157	116	3032A	504A	1104	
T-E1171	116	3032A	504A	1104	
T-E1176	116	3119/113	504A	1104	
T-E1177	110MP	3087	1N34AS	1123(2)	
T-G1138			1N34AS	1123	
T-G5033			18	2030	
T-G5055			18	2030	
T-H1S557		3016	504A	1104	
T-H2SC313	108	3019	86	2038	
T-H2SC387	108	3019	86	2038	
T-H2SC536	123A	3444	20	2051	
T-H2SC693	123A	3444	20	2051	
T-H2SC715	123A	3444	20	2051	
T-HS6105			504A	1104	
T-HSG105		3016	504A	1104	
T-HU60U		3004	21	2034	
T-Q5019	163A	3439	36		
T-Q5020		3005	1	2007	
T-Q5021		3006	51	2004	
T-Q5022		3006	51	2004	
T-Q5023			2	2007	
T-Q5025		3004	2	2007	
T-Q5026		3004	2	2007	
T-Q5027		3004	53	2007	
T-Q5028		3034	16	2006	
T-Q5030		3035	25		
T-Q5031		3011	8	2002	
T-Q5032		3011	18	2030	
T-Q5034		3006	51	2004	
T-Q5035		3006	51	2004	
T-Q5036		3009	16	2006	
T-Q5038		3006	51	2004	
T-Q5039		3010	8	2002	
T-Q5049		3018	11	2015	
T-Q5050		3010	8	2002	
T-Q5053	123A	3444	20	2051	
T-Q5053C	123A	3444	20	2051	
T-Q5055	108	3117	86	2038	
T-Q5057	124	3021	12		
T-Q5063		3020	18	2030	
T-Q5071	161	3018	39	2015	
T-Q5073	123A	3444	20	2051	
T-Q5075	124	3021	12		
T-Q5077	159	3114/290	82	2032	
T-Q5078	233	3018	210	2009	
T-Q5079	107	3018	11	2015	
T-Q5080	175	3021/124	246	2020	
T-Q5081	128	3024	243	2030	
T-Q5082	154	3045/225	40	2012	
T-Q5083	164	3079	37		
T-Q5084	165	3111	38		
T-Q5086	161	3018	39	2015	
T-Q5087	159	3114/290	82	2032	
T-Q5093	199	3124/289	62	2010	
T-Q5099	128	3024	243	2030	
T-Q5104	124	3021	12		
T-Q5105	130	3027	14	2041	
T-Q5106	107	3018	11	2015	
T-Q7037	1004	3102/710			
T-R9S	139A	3059/138A	ZD-9.1	562	
T0-101	100	3005	1	2007	
T0-102	100	3005	1	2007	
T00014	102A		53	2007	
T0003	102A	3003	53	2005	
T0004	102A	3003	53	2005	
T0005	102A	3003	53	2005	

Industry Standard No.	ECG	SK	GE	RS 276-	MOTOR.
T0012	102A	3003	53	2007	
T0014	102A		53	2007	
T0015	102A	3003	53	2007	
T0031	102A	3123	53	2007	
T0033	102A	3003	53	2007	
T0038	102A	3004	53	2007	
T0039	102A	3004	53	2007	
T0040	102A	3004	53	2007	
T0041	102A	3004	53	2007	
T0051	102A	3123	53	2007	
T01-013	123A	3444	20	2051	
T01-014	123A	3444	20	2051	
T01-022	128		243	2030	
T01-023	159		82	2032	
T01-030	152	3893	66	2048	
T01-044	312			2028	
T01-047	199	3245	62	2010	
T01-101	123A	3444	20	2051	
T01-104	123A	3444	20	2051	
T01-105	123A	3444	20	2051	
T01003	314	3898			
T0101	100	3721	1	2007	
T0102	100	3721	1	2007	
T0150	116		504A	1104	
T02003	314	3898			
T0253	314	3898			
T03003	314	3898			
T03323	100	3721	1	2007	
T04003	314	3898			
T0503	314	3898			
T065	116		504A	1104	
T075	116		504A	1104	
TOA1748V	1171	3565			
TOA2748V	1171	3565			
T1-1A6	123A	3444	20	2051	
T1-503	159	3114/290	82	2032	
T1-503A	159	3114/290	82	2032	
T1-741	312	3448	FET-1	2035	
T1-743	159	3114/290	82	2032	
T1-743A	159	3114/290	82	2032	
T1-744	159	3114/290	82	2032	
T1-744A	159	3114/290	82	2032	
T1-752	159	3114/290	82	2032	
T1-752A	159	3114/290	82	2032	
T1-906	159	3114/290	82	2032	
T100	102A		53	2007	
T1000	158	3033	53	2007	1N4007
T1000X	125		510,531	1114	
T1001	102A	3004	53	2007	
T10010	102A	3004	53	2007	
T1002	102A		53	2007	
T1002A	102A		53	2007	
T1003	102A	3005	53	2007	
T1003-521	108	3039/316	86	2038	
T1003521	108		86	2038	
T1004	102A		53	2007	
T1004671	123A	3444	20	2051	
T1005	102A	3004	53	2007	
T1006	102A	3004	53	2007	
T1007	102A	3024/128	53	2007	
T1007(ZENITH)	199		62	2010	
T1008	102A	3004	53	2007	
T1008-834	123A	3444	20	2051	
T10085	102A		53	2007	
T1008834	123A	3444	20	2051	
T1009	102A		53	2007	
T101	102A	3717/121	53	2007	
T1010	102A	3123	53	2007	
T1011	160	3005	245	2004	
T1012	160	3005	245	2004	
T1013	102A	3003	53	2007	
T10144	116	3031A	504A	1104	
T10175	116	3311	504A	1104	
T10185	116		504A	1104	
T10195	166	9075	BR-600	1152	
T1023	102A	3123	53	2007	
T1028	160	3008	245	2004	
T102A	102A		53	2007	
T1032	160	3006	245	2004	
T1033	160	3007	245	2004	
T1034	160	3008	245	2004	
T1036	102A		53	2007	
T1037	102A		53	2007	
T1038	160	3008	245	2004	
T1040	104	3009	16	2006	
T1041	104	3009	16	2006	
T1042	102A	3004	53	2007	
T1043	102A	3004	53	2007	
T10453	116	3017B/117	504A	1104	
T1046	102A	3004	53	2007	
T1047	102A	3004	53	2007	
T1076	102A		53	2007	
T108	102A		53	2007	
T1085	116		504A	1104	
T109	102A		53	2007	
T11	109	3087		1123	
T112			82	2032	
T1145A0					2N1595
T1145A1					2N1596
T1145A2					2N1597
T1145A3					2N1598
T1145A4					2N1599
T116	102A	3008	53	2007	
T11618	102A	3005	53	2007	
T1166	160	3008	245	2004	
T1167	121	3009	239	2006	
T1168	121		239	2006	
T12	109	3100/519		1123	
T12-1	177	3100/519	300	1122	
T12-2	177	3100/519	300	1122	
T1202	102A	3039/316	53	2007	
T1202(GE)	161	3716	39	2015	
T1203	102A		53	2007	
T1208	312	3448	FET-1	2035	
T1224	160	3006	245	2004	
T1225	160	3006	245	2004	
T1232	160	3005	245	2004	
T1233	160	3005	245	2004	
T1250	160	3006	245	2004	
T1251	100	3721	1	2007	
T126	102A		53	2007	
T127	102A		53	2007	
T1275	129	3025	244	2027	
T1276	129	3025	244	2027	
T1289	100	3005	1	2007	
T129	102A	3123	53	2007	
T1291	100	3005	1	2007	
T1298	160	3005	245	2004	
T1299	160	3005	245	2004	
T12A	177	3100/519	300	1122	
T12A6F					SPECIAL
T12B	177	3100/519	300	1122	
T12C	177	3100/519	300	1122	
T12G	109	3087		1123	
T12N100COB	5542	6642			
T13	109	3091		1123	
T130	102A	3123	53	2007	
T1300	102A	3123	53	2007	
T13000	102A	3004	53	2007	
T13015	128	3024	243	2030	
T13029	121	3717	239	2006	
T1305	160	3005	245	2004	
T1306	160	3005	245	2004	
T131	158	3721/100	1	2007	
T1310	102A	3004	53	2007	
T1312	100	3005	1	2007	
T1314	160	3008	245	2004	
T1322	100	3005	1	2007	
T1326	100	3005	1	2007	
T1327	102A	3005	53	2005	
T1328	102A	3005	53	2005	
T1334	102A	3005	53	2007	
T1340A31	123A	3124/289	20	2051	
T1340A3H	128	3024	243	2030	
T1340A3I	123A	3444	20	2051	
T1340A3J	123A	3444	20	2051	
T1340A3K	123A	3444	20	2051	
T1341A3K	199	3122	62	2010	
T1342	102A		53	2005	
T1346	102A	3005	53	2007	
T1352	102A	3004	53	2007	
T1363	102A		53	2007	
T1364	102A		53	2007	
T1366	104	3009	16	2006	
T1366A	104	3009	16	2006	
T1367	104	3009	16	2006	
T1367A	104	3009	16	2006	
T1368	104	3009	16	2006	
T1368A	104	3009	16	2006	
T1369	104	3009	16	2006	
T1369A	104		16	2006	
T1370	104	3009	16	2006	
T1370A	104	3009	16	2006	
T1387	160	3006	245	2004	
T1388	126	3006/160	245	2004	
T1389	160	3006	245	2004	
T139	104	3020/123	16	2006	
T1390	160	3006	245	2004	
T1391	160	3006	245	2004	
T13G	109	3091		1123	
T14	109	3091		1123	
T1400	160	3006	245	2004	
T1401	160	3006	245	2004	
T1402	199		245	2004	
T1403	160	3006	245	2004	
T1408	229	3018	61	2038	
T1413	123A	3444	20	2051	
T1414	123A	3444	20	2051	
T1415	123A	3444	20	2051	
T1416	123A	3444	20	2051	
T1417	123A	3444	20	2051	
T142	121	3009	239	2006	
T143	123A	3444	20	2051	
T1450	116		504A	1104	
T1454	160	3005	245	2004	
T1459	160	3005	245	2004	
T1460	126		52	2024	
T1461	160		245	2004	
T1474	100	3005	1	2007	
T1486	152	3893	66	2048	
T1487	152	3893	66	2048	
T1495	123A	3444	20	2051	
T14G	109	3091		1123	
T15.1N100COB	5542	6642			
T151	177		300	1122	
T1510	100	3005	1	2007	
T152	177		300	1122	
T152148	100	3005	1	2007	
T1524	126	3008	52	2024	
T1524BRN	126	3006/160	52	2024	
T1524BRN/RED	126		52	2024	
T153	177		300	1122	
T154	177		300	1122	
T1546	102A	3004	53	2007	
T1548	160	3008	245	2004	
T155	177		300	1122	
T1559	102A		53	2007	
T156	116		504A	1104	
T157	123A	3444	20	2051	
T1573	102A	3004	53	2007	
T1574	102A	3004	53	2007	
T1577	102A	3004	53	2007	
T158	123A	3444	20	2051	
T1583	102A	3004	53	2007	
T159	116		504A	1104	
T1593	102A		53	2007	
T1594	102A		53	2007	
T1595	102A		53	2007	
T1596	102A		53	2007	
T1597	102A		53	2007	
T1598	102A		53	2007	
T1599	102A		53	2007	
T16	177	3100/519	300	1122	
T160	102A		53	2007	
T1601	121	3009	239	2006	
T1602	159	3009	239	2006	
T1618	160	3008	245	2004	
T163	160		245	2004	
T164213	128	3024	243	2030	
T1642B	123A	3444	20	2051	
T1654	126	3006/160	52	2024	
T1654BLU	126		52	2024	
T1657	160	3006	245	2004	
T1690	160	3006	245	2004	
T1691	160	3006	245	2004	
T1692	160	3006	245	2004	
T17	109	3087		1123	
T17-A	138A	3059	ZD-7.5		
T170	123A	3444	20	2051	
T1706	128	3024	243	2030	
T1706A	128	3024	243	2030	
T1706B	128	3024	243	2030	
T1706C	128	3024	243	2030	
T171	123A	3444	20	2051	
T1731	222	3065	FET-4	2036	
T1737	160	3006	245	2004	
T1738	160	3006	245	2004	
T1740	102A	3003	53	2007	
T1746	123A	3444	20	2051	
T1746A	123A	3444	20	2051	
T1746B	123A	3444	20	2051	
T1746C	123A	3444	20	2051	
T1748	123A	3444	20	2051	
T1748A	123A	3444	20	2051	
T1748B	123A	3444	20	2051	
T1748C	123A	3444	20	2051	
T1788	160	3008	245	2004	
T17A	138A		ZD-7.5		

Industry Standard No.	ECG	SK	GE	RS 276-	MOTOR.
T18	109	3087		1123	
T1802	123A	3444	20	2051	
T1802A	123A	3444	20	2051	
T1802B	123A	3444	20	2051	
T1804	123A	3444	20	2051	
T1805	123A	3444	20	2051	
T1808	129	3025	244	2027	
T1808A	129	3025	244	2027	
T1808B	129	3025	244	2027	
T1808C	129	3025	244	2027	
T1808D	129	3025	244	2027	
T1808E	129	3025	244	2027	
T1810	123A		20	2051	
T1810B	123A	3020/123	20	2051	
T1811	128		243	2030	
T1811E	128	3024	243	2030	
T1811G	128	3024	243	2030	
T1814	160	3006	245	2004	
T1831	160	3007	245	2004	
T185	123A	3444	20	2051	
T1877	100	3005	1	2007	
T1902	100	3005	1	2007	
T1903	102A	3004	53	2007	
T1904	102A	3005	53	2007	
T1909	123A	3018	20	2051	
T1930				2005	
T1961	102A	3005	53	2007	
T1A	704	3023	IC-205		
T1B	703A	3157	IC-12		
T1C116A	5462	3574			C122A1
T1C116B	5463	3574			C122B1
T1C116C	5465	3573/5466			C122C1
T1C116D	5465	3573/5466			C122D1
T1C116E		3573			C122E1
T1C116F	5461	3574			C122F1
T1C116M		3573			C122M1
T1C126A	5462	3574			2N6395
T1C126B	5463	3574			2N6396
T1C126C	5465				MCR220-5
T1C126D	5465				2N6397
T1C126E	5466				MCR220-7
T1C126F	5461	3574			2N6394
T1C126M	5466				2N6398
T1C20					2N5567
T1C205A	5650				T2301PA
T1C205B	5651				T2301PB
T1C205D	5652				T2301PD
T1C206A					2N6070A
T1C206B					2N6071A
T1C206D					2N6073A
T1C21					2N5568
T1C215A					2N6070A
T1C215B					2N6071A
T1C215D					2N6073A
T1C216A					MAC228A3
T1C216B					MAC228A4
T1C216D					MAC228A6
T1C22	5667A				2N5569
T1C226B	56004	3658			
T1C226D	56006	3659			
T1C23	5667A				2N5570
T1C236B	56004	3658			
T1C236D	56006	3659			
T1C35					SPECIAL
T1C36					SPECIAL
T1C39A					2N6238
T1C39B					2N6239
T1C39C					MCR100-5
T1C39D					2N6240
T1C39E					MCR100-7
T1C39F					2N6237
T1C39Y					2N6236
T1C44	5400	3950			2N5060
T1C45	5401	3597/5455			2N5061
T1C46	5402	3597/5455			2N5062
T1C47	5404	3627			2N5064
T1C54	6407	9083A			
T1C57	6406	9082A			
T1C60	5400	3950			2N5060
T1C61	5401				2N5061
T1C62	5402				2N5062
T1C63	5404	3627			2N5063
T1C64	5404	3627			2N5064
T1C67					SPECIAL
T1C68					SPECIAL
T1E	722	3159/718	IC-9		
T1F	748	3236			
T1G	109			1123	
T1H	710	3102	IC-89		
T1J	718	3159	IC-8		
T1J6G	116	3017B/117	504A	1104	
T1K	798	3216	IC-234		
T1M	717		IC-209		
T1N	718	3159	IC-8		
T1P125	262	3221			
T1P140	270	3180/263			
T1P27	157	3747			
T1P29	152	3041	66	2048	
T1P2955	393		74	2043	
T1P29B	291	3220/198			
T1P29C	291	3220/198			
T1P29X	152		66	2048	
T1P3055	392		14	2041	
T1P31	152	3192/186	57	2017	
T1P31A	291		57	2017	
T1P33	390		55	2020	
T1P33A	390		55	2020	
T1P34	391	3189A/183			
T1P34A	391	3189A/183			
T1P34B	391	3189A/183			
T1P36A	393		56	2027	
T1P41	196	3188A/182			
T1P41A	196	3188A/182			
T1P41B	196	3188A/182			
T1P42	197	3189A/183			
T1P42A	197	3189A/183			
T1P42B	197	3189A/183			
T1P47	198	3220			
T1P48	198	3220			
T1P49	198	3220			
T1P645	271	3183A			
T1S-18	229	3019	61	2038	
T1S-97	199	3220/198			
T1S03	159	3114/290	82	2032	
T1S03A	159	3114/290	82	2032	
T1S04	159	3114/290	82	2032	
T1S04A	159	3114/290	82	2032	
T1S14	312	3112		2028	
T1S18	229	3018	61	2038	
T1S34	312	3116	FET-1	2035	
T1S37A	159	3114/290	82	2032	
T1S38	159	3025/129	82	2032	
T1S38A	159	3114/290	82	2032	
T1S53	159	3114/290	82	2032	
T1S53A	159	3114/290	82	2032	
T1S54	159	3114/290	82	2032	
T1S55	123A	3122	20	2051	
T1S58	457	3448	FET-1	2035	
T1S59	312	3448	FET-1	2035	
T1S5A	159	3114/290	82	2032	
T1S61A	159	3114/290	82	2032	
T1S86	107		60	2015	
T1S88	312	3448	FET-1	2035	
T1S91	159	3114/290	82	2032	
T1S91A	159	3114/290	82	2032	
T1S92	123A		S92		
T1S93	159	3114/290	82	2032	
T1S93A	159	3114/290	82	2032	
T1S94	108	3018	86	2038	
T1S95			10	2051	
T1S97	107		S97		
T1S98	108	3018	86	2038	
T1SDS1K			504A	1104	
T1SZ8	108	3018	86	2038	
T1V	917		IC-258		
T1X-M14	161	3018	39	2015	
T1X-M15	161	3018	39	2015	
T1X-M16	161	3716	39	2015	
T1XM05			245	2004	
T1XM15	108		86	2038	
T1XM17	108		86	2038	
T1Y	917		IC-258		
T1Z	722	3161	IC-9		
T20	109	3087		1123	
T200	116		504A	1104	
T2015	160	3006	245	2004	
T2016	160	3006	245	2004	
T2017	160	3006	245	2004	
T2019	160	3006	245	2004	
T201AV	1171	3565			
T2020	160	3008	245	2004	
T2021	160	3006	245	2004	
T2022	160	3006	245	2004	
T2024	160	3008	245	2004	
T2025	160	3008	245	2004	
T2026	160	3008	245	2004	
T2028	160		245	2004	
T2029	160		245	2004	
T2030	160		245	2004	
T2038	100	3005	1	2007	
T2039	100	3005	1	2007	
T2040	100	3005	1	2007	
T2062	783	3215	IC-225		
T2091	100	3005	1	2007	
T20A6F					SPECIAL
T20G	109	3087		1123	
T21	109	3087		1123	
T2122	100	3005	1	2007	
T21237	109			1123	
T21238	109	3087		1123	
T21271	109	3087		1123	
T21312	116	3311	504A	1104	
T21313	109			1123	
T21333	116	3311	504A	1104	
T21334	139A		ZD-9.1	562	
T21507	116		504A	1104	
T2159	102A	3005	53	2007	
T21600	177		300	1122	
T21602	116	3031A	504A	1104	
T21638	117		504A	1104	
T21639	139A		ZD-9.1	562	
T21649	116	3031A	504A	1104	
T21679	116		504A	1104	
T2172	100	3005	1	2007	
T2173	100	3005	1	2007	
T2191	160		245	2004	
T21G	109	3091		1123	
T22	109	3087		1123	
T222	162		35		
T225	162		35		
T2256	100	3005	1	2007	
T2257	100	3005	1	2007	
T2258	100	3005	1	2007	
T2259	100	3005	1	2007	
T2260	100	3005	1	2007	
T2261	100	3005	1	2007	
T22G	109	3091		1123	
T23	109	3087		1123	
T23-93	196	3054	241	2020	
T23-94	128	3024	243	2030	
T2300A	5650	3506/5642			T2300PA5
T2300B	5651	3506/5642			T2300PB5
T2300C	5652				T2300PC5
T2300D	5652	3506/5642			T2300PD5
T2300E					T2300PE5
T2300F	5650				T2300PF5
T2300M					T2300PM5
T2300PA	5650				T2300PA
T2300PB	5651				T2300PB
T2300PC	5652				T2300PC
T2300PD	5652				T2300PD
T2300PE					T2300PE
T2300PF	5650				T2300PF
T2300PM					T2300PM
T2301A	5650	3506/5642			T2301PA5
T2301B	5651	3506/5642			T2301PB5
T2301C					T2301PC5
T2301D	5652	3506/5642			T2301PD5
T2301E					T2301PE5
T2301F					T2301PF5
T2301M					T2301PM5
T2301PA	5650				T2301PA
T2301PB	5651				T2301PB
T2301PC					T2301PC
T2301PD	5652				T2301PD
T2301PE					T2301PE
T2301PF					T2300PF
T2301PM					T2301PM
T2302A	5650	3583/5641			T2302PA5
T2302B	5651	3583/5641			T2302PB5
T2302C	5652				T2302PC5
T2302D	5652	3506/5642			T2302PD5
T2302E					T2302PE5
T2302F					T2302PF5
T2302M					T2302PM5
T2302PA	5650				T2302PA
T2302PB	5651				T2302PB
T2302PC					T2302PC
T2302PD	5652				T2300PD
T2302PE					T2302PE
T2302PF					T2302PF
T2302PM					T2302PM
T2303F	5640				SC136A
T2306A		3583			SC136A
T2306B		3583			SC136B
T2306D		3506			SC136D
T2310A		3506			T2300PA
T2310B		3506			T2300PB

Industry Standard No.	ECG	SK	GE	RS 276-	MOTOR.
T2310D		3506			T2300PD
T2310F					T2300PF
T2311A		3506			T2301PA
T2311B		3506			T2301PB
T2311D		3506			T2301PD
T2311F					T2301PF
T2312A		3506			T2302A
T2312B		3506			T2302B
T2312D		3506			T2302D
T2312F					T2302F
T2313A		3583			SC136A
T2313B		3583			SC136B
T2313D		3583			SC136D
T2313F					SC136A
T2313M		3519			SC136M
T2320A					T2320A
T2320B					T2320B
T2320C					T2320C
T2320D					T2320D
T2320E					T2320E
T2320F					T2320F
T2322	126	3008	52	2024	
T2322A					T2322A
T2322B					T2322B
T2322C					T2322C
T2322D					T2322D
T2322E					T2322E
T2322F					T2322F
T2323	126	3008	52	2024	
T2323A					T2323A
T2323B					T2323B
T2323C					T2323C
T2323D					T2323D
T2323E					T2323E
T2323F					T2323F
T2324	126	3008	52	2024	
T2327A					T2327A
T2327B					T2327B
T2327C					T2327C
T2327D					T2327D
T2327E					T2327E
T2327F					T2327F
T2357	106		21	2034	
T235A013-2	123A	3444	20	2051	
T2364	160		245	2004	
T237	123A	3444	20	2051	
T2379	160	3006	245	2004	
T2384	160	3006	245	2004	
T23G	109	3087		1123	
T2439	102A		53	2007	
T2440	102A		53	2007	
T2441	102A		53	2007	
T2446	123A	3444	20	2051	
T246	129	3025	244	2027	
T247	128	3024	243	2030	
T24G	109	3091		1123	
T2500A					T2500A
T2500B	5645	3658/56004			T2500B
T2500C					T2500C
T2500D	5645	3659/56006			T2500D
T2500E					T2500E
T2500M					T2500M
T2500N					T2500N
T2500S					T2500S
T2506B					T2500B
T2506D					T2500D
T2515	102A		53	2007	
T2517	102A		53	2007	
T253	160	3006	245	2004	
T253(SEARS)	160		245	2004	
T255	123A	3444	20	2051	
T256	123A	3444	20	2051	
T257	124		12		
T261	124		12		
T2634	108	3018	86	2038	
T26G	109	3087		1123	
T2700B	5624	3612			
T2700D	5635	3507			
T271	175		246	2020	
T2710B	5624	3507			
T2710D	5626	3507			
T276	159	3025/129	82	2032	
T277	123A	3444	20	2051	
T278	160		245	2004	
T2788	160		245	2004	
T279	160		245	2004	
T27G	109	3091		1123	
T280	160	3006	245	2004	
T280(SEARS)	126		52	2024	
T2800B	5645	3658/56004			T2800B
T2800C	5645	3659/56006			T2800C
T2800D	5645	3659/56006			T2800D
T2800E	5645	3660/56008			T2800E
T2800M	5645	3660/56008			T2800M
T2801B	5633	3938			T2801B
T2801C	5634				T2801C
T2801D	5635	3533			T2801D
T2801DF	5636	3939			
T2801E	5636	3939			T2801E
T2801M	5637		SC141M		T2801M
T2802B	5633	3938			T2802B
T2802C	5634	3533/5635			T2802C
T2802D	5635	3533			T2802D
T2802E	5636	3939			T2802E
T2802M	5637		SC143M		T2802M
T2806B		3938			T2800B
T2806C					T2800C
T2806D					T2800D
T2806M					T2800M
T2806N					T2800N
T281	160	3008	245	2004	
T281(SEARS)	126		52	2024	
T282	160	3008	245	2004	
T282(SEARS)	102A		53	2007	
T2850A	5645	3658/56004			2N6346A
T2850B	5645	3658/56004			2N6346A
T2850D	5645	3659/56006			2N6347A
T2850E					T2850E
T2851B	5645				T2801B
T2851C	5645				T2801C
T2851D	5645				T2801D
T2851E	5645				T2801E
T2856B					T2800B
T2856C					T2800C
T2856D					T2800D
T287	157	3747	232		
T2878	160		245	2004	
T2896	160		245	2004	
T291	123A	3444	20	2051	
T2945	160		245	2004	
T2946	160		245	2004	
T2A	722		IC-9		
T2A(I.C.)	798		IC-234		
T2A(MOTOROLA)	722		IC-9		

Industry Standard No.	ECG	SK	GE	RS 276-	MOTOR.
T2A(TRANSISTOR)	172A		64		
T2C	722	3161	IC-9		
T2D	723		IC-15		
T2F	718	3159	IC-8		
T2G	723		IC-15		
T2G(DIODE)	109			1123	
T2G(I.C.)	723		IC-15		
T2J	708	3159/718	IC-10		
T2M	781	3169	IC-223		
T2R	778A		IC-220		
T2T	788	3829	IC-229		
T2T-1	788	3829			
T2T-2	788	3829	IC-229		
T2V	736		IC-17		
T3/2	116	3311	504A	1104	
T300	116		504A	1104	
T3005	102A	3005	53	2007	
T30155	117	3017B	504A	1104	
T30155-001	116	3311	504A	1104	
T30155-1	116	3311	504A	1104	
T301AV	1171	3565			
T308	108	3039/316	86	2038	
T309	234		65	2050	
T309K	309K	3629			
T30A6F					SPECIAL
T327	123A	3444	20	2051	
T327-2	123A	3444	20	2051	
T328	123A	3444	20	2051	
T3321	102A		53	2007	
T3322	102A		53	2007	
T334-2	129		244	2027	
T336-2	128		243	2030	
T339	123A	3444	20	2051	
T33Z1	102A	3004	53	2007	
T33Z3	102A	3004	53	2007	
T340	159		82	2032	
T342	123A	3444	20	2051	
T344	184	3054/196	57	2017	
T345	185	3083/197	58	2025	
T346			1	2007	
T348	160	3008	245	2004	
T3530	108	3019	86	2038	
T3535	108	3117	86	2038	
T3536	108	3018	86	2038	
T3539	108	3018	86	2038	
T3565	123A	3444	20	2051	
T3568	107	3039/316	11	2015	
T3568(RCA)	108		86	2038	
T3570	159	3025/129	82	2032	
T3570A	159	3114/290	82	2032	
T35A-5	123A	3020/123	20	2051	
T3601	123A	3444	20	2051	
T3601(RCA)	107		11	2015	
T367	160	3006	245	2004	
T368	160	3006	245	2004	
T370	179	3034	76		
T373	160	3006	245	2004	
T374	160	3006	245	2004	
T381	161	3716	39	2015	
T381(SEARS)	161	3716	39	2015	
T386	123A	3444	20	2051	
T386(SEARS)	108		86	2038	
T3889					SPECIAL
T3890					SPECIAL
T3891					SPECIAL
T3892					SPECIAL
T3893					SPECIAL
T3899					SPECIAL
T39	102A	3123	82	2007	
T3900					SPECIAL
T3901					SPECIAL
T3902					SPECIAL
T3903					SPECIAL
T3909					SPECIAL
T3910					SPECIAL
T3911					SPECIAL
T3912					SPECIAL
T3913					SPECIAL
T396	185	3083/197	58	2025	
T399	123A	3444	20	2051	
T3A	743		IC-214		
T3B	941M		IC-265	007	
T3G	109			1123	
T40	106	3984	21	2034	
T400	116		504A	1104	
T400001606	5521	6621			
T400001608	5521	6621			
T400002206	5541	6615/5515			
T400002208	5541	6615/5515	C228F		
T400011606	5522	6622			
T400011608	5522	6622			
T400012206	5542	6615/5515			
T400012208	5542	6615/5515	C228A		
T400021606	5524	6624			
T400021608	5524	6624			
T400022208	5543		C228B		
T400031606	5526	6627/5527			
T400031608	5526	6627/5527			
T400032208	5544		C228C		
T400041606	5527	6627			
T400041608	5527	6627			
T400042208	5545		C228D		
T400051606	5528	6629/5529			
T400051608	5528	6629/5529			
T400052208	5546		C228E		
T400061006	5529	6629			
T400061008	5529	6629			
T400061606	5529	6629			
T400061608	5529	6629			
T400062208	5547		C228M		
T400081006	5531	6631			
T400081008	5531	6631			
T400081606	5531	6631			
T400081608	5531	6631			
T407185152			22	2032	
T4100D					T4100B
T4100C					T4100C
T4100D					T4100D
T4100E					T4100E
T4100M			SC251M		T4100M
T4101B					T4101B
T4101C					T4101C
T4101D					T4101D
T4101E					T4101E
T4101M					T4101M
T4106B					2N5571
T4106D					2N5572
T4106M					T4100M
T4107B					2N5567
T4107D					2N5568
T4107M					T4101M
T4110B					T4110B
T4110C					T4110C
T4110D					T4110D

Industry Standard No.	ECG	SK	GE	RS 276-	MOTOR.
T4110E					T4110E
T4110M	5677	3520			T4110M
T4111B					T4111B
T4111C					T4111C
T4111D					T4111D
T4111E					T4111E
T4111M	5677	3520			T4111M
T4116B		3508			2N5573
T4116D		3508			2N5574
T4116M					T4110M
T4117B		3508			2N5569
T4117D					2N5570
T4117M					T4111M
T4120A					T4120A
T4120B					T4120B
T4120C					T4120C
T4120D			SC250D3		T4120D
T4120E					T4120E
T4120M			SC250M3		T4120M
T4120N					T4120N
T4120S					T4120S
T4121A					T4121A
T4121B					T4121B
T4121C					T4121C
T4121D					T4121D
T4121E					T4121E
T4121M	5677				T4121M
T4121N					T4121N
T4121S					T4121S
T4126B					T4120B
T4126D					T4120D
T4126M					T4120M
T4127B					T4121B
T4127D					T4121D
T4127M					T4121M
T4130A					SPECIAL
T4130B			SC250B8		SPECIAL
T4130C					SPECIAL
T4130D			SC250D8		SPECIAL
T4130E					SPECIAL
T4130F					SPECIAL
T4130M			SC250M8		SPECIAL
T4131A					SPECIAL
T4131B					SPECIAL
T4131C					SPECIAL
T4131D					SPECIAL
T4131E					SPECIAL
T4131F					SPECIAL
T4131M					SPECIAL
T4140A					SPECIAL
T4140B	56014		SC250B4		SPECIAL
T4140C					SPECIAL
T4140D	56014		SC250D4		SPECIAL
T4140E					SPECIAL
T4140F	56014				SPECIAL
T4140M			SC250M4		SPECIAL
T4141A					SPECIAL
T4141B					SPECIAL
T4141C					SPECIAL
T4141D					SPECIAL
T4141E					SPECIAL
T4141F					SPECIAL
T4141M					SPECIAL
T4150A					SPECIAL
T4150B	56014		SC250B6		SPECIAL
T4150C					SPECIAL
T4150D	56014		SC250D6		SPECIAL
T4150E					SPECIAL
T4150F	56014				SPECIAL
T4150M			SC250M6		SPECIAL
T4151A					SPECIAL
T4151B					SPECIAL
T4151C					SPECIAL
T4151D					SPECIAL
T4151E					SPECIAL
T4151F					SPECIAL
T4151M					SPECIAL
T416-16(SEARS)	123A		20	2051	
T417	123A	3444	20	2051	
T4205L1			243	2030	
T422	188		226	2018	
T423	189		218	2026	
T42692-001	116	3031A	504A	1104	
T42692-1R	116		504A	1104	
T449	160	3006	245	2004	
T449(SEARS)	126		52	2024	
T45	102A	3123	53	2007	
T450	116		504A	1104	
T452	128	3024	243	2030	
T457-16	123A	3444	20	2051	
T457-16(SEARS)	123A		20	2051	
T458-16	123A	3444	20	2051	
T459	129	3025	244	2027	
T459(SEARS)	123A		20	2051	
T4590	110MP	3709	504A	1123(2)	
T46	102A	3123	53	2007	
T460	123A	3444	20	2051	
T460(SEARS)	159		82	2032	
T461-16	123A	3444	20	2051	
T461-16(SEARS)	123A		20	2051	
T462	123A		20	2051	
T462(SEARS)	123A		20	2051	
T464	222		FET-4	2036	
T47	102A	3123	53	2007	
T4700B		3611	SC151B2		MAC15A4
T4700D		3507	SC151D2		MAC15A6
T4700E					MAC15A7
T4700F					MAC15A2
T4706B					MAC15A4
T4706D	5675				MAC15A6
T472	123A	3444	20	2051	
T472(SEARS)	123A		20	2051	
T475	129		244	2027	
T475(SEARS)	159		82	2032	
T48	102A	3123	53	2007	
T481(SEARS)	154		40	2012	
T482(SEARS)	159		82	2032	
T483(SEARS)	123A		20	2051	
T484(SEARS)	123A		20	2051	
T485(SEARS)	123A		20	2051	
T486(SEARS)	123A		20	2051	
T4F	1115	3184			
T50	102A		53	2007	
T500	116		504A	1104	
T50339A	102A	3004	53	2007	
T50631	102A		53	2007	
T50818	126	3008	52	2024	
T50931B	102	3722	2	2007	
T50933B	158		53	2007	
T50944	100	3005	1	2007	
T513			FET-4	2036	
T51573A	102	3722	2	2007	
T52054	126		52	2024	
T52147	100	3005	1	2007	

Industry Standard No.	ECG	SK	GE	RS 276-	MOTOR.
T52147Z	100	3005	1	2007	
T52148Z	100	3721	1	2007	
T52149	100	3721	1	2007	
T52149Z	100	3721	1	2007	
T52150	102A		53	2007	
T52150Z	102A		53	2007	
T52151	102A		53	2007	
T52151Z	102A		53	2007	
T52159	102A	3004	53	2007	
T550	116		504A	1104	
T576-1	161	3117	39	2015	
T59	102A		53	2007	
T59235A	123A	3020/123	20	2051	
T59247	102A	3004	53	2007	
T59249	102A	3004	53	2007	
T59276	101	3861	8	2002	
T59277	101	3861	8	2002	
T597-1	159		82	2032	
T60	102A		53	2007	
T600	116		504A	1104	
T6000A					MAC15A3
T6000B		3938			MAC15A4
T6000C		3533			MAC15A5
T6000D		3533			MAC15A6
T6000E		3939			MAC15A7
T6000F		3937			MAC15A2
T6000M					MAC15A8
T6001A					MAC15A3
T6001B					MAC15A4
T6001C					MAC15A5
T6001D					MAC15A6
T6001E					MAC15A7
T6001F					MAC15A2
T6001M					MAC15A8
T6006A					MAC15A3
T6006B		3938			MAC15A4
T6006C		3533			MAC15A5
T6006D		3533			MAC15A6
T6006E		3939			MAC15A7
T6006F					MAC15A2
T6006M					MAC15A8
T6028	126		52	2024	
T6029	126		52	2024	
T6030	126		52	2024	
T6031	126		52	2024	
T6032	126		52	2024	
T6058	160		245	2004	
T61	102A	3004	53	2007	
T611-1	152	3054/196	66	2048	
T611-1(RCA)	184		57	2017	
T612-1	152	3054/196	66	2048	
T612-1(RCA)	184		57	2017	
T615A002	123A	3444	20	2051	
T615A006-1	123A	3444	20	2051	
T6400A					T6400A
T6400B					T6400B
T6400C					T6400C
T6400D					T6400D
T6400E					T6400E
T6400M					T6400M
T6400N					T6400N
T6400S					T6400S
T6401A					T6401A
T6401B					T6401B
T6401C					T6401C
T6401D					T6401D
T6401E					T6401E
T6401M					T6401M
T6401N					T6401N
T6401S					T6401S
T6402E					2N5443
T6402F					2N5441
T6406B					2N5441
T6406D					2N5442
T6406E					2N5443
T6406M					2N5443
T6407B					T6401B
T6407D					T6401D
T6407E					T6401E
T6407M					T6401M
T6410A					T6410A
T6410B					T6410B
T6410C					T6410C
T6410D					T6410D
T6410E					T6410E
T6410M		3521			T6410M
T6410N		3509			T6410N
T6410S					T6410S
T6411A					T6411A
T6411B	5693	3652			T6411B
T6411C					T6411C
T6411D	5695	3509			T6411D
T6411E					T6411E
T6411M	5697	3522			T6411M
T6411N					T6411N
T6411S					T6411S
T6412E					2N5446
T6412F					2N5444
T6420A					T6420A
T6420B	56022	3661			T6420B
T6420C	56024				T6420C
T6420D	56024	3662			T6420D
T6420E	56026				T6420E
T6420M	56026	3663			T6420M
T6420N					T6420N
T6420S					T6420S
T6421A					T6421A
T6421B	56022				T6421B
T6421C	56024				T6421C
T6421D	56024				T6421D
T6421E	56026				T6421E
T6421M	56026				T6421M
T6421N					T6421N
T6421S					T6421S
T6426B					T6420B
T6426D					T6420D
T6426M					T6420M
T6427B					T6421B
T6427D					T6421D
T6427M					T6421M
T6430B					SPECIAL
T6430D					SPECIAL
T6430M					SPECIAL
T6431A					SPECIAL
T6431B					SPECIAL
T6431C					SPECIAL
T6431D					SPECIAL
T6431E					SPECIAL
T6431F					SPECIAL
T6431M					SPECIAL
T6440A					SPECIAL
T6440B					SPECIAL
T6440C					SPECIAL
T6440D					SPECIAL

Industry Standard No.	ECG	SK	GE	RS 276-	MOTOR.
T6440E					SPECIAL
T6440F					SPECIAL
T6440M					SPECIAL
T6441A					SPECIAL
T6441B	56014				SPECIAL
T6441C					SPECIAL
T6441D	56014				SPECIAL
T6441E					SPECIAL
T6441F	56014				SPECIAL
T6441M					SPECIAL
T6450A					SPECIAL
T6450B					SPECIAL
T6450C					SPECIAL
T6450D					SPECIAL
T6450E					SPECIAL
T6450F					SPECIAL
T6450M					SPECIAL
T6451A					SPECIAL
T6451B					SPECIAL
T6451C					SPECIAL
T6451D	56014				SPECIAL
T6451E					SPECIAL
T6451F	56014				SPECIAL
T6451M					SPECIAL
T650	116	3122	504A	1104	
T650(TRANSISTOR)	123A		20	2051	
T6565	123A		20	2051	
T7	109			1123	
T700-709			1N295	1123	
T72	102A	3004	53	2007	
T74	102A	3004	53	2007	
T7400B1	7400	7400		1801	
T7402B1	7402	7402		1811	
T7403B1	7403	7403			
T7403D1	7403	7403			
T7404B1	7404	7404		1802	
T7405B1	7405	7405			
T7406B1	7406	7406		1821	
T7407B1	7407	7407			
T7407D1	7407	7407			
T7409B1	7409	7409			
T7409D1	7409	7409			
T74107B1	74107	74107			
T7410B1	7410	7410		1807	
T7410D1	7410			1807	
T7410D2				1807	
T74122B1	74122	74122			
T74122D1	74122	74122			
T7416B1	7416	7416			
T7417B1	7417	7417			
T7417D1	7417	7417			
T74180B1	74180	74180			
T74180D1	74180	74180			
T74193B1	74193	74193		1820	
T7420B1	7420	7420		1809	
T7420D1	7420			1809	
T7420D2				1809	
T7426B1	7426	7426			
T7426D1	7426	7426			
T7428B1	7428	7428			
T7428D1	7428	7428			
T7430B1	7430	7430			
T7433B1	7433	7433			
T7433D1	7433	7433			
T7441AB1	7441	7441		1804	
T7441AD1	7441	7441			
T7443B1	7443	7443			
T7443D1	7443	7443			
T7444B1	7444	7444			
T7444D1	7444	7444			
T7450B1	7450	7450			
T7450D1	7450	7450			
T7451B1	7451	7451		1825	
T7451D1	7451			1825	
T7451D2				1825	
T7453B1	7453	7453			
T7453D1	7453	7453			
T7454B1	7454	7454			
T7460B1	7460	7460			
T7460D1	7460	7460			
T7472B1	7472	7472			
T7472D1	7472	7472			
T7473B1	7473	7473		1803	
T7474B1	7474	7474		1818	
T7475B1	7475	7475		1806	
T7476B1	7476	7476		1813	
T7481B1	7481	7481			
T7481D1	7481	7481			
T7483B1	7483	7483			
T7483D1	7483	7483			
T7486B1	7486	7486		1827	
T7486D1	7486			1827	
T7486D2				1827	
T7490B1	7490	7490		1808	
T7493B1	7493A	7493			
T76	123A	3122	53	2051	
T77	102A		53	2007	
T78	102A		53	2007	
T8/2	116	3311	504A	1104	
T800	125	3032A	510,531	1114	1N4006
T800X	125	3032A	510,531	1114	
T81	103A		59	2002	
T811	160	3008	245	2004	
T814	158	3004/102A	53	2007	
T815	158	3004/102A	53	2007	
T82	102A	3004	53	2007	
T83	102A		53	2007	
T84	102A	3004	53	2007	
T841	181	3036	75	2041	
T842	181	3036	75	2041	
T843	181	3036	75	2041	
T844	181	3036	75	2041	
T87	102A		53	2007	
T89412	911		IC-253		
T8G	109	3087		1123	
T9	109	3087		1123	
T900			82	2032	
T9011A1C	123A	3444	20	2051	
T9011A1G	123A	3444	20	2051	
T9011AZ	123A	3444	20	2051	
T9011CD	108		86	2038	
T9011EF	108		86	2038	
T9011G	108		86	2038	
T9011G(CD)			20	2051	
T9011G(EF)			20	2051	
T9011GEF	108		86	2038	
T9011GH	108		86	2038	
T9011H	108		86	2038	
T9011H(EF)			20	2051	
T9011HEF	108		86	2038	
T9011I	108		86	2038	
T9011I(EF)			20	2051	
T9011J	108		86	2038	
T9011J(GH)			20	2051	
T9016F	108		86	2038	
T9016H	108	3122	86	2038	
T9418	293	3137/192A	47	2030	
T9423	229	3018	61	2038	
T9468	294	3138/193A	48		
T95	102A		53	2007	
T9631	123AP	3018	243	2030	
T9681	159	3114/290	82	2032	
T99	102A		53	2007	
T9G	109	3087		1123	
TA-1	105	3012			
TA-1575	102A	3004	53	2007	
TA-1575B	100	3005	1	2007	
TA-1614	121	3009	239	2006	
TA-1620A	103A		59	2002	
TA-1620B	103A		59	2002	
TA-1628	160	3007	245	2004	
TA-1650A	126	3008	52	2024	
TA-1655B	100	3005	1	2007	
TA-1658	160	3007	245	2004	
TA-1659	160	3007	245	2004	
TA-1660	160	3007	245	2004	
TA-1662	160	3007	245	2004	
TA-1682	121	3009	239	2006	
TA-1682A	121	3009	239	2006	
TA-1697	102A	3003	53	2007	
TA-1704	100	3005	1	2007	
TA-1705	121	3009	239	2006	
TA-1706	102A	3003	53	2007	
TA-1730	102A	3004	53	2007	
TA-1731	160	3007	245	2004	
TA-1755	126	3008	52	2024	
TA-1756	126	3008	52	2024	
TA-1757	160	3007	245	2004	
TA-1759	101	3011	8	2002	
TA-1763	100	3005	1	2007	
TA-1763A	100	3005	1	2007	
TA-1765	121	3009	239	2006	
TA-1766	121	3009	239	2006	
TA-1767	101	3011	8	2002	
TA-1771	101	3011	8	2002	
TA-1772	101	3011	8	2002	
TA-1773	121	3009	239	2006	
TA-1778	100	3005	1	2007	
TA-1782	100	3005	1	2007	
TA-1783	100	3005	1	2007	
TA-1794	121	3009	239	2006	
TA-1796	160	3007	245	2004	
TA-1797	160	3007	245	2004	
TA-1798	160	3007	245	2004	
TA-1828	160	3007	245	2004	
TA-1829	160	3007	245	2004	
TA-1830	126	3008			
TA-1830+			52	2024	
TA-1846	160	3006	245	2004	
TA-1847	160	3006	245	2004	
TA-1860	160	3006	245	2004	
TA-1861	160	3006	245	2004	
TA-1881	121	3009	239	2006	
TA-1890	121	3009	239	2006	
TA-1891	121	3009	239	2006	
TA-2	121	3009	239	2006	
TA-3	105	3012	4		
TA-4	102A		53	2007	
TA-4846				2004	
TA-5	126		52	2024	
TA-6	123A	3444	20	2051	
TA-7	108	3444/123A	86	2038	
TA-7045-M	724		IC-86		
TA-7045M	724		IC-86		
TA-7051P	710	3102	IC-89		
TA-7061AP	1100		IC-92		
TA-7062P	1102		IC-93		
TA-7063P	1103	3281	IC-94		
TA-7069P	1101		IC-95		
TA-7120P	1087		IC-103		
TA-7155	196	3054	241	2020	
TA-7205P			IC-179	705	
TA05	6408	9084A			
TA10					1N4002
TA100	116	3017B/117	504A	1104	1N4007
TA1000	125		510,531	1114	1N4007
TA1062	116	3017B/117	504A	1104	
TA1063	116	3017B/117	504A	1104	
TA1064	116	3017B/117	504A	1104	
TA117	704	3023	IC-205		
TA120					MR1-1200
TA1222	5511	3683			
TA1225	5512	3502/5513			
TA1575	102A		53	2007	
TA1575B	102A		53	2007	
TA1614	121	3717	239	2006	
TA1620A	103		8	2002	
TA1620B	103		8	2002	
TA1628	160		245	2004	
TA1650A	160		245	2004	
TA1655B	102A		53	2007	
TA1658	160		245	2004	
TA1659	160		245	2004	
TA1660	160		245	2004	
TA1662	160		245	2004	
TA1682	121	3717	239	2006	
TA1682A	121	3717	239	2006	
TA1697	102A		53	2007	
TA1704	100	3721	1	2007	
TA1705	121	3717	239	2006	
TA1706	102A		53	2007	
TA1730	102A		53	2007	
TA1731	160		245	2004	
TA1755	160		245	2004	
TA1756	160		245	2004	
TA1757	160		245	2004	
TA1759	101		8	2002	
TA1763	100	3721	1	2007	
TA1763A	100	3721	1	2007	
TA1765	121	3717	239	2006	
TA1766	121	3717	239	2006	
TA1767	101		8	2002	
TA1771	101		8	2002	
TA1772	101		8	2002	
TA1773	121	3717	239	2006	
TA1778	100	3721	1	2007	
TA1782	100	3721	1	2007	
TA1783	100	3721	1	2007	
TA1794	121	3717	239	2006	
TA1796	160		245	2004	
TA1797	160		245	2004	
TA1798	160		245	2004	
TA1828	160	3007	245	2004	
TA1829	160	3006	245	2004	
TA1830	160		245	2004	
TA1846	160		245	2004	
TA1847	160		245	2004	
TA1860	160	3006	245	2004	

Industry Standard No.	ECG	SK	GE	RS 276-	MOTOR.
TA1861	160		245	2004	
TA1881	121	3717	239	2006	
TA1890	121	3717	239	2006	
TA1891	121	3717	239	2006	
TA1928A	127	3035	25		
TA198030-4	123A		20	2051	
TA198035-1	128		243	2030	
TA198036-2	159		82	2032	
TA198785-2	519		514	1122	
TA1990		3006	50	2004	
TA1Z			18	2030	
TA20					1N4003
TA200	116	3017B/117	504A	1104	1N4003
TA2083	127	3034	25		
TA2188	127	3034	25		
TA2301	121	3009	239	2006	
TA2322	160		245	2004	
TA2401	108		86	2038	
TA2402	175	3026	246	2020	
TA2402A	175	3026	246	2020	
TA2503	108	3039/316	86	2038	
TA2509	124	3021	12		
TA2509A	124	3021	12		
TA2554	161	3716	39	2015	
TA2555	161	3716	39	2015	
TA2577A	130	3027	14	2041	
TA2644	222	3065		2036	
TA2653	230	3502/5513			
TA2654	230	3502/5513			
TA2655	230	3502/5513			
TA2672	121	3014	239	2006	
TA2700	124		12		
TA2710	278	3024/128	261		
TA2773	5513	3502			
TA2800	278	3218	261		
TA2819	191	3053	249		
TA2840	220			2036	
TA2888	5455	3597	MR-5	1067	
TA2889	5455	3597	MR-5	1067	
TA2892A	5640	3506/5642			
TA2893A	5651	3506/5642			
TA2894A	5652	3507			
TA2911	152	3054/196	66	2048	
TA2A			20	2051	
TA300	116	3017B/117	504A	1104	1N4004
TA40		3080			1N4004
TA400	116	3031A	504A	1104	1N4004
TA5					1N4001
TA50	116	3017B/117	504A	1104	1N4001
TA500	116	3017B/117	504A	1104	1N4005
TA5274	711		IC-207		
TA5274(RCA)	711		IC-207		
TA5628	788	3829			
TA5649	714	3075	IC-4		
TA5649(RCA)	714		IC-4		
TA5649A		3075	IC-4		
TA5702	715		IC-6		
TA5702(RCA)	715		IC-6		
TA5814	712	3072	IC-2		
TA5814(RCA)	712		IC-2		
TA5912	790	3077	IC-5		
TA5912(RCA)	790		IC-5		
TA6	123A		20	2051	
TA60		3080			1N4005
TA600	116	3017B/117	504A	1104	1N4005
TA6200	128	3024	243	2030	
TA6220	708	3135/709	IC-10		
TA6243	731		IC-13		
TA6243(RCA)	731		IC-13		
TA6319(RCA)	797		IC-233		
TA6404	798	3216	IC-234		
TA6405	738	3167	IC-29		
TA6523	749		IC-97		
TA7	123A		20	2051	
TA7006P	1149		IC-285		
TA7027M	726	3129			
TA7028M	1188	3022	IC-82		
TA7031M	901	3549	IC-180		
TA7038M	704		IC-205		
TA7045	724	3525	IC-86		
TA7045M	724	3525	IC-86		
TA7046P	1108		IC-87		
TA7047		3024	IC-289		
TA7050M	711	3070	IC-207		
TA7050P	1061	3228			
TA7051/01			511	1114	
TA7051P	710	3102	IC-89		
TA7054P	1106		IC-90		
TA7055P	1205	3482	IC-178		
TA7057M	905		IC-270		
TA7060	1104	3225	IC-91		
TA7060P	1104	3225	IC-91		
TA7060P-10	1104	3225	IC-91		
TA7060PR	1104	3225	IC-91		
TA7060PRW	1104	3225	IC-91		
TA7060PW	1104	3225	IC-91		
TA7061	1100	3223	IC-92		
TA7061(AP)	1100	3223	IC-92		
TA7061A	1100	3223			
TA7061AP	1100	3223	IC-92		
TA7061B	1100	3223			
TA7061P	1100	3223	IC-92		
TA7062P	1102	3224	IC-93		
TA7063	1103	3281	IC-94		
TA70639	1052	3249	IC-135		
TA7063P	1103	3281	IC-94		
TA7063P-B	1103	3281	IC-94		
TA7063P-C	1103	3281	IC-94		
TA7063P-D	1103	3281	IC-94		
TA7063P-O	1103	3281	IC-94		
TA7064P	1235	3637			
TA7064P-JA	1235	3637			
TA7068	130		14	2041	
TA7069	130		14	2041	
TA7069P	1101		IC-95		
TA7070P	1004	3365	IC-149		
TA7070PFA-1	1004	3365	IC-149		
TA7070PGL	1004	3365	IC-149		
TA7071P	712	3072	IC-148		
TA7072P	748	3236	IC-183		
TA7073AP	748	3236			
TA7073P	748	3236	IC-183		
TA7074P	749	3168	IC-97		
TA7074PGL	749	3168	IC-97		
TA7075	1080	3284	IC-98		
TA7075P	1080	3284	IC-98		
TA7076P	1109	3711	IC-99		
TA7076P(FA-1)	1109	3711	IC-99		
TA7076P(FA-6)	1109	3711	IC-99		
TA7076P(FA-7)	1109	3711	IC-99		
TA7086M	309K	3629			
TA7092AP	1023		IC-272		
TA7092AP-C	1023		IC-272		
TA7092P	1107		IC-276		
TA7092P-A	1107		IC-276		
TA7092P-B	1107		IC-276		
TA7092P-C	1107		IC-276		
TA7092P-D	1107		IC-276		
TA7092P-H	1107		IC-276		
TA7102P	1105	3285	IC-101		
TA7102P(FA-1)	1105	3285	IC-101		
TA7102P(FA-2)	1105	3285	IC-101		
TA7106P	1149		IC-285		
TA7117P	1022		IC-102		
TA7120	1087	3477	IC-103		
TA7120B	1087	3477	IC-103		
TA7120P	1087	3477	IC-103		
TA7120P-C	1087		IC-103		
TA7120P-D	1087		IC-103		
TA7120P-E	1087		IC-103		
TA7122		3512	IC-104		
TA7122AP	1085	3476	IC-104		
TA7122AP-C	1085	3476			
TA7122AP-D	1085	3476	IC-104		
TA7122APC	1085	3476			
TA7122AR	1085	3476	IC-104		
TA7122P-B	1085	3476	IC-104		
TA7124P	1128	3488	IC-105		
TA7127P	1200	3714			
TA7130P	1234	3487	IC-326		
TA7130P-B	1234		IC-181		
TA7130PB	1234	3487	IC-181		
TA7130PC	1234	3487	IC-181		
TA7134		3103A	257		
TA7137			66	2048	
TA7145P	1134	3489	IC-106		
TA7146P	1133	3490	IC-107		
TA7148P	1131	3286	IC-109		
TA7149	221	3050		2036	
TA7149P	1132	3287	IC-110		
TA7150	221	3050		2036	
TA7150P	1130	3478	IC-111		
TA7151	221	3050		2036	
TA7152P		3294	IC-112		
TA7155	196	3054	241	2020	
TA7156	152	3054/196	66	2048	
TA7157P	1206	3160/801	IC-35		
TA7159P	1243	3731	IC-113		
TA7167	1200	3714			
TA7167P	1200	3714			
TA7176AP	712	3072	IC-148		
TA7176P	1236	3072/712	IC-148		
TA7176PFA-1	712	3072	IC-148		
TA7176PG	1236	3072/712			
TA7189	222	3050/221		2036	
TA7192P	1196	3725			
TA7199		3027	255	2041	
TA7200	390		255	2041	
TA7201	390	3036	255	2041	
TA7202	390		255	2041	
TA7202P	1259	3915			
TA7203	1154	3230	IC-287		
TA7203P	1154	3230	IC-287		
TA7204	1153	3282			
TA7204P	1153	3282	IC-182		
TA7205	1155	3231	IC-179	705	
TA7205A	1155	3231	IC-179	705	
TA7205AP	1155	3231	IC-179	705	
TA7205P	1155	3231	IC-179	705	
TA7209P	1222	3730			
TA7214P	1273	3916			
TA7222P	1278	3726			
TA7262	152	3050/221	66	2048	
TA7262(RCA)	221			2036	
TA7274	222	3065		2036	
TA7292	154	3040	40	2012	
TA7293	154	3040	40	2012	
TA7303		3039	86	2038	
TA7310	1192	3445			
TA7310A	1192	3445			
TA7310P	1192	3445			
TA7310P-O	1192	3445			
TA7311	182	3054/196	55		
TA7312	182	3054/196	55		
TA7313	182	3054/196	55		
TA7314	182	3054/196	55		
TA7315	182	3054/196	55		
TA7316	182	3054/196	55		
TA7318	182	3054/196	55		
TA7319	108	3039/316	86	2038	
TA7362	196	3054	241	2020	
TA7363	152	3054/196	66	2048	
TA7374	222	3065		2036	
TA7399	222	3050/221		2036	
TA7404	5463	3572			
TA7405	5465	3687			
TA7461	5697	3521/5686			
TA7462	5697	3521/5686			
TA7500	5640	3519/5643			
TA7501	5641	3519/5643			
TA7502	5642	3519/5643			
TA7502M	909		IC-249		
TA7503	5643	3519			
TA7504P	941M		IC-265	007	
TA7506M	1171	3565			
TA7506P	975	3641			
TA7520	185		58	2025	
TA7554	152	3578	66	2048	
TA7555	152	3578	66	2048	
TA7556	153	3179A	69	2049	
TA7557	153	3179A	69	2049	
TA7579	5643	3519			
TA7580	5643	3519			
TA7581	5643	3519			
TA7582	5643	3519			
TA7584	5697	3521/5686			
TA7669	222	3065		2036	
TA7684	222			2036	
TA7739	228A	3104A	257		
TA7740	228A	3103A/396	257		
TA7741	187	3083/197	29	2025	
TA7742	197	3083	250	2027	
TA7743	197	3084	250	2027	
TA7782	196	3054	241	2020	
TA7783	196	3054	241	2020	
TA7784	196	3054	241	2020	
TA78005M	309K	3629			
TA78005P	960	3591	VR-102		
TA78012P	966	3592			
TA78015P	968	3593			
TA7802	116	3311	504A	1104	
TA7803	116	3031A	504A	1104	
TA7804	116	3032A	504A	1104	
TA7805	125	3032A	510,531	1114	
TA7806	125	3080	510,531	1114	
TA7996	116	3311	504A	1104	
TA80		3080			1N4006
TA800	125	3032A	510,531	1114	1N4006

Industry Standard No.	ECG	SK	GE	RS 276-	MOTOR.
TA8158	230	3042			
TA8159	230	3857/231			
TA8160	515	3314			
TA8161	515	3314			
TA8162	515	3314			
TA8210	197	3083	250	2027	
TA8211	197	3084	250	2027	
TA8212	197	3083	250	2027	
TA8231	196	3054	241	2020	
TA8232	196	3054	241	2020	
TA8233	196	3054	241	2020	
TA8242	222	3065		2036	
TA8532787	519		514	1122	
TAA320	312	3112		2028	
TAA380	704		IC-205		
TAA521	909	3590	IC-249		
TAA521/709	909	3590	IC-249		
TAA521A	909D		IC-250		
TAA521A/709C	909D		IC-250		
TAA522	909		IC-249		
TAA522/709	909		IC-249		
TAA630					MC1327P
TAC-047	123A	3020/123	20	2051	
TAC047	123A	3444	20	2051	
TAG10-100R	5491	6791			
TAG10-200R	5492	6792			
TAG10-300R	5494	6794			
TAG10-400R	5494	6794			
TAG10-500R	5496	6796			
TAG10-50R	5491	6791			
TAG10-600R	5496	6796			
TAG20-100R	5542	6615/5515			
TAG20-50R	5541	6615/5515			
TAG255-200	56004	3658			
TAG255-400	56006	3659			
TAG255-600	56008	3660			
TAG3-100R	5511	3683			
TAG3-200R	5511	3683			
TAG3-300R	5512	3684			
TAG3-400R	5512	3684			
TAG3-50R	5511	3683			
TAG7-100R	5491	6791			
TAG7-300R	5494	6794			
TAG7-400R	5494	6794			
TAG7-500R	5496	6796			
TAG7-50R	5491	6791			
TAG7-600R	5496	6796			
TB-1				1152	
TB-2				1172	
TB-4				1173	
TB100	5802	9005		1143	
TB200	5802	9005		1143	
TB5	5800	9003		1142	
TBA1190Z		3832			TBA1190Z
TBA120S					MC1358P
TBA221	941	3514	IC-263	010	
TBA221/741C	941	3514	IC-263	010	
TBA221A	941D		IC-264		
TBA221A/741C	941D	3552/941M	IC-264		
TBA222	941	3514	IC-263	010	
TBA222/741	941	3514	IC-263	010	
TBA231	725	3162	IC-19		
TBA281	923	3164	IC-259		
TBA281/723	923	3164	IC-259		
TBA281/723C	923	3164	IC-259		
TBA440					MC1352P
TBA520					MC1327P
TBA641B11	OBS-NLA		IC-277		
TBA800	1116	3969	IC-279		
TBA810ACB	1115	3184			
TBA810ACBS	1115	3184			
TBA810AD	1115	3184			
TBA810ADS	1115	3184			
TBA810AP	1115	3184			
TBA810APS	1115	3184			
TBA810AS	1115	3184	IC-278		
TBA810DS	1115	3184	IC-278		
TBA810S	1115	3917	IC-278		
TBA810S-H		3184	IC-278		
TBA820	1117		IC-280		
TBA820L	1117		IC-280		
TBA920					MC1391P
TBA920S					MC1391P
TBA940					MC1344P
TBA950					MC1344P
TBA990					MC1327P
TBB0747A	947D	3556			
TBB0748	1171	3565			
TBB0748B	975	3641			
TBC0748	1171	3565			
TBRC147B			10	2051	
TC-0-2P11/1			511	1114	
TC-0.09M21/5			6GD1	1104	
TC-0.2	506	3311	511	1114	
TC-0.2P11/1	506	3130	511	1114	
TC-0918	108	3019	86	2038	
TC-136		3016	504A	1104	
TC-6(ELCOM)	703A		IC-12		
TC0.09M21/3	114		6GD1	1104	
TC0.09M22/1	166	9007/5804		1144	
TC0.1P	506		511	1114	
TC0.2P11/2	116		504A	1104	
TC02P112	116		504A	1104	
TC02P12	116		504A	1104	
TC0914	108	3039/316	86	2038	
TC0918	108		86	2038	
TC0P11/2	116		504A	1104	
TC100	5802	9005		1143	
TC106A1	5454	6754		1067	
TC106A2	5454	6754		1067	
TC106A3	5454	6754		1067	
TC106A4	5454	6754		1067	
TC106B1	5455	3597		1067	
TC106B2	5455	3597	C106B2		
TC106B3	5455	3597		1067	
TC106B4	5455	3597		1067	
TC106C1	5456			1020	
TC106C2	5456			1020	
TC106C3	5456			1020	
TC106C4	5456			1020	
TC106D1	5457	3598			
TC106D2	5457	3598			
TC106D3	5457	3598			
TC106D4	5457	3598			
TC106F1	5453	6753		1067	
TC106F2	5453	6753		1067	
TC106F3	5453	6753		1067	
TC106F4	5453	6753		1067	
TC106Q1	5452	6752		1067	
TC106Q2	5452	6752		1067	
TC106Q3	5452	6752		1067	
TC106Q4	5452	6752		1067	
TC106Y1	5452	6752		1067	
TC106Y2	5452	6752		1067	
TC106Y3	5452	6752		1067	
TC106Y4	5452	6752		1067	
TC10LB2				1067	
TC110A5B	151A	3099	ZD-110		
TC136	116		504A	1104	
TC200	5802	9005		1143	
TC2369A	108	3039/316	86	2038	
TC2483	108	3039/316	86	2038	
TC2484	108	3039/316	86	2038	
TC27A5A	146A	3064	ZD-27		
TC27A5A-5	146A	3064	ZD-27		
TC3001	1192	3445			
TC311200600	109	3087		1123	
TC3112006000	109	3091		1123	
TC3112319300	116	3017B/117	504A	1104	
TC3123036722	123A	3444	20	2051	
TC3123036900	123A	3444	20	2051	
TC3123037111	123A	3444	20	2051	
TC3123037222	123A	3444	20	2051	
TC3123037412	123A	3444	20	2051	
TC3123041557	102A	3004	53	2007	
TC312307222			10	2051	
TC33A5A	147A	3095	ZD-33		
TC4001BP	4001B	4001			
TC4001P	4001B	4001	4001	2401	
TC4002BP	4002B	4002			
TC4006B	4006B	4006			
TC4007	4007	4007			
TC4008BP	4008B	4008			
TC4009UBP	4049	4009			
TC4010BP	4050B	4010			
TC4011	4011B	4011			
TC4011BP	4011B	4011	IC-4011	2411	
TC4011P	4011B	4011	4011	2411	
TC4012BP	4012B	4012			
TC4013	4013B	4013			
TC4013BP	4013B	4013			
TC4013P	4013B	4013			
TC4015BP	4015B	4015			
TC40161BP	40161B	40161			
TC4017BP	4017B	4017			
TC4018BP	4018B	4018			
TC4019BP	4019B	4019			
TC4021BP	4021B	4021			
TC4021P	4021B	4021			
TC4023BP	4023B	4023			
TC4024BP	4024B	4024			
TC4025BP	4025B	4025			
TC4027	4027B	4027			
TC4027BP	4027B	4027			
TC4028BP	4028B	4028			
TC4029BP	4029B	4029			
TC4030BP	4030B	4030			
TC4035	4035B	4035			
TC4035BP	4035B	4035			
TC4042BP	4042B	4042			
TC4043BP	4043B	4043			
TC4044BP	4044B	4044			
TC4047BP	4047B	4047			
TC4050BP	4050B	4050			
TC4051BP	4051B	4051			
TC4053B	4053B	4053			
TC4055BP	4055B	4055			
TC4066	4066B	4066			
TC4066BP	4066B	4066			
TC4072	4072B	4072			
TC4075	4075B	4075			
TC4075BP	4075B	4075			
TC4081		4081	4081		
TC4081BP	4081B	4081			
TC4093BP	4093B	4093			
TC4511BP	4511B	4511			
TC4520BP	4520B	4520			
TC4555BP	4555B	4555			
TC4556BP	4556B	4556			
TC50	5800	9003		1142	
TC5080P	1207	3713			
TC5081P	1208	3712			
TC5082D	1197	3733			
TC5082L	1197	3733			
TC5082P	1197	3733			
TC5082P-L	1197	3733			
TC5501					MCM5101
TC5508	6508				MCM6508
TC5516					MCM4016
TC62A5B	149A		ZD-62		
TC82A5B	150A	3098	ZD-82		
TC9100C	1167	3732			
TC9100P	1167	3732	IC-317		
TCH98	159		82	2032	
TCH99			82	2032	
TCH99B			82	2032	
TCO.09M22/1			BR-600	1152	
TCR1005	5483	3942			
TCR13	5483	3942			
TCR1505	5483	3942			
TCR1510	5483	3942			
TCR18	5483	3942			
TCR2005	5483	3942			
TCR2010	5483	3942			
TCR23	5483	3942			
TCR2505	5484	6754/5454			
TCR2510	5484	3943/5485			
TCR28	5483	3942			
TCR3005	5484	3943/5485			
TCR3010	5484	3943/5485			
TCR3505	5485	3943			
TCR3510	5485	3943			
TCR38	5484	3943/5485			
TCR4005	5485	3943			
TCR4010	5485	3943			
TCR48	5485	3943			
TCR52	5483	3942			
TCR53	5483	3942			
TCR54	5484	3943/5485			
TCR55	5484	3943/5485			
TCR56	5485	3943			
TCR72	5483	3942			
TCR732	5483	3942			
TCR733	5484	3943/5485			
TCR734	5485	3943			
TCR74	5484	3943/5485			
TCR744	5483	3942			
TCR745	5483	3942			
TCR746	5484	3943/5485			
TCR747	5484	3943/5485			
TCR748	5485	3943			
TCR75	5484	3943/5485			
TCS100			63	2030	MPSA05
TCS101	383		67	2023	MPSA55
TCS102			63	2030	MPSA05
TCS103	383		67	2023	MPSA55
TD-13	551	3843	511	1114	
TD-15	551	3998/506	511	1114	

Industry Standard No.	ECG	SK	GE	RS 276-	MOTOR.
TD-15-BL	177	3100/519	300	1122	
TD-15H	551	3998/506			
TD100			210	2051	
TD101		3444	20	2051	
TD102		3444	20	2009	
TD1401	7400	7400		1801	
TD1401P	7400	7400		1801	
TD1402	7410	7410		1807	
TD1402P	7410	7410		1807	
TD1403	7420	7420		1809	
TD1403P	7420	7420		1809	
TD1404	7430	7430			
TD1404P	7430	7430			
TD1405	7440	7440			
TD1405P	7440	7440			
TD1409	7473	7473		1803	
TD1409P	7473	7473		1803	
TD1419	7451			1825	
TD1419P	7451			1825	
TD15	551		511	1114	
TD15M			511	1114	
TD200			210	2051	
TD2001P	9661				MC661P
TD2002P	9670				MC670P
TD2003P	9668				MC668P
TD2004P	9664				MC664P
TD2005P	9663				MC663P
TD2006P	9669				MC669P
TD2008P	9660				MC660P
TD2009P	9671				MC671P
TD201		3444	20	2009	
TD2010P	9672				MC672P
TD2011P	9662				MC662P
TD2012P	9680				MC697P
TD2013P	9677				MC680P
TD2015P	9667				MC667P
TD2016P					MC665P
TD2017P					MC691P
TD2018P					MC676P
TD202		3444	20	2009	
TD2219			210	2009	
TD250			210	2051	
TD2905		3118	82	2032	
TD3400A	7400	7400		1801	
TD3400AP	7400	7400		1801	
TD3400P	7400	7400		1801	
TD3401A	7401	7401			
TD3401AP	7401	7401			
TD3402A	7402	7402		1811	
TD3402AP	7402	7402		1811	
TD3403AP	7403	7403			
TD3404A	7404	7404		1802	
TD3404AP	7404	7404		1802	
TD3405A	7405	7405			
TD3405AP	7405	7405			
TD3407AP	7407	7407			
TD3409A	7409	7409			
TD3409AP	7409	7409			
TD3410A	7410	7410		1807	
TD3410AP	7410	7410		1807	
TD3410P	7410	7410		1807	
TD34121A	74121	74121			
TD34121AP	74121	74121			
TD3417AP	7417	7417			
TD34192A	74192	74192		1831	
TD34192AP	74192	74192		1831	
TD3420A	7420	7420		1809	
TD3420AP	7420	7420		1809	
TD3420P	7420			1809	
TD3426AP	7426	7426			
TD3430A	7430	7430			
TD3430AP	7430	7430			
TD3430P	7430	7430			
TD3440A	7440	7440			
TD3440AP	7440	7440			
TD3440P	7440	7440			
TD3441A	7441	7441		1804	
TD3441AP	7441	7441		1804	
TD3442A	7442	7442			
TD3442AP	7442	7442			
TD3447A	7447	7447		1805	
TD3447AP	7447	7447		1805	
TD3450A	7450	7450			
TD3450AP	7450	7450			
TD3450P	7450	7450			
TD3451A	7451	7451		1825	
TD3451AP	7451	7451		1825	
TD3451P	7451	7451		1825	
TD3460A	7460	7460			
TD3460AP	7460	7460			
TD3460P	7460	7460			
TD3472A	7472	7472			
TD3472AP	7472	7472			
TD3473A	7473	7473		1803	
TD3473AP	7473	7473	7473	1803	
TD3474A	7474	7474		1818	
TD3474AP	7474	7474		1818	
TD3474P	7474	7474		1818	
TD3475A	7475	7475		1806	
TD3475AP	7475	7475		1806	
TD3480A	7480	7480			
TD3480AP	7480	7480			
TD3483P	7483	7483			
TD3490A	7490	7490		1808	
TD3490AP	7490	7490		1808	
TD3490P	7490	7490		1808	
TD3492A	7492	7492		1819	
TD3492AP	7492	7492		1819	
TD3492P	7492			1819	
TD3493BP	7493A	7493			
TD3493P	7493A	7493			
TD3495A	7495	7495			
TD3495AP	7495	7495			
TD3503A	74164	74164			
TD3503AP	74164	74164			
TD3F800H	310	3856			
TD3F800R	308	3855			
TD400		3114	82	2032	
TD401		3114	82	2032	
TD402		3114	82	2032	
TD500		3114	82	2032	
TD501		3114	82	2032	
TD502		3114	82	2032	
TD550		3114	82	2032	
TD81515	177	3100/519	300	1122	
TD81518			300	1122	
TD960016-1M	177	3100/519	300	1122	
TDA1190Z	1231	3832	IC-333		TDA1190Z
TDA1200	788	3829	IC-229		
TDA1330	747	3279			
TDA1330P	747	3279	IC-218		
TDA1352	749	3171/744			
TDA1405	960	3591			
TDA1412	966	3592			

Industry Standard No.	ECG	SK	GE	RS 276-	MOTOR.
TDA1415	968	3593			
TDA2002	1232	3852	IC-334	703	TDA2002
TDA3081B	916	3550			
TDA3081N	916	3550			
TDB0555B	955M	3564			
TDB0556A	978	3689			
TDB0723A	923D		IC-260	1740	
TDB2912SP	967	3673			
TDB2915SP	969	3674			
TDB7805	309K	3629			
TDB7805T	960	3591			
TDB7806T	962	3669			
TDB7808T	964	3630			
TDB7812T	966	3592			
TDB7815T	968	3593			
TE-1011			504A	1104	
TE-1014			1N60	1123	
TE-1029			300	1122	
TE-1031			1N60	1123	
TE-1050			504A	1104	
TE-500-E	312	3448	FET-1	2035	
TE1010	116		504A	1104	
TE1011	116		504A	1104	
TE1014	109	3087		1123	
TE1024	116		504A	1104	
TE1024C	116		504A	1104	
TE1024D	116		504A	1104	
TE1029	116		504A	1104	
TE1031	109	3087		1123	
TE1042	116		504A	1104	
TE1050	116		504A	1104	
TE1061	113A		6GC1		
TE1061A	113A		6GC1		
TE1064	125		510,531	1114	
TE1068	142A		ZD-12	563	
TE1077	142A		ZD-12	563	
TE1078	116		504A	1104	
TE1080	116		504A	1104	
TE1086	113A		6GC1		
TE1088	116		504A	1104	
TE1089	116		504A	1104	
TE1090	116		504A	1104	
TE1095	118		CR-1		
TE1097	116		504A	1104	
TE1098	109			1123	
TE1105	109			1123	
TE1108	116		504A	1104	
TE1114	118		CR-1		
TE1420	123A	3444	20	2051	
TE1990	128		243	2030	
TE2369	123A	3444	20	2051	
TE2484	108	3244	86	2038	
TE2711	199	3245	62	2010	
TE2712	199	3245	62	2010	
TE2713	172A	3156	64		
TE2714	172A	3156	64		
TE2715	108	3245/199	86	2038	
TE2716	108	3245/199	86	2038	
TE2921	199	3245	62	2010	
TE2922	199	3245	62	2010	
TE2923	199	3245	62	2010	
TE2924	199	3245	62	2010	
TE2925	199	3245	62	2010	
TE2926	199	3245	62	2010	
TE3390	199	3245	62	2010	
TE3391	199	3245	62	2010	
TE3391A	199	3245	62	2010	
TE3392	199	3245	62	2010	
TE3393	199	3245	62	2010	
TE3394	199	3245	62	2010	
TE3395	199	3245	62	2010	
TE3396	199	3245	62	2010	
TE3397	199	3245	62	2010	
TE3398	199	3245	62	2010	
TE3414	123A	3444	20	2051	
TE3415	123A	3444	20	2051	
TE3416	128		243	2030	
TE3417	192		63	2030	
TE3605	123A	3444	20	2051	
TE3605A	123A	3444	20	2051	
TE3606		3122	20	2051	
TE3606A		3122	20	2051	
TE3607	123A	3444	20	2051	
TE3702	193		67	2023	
TE3703	193		67	2023	
TE3704	123A	3444	20	2051	
TE3705	123A	3444	20	2051	
TE3707	108	3245/199	86	2038	
TE3708	108	3245/199	86	2038	
TE3709	108	3245/199	86	2038	
TE3710	108	3245/199	86	2038	
TE3711	108	3245/199	86	2038	
TE3843	199	3245	62	2010	
TE3844	199	3245	62	2010	
TE3845	199	3245	62	2010	
TE3854	199	3245	62	2010	
TE3854A	199	3245	62	2010	
TE3855	199	3245	62	2010	
TE3855A	199	3245	62	2010	
TE3859	199	3245	62	2010	
TE3859A	192	3245/199	63	2030	
TE3860	199	3245	62	2010	
TE3900	199	3245	62	2010	
TE3900A	199	3245	62	2010	
TE3901	199	3245	62	2010	
TE3903	123A	3444	20	2051	
TE3904	123A	3444	20	2051	
TE3905	159		82	2032	
TE3906	123A	3444	20	2051	
TE4123	123A	3444	20	2051	
TE4124	123A	3444	20	2051	
TE4125	159		82	2032	
TE4126	159		82	2032	
TE4256	199	3245	62	2010	
TE4424	123A	3444	20	2051	
TE4425	192		63	2030	
TE4951	123A	3444	20	2051	
TE4952	123A	3444	20	2051	
TE4953	123A	3444	20	2051	
TE4954	123A	3444	20	2051	
TE500	312	3116	FET-2	2035	
TE5086	193		67	2023	
TE5087	193		67	2023	
TE5088	199	3245	62	2023	
TE5089	199	3245	62	2023	
TE5249	199	3245	62	2023	
TE5309	199	3122	62	2023	
TE5309A	123A	3444	20	2051	
TE5310	199	3122	62	2010	
TE5311	199	3122	62	2010	
TE5311A	123A	3444	20	2051	
TE5365	159		82	2032	
TE5366	159		82	2032	
TE5367	193		67	2023	

Industry Standard No.	ECG	SK	GE	RS 276-	MOTOR.
TE5368	123A	3444	20	2051	
TE5369	123A	3444	20	2051	
TE5370	123A	3444	20	2051	
TE5371	123A	3444	20	2051	
TE5376	123A	3444	20	2051	
TE5377	123A	3444	20	2051	
TE5378	159		82	2032	
TE5379	159		82	2032	
TE5447	159		82	2032	
TE5448	159		82	2032	
TE5449	123A	3444	20	2051	
TE5450	123A	3444	20	2051	
TE5451	123A	3444	20	2051	
TE697	123A	3444	20	2051	
TE706	108		86	2038	
TEG0129			39	2015	
TEH0143			40	2012	
TEH0147			20	2051	
TF-30	102A	3004	53	2007	
TF-65	102A	3004	53	2007	
TF-66	102A	3004	53	2007	
TF-78	123A	3444	20	2051	
TF-80/30	121	3009	239	2006	
TF101-A	128		243	2030	
TF101-B	128		243	2030	
TF101-D	128		243	2030	
TF20	116	3311	504A	1104	
TF21	116	3311	504A	1104	
TF22	116	3311	504A	1104	
TF23	116	3311	504A	1104	
TF30	102A		53	2007	
TF34	177	3100/519	300	1122	
TF44	177	3100/519	300	1122	
TF44J	177	3100/519	300	1122	
TF49	102A		53	2007	
TF65	102A		53	2007	
TF65/30	102A		53	2007	
TF65/M	102A		53	2007	
TF65/S/30	102A		53	2007	
TF65M	158		53	2007	
TF66	102A	3123	53	2007	
TF66/30	102A		53	2007	
TF66/60	102A		53	2007	
TF7	124		12		
TF70	101	3861	8	2002	
TF71	101	3861	8	2002	
TF72	101	3861	8	2002	
TF75	102A	3004	53	2007	
TF77	102A	3004	53	2007	
TF78	104	3719	16	2006	
TF78/30	104	3009	16	2006	
TF78/30Z	121	3009	239	2006	
TF78/60	121	3009	239	2006	
TF80/30	121	3717	239	2006	
TF80/302	102A	3004	53	2007	
TF80/30Z	121	3717	239	2006	
TFR-120	116	3016	504A	1104	
TFR105	125		510,531	1114	1N3879
TFR110	125		510,531	1114	1N3880
TFR120	116		504A	1104	1N3881
TFR1205					1N3889
TFR1210					1N3890
TFR1220					1N3891
TFR1240					1N3893
TFR140	125		510,531	1114	1N3883
TFR305					1N3879
TFR310					1N3880
TFR320					1N3881
TFR340					1N3883
TFR605					1N3879
TFR610					1N3880
TFR620					1N3881
TFR640					1N3883
TG-11	116	3016	504A	1104	
TG-12	116	3016	504A	1104	
TG-21	116	3016	504A	1104	
TG-22	116	3016	504A	1104	
TG-28	109	3091		1123	
TG-31	116	3016	504A	1104	
TG-32	116	3016	504A	1104	
TG-41	116	3016	504A	1104	
TG-42	116	3016	504A	1104	
TG-48	109	3004/102A		1123	
TG-51	116	3016	504A	1104	
TG-52	116	3016	504A	1104	
TG-61	116	3017B/117	504A	1104	
TG-62	116	3017B/117	504A	1104	
TG11	116		80		
TG12	116		504A	1104	
TG20A	116	3016	504A	1104	
TG21	116		504A	1104	
TG22	116		504A	1104	
TG2SA201	126	3005	52	2024	
TG2SA201(C)			50	2004	
TG2SA201-O	126		52	2024	
TG2SA201-N	126		52	2024	
TG2SA201C	126	3005	52	2024	
TG2SA608	159	3114/290	82	2032	
TG2SA608-D	290A		82	2032	
TG2SA608-E	290A		82	2032	
TG2SA608C	159	3114/290	82	2032	
TG2SC1025	175		246	2020	
TG2SC1025D	175	3131A/369	246	2020	
TG2SC1046N-A	165	3115	38		
TG2SC1175	289A	3122	210	2051	
TG2SC1175(C)			20	2051	
TG2SC1175-			210	2051	
TG2SC1175-C	123A		210		
TG2SC1175-D	123A		210	2051	
TG2SC1175-E	123A		210	2051	
TG2SC1175C	123A	3122	243	2030	
TG2SC1293	107	3018	61	2009	
TG2SC1293(A)			61	2009	
TG2SC1293-			61	2009	
TG2SC1293-A-A	233	3132	61	2009	
TG2SC1293-B-A	233	3132	61	2009	
TG2SC1293-C-A	233	3132	61	2009	
TG2SC1293-D-A	233	3132	61	2009	
TG2SC1293A	161	3018	61	2009	
TG2SC1295	165	3115	38		
TG2SC1295-O	165		38		
TG2SC1295-O-A	165		38		
TG2SC12950	165	3115	38		
TG2SC1755-C-A	198		264		
TG2SC1756	198	3220	251		
TG2SC1756-C	198		251		
TG2SC1756-C-A	198		264		
TG2SC1756-C-B	198		264		
TG2SC1756-D	198		251		
TG2SC1756-D-A	198		264		
TG2SC1756-D-B	198		264		
TG2SC1756-D-R	198		251		
TG2SC1756-E	198		251		
TG2SC1756-E-A	198		264		
TG2SC1756-E-B	198		264		
TG2SC2057-C	319P		39	2016	
TG2SC2057-D	229		39	2016	
TG2SC2057C	319P	3246/229	39	2016	
TG2SC2228	287	3866	220		
TG2SC536	199	3122	212	2010	
TG2SC536(C)			212	2010	
TG2SC536(E)			212	2010	
TG2SC536-D-A	199	3122	212	2010	
TG2SC536-D-B	199	3122	212	2010	
TG2SC536-E	199		212	2010	
TG2SC536-E-A	199	3122	212	2010	
TG2SC536-E-B	199	3122	212	2010	
TG2SC536-F	199	3122	212	2010	
TG2SC536-F-A	199	3122	212	2010	
TG2SC536C	199	3122	212	2010	
TG2SC536E	199	3122	212	2010	
TG2SC65	154		40	2012	
TG2SC65(Y)			18	2030	
TG2SC65Y	154	3044	40	2012	
TG2SC927	161	3117	39	2015	
TG2SC927(C)			39	2015	
TG2SC927-C-A	319P	3018	283	2016	
TG2SC927-D-A	319P	3018	283	2016	
TG2SC927-DD-A	161	3716			
TG2SC927-E-A	319P	3018	283	2016	
TG2SC927A	107	3117	11	2015	
TG2SC927C	107	3117	11	2015	
TG2SD24	124		12		
TG2SD24Y	124	3021	12		
TG2SD386Y-D	375	3104A			
TG2SD386Y-D-A	198	3104A	251		
TG2SD386Y-E	375	3104A			
TG2SD386Y-E-A	198	3104A	251		
TG31	116		504A	1104	
TG32	116		504A	1104	
TG41	116		504A	1104	
TG42	116		504A	1104	
TG48	158		53	2007	
TG51	116		504A	1104	
TG52	116		504A	1104	
TG61	116		504A	1104	
TG62	116		504A	1104	
TGB0331			210	2051	
TGZSA608(C)			21	2034	
TH-1S750	112	3089	1N82A		
TH-H2SC313	161	3117	39	2015	
TH1000	125		510,531	1114	
TH1S557	116	3017B/117	504A	1104	
TH2SC536	123A	3444	20	2051	
TH2SC693	123A	3444	20	2051	
TH2SC715	123A	3444	20	2051	
TH400	116	3017B/117	504A	1104	
TH50	116		504A	1104	
TH600	116	3017B/117	504A	1104	
TH7251	196		241	2020	
TH800	125	3032A	510,531	1114	
TH801	116	3017B/117	504A	1104	
TH802	116	3017B/117	504A	1104	
TH803	116	3031A	504A	1104	
TH804	116	3017B/117	504A	1104	
TH805	116	3017B/117	504A	1104	
TH806	116	3017B/117	504A	1104	
TH808	125	3032A	510,531	1114	
TH810	125		510,531	1114	
THSG105	116	3017B/117	504A	1104	
THU60U	102A		53	2007	
TI-132	5483	3942			
TI-136	5483	3942			
TI-1A6	131		44	2006	
TI-266A	121	3009	239	2006	
TI-269	121	3009	239	2006	
TI-3016	107		11	2015	
TI-3029			239	2006	
TI-3030	179	3642	76		
TI-3031	179	3642	76		
TI-338	160	3006	245	2004	
TI-363	100	3005	1	2007	
TI-364	100	3721	1	2007	
TI-365A	166	9075		1152	
TI-366	121	3009	239	2006	
TI-366A	121	3009	239	2006	
TI-367	121	3009	239	2006	
TI-367A	121	3009	239	2006	
TI-368	121	3009	239	2006	
TI-368A	121	3009	239	2006	
TI-369	121	3009	239	2006	
TI-369A	121	3009	239	2006	
TI-370	121	3009	239	2006	
TI-370A	121	3009	239	2006	
TI-387	160	3006	245	2004	
TI-388	160	3006	245	2004	
TI-389			51	2004	
TI-400	160	3006	245	2004	
TI-401	160	3006	245	2004	
TI-402				2005	
TI-403	160	3006	245	2004	
TI-407	161	3019	39	2015	
TI-408	161	3019	39	2015	
TI-409	161	3019	39	2015	
TI-40A1	5483	3942			
TI-40A2	5483	3942			
TI-410	108	3019	86	2038	
TI-412	123A	3122	20	2051	
TI-413	123A		20	2051	
TI-414	123A	3122	20	2051	
TI-415	199	3020/123	62	2010	
TI-416	199	3020/123	62	2010	
TI-417	108	3019	86	2038	
TI-418	199	3019	62	2010	
TI-419	199	3019	62	2010	
TI-420	108		86	2038	
TI-421	199		62	2010	
TI-422	123A		20	2051	
TI-423	123A		20	2051	
TI-424	128		243	2030	
TI-425	128		243	2030	
TI-428	159		82	2032	
TI-429	159		82	2032	
TI-430	108		86	2038	
TI-431	108		86	2038	
TI-432	123A		20	2051	
TI-433	123A		20	2051	
TI-474	108		86	2038	
TI-475	128		243	2030	
TI-480	128		243	2030	
TI-481			81	2051	
TI-482	128		243	2030	
TI-483	128		243	2030	
TI-484	128		243	2030	
TI-485	192		63	2030	
TI-490	108		86	2038	
TI-492	161	3020/123	39	2015	
TI-493	161	3020/123	39	2015	

Industry Standard No.	ECG	SK	GE	RS 276-	MOTOR.
TI-494	161	3020/123	39	2015	
TI-495	108	3020/123	86	2038	
TI-496	128		243	2030	
TI-503	159	3114/290	82	2032	
TI-51	177		300	1122	
TI-53	116	3311	504A	1104	
TI-55	116	3016	504A	1104	
TI-56	5800			1142	
TI-71	116	3017B/117	504A	1104	
TI-714	123A	3122	20	2051	
TI-714A	123A	3122	20	2051	
TI-722	154		40	2012	
TI-741	312		FET-1	2035	
TI-743	159	3114/290	82	2032	
TI-744	159	3114/290	82	2032	
TI-751	123A	3122	20	2051	
TI-752	159	3114/290	82	2032	
TI-7A	131		44	2006	
TI-806G	123A	3122	20	2051	
TI-890	159		82	2024	
TI-891	193		67	2023	
TI-905	159		82	2032	
TI-906	159	3114/290	82	2032	
TI-907	123A	3122	20	2051	
TI-908	123A	3122	20	2051	
TI-92	123A		62	2010	
TI-UG-1888	177		300	1122	
TI-UG1888	177		300	1122	
TI136	5483	3942			
TI137	5484	3943/5485			
TI138	5485	3943			
TI152	116		504A	1104	
TI1A6	123A	3444	20	2051	
TI24A	123A	3444	20	2051	
TI24B	123A	3444	20	2051	
TI25A	108		86	2038	
TI25B	108		86	2038	
TI266A	121	3717	239	2006	
TI269	121	3717	239	2006	
TI3010	160		245	2004	
TI3011	160		245	2004	
TI3012	121	3009	239	2006	
TI3015	128	3024	243	2030	
TI3016	108	3039/316	86	2038	
TI3027	121	3009	239	2006	
TI3028	121	3009	239	2006	
TI3029	121	3014	239	2006	
TI338	160		245	2004	
TI363	160		245	2004	
TI364	160		245	2004	
TI365	160		245	2004	
TI366	121	3717	239	2006	
TI366A	121	3717	239	2006	
TI367	121	3717	239	2006	
TI367A	121	3717	239	2006	
TI368	121	3717	239	2006	
TI368A	121	3717	239	2006	
TI369	121	3717	239	2006	
TI369A	121	3717	239	2006	
TI370	104	3719	16	2006	
TI370A	104	3719	16	2006	
TI387	160		245	2004	
TI388	160		245	2004	
TI389	160		245	2004	
TI390	160		245	2004	
TI391	160		245	2004	
TI393	160		245	2004	
TI395	160		245	2004	
TI396	160		245	2004	
TI397	160		245	2004	
TI398	160		245	2004	
TI399	160		245	2004	
TI400	160	3006	245	2004	
TI401	160	3006	245	2004	
TI402	160	3006	245	2004	
TI403	160	3006	245	2004	
TI407	108		86	2038	
TI408	108		86	2038	
TI409	108		86	2038	
TI40A2	5483	3942			
TI40A3	5484	3943/5485			
TI40A4	5485	3943			
TI410	108		86	2038	
TI411	123A	3444	20	2051	
TI415	123A	3444	20	2051	
TI416	123A	3444	20	2051	
TI417	123A	3444	20	2051	
TI418	123A	3444	20	2051	
TI419	123A	3444	20	2051	
TI42	6407	9083A			
TI420	123A	3444	20	2051	
TI422	123A	3444	20	2051	
TI424	123A	3444	20	2051	
TI430	123A	3444	20	2051	
TI431	108	3039/316	86	2038	
TI432	123A	3444	20	2051	
TI480	123A	3444	20	2051	
TI481	128	3444/123A	20	2051	
TI482	123A	3444	20	2051	
TI483	123A	3444	20	2051	
TI484	123A	3444	20	2051	
TI485	123A	3444	20	2051	
TI486	152	3893	66	2048	
TI487	152	3893	66	2048	
TI492	123A	3444	20	2051	
TI493	123A	3444	20	2051	
TI494	123A	3444	20	2051	
TI495	123A	3444	20	2051	
TI496	123A	3444	20	2051	
TI503	159		82	2032	
TI51	177		300	1122	
TI52	116		504A	1104	
TI53	116		504A	1104	
TI54	116	3311	504A	1104	
TI54A	123A	3444	20	2051	
TI54B	123A	3444	20	2051	
TI54C	123A	3444	20	2051	
TI54D	199		62	2010	
TI54E	123A	3444	20	2051	
TI55	116		504A	1104	
TI56	116	9003/5800	504A	1104	
TI57	116	3311	504A	1104	
TI58	116	3031A	504A	1104	
TI59	116	3031A	504A	1104	
TI60	116	3031A	504A	1104	
TI64213	128	3024	243	2030	
TI642B			10	2051	
TI71	116		504A	1104	
TI714	123A	3444	20	2051	
TI722	171		27		
TI741	312	3448	FET-1	2035	
TI743	159		82	2032	
TI744	159		82	2032	
TI751	123A	3444	20	2051	
TI752	159		82	2032	
TI8003B				2051	
TI802B	123A	3444	20	2051	
TI803B	123A	3444	20	2051	
TI808E	129		244	2027	
TI810B	123A	3444	20	2051	
TI811G	128	3024	243	2030	
TI904	123A	3444	20	2051	
TI907	123A	3444	20	2051	
TI908	123A	3444	20	2051	
TIA-01	102A	3004	53	2007	
TIA01	102A		53	2007	
TIA02	160	3123	245	2004	
TIA03	100	3005	1	2007	
TIA04	102	3004	2	2007	
TIA042	5483	3942			
TIA05	100	3005	1	2007	
TIA05A	100	3721	1	2007	
TIA06	123A	3444	20	2051	
TIA102	123A	3444	20	2051	
TIA35	176	3123	80		
TIC106A	5454	6754		1067	
TIC106B	5455	3597		1067	
TIC106C	5456			1020	
TIC106F	5453	6753		1067	
TIC106Y	5452	6752		1067	
TIC116A	5462	3686			
TIC116C	5465	3687			
TIC116D	5465	3687			
TIC116F	5461	3685			
TIC126A		3574	C126A		
TIC126B		3574	C126B		
TIC126C		3575	C126C		
TIC126D		3575	C126D		
TIC126E		3576	C126E		
TIC126M		3576	C126M		
TIC22	5667A	3632/5624			
TIC220B					2N5567
TIC220D					2N5568
TIC220E					T4101M
TIC221B					T4121B
TIC221D					T4121D
TIC221E					T4121M
TIC222B	5673	3520/5677			2N5569
TIC222D	5675	3520/5677			2N5570
TIC222E	5676	3520/5677			T4111M
TIC226B	56004	3658			SC141B
TIC226D	56006	3659			SC141D
TIC23	5667A	3633/5626			
TIC230B					2N5567
TIC230D					2N5568
TIC230E					T4101M
TIC231B					T4121B
TIC231D					T4121D
TIC231E					T4121M
TIC232B	5673	3520/5677			2N5569
TIC232D	5675	3520/5677			2N5570
TIC232E	5676	3520/5677			T4111M
TIC236B	56004	3658	SC149B		2N6342A
TIC236D	56006	3659	SC149D		2N6343A
TIC240B					2N5571
TIC240D					2N5572
TIC240E					T4100M
TIC241B					T4120B
TIC241D					T4120D
TIC241E					T4120M
TIC2420	5677	3520			
TIC242B	5673	3520/5677			2N5573
TIC242D	5675				2N5574
TIC242E	5676	3520/5677			T4110M
TIC246B		3938	SC151B		MAC15-4
TIC246D		3533	SC151D		MAC15-6
TIC250B					S6401B
TIC250D					S6401D
TIC250E					S6401E
TIC250M					S6401M
TIC252B	5683	3521/5686			T6411B
TIC252D	5685	3521/5686			T6411D
TIC252E	5686	3521			T6411E
TIC252M	5687				T6411M
TIC253B					MAC20-4
TIC253D					MAC20-6
TIC253E					MAC20-7
TIC253M					MAC20-8
TIC260B					T6401B
TIC260D					T6401D
TIC260E					T6401E
TIC260M					T6401M
TIC262B	5683	3521/5686			
TIC262D	5685	3521/5686			T6411D
TIC262E	5686	3521			T6411E
TIC262M	5687				T6411M
TIC263B					MAC25-4
TIC263D					MAC25-5
TIC263E					MAC25-7
TIC263M					MAC25-8
TIC270B					2N5441
TIC270D					2N5442
TIC270E					2N5443
TIC270M					2N5444
TIC272B	5693	3652			2N5444
TIC272D	5695	3509			2N5445
TIC272E	5697	3522			2N5446
TIC272M	5697	3522			2N5446
TIC44	5400	3950	MR-5		
TIC45	5401	3638/5402	MR-5		
TIC46	5402	3638	MR-5		
TIC47	5404	3627	MR-5		
TIC54	6407	9083A			
TIC56	6407	3523			
TIC57	6406	9082A			
TIC60	5400	3950			
TIC61	5401	3638/5402			
TIC62	5402	3638			
TIC63	5403	3627/5404			
TIC64	5404	3627			
TIE	718		IC-8		
TIJ	718	3159	IC-8		
TIL111	3042		H11A4		TIL111
TIL112			H11A5		TIL112
TIL113	3041		H11B2		TIL113
TIL114	3042		H11A3		TIL114
TIL115			H11A3		TIL115
TIL116	3043		H11A3		TIL116
TIL117			H11A1		TIL117
TIL118			H11A5		TIL118
TIL119			H11B2		TIL119
TIL26					MLED60
TIL31	3028		LED55B		MLED930
TIL33			LED55B		MLED930
TIL34	3028		LED56		MLED930
TIL63					MRD310
TIL64	3032				MRD310
TIL65					MRD310

Industry Standard No.	ECG	SK	GE	RS 276-	MOTOR.
TIL66	3032				MRD300
TIL67	3032				MRD300
TIL78					MRD450
TIL81	3032		L14G1		MRD300
TIM-01	160		245	2004	
TIM-10	160		245	2004	
TIM-11	160		245	2004	
TIN	718		IC-8		
TIP-14	152	3054/196	66	2048	
TIP-24	196	3054	241	2020	
TIP-29	291		23	2020	
TIP-30	292			2049	
TIP-31	152	3054/196	241	2020	
TIP-31A	152	3054/196	241	2048	
TIP-31B	291	3054/196	241	2020	
TIP100		3180			TIP100
TIP101		3180			TIP101
TIP102					TIP102
TIP105					TIP105
TIP106					TIP106
TIP107					TIP107
TIP110	261				TIP110
TIP111	261		D44E3		TIP111
TIP112	261				TIP112
TIP115	262				TIP115
TIP116	262		D45E3		TIP116
TIP117	262				TIP117
TIP120	261	3896			TIP120
TIP121	261	3896			TIP121
TIP122		3978			TIP122
TIP125	262	3897			TIP125
TIP126	262	3897			TIP126
TIP127					TIP127
TIP14	184		57	2017	
TIP140	270	3935			TIP140
TIP141	270	3935			TIP141
TIP142	270	3935			TIP142
TIP145	271	3936			TIP145
TIP146	271	3936			TIP146
TIP147	271	3936			TIP147
TIP150					MJE13006
TIP151					MJE13007
TIP152					MJE13007
TIP160					MJE5740
TIP161					MJE5741
TIP162					MJE5742
TIP24	152		66	2048	
TIP27	157	3103A/396	232		
TIP29	291	3893/152	66	2048	TIP29
TIP2955	393	3961	74	2043	MJE2955T
TIP29A	291	3054/196	66	2048	TIP29A
TIP29B	291		251		TIP29B
TIP29C	291		251		TIP29C
TIP29XA	196	3054	241	2020	
TIP30	292	3083/197	69	2049	TIP30
TIP303					2N6544
TIP304					2N6544
TIP305					2N6545
TIP3055	392				MJE3055T
TIP306					2N6545
TIP309					BU208
TIP30A	292		69	2049	TIP30A
TIP30B	292	3200/189	218	2026	TIP30B
TIP30C	292				TIP30C
TIP31	152	3054/196	66	2048	TIP31
TIP310					BU208
TIP31A	152	3054/196	66	2048	TIP31A
TIP31B	291		57	2017	TIP31B
TIP31C	291				TIP31C
TIP32	153	3083/197	69	2049	TIP32
TIP32A	153		69	2049	TIP32A
TIP32B	292		250	2027	TIP32B
TIP32C	292				TIP32C
TIP33	390	3958			TIP41
TIP33A	390	3958			TIP41A
TIP33B	390	3958			TIP41B
TIP33C	390	3958			TIP41C
TIP34	391	3959	56	2027	TIP42
TIP34A	391	3959	56	2027	TIP42A
TIP34B	391	3959	56	2027	TIP42B
TIP34C	391	3959			TIP42C
TIP35	392	3960			TIP35
TIP35A	392	3960			TIP35A
TIP35B	392	3960			TIP35B
TIP35C	392	3960			TIP35C
TIP35D					2N6339
TIP35E					2N6340
TIP36	393	3961			TIP36
TIP36A	393	3961			TIP36A
TIP36B	393	3961			TIP36B
TIP36C	393	3961			TIP36C
TIP3A	197	3083	250	2027	
TIP4	196		241	2020	
TIP41	331		241	2020	TIP41
TIP41A	331		241	2020	TIP41A
TIP41B	331		241	2020	TIP41B
TIP41C	331				TIP41C
TIP42	332		250	2027	TIP42
TIP42A	332		250	2027	TIP42A
TIP42B	332		250	2027	TIP42B
TIP42C	332				TIP42C
TIP47	198	3219	251		TIP47
TIP48	198	3219	251		TIP48
TIP49	198	3220	251		TIP49
TIP50	198	3220			TIP50
TIP501	282				2N4877
TIP502	285				2N4877
TIP503	175		246	2020	2N5050
TIP504					2N5051
TIP505	95$				2N5050
TIP506	95$				2N5051
TIP507					2N6211
TIP508					2N6211
TIP509	328	3895			MJ4247
TIP51	394	3983			2N6306
TIP510	328	3895			MJ4248
TIP511					MJ4247
TIP512					MJ4248
TIP513					MJ15012
TIP514					MJ3238
TIP515	328				2N6339
TIP516	327	3895/328			2N6341
TIP517					2N6339
TIP518					2N6341
TIP519					MJ4238
TIP52	394	3983			2N6307
TIP520					MJ4238
TIP521					2N6211
TIP522					2N6211
TIP523					MJ15012
TIP524					2N6497
TIP525	385				MJ15011
TIP526	73$				MJ15011
TIP527					MJ15012
TIP528					MJ15012
TIP529					2N6542
TIP53	394	3983			2N6308
TIP530	384				2N6235
TIP531	386				2N6546
TIP532	386				2N6547
TIP533					2N6546
TIP534					2N6457
TIP535	385				2N6544
TIP536	385				2N6544
TIP537	386				2N6545
TIP538	386				2N6249
TIP539	386				2N6546
TIP54	394	3983			2N6545
TIP540	386				2N6547
TIP541	282	3748			
TIP542					2N5347
TIP543					2N5347
TIP544	285		74	2043	2N6226
TIP545	285				2N6227
TIP546	285				2N6228
TIP550		3111			BU205
TIP551		3115			BU205
TIP552		3115			BU207
TIP553		3115			BU208
TIP554	385				2N6306
TIP555	385				2N6307
TIP556					2N6545
TIP558	283				2N6544
TIP559	283				2N6544
TIP55A					MJE13008
TIP560	283				2N6545
TIP561	283				2N6545
TIP562	283				2N6546
TIP563	283				2N6547
TIP564					MJ13014
TIP565					MJ13015
TIP56A					MJE13008
TIP575					MJ410
TIP575A					MJ3029
TIP575B					2N6542
TIP575C					2N6543
TIP57A					MJE13009
TIP58A					MJE13009
TIP600					2N6384
TIP601					2N6385
TIP602					2N6059
TIP605					2N6649
TIP606					2N6650
TIP607					2N6052
TIP61					TIP61
TIP61A					TIP61A
TIP61B					TIP61B
TIP61C					TIP61C
TIP62					TIP62
TIP620	243				2N6055
TIP621	243				2N6056
TIP622	243				2N6578
TIP625	244				2N6053
TIP626	244				2N6054
TIP627	244				2N6052
TIP62A					TIP62A
TIP62B					TIP62B
TIP62C					TIP62C
TIP640	270	3858/251			MJ3000
TIP641	270	3858/251			MJ3001
TIP642	270	3858/251			2N6578
TIP645	271	3859/252			MJ2500
TIP646	271	3859/252			MJ2501
TIP647	271	3859/252			2N6052
TIP660					MJ10002
TIP661					MJ10002
TIP662					MJ10002
TIP663					MJ10004
TIP664					MJ10004
TIP665					MJ10005
TIP666	97				MJ10006
TIP667	97				MJ10006
TIP668	98				MJ10007
TIP69					BU205
TIP70					BU205
TIP71					BU205
TIP72					BU205
TIP73					2N6486
TIP73A					2N6487
TIP73B					2N6488
TIP74					2N6489
TIP74A					2N6490
TIP74B					2N6491
TIP75					MJE13005
TIP75A					MJE13004
TIP75B					MJE13004
TIP75C					MJE13005
TIP8350					2N6648
TIP8350A					2N6649
TIP8350B					2N6650
TIR01	116	3017B/117	504A	1104	
TIR02	116	3017B/117	504A	1104	
TIR03	116	3031A	504A	1104	
TIR04	116	3031A	504A	1104	
TIR05	116	3017B/117	504A	1104	
TIR06	116	3017B/117	504A	1104	
TIR07	125	3032A	510,531	1114	
TIR08	125	3032A	510,531	1114	
TIR09	125	3033	510,531	1114	
TIR10	125	3033	510,531	1114	
TIR101A					SPECIAL
TIR101B					SPECIAL
TIR101C					SPECIAL
TIR101D					SPECIAL
TIR102A					SPECIAL
TIR102B					SPECIAL
TIR102C					SPECIAL
TIR102D					SPECIAL
TIR201A					SPECIAL
TIR201B					SPECIAL
TIR201C					SPECIAL
TIR201D					SPECIAL
TIR202A					SPECIAL
TIR202B					SPECIAL
TIR202C					SPECIAL
TIR202D					SPECIAL
TIS-03	159	3114/290	82	2032	
TIS-125			17	2051	
TIS-18			60	2015	
TIS-1B			11	2015	
TIS-62			20	2051	
TIS-88	312	3116	FET-1	2035	
TIS-94	199		62	2010	
TIS-97	199		62	2010	
TIS03	159		82	2032	
TIS04	159	3114/290	82	2032	
TIS05					2N5465
TIS100	287	3045/225	40	2012	MPSA43

Industry Standard No.	ECG	SK	GE	RS 276-	MOTOR.
TIS101	287	3045/225	40	2012	MPS5551
TIS102	154	3045/225	40	2012	
TIS103	154	3045/225	40	2012	
TIS104	159		82	2032	MPSH24
TIS105	229		86	2038	MPS918
TIS106	194		220		
TIS107	128		243	2030	
TIS108	229	3039/316	11	2015	MPSH32
TIS109	195A	3765	46		MPS2222
TIS110	128	3765/195A	46		MPS2222
TIS111	128	3765/195A	46		MPS2222A
TIS112	159		82	2032	MPS2907
TIS113	123A	3444	20	2051	
TIS114	123A	3444	20	2051	
TIS125	123AP		61	2038	MPSH08
TIS126	229				MPSH34
TIS128			21	2034	MPSH83
TIS129	229				MPSH10
TIS133	123AP		47	2030	2N3724
TIS134	123AP		47	2030	2N3724
TIS135	128				2N3725
TIS136	128				2N3725
TIS137	159				MPSH54
TIS138	288		21	2034	MPSH54
TIS14	312	3112	FET-1	2028	2N4220A/2A
TIS18	107	3039/316	39	2015	
TIS22	123A	3444	20	2051	
TIS23	123A	3444	20	2051	
TIS24	108	3039/316	86	2038	
TIS34	312	3448	FET-1	2035	2N5486
TIS37	159	3025/129	82	2032	
TIS38	159	3114/290	82	2032	
TIS412	107	3039/316	11	2015	
TIS42	312	3448	FET-1	2035	2N4092
TIS44	123A	3444	20	2051	
TIS45	123A	3444	20	2051	
TIS46	123A	3444	20	2051	
TIS47	123A	3444	20	2051	
TIS48	123A	3444	20	2051	
TIS49	123A	3444	20	2051	
TIS50	159	3122	82	2032	MPS2369
TIS51	123A	3444	20	2051	MPS2369
TIS52	123A	3444	20	2051	
TIS53	159	3114/290	82	2032	MPS3640
TIS54	159	3114/290	82	2032	MPS3640
TIS55	123A	3444	20	2051	
TIS56	161	3444/123A	20	2051	MPSH32
TIS57	161	3444/123A	20	2051	MPSH32
TIS58	312	3448	FET-1	2035	2N5458
TIS59	312	3448	FET-1	2035	2N5670
TIS60	289A	3124/289	20	2051	MPS6531
TIS60A	128	3124/289	243	2030	
TIS60B	128	3124/289	243	2030	
TIS60C	128	3124/289	243	2030	
TIS60D	128	3124/289	243	2030	
TIS60E	128	3124/289	243	2030	
TIS60M	128	3124/289	243	2030	
TIS61	159	3114/290	82	2032	MPS6534
TIS61A	129	3114/290	244	2027	
TIS61B	129	3114/290	244	2027	
TIS61C	129	3114/290	244	2027	
TIS61D	129	3114/290	244	2027	
TIS61E	129	3114/290	244	2027	
TIS61M	129	3114/290	244	2027	
TIS62	108	3452	86	2038	MPS918
TIS62A	108	3452			
TIS63	108	3452	86	2038	MPS918
TIS63A	108	3452			
TIS64	108	3452	86	2038	MPS918
TIS64A	108	3452			
TIS71	123A	3444	20	2051	
TIS72	123A	3444	20	2051	
TIS73					2N5638
TIS74					2N5639
TIS75					2N5640
TIS78	312	3112		2028	
TIS79	312	3112		2028	
TIS82	186	3192	28	2017	
TIS83	123AP	3444/123A	20	2051	
TIS84	229	3039/316	86	2038	MPSH32
TIS85	108	3039/316	86	2038	
TIS86	229	3039/316	11	2015	MPSH24
TIS87	229	3039/316	11	2015	MPSH34
TIS88	312	3448	FET-1	2035	2N5486
TIS90	289A	3444/123A	20	2051	MPS8092
TIS90-2	123A	3444	20	2051	
TIS90M					2N4401
TIS91	159	3114/290	82	2032	MPS8093
TIS92	123A	3444	20	2051	MPS8092
TIS92-BLU	123A	3444	20	2051	
TIS92-GRN	123A	3444	20	2051	
TIS92-GRY	123A	3444	20	2051	
TIS92-VIO	123A		20	2051	
TIS92-YEL	123A	3444	20	2051	
TIS92M	128	3024	243	2030	2N4401
TIS93	159	3114/290	82	2032	MPS8093
TIS93-BLU	159		82	2032	
TIS93-GRN	159		82	2032	
TIS93-GRY	159		82	2032	
TIS93-VIO	159		82	2032	
TIS93-YEL	159		82	2032	
TIS93M	129	3025	244	2027	2N4403
TIS94	123A	3444	20	2051	MPS6521
TIS94(APAMP)			10	2051	
TIS94(XSTR)			10	2051	
TIS95	293	3444/123A	20	2051	MPS8098
TIS96	293	3444/123A	20	2051	MPS8099
TIS97	123AP	3039/316	11	2015	MPS8097
TIS98	128	3122	11	2015	MPS8098
TIS98A	108	3018	86	2038	
TIS99	128	3039/316	11	2015	MPS8099
TIS991				2032	
TIS992				2051	
TIS993				2032	
TISM74LS00N				1900	
TIU3304	912		IC-172		
TIX-M01	160		245	2004	
TIX-M02	160		245	2004	
TIX-M03	160		245	2004	
TIX-M04	160		245	2004	
TIX-M05	160		245	2004	
TIX-M06	160		245	2004	
TIX-M07	160		245	2004	
TIX-M08	160		245	2004	
TIX-M101	160		245	2004	
TIX-M11	160		245	2004	
TIX-M14	161	3006/160	39	2015	
TIX-M15	161	3006/160	39	2015	
TIX-M16	161	3006/160	39	2015	
TIX-M17	160		245	2004	
TIX-M201	160		245	2004	
TIX-M202	160		245	2004	
TIX-M203	160		245	2004	
TIX-M204	160		245	2004	

Industry Standard No.	ECG	SK	GE	RS 276-	MOTOR.
TIX-M205	160		245	2004	
TIX-M206	160		245	2004	
TIX-M207	160		245	2004	
TIX1392	195A		46		
TIX1393	195A		46		
TIX3016	160		245	2004	
TIX3016A	160		245	2004	
TIX3032	160		245	2004	
TIX316	160		245	2004	
TIX440	5940		5048		
TIX441	5944		5048		
TIX442	5944		5048		
TIX712	123A	3444	20	2051	
TIX804	159		82	2032	
TIX805	159		82	2032	
TIX876	108		86	2038	
TIX880	108		86	2038	
TIX888	195A	3765	46		
TIX890	159		82	2032	
TIX891	159		82	2032	
TIX895	100	3721	1	2007	
TIX896	101	3861	8	2002	
TIX90	102A		53	2007	
TIX91	160		245	2004	
TIX92	160		245	2004	
TIXA-03	100	3005	1	2007	
TIXA-04	100	3005	1	2007	
TIXA-05	100	3005	1	2007	
TIXA01	100	3005	1	2007	
TIXA02	100	3005	1	2007	
TIXA03	100	3123	1	2007	
TIXA04	100	3123	1	2007	
TIXA05	100	3123	1	2007	
TIXD747	134A	3055	ZD-3.6		
TIXD753	137A	3058	ZD-6.2	561	
TIXD758	140A	3061	ZD-10	562	
TIXM-201	160	3006	245	2004	
TIXM-203	160	3006	245	2004	
TIXM-205	160	3006	245	2004	
TIXM-206	160	3006	245	2004	
TIXM01	160		245	2004	
TIXM02	160		245	2004	
TIXM03	160		245	2004	
TIXM04	160		245	2004	
TIXM05	160		245	2004	
TIXM06	160		245	2004	
TIXM07	160		245	2004	
TIXM08	160		245	2004	
TIXM10	160		245	2004	
TIXM101	160		245	2004	
TIXM103	160		245	2004	
TIXM104	160		245	2004	
TIXM105	160		245	2004	
TIXM106	160		245	2004	
TIXM107	160		245	2004	
TIXM108	160		245	2004	
TIXM11	160		245	2004	
TIXM13	160		245	2004	
TIXM14	160	3716/161	245	2004	
TIXM15	160		245	2004	
TIXM16	160		245	2004	
TIXM17	160		245	2004	
TIXM18	160		245	2004	
TIXM19	160		245	2004	
TIXM201	160		245	2004	
TIXM202	160	3006	245	2004	
TIXM203	160		245	2004	
TIXM204	160	3006	245	2004	
TIXM205	160		245	2004	
TIXM206	160		245	2004	
TIXM207	160		245	2004	
TIXS09	108	3039/316	86	2038	
TIXS10	108	3039/316	86	2038	
TIXS11					2N4352
TIXS12	123A	3444	20	2051	
TIXS13	123A	3444	20	2051	
TIXS28	108	3039/316	86	2038	
TIXS29	108	3039/316	86	2038	
TIXS30	108	3039/316	86	2038	
TIXS31	108	3039/316	86	2038	
TIXS42					MFE2004
TIXS67					2N4352
TIXS78					3N126
TIXS79					3N126
TJ-5A	116	3016	504A	1104	
TJ10A	116	3016	504A	1104	
TJ15A	116	3016	504A	1104	
TJ20A	116		504A	1104	
TJ25A	116	3016	504A	1104	
TJ30A	116	3016	504A	1104	
TJ35A	116	3016	504A	1104	
TJ40A	116	3016	504A	1104	
TJ5A	116		504A	1104	
TJ60A	116	3017B/117	504A	1104	
TK-10	116	3016	504A	1104	
TK-23C	102A	3004	53	2007	
TK-30	116	3016	504A	1104	
TK-40	116	3004/102A	504A	1104	
TK-40C	102A	3004	53	2007	
TK-41	116		504A	1104	
TK-41C	102A	3004	53	2007	
TK-42C	102A	3004	53	2007	
TK-45C	102A	3004	53	2007	
TK-705M	113A		6GC1		
TK10	116		504A	1104	1N4002
TK1000	125	3033	510,531	1114	
TK11	116	3017B/117	504A	1104	1N4002
TK1228-001	160	3006	245	2004	
TK1228-1001	126	3008	52	2024	
TK1228-1002	102A		53	2007	
TK1228-1003	102A		53	2007	
TK1228-1004	102A		53	2007	
TK1228-1005	102A		53	2007	
TK1228-1006	102A		53	2007	
TK1228-1007	102A		53	2007	
TK1228-1008	123	3020	20	2051	
TK1228-1009	123	3020	20	2051	
TK1228-1010	123A	3444	20	2051	
TK1228-1011	123A	3444	20	2051	
TK1228-1012	123A	3444	20	2051	
TK20	116	3016	504A	1104	1N4003
TK21	116	3017B/117	504A	1104	1N4003
TK23C	102A		53	2007	
TK30	116		504A	1104	1N4004
TK30551			75	2041	
TK30552			75	2041	
TK30555			75	2041	
TK30556			75	2041	
TK30557			75	2041	
TK30560			75	2041	
TK33C	101	3861	8	2002	
TK40	102	3004	2	2007	1N4004
TK400	116	3017B/117	504A	1104	
TK40A	102A	3004	53	2007	

Industry Standard No.	ECG	SK	GE	RS 276-	MOTOR.
TK40C	102	3004	2	2007	
TK40CA	102A	3004	53	2007	
TK41	102	3004	2	2007	1N4004
TK41A	102A	3004	53	2007	
TK41C	160		245	2004	
TK42	102	3004	2	2007	
TK42A	102A	3004	53	2007	
TK42C	160		245	2004	
TK45C	102	3004	2	2007	
TK49C	102A	3010	53	2007	
TK5	116	3311	504A	1104	1N4001
TK50	116	3017B/117	504A	1104	1N4005
TK60	116	3017B/117	504A	1104	1N4005
TK600	116	3017B/117	504A	1104	
TK61	116	3017B/117	504A	1104	1N4005
TK705M	113A		6GC1		
TK800	125	3032A	510,531	1114	
TK9201	181		75	2041	
TKF10	116	3016	504A	1104	1N4934
TKF100	125	3033	510,531	1114	MR817
TKF20	116	3016	504A	1104	1N4935
TKF40	116	3017B/117	504A	1104	1N4936
TKF5	116	3017B/117	504A	1104	1N4933
TKF50					1N4937
TKF60	116	3017B/117	504A	1104	1N4937
TKF80	125	3032A	510,531	1114	MR817
TL022CJG					LM358J
TL022CL					LM358H
TL022CP					LM358N
TL022MJG					LM158J
TL022ML					LM158H
TL044CJ					LM324J
TL044CN					LM324N
TL044MJ					LM124J
TL071ACJG					MC34001BU
TL071ACL					MC34001BG
TL071ACP					MC34001BP
TL071BCJG					MC34001AU
TL071BCP					MC34001AP
TL071CJG					MC34001U
TL071CL					MC34001G
TL072ACJG					MC34002BU
TL072ACL					MC34002BG
TL072ACP					MC34002BP
TL072BCJG					MC34002AU
TL072BCL					MC34002AG
TL072BCP					MC34002AP
TL072CJG					MC34002U
TL072CL					MC34002G
TL072CP					MC34002P
TL074ACJ					MC34004BL
TL074ACN					MC34004BP
TL074BCJ					MC34004A
TL074BCL					MC34001AG
TL074BCN					MC34004AP
TL074CJ					MC34004L
TL074CN					MC34004P
TL081ACJG					MC34001BU
TL081ACL					MC34001BG
TL081ACP					MC34001BP
TL081BCJG					MC34001AU
TL081BCL					MC34001AG
TL081BCP					MC34001AP
TL081CJG					MC34001U
TL081CL					MC34001G
TL081CP					MC34001P
TL082ACJG					MC34002BU
TL082ACL					MC34002BG
TL082ACP					MC34002BP
TL082BCJG					MC34002AU
TL082BCL					MC34002AG
TL082BCP					MC34002AP
TL082CJG					MC34002U
TL082CL					MC34002G
TL082CP					MC34002P
TL084ACJ					MC34004BL
TL084ACN					MC34004BP
TL084BCJ					MC34004AL
TL084BCN					MC34004AP
TL084CJ					MC34004L
TL084CN					MC34004P
TL1	116	3311	504A	1104	
TL11		3016	504A	1104	
TL12	116	3016	504A	1104	
TL2		3017B	504A	1104	
TL21		3016	504A	1104	
TL22		3017B	504A	1104	
TL31		3017B	504A	1104	
TL32		3031A	504A	1104	
TL41		3031A	504A	1104	
TL42		3017B	504A	1104	
TL494CJ					TL494CJ
TL494CN					TL494CN
TL495CJ					TL495CJ
TL495CN					TL495CN
TL497CJ					MC3420L
TL497CN					MC3420P
TL497MJ					MC3520L
TL51		3017B	504A	1104	
TL61		3017B	504A	1104	
TL660L					MC660L
TL660P					MC660P
TL661L					MC661L
TL661P					MC661P
TL662L					MC662L
TL662P					MC662P
TL664L					MC664L
TL664P					MC664P
TL665L					MC665L
TL665P					MC665P
TL666L					MC666L
TL666P					MC666P
TL668L					MC668L
TL668P					MC668P
TL670L					MC670L
TL670P					MC670P
TL671L					MC671L
TL671P					MC671P
TL672L					MC672L
TL672P					MC672P
TL681L					MC681L
TL681P					MC681P
TL7400N	7400	7400		1801	
TL7401N	7401	7401			
TL7402N	7402	7402		1811	
TL7403N	7403	7403			
TL7404N	7404	7404		1802	
TL7405N	7405	7405			
TL7406N	7406	7406		1821	
TL7407N	7407	7407			
TL7408N	7408	7408			
TL7409N	7409	7409			
TL74107N	74107	74107			
TL7410N	7410	7410		1807	

Industry Standard No.	ECG	SK	GE	RS 276-	MOTOR.
TL74110	74110	74110			
TL74111	74111	74111			
TL74121N	74121	74121			
TL74122N	74122	74122			
TL74123N	74123	74123		1817	
TL74125	74125	74125			
TL74126	74126	74126			
TL7412N	7412	7412			
TL74132	74132	74132			
TL7413N	7413	7413		1815	
TL74141N	74141	74141			
TL74145N	74145	74145		1828	
TL74150N	74150	74150		1829	
TL74151N	74151	74151			
TL74153N	74153	74153			
TL74154N	74154	74154		1834	
TL74155N	74155	74155			
TL74156N	74156	74156			
TL74162N	74162	74162			
TL74163N	74163	74163			
TL74164N	74164	74164			
TL74165N	74165	74165			
TL74166N	74166	74175			
TL7416N	7416	7416			
TL74180N	74180	74180			
TL74181N	74181	74181			
TL74190N	74190	74190			
TL74191N	74191	74191			
TL74192N	74192	74192		1831	
TL74193N	74193	74193		1820	
TL74196N	74196	74196		1833	
TL74198N	74198	74198			
TL74199N	74199	74199			
TL7420N	7420	7420		1809	
TL7422	7422	7422			
TL7425N	7425	7425			
TL7426N	7426	7426			
TL7428	7428	7428			
TL7430N	7430	7430			
TL7433	7433	7433			
TL7437N	7437	7437			
TL7438N	7438	7438			
TL7440N	7440	7440			
TL7442N	7442	7442			
TL7443N	7443	7443			
TL7444N	7444	7444			
TL7445N	7445	7445			
TL7446AN	7446	7446			
TL7447AN	7447	7447		1805	
TL7448N	7448	7448		1816	
TL7450N	7450	7450			
TL7451N	7451	7451		1825	
TL7453N	7453	7453			
TL7454N	7454	7454			
TL7460N	7460	7460			
TL7470N	7470	7470			
TL7472N	7472	7472			
TL7473N	7473	7473		1803	
TL7474N	7474	7474		1818	
TL7475N	7475	7475		1806	
TL7476N	7476	7476		1813	
TL7480N	7480	7480			
TL7481	7481	7481			
TL7481N	7481	7481			
TL7483N	7483	7483			
TL7485N	7485	7485		1826	
TL7486N	7486	7486		1827	
TL7489N	7489	7489			
TL7490N	7490	7490		1808	
TL7492N	7492	7492		1819	
TL7493N	7493A	7493			
TL7494N	7494	7494			
TL7495AN	7495	7495			
TL7496N	7496	7496			
TL7497N	7497	7497			
TLP501					4N27
TLP503					4N25
TLP504					4N25
TM-22	106		21	2034	
TM-33	116	3016	504A	1104	
TM-43	116	3016	504A	1104	
TM1	5870		5032		
TM100	100	3721	1	2007	
TM1003	1003	3288			
TM1004	1080	3365/1004			
TM1005	1005	3723			
TM101	101		8	2002	
TM102	102	3722	2	2007	
TM102A	102A	3004	52	2007	
TM103	103		8	2002	
TM103A	103A	3835	59	2002	
TM104		3051	512		MR1130
TM104MP	104MP	3720	16	2006(2)	
TM105		3051	512		MR1130
TM1058	1058		IC-49		
TM106	106	3118	21	2034	MR1130
TM107	107	3293	11	2015	
TM108	108	3452	11	2038	
TM1080	1080	3284			
TM109		3090	1N34AS	1123	
TM11	5872		5032		
TM110		3709		1123(2)	
TM1115		3184	IC-278		
TM1116		3969	IC-279		
TM112		3089	1N82A		
TM1131	1131		IC-109		
TM1132	1132		IC-110		
TM1133	1133		IC-107		
TM114			6GD1	1104	
TM1155	1155	3231	IC-179	705	
TM116		3313	504A	1104	
TM1160	1160	3243	IC-196		
TM1166		3827	IC-316	704	
TM1167	1167	3732	IC-317		
TM117			504A	1104	
TM1178	1178	3480			
TM1183	1183	3475/1050			
TM12	5872		5032		
TM121	121	3717	76	2006	
TM1217	1217	3700			
TM121MP	121MP	3718	76	2006(2)	
TM123	123	3020	20	2051	
TM123A	123A	3444		2051	
TM123AP		3854		2051	
TM124	124	3271			
TM125		3081		1114	
TM126	126			2024	
TM127	127	3764			
TM128	128	3024		2030	
TM129	129	3025		2027	
TM13	5834		5048		
TM130	130	3027		2041	
TM130MP	130MP			2041(2)	
TM131	131	3198		2006	

Industry Standard No.	ECG	SK	GE	RS 276-	MOTOR.
TM131MP	131MP			2006(2)	
TM132				2035	
TM133				2028	
TM136		3057		561	
TM137		3058		561	
TM137A		3058		561	
TM139				562	
TM139A		3060		562	
TM14	5874		5032		
TM140		3061		562	
TM142		3062		563	
TM142A				563	
TM143				563	
TM144				564	
TM145		3063		564	
TM145A				564	
TM15	5874		5032		
TM152	152	3893		2048	
TM153	153	3274		2049	
TM154	154	3044		2012	
TM157	157	3747			
TM158	158			2007	
TM159	159	3466		2032	
TM16	5874		5032		
TM160	160	3006		2004	
TM161	161	3716		2015	
TM1613	123A		47	2030	
TM1614	159		82	2032	
TM163A	163A	3439			
TM165	165	3115			
TM166		3647		1152	
TM167		3647		1172	
TM168				1173	
TM17	5944		5048		
TM171	171	3201			
TM1711	123A		47	2030	
TM1712	159		82	2032	
TM175	175	3261		2020	
TM177		9091		1122	
TM178MP				1122(2)	
TM18	5944		5048		
TM180	180			2043	
TM181	181			2041	
TM183	183			2027	
TM184	184	3190		2017	
TM185	185			2025	
TM186	186			2017	
TM186A	186A			2052	
TM187	187			2025	
TM188	188			2018	
TM189	189	3200		2026	
TM19	5916		5048		
TM191	191	3232			
TM192	192			2030	
TM193	193			2023	
TM196	196			2020	
TM197	197			2027	
TM199	199	3245		2010	
TM2	5830		5048		
TM2007	5483	3942			
TM21	5874		5032		
TM210	210	3202		2018	
TM211	211			2026	
TM219	219			2043	
TM22	5874	3984/106	5032		
TM220				2036	
TM221				2036	
TM222		3065		2036	
TM223	390			2041	
TM225	225	3045			
TM226	226			2025	
TM226MP	226MP			2025(2)	
TM229	229	3246		2038	
TM23	5834		5048		
TM233	233	3854/123AP		2009	
TM234	234	3247		2050	
TM235	235	3197			
TM236	236	3239		2053	
TM238	238	3710			
TM24	5944		5048		
TM241	241			2020	
TM242	242			2027	
TM25	5835		5048		
TM26	5944		5048		
TM2613	123A	3444	20	2051	
TM2614	159		82	2032	
TM27	5944		5048		
TM2711	123A	3444	20	2051	
TM2712	159		82	2032	
TM28	5944		5048		
TM280	280			2041	
TM281	281			2043	
TM282	282	3748			
TM283	283	3467			
TM284	284			2047	
TM287	287	3433			
TM289	289A	3124/289		2038	
TM29	5916		5048		
TM290	290A	3114/290		2032	
TM291	291	3440			
TM292	292	3441			
TM293	293	3849		2030	
TM293MP	293MP			2030(2)	
TM294	294	3841			
TM295	295	3253			
TM297	297	3449		2030	
TM298	298	3450			
TM3	5874		5032		
TM300	300	3004/102A			
TM3007	5484	3943/5485			
TM31	5838		5048		
TM312				2035	
TM313	313			2016	
TM315	315	3250			
TM316	316	3039			
TM319	319P			2016	
TM32	5874		5032		
TM321	321	3844			
TM33	116		504A	1104	
TM34	5874		5032		
TM35	5874		5032		
TM36	5874		5032		
TM37	5874		5032		
TM38	5874		5032		
TM39	5874		5032		
TM4	5874		5032		
TM4007	5485	3943			
TM4016B		4016	4016		
TM41	5878		5036		
TM42	5878		5036		
TM43	116		504A	1104	
TM44	5874		5032		
TM45	5874		5032		
TM46	5874		5032		

Industry Standard No.	ECG	SK	GE	RS 276-	MOTOR.
TM47	5874		5032		
TM48	5874		5032		
TM49	5874		5032		
TM5	5830		5048		
TM5007	5486	3944/5487			
TM5016				562	
TM5022		3788		563	
TM5023				564	
TM506		3998		1114	
TM5070		9021		561	
TM5071				561	
TM5072		3136		562	
TM5073		3749		562	
TM5074				563	
TM5075		3751		564	
TM51	5882		5040		
TM519		3100		1122	
TM52	5882		5040		
TM5400	5400	3950			
TM5414				1067	
TM5452				1067	
TM5454				1067	
TM5455				1067	
TM55	5882		5040		
TM56	5882	3500	5040		
TM5800				1142	
TM5802				1143	
TM5804				1144	
TM6	5874		5032		
TM6007	5487	3944			
TM61	156	3051	512		
TM62	116	3051/156	504A	1104	
TM63	116	3017B/117	504A	1104	
TM6407				1050	
TM65	116	3051/156	504A	1104	
TM66	116	3017B/117	504A	1104	
TM7	5940		5048		
TM703A		3157	IC-12		
TM712	712	3072	IC-148		
TM714	714	3075			
TM738	738	3167			
TM74					MR1128
TM744		3171	IC-24	2022	
TM75					MR1128
TM76					MR1128
TM78					MR1128
TM788	788	3147			
TM79					MR1128
TM791	791	3149			
TM8	5940		5048		
TM801	801	3160			
TM84	156	3051	512		MR1128
TM85	156	3051	512		MR1128
TM86	125	3032A	510,531	1114	MR1128
TM88					MR1128
TM89					MR1128
TM9	5912		5048		
TM915				017	
TM941			IC-263	010	
TM941M			IC-265	007	
TM955M		3564	IC-269	1723	
TMD01	135A	3056	ZD-5.1		
TMD01A	135A	3056	ZD-5.1		
TMD02	136A	3057	ZD-5.6	561	
TMD02A	136A	3057	ZD-5.6	561	
TMD03	137A	3058	ZD-6.2	561	
TMD03A	137A	3058	ZD-6.2	561	
TMD04	5014A	3780	ZD-6.8	561	
TMD04A	5014A	3780	ZD-6.8	561	
TMD06	5016A	3782	ZD-8.2	562	
TMD06A	5016A	3782	ZD-8.2	562	
TMD07	139A	3060	ZD-9.1	562	
TMD07A	139A	3059/138A	ZD-9.1	562	
TMD08	140A	3061	ZD-10	562	
TMD08A	140A	3061	ZD-10	562	
TMD10	142A		ZD-12	563	
TMD10A	142A		ZD-12	563	
TMD41	116	3311	504A	1104	
TMD42	116	3311	504A	1104	
TMD45	116	3311	504A	1104	
TMM314	2114				MCM2114
TMM322					MCM2708
TMM323					MCM2716
TMM415					MCM4027
TMM416					MCM4116
TMM4164					MCM6665
TMS2114					MCM2114
TMS2147	2147				MCM2147
TMS2516					MCM2716
TMS2532					MCM2532
TMS2708					MCM2708
TMS2716					MCM2716
TMS4027					MCM4027
TMS4044					MCM6641
TMS4732					MCM68A332
TMS4764					MCM68365
TMT-1543	123A	3020/123	20	2051	
TMT-2427	108	3019	86	2038	
TMT1543	123A	3444	20	2051	
TMT2427	108		86	2038	
TMT696	108	3039/316	86	2038	
TMT697	108	3039/316	86	2038	
TMT839	108	3039/316	86	2038	
TMT840	108	3039/316	86	2038	
TMT841	108	3039/316	86	2038	
TMT842	108	3039/316	86	2038	
TMT843	108	3039/316	86	2038	
TN-3200				2011	
TN061690	159	3114/290	82	2032	
TN061703	159	3114/290	82	2032	
TN237	123A	3444	20	2051	
TN2SA733-Q	290A	3114/290	48		
TN2SA733-Q-A	294		48		
TN2SA733-Q-B	294		48		
TN2SA733-R	290A	3114/290	48		
TN2SA733-R-A	294		48		
TN2SA733Q	294		48		
TN2SA733R	294		48		
TN2SC1507	198	3220	251		
TN2SC1507-K-A	198	3220	251		
TN2SC1507-L	198		251		
TN2SC1507-L-A	198	3220	251		
TN2SC1507-L-B	198	3220			
TN2SC1507-M	198		251		
TN2SC1507-M-A	198	3220	251		
TN2SC1507L	198		251		
TN2SC1520-1	198	3220	251		
TN2SC1520-K-1	198		251		
TN2SC1520-K-1A	198	3220	251		
TN2SC1520-L-1	198		251		
TN2SC1520-L-1A	198		251		
TN2SC1520-L-3A	198		251		
TN2SC1520-M-1	198		251		
TN2SC1520-M-1A	198	3220	251		

Industry Standard No.	ECG	SK	GE	RS 276-	MOTOR.
TN2SC1520-M-3A	198	3220	251		
TN2SC1941	287	3866	220		
TN2SC945-Q	289A	3124/289	212	2010	
TN2SC945-R	199	3124/289	212	2010	
TN2SC945R	289A		212	2010	
TN3200			39	2015	
TN4117					MFE2093
TN4117A					MFE2093
TN4118					MFE2094
TN4118A					MFE2094
TN4119,A					MFE2095
TN41A	6402	3628			
TN421			243	2030	
TN4338					2N4220A
TN4339					2N4220A
TN4340					2N4220A
TN4341					2N4220A
TN53	123A	3444	20	2051	
TN55	123A	3444	20	2051	
TN56	123A	3444		2051	
TN59	123A	3444	20	2051	
TN591	102A		53	2007	
TN60	123A	3444	20	2051	
TN61	123A	3444	20	2051	
TN62	123A	3444	20	2051	
TN63	123A	3444	20	2051	
TN64	123A	3444	20	2051	
TN79	192		63	2030	
TN80	123A	3444	20	2051	
TN81	192		63	2030	
TNC61688	172A	3156	64		
TNC61689	123A	3444	20	2051	
TNC61690	159		82	2032	
TNC61702	123A	3444	20	2051	
TNC61703	159		82	2032	
TNJ-60066	108	3019	86	2038	
TNJ-60067	160	3006	245	2004	
TNJ-60068	160	3006	245	2004	
TNJ-60069	160	3006	245	2004	
TNJ-60070	102A	3004	53	2007	
TNJ-60071	160	3006	245	2004	
TNJ-60073	160	3006	245	2004	
TNJ-60074	102A	3004	53	2007	
TNJ-60075	127	3034	25		
TNJ-60076	123A	3020/123	20	2051	
TNJ-60077	160	3006	245	2004	
TNJ-60079	102A	3004	53	2007	
TNJ-60080	127	3034	25		
TNJ-60279	160	3006	245	2004	
TNJ-60280	160	3006	245	2004	
TNJ-60281	160	3006	245	2004	
TNJ-60282	102A	3004	53	2007	
TNJ-60283	102A	3004	53	2007	
TNJ-60362	160	3008	245	2004	
TNJ-60363	160	3008	245	2004	
TNJ-60364	160	3008	245	2004	
TNJ-60365	160	3004/102A	245	2004	
TNJ-60604		3018	17	2051	
TNJ-60605	107	3018	11	2015	
TNJ-60606	123A	3444	20	2051	
TNJ-60607	123A	3444	20	2051	
TNJ-60608	160	3008	245	2004	
TNJ-60610	100	3004/102A	1	2007	
TNJ-60611	100	3004/102A	1	2007	
TNJ-60612	100	3004/102A	1	2007	
TNJ-60728	102A	3004	53	2007	
TNJ1034	199		62	2010	
TNJ1036			20	2051	
TNJ1173	161	3716	39	2015	
TNJ60063	160		245	2004	
TNJ60064	160		245	2004	
TNJ60065	160		245	2004	
TNJ60066	108		86	2038	
TNJ60067	160		245	2004	
TNJ60068	160		245	2004	
TNJ60069	160		245	2004	
TNJ60069(2SC74)	107		11	2015	
TNJ60070	102A		53	2007	
TNJ60071	160		245	2004	
TNJ60072	154		40	2012	
TNJ60073	160		245	2004	
TNJ60074	102A		53	2007	
TNJ60075	127		25		
TNJ60076	123A	3444	20	2051	
TNJ60077	160		245	2004	
TNJ60078	162		35		
TNJ60079	102A		53	2007	
TNJ60080	127	3764	25		
TNJ60279	160		245	2004	
TNJ60280	160		245	2004	
TNJ60281	160		245	2004	
TNJ60282	102A		53	2007	
TNJ60283	102A		53	2007	
TNJ60362	126		52	2024	
TNJ60363	126		52	2024	
TNJ60364	126		52	2024	
TNJ60365	102A		53	2007	
TNJ60447	107	3039/316	11	2015	
TNJ60448	107	3039/316	11	2015	
TNJ60449	107	3039/316	11	2015	
TNJ60450	160		225	2004	
TNJ60451	175	3131A/369	246	2020	
TNJ60453	175	3131A/369	246	2020	
TNJ60454	121	3009	239	2006	
TNJ60455	162		35		
TNJ60456	160		245	2004	
TNJ60604	107		11	2015	
TNJ60605	107		11	2015	
TNJ60606	107		11	2015	
TNJ60607	107		11	2015	
TNJ60608	100	3721	1	2007	
TNJ60610	102A		53	2007	
TNJ60611	102A	3123	53	2007	
TNJ60612	158		53	2007	
TNJ60728	102A		53	2007	
TNJ61217	107	3018	11	2015	
TNJ61218	107	3018	11	2015	
TNJ61219	123A	3444	20	2051	
TNJ61220	123A	3444	20	2051	
TNJ61221		3004	2	2007	
TNJ61222	102A	3004	53	2007	
TNJ61223	176	3845	80		
TNJ61282	102A	3004	53	2007	
TNJ61671	101	3861	8	2002	
TNJ61671(2SC688)	107		11	2015	
TNJ61671(2SD72)	103A		59	2002	
TNJ61672(2SK25)	312	3448	FET-1	2035	
TNJ61672(RECT)			504A	1104	
TNJ61673	312	3112		2028	
TNJ61673(2SB186)	102A		53	2007	
TNJ61673(2SK24)	312			2028	
TNJ61674	158	3004/102A	53	2007	
TNJ61679	107		11	2015	
TNJ6172(2SC722)	107		11	2015	

Industry Standard No.	ECG	SK	GE	RS 276-	MOTOR.
TNJ61729	108	3039/316	86	2038	
TNJ61730	107	3039/316	11	2015	
TNJ61731	107	3039/316	11	2015	
TNJ61734	103A	3835	59	2002	
TNJ70450	186	3192	28	2017	
TNJ70478	108	3018	86	2038	
TNJ70478-1	108	3018	86	2038	
TNJ70479	108	3122	86	2038	
TNJ70479-1	123A	3444	20	2051	
TNJ70480	108	3018	86	2038	
TNJ70481	193	3114/290	67	2023	
TNJ70482	128	3024	243	2030	
TNJ70483	131	3052	44	2006	
TNJ70484	108	3018	86	2038	
TNJ70537	123A	3444	20	2051	
TNJ70539	123A	3444	20	2051	
TNJ70540	184	3054/196	57	2017	
TNJ70541	131		44	2006	
TNJ70634	102	3722	2	2007	
TNJ70635	102	3722	2	2007	
TNJ70637	123A	3444	20	2051	
TNJ70638	123A	3444	20	2051	
TNJ70639	123A	3444	20	2051	
TNJ70640	123A	3444	20	2051	
TNJ70641	160		245	2004	
TNJ70688	158		53	2007	
TNJ70691	199	3124/289	62	2010	
TNJ71034	199	3124/289	62	2010	
TNJ71035	128	3024	243	2030	
TNJ71036	123A	3444	20	2051	
TNJ71037	159	3122	82	2032	
TNJ71173	108	3019	86	2038	
TNJ71234	128	3024	243	2030	
TNJ71248	126	3006/160	52	2024	
TNJ71252	192	3054/196	63	2030	
TNJ71271	199	3124/289	62	2010	
TNJ71277	199	3124/289	62	2010	
TNJ71498	108	3018	86	2038	
TNJ71629	107		11	2015	
TNJ71773	159	3114/290	82	2032	
TNJ71774	159	3114/290	82	2032	
TNJ71937	107	3018	11	2015	
TNJ71963	107	3018	11	2015	
TNJ71964	161	3018	39	2015	
TNJ71965	199	3018	62	2010	
TNJ72146	164	3133	37		
TNJ72147	124	3021	12		
TNJ72148	130	3027	14	2041	
TNJ72149	165	3111	38		
TNJ72150	161	3039/316	39	2015	
TNJ72151	161	3117	39	2015	
TNJ72152	193		67	2023	
TNJ72153	124	3021	12		
TNJ72154	159	3114/290	82	2032	
TNJ72275	161	3039/316	39	2015	
TNJ72276	233	3132	210	2009	
TNJ72277	107	3018	11	2015	
TNJ72278	102A	3005	53	2007	
TNJ72279	107	3018	11	2015	
TNJ72280	123A	3444	20	2051	
TNJ72281	123A	3444	20	2051	
TNJ72282	154		40	2012	
TNJ72283	102A	3004	53	2007	
TNJ72284	103A	3010	59	2002	
TNJ72285	102A		53	2007	
TNJ72286	175		246	2020	
TNJ72287	102A	3006/160	53	2007	
TNJ72288	128	3024	243	2030	
TNJ72289	102A	3004	53	2007	
TNJ72318	121	3009	239	2006	
TNJ72319	163A	3111	36		
TNJ72320	162	3027/130	35		
TNJ72368	161	3039/316	39	2015	
TNJ72701	161	3039/316	39	2015	
TNJ72774	187	3193	29	2025	
TNJ72775	186	3192	28	2017	
TNJ72783	123A	3444	20	2051	
TNJ72784	123A	3444	20	2051	
TNT-839	108	3018	86	2038	
TNT-840	108	3018	86	2038	
TNT-841	108	3018	86	2038	
TNT-843	123A	3018	20	2051	
TNT839	108		86	2038	
TNT840	108		86	2038	
TNT841	108		86	2038	
TNT842	123A	3444	20	2051	
TNT843	108	3444/123A	86	2038	
TO-003	160		245	2004	
TO-004	160		245	2004	
TO-005	102A		53	2007	
TO-012	121	3009	239	2006	
TO-014	102A		53	2007	
TO-015	121	3009	239	2006	
TO-033	123A	3444	20	2051	
TO-038	123A	3444	20	2051	
TO-039	123A	3444	20	2051	
TO-040	123A	3444	20	2051	
TO-041	102A		53	2007	
TO-101	100	3721	1	2007	
TO-102	100	3721	1	2007	
TO-103	102A		53	2007	
TO-104	102A		53	2007	
TO1-101	123A	3444	20	2051	
TO1-104	123A	3444	20	2051	
TO1-105	123A	3444	20	2051	
TO1003	314	3898			
TO101	102A	3005	53	2007	
TO102	102A	3005	53	2007	
TO103	102A	3003	53	2007	
TO104	102A	3003	53	2007	
TO2003	314	3898			
TO253	314	3898			
TO3003	314	3898			
TO4003	314	3898			
TO503	314	3898			
TOA1748V	1171	3645			
TOA2748V	1171	3645			
TOB	703A		IC-12		
TP101	116	3016	504A	1104	
TP20-0284			ZD-6.2	561	
TP201	116	3016	504A	1104	
TP300	5874		5032		
TP302	116	3017B/117	504A	1104	
TP312					BFR96
TP34	109	3087		1123	
TP34A	109	3087		1123	
TP3638	159	3247/234	82	2032	
TP3638A	193	3247/234	67	2023	
TP390					BFW92A
TP393					BFR91
TP394					BFR96
TP400	5874		5032		
TP4000		4000	4000		
TP4001		4001	4001	2401	

Industry Standard No.	ECG	SK	GE	RS 276-	MOTOR.
TP4002		4002	4002		
TP4007		4007	4007		
TP4008AN	4008B	4008			
TP4008BN	4008B	4008			
TP4009		4009	4009		
TP4011		4011	4011	2411	
TP4012		4012	4012	2412	
TP4013		4013	4013	2413	
TP4013BN	4013B	4013			
TP4014AN	4014B	4014			
TP4015		4015	4015		
TP4016		4016	4016		
TP4016A	4016B	4016			
TP4017		4017	4017	2417	
TP4018	4018B	4022			
TP4018AN	4018B	4022			
TP4018BN	4018B	4022			
TP4019		4019	4019		
TP402	116	3017B/117	504A	1104	
TP4020			4020	2420	
TP4021			4021	2421	
TP4022AN	4022B	4022			
TP4023		4023	4023	2423	
TP4024		4024	4024		
TP4025		4025	4025		
TP4027		4027	4027	2427	
TP4028AN	4028B	4028			
TP4028BN	4028B	4028			
TP4035	4035B	4035			
TP4042		4042	4042		
TP4043AN	4043B	4043			
TP4043BN	4043B	4043			
TP4044AN	4044B	4044			
TP4044BN	4044B	4044			
TP4049		4049	4049	2449	
TP4050		4050	4050	2450	
TP4051		4051	4051	2451	
TP4053	4053B	4053			
TP4053BN	4053B	4053			
TP4067-409	116		504A	1104	
TP4067-410	199	3245	62	2010	
TP4067-411	199	3245	62	2010	
TP4072	4072B	4072			
TP4081		4081	4081		
TP4123	123A	3444	20	2051	
TP4124	123A	3444	20	2051	
TP4125	159		82	2032	
TP4126	159		82	2032	
TP4257	159	3118	82	2032	
TP4258	159	3118	82	2032	
TP4274	101	3861	8	2002	
TP4275	108		86	2038	
TP4511	4511B	4511			
TP4518			4518	2490	
TP491					BFR91
TP5114					2N3993
TP5115					2N3993
TP5116					2N3993
TP5142	159			2032	
TPR10					MRF1015MB
TPR150					MRF1150MB
TPR50					MRF1090MB
TPS6512	123A		20	2051	
TPS6513	123A		20	2051	
TPS6514	123A	3444	20	2051	
TPS6515	123A	3444	20	2051	
TPS6516	159	3247/234	82	2032	
TPS6517	159	3247/234	82	2032	
TPS6518	159	3247/234	82	2032	
TPS6519	159	3247/234	82	2032	
TPS6520	123A	3444	20	2051	
TPS6521	123A	3444	20	2051	
TPS6522	159	3247/234	82	2032	
TPS6523	159	3247/234	82	2032	
TQ-5034	160		245	2004	
TQ-5051	158		53	2007	
TQ-5052	123A		20	2051	
TQ-5053	123A		20	2051	
TQ-5054	123A		20	2051	
TQ-5055	128		243	2030	
TQ-5060	123A		20	2051	
TQ-5061	158		53	2007	
TQ-5063	171	3201	27		
TQ-5064	121	3717	239	2006	
TQ-63	159		82	2032	
TQ-64	159		82	2032	
TQ-PD-3055	130		14	2041	
TQ1	108	3018	86	2038	
TQ2	108	3018	86	2038	
TQ3	108	3018	86	2038	
TQ4	123A	3444	20	2051	
TQ5	108	3018	86	2038	
TQ5020	100	3119/113	1	2007	
TQ5020(SANYO)	113A		6GC1		
TQ5021	160		245	2004	
TQ5022	160		245	2004	
TQ5023	102A		53	2007	
TQ5025	102A		53	2007	
TQ5026	102A		53	2007	
TQ5027			2	2007	
TQ5028	127	3764	25		
TQ5030	179	3642	76		
TQ5031	101	3861	8	2002	
TQ5032	101	3861	8	2002	
TQ5034	160		245	2004	
TQ5035	160		245	2004	
TQ5036	121	3009	239	2006	
TQ5038	160		245	2004	
TQ5039	101	3011	8	2002	
TQ5044	103A		59	2002	
TQ5049	108	3039/316	86	2038	
TQ5050	103A	3835	59	2002	
TQ5051	102A		53	2007	
TQ5052	123A	3444	20	2051	
TQ5053	123A	3444	20	2051	
TQ5054	123A	3444	20	2051	
TQ5055	128	3024	243	2030	
TQ5060	123A	3444	20	2051	
TQ5061	102A		53	2007	
TQ5062	103A	3835	59	2002	
TQ5063	154		40	2012	
TQ5064	121	3009	239	2006	
TQ5081			18	2030	
TQ53	193		67	2023	
TQ54	193		67	2023	
TQ55	193		67	2023	
TQ57	193		67	2023	
TQ58	193		67	2023	
TQ59	193		67	2023	
TQ59A	193		67	2023	
TQ6	108	3018	86	2038	
TQ60	193		67	2023	
TQ60A	193		67	2023	
TQ61	159		82	2032	
TQ61A	159		82	2032	
TQ62	159		82	2032	
TQ62A	159		82	2032	
TQ63	129		244	2027	
TQ63A	129		244	2027	
TQ64	129		244	2027	
TQ64A	129		244	2027	
TQ7	108	3018	86	2038	
TQ8	108	3018	86	2038	
TQ9	108	3018	86	2038	
TQPD3053	128		243	2030	
TQSA0-222	102	3722	2	2007	
TR 19A	199			2010	
TR-01	121	3009	239	2006	
TR-01(PENNCREST)	128		243	2030	
TR-01045	182	3188A			
TR-016	108		86	2038	
TR-01B	108	3009	86	2038	
TR-01B(PENNCREST)	123A		20	2051	
TR-01C	121	3009	239	2006	
TR-01C(PENNCREST)	123A		20	2051	
TR-01E	128	3009	243	2030	
TR-01E(PENNCREST)	128		243	2030	
TR-01MP			16	2006	
TR-02	104	3719	16	2006	
TR-02E	116	3017B/117	504A	1104	
TR-03	105		4		
TR-03(PENNCREST)	175		246	2020	
TR-04	102A		53	2007	
TR-04C	129		244	2027	
TR-04C(PENNCREST)	129		244	2027	
TR-05	100	3721	1	2007	
TR-06	100	3004/102A	1	2007	
TR-06(PENNCREST)	123A		20	2051	
TR-07	160		245	2004	
TR-07(PENNCREST)	128		243	2030	
TR-08	101	3861	8	2002	
TR-08(PENNCREST)	129		244	2027	
TR-08C	101	3861	8	2002	
TR-09	101	3861	8	2002	
TR-09C	101	3861	8	2002	
TR-10	101		8	2002	
TR-1000-2	159		82	2032	
TR-1000-3	128		243	2030	
TR-1000-7	130		14	2041	
TR-1030-1	159	3114/290	82	2032	
TR-1030-2	159	3114/290	82	2032	
TR-1032-1		3114	21	2034	
TR-1032-2	159	3114/290	82	2032	
TR-1033-1	123A	3444	20	2051	
TR-1033-2	123A	3444	20	2051	
TR-1033-3			20	2051	
TR-1036-1	183	3189A	56	2027	
TR-1036-2	183	3189A	56	2027	
TR-1036-3	187	3193	29	2025	
TR-1037-1	182	3188A	55		
TR-1037-2	182	3188A	55		
TR-1037-3	186	3192	28	2017	
TR-1038-4				2043	
TR-1039-4	130		14	2041	
TR-10C	101		8	2002	
TR-11	126		52	2024	
TR-11C	160		245	2004	
TR-12	160		245	2004	
TR-12C	160		245	2004	
TR-13	160		243	2030	
TR-13(RF)			245	2004	
TR-1347	123A	3444	20	2051	
TR-13C	160		245	2004	
TR-14	102A		53	2007	
TR-14001-2	5130A		5ZD-15		
TR-14001-4	5135A		5ZD-20		
TR-14002-10	147A	3095	ZD-33		
TR-14C	102A		53	2007	
TR-15	102A		53	2007	
TR-157(OLSON)	158		53	2007	
TR-158(OLSON)	158		53	2007	
TR-159(OLSON)	101		8	2002	
TR-15C	102A		53	2007	
TR-16	104	3719	16	2006	
TR-160(OLSON)	101		8	2002	
TR-161(OLSON)	160		245	2004	
TR-162(OLSON)	123A		20	2051	
TR-163(OLSON)	108		86	2038	
TR-164(OLSON)	128		243	2030	
TR-165(OLSON)	129		244	2027	
TR-166(OLSON)	160		245	2004	
TR-167(OLSON)	159		82	2032	
TR-168(OLSON)	126		52	2024	
TR-169(OLSON)	102A		53	2007	
TR-16C	104	3719	16	2006	
TR-17	160		245	2004	
TR-170(OLSON)	158		53	2007	
TR-171	161	3716	39	2015	
TR-172(OLSON)	104	3719	16	2006	
TR-174(OLSON)	105		4		
TR-175(OLSON)	124		12		
TR-176(OLSON)	181		75	2041	
TR-178(OLSON)	121	3717	239	2006	
TR-17A	160		245	2004	
TR-17C	160		245	2004	
TR-18	160		245	2004	
TR-180(OLSON)	175		246	2020	
TR-182(OLSON)	185		58	2025	
TR-183(OLSON)	127	3764	25		
TR-184(OLSON)	131		44	2006	
TR-185(OLSON)	150A		ZD-82		
TR-186(OLSON)	162		35		
TR-187(OLSON)	163A		36		
TR-188(OLSON)	175		246	2020	
TR-18C	160		245	2004	
TR-19	152	3025/129	82	2032	
TR-1993	199	3245	62	2010	
TR-19A	159	3114/290	82	2032	
TR-1R26	160		245	2004	
TR-1R31	108		86	2038	
TR-1R33	123A	3444	20	2051	
TR-1R35	108	3018	86	2038	
TR-20	159	3114/290	82	2032	
TR-20A	159	3114/290	82	2032	
TR-21	123A	3018	20	2051	
TR-21-6	123A	3444	20	2051	
TR-21C	123A	3444	20	2051	
TR-22	123A	3444	20	2051	
TR-22C	123A	3444	20	2051	
TR-23	124	3021	12		
TR-23C	124	3021	12		
TR-24	108	3018	86	2038	
TR-24(PHILCO)	123A		20	2051	
TR-25	128	3024	243	2030	
TR-26	130		14	2041	
TR-28	129	3025	66	2027	
TR-2880	116	3016	504A	1104	
TR-29	219		74	2043	

Industry Standard No.	ECG	SK	GE	RS 276-	MOTOR.
TR-2N2641C	102A		53	2007	
TR-2R26	160		245	2004	
TR-2R31	108		86	2038	
TR-2R33	108		86	2038	
TR-2R35	108	3018	86	2038	
TR-2SC367	123A	3444	20	2051	
TR-2SC371	108		86	2038	
TR-2SC372	108		86	2038	
TR-2SC373	123A	3444	20	2051	
TR-2SC384	108		86	2038	
TR-2SC482	128	3024	243	2030	
TR-2SC735	123A	3444	20	2051	
TR-2SK55	312	3116			
TR-3			3	2006	
TR-30	159	3114/290	82	2032	
TR-30A	159	3114/290	82	2032	
TR-31			82	2032	
TR-31B	128		243	2030	
TR-320	102A	3004	53	2007	
TR-320A	102A	3004	53	2007	
TR-321	102A	3004	53	2007	
TR-321A	102A	3004	53	2007	
TR-323A	102A	3004	53	2007	
TR-33		3018	47	2030	
TR-36			75	2041	
TR-36MP			75	2041	
TR-3R26	160		245	2004	
TR-3R31	108		86	2038	
TR-3R33	108			2038	
TR-3R35	108	3018	86	2038	
TR-3R38	123A	3444	20	2051	
TR-3R03			86	2038	
TR-43B	121	3717	239	2006	
TR-482A	102A	3004	53	2007	
TR-4R26	160		245	2004	
TR-4R31	128	3024	243	2030	
TR-4R33	123A	3444	20	2051	
TR-4R35	108	3018	86	2038	
TR-4R38	159	3114/290	82	2032	
TR-5	121	3717	239	2006	
TR-50				2006	
TR-508A	102A	3004	53	2007	
TR-51			17	2051	
TR-53			18	2030	
TR-55			28	2017	
TR-56			29	2025	
TR-57			246	2020	
TR-59			14	2041	
TR-5R26	102A		53	2007	
TR-5R31	128	3024	243	2030	
TR-5R33	123A	3444	20	2051	
TR-5R35	123A	3444	20	2051	
TR-5R38	123A	3444	20	2051	
TR-62			20	2051	
TR-65	195A	3048/329	46		
TR-6R26	102A		53	2007	
TR-6R33	123A	3444	20	2051	
TR-6R35	159	3114/290	82	2032	
TR-6R35A	159	3114/290	82	2032	
TR-70			61	2038	
TR-75	142A		ZD-12	563	
TR-76			57	2017	
TR-77	116	3311	504A	1104	
TR-77(XSTR)			58	2025	
TR-7GS	138A	3059	ZD-7.5		
TR-7R31	128	3024	245	2030	
TR-7R35	123A	3444	20	2051	
TR-7SA	138A		ZD-7.5		
TR-7SB		3059	ZD-12	563	
TR-80			86	2038	
TR-8001	160	3006	245	2004	
TR-8002	160	3006	245	2004	
TR-8003	160	3006	245	2004	
TR-8004	123A	3018	20	2051	
TR-8004-4	108		86	2038	
TR-8004-5	108	3018	86	2038	
TR-8005	124	3021	12		
TR-8006	121	3009	239	2006	
TR-8007	159	3004/102A	82	2032	
TR-8007(FISHER)	159		82	2032	
TR-8010	107		11	2015	
TR-8014	123A	3020/123	20	2051	
TR-8018	181	3036	75	2041	
TR-8019	153		69	2049	
TR-8020	129	3025	244	2027	
TR-8021	128	3024	243	2030	
TR-8022	190	3024/128	217		
TR-8023	128	3024	243	2030	
TR-8024	128	3024	243	2030	
TR-8025	123A	3020/123	20	2051	
TR-8026	159	3114/290	82	2032	
TR-8027	312	3116	FET-1	2035	
TR-8028	108	3018	86	2038	
TR-8029	108	3018	86	2038	
TR-8030	108	3018	86	2038	
TR-8031	108	3018	86	2038	
TR-8032	108	3018	86	2038	
TR-8034	199	3018	62	2010	
TR-8035	123A		20	2051	
TR-8036	128	3122	243	2030	
TR-8037	159		82	2032	
TR-8038	108		86	2038	
TR-8039	123A		20	2051	
TR-8040	199	3245	62	2010	
TR-8042	123A	3122	20	2051	
TR-8043	107		11	2015	
TR-81				2020	
TR-81MP				2020	
TR-83			86	2038	
TR-84			53	2007	
TR-85			53	2007	
TR-86			20	2051	
TR-87			18	2030	
TR-88			244	2027	
TR-8R35	123A	3444	20	2051	
TR-9100-18	123A	3444	20	2051	
TR-95	139A		ZD-9.1	562	
TR-95(B)	139A		ZD-9.1	562	
TR-95(IR)				2011	
TR-95(XSTR)			39	2015	
TR-95B			ZD-9.1	562	
TR-9GS	139A		ZD-9.1	562	
TR-9S	139A		ZD-9.1	562	
TR-9SA	139A		ZD-9.1	562	
TR-9SB	139A		ZD-9.1	562	
TR-BC147B	123A	3444	20	2051	
TR-BC149C	199	3122	62	2010	
TR-BRC149C	123A	3444	20	2051	
TR-C44	100	3005	1	2007	
TR-C44A	100	3005	1	2007	
TR-C45	102	3004	53	2007	
TR-C45A	100	3005	1	2007	
TR-C70	102		53	2007	
TR-C71	102		53	2007	
TR-C72	102		53	2007	
TR-C72(DIODE)	178MP		300(2)	1122(2)	
TR-FE100			FET-2	2035	
TR-FET-1	312	3448	FET-1	2035	
TR-RR38	123A	3444	20	2051	
TR-TR38	123A	3444	20	2051	
TR-U1650E	312	3448	FET-1	2035	
TR-U1650E-1			FET-2	2035	
TR-U1835E	312	3448	FET-1	2035	
TRO-1057-1	182	3188A	55		
TRO-1057-3	182	3188A	55		
TRO-1057-4	182	3188A	55		
TRO-1058-1	182		55		
TRO-1058-5	182		55		
TRO-2012			50	2004	
TRO-2028-5				2027	
TRO-2057-1	183		56	2027	
TRO-2057-3	183		56	2027	
TRO-2057-4	183		56	2027	
TRO-2058-1	183		56	2027	
TRO-2058-5	183		56	2027	
TRO-9005	703A		IC-12		
TRO-9006	703A		IC-12		
TRO-9007	750		IC-219		
TRO-9010	746		IC-217		
TRO-9011	720		IC-7		
TRO-9017	722		IC-9		
TRO-9018	788		IC-229		
TR0055	159	3114/290	82	2032	
TR01			16	2006	
TR01014	199		62	2010	
TR01015		3122	85	2010	
TR01026	108	3018	86	2038	
TR01037	123A	3444	20	2051	
TR01040	199	3122	62	2010	
TR01042	161	3039/316	39	2015	
TR01045		3041	57	2017	
TR01053-1	159	3114/290	82	2032	
TR01054-7	128		243	2030	
TR01056-5		3041	57	2017	
TR01057-3	184		57	2017	
TR01060-7	181		75	2041	
TR010602-1	108	3020/123	86	2038	
TR01062-1	123AP	3024/128		2030	
TR01062-7		3124	47	2030	
TR01065		3010	59	2002	
TR01073		3039	20	2051	
TR01074		3039	20	2051	
TR01C	121	3009	239	2006	
TR01MP			16	2006	
TR02	121	3009	16,237	2006	
TR02012	126	3006/160	52	2024	
TR02020-2	234	3118	65	2050	
TR02051-1	159	3114/290	82	2032	
TR02051-3	159		82	2032	
TR02051-5	159		82	2032	
TR02051-6	159		82	2032	
TR02053-5	189		218	2026	
TR02053-7	189	3200	218	2026	
TR02054-7	129		244	2027	
TR02057-3	185		58	2025	
TR02060-7	180		74	2043	
TR02062-1	159	3114/290	82	2032	
TR02062-6	159	3114/290	82	2032	
TR02063-1	126	3003	52	2024	
TR02063-8	159	3114/290	82	2032	
TR0259-6	219		74	2043	
TR02C	121	3009	239	2006	
TR03	105	3012	240		
TR03C	105	3012	4		
TR04	101	3003	2	2007	
TR04C	102A	3003	53	2007	
TR05	101	3004/102A	1,2	2007	
TR0573486	233	3124/289	61	2038	
TR0573491	123A	3124/289	210	2051	
TR0573507	229	3018	61	2038	
TR0575002	109	3087	1N34AS	1123	
TR0575005	110MP	3088	1N60	1123	
TR05C	100	3004/102A	1	2007	
TR06		3004	51	2004	
TR06011	312	3448	FET-1	2035	
TR06014	312	3448	FET-1	2035	
TR06C	160	3004/102A	245	2004	
TR07	101	3007	245	2004	
TR07C	100	3007	1	2007	
TR08	101	3862/103	8	2002	
TR08004	222	3050/221	FET-4	2036	
TR08C	101		8	2002	
TR09	103	3862	8	2002	
TR09004	718		IC-8		
TR09005	703A		IC-12		
TR09006	703A		IC-12		
TR09007	750	3280	IC-219		
TR09010	746		IC-27		
TR09011	720		IC-7		
TR09012	703A		IC-12		
TR09017	722		IC-9		
TR09018	788	3147	IC-229		
TR09019	801		IC-35		
TR09022	1090	3291			
TR09024	1090	3291			
TR09027	801	3160	IC-35		
TR09029	708		IC-10		
TR09032	799		IC-34		
TR09033	802	3277	IC-235		
TR09034	803	3278	IC-237		
TR09C	101		8	2002	
TR10	101	3011	8	2002	
TR100	5982	3609/5986	5096		1N249B
TR1000	129	3114/290	244	2027	
TR1000A	159	3114/290	82	2032	
TR1001	128	3024	243	2030	
TR1002	129	3025	244	2027	
TR1003	128	3024	243	2030	
TR1004	129	3025	244	2027	
TR1005	128	3024	243	2030	
TR1007		3027	14	2041	
TR1009	162	3027/130	35		
TR1009A	181		75	2041	
TR1011	123A	3444	20	2051	
TR1012	129		244	2027	
TR1015T	5542	6615/5515			
TR1025	390		255	2041	
TR103	6054	3608/5990	5100		1N1184
TR1030	159	3114/290	82	2032	
TR1030-1	159		82	2032	
TR1030-2	159		82	2032	
TR1030A	159	3114/290	82	2032	
TR1031	123A	3444	20	2051	
TR1031-1	194		220		
TR1031-2	194		220		
TR1032	159	3114/290	82	2032	
TR1032-1	159		82	2032	
TR1032A	159	3114/290	82	2032	

Industry Standard No.	ECG	SK	GE	RS 276-	MOTOR.
TR1033	123A	3444	20	2051	
TR1033-1	190		217		
TR1033-3	123A	3232/191	217		
TR1033-4	190		217		
TR1033-5	190		217		
TR1033-6	190		217		
TR1034	159	3114/290	82	2032	
TR1034A	159	3114/290	82	2032	
TR1036	179	3642	76		
TR1036-2	183	3189A	56	2027	
TR1036-3	183		56	2027	
TR1037-1	183		56	2027	
TR1037-2	182		55		
TR1037-3	182		55		
TR1038	179	3642	76		
TR1038-4	219		74	2043	
TR1038-5	281	3189A/183	56	2027	
TR1038-6	183	3189A	56	2027	
TR1039-4	130		14	2041	
TR1039-5	280		76		
TR1039-6	130		14	2041	
TR104			1	2007	
TR105			50	2004	
TR105(SPRAGUE)	199		62	2010	
TR106(SPRAGUE)	234		65	2050	
TR1077	130	3027	14	2041	
TR109			1	2007	
TR10C	101	3011	8	2002	
TR11	100	3721	1	2007	
TR112	108	3018	86	2038	
TR1120	5870		5032		MR1120
TR1121	5872		5032		MR1121
TR1122	5874		5032		MR1122
TR1123	5878		5036		MR1123
TR1124	5878		5036		MR1124
TR1125	5882	3500	5040		MR1125
TR1126	5882	3500	5040		MR1126
TR1128	5886		5044		MR1128
TR1130	5890	7090	5044		MR1130
TR11C	160	3006	245	2004	
TR12	160	3787/5021A	50,245	2004	
TR12001-4	109	3087		1123	
TR123				2005	
TR125B	145A		ZD-15	564	
TR12C	160	3787/5021A	245	2004	
TR128	5022A		ZD-15	563	
TR12SA	5021A	3787		563	
TR12SB	5022A	3787/5021A	ZD-15	563	
TR12SC	5023A	3789		564	
TR13	160	3788/5022A	245	2004	
TR139			50	2004	
TR13C	160	3788/5022A	245	2004	
TR14	102A	3004	53	2004	
TR14001-1	5081A	3151	ZD-24		
TR14002-12	145A	3063	ZD-15	564	
TR14002-13	5079A		ZD-20		
TR14002-6	142A	3062	ZD-12	563	
TR1490	130	3027	14	2041	
TR1491	130	3027	14	2041	
TR1492	130	3027	14	2041	
TR1493	130	3027	14	2041	
TR14C	102A	3004	53	2007	
TR15	102A	3004	53	2007	
TR150	5896	7096	5064		1N250B
TR151	5986		5048		1N3210
TR1512-80	108	3019	86	2038	
TR152	5916	3609/5986	5096		1N250B
TR153	6054	3608/5990	5100		1N1186A
TR1591	175	3026	246	2020	
TR1593	175	3026	246	2020	
TR15C	102A	3004	53	2007	
TR16	127	3034	3,25		
TR1605LP	124	3021	12		
TR167	101	3861	8	2002	
TR16C	121	3009	239	2006	
TR16X2	104MP		13MP		
TR17	160	3006	50,245	2004	
TR18	160	3006	245	2004	
TR182	101	3861	8	2002	
TR183	101	3861	8	2002	
TR184	101	3861	8	2002	
TR18C	160	3006	245	2004	
TR19			82	2032	
TR193	101	3861	8	2002	
TR194	101	3861	8	2002	
TR1993-2	123A	3444	20	2051	
TR19A		3245	62	2010	
TR1N4002	116	3311	504A	1104	
TR1R26	160	3006	245	2004	
TR20	159	3025/129	48,82	2005	
TR200	5986	3609	5096		1N1194
TR203	6054		5100		1N1188A
TR2083-40		3126	90		
TR2083-41		3088	1N60	1123	
TR2083-42	177	3100/519	300	1122	
TR2083-44		3100	300	1122	
TR2083-70		3116	FET-2	2035	
TR2083-71		3018	17	2051	
TR2083-72		3122	20	2051	
TR2083-73		3122	20	2051	
TR2083-74		3124	60	2015	
TR2083-75		3004	2	2007	
TR209T2			ZD-8.7	562	
TR21	123A	3444	20,47	2051	
TR211	101	3861	8	2002	
TR212	101	3861	8	2002	
TR213	101	3861	8	2002	
TR214	101	3861	5	2002	
TR215			1	2007	
TR216	101	3861	8	2002	
TR217			1	2007	
TR218			50	2004	
TR21C	108	3018	86	2038	
TR22	129	3020/123	244	2027	
TR228735045311	108	3018	86	2038	
TR228735046011	107	3018	11	2015	
TR228735048617	108	3018	86	2038	
TR228735048618	108	3018	86	2038	
TR228735120325	312	3448	FET-1	2035	
TR228736002003	109	3087		1123	
TR228736002004	110MP	3709	1N34AS	1123(2)	
TR228736003026	140A	3061	ZD-10	562	
TR22873700104		9014	IC-7		
TR228779905696	1003		IC-43		
TR22A	129	3025	244	2027	
TR22C	108	3018	86	2038	
TR23	128	3021/124	243	2030	
TR2320063	289A	3124/289	210	2051	
TR2327031	125	3311	504A	1104	
TR2327041	116	3311	504A	1104	
TR2327203	152	3054/196	57	2017	
TR2327293	289A	3122	210	2051	
TR2327312	1039		IC-59		
TR2327333	289A	3122	210	2051	

Industry Standard No.	ECG	SK	GE	RS 276-	MOTOR.
TR2327363	199	3122	210	2051	
TR2327393	185	3179A	58	2025	
TR2327411	1041		IC-160		
TR2327422	720	9014	IC-7		
TR2327431	222	3050/221	FET-4	2036	
TR2327443	199	3124/289	62	2010	
TR2327444	199	3124/289	62	2010	
TR2327574	390	3027/130	75	2041	
TR2327603	184		55		
TR2327607	184		215		
TR2327723	153	3083/197	69	2049	
TR2327733	138A		ZD-7.5		
TR2327743	129	3114/290	82	2032	
TR2327841	219	3173	74	2043	
TR2327852	175	3026	246	2020	
TR2337011	177	3100/519	300	1122	
TR2337063	5080A		ZD-22		
TR2337103	144A			564	
TR2337123	137A		ZD-6.2	561	
TR2367171	801	3160	IC-35		
TR23A	129	3024/128	244	2027	
TR23C	124	3021	12		
TR24	108	3039/316	210	2038	
TR25	128	3024	45,243	2030	
TR251	5990	3608	5100		1N3211
TR252	5988	3608/5990	5100		1N3211
TR253	6060	3608/5990	5100		1N1188A
TR26	130	3027	14	2041	
TR26-1	175	3026	246	2020	
TR262-2	124	3021	12		
TR266-2	124	3021	12		
TR269				2005	
TR26C	130	3027	14	2041	
TR271TR26	130	3027	14	2041	
TR28		3025	82	2032	
TR2880	116		504A	1104	
TR29	219	3173	74	2043	
TR2A	116		504A	1104	
TR2N2614C	102A		53	2007	
TR2R26	160	3006	245	2004	
TR2S1570LH	123A	3124/289	212	2010	
TR2SA763	234	3247	65	2050	
TR2SC1342	229	3018	61	2038	
TR2SC3677	123A		20	2051	
TR2SC371	108	3018	86	2038	
TR2SC372	108	3018	86	2038	
TR2SC373	123A	3020/123	20	2051	
TR2SC384	108	3018	86	2038	
TR2SC482	128	3024	243	2030	
TR2SC535	229	3018	86	2038	
TR2SC671	195A	3765	46		
TR2SC735	123A	3020/123	20	2051	
TR2SD330E	152	3054/196	66	2048	
TR2SK55	312		FET-2	2035	
TR30	159	3114/290	82	2032	
TR300	5990	3608	5100		1N3211
TR301	154	3045/225	40	2012	1N3211
TR302	123A	3444	20	2051	1N3211
TR303	6060	3608/5990	5100		1N1187
TR31	159	3046	48,82,	2032	
TR310011	102A	3004	53	2007	
TR310012	102A	3004	53	2007	
TR310015	100	3005	1	2007	
TR310017	102A	3004	53	2007	
TR310018	102A	3004	53	2007	
TR310019	160	3008	245	2004	
TR3100227		3004	2	2007	
TR310025	160	3004/102A	245	2004	
TR310026	102A	3004	53	2007	
TR310065	160	3008	245	2004	
TR310068	160	3006	245	2004	
TR310069	160	3007		2004	
TR310075	102A	3004	53	2007	
TR310107	102A	3004	53	2007	
TR310123	160	3006	245	2004	
TR310124	160	3006	245	2004	
TR310125	102A	3004	53	2007	
TR310136	102A	3004	53	2007	
TR310139	160	3006	245	2004	
TR310147	160	3006	245	2004	
TR310149	102A	3004	53	2007	
TR310150	160	3006	245	2004	
TR310153	102A	3004	53	2007	
TR310155	160	3008	245	2004	
TR310156	160	3008	245	2004	
TR310157	160	3008	245	2004	
TR310158	160	3008	245	2004	
TR310159	102A	3004	53	2007	
TR310160	103A	3835	59	2002	
TR310161	100	3005	1	2007	
TR310164	102A	3004	53	2007	
TR310193	160	3007	245	2004	
TR310224	160		245	2004	
TR310225	102	3004	2	2007	
TR310227	102A		53	2007	
TR310230	108	3018	86	2038	
TR310231	123A	3444	20	2051	
TR310232	160	3007	245	2004	
TR310235	102A	3004	53	2007	
TR310236	103A	3010	59	2002	
TR310243	123A	3444	20	2051	
TR310244	108	3018	86	2038	
TR310245	123A	3444	20	2051	
TR310249	108	3039/316	86	2038	
TR310250	108	3018	86	2038	
TR310251	102A	3008	53	2007	
TR310252	102A	3004	53	2007	
TR310255	102A	3008	53	2007	
TR32	194	3275	220		
TR320	102	3123	2	2007	
TR320007	110MP	3091		1123(2)	
TR320008	109	3087		1123	
TR320020	116		504A	1104	
TR320022	116		504A	1104	
TR320039	109	3087	510,531	1123	
TR320041	109	3087	510,531	1123	
TR320048	109	3087	510,531	1123	
TR320A	102	3004	2	2007	
TR320AN	102A	3004	53	2007	
TR321	102	3123	2	2007	
TR321(HFGH1)	100	3721	1	2007	
TR321A		3004	2	2007	
TR323	102	3004	2	2007	
TR323A	102	3004	2	2007	
TR323AN	102A	3004	53	2007	
TR33	199	3122	47,62	2010	
TR330027	116	3031A	504A	1104	
TR330028	140A	3061	ZD-10	562	
TR331	160		245	2004	
TR332	102	3123	2	2007	
TR333	121	3009	239	2006	
TR334	105	3012	4		
TR335	101	3861	8	2002	
TR336	101	3861	8	2002	

Industry Standard No.	ECG	SK	GE	RS 276-	MOTOR.
TR337	101	3861	8	2002	
TR338	103	3862	8	2002	
TR34	127	3123	25		
TR35	179	3642	76		
TR351	5990	3608	5100		1N3212
TR35144	121	3014	239	2006	
TR35144A	121	3014	239	2006	
TR352	5990	3608	5100		1N1196
TR353	6060	3608/5990	5100		1N1188A
TR35524	121	3717	239	2006	
TR36	181	3018	75	2041	
TR36643	128		243	2030	
TR37	218		234		
TR38	108	3018	86	2038	
TR381			53	2007	
TR38117	102A	3004	53	2007	
TR382			53	2005	
TR383	102	3123	53	2005	
TR383(HGFH-2)	102	3722	2	2007	
TR39453	221			2036	
TR3R26	160	3006	245	2004	
TR400	5990	3608	5100		1N1196
TR401	5990	3608	5100		1N3212
TR4010	199	3245	62	2010	
TR4010-2	123A	3444	20	2051	
TR402	5990	3608	5100		1N1196
TR403	6060	3608/5990	5100		1N1188A
TR40603	221			2036	
TR4104-2327421	720	9014	IC-7		
TR43	102A		53	2007	
TR44	102A		53	2007	
TR45	102A	3123	53	2007	
TR48	109	3087		1123	
TR482	100	3004/102A	1	2007	
TR482A	100	3721	1	2007	
TR4R26	160	3006	245	2004	
TR4S	5008A	3774	ZD-4.3		
TR50	131	3052	44	2006	1N248B
TR501	5952	3585			
TR502	5992	3501/5994	5104		
TR503					1N1189
TR508	102	3004	2	2007	
TR508A	102	3004	2	2007	
TR508AN	102A	3004	53	2007	
TR51	160		17	2004	
TR515T	5541	6615/5515			
TR52	160		82	2004	
TR53	100	3122	18	2007	1N1183A
TR5320326	229	3018	86	2038	
TR54	102A	3118	48	2007	
TR55	100	3776/5010A	28	2007	
TR5528	312	3448	FET-1	2035	
TR56	104	3009	29	2006	
TR57	105	3054/196	246		
TR59			14	2041	
TR5R26	102	3004	2	2007	
TR5S	5010A		ZD-5.1		
TR600					1N1198
TR601	123A	3444	20	2051	1N1198
TR6010	5529	6629			
TR602	5994	3501	5104		1N1198
TR603					1N1190
TR62	160		20	2004	
TR63			46	2024	
TR64			46	2007	
TR65			46	2007	
TR650	102	3004	2	2007	
TR650A	102A	3004	53	2007	
TR653	102	3004	2	2007	
TR653A	102A	3004	53	2007	
TR67	165	3111	36		
TR68	162	3079	38		
TR6R26	102	3004	2	2007	
TR6S	5070A	9021	ZD-6.2	561	
TR70			61	2038	
TR7010	5530	6648/5548			
TR71	102A	3004	53	2007	
TR72	102A	3004	217	2007	
TR721	102	3004	2	2007	
TR721A	102A	3004	53	2007	
TR722	102	3004	2	2007	
TR722A	102A	3004	53	2007	
TR758A			1	2007	
TR759			1	2007	
TR76		3190	57	2017	
TR760	100	3007	52	2024	
TR761	126	3007	52	2024	
TR762	126	3007	52	2024	
TR763	102	3004	2	2007	
TR763A	102A	3004	53	2007	
TR764			1	2007	
TR77			58	2004	
TR792			1	2007	
TR7SA	5014A	3780	ZD-6.8	561	
TR7SB	5071A	3334	ZD-6.8	561	
TR80			86	2038	
TR8001	160	3006	245	2004	
TR8002	160	3006	245	2004	
TR8003	160	3006	245	2004	
TR8004	108	3444/123A	86	2038	
TR8004-4	123A	3444	20	2051	
TR8005	124	3021	12		
TR8006	121	3717	239	2006	
TR8007	102	3722	2	2007	
TR8007A	159	3114/290	82	2032	
TR801			1	2007	
TR8010		3444	20	2051	
TR8014	123A	3444	20	2051	
TR8018	181		75	2041	
TR8019	153		69	2049	
TR802			1	2007	
TR8020	159		82	2032	
TR8021	123A	3444	20	2051	
TR8022	190		217		
TR8023	190		217		
TR8024	190		217		
TR8025	123A	3444	20	2051	
TR8026A	159	3114/290	82	2032	
TR8028	123A	3444	20	2051	
TR8029	123A	3444	20	2051	
TR803				2005	
TR8030	123A	3444	20	2051	
TR8031	123A	3444	20	2051	
TR8034	123A	3444	20	2051	
TR8035	123A	3444	20	2051	
TR8036	128		243	2030	
TR8037	129		244	2027	
TR8038			20	2051	
TR8039	123A	3444	20	2051	
TR8040	123A	3444	20	2051	
TR8042	108	3444/123A	86	2038	
TR8043	123A	3444	20	2051	
TR8055	159	3114/290	82	2032	
TR81	102A	3004	53	2007	
TR83			86	2038	
TR8330			17	2051	
TR84			53	2007	
TR85			53	2007	
TR86			20	2051	
TR87			18,243	2004	
TR88			244	2004	
TR9100	123A	3444	20	2051	
TR9S	139A	3782/5016A	ZD-9.1	562	
TR9SA	5016A	3784/5018A	ZD-8.2	562	
TR9SB	139A	3784/5018A	ZD-9.1	562	
TRA-10R	126	3008	52	2024	
TRA-11R	126	3008	52	2024	
TRA-12R	126	3008	52	2024	
TRA-2	126	3008	52	2024	
TRA-22	126	3006/160	52	2024	
TRA-22A	126	3008	52	2024	
TRA-22B	126	3008	52	2024	
TRA-23	126	3008	52	2024	
TRA-23A	126	3008	52	2024	
TRA-23B	126	3008	52	2024	
TRA-24	126	3008	52	2024	
TRA-24A	126	3008	52	2024	
TRA-24B	126	3008	52	2024	
TRA-32	102A	3004	53	2007	
TRA-33	102A	3004	53	2007	
TRA-34	123A	3020/123	20	2051	
TRA-36	123A	3020/123	20	2051	
TRA-4	123A	3020/123	20	2051	
TRA-4A	123A	3020/123	20	2051	
TRA-4B	123A	3020/123	20	2051	
TRA-7R	121	3009	239	2006	
TRA-7RM	121	3009	239	2006	
TRA-8R	121	3009	239	2006	
TRA-9R	123A	3020/123	20	2051	
TRA10R	160		245	2004	
TRA11R	160		245	2004	
TRA12R	160		245	2004	
TRA22A	160		245	2004	
TRA22B	160		245	2004	
TRA23	160		245	2004	
TRA23A	160		245	2004	
TRA23B	160		245	2004	
TRA24			51	2004	
TRA24A			51	2004	
TRA24C	160		245	2004	
TRA32	102A		53	2007	
TRA33	102A		53	2007	
TRA34	123A	3444	20	2051	
TRA36	123A	3444	20	2051	
TRA4	123A	3444	20	2051	
TRA4A	123A	3444	20	2051	
TRA4B	123A	3444	20	2051	
TRA705	5514	3613			
TRA705D	5491	6791			
TRA71	5514	3613			
TRA71D	5491	6791			
TRA72	5514	3613			
TRA72D	5492	6792			
TRA74	5515	6615			
TRA74D	5494	6794			
TRA7R	121	3717	239	2006	
TRA8R	121	3717	239	2006	
TRA9R	123A	3444	20	2051	
TRAPLC1013	186	3192	28	2017	
TRAPLC711	123A	3444	20	2051	
TRAPLC871	123A	3444	20	2051	
TRAPLC871A	123A	3444	20	2051	
TRBC147B	123A	3444	20	2051	
TRC-P4	113A		6GC1		
TRC44	100	3721	1	2007	
TRC44A	100	3721	1	2007	
TRC45	100	3721	1	2007	
TRC45A	100	3721	1	2007	
TRC70	102	3722	2	2007	
TRC71	102	3722	2	2007	
TRC72	102	3722	2	2007	
TRHA1151	1237	3707	IC-154		
TRLA3350		3700	IC-76		
TRM13			50	2004	
TRM14			50	2004	
TRM15			1	2007	
TRM16			52	2024	
TRM17			52	2024	
TRM21			1	2007	
TRM81			50	2004	
TRO-9004	718		IC-8		
TRO-9005	703A	3157	IC-12		
TRO-9006	703A	3157	IC-12		
TRO-9011	720	9014	IC-7		
TRO1026			20	2051	
TRO1037			62	2010	
TRO1042			39	2015	
TRO1053-1			21	2034	
TRO1054-1				2030	
TRO1054-7	128		243	2030	
TRO10602-1			20	2051	
TRO2012			50	2004	
TRO2051-1			21	2034	
TRO2054-1				2027	
TRO2062-1			21	2034	
TRO31	176	3123	80		
TRO6011			FET-2	2035	
TRO9004		3159	IC-8		
TRPLC711		3018	62	2010	
TRR6	137A	3058	ZD-6.2	561	
TRS100	154	3045/225	40	2012	
TRS1005	124	3021	12		
TRS1005LP	124	3021	12		
TRS100A	154		222		
TRS100HC	282	3748			
TRS101	154	3045/225	40	2012	
TRS120	154	3045/225	40	2012	
TRS1204	198	3220	251		
TRS1205	124	3021	12		
TRS1205LP	124	3021	12		
TRS125HC	282	3748			
TRS140	154	3045/225	40	2012	
TRS1404	198	3220	251		
TRS1405	124	3021	12		
TRS1405LP	124	3021	12		
TRS140HP	171		27		
TRS140MP	198	3220	251		
TRS160	154	3045/225	40	2012	
TRS1604	198	3220	251		
TRS1605	124	3021	12		
TRS160HP	171		27		
TRS160MP	198	3220	251		
TRS180	154	3045/225	40	2012	
TRS1804	198	3220	251		
TRS1805	124	3021	12		
TRS1805LP	124	3021	12		
TRS180HP	171		27		
TRS180MP	198	3220	251		

Industry Standard No.	ECG	SK	GE	RS 276-	MOTOR.
TRS200	154	3045/225	40	2012	
TRS2004	198	3220	251		
TRS2005	124	3021	12		
TRS2005LP	124	3021	12		
TRS2006	198	3220	251		
TRS200HP	171		27		
TRS200MP	198	3220	251		
TRS225	154	3045/225	40	2012	
TRS2254	198	3220	251		
TRS2255	124	3021	12		
TRS2255LP	124	3021	12		
TRS225HP	171		27		
TRS225MP	198	3220	251		
TRS250	154	3045/225	40	2012	
TRS2504	198	3220	251		
TRS2505	124	3021	12		
TRS2505LP	124	3021	12		
TRS250HP	171		27		
TRS250MP	198	3220	251		
TRS275	154	3045/225	40	2012	
TRS2754	198	3220	251		
TRS2755	124	3021	12		
TRS2755LP	124	3021	12		
TRS275HP	171		27		
TRS275MP	198	3220	251		
TRS2804S	198	3220	251		
TRS2805S	198	3220	251		
TRS3006	198	3220	251		
TRS301	154	3045/225	40	2012	
TRS3011	154	3045/225	40	2012	
TRS3012	154	3045/225	40	2012	
TRS3014	171	3201	27		
TRS3015	198	3220	251		
TRS3015LP	124	3021	12		
TRS3016LC	124	3021	12		
TRS301HP	171		27		
TRS301LC	198	3220	251		
TRS301MP	198	3220	251		
TRS3204S	198	3220	251		
TRS3205S	198	3220	251		
TRS325	396		222		
TRS3254	198	3220	251		
TRS3255	198	3220	251		
TRS325MP	198		251		
TRS350	396		32		
TRS3742	198	3220	251		
TRS375	396		32		
TRS401	396		32		
TRS4016LC	124	3021	12		
TRS425	396		32		
TRS450	396		32		
TRS451	396		32		
TRS4926	198	3220	251		
TRS4927	198	3220	251		
TRS5016LC	124	3021	12		
TRS6016LC	124	3021	12		
TRSR3AM	125	3051/156	510	1114	
TRW2001					MRF2001
TRW2003					MRF2003
TRW2005					MRF2005
TRW2010					MRF2010
TRWZ130	5022A	3788			
TS-05					1N4001
TS-1	102A	3004	53	2007	1N4002
TS-1007	102A		53	2007	
TS-1193-736	130		14	2041	
TS-1266	102A		53	2007	
TS-13	102A	3004	53	2007	
TS-14	102A	3004	53	2007	
TS-15	102A	3004	53	2007	
TS-162	102A		53	2007	
TS-163	102A		53	2007	
TS-164	102	3722	2	2007	
TS-165	102A		53	2007	
TS-1657	121	3717	239	2006	
TS-166	102A		53	2007	
TS-1727	102A		53	2007	
TS-1728	102A		53	2007	
TS-173	121	3009	239	2006	
TS-176	121	3717	239	2006	
TS-2	102A	3017B/117	53	2007	1N4003
TS-21756640	159		82	2032	
TS-2A	116	3016	504A	1104	
TS-3	102A	3004	53	2007	
TS-337	142A	3062	ZD-12	563	
TS-4					1N4004
TS-6					1N4005
TS-601	100	3005	1	2007	
TS-602	100	3005	1	2007	
TS-603	102A		53	2007	
TS-604	102A		53	2007	
TS-609	105	3012	4		
TS-610	104	3009	16	2006	
TS-612	104	3009	16	2006	
TS-613	104	3009	16	2006	
TS-614	104	3009	16	2006	
TS-615	160		245	2004	
TS-616	102A		53	2007	
TS-617	102A		53	2007	
TS-618	102A		53	2007	
TS-620	160		245	2004	
TS-621	160		245	2004	
TS-627	102A		53	2007	
TS-627A	160		245	2004	
TS-627B	160		245	2004	
TS-629	102A		53	2007	
TS-630	160		245	2004	
TS-672A	160		245	2004	
TS-672B	160		245	2004	
TS-673A	160		245	2004	
TS-673B	160		245	2004	
TS-739	102A		53	2007	
TS-739B	102A		53	2007	
TS-740	102A		53	2007	
TS-765	102A		53	2007	
TS-8					1N4006
TS05	116	3017B/117	504A	1104	
TS1	116		504A	1104	
TS10					1N4934
TS1007	158	3004/102A	53	2007	
TS1266	158	3123	53	2007	
TS13	102	3722	2	2007	
TS14	102	3722	2	2007	
TS15	102	3722	2	2007	
TS1541	176	3123	80		
TS162	158	3123	53	2007	
TS163	158	3123	53	2007	
TS164	158	3004/102A	53	2007	
TS165	158	3004/102A	53	2007	
TS1657	121	3009	239	2006	
TS166	158	3004/102A	53	2007	
TS1727	158	3004/102A	53	2007	
TS1728	158	3004/102A	53	2007	
TS173	121	3014	239	2006	
TS176	121	3009	239	2006	
TS1792	102A		53	2007	
TS2	116		504A	1104	
TS20					1N4935
TS2218	128		63	2030	
TS2219	128		63	2030	
TS2221	123A	3444	20	2051	
TS2222	123A	3444	20	2051	
TS2776			40	2012	
TS2904	159		82	2032	
TS2905	159		82	2032	
TS2906	159		82	2032	
TS2907	159		82	2032	
TS2A	116		504A	1104	
TS3	125		510,531	1114	1N4933
TS4	5804	9007	504A	1144	
TS40					1N4936
TS5				1104	1N4933
TS50					1N4937
TS6	116	3017B/117	504A	1104	
TS60					1N4937
TS601			53	2007	
TS602			53	2007	
TS603	158		53	2007	
TS604	158		53	2007	
TS609	105		4		
TS610	121	3014	239	2006	
TS612	121	3009	239	2006	
TS613	121	3014	239	2006	
TS614	121	3009	239	2006	
TS615			1	2007	
TS616	158	3004/102A	53	2007	
TS617	158	3004/102A	53	2007	
TS618	158	3004/102A	53	2007	
TS619	158	3004/102A	53	2005	
TS620		3008	53	2005	
TS621		3008	53	2007	
TS627	158	3004/102A	53	2007	
TS627A		3123	2	2007	
TS627B		3123	2	2007	
TS629	158	3004/102A	53	2007	
TS630	160	3123	245	2004	
TS669A	100	3721	1	2007	
TS669B	100	3721	1	2007	
TS669C	160		245	2004	
TS669D	100	3721	1	2007	
TS669E	100	3721	1	2007	
TS669F	100	3721	1	2007	
TS672A		3005	2	2007	
TS672B	160	3005	245	2004	
TS673A	160	3005	245	2004	
TS673B	160	3005	245	2004	
TS67BB	176	3123	80		
TS739	158	3004/102A	53	2007	
TS739B	158		53	2007	
TS740	158	3004/102A	53	2007	
TS765	158	3123	53	2007	
TS8	125	3032A	510,531	1114	
TS80					MR817
TS9013	108	3122	86	2038	
TS97-1	159		82	2032	
TSB-1000	116	3017B/117	504A	1104	
TSB-245	116	3016	504A	1104	
TSB245	116		504A	1104	
TSC-499	199	3245	62	2010	
TSC-722	128		243	2030	
TSC136	177	3100/519	300	1122	
TSC159	116	3311	504A	1104	
TSC499	123A	3444	20	2051	
TSC614	108	3018	86	2038	
TSC695	123A	3444	20	2051	
TSC722	128		243	2030	
TSC767	199	3124/289	62	2010	
TSS-616-1	358		42		
TST705899A	123A	3444	20	2051	
TSTD-15	506	3130	511	1114	
TSV					1N4933
TSV-1000			504A	1104	
TSW100C	5402	3638			
TSW200C	5404	3627			
TSW20C	5400	3950			
TSW21C	5401	3638/5402			
TSW22C	5402	3638			
TSW23C	5404	3627			
TSW30C	5400	3950			
TSW5060	5400	3950			
TSW5061	5401	3638/5402			
TSW5062	5402	3638			
TSW5063	5403	3627/5404			
TSW5064	5404	3627			
TSW5065	5405	3951			
TSW5066	5405	3951			
TSW60C	5401	3638/5402			
TT-1083	121	3009	239	2006	
TT-1097	123A	3020/123	20	2051	
TT-204	108	3018	86	2038	
TT-204A	108	3018	86	2038	
TT-204AB	108	3018	86	2038	
TT-204B	108	3018	86	2038	
TT-204C	108	3018	86	2038	
TT1083	121	3717	239	2006	
TT1097	123A	3444	20	2051	
TT204	108		86	2038	
TT204A	108		86	2038	
TT204AB	108		86	2038	
TT204B	108		86	2038	
TT204C	108		86	2038	
TT2SA495-O	290A	3466/159	221	2032	
TT2SA495-O-A	159		82	2032	
TT2SA495-O-A	159		82	2032	
TT2SA495-Y	290A	3466/159	221	2032	
TT2SA495-Y-A	159		82	2032	
TT2SC983-O	287	3044/154			
TT2SC983-O-A	198	3044/154	251		
TT2SC983-O-A	171	3044/154	27		
TT2SC983-R-A	198	3044/154	251		
TT2SC983-Y	287	3044/154			
TT2SC983-Y-A	198	3044/154	251		
TT66X26	116	3017B/117	504A	1104	
TU000				2007	
TUS-185D	116	3031A	504A	1104	
TUS34	312	3448	FET-1	2035	
TV-109			23	2020	
TV-112	196	3054	241	2020	
TV-114	127	3764	25		
TV-115	152		66	2048	
TV-116	153		69	2049	
TV-117	152	3054/196	66	2048	
TV-118	165		38		
TV-119	165	3111	38		
TV-120	157	3104A	232		
TV-121	164		37		
TV-124	165		38		
TV-125	165	3111	38		

Industry Standard No.	ECG	SK	GE	RS 276-	MOTOR.
TV-15	108	3018	86	2038	
TV-15A			60	2015	
TV-15B	108		86	2038	
TV-16			39	2015	
TV-17		3444	20	2051	
TV-18	123A	3444	20	2051	
TV-19	154	3040	40	2012	
TV-20	233	3018	210	2009	
TV-21	123A	3444	20	2051	
TV-215	513	3443	513		
TV-22	108		86	2038	
TV-23	123A	3444	20	2051	
TV-24104	116	3016	504A	1104	
TV-24125	116	3016	504A	1104	
TV-24191			504A	1104	
TV-24232			504A	1104	
TV-24266	116		504A	1104	
TV-24399			39	2015	
TV-2496	116	3016	504A	1104	
TV-25	191	3232	249		
TV-26	128	3024	243	2030	
TV-27	103A		59	2002	
TV-28	128	3024	243	2030	
TV-29	129		244	2027	
TV-32	123AP	3018	86	2038	
TV-33	319P	3117	283	2016	
TV-34	319P	3117	283	2016	
TV-35	319P	3132	283	2016	
TV-36	161	3132	39	2015	
TV-37	161	3444/123A	39	2015	
TV-38	161	3444/123A	39	2015	
TV-39	161	3444/123A	39	2015	
TV-40	123A	3444	20	2051	
TV-41	128	3024	243	2030	
TV-42			20	2051	
TV-43	128		243	2030	
TV-44	159	3025/129	82	2032	
TV-45	128	3024	243	2030	
TV-46	123A	3444	20	2051	
TV-47		3114	22	2032	
TV-47A	159	3114/290	82	2032	
TV-48	229	3444/123A	61	2038	
TV-49	154	3024/128	40	2012	
TV-50	319P	3117	283	2016	
TV-51	123A	3444	20	2051	
TV-52	123A	3444	20	2051	
TV-53	123A	3444	20	2051	
TV-54	233	3117	39	2009	
TV-55	233	3039/316		2009	
TV-56	123A	3444	20	2051	
TV-57	159	3122	82	2032	
TV-58	123A	3444	20	2051	
TV-59	128	3024	243	2030	
TV-6	123A	3444	20	2051	
TV-60	123A	3444	20	2051	
TV-61	102	3004	2	2007	
TV-61A	102A	3004	53	2007	
TV-62			20	2051	
TV-65	123A	3444	20	2051	
TV-66	123A	3444	20	2051	
TV-67	128	3024	243	2030	
TV-68	123A	3444	20	2051	
TV-7	108	3124/289	86	2038	
TV-70	154	3044	40	2012	
TV-70B	128	3044/154	243	2030	
TV-71		3444	21	2034	
TV-73	186	3192	28	2017	
TV-74	187	3193	29	2025	
TV-75	186	3192	28	2017	
TV-77	233			2009	
TV-82	186	3192	28	2017	
TV-83	312	3448	FET-1	2035	
TV-84	123A	3444	20	2051	
TV-85	171		27		
TV-87	159	3114/290	82	2032	
TV-92	123A	3444	20	2051	
TV-93	159	3114/290	82	2032	
TV1000	108	3039/316	86	2038	
TV106	127	3764	25		
TV108	163A	3111	36		
TV109	175	3026	246	2020	
TV110		3035	25		
TV111	127	3035	25		
TV112	196	3026	241	2020	
TV113	124	3021	12		
TV114	127	3035	25		
TV115	108	3893/152	86	2038	
TV116	197	3018	250	2027	
TV117	196		241	2020	
TV118	163A	3111	36		
TV121	162	3133/164	35		
TV122	124	3021	12		
TV124	165	3111	38		
TV13-11K60	502	3067	CR-4		
TV13-12K60	503	3068	CR-5		
TV15	161	3132	39	2015	
TV15A	161	3018	39	2015	
TV15B	161	3018	39	2015	
TV15C	161	3018	39	2015	
TV16	108	3018	86	2038	
TV17	123A	3018	20	2051	
TV17A	108	3018	86	2038	
TV18	108	3018	86	2038	
TV18B			20	2051	
TV19	154	3040	220	2012	
TV20	161	3716	39	2015	
TV20-10K80	505	3108	CR-7		
TV20-S	505	3108	CR-7		
TV21	192	3020/123	63	2030	
TV215(I.R.)	513		513		
TV22	191	3018	249		
TV23	128	3024	243	2030	
TV2403	108	3039/316	86	2038	
TV2404	108	3039/316	86	2038	
TV241013	110MP	3088	1N60	1123	
TV24102	108	3018	86	2038	
TV24103	112	3089	1N82A		
TV24103A	112	3089	1N82A		
TV24103B	112	3089	1N82A		
TV24103C	112	3089	1N82A		
TV24103D	112	3089	1N82A		
TV24104	116		504A	1104	
TV241073	116	3017B/117	504A	1104	
TV241074	116	3017B/117	504A	1104	
TV241077	123A	3444	20	2051	
TV241078	123A	3444	20	2051	
TV24115	102A	3004	53	2007	
TV24122	110MP	3709	1N60	1123(2)	
TV24125	116		504A	1104	
TV24136	116	3016	504A	1104	
TV24137	160		245	2004	
TV24142	127	3035	25		
TV24143	103A	3010	59	2002	
TV24148	108	3039/316	86	2038	
TV24152	100	3005	1	2007	
TV24154	102A	3004	53	2007	
TV24155	116	3016	504A	1104	
TV24156	102A	3004	53	2007	
TV24157			52	2024	
TV24158	160	3006	245	2004	
TV24159	112	3089	1N82A		
TV24160	161	3018	39	2015	
TV24161	161	3018	39	2015	
TV24162	127	3034	25		
TV24163	127	3034	25		
TV24164	154	3045/225	40	2012	
TV24166	160		245	2004	
TV24167	506	3031A	511	1114	
TV24169	506	3017B/117	511	1114	
TV24172	160	3006	245	2004	
TV24182	112	3089	1N82A		
TV24189	102A	3004	53	2007	
TV2419	116	3017B/117	504A	1104	
TV24190	113A		6GC1		
TV24191	116	3016	504A	1104	
TV24193	116	3017B/117	504A	1104	
TV24194	102A	3004	53	2007	
TV24200	116	3016	504A	1104	
TV24203	108	3018	86	2038	
TV24204	108	3018	86	2038	
TV24209	161	3018	39	2015	
TV24210	107	3018	11	2015	
TV24211	175	3021/124	246	2020	
TV24214	159	3009	82	2032	
TV24214A	159	3114/290	82	2032	
TV24215	123A	3444	20	2051	
TV24216	123A	3444	20	2051	
TV24217	506	3125	511	1114	
TV24219	506	3125	511	1114	
TV24221	125		510,531	1114	
TV24222	116	3017B/117	504A	1104	
TV24224	116	3016	504A	1104	
TV24225	156		512		
TV24226	113A	3119/113	6GC1		
TV24229	160		245	2004	
TV24230	160		245	2004	
TV24232	116		504A	1104	
TV24234	116	3017B/117	504A	1104	
TV24237	120	3017B/117	CR-3		
TV24239	160		245	2004	
TV24266	116	3032A	504A	1104	
TV24272	138A		ZD-7.5		
TV24273	110MP		1N60	1123	
TV24278	116(2)	3311	504A	1104	
TV2428	102A	3004	53	2007	
TV24281	123A	3444	20	2051	
TV24282	116	3311	504A	1104	
TV24283	116		504A	1104	
TV24285	166		504A	1152	
TV2429	102A	3004	53	2007	
TV24292	116		504A	1104	
TV24298	116	3032A	504A	1104	
TV24313	108	3039/316	86	2038	
TV24337	121	3009	239	2006	
TV2434	102A	3005	53	2007	
TV24341	127	3034	25		
TV24351	160	3006	245	2004	
TV24363	159	3114/290	82	2032	
TV24363A	159	3114/290	82	2032	
TV24370	102A	3004	53	2007	
TV24372	123A	3444	20	2051	
TV24380	107	3018	11	2015	
TV24382	107	3117	11	2015	
TV24383	107	3018	11	2015	
TV24385	107		11	2015	
TV24387	108	3019	86	2038	
TV24399	161	3716	39	2015	
TV24435	154	3040	40	2012	
TV24436	161	3132	39	2015	
TV24437	161	3117	39	2015	
TV24438	107		11	2015	
TV24453	123A	3444	20	2051	
TV24454	123A	3444	20	2051	
TV24458	123A	3444	20	2051	
TV24468	127	3035	25		
TV24487	175		246	2020	
TV24495	159		82	2032	
TV24495A	159	3114/290	82	2032	
TV24499	154		40	2012	
TV24540	503	3035	CR-5		
TV2455	160	3006	245	2004	
TV24554	177	3311	300	1122	
TV24568	127	3035	25		
TV24571	161	3117	39	2015	
TV24573	108	3039/316	86	2038	
TV24574	108	3039/316	86	2038	
TV24576	123A	3444	20	2051	
TV24582	116	3311	504A	1104	
TV24586	116	3031A	504A	1104	
TV24589	108	3039/316	86	2038	
TV24599	102A		53	2007	
TV24617	506	3032A	511	1114	
TV24648	506	3125	511	1114	
TV24650	503		CR-5		
TV24651	5800	9003	504A	1142	
TV24655	123A	3444	20	2051	
TV24678	121	3717	239	2006	
TV24684	108		86	2038	
TV2479	160	3006	245	2004	
TV24803	116	3017B/117	504A	1104	
TV24806	107	3019	11	2015	
TV24848			21	2034	
TV24941	116	3130	504A	1104	
TV24942	125		510,531	1114	
TV24945	102A		53	2007	
TV2496	116		504A	1104	
TV24979	116		504A	1104	
TV24981	5071A	3334	ZD-6.8	561	
TV24983	103A	3835	59	2002	
TV24984	102A		53	2007	
TV24985	115		6GX1		
TV24987	113A		6GC1		
TV25		3045	18	2030	
TV26	192	3024/128	63	2030	
TV27	103	3862	8	2002	
TV28	192	3020/123	63	2030	
TV29	129	3025	244	2027	
TV2SB126	127	3764	25		
TV2SB126P	121	3009	239	2006	
TV2SB126V	127		25		
TV2SB448	127	3764	25		
TV2SC1505	198	3220	251		
TV2SC1507	198	3220	251		
TV2SC208	123A	3444	20	2051	
TV32	108	3018	86	2038	
TV33	161	3018	39	2015	
TV34	506		511	1114	
TV34232	116	3031A	504A	1104	

Industry Standard No.	ECG	SK	GE	RS 276-	MOTOR.
TV35	161	3018	39	2015	
TV36		3018	20	2051	
TV37	319P	3018	20	2009	
TV38	123A	3018	20	2051	
TV39	123A	3018	20	2051	
TV4	116	3017B/117	504A	1104	
TV40		3018	20	2051	
TV41	128	3024	243	2030	
TV4152	100	3005	1	2007	
TV42	128	3020/123	243	2030	
TV43		3018	20	2051	
TV44	159	3018	82	2032	
TV45	128	3024	243	2030	
TV46	123A	3018	20	2051	
TV47	158		53	2007	
TV48	123A		20	2051	
TV49	191	3232	249		
TV50	161	3039/316	39	2015	
TV51	128	3024	243	2030	
TV52	128	3024	243	2030	
TV53	128	3024	243	2030	
TV54	161	3039/316	39	2015	
TV55	108		86	2038	
TV56	123A	3018	20	2051	
TV57	107	3020/123	11	2015	
TV57A	123A	3020/123	20	2051	
TV58	107	3020/123	11	2015	
TV58A	123A	3020/123	20	2051	
TV59	128	3020/123	243	2030	
TV59A	123A	3020/123	20	2051	
TV6.5	118		CR-1	1104	
TV60	107	3020/123	11	2015	
TV6080	128		243	2030	
TV60A	123A	3020/123	20	2051	
TV61	158	3004/102A	53	2007	
TV62	172A	3156	64		
TV65	123A		20	2051	
TV695	118		CR-1		
TV70	154		40	2012	
TV71	123A		20	2051	
TV72	159		82	2032	
TV73	210		252	2018	
TV75	210		252	2018	
TV8	125	3032A	510,531	1114	
TV80	312	3116	FET-1	2035	
TV81	233	3117	210	2009	
TV83			FET-2	2035	
TV85		3054	27		
TV92	123A		20	2051	
TVC-3	116	3110/120	504A	1104	
TVC-6	120		CR-3		
TVC-MK3800	1136		IC-284		
TVC6	120		CR-3		
TVCM-1	705A	3134	IC-3		
TVCM-10	722	3161	IC-9		
TVCM-11	712	3072	IC-2		
TVCM-12	719		IC-5		
TVCM-13	721		IC-14		
TVCM-14		3163	IC-211		
TVCM-15	731	3170	IC-13		
TVCM-16	723	3144	IC-15		
TVCM-18	737	3375	IC-16		
TVCM-19	744		IC-24	2022	
TVCM-2	790	3077	IC-5		
TVCM-20	736		IC-17		
TVCM-21	739	3235	IC-30		
TVCM-22	790		IC-18		
TVCM-26	804		IC-27		
TVCM-27	738	3167	IC-29		
TVCM-28	742		IC-213		
TVCM-29	743		IC-214		
TVCM-3	707	3134/705A			
TVCM-30		3141	IC-225		
TVCM-33	729	3074			
TVCM-34	791	3149	231		
TVCM-35	740A	3328	IC-31		
TVCM-39	788	3147	IC-229		
TVCM-4	708	3135/709	IC-10		
TVCM-49		3254	IC-226		
TVCM-5	709	3135	IC-11		
TVCM-500	7400		7400		
TVCM-502	7474		IC-193	1818	
TVCM-503	7447			1805	
TVCM-505	7490			1808	
TVCM-551		3512	46		
TVCM-6	718	3358/1006	IC-8		
TVCM-65	804	3455			
TVCM-66		3543	IC-172		
TVCM-7	720	9014	IC-7		
TVCM-8	714	3075	IC-4		
TVCM-81	1155		IC-179	705	
TVCM-82		3249	IC-135		
TVCM-9	715	3076	IC-6		
TVCM-S	709		IC-11		
TVDS1M	116		504A	1104	
TVH-526	114		6GD1	1104	
TVH538	113A		6GC1		
TVM-108	501B		520		
TVM-511	116		504A	1104	
TVM-526	113A	3119/113	6GC1		
TVM-530	178MP		300(2)	1122(2)	
TVM-537	119	3109	CR-2		
TVM-778	500A		517		
TVM-EH2C	116	3017B/117	504A	1104	
TVM-EH2C11	116		504A	1104	
TVM-EH2C11/1			504A	1104	
TVM-EH2C11/1+12/1			504A	1104	
TVM-EH2C12/1			504A	1104	
TVM-K-112C	178MP		300(2)	1122(2)	
TVM-K112C	113A		6GC1		
TVM-LOO.09M1115	119		CR-2		
TVM-M204B	116	3016	504A	1104	
TVM-PH9D22/1	116	3016	504A	1104	
TVM-PH9D522/11	116(4)		504A(4)	1104	
TVM-PT6D22/1	166	3647/167	504A	1152	
TVM-TCO.2P	114		6GD1		
TVM-TCO.2P11/2	116	3100/519	504A	1104	
TVM-TK-705M	113A		6GC1		
TVM-TK705M	113A		6GC1		
TVM/HS15/16	502		CR-4		
TVM108	501B	3069/501A	520		
TVM153	522	3303			
TVM35	116	3032A	504A	1104	
TVM511	116(3)	3032A	504A(3)	1104	
TVM526	114	3120	6GD1		
TVM529	166	9075	BR-600	1152	
TVM531	503	3068	CR-5		
TVM533	504	3068/503	CR-5		
TVM535	506	3125	511	1114	
TVM537	506	3125	511	1114	
TVM538	113A		6GC1		
TVM540	505	3108	CR-7		
TVM546	507	3311	511		
TVM550	116	3032A		1104	

Industry Standard No.	ECG	SK	GE	RS 276-	MOTOR.
TVM553	502		CR-4		
TVM554	113A	3119/113	6GC1		
TVM554A	178MP	3119/113	300(2)	1122(2)	
TVM56	116	3031A	504A	1104	
TVM563	116	3017B/117		1104	
TVM567	502	3067	CR-4		
TVM569	503	3068	CR-5		
TVM778	500A	3304	517		
TVMHS151B	116		504A	1104	
TVMK112C	113A		6GC1		
TVML00.09D1115				1104	
TVML00.09M1115	116		504A	1104	
TVMM204B	116		504A	1104	
TVMPH9DZ2/1	116		504A	1104	
TVMTC00921-3	114	3120	6GD1	1104	
TVS-0A70	109	3091		1123	
TVS-0A81	109			1123	
TVS-0A90	109	3091		1123	
TVS-0A91	109	3091		1123	
TVS-0A95	109			1123	
TVS-0V-02	506		511	1114	
TVS-10D8	125			1104	
TVS-1303	159		82		
TVS-182G	112	3089	1N82A		
TVS-185D	116		504A	1104	
TVS-1N82G	112	3089	1N82A		
TVS-1S1211	177	3100/519	300	1122	
TVS-1S18	108		86	2038	
TVS-1S1850	116	3031A	504A	1104	
TVS-1S1893	5800	9003		1142	
TVS-1S1906	116	3031A	504A	1104	
TVS-1S1922	5800	9003		1142	
TVS-1S750	112	3089	1N82A		
TVS-2B-2C	506	3125	511	1114	
TVS-2CS645A	123A		20	2051	
TVS-2S172F			2	2007	
TVS-2S288A	107	3039/316	11	2015	
TVS-2SA103	160	3006	245	2004	
TVS-2SA171	100	3005	1	2007	
TVS-2SA385	102A	3006/160	53	2007	
TVS-2SA385A	102A	3006/160	53	2007	
TVS-2SA385L	102A		53	2007	
TVS-2SA564	159	3114/290	82	2032	
TVS-2SA564A	159	3114/290	82	2032	
TVS-2SA564P	159	3114/290	82	2032	
TVS-2SA71B	102A	3004	53	2007	
TVS-2SB126	121	3009	239	2006	
TVS-2SB126F	121	3009	239	2006	
TVS-2SB126V	127	3034	25		
TVS-2SB171	102A	3004	53	2007	
TVS-2SB171A	158		53	2007	
TVS-2SB172	102A	3004	53	2007	
TVS-2SB172A	100		1	2007	
TVS-2SB172F	102A	3004	53	2007	
TVS-2SB176	102A	3004	53	2007	
TVS-2SB234	102A		53	2007	
TVS-2SB324	158		53	2007	
TVS-2SB448	127	3035	25		
TVS-2SB449F	121	3717	239	2006	
TVS-2SC183P			11	2015	
TVS-2SC183Q			11	2015	
TVS-2SC185A	108	3039/316	86	2038	
TVS-2SC206			20	2051	
TVS-2SC208	108	3444/123A	86	2038	
TVS-2SC208A			20	2051	
TVS-2SC287A	108		86	2038	
TVS-2SC313	161	3039/316	39	2015	
TVS-2SC429A	107	3039/316	11	2015	
TVS-2SC446	108	3018	86	2038	
TVS-2SC466	161	3039/316	39	2015	
TVS-2SC469A	107		11	2015	
TVS-2SC526	154	3045/225	40	2012	
TVS-2SC538	123A	3444	20	2051	
TVS-2SC538A	123A	3444	20	2051	
TVS-2SC562	161	3444/123A	39	2015	
TVS-2SC563	161	3018	39	2015	
TVS-2SC563A	161	3018	39	2015	
TVS-2SC564	159		82	2032	
TVS-2SC58	154	3040	40	2012	
TVS-2SC582	128	3024	243	2030	
TVS-2SC582A	128	3024	243	2030	
TVS-2SC58A	154	3040	40	2012	
TVS-2SC605	108		86	2038	
TVS-2SC606	108		86	2038	
TVS-2SC644	107	3444/123A	11	2015	
TVS-2SC645	107	3444/123A	11	2015	
TVS-2SC645A	107	3444/123A	11	2015	
TVS-2SC645B	107	3444/123A	11	2015	
TVS-2SC645C	107	3044/154	11	2015	
TVS-2SC646	181		75	2041	
TVS-2SC683	161	3018	39	2015	
TVS-2SC683V	161	3018	39	2015	
TVS-2SC684	161	3444/123A	39	2015	
TVS-2SC687	162		35		
TVS-2SC762	161	3018	39	2015	
TVS-2SC828	123A	3444	20	2051	
TVS-2SC828A	123A	3444	20	2051	
TVS-2SC828Q	123A	3444	20	2051	
TVS-2SC840	175	3021/124	246	2020	
TVS-2SC840A	175	3021/124	246		
TVS-2SC901	163A	3238/799	36		
TVS-2SC948	161	3132	39	2011	
TVS-2SD226A	175	3131A/369	246		
TVS-828A	123A	3444	20		
TVS-82G	112	3089	1N82A		
TVS-AN227	1063		IC-53		
TVS-AN241	712		IC-2		
TVS-AN241D	712		IC-2		
TVS-CS1255H	123A	3020/123	20		
TVS-CS1255HF	123A	3122	20		
TVS-CS1256HG	192	3195/311	63		
TVS-CS1303	159	3114/290	82		
TVS-CS1303A	159	3114/290	82		
TVS-DG-1N-R	116	3031A	504A	1104	
TVS-DG1NR	506	3125	511		
TVS-DS-1K	116	3016	504A	1104	
TVS-DS-1M	116	3017B/117	504A	1104	
TVS-DS1K	116	3032A	504A	1104	
TVS-DS1M	116		504A	1104	
TVS-DS2K	116		504A	1104	
TVS-ET1P	116		504A	1104	
TVS-FR-1P	116		504A	1104	
TVS-FR-1P(FR1P)	116		504A	1104	
TVS-FR-2M				1104	
TVS-FR-2PC	116	3031A	504A	1104	
TVS-FR1-PC	116	3031A	504A	1104	
TVS-FR10	125	3033	510,531	1114	
TVS-FR1MD	116		504A	1104	
TVS-FR1P	506	3125	511		
TVS-FR1PC	116		504A	1104	
TVS-FR2M	116		504A	1104	
TVS-FR2P	506	3031A	511		
TVS-FR2PC	156		512		
TVS-FT-1N	125	3033	510,531	1114	

Industry Standard No.	ECG	SK	GE	RS 276-	MOTOR.
TVS-PT-1P	116	3033	504A	1104	
TVS-PT10	125	3033	510,531	1114	
TVS-PT1N	506	3125	511		
TVS-PT1P	506	3125	511		
TVS-PT1PC	506	3125	511		
TVS-PU1N	116	3031A	504A	1104	
TVS-H-339W	505	3108	CR-7		
TVS-H-399W	505	3108	CR-7		
TVS-HF-SD-12	116		504A	1104	
TVS-HF-SD-1C	506		511		
TVS-HF-SD1Z	506	3031A	511		
TVS-HF-SD12C	119		CR-2		
TVS-HFSD1Z	116		504A	1104	
TVS-HS25/16	503		CR-5		
TVS-HS7/1				1104	
TVS-K112C	113A	3119/113	6GC1		
TVS-KC2-LP	116		504A	1104	
TVS-KC20P12/1	116		504A	1104	
TVS-KC20P12/1	116		504A	1104	
TVS-KC2DP12/2	116		504A	1104	
TVS-MPC23C	1046	3471	IC-118		
TVS-OA70	109			1123	
TVS-OA81	177	3100/519	300	1122	
TVS-OA90	109	3087		1123	
TVS-OA91	109	3087		1123	
TVS-OA95	177	3087	300	1122	
TVS-OV-02	116		504A	1104	
TVS-PC02P11/2	116	3016	504A	1104	
TVS-PCD2P11/2	116		504A	1104	
TVS-RD(M)(P)D	142A		ZD-12	563	
TVS-RD-13A				563	
TVS-RD11A	5074A	3061/140A	ZD-11.0	563	
TVS-RD13A	142A	3750/143A	ZD-12	563	
TVS-RD13D	142A	3062	ZD-12	563	
TVS-RD13M	142A	3062	ZD-12	563	
TVS-RD13P	142A	3062	ZD-12	563	
TVS-RD29AN	147A	3095	ZD-33		
TVS-RD5A	135A		ZD-5.1		
TVS-RD7A	138A	3059	ZD-7.5		
TVS-S1B02-03C	116	3106/5304	504A	1104	
TVS-S1B02-03CR	116	3106/5304	504A	1104	
TVS-S8-2C	125	3080	510,531	1114	
TVS-S82	112	3089	1N82A		
TVS-SB-20	125		510,531	1114	
TVS-SB-2C	506		511		
TVS-SD-1				1104	
TVS-SD-1B	125	3033	510,531	1114	
TVS-SD-1B(BOOST)	506		511		
TVS-SD1A	116		504A	1104	
TVS-SD1B	125		510,531	1114	
TVS-SD82	112		1N82A		
TVS-SD82A	112	3089	1N82A		
TVS-SF-1	506	3033	511		
TVS-SN76665N	712		IC-2		
TVS-SPN01				1104	
TVS-TCO.09M11/10				1104	
TVS-TC009M11/10	116		504A	1104	
TVS-TC009M21/3	114	3120	6GD1		
TVS-TIS18	108	3039/316	86		
TVS-UFSD-1	116	3031A	504A	1104	
TVS-UPC23C	1046	3471	IC-118		
TVS-ZB1-11	141A	3092	ZD-11.5		
TVS-ZB1-15	145A	3063	ZD-15	564	
TVS010D1			504A	1104	
TVSOA70	109			1123	
TVSOA71	116		504A	1104	
TVSOA90	109	3087		1123	
TVSOA91	109			1123	
TVS10D	116	9000/552	504A	1104	
TVS10D1	506	9000/552	511		
TVS10D2	552	9000	504A		
TVS10D8	125	3017B/117	504A	1104	
TVS10DC4	116	3016	504A		
TVS10DC4R	116	3016	504A		
TVS1850	116		504A	1104	
TVS1N4002	116		504A	1104	
TVS1N4741		3092	ZD-11	563	
TVS1N4741A	141A	3092	ZD-11.5		
TVS1N741A	142A		ZD-12	563	
TVS1N741H	142A	3062	ZD-12	563	
TVS1P20		3032A	504A	1104	
TVS1P80	125	3080	510,531	1114	
TVS1S1850	116		504A	1104	
TVS1S1906	116	3032A	504A	1104	
TVS1S1922G			504A	1104	
TVS1S1950	116(2)	3311	504A(2)	1104	
TVS1S2076	177	3100/519	514	1122	
TVS1S750	112		1N82A		
TVS1S954	177		514	1122	
TVS25126F	121	3717	239		
TVS25A103	160		245	2005	
TVS25B172	102	3722	2		
TVS25B448	127	3764	25		
TVS25C208	123A		20		
TVS25C562	123A		20		
TVS25C645	123A		20		
TVS2CS1256HG			67	2023	
TVS2SA171			2	2007	
TVS2SA483			69	2049	
TVS2SA543	192	3025/129	63	2030	
TVS2SA546	129	3025	244	2027	
TVS2SA546B				2027	
TVS2SA546E				2027	
TVS2SA546H				2027	
TVS2SA564	159	3025/129	82	2032	
TVS2SA564-O	159	3114/290	82	2032	
TVS2SA564A	159		82	2032	
TVS2SA564C	159		82	2032	
TVS2SA564P	159	3114/290	82	2032	
TVS2SA564PY	159		82	2032	
TVS2SA564Q	159	3114/290	82	2032	
TVS2SA607	159	3118	82	2032	
TVS2SA609	159	3118	82	2032	
TVS2SA71B			2	2007	
TVS2SB126	121	3717	239	2006	
TVD2SB171	158		53	2007	
TVS2SB171A	102A	3004	53	2007	
TVS2SB171B	102A	3004	53	2007	
TVS2SB172			53	2007	
TVS2SB172A			2	2007	
TVS2SB324	102A	3004	53	2007	
TVS2SB449	121	3009	239	2006	
TVS2SB449(F)			16	2006	
TVS2SB449F		3009	25		
TVS2SB546	292	3083/197	69		
TVS2SC1255	128	3024	243		
TVS2SC1255HF	128	3024	243		
TVS2SC1256	129	3025	244		
TVS2SC1256HG	129	3025	244		
TVS2SC1505	198	3219	251		
TVS2SC1507	198		251		
TVS2SC1520	198	3103A/396	251		
TVS2SC1629A	238		75	2041	
TVS2SC1828	175	3131A/369			
TVS2SC288A	108		86	2038	
TVS2SC466	108		86	2038	
TVS2SC538	108	3018	86	2038	
TVS2SC538A	123A		20	2051	
TVS2SC562	161	3039/316	39	2015	
TVS2SC563	161		39	2015	
TVS2SC564	159		82	2032	
TVS2SC564-3			82	2032	
TVS2SC564-O	159			2032	
TVS2SC5640	159		82		
TVS2SC5640	159	3025/129	82	2032	
TVS2SC564R	159	3025/129	82	2032	
TVS2SC58A		3044	20	2051	
TVS2SC644	123A		20	2051	
TVS2SC645	108		86	2038	
TVS2SC645A	108		86	2038	
TVS2SC645B	123A		20	2051	
TVS2SC645C	108		86	2038	
TVS2SC647	130	3561	14	2041	
TVS2SC647-O	165	3561	38		
TVS2SC647-P	165	3561	38		
TVS2SC647A	101	3561	8	2002	
TVS2SC647B	165	3561	38		
TVS2SC647C	165	3561	38		
TVS2SC647D	165	3561	38		
TVS2SC647E	165	3561	38		
TVS2SC647Q	165	3561	38		
TVS2SC647R	165	3561	38		
TVS2SC683	108		86	2038	
TVS2SC684	123A	3716/161	20	2051	
TVS2SC696	128	3024	243	2030	
TVS2SC762	108		86	2038	
TVS2SC828	123A	3020/123	20	2051	
TVS2SC828(Q)			20	2051	
TVS2SC828A	123A		20	2051	
TVS2SC828P	123A	3020/123	20	2051	
TVS2SC828Q	108	3122	86	2038	
TVS2SC828R	123A	3020/123	20	2051	
TVS2SC829(B)			20	2051	
TVS2SC829B	107		11	2015	
TVS2SC840	124		12		
TVS2SC840A	124		12		
TVS2SC840C	175		246		
TVS2SC901	162		35		
TVS2SC920-OQ	108	3018	86		
TVS2SC968	123	3020	20		
TVS2SD201M	130	3027			
TVS2SD226	175	3026	246		
TVS2SD226-O	175	3026	246		
TVS2SD226A	175	3026	246		
TVS2SD226B	175	3026	246		
TVS2SD226C	175	3026	246		
TVS2SD226D	175	3026	246		
TVS2SD2260	175	3026			
TVS2SD226P	175	3026	246		
TVS2SD401	291	3054/196			
TVS2SD968	128	3024	243		
TVS5-B-HW5	506		511		
TVS550	116		504A	1104	
TVS5B-2-H5W	506	3043	511		
TVS5N76642N	709		IC-11		
TVS828A	123A		20		
TVSAFS01	506	3010	511		
TVSAFSD1	506		511		
TVSAN155	147A	3095	ZD-33		
TVSAN220	711	3070	IC-207		
TVSAN225	739		IC-30		
TVSAN241	712	3072	IC-2		
TVSAN241D	712	3072	IC-2		
TVSAW01-11	5074A	3092/141A	ZD-11.0	563	
TVSB01-02	552	3311	504A	1104	
TVSB01-2	177	3311	300	1122	
TVSB24-06C	552	9000	509	1114	
TVSB24-06D	552	9000	509		
TVSBAX-13	177	3100/519	300	1122	
TVSBAX13	177	3100/519	300	1122	
TVSBB10	552	3033	510,531	1114	
TVSBB2	506	3311	509	1114	
TVSBB2A	522	3032A			
TVSBBE	116		504A	1104	
TVSC0410	125	3032A	509	1114	
TVSC1255H	128		243		
TVSCA3126	797		IC-233		
TVSCS1255HF	192		63		
TVSCS1256HG	193		63		
TVSD1K	116		504A	1104	
TVSD52K	116		504A	1104	
TVSDG1NR	116		504A	1104	
TVSDS-1M			504A	1104	
TVSDS-2K	125	3032A	531	1114	
TVSDS1K			504A	1104	
TVSDS1M	156	3051	512		
TVSDS2K	156	3017B/117	512		
TVSDS2M	116	3311	504A	1104	
TVSEA01-07R		3063		564	
TVSEA06	116(4)	3016			
TVSEQA01-05T	5010A	3057/136A	ZD-5.6	561	
TVSEQA01-06S	5075A		ZD-16	564	
TVSEQA01-125	142A	3062	ZD-12	563	
TVSEQB01-12	142A	3062	ZD-12	563	
TVSEQB01-15	145A	3063	ZD-15	564	
TVSEQB01-15Z				564	
TVSERB-06B				1104	
TVSERB24-04D	506	3311	504A	1104	
TVSERB24-06	506	3311	504A		
TVSERB24-06A	506	3016	504A		
TVSERB24-06B	506	3016	504A		
TVSESA06	116	3016	504A	1104	
TVSPG2PC	506	3130	511		
TVSFR1P	116	3031A	504A	1104	
TVSFR1PC	116		504A	1104	
TVSFR2-005	506	3032A	511		
TVSFR2-02	506	3130	511		
TVSFR2-02C	506	3130	511		
TVSFR2-02G	506	3130	511		
TVSFR2-04	506	3032A	511		
TVSFR2-06	116	3311	504A	1104	
TVSFR2-10	506	3032A	511		
TVSFR2PC		3031A	504A	1104	
TVSFT-1P	506		511		
TVSFT1M	506	3130	511		
TVSFT1P	116	3311	504A	1104	
TVSFU1N	116		504A	1104	
TVSFUIN			504A	1104	
TVSGP2-354	177	3100/519	300	1122	
TVSHA-1	506		511		
TVSHA-1/HP-1	506		511		
TVSHA1151	1237		IC-154		
TVSHF-1	506	3130	511		
TVSHF-1/HF-1	506		511		
TVSHF1	506	3130	511		
TVSHFD1Z	506		511		
TVSHFSD-1/HF●1	506		511		
TVSHFSD-1A	116	3311	504A	1104	

Industry Standard No.	ECG	SK	GE	RS 276-	MOTOR.
TVSHFSD-1C	506	3032A	511		
TVSHFSD-1Z	506	3130	511		
TVSHFSD1	506	3130	511		
TVSHFSD12C	116		504A	1104	
TVSHFSD1Z	506		511		
TVSHS7/1	118		CR-1		
TVSIS1850	116		504A	1104	
TVSJA1200	123A	3444	20		
TVSJL41A	116	3311	504A	1104	
TVSJL41AM				1104	
TVSK112C	113A		6GC1		
TVSKB462F	607	3864/605			
TVSM1-02	116		504A	1104	
TVSM51247	1173		IC-302		
TVSM51247P	1173		IC-302		
TVSMA26	177	3100/519	300	1122	
TVSMPC-23C	1046		IC-118		
TVSMPC574J	615	3095/147A	ZD-33		
TVSMPC595C	1186	3168/749	IC-175		
TVSMPC596C	1187	3742			
TVSMPC596C2	1187	3742	IC-176		
TVSMR-1M	125	3032A	509	1114	
TVSMR1C	125	3311	510,531	1114	
TVSOA71	116	3031A	504A	1104	
TVSOA90				1123	
TVSPCD2P11/2	116		504A	1104	
TVSQ01-06SB	5070A	9021			
TVSQA01-06SB	5070A	9021			
TVSQA01-07BE				561	
TVSQA01-07R	145A	3063	ZD-15	564	
TVSQA01-07RE	5071A	3334	ZD-6.8	561	
TVSQA01-11.5E				563	
TVSQA01-115E	5074A	3139			
TVSQA01-11SE	5074A	3139	ZD-11	563	
TVSQA01-12S	142A	3062	ZD-12	563	
TVSQA01-14RD	144A	3094	ZD-14		
TVSQA01-15RB	145A		ZD-15	564	
TVSQA01-15S			ZD-15	564	
TVSQA01-25A	5080A	3336	ZD-22		
TVSQA01-25R	5080A	3151/5081A	ZD-22		
TVSQA01-25RA	5081A	3151	ZD-24		
TVSQB01-15ZB	145A	3063	ZD-15	564	
TVSQB01-18	5077A	3752			
TVSRA-1Z	506	3843	504A	1104	
TVSRA-26		3061	ZD-10	562	
TVSRA1A	177	3100/519	300	1122	
TVSRA1Z	506	3843	511		
TVSRA1ZC	506		511		
TVSRC2	525	3925			
TVSRD11	140A	3061	ZD-10	562	
TVSRD11A	140A	3061	ZD-10	562	
TVSRD12EBH		3062	ZD-12	563	
TVSRD13AL	143A	3750	ZD-13	563	
TVSRD27EB	146A		ZD-27		
TVSRD4AM	134A	3055	ZD-3.6		
TVSRD4R7EB	5009A	3775			
TVSRD5A	135A	3056	ZD-5.1		
TVSRD6A	134A		ZD-3.6		
TVSRD9AL	139A	3060	ZD-9.1	562	
TVSRF-1A			511	1114	
TVSRM25	5089A	3341	ZD-51		
TVSRM26V	5089A		ZD-10	562	
TVSRMP5020	116		504A	1104	
TVSS-34	506	3130	511		
TVSS15	506	3032A	511		
TVSS1D30-15	506	3080	511		
TVSS1P20	116	3032A	504A	1104	
TVSS1R20	552	3311	504A	1122	
TVSS1R80	552	3017B/117	511	1104	
TVSS1R8D	116		504A	1104	
TVSS3-2	116	3032A	504A	1104	
TVSS34	506	3130	511		
TVSS34RECT	116		504A	1104	
TVSS3G4	125	3081	510,531	1114	
TVSS4C	116	3032A	504A	1104	
TVSSA-2				1104	
TVSSA-2B	125	3032A	510,531	1114	
TVSSA-2H	117	3051/156	504A	1104	
TVSSA2B	116	3032A	504A	1104	
TVSSA2H	156		512		
TVSSA71B	102	3722	2		
TVSSB-2	506	3311	511		
TVSSB-2T	117		504A	1104	
TVSSB2-C	506		511		
TVSSB2C	506	3130	511		
TVSSD-1Y	116		504A		
TVSSD1A	116		504A	1104	
TVSSD82A	112		1N82A		
TVSSF-1	125		510,531	1114	
TVSSID30-15	125		510,531	1114	
TVSSJE5472-1	157	3054/196	232		
TVSSM1-02	116	3311			
TVSSN76642N	709	3135	IC-11		
TVSSN76665	712	3072	IC-2		
TVSSN7666N	712		IC-148		
TVSSV02	116	3311	504A	1104	
TVSSV04	177(4)		300(4)		
TVSTD15		3080	511		
TVSTD15M	551	3130	511		
TVSUF2	506	3017B/117		1104	
TVSUFSD-1	506	3130	511		
TVSUFSD1	506		511		
TVSUFSD1F	116	3016	504A	1104	
TVSUP574J		3095	ZD-33		
TVSUPC23C	1046	3471	IC-118		
TVSUPC574J	147A		ZD-33		
TVSUPC595C	1186	3171/744	IC-175		
TVSUPC596C2	1187	3742	IC-176		
TVSVD1221L	605	3864			
TVSVPC23C	1046		IC-118		
TVSW04	166	9075	BR-600	1152	
TVSW04M	166	9075	BR-600	1152	
TVSWF2	116	3017B/117	504A	1104	
TVSWZ061	5013A	3779			
TVSWZ270	146A		ZD-27		
TVSYZ-080	5072A		ZD-8.2	562	
TVSYZ080	5072A	3136	ZD-8.2	562	
TVSZB1-11	5074A	3092/141A	ZD-11.0	563	
TVSZB1-15	5075A	3751	ZD-16	564	
TVSZB1-23	5081A		ZD-24		
TVSZB1-27	146A	3064	ZD-27		
TVSZB1-29	146A	3064	ZD-27		
TVSZB1-6	5070A		ZD-6.2	561	
TW10	116	3016	504A	1104	1N4002
TW100	125	3033	510,531	1114	1N4007
TW10N400CZ	5685	3521/5686			
TW10N500CZ	5686	3521			
TW10N600CZ	5697	3522			
TW120					MR1-1200
TW12N400CX	5695	3509			
TW12N600CX	5697	3522			
TW135	106	3984	21		
TW20	116	3017B/117	504A	1104	1N4003
TW3	116	3311	504A	1104	
TW30	116	3017B/117	504A	1104	1N4004
TW38	156		512		
TW40	116	3017B/117	504A	1104	1N4004
TW5	116	3017B/117	504A	1104	1N4001
TW50	116	3017B/117	504A	1104	1N4005
TW60	116	3017B/117	504A	1104	1N4005
TW80	125	3032A	510,531	1114	1N4006
TW8N400CZ	5685	3521/5686			
TW8N500CZ	5686	3521			
TW8N600CZ	5697	3522			
TWV	116	3311	504A	1104	
TX-100-1	123A	3020/123	20		
TX-100-2	123A	3020/123	20		
TX-100-3	124	3021	12		
TX-100-4	128	3024	243		
TX-1005	175	3021/124			
TX-101-11	124	3021	12		
TX-101-12	123A	3020/123	20		
TX-101-8	124	3021	12		
TX-102-4	124	3021	12		
TX-104-3	158	3004/102A	53		
TX-106-1	158		53		
TX-107-1	123A	3020/123	20		
TX-107-10	123A	3020/123	20		
TX-107-12	107	3020/123	11		
TX-107-13	124	3020/123	12		
TX-107-16	123A	3020/123	20		
TX-107-3	107	3020/123	11		
TX-107-4	123A	3020/123	20		
TX-107-5	123A	3020/123	20		
TX-107-6	123A	3020/123	20		
TX-108-1	123A	3020/123	20		
TX-111-1	124	3021	12		
TX-112-1	123A	3020/123	20		
TX-119-1	123A	3020/123	20		
TX-122-1	159	3025/129			
TX-123	172A	3156			
TX-124	128	3024			
TX-125	129	3025			
TX-126-1	103A	3835			
TX-134-1	159	3114/290			
TX-134-1A	159	3114/290			
TX-135	128	3024			
TX-136	129	3025			
TX-139	129	3025			
TX-140	128	3024			
TX-141	123A	3024/128			
TX-145	5452	6752		1067	
TX-1838	162	3079			
TX100	317		281		
TX100-1	123A	3444	20		
TX100-2	123A	3444	20		
TX100-3	123A	3444	20		
TX100-4	128	3024	243		
TX100-5	124	3021	12		
TX101-11	124	3021	12		
TX101-12	123A	3444	20		
TX101-8	124	3021	12		
TX101-9	128	3024	243		
TX102-1	123A	3444	20		
TX102-2	123A	3444	20		
TX102-4	124	3021	12		
TX103-1	104	3009	16		
TX104-3	158		53		
TX105-4	128		243		
TX106-1	102A		53		
TX107-1	123A	3444	20		
TX107-10	123A	3444	20		
TX107-12	123A	3444	20		
TX107-13	124	3021	12		
TX107-16	123A	3444	20		
TX107-3	123A	3444	20		
TX107-4	123A	3444	20		
TX107-5	123A	3444	20		
TX107-6	123A	3444	20		
TX108-1	123A	3444	20		
TX111-1	124	3021	12		
TX112-1	123A	3444	20		
TX120	107		11		
TX124-1	123A	3444			
TX128-1	128	3444/123A			
TX134-1	129	3025			
TX138	128	3444/123A			
TX139	129	3025			
TX140	128	3024			
TX141	107	3444/123A			
TX1N3190	116	3016	504A	1104	
TX1N3191	116	3017B/117	504A	1104	
TX1N645	116	3016	504A	1104	
TX1N647	116	3017B/117	504A	1104	
TX27A	146A	3064	ZD-27		
TX50	318		282		
TY-107-12	123A	3444			
TY-107-4	123A	3444			
TZ-15A	145A	3063		564	
TZ-15AB	145A	3063		564	
TZ-15BC	145A	3063		564	
TZ-15C	145A	3063		564	
TZ-8	121	3009			
TZ110	151A	3099			
TZ110A	151A	3099			
TZ110B	151A	3099			
TZ110C	151A	3099			
TZ1151	199	3124/289			
TZ1152	199	3124/289			
TZ1153	177	3100/519		1122	
TZ1180	123A	3444			
TZ1182	123A	3444			
TZ12	142A	3062		563	
TZ12A	142A	3062		563	
TZ12B	142A	3062		563	
TZ12C	142A	3062		563	
TZ15	145A	3063		564	
TZ15A	145A	3063		564	
TZ15B	145A	3063		564	
TZ15C	145A	3063		564	
TZ27	146A	3064			
TZ27-A	146A	3064			
TZ27A	146A	3064			
TZ27AA	146A	3064			
TZ27B	146A	3064			
TZ27BA	146A	3064			
TZ27C	146A	3064			
TZ27CJ	146A	3064			
TZ33	147A	3095			
TZ33A	147A	3095			
TZ33B	147A	3095			
TZ33C	147A	3095			
TZ5.6	136A			561	
TZ551	159	3114/290			
TZ581	211	3203			
TZ582	211	3203			
TZ6	121	3009			
TZ62	149A	3097			

Industry Standard No.	ECG	SK	GE	RS 276-	MOTOR.
TZ62A	149A	3097			
TZ62B	149A	3097			
TZ62C	149A	3097			
TZ7	121	3009			
TZ8	121	3717			
TZ8.2	139A	3136/5072A		562	
TZ82	150A	3098			
TZ82A	150A	3098			
TZ82B	5153A	3098/150A			
TZ82C	150A	3098			
TZ9.1	139A	3059/138A		562	
TZ9.1A	139A	3059/138A		562	
TZ9.1B	139A	3059/138A		562	
TZ9.1C	139A	3066/118		562	
U	116			1104	
U-2400-03	116	3016		1104	
U-422				1104	
U-633	116	3016		1104	
U0-5E		3100		1122	
U04	190	3103A/396			
U05B	156	3051		1143	
U05C	156	3051		1143	
U05E	156	3032A		1144	
U05G	558	3314			
U06C	506	3843		1143	
U06E	116			1144	
U07L	558			1114	
U10	191	3232			
U110					2N5474
U112					2N5476
U1177	312	3112			2N4220A
U1178	312	3112			MFE2093/4
U1179					MFE2093
U1180	456	3112			2N4221A
U1181	312	3112			2N4220A
U1182					MFE2093
U119	125			1114	
U120	125			1114	
U1277	312	3112			MFE2095
U1278	312	3112			MFE2094
U1279	312	3112			MFE2093
U1280	312	3112			MFE2093/5
U1281	459				2N3822
U1282	459	3448			2N3822
U1283	459	3112			2N3822
U1284	459	3112			2N3821/2
U1285	312	3112			MFE2093
U1286	312	3112			2N3821
U1287					2N4092
U13033801	116	3017B/117		1104	
U1321					2N4221
U1322	456	3112			2N4221A/2A
U1323	312	3112			2N4221A
U1324	312	3112			2N4220A
U1325	312	3112			2N4220A
U133					2N5475
U146					2N3909
U147					2N3909
U148					2N3909
U149					2N3330
U1585E	123A	3018		2051	
U1585F	123A	3018			
U1585G	123A	3018			
U1585H	123A	3018			
U1650E	312	3448			
U168					2N3330
U1714	312	3112			2N4220A
U1715	312	3112			2N4392
U182					2N4092
U183					MFE2000
U1835E	312	3448			
U1837E	312	3116			2N5486
U184					MFE2001
U1897E	467				2N5638
U1898E					2N5639
U1899E					2N5640
U197	312	3112			
U198					2N3437
U199	466				2N3436
U1994E	312	3116			2N5486
U18102	116			1104	
U200					MFE2004
U201	466				MFE2005
U202					MFE2006
U2047E	312	3448			
U2091858-11	121	3009			
U212	116	3017B/117		1104	
U212-25	116	3017B/117		1104	
U213	116	3016		1104	
U214	116	3016		1104	
U221					2N4092
U222	466				2N4856
U235	312	3112			
U240					MFE2012
U2400-03	116			1104	
U241					MFE2012
U242					MFE2012
U243					MFE2011
U244					MFE2012
U266					MFE2012
U273,A					2N4220
U274,A					2N4220
U275					2N4220
U275A					2N4220
U2848-1	123A	3020/123			
U290					MFE2012
U291					MFE2012
U2N34	102	3004			
U2N34A	102A	3004			
U2N474A	123	3122			
U2N96	102	3004			
U2N96A	102A	3004			
U2S93	126	3006/160			
U2SB267	102A	3004			
U2T85	103A	3010			
U300					2N3993
U3001					2N3365
U3002					2N3366
U3003					2N3367
U301					2N3993
U3010					2N3365
U3011					2N3366
U3012	312	3112			2N3367
U304					2N3993
U305					2N3993
U306					2N3993
U308	452				U308
U309	452				U309
U310					U310
U311	452				2N4416
U312	452				2N4416
U314					2N5484
U315					2N5486
U320					2N4391

Industry Standard No.	ECG	SK	GE	RS 276-	MOTOR.
U321					2N4391
U322					2N4391
U361	125			1114	
U3I109659					MC1806F
U3I180051					MC1900F
U3I180059					MC1800F
U3I180151					MC1901F
U3I180159					MC1801F
U3I180251					MC1902F
U3I180259					MC1802F
U3I180351					MC1903F
U3I180359					MC1803F
U3I180451					MC1904F
U3I180459					MC1804F
U3I180551					MC1905F
U3I180559					MC1805F
U3I180651					MC1906F
U3I180751					MC1907F
U3I180759					MC1807F
U3I180851					MC1908F
U3I180859					MC1808F
U3I180951					MC1909F
U3I180959					MC1809F
U3I181051					MC1910F
U3I181059					MC1810F
U3I181151					MC1911F
U3I181159					MC1811F
U3I181251					MC1912F
U3I181259					MC1812F
U3I181451					MC1914F
U3I181459					MC1814F
U3I181651					MC1916F
U3I181659					MC1816F
U3I909351					MC953F
U3I909359					MC953F
U3I909451					MC956F
U3I909459					MC956F
U3I909751					MC955F
U3I909759					MC955F
U3I909951					MC952F
U3I909959					MC952F
U3I913551					MC935F
U3I913559					MC935F
U3I915751					MC957F
U3I915759					MC957F
U3I915851					MC958F
U3I915859					MC958F
U3I993051					MC930F
U3I993059					MC930F
U3I993251					MC932F
U3I993259					MC932F
U3I993351					MC933F
U3I993551					MC940F
U3I993559					MC940F
U3I993651					MC936F
U3I993659					MC936F
U3I993751					MC937F
U3I993759					MC937F
U3I994451					MC944F
U3I994459					MC944F
U3I994551					MC945F
U3I994559					MC945F
U3I994651					MC946F
U3I994659					MC946F
U3I994851					MC948F
U3I994859					MC948F
U3I994951					MC949F
U3I994959					MC949F
U3I995051					MC950F
U3I995059					MC950F
U3I995151					MC951F
U3I995159					MC951F
U3I996151					MC961F
U3I996159					MC961F
U3I996251					MC962F
U3I996259					MC962F
U3I996351					MC963F
U3I996359					MC963F
U460A	107	3018			
U460B	107	3018			
U51	189	3025/129			
U535/7825B(ZENER				561	
U535A	229	3246			
U535B	229	3246			
U535B/7825B	229	3246			
U535C	107	3293			
U535M	107	3293			
U574	147A	3095			
U5A7064354	780	3141			
U5B770939X	909	3590			
U5B7741393	941	3514			
U5D7703312	703A	3157			
U5D7703393	703A	3157			
U5D7703394	703A	3157			
U5D770339X	703A	3157			
U5E7064394	780	3141			
U5F7747393	947	3514/941			
U5R7723393	923	3164			
U633	116			1104	
U6A180051					MC1900L
U6A180059					MC1800L
U6A180151					MC1901L
U6A180159					MC1801L
U6A180251					MC1902L
U6A180259					MC1802L
U6A180351					MC1903L
U6A180359					MC1803L
U6A180451					MC1904L
U6A180459					MC1804L
U6A180551					MC1905L
U6A180559					MC1805L
U6A180651					MC1906L
U6A180659					MC1806L
U6A180751					MC1907L
U6A180759					MC1807L
U6A180851					MC1908L
U6A180859					MC1000L
U6A180951					MC1909L
U6A180959					MC1809L
U6A181051					MC1910L
U6A181059					MC1810L
U6A181151					MC1911L
U6A181159					MC1811L
U6A181251					MC1912L
U6A181259					MC1812L
U6A181451					MC1914L
U6A181459					MC1814L
U6A181651					MC1916L
U6A181659					MC1816L
U6A7136354	737	3375			
U6A7720354	744			2022	
U6A7720395	744			2022	
U6A7723393	923D	3165			
U6A7729394	720	9014			

Industry Standard No.	ECG	SK	GE	RS 276-	MOTOR.
U6A7732394	718	3159			
U6A7739394	725	3162			
U6A7742393	979	3886			
U6A7746394	790	3077			
U6A7767394	722	3161			
U6A7781394	715	3076			
U6A909351					MC953L
U6A909359					MC853L
U6A909451					MC956L
U6A909459					MC856L
U6A909751					MC955L
U6A909759					MC855L
U6A909951					MC952L
U6A909959					MC852L
U6A913551					MC935L
U6A913559					MC835L
U6A915751					MC957L
U6A915759					MC857L
U6A915851					MC958L
U6A915859					MC858L
U6A993051					MC930L
U6A993059					MC830L
U6A993251					MC932L
U6A993259					MC832L
U6A993351					MC933L
U6A993359					MC833L
U6A993551					MC940L
U6A993559					MC940F
U6A993651					MC936L
U6A993659					MC836L
U6A993751					MC937L
U6A993759					MC837L
U6A994451					MC944L
U6A994459					MC844L
U6A994551					MC945L
U6A994559					MC845L
U6A994651					MC946L
U6A994659					MC846L
U6A994851					MC948L
U6A994859					MC848L
U6A994951					MC949L
U6A994959					MC849L
U6A995051					MC950L
U6A995059					MC850L
U6A995151					MC951L
U6A995159					MC851L
U6A996151					MC961L
U6A996159					MC861L
U6A996251					MC962L
U6A996259					MC862L
U6A996351					MC963L
U6A996359					MC863L
U6AH00059X	74H00	74H00			
U6AH00459X	74H04	74H04			
U6AN00059X	7400	7400			
U6AN00159X	7401	7401			
U6AN00259X	7402	7402			
U6AN00359X	7403	7403			
U6AN00459X	7404	7404			
U6AN00559X	7405	7405			
U6AN00659X	7406	7406			
U6AN00759X	7407	7407			
U6AN00859X	7408	7408			
U6AN00959X	7409	7409			
U6AN01059X	7410	7410			
U6AN01159X	7411	7411			
U6AN01259X	7412	7412			
U6AN01359X	7413	7413			
U6AN01459X	7414	7414			
U6AN01659X	7416	7416			
U6AN01759X	7417	7417			
U6AN02059X	7420	7420			
U6AN02559X	7425	7425			
U6AN02659X	7426	7426			
U6AN02759X	7427	7427			
U6AN03059X	7430	7430			
U6AN03259X	7432	7432			
U6AN03759X	7437	7437			
U6AN03859X	7438	7438			
U6AN04059X	7440	7440			
U6AN05059X	7450	7450			
U6AN05159X	7451	7451			
U6AN05359X	7453	7453			
U6AN05459X	7454	7454			
U6AN06059X	7460	7460			
U6AN07059X	7470	7470			
U6AN07259X	7472	7472			
U6AN07359X	7473	7473			
U6AN07459X	7474	7474			
U6AN08059X	7480	7480			
U6AN09059X	7490	7490			
U6AN09359X	7493A	7493			
U6AN09559X	7495	7495			
U6AN10759X	74107	74107			
U6AN12259X	74122	74122			
U6AN13259X	74132	74132			
U6AN17659X	74176	74197			
U6AN17759X	74177	74177			
U6AN18059X	74180	74180			
U6AN19659X	74196	74196			
U6AS00059X	74S00	74S00			
U6AS00459X	74S04	74S04			
U6B7414159X	74141	74141			
U6B7780394	714	3075			
U6BA01559X	7441	7441			
U6BN02359X	7423	7423			
U6BN04259X	7442	7442			
U6BN04359X	7443	7443			
U6BN04459X	7444	7444			
U6BN07559X	7475	7475			
U6BN07659X	7476	7476			
U6BN08359X	7483	7483			
U6BN08559X	7485	7485			
U6BN09459X	7494	7494			
U6BN15359X	74153	74153			
U6BN15559X	74155	74155			
U6BN15659X	74156	74156			
U6BN17559X	74175	74175			
U6E7729394	720	9014			
U6E7739393	725	3162			
U6NA01159X	74154	74154			
U6NA04159X	74181	74181			
U6NN15059X	74150	74150			
U6NN19859X	74198	74198			
U6NN19959X	74199	74199			
U7AN16459X	74164	74164			
U7AN17859X	74178	74178			
U7BA04559X	7445	7445			
U7BN046591	7446	7446			
U7BN04859X	7448	7448			
U7BN09659X	7496	7496			
U7BN12359X	74123	74123			
U7BN14159X	74141	74141			
U7BN14559X	74145	74145			

Industry Standard No.	ECG	SK	GE	RS 276-	MOTOR.
U7BN15159X	74151	74151			
U7BN16659X	74166	74175			
U7BN17059X	74170	74170			
U7BN17459X	74174	74174			
U7BN17959X	74179	74179			
U7BN19159X	74191	74191			
U7BN19259X	74192	74192			
U7F7065354	712	3072			
U7F7065394	712	3072			
U7F7729394	720	9014			
U7F7732394	718	3159			
U7F77463394	790	3077			
U7F7746394	790	3077			
U7F7767394	722	3161			
U7F77803394	714	3075			
U7F7780394	714	3075			
U7F7781394	715	3076			
U8F7737394	705A	3134			
U8F7746394	705A	3134			
U9A180059					MC1800P
U9A180159					MC1801P
U9A180259					MC1802P
U9A180359					MC1803P
U9A180459					MC1804P
U9A180559					MC1805P
U9A180659					MC1806P
U9A180759					MC1807P
U9A180859					MC1808P
U9A180959					MC1809P
U9A181059					MC1810P
U9A181159					MC1811P
U9A181259					MC1812P
U9A181459					MC1814P
U9A181659					MC1816P
U9A7136354	737	3375			
U9A74259X	979	3886			
U9A7720354	744			2022	
U9A7720395	744			2022	
U9A7723393	923D	3165			
U9A7729394	720	9014			
U9A7746394	790	3077			
U9A909359					MC853P
U9A909459					MC856P
U9A909759					MC855P
U9A909959					MC852P
U9A913559					MC835P
U9A915759					MC857P
U9A915859					MC858P
U9A993059					MC830P
U9A993259					MC832P
U9A993359					MC833P
U9A993559					MC840P
U9A993659					MC836P
U9A993759					MC837P
U9A994459					MC844P
U9A994559					MC845P
U9A994659					MC846P
U9A994859					MC848P
U9A994959					MC849P
U9A995059					MC850P
U9A995159					MC851P
U9A996159					MC861P
U9A996259					MC862P
U9A996359					MC863P
U9A9N0759X	7407	7407			
U9AH00059X	74H00	74H00			
U9AH00459X	74H04	74H04			
U9AN00059X	7400	7400			
U9AN00159X	7401	7401			
U9AN00259X	7402	7402			
U9AN00359X	7403	7403			
U9AN00459X	7404	7404			
U9AN00559X	7405	7405			
U9AN00659X	7406	7406			
U9AN00759X	7407	7407			
U9AN00859X	7408	7408			
U9AN00959X	7409	7409			
U9AN01059X	7410	7410			
U9AN01159X	7411	7411			
U9AN01259X	7412	7412			
U9AN01359X	7413	7413			
U9AN01459X	7414	7414			
U9AN01659X	7416	7416			
U9AN01759X	7417	7417			
U9AN02059X	7420	7420			
U9AN02559X	7425	7425			
U9AN02659X	7426	7426			
U9AN02759X	7427	7427			
U9AN03059X	7430	7430			
U9AN03259X	7432	7432			
U9AN03759X	7437	7437			
U9AN03859X	7438	7438			
U9AN04059X	7440	7440			
U9AN05059X	7450	7450			
U9AN05159X	7451	7451			
U9AN05359X	7453	7453			
U9AN05459X	7454	7454			
U9AN06059X	7460	7460			
U9AN07059X	7470	7470			
U9AN07259X	7472	7472			
U9AN07359X	7473	7473			
U9AN07459X	7474	7474			
U9AN08059X	7480	7480			
U9AN08659X	7486	7486			
U9AN09059X	7490	7490			
U9AN09359X	7493A	7493			
U9AN09559X	7495	7495			
U9AN10759X	74107	74107			
U9AN12259X	74122	74122			
U9AN13259X	74132	74132			
U9AN16459X	74164	74164			
U9AN17659X	74176	74197			
U9AN17759X	74177	74177			
U9AN17859X	74178	74178			
U9AN18059X	74180	74180			
U9AN19659X	74196	74196			
U9AS00059X	74S00	74S00			
U9AS00459X	74S04	74S04			
U9B7780394	714	3075			
U9BA01559X	7441	7441			
U9BN02359X	7423	7423			
U9BN04259X	7442	7442			
U9BN04359X	7443	7443			
U9BN04459X	7444	7444			
U9BN046591	7446	7446			
U9BN04859X	7448	7448			
U9BN07559X	7475	7475			
U9BN07659X	7476	7476			
U9BN08359X	7483	7483			
U9BN08559X	7485	7485			
U9BN09459X	7494	7494			
U9BN09659X	7496	7496			
U9BN12359X	74123	74123			
U9BN14159X	74141	74141			

Industry Standard No.	ECG	SK	GE	RS 276-	MOTOR.
U9BN14559X	74145	74145			
U9BN15159X	74151	74151			
U9BN15559X	74155	74155			
U9BN15659X	74156	74156			
U9BN16659X	74166	74175			
U9BN17059X	74170	74170			
U9BN17459X	74174	74174			
U9BN17559X	74175	74175			
U9BN17959X	74179	74179			
U9BN19159X	74191	74191			
U9BN19259X	74192	74192			
U9BN19359X	74193	7493			
U9C7065354	712	3072			
U9NA04159X	74181	74181			
U9NN15059X	74150	74150			
U9NN19859X	74198	74198			
U9NN19959X	74199	74199			
U9T7101392	975	3641			
U9T7741392	976	3596			
UA1312PC	799	3238			
UA1314PC	802	3277			
UA1315PC	803	3278			
UA1391TC	815	3255			
UA1748G	1171	3565			
UA2002	1232	3852			
UA201AH	1171	3565			
UA201H	1171	3565			
UA209KM	309K	3629			
UA2136DC	737	3375			
UA2136P	737	3375			
UA2136PC	737	3375			
UA3018AHM	904	3542			
UA301AH	1171	3565			
UA301AHC	1171	3565			
UA3046DC	912	3543			
UA3054DC	917	3544			
UA3064	780	3141			
UA3064HC	780	3141			
UA3064PC	783	3215			
UA3065	712	3072			
UA307T	976	3596			
UA307TC	976	3596			
UA3086DC	912	3543			
UA309KC	309K	3629			
UA3401PC	992	3688			
UA555TC	955M	3564			
UA556PC	978	3689			
UA67	722	3161			
UA703	703A	3157			
UA703C	703A	3157			
UA703E	703A	3157			
UA703HC	703A	3157			
UA705	804	3455			
UA709A	909D	3590			
UA709CT	909	3590			
UA709DC	909D	3590			
UA709PC	909D	3590			
UA710HC	910	3553			
UA720	744	3171		2022	
UA720DC	744			2022	
UA720PC	744			2022	
UA723CA	923D	3165			
UA723CL	923	3164			
UA723DC	923D	3165			
UA723HC	923	3164			
UA723HM	923D	3164/923			
UA723PC	923D	3165			
UA729	720	9014			
UA732	718	3159			
UA732PC	718	3159			
UA737E	705A	3134			
UA739	725	3162			
UA739C	725	3162			
UA741	941M	3514/941			
UA741CA	941D	3552/941M			
UA741CP	941M	3552			
UA741CT	941	3514			
UA741CV	941M	3552			
UA741TC		3514		007	
UA742	979	3886			
UA742PC	979	3886			
UA746	707	3134/705A			
UA746C	790	3134/705A			
UA746PC	790	3077			
UA747CA	947D	3556			
UA747CF	947D	3556			
UA747CJ	947D	3556			
UA747CK	947	3526			
UA747CN	947D	3556			
UA747DC	947D	3556			
UA748HC	1171	3565			
UA748TC	975	3641			
UA748V	975	3641			
UA753	736	3163			
UA758	743	3172			
UA767	722	3161			
UA767PC	722	3161			
UA780	714	3075			
UA7805C	960	3462/977			
UA7805CDA	309K	3629			
UA7805KC	309K	3629			
UA7805UC	960	3462/977			
UA7806CKC	962	3669			
UA7808C	964	3630			
UA7808CKC	964	3630			
UA7808CU	964	3630			
UA7808UC	964	3630			
UA780C	714	3075			
UA780DC	714	3075			
UA780PC	714	3075			
UA781	715	3076			
UA7812C	966	3592			
UA7812CKC	966	3592			
UA7812UC	966	3592			
UA7815C	968	3593			
UA7815CKC	968	3593			
UA7815CU	968	3593			
UA7815UC	968	3593			
UA781C	715	3076			
UA781DC	715	3076			
UA781PC	715	3076			
UA7824C	972	3670			
UA7824UC	972	3670			
UA787	797	3158			
UA787PC	797	3158			
UA78L	981	3724			
UA78L05	977	3462			
UA78L05AC	977	3462			
UA78L05AWC	977	3462			
UA78L05CLP	977	3462			
UA78L05S	977	3462			
UA78L05WC	977	3462			
UA78L062WV	988	3973			
UA78L08	981	3724			
UA78L08AC	981	3724			
UA78L08C	981	3724			
UA78L08CLP	981	3724			
UA78L08WC	981	3724			
UA78L82	981	3724			
UA78M05UC	960	3591			
UA78M06CKC	962	3669			
UA78M06UC	962	3669			
UA78M08CKC	964	3630			
UA78M08CKD	964	3630			
UA78M08UC	964	3630			
UA78M12CKC	966	3592			
UA78M12CKD	966	3592			
UA78M12UC	966	3592			
UA78M15CKC	968	3593			
UA78M15CKD	968	3593			
UA78M24CKD	972	3670			
UA78M24UC	972	3670			
UA7905C	961	3671			
UA7905UC	961	3671			
UA7906CKC	963	3672			
UA7906UC	963	3672			
UA7912CKC	967	3673			
UA7912UC	967	3673			
UA7915CKC	969	3674			
UA7915UC	969	3674			
UA7924CKC	971	3675			
UA7924UC	971	3675			
UA796DC	973D	3892			
UA796HC	973	3233			
UA796PC	973D	3892			
UA79M05AUC	961	3671			
UA79M06AUC	963	3672			
UA79M06CKC	963	3672			
UA79M12AUC	967	3673			
UA79M12CKC	967	3673			
UA79M12CKD	967	3673			
UA79M15AUC	969	3674			
UA79M15CKC	969	3674			
UA79M15CKD	969	3674			
UA79M24AUC	971	3675			
UA79M24CKC	971	3675			
UA79M24CKD	971	3675			
UBB7703393X	703A	3157			
UBFY11	108	3039/316			
UC100	312	3112			2N4221A
UC105	312	3112			
UC110	312	3112			2N4220A
UC1100	159	3114/290			
UC1100A	159	3114/290			
UC115	312	3112			
UC120	312	3112			2N4220A
UC125	312	3112			
UC130					2N4220A
UC150	452				2N4416
UC150W	452				2N4416
UC155	312	3116			2N4416
UC155W					2N5640
UC1700					3N155A
UC20	312	3112			2N5486
UC200					2N4393
UC201	312	3116			2N4093
UC21					2N5486
UC210	312	3448			2N4416
UC2136	312	3112			
UC2138	312	3112			
UC2139	312	3112			
UC2147	461	3112			
UC2148	312	3112			
UC2149	312	3112			
UC22					2N5486
UC220	312	3112			2N4221
UC23					2N5486
UC240	459				2N4221A
UC241		3112			2N3822
UC250					2N4091
UC251					2N4093
UC300					2N5267
UC310					2N5265
UC320					2N5265
UC330					2N5473
UC340					2N5265
UC400					MFE4012
UC401					2N3993
UC410					MFE4009
UC420					MFE4001
UC450					2N3993
UC451					2N3994
UC4741C	941	3514			
UC547C	123A	3124/289			
UC588	312	3448			
UC701	312				2N4220
UC703	312				2N4220
UC704	456				2N4221A
UC705	456				2N4221A
UC707	466				MFE2000
UC714	312	3116			2N3437
UC714A					2N5484
UC734	312	3448			2N4416
UC734E	312	3448			2N5486
UC751	312				2N4091
UC752	312				2N4091
UC753	312				2N4091
UC754					2N4091
UC755					2N4091
UC756	312	3112			2N4091
UC805	460				2N3330
UC807					2N3972
UC814					2N5265
UC851					2N5265
UC853					2N5265
UC854					2N5265
UC855					2N5265
UD5B	156	3051			
UDN-6118A	2021	9020			
UDN 6128A	2022	9019			
UDN-6144A					MC3490P
UDN-6164A					MC3490P
UDN-6184A					MC3490P
UDN-7183A					MC3491P
UDN-7184A					MC3491P
UDN-7186A					MC3491P
UDN-G118A	2021	9020			
UDN5711M					MC1471P1
UDN5712M					MC1472P1
UDN5713M					MC1473P1
UDN5714M					MC1474P1
UDZ8718	5077A	3752			
UES1001					SPECIAL
UES1002					SPECIAL
UES1003					SPECIAL
UES1101					SPECIAL
UES1102					SPECIAL
UES1103					SPECIAL

Industry Standard No.	ECG	SK	GE	RS 276-	MOTOR.
UES1301					SPECIAL
UES1302					SPECIAL
UES1303					SPECIAL
UES2401					SPECIAL
UES2402					SPECIAL
UES2403					SPECIAL
UES2601					SPECIAL
UES2602					SPECIAL
UES2603					SPECIAL
UES701					SPECIAL
UES702					SPECIAL
UES703					SPECIAL
UES801					SPECIAL
UES802					SPECIAL
UES803					SPECIAL
UF-01	506	3016			
UF-1	558	3515		1104	
UF-1A	558			1114	
UF-1B	558	3515			
UF-1C	558	3515		1114	
UF-2	558	3843			
UF-SD1	109	3087		1123	
UF01	116	3843		1114	
UF2	552			1104	
UFSD-1	116	3031A		1103	
UFSD-1B	125			1114	
UFSD-1A	116			1104	
UFSD-1C	506			1114	
UFSD1	506	3130			
UFSD1A	506	3130			
UFSD1F				1104	
UG1888	177	3100/519		1122	
UGH7812	966	3592			
UGJ7109393	309K	3629			
UHD-490					MC3494P
UHD-491					MC3494P
UHP-490					MC3494P
UHP-491					MC3494P
UHP-495					MC3490P
ULN-2001A	2011	3975			
ULN-2002A	2012	9092			
ULN-2003A	2013	9093			
ULN-2004A	2014	9094			
ULN-2005A	2015	9095			
ULN-2011A	2011	3975			
ULN-2012A	2012	9092			
ULN-2013A	2013	9093			
ULN-2014A	2014	9094			
ULN-2015A	2015	9095			
ULN-2081	916	3550			
ULN-2081A	916	3550			
ULN-2082A	2023	3694			
ULN-2801A	2016	9077			
ULN-2802A	2017	9078			
ULN-2803A	2018	9079			
ULN-2804A	2019	9080			
ULN-2805A	2020	9081			
ULN-2811A	2016	9077			
ULN-2812A	2017	9078			
ULN-2813A	2018	9079			
ULN-2814A	2019	9080			
ULN-2815A	2020	9081			
ULN2001A					MC1411
ULN2002A					MC1412
ULN2003A					MC1413
ULN2004A					MC1416
ULN2111	708	3135/709			
ULN2111A	708	3135/709			MC1357P
ULN2111N	708	3135/709			MC1357PQ
ULN2113	709	3375/737			
ULN2113A	709	3375/737			MC1357P
ULN2113N	709	3375/737			MC1357P
ULN2114	790	3077			
ULN2114A	790	3077			MC1323P
ULN2114K	707				MC1323P
ULN2114N	790	3077			MC1323P
ULN2114W	705A	3134			
ULN2120A	718	3159			MC1310P
ULN2120N	718	3159			
ULN2121A	719				MC1310P
ULN2122A	720	9014			MC1310P
ULN2122N	720	9014			
ULN2124	714	3075			
ULN2124A	714	3075			MC1399P
ULN2124N	714	3075			
ULN2125A	731	3170			MC1344P
ULN2125N	731	3170			
ULN2127	715	3076			
ULN2127A	715	3076			MC1399P
ULN2127N	715	3076			
ULN2128A	722	3161			MC1310P
ULN2128N	722	3161			
ULN2129A	723	3144			
ULN2129N	723	3144			
ULN2134A	806	3734			
ULN2136A	737	3375			MC1356P
ULN2136N	737	3375			
ULN2137A	744	3171		2022	
ULN2137N	744			2022	
ULN2139D					MC1439G
ULN2139G					MC1439G
ULN2139H					MC1439P2
ULN2139M					MC1439P1
ULN2151D					MC1741CG
ULN2151G					MC1741CF
ULN2151H					MC1741CP2
ULN2151M					MC1741CP1
ULN2156D					MC1456G
ULN2156G					MC1456G
ULN2156H					MC1456G
ULN2156M					MC1456G
ULN2157A					MC1458P2
ULN2157H					MC1458P2
ULN2157K					MC1458G
ULN2165	712	3072			
ULN2165A	712	3072			MC1358P
ULN2165N	712	3072			MC1358PQ
ULN2208M	805	3163/736			
ULN2209A					MC1356P
ULN2209M	736	3163			
ULN2210A	801	3160			MC1310P
ULN2210N	801	3160			
ULN2211B	742	3453			
ULN2211P	742	3453			
ULN2212A	807	3451			
ULN2212B	807	3451			
ULN2212P	807	3451			
ULN2224A	1062	3235/739			MC1324P
ULN2226A	739	3235			
ULN2226N	739	3235			
ULN2228A	790	3454			MC1323P
ULN2228N	790	3077			
ULN2244	743	3172			
ULN2244A	743	3172			MC1310P

Industry Standard No.	ECG	SK	GE	RS 276-	MOTOR.
ULN2244N	743	3172			
ULN2261A	818	3207			
ULN2262A	797	3158			MC1399P
ULN2262N	797	3158			
ULN2264A	783	3215			MC1364P
ULN2264N	783	3215			
ULN2266A	728	3073			
ULN2266N	728	3073			
ULN2267A	729	3074			MC1323P
ULN2268A	982	3205			
ULN2269A	791	3149			
ULN2269N	791	3149			
ULN2274B	804	3455			
ULN2274P	804	3455			
ULN2275P	804	3455			
ULN2275Q	804	3455			
ULN2276P	804	3455			
ULN2276Q	804	3455			
ULN2277P	804	3455			
ULN2277Q	804	3455			
ULN2278B	804	3455			
ULN2278P	804	3455			
ULN2278Q	804	3455			
ULN2280A	740A	3328			
ULN2280B	740A	3328			
ULN2289A	788	3829			
ULN2289N	788	3829			
ULN2298A	738	3167			MC1398P
ULN2298N	738	3167			
ULN2741D	941	3514			MC1741CG
ULN2747A	947D				MC1747CL
ULN3262A	797	3158			
ULN3701Z	1232	3852			
ULS2139D					MC1539G
ULS2139G					MC1539G
ULS2139H					MC1539L
ULS2139M					MC1439P1
ULS2151D					MC1741G
ULS2151G					MC1741F
ULS2151H					MC1741L
ULS2151M					MC1741CP1
ULS2156D					MC1556G
ULS2156G					MC1556G
ULS2156H					MC1556G
ULS2156M					MC1556G
ULS2157A					MC1558L
ULS2157H					MC1558L
ULS2157K					MC1558G
ULT3730	127	3764			
ULT3731	127	3764			
ULX2277	804	3455			
ULX2298N	738	3167			
UMT1008					MJ13014
UMT1009					MJ13015
UMT1203					MJE13004
UMT1204					MJE13005
UMT3202	124	3021			
UO-5E	156			1122	
UO5E	156			1104	
UO6C	506			1104	
UP12217	186	3192			
UP12218	186	3192			
UP14046	186	3192			
UP14047	186	3192			
UP32C	1050	3475			
UPB201C	7400	7400			
UPB201D	7400	7400			
UPB202C	7410	7410		1807	
UPB202D	7410	7410		1807	
UPB203C	7420	7420			
UPB203D	7420	7420		1809	
UPB2047D	7447	7447			
UPB204C	7430	7430			
UPB204D	7430	7430			
UPB205C	7440	7440			
UPB205D	7440	7440			
UPB206C	7450	7450			
UPB206D	7450	7450			
UPB207C	7451	7451			
UPB207D	7451	7451		1825	
UPB2085D	7485	7485		1826	
UPB2086D	7486	7486		1827	
UPB208D	7453	7453			
UPB209D	7454	7454			
UPB210C	7460	7460			
UPB210D	7460	7460			
UPB211C	7472	7472			
UPB211D	7470	7470			
UPB212C	7473	7473			
UPB212D	7472	7472			
UPB214C	7474	7474			
UPB214D	7474	7474			
UPB2150D	74150	74150		1829	
UPB2151D	74151	74151			
UPB2154D	74154	74154			
UPB215C	7401	7401			
UPB215D	7401	7401			
UPB2161D	74161	74161			
UPB2170D	74170	74170			
UPB217C	7475	7475			
UPB217D	7475	7475			
UPB2180D	74180	74180			
UPB2181D	74181	74181			
UPB218C	7441	7441			
UPB218D	7441	7441			
UPB2192D	74192	74192			
UPB2193D	74193	74193			
UPB2195D	74195	74195			
UPB2198D	74198	74198			
UPB219C	7490	7490			
UPB219D	7490	7490			
UPB222C	7492	7492			
UPB222D	7492	7492			
UPB223C	7493A	7493			
UPB223D	7493A	7493			
UPB224C	7476	7476			
UPB224D	7476	7476			
UPB225C	7473	7473			
UPB225D	7473	7473			
UPB226C	7495	7495			
UPB226D	7495	7495			
UPB227D	7442	7442			
UPB230D	7483	7483			
UPB232C	7402	7402			
UPB232D	7402	7402			
UPB234C	7408	7408			
UPB234D	7408	7408			
UPB235D	7404	7404			
UPB236D	7405	7405			
UPB237D	7437	7437			
UPB238D	7438	7438			
UPB7400C	7400	7400			
UPB7402C	7402	7402			
UPB7404C	7404	7404			

Industry Standard No.	ECG	SK	GE	RS 276-	MOTOR.
UPB7405C	7405	7405			
UPB74107C	74107	74107			
UPB7410C	7410	7410		1807	
UPB74123C	74123	74123		1817	
UPB7413C	7413	7413		1815	
UPB74141D	74141	74141			
UPB74150C	74150	74150		1829	
UPB74151C	74151	74151			
UPB74153C	74153	74153			
UPB74154C	74154	74154			
UPB74155C	74155	74155			
UPB74156C	74156	74156			
UPB74161C	74161	74161			
UPB74164C	74164	74164			
UPB74170D	74170	74170			
UPB74175C	74175	74175			
UPB74180C	74180	74180			
UPB74181D	74181	74181			
UPB74192C	74192	74192			
UPB74193C	74193	74193			
UPB74195C	74195	74195			
UPB74198D	74198	74198			
UPB7420C	7420	7420		1809	
UPB7430C	7430	7430			
UPB7437C	7437	7437			
UPB7438C	7438	7438			
UPB7440C	7440	7440			
UPB7442C	7442	7442			
UPB7445C	7445	7445			
UPB7447C	7447	7447			
UPB7450C	7450	7450			
UPB7451C	7451	7451		1825	
UPB7453C	7453	7453			
UPB7454C	7454	7454			
UPB7460C	7460	7460			
UPB7473C	7473	7473			
UPB7474C	7474	7474			
UPB7476C	7476	7476			
UPB7480C	7480	7480			
UPB7485C	7485	7485		1826	
UPB7486				1827	
UPB7486C	7486	7486			
UPC1001H	1127	3243/1160			
UPC1020	1160	3243			
UPC1020H	1160	3243			
UPC1025H	1081A*	3474/1081A			
UPC1026C	1226	3763			
UPC1031H	1245	3878			
UPC1031H2	1245	3878			
UPC1032H	1170	3874			
UPC1156	1194	3484			
UPC1156H	1194	3484			
UPC1156H2	1194	3484			
UPC1181H	1285	3922			
UPC1182H	1286	3923			
UPC1353C	1246	3879			
UPC1380C	1196	3725			
UPC143-05	960	3591			
UPC14305	977	3591/960			
UPC14305H	960	3591			
UPC14308	964	3630			
UPC14308H	964	3630			
UPC14312	966	3592			
UPC14312H	966	3592			
UPC14315	968	3593			
UPC14315H	968	3593			
UPC1458C	778A	3465			
UPC157A	1171	3565			
UPC2002	1232	3852			
UPC206	1075A	3877			
UPC20C	1075A	3877			
UPC20C1	1075A	3877			
UPC20C2	1075A	3877			
UPC22C	1078	3498			
UPC23C	1046	3471			
UPC301AM	975	3641			
UPC30C	1049	3470			
UPC324C	987	3643			
UPC32C	1050	3475			
UPC544C	1006	3358			
UPC554	1142	3485			
UPC554C	1142	3485			
UPC555	703A	3157			
UPC555A	703A	3157			
UPC555C	703A	3157			
UPC555H	1092	3472			
UPC562C	1050	3475			
UPC566H	1052	3249			
UPC566H-B	1052	3249			
UPC566H-L	1052	3249			
UPC566H-M	1052	3249			
UPC566H-N	1052	3249			
UPC566H2	1052	3249			
UPC566H3	1052	3249			
UPC571C	1078	3498			
UPC573C	1184	3486			
UPC575C	1141	3473/1140			
UPC575C2	1140	3473			
UPC576H	1185	3468			
UPC577A	1082	3461			
UPC577H	1082	3461			
UPC577N	1082	3461			
UPC580C	1196	3725			
UPC592A2	1195	3469			
UPC592H2	1195	3469			
UPC595C	1186	3168/749			
UPC595C2	1187	3168/749			
UPC596	1187	3742			
UPC596C	1187	3742			
UPC596C2	1187	3742			
UPC741C	941D	3552/941M			
UPC767PC	722	3161			
UPC78L05	977	3462			
UPC78L08	981	3630/964			
UPC81C	1075A	3877			
UPC861C	1254	3880			
UPD4001C	4001B	4001			
UPD4002C	4002B	4002			
UPD4011C	4011B	4011		2411	
UPD4012C	4012B	4012			
UPD4013C	4013B	4013			
UPD4014C	4014B	4014			
UPD4015C	4015B	4015			
UPD4017C	4017B	4017			
UPD4020C	4020B	4020			
UPD4021C	4021B	4021			
UPD4023C	4023B	4023			
UPD4025C	4025B	4025			
UPD4027C	4027B	4027			
UPD4028C	4028B	4028			
UPD4029C	4029B	4029			
UPD4030C	4030B	4030			
UPD4035C	4035B	4035			
UPD4040C	4040B	4040			
UPD4042C	4042B	4042			
UPD4049C	4049	4049			
UPD4050C	4050B	4050			
UPD4066C	4066B	4066			
UPD4069C	4069	4069			
UPD4071C	4071B	4071			
UPD4081C	4081B	4081			
UPD4511	4511B	4511			
UPD4518	4518B	4518			
UPD4520C	4520B	4520			
UPD4555C	4555B	4555			
UPD4556C	4556B	4556			
UPD861C	1254	3880			
UPD861CE	1254	3880			
UPI1301				2005	
UPI1303	102			2007	
UPI1305	102			2007	
UPI1307	102			2007	
UPI1309	102			2007	
UPI1345	100			2007	
UPI1347	100	3721		2007	
UPI1352	102A			2005	
UPI1353	102A	3722/102		2007	
UPI1613	128			2030	
UPI2217	128			2030	
UPI2218	128			2030	
UPI2222	123A	3444		2051	
UPI2222B	123A			2051	
UPI2222P	123AP			2051	
UPI404				2005	
UPI4046	123A			2030	
UPI4046-46	123A	3444		2051	
UPI4047-46	123A			2051	
UPI706	123A			2038	
UPI706A	123A			2038	
UPI706B	123A			2038	
UPI718A	123A			2051	
UPI956	128			2051	
UPT1131	328	3895			
UPT1132	328	3895			
UPT1133	328	3895			
UPT1134	328	3895			
UPT1135	328	3895			
UPT121				2053	
UPT214	282	3748			
UPT215	282	3748			
UPT221		3131A		2020	
UPT410	284	3836			
UPT611	282	3748		2053	
UPT612	282	3748			
UPT613	282	3748			
UPT614	282	3748			
UPT615	282	3748			
UPT621				2053	
UPT731	385	3946			
UPT732	385	3946			
UPT733	385	3946			
UPT734	385	3946			
UPT735	385	3946			
UPT833	385	3946			
UPT834	385	3946			
UPT835	385	3946			
UPT931	327	3945			
UR105	116	3017B/117		1104	
UR110	116	3017B/117		1104	
UR115	116	3017B/117		1104	
UR120	116	3017B/117		1104	
UR123VA	5890	7090			
UR125	116	3031A		1104	
UR205	5800	9003		1142	
UR210	5801	9004		1142	
UR215	5802	9005		1143	
UR220	5802	9005		1143	
UR225	5804	9007		1144	
US123VA	5890	7090			
US15/1	502	3067			
US15/1B	502	3067			
US15/1S	502	3067			
US15/3	505	3108			
US1555	177	3100/519		1122	
US20-1B	505	3067/502			
US20/16	504	3108/505			
US25/1AS	503	3068			
US25/1CS	504	3108/505			
US7400A	7400	7400		1801	
US7400J	7400	7400		1801	
US7401A	7401	7401			
US7401J	7401	7401			
US7402A	7402	7402		1811	
US7402J	7402	7402		1811	
US7403A	7403	7403			
US7403J	7403	7403			
US7404A	7404	7404		1802	
US7404J	7404	7404		1802	
US7405A	7405	7405			
US7405J	7405	7405			
US7408A	7408	7408		1822	
US7408J	7408	7408		1822	
US7409A	7409	7409			
US7409J	7409	7409			
US74107A	74107	74107			
US7410A	7410	7410		1807	
US7410J	7410	7410		1807	
US7411A	7411	7411			
US7411J	7411	7411			
US74121A	74121	74121			
US74121J	74121	74121			
US74145A	74145	74145		1828	
US74153A	74153	74153			
US74154A	74154			1834	
US74180A	74180	74180			
US74180J	74180	74180			
US7420A	7420	7420		1809	
US7420J	7420	7420		1809	
US7426A	7426	7426			
US7427A	7427	7427		1803	
US7430A	7430	7430			
US7430J	7430	7430			
US7432A	7432	7432		1824	
US7432J	7432	7432		1824	
US7438A	7438	7438			
US7438J	7438	7438			
US7440A	7440	7440			
US7440J	7440	7440			
US7441A	7441	7441		1804	
US7442A	7442	7442			
US7443A	7443	7443			
US7444A	7444	7444			
US7445A	7445	7445			
US7446A	7446	7446			
US7447A	7447	7447		1805	
US7448A	7448	7448		1816	
US7450A	7450	7450			
US7450J	7450	7450			

Industry Standard No.	ECG	SK	GE	RS 276-	MOTOR.
US7451A	7451	7451		1825	
US7451J	7451	7451		1825	
US7453A	7453	7453			
US7453J	7453	7453			
US7454A	7454	7454			
US7454J	7454	7454			
US7460A	7460	7460			
US7460J	7460	7460			
US7470A	7470	7470			
US7470J	7470	7470			
US7472A	7472	7472			
US7472J	7472	7472			
US7473A	7473	7473		1803	
US7473J	7473	7473		1803	
US7474A	7474	7474		1818	
US7474J	7474	7474		1818	
US7475A	7475	7475		1806	
US7475J	7475	7475		1806	
US7476A	7476	7476		1813	
US7480A	7480	7480			
US7483A	7483	7483			
US7486A	7486	7486		1827	
US7486J	7486	7486		1827	
US7490A	7490	7490		1808	
US7490J	7490	7490		1808	
US7492A	7492	7492		1819	
US7492J	7492	7492		1819	
US7493A	7493A	7493			
US7493J	7493A	7493			
US74H00A	74H00	74H00			
US74H00J	74H00	74H00			
US74H04A	74H04	74H04			
US74H04J	74H04	74H04			
USD320C					MBR3020CT
USD335C					MBR3035CT
USD345C					MBR3045CT
USD420					MBR3520
USD435					MBR3535
USD445					MBR3545
USD520					MBR7520
USD535					MBR7535
USD545					MBR7545
USFD-1	116	3031A		1104	
USFD-1A	116	3031A		1104	
UT-112	116			1104	
UT-16	116	3016		1104	
UT-234	116			1104	
UT-238	116			1104	
UT11	116	3311		1142	
UT111	116	9003/5800			1N4001
UT112	116	9004/5801		1142	1N4002
UT113	116	9005/5802		1143	1N4003
UT114	116	9006/5803		1144	1N4004
UT115	116	9007/5804		1144	1N4004
UT116	116	3017B/117		1104	
UT117	116	9008/5805			1N4005
UT118	125	3848/5806			1N4005
UT119		9009			1N4006
UT12	5802	9005		1143	
UT120	125	9010/5809			
UT13	5802	9005		1143	
UT14	116	3017B/117		1104	
UT15	116	3311		1104	
UT16	116			1104	
UT17	116	3016		1104	
UT18	116	3016		1104	
UT2005	5800	9003		1142	MR501
UT2010	5801	9004		1142	MR501
UT2020	5802	9005		1143	MR502
UT2040	5804	9007		1144	MR504
UT2060	5806	3848			MR506
UT2080	156	3051			
UT21	116	3311		1104	
UT211	5803	9006		1144	1N4004
UT212	5803	9006		1144	1N4004
UT213	5804	9007		1144	1N4004
UT214	5805	9008			1N4005
UT215	5806	3848			1N4005
UT22	116	3016		1104	
UT221	116	3016		1104	
UT222	116	3017B/117		1104	
UT223	116	3017B/117		1104	
UT224	116	3016		1104	
UT225	116	3031A		1104	1N4005
UT226	116	3031A		1104	
UT227	116	3031A		1104	
UT228	116	3017B/117		1104	
UT229	116	3017B/117		1104	
UT23	116	3017B/117		1104	
UT231	116	3017B/117		1104	
UT232	116	3017B/117		1104	
UT233	116	3017B/117		1104	
UT234	5802	9005		1143	1N4003
UT235	5804	9007		1144	1N4004
UT236	5801	9004		1142	1N4002
UT237	5805	9008			1N4005
UT238	5806	3848		1104	
UT24	116	3017B/117		1104	
UT242	5802	9005		1143	1N4003
UT244	5804	9007		1144	1N4004
UT245	5805	9008			1N4005
UT247	5806	3848			1N4005
UT249	5801	9004		1142	1N4002
UT25	116	3031A		1104	
UT251	5801	9004		1142	1N4002
UT252	5802	9005		1143	1N4003
UT254	5804	9007		1144	1N4004
UT255	5805	9008			
UT257	5806	3848			1N4005
UT258	5808	9009			1N4006
UT26	116	3031A		1104	
UT261	5801	9004		1142	MR501
UT262	5802	9005		1143	MR502
UT264	5804	9007		1144	MR504
UT265	5805	9008			MR506
UT267	5806	3848			MR506
UT268	5808	9009			MR508
UT27	116	3017B/117		1104	
UT3005	5800	9003		1142	
UT3010	5801	9004		1142	
UT3020	5802	9005		1143	
UT3040	5804	9007		1144	
UT3060	5806	3848			
UT3080	5808	9009			
UT338					1N4005
UT345	125	3033		1114	
UT347	5809	9010			1N4007
UT361	5808	9009			1N4006
UT362	5808	9009			1N4006
UT363	5809	9010			1N4007
UT364	5809	9010			1N4007
UT4005					MR501
UT4010					MR501
UT4020					MR502

Industry Standard No.	ECG	SK	GE	RS 276-	MOTOR.
UT4040					MR504
UT4060					MR506
UTR01					1N4933
UTR02					1N4933
UTR10					1N4934
UTR11					1N4934
UTR12					1N4934
UTR20					1N4935
UTR21					1N4935
UTR22					1N4935
UTR2305					MR850
UTR2310					MR851
UTR2320					MR852
UTR2340					MR854
UTR2350					MR856
UTR2360					MR856
UTR3305					MR850
UTR3310					MR851
UTR3320					MR852
UTR3340					MR854
UTR3350					MR856
UTR40					1N4936
UTR41					1N4936
UTR42					1N4936
UTR4305					MR850
UTR4310					MR851
UTR4320					MR852
UTR4340					MR854
UTR4350					MR856
UTR4360					MR856
UTR50					1N4937
UTR51					1N4937
UTR52					1N4937
UTR60					1N4937
UTR61					1N4937
UTR62					1N4937
UTRA-7RM	121	3009		2006	
UTX3105					MR850
UTX3110					MR851
UTX3115					MR852
UTX3120					MR852
UTX4105					MR850
UTX4110					MR851
UTX4115					MR852
UTX4120					MR852
UU/J				1104	
UZ-12B	142A	3787/5021A		563	
UZ-15C	145A	3063		564	
UZ-9.1F	139A	3060			
UZ110	5156A	3422			
UZ111	5157A	3423			
UZ112	5158A	3424			
UZ113	5159A	3425			
UZ114	5160A	3426			
UZ115	5161A	3427			
UZ116	5162A	3428			
UZ117	5163A	3429			
UZ118	5164A	3430			
UZ119	5165A	3431			
UZ120	5166A	3432			5M200ZS5
UZ122					5M110ZSB5
UZ13B	142A	3093		563	
UZ140					5M200ZSB5
UZ15	145A	3063		564	
UZ210	5156A	3422			
UZ211	5157A	3423			
UZ212	5158A	3424			
UZ213	5159A	3425			
UZ214	5160A	3426			
UZ215	5161A	3427			
UZ216	5162A	3428			
UZ217	5163A	3429			
UZ218	5164A	3430			
UZ219	5165A	3431			
UZ220	5166A	3432			5M200ZS10
UZ222					5M100ZSB10
UZ240					5M200ZSB10
UZ3.6	134A	3055			
UZ3016,A,B					1N3016A,B
UZ3051,A,B					1N3051A,B
UZ310	5156A	3422			
UZ311	5157A	3423			
UZ312	5158A	3424			
UZ313	5159A	3425			
UZ314	5160A	3426			
UZ315	5161A	3427			
UZ316	5162A	3428			
UZ317	5163A	3429			
UZ318	5164A	3430			
UZ319	5165A	3431			
UZ320	5166A	3432			
UZ3235,A,B					1N5235,A,B
UZ3281,A,B					1N5281,A,B
UZ3470,A,B					1N2970A,B
UZ3515,A,B					1N3015A,B
UZ4116,A,B					1N5384A,B
UZ4706,A,B					1N5342A,B
UZ4712	5127A	3393			
UZ4736,A					1N4736,A
UZ4764,A					1N4764,A
UZ4812	5127A	3393			
UZ5.1	135A	3056			
UZ5.1B	135A	3056			
UZ5110	5156A	3422			
UZ5111	5157A	3423			
UZ5112	5158A	3424			
UZ5113	5159A	3425			
UZ5114	5160A	3426			
UZ5115	5161A	3427			
UZ5116	5162A	3428			
UZ5117	5163A	3429			
UZ5118	5164A	3430			
UZ5119	5165A	3431			
UZ5120	5166A	3432			5M200ZS5
UZ5122					5M110ZSB5
UZ5140					5M200ZSB5
UZ5210	5156A	3422			
UZ5211	5157A	3423			
UZ5212	5158A	3424			
UZ5213	5159A	3425			
UZ5214	5160A	3426			
UZ5215	5161A	3427			
UZ5216	5162A	3428			
UZ5217	5163A	3429			
UZ5218	5164A	3430			
UZ5219	5165A	3431			
UZ5220	5166A	3432			5M200ZS10
UZ5222					5M110ZSB10
UZ5240					5M200ZSB10
UZ5310	5156A	3422			
UZ5311	5157A	3423			
UZ5312	5158A	3424			
UZ5313	5159A	3425			
UZ5314	5160A	3426			

Industry Standard No.	ECG	SK	GE	RS 276-	MOTOR.
UZ5315	5161A	3427			
UZ5316	5162A	3428			
UZ5317	5163A	3429			
UZ5318	5164A	3430			
UZ5319	5165A	3431			
UZ5320	5166A	3432			
UZ5706	5120A	3386			5M6.8ZS5
UZ5707	5121A	3387			
UZ5708	5122A	3388			
UZ5709	5124A	3390			
UZ5710	5125A	3391			
UZ5712	5127A	3393			
UZ5713	5128A	3394			
UZ5714	5129A	3395			
UZ5715	5130A	3396			
UZ5716	5131A	3397			
UZ5718	5133A	3399			
UZ5720	5135A	3401			
UZ5722	5136A	3402			
UZ5724	5137A	3403			
UZ5727	5139A	3405			
UZ5730	5141A	3407			
UZ5733	5142A	3408			
UZ5736	5143A	3409			
UZ5740	5144A	3410/5144			
UZ5750	5147A	3413			
UZ5756	5148A	3414			
UZ5760	5149A	3415			
UZ5775	5152A	3418			
UZ5780	5153A	3419			
UZ5790	5155A	3421			
UZ5806	5120A	3386			5M6.8ZS10
UZ5807	5121A	3387			
UZ5808	5122A	3388			
UZ5809	5124A	3390			
UZ5810	5125A	3391			
UZ5812	5127A	3393			
UZ5813	5128A	3394			
UZ5814	5129A	3395			
UZ5815	5130A	3396			
UZ5816	5131A	3397			
UZ5818	5133A	3399			
UZ5820	5135A	3401			
UZ5822	5136A	3402			
UZ5824	5137A	3403			
UZ5827	5139A	3405			
UZ5830	5141A	3407			
UZ5833	5142A	3408			
UZ5836	5143A	3409			
UZ5840	5144A	3410			
UZ5850	5147A	3413			
UZ5856	5148A	3414			
UZ5860	5149A	3415			
UZ5875	5152A	3418			
UZ5880	5153A	3419			
UZ5890	5155A	3421			
UZ5906	5120A	3386			
UZ5907	5121A	3387			
UZ5908	5122A	3388			
UZ5909	5124A	3390			
UZ5910	5125A	3391			
UZ5912	5127A	3393			
UZ5913	5128A	3394			
UZ5914	5129A	3395			
UZ5915	5130A	3396			
UZ5916	5131A	3397			
UZ5918	5133A	3399			
UZ5920	5135A	3401			
UZ5922	5136A	3402			
UZ5924	5137A	3403			
UZ5927	5139A	3405			
UZ5930	5141A	3407			
UZ5933	5142A	3408			
UZ5936	5143A	3409			
UZ5940	5144A	3410			
UZ5950	5147A	3413			
UZ5956	5148A	3414			
UZ5960	5149A	3415			
UZ5975	5152A	3418			
UZ5980	5153A	3419			
UZ5990	5155A	3421			
UZ6.2	5013A	3779		561	
UZ6.2B	5012A	3058/137A		561	
UZ7.5	138A	3059			
UZ706	5120A	3386			5M6.8ZS5
UZ707	5121A	3387			
UZ708	5122A	3388			
UZ709	5124A	3390			
UZ710	5125A	3391			
UZ7110					10M100Z5
UZ712	5127A	3393			
UZ713	5128A	3394			
UZ714	5129A	3395			
UZ715	5130A	3396			
UZ716	5131A	3397			
UZ718	5133A	3399			
UZ720	5135A	3401			
UZ7210					10M100Z10
UZ722	5136A	3402			
UZ724	5137A	3403			
UZ727	5139A	3405			
UZ730	5141A	3407			
UZ733	5142A	3408			
UZ736	5143A	3409			
UZ740	5144A	3410			
UZ750	5147A	3413			
UZ756	5148A	3414			
UZ760	5149A	3415/5148A			
UZ7706					10M6.8Z5
UZ775	5152A	3418			
UZ780	5153A	3419			
UZ7806					10M6.8Z10
UZ790	5155A	3421			
UZ8.2B	5072A			562	
UZ806	5120A	3386			5M6.8ZS10
UZ807	5121A	3387			
UZ808	5122A	3388			
UZ809	5124A	3390			
UZ810	5125A	3391			
UZ8110	5096A	3346			
UZ8111	151A	3099			
UZ8112	5097A	3347			
UZ8113	5098A	3348			
UZ8114	5099A	3349			
UZ8115	5100A	3350			
UZ8116	5101A	3351			
UZ8117	5102A	3352			
UZ8118	5103A	3353			
UZ8119	5104A	3354			
UZ812	5127A	3393			
UZ8120	5105A	3355			1M200ZS5
UZ813	5128A	3394			
UZ814	5129A	3395			
UZ815	5130A	3396			
UZ816	5131A	3397			
UZ818	5133A	3399			
UZ820	5135A	3401			
UZ8210	5096A	3346			
UZ8211	151A	3099			
UZ8212	5097A	3347			
UZ8217	5102A	3352			
UZ8218	5103A	3353			
UZ8219	5104A	3354			
UZ822	5136A	3402			
UZ8220	5105A	3355			1M200ZS10
UZ824	5137A	3403			
UZ827	5139A	3405			
UZ830	5141A	3407			
UZ833	5142A	3408			
UZ836	5143A	3409			
UZ840	5144A	3410			
UZ850	5147A	3413			
UZ856	5148A	3414			
UZ860	5149A	3415			
UZ8706	5071A	3334		561	1M6.8ZS5
UZ8707	138A	3059			
UZ8708	5072A	3136		562	
UZ8709	139A	3060		562	
UZ8710	140A	3061		562	
UZ8712	142A	3062		563	
UZ8713	143A	3750		563	
UZ8714	144A	3060/139A		562	
UZ8715	145A	3063		564	
UZ8716	5075A	3751		564	
UZ8718	5077A	3752			
UZ8720	5079A	3335			
UZ8722	5080A	3336			
UZ8724	5081A	3151			
UZ8727	146A	3064			
UZ8730	5084A	3755			
UZ8733	147A	3095			
UZ8736	5085A	3337			
UZ8740	5086A	3338			
UZ875	5152A	3418			
UZ8750	5089A	3341			
UZ8756	5090A	3342			
UZ8760	5091A	3097/149A			
UZ8775	5093A	3344			
UZ8780	150A	3098			
UZ8790	5095A	3345			
UZ880	5153A	3419			
UZ8806	5071A	3334		561	1M6.8ZS10
UZ8807	138A	3059			
UZ8808	5072A	3136		562	
UZ8809	139A	3060		562	
UZ8810	140A	3061		562	
UZ8812	142A	3062		563	
UZ8813	143A	3750		563	
UZ8814	139A			562	
UZ8815	145A	3063		564	
UZ8816	5075A	3751		564	
UZ8818	5077A	3752			
UZ8820	5079A	3335			
UZ8822	5080A	3336			
UZ8824	5081A	3151			
UZ8827	146A	3064			
UZ8830	5084A	3755			
UZ8833	147A	3095			
UZ8836	5085A	3337			
UZ8840	5086A	3338			
UZ8850	5089A	3341			
UZ8856	5090A	3342			
UZ8860	5091A	3097/149A			
UZ8875	5093A	3344			
UZ8880	150A	3098			
UZ8890	5095A	3345			
UZ890	5155A	3421			
UZ9.1	139A	3066/118		562	
UZ9.1B	139A	3060		562	
UZ906	5120A	3386			
UZ907	5121A	3387			
UZ908	5122A	3388			
UZ909	5124A	3390			
UZ910	5125A	3391			
UZ912	5127A	3393			
UZ913	5128A	3394			
UZ914	5129A	3395			
UZ915	5130A	3396			
UZ916	5131A	3397			
UZ918	5133A	3399			
UZ920	5135A	3401			
UZ922	5136A	3402			
UZ924	5137A	3403			
UZ927	5139A	3405			
UZ930	5141A	3407			
UZ933	5142A	3408			
UZ936	5143A	3409			
UZ940	5144A	3410			
UZ950	5147A	3413			
UZ956	5148A	3414			
UZ960	5149A	3415			
UZ975	5152A	3418			
UZ980	5153A	3419			
UZ990	5155A	3421			
UZ9B				562	
V-06B	116	3311		1104	
V-06C	116			1104	
V-1	173BP	3999			
V-10916-3	109	3091		1123	
V-11N				1114	
V-1266	145A	3063		564	
V-210C	109	3091		1123	
V-270D1	116	3016		1104	
V-442	116	3016		1104	
V-8634-3	112	3089			
V0-3C	125	3017B/117		1104	
V0-6	116	3311		1104	
V0-6A	125			1114	
V0-6B	116	3311		1104	
V0-6C		3311		1104	
V0-6C-401	116	3311		1104	
V01C				1143	
V01E				1144	
V01G	125			1114	
V03	125	3051/156		1114	
V03-C	125	3311		1114	
V03-E	125			1104	
V03C	125	3032A		1104	
V03E	125	3051/156		1114	
V03E(HITACHI)	506			1114	
V03G	116			1104	
V030	125			1114	
V05B	156	3051			
V05E				1104	
V06	116	3017B/117		1104	
V06(DIO)				1104	
V06(RECT)				1104	
V06-C	116	3311		1104	

Industry Standard No.	ECG	SK	GE	RS 276-	MOTOR.
V06-E	125	3313/116		1104	
V06-G	125	3017B/117		1114	
V06A	116	3017B/117		1104	
V06B	125	3311		1104	
V06BX4	116	3016		1104	
V06C	116	3311		1114	
V06C(BOOST)	506			1114	
V06C(HITACHI)	506			1114	
V06CS	125	3312			
V06E	116	3312		1104	
V06G	125	3032A		1104	
V07E				1103	
V07G				1104	
V08E				1103	
V08G				1104	
V09	506	3032A		1114	
V09-E	506	3016		1104	
V09A	506	3032A		1114	
V09B	506	3032A		1114	
V09C	552	3081/125		1114	
V09E	552	3843		1104	
V09G	552	3311		1104	
V0C236-001	221			2036	
V0C236-00I	221			2036	
V0G				1104	
V10	611				1N5443A
V10/15A	102	3004		2007	
V10/15A18	102A	3004		2007	
V10/1S	126	3005		2024	
V10/1SJ	126	3006/160		2024	
V10/2S	100	3005		2007	
V10/2SJ	100	3005		2007	
V10/30A	102	3004		2007	
V10/30A18	102A	3004		2007	
V10/50A	102	3004		2007	
V10/50A18	102A	3004		2007	
V100					MV840
V1000					MR331
V100E					1N5476A
V10158	116	3031A		1104	
V10916-1				1123	
V10916-3	109	3088		1123	
V10A	611	3324			
V10E					1N5140
V1112	177	3100/519		1122	
V11189-1	116			1104	
V115	109	3087		1123	
V117	109	3087		1123	
V118	107	3039/316		2015	
V119	123A	3444		2051	
V11J	506	3016		1104	
V11L	506	3017B/117		1104	
V11N	525	9000/552		1114	
V12	612	3325			1N5444A
V120	160			2004	
V120PH	108	3039/316		2038	
V120RH	128	3024		2030	
V126	145A	3063		564	
V129	107	3039/316		2015	
V12A	612	3325			
V12E					1N5141
V13/11	100	3004/102A		2007	
V135	109	3087		1123	
V139	128	3024		2030	
V143	107	3039/316		2015	
V144	124	3021			
V145	121	3009		2006	
V146	128	3024		2030	
V148	117	3017B		1104	
V15					MV830
V15/10DP	121	3009		2006	
V15/20DP	121	3009		2006	
V15/20R	126	3006/160		2024	
V15/30DP	121	3009		2006	
V152	159	3114/290		2032	
V152A	121	3114/290		2006	
V154	128	3024		2030	
V159	312	3112		2028	
V15920	116			1104	
V15C200/80-V&				1104	
V15C200/80-VF	116			1104	
V15E					1N5142
V160	142A	3062		563	
V162	159	3114/290		2032	
V162A	121	3114/290		2006	
V1650E-1	312			2028	
V1650E-4	312	3112		2028	
V166	128	3024		2030	
V167	134A	3055			
V169	123A	3444		2051	
V171	117	3017B		1104	
V172	128	3024		2030	
V176	184	3054/196		2017	
V177	128	3024		2030	
V178	703A	3157			
V17L	116	3017B/117		1104	
V180	129	3025		2027	
V182	725	3162			
V183	312	3448		2035	
V1833E	312	3112		2028	
V184	703A	3157			
V19C	552	3318			
V19E	552	3081/125			
V20					1N5447A
V205	160	3006		2004	
V20E					1N5467A
V210C	116			1104	
V220	108	3039/316		2038	
V221	108	3039/316		2038	
V222	108	3039/316		2038	
V27					MV833
V270-D1	116			1104	
V27E					1N5145
V297	123A	3444		2051	
V30/10DP	121	3009		2006	
V30/20DP	121	3009		2006	
V30/30DP	104	3009		2006	
V3074A20	116	3016		1104	
V3074A21	116	3016		1104	
V30N		3998		1114	
V33	614	3327			MV834
V330				1141	MR500
V330X					MR850
V331				1143	MR501
V3310					MR510
V331X					MR851
V332				1143	MR502
V332X					MR852
V334				1144	MR504
V334X					MR854
V336					MR506
V336X					MR856
V338					MR508
V33A	614	3327			

Industry Standard No.	ECG	SK	GE	RS 276-	MOTOR.
V33E					1N5146
V33EG	614	3327			
V33G	614	3327			
V39					MV835
V39E					1N5147
V405	108	3039/316		2038	
V409	152	3893		2048	
V410	159	3114/290		2032	
V410A	159	3025/129		2027	
V413	6408	9084A			
V413K	6408	9084A			
V413L	6407	9084A/6408			
V413M	6407	9084A/6408			
V413N	6408	9084A			
V415	108			2038	
V417	108			2038	
V435	108	3039/316		2038	
V435A	159			2032	
V442	116			1104	
V47					MV836
V47E					1N5148
V500					MR328
V50260-10	177	3100/519		1122	
V50260-16	109	3087		1123	
V50260-36	177	3100/519		1122	
V50A260-36	109	3087		1123	
V51	102A			2007	
V56					MV857
V56E					1N5473A
V58	160			2004	
V6/2R	126	3005		2024	
V6/2RC	102A	3005		2007	
V6/2RJ	102A			2007	
V6/4R	126	3005		2024	
V6/4RC	102A	3005		2007	
V6/4RJ	102A			2007	
V6/8R	126	3005		2024	
V6/8RJ	102A			2007	
V6/RC	102A			2007	
V60/10DP	121	3009		2006	
V60/10P	121	3009		2006	
V60/20DP	121	3014		2006	
V60/20P	121	3009		2006	
V60/30DP	121	3009		2006	
V60/30P	121	3009		2006	
V600					MR328
V654	193			2023	
V655	193	3118		2023	
V66	116			1104	
V68					MV838
V68E					1N5474A
V6C	116	3017B/117		1104	
V7					1N5441A
V74	109	3087		1123	
V741	193			2023	
V75	160			2004	
V761	159			2032	
V763	159			2032	
V78	117			1104	
V7E					1N5138
V800					MR330
V82					MV839
V82E					1N5457A
V900					1N5456A
V900E					1N5476A
V907					1N5441A
V907E					1N5139
V910	611	3324			1N5443A
V910A	611	3324			
V910E					1N5140
V912	612	3325			1N5444A
V912A	612	3325			
V915					1N5445A
V920					1N5447A
V927					1N5449A
V933	614	3327			1N5450A
V933A	614	3327			
V939					1N5451A
V9446-4	116			1104	
V956					1N5453A
V968					1N5454A
V982					1N5455A
V996					2N6603
VA127	614	3327			
VA128	614	3327			
VA132	611	3324			
VA135	611	3324			
VA157	611	3324			
VA160	614	3327			
VA5139	610	3323			
VA5139A	610	3323			
VAMV-1	116	3031A		1104	
VAR	177	3100/519		1122	
VAR-1R2	177			1122	
VARIST-5				1104	
VB-11	116			1104	
VB-400	116	3016		1104	
VB-600	116	3017B/117		1104	
VB100	116	3016		1104	
VB11	102A			2007	
VB15	525	3925			
VB300	116	3016		1104	
VB400	116			1104	
VB500	116	3016		1104	
VB600	116			1104	
VB600A	116	3017B/117		1104	
VBH600	116	3017B/117		1104	
VC10	611	3324			
VC10A	611	3324			
VC12	612	3325			
VC12A	612	3325			
VC210	611	3324			
VC210A	611	3324			
VC212	612	3325			
VC212A	612	3325			
VC233	614	3327			
VC233A	614	3327			
VC309	611	3324			
VC309A	611	3324			
VC321	613	3326			
VC321A	613	3326			
VC33	614	3327			
VC332	614	3327			
VC332A	614	3327			
VC33A	614	3327			
VC408	610	3323			
VC408A	610	3323			
VC409	611	3324			
VC409A	611	3324			
VC410	611	3324			
VC410A	611	3324			
VC421	613	3326			
VC421A	613	3326			
VC422	613	3326			

Industry Standard No.	ECG	SK	GE	RS 276-	MOTOR.
VC422A	613	3326			
VC432	614	3327			
VC432A	614	3327			
VC433	614	3327			
VC433A	614	3327			
VC610	611	3324			
VC610A	611	3324			
VC611	611	3324			
VC611A	611	3324			
VC612	612	3325			
VC612A	612	3325			
VC613	612	3325			
VC613A	612	3325			
VC622	613	3326			
VC622A	613	3326			
VC623	613	3326			
VC623A	613	3326			
VC633	614	3327			
VC633A	614	3327			
VC634	614	3327			
VC634A	614	3327			
VC6E	116			1104	
VC822	611	3324			
VC822A	611	3324			
VC823	612	3325			
VC823A	612	3325			
VC827	614	3327			
VC827A	614	3327			
VC841	610	3323			
VC841A	610	3323			
VD-1121				1122	
VD-1122	601			1104	
VD-1123	177			1122	
VD-1124	601			1122	
VD-121C	116	3017B/117		1104	
VD-1222				1104	
VD-S-7-013626-001	519			1122	
VD10E11F				1104	
VD11	109	3091		1123	
VD1120	601	3463		1104	
VD1121	601	3463		1122	
VD1122	601	3463		1104	
VD1123	601	3463		1122	
VD1124	601	3463		1122	
VD1127	177	3100/519		1122	
VD1129	601	3463			
VD1150L	601	3463			
VD1150M	601	3463		1122	
VD1150M				1122	
VD12	109	3087		1123	
VD1210	605	3864			
VD1210L	605	3864		1122	
VD1211	605	3864			
VD1212	605	3864		1122	
VD1213	605	3864		1104	
VD1220	605	3864			
VD1221	605	3864			
VD1221L	605	3864			
VD1222	605	3864			
VD1223	605	3864			
VD1251L	601	3864/605			
VD1252L	605	3864			
VD13	109	3004/102A		1123	
VD1E1////-1				1104	
VD6	116	3017B/117		1104	
VE18	5304				MDA920A3
VE28	5304				MDA920A4
VE48	5304				MDA920A6
VE67	5305	3676			
VE68	5305	3676			MDA920A7
VE88	5306	3677			
VF-1043	614	3327			
VF-2	125	3081			
VF-SD1A	506			1114	
VFA-2745C	116	3016		1104	
VFA2745C	116			1104	
VFG-274513	104	3009		2006	
VFG2745B	104	3014		2006	
VFL-2744K	126	3006/160		2024	
VFL2744K	160			2004	
VFP-2746C	104	3009		2006	
VFP-6537C	121	3009		2006	
VFP2746C	121	3014		2006	
VFP6357C				2006	
VFP6537C	121	3014		2006	
VFQ-2745F	100	3005		2007	
VFQ2745F	102A			2007	
VFS-2745	102A			2007	
VFS-2745J	102A	3005		2007	
VFS2745	158			2007	
VFS2745J	158			2007	
VFS5K	126	3008		2024	
VFT-2745H	102A	3004		2007	
VFT2745H				2007	
VFU-2746B	121	3009		2006	
VFU2746B	121	3014		2006	
VFU65326B	121	3717		2006	
VFW-27450	126	3006/160		2024	
VFW2745D	160			2004	
VFY-2745E	100	3005		2007	
VFY2745E				2007	
VG310	611	3324			
VG310A	611	3324			
VG322	613	3326			
VG322A	613	3326			
VG333	614	3327			
VG333A	614	3327			
VH10	611	3324			
VH148	5312	3985			
VH247	5313	3986			
VH248	5313	3986			
VH33	614	3327			
VH447	5314	3987			
VH448	5314	3987			
VH647	5315	3988			
VH648	5315	3988			
VH7084	614	3327			
VH848	5316	3989			
VH910	611	3324			
VH933	614	3327			
VHCBB1096//-1	614	3327			
VHCBB109G-1	610	3323			
VHD1N34A///-1	110MP	3087		1123	
VHD1N34A///1				1123	
VHD1N60-1	109	3087		1123	
VHD1N60////-1	109			1123	
VHD1S1553//-1	177	3175		1122	
VHD1S1555-R-1	177	3100/519		1122	
VHD1S1555//1A	177	3175		1122	
VHD1S1834				1104	
VHD1S1834//-1	506	3843		1114	
VHD1S1885-1	116			1104	
VHD1S1885//-1	116	3311		1104	
VHD1S1887//-1				1104	
VHD1S2076-//-1		3100		1122	
VHD1S2076-1	177	3100/519		1122	
VHD1S2230//1E	116			1114	
VHD1S34///-1	109	3087		1123	
VHEHZ6B2///1A	5012A	3057/136A			
VHEHZ6B3///1A	136A	3057		561	
VHERD27ED//-1	146A	3064			
VHERD6R8EE/-1	5071A	3334		561	
VHEWZ-100-1F	140A	3061		562	
VHEWZ-100//1F	140A	3061		562	
VHEWZ-10001F	5096A	3346			
VHEWZ-1001E	5096A	3346			
VHEWZ-1001F	177	3061/140A		562	
VHEXZ-090-1	139A	3097/149A		562	
VHF1S2076-1	177	3100/519			
VHS2SP1422/1E	230	3042			
VHS2SP656//1E	5402	3638			
VHSCR3AM28LB1	5457	3598			
VHSS6089///1E	230	3042			
VHSS61420LB1E	5457	3598			
VHV1S1209-1	177	3100/519		1122	
VJ148	5312	3985			
VJ248	5313	3986			
VJ448	5314	3987			
VJ648	5315	3988			
VJ848	5316	3989			
VL/8RJ	100	3721		2007	
VL18RJ	126			2024	
VM-20233	124	3021			
VM-30203	121	3009		2006	
VM-30209	123A	3020/123		2051	
VM-30234	124	3021			
VM-30241	123A	3020/123		2051	
VM-30242	123A	3020/123		2051	
VM-30244	102A	3004		2007	
VM-30245	124	3021			
VM-PH11D522/1	116			1104	
VM-PH90522/1				1104	
VM-PH9D522/1	116	3017B/117		1104	
VM-TCO.2P11/2				1104	
VM-TCO2P11/2	116	3017B/117		1104	
VM08	5332			1151	
VM18	5332			1152	
VM204	610	3323			
VM204A	610	3323			
VM205	610	3323			
VM205A	610	3323			
VM210	610	3323			
VM210A	610	3323			
VM211	610	3323			
VM211A	610	3323			
VM216	610	3323			
VM216A	610	3323			
VM217	610	3323			
VM217A	610	3323			
VM222	610	3323			
VM222A	610	3323			
VM223	610	3323			
VM223A	610	3323			
VM228	610	3323			
VM228A	610	3323			
VM229	610	3323			
VM229A	610	3323			
VM234	610	3323			
VM234A	610	3323			
VM235	610	3323			
VM235A	610	3323			
VM25	5332			1151	
VM28	5332			1172	
VM30203	121	3717		2006	
VM30209	123A	3444		2051	
VM30233	124	3021			
VM30234	124	3021			
VM30241	123A	3444		2051	
VM30242	123A	3444		2051	
VM30244	102A			2007	
VM30245	124	3021			
VM304	611	3324			
VM304A	611	3324			
VM305	611	3324			
VM305A	611	3324			
VM310	611	3324			
VM310A	611	3324			
VM311	611	3324			
VM311A	611	3324			
VM316	611	3324			
VM316A	611	3324			
VM317	611	3324			
VM317A	611	3324			
VM328	611	3324			
VM328A	611	3324			
VM329	611	3324			
VM329A	611	3324			
VM334	611	3324			
VM334A	611	3324			
VM335	611	3324			
VM335A	611	3324			
VM48	5332			1173	
VM504	613	3326			
VM504A	613	3326			
VM505	613	3326			
VM505A	613	3326			
VM510	613	3326			
VM510A	613	3326			
VM511	613	3326			
VM511A	613	3326			
VM516	613	3326			
VM516A	613	3326			
VM517	613	3326			
VM517A	613	3326			
VM522	613	3326			
VM522A	613	3326			
VM523	613	3326			
VM523A	613	3326			
VM528	613	3326			
VM528A	613	3326			
VM529	613	3326			
VM529A	613	3326			
VM534	613	3326			
VM534A	613	3326			
VM535	613	3326			
VM535A	613	3326			
VM604	614	3327			
VM604A	614	3327			
VM605	614	3327			
VM605A	614	3327			
VM610	614	3327			
VM610A	614	3327			
VM611	614	3327			
VM611A	614	3327			
VM616	614	3327			
VM616A	614	3327			
VM617	614	3327			
VM617A	614	3327			

Industry Standard No.	ECG	SK	GE	RS 276-	MOTOR.
VM622	614	3327			
VM622A	614	3327			
VM623	614	3327			
VM623A	614	3327			
VM628	614	3327			
VM628A	614	3327			
VM629	614	3327			
VM629A	614	3327			
VM634	614	3327			
VM634A	614	3327			
VM635	614	3327			
VM635A	614	3327			
VMPH11D522-1	116			1104	
VMT-01	532	3301/531			
VMT-02	531	3301			
VMT-03	500A	3304			
VMT-03-01	500A	3304			
VMT-03-02	500A	3304			
VMT-04	521	3304/500A			
VMT-05	522	3303			
VMT-05-01	522	3303			
VMT-05-02	522	3304/500A			
VMT-05-03	522	3303			
VMT-06	537	3302/533			
VMT-07	533	3302			
VMT-08	523	3306			
VMT-09	529	3307			
VMT-09-01	529	3307			
VMT-10	536A	3900			
VMT-11	534	3305			
VMT-12	538	3310			
VMT-13	539	3309			
VMT-14	539	3309			
VMT-15	530	3308			
VO-3C	125			1114	
VO-5E	116			1104	
VO-6A	116			1104	
VO-6C	116			1104	
VO-9C	506			1114	
VO3-C	125	3017B/117		1104	
VO3-G				1114	
VO3C	125			1114	
VO3G	125			1104	
VO6-C	116	3311		1104	
VO6B	116			1104	
VO6C	116			1104	
VO9C				1114	
VO9E	506			1104	
VOG	116			1104	
VP20A					MHW612
VP20B					MHW612
VR-14	145A			564	
VR-9.1				562	
VR10	140A			562	
VR10A	140A			562	
VR10B	140A			562	
VR11	141A	3092			
VR110	151A	3099			
VR110A	151A	3099			
VR110B	151A	3099			
VR11A	141A	3092			
VR11B	5074A	3092/141A			
VR12	142A	3062		563	
VR12A	142A	3062		563	
VR12B	142A	3062		563	
VR13	143A	3750		563	
VR13A	143A	3750		563	
VR13B	143A	3750		563	
VR14	144A	3094		564	
VR14A	144A	3094		564	
VR14B	144A	3094		564	
VR15	145A			564	
VR15A	145A			564	
VR15B	145A			564	
VR16B	5075A	3751			
VR18B	5077A	3752			
VR200					1M200ZS10
VR28B	5083A	3754			
VR30B	5084A	3755			
VR33	147A	3095			
VR33A	147A	3095			
VR33B	147A	3095			
VR5.6	136A	3057		561	
VR5.6A	136A	3057		561	
VR5.6B	136A	3057		561	
VR56	148A	3096			
VR56A	148A	3096			
VR56B	5090A	3096/148A			
VR6.2					1M6.2ZS10
VR62	149A	3097			
VR62A	149A	3097			
VR62B	149A	3097			
VR8.2B	5072A	3136			
VR82	150A	3098			
VR82A	150A	3098			
VR82B	150A	3098			
VR9.1	139A	3059/138A		562	
VR9.1A	139A	3059/138A		562	
VR9.1B	139A	3059/138A		562	
VR9B	139A	3059/138A		562	
VS-0A70	109			1123	
VS-1	116	3016		1104	
VS-102	116	3016		1104	
VS-2SA103	160	3007		2004	
VS-2SA288A				2051	
VS-2SA358	126	3006/160		2024	
VS-2SA378	126	3006/160		2024	
VS-2SA379	126	3006/160		2024	
VS-2SA385	126	3008		2024	
VS-2SA385L	160	3006		2004	
VS-2SA71	160	3006		2004	
VS-2SA71B	160	3006		2004	
VS-2SA71BS	160	3006		2004	
VS-2SB126	121	3034		2006	
VS-2SB126F	127	3009			
VS-2SB126V	127	3034			
VS-2SB128V	127	3034			
VS-2SB171	102A	3004		2007	
VS-2SB172	102A			2007	
VS-2SB172FN	102A	3004		2007	
VS-2SB176	102A	3004		2007	
VS-2SB178	102A	3004		2007	
VS-2SB178A	102A	3004		2007	
VS-2SB324	102A	3004		2007	
VS-2SB448	127	3035			
VS-2SC-458	123A	3020/123		2051	
VS-2SC206	108	3019		2038	
VS-2SC208	108	3018		2038	
VS-2SC288A	108	3019		2038	
VS-2SC324				2038	
VS-2SC324H	123A	3122		2051	
VS-2SC371-R-1				2038	
VS-2SC385L	160	3006		2004	
VS-2SC41				2041	
VS-2SC446	161	3039/316		2015	
VS-2SC458	123A			2051	
VS-2SC466				2038	
VS-2SC538	123A	3020/123		2051	
VS-2SC563	108	3018		2038	
VS-2SC58	154	3041		2012	
VS-2SC58A	154	3041		2012	
VS-2SC58B	154	3041		2012	
VS-2SC58C	154	3041		2012	
VS-2SC645	108	3019		2038	
VS-2SC645A	108	3018		2038	
VS-2SC645B				2038	
VS-2SC645C				2038	
VS-2SC683	161	3039/316		2015	
VS-2SC683Y				2038	
VS-2SC684	108	3019		2038	
VS-2SC762	108	3018		2038	
VS-CS1255H	128	3024		2030	
VS-CS1255HF	128	3024		2030	
VS-CS1256HG	129	3025		2027	
VS-DG1N				1104	
VS-DG1NR	116	3017B/117		1104	
VS-FR-1	116	3016		1104	
VS-FR-1P	116	3016		1104	
VS-FR-1U				1104	
VS-FR1	116	3017B/117		1104	
VS-FR1P	116	3017B/117		1104	
VS-FT-1N	116	3017B/117		1104	
VS-PC02P11/2				1104	
VS-PH9D522/1	116	3016		1104	
VS-RD11AM				563	
VS-SD-1B	506	3016		1114	
VS-SD-1Z	116	3017B/117		1104	
VS-SD1				1114	
VS-SD1B				1114	
VS-TCO-2P11/2	116			1104	
VS-TCO.2P11/2				1104	
VS-TC02P11/2	116	3016		1104	
VS1	116			1104	
VS120	116	3017B/117		1104	
VS148	5312	3985			
VS202	116	3017B/117		1104	
VS244					MDA920A4
VS247	5313	3986			
VS248	5313	3986			MDA920A4
VS2A71BS				2004	
VS2SA103	160			2004	
VS2SA128V	127	3764			
VS2SA378	160			2004	
VS2SA379	160			2004	
VS2SA385	102	3722		2007	
VS2SA448	127	3764			
VS2SA495-0/1E	159	3114/290		2032	
VS2SA495-0/-1				2032	
VS2SA495-Y/1E	159	3114/290		2032	
VS2SA562-0/1E		3114		2032	
VS2SA562-Y/1E	290A	3114/290		2032	
VS2SA673-B/1E	290A	3114/290		2032	
VS2SA673-C/1A	290A	3114/290		2032	
VS2SA673-C/1E	290A	3114/290		2032	
VS2SA71B	160			2004	
VS2SA71BS	160			2004	
VS2SA740///1E	292	3930			
VS2SA844-D/-1	129	3114/290		2027	
VS2SA844-D/-2	159	3114/290			
VS2SA854-Q/1E	290A	3247/234		2050	
VS2SB126	121	3717		2006	
VS2SB126F	121	3717		2006	
VS2SB126V	121	3717		2006	
VS2SB171	102	3722		2007	
VS2SB172	102	3722		2007	
VS2SB172F				2007	
VS2SB172FN				2007	
VS2SB176	102	3722		2007	
VS2SB178	102	3722		2007	
VS2SB178A	102	3722		2007	
VS2SB324	123A	3444		2051	
VS2SB561-C/-1	290A	3114/290		2032	
VS2SB568///-1	292	3930			
VS2SC1166-0-1	297	3122			
VS2SC1166-0-1	297			2051	
VS2SC1166-Y-1	297	3137/192A		2030	
VS2SC1166Y-1	297			2051	
VS2SC1172-/1E	238	3710			
VS2SC1173-Y-3	152	3054/196		2048	
VS2SC1174-/1E	238	3710			
VS2SC1213-C/1A	289A			2038	
VS2SC1213-C1A	289A	3122			
VS2SC1213AC/1A		3024		2038	
VS2SC1237-1	236	3197/235		2053	
VS2SC1237-1F	236	3197/235			
VS2SC1335D/-1	199			2010	
VS2SC1335D/-1E		3122		2010	
VS2SC1335D/1	199			2051	
VS2SC1447LB1E	198	3219			
VS2SC1514//-1	198	3219			
VS2SC1514BK1E	376	3219			
VS2SC1569K/1E	376	3219			
VS2SC1627A				2030	
VS2SC1627Y				2030	
VS2SC16742/-1				2038	
VS2SC1674L/-1	229			2038	
VS2SC1675M-1	229	3124/289		2038	
VS2SC1681G/1E	199	3245		2010	
VS2SC1723S/-1	198	3219			
VS2SC1741-1	129			2030	
VS2SC1815Y/-1	123AP	3245/199			
VS2SC1829-/1E	162	3836/284			
VS2SC1855//-1	108	3018		2038	
VS2SC1890A/1E	375			2010	
VS2SC1894//1E	238	3111			
VS2SC1906//1E	107	3293			
VS2SC1921//1E	191	3433/287		2012	
VS2SC1942//1E	389	3710/238			
VS2SC1983//-1		3041		2048	
VS2SC206	123A	3444		2051	
VS2SC2068LB1E	198	3201/171			
VS2SC2073//1E	375	3929			
VS2SC208	123A	3444		2051	
VS2SC22290/1E	287	3244			
VS2SC2231Y/1E	198	3219			
VS2SC2371K/1E	287	3244			
VS2SC2481Y/1E	373	9041			
VS2SC2482//-1	287	3244			
VS2SC2610K/1E	287	3244			
VS2SC288A	123A	3444		2051	
VS2SC324H	123A	3444		2051	
VS2SC371-R-1	199	3122		2010	
VS2SC372-Y/1E	289A	3245/199		2051	
VS2SC373-//1E	199	3122		2010	
VS2SC373//-1E				2010	
VS2SC373G-1	199	3122		2010	
VS2SC374-B-1	199	3122		2010	
VS2SC383-W/1E	199			2010	

Industry Standard No.	ECG	SK	GE	RS 276-	MOTOR.
VS2SC385L	160			2004	
VS2SC388A-/1A	233	3132			
VS2SC394-0-1	229			2038	
VS2SC394-0-1				2015	
VS2SC394-Y-1	229	3018		2038	
VS2SC454-B/1E	289A	3124/289		2038	
VS2SC454-C/1A	289A			2038	
VS2SC454-C/1E	289A	3124/289		2038	
VS2SC454-C/3A	289A	3124/289		2038	
VS2SC458	123A	3124/289		2051	
VS2SC458-C/1E	289A	3124/289		2038	
VS2SC460B-1	233	3018		2038	
VS2SC481-1	195A	3048/329			
VS2SC538		3444		2051	
VS2SC645A	123A	3444		2051	
VS2SC681A-Y2E	163A	3439			
VS2SC684	107			2015	
VS2SC717///-1	161			2015	
VS2SC732-V1F	199	3122		2010	
VS2SC732-VIF				2051	
VS2SC733B-1	199	3122		2010	
VS2SC735-Y-1	289A	3138/193A		2038	
VS2SC735Y-1	289A	3122			
VS2SC784-R1F	229	3122		2038	
VS2SC784R1F	229			2015	
VS2SC945LK-1	199			2010	
VS2SC945LP-1	199	3124/289		2010	
VS2SC983-0/2E	198	3044/154			
VS2SD2271-1	123A	3124/289			
VS2SD227V-1	123A			2051	
VS2SD467-C/-1	289A	3024/128		2030	
VS2SD471-S/-1				2038	
VS2SD476-C/-1		3054		2020	
VS2SD478YL2E	291	3929			
VS2SD604-//1E	162	3560			
VS2SD666-C/-1	194	3275			
VS2SD724///1A	375	3929			
VS2SK49F-1	312	3116		2035	
VS444					MDA920A6
VS447	5314	3987			
VS448	5314	3987			MDA920A6
VS500	5952	3585			
VS600	5952	3585			
VS644					MDA920A7
VS647	5315	3988			
VS648	5315	3988			MDA920A7
VS848	5316	3989			
VS9-0001-911	116			1104	
VS9-0002-911	116			1104	
VS9-0003-911	125			1114	
VS9-0003-913	199	3245		2010	
VS9-0004-911	125			1114	
VS9-0004-923	234			2050	
VS9-0005-911	116			1104	
VS9-0005-913	123A	3444		2051	
VS9-0006-911	116			1104	
VS9-0006-913	123A	3444		2051	
VS9-0007-911	116			1104	
VS9-0008-911	177			1122	
VS9-0008-923	234			2050	
VS9-0014-911	177			1122	
VSCS1255H	128			2030	
VSCS1256HG	159			2032	
VSF2745	100	3005		2007	
VSFR1	116			1104	
VSFR1P	116			1104	
VSFT1N	116			1104	
VSG-20024	116	3016		1104	
VSK1520					MBR3520
VSK1530					MBR3535
VSK1540					MBR3545
VSK3020S					MBR3520
VSK3020T					MBR3020CT
VSK3030S					MBR3535
VSK3030T					MBR3035CT
VSK3040S					MBR3545
VSK3040T					MBR3045CT
VSK4020					MBR6020
VSK4030					MBR6035
VSK4040					MBR6045
VSK51					SD51
VSK51B					SD5171
VSK520					1N5823
VSK530					1N5824
VSK540					1N5825
VSOA70				1104	
VSSD1B	116			1104	
VSSD1Z	116			1104	
VSTC02P11/2	116			1104	
VT200-S					MDA3502
VT200-T					MDA3502
VT400-S					MDA3504
VT400-T					MDA3504
VT600-S					MDA3506
VT600-T					MDA3506
VX3375	186	3192		2017	
VX3733	152			2048	
VZ-050	135A	3056			
VZ-052	5010A	3776			
VZ-054	5011A			561	
VZ-056	5011A	3057/136A		561	
VZ-058	5012A			561	
VZ-061	5012A	3058/137A		561	
VZ-063	5013A	3779		561	
VZ-065	5014A	3780		561	
VZ-067	5014A	3780		561	
VZ-069	5014A	3780		561	
VZ-071	5015A	3781			
VZ-073	5015A	3781			
VZ-075	5015A	3781			
VZ-077	5015A	3781			
VZ-079		3782		562	
VZ-081	5016A	3782		562	
VZ-083	5016A	3782		562	
VZ-085	5017A	3783		562	
VZ-088	5017A	3783		562	
VZ-090	5018A	3784		562	
VZ-092	5018A	3784		562	
VZ-094	5018A	3784		562	
VZ-096	5019A	3785		562	
VZ-098	5019A	3785		562	
VZ-100	5019A	3785		562	
VZ-105		3786		563	
VZ-110	5020A	3786		563	
VZ-115	141A	3092			
VZ-120	5021A	3787		563	
VZ-125	5022A	3093		563	
VZ-130	5022A	3788		563	
VZ-135	5023A	3789		564	
VZ-140	5023A	3789		564	
VZ-145	5024A	3790		564	
VZ-150	5024A	3790		564	
VZ-157	5025A	3791		564	
VZ-162	5025A	3791		564	
VZ-167	5026A	3792			
VZ-172	5026A	3792			
VZ-177	5027A	3145			
VZ-182	5027A	3793			
VZ-192	5028A	3794			
VZ-197	5029A	3795			
W-005	166	9075		1152	
W-06A				1104	
W/6A				1104	
WO-61	137A	3057/136A		561	
WO-6A				1104	
W005					MDA100A
W005M	5304	3647/167		1104	MDA100A
W010M	5307	3107			
W02	5304	3105			MDA102A
W02M	5304				MDA102A
W03A	116			1011	
W03B	116	3311		1104	
W03C	116	3017B/117			
W04	167	3106/5304		1172	MDA104A
W04M	5304				MDA104A
W06	116	3676/5305		1104	MDA106A
W06A	116	3313		1104	
W06B	116	3313		1104	
W06C	116	3313		1104	
W06M	5305	3676			MDA106A
W08		3677			MDA108A
W08M	5306	3677			MDA108A
W1	100	3005		2007	
W10	123	3122		2051	MDA110A
W106A1				1067	
W106C1				1020	
W106Y1				1067	
W10M	5307				MDA110A
W11	124	3021			
W12	160	3006		2004	
W13	127	3035			
W14	128	3104A		2030	
W15	129	3025		2027	
W16	130	3027		2041	
W16MP	130MP	3029		2041(2)	
W17	131			2006	
W17MP	131MP	3840		2006(2)	
W18	152	3893		2048	
W19	153	3083/197		2049	
W1A	312	3112			
W1E	220	3990		2036	
W1P	312	3448		2035	
W1R	177			1122	
W1R(DIODE)	177			1122	
W1R(SCR)	5404	3627			
W1U	222	3050/221		2036	
W1U-1	222			2036	
W1W	5444	3572/5463			
W2	101	3861		2002	
W20	123A	3444		2051	
W2058	5402	3638			
W21	159	3114/290		2032	
W22	160			2004	
W23	161	3018		2015	
W24	123A	3444		2051	
W25	158	3004/102A		2007	
W26	157	3103A/396			
W27	155	3839			
W28	154	3044		2012	
W29	107	3039/316		2015	
W3	102A	3004		2007	
W3D	5633	3533/5635			
W4	103A	3010		2002	
W4002	116			1104	
W5	104	3009		2006	
W6	105	3012			
W7	106	3118		2034	
W8	108	3019		2038	
W9	121	3009		2006	
W9MP	121MP	3013		2006(2)	
WA-26	177			1122	
WAOA-90				1122	
WC-14020	116	3016		1104	
WC-14027	116	3016		1104	
WC120	116	3311		1104	
WC14020	116			1104	
WC14027	116			1104	
WC19862	103A	3010		2002	
WC19862A	158	3004/102A		2007	
WC19863	102A	3004		2007	
WC19864	102A	3004		2007	
WC19865	116	3017B/117		1104	
WD001	116	3311		1104	
WD002	116	3311		1104	
WD003	116	3311		1104	
WD004	116	3311		1104	
WD005	116	3311		1104	
WD006	116	3311		1104	
WD007	116	3311		1104	
WD008	116	3311		1104	
WD009	116	3311		1104	
WD010	116	3311		1104	
WD011	116	3311		1104	
WD012	116	3311		1104	
WD013	116	3031A		1104	
WD014	116	3031A		1104	
WD015	116	3031A		1104	
WD1	109	3087		1123	
WD10	113A	3119/113			
WD11	114	3120			
WD12	115	3121			
WD2	112	3089			
WD20	118	3066			
WD21	119	3109			
WD4	177	3100/519		1122	
WD90	139A	3060		562	
WE081	5072A	3136		562	
WEP1008				561	
WEP101		3061		562	
WEP1010				562	
WEP103		3058		561	
WEP104		3060		562	
WEP105		3062		563	
WEP1060				1122	
WEP1082				1104	
WEP1083				1104	
WEP1096				2017	
WEP1112		3062		563	
WEP1307				2053	
WEP134		3090		1123	
WEP156		3311		1104	
WEP158		3313		1104	
WEP158A		3313		1114	
WEP160		3080		1114	
WEP165A		3120		1104	
WEP170		3081		1114	
WEP1717		3132		2051	
WEP172		3998		1114	

Industry Standard No.	ECG	SK	GE	RS 276-	MOTOR.
WEP186		3125		1114	
WEP1860				2007	
WEP187				2007	
WEP189	HIDIV-1	3868/DIV-1			
WEP190	HIDIV-2	3869/DIV-2			
WEP191	HIDIV-4	3871/DIV-4			
WEP193	HIDIV-3	3870/DIV-3			
WEP1945		3124		2010	
WEP2				2007	
WEP230		3719		2006	
WEP230MP		3720		2006	
WEP232		3717		2006	
WEP241		3261		2020	
WEP242		3025		2027	
WEP245				2017	
WEP246		3083		2025	
WEP247		3036		2041	
WEP247MP		3037		2041	
WEP250		3003		2007	
WEP253		3004		2007	
WEP254		3004		2007	
WEP3		3006		2004	
WEP310				2029	
WEP367				2006	
WEP371				2038	
WEP373				2010	
WEP380				2038	
WEP394				2038	
WEP403				2010	
WEP405				2007	
WEP454				2051	
WEP458		3122		2051	
WEP460		3122		2038	
WEP474				2025	
WEP474MP				2025	
WEP481				2006	
WEP50		3122		2051	
WEP51		3114		2034	
WEP52		3118		2034	
WEP53		3020		2030	
WEP535		3293		2038	
WEP536		3124		2010	
WEP537				2010	
WEP538				2010	
WEP54				2051	
WEP55				2051	
WEP56		3452		2015	
WEP603		3057		561	
WEP605		3093		563	
WEP606		3094		564	
WEP607		3063		564	
WEP624		3719		2006	
WEP628		3719		2006	
WEP628MP		3720		2006	
WEP630				2007	
WEP631		3004		2007	
WEP632		3004		2007	
WEP633				2030	
WEP634				2010	
WEP635		3006		2024	
WEP637		3007		2004	
WEP641		3011		2002	
WEP641A		3010		2002	
WEP641B		3835		2002	
WEP642		3052		2006	
WEP642MP				2006	
WEP643		3052		2006	
WEP644				2010	
WEP700		3084		2025	
WEP701		3253		2017	
WEP703		3026		2020	
WEP704		3027		2041	
WEP709		3132		2038	
WEP710		3444		2016	
WEP712		3044		2012	
WEP715		3114		2034	
WEP716		3114		2034	
WEP717		3114		2032	
WEP719		3132		2038	
WEP720		3018		2038	
WEP723		3122		2051	
WEP724		3122		2051	
WEP728		3122		2051	
WEP729		3122		2051	
WEP735		3122		2051	
WEP735A				2051	
WEP736		3444		2051	
WEP7400	7400			1801	
WEP7402				1811	
WEP7408				1822	
WEP74145		74145		1828	
WEP7473				1803	
WEP7474	7474			1818	
WEP7476				1813	
WEP7490		7490		1808	
WEP772				2038	
WEP773		3849		2051	
WEP784		3132		2015	
WEP801		3112		2028	
WEP802		3116		2028	
WEP828		3444		2038	
WEP829		3444		2051	
WEP838				2051	
WEP900				2052	
WEP903				2036	
WEP904				2036	
WEP905				2036	
WEP906				2051	
WEP907				2050	
WEP908				2017	
WEP910		3124		2038	
WEP911		3114		2032	
WEP912		3849		2030	
WEP920				2035	
WEP921				2016	
WEP924				2016	
WEP925		3100		1122	
WEP933				007	
WEP949		3231		705	
WEP956				2038	
WEPS1000-1				1114	
WEPS3023				2018	
WEPS3027				2026	
WEPS3031		3083		2026	
WEPS5003				2017	
WEPS5005				2027	
WEPS5007				2025	
WEPS7000				2041	
WEPS7001				2043	
WP1	312	3112		2028	
WP2	312	3448		2035	
WG-1010-A	177			1122	
WG-1010A	177	3100/519		1122	
WG-1010AS	177	3100/519			

Industry Standard No.	ECG	SK	GE	RS 276-	MOTOR.
WG-1012	177	3100/519		1122	
WG-101DA				1122	
WG-10AS	177			2035	
WG-599	177	3100/519		1122	
WG-713	177	3100/519		1122	
WGOA-90	177			1122	
000WG1010	177	3100/519	300	1122	
WG1010A	177	3100/519		1122	
WG1010B	177	3100/519		1122	
WG1012	177	3100/519		1122	
WG1014A	177	3100/519		1122	
WG1021	177	3100/519		1122	
WG599	177			1122	
WG713	177	3100/519		1122	
WG714	177	3100/519		1122	
WG851	519			1122	
WG91	139A	3060		562	
WH1012		3100		1122	
WIE	220			2036	
WK5457		3112			2N5457
WK5458	105	3112			2N5458
WK5459	105	3112			2N5459
WL-125	143A	3093		563	
WL-125A	143A	3093		563	
WL-125B	143A	3093		563	
WL-125C	143A	3093		563	
WL-125D	143A	3093		563	
WL-130	143A	3750		563	
WL-130A	143A	3750		563	
WL-130B	143A	3750		563	
WL-130C	143A	3750		563	
WL-130D	143A	3750		563	
WL-140	144A	3094		564	
WL-140A	144A	3094		564	
WL-140B	144A	3094		564	
WL-140C	144A	3094		564	
WL-140D	144A	3094		564	
WL-150	145A	3063		564	
WL-150A	145A	3063		564	
WL-150B	145A	3063		564	
WL-150C	145A	3063		564	
WL-150D	145A	3063		564	
WL-157	5075A	3751		564	
WL-157A	5075A	3751		564	
WL-157B	5075A	3751		564	
WL-157C	5075A	3751		564	
WL-157D	5075A	3751		564	
WL-162	5075A	3751		564	
WL-162A	5075A	3751		564	
WL-162B	5075A	3751		564	
WL-162C	5075A	3142		564	
WL-162D	5075A	3142		564	
WL-167	5075A	3751		564	
WL-167A	5075A	3751		564	
WL-167B	5075A	3751		564	
WL-167C	5075A	3751		564	
WL-167D	5075A	3751		564	
WL-172	5077A	3145			
WL-172A	5077A	3145			
WL-172B	5077A	3145			
WL-172C	5077A	3145			
WL-172D	5077A	3145			
WL-177	5077A	3145			
WL-177A	5077A	3145			
WL-177B	5077A	3145			
WL-177C	5077A	3145			
WL-177D	5077A	3145			
WL-182	5077A	3752			
WL-182A	5077A	3752			
WL-182B	5077A	3752			
WL-182C	5077A	3752			
WL-182D	5077A	3752			
WL-230	5081A	3151			
WL-230A	5081A	3151			
WL-230B	5081A	3151			
WL-230C	5081A	3151			
WL-230D	5081A	3151			
WL-240	5081A	3151			
WL-240A	5081A	3151			
WL-240B	5081A	3151			
WL-240C	5081A	3151			
WL-240D	5081A	3151			
WL-250	5081A	3151			
WL-250A	5081A	3151			
WL-250B	5081A	3151			
WL-250C	5081A	3151			
WL-250D	5081A	3151			
WL-260	146A	3148			
WL-260A	146A	3148			
WL-260B	146A	3148			
WL-260C	146A	3148			
WL-260D	146A	3148			
WL-270	146A	3064			
WL-270A	146A	3064			
WL-270B	146A	3064			
WL-270C	146A	3064			
WL-270D	146A	3064			
WL-290	5084A	3150			
WL-290A	5084A	3150			
WL-290B	5084A	3150			
WL-290C	5084A	3150			
WL-290D	5084A	3150			
WL-300	5084A	3755			
WL-300A	5084A	3755			
WL-300B	5084A	3755			
WL-300C	5084A	3755			
WL-300D	5084A	3755			
WL-330	147A	3095			
WL-330A	147A	3095			
WL-330B	147A	3095			
WL-330C	147A	3095			
WL-330D	147A	3095			
WL-340	147A	3095			
WL-340A	147A	3095			
WL-340B	147A	3095			
WL-340C	147A	3095			
WL-340D	147A	3095			
WL02M					MDA102A
WL04M					MDA104A
WL06M		3676			MDA106A
WL08M		3677			MDA108A
WM-050	135A	3056			
WM-050A	135A	3056			
WM-050B	135A	3056			
WM-050C	135A	3056			
WM-050D	135A	3056			
WM-052	135A	3056			
WM-052A	135A	3056			
WM-052B	135A	3056			
WM-052C	135A	3056			
WM-052D	135A	3056			
WM-054	136A	3057		561	
WM-054A	136A	3057		561	
WM-054B	136A	3057		561	

Industry Standard No.	ECG	SK	GE	RS 276-	MOTOR.
WM-054C	136A	3057		561	
WM-054D	136A	3057		561	
WM-056	136A	3057		561	
WM-056A	136A	3057		561	
WM-056B	136A	3057		561	
WM-056C	136A	3057		561	
WM-056D	136A	3057		561	
WM-058	136A	3057		561	
WM-058A	136A	3057		561	
WM-058B	136A	3057		561	
WM-058C	136A	3057		561	
WM-058D	136A	3057		561	
WM-061	137A	3058		561	
WM-061A	137A	3058		561	
WM-061B	137A	3058		561	
WM-061C	137A	3058		561	
WM-061D	137A	3058		561	
WM-063	137A	3058		561	
WM-063A	137A	3058		561	
WM-063B	137A	3058		561	
WM-063C	137A	3058		561	
WM-063D	137A	3058		561	
WM-065	137A	3058		561	
WM-065A	137A	3058		561	
WM-065B	137A	3058		561	
WM-065C	137A	3058		561	
WM-065D	137A	3058		561	
WM-071	138A	3059			
WM-071A	138A	3059			
WM-071B	138A	3059			
WM-071C	138A	3059			
WM-071D	138A	3059			
WM-073	138A	3059			
WM-073A	138A	3059			
WM-073B	138A	3059			
WM-073C	138A	3059			
WM-073D	138A	3059			
WM-075	138A	3059			
WM-075A	138A	3059			
WM-075B	138A	3059			
WM-075C	138A	3059			
WM-075D	138A	3059			
WM-077	138A	3059			
WM-077A	138A	3059			
WM-077B	138A	3059			
WM-077C	138A	3059			
WM-077D	138A	3059			
WM-079	5072A	3136		562	
WM-079A	5072A	3136		562	
WM-079B	5072A	3136		562	
WM-079C	5072A	3136		562	
WM-079D	5072A	3136		562	
WM-081	5072A	3136		562	
WM-081A	5072A	3136		562	
WM-081B	5072A	3136		562	
WM-081C	5072A	3136		562	
WM-081D	5072A	3136		562	
WM-083	5072A	3136		562	
WM-083A	5072A	3136		562	
WM-083B	5072A	3136		562	
WM-083C	5072A	3136		562	
WM-083D	5072A	3136		562	
WM-085	5072A	3136		562	
WM-085A	5072A	3136		562	
WM-085B	5072A	3136		562	
WM-085C	5072A	3136		562	
WM-085D	5072A	3136		562	
WM-088	139A	3060		562	
WM-088A	139A	3060		562	
WM-088B	139A	3060		562	
WM-088C	139A	3060		562	
WM-088D	139A	3060		562	
WM-090	139A	3060		562	
WM-090A	139A	3060		562	
WM-090B	139A	3060		562	
WM-090C	139A	3060		562	
WM-090D	139A	3060		562	
WM-092	139A	3060		562	
WM-092A	139A	3060		562	
WM-092B	139A	3060		562	
WM-092C	139A	3060		562	
WM-092D	139A	3060		562	
WM-094A	139A	3060		562	
WM-094B	139A	3060		562	
WM-094C	139A	3060		562	
WM-094D	139A	3060		562	
WM-098	140A	3061		562	
WM-098A	140A	3061		562	
WM-098B	140A	3061		562	
WM-098C	140A	3061		562	
WM-098D	140A	3061		562	
WM-100	140A	3061		562	
WM-100A	140A	3061		562	
WM-100B	140A	3061		562	
WM-100C	140A	3061		562	
WM-100D	140A	3061		562	
WM-110	5074A	3139		563	
WM-110A	5074A	3139		563	
WM-110B	5074A	3139		563	
WM-110C	5074A	3139		563	
WM-110D	5074A	3139		563	
WM-115	141A	3092			
WM-115A	141A	3092			
WM-115B	141A	3092			
WM-115C	141A	3092			
WM-115D	5074A	3092/141A			
WM-120	142A	3062		563	
WM-120A	142A	3062		563	
WM-120B	142A	3062		563	
WM-120C	142A	3062		563	
WM-120D	142A	3062		563	
WM-125	143A	3093		563	
WM-125A	143A	3093		563	
WM-125B	143A	3093		563	
WM-125C	143A	3093		563	
WM-125D	143A	3093		563	
WM-130	143A	3750		563	
WM-130A	143A	3750		563	
WM-130B	143A	3750		563	
WM-130C	143A	3750		563	
WM-130D	143A	3750		563	
WM-140	144A	3094		564	
WM-140A	144A	3094		564	
WM-140B	144A	3094		564	
WM-140C	144A	3094		564	
WM-140D	144A	3094		564	
WM-150	145A	3063		564	
WM-150A	145A	3063		564	
WM-150B	145A	3063		564	
WM-150C	145A	3063		564	
WM-150D	145A	3063		564	
WM-157	5075A	3751		564	
WM-157A	5075A	3751		564	
WM-157B	5075A	3751		564	

Industry Standard No.	ECG	SK	GE	RS 276-	MOTOR.
WM-157C	5075A	3751		564	
WM-157D	5075A	3751		564	
WM-162	5075A	3142		564	
WM-162A	5075A	3142		564	
WM-162B	5075A	3142		564	
WM-162C	5075A	3142		564	
WM-162D	5075A	3142		564	
WM-167	5075A	3751		564	
WM-167A	5075A	3751		564	
WM-167B	5075A	3751		564	
WM-167C	5075A	3751		564	
WM-167D	5075A	3751		564	
WM-172	5077A	3145			
WM-172A	5077A	3145			
WM-172B	5077A	3145			
WM-172C	5077A	3145			
WM-172D	5077A	3145			
WM-177	5077A	3145			
WM-177A	5077A	3145			
WM-177B	5077A	3145			
WM-177C	5077A	3145			
WM-177D	5077A	3145			
WM-182	5077A	3752			
WM-182A	5077A	3752			
WM-182B	5077A	3752			
WM-182C	5077A	3752			
WM-182D	5077A	3752			
WM-230	5081A	3151			
WM-230A	5081A	3151			
WM-230B	5081A	3151			
WM-230C	5081A	3151			
WM-230D	5081A	3151			
WM-240	5081A	3151			
WM-240A	5081A	3151			
WM-240B	5081A	3151			
WM-240C	5081A	3151			
WM-240D	5081A	3151			
WM-250	5081A	3151			
WM-250A	5081A	3151			
WM-250B	5081A	3151			
WM-250C	5081A	3151			
WM-250D	5081A	3151			
WM-260	146A	3148			
WM-260A	146A	3148			
WM-260B	146A	3148			
WM-260C	146A	3148			
WM-260D	146A	3148			
WM-270	146A	3064			
WM-270A	146A	3064			
WM-270B	146A	3064			
WM-270C	146A	3064			
WM-270D	146A	3064			
WM-290	5084A	3150			
WM-290A	5084A	3150			
WM-290B	5084A	3150			
WM-290C	5084A	3150			
WM-290D	5084A	3150			
WM-300	5084A	3755			
WM-300A	5084A	3755			
WM-300B	5084A	3755			
WM-300C	5084A	3755			
WM-300D	5084A	3755			
WM-330	147A	3095			
WM-330A	147A	3095			
WM-330B	147A	3095			
WM-330C	147A	3095			
WM-330D	147A	3095			
WM-340	147A	3095			
WM-340A	147A	3095			
WM-340B	147A	3095			
WM-340C	147A	3095			
WM-340D	147A	3095			
WN-050	135A	3056			
WN-050A	135A	3056			
WN-050B	135A	3056			
WN-050C	135A	3056			
WN-050D	135A	3056			
WN-052	135A	3056			
WN-052A	135A	3056			
WN-052B	135A	3056			
WN-052C	135A	3056			
WN-052D	135A	3056			
WN-054	136A	3057		561	
WN-054A	136A	3057		561	
WN-054B	136A	3057		561	
WN-054C	136A	3057		561	
WN-054D	136A	3057		561	
WN-056	136A	3057		561	
WN-056A	136A	3057		561	
WN-056B	136A	3057		561	
WN-056C	136A	3057		561	
WN-056D	136A	3057		561	
WN-058	136A	3057		561	
WN-058A	136A	3057		561	
WN-058B	136A	3057		561	
WN-058C	136A	3057		561	
WN-058D	136A	3057		561	
WN-061	137A	3058		561	
WN-061A	137A	3058		561	
WN-061B	137A	3058		561	
WN-061C	137A	3058		561	
WN-061D	137A	3058		561	
WN-063	137A	3058		561	
WN-063A	137A	3058		561	
WN-063B	137A	3058		561	
WN-063C	137A	3058		561	
WN-063D	137A	3058		561	
WN-065	137A	3058		561	
WN-065A	137A	3058		561	
WN-065B	137A	3058		561	
WN-065C	137A	3058		561	
WN-065D	137A	3058		561	
WN-071	138A	3059			
WN-071A	138A	3059			
WN-071B	138A	3059			
WN-071C	138A	3059			
WN-071D	138A	3059			
WN-073	138A	3059			
WN-073A	138A	3059			
WN-073B	138A	3059			
WN-073C	138A	3059			
WN-073D	138A	3059			
WN-075	138A	3059			
WN-075A	138A	3059			
WN-075B	138A	3059			
WN-075C	138A	3059			
WN-075D	138A	3059			
WN-077	138A	3059			
WN-077A	138A	3059			
WN-077B	138A	3059			
WN-077C	138A	3059			
WN-077D	138A	3059			
WN-079	5072A	3136		562	
WN-079A	5072A	3136		562	

Industry Standard No.	ECG	SK	GE	RS 276-	MOTOR.
WN-079B	5072A	3136		562	
WN-079C	5072A	3136		562	
WN-079D	5072A	3136		562	
WN-081	5072A	3136		562	
WN-081A	5072A	3136		562	
WN-081B	5072A	3136		562	
WN-081C	5072A	3136		562	
WN-081D	5072A	3136		562	
WN-083	5072A	3136		562	
WN-083A	5072A	3136		562	
WN-083B	5072A	3136		562	
WN-083C	5072A	3136		562	
WN-083D	5072A	3136		562	
WN-085	5072A	3136		562	
WN-085A	5072A	3136		562	
WN-085B	5072A	3136		562	
WN-085C	5072A	3136		562	
WN-085D	5072A	3136		562	
WN-088	139A	3060		562	
WN-088A	139A	3060		562	
WN-088B	139A	3060		562	
WN-088C	139A	3060		562	
WN-088D	139A	3060		562	
WN-090	139A	3060		562	
WN-090A	139A	3060		562	
WN-090B	139A	3060		562	
WN-090C	139A	3060		562	
WN-090D	139A	3060		562	
WN-092	139A	3060		562	
WN-092A	139A	3060		562	
WN-092B	139A	3060		562	
WN-092C	139A	3060		562	
WN-092D	139A	3060		562	
WN-094	139A	3060		562	
WN-094A	139A	3060		562	
WN-094B	139A	3060		562	
WN-094C	139A	3060		562	
WN-094D	139A	3060		562	
WN-098	140A	3061		562	
WN-098A	140A	3061		562	
WN-098B	140A	3061		562	
WN-098C	140A	3061		562	
WN-098D	140A	3061		562	
WN-100	140A	3061		562	
WN-100A	140A	3061		562	
WN-100B	140A	3061		562	
WN-100C	140A	3061		562	
WN-100D	140A	3061		562	
WN-110	5074A	3139		563	
WN-110A	5074A	3139		563	
WN-110B	5074A	3139		563	
WN-110C	5074A	3139		563	
WN-110D	5074A	3139		563	
WN-115	141A	3092			
WN-115A	141A	3092			
WN-115B	141A	3092			
WN-115C	141A	3092			
WN-115D	141A	3092			
WN-120	142A	3062		563	
WN-120A	142A	3062		563	
WN-120B	142A	3062		563	
WN-120C	142A	3062		563	
WN-120D	142A	3062		563	
WN-125	143A	3093		563	
WN-125A	143A	3093		563	
WN-125B	143A	3093		563	
WN-125C	143A	3093		563	
WN-125D	143A	3093		563	
WN-130	143A	3750		563	
WN-130A	143A	3750		563	
WN-130B	143A	3750		563	
WN-130C	143A	3750		563	
WN-130D	143A	3750		563	
WN-140	144A	3094		564	
WN-140A	144A	3094		564	
WN-140B	144A	3094		564	
WN-140C	144A	3094		564	
WN-140D	144A	3094		564	
WN-150	145A	3063		564	
WN-150A	145A	3063		564	
WN-150B	145A	3063		564	
WN-150C	145A	3063		564	
WN-150D	145A	3063		564	
WN-157	5075A	3751		564	
WN-157A	5075A	3751		564	
WN-157B	5075A	3751		564	
WN-157C	5075A	3751		564	
WN-157D	5075A	3751		564	
WN-162	5075A	3142		564	
WN-162A	5075A	3142		564	
WN-162B	5075A	3142		564	
WN-162C	5075A	3142		564	
WN-162D	5075A	3142		564	
WN-167	5075A	3751		564	
WN-167A	5075A	3751		564	
WN-167B	5075A	3751		564	
WN-167C	5075A	3751		564	
WN-167D	5075A	3751		564	
WN-170D				564	
WN-177	5077A	3145			
WN-177A	5077A	3145			
WN-177B	5077A	3145			
WN-177C	5077A	3145			
WN-177D	5077A	3145			
WN-182	5077A	3752			
WN-182A	5077A	3752			
WN-182B	5077A	3752			
WN-182C	5077A	3752			
WN-182D	5077A	3752			
WN-230	5081A	3151			
WN-230A	5081A	3151			
WN-230B	5081A	3151			
WN-230C	5081A	3151			
WN-230D	5081A	3151			
WN-240	5081A	3151			
WN-240A	5081A	3151			
WN-240B	5081A	3151			
WN-240C	5081A	3151			
WN-240D	5081A	3151			
WN-250	5081A	3151			
WN-250A	5081A	3151			
WN-250B	5081A	3151			
WN-250C	5081A	3151			
WN-250D	5081A	3151			
WN-260	146A	3148			
WN-260A	146A	3148			
WN-260B	146A	3148			
WN-260C	146A	3148			
WN-260D	146A	3148			
WN-270	146A	3064			
WN-270A	146A	3064			
WN-270B	146A	3064			
WN-270C	146A	3064			
WN-270D	146A	3064			
WN-290	5084A	3150			
WN-290A	5084A	3150			
WN-290B	5084A	3150			
WN-290C	5084A	3150			
WN-290D	5084A	3150			
WN-300	5084A	3755			
WN-300A	5084A	3755			
WN-300B	5084A	3755			
WN-300C	5084A	3755			
WN-300D	5084A	3755			
WN-330	147A	3095			
WN-330A	147A	3095			
WN-330B	147A	3095			
WN-330C	147A	3095			
WN-330D	147A	3095			
WN-340	147A	3095			
WN-340A	147A	3095			
WN-340B	147A	3095			
WN-340C	147A	3095			
WN-340D	147A	3095			
WO-050	135A	3056			
WO-050A	135A	3056			
WO-050B	135A	3056			
WO-050C	135A	3056			
WO-050D	135A	3056			
WO-052	135A	3056			
WO-052A	135A	3056			
WO-052B	135A	3056			
WO-052C	135A	3056			
WO-052D	135A	3056			
WO-054	136A	3057		561	
WO-054A	136A	3057		561	
WO-054B	136A	3057		561	
WO-054C	136A	3057		561	
WO-054D	136A	3057		561	
WO-056	136A	3057		561	
WO-056A	136A	3057		561	
WO-056B	136A	3057		561	
WO-056C	136A	3057		561	
WO-056D	136A	3057		561	
WO-058	136A	3057		561	
WO-058A	136A	3057		561	
WO-058B	136A	3057		561	
WO-058C	136A	3057		561	
WO-058D	136A	3057		561	
WO-061	137A	3058		561	
WO-061A	137A	3058		561	
WO-061B	137A	3058		561	
WO-061C	137A	3058		561	
WO-061D	137A	3058		561	
WO-063	137A	3058		561	
WO-063A	137A	3058		561	
WO-063B	137A	3058		561	
WO-063C	137A	3058		561	
WO-063D	137A	3058		561	
WO-065	137A	3058		561	
WO-065A	137A	3058		561	
WO-065B	137A	3058		561	
WO-065C	137A	3058		561	
WO-065D	137A	3058		561	
WO-071	138A	3059			
WO-071A	138A	3059			
WO-071B	138A	3059			
WO-071C	138A	3059			
WO-071D	138A	3059			
WO-073	138A	3059			
WO-073A	138A	3059			
WO-073B	138A	3059			
WO-073C	138A	3059			
WO-073D	138A	3059			
WO-075	138A	3059			
WO-075A	138A	3059			
WO-075B	138A	3059			
WO-075C	138A	3059			
WO-075D	138A	3059			
WO-077	138A	3059			
WO-077A	138A	3059			
WO-077B	138A	3059			
WO-077C	138A	3059			
WO-077D	138A	3059			
WO-079	5072A	3136		562	
WO-079A	5072A	3136		562	
WO-079B	5072A	3136		562	
WO-079C	5072A	3136		562	
WO-079D	5072A	3136		562	
WO-081	5072A	3136		562	
WO-081A	5072A	3136		562	
WO-081B	5072A	3136		562	
WO-081C	5072A	3136		562	
WO-081D	5072A	3136		562	
WO-083	5072A	3136		562	
WO-083A	5072A	3136		562	
WO-083B	5072A	3136		562	
WO-083C	5072A	3136		562	
WO-083D	5072A	3136		562	
WO-085	5072A	3136		562	
WO-085A	5072A	3136		562	
WO-085B	5072A	3136		562	
WO-085C	5072A	3136		562	
WO-085D	5072A	3136		562	
WO-088	139A	3060		562	
WO-088A	139A	3060		562	
WO-088B	139A	3060		562	
WO-088C	139A	3060		562	
WO-088D	139A	3060		562	
WO-090	139A	3060		562	
WO-090A	139A	3060		562	
WO-090B	139A	3060		562	
WO-090C	139A	3060		562	
WO-090D	139A	3060		562	
WO-092	139A	3060		562	
WO-092A	139A	3060		562	
WO-092B	139A	3060		562	
WO-092C	139A	3060		562	
WO-092D	139A	3060		562	
WO-094	139A	3060		562	
WO-094A	139A	3060		562	
WO-094B	139A	3060		562	
WO-094C	139A	3060		562	
WO-094D	139A	3060		562	
WO-098	140A	3061		562	
WO-098A	140A	3061		562	
WO-098B	140A	3061		562	
WO-098C	140A	3061		562	
WO-098D	140A	3061		562	
WO-100A	140A	3061		562	
WO-100B	140A	3061		562	
WO-100C	140A	3061		562	
WO-100D	140A	3061		562	
WO-110	5074A	3139		563	
WO-110A	5074A	3139		563	
WO-110B	5074A	3139		563	
WO-110D	5074A	3139		563	
WO-115	141A	3092			

Industry Standard No.	ECG	SK	GE	RS 276-	MOTOR.
WO-115A	141A	3092			
WO-115B	141A	3092			
WO-115C	141A	3092			
WO-120	142A	3062		563	
WO-120A	142A	3062		563	
WO-120B	142A	3062		563	
WO-120C	142A	3062		563	
WO-120D	142A	3062		563	
WO-125	143A	3093		563	
WO-125A	143A	3093		563	
WO-125B	143A	3093		563	
WO-125C	143A	3093		563	
WO-125D	143A	3093		563	
WO-130A	143A	3750		563	
WO-130B	143A	3750		563	
WO-130C	143A	3750		563	
WO-130D	143A	3750		563	
WO-140	144A	3094		564	
WO-140A	144A	3094		564	
WO-140B	144A	3094		564	
WO-140C	144A	3094		564	
WO-140D	144A	3094		564	
WO-150	145A	3063		564	
WO-150A	145A	3063		564	
WO-150B	145A	3063		564	
WO-150C	145A	3063		564	
WO-150D	145A	3063		564	
WO-157	5075A	3751		564	
WO-157A	5075A	3751		564	
WO-157B	5075A	3751		564	
WO-157C	5075A	3751		564	
WO-157D	5075A	3751		564	
WO-162	5075A	3142		564	
WO-162A	5075A	3142		564	
WO-162B	5075A	3142		564	
WO-162C	5075A	3142		564	
WO-162D	5075A	3142		564	
WO-167	5075A	3751		564	
WO-167A	5075A	3751		564	
WO-167B	5075A	3751		564	
WO-167C	5075A	3751		564	
WO-167D	5075A	3751		564	
WO-172	5077A	3145			
WO-172A	5077A	3145			
WO-172B	5077A	3145			
WO-172C	5077A	3145			
WO-172D	5077A	3145			
WO-177	5077A	3145			
WO-177A	5077A	3145			
WO-177B	5077A	3145			
WO-177C	5077A	3145			
WO-177D	5077A	3145			
WO-182	5077A	3752			
WO-182A	5077A	3752			
WO-182B	5077A	3752			
WO-182C	5077A	3752			
WO-182D	5077A	3752			
WO-230	5081A	3151			
WO-230A	5081A	3151			
WO-230B	5081A	3151			
WO-230C	5081A	3151			
WO-230D	5081A	3151			
WO-240	5081A	3151			
WO-240A	5081A	3151			
WO-240B	5081A	3151			
WO-240C	5081A	3151			
WO-240D	5081A	3151			
WO-250	5081A	3151			
WO-250A	5081A	3151			
WO-250B	5081A	3151			
WO-250C	5081A	3151			
WO-250D	5081A	3151			
WO-260	146A	3148			
WO-260A	146A	3148			
WO-260B	146A	3148			
WO-260C	146A	3148			
WO-260D	146A	3148			
WO-270	146A	3064			
WO-270A	146A	3064			
WO-270B	146A	3064			
WO-270C	146A	3064			
WO-270D	146A	3064			
WO-290	5084A	3150			
WO-290A	5084A	3150			
WO-290B	5084A	3150			
WO-290C	5084A	3150			
WO-290D	5084A	3150			
WO-300	5084A	3755			
WO-300A	5084A	3755			
WO-300B	5084A	3755			
WO-300C	5084A	3755			
WO-300D	5084A	3755			
WO-330	147A	3095			
WO-330A	147A	3095			
WO-330B	147A	3095			
WO-330C	147A	3095			
WO-330D	147A	3095			
WO-340	147A	3095			
WO-340A	147A	3095			
WO-340B	147A	3095			
WO-340C	147A	3095			
WO-340D	147A	3095			
WO-61	137A			561	
WO-6A	116	3031A		1104	
WO110C	5074A			563	
WO130	143A			563	
WO6A	125			1104	
WO6B	116			1104	
WO6C				1102	
WP-050	135A	3056			
WP-050A	135A	3056			
WP-050B	135A	3056			
WP-050C	135A	3056			
WP-050D	135A	3056			
WP-052	135A	3056			
WP-052A	135A	3056			
WP-052B	135A	3056			
WP-052C	135A	3056			
WP-052D	135A	3056		561	
WP-054	136A	3057		561	
WP-054A	136A	3057		561	
WP-054B	136A	3057		561	
WP-054C	136A	3057		561	
WP-054D	136A	3057		561	
WP-056	136A	3057		561	
WP-056A	136A	3057		561	
WP-056B	136A	3057		561	
WP-056C	136A	3057		561	
WP-056D	136A	3057		561	
WP-058	136A	3057		561	
WP-058A	136A	3057		561	
WP-058B	136A	3057		561	
WP-058C	136A	3057		561	
WP-058D	136A	3057		561	
WP-061	137A	3058		561	
WP-061A	137A	3058		561	
WP-061B	137A	3058		561	
WP-061C	137A	3058		561	
WP-061D	137A	3058		561	
WP-063	137A	3058		561	
WP-063A	137A	3058		561	
WP-063B	137A	3058		561	
WP-063C	137A	3058		561	
WP-063D	137A	3058		561	
WP-065	137A	3058		561	
WP-065A	137A	3058		561	
WP-065B	137A	3058		561	
WP-065C	137A	3058		561	
WP-065D	137A	3058		561	
WP-071	138A	3059			
WP-071A	138A	3059			
WP-071B	138A	3059			
WP-071C	138A	3059			
WP-071D	138A	3059			
WP-073	138A	3059			
WP-073A	138A	3059			
WP-073B	138A	3059			
WP-073C	138A	3059			
WP-073D	138A	3059			
WP-075	138A	3059			
WP-075A	138A	3059			
WP-075B	138A	3059			
WP-075C	138A	3059			
WP-075D	138A	3059			
WP-077	138A	3059			
WP-077A	138A	3059			
WP-077B	138A	3059			
WP-077C	138A	3059			
WP-077D	138A	3059			
WP-079	5072A	3136		562	
WP-079A	5072A	3136		562	
WP-079B	5072A	3136		562	
WP-079C	5072A	3136		562	
WP-079D	5072A	3136		562	
WP-081	5072A	3136		562	
WP-081A	5072A	3136		562	
WP-081B	5072A	3136		562	
WP-081C	5072A	3136		562	
WP-081D	5072A	3136		562	
WP-083	5072A	3136		562	
WP-083A	5072A	3136		562	
WP-083B	5072A	3136		562	
WP-083C	5072A	3136		562	
WP-083D	5072A	3136		562	
WP-085	5072A	3136		562	
WP-085A	5072A	3136		562	
WP-085B	5072A	3136		562	
WP-085C	5072A	3136		562	
WP-085D	5072A	3136		562	
WP-088	139A	3060		562	
WP-088A	139A	3060		562	
WP-088B	139A	3060		562	
WP-088C	139A	3060		562	
WP-088D	139A	3060		562	
WP-090	139A	3060		562	
WP-090A	139A	3060		562	
WP-090B	139A	3060		562	
WP-090C	139A	3060		562	
WP-090D	139A	3060		562	
WP-092	139A	3060		562	
WP-092A	139A	3060		562	
WP-092B	139A	3060		562	
WP-092C	139A	3060		562	
WP-092D	139A	3060		562	
WP-094	139A	3060		562	
WP-094A	139A	3060		562	
WP-094B	139A	3060		562	
WP-094C	139A	3060		562	
WP-094D	139A	3060		562	
WP-098	140A	3061		562	
WP-098A	140A	3061		562	
WP-098B	140A	3061		562	
WP-098C	140A	3061		562	
WP-098D	140A	3061		562	
WP-100	140A	3061		562	
WP-100A	140A	3061		562	
WP-100B	140A	3061		562	
WP-100C	140A	3061		562	
WP-100D	140A	3061		562	
WP-110	5074A	3139		563	
WP-110A	5074A	3139		563	
WP-110B	5074A	3139		563	
WP-110C	5074A	3139		563	
WP-110D	5074A	3139		563	
WP-115	141A	3092			
WP-115A	141A	3092			
WP-115B	141A	3092			
WP-115C	141A	3092			
WP-115D	141A	3092			
WP-120	142A	3062		563	
WP-120A	142A	3062		563	
WP-120B	142A	3062		563	
WP-120C	142A	3062		563	
WP-120D	142A	3062		563	
WP-125	143A	3093		563	
WP-125A	143A	3093		563	
WP-125B	143A	3093		563	
WP-125C	143A	3093		563	
WP-125D	143A	3093		563	
WP-130	143A	3750		563	
WP-130A	143A	3750		563	
WP-130B	143A	3750		563	
WP-130C	143A	3750		563	
WP-130D	143A	3750		563	
WP-140	144A	3094		564	
WP-140A	144A	3094		564	
WP-140B	144A	3094		564	
WP-140C	144A	3094		564	
WP-140D	144A	3094		564	
WP-150	145A	3063		564	
WP-150A	145A	3063		564	
WP-150B	145A	3063		564	
WP-150C	145A	3063		564	
WP-150D	145A	3063		564	
WP-157	5075A	3751		564	
WP-157A	5075A	3751		564	
WP-157B	5075A	3751		564	
WP-157C	5075A	3751		564	
WP-157D	5075A	3751		564	
WP-162	5075A	3142		564	
WP-162A	5075A	3142		564	
WP-162B	5075A	3142		564	
WP-162C	5075A	3142		564	
WP-162D	5075A	3142		564	
WP-167	5075A	3751		564	
WP-167A	5075A	3751		564	
WP-167B	5075A	3751		564	
WP-167C	5075A	3751		564	

Industry Standard No.	ECG	SK	GE	RS 276-	MOTOR.
WP-167D	5075A	3751		564	
WP-172	5077A	3751/5075A			
WP-172A	5077A	3751/5075A			
WP-172B	5077A	3751/5075A			
WP-172C	5077A	3751/5075A			
WP-172D	5077A	3751/5075A			
WP-177	5077A	3145			
WP-177A	5077A	3145			
WP-177B	5077A	3145			
WP-177C	5077A	3145			
WP-177D	5077A	3145			
WP-182	5077A	3752			
WP-182A	5077A	3752			
WP-182B	5077A	3752			
WP-182C	5077A	3752			
WP-182D	5077A	3752			
WP-230	5081A	3151			
WP-230A	5081A	3151			
WP-230B	5081A	3151			
WP-230C	5081A	3151			
WP-230D	5081A	3151			
WP-240	5081A	3151			
WP-240A	5081A	3151			
WP-240B	5081A	3151			
WP-240C	5081A	3151			
WP-240D	5081A	3151			
WP-250	5081A	3151			
WP-250A	5081A	3151			
WP-250B	5081A	3151			
WP-250C	5081A	3151			
WP-250D	5081A	3151			
WP-260	146A	3148			
WP-260A	146A	3148			
WP-260B	146A	3148			
WP-260C	146A	3148			
WP-260D	146A	3148			
WP-270	146A	3064			
WP-270A	146A	3064			
WP-270B	146A	3064			
WP-270C	146A	3064			
WP-270D	146A	3064			
WP-290	5084A	3150			
WP-290A	5084A	3150			
WP-290B	5084A	3150			
WP-290C	5084A	3150			
WP-290D	5084A	3150			
WP-300	5084A	3755			
WP-300A	5084A	3755			
WP-300B	5084A	3755			
WP-300C	5084A	3755			
WP-300D	5084A	3755			
WP-330	147A	3095			
WP-330A	147A	3095			
WP-330B	147A	3095			
WP-330C	147A	3095			
WP-330D	147A	3095			
WP-340	147A	3095			
WP-340A	147A	3095			
WP-340B	147A	3095			
WP-340C	147A	3095			
WP-340D	147A	3095			
WQ-050	135A	3056			
WQ-050A	135A	3056			
WQ-050B	135A	3056			
WQ-050C	135A	3056			
WQ-050D	135A	3056			
WQ-052	135A	3056			
WQ-052A	135A	3056			
WQ-052B	135A	3056			
WQ-052C	135A	3056			
WQ-052D	135A	3056			
WQ-054	136A	3057		561	
WQ-054A	136A	3057		561	
WQ-054B	136A	3057		561	
WQ-054C	136A	3057		561	
WQ-054D	136A	3057		561	
WQ-056	136A	3057		561	
WQ-056A	136A	3057		561	
WQ-056B	136A	3057		561	
WQ-056C	136A	3057		561	
WQ-056D	136A	3057		561	
WQ-058	136A	3057		561	
WQ-058A	136A	3057		561	
WQ-058B	136A	3057		561	
WQ-058C	136A	3057		561	
WQ-058D	136A	3057		561	
WQ-061	137A	3058		561	
WQ-061A	137A	3058		561	
WQ-061B	137A	3058		561	
WQ-061C	137A	3058		561	
WQ-061D	137A	3058		561	
WQ-063	137A	3058		561	
WQ-063A	137A	3058		561	
WQ-063B	137A	3058		561	
WQ-063C	137A	3058		561	
WQ-063D	137A	3058		561	
WQ-065	137A	3058		561	
WQ-065A	137A	3058		561	
WQ-065B	137A	3058		561	
WQ-065C	137A	3058		561	
WQ-065D	137A	3058		561	
WQ-071	138A	3059			
WQ-071A	138A	3059			
WQ-071B	138A	3059			
WQ-071C	138A	3059			
WQ-071D	138A	3059			
WQ-073	138A	3059			
WQ-073A	138A	3059			
WQ-073B	138A	3059			
WQ-073C	138A	3059			
WQ-073D	138A	3059			
WQ-075	138A	3059			
WQ-075A	138A	3059			
WQ-075B	138A	3059			
WQ-075C	138A	3059			
WQ-075D	138A	3059			
WQ-077	138A	3059			
WQ-077A	138A	3059			
WQ-077B	138A	3059			
WQ-077C	138A	3059			
WQ-077D	138A	3059			
WQ-079	5072A	3136		562	
WQ-079A	5072A	3136		562	
WQ-079B	5072A	3136		562	
WQ-079C	5072A	3136		562	
WQ-079D	5072A	3136		562	
WQ-081	5072A	3136		562	
WQ-081A	5072A	3136		562	
WQ-081B	5072A	3136		562	
WQ-081C	5072A	3136		562	
WQ-081D	5072A	3136		562	
WQ-083	5072A	3136		562	
WQ-083A	5072A	3136		562	
WQ-083B	5072A	3136		562	
WQ-083C	5072A	3136		562	
WQ-083D	5072A	3136		562	
WQ-085	5072A	3136		562	
WQ-085A	5072A	3136		562	
WQ-085B	5072A	3136		562	
WQ-085C	5072A	3136		562	
WQ-085D	5072A	3136		562	
WQ-088	139A	3060		562	
WQ-088A	139A	3060		562	
WQ-088B	139A	3060		562	
WQ-088C	139A	3060		562	
WQ-088D	139A	3060		562	
WQ-090	139A	3060		562	
WQ-090A	139A	3060		562	
WQ-090B	139A	3060		562	
WQ-090C	139A	3060		562	
WQ-090D	139A	3060		562	
WQ-092	139A	3060		562	
WQ-092A	139A	3060		562	
WQ-092B	139A	3060		562	
WQ-092C	139A	3060		562	
WQ-092D	139A	3060		562	
WQ-094	139A	3060		562	
WQ-094A	139A	3060		562	
WQ-094B	139A	3060		562	
WQ-094C	139A	3060		562	
WQ-094D	139A	3060		562	
WQ-098	140A	3061		562	
WQ-098A	140A	3061		562	
WQ-098B	140A	3061		562	
WQ-098C	140A	3061		562	
WQ-098D	140A	3061		562	
WQ-100	140A	3061		562	
WQ-100A	140A	3061		562	
WQ-100B	140A	3061		562	
WQ-100C	140A	3061		562	
WQ-100D	140A	3061		562	
WQ-110	5074A	3139		563	
WQ-110A	5074A	3139		563	
WQ-110B	5074A	3139		563	
WQ-110C	5074A	3139		563	
WQ-110D	5074A	3139		563	
WQ-115	141A	3092			
WQ-115A	141A	3092			
WQ-115B	141A	3092			
WQ-115C	141A	3092			
WQ-115D	141A	3092			
WQ-120	142A	3062		563	
WQ-120A	142A	3062		563	
WQ-120B	142A	3062		563	
WQ-120C	142A	3062		563	
WQ-120D	142A	3062		563	
WQ-125	143A	3093		563	
WQ-125A	143A	3093		563	
WQ-125B	143A	3093		563	
WQ-125C	143A	3093		563	
WQ-125D	143A	3093		563	
WQ-130	143A	3750		563	
WQ-130A	143A	3750		563	
WQ-130B	143A	3750		563	
WQ-130C	143A	3750		563	
WQ-130D	143A	3750		563	
WQ-140	144A	3094		564	
WQ-140A	144A	3094		564	
WQ-140B	144A	3094		564	
WQ-140C	144A	3094		564	
WQ-140D	144A	3094		564	
WQ-150	145A	3063		564	
WQ-150A	145A	3063		564	
WQ-150B	145A	3063		564	
WQ-150C	145A	3063		564	
WQ-150D	145A	3063		564	
WQ-157	5075A	3751		564	
WQ-157A	5075A	3751		564	
WQ-157B	5075A	3751		564	
WQ-157C	5075A	3751		564	
WQ-157D	5075A	3751		564	
WQ-162	5075A	3142		564	
WQ-162A	5075A	3142		564	
WQ-162B	5075A	3142		564	
WQ-162C	5075A	3142		564	
WQ-162D	5075A	3142		564	
WQ-167	5075A	3751		564	
WQ-167A	5075A	3751		564	
WQ-167B	5075A	3751		564	
WQ-167C	5075A	3751		564	
WQ-167D	5075A	3751		564	
WQ-172	5077A	3145			
WQ-172A	5077A	3145			
WQ-172B	5077A	3145			
WQ-172C	5077A	3145			
WQ-172D	5077A	3145			
WQ-177	5077A	3145			
WQ-177A	5077A	3145			
WQ-177B	5077A	3145			
WQ-177C	5077A	3145			
WQ-177D	5077A	3145			
WQ-182	5077A	3752			
WQ-182A	5077A	3752			
WQ-182B	5077A	3752			
WQ-182C	5077A	3752			
WQ-182D	5077A	3752			
WQ-230	5081A	3151			
WQ-230A	5081A	3151			
WQ-230B	5081A	3151			
WQ-230C	5081A	3151			
WQ-230D	5081A	3151			
WQ-240	5081A	3151			
WQ-240A	5081A	3151			
WQ-240B	5081A	3151			
WQ-240C	5081A	3151			
WQ-240D	5081A	3151			
WQ-250	5081A	3151			
WQ-250A	5081A	3151			
WQ-250B	5081A	3151			
WQ-250C	5081A	3151			
WQ-250D	5081A	3151			
WQ-260	146A	3148			
WQ-260A	146A	3148			
WQ-260B	146A	3148			
WQ-260C	146A	3148			
WQ-260D	146A	3148			
WQ-270	146A	3064			
WQ-270A	146A	3064			
WQ-270B	146A	3064			
WQ-270C	146A	3064			
WQ-270D	146A	3064			
WQ-290	5084A	3150			
WQ-290A	5084A	3150			
WQ-290B	5084A	3150			
WQ-290C	5084A	3150			
WQ-290D	5084A	3150			
WQ-300	5084A	3755			
WQ-300A	5084A	3755			

Industry Standard No.	ECG	SK	GE	RS 276-	MOTOR.
WQ-300B	5084A	3755			
WQ-300C	5084A	3755			
WQ-300D	5084A	3755			
WQ-330	147A	3095			
WQ-330A	147A	3095			
WQ-330B	147A	3095			
WQ-330C	147A	3095			
WQ-330D	147A	3095			
WQ-340	147A	3095			
WQ-340A	147A	3095			
WQ-340B	147A	3095			
WQ-340C	147A	3095			
WQ-340D	147A	3095			
WR-006	5800	9003		1142	
WR-013	116			1104	
WR-030	5800			1142	
WR-040	5800			1142	
WR-050	135A	3056			
WR-050A	135A	3056			
WR-050B	135A	3056			
WR-050C	135A	3056			
WR-050D	135A	3056			
WR-052	135A	3056			
WR-052A	135A	3056			
WR-052B	135A	3056			
WR-052C	135A	3056			
WR-052D	135A	3056			
WR-054	136A	3057		561	
WR-054A	136A	3057		561	
WR-054B	136A	3057		561	
WR-054C	136A	3057		561	
WR-054D	136A	3057		561	
WR-056	136A	3057		561	
WR-056A	136A	3057		561	
WR-056B	136A	3057		561	
WR-056C	136A	3057		561	
WR-056D	136A	3057		561	
WR-058	136A	3057		561	
WR-058A	136A	3057		561	
WR-058B	136A	3057		561	
WR-058C	136A	3057		561	
WR-058D	136A	3057		561	
WR-061	137A	3058		561	
WR-061A	137A	3058		561	
WR-061B	137A	3058		561	
WR-061C	137A	3058		561	
WR-061D	137A	3058		561	
WR-063	137A	3058		561	
WR-063A	137A	3058		561	
WR-063B	137A	3058		561	
WR-063C	137A	3058		561	
WR-063D	137A	3058		561	
WR-065	137A	3058		561	
WR-065A	137A	3058		561	
WR-065B	137A	3058		561	
WR-065C	137A	3058		561	
WR-065D	137A	3058		561	
WR-071	138A	3059			
WR-071A	138A	3059			
WR-071B	138A	3059			
WR-071C	138A	3059			
WR-071D	138A	3059			
WR-073	138A	3059			
WR-073A	138A	3059			
WR-073B	138A	3059			
WR-073C	138A	3059			
WR-073D	138A	3059			
WR-075	138A	3059			
WR-075A	138A	3059			
WR-075B	138A	3059			
WR-075C	138A	3059			
WR-075D	138A	3059			
WR-077	138A	3059			
WR-077A	138A	3059			
WR-077B	138A	3059			
WR-077C	138A	3059			
WR-077D	138A	3059			
WR-079	5072A	3136		562	
WR-079A	5072A	3136		562	
WR-079B	5072A	3136		562	
WR-079C	5072A	3136		562	
WR-079D	5072A	3136		562	
WR-081	5072A	3136		562	
WR-081A	5072A	3136		562	
WR-081B	5072A	3136		562	
WR-081C	5072A	3136		562	
WR-081D	5072A	3136		562	
WR-083	5072A	3136		562	
WR-083A	5072A	3136		562	
WR-083B	5072A	3136		562	
WR-083C	5072A	3136		562	
WR-083D	5072A	3136		562	
WR-085	5072A	3136		562	
WR-085A	5072A	3136		562	
WR-085B	5072A	3136		562	
WR-085C	5072A	3136		562	
WR-085D	5072A	3136		562	
WR-088	139A	3060		562	
WR-088A	139A	3060		562	
WR-088B	139A	3060		562	
WR-088C	139A	3060		562	
WR-088D	139A	3060		562	
WR-090	139A	3060		562	
WR-090A	139A	3060		562	
WR-090B	139A	3060		562	
WR-090C	139A	3060		562	
WR-090D	139A	3060		562	
WR-092	139A	3060		562	
WR-092A	139A	3060		562	
WR-092B	139A	3060		562	
WR-092C	139A	3060		562	
WR-092D	139A	3060		562	
WR-094	139A	3060		562	
WR-094A	139A	3060		562	
WR-094B	139A	3060		562	
WR-094C	139A	3060		562	
WR-094D	139A	3060		562	
WR-098	140A	3061		562	
WR-098A	140A	3061		562	
WR-098B	140A	3061		562	
WR-098C	140A	3061		562	
WR-098D	140A	3061		562	
WR-100	140A	3061		562	
WR-100A	140A	3061		562	
WR-100B	140A	3061		562	
WR-100C	140A	3061		562	
WR-100D	140A	3061		562	
WR-110	5074A	3139		563	
WR-110A	5074A	3139		563	
WR-110B	5074A	3139		563	
WR-110C	5074A	3139		563	
WR-110D	5074A	3139		563	
WR-115	141A	3092			
WR-115A	141A	3092			
WR-115B	141A	3092			
WR-115C	141A	3092			
WR-115D	141A	3092			
WR-120	142A	3062		563	
WR-120A	142A	3062		563	
WR-120B	142A	3062		563	
WR-120C	142A	3062		563	
WR-120D	142A	3062		563	
WR-125	143A	3093		563	
WR-125A	143A	3093		563	
WR-125B	143A	3093		563	
WR-125C	143A	3093		563	
WR-125D	143A	3093		563	
WR-130	143A	3750		563	
WR-130A	143A	3750		563	
WR-130B	143A	3750		563	
WR-130C	143A	3750		563	
WR-130D	143A	3750		563	
WR-140	144A	3094		564	
WR-140A	144A	3094		564	
WR-140B	144A	3094		564	
WR-140C	144A	3094		564	
WR-140D	144A	3094		564	
WR-150	145A	3063		564	
WR-150A	145A	3063		564	
WR-150B	145A	3063		564	
WR-150C	145A	3063		564	
WR-150D	145A	3063		564	
WR-157	5075A	3751		564	
WR-157A	5075A	3751		564	
WR-157B	5075A	3751		564	
WR-157C	5075A	3751		564	
WR-157D	5075A	3751		564	
WR-162	5075A	3142		564	
WR-162A	5075A	3142		564	
WR-162B	5075A	3142		564	
WR-162C	5075A	3142		564	
WR-162D	5075A	3142		564	
WR-167	5075A	3142		564	
WR-167A	5075A	3142		564	
WR-167B	5075A	3142		564	
WR-167C	5075A	3142		564	
WR-167D	5075A	3142		564	
WR-172	5077A	3145			
WR-172A	5077A	3145			
WR-172B	5077A	3145			
WR-172C	5077A	3145			
WR-172D	5077A	3145			
WR-177	5077A	3145			
WR-177A	5077A	3145			
WR-177B	5077A	3145			
WR-177C	5077A	3145			
WR-177D	5077A	3145			
WR-182	5077A	3752			
WR-182A	5077A	3752			
WR-182B	5077A	3752			
WR-182C	5077A	3752			
WR-182D	5077A	3752			
WR-200	116	3016		1104	
WR-230	5081A	3151			
WR-230A	5081A	3151			
WR-230B	5081A	3151			
WR-230C	5081A	3151			
WR-230D	5081A	3151			
WR-240	5081A	3151			
WR-240A	5081A	3151			
WR-240B	5081A	3151			
WR-240C	5081A	3151			
WR-240D	5081A	3151			
WR-250	5081A	3151			
WR-250A	5081A	3151			
WR-250B	5081A	3151			
WR-250C	5081A	3151			
WR-250D	5081A	3151			
WR-260	146A	3148			
WR-260A	146A	3148			
WR-260B	146A	3148			
WR-260C	146A	3148			
WR-260D	146A	3148			
WR-270	146A	3064			
WR-270A	146A	3064			
WR-270B	146A	3064			
WR-270C	146A	3064			
WR-270D	146A	3064			
WR-290	5084A	3150			
WR-290A	5084A	3150			
WR-290B	5084A	3150			
WR-290C	5084A	3150			
WR-290D	5084A	3150			
WR-300	5084A	3755			
WR-300A	5084A	3755			
WR-300B	5084A	3755			
WR-300C	5084A	3755			
WR-300D	5084A	3755			
WR-330	147A	3095			
WR-330A	147A	3095			
WR-330B	147A	3095			
WR-330C	147A	3095			
WR-330D	147A	3095			
WR-340	147A	3095			
WR-340A	147A	3095			
WR-340B	147A	3095			
WR-340C	147A	3095			
WR-340D	147A	3095			
WR006	116			1104	
WR011	166	9075		1152	
WR013	116(2)			1104	
WR030	166	9075		1152	
WR040	166	9075		1152	
WR10	5492	6792			
WR100	116	3017B/117		1104	
WR2	156	3033			
WR200	116			1104	
WR3	5809	9010			
WR300	116	3016		1104	
WR400	116	3016		1104	
WR50	5980	3610			
WR51	5982	3609/5986			
WR52	5986	3609			
WR53	5990	3608			
WR54	5994	3501			
WR60	5980	3610			
WR61	5982	3609/5986			
WR62	5986	3609			
WR63	5990	3608			
WRE-981	116	3016		1104	
WRE981	116			1104	
WRR1952	123A	3444		2051	
WRR1953	123A	3444		2051	
WRR1954	123A	3444		2051	
WRR1955	5804	9007		1144	
WRR1956	5804	9007		1144	
WRT1114	102	3722		2007	
WS-050	135A	3056			

Industry Standard No.	ECG	SK	GE	RS 276-	MOTOR.
WS-050A	135A	3056			
WS-050B	135A	3056			
WS-050C	135A	3056			
WS-050D	135A	3056			
WS-052	135A	3056			
WS-052A	135A	3056			
WS-052B	135A	3056			
WS-052C	135A	3056			
WS-052D	135A	3056			
WS-054	136A	3057		561	
WS-054A	136A	3057		561	
WS-054B	136A	3057		561	
WS-054C	136A	3057		561	
WS-054D	136A	3057		561	
WS-056	136A	3057		561	
WS-056A	136A	3057		561	
WS-056B	136A	3057		561	
WS-056C	136A	3057		561	
WS-056D	136A	3057		561	
WS-058	136A	3057		561	
WS-058A	136A	3057		561	
WS-058B	136A	3057		561	
WS-058C	136A	3057		561	
WS-058D	136A	3057		561	
WS-061	137A	3058		561	
WS-061A	137A	3058		561	
WS-061B	137A	3058		561	
WS-061C	137A	3058		561	
WS-061D	137A	3058		561	
WS-063	137A	3058		561	
WS-063A	137A	3058		561	
WS-063B	137A	3058		561	
WS-063C	137A	3058		561	
WS-063D	137A	3058		561	
WS-065	137A	3058		561	
WS-065A	137A	3058		561	
WS-065B	137A	3058		561	
WS-065C	137A	3058		561	
WS-065D	137A	3058		561	
WS-071	138A	3059			
WS-071A	138A	3059			
WS-071B	138A	3059			
WS-071C	138A	3059			
WS-071D	138A	3059			
WS-073	138A	3059			
WS-073A	138A	3059			
WS-073B	138A	3059			
WS-073C	138A	3059			
WS-073D	138A	3059			
WS-075	138A	3059			
WS-075A	138A	3059			
WS-075B	138A	3059			
WS-075C	138A	3059			
WS-075D	138A	3059			
WS-077	138A	3059			
WS-077A	138A	3059			
WS-077B	138A	3059			
WS-077C	138A	3059			
WS-077D	138A	3059			
WS-079	5072A	3136		562	
WS-079A	5072A	3136		562	
WS-079B	5072A	3136		562	
WS-079C	5072A	3136		562	
WS-079D	5072A	3136		562	
WS-081	5072A	3136		562	
WS-081A	5072A	3136		562	
WS-081B	5072A	3136		562	
WS-081C	5072A	3136		562	
WS-081D	5072A	3136		562	
WS-083	5072A	3136		562	
WS-083A	5072A	3136		562	
WS-083B	5072A	3136		562	
WS-083C	5072A	3136		562	
WS-083D	5072A	3136		562	
WS-085	5072A	3136		562	
WS-085A	5072A	3136		562	
WS-085B	5072A	3136		562	
WS-085C	5072A	3136		562	
WS-085D	5072A	3136		562	
WS-088	139A	3060		562	
WS-088A	139A	3060		562	
WS-088B	139A	3060		562	
WS-088C	139A	3060		562	
WS-088D	139A	3060		562	
WS-090	139A	3060		562	
WS-090A	139A	3060		562	
WS-090B	139A	3060		562	
WS-090C	139A	3060		562	
WS-090D	139A	3060		562	
WS-092	139A	3060		562	
WS-092A	139A	3060		562	
WS-092B		3060		562	
WS-092C	139A	3060		562	
WS-092D	139A	3060		562	
WS-094	139A	3060		562	
WS-094A	139A	3060		562	
WS-094B	139A	3060		562	
WS-094C	139A	3060		562	
WS-094D	139A	3060		562	
WS-098	140A	3061		562	
WS-098A	140A	3061		562	
WS-098B	140A	3061		562	
WS-098C	140A	3061		562	
WS-098D	140A	3061		562	
WS-100	140A	3061		562	
WS-100A	140A	3061		562	
WS-100B	140A	3061		562	
WS-100C	140A	3061		562	
WS-100D	140A	3061		562	
WS-110	5074A	3139		563	
WS-110A	5074A	3139		563	
WS-110B	5074A	3139		563	
WS-110C	5074A	3139		563	
WS-110D	5074A	3139		563	
WS-115	141A	3092			
WS-115A	141A	3092			
WS-115B	141A	3092			
WS-115C	141A	3092			
WS-115D	141A	3092			
WS-120	142A	3062		563	
WS-120A	142A	3062		563	
WS-120B	142A	3062		563	
WS-120C	142A	3062		563	
WS-120D	142A	3062		563	
WS-125	143A	3093		563	
WS-125A	143A	3093		563	
WS-125B	143A	3093		563	
WS-125C	143A	3093		563	
WS-125D	143A	3093		563	
WS-130	143A	3750		563	
WS-130A	143A	3750		563	
WS-130B	143A	3750		563	
WS-130C	143A	3750		563	
WS-130D	143A	3750		563	
WS-140	144A	3094		564	
WS-140A	144A	3094		564	
WS-140B	144A	3094		564	
WS-140C	144A	3094		564	
WS-140D	144A	3094		564	
WS-150	145A	3063		564	
WS-150A	145A	3063		564	
WS-150B	145A	3063		564	
WS-150C	145A	3063		564	
WS-150D	145A	3063		564	
WS-157	5075A	3751		564	
WS-157A	5075A	3751		564	
WS-157B	5075A	3751		564	
WS-157C	5075A	3751		564	
WS-157D	5075A	3751		564	
WS-162	5075A	3142		564	
WS-162A	5075A	3142		564	
WS-162B	5075A	3142		564	
WS-162C	5075A	3142		564	
WS-162D	5075A	3142		564	
WS-167	5075A	3751		564	
WS-167A	5075A	3751		564	
WS-167B	5075A	3751		564	
WS-167C	5075A	3751		564	
WS-167D	5075A	3751		564	
WS-172	5077A	3145			
WS-172A	5077A	3145			
WS-172B	5077A	3145			
WS-172C	5077A	3145			
WS-172D	5077A	3145			
WS-177	5077A	3145			
WS-177A	5077A	3145			
WS-177B	5077A	3145			
WS-177C	5077A	3145			
WS-177D	5077A	3145			
WS-182	5077A	3752			
WS-182A	5077A	3752			
WS-182B	5077A	3752			
WS-182C	5077A	3752			
WS-182D	5077A	3752			
WS-230	5081A	3151			
WS-230A	5081A	3151			
WS-230B	5081A	3151			
WS-230C	5081A	3151			
WS-230D	5081A	3151			
WS-240	5081A	3151			
WS-240A	5081A	3151			
WS-240B	5081A	3151			
WS-240C	5081A	3151			
WS-240D	5081A	3151			
WS-250	5081A	3151			
WS-250A	5081A	3151			
WS-250B	5081A	3151			
WS-250C	5081A	3151			
WS-250D	5081A	3151			
WS-260	146A	3148			
WS-260A	146A	3148			
WS-260B	146A	3148			
WS-260C	146A	3148			
WS-260D	146A	3148			
WS-270	146A	3064			
WS-270A	146A	3064			
WS-270B	146A	3064			
WS-270C	146A	3064			
WS-270D	146A	3064			
WS-290	5084A	3150			
WS-290A	5084A	3150			
WS-290B	5084A	3150			
WS-290C	5084A	3150			
WS-290D	5084A	3150			
WS-300	5084A	3755			
WS-300A	5084A	3755			
WS-300B	5084A	3755			
WS-300C	5084A	3755			
WS-300D	5084A	3755			
WS-330	147A	3095			
WS-330A	147A	3095			
WS-330B	147A	3095			
WS-330C	147A	3095			
WS-330D	147A	3095			
WS-340	147A	3095			
WS-340A	147A	3095			
WS-340B	147A	3095			
WS-340C	147A	3095			
WS-340D	147A	3095			
WS100	177	3100/519		1122	
WS100A	177	3100/519		1122	
WS100B	177	3100/519		1122	
WS100C	177	3100/519		1122	
WS20-1B	503	3068			
WS200	177	3100/519		1122	
WS200A	177	3100/519		1122	
WS200B	177	3100/519		1122	
WS200C	177	3100/519		1122	
WS300	177	3100/519		1122	
WS300A	177	3100/519		1122	
WS300B	177	3100/519		1122	
WS300C	177	3100/519		1122	
WSD002C	177	3100/519		1122	
WT-050	135A	3056			
WT-050A	135A	3056			
WT-050B	135A	3056			
WT-050C	135A	3056			
WT-050D	135A	3056			
WT-052	135A	3056			
WT-052A	135A	3056			
WT-052B	135A	3056			
WT-052C	135A	3056			
WT-052D	135A	3056			
WT-054	136A	3057		561	
WT-054A	136A	3057		561	
WT-054B	136A	3057		561	
WT-054C	136A	3057		561	
WT-054D	136A	3057		561	
WT-056	136A	3057		561	
WT-056A	136A	3057		561	
WT-056B	136A	3057		561	
WT-056C	136A	3057		561	
WT-056D	136A	3057		561	
WT-058	136A	3057		561	
WT-058A	136A	3057		561	
WT-058B	136A	3057		561	
WT-058C	136A	3057		561	
WT-058D	136A	3057		561	
WT-061	137A	3058		561	
WT-061A	137A	3058		561	
WT-061B	137A	3058		561	
WT-061C	137A	3058		561	
WT-061D	137A	3058		561	
WT-063	137A	3058		561	
WT-063A	137A	3058		561	
WT-063B	137A	3058		561	
WT-063C	137A	3058		561	
WT-063D	137A	3058		561	

Industry Standard No.	ECG	SK	GE	RS 276-	MOTOR.
WT-065	137A	3058		561	
WT-065A	137A	3058		561	
WT-065B	137A	3058		561	
WT-065C	137A	3058		561	
WT-065D	137A	3058		561	
WT-071	138A	3059			
WT-071A	138A	3059			
WT-071B	138A	3059			
WT-071C	138A	3059			
WT-071D	138A	3059			
WT-073	138A	3059			
WT-073A	138A	3059			
WT-073B	138A	3059			
WT-073C	138A	3059			
WT-073D	138A	3059			
WT-075	138A	3059			
WT-075A	138A	3059			
WT-075B	138A	3059			
WT-075C	138A	3059			
WT-075D	138A	3059			
WT-077	138A	3059			
WT-077A	138A	3059			
WT-077B	138A	3059			
WT-077C	138A	3059			
WT-077D	138A	3059			
WT-079	5072A	3136		562	
WT-079A	5072A	3136		562	
WT-079B	5072A	3136		562	
WT-079C	5072A	3136		562	
WT-079D	5072A	3136		562	
WT-081	5072A	3136		562	
WT-081A	5072A	3136		562	
WT-081B	5072A	3136		562	
WT-081C	5072A	3136		562	
WT-081D	5072A	3136		562	
WT-083	5072A	3136		562	
WT-083A	5072A	3136		562	
WT-083B	5072A	3136		562	
WT-083C	5072A	3136		562	
WT-083D	5072A	3136		562	
WT-085	5072A	3136		562	
WT-085A	5072A	3136		562	
WT-085B	5072A	3136		562	
WT-085C	5072A	3136		562	
WT-085D	5072A	3136		562	
WT-088	139A	3060		562	
WT-088A	139A	3060		562	
WT-088B	139A	3060		562	
WT-088C	139A	3060		562	
WT-088D	139A	3060		562	
WT-090	139A	3060		562	
WT-090A	139A	3060		562	
WT-090B	139A	3060		562	
WT-090C	139A	3060		562	
WT-090D	139A	3060		562	
WT-092	139A	3060		562	
WT-092A	139A	3060		562	
WT-092B	139A	3060		562	
WT-092C	139A	3060		562	
WT-092D	139A	3060		562	
WT-094	139A	3060		562	
WT-094A	139A	3060		562	
WT-094B	139A	3060		562	
WT-094C	139A	3060		562	
WT-094D	139A	3060		562	
WT-098	140A	3061		562	
WT-098A	140A	3061		562	
WT-098B	140A	3061		562	
WT-098C	140A	3061		562	
WT-098D	140A	3061		562	
WT-100	140A	3061		562	
WT-100A	140A	3061		562	
WT-100B	140A	3061		562	
WT-100C	140A	3061		562	
WT-100D	140A	3061		562	
WT-10C				563	
WT-110	5074A	3139		563	
WT-110A	5074A	3139		563	
WT-110B	5074A	3139		563	
WT-110C	5074A	3139		563	
WT-110D	5074A	3139		563	
WT-115	141A	3092			
WT-115A	141A	3092			
WT-115B	141A	3092			
WT-115C	141A	3092			
WT-115D	141A	3092			
WT-120	142A	3062		563	
WT-120A	142A	3062		563	
WT-120B	142A	3062		563	
WT-120C	142A	3062		563	
WT-120D	142A	3062		563	
WT-125	143A	3093		563	
WT-125A	143A	3093		563	
WT-125B	143A	3093		563	
WT-125C	143A	3093		563	
WT-125D	143A	3093		563	
WT-130	143A	3750		563	
WT-130A	143A	3750		563	
WT-130B	143A	3750		563	
WT-130C	143A	3750		563	
WT-130D	143A	3750		563	
WT-140	144A	3094		564	
WT-140A	144A	3094		564	
WT-140B	144A	3094		564	
WT-140C	144A	3094		564	
WT-140D	144A	3094		564	
WT-150	145A	3063		564	
WT-150A	145A	3063		564	
WT-150B	145A	3063		564	
WT-150C	145A	3063		564	
WT-150D	145A	3063		564	
WT-157	5075A	3751		564	
WT-157A	5075A	3751		564	
WT-157B	5075A	3751		564	
WT-157C	5075A	3751		564	
WT-157D	5075A	3751		564	
WT-162	5075A	3142		564	
WT-162A	5075A	3142		564	
WT-162B	5075A	3142		564	
WT-162C	5075A	3142		564	
WT-162D	5075A	3142		564	
WT-167	5075A	3751		564	
WT-167A	5075A	3751		564	
WT-167B	5075A	3751		564	
WT-167C	5075A	3751		564	
WT-167D	5075A	3751		564	
WT-16X9	114			1104	
WT-172	5077A	3145			
WT-172A	5077A	3145			
WT-172B	5077A	3145			
WT-172C	5077A	3145			
WT-172D	5077A	3145			
WT-177	5077A	3145			
WT-177A	5077A	3145			
WT-177B	5077A	3145			
WT-177C	5077A	3145			
WT-177D	5077A	3145			
WT-182	5077A	3752			
WT-182A	5077A	3752			
WT-182B	5077A	3752			
WT-182C	5077A	3752			
WT-182D	5077A	3752			
WT-230	5081A	3151			
WT-230A	5081A	3151			
WT-230B	5081A	3151			
WT-230C	5081A	3151			
WT-230D	5081A	3151			
WT-240	5081A	3151			
WT-240A	5081A	3151			
WT-240B	5081A	3151			
WT-240C	5081A	3151			
WT-240D	5081A	3151			
WT-250	5081A	3151			
WT-250A	5081A	3151			
WT-250B	5081A	3151			
WT-250C	5081A	3151			
WT-250D	5081A	3151			
WT-260	146A	3148			
WT-260A	146A	3148			
WT-260B	146A	3148			
WT-260C	146A	3148			
WT-260D	146A	3148			
WT-270	146A	3064			
WT-270A	146A	3064			
WT-270B	146A	3064			
WT-270C	146A	3064			
WT-270D	146A	3064			
WT-290	5084A	3150			
WT-290A	5084A	3150			
WT-290B	5084A	3150			
WT-290C	5084A	3150			
WT-290D	5084A	3150			
WT-300	5084A	3755			
WT-300A	5084A	3755			
WT-300B	5084A	3755			
WT-300C	5084A	3755			
WT-300D	5084A	3755			
WT-330	147A	3095			
WT-330A	147A	3095			
WT-330B	147A	3095			
WT-330C	147A	3095			
WT-330D	147A	3095			
WT-340	147A	3095			
WT-340A	147A	3095			
WT-340B	147A	3095			
WT-340C	147A	3095			
WT-340D	147A	3095			
WT16X9	114	3120			
WT5100					MJ13015
WT5200					2N6547
WTV-BMC	102A			2007	
WTV-L6	101			2002	
WTV129PWR	105	3012			
WTV12MC	160	3005		2004	
WTV12PWR	121	3014		2006	
WTV15MG	102A	3004		2007	
WTV15VMG	102A	3123		2007	
WTV199PWR	121	3009		2006	
WTV20MC	160	3005		2004	
WTV20MG	102A			2007	
WTV20VH6	102A	3123		2007	
WTV20VHG	102A			2007	
WTV20VMG	102A	3004		2007	
WTV25PWR	121	3009		2006	
WTV299PWR	105	3012			
WTV30LVG	176	3123			
WTV30VH6	158			2007	
WTV30VHG	102A	3004		2007	
WTV30VLG	102A	3004		2007	
WTV30VMG	102A	3004		2007	
WTV3MC	160	3123		2004	
WTV40PWR	121	3014		2006	
WTV6MC	160	3005		2004	
WTV6PWR	121	3014		2006	
WTV99PWR	121	3014		2006	
WTVAT6	102A	3004		2007	
WTVAT6A	160	3005		2004	
WTVB5	160	3005		2004	
WTVB5A	102A	3004		2007	
WTVB6	100	3123		2007	
WTVB6A	102	3722		2007	
WTVBA6	160	3005		2004	
WTVBA6A	102A	3004		2007	
WTVBB6	160	3005		2004	
WTVBB6A	102A	3123		2007	
WTVBMC	158	3005		2007	
WTVL6	103	3862		2002	
WTVSA7	101	3861		2002	
WTVSK7	101	3861		2002	
WTVSQ7	101	3861		2002	
WU-050	135A	3056			
WU-050A	135A	3056			
WU-050B	135A	3056			
WU-050C	135A	3056			
WU-050D	135A	3056			
WU-052	135A	3056			
WU-052A	135A	3056			
WU-052B	135A	3056			
WU-052C	135A	3056			
WU-052D	135A	3056			
WU-054	136A	3057		561	
WU-054A	136A	3057		561	
WU-054B	136A	3057		561	
WU-054C	136A	3057		561	
WU-054D	136A	3057		561	
WU-056	136A	3057		561	
WU-056A	136A	3057		561	
WU-056B	136A	3057		561	
WU-056C	136A	3057		561	
WU-056D	136A	3057		561	
WU-058	136A	3057		561	
WU-058A	136A	3057		561	
WU-058B	136A	3057		561	
WU-058C	136A	3057		561	
WU-058D	136A	3057		561	
WU-061	137A	3058		561	
WU-061A	137A	3058		561	
WU-061B	137A	3058		561	
WU-061C	137A	3058		561	
WU-061D	137A	3058		561	
WU-063	137A	3058		561	
WU-063A	137A	3058		561	
WU-063B	137A	3058		561	
WU-063C	137A	3058		561	
WU-063D	137A	3058		561	
WU-065	137A	3058		561	
WU-065A	137A	3058			
WU-065B	137A	3058		561	

Industry Standard No.	ECG	SK	GE	RS 276-	MOTOR.
WU-065C	137A	3058		561	
WU-065D	137A	3058		561	
WU-071	138A	3059			
WU-071A	138A	3059			
WU-071B	138A	3059			
WU-071C	138A	3059			
WU-071D	138A	3059			
WU-073	138A	3059			
WU-073A	138A	3059			
WU-073B	138A	3059			
WU-073C	138A	3059			
WU-073D	138A	3059			
WU-075	138A	3059			
WU-075A	138A	3059			
WU-075B	138A	3059			
WU-075C	138A	3059			
WU-075D	138A	3059			
WU-077	138A	3059			
WU-077A	138A	3059			
WU-077B	138A	3059			
WU-077C	138A	3059			
WU-077D	138A	3059			
WU-079	5072A	3136		562	
WU-079A	5072A	3136		562	
WU-079B	5072A	3136		562	
WU-079C	5072A	3136		562	
WU-079D	5072A	3136		562	
WU-081	5072A	3136		562	
WU-081A	5072A	3136		562	
WU-081B	5072A	3136		562	
WU-081C	5072A	3136		562	
WU-081D	5072A	3136		562	
WU-083	5072A	3136		562	
WU-083A	5072A	3136		562	
WU-083B	5072A	3136		562	
WU-083C	5072A	3136		562	
WU-083D	5072A	3136		562	
WU-085	5072A	3136		562	
WU-085A	5072A	3136		562	
WU-085B	5072A	3136		562	
WU-085C	5072A	3136		562	
WU-085D	5072A	3136		562	
WU-088	139A	3060		562	
WU-088A	139A	3060		562	
WU-088B	139A	3060		562	
WU-088C	139A	3060		562	
WU-088D	139A	3060		562	
WU-090	139A	3060		562	
WU-090A	139A	3060		562	
WU-090B	139A	3060		562	
WU-090C	139A	3060		562	
WU-090D	139A	3060		562	
WU-092	139A	3060		562	
WU-092A	139A	3060		562	
WU-092B	139A	3060		562	
WU-092C	139A	3060		562	
WU-092D	139A	3060		562	
WU-094	139A	3060		562	
WU-094A	139A	3060		562	
WU-094B	139A	3060		562	
WU-094C	139A	3060		562	
WU-094D	139A	3060		562	
WU-098	140A	3061		562	
WU-098A	140A	3061		562	
WU-098B	140A	3061		562	
WU-098C	140A	3061		562	
WU-098D	140A	3061		562	
WU-100	140A	3061		562	
WU-100A	140A	3061		562	
WU-100B	140A	3061		562	
WU-100C	140A	3061		562	
WU-100D	140A	3061		562	
WU-110	5074A	3139		563	
WU-110A	5074A	3139		563	
WU-110B	5074A	3139		563	
WU-110C	5074A	3139		563	
WU-110D	5074A	3139		563	
WU-115	141A	3092			
WU-115A	141A	3092			
WU-115B	141A	3092			
WU-115C	141A	3092			
WU-115D	141A	3092			
WU-120	142A	3062		563	
WU-120A	142A	3062		563	
WU-120C	142A	3062		563	
WU-120D	142A	3062		563	
WU-125	142A	3093		563	
WU-125A	142A	3093		563	
WU-125B	142A	3093		563	
WU-125C	142A	3093		563	
WU-125D	142A	3093		563	
WU-130	143A	3750		563	
WU-130A	143A	3750		563	
WU-130B	143A	3750		563	
WU-130C	143A	3750		563	
WU-130D	143A	3750		563	
WU-140	144A	3094		564	
WU-140A	144A	3094		564	
WU-140B	144A	3094		564	
WU-140C	144A	3094		564	
WU-140D	144A	3094		564	
WU-150	145A	3063		564	
WU-150A	145A	3063		564	
WU-150B	145A	3063		564	
WU-150C	145A	3063		564	
WU-150D	145A	3063		564	
WU-157	5075A	3751		564	
WU-157A	5075A	3751		564	
WU-157B	5075A	3751		564	
WU-157C	5075A	3751		564	
WU-157D	5075A	3751		564	
WU-162	5075A	3142		564	
WU-162A	5075A	3142		564	
WU-162B	5075A	3142		564	
WU-162C	5075A	3142		564	
WU-162D	5075A	3142		564	
WU-167	5075A	3751		564	
WU-167A	5075A	3751		564	
WU-167B	5075A	3751		564	
WU-167C	5075A	3751		564	
WU-167D	5075A	3751		564	
WU-172	5077A	3145			
WU-172A	5077A	3145			
WU-172B	5077A	3145			
WU-172C	5077A	3145			
WU-172D	5077A	3145			
WU-177	5077A	3145			
WU-177A	5077A	3145			
WU-177B	5077A	3145			
WU-177C	5077A	3145			
WU-177D	5077A	3145			
WU-182	5077A	3752			
WU-182A	5077A	3752			
WU-182B	5077A	3752			
WU-182C	5077A	3752			
WU-182D	5077A	3752			
WU-20B	142A			563	
WU-230	5081A	3151			
WU-230A	5081A	3151			
WU-230B	5081A	3151			
WU-230C	5081A	3151			
WU-230D	5081A	3151			
WU-240	5081A	3151			
WU-240A	5081A	3151			
WU-240B	5081A	3151			
WU-240C	5081A	3151			
WU-240D	5081A	3151			
WU-250	5081A	3151			
WU-250A	5081A	3151			
WU-250B	5081A	3151			
WU-250C	5081A	3151			
WU-250D	5081A	3151			
WU-260	146A	3148			
WU-260A	146A	3148			
WU-260B	146A	3148			
WU-260C	146A	3148			
WU-260D	146A	3148			
WU-270	146A	3064			
WU-270A	146A	3064			
WU-270B	146A	3064			
WU-270C	146A	3064			
WU-270D	146A	3064			
WU-290	5084A	3150			
WU-290A	5084A	3150			
WU-290B	5084A	3150			
WU-290C	5084A	3150			
WU-290D	5084A	3150			
WU-300	5084A	3755			
WU-300A	5084A	3755			
WU-300B	5084A	3755			
WU-300C	5084A	3755			
WU-300D	5084A	3755			
WU-330	147A	3095			
WU-330A	147A	3095			
WU-330B	147A	3095			
WU-330D	147A	3095			
WU-340	147A	3095			
WU-340A	147A	3095			
WU-340B	147A	3095			
WU-340C	147A	3095			
WU-340D	147A	3095			
WU065A				561	
WU2N1307	140A	3004/102A		562	
WV-050	135A	3056			
WV-050A	135A	3056			
WV-050B	135A	3056			
WV-050C	135A	3056			
WV-050D	135A	3056			
WV-052	135A	3056			
WV-052A	135A	3056			
WV-052B	135A	3056			
WV-052C	135A	3056			
WV-052D	135A	3056			
WV-054	136A	3057		561	
WV-054A	136A	3057		561	
WV-054B	136A	3057		561	
WV-054C	136A	3057		561	
WV-054D	136A	3057		561	
WV-056	136A	3057		561	
WV-056A	136A	3057		561	
WV-056B	136A	3057		561	
WV-056C	136A	3057		561	
WV-056D	136A	3057		561	
WV-058	136A	3057		561	
WV-058A	136A	3057		561	
WV-058B	136A	3057		561	
WV-058C	136A	3057		561	
WV-058D	136A	3057		561	
WV-061	137A	3058		561	
WV-061A	137A	3058		561	
WV-061B	137A	3058		561	
WV-061C	137A	3058		561	
WV-061D	137A	3058		561	
WV-063	137A	3058		561	
WV-063A	137A	3058		561	
WV-063B	137A	3058		561	
WV-063C	137A	3058		561	
WV-063D	137A	3058		561	
WV-065	137A	3058		561	
WV-065A	137A	3058		561	
WV-065B	137A	3058		561	
WV-065C	137A	3058		561	
WV-065D	137A	3058		561	
WV-071	138A	3059			
WV-071A	138A	3059			
WV-071B	138A	3059			
WV-071C	138A	3059			
WV-071D	138A	3059			
WV-073	138A	3059			
WV-073A	138A	3059			
WV-073B	138A	3059			
WV-073C	138A	3059			
WV-073D	138A	3059			
WV-075	138A	3059			
WV-075A	138A	3059			
WV-075B	138A	3059			
WV-075C	138A	3059			
WV-075D	138A	3059			
WV-077	138A	3059			
WV-077A	138A	3059			
WV-077B	138A	3059			
WV-077C	138A	3059			
WV-077D	138A	3059			
WV-079	5072A	3136		562	
WV-079A	5072A	3136		562	
WV-079B	5072A	3136		562	
WV-079C	5072A	3136		562	
WV-079D	5072A	3136		562	
WV-081	5072A	3136		562	
WV-081A	5072A	3136		562	
WV-081B	5072A	3136		562	
WV-081C	5072A	3136		562	
WV-081D	5072A	3136		562	
WV-083	5072A	3136		562	
WV-083A	5072A	3136		562	
WV-083B	5072A	3136		562	
WV-083C	5072A	3136		562	
WV-083D	5072A	3136		562	
WV-085	5072A	3136		562	
WV-085A	5072A	3136		562	
WV-085B	5072A	3136		562	
WV-085C	5072A	3136		562	
WV-085D	5072A	3136		562	
WV-088	139A	3060		562	
WV-088A	139A	3060		562	
WV-088B	139A	3060		562	
WV-088C	139A	3060		562	
WV-088D	139A	3060		562	

Industry Standard No.	ECG	SK	GE	RS 276-	MOTOR.
WV-090	139A	3060		562	
WV-090A	139A	3060		562	
WV-090B	139A	3060		562	
WV-090C	139A	3060		562	
WV-090D	139A	3060		562	
WV-092	139A	3060		562	
WV-092A	139A	3060		562	
WV-092B	139A	3060		562	
WV-092C	139A	3060		562	
WV-092D	139A	3060		562	
WV-094	139A	3060		562	
WV-094A	139A	3060		562	
WV-094B	139A	3060		562	
WV-094C	139A	3060		562	
WV-094D	139A	3060		562	
WV-098	140A	3061		562	
WV-098A	140A	3061		562	
WV-098B	140A	3061		562	
WV-098C	140A	3061		562	
WV-098D	140A	3061		562	
WV-100	140A	3061		562	
WV-100A	140A	3061		562	
WV-100B	140A	3061		562	
WV-100C	140A	3061		562	
WV-100D	140A	3061		562	
WV-110	5074A	3139		563	
WV-110A	5074A	3139		563	
WV-110B	5074A	3139		563	
WV-110C	5074A	3139		563	
WV-110D	5074A	3139		563	
WV-115	141A	3092			
WV-115A	141A	3092			
WV-115B	141A	3092			
WV-115C	141A	3092			
WV-115D	141A	3092			
WV-120	142A	3062		563	
WV-120A	142A	3062		563	
WV-120B	142A	3062		563	
WV-120C	142A	3062		563	
WV-120D	142A	3062		563	
WV-125	143A	3093		563	
WV-125A	143A	3093		563	
WV-125B	143A	3093		563	
WV-125C	143A	3093		563	
WV-125D	143A	3093		563	
WV-130	143A	3750		563	
WV-130A	143A	3750		563	
WV-130B	143A	3750		563	
WV-130C	143A	3750		563	
WV-130D	143A	3750		563	
WV-140	144A	3094		564	
WV-140A	144A	3094		564	
WV-140B	144A	3094		564	
WV-140C	144A	3094		564	
WV-140D	144A	3094		564	
WV-150	146A	3063/145A			
WV-150A	146A	3063/145A			
WV-150B	146A	3063/145A			
WV-150C	146A	3063/145A			
WV-150D	146A	3063/145A			
WV-157	5075A	3751		564	
WV-157A	5075A	3751		564	
WV-157B	5075A	3751		564	
WV-157C	5075A	3751		564	
WV-157D	5075A	3751		564	
WV-162	5075A	3142		564	
WV-162A	5075A	3142		564	
WV-162B	5075A	3142		564	
WV-162C	5075A	3142		564	
WV-162D	5075A	3142		564	
WV-167	5075A	3751		564	
WV-167A	5075A	3751		564	
WV-167B	5075A	3751		564	
WV-167C	5075A	3751		564	
WV-167D	5075A	3751		564	
WV-172	5077A	3145			
WV-172A	5077A	3145			
WV-172B	5077A	3145			
WV-172C	5077A	3145			
WV-172D	5077A	3145			
WV-177	5077A	3145			
WV-177A	5077A	3145			
WV-177B	5077A	3145			
WV-177C	5077A	3145			
WV-177D	5077A	3145			
WV-182	5077A	3752			
WV-182A	5077A	3752			
WV-182B	5077A	3752			
WV-182C	5077A	3752			
WV-182D	5077A	3752			
WV-230	5081A	3151			
WV-230A	5081A	3151			
WV-230B	5081A	3151			
WV-230C	5081A	3151			
WV-230D	5081A	3151			
WV-240	5081A	3151			
WV-240A	5081A	3151			
WV-240B	5081A	3151			
WV-240C	5081A	3151			
WV-240D	5081A	3151			
WV-250	5081A	3151			
WV-250B	5081A	3151			
WV-250C	5081A	3151			
WV-250D	5081A	3151			
WV-260	146A	3148			
WV-260A	146A	3148			
WV-260B	146A	3148			
WV-260C	146A	3148			
WV-260D	146A	3148			
WV-270	146A	3064			
WV-270A	146A	3064			
WV-270B	146A	3064			
WV-270C	146A	3064			
WV-270D	146A	3064			
WV-290	5084A	3150			
WV-290A	5084A	3150			
WV-290B	5084A	3150			
WV-290C	5084A	3150			
WV-290D	5084A	3150			
WV-300	5084A	3755			
WV-300A	5084A	3755			
WV-300B	5084A	3755			
WV-300C	5084A	3755			
WV-300D	5084A	3755			
WV-330	147A	3095			
WV-330A	147A	3095			
WV-330B	147A	3095			
WV-330D	147A	3095			
WV-340	147A	3095			
WV-340A	147A	3095			
WV-340B	147A	3095			
WV-340C	147A	3095			
WV-340D	147A	3095			
WW-050	135A	3056			
WW-050A	135A	3056			
WW-050B	135A	3056			
WW-050C	135A	3056			
WW-050D	135A	3056			
WW-052	135A	3056			
WW-052A	135A	3056			
WW-052B	135A	3056			
WW-052C	135A	3056			
WW-052D	135A	3056			
WW-054	136A	3057		561	
WW-054A	136A	3057		561	
WW-054B	136A	3057		561	
WW-054C	136A	3057		561	
WW-054D	136A	3057		561	
WW-056	136A	3057		561	
WW-056A	136A	3057		561	
WW-056B	136A	3057		561	
WW-056C	136A	3057		561	
WW-056D	136A	3057		561	
WW-058	136A	3057		561	
WW-058A	136A	3057		561	
WW-058B	136A	3057		561	
WW-058C	136A	3057		561	
WW-058D	136A	3057		561	
WW-061	137A	3058		561	
WW-061A	137A	3058		561	
WW-061B	137A	3058		561	
WW-061C	137A	3058		561	
WW-061D	137A	3058		561	
WW-063	137A	3058		561	
WW-063A	137A	3058		561	
WW-063B	137A	3058		561	
WW-063C	137A	3058		561	
WW-063D	137A	3058		561	
WW-065	137A	3058		561	
WW-065A	137A	3058		561	
WW-065B	137A	3058		561	
WW-065C	137A	3058		561	
WW-065D	137A	3058		561	
WW-071	138A	3059			
WW-071A	138A	3059			
WW-071B	138A	3059			
WW-071C	138A	3059			
WW-071D	138A	3059			
WW-073	138A	3059			
WW-073A	138A	3059			
WW-073B	138A	3059			
WW-073C	138A	3059			
WW-073D	138A	3059			
WW-075	138A	3059			
WW-075A	138A	3059			
WW-075B	138A	3059			
WW-075C	138A	3059			
WW-075D	138A	3059			
WW-077	138A	3059			
WW-077A	138A	3059			
WW-077B	138A	3059			
WW-077C	138A	3059			
WW-077D	138A	3059			
WW-079	5072A	3136		562	
WW-079A	5072A	3136		562	
WW-079B	5072A	3136		562	
WW-079C	5072A	3136		562	
WW-079D	5072A	3136		562	
WW-081	5072A	3136		562	
WW-081A	5072A	3136		562	
WW-081B	5072A	3136		562	
WW-081C	5072A	3136		562	
WW-081D	5072A	3136		562	
WW-083	5072A	3136		562	
WW-083A	5072A	3136		562	
WW-083B	5072A	3136		562	
WW-083C	5072A	3136		562	
WW-083D	5072A	3136		562	
WW-085	5072A	3136		562	
WW-085A	5072A	3136		562	
WW-085B	5072A	3136		562	
WW-085C	5072A	3136		562	
WW-085D	5072A	3136		562	
WW-088	139A	3060		562	
WW-088A	139A	3060		562	
WW-088B	139A	3060		562	
WW-088C	139A	3060		562	
WW-088D	139A	3060		562	
WW-090	139A	3060		562	
WW-090A	139A	3060		562	
WW-090B	139A	3060		562	
WW-090C	139A	3060		562	
WW-090D	139A	3060		562	
WW-092	139A	3060		562	
WW-092A	139A	3060		562	
WW-092B	139A	3060		562	
WW-092C	139A	3060		562	
WW-092D	139A	3060		562	
WW-094	139A	3060		562	
WW-094A	139A	3060		562	
WW-094B	139A	3060		562	
WW-094C	139A	3060		562	
WW-094D	139A	3060		562	
WW-098	140A	3061		562	
WW-098A	140A	3061		562	
WW-098B	140A	3061		562	
WW-098C	140A	3061		562	
WW-098D	140A	3061		562	
WW-100	140A	3061		562	
WW-100A	140A	3061		562	
WW-100B	140A	3061		562	
WW-100C	140A	3061		562	
WW-100D	140A	3061		562	
WW-110	5074A	3139		563	
WW-110A	5074A	3139		563	
WW-110B	5074A	3139		563	
WW-110C	5074A	3139		563	
WW-110D	5074A	3139		563	
WW-115	142A	3092/141A		563	
WW-115A	142A	3092/141A		563	
WW-115B	142A	3092/141A		563	
WW-115C	142A	3092/141A		563	
WW-115D	142A	3092/141A		563	
WW-120	142A	3062		563	
WW-120A	142A	3062		563	
WW-120B	142A	3062		563	
WW-120C	142A	3062		563	
WW-120D	142A	3062		563	
WW-125	144A	3093		564	
WW-125A	144A	3093		564	
WW-125B	144A	3093		564	
WW-125C	144A	3093		564	
WW-125D	144A	3093		564	
WW-130	144A	3093		564	
WW-130A	144A	3093		564	
WW-130B	144A	3093		564	
WW-130C	144A	3093		564	
WW-130D	144A	3093		564	

Industry Standard No.	ECG	SK	GE	RS 276-	MOTOR.
WW-140	144A	3094		564	
WW-140A	144A	3094		564	
WW-140B	144A	3094		564	
WW-140C	144A	3094		564	
WW-140D	144A	3094		564	
WW-150	145A	3063		564	
WW-150A	145A	3063		564	
WW-150B	145A	3063		564	
WW-150C	145A	3063		564	
WW-150D	145A	3063		564	
WW-157	5075A	3751		564	
WW-157A	5075A	3751		564	
WW-157B	5075A	3751		564	
WW-157C	5075A	3751		564	
WW-157D	5075A	3751		564	
WW-162	5075A	3142		564	
WW-162A	5075A	3142		564	
WW-162B	5075A	3142		564	
WW-162C	5075A	3142		564	
WW-162D	5075A	3142		564	
WW-167	5075A	3751		564	
WW-167A	5075A	3751		564	
WW-167B	5075A	3751		564	
WW-167C	5075A	3751		564	
WW-167D	5075A	3751		564	
WW-172	5077A	3145			
WW-172A	5077A	3145			
WW-172B	5077A	3145			
WW-172C	5077A	3145			
WW-172D	5077A	3145			
WW-177	5077A	3145			
WW-177A	5077A	3145			
WW-177B	5077A	3145			
WW-177C	5077A	3145			
WW-177D	5077A	3145			
WW-182	5077A	3145			
WW-182A	5077A	3145			
WW-182B	5077A	3145			
WW-182C	5077A	3145			
WW-182D	5077A	3145			
WW-230	5081A	3151			
WW-230A	5081A	3151			
WW-230B	5081A	3151			
WW-230C	5081A	3151			
WW-230D	5081A	3151			
WW-240	5081A	3151			
WW-240A	5081A	3151			
WW-240B	5081A	3151			
WW-240C	5081A	3151			
WW-240D	5081A	3151			
WW-250	5081A	3151			
WW-250A	5081A	3151			
WW-250B	5081A	3151			
WW-250C	5081A	3151			
WW-250D	5081A	3151			
WW-260	146A	3148			
WW-260A	146A	3148			
WW-260B	146A	3148			
WW-260C	146A	3148			
WW-260D	146A	3148			
WW-270	146A	3064			
WW-270A	146A	3064			
WW-270B	146A	3064			
WW-270C	146A	3064			
WW-270D	146A	3064			
WW-290	5084A	3150			
WW-290A	5084A	3150			
WW-290B	5084A	3150			
WW-290C	5084A	3150			
WW-290D	5084A	3150			
WW-300	5084A	3755			
WW-300A	5084A	3755			
WW-300B	5084A	3755			
WW-300C	5084A	3755			
WW-300D	5084A	3755			
WW-330	147A	3095			
WW-330A	147A	3095			
WW-330B	147A	3095			
WW-330C	147A	3095			
WW-330D	147A	3095			
WW-340	147A	3095			
WW-340A	147A	3095			
WW-340B	147A	3095			
WW-340C	147A	3095			
WW-340D	147A	3095			
WX-050	135A	3056			
WX-050A	135A	3056			
WX-050B	135A	3056			
WX-050C	135A	3056			
WX-050D	135A	3056			
WX-052	135A	3056			
WX-052A	135A	3056			
WX-052B	135A	3056			
WX-052C	135A	3056			
WX-052D	135A	3056			
WX-054	136A	3057		561	
WX-054A	136A	3057		561	
WX-054B	136A	3057		561	
WX-054C	136A	3057		561	
WX-054D	136A	3057		561	
WX-056	136A	3057		561	
WX-056A	136A	3057		561	
WX-056B	136A	3057		561	
WX-056C	136A	3057		561	
WX-056D	136A	3057		561	
WX-058	136A	3057		561	
WX-058A	136A	3057		561	
WX-058B	136A	3057		561	
WX-058C	136A	3057		561	
WX-058D	136A	3057		561	
WX-061	137A	3058		561	
WX-061A	137A	3058		561	
WX-061B	137A	3058		561	
WX-061C	137A	3058		561	
WX-061D	137A	3058		561	
WX-063	137A	3058		561	
WX-063A	137A	3058		561	
WX-063B	137A	3058		561	
WX-063C	137A	3058		561	
WX-063D	137A	3058		561	
WX-065	137A	3058		561	
WX-065A	137A	3058		561	
WX-065B	137A	3058		561	
WX-065C	137A	3058		561	
WX-065D	137A	3058		561	
WX-071	138A	3059			
WX-071A	138A	3059			
WX-071B	138A	3059			
WX-071C	138A	3059			
WX-071D	138A	3059			
WX-073	138A	3059			
WX-073A	138A	3059			
WX-073B	138A	3059			
WX-073C	138A	3059			
WX-073D	138A	3059			
WX-075	138A	3059			
WX-075A	138A	3059			
WX-075B	138A	3059			
WX-075C	138A	3059			
WX-075D	138A	3059			
WX-077	138A	3059			
WX-077A	138A	3059			
WX-077B	138A	3059			
WX-077C	138A	3059			
WX-077D	138A	3059			
WX-079	5072A	3136		562	
WX-079A	5072A	3136		562	
WX-079B	5072A	3136		562	
WX-079C	5072A	3136		562	
WX-079D	5072A	3136		562	
WX-081	5072A	3136		562	
WX-081A	5072A	3136		562	
WX-081B	5072A	3136		562	
WX-081C	5072A	3136		562	
WX-081D	5072A	3136		562	
WX-083	5072A	3136		562	
WX-083A	5072A	3136		562	
WX-083B	5072A	3136		562	
WX-083C	5072A	3136		562	
WX-083D	5072A	3136		562	
WX-085	5072A	3136		562	
WX-085A	5072A	3136		562	
WX-085B	5072A	3136		562	
WX-085C	5072A	3136		562	
WX-085D	5072A	3136		562	
WX-088	139A	3060		562	
WX-088A	139A	3060		562	
WX-088B	139A	3060		562	
WX-088C	139A	3060		562	
WX-088D	139A	3060		562	
WX-090	139A	3060		562	
WX-090A	139A	3060		562	
WX-090B	139A	3060		562	
WX-090C	139A	3060		562	
WX-090D	139A	3060		562	
WX-092	139A	3060		562	
WX-092A	139A	3060		562	
WX-092B	139A	3060		562	
WX-092C	139A	3060		562	
WX-092D	139A	3060		562	
WX-094	139A	3060		562	
WX-094A	139A	3060		562	
WX-094B	139A	3060		562	
WX-094C	139A	3060		562	
WX-094D	139A	3060		562	
WX-098	140A	3061		562	
WX-098A	140A	3061		562	
WX-098B	140A	3061		562	
WX-098C	140A	3061		562	
WX-098D	140A	3061		562	
WX-100	140A	3061		562	
WX-100A	140A	3061		562	
WX-100B	140A	3061		562	
WX-100C	140A	3061		562	
WX-100D	140A	3061		562	
WX-110	141A	3139/5074A			
WX-110A	141A	3139/5074A			
WX-110B	141A	3139/5074A			
WX-110C	141A	3139/5074A			
WX-110D	141A	3139/5074A			
WX-115	141A	3092			
WX-115A	141A	3092			
WX-115B	141A	3092			
WX-115C	141A	3092			
WX-115D	141A	3092			
WX-120	142A	3062		563	
WX-120A	142A	3062		563	
WX-120B	142A	3062		563	
WX-120C	142A	3062		563	
WX-120D	142A	3062		563	
WX-125	143A	3093		563	
WX-125A	143A	3093		563	
WX-125B	143A	3093		563	
WX-125C	143A	3093		563	
WX-125D	143A	3093		563	
WX-130	143A	3750		563	
WX-130A	143A	3750		563	
WX-130B	143A	3750		563	
WX-130C	143A	3750		563	
WX-130D	143A	3750		563	
WX-140	144A	3094		564	
WX-140A	144A	3094		564	
WX-140B	144A	3094		564	
WX-140C	144A	3094		564	
WX-140D	144A	3094		564	
WX-150	145A	3063		564	
WX-150A	145A	3063		564	
WX-150B	145A	3063		564	
WX-150C	145A	3063		564	
WX-150D	145A	3063		564	
WX-157	5075A	3751		564	
WX-157A	5075A	3751		564	
WX-157B	5075A	3751		564	
WX-157C	5075A	3751		564	
WX-157D	5075A	3751		564	
WX-162	5075A	3142		564	
WX-162A	5075A	3142		564	
WX-162B	5075A	3142		564	
WX-162C	5075A	3142		564	
WX-162D	5075A	3142		564	
WX-167	5075A	3751		564	
WX-167A	5075A	3751		564	
WX-167B	5075A	3751		564	
WX-167C	5075A	3751		564	
WX-167D	5075A	3751		564	
WX-172	5077A	3145			
WX-172A	5077A	3145			
WX-172B	5077A	3145			
WX-172C	5077A	3145			
WX-172D	5077A	3145			
WX-177	5077A	3145			
WX-177A	5077A	3145			
WX-177B	5077A	3145			
WX-177C	5077A	3145			
WX-177D	5077A	3145			
WX-182A	5077A	3752			
WX-182C	5077A	3752			
WX-182D	5077A	3752			
WX-230	5081A	3151			
WX-230A	5081A	3151			
WX-230B	5081A	3151			
WX-230C	5081A	3151			
WX-230D	5081A	3151			
WX-240	5081A	3151			
WX-240A	5081A	3151			
WX-240B	5081A	3151			
WX-240C	5081A	3151			
WX-240D	5081A	3151			

Industry Standard No.	ECG	SK	GE	RS 276-	MOTOR.
WX-250	5081A	3151			
WX-250A	5081A	3151			
WX-250B	5081A	3151			
WX-250C	5081A	3151			
WX-250D	5081A	3151			
WX-260	146A	3148			
WX-260A	146A	3148			
WX-260B	146A	3148			
WX-260C	146A	3148			
WX-260D	146A	3148			
WX-270	146A	3064			
WX-270A	146A	3064			
WX-270B	146A	3064			
WX-270C	146A	3064			
WX-270D	146A	3064			
WX-290	5084A	3150			
WX-290A	5084A	3150			
WX-290B	5084A	3150			
WX-290C	5084A	3150			
WX-290D	5084A	3150			
WX-300	5084A	3755			
WX-300A	5084A	3755			
WX-300B	5084A	3755			
WX-300C	5084A	3755			
WX-300D	5084A	3755			
WX-330	147A	3092/141A			
WX-330A	147A	3092/141A			
WX-330B	147A	3092/141A			
WX-330C	147A	3092/141A			
WX-330D	147A	3092/141A			
WX-340	147A	3092/141A			
WX-340A	147A	3092/141A			
WX-340B	147A	3092/141A			
WX-340C	147A	3092/141A			
WX-340D	147A	3092/141A			
WX6	109			1123	
WY-050	135A	3056			
WY-050A	135A	3056			
WY-050B	135A	3056			
WY-050C	135A	3056			
WY-050D	135A	3056			
WY-052	135A	3056			
WY-052A	135A	3056			
WY-052B	135A	3056			
WY-052C	135A	3056			
WY-052D	135A	3056			
WY-054	136A	3057		561	
WY-054A	136A	3057		561	
WY-054B	136A	3057		561	
WY-054C	136A	3057		561	
WY-054D	136A	3057		561	
WY-056	136A	3057		561	
WY-056A	136A	3057		561	
WY-056B	136A	3057		561	
WY-056C	136A	3057		561	
WY-056D	136A	3057		561	
WY-058	136A	3057		561	
WY-058A	136A	3057		561	
WY-058B	136A	3057		561	
WY-058C	136A	3057		561	
WY-058D	136A	3057		561	
WY-061	137A	3058		561	
WY-061A	137A	3058		561	
WY-061B	137A	3058		561	
WY-061C	137A	3058		561	
WY-061D	137A	3058		561	
WY-063	137A	3058		561	
WY-063A	137A	3058		561	
WY-063B	137A	3058		561	
WY-063C	137A	3058		561	
WY-063D	137A	3058		561	
WY-065	137A	3058		561	
WY-065A	137A	3058		561	
WY-065B	137A	3058		561	
WY-065C	137A	3058		561	
WY-065D	137A	3058		561	
WY-071	138A	3059			
WY-071A	138A	3059			
WY-071B	138A	3059			
WY-071C	138A	3059			
WY-071D	138A	3059			
WY-073	138A	3059			
WY-073A	138A	3059			
WY-073B	138A	3059			
WY-073C	138A	3059			
WY-073D	138A	3059			
WY-075	138A	3059			
WY-075A	138A	3059			
WY-075B	138A	3059			
WY-075C	138A	3059			
WY-075D	138A	3059			
WY-077	138A	3059			
WY-077A	138A	3059			
WY-077B	138A	3059			
WY-077C	138A	3059			
WY-077D	138A	3059			
WY-079	5072A	3136		562	
WY-079A	5072A	3136		562	
WY-079B	5072A	3136		562	
WY-079C	5072A	3136		562	
WY-079D	5072A	3136		562	
WY-081	5072A	3136		562	
WY-081A	5072A	3136		562	
WY-081B	5072A	3136		562	
WY-081C	5072A	3136		562	
WY-081D	5072A	3136		562	
WY-083	5072A	3136		562	
WY-083A	5072A	3136		562	
WY-083B	5072A	3136		562	
WY-083C	5072A	3136		562	
WY-083D	5072A	3136		562	
WY-085	5072A	3136		562	
WY-085A	5072A	3136		562	
WY-085B	5072A	3136		562	
WY-085C	5072A	3136		562	
WY-085D	5072A	3136		562	
WY-088	139A	3060		562	
WY-088A	139A	3060		562	
WY-088B	139A	3060		562	
WY-088C	139A	3060		562	
WY-088D	139A	3060		562	
WY-090	139A	3060		562	
WY-090A	139A	3060		562	
WY-090B	139A	3060		562	
WY-090C	139A	3060		562	
WY-090D	139A	3060		562	
WY-092	139A	3060		562	
WY-092A	139A	3060		562	
WY-092B	139A	3060		562	
WY-092C	139A	3060		562	
WY-092D	139A	3060		562	
WY-094	139A	3060		562	
WY-094A	139A	3060		562	
WY-094B	139A	3060		562	
WY-094C	139A	3060		562	
WY-094D	139A	3060		562	
WY-098	140A	3061		562	
WY-098A	140A	3061		562	
WY-098B	140A	3061		562	
WY-098C	140A	3061		562	
WY-098D	140A	3061		562	
WY-100	140A	3061		562	
WY-100A	140A	3061		562	
WY-100B	140A	3061		562	
WY-100C	140A	3061		562	
WY-100D	140A	3061		562	
WY-110	5074A	3139		563	
WY-110A	5074A	3139		563	
WY-110B	5074A	3139		563	
WY-110C	5074A	3139		563	
WY-110D	5074A	3139		563	
WY-115	141A	3092			
WY-115A	141A	3092			
WY-115B	141A	3092			
WY-115C	141A	3092			
WY-115D	141A	3092			
WY-120	142A	3062		563	
WY-120A	142A	3062		563	
WY-120B	142A	3062		563	
WY-120C	142A	3062		563	
WY-120D	142A	3062		563	
WY-125	143A	3093		563	
WY-125A	143A	3093		563	
WY-125B	143A	3093		563	
WY-125C	143A	3093		563	
WY-125D	143A	3093		563	
WY-130	143A	3750		563	
WY-130A	143A	3750		563	
WY-130B	143A	3750		563	
WY-130C	143A	3750		563	
WY-130D	143A	3750		563	
WY-140	144A	3094		564	
WY-140A	144A	3094		564	
WY-140B	144A	3094		564	
WY-140C	144A	3094		564	
WY-140D	144A	3094		564	
WY-150	145A	3063		564	
WY-150A	145A	3063		564	
WY-150B	145A	3063		564	
WY-150C	145A	3063		564	
WY-150D	145A	3063		564	
WY-157	5071A	3751/5075A		561	
WY-157A	5071A	3751/5075A		561	
WY-157B	5075A	3751		564	
WY-157C	5075A	3751		564	
WY-157D	5075A	3751		564	
WY-162	5075A	3142		564	
WY-162A	5075A	3142		564	
WY-162B	5075A	3142		564	
WY-162C	5075A	3142		564	
WY-162D	5075A	3142		564	
WY-167	5075A	3751		564	
WY-167A	5075A	3751		564	
WY-167B	5075A	3751		564	
WY-167C	5075A	3751		564	
WY-167D	5075A	3751		564	
WY-172	5077A	3145			
WY-172A	5077A	3145			
WY-172B	5077A	3145			
WY-172C	5077A	3145			
WY-172D	5077A	3145			
WY-177	5077A	3145			
WY-177A	5077A	3145			
WY-177B	5077A	3145			
WY-177C	5077A	3145			
WY-177D	5077A	3145			
WY-182	5077A	3752			
WY-182A	5077A	3752			
WY-182B	5077A	3752			
WY-182C	5077A	3752			
WY-182D	5077A	3752			
WY-230	5081A	3151			
WY-230A	5081A	3151			
WY-230B	5081A	3151			
WY-230C	5081A	3151			
WY-230D	5081A	3151			
WY-240	5081A	3151			
WY-240A	5081A	3151			
WY-240B	5081A	3151			
WY-240C	5081A	3151			
WY-250	5081A	3151			
WY-250B	5081A	3151			
WY-250C	5081A	3151			
WY-250D	5081A	3151			
WY-260	146A	3148			
WY-260A	146A	3148			
WY-260B	146A	3148			
WY-260C	146A	3148			
WY-270	146A	3064			
WY-270A	146A	3064			
WY-270B	146A	3064			
WY-270C	146A	3064			
WY-270D	146A	3064			
WY-290	5084A	3150			
WY-290A	5084A	3150			
WY-290B	5084A	3150			
WY-290C	5084A	3150			
WY-290D	5084A	3150			
WY-300	5084A	3755			
WY-300A	5084A	3755			
WY-300B	5084A	3755			
WY-300C	5084A	3755			
WY-300D	5084A	3755			
WY-330	147A	3095			
WY-330A	147A	3095			
WY-330B	147A	3095			
WY-330C	147A	3095			
WY-330D	147A	3095			
WY-340	147A	3095			
WY-340A	147A	3095			
WY-340B	147A	3095			
WY-340C	147A	3095			
WY-340D	147A	3095			
WZ-050	5010A	3776			
WZ-050A	135A	3776/5010A			
WZ-052	5010A	3776		561	
WZ-052A	5010A	3776			
WZ-054	5011A	3057/136A		561	
WZ-054A	136A	3057		561	
WZ-056	5011A	3057/136A		561	
WZ-056A	136A	3057		561	
WZ-058	136A	3057		561	
WZ-058A	136A	3057		561	
WZ-061	5013A			561	
WZ-061A	137A			561	
WZ-063	5013A	3779		561	
WZ-063A	137A	3779/5013A		561	
WZ-065	5014A	3780		561	

Industry Standard No.	ECG	SK	GE	RS 276-	MOTOR.
WZ-065A	137A	3780/5014A		561	
WZ-067	5014A	3780		561	
WZ-069	5014A	3780		561	
WZ-071	5015A	3781			
WZ-071A	138A	3781/5015A			
WZ-073	5015A	3781			
WZ-073A	138A	3781/5015A			
WZ-075	5015A	3781			
WZ-077	5015A	3781			
WZ-077A	138A	3781/5015A			
WZ-079		3782		562	
WZ-079A	5072A	3782/5016A		562	
WZ-081	5072A	3136		562	
WZ-081A	5072A	3136		562	
WZ-083	5016A	3782		562	
WZ-083A	5072A	3782/5016A		562	
WZ-085	5017A	3749/5073A		562	
WZ-085A	5072A	3749/5073A		562	
WZ-088	5017A	3783		562	
WZ-088A	140A	3783/5017A		562	
WZ-090	139A	3060		562	
WZ-092	139A	3060		562	
WZ-092A	139A	3060		562	
WZ-094	139A	3060		562	
WZ-094A		3060		562	
WZ-096	5019A	3785		562	
WZ-098	5019A	3785		562	
WZ-098A	140A	3785/5019A		562	
WZ-100	140A	3061		562	
WZ-100A	140A	3061		562	
WZ-105		3786		563	
WZ-110	5074A	3139		563	
WZ-110A	5074A	3139		563	
WZ-115	5074A	3092/141A		563	
WZ-115A	5074A	3092/141A			
WZ-120	5021A	3787		563	
WZ-125	5022A	3093		563	
WZ-130	5022A	3788		563	
WZ-135	5023A	3789		564	
WZ-140	5023A	3789		564	
WZ-145	5024A	3790		564	
WZ-150	145A	3063		564	
WZ-157	5025A	3791		564	
WZ-162	5025A	3791		564	
WZ-167	5026A	3792			
WZ-172	5026A	3792			
WZ-177	5077A	3145			
WZ-182	5027A	3793			
WZ-187	5028A	3794			
WZ-192	5028A	3794			
WZ-197	5029A	3795			
WZ-210	5080A	3336			
WZ-220	5030A	3796			
WZ-230	5031A	3797			
WZ-240	5081A	3151			
WZ-250	5032A	3798			
WZ-260	5033A	3799			
WZ-270	146A	3064			
WZ-280	5034A	3800			
WZ-290	5035A	3801			
WZ-300	5035A	3801			
WZ-320	5036A	3802			
WZ-330	5036A	3802			
WZ-340	5036A	3802			
WZ-350	5037A	3803			
WZ-538	145A	3063		564	
WZ-90	139A			562	
WZ-920	145A	3063		564	
WZ.061	137A			561	
WZ050	5010A			561	
WZ052	5010A			561	
WZ060	136A	3057		561	
WZ061	5013A	3779		561	
WZ065	137A	3058		561	
WZ069	5071A			561	
WZ081	5072A	3136		562	
WZ085	5073A			562	
WZ09	5018A	3060/139A			
WZ090	139A	3060	ZD-9.1	562	
WZ090A	139A			562	
WZ092		3060		562	
WZ094	139A			562	
WZ096	140A	3061		562	
WZ10	140A	3061		562	
WZ100	140A			562	
WZ11.5	141A	3092			
WZ110	5074A	3092/141A			
WZ12	142A	3062		563	
WZ12.8	143A			563	
WZ120	142A	3787/5021A		563	
WZ130	143A			563	
WZ13B	143A	3093		563	
WZ14	144A	3094		564	
WZ140	144A			564	
WZ15	145A	3063		564	
WZ150	145A	3063		564	
WZ177	5076A	9023/5078A			
WZ192	5078A	9023			
WZ27	146A	3064			
WZ33	147A	3095			
WZ5	135A	3056			
WZ5.6	136A			561	
WZ522	135A	3056			
WZ523	136A	3057		561	
WZ533	140A	3061		562	
WZ535	142A	3062		563	
WZ537	144A	3094		564	
WZ538	145A	3063		564	
WZ542	5027A	3793			
WZ544	5029A	3795			
WZ6.2	137A			561	
WZ62	149A	3097			
WZ7.5	138A	3059			
WZ82	150A	3098			
WZ9.1	139A	3060		562	
WZ905	135A	3056			
WZ916	141A	3092			
WZ917	142A	3062		563	
WZ919	144A	3094		564	
WZ920	145A	3063		564	
X-19031-A	519			1122	
X-23305-3	5804	3755/5084A		1144	
X-23305-4	5804	3755/5084A		1144	
X-439623	912	3543			
X-78	102A	3123		2007	
X092	139A	3060		562	
X1-548				2041	
X1-549				2020	
X1005	121	3009		2006	
X1022220-1	177			1122	
X10G1829				2051	
X137	105	3012			
X16	109			1123	
X16A1938	123A	3444		2051	

Industry Standard No.	ECG	SK	GE	RS 276-	MOTOR.
X16A545-7	123A	3444		2051	
X16E3860	123A	3444		2051	
X16E3890	192			2030	
X16E3960	123A	3444		2051	
X16N1485	123A	3444		2051	
X18	109	3087		1123	
X19001-A	123A	3444		2051	
X19001-B	128			2030	
X19001-C	128			2030	
X19001-D	199	3245		2010	
X194-3005829A	130			2041	
X1C1644	102A	3004		2007	
X253223	124	3021			
X29A829	159			2032	
X300-1300A	128			2030	
X32A1389				2032	
X32C4211				2051	
X32C4293				2004	
X32C4296				2051	
X32C5099	128			2030	
X32C5111	128			2030	
X32C5198				2051	
X32C6105				2051	
X32D5422				2051	
X32M5026				2051	
X32P5660				2038	
X330302	142A	3062		563	
X34E1226				2032	
X34E2111				2051	
X42	160	3123		2004	
X43C248				2026	
X44C358				2048	
X45C-H06	102A	3004		2007	
X45C359				2049	
X5M6	116			1104	
X6584-C	123A	3444		2051	
X713	184			2017	
X72A42416	177			1122	
X7338934	175			2020	
X7338935	191	3232			
X735-40C	196	3054		2020	
X735-41	123A	3444		2051	
X78	100	3005		2007	
X908922-001	724	3525			
X925940-5018	177			1122	
X925940-501B	177			1122	
XA-1071	123A	3444		2051	
XA-1072	159			2032	
XA-1078	130			2041	
XA-1095	128			2030	
XA-1139	123A	3444		2051	
XA-1140	159			2032	
XA-1160	152			2048	
XA-1161	130			2041	
XA-1164	234			2050	
XA-495	159	3025/129		2032	
XA-495C	129	3114/290		2027	
XA101	160	3005		2004	
XA1018	128			2030	
XA102	160	3005		2004	
XA103	160	3005		2004	
XA104	160	3005		2004	
XA111	160	3005		2004	
XA112	160	3005		2004	
XA1199	197	3083		2027	
XA121	116	3007		1104	
XA122	102A	3007		2007	
XA123	160	3007		2004	
XA124	160	3007		2004	
XA126	160	3007		2004	
XA131	160	3006		2004	
XA141	160			2004	
XA142	160			2004	
XA143	160			2004	
XA151	158			2007	
XA152	158			2007	
XA161	160			2004	
XA492D	129			2027	
XA494	159	3025/129		2032	
XA495	159	3004/102A		2032	
XA495(C)	159	3025/129		2032	
XA495AC	159	3025/129		2032	
XA495C	159	3025/129		2032	
XA495D	129	3114/290		2027	
XA701	101	3861		2002	
XA702	101	3861		2002	
XA703	101	3861		2002	
XAA104	7402	7402		1811	
XAA105	7430	7430			
XAA106	7440	7440			
XAA107	7473	7473		1803	
XAA108	7476	7476		1813	
XAA109	7490	7490		1808	
XB-5	121	3717		2006	
XB-7	121	3717		2006	
XB1	102	3004		2007	
XB10	160	3005		2004	
XB102	102A	3004		2007	
XB103	102A	3004		2007	
XB104	102A	3004		2007	
XB112	102A	3004		2007	
XB113	102A	3004		2007	
XB114	102A	3004		2007	
XB12	128			2030	
XB13	102A			2007	
XB14	121	3009		2006	
XB152	145A	3063		564	
XB1A		3004		2007	
XB2	102	3004		2007	
XB2A	102A	3004		2007	
XB3	102	3004		2007	
XB3A	102A	3004		2007	
XB3B	102	3004		2007	
XB3BN	102A	3004		2007	
XB3C	102	3004		2007	
XB3C-1	102A	3004		2007	
XB4	103	3862		2002	
XB4-1	102A	3004		2007	
XB401	186	3192		2017	
XB404	152	3893		2048	
XB408	152	3893		2048	
XB476	152	3893		2048	
XB5	104	3009		2006	
XB7	104	3009		2006	
XB8	160	3006		2004	
XB9	160	3005		2004	
XC-1312A	799	3238			
XC101	102A	3004		2007	
XC121	102A	3004		2007	
XC131	102A			2007	
XC1312	799	3238			
XC1312A	799	3238			
XC1312P	799	3238			
XC141	121	3009		2006	

Industry Standard No.	ECG	SK	GE	RS 276-	MOTOR.
XC142	121	3009		2006	
XC155	104	3009		2006	
XC156	104	3014		2006	
XC171	102A	3004		2007	
XC371	107	3039/316		2015	
XC372	123A	3444		2051	
XC373	123A	3444		2051	
XC374	123A	3444		2051	
XC70	5071A			561	
XC723	130			2041	
XD-2581-42	5635	3533			
XD2A	110MP	3709		1123(2)	
XEJ040017	123A	3444		2051	
XF-14B217				2007	
XG1	160	3006		2004	
XG10	160	3006		2004	
XG11	160	3006		2004	
XG12	160	3006		2004	
XG2	160	3006		2004	
XG24	160	3006		2004	
XG28	103A	3835		2002	
XG29	103A	3835		2002	
XG3	160	3006		2004	
XG30	123A	3444		2051	
XG32	158	3004/102A		2007	
XG33	103A	3010		2002	
XG5	160	3006		2004	
XG8	102A	3004		2007	
XHPF1299	1232	3852			
XI-548	130	3027		2041	
XI-549	175	3026		2020	
XJ13	102	3004		2007	
XJ13-1	102A	3004		2007	
XJ71	160	3008		2004	
XJ72	160	3006		2004	
XJ73	160	3006		2004	
XN-400-318-P1	123A	3444		2051	
XN-400-319-P2	159			2032	
XN12A	104	3009		2006	
XN12B	104	3009		2006	
XN12C	104	3009		2006	
XN12E	104	3009		2006	
XN12F	104	3719		2006	
XNC101	103	3862		2002	
XP-2581-82	5635	3533			
XPS30800	124	3021			
XR2001CP	2011	3975			
XR2002CP	2012	9092			
XR2003CP	2013	9093			
XR2201CP	2011	3975			
XR2202CP	2012	9092			
XR2203CP	2013	9093			
XR2204CP	2014	9094			
XR556V	978	3689			
XR6118CP	2021	9020			
XS-10	116	3016		1104	
XS-31	116	3016		1104	
XS1	108	3039/316		2038	
XS10	116			1104	
XS101	158			2007	
XS104	158			2007	
XS121	158			2007	
XS14	108	3039/316		2038	
XS15	108	3039/316		2038	
XS16	116	3017B/117		1104	
XS16A	116	3031A		1104	
XS17	116	3016		1104	
XS17A	116	3016		1104	
XS18	116	3017B/117		1104	
XS19	159	3025/129		2032	
XS2	108	3039/316		2038	
XS21	123A	3444		2051	
XS22	123A	3444		2051	
XS22(RECT.)	116			1104	
XS23	116	3017B/117		1104	
XS23A	116	3017B/117		1104	
XS26	103A	3010		2002	
XS3	108	3039/316		2038	
XS30	128	3024		2030	
XS31	116			1104	
XS36	161	3132		2015	
XS37	161	3132		2015	
XS38	161	3132		2015	
XS39	161	3132		2015	
XS4	108	3039/316		2038	
XS40	107	3039/316		2015	
XS40A	116	3017B/117		1104	
XS6	108	3039/316		2038	
XS9.1B	139A			562	
XT-548A	130	3027		2041	
XT-549A	175	3026		2020	
XT100	158			2007	
XT15X3	108	3019		2038	
XT200	158			2007	
XT300	126			2024	
XT400	126			2024	
XU604	116	3016		1104	
XV604	125			1114	
XZ-049	135A	3056			
XZ-051	135A	3056			
XZ-064				562	
XZ-070	5071A	3334		561	
XZ-072	138A	3059			
XZ-076	138A	3059			
XZ-082	5072A	3136		562	
XZ-086	139A			562	
XZ-090	5018A			562	
XZ-092	139A	3060		562	
XZ-096	140A	3061		562	
XZ-100	140A	3061		562	
XZ-122	142A	3062		563	
XZ-152	145A	3063		564	
XZ055	136A	3057		561	
XZ058	5070A	9021			
XZ064	5072A	3060/139A		562	
XZ070	5071A	3334		561	
XZ084	5072A	3059/138A		562	
XZ086	139A	3749/5073A			
XZ090	139A	3060		562	
XZ092	139A			562	
XZ098	140A	3061		562	
XZ102	140A			562	
XZ070	5071A			561	
Y0-6A				1104	
Y100	114	3120		1104	
Y363	102A	3004		2007	
Y410	121	3114/290		2006	
Y49001-21	123A	3444		2051	
Y56001-21	519			1122	
Y56001-86	123A	3444		2051	
Y56601-08	123A	3444		2051	
Y56601-44	234			2050	
Y56601-45	123A	3444		2051	
Y56601-46	129			2027	
Y56601-47	128			2030	
Y56601-48	234			2050	
Y56601-49	123A	3444		2051	
Y56601-50	159			2032	
Y56601-51	123A	3444		2051	
Y56601-63	159			2032	
Y56601-73	123A	3444		2051	
Y56601-74	159			2032	
Y56601-75	123A	3444		2051	
Y56601-76	129			2027	
Y56601-79	159			2032	
Y56601-80	123A	3444		2051	
Y56601-82	159			2032	
Y56601-84	159			2032	
Y56601-86-AD	123A	3444		2051	
Y56601-93	123A	3444		2051	
Y633	158			2007	
YAAD001	109	3136/5072A		562	
YAAD004	116	3031A		1104	
YAAD007	116			1104	
YAAD009	109	3088		1123	
YAAD010	177	3100/519		1122	
YAAD017	5073A	3749		562	
YAAD018	177	3100/519		1122	
YAAD019	116	3311		1104	
YAAD020	116	3311		1104	
YAAD021	5072A	3136		562	
YAAD022	116	3311		1104	
YAAM003	1006	3358			
YAAN2SC1096	186	3248		2017	
YAAN2SC1096K	186	3248		2017	
YAAN2SC1096L	186	3248		2017	
YAAN2SC1096M	186	3248		2017	
YAAN2SC1096N	186	3248		2017	
YAAN2SD141	175	3538		2020	
YAANL2SC1096	186	3248		2017	
YAANL2SC1096K	186A	3248		2052	
YAANL2SC1096KLMN	186A			2052	
YAANL2SC1096L	186A	3248		2052	
YAANL2SC1096M	186A	3248		2052	
YAANL2SC1096N		3248		2052	
YAANL2SC1096Q				2017	
YAANL2SC1096R				2017	
YAANZ	186	3192		2017	
YAANZ2S1096	186	3192		2017	
YAANZ2SC1096	186	3248		2017	
YAANZ2SC1096-BLUE	186			2017	
YAANZ2SC1096-RED	186			2017	
YAANZ2SC1096K	186	3248		2017	
YAANZ2SC1096L	186	3248		2017	
YAANZ2SC1096M	186	3248		2017	
YAANZSC1096	184			2017	
YANZ2SC1096				2017	
YANZ2SC1096K				2017	
YANZ2SC1096L				2017	
YANZ2SC1096M				2017	
YBAD009	116	3311		1104	
YD1121	177	3100/519		1122	
YEAD010	5072A	3136		562	
YEAD014	5072A	3136		562	
YEAD015	139A	3060		562	
YEAD024	5072A	3136		562	
YEAD029	614	3327			
YEAD030	116	3311		1104	
YEAD032	110MP	3087		1123	
YEAD1N60P	109	3088		1123	
YEAD1SV53	614	3327			
YEADAW01-06	137A			561	
YEADUPC577H	1082	3461			
YEAM53274P	7474			1818	
YEAMM5152L	1170	3745			
YEAMNJM78L05	977	3462			
YEAMUPC577H	1082	3461			
YEAN2SC941	123A	3122		2051	
YEAN3SK39Q	222	3050/221		2036	
YEAUPC577H	1082	3461			
YOD4	117			1104	
YR-011	116	3031A		1104	
YR011	116			1104	
YSG-20024				1104	
YSG-V139-2-2	109	3017B/117		1123	
YSG-V139-22	109			1123	
YSG-V47-1-3	116	3016		1104	
YSG-V47-7-51-1	116	3031A		1104	
YSG-V47-7-51-2	116			1104	
YSG-V81-2-3	178MP			1122(2)	
YV1	102A	3004		2007	
YV1A	102A			2007	
YV2	102A	3004		2007	
YZ-080	5072A	3136		562	
YZ037	5067A	3331			
YZ058	136A	9021/5070A		561	
YZ060	5071A	3334		561	
YZ063	137A	3779/5013A		561	
YZ284	5084A	3755			
Z-1006	136A	3777/5011A		561	
Z-1010	138A	3059		562	
Z-1012	5019A	3785		562	
Z-1014	142A			563	
Z-1016	145A	3790/5024A		564	
Z-1016A	145A	3790/5024A		564	
Z-1024	147A	3802/5036A			
Z-1102	5069A	3056/135A			
Z-1104	136A			561	
Z-1112C	142A	3062		563	
Z-1114	145A			564	
Z-1122	147A	3095			
Z-1128	5090A	3096/148A			
Z-1132	150A	3098			
Z-1140	136A	3057		561	
Z-1145	136A	3057		561	
Z-1150	136A	3057		561	
Z-1155	136A	3057		561	
Z-1160	136A	3057		561	
Z-1165	136A	3057		561	
Z-1170	136A	3057		561	
Z-1240	136A	3777/5011A		561	
Z-1245	136A	3777/5011A		561	
Z-1250	136A	3057		561	
Z-1255	136A	3057		561	
Z-1260	136A	3057		561	
Z-1265	136A	3057		561	
Z-1270	136A	3057		561	
Z-15	145A	3063		564	
Z-1540	136A	3057		561	
Z-1545	136A	3057		561	
Z-1550	136A	3777/5011A		561	
Z-1555	136A	3057		561	
Z-1560	136A	3057		561	
Z-1565	136A	3057		561	
Z-1570	136A	3057		561	
Z-15K	145A	3063		564	
Z-175-011	177			1122	
Z-28058-1	123A			2051	

Industry Standard No.	ECG	SK	GE	RS 276-	MOTOR.
Z-5140	136A	3057		561	
Z-5145	136A	3057		561	
Z-5150	136A	3057		561	
Z-5155	136A	3057		561	
Z-5160	136A	3057		561	
Z-5165	136A	3057		561	
Z-5170	136A	3057		561	
Z-5240	136A	3057		561	
Z-5245	136A	3057		561	
Z-5250	136A	3057		561	
Z-5255	136A	3057		561	
Z-5260	136A	3057		561	
Z-5265	136A	3057		561	
Z-5270	136A	3057		561	
Z-5540	136A	3057		561	
Z-5545	136A	3057		561	
Z-5550	136A	3057		561	
Z-5555	136A	3057		561	
Z-5560	136A	3057		561	
Z-5565	136A	3057		561	
Z-5570	136A	3057		561	
Z-963B	142A			563	
Z0-12	142A	3062		563	
Z0-12A	142A	3062		563	
Z0-12B	142A	3062		563	
Z0-27	146A	3064			
Z0-27A	146A	3064			
Z0-27B	146A	3064			
Z01-13	143A	3093		563	
Z0211	135A	3056			
Z0212	136A	3057		561	
Z0214	137A	3058		561	
Z0216	138A	3059			
Z0217	5072A	3136		562	
Z0219	139A	3060		562	
Z0220	140A	3061		562	
Z0222	142A	3062		563	
Z0225	145A	3063		564	
Z0228	5077A	3752			
Z0234	146A	3064			
Z0402	134A	3055			
Z0406	135A	3056			
Z0407	136A	3057		561	
Z0408	137A	3058		561	
Z0410	138A	3059			
Z0411	5072A	3136		562	
Z0412	139A	3060		562	
Z0413	140A	3061		562	
Z0414	5074A	3139		563	
Z0415	142A	3062		563	
Z0416	143A	3750		563	
Z0417	144A	3094		564	
Z0418	145A	3063		564	
Z0419	5075A	3751		564	
Z0420	5077A	3145			
Z0423	5081A	3151			
Z0424	146A	3064			
Z0426	147A	3095			
Z0427	5085A	3337			
Z0432	148A	3096			
Z0433	149A	3097			
Z0436	150A	3098			
Z0439	151A	3099			
Z0A3.3	134A	3055			
Z0A3.9	134A	3055			
Z0B-12	142A	3062		563	
Z0B-15	145A	3063		564	
Z0B10	140A	3061		562	
Z0B12	142A	3062		563	
Z0B15	145A	3063		564	
Z0B16	5075A	3751			
Z0B18	5077A	3752			
Z0B2.7	5063A	3837			
Z0B20	5079A	3335			
Z0B22	5080A	3336			
Z0B27	146A	3064			
Z0B3.0	5065A	3838			
Z0B3.3	5066A	3055/134A			
Z0B3.9	5067A	3055/134A			
Z0B4.7	5069A	3056/135A			
Z0B5.1	5116A	3056/135A			
Z0B5.6	136A			561	
Z0B6.2	137A			561	
Z0B6.8	5071A	3334		561	
Z0B8.2	5072A			562	
Z0B9.1	139A	3059/138A		562	
Z0C-15	145A	3063		564	
Z0C10	140A	3061		562	
Z0C12	142A	3062		563	
Z0C15	145A	3063		564	
Z0C18	5027A	3793			
Z0C20	5029A	3795			
Z0C22	5030A	3796			
Z0C27	146A	3064			
Z0C27A	146A	3064			
Z0C3.3	134A	3055			
Z0C3.9	134A	3055			
Z0C4.7	135A	3056			
Z0C5.1	135A	3056			
Z0C5.6	136A			561	
Z0C6.2	137A			561	
Z0C6.8	5014A	3334/5071A		561	
Z0C8.2	5016A			562	
Z0C9.1	139A	3059/138A		562	
Z0D-15	145A			564	
Z0D10	140A	3061		562	
Z0D12	142A	3062		563	
Z0D15	145A	3063		564	
Z0D20	5029A	3795			
Z0D22	5030A	3796			
Z0D27	5033A	3799			
Z0D27B	146A	3064			
Z0D3.3	134A	3055			
Z0D3.9	134A	3055			
Z0D4.7	135A	3056			
Z0D5.1	135A	3056			
Z0D5.6	136A	3057		561	
Z0D6.2	137A	3058		561	
Z0D8.2	5016A			562	
Z0D9.1	139A	3059/138A		562	
Z10	140A	3061		562	
Z1000	5005A	3771			
Z10003	923D			1740	
Z1002	5007A	3773			
Z1004	5009A	3775			
Z1006				561	
Z1008	5014A	3780		561	
Z101	5072A	3136		562	
Z1010	5016A	3782		562	
Z1012	5019A			562	
Z1014	5021A	3787		563	
Z1016	5024A			564	
Z1018	5027A	3793			
Z1020	5030A	3796			
Z1022	5033A	3799			
Z10K	140A	3061		562	
Z1100	5067A	3331			
Z1100-C	5067A	3331			
Z1102	5069A	3330/5066A			
Z1102-C	5069A	3330/5066A			
Z1103	135A	3056			
Z1104	136A	3057		561	
Z1104-C	136A			561	
Z1106	5071A	3334		561	
Z1106-C	5071A	3334		561	
Z1107	138A	3059			
Z1108	5072A	3136		562	
Z1108-C	5072A			562	
Z1109	139A	3060		562	
Z1110	140A	3061		562	
Z1110-C	140A			562	
Z1112	142A	3062		563	
Z1112-C	142A			563	
Z1112C		3062		563	
Z1114	145A	3063		564	
Z1114-C	145A			564	
Z1116	5077A	3752			
Z1116-C	5077A	3752			
Z111Z	142A			563	
Z1120	146A	3064			
Z1130	149A	3097			
Z1134	151A	3099			
Z1140	136A			561	
Z1145	136A			561	
Z1150	136A			561	
Z1155	136A			561	
Z1160	136A			561	
Z1165	136A			561	
Z1170	136A			561	
Z12	142A	3787/5021A			
Z12.0	5021A			563	
Z1202	5069A	3330/5066A			
Z1204				561	
Z1206				561	
Z1208				562	
Z1209				562	
Z1210		3061		562	
Z1212	142A	3062		563	
Z1214	145A	3063		564	
Z1216	5077A	3752			
Z1220	146A	3064			
Z1222	147A	3095			
Z1226	148A	3096			
Z1240				561	
Z1245				561	
Z1250	136A			561	
Z1255	136A			561	
Z1260	136A			561	
Z1270	136A			561	
Z12A	142A	3787/5021A		563	
Z12K	142A	3062		563	
0Z12T10	142A	3062	ZD-12	563	
0Z12T5	142A	3062	ZD-12	563	
Z13	143A	3750			
Z15	145A	3063		564	
Z1540	136A			561	
Z1545	136A			561	
Z1550				561	
Z1555	136A			561	
Z1560	136A			561	
Z1565	136A			561	
Z1570	136A			561	
Z15K	145A	3063		564	
Z18	5077A	3752			
Z1A103-018	125	3032A		1114	
Z1A3.3	134A	3055			
Z1A3.9	134A	3055			
Z1A4.7	5009A	3775			
Z1A8.2	5072A	3136			
Z1A8.2A	5072A	3136		562	
Z1A8.2B	5072A	3136		562	
Z1AB.2				562	
Z1B-15	145A	3063		564	
Z1B10	140A	3061		562	
Z1B12	142A	3062		563	
Z1B13	143A	3750			
Z1B15	145A	3063		564	
Z1B16	5075A	3751			
Z1B18	5077A	3752			
Z1B2.7	5063A	3837			
Z1B20	5079A	3335			
Z1B22	5080A	3336			
Z1B27	146A	3064			
Z1B27A	146A	3064			
Z1B3.0	5065A	3838			
Z1B3.9	5067A	3055/134A			
Z1B30	5084A	3755			
Z1B4.7	5069A	3056/135A			
Z1B5.1	135A	3056			
Z1B5.6	136A	3057		561	
Z1B6.2	137A	3058		561	
Z1B6.8	5071A	3334		561	
Z1B7.5	138A	3059			
Z1B8.2	5072A			562	
Z1B9.1	139A	3059/138A		562	
Z1C-15	145A	3063		564	
Z1C10	140A	3061		562	
Z1C12	142A	3062		563	
Z1C15	145A	3063		564	
Z1C18	5027A	3793			
Z1C20	5029A	3795			
Z1C22	5030A	3796			
Z1C27	146A	3064			
Z1C27B	146A	3064			
Z1C3.3	134A	3055			
Z1C3.9	134A	3055			
Z1C5.1	135A	3056			
Z1C5.6	135A	3056			
Z1C6.2	137A	3058		561	
Z1C8.2	5016A	3782		562	
Z1C9.1		3784		562	
Z1D-15	145A	3063		564	
Z1D10	140A	3061		562	
Z1D12	142A	3062		563	
Z1D15	145A	3063		564	
Z1D18	5027A	3793			
Z1D20	5029A	3795			
Z1D22	5030A	3796			
Z1D27	5033A	3799			
Z1D27A	146A	3799/5033A			
Z1D3.3	134A	3055			
Z1D3.9	5007A	3773			
Z1D4.7	135A	3056			
Z1D5.1	135A	3056			
Z1D5.6	136A	3057		561	
Z1D6.2	137A	3058		561	
Z1D7.5	138A	3059			
Z1D8.2	5016A	3782		562	

Industry Standard No.	ECG	SK	GE	RS 276-	MOTOR.
Z1D9.1		3784		562	
Z20	104	3009		2006	
Z22	5080A	3336			
Z2525	5136A	3402			
Z2530	5141A	3407			
Z2548	5159A	3425			
Z27	146A	3064			
Z27-A	146A	3064			
Z2A-150F	145A	3063		564	
Z2A120F	142A	3062		563	
Z2A150F	145A	3063		564	
Z2A33F	5066A	3330			
Z2A36F	134A	3055			
Z2A51F	135A	3056			
Z2A56CF	136A			561	
Z2A56F	136A	3057		561	
Z2A62F	137A	3058		561	
Z2A82F	5072A	3136		562	
Z2A91F	139A	3066/118		562	
Z2B10	5125A	3391			
Z2B12	5127A	3393			
Z2B15	5130A	3396			
Z2B18	5133A	3399			
Z2B22	5136A	3402			
Z2B3.9	5113A	3379			
Z2B33	5142A	3408			
Z2B4.7	5115A	3381			
Z2B5.6	5117A	3383			
Z2B6.2	5119A	3385			
Z2B7.5	5121A	3387			
Z2B9.1	5124A	3390			
Z2C10	5125A	3391			
Z2C15	5130A	3396			
Z2C18	5133A	3399			
Z2C22	5136A	3402			
Z2C3.9	5113A	3379			
Z2C33	5142A	3408			
Z2C4.7	5115A	3381			
Z2C5.6	5117A	3383			
Z2C6.2	5119A	3385			
Z2C7.5	5121A	3387			
Z2C9.1	5124A	3390			
Z2D10	5125A	3391			
Z2D12	5127A	3393			
Z2D15	5130A	3396			
Z2D18	5133A	3399			
Z2D22	5136A	3402			
Z2D27	5139A	3405			
Z2D3.9	5113A	3379			
Z2D33	5142A	3408			
Z2D5.6	5117A	3383			
Z2D6.2	5119A	3385			
Z2D7.5	5121A	3387			
Z2D9.1	5124A	3390			
Z3.3	5066A	3055/134A			
Z3.6	134A	3055			
Z3.9	134A	3055			
Z33	147A	3095			
Z330611	116	3031A		1104	
Z3B1000CF	5156A	3422			
Z3B100CF	5125A	3391			
Z3B120CF	5127A	3393			
Z3B150CF	5130A	3396			
Z3B180CF	5133A	3399			
Z3B220CF	5136A	3402			
Z3B270CF	5139A	3405			
Z3B330CF	5142A	3408			
Z3B47CF	5115A	3381			
Z3B560CF	5148A	3414			
Z3B56CF	5117A	3383			
Z3B62CF	5119A	3385			
Z3B75CF	5121A	3387			
Z3B820CF	5153A	3419			
Z3B91CF	5124A	3390			
Z4.7	5069A	3056/135A			
Z4743A	143A	3750			
Z4745A	5075A	3751			
Z4746A	5077A	3752			
Z4751A	5084A	3755			
Z4A8.2	5072A	3136		562	
Z4B-15	145A	3063		564	
Z4B12	142A	3062		563	
Z4B13	143A	3750			
Z4B15	145A	3063		564	
Z4B18	5077A	3752			
Z4B27	146A	3064			
Z4B27-9	146A	3064			
Z4B3.0	5065A	3838			
Z4B30	5084A	3755			
Z4B33	147A	3095			
Z4B5.1	5116A	3056/135A			
Z4B5.6	136A			561	
Z4B9.1	139A	3059/138A		562	
Z4C3.6	134A	3055			
Z4C5.6	136A	3057		561	
Z4D3.6	134A	3055			
Z4D5.6	136A	3057		561	
Z4MW333	123A	3444		2051	
Z4X-15	145A	3063		564	
Z4X11	141A	3092			
Z4X12	142A	3062		563	
Z4X13	143A	3750		563	
Z4X14	144A	3094		564	
Z4X14B	144A	3094		564	
Z4X14B,A					1N14Z10,5
Z4X15	145A	3063		564	
Z4X5.1A	135A	3056			
Z4X5.1B	135A	3056			
Z4X5.6	136A	3057		561	
Z4X9.1	139A	3059/138A		562	
Z4XL12	142A	3062		563	
Z4XL12B	142A	3062		563	
Z4XL14	144A	3094		564	
Z4XL14B	144A	3094		564	
Z4XL6.2	137A	3058		561	
Z4XL6.2B	137A	3058		561	
Z4XL7.5	138A	3059			
Z4XL75	138A	3059			
Z4XL9.1	139A	3059/138A		562	
Z4XL9.1B	139A	3066/118		562	
Z5.6	136A	3057		561	
Z5140	136A			561	
Z5145	136A			561	
Z5150	136A			561	
Z5155	136A			561	
Z5160	136A			561	
Z5165	136A			561	
Z5170	136A			561	
Z5240	136A			561	
Z5250	136A			561	
Z5255	136A			561	
Z5260	136A			561	
Z5265	136A			561	
Z5270	136A			561	
Z537	147A	3095			
Z5540	136A			561	
Z5545	136A			561	
Z5550	136A			561	
Z5555	136A			561	
Z5560	136A			561	
Z5565	136A			561	
Z5570	136A			561	
Z5A3.3	5005A	3771			
Z5A3.9	134A	3055			
Z5B-15	145A	3063		564	
Z5B10	140A	3061		562	
Z5B12	142A	3062		563	
Z5B13	143A	3750			
Z5B15	145A	3063			
Z5B16	5075A	3751			
Z5B18	5077A	3752			
Z5B2	142A	3062		563	
Z5B20	5079A	3335			
Z5B22	5080A	3336			
Z5B3.0	5065A	3838			
Z5B3.3	5066A	3055/134A			
Z5B3.6	134A	3055			
Z5B3.9	5067A	3055/134A			
Z5B30	5084A	3755			
Z5B4.7	5069A	3056/135A			
Z5B5.1	135A	3056			
Z5B5.6	136A	3057		561	
Z5B6.2	137A	3058		561	
Z5B6.8	5071A	3334		561	
Z5B8.2	5072A			562	
Z5B9.1	139A	3059/138A		562	
Z5C-15	145A	3063		564	
Z5C10	140A	3061		562	
Z5C12	142A	3062		563	
Z5C15	145A	3063		564	
Z5C18	5027A	3793			
Z5C20	5029A	3795			
Z5C22	5030A	3796			
Z5C3.3	134A	3055			
Z5C3.9	134A	3055			
Z5C4.7	135A	3056			
Z5C5.1	135A	3056			
Z5C5.6	136A	3057		561	
Z5C6.2	137A	3058		561	
Z5C6.8	5014A	3780		561	
Z5C8.2	5016A	3782		562	
Z5C9.1	139A	3059/138A		562	
Z5D-15	145A	3063		564	
Z5D10	140A	3061		562	
Z5D12	142A	3062		563	
Z5D15	145A	3063		564	
Z5D18	5027A	3793			
Z5D20	5029A	3795			
Z5D22	5030A	3796			
Z5D3.3	134A	3055			
Z5D3.9	134A	3055			
Z5D4.7	135A	3056			
Z5D5.1	135A	3056			
Z5D5.6	136A	3057		561	
Z5D560CF	5029A	3795			
Z5D6.2	137A	3058		561	
Z5D6.8	5014A	3780		561	
Z5D8.2	5016A	3782		562	
Z5D9.1	139A	3059/138A		562	
Z6	5070A	9021		561	
Z6.2	137A	3058		561	
Z6.8	5071A	3334		561	
Z694	139A	3059/138A		562	
Z714	139A	3059/138A		562	
Z8.2	5072A			562	
Z8.2A	5072A	3136		562	
Z8.2B	5072A	3136		562	
Z8.2C	5072A	3136		562	
Z8.2D	5072A	3136		562	
Z801	137A	3058		561	
ZA-15B				564	
ZA100962B	108	3039/316		2038	
ZA105604	102A	3003		2007	
ZA11	141A	3092			
ZA110	151A	3099			
ZA110A	151A	3099			
ZA110B	151A	3099			
ZA11A	141A	3092			
ZA11B	5074A	3092/141A			
ZA12	142A	3062		563	
ZA12A	142A			563	
ZA12B	142A	3062		563	
ZA13	143A	3750		563	
ZA13A	143A	3750		563	
ZA13B	143A	3750		563	
ZA15	145A	3063		564	
ZA150	166	9075		1152	
ZA15A	145A	3063		564	
ZA15B	145A	3063		564	
ZA15V	145A			564	
ZA16B	5075A	3751			
ZA18B	5077A	3752			
ZA27	146A	3064			
ZA27A	146A	3064			
ZA27B	146A	3064		564	
ZA29312	519			1122	
ZA30B	5084A	3755			
ZA33	147A	3095			
ZA33A	147A	3095			
ZA33B	147A	3095			
ZA56	148A	3096			
ZA56A	148A	3096			
ZA56B	5090A	3096/148A			
ZA62	149A	3097			
ZA62A	149A	3097			
ZA62B	149A	3097			
ZA8.2	5072A	3136		562	
ZA8.2A	5072A	3136		562	
ZA8.2B	5072A	3136		562	
ZA82	150A	3098			
ZA82A	150A	3098			
ZA82B	150A	3098			
ZA9.1	139A	3059/138A		562	
ZA9.1A	139A	3059/138A		562	
ZA9.1B	139A	3059/138A		562	
ZAC10	5125A	3391			
ZAC100	5156A	3422			
ZAC100A	5156A	3422			
ZAC100B	5156A	3422			
ZAC10A	5125A	3391			
ZAC10B	5125A	3391			
ZAC12	5127A	3393			
ZAC120	5158A	3424			
ZAC120A	5158A	3424			
ZAC120B	5158A	3424			
ZAC12A	5127A	3393			
ZAC12B	5127A	3393			
ZAC15	5130A	3396			

Industry Standard No.	ECG	SK	GE	RS 276-	MOTOR.
ZAC15A	5130A	3396			
ZAC15B	5130A	3396			
ZAC18	5133A	3399			
ZAC18A	5133A	3399			
ZAC18B	5133A	3399			
ZAC22	5136A	3402			
ZAC22A	5136A	3402			
ZAC22B	5136A	3402			
ZAC27	5139A	3405			
ZAC27A	5139A	3405			
ZAC27B	5139A	3405			
ZAC33	5142A	3408			
ZAC33A	5142A	3408			
ZAC33B	5142A	3408			
ZAC56	5148A	3414			
ZAC56A	5148A	3414			
ZAC56B	5148A	3414			
ZAC7.5	5121A	3387			
ZAC7.5A	5121A	3387			
ZAC7.5B	5121A	3387			
ZAC82	5153A	3419			
ZAC82A	5153A	3419			
ZAC82B	5153A	3419			
ZAC9.1	5124A	3390			
ZAC9.1A	5124A	3390			
ZAC9.1B	5124A	3390			
ZAG-9673	159			2032	
ZAR110	5809	9010			
ZAR210	5890	7090			
ZAR610	5809	9010			
ZAR710	5809	9010			
ZAX9.1	118	3066			
ZB-1	141A	3092			
ZB-1-15A	118	3063/145A			
ZB-1-9.5	139A			562	
ZB-11	5074A			563	
ZB-15A	145A	3063		564	
ZB-31.12				563	
ZB1-09				562	
ZB1-1	5074A			563	
ZB1-10	142A	3061/140A		563	
ZB1-100	5096A	3346			
ZB1-10A				562	
ZB1-11	141A	3092		563	
ZB1-110	151A	3099			
ZB1-12	142A	3062		563	
ZB1-13	143A	3750		563	
ZB1-14	144A	3094		564	
ZB1-15	145A	3063		564	
ZB1-16	5075A	3751		564	
ZB1-16V	5075A	3751		564	
ZB1-18	5077A	3752			
ZB1-19	5078A	9023			
ZB1-35	147A	3095			
ZB1-6	5070A	9021		561	
ZB1-7				561	
ZB1-8	5072A	3059/138A		561	
ZB1-9	139A	3060		562	
ZB1-9V	139A	3060		562	
ZB10	140A	3061		562	
ZB100	5096A	3346			
ZB10A	140A	3061		562	
ZB10B	140A	3061		562	
ZB10X	140A	3061		562	
ZB11	5074A			563	
ZB110A	151A	3099			
ZB11A	5074A	3139		563	
ZB11B	5074A	3139		563	
ZB12	142A	3062		563	
ZB120	5097A	3347			
ZB12A	142A	3062		563	
ZB12B	142A	3062		563	
ZB13	143A	3750		563	
ZB13A	145A	3750/143A		564	
ZB13B	143A	3750		563	
ZB15	145A	3063		564	
ZB15A	145A	3063		564	
ZB15B	145A	3063		564	
ZB16	5075A	3751		564	
ZB16A	5075A	3751		564	
ZB16B	5075A	3751		564	
ZB18	5027A	3752/5077A			
ZB18A	5077A	3752			
ZB18B	5077A	3752			
ZB20	5029A	3795			
ZB200	5105A	3355			
ZB205				561	
ZB206				561	
ZB209				562	
ZB210				562	
ZB212				563	
ZB215				564	
ZB22	5030A	3796			
ZB22B	5080A	3336			
ZB24	5081A	3151			
ZB24A	5081A	3151			
ZB24B	5081A	3151			
ZB27	146A	3064			
ZB27-3	146A	3064			
ZB27A	146A	3064			
ZB27A-3	146A	3064			
ZB27B	146A	3064			
ZB27B-3	146A	3064			
ZB3.3	134A	3055			
ZB30	5084A	3755			
ZB30A	5084A	3755			
ZB30B	5084A	3755			
ZB33	147A	3095			
ZB33A	147A	3095			
ZB33B	147A	3095			
ZB36	5085A	3337			
ZB39	5144A	3338/5086A			
ZB4.7	135A	3056			
ZB5.1	135A	3056			
ZB5.6	136A	3057		561	
ZB5.6A	136A	3057		561	
ZB5.6B	136A	3057		561	
ZB56A	5090A	3342			
ZB6.2	137A	3058		561	
ZB6.2A	137A	3058		561	
ZB6.2B	137A	3058		561	
ZB6.8	5014A	3780		561	
ZB6.8A				561	
ZB6.8B	5071A			561	
ZB62	149A	3097			
ZB62A	149A	3097			
ZB62B	149A	3097			
ZB7.5A	138A	3059			
ZB8.2	5016A	3782		562	
ZB8.2A		3782		562	
ZB8.2B	5072A			562	
ZB82	150A	3098			
ZB82A	150A	3098			
ZB82B	150A	3098			

Industry Standard No.	ECG	SK	GE	RS 276-	MOTOR.
ZB9.1	139A			562	
ZB9.1A	139A	3060		562	
ZB9.1B	139A	3060		562	
ZB91B	5095A	3345			
ZBC12	5127A	3393			
ZBC15	5130A	3396			
ZBC18	5133A	3397/5131A			
ZBC22	5136A	3402			
ZBC7.5	5121A	3387			
ZBI-09	139A	3060		562	
ZBI-13				561	
ZBU-19	5079A	3335			
ZC-015	145A	3063		564	
ZC-140		3062		2048	
ZC012	142A	3062		563	
ZC013	143A	3750			
ZC015	145A	3063		564	
ZC016	5075A	3751			
ZC018	5077A	3752			
ZC027	146A	3064			
ZC027-3	146A	3064			
ZC030	5084A	3755			
ZC033	147A	3095			
ZC10	5125A	3391			
ZC100	5156A	3422			
ZC105				561	
ZC106				561	
ZC109				562	
ZC110				562	
ZC112				563	
ZC115				564	
ZC12	5127A	3393			
ZC18	5133A	3399			
ZC1N34A	109			1123	
ZC1S358B	177			1122	
ZC2010	5125A	3391			
ZC2015	5130A	3396			
ZC2022	5136A	3402			
ZC2027	5139A	3405			
ZC2033	5142A	3408			
ZC22	5136A	3402			
ZC27	5139A	3405			
ZC2SA101	160			2004	
ZC2SA101BA	160			2004	
ZC2SA102	778A			2004	
ZC2SA102CA	160			2004	
ZC2SA103	160			2004	
ZC2SA103CA	160			2004	
ZC2SA377	160			2004	
ZC2SA70	160			2004	
ZC2SA700A	160			2004	
ZC2SA700B	160			2004	
ZC2SA71	160			2004	
ZC2SA71A	160			2004	
ZC2SB172	158			2007	
ZC2SB172A	158			2007	
ZC33	5142A	3414/5148A			
ZC7.5	5121A	3387			
ZC82	5153A	3419			
ZC9.1	5124A	3390			
ZCC100	5156A	3422			
ZCC12	5127A	3393			
ZCC120	5158A	3424			
ZCC15	5130A	3396			
ZCC22	5136A	3402			
ZCC27	5139A	3405			
ZCC33	5142A	3408			
ZCC56	5148A	3414			
ZCC7.5	5121A	3387			
ZCC82	5153A	3419			
ZCOM-5683-0	116			1104	
ZCOM3679	116			1104	
ZCSR16C	5524	6624			
ZCSR16D	5526	6627/5527			
ZCSR16E	5527	6627			
ZCSR16F	5528	6629/5529			
ZCSR16G	5529	6629			
ZCSR16J	5531	6631			
ZD-015	145A	3063		564	
ZD-15	145A	3063		564	
ZD-15A	145A	3063		564	
ZD-15B	145A	3063		564	
ZD012	142A	3062		563	
ZD015	145A	3063		564	
ZD027	146A	3064			
ZD033	147A	3095			
ZD10B	140A			562	
ZD110	5157A	3099/151A			
ZD110A	151A	3099			
ZD110B	151A	3099			
ZD12	5127A	3062/142A		563	
ZD12A	142A	3062		563	
ZD12B	142A	3062		563	
ZD13B	143A	3750			
ZD15	5130A	3063/145A		564	
ZD15A	145A	3063		564	
ZD15B	145A	3063		564	
ZD16B	5075A	3751			
ZD2010	5125A	3391			
ZD2012	5127A	3392/5126A			
ZD2015	5130A	3396			
ZD2018	5133A	3399			
ZD2027	5139A	3405			
ZD2028	5136A	3406/5140A			
ZD2033	5142A	3408			
ZD27	5139A	3064/146A			
ZD27-3	146A	3064			
ZD27A	146A	3064			
ZD27B	146A	3064			
ZD3.6B	134A	3055			
ZD30B	5084A	3755			
ZD33	5142A	3095/147A			
ZD33A	147A	3095			
ZD33B	147A	3095			
ZD4.7	5115A	3381			
ZD5.1A	135A	3056			
ZD5.1B	5116A	3056/135A			
ZD5.6A	136A	3057		561	
ZD5.6B	136A	3057		561	
ZD56A	136A			561	
ZD56B	5090A			561	
ZD6.2A	137A	3058		561	
ZD6.2B	137A	3058		561	
ZD62	5150A	3097/149A			
ZD62A	149A	3097			
ZD62B	149A	3097			
ZD7.5	5121A	3059/138A			
ZD8.2	5122A	3136/5072A			
ZD8.2B	5072A	3136			
ZD82	150A	3098			
ZD82A	150A	3098			
ZD82B	150A	3098			
ZD9.1	5124A	3059/138A		562	
ZD9.1A	139A	3059/138A		562	

Industry Standard No.	ECG	SK	GE	RS 276-	MOTOR.
ZD9.1B	139A	3059/138A		562	
ZDP-D22-69 54	129			2027	
ZDP-D22-69-54				2027	
ZDT	123A	3444		2051	
ZDT-30	108	3019		2038	
ZDT-31	108	3019		2038	
ZDT10	108			2038	
ZDT11	108			2038	
ZDT20	108			2038	
ZDT21	108			2038	
ZDT30	108			2038	
ZDT31	108			2038	
ZE-1.5	109	3087		1123	
ZE-15	145A	3063		564	
ZE-15A	145A	3063		564	
ZE-15B	145A	3063		564	
ZE10	140A	3061		562	
ZE110	151A	3099			
ZE110A	151A	3099			
ZE110B	151A	3099			
ZE12	142A	3062		563	
ZE12A	142A	3062		563	
ZE12B	142A	3062		563	
ZE15	145A	3063		564	
ZE15A	145A	3063		564	
ZE15B	145A	3063		564	
ZE22	5030A	3796			
ZE27	146A	3064			
ZE27A	146A	3064			
ZE27B	146A	3064			
ZE33	147A	3095			
ZE33A	147A	3095			
ZE33B	147A	3095			
ZE4.7	5009A	3775			
ZE5.6	136A	3057		561	
ZE6.8	5014A	3780		561	
ZE62	149A	3097			
ZE62A	149A	3097			
ZE62B	149A	3097			
ZE8.2	5016A	3782		562	
ZE82	150A	3098			
ZE82A	150A	3098			
ZE9.1				562	
ZEC12	142A	3062		563	
ZEC27	146A	3064			
ZEC5.6	136A	3057		561	
ZEN100	123A	3444		2051	
ZEN101	129	3118		2032	
ZEN102	123A	3444		2051	
ZEN103	123A	3444		2051	
ZEN104	108	3019		2038	
ZEN105	161	3039/316		2015	
ZEN106	159			2032	
ZEN107	159	3114/290		2032	
ZEN108	108	3018		2038	
ZEN109	108	3039/316		2038	
ZEN110	123A	3444		2051	
ZEN111	123A	3444		2051	
ZEN112	123A	3444		2051	
ZEN113	123A	3444		2051	
ZEN114	123A	3444		2051	
ZEN115	123A	3444		2051	
ZEN116	199	3245		2010	
ZEN117	229	3122		2038	
ZEN118	108	3122		2038	
ZEN119	123A	3444		2051	
ZEN120	123A	3444		2051	
ZEN121	229	3245/199		2038	
ZEN122	159	3247/234		2032	
ZEN123	312	3116		2035	
ZEN124	221	3050		2036	
ZEN125	194	3244			
ZEN126		3122		2051	
ZEN127	123A			2051	
ZEN128	172A	3156			
ZEN129	6409			2029	
ZEN200	124	3021			
ZEN201	225	3104A			
ZEN202	184	3190		2017	
ZEN203	185	3191		2025	
ZEN204	162	3559			
ZEN205	154	3104A		2012	
ZEN206	165	3115			
ZEN207	282	3190/184			
ZEN208	191	3232			
ZEN209	182	3188A			
ZEN210	184	3190		2017	
ZEN211	185	3191		2025	
ZEN300	160			2004	
ZEN301	160			2004	
ZEN302	158	3004/102A		2007	
ZEN303	102	3722		2007	
ZEN304	158			2007	
ZEN305	158			2007	
ZEN306	102	3722		2007	
ZEN307	158	3004/102A		2007	
ZEN308	158	3004/102A		2007	
ZEN309	102A			2007	
ZEN310	102A			2007	
ZEN311	126	3006/160		2024	
ZEN312	126	3006/160		2024	
ZEN313	126	3007		2024	
ZEN314	126	3007		2024	
ZEN315	101	3861		2002	
ZEN325	104	3009		2006	
ZEN326	121	3717		2006	
ZEN327	213	3012/105			
ZEN328	127	3035			
ZEN330	121	3717		2006	
ZEN331	121	3717		2006	
ZEN401	5806	3848			
ZEN403	5878	3602			
ZEN430	109	3087		1123	
ZEN431	112	3089			
ZEN432	178MP			1122(2)	
ZEN433	167	3647		1172	
ZEN434		3105		1172	
ZEN500	5013A	3779		561	
ZEN501	138A	3059			
ZEN502	5021A	3787		563	
ZEN503	5027A	3793			
ZEN504	5033A	3799			
ZEN505	138A	3059			
ZEN506	143A	3750		563	
ZEN507	144A	3094		564	
ZEN508	145A	3063		564	
ZEN512	147A	3095			
ZEN513	5098A	3348			
ZEN514	5099A	3349			
ZEN515	5085A	3337			
ZEN601	760	3157/703A			
ZEN602	722	3161			
ZEN603	790	3077			
ZEN604	708	3135/709			
ZEN605	712	3072			
ZEN607	714	3075			
ZEN608	715	3076			
ZEN609	739	3235			
ZENER-122	142A	3062		563	
ZENER-132	5082A	3753			
ZENER-136	5078A	9023			
ZF-15	145A	3063		564	
ZF-15A	145A	3063		564	
ZF-15B	145A	3063		564	
ZF-16	145A			564	
ZF-8.2	5072A			562	
ZF10	140A	3061		562	
ZF110				562	
ZF112				563	
ZF113				563	
ZF115				564	
ZF116				564	
ZF12	142A	3062		563	
ZF12A	142A	3062		563	
ZF12B	142A	3062		563	
ZF13	143A	3750		563	
ZF13P	143A	3750			
ZF15	145A	3063		564	
ZF15.6R				561	
ZF15A	145A	3063		564	
ZF15B	145A	3063		564	
ZF16	5075A	3751		564	
ZF16.2				561	
ZF16P	5075A	3751			
ZF18	5077A	3752			
ZF18P	5077A	3752			
ZF19.1				562	
ZF2.7	5063A	3837			
ZF2.7P	5063A	3837			
ZF20	5079A	3335			
ZF210				562	
ZF211				563	
ZF212				563	
ZF213				563	
ZF215				564	
ZF216				564	
ZF22	5080A	3336			
ZF25.6R				561	
ZF26.2				561	
ZF26.8				561	
ZF27	146A	3064			
ZF27A	146A	3064			
ZF27B	146A	3064			
ZF29.1				562	
ZF3.0	5065A	3838			
ZF3.0P	5065A	3838			
ZF3.3	5066A	3330			
ZF3.3P	5066A	3330			
ZF3.6	134A	3055			
ZF30	5084A	3755			
ZF30P	5084A	3755			
ZF33	147A	3095			
ZF33A	147A	3095			
ZF33B	147A	3095			
ZF4.3	5068A	3332			
ZF4.7	5069A	3056/135A			
ZF413	5068A	3332			
ZF43	5068A	3332			
ZF5.1	135A	3056			
ZF5.6	136A	3057		561	
ZF6.2	137A	3058		561	
ZF6.8	5071A	3334		561	
ZF62	149A	3097			
ZF62A	149A	3097			
ZF62B	149A	3097			
ZF8.2	5072A	3136		562	
ZF8.2P	5072A	3136		562	
ZF82	150A	3098			
ZF82A	150A	3098			
ZF82B	150A	3098			
ZF8A	5072A	3136		562	
ZF9.1	139A			562	
ZG-15	145A	3063		564	
ZG-15B	145A	3063		564	
ZG10	140A	3061		562	
ZG100119				564	
ZG12	142A	3062		563	
ZG15	145A	3063		564	
ZG15A	145A	3063		564	
ZG15B	145A	3063		564	
ZG18	5027A	3793			
ZG2.7	5002A	3768			
ZG22	5030A	3796			
ZG27	146A	3064			
ZG27A	146A	3064			
ZG27B	146A	3064			
ZG3.3	5005A	3771			
ZG3.9	134A	3055			
ZG33	147A	3095			
ZG33A	147A	3095			
ZG33B	147A	3095			
ZG4.7	135A	3056			
ZG5.6	136A	3057		561	
ZG6.8	5014A	3780		561	
ZG62	149A	3097			
ZG62A	149A	3097			
ZG62B	149A	3097			
ZG8.2	5016A	3782		562	
ZG82	150A	3098			
ZG82A	150A	3098			
ZG82B	150A	3098			
ZG9.1	139A			562	
ZH-15	145A	3063		564	
ZH-15A	145A	3063		564	
ZH-15B	145A	3063		564	
ZH105				561	
ZH106				561	
ZH109				562	
ZH11	141A	3092			
ZH110	151A	3099		562	
ZH110A	151A	3099			
ZH110B	151A	3099			
ZH112				563	
ZH115				564	
ZH11A	141A	3092			
ZH11B	5074A	3092/141A			
ZH12	142A	3062		563	
ZH12A	142A	3062		563	
ZH12B	142A	3062		563	
ZH13	143A	3750		563	
ZH13A	143A	3750		563	
ZH13B	143A	3750		563	
ZH15	145A	3063		564	
ZH15A	145A	3063		564	
ZH15B	145A	3063		564	
ZH16B	5075A	3751			
ZH18B	5077A	3752			
ZH27	146A	3064			

Industry Standard No.	ECG	SK	GE	RS 276-	MOTOR.
ZH27A	146A	3064			
ZH27B	146A	3064			
ZH30B	5084A	3755			
ZH33	147A	3095			
ZH33A	147A	3095			
ZH33B	147A	3095			
ZH56	148A	3096			
ZH56A	148A	3096			
ZH56B	5090A	3096/148A			
ZH62	149A	3097			
ZH62A	149A	3097			
ZH62B	149A	3097			
ZH8.2B	5072A	3136			
ZH82	150A	3098			
ZH82A	150A	3098			
ZH82B	150A	3098			
ZH9.1	139A	3059/138A		562	
ZH9.1A	139A	3066/118		562	
ZH9.1B	139A	3059/138A		562	
ZJ-15	145A	3063		564	
ZJ-15A	145A	3063		564	
ZJ-15B	145A	3063		564	
ZJ110	151A	3099			
ZJ110A	151A	3099			
ZJ110B	151A	3099			
ZJ12	142A	3062		563	
ZJ12A	142A	3062		563	
ZJ12B	142A	3062		563	
ZJ13	158			2007	
ZJ15	145A	3063		564	
ZJ15A	145A	3063		564	
ZJ15B	145A	3063		564	
ZJ226	5496	6796			
ZJ230	5527	6627			
ZJ252B	116	3016		1104	
ZJ27	146A	3064			
ZJ27A	146A	3064			
ZJ27B	146A	3064			
ZJ33	147A	3095			
ZJ33A	147A	3095			
ZJ33B	147A	3095			
ZJ39A	5531	6631			
ZJ39L	5531	6631			
ZJ40	108	3018		2038	
ZJ54A	5487	3944			
ZJ54L	5487	3944			
ZJ62	149A	3097			
ZJ62A	149A	3097			
ZJ62B	149A	3097			
ZJ72	160	3008		2004	
ZJ73	160	3008		2004	
ZJ82	150A	3098			
ZJ82A	150A	3098			
ZJ82B	150A	3098			
ZL6	5012A	3778		561	
ZM10				562	
ZM10A		3785		562	
ZM10B	140A	3785/5019A		562	
ZM11				563	
ZM11A				563	
ZM11B	5074A	3786/5020A		563	
ZM12				563	
ZM12A				563	
ZM12B	142A	3787/5021A		563	
ZM13				563	
ZM13A				563	
ZM13B	143A	3788/5022A		563	
ZM14				564	
ZM14A				564	
ZM14B	144A	3789/5023A		564	
ZM15				564	
ZM15A				564	
ZM15B	5130A	3790/5024A		564	
ZM16				564	
ZM16A				564	
ZM16B	5075A	3791/5025A		564	
ZM17B	5076A	9022			
ZM18B	5077A	3793/5027A			
ZM19B	5078A	9023			
ZM20B	5079A	3795/5029A			
ZM22B	5080A	3796/5030A			
ZM24B	5081A	3797/5031A			
ZM25B	5082A	3753			
ZM27B	146A	3799/5033A			
ZM30B	5084A	3755			
ZM33	147A	3802/5036A			
ZM33B	147A	3802/5036A			
ZM5.6	136A	3777/5011A		561	
ZM5.6A				561	
ZM5.6B	136A	3777/5011A		561	
ZM6				561	
ZM6.2				561	
ZM6.2A				561	
ZM6.2B	137A	3779/5013A		561	
ZM6.8				561	
ZM6.8A				561	
ZM6.8B	5071A	3780/5014A		561	
ZM6A				561	
ZM6B		3778		561	
ZM7.5B	138A	3781/5015A			
ZM8.2				562	
ZM8.2A	5072A	3782/5016A		562	
ZM8.2B	5072A	3782/5016A		562	
ZM8.2C	5072A	3782/5016A		562	
ZM8.2D	5072A	3782/5016A		562	
ZM8.7				562	
ZM8.7A				562	
ZM8.7B		3783		562	
ZM9.1				562	
ZM9.1A				562	
ZM9.1B	139A	3784/5018A		562	
ZMZ205	135A	3056			
ZN 35024712	197			2027	
ZN-35024712				2027	
ZN35050411	198	3220			
ZN37003008	738	3167			
ZN37007003	1109	3711			
ZN37011002	749	3168			
ZN7400E	7400			1801	
ZN7401E	7401	7401			
ZN7402E	7402	7402		1811	
ZN7404E	7404	7404		1802	
ZN7405E	7405	7405			
ZN74107E	74107	74107			
ZN7410E	7410	7410		1807	
ZN7410F				1807	
ZN7420E	7420	7420		1809	
ZN7420F				1809	
ZN7427E				1823	
ZN7427J				1823	
ZN7430E	7430	7430			
ZN7432E				1824	
ZN7432J				1824	
ZN7440E	7440	7440			

Industry Standard No.	ECG	SK	GE	RS 276-	MOTOR.
ZN7441AE	7441	7441		1804	
ZN7451E	7451			1825	
ZN7451F				1825	
ZN7454E	7454	7454			
ZN7473E	7473	7473		1803	
ZN7474E	7474	7474		1818	
ZN7475E	7475	7475		1806	
ZN7476E	7476	7476		1813	
ZN7486E				1827	
ZN7486J				1827	
ZO-110	151A	3099			
ZO-110A	151A	3099			
ZO-110B	151A	3099			
ZO-12	142A	3062		563	
ZO-12A	142A	3062		563	
ZO-12B	142A	3062		563	
ZO-15	145A	3063		564	
ZO-15A	145A	3063		564	
ZO-15AC	145A	3063		564	
ZO-15B	145A	3063		564	
ZO-15BY	145A	3063		564	
ZO-27	146A	3064			
ZO-27A	146A	3064			
ZO-27B	146A	3064			
ZO-33	147A	3095			
ZO-33A	147A	3095			
ZO-33B	147A	3095			
ZO-62	149A	3097			
ZO-62A	149A	3097			
ZO-62B	149A	3097			
ZO-82	150A	3098			
ZO-82A	150A	3098			
ZO-82B	150A	3098			
ZOB-15	145A	3063		564	
ZOB110	151A	3099			
ZOB12	142A	3062		563	
ZOB15	145A	3063		564	
ZOB27	146A	3064			
ZOB33	147A	3095			
ZOB5.6	136A	3057		561	
ZOB6.2	137A	3058		561	
ZOB62	149A	3097			
ZOB82	150A	3098			
ZOD-15	145A	3063		564	
ZOD110	151A	3099			
ZOD15	145A	3063		564	
ZOD27	146A	3064			
ZOD33	147A	3095			
ZOD6.2	137A	3058		561	
ZOD6.8	5071A			561	
ZOD62	149A	3097			
ZOD82	150A	3098			
ZP-15	145A			564	
ZP-15A	145A			564	
ZP-15B	145A			564	
ZP10	140A	3061		562	
ZP110	151A	3099			
ZP110A	151A	3099			
ZP110B	151A	3099			
ZP12	142A	3062		563	
ZP12A	142A	3062		563	
ZP12B	142A	3062		563	
ZP15	145A	3063		564	
ZP15A	145A	3063		564	
ZP15B	145A	3063		564	
ZP18	5027A	3793			
ZP2.7	5002A	3768			
ZP20	5029A	3795			
ZP22	5030A	3796			
ZP27	146A	3064			
ZP27A	146A	3064			
ZP27B	146A	3064			
ZP3.3	5005A	3771			
ZP3.9	5007A	3773			
ZP33	147A	3095			
ZP33A	147A	3095			
ZP33B	147A	3095			
ZP4.3	5068A	3332			
ZP4.7	5009A	3775			
ZP5.1	135A	3056			
ZP5.6	136A	3057		561	
ZP6.2	137A	3058		561	
ZP6.8	5014A	3780		561	
ZP62	149A	3097			
ZP62A	149A	3097			
ZP62B	149A	3097			
ZP8.2	5016A	3782		562	
ZP82	150A	3098			
ZP82A	150A	3098			
ZP82B	150A	3098			
ZP9.1	139A	3059/138A		562	
ZP9.1A	139A			562	
ZP9.1B	139A			562	
ZPD13	143A	3750			
ZPD16	5075A	3751			
ZPD18	5077A	3752			
ZPD2.7	5063A	3837			
ZPD30	5084A	3755			
ZPD6.2	137A			561	
ZPD618	5071A	3334		561	
ZPD8.2	5072A	3136		562	
ZPD9.1	139A			562	
ZQ-15	145A	3063		564	
ZQ-15A	145A	3063		564	
ZQ-15B	145A	3063		564	
ZQ-6	5070A	3058/137A		561	
ZQ110	151A	3099			
ZQ110A	151A	3099			
ZQ110B	151A	3099			
ZQ12	142A	3062		563	
ZQ12A	142A	3062		563	
ZQ12B	142A	3062		563	
ZQ15	145A	3063		564	
ZQ15A	145A	3063		564	
ZQ15B	145A	3063		564	
ZQ15X	145A	3063		564	
ZQ27	146A	3064			
ZQ27A	146A	3064			
ZQ27B	146A	3064			
ZQ33	147A	3095			
ZQ33A	147A	3095			
ZQ33B	147A	3095			
ZQ62	149A	3097			
ZQ62A	149A	3097			
ZQ62B	149A	3097			
ZQ82	150A	3098			
ZQ82A	150A	3098			
ZQ82B	150A	3098			
ZQ9.1	139A	3059/138A		562	
ZQ9.1A	139A	3060		562	
ZQ9.1B	139A	3066/118		562	
ZR-1025	116	3016		1104	
ZR-1031	116	3016		1104	
ZR-1035	116	3016		1104	

Industry Standard No.	ECG	SK	GE	RS 276-	MOTOR.
ZR-1076	116	3016		1104	
ZR-10B				1104	
ZR-500	116	3016		1104	
ZR-590	116			1104	
ZR-590A	116	3016		1104	
ZR-61	116	3016		1104	
ZR-63	116	3016		1104	
ZR10	5800	9003		1142	
ZR1025	116			1104	
ZR1031	116			1104	
ZR1035	116			1104	
ZR1076	116			1104	
ZR11	5801	9004		1142	
ZR12	5802	9005		1143	
ZR13	5804	9007		1144	
ZR14	5804	9007		1144	
ZR15	116	3017B/117		1104	
ZR205				561	
ZR206	5882			561	
ZR209				562	
ZR209T1				562	
ZR209T3				562	
ZR210				562	
ZR212				563	
ZR215				564	
ZR500	116			1104	
ZR50B793-1	142A	3062		563	
ZR50B921-2	138A	3059			
ZR50B921-3	145A			564	
ZR590				1104	
ZR590A	116			1104	
ZR60	116	3017B/117		1104	
ZR601	5801	9004		1142	
ZR602	5802	9005		1143	
ZR604	5804	9007		1144	
ZR606	5806	3848			
ZR608	156	3051			
ZR61	116			1104	
ZR62	116	3016		1104	
ZR63	116			1104	
ZR64	116	3017B/117		1104	
ZR66	116	3017B/117		1104	
ZR68	156	3051			
ZS-10B	116	3016		1104	
ZS-15A	145A	3063		564	
ZS-15B	145A	3063		564	
ZS-20A	116	3016		1104	
ZS-20B	116	3016		1104	
ZS-21	116	3311		1104	
ZS-23	116	3016		1104	
ZS-24	116	3031A		1104	
ZS-25	116	3016		1104	
ZS-30A	116	3016		1104	
ZS-30B	116	3016		1104	
ZS-31A	116	3016		1104	
ZS-31B	116	3016		1104	
ZS-32A	116	3311		1104	
ZS-32B	116	3311		1104	
ZS-33A	116	3031A		1104	
ZS-33B	116	3031A		1104	
ZS-34A	116	3031A		1104	
ZS-34B	116	3031A		1104	
ZS-50	116	3016		1104	
ZS-52	116	3311		1104	
ZS-53	116	3031A		1104	
ZS-71	116	3016		1104	
ZS-73	116	3016		1104	
ZS100	116	3311		1104	
ZS101	116	3311		1104	
ZS102	116	3311		1104	
ZS103	116	3031A		1104	
ZS104	116	3031A		1104	
ZS108	116	3032A		1104	
ZS10A	116	3016		1104	
ZS10B	116			1104	
ZS110	151A	3099			
ZS110A	151A	3099			
ZS110B	151A	3099			
ZS12	142A	3062		563	
ZS120	116	3311		1104	
ZS121	116	3311		1104	
ZS122	116	3311		1104	
ZS123	116	3031A		1104	
ZS124	116	3031A		1104	
ZS12A	142A	3062		563	
ZS12B	142A	3062		563	
ZS142	177	3100/519		1122	
ZS15	145A	3063		564	
ZS15A	145A	3063		564	
ZS15B	145A	3063		564	
ZS173	116	3031A		1104	
ZS174	116	3031A		1104	
ZS174B	116	3031A		1104	
ZS20A	116			1104	
ZS20B	116			1104	
ZS21	116			1104	
ZS22	116	3031A		1104	
ZS23	116			1104	
ZS24	116			1104	
ZS25	116			1104	
ZS27	146A	3064			
ZS270	5800	9003		1142	
ZS271	5801	9004		1142	
ZS272	5802	9005		1143	
ZS274	5804	9007		1144	
ZS276	5806	3848			
ZS278	156	3051			
ZS27A	146A	3064			
ZS27B	146A	3064			
ZS30	116	3311		1104	
ZS30A	116			1104	
ZS30B	116			1104	
ZS31	116	3311		1104	
ZS31A	116			1104	
ZS31B	116			1104	
ZS32	116	3311		1104	
ZS32A	116			1104	
ZS32B	116			1104	
ZS33	147A	3031A			
ZS33A	147A	3095			
ZS33B	147A	3095			
ZS34	116	3031A		1104	
ZS34A	116			1104	
ZS34B	116			1104	
ZS38	158			2007	
ZS40	177	3100/519		1122	
ZS4F	177	3100/519		1122	
ZS5.6	136A	3057		561	
ZS5.6A	136A	3057		561	
ZS5.6B	136A	3057		561	
ZS50	116			1104	
ZS51	116	3311		1104	
ZS52	116			1104	
ZS53	116			1104	
ZS56	158			2007	
ZS62	149A	3097			
ZS62A	149A	3097			
ZS62B	149A	3097			
ZS7	116	3311		1104	
ZS70	116	3017B/117		1104	
ZS701	5802	9005		1143	
ZS702	5802	9005		1143	
ZS71	116			1104	
ZS72	116	3016		1104	
ZS73	116			1104	
ZS74	116	3017B/117		1104	
ZS74B	116	3017B/117		1104	
ZS76	116	3017B/117		1104	
ZS78	125	3032A		1114	
ZS78A	125	3032A		1114	
ZS78B	125	3032A		1114	
ZS8	116	3311		1104	
ZS82	150A	3098			
ZS82A	150A	3098			
ZS82B	150A	3098			
ZS9.1	139A	3060		562	
ZS9.1A	139A	3066/118		562	
ZS9.1B	139A	3066/118		562	
ZS90	116	3311		1104	
ZS91	116	3311		1104	
ZS92	116	3311		1104	
ZS94	116	3031A		1104	
ZSC535B	108	3018		2038	
ZSD235Y	196	3054		2020	
ZSP88.2	5072A			562	
ZT-110	123A	3444		2051	
ZT-15	145A	3063		564	
ZT-15A	145A	3063		564	
ZT-15B	145A	3063		564	
ZT-62	123A	3444		2051	
ZT-82	123A	3444		2051	
ZT110	151A	3099			
ZT110A	151A	3099			
ZT110B	151A	3099			
ZT111	123A	3444		2051	
ZT112	123A	3444		2051	
ZT113	123A	3444		2051	
ZT114	123A	3444		2051	
ZT116	123A	3444		2051	
ZT117	123A	3444		2051	
ZT118	123A	3444		2051	
ZT119	123A	3444		2051	
ZT12	142A	3062		563	
ZT12A	142A	3062		563	
ZT12B	142A	3062		563	
ZT131	159	3114/290		2032	
ZT131A	159	3114/290		2032	
ZT1420	123A	3444		2051	
ZT1479	128	3024		2030	
ZT1481	128	3024		2030	
ZT1483	152	3893		2048	
ZT1484	152	3893		2048	
ZT1485	152	3893		2048	
ZT1486	152	3893		2048	
ZT1487	130	3027		2041	
ZT1488	130	3027		2041	
ZT1489	130	3027		2041	
ZT1490	130	3027		2041	
ZT15	145A	3063		564	
ZT1511	182	3188A			
ZT1512	182	3188A			
ZT1513	182	3188A			
ZT152	159	3114/290		2032	
ZT152A	159	3114/290		2032	
ZT153	159	3114/290		2032	
ZT153A	159	3114/290		2032	
ZT154	159	3114/290		2032	
ZT154A	159	3114/290		2032	
ZT15A	145A	3063		564	
ZT15B	145A	3063		564	
ZT1613	128	3024		2030	
ZT1700	128	3024		2030	
ZT1701	152	3893		2048	
ZT1702	130	3027		2041	
ZT1708	123A	3444		2051	
ZT1711	128	3444/123A		2051	
ZT180	159	3114/290		2032	
ZT180A	159	3114/290		2032	
ZT181	159	3114/290		2032	
ZT181A	159	3114/290		2032	
ZT182	159	3114/290		2032	
ZT182A	159	3114/290		2032	
ZT183	159	3114/290		2032	
ZT183A	159	3114/290		2032	
ZT184	159	3114/290		2032	
ZT184A	159	3114/290		2032	
ZT187	159	3114/290		2032	
ZT187A	159	3114/290		2032	
ZT189	159			2032	
ZT190	128	3024		2030	
ZT191	128	3024		2030	
ZT192	128	3024		2030	
ZT193	128	3024		2030	
ZT20	123A	3444		2051	
ZT20-1	123A	3444		2051	
ZT20-12	123A	3444		2051	
ZT20-55	123A	3444		2051	
ZT202	123A	3444		2051	
ZT203	123A	3444		2051	
ZT204	123A	3444		2051	
ZT20A	123A	3444		2051	
ZT20B	123A	3444		2051	
ZT20C	123A	3444		2051	
ZT21	123A	3444		2051	
ZT21-1	123A	3444		2051	
ZT21-12	123A	3444		2051	
ZT21-55	123A	3444		2051	
ZT210	129	3024/128		2030	
ZT2102	128	3722/102		2007	
ZT211	129	3024/128		2030	
ZT21A	123A	3444		2051	
ZT21B	123A	3444		2051	
ZT21C	123A	3444		2051	
ZT22	123A	3444		2051	
ZT22-1	123A	3444		2051	
ZT22-12	123A	3444		2051	
ZT22-55	123A	3444		2051	
ZT2205	123A	3444		2051	
ZT2206	123A	3444		2051	
ZT2270	128			2030	
ZT22A	123A	3444		2051	
ZT22B	123A	3444		2051	
ZT22C	123A	3444		2051	
ZT23	123A	3444		2051	
ZT23-1	123A	3444		2051	
ZT23-12	123A	3444		2051	
ZT2368	123A	3452/108		2038	
ZT2369	123A	3452/108		2038	

Industry Standard No.	ECG	SK	GE	RS 276-	MOTOR.
ZT2369A	123A	3452/108		2038	
ZT23A	123A	3444		2051	
ZT23B	123A	3444		2051	
ZT23C	123A	3444		2051	
ZT24	123A	3444		2051	
ZT24-1	123A	3444		2051	
ZT24-12	123A	3444		2051	
ZT24-55	123A	3444		2051	
ZT2475	108	3452		2038	
ZT2476	123A	3444		2051	
ZT2477	123A	3444		2051	
ZT24A	123A	3444		2051	
ZT24B	123A	3444		2051	
ZT24C	123A	3444		2051	
ZT2631	282	3748			
ZT27	146A	3064			
ZT2708	161	3452/108		2038	
ZT27A	146A	3064			
ZT27B	146A	3064			
ZT280	159	3114/290		2032	
ZT280A	159	3114/290		2032	
ZT281	159	3114/290		2032	
ZT281A	159	3114/290		2032	
ZT282	159	3114/290		2032	
ZT282A	159	3114/290		2032	
ZT283	159	3114/290		2032	
ZT283A	159	3114/290		2032	
ZT284	159	3114/290		2032	
ZT284A	159	3114/290		2032	
ZT2857	108	3452		2038	
ZT286	129	3025		2027	
ZT287	159	3114/290		2032	
ZT2876	152	3893		2048	
ZT287A	159	3114/290		2032	
ZT2938	123A	3452/108		2038	
ZT3053	192			2030	
ZT323.55		3444		2051	
ZT3269A	108	3452		2038	
ZT33	147A	3095			
ZT3375	186	3192		2017	
ZT33A	147A	3095			
ZT33B	147A	3095			
ZT3440	198	3220			
ZT3512	128	3024		2030	
ZT3600	108	3452		2038	
ZT3866	128	3024		2030	
ZT40	123A	3444		2051	
ZT402	123A	3444		2051	
ZT403	123A	3444		2051	
ZT404	123A	3444		2051	
ZT406	123A	3444		2051	
ZT41	123A	3444		2051	
ZT42	123A	3444		2051	
ZT43	123A	3444		2051	
ZT44	123A	3444		2051	
ZT5.6	136A	3057		561	
ZT5.6A	136A	3057		561	
ZT5.6B	136A	3057		561	
ZT50	123A	3444		2051	
ZT6.2	137A	3058		561	
ZT6.2A	137A	3058		561	
ZT6.2B	137A	3058		561	
ZT60	123A	3444		2051	
ZT60-1	123A	3444		2051	
ZT60-12	123A	3444		2051	
ZT60-55	123A	3444		2051	
ZT600	186	3192		2017	
ZT60A	123A	3444		2051	
ZT60B	123A	3444		2051	
ZT60C	123A	3444		2051	
ZT61	123A	3444		2051	
ZT61-1	123A	3444		2051	
ZT61-12	123A	3444		2051	
ZT61-55	123A	3444		2051	
ZT61A	123A	3444		2051	
ZT61B	123A	3444		2051	
ZT61C	123A	3444		2051	
ZT62	149A	3122			
ZT62-1	123A	3444		2051	
ZT62-12	123A	3444		2051	
ZT62-15	123A	3444		2051	
ZT62A	149A	3097			
ZT62B	149A	3097			
ZT62C	123A	3444		2051	
ZT63	123A	3444		2051	
ZT63-1	123A	3444		2051	
ZT63-12	123A	3444		2051	
ZT63-55	123A	3444		2051	
ZT63A	123A	3444		2051	
ZT63B	123A	3444		2051	
ZT63C	123A	3444		2051	
ZT64	123A	3444		2051	
ZT64-1	123A	3444		2051	
ZT64-12	123A	3444		2051	
ZT64-5	123A	3444		2051	
ZT64-55	123A	3444		2051	
ZT64A	123A	3444		2051	
ZT64B	123A	3444		2051	
ZT64C	123A	3444		2051	
ZT66	123A	3122		2051	
ZT66A	128	3024		2030	
ZT66B	128	3024		2030	
ZT66C	128	3024		2030	
ZT68	123A	3444		2051	
ZT696	123A	3444		2051	
ZT697	123A	3444		2051	
ZT706	123A	3444		2051	
ZT706A	123A	3444		2051	
ZT708	123A	3444		2051	
ZT709	108	3452		2038	
ZT80	123A	3444		2051	
ZT81	123A	3444		2051	
ZT82	123A	3122			
ZT82(TRANSISTOR)	123A			2051	
ZT83	123A	3444		2051	
ZT84	123A	3444		2051	
ZT86	128	3444/123A		2051	
ZT87	123A	3444		2051	
ZT88	128	3444/123A		2051	
ZT89	123A	3444		2051	
ZT9.1	139A	3060		562	
ZT9.1A	139A	3060		562	
ZT9.1B	139A	3066/118		562	
ZT90	128	3024		2030	
ZT917	108	3452		2038	
ZT918	108	3452		2038	
ZT93	128	3024		2030	
ZT930	199	3245		2010	
ZT94	128	3024		2030	
ZTK-33	147A	3095			
ZTR-1N60	109	3088		1123	
ZTR-B54	102A	3004		2007	
ZTR-B56	102A	3004		2007	
ZTR-W06B	116	3100/519		1104	

Industry Standard No.	ECG	SK	GE	RS 276-	MOTOR.
ZTR-W06C	116	3311		1104	
ZTR-W06B	116			1104	
ZTX107	123AP	3245/199		2010	
ZTX108	123AP	3245/199		2010	
ZTX109	123AP	3245/199		2010	
ZTX300	123AP	3444/123A		2051	
ZTX301	123AP	3444/123A		2051	
ZTX302	123AP	3244		2030	
ZTX303	123AP	3244		2030	
ZTX304	123AP	3444/123A		2051	
ZTX310	123AP	3444/123A		2051	
ZTX311	123AP	3444/123A		2051	
ZTX312	123AP	3444/123A		2051	
ZTX320	108	3452		2038	
ZTX321	108	3452		2038	
ZTX330	123AP	3444/123A		2051	
ZTX331	123AP	3244		2030	
ZTX341	171	3201			
ZTX342	171	3201			
ZTX360	123AP			.2018	
ZTX3IU				2038	
ZTX500	129			2027	
ZTX501	159			2032	
ZTX502	159			2032	
ZTX503	159			2032	
ZTX504	159			2032	
ZTX510	159			2032	
ZTX530	159	3114/290		2032	
ZTX530A	159	3114/290		2032	
ZTX530B	159	3114/290		2032	
ZTX530C	159	3114/290		2032	
ZTX530D	159	3114/290		2032	
ZTX531	159	3114/290		2032	
ZTX531A	159	3114/290		2032	
ZTX531B	159	3114/290		2032	
ZU-15	145A	3063		564	
ZU-15A	145A	3063		564	
ZU-15B	145A	3063		564	
ZU110	151A	3099			
ZU110A	151A	3099			
ZU110B	151A	3099			
ZU12	142A	3062		563	
ZU12A	142A	3062		563	
ZU12B	142A	3062		563	
ZU15	145A	3063		564	
ZU15A	145A	3063		564	
ZU15B	145A	3063		564	
ZU27	146A	3064			
ZU27A	146A	3064			
ZU27B	146A	3064			
ZU33	147A	3095			
ZU33A	147A	3095			
ZU33B	147A	3095			
ZU62	149A	3097			
ZU62A	149A	3097			
ZU62B	149A	3097			
ZU82	150A	3098			
ZU82A	150A	3098			
ZU82B	150A	3098			
ZV-15	145A	3063		564	
ZV-15A	145A	3063		564	
ZV-15B	145A	3063		564	
ZV110	151A	3099			
ZV110A	151A	3099			
ZV110B	151A	3099			
ZV12	142A	3062		563	
ZV12A	142A	3062		563	
ZV12B	142A	3062		563	
ZV15	145A	3063		564	
ZV15A	145A	3063		564	
ZV15B	145A	3063		564	
ZV27	146A	3064			
ZV27A	146A	3064			
ZV27B	146A	3064			
ZV33	147A	3095			
ZV33A	147A	3095			
ZV33B	147A	3095			
ZV62	149A	3097			
ZV62A	149A	3097			
ZV62B	149A	3097			
ZV82	150A	3098			
ZV82A	150A	3098			
ZV82..	150A	3098			
ZV9.1	139A	3099/118		562	
ZV9.1A	139A	3060		562	
ZV9.1B	139A	3060		562	
ZW0-9.1	139A			562	
ZW2	116	3311		1104	
ZW2.7	5063A	3837			
ZW3.3	5066A	3330			
ZW9.1	139A	3060		562	
ZX070	138A	3059			
ZX12	5188A			563	
ZX9.1	139A	3060		562	
ZXL-14	144A	3094		564	
ZXY-14	144A			564	
ZXY14B	144A	3094		564	
ZY-15	145A	3063		564	
ZY-15A	145A	3063		564	
ZY-15B	145A	3063		564	
ZY110	5157A	3099/151A			
ZY110A	151A	3099			
ZY110B	151A	3099			
ZY12	5127A	3062/142A		563	
ZY12A	142A	3062		563	
ZY12B	142A	3062		563	
ZY15	5130A	3063/145A		564	
ZY15A	145A	3063		564	
ZY15B	145A	3063		564	
ZY27	5139A	3064/146A			
ZY27A	146A	3064			
ZY27B	146A	3064			
ZY33	5142A	3095/147A			
ZY33A	147A	3095			
ZY33B	147A	3095			
ZY62	5150A	3097/149A			
ZY62A	149A	3097			
ZY62B	149A	3097			
ZY82	150A	3098			
ZY82A	150A	3098			
ZY82B	150A	3098			
ZZ-15	145A	3063		564	
ZZ10	140A	3061		562	
ZZ110	151A	3099			
ZZ12	142A	3062		563	
ZZ13	143A	3750			
ZZ15	145A	3063		564	
ZZ18	5027A	3793			
ZZ22	5030A	3796			
ZZ27	146A	3064			
ZZ3.3	5066A	3055/134A			
ZZ3.6	134A	3055			
ZZ3.9	5067A	3055/134A			
ZZ33	147A	3095			

Industry Standard No.	ECG	SK	GE	RS 276-	MOTOR.
ZZ4.1	135A	3056			
ZZ5.1	135A	3056			
ZZ5.6	136A	3057		561	
ZZ6.2	137A	3058		561	
ZZ6.8	5071A	3334		561	
ZZ62	149A	3097			
ZZ8.2	5072A	3136		562	
ZZ82	150A	3098			
ZZ9.1	139A	3060		562	
]A0801ADM					DAC-08AQ
]A0801AFM					DAC-08AQ
]A0801CDC					DAC-08CQ
]A0801CPC					DAC-08CP
]A0801DM					DAC-08Q
]A0801EDC					DAC-08EQ
]A0801EPC					DAC-08EP
]A0801FM					DAC-08Q
]A0801HDC					DAC-08HQ
]A0801HPC					DAC-08HP
]A0802DC-1					MC1408L8
]A0802DC-2					MC1408L7
]A0802DC-3					MC1408L6
]A0802DM-1					MC1508L8
]A0802PC-1					MC1408P8
]A0802PC-2					MC1408P7
]A0802PC-3					MC1408P6
]A101AD					LM101AJ
]A101AF					LM101AJ
]A101AH					LM101AH
]A101D					LM101AJ
]A101F					LM101AJ
]A101H					LM101AH
]A104HM					LM104H
]A105HM					LM105H
]A107H					LM107H
]A108AD					LM108AJ
]A108AF					LM108AF
]A108AH					LM108AH
]A108D					LM108J
]A108F					LM108F
]A108H					LM108H
]A109KM					LM109K
]A1312PC					MC1312P
]A1314PC					MC1314P
]A1315PC					MC1315P
]A1391PC					MC1391P
]A1394PC					MC1394P
]A1458CHC					MC1458CG
]A1458CP					MC1458CP1
]A1458CRC					MC1458CU
]A1458CTC					MC1458CP1
]A1458E					MC1458G
]A1458HC					MC1558G
]A1458P					MC1458P1
]A1458RC					MC1458U
]A1458TC					MC1458P1
]A1558E					MC1558G
]A1558HM					MC1558G
]A201AD					LM201AJ
]A201AF					LM201AJ
]A201AH					LM201AH
]A201D					LM201AJ
]A201F					LM201AJ
]A201H					LM201AH
]A207H					LM207H
]A208AD					LM208AJ
]A208AF					LM208AF
]A208AH					LM208AH
]A208D					LM208J
]A208F					LM208F
]A208H					LM208H
]A209KM					LM209K
]A2136PC					MC1356P
]A2240DC					MC1455U
]A2240DM					MC1555G
]A2240PC					MC1455P1
]A301AD					LM301AJ
]A301AH					LM301AH
]A301AT					LM301AN
]A3026HM					CA3054
]A3045					MC3346P
]A3046DC					MC3346P
]A304HC					LM304H
]A3054DC					CA3054P
]A305HC					LM305H
]A3064PC					MC1364P
]A3065PC					MC1358P
]A307H					LM307H
]A307T					LM307N
]A3086DM					MC3386P
]A308AD					LM308AJ
]A308AH					LM308AH
]A308D					LM308J
]A308H					LM308H
]A309KC					LM309K
]A311T					LM311N
]A3301P					MC3301P
]A3302P					MC3302P
]A3303P					MC3303P
]A3401P					MC3401P
]A3403D					MC3403L
]A3403P					MC3403P
]A376TC					LM305H
]A4136DC					MC4741CL
]A4136DM					MC4741L
]A4136PC					MC4741CP
]A4558HC					MC4558CG
]A4558HM					MC4558G
]A4558TC					MC4558CP1
]A555HC					MC1455G
]A555HM					MC1555G
]A555TC					MC1455P1
]A556DC					MC3456L
]A556DM					MC3556L
]A556PC					MC3456P
]A702DC					MC1712CL
]A702DM					MC1712L
]A702FM					MC1712F
]A702HC					MC1712CG
]A702HM					MC1712G
]A702MJ					MC1712L
]A702ML					MC1712G
]A709ADM					MC1709AL
]A709AFM					MC1709AF
]A709AHM					MC1709AG
]A709AMJ					MC1709AL
]A709AMJG					MC1709AU
]A709AML					MC1709AG
]A709CJ					MC1709CL
]A709CJG					MC1709CU
]A709CL					MC1709CG
]A709CN					MC1709CP2
]A709CP					MC1709CP1
]A709DC					MC1709CL
]A709DM					MC1709L
]A709FM					MC1709F
]A709HC					MC1709CG
]A709HM					MC1709G
]A709MJ					MC1709L
]A709MJG					MC1709U
]A709ML					MC1709G
]A709PC					MC1709CP2
]A709TC					MC1709CP1
]A710DC					MC1710CL
]A710DM					MC1710L
]A710HC					MC1710CG
]A710HM					MC1710G
]A710PC					MC1710CP
]A711DC					MC1711CL
]A711DM					MC1711L
]A711HC					MC1711CG
]A711HM					MC1711G
]A711PC					MC1711CP
]A715DC					MC1741SCL
]A715DM					MC1741SL
]A715HC					MC1741SCG
]A715HM					MC1741SG
]A723CJ					MC1723CL
]A723CL					MC1723CG
]A723CN					MC1723CP
]A723DC					]A723DC
]A723DM					MC1723L
]A723HC					]A723HC
]A723HM					MC1723G
]A723MJ					MC1723L
]A723ML					MC1723G
]A723PC					]A723PC
]A725AHM					LM108AH
]A725EHC					LM308AH
]A725HC					LM308AH
]A725HM					LM108AH
]A727HC					MC1420G
]A727HM					MC1520G
]A730HC					MC1420G
]A730HM					MC1520G
]A732DC					MC1310P
]A732PC					MC1310P
]A733CJ					MC1733CL
]A733CL					MC1733CG
]A733CN					MC1733CP
]A733DC					MC1733CL
]A733DM					MC1733L
]A733FM					MC1733F
]A733HC					MC1733CG
]A733HM					MC1733G
]A733MJ					MC1733L
]A733ML					MC1733G
]A734DC					LM311J
]A734DM					LM311J
]A734HC					LM311H
]A734HM					LM311H
]A740HC					LF355H
]A740HM					LF155H
]A741ADM					MC1741L
]A741AFM					MC1741F
]A741AHM					MC1741G
]A741CJ					MC1741CL
]A741CJG					MC1741CU
]A741CL					MC1741CG
]A741CN					MC1741CP2
]A741CP					MC1741CP1
]A741DC					]A741DC
]A741DM					MC1741L
]A741EDC					MC1741L
]A741EHC					MC1741G
]A741FC					MC1741CF
]A741FM					MC1741F
]A741HC					]A741HC
]A741HM					MC1741G
]A741MJ					MC1741L
]A741MJG					MC1741U
]A741ML					MC1741G
]A741PC					MC1741CP2
]A741RC					MC1741CU
]A741RM					MC1741U
]A741TC					]A741TC
]A742DC					CA3059
]A746DC					MC1323P
]A746HC					MC1323P
]A747ADM					MC1747L
]A747AHM					MC1747G
]A747CJ					MC1741CL
]A747CL					MC1747CG
]A747CN					MC1747CP2
]A747DC					MC1747CL
]A747DM					MC1747L
]A747EDC					MC1747CCBM
]A747EHC					MC1747CICM
]A747HC					MC1747CG
]A747HM					MC1747G
]A747MJ					MC1747L
]A747ML					MC1747G
]A747PC					MC1747CP2
]A748AFM					MC1748F
]A748AHM					MC1748G
]A748CJ					MC1748CL
]A748CJG					MC1748CU
]A748CL					MC1748CG
]A748CN					MC1748CP2
]A748CP					MC1748CP1
]A748DC					MC1748CL
]A748DM					MC1748L
]A748FM					MC1748F
]A748HC					MC1748CG
]A748HM					MC1748G
]A748MJ					MC1748L
]A748MJG					MC1748U
]A748ML					MC1748G
]A748TC					MC1748CP1
]A749DC					MC1435L
]A749DHC					MC1435G
]A749DM					MC1535L
]A749HC					MC1435G
]A753TC					MC1356P
]A754HC					MC1355P
]A754TC					MC1355P
]A757DC					MC1350P
]A757DM					MC1350P
]A758DC					MC1310P
]A758PC					MC1310P
]A767DC					MC1310P
]A767PC					MC1310P
]A772					MC1741S
]A775DC					LM339J
]A775DM					LM339J
]A775PC					LM339N
]A776DC					MC1776CG
]A776DM					MC1776G

Industry Standard No.	ECG	SK	GE	RS 276-	MOTOR.
]A776HC					MC1776CG
]A776HM					MC1776G
]A776TC					MC1776CP1
]A777CJ					LM308AJ-8
]A777CJG					LM308AJ-8
]A777CL					LM308AH
]A777CN					LM308AN
]A777CP					LM308AN
]A777DC					LM308AJ-8
]A777HC					LM308AH
]A777MJ					LM108AJ-8
]A777MJG					LM108AJ-8
]A777ML					LM108AH
]A777TC					LM308AN
]A7805CKC					MC7805CT
]A7805KC					MC7805CK
]A7805KM					MC7805K
]A7805UC					MC7805CT
]A7806CKC					MC7806CT
]A7806KC					MC7806CK
]A7806KM					MC7806K
]A7806UC					MC7806CT
]A7808CKC					MC7808CT
]A7808KC					MC7808CK
]A7808KM					MC7808K
]A7808UC					MC7808CT
]A780DC					MC1399P
]A780PC					MC1399P
]A7812CKC					MC7812CT
]A7812KC					MC7812CK
]A7812KM					MC7812K
]A7812UC					MC7812CT
]A7815CKC					MC7815CT
]A7815KC					MC7815CK
]A7815KM					MC7815K
]A7815UC					MC7815CT
]A7818CKC					MC7818CT
]A7818KC					MC7818CK
]A7818KM					MC7818K
]A7818UC					MC7815CT
]A781DC					MC1399P
]A781PC					MC1399P
]A7824CKC					MC7824CT
]A7824KC					MC7824CK
]A7824KM					MC7824K
]A7824UC					MC7824CT
]A786DC					MC1327P
]A787PC					MC1399P
]A78GHM					LM117K
]A78GKC					LM117K
]A78GKM					LM117K
]A78GU1C					LM317T
]A78H05KC					MC7805CK
]A78L02ACJG					MC78L02ACG
]A78L05ACJG					MC78L05ACG
]A78L05ACLP					MC78L05ACP
]A78L05AHC					MC78L05ACG
]A78L05AWC					MC78L05ACP
]A78L05CJG					MC78L05CG
]A78L05CLP					MC78L05CP
]A78L05HC					MC78L05CG
]A78L05WC					MC78L05CP
]A78L06ACJG					MC78L06ACG
]A78L06ACLP					MC78L06ACP
]A78L06CJG					MC78L06CG
]A78L06CLP					MC78L06CP
]A78L08ACJG					MC78L08ACG
]A78L08ACLP					MC78L08ACP
]A78L08CJG					MC78L08CG
]A78L08CLP					MC78L08CP
]A78L12ACJG					MC78L12ACG
]A78L12ACLP					MC78L12ACP
]A78L12AHC					MC78L12ACG
]A78L12AWC					MC78L12ACP
]A78L12CJG					MC78L12CG
]A78L12CLP					MC78L12CP
]A78L12HC					MC78L12CG
]A78L12WC					MC78L12CP
]A78L15ACJG					MC78L15ACG
]A78L15ACLP					MC78L15ACP
]A78L15AHC					MC78L15ACG
]A78L15AWC					MC78L15ACP
]A78L15CJG					MC78L15CG
]A78L15CLP					MC78L15CP
]A78L15HC					MC78L15CG
]A78L15WC					MC78L15CP
]A78L26AWC					MC7802ACP
]A78M05CKC					MC78M05CT
]A78M05HC					MC78M05CG
]A78M05HM					MC78M05CG
]A78M05UC					MC78M05CT
]A78M06CKC					MC78M06CT
]A78M06HC					MC78M06CG
]A78M06HM					MC78M06CG
]A78M06UC					MC78M06CT
]A78M08CKC					MC78M08CT
]A78M08HC					MC78M08CG
]A78M08HM					MC78M08CG
]A78M08UC					MC78M08CT
]A78M12CKC					MC78M12CT
]A78M12HC					MC78M12CG
]A78M12HM					MC78M12CG
]A78M12UC					MC78M12CT
]A78M15CKC					MC78M15CT
]A78M15HC					MC78M15CG
]A78M15HM					MC78M15CG
]A78M15UG					MC78M15CT
]A78M18HC					MC78M18CG
]A78M18HM					MC78M18CG
]A78M18UG					MC78M18CT
]A78M20CKC					MC78M20CT
]A78M20HC					MC78M20CG
]A78M20HM					MC78M20CG
]A78M20UG					MC78M20CT
]A78M24CKC					MC78M24CT
]A78M24HC					MC78M24CG
]A78M24HM					MC78M24CG
]A78M24UC					MC78M24CT
]A78MGHC					LM317H
]A78MGT2C					LM317T
]A78MGU1C					LM317T
]A7902KC					MC7902K
]A7902KM					MC7902K
]A7902UC					MC7902CT
]A7905KC					MC7905CK
]A7905KM					MC7905CK
]A7905UC					MC7905CT
]A7906KC					MC7906CK
]A7906KM					MC7906CK
]A7906UC					MC7906CT
]A7908KC					MC7908CK
]A7908KM					MC7908CK
]A7908UC					MC7908CT
]A7912KC					MC7912CK
]A7912KM					MC7912CK
]A7912UC					MC7912CT
]A7915KC					MC7915CK
]A7915KM					MC7915CK
]A7915UC					MC7915CT
]A7918CKC					MC7918CT
]A7918KC					MC7918CK
]A7918KM					MC7918CK
]A7918UC					MC7918CT
]A791KC					MC1438R
]A791KM					MC1538R
]A791P5					MC1438R
]A7924CKC					MC7924CT
]A7924KC					MC7924CK
]A7924KM					MC7924CK
]A7924UC					MC7924CT
]A796DC					MC1496L
]A796DM					MC1596L
]A796HC					MC1496G
]A796HM					MC1596G
]A798HC					MC3458G
]A798HM					MC3558G
]A798RC					MC3458U
]A798RM					MC3558U
]A798TC					MC3458P1
]A799HC					MC1741G
]A799HM					MC1741G
]A79L05AHC					MC79L05ACG
]A79L05AWC					MC79L05ACP
]A79L05HC					MC79L05CG
]A79L05WC					MC79L05CP
]A79L12AHC					MC79L12ACG
]A79L12AWC					MC79L12ACP
]A79L12HC					MC79L12CG
]A79L12WC					MC79L12CP
]A79L15AHC					MC79L15ACG
]A79L15AWC					MC79L15ACP
]A79L15HC					MC79L15CG
]A79L15WC					MC79L15CP
]A79M05AHM					MC7905CK
]A79M05AUC					MC7905CT
]A79M05CKC					MC7905CT
]A79M05HM					MC7905CK
]A79M05UC					MC7905CT
]A79M06AHM					MC7906CK
]A79M06AUC					MC7906CT
]A79M06CKC					MC7906CT
]A79M06HM					MC7906CK
]A79M06UC					MC7906CT
]A79M08AHM					MC7908CK
]A79M08AUC					MC7908CT
]A79M08CKC					MC7908CT
]A79M08HM					MC7908CK
]A79M08UC					MC7908CT
]A79M12AHM					MC7912CK
]A79M12AUC					MC7912CT
]A79M12CKC					MC7912CT
]A79M12HM					MC7912CK
]A79M12UC					MC7912CT
]A79M15AHM					MC7915CK
]A79M15AUC					MC7915CT
]A79M15CKC					MC7915CT
]A79M15HM					MC7915CK
]A79M15UC					MC7915CT
]A79M18AHM					MC7918CK
]A79M18AUC					MC7918CT
]A79M18HM					MC7918CK
]A79M18UC					MC7918CT
]A79M24AHM					MC7924CK
]A79M24AUC					MC7924CT
]A79M24HM					MC7924CK
]A79M24UC					MC7924CT
]A8T13DC					MC8T13L
]A8T13PC					MC8T13P
]A8T14DC					MC8T14L
]A8T14PC					MC8T14P
]A8T23DC					MC8T23L
]A8T23PC					MC8T23P
]A8T24DC					MC8T24L
]A8T24PC					MC8T24P
]AF155AHM					LF155AH
]AF155HM					LF155H
]AF156AHM					LF156AH
]AF156HM					LF156H
]AF157AHM					LF157AH
]AF157HM					LF157H
]AF355AHC					LF355AH
]AF355HC					LF355H
]AF356AHC					LF356AH
]AF356HC					LF356H
]AF357AHC					LF357AH
]AF357HC					LF357H
]PD2114L					MCM21L14
]PD2118					MCM4517
]PD2147					MCM2147
]PD2332					MCM68A332
]PD2364					MCM68A364
]PD2716					MCM2716
]PD4104					MCM66L41
]PD414A					MCM4027A
]PD416					MCM4116
]PD4164					MCM6665
]PD443					MCM6508
001	6400A	3466/159	300	2032	
1 735 0060	159	3715			
1 735 0061	159	3715			
1 735 0066	159	3715			
1&12/1			504A	1104	
1+12/1			504A	1104	
1,000,111-00	159		ZD-82	2032	
001-00	123A	3444	20	2051	
1-000-099-00	7490			1808	
001-0000-00	109	3090	1N34AS	1123	
1-0002-0001	519		514	1122	
1-0002-001				1122	
1-0006-0021	123A	3444	20	2051	
1-0006-0022	123A	3444	20	2051	
1-0006-0023	159	3715	82	2032	
1-001-003-15	123A	3444	20	2051	
1-001/2207	177	3100/519	300	1122	
001-0010-00	109	3087		1123	
1-002	130	3715	14	2041	
001-0020-00	110MP	3709		1123(2)	
001-0022-00	109	3090		1123	
1-003	130	3715	14	2041	
001-0036	720	9014	IC-7		
001-0072-00	125	3031A	510,531	1114	
001-0077-00	116	3031A	504A	1104	
001-0081	109	3088		1123	
001-0081-00	117	3017B	504A	1104	
001-0082-00	5071A	3334	ZD-6.8	561	
001-0085-00		3126	90		
001-0091	1140	3473	IC-138		
001-0095-00	177	3175	300	1122	

Industry Standard No.	ECG	SK	GE	RS 276-	MOTOR.
001-0095-02	519	3100			
001-0099-00	138A	3059			
001-0099-01	138A	3059	ZD-6.8	561	
001-0099-02	138A	3059	ZD-7.5		
001-0101-01	5073A	3749	ZD-9.1	562	
01-010473	153	3083/197	69	2049	
01-010495	290A	3114/290	221	2032	
01-010562	290A		82	2032	
01-010564	234	3114/290	65	2050	
01-010628	159	3114/290	82		
01-010673	290A	3114/290	269	2032	
01-010719	290A	3114/290	269	2032	
01-010733	159	3138/193A	38		
01-010844	159	3114/290	244	2050	
001-01101-0	101	3861	8	2002	
001-011010	101	3861	8	2002	
001-0112-00	519	3100		1122	
001-012010	102	3004			
001-012011	102	3004	53	2007	
001-01202-1	100	3005	1	2007	
001-012020	100	3123			
001-012021	100	3004/102A	53	2007	
001-01203-1	100	3005	1	2007	
001-012030	100	3123			
001-012031	100	3004/102A	53	2007	
001-01204-0	104	3009	16	2006	
001-012040	104	3009	16	2006	
001-01205-0	104	3009	16	2006	
001-01205-1	104	3719	16	2006	
001-012050	179	3009	16	2006	
001-012051	104	3009	16	2006	
001-012052	104	3009			
001-012053	104	3009			
001-01206-0	102	3722	2	2007	
001-012060	102	3004	53	2007	
001-0125-00	177	3175	300	1122	
001-01250			3	2006	
001-01251			3	2006	
001-01252			3	2006	
001-01253			3	2006	
001-01254			3	2006	
001-01255			3	2006	
001-01256			3	2006	
001-01257			3	2006	
001-01258			3	2006	
001-01259			3	2006	
001-0127-00	140A	3061	ZD-10	562	
001-0130-00	613	3126	90		
1-014/2207	177	3100/519	300	1122	
001-01501-0	109	3087		1123	
001-015010	109	3087		1123	
001-015011	109	3087	1N34AS	1123	
001-0151-00	177	3175	300	1122	
001-0151-01	177	3175	300	1122	
001-0152-00	5072A	3136		562	
001-0153-00	116	3313		1104	
1-016		3087		1123	
001-0160-00	614	3327	90		
001-0161-00	5072A	3136	ZD-8.2	562	
001-0163-02	135A	3056	ZD5.1		
001-0163-04	136A	3057	ZD-5.6	561	
001-0163-13	5072A	3136	ZD-8.7	562	
001-0163-15	139A	3060	ZD-9.1	562	
1-017	109	3087		1123	
001-0176-00	614	3327	90		
001-02	159	3466	82	2032	
001-02010	158	3004/102A	82	2007	
001-02011	123	3122	20	2051	
001-02020	123A	3444	20	2051	
01-020562	294		48		
01-020566	197	3083	250	2027	
001-02101-0	123A	3444	20	2051	
001-02101-1	123A	3444	20	2051	
001-021010	289A	3020/123	20	2051	
001-021011	289A	3124/289			
001-02102-0	123A	3444	20	2051	
001-021020	289A	3444/123A	20	2051	
001-02103-0	123A	3444	20	2051	
001-021030	289A	3444/123A	20	2051	
001-02104-0	123A	3444	20	2051	
001-021040	289A	3020/123	20	2051	
001-02105-0	123A	3444	20	2051	
001-021050	289A	3020/123	20	2051	
001-02106-0	123A	3444	20	2051	
001-021060	289A	3020/123	20	2051	
001-02107-0	123A	3444	20	2051	
001-021070	199	3444/123A	20	2051	
001-02108-0	123A	3444	20	2051	
001-021080	289A	3444/123A	20	2051	
001-02109-0	123A	3444	20	2051	
001-021090	289A	3020/123	20	2051	
001-02110-0	123A	3444	20	2051	
001-021100	194	3044/154	40	2012	
001-02111-0			20	2051	
001-02111-1	123A	3444	20	2051	
001-021110	128	3024	20	2051	
001-021111	128	3024	20	2051	
001-02113-2	123A	3444	20	2051	
001-02113-3	123A	3444	20	2051	
001-02113-4	123A	3444	20	2051	
001-02113-5	123A	3444	20	2051	
001-021130	289A	3124/289			
001-021131	289A	3124/289			
001-021132	289A	3444/123A	20	2051	
001-021133	289A	3020/123	20	2051	
001-021134	289A	3444/123A	20	2051	
001-021135	289A	3444/123A	20	2051	
001-021136	199	3124/289			
001-02114-0	198	3220	251		
001-021140	396	3021/124	12		
001-02115-0	124	3021	12		
001-021150	124	3021	12		
001-021151	124	3021	12		
001-02116-0	124	3021	12		
001-021160	124	3021	12		
001-021161	124	3021			
001-021162	124	3021	12		
001-021163	124	3021			
001-02117-2	193	3138	67	2023	
001-021170	159	3247/234	65	2050	
001-021171	159	3247/234	65	2050	
001-021172	159	3247/234	65	2050	
001-021173	159	3247/234	65	2050	
001-021180	130	3027			
001-02119-0	128	3024	243	2030	
001-021190	128	3024	243	2030	
001-021200	130	3027			
001-02121-0	123A	3444	20	2051	
001-021210	123A	3124/289	20	2051	
001-021211	123A	3124/289			
001-021218	123A	3124/289			
001-021221	225		243	2030	
001-021230	192A	3124/289			
001-021231	192A	3124/289			
001-021232	192A	3124/289	20	2051	
001-021270	130	3027	14	2041	
001-021280	181	3036	75	2041	
001-021290	128	3024	243	2030	
001-02201-0	159	3715	82	2032	
001-022010	159	3466	82	2032	
001-022020	290A	3466/159	82	2032	
001-022030	192A	3124/289			
001-022031	192A	3124/289			
001-022032	193A	3114/290			
001-022050	6402	3628			
001-023027	146A	3064			
001-02303-0	135A	3056	ZD-5.1		
001-02303-3	137A	3058	ZD-6.2	561	
001-02303-4	145A	3063	ZD-15	564	
001-023030	135A	3333/5069A	ZD-4.7		
001-023031	5080A	3336	ZD-22		
001-023032	5079A	3335	ZD-20		
001-023033	137A	3057/136A	ZD-5.6	561	
001-023034	145A	3063	ZD-15	564	
001-023035	140A	3061	ZD-10	562	
001-023036	5067A	3333/5069A	ZD-4.7		
001-023037	5069A	3062/142A	ZD-12	563	
001-023038	5075A	3751	ZD-16	564	
001-023041	5130A	3396	5ZD-15		
001-023042	5121A	3386/5120A	5ZD-6.8		
001-024010	116	3016			
001-024020	116	3017B/117			
001-024030	116	3017B/117			
001-024040	5986	3501/5994			
001-02405-0	116	3017B/117	504A	1104	
001-02405-1	116	3017B/117	504A	1104	
001-02405-2	116	3017B/117	504A	1104	
001-024050	116	3016	504A	1104	
001-024051	116	3016	504A	1104	
001-024052	116	3016	504A	1104	
001-02406-0	156	3051	512		
001-02406-1	156	3051	512		
001-024060	5801	3501/5994			
001-024061	5801	3051/156	512		
001-024070	5980	3610	5096		
001-024080	289A	3124/289			
001-02601-0	116	3017B/117	504A	1104	
001-026010	177	3100/519	514	1122	
001-02603-0	116	3017B/117	504A	1104	
001-026030	177	3175	300	1122	
001-026050	6402	3628			
001-026060	519	3170/731	300	1122	
001-02701-1			FET-1	2028	
001-027010	326	3116			
001-02702-0			FET-1	2028	
001-027020	326	3116			
001-02703-0			FET-2	2035	
001-027030	312	3834/132	FET-1	2035	
001-028-000	909		249		
001-03	159	3466	82	2032	
01-030372	123AP	3245/199			
01-030373	199	3245	62	2010	
01-030380	229		61	2038	
01-030388	233	3132			
01-030394	229	3018	61	2038	
01-030454	289A	3124/289	210		
01-030458	123A	3124/289	210	2009	
01-030495	295		270		
01-030509	289A	3124/289	269	2032	
01-030536	199	3122	212	2032	
01-030643		3115	259		
01-030668	107	3018			
01-030682	161	3018	61	2015	
01-030710	123AP	3356	211	2009	
01-030711	199		62	2010	
01-030732	199		62	2010	
01-030733	159	3245/199	48	2027	
01-030734	128	3122	212	2030	
01-030735	289A	3122	210	2038	
01-030763	229	3122	61	2038	
01-030784	229		62	2010	
01-030785	229	3246			
01-030828	199	3122	61		
01-030829	123AP	3122	20	2051	
01-030900	199	3124/289	62	2010	
01-030930	107	3356	61	2038	
01-030945	123AP	3124/289	212	2010	
01-030983	287	3244	27		
01-031018	299	3298	236		
01-031047	229		60	2038	
01-031096	186A		28	2052	
01-031166	297	3122	81	2030	
01-031173	152	3054/196	215	2053	
01-031175	123A	3444	210	2051	
01-031213	289A	3122	268		
01-031239	237		46		
01-031293	233	3132	61	2009	
01-031306	235	3239/236	215		
01-031317	289A		81	2038	
01-031318	297	3122	210	2030	
01-031327	199		85	2010	
01-031359	229		212	2038	
01-031360		3132	11	2015	
01-031364	123A	3124/289	210	2051	
01-031446		3220	251		
01-031507	198	3219	251		
01-031514	376	3219	251		
01-031520		3220	251		
01-031550		3747	232		
01-031674	229		61	2038	
01-031675	229	3124/289	213	2038	
01-031678	235		215		
01-031685		3124	62	2010	
01-031686		3246	61	2038	
01-031687	123A		20	2051	
01-031688		3246	81	2051	
01-031756	198	3219	251		
01-031760	306	3251	276		
01-031815	123AP	3854	62	2010	
01-031846	306	3251	336		
01-031855	108	3246/229	61		
01-031906	108	3293/107	86		
01-031909	235	3197	215		
01-031921	287	3433	222		
01-031923	229	3132			
01-031957	295	3253	270		
01-031964	236	3856/310	215		
01-031973	340	3250/315			
01-032057	319P	3246/229	39	2016	
01-032076	123AP			2051	
01-032078	235	3197			
01-032092	235		215		
01-032314	295	3253			
1-034/2207	159	3466	82	2032	
1-035/2207	128	3124/289	243	2030	
1-037/2207	109	3088		1123	
001-04	159	3466	82	2032	
01-040201		3027	75	2041	

Industry Standard No.	ECG	SK	GE	RS 276-	MOTOR.
01-040234	152	3893			
01-040243	152	3054/196	66	2048	
01-040299		3115	259		
01-040313	196		241	2020	
01-040389		3054	66	2048	
01-040471	293	3849			
01-040476	196	3054	241		
01-040724	291	3440			
1-041/2207	108	3452	86	2038	
1-042/2207	123A	3444	20	2051	
1-043/2207	159	3466	82	2032	
1-044/2207	123A	3444	20	2051	
001-044272-002	289A	3124/289			
001-044273-002	128	3024			
001-044275-002	175	3024/128			
001-044277-002	116	3311			
001-044672-001	102A	3004			
001-044673-001	102A	3004			
001-044674-001	289A	3124/289			
001-044676-001	177	3100/519			
001-044677-001	199	3018			
01-070019	453	3834/132			
01-070030	312		FET-1	2035	
01-080045	222	3065	FET-4	2036	
01-080059	222	3065			
1-101	116	3017B/117	504A	1104	
01-117005	108			2038	
01-117006	108			2038	
01-119185-01		3328	IC-31		
01-119185-02		3328	IC-31		
01-119185-03		3328	IC-31		
01-1201-0	102A	3004	53	2007	
01-1205-0P	179		76		
01-121365	712	3072	IC-148		
1-12689	109	3088	1N60	1123	
1-13989	519		514	1122	
01-1501	177	3175	300	1122	
1-16549	177	3100/519	300	1122	
1-18341		3126	90		
1-20-001-890	116		504A	1104	
01-201-0				2051	
1-20363	139A		ZD-9.1	562	
1-20398	140A	3061	ZD-10	562	
1-21-100	100	3005	1	2007	
1-21-102	100	3005	1	2007	
1-21-103	100	3005	1	2007	
1-21-104	100	3005	1	2007	
1-21-105	100	3005	1	2007	
1-21-106	102A	3004	53	2007	
1-21-107	102A	3004	53	2007	
1-21-120	102A	3004	53	2007	
1-21-128	100	3005	1	2007	
1-21-135	160	3006	245	2004	
1-21-137	160	3006	245	2004	
1-21-138	126	3008	52	2024	
1-21-139	160	3006	245	2004	
1-21-148	102A	3004	53	2007	
1-21-150	160	3007	245	2004	
1-21-157	160	3006	245	2004	
1-21-161	100	3005	1	2007	
1-21-162	100	3005	1	2007	
1-21-164	102A	3004	53	2007	
1-21-179	100	3005	1	2007	
1-21-180	100	3005	1	2007	
1-21-184	102A	3004	53	2007	
1-21-186	100	3005	1	2007	
1-21-189	126	3008	52	2024	
1-21-190	160	3007	245	2004	
1-21-191	102A	3004	53	2007	
1-21-192	102A	3004	53	2007	
1-21-225	102A	3004	53	2007	
1-21-226	102A	3004	53	2007	
1-21-227	102A	3004	53	2007	
1-21-228	160	3006	245	2004	
1-21-229	160	3006	245	2004	
1-21-230	160	3006	245	2004	
1-21-231	160	3006	245	2004	
1-21-232	102A	3004	53	2007	
1-21-233	160	3006	245	2004	
1-21-234	100	3005	1	2007	
1-21-235	100	3005	1	2007	
1-21-236	100	3005	1	2007	
1-21-240	100	3005	1	2007	
1-21-241	100	3005	1	2007	
1-21-242	126	3008	52	2024	
1-21-243	126	3008	52	2024	
1-21-244	126	3008	52	2024	
1-21-246	102A	3004	53	2007	
1-21-254	100	3005	1	2007	
1-21-256	160	3007	245	2004	
1-21-257	126	3008	52	2024	
1-21-258	160	3007	245	2004	
1-21-259	160	3007	245	2004	
1-21-260	160	3007	245	2004	
1-21-266	102A	3004	53	2007	
1-21-267	102A	3004	53	2007	
1-21-270	121	3009	239	2006	
1-21-271	121	3009	239	2006	
1-21-272	102A	3004	53	2007	
1-21-273	100	3005	1	2007	
1-21-274	102A	3003	53	2007	
1-21-275	100	3005	1	2007	
1-21-276	123A	3444	20	2051	
1-21-277	123A	3444	20	2051	
1-21-278	123A	3444	20	2051	
1-21-279	123A	3444	20	2051	
1-21-289	100	3005	1	2007	
1-21-73	100	3005	1	2007	
1-21-74	100	3005	1	2007	
1-21-75	100	3005	1	2007	
1-21-76	100	3005	1	2007	
1-21-78	100	3005	1	2007	
1-21-83	100	3005	1	2007	
1-21-91	100	3005	1	2007	
1-21-92	100	3005	1	2007	
1-21-93	100	3005	1	2007	
1-21-95	102A	3004	53	2007	
1-21-96	102A	3004	53	2007	
1-210	136A	3057	ZD-5.6	561	
01-2101	123A	3444	20	2051	
01-2101-0	123A	3444	20	2051	
001-21011	123A	3444	20	2051	
01-2102	123A	3444	20	2051	
01-2104	123A	3444	20	2051	
01-2105	123A	3444	20	2051	
01-2106	123A	3444	20	2051	
01-2107	123A	3444	20	2051	
01-2108	123A	3444	20	2051	
01-2109	123A	3444	20	2051	
01-2110	128	3024	243	2030	
01-2111	128	3024	243	2030	
01-2114	128	3024	243	2030	
1-2114-0	128		243	2030	
001-223027	146A	3064			

Industry Standard No.	ECG	SK	GE	RS 276-	MOTOR.
001-223034	145A	3063			
001-226010	177	3100/519	514	1122	
01-226030	177	3016			
01-2303-0	5080A	3336	ZD-22		
01-2303-1	5080A	3336	ZD-22		
01-2303-3	5071A	3334	ZD-6.8		
01-2303-33				561	
01-2405-0	116	3175	504A	1104	
01-2405-1	116	3017B/117	504A	1104	
01-2405-2	116	3017B/117	504A	1104	
01-2406-1	116	3017B/117	504A	1104	
01-2601	177	3175	300	1122	
01-2601-0	177	3175	300	1122	
01-2603-0	177	3175	300	1122	
01-30828			61	2038	
01-30829			20	2051	
01-349418	123AP	3854	20	2051	
01-349426	229	3246			
01-349634	123AP	3854			
01-349681	159	3466			
1-425-636	109	3091		1123	
1-506-319	156	3051	512		
1-52221011	102A	3004	53	2007	
1-522210111	102A	3004	53	2007	
1-522210131	160	3006	245	2004	
1-522210300	160	3006	245	2004	
1-522210921	160	3007	245	2004	
1-522211021	160	3006	245	2004	
1-522211200	102A	3004	53	2007	
1-522211328	102A	3004	53	2007	
1-522211921	160	3006	245	2004	
1-522214400	160	3006	245	2004	
1-522214411	160	3006	245	2004	
1-522214435	160	3006	245	2004	
1-522214821	160	3006	245	2004	
1-522214831	160	3006	245	2004	
1-522216500	102A	3004	53	2007	
1-522216600	160	3007	245	2004	
1-522217400	160	3006	245	2004	
1-522223720	123A	3444	20	2051	
1-530-012-11	116	3031A	504A	1104	
1-531-024	167	3647	512	1172	
1-531-027	116	3081/125	504A	1104	
1-531-028	503	3068	CR-5		
1-531-028-21	503		CR-5		
1-531-036-11		3066	CR-1		
1-531-052		3067	CR-4		
1-531-052(SONY)	502		CR-4		
1-531-055		3067	CR-4		
1-531-055(SONY)	502		CR-4		
1-531-105	116	3017B/117	504A	1104	
1-531-105-11		3016	504A	1104	
1-531-105-13		3017B	504A	1104	
1-531-10513	116		504A	1104	
1-531-106			504A	1104	
1-531-106-13	116	3017B/117	504A	1104	
1-531-106-17	116	3016	504A	1104	
1-531-5-11	116		504A	1104	
001-533-00	159	3466	82	2032	
1-534-105-13	116	3016	504A	1104	
1-534-106-13			504A	1104	
01-571591	159	3466	82		
01-571751	159	3466	82		
01-571794	161	3246/229	39	2015	
01-571804	233	3132	39		
01-571811		3444	20	2051	
01-571821	123AP	3444/123A	20	2051	
01-571831	154	3433/287			
01-571921	124	3131A/369	12		
01-571941	123A	3433/287	40	2012	
01-572088	171	3201	27		
01-572588	159	3466	82	2032	
01-572631		3710	259		
01-572774	153	3083/197	69	2049	
01-572784	152	3054/196	66	2048	
01-572791	152	3054/196	66	2048	
01-572811	290A	3114/290	269	2032	
01-572814	289A	3124/289	268	2038	
01-572831	154	3433/287	40	2012	
01-572861	152	3054/196	66	2048	
01-57291	152		66	2048	
1-590-1261	7440	7440			
1-6147191229	123A	3444	20	2051	
1-6171191368	123A	3444	20	2051	
1-6207190405	102A	3004	53	2007	
1-6501190016	116	3031A	504A	1104	
01-680815	123AP	3124/289		2051	
01-690261	192	3137			
01-690471	297	3449			
01-690564	298	3450			
01-690643	193A	3138			
01-690733	290A	3114/290			
01-691674	107	3132			
1-800-662-11	1185		IC-139		
1-801-003	123A	3018	20	2051	
1-801-003-12	108	3452	86	2038	
1-801-003-13	108	3452	86	2038	
1-801-003-14	108	3452	86	2038	
1-801-003-15	108	3452	86	2038	
1-801-004	123A	3444	20	2051	
1-801-004-17	123A	3444	20	2051	
1-801-005	158	3004/102A	53	2007	
1-801-005-23	102A	3004	53	2007	
1-801-006	158	3004/102A	53	2007	
1-801-006-12	102A	3004	53	2007	
1-801-006-14	102A	3004	53	2007	
1-801-301-13	124	3021	12		
1-801-301-14	124	3021	12		
1-801-301-15	124	3021	12		
1-801-304-15	107	3039/316	11	2015	
1-801-305-13	108	3452	86	2038	
1-801-306	108	3452	86	2038	
1-801-306-13	108	3452	86	2038	
1-801-306-14	108	3452	86	20388	
1-801-306-15	108	3452	86	2038	
1-801-308	102A	3004	53	2007	
1-801-308-24	102A	3004	53	2007	
1-801-309	103A	3018	59	2002	
1-801-310	158	3004/102A	53	2007	
1-801-314	123A	3444	20	2051	
1-801-314-15	123A	3444	20	2051	
1-801-314-16	123A	3444	20	2051	
1-8259	116	3016	504A	1104	
01-9011-5/2221-3			20	2051	
01-9011-52221-3	123A	3444	20	2051	
01-9013-7/2221-3			810	2051	
01-9013-72221-3	123A	3444	20	2051	
01-9014-2/2221-3			10	2051	
01-9014-22221-3	123A	3444	20	2051	
01-9016-4/2221-3			20	2051	
01-9016-42221-3	108	3452	86	2038	
01-9018-6/2221-3			17	2051	
01-9018-62221-3	108	3452	86	2038	
1-DI-007	177		300	1122	

Industry Standard No.	ECG	SK	GE	RS 276-	MOTOR.
1-DI-009	109			1123	
1-RE-004	116		504A	1104	
1-TR-046	108		86	20388	
1-TR-048	107		11	2015	
1-TR-111	102A		53	2007	
1-TR-112	103A	3835	59	2002	
1.5B05	156	3051	512		
1.5BZ100D	5156A	3422			
1.5BZ110B	5157A	3423			
1.5BZ120	5158A	3424			
1.5BZ120C	5158A	3424			
1.5BZ130D	5159A	3425			
1.5BZ140B	5160A	3426			
1.5BZ150	5161A	3427			
1.5BZ150C	5161A	3427			
1.5C05	156	3051	512		
1.5DKZ10	5125A	3391			
1.5DKZ10A	5125A	3391	5ZD-10		
1.5DKZ10B	5125A	3391			
1.5DKZ11	5126A	3392			
1.5DKZ11A	5126A	3392	5ZD-11		
1.5DKZ11B	5126A	3392			
1.5DKZ12	5127A	3393			
1.5DKZ12A	5127A	3393	5ZD-12		
1.5DKZ12B	5127A	3393			
1.5DKZ13	5128A	3394			
1.5DKZ13A	5128A	3394	5ZD-13		
1.5DKZ13B	5128A	3394			
1.5DKZ15	5130A	3396			
1.5DKZ15A	5130A	3396	5ZD-15		
1.5DKZ15B	5130A	3396			
1.5DKZ16	5131A	3397			
1.5DKZ16A	5131A	3397	5ZD-16		
1.5DKZ16B	5131A	3397			
1.5DKZ18	5133A	3399			
1.5DKZ18A	5133A	3399	5ZD-18		
1.5DKZ18B	5133A	3399			
1.5DKZ20	5135A	3401	5ZD-20		
1.5DKZ20A	5135A	3401			
1.5DKZ20B	5135A	3401			
1.5DKZ22	5136A	3402			
1.5DKZ22A	5136A	3402	5ZD-22		
1.5DKZ22B	5136A	3402			
1.5DKZ24	5137A	3403			
1.5DKZ24A	5137A	3403	5ZD-24		
1.5DKZ24B	5137A	3403			
1.5DKZ27	5139A	3405			
1.5DKZ27A	5139A	3405	5ZD-27		
1.5DKZ27B	5139A	3405			
1.5DKZ30	5141A	3407	5ZD-30		
1.5DKZ30A	5141A	3407			
1.5DKZ30B	5141A	3407			
1.5DKZ33	5142A	3408			
1.5DKZ33A	5142A	3408			
1.5DKZ33B	5142A	3408	5ZD-33		
1.5DKZ36	5143A	3409			
1.5DKZ36A	5143A	3409			
1.5DKZ36B	5143A	3409			
1.5DKZ39	5144A	3410			
1.5DKZ39A	5144A	3410			
1.5DKZ39B	5144A	3410			
1.5DKZ43	5145A	3411			
1.5DKZ43A	5145A	3411			
1.5DKZ43B	5145A	3411			
1.5DKZ47	5146A	3412			
1.5DKZ47A	5146A	3412			
1.5DKZ47B	5146A	3412			
1.5DKZ51	5147A	3413			
1.5DKZ51A	5147A	3413			
1.5DKZ51B	5147A	3413			
1.5DKZ56	5148A	3414			
1.5DKZ56A	5148A	3414			
1.5DKZ56B	5148A	3414			
1.5DKZ6.8	5120A	3386	5ZD-6.8		
1.5DKZ6.8A	5120A	3386			
1.5DKZ6.8B	5120A	3386			
1.5DKZ62	5150A	3416			
1.5DKZ62A	5150A	3416			
1.5DKZ62B	5150A	3416			
1.5DKZ68	5151A	3417			
1.5DKZ68A	5151A	3417			
1.5DKZ68B	5151A	3417			
1.5DKZ7.5	5121A	3387			
1.5DKZ7.5A	5121A	3387	5ZD-7.5		
1.5DKZ75	5152A	3418			
1.5DKZ75A	5152A	3418			
1.5DKZ75B	5152A	3418			
1.5DKZ8.2	5122A	3388	5ZD-8.2		
1.5DKZ8.2A	5122A	3388			
1.5DKZ8.2B	5122A	3388			
1.5DKZ9.1	5124A	3390			
1.5DKZ9.1A	5124A	3390	5ZD-9.1		
1.5DKZ9.1B	5124A	3390			
1.5E05	5800	9003		1142	
1.5E1	5801	9004		1142	
1.5E12	506	3843	511	1114	
1.5E140	5160A	3426			
1.5E2	5802	9005		1143	
1.5E4	5804	9007		1144	
1.5E5	5805	9008			
1.5E6	5806	3848			
1.5J05	5800	9003		1142	
1.5J1	5801	9004		1142	
1.5J12	506	3843	511	1114	
1.5J2	5802	9005		1143	
1.5J3	5803	9006		1144	
1.5J4	5804	9007		1144	
1.5J5	5805	9008			
1.5J6	5806	3848			
1.5JZ10	5125A	3391			
1.5JZ100	5156A	3422			
1.5JZ100A	5156A	3422			
1.5JZ100B	5156A	3422			
1.5JZ100C	5156A	3422			
1.5JZ10A	5125A	3391			
1.5JZ10B	5125A	3391			
1.5JZ10C	5125A	3391			
1.5JZ10D	5125A	3391			
1.5JZ11	5126A	3392			
1.5JZ110	5157A	3423			
1.5JZ110A	5157A	3423			
1.5JZ110C	5157A	3423			
1.5JZ110D	5157A	3423			
1.5JZ11A	5126A	3392	5ZD-11		
1.5JZ11B	5126A	3392			
1.5JZ11C	5126A	3392			
1.5JZ11D	5126A	3392			
1.5JZ12	5127A	3393			
1.5JZ120A	5158A	3424			
1.5JZ120B	5158A	3424			
1.5JZ120D	5158A	3424			
1.5JZ12A	5127A	3393	5ZD-12		
1.5JZ12B	5127A	3393			
1.5JZ12C	5127A	3393			
1.5JZ12D	5127A	3393			
1.5JZ13	5128A	3394			
1.5JZ130	5159A	3425			
1.5JZ130A	5159A	3425			
1.5JZ130B	5159A	3425			
1.5JZ130C	5159A	3425			
1.5JZ13A	5128A	3394	5ZD-13		
1.5JZ13B	5128A	3394			
1.5JZ13C	5128A	3394			
1.5JZ13D	5128A	3394			
1.5JZ14	5129A	3395			
1.5JZ140	5160A	3426			
1.5JZ140A	5160A	3426			
1.5JZ140C	5160A	3426			
1.5JZ140D	5160A	3426			
1.5JZ14A	5129A	3395	5ZD-14		
1.5JZ14B	5129A	3395			
1.5JZ14C	5129A	3395			
1.5JZ14D	5129A	3395			
1.5JZ15	5130A	3396			
1.5JZ150A	5161A	3427			
1.5JZ150B	5161A	3427			
1.5JZ150D	5161A	3427			
1.5JZ15A	5130A	3396	5ZD-15		
1.5JZ15B	5130A	3396			
1.5JZ15C	5130A	3396			
1.5JZ15D	5130A	3396			
1.5JZ16	5131A	3397			
1.5JZ160	5162A	3428			
1.5JZ160A	5162A	3428			
1.5JZ160B	5162A	3428			
1.5JZ160C	5162A	3428			
1.5JZ160D	5162A	3428			
1.5JZ16A	5131A	3397	5ZD-16		
1.5JZ16B	5131A	3397			
1.5JZ16C	5131A	3397			
1.5JZ16D	5131A	3397			
1.5JZ17	5132A	3398			
1.5JZ17A	5132A	3398	5ZD-17		
1.5JZ17B	5132A	3398			
1.5JZ17C	5132A	3398			
1.5JZ17D	5132A	3398			
1.5JZ18	5133A	3399			
1.5JZ18A	5133A	3399	5ZD-18		
1.5JZ18B	5133A	3399			
1.5JZ18C	5133A	3399			
1.5JZ18D	5133A	3399			
1.5JZ19	5134A	3400			
1.5JZ19A	5134A	3400	5ZD-19		
1.5JZ19C	5134A	3400			
1.5JZ19D	5134A	3400			
1.5JZ20	5135A	3401	5ZD-20		
1.5JZ200	5166A	3432			
1.5JZ200A	5166A	3432			
1.5JZ200B	5166A	3432			
1.5JZ200C	5166A	3432			
1.5JZ200D	5166A	3432			
1.5JZ20A	5135A	3401			
1.5JZ20B	5135A	3401			
1.5JZ20C	5135A	3401			
1.5JZ20D	5135A	3401			
1.5JZ22	5136A	3402			
1.5JZ22A	5136A	3402	5ZD-22		
1.5JZ22B	5136A	3402			
1.5JZ22C	5136A	3402			
1.5JZ22D	5136A	3402			
1.5JZ24	5137A	3403			
1.5JZ24A	5137A	3403	5ZD-24		
1.5JZ24B	5137A	3403			
1.5JZ24C	5137A	3403			
1.5JZ24D	5137A	3403			
1.5JZ25	5138A	3404	5ZD-25		
1.5JZ25A	5138A	3404			
1.5JZ25B	5138A	3404			
1.5JZ25D	5138A	3404			
1.5JZ27	5139A	3405			
1.5JZ27A	5139A	3405	5ZD-27		
1.5JZ27B	5139A	3405			
1.5JZ27C	5139A	3405			
1.5JZ27D	5139A	3405			
1.5JZ30	5141A	3407			
1.5JZ30A	5141A	3407			
1.5JZ30B	5141A	3407			
1.5JZ30C	5141A	3407			
1.5JZ30D	5141A	3407			
1.5JZ33	5142A	3408			
1.5JZ33A	5142A	3408	5ZD-33		
1.5JZ33B	5142A	3408			
1.5JZ33C	5142A	3408			
1.5JZ33D	5142A	3408			
1.5JZ36	5143A	3409			
1.5JZ36A	5143A	3409			
1.5JZ36B	5143A	3409			
1.5JZ36C	5143A	3409			
1.5JZ36D	5143A	3409			
1.5JZ39	5144A	3410			
1.5JZ39A	5144A	3410			
1.5JZ39B	5144A	3410			
1.5JZ39C	5144A	3410			
1.5JZ39D	5144A	3410			
1.5JZ43	5145A	3411			
1.5JZ43A	5145A	3411			
1.5JZ43B	5145A	3411			
1.5JZ43C	5145A	3411			
1.5JZ43D	5145A	3411			
1.5JZ47	5146A	3412			
1.5JZ47A	5146A	3412			
1.5JZ47B	5146A	3412			
1.5JZ47C	5146A	3412			
1.5JZ47D	5146A	3412			
1.5JZ51	5147A	3413			
1.5JZ51A	5147A	3413			
1.5JZ51B	5147A	3413			
1.5JZ51C	5147A	3413			
1.5JZ51D	5147A	3413			
1.5JZ56	5148A	3414			
1.5JZ56A	5148A	3414			
1.5JZ56B	5148A	3414			
1.5JZ56C	5148A	3414			
1.5JZ56D	5148A	3414			
1.5JZ6.8	5120A	3386	5ZD-6.8		
1.5JZ6.8A	5120A	3386			
1.5JZ6.8B	5120A	3386			
1.5JZ6.8C	5120A	3386			
1.5JZ6.8D	5120A	3386			
1.5JZ62	5150A	3416			
1.5JZ62A	5150A	3416			
1.5JZ62B	5150A	3416			
1.5JZ62C	5150A	3416			
1.5JZ62D	5150A	3416			
1.5JZ68A	5151A	3417			
1.5JZ68B	5151A	3417			
1.5JZ68C	5151A	3417			
1.5JZ7.5	5121A	3387			

Industry Standard No.	ECG	SK	GE	RS 276-	MOTOR.
1.5JZ7.5A	5121A	3387	5ZD-7.5		
1.5JZ7.5B	5121A	3387			
1.5JZ7.5C	5121A	3387			
1.5JZ7.5D	5121A	3387			
1.5JZ75	5152A	3418			
1.5JZ75A	5152A	3418			
1.5JZ75B	5152A	3418			
1.5JZ75C	5152A	3418			
1.5JZ75D	5152A	3418			
1.5JZ8.2	5122A	3388	5ZD-8.2		
1.5JZ8.2A	5122A	3388			
1.5JZ8.2B	5122A	3388			
1.5JZ8.2C	5122A	3388			
1.5JZ8.2D	5122A	3388			
1.5JZ82	5153A	3419			
1.5JZ82A	5153A	3419			
1.5JZ82B	5153A	3419			
1.5JZ82C	5153A	3419			
1.5JZ82D	5153A	3419			
1.5JZ9.1	5124A	3390			
1.5JZ9.1A	5124A	3390	5ZD-9.1		
1.5JZ9.1B	5124A	3390			
1.5JZ9.1C	5124A	3390			
1.5JZ9.1D	5124A	3390			
1.5JZ91	5155A	3421			
1.5JZ91A	5155A	3421			
1.5JZ91B	5155A	3421			
1.5JZ91C	5155A	3421			
1.5JZ91D	5155A	3421			
1.5M140Z	5160A	3426			
1.5M14Z	5129A	3395	5ZD-14		
1.5M14Z10	5129A	3395			
1.5M17Z	5132A	3398			
1.5M17Z10	5132A	3398			
1.5M19Z	5134A	3400	5ZD-19		
1.5M19Z10	5134A	3400			
1.5M25Z	5138A	3404	5ZD-25		
1.5M25Z10	5138A	3404			
1.5R10	5125A	3391			
1.5R100	5156A	3422			
1.5R100A	5156A	3422			
1.5R100B	5156A	3422			
1.5R10A	5125A	3391			
1.5R10B	5125A	3391			
1.5R11	5126A	3392			
1.5R110	5157A	3423			
1.5R110A	5157A	3423			
1.5R110B	5157A	3423			
1.5R11A	5126A	3392			
1.5R11B	5126A	3392			
1.5R12	5127A	3393			
1.5R120	5158A	3424			
1.5R120A	5158A	3424			
1.5R120B	5158A	3424			
1.5R12A	5127A	3393			
1.5R12B	5127A	3393			
1.5R13	5128A	3394			
1.5R130	5159A	3425			
1.5R130A	5159A	3425			
1.5R130B	5159A	3425			
1.5R13A	5128A	3394			
1.5R13B	5128A	3394			
1.5R140	5160A	3426			
1.5R140A	5160A	3426			
1.5R140B	5160A	3426			
1.5R15	5130A	3396			
1.5R150	5161A	3427			
1.5R150A	5161A	3427			
1.5R150B	5161A	3427			
1.5R15A	5130A	3396			
1.5R15B	5130A	3396			
1.5R16	5131A	3397			
1.5R160	5162A	3428			
1.5R160A	5162A	3428			
1.5R160B	5162A	3428			
1.5R16A	5131A	3397			
1.5R16B	5131A	3397			
1.5R18	5133A	3399			
1.5R180	5164A	3430			
1.5R180A	5164A	3430			
1.5R180B	5164A	3430			
1.5R18A	5133A	3399			
1.5R18B	5133A	3399			
1.5R20	5135A	3401			
1.5R200	5166A	3432			
1.5R200A	5166A	3432			
1.5R200B	5166A	3432			
1.5R20A	5135A	3401			
1.5R20B	5135A	3401			
1.5R22	5136A	3402			
1.5R22A	5136A	3402			
1.5R22B	5136A	3402			
1.5R24	5137A	3403			
1.5R24A	5137A	3403			
1.5R24B	5137A	3403			
1.5R27	5139A	3405			
1.5R27A	5139A	3405			
1.5R27B	5139A	3405			
1.5R30	5141A	3407			
1.5R30A	5141A	3407			
1.5R30B	5141A	3407			
1.5R33	5142A	3408			
1.5R33A	5142A	3408			
1.5R33B	5142A	3408			
1.5R36	5143A	3409			
1.5R36A	5143A	3409			
1.5R36B	5143A	3409			
1.5R39A	5144A	3410			
1.5R39B	5144A	3410			
1.5R43	5145A	3411			
1.5R43A	5145A	3411			
1.5R43B	5145A	3411			
1.5R47	5146A	3412			
1.5R47A	5146A	3412			
1.5R47B	5146A	3412			
1.5R56	5148A	3414			
1.5R56A	5148A	3414			
1.5R56B	5148A	3414			
1.5R62	5150A	3416			
1.5R62A	5150A	3416			
1.5R62B	5150A	3416			
1.5R68	5151A	3417			
1.5R68A	5151A	3417			
1.5R68B	5151A	3417			
1.5R8.2	5122A	3388			
1.5R8.2A	5122A	3388			
1.5R8.2B	5122A	3388			
1.5R9.1	5124A	3390			
1.5R9.1A	5124A	3390			
1.5R9.1B	5124A	3390			
1.5R91	5155A	3421			
1.5R91A	5155A	3421			
1.5R91B	5155A	3421			
1.5Z10	5125A	3391			

Industry Standard No.	ECG	SK	GE	RS 276-	MOTOR.
1.5Z100	5156A	3422			
1.5Z100A	5156A	3422			
1.5Z100B	5156A	3422			
1.5Z100C	5156A	3422			
1.5Z10A	5125A	3391			
1.5Z10B	5125A	3391			
1.5Z10C	5125A	3391			
1.5Z10D	5125A	3391			
1.5Z11	5126A	3392			
1.5Z110	5157A	3423			
1.5Z110A	5157A	3423			
1.5Z110B	5157A	3423			
1.5Z110C	5157A	3423			
1.5Z110D	5157A	3423			
1.5Z11A	5126A	3392			
1.5Z11B	5126A	3392			
1.5Z11C	5126A	3392			
1.5Z11D	5126A	3392			
1.5Z12	5127A	3393			
1.5Z120	5158A	3424			
1.5Z120A	5158A	3424			
1.5Z120B	5158A	3424			
1.5Z120C	5158A	3424			
1.5Z120D	5158A	3424			
1.5Z12A	5127A	3393			
1.5Z12B	5127A	3393			
1.5Z12C	5127A	3393			
1.5Z12D	5127A	3393			
1.5Z13	5128A	3394			
1.5Z130	5159A	3425			
1.5Z130A	5159A	3425			
1.5Z130B	5159A	3425			
1.5Z130C	5159A	3425			
1.5Z130D	5159A	3425			
1.5Z13A	5128A	3394			
1.5Z13B	5128A	3394			
1.5Z13C	5128A	3394			
1.5Z13D	5128A	3394			
1.5Z14	5129A	3395			
1.5Z140	5160A	3426			
1.5Z140A	5160A	3426			
1.5Z140B	5160A	3426			
1.5Z140C	5160A	3426			
1.5Z140D	5160A	3426			
1.5Z14A	5129A	3395	5ZD-14		
1.5Z14B	5129A	3395			
1.5Z14C	5129A	3395			
1.5Z14D	5129A	3395			
1.5Z15	5130A	3396			
1.5Z150	5161A	3427			
1.5Z150A	5161A	3427			
1.5Z150B	5161A	3427			
1.5Z150C	5161A	3427			
1.5Z150D	5161A	3427			
1.5Z15A	5130A	3396			
1.5Z15B	5130A	3396			
1.5Z15C	5130A	3396			
1.5Z16	5131A	3397			
1.5Z160	5162A	3428			
1.5Z160A	5162A	3428			
1.5Z160B	5162A	3428			
1.5Z160C	5162A	3428			
1.5Z160D	5162A	3428			
1.5Z16A	5131A	3397			
1.5Z16B	5131A	3397			
1.5Z16C	5131A	3397			
1.5Z16D	5131A	3397			
1.5Z17	5132A	3398			
1.5Z17A	5132A	3398			
1.5Z17B	5132A	3398			
1.5Z17C	5132A	3398			
1.5Z17D	5132A	3398	5ZD-17		
1.5Z18	5133A	3399			
1.5Z180	5164A	3430			
1.5Z180A	5164A	3430			
1.5Z180B	5164A	3430			
1.5Z180C	5164A	3430			
1.5Z180D	5164A	3430			
1.5Z18A	5133A	3399			
1.5Z18B	5133A	3399			
1.5Z18C	5133A	3399			
1.5Z18D	5133A	3399			
1.5Z19	5134A	3400			
1.5Z19A	5134A	3400			
1.5Z19B	5134A	3400			
1.5Z19C	5134A	3400			
1.5Z19D	5134A	3400			
1.5Z20	5135A	3401			
1.5Z200	5166A	3432			
1.5Z200A	5166A	3432			
1.5Z200B	5166A	3432			
1.5Z200C	5166A	3432			
1.5Z200D	5166A	3432			
1.5Z20A	5135A	3401			
1.5Z20B	5135A	3401			
1.5Z20C	5135A	3401			
1.5Z20D	5135A	3401	5ZD-20		
1.5Z22	5136A	3402			
1.5Z22A	5136A	3402			
1.5Z22B	5136A	3402			
1.5Z22C	5136A	3402			
1.5Z22D	5136A	3402	5ZD-22		
1.5Z24	5137A	3403			
1.5Z24A	5137A	3403			
1.5Z24B	5137A	3403			
1.5Z24C	5137A	3403			
1.5Z24D	5137A	3403	5ZD-24		
1.5Z27	5139A	3405			
1.5Z27A	5139A	3405	5ZD-27		
1.5Z27B	5139A	3405			
1.5Z27C	5139A	3405			
1.5Z27D	5139A	3405			
1.5Z30	5141A	3407			
1.5Z30A	5141A	3407			
1.5Z30B	5141A	3407			
1.5Z30C	5141A	3407			
1.5Z30D	5141A	3407			
1.5Z33	5142A	3408			
1.5Z33A	5142A	3408			
1.5Z33B	5142A	3408			
1.5Z33C	5142A	3408			
1.5Z33D	5142A	3408			
1.5Z36	5143A	3409			
1.5Z36A	5143A	3409			
1.5Z36B	5143A	3409			
1.5Z36C	5143A	3409			
1.5Z36D	5143A	3409			
1.5Z39	5144A	3410			
1.5Z39A	5144A	3410			
1.5Z39B	5144A	3410			
1.5Z39D	5144A	3410			
1.5Z43	5145A	3411			
1.5Z43A	5145A	3411			
1.5Z43B	5145A	3411			

Industry Standard No.	ECG	SK	GE	RS 276-	MOTOR.
1.5Z43C	5145A	3411			
1.5Z43D	5145A	3411			
1.5Z47	5146A	3412			
1.5Z47A	5146A	3412			
1.5Z47B	5146A	3412			
1.5Z47C	5146A	3412			
1.5Z47D	5146A	3412			
1.5Z51	5147A	3413			
1.5Z51A	5147A	3413			
1.5Z51B	5147A	3413			
1.5Z51C	5147A	3413			
1.5Z51D	5147A	3413			
1.5Z56	5148A	3414			
1.5Z56A	5148A	3414			
1.5Z56B	5148A	3414			
1.5Z56C	5148A	3414			
1.5Z6.8	5120A	3386	5ZD-6.8		
1.5Z6.8A	5120A	3386			
1.5Z6.8B	5120A	3386			
1.5Z6.8C	5120A	3386			
1.5Z62	5150A	3416			
1.5Z62A	5150A	3416			
1.5Z62B	5150A	3416			
1.5Z62C	5150A	3416			
1.5Z62D	5150A	3416			
1.5Z68	5151A	3417			
1.5Z68A	5151A	3417			
1.5Z68B	5151A	3417			
1.5Z68C	5151A	3417			
1.5Z68D	5151A	3417			
1.5Z7.5	5121A	3387			
1.5Z7.5A	5121A	3387			
1.5Z7.5B	5121A	3387			
1.5Z7.5C	5121A	3387			
1.5Z7.5D	5121A	3387			
1.5Z75	5152A	3418			
1.5Z75A	5152A	3418			
1.5Z75B	5152A	3418			
1.5Z75C	5152A	3418			
1.5Z75D	5152A	3418			
1.5Z8.2	5122A	3388			
1.5Z8.2A	5122A	3388	5ZD-8.2		
1.5Z8.2B	5122A	3388			
1.5Z8.2C	5122A	3388			
1.5Z8.2D	5122A	3388			
1.5Z82	5153A	3419			
1.5Z82A	5153A	3419			
1.5Z82B	5153A	3419			
1.5Z82C	5153A	3419			
1.5Z82D	5153A	3419			
1.5Z9.1	5124A	3390			
1.5Z9.1A	5124A	3390			
1.5Z9.1B	5124A	3390			
1.5Z9.1C	5124A	3390			
1.5Z9.1D	5124A	3390			
1.5Z91	5155A	3421			
1.5Z91A	5155A	3421			
1.5Z91B	5155A	3421			
1.5Z91C	5155A	3421			
1.5Z91D	5155A	3421			
1/2Z100T5		3816	ZD-100		
1/2Z105T5			ZD-10	562	
1/2Z10T5		3785	ZD-10	562	
1/2Z11T5		3786	ZD-11	563	
1/2Z12T5		3062	CR-6		
1/2Z13T5		3788	ZD-13	563	
1/2Z15T5		3790	ZD-15	564	
1/2Z16T5		3791		564	
1/2Z18T5		3793	ZD-18.		
1/2Z20T5		3795	ZD-20		
1/2Z22T5		3796	ZD-22		
1/2Z27T5		3799	ZD-27		
1/2Z2TT5		3064	ZD-27		
1/2Z3.3T5		3771	ZD-3.3		
1/2Z3.6T5		3772	ZD-3.6		
1/2Z3.9T5		3773	ZD-3.9		
1/2Z30T5		3801	ZD-30		
1/2Z33T5		3061	ZD-10	562	
1/2Z36T5		3803	ZD-36		
1/2Z39T5		3804	ZD-39		
1/2Z4.3T5		3774	ZD-4.3		
1/2Z4.7T5		3775	ZD-4.7		
1/2Z43T5		3805	ZD-43		
1/2Z47T5		3806	ZD-47		
1/2Z5.1T5		3776	ZD-5.1		
1/2Z5.6T5		3777	ZD-5.6	561	
1/2Z51T5		3807	ZD-51		
1/2Z56T5		3808	ZD-56		
1/2Z6.2T5		3779	ZD-6.2	561	
1/2Z6.5T5			ZD-5.6	561	
1/2Z6.8T5		3780	ZD-6.8	561	
1/2Z62T5		3810	ZD-62		
1/2Z68T5		3811	ZD-68		
1/2Z7.5T5		3781	ZD-7.5		
1/2Z75T5		3812	ZD-75		
1/2Z8.2T5		3782	ZD-8.2	562	
1/2Z82T5		3813	ZD-8.2	562	
1/2Z9.1T5		3784	ZD-9.1	562	
1/2Z91T5		3815	ZD-91		
1/4A10		3061	ZD-10	562	
1/4A10A		3061	ZD-10	562	
1/4A10B		3061	ZD-10	562	
1/4A110		3099	ZD-110		
1/4A110A		3099	ZD-110		
1/4A110B		3099	ZD-110		
1/4A12		3062	ZD-12	563	
1/4A12A		3062	ZD-12	563	
1/4A12B		3094	ZD-12	563	
1/4A14		3094		564	
1/4A14A		3094	ZD-14	564	
1/4A14B		3094	ZD-14	564	
1/4A15		3063	ZD-15	564	
1/4A15A		3063	ZD-15	564	
1/4A15B		3063	ZD-15	564	
1/4A27		3064	ZD-27		
1/4A27A		3064	ZD-27		
1/4A27B		3064	ZD-27		
1/4A33		3095	ZD-33		
1/4A33A		3095	ZD-33		
1/4A33B		3095	ZD-33		
1/4A62		3097	ZD-62		
1/4A62A		3097	ZD-62		
1/4A82		3098	ZD-82		
1/4A82A		3098	ZD-82		
1/4A82B		3098	ZD-82		
1/4A9.1		3060	ZD-9.1	562	
1/4A9.1A		3060	ZD-9.1	562	
1/4A9.1B		3060	ZD-9.1	562	
1/4AZ3.6D10		3055	ZD-3.6		
1/4AZ3.6D5		3055	ZD-3.6		
1/4AZ5.1D		3056	ZD-5.1		
1/4AZ5.1D10		3056	ZD-5.1		
1/4AZ5.1D5		3056	ZD-5.1		
1/4AZ5.6D		3057	ZD-5.6	561	

Industry Standard No.	ECG	SK	GE	RS 276-	MOTOR.
1/4AZ5.6D10		3057	ZD-5.6	561	
1/4AZ5.6D5		3057	ZD-5.6	561	
1/4AZ6.2D5		3058	ZD-6.2	561	
1/4LZ3.6D		3055	2	2007	
1/4LZ3.6D5		3055	2	2007	
1/4LZ5.6D		3057	ZD-5.6	561	
1/4LZ5.6D10		3057	ZD-5.6	561	
1/4LZ5.6D5		3057	ZD-5.6	561	
1/4LZ6.2D		3058	ZD-6.2	561	
1/4LZ6.2D5		3058	ZD-6.2	561	
1/4M10Z		3061	ZD-10-4		
1/4M10Z10		3061	ZD-10-4		
1/4M120Z10		3818	ZD-120		
1/4M150Z		3821	ZD-150		
1/4M150Z10		3821	ZD-150		
1/4M150Z5		3821	ZD-150		
1/4M15Z		3063	ZD-15	564	
1/4M15Z10		3063	ZD-15	564	
1/4M15Z5		3063	ZD-15	564	
1/4M160Z		3822	ZD-160		
1/4M160Z10		3822	ZD-160		
1/4M160Z5		3822	ZD-160		
1/4M16Z5	5075A	3751			
1/4M175Z		3823	ZD-170		
1/4M175Z10		3823	ZD-170		
1/4M175Z5		3823	ZD-170		
1/4M180Z		3824	ZD-180		
1/4M180Z10		3824	ZD-180		
1/4M180Z5		3824	ZD-180		
1/4M200Z		3826	ZD-200		
1/4M200Z10		3826	ZD-200		
1/4M200Z5	5105A	3826/5060A	ZD-200		
1/4M20Z10		3795	ZD-20		
1/4M22Z		3796	ZD-22		
1/4M25Z10		3798	ZD-25		
1/4M25Z5		3798	ZD-25		
1/4M27Z		3064	ZD-27		
1/4M27Z10		3064	ZD-27		
1/4M27Z5		3064	ZD-27		
1/4M3.3AZ		3771	ZD-3.3		
1/4M3.3AZ10		3771	ZD-3.3		
1/4M3.3AZ5		3771	ZD-3.3		
1/4M3.6AZ10		3055	ZD-3.6		
1/4M3.6AZ5		3055	ZD-3.6		
1/4M3.9AZ		3773	ZD-3.9		
1/4M3.9AZ10		3773	ZD-3.9		
1/4M3.9AZ5		3331	ZD-3.9		
1/4M4.3AZ		3774	ZD-4.3		
1/4M4.3AZ10		3774	ZD-4.3		
1/4M4.3AZ5		3774	ZD-4.3		
1/4M4.7AZ		3775	ZD-4.7		
1/4M4.7AZ10		3775	ZD-4.7		
1/4M4.7AZ5	5069A	3775/5009A	ZD-4.7		
1/4M5.1AZ		3776	ZD-5.1		
1/4M5.1AZ10		3776	ZD-5.1		
1/4M5.6AZ		3057	ZD-5.6	561	
1/4M5.6AZ10		3057	ZD-5.6	561	
1/4M5.6AZ25			ZD-5.6	561	
1/4M5.6AZ5		3057	ZD-5.6	561	
1/4M6.2AZ5		3058	ZD-6.2.	561	
1/4M6.8AZ		3780	ZD-6.8	561	
1/4M6.8AZ10		3780	ZD-6.8	561	
1/4M6.8AZ5		3780	ZD-6.8	561	
1/4M6.8Z		3780	ZD-6.8	561	
1/4M6.8Z10		3780	ZD-6.8	561	
1/4M6.8Z5	5071A	3780/5014A	ZD-6.8	561	
1/4M62Z		3097	ZD-62		
1/4M62Z5		3097	ZD-62		
1/4M7.5Z		3781	ZD-7.5		
1/4M7.5Z10		3781	ZD-7.5		
1/4M8/27			ZD-8.2	562	
1/4M9.1Z		3784	ZD-9.1	562	
1/4M9.1Z10		3784	ZD-9.1	562	
1/4M9.1Z5			ZD-9.1	562	
1/4Z110D		3099	ZD-110		
1/4Z110D10		3099	ZD-110		
1/4Z110D5		3099	ZD-110		
1/4Z12T5			ZD-12	563	
1/4Z14D		3094	ZD-14	564	
1/4Z14D10		3094	ZD-14	564	
1/4Z14D5		3094	ZD-14	564	
1/4Z15D10		3063	ZD-15	564	
1/4Z15D5		3063	ZD-15	564	
1/4Z15T5		3063	ZD-15	564	
1/4Z27D		3064	ZD-27		
1/4Z27D10		3064	ZD-27		
1/4Z27D5		3064	ZD-27		
1/4Z27T5		3799	ZD-27		
1/4Z33D		3802	ZD-33		
1/4Z33D10		3802	ZD-33		
1/4Z33D5		3802	ZD-33		
1/4Z33T5		3802	ZD-33		
1/4Z5.1T5		3776	ZD-5.1		
1/4Z5.6T5		3777	ZD-5.6	561	
1/4Z6.2T5		3779	ZD-6.2	561	
1/4Z62D		3097	ZD-62		
1/4Z62D10		3097	ZD-62		
1/4Z62D5		3097	ZD-62		
1/4Z62T5		3810	ZD-62		
1/4Z82D		3098	ZD-82		
1/4Z82D10		3098	ZD-82		
1/4Z82T5		3813	ZD-82		
1A0013	282		62	2010	
1A0013(YAMAHA)	175		246	2020	
1A0020	313	3444/123A	20	2051	
1A0021	289A	3444/123A	20	2051	
1A0022	289A	3444/123A	20	2051	
1A0024	199	3444/123A	20	2051	
1A0025	123AP	3444/123A	20	2051	
1A0027	175	3024/128	246	2020	
1A0027(YAMAHA)	175		246	2020	
1A0029	313	3444/123A	20	2051	
1A0032	289A	3444/123A	20	2051	
1A0033	289A	3444/123A	20	2051	
1A0034	123A	3444	20	2051	
1A0035	123A	3444	20	2051	
1A0037	289A	3444/123A	20	2051	
1A0038	124	3021	12		
1A0043	123A	3444	20	2051	
1A0044	107	3444/123A	20	2051	
1A0045	289A	3124/289	17	2051	
1A0046	291	3893/152	66	2048	
1A0048	175	3024/128	28	2020	
1A0048(YAMAHA)	175		246	2020	
1A0051	289A	3444/123A	20	2051	
1A0055	102A	3003	53	2007	
1A0056	102A	3003	53	2007	
1A0058	152	3893	66	2048	
1A0059	291	3893/152	66	2048	
1A0063	123A	3444	20	2051	
1A0066	128	3024	243	2030	
1A0067	289A	3444/123A	20	2051	
1A0070	123AP	3444/123A	20	2051	
1A0076	123AP	3444/123A	20	2051	

Industry Standard No.	ECG	SK	GE	RS 276-	MOTOR.
1A0077	123AP	3444/123A	20	2051	
1A0078	123AP	3444/123A	20	2051	
1A0079	123A	3444	20	2051	
1A0080	123A	3444	20	2051	
1A0081	123A	3444	20	2051	
1A0083	289A	3444/123A	20	2051	
1A0084	289A	3444/123A	20	2051	
1A1	116	3313			
1A100V	116	3311			
1A10425	116	3311	504A	1104	
1A10952			504A	1104	
1A10M	140A	3061	ZD-10	562	
1A10MA	140A	3061	ZD-10	562	
1A110M	151A	3099	ZD-110		
1A110MA	151A	3099	ZD-110		
1A11184	116	3017B/117	504A	1104	
1A1123100-1	181		75	2041	
1A11306	109	3087		1123	
1A11671	116	3017B/117	504A	1104	
1A12214	116	3031A	504A	1104	
1A12407	116	3031A	504A	1104	
1A12688	139A	3060	ZD-9.1	562	
1A12689	109	3088		1123	
1A12690	116	3017B/117	504A	1104	
1A12M	142A	3062	ZD-12	563	
1A12MA	142A	3062	ZD-12	563	
1A13219	116	3017B/117	504A	1104	
1A13719	116	3017B/117	504A	1104	
1A13720	116	3017B/117	504A	1104	
1A13M	143A	3750	ZD-13	563	
1A13MA	143A	3750	ZD-13	563	
1A14384	109	3087		1123	
1A15790	116	3031A	504A	1104	
1A15M	145A	3063	ZD-15	564	
1A15MA	145A	3063	ZD-15	564	
1A16549	177	3100/519	300	1122	
1A16550	116	3311	504A	1104	
1A16551	177	3100/519	300	1122	
1A27M	146A	3064	ZD-27		
1A27MA	146A	3064	ZD-27		
1A33M	147A	3095	ZD-33		
1A33MA	147A	3095	ZD-33		
1A34	128		63	2030	
1A34(R)			63	2030	
1A348	128		ZD-243	2030	
1A348(R)			63	2030	
1A348R	128	3024	ZD-243	2030	
1A34R	128	3024	63	2030	
1A35	519		514	1122	
1A38			63	2030	
1A38(R)			63	2030	
1A38R			63	2030	
1A4757-1	123A	3444	20	2051	
1A50	116	3031A	504A	1104	
1A52	5000A	3766			
1A54	5004A	3770			
1A55		3771	ZD-3.3		
1A56		3772	ZD-3.6		
1A56M	148A	3096	ZD-55		
1A56MA	148A	3096	ZD-55		
1A57		3773	ZD-3.9		
1A58	5008A	3774	ZD-4.3		
1A59		3775	ZD-4.7		
1A60		3776	ZD-5.1		
1A62		3779	ZD-6.2	561	
1A62M	149A	3779/5013A	ZD-62		
1A62MA	149A	3779/5013A	ZD-62		
1A69425-1	519		514	1122	
1A7.5M	138A	3059	ZD-7.5		
1A7.5MA	138A	3059	ZD-7.5		
1A8-1A82	160		ZD-245	2004	
1A82M	150A	3098	ZD-82		
1A82MA	150A	3098	ZD-82		
1A9.1M	139A	3060	ZD-9.1	562	
1A9.1MA	139A	3060	ZD-9.1	562	
1A99812-1001	519		514	1122	
1AC110	151A	3099	ZD-110		
1AC110A	151A	3099	ZD-110		
1AC110B	151A	3099	ZD-110		
1AC12	142A	3062	ZD-12	563	
1AC12A	142A	3062	ZD-12	563	
1AC12B	142A	3062	ZD-12	563	
1AC15	145A	3063	ZD-15	564	
1AC15A	145A	3063	ZD-15	564	
1AC15B	145A	3063	ZD-15	564	
1AC27	146A	3064	ZD-27		
1AC27A	146A	3064	ZD-27		
1AC27B	146A	3064	ZD-27		
1AC33	147A	3095	ZD-33		
1AC33A	147A	3095	ZD-33		
1AC33B	147A	3095	ZD-33		
1AC62	149A	3097	ZD-62		
1AC62A	149A	3097	ZD-62		
1AC62B	149A	3097	ZD-62		
1AC82	150A	3098	ZD-82		
1AC82A	150A	3098	ZD-82		
1AC82B	150A	3098	ZD-82		
1AMH2	525	3925			
1AS027	156	3051	512		
1AS029	156	3051	512		
1B	504		CR-6		
1B-2C1	116	3199/188	504A(2)		
1B05	166			1152	
1B05J05	166	9075		1152	
1B05J20	116	3016	504A	1104	
1B05J40	116	3031A	504A	1104	
1B08T05	166	9075		1152	
1B08T10	166	9075		1152	
1B08T20	167	3647		1172	
1B1	166	9075		1152	
1B10J20	116	3016	504A	1104	
1B12421	5882		5040		
1B2	167	3647	504A	1172	
1B2992	519		514	1122	
1B2C1	116(2)	3119/113	6GC1		
1B2Z1	116(2)	3311			
1B3096-1	159	3466	82	2032	
1B4	168	3648		1173	
1B6	169	3678			
1B759			ZD-12	563	
1BZ61		3311		1102	
1C0009	116	3017B/117	504A	1104	
1C0017	177	3017B/117	504A	1104	
1C0020	138A	3059	ZD-7.5		
1C0025	116	3017B/117	504A	1104	
1C0026	116	3017B/117	504A	1104	
1C0029	109	3087		1123	
1C0031	116	3017B/117	504A	1104	
1C0038	137A	3334/5071A	ZD-6.8	561	
1C0039	109	3087		1123	
1C05	116(2)	3311	504A(2)	1104	
1C1	116(2)	3017B/117	504A(2)	1104	
1C10	125(2)	3033	510,531	1114	
1C110Z	151A	3099	ZD-110		

Industry Standard No.	ECG	SK	GE	RS 276-	MOTOR.
1C110ZA	151A	3099	ZD-110		
1C12Z	142A	3062	ZD-12	563	
1C12ZA	142A	3062	ZD-12	563	
1C15Z	145A	3063	ZD-15	564	
1C15ZA	145A	3063	ZD-15	564	
1C2	116(2)	3016	504A(2)	1104	
1C21	782	3077/790			
1C27Z	146A	3064	ZD-27		
1C27ZA	146A	3064	ZD-27		
1C33Z	147A	3095	ZD-33		
1C33ZA	147A	3095	ZD-33		
1C3576	123A		20	2051	
1C4	116(2)	3017B/117	504A(2)	1104	
1C6	711	3017B/117	504A(2)	1104	
1C62Z	149A	3097	ZD-62		
1C62ZA	149A	3097	ZD-62		
1C73045	235	3041	215		
1C744005	161	3716			
1C744006	293	3849			
1C744009	236	3239			
1C8	748	3033	510,513	1114	
1C82Z	150A	3098	ZD-82		
1C82ZA	150A	3098	ZD-82		
1CL-201A-TY	1171	3565			
1CL-301A-TY	1171	3565			
1CL-748C-TY	1171	3565			
1D08	145A	3063	ZD-15	564	
1D098-001V-022	177		300	1122	
1D1	116(2)	3017B/117	504A(2)	1104	
1D10	125(2)	3033	510,531	1114	
1D100	5400	3950			
1D101	5401	3638/5402			
1D102	5402	3638			
1D103	5403	3627/5404			
1D104	5404	3627			
1D105	5405	3951			
1D106	5405	3951			
1D2	116(2)	3016	504A(2)	1104	
1D261	116	3311	504A	1104	
1D27Z5	146A	3064	ZD-27		
1D2Z1	116	3311	504A	1104	
1D3.3	5066A	3330	ZD-3.3		
1D3.3A	5066A	3330	ZD-3.3		
1D3.3B	5066A	3330	ZD-3.3		
1D3.6	134A	3055	ZD-3.6		
1D3.6A	134A	3055	ZD-3.6		
1D3.6B	134A	3055	ZD-3.6		
1D3.9	5067A	3331	ZD-3.9		
1D3.9A	5067A	3331	ZD-3.9		
1D3.9B	5067A	3331	ZD-3.9		
1D4.3	5068A	3332	ZD-4.3		
1D4.3A	5068A	3332	ZD-4.3		
1D4.3B	5068A	3332	ZD-4.3		
1D4.7	5069A	3333	ZD-4.7		
1D4.7B	5069A	3333	ZD-4.7		
1D5.1	135A		ZD-5.1		
1D5.1A	135A	3056	ZD-5.1		
1D5.1B	5116A	3056/135A	ZD-5.1		
1D5.6	136A	3057	ZD-5.6	561	
1D5.6A	136A	3057	ZD-5.6	561	
1D5.6B	136A	3057	ZD-5.6	561	
1D6	116(2)		504A(2)	1104	
1D6.2	137A	3058	ZD-6.2	561	
1D6.2A	137A	3058	ZD-6.2	561	
1D6.2B	137A	3058	ZD-6.2	561	
1D6.2SA	137A	3058	ZD-6.2	561	
1D6.2SB	137A	3058	ZD-6.2	561	
1D8	125(2)	3032A	510,531	1114	
1DA1Z18B	5077A	3752			
1DA1Z8.2B	5072A	3136			
1DAZ229B	5078A	9023			
1DAZ3401B	5072A	3136			
1DAZ3405B	5077A	3752			
1DC1	116		504A	1104	
1DG2	109	3088		1123	
1DZ12	142A	3062	ZD-12	563	
1DZ61		3311		1102	
1E05	116	3311	504A	1104	
1E1	116	3016	504A	1104	
1E10	125	3033	510,531	1114	
1E10Z	140A	3061	ZD-10	562	
1E10Z10	140A		ZD-10	562	
1E110Z	151A	3099	ZD-110		
1E110Z10	151A	3099	ZD-110		
1E110Z5	151A	3099	ZD-110		
1E11Z	141A	3092	ZD-11.5		
1E11Z10	141A	3092	ZD-11.5		
1E12Z	142A	3062	ZD-12	563	
1E12Z10	142A	3062	ZD-12	563	
1E12Z5	142A	3062	ZD-12	563	
1E14Z	144A	3094	ZD-14	564	
1E14Z10	144A	3094	ZD-14	564	
1E14Z5	144A	3094	ZD-14	564	
1E15Z	145A	3063	ZD-15	564	
1E15Z10	145A	3063	ZD-15	564	
1E15Z5	145A	3063	ZD-15	564	
1E2	116	3016	504A	1104	
1E27Z	146A	3064	ZD-27		
1E27Z10	146A	3064	ZD-27		
1E27Z5	146A		ZD-27		
1E2Z5	142A		ZD-12	563	
1E3	116	3016	504A	1104	
1E33Z	147A	3095	ZD-33		
1E33Z10	147A	3095	ZD-33		
1E33Z5	147A	3095	ZD-33		
1E4	116	3016	504A	1104	
1E5	116	3016	504A	1104	
1E535A	229	3018	61	2038	
1E535A/7825B	229	3018	61	2038	
1E535B	107	3018	11	2015	
1E6	116	3017B/117	504A	1104	
1E62Z	149A	3097	ZD-62		
1E62Z10	149A	3097	ZD-62		
1E62Z5	149A	3097	ZD-62		
1E7	125	3032A	510,531	1114	
1E7.5Z	138A		ZD-7.5		
1E7.5Z10	138A		ZD-7.5		
1E703E	703A	3157	IC-12		
1E8	125	3032A	510,531	1114	
1E82Z	150A	3098	ZD-82		
1E82Z10	150A	3098	ZD-82		
1E82Z5	150A	3098	ZD-82		
1E9.1Z	139A	3060	ZD-9.1	562	
1E9.1Z10	139A	3060	ZD-9.1	562	
1EA10A	117	3016	504A	1104	
1EA20A	117	3017B	504A	1104	
1EA30A	117	3017B	504A	1104	
1EA40A	117	3017B	504A	1104	
1EA50A	117	3017B	504A	1104	
1EA60A	117	3017B	504A	1104	
1EB100A	5890	7090	5044		
1EB10A	5872		5032		
1EB20A	5874		5032		
1EB30A	5876		5036		

Industry Standard No.	ECG	SK	GE	RS 276-	MOTOR.
1EB40A	5878		5036		
1EB60A	5882		5040		
1EB70A	5886		5044		
1EB80A	5886		5044		
1ET02	116	3311	504A	1104	
1ET05	116	3017B/117	504A	1104	
1ET1	116	3017B/117	504A	1104	
1ET10	125	3033	510,531	1114	
1ET2	116	3017B/117	504A	1104	
1ET3	116	3031A	504A	1104	
1ET4	116	3031A	504A	1104	
1ET5	116	3017B/117	504A	1104	
1ET6	116	3017B/117	504A	1104	
1ET7	125	3032A	510,531	1114	
1ET8	125	3032A	510,531	1114	
1EZ-5Z10	138A	3059	ZD-7.5		
1EZ110D5	151A		ZD-110		1M110ZS5
1EZ12	142A	3062	ZD-12	563	
1EZ120D5	5097A				1M120ZS5
1EZ130D5	5098A	3348	ZD-130		1M130ZS5
1EZ140D5	5099A	3349	ZD-140		1M140ZS5
1EZ15	145A	3063	ZD-15	564	
1EZ150D5	5100A				1M150ZS5
1EZ160D5	5101A	3351	ZD-160		1M160ZS5
1EZ170D5	5102A				1M170ZS5
1EZ17D5	5076A	9022	ZD-17		
1EZ180D5	5103A	3353	ZD-180		1M180ZS5
1EZ190D5	5104A	3354			1M190ZS5
1EZ19D5	5078A	9023			
1EZ200D5	5105A	3354/5104A			1M200ZS5
1EZ27	146A	3064	ZD-27		
1EZ3.3D5		3330	ZD-3.3		
1EZ3.6	134A	3055	ZD-3.6		
1EZ3.9D5		3331	ZD-3.9		
1EZ4.3D5		3332	ZD-4.3		
1EZ43D5		3339	ZD-43		
1EZ5.1	135A	3056	ZD-5.1		
1EZ5.6	136A	3057	ZD-5.6	561	
1EZ51D5		3341	ZD-51		
1EZ7.5Z	138A	3059	ZD-7.5		
1EZ9.1	139A	3066/118	ZD-9.1	562	
01F	703A	3157	IC-12		
01F-SL8020	703A	3157	IC-12		
1F05	116		504A	1104	
1F14A	116	3311	504A	1104	
1F2	116	3016	504A	1104	
1F8	125	3032A	510,531	1114	
1FM2	116	3311	504A	1104	
1G01			1N34AS	1123	
1G02	109	3087		1123	
1G100	105	3012	4		
1G25			1N34AS	1123	
1G2C1	116	3081/125	510	1114	
1G2Z1	116	3081/125	510	1114	
1G8	125	3033	510,531	1114	
1G86			1N34AS	1123	
1GA	116	3311	504A	1104	
1GD10			1N34AS	1123	
1GD2	109	3087		1123	
1GD4	109	3087		1123	
1GD5X			1N34AS	1123	
1GD6	109	3087		1123	
1GZ61		3031A		1103	
1HC-100F				1114	
1HC-100R				1114	
1HC-10F				1102	
1HC-10R				1102	
1HC-15F				1102	
1HC-15R				1102	
1HC-20F				1102	
1HC-20R				1102	
1HC-25F				1103	
1HC-25R				1103	
1HC-30F				1103	
1HC-30R				1103	
1HC-40F				1103	
1HC-40R				1103	
1HC-50F				1104	
1HC-50R				1104	
1HC-60F				1104	
1HC-60R				1104	
1HY100	125	3032A	510,531	1114	
1HY40	116	3031A	504A	1104	
1HY50	116	3016	504A	1104	
1HY80	125	3032A	510,531	1114	
1J1	123A	3444	20	2051	
1JH11F		3313	504A	1104	
1JZ61	125	3017B/117	504A	1104	
01K-5.4E			ZD-5.6	561	
01K-5.8E			ZD-5.6	561	
01K-6.5E			ZD-6.2	561	
1K110	109			1123	
1K188FM				1123	
1K188FM-1				1123	
1K2	5802	9005		1143	
1K261	109	3087	1N34AS	1123	
1K34A	109			1123	
01K6.5E			ZD-6.2	561	
1K60	109	3088		1123	
1K60A	109	3088		1123	
1K90				1123	
1LE11	156	3033	512		
1LF11	156	3051	512		
1M100Z	5096A	3346	ZD-100		
1M100Z10	5096A	3346	ZD-100		
1M100Z5	5096A	3346	ZD-100		
1M100ZS10	5096A	3346	ZD-100		
1M100ZS5	5096A	3346	ZD-100		
1M10Z	140A		ZD-10	562	
1M10Z10	140A		ZD-10	562	
1M10Z5	140A		ZD-10	562	
1M10ZS10	140A		ZD-10	562	
1M10ZS5	140A		ZD-10	562	
1M110Z	151A	3099	ZD-110		
1M110Z10	151A	3099	ZD-110		
1M110Z3	151A	3099	ZD-110		
1M110Z5	151A	3099	ZD-110		
1M110ZS10	151A	3099	ZD-110		
1M110ZS5	151A	3099	ZD-110		
1M11Z	5074A		ZD-11.0	563	
1M11Z10	5074A		ZD-11.0	563	
1M11Z5	5074A		ZD-11.0	563	
1M11ZS10	5074A		ZD-11.0	563	
1M11ZS5	5074A		ZD-11.0	563	
1M120Z	5097A	3347	ZD-120		
1M120Z10	5097A	3347	ZD-120		
1M120Z3	5097A	3347	ZD-120		
1M120Z5	5097A	3347	ZD-120		
1M120ZS10	5097A	3347	ZD-120		
1M120ZS5	5097A	3347	ZD-120		
1M12Z	142A	3062	ZD-12	563	
1M12Z10	142A	3062	ZD-12	563	
1M12Z5	142A	3062	ZD-12	563	
1M12ZS10	142A	3062	ZD-12	563	
1M12ZS5	142A	3062	ZD-12	563	
1M130Z	5098A	3348	ZD-130		
1M130Z10	5098A	3348	ZD-130		
1M130Z3	5098A	3348	ZD-130		
1M130Z5	5098A	3348	ZD-130		
1M130ZS10	5098A	3348	ZD-130		
1M130ZS5	5098A	3348	ZD-130		
1M13Z5	143A	3750			
1M13ZS10	143A	3750	ZD-13	563	
1M13ZS5	143A	3750	ZD-13	563	
1M140Z	5099A	3349			
1M140Z10	5099A	3349	ZD-140		
1M140Z5	5099A	3349	ZD-140		
1M140ZS5		3349	ZD-140		
1M14Z	144A	3094	ZD-14	564	
1M14Z10	144A	3094	ZD-14	564	
1M14Z5	144A	3094	ZD-14	564	
1M150Z	5100A	3350	ZD-150		
1M150Z10	5100A	3350	ZD-150		
1M150Z3	5100A	3350	ZD-150		
1M150Z5	5100A	3350	ZD-150		
1M150ZS10	5100A	3350	ZD-150		
1M150ZS5	5100A	3350	ZD-150		
1M15Z	145A	3063	ZD-15	564	
1M15Z10	145A	3063	ZD-15	564	
1M15Z5	145A	3063	ZD-15	564	
1M15ZS10	145A		ZD-15	564	
1M15ZS5	145A		ZD-15	564	
1M160Z	5101A	3351	ZD-160		
1M160Z10	5101A	3351	ZD-160		
1M160Z3	5101A	3351	ZD-160		
1M160Z5	5101A	3351	ZD-160		
1M160ZS10	5101A	3351	ZD-160		
1M160ZS5	5101A	3351	ZD-160		
1M16Z	5075A	3751	ZD-16	564	
1M16Z5	5075A	3751	ZD-16	564	
1M16ZS10	5075A	3751	ZD-16	564	
1M16ZS5	5075A	3751	ZD-16	564	
1M170ZS5		3352	ZD-170		
1M17Z	5076A	9022	ZD-17		
1M17Z10	5076A	9022	ZD-17		
1M17Z5	5076A	9022	ZD-17		
1M180Z	5103A	3353	ZD-180		
1M180Z10	5103A	3353	ZD-180		
1M180Z3	5103A	3353	ZD-180		
1M180Z5	5103A	3353	ZD-180		
1M180ZS10	5103A	3353	ZD-180		
1M180ZS5	5103A	3353	ZD-180		
1M18Z	5077A	3752	ZD-18		
1M18Z10	5077A	3752	ZD-18		
1M18Z5	5077A	3752	ZD-18		
1M18ZS10	5077A	3752	ZD-18		
1M18ZS5	5077A	3752	ZD-18		
1M190ZS5		3354	ZD-190		
1M19Z	5078A		ZD-19		
1M19Z10	5078A	9023	ZD-19		
1M19Z5	5078A	9023	ZD-19		
1M200Z	5105A	3355	ZD-200		
1M200Z10	5105A	3355	ZD-200		
1M200Z3	5105A	3355	ZD-200		
1M200Z5	5105A	3355	ZD-200		
1M200ZS10	5105A	3355	ZD-200		
1M200ZS5	5105A	3355	ZD-200		
1M2086			504A	1104	
1M20Z	5079A	3335	ZD-20		
1M20Z10	5079A	3335	ZD-20		
1M20Z20	5079A	3335	ZD-20		
1M20Z3	5079A	3335	ZD-20		
1M20Z5	5079A	3335	ZD-20		
1M20ZS10	5079A	3335	ZD-20		
1M20ZS5	5097A	3335/5079A	ZD-20		
1M22Z	5080A	3336	ZD-22		
1M22Z10	5080A	3336	ZD-22		
1M22Z5	5080A	3336	ZD-22		
1M22ZS10	5080A	3336	ZD-22		
1M22ZS5	5080A	3336	ZD-22		
1M24Z	5081A	3151	ZD-24		
1M24Z10	5081A	3151	ZD-24		
1M24Z5	5081A	3151	ZD-24		
1M24ZS10	5081A	3151	ZD-24		
1M24ZS5	5081A	3151	ZD-24		
1M25Z	5082A	3753	ZD-25		
1M25Z10	5082A	3753	ZD-25		
1M25Z5	5082A	3753	ZD-25		
1M27Z	146A	3064	ZD-27		
1M27Z10	146A	3064	ZD-27		
1M27Z5	146A	3064	ZD-27		
1M27ZS10	146A	3064	ZD-27		
1M27ZS5	146A	3064	ZD-27		
1M3.34Z3		3330	ZD-3.3		
1M3.3ZS10	5066A	3330	ZD-3.3		
1M3.3ZS5	5066A	3330	ZD-3.3		
1M3.6AZ3	134A		ZD-3.6		
1M3.6ZS10	134A		ZD-3.6		
1M3.6ZS5	134A		ZD-3.6		
1M3.9ZS10	5067A	3331	ZD-3.9		
1M3.9ZS5	5067A	3331	ZD-3.9		
1M30Z	5084A	3755	ZD-30		
1M30Z10	5084A	3755	ZD-30		
1M30Z5	5084A	3755	ZD-30		
1M30ZS10	5084A	3755	ZD-30		
1M30ZS5	5084A	3755	ZD-30		
1M33Z	147A	3095	ZD-33		
1M33Z10	147A	3095	ZD-33		
1M33Z5	147A	3095	ZD-33		
1M33ZS5	147A	3095	ZD-33		
1M353					1N1204C
1M36Z	5085A	3337	ZD-36		
1M36Z10	5085A	3337	ZD-36		
1M36Z5	5085A	3337	ZD-36		
1M36ZS10	5085A	3337	ZD-36		
1M36ZS5	5085A	3337	ZD-36		
1M383T				703	
1M39Z	5086A	3338	ZD-39		
1M39Z10	5086A	3338	ZD-39		
1M39Z5	5086A	3338	ZD-39		
1M39ZS10	5086A	3338	ZD-39		
1M39ZS5	5086A	3338	ZD-39		
1M4.3ZS10	5068A	3332	ZD-4.3		
1M4.3ZS5	5068A	3332	ZD-4.3		
1M4.7ZS10	5069A	3333	ZD-4.7		
1M4.7ZS5	5069A	3333	ZD-4.7		
1M43Z	5087A	3339	ZD-43		
1M43Z10	5087A	3339	ZD-43		
1M43Z5	5087A	3339	ZD-43		
1M43ZS10	5087A	3339	ZD-43		
1M43ZS5	5087A	3339	ZD-43		
1M47Z	5088A	3340	ZD-47		
1M47Z10	5088A	3340	ZD-47		
1M47Z5	5088A	3340	ZD-47		
1M47ZS10	5088A	3340	ZD-47		
1M47ZS5	5088A	3340	ZD-47		
1M5.1AZ3	135A		ZD-5.1		
1M5.1ZS10	135A		ZD-5.1		

Industry Standard No.	ECG	SK	GE	RS 276-	MOTOR.
1M5.1ZS5	5116A		ZD-5.1		
1M5.6AZ10				561	
1M5.6ZS10	136A		ZD-5.6	561	
1M5.6ZS5	136A		ZD-5.6	561	
1M51Z	5089A	3341	ZD-51		
1M51Z10	5089A	3341	ZD-51		
1M51Z5	5089A	3341	ZD-51		
1M51ZS10	5089A	3341	ZD-51		
1M51ZS5	5089A	3341	ZD-51		
1M56Z	5090A	3342	ZD-56		
1M56Z10	5090A	3342	ZD-56		
1M56Z5	5090A	3342	ZD-56		
1M56ZS10	5090A	3342	ZD-56		
1M56ZS5	5090A	3342	ZD-56		
1M6.2AZ10				561	
1M6.2ZS10	137A	3334/5071A	ZD-6.2	561	
1M6.2ZS5	137A		ZD-6.2	561	
1M6.8AZ10				561	
1M6.8Z	5071A	3334	ZD-6.8	561	
1M6.8Z10	5071A	3334	ZD-6.8	561	
1M6.8Z5	5071A	3334	ZD-6.8	561	
1M6.8ZS10	5071A	3334	ZD-6.8	561	
1M6.8ZS5	5071A	3334	ZD-6.8	561	
1M62Z	149A	3097	ZD-62		
1M62Z10	149A	3097	ZD-62		
1M62Z5	149A	3097	ZD-62		
1M62ZS10	149A	3097	ZD-62		
1M62ZS5	149A	3097	ZD-62		
1M68Z	5092A	3343	ZD-68		
1M68Z10	5092A	3343	ZD-68		
1M68Z3	5092A	3343	ZD-68		
1M68Z5	5092A	3343	ZD-68		
1M68ZS10	5092A	3343	ZD-68		
1M68ZS5	5092A	3343	ZD-68		
1M7.5Z	138A		ZD-7.5		
1M7.5Z10	138A		ZD-7.5		
1M7.5Z5	138A		ZD-7.5		
1M7.5ZS10	138A		ZD-7.5		
1M7.5ZS5	138A		ZD-7.5		
1M753A	137A	3058	ZD-6.2	561	
1M75Z	5093A	3344	ZD-75		
1M75Z10	5093A	3344	ZD-75		
1M75Z3	5093A	3344	ZD-75		
1M75Z5	5093A	3344	ZD-75		
1M75ZS10	5093A	3344	ZD-75		
1M75ZS5	5093A	3344	ZD-75		
1M8.2Z	5072A		ZD-8.2	562	
1M8.2Z10	5072A		ZD-8.2	562	
1M8.2Z5	5072A		ZD-8.2	562	
1M8.2ZS10	5072A		ZD-8.2	562	
1M8.2ZS5	5072A		ZD-8.2	562	
1M82Z	150A	3098	ZD-82		
1M82Z10	150A	3098	ZD-82		
1M82Z5	150A	3098	ZD-82		
1M82ZS10	150A	3098	ZD-82		
1M82ZS5	150A	3098	ZD-82		
1M8513A	116	3017B/117	504A	1104	
1M9.1Z	139A		ZD-9.1	562	
1M9.1Z10	139A		ZD-9.1	562	
1M9.1Z5	139A		ZD-9.1	562	
1M9.1ZS10	139A		ZD-9.1	562	
1M9.1ZS5	139A		ZD-9.1	562	
1M91Z	5095A	3345	ZD-91		
1M91Z10	5095A	3345	ZD-91		
1M91Z3	5095A	3345	ZD-91		
1M91Z5	5095A	3345	ZD-91		
1M91ZS10	5095A	3345	ZD-91		
1M91ZS5	5095A	3345	ZD-91		
1MA4			504A	1104	
1MZ10	5075A	3751	ZD-16	564	
1N-22			1N60	1123	
1N-4002	116		504A	1104	
1N10	109	3087		1123	
1N100	109	3087		1123	
1N1008	116	3017B/117	504A	1104	
1N100A	109	3087		1123	
1N1028	116	3016	504A	1104	
1N1029	116	3016	504A	1104	
1N103	109	3091		1123	
1N1030	116	3016	504A	1104	
1N1031	116	3016	504A	1104	
1N1032	116	3016	504A	1104	
1N1033	116	3016	504A	1104	
1N1034	5830		5005		
1N1035	5833	3017B/117	5005		
1N1036	5834	3017B/117	5005		
1N1037	5834	3031A	5005		
1N1038	5837	3017B/117	5005		
1N1039	5838	3031A	5005		
1N104	109	3091		1123	
1N1040	5834		5005		
1N1041	5834	3016	5005		
1N1042	5834	3016	5005		
1N1043	5834	3016	5005		
1N1044	5837	3031A	5005		
1N1045	5838	3031A	5005		
1N1046	5831	3311	5005	1104	
1N1047	5834	3016	5005	1104	
1N1048	5834	3016	5005	1104	
1N1049	5834	3016	5005	1104	
1N105	109	3091		1123	
1N1050	5837	3031A	5005	1104	
1N1051	5838		5005	1104	
1N1052	116	3016	504A	1104	
1N1053	116	3016	504A	1104	
1N1054	166	9075	504A	1152	
1N1055	5802	9005	504A	1143	
1N1056	156	3016	512		
1N1057	156	3016	512		
1N1058	5850		5017		
1N1059	5854		5017		
1N1060	5854		5017		
1N1061	5854		5017		
1N1062	5858		5021		
1N1063	5858		5021		
1N1064	5850		5017		
1N1065	5854		5017		
1N1066	5854		5016		
1N1067	5854		5016		
1N1068	5856		5020		
1N107	109	3087		1123	
1N1070	5850		5017		
1N1071	5854		5017		
1N1072	5854		5017		
1N1073	5854		5017		
1N1074	5858		5021		
1N1075	5858		5021		
1N1077	5895		5065		
1N1078	5944		5048		
1N1079	5897		5065		
1N108	109	3087		1123	
1N1081	116	3016	504A	1104	
1N1081A	116		504A	1104	
1N1082	116	3016	504A	1104	
1N1082A			504A	1104	
1N1083	116	3016	504A	1104	
1N1083A			504A	1104	
1N1084	116	3016	504A	1104	
1N1084A			504A	1104	
1N109	109	3087		1123	
1N1093	109	3087		1123	
1N1095	116	3016	510,531	1114	1N4005
1N1096	116	3033	510,531	1114	1N4005
1N10D-4F	116		504A	1104	
1N110	112		1N82A		
1N1100	116	3016	504A	1104	1N4002
1N1101	116	3016	504A	1104	1N4003
1N1102	116	3016	504A	1104	1N4004
1N1103	116	3016	504A	1104	1N4004
1N1104	116	3016	504A	1104	1N4005
1N1104R	5987		5097		
1N1105	116	3033	504A	1104	1N4006
1N1108			509	1114	
1N111	109	3087		1123	
1N1115	5832	3500/5882	5004		1N1200C
1N1116	5834	3500/5882	5004		1N1202C
1N1117	5836	3500/5882	5004		1N1204C
1N1118	5838	3500/5882	5004		1N1204C
1N1119	5840	3500/5882	5008	1104	1N1206C
1N112	109	3087		1123	
1N1120	5842	7042	5008	1104	1N1206C
1N1122A	116	3017B/117	504A	1104	
1N1124	5834	3500/5882	5004		
1N1124,A					MR1122
1N1124A	5834	3500/5882	5016		
1N1124AR	5855		5017		
1N1124R	5835		5005		
1N1125	5836	3500/5882	5004		
1N1125,A					MR1124
1N1125A	5838	3500/5882	5020		
1N1125AR	5859		5021		
1N1125R	5837		5005		
1N1126	5838	3500/5882	5004		
1N1126,A					MR1124
1N1126A	5838	3500/5882	5020		
1N1126AR	5859		5021		
1N1126R	5839		5005		
1N1127	5840	3500/5882	5008		
1N1127,A					MR1126
1N1127A	5842	3500/5882	5024		
1N1127AR	5863		5025		
1N1127R	5841	3517/5883	5009		
1N1128	5842	7042	5008		
1N1128,A					MR1126
1N1128A	5842	3500/5882	5024		
1N1128AR	5863		5025		
1N1128R	5843	3517/5883	5009		
1N113	109	3087		1123	
1N114	109	3087		1123	
1N115	109	3087		1123	
1N116	109	3087		1123	
1N1169	116	3016	504A	1104	
1N1169,A					1N4004
1N1169A	116	3016	504A	1104	
1N116A	109	3087		1123	
1N117	109	3087		1123	
1N117A	109	3087		1123	
1N118	109	3087		1123	
1N1183	5980	3610	5096		
1N1183A	5980	3610	5096		
1N1183AR	5981		5097		
1N1183R	5981	3698/5987	5097		
1N1183RA	5981	3698/5987			
1N1184	5982	3609/5986	5096		
1N1184A	5982	3609/5986	5096		
1N1184AR	5983		5097		
1N1184R	5983	3698/5987	5097		
1N1184RA	5987	3698			
1N1185	5986	3609	5096		
1N1185A	5986	3609	5096		
1N1185AR	5987		5097		
1N1185R	5987	3698	5097		
1N1185RA	5987	3698			
1N1186	5986	3609	5096		
1N1186A	5986	3609	5096		
1N1186AR	5987	3518/5995	5097		
1N1186R	5987	3698	5097		
1N1186RA	5987	3698			
1N1187	5990	3608	5097		
1N1187A	5988	3608/5990	5100		
1N1187AR	5989		5101		
1N1187R	5989	3518/5995	5101		
1N1187RA	5989	3518/5995			
1N1188	5990	3608	5100		
1N1188A	5990	3608	5100		
1N1188AR	5991		5101		
1N1188R	5991	3518/5995	5101		
1N1188RA	5991	3518/5995			
1N1189	5992	3501/5994	5104		
1N1189A	5992	3501/5994	5104		
1N1189AR	5993		5105		
1N1189R	5993	3518/5995	5105		
1N1189RA	5993	3518/5995			
1N118A	109	3087		1123	
1N119	109	3087	5008	1123	
1N1190	5994	3501	5104		
1N1190A	5994	3501	5104		
1N1190AR	5995		5105		
1N1190R	5995	3518	5105		
1N1190RA	5995	3518			
1N1191	5980	3610	5096		
1N1191A	5980	3610	5096		
1N1191AR	5981		5033		
1N1191R	5981	3698/5987	5097		
1N1191RA	5981	3698/5987			
1N1192	5982	3609/5986	5096		
1N1192A	5982	3609/5986	5096		
1N1192AR	5983		5097		
1N1192R	5983	3698/5987	5097		
1N1192RA	5987	3698			
1N1193	5986	3609	5096		
1N1193A	5986	3609	5096		
1N1193AR	5987		5097		
1N1193R	5987	3698	5097		
1N1193RA	5987	3698			
1N1194	5986	3609	5096		
1N1194A	5986	3609	5096		
1N1194AR	5987		5097		
1N1194R	5987	3698	5097		
1N1194RA	5987	3698			
1N1195	5988	3608/5990	5100		
1N1195A	5988	3608/5990	5100		
1N1195AR	5989		5101		
1N1195R	5989	3518/5995	5101		
1N1195RA	5991	3518/5995			
1N1196	5990	3608	5100		
1N1196A	5990	3608	5100		
1N1196AR	5991		5101		

Industry Standard No.	ECG	SK	GE	RS 276-	MOTOR.
1N1196R	5991	3518/5995	5101		
1N1196RA	5991	3518/5995			
1N1197	5992	3501/5994	5104		
1N1197A	5992	3501/5994	5104		
1N1197AR	5993		5105		
1N1197R	5993	3518/5995	5105		
1N1197RA	5995	3518			
1N1198	5994	3501			
1N1198A	5994	3501	5104		
1N1198AR	5995		5105		
1N1198R	5995	3518	5105		
1N1198RA	5995	3518			
1N1199	5870		5032		
1N1199A	5870	3604	5032		
1N1199AR	5871	3517/5883	5033		
1N1199B	5870		5032		
1N1199R	5871	3517/5883	5033		
1N1199RA	5871	3517/5883			
1N119A	109	3087		1123	
1N120	109	3087		1123	
1N1200	5872		5032		
1N1200A	5872	3500/5882	5032		
1N1200AR	5873	3517/5883	5033		
1N1200B	5872		5032		
1N1200BR	5873	3517/5883	5033		
1N1200R	5873	3517/5883	5033		
1N1200RA	5875	3517/5883			
1N1201	5874		5032		
1N1201A	5874		5032		
1N1201AR	5875	3517/5883	5033		
1N1201B	5874		5032		
1N1201BR	5875	3517/5883	5033		
1N1201R	5875	3517/5883	5033		
1N1201RA	5875	3517/5883			
1N1202	5874		5032		
1N1202A	5874	3603	5032		
1N1202AR	5875	3517/5883	5033		
1N1202B	5874		5032		
1N1202BR	5875	3517/5883	5033		
1N1202R	5875	3517/5883	5033		
1N1202RA	5875	3517/5883			
1N1203	5876		5036		
1N1203A	5876	3500/5882	5036		
1N1203AR	5877	3517/5883	5037		
1N1203B	5876		5036		
1N1203BR	5877	3517/5883	5037		
1N1203R	5877	3517/5883	5037		
1N1203RA	5877	3517/5883			
1N1204	5878		5036		
1N1204A	5878	3602	5036		
1N1204AR	5879	3517/5883	5037		
1N1204B	5878		5036		
1N1204BR	5879	3517/5883	5037		
1N1204R	5879	3517/5883	5037		
1N1204RA	5879	3517/5883			
1N1205	5880		5040		
1N1205A	5880	3500/5882	5040		
1N1205AR	5881	3517/5883	5041		
1N1205B	5880		5040		
1N1205BR	5881	3517/5883	5041		
1N1205R	5881	3517/5883	5041		
1N1205RA	5881	3517/5883			
1N1206	5882		5040		
1N1206A	5882	3500	5040		
1N1206AR	5883	3517	5041		
1N1206B	5882		5040		
1N1206BR	5883	3517	5041		
1N1206R	5883	3517	5041		
1N1206RA	5883	3517			
1N120A	109	3087		1123	
1N1217	116	3081/125	504A	1104	
1N1217,A,B					1N4001
1N1217A	116	3081/125	504A	1104	
1N1217B	116	3017B/117	504A	1104	
1N1218	116	3016	504A	1104	
1N1218,A,B					1N4002
1N1218A	116	3016	504A	1104	
1N1218B	116	3017B/117	504A	1104	
1N1219	116	3016	504A	1104	
1N1219,A,B					1N4003
1N1219A	116	3016	504A	1104	
1N1219B	116	3017B/117	504A	1104	
1N1220	116	9005/5802	504A	1143	
1N1220,A,B					1N4003
1N1220A	116	9005/5802	504A	1143	
1N1220B	5802	9005	504A	1143	
1N1221	116	9006/5803	504A	1144	
1N1221,A,B					1N4004
1N1221A	116	9006/5803	504A	1144	
1N1221B	5803	9006	504A	1144	
1N1222	5804	9007	504A	1144	
1N1222,A,B					1N4004
1N1222A	5806	9007/5804	504A	1104	
1N1222B	117		504A	1104	
1N1223	5806	9008/5805	504A	1104	
1N1223,A,B					1N4005
1N1223A	5806	9008/5805	504A	1104	
1N1223B	5805	9008	504A	1104	
1N1224	116	3848/5806	504A	1104	
1N1224,A,B					1N4005
1N1224A	116	3848/5806	509	1114	
1N1224B	5806	7048/5848	509	1114	
1N1225	125	3033	512		
1N1225,A,B					1N4006
1N1225A	125	3033	510,531	1114	
1N1225B	156	7048/5848	512		
1N1226	125	3033	512		
1N1226,A,B					1N4006
1N1226A	125	3032A	510,531	1114	
1N1226B	156	3033	512		
1N1227	5830		5004		
1N1227,A,B					1N1199C
1N1227A	5830		5004		
1N1227B	5830		5004		
1N1228	5832		5004		
1N1228,A,B					1N1200C
1N1228A	5832		5004		
1N1228B	5832		5004		
1N1229	5834		5004		
1N1229,A,B					1N1202C
1N1229A	5834		5004		
1N1229B	5834		5004		
1N122B				1104	
1N1230	5834		5004		
1N1230,A,B					1N1202C
1N1230A	5834		5004		
1N1230B	5834		5004		
1N1231	5836		5004		
1N1231,A,B					1N1204C
1N1231A	5836		5004		
1N1231B	5836		5004		
1N1232	5838		5004		
1N1232,A,B					1N1204C
1N1232A	5838		5004		
1N1232B	5838		5004		
1N1233	5840	3584/5862	5008		
1N1233,A,B					1N1206C
1N1233A	5840	3584/5862	5008		
1N1233B	5840	3584/5862	5008		
1N1234	5842	7042	5008		
1N1234,A,B					1N1206C
1N1234A	5842	7042	5008		
1N1234B	5842	7042	5008		
1N1235	5846	3516	5012		
1N1235,A,B					MR1128
1N1235A	5846	3516	5012		
1N1235B	5846	3516	5012		
1N1236	5846	3516	5012		
1N1236,A,B					MR1128
1N1236A	5846	3516			
1N1236B	5846	3516	5012		
1N124	112		1N82A		
1N124A	112		1N82A		
1N125	109	3087		1123	
1N1251	116	3016	504A	1104	1N4001
1N1252	116	3016	504A	1104	1N4002
1N1253	116	3016	504A	1104	1N4003
1N1254	116	3016	504A	1104	1N4004
1N1254(WEST)	5980		5096		
1N1255	116	3016	504A	1104	
1N1255(WEST)	5982		5096		
1N1255,A					1N4004
1N1255A	116	3031A	504A	1104	
1N1256	116	3016	504A	1104	1N4005
1N1257	116	3017B/117	504A	1104	1N4005
1N1257(WEST)	5988		5100		
1N1258	125	3032A	510,531	1114	1N4006
1N1259	125	3032A	510,531	1114	1N4006
1N1259(WEST)	5992		5104		
1N126	109	3087		1123	
1N1260	125	3033	510,531	1114	1N4007
1N1261	125	3033	510,531	1114	1N4007
1N126A	109	3087		1123	
1N128	109	3087		1123	
1N128A	109			1123	
1N1301	5980	3610	5096		1N1183A
1N1302	5982	3609/5986	5096		1N1184A
1N1304	5986	3609	5096		1N1186A
1N1306	5988	3608/5990	5100		1N1188A
1N1313	5017A	3783	ZD-9.1	562	1N4102
1N1313A	5017A	3783	ZD-9.1	562	1N4102
1N1314		3786	ZD-11	563	
1N1314A		3786	ZD-11	563	1/4M10.5Z5
1N1315	5022A	3788	ZD-13	563	
1N1315A	5022A	3788	ZD-13	563	
1N1316	5025A	3791	ZD-15	564	
1N1316A	5025A	3791	ZD-16	564	
1N1317	5028A	3794	ZD-19		1N4113
1N1317A	5028A	3794	ZD-19		1N4113
1N1318	5031A	3797	ZD-24		
1N1318A	5031A	3797	ZD-24		1/4M23.5Z5
1N1319	5034A	3800	ZD-28		
1N1319A	5034A	3800	ZD-28		1/4M28.5Z5
1N132	109	3091		1123	
1N1320	5036A	3802	ZD-33		
1N1320A	5036A	3802	ZD-33		1/4M34.5Z5
1N1321	5038A	3804	ZD-39		1/4M41Z10
1N1321A	5038A	3804	ZD-39		1/4M41Z5
1N1322	5040A	3806	ZD-47		
1N1322A	5040A	3806	ZD-47		1/4M48.5Z5
1N1323	5043A	3809	ZD-60		1/4M58Z10
1N1323A	5043A	3809	ZD-60		1/4M58Z5
1N1324			ZD-68		1/4M71Z10
1N1324A			ZD-68		1/4M71Z5
1N1325	5048A	3814	ZD-87		
1N1325A	5094A	9024	ZD-87		1/4M87.5Z5
1N1326			ZD-110		.4M105Z10
1N1326A			ZD-110		.4M105Z5
1N1327	5053A	3819	ZD-130		
1N1327A			ZD-130		.4M127.5Z5
1N133	109	3091		1123	
1N1337-5	116	3017B/117	504A	1104	
1N134	109	3087		1123	
1N1341	5850	3601	5016		
1N1341A	5850	3601	5016		
1N1341AR	5851	3517/5883	5017		
1N1341B	5850	3601	5016		
1N1341BR	5851	3517/5883	5017		
1N1341R	5851	3517/5883	5017		
1N1341RB	5851	3517/5883			
1N1342	5852	3500/5882	5016		
1N1342A	5852	3500/5882	5016		
1N1342AR	5852		5017		
1N1342B	5852	3500/5882	5016		
1N1342BR	5852	3517/5883	5017		
1N1342R	5853	3517/5883	5017		
1N1342RB	5853	3517/5883			
1N1343	5854	3584/5862	5016		
1N1343A	5854	3500/5882	5016		
1N1343AR	5855	3517/5883	5017		
1N1343B	5854	3500/5882	5016		
1N1343BR	5855	3517/5883	5017		
1N1343R	5855	3517/5883	5017		
1N1344	5854	3600	5016		
1N1344A	5854	3600	5016		
1N1344AR	5855	3517/5883	5017		
1N1344B	5854	3600	5016		
1N1344BR	5855	3517/5883	5017		
1N1344R	5855	3517/5883	5017		
1N1344RB	5855	3517/5883			
1N1345	5856	3500/5882	5020		
1N1345A	5856	3500/5882	5020		
1N1345AR	5857	3517/5883	5021		
1N1345B	5856	3500/5882	5020		
1N1345BR	5857	3517/5883	5021		
1N1345R	5857	3517/5883	5021		
1N1345RB	5857	3517/5883			
1N1346	5858	3599	5020		
1N1346A	5858	3599	5020		
1N1346AR	5859	3517/5883	5021		
1N1346B	5858	3599	5020		
1N1346BR	5859	3517/5883	5021		
1N1346R	5859	3517/5883	5021		
1N1346RB	5859	3517/5883			
1N1347	5860	3500/5882	5024		
1N1347A	5860	3500/5882	5024		
1N1347AR	5861	3517/5883	5025		
1N1347B	5861	3500/5882	5024		
1N1347BR	5861	3517/5883	5025		
1N1347R	5861	3517/5883	5025		
1N1347RB	5863	3517/5883			
1N1348	5862	3500/5882	5024		
1N1348A	5862	3500/5882	5024		
1N1348AR	5863	3517/5883	5025		
1N1348B	5862	3500/5882	5024		
1N1348BR	5863	3517/5883	5025		
1N1348R	5863	3517/5883	5025		
1N1348RB	5863	3517/5883			

Industry Standard No.	ECG	SK	GE	RS 276-	MOTOR.
1N135	177	3100/519	300	1122	
1N1351	5186A				1N2974A
1N1351A	5186A				1N2974B
1N1352	5187A				1N2975A
1N1352A	5187A				1N2975B
1N1353	5188A				1N2976A
1N1353A	5188A				1N2976B
1N1354	5189A				1N2977A
1N1356	5192A				1N2980A
1N1356A	5192A				1N2980B
1N1357	5194A				1N2982A
1N1357A	5194A				1N2982B
1N1358	5196A				1N2984A
1N1358A	5196A				1N2984B
1N1359	5197A				1N2985A
1N1359A	5197A				1N2985B
1N136			1N295	1123	
1N1360	5198A				1N2986A
1N1360A	5198A				1N2986B
1N1361	5200A				1N2988A
1N1361A	5200A				1N2988B
1N1362	5202A				1N2989A
1N1362A	5202A				1N2989B
1N1363	5203A				1N2990A
1N1363A	5203A				1N2990B
1N1364	5204A				1N2991A
1N1364A	5204A				1N2991B
1N1365	5205A				1N2992A
1N1365A	5205A				1N2992B
1N1366	5206A				1N2993A
1N1366A	5206A				1N2993B
1N1367	5208A				1N2995A
1N1367A	5208A				1N2995B
1N1368	5210A				1N2997A
1N1368A	5210A				1N2997B
1N1369	5212A				1N2999A
1N1369A	5212A				1N2999B
1N137			1N295	1123	
1N1370	5214A				1N3000A
1N1370A	5214A				1N3000B
1N1371	5215A				1N3001A
1N1371A	5215A				1N3001B
1N1372	5216A				1N3002A
1N1372A	5216A				1N3002B
1N1373	5217A				1N3003A
1N1373A	5217A				1N3003B
1N1374	5219A				1N3004A
1N1374A	5219A				1N3004B
1N1375	5220A				1N3005A
1N1375A	5220A				1N3005B
1N137A	177	3100/519	300	1122	
1N137B	177	3100/519	300	1122	
1N138	177	3100/519	300	1122	
1N138A	177	3100/519	300	1122	
1N138B	177	3100/519	300	1122	
1N139	109	3087		1123	
1N1391	109			1123	
1N1396	6154		5128		
1N1399	6154	7226/6026	5128		
1N140	109	3087		1123	
1N1401	6060	7260	5132		
1N1406	116	3017B/117	504A	1104	
1N1407	125	3033	510,531	1114	
1N1408	125	3033	510,531	1114	
1N1409	506	3843	511	1114	
1N1415	116	3016	504A	1104	
1N1416	5183A				1N2972B
1N1417	5188A				1N2976B
1N1418	5191A				1N2979B
1N1419	5194A				1N2982B
1N142	109	3087		1123	
1N1420	5197A				1N2985B
1N1421	5200A				1N2988B
1N1422	5215A				1N3001B
1N1423	5220A				1N3005B
1N1424	5226A				1N3011B
1N1425	5072A	3136	ZD-8.2	562	1N4738A
1N1426	142A	3062	ZD-12	563	1N4742A
1N1427	145A	3063	ZD-15	564	1N4744A
1N1428	5077A	3752	ZD-18		1N4746A
1N1429	5080A	3336	ZD-22		1N4748A
1N143	109	3087		1123	
1N1430	146A	3064	ZD-27		1N4750A
1N1431	5092A	3343	ZD-68		1N4760A
1N1432	5096A	3346	ZD-100		1N4764A
1N1433	5100A	3350	ZD-150		1M150ZS5
1N1434	5980	3610	5096		1N1183A
1N1435	5982	3609/5986	5096		1N1184A
1N1436	5986	3609	5096		1N1186A
1N1437	5990	3608	5100		1N1188A
1N1438	5994	3501	5104		1N1190A
1N1439	116	3016	504A	1104	
1N144	109	3087		1123	
1N1440	116	3016	504A	1104	
1N1441	116	3016	504A	1104	
1N1442	116	3016	504A	1104	
1N1443	551	3033	510,531	1114	
1N1443,A,B					1N4007
1N1443A	551	3033	510,531	1114	
1N1443B	551		5012		
1N1444	5848	7048	5012		
1N1444,A,B					MR1130
1N1444A	5848	7048	5012		
1N1444B	5848	7048	5012		
1N1446	5834		5004		
1N1447	5834		5004		
1N1448	5838	3100/519	5004	1122	
1N1449	5838		5004		
1N145	109	3087		1123	
1N1450	5834		5004		
1N1451	5834		5004		
1N1452	5838		5004		
1N1453	5838		5004		
1N147	112		1N82A		
1N147A	112		1N82A		
1N148	109	3087		1123	
1N1482	5176A				1N3995A
1N1483	5180A				1N3998A
1N1484	5069A	3330/5066A	ZD-4.7		1N4732A
1N1485	137A	3058	ZD-6.2	561	1N4735A
1N1486	116	3016	504A	1104	1N4005
1N1487	116	3016	504A	1104	1N4002
1N1488	116	3016	504A	1104	1N4003
1N1489	116	3016	504A	1104	1N4004
1N1490	116	3016	504A	1104	1N4004
1N1491	116	3016	504A	1104	1N4005
1N1492	116	3033	504A	1104	1N4005
1N1507	5067A	3331	ZD-3.9		1N4730
1N1507A	5067A	3331	ZD-3.9		1N4730A
1N1508	5069A	3330/5066A	ZD-4.7		1N4732
1N1508A	5069A	3330/5066A	ZD-4.7		1N4732A
1N1509	136A		ZD-5.6	561	1N4734
1N1509A	136A	3057	ZD-5.6	561	1N4734A
1N151	116	3017B/117	504A	1104	
1N1510	5071A	3334	ZD-6.8	561	1N4736
1N1510A	5071A	3334	ZD-6.8	561	1N4736A
1N1511	5072A	3136	ZD-8.2	562	1N4738
1N1511A	5072A	3136	ZD-8.2	562	1N4738A
1N1512	140A		ZD-10	562	1N4740
1N1512A	140A		ZD-10	562	1N4740A
1N1513	142A	3062	ZD-12	563	1N4742
1N1513A	142A	3062	ZD-12	563	1N4742A
1N1514	145A	3063	ZD-15	564	1N4744
1N1514A	145A	3063	ZD-15	564	1N4744A
1N1515	5077A	3752	ZD-18		1N4746
1N1515A	5077A	3752	ZD-18		1N4746A
1N1516	5080A	3336	ZD-22		1N4748
1N1516A	5080A	3336	ZD-22		1N4748A
1N1517	146A	3064	ZD-27		1N4750
1N1517A	146A	3064	ZD-27		1N4750A
1N1518	5067A	3331	ZD-3.9		1N4730
1N1518A	5067A	3331	ZD-3.9		1N4730A
1N1519	5069A	3333	ZD-4.7		1N4732
1N1519A	5069A	3333	ZD-4.7		1N4732A
1N152	116	3017B/117	504A	1104	
1N1520	136A		ZD-5.6	561	1N4734
1N1520A	136A	3057	ZD-5.6	561	1N4734A
1N1521	5071A	3334	ZD-6.8	561	1N4736
1N1521A	5071A	3334	ZD-6.8	561	1N4736A
1N1522	5072A	3136	ZD-8.2	562	1N4738
1N1522A	5072A	3136	ZD-8.2	562	1N4738A
1N1523	140A		ZD-10	562	1N4740
1N1523A	140A		ZD-10	562	1N4740A
1N1524	142A	3062	ZD-12	563	1N4742
1N1524A	142A	3062	ZD-12	563	1N4742A
1N1525	145A	3063	ZD-15	564	1N4744
1N1525A	145A	3063	ZD-15	564	1N4744A
1N1526	5077A	3752	ZD-18		1N4746
1N1526A	5077A	3752	ZD-18		1N4746A
1N1527	5080A	3336	ZD-22		1N4748
1N1527A	5080A	3336	ZD-22		1N4748A
1N1528	146A	3064	ZD-27		1N4750
1N1528A	146A	3064	ZD-27		1N4750A
1N153	116	3017B/117	504A	1104	
1N1530		3749	ZD-9.1	562	1N3156
1N1530A		3749	ZD-9.1	562	1N3157
1N1537	5830	3500/5882	5004		1N1199C
1N1537R	5831	3517/5883	5005		
1N1538	5834	3500/5882	MR-1		1N1200C
1N1538R	5835	3517/5883	5005		
1N1539	5834	3500/5882	5004		1N1202C
1N1539R	5835	3517/5883	5005		
1N1540	5834	3500/5882	5004		1N1202C
1N1540R	5835	3517/5883	5005		
1N1541	5838	3500/5882	5004		1N1204C
1N1541R	5837	3517/5883	5005		
1N1542	5838	3500/5882	5004		1N1204C
1N1542R	5839		5005		
1N1543	5842	3500/5882	5008	1104	1N1206C
1N1543R	5841	3517/5883	5009		
1N1544	5842	7042	5008	564	1N1206C
1N1544A				564	
1N1544R	5843	3517/5883	5009		
1N1551	5832	3517/5883	5004		1N1200C
1N1552	5834		5004		1N1202C
1N1553	5836		5004	1104	1N1204C
1N1554	5838		5004	1104	1N1204C
1N1555	5840	3584/5862	5008		1N1206C
1N1556	5801	9004	504A	1142	1N4002
1N1557	5802	9005	504A	1143	1N4003
1N1558	5803	9006	504A	1144	1N4004
1N1559	5804	9007	504A	1144	1N4004
1N1560	5805	9008	504A	1104	1N4005
1N1561	109	3087		1123	
1N1562	109	3087		1123	
1N1563	5801	9004	504A	1142	
1N1563A	116	9004/5801	504A	1104	
1N1564	5802	9005	504A	1143	
1N1564A	116	9005/5802	504A	1104	
1N1565	5803	9006	504A	1144	
1N1565A	116	9006/5803	504A	1104	
1N1566	5804	9007	504A	1144	
1N1566A	116	9007/5804	504A	1104	
1N1567	5805	9008	504A	1104	
1N1567A	116	9008/5805	504A	1104	
1N1568	5806	3848	504A	1104	
1N1568A	116	3848/5806	504A	1104	
1N1575		3812	ZD-75		
1N158	116	3017B/117	504A	1104	
1N1581	5830		5004		1N1199C
1N1581R	5831	3517/5883	5005		
1N1582	5832		5004		1N1200C
1N1582R	5835	3517/5883	5005		
1N1583	5834		5004		1N1202C
1N1583R	5835	3517/5883	5005		
1N1584	5836		5004		1N1204C
1N1584R	5837	3517/5883	5005		
1N1585	5838		5004		1N1204C
1N1585R	5839	3517/5883	5005		
1N1586	5840	3584/5862	5008		1N1206C
1N1586R	5841	3517/5883	5009		
1N1587	5842	7042	5008		1N1206C
1N1587R	5843	3517/5883	5009		
1N1588	5174AK				1N3993A
1N1588A	5174AK				1N3993A
1N1589	5176AK				1N3995A
1N1589A	5176AK				1N3995A
1N1590	5178AK				1N3997A
1N1590A	5177AK				1N3997A
1N1591	5181AK				1N2970RA
1N1591A	5181AK				1N2970RB
1N1592	5183AK				1N2972RA
1N1592A	5183AK				1N2972RB
1N1593	5186AK				1N2974RA
1N1593A	5186AK				1N2974RB
1N1594	5188AK				1N2976RA
1N1594A	5188AK				1N2976RB
1N1595	5191AK				1N2979RA
1N1595A	5191AK				1N2979RB
1N1596	5194AK				1N2982RA
1N1596A	5194AK				1N2982RB
1N1597	5197AK		509		
1N1597A	5197AK				1N2985RB
1N1598	5200AK				1N2988RA
1N1598A	5200AK				1N2988RB
1N1599	5174AK				1N3993A
1N1599A					1N3993A
1N1600	5176AK				1N3995A
1N1600A	5176AK				1N3995A
1N1601	5178AK				1N3997A
1N1601A	5177AK				1N3997A
1N1602	5181AK				1N2970RA
1N1602A	5181AK				1N2970RB
1N1603	5183AK				1N2972RA
1N1603A	5183AK				1N2972RB
1N1604	5186AK				1N2974RA
1N1604A	5186AK				1N2974RB
1N1605	5188AK				1N2976RA

Industry Standard No.	ECG	SK	GE	RS 276-	MOTOR.
1N1605A	5188AK				1N2976RB
1N1606	5191AK				1N2979RA
1N1606A	5191AK				1N2979RB
1N1607	5194AK				1N2982RA
1N1607A	5194AK				1N2982RB
1N1608	5197AK				1N2985RA
1N1608A	5197AK				1N2985RB
1N1609	5200AK				1N2988RA
1N1609A	5200AK				1N2988RB
1N1612	5874	3500/5882	5016		
1N1612A	5850		5016		
1N1612AR	5851		5017		
1N1612R	5851	3517/5883	5017		
1N1613	5852	3500/5882	5016		
1N1613A	5852		5016		
1N1613AR	5853		5017		
1N1613R	5853	3517/5883	5017		
1N1614	5854	3500/5882	5016		
1N1614A	5854		5016		
1N1614AR	5855		5017		
1N1614R	5855	3517/5883	5017		
1N1615	5878	3500/5882	5020		
1N1615A	5858		5020		
1N1615AR	5859		5021		
1N1615R	5859	3517/5883			
1N1616	5882	3500	5025		
1N1616A	5862		5024		
1N1616AR	5863		5025		
1N1616R	5883	3517			
1N1617	116	9004/5801	504A	1142	
1N1618	116	9005/5802	504A	1143	
1N1619	116	9006/5803	504A	1144	
1N1620	116	9007/5804	504A	1144	
1N1621	5872		5032		
1N1622	5874	3500/5882	5032		
1N1623	5876	3500/5882	5036		
1N1624	5878	3500/5882	5036		
1N1630	177		300	1122	
1N1638	177		300	1122	
1N1644	5800	9003	504A	1142	1N4001
1N1645	5801	9004	504A	1142	1N4003
1N1646	5802	9005	504A	1143	1N4003
1N1647	5802	9005	504A	1143	1N4004
1N1648	5803	9006	504A	1144	1N4004
1N1649	5803	9006	504A	1144	1N4004
1N1650	5804	9007	504A	1144	1N4004
1N1651	5804	9007	504A	1144	1N4005
1N1652	5805	9008	504A	1104	1N4005
1N1653	5806	3848	504A	1104	1N4005
1N169	116	3017B/117	504A	1104	
1N1692	116	9004/5801	504A	1142	1N4002
1N1693	116	9005/5802	504A	1143	1N4003
1N1694	116	9006/5803	504A	1144	1N4004
1N1695	116	9007/5804	504A	1144	1N4004
1N1696	116	9008/5805	504A	1104	1N4005
1N1697	116	3848/5806	504A	1104	1N4005
1N1701	116	9003/5800	504A	1142	1N4001
1N1702	116	9004/5801	504A	1142	1N4002
1N1703	116	9005/5802	504A	1143	1N4003
1N1704	116	9006/5803	504A	1144	1N4004
1N1705	116	9007/5804	504A	1144	1N4004
1N1706	116	9008/5805	504A	1104	1N4005
1N1707	116	9003/5800	504A	1142	1N4001
1N1708	116	9004/5801	504A	1142	1N4002
1N1709	116	9005/5802	504A	1143	1N4003
1N1710	116	9006/5803	504A	1144	1N4004
1N1711	116	9007/5804	504A	1144	1N4004
1N1712	116	9008/5805	504A	1104	1N4005
1N1713			ZD-12	563	
1N172	112		1N82A		
1N173	112		1N82A		
1N1730	125	3033	510,531	1114	
1N1730,A					MR250-1
1N1730A	125	3033	510,531	1114	
1N1731A	525	3925			
1N1735			ZD-6.2	561	1N823
1N1736		3787	ZD-12	563	1N941A
1N1736A		3787	ZD-12	563	1N942A
1N1737		9023	ZD-19		
1N1737A		9023	ZD-19		
1N1738		3753	ZD-25		
1N1739		3755	ZD-30		
1N1739A		3755	ZD-30		
1N1743					1N2974A
1N1744	140A		ZD-10	562	1N4740
1N1763	116	9007/5804	504A	1144	1N4004
1N1763A	5804	9007	504A	1144	
1N1764	116	9008/5805	504A	1104	1N4005
1N1764A	5805	9008	504A	1104	
1N1765	135A	3057/136A	ZD-5.1	561	1N4734
1N1765A	136A	3057	ZD-5.6	561	1N4734A
1N1766	137A	3058	ZD-6.2	561	1N4735
1N1766A	137A	3058	ZD-6.2	561	1N4735A
1N1767	5071A	3334	ZD-6.8	561	1N4736
1N1767A	5071A			561	1N4736A
1N1768	138A		ZD-7.5		1N4737
1N1768A	138A				1N4737A
1N1769	5072A	3136	ZD-8.2	562	1N4738
1N1769A	5072A	3136		562	1N4738A
1N1770	139A	3060	ZD-9.1	562	1N4739
1N1770A	139A	3060	ZD-9.1	562	1N4739A
1N1771	140A		ZD-10	562	1N4740
1N1771A	140A			562	1N4740A
1N1772	5074A		ZD-11.0	563	1N4741
1N1772A	5074A			563	1N4741A
1N1773	142A	3062	ZD-12	563	1N4742
1N1773A	142A	3062	ZD-12	563	1N4742A
1N1774	143A	3750	ZD-13	563	1N4743
1N1774A	143A	3750		563	1N4743A
1N1775	145A	3063	ZD-15	564	1N4744
1N1775A	145A	3063	ZD-15	564	1N4744A
1N1776	5075A	3751	ZD-16	564	1N4745
1N1776A	5075A	3751		564	1N4745A
1N1777	5077A	3752	ZD-18		1N4746
1N1777A	5077A	3752			1N4746A
1N1778	5079A	3335	ZD-20		1N4747
1N1778A	5079A				1N4747A
1N1779	5070A	9021	ZD-6.2	561	1N4748
1N1779A	5079A			561	1N4748A
1N1780	5081A		ZD-24		1N4749
1N1780A	5079A				1N4749A
1N1781	146A	3064	ZD-27		1N4750
1N1781A	146A	3064	ZD-27		1N4750A
1N1782	5084A	3755	ZD-30		1N4751
1N1782A	5084A	3755			1N4751A
1N1783	143A	3750	ZD-13	563	1N4752
1N1783A	147A	3750/143A	ZD-33		1N4752A
1N1784	5085A	3337	ZD-36		1N4753
1N1784A	5085A				1N4753A
1N1785	5086A	3338	ZD-39		1N4754
1N1785A	5086A				1N4754A
1N1786	5087A	3339	ZD-43		1N4755
1N1786A	5087A	3339	ZD-43		1N4755A
1N1787	5088A	3340	ZD-47		1N4756
1N1787A	5088A				1N4756A
1N1788	5089A	3341	ZD-51		1N4757
1N1788A	5089A	3341	ZD-51		1N4757A
1N1789	5090A	3342	ZD-56		1N4758
1N1789A	5090A	3342	ZD-56		1N4758A
1N1790	149A		ZD-62		1N4759
1N1790A	149A				1N4759A
1N1791	5092A	3343	ZD-68		1N4760
1N1791A	5092A	3343	ZD-68		1N4760A
1N1792	5093A	3344	ZD-75		1N4761
1N1792A	5093A	3344	ZD-75		1N4761A
1N1793	150A	3098	ZD-82		1N4762
1N1793A	150A	3098	ZD-82		1N4762A
1N1794	5095A	3345	ZD-91		1N4763
1N1794A	5095A	3345	ZD-91		1N4763A
1N1795	5096A	3346	ZD-100		1N4764
1N1795A	5096A	3346	ZD-100		1N4764A
1N1796	151A	3099	ZD-110		1M110ZS10
1N1796A	151A	3099	ZD-110		1M110ZS5
1N1797	5097A	3347	ZD-120		1M120ZS10
1N1797A	5097A				1M120ZS5
1N1798	5098A	3348	ZD-130		1M130ZS10
1N1798A	5098A	3348	ZD-130		1M130ZS5
1N1799	5100A	3350	ZD-150		1M150ZS10
1N1799A	5100A				1M150ZS5
1N1800	5101A	3351	ZD-160		1M160ZS10
1N1800A	5101A	3351	ZD-160		1M160ZS5
1N1801	5103A	3353	ZD-180		1M180ZS10
1N1801A	5103A	3353	ZD-180		1M180ZS5
1N1802	5105A	3355	ZD-200		1M200ZS10
1N1802A	5105A				1M200ZS5
1N1803	5178A				1N3997RA
1N1803A	5178A				1N3997RA
1N1804	5180A				1N3998RA
1N1804A	5180A				1N3998RA
1N1805	5181A				1N2970A
1N1805A	5181A				1N2970B
1N1806	5182A				1N2971A
1N1806A	5182A				1N2971B
1N1807	5183A				1N2972A
1N1807A	5183A				1N2972B
1N1808	5185A				1N2973A
1N1808A	5185A				1N2973B
1N1809	5222A				1N3007A
1N1809A	5222A				1N3007B
1N1810	5223A				1N3008A
1N1810A	5223A				1N3008B
1N1811	5224A				1N3009A
1N1811A	5224A				1N3009B
1N1812	5226A				1N3011A
1N1812A	5226A				1N3011B
1N1813	5227A				1N3012A
1N1813A	5227A				1N3012B
1N1814	5230A				1N3014A
1N1814A	5230A				1N3014B
1N1815	5232A				1N3015A
1N1815A	5232A				1N3015B
1N1816	5189A				1N2977A
1N1816A	5189A				1N2977B
1N1816C					10M13ZZ10
1N1816CA	5189A				10M13ZZ5
1N1817	5191A				1N2979A
1N1817A	5191A				1N2979B
1N1817C					10M15ZZ10
1N1817CA					10M15ZZ5
1N1818	5192A				1N2980A
1N1818A	5192A				1N2980B
1N1818C					10M16ZZ10
1N1818CA					10M16ZZ5
1N1819	5194A				1N2982A
1N1819A	5194A				1N2982B
1N1819C					10M18ZZ10
1N1819CA					10M18ZZ5
1N1820	5196A	3057/136A	ZD-5.6		1N2984A
1N1820A	5196A				1N2984B
1N1820C					10M20ZZ10
1N1820CA					10M20ZZ5
1N1821	5197A				1N2985A
1N1821A	5197A				1N2985B
1N1821C					10M22ZZ10
1N1821CA					10M22ZZ5
1N1822	5198A				1N2986A
1N1822A	5198A				1N2986B
1N1822C					10M24ZZ10
1N1822CA					10M24ZZ5
1N1823	5200A				1N2988A
1N1823A	5200A				1N2988B
1N1823C					10M27ZZ10
1N1823CA					10M27ZZ5
1N1824	5202A				1N2989A
1N1824A	5202A				1N2989B
1N1824C					10M30ZZ10
1N1824CA					10M30ZZ5
1N1825	5203A				1N2990A
1N1825A	5203A				1N2990B
1N1825C					10M33ZZ10
1N1825CA					10M33ZZ5
1N1826	5204A				1N2991A
1N1826A	5204A				1N2991B
1N1826C					10M36ZZ10
1N1826CA					10M36ZZ5
1N1827	5205A				1N2992A
1N1827A					1N2992B
1N1827C					10M39ZZ10
1N1827CA					10M39ZZ5
1N1828	5206A				1N2993A
1N1828A	5206A				1N2993B
1N1828C					10M43ZZ10
1N1828CA					10M43ZZ5
1N1829	5208A				1N2995A
1N1829A	5208A				1N2995B
1N1829C					10M47ZZ10
1N1829CA					10M47ZZ5
1N1830	5210A				1N2997A
1N1830A	5210A				1N2997B
1N1830C					10M51ZZ10
1N1830CA					10M51ZZ5
1N1831	5212A	3064/146A	ZD-27		1N2999A
1N1831A	5212A	3064/146A	ZD-27		1N2999B
1N1831C					10M56ZZ10
1N1831CA					10M56ZZ5
1N1832	5214A				1N3000A
1N1832A	5214A				1N3000B
1N1832C					10M62ZZ10
1N1832CA					10M62ZZ5
1N1833	5215A				1N3001A
1N1833A	5215A				1N3001B
1N1833C					10M68ZZ10
1N1833CA					10M68ZZ5
1N1834	5216A				1N3002A
1N1834A	5216A				1N3002B
1N1834C					10M75ZZ10
1N1834CA					10M75ZZ5
1N1835	5217A				1N3003A

Industry Standard No.	ECG	SK	GE	RS 276-	MOTOR.
1N1835A	5217A				1N3003B
1N1835C					10M82ZZ10
1N1835CA					10M82ZZ5
1N1836	5219A				1N3004A
1N1836A	5219A				1N3004B
1N1836C					10M91ZZ10
1N1836CA					10M91ZZ5
1N1839	177		300	1122	
1N1840	177		300	1122	
1N1841	177		300	1122	
1N1842	178MP		300(2)	1122	
1N1843	177		300	1122	
1N1844	177		300	1122	
1N1845	177		300	1122	
1N1846	177		300	1122	
1N1847	177		300	1122	
1N1875	5072A	3136	ZD-8.2	562	
1N1875A	5072A	3136	ZD-8.2	562	
1N1875B	5072A		ZD-8.2	562	
1N1876	140A		ZD-10	562	1N4740
1N1876A	140A		ZD-10	562	
1N1877	142A	3062	ZD-12	563	1N4742
1N1877A	142A	3062	ZD-12	563	
1N1878	145A	3063	ZD-15	564	1N4744
1N1878A	145A	3063	ZD-15	564	
1N1879	5077A	3752	ZD-18		1N4746
1N1879A	5077A	3752	ZD-18		
1N1880	5080A	3336	ZD-22		1N4748
1N1880A	5080A	3336	ZD-22		
1N1881	146A	3064	ZD-27		1N4750
1N1881A	146A	3064	ZD-27		
1N1882	147A	3095	ZD-33		1N4752
1N1882A	147A	3095	ZD-33		
1N1883	5086A	3338	ZD-39		1N4754
1N1883A	5086A	3338	ZD-39		
1N1884	5088A	3340	ZD-47		1N4756
1N1884A	5088A	3340	ZD-47		
1N1885	5090A	3342	ZD-56		1N4758
1N1885A	5090A	3342	ZD-56		
1N1886	5092A	3343	ZD-68		1N4760
1N1886A	5092A	3343	ZD-68		
1N1887	150A	3098	ZD-82		1N4762
1N1887A	150A	3098	ZD-82		
1N1888	5096A	3346	ZD-100		1N4764
1N1888A	5096A	3346	ZD-100		
1N1889	5097A	3347	ZD-120		1M120ZS10
1N1890	5100A	3350	ZD-150		1M150ZS10
1N1891	5183A				1N2972A
1N1892	5186A				1N2974A
1N1893					1N2976A
1N1894	5191A				1N2979A
1N1895	5194A				1N2982A
1N1896	5197A				1N2985A
1N1897	5200A				1N2988A
1N1898	5203A				1N2990A
1N1899	5205A				1N2992A
1N1900	5208A				1N2995A
1N1901	5212A				1N2999A
1N1902	5215A				1N3001A
1N1903	5217A				1N3003A
1N1904	5221A				1N3005A
1N1905	5223A				1N3008A
1N1906	5226A				1N3011A
1N1907	116	9003/5800	504A	1142	1N4001
1N1908	116	9004/5801	504A	1142	1N4002
1N1909	116	9005/5802	504A	1143	1N4003
1N191	109	3087		1123	
1N1910	5803	9006	504A	1144	1N4004
1N1911	116	9007/5804	504A	1144	1N4004
1N1912	116	9008/5805	504A	1104	1N4005
1N1913	116	3848/5806	504A	1104	1N4005
1N1914	125	9009/5808	509	1114	1N4006
1N1915	125	9009/5808	509	1114	1N4006
1N1916	125	9010/5809	509	1114	1N4007
1N1917	5850		5016		
1N192	109	3087		1123	
1N1927	5007A	3773	ZD-3.9		1N5228A
1N1928	5009A	3775	ZD-4.7		1N5230A
1N1929	5011A	3777	ZD-5.6	561	1N5232A
1N1929A	136A	3777/5011A	ZD-5.6	561	
1N1929B		3777	ZD-5.6	561	
1N193		3017B	504A	1104	
1N1930	5014A	3780	ZD-6.8	561	1N5235A
1N1931	5016A	3782	ZD-8.2	562	1N5237A
1N1932	5019A	3785	ZD-10	562	1N5240A
1N1932A	140A	3785/5019A	ZD-10	562	
1N1932B		3785	ZD-10	562	
1N1933	5021A	3787	ZD-12	563	1N5242A
1N1934	5024A	3790	ZD-15	564	1N5245A
1N1934A	145A	3790/5024A	ZD-15	564	
1N1934B		3790	ZD-15	564	
1N1935	5027A	3793	ZD-18		1N5248A
1N1935A	5077A	3793/5027A			
1N1936	5030A	3796	ZD-22		1N5251A
1N1937	5033A	3799	ZD-27		1N5254A
1N1937A	146A	3799/5033A	ZD-27		
1N1937B		3064	ZD-27		
1N1938	5036A	3802	ZD-33		1N5257A
1N1938A	147A	3802/5036A	ZD-33		
1N1938B		3802	ZD-33		
1N1939	5038A	3804	ZD-39		1N5259A
1N194	177	3100/519	300	1122	
1N1940	5040A	3806	ZD-47		1N5261A
1N1941	5042A	3808	ZD-56		1N5263A
1N1942	5042A	3808	ZD-56		1N5266A
1N1943	5047A	3813	ZD-82		1N5268A
1N1943A	150A	3813/5047A	ZD-82		
1N1943B		3813	ZD-82		
1N1944	5050A	3816	ZD-100		1N5271A
1N1945	5052A	3818	ZD-120		1N5273A
1N1946	5055A	3821	ZD-150		1N5276A
1N1947	5058A	3824	ZD-180		1N5279A
1N194A	177	3100/519	300	1122	
1N195	177	3100/519	300	1122	
1N1954	5007A	3773	ZD-3.9		1N5228A
1N1954A		3773	ZD-3.9		
1N1954B		3773	ZD-3.9		
1N1955	5009A	3775	ZD-4.7		1N5230A
1N1956	5011A	3777	ZD-5.6	561	1N5232A
1N1957	5014A	3780	ZD-6.8	561	1N5235A
1N1958	5016A	3782	ZD-8.2	562	1N5237A
1N1959	5019A	3785	ZD-10	562	1N5240A
1N196	177	3100/519	300	1122	
1N1960	5021A	3062/142A	ZD-12	563	1N5242A
1N1960A	142A	3062	ZD-12	563	
1N1960B		3062	ZD-12	563	
1N1961	5023A	3789	ZD-14	564	1N5245A
1N1962	5027A	3793	ZD-18		1N5248A
1N1962A	5077A	3793/5027A			
1N1963	5030A	3796	ZD-22		1N5251A
1N1964	5033A	3799	ZD-27		1N5254A
1N1964A	146A	3799/5033A	ZD-27		
1N1964B		3064	ZD-27		
1N1965	5036A	3802	ZD-33		1N5257A

Industry Standard No.	ECG	SK	GE	RS 276-	MOTOR.
1N1965A	147A	3802/5036A	ZD-33		
1N1965B		3802	ZD-33		
1N1966	5038A	3804	ZD-39		1N5259A
1N1967	5040A	3806	ZD-47		1N5261A
1N1968	5042A	3808	ZD-56		1N5263A
1N1969	5045A	3811	ZD-68		1N5266A
1N1970	5047A	3813	ZD-82		1N5268A
1N1971	5050A	3816	ZD-100		1N5271A
1N1972		3818	ZD-120		1N5273A
1N1973		3821	ZD-150		1N5276A
1N1974		3824	ZD-180		1N5279A
1N198	109	3087		1123	
1N1981	5007A	3773	ZD-3.9		1N5228A
1N1981A	5067A	3773/5007A	ZD-3.9		
1N1981B		3773	ZD-3.9		
1N1982	5009A	3775	ZD-4.7		1N5230A
1N1983	5011A	3777	ZD-5.6	561	1N5232A
1N1983A	136A	3777/5011A	ZD-5.6	561	
1N1983B		3777	ZD-5.6	561	
1N1984	5014A	3780	ZD-6.8	561	1N5235A
1N1985	5016A	3782	ZD-8.2	562	1N5237A
1N1985A	5072A	3782/5016A			
1N1986	5019A	3785	ZD-10	562	1N5240A
1N1986A	140A	3785/5019A	ZD-10	562	
1N1986B		3785	ZD-10	562	
1N1987	5021A	3787	ZD-12	563	1N5242A
1N1987A	142A	3787/5021A			
1N1988	5024A	3790	ZD-15	564	1N5245A
1N1988A	145A	3790/5024A	ZD-15	564	
1N1988B		3790	ZD-15	564	
1N1989	5027A	3793	ZD-18		1N5248A
1N198A	109	3087		1123	
1N198B	109	3087		1123	
1N1990	5030A	3796	ZD-22		1N5251A
1N1991	5033A	3799	ZD-27		1N5254A
1N1991A	146A	3799/5033A	ZD-27		
1N1991B	146A	3799/5033A	ZD-27		
1N1992	5036A	3802	ZD-33		1N5257A
1N1992A	147A	3802/5036A	ZD-33		
1N1992B	147A	3802/5036A	ZD-33		
1N1993	5038A	3804	ZD-39		1N5259A
1N1994	5040A	3806	ZD-47		1N5261A
1N1995	5042A	3808	ZD-56		1N5263A
1N1996	5045A	3811	ZD-68		1N5266A
1N1997	5047A	3813	ZD-82		1N5268A
1N1997A	150A	3098			
1N1998	5050A	3816	ZD-100		1N5271A
1N1999	5052A	3818	ZD-120		1N5273A
1N1A13MA	143A	3750	ZD-13	563	
1N200	177	3100/519	300	1122	
1N2000	5055A	3821	ZD-150		1N5276A
1N2001	5058A	3824	ZD-180		1N5279A
1N2001A	5103A	3824/5058A	ZD-180		
1N2001B		3824	ZD-180		
1N2008	5220A				1N3005A
1N2008C					10M100ZZ10
1N2008CA					10M100ZZ5
1N2009	5222A				1N3007A
1N2009C					10M110ZZ10
1N2009CA					10M110ZZ5
1N201	178MP		300(2)	1122	
1N2010	5223A				1N3008A
1N2010C					10M120ZZ10
1N2010CA					10M120ZZ5
1N2011	5224A				1N3009A
1N2011C					10M130ZZ10
1N2011CA					10M130ZZ5
1N2012	5226A				1N3011A
1N2012A,AR					1N3011B
1N2012C					10M150ZZ10
1N2012CA					10M150ZZ5
1N2013	5800	9003	504A	1142	1N4001
1N2014	5801	9004	504A	1142	1N4002
1N2015	5802	9005	504A	1143	1N4003
1N2016	5802	9005	504A	1143	1N4003
1N2017	5803	9006	504A	1144	1N4004
1N2018	5803	9006	504A	1144	1N4004
1N2019	5804	9007	504A	1144	1N4004
1N202	177	3100/519	300	1122	
1N2020	5804	9007	504A	1144	1N4004
1N2021	5944		5048		1N1186
1N2022	5946				1N1188
1N2023	5946				1N1188
1N2024	5948				1N1188
1N2025	5948				1N1188
1N2026	5830	3051/156	5004		1N1199C
1N2027	5834	3051/156	5004		1N1202C
1N2028	5836	3051/156	5004		1N1204C
1N2029	5838	3051/156	5004		1N1204C
1N203	177	3100/519	300	1122	
1N2030	5840	3051/156	5008		1N1206C
1N2031	5842	7042	5008		1N1206C
1N2032	5069A	3330/5066A	ZD-4.7		1N4732
1N2032-2	135A		ZD-5.1		
1N2032A	5069A	3330/5066A	ZD-4.7		
1N2033	5070A		ZD-6.2	561	1N4734
1N2033-2	5070A		ZD-6.2	561	
1N2033A	136A		ZD-5.6	561	
1N2034	5071A	3334	ZD-6.8	561	1N4736
1N2034-2	5071A	3334	ZD-6.8	561	
1N2034-3	138A		ZD-7.5	561	
1N2034A	5071A	3334	ZD-6.8	561	
1N2035	139A	3749/5073A			1N4739
1N2035-1	5073A		ZD-9.1	562	
1N2035-12		3749	ZD-9.1	562	
1N2035A	5072A		ZD-8.2	562	
1N2036	140A	3017B/117	ZD-10	562	1N4740
1N2036-2	5074A		ZD-11.0	563	
1N2037	143A	3750	ZD-13	563	1N4743
1N2037-1	142A	3062	ZD-12	563	
1N2037-2	143A	3750	ZD-13	563	
1N2037-3	144A	3094	ZD-14	564	
1N2037A	142A		ZD-12	563	
1N2038	5075A	3751	ZD-16	564	1N4745
1N2038-1	145A	3063	ZD-15	564	
1N2038-2	5075A	3751	ZD-16	564	
1N2038-3	5076A	9022	ZD-17		
1N2038A	5075A		ZD-15	564	
1N2039	5078A	9023	ZD-19		1N4747
1N2039-1	5077A	3752			
1N2039-2	5078A	9023	ZD-19		
1N2039-3	5079A	3335	ZD-20		
1N2039A	5078A	3752/5077A	ZD-18		
1N203CA			ZD-10	562	
1N204	177	3100/519	300	1122	
1N2040	5081A		ZD-24		1N4749
1N2040-2	5081A		ZD-24		
1N2040-3	5082A	3753	ZD-25		
1N2040-4	5082A	3754/5083A	ZD-28		
1N2040A	5081A	3753/5082A	ZD-22		
1N2041	5176AK	3056/135A	ZD-5.1		1N3995A
1N2041-2	5177AK	3056/135A	ZD-5.1		
1N2041B	5177AK		ZD-5.0		
1N2042	5178AK				1N3997A

Industry Standard No.	ECG	SK	GE	RS 276-	MOTOR.
1N2043	5181AK				1N2970RA
1N2044	5184AK				1N2973RA
1N2044A	5183AK		ZD-8.0		
1N2045	5186AK				1N2974RB
1N2046	5189AK	3093	ZD-12		1N2977RA
1N2047	5192AK				1N2980RA
1N2047B	5192AK		ZD-16		
1N2048	5195AK				1N2983RA
1N2048B	5195AK		ZD-19		
1N2049	5198AK				1N2986RA
1N2049C	5200AK	3753/5082A	ZD-25		
1N205	177	3100/519	300	1122	
1N2054	6354	6554			
1N2055	6354	6554			
1N2056	6354	6554			
1N2057	6354	6554			
1N2058	6354	6554			
1N2059	6354	6554			
1N206	178MP		300(2)	1122	
1N2060	6354	6554			
1N2061	6354	6554			
1N2062	6356	6556			
1N2063	6356	6556			
1N2064	6356	6556			
1N2065	6358	6558			
1N2066	6358	6558			
1N2067	6358	6558			
1N2068	6358	6558			
1N2069	116	3017B/117	504A	1104	
1N2069,A					1N4003
1N2069A	116	3311	504A	1104	
1N207	178MP		300	1122	
1N2070	116	3016	504A	1104	
1N2070,A					1N4004
1N2070A	116	3031A	504A	1104	
1N2071	116	3017B/117	504A	1104	
1N2071,A					1N4005
1N2071A	116	3017B/117	504A	1104	
1N2072	116	3016	504A	1104	1N4001
1N2073	116	3016	504A	1104	1N4002
1N2074	116	3016	504A	1104	1N4003
1N2075	116	3016	504A	1104	1N4003
1N2075K	177	3100/519	300	1122	
1N2076	116	3016	504A	1104	1N4004
1N2077	116	3016	504A	1104	1N4004
1N2078	116	3016	504A	1104	1N4004
1N2079	116	3016	504A	1104	1N4005
1N208	177		300	1122	
1N2080	116	3016	504A	1104	1N4001
1N2081	116	3016	504A	1104	1N4002
1N2082	116	3016	504A	1104	1N4003
1N2083	116	3016	504A	1104	1N4004
1N2084	116	3016	504A	1104	1N4004
1N2085	116	3016	504A	1104	1N4005
1N2086	116	3017B/117	504A	1104	1N4005
1N2088	116	3016	504A	1104	
1N2089	116	3017B/117	504A	1104	
1N209	178MP		300(2)	1122	
1N2090	116	3016	504A	1104'	
1N2091	116	3016	504A	1104	
1N2092	116	3016	504A	1104	
1N2093	116	3016	504A	1104	
1N2094	116	3016	504A	1104	
1N2095	116	3016	504A	1104	
1N2096	116	3017B/117	504A	1104	
1N210	177	3100/519	300	1122	
1N2103	116	3016	504A	1104	1N4001
1N2104	116	3016	504A	1104	1N4002
1N2105	116	3016	504A	1104	1N4003
1N2106	116	3016	504A	1104	1N4004
1N2107	116	3016	504A	1104	1N4004
1N2108	116	3016	504A	1104	1N4005
1N2109	5830		5004		
1N211	177	3100/519	300	1122	
1N2115	116	3031A	504A	1104	
1N2116	116	3031A	504A	1104	1N4004
1N2117	116	3017B/117	504A	1104	1N4006
1N212	177	3100/519	300	1122	
1N2128	6020	7220	5128		
1N2128A	6020	7220	5128		
1N2128AR	6021		5129		
1N2128R	6021		5129		
1N2129	6022		5128		
1N2129A	6022		5128		
1N2129AR	6027		5129		
1N2129R	6023		5129		
1N213	177	3100/519	300	1122	
1N2130	6026	7226	5128		
1N2130A	6026	7226	5128		
1N2130AR	6027		5129		
1N2130R	6027	7227	5129		
1N2130RA	6027	7227			
1N2131	6026	7226	5128		
1N2131A	6026	7226	5128		
1N2131AR	6027		5129		
1N2131R	6027	7227	5129		
1N2131RA	6027	7227			
1N2132	6023		5132		
1N2132A	6030		5132		
1N2132AR	6035		5133		
1N2132R	6031		5133		
1N2133	6030		5132		
1N2133A	6034		5132		
1N2133AR	6035		5133		
1N2133R	6031		5133		
1N2134	6034	7234	5132		
1N2134A	6034	7234	5132		
1N2134AR	6035		5133		
1N2134R	6035		5133		
1N2135	6034	7234	5132		
1N2135A	6034	7234	5132		
1N2135AR	6035		5133		
1N2135R	6035		5133		
1N2136	6038		5136		
1N2136A	6038		5136		
1N2136AR	6041		5137		
1N2136R	6039		5137		
1N2137	6038		5136		
1N2137A	6038		5136		
1N2137AR	6039		5137		
1N2137R	6039		5137		
1N2138	6040	7240	5136		
1N2138A	6040	7240	5136		
1N2138AR	6041		5137		
1N2138R	6041		5137		
1N2139	6040	7240	5136		
1N2139A	6040	7240	5136		
1N2139AR	6041		5137		
1N2139R	6041		5137		
1N214	177		300	1122	
1N2147	5850		5016		
1N2147A	5850		5016		
1N2148	5852		5016		
1N2148A	5852		5016		

Industry Standard No.	ECG	SK	GE	RS 276-	MOTOR.
1N2149	5854		5016		
1N2149A	5854		5016		
1N215		3100	300	1122	
1N2150	5856		5020		
1N2150A	5856		5020		
1N2151	5862		5024		
1N2151A	5858		5020		
1N2152A	5860		5024		
1N2153	5862		5024		
1N2153A	5862		5024		
1N2154	5980	3610	5096		1N1183
1N2154R	5981	3698/5987	5097		
1N2155	5982	3609/5986	5096		1N1184
1N2155R	5983	3698/5987	5097		
1N2156	5986	3501/5994	MR-2		1N1186
1N2156R	5987	3698			
1N2157	5988	3608/5990	5100		1N1188
1N2157R	5989	3518/5995	5101		
1N2158	5990	3608	5100		1N1188
1N2158R	5991	3518/5995	5101		
1N2159	5992	3501/5994	5104		1N1190
1N2159R	5993	3518/5995	5105		
1N216	177	3100/519	300	1122	
1N2160	5994	3501	5104		1N1190
1N2160R	5995	3518	5105		
1N2163			ZD-9.1	562	
1N2163A			ZD-9.1	562	
1N2164			ZD-9.1	562	
1N2164A			ZD-9.1	562	
1N2165			ZD-9.1	562	
1N2165A			ZD-9.1	562	
1N2166			ZD-9.1	562	
1N2166A			ZD-9.1	562	
1N2167			ZD-9.1	562	
1N2167A			ZD-9.1	562	
1N2168			ZD-9.1	562	
1N2168A			ZD-9.1	562	
1N2169			ZD-9.1	562	
1N2169A			ZD-9.1	562	
1N217	177	3100/519	300	1122	
1N2170			ZD-9.1	562	
1N2170A			ZD-9.1	562	
1N2171			ZD-9.1	562	
1N2171A			ZD-9.1	562	
1N2172	6154	7354			
1N2172R	6155	7355			
1N2173	6154	7354			
1N2173R	6155	7355			
1N2174	6154	7354			
1N2174R	6155	7355			
1N218	177	3100/519	300	1122	
1N2181			504A	1104	
1N22	109	3087	1N82A	1123	
1N2214			ZD-5.6	561	
1N2216	5830		5004		1N1199C
1N2217	5830		5004	1104	
1N2218	5840	3500/5882	5008	1104	1N1206C
1N2219	5842	3500/5882	5008	1104	
1N2220	5842	7042	5008	1104	1N1206C
1N2221	5842	7042	5008	1104	
1N2222	5846	3516	5012	1114	
1N2222,A					MR1128
1N2222A	5846	3516	5012	1114	
1N2223	5846	3516	5012	1114	
1N2223A	5846	3516	5012	1114	
1N2224	5848		5012	1114	
1N2224,A					MR1130
1N2224A	5848		5012	1114	
1N2225	5848		5012	1114	
1N2225A	5848		5012	1114	
1N2226,A					SPECIAL
1N2228	5850		5016		
1N2228,A					1N1199C
1N2228A	5850		5016		
1N2229	5850		5016		
1N2229A	5850		5016		
1N2230	5854		5016		
1N2230,A					1N1202C
1N2230A	5854		5016		
1N2231	5854		5016		
1N2231A	5854		5016		
1N2232	5856	3500/5882	5020		
1N2232,A					1N1204C
1N2232A	5856	3500/5882	5020		
1N2233	5858	3500/5882	5020		
1N2233A	5858	3500/5882	5020		
1N2234	5858	3500/5882	5016		
1N2234,A					1N1204C
1N2234A	5858	3500/5882	5016		
1N2235	5858	3500/5882	5020		
1N2235A	5858	3500/5882	5020		
1N2236	5860	3500/5882	5024		
1N2236,A					1N1206C
1N2236A	5860	3500/5882	5024		
1N2237	5860	3500/5882	5024		
1N2237A	5860	3500/5882	5024		
1N2238	5862	3500/5882	5024		
1N2238,A					1N1206C
1N2238A	5862	3500/5882	5024		
1N2239	5862	3500/5882	5024		
1N2239A	5862	3500/5882	5024		
1N2240	5866		5028		
1N2240,A					MR1128
1N2240A	5866		5028		
1N2241	5866		5028		
1N2241A	5866		5028		
1N2242	5868	7068	5028		
1N2242,A					MR1130
1N2242A	5868	7068	5028		
1N2243	5868	7068	5028		
1N2243A	5868	7068	5028		
1N2244,A					SPECIAL
1N2246	5870	3500/5882	5032		
1N2246A	5870	3500/5882	5032		1N1199C
1N2247	5870		5032		
1N2247A	5870		5032		
1N2248	5872	3500/5882	5032		
1N2248A	5872	3500/5882	5032		1N1200C
1N2249	5872		5032		
1N2249A	5872		5032		
1N225		3783	ZD-9.1	562	
1N2250	5874	3500/5882	5032		
1N2250A	5874	3500/5882	5032		1N1202C
1N2251	5874		5032		
1N2251A	5874		5032		
1N2252	5876	3500/5882	5036		
1N2252A	5876	3500/5882	5036		1N1204C
1N2253	5876	3500/5882	5036		
1N2253A	5876	3500/5882	5036		
1N2254	5878	3500/5882	5036		
1N2254A	5878	3500/5882	5036		1N1204C
1N2255	5878	3500/5882	5036		
1N2255A	5878	3500/5882	5036		
1N2256	5880	3500/5882	5040		

Industry Standard No.	ECG	SK	GE	RS 276-	MOTOR.
1N2256A	5880	3500/5882	5040		1N1206C
1N2257	5880	3500/5882	5040		
1N2257A	5880	3500/5882	5040		
1N2258	5882	3500	5040		
1N2258A	5882	3500	5040		1N1206C
1N2259	5882	3500	5040		
1N2259A	5882	3500	5040		
1N225A		3783	ZD-9.1	562	
1N226		3785	ZD-10	562	
1N2260	5886		5044		
1N2260A	5886		5044		MR1128
1N2261	5886		5044		
1N2261A	5886		5044		
1N2262	5890	7090	5044		
1N2262A	5890	7090	5044		MR1130
1N2263	5890	7090	5044		
1N2263A	5890	7090	5044		
1N2266	5830		5004		1N1199C
1N2267	5830		5004	1104	
1N2268	5840	3500/5882	5008	1104	1N1206C
1N2269	5840	3500/5882	5008	1104	
1N226A		3785	ZD-10	562	
1N227		3750	ZD-13	563	
1N2270	5842	7042	5008	1104	1N1206C
1N2271	5842	7042	5008	1104	
1N2272	5912		MR-1		
1N2273	5944		MR-1		
1N2274	5944		MR-1		
1N2275	5948		MR-1		
1N2276	5948		MR-1		
1N227A		3750	ZD-13	563	
1N228		3791	ZD-15	564	
1N2282	5988	3608/5990	5100		1N1188
1N2283	5990	3608	5100		1N1188
1N2284	5992	3501/5994	5104		1N1190
1N2285	5994	3501	5104		1N1190
1N2286	5998		5108		1N3766
1N2287	6002	7202	5108		1N3768
1N2289	5830		5004	1104	
1N2289A	5830		5004	1104	
1N228A		3791		564	
1N229		3794	ZD-19		
1N2290	5830		5004		
1N2290A	5830		5004	1104	
1N2291	5834		5004		
1N2291A	5836		5004		
1N2292	5836	3500/5882	5004		
1N2292A	5836	3500/5882	5004		
1N2293	5838	3500/5882	5004		
1N2293A	5838	3500/5882	5004		
1N229A		9023	ZD-19		
1N230		3797	ZD-24		
1N230A		3797	ZD-24		
1N231		3800	ZD-28		
1N231A		3800	ZD-28		
1N2323	116		504A	1104	
1N2326	604			1123	
1N2327	506	3843	511	1114	
1N234		3806	ZD-47		
1N2348	5830		5004		MR1120
1N2349	5832		5004		MR1121
1N234A		3806	ZD-47		
1N235		3809	ZD-60		
1N2350	5834		5004		MR1122
1N2357	506	3843	511	1114	MR1-1400
1N2358					MR1-1600
1N2359					MR1-1600
1N237		3814	ZD-87		
1N2372	5848	7048	5012	1114	
1N2373	5806	3848	504A	1104	
1N2374	5808	9009	509	1114	
1N2387	5084A	3755	ZD-30		1N4751
1N239		3819	ZD-130		
1N2390	5800	9003		1142	
1N2391	5801	9004		1142	
1N2392	5802	9005		1143	
1N2393	5803	9006		1144	
1N2394	5804	9007		1144	
1N2395	5805	9008	504A	1104	
1N2396	5806	3848	504A	1104	
1N2397	5808	9009	509	1114	
1N2398	5808	9009	509	1114	
1N2399	5800	9003		1142	
1N23WF			1N82A	1104	
1N2400	5801	9004		1142	
1N2401	5802	9005		1143	
1N2402	5803	9006		1144	
1N2403	5804	9007		1144	
1N2404	5805	9008	504A	1104	
1N2405	5806	3848	504A	1104	
1N2406	5808	9009	509	1114	
1N2407	5808	9009	509	1114	
1N2408	5800	9003		1142	
1N2409	5801	9004		1142	
1N2410	5802	9005		1143	
1N2411	5803	9006		1144	
1N2412	5804	9007		1144	
1N2413	5805	9008	504A	1104	
1N2414	5806	3848	504A	1104	
1N2415	5808	9009	509	1114	
1N2416	5808	9009	509	1114	
1N2422			504A	1104	
1N2423			504A	1104	
1N2424			509	1114	
1N2425			509	1114	
1N2426	6154	7354			
1N2426R	6155	7355			
1N2427	6154	7354			
1N2427R	6155	7355			
1N2428	6154	7354			
1N2428R	6155	7355			
1N2429	6154	7354			
1N2429R	6155	7355			
1N2430	6154	7354			
1N2430R	6155	7355			
1N2431	6154	7354			
1N2431R	6155	7355			
1N2432	6154	7354			
1N2432R	6155	7355			
1N2433	6154	7354			
1N2433R	6155	7355			
1N2434	6156	7356			
1N2435	6156	7356			
1N2436	6154	7354			
1N2436R	6155	7355			
1N2437	6154	7354			
1N2437R	6155	7355			
1N2438	6154	7354			
1N2438R	6155	7355			
1N2439	6154	7354			
1N2439R	6155	7355			
1N2440	6154	7354			
1N2440R	6155	7355			
1N2443	6154	7354			
1N2443R	6155	7355			
1N2444	6156	7356			
1N2445	6156	7356			
1N2446	6020	7220	5128		1N1183
1N2447	6022		5128		1N1184
1N2448	6026	7226	5128		1N1186
1N2449	6026		MR-2		1N1186
1N2450	6030		5132		1N1188
1N2451	6030		5132		1N1188
1N2452	6034	7234	5132		1N1188
1N2453	6034	7234	5132		1N1188
1N2454	6038		5136		1N1190
1N2455	6040	7240	5136		1N1190
1N2456	6042		5140		1N3766
1N2457	6042		5140		1N3766
1N2458	6020	7220	5128		1N1183.
1N2459	6022		5128		1N1184
1N2460	6026	7226	5128		1N1186
1N2461	6026	7226	5128		1N1186
1N2462	6030		5132		1N1188
1N2463	6030		5132		1N1188
1N2464	6034	7234	5132		1N1188
1N2465	6034	7234	5132		1N1188
1N2466	6038		5136		1N1190
1N2467	6040	7240	5136		1N1190
1N2468	6042		5140		1N3766
1N2469	6042		5140		1N3766
1N2473	177		300	1122	
1N248	5980	3610	5048		
1N2482	116	9005/5802	504A	1143	1N4003
1N2483	116	9007/5804	504A	1144	1N4004
1N2484	116	3848/5806	504A	1104	1N4005
1N2485	116	9005/5802	504A	1143	1N5393
1N2486	116	3848/5806	504A	1104	1N5395
1N2487	116	9007/5804	504A	1144	1N5395
1N2488	116	9008/5805	504A	1104	1N5397
1N2489	116	3848/5806	504A	1104	1N5397
1N248A	5980	3610	5096		
1N248AR	5981		5097		
1N248B	5980	3610	5096		
1N248BR	5981		5097		
1N248C	5986	3610/5980	5096		
1N248CR	5981		5097		
1N248R	5981	3698/5987			
1N249	5986	3609	5048		
1N2491	5850		5016		1N1199C
1N2492	5852		5016		1N1200C
1N2493	5854		5016		1N1202C
1N2494	5856		5020		1N1204C
1N2495	5858		5020		1N1204C
1N2496	5860		5024		1N1206C
1N2497	5862	3500/5882	5024		1N1206C
1N2498					1N2974A
1N2498A	5186A				1N2974B
1N2498C					10M10Z10
1N2498CA					10M10Z5
1N2499	5187A				1N2975A
1N2499A	5187A				1N2975B
1N2499C					10M11Z10
1N2499CA					10M11Z5
1N249A	5982	3609/5986	5096		
1N249AR	5983		5097		
1N249B	5982	3609/5986	5096		
1N249BR	5983		5097		
1N249C	5982	3609/5986	5096		
1N249CR	5983		5097		
1N249R	5987	3698			
1N250	5944	3609/5986	5048		
1N2500	5188A				1N2976A
1N2500A	5188A				1N2976B
1N2500C					10M12Z10
1N2500CA					10M12Z5
1N2501	5808	9009	509	1114	1N4006
1N2502	5809	9010	509	1114	1N4007
1N2503	506	3843	511	1114	MR1-1200
1N2504	506	3843	511	1114	MR1-1600
1N2505	5808	9009	509	1114	1N4006
1N2506	5809	9010	509	1114	1N4007
1N2507	506	3843	511	1114	MR1-1200
1N2508	506	3843	511	1114	MR1-1600
1N250A	5986	3609	5096		
1N250AR	5987		5097		
1N250B	5986	3609	5096		
1N250BR	5987		5097		
1N250C	5990	3609/5986	5096		
1N250CR	5987		5097		
1N250R	5945	3698/5987			
1N250RA	5987	3698			
1N250RB	5987	3698			
1N250RC	5991	3698/5987			
1N251	177	3100/519	300	1122	
1N2512	5832		5004		1N1200C
1N2513	5834		5004		1N1202C
1N2514	5836		5004		1N1204C
1N2515	5838		5004		1N1204C
1N2516	5860		5024		1N1206C
1N2517	5842	7042	5008		1N1206C
1N2518	5834		5004		
1N2519	5854		5016		
1N251A	177	3100/519	300	1122	
1N252	177		300	1122	
1N2520	5836		5004		
1N2521	5838		5004		
1N2522	5860		5024		
1N2523	5842	7042	5008		
1N2524	5830	3081/125	5004		
1N2525	5834		5004		
1N2526	5832		5004		
1N2527	5836		5004		
1N2528	5838		5004		
1N2529	5840	3584/5862	5008		
1N252A	177	3100/519	300	1122	
1N253	5834	3500/5882	5004		1N1200C
1N2530	5842	7042	5008		
1N2531	5846	3516	5012		
1N2532	5846	3516	5012		
1N2533	5848		5012		
1N2534	5848		5012		
1N2535	5830	3081/125	5004		
1N2536	5834		5004		
1N2537	5834		5004		
1N2538	5836		5004		
1N2539	5838		5004		
1N254	5834	3500/5882	5004		1N1202C
1N2540	5840	3584/5862	5008		
1N2541	5842	7042	5008		
1N2542	5846	3516	5012		
1N2543	5846	3516	5012		
1N2544	5848		5012		
1N2545	5848		5012		
1N2547	5832		5004		
1N2548	5834		5004		
1N2549	5836		5004		
1N255	5838	3500/5882	5004	1104	1N1204C

Industry Standard No.	ECG	SK	GE	RS 276-	MOTOR.
1N2550	5838		5004		
1N2551	5840	3584/5862	5008		
1N2552	5842	7042	5008		
1N2553	5846	3516	5012		
1N2554	5846	3516	5012		
1N2555	5848		5012		
1N2556	5848		5012		
1N2557	5866		5028		
1N2558	5866		5028		
1N2559	5868	7068	5028		
1N256	5842	7042	5008	1104	1N1206C
1N2560	5868	7068	5028		
1N2561	5866		5028		
1N2562	5866		5028		
1N2563	5868	7068	5028		
1N2564	5868	7068	5028		
1N2565	5850		5016		
1N2566	5852		5016		
1N2567	5854		5016		
1N2568	5856		5020		
1N2569	5858		5020		
1N2570	5860		5024		
1N2571	5862		5024		
1N2572	5866		5028		
1N2573	5866		5028		
1N2574	5868	7068	5028		
1N2575	5868	7068	5028		
1N2576	5870		5032		
1N2577	5872		5032		
1N2578	5874		5032		
1N2579	5876		5036		
1N2580	5878		5036		
1N2581	5880		5040		
1N2582	5882		5040		
1N2583	5886		5044		
1N2584	5886		5044		
1N2585	5890	7090	5044		
1N2586	5890	7090	5044		
1N2587	5870		5032		
1N2588	5872		5032		
1N2589	5874		5032		
1N2590	5876		5036		
1N2591	5878		5036		
1N2592	5880		5040		
1N2593	5882		5040		
1N2594	5886		5044		
1N2595	5886		5044		
1N2596	5890	7090	5044		
1N2597	5890	7090	5044		
1N2598	5870		5032		
1N2599	5872		5032		
1N2600	5874		5032		
1N2601	5876		5036		
1N2602	5878		5036		
1N2603	5880		5040		
1N2604	5882		5040		
1N2605	5886		5044		
1N2606	5886		5044		
1N2607	5890	7090	5044		
1N2608	5890	7090	5044		
1N2609	116	9003/5800	504A	1142	1N4001
1N2610	116	9004/5801	504A	1142	1N4002
1N2611	116	9005/5802	504A	1143	1N4003
1N2612	116	9006/5803	504A	1144	1N4004
1N2613	116	9007/5804	504A	1144	1N4004
1N2614	116	9008/5805	504A	1103	1M4005
1N2615	116	3848/5806	504A	1104	1N4005
1N2616	125	9009/5808	509		1N4006
1N2617	125	9010/5809	509	1114	1N4007
1N2618	506	3843	511	1114	MR1-1200
1N2619	506	3843	511	1114	MR1-1600
1N2620			ZD-9.1	562	
1N2620A			ZD-9.1	562	
1N2620B			ZD-9.1	562	
1N2621			ZD-9.1	562	
1N2621A			ZD-9.1	562	
1N2621B			ZD-9.1	562	
1N2622			ZD-9.1	562	
1N2622A			ZD-9.1	562	
1N2622B			ZD-9.1	562	
1N2623			ZD-9.1	562	
1N2623A			ZD-9.1	562	
1N2623B			ZD-9.1	562	
1N2624			ZD-9.1	562	
1N2624A			ZD-9.1	562	
1N2624B			ZD-9.1	562	
1N2625					1N937
1N2625A					1N937A
1N2625B					1N937B
1N2626					1N938
1N2626A					1N938A
1N2626B					1N938B
1N2638			504A	1104	
1N265	109	3087		1123	
1N2650			504A	1104	
1N2653			509	1114	
1N266	109	3017B/117		1123	
1N267	109	3087		1123	
1N268	109	3087		1123	
1N270	109	3087		1123	
1N2702	5834		5004		
1N273	109	3087		1123	
1N2750	5803		5004		
1N276	109	3087		1123	
1N2765		3780	ZD-6.8	561	1N823A
1N2765A		3780	ZD-6.8	561	1N825A
1N2766		3789	ZD-14	564	1N1736A
1N2766A		3789	ZD-14	564	1N1736A
1N2767		3335	ZD-20		
1N2767A		3335	ZD-20		
1N277	109	3087		1123	
1N2772	5808	9009	509	1114	
1N2773	5808	9009	509	1114	
1N2774	5809	9010	509	1114	
1N2775	5809	9010	509	1114	
1N2776	506	3843	511	1114	
1N2777	506	3843	511	1114	
1N2778	506	3843	511	1114	
1N2779	506	3843	511	1114	
1N278	109	3087		1123	
1N2783	5150A				1N3000A
1N2786	5944		5048		1N1186
1N2787	5948				1N1188
1N2788	6026	7226	5128		1N1186
1N2788R	6027	7227	5129		
1N2789	6034	7234	5132		1N1188
1N2789R	6035		5133		
1N279	109	3087		1123	
1N2790		3749	ZD-9.1	562	1N3156
1N2793	5940		5048		
1N2794	5942		5048		
1N2795	5944		5048		
1N2796	5944		5048		
1N2797	5946		MR-2		
1N2798	5946		MR-2		
1N2799	5948		MR-2		
1N2800	5948		MR-2		
1N2801	109	3087		1123	
1N281	109	3087		1123	
1N282	109			1123	
1N283	109	3087		1123	
1N2847	5832		5004		
1N2848	5834		5004		
1N2849	5836		5004		
1N285	109	3087		1123	
1N2850	5838	3017B/117	5004	1104	
1N2851	5840	3584/5862	5008	1104	
1N2852	5842	7042	5008	1114	
1N2858	116	9003/5800	504A	1142	
1N2858,A					1N5391
1N2858A	5800	9003	504A	1142	
1N2859	116	9004/5801	504A	1142	
1N2859,A					1N5392
1N2859A	5801	9004	504A	1142	
1N2860	116	9005/5802	504A	1143	
1N2860,A					1N5393
1N2860A	5802	9005	504A	1143	
1N2861	116	9006/5803	504A	1144	
1N2861,A					1N5395
1N2861A	5803	9006	504A	1144	
1N2862	116	9007/5804	504A	1144	
1N2862,A					1N5395
1N2862A	5804	9007	504A	1144	
1N2863	116	9008/5805	504A	1104	
1N2863,A					1N5397
1N2863A	5805	9008	504A	1104	
1N2864	116	3848/5806	504A	1104	
1N2864,A					1N5397
1N2864A	5806	3848	504A	1104	
1N2865	5808	9009	509	1114	1N4007
1N2866	5809	9010	509	1114	MR1-1600
1N2867	5808	9009	509	1114	
1N2868	5809	9010	509	1114	
1N287	109	3087		1123	
1N2878	125	9009/5808	509	1114	
1N2879	125	9009/5808	509	1114	
1N288	109	3087		1123	
1N2880	125	9010/5809	509	1114	
1N2881	125	9010/5809	509	1114	
1N2882	125	9010/5809	509	1114	
1N2883	125	9010/5809	509	1114	
1N2884	506	3843	511	1114	
1N2885	506	3843	511	1114	
1N289	109	3087		1123	
1N290	109	3087		1123	
1N292	109	3087		1123	
1N2937					1N2996A
1N294	109	3087		1123	
1N294A	109			1123	
1N295	109	3087	1N295	1123	
1N295A	109	3091		1123	
1N295B	109			1123	
1N295X	109	3087		1123	
1N296	109	3087		1123	
1N297	109	3087		1123	
1N297A	109	3087		1123	
1N298	109			1123	
1N298A	109	3087		1123	
1N299	112		1N82A		
1N299B	112		1N82A		
1N300	177	3100/519	300	1122	
1N3005	5220A	3346/5096A			
1N300A	177	3100/519	300	1122	
1N300B	177	3100/519	300	1122	
1N301	177	3100/519	300	1122	
1N3016	5071A	3334	ZD-6.8	561	
1N3016A	5071A	3334	ZD-6.8	561	
1N3016B	5071A	3334	ZD-6.8	561	
1N3017	138A		ZD-7.5		
1N3017A	138A		ZD-7.5		
1N3017B	138A	3059	ZD-7.5		
1N3018	5072A	3136	ZD-8.2	562	
1N3018A	5072A	3136	ZD-8.2	562	
1N3018B	5072A	3136	ZD-8.2	562	
1N3019	139A		ZD-9.1	562	
1N3019A	139A	3060	ZD-9.1	562	
1N3019B	139A	3066/118	ZD-9.1	562	
1N301A	177	3100/519	300	1122	
1N301B	177	3100/519	300	1122	
1N3020	140A		ZD-10	562	
1N3020A	140A		ZD-10	562	
1N3020B	140A	3061	ZD-10	562	
1N3021	5074A		ZD-11.0	563	
1N3021A	5074A		ZD-11.0	563	
1N3021B	5074A	3139	ZD-11	563	
1N3022	142A	3062	ZD-12	563	
1N3022A	142A	3062	ZD-12	563	
1N3022B	142A	3062	ZD-12	563	
1N3023	143A	3750	ZD-13	563	
1N3023A	143A	3750	ZD-13	563	
1N3023B	143A	3750	ZD-13	563	
1N3024	145A	3063	ZD-15	564	
1N3024A	145A	3063	ZD-15	564	
1N3024B	145A	3063	ZD-15	564	
1N3025	5075A	3751	ZD-16	564	
1N3025A	5075A	3751	ZD-16	564	
1N3025B	5075A	3751	ZD-16	564	
1N3026	5077A	3752	ZD-18		
1N3026A	5077A	3752	ZD-18		
1N3026B	5077A	3752	ZD-18		
1N3027	5079A	3335	ZD-20		
1N3027A	5079A	3335	ZD-20		
1N3027B	5079A	3335	ZD-20		
1N3028	5080A	3336	ZD-22		
1N3028A	5080A	3336	ZD-22		
1N3028B	5080A	3336	ZD-22		
1N3029	5081A		ZD-24		
1N3029A	5081A		ZD-24		
1N3029B	5081A	3151	ZD-24		
1N303	177	3100/519	300	1122	
1N3030	146A	3064	ZD-27		
1N3030A	146A	3064	ZD-27		
1N3030B	146A	3064	ZD-27		
1N3031	5084A	3755	ZD-30		
1N3031A	5084A	3755	ZD-30		
1N3031B	5084A	3755	ZD-30		
1N3032	147A	3095	ZD-33		
1N3032A	147A	3095	ZD-33		
1N3032B	147A	3095	ZD-33		
1N3033	5085A	3337	ZD-36		
1N3033A	5085A	3337	ZD-36		
1N3033B	5085A	3337	ZD-36		
1N3034	5086A	3338	ZD-39		
1N3034A	5086A	3338	ZD-39		
1N3034B	5086A	3338	ZD-39		
1N3035	5087A	3339	ZD-43		
1N3035A	5087A	3339	ZD-43		
1N3035B	5087A	3339	ZD-43		

Industry Standard No.	ECG	SK	GE	RS 276-	MOTOR.
1N3036	5088A	3340	ZD-47		
1N3036A	5088A	3340	ZD-47		
1N3036B	5088A	3340	ZD-47		
1N3037	5089A	3341	ZD-51		
1N3037A	5089A	3341	ZD-51		
1N3037B	5089A	3341	ZD-51		
1N3038	5090A	3342	ZD-56		
1N3038A	5090A	3342	ZD-56		
1N3038B	5090A	3342	ZD-56		
1N3039	149A	3097	ZD-62		
1N3039A	149A	3097	ZD-62		
1N3039B	149A	3097	ZD-62		
1N303A	177	3100/519	300	1122	
1N303B	177	3100/519	300	1122	
1N304	109	3087		1123	
1N3040	5092A	3343	ZD-68		
1N3040A	5092A	3343	ZD-68		
1N3040B	5092A	3343	ZD-68		
1N3041	5093A	3344	ZD-75		
1N3041A	5093A	3344	ZD-75		
1N3041B	5093A	3344	ZD-75		
1N3042	150A	3098	ZD-82		
1N3042A	150A	3098	ZD-82		
1N3042B	150A	3098	ZD-82		
1N3043	5095A	3345	ZD-91		
1N3043A	5095A	3345	ZD-91		
1N3043B	5095A	3345	ZD-91		
1N3044	5096A	3346	ZD-100		
1N3044A	5096A	3346	ZD-100		
1N3044B	5096A	3346	ZD-100		
1N3045	151A	3099	ZD-110		
1N3045A	151A	3099	ZD-110		
1N3045B	151A	3099	ZD-110		
1N3046	5097A	3347	ZD-120		
1N3046A	5097A	3347	ZD-120		
1N3046B	5097A	3347	ZD-120		
1N3047	5098A	3348	ZD-130		
1N3047A	5098A	3348	ZD-130		
1N3047B	5098A	3348	ZD-130		
1N3048	5100A	3350	ZD-150		
1N3048A	5100A	3350	ZD-150		
1N3048B	5100A	3350	ZD-150		
1N3049	5101A	3351	ZD-160		
1N3049A	5101A	3351	ZD-160		
1N3049B	5101A	3351	ZD-160		
1N305	109	3087		1123	
1N3050	5103A	3353	ZD-180		
1N3050A	5103A	3353	ZD-180		
1N3050B	5103A	3353	ZD-180		
1N3051	5105A	3355	ZD-200		
1N3051A	5105A	3355	ZD-200		
1N3051B	5105A	3355	ZD-200		
1N306	109	3087		1123	
1N3062	519	3100	514	1122	
1N3063	177	9091	300	1122	
1N3064	177	9091	5017	1122	
1N3065	177	9091	300	1122	
1N3066	177	9091	300	1122	
1N3067	177	9091	300	1122	
1N3068	177	9091	300	1122	
1N3069	177	9091	300	1122	
1N3069M	177	9091	300	1122	
1N307	109	3087		1123	
1N3070	177	9091	300	1122	
1N3071	177	9091	300	1122	
1N3072	5800	9003	504A	1142	1N4001
1N3073	5801	9004	504A	1142	1N4002
1N3074	5802	9005	504A	1143	1N4003
1N3075	5802	9005	504A	1143	1N4003
1N3076	5803	9006	504A	1144	1N4004
1N3077	5803	9006	504A	1144	1N4004
1N3078	5804	9007	504A	1144	1N4004
1N3079	5804	9007	504A	1144	1N4004
1N308	109	3087		1123	
1N3080	5808	9009	504A	1104	1N4005
1N3081	5806	3848	504A	1104	1N4005
1N3082	5802	9005	504A	1143	1N5393
1N3083	5804	9007	504A	1144	1N5395
1N3084	5806	3848	504A	1104	1N5397
1N309	109	3087		1123	
1N3094	116	3017B/117	504A	1104	
1N3098	5097A	3347	ZD-120		
1N3098,A					1N3046A
1N3098A	5097A	3347	ZD-120		
1N3099	5100A	3350	ZD-150		
1N3099,A					1N3048A
1N310	109	3087		1123	
1N3100	5103A	3353	ZD-180		
1N3100,A					1N3050A
1N3100A	5103A	3353	ZD-180		
1N3101,A					1N3051A
1N3102,A					1N3008A
1N3103,A					1N3011A
1N3104,A					1N3014A
1N3105,A					1N3015A
1N3106	5806	3848	504A	1104	1N4006
1N3107	5809	9010	509	1114	MR1-1200
1N3108			ZD-9.1	562	
1N3110	109	3087		1123	
1N3112	128	3024	243	2030	1N4737A
1N312	109	3087		1123	
1N3121	109	3087		1123	
1N3122	109	3091		1123	
1N3123	177	3100/519	300	1122	
1N3124	177	3100/519	300	1122	
1N3125	109	3087		1123	
1N314	109			1123	
1N3142	6026	7226	5128		
1N3146	109	3087		1123	
1N3147	177	3100/519	300	1122	
1N3148		3783	ZD-9.1	562	1N3155A
1N315	116	3031A	504A	1104	
1N3154			ZD-8.2	562	
1N3154A			ZD-8.2	562	1N2977B
1N3155			ZD-8.2	562	
1N3155A			ZD-8.2	562	
1N3156			ZD-8.2	562	
1N3156A			ZD-8.2	562	
1N3157			ZD-8.2	562	
1N3157A			ZD-8.2	562	
1N315A	116	3017B/117	504A	1104	
1N316	116	3016	504A	1104	
1N316,A					1N4001
1N3160	116	3017B/117	504A	1104	
1N3161	6354	6554			
1N3162	6354	6554			
1N3163	6354	6554			
1N3164	6354	6554			
1N3165	6354	6554			
1N3166	6354	6554			
1N3167	6354	6554			
1N3168	6354	6554			
1N3169	6356	6556			
1N316A	116	3016	504A	1104	

Industry Standard No.	ECG	SK	GE	RS 276-	MOTOR.
1N317	116	3016	504A	1104	
1N317,A					1N4002
1N3170	6356	6556			
1N3171	6358	6558			
1N3172	6358	6558			
1N3173	6358	6558			
1N317A	116	3016	504A	1104	
1N318	116	3016	504A	1104	
1N318,A					1N4003
1N3181			ZD-7.5		1N5237A
1N3182	614	3126	90		1N3182
1N3183			504A	1104	
1N3184			504A	1104	
1N3185			509	1114	
1N3186			509	1114	
1N3189	116	3016	504A	1104	1N4003
1N318A	116	3016	504A	1104	
1N319	116	3016	504A	1104	
1N319,A					1N4004
1N3190	116	3016	504A	1104	1N4004
1N3191	116	3017B/117	504A	1104	1N4005
1N3191B		3351	ZD-160		
1N3192					SPECIAL
1N3193	116	3017B/117	504A	1104	1N4003
1N3194	116	3017B/117	504A	1104	1N4004
1N3195	116	3017B/117	504A	1104	1N4005
1N3196	125	3032A	510,531	1114	1N4006
1N3197	177	3100/519	300	1122	
1N3198					1N5221B
1N3199		3782	ZD-8.2	562	1N3155
1N319A	116	3016	504A	1104	
1N31A	109			1123	
1N320	116	3016	504A	1104	
1N320,A					1N4005
1N3200					1N3156
1N3201					1N3156
1N3202					1N3157
1N3203	116	3016	504A	1104	
1N3204	109	3087		1123	
1N3206	177	3100/519	300	1122	
1N3207	177	3100/519	300	1122	
1N3208	5940	3501/5994	5048		
1N3208R	5941	3518/5995			
1N3209	5942	3501/5994	5048		
1N3209R	5943	3518/5995			
1N320A	116	3016	504A	1104	
1N321	116	3017B/117	504A	1104	
1N321,A					1N4007
1N3210	5944	3501/5994	5048		
1N3210R	5945	3518/5995			
1N3211	5946	3501/5994	MR-2		
1N3211R	5947	3518/5995			
1N3212	5948		MR-2		
1N3212R	5949	3518/5995			
1N3213	5924	3501/5994			
1N3214	5924	3501/5994			
1N3214R	5925	7153/5953			
1N321A	125	3033	510,531	1114	
1N322	125	3033	510,531	1114	
1N322,A					1N4007
1N3221	125		510,531	1114	
1N3223	177	3100/519	300	1122	
1N3227	5801	9004	504A	1142	
1N3228	5802	9005	504A	1143	
1N3229	5805	9008	504A	1104	
1N322A	125	3033	510,531	1114	
1N323	116	3016	504A	1104	
1N323,A					1N4001
1N3230	5806	3848	504A	1104	
1N3231	5808	9009	509	1114	
1N3232	5809	9010	509	1114	
1N3233	506	3843	511	1114	
1N3237	5800	9003	504A	1142	
1N3238	5801	9004	504A	1142	
1N3239	5802	9005	504A	1143	
1N323A	116	3016	504A	1104	
1N324	116	3016	504A	1104	
1N324,A					1N4002
1N3240	5804	9007	504A	1144	
1N3241	5806	3848	504A	1104	
1N3242	5808	9009	509	1114	
1N3243	5809	9010	509	1114	
1N3244	506	3843	511	1114	
1N3246	5800	9003	504A	1142	
1N3247	5801	9004	504A	1142	
1N3248	5802	9005	504A	1143	
1N3249	5804	9007	504A	1144	
1N324A	116	3016	504A	1104	
1N325	116	3016	504A	1104	
1N325,A					1N4003
1N3250	5806	3848	504A	1104	
1N3251	5808	9009	509	1114	
1N3252	5809	9010	509	1114	
1N3253	116	9005/5802	504A	1143	1N4003
1N3254	116	9007/5804	504A	1144	1N4004
1N3255	116	3848/5806	504A	1104	1N4005
1N3256	125	9009/5808	509	1114	1N4006
1N3257	177	3016	300	1122	
1N3258	177	3016	300	1122	
1N325A	116	3016	504A	1104	
1N326	116	3016	504A	1104	
1N326,A					1N4004
1N3260	6354	6554			
1N3261	6354	6554			
1N3262	6354	6554			
1N3263	6354	6554			
1N3264	6354	6554			
1N3265	6354	6554			
1N3266	6354	6554			
1N3267	6354	6554			
1N3268	6356	6556			
1N3269	6356	6556			
1N326A	116	3016	504A	1104	
1N327	116	3016	504A	1104	
1N327,A					1N4006
1N3270	6358	6558			
1N3270R	6359	6559			
1N3271	6358	6558			
1N3271R	6359	6559			
1N3272	6358	6558			
1N3272R	6359	6559			
1N3273	6358	6558			
1N3273R	6359	6559			
1N3277	116	3017B/117	504A	1104	
1N3278	116	3017B/117	504A	1104	
1N3279	116	3017B/117	504A	1104	
1N327A	116	3016	504A	1104	
1N328	125	3033	510,531	1114	
1N328,A					1N4007
1N3280	125	3032A	510,531	1114	
1N3281	125	3033	510,531	1114	
1N3282	125	3033	510,531	1114	
1N3283	525	3925			
1N3287	109	3091		1123	

Industry Standard No.	ECG	SK	GE	RS 276-	MOTOR.
1N3287N	109	3087		1123	
1N3287W	109	3091		1123	
1N3288	6154	7354			MR1215FL
1N3288R	6155	7355			
1N3289	6154	7354			MR1215FL
1N3289R	6155	7355			
1N328A	125	3033	510,531	1114	
1N329	125	3033	510,531	1114	
1N329,A					1N4007
1N3290	6154	7354			MR1215FL
1N3290R	6155	7355			
1N3291	6154	7354			MR1219SL
1N3291R	6155	7355			
1N3292	6156	7356	1N3292		MR1219SL
1N3292A	6156	7356			
1N3293	6156	7356			MR1219SL
1N3294	6158	7358			
1N3294R	6159	7359			
1N3295	6158	7358	1N3295		
1N3295R	6159	7359	1N3295R		
1N3298	116	3016	504A	1104	
1N329A	125	3033	510,531	1114	
1N330	178MP			1122	
1N3301	6407	3523	IC-38		
1N3301A	6407	3523			
1N331	177	3100/519	300	1122	
1N332	5838	3500/5882	5004		1N1204C
1N332R	5839		5005		
1N333	5838	3500/5882	MR-1		1N1204C
1N333R	5839		5005		
1N334	116	3500/5882	5004		1N1204C
1N334R	5837		5005		
1N335	5838	3500/5882	MR-1		1N1204C
1N335R	5837		5005		
1N336	5834	3500/5882	5004		1N1202C
1N336R	5835		5005		
1N337	5834	3500/5882	5004		1N1202C
1N337R	5835		5005		
1N338	5834	3017B/117	5004		1N1200C
1N338R	5833		5005		
1N339	5834	3500/5882	5004		1N1200C
1N3396		3771	ZD-3.3		
1N33974			ZD-9.1	562	
1N3399			ZD-5.6	561	
1N339R	5833		5005		
1N34	109	3087	1N34AS	1123	
1N340	5834	3500/5882	5004		1N1200C
1N3400		3780	ZD-6.8	561	
1N3401		3782	ZD-8.2	562	
1N3402		3785	ZD-10	562	
1N3403		3787	ZD-12	563	
1N3404		3790	ZD-15	564	
1N3405		3793	ZD-18		
1N3406		3796	ZD-22		
1N3407		3799	ZD-27		
1N3408		3802	ZD-33		
1N3409		3804	ZD-39		
1N340R	5833		5005		
1N341	5838	3500/5882	5004		1N1204C
1N3410		3806	ZD-47		
1N3411		3779	ZD-6.2	561	1N5234A
1N3412		3780	ZD-6.8	561	1N5235A
1N3413		3781	ZD-7.5		1N5236A
1N3414		3782	ZD-8.2	562	1N5237A
1N3415		3785	ZD-10	562	1N5240A
1N3416		3787	ZD-12	563	1N5242A
1N3417		3790	ZD-15	564	1N5245A
1N3418		3793	ZD-18		1N5248A
1N3419		3796	ZD-22		1N5251A
1N341R	5839		5005		
1N342	5838	3500/5882	5004		1N1204C
1N3420		3799	ZD-27		1N5254A
1N3421		3801	ZD-30		1N5256A
1N3422		3802	ZD-33		1N5257A
1N3423		3804	ZD-39		1N5259A
1N3424		3806	ZD-47		1N5261A
1N3425		3808	ZD-56		1N5263A
1N3426		3811	ZD-68		1N5266A
1N3427		3813	ZD-82		1N5268A
1N3428		3816	ZD-100		1N5271A
1N3429		3818	ZD-120		1N5273A
1N342R	5839		5005		
1N343	5838	3500/5882	5004		1N1204C
1N3430		3821	ZD-150		1N5276A
1N3431		3824	ZD-180		1N5279A
1N3432					1N5281A
1N3433	5072A		ZD-8.2	562	1N4738
1N3434	140A		ZD-10	562	1N4740
1N3435	142A		ZD-12	563	1N4742
1N3436	145A		ZD-15	564	1N4744
1N3437	5077A	3752	ZD-18		1N4746
1N3438	5080A	3336	ZD-22		1N4748
1N3439	146A		ZD-27		1N4750
1N343R	5837		5005		
1N344	5838	3500/5882	5004		1N1204C
1N3440	147A		ZD-33		1N4752
1N3441	5086A	3338	ZD-39		1N4754
1N3442	5088A	3340	ZD-47		1N4756
1N3443	137A		ZD-6.2	561	1N4735
1N3444	5071A	3334	ZD-6.8	561	1N4736
1N3445	5072A		ZD-8.2	562	1N4738
1N3446	140A		ZD-10	562	1N4740
1N3447	142A		ZD-12	563	1N4742
1N3448	145A		ZD-15	564	1N4744
1N3449	5077A	3752	ZD-18		1N4746
1N344R	5837		5005		
1N345	5834	3500/5882	5004		1N1202C
1N3450	5080A	3336	ZD-22		1N4748
1N3451	146A		ZD-27		1N4750
1N3452	5084A	3755	ZD-30		1N4751
1N3453	147A		ZD-33		1N4752
1N3454	5086A	3338	ZD-39		1N4754
1N3455	5088A	3340	ZD-47		1N4756
1N3456	5090A	3342	ZD-56		1N4758
1N3457	5092A	3343	ZD-68		1N4760
1N3458	150A		ZD-82		1N4762
1N3459	5096A	3346	ZD-100		1N4764
1N345R	5835		5005		
1N346	5834	3500/5882	5004		1N1202C
1N3460	5097A	3347	ZD-120		1M120ZS10
1N3461	5100A	3350	ZD-150		1M150ZS10
1N3462	5103A	3353	ZD-180		1M180ZS10
1N3463					1M200ZS5
1N3465	109	3087		1123	
1N3466	109	3087		1123	
1N3467	109	3091		1123	
1N3468	109	3091		1123	
1N3469	109	3087		1123	
1N346R	5835		5005		
1N347	5834	3017B/117	MR-1		1N1200C
1N3470	109	3087		1123	
1N3471	177	3100/519	300	1122	
1N3477					1N5221A
1N3477A					1N5221B

Industry Standard No.	ECG	SK	GE	RS 276-	MOTOR.
1N347R	5833		5005		
1N348	5834	3500/5882	5004	1104	1N1200C
1N3482			1N295	1123	
1N3483	109	3091		1123	
1N3484	109	3087		1123	
1N3486	5801	9004	509	1142	1N4007
1N3487	506	3843	511	1114	MR1-1200
1N348R	5833		5005		
1N349	5834	3500/5882	5004		1N1200C
1N3493	5966		504A	1104	
1N3496		3058	ZD-6.2	561	1N823
1N3497		3058	ZD-6.2	561	1N825
1N3498		3058	ZD-6.2	561	1N827
1N3499		3058	ZD-6.2	561	1N829
1N349R	5833		5005		
1N34A	109	3087	1N34AS	1123	
1N34A-Z	109	3087	1N34AS	1123	
1N34AM	109			1123	
1N34AS	109	3087	1N34AS	1123	
1N34G	109	3087	1N34AS	1123	
1N34GA	109	3087	1N34AS	1123	
1N34M	110MP	3087	1N34AS	1123(2)	
1N34Z			1N34AS	1123	
1N35	109	3087		1123	
1N350	177	3016	300	1122	1N1200C
1N3500		3058	ZD-6.2	561	1N821
1N3501					MZ640
1N3502					MZ620
1N3503					MZ610
1N3504					MZ605
1N3506		3330	ZD-3.3		1N5226B
1N3507		3055	ZD-3.6		1N5227B
1N3508		3331	ZD-3.9		1N5228B
1N3509		3332	ZD-4.3		1N5229B
1N351	177	3016	300	1122	1N1202C
1N3510		3333	ZD-4.7		1N5230B
1N3511			ZD-5.1		1N5231B
1N3512		3057	ZD-5.6	561	1N5232B
1N3513		3058	ZD-6.2	561	1N5234B
1N3514		3334	ZD-6.8	561	1N5235B
1N3515			ZD-7.5		1N5236B
1N3516		3136	ZD-8.2	562	1N5237B
1N3517			ZD-9.1	562	1N5239B
1N3518		3061	ZD-10	562	1N5240B
1N3519			ZD-11	563	1N5241B
1N352	177	3016	300	1122	1N1204C
1N3520		3062	ZD-12	563	1N5242B
1N3521		3750	ZD-13	563	1N5243B
1N3522		3063	ZD-15	564	1N5245B
1N3523		3751		564	1N5246B
1N3524		3752	ZD-18		1N5248B
1N3525		3335	ZD-20		1N5250B
1N3526		3336	ZD-22		1N5251B
1N3527			ZD-24		1N5252B
1N3528		3064	ZD-27		1N5254B
1N3529		3755	ZD-30		1N5256B
1N353	178MP	3016	300(2)	1122	
1N3530		3095	ZD-33		1N5257B
1N3531		3337	ZD-36		1N5258B
1N3532		3338	ZD-39		1N5259B
1N3533		3339	ZD-43		1N5260B
1N3534		3340	ZD-47		1N5261B
1N3537		3062	ZD-12	563	
1N354	116	3016	504A	1104	1N1206C
1N3544	116	9004/5801	504A	1142	
1N3545	116	9005/5802	504A	1143	
1N3546	116	9006/5803	504A	1144	
1N3547	116	9007/5804	504A	1144	
1N3548	116	9008/5805	504A	1104	
1N3549	116	3848/5806	504A	1104	
1N355	109	3087		1123	1N1206C
1N3550	177	3100/519	300	1122	
1N3551					MV1642
1N3552					1N5447A
1N3553		3779	ZD-6.2	561	1N821
1N3554	612	3325			1N5141A
1N3555					1N5144
1N3556					1N5148
1N3557					1N5144
1N3563	5809	9010	509	1114	1N4007
1N3564	109	3087		1123	
1N3565	5830		5004		
1N3569	5852		5016		MR1121
1N3570	5854		5016		MR1122
1N3571	5856		5020		MR1124
1N3572	5858		5020		MR1124
1N3573	5860		5024		MR1126
1N3574	5862		5024		MR1126
1N3575	177	3100/519	300	1122	
1N3576	177	3100/519	300	1122	
1N3577	177	3100/519	300	1122	
1N3580		3092	ZD-11.5		1N941
1N3580A		3092	ZD-11.5		1N941A
1N3580B		3092	ZD-11.5		1N941B
1N3581		3092	ZD-11.5		1N942
1N3581A		3092	ZD-11.5		1N942A
1N3581B		3092	ZD-11.5		1N942B
1N3582			ZD-11.5		1N943
1N3582A		3092	ZD-11.5		1N943A
1N3582B		3092	ZD-11.5		1N943B
1N3583			ZD-12	563	1N944
1N3583A			ZD-12	563	1N944A
1N3583B			ZD-12	563	1N944B
1N3584		3092	ZD-11.5		1N945
1N3584A		3092	ZD-11.5		1N945A
1N3584B		3092	ZD-11.5		1N945B
1N359	117	3016	504A	1104	
1N359,A					1N4001
1N3592	109	3091		1123	
1N3593	177	3100/519	300	1122	
1N3594	177	3100/519	300	1122	
1N3595	177	3100/519	300	1122	
1N3596			300	1122	
1N3598	177	3100/519	300	1122	
1N3599	177	3100/519	300	1122	
1N359A	116	3016	504A	1104	
1N36	109	3087		1123	
1N360	116	3016	504A	1104	
1N360,A					1N4002
1N3600	519	3100	514	1122	
1N3601	177	3100/519	300	1122	
1N3602	177	3100/519	300	1122	
1N3603	177	3100/519	300	1122	
1N3604	177	3100/519	300	1122	
1N3605	177	3100/519	300	1122	
1N3606	177	3100/519	300	1122	
1N3607	177	3100/519	300	1122	
1N3608	177	3100/519	300	1122	
1N3609	177	3100/519	300	1122	
1N360A	116	3016	504A	1104	
1N361	116	3016	504A	1104	
1N361,A					1N4003
1N3610		7096	5064		
1N3611	116	9005/5802		1143	1N4003

Industry Standard No.	ECG	SK	GE	RS 276-	MOTOR.
1N3612	116	9007/5804	504A	1144	1N4004
1N3613	116	3848/5806	504A	1104	1N4005
1N3614	125	9010/5809	504A	1104	1N4006
1N3615	5892	3017B/117	5064	1104	MR2000S
1N3615R	5893		5065		
1N3616	5894		5064		MR2001S
1N3616R	5895		5065		
1N3617	5896	7096	5064		MR2002S
1N3617R	5897		5065		
1N3618	5916				MR2002S
1N3618R	5897		5065		
1N3619	5900	7100	5069		MR2004S
1N361A	116	3016	504A	1104	
1N362	116	3016	504A	1104	
1N362,A					1N4004
1N3620	5900	7100	5068		MR2004S
1N3620R	5901		5069		
1N3621	5902		5072		MR2006S
1N3621R	5903		5073		
1N3622	5904	7104	5072		MR2006S
1N3622R	5925		5073		
1N3623	5908		5076		MR2008S
1N3623R	5909		5077		
1N3624	5910	7110	5076		MR2010S
1N3624R	5911	7111	5077		
1N3625	177		300	1122	
1N3627					1N5447A
1N3628					1N5452A
1N362A	116	3016	504A	1104	
1N363	116	3016	504A	1104	
1N363,A					1N4006
1N3639	116	9005/5802	504A	1143	1N5393
1N363A	116	3016	504A	1104	
1N364	125	3033	510,531	1114	
1N364,A					1N4007
1N3640	116	9007/5804	504A	1144	1N5395
1N3641	116	3848/5806	504A	1104	1N5397
1N3642	125	9009/5808	504A	1104	1N5398
1N3649	5866	7068/5868	5028		MR1128
1N3649R	5867		5029		
1N364A	125	3033	510,531	1114	
1N365	125	3033	510,531	1114	
1N365,A					1N4007
1N3650	5868	7068	5028		MR1128
1N3653	177	3100/519	300	1122	
1N3654	177	3100/519	300	1122	
1N3656	5802	9005	504A	1143	
1N3657	5804	9007	504A	1144	
1N3658	5806	3848	504A	1104	
1N365A	125	3033	510,531	1114	
1N3666	177	3100/519	300	1122	
1N3668	177	3100/519	300	1122	
1N3669	116	3016	504A	1104	
1N367	109	3087		1123	
1N3670	5886		5044		
1N3670,A					MR1128
1N3670A	5886		5044		
1N3670R	5887		5045		
1N3671	5886		5044		
1N3671,A					MR1128
1N3671A	5886		5044		
1N3671AR	5887		5045		
1N3671R	5887		5045		
1N3672	5890	7090	5044		
1N3672,A					MR1130
1N3672A	5890	7090	5044		
1N3672AR	5891	7091	5045		
1N3672R	5891	7091	5045		
1N3673	5890	7090	5044		
1N3673,A					MR1130
1N3673A	5890	7090	5044		
1N3673AR	5891	7091	5045		
1N3673R	5891	7091	5045		
1N3675	5071A	3334	ZD-6.8	561	1N4736
1N3675A	5071A	3334	ZD-6.8	561	1N4736
1N3675B	5071A				1N4736A
1N3676	138A		ZD-7.5		1N4737
1N3676A	138A		ZD-7.5		1N4737
1N3676B	138A				1N4737A
1N3677	5072A	3136	ZD-8.2	562	1N4738
1N3677A	5072A	3136	ZD-8.2	562	1N4738
1N3677B	5072A	3136			1N4738A
1N3678	139A	3060	ZD-9.1	562	1N4739
1N3678A	139A		ZD-9.1	562	1N4739
1N3678B	139A				1N4739A
1N3679	140A		ZD-10	562	1N4740
1N3679A	140A		ZD-10	562	1N4740
1N3679B	140A				1N4740A
1N367B	109	3087		1123	
1N368		3017B	504A	1104	
1N3680	5074A		ZD-11.0	563	1N4741
1N3680A	5074A		ZD-11.0	563	1N4741
1N3680B	5074A				1N4741A
1N3681	142A	3062	ZD-12	563	1N4742
1N3681A	142A	3062	ZD-12	563	1N4742
1N3681B	142A	3062	ZD-12	563	1N4742A
1N3682	143A	3750	ZD-13	563	1N4743
1N3682A	143A	3750	ZD-13	563	1N4743
1N3682B	143A	3750			1N4743A
1N3683	145A	3063	ZD-15	564	1N4744
1N3683A	145A	3063	ZD-15		1N4744
1N3683B	145A	3063	ZD-15	564	1N4744A
1N3684	5075A	3751	ZD-16	564	1N4745
1N3684A	5075A	3751	ZD-16	564	1N4745
1N3684B	5075A				1N4745A
1N3685	5077A	3752	ZD-18		1N4746
1N3685A	5077A	3752	ZD-18		1N4746
1N3685B	5077A	3752			1N4746A
1N3686	5079A	3335	ZD-20		1N4747
1N3686A	5079A	3335	ZD-20		1N4747
1N3686B	5079A				1N4747A
1N3687	5080A	3336	ZD-22		1N4748
1N3687A	5080A	3336	ZD-22		1N4748
1N3687B	5080A				1N4748A
1N3688	5081A		ZD-24		1N4749
1N3688A	5081A		ZD-24		1N4749
1N3688B	5081A				1N4749A
1N3689	146A	3064	ZD-27		1N4750
1N3689A	146A	3064	ZD-27		1N4750
1N3689B	146A	3064	ZD-27		1N4750A
1N3690	5084A	3755	ZD-30		1N4751
1N3690A	5084A	3755	ZD-30		1N4751
1N3690B	5084A	3755			1N4751A
1N3691	147A	3095	ZD-33		1N4752
1N3691A	147A	3095	ZD-33		1N4752
1N3691B	147A	3095	ZD-33		1N4752A
1N3692	5085A	3337	ZD-36		1N4753
1N3692A	5085A	3337	ZD-36		1N4753
1N3692B	5085A				1N4753A
1N3693	5086A	3338	ZD-39		1N4754
1N3693A	5086A	3338	ZD-39		1N4754
1N3693B	5086A				1N4754A
1N3694	5087A	3339	ZD-43		1N4755
1N3694A	5087A	3339	ZD-43		1N4755
1N3694B	5087A				1N4755A
1N3695	5088A	3340	ZD-47		1N4756
1N3695A	5088A	3340	ZD-47		1N4756
1N3695B	5088A				1N4756A
1N3696	5089A	3341	ZD-51		1N4757
1N3696A	5089A	3341	ZD-51		1N4757
1N3696B	5089A	3341	ZD-51		1N4757A
1N3697	5090A	3342	ZD-56		1N4758
1N3697A	5090A	3342	ZD-56		1N4758
1N3697B	5090A				1N4758A
1N3698	149A	3097	ZD-62		1N4759
1N3698A	149A	3097	ZD-62		1N4759
1N3698B	149A	3097	ZD-62		1N4759A
1N3699	5092A	3343	ZD-68		1N4760
1N3699A	5092A	3343	ZD-68		1N4760
1N3699B	5092A				1N4760A
1N370					1N5221B
1N3700	5093A	3344	ZD-75		1N4761
1N3700A	5093A	3344	ZD-75		1N4761
1N3700B	5093A				1N4761A
1N3701	150A	3098	ZD-82		1N4762
1N3701A	150A	3098	ZD-82		1N4762
1N3701B	150A	3098	ZD-82		1N4762A
1N3702	5095A	3345	ZD-91		1N4763
1N3702A	5095A	3345	ZD-91		1N4763
1N3702B	5095A				1N4763A
1N3703	5096A	3346	ZD-100		1N4764
1N3703A	5096A	3346	ZD-100		1N4764
1N3703B	5096A				1N4764A
1N3704	151A	3099	ZD-110		1M110ZS10
1N3704A	151A	3099	ZD-110		1M110ZS10
1N3704B	151A	3099	ZD-110		1M110ZS5
1N3705	5097A	3347	ZD-120		1M120ZS10
1N3705A	5097A	3347	ZD-120		1M120ZS10
1N3705B	5097A				1M120ZS5
1N3706	5098A	3348	ZD-130		1M130ZS10
1N3706A	5098A	3348	ZD-130		1M130ZS10
1N3706B	5098A				1M130ZS5
1N3707	5100A	3350	ZD-150		1M150ZS10
1N3707A	5100A	3350	ZD-150		1M150ZS10
1N3707B	5100A				1M150ZS5
1N3708	5101A	3351	ZD-160		1M160ZS10
1N3708A	5101A	3351	ZD-160		1M160ZS10
1N3708B	5101A				1M160ZS5
1N3709	5103A	3353	ZD-180		1M180ZS10
1N3709A	5103A	3353	ZD-180		1M180ZS10
1N3709B	5103A				1M180ZS5
1N371	5000A	3766			1N5221A
1N3710	5105A	3355	ZD-200		1M200ZS10
1N3710A	5105A	3355	ZD-200		1M200ZS10
1N3710B	5105A				1M200ZS5
1N372		3838			1N5225A
1N3722	177	3100/519	300	1122	
1N3723		9010	509	1114	
1N3724	506	3843	511	1114	
1N3725	506	3843	511	1114	
1N3729		3843	511	1114	
1N373		3055	ZD-3.6		1N5227A
1N3731		3843	511	1114	
1N3732	135A		ZD-5.1		
1N3735	6354	6554			
1N3735R	6355	6555			
1N3736	6354	6554			
1N3736R	6355	6555			
1N3738	6354	6554			
1N3738R	6355	6555			
1N3739	6356	6556			
1N374	5008A	3774	ZD-4.3		1N5229A
1N3740	6356	6556			
1N3741	6358	6558			
1N3741R	6359	6559			
1N3742	6358	6558			
1N3742R	6359	6559			
1N3743	6104	6560			
1N3743R	6105	6561			
1N3748	5802	9005	504A	1143	
1N3749	5804	9007	504A	1144	
1N375	5008A	3774			1N5230A
1N3750	5806	3848	ZD-4.3		
1N3751	5808	9009	509	1114	
1N3752	5809	9010	509	1114	
1N3753	109	3087		1123	
1N3754	116	9004/5801	1N34AS	1142	
1N3755	5802	9005	504A	1143	
1N3756	5804	9007	504A	1144	
1N3757	5802	9005	504A	1143	
1N3758	5804	9007	504A	1144	
1N3759	5806	3848	504A	1104	
1N376			ZD-5.1		1N5233A
1N3760	5808	9009	509	1114	
1N3761	5809	9010	509	1114	
1N3765	5998		5108		
1N3765R	5999		5109		
1N3766	5998		5108		
1N3766R	5999		5109		
1N3767R	6003	7203	5109		
1N3768	6002	7202	5108		
1N3768R	6003	7203	5109		
1N377		9021		561	1N5236A
1N3773	109			1123	
1N3776			ZD-10	562	
1N3777	5998		5108		
1N3779		3780			1N821A
1N378	5014A	3780	ZD-6.8	561	1N5238A
1N3780		3780			1N821A
1N3781		3780			1N823A
1N3782		3780			1N825A
1N3783		3780			1N827A
1N3784		3780			1N829A
1N3785	5120A	3386	5ZD-6.8		
1N3785A	5120A	3386	5ZD-6.8		
1N3785B	5120A		5ZD-6.8		
1N3786	5121A	3387	5ZD-7.5		
1N3786A	5121A	3387	5ZD-7.5		
1N3786B	5121A		5ZD-7.5		
1N3787	5122A	3388	5ZD-8.2		
1N3787A	5122A	3388	5ZD-8.2		
1N3787B	5122A		5ZD-8.2		
1N3788	5124A	3390	5ZD-9.1		
1N3788A	5124A	3390	5ZD-9.1		
1N3788B	5124A		5ZD-9.1		
1N3789	5125A	3391	5ZD-10		
1N3789A	5125A	3391	5ZD-10		
1N3789B	5125A		5ZD-10		
1N379	178MP		300(2)	1122	1N5240A
1N3790	5126A	3392	5ZD-11		
1N3790A	5126A	3392	5ZD-11		
1N3790B	5126A		5ZD-11		
1N3791	5127A	3393	5ZD-12		
1N3791A	5127A	3393	5ZD-12		
1N3791B	5127A		5ZD-12		
1N3792	5128A	3394	5ZD-13		
1N3792A	5128A	3394	5ZD-13		
1N3792B	5128A		5ZD-13		

Industry Standard No.	ECG	SK	GE	RS 276-	MOTOR.
1N3793	5130A	3396	ZD-15		
1N3793A	5130A	3396	5ZD-15		
1N3793B	5130A		5ZD-15		
1N3794	5131A	3397	5ZD-16		
1N3794A	5131A	3397	5ZD-16		
1N3794B	5131A		5ZD-16		
1N3795	5133A	3399	5ZD-18		
1N3795A	5133A	3399	5ZD-18		
1N3795B	5133A		5ZD-18		
1N3796	5135A	3401	5ZD-20		
1N3796A	5135A	3401	5ZD-20		
1N3796B	5135A		5ZD-20		
1N3797	5136A	3402	5ZD-22		
1N3797A	5136A	3402	5ZD-22		
1N3797B	5136A		5ZD-22		
1N3798	5137A	3403	5ZD-24		
1N3798A	5137A	3403	5ZD-24		
1N3798B	5137A		5ZD-24		
1N3799	5139A	3405	5ZD-27		
1N3799A	5139A	3405	5ZD-27		
1N3799B	5139A		5ZD-27		
1N38	109	3087	1N91	1123	
1N380	177	3100/519	300	1122	1N5243A
1N3800	5141A	3407	5ZD-30		
1N3800A	5141A	3407	5ZD-30		
1N3800B	5141A		5ZD-30		
1N3801	5142A	3408	5ZD-33		
1N3801A	5142A	3408	5ZD-33		
1N3801B	5142A		5ZD-33		
1N3802	5143A	3409			
1N3802A	5143A	3409			
1N3803	5144A	3410			
1N3803A	5144A	3410			
1N3804	5145A	3411			
1N3804A	5145A	3411			
1N3805	5146A	3412			
1N3805A	5146A	3412			
1N3806	5147A	3413			
1N3806A	5147A	3413			
1N3807	5148A	3414			
1N3807A	5148A	3414			
1N3808	5150A	3416			
1N3808A	5150A	3416			
1N3809	5151A	3417			
1N3809A	5151A	3417			
1N381	177	3100/519	300	1122	1N5246A
1N3810	5152A	3418			
1N3810A	5152A	3418			
1N3811	5153A	3419			
1N3811A	5153A	3419			
1N3812	5155A	3421			
1N3812A	5155A	3421			
1N3813	5156A	3422			
1N3813A	5156A	3422			
1N3814	5157A	3423			
1N3814A	5157A	3423			
1N3815	5158A	3424			
1N3815A	5158A	3424			
1N3816	5159A	3425			
1N3816A	5159A	3425			
1N3817	5161A	3427			
1N3817A	5161A	3427			
1N3818	5162A	3428			
1N3818A	5162A	3428			
1N3819	5164A	3430			
1N3819A	5164A	3430			
1N382	177	3100/519	300	1122	1N5249A
1N3820	5166A	3432			
1N3820A	5166A	3432			
1N3821	5066A	3330	ZD-3.3		
1N3821A	5066A	3330	ZD-3.3		
1N3822	134A	3055	ZD-3.6		
1N3822A	134A		ZD-3.6		
1N3823	5067A	3331	ZD-3.9		
1N3823A	5067A	3331	ZD-3.9		
1N3824	5068A	3332	ZD-4.3		
1N3824A	5068A	3332	ZD-4.3		
1N3825	5069A	3333	ZD-4.7		
1N3825A	5069A	3333	ZD-4.7		
1N3826	135A	3056	ZD-5.1		
1N3826A	135A	3056	ZD-5.1		
1N3827	136A		ZD-5.6	561	
1N3827A	136A		ZD-5.6	561	
1N3828	137A		ZD-6.2	561	
1N3828A	137A		ZD-6.2	561	
1N3829	5071A	3334	ZD-6.8	561	
1N3829A	5071A	3334	ZD-6.8	561	
1N383	177	3100/519	300	1122	1N5252A
1N3830	138A		ZD-7.5		
1N3830A	138A		ZD-7.5		
1N384	178MP		300(2)	1122	1N5255A
1N385	178MP		300(2)	1122	1N5258A
1N3855A	139A	3060	ZD-9.1	562	
1N386	177	3100/519	300	1122	1N5260A
1N3864	177	3100/519	300	1122	
1N3866	5802	9005	504A	1143	1N4003
1N3867	5804	9007	504A	1144	1N4004
1N3868	5806	3848	504A	1104	1N4005
1N3869	5809	9010	509	1114	1N4007
1N387	178MP		300(2)	1122	1N5261A
1N3870					MR1-1600
1N3872	177	3100/519	300	1122	
1N3873	177	3100/519	300	1122	
1N388	178MP		300(2)	1122	
1N3889R	5819		5029		
1N389	177	3100/519	300	1122	
1N3895	116	3016	504A	1104	
1N3898					1N5221B
1N38A	109	3087	1N91	1123	
1N38B	109	3087	1N91	1123	
1N390	177	3100/519	300	1122	
1N391	177	3100/519	300	1122	
1N3915	506	3843	511	1114	
1N3919	5868	7068	5028		
1N392	177	3100/519	300	1122	
1N3924	5890	7090	5044		MR1130
1N3929	125	3033	510,513	1114	MR1-1000
1N393	177	3100/519	300	1122	
1N3930					MR1-1600
1N3938	5802	9005	504A	1143	SPECIAL
1N3939	5804	9007	504A	1144	SPECIAL
1N394	177	3100/519	300	1122	
1N3940	5806	3848	504A	1104	SPECIAL
1N3941	5808	9009			
1N3942	5809	9010			
1N3945					MV1632
1N3946					1N5457A
1N3947	137A				1N5474A
1N3949	5196A				1N2984B
1N3950	5135A		5ZD-20		1N3796B
1N3951	5138A		5ZD-25		1.5M25Z5
1N3952	5802	9005	504A	1143	
1N3953	177		300	1122	
1N3954	177		300	1122	
1N3956	177	3100/519	300	1122	
1N3957	5809	9010	509	1114	
1N3968	6026	7226	5128		
1N3968R	6027	7227	5129		
1N3969	6034	7234	5132		
1N3969R	6035		5133		
1N3970	6040	7240	5136		
1N3970R	6041		5137		
1N3971	6042		5140		
1N3971R	6043		5142		
1N3972	6154	7354			
1N3973	6154	7354			
1N3974	6156	7356			
1N3975	6158	7358			
1N3976	6354	6554			
1N3977	6354	6554			
1N3978	6356	6556			
1N3981	5802	9005		1143	1N4003
1N3982	5804	9007		1144	1N4004
1N3983	5806	3848			1N4005
1N3984	5178AK				1N3997A
1N3985	5179AK				1N3998A
1N3986	5180AK				1N3998A
1N3987	5866	3500/5882	5028		MR1128
1N3987R	5867		5029		
1N3988	5866		5028		MR1128
1N3988R	5867		5029		
1N3989	5868	7068	5028		MR1130
1N3989R	5869	7069	5029		
1N3990	5868	7068	5028		MR1130
1N3990R	5869	7069	5029		
1N3991	109	3087		1123	
1N40	109	3087		1123	
1N400	116	3016	504A	1104	
1N4001	125	3311	504A	1104	
1N4002	116	3311	504A	1104	
1N4003	116	3311	504A	1104	
1N4003GP	116	3311	504A	1104	
1N4004	116	3312	504A	1104	
1N4005	116	3313	504A	1104	
1N4006	125	3032A	510,531	1114	
1N4007	125	3080	510,531	1114	
1N4009	177	3016	300	1122	
1N400B	116	3016	504A	1104	
1N4010			ZD-6.2	561	1N821
1N4011	125	3033	510,531	1114	
1N4012	5886	3500/5882	5044		MR1128
1N4012R	5887		5045		
1N4013	5886		5044		MR1128
1N4013R	5887		5045		
1N4014	5890	7090	5044	562	MR1130
1N4014R	5891	7091	5045		
1N4015	5890	7090	5044		MR1130
1N4015R	5891	7091	5045		
1N4016	5183A				1N2972A
1N4016A	5183A				1N2972A
1N4016B	5183A				1N2972B
1N4017	5185A				1N2973A
1N4017A	5185A				1N2973A
1N4017B	5185A				1N2973B
1N4018	5186A				1N2974A
1N4018A	5186A				1N2974A
1N4018B	5186A				1N2974B
1N4019	5187A				1N2975A
1N4019A	5187A				1N2975A
1N4019B	5187A				1N2975B
1N4020	5188A				1N2976A
1N4020A	5188A				1N2976A
1N4020B	5188A				1N2976B
1N4021	5189A				1N2977A
1N4021A	5189A				1N2977A
1N4021B	5189A				1N2977B
1N4022	5191A				1N2979A
1N4022A	5191A				1N2979A
1N4022B	5191A				1N2979B
1N4023	5192A				1N2980A
1N4023A	5192A				1N2980A
1N4023B	5192A				1N2980B
1N4024	5194A				1N2982A
1N4024A	5194A				1N2982A
1N4024B	5194A				1N2982B
1N4025	5196A				1N2984A
1N4025A	5196A				1N2984A
1N4025B	5196A				1N2984B
1N4026	5197A				1N2985A
1N4026A	5197A				1N2985A
1N4026B	5197A				1N2985B
1N4027	5198A				1N2986A
1N4027A	5198A				1N2986A
1N4027B	5198A				1N2986B
1N4028	5200A				1N2988A
1N4028A	5200A				1N2988A
1N4028B	5200A				1N2988B
1N4029	5202A				1N2989A
1N4029A	5202A				1N2989A
1N4029B					1N2989B
1N4030	5203A				1N2990A
1N4030A	5203A				1N2990A
1N4030B	5203A				1N2990B
1N4031	5204A				1N2991A
1N4031A	5204A				1N2991A
1N4031B	5204A				1N2991B
1N4032	5205A				1N2992A
1N4032A	5205A				1N2992A
1N4032B	5205A				1N2992B
1N4033	5206A				1N2993A
1N4033A	5206A				1N2993A
1N4033B					1N2993B
1N4034	5208A				1N2995A
1N4034A	5208A				1N2995A
1N4034B	5208A				1N2995B
1N4035	5210A				1N2997A
1N4035A	5210A				1N2997A
1N4035B	5210A				1N2997B
1N4036	5212A				1N2999A
1N4036A	5212A				1N2999A
1N4036B	5212A				1N2999B
1N4037	5214A				1N3000A
1N4037A	5214A				1N3000A
1N4037B	5214A				1N3000B
1N4038	5215A				1N3001A
1N4038A	5215A				1N3001A
1N4038B					1N3001B
1N4039	5216A				1N3002A
1N4039A	5216A				1N3002A
1N4039B	5216A				1N3002B
1N4040	5217A				1N3003A
1N4040A	5217A				1N3003A
1N4040B					1N3003B
1N4041	5219A				1N3004A
1N4041A	5219A				1N3004A
1N4041B					1N3004B
1N4042	5220A				1N3005A
1N4042A	5220A				1N3005A

Industry Standard No.	ECG	SK	GE	RS 276-	MOTOR.
1N4042B	5220A				1N3005B
1N4043	177	3016	300	1122	
1N4044	6354	3095/147A			
1N4044R	6355	6555			
1N4045	6354	6554	ZD-5.0		
1N4045R	6355	6555			
1N4046	6354	6554			
1N4046R	6355	6555			
1N4047	6354	6554			
1N4047R	6355	6555			
1N4048	6354	6554	300	1122	
1N4048R	6355	6555			
1N4049	6354	6554			
1N4049R	6355	6555			
1N4050	6354	6554			
1N4050R	6355	6555			
1N4052	6356	6556			
1N4053	6358	6558			
1N4053R	6359	6559			
1N4054	6358	6558			
1N4054R	6359	6559			
1N4055	6358	6558			
1N4055R	6359	6559			
1N4056	6358	6558			
1N4056R	6359	6559			
1N4057		3393	5ZD-12		
1N4057A		3393	5ZD-12		
1N4058		3396	5ZD-15		
1N4058A		3396	5ZD-15		
1N4059		3398	5ZD-17		
1N4059A		3398	5ZD-17		
1N4060		3400	5ZD-19		
1N4060A		3400	5ZD-19		
1N4061		3401	5ZD-20		
1N4061A		3401	5ZD-20		
1N4062		3402	5ZD-22		
1N4062A		3402	5ZD-22		
1N4063		3405	5ZD-27		
1N4063A		3405	5ZD-27		
1N4064		3045	5ZD-30		
1N4064A		3407	5ZD-30		
1N4065		3408	5ZD-33		
1N4065A		3408	5ZD-33		
1N4085					1N5142
1N4087	177	3100/519	300	1122	
1N4088	109	3091		1123	
1N4089	5804	9007	504A	1144	
1N4091					1N5461A
1N4092	178MP		300(2)	1122	
1N4093	178MP		300(2)	1122(2)	
1N4094					1N2624B
1N4095	5010A	3785/5019A	ZD-10	562	1N5231A
1N4096	5155A	3421			1N4763A
1N4097	5156A	3422			1N4764A
1N4098	5161A	3427			1M150ZS5
1N4099		3334	ZD-6.8	561	
1N41	109	3087		1123	
1N4101		3136	ZD-8.2	562	
1N4102		3749	ZD-9.1	562	
1N4103		3060	ZD-9.1	562	
1N4103A		3066		562	
1N4104		3061	ZD-10	562	
1N4105			ZD-11	563	
1N4106		3062	ZD-12	563	
1N4106A		3062	ZD-12	563	
1N4107		3750	ZD-13	563	
1N4108		3094	ZD-14	564	
1N4109		3063	ZD-15	564	
1N4110		3751	ZD-16	564	
1N4111		9022	ZD-17		
1N4112		3752	ZD-18		
1N4113		9023	ZD-19		
1N4114		3335	ZD-20		
1N4115		3336	ZD-22		
1N4117		3798	ZD-25		
1N4118		3064	ZD-27		
1N4119		3754	ZD-28		
1N411B	6154	7354			
1N4120		3755	ZD-30		
1N4121		3095	ZD-33		
1N4121A		3095	ZD-33		
1N4121B		3095	ZD-33		
1N4122		3337	ZD-36		
1N4123		3338	ZD-39		
1N4124		3339	ZD-43		
1N4125		3340	ZD-47		
1N4126		3341	ZD-51		
1N4127		3342	ZD-56		
1N4129		3097	ZD-62		
1N412B	6154	7354			
1N4130		3343	ZD-68		
1N4131		3344	ZD-75		
1N4132		3098	ZD-82		
1N4133		9024	ZD-87		
1N4134		3345	ZD-91		
1N4135		3346	ZD-100		
1N4136	6054	7254	5128		
1N4137	6060	7260	5132		
1N4138	6064	7264	5136		
1N4139	5800	9003	504A	1142	
1N413B	6154	7354			
1N4140	5801	9004		1142	
1N4141	5802	9005	504A	1143	
1N4142	5804	9007		1144	
1N4143	5806	3848			
1N4144	5808	9009			
1N4145	5809	9010			
1N4146	506	3843	511	1114	
1N4147	177	3100/519	300	1122	
1N4148	519	3100	514	1122	
1N4149	519	3100	514	1122	
1N414B	177	3100/519	300	1122	
1N4150	177	3100/519	300	1122	
1N4151	177	3100/519	300	1122	
1N4152	177	3100/519	300	1122	
1N4153	519	3100	514	1122	
1N4154	519	3100	514	1122	
1N4155	506	3843	511	1114	
1N4156	177		300	1122	
1N4157	177		300	1122	
1N4158	5071A	3334	ZD-6.8	561	1N4736
1N4158A	5071A	3334	ZD-6.8	561	1N4736
1N4158B	5071A				1N4736A
1N4159	138A		ZD-7.5		1N4737
1N4159A	138A		ZD-7.5		1N4737
1N4159B	138A				1N4737A
1N4160	5072A	3136	ZD-8.2	562	1N4738
1N4160A	5072A	3136	ZD-8.2	562	1N4738
1N4160B	5072A	3136			1N4738A
1N4161	139A	3066/118	ZD-9.1	562	1N4739
1N4161A	139A	3060	ZD-9.1	562	1N4739
1N4161B	139A	3060	ZD-9.1	562	1N4739A
1N4162	140A		ZD-10	562	1N4740
1N4162A	140A		ZD-10	562	1N4740

Industry Standard No.	ECG	SK	GE	RS 276-	MOTOR.
1N4162B	140A				1N4740A
1N4163	5074A		ZD-11.0	563	1N4741
1N4163A	5074A		ZD-11.0	563	1N4741
1N4163B	5074A				1N4741A
1N4164	142A	3062	ZD-12	563	1N4742
1N4164A	142A	3062	ZD-12	563	1N4742
1N4164B	142A	3062	ZD-12	563	1N4742A
1N4165	143A	3750	ZD-13	563	1N4743
1N4165A	143A	3750	ZD-13	563	1N4743
1N4165B	143A	3750			1N4743A
1N4166	145A	3063	ZD-15	564	1N4744
1N4166A	145A	3063	ZD-15	564	1N4744
1N4166B	145A	3063	ZD-15	564	1N4744A
1N4167	5075A	3751	ZD-16	564	1N4745
1N4167A	5075A	3751	ZD-16	564	1N4745
1N4167B	5075A	3751			1N4745A
1N4168	5077A	3752	ZD-18		1N4746
1N4168A	5077A	3752	ZD-18		1N4746
1N4168B	5077A	3752			1N4746A
1N4169	5079A	3335	ZD-20		1N4747
1N4169A	5079A	3335	ZD-20		1N4747
1N4169B	5079A				1N4747A
1N417	109	3087		1123	
1N4170	5080A	3336	ZD-22		1N4748
1N4170A	5080A	3336	ZD-22		1N4748
1N4170B	5080A				1N4748A
1N4171	5081A		ZD-24		1N4749
1N4171A	5081A		ZD-24		1N4749
1N4171B	5081A				1N4749A
1N4172	146A	3064	ZD-27		1N4750
1N4172A	146A	3064	ZD-27		1N4750
1N4172B	146A	3064	ZD-27		1N4750A
1N4173	5084A	3755	ZD-30		1N4751
1N4173A	5084A	3755	ZD-30		1N4751
1N4173B	5084A	3755			1N4751A
1N4174	147A	3095	ZD-33		1N4752
1N4174A	147A	3095	ZD-33		1N4752
1N4174B	147A	3095	ZD-33		1N4752A
1N4175	5085A	3337	ZD-36		1N4753
1N4175A	5085A	3337	ZD-36		1N4753
1N4175B	5085A				1N4753A
1N4176	5086A	3338	ZD-39		1N4754
1N4176A	5086A	3338	ZD-39		1N4754
1N4176B	5086A				1N4754A
1N4177	5087A	3339	ZD-43		1N4755
1N4177A	5087A	3339	ZD-43		1N4755
1N4177B	5087A	3339	ZD-43		1N4755A
1N4178	5088A	3340	ZD-47		1N4756
1N4178A	5088A	3340	ZD-47		1N4756
1N4178B	5088A				1N4756A
1N4179	5089A				1N4757
1N4179A	5089A	3341	ZD-51		1N4757
1N4179B	5089A	3341	ZD-51		1N4757A
1N418	109	3087		1123	
1N4180	5090A	3342	ZD-56		1N4758
1N4180A	5090A	3342	ZD-56		1N4758
1N4180B	5090A		ZD-56		1N4758A
1N4181	149A	3097	ZD-62		1N4759
1N4181A	149A	3097	ZD-62		1N4759
1N4181B	149A	3097	ZD-62		1N4759A
1N4182	5092A	3343	ZD-68		1N4760
1N4182A	5092A	3343	ZD-68		1N4760
1N4182B	5092A	3343	ZD-68		1N4760A
1N4183	5093A	3344	ZD-75		1N4761
1N4183A	5093A	3344	ZD-75		1N4761
1N4183B	5093A	3344	ZD-75		1N4761A
1N4184	150A	3098	ZD-82		1N4762
1N4184A	150A	3098	ZD-82		1N4762
1N4184B	150A	3098	ZD-82		1N4762A
1N4185	5095A	3345	ZD-91		1N4763
1N4185A	5095A	3345	ZD-91		1N4763
1N4185B	5095A	3345	ZD-91		1N4763A
1N4186	5096A	3346	ZD-100		1N4764
1N4186A	5096A	3346	ZD-100		1N4764
1N4186B	5096A	3346	ZD-100		1N4764A
1N4187	151A	3099	ZD-110		
1N4187A	151A	3099	ZD-110		
1N4187B	151A	3099	ZD-110		
1N4188	5097A	3347	ZD-120		
1N4188A	5097A	3347	ZD-120		
1N4189	5098A	3348	ZD-130		
1N4189A	5098A	3348	ZD-130		
1N4189B	5098A	3348	ZD-130		
1N419	109	3087		1123	
1N4190	5100A	3350	ZD-150		
1N4190A	5100A	3350	ZD-150		
1N4191	5101A	3351	ZD-160		
1N4191A	5101A	3351	ZD-160		
1N4192	5103A	3353	ZD-180		
1N4192A	5103A	3353	ZD-180		
1N4192B	5103A	3353	ZD-180		
1N4193	5105A	3355	ZD-200		
1N4193A	5105A	3355	ZD-200		
1N4194	5181A				1N2970A
1N4194A	5181A				1N2970A
1N4194B					1N2970B
1N4195	5182A				1N2971A
1N4195A	5182A				1N2971A
1N4195B					1N2971B
1N4196	5183A				1N2972A
1N4196A	5183A				1N2972A
1N4196B					1N2972B
1N4197	5185A				1N2973A
1N4197A	5185A				1N2973A
1N4197B					1N2973B
1N4198	5186A		1N34AS	1123	1N2974A
1N4198A	5186A				1N2974A
1N4198B					1N2974B
1N4199	5187A				1N2975A
1N4199A	5187A				1N2975A
1N4199B					1N2975B
1N42			1N60	1123	
1N4200	5188A				1N2976A
1N4200A	5188A				1N2976A
1N4200B					1N2976B
1N4201	5189A				1N2977A
1N4201A	5189A				1N2977A
1N4201B					1N2977B
1N4202	5190A				1N2978A
1N4202A	5190A				1N2978A
1N4202B					1N2978B
1N4203	5191A				1N2979A
1N4203A	5191A				1N2979A
1N4203B					1N2979B
1N4204	5192A				1N2980A
1N4204A	5192A				1N2980A
1N4204B					1N2980B
1N4205	5193A				1N2981A
1N4205A	5193A				1N2981A
1N4205B					1N2981B
1N4206	5194A				1N2982A
1N4206A	5194A				1N2982A
1N4206B					1N2982B
1N4207	5195A				1N2983A

Industry Standard No.	ECG	SK	GE	RS 276-	MOTOR.
1N4207A	5195A				1N2983A
1N4207B					1N2983B
1N4208	5196A				1N2984A
1N4208A	5196A				1N2984A
1N4208B					1N2984B
1N4209	5197A				1N2985A
1N4209A	5197A				1N2985A
1N4209B					1N2985B
1N4210	5198A				1N2986A
1N4210A	5198A				1N2986A
1N4210B					1N2986B
1N4211	5199A				1N2987A
1N4211A	5199A				1N2987A
1N4211B					1N2987B
1N4212	5200A				1N2988A
1N4212A	5200A				1N2988A
1N4212B					1N2988B
1N4213	5202A				1N2989A
1N4213A	5202A				1N2989A
1N4213B					1N2989B
1N4214	5203A				1N2990A
1N4214A	5203A				1N2990A
1N4214B					1N2990B
1N4215	5204A				1N2991A
1N4215A	5204A				1N2991A
1N4215B					1N2991B
1N4216	5205A				1N2992A
1N4216A	5205A				1N2992A
1N4216B					1N2992B
1N4217	5206A				1N2993A
1N4217A	5206A				1N2993A
1N4217B					1N2993B
1N4218	5207A				1N2994A
1N4218A	5207A				1N2994A
1N4218B					1N2994B
1N4219	5208A				1N2995A
1N4219A	5208A				1N2995A
1N4219B					1N2995B
1N4220	5209A				1N2996A
1N4220A	5209A				1N2996A
1N4220B					1N2996B
1N4221	5210A				1N2997A
1N4221A	5210A				1N2997A
1N4221B					1N2997B
1N4222	5211A				1N2998A
1N4222A	5211A				1N2998A
1N4222B					1N2998B
1N4223	5212A				1N2999A
1N4223A	5212A				1N2999A
1N4223B					1N2999B
1N4224	5214A				1N3000A
1N4224A	5214A				1N3000A
1N4224B					1N3000B
1N4225	5215A				1N3001A
1N4225A	5215A				1N3001A
1N4225B					1N3001B
1N4226	5216A	3017B/117			1N3002A
1N4226A	5216A				1N3002A
1N4226B					1N3002B
1N4227	5217A				1N3003A
1N4227A	5217A				1N3003A
1N4227B					1N3003B
1N4228	5219A				1N3004A
1N4228A	5219A				1N3004A
1N4228B					1N3004B
1N4229	5220A				1N3005A
1N4229A	5220A				1N3005A
1N4229B					1N3005B
1N4230	5221A				1N3006A
1N4230A	5221A				1N3006A
1N4230B					1N3006B
1N4231	5222A				1N3007A
1N4231A	5222A				1N3007A
1N4231B					1N3007B
1N4232	5223A				1N3008A
1N4232A	5223A				1N3008A
1N4232B					1N3008B
1N4233	5224A				1N3009A
1N4233A	5224A				1N3009A
1N4233B					1N3009B
1N4234	5225A				1N3010A
1N4234A	5225A				1N3010A
1N4234B					1N3010B
1N4235	5226A				1N3011A
1N4235A	5226A				1N3011A
1N4235B					1N3011B
1N4236	5227A				1N3012A
1N4236A	5227A				1N3012A
1N4236B					1N3012B
1N4237	5229A				1N3013A
1N4237A	5229A				1N3013A
1N4237B					1N3013B
1N4238	5230A				1N3014A
1N4238A	5230A				1N3014A
1N4238B					1N3014B
1N4239	5232A				1N3015A
1N4239A	5232A				1N3015A
1N4239B					1N3015B
1N4240	5177A		ZD-5.1		
1N4242	177	3100/519	300	1122	
1N4243	177	3100/519	300	1122	
1N4245	116	9005/5802	504A	1143	1N4003
1N4246	116	9007/5804	504A	1144	1N4004
1N4247	116	3848/5806	504A	1104	1N4005
1N4248	125	9009/5808	509	1114	1N4006
1N4249	125	9009/5808	509	1114	1N4007
1N4250	125	3032A	510,531	1114	
1N4251	125	3033	510,531	1114	
1N4252	506	3843	511	1114	
1N4258	5181A				1N2970A
1N4258A	5181A				1N2970A
1N4258B					1N2970B
1N4259	5182A				1N2971A
1N4259A	5182A				1N2971A
1N4259B					1N2971B
1N4260	5183A				1N2972A
1N4260A	5183A				1N2972A
1N4260B					1N2972B
1N4261	5185A				1N2973A
1N4261A	5185A				1N2973A
1N4261B					1N2973B
1N4262	5186A				1N2974A
1N4262A	5186A				1N2974A
1N4262B					1N2974B
1N4263	5187A				1N2975A
1N4263A	5187A				1N2975A
1N4263B	5187A				1N2975B
1N4264	5188A				1N2976A
1N4264A	5188A				1N2976A
1N4264B					1N2976B
1N4265	5189A				1N2977A
1N4265A	5189A				1N2977A
1N4265B					1N2977B
1N4266	5191A				1N2979A
1N4266A	5191A				1N2979A
1N4266B					1N2979B
1N4267	5192A				1N2980A
1N4267A	5192A				1N2980A
1N4267B					1N2980B
1N4268	5194A				1N2982A
1N4268A	5194A				1N2982A
1N4268B					1N2982B
1N4269	5196A				1N2984A
1N4269A	5196A				1N2984A
1N4269B					1N2984B
1N4270	5197A				1N2985A
1N4270A	5197A				1N2985A
1N4270B					1N2985B
1N4271	5198A				1N2986A
1N4272A	5200A				1N2988A
1N4272B					1N2988B
1N4273	5202A				1N2989A
1N4273A	5202A				1N2989A
1N4273B					1N2989B
1N4274	5203A				1N2990A
1N4274A	5203A				1N2990A
1N4274B					1N2990B
1N4275	5204A				1N2991A
1N4275A	5204A				1N2991A
1N4275B					1N2991B
1N4276	5205A				1N2992A
1N4276A	5205A				1N2992A
1N4276B					1N2992B
1N4277	5206A				1N2993A
1N4277A	5206A				1N2993A
1N4277B					1N2993B
1N4278	5208A				1N2995A
1N4278A	5208A				1N2995A
1N4278B					1N2995B
1N4279	5210A				1N2997A
1N4279A	5210A				1N2997A
1N4279B					1N2997B
1N4280	5212A				1N2999A
1N4280A	5212A				1N2999A
1N4280B					1N2999B
1N4281	5214A				1N3000A
1N4281A	5214A				1N3000A
1N4281B					1N3000B
1N4282	5215A				1N3001A
1N4282A	5215A				1N3001A
1N4282B					1N3001B
1N4283	5216A				1N3002A
1N4283A	5216A				1N3002A
1N4283B					1N3002B
1N4284	5217A				1N3003A
1N4284A	5217A				1N3003A
1N4284B					1N3003B
1N4285	5219A				1N3004A
1N4285A	5219A				1N3004A
1N4285B					1N3004B
1N4286	5220A				1N3005A
1N4286A	5220A				1N3005A
1N4286B					1N3005B
1N4287	5222A				1N3007A
1N4287A	5222A				1N3007A
1N4287B					1N3007B
1N4288	5223A				1N3008A
1N4288A	5223A				1N3008A
1N4288B					1N3008B
1N4289	5224A				1N3009A
1N4289A	5224A				1N3009A
1N4289B					1N3009B
1N429			ZD-6.2	561	
1N4290	5226A				1N3011A
1N4290A	5226A				1N3011A
1N4290B					1N3011B
1N4291	5227A				1N3012A
1N4291A	5227A				1N3012A
1N4291B					1N3012B
1N4292	5230A				1N3014A
1N4292A	5230A				1N3014A
1N4292B					1N3014B
1N4293	5232A				1N3015A
1N4293A					1N3015A
1N4293B	5232A				1N3015B
1N4295		3061	ZD-10	562	
1N4295A		3061	ZD-10	562	
1N4296			ZD-10	562	
1N4296A			ZD-10	562	
1N4297		3749	ZD-9.1	562	
1N4297A		3749	ZD-9.1	562	
1N4298		3749	ZD-9.1	562	
1N4298A		3749	ZD-9.1	562	
1N4299		3092	ZD-11.0	563	
1N4299A		3092	ZD-11.0	563	
1N4299B		3092	ZD-11.5		
1N43	109	3087		1123	
1N430		3782	ZD-8.2	562	1N3156
1N4300			ZD-11.0	563	
1N4300A			ZD-11.0	563	
1N4301			ZD-11.0	563	
1N4301A			ZD-11.0	563	
1N4302		3749	ZD-9.1	562	
1N4302A		3749	ZD-9.1	562	
1N4303		3092	ZD-11.0	563	
1N4303A			ZD-11.0	563	
1N4304			ZD-11.0	563	
1N4304A			ZD-11.0	563	
1N4305	519	3100	514	1122	
1N4306			300	1122	
1N4307			300	1122	
1N4308	177	3100/519	300	1122	
1N4309	177	3100/519	300	1122	
1N430A		3782	ZD-8.2	562	1N3157
1N430B			ZD-8.2	562	1N3157A
1N431	177	3100/519	300	1122	
1N4311	177	3100/519	300	1122	
1N4315	177	3100/519	300	1122	
1N4318	177	3100/519	300	1122	
1N432	177	3100/519	300	1122	
1N4321	5089A	3341	ZD-51		5M50Z810
1N4322	177	3100/519	300	1122	
1N4323	5071A	3334	ZD-6.8	561	1N4736
1N4323A	5071A	3334	ZD-6.8	561	1N4736
1N4323B	5071A				1N4736A
1N4324	138A		ZD-7.5		1N4737
1N4324A	138A		ZD-7.5		1N4737
1N4324B	138A				1N4737A
1N4325	5072A	3136	ZD-8.2	562	1N4738
1N4325A	5072A	3136	ZD-8.2	562	1N4738
1N4325B	5072A	3136			1N4738A
1N4326	139A	3060	ZD-9.1	562	1N4739
1N4326A	139A	3060	ZD-9.1	562	1N4739
1N4326B	139A	3060	ZD-9.1	562	1N4739A
1N4327	140A		ZD-10	562	1N4740
1N4327A	140A		ZD-10	562	1N4740
1N4327B	140A				1N4740A
1N4328	5074A		ZD-11.0	563	1N4741

Industry Standard No.	ECG	SK	GE	RS 276-	MOTOR.
1N4328A	5074A		ZD-11.0	563	1N4741
1N4328B	5074A				1N4741A
1N4329	142A	3062	ZD-12	563	1N4742
1N4329A	142A	3062	ZD-12	563	1N4742
1N4329B	142A	3062	ZD-12	563	1N4742A
1N432A	177	3100/519	300	1122	
1N432B	177	3100/519	300	1122	
1N433	177	3100/519	300	1122	
1N4330	143A	3750	ZD-13	563	1N4743
1N4330A	143A	3750	ZD-13	563	1N4743
1N4330B	143A	3750			1N4743A
1N4331	145A	3063	ZD-15	564	1N4744
1N4331A	145A	3063	ZD-15	564	1N4744
1N4331B	145A	3063	ZD-15	564	1N4744A
1N4332	5075A	3751	ZD-16	564	1N4745
1N4332A	5075A	3751	ZD-16	564	1N4745
1N4332B	5075A	3751			1N4745A
1N4333	5077A	3752	ZD-18		1N4746
1N4333A	5077A	3752	ZD-18		1N4746
1N4333B	5077A	3752			1N4746A
1N4334	5079A	3335	ZD-20		1N4747
1N4334A	5079A	3335	ZD-20		1N4747
1N4334B	5079A				1N4747A
1N4335	5080A	3336	ZD-22		1N4748
1N4335A	5080A	3336	ZD-22		1N4748
1N4335B	5080A				1N4748A
1N4336	5081A		ZD-24		1N4749
1N4336A	5081A		ZD-24		1N4749
1N4336B	5081A				1N4749A
1N4337	146A	3064	ZD-27		1N4750
1N4337A	146A	3064	ZD-27		1N4750
1N4337B	146A	3064	ZD-27		1N4750A
1N4338	5084A	3755	ZD-30		1N4751
1N4338A	5084A	3755	ZD-30		1N4751
1N4338B	5084A	3755			1N4751A
1N4339	147A	3064/146A	ZD-33		1N4752
1N4339A	147A	3095	ZD-33		1N4752
1N4339B	147A	3095	ZD-33		1N4752A
1N433A	177	3100/519	300	1122	
1N433B	177	3100/519	300	1122	
1N434	177	3100/519	300	1122	
1N4340	5085A	3337	ZD-36		1N4753
1N4340A	5085A	3337	ZD-36		1N4753
1N4340B	5085A				1N4753A
1N4341	5086A	3338	ZD-39		1N4754
1N4341A	5086A	3338	ZD-39		1N4754
1N4341B	5086A				1N4754A
1N4342	5087A	3339	ZD-43		1N4755
1N4342A	5087A	3339	ZD-43		1N4755
1N4342B	5087A	3339	ZD-43		1N4755A
1N4343	5088A	3340	ZD-47		1N4756
1N4343A	5088A	3340	ZD-47		1N4756
1N4343B	5088A				1N4756A
1N4344	5089A	3341	ZD-51		1N4757
1N4344A	5089A	3341	ZD-51		1N4757
1N4344B	5089A	3341	ZD-51		1N4757A
1N4345	5090A	3342	ZD-56		1N4758
1N4345A	5090A	3342	ZD-56		1N4758
1N4345B	5090A	3342	ZD-56		1N4758A
1N4346	149A	3097	ZD-62		1N4759
1N4346A	149A	3097	ZD-62		1N4759
1N4346B	149A	3097	ZD-62		1N4759A
1N4347	5092A	3343	ZD-68		1N4760
1N4347A	5092A	3343	ZD-68		1N4760
1N4347B	5092A	3343	ZD-68		1N4760A
1N4348	5093A	3344	ZD-75		1N4761
1N4348A	5093A	3344	ZD-75		1N4761
1N4348B	5093A	3344	ZD-75		1N4761A
1N4349	150A	3098	ZD-82		1N4762
1N4349A	150A	3098	ZD-82		1N4762
1N4349B	150A	3098	ZD-82		1N4762A
1N434A	177	3340/5088A	300	1122	
1N434B	177	3100/519	300	1122	
1N435	109	3087		1123	
1N4350	5095A	3345	ZD-91		1N4763
1N4350A	5095A	3345	ZD-91		1N4763
1N4350B	5095A	3345	ZD-91		1N4763A
1N4351	5096A	3346	ZD-100		1N4764
1N4351A	5096A	3346	ZD-100		1N4764
1N4351B	5096A	3346	ZD-100		1N4764A
1N4352	151A	3099	ZD-110		1M110ZS10
1N4352A	151A	3099	ZD-110		1M110ZS10
1N4352B	151A	3099	ZD-110		1M110ZS5
1N4353	5097A	3347	ZD-120		1M120ZS10
1N4353A	5097A	3347	ZD-120		1M120ZS10
1N4353B	5097A				1M120ZS5
1N4354	5098A	3348	ZD-130		1M130ZS10
1N4354A	5098A	3348	ZD-130		1M130ZS10
1N4354B	5098A	3348	ZD-130		1M130ZS5
1N4355	5100A	3350	ZD-150		1M150ZS10
1N4355A	5100A	3350	ZD-150		1M150ZS10
1N4355B	5100A				1M150ZS5
1N4356	5101A	3351	ZD-160		1M160ZS10
1N4356A	5101A	3351	ZD-160		1M160ZS10
1N4356B	5101A	3351	ZD-160		1M160ZS5
1N4357	5103A	3353	ZD-180		1M180ZS10
1N4357A	5103A	3353	ZD-180		1M180ZS10
1N4357B	5103A	3353	ZD-180		1M180ZS5
1N4358	5105A	3355	ZD-200		1M200ZS10
1N4358A	5105A	3355	ZD-200		1M200ZS10
1N4358B	5105A				1M200ZS5
1N4360	5000A				1N4370A
1N4361	125		510,531	1114	
1N4363	177	3100/519	300	1122	
1N4364	116	9004/5801	504A	1142	1N4002
1N4365	116	9005/5802	504A	1143	1N4003
1N4366	116	9006/5803	504A	1144	1N4004
1N4367	116	9007/5804	504A	1144	1N4004
1N4368	116	9008/5805	504A	1104	1N4005
1N4369	116	3848/5806	504A	1104	1N4005
1N4370	5000A	3766			
1N4370A	5000A	3766			
1N4371	5002A	3768			
1N4371A	5002A	3768			
1N4372	5004A	3770			
1N4372A	5004A	3770			
1N4374					MR1-1600
1N4375	177	3100/519	300	1122	
1N4376			300	1122	
1N4382	177	3100/519	300	1122	
1N4383	116	9005/5802	504A	1143	
1N4384	116	9007/5804	504A	1144	
1N4385	116	3848/5806	504A	1104	
1N4389	178MP		300(2)	1122	
1N4392	177	3100/519	300	1122	
1N4395	177	3100/519	300	1122	
1N4395A	177	3100/519	300	1122	
1N44	109	3087		1123	
1N440	116	3016	504A	1104	
1N440,B					1N4002
1N4400	5071A	3334	ZD-6.8	561	1N4736
1N4400A	5071A	3334	ZD-6.8	561	
1N4401	138A		ZD-7.5		1N4737
1N4401A	138A		ZD-7.5		

Industry Standard No.	ECG	SK	GE	RS 276-	MOTOR.
1N4401B	138A		ZD-7.5		
1N4402	5072A	3311	ZD-8.2	562	1N4738
1N4402A	5072A	3136	ZD-8.2	562	
1N4402B	5072A	3136			
1N4403	139A	3066/118	ZD-9.1	562	1N4739
1N4403A	139A	3066/118	ZD-9.1	562	
1N4403B	139A	3066/118	ZD-9.1	562	
1N4404	140A		ZD-10	562	1N4740
1N4404A	140A		ZD-10	562	
1N4405	5074A		ZD-11.0	563	1N4741
1N4405A	5074A		ZD-11.0	563	
1N4406	142A	3062	ZD-12	563	1N4742
1N4406A	142A	3062	ZD-12	563	
1N4406B	142A	3062	ZD-12	563	
1N4407	143A	3750	ZD-13	563	1N4743
1N4407A	143A	3750	ZD-13	563	
1N4407B	143A	3750			
1N4408	145A	3063	ZD-15	564	1N4744
1N4408A	145A	3063	ZD-15	564	
1N4408B	145A	3063	ZD-15	564	
1N4409	5075A	3751	ZD-16	564	1N4745
1N4409A	5075A	3751	ZD-16	564	
1N440B	116	3016	504A	1104	
1N441	116	3016	504A	1104	
1N441,B					1N4003
1N4410	5077A	3752	ZD-18		1N4746
1N4410A	5077A	3752	ZD-18		
1N4410B	5077A	3752			
1N4411	5079A	3335	ZD-20		1N4747
1N4411A	5079A	3335	ZD-20		
1N4412	5080A	3336	ZD-22		1N4748
1N4412A	5080A	3336	ZD-22		
1N4413	5081A		ZD-24		1N4749
1N4413A	5081A		ZD-24		
1N4414	146A	3064	ZD-27		1N4750
1N4414A	146A	3064	ZD-27		
1N4414B	146A	3064	ZD-27		
1N4415	5084A	3755	ZD-30		1N4751
1N4415A	5084A	3755	ZD-30		
1N4416	147A	3095	ZD-33		1N4752
1N4416A	147A	3095	ZD-33		
1N4416B	147A	3095	ZD-33		
1N4417	5085A	3337	ZD-36		1N4753
1N4417A	5085A	3337	ZD-36		
1N4418	5086A	3338	ZD-39		1N4754
1N4418A	5086A	3338	ZD-39		
1N4419	5087A	3339	ZD-43		1N4755
1N4419A	5087A	3339	ZD-43		
1N4419B	5087A	3339	ZD-43		
1N441B	116	3016	504A	1104	
1N442	116	3016	504A	1104	
1N442,B					1N4004
1N4420	5088A	3340	ZD-47		1N4756
1N4420A	5088A	3340	ZD-47		
1N4421	5089A	3341	ZD-51		1N4757
1N4421A	5089A	3341	ZD-51		
1N4422	5090A	3342	ZD-56		1N4758
1N4422A	5090A	3342	ZD-56		
1N4422B	5090A	3342	ZD-56		
1N4423	149A	3097	ZD-62		1N4759
1N4423A	149A	3097	ZD-62		
1N4423B	149A	3097	ZD-62		
1N4424	5092A	3343	ZD-68		1N4760
1N4424A	5092A	3343	ZD-68		
1N4424B	5092A	3343	ZD-68		
1N4425	5093A	3344	ZD-75		1N4761
1N4425A	5093A	3344	ZD-75		
1N4425B	5093A	3344	ZD-75		
1N4426	150A	3098	ZD-82		1N4762
1N4426A	150A	3098	ZD-82		
1N4426B	150A	3098	ZD-82		
1N4427	5095A	3345	ZD-91		1N4763
1N4427A	5095A	3345	ZD-91		
1N4428	5096A	3346	ZD-100		1N4764
1N4428A	5096A	3346	ZD-100		
1N4428B	5096A	3346	ZD-100		
1N4429	151A	3099	ZD-110		1M110ZS10
1N4429A	151A	3099	ZD-110		
1N4429B	151A	3099	ZD-110		
1N442B	116	3016	504A	1104	
1N443	116	3016	504A	1104	
1N443,B					1N4004
1N4430	5097A	3347	ZD-120		1M120ZS10
1N4430A	5097A	3347	ZD-120		
1N4431	5098A	3348	ZD-130		1M130ZS10
1N4431A	5098A	3348	ZD-130		
1N4431B	5098A	3348	ZD-130		
1N4432	5100A	3350	ZD-150		1M150ZS10
1N4432A	5100A	3350	ZD-150		
1N4433	5101A	3351	ZD-160		1M160ZS10
1N4433A	5101A	3351	ZD-160		
1N4434	5103A	3353	ZD-180		1M180ZS10
1N4434A	5103A	3353	ZD-180		
1N4434B	5103A	3353	ZD-180		
1N4435	5105A	3355	ZD-200		1M200ZS10
1N4435A	5105A	3355	ZD-200		
1N443B	116	3016	504A	1104	
1N444	116	3016	504A	1104	
1N444,B					1N4005
1N4444			300	1122	
1N4446	519	3100	514	1122	
1N4447	519	3100	514	1122	
1N4448	519	3100	514	1122	
1N4449	519	3100	514	1122	
1N444B	116	3100/519	504A	1104	
1N445	116	3033	504A	1104	
1N445,B					1N4005
1N4450	177	3100/519	300	1122	
1N4451			300	1122	
1N4453	177	3100/519	300	1122	
1N4454	519	3100	514	1122	
1N4455	177	3100/519	300	1122	
1N4458	5890	7090	5028		
1N4459	5868	7068	5028		
1N445B	116	3033	504A	1104	
1N4460	5119A	3385			1N4735A
1N4461	5120A	3386	5ZD-6.8		1N4736A
1N4462	5121A	3387	5ZD-7.5		1N4737A
1N4463	5122A	3388	5ZD-8.2		1N4738A
1N4464	5124A	3390	5ZD-9.1		1N4739A
1N4465	5125A	3391	5ZD-10		1N4740A
1N4466	5126A	3392	5ZD-11		1N4741A
1N4467	5127A	3393	5ZD-12		1N4742A
1N4468	5128A	3394	5ZD-13		1N4743A
1N4469	5130A	3396	5ZD-15		1N4744A
1N447	109	3087		1123	
1N4470	5131A	3397	5ZD-16		1N4745A
1N4471	5133A	3399	5ZD-18		1N4746A
1N4472	5135A	3401	5ZD-20		1N4747A
1N4473	5136A	3402	5ZD-22		1N4748A
1N4474	5137A	3403	5ZD-24		1N4749A
1N4475	5139A	3405	5ZD-27		1N4750A
1N4476	5141A	3407	5ZD-30		1N4751A
1N4477	5142A	3408	5ZD-33		1N4752A

Industry Standard No.	ECG	SK	GE	RS 276-	MOTOR.
1N4478	5143A	3409			1N4753A
1N4479	5144A	3410			1N4754A
1N448	116	3017B/117	504A	1104	
1N4480	5145A	3411			1N4755A
1N4481	5146A	3412			1N4756A
1N4482	5147A	3413			1N4757A
1N4483	5148A	3414			1N4758A
1N4484	5150A	3416			1N4759A
1N4485	5151A	3417			1N4760A
1N4486	5152A	3418			1N4761A
1N4487	5153A	3419			1N4762A
1N4488	5155A	3421			1N4763A
1N4489	5156A	3422			1N4764A
1N449	109	3016		1123	
1N4490	5157A	3423			1M110ZS5
1N4491	5158A	3424			1M120ZS5
1N4492	5159A	3425			1M130ZS5
1N4493	5161A	3427			1M150ZS5
1N4494	5162A	3428			1M160ZS5
1N4495	5164A	3430			1M180ZS5
1N4496	5166A	3432			1M200ZS5
1N4499	137A	3058	ZD-6.2	561	1N4735A
1N45	109	3087		1123	
1N450	116	3017B/117	504A	1104	
1N4501		3848	ZD-6.8	561	
1N4502	109	3848		1123	
1N4503	5142A	3408			1N4752
1N4504	5166A	3432			1N5388A
1N4506	5874		5032		SPECIAL
1N4507	5878		5036		SPECIAL
1N4508	5882		5040		SPECIAL
1N4509	5886		5044		
1N451	116	3016	504A	1104	
1N4510	5890	7090	5044		
1N4514	5808	9009	509	1114	
1N4517	5802	9005		1143	
1N452	109	3087		1123	
1N4523	109			1123	
1N4525	5986	3609	5096		
1N4526	5990	3608	5100		
1N4527	5994	3501	5104		
1N4528	5998		5108		
1N4529	6002	7202	5108		
1N453	116	3779/5013A	504A	1104	
1N4531	177	3100/519	300	1122	
1N4532	177	3100/519	300	1122	
1N4533	177	3100/519	300	1122	
1N4534	177	3100/519	300	1122	
1N4535		3330	ZD-3.3		
1N4536	177	3100/519	300	1122	
1N454	109	3087		1123	
1N4547	177	3100/519	300	1122	
1N4548	177	3100/519	300	1122	
1N455	109	3087		1123	
1N456	177	3100/519	300	1122	
1N4565			ZD-6.2	561	
1N4565A			ZD-6.2	561	
1N4566			ZD-6.2	561	
1N4566A			ZD-6.2	561	
1N4567			ZD-6.2	561	
1N4567A			ZD-6.2	561	
1N4568			ZD-6.2	561	
1N4568A			ZD-6.2	561	
1N4569			ZD-6.2	561	
1N4569A			ZD-6.2	561	
1N456A	177	3016	300	1122	
1N457	177	3016	300	1122	
1N4570			ZD-6.2	561	
1N4570A			ZD-6.2	561	
1N4571			ZD-6.2	561	
1N4571A			ZD-6.2	561	
1N4572			ZD-6.2	561	
1N4572A			ZD-6.2	561	
1N4573			ZD-6.2	561	
1N4573A			ZD-6.2	561	
1N4574			ZD-6.2	561	
1N4574A			ZD-6.2	561	
1N4575			ZD-6.2	561	
1N4575A			ZD-6.2	561	
1N4576			ZD-6.2	561	
1N4576A			ZD-6.2	561	
1N4577			ZD-6.2	561	
1N4577A			ZD-6.2	561	
1N4578			ZD-6.2	561	
1N4578A			ZD-6.2	561	
1N4579			ZD-6.2	561	
1N4579A			ZD-6.2	561	
1N457A	177	3016	300	1122	
1N457M	177	3100/519	300	1122	
1N458	177	3016	300	1122	
1N4580			ZD-6.2	561	
1N4580A			ZD-6.2	561	
1N4581			ZD-6.2	561	
1N4581A			ZD-6.2	561	
1N4582			ZD-6.2	561	
1N4582A			ZD-6.2	561	
1N4583		3092	ZD-11.5	561	
1N4583A		3092	ZD-11.5	561	
1N4583B		3092	ZD-6.2	561	
1N4584			ZD-6.2	561	
1N4584A			ZD-6.2	561	
1N4585	5808	9009	509	1114	
1N4586	5808	9009	504A	1104	
1N458A	177	3016	300	1122	
1N458M	177	3100/519	300	1122	
1N459	177	3016	300	1122	
1N4593	6158	7358			
1N4593R	6159	7359			
1N4594	6158	7358			
1N459A	177	3016	300	1122	
1N459M	177	3100/519	300	1122	
1N45A	109	3087		1123	
1N46	109	3087		1123	
1N460	177	3016	300	1122	
1N4606			300	1122	
1N4607			300	1122	
1N4608			300	1122	
1N4609	613	3326			
1N4609A	613	3326			
1N460A	177	3016	300	1122	
1N460B	177	3100/519	300	1122	
1N461	177	3016	300	1122	
1N4610			300	1122	
1N4611			ZD-6.6	561	1N4576A
1N4611A					1N4577A
1N4611B					1N4578A
1N4611C					1N4579A
1N4612			ZD-6.6	561	1N4581A
1N4612A			ZD-6.6	561	1N4582A
1N4612B			ZD-6.6	561	1N4583A
1N4612C			ZD-6.6	561	1N4584A
1N4613			ZD-6.6	561	1N4581A
1N4613A					1N4582A
1N4613B					1N4583A
1N4613C					1N4584A
1N461A	177	3016	300	1122	
1N462	177	3016	300	1122	
1N4620		3330	ZD-3.3		
1N4622		3331	ZD-3.9		
1N4623		3332	ZD-4.3		
1N4624		3333	ZD-4.7		
1N4625		3776	ZD-5.1		
1N4626			ZD-5.6	561	
1N4627			ZD-6.2	561	
1N4628	5071A	3334	ZD-6.8	561	1N4736A
1N4629	138A		ZD-7.5		1N4737A
1N462A	177	3016	300	1122	
1N463	177	3016	300	1122	
1N4630	5072A	3136	ZD-8.2	562	1N4738A
1N4631	139A		ZD-9.1	562	1N4739A
1N4632	140A		ZD-10	562	1N4740A
1N4633	5074A		ZD-11.0	563	1N4741A
1N4634	142A	3062	ZD-12	563	1N4742A
1N4635	143A	3750	ZD-13	563	1N4743A
1N4636	145A	3063	ZD-15	564	1N4744A
1N4637	5075A	3751	ZD-16	564	1N4745A
1N4638	5077A	3752	ZD-18		1N4746A
1N4639	5079A	3335	ZD-20		1N4747A
1N463A	177	3016	300	1122	
1N464	177	3016	300	1122	
1N4640	5080A	3336	ZD-22		1N4748A
1N4641	5081A		ZD-24		1N4749A
1N4642	146A	3064	ZD-27		1N4750A
1N4643	5084A	3755	ZD-30		1N4751A
1N4644	147A		ZD-33		1N4752A
1N4645	5085A	3337	ZD-36		1N4753A
1N4646	5086A	3338	ZD-39		1N4754A
1N4647	5087A	3339	ZD-43		1N4755A
1N4648	5088A	3340	ZD-47		1N4756A
1N4649		3330	ZD-3.3		1N4728A
1N464A	177	3016	300	1122	
1N465	5002A	3768			1N5223A
1N4650		3055	ZD-3.6		1N4729A
1N4651		3331	ZD-3.9		1N4730A
1N4652		3332	ZD-4.3		1N4731A
1N4653		3333	ZD-4.7		1N4732A
1N4654		3056	ZD-5.1		1N4733A
1N4655		3057	ZD-5.6	561	1N4734A
1N4656		3058	ZD-6.2	561	1N4735A
1N4657		3334	ZD-6.8	561	1N4736A
1N4658			ZD-7.5		1N4737A
1N4659		3136	ZD-8.2	562	1N4738A
1N465A	5002A	3768			1N5223B
1N466	5006A	3772	ZD-3.6		1N5226A
1N4660		3060	ZD-9.1	562	1N4739A
1N4661			ZD-10	562	1N4740A
1N4662			ZD-11.0	563	1N4741A
1N4663		3062	ZD-12	563	1N4742A
1N4664		3750	ZD-13	563	1N4743A
1N4665		3063	ZD-15	564	1N4744A
1N4666		3751	ZD-16	564	1N4745A
1N4667		3752	ZD-18		1N4746A
1N4668		3335	ZD-20		1N4747A
1N4669		3336	ZD-22		1N4748A
1N466A	5066A		ZD-3.6		1N5226B
1N467	5007A	3773	ZD-3.9		1N5228B
1N4670			ZD-24		1N4749A
1N4671		3064	ZD-27		1N4750A
1N4672		3755	ZD-30		1N4751A
1N4673		3095	ZD-33		1N4752A
1N4674		3337	ZD-36		1N4753A
1N4675		3338	ZD-39		1N4754A
1N4676		3339	ZD-43		1N4755A
1N4677		3340	ZD-47		1N4756A
1N467A	5007A	3773	ZD-3.9		1N5228B
1N468	5009A	3775	ZD-4.7		1N5230A
1N4681	5000A	3766			
1N4682	5063A	3837			
1N4683	5065A	3838			
1N4684	5066A	3330	ZD-3.3		
1N4685	134A		ZD-3.6		
1N4686	5067A	3331	ZD-3.9		
1N4687	5068A	3332	ZD-4.3		
1N4688	5069A	3333	ZD-4.7		
1N4689	5010A	3776	ZD-5.1		
1N468A	5009A	3775	ZD-4.7		1N5230B
1N468B		3775	ZD-5.1		
1N469	5014A	3780	ZD-6.8	561	1N5232B
1N4690	136A	3057	ZD-5.6	561	
1N4691	137A		ZD-6.2	561	
1N4692	5071A	3334	ZD-6.8	561	
1N4693	5015A	3781	ZD-7.5		
1N4694	5072A	3136	ZD-8.2	562	
1N4695	5073A	3749	ZD-9.1	562	
1N4697	140A		ZD-10	562	
1N4698	5074A		ZD-11	563	
1N4699	142A		ZD-12	563	
1N469A	5014A	3780	ZD-6.8	561	1N5232B
1N46A	109	3088		1123	
1N47	109	3087		1123	
1N470	5014A	3780	ZD-6.8	561	1N5235B
1N4700	143A	3750	ZD-13	563	
1N4701	144A	3094	ZD-14	564	
1N4702	145A	3063	ZD-15	564	
1N4703	5075A	3751	ZD-16	564	
1N4704	5076A	9022	ZD-17		
1N4705	5077A	3752	ZD-18		
1N4706	5078A		ZD-19		
1N4707	5079A	3335	ZD-20		
1N4708	5080A	3336	ZD-22		
1N4709	5081A		ZD-24		
1N470A	5014A	3780	ZD-6.8	561	1N5235B
1N470B		3780	ZD-6.8	561	
1N471		3055	ZD-3.6		
1N4710	5032A	3798	ZD-25		
1N4711	146A	3064	ZD-27		
1N4712	5083A	3754	ZD-28		
1N4713	5084A	3755	ZD-30		
1N4714	147A	3095	ZD-33		
1N4715	5085A	3337	ZD-36		
1N4716	5086A	3338	ZD-39		
1N4717	5087A	3339	ZD-43		
1N4718	507	3051/156	504A		
1N4719	5800	9003	512	1142	
1N4719R			512	1141	
1N472		3774	ZD-4.3		
1N4720	5801	9004	504A	1142	
1N4720R			512	1143	
1N4721	5802	9005	512	1143	
1N4721R	5802		512	1143	
1N4722	5804	9007	512	1144	
1N4722R			512	1144	
1N4723	5806	3848	512		
1N4724	5808	9009	512		
1N4725	5809	9010	512		
1N4726	177	3100/519	300	1122	
1N4727	177	3100/519	300	1122	
1N4728	5066A	3330	ZD-3.3		

Industry Standard No.	ECG	SK	GE	RS 276-	MOTOR.
1N4728A	5066A	3330	ZD-3.3		
1N4729	134A	3055	ZD-3.6		
1N4729A	134A	3055	ZD-3.6		
1N472A		3774	ZD-4.3		
1N472B		3774	ZD-4.3		
1N473		3056	ZD-5.1		
1N4730	5067A	3331	ZD-3.9		
1N4730A	5067A	3331	ZD-3.9		
1N4731	5068A	3332	ZD-4.3		
1N4731A	5068A	3332	ZD-4.3		
1N4732	5069A	3333	ZD-4.7		
1N4732A	5069A	3333	ZD-4.7		
1N4733	135A	3056	ZD-5.1		
1N4733A	135A	3056	ZD-5.1		
1N4734	136A	3057	ZD-5.6	561	
1N4734A	136A	3057	ZD-5.6	561	
1N4735	137A	3058	ZD-6.2	561	
1N4735A	137A	3058	ZD-6.2	561	
1N4736	5071A	3334	ZD-6.8	561	
1N4736A	5071A	3334	ZD-6.8	561	
1N4737	138A	3059	ZD-7.5		
1N4737A	138A	3059	ZD-7.5		
1N4738	5072A	3136	ZD-8.2	562	
1N4738A	5072A	3136	ZD-7.5	562	
1N4739	139A	3060	ZD-9.1	562	
1N4739A	139A	3060	ZD-9.1	562	
1N473A		3056	ZD-5.1		
1N473B		3056	ZD-5.1		
1N474		3778	ZD-6.0	561	
1N4740	140A	3061	ZD-10	562	
1N4740A	140A	3061	ZD-10	562	
1N4741	5074A	3139	ZD-11.0	563	
1N4741A	5074A	3139	ZD-11	563	
1N4742	142A	3062	ZD-12	563	
1N4742A	142A	3062	ZD-12	563	
1N4743	143A	3750	ZD-12	563	
1N4743A	143A	3750	ZD-12	563	
1N4744	145A	3063	ZD-15	564	
1N4744A	145A	3063	ZD-15	564	
1N4745	5075A	3751	ZD-16	564	
1N4745A	5075A	3751	ZD-16	564	
1N4746	5077A	3752	ZD-18		
1N4746A	5077A	3752	ZD-18		
1N4747	5079A	3335	ZD-20		
1N4747A	5079A	3335	ZD-20		
1N4748	5080A	3336	ZD-22		
1N4748A	5080A	3336	ZD-22		
1N4749	5081A	3151	ZD-24		
1N4749A	5081A	3151	ZD-24		
1N474A		3778	ZD-6.0	561	
1N474B		3778	ZD-6.0	561	
1N475		3780	ZD-6.8	561	
1N4750	146A	3064	ZD-27		
1N4750A	146A	3064	ZD-27		
1N4751	5084A	3755	ZD-30		
1N4751A	5084A	3755	ZD-30		
1N4752	147A	3095	ZD-33		
1N4752A	147A	3095	ZD-33		
1N4753	5085A	3337	ZD-36		
1N4753A	5085A	3337	ZD-36		
1N4754	5086A	3338	ZD-39		
1N4754A	5086A	3338	ZD-39		
1N4755	5087A	3339	ZD-43		
1N4755A	5087A	3339	ZD-43		
1N4756	5088A	3340	ZD-47		
1N4756A	5088A	3340	ZD-47		
1N4757	5089A	3341	ZD-51		
1N4757A	5089A	3341	ZD-51		
1N4758	5090A	3342	ZD-56		
1N4758A	5090A	3342	ZD-56		
1N4759	149A	3097	ZD-62		
1N4759A	149A	3097	ZD-62		
1N475A		3780	ZD-6.8	561	
1N475B		3780	ZD-6.8	561	
1N476	109	3087		1123	
1N4760	5092A	3343	ZD-68		
1N4760A	5092A	3343	ZD-68		
1N4761	5093A	3344	ZD-75		
1N4761A	5093A	3344	ZD-75		
1N4762	150A	3098	ZD-82		
1N4762A	150A	3098	ZD-82		
1N4763	5095A	3345	ZD-91		
1N4763A	5095A	3345	ZD-91		
1N4764	5096A	3346	ZD-100		
1N4764A	5096A	3346	ZD-100		
1N4765			ZD-9.1	562	
1N4765A			ZD-9.1	562	
1N4766			ZD-9.1	562	
1N4766A			ZD-9.1	562	
1N4767			ZD-9.1	562	
1N4767A			ZD-9.1	562	
1N4768			ZD-9.1	562	
1N4768A			ZD-9.1	562	
1N4769			ZD-9.1	562	
1N4769A			ZD-9.1	562	
1N477	109	3087		1123	
1N4770		3784	ZD-9.1	562	
1N4770A		3784	ZD-9.1	562	
1N4771			ZD-9.1	562	
1N4771A			ZD-9.1	562	
1N4772			ZD-9.1	562	
1N4772A			ZD-9.1	562	
1N4773			ZD-9.1	562	
1N4773A			ZD-9.1	562	
1N4774			ZD-9.1	562	
1N4774A			ZD-9.1	562	
1N4775			ZD-8.7	562	
1N4775A			ZD-8.7	562	
1N4776			ZD-8.7	562	
1N4776A			ZD-8.7	562	
1N4777			ZD-8.7	562	
1N4777A			ZD-8.7	562	
1N4778			ZD-8.7	562	
1N4778A			ZD-8.7	562	
1N4779			ZD-8.7	562	
1N4779A			ZD-8.7	562	
1N478	109	3087		1123	
1N4780			ZD-8.7	562	
1N4780A			ZD-8.7	562	
1N4781			ZD-8.7	562	
1N4781A			ZD-8.7	562	
1N4782			ZD-8.7	562	
1N4782A			ZD-8.7	562	
1N4783			ZD-8.7	562	
1N4783A			ZD-8.7	562	
1N4784			ZD-8.7	562	
1N4784A			ZD-8.7	562	
1N4785	127	3113/516			
1N4786	610	3323			1N5441A
1N4786A	610	3323			
1N4787					1N5442A
1N4788	611	3324			1N5443A
1N4788A	611	3324			
1N4789	612	3325			1N5444A
1N4789A	612	3325			
1N479	109	3087		1123	
1N4790					1N5445A
1N4791					1N5446A
1N4792	613	3326			1N5448A
1N4792A	613	3326			
1N4793					1N5449A
1N4794	614	3327			1N5450A
1N4794A	614	3327			
1N4795					1N5451A
1N4796					1N5452A
1N4797					1N5453A
1N4798					1N5454A
1N4799					1N5455A
1N48	109	3087		1123	
1N480	109			1123	
1N4800					1N5446A
1N4801	610	3323			1N5139
1N4801A	610	3323			
1N4802					1N5462A
1N4803					1N5140
1N4804					1N5141
1N4806					1N5143
1N4807					1N5144
1N4808					1N5145
1N4809					1N5146
1N480B	116		504A	1104	
1N4810					1N5147
1N4811					1N5148
1N4812					1N5148
1N4813					1N5454A
1N4814					1N5455A
1N4815					1N5456A
1N4816	5800	9003	504A	1142	1N5391
1N4816GP					1N5391
1N4817	5801	9004	510	1114	1N5392
1N4817GP					1N5392
1N4818	125	3081	510,531	1114	1N5393
1N4818GP					1N5393
1N4819	5803	9006	510	1144	1N5395
1N4819GP					1N5395
1N482	177	3016	300	1122	
1N4820	5804	9007	504A	1144	1N5395
1N4820GP					1N5395
1N4821	5805	9008	504A	1104	1N5396
1N4821GP					1N5396
1N4822	5806	3848	504A	1104	1N5397
1N4822GP					1N5397
1N4826	506	3843	511	1114	
1N4827	177	3100/519	300	1122	
1N4828	177		300	1122	
1N4829	177		300	1122	
1N482A	177	3016	300	1122	
1N482B	177	3016	300	1122	
1N482C	177		300	1122	
1N483	177	3016	300	1122	
1N4830	177		300	1122	
1N4831	5124A		ZD-9.1	562	1N4739
1N4831A	5124A		ZD-9.1	562	1N4739
1N4831B					1N4739A
1N4832	5125A				1N4740
1N4832A	5125A		ZD-10	562	1N4740
1N4832B	5125A				1N4740A
1N4833	5126A		ZD-11.0	563	1N4741
1N4833A	5126A		ZD-11.0	563	1N4741
1N4833B	5126A				1N4741A
1N4834	5127A		ZD-12	563	1N4742
1N4834A	5127A		ZD-12	563	1N4742
1N4834B	5127A				1N4742A
1N4835	5128A	3750/143A	ZD-13	563	1N4743
1N4835A	5128A	3750/143A	ZD-13	563	1N4743
1N4835B	5128A				1N4743A
1N4836	5130A		ZD-15	564	1N4744
1N4836A	5130A		ZD-15	564	1N4744
1N4836B	5130A				1N4744A
1N4837	5131A	3751/5075A	ZD-16	564	1N4745
1N4837A	5131A	3751/5075A	ZD-16	564	1N4745
1N4837B	5131A				1N4745A
1N4838	5133A	3752/5077A	ZD-18		1N4746
1N4838A	5133A		ZD-18		1N4746
1N4838B	5133A				1N4746A
1N4839	5135A	3335/5079A	ZD-20		1N4747
1N4839A	5135A	3335/5079A	ZD-20		1N4747
1N4839B	5135A				1N4747A
1N483A	177	3016	300	1122	
1N483AM	177	3100/519	300	1122	
1N483B	177	3016	300	1122	
1N483BM	177	3100/519	300	1122	
1N483C	177	3100/519	300	1122	
1N484	177	3016	300	1122	
1N4840	5136A	3336/5080A	ZD-22		1N4748
1N4840A					1N4748
1N4840B	5136A				1N4748A
1N4841	5137A		ZD-24		1N4749
1N4841A	5137A		ZD-24		1N4749
1N4841B	5137A				1N4749A
1N4842	5139A	3064/146A	ZD-27		1N4750
1N4842A	5139A	3064/146A	ZD-27		1N4750
1N4842B	5139A	3064/146A	ZD-27		1N4750A
1N4843	5141A	3755/5084A	ZD-30		1N4751
1N4843A	5141A	3755/5084A	ZD-30		1N4751
1N4843B	5141A				1N4751A
1N4844	5142A	3095/147A	ZD-33		1N4752
1N4844A	5142A	3095/147A	ZD-33		1N4752
1N4844B	5142A	3095/147A	ZD-33		1N4752A
1N4845	5143A	3337/5085A	ZD-36		1N4753
1N4845A	5143A	3337/5085A	ZD-36		1N4753
1N4845B	5143A				1N4753A
1N4846	5144A	3338/5086A	ZD-39		1N4754
1N4846A	5144A	3338/5086A	ZD-39		1N4754
1N4846B	5144A				1N4754A
1N4847	5145A	3339/5087A	ZD-43		1N4755
1N4847A	5145A	3339/5087A	ZD-43		1N4755
1N4847B	5145A				1N4755A
1N4848	5146A	3340/5088A	ZD-47		1N4756
1N4848A	5146A	3340/5088A	ZD-47		1N4756
1N4848B	5146A				1N4756A
1N4849	5147A	3341/5089A	ZD-51		1N4757
1N4849A	5147A	3341/5089A	ZD-51		1N4757
1N4849B	5147A				1N4757A
1N484A	177	3016	300	1122	
1N484B	177	3016	300		
1N484C	177	3100/519	300	1122	
1N485	177	3311	300	1122	
1N4850	5148A	3342/5090A	ZD-56		1N4758
1N4850A	5148A	3342/5090A	ZD-56		1N4758
1N4850B	5148A				1N4758A
1N4851	5150A		ZD-62		1N4759
1N4851A	5150A		ZD-62		1N4759
1N4851B					1N4759A
1N4852	5151A	3343/5092A	ZD-68		1N4760
1N4852A	5151A	3343/5092A	ZD-68		1N4760
1N4852B	5151A				1N4760A
1N4853	5152A	3344/5093A	ZD-75		1N4761

Industry Standard No.	ECG	SK	GE	RS 276-	MOTOR.
1N4853A	5152A	3344/5093A	ZD-75		1N4761
1N4853B	5152A				1N4761A
1N4854	5153A	3098/150A	ZD-82		1N4762
1N4854A	5153A	3098/150A	ZD-82		1N4762
1N4854B	5153A	3098/150A	ZD-82		1N4762A
1N4855	5155A	3345/5095A	ZD-91		1N4763
1N4855A	5155A	3345/5095A	ZD-91		1N4763
1N4855B	5155A				1N4763A
1N4856	5156A	3346/5096A	ZD-100		1N4764
1N4856A	5156A	3346/5096A	ZD-100		1N4764
1N4856B	5156A				1N4764A
1N4857	5157A	3099/151A	ZD-110		1M110ZS10
1N4857A	5157A	3099/151A	ZD-110		1M110ZS10
1N4857B	5157A	3099/151A	ZD-110		1M110ZS5
1N4858	5158A	3347/5097A	ZD-120		1M120ZS10
1N4858A	5097A	3347	ZD-120		1M120ZS10
1N4858B	5158A				1M120ZS5
1N4859	5098A	3348	ZD-130		1M130ZS10
1N4859A	5098A	3348	ZD-130		1M130ZS10
1N4859B	5159A				1M130ZS5
1N485A	177	3311	300	1122	
1N485B	177	3311	300	1122	
1N485C	177	3311	300	1122	
1N486	177	3016	300	1122	
1N4860	5100A	3350	ZD-150		1M150ZS10
1N4860A	5100A	3350	ZD-150		1M150ZS10
1N4860B	5161A				1M150ZS5
1N4861	177	3100/519	300	1122	
1N4862	177	3100/519	300	1122	
1N4863	177	3100/519	300	1122	
1N4864			300	1122	
1N486A	177	3016	300	1122	
1N486B	177	3031A	300	1122	
1N487	506	3843	511	1114	
1N4878	6154	7354			
1N4879	6354	6554			
1N487A	506	3843	511	1114	
1N487B	506	3843	511	1114	
1N488	506	3843	511	1114	
1N4880	6354	6554			
1N4881	5135A	3401			1N4747
1N4882	5143A	3409			1N4753
1N4883	5127A	3393			1N4742A
1N4884	5135A	3401			1N4747A
1N4889	5150A	3416			1N3000B
1N488A	506	3843	511	1114	
1N488B	506	3843	511	1114	
1N4890					MZ640
1N4890A					MZ640
1N4891					MZ640
1N4891A					MZ640
1N4892					MZ620
1N4892A					MZ620
1N4893					MZ620
1N4893A					MZ620
1N4894					MZ610
1N4894A					MZ610
1N4895					MZ610
1N4895A					MZ610
1N4896		3062	ZD-12	563	
1N4896A		3093	ZD-12	563	
1N4897		3093	ZD-13	563	
1N4897A		3093	ZD-13	563	
1N4898		3093	ZD-13	563	
1N4898A		3093	ZD-13	563	
1N4899		3093	ZD-13	563	
1N4899A		3093	ZD-13	563	
1N48A	109	3087		1123	
1N49		3087	1N34AS	1123	
1N4900		3093	ZD-13	563	
1N4900A		3093	ZD-13	563	
1N4901		3093	ZD-13	563	
1N4901A		3093	ZD-13	563	
1N4902		3093	ZD-13	563	
1N4902A		3093	ZD-13	563	
1N4903		3093	ZD-13	563	
1N4903A		3093	ZD-13	563	
1N4904		3093	ZD-13	563	
1N4904A		3093	ZD-13	563	
1N4905		3093	ZD-13	563	
1N4905A		3093	ZD-13	563	
1N4906		3093	ZD-13	563	
1N4906A		3093	ZD-13	563	
1N4907		3093	ZD-13	563	
1N4907A		3093	ZD-13	563	
1N4908		3093	ZD-13	563	
1N4908A		3093	ZD-13	563	
1N4909		3093	ZD-13	563	
1N4909A		3093	ZD-13	563	
1N4910		3093	ZD-13	563	
1N4910A		3093	ZD-13	563	
1N4911		3093	ZD-13	563	
1N4911A		3093	ZD-13	563	
1N4912		3093	ZD-13	563	
1N4912A		3093	ZD-13	563	
1N4913		3093	ZD-13	563	
1N4913A		3093	ZD-13	563	
1N4914		3093	ZD-13	563	
1N4914A		3093	ZD-13	563	
1N4915		3093	ZD-13	563	
1N4915A		3093	ZD-13	563	
1N492	141A	3092	ZD-11.5		
1N492A	141A	3092	ZD-11.5		
1N492B	141A	3092	ZD-11.5		
1N493	141A	3092	ZD-11.5		
1N4933	507	3100/519	504A		
1N4933GP					1N4933
1N4934	506	3843	511	1114	
1N4934GP					1N4934
1N4935	506	3843	511	1114	
1N4935GP					1N4935
1N4936	506	3843	511	1114	
1N4936GP					1N4936
1N4937	116	3125	511	1114	
1N4937GP					1N4937
1N4938	177		300	1122	
1N493A	141A	3092	ZD-11.5		
1N493B	141A	3092	ZD-11.5		
1N4942	552	3051/156			1N4935
1N4943		3051			1N4936
1N4944	552	3051/156			1N4936
1N4945	506	3051/156	511	1114	1N4937
1N4946	552	3051/156	511	1114	1N4937
1N4947	506	3051/156	511	1114	MR817
1N4948	506	3051/156	511	1114	MR818
1N4949	177	3100/519	300	1122	
1N4950	177		300	1122	
1N4951	178MP		300(2)	1122(2)	
1N4952	178MP		300(2)	1122(2)	
1N4954	5120A		5ZD-6.8		1N5342B
1N4955	5121A	3387	5ZD-7.5		1N5343B
1N4956	5122A	3388	5ZD-8.2		1N5344B
1N4957	5124A	3390	5ZD-9.1		1N5346B
1N4958	5125A	3391	5ZD-10		1N5347B
1N4959	5126A	3392	5ZD-11		1N5348B
1N4960	5127A	3393	5ZD-12		1N5349B
1N4960D	5127A	3393			
1N4961	5128A	3394	5ZD-13		1N5350B
1N4962	5130A	3396			1N5352B
1N4963	5131A	3397	5ZD-16		1N5353B
1N4964	5133A	3399	5ZD-18		1N5355B
1N4965	5135A	3401	5ZD-20		1N5357B
1N4966	5136A	3402	5ZD-22		1N5358B
1N4967	5137A	3403	5ZD-24		1N5359B
1N4968	5139A	3405	5ZD-27		1N5361B
1N4969	5141A	3407	5ZD-30		1N5363B
1N497	109	3087		1123	
1N4970	5142A	3408	5ZD-33		1N5364B
1N4971	5143A	3409			1N5365B
1N4972	5144A	3410			1N5366B
1N4973	5145A	3411			1N5367B
1N4974	5146A	3412			1N5368B
1N4975	5147A	3413			1N5369B
1N4976	5148A	3414			1N5370B
1N4977	5150A	3416			1N5372B
1N4978	5151A	3417			1N5373B
1N4979	5152A	3418			1N5374B
1N498	109	3087		1123	
1N4980	5153A	3419			1N5375B
1N4981	5155A	3421			1N5377B
1N4982	5156A	3422			1N5378B
1N4983	5157A	3423			1N5379B
1N4984	5158A	3424			1N5380B
1N4985	5159A	3425			1N5381B
1N4986	5161A	3427			1N5383B
1N4987	5162A	3428			1N5384B
1N4988	5164A	3430			1N5386B
1N4989	5166A	3432			1N5388B
1N4997	5800	9003	512	1142	
1N4997R			512	1141	
1N4998	5801	9004	512	1142	
1N4998R			512	1143	
1N4999	5802	9005	512	1143	
1N4999R			512	1143	
1N50	109	3087		1123	
1N500	109	3087		1123	
1N5000	5804	9007	512	1144	
1N5000R			512	1144	
1N5001	5806	3848	512		
1N5002	5808	9009	512		
1N5003	5809	9010	512		
1N5004	5801	9004	504A	1142	1N5392
1N5005	5802	9005	504A	1143	1N5393
1N5006	5804	9007	504A	1144	1N5395
1N5007	5806	3848	504A	1104	1N5397
1N5008	5111A	3377			1N4728
1N5008A	5111A	3377			1N4728A
1N5009	5112A	3378			1N4729
1N5009A	5112A	3378			1N4729A
1N5010	5113A	3379			1N4730
1N5010A	5113A	3379			1N4730A
1N5011	5114A	3380			1N4731
1N5011A	5114A				1N4731A
1N5012	5115A	3381			1N4732
1N5012A	5115A	3381			1N4732A
1N5013	5116A	3382			1N4733
1N5013A	5116A				1N4733A
1N5014	5117A	3383			1N4734
1N5014A	5117A				1N4734A
1N5015	5119A	3385			1N4735
1N5015A	5119A				1N4735A
1N5016	5120A	3386	5ZD-6.8		1N4736
1N5016A	5120A				1N4736A
1N5017	5121A	3387	ZD-7.5		1N4737
1N5017A	5121A				1N4737A
1N5018	5122A	3388	5ZD-8.2		1N4738
1N5018A	5122A				1N4738A
1N5019	5124A	3390			1N4739
1N5019A	5124A	3390			1N4739A
1N5020	5125A	3391	5ZD-10		1N4740
1N5020A					1N4740A
1N5021	5126A	3392	5ZD-11		1N4741
1N5021A	5126A				1N4741A
1N5022	5127A	3393	5ZD-12		1N4742
1N5022A	5127A				1N4742A
1N5023	5128A	3394	5ZD-13		1N4743
1N5023A	5128A				1N4743A
1N5024	5129A	3395	5ZD-14		1M14ZS10
1N5024A	5129A				1M14ZS5
1N5025	5130A	3396	5ZD-15		1N4744
1N5025A	5130A				1N4744A
1N5026	5131A	3397	5ZD-16		1N4745
1N5026A	5131A				1N4745A
1N5027	5132A	3398	5ZD-17		1M17ZS10
1N5027A	5132A				1M17ZS5
1N5028	5133A	3399	5ZD-18		1N4746
1N5028A					1N4746A
1N5029	5134A	3400	5ZD-19		1M19ZS10
1N5029A	5134A				1M19ZS5
1N503	116	3311	504A	1104	
1N5030	5135A	3401	5ZD-20		1N4747
1N5030A	5135A				1N4747A
1N5031	5136A	3402	5ZD-22		1N4748
1N5031A	5136A				1N4748A
1N5032	5137A	3403	5ZD-24		1N4749
1N5032A	5137A				1N4749A
1N5033	5138A	3404	5ZD-25		1M25ZS10
1N5033A	5138A				1M25ZS5
1N5034	5139A	3405	5ZD-27		1N4750
1N5034A	5139A				1N4750A
1N5035	5141A	3407	5ZD-30		1N4751
1N5035A	5141A				1N4751A
1N5036	5142A	3408			1N4752
1N5036A	5142A				1N4752A
1N5037	5143A	3409			1N4753
1N5037A	5143A				1N4753A
1N5038	5144A	3410			1N4754
1N5038A	5144A				1N4754A
1N5039	5145A	3411			1N4755
1N5039A	5145A				1N4755A
1N504	116	3017B/117	504A	1104	
1N5040	5146A	3412			1M45ZS10
1N5040A	5146A	3412			1M45ZS5
1N5041	5146A	3412			1N4756
1N5041A	5146A				1N4756A
1N5042	5147A	3413			1M50ZS10
1N5042A					1M50ZS5
1N5043	5147A	3413			1N4757
1N5043A	5147A				1N4757A
1N5044	5147A	3413			1M52ZS10
1N5044A	5147A	3413			1M52ZS5
1N5045	5148A	3414			1N4758
1N5045A	5148A				1N4758A
1N5046	5150A	3416			1N4759
1N5046A	5150A				1N4759A
1N5047	5151A	3417			1N4760
1N5047A	5151A				1N4760A
1N5048	5152A	3418			1N4761

Industry Standard No.	ECG	SK	GE	RS 276-	MOTOR.
1N5048A	5152A				1N4761A
1N5049	5153A	3419			1N4762
1N5049A					1N4762A
1N505	116	3017B/117	504A	1104	
1N5050	5155A	3421			1N4763
1N5050A	5155A				1N4763A
1N5051	5156A	3422			1N4764
1N5051A	5156A				1N4764A
1N5052	5808	9009	509	1114	1N5398
1N5053	5808	9009	509	1114	1N5398
1N5054	125	3081	510,531	1114	1N5399
1N5054A	125	3081	510,531	1114	
1N5055	5801	9004	504A	1142	1N4934
1N5056	5802	9005	504A	1143	1N4935
1N5057	5803	9006	504A	1144	1N4936
1N5058	5804	9007	504A	1144	1N4937
1N5059	125	9005/5802	504A	1143	
1N5059GP					MR5059
1N506	116		504A	1104	
1N5060	125	9007/5804	504A	1144	
1N5060GP					MR5060
1N5061	125	3848/5806	504A		
1N5061GP					MR5061
1N5062	125	9009/5808	300		
1N5062(SEARS)	177		300	1122	
1N5062GP					MR5062
1N5063	5120A	3386			1N4736A
1N5064	5121A	3387			1N4737A
1N5065	5122A	3388			1N4738A
1N5066	5124A	3390			1N4739A
1N5067	5125A	3391			1N4740A
1N5068	5126A	3392			1N4741A
1N5069	5128A	3394	5ZD-13		1N4743A
1N507	116	3031A	504A	1104	
1N5070	5129A	3395	5ZD-14		1M14ZS5
1N5071	5130A	3396			1N4744A
1N5072	5131A	3397	5ZD-16		1N4745A
1N5073	5133A	3399	5ZD-18		1N4746A
1N5074	5136A	3402	5ZD-22		1N4748A
1N5075	5137A		5ZD-24		1N4749A
1N5076	5139A	3405	5ZD-27		1N4750A
1N5077	5141A	3407	5ZD-30		1N4751A
1N5078	5142A	3408	5ZD-33		1N4752A
1N5079	5143A	3409			1N4753A
1N508	116	3017B/117	504A	1104	
1N5080	5144A	3410			1N4754A
1N5081	5144A	3410			1M40ZS5
1N5082	5145A	3411			1N4755A
1N5083	5146A	3412			1M45ZS5
1N5084	5146A	3412			1N4756A
1N5085	5147A	3413			1M50ZS5
1N5086	5147A	3413			1N4757A
1N5087	5148A	3414			1N4758A
1N5088	5149A	3415			1M60ZS5
1N5089	5150A	3416			1N4759A
1N509	125	3033	510,531	1114	
1N5090	5151A	3417			1N4760A
1N5091					1M70ZS5
1N5092	5152A	3418			1N4761A
1N5093	5153A	3419			1M80ZS5
1N5094	5153A	3420/5154A			1N4762A
1N5095	5155A	3421			1N4763A
1N5096	5157A	3423			1M110ZS5
1N5097	5158A	3424			1M120ZS5
1N5098	5159A	3425			1M130ZS5
1N5099	5160A	3426			1M140ZS5
1N51	109	3087		1123	
1N510	125	3033	510,531	1114	
1N5100	5162A	3428			1M160ZS5
1N5101	5163A	3429			1M170ZS5
1N5102	5164A	3430			1M180ZS5
1N5103	5165A	3431			1M190ZS5
1N5104	5166A	3432			1M200ZS5
1N5105					1M110ZSB5
1N5106					1M120ZSB5
1N5107					1M130ZSB5
1N5108					1M135ZSB5
1N5109					1M140ZSB5
1N511	116	3311	504A	1104	
1N5110					1M150ZSB5
1N5111					1M160ZSB5
1N5112					1M165ZSB5
1N5113					1M170ZSB5
1N5114					1M180ZSB5
1N5115					1M190ZSB5
1N5116					1M195ZSB5
1N5118	5129A	3395	5ZD-14		1N5341B
1N5119	5144A	3410			
1N512	116	3017B/117	504A	1104	
1N5120	5146A	3412			
1N5121	5147A	3413			
1N5122	5149A	3415			1N5371B
1N5123	5151A	3417			
1N5124	5153A	3419			
1N5125	5155A	3421			
1N5126	5160A	3426			1N5382B
1N5127	5163A	3429			1N5385B
1N5128	5165A	3431			1N5387B
1N513	116	3017B/117	504A	1104	
1N5139	610	3323			
1N5139,A					1N5139,A
1N5139A	610	3323			
1N514	116	3017B/117	504A	1104	
1N5140	611	3324			
1N5140,A					1N5140,A
1N5140A	611	3324			
1N5141	612	3325			
1N5141,A					1N5141,A
1N5141A	612	3325			
1N5142,A					1N5142,A
1N5143,A					1N5143,A
1N5144	613	3326			
1N5144,A					1N5144,A
1N5144A	613	3326			
1N5145,A					1N5145,A
1N5146,A					1N5146,A
1N5147,A					1N5147,A
1N5148,A					1N5148,A
1N515	116	3017B/117	504A	1104	
1N5154B		3064	ZD-27		
1N516	116	3017B/117	504A	1104	
1N517	125	3032A	510,531	1114	
1N5170	156	3081/125	512		1N5391
1N5171	5800	9003		1142	1N5391
1N5172	156	3051	512		1N5392
1N51/3	5803	9006		1144	1N5395
1N5174	5804	9007		1144	1N5395
1N5175	5805	9008			1N5397
1N5176	5806	3848			1N5397
1N5177	156	3051	512		1N5398
1N5178	156	3051	512		1N5399
1N5179	178MP		300(2)	1122	
1N518	125	3033	510,531	1114	
1N5185,GP					MR850

Industry Standard No.	ECG	SK	GE	RS 276-	MOTOR.
1N5186,GP					MR851
1N5187,GP					MR852
1N5188	506	3848/5806	511	1114	
1N5188,GP					MR854
1N5189	506	3848/5806	511	1114	
1N5189,GP					MR856
1N519	116	3311	504A	1104	
1N5190	506	3051/156	511	1114	
1N5190,GP					MR856
1N5194	177	3100/519	300	1122	
1N5195	177	3100/519	300	1122	
1N5196	177	3100/519	300	1122	
1N5197	5800	9003	512	1142	MR500
1N5198	117	3051/156	504A	1104	MR501
1N5199	5802	9005	512	1143	MR502
1N52	109	3087		1123	
1N520	116	3017B/117	504A	1104	
1N5200	5804	3051/156	512	1144	MR504
1N5201	5806	3848	512		MR506
1N5206	5806	3848			1N4936
1N5208	177	3100/519	300	1122	
1N5209	177		300	1122	
1N521	116	3017B/117	504A	1104	
1N5210	177	3100/519	300	1122	
1N5211	116	3311	504A	1104	
1N5212	116	3031A	504A	1104	
1N5213	116	3017B/117	504A	1104	
1N5214	125	3032A	510,531	1114	
1N5215	116	3311	504A	1104	
1N5216	116	3031A	504A	1104	
1N5217	116	3017B/117	504A	1104	
1N5218	125	3032A	510,531	1114	
1N5219	177	3100/519	300	1122	
1N522	116	3031A	504A	1104	
1N5220	177	3100/519	300	1122	
1N5221	5000A	3766			
1N5221A	5000A	3766			
1N5221B	5000A	3766			
1N5222	5001A	3767			
1N5222A	5001A	3767			
1N5223	5002A	3768			
1N5223A	5002A	3768			
1N5223B	5063A	3768/5002A			
1N5224	5003A	3769			
1N5224A	5003A	3769			
1N5224B	5003A	3982/5064A			
1N5225	5004A	3770			
1N5225A	5004A	3770			
1N5225B	5065A	3770/5004A			
1N5226	5005A	3771	ZD-3.3		
1N5226A	5005A	3771	ZD-3.3		
1N5226B	5005A	3771	ZD-3.3		
1N5227	5006A	3772	ZD-3.6		
1N5227A	5006A	3772	ZD-3.6		
1N5227B	134A	3772/5006A	ZD-3.6		
1N5228	5007A	3773	ZD-3.9		
1N5228A	5007A	3773	ZD-3.9		
1N5228B	5007A	3773	ZD-3.9		
1N5229	5008A	3774	ZD-4.3		
1N5229A	5008A	3784/5018A	ZD-4.3		
1N5229B	5068A	3784/5018A	ZD-4.3		
1N523	116	3017B/117	504A	1104	
1N5230	5009A	3775	ZD-4.7		
1N5230A	5009A	3775	ZD-4.7		
1N5230B	5069A	3775/5009A	ZD-5.1		
1N5231	5010A	3776	ZD-5.1		
1N5231A	5010A	3776	ZD-5.1		
1N5231B	5010A	3776	ZD-5.1		
1N5232	5011A	3777	ZD-5.6	561	
1N5232A	5011A	3777	ZD-5.6	561	
1N5232B	136A	3777/5011A	ZD-5.6	561	
1N5233	5012A	3778	ZD-6.0	561	
1N5233A	5012A	3778	ZD-6.0	561	
1N5233B	5012A	3778	ZD-6.0	561	
1N5234	5013A	3779	ZD-6.2	561	
1N5234A	5013A	3779	ZD-6.2	561	
1N5234B	137A	3777/5011A	ZD-6.2	561	
1N5235	5014A	3780	ZD-6.8	561	
1N5235A	5014A	3780	ZD-6.8	561	
1N5235B	5071A	3780/5014A	ZD-6.8	561	
1N5236	5015A	3781	ZD-7.5		
1N5236A	5015A	3781	ZD-7.5		
1N5236B	5015A	3781	ZD-7.5		
1N5237	5016A	3782	ZD-8.2	562	
1N5237A	5016A	3782	ZD-8.2	562	
1N5237B	5016A	3782	ZD-8.2	562	
1N5238	5017A	3783	ZD-8.7	562	
1N5238A	5017A	3783	ZD-9.1	562	
1N5238B	5073A	3783/5017A	ZD-8.7	562	
1N5239	5018A	3784	ZD-9.1	562	
1N5239A	5018A	3784	ZD-9.1	562	
1N5239B	139A	3784/5018A	ZD-9.1	562	
1N524	116	3017B/117	504A	1104	
1N5240	5019A	3785	ZD-10	562	
1N5240A	5019A	3785	ZD-10	562	
1N5240B	140A	3785/5019A	ZD-10	562	
1N5241	5020A	3786	ZD-11	563	
1N5241A	5020A	3786	ZD-11	563	
1N5241B	5074A	3786/5020A	ZD-11	563	
1N5242	5021A	3787	ZD-12	563	
1N5242A	5021A	3787	ZD-12	563	
1N5242B	142A	3787/5021A	ZD-12	563	
1N5243	5022A	3788	ZD-13	563	
1N5243A	5022A	3788	ZD-13	563	
1N5243B	143A	3788/5022A	ZD-13	563	
1N5244	5023A	3789	ZD-14	564	
1N5244A	5023A	3789	ZD-14	564	
1N5244B	144A	3789/5023A	ZD-14	564	
1N5245	5024A	3790	ZD-15	564	
1N5245A	5024A	3790	ZD-15	564	
1N5245B	145A	3790/5024A	ZD-15	564	
1N5246	5025A	3791	ZD-16	564	
1N5246A	5025A	3791	ZD-16	564	
1N5246B	5075A	3751	ZD-16	564	
1N5247	5026A	3792	ZD-17		
1N5247A	5026A	3792	ZD-17		
1N5247B	5076A	3792/5026A	ZD-17		
1N5248	5027A	3793	ZD-18		
1N5248A	5027A	3793	ZD-18		
1N5248B	5077A	3793/5027A	ZD-18		
1N5249	5028A	3794	ZD-19		
1N5249A	5028A	3794	ZD-19		
1N5249B	5078A	3794/5028A	ZD-19		
1N525	125	3033	510,531	1114	
1N5250	5029A	3795	ZD-20		
1N5250A	5029A	3795	ZD-20		
1N5250B	5079A	3795/5029A	ZD-20		
1N5251	5030A	3796	ZD-22		
1N5251A	5030A	3796	ZD-22		
1N5251B	5030A	3796	ZD-22		
1N5252	5031A	3797	ZD-24		
1N5252A	5031A	3797	ZD-24		
1N5252B	5081A	3797/5031A	ZD-24		
1N5253	5032A	3798	ZD-25		

Industry Standard No.	ECG	SK	GE	RS 276-	MOTOR.
1N5253A	5032A	3798	ZD-25		
1N5253B	5082A	3798/5032A	ZD-25		
1N5254	5033A	3799	ZD-27		
1N5254A	5033A	3799	ZD-27		
1N5254B	146A	3799/5033A	ZD-27		
1N5255	5034A	3800	ZD-28		
1N5255A	5034A	3800	ZD-28		
1N5255B	5083A	3800/5034A	ZD-28		
1N5256	5035A	3801	ZD-30		
1N5256A	5035A	3801	ZD-30		
1N5256B	5084A	3801/5035A	ZD-30		
1N5257	5036A	3802	ZD-33		
1N5257A	5036A	3802	ZD-33		
1N5257B	147A	3802/5036A	ZD-33		
1N5258	5037A	3803	ZD-36		
1N5258A	5037A	3803	ZD-36		
1N5258B	5085A	3803/5037A	ZD-36		
1N5259	5038A	3804	ZD-39		
1N5259A	5038A	3804	ZD-39		
1N5259B	5086A	3804/5038A	ZD-39		
1N526	125	3033	510,531	1114	
1N5260	5039A	3805	ZD-43		
1N5260A	5039A	3805	ZD-43		
1N5260B	5087A	3805/5039A	ZD-43		
1N5261	5040A	3806	ZD-47		
1N5261A	5040A	3806	ZD-47		
1N5261B	5088A	3806/5040A	ZD-47		
1N5262	5041A	3807	ZD-51		
1N5262A	5041A	3807	ZD-51		
1N5262B	5089A	3807/5041A	ZD-51		
1N5263	5042A	3808	ZD-56		
1N5263A	5042A	3808	ZD-56		
1N5263B	5090A	3808/5042A	ZD-55		
1N5264	5043A	3809	ZD-60		
1N5264A	5043A	3809	ZD-60		
1N5264B	5091A	3809/5043A	ZD-60		
1N5265	5044A	3810	ZD-62		
1N5265A	5044A	3810	ZD-62		
1N5265B	149A	3810/5044A	ZD-62		
1N5266	5045A	3811	ZD-68		
1N5266A	5045A	3811	ZD-68		
1N5266B	5092A	3811/5045A	ZD-68		
1N5267	5046A	3812	ZD-75		
1N5267A	5046A	3812	ZD-75		
1N5267B	5093A	3812/5046A	ZD-75		
1N5268	5047A	3813	ZD-82		
1N5268A	5047A	3813	ZD-82		
1N5268B	150A	3813/5047A	ZD-82		
1N5269	5048A	3814	ZD-87		
1N5269A	5048A	3814	ZD-87		
1N5269B	5094A	3814/5048A	ZD-87		
1N527	109	3087		1123	
1N5270	5049A	3815	ZD-91		
1N5270A	5049A	3815	ZD-91		
1N5270B	5095A	3815/5049A	ZD-91		
1N5271	5050A	3816	ZD-100		
1N5271A	5050A	3816	ZD-100		
1N5271B	5096A	3816/5050A	ZD-100		
1N5272	5051A	3817	ZD-110		
1N5272A	5051A	3817	ZD-110		
1N5272B	151A	3817/5051A	ZD-110		
1N5273	5052A	3818	ZD-120		
1N5273A	5052A	3818	ZD-120		
1N5273B	5097A	3818/5052A	ZD-120		
1N5274	5053A	3819	ZD-130		
1N5274A	5053A	3819	ZD-130		
1N5274B	5098A	3819/5053A	ZD-130		
1N5275	5054A	3820	ZD-140		
1N5275A	5054A	3820	ZD-140		
1N5275B	5099A	3820/5054A	ZD-140		
1N5276	5055A	3821	ZD-150		
1N5276A	5055A	3821	ZD-150		
1N5276B		3821	ZD-150		
1N5277	5056A	3822	ZD-160		
1N5277A	5056A	3822	ZD-160		
1N5277B	5101A	3822/5056A	ZD-160		
1N5278	5057A	3823	ZD-170		
1N5278A	5057A	3823	ZD-170		
1N5278B	5102A	3823/5057A	ZD-170		
1N5279	5058A	3824	ZD-180		
1N5279A	5058A	3824	ZD-180		
1N5279B	5103A	3824/5058A	ZD-180		
1N5280	5059A	3825	ZD-190		
1N5280A	5059A	3825	ZD-190		
1N5280B	5104A	3825/5059A	ZD-190		
1N5281	5060A	3826	ZD-200		
1N5281A	5060A	3826	ZD-200		
1N5281B	5105A	3826/5060A	ZD-200		
1N5282	519	3100	300	1122	
1N52A	109	3087		1123	
1N530	116	3016	504A	1104	1N4002
1N531	116	3016	504A	1104	1N4003
1N5315	177	3100/519	300	1122	
1N5316	177	3100/519	300	1122	
1N5317	177	3100/519	300	1122	
1N5318	177	3100/519	300	1122	
1N5319	177	3100/519	300	1122	
1N532	116	3031A	504A	1104	1N4004
1N5320	177	3311	300	1122	
1N533	116	3016	504A	1104	1N4004
1N5333	5111A	3377	5ZD-3.3		
1N5333A	5111A		5ZD-3.3		
1N5333B	5111A		5ZD-3.3		
1N5334	5112A	3378	5ZD-3.6		
1N5334A	5112A	3378	5ZD-3.6		
1N5334B	5112A		5ZD-3.6		
1N5335	5113A	3379	5ZD-3.9		
1N5335A	5113A	3379	5ZD-3.9		
1N5335B	5113A		5ZD-3.9		
1N5336	5114A	3380	5ZD-4.3		
1N5336A	5114A	3380	5ZD-4.3		
1N5336B	5114A		5ZD-4.3		
1N5337	5115A	3381	5ZD-4.7		
1N5337A	5115A	3381	5ZD-4.7		
1N5337B	5115A		5ZD-4.7		
1N5338	5116A	3382	5ZD-5.1		
1N5338A	5116A	3382	5ZD-5.1		
1N5338B	5116A		5ZD-5.1		
1N5339	5117A	3383	5ZD-5.6		
1N5339A	5117A	3383	5ZD-5.6		
1N5339B	5117A		5ZD-5.6		
1N534	116	3016	504A	1104	1N4005
1N5340	5118A	3384	5ZD-6.0		
1N5340A	5118A	3384	5ZD-6.0		
1N5340B	5118A		5ZD-6.0		
1N5341	5119A	3385	5ZD-6.2		
1N5341A	5119A	3385	5ZD-6.2		
1N5341B	5119A		5ZD-6.2		
1N5342	5120A	3386	5ZD-6.8		
1N5342A	5120A	3386	5ZD-6.8		
1N5342B	5120A		5ZD-6.8		
1N5343	5121A	3387	5ZD-7.5		
1N5343A	5121A	3387	5ZD-7.5		
1N5343B	5121A		5ZD-7.5		

Industry Standard No.	ECG	SK	GE	RS 276-	MOTOR.
1N5344	5122A	3388	5ZD-8.2		
1N5344A	5122A	3388	5ZD-8.2		
1N5344B	5122A		5ZD-8.2		
1N5345	5123A	3389	5ZD-8.7		
1N5345A	5123A	3389	5ZD-8.7		
1N5345B	5123A		5ZD-8.7		
1N5346	5124A	3390	5ZD-9.1		
1N5346A	5124A	3390	5ZD-9.1		
1N5346B	5124A		5ZD-9.1		
1N5347	5125A	3391	5ZD-10		
1N5347A	5125A	3391	5ZD-10		
1N5347B	5125A		5ZD-10		
1N5348	5126A	3392	5ZD-11		
1N5348A	5126A	3392	5ZD-11		
1N5348B	5126A	3392	ZD-10	562	
1N5349	5127A	3393	5ZD-12		
1N5349A	5127A	3393	5ZD-12		
1N5349B	5127A		5ZD-12		
1N535	116	3017B/117	504A	1104	1N4005
1N5350	5128A	3394	5ZD-13		
1N5350A	5128A	3394	5ZD-13		
1N5350B	5128A		5ZD-13		
1N5351	5129A	3395	5ZD-14		
1N5351A	5129A	3395	5ZD-14		
1N5351B	5129A		5ZD-14		
1N5352	5130A	3396	5ZD-15		
1N5352A	5130A	3396	5ZD-15		
1N5352B	5130A		5ZD-15		
1N5353	5131A	3397	5ZD-16		
1N5353A	5131A	3397	5ZD-16		
1N5353B	5131A		5ZD-16		
1N5354	5132A	3398	5ZD-17		
1N5354A	5132A	3398	5ZD-17		
1N5354B	5132A		5ZD-17		
1N5355	5133A	3399	5ZD-18		
1N5355A	5133A	3399	5ZD-18		
1N5355B	5133A		5ZD-18		
1N5356	5134A	3400	5ZD-19		
1N5356A	5134A	3400	5ZD-19		
1N5356B	5134A		5ZD-18		
1N5357	5135A	3401	5ZD-20		
1N5357A	5135A	3401	5ZD-20		
1N5357B	5135A		5ZD-20		
1N5358	5136A	3402	5ZD-22		
1N5358A	5136A	3402	5ZD-22		
1N5358B	5136A		5ZD-22		
1N5359	5137A	3403	5ZD-24		
1N5359A	5137A	3403	5ZD-24		
1N5359B	5137A		5ZD-24		
1N536	116	3016	504A	1104	1N4001
1N5360	5138A	3753/5082A	5ZD-25		
1N5360A	5138A	3753/5082A	5ZD-25		
1N5360B	5138A	3404	5ZD-25		
1N5361	5139A	3405	5ZD-27		
1N5361A	5139A	3405	5ZD-27		
1N5361B	5139A		5ZD-27		
1N5362	5140A	3406	5ZD-28		
1N5362A	5140A	3406	5ZD-28		
1N5362B	5140A		5ZD-28		
1N5363	5141A	3407	5ZD-30		
1N5363A	5141A	3407	5ZD-30		
1N5363B	5141A		5ZD-30		
1N5364	5142A	3408	5ZD-33		
1N5364A	5142A	3408	5ZD-33		
1N5364B	5142A		5ZD-33		
1N5365	5143A	3409			
1N5365A	5143A	3409			
1N5366	5144A	3410			
1N5366A	5144A	3410			
1N5367	5145A	3411			
1N5367A	5145A	3411			
1N5368	5146A	3412			
1N5368A	5146A	3412			
1N5369	5147A	3413			
1N5369A	5147A	3413			
1N537	116	3017B/117	504A	1104	1N4002
1N5370	5148A	3414			
1N5370A	5148A	3414			
1N5371	5149A	3415			
1N5371A	5149A	3415			
1N5372	5150A	3416			
1N5372A	5150A	3416			
1N5373	5151A	3417			
1N5373A	5151A	3417			
1N5374	5152A	3418			
1N5374A	5152A	3418			
1N5375	5153A	3419			
1N5375A	5153A	3419			
1N5376	5154A	3420			
1N5376A	5154A	3420			
1N5377	5155A	3421			
1N5377A	5155A	3421			
1N5378	5156A	3422			
1N5378A	5156A	3422			
1N5379	5157A	3423			
1N5379A	5157A	3423			
1N538	116	3016	504A	1104	1N4003
1N5380	5158A	3424			
1N5380A	5158A	3424			
1N5381	5159A	3425			
1N5381A	5159A	3425			
1N5382	5160A	3426			
1N5382A	5160A	3426			
1N5383	5161A	3427			
1N5383A	5161A	3427			
1N5384	5162A	3428			
1N5384A	5162A	3428			
1N5385	5163A	3429			
1N5385A	5163A	3429			
1N5386	5164A	3430			
1N5386A	5164A	3430			
1N5387	5165A	3431			
1N5387A	5165A	3431			
1N5388	5166A	3432			
1N5388A	5166A	3432			
1N539	116	3016	504A	1104	1N4004
1N5391	5800	9003		1142	
1N5391GP		9003			1N5391
1N5392	125	9004/5801	510	1114	
1N5392GP	5801	9004	510	1114	1N5392
1N5393	5802	9005	504A	1143	
1N5393GP		9005			1N5393
1N5394	5803	9006		1144	
1N5394GP		9006			1N5394
1N5395	5804	9007		1144	
1N5395GP		9007			1N5395
1N5396	5805	9008		1104	
1N5396GP		9008			1N5396
1N5397	5806	3848		1104	
1N5397GP		3848			1N5397
1N5398	5808	9009		1114	
1N5398GP		9009			1N5398
1N5399	5809	9010			
1N5399GP		9010			1N5399

Industry Standard No.	ECG	SK	GE	RS 276-	MOTOR.
1N54	109	3087		1123	
1N540	116	3016	504A	1104	1N4004
1N5400	156	9003/5800		1142	
1N5401	5801	9004			
1N5402	5802	9005		1143	
1N5403	5803	9006		1144	
1N5404	5804	9007		1144	
1N5405	5805	9008			
1N5406	5806	3848			
1N5407	5808	9009			
1N5408	5809	9010			
1N5409	5988	3608/5990			
1N5409R	5989	3518/5995			
1N541	110MP	3709	1N34AS	1123(2)	
1N5411	6408	3523/6407			
1N5412	117	3100/519		1104	
1N5413	177			1122	
1N5414	177	3100/519		1122	
1N5415	507	3051/156			MR850
1N5416		3051			MR851
1N5417		3051			MR852
1N5418	506	3051/156		1114	MR854
1N5419	506	3051/156		1114	MR856
1N542	110MP	3709	1N34AS	1123(2)	
1N5420		3051			MR856
1N5426	177	3100/519	300	1122	
1N542A			1N60	1123	
1N542MP		3709	1N60	1123	
1N543	506	3843	511	1114	
1N5433	506	3175	511	1114	
1N5434	506	3175	511	1114	
1N5435	5886		5044		
1N543A	506	3843	511	1114	
1N5441	610	3323			
1N5441A	610	3323			
1N5443	611	3324			
1N5443A	611	3324			
1N5444	612	3325			
1N5444A	612	3325			
1N5448	613	3326			
1N5448A	613	3326			
1N5450	614	3327			
1N5450A	614	3327	90		
1N5461	610	3323			
1N5461A	610	3323			
1N5463	611	3324			
1N5463A	611	3324			
1N5464	612	3325			
1N5464A	612	3325			
1N5468	613	3326			
1N5468A	613	3326			
1N547	116	3033	504A	1104	1N4005
1N5470	614	3327			
1N5470A	614	3327	90		
1N548	125		510,531	1114	
1N549	506	3843	511	1114	
1N54A	109	3087		1123	
1N54G	109	3087		1123	
1N54GA	109	3087		1123	
1N550	5834	3016	5004	1104	
1N550R	5833		5005		
1N551	5834	3017B/117	5004		
1N5518	5005A	3771	ZD-3.3		
1N5518A	5005A	3771	ZD-3.3		
1N5518B	5066A		ZD-3.3		
1N5519	5006A	3772	ZD-3.6		
1N5519A	5006A	3772	ZD-3.6		
1N5519B	134A		ZD-3.6		
1N551R	5835		5005		
1N552	5838	3017B/117	5004		
1N5520	5007A	3773	ZD-3.9		
1N5520A	5007A	3773	ZD-3.9		
1N5520B	5067A		ZD-3.9		
1N5521	5008A	3774	ZD-4.3		
1N5521A	5008A	3774	ZD-4.3		
1N5521B	5068A		ZD-4.3		
1N5522	5009A	3333/5069A	ZD-4.7		
1N5522A	5009A	3333/5069A	ZD-4.7		
1N5522B	5069A		ZD-4.7		
1N5523	5010A	3776	ZD-5.1		
1N5523A	5010A	3776	ZD-5.1		
1N5524	5011A	3777	ZD-5.6	561	
1N5524A	5011A	3777	ZD-5.6	561	
1N5524B	136A		ZD-5.6	561	
1N5525	5013A	3779	ZD-6.2	561	
1N5525A	5013A	3779	ZD-6.2	561	
1N5525B	137A		ZD-6.2	561	
1N5526	5014A	3780	ZD-6.8	561	
1N5526A	5014A	3780	ZD-6.8	561	
1N5526B	5071A		ZD-6.8	561	
1N5527	5015A	3781	ZD-7.5		
1N5527A	5015A	3781	ZD-7.5		
1N5527B	138A		ZD-7.5		
1N5528	5016A	3782	ZD-8.2	562	
1N5528A	5016A	3782	ZD-8.2	562	
1N5528B	5072A	3782/5016A	ZD-8.2	562	
1N5529	5018A	3784	ZD-9.1	562	
1N5529A	5018A	3784	ZD-9.1	562	
1N5529B	139A		ZD-9.1	562	
1N552R	5837		5005		
1N553	5838	3031A	5004		
1N5530	5019A	3785	ZD-10	562	
1N5530A	5019A	3785	ZD-10	562	
1N5530B	140A		ZD-10	562	
1N5531			ZD-11	563	
1N5531A	5020A		ZD-11	563	
1N5531B	5074A		ZD-11	563	
1N5532	5021A		ZD-12	563	
1N5532A	5021A		ZD-12	563	
1N5532B	142A		ZD-12	563	
1N5533	5022A	3788	ZD-13	563	
1N5533A	5022A	3788	ZD-13	563	
1N5533B	143A	3750	ZD-13	563	
1N5534	5023A	3789	ZD-14	564	
1N5534A	5023A	3789	ZD-14	564	
1N5534B	144A			564	
1N5535	5024A	3790	ZD-15	564	
1N5535A	5024A	3790	ZD-15	564	
1N5535B	145A		ZD-15	564	
1N5536	5025A	3791	ZD-16	564	
1N5536A	5025A	3791	ZD-16	564	
1N5536B	5075A	3751	ZD-16	564	
1N5537	5026A	3792	ZD-17		
1N5537A	5026A	3792	ZD-17		
1N5537B	5076A	9022	ZD-17		
1N5538	5027A	3793	ZD-18		
1N5538A	5027A	3793	ZD-18		
1N5538B	5077A	3752	ZD-18		
1N5539A	5028A		ZD-19		
1N5539B	5078A	9023	ZD-19		
1N553R	5839		5005		
1N554	5842	3584/5862	5008	1104	
1N5540	5029A	3795	ZD-20		
1N5540A	5029A	3795	ZD-20		
1N5540B	5079A		ZD-20		
1N5541	5030A	3796	ZD-22		
1N5541A	5030A	3796	ZD-22		
1N5541B	5080A		ZD-22		
1N5542	5031A	3797	ZD-24		
1N5542A	5031A	3797	ZD-24		
1N5542B	5081A		ZD-24		
1N5543	5032A	3798	ZD-25		
1N5543A	5032A	3798	ZD-25		
1N5543B	5082A	3753	ZD-25		
1N5544	5034A	3800	ZD-28		
1N5544A	5034A	3800	ZD-28		
1N5544B	5083A	3754			
1N5545	5035A	3801	ZD-30		
1N5545A	5035A	3801	ZD-30		
1N5545B	5084A	3801/5035A	ZD-30		
1N5546	5036A	3802	ZD-33		
1N5546A	5036A	3802	ZD-33		
1N554R	5841	3517/5883	5009		
1N555	5842	7042	5008	1104	
1N5552	506	3051/156	511	1114	
1N5553	506	3051/156	511	1114	
1N5554	506	3051/156	511	1114	
1N5555					1N6283
1N5556					1N6283A
1N5557					1N6289A
1N5558					1N6303A
1N5559	5071A	3334	ZD-6.8	561	
1N5559A	5071A	3334	ZD-6.8	561	
1N555R	5843	3517/5883	5009		
1N5560	138A		ZD-7.5		
1N5560A	138A		ZD-7.5		
1N5561	5072A	3136	ZD-8.2	562	
1N5561A	5072A	3136	ZD-8.2	562	
1N5561B	5072A	3136			
1N5562	139A		ZD-9.1	562	
1N5562A	139A		ZD-9.1	562	
1N5563	140A		ZD-10	562	
1N5563A	140A		ZD-10	562	
1N5564	5074A		ZD-11.0	563	
1N5564A	5074A		ZD-11.0	563	
1N5565	142A		ZD-12	563	
1N5565A	142A		ZD-12	563	
1N5566	143A	3750	ZD-13	563	
1N5566A	143A	3750	ZD-13	563	
1N5566B	143A	3750			
1N5567	145A		ZD-15	564	
1N5567A	145A		ZD-15	564	
1N5568	5075A	3751	ZD-16	564	
1N5568A	5075A	3751	ZD-16	564	
1N5568B	5075A	3751			
1N5569	5077A	3752	ZD-18		
1N5569A	5077A	3752	ZD-18		
1N5569B	5077A	3752			
1N5570	5079A	3335	ZD-20		
1N5570A	5079A	3335	ZD-20		
1N5571	5080A	3336	ZD-22		
1N5571A	5080A	3336	ZD-22		
1N5572	5081A		ZD-24		
1N5572A	5081A		ZD-24		
1N5573	146A		ZD-27		
1N5573A	146A		ZD-27		
1N5574	5084A	3755	ZD-30		
1N5574A	5084A	3755	ZD-30		
1N5574B	5084A	3755			
1N5575	147A		ZD-33		
1N5575A	147A		ZD-33		
1N5576	5085A	3337	ZD-36		
1N5576A	5085A	3337	ZD-36		
1N5577	5086A	3338	ZD-39		
1N5577A	5086A	3338	ZD-39		
1N5578	5087A	3339	ZD-43		
1N5578A	5087A	3339	ZD-43		
1N5579	5088A	3340	ZD-47		
1N5579A	5088A	3340	ZD-47		
1N5580	5089A	3341	ZD-51		
1N5580A	5089A	3341	ZD-51		
1N5580B	5089A	3341	ZD-51		
1N5581	5090A	3342	ZD-56		
1N5581A	5090A	3342	ZD-56		
1N5582	149A		ZD-62		
1N5582A	149A		ZD-62		
1N5583	5092A	3343	ZD-68		
1N5583A	5092A	3343	ZD-68		
1N5583B		3343	ZD-68		
1N5584	5093A	3344	ZD-75		
1N5584A	5093A	3344	ZD-75		
1N5584B	5093A	3344	ZD-75		
1N5585	150A		ZD-82		
1N5585A	150A		ZD-82		
1N5586	5095A	3345	ZD-91		
1N5586A	5095A	3345	ZD-91		
1N5586B	5095A	3345	ZD-91		
1N5587	5096A	3346	ZD-100		
1N5587A	5096A	3346	ZD-100		
1N5587B	5096A	3346	ZD-100		
1N5588	151A		ZD-110		
1N5588A	151A		ZD-110		
1N5588B	151A		ZD-110		
1N5589	5097A	3347	ZD-120		
1N5589A	5097A	3347	ZD-120		
1N5590	5098A	3348	ZD-130		
1N5590A	5098A	3348	ZD-130		
1N5590B	5098A	3348	ZD-130		
1N5591	5100A	3350	ZD-150		
1N5591A	5100A	3350	ZD-150		
1N5592	5101A	3351	ZD-160		
1N5592A	5101A	3351	ZD-160		
1N5592B	5101A	3351	ZD-160		
1N5593	5103A	3353	ZD-180		
1N5593A	5103A	3353	ZD-180		
1N5593B	5103A	3353	ZD-180		
1N5594	5105A	3355	ZD-200		
1N5594A	5105A	3355	ZD-200		
1N5594B	5105A	3355	ZD-200		
1N56	109	3087		1123	
1N560	125	3033	510,531	1114	1N4006
1N5605	177	3100/519	300	1122	
1N5606	177		300	1122	
1N5607	177	3100/519	300	1122	
1N561	125	3033	510,531	1114	1N4007
1N5610	177	3100/519	300	1122	
1N5614GP					1N4003
1N5615GP					1N4935
1N5616	506	3125	511	1114	
1N5616GP					1N4004
1N5617	552	3125	511	1114	
1N5617GP					1N4936
1N5618	506	3125	511	1114	
1N5618GP					1N4005
1N5619	552	3125	511	1114	
1N5619GP					1N4937
1N562	5846	3516	5012	1114	MR1128
1N5620	506	3125	511	1114	

Industry Standard No.	ECG	SK	GE	RS 276-	MOTOR.
1N5620GP					1N4006
1N5621	506	3125	511	1114	
1N5621GP					MR817
1N5622GP					1N4007
1N5623GP					MR818
1N5624	5802	3848/5806	1N5624	1143	
1N5624,GP					MR502
1N5625		3848	1N5625	1144	
1N5625,GP					MR504
1N5626		3848	1N5626		
1N5626,GP					MR506
1N5627		3051	1N5627		
1N5627GP		3051			MR508
1N5629					1N6267
1N5629A					1N6267A
1N562R	5847		5013		
1N563	5848	3033	5012	1114	MR1130
1N5630					1N6268
1N5630A					1N6268A
1N5631					1N6269
1N5631A					1N6269A
1N5632					1N6270
1N5632A					1N6270A
1N5633					1N6271
1N5633A					1N6271A
1N5634					1N6272
1N5634A					1N6272A
1N5635					1N6273
1N5635A					1N6273A
1N5636					1N6274
1N5636A					1N6274A
1N5637					1N6275
1N5637A					1N6275A
1N5638					1N6276
1N5638A					1N6276A
1N5639					1N6277
1N5639A					1N6277A
1N563R	5849	7049	5013		
1N564	109	3087		1123	
1N5640					1N6278
1N5640A					1N6278A
1N5641					1N6279
1N5641A					1N6279A
1N5642					1N6280
1N5642A					1N6280A
1N5643					1N6281
1N5643A					1N6281A
1N5644					1N6282
1N5644A					1N6282A
1N5645					1N6283
1N5645A					1N6283A
1N5646					1N6284
1N5646A					1N6284A
1N5651A					1N6289A
1N5652					1N6290
1N5652A					1N6290A
1N5653					1N6291
1N5653A					1N6291A
1N5654					1N6292
1N5654A					1N6292A
1N5655					1N6293
1N5655A					1N6293A
1N5656					1N6294
1N5656A					1N6294A
1N5657					1N6295
1N5657A					1N6295A
1N5658					1N6296
1N5658A					1N6296A
1N5659					1N6297
1N5659A					1N6297A
1N5660					1N6298
1N5660A					1N6298A
1N5661					1N6299
1N5661A					1N6299A
1N5662					1N6300
1N5662A					1N6300A
1N5663					1N6301
1N5663A					1N6301A
1N5664					1N6302
1N5664A					1N6302A
1N5665					1N6303
1N5665A					1N6303A
1N568	177		300	1122	
1N5681	610	3323			1N5461A
1N5681A	610	3323			
1N5682					1N5462A
1N5683	611	3324			1N5463A
1N5683A	611	3324			
1N5684	612	3325			1N5464A
1N5684A	612	3325			
1N5685					1N5465A
1N5686					1N5457A
1N5687	613	3326			1N5458A
1N5687A	613	3326			
1N5688					1N5469A
1N5689	614	3327			1N5470A
1N5689A	614	3327			
1N569	109	3091		1123	
1N5690					1N5471A
1N5691					1N5472A
1N5692					1N5473A
1N5693					1N5474A
1N5694					1N5475A
1N5695					1N5476A
1N5696	610	3323			1N5461A
1N5696A	610	3323			
1N5697					1N5462A
1N5698					1N5463A
1N5699	612	3325			1N5464A
1N5699A	612	3325			
1N56A	109	3087		1123	
1N57	109	3087		1123	
1N5700					1N5465A
1N5701					1N5467A
1N5702	613	3326			1N5468A
1N5702A	613	3326			
1N5703					1N5469A
1N5704					1N5470A
1N5705					1N5471A
1N5706					1N5472A
1N5707					1N5473A
1N5708					1N5474A
1N5709					1N5475A
1N571	109	3091		1123	
1N5710					1N5476A
1N5711	177	3100/519	300	1122	MBD701/702
1N5712	177	3100/519	300	1122	MBD201
1N5713	177	3100/519	300	1122	MBD201
1N5719	177	3100/519	300	1122	
1N5728	5009A	3775	ZD-4.7		1N5230B
1N5729	5010A	3776	ZD-5.1		1N5231B
1N5730	5011A	3777	ZD-5.6	561	1N5232B
1N5731	5013A	3779	ZD-6.2	561	1N5234B
1N5732	5014A	3334/5071A	ZD-6.8	561	

Industry Standard No.	ECG	SK	GE	RS 276-	MOTOR.
1N5732B	5071A				1N5235B
1N5733	5015A	3781	ZD-7.5		
1N5733B	138A				1N5236B
1N5734	5016A	3782	ZD-8.2	562	
1N5734B	5072A	3782/5016A	ZD-8.2	562	1N5237B
1N5735	5018A	3784	ZD-9.1	562	
1N5735B	139A				1N5239B
1N5736	5019A	3785	ZD-10	562	
1N5736B	140A				1N5240B
1N5737	5020A	3786	ZD-11	563	
1N5738	5021A	3787	ZD-12	563	
1N5738B	142A				1N5242B
1N5739	5022A	3788	ZD-13	563	
1N5739B	143A	3788/5022A			1N5243B
1N5740	5024A	3790	ZD-15	564	
1N5740B	145A				1N5245B
1N5741	5025A	3791		564	
1N5741B	5075A	3751			1N5246B
1N5742	5027A	3793	ZD-18		
1N5742B	5077A	3752			1N5248B
1N5743	5029A	3795	ZD-20		
1N5743B	5079A				1N5250B
1N5744	5030A	3796	ZD-22		
1N5744B	5080A				1N5251B
1N5745	5031A	3797	ZD-24		
1N5745B	5081A				1N5252B
1N5746	5033A	3799	ZD-27		
1N5746B	146A		ZD-27		1N5254B
1N5746C		3799	ZD-27		
1N5746D		3799	ZD-27		
1N5747	5035A	3801	ZD-30		
1N5747B	5084A	3755			1N5256B
1N5748	5036A	3802	ZD-33		
1N5748B	147A				1N5257B
1N5749	5037A	3803	ZD-36		1N5258B
1N5750	5038A	3804	ZD-39		1N5259B
1N5751	5039A	3805	ZD-43		1N5760B
1N5752	5040A	3806	ZD-47		1N5261B
1N5753	5041A	3807	ZD-51		1N5262B
1N5754	5042A	3808	ZD-56		
1N5755	5044A	3810	ZD-62		
1N5756	5045A	3811	ZD-68		
1N5758	6406	9082A			1N5758
1N5758A	6406	9082A			1N5758A
1N5759	6407	9083A			
1N5759A	6407	9083A			
1N5760	6407	9083A		1050	1N5760
1N5760A	6408	9084A		1050	
1N5761	6408	9084A		1050	1N5761
1N5761A	6407	9083A		1050	1N5761A
1N5762					1N5762
1N5762A					1N5762A
1N5765					MBD301
1N5766					MBD301
1N5767	177		300	1122	MBD201
1N5779					MBS4991
1N5780					MBS4991
1N5781					MBS4991
1N5782					MBS4991
1N5783					MBS4991
1N5784					MBS4991
1N5785					MBS4991
1N5786					MBS4991
1N5787					MBS4991
1N5788					MBS4991
1N5789					MBS4991
1N5790					MBS4991
1N5791					MBS4991
1N5792					MBS4991
1N5793					MBS4991
1N57A	109	3087		1123	
1N58	109	3087		1123	
1N5812					SPECIAL
1N5813					SPECIAL
1N5814					SPECIAL
1N5815					SPECIAL
1N5816					SPECIAL
1N5837					1N4370
1N5838					.5M2.5AZ10
1N5839					1N4371
1N5839B	5063A	3837			
1N5840					.5M2.8AZ10
1N5841					1N4372
1N5841B	5065A	3838			
1N5842			ZD-3.3		1N746
1N5842B	5066A		ZD-3.3		
1N5843			ZD-3.6		1N747
1N5843B	134A		ZD-3.6		
1N5844			ZD-3.9		1N748
1N5844B	5067A		ZD-3.9		
1N5845			ZD-4.3		1N749
1N5845B	5068A	3332	ZD-4.3		
1N5846			ZD-4.7		1N750
1N5846B	5069A	3752/5077A	ZD-4.7		
1N5847			ZD-5.1		1N751
1N5848			ZD-5.6	561	1N752
1N5848A			ZD-5.6	561	
1N5848B	136A		ZD-5.6	561	
1N5849			ZD-6.2	561	5M6.0AZ10
1N5849A			ZD-6.2	561	
1N5849B	5070A	9021	ZD-6.2	561	
1N5850			ZD-6.2	561	1N753
1N5850A			ZD-6.2	561	
1N5850B	137A		ZD-6.2	561	
1N5851			ZD-6.8	561	1N754
1N5851A			ZD-6.8	561	
1N5851B	5071A	3334	ZD-6.8	561	
1N5852	138A		ZD-7.5		1N755
1N5852A	138A		ZD-7.5		
1N5852B	138A		ZD-7.5		
1N5853	5016A		ZD-8.2	562	1N756
1N5853A			ZD-8.2	562	
1N5853B	5072A		ZD-8.2	562	
1N5854			ZD-8.7	562	.5M8.7AZ10
1N5854A			ZD-8.7	562	
1N5854B	5073A	3749	ZD-8.7	562	
1N5855			ZD-9.1	562	1N757
1N5855A	139A		ZD9.1	562	
1N5855B	139A		ZD-9.1	562	
1N5856			ZD-10	562	1N758
1N5856A			ZD-10	562	
1N5856B	140A	3061	ZD-10	562	
1N5857			ZD-11	563	.5M11AZ10
1N5857A			ZD-11	563	
1N5857B	5074A		ZD-11	563	
1N5858			ZD-12	563	1N759
1N5858A	142A	3062	ZD-12	563	
1N5858B	142A		ZD-12	563	
1N5859			ZD-13	563	1N964A
1N5859A			ZD-13	563	
1N5859B	143A	3750	ZD-13	563	
1N5860				564	.5M14Z10
1N5860A			ZD-14	564	
1N5860B	144A			564	

Industry Standard No.	ECG	SK	GE	RS 276-	MOTOR.
1N5861			ZD-15	564	1N965A
1N5861A			ZD-15	564	
1N5861B	145A		ZD-15	564	
1N5862			ZD-16	564	1N966A
1N5862A			ZD-16	564	
1N5862B	5075A		ZD-16	564	
1N5863			ZD-17		.5M17Z10
1N5863B	5076A	9022	ZD-17		
1N5864			ZD-18		1N967A
1N5864B	5077A		ZD-18		
1N5865			ZD-19		.5M19Z10
1N5865B	5078A	9023	ZD-19		
1N5866			ZD-20		1N968A
1N5866B	5079A		ZD-20		
1N5867			ZD-22		1N969A
1N5867B	5080A		ZD-22		
1N5868			ZD-24		1N970A
1N5868B	5081A		ZD-24		
1N5869			ZD-25		.5M25Z10
1N5869B	5082A	3753	ZD-25		
1N5870			ZD-27		1N971A
1N5870B	146A		ZD-27		
1N5871			ZD-27		.5M28Z10
1N5871B	5083A	3754	ZD-28		
1N5872			ZD-30		1N972A
1N5872B	5084A	3755	ZD-30		
1N5873			ZD-33		1N973A
1N5873B	147A		ZD-33		
1N5874			ZD-36		1N974A
1N5874B	5085A		ZD-36		
1N5875			ZD-39		1N975A
1N5875B	5086A		ZD-39		
1N5876			ZD-43		1N976A
1N5876B	5087A		ZD-43		
1N5877			ZD-47		1N977A
1N5877B	5088A		ZD-47		
1N5878			ZD-51		1N978A
1N5878B	5089A		ZD-51		
1N5879			ZD-56		1N979A
1N5879B	5090A		ZD-56		
1N5880			ZD-60		.5M60Z10
1N5880B	5091A		ZD-60		
1N5881			ZD-62		1N980A
1N5881B	149A		ZD-62		
1N5882			ZD-68		1N981A
1N5882B	5092A		ZD-68		
1N5883			ZD-75		1N982A
1N5883B	5093A		ZD-75		
1N5884			ZD-82		1N983A
1N5884B	150A		ZD-82		
1N5885			ZD-87		.5M87Z10
1N5885B	5094A	9024	ZD-87		
1N5886			ZD-91		1N984A
1N5886B	5095A		ZD-91		
1N5887			ZD-100		1N985A
1N5887B	5096A		ZD-100		
1N5888			ZD-110		1N986A
1N5889			ZD-120		1N987A
1N5889A			ZD-120	562	
1N5889B	5097A		ZD-120		
1N5890	5098A	3348	ZD-130		1N988A
1N5890A	5098A	3348	ZD-130		
1N5890B	5098A		ZD-130		
1N5891	5099A	3349	ZD-140		.4M140Z10
1N5891A	5099A	3349	ZD-140		
1N5891B	5099A	3349	ZD-140		
1N5892			ZD-150		1N989A
1N5892B	5100A		ZD-150		
1N5893			ZD-160		1N990A
1N5893B	5101A		ZD-160		
1N5894			ZD-170		.4M170Z10
1N5894B	5102A	3352	ZD-170		
1N5895			ZD-180		1N991A
1N5895B	5103A		ZD-180		
1N5896			ZD-190		.4M190Z10
1N5896A		3354	ZD-190		
1N5896B	5104A	3354	ZD-190		
1N5897			ZD-190		1N992A
1N5897B	5105A	3355	ZD-200		
1N58A	109	3087		1123	
1N5913B	5111A		5ZD-3.3		
1N5914B	5112A		5ZD-3.6		
1N5915B	5113A		5ZD-3.9		
1N5916B	5114A		5ZD-4.3		
1N5917B	5115A		5ZD-4.7		
1N5918B	5116A		5ZD-5.1		
1N5919B	5117A		5ZD-5.6		
1N5920B	5119A		5ZD-6.2		
1N5921B	5120A		5ZD-6.8		
1N5922B	5121A		5ZD-7.5		
1N5923B	5122A		5ZD-8.2		
1N5924B	5124A		5ZD-9.1		
1N5925B	5125A		5ZD-10		
1N5926B	5126A		5ZD-11		
1N5927B	5127A		5ZD-12		
1N5928B	5128A		5ZD-13		
1N5929B	5130A		5ZD-15		
1N5930B	5131A		5ZD-16		
1N5931B	5133A		5ZD-18		
1N5932B	5135A		5ZD-20		
1N5933B	5136A		5ZD-22		
1N5934B	5137A		5ZD-24		
1N5935B	5139A		5ZD-27		
1N5936B	5141A		5ZD-30		
1N5937B	5142A		5ZD-33		
1N596	116	3017B/117	504A	1104	1N4005
1N5968	611	3324			
1N5968A	611	3324			
1N597	125	3032A	510,531	1114	1N4006
1N598	125	3033	510,531	1114	1N4007
1N5985					1N5221
1N5986					1N5223
1N5986B	5063A	3837			
1N5987					1N5225
1N5987B	5065A	3838			
1N5988					1N5226
1N5988B	5066A	3330	ZD-3.3		
1N5989					1N5227
1N599	116	3016	504A	1104	
1N599,A					1N4001
1N5990					1N5228
1N5991					1N5229
1N5992					1N5230
1N5993					1N5231
1N5994					1N5232
1N5995					1N5234
1N5996					1N5235
1N5997					1N5236
1N5998					1N5237
1N5999					1N5239
1N599A	116	3016	504A	1104	
00001N60	109	3088		1123	
1N60(TV)(FA-1)	109		1N60	1123	
1N60-1			1N60	1123	

Industry Standard No.	ECG	SK	GE	RS 276-	MOTOR.
1N60-5	110MP	3088	1N60	1123(2)	
1N60-FA1			1N60	1123	
1N60-M3	109	3088	1N60	1123	
1N60-P	109		1N60	1123	
1N60-S	109		1N60	1123	
1N60-T	109	3088	1N60	1123	
1N60-TF1	109	3088			
1N60-Z	109	3088	1N60	1123	
1N60/3490	109		1N60	1123	
1N60/4454C	109			1123	
1N60/7825B	110MP	3088	1N60	1123	
1N600	116	3016	504A	1104	
1N600,A					1N4002
1N6000					1N5240
1N6001					1N5241
1N6002					1N5242
1N6003					1N5243
1N6003B	143A	3750			
1N6004					1N5245
1N6005					1N5246
1N6006					1N5248
1N6006B	5077A	3752			
1N6007					1N5250
1N6008					1N5251
1N6009					1N5252
1N600A	116	3016	504A	1104	
1N601	116	3016	504A	1104	
1N601,A					1N4003
1N6010					1N5254
1N6011	5035A				1N5256
1N6012					1N5257
1N6013					1N5258
1N6014					1N5259
1N6015					1N5260
1N6016					1N5261
1N6017					1N5262
1N6018					1N5263
1N6019					1N5265
1N601A	116	3016	504A	1104	
1N602	116	3016	504A	1104	
1N602,A					1N4003
1N6020					1N5266
1N6021					1N5267
1N6022					1N5268
1N6023					1N5270
1N6024					1N5271
1N6025					1N5272
1N6026					1N5273
1N6027					1N5274
1N6028					1N5276
1N6029					1N5277
1N602A	116	3016	504A	1104	
1N603	116	3016	504A	1104	
1N603,A					1N4004
1N6030					1N5279
1N6031					1N5281
1N6031B	5105A	3355	ZD-200		
1N603A	116	3016	504A	1104	
1N604	116	3016	504A	1104	
1N604,A					1N4004
1N604A	116	3016	504A	1104	
1N605	116	3016	504A	1104	
1N605,A					1N4005
1N605A	116	3016	504A	1104	
1N606	116	3033	504A	1104	
1N606,A					1N4005
1N606A	116	3017B/117	504A	1104	
1N607	5830	3500/5882	5004		
1N607,A					1N1199C
1N607A	5830	3500/5882	5004		
1N608	5834	3500/5882	5004		
1N608,A					1N1200C
1N608A	5834	3500/5882	5004		
1N609	5834	3500/5882	5004		
1N609,A					1N1202C
1N609A	5834	3500/5882	5004		
1N60A	109	3088	1N60	1123	
1N60AM	109	3088	1N60	1123	
1N60B			1N60	1123	
1N60C	109	3088	1N60	1123	
1N60D	109	3088	1N60	1123	
1N60F	109	3088	1N60	1123	
1N60FA1			1N60	1123	
1N60FD1	109	3088	1N60	1123	
1N60FM	109	3088	1N60	1123	
1N60FM1	109	3088			
1N60FMX	109		1N60	1123	
1N60G	109	3088	1N60	1123	
1N60GA	109	3088	1N60	1123	
1N60GB	109	3088	1N60	1123	
1N60M	109		1N34AS	1123	
1N60M3			1N60	1123	
1N60MP	110MP	3709	1N60	1123(2)	
1N60P	109	3088	1N60	1123	
1N60R			1N60	1123	
1N60S	109	3088	1N60	1123	
1N60SD60	109		1N60	1123	
1N60T			1N60	1123	
1N60TV	109	3088	1N60	1123	
1N60TV-TOGL			1N60	1123	
1N60TVGL	109	3088	1N60	1123	
1N60Z			1N60	1123	
1N610	5834	3500/5882	5004		
1N610,A					1N1202C
1N610A	5834	3500/5882	5004		
1N611	5838	3500/5882	5004		
1N611,A					1N1204C
1N611A	5838	3500/5882	5004		
1N612	5838	3500/5882	5004		
1N612,A					1N1204C
1N612A	5838	3500/5882	5004		
1N613	5842	3500/5882	5008	1104	
1N613,A					1N1206C
1N613A	5842	3500/5882	5008	1104	
1N614	5842	7042	5008	1104	
1N614,A					1N1206C
1N614A	5842	7042	5008	1104	
1N616	5861	3087		1123	
1N617	109	3087		1123	
1N618	109	3087		1123	
1N619	177		300	1122	
1N62	109	3087		1123	
1N622	177	3100/519	300	1122	
1N625	177	3311	300	1122	
1N625A	177	3100/519	300	1122	
1N625M	177	3100/519	300	1122	
1N626	177	3311	300	1122	
1N6263					MBD701
1N626A	177	3100/519	300	1122	
1N626M	177	3100/519	300	1122	
1N627	177	3100/519	300	1122	
1N627A	177	3100/519	300	1122	
1N628	177	3100/519	300	1122	
1N628A	177	3100/519	300	1122	

Industry Standard No.	ECG	SK	GE	RS 276-	MOTOR.
1N629	177	3100/519	300	1122	
1N629A	177	3100/519	300	1122	
1N63	109	3087		1123	
1N631	109	3087		1123	
1N632	109	3087		1123	
1N633	177	3100/519	300	1122	
1N636	109	3087		1123	
1N636A			1N34AS	1123	
1N63A	109	3087		1123	
1N64	109	3091		1123	
1N643	177	3100/519	300	1122	
1N643A	177	3016	300	1122	
1N645	116	3311	504A	1104	
1N645A	116	3311	504A	1104	
1N645B	116	3311	504A	1104	
1N646	116	3017B/117	504A	1104	
1N647	116	3017B/117	504A	1104	
1N648	116	3017B/117	504A	1104	
1N649	116	3017B/117	504A	1104	
1N64A	109	3091		1123	
1N64B	109	3091		1123	
1N64G	109	3087		1123	
1N64GA	109	3091		1123	
1N64P			1N60	1123	
1N65	109	3087		1123	
1N658	177	3016	300	1122	
1N658A	177	3100/519	300	1122	
1N658M	177	3100/519	300	1122	
1N659	177	3311	300	1122	
1N659/A			300	1122	
1N659A	177	3100/519	300	1122	
1N65A	109	3087		1123	
1N66	109	3087		1123	
1N660	177	3100/519	300	1122	
1N660A	177	3100/519	300	1122	
1N661	177	3100/519	300	1122	
1N661A	177	3100/519	300	1122	
1N662	177	3100/519	300	1122	
1N662A	177	3100/519	300	1122	
1N663	177	3016	300	1122	
1N663A	177	3100/519	300	1122	
1N663M	177	3100/519	300	1122	
1N664	5072A	3136	ZD-8.2	562	1N5237A
1N665	142A	3062	ZD-12	563	1N5242A
1N666	145A	3063	ZD-15	564	1N5245B
1N667	5077A	3752	ZD-18		1N5248A
1N668	5080A	3336	ZD-22		1N5251A
1N669	146A	3064	ZD-27		1N5254A
1N66A	109	3087		1123	
1N67	109	3087		1123	
1N670	5092A	3343	ZD-68		1N5266B
1N671	5096A	3346	ZD-100		1N5271A
1N672	5100A	3350	ZD-150		1N5276A
1N673	116	3016	504A	1104	
1N674	5069A	3333	ZD-4.7		1N5230A
1N675	137A	3058	ZD-6.2	561	1N5234B
1N676	116	3017B/117	504A	1104	
1N677	116	3017B/117	504A	1104	
1N678	116	3017B/117	504A	1104	
1N679	116	3017B/117	504A	1104	
1N67A	109	3087	1N34AS	1123	
1N67D			1N60	1123	
1N68	109	3087		1123	
1N681	116	3017B/117	504A	1104	
1N682	116	3017B/117	504A	1104	
1N683	116	3017B/117	504A	1104	
1N684	116	3017B/117	504A	1104	
1N685	116	3017B/117	504A	1104	
1N686	116	3017B/117	504A	1104	
1N687	116	3017B/117	504A	1104	
1N689	116	3017B/117	504A	1104	
1N68A	109	3087		1123	
1N69	109	3087		1123	
1N690	177	3100/519	300	1122	
1N692	116		504A	1104	
1N693	506	3125	511	1114	
1N695	109	3087		1123	
1N695A	109	3087		1123	
1N696	177	3100/519	300	1122	
1N697	177	3100/519	300	1122	
1N698	109	3087		1123	
1N69A	109	3087		1123	
1N70	109	3087		1123	
1N701	5019A	3785	ZD-10	562	
1N702	5002A	3768			
1N702A	5002A	3768			
1N703	5006A	3772	ZD-3.6		
1N703A	5006A	3772	ZD-3.6		
1N704		3773	ZD-3.9		
1N704A		3773	ZD-3.9		
1N705	5009A	3775	ZD-4.7		
1N705A	5009A	3775	ZD-4.7		
1N706	5012A	3778	ZD-6.0	561	
1N706A	5012A	3778	ZD-6.0	561	
1N707	5014A	3780	ZD-6.8	561	
1N707A	5014A	3780	ZD-6.8	561	
1N707A(ZENER)	5014A		ZD-6.8	561	
1N708	5011A	3777	ZD-5.6	561	
1N708A	136A	3777/5011A	ZD-5.6	561	
1N709	137A	3058	ZD-6.2	561	
1N70980			300	1122	
1N709A	137A	3058	ZD-6.2	561	
1N709B	137A	3058	ZD-6.2	561	
1N70A	109	3087		1123	
1N71	109	3087		1123	
1N710	5014A	3780	ZD-6.8	561	
1N710A	5071A	3780/5014A	ZD-6.8	561	
1N711	5015A	3781	ZD-7.5		
1N711A	138A		ZD-7.5		
1N712	5016A	3782	ZD-8.2	562	
1N712A	5072A	3782/5016A	ZD-8.2	562	
1N713	5018A	3784	ZD-9.1	562	
1N713A	139A	3784/5018A	ZD-9.1	562	
1N713B	139A	3784/5018A	ZD-9.1	562	
1N714	5019A	3785	ZD-10	562	
1N714A	140A	3785/5019A	ZD-10	562	
1N714B	5019A	3785	ZD-10	562	
1N715	5020A	3786	ZD-11	563	
1N715A	5020A		ZD-11	563	
1N716	5021A	3787	ZD-12	563	
1N716(ZENER)	5021A	3787	ZD-12	563	
1N716A	142A	3787/5021A	ZD-12	563	
1N716B	142A	3787/5021A	ZD-12	563	
1N717	5022A	3788	ZD-13	563	
1N717A	143A	3788/5022A	ZD-13	563	
1N718	5024A	3790	ZD-15	564	
1N718A	145A	3790/5024A	ZD-15	564	
1N718B	145A	3790/5024A	ZD-15	564	
1N719	5025A	3791	ZD-15	564	
1N719A	5075A	3751		564	
1N72	109	3087		1123	
1N720	5027A	3793	ZD-18		
1N720A	5077A	3793/5027A	ZD-18		
1N720B	5027A	3793	ZD-18		
1N721	5029A	3795	ZD-20		
1N721A	5079A	3795/5029A	ZD-20		
1N722	5030A	3796	ZD-22		
1N722A	5080A	3796/5030A	ZD-22		
1N723	5031A	3797	ZD-24		
1N723A	5081A		ZD-24		
1N724	5033A	3799	ZD-27		
1N724A	146A	3799/5033A	ZD-27		
1N724B	146A	3799/5033A	ZD-27		
1N725	5035A	3755/5084A	ZD-30		
1N725A	5084A	3755	ZD-30		
1N726	5036A	3802	ZD-33		
1N726A	147A	3802/5036A	ZD-33		
1N726B	5036A		ZD-33		
1N727	5037A	3803	ZD-36		
1N727A	5085A	3803/5037A	ZD-36		
1N728	5086A	3804/5038A	ZD-39		
1N728A	5038A	3804	ZD-39		
1N729	5039A	3805	ZD-43		
1N729A	5087A	3805/5039A	ZD-43		
1N72G	109	3087		1123	
1N73	109	3087	1N60	1123	
1N730	5040A	3806	ZD-47		
1N730A	5088A	3806/5040A	ZD-47		
1N731	5041A	3807	ZD-51		
1N731A	5089A	3807/5041A	ZD-51		
1N731B	5041A	3807	ZD-51		
1N732	5042A	3808	ZD-56		
1N732A	5090A	3808/5042A	ZD-56		
1N732B	5042A	3808	ZD-56		
1N733	149A	3097	ZD-62		
1N733A	149A	3097	ZD-62		
1N734	5045A	3811	ZD-68		
1N734A	5092A	3811/5045A	ZD-68		
1N734B	5045A	3811	ZD-68		
1N735	5046A	3812	ZD-75		
1N735A	5093A	3812/5046A	ZD-75		
1N735B	5046A	3812	ZD-75		
1N736	150A	3098	ZD-82		
1N736A	150A	3098	ZD-82		
1N737	5049A	3815	ZD-91		
1N737A	5095A	3815/5049A	ZD-91		
1N737B	5049A	3815	ZD-91		
1N738	5050A	3816	ZD-100		
1N738A	5096A	3816/5050A	ZD-100		
1N738B	5050A	3816	ZD-100		
1N739	151A	3099	ZD-110		
1N739A	151A	3099	ZD-110		
1N74	109	3087		1123	
1N740	5052A	3818	ZD-120		
1N741	5053A	3819	ZD-130		
1N741A	5098A	3821/5055A	ZD-130		
1N741B	5055A	3821	ZD-150		
1N742	5055A	3821	ZD-150		
1N742A	5100A	3821/5055A	ZD-150		
1N742B	5055A	3821	ZD-150		
1N743	5056A	3822	ZD-160		
1N743A	5101A	3822/5056A	ZD-160		
1N744	5058A	3824	ZD-180		
1N744A	5103A	3824/5058A	ZD-180		
1N744B	5058A	3824	ZD-180		
1N745	5060A	3826	ZD-200		
1N745A	5105A	3826/5060A	ZD-200		
1N745B	5060A	3826	ZD-200		
1N746	5005A	3771	ZD-3.3		
1N746A	5066A	3771/5005A	ZD-3.3		
1N747	5006A	3772	ZD-3.6		
1N747A	134A	3772/5006A	ZD-3.6		
1N747B	134A	3772/5006A	ZD-3.6		
1N748	5007A	3773	ZD-3.9		
1N748A	5067A	3773/5007A	ZD-3.9		
1N749	5008A	3774	ZD-4.3		
1N749A	5008A	3774	ZD-4.3		
1N750	5009A	3775	ZD-4.7		
1N750A	5069A	3775/5009A	ZD-4.7		
1N751	5010A	3776	ZD-5.1		
1N751A	5010A	3776	ZD-5.1		
1N752	136A	3777/5011A	ZD-5.6	561	
1N752A	5011A	3777	ZD-5.6	561	
1N753	5013A	3779	ZD-6.2	561	
1N753A	137A	3779/5013A	ZD-6.2	561	
1N754	140A	3334/5071A	ZD-10	562	
1N754A	5014A	3334/5071A	ZD-6.8	561	
1N755	5015A	3781	ZD-7.5		
1N755A	5015A	3781	ZD-7.5		
1N756	5072A	3136	ZD-8.2	562	
1N756A	5072A	3136	ZD-8.2	562	
1N757	5018A	3784	ZD-9.1	562	
1N757A	139A	3784/5018A	ZD-9.1	562	
1N758	5019A	3785	ZD-10	562	
1N758A	140A	3785/5019A	ZD-10	562	
1N759	5021A	3787	ZD-12	563	
1N759A	142A	3787/5021A	ZD-12	563	
1N75A	109	3087		1123	
1N76	109	3087		1123	
1N760	177	3091	300	1122	
1N761	5009A	3775	ZD-4.7		
1N761-1	5009A	3775	ZD-4.7		
1N761-2	135A		ZD-5.1		
1N761A	135A	3775/5009A	ZD-5.1		
1N762	5012A	3778	ZD-6.0	561	
1N762-1	5011A	3777	ZD-5.6	561	
1N762-2	5070A	9021	ZD-6.0	561	
1N762A	136A		ZD-5.6	561	
1N763		3781	ZD-7.5		
1N763-1	5014A	3781/5015A	ZD-6.8	561	
1N763-2	5014A	3781/5015A	ZD-6.8	561	
1N763-3	138A		ZD-7.5		
1N764	5073A	3749	ZD-9.1	562	
1N764-1	5016A	3782	ZD-8.2	562	
1N764-3	5018A	3784	ZD-9.1	562	
1N765		3062	ZD-12	563	
1N765-1	140A	3005	ZD-10	562	
1N765-2	5074A		ZD-11	563	
1N765A	140A		ZD-10	562	
1N766	5022A	3788	ZD-13	563	
1N766-1	142A		ZD-12	563	
1N766-2	143A	3750	ZD-13	563	
1N766-3	144A	3094	ZD-14	564	
1N766A	143A	3788/5022A	ZD-13	563	
1N767	5025A	3791	ZD-15	564	
1N767-1	145A	3791/5025A	ZD-15	564	
1N767-2	5075A	3751			
1N767-22				564	
1N767-3	5076A	9022	ZD-17		
1N768	5028A	3794	ZD-19		
1N768-1	5077A	3752	ZD-18		
1N768-2	5078A	9023	ZD-19		
1N768-3	5079A	3335	ZD-20		
1N768A	5078A	9023	ZD-19		
1N769	5081A		ZD-24		
1N769-1	5080A	3798/5032A	ZD-22		
1N769-2	5081A		ZD-24		
1N769-3	5032A	3798	ZD-25		

Industry Standard No.	ECG	SK	GE	RS 276-	MOTOR.
1N769-4	5083A	3754	ZD-28		
1N76A	109	3087		1123	
1N76C	109	3087		1123	
1N76G	109	3087		1123	
1N770	109	3087		1123	
1N771	109	3087		1123	
1N771A	109	3087		1123	
1N771B	109	3087		1123	
1N772A	109	3087		1123	
1N773	109	3087		1123	
1N773A	109	3087		1123	
1N774	109	3087		1123	
1N774A	109	3087		1123	
1N775	109	3087		1123	
1N776	109	3087		1123	
1N777	109	3087		1123	
1N778	177	3100/519	300	1122	
1N779	177	3100/519	300	1122	
1N781	109	3087		1123	
1N788	177	3100/519	300	1122	
1N789	177	3100/519	300	1122	
1N789M	177	3100/519	300	1122	
1N790	177	3100/519	300	1122	
1N790M	177	3100/519	300	1122	
1N791	177	3100/519	300	1122	
1N791M	177	3100/519	300	1122	
1N792	177	3100/519	300	1122	
1N792M	177	3100/519	300	1122	
1N793	177	3100/519	300	1122	
1N793M	177	3100/519	300	1122	
1N794	177	3100/519	300	1122	
1N795	177	3100/519	300	1122	
1N796	177	3100/519	300	1122	
1N797	177	3100/519	300	1122	
1N798	177	3100/519	300	1122	
1N799	177	3100/519	300	1122	
1N800	177	3100/519	300	1122	
1N801	177	3100/519	300	1122	
1N802	177	3100/519	300	1122	
1N803	177	3100/519	300	1122	
1N804	177	3100/519	300	1122	
1N805	109	3087		1123	
1N806	177	3100/519	300	1122	
1N807	177	3100/519	300	1122	
1N808	177	3100/519	300	1122	
1N809	177	3100/519	300	1122	
1N81	109	3087		1123	
1N810	177	3100/519	300	1122	
1N811	177	3100/519	300	1122	
1N811M	177	3100/519	300	1122	
1N812	177	3100/519	300	1122	
1N812M	177	3100/519	300	1122	
1N813	177	3100/519	300	1122	
1N813M	177	3100/519	300	1122	
1N814	177	3100/519	300	1122	
1N814M	177	3100/519	300	1122	
1N815	177	3100/519	300	1122	
1N815M	177	3100/519	300	1122	
1N816	5070A	9021	ZD-6.2	561	
1N817	177	3100/519	300	1122	
1N818	177	3100/519	300	1122	
1N818M	177	3100/519	300	1122	
1N819	116	3016	504A	1104	
1N81A	109	3087		1123	
1N82	112		1N82A		
1N821		3058	ZD-6.2	561	
1N821A		3058	ZD-6.2	561	
1N822		3058	ZD-6.2	561	
1N823		3058	ZD-6.2	561	
1N823A		3058	ZD-6.2	561	
1N824		3058	ZD-6.2	561	
1N825		3058	ZD-6.2	561	
1N825A			ZD-6.2	561	
1N826		3780	ZD-6.8	561	1N825
1N826A		3780	ZD-6.8	561	
1N827		3058	ZD-6.2	561	
1N827A		3058	ZD-6.2	561	
1N828		3780	ZD-6.8	561	1N827
1N828A		3780	ZD-6.8	561	
1N829		3058	ZD-6.2	561	
1N829A		3058	ZD-6.2	561	
1N82A	112	3089	1N82A		
1N82AG	112	3089	1N82A		
1N82D	112		1N82A		
1N82G	112	3089	1N82A		
1N835	109	3087		1123	
1N837	177	3100/519	300	1122	
1N837A	177	3100/519	300	1122	
1N838	177	3100/519	300	1122	
1N839	177	3100/519	300	1122	
1N84	109	3087		1123	
1N840	177	3100/519	300	1122	
1N840M	177	3100/519	300	1122	
1N841	177	3100/519	300	1122	
1N842	177	3100/519	300	1122	
1N843	177	3100/519	300	1122	
1N844	177	3100/519	300	1122	
1N845	177	3100/519	300	1122	
1N846	116	3016	504A	1104	
1N847	116	3016	504A	1104	
1N848	116	3016	504A	1104	
1N849	116	3016	504A	1104	
1N850	116	3016	504A	1104	
1N851	116	3016	504A	1104	
1N852	116	3017B/117	504A	1104	
1N853	125	3032A	510,531	1114	
1N854	125	3032A	510,531	1114	
1N855	125	3033	510,531	1114	
1N856	125	3033	510,531	1114	
1N857	116	3016	504A	1104	
1N858	116	3016	504A	1104	
1N859	116	3016	504A	1104	
1N86	109	3087		1123	
1N860	116	3016	504A	1104	
1N861	116	3016	504A	1104	
1N862	116	3016	504A	1104	
1N863	116	3017B/117	504A	1104	
1N864	125	3033	510,531	1114	
1N865	125	3033	510,531	1114	
1N866	125	3033	510,531	1114	
1N867	125	3033	510,531	1114	
1N868	116	3016	504A	1104	
1N869	116	3016	504A	1104	
1N86AG	109			1123	
1N87	109	3091		1123	
1N870	116	3016	504A	1104	
1N871	116	3016	504A	1104	
1N872	116	3016	504A	1104	
1N873	116	3016	504A	1104	
1N874	116	3017B/117	504A	1104	
1N875	125	3032A	510,531	1114	
1N876	125	3033	510,531	1114	
1N877	125	3033	510,531	1114	
1N878	125	3033	510,531	1114	

Industry Standard No.	ECG	SK	GE	RS 276-	MOTOR.
1N879	116	3016	504A	1104	
1N87A	109	3091		1123	
1N87G	109	3091		1123	
1N87GA	109	3087		1123	
1N87S	109	3091		1123	
1N87T	109	3091		1123	
1N88	109	3087		1123	
1N880	116	3016	504A	1104	
1N881	116	3311	504A	1104	
1N881B			504A	1104	
1N882	116	3017B/117	504A	1104	
1N883	116	3016	504A	1104	
1N884	116	3016	504A	1104	
1N885	116	3017B/117	504A	1104	
1N886	125	3033	510,531	1114	
1N887	125	3033	510,531	1114	
1N888	125	3033	510,531	1114	
1N889	125	3033	510,531	1114	
1N89	109	3087		1123	
1N890	177	3100/519	300	1122	
1N891	177	3100/519	300	1122	
1N892	177	3100/519	300	1122	
1N893	506	3843	511	1114	
1N897	177	3100/519	300	1122	
1N898	177	3100/519	300	1122	
1N899	177	3100/519	300	1122	
1N90	109	3087		1123	
1N900	177	3100/519	300	1122	
1N901	177	3100/519	300	1122	
1N902	177	3100/519	300	1122	
1N903	177	3100/519	300	1122	
1N903/A			300	1122	
1N903A	177	3100/519	300	1122	
1N903AM	177	3100/519	300	1122	
1N903M	177	3100/519	300	1122	
1N904	177	3100/519	300	1122	
1N904/A			300	1122	
1N904A	177	3100/519	300	1122	
1N904AM	177	3100/519	300	1122	
1N904M	177	3100/519	300	1122	
1N905	177	3100/519	300	1122	
1N905/A			300	1122	
1N905A	177	3100/519	300	1122	
1N905AM	177	3100/519	300	1122	
1N905M	177	3100/519	300	1122	
1N906	177	3100/519	300	1122	
1N906/A			300	1122	
1N906A	177	3100/519	300	1122	
1N906AM	177	3100/519	300	1122	
1N906M	177	3100/519	300	1122	
1N907	177	3100/519	300	1122	
1N907/A			300	1122	
1N907A	177	3100/519	300	1122	
1N907AM	177	3100/519	300	1122	
1N907M	177	3100/519	300	1122	
1N908	177	3100/519	300	1122	
1N908/A			300	1122	
1N908A	177	3100/519	300	1122	
1N908AM	177	3100/519	300	1122	
1N908M	177	3100/519	300	1122	
1N909	109	3087		1123	
1N90G	109	3087		1123	
1N90GA	109	3087		1123	
1N91	116	3311	504A	1104	
1N910	109	3087		1123	
1N911	109	3087		1123	
1N914	177	3100/519	514	1122	
1N914/A/B			514	1122	
1N914A	178MP	3100/519	514	1122	
1N914B	177	3100/519	514	1122	
1N914M	177	3100/519	514	1122	
1N915	177	3100/519	300	1122	
1N916	519	3017B/117	504A	1122	
1N916A	177	3100/519	300	1122	
1N916B	177	3100/519	300	1122	
1N917	177	3100/519	300	1122	
1N919	177	3311	300	1122	
1N92	116	3017B/117	504A	1104	
1N920	177	3100/519	300	1122	
1N921	177	3100/519	300	1122	
1N922	177	3100/519	300	1122	
1N923	506	3843	511	1114	
1N924	177	3100/519	300	1122	
1N925	177	3100/519	300	1122	
1N926	177	3100/519	300	1122	
1N927	177	3100/519	300	1122	
1N928	177	3100/519	300	1122	
1N929	178MP		300(2)	1122	
1N93	116	3017B/117	504A	1104	
1N930	177	3100/519	300	1122	
1N931	177	3100/519	300	1122	
1N932	177	3100/519	300	1122	
1N933	177	3100/519	300	1122	
1N934	177	3100/519	300	1122	
1N935			ZD-9.1	562	
1N935A			ZD-9.1	562	
1N935B			ZD-9.1	562	
1N936			ZD-9.1	562	
1N936A			ZD-9.1	562	
1N936B			ZD-9.1	562	
1N937			ZD-9.1	562	
1N937A			ZD-9.1	562	
1N937B			ZD-9.1	562	
1N938			ZD-9.1	562	
1N938A			ZD-9.1	562	
1N938B			ZD-9.1	562	
1N939			ZD-9.1	562	
1N939A			ZD-9.1	562	
1N939B			ZD-9.1	562	
1N93A	116	3017B/117	504A	1104	
1N941		3092	ZD-11.5	563	
1N941A		3092	ZD-11.5	563	
1N941B		3092	ZD-11.5	563	
1N942		3092	ZD-11.5	563	
1N942A			ZD-12	563	
1N942B			ZD-12	563	
1N943			ZD-12	563	
1N943A			ZD-12	563	
1N943B			ZD-12	563	
1N944			ZD-12	563	
1N944A			ZD-12	563	
1N944B			ZD-12	563	
1N945		3092	ZD-11.5	563	
1N945A		3092	ZD-11.5	563	
1N945B		3092	ZD-11.5	563	
1N946		3092	ZD-11.5	563	
1N946A		3092	ZD-11.5	563	
1N946B		3092	ZD-11.5	563	
1N947	116	3017B/117	504A	1104	
1N948	177	3100/519	300	1122	
1N949	109	3087		1123	
1N95	109	3087		1123	
1N957	5014A	3780	ZD-6.8	561	
1N957A	5014A	3780	ZD-6.8	561	

Industry Standard No.	ECG	SK	GE	RS 276-	MOTOR.
1N957B	5071A		ZD-6.8	561	
1N958	5015A	3781	ZD-7.5		
1N958A	5015A	3781	ZD-7.5		
1N958B	138A		ZD-7.5		
1N959	5016A	3782	ZD-8.2	562	
1N959A	5016A	3782	ZD-8.2	562	
1N959B	5072A	3782/5016A	ZD-8.2	562	
1N96	109	3087		1123	
1N960	139A	3784/5018A	ZD-9.1	562	
1N960A	139A	3784/5018A	ZD-9.1	562	
1N960B	139A	3784/5018A	ZD-9.1	562	
1N961	140A		ZD-10	562	
1N961A	140A		ZD-10	562	
1N961B	140A	3060/139A	ZD-10	562	
1N962	5020A	3786	ZD-11	563	
1N962A	5020A	3786	ZD-11	563	
1N962B	5020A	3786	ZD-11.0	563	
1N963	5021A	3787	ZD-12	563	
1N963A	5021A	3787	ZD-12	563	
1N963B	142A	3787/5021A	ZD-12	563	
1N964	5022A	3788	ZD-13	563	
1N964A	5022A	3788	ZD-13	563	
1N964B	5022A	3788	ZD-13	563	
1N964B-F	143A	3788/5022A			
1N965	5024A	3790	ZD-15	564	
1N965A	5024A	3790	ZD-15	564	
1N965B	145A	3790/5024A	ZD-15	564	
1N966	5025A	3791	ZD-15	564	
1N966A	5025A	3791	ZD-16	564	
1N966B	5075A	3751	ZD-16	564	
1N966B-F	5075A	3751			
1N967	5027A	3793	ZD-18		
1N967A	5027A	3793	ZD-18		
1N967B	5027A	3793	ZD-18		
1N967B-F	5077A	3793/5027A			
1N968	5029A	3795	ZD-20		
1N968A	5029A	3795	ZD-20		
1N968B	5029A	3795	ZD-20		
1N969	5080A	3796/5030A	ZD-22		
1N969A	5030A	3796	ZD-22		
1N969B	5080A	3796/5030A	ZD-22		
1N96A	109	3087		1123	
1N97	109	3087		1123	
1N970	5031A	3797	ZD-24		
1N970A	5031A	3797	ZD-24		
1N970B	5031A		ZD-24		
1N971	146A	3799/5033A	ZD-27		
1N971A	146A	3799/5033A	ZD-27		
1N971B	146A	3799/5033A	ZD-27		
1N972	5035A	3801	ZD-30		
1N972A	5035A	3801	ZD-30		
1N972B	5084A	3801/5035A	ZD-30		
1N972B-F	5084A	3801/5035A			
1N973	5036A	3802	ZD-33		
1N973A	5036A	3802	ZD-33		
1N973B	147A	3802/5036A	ZD-33		
1N974	5037A	3803	ZD-36		
1N974A	5037A	3803	ZD-36		
1N974B	5085A	3803/5037A	ZD-36		
1N975	5038A	3804	ZD-39		
1N975A	5038A	3804	ZD-39		
1N975B	5086A	3804/5038A	ZD-39		
1N976	5039A	3805	ZD-43		
1N976A	5039A	3805	ZD-43		
1N976B	5087A	3805/5039A	ZD-43		
1N977	5040A	3806	ZD-47		
1N977A	5040A	3806	ZD-47		
1N977B	5088A	3806/5040A	ZD-47		
1N978	5041A	3807	ZD-51		
1N978A	5041A	3807	ZD-51		
1N978B	5089A	3807/5041A	ZD-51		
1N979	5042A	3808	ZD-56		
1N979A	5042A	3808	ZD-56		
1N979B	5090A	3808/5042A	ZD-56		
1N97A	109	3087		1123	
1N98	109	3087		1123	
1N980	5044A	3810	ZD-62		
1N980A	5044A	3810	ZD-62		
1N980B	149A	3810/5044A	ZD-62		
1N981	5045A	3811	ZD-68		
1N981A	5045A	3811	ZD-68		
1N981B	5092A	3811/5045A	ZD-68		
1N982	5046A	3812	ZD-75		
1N982A	5046A	3812	ZD-75		
1N982B	5093A		ZD-75		
1N983	5047A	3813	ZD-82		
1N983A	5047A	3813	ZD-82		
1N983B	150A	3813/5047A	ZD-82		
1N984	5049A	3815	ZD-91		
1N984A	5049A	3815	ZD-91		
1N984B		3815	ZD-91		
1N985	5050A	3816	ZD-100		
1N985A	5050A	3816	ZD-100		
1N985B	5096A		ZD-100		
1N986	5051A	3817	ZD-110		
1N986A	5051A	3817	ZD-110		
1N986B	151A	3817/5051A	ZD-110		
1N987	5052A	3818	ZD-120		
1N987A	5052A	3818	ZD-120		
1N987B	5097A	3818/5052A	ZD-120		
1N988	5053A	3819	ZD-130		
1N988A	5053A	3819	ZD-130		
1N988B	5053A	3819	ZD-130		
1N989	5055A	3821	ZD-150		
1N989A	5077A	3821/5055A	ZD-150		
1N989B	5100A	3821/5055A	ZD-130		
1N98A	109	3087		1123	
1N99	109	3087		1123	
1N990	5056A	3822	ZD-160		
1N990A	5056A	3822	ZD-160		
1N990B	5101A	3822/5056A	ZD-160		
1N991	5058A	3824	ZD-180		
1N991A	5058A	3824	ZD-180		
1N991B	5103A	3824/5058A	ZD-180		
1N992	5060A	3826	ZD-200		
1N992A	5060A	3826	ZD-200		
1N992B	5105A	3826/5060A	ZD-200		
1N993	177	3100/519	300	1122	
1N994	109	3087		1123	
1N995	177	3100/519	300	1122	
1N995M			1N60	1123	
1N996	109	3087		1123	
1N997	177	3100/519	300	1122	
1N998	116		504A	1104	
1N999	177	3100/519	300	1122	
1N99A	109	3351/5101A		1123	
1NA4	109	3087		1123	
1NA4G	109	3087		1123	
1NC61684	116	3017B/117	504A	1104	
1NE11	156	3033	512	1114	
1NJ27	125		510,531	1114	
1NJ33233	109	3087		1123	
1NJ60284	109	3087		1123	
1NJ61224	109	3087		1123	

Industry Standard No.	ECG	SK	GE	RS 276-	MOTOR.
1NJ61225	136A	3057	ZD-5.6	561	
1NJ61433	177	3100/519	300	1122	
1NJ61675	109	3087		1123	
1NJ61676	116	3016	504A	1104	
1NJ61677	177	3100/519	300	1122	
1NJ61725	177	3100/519	300	1122	
1NJ61726	116		504A	1104	
1NJ70972	113A	3119/113	6GC1		
1NJ70973	109	3088		1123	
1NJ70976	506	3843	511	1114	
1NJ70980	177	3016	300	1122	
1NJ71126	507	3016	504A	1104	
1NJ71185	109	3087		1123	
1NJ71186	507	3016	504A	1104	
1NJ71224	177	3100/519	300	1122	
1NZ61		3080		1114	
1P100	5400	3950			
1P101	5401	3638/5402			
1P102	5402	3638			
1P103	5403	3627/5404			
1P104	5404	3627			
1P541	109	3091		1123	
1P542	109	3091		1123	
1P643	5800	9003		1142	
1P644	5801	9004		1142	
1P645	5803	9006		1144	
1P646	5803	9006		1144	
1P647	5804	9007		1144	
1P649	5806	3848			
1R0	177		300	1122	
1R01	116	3017B/117	504A	1104	
1R0E			300	1122	
1R0F			504A	1104	
1R0H			504A	1104	
1R10	140A	3061	ZD-10	562	
1R100	5096A	3346	ZD-100		
1R100A	5096A	3346	ZD-100		
1R100B	5096A	3346	ZD-100		
1R10A	140A	3061	ZD-10	562	
1R10B	140A		ZD-10	562	
1R10D3K	177	3100/519	300	1122	
1R10D3L	506	3843	511	1114	
1R11	5074A	3092/141A	ZD-11.0	563	
1R110	151A	3099	ZD-110		
1R110A	151A	3099	ZD-110		
1R110B	151A	3099	ZD-110		
1R11A	5074A		ZD-11.0	563	
1R11B	5074A		ZD-11.0	563	
1R12	142A	3062	ZD-12	563	
1R120	5097A	3347	ZD-120		
1R120A	5097A	3347	ZD-120		
1R120B	5097A	3347	ZD-120		
1R12A	142A	3062	ZD-12	563	
1R12B	142A	3062	ZD-12	563	
1R13	143A	3750	ZD-13	563	
1R130	5098A	3348	ZD-130		
1R130A	5098A	3348	ZD-130		
1R130B	5098A	3348	ZD-130		
1R13A	143A	3750	ZD-13	563	
1R13B	143A	3750	ZD-13	563	
1R140	5099A	3349	ZD-140		
1R140A	5099A	3349	ZD-140		
1R140B	5099A	3349	ZD-140		
1R15	145A	3063	ZD-15	564	
1R150	5100A	3350	ZD-150		
1R150A	5100A	3350	ZD-150		
1R150B	5100A	3350	ZD-150		
1R15A	145A	3063	ZD-15	564	
1R15B	145A	3063	ZD-15	564	
1R16	5075A	3751	ZD-16	564	
1R160	5101A	3351	ZD-160		
1R160A	5101A	3351	ZD-160		
1R160B	5101A	3351	ZD-160		
1R161R16B	5075A	3751			
1R16A	5075A	3751	ZD-16	564	
1R16B	5075A	3751	ZD-16	564	
1R18	5077A	3752	ZD-18		
1R180	5103A	3353	ZD-180		
1R180A	5103A	3353	ZD-180		
1R180B	5103A	3353	ZD-180		
1R18A	5077A	3752	ZD-18		
1R18B	5077A	3752	ZD-18		
1R1D	116	3017B/117	504A	1104	
1R1K	116		504A	1104	
1R2	177	3100/519	300	1122	
1R20	5079A	3335	ZD-20		
1R200	5105A	3355	ZD-200		
1R200A	5105A	3355	ZD-200		
1R200B	5105A	3355	ZD-200		
1R20A	5079A	3335	ZD-20		
1R20B	5079A	3335	ZD-20		
1R22	5080A	3336	ZD-22		
1R22A	5080A	3336	ZD-22		
1R22B	5080A	3336	ZD-22		
1R24	5081A		ZD-24		
1R24A	5081A		ZD-24		
1R24B	5081A		ZD-24		
1R27	146A	3064	ZD-27		
1R27A	146A	3064	ZD-27		
1R27B	146A	3064	ZD-27		
1R2A	116	3311	504A	1104	
1R2D	116	3311	504A	1104	
1R2E	116	3311	504A	1104	
1R30	5084A	3755	ZD-30		
1R30A	5084A	3755	ZD-30		
1R30B	5084A	3755	ZD-30		
1R31	506	3843	511	1114	
1R33	147A	3095	ZD-33		
1R33A	147A	3095	ZD-33		
1R33B	147A	3095	ZD-33		
1R36	5085A	3337	ZD-36		
1R36A	5085A	3337	ZD-36		
1R36B	5085A	3337	ZD-36		
1R39	5085A	3338/5086A	ZD-36		
1R39A	5086A	3338	ZD-39		
1R39B	5086A	3338	ZD-39		
1R3A	177	3017B/117	300	1122	
1R3B	506	3843	511	1114	
1R3D	116		504A	1104	
1R3G	116	3016	504A	1104	
1R3J	116	3031A	504A	1104	
1R4	177	3100/519	300	1122	
1R43	5087A	3339	ZD-43		
1R43A	5087A	3339	ZD-43		
1R43B	5087A	3339	ZD-43		
1R47	5088A	3340	ZD-47		
1R47A	5088A	3340	ZD-47		
1R47B	5088A	3340	ZD-47		
1R51	5089A	3341	ZD-51		
1R51A	5089A	3341	ZD-51		
1R51B	5089A	3341	ZD-51		
1R56	5090A	3342	ZD-56		
1R56A	5090A	3342	ZD-56		
1R56B	5090A	3342	ZD-56		

Industry Standard No.	ECG	SK	GE	RS 276-	MOTOR.
1R5A	116		504A	1104	
1R5B	116	3311	504A	1104	
1R5BZ61	156	3081/125		1143	
1R5DZ61		3081		1143	
1R5G	116	3311	504A	1104	
1R5GZ61	125	3081	510,531	1114	
1R5GZ61FA-1	125		510	1114	
1R5H	116	3016	504A	1104	
1R5TH61	506	3125	511	1114	
1R62	149A	3097	ZD-62		
1R62A	149A	3097	ZD-62		
1R62B	149A	3097	ZD-62		
1R68	5092A	3343	ZD-68		
1R68A	5092A	3343	ZD-68		
1R68B	5092A	3343	ZD-68		
1R75	5093A	3344	ZD-75		
1R75A	5093A	3344	ZD-75		
1R75B	5093A	3344	ZD-75		
1R8.2	5072A	3136	ZD-8.2	562	
1R8.2A	5072A	3136	ZD-8.2	562	56K
1R8.2B	5072A	3136	ZD-8.2	562	
1R82	150A	3098	ZD-82		
1R82A	150A	3098	ZD-82		
1R82B	150A	3098	ZD-82		
1R9	116		504A	1104	
1R9.1	131	3060/139A	44	2006	
1R9.1A	139A	3060	ZD-9.1	562	
1R9.1B	139A	3066/118	ZD-9.1	562	
1R90	116		504A	1104	
1R91	116	3311	ZD-91		
1R91A	5095A	3345	ZD-91		
1R91B	5095A	3345	ZD-91		
1R96	125		510,531	1114	
1R9H	116	3031A	504A	1104	
1R9I		3017B		1104	
1R9J	116	3311	504A	1104	
1R9L			504A	1104	
1R9U		3311	504A	1104	
1RC10	5454	3597/5455			
1RC20	5455	3597		1067	
1RC30	5456	3598/5457			
1RC40	5457	3598			
1RC5	5453	3597/5455			
1RLD	166		504A(4)	1104	
1ROE	177	3017B/117		1122	
1ROF	116	3017B/117		1104	
1ROH	116	3311		1104	
1S-1209			504A	1104	
1S-1555-Z			300	1122	
1S-1555V	177		300	1122	
1S-180	177		300	1122	
1S-188			1N60	1123	
1S-188AM	109			1123	
1S-2144Z	177		300	1122	
1S-2472	519	3100	300	1122	
1S-331			ZD-6.2	561	
1S-446D	109			1123	
1S005	116	3017B/117	504A	1104	
1S020	5801			1142	
1S020(RECT)	5801	9004			
1S021	5802	9005		1143	
1S023	5804	9007		1144	
1S025	5806	3848			
1S027	5808	9009			
1S030	5800	9003		1142	
1S031	5801	9004	504A	1104	
1S032	5802	9005	504A	1143	
1S034	5804	9007		1104	
1S036	5806	3848			
1S038	5808	9009		1114	
1S054		3031A		1104	
1S058		3032A		1114	
1S1-0336-00-A	130		14	2041	
1S100	116	3016	504A	1104	
1S1000	112		1N82A		
1S1000A	112		1N82A		
1S1001	177		300	1122	
1S1004	116	3311	504A	1104	
1S1006				1123	
1S1007	109	3100/519		1123	
1S1007S	109	3087		1123	
1S1008				1123	
1S1009				1123	
1S100R			504A	1104	
1S101	116	3016	504A	1104	
1S1010				1123	
1S1012				1103	
1S1013				1104	
1S1015				1114	
1S101R			504A	1104	
1S102	116	3016	504A	1104	
1S1021				1103	
1S1022				1104	
1S1024				1114	
1S1027				1103	
1S1028				1104	
1S102R			504A	1104	
1S103	116	3016	504A	1104	
1S1030				1114	
1S1039				1144	
1S103R			504A	1104	
1S104	116	3016	504A	1104	
1S1043				1103	
1S1044				1104	
1S1046				1114	
1S104R			504A	1104	
1S105	116	3017B/117	504A	1104	
1S1052	177	3100/519	300	1122	
1S1053	177	3100/519	300	1122	
1S105R			504A	1104	
1S106	116	3032A	504A	1104	
1S1061	116	3032A	504A	1104	
1S1062	116	3017B/117	504A	1104	
1S1063	116	3017B/117	504A	1104	
1S1064	116	3017B/117	504A	1104	
1S1065	125	3032A	510,531	1114	
1S1066	125	3033	510,531	1114	
1S106A	116	3016	504A	1104	
1S106R			504A	1104	
1S107	125	3032A	510,531	1114	
1S1071	156		512	1143	
1S1072	5802	9005	504A	1143	
1S1073	5804	9007		1144	
1S1074	5806	3848			
1S1075	5808	9009			
1S1076	156	3051	512		
1S107R			509	1114	
1S108	125	3033	510,531	1114	
1S108R			509	1114	
1S109	125	3033	510,531	1114	
1S1095				562	
1S1096	116	3017B/117	504A	1104	
1S1097			ZD-13	563	
1S1098				564	

Industry Standard No.	ECG	SK	GE	RS 276-	MOTOR.
1S109R			509	1114	
1S11			1N34AS	1123	
1S110	116	3016	504A	1104	
1S110A	116	3016	504A	1104	
1S111	116	3016	504A	1104	
1S112	116	3016	504A	1104	
1S1121				1152	
1S1122				1172	
1S1123				1173	
1S1124	177			1173	
1S113	116	3016	504A	1104	
1S113A	116	3016	504A	1104	
1S114	116	3031A	504A	1104	
1S1146				1011	
1S1147				1122	
1S115	116	3017B/117	504A	1104	
1S1155	177		300	1122	
1S1156				562	
1S116	116	3017B/117	504A	1104	
1S1162				561	
1S1164			ZD-8.2	562	
1S1165			ZD-10	562	
1S1166			ZD-13	563	
1S1167			ZD-16	564	
1S117	125	3032A	510,531	1114	
1S1173			ZD-6.0	561	
1S1174			ZD-6.8	561	
1S1175			ZD-8.7	562	
1S1176			ZD-11	563	
1S1177			ZD15	564	
1S1178			ZD-16	564	
1S118			504A	1104	
1S119	116	3311	504A	1104	
1S1191			ZD-16	564	
1S11941	177		300	1122	
1S12	177	3087	1N60	1123	
1S120	116	3017B/117	504A	1104	
1S1201			ZD-6.2	561	
1S1201(H)				561	
1S1202			ZD-6.8	561	
1S1207		3100	300	1122	
1S1208				1102	
1S1209	601	3463	300		
1S121	116	3017B/117	504A	1104	
1S121(RECT)	116		504A	1104	
1S1210	601	3175	300	1122	
1S1211	601	3463			
1S1212	601		300	1122	
1S1212A	601	3100/519	300	1122	
1S1213	177	3100/519	300	1122	
1S1214	177	3100/519	300	1122	
1S1215	177	3100/519	300	1122	
1S1216	177	3100/519	300	1122	
1S1217	177	3100/519	300	1122	
1S1218	177	3100/519	300	1122	
1S1218GR	177		300	1122	
1S1219	177	3100/519	300	1122	
1S1219H				1122	
1S122	116	3017B/117	504A	1104	
1S1220	177	3100/519	300	1122	
1S1221	116	3017B/117	504A	1104	
1S1222	116	3017B/117	504A	1104	
1S1223	125	3033	510,531	1114	
1S1224	116	3016	504A	1104	
1S1225	125	3033	510,531	1114	
1S1225A	125		510,531	1114	
1S1226	506	3125	511	1114	
1S123	116	3016	504A	1104	
1S1230	116	3017B/117	504A	1104	
1S1230H				1103	
1S1231	116	3017B/117	504A	1104	
1S1231H				1104	
1S1232	116	3017B/117	504A	1104	
1S1232H				1104	
1S1232N				1104	
1S1233	125	3033	510,531	1114	
1S1234	125	3033	510,531	1114	
1S1234H				1114	
1S1234N				1114	
1S1236				1143	
1S1237	506	3843	511	1114	
1S1238	506	3843	511	1114	
1S1239				1123	
1S124	116	3016	504A	1104	
1S1244				1123	
1S125	116	3016	504A	1104	
1S1255A	125	3033	510,531	1114	
1S126	116	3017B/117	504A	1104	
1S127	109	3087		1123	
1S128			1N60	1123	
1S128FM	110MP	3709		1123(2)	
1S129			1N34AS	1123	
1S1295				1102	
1S1296				1102	
1S1297				1102	
1S1298				1102	
1S1299				1102	
1S13	109	3087		1123	
1S130	177	3100/519	300	1122	
1S130(ZENER)	139A	3060	ZD-9.1	562	
1S1301				1122	
1S1302	177	3100/519	300	1122	
1S1303	177	3100/519	300	1122	
1S1305	177		300	1122	
1S131	177	3100/519	300	1122	
1S1314				1011	
1S1315	611	3324			
1S1316				1011	
1S1317				1102	
1S1318				1102	
1S1319				1103	
1S132	177	3100/519	300	1122	
1S1320				1103	
1S1329B		3126	90		
1S133			ZD-8.7	562	
1S134	116	3017B/117	ZD-4.7	1104	
1S1341	116	3017B/117	504A	1104	
1S1342	116	3017B/117	504A	1104	
1S1343	116	3017B/117	504A	1104	
1S1344	116	3017B/117	504A	1104	
1S1345	125	3017B/117	510,531	1114	
1S1346	116	3017B/117	504A	1104	
1S1347	125	3033	510,531	1114	
1S1348	125	3033	510,531	1114	
1S1349	506	3843	511	1114	
1S135	141A	3092	ZD-11.5		
1S1351	5832		5004		
1S1352	5834	3126	5004		
1S1353	5836		5004		
1S1354	5838		5004		
1S1355	5842	7042	5008		
1S1356	5846	3516	5012		
1S1357	5848	7048	5012		
1S1359				1102	

Industry Standard No.	ECG	SK	GE	RS 276-	MOTOR.
1S1359A				1102	
1S136	116	3093	ZD-6.2	561	
1S136(RECT)			504A	1104	
1S136(ZENER)	137A		ZD-6.2	561	
1S1360				1103	
1S1360A				1103	
1S1361				1104	
1S1361A				1104	
1S1363				1114	
1S1363A				1114	
1S1365				1102	
1S1365A				1102	
1S1366				1103	
1S1366A				1103	
1S1367				1104	
1S1367A				1104	
1S1369				1114	
1S1369A				1114	
1S136Q				1103	
1S137	145A	3063	ZD-15	564	
1S1374			ZD-5.6	561	56J
1S1374A			ZD-5.6	561	
1S1375			ZD-6.2	561	
1S1375A			ZD-6.2	561	
1S1376			ZD-6.8	561	
1S1376A			ZD-6.8	561	
1S1377			ZD-8.7	562	
1S1377A			ZD-8.7	562	
1S1378			ZD-10	562	
1S1378A				562	
1S1379			ZD-13	563	
1S1379A			ZD-13	563	
1S138	125	3033	510,531	1114	
1S138(RECT)				1114	
1S138(ZENER)				562	
1S1380			ZD-16	564	
1S1380A			ZD-16	564	
1S1389			ZD-5.6	561	
1S1389A			ZD-5.6	561	
1S139			ZD-9.1	562	
1S1390			ZD-6.2	561	
1S1390A			ZD-6.2	561	
1S1392			ZD-8.7	562	
1S1392A			ZD-8.7	562	
1S1393			ZD-10	562	
1S1393A			ZD-10	562	
1S1394			ZD-13	563	
1S1394A			ZD-13	563	
1S1395			ZD-16	564	
1S1395A			ZD-16	564	
1S14			1N34AS	1123	
1S140			ZD-11	563	
1S1405			ZD-5.6	561	
1S1405R			ZD-5.6	561	
1S1406			ZD-6.2	561	
1S1406R			ZD-6.2	561	
1S1408			ZD-8.7	562	
1S1408R			ZD-8.7	562	
1S1409			ZD-10	562	
1S1409R			ZD-10	562	
1S141			ZD-12	563	
1S1410			ZD-14	564	
1S1410R			ZD-14	563	
1S1411			ZD-16	564	
1S1411R			ZD-16	564	
1S1417	5999		5109		
1S1418	6003	7203	5109		
1S142			ZD-15	564	
1S1420H	177	3100/519	300	1122	
1S143			ZD-16	564	
1S144			300	1122	
1S145	611	3324	90		
1S146	5832	3016	5004	1102	
1S147	116	3016	504A	1104	
1S1471				1114	
1S1472	117		504A	1104	
1S1473	177	3100/519	300	1122	
1S1474				1143	
1S1475				1144	
1S148	116	3016	504A	1104	
1S1486				1102	
1S1487				1102	
1S1488				1143	
1S1489				1144	
1S149	116	3016	504A	1104	
1S15	109	3091		1123	
1S150	116	3016	504A	1104	
1S1500	610	3323			
1S1501	611	3324	90		
1S1502	611	3324	90		
1S1503	614	3126	90		
1S151			MR-1	1114	
1S1514	177	3016	300	1122	
1S1515	177	3016	300	1122	
1S1516	177	3016	300	1122	
1S1517	506	3843	511	1114	
1S1517A	506	3843	511	1114	
1S1532	177	3100/519	300	1122	
1S1533	177	3126	90		
1S1544	553	3100/519	300	1122	
1S1544A				1122	
1S1545	553	3100/519	300	1122	
1S1546				1102	
1S1547				1102	
1S1548				1103	
1S155	177	3100/519	300	1122	
1S155-1	177	3100/519	300	1122	
1S1553	519	3100	300	1122	
1S1553(TV)	177	3100/519			
1S1554	177	3311	300	1122	
00001S1555	177	3175	300	1122	
1S1555(TV)	177	3100/519			
1S1555-1			300	1122	
1S1555-S	177	3100/519	300	1122	
1S1555-Z	177		300	1122	
1015551	177		300	1122	
1S1555FA1			300	1122	
1S1555V		3100	300	1122	
1S1555Z			300	1122	
1S1558	614	3126	90		
1S1575	5878		5036		
1S1576	5882		5040		
1S1577	5886		5044		
1S1578	5890	7090	5044		
1S1579	178MP	3119/113	300(2)	1122(2)	
1S1580	177		300	1122	
1S1582	610	3323			
1S1585	177	3100/519	300	1122	
1S1586	177	3100/519	300	1122	
1S1587	177	3175	300	1122	
1S1588	519	3100	514	1122	
1S1588V	519		514	1122	
1S1589	177	3087	300	1122	
1S1597				1011	
1S1598				1011	
1S1599				1102	
1S16			1N34AS	1123	
1S160	5870		5032		
1S1600				1102	
1S1601				1103	
1S1602				1103	
1S1604				1102	
1S1608				1114	
1S161	5872		5032		
1S1614	6068		5140		
1S1615	6072	7272	5140		
1S162	5874		5032		
1S1621	177	3016	300	1122	
1S1621-O	177		300	1122	
1S1621-R	177		300	1122	
1S1621-Y	177		300	1122	
1S1622	116	3016	504A	1104	
1S1623	116	3016	504A	1104	
1S1624	116	3016	504A	1104	
1S1625	116	3017B/117	504A	1104	
1S1626	5866		5028		
1S1627	5868	7068	5028		
1S1629	5846	3516	5012		
1S163	5876		5036		
1S1630	5848	7048	5012		
1S164	5878		5036		
1S1643R	6355	7227/6027	5129		
1S1644R	6355		5133		
1S1648		3126	90		
1S165	5880		5040		
1S1650	177	3100/519	300	1122	
1S1651	177	3100/519	300	1122	
1S1652	5944		5048		
1S1652R	5875		5033		
1S1653R	5877		5037		
1S1654	5986	3609	5096		
1S1654R	5987	3698	5097		
1S1655	5988	3608/5990	5100		
1S1655R		3518	5101		
1S1658	614	3327	90		
1S1658FA-2		3126	90		
1S1658FA-3	614	3327	90		
1S166	5882		5040		
1S1660	5834	3051/156	5004		
1S1660R	5835		5005		
1S1661	5836	3051/156	5004		
1S1661R	5837	3051/156	5005		
1S1662	5854		5016		
1S1662R	5855		5017		
1S1663	5856		5020		
1S1663R	5859		5021		
1S1664	116	3031A	504A	1104	
1S1665	117		504A	1104	
1S1666	116	3311	504A	1104	
1S1667				1104	
1S1668	116	3032A	504A	1104	
1S1668F	116	3032A	504A	1104	
1S1669				1114	
1S1670				1152	
1S1671				1172	
1S1672				1173	
1S1678				1102	
1S1679				1102	
1S1680				1103	
1S1681				1104	
1S1683				1114	
1S1684				1102	
1S1685				1102	
1S1686				1103	
1S1687				1104	
1S1689				1114	
1S1690	109	3100/519	300	1123	
1S1690A				1011	
1S1691	116	3016	504A	1104	
1S1691A				1011	
1S1692	117		504A	1104	
1S1692A				1102	
1S1693	116	3017B/117	504A	1104	
1S1693A				1102	
1S1694	117	3017B	504A	1104	
1S1694A				1102	
1S1695	125	3033	510,531	1114	
1S1695A				1102	
1S1696				1102	
1S1696A				1102	
1S1697	116	3017B/117	504A	1104	
1S1698	116	3017B/117	504A	1104	
1S17	177	3100/519	1N91		
1S1701	178MP	3087	300(2)	1122(2)	
1S1710				1122	
1S1711				1122	
1S1712				1122	
1S1712A				1122	
1S1713				1122	
1S1714				1122	
1S1715	137A	3058	ZD-6.2	561	
1S1717	5073A	3749	ZD-9.1	562	
1S17176	138A		ZD-7.5		
1S1717L	139A	3060	ZD-9.1	562	
1S1718	139A	3061/140A	ZD-9.1	562	
1S1718/4454C	140A		ZD-10	562	
1S1722	112		1N82A		
1S1723				1102	
1S1724				1102	
1S1724R				1102	
1S1725				1102	
1S1725R				1102	
1S1726				1103	
1S1726R				1103	
1S1727				1103	
1S1727R				1103	
1S1728	139A	3060	ZD-9.1	562	
1S1728R				1103	
1S1729				1104	
1S1729R				1104	
1S1730				1104	
1S1730R				1104	
1S1732				1114	
1S1732R				1114	
1S1735			ZD-6.2	561	
1S1736			ZD-7.5	561	
1S1737			ZD-8.7	562	
1S1738			ZD-11	563	
1S1739			ZD-12	563	
1S1740			ZD-15	564	
1S1758	613	3326			
1S1759	613	3326			
1S1767			ZD-6.0	561	
1S1768			ZD-7.5	561	
1S1769			ZD-8.7	562	
1S1770			ZD-11	563	
1S1771			ZD-13	563	
1S1772			ZD-15	564	

Industry Standard No.	ECG	SK	GE	RS 276-	MOTOR.
1S1781			ZD-8.7	562	
1S1782			ZD-8.7	562	
1S1783			ZD-8.7	562	
1S17D1	109	3087		1123	
1S18	177	3100/519	1N91		
1S180	177	3100/519	300	1122	
1S180-M				1011	
1S180/5GB	116	3016	504A	1104	
1S1805	611	3324			
1S1805A	611	3324			
1S1806	611	3324			
1S1806A	611	3324			
1S180B			300	1122	
1S181	177	3016	300	1122	
1S181-M				1102	
1S181FA			300	1122	
1S182	177	3100/519	300	1122	
1S182-M				1102	
1S1825	177	3100/519	300	1122	
1S1829	125	3051/156	510,531	1114	
1S183	116		504A	1104	
1S183-M				1103	
1S1830		3080		1114	
1S1832	506	3843	511	1114	
1S1834	552	3843	511	1114	
1S1835	552	3843	511	1114	
1S184				1122(2)	
1S1841	5990	3608	5100		
1S1842	5994	3501	5104		
1S1843	5998		5108		
1S1844	6002	7202	5108		
1S1845				1102	
1S1845A				1103	
1S1845B				1104	
1S1845D				1114	
1S1846				1102	
1S1846A				1103	
1S1846B				1104	
1S1846D				1114	
1S1848			504A	1104	
00001S1849	116(2)	3311	504A(2)	1104	
00001S1849R	116(2)	9002	504A(2)	1104	
1S185	109	3311		1123	
1S1850	116(2)	3017B/117	504A(2)	1104	
1S1850R	116(2)	3311		1104	
1S1851	116	3031A	504A	1104	
1S1851R	116	3031A	504A	1104	
1S1855	506	3843	511	1114	
1S1850R	116		504A	1104	
1S186	109	3087		1123	
1S186(FM)	109			1123	
1S186FM	110MP	3709	1N34AS	1123(2)	
1S187	109	3087		1123	
1S187S				1123	
0001S188	109	3087	1N34AS	1123	
1S1881AM	116	3311	504A	1104	
1S1882	506	3843	511	1114	
1S1885	116	3311	504A	1104	
1S1885-3	116	3311	504A	1104	
1S1886	116	3311	504A	1104	
1S1887	116	3016	504A	1104	
1S1888	116	3313	504A	1104	
1S188A	109	3087	1N60	1123(2)	
1S188AM	109	3087		1123	
1S188AR			1N34AS	1123(2)	
1S188F	110MP	3091			
1S188FM	109	3087		1123	
1S188FM-1	109	3088	1N34AS	1123(2)	
1S188FM1	109		1N34AS	1123(2)	
1S188FM1A	109	3087	1N34AS	1123(2)	
1S188FM2	109	3087		1123(2)	
1S188FMA	109	3087	1N34AS	1123(2)	
1S188FMI	110MP	3091	1N60	1123(2)	
1S188G	109	3087		1123(2)	
1S188MPX	110MP	3091	1N34AS	1123(2)	
1S188P	110MP	3091		1123(2)	
1S188PM	109			1123(2)	
1S188S	109			1123(2)	
1S188TV	109	3087	1N34AS	1123(2)	
1S189	109	3087		1123	
1S1890	5806	3848		1104	
1S1891	156	3051	512		
1S1892	156	3051	512	1114	
1S1893	610	3323			
1S1895	612	3325	90		
1S19			1N60	1123	
1S190	5010A	3776	ZD-5.1		
1S1906	116		504A	1104	
1S191	5012A	3778	ZD-6.0	561	
1S1914				1143	
1S1915				1143	
1S1916				1144	
1S192	5014A	3780	ZD-6.8	561	
1S1920	506	3125	511	1114	
1S1921A	506	3125	511	1114	
1S1921B	506	3125	511	1114	
1S1921C	506	3125	511	1114	
1S1921D	506	3125	511	1114	
1S1921E	506	3125	511	1114	
1S1921F	506	3125	511	1114	
1S1922	112	3848	1N82A		
1S1923	611	3324	90		
1S1923A	611		90		
1S1924	612	3325	90		
1S1925	112	3089	1N82A		
1S1926	112	3089	1N82A		
1S1926K	112	3089	1N82A		
1S1928		3089	1N82A		
1S193	5072A	3031A	ZD-8.2	562	
1S194	5018A	3784	ZD-9.1	562	
1S1941	116	3311	504A	1104	
1S19413	116		504A	1104	
1S1942	116	3311	504A	1104	
1S1943	116	3031A	504A	1104	
1S1944	156	3017B/117	512	1104	
1S1948				1102	
1S1949				1103	
1S195	140A	3061	ZD-10-4	562	
1S1950			504A	1104	
1S1951	147A		ZD-33		
1S1956	5070A	9021	ZD-6.2	561	
1S1957	5071A		ZD-6.8	561	
1S1958	5072A		ZD-8.2	562	
1S1959	139A		ZD-9.1	562	
1S196	5074A		ZD-11	563	
1S1960	140A		ZD-10	562	
1S1961	5074A		ZD-11.0	563	
1S1962	142A		ZD-12	563	
1S1963	144A		ZD-14	564	
1S1964	145A		ZD-15	564	
1S1965	5075A	3751	ZD-16	564	
1S1966	5077A	3752	ZD-18		
1S1967	5079A	3335	ZD-20		
1S1968	5080A	3336	ZD-22		

Industry Standard No.	ECG	SK	GE	RS 276-	MOTOR.
1S1969	5081A		ZD-24		
1S197	142A		ZD-12	563	
1S1970	146A		ZD-27		
1S1971	5084A	3755	ZD-30		
1S1973				1122	
1S198	144A		ZD-14	564	
1S199	5077A	3752	ZD-18		
1S1991				1143	
1S1992		3100	300	1122	
1S1993	177	3100/519	300	1122	
1S1994	177	3100/519	300	1122	
1S1995	177	3100/519	300	1122	
1S1Z09	116	3311	504A	1104	
1S20	109	3016		1123	
1S200				1122(2)	
1S2030A	5065A	3838			
1S2031		3773	ZD-3.9		
1S2033	5005A	3771	ZD-3.3		
1S2033A	5066A	3771/5005A	ZD-3.3		
1S2036	5006A	3772	ZD-3.6		
1S2036A	134A	3772/5006A	ZD-3.6		
1S2039A	5067A	3773/5007A	ZD-3.9		
1S2039B		3773	ZD-3.9		
1S204	116	3031A	504A	1104	
1S2043	5008A	3774	ZD-4.3		
1S2043A	5068A		ZD-4.3		
1S2047	5009A	3775	ZD-4.7		
1S2047A	5069A	3775/5009A	ZD-4.7		
1S2047B		3775	ZD-4.7		
1S205	116	3017B/117	504A	1104	
1S2051	5010A	3776	ZD-5.1		
1S2051A	5010A	3776	ZD-5.1		
1S2051B		3776	ZD-5.1		
1S2056	5011A	3777	ZD-5.6	561	
1S2056A	136A	3777/5011A	ZD-5.6	561	
1S2056B		3777	ZD-5.6	561	
1S206	116	3017B/117	504A	1104	
1S2062	5013A	3779	ZD-6.2	561	
1S2062A	137A		ZD-6.2	561	
1S2062B		3779	ZD-6.2	561	
1S2067		3783	ZD-9.1	562	
1S2068		3783	ZD-9.1	562	
1S2069		3783	ZD-9.1	562	
1S207	116	3017B/117	504A	1104	
1S2070				1122	
1S2071				1122	
1S2074H	177	3100/519	300	1122	
1S2075		3100	514	1122	
1S20751C	177	3100/519	300	1122	
1S2075A	138A		ZD-7.5		
1S2075A(ZENER)	5015A	3781			
1S2075K	519	3100	514	1122	
1S2076	177	3175	300	1122	
1S2076-27	519		514	1122	
1S2076-TF1	519		514	1122	
1S2076-TFI	177		514	1122	
1S2076A	519	3175	514	1122	
1S2076A-07	519		300	1122	
1S208	116	3017B/117	504A	1104	
1S208/2SJ2A	117		504A	1104	
1S2080	116			1102	
1S2081	116			1103	
1S2082	5016A	3782	ZD-8.2	562	
1S2082A	5072A	3782/5016A	ZD-8.2	562	
1S2085	612	3325	90		
1S2085A	612	3325	90		
1S2086	612	3325			
1S2086A	612	3325			
1S2087	612	3325	90		
1S2087A	612	3325	90		
1S2088	612	3325			
1S2088A	612	3325			
1S2089	611	3324			
1S209	116	3017B/117	504A	1104	
1S209/2SJ4A	117		504A	1104	
1S2090	610	3323	1N82A		
1S2091	177	3100/519	300	1122	
1S2091(DIODE)	177	3100/519	300	1122	
1S2091(ZENER)	5018A		ZD-9.1	562	
1S2091-BK	177		300	1122	
1S2091-BL	177		300	1122	
1S2091-W	177		300	1122	
1S2091A	5018A		ZD-9.1	562	
1S2091BK			300	1122	
1S2091BL			300	1122	
1S2091W			300	1122	
1S2092	177	3100/519	300	1122	
1S2093	6408			1050	
1S2094	611	3324			
1S2097	177	3100/519	300	1122	
1S2098	177	3100/519	300	1122	
1S2099				1122	
1S210	125		510,531	1114	
1S210/2SJ6A	125		510,531	1114	
1S2100	5019A	3785	ZD-10	562	
1S2100A	140A	3785/5019A	ZD-10	562	
1S211	125		510,531	1114	
1S211/2SJ8A	125		510,531	1114	
1S2110	5020A	3786	ZD-11	563	
1S2110A	5020A	3786	ZD-11	563	
1S2111		3057	ZD-5.6	561	
1S2111A	136A	3057	ZD-5.6	561	
1S2112		3058	ZD-6.2	561	
1S2112A	137A	3058	ZD-6.0	561	
1S2113			ZD-6.8	561	
1S2113A	5071A	3334	ZD-6.8	561	
1S2114A	138A	3059	ZD-7.5		
1S2115		3136	ZD-8.2	562	
1S2115A	5072A	3136	ZD-8.2	562	
1S2116		3060	ZD-9.1	562	
1S2116A	139A	3060	ZD-9.1	562	
1S2117	140A	3061	ZD-10	562	
1S2117A	140A	3061	ZD-10	562	
1S2118		3139	ZD-11	563	
1S2118A	5074A	3092/141A	ZD-11	563	
1S2119		3062	ZD-12	563	
1S2119A	5021A	3062/142A	ZD-12	563	
1S211A	136A		ZD-5.6	561	
1S212		3783	ZD-9.1	562	
1S2120	5021A	3787	ZD-12	563	
1S2120A	142A	3787/5021A	ZD-12	563	
1S2121		3063	ZD-15	564	
1S2121A	145A	3063	ZD-15	564	
1S2122		3751	ZD-17	564	
1S2122A	5075A	3751	ZD-16	564	
1S2123		3752	ZD-18		
1S2123A	5077A	3752	ZD-18		
1S2124				1103	
1S2125			ZD-6.0	561	
1S2126			ZD-8.7	562	
1S2127			ZD-12	563	
1S2128			ZD-6.0	561	
1S2129			ZD-8.7	562	
1S213		3783	ZD-9.1	562	

Industry Standard No.	ECG	SK	GE	RS 276-	MOTOR.
1S2130	5022A	3788	ZD-13	563	
1S2130A	5022A	3788	ZD-13	563	
1S2134	519	3100		1122	
1S2136				561	
1S2137			ZD-8.7	562	
1S2138			ZD-12	563	
1S2139	611	3324	90		
1S2139A	610	3323	90		
1S2139B	611	3324			
1S2139C	612	3325			
1S213C		3783	90		
1S214		3783	ZD-9.1	562	
1S2140				1011	
1S2142	611	3324			
1S2143	611	3324			
1S2144	177	3100/519	300	1122	
1S2144A	177	3100/519	300	1122	
1S2144Z	177	3175	300	1122	
1S2145				1122	
1S2147	610	3323			
1S2147A	612	3325			
1S2147B	612	3325			
1S2147C	612	3325			
1S215	140A		ZD-10	562	
1S2150	5024A	3790	ZD-15	564	
1S2150A	145A	3790/5024A	ZD-15	564	
1S216	140A		ZD-10	562	
1S2160,A					.5M16Z10,5
1S2160A	5075A	3751		564	
1S217			ZD-9.1	562	
1S218				1122	
1S2180	610	3793/5027A	ZD-18		
1S2180A	5077A	3752	ZD-18		
1S2186	553	3100/519	300	1122	
1S2186GR	177	3100/519	300	1122	
1S2188	611	3324			
1S2189		3126	90		
1S2190		3136	ZD-8.2	562	
1S2191		3136	ZD-8.2	562	
1S2192		3136	ZD-8.2	562	
1S2193				562	
1S2194			ZD-6.0	561	
1S2195			ZD-8.7	562	
1S2196			ZD-12	563	
1S2198	112	3089	1N82A		
1S220	5068A	3332	ZD-4.3		
1S2200	5029A	3795	ZD-20		
1S2200A	5079A	3795/5029A	ZD-20		
1S2207	616	3327/614	90		
1S2207-9	611	3324			
1S2208	616	3324/611	90		BB105GM
1S2209	616	3324/611			
1S221	136A	3057	ZD-5.6	561	
1S2210				1122(2)	
1S2211				1122(2)	
1S2212				1122	
1S222	5071A	3334	ZD-6.8	561	MPN3401
1S2220	5030A	3796	ZD-22		
1S2220A	5080A	3796/5030A	ZD-22		
1S2226				1102	
1S2227		3925		1102	
1S2228				1103	
1S2229				1104	
1S223	138A	3059	ZD-7.5		
1S2230	116	3017B/117	504A	1104	
1S2231				1114	
1S2233	5806	3848			
1S2234	5808	9009			
1S2236	611	3324	90		
1S2238				1103	
1S2239	610	3323		1104	
1S224	5073A	3749	ZD-9.1	562	
1S2240	5031A	3797	ZD-24		
1S2240A	5081A		ZD-24		
1S2241				1103	
1S2242				1104	
1S2244	552			1102	
1S2245	552			1103	
1S2246	552			1104	
1S225	140A	3060/139A	ZD-10	562	
1S226	5074A	3092/141A	ZD-11.0	563	
1S2267	612	3325	90		
1S2268	612	3325	90		
1S2269				1103	
1S227	142A	3062	ZD-12	563	
1S2270	5033A	3799	ZD-27	1102	
1S2270A	146A		ZD-27	1102	
1S2271				1103	
1S2272				1104	
1S2274				1114	
1S2276	177	3100/519	300	1122	
1S2277	552			1102	
1S2278	552			1103	
1S2279	552			1104	
1S228	143A	3750	ZD-13	563	
1S2281	558			1114	
1S229	144A		ZD-14	564	
1S230	145A	3063	ZD-15	564	
1S2300	5035A	3801	ZD-30		
1S2300A	5084A	3755	ZD-30		
1S2306	506	3843	511	1114	
1S2307	506	3843	511	1114	
1S2308	506	3843	511	1114	
1S2309	506	3843	511	1114	
1S231	5075A	3751	ZD-16	564	
1S2310	116	3016	504A	1104	
1S2311	125		510,531	1114	
1S2312	125		510,531	1114	
1S2313	116		504A	1104	
1S2314	125		510,531	1114	
1S2315	125		510,531	1114	
1S2316	552			1103	
1S2317	552			1104	
1S2319	558			1114	
1S232	5076A	9022	ZD-17		
1S2324	558			1114	
1S2329			CR-2	1103	
1S233	5077A	3752	ZD-18		
1S2330	5036A	3802	ZD-33		
1S2330A	147A		ZD-33		
1S2336	611	3324	90		
1S2339	610	3323			
1S2339C	611	3323/610			
1S234	5078A	9023	ZD-19		
1S2340	611	3324			
1S235	5079A	3335	ZD-20		
1S2350	5804	9007		1144	
1S2351	116		504A	1104	
1S2352	116		504A	1104	
1S2353	125		510,531	1114	
1S2354	125		510,531	1114	
1S2356	116		504A	1104	
1S2357	116		504A	1104	
1S2358				1102	
1S2359				1103	
1S236	5080A	3336	ZD-22		
1S2361	116		504A	1104	
1S2362	116		504A	1104	
1S2363	116		504A	1104	
1S2364	125		510,531	1114	
1S2365	125		510,531	1114	
1S2367	116	3081/125	511		
1S237	5081A	3062/142A	ZD-24		
1S2371	166	3311	504A	1152	
1S2371A	116(4)			1152	
1S2372	116		504A	1104	
1S2372A				1172	
1S2373	116		504A	1104	
1S2373A				1173	
1S2374	116		504A	1104	
1S2375	116		504A	1104	
1S2375A				1173	
1S2376	116		504A	1104	
1S2377	156		512		
1S2378	5834		5004		
1S2378R	5835		5005		
1S2379	5804	9007		1144	
1S238	5082A	3753	ZD-25		
1S2380	5806	3848			
1S239	146A	3064	ZD-27		
1S2390				1011	
1S2391				1102	
1S2392				1102	
1S2393				1103	
1S2394				1103	
1S2395				1104	
1S2396				1104	
1S2399				1114	
1S240	5084A		ZD-30		
1S2400				1114	
1S2401	116		504A	1104	
1S2402	116		504A	1104	
1S2403	125		510,531	1114	
1S2404	116		504A	1104	
1S2405	125		510,531	1114	
1S2406	125		510,531	1114	
1S2408	5800	9003		1142	
1S2409	156		512	1143	
1S241	147A	3095	ZD-33		
1S2410	5802	9005		1143	
1S2411	5803	9006		1144	
1S2412	5804	9007		1144	
1S2413	5805	9008			
1S2414	5806	3848			
1S2415	5800	9003		1142	
1S2416	5801	9004		1142	
1S2417	5802	9005		1143	
1S2418	5803	9006		1144	
1S2419	5804	9007		1144	
1S242	5085A	3337	ZD-36		
1S2420	5805	9008			
1S2421	5806	3848			
1S243	5086A	3338	ZD-39		
1S2434	5870		5032		
1S2435	5872		5032		
1S2437	5876		5036		
1S2438	5878		5036		
1S2439	5880		5040		
1S2439R	5881		5041		
1S244	5087A	3339	ZD-43		
1S2446	5980	3610	5096		
1S2447	5986	3609	5096		
1S2448	5986	3609	5096		
1S2449	5988	3608/5990	5100		
1S2451	5992	3501/5994	5104		
1S2452			ZD-6.2	561	
1S2453			ZD-6.2	561	
1S2454			ZD-6.2	561	
1S2455				1143	
1S2456				1143	
1S2457				1144	
1S246	5088A	3340	ZD-47		
1S2460	177	3100/519	300	1122	
1S2461	177	3031A	300	1122	
1S2462	116	9091/177	300	1122	
1S2463		3515		1103	
1S247	5089A	3341	ZD-51		
1S2471	519	3100	514	1122	
1S2472	519	3100	514	1122	
1S2473	519	3100	514	1122	
1S2473-T72	519	3100	514	1122	
1S2473H	519	3100	514	1122	
1S2473HC	519	3100	514	1122	
1S2473K	519	3100	514	1122	
1S2473T	177	3100/519	514	1122	
1S2473VE	519	3100	514	1122	
1S2479			1N60	1123	
1S248	5089A	3341	ZD-51		
1S2488			ZD-5.6	561	
1S2489			ZD-6.2	561	
1S249	5090A	3342	ZD-56		
1S2490			ZD-6.8	561	
1S2492				562	
1S2493			ZD-9.1	562	
1S2494			ZD-10	562	
1S2495				563	
1S2496			ZD-12	563	
1S2497			ZD-13	563	
1S2498			ZD-14	564	
1S2499			ZD-16	564	
1S250	149A	3097	ZD-62		
1S251	5092A	3343	ZD-68		
1S252	5093A	3344	ZD-75		
1S253	150A	3098	ZD-82		
1S254	5095A	3345	ZD-91		
1S2544			ZD-5.6	561	
1S2545			ZD-5.6	561	
1S2546			ZD-6.0	561	
1S2547			ZD-6.2	561	
1S2549				562	
1S255	5096A	3346	ZD-100		
1S2550			ZD-9.1	562	
1S2551			ZD-10	562	
1S2552			ZD-11	563	
1S2553			ZD-12	563	
1S2554			ZD-13	563	
1S2555			ZD-14	564	
1S2556			ZD-16	564	
1S257	151A	3099	ZD-110		
1S258	5097A	3347	ZD-120		
1S2588	519	3100			
1S2589				1011	
1S259	5098A	3348	ZD-130		
1S2590				1102	
1S2591				1102	
1S2593	558			1114	
1S2596				1143	
1S2597		3100	300	1122	

Industry Standard No.	ECG	SK	GE	RS 276-	MOTOR.
1S260	5099A	3349	ZD-140		
1S2602				1102	
1S2603				1103	
1S2604				1103	
1S2605				1104	
1S2606	116		504A	1104	
1S2608	125		510,531	1114	
1S2609				1114	
1S261	5100A	335//5100A	ZD-150		
1S2610	125		510,531	1114	
1S2638	611	3327/614	90		
1S2638VVC		3126	90		
1S268	5187AK		504A		
1S2686	611	3324			
1S2687	612	3325	90		
1S2687AA	612	3325			
1S2687D	612	3325	90		
1S2688	614	3327	90		
1S2688B	614	3327			
1S2688C		3327	90		
1S2688EA	614	3327			
1S2688EB	613	3325/612			
1S2688EP	613	3326			
1S2689	614	3327	90		
1S268AA		3311	504A	1104	
1S268T	116	3311	504A	1104	
1S2692	555	3319	514	1122	
1S2692AB		3319	514	1122	
1S2692B	177	3319			
1S2711	506	3843	511	1114	
1S2746	515	9098			
1S2756	506	3843	511	1114	
1S2757				1104	
1S2760				1141	
1S2761				1143	
1S2762				1143	
1S2763				1144	
1S2764				1144	
1S2769			ZD-6.2	561	
1S2770			ZD-6.2	561	
1S2771			ZD-6.2	561	
1S2772				561	
1S2773				561	
1S2774			ZD-6.2	561	
1S2775				1102	
1S2775FA-1	552	3311	504A	1104	
1S2776				1103	
1S2777				1104	
1S2778				1143	
1S2779				1144	
1S2787	611	3324			
1S2788	177	3100/519	300	1122	
1S2788B	177	3100/519	300	1122	
1S2789	616	3325/612			
1S2790	613	3327/614			
1S2790W	613	3327/614			
1S2790WT	613	3327/614			
1S2827A				1143	
1S2828A				1144	
1S2832				1102	
1S3006	5071A	3334	ZD-6.8	561	
1S3007	138A		ZD-7.5		
1S3008	5072A	3136	ZD-8.2	562	
1S3009	139A		ZD-9.1	562	
1S301.5	145A		ZD-15	564	
1S3010	140A		ZD-10	562	
1S3011	5074A		ZD-11.0	563	
1S3012	142A		ZD-12	563	56L
1S3013	143A	3750	ZD-13	563	
1S3016		3311	ZD-15	564	
1S3016(RECT)	116		504A	1104	
1S3016(ZENER)	5075A	3751	ZD-16	564	
1S3016R		3100	504A	1104	
1S3018	5077A	3752	ZD-18		
1S3020	5079A	3335	ZD-20		
1S3022	5080A	3336	ZD-22		
1S3024	5081A		ZD-24		
1S3027	146A		ZD-27		
1S3030	5084A	3755	ZD-30		
1S3033	147A		ZD-33		
1S3036	5085A	3337	ZD-36		
1S3039	5086A	3338	ZD-39		
1S304				562	
1S3043	5087A	3339	ZD-43		
1S3047	5088A	3340	ZD-47		
1S3051	5089A	3341	ZD-51		
1S3056	5090A	3342	ZD-56		
1S306	177	3100/519	300	1122	
1S3062	149A		ZD-62		
1S3068	5092A	3343	ZD-68		
1S306M				1122	
1S307	177	3100/519	300	1122	
1S3075	5093A	3344	ZD-75		
1S3076	177		300	1122	
1S3082	150A		ZD-82		
1S309	116	3016	504A	1104	
1S3091	5095A	3345	ZD-91		
1S310	116	3017B/117	504A	1104	
1S3100	5096A	3346	ZD-100		
1S310H				1011	
1S311	117	3017B	504A	1104	
1S3110	151A		ZD-110		
1S311H				1102	
1S311N				1102	
1S312	116	3017B/117	504A	1104	
1S312(HITACHI)	506		511	1114	
1S3120	5097A	3347	ZD-120		
1S312A				1102	
1S312H				1102	
1S313	116	3016	504A	1104	
1S3130	5098A		ZD-130		
1S313H				1103	
1S314	116	3017B/117	504A	1104	
1S314H				1103	
1S314N				1103	
1S315	116	3032A	504A	1104	
1S315(HITACHI)	506		511	1114	
1S3150	5100A		ZD-150		
1S315H				1104	
1S3160	5101A		ZD-160		
1S318	109	3087		1123	
1S3180	5103A	3353	ZD-180		
1S319				1123	
1S3195			504A	1104	
1S32	109	3087	1N34AS	1123	
1S320	135A	3056	ZD-5.1		
1S3200	5105A		ZD-200		
1S322	177	3100/519	300	1122	
1S323				1123	
1S324				1122	
1S325				1122	
1S326				1102	
1S327				1103	
1S328				1104	

Industry Standard No.	ECG	SK	GE	RS 276-	MOTOR.
1S33	109	3087	1N34AS	1123	
1S330	5010A	3056/135A	ZD-5.1		
1S3305	5010A	3776	ZD-5.1		
1S330A	5010A	3056/135A	ZD-5.1		
1S330A1	135A		ZD-5.1		
1S331	137A	9021/5070A	ZD-6.2	561	
1S331A	137A		ZD-6.2	561	
1S331AZ	137A	9021/5070A	ZD-6.2	561	
1S332	138A	3059	ZD-7.5	561	
1S332M	5071A	3334	ZD-6.8	561	
1S333	5072A	3136	ZD-8.2	562	
1S3332	5071A	3334	ZD-6.8	561	
1S333Y	5072A	3136	ZD-8.2	562	
000001S334	139A	3060	ZD-9.1	562	
1S334K	139A		ZD-9.1	562	
1S334M			ZD-9.1	562	
1S334N	139A	3060	ZD-9.1	562	
1S335	140A	3785/5019A	ZD-10	562	
1S335S	5019A	3785			
1S336	5074A	3062/142A	ZD-11	563	
1S337	142A	3062	ZD-12	563	
1S337-Y	142A	3062	ZD-12	563	
1S337A	142A	3062	ZD-12	563	
1S337E	142A		ZD-12	563	
1S337Y	142A		ZD-12	563	
1S338	5023A	3789	ZD-14	564	
1S338Q	143A		ZD-13	563	
1S338U	144A	3789/5023A	ZD-14	564	
1S339	5027A	3793	ZD-18		
1S34	109	3087	1N34AS	1123	
1S341W1		3126	90		
1S341W2		3126	90		
1S341W3		3126	90		
1S347	5070A	9021	ZD-6.2	561	
1S34A	109	3087	1N34AS	1123	
1S34S	109	3087	1N34AS	1123	
1S35	109	3016		1123	
1S351		3325	90		
1S351B	612	3325			
1S351M	611	3325/612	90		
1S351R	613	3326	90		
1S351S	611	3324	90		
1S351W		3326	90		
1S352	612	3325	90		
1S352M	612	3325	90		
1S352N	612	3325			
1S352R		3126	90		
1S353	613	3326			
1S353M	614	3327			
1S354	109	3087		1123	
1S355	109	3087		1123	
1S356				1123	
1S357	109	3087		1123	
1S358	116	3100/519	504A	1104	
1S358(S)	116	3017B/117	504A	1104	
1S3585	109			1123	
1S358S	116	3017B/117	504A	1104	
1S359				1103	
1S36			504A	1104	
1S367				1102	
1S368				1102	
1S369				1103	
1S37	5070A	3058/137A	ZD-6.2	561	
1S370				1103	
1S371				1104	
1S372				1104	
1S373				1102	
1S374				1102	
1S375				1103	
1S376				1104	
1S377				1104	
1S378				1102	
1S379				1102	
1S38	177	3100/519	509	1122	
1S380				1103	
1S381				1104	
1S382				1104	
1S39			509	1114	
1S390			504A	1104	
1S393				1103	
1S395	116	3017B/117	504A	1104	
1S396	116	3017B/117	504A	1104	
1S397	125	3033	510,531	1114	
1S398	125	3033	510,531	1114	
1S399	116	3031A	504A	1104	
1S40	5801	9004	504A	1142	
1S400	116	3017B/117	504A	1104	
1S4006	5120A	3386			
1S4006A	5120A	3386			
1S4007	5121A	3387			
1S4007A	5121A	3387			
1S4008	5122A	3388			
1S4008A	5122A	3388	5ZD-8.2		
1S4009	5124A	3390			
1S4009A	5124A	3390			
1S401	125	3032A	510,531	1114	
1S4010	5125A	3391			
1S4010A	5125A	3391			
1S4011	5126A	3392			
1S4011A	5126A	3392	5ZD-11		
1S4012	5127A	3393			
1S4012A	5127A	3393	5ZD-12		
1S4013	5128A	3394			
1S4013A	5128A	3394	5ZD-13		
1S4015	5130A	3396	5ZD-15		
1S4015A	5130A	3396			
1S4016	5131A	3397			
1S4016A	5131A	3397	5ZD-16		
1S4018	5133A	3399	5ZD-18		
1S4018A	5133A	3399			
1S402	125	3033	510,531	1114	
1S4020	5135A	3401			
1S4020A	5135A	3401			
1S4022	5136A	3402	5ZD-22		
1S4022A	5136A	3402			
1S4024	5137A	3403			
1S4024A	5137A	3403	5ZD-24		
1S4027	5139A	3405	5ZD-27		
1S4027A	5139A	3405			
1S4030	5141A	3407			
1S4030A	5141A	3407	5ZD-30		
1S4033	5142A	3408			
1S4033A	5142A	3408	5ZD-33		
1S4036	5143A	3409			
1S4036A	5143A	3409			
1S4039	5144A	3410			
1S4039A	5144A	3410			
1S404				1104	
1S4043	5145A	3411			
1S4043A	5145A	3411			
1S4047	5146A	3412			
1S4047A	5146A	3412			
1S4051	5147A	3413			
1S4051A	5147A	3413			

Industry Standard No.	ECG	SK	GE	RS 276-	MOTOR.
1S4056	5148A	3414			
1S4056A	5148A	3414			
1S406				1114	
1S4062	5150A	3416			
1S4062A	5150A	3416			
1S4068	5151A	3417			
1S4068A	5151A	3417			
1S4075	5152A	3418			
1S4082	5153A	3419			
1S4082A	5153A	3419			
1S4091	5155A	3421			
1S4091A	5155A	3421			
1S41	5002A	3768	504A	1102	
1S410	5832		5004		
1S4100	5156A	3422			
1S4100A	5156A	3422			
1S411	5834		5004		
1S4110	5157A	3423			
1S4110A	5157A	3423			
1S4120	5158A	3424			
1S4120A	5158A	3424			
1S413	5838		5004		
1S4130	5159A	3425			
1S4130A	5159A	3425			
1S415	5842	7042	5008		
1S4150	5161A	3427			
1S4150A	5161A	3427			
1S4160	5162A	3428			
1S4160A	5162A	3428			
1S417	5846	3516	5012		
1S4180	5164A	3430			
1S4180A	5164A	3430			
1S419	5848	7048	5012		
1S42	5803	9006	504A	1144	
1S420	5872		5032		
1S4200	5166A	3432			
1S421	5874		5032		
1S423	5878		5036		
1S425	5882		5040		
1S426	109	3088		1123	
1S4266			1N60	1123	
1S4266FM	109	3088		1123	
1S426FM	109			1123	
1S426G	109	3087		1123	
1S426GFM		3087	1N34AS	1123	
1S427	5886		5044	564	
1S428	109	3087		1123	
1S43	5804	9007	504A	1144	
1S430	5800	9003	504A	1142	
1S431	5801	9004	504A	1142	
1S432	5802	9005	504A	1143	
1S434	5804	9007		1144	
1S436	5806	3848			
1S438	5808	9009			
1S44	5805	9008	504A	1104	
1S441	109	3087		1123	
1S441(RECT)	5830		5004		
1S442		3087		1123	
1S442(RECT)	5830		5004		
1S443(RECT)	5830		5004		
1S444	177	3100/519	300	1122	
1S444(RECT)	5832		5004		
1S445(RECT)	5832		5004		
1S446	109	3087		1123	
1S446(RECT)	5834		5004		
00001S4460	109	3090		1123	
00001S446D	109	3088		1123	
1S447	109	3087		1123	
1S447P	109	3087		1123	
1S448	109	3087		1123	
1S449	109	3087		1123	
1S45	5806	3848	504A	1104	
1S451	109	3087		1123	
1S452	109	3087		1123	
1S453	109	3087		1123	
1S454	109	3087		1123	
1S455	109			1123	
1S456	116	3017B/117	504A	1104	
1S457	116	3016	504A	1104	
1S458	116	3017B/117	504A	1104	
1S459	116	3017B/117	504A	1104	
1S46	5808	3090/109	504A		
1S460				1122	
1S461				1122	
1S465				1122	
1S466	109	3087		1123	
1S467R	109	3087		1123	
1S47	5008A	3774	509	1114	
1S470	5012A	3778	ZD-6.0	561	
1S471	5014A	3780	ZD-6.8	561	
1S472	5018A	3784	ZD-9.1	562	
1S473	5020A	3100/519	ZD-11	563	
1S4735	5010A	3776	ZD-5.1		
1S474	5023A	3789	ZD-14	564	
1S475	5027A	3793	ZD-18		
1S476	5030A	3796	ZD-22		
1S477	5034A	3800	ZD-28		
1S478	5037A	3803	ZD-36		
1S479	5040A	3806	ZD-47		
1S48	612	3325	90		
1S480	5070A	9021	ZD-6.2	561	
1S481	5071A	3334	ZD-6.8	561	
1S482	139A		ZD-9.1	562	
1S483	5074A		ZD-11.0	563	
1S484	144A		ZD-14	564	
1S485	5077A	3752	ZD-18		
1S487	5083A	3754	ZD-28		
1S488	5085A	3337	ZD-36		
1S489	5088A	3340	ZD-47		
1S48A	613	3326			
00001S49	116	3017B/117	90	1104	
1S499	5146A	3412			
1S50	110MP	3709	1N295	1123(2)	
1S500				1122	
1S501				1122	
1S501M				1122	
1S52			ZD-5.6	561	
1S53			ZD-6.6	561	
1S535		3126	90		
1S5351		3126	90		
1S54	117		504A	1104	
1S542			1N60	1123	
1S543	112		1N82A		
1S544	5886		5044		
1S545	5890	7090	5044		
1S5454	109	3087		1123	
1S548				1122	
1S55			ZD-9.1	562	
1S550				562	
1S551				562	
1S553	612	3126	90		
1S554	613	3126	90		
1S555	613	3326	300	1122	
1S556	614	3126			
1S557	125	3016	510,531	1114	
1S557H				1114	
1S558	116	3017B/117	504A	1104	
1S559	116	3017B/117	504A	1104	
1S56			ZD-9.1	562	
1S560	177		300	1122	
1S560H				1122	
1S58	117		1N34AS	1123	
1S588	116	3016	504A	1104	
1S589			1N34AS	1123	
1S597		3063	ZD-15	564	
1S60	110MP	3709	1N60	1123(2)	
1S60(RECT)			504A	1104	
1S60P	109	3088		1123	
1S61			504A	1104	
1S615				561	
1S617			ZD-7.5	562	
1S618			ZD-10	562	
1S619			ZD-13	563	
1S62			504A	1104	
1S620			ZD-16	564	
1S63			504A	1104	
1S635			ZD-6.0	561	
1S636			ZD-6.8	561	
1S64			504A	1104	
1S642	177	3100/519	300	1122	
1S65			504A	1104	
1S66			504A	1104	
1S661				1103	
1S663				1114	
1S685	116	3016	504A	1104	
1S686	125	3033	510,531	1114	
1S687	125	3033	510,531	1114	
1S692	136A	3057	ZD-5.6	561	
1S693			300	1122	
1S694				562	
1S695			ZD-10	562	
1S696	142A	3062	ZD-12	563	
1S697	145A	3063	ZD-15	564	
1S700	146A	3064	ZD-27		
1S7030A	5004A	3770			
1S7033	5005A	3771	ZD-3.3		
1S7033A	5005A	3771	ZD-3.3		
1S7033B	5005A	3771	ZD-3.3		
1S7036	5006A	3772	ZD-3.6		
1S7036A	5006A	3772	ZD-3.6		
1S7036B	5006A	3772	ZD-3.6		
1S7039	5007A	3773	ZD-3.9		
1S7039A	5007A	3773	ZD-3.9		
1S7039B	5007A	3773	ZD-3.9		
1S7043	5008A	3774	ZD-4.3		
1S7043A	5008A	3332/5068A	ZD-4.3		
1S7043B	5008A	3332/5068A	ZD-4.3		
1S7047	5009A	3775	ZD-4.7		
1S7047A	5009A	3775	ZD-4.7		
1S7047B	5009A	3775	ZD-4.7		
1S7051	5010A	3776	ZD-5.1		
1S7051A	5010A	3776	ZD-5.1		
1S7051B	5010A	3776	ZD-5.1		
1S7056	5011A	3777	ZD-5.6	561	
1S7056A	5011A	3777	ZD-5.6	561	
1S7056B	5011A	3777	ZD-5.6	561	
1S7062	5013A	3779	ZD-6.2	561	
1S7062A	5013A	3779	ZD-6.2	561	
1S7062B	5013A	3779	ZD-6.2	561	
1S7068	5014A	3780	ZD-6.8	561	
1S7068A	5014A	3780	ZD-6.8	561	
1S7068B	5014A	3780	ZD-6.8	561	
1S7075	5015A	3781	ZD-7.5		
1S7075A	5015A	3781	ZD-7.5		
1S7075B	5015A	3781	ZD-7.5		
1S7082	5016A	3782	ZD-8.2	562	
1S7082A	5016A	3782	ZD-8.2	562	
1S7082B	5016A	3782	ZD-8.2	562	
1S7091	5018A	3784	ZD-9.1	562	
1S7091A	5018A	3784	ZD-9.1	562	
1S7091B	5018A	3784	ZD-9.1	562	
1S71	116	3016	504A	1104	
1S7100	5019A	3785	ZD-10	562	
1S7100A	5019A	3785	ZD-10	562	
1S7100B	5019A	3785	ZD-10	562	
1S7110	5020A	3786	ZD-11	563	
1S7110A	5020A	3786	ZD-11	563	
1S7110B	5020A	3786	ZD-11	563	
1S7120	5021A	3787	ZD-12	563	
1S7120A	5021A	3787	ZD-12	563	
1S7120B	5021A	3787	ZD-12	563	
1S7150	5024A	3790	ZD-15	564	
1S7150A	5024A	3790	ZD-15	564	
1S7150B	5024A	3790	ZD-15	564	
1S7160,A					.5M16Z10.5
1S7160A	5025A	3791		564	
1S72	109	3017B/117		1123	
1S73	109			1123	
1S731				1122	
1S73A			1N34A	1123	
1S74	109	3087		1123	
1S744				1123	
1S745				1123	
1S746				1123	
1S747				1123	
1S75	109	3087		1123	
1S750	112	3089	1N82A		
1S750GR	112		1N82A		
1S752	5002A	3768			
1S753	5006A	3772	ZD-3.6		
1S754	5008A	3774	ZD-4.3		
1S755	5009A	3775	ZD-4.7		
1S756	5012A	3778	ZD-6.0	561	
1S756(H)				561	
1S757	5014A	3780	ZD-6.8	561	
1S757A	139A	3780/5014A	ZD-9.1	562	
1S758	5017A	3783	ZD-9.1	562	
1S758(H)				562	
1S758GS	139A	3783/5017A	ZD-9.1	562	
1S758H			ZD-8.7	562	
1S759	5019A	3785	ZD-10	562	
1S759(H)				563	
1S759H			ZD-10	562	
1S76	109	3087		1123	
1S760	5022A	3788	ZD-13	563	
1S760(H)				563	
1S760H			ZD-13	563	
1S761	5025A	3791		564	
1S761(H)				564	
1S762	5028A	3794	ZD-19		
1S763	5031A	3797	ZD-24		
1S764	5033A	3799	ZD-27		
1S766	611	3324			
1S767	611	3324			
1S767A	611	3324			
1S769	611	3324			
1S77			1N34AS	1123	
1S773				1122	

Industry Standard No.	ECG	SK	GE	RS 276-	MOTOR.
1S773A				1122	
1S775				1122	
1S776				1122	
1S77H				1123	
1S78	109	3087		1123	
1S788	109	3087		1123	
1S78H				1123	
1S79	109	3087		1123	
1S79H				1123	
1S80	109	3087		1123	
1S801				1122	
1S802				1122	
1S803				1122	
1S804				1122	
1S805				1122	
1S806				1122	
1S807				1122	
1S808				1122	
1S81	116	3016	504A	1104	
1S811				1103	
1S812				1103	
1S814				1114	
1S816				1103	
1S817				1103	
1S819				1114	
1S82	109	3091		1123	
1S83	116	3017B/117	504A	1104	
1S84	177	3100/519	300	1122	
1S841				1102	
1S842				1102	
1S843				1103	
1S844	116	3017B/117	504A	1104	
1S844N				1103	
1S845				1104	
1S846	116	3017B/117	504A	1104	
1S846N				1104	
1S848	125	3032A	510,531	1114	
1S849	116	3311	504A	1104	
1S849,R	116		504A	1104	
1S849R	116(2)	3311	504A(2)	1114	
1S84H				1102	
1S85	612	3126	90		
1S850	125	3033	510,531	1114	
1S850N				1114	
1S851			1N60	1123	
1S853		3126	90	1144	
1S854				1144	
1S85H	612	3126			
1S85M		3126	90		
1S85V		3126	504A	1104	
1S85W		3126	90		
1S85WR		3126	90		
1S85WT		3126	90		
1S85Y		3126	90		
1S86	612	3126	90		
1S87			1N60	1123	
1S871				1103	
1S871A				1104	
1S871C				1114	
1S872				1103	
1S872A				1104	
1S872C				1114	
1S88	109			1123	
1S881				1103	
1S881A				1104	
1S881C				1114	
1S882				1103	
1S882A				1104	
1S882C				1114	
1S885	116	3311	504A	1104	
1S89	177	3100/519	300	1122	
1S891	611	3324			
1S892	612	3325			
1S893	613	3326	90		
1S90	116	3016	504A	1104	
1S90R			300	1122	
1S91	116	3017B/117	504A	1104	
1S91R			504A	1104	
1S92	116	3016	504A	1104	
1S920	177	3100/519	300	1122	
1S921	177	3100/519	300	1122	
1S921(RECT)	5802			1143	
1S922				1144	
1S922(DIODE)	177		300	1122	
1S922(RECT)	5804			1144	
1S923	177	3848/5806	300	1122	
1S924	5808	9009			
1S925	5809	9010			
1S926	5874		5032		
1S926G	109			1123	
1S927	5878		5036		
1S928	5882		5040		
1S929	5886		5044		
1S92R			504A	1104	
1S93	116	3017B/117	504A	1104	
1S93/SGJ	116	3017B/117	504A	1104	
1S930	5890	7090	5044		
1S936	5986	3609	5096		
1S939	6158		5108		
1S93R			504A	1104	
1S94	116	3017B/117	504A	1104	
1S940	6158	3311	5108		
1S941	177	3311	300	1122	
1S941/4454C	116		504A	1104	
1S9413	116	3031A	504A	1104	
1S942	177	3016	300	1122	
1S943			504A	1104	
1S946				1102	
1S947				1103	
1S948				1104	
1S94R			504A	1104	
1S95	116	3017B/117	504A	1104	
1S950				1114	
1S951	177	3100/519	300	1122	
1S952	177	3100/519	300	1122	
1S953	177		514	1122	
1S954	519	3100	514	1122	
1S955	177	3100/519	300	1122	
1S956				1143	
1S957				1144	
1S95R			504A	1104	
1S96	125	3032A	510,531	1114	
1S963	117	3017B	504A	1104	
1S96R			509	1114	
1S97	125	3032A	510,531	1114	
1S977				1122	
1S978				1122	
1S97R			509	1114	
1S98	125	3033	510,531	1114	
1S981				1122	
1S982				1122	
1S983	177	3100/519	300	1122	
1S984				1122	
1S985				1122	
1S986				1122	
1S987				1122	
1S988				1122	
1S98R			509	1114	
1S99	125	3033	510,531	1114	
000001S990	177	3175	300	1122	
1S990(ZENER)			ZD-15	564	
1S990-AM	177		300	1122	
1S9908			1N34AS	1123	
1S990A	177	3175	300	1122	
1S990AM	109			1123	
1S990S	145A	3175	ZD-15	564	
1S990S(DIODE)	177	3175	300	1122	
1S990S(ZENER)	145A		ZD-15	564	
1S993	5004A	3770			
1S994	5067A	3331	ZD-3.9		
1S994A	177	3100/519	300	1122	
1S994S	5069A	3330/5066A	ZD-4.7		
1S994Y	138A		ZD-7.5		
1S995	611	3324			
1S99A	177	3100/519	300	1114	
1S99R			509	1114	
1SB01-02	116	3311	504A	1104	
1SB50S	614	3327			
1SB69-06	612	3325			
1SC2367	177		300	1122	
1SC683			60	2015	
1SD-2	116		504A	1104	
1SD1212		3311	504A	1104	
1SD2	116	3311	504A	1104	
1SIZ09	116		504A	1104	
1SL1885	116	3017B/117	504A	1104	
1SN1835			CR-2	1114	
1SO30	116	3017B/117	504A	1104	
1SO31	116	3017B/117	504A	1104	
1SO32	116	3017B/117	504A	1104	
1SO34	116	3017B/117	504A	1104	
1SO36	116	3017B/117	504A	1104	
1SO38	125	3033	510,531	1114	
1SO54	116	3017B/117	504A	1104	
1SO58	125	3033	510,531	1114	
1SR11-100				1102	
1SR11-200				1102	
1SR11-400				1103	
1SR11-600				1104	
1SR12-100				1102	
1SR12-1000				1114	
1SR12-200				1102	
1SR12-400				1103	
1SR12-50				1011	
1SR12-600				1104	
1SR13-100				1102	
1SR13-1000				1114	
1SR13-200				1102	
1SR13-400				1103	
1SR13-50				1011	
1SR13-600				1104	
1SR14-100				1102	
1SR14-200				1102	
1SR14-400				1103	
1SR14-50				1011	
1SR15-100				1143	
1SR15-200				1143	
1SR15-400				1144	
1SR15-50				1141	
1SR16-100				1143	
1SR16-200				1143	
1SR16-400				1144	
1SR16-50				1141	
1SR17-100				1143	
1SR17-200				1143	
1SR17-400				1144	
1SR17-50				1141	
1SR1FM4			504A	1104	
1SR1K	116	3311	504A	1104	
1SR30-200				1102	
1SR30-400				1103	
1SR30-600				1104	
1SR31-1000				1114	
1SR55-100				1102	
1SR55-200				1102	
1SR55-50				1011	
1SR70-100				1152	
1SR70-200				1172	
1SR70-400				1173	
1SR77-1000				1114	
1SR77-200				1102	
1SR77-400				1103	
1SR78-200				1102	
1SR78-400				1103	
1SR79-1000				1114	
1SRBA20Z				1172	
1SS16	112	3089	1N82A		
1SS24				1122(2)	
1SS25				1122(2)	
1SS26				1122(2)	
1SS27				1122(2)	
1SS28				1122(2)	
1SS29				1122(2)	
1SS30				1122	
1SS31				1122	
1SS49				1122	
1SS50				1122	
1SS51				1122	
1SS53	177	3100/519	300	1122	
1SS54	116	3311	504A	1104	
1SS55	519		514	1122	
1SS68				1122	
1SS81	177	9091	300	1122	
1SV34				1122	
1SV35				1122	
1SV50	614	3327	90		
1SV53	614	3126	90		
1SV68	614	3327	90		
1SV69-06	612	3325			
1SV70-06	612	3325			
1SV77	611	3324			
1SZ25				562	
1SZ26				561	
1SZ27				562	
1SZ28				564	
1SZ39				561	
1SZ40-11				563	
1SZ40-12				563	
1SZ40-13				563	
1SZ40-15				564	
1SZ40-16				564	
1SZ50				561	
1SZ51				561	
1SZ52				561	
1SZ53				561	
1SZ56-08				562	
1T010	5645	3658/56004			
1T015	56004#	3658/56004			

Industry Standard No.	ECG	SK	GE	RS 276-	MOTOR.
1T06	5645	3658/56004			
1T08	5645	3658/56004			
1T10	5645	3061/140A	ZD-10	562	
1T100					1M100Z
1T10B	140A	3061	ZD-10	562	
1T11.5	141A	3092	ZD-11.5		
1T11.5B	141A	3092	ZD-11.5		
1T110	5645		ZD-110		
1T110B	151A		ZD-110		
1T12	142A	3062	ZD-12	563	
1T12.8	143A	3093	ZD-13	563	
1T12.8B	143A	3093	ZD-13	563	
1T12A		3062	ZD-12	563	
1T12B	142A	3062	ZD-12	563	
1T13	112	3093	1N82A		
1T1331	110MP		1N82A		
1T13A	112	3093	1N82A		
1T13B	112	3093	1N82A		
1T14	144A	3094	ZD-14	564	
1T14B	144A	3094	ZD-14	564	
1T15	145A	3063	ZD-15	564	
1T15B	145A	3063	ZD-15	564	
1T16	5645	3658/56004			
1T18	5645	3658/56004			
1T188	109	3087		1123	
1T19-15B	145A	3063	ZD-15	564	
1T2011	116	3016	504A	1104	
1T2012	116	3016	504A	1104	
1T2013	116	3016	504A	1104	
1T2014	116	3017B/117	504A	1104	
1T2015	116	3017B/117	504A	1104	
1T2016	116	3017B/117	504A	1104	
1T205A			51	2004	
1T210	5645	3658/56004			
1T213	109	3087		1123	
1T215	56004#	3658/56004			
1T22	109	3016		1123	
1T22A	109	3087	1N60	1123	
1T22AJ	109	3087		1123	
1T22B	109	3087		1123	
1T22G	109	3087		1123	
1T23	109	3091		1123	
1T231	109	3087		1123	
1T236	109	3087		1123	
1T23A	109			1123	
1T23B	109			1123	
1T23G	109	3091		1123	
1T23J	109	3087		1123	
1T23M	109	3087		1123	
1T23S	110MP	3088	1N60	1123(2)	
1T240	109	3126		1123	
1T240A	109			1123	
1T243	177	3126	300	1122	
1T243M	139A	3100/519	ZD-9.1	562	
1T26	5645			1123	
1T26-2	109			1123	
1T261	109	3088	1N34AS	1123	
1T262	109	3091		1123	
1T263	5089A	3096/148A	ZD-51		
1T264	5089A		51	2005	
1T27	146A	3064	ZD-27		
1T27B	146A	3064	ZD-27		
1T28	5645	3658/56004			
1T2C1	116(2)		504A	1104	
1T2Z1	116(2)		504A	1104	
1T310	5645	3659/56006			
1T315	56006#	3659/56006			
1T33	147A	3095	ZD-33		
1T33B	147A	3095	ZD-33		
1T36	5645	3659/56006			
1T378	116	3311	504A	1104	
1T38	5645	3659/56006			
1T40	177	3100/519	300	1122	
1T410	5645	3659/56006			
1T415	56006#	3659/56006			
1T46	5645	3659/56006			
1T48	5645	3659/56006			
1T495	158		53	2007	
1T5.6	136A	3057	ZD-5.6	561	1M5.6AZ
1T5.6B	136A	3057	ZD-5.6	561	
1T501	116	3017B/117	504A	1104	
1T502	116	3017B/117	504A	1104	
1T503	116	3031A	504A	1104	
1T504	116	3016	504A	1104	
1T505	116	3017B/117	504A	1104	
1T506	116	3017B/117	504A	1104	
1T507	125	3032A	510,531	1114	
1T508	125	3032A	510,531	1114	
1T509	125	3033	510,531	1114	
1T510	5645	3033	510,531	1114	
1T515	56008#	3660/56008			
1T55	148A	3096	ZD-55		
1T55B	148A	3096	ZD-55		
1T58	5645	3660/56008			
1T6.2	137A	3058	ZD-6.2	561	
1T6.2B	137A	3058	ZD-6.2	561	
1T610	5645	3660/56008			
1T615	56008#	3660/56008			
1T62	149A	3097	ZD-62		
1T62B	149A	3097	ZD-62		
1T66	5645	3660/56008			
1T68	5645	3660/56008			
1T7.5	138A	3059	ZD-7.5		
1T7.5B	138A	3059	ZD-7.5		
1T82	150A	3098	ZD-82		
1T82B	150A	3098	ZD-82		
1T9.1	139A	3060	ZD-9.1	562	
1T9.1A		3066	ZD-9.1	562	
1T9.1B	139A	3060	ZD-9.1	562	
1TA10	140A	3061	ZD-10	562	
1TA10A	140A	3061	ZD-10	562	
1TA11.5	141A	3092	ZD-11.5		
1TA11.5A	141A	3092	ZD-11.5		
1TA110	151A		ZD-110		
1TA110A	151A		ZD-110		
1TA12	142A		ZD-12	563	
1TA12.8	143A	3093	ZD-13	563	
1TA12.8A	143A	3093	ZD-13	563	
1TA12A	142A		ZD-12	563	
1TA14	144A	3094	ZD-14	564	
1TA14A	144A	3094	ZD-14	564	
1TA15	145A	3063	ZD-15	564	
1TA15A	145A	3063	ZD-15	564	
1TA27	146A	3064	ZD-27		
1TA27A	146A	3064	ZD-27		
1TA33	147A	3095	ZD-33		
1TA33A	147A	3095	ZD-33		
1TA5.6	136A	3057	ZD-5.6	561	
1TA5.6A	136A	3057	ZD-5.6	561	
1TA55	148A	3096	ZD-55		
1TA55A	148A	3096	ZD-55		
1TA6.2	137A	3058	ZD-6.2	561	
1TA6.2A	137A	3058	ZD-6.2	561	
1TA6.2B			ZD-6.2	561	

Industry Standard No.	ECG	SK	GE	RS 276-	MOTOR.
1TA62	149A	3097	ZD-62		
1TA62A	149A	3097	ZD-62		
1TA7.5	138A	3059	ZD-7.5		
1TA7.5A	138A	3059	ZD-7.5		
1TA82	150A	3098	ZD-82		
1TA82A	150A	3098	ZD-82		
1TA9.1	139A	3060	ZD-9.1	562	
1TA9.1A	139A	3060	ZD-9.1	562	
1TB06	116	3016	504A	1104	
1TH61	525	3925	530	1114	
1TS05	116	3017B/117	504A	1104	
1TT-410		3126	90		
1TV24237	120	3110	CR-3		
1TW85B		3126	90		
1TWC8H			ZD-8.7	562	
1TWC8L			ZD-8.7	562	
1TWC8M			ZD-8.7	562	
1U585F	123A	3444	20	2051	
1U585F/7825B	123A	3444	20	2051	
1V3074A20	116	3016	504A	1104	
1V3074A21	116	3016	504A	1104	
1V68611A47	159	3114/290	82	2032	
1V68611A47A	159	3114/290	82	2032	
1V9002	109	3087		1123	
1VA10	116	3016	504A	1104	
1W01-08J	5074A		ZD-11.0	563	
1W11700	159	3114/290	82	2032	
1W11700A	159	3114/290	82	2032	
1W11702	159	3715	82	2032	
1W11702A	159	3715	82	2032	
1W11706	312	3112		2028	
1W11708-1	123A	3024/128			
1W11708-2	123A	3024/128			
1W11711	159	3114/290	82	2032	
1W1706			FET-1	2028	
1W8358	123A		20	2051	
1W8537	159	3715	82	2032	
1W8995A	128		243	2030	
1W9148	159	3715	82	2032	
1W9640	159	3715	82	2032	
1W9640A	159	3715	82	2032	
1W9723	123A	3444	20	2051	
1W9728	159	3114/290	82	2032	
1W9728A	159	3114/290	82	2032	
1W9782	159	3114/290	82	2032	
1W9782A	159	3114/290	82	2032	
1W9787	123A	3444	20	2051	
1W9810	159	3114/290	82	2032	
1W9810A	159	3114/290	82	2032	
1W9810S	159	3114/290	82	2032	
1W9810SA	159	3114/290	82	2032	
1WB-11D			ZD-11	563	
1WB-13D			ZD-13	563	
1WB-15D			ZD-15	564	
1WB-6A			ZD-5.6	561	
1WB-7A			ZD-6.8	561	
1WB-9D			ZD-8.7	562	
1WM6	116	3017B/117	504A	1104	
1WS1	156		512		
1WS10	156	3051	512		
1WS2	156		512		
1WS4	5804	9007		1144	
1WS6	5806	3848			
1WS8	156		512		
1X9179	177	3100/519	300	1122	
1X9805	177	3100/519	300	1122	
1X9809	177	3100/519	300	1122	
1Z10	140A	3061	ZD-10	562	
1Z100	5096A	3346	ZD-100		
1Z100A	5096A	3346	ZD-100		
1Z10A	140A		ZD-10	562	
1Z10T10	140A		ZD-10	562	
1Z10T20	140A		ZD-10	562	
1Z10T5	140A		ZD-10	562	
1Z11	5074A	3139	ZD-11.0	563	
1Z110	151A	3099	ZD-110		
1Z110A	151A	3099	ZD-110		
1Z110B	151A	3099	ZD-110		
1Z110C	151A	3099	ZD-110		
1Z110D	151A	3099	ZD-110		
1Z110D10	151A	3099	ZD-110		
1Z110D5	151A	3099	ZD-110		
1Z11A	5074A		ZD-11.0	563	
1Z11T10	5074A		ZD-11.0	563	
1Z11T20	5074A		ZD-11.0	563	
1Z11T5			ZD-11.0	563	
1Z12	142A	3062	ZD-12	563	
1Z12A	142A	3062	ZD-12	563	
1Z12B	142A	3062	ZD-12	563	
1Z12C	142A	3062	ZD-12	563	
1Z12T10	142A	3062	ZD-12	563	
1Z12T20	142A	3062	ZD-12	563	
1Z12T5	142A	3062	ZD-12	563	
1Z13	143A	3750	ZD-13	563	
1Z13A	143A	3750	ZD-13	563	
1Z13T10	143A	3750	ZD-13	563	
1Z13T20	143A	3750	ZD-13	563	
1Z13T5	143A	3750	ZD-13	563	
1Z15	145A	3063	ZD-15	564	
1Z15A	145A	3063	ZD-15	564	
1Z15B	145A	3063	ZD-15	564	
1Z15C	145A	3063	ZD-15	564	
1Z15D	145A	3063	ZD-15	564	
1Z15D10	145A	3063	ZD-15	564	
1Z15D5	145A	3063	ZD-15	564	
1Z15T10	145A	3063	ZD-15	564	
1Z15T20	145A	3063	ZD-15	564	
1Z15T5	145A		ZD-15	564	
1Z16	5075A	3751	ZD-16	564	
1Z16A	5075A	3751	ZD-16	564	
1Z16T10	5075A	3751	ZD-16	564	
1Z16T20	5075A	3751	ZD-16	564	
1Z16T5	5075A	3751	ZD-16	564	
1Z18	5077A	3752	ZD-18		
1Z18A	5077A	3752	ZD-18		
1Z18T10	5077A	3752	ZD-10		
1Z18T20	5077A	3752	ZD-18		
1Z18T5	5077A	3752	ZD-18		
1Z20	5079A	3335	ZD-20		
1Z20A	5079A	3335	ZD-20		
1Z20T10	5079A	3335	ZD-20		
1Z20T20	5079A	3335	ZD-20		
1Z20T5	5079A	3335	ZD-20		
1Z22	5080A	3336	ZD-22		
1Z22A	5080A	3336	ZD-22		
1Z22T10	5080A	3336	ZD-22		
1Z22T20	5080A	3336	ZD-22		
1Z22T5	5080A	3336	ZD-22		
1Z24	5081A	3151	ZD-24		
1Z24A	5081A	3151	ZD-24		
1Z24T10	5081A	3151	ZD-24		
1Z24T20	5081A	3151	ZD-24		
1Z24T5	5081A	3151	ZD-24		
1Z27	146A	3064	ZD-27		

Industry Standard No.	ECG	SK	GE	RS 276-	MOTOR.
1Z27A	146A	3064	ZD-27		
1Z27B	146A	3064	ZD-27		
1Z27C	146A	3064	ZD-27		
1Z27D	146A	3064	ZD-27		
1Z27D10	146A	3064	ZD-27		
1Z27D5	146A	3064	ZD-27		
1Z27T10	146A	3064	ZD-27		
1Z27T20	146A	3064	ZD-27		
1Z27T5	146A	3064	ZD-27		
1Z3.3	5065A	3838			
1Z3.3A	5066A	3330	ZD-3.3		
1Z3.3T10.5	5066A	3330	ZD-3.3		
1Z3.6	134A	3055	ZD-3.6		
1Z3.6A	134A	3055	ZD-3.6		
1Z3.6T5	134A	3055			
1Z3.9	5067A	3331	ZD-3.9		
1Z3.9A	5067A	3331	ZD-3.9		
1Z30	5084A	3755	ZD-30		
1Z30A	5084A	3755	ZD-30		
1Z30T10	5084A	3755	ZD-30		
1Z30T20	5084A	3755	ZD-30		
1Z30T20,10,5					1M30Z,10,5
1Z30T5	5084A	3755	ZD-30		
1Z33	147A	3095	ZD-33		
1Z33A	147A	3095	ZD-33		
1Z33B	147A	3095	ZD-33		
1Z33C	147A	3095	ZD-33		
1Z33D	147A	3095	ZD-33		
1Z33D10	147A	3095	ZD-33		
1Z33D5	147A	3095	ZD-33		
1Z33T10	147A	3095	ZD-33		
1Z33T20	147A	3095	ZD-33		
1Z33T5	147A	3095	ZD-33		
1Z36	5085A	3337	ZD-36		
1Z36A	5085A	3337	ZD-36		
1Z39	5086A	3338	ZD-39		
1Z39A	5086A	3338	ZD-39		
1Z4.3	5068A	3332	ZD-4.3		
1Z4.3A	5068A	3332	ZD-4.3		
1Z4.7	5069A	3330/5066A	ZD-4.7		
1Z4.7A	5069A	3330/5066A	ZD-4.7		
1Z43	5087A	3339	ZD-43		
1Z43A	5087A	3339	ZD-43		
1Z47	5088A	3340	ZD-47		
1Z47A	5088A	3340	ZD-47		
1Z5.1	135A		ZD-5.1		
1Z5.1A	135A		ZD-5.1		
1Z5.1T10	135A	3056	ZD-5.1		
1Z5.1T5	5116A		ZD-5.1		
1Z5.6	136A		ZD-5.6	561	
1Z5.6A	136A		ZD-5.6	561	
1Z51	5089A	3341	ZD-51		
1Z51A	5089A	3341	ZD-51		
1Z55	148A	3096	ZD-55		
1Z55A	148A	3096	ZD-55		
1Z55B	148A	3096	ZD-55		
1Z56	5090A	3342	ZD-56		
1Z56A	5090A	3342	ZD-56		
1Z6.2	137A		ZD-6.2	561	
1Z6.2A			ZD-6.2	561	
1Z6.2T10	137A	3058	ZD-6.2	561	
1Z6.2T5	137A	3058	ZD-6.2	561	
1Z6.8	5071A	3334	ZD-6.8	561	
1Z6.8A	5071A	3334	ZD-6.8	561	
1Z62	149A	3097	ZD-62		
1Z62A	149A	3097	ZD-62		
1Z62B	149A	3097	ZD-62		
1Z62C	149A	3097	ZD-62		
1Z62D	149A	3097	ZD-62		
1Z62D10	149A	3097	ZD-62		
1Z62D5	149A	3097	ZD-62		
1Z68	5092A	3343	ZD-6.8	561	
1Z68A	5092A	3343	ZD-6.8	561	
1Z7.5	138A	3059	ZD-7.5		
1Z7.5A	138A		ZD-7.5		
1Z72	5093A	3344	ZD-75		
1Z75	5093A	3344	ZD-75		
1Z75A	5093A	3344	ZD-75		
1Z8.2	5072A	3136	ZD-8.2	562	
1Z8.2A	137A	3136/5072A	ZD-8.2	562	
1Z8.2T10	5072A	3136	ZD-8.2	562	
1Z8.2T20	5072A	3136	ZD-8.2	562	
1Z8.2T5	5072A	3136	ZD-8.2	562	
1Z8.3			ZD-8.2	562	
1Z82	150A	3098	ZD-82		
1Z82A	150A	3098	ZD-82		
1Z82B	150A	3098	ZD-82		
1Z82C	150A	3098	ZD-82		
1Z82D	150A	3098	ZD-82		
1Z82D10	150A	3098	ZD-82		
1Z82D5	150A	3098	ZD-82		
1Z9.1	139A	3060	ZD-9.1	562	
1Z9.1A	139A	3060	ZD-9.1	562	
1Z9.1B	139A	3060	ZD-9.1	562	
1Z9.1C	139A	3066/118	ZD-9.1	562	
1Z9.1D10	139A	3060	ZD-9.1	562	
1Z9.1D15			ZD-9.1	562	
1Z9.1D5	139A	3060	ZD-9.1	562	
1Z9.1T10	139A	3066/118	ZD-9.1	562	
1Z9.1T20	139A	3066/118	ZD-9.1	562	
1Z9.1T5	139A	3060	ZD-9.1	562	
1Z91	5095A	3345	ZD-91		
1Z91.1D10			ZD-9.1	562	
1Z91.D10			ZD-9.1	562	
1Z91A	5095A	3345	ZD-91		
1ZB12	142A	3062	ZD-12	563	
1ZB12B	142A	3062	ZD-12	563	
1ZB15	145A	3063	ZD-15	564	
1ZB15B	145A	3063	ZD-15	564	
1ZB27	146A	3064	ZD-27		
1ZB27B	146A	3064	ZD-27		
1ZB33	147A	3095	ZD-27		
1ZB33B	147A	3095	ZD-27		
1ZB82	150A	3098	ZD-82		
1ZB82B	150A	3098	ZD-82		
1ZC10T5	140A		ZD-10	562	
1ZC11T10	5074A		ZD-11.0	563	
1ZC11T5	5074A		ZD-11.0	563	
1ZC12	142A	3062	ZD-12	563	
1ZC12T10	142A		ZD-12	563	
1ZC12T20	142A		ZD-12	563	
1ZC12T5	142A		ZD-12	563	
1ZC13T10	143A		ZD-13	563	
1ZC13T5	143A		ZD-13	563	
1ZC15	145A	3063	ZD-15	564	
1ZC15T10	145A	3063	ZD-15	564	
1ZC15T5	145A	3063	ZD-15	564	
1ZC16T10	5075A	3142	ZD-16	564	
1ZC16T5	5075A	3751	ZD-16	564	
1ZC18T10	5077A	3752	ZD-18		
1ZC18T5	5077A	3752	ZD-18		
1ZC20T10	5079A	3335	ZD-20		
1ZC20T5	5079A	3335	ZD-20		
1ZC22T10	5080A	3336	ZD-22		
1ZC22T5	5080A	3336	ZD-22		
1ZC24T10	5081A	3151	ZD-24		
1ZC24T5	5081A	3151	ZD-24		
1ZC27	146A	3064	ZD-27		
1ZC27T10	146A	3064	ZD-27		
1ZC27T5	146A	3064	ZD-27		
1ZC3.6	134A	3055	ZD-3.6		
1ZC30T10	5084A	3755	ZD-30		
1ZC30T5	5084A	3755	ZD-30		
1ZC4.3T10.5	5068A	3332	ZD-4.3		
1ZC5.1		3056	ZD-5.1		
1ZC5.6	136A	3057	ZD-5.6	561	
1ZC5.6T10.5	136A		ZD-5.6	561	
1ZC6.2	137A	3058	ZD-6.2	561	
1ZC8.2T10	5072A	3136	ZD-8.2	562	
1ZC8.2T20	5072A		ZD-8.2	562	
1ZC8.2T5	5072A	3136	ZD-8.2	562	
1ZC9.1	139A	3060	ZD-9.1	562	
1ZC9.1T10	139A		ZD-9.1	562	
1ZC9.1T5	139A		ZD-9.1	562	
1ZCT10	140A	3061	ZD-10	562	
1ZD120	5097A		ZD-120		
1ZD8-2			ZD-8.2	562	
1ZD8-5			ZD-8.2	562	
1ZD8.2			X11	562	
1ZD8.2V			ZD-8.2	562	
1ZD825	5072A	3136	ZD-8.2	562	
1ZD82S			ZD-8.2	562	
1ZD892			X11	562	
1ZF12T10	142A	3062	ZD-12	563	
1ZF12T20	142A	3062	ZD-12	563	
1ZF12T5		3062	ZD-12	563	
1ZF14T10	144A	3094	ZD-14	564	
1ZF14T20	144A	3094	ZD-14	564	
1ZF14T5	144A	3094	ZD-14	564	
1ZF15T10	145A	3063	ZD-15	564	
1ZF15T20	145A	3063	ZD-15	564	
1ZF15T5	145A	3063	ZD-15	564	
1ZF27T10	146A	3064	ZD-27		
1ZF27T20	146A	3064	ZD-27		
1ZF27T5	146A	3064	ZD-27		
1ZF33T10	147A	3095	ZD-33		
1ZF33T20	147A	3095	ZD-33		
1ZF33T5	147A	3095	ZD-33		
1ZF5.1T10	135A	3056	ZD-5.1		
1ZF5.1T20	135A	3056	ZD-5.1		
1ZF5.1T5	135A	3056	ZD-5.1		
1ZF5.6T10	136A	3057	ZD-5.6	561	
1ZF5.6T20	136A	3057	ZD-5.6	561	
1ZF5.6T5	136A	3057	ZD-5.6	561	
1ZF9.1T10	139A	3060	ZD-9.1	562	
1ZF9.1T20	139A	3060	ZD-9.1	562	
1ZF9.1T5	139A	3066/118	ZD-9.1	562	
1ZM10T10	140A	3061	ZD-10	562	
1ZM10T20	140A	3061	ZD-10	562	
1ZM10T5	140A	3061	ZD-10	562	
1ZM12T10	142A	3062	ZD-12	563	
1ZM12T20	142A	3062	ZD-12	563	
1ZM12T5	142A	3062	ZD-12	563	
1ZM13T5	143A	3750			
1ZM15T10	145A	3063	ZD-15	564	
1ZM15T20	145A	3063	ZD-15	564	
1ZM15T5	145A	3063	ZD-15	564	
1ZM16T5	5075A	3751			
1ZM18T5	5077A	3752			
1ZM3.6T10	134A	3055	ZD-3.6		
1ZM3.6T20	134A	3055	ZD-3.6		
1ZM3.6T5	134A	3055	ZD-3.6		
1ZM5.1T10	135A	3056	ZD-5.1		
1ZM5.1T20	135A	3056	ZD-5.1		
1ZM5.1T5	5116A	3056/135A	ZD-5.1		
1ZM5.6T10	136A	3057	ZD-5.6	561	
1ZM5.6T20	136A	3057	ZD-5.6	561	
1ZM5.6T5	136A	3057	ZD-5.6	561	
1ZM6.2	137A	3058	ZD-6.2	561	
1ZM6.2T10	137A	3058	ZD-6.2	561	
1ZM6.2T20	137A	3058	ZD-6.2	561	
1ZM6.2T5	137A	3058	ZD-6.2	561	
1ZM7.5T10	138A	3059	ZD-7.5		
1ZM7.5T20	138A	3059	ZD-7.5		
1ZM7.5T5	138A	3059			
1ZM75T5	5093A		ZD-7.5		
1ZM9.1T10	139A	3060	ZD-9.1	562	
1ZM9.1T20	139A	3060	ZD-9.1	562	
1ZM9.1T5	139A	3060	ZD-9.1	562	
1ZS100					1M100ZS
1ZS13A	143A	3750			
1ZS16A	5075A	3751			
1ZS18A	5077A	3752			
1ZS3.3					1M3.3ZS
1ZS30A	5084A	3755			
1ZT110A	151A		ZD-110		
1ZT12	142A	3062	ZD-12	563	
1ZT120A	5097A		ZD-120		
1ZT12B	142A	3062	ZD-12	563	
1ZT130A	5098A		ZD-130		
1ZT15	145A	3063	ZD-15	564	
1ZT150A	5100A		ZD-150		
1ZT15B	145A	3063	ZD-15	564	
1ZT160A	5101A		ZD-160		
1ZT180A	5103A		ZD-180		
1ZT200A	5105A		ZD-200		
1ZT27	146A	3064	ZD-27		
1ZT27B	146A	3064	ZD-27		
1ZT33	147A	3095	ZD-33		
1ZT33B	147A	3095	ZD-33		
1ZT82	150A	3098	ZD-82		
1ZT82B	150A	3098	ZD-82		
2	198	3102/710	IC-89		
002(MCINTOSH)	710		IC-89		
002-005100	102A	3004	53	2007	
002-006300	126	3008	52	2024	
002-006500	123	3020	20	2051	
002-006600	102A	3004	53	2007	
002-006800	102A	3004	53	2007	
002-006900	102A	3004	53	2007	
002-007000	104	3009	16	2006	
002-007100	160	3007	245	2004	
002-007200	160	3006	245	2004	
002-007300	102A	3004	53	2007	
002-007400	160	3007	245	2004	
002-008100	104	3009	16	2006	
002-008300			10	2051	
002-00840			2	2007	
002-008400	102A	3004	53	2007	
002-008800	121	3717	239	2006	
002-00900			51	2004	
002-009000			9	2004	
002-009100	124	3021	12		
002-009500	123A	3444	20	2051	
002-009501	123A	3444	20	2051	
002-009502	123A	3444	20	2051	
002-009600	108	3452	86	2038	
002-009601	108	3452	86	2038	

Industry Standard No.	ECG	SK	GE	RS 276-	MOTOR.
002-009700			16	2006	
002-009701	121	3717	239	2006	
002-009800	159	3715	82	2032	
002-009900	123A	3444	20	2051	
002-010100	121	3009	239	2006	
02-010103	1192	3445			
002-010300	159	3715	82	2032	
002-010300-6	129	3025	244	2027	
002-010300A	159	3715	82	2032	
02-010331		3728	IC-66		
002-010400	123A	3444	20	2051	
002-0105-00	106	3984	21	2034	
002-010500	159	3715	82	2032	
002-010500A	159	3715	82	2032	
002-010600	128	3024	243	2030	
02-010612	1249	3963			
002-010700	129	3025	244	2027	
002-010800	123A	3444	20	2051	
002-010900	159	3715	82	2032	
002-010900A	159	3715	82	2032	
002-011000	102A	3004	53	2007	
002-011400	161	3716	39	2015	
002-011500	161	3716	39	2015	
002-011600	126		52	2024	
002-011700	103A	3835	59	2002	
002-011800	176	3845	80		
002-011900	102A	3003	53	2007	
002-012000	123	3124/289	20	2051	
002-012100	185	3191	58	2025	
002-012200	184	3190	57	2017	
002-012300	153	3083/197	69	2049	
002-012400	152	3893	66	2048	
002-012500	128	3024	243	2030	
002-012600	129	3025	244	2027	
002-012700	121	3009	239	2006	
002-012700-12	127	3764	25		
002-012800	159	3715	82	2032	
002-012800A	159	3715	82	2032	
002-03	123A	3444	20	2051	
02-091128	712	3072	IC-148		
02-091366	1261	3872			
2-0A90		3088	1N34AS	1123	
002-1(SYLVANIA)	196		241	2020	
02-1001-1/2221-3			1N60	1123	
02-1001-1221-3	109	3087		1123	
02-1006-2/2221-3	177		300	1122	
002-104-000	199	3245	62	2010	
02-1078-01	123A	3444	20	2051	
002-11700	103A	3835	59	2002	
002-11800	102A	3004	53	2007	
002-11900	102A	3004	53	2007	
002-12000	123A	3444	20	2051	
02-121201	1003		IC-43		
02-121365	712	3072	IC-148		
02-124400	1193	3701			
02-124422	1155	3231			
02-161521	1170	3745			
02-165143	138A	3072/712	IC-148		
02-173710	1194	3484			
2-1K60	177	3100/519	300	1122	
002-2(SYLVANIA)	196		241	2020	
02-257130	1234	3487	IC-181		
02-257205	1155	3231	IC-179	705	
02-257310	1192	3445			
02-263001	1192	3445			
002-3(SYLVANIA)	241	3188A/182	57	2020	
02-3002-2/2221-3	116		504A	1104	
02-300577	1082	3461			
02-310023		3471	IC-118		
02-33379-6			18	2030	
02-341125	712	3072			
02-341128	712	3072			
02-341358	712	3072			
02-343065	712	3072	IC-148		
2-36	103	3862	8	2002	
02-360002	1167	3732	IC-317		
02-373001	1192	3445			
002-4(SYLVANIA)	196		241	2020	
02-403065	712	3072	IC-148		
02-437205	1155	3231			
02-437310	1192	3445			
02-5004	5697	3522			
2-5004-14	5697	3522			
2-5004-6	5695	3509			
2-5184-11	5675	3508			
02-537666	712	3072			
02-561171	791	3149	IC-231		
02-561201	783	3215	IC-225		
02-561410	714	3075	IC-4		
2-5SR	116	3311	504A	1104	
2-64701508	125	3032A	510,531	1114	
02-781050	977	3462	VR-100		
02-781060	988	3973			
2-8454-031	123A	3444	20	2051	
002-9501	199	3245	62	2010	
002-9502	199	3245	62	2010	
002-9502-12	199	3245	62	2010	
002-9601	108	3452	86	2038	
002-9601-12	108	3452	86	2038	
002-9700	121	3009	239	2006	
002-9800-12	129	3025	244	2027	
002-9800-A	159	3715	82	2032	
2-G-3055	130		14	2041	
2-OA90			1N60	1123	
2.01.03.02	159		82	2032	
2.4341.0018			3	2006	
2/A4					1N4004
2A	126	3006/160	52	2024	
2A100	156	3051	512		
2A119	110MP	3709	1N34AS	1123(2)	
2A12	129		244	2027	
2A18	5008A	3774	ZD-4.3		
2A200	156	3051	512		
2A21	5010A		ZD-5.1		
2A22	5010A		ZD-5.1		
2A25(ZENER)	5006A	3772	ZD-3.6		
2A28(ZENER)	5004A	3770			
2A30	156	3081/125	512		
2A43	5003A	3769			
2A44	5005A	3330/5066A	ZD-3.3		
2A47	5008A	3774	ZD-4.3		
2A64	5036A	3802	ZD-33		
2AA113	116(2)	3016	504A(2)	1104	
2AA119	110MP	3090/109	1N60	1123	
2AC128		3123	52	2024	
2AC187	176(2)		80(2)		
2AC188	158(2)		53(2)		
2AD140	121	3009	239	2006	
2AD149	121MP	3014	239(2)	2006(2)	
2AF1					MR501
2AF10					MR510
2AF2					MR502
2AF3					MR504
2AF4					MR504
2AF6					MR506
2AF8					MR508
2AFR1					MR851
2AFR2					MR852
2AFR2PF	156		512		
2AFR3					MR854
2AFR4					MR854
2AFR6					MR856
2AG	161	3018	39	2015	
2AH	161	3018	39	2015	
2AM-1	5452	6752			
2AT-16	503		CR-5		
2B	100	3005	1	2007	
2B-2	506		511	1114	
2B2DM				1152	
2B4DM				1172	
2B6DM				1173	
2B8DM				1173	
2C	100	3005	1	2007	
2C291(TRANSISTOR)	128		243	2030	
2CD198S	130		14	2041	
2CS537FC			62	2010	
2CS900	123A	3444	20	2051	
2CY34		3025	221		
2CY38			82	2032	
2CY39			82	2032	
2D	100	3005	1	2007	
2D-02	116	3311	504A	1104	
2D001	121	3717	239	2006	
2D002	123AP	3444/123A	20	2051	
2D002-168	123A	3444	20	2051	
2D002-169	123A	3444	20	2051	
2D002-170	123AP	3444/123A	20	2051	
2D002-171	171	3444/123A	20	2051	
2D002-175	123AP	3444/123A	20	2051	
2D002-41	123AP	3444/123A	20	2051	
2D004	121	3009	239	2006	
2D004-9	121	3009	16	2006	
2D010	130	3027	14	2041	
2D013	102	3004	53	2007	
2D013-109	102	3004	53	2007	
2D013-13	102	3004	53	2007	
2D013-160	102	3004	53	2007	
2D013-54	102	3004	53	2007	
2D015	121	3009	239	2006	
2D016	102	3004	53	2007	
2D016-45	102	3009	53	2007	
2D016-54	102	3009	53	2007	
2D017	159	3444/123A	20	2051	
2D017-165	159	3466	82	2032	
2D017-166	159	3466	82	2032	
2D017-167	159	3466	82	2032	
2D017-169	159	3466	82	2032	
2D020	194	3275	220		
2D020-173	194	3275	220		
2D020-174	194	3275	220		
2D021	100	3009	53	2007	
2D021-11	100	3009	53	2007	
2D021-56	100	3009	53	2007	
2D021-8	100	3009	53	2007	
2D022	6400A			2029	
2D022-211	6400A			2029	
2D023	102	3009	53	2007	
2D026	123AP	3444/123A	20	2051	
2D026-274	123AP	3444/123A	20	2051	
2D027	159	3114/290	82	2032	
2D030	6410			2029	
2D031	457	3112	FET-1	2028	
2D033	289A	3444/123A	20	2051	
2D036	102A	3004	53	2007	
2D038	313	3444/123A	20	2051	
2D039	102A	3004	53	2007	
2D040	396	3044/154			
2D165	177	3100/519	300	1122	
002D235RY	152		66	2048	
2D41	5400	3950			
2D7A	5096A	3346	ZD-100		
2D021-56			53	2007	
2D023			53	2007	
2D033			17	2051	
2E	126	3006/160	52	2024	
2E05	116	3081/125	512		
2E1	116		512		
2E10	156	3051	512		
2E1A20A22AAB	128		243	2030	
2E2	156		512		
2E4	116	3016	512		
2E6	156	3051	512		
2E8	156	3051	512		
2EM015	156	3081/125	512		
2F	160		245	2004	
2F05	5800	9003		1142	
2F2CN1	156		512		
2F2CN5	156		512		
2F4	116	3017B/117	504A	1104	
2F5CN1			509	1114	
2FB050	5304				MDA200
2FB050R					SPECIAL
2FB100	5304				MDA201
2FB1000	5307	3107			MDA210
2FB100R					SPECIAL
2FB200	5304				MDA202
2FB200R					SPECIAL
2FB400	5304				MDA204
2FB400R					SPECIAL
2FB600	5305	3676			MDA206
2FB600R					SPECIAL
2FB800	5306	3677			MDA208
2G	126	3006/160	52	2024	
2G101	126		245	2004	
2G102	160		245	2004	
2G1024	100	3005	1	2007	
2G1025	100	3005	1	2007	
2G1026	100	3005	1	2007	
2G1027	102	3004	2	2007	
2G103	160		245	2004	
2G104	160		245	2004	
2G106	160		245	2004	
2G108	102	3123	245	2004	
2G109	160	3123	245	2004	
2G110	160		245	2004	
2G13	116	3016	504A	1104	
2G138	100	3005	1	2007	
2G139	100	3005	1	2007	
2G140	100	3005	1	2007	
2G201	160	3123	245	2004	
2G202	160	3123	245	2004	
2G210	121	3014	239	2006	
2G220	121	3009	239	2006	
2G222	121	3014	239	2006	
2G223	104	3009	16	2006	
2G224	121	3014	239	2006	
2G225	121	3009	239	2006	
2G226	179	3642	76		

Industry Standard No.	ECG	SK	GE	RS 276-	MOTOR.
2G227	179	3642	76		
2G228	179	3642	76		
2G240	121	3014	239	2006	
2G270	102	3004	2	2007	
2G271	102	3004	2	2007	
2G301	160		245	2004	
2G302	102	3004	2	2007	
2G303	102A		53	2007	
2G304	102A		53	2007	
2G3055	130		14	2041	
2G306	102	3004	2	2007	
2G308	102A	3004	53	2007	
2G309	102A		53	2007	
2G319	102	3005	1	2007	
2G320	102	3005	1	2007	
2G321	102	3004	2	2007	
2G322	102	3004	2	2007	
2G323	102	3004	2	2007	
2G324	102	3004	2	2007	
2G339	103	3862	8	2002	
2G339A	101	3862/103	8	2002	
2G344	102A	3006/160	52	2024	
2G345	102A	3005	245	2004	
2G371	102A	3005	53	2007	
2G371A	102A	3005	53	2007	
2G374	102A	3005	53	2007	
2G374A	102A	3004	53	2007	
2G376	102A		53	2007	
2G377	102A		53	2007	
2G381	102A	3004	53	2007	
2G381A	102A	3004	53	2007	
2G382	102A		245	2004	
2G383	100	3005	1	2007	
2G384	102	3004	2	2007	
2G385	102	3004	2	2007	
2G386	102	3004	2	2007	
2G387	102	3004	2	2007	
2G394	100	3005	1	2007	
2G395	102	3123	53	2007	
2G396	100	3005	1	2007	
2G397	100	3005	1	2007	
2G401	126		245	2004	
2G402	126		245	2004	
2G403	160		245	2004	
2G404	160		245	2004	
2G413	160		245	2004	
2G414	126		245	2004	
2G415	126		245	2004	
2G416	126		245	2004	
2G417	126		245	2004	
2G508	102	3004	2	2007	
2G509	102	3004	2	2007	
2G524	100	3005	1	2007	
2G525	100	3005	1	2007	
2G526	100	3005	1	2007	
2G527	100	3005	1	2007	
2G577	100	3005	1	2007	
2G601	100	3006/160	52	2024	
2G602	100	3006/160	52	2024	
2G603	100	3005	1	2007	
2G604	100	3005	1	2007	
2G605	100	3123	1	2007	
2G8	116	3016	504A	1104	
2G805	116	3016	504A	1104	
2GA	116	3311	504A	1104	
2H1254	159		21	2034	
2H1255	159		21	2034	
2H1256	106		21	2034	
2H1257	106		21	2034	
2H1258	159		21	2034	
2H1259	159		21	2034	
2HR3J	125	3032A	510,531	1114	
2HR3M	125	3033	510,531	1114	
2HT-16		3068	CR-4		
2HT-6	505	3108	CR-7		
2HT16	504		CR-6		
2HT6	505		CR-7		
2J72	160		245	2004	
2J73	160		245	2004	
2J8138			504A	1104	
2JMW961			20	2051	
2K02	125	3033	510,531	1114	
2K48	160		245	2004	
2KBB10R					MDA970A2
2KBB20R					MDA970A3
2KBB40R					MDA970A4
2KBP002					MDA202
2KBP004					MDA204
2KBP005					MDA200
2KBP006					MDA206
2KBP008					MDA208
2KBP010					MDA210
2KBP02	167	3647			
2L15	525	3925			
2M1127	102	3722	2	2007	
2M1303			53	2007	
2M1305			51	2004	
2M214	5073A	3749	ZD-9.1	562	
2M3574					2N5265
2M6219			214	2038	
2M78	101	3861	8	2002	
2MA	119	3109	CR-2		
2MA509	298	3450			
2MC	126	3006/160	52	2024	
2MN6374			214	2038	
2MW665	109	3088		1123	
2MW742	137A	3058	ZD-6.2	561	
2N100	101	3861	8	2002	
2N1000	101	3861	8	2002	
2N1003	126	3006/160	52	2024	
2N1004	126		52	2024	
2N1005	108	3452	86	2038	2N2242
2N1006	123	3444	20	2051	2N2242
2N1007	104	3009	16	2006	
2N1008	102	3004	2	2007	
2N1008A	102	3004	2	2007	
2N1008B	102	3004	2	2007	
2N1009	102	3003	2	2007	
2N1010	103A	3011	59	2002	
2N1010#				2002	
2N1011	179	3009	239	2006	
2N1012	101	3861	8	2002	
2N1014	179	3009	239	2006	
2N1017	160	3011	245	2004	
2N1018	160	3011	245	2004	
2N102	101	3011A	8	2002	
2N1020	121	3009	239	2006	
2N1021	179	3009	239	2006	
2N1021A	179	3009	239	2006	
2N1022	179	3009	239	2006	
2N1022A	179	3009	239	2006	
2N1023	160	3006	245	2004	
2N1024	106	3984	21	2034	2N3250
2N1025	159	3466	82	2032	2N3250

Industry Standard No.	ECG	SK	GE	RS 276-	MOTOR.
2N1026	159	3466	82	2032	2N3250
2N1026A	159	3466	82	2032	
2N1027	159	3466	82	2032	2N3250
2N1028	159	3466	82	2032	
2N1029	179	3009	239	2006	
2N1029A	179	3009	239	2006	
2N1029B	179	3009	239	2006	
2N1029C	179	3009	239	2006	
2N103	101	3861	8	2002	
2N1030	179	3009	239	2006	
2N1030A	179	3009	239	2006	
2N1030B	179	3009	239	2006	
2N1030C	179	3009	239	2006	
2N1031	179	3009	239	2006	
2N1031A	179	3009	239	2006	
2N1031B	179	3009	239	2006	
2N1031C	179	3009	239	2006	
2N1032	179	3009	239	2006	
2N1032A	179	3009	239	2006	
2N1032B	179	3009	239	2006	
2N1032C	179	3009	239	2006	
2N1033	121	3009	239	2006	
2N1034	129	3114/290	244	2027	
2N1035	129	3114/290	244	2027	
2N1036	129	3114/290	244	2027	
2N1037	129	3114/290	244	2027	
2N1038	176	3845	80		
2N1039			3	2006	
2N104	102A	3004	53	2007	
2N1040	104	3009	16	2006	
2N1041		3009	3	2006	
2N1042	126		52	2024	
2N1043	126		52	2024	
2N1044	102	3004	2	2007	
2N1045	102	3004	2	2007	
2N1046	179	3009	239	2006	
2N1046A		3009	25		
2N1046B	121	3009	239	2006	
2N1047			66	2048	
2N1049			66	2048	
2N105	102A	3004	53	2007	
2N1051	128	3444/123A	20	2051	2N2218
2N1052	154	3045/225	40	2012	
2N1053	154	3045/225	40	2012	
2N1054	154	3045/225	40	2012	
2N1055	128		222		
2N1056	102	3004	2	2007	
2N1057	102	3004	2	2007	
2N1058	101	3011	59	2002	
2N1059	103A	3010	59	2002	
2N1059-1			8	2002	
2N106	102A	3003	53	2004	
2N106#				2004	
2N1060	108	3452	86	2038	
2N1065	160	3008	245	20044	
2N1066	160	3006	245	2004	
2N1067			28	2017	
2N1068			28	2017	
2N1069	130	3027	14	2041	
2N107	100	3003	53	2004	
2N1070	130	3027	14	2041	
2N1072			14	2041	
2N1073	179	3034	239	2006	
2N1073A	179	3034	239	2006	
2N1073B	127	3034	25		
2N1074					2N2218
2N1075					2N2218
2N1076					2N2218
2N1077	123A				2N2218
2N1078		3009	54		
2N108	102A	3003	2	2007	
2N1081	128	3444/123A	20	2051	2N2221
2N1082	123	3444	20	2051	2N2221
2N1085			214	2038	
2N1086	101	3861	8	2002	
2N1086A	101	3861	8	2002	
2N1087	101	3861	8	2002	
2N109	102A	3004	53	2007	
2N109/5	102	3004	2	2007	
2N1090	101	3861	8	2002	
2N1091	101	3861	8	2002	
2N1092	128	3024	243	2030	
2N1093	102		245	2004	
2N1094	160		245	2004	
2N1095	103	3862	8	2002	
2N1096	103	3862	8	2002	
2N1097	102	3003	2	2007	
2N1098	102	3003	2	2007	
2N1099	105	3012	4		
2N109BLU	102A	3123	53	2007	
2N109GRN	102A	3123	53	2007	
2N109M1	102A		53	2007	
2N109M2	102A		53	2007	
2N109WHT	102A		53	2007	
2N109YEL	102A		53	2007	
2N1100	105	3012	4		
2N1101	103	3010	59	2002	
2N1102	103A	3010	59	2002	
2N1102/5	101	3011	8	2002	
2N1103	123A	3444	20	2051	2N2221
2N1104	123A	3124/289	20	2051	2N2221
2N1107	126	3005	52	2024	
2N1108	126	3008	52	2024	
2N1108#				2005	
2N1108RED	126	3006/160	52	2024	
2N1109	126	3007	52	2024	
2N1109#				2005	
2N111	102	3005	1	2007	
2N1110	126	3008	52	2024	
2N1110#				2005	
2N1111	126	3007	52	2024	
2N1111A	126	3006/160	52	2024	
2N1111B	126	3006/160	52	2024	
2N1111M1			52	2024	
2N1111RED	126	3006/160	52	2024	
2N1112	101	3861	8	2002	
2N1114	101	3861	8	2002	
2N1115	102A	3005	245	2004	
2N1115A	160	3005	245	2004	
2N1116	128	3444/123A	20	2051	
2N1117	128	3444/123A	20	2051	
2N1118	159	3466	82	2032	
2N1118A	159	3466	82	2032	
2N1119	159	3466	82	2032	
2N111A	102	3005	1	2007	
2N111B	126	3006/160	52	2024	
2N111M1	126	3006/160		2024	
2N111M2	126		52	2024	
2N112	126	3005	8	2024	
2N1120	179	3009	239	2006	
2N1121	101	3861	8	2002	
2N1122	160	3005	245	2004	
2N1122A	160	3005	245	2004	
2N1123			53	2007	

Industry Standard No.	ECG	SK	GE	RS 276-	MOTOR.
2N1124	102	3004	2	2007	
2N1125	102	3004	2	2007	
2N1126	102	3004	2	2007	
2N1127		3004	53	2007	
2N1128	176	3004/102A	2	2007	
2N1129		3004	53	2007	
2N112A	102	3005	1	2007	
2N112M1	126	3006/160	52	2024	
2N113		3004	1	2007	
2N1130	102A	3004	53	2007	
2N1131	176	3466/159	82	2032	2N2905
2N1131/46	159		221		
2N1131/51			21	2034	
2N1131A	159	3466	82	2032	2N2905A
2N1131A/51			21	2034	
2N1132	159	3466	82	2032	2N1132
2N1132/46	159	3466	82	2032	
2N1132/51			21	2034	
2N1132A	159	3466	82	2032	2N1132A
2N1132A/51			21	2034	
2N1132A46	159		21	2034	
2N1132B	159	3466	82	2032	
2N1132B/51			67	2023	
2N1132B46	288		67	2023	
2N1135	159	3466	82	2032	2N2369
2N1135A	159	3466	82	2032	2N2369
2N1136	104	3009	239	2006	
2N1136A	104	3009	239	2006	
2N1136B	179	3009	25		
2N1136C			16	2006	
2N1137	104	3009	239	2006	
2N1137A	179	3009	239	2006	
2N1137B	179	3009	25		
2N1138	179	3009	239	2006	
2N1138A	179	3009	239	2006	
2N1138B		3009	16	2006	
2N1139	123	3444	20	2051	2N742
2N114	100	3005	1	2007	
2N1140	123A	3444	20	2051	
2N1141			245	2004	
2N1141A			245	2004	
2N1142			245	2004	
2N1142A			245	2004	
2N1143			245	2004	
2N1143A			245	2004	
2N1144	102	3003	245	2004	
2N1145	102	3003	245	2004	
2N1146	179	3009	16	2006	
2N1146A	179	3009	16	2006	
2N1146B	179	3009	16	2006	
2N1146C	179	3009	16	2006	
2N1147	179	3009	16	2006	
2N1147A	179	3009	16	2006	
2N1147B	179	3009	16	2006	
2N1147C	179	3009	16	2006	
2N1149	123AP	3124/289	20	2051	2N2221
2N115			3	2006	
2N1150	123AP	3020/123	20	2051	2N2221
2N1150/904			17	2051	
2N1151	123AP	3124/289	20	2051	2N2221
2N1151/904A			17	2051	
2N1152	123AP	3124/289	20	2051	2N2221
2N1153	123AP	3124/289	20	2051	2N2221
2N1153/910			17	2051	
2N1154	194	3275	47	2030	2N2221
2N1154/951			47	2030	
2N1155	194	3275	18	2030	2N2221
2N1155/952			47	2030	
2N1156	194	3275	18	2030	2N2221
2N1156/953			18	2030	
2N1157	213	3012/105	4		
2N1157A	213	3012/105	4		
2N1158	126		245	2004	
2N1158A	126		245	2004	
2N1159	179	3009	239	2006	
2N1160	179	3009	239	2006	
2N1162	179	3009	16	2006	
2N1162A	179	3009	16	2006	
2N1163	179	3009	239	2006	
2N1163A	179	3009	239	2006	
2N1164	179	3009	16	2006	
2N1164A	179	3009	16	2006	
2N1165	104	3009	16	2006	
2N1165A	179	3009	239	2006	
2N1166	179	3009	16	2006	
2N1166A	179	3009	239	2006	
2N1167		3009	25		
2N1167A		3009	25		
2N1168	121	3009	239	2006	
2N1169	103	3862	8	2002	
2N117	123AP	3124/289	20	2051	
2N117#				2011	
2N1170	103	3862	8	2002	
2N1171	100	3005	1	2007	
2N1172		3009	16	2006	
2N1173	103A	3010	59	2002	
2N1173W	103A	3010	59	2002	
2N1174	102	3010	2	2007	
2N1174W	176	3123	80		
2N1175	102	3004	2	2007	
2N1175A	102	3004	2	2007	
2N1176	100	3005	1	2007	
2N1176A	100	3005	1	2007	
2N1176B	100	3005	1	2007	
2N1177	160	3007	245	2004	
2N1178	160	3007	245	2004	
2N1179	160	3007	245	20041	
2N118	123AP	3124/289	20	2051	
2N118#				2011	
2N1180	126	3006/160	52	2024	
2N1182	179	3014	239	2006	
2N1183	176	3123	80		
2N1183A		3123	80		
2N1183B	176		80		
2N1184	176	3123	80		
2N1184A	176		80		
2N1184B	176		80		
2N1185	100	3005	1	2007	
2N1186	100	3005	1	2007	
2N1187	100	3005	1	2007	
2N1188	100	3005	1	2007	
2N1189	102	3004	2	2007	
2N118A	123AP	3124/289	20	2051	
2N118A#				2011	
2N119	123AP	3124/289	20	2051	
2N119#				2011	
2N1190	102	3004	2	2007	
2N1191	102	3004	2	2007	
2N1192	102	3004	2	2007	
2N1193	102	3004	2	2007	
2N1194	102	3004	2	2007	
2N1195			245	2004	
2N1196	159	3715	82	2032	
2N1197	129	3025	244	2027	
2N1198	101	3862/103	8	2002	
2N1199	123	3444	20	2051	2N834
2N1199A	123	3444	20	2051	2N834
2N120	123AP	3124/289	20	2051	
2N120#				2011	
2N1200	123	3444	20	2051	
2N1201	123	3124/289	20	2051	
2N1202	105	3012	4		
2N1203	105	3012	4		
2N1204	126	3006/160	52	2024	
2N1204A	126	3006/160	52	2024	
2N1205	123	3124/289	20	2051	
2N1206			63	2030	2N3020
2N1207	154	3045/225	40	2012	2N3500
2N1213	126	3006/160	52	2024	
2N1214	126	3006/160	52	2024	
2N1215	126	3006/160	52		
2N1216	126	3006/160	52	2024	
2N1217	101	3862/103	8	2002	
2N1218		3717	239	2006	
2N1219	159	3466	82	2032	2N3250
2N1220	159	3466	82	2032	
2N1221	159	3466	82	2032	
2N1222	159	3466	82	2032	
2N1223	159	3466	82	2032	2N3250
2N1224	126	3007	52	2024	
2N1225	126	3006/160	52	2024	
2N1226	126	3008	52	2024	
2N1227	104	3014	16	2006	
2N1227-3	104	3014	16	2006	
2N1227-4	104	3014	16	2006	
2N1227-4A			16	2006	
2N1227-4R	104	3014	16	2006	
2N1227A	104	3014	16	2006	
2N1228	129	3025	244	2027	2N2904
2N1229	129	3025	244	2027	2N2904
2N123	100	3005	1	2007	
2N123/5			51	2004	
2N1230	129	3025	244	2027	2N2904
2N1231	129	3025	244	2027	2N2904
2N1232	129	3025	244	2027	2N2905A
2N1233	129	3025	244	2027	2N2905A
2N1234	129	3025	244	2027	2N3495
2N1235					2N5759
2N1238	106	3984	21	2034	2N3467
2N1239	106	3984	21	2034	2N3467
2N123A	126	3006/160	52	2024	
2N123A5			51	2004	
2N124	101	3861	8	2002	
2N1240	106	3984	21	2034	2N3467
2N1241	106	3984	21	2034	2N3467
2N1242		3984	67	2023	2N3763
2N1243			67	2023	2N3763
2N1245	104	3009	16	2006	
2N1246	104	3009	16	2006	
2N1247	123	3124/289	20	2051	2N2222
2N1248	123	3124/289	20	2051	2N2222
2N1249	123	3124/289	20	2051	2N2222
2N125	101	3861	8	2002	
2N1251	103	3010	59	2002	
2N1252	123	3024/128	243	2030	2N2845
2N1253	123	3024/128	243	2030	2N2845
2N1253A			63	2030	
2N1254	129	3025	244	2027	2N869A
2N1255	129	3025	244	2027	2N869A
2N1256	159	3025/129	244	2027	2N869A
2N1257	159	3025/129	244	2027	2N869A
2N1258	129	3025	244	2027	2N869A
2N1259	129	3025	244	2027	2N869A
2N126	101	3011	59	2002	
2N1261	105	3012	4		
2N1262	105	3012	4		
2N1263	105	3012	4		
2N1264	100	3006/160	1	2007	
2N1265	102	3003	2	2007	
2N1265/5			1	2007	
2N1265A	102	3004	2	2007	
2N1266	126	3005	53	2004	
2N1266#				2004	
2N1267	123	3124/289	20	2051	2N2481
2N1268	123		212	2010	2N2481
2N1269	123		212	2010	2N2481
2N127	101	3861	8	2002	
2N1270	123		212	2010	2N2481
2N1271	123		212	2010	2N2481
2N1272	123		212	2010	2N2481
2N1273	102	3004	2	2007	
2N1273BLU	102	3004	2	2007	
2N1273GRN	102	3004	2	2007	
2N1273ORN	102	3004	2	2007	
2N1273RED	102	3004	2	2007	
2N1273YEL	102	3004	2	2007	
2N1274	102	3845/176	2	2007	
2N1274BLU	102	3845/176	2	2007	
2N1274BRN	102	3845/176	2	2007	
2N1274GRN	102	3845/176	2	2007	
2N1274ORN	102	3845/176	2	2007	
2N1274PUR	102	3845/176	2	2007	
2N1274RED	102	3845/176	2	2007	
2N1274V10	176	3845			
2N1275	129	3025	244	2027	
2N1276	123A	3124/289	20	2051	2N2501
2N1277	123A	3124/289	20	2051	2N2501
2N1278	123A	3124/289	20	2051	2N2501
2N1279	123A	3124/289	20	2051	2N2501
2N128	126	3008	52	2024	
2N1280	100	3005	1	2007	
2N1281	100	3005	1	2007	
2N1282	100	3005	245	2004	
2N1284	100	3005	1	2007	
2N1285	160	3006	245	2004	
2N1287	102	3004	2	2007	
2N1287A	102	3004	2	2007	
2N1288	101	3861	8	2002	
2N1289	101	3861	8	2002	
2N129	126	3008	52	2024	
2N1291		3009	16	2006	
2N1292	121	3009	239	2006	
2N1293	121	3009	239	2006	
2N1294		3014	239	2006	
2N1295	179	3009	239	2006	
2N1296	121	3014	239	2006	
2N1297	179	3009	239	2006	
2N1299	101	3861	8	2002	
2N130	102	3003	2	2007	
2N1300	126		245	2004	
2N1301	160	3006	245	2004	
2N1302	101	3011	8	2002	
2N1303	102	3721/100	53	2007	
2N1304	101	3861	8	2002	
2N1305	102	3721/100	53	2007	
2N1306	101	3862/103	8	2002	
2N1307	100	3722/102	53	2007	
2N1308	101	3861	8	2002	

Industry Standard No.	ECG	SK	GE	RS 276-	MOTOR.
2N1309	100	3721	1	2007	
2N1309A	160		245	2004	
2N130A	102	3004	2	2007	
2N131	102	3003	2	2007	
2N131#				2005	
2N1310	101	3861	8	2002	
2N1311	101	3861	8	2002	
2N1312	103	3862	8	2002	
2N1313	126	3008	52	2024	
2N1314	104	3009	16	2006	
2N1314R	104	3009	16	2006	
2N1315	176	3123	80		
2N1316	100	3005	1	2007	
2N1317	100	3005	1	2007	
2N1318	100	3005	1	2007	
2N1319	100	3005	1		
2N131A	102	3003	2	2007	
2N132	102	3003	2	2007	
2N132#				2004	
2N132A	102A	3003	2	2007	
2N133	102	3003	2	2007	
2N133#				2005	
2N1335	154		18	2030	2N3019
2N1336	154		18	2030	2N3019
2N1337	154		18	2030	2N3019
2N1338	128		20	2051	2N2219
2N1339	154		18	2030	2N3019
2N133A	102A	3003	53	2007	
2N1340	154		18	2030	2N3019
2N1341	154		18	2030	2N3019
2N1342	154				2N3019
2N1343	100	3005	1	2007	
2N1344	100	3005	1	2007	
2N1345	100	3005	1	2007	
2N1346	100	3005	1	2007	
2N1347	100	3005	1	2007	
2N1348	102	3004	2	2007	
2N1349	100	3005	1	2007	
2N135	100	3005	1	2007	
2N135#				2005	
2N1350	100	3005	1	2007	
2N1351	100	3005	1	2007	
2N1352	102	3003	2	2007	
2N1353	102	3003	2	2007	
2N1354	100	3005	1	2007	
2N1355	100	3005	1	2007	
2N1356	100	3005	1	2007	
2N1357	100	3005	1	2007	
2N1358	213	3012/105	4		
2N1358A	213	3012/105	4		
2N1358M	213	3012/105	4		
2N1359	121	3009	239	2006	
2N136	100	3005	1	2007	
2N136#				2005	
2N1360	179	3009	239	2006	
2N1361	100	3009	1	2007	
2N1361A	100	3005	2	2007	
2N1362	179	3009	239	2006	
2N1363	179	3009	239	2006	
2N1364	121	3009	239	2006	
2N1365	121	3009	239	2006	
2N1366	101	3861	8	2002	
2N1367	101	3861	8	2002	
2N137	100	3005	1	2007	
2N137#				2005	
2N1370	102	3004	2	2007	
2N1371	102	3004	2	2007	
2N1372	102	3004	2	2007	
2N1373	102	3004	2	2007	
2N1374	102	3004	2	2007	
2N1375	102	3004	2	2007	
2N1376	102	3004	2	2007	
2N1377	102	3004	2	2007	
2N1378	102	3004	2	2007	
2N1379	102	3004	2	2007	
2N138	126	3003	53	2007	
2N1380	102	3004	2	2007	
2N1381	102	3004	2	2007	
2N1382	102	3004	2	2007	
2N1383	102	3004	2	2007	
2N1384	126	3008	52	2024	
2N1385	160		245	2004	
2N1386	123	3124/289	20	2051	2N2222
2N1387	123	3124/289	20	2051	2N2222
2N1388	123A	3124/289	20	2051	2N2222
2N1389	123A	3124/289	20	2051	2N2222
2N138A	102A	3003	53	2007	
2N138B	102A	3003	53	2007	
2N139	126	3005	52	2024	
2N1390	123		212	2010	2N2222
2N1391	103	3862	8	2002	
2N1392	102	3004	2	2007	
2N1393	100	3005	1	2007	
2N1394			51	2004	
2N1395	100	3006/160	1	2007	
2N1396	126	3006/160	52	2024	
2N1397	126	3006/160	52	2024	
2N1398	126	3006/160	52	2024	
2N1399	126	3006/160	52	2024	
2N140	126	3005	52	2024	
2N1400	126	3006/160	52	2024	
2N1401	126	3006/160	52	2024	
2N1401A	126	3006/160	52	2024	
2N1402	126	3006/160	52	2024	
2N1403	160	3123	245	2004	
2N1404	102	3722	2	2007	
2N1404A	100		245	2004	
2N1405	160		245	2004	
2N1406	160		245	2004	
2N1407	160		245	2004	
2N1408	160		245	2004	
2N1409	123		245	2004	2N5859
2N1409A	123A		47	2030	2N5859
2N140M1	126		52	2024	
2N140M2	126	3006/160	52	2024	
2N141			3	2006	
2N1410	128		245	2004	2N5859
2N1410A	123A		47	2030	2N5859
2N1411	160		245	2004	
2N1412	213		4		
2N1413	102	3004	2	2007	
2N1414	102	3004	2	2007	
2N1415	102	3004	2	2007	
2N1416	102A		53	2007	
2N1417	123	3444	20	2051	
2N1418	123	3444	20	2051	
2N1419	179	3009	16	2006	
2N1420	128	3444/123A	20	2051	2N2219
2N1420A	128	3444/123A	20	2051	
2N1422	130		14	2041	
2N1423	130		14	2041	
2N1425	126	3008	52	2024	
2N1426	126	3008	52	2024	
2N1427	126	3003	52	2024	
2N1428	159	3114/290	82	2032	2N869A
2N1429	129	3025	244	2027	2N869A
2N143			3	2006	
2N1430	121	3014	239	2006	
2N1431	103	3010	59	2002	
2N1432	102	3004	2	2007	
2N1433	105	3012	4		
2N1434	105	3012	4		
2N1435	105	3012	4		
2N1436	160		245	2004	
2N1437			3	2006	
2N1438			3	2006	
2N1439	159	3466	82	2032	2N2907A
2N1440	159	3466	82	2032	2N2907A
2N1441	159	3466	82	2032	2N2907A
2N1442	159	3466	82	2032	2N2907A
2N1443	159	3466	82	2032	2N2907A
2N1444	128		63	2030	2N5859
2N1445	396		32		
2N1446	102	3004	2	2007	
2N1447	102	3004	2	2007	
2N1448	102	3004	2	2007	
2N1449	102	3004	2	2007	
2N145	101	3861	8	2002	
2N1450	126	3004/102A	2	2007	
2N1451	102	3004	2	2007	
2N1452	102	3004	2	2007	
2N146	101	3861	8	2002	
2N1469	100	3005	1	2007	
2N147	101	3861	8	2002	
2N1470	100	3005	1	2007	
2N1471	100	3005	1	2007	
2N1472	128	3024	243	2030	2N834
2N1473	101	3861	8	2002	
2N1474	159	3466	82	2032	2N2906A
2N1474A	159	3466	82	2032	2N2906A
2N1475	159	3466	82	2032	2N2906A
2N1476			221		2N4928
2N1477			221		2N4928
2N1478	102	3722	2	2007	
2N1479	324	3024/128	243	2030	2N4237
2N148	101	3124/289	20	2051	
2N1480	324	3024/128	243	2030	2N4238
2N1481	324	3024/128	243	2030	2N4237
2N1482	324	3024/128	243	2030	2N4238
2N1483		3530	66	2048	TIP41
2N1484		3530	66	2048	TIP41A
2N1485		3530	66	2048	TIP41
2N1486		3530	66	2048	TIP41A
2N1487	280	3297	262	2041	2N5877
2N1488	280	3297	262	2041	2N5878
2N1489	280	3297	262	2041	2N5877
2N148A	103A	3124/289	59	2002	
2N148B	103A	3010	59	2002	
2N148C			20	2051	
2N148C/D	103A	3011	20	2051	
2N148D			20	2051	
2N149	103A	3835	59	2002	
2N1490	283	3510	14	2041	2N5878
2N1491	128	3765/195A	46		MRF531
2N1492	128	3765/195A	46		
2N1493	154	3045/225	40	2012	
2N1494			51	2004	
2N1495	160		245	2004	
2N1499	126	3006/160	52	2024	
2N1499A	126	3006/160	52	2024	
2N1499B	126	3006/160	52	2024	
2N149A	103A	3835	59	2002	
2N149B	158	3835/103A	53	2007	
2N150	101	3861	8	2002	
2N1500	126	3721/100	52	2024	
2N1500/18	160		245	2004	
2N1501	104	3012/105	16	2006	
2N1502	104	3012/105	16	2006	
2N1505	123A	3452/108	86	2038	2N2219A
2N1506	123A	3452/108	86	2038	2N4237
2N1506A	128	3452/108	86	2038	2N4239
2N1507	128	3452/10R	86	2038	
2N1508					2N3019
2N1509					2N3019
2N150A	101	3861	8	2002	
2N151	102	3004	2	2007	
2N1510	101	3861	8	2002	
2N1515	126	3006/160	245	2004	
2N1516	126	3006/160	245	2004	
2N1517	126	3006/160	245	2004	
2N1517A	160		245	2004	
2N1518	213	3012/105	254		
2N1519	213	3012/105	254		
2N1520	213	3012/105	254		
2N1521	213	3012/105	254		
2N1522	213	3012/105	242		
2N1523	105	3012	4		
2N1524	126	3007	245	2004	
2N1524-1	160	3007	245	2004	
2N1524-2	160	3007	245	2004	
2N1524/33	160		245	2004	
2N1525	160	3008	245	2004	
2N1526	126	3007	52	2024	
2N1526/33	160		245	2004	
2N1527	126	3008	245	2004	
2N1528	123	3444	20	2051	
2N1529	179	3009	239	2006	
2N1529A	179	3009	239	2006	
2N1530	179	3009	239	2006	
2N1530A	121	3009	239	2006	
2N1531	179	3009	239	2006	
2N1531A	179	3009	239	2006	
2N1532	179	3009	239	2006	
2N1532A		3009	239	2006	
2N1533	121	3009	239	2006	
2N1534	179	3009	239	2006	
2N1534A	179	3009	239	2006	
2N1535		3009	239	2006	
2N1535A	121	3009	239	2006	
2N1535B	121	3009	239	2006	
2N1536	179	3009	239	2006	
2N1536A	121	3009	239	2006	
2N1537	179	3009	239	2006	
2N1537A	179	3009	239	2006	
2N1538	121	3009	239	2006	
2N1538A	121	3009	239	2006	
2N1539	179	3009	239	2006	
2N1539A	179	3009	239	2006	
2N1540	179	3717/121	239	2006	
2N1540A	179	3717/121	239	2006	
2N1541	179	3717/121	239	2006	
2N1541A	179	3717/121	239	2006	
2N1542	179	3717/121	239	2006	
2N1542A	179	3717/121	239	2006	
2N1543	179	3717/121	239	2006	
2N1543A	179	3717/121	239	2006	
2N1544	179	3009	239	2006	
2N1544A	179	3009	239	2006	

Industry Standard No.	ECG	SK	GE	RS 276-	MOTOR.
2N1545	179	3717/121	239	2006	
2N1545A	179	3717/121	239	2006	
2N1546	179	3717/121	239	2006	
2N1546A	179	3717/121	239	2006	
2N1547	179	3717/121	239	2006	
2N1547A	179	3717/121	239	2006	
2N1548	179	3717/121	239	2006	
2N1548A	179	3717/121	239	2006	
2N1549	179	3009	239	2006	
2N1549A	179	3009	239	2006	
2N155	121	3009	239	2006	
2N1550	179	3717/121	239	2006	
2N1550A	179	3717/121	239	2006	
2N1551	179	3717/121	239	2006	
2N1551A	179	3717/121	239	2006	
2N1552	179	3717/121	239	2006	
2N1552A	179	3717/121	239	2006	
2N1553	179	3009	239	2006	
2N1553A	179	3009	239	2006	
2N1554	179	3717/121	239	2006	
2N1554A	179	3717/121	239	2006	
2N1555	179	3717/121	239	2006	
2N1555A	179	3717/121	239	2006	
2N1556	179	3717/121	239	2006	
2N1556A	179	3717/121	239	2006	
2N1557	179	3009	239	2006	
2N1557A	179	3009	239	2006	
2N1558	179	3717/121	239	2006	
2N1558A	179	3717/121	239	2006	
2N1559	179	3717/121	239	2006	
2N1559A	179	3717/121	239	2006	
2N156		3009	239	2006	
2N1560	179	3717/121	239	2006	
2N1560A	179	3717/121	239	2006	
2N1561	102		245	2004	
2N1562	102		245	2004	
2N1564	128	3479	81	2051	
2N1565	128	3479	81	2051	
2N1566	128	3479	81	2051	
2N1566A	128	3479	81	2051	2N2219
2N157	121	3009	239	2006	
2N1570	100	3005	1	2007	
2N1572	154	3045/225	40	2012	
2N1573	154	3045/225	40	2012	
2N1574	154	3045/225	40	2012	
2N157A	121	3009	239	2006	
2N158		3009	3	2006	
2N1581	100	3005	1	2007	
2N1583	100	3005	1	2007	
2N1584	100	3005	1	2007	
2N1585	101	3861	8	2002	
2N1586	107	3124/289	11	2015	2N706A
2N1587	123AP	3124/289	11	2051	2N2501
2N1588	123AP	3444/123A	20	2051	2N2221
2N1589	107	3124/289	11	2015	2N834
2N158A		3009	3	2006	
2N159	102	3004	2	2007	
2N1590	123AP	3124/289	11	2051	2N2501
2N1591	123AP	3444/123A	20	2051	2N2221
2N1592	107	3124/289	11	2015	2N2222
2N1593	123AP	3124/289	11	2051	2N2222
2N1594	123AP	3444/123A	20	2051	2N2222
2N1595	5408	3577	MR-5	1067	2N1595
2N1596	5408	3577	MR-5	1067	2N1596
2N1597	5408	3577	MR-5	1067	2N1597
2N1598	5409	3578			2N1598
2N1599	5409	3578			2N1599
2N160	123AP	3124/289	20	2051	
2N1600	5470	3577			2N4168
2N1601	5471	3577			2N4169
2N1602	5472	3577			2N417C
2N1603	5474	3578			2N4171
2N1604	5474	3578			2N4172
2N1605	101	3861	8	2002	
2N1605-A				2002	
2N1605A	101	3861	8	2002	
2N1606	129	3114/290	244	2027	2N3546
2N1607	159	3466	82	2032	2N3546
2N1608	159	3466	82	2032	2N3546
2N1609			16	2006	
2N160A	123AP	3124/289	20	2051	
2N161	123AP	3124/289	20	2051	
2N1610			16	2006	
2N1611			3	2006	
2N1612			3	2006	
2N1613	128	3024	243	2030	2N1613
2N1613/46	123A	3479	81	2051	
2N1613A			235	2012	
2N1613B			235	2012	
2N1613L	128		243	2030	
2N1613S	128		243	2030	
2N1614	158		53	2007	
2N1615	128		222		2N3500
2N161A	123AP	3124/289	20	2051	
2N162	123AP	3039/316	11	2015	2N2221
2N1622	101	3861	8	2002	
2N1623	129	3114/290	244	2027	2N2906
2N1624	101	3861	6	2002	
2N1625			50	2004	
2N162A	107	3039/316	11	2015	2N2221
2N163	123AP	3124/289	20	2051	2N2221
2N1631	126	3008	52	2024	
2N1632	126	3008	52	2024	
2N1633	126	3008	245	2004	
2N1634	126	3008	245	2004	
2N1635	126	3008	52	2024	
2N1636	126	3008	52	2024	
2N1637	126	3008	52	2024	
2N1637/33	160	3123	245	2004	
2N1638	126	3007	245	2004	
2N1638/33			52	2024	
2N1639	126	3008	245	2004	
2N1639/33	160	3123	245	2004	
2N163A	123AP	3124/289	20	2051	2N2221
2N164	101	3861	8	2002	
2N1640	106	3118	21	2034	
2N1641	106	3118	21	2034	
2N1642	106	3118	21	2034	
2N1643	159	3466	82	2032	
2N1644	128	3444/123A	20	2051	
2N1644A	123A	3444	20	2051	
2N1646	160		245	2004	
2N164A	101	3861	8	2002	
2N165	101	3861	8	2002	
2N1651	179	3642	76		
2N1652	179	3014	239	2006	
2N1653	121	3009	239	2006	
2N1654	129		220		MM3006
2N1655	288		220		MM3007
2N1656	288		220		
2N166	101	3861	8	2002	
2N1662			52	2024	
2N1663	123	3444	20	2051	
2N1664	100	3005	1	2007	
2N1665	160		245	2004	
2N1666	179	3009	239	2006	
2N1667	121	3009	239	2006	
2N1668	104	3009	16	2006	
2N1669	104	3009	16	2006	
2N167	101	3861	8	2002	
2N1670	160	3005	245	2004	
2N1671	6400A		2N1671	2029	
2N1672	103A	3835	59	2002	
2N1672A	103A	3835	59	2002	
2N1673	160	3005	53	2007	
2N1674	123A	3444	20	2051	
2N1676	159	3466	82	2032	
2N1677	159	3466	82	2032	
2N1678	160	3008	245	2004	
2N167A	101	3861	8	2002	
2N168	101	3861	8	2002	
2N1681	102		53	2007	
2N1682	123	3124/289	20	2051	
2N1683	126	3005	1	2007	
2N1684	100	3005	1	2007	
2N1685	101	3861	8	2002	
2N168A	101	3861	8	2002	
2N169	101	3861	8	2002	
2N1694	101	3861	8	2002	
2N1699	126		52	2024	
2N169A	101	3861	8	2002	
2N170	101	3861	8	2002	
2N1700	128	3512	243	2030	
2N1701		3530	66	2048	
2N1702	283	3510	14	2041	2N5877
2N1703	130		14	2041	
2N1704	123A	3444	20	2051	
2N1705	102	3004	2	2007	
2N1706	102	3004	2	2007	
2N1707	102	3004	2	2007	
2N1708	123A	3444	20	2051	2N3508
2N1708A	123A	3444	20	2051	
2N1709			66	2048	
2N171			8	2002	
2N1710			28	2017	
2N1711	128	3024	243	2030	2N1711
2N1711/46	123A	3479	81	2051	
2N1711A	128		243	2030	
2N1711L	128		243	2030	
2N1711S	128		243	2030	
2N1713	126		245	2004	
2N1714	128		63	2030	2N4237
2N1715	396		32		2N5681
2N1716	128		63	2030	2N4237
2N1717	396		32		2N5681
2N1718			66	2048	
2N172	101	3861	8	2002	
2N1720			66	2048	
2N1721			D42R1		2N5681
2N1725	73		D44H11		
2N1726	126	3006/160	245	2004	
2N1727	126	3006/160	245	2004	
2N1728	126	3006/160	245	2004	
2N1729	100	3005	1	2007	
2N173	213	3012/105	4		
2N1730	101	3861	8	2002	
2N1731	100	3005	1	2007	
2N1732	101	3861	8	2002	
2N174	213	3012/105	4		
2N1742	126	3006/160	245	2004	
2N1743	126	3005	1	2007	
2N1744	126	3005	1	2007	
2N1745	126	3006/160	245	2004	
2N1746	126	3006/160	245	2004	
2N1747	126	3011	245	2004	
2N1748	126	3006/160	245	2004	
2N1748A	126	3006/160	245	2004	
2N1749	160	3006	245	2004	
2N174A	213	3012/105	4		
2N174RED	213	3123	4		
2N175	102A	3003	53	2007	
2N175#				2004	
2N1750	160	3007	245	2004	
2N1751	179	3642	76		
2N1752	126		245	2004	
2N1753	102A		245	2004	
2N1754	126	3005	245	2004	
2N1755	104	3009	16	2006	
2N1756	104	3009	16	2006	
2N1757	104	3009	16	2006	
2N1758	104	3009	16	2006	
2N1759	104	3009	16	2006	
2N176	104	3009	16	2006	
2N176(WHT/BRN)	121	3717			
2N176(WHT/GRN)	121	3717			
2N176(WHT/RED)	121	3717			
2N176-1	104	3009	16	2006	
2N176-1BLU	121	3717	239	2006	
2N176-1WHT	121		239	2006	
2N176-1YEL	121	3009	239	2006	
2N176-3PUR	121	3009	239	2006	
2N176-4PUR	121	3009	239	2006	
2N176-5WHT	121	3009	239	2006	
2N176-6WHT	121	3009	239	2006	
2N1760	104	3009	16	2006	
2N1761	104	3009	16	2006	
2N1762	104	3009	16	2006	
2N1763	123	3124/289	20	2051	
2N1764	128	3024	243	2030	2N2369A
2N1768			66	2048	
2N1769			66	2048	
2N176A	104	3009	16	2006	
2N176BLK	121	3717	239	2006	
2N176BLU	121	3717	239	2006	
2N176G	104	3009	16	2006	
2N176GRN	121	3717	239	2006	
2N176M	121	3717	239	2006	
2N176PUR	121	3717	239	2006	
2N176RED	121	3717	239	2006	
2N176W	104	3009	16	2006	
2N176WHT	121	3717	239	2006	
2N176YEL	121	3717	239	2006	
2N1770	5480				2N4167
2N1771	5481				2N4168
2N1771A	5481				2N4168
2N1772	5482	3942/5483			2N4169
2N1772A	5482	3942/5483			2N4169
2N1773	5483	3942			2N4170
2N1773A	5483	3942			2N4170
2N1774	5483	3942			2N4170
2N1774A	5483	3942			2N4170
2N1775	5484	3943/5485			2N4171
2N1775A	5484	3943/5485			2N4171
2N1776	5484	3943/5485			2N4171
2N1776A	5484	3943/5485			2N4171
2N1777	5485	3943			2N4172
2N1777A	5485	3943			2N4172
2N1778	5487				2N4173

Industry Standard No.	ECG	SK	GE	RS 276-	MOTOR.
2N1778A	5486	3944/5487			2N4173
2N1779	101	3861	8	2002	
2N178	104	3009	16	2006	
2N1780	101	3861	8	2002	
2N1781	101	3011	8	2002	
2N1782	160	3005	245	2004	
2N1783	101	3861	8	2002	
2N1784	160	3005	245	2004	
2N1785	126	3006/160	245	2004	
2N1786	126	3006/160	245	2004	
2N1787	126	3006/160	245	2004	
2N1788	160	3006	245	2004	
2N1789	160	3006	245	2004	
2N179	104	3009	16	2006	
2N1790	160		245	2004	
2N180	102A	3004	53	2007	
2N1800			53	2007	
2N1808·	101	3861	8	2002	
2N181	102A	3004	53	2007	
2N182	103A	3861/101	8	2002	
2N182#			5	2002	
2N183	103A	3861/101	8	2002	
2N183#			5	2002	
2N1837	128		81	2051	
2N1838	128	3444/123A	20	2051	2N2218
2N1839	128	3444/123A	20	2051	
2N184	101	3861	8	2002	
2N1840	123A	3444	20	2051	
2N1842	5500	3579/5504	MR-3		2N1842
2N1842A	5500	3579/5504	MR-3		2N1842A
2N1842B		3579	MR-3		
2N1843	5501	3579/5504	MR-3		2N1843
2N1843A	5501	3579/5504	MR-3		2N1843A
2N1843B	5501	3579/5504	MR-3		
2N1844	5502	3579/5504	MR-3		2N1844
2N1844A	5502	3579/5504	MR-3		2N1844A
2N1844B	5502	3579/5504	MR-3		
2N1845	5503	3579/5504	MR-3		2N1845
2N1845A	5503	3579/5504	MR-3		2N1845A
2N1845B	5504	3579	MR-3		
2N1846	5504	3579	MR-3		2N1846
2N1846A	5504	3579	MR-3		2N1846A
2N1846B	5504	3579	MR-3		
2N1847	5505	3580/5507	MR-3		2N1847
2N1847A	5505	3580/5507	MR-3		2N1847A
2N1847B	5505	3580/5507	MR-3		
2N1848	5506	3580/5507	MR-3		2N1848
2N1848A	5506	3580/5507	MR-3		2N1848A
2N1848B	5506	3580/5507	MR-3		
2N1849	5507	3580	MR-3		2N1849
2N1849A	5507	3580	MR-3		2N1849A
2N1849B	5507	3580	MR-3		
2N185	126	3004/102A	53	2007	
2N1850	5508	3504			2N1850
2N1850A	5508	3504			2N1850A
2N1850B	5508	3504			
2N1853	126	3005	1	2007	
2N1853/18	160	3123	245	2004	
2N1854	126	3004/102A	2	2007	
2N1858	103	3862	8	2002	
2N185BLU	102A	3123	53	2007	
2N186	102A	3004	53	2007	
2N1864	126	3008	245	2004	
2N1865	126	3008	245	2004	
2N1866	160	3008	245	2004	
2N1867	160	3008	245	2004	
2N1868	126		245	2004	
2N1869					2N4212
2N1869A					MCR1906-1
2N186A	102A	3004	53	2007	
2N187	102A	3004	53	2007	
2N1870					2N4212
2N1870A					MCR1906-1
2N1871					2N4213
2N1871A					MCR1906-2
2N1872			50	2004	2N4214
2N1872A					MCR1906-3
2N1873			50	2004	2N4215
2N1873A					MCR1906-4
2N1874			50	2004	2N4216
2N1874A					MCR1906-4
2N1875			50	2004	2N2322
2N1875A					MCR1906-1
2N1876					2N2322
2N1876A					MCR1906-1
2N1877					2N2323
2N1877A					MCR1906-2
2N1878					2N2324
2N1878A					MCR1906-3
2N1879					2N2326
2N1879A					MCR1906-4
2N187A	102A	3004	2	2007	
2N188	102A	3004	2	2007	
2N1880					2N2326
2N1880A					MCR1906-4
2N1881					MCR1906-1
2N1882					MCR1906-2
2N1883					MCR1906-3
2N1884					MCR1906-4
2N1885					MCR1906-4
2N1886			66	2048	
2N1889	128		243	2030	
2N188A	102A	3004	2	2007	
2N189	102	3004	2	2007	
2N189#			53	2007	
2N189#				2007	
2N1890	128		243	2030	2N1890
2N1891	101	3861	8	2002	
2N1892	102	3861/101	8	2002	
2N1893	154	3024/128	264		2N1893
2N1893L	128		264		
2N1893S	128		264		
2N190	102	3004	2	2007	
2N1905	127	3014	239	2006	
2N1906	179	3642	76		
2N1907	179	3642	76		
2N1907A	179	3642	76		
2N1908	179	3642	76		
2N1908A	179	3642	76		
2N191	102	3004	2	2007	
2N1917	288	3466/159	82	2032	
2N1918	288	3466/159	82	2032	
2N1919	159	3466	82	2032	
2N192	102	3004	2	2007	
2N1920	159	3466	82	2032	
2N1921	159	3466	82	2032	
2N1922	129		244	2027	
2N1923	128				2N3498
2N1924	102A	3004	53	2007	
2N1925	102A	3004	53	2007	
2N1926	102A	3004	53	2007	
2N1929	5400	3950			
2N193	103	3011	59	2002	
2N1931	5483	3942			

Industry Standard No.	ECG	SK	GE	RS 276-	MOTOR.
2N1932	5483	3942			
2N1933	5483	3942			
2N194	103	3011	59	2002	
2N1940	100	3005	1	2007	
2N1941	128		214	2038	
2N1943	128		18	2030	2N3020
2N1944	123A	3444	20	2051	2N2219A
2N1945	123A	3444	20	2051	2N2219A
2N1946	123A	3444	20	2051	2N2219A
2N1947	123A	3444	20	2051	
2N1948	123A	3444	20	2051	
2N1949	123A	3444	20	2051	
2N194A	103	3011	59	2002	
2N195	102	3004	2	2007	
2N1950	123A	3444	20	2051	
2N1951	123A	3444	20	2051	
2N1952	123A	3444	20	2051	
2N1953	128	3444/123A	20	2051	2N2218
2N1954	102A	3005	53	2007	
2N1955	102A	3005	53	2007	
2N1956	102A	3005	53	2007	
2N1957	102A	3005	53	2007	
2N1958	128	3444/123A	20	2051	
2N1958A	128		47	2030	
2N1958A/51			47	2030	
2N1959	128	3444/123A	20	2051	2N5859
2N1959A	128		47	2030	
2N1959A/51			47	2030	
2N196	102	3004	2	2007	
2N1960	160		245	2004	
2N1960/46	160	3123	245	2004	
2N1961	160	3123	245	2004	
2N1961/46		3123	2	2007	
2N19616	160		245	2004	
2N1962	123A	3444	20	2051	
2N1962/46			63	2030.	
2N1963	123A	3444	20	2051	
2N1964	123A	3444	20	2051	
2N1964/46			63	2030	
2N1965	123A	3444	20	2051	
2N1965/46			63	2030	
2N1969	100		53	2007	
2N197	102	3004	2	2007	
2N1970	213	3012/105	4		
2N1971	121	3009	239	2006	
2N1972	128	3444/123A	20	2051	
2N1973	128	3444/123A	20	2051	2N3019
2N1974	128	3444/123A	20	2051	2N3020
2N1975	128	3444/123A	20	2051	2N3020
2N198	102	3004	2	2007	
2N1980	213	3012/105	4		
2N1981	213	3012/105	4		
2N1982	213	3012/105	4		
2N1983	128	3444/123A	20	2051	2N2218
2N1984	128	3444/123A	20	2051	2N2219
2N1985	128	3444/123A	20	2051	2N1984
2N1986	128	3444/123A	20	2051	2N2219A
2N1987	128	3444/123A	20	2051	2N2218A
2N1988	128	3444/123A	20	2051	2N3020
2N1989	128	3452/108	86	2038	2N3020
2N199	102	3004	2	2007	
2N1990	128		243	2030	2N1990
2N1990R	128		220		
2N1990S	128		222		
2N1991	159	3466	82	2032	2N1991
2N1992	123A	3444	20	2051	
2N1993	103	3011	8	2002	
2N1994	101	3861	8	2002	
2N1995	101	3861	8	2002	
2N1996	101	3861	8	2002	
2N1997	100	3005	1	2007	
2N1998	100	3005	1	2007	
2N1999	160	3005	245	2004	
2N200	102	3722	2	2007	
2N2000	102A		53	2007	
2N2001	102		53	2007	
2N2002	159	3466	82	2032	
2N2003	159	3466	82	2032	
2N2004	159	3466	82	2032	
2N2005	159	3466	82	2032	
2N2006	159	3466	82	2032	
2N2007	159	3466	82	2032	
2N2008	154		220		2N3114
2N2009			51	2004	2N4212
2N2010					2N4213
2N2011					2N4214
2N2012					2N4216
2N2013					2N4218
2N2014					2N4219
2N2017	128	3024	243	2030	
2N2022	160		245	2004	
2N2032	108	3452	86	2038	
2N2033	282		66	2048	2N4238
2N2034	282		66	2048	
2N2035			66	2048	
2N2036			66	2048	
2N2038	128	3444/123A	20	2051	2N4237
2N2039	128	3444/123A	20	2051	2N4238
2N204	102	3004	2	2007	
2N2040	128	3444/123A	20	2051	2N4237
2N2041	128	3444/123A	20	2051	2N4238
2N2048	126		245	2004	
2N2048A	126		245	2004	
2N2049	128	3444/123A	20	2051	
2N205	102	3004	2	2007	
2N2059	126		245	2004	
2N206	102A	3004	2	2007	
2N2060			222		2N2060
2N2060A					2N2060A
2N2060B					2N2060A
2N2060JAN					2N2060JAN
2N2060JTX					2N2060JTX
2N2060TXV					2N2060TXV
2N2061	104	3009	16	2006	
2N2061A	104	3009	239	2006	
2N2062	104	3009	16	2006	
2N2062A	104	3009	16	2006	
2N2063	104	3009	16	2006	
2N2063A	104	3009	16	2006	
2N2064	104	3009	16	2006	
2N2064A	104	3009	16	2006	
2N2065	104	3009	16	2006	
2N2065A	104	3009	16	2006	
2N2066	104	3009	16	2006	
2N2066A	104	3014	16	2006	
2N2067	104	3009	16	2006	
2N2067B	104	3009	16	2006	
2N2067G	104	3009	16	2006	
2N2067W	104	3009	16	2006	
2N2069	179	3009	16	2006	
2N207	102	3003	2	2007	
2N2070	179	3009	16	2006	
2N2071			3	2006	
2N2072			3	2006	

Industry Standard No.	ECG	SK	GE	RS 276-	MOTOR.
2N2075	213	3012/105	4		
2N2075A	213	3012/105	4		
2N2076	213	3012/105	4		
2N2076A	213	3012/105	4		
2N2077	213	3012/105	4		
2N2077A	213	3012/105	240		
2N2078	213	3012/105	4		
2N2078A	213	3012/105	4		
2N2079	213	3012/105	4		
2N2079A	213	3012/105	4		
2N207A	102	3003	2	2007	
2N207B	102	3003	2	2007	
2N207B#				2005	
2N207B#				2005	
2N207BLU	102	3004	2	2007	
2N2080	213	3012/105	4		
2N2080A	213	3012/105	4		
2N2081	213	3012/105	4		
2N2081A	213	3012/105	240		
2N2082	213	3012/105	4		
2N2082A	213	3012/105	4		
2N2083	126		245	2004	
2N2084	160	3006	245	2004	
2N2085	101	3861	8	2002	
2N2086	154		18	2030	2N3020
2N2087		3479	81	2051	2N3020
2N2089	126	3006/160	52	2024	
2N2090	126	3006/160	52	2024	
2N2091	126	3006/160	245	2004	
2N2092	126	3007	52	2024	
2N2093	126	3008	245	2004	
2N2094	128	3444/123A	20	2051	
2N2094A	128	3444/123A	20	2051	
2N2095	158	3444/123A	20	2051	
2N2095A	128	3444/123A	20	2051	
2N2096	158	3444/123A	20	2051	
2N2096A	123A	3444	20	2051	
2N2097			51	2004	
2N2097A	123A	3444	20	2051	
2N2098	160		245	2004	
2N2099	126		245	2004	
2N21			53	2007	
2N2100	160		245	2004	
2N2102	128	3024	243	2030	2N2102
2N2102A			235	2012	
2N2102L	128		243	2030	
2N2102S	128		243	2030	
2N2104	159	3114/290	82	2032	
2N2105	159	3114/290	82	2032	
2N2106	128	3024	243	2030	2N4238
2N2107	128	3024	243	2030	2N4238
2N2108	128	3024	243	2030	2N5681
2N211	103	3011	59	2002	
2N212	103	3011	59	2002	
2N2121	159	3466	82	2032	
2N213	103	3862	59	2002	
2N2137	104	3009	16	2006	
2N2137A	104	3009	16	2006	
2N2138	104	3009	16	2006	
2N2138A	104	3009	16	2006	
2N2139	104	3009	16	2006	
2N2139A	104	3009	16	20066	
2N213A	103A	3862/103	59	2002	
2N214	103A	3010	59	2002	
2N2140	104	3009	16	2006	
2N2140A	104	3009	16	2006	
2N2141	104	3009	16	2006	
2N2141A	104	3009	16	2006	
2N2142	104	3009	16	2006	
2N2142A	104	3009	16	2006	
2N2143	104	3009	16	2006	
2N2143A	104	3009	16	2006	
2N2144	104	3009	16	2006	
2N2144A	104	3009	16	2006	
2N2145	104	3009	16	2006	
2N2145A	104	3009	16	2006	
2N2146	104	3009	16	2006	
2N2146A	104	3009	16	2006	
2N2147	121	3014	16	2006	
2N2148	121	3014	239	2006	
2N214A	103	3010	8	2002	
2N215	102A	3004	53	2007	
2N2150	75		D44Q1		
2N2152	213		254		
2N2152A	213	3012/105	254		
2N2153	213	3012/105	254		
2N2153A	213	3012/105	254		
2N2154	213		254		
2N2154A	213	3012/105	254		
2N2155	213	3012/105	4		
2N2155A	213	3012/105	4		
2N2156	213		254		
2N2156A	213	3012/105	254		
2N2157	213	3012/105	254		
2N2157A	213	3012/105	254		
2N2158	213	3012/105	254		
2N2158A	213	3012/105	254		
2N2159	213	3012/105	4		
2N2159A	213	3012/105	4		
2N216	103	3011	59	2002	
2N2160	6400A		2N2160	2029	2N4853
2N2161	128	3444/123A	20	2051	
2N2162	159	3466	82	2032	
2N2163	159	3466	82	2032	
2N2164	159	3466	82	2032	
2N2165	159	3466	82	2032	
2N2166	159	3466	82	2032	
2N2167	159	3466	82	2032	
2N2168	126		245	2004	
2N2169	126		51	2005	
2N217	102A	3004	53	2007	
2N2170	126		245	2004	
2N2171	100	3005	1	2007	
2N2172	100	3005	1	2007	
2N2173	176		53	2007	
2N2175	106	3025/129	21	2034	
2N2176	106	3984	21	2034	
2N2177	106	3025/129	21	2034	
2N2178	159	3984/106	82	2032	
2N217A	102A		53	2007	
2N217RED	102A		53	2007	
2N217WHT	102A		53	2007	
2N217YEL	102A		53	2007	
2N218	126	3005	245	2004	
2N2180	126		52	2024	
2N2181	159	3466	82	2032	2N2945
2N2182	159	3466	82	2032	2N2945
2N2183	159	3466	82	2032	2N2944
2N2184	159	3466	82	2032	
2N2185	159	3466	82	2032	
2N2186	159	3466	82	2032	
2N2187	159	3466	82	2032	
2N2188	126	3006/160	245	2004	
2N2189	126	3006/160	245	2004	

Industry Standard No.	ECG	SK	GE	RS 276-	MOTOR.
2N219	160	3005	245	2004	
2N2190	160	3006	245	2004	
2N2191	160	3006	245	2004	
2N2192	128	3039/316	243	2030	2N3019
2N2192A	128	3039/316	243	2030	2N3019
2N2192B	128		243	2030	2N3019
2N2193	128	3039/316	243	2030	2N3020
2N2193A	128	3452/108	86	2038	2N2193A
2N2193B	128	3452/108	86	2038	2N3020
2N2194	128	3444/123A	20	2051	2N2219A
2N2194A	128	3444/123A	20	2051	2N2219A
2N2194B	128	3444/123A	20	2051	2N2219
2N2195	128	3452/108	86	2038	2N2219
2N2195A	128	3024	243	2030	2N2219
2N2195B	128	3452/108	86	2038	2N2219
2N2196	224	3039/316	46		
2N2197	108	3452	86	2038	
2N2198	128	3124/289	20	2051	
2N2199	126		245	2004	
2N22			53	2007	
2N220	102A	3004	53	2007	
2N220#				2004	
2N220#				2004	
2N2200	126		245	2004	
2N2201					2N5681
2N2202					2N5681
2N2204	124	3021	12		2N5681
2N2205	123A	3444	20	2051	2N2369
2N2206	123A	3444	20	2051	
2N2207	126	3006/160	52	2024	
2N2208	126	3006/160	52	2024	
2N2209	102	3005	1	2007	
2N2210	105	3012	4		
2N2212	121	3009	239	2006	
2N2217	128	3444/123A	20	2051	2N2218
2N2217/51			20	2051	
2N2218	128	3124/289	20	2051	2N2218
2N2218/51			20	2051	
2N2218A	128	3124/289	20	2051	
2N2219	128		20	2051	2N2219
2N2219/51			20	2051	
2N2219A	128	3444/123A	20	2051	2N2219A
2N222	102A	3444/123A	53	2015	
2N2220	123A	3444	20	2051	2N2221
2N2221	123A	3444	20	2051	2N2221
2N2221A	123A	3444	20	2051	2N2221A
2N2222	123A	3444	20	2051	2N2222
2N2222A	123A	3444	20	2051	2N2222A
2N2223			222		2N2223
2N2223A			222		2N2223A
2N2224	123A		47	2030	2N2218
2N2225	126		245	2004	
2N222A	102A		20	2051	
2N223	102A	3003	53	2007	
2N2234	123	3124/289	20	2051	
2N2235	123	3124/289	20	2051	
2N2236	123A	3444	20	2051	2N2218
2N2237	123A	3444	20	2051	2N2218
2N2238	160	3006	245	2004	
2N2239			63	2030	
2N224	102A	3004	53	2007	
2N224#				2005	
2N224#				2005	
2N2240	123	3444	20	2051	
2N2241	123	3444	20	2051	
2N2242	123A	3444	20	2051	2N708
2N2243	128		32		2N3020
2N2243A	128		32		2N3020
2N2244	123A	3444	20	2051	
2N2245	123A	3444	20	2051	
2N2246	123A	3444	20	2051	
2N2247	123A	3444	20	2051	
2N2248	123A	3444	20	2051	
2N2249	123A	3444	20	2051	
2N225	102A	3004	53	2007	
2N2250	123A	3444	20	2051	
2N2251	286	3444/123A	20	2051	
2N2252	286	3444/123A	20	2051	
2N2253	123A	3444	20	2051	
2N2254	123A	3444	20	2051	
2N2255	123A	3444	20	2051	
2N2256	123A	3444	20	2051	2N706
2N2257	123A	3444	20	2051	
2N2258	126		245	2004	MM2258
2N2259	126		245	2004	MM2259
2N226	102A	3004	53	2007	
2N226#				2005	
2N2260					MM2260
2N2266	105	3012	4		
2N2267	105	3012	4		
2N2268	105	3012	4		
2N2269	105	3012	4		
2N227	102A	3004	53	2007	
2N2270	128	3263	243	2030	2N2270
2N2270L	128		243	2030	
2N2270S	128		243	2030	
2N2271	102	3004	2	2007	
2N2272	123A	3444	20	2051	
2N2273	126		245	2004	
2N2274	159	3466	82	2032	
2N2275	159	3466	82	2032	
2N2276	159	3466	82	2032	
2N2277	159	3466	82	2032	
2N2278	159	3466	82	2032	
2N2279	159	3466	82	2032	
2N2279/51			63	2030	
2N228	103A	3010	59	2002	
2N2280	159	3466	82	2032	
2N2281	159	3466	82	2032	
2N2282	104	3009	16	2006	
2N2285	179	3642	76		
2N2286	179	3642	76		
2N2287	121	3642/179	239	2006	
2N2288	179	3009	239	2006	
2N2289	179	3009	239	2006	
2N228A			53	2007	
2N229	103	3010	59	2002	
2N2290	179	3009	25		
2N2291	179	3009	239	2006	
2N2292	179	3009	239	2006	
2N2293	121	3009	239	2006	
2N2294	121	3009	239	2006	
2N2295	121	3009	239	2006	
2N2296	121	3009	239	2006	
2N2297	128	3104A	243	2030	2N2297
2N2299	159	3715	82	2032	
2N23	102	3004	2	2007	
2N230	104	3009	16	2006	
2N2303	159	3466	82	2032	
2N2303/46		3114	221	2032	
2N2305	130	3027	14	2041	
2N2309	123	3444	20	2051	
2N231	126	3008	51	2004	
2N231-YEL-RED	126	3006/160	52	2024	

Industry Standard No.	ECG	SK	GE	RS 276-	MOTOR.
2N2310	287	3444/123A	20	2051	
2N2311	287		222		
2N2312	287	3444/123A	20	2051	2N3020
2N2313	287		222		
2N2314	123A	3444	20	2051	2N2221A
2N2315	123A	3444	20	2051	2N2221A
2N2316	287		222		2N3020
2N2317	128	3479	81	2051	2N2193A
2N2318	123A	3444	20	2051	
2N2319	123A	3444	20	2051	2N3508
2N231BLU	126	3006/160	52	2024	
2N231RED			52	2024	
2N231YEL	126	3006/160	52	2024	
2N232	126	3008	245	2004	
2N2320	123A	3444	20	2051	
2N2322			MR-5	1067	2N2322
2N2323			MR-5	1067	2N2323
2N2324			2N2324	1067	2N2324
2N2325			MR-5	1067	2N2325
2N2326			MR-5	1067	2N2326
2N2327				1020	2N2327
2N2328					2N2328
2N2329			2N2329		2N2329
2N233	103	3011	59	2002	
2N233#				2002	
2N2330	123	3124/289	20	2051	
2N2331	123A	3444	20	2051	
2N2332	159	3466	82	2032	
2N2333	159	3466	82	2032	
2N2334	159	3466	82	2032	
2N2335	159	3466	82	2032	
2N2336	159	3466	82	2032	
2N2337	159	3466	82	2032	
2N2339			66	2048	
2N233A	103	3011	59	2002	
2N233A#				2002	
2N234	121	3009	239	2006	
2N2344					MCR1906-1
2N2345			8	2002	MCR1906-2
2N2346					MCR1906-3
2N2347					MCR1906-4
2N2348					MCR1906-4
2N2349	123A	3444	20	2051	
2N234A	121	3009	239	2006	
2N235	121	3009	239	2006	
2N2350	128		47	2030	
2N2350A	128		47	2030	
2N2351	128		18	2030	2N3737
2N2351A	128				2N3737
2N2352	128		47	2030	
2N2352A	128		47	2030	
2N2353	128	3444/123A	20	2051	
2N2353A	128	3444/123A	20	2051	
2N2354	103	3862	8	2002	
2N2357	121	3009	239	2006	
2N2358	121	3009	239	2006	
2N235A	121	3009	239	2006	
2N235B	121	3009	239	2006	
2N236	121	3009	239	2006	
2N2360	160		245	2004	
2N2361	160		245	2004	
2N2362	160		245	2004	
2N2363	160	3006	52	2024	
2N2364	396		222		
2N2364A	396		222		
2N2368	123A	3444	20	2051	2N2368
2N2368/51			63	2030	
2N2368A	123A	3444	20	2051	
2N2369	123A	3444	20	2051	2N2369
2N2369/51			63	2030	
2N2369A	311	3444/123A	20	2051	2N2369A
2N236A	121	3009	239	2006	
2N236B	121	3009	239	2006	
2N237	102	3004	2	2007	
2N2370	159	3466	82	2032	
2N2371	159	3466	82	2032	
2N2372	159	3466	82	2032	
2N2373	159	3466	82	2032	
2N2374	102	3004	2	2007	
2N2375	102	3004	2	2007	
2N2376	102	3004	2	2007	
2N2377	159	3466	82	2032	
2N2378	159	3466	82	2032	
2N238	102A	3004	53	2004	
2N238-ORN	102A		53	2007	
2N2380	128	3479	81	2051	
2N2380A	128	3479	81	2051	
2N2381	126		245	2004	
2N2382	160		245	2004	
2N2383			14	2041	
2N2384			14	2041	
2N2386	460				2N2608
2N2387		3444	20	2051	
2N2388		3444	20	2051	
2N2389	123	3124/289	20	2051	2N2193A
2N238D	102A		53	2007	
2N238E	102A		53	2007	
2N238F	102A		53	2007	
2N2390	123	3124/289	20	2051	2N3019
2N2391				2024	2N3250
2N2392				2024	2N3250
2N2393	159	3466	82	2032	2N2905
2N2394	159	3466	82	2032	2N2905
2N2395	106	3984	21	2034	2N2219
2N2396	123	3124/289	20	2051	2N2219
2N2397	123	3124/289	20	2051	2N2369A
2N2398	160		245	2004	
2N2399	160		245	2004	
2N24	102	3004	2	2007	
2N240	126	3008	245	2004	
2N2400	126		245	2004	
2N2401	126		245	2004	
2N2402	126		245	2004	
2N2403			63	2030	
2N2404			63	2030	
2N2405	282	3024/128	264		2N2405
2N2405L	282		264		
2N2405S	128		264		
2N2406	176	3123	80		
2N241	102A	3004	2	2007	
2N2410	128		47	2030	2N5859
2N2410/51			47	2030	
2N2411	159	3466	82	2024	
2N2412	159	3466	82	2024	
2N2413	123A	3444	20	2051	
2N2414					2N2223A
2N2415	160		245	2004	
2N2416	160		245	2004	
2N2417	123A	3444	20	2051	
2N241A	102A	3004	2	2007	
2N242	104	3009	16	2006	
2N2423	179	3009	239	2006	
2N2424	159	3466	82	2032	2N3250
2N2425	159	3466	82	2032	2N3250A

Industry Standard No.	ECG	SK	GE	RS 276-	MOTOR.
2N2426	101	3861	8	2002	
2N2427	123A	3444	20	2051	
2N2428	102A	3004	53	2007	
2N2429	102A	3004	53	2007	
2N243			47	2030	
2N2430	103A	3010	59	2002	
2N2431	158	3004/102A	53	2007	
2N2431B			2	2007	
2N2432	123A	3444	20	2051	
2N2432A			212	2010	
2N2433	128		27		
2N2435	287		27		
2N2437	287		27		2N3020
2N2438	287		27		2N3019
2N2439	287		27		2N3019
2N244			47	2030	
2N2443	154		243	2030	
2N2446	179	3009	239	2006	
2N2447	102A	3004	53	2007	
2N2448	102A	3004	53	2007	
2N2449	102A	3004	53	2007	
2N245			63	2030	
2N2450	102A	3004	53	2007	
2N2451	126	3006/160	52	2024	
2N2453	81				2N2453
2N2453A					2N2453A
2N2455	160		245	2004	
2N2456	160		52	2024	
2N246			63	2030	
2N2463	128		27		
2N2465	128		54		
2N2466	128	3444/123A	20	2051	
2N2468	102	3004	2	2007	
2N2469	102	3004	2	2007	
2N247	126	3007	52	2024	
2N247/33	126	3006/160	52	2024	
2N2472			20	2051	2N3500
2N2473			20	2051	2N3500
2N2474		3466	221	2032	
2N2475	108	3444/123A	61	2038	
2N2475/46			61	2038	
2N2475/51			61	2038	
2N2476	123A	3195/311	20	2051	2N5859
2N2477	123A	3195/311	20	2051	2N5859
2N2478			18	2030	
2N2479	128	3479	81	2051	2N2218
2N248	126	3007	52	2024	
2N2480	81				2N2480
2N2480A	81				2N2480A
2N2481	123A	3444	20	2051	2N2481
2N2482	103	3862	8	2002	
2N2483	123A	3444	20	2051	2N2484
2N2484	123A	3444	20	2051	2N2484
2N2484A			63	2030	
2N2487	126	3006/160	52	2024	
2N2488	126	3006/160	52	2024	
2N2489	126	3006/160	52	2024	
2N2489A	128		214	2038	
2N249	102A	3005	1	2007	
2N2490	213	3012/105	4		
2N2491	213	3012/105	4		
2N2492	213	3012/105	4		
2N2493	213	3012/105	4		
2N2494	126	3006/160	52	2024	
2N2495	160	3006	245	2004	
2N2496	160	3006	245	2004	
2N2497	326				2N5267
2N2498	460				2N3330
2N2499					2N3909A
2N25	102	3004	2	2007	
2N250	104	3009	16	2006	
2N2500					2N5267/8
2N2501	123A	3444	20	2051	2N2501
2N2509	154	3045/225	40	2012	2N720A
2N250A	104	3009	16	2006	
2N251	121	3009	239	2006	
2N2510	128	3045/225	40	2012	2N930
2N2511	128		40	2012	2N930A
2N2512	160		245	2004	
2N2514	123A	3479	81	2051	
2N2515	123A	3479	81	2051	
2N2517	287		222		
2N251A	121	3009	239	2006	
2N252	126	3005	52	2024	
2N252#				2004	
2N2520	287	3444/123A	20	2051	
2N2521	287	3444/123A	20	2051	
2N2522	287	3444/123A	20	2051	
2N2523	123A	3444	20	2051	
2N2524	123A	3444	20	2051	2N3485A
2N2526	179	3764/127	25		
2N2527	127	3764	25		
2N2528	127	3764	25		
2N2529	123A	3444	20	2051	2N930
2N253	101	3861	8	2002	
2N2530	123A	3444	20	2051	2N930
2N2531	123A	3444	20	2051	2N930
2N2532	123A	3444	20	2051	2N930
2N2533	123A	3444	20	2051	2N930
2N2534	123A	3444	20	2051	2N930
2N2537	128		47	2030	2N5859
2N2538	128		18	2030	2N5859
2N2539	123A	3444	20	2051	2N2540
2N254	101	3861	8	2002	
2N2540	123A	3765/195A	20	2051	2N2540
2N2540A	195A	3765	46		
2N2541	176	3845	80		
2N255	104	3009	16	2006	
2N2551	288		222		
2N255A	104	3009	16	2006	
2N256	104	3009	16	2006	
2N2564	102	3004	2	2007	
2N2565	102	3004	2	2007	
2N2569	123A	3444	20	2051	
2N256A	104	3009	16	2006	
2N257	104	3009	16	2006	
2N2570	123A	3444	20	2051	
2N2571	123A	3444	20	2051	
2N2572	123A	3444	20	2051	
2N2573					2N2573
2N2574					2N2574
2N2575					2N2575
2N2576					2N2576
2N2577					2N2577
2N2578					2N2578
2N2579					2N2579
2N257A	104	3009	16	2006	
2N257B	104	3009	16	2006	
2N257G	104	3009	16	2006	
2N257W	104	3009	16	2006	
2N2586	123A	3444	20	2051	2N930A
2N2587	160		245	2004	
2N2588	126		52	2024	
2N2590	129		223		

Industry Standard No.	ECG	SK	GE	RS 276-	MOTOR.
2N2591	129		21	2034	
2N2592	129		21	2034	
2N2593	129		21	2034	
2N2594	128	3024	243	2030	
2N2595	159	3466	82	2032	
2N2596	159	3466	82	2032	
2N2597	159	3466	82	2032	
2N2598	288	3466/159	82	2032	
2N2599	288	3466/159	82	2032	
2N2599A	288	3466/159	223		
2N26	102	3004	2	2007	
2N2600	288	3466/159	82	2032	2N3497
2N2600A	288	3466/159			2N3497
2N2601	159	3466	82	2032	
2N2602	159	3466	82	2032	
2N2603	159	3466	82	2032	
2N2604	159	3466	82	2032	2N2604
2N2605	159	3466	82	2032	2N2605
2N2605A	159	3466	82	2032	
2N2606					2N5474
2N2607					2N5475
2N2608					2N2608
2N2609	460				2N2609
2N2610	107	3444/123A	20	2051	2N930
2N2612	179	3009	16	2006	
2N2613	102A	3004	53	2007	
2N2614	102A	3004	53	2007	
2N2615	108	3018	61	2038	
2N2616	108	3452	86	2038	
2N2617	123AP	3444/123A	20	2051	
2N2618	154	3045/225	40	2012	
2N2618/46			20	2051	
2N2619	5476				2N4174
2N262	100	3005	1	2007	
2N2621	126	3123	245	2004	
2N2622	126	3123	245	2004	
2N2623	126	3123	245	2004	
2N2624	126	3123	245	2004	
2N2625	126	3123	245	2004	
2N2626	126	3122	245	2004	
2N2627	126	3123	245	2004	
2N2628	126	3123	245	2004	
2N2629	126	3123	245	2004	
2N263	123AP	3452/108	86	2051	
2N2630	126	3123	245	2004	
2N2631	312				2N3553
2N2632	74		66	2048	
2N2635	126		245	2004	
2N2636	179	3642	76		
2N2637	179	3642	76		
2N2638	179	3642	76		
2N2639	81$	3039/316	212	2010	2N2639
2N2639JAN					2N2639
2N2639JTX					2N2639
2N2639TXV					2N2639
2N264	123AP	3452/108	86	2051	
2N2640	81$	3039/316	212	2010	2N2640
2N2641	81	3039/316	212	2010	2N2641
2N2642	81$	3039/316	212	2010	2N2642
2N2642JAN					2N2642
2N2642JTX					2N2642
2N2642TXV					2N2642
2N2643	81$	3039/316	212	2010	2N2643
2N2644	81	3039/316	212	2010	2N2644
2N2645	123A	3444	20	2051	
2N2646	6401		2N2646	2029	2N2646
2N2647	6409		2N2647	2029	2N2647
2N2648	100	3845/176	80		
2N265	102	3003	53	2007	
2N2651	123A	3452/108	86	2038	
2N2652			220		2N2652
2N2652A			220		2N2652A
2N2653	5474	3578			2N4172
2N2654	160	3006	245	2004	
2N2655			214	2038	
2N2656	123A	3444	20	2051	2N2222
2N2657			66	2048	2N4238
2N2658					2N4239
2N266	102	3004	2	2007	
2N267	160	3005	245	2004	
2N2671	160	3007	245	2004	
2N2672	176	3007	245	2004	
2N2672A	160		245	2004	
2N2672BLK	160	3123	245	2004	
2N2672GRN	160	3123	245	2004	
2N2673	123A	3444	20	2051	
2N2674	123A	3444	20	2051	
2N2675	123A	3444	20	2051	
2N2676	123A	3444	20	2051	2N2222A
2N2677	123A	3444	20	2051	2N2221A
2N2678	123A	3444	20	2051	2N2221A
2N2679	5400	3950			MCR102
2N268	179	3009	239	2006	
2N2680	5401	3638/5402			MCR103
2N2680A	5401	3638/5402			
2N2681					MCR104
2N2682					MCR120
2N2683					MCR102
2N2684					MCR103
2N2685					MCR104
2N2686					MCR120
2N2687	5400	3950			MCR102
2N2688	5401	3638/5402			MCR103
2N2689	5402	3638			MCR104
2N268A	179	3009	239	2006	
2N269	102A	3005	245	2004	
2N2690	5404	3627			MCR120
2N2691	179	3642	76		
2N2691A	179	3642	76		
2N2692	123A	3444	20	2051	
2N2693	123A	3444	20	2051	2N930
2N2694			212	2010	
2N2695	159	3466	82	2032	2N5582
2N2696	159	3466	82	2032	2N2907
2N2699	101	3861	8	2002	
2N27	100	3005	1	2007	
2N270	102	3004	2	2007	
2N270-5B	102	3722	2	2007	
2N2706	158	3004/102A	53	2007	
2N2707	158	3010	53	2007	
2N2708	316	3452/108	86	2038	
2N2709	159	3715	82	2032	2N2800
2N270A	102	3004	2	2007	
2N271	100	3005	1	2007	
2N2710	311	3452/108	86	2038	2N3014
2N2711	108	3452	86	2038	
2N2711#			2N2711	2033	
2N2712	289A	3452/108	86	2038	
2N2712BLUE			86	2038	
2N2713	289A	3444/123A	20	2051	
2N2714	289A	3444/123A	20	2051	
2N2715	107	3452/108	86	2038	
2N2716	107	3452/108	86	2038	
2N2717	126		245	2004	
2N2719	123A	3444	20	2051	
2N271A	100	3005	1	2007	
2N272	102A	3003	2	2007	
2N2720	81$		20	2051	2N2720
2N2721	81$		20	2051	2N2721
2N2722	81		20	2051	2N2722
2N2723		3156	D40C7		2N2723
2N2724		3156	D40C7		2N2723
2N2725		3156			2N2723
2N2726			235	2012	2N3440
2N2728	105	3012	4		
2N2729	311	3452/108	86	2038	
2N273	102A	3004	1	2007	
2N2730	105	3012	4		
2N2731	105	3012	4		
2N2732	105	3012	4		
2N274	160	3007	245	2004	
2N274BLU	160		245	2004	
2N274WHT	160		245	2004	
2N275	126		52	2024	
2N275W			16	2006	
2N275W				2006	
2N276	160		245	2004	
2N277	213	3012/105	4		
2N278	213	3012/105	4		
2N2783	126		52	2024	
2N2784	108	3452	86	2038	
2N2784/52	108	3452	86	2038	
2N2784/TNT			20	2051	
2N2785					2N2785
2N2786	160		245	2004	
2N2786A	160		245	2004	
2N2787	128		20	2051	
2N2788	128		18	2030	2N2219
2N2789	128		18	2030	
2N279	102A	3004	53	2007	
2N2790	123A		20	2051	2N697
2N2791	123A		20	2051	2N2221A
2N2792	123A		20	2051	2N2222
2N2793	105	3012	4		
2N2795	126		245	2004	
2N2796	126		245	2004	
2N2797	160		245	2004	
2N2798	160		245	2004	
2N2799	160		245	2004	
2N28	101	3861	8	2002	
2N280	102A	3004	53	2005	
2N2800	129	3466/159	82	2032	2N2800
2N2800/46	159	3466	82	2032	
2N2800/51			21	2034	
2N2801	159	3466	82	2032	2N2905
2N2801/46	129	3466/159	48		
2N2801/51			21	2034	
2N2802	82$	3118	21	2034	2N4937
2N2803	82$	3118	21	2034	2N4938
2N2804	82	3118	21	2034	2N4939
2N2805	82$	3118	21	2034	2N4937
2N2806	82$	3118	21	2034	2N4938
2N2807	82	3118	21	2034	2N4939
2N281	102A	3003	53	2007	
2N282	102	3004	2	2007	
2N2828			66	2048	
2N2829			66	2048	
2N283	102A	3003	53	2007	
2N2831	123A	3444	20	2051	
2N2832	179	3717/121	239	2006	
2N2835	131		44	2006	
2N2836	121	3009	239	2006	
2N2837	159	3466	82	2032	2N2907
2N2838	159	3466	82	2032	2N2907
2N284	102A	3004	53	2007	
2N2841					2N5472
2N2842					2N5473
2N2843					2N5475
2N2844					2N2608
2N2845	123A	3444	20	2051	
2N2846	123A	3024/128	243	2030	2N5859
2N2847	123A	3444	20	2051	
2N2848	123A	3024/128	243	2030	2N5859
2N2849	282	3024/128	243	2030	
2N284A	102A	3004	53	2007	
2N285	104	3009	16	2006	
2N2850	282	3024/128	243	2030	
2N2851	282	3024/128	243	2030	
2N2852	282	3024/128	243	2030	
2N2853	282	3024/128	243	2030	
2N2853-1	282		252	2018	
2N2854	282	3024/128	243	2030	
2N2854-1	282		252	2018	
2N2855	282	3024/128	243	2030	
2N2855-1	282		252	2018	
2N2856	282	3024/128	243	2030	
2N2856-1	282		252	2018	
2N2857	316	3039	280		2N2857
2N285A	104	3009	16	2006	
2N285B	104	3009	16	2006	
2N286	160		245	2004	
2N2860	160		245	2004	
2N2861	159	3466	82	2024	
2N2862	159	3466	82	2024	
2N2863	128		47	2030	2N2219
2N2864	128		47	2030	2N2219
2N2865	161	3019	39	2015	
2N2868	128	3444/123A	20	2051	2N3253
2N2869	179	3009	239	2006	
2N2869/2N301	179	3009	239	2006	
2N2870	121	3009	239	2006	
2N2873	160		245	2004	
2N2875			69	2049	
2N2876			216	2053	2N3375
2N2877	75		66	2048	
2N2878	75		66	2048	
2N2881					2N4235
2N2883	123A	3452/108	86	2038	
2N2884	123A	3452/108	86	2038	
2N2885				2014	
2N2886			47	2030	
2N2888	5524	6624			2N1846
2N2889	5525	6627/5527			2N1847
2N289			51	2004	
2N289				2004	
2N2890	128		18	2030	
2N2891	128		18	2030	
2N2894	159	3466	82	2024	2N2894
2N2894A	159	3466	82	2024	2N3546
2N2895	128	3024	243	2030	2N2895
2N2896	282				2N2896
2N2897	128	3024	243	2030	2N2897
2N29	101	3861	8	2002	
2N290	213	3012/105	16	2006	
2N2900	128		47	2030	
2N2901			11	2015	
2N2903	81				2N2903
2N2903A	81				2N2903A

Industry Standard No.	ECG	SK	GE	RS 276-	MOTOR.
2N2904	129	3025	244	2027	2N2904
2N2904A	129	3025	244	2027	2N2904A
2N2905	129	3025	244	2027	2N2905
2N2905A	129	3025	244	2027	2N2905A
2N2906	159	3466	82	2032	2N2906
2N2906A	159	3466	82	2032	2N2906A
2N2907	159	3466	82	2032	2N2907
2N2907A	159	3466	82	2032	2N2907A
2N2909	128		27		2N2221A
2N291	102A	3004	2	2007	
2N291#				2005	
2N291#				2005	
2N2910			20	2051	MD2219
2N2911					2N5682
2N2913	81		20	2051	2N2913
2N2914			20	2051	2N2914
2N2915	81				2N2915
2N2915A	81				2N2915
2N2916					2N2916
2N2916A					2N2916
2N2917	81		20	2051	2N2917
2N2918			20	2051	2N2918
2N2919	81$				2N2919
2N2919A	81$				2N2919
2N292	101	3861	8	2002	
2N2920					2N2920
2N2920A					2N2920
2N2921	289A	3452/108	86	2038	
2N2922	289A	3452/108	86	2038	
2N2923	289A	3444/123A	20	2051	
2N2924	289A	3444/123A	20	2051	
2N2925	289A	3245/199	20	2051	
2N2926	289A	3444/123A	20	2051	
2N2926-6	123A	3444	20	2051	
2N2926-BRN			212	2010	
2N2926-GRN			212	2010	
2N2926-ORG			212	2010	
2N2926-RED			212	2010	
2N2926-YEL			212	2010	
2N29260RN			20	2051	
2N2926G	123A	3444	20	2051	
2N2926GRN	123A	3444	20	2051	
2N29260			63	2030	
2N29260RN	123A	3444		2051	
2N2926R			63	2030	
2N2926Y			10	2051	
2N2927	159	3715	82	2032	2N2904
2N2927/46	159	3715	82	2032	
2N2927/51			21	2034	
2N2928	160	3123	245	2004	
2N2929	160		245	2004	
2N292A	101	3861	8	2002	
2N293	101	3861	8	2002	
2N2930	102	3010	1	2007	
2N2931	199	3444/123A	20	2051	
2N2932	199	3444/123A	20	2051	
2N2933	199	3444/123A	20	2051	
2N2934	199	3444/123A	20	2051	
2N2935	199	3444/123A	20	2051	
2N2936			20	2051	2N2920
2N2937			20	2051	2N2920
2N2938	123A	3444	20	2051	
2N2941	396		D44R1		
2N2942	160		245	2004	
2N2943	160		245	2004	
2N2944	159	3466	82	2032	2N2944
2N2944A	159	3466	82	2032	2N2945A
2N2945	159	3466	82	2032	2N2945
2N2945A	159	3466	82	2032	2N2945A
2N2946	159	3466	82	2032	2N2946
2N2946A	159	3466	82	2032	2N2946A
2N2947			66	2048	2N5641
2N2948			19	2041	
2N2949	195A	3765	46		
2N2951	123A	3444	20	2051	
2N2952	123A	3444	20	2051	2N1711
2N2953	102A	3004	2	2007	
2N2954			20	2051	2N834
2N2955	126		245	2004	
2N2956	126		245	2004	
2N2957	126		245	2004	
2N2958	123A	3444	20	2051	2N2219
2N2959	123A	3444	20	2051	2N2959
2N296	104	3020/123	16	2006	
2N2960	123A	3444	20	2051	
2N2961	123A	3444	20	2051	
2N2966	100	3721	1	2007	
2N2968	106	3025/129	21	2034	
2N2968A	312			2028	
2N2969	106	3984	21	2034	
2N297	121	3009	239	2006	
2N2970	106	3025/129	21	2034	2N3250
2N2971	106	3984	21	2034	2N3250
2N2972	81				2N2913
2N2973					2N2914
2N2974	81		20	2051	2N2915
2N2975					2N2916
2N2976	81				2N2917
2N2977					2N2918
2N2978	81$				2N2919
2N2979					2N2920
2N2979A					2N2979
2N297A	121	3009	239	2006	
2N2980					2N2060
2N2981					2N2060
2N2982					2N2060
2N2984					2N5682
2N2987			32		2N4238
2N2988			32		2N5681
2N299	126	3006/160	245	2004	
2N2990			32		2N5682
2N2996	160	3006	245	2004	
2N2997	160		245	2004	
2N2998	160		245	2004	
2N2999	160		245	2004	
2N29A				2002	
2N2G374			2	2007	
2N2J324	128	3024	243	2030	
2N2J374			63	2030	
2N30	100	3005	1	2007	
2N300	160	3006	245	2004	
2N3001					MCR102
2N3002					MCR103
2N3003					MCR104
2N3004					MCR120
2N3005	5400	3950			MCR102
2N3006	5401	3638/5402			MCR103
2N3007	5402	3638			MCR104
2N3008	5404	3627			MCR120
2N3009	123A	3444	20	2051	2N3009
2N301	121	3009	239	2006	
2N3010	108	3452	86	2038	MM1505
2N3011	123A	3444	20	2051	2N3011
2N3012	159	3715	82	2032	2N3012
2N3013	123A	3444	20	2051	2N3013
2N3014	123A	3444	20	2051	2N3014
2N3015	123A		20	2051	2N5859
2N3019	396	3748/282	243	2030	2N3019
2N301A	121	3009	239	2006	
2N301B	121	3009	239	2006	
2N301G		3009	52	2024	
2N301W	121	3009	239	2006	
2N302	102	3003	1	2007	
2N3020	128		243	2030	2N3020
2N3021			69	2049	
2N3022			69	2049	
2N3023			69	2049	
2N3024			69	2049	
2N3025			69	2049	
2N3026			69	2049	
2N3027	5400	3950			MCR102
2N3028	5401	3638/5402			MCR103
2N3029	5402	3638			MCR104
2N303	102	3003	2	2007	
2N3030	5400	3950			MCR102
2N3031	5401	3638/5402			MCR103
2N3032	5402	3638			MCR104
2N3035	108	3452	86	2038	
2N3036	128		18	2030	2N3053
2N3038			18	2030	
2N3039	129		82	2032	
2N3040	129		82	2032	
2N3043			17	2051	2N3043
2N3044			17	2051	2N3044
2N3045			17	2051	2N3045
2N3046			17	2051	2N3046
2N3047			17	2051	2N3047
2N3048			17	2051	2N3048
2N3049			21	2034	MD3250AF
2N3050			21	2034	MD3250AF
2N3051			21	2034	MD3250AF
2N3052			63	2030	MD2369F
2N3053	128	3024	243	2030	2N3053
2N3053/40053			63	2030	
2N3053A		3024			2N3053A
2N3054	384	3026	246	2020	
2N3055	130	3027	14	2041	
2N3055-1	130		14	2041	
2N3055-10			14	2041	
2N3055-2			14	2041	
2N3055-3			75	2041	
2N3055-4			14	2041	
2N3055-5			14	2041	
2N3055-6			75	2041	
2N3055-7			75	2041	
2N3055-8			75	2041	
2N3055-9			14	2041	
2N3055S			14	2041	
2N3056	128		18	2030	
2N3057	287		18	2030	
2N3058	159	3466	82	2032	
2N3059	159	3466	82	2032	
2N306	103	3010	59	2002	
2N306#				2002	
2N3060	159	3466	82	2032	
2N3061			82	2032	
2N3062	159	3466	82	2032	
2N3066	312	3112	FET-1	2028	2N4339
2N3067	312	3112	FET-1	2028	2N4339
2N3068	312	3112		2028	2N4338
2N3068A					2N3367
2N3069			FET-1	2028	2N3822
2N306A	103A	3010	59	2002	
2N307	104	3009	16	2006	
2N3070	312	3112	FET-1	2028	2N3821
2N3071	312	3112	FET-1	2028	2N4338
2N3072	159	3466	82	2032	2N3763
2N3073	159	3466	82	2032	2N3073
2N3074	160	3006	245	2004	
2N3075	100	3006/160	1	2007	
2N3077	128	3124/289	243	2030	
2N3078	128	3124/289	243	2030	
2N307A	104	3009	239	2006	
2N307B	104	3009	239	2006	
2N308	100	3007	1	2007	
2N308#				2004	
2N3081	159	3715	82	2032	
2N3081/46			67	2023	
2N3082	107	3039/316	11	2015	
2N3083	107	3039/316	11	2015	
2N3084	312	3112	FET-1	2028	2N3821
2N3085	312	3112	FET-1	2028	2N3821
2N3086	312	3112	FET-1	2028	2N3821
2N3087	312	3112	FET-1	2028	2N3821
2N3088	312	3112	FET-1	2028	2N4339
2N3088A	312	3112		2028	2N3821
2N3089	312	3112	FET-1	2028	2N4339
2N3089A	312	3112		2028	2N3821
2N309	100	3007	1	2007	
2N309#				2004	
2N31	100	3005	1	2007	
2N310	102	3007	2	2007	
2N3107	128		243	2030	2N3019
2N3108	128		243	2030	2N3020
2N3109	128		243	2030	2N3499
2N311	100	3005	1	2007	
2N3110	128		243	2030	2N3020
2N3112					2N5471/3
2N3113					2N5471/3
2N3114	154	3104A	40	2012	2N3114
2N3115	123A	3444	20	2051	2N2221
2N3116	123A	3444	20	2051	2N2222
2N3117	199	3124/289	243	2030	2N930A
2N3118			18	2030	2N3553
2N3119		3512	27		MRF531
2N312	101	3861	8	2002	
2N3120	129	3025	244	2027	2N3764
2N3121	159	3466	82	2032	2N3468
2N3122	128	3444/123A	20	2051	
2N3125	179	3014	239	2006	
2N3126			3	2006	
2N3127	160		245	2004	
2N3128	123AP	3444/123A	20	2051	
2N3129	123AP	3444/123A	20	2051	
2N313	101	3861	8	2002	
2N3130	123AP	3444/123A	20	2051	
2N3132	179	3009	239	2006	
2N3133	129	3025	244	2027	2N3133
2N3134	129	3025	244	2027	2N2905
2N3135	159	3466	82	2032	2N3135
2N3136	159	3466	82	2032	2N2907
2N3137	311	3039/316	11	2015	2N3137
2N3138			66	2048	
2N314	101	3861	8	2002	
2N3140			66	2048	
2N3142			66	2048	
2N3144			66	2048	
2N3146			3	2006	

Industry Standard No.	ECG	SK	GE	RS 276-	MOTOR.
2N3147			3	2006	
2N3148	160		245	2004	
2N315	100	3005	1	2007	
2N3153	160		245	2004	
2N315A	102	3004	1	2007	
2N315B	102	3005	1	2007	
2N316	100	3005	245	2004	
2N316A	102	3004	2	2007	
2N317	100	3005	1	2007	
2N3171			74	2043	2N3789
2N3172			74	2043	2N3789
2N3173			74	2043	2N3790
2N3174			74	20433	2N6226
2N317A	100	3005	1	2007	
2N318	126		52	2024	
2N3183			74	2043	2N3789
2N3184			74	2043	2N3789
2N3185			74	2043	2N3790
2N3186			74	2043	2N6226
2N319	102	3003	2	2007	
2N3195			74	2043	2N3789
2N3196			74	2043	2N3789
2N3197			74	2043	2N3790
2N3198			74	2043	2N6226
2N3199			69	2049	
2N32	102	3004	2	2007	
2N320	102A	3003	53	2007	
2N3200			69	2049	
2N3202	129	3258	244	2027	2N3719
2N3203	129	3259	244	2027	2N3720
2N3204					2N6303
2N3205			69	2049	
2N3206			69	2049	
2N3208	129	3258	244	2027	
2N3209	106	3118	21	2034	2N869A
2N321	102	3003	2	2007	
2N3210	123A	3444	20	2051	2N708
2N3211	123A	3444	20	2051	2N3013
2N3212	121	3009	239	2006	
2N3213	121	3009	239	2006	
2N3214	121	3009	239	2006	
2N3215	121	3009	239	2006	
2N3216	100		245	2004	
2N3217	159	3466	82	2032	
2N3218	159	3466	82	2032	
2N3219	159	3466	82	2032	
2N322	102	3003	2	2007	
2N3220					MM3220
2N3224	397		21	2034	
2N3226	130		14	2041	
2N3227	311	3452/108	86	2038	2N3227
2N3228	5511	3683	MR-4		S2800B
2N3229			66	2048	
2N323	102	3004	2	2007	
2N3232	130	3027	14	2041	2N5877
2N3233	130	3027	14	2041	2N5632
2N3234	280		14	2041	2N5760
2N3235		3027	77	2007	2N3055
2N3236			75	2041	2N5632
2N3237			75	2041	2N5302
2N3238			75	2041	2N5882
2N3239			75	2041	2N5882
2N324	102	3003	2	2007	
2N3240					2N5882
2N3241	123A	3854	20	2051	
2N3241A	123A	3854	20	2051	
2N3242	123A	3444	20	2051	
2N3242A	123A	3444	20	2051	
2N3244	129		67	2023	2N3244
2N3245	129	3118	21	2034	2N3245
2N3246	123A	3124/289	20	2051	2N3964
2N3247	123A	3444	20	2051	2N930A
2N3248	159	3466	82	2032	2N3546
2N3249	159	3466	82	2032	2N3249
2N325	104	3009	16	2006	
2N3250	159	3466	82	2032	2N3250
2N3250A	159	3466	82	2032	2N3250A
2N3251	159	3466	82	2032	2N3251
2N3251A	159	3466	82	2032	2N3251A
2N3252			63	2030	2N3252
2N3253			216	2053	2N3253
2N3254					MCR102
2N3255					MCR102
2N3256					MCR103
2N3257					MCR102
2N3258					MCR102
2N3259	5401	3638/5402			MCR103
2N3261	123A	3444	20	2051	
2N3262	282	3024/128	243	2030	
2N3267	160		245	2004	
2N3268	123A	3124/289	20	2051	
2N3269					2N4169
2N327	100	3005	1	2007	2N2906
2N3270					2N4170
2N3271					2N4171
2N3272					2N4172
2N3273					2N2324
2N3274					2N2326
2N3275					2N2328
2N3276					2N2329
2N3277					2N5265
2N3278					2N5265
2N3279	160		245	2004	
2N327A	100	3114/290	1	2007	2N2906
2N327B	129	3025	244	2027	2N2906
2N328	129	3114/290	244	2027	
2N3280	160		245	2004	
2N3281	160		245	2004	
2N3282	160		245	2004	
2N3283	160		245	2004	
2N3284	160		245	2004	
2N3285	160		245	2004	
2N3286	160		245	2004	
2N3287	161	3018	39	2015	2N3287
2N3288	161	3018	39	2015	2N3287
2N3289	161	3019	39	2015	2N3287
2N328A	129	3114/290	244	2027	
2N328B	129	3025	244	2027	
2N329	129	3025	244	2027	
2N3290	161	3019	39	2015	2N3287
2N3291	161	3019	39	2015	
2N3292	161	3019	39	2015	
2N3293	161	3019	39	2015	2N3287
2N3294	161	3019	39	2015	
2N3295			63	2030	
2N3296			28	2017	2N5641
2N3297			19	2041	
2N3298			10	2051	2N2369
2N3299	123A	3195/311	47	2030	2N3299
2N3299S	311	3195			
2N329A	129	3025	244	2027	
2N329B	129	3025	244	2027	
2N32A	102	3004	2	2007	
2N33	126	3006/160	52	2024	
2N330	129	3025	244	2027	2N2906
2N3300	123A	3195/311	20	2051	2N3300
2N3300S	311	3195			
2N3301	123A	3444	20	2051	2N3301
2N3302	123A	3444	20	2051	2N3302
2N3303					2N3303
2N3304	106	3984	21	2034	2N4209
2N3305	159	3466	82	2032	
2N3306	159	3466	82	2032	
2N3307	159	3466	82	2032	2N3307
2N3308	159	3466	82	2032	2N3308
2N3309A					2N3553
2N330A	129	3025	244	2027	2N2906
2N331	160	3008	245	2004	
2N3310	123A		20	2051	
2N3311	213	3012/105	4		
2N3312	213	3012/105	4		
2N3313	213	3012/105	4		
2N3314	213	3012/105	4		
2N3315	213	3012/105	4		
2N3317	159	3466	82	2032	
2N3318	159	3466	82	2032	
2N3319	159	3466	82	2032	
2N331M		3123	2	2007	
2N332	123A	3124/289	20	2051	2N2221
2N3320	126		245	2004	
2N3321	126		245	2004	
2N3322	126		245	2004	
2N3323	126		245	2004	
2N3324	126		245	2004	
2N3325	126		245	2004	
2N3326	128	3444/123A	20	2051	
2N3328			50	2004	2N5265
2N3329	460				2N3330
2N332A	123A	3124/289	20	2051	2N2218
2N333	123A	3124/289	20	2051	2N2221
2N3330,J					JAN2N3330
2N3331	460				MFE4012
2N3332	460				2N3330
2N3337	161	3716	39	2015	
2N3338	161	3117	39	2015	
2N3339	161	3117	39	2015	
2N333A	123A	3124/289	20	2051	2N2218
2N334	123A	3124/289	20	2051	2N2221
2N3340	123A	3444	20	2051	
2N3341	159	3466	82	2032	
2N3342	106	3984	21	2034	
2N3343	129		244	2027	
2N3344	129	3025	244	2027	
2N3345	129	3025	244	2027	
2N3346	159	3466	82	2032	
2N3347	82				2N3726/27
2N3348	82				2N3726/27
2N3349	82				2N3726/27
2N334A	123A	3124/289	20	2051	2N2218
2N334B					2N2218
2N335	123A	3124/289	20	2051	2N2221
2N3350	82				2N3726/27
2N3351	82				2N3726/27
2N3352	82				2N3726/27
2N335A	123	3124/289	20	2051	2N2218
2N335B	123A	3124/289	20	2051	2N2218
2N336	123A	3124/289	20	2051	2N2221
2N3365			FET-1	2028	
2N3366		3112	FET-1	2028	
2N3367		3112	FET-1	2028	
2N3368		3112	FET-1	2028	2N4341
2N33681				2015	
2N33682				2038	
2N3369		3112	FET-1	2028	2N4339
2N336A	123A	3124/289	20	2051	2N2218
2N337	123A	3039/316	11	2015	2N2221
2N3370		3112	FET-1	2028	2N4338
2N3371	160		245	2004	
2N3374			63	2030	
2N3375			28	2017	2N3375
2N3376	460				2N3330
2N3377	460				2N2608
2N3378	460				2N3330
2N3379	460				2N2608
2N337A	123A	3452/108	86	2038	2N2218
2N338	123A	3124/289	39	2015	2N2221
2N3380					2N3307
2N3381					2N3909A
2N3382					2N3394
2N3383					2N3394
2N3384					2N3394
2N3385					2N3393
2N3386					2N3993
2N3387					2N3993
2N3388	154	3045/225	40	2012	
2N3389	154	3045/225	40	2012	
2N338A	123A	3124/289	20	2051	2N2218
2N339	128		63	2030	
2N3390	199	3444/123A	20	2051	
2N3390-U29			212	2010	
2N3391	199	3444/123A	20	2051	
2N3391-U29			212	2010	
2N3391A	199	3444/123A	20	2051	
2N3392	289A	3444/123A	20	2051	
2N3392-U29			212	2010	
2N3393	289A	3444/123A	20	2051	
2N3393-U29			212	2010	
2N3394	289A	3444/123A	20	2051	
2N3394-U29			212	2010	
2N3395	289A	3444/123A	20	2051	
2N3395-WHT		3444	212	2010	
2N3395-YEL			212	2010	
2N3396	289A	3444/123A	20	2051	
2N3396-ORG			212	2010	
2N3396-WHT		3444	212	2010	
2N3396-YEL			212	2010	
2N3397	289A	3444/123A	20	2051	
2N3397-ORG			212	2010	
2N3397-RED			212	2010	
2N3397-WHT		3444	212	2010	
2N3397-YEL			212	2010	
2N3398	199	3444/123A	20	2051	
2N3398-BLU		3444	212	2010	
2N3398-ORG			212	2010	
2N3398-RED			212	2010	
2N3398-WHT		3444	212	2010	
2N3399	160		245	2004	
2N339A	128		63	2030	
2N34	102A	3004	53	2007	
2N340	128		235	2012	
2N3400	126	3123	245	2004	
2N3401	159	3715	82	2032	
2N3402	192A	3124/289	81	2051	
2N3403	192A	3124/289	63	2030	
2N3404	192A	3124/289	63	2030	
2N3405	192A	3124/289	47	2030	
2N3407	161	3019	86	2038	
2N3409			20	2051	MD3409

Industry Standard No.	ECG	SK	GE	RS 276-	MOTOR.
2N340A	128		235	2012	
2N341	154		235	2012	
2N3410			20	2051	MD3410
2N3411			20	2051	MD3410
2N3412	126		245	2004	
2N3413	288		221		
2N3414	289A	3444/123A	20	2051	
2N3415	123A	3444	20	2051	
2N3416	289A	3444/123A	20	2051	
2N3417	192A	3444/123A	20	2051	
2N3418			28	2017	2N5336
2N3419					2N5336
2N341A	154		235	2012	
2N342	128		63	2030	
2N3420			28	2017	2N5336
2N3421					2N5336
2N3423			20	2051	MD918
2N3424			20	2051	MD918
2N3425					2N3425
2N3426	195A	3765	47	2030	
2N3427	102	3004	2	2007	
2N3428	102	3004	2	2007	
2N342A	128		235	2012	
2N342B	128		18	2030	
2N343	128		63	2030	
2N3435			63	2030	
2N3436	312	3112	FET-1	2028	2N3436
2N3437	312	3112	FET-1	2028	2N3437
2N3438	312	3112	FET-1	2028	2N3438
2N3439	396	3103A	243	2030	2N3439
2N343A	128	3024	243	2030	
2N343B	128		63	2030	
2N344	126	3006/160	52	2024	
2N3440	396	3044/154	12		2N3440
2N3441	384	3538	267		2N3441
2N3442	284	3621	265	2047	2N3442
2N3443	126		245	2004	
2N3444					2N3444
2N3445	385	3027/130	14	2041	2N3445
2N3446	385	3027/130	14	2041	2N3446
2N3447	281	3027/130	14	2041	2N3447
2N3448	130	3027	14	2041	2N3448
2N3449	126		245	2004	
2N345	126	3006/160	52	2024	
2N3451	159	3466	82	2032	
2N3452	312	3112	FET-1	2028	2N3821
2N3453	312	3112	FET-1	2028	2N4119
2N3454	312	3112	FET-1	2028	2N4118
2N3455	312	3112	FET-1	2028	2N3821
2N3456	312	3112	FET-1	2028	2N4119
2N3457		3112	FET-1	2028	2N4118
2N3458			FET-1	2028	2N4341
2N3459			FET-1	2028	2N4339
2N346	126	3006/160	52	2024	
2N3460	312	3112	FET-1	2028	2N4338
2N3462	123A	3444	20	2051	
2N3463	123A	3444	20	2051	
2N3464	159	3466	82	2032	
2N3465	312	3112	FET-1	2028	2N4220A
2N3466	312	3112	FET-1	2028	2N4220A
2N3467	129		244	2027	2N3467
2N3467JAN					2N3467JAN
2N3467JTX					2N3467JTX
2N3467JTXV					2N3467JTXV
2N3468	129		244	2027	2N3468
2N3468JAN					2N3468JAN
2N3468JTX					2N3468JTX
2N3468JTXV					2N3468JTXV
2N3469	128	3024	243	2030	
2N3478	316	3039	280		2N5179
2N3483	175		246	2020	
2N3485	159	3466	82	2032	2N3485
2N3485A	159	3466	82	2032	2N3485A
2N3486	159	3466	82	2032	2N3486
2N3486A	159	3466	82	2032	2N3486A
2N3493	161	3452/108	86	2015	
2N3494	129	3025	244	2027	2N3494
2N3495	288				2N3495
2N3496	159	3466	82	2032	2N3496
2N3497	288				2N3497
2N3498	128		20	2051	2N3498
2N3499			18	2030	2N3499
2N34A	102A		53	2007	
2N35	101	3010	59	2002	
2N350	104	3009	239	2006	
2N3500		3104A	32		2N3500
2N3501		3104A	18	2030	2N3501
2N3502	129	3025	244	2027	2N3502
2N3503	129	3025	244	2027	2N3503
2N3504	159	3466	82	2032	2N3504
2N3505	159	3466	82	2032	2N3505
2N3506	282	3024/128	243	2030	2N3506
2N3507	282	3024/128	243	2030	2N3507
2N3508	123A	3452/108	86	2038	2N3508
2N3509	123A	3452/108	86	2038	2N3509
2N350A		3009	239	2006	
2N351	104	3009	239	2006	
2N3510	123A	3122	20	2051	2N3510
2N3511	123A	3452/108	86	2038	2N3511
2N3512	123A		47	2030	2N5859
2N3513					2N2060
2N3514					MD2219AF
2N3515					MD2219AF
2N3516					2N2060
2N3517					MD2219AF
2N3518					MD2219AF
2N3519					2N3043
2N351A	104	3009	239	2006	
2N352	121	3009	239	2006	
2N3520					2N3043
2N3521					2N2919
2N3522					2N2919
2N3523					2N3043
2N3524					2N3043
2N3525	5512	3684	MR-4		S2800D
2N3525A			28	2017	
2N3526	154		40	2012	
2N3527	159	3466	82	2032	
2N3528	5641	3850	MR-5	1067	S2800B
2N3529		3851			S2800D
2N353	121	3009	239	2006	
2N354					2N2906
2N3542			20	2051	
2N3544	311	3452/108	86	2038	2N869A
2N3545	159	3466	82	2032	2N869A
2N3546	159	3466	20	2032	2N3546
2N3547	159	3466	82	2032	2N2907A
2N3548	159	3466	82	2032	2N2907
2N3549	159	3466	82	2032	2N2907A
2N3550	159	3466	82	2032	2N2907
2N3553			285		2N3553
2N3554	195A	3197/235	215		
2N3555					2N4212
2N3556					2N4213

Industry Standard No.	ECG	SK	GE	RS 276-	MOTOR.
2N3557					2N4214
2N3558					2N4216
2N3559					2N4212
2N356	101	3861	8	2002	
2N3560					2N4213
2N3561					2N4214
2N3562	108	3452	86	2038	2N4216
2N3563	108	3018	61	2038	
2N3563-1	108	3018	61	2038	
2N3564	123A	3018	61	2051	
2N3565	123A	3444	20	2051	
2N3566	123A	3444	20	2051	
2N3567	123A	3024/128	243	2030	
2N3568	128	3024	243	2030	
2N3569	128	3024	243	2030	
2N356A	103	3861/101	8	2002	
2N357	101	3861	8	2002	
2N3570	316	3452/108	86	2038	2N5032
2N3571	316	3452/108	86	2038	2N5032
2N3572	316	3452/108	86	2038	2N5032
2N3573					2N5265
2N3575					2N5265
2N3576	106	3984	21	2034	2N869A
2N3578					2N2608
2N3579	159	3466	82	2032	2N3799
2N357A	103	3861/101	8	2002	
2N358	101	3861	8	2002	
2N3580	159	3466	82	2032	2N3799
2N3581	159	3466	82	2032	2N3799
2N3582	159	3466	82	2032	2N3799
2N3583	286	3271	267		2N3583
2N3584	384	3271	267		2N3584
2N3585	286	3261/175	12		2N3585
2N3586					2N5794
2N3587	81				2N5794
2N3588	160	3006	245	2004	
2N358A	103	3861/101	8	2002	
2N359	102	3004	2	2007	
2N36	102A	3003	53	2004	
2N36#				2004	
2N360	102	3004	2	2007	
2N3600	161	3452/108	86	2038	2N5179
2N3605	289A	3444/123A	20	2051	
2N3605A	289A		20	2051	
2N3606	289A	3444/123A	20	2051	
2N3606A	289A		20	2051	
2N3607	289A	3444/123A	20	2051	
2N3608			FET-1	2028	2N2608
2N361	102	3004	2	2007	
2N3610					2N4352
2N3611	179	3009	239	2006	
2N3612	179	3009	239	2006	
2N3613	179	3009	239	2006	
2N3614	179	3009	239	2006	
2N3615	179	3009	239	2006	
2N3616	179	3009	239	2006	
2N3617	179	3009	239	2006	
2N3618	179	3009	239	2006	
2N3619			28	2017	
2N362	102	3004	2	2007	
2N3620			28	2017	
2N3621	72		66	2048	
2N3622	73		66	2048	
2N3623			28	2017	
2N3624			28	2017	
2N3625	72		66	2048	
2N3626			66	2048	
2N3627			66	2048	
2N3628			28	2017	
2N3629	72		66	2048	
2N362B	102	3004	2	2007	
2N363	102	3004	2	2007	
2N3630			66	2048	
2N3631	462				2N3797
2N3632			66	2048	2N3632
2N3633	278	3452/108	86	2038	
2N3633/46			11	2015	
2N3633/52			63	2030	
2N3633/TNT			11	2015	
2N3634					2N3634
2N3635					2N3635
2N3636					2N3636
2N3637					2N3637
2N3638	159	3466	82	2032	
2N3638A	159	3466	82	2032	
2N3639	159	3466	82	2032	
2N364	103	3862	8	2002	
2N364#				2002	
2N364#				2002	
2N3640	159	3466	82	2032	2N4091
2N3641	123A	3444	20	2051	
2N3642	123A	3444	243	2051	
2N3643	123A	3444	20	2051	
2N3644	159	3466	82	2032	2N1843
2N3645	129	3025	244	2027	
2N3646	123A	3444	20	2051	
2N3647	123A		20	2051	2N3647
2N3648	123A	3039/316	86	2038	2N3648
2N365	103	3862	8	2002	
2N365#				2002	
2N365#				2002	
2N3650					2N1844
2N3651					2N1846
2N3652					2N1848
2N3653					2N1849
2N3654	5501				2N1843
2N3655					2N1844
2N3656					2N1846
2N3657			2N3657		2N1848
2N3658					2N1849
2N366	105	3862	8	2002	
2N366#				2002	
2N366#				2002	
2N3660	323		84		2N4234
2N3661	323				2N4235
2N3662	107	3039/316	11	2015	
2N3663	107	3018	11	2015	
2N3665	128		243	2030	2N3020
2N3666	128		243	2030	2N3019
2N3667	130	3036	14	2041	2N5881
2N3668		3527			2N3668
2N3669		3527			2N3669
2N367	102	3004	2	2007	
2N3670		3898			2N3670
2N3671	129	3025	244	2027	2N2905
2N3672	159	3466	82	2032	
2N3673	159	3466	82	2032	2N5582
2N3675			66	2048	
2N3677	129	3025	244	2027	2N3677
2N3678	128	3444/123A	20	2051	2N3020
2N368	102	3004	2	2007	
2N3680					2N2920
2N3681	161		11		
2N3682	108	3039/316	86		

Industry Standard No.	ECG	SK	GE	RS 276-	MOTOR.
2N3683	161		86	2038	
2N3684	459	3112	FET-1	2028	2N3822
2N3684A	312	3112		2028	
2N3685	312	3112	FET-1	2028	2N3821
2N3685A	312	3112		2028	
2N3686	312	3112	FET-1	2028	2N3821
2N3686A	312	3112		2028	
2N3687	312	3112	FET-1	2028	2N4338
2N3687A	311	3112	277		
2N3688	123A	3452/108	86	2051	
2N3689	123A	3452/108	86	2051	
2N369	102	3004	2	2007	
2N3690	123A	3452/108	86	2051	
2N3691	123A	3452/108	86	2051	
2N3692	123A	3452/108	86	2051	
2N3693	123A	3444	20	2051	
2N3694	123A	3444	20	2051	
2N3695					MFE4009
2N3696					MFE4009
2N3697					MFE4007
2N3698					2N5265
2N37	126	3003	53	2004	
2N370	126	3007	52	2024	
2N370/33	126	3007	52	2024	
2N3700	154	3045/225	40	2012	2N3700
2N3701	154		40	2012	2N3019
2N3702	159	3466	82	2032	2N3250
2N3703	159	3466	82	2032	2N3251
2N3704	297	3444/123A	20	2051	2N2222A
2N3705	123AP	3444/123A	20	2051	2N2222A
2N3706	123AP	3444/123A	20	2051	2N930
2N3707	289A	3444/123A	20	2051	
2N3708	289A	3444/123A	20	2051	
2N3708-BLU			212	2010	
2N3708-BRN			212	2010	
2N3708-GRN			212	2010	
2N3708-ORG			212	2010	
2N3708-RED			212	2010	
2N3708-VIO			212	2010	
2N3708-YEL			212	2010	
2N3709	107	3444/123A	20	2051	
2N370A	126	3007	52	2024	
2N371	126	3007	52	2024	
2N371/33	126	3007	52	2024	
2N3710	107	3444/123A	20	2051	
2N3711	289A	3444/123A	20	2051	
2N3712	154	3104A	40	2012	2N3712
2N3712S	154	3104A			
2N3713	385	3945/327	14	2041	2N3713
2N3714	385	3945/327	14	2041	2N3714
2N3715	385	3945/327	14	2041	2N3715
2N3715JAN					2N3715JAN
2N3715JTX					2N3715JTX
2N3715JTXV					2N3715JTXV
2N3716	385	3945/327	265	2047	2N3716
2N3716HS	385	3945/327			
2N3716JAN					2N3716JAN
2N3716JTX					2N3716JTX
2N3716JTXV					2N3716JTXV
2N3719	129	3025	244	2027	2N3719
2N372	126	3007	52	2024	
2N372/33	126	3007	52	2024	
2N3720	129	3025	244	2027	2N3720
2N3721	289A	3124/289	11	2015	
2N3722	128		D41E7		
2N3723	128	3444/123A	20	2051	
2N3724	311		47	2030	2N3724
2N3724A			215		2N3734
2N3725	123A		18	2030	2N3725
2N3725A			D40E7		2N3725A
2N3726	82				MM3726
2N3727	82				2N3727
2N3728					2N2903
2N3729					2N2903A
2N373	126	3007	52	2024	
2N3730	127	3034	25		
2N3731	127	3764	25		
2N3732	127	3034	25		
2N3733					2N3632
2N3734			215		2N3734
2N3735					2N3735
2N3735JAN					2N3735JAN
2N3735JTX					2N3735JTX
2N3735JTXV					2N3735JTXV
2N3736			20	2051	2N3736
2N3737					2N3737
2N3738	286	3271	12		2N3738
2N3739	286	3261/175	12		2N3739
2N3739JAN					2N3739JAN
2N3739JTX					2N3739JTX
2N3739JTXV					2N3739JTXV
2N374	126	3007	52	2024	
2N3740			234		2N3740
2N3740A			69	2049	2N3740A
2N3740JAN					2N3740JAN
2N3740JTX					2N3740JTX
2N3740JTXV					2N3740JTXV
2N3741	292	3441	234		2N3741
2N3741A	292		234		2N3741A
2N3741JAN					2N3741JAN
2N3741JTX					2N3741JTX
2N3741JTXV					2N3741JTXV
2N3742	154		40	2012	2N3742
2N3743	154		228	2012	2N3743
2N3744			66	2048	
2N3745			66	2048	
2N3747			66	2048	
2N3748			66	2048	
2N375	104	3009	16	2006	
2N3753	5491	6791			
2N376	104	3009	16	2006	
2N3762	129	3025	244	2027	2N3762
2N3762JAN					2N3762JAN
2N3762JTX					2N3762JTX
2N3762JTXV					2N3762JTXV
2N3763	129	3025	244	2027	2N3763
2N3763JAN					2N3763JAN
2N3763JTX					2N3763JTX
2N3763JTXV					2N3763JTXV
2N3764	129	3025	244	2027	2N3764
2N3765	294	3025/129	244	2027	2N3765
2N3766	384		246	2020	2N3766
2N3766JAN					2N3766JAN
2N3766JTX					2N3766JTX
2N3766JTXV					2N3766JTXV
2N3767	384	3538	246	2020	2N3767
2N3767JAN					2N3767JAN
2N3767JTX					2N3767JTX
2N3767JTXV					2N3767JTXV
2N376A	104	3009	16	2006	
2N377	101	3861	8	2002	
2N3770	160		245	2004	
2N3771	181	3036	75	2041	2N3771
2N3772	284	3036	75	2041	2N3772

Industry Standard No.	ECG	SK	GE	RS 276-	MOTOR.
2N3773	284	3260	265	2047	2N3773
2N3774	129	3258	244	2027	2N4234
2N3775	129	3259	244	2027	2N4235
2N3776	129	3025	244	2027	2N4236
2N3777	323				2N5679
2N3778	129	3258	244	2027	2N4234
2N3779	129	3259	246	2020	2N4235
2N377A	101	3861	8	2002	
2N378	104	3009	16	2006	
2N3780	129	3025	244	2027	2N4236
2N3781	322	3025/129	244	2027	2N5679
2N3782	129	3258	244	2027	2N4234
2N3783	160		245	2004	
2N3784	160		245	2004	
2N3785	160		245	2004	
2N3788			36		2N6542
2N3789	219		74	2043	2N3789
2N379	179	3009	239	2006	
2N3790	219	3173	74	2043	2N3790
2N3791	219	3173	74	2043	2N3791
2N3791JAN					2N3791JAN
2N3791JTX					2N3791JTX
2N3791JTXV					2N3791JTXV
2N3792	219	3437/180	74	2043	2N3792
2N3792JAN					2N3792JAN
2N3792JTX					2N3792JTX
2N3792JTXV					2N3792JTXV
2N3793	123AP	3444/123A	20	2051	
2N3794	123AP	3444/123A	20	2051	
2N3796	462				2N3796
2N3797	462				2N3797
2N3798	159	3466	82	2032	2N3798
2N3798A	159	3114/290			2N3798
2N3799	159	3466	82	2032	2N3799
2N3799A	159	3114/290			2N3799
2N38	126	3003	53	2004	
2N38#				2004	
2N380	104	3009	16	2006	
2N3800	82				2N3806
2N3801					2N3807
2N3802					2N3808
2N3803					2N3809
2N3804					2N3810
2N3804A					2N3810A
2N3805					2N3811
2N3805A					2N3811A
2N3806	82				2N3811A
2N3807					2N3811A
2N3808	82				2N3811A
2N3809					2N3811A
2N381	102	3004	2	2007	
2N3810	82				2N3810
2N3810A	82				2N3810A
2N3810JAN					2N3810JAN
2N3810JTX					2N3810JTX
2N3810TXV					2N3810TXV
2N3811					2N3811
2N3811A					2N3811A
2N3811JAN					2N3811JAN
2N3811JTX					2N3811JTX
2N3811TXV					2N3811TXV
2N3812		3114			2N3812
2N3813		3114			2N3813
2N3814		3114			2N3814
2N3815		3114			2N3815
2N3816		3114			2N3816
2N3816A					2N3816
2N3817					2N3817
2N3817A					2N3817A
2N3818			66	2048	2N3632
2N3819	312	3448	FET-1	2035	MPF108/112
2N382	102	3004	2	2007	
2N3820		3746			MPF161
2N3821	312	3112		2028	2N3821
2N3822	459	3112	FET-1	2028	2N3822
2N3823	220	3116	FET-1	2028	2N3823
2N3824	459				2N3824
2N3825	289A	3039/316	11	2015	
2N3826	289A	3018	11	2015	
2N3827	289A	3018	61	2038	
2N3828	123AP	3444/123A	20	2051	
2N3829	159	3466	82	2032	2N3250
2N383	102	3004	2	2007	
2N3830	128	3748/282	D40E7		2N2193A
2N3831	282	3748	215		2N2193A
2N3832	161	3452/108	86	2038	
2N3838			21	2034	2N3838
2N3838JAN					2N3838
2N3838JTX					2N3838
2N3838TXV					2N3838
2N3839	316		280		2N3839
2N384	160	3007	245	2004	
2N384/33	160	3007	245	2004	
2N3840	159	3466	82	2032	
2N3841	288	3466/159	221	2032	
2N3842		3466	221	2032	
2N3843	289A	3444/123A	20	2051	
2N3843A	289A	3444/123A	20	2051	
2N3844	289A	3444/123A	20	2051	
2N3844A	289A	3444/123A	20	2051	
2N3845	289A	3039/316	11	2015	
2N3845A	289A	3039/316	11	2015	
2N3846	107	3039/316	11	2015	
2N385	101	3861	8	2002	
2N3852	74$		28	2017	
2N3853	74$		28	2017	
2N3854	289A	3452/108	86	2038	
2N3854A	289A	3452/108	86	2038	
2N3855	289A	3018	11	2015	
2N3855A	107	3039/316	11	2015	
2N3856	289A	3039/316	11	2015	
2N3856A	289A	3039/316	11	2015	
2N3857	159	3466	82	2032	
2N3858	289A	3444/123A	20	2051	
2N3858A	289A	3444/123A	20	2051	
2N3859	289A	3444/123A	20	2051	
2N3859A	289A	3444/123A	20	2051	
2N385A	101	3861	8	2002	
2N386	104	3009	16	2006	
2N3860	289A	3039/316	11	2015	
2N3862	311	3444/123A	20	2051	
2N3863	130	3036	14	2041	2N3715
2N3864	130	3036	14	2041	2N5632
2N3865					2N5634
2N3866	311	3195	277		2N3866
2N3866A	311		277		
2N3867	129	3025	244	2027	2N3867
2N3867JAN					2N3867JAN
2N3867JTX					2N3867JTX
2N3867JTXV					2N3867JTXV
2N3868	129	3025	244	2027	2N3868
2N3868JAN					2N3868JAN
2N3868JTX					2N3868JTX
2N3868JTXV					2N3868JTXV

Industry Standard No.	ECG	SK	GE	RS 276-	MOTOR.
2N3869	311		20	2051	
2N387	104	3009	16	2006	
2N3870	5517	3615			2N3870
2N3871	5517	3615			2N3871
2N3872	5518	3653			2N3872
2N3873	5519	3654			2N3873
2N3877	289A	3024/128	243	2030	
2N3877A	289A		81	2051	
2N3878	384	3562	246	2020	2N5428
2N3879		3562	23	2020	2N5430
2N388	101	3861	8	2002	
2N3880	316	3122	20	2051	2N5032
2N3881	128		47	2030	
2N3882					2N4352
2N3883	126		245	2004	
2N3885			20	2051	
2N388A	101	3861	8	2002	
2N3896	5522	6642/5542	MR-3		2N3896
2N3897	5524	3581/5543	MR-3		2N3897
2N3898	5527	3582/5545	MR-3		2N3898
2N3899	5547	3505			2N3899
2N38A	102A	3003	53	2007	
2N39	102	3004	2	2007	
2N3900	289A	3124/289	20	2051	
2N3900A	289A	3444/123A	20	2051	
2N3901	199	3124/289	20	2051	
2N3902	283		36		2N3902
2N3903	123AP	3444/123A	20	2051	
2N3904	123AP	3444/123A	20	2051	
2N3905	159	3466	82	2032	
2N3906	159	3466	82	2032	
2N3907	81				2N2919
2N3908	81$				2N2919
2N3910			82	2032	
2N3911		3114	82	2032	
2N3912		3114	82	2032	
2N3913	159		82	2032	
2N3914	159	3466	82	2032	
2N3915	106	3984	21	2034	
2N3917			19	2041	
2N3918			19	2041	
2N3919	328	3895	28	2017	
2N392	104	3009	16	2006	
2N3920	328	3895			
2N3923	154	3045/225	40	2012	2N3500
2N3924	123		20	2051	
2N3925			28	2017	2N5589
2N3926	475		28	2017	2N3926
2N3927	476		66	2048	2N3927
2N393	126	3008	52	2024	
2N3930	159	3466	82	2032	
2N3931	159	3466	82	2032	
2N3932	161		86	2015	
2N3933	161	3452/108	86	2015	
2N3936					2N4169
2N3937					2N4170
2N3938					2N4171
2N3939					2N4172
2N394	100	3005	1	2007	
2N3940					2N4173
2N3945			215		2N4404
2N3946	123A		210	2051	2N3946
2N3947	123A		20	2051	2N3947
2N3948	311	3024/128	243	2030	2N3948
2N394A	100	3005	1	2007	
2N395	100	3005	1	2007	
2N3950					2N3950
2N3953	316		86	2038	
2N3959	278				2N3959
2N396	100	3005	1	2007	
2N3960	278				2N3960
2N3961			28	2017	2N5641
2N3962	159	3466	82	2032	2N3962
2N3963	159	3466	82	2032	2N3963
2N3964	159	3466	82	2032	2N3964
2N3965		3466	82	2032	2N3965
2N3966			FET-1	2028	2N3966
2N3967	312	3112	FET-1	2028	2N4222A
2N3967A		3112		2028	2N4222A
2N3968	312	3112	FET-1	2028	2N4221A
2N3968A		3112		2028	2N4221A
2N3969		3112	FET-1	2028	2N4220A
2N3969A	312	3112		2028	2N4220A
2N396A	100	3005	1	2007	
2N397	100	3005	1	2007	
2N3970	466				2N3970
2N3971	466				2N3971
2N3972					2N3972
2N3973	293	3122	210	2051	
2N3974	293	3122	210	2051	
2N3975	293	3444/123A	20	2051	
2N3976	293	3444/123A	20	2051	
2N3977	159	3466	82	2032	2N2944
2N3978	159	3466	82	2032	
2N3979	159	3466	82	2032	
2N3980					2N3980
2N3981			63	2030	
2N3982			63	2030	
2N3983	107	3452/108	86	2038	
2N3984	107	3452/108	86	2038	
2N3985	107	3452/108	86	2038	
2N399	104	3009	16	2006	
2N3993					2N3993
2N3994					2N3994
2N3995	160		245	2004	
2N3996					2N5347
2N3997					2N5347
2N3998	75				2N5347
2N3999	75$				2N5347
2N40	102	3004	2	2007	
2N400	104	3009	16	2006	
2N4000			32		2N5347
2N4001	282	3748	32		2N6056
2N4002	70				2N6274
2N4003	70				2N6274
2N4006			22	2032	
2N4007			22	2032	
2N4008			82	2032	
2N401	104	3009	16	2006	
2N4012			28	2017	2N4012
2N4013	123A	3444	20	2051	2N4013
2N4014	123A	3444	20	2051	2N4014
2N4015	82				2N4015
2N4016	82				2N4016
2N4017					2N3811
2N4018	82				2N3806
2N4019					2N3807
2N402	102	3003	2	2007	
2N4020					2N3811
2N4021	82				2N3811
2N4022					2N3811
2N4023					2N3810
2N4024	82				2N3810
2N4025					2N3810

Industry Standard No.	ECG	SK	GE	RS 276-	MOTOR.
2N4026	129	3025	244	2027	2N4026
2N4027	129	3025	244	2027	2N4027
2N4028	129	3025	244	2027	2N4028
2N4029	129	3025	244	2027	2N4029
2N403	100	3003	2	2007	
2N4030	129	3025	244	2027	2N4030
2N4031	129	3025	244	2027	2N4031
2N4032	129	3025	244	2027	2N4032
2N4033	129	3025	244	2027	2N4033
2N4034	106	3984	21	2034	2N3250
2N4035	106	3984	21	2034	2N3251
2N4036	129	3025	244	2027	2N4036
2N4037	129	3262	244	2027	2N4037
2N4038	465				2N4351
2N4039	465				2N4351
2N404	100	3004/102A	2	2007	
2N4040			28	2017	2N5636
2N4041			28	2017	2N5635
2N4042					2N2920
2N4043					2N2920
2N4044					2N2920
2N4045					2N2920
2N4046	123A		47	2030	2N5859
2N4047	128		D40E7		2N5861
2N404A	102	3004	2	2007	
2N405	102A	3003	2	2007	
2N4054	157	3103A/396	232		
2N4055	157	3103A/396	232		
2N4056	157	3103A/396	232		
2N4057	157	3103A/396	232		
2N4058	106	3118	21	2034	
2N4059	106	3118	21	2034	
2N406	102A	3004	53	2007	
2N4060	106	3118	21	2034	
2N4061	106	3118	21	2034	
2N4062	106	3118	21	2034	
2N4063	396/427	3045/225	20	2051	2N3439
2N4064	396/427	3444/123A	20	2051	2N3440
2N4065					3N156
2N4068	154	3045/225	40	2012	
2N406BLU	102A		53	2007	
2N406BRN	102A		53	2007	
2N406GRN	102A		53	2007	
2N406GRN-YEL	102A		53	2007	
2N4060RN	102A		53	2007	
2N406RED	102A		53	2007	
2N407	126	3003	53	2007	
2N4070		3561			2N6306
2N4071	385	3945/327			2N6306
2N4072	123A	3452/108	86	2038	2N4072
2N4073	311	3452/108	86	2038	2N4073
2N4074	123A	3444	20	2051	
2N4077	155	3839	43		
2N4078	131	3052	44	2006	
2N407BLK	102A		53	2007	
2N407GRN	102A		53	2007	
2N407J	102A		53	2007	
2N407RED	102A		53	2007	
2N407WHT	102A		53	2007	
2N407YEL	102A		53	2007	
2N408	102A	3003	53	2007	
2N4080	395	3118	21	2034	
2N4081			60	2015	
2N4086		3124	212	2010	
2N4087		3245	212	2010	
2N4087A		3245	212	2010	
2N4088					MPF161
2N4089					MPF161
2N408J	102A		53	2007	
2N408WHT	102A		53	2007	
2N409	126	3005	52	2024	
2N409#				2005	
2N409#				2005	
2N4090					MPF161
2N4091,J					2N4091,J
2N4092,J					2N4092,J
2N4093			334	2035	
2N4093,J					2N4093,J
2N4094					2N4856
2N4095					2N4857
2N4096					MCR103
2N4097					MCR104
2N4098					MCR120
2N41	102	3003	2	2007	
2N410	126	3005	52	2024	
2N410#				2005	
2N410#				2005	
2N4100			210	2051	
2N4101	5513	3502			S2800M
2N4102		3851			S2800M
2N4103		3527			2N4103
2N4104		3124	20	2051	
2N4105		3010	8	2002	
2N4106	158	3004/102A	2	2007	
2N4107			53	2007	
2N4108					MCR103
2N4109					MCR104
2N411	126	3005	52	2024	
2N411#				2005	
2N411#				2005	
2N4110					MCR120
2N4111	328	3895	19	2041	2N3715
2N4112			19	2041	
2N4113	328	3895	19	2041	2N3716
2N4114			19	2041	
2N4115					2N5347
2N4116					2N5347
2N4117	312	3112	FET-1	2028	2N4117
2N4117A	312	3112		2028	2N4117A
2N4118	312	3112	FET-1	2028	2N4118
2N4118A		3112			2N4118A
2N4119	312	3112	FET-1	2028	2N4119
2N4119A	312	3112		2028	2N4119A
2N412	126	3005	52	2024	
2N412#				2005	
2N412#				2005	
2N4120					3N156
2N4121	159	3715	82	2032	
2N4122		3118	21	2034	
2N4123	123AP	3444/123A	20	2051	
2N4124	123AP	3444/123A	20	2051	
2N4125	159	3466	82	2032	
2N4126	159	3715	82	2032	
2N4127	350	3176	66	2048	2N6081
2N4128	351	3177	66	2048	2N6082
2N413	100	3005	1	2007	
2N4130		3036	14	2041	MRF463
2N4134	161	3452/108	86	2038	
2N4135	161	3452/108	86	2038	
2N4137	311	3444/123A	20	2051	
2N4138	123A		212	2010	
2N4139					2N4222
2N413A	100	3005	1	2007	
2N414	100	3005	1	2007	

Industry Standard No.	ECG	SK	GE	RS 276-	MOTOR.
2N4140	123A	3122	210	2051	
2N4141	123A	3122	62	2051	
2N4142	159	3114/290	21	2034	
2N4143	159	3715	48		
2N4144	5400	3950			MCR101
2N4145	5400	3950			MCR101
2N4146	5401	3638/5402			MCR102
2N4147	5402	3638			MCR103
2N4148	5403	3627/5404			MCR104
2N4149	5404	3627	82	2032	MCR120
2N414A	100	3005	1	2007	
2N414B	102	3005	1	2007	
2N414C	102	3005	1	2007	
2N415	100	3005	1	2007	
2N415A	100	3005	1	2007	
2N416	100	3005	1	2007	
2N4167	5480		MR-4		2N4167
2N4168	5481		MR-4		2N4168
2N4169	5482		MR-4		2N4169
2N417	100	3005	1	2007	
2N4170	5483	3942	MR-4		2N4170
2N4171	5484	3943/5485	MR-4		2N4171
2N4172	5485	3943	MR-4		2N4172
2N4173	5486	3944/5487			2N4173
2N4174	5487	3944			2N4174
2N4175	5481		MR-4		
2N4176	5480		MR-4		
2N4177	5482		MR-4		
2N4178	5483	3942	MR-4		
2N4179	5484	3943/5485	MR-4		
2N418	179	3009	239	2006	
2N4180	5485	3943	MR-4		
2N4181	5486	3944/5487			
2N4182	5487	3944			
2N4183			MR-4		2N4183
2N4184			MR-4		2N4184
2N4185			MR-4		2N4185
2N4186			MR-4		2N4186
2N4187			MR-4		2N4187
2N4188			MR-4		2N4188
2N4189					2N4189
2N419	104	3009	16	2006	
2N4190					2N4190
2N4192			MR-4		MRF531
2N4193			MR-4		MRF531
2N4199					2N4199
2N42	102	3004	2	2007	
2N420	121	3009	239	2006	
2N4200					2N4200
2N4201					2N4201
2N4202					2N4202
2N4203					2N4203
2N4204					2N4204
2N4207		3466	82	2032	2N4208
2N4208	159	3466	82	2032	2N4208
2N4209	159	3466	82	2032	2N4209
2N420A	121	3009	239	2006	
2N4212					2N4212
2N4213					2N4213
2N4214					2N4214
2N4215					2N4215
2N4216					2N4216
2N4217					2N4217
2N4218					2N4218
2N4219					2N4219
2N422	102	3003	2	2007	
2N4220	456	3977	FET-1	2028	2N4220
2N4220A		3977			2N4220A
2N4221	456	3977	82	2032	2N4221
2N4221A	456	3977			2N4221A
2N4222	312	3112	FET-1	2028	2N4222
2N4222A	312	3112		2028	2N4222A
2N4223	312	3112	FET-2	2028	2N4223
2N4224		3116	FET-1	2028	2N4224
2N4225			28	2017	
2N4226			28	2017	
2N4227	123A	3444	210	2051	
2N4228	159	3466	21	2034	
2N422A	102	3722	2	2007	
2N4231	175	3272	246	2020	2N4231A
2N4231A					2N4231A
2N4232	175	3131A/369	246	2020	2N4232A
2N4232A					2N4232A
2N4233	175	3131A/369	246	2020	2N4233A
2N4233A		3131A	233		2N4233A
2N4234	129	3258	244	2027	2N4234
2N4235	129	3259	244	2027	2N4235
2N4236	129		244	2027	2N4236
2N4237	128	3024	243	2030	2N4237
2N4238	128	3024	32		2N4238
2N4239	128	3024	32		2N4239
2N4240	286	3021/124	12		2N4240
2N4241	121	3014	239	2006	
2N4244			16	2006	
2N4247			16	2006	
2N4248	159	3466	82	2032	
2N4249	159	3466	82	2032	
2N425	102	3005	1	2007	
2N4250	159	3466	82	2032	
2N4250A	159	3466	82	2032	
2N4251	278	3452/108	86	2038	
2N4252	161		86	2038	
2N4253	161		86	2038	
2N4254	107	3039/316	11	2015	
2N4255	107	3018	11	2015	
2N4256	289A	3444/123A	20	2051	
2N4257	159	3466	82	2032	MM4257
2N4257A	159	3466	82	2032	
2N4258		3466	82	2032	MM4258
2N4258A		3466	82	2032	
2N4259	161	3444/123A	20	2051	
2N426	100	3005	1	2007	
2N4260	295				2N4260
2N4261	395				2N4261
2N4264	123AP		20	2051	
2N4265	123AP		20	2051	
2N4267					2N4352
2N4268					MFE3003
2N4269	154	3045/225	40	2012	
2N427	100	3005	1	2007	
2N4270	154	3045/225	40	2012	
2N4271			32		2N5682
2N4272					2N5682
2N4274	123A	3452/108	86	2051	
2N4275	123A	3452/108	86	2051	
2N427A	102A	3123	53	2007	
2N428	100	3005	1	2007	
2N4284		3466	221	2032	
2N4285		3466	221	2032	
2N4286	289A	3444/123A	20	2051	
2N4287	289A	3444/123A	20	2051	
2N4288	159	3466	82	2032	
2N4289	159	3466	82	2032	
2N428A	100	3005	1	2007	

Industry Standard No.	ECG	SK	GE	RS 276-	MOTOR.
2N4290	159	3466	82	2032	
2N4291	159	3466	82	2032	
2N4292	108		86	2038	
2N4293	108		86	2038	
2N4294			20	2051	
2N4295			60	2015	
2N4296	286	3021/124	12		2N3738
2N4297	286	3021/124	12		2N3738
2N4298	124	3021	12		2N6235
2N4299	124	3021	12		2N6235
2N43	102	3003	2	2007	
2N4300					2N5337
2N4301	738				2N5337
2N4302	457	3112	FET-1	2028	2N5457
2N4303	457	3112	FET-1	2028	2N5458
2N4304	457	3112	FET-1	2028	2N5458
2N4305					2N5337
2N4307			66	2048	2N5337
2N4308			66	2048	
2N4309					2N5339
2N431	123	3124/289	20	2051	
2N4311			66	2048	2N5337
2N4312			66	2048	
2N4313	159	3118	21	2034	
2N4314	129	3025	244	2027	2N3868
2N432	123	3124/289	20	2051	
2N433	123	3124/289	20	2051	
2N4332					MCR102
2N4333					MCR103
2N4334					MCR104
2N4335					MCR115
2N4336					MCR120
2N4338	312	3112	FET-1	2028	2N4338
2N4339		3112	FET-1	2028	2N4339
2N4340	459$	3977/456	FET-1	2028	2N4340
2N4341			FET-1	2028	
2N4342					2N4342
2N4343					2N4342
2N4346	127	3035	25		
2N4347	280	3297	262	2041	2N4347
2N4348	280	3297	262	2041	2N5630
2N4349		3197	215		
2N435			53	2007	
2N435				2007	
2N4350			28	2017	
2N4351	465				2N4351
2N4352	464				2N4352
2N4353					2N4352
2N4354	159	3466	82	2032	
2N4355	159	3466	82	2032	
2N4356	159	3466	82	2032	
2N4359	159	3466	82	2032	2N3799
2N4360	326	3746			2N4360
2N438	103	3861/101	8	2002	
2N4381	460				2N2609
2N4382					2N3994
2N4383	128		47	2030	
2N4384			47	2030	2N2906
2N4385	128		47	2030	
2N4386			47	2030	2N2906
2N4387	153	3257	69	2049	2N4898
2N4388	153	3083/197	69	2049	2N4898
2N4389	159	3118	21	2034	
2N438A	103	3861/101	1	2007	
2N439	103	3011	8	2002	
2N4391	466				2N4391
2N4392	466				2N4392
2N4393					2N4393
2N4395	130	3027	14	2041	
2N4396	130	3027	14	2041	
2N4397			60	2015	
2N4398	180	3437			2N4398
2N4399					2N4399
2N4399JAN					2N4399JAN
2N4399JTX					2N4399JTX
2N4399JTXV					2N4399JTXV
2N439A	103	3011	1	2007	
2N43A	102	3004	2	2007	
2N44	102	3004	2	2007	
2N440	101	3011	8	2002	
2N4400	123AP	3444/123A	20	2051	
2N4401	123AP	3444/123A	20	2051	
2N4402	159	3466	82	2032	
2N4403	159	3466	82	2032	
2N4404	129	3025	244	2027	2N4404
2N4405	129	3025	244	2027	2N4405
2N4406	129	3025	244	2027	2N4406
2N4407	129	3025	244	2027	2N4407
2N4409	194		220		
2N440A	103	3011	1	2007	
2N441	213	3012/105	4		
2N4410	194	3045/225	40	2012	
2N4411	159	3466	82	2032	
2N4412	129	3025	244	2027	
2N4412A	129	3025	244	2027	
2N4413	159	3466	82	2032	
2N4413A	159	3466	82	2032	
2N4414	129	3025	244	2027	
2N4414A	129	3025	244	2027	
2N4415	159	3466	82	2032	
2N4415A	159	3466	82	2032	
2N4416	452	3112	FET-2	2028	2N4416
2N4416A	452	3112		2028	2N4416A
2N4417	452	3112	FET-2	2028	2N4416
2N4418	123AP	3452/108	86	2038	
2N4419	123AP	3452/108	86	2038	
2N441BLU	213		4		
2N442	213	3012/105	4		
2N4420	123AP		20	2051	
2N4421	123AP	3118	20	2051	
2N4422	123AP		20	2051	
2N4423			21	2034	
2N4424	289A	3444/123A	20	2051	
2N4425	192A	3124/289	63	2030	
2N4427	346	9038	243	2030	2N4427
2N4428			28	2017	2N4428
2N4429			28	2017	
2N443	213	3012/105	4		
2N4430			28	2017	
2N4432	123A	3195/311	20	2051	
2N4432A	123A	3195/311	20	2051	
2N4433	161	3018	11	2015	
2N4434	161	3117	39	2015	
2N4435	161	3117	39	2015	
2N4436	123A	3444	20	2051	
2N4437	123A	3195/311	277	2051	
2N444	101	3861	8	2002	
2N444#				2002	
2N444#				2002	
2N4440			28	2017	2N4440
2N4441	5442	3634/5444	MR-4		2N4441
2N4442	5444	3634			2N4442
2N4443	5446	3635	MR-4		2N4443
2N4444	5448	3636			2N4444

Industry Standard No.	ECG	SK	GE	RS 276-	MOTOR.
2N4445					MFE2012
2N4446					MFE2012
2N4447					MFE2012
2N4448					MFE2012
2N4449	311	3452/108	86	2038	2N4449
2N444A	103	3861/101	8	2002	
2N445	101	3861	8	2002	
2N445#				2002	
2N445#				2002	
2N4450	123A	3444	20	2051	2N5582
2N4451	106	3984	21	2034	
2N4452	159	3466	82	2032	
2N4453	106	3118	21	2034	2N4453
2N445A	103	3861/101	8	2002	
2N446	101	3861	8	2002	
2N446A	103	3861/101	8	2002	
2N447	101	3861	8	2002	
2N447#				2002	
2N447A	103	3861/101	8	2002	
2N447B	103	3861/101	8	2002	
2N448	101	3011	8	2002	
2N449	101	3011	8	2002	
2N44A	102A	3004	2	2007	
2N45	102A		53	2007	
2N450	102A	3005	1	2007	
2N456	104	3009	16	2006	
2N456A	104	3009	16	2006	
2N456B	104	3009	16	2006	
2N457	104	3719	16	2006	
2N457A	104	3719	16	2006	
2N457B	104	3719	16	2006	
2N458	104	3009	16	2006	
2N458A	104	3009	16	2006	
2N458B	121	3009	239	2006	
2N459	179	3012/105	239	2006	
2N459A	179	3009	239	2006	
2N45A	102A		53	2007	
2N46	102	3004	2	2007	
2N460	102	3004	2	2007	
2N461	102	3004	2	2007	
2N462	102	3004	2	2007	
2N464	102	3004	2	2007	
2N465	102	3004	2	2007	
2N466	102	3003	2	2007	
2N467	102	3003	2	2007	
2N468	102	3004	2	2007	
2N469			53	2007	
2N47	102	3004	2	2007	
2N470	123	3124/289	20	2051	2N2221
2N471	123		60	2015	2N2221
2N471A	123	3124/289	20	2051	2N2221
2N472	123A	3444	20	2051	2N2221
2N472A	123A	3444	20	2051	2N2221
2N473	123	3124/289	20	2051	2N2221
2N474	123	3444	20	2051	2N2221
2N474A	123	3444	20	2051	2N2221
2N475	123A	3444	20	2051	2N2221
2N475A	123A	3444	20	2051	2N2221
2N476	123	3124/289	20	2051	2N2221
2N476A			11	2015	
2N477	123	3124/289	20	2051	2N2221
2N478	123	3124/289	20	2051	2N2221
2N479	123	3124/289	20	2051	2N2221
2N479A	123	3124/289	20	2051	2N2221
2N48	102	3004		2007	
2N480	123A	3124/289	20	2051	2N2221
2N480A	123A	3444	20	2051	2N2221
2N480B	128	3444/123A	20	2051	
2N481	100	3005	1	2007	
2N482	100	3005	1	2007	
2N483	100	3005	1	2007	
2N483-6M	102	3004	2	2007	
2N483B	102	3004	2	2007	
2N484	100	3005	1	2007	
2N485	100	3005	1	2007	
2N4851	6401				2N4851
2N4852	6409				2N4852
2N4853	6409				2N4853
2N4854					2N4854
2N4854JAN					2N4854JAN
2N4854JTX					2N4854JTX
2N4854TXV					2N4854TXV
2N4855					2N4855
2N4856,J					2N4856,J
2N4856A	466				2N4856A
2N4857,J					2N4857,J
2N4857A	466				2N4857A
2N4858,J					2N4858,J
2N4858A					2N4858A
2N4859,J					2N4859,J
2N486	100	3005	1	2007	
2N4860,J					2N4860,J
2N4860A	466				2N4860A
2N4861,J					2N4861,J
2N4861A	466				2N4861A
2N4867	312	3112	FET-1	2028	2N4220A
2N4867A	312	3112		2028	2N4220A
2N4868	312	3112	FET-1	2028	2N4220A
2N4868A	312	3112		2028	2N4220A
2N4869	459	3112	FET-1	2028	2N4221
2N4869A	459	3112		2028	2N4221A
2N486B	100	3005	1	2007	
2N487	100	3004/102A	2	2007	
2N4870	6410		2N2160		2N4870
2N4871	6410		2N2160		2N4871
2N4872	106	3118	21	2034	2N4208
2N4873	311	3452/108	86	2038	
2N4877	282	3748	66	2048	2N5337
2N4878					2N2060
2N4879					2N2060
2N4888		3715	221		
2N4889	159	3715	221		
2N4890	129		244	2027	2N4890
2N4891					MU4891
2N4892					MU4892
2N4893	6401				MU4893
2N4894					MU4894
2N4898	218	3257	234		2N4898
2N4899	218	3083/197	234		2N4899
2N48A	102	3004		2007	
2N49	102	3004		2007	
2N4900	218		234		2N4900
2N4901	219		234		2N4901
2N4902	219		234		2N4902
2N4903	219		234		2N4903
2N4904	281	3173/219	234		2N4904
2N4905	281	3173/219	234		2N4905
2N4906	281	3437/180	234		2N4906
2N4907	219		234		2N3791
2N4908	219		234		2N3791
2N4909	219		234		2N3792
2N4910	384	3131A/369	246	2020	2N3054
2N4911	384	3131A/369	246	2020	2N3054
2N4912	384	3131A/369	246	2020	2N4912
2N4913	280	3027/130	14	2041	2N4913
2N4914	280	3027/130	14	2041	2N4914
2N4915	280	3027/130	14	2041	2N4915
2N4916	159	3715	82	2032	
2N4917	159	3466	82	2032	
2N4918	185	3191	58	2025	2N4918
2N4919	185	3191	58	2025	2N4919
2N4920	185S	3191/185	58	2025	2N4920
2N4921	184	3190	57	2017	2N4921
2N4922	184	3190	57	2017	2N4922
2N4923	184	3190	57	2017	2N4923
2N4924	282	3045/225	40	2012	2N4924
2N4925	154	3045/225	40	2012	2N4925
2N4926	154	3045/225	40	2012	2N4926
2N4927	154	3045/225	40	2012	2N4927
2N4928	129	3025	244	2027	2N4928
2N4929			228		2N4929
2N4930	397		228		2N4930
2N4930JAN					2N4930JAN
2N4930JTX					2N4930JTX
2N4931	397		228		2N4931
2N4931JAN					2N4931JAN
2N4931JTX					2N4931JTX
2N4931S	397		228		
2N4932					2N3926
2N4933					2N3927
2N4934	316	3452/108	86	2038	
2N4935	311	3452/108	86	2038	
2N4936	311	3452/108	86	2038	
2N4937			21	2034	2N4937
2N4938			21	2034	2N4938
2N4939					2N4939
2N4940			21	2034	2N4040
2N4941			21	2034	2N4941
2N4942					2N4942
2N4943	128	3024	243	2030	
2N4944	128	3024	243	2030	
2N4945	194	3275	220		
2N4946	128	3024	243	2030	
2N4948	6409				2N4948
2N4949					2N4949
2N495	106	3114/290	21	2034	
2N495/18	106	3984	21	2034	
2N4950			20	2051	
2N4951	123AP	3444/123A	20	2051	
2N4952	123AP	3444/123A	20	2051	
2N4953	123AP	3444/123A	20	2051	
2N4954	289A	3444/123A	20	2051	
2N4955					2N2916
2N4956					2N2918
2N4957	395				2N4957
2N4958	395				2N4959
2N4959	395				2N4959
2N496	106	3114/290	21	2034	
2N4960	128	3024	243	2030	2N3499
2N4961	128	3024	243	2030	2N3499
2N4962	128	3024	243	2030	2N3499
2N4963	128	3024	243	2030	2N3499
2N4964	128	3024	243	2030	
2N4965	159	3466	82	2032	
2N4966	123A	3444	20	2051	
2N4967	199	3444/123A	20	2051	
2N4968	123A	3444	20	2051	
2N4969	123A	3444	20	2051	
2N497	128		243	2030	2N3020
2N4970	123A	3444	20	2051	
2N4971	159	3466	82	2032	
2N4971A		3114	89	2023	
2N4972	159	3466	82	2032	
2N4972A		3114	89	2023	
2N4974					2N6521
2N4975					2N6522
2N4976			28	2017	
2N4977					MFE2012
2N4978					MFE4856
2N4979					MFE4857
2N497A	128		243	2030	
2N498	128		243	2030	2N3498
2N4980			82	2032	
2N4981			82	2032	
2N4982	159	3466	82	2032	
2N4982A		3114	89	2023	
2N4987	6404		2N4987		
2N499	126	3006/160	52	2024	
2N4990	6404	3638/5402	2N4990		
2N4991	6403				MBS4991
2N4992	6403		2N4992		MBS4992
2N4993	6403				MBS4991
2N4994	123AP	3444/123A	20	2051	
2N4995	123AP	3444/123A	20	2051	
2N4996	108	3452	86	2038	
2N4997	108	3452	86	2038	
2N4998					2N5347
2N4999					2N6187
2N499A	126	3006/160	52	2024	
2N50	102	3004		2007	
2N500	126	3008	52	2024	
2N5000					2N5347
2N5001					2N6187
2N5002					2N5347
2N5003					2N6187
2N5004					2N5347
2N5005					2N6187
2N500BLU	126	3006/160	52	2024	
2N500RED	126	3006/160	52	2024	
2N500WHT	126	3006/160	52	2024	
2N501	126	3006/160	52	2024	
2N501/18	126	3006/160	52	2024	
2N5016					2N5016
2N5018					2N3993
2N5019					2N3994
2N501A	126	3006/160	52	2024	
2N502	126	3006/160	52	2024	
2N5020					2N5265
2N5021					2N2608
2N5022	129	3025	244	2027	2N5022
2N5023	129	3025	244	2027	2N5023
2N5024	316	3039	86	2038	
2N5027			210	2051	
2N5028			210	2051	
2N502A	126	3006/160	52	2024	
2N502B	126	3006/160	52	2024	
2N502BRN				2005	
2N502BRN				2005	
2N502GRN				2005	
2N502GRN				2005	
2N502RED				2005	
2N502RED				2005	
2N503	126	3005	1	2007	
2N5030			20	2051	
2N5031	161	3039/316	11	2015	2N5031
2N5032	161	3039/316	11	2015	2N5032
2N5033					2N5460
2N5034	390	3036	255	2041	2N3055

Industry Standard No.	ECG	SK	GE	RS 276-	MOTOR.
2N5035	390		255	2041	2N3055
2N5036	390	3027/130	255	2041	2N3055
2N5037	390		255	2041	2N3055
2N5038	327	3561			2N5038
2N5038-1	327	3561			
2N5039	327	3561			2N5039
2N5039-1	327	3561			
2N504	126	3008	52	2024	
2N5040	129		244	2027	
2N5041	129		244	2027	
2N5042	129		244	2027	
2N5045	312	3112	FET-1	2028	
2N5046	312	3112	FET-1	2028	
2N5047	312	3112	FET-1	2028	
2N505	100	3005	2	2007	
2N5050	384	3131A/369	267		2N5050
2N5051	384	3440/291			2N5051
2N5052	384	3131A/369	267		2N5052
2N5053	316	3452/108	86	2038	2N6305
2N5054	316	3452/108	86	2038	2N6304
2N5055		3118	89	2023	
2N5056					2N4209
2N5057					2N4209
2N5058	154	3045/225	40	2012	2N5058
2N5058S			214	2038	
2N5059	154	3045/225	40	2012	2N5059
2N5059S			214	2038	
2N506	126	3004/102A	53	2007	
2N5060	5400	3950	MR-5		2N5060
2N5061	5401	3638/5402	MR-5		2N5061
2N5062	5402	3638	MR-5		2N5062
2N5063	5403	3627/5404	MR-5		2N5063
2N5064	5404	3627			2N5064
2N5066			210	2051	
2N5067	280	3027/130	262	2041	2N5067
2N5068	280	3027/130	262	2041	2N5068
2N5069	280		262	2041	2N5069
2N507	103A	3010	59	2002	
2N5070					2N5070
2N5071					2N5071
2N5073	154		222		
2N5078	452		FET-2	2035	U308
2N5078A		3116	FET-2	2035	
2N5079			28	2017	
2N508	102	3003	2	2007	
2N5080			28	2017	
2N5081	123A	3444	20	2051	
2N5082	123A	3444	20	2051	
2N5083					2N5347
2N5084					2N5347
2N5085					2N5347
2N5086	159	3466	82	2032	
2N5086A		3114	89	2023	
2N5087	159	3466	82	2032	
2N5087A		3114	89	2023	
2N5088	123AP	3250/315	20	2051	
2N5089	123AP	3444/123A	20	2051	
2N508A	102	3004	2	2007	
2N509	160		245	2004	
2N5090					2N5090
2N51	102	3004		2007	
2N5102					2N5071
2N5103			17	2051	2N4416
2N5103A		3116	FET-2	2035	
2N5104					2N4416
2N5104A		3116	FET-2	2035	
2N5105					2N4416
2N5105A		3116	FET-2	2035	
2N5106			47	2030	
2N5107	123A	3444	20	2051	
2N5108	278	3024/128	261		2N5108
2N5109	278	3218	261		2N5109
2N511	104	3009	16	2006	
2N5110	129	3258	244	2027	
2N5111	129	3025	244	2027	
2N5112			69	2049	
2N5114					2N3993
2N5115					2N3993
2N5116					2N3994
2N5117					2N5796
2N5118					2N5796
2N5119					2N5796
2N511A	104	3009	16	2006	
2N511B	104	3009	16	2006	
2N512	104	3009	16	2006	
2N5120			18	2030	2N5796
2N5121					2N5796
2N5122					2N5796
2N5123					2N5796
2N5124					2N5796
2N5125					2N5796
2N5126	161	3124/289	39	2015	
2N5127	123A	3444	20	2051	
2N5128	123A	3444	20	2051	
2N5129	123A	3024/128	243	2030	
2N512A	104	3009	16	2006	
2N512B	104	3009	16	2006	
2N513	104	3009	16	2006	
2N5130	161	3018	39	2015	
2N5131	123AP	3452/108	86	2038	
2N5132	161	3124/289	11	2015	
2N5133	161	3018	39	2015	
2N5134	123A	3444	20	2051	
2N5135	123	3124/289	243	2030	
2N5136	123	3124/289	243	2030	
2N5137	123A	3444	20	2051	
2N5138	159	3715	82	2032	
2N5138A		3715	89	2023	
2N5139	159	3715	82	2032	
2N5139A		3715	89	2023	
2N513A	104	3009	16	2006	
2N513B	104	3009	16	2006	
2N514	179	3009	239	2006	
2N5140	106	3118	21	2034	
2N5141	106	3118	21	2034	
2N5142	159	3466	82	2032	
2N5142A		3114	89	2023	
2N5143	159	3466	82	2032	
2N5143A		3114	89	2023	
2N5144			20	2051	
2N5146					2N5146
2N5147					2N6191
2N5148					2N5336
2N5149					2N6191
2N514A	179	3009	239	2006	
2N514B	179	3009	239	2006	
2N515	103	3011	59	2002	
2N5150					2N5336
2N5151					2N6191
2N5152					2N5337
2N5153					2N6191
2N5154					2N5337
2N5156	179		25		
2N5157	283				2N6545
2N5158					MFE2012
2N5159					MFE2012
2N516	103	3010	59	2002	
2N5160	129	3025	291	2027	2N5160
2N5161			69	2049	2N5161
2N5162					2N5162
2N5163	312	3116	FET-1	2035	2N5458
2N5163A		3116	FET-2	2035	
2N5164	5514	3613			2N5164
2N5165	5514	3613			2N5165
2N5166	5515	6615			2N5166
2N5167	5516	6616			2N5167
2N5168	5521	6621	MR-3		2N5168
2N5169	5524	3504/5508	MR-3		2N5169
2N517	103	3011	59	2002	
2N5170		3504	MR-3		2N5170
2N5171	5529	6629			2N5171
2N5172	289A	3124/289	20	2051	
2N5174	289A	3045/225	40	2012	
2N5175	194	3045/225	40	2012	
2N5176	154	3045/225	40	2012	
2N5179	316	3039	280		2N5179
2N518	100	3005	1	2007	
2N5180	161	3018	86	2038	2N5179
2N5181	161	3716	39	2015	
2N5182	161	3716	39	2015	
2N5183	123	3124/289	20	2051	
2N5184	154	3040	40	2012	
2N5185		3040	235	2012	
2N5186	123A	3444	20	2051	
2N5187	123A	3444	20	2051	
2N5188	128	3024	243	2030	
2N5189	282	3529	264		2N2905A
2N519	100	3005	1	2007	
2N5190	184	3190	57	2017	2N5190
2N5191	184	3054/196	57	2017	2N5191
2N5192	184	3054/196	57	2017	2N5192
2N5193	185	3083/197	58	2025	2N5193
2N5194	185	3083/197	58	2025	2N5194
2N5195	185	3083/197	58	2025	2N5195
2N519A	100	3005	1	2007	
2N52	102	3004		2007	
2N520	100	3005	1	2007	
2N5200	278	3452/108	86	2038	
2N5201	278	3020/123	1	2007	
2N5202	384	3622	246	2020	2N5428
2N5204	5547				2N3899
2N5205					MCR3935-10
2N5208	106	3118	21	2034	
2N5209	123AP	3444/123A	20	2051	
2N520A	100	3005	1	2007	
2N521	100	3005	1	2007	
2N5210	123AP	3444/123A	20	2051	
2N5211	128	3024	243	2030	
2N5219	123AP	3444/123A	20	2051	
2N521A	100	3005	1	2007	
2N522	100	3003	1	2007	
2N5220	123AP	3122	268	2038	
2N5221	290A	3114/290	269	2032	
2N5222	229	3039/316	11		
2N5223	123AP	3444/123A	20	2051	
2N5224	123AP	3444/123A	20	2051	
2N5225	123AP	3444/123A	20	2051	
2N5226	159	3466	82	2032	
2N5226A		3114	89	2023	
2N5227	159	3466	82	2032	
2N5227A		3114	89	2023	
2N5228	106	3984	21	2034	
2N5229			22	2032	2N5229
2N522A	100	3005	1	2007	
2N523	100	3003	1	2007	
2N5230	161	3716	82	2032	2N5230
2N5231			82	2032	2N5231
2N5232	199	3444/123A	20	2051	
2N5232A	199	3444/123A	20	2051	
2N5233	297	3024/128	243	2030	
2N5234	297	3024/128	243	2030	
2N5235	297	3024/128	243	2030	
2N5239	94	3268	36		2N6306
2N523A	100	3005	1	2007	
2N524	102	3004	2	2007	
2N5240	94	3467/283	36		2N6544
2N5241	283	3467	36		2N5241
2N5242	129		244	2027	
2N5243	129		244	2027	
2N5245	312	3834/132	FET-2	2035	2N5245
2N5245A		3834	FET-2	2035	
2N5246	312	3834/132	FET-2	2035	2N5246
2N5246A		3834	FET-2	2035	
2N5247	312	3834/132	FET-2	2035	2N5247
2N5247A		3834	FET-2	2035	
2N5248	312	3834/132	FET-1	2035	MPF102
2N5248A		3834	FET-2	2035	
2N5249	199	3024/128	212	2010	
2N5249A	199	3024/128	62	2010	
2N524A	102	3004	2	2007	
2N525	102	3004	2	2007	
2N5254					2N3806
2N5255					2N3808
2N5258			FET-2	2035	
2N525A	102	3004	2	2007	
2N526	102	3004	2	2007	
2N5262	282	3529	264		MRF531
2N5264	385		35		2N6249
2N5265					2N5265
2N5266					2N5266
2N5267		3112	FET-1	2028	2N5267
2N5268		3112	FET-1	2028	2N5268
2N5269		3112	FET-1	2028	2N5269
2N526A	102	3004	2	2007	
2N527	102	3004	2	2007	
2N5270		3112	FET-1	2028	2N5270
2N5271					2N5271
2N5272	311	3452/108	86	2038	
2N5273	5683				2N6160
2N5274	5685	3521/5686			2N6161
2N5275	5697	3522			2N6162
2N5277	312	3112	FET-1	2028	2N3822
2N5278					2N5364
2N5278A		3116	FET-2	2035	
2N5279			32		2N3742
2N527A	102	3004	2	2007	
2N5284					2N5347
2N5285					2N5347
2N5286					2N6189
2N5287					2N6189
2N529	100	3005	1	2007	
2N529/N			5	2002	
2N529/P			50	2004	
2N5292	311	3452/108	86	2038	
2N5293	152	3054/196	66	2048	2N6123
2N5294	196	3054	66	2048	2N6123
2N5295	152	3054/196	66	2048	2N6121
2N5296	152	3054/196	66	2048	2N6121

Industry Standard No.	ECG	SK	GE	RS 276-	MOTOR.
2N5297	196	3054	66	2048	2N6122
2N5298	152	3054/196	66	2048	2N6122
2N53	102	3004		2007	
2N530	100	3005	1	2007	
2N530/N			5	2002	
2N530/P			50	2004	
2N5300			18	2030	
2N5301	181		75	2041	2N5301
2N5302	181		75	2041	2N5302
2N5302JAN					2N5302JAN
2N5302JTX					2N5302JTX
2N5302JTXV					2N5302JTXV
2N5303	181		75	2041	2N5303
2N5303JAN					2N5303JAN
2N5303JTX					2N5303JTX
2N5303JTXV					2N5303JTXV
2N5304			214	2038	
2N5305	172A	3156	64		
2N5306	172A	3156	64		
2N5307	172A	3156	64		
2N5308	172A	3156	64		
2N5308A	172A		2N5308A		
2N5309	289A	3024/128	243	2030	
2N531	100	3005	1	2007	
2N531/N			5	2002	
2N531/P		3123	80		
2N5310	289A	3024/128	243	2030	
2N5311	289A	3024/128	243	2030	
2N532	100	3005	1	2007	
2N532/N			5	2002	
2N532/P			50	2004	
2N5320	282	3512	46	2030	
2N5320HS	192	3024/128	63	2030	
2N5321	282	3765/195A	46		
2N5322	129	3513	46	2027	
2N5326					2N5347
2N533	100	3005	1	2007	
2N533/N			5	2002	
2N533/P			50	2004	
2N5333					2N5680
2N5334			66	2048	2N4877
2N5335					2N5337
2N5336					2N5336
2N5337					2N5337
2N5338					2N5338
2N5339					2N5339
2N534	102A	3123	53	2007	
2N5344		3624	D44R1		2N5344
2N5345			D44R3		2N5345
2N5346	96				2N5346
2N5347	96				2N5347
2N5348	96				2N5348
2N5349	96				2N5349
2N535	102A	3003	53	2007	
2N5352	106	3984	21	2034	
2N5354	290A	3466/159	82	2032	
2N5354A		3114	89	2023	
2N5355	290A	3466/159	82	2032	
2N5355A		3114	89	2023	
2N5356	159	3466	82	2032	
2N5356A		3114	89	2023	
2N5358		3112	FET-1	2028	2N5358
2N5359		3112	FET-1	2028	2N5359
2N535A	102A	3003	53	2007	
2N535B	102A	3003	53	2007	
2N536	102A	3003	53	2007	
2N5360			FET-1	2028	2N5360
2N5360A		3116	FET-2	2035	
2N5361		3116	FET-1	2028	2N5361
2N5362		3116	FET-1	2028	2N5362
2N5363		3116	FET-1	2028	2N5363
2N5364		3116	FET-1	2028	2N5364
2N5365	159	3466	82	2032	
2N5365A		3114	89	2023	
2N5366	159	3466	82	2032	
2N5366A		3114	89	2023	
2N5367	159	3466	82	2032	
2N5367A		3114	89	2023	
2N5368	123AP	3444/123A	20	2051	
2N5369	123AP	3444/123A	20	2051	
2N537	160		245	2004	
2N5370	123AP	3444/123A	20	2051	
2N5371	123AP	3444/123A	20	2051	
2N5372	159	3466	82	2032	
2N5372A		3114	89	2023	
2N5373	159	3466	82	2032	
2N5373A		3114	89	2023	
2N5374	159	3466	82	2032	
2N5374A		3114	89	2023	
2N5375	159	3466	82	2032	
2N5375A		3114	89	2023	
2N5378	159	3466	82	2032	
2N5378A		3114	89	2023	
2N5379	159	3466	82	2032	
2N5379A		3114	89	2023	
2N538	121	3009	239	2006	
2N5380	123AP	3444/123A	20	2051	
2N5381	123AP		20	2051	
2N5382	159	3466	82	2032	
2N5382A		3114	89	2023	
2N5383	159	3466	82	2032	
2N5383A		3114	89	2023	
2N5384					2N6187
2N5385					2N6187
2N5386					2N5038
2N5387					2N6546
2N5388					2N6546
2N5389					2N6546
2N538A	121	3009	239	2006	
2N539	104	3009	16	2006	
2N5391		3112	FET-1	2028	2N4220A
2N5392		3116	FET-1	2028	2N4220A
2N5393		3116	FET-1	2028	2N4221A
2N5394		3116	FET-1	2028	2N4221A
2N5395		3116	FET-1	2028	2N4221A
2N5396		3116	FET-1	2028	2N4393
2N5397	452	3116	FET-2	2035	U309
2N5398	452	3116	FET-2	2035	U309
2N5399	311	3452/108	86	2038	
2N539A	104	3009	16	2006	
2N54	102	3004		2007	
2N540	104	3009	16	2006	
2N5400	288	3715	223		
2N5401	288	3715	223		
2N5404					2N6191
2N5405					2N6192
2N5406					2N6191
2N5407					2N6193
2N5408					2N6187
2N5409					2N6189
2N540A	104	3009	16	2006	
2N541	123	3444	20	2051	
2N5410					2N6187
2N5411					2N6189

Industry Standard No.	ECG	SK	GE	RS 276-	MOTOR.
2N5415	397	3053			2N5415
2N5416	397	3528			2N5416
2N5417	123A		20	2051	
2N5418	289A	3444/123A	20	2051	
2N5419	289A	3444/123A	20	2051	
2N542	128	3444/123A	20	2051	
2N5420	289A	3444/123A	20	2051	
2N5421	346	3024/128	243	2030	2N4427
2N5422			28	2017	MRF606
2N5423			28	2017	2N3926
2N5424			66	2048	2N3927
2N5427					2N5427
2N5428					2N5428
2N5429					2N5429
2N542A	123	3124/289	20	2051	
2N543	123A	3444	20	2051	
2N5430					2N5430
2N5431					2N5431
2N5432					MFE2012
2N5433					MFE2012
2N5434					MFE2012
2N543A	123A	3444	20	2051	
2N544	126	3008	52	2024	
2N544/33	126	3008	52	2024	
2N5441					2N5441
2N5442					2N5442
2N5443					2N5443
2N5444	5693	3652			2N5444
2N5445	5695	3509			2N5445
2N5446	5697	3522			2N5446
2N5447	159	3466	82	2032	2N5556
2N5447A		3114	89	2023	
2N5448	159	3466	82	2032	
2N5448A		3114	89	2023	
2N5449	123A	3024/128	243	2030	
2N545	128		63	2030	
2N5450	128	3024	243	2030	
2N5451	128	3024	243	2030	
2N5452	461	3112	FET-1	2028	
2N5453	461	3112	FET-1	2028	
2N5454	461	3112	FET-1	2028	
2N5456			20	2051	
2N5457	457	3116	FET-1	2028	2N5457
2N5458	457	3112	FET-1	2028	2N5458
2N5459	459	3116	FET-1	2028	2N5459
2N546	123A	3124/289	20	2051	
2N5460	326	3746			MRF454
2N5461	326	3746			2N5461
2N5462	326	3746			2N5462
2N5463	326	3746			2N5463
2N5464	326				2N5464
2N5465					2N5465
2N5466	283		36		2N6545
2N5467	283		38		2N6545
2N547	128	3124/289	20	2051	
2N5471			D40E1		2N5471
2N5472					2N5472
2N5473					2N5473
2N5474					2N5474
2N5475					2N5475
2N5476					2N5476
2N5477	74				2N5347
2N5478	74				2N5347
2N5479	74				2N5349
2N548	123A	3124/289	20	2051	
2N5480	74				2N5349
2N5481			28	2017	
2N5482			28	2017	
2N5483			66	2048	
2N5484	312	3116	FET-2	2035	2N5484
2N5485	451	3116	FET-2	2035	2N5485
2N5485-1	451		FET-2	2035	
2N5486	312	3116	FET-2	2035	2N5486
2N5487-1			214	2038	
2N5487-3			214	2038	
2N5489			28	2017	
2N549	128	3124/289	20	2051	
2N5490	196	3893/152	241	2020	2N6290
2N5491	196	3893/152	241	2020	2N6290
2N5492	196	3893/152	66	2048	2N6292
2N5493	196	3893/152	241	2020	2N6292
2N5494	196	3893/152	241	2020	2N6290
2N5495	196	3893/152	241	2020	2N6290
2N5496	196	3054	241	2020	2N6292
2N5497	196	3054	241	2020	2N6292
2N55	102	3004		2007	
2N550	123A	3124/289	20	2051	
2N5508					2N5428
2N551	123A	3124/289	20	2051	
2N552	123	3124/289	20	2051	
2N5525	172A	3156	64		
2N553	121	3009	239	2006	
2N5539	70				2N6379
2N554	121	3009	239	2006	
2N5543	459	3112	FET-1	2028	2N3822
2N5544	459	3112	FET-1	2028	2N3822
2N5548					MFE3003
2N5549	312				2N4093
2N555	121	3009	239	2006	
2N5550	287	3275/194	222		
2N5551	287	3433	222		
2N5555	312		FET-2	2035	
2N5556		3112	FET-1	2028	
2N5557		3112	FET-1	2028	
2N5558		3112	FET-1	2028	
2N5559					2N5633
2N556	101	3861	8	2002	
2N5561		3112	FET-1	2028	
2N5562		3112	FET-1	2028	
2N5563		3112	FET-1	2028	
2N5567					2N5567
2N5568					2N5568
2N5569	5673	3508/5675			2N5569
2N557	101	3861	8	2002	
2N5570	5675	3508			2N5570
2N5571					2N5571
2N5572					2N5572
2N5573	5673	3508/5675			2N5573
2N5574	5675	3508			2N5574
2N5575					2N5685
2N5578					2N5685
2N558	101	3861	8	2002	
2N5581			20	2051	2N5581
2N5582			20	2051	2N5582
2N5583					2N5583
2N5589	347		286		2N5589
2N559	160		245	2004	
2N5590	349	3176/350	295		2N5590
2N5591	351	3177	288		2N5591
2N5592	459	3116	FET-2	2035	
2N5593	459	3116	FET-2	2035	
2N5594	459	3116	FET-2	2035	
2N5597			69	2049	
2N5598	175	3131A/369	246	2020	2N5428

Industry Standard No.	ECG	SK	GE	RS 276-	MOTOR.
2N56	102	3004		2007	
2N560	128		20	2051	
2N5600	175	3131A/369	246	2020	2N5428
2N5601			246	2020	
2N5602	175	3131A/369	246	2020	2N5428
2N5604	175	3131A/369	246	2020	2N5430
2N5606			66	2048	2N5428
2N5608			28	2017	
2N561	179	3009	239	2006	
2N5610			28	2017	2N5428
2N5612			28	2017	2N5430
2N5614	328	3895	14	2041	2N3448
2N5616	328	3895	75	2041	2N3448
2N5618	328	3895	3	2006	2N3448
2N5620	328	3895			
2N5621	180		74	2043	
2N5622	328	3895	14	2041	
2N5623	180		74	2043	
2N5624	328	3895			
2N5625	180		74	2043	
2N5626	328	3895			
2N5628	328	3895			
2N5629	284	3836	75	2041	2N5629
2N563	102A	3004	2	2007	
2N5630	284	3621	265	2047	2N5630
2N5631	284	3836			2N5631
2N5632	385	3260	75	2041	2N5632
2N5633	385	3438	35		2N5633
2N5634	385	3260	265	2047	2N5634
2N5635					2N5635
2N5636					2N5636
2N5637			66	2048	2N5637
2N564	102	3004	2	2007	
2N5641	357				2N5641
2N5642	359		66	2048	2N5642
2N5643	360				2N5643
2N5644	362		28	2017	2N5644
2N5645	363		28	2017	2N5645
2N5646	364		66		2N5646
2N5647					MFE2093
2N5648		3112	FET-1	2028	MFE2094
2N5649		3112	FET-1	2028	MFE2094
2N565	102A	3004	2	2007	
2N5650	316	3117	39	2015	
2N5651	316	3117	39	2015	2N6235
2N5652	316	3039	39	2015	
2N5653					2N5653
2N5654			279		2N5654
2N5655	157	3103A/396	232		2N5655
2N5656	157	3103A/396	232		2N5656
2N5657					2N5657
2N566	102	3004	2	2007	
2N5660	369	3131A	32		2N6233
2N5664					2N6233
2N5665					2N6235
2N5668	457	3116	FET-1	2028	2N5668
2N5669		3116	FET-2	2035	2N5669
2N567	102A	3835/103A	59	2002	
2N5670		3116	FET-2	2035	2N5670
2N5671	327	3945			2N6338
2N5672	327	3945			2N6339
2N5678					2N6378
2N5679	323	3528/397			2N5679
2N568	102	3004	2	2007	
2N5680	323	3528/397			2N5680
2N5681	128	3748/282	264		2N5681
2N5682	282	3024/128	264		2N5682
2N5683					2N5683
2N5683JAN					2N5683JAN
2N5683JTX					2N5683JTX
2N5683JTXV					2N5683JTXV
2N5684					2N5684
2N5684JAN					2N5684JAN
2N5684JTX					2N5684JTX
2N5684JTXV					2N5684JTXV
2N5685	387				2N5685
2N5685JAN					2N5685JAN
2N5685JTX					2N5685JTX
2N5685JTXV					2N5685JTXV
2N5686					2N5686
2N5686JAN					2N5686JAN
2N5686JTX					2N5686JTX
2N5686JTXV					2N5686JTXV
2N5687	346		219		2N4427
2N5688			28	2017	MRF475
2N5689	337		66	2048	2N5847
2N569	102A	3003	2	2007	
2N5690	338		66	2048	2N5848
2N5691					2N5849
2N5697	486		28	2017	MRF515
2N5698	362		28	2017	2N5944
2N5699			28	2017	2N5847
2N57				2006	
2N570	102	3003	2	2007	
2N5700			66	2048	
2N5701			66	2048	
2N5702	311		320		
2N5703			28	2017	
2N5704			66	2048	
2N5705			66	2048	
2N5707					MRF401
2N5708					MRF466
2N571	102A	3003	1	2007	
2N5710			219		2N4073
2N5711	357		D40E1		2N5641
2N5712	359		66	2048	2N5642
2N5713	359		66	2048	2N5642
2N5714					2N5643
2N5716		3112	FET-1	2028	
2N5717		3112	FET-1	2028	
2N5718	457	3112	FET-1	2028	
2N5719					MCR103
2N572	100	3003	1	2007	
2N5720					MCR104
2N5721					MCR120
2N5722					MCR100-5
2N5723					MCR100-6
2N5724					MCR103
2N5725					MCR104
2N5726					MCR120
2N5727					MCR100-5
2N5728					MCR100-6
2N5729					2N5337
2N573	102	3004	2	2007	
2N5730			19	2041	2N5347
2N5733	70				2N6274
2N5734					2N6338
2N5735			210	2051	
2N5736			210	2051	
2N5737					2N5878
2N5738	281		74	2043	2N6229
2N5739					2N5878
2N573BRN	102	3004	2	2007	
2N5730RN	102	3004	2	2007	
2N573RED	102	3004	2	2007	
2N574		3004	4		
2N5740					2N6229
2N5741	180		74	2043	2N5883
2N5742	180		74	2043	2N6029
2N5743					2N5883
2N5744					MJ4502
2N5745	180		74	2043	2N5745
2N5745JAN					2N5745JAN
2N5745JTX					2N5745JTX
2N5745JTXV					2N5745JTX'
2N574A		3004	4		
2N575	213	3012/105	4		
2N5754	5640	3506/5642			SC136A
2N5755	5641	3506/5642			SC136B
2N5756	5642	3506			SC136C
2N5757	5643	3519			SC136D
2N5758	385	3438	35		2N5758
2N5759	385	3438	35		2N5759
2N575A	213	3012/105	4		
2N576	101	3861	8	2002	
2N5760	385	3438	35		2N5760
2N5763			21	2034	2N2907A
2N5764			28	2017	
2N5765			66	2048	
2N5766			28	2017	
2N5767			28	2017	
2N5768			66	2048	
2N5769	123AP	3444/123A	20	2051	MPS2369
2N576A	101	3861	8	2002	
2N5770	108		214	2038	
2N5771			221		MPS2369
2N5772	123AP		214	2038	MPS3646
2N5773					MRF5174
2N5774					2N5636
2N5775					MRF5177
2N5777	3035		2N5777		2N5777
2N5778	3035		2N5778		2N5778
2N5779	3035		2N5779		2N5779
2N578	100	3005	1	2007	
2N5780	3035		2N5780		2N5780
2N5782		3025	29	2025	
2N5783		3258	29	2025	
2N5784	282	3024/128	18	2030	
2N5785	282	3024/128	216	2053	
2N5786		3024	46		
2N5787	5400	3950			
2N5788	5401	3638/5402			
2N5789	5402	3638			
2N579	100	3005	1	2007	
2N5790	5404	3627			
2N5793					2N5793
2N5794	81				2N5794
2N5795					2N5795
2N5796	82				2N5796
2N58	102			2007	
2N580	100	3005	1	2007	
2N5804	283	3559/162	36		2N6306
2N5805	283	3559/162	36		2N6542
2N5806	5543				2N6160
2N5807	5545				2N6161
2N5808	5546				2N6162
2N5809	5547				2N6162
2N581	100	3005	1	2007	
2N5810	123AP		20	2051	
2N5811	159		48		
2N5812	123AP		47	2030	
2N5814	128		20	2051	
2N5815		3118	48		
2N5816	128		20	2051	
2N5818	128	3444/123A	20	2051	
2N582	100	3003	1	2007	
2N5820	128	3024	243	2030	
2N5820HS	192		63	2030	
2N5821	129	3025	244	2027	
2N5821HS	193		67	2023	
2N5822	128		63	2030	
2N5823			67	2023	
2N5824	123AP		62	2010	
2N5825	123AP		62	2010	
2N5826	123AP		62	2010	
2N5827	123AP		62	2010	
2N5827A	123AP		62	2010	
2N5828	123AP		62	2010	
2N5828A	123AP		62	2010	
2N5829	395				2N5829
2N583	100	3005	52	2024	
2N5830	287	3444/123A	20	2051	
2N5830A		3018	61	2038	
2N5834	129		221		2N3553
2N5835					2N5835
2N5836					2N5836
2N5837					2N5837
2N5838	94	3439/163A	36		2N5838
2N5839	94	3439/163A	36		2N5839
2N584	102A	3003	52	2024	
2N5840	389	3439/163A	36		2N5840
2N5841					2N5841
2N5842					2N5842
2N5843					MD3250A
2N5844					MD3251A
2N5845	123AP		20	2051	
2N5845A	123AP		20	2051	
2N5846	282	3748			MRF231
2N5847	337				2N5847
2N5848	338				2N5849
2N5849	339				2N5849
2N585	103A	3861/101	8	2002	
2N5852	161		62	2010	
2N5855		3466	67	2023	
2N5856	128	3854/123AP	63	2030	
2N5857	159		21	2034	
2N5859			18	2030	2N5859
2N586	100	3005	1	2007	
2N5860					2N3725
2N5861					2N5861
2N5862					2N5862
2N5864	323				2N4404
2N5865	129		244	2027	2N4406
2N5867		3173			2N3789
2N5868		3173			2N3790
2N5869	283	3027/130			2N3713
2N587	101	3861	8	2002	
2N5870	283	3027/130			2N3714
2N5871	281	3173/219			2N3789
2N5872		3173			2N3790
2N5873		3027			2N3713
2N5874	280	3563			2N3714
2N5875		3173			2N5875
2N5876	285	3846	266		2N5876
2N5877		3027			2N5877
2N5878	284	3621	265	2047	2N5878
2N5879	285	3173/219	74	2043	2N5879
2N588	126	3006/160	52	2024	

Industry Standard No.	ECG	SK	GE	RS 276-	MOTOR.
2N5880	285	3437/180	74	2043	2N5880
2N5881	328	3024/128	243	2030	2N5881
2N5882	328	3024/128	243	2030	2N5882
2N5883	180	3437	74	2043	2N5883
2N5884	180	3437	74	2043	2N5884
2N5885	181	3535	75	2041	2N5885
2N5886	181	3535	75	2041	2N5886
2N5887			30	2006	
2N5888			30	2006	
2N5889			30	2006	
2N588A	126	3006/160	52	2024	
2N589	121	3009	239	2006	
2N5893	131		44	2006	
2N5897	131	3052	44	2006	
2N59	100	3003		2007	
2N5901			30	2006	
2N591	102A	3004	53	2007	
2N591-6M	126	3006/160	52	2024	
2N591/5	102A		53	2007	
2N5910			22	2032	MPS2369
2N5913	472		285		MRF607
2N5914	362		284		2N5944
2N5915	364				2N5946
2N5916			231		MRF5177
2N5917			284		MRF5174
2N5918					MRF321
2N5919A					MRF323
2N591A	102A		53	2007	
2N592	100	3005	1	2007	
2N5929					2N6338
2N593	100	3005	1	2007	
2N5930		3619			2N6338
2N5931					2N6341
2N5932					2N6338
2N5933		3619			2N6338
2N5935					2N6341
2N5936					2N6338
2N5937					2N6341
2N594	101	3861	8	2002	
2N5941	471$				MRF466
2N5942	471$				MRF463
2N5943	77				2N5943
2N5944	362				2N5944
2N5945	363				2N5945
2N5946	364				2N5946
2N5947	76				2N5947
2N5949		3116	FET-2	2035	2N5638
2N595	101	3861	8	2002	
2N5950		3116	FET-2	2035	2N5639
2N5951		3116	FET-2	2035	2N5640
2N5952		3116	FET-2	2035	2N5640
2N5953	312	3116	FET-2	2035	2N5640
2N5954		3085	56	2027	2N6318
2N5955		3085	56	2027	2N6317
2N5956		3257	56	2027	2N6317
2N596	101	3861	8	2002	
2N5961	123AP		81	2051	
2N5964			18	2030	
2N5965	154		18	2030	
2N597	100	3005	1	2007	
2N5970	284		19	2041	2N5882
2N5971	284				2N5882
2N5972	284				MJ15003
2N5974					2N5974
2N5975					2N5975
2N5976					2N5976
2N5977					2N5977
2N5978					2N5978
2N5979					2N5979
2N598	100	3005	1	2007	
2N5980					2N5980
2N5981		3189A			2N5981
2N5982		3189A			2N5982
2N5983					2N5983
2N5984					2N5984
2N5985		3188A			2N5985
2N5986	183	3189A			2N5986
2N5987	183				2N5987
2N5988					2N5988
2N5989					2N5989
2N599	100	3005	1	2007	
2N5990					2N5990
2N5991					2N5991
2N5992					MRF232
2N5993					MRF234
2N5994	360				2N5643
2N5995					MRF212
2N5996	350				2N6081
2N5998	289A		210	2051	
2N5999	290A		82	2032	
2N59A	102	3003		2007	
2N59B	102	3004		2007	
2N59C	102	3003		2007	
2N59D	102	3004		2007	
2N60	100	3003	2	2007	
2N600	100	3005	1	2007	
2N6000	123AP		210	2051	2N4401
2N6001			48		2N4403
2N6002	123AP		47	2030	2N4401
2N6003		3466	82	2032	2N4403
2N6004			210	2009	2N4401
2N6005	159		48		2N4403
2N6006	123AP		47	2030	2N4401
2N6007			48		2N4003
2N6008	289A		210	2051	MPS6515
2N601			51	2004	
2N6010	128		47	2030	2N4401
2N6011			48		2N4403
2N6012	128		S6012		2N4401
2N6013			S6013		2N4403
2N6014	128		63	2030	MPS8098
2N6015			67	2023	2N4403
2N6016	128		S6016		
2N6017			S6017		MPS8598
2N602	126	3008	52	2024	
2N6021	153		69	2049	2N6126
2N6022	292		69	2049	2N6126
2N6023	292		69	2049	2N6124
2N6024	292		69	2049	2N6124
2N6025	292		74	2043	2N6125
2N6026	292		69	2049	2N6125
2N6027	6402	3628	X17		2N6027
2N6028	6402		2N6028		2N6028
2N6029	285	3846	266		2N6029
2N602A	126	3008	52	2024	
2N603	126	3008	52	2024	
2N6030	285	3846	266		2N6030
2N6031	285	3846	266		2N6031
2N6032		3619			2N6275
2N6033		3619			2N6277
2N6034	254	3997			2N6034
2N6035	254	3997			2N6035
2N6036	254		250	2027	2N6036
2N6037	253	3180/263	57	2017	2N6037
2N6038	253		57	2017	2N6038
2N6039	253	3996	64		2N6039
2N603A	126	3008	52	2024	
2N604	126	3008	52	2024	
2N6040	260	3897/262	221		2N6040
2N6041	260	3897/262			2N6041
2N6042	262		221		2N6042
2N6043	259	3896/261	214		2N6043
2N6044	259	3896/261	214		2N6044
2N6045	261		214	2038	2N6045
2N6049	218		234		2N6049
2N604A	126	3006/160	52	2024	
2N605	126	3008	52	2024	
2N6050	248	3949			2N6050
2N6051	248	3949			2N6051
2N6051JAN					2N6051JAN
2N6051JTX					2N6051JTX
2N6051JTXV					2N6051JTXV
2N6052	248	3949			2N6052
2N6052JAN					2N6052JAN
2N6052JTX					2N6052JTX
2N6052JTXV					2N6052JTXV
2N6053	244	3183A			2N6053
2N6054	244	3183A			2N6054
2N6055	243	3182			2N6055
2N6056	243	3182			2N6056
2N6057	247	3948			2N6057
2N6058	247	3948			2N6058
2N6058JAN					2N6058JAN
2N6058JTX					2N6058JTX
2N6058JTXV					2N6058JTXV
2N6059	247	3948			2N6059
2N6059JAN					2N6059JAN
2N6059JTX					2N6059JTX
2N6059JTXV					2N6059JTXV
2N606	126	3008	52	2024	
2N6067	159		82	2032	
2N6068	5600	3664/5602		1001	2N6068
2N6068A					2N6068A
2N6068B					2N6068B
2N6069	5601	3664/5602		1001	2N6069
2N6069A					2N6069A
2N6069B					2N6069B
2N607	126	3008	52	2024	
2N6070	5602	3664		1001	2N6070
2N6070A					2N6070A
2N6070B					2N6070B
2N6071	5603	3665		1001	2N6071
2N6071A					2N6071A
2N6071B					2N6071B
2N6072	5604	3666/5605		1000	2N6072
2N6072A					2N6072A
2N6072B					2N6072B
2N6073	5605	3666		1000	2N6073
2N6073A				1000	2N6073A
2N6073B					2N6073B
2N6074	5606				2N6074
2N6074A					2N6074A
2N6074B					2N6074B
2N6075	5607				2N6075
2N6075A					2N6075A
2N6075B					2N6075B
2N6076	159	3466	82	2032	
2N6077	384	3894			2N6077
2N6078	384	3894			2N6078
2N6079	384	3894			2N6235
2N608	126	3008	52	2024	
2N6080	348				2N6080
2N6081	350	3176	295		2N6081
2N6082	351		296		2N6082
2N6083	351		296		2N6083
2N6084	320		284		2N6084
2N6085					2N2920
2N6086					2N2920
2N6087					2N2920
2N6088					2N2920
2N6089					2N2920
2N609	102	3004	2	2007	
2N6090					2N2920
2N6091					2N2920
2N6092					2N2920
2N6093	471				MRF464
2N6094	353		321		2N6094
2N6095	354		321		2N6095
2N6096	355		321		2N6096
2N6097	356		321		2N6097
2N6098	377	3534	D44H7		2N6487
2N6099	331	3534	D44H7		2N6487
2N60A	102	3003	2	2007	
2N60B	102	3003	2	2007	
2N60C	102	3003	2	2007	
2N60R	102	3004	2	2007	
2N61	100	3003	2	2007	
2N610	102	3004	2	2007	
2N6100	331	3534	D44H10		2N6487
2N6101	331	3534	D44H10		2N6488
2N6102	331	3620	D44H4		2N6488
2N6103	331	3620	D44H4		2N6486
2N6104					MRF5177
2N6105					MRF5177A
2N6106	197	3083	250	2027	2N6107
2N6107	197	3083	250	2027	2N6107
2N6108	197	3083	250	2027	2N6109
2N6109	153	3083/197	250	2027	2N6109
2N611	102	3004	2	2007	
2N6110	197	3084	250	2027	2N6111
2N6111	197	3084	250	2027	2N6111
2N6112			62	2010	
2N6116					2N6116
2N6117					2N6117
2N6118					2N6118
2N612	102	3004	2	2007	
2N6121	377	3054/196	66	2048	2N6121
2N6122	377	3054/196	66	2048	2N6122
2N6123	377	3054/196			2N6123
2N6124	378	3083/197	69	2049	2N6124
2N6125	378	3083/197	69	2049	2N6125
2N6126	378		69	2049	2N6126
2N6127					2N6436
2N6128	72$				2N6338
2N6129		3893			2N6129
2N613	102	3004	2	2007	
2N6130		3893			2N6130
2N6131					2N6131
2N6132	197	3274/153	250	2027	2N6132
2N6133	197	3274/153	250	2027	2N6133
2N6134	197		250	2027	2N6134
2N6135	76				MRF511
2N6136					2N6136
2N6137					2N6116
2N6138					2N6116
2N614	100	3005	1	2007	
2N6145					2N6145
2N6146	5675				2N6146

Industry Standard No.	ECG	SK	GE	RS 276-	MOTOR.
2N6147					2N6147
2N615	100	3005	1	2007	
2N6151	5614				2N6151
2N6152	5616				2N6152
2N6153	5618				2N6153
2N6154	5624	3632			2N6154
2N6155	5626	3633			2N6155
2N6156	5628				2N6156
2N6157					2N6157
2N6158					2N6158
2N6159					2N6159
2N616	100	3005	1	2007	
2N6160	5693	3652			2N6160
2N6161	5695	3509			2N6161
2N6162	5697	3522			2N6162
2N6163	56022	3661			2N6163
2N6164	56024	3509/5695			2N6164
2N6165	56026	3663			2N6165
2N6166					2N6166
2N6167	5562				2N6167
2N6168	5562	3655			2N6168
2N6169	5564				2N6169
2N617	100	3005	1	2007	
2N6170	5566	3657			2N6170
2N6171					2N6171
2N6172	5562				2N6172
2N6173	5564				2N6173
2N6174	5566				2N6174
2N6175	228A	3104A	257		MPSU10
2N6176	228A	3103A/396	232		MPSU10
2N6177	228A	3103A/396	257		2N6559
2N6178		3024	D44C10		MPSU06
2N6179		3024			MPSU05
2N618		3009	16	2006	
2N6180		3025	D45C10		MPSU56
2N6181		3025			MPSU55
2N6186					2N6186
2N6187					2N6187
2N6188					2N6188
2N6189					2N6189
2N619	123A	3452/108	86	2038	
2N6190					2N6190
2N6191					2N6191
2N6192					2N6192
2N6193					2N6193
2N6197					2N3553
2N6198	357				2N5641
2N6199	359				2N5642
2N61A	102	3003	2	2007	
2N61B	102	3003	2	2007	
2N61C	102	3003	2	2007	
2N62	102A	3004	53	2007	
2N620	123A	3452/108	86	2038	
2N6200	360				2N5643
2N6201					2N6166
2N6202					MRF5174
2N6203					MRF5175
2N6204					2N5637
2N6205					MRF5177
2N621	123A	3452/108	86	2038	
2N6211	162	3624	35		2N6211
2N6212		3623	35		2N6212
2N6213	162	3623	35		2N6213
2N6214	167	3623	221	1172	
2N6218			214	2038	
2N622	123A		20	2051	
2N6220			214	2038	
2N6221			214	2038	
2N6222	159		214	2038	
2N6223		3466	221		
2N6224	159		214		
2N6226	285		263	2043	2N6226
2N6227	285		263	2043	2N6227
2N6228	285		263	2043	2N6228
2N6229	285		263	2043	2N6229
2N623	126		52	2024	
2N6230	285				2N6230
2N6231	285		263	2043	2N6231
2N6232			214	2038	
2N6232-4			214	2038	
2N6233			246	2020	2N6233
2N6234					2N6234
2N6235					2N6235
2N6236	5421	3570		1067	2N6236
2N6237	5422	3570		1067	2N6237
2N6238	5423	3570		1067	2N6238
2N6239	5414	3954			2N6239
2N624	126	3008	52	2024	
2N6240	5415	3955			2N6240
2N6241	5416	3956			2N6241
2N6242					MJ13015
2N6243					MJ13334
2N6244					MJ13333
2N6245					MJ13334
2N6246	219	3173	74	2043	2N5879
2N6247	219	3173	74	2043	2N5880
2N6248		3437			MJ15016
2N6249	97	3559/162			2N6249
2N625	103	3862	8	2002	
2N6250	97	3559/162			2N6250
2N6251	97	3559/162			2N6251
2N6253	130	3027	14	2041	2N5877
2N6254	130	3511	75	2041	2N5878
2N6255			219		2N6255
2N6256					2N6256
2N6257	181	3036	75	2041	2N6257
2N6258	387		75	2041	2N5686
2N6259	284	3535/181	265	2047	2N5631
2N626			21	2034	
2N6260	175	3026	246	2020	2N4231A
2N6261	384	3026	246	2020	2N4233A
2N6262	284	3621	265	2047	2N5760
2N6263	175	3538	246	2020	2N5050
2N6264	175	3538	267		2N5051
2N627	179	3009	239	2006	
2N6270					2N6338
2N6271		3619			2N6338
2N6272					2N6338
2N6273					2N6338
2N6274		9040			2N6274
2N6275		9040			2N6275
2N6276		9040			2N6276
2N6277	387	9040			2N6277
2N6278	70				2N6274
2N6279	70				2N6275
2N628	179	3009	239	2006	
2N6280	70				2N6276
2N6281	70				2N6277
2N6282	251	3858			2N6282
2N6283	251	3858			2N6283
2N6283JAN					2N6283JAN
2N6283JTX					2N6283JTX
2N6283JTXV					2N6283JTXV
2N6284	251	3858			2N6284
2N6284JAN					2N6284JAN
2N6284JTX					2N6285JTX
2N6284JTXV					2N6284JTXV
2N6285	250	3859/252			2N6285
2N6286	252	3859			2N6286
2N6286JAN					2N6286JAN
2N6286JTX					2N6286JTX
2N6286JTXV					2N6286JTXV
2N6287	252	3859			2N6287
2N6287JAN					2N6287JAN
2N6287JTX					2N6287JTX
2N6287JTXV					2N6287JTXV
2N6288	152	3256	28	2017	2N6288
2N6289	152	3054/196	28	2017	2N6288
2N629	179	3009	239	2006	
2N6290	152	3054/196	28	2017	2N6290
2N6291	196	3054	28	2017	2N6290
2N6292	196	3054	241	2020	2N6292
2N6293	196	3054	28	2017	2N6292
2N6294	274	3182/243			2N6294
2N6295	274	3182/243			2N6295
2N6296	275	3181A/264			2N6296
2N6297	275	3183A			2N6297
2N6298	262	3859/252			2N6298
2N6299	262	3859/252			2N6299
2N63	102A	3003	53	2007	
2N63#				2004	
2N630	179	3642	76		
2N6300	261	3182/243			2N6300
2N6301	261	3182/243			2N6301
2N6302	284	3836			2N5630
2N6303					2N6303
2N6304	316	3039	280		2N6304
2N6305	316	3039	280		2N6305
2N6306	283	3438	36		2N6306
2N6307	283	3439/163A	36		2N6307
2N6308	283	3439/163A	36		2N6308
2N631	102	3003	2	2007	
2N6312		3257	234		2N6312
2N6313			234		2N6313
2N6314			234		2N6314
2N6315					2N6315
2N6316					2N6316
2N6317					2N6317
2N6318					2N6318
2N632	102	3004	2	2007	
2N6322					MJ10015
2N6323					MJ10015
2N6324					MJ10015
2N6325					MJ10015
2N6326	181		75	2041	2N6326
2N6327	181		75	2041	2N6327
2N6328	181		75	2041	2N6328
2N6329	180	3437	74	2043	2N6329
2N633	102	3004	2	2007	
2N6330		3437	74	2043	2N6330
2N6331		3437	74	2043	2N6331
2N6332					2N6236
2N6333					2N6237
2N6334					2N6238
2N6335					2N6239
2N6336					MCR106-5
2N6337					2N6240
2N6338	327	3945			2N6338
2N6339	327	3945			2N6339
2N633B	102	3014	2	2007	
2N634	101	3861	8	2002	
2N6340	327	3945			2N6340
2N6341	327	3945			2N6341
2N6342	56004	3658			2N6342
2N6342A	56004	3658			2N6342A
2N6343	56006	3659			2N6343
2N6343A	56006	3659			2N6343A
2N6344	56008	3660			2N6344
2N6344A	56008	3660			2N6344A
2N6345	56010				2N6345
2N6345A	56010				2N6345A
2N6346	56004	3658			2N6346
2N6346A	56004	3658			2N6346A
2N6347	56006	3659			2N6347
2N6347A	56006	3659			2N6347A
2N6348	56008	3660			2N6348
2N6348A	56008	3660			2N6348A
2N6349	56010				2N6349
2N6349A	56010				2N6349A
2N634A	101	3861	8	2002	
2N635	101	3861	8	2002	
2N6354	328	3561	265	2047	2N6339
2N6355	251	3858	75		2N6057
2N6356	251	3858	64		2N6057
2N6357	247	3948	75		2N6058
2N6358	251	3858			2N6058
2N6359	284	3836	75	2041	2N5885
2N635A	101	3861	8	2002	
2N636	101	3861	8	2002	
2N6360	284	3836			2N5629
2N6361					MRF325
2N6362					2N6439
2N6365			51	2004	
2N6365A			51	2004	
2N6366					MRF231
2N6367			214	2038	MRF433
2N6368			293		MRF460
2N636A	101	3861	8	2002	
2N637	104	3009	16	2006	
2N6370			214	2038	
2N6371	130	3511	14	2041	2N6569
2N6372	384	3894	214	2038	2N6316
2N6373	384	3894	214	2038	2N6315
2N6374	384	3894			2N6315
2N6377					2N6377
2N6378					2N6378
2N6379					2N6379
2N637A	104	3009	16	2006	
2N637B	104	3009	16	2006	
2N638	104	3009	16	2006	
2N6380					2N6377
2N6381					2N6378
2N6382					2N6379
2N6383	245	3182/243			2N6383
2N6384	245	3182/243			2N6384
2N6385	245	3182/243			2N6385
2N6386	263	3180			2N6386
2N6387	263	3180			2N6387
2N6388	263	3180			2N6388
2N6389	316	3039	280		
2N638A	104	3009	16	2006	
2N638B	104	3009	16	2006	
2N639	104	3009	16	2006	
2N6394	5461	3574			2N6394
2N6395		3574			2N6395
2N6396		3574			2N6396
2N6397		3575			2N6397
2N6398		3576			2N6398

Industry Standard No.	ECG	SK	GE	RS 276-	MOTOR.
2N6399	5558				2N6399
2N639A	104	3009	16	2006	
2N639B	104	3009	16	2006	
2N64	102A	3003	53	2007	
2N64#				2004	
2N640	126	3008	52	2024	
2N6400		3574			2N6400
2N6401		3574			2N6401
2N6402		3574			2N6402
2N6403		3575			2N6403
2N6404		3576			2N6404
2N6405					2N6405
2N6406					MJE171
2N6407					MJE172
2N6408					2N6408
2N6409					2N6409
2N641	126	3008	52	2024	
2N6410					2N6410
2N6411					MJE210
2N6412					MJE180
2N6413					MJE181
2N6414					MJE170
2N6415					MJE171
2N6416					MJE243
2N6417					MJE243
2N6418					MJE253
2N6419					MJE253
2N641REDM/F	126		52	2024	
2N642	126	3007	52	2024	
2N6420					2N6420
2N6421					2N6421
2N6422					2N6422
2N6423					2N6423
2N6424					2N6424
2N6425	124	3021			2N6425
2N643	126	3007	52	2024	
2N6430	396				2N6430
2N6431	396				2N6431
2N6432	288				2N6432
2N6433	288				2N6433
2N6436					2N6436
2N6437					2N6437
2N6438					2N6438
2N6439					2N6439
2N644	126	3007	52	2024	
2N6441					2N2920
2N6442					2N2920
2N6443					2N2920
2N6444					2N2920
2N6445					2N2920
2N6446					2N2920
2N6447					2N2920
2N6448					2N2920
2N645	126	3007	52	2024	
2N6455					MRF406
2N6456					MRF406
2N6457					MRF460
2N6458	328				MRF460
2N6459	335				MRF454
2N646	103A	3862/103	8	2002	
2N6465		3626			MJ3247
2N6466		3626			MJ3247
2N6467	292	3625/218			MJ3237
2N6468	292	3625/218			MJ3237
2N6469	219	3173			2N5879
2N647	103A	3835	59	2002	
2N647/22	103A	3835	59	2002	
2N6470		3270			2N5881
2N6471		3563			2N5881
2N6472		3563			2N5882
2N6473	291	3440			FT317
2N6474	291	3440			FT317A
2N6475	292	3441			FT417
2N6476	292	3441			FT417A
2N6477	379	3929			FT317A
2N6478	379	3929			FT317B
2N648	103	3862	8	2002	
2N6486	331	3188A/182			2N6486
2N6487	331	3188A/182			2N6487
2N6488	332	3188A/182			2N6488
2N6489	332	3189A/183			2N6489
2N649	103A	3010	59	2002	
2N649/22			8	2002	
2N649/22				2002	
2N649/5	103A		59	2002	
2N6490	332	3189A/183			2N6490
2N6491	332	3189A/183			2N6491
2N6492	243	3182			2N6055
2N6493	243	3182			2N6056
2N6494	243	3182			2N6056
2N6495					2N6316
2N6496	328	3895	265	2047	2N6339
2N6497	379				2N6497
2N6498	379				2N6498
2N6499	379				2N6499
2N65	102	3003	2	2007	
2N65#				2004	
2N650	102	3004	2	2007	
2N6500	175	3562	246	2020	2N5430
2N6501					2N6501
2N6502					2N6502
2N6503					2N6503
2N6504	5550				2N6504
2N6505	5552				2N6505
2N6506	5552				2N6506
2N6507	5554				2N6507
2N6508	5556				2N6508
2N6509	5558				2N6509
2N650A	102	3004	2	2007	
2N651	102	3004	2	2007	
2N6510	283	3467	36		2N6306
2N6511	283	3467	36		2N6306
2N6512	283	3467	36		2N6544
2N6513	283	3467	36		2N6545
2N6514	283	3467	36		2N6544
2N6516	287	3433			
2N651A	102	3004	2	2007	
2N652	102	3004	2	2007	
2N6521					2N2920
2N6522					2N2920
2N652A	102	3004	2	2007	
2N653	102	3004	53	2007	
2N6530	263	3180			TIP101
2N6531	261				TIP102
2N6532	261				TIP102
2N6534					2N6301
2N6535					TIP102
2N6536					TIP102
2N654	102	3004	2	2007	
2N6542	283				2N6542
2N6543	283				2N6543
2N6544	283	3467			2N6544
2N6545		3710			2N6545
2N6546					2N6546

Industry Standard No.	ECG	SK	GE	RS 276-	MOTOR.
2N6546JAN					2N6546JAN
2N6546JTX					2N6546JTX
2N6546JTXV					2N6546JTXV
2N6547	386				2N6547
2N6547JAN					2N6547JAN
2N6547JTX					2N6547JTX
2N6547JTXV					2N6547JTXV
2N6548					2N6548
2N6549					2N6549
2N655	102	3004	2	2007	
2N6551	186	3178A			2N6551
2N6552		3178A			2N6552
2N6553		3178A			2N6553
2N6554	186	3203/211			2N6554
2N6555					2N6555
2N6556					2N6556
2N6557	171	3232/191			2N6557
2N6558	171	3232/191	27		2N6558
2N6559					2N6559
2N655A	128		243	2030	
2N655GRN	102	3004	2	2007	
2N655RED	102	3004	2	2007	
2N656	128	3024	243	2030	2N656
2N6569	130	3027			2N6569
2N656A		3024	18	2030	
2N657	128	3024	243	2030	2N657
2N6573	283				2N6546
2N6574	386				2N6546
2N6575	386				2N6547
2N6576	249				2N6576
2N6577	249				2N6577
2N6578					2N6578
2N6579					MJ13014
2N658	100	3845/176	80		
2N6580					MJ13015
2N6581					MJ13334
2N6582	385				2N6308
2N6583	385				2N6545
2N6584					MJ13334
2N659	100	3845/176	80		
2N6591					2N6591
2N6592					2N6592
2N6593					2N6593
2N6594	219	3173			2N6594
2N6595					MRF904
2N6596					MRF904
2N6597					MRF914
2N6598					MRF914
2N6599					MRF965
2N66			16	2006	
2N660	176	3845	80		
2N6600					MRF965
2N6605					2N1595
2N6606					2N1595
2N6607					2N1596
2N6608					2N1597
2N6609		9032			2N6609
2N661	176	3845	80		
2N662	100	3845/176	80		
2N663	121	3009	239	2006	
2N6648	246	3183A			2N6648
2N6649	246	3183A			2N6649
2N665	121	3009	239	2006	
2N6650	246	3183A			2N6650
2N6653					MJ13332
2N6654					MJ13332
2N6655					MJ13333
2N6666	264	3181A			2N6666
2N6667	264	3181A			2N6667
2N6668	264	3181A			2N6668
2N6669					MJE15028
2N6671					2N6544
2N6672					2N6545
2N6673					2N6545
2N6674					MJ13014
2N6675					MJ13015
2N6676					MJ13332
2N6677	386				MJ13332
2N6678	386				MJ13333
2N6669	104	3009	239	2006	
2N667			16	2006	
2N670			53	2007	
2N670				2007	
2N671			53	2007	
2N671				2007	
2N672	176	3845	80		
2N673			53	2007	
2N673				2007	
2N674	100	3005	1	2007	
2N675			53	2007	
2N675				2007	
2N677	179	3009	76		
2N677A	179	3009	76		
2N677B	179	3009	76		
2N677C	179	3009	239	2006	
2N678	179	3009	76		
2N678A	179	3009	76		
2N678B	179	3009	239	2006	
2N678C	179	3009	239	2006	
2N679	101	3861	8	2002	
2N68			16	2006	
2N680	102A	3004	52	2024	
2N681	5520	6621/5521	MR-3		2N681
2N682	5521	6621	MR-3		2N682
2N682A	5521	6621			
2N683	5522	6622	MR-3		2N683
2N683A	5522	6622			
2N684	5523	6624/5524	MR-3		2N684
2N684A	5523	6624/5524			
2N685	5524	6624	MR-3		2N685
2N686	5525	6627/5527	MR-3		2N686
2N686A	5525	6627/5527			
2N687	5526	6627/5527	MR-3		2N687
2N688	5527	6627	MR-3		2N688
2N688A	5527	6627			
2N689	5528	6629/5529	2N689		2N689
2N689A	5528	6629/5529			
2N690	5529	6629	2N690		2N690
2N691	5530				2N691
2N692	5531	6631			2N692
2N694	126	3008	52	2024	
2N695	126	3006/160	52	2024	
2N696	123	3124/289	20	2051	2N697
2N696A	128		47	2030	
2N697	128	3024	243	2030	2N697
2N697A	128	3024	20	2051	
2N697L	128		243	2030	
2N697S	128		243	2030	
2N698	396		18	2030	2N3020
2N699	396	3024/128	222		2N699
2N699A		3024	18	2030	
2N699A,B	396		40		
2N699AB				2012	
2N699AB				2012	

Industry Standard No.	ECG	SK	GE	RS 276-	MOTOR.
2N699B		3024	18	2030	
2N699L	128		264		
2N699S	128		264		
2N700	160		245	2004	
2N700/18	160		245	2004	
2N700A	160		245	2004	
2N700A/18	160		245	2004	
2N701	123	3124/289	20	2051	
2N702	123A	3452/108	86	2038	2N706A
2N703	123A	3452/108	86	2038	2N703
2N705	126		245	2004	
2N705A	126		245	2004	
2N706	123A	3452/108	86	2038	2N706
2N706/46	123A		210	2051	
2N706/51			210	2051	
2N706A	123A	3452/108	86	2038	2N706A
2N706A/46			210	2051	
2N706A/51			210	2051	
2N706B	123A	3452/108	86	2038	2N706B
2N706B/46			210	2051	
2N706B/51			210	2051	
2N706C	123A	3452/108	86	2038	
2N706M-JAN			63	2030	
2N707	311	3452/108	86	2038	
2N707A	123A	3124/289	20	2051	
2N708	123A	3444	20	2051	2N708
2N708/46	123A		17	2051	
2N708/51			61	2038	
2N708/TNT			11	2015	
2N708A	123A	3444	20	2051	
2N709	108	3444/123A	86	2038	MM1748
2N709/46	278		63	2030	
2N709/51			63	2030	
2N709/52	108	3452	86	2038	
2N709/TNT			11	2015	
2N709A	108	3452	86	2038	
2N709A/51			63	2030	
2N709A46	278		210	2051	
2N71	102	3004	2	2007	
2N710	126		245	2004	
2N710A	126		245	2004	
2N711			245	2004	
2N711A	126		245	2004	
2N711B	126		245	2004	
2N715	123A	3124/289	20	2051	2N2221
2N716			20	2051	2N2221
2N717	123A	3452/108	86	2038	2N2221
2N717A	108	3452	86	2038	
2N718	123A	3452/108	86	2038	2N718
2N718A	123A	3452/108	86	2038	2N718A
2N719	194	3275	220		2N2895
2N719A	194	3275	220		2N2895
2N72	126	3006/160	52	2024	
2N720	194		243	2030	2N720A
2N720A	194		243	2030	2N720A
2N721	159	3466	82	2032	2N2906
2N721A	159	3466	82	2032	
2N722	159	3466	82	2032	2N2906
2N722A	159	3466	82	2032	
2N723	106		21	2034	
2N725	126	3008	245	2004	
2N726	106	3984	21	2034	2N869A
2N727	106	3984	21	2034	
2N728	123	3124/289	20	2051	
2N729	123A	3124/289	20	2051	
2N73	102	3004	2	2007	
2N730	128		20	2051	
2N731	128		20	2051	2N2221
2N734	128	3479	81	2051	
2N734A		3479	81	2051	
2N735	128	3479	81	2051	2N720A
2N735A	128	3479	81	2051	
2N736	128	3479	81	2051	2N736
2N736A	128	3479	81	2051	
2N736B	128	3479	81	2051	
2N738	154	3045/225	40	2012	
2N739	154	3045/225	40	2012	2N740
2N739A		3045	40	2012	
2N74	102	3004	2	2007	
2N740	154	3045/225	40	2012	2N740
2N740A	126	3045/225	40	2012	
2N741	126	3006/160	245	2004	
2N741A	160		245	2004	
2N742	123A	3124/289	20	2051	
2N742A	123A	3124/289	20	2051	
2N743	123A	3452/108	86	2038	2N2368
2N743/46			63	2030	
2N743/51			63	2030	
2N743A	311	3452/108	86	2038	
2N744	123A	3452/108	86	2038	2N2369
2N744/46			63	2030	
2N744/51			63	2030	
2N745	123A	3444	20	2051	
2N746	123A	3444	20	2051	
2N747	123A	3444	20	2051	
2N748	123A	3444	20	2051	
2N749	123A	3444	20	2051	
2N75	126	3006/160	52	2024	
2N750	194	3444/123A	20	2051	
2N751	123A	3444	20	2051	
2N752	128	3452/108	86	2038	
2N753	123A	3444	20	2051	2N2369A
2N753/46	123A		210	2051	
2N753/51			63	2030	
2N754	123A	3444	20	2051	
2N755	128		220		
2N756	123A	3452/108	86	2038	
2N756A	123A	3124/289	20	2051	
2N757	123A	3452/108	86	2038	
2N757A	123A	3124/289	20	2051	
2N758	123A	3452/108	86	2038	
2N758A	123A	3124/289	20	2051	
2N758B	123A	3124/289	20	2051	
2N759	123A	3452/108	86	2038	
2N759A	123A	3124/289	20	2051	
2N759B	123A	3124/289	20	2051	
2N76	102	3004	2	2007	
2N760	123A	3452/108	86	2038	2N3700
2N760A	123A	3124/289	20	2051	2N3700
2N760B	123A	3124/289	20	2051	
2N761	123A	3452/108	86	2038	
2N762	123A	3452/108	86	2038	
2N768	126		245	2004	
2N769	126		245	2004	
2N77	102A	3004	1	2007	
2N770	123	3124/289	20	2051	2N3014
2N771	123	3124/289	20	2051	2N3014
2N772	123	3124/289	20	2051	2N3014
2N773	123	3124/289	20	2051	2N3014
2N774	123	3124/289	20	2051	2N3014
2N775	123	3124/289	20	2051	2N3014
2N776	123	3124/289	20	2051	2N3014
2N777	123	3124/289	20	2051	2N3014
2N778	123	3124/289	20	2051	2N3014
2N779	126		245	2004	
2N779A	126		245	2004	
2N779B	160		245	2004	
2N78	101	3861	8	2002	
2N780	123A	3039/316	11	2015	
2N781	126	3004/102A	245	2004	
2N782	126	3123	245	2004	
2N783	123A	3122	20	2051	2N834
2N784	123A	3122	20	2051	
2N784/51			63	2030	
2N784A	123A	3444	20	2051	
2N784A/46			210	2051	
2N784A/51			210	2051	
2N789	123	3124/289	20	2051	2N3946
2N78A	101	3861	8	2002	
2N79	102	3004	2	2007	
2N790	123	3124/289	20	2051	2N3946
2N791	123	3124/289	20	2051	2N3946
2N792	123	3124/289	20	2051	2N3946
2N793	123	3124/289	20	2051	2N3946
2N794	126		245	2004	
2N795	126		245	2004	
2N796	126		245	2004	
2N797	103	3862	8	2002	
2N799	102	3004	2	2007	
2N80	102	3004	2	2007	
2N800	102A	3123	53	2007	
2N801	100	3005	1	2007	
2N802	100	3005	1	2007	
2N803	158	3005	53	2007	
2N804	158	3005	53	2007	
2N805	158	3005	53	2007	
2N806	158	3005	53	2007	
2N807	102A	3005	53	2007	
2N808	102A	3005	53	2007	
2N809	100	3005	1	2007	
2N81	102	3003	53	2007	
2N810	100	3005	1	2007	
2N811	100	3005	1	2007	
2N812	100	3005	1	2007	
2N813	102A	3005	53	2007	
2N814	102A	3005	53	2007	
2N815	102A	3011	53	2007	
2N816	102A	3011	53	2007	
2N817	158	3011	53	2007	
2N818	158	3011	53	2007	
2N819	102A	3011	53	2007	
2N82	102	3003	2	2007	
2N820	102A	3011	53	2007	
2N821	101	3861	8	2002	
2N822	101	3861	8	2002	
2N823	101	3861	8	2002	
2N824	101	3861	8	2002	
2N825	102	3011	2	2007	
2N826	102	3004	2	2007	
2N827	126		245	2004	
2N828	126		245	2004	
2N828A	126		245	2004	
2N829	126		245	2004	
2N82AG	112		1N82A		
2N83	100	3005	1	2007	
2N834	123A	3452/108	86	2038	2N834
2N834/46			20	2051	
2N834/51			63	2030	
2N834A	123A	3444	20	2051	
2N835	123A	3444	20	2051	2N706A
2N835/46			20	2051	
2N835/51			63	2030	
2N837	126		245	2004	
2N838	126		245	2004	
2N839	123A	3444	20	2051	
2N84			51	2004	
2N840	123A	3444	20	2051	2N720A
2N841	123A	3452/108	86	2038	2N720A
2N841/46			212	2010	
2N842	123A	3444	20	2051	2N2221
2N843	123A	3452/108	86	2038	2N2222
2N844	123A	3444	20	2051	2N2221
2N845	128		220		2N2222
2N846	160		245	2004	
2N846A	126		245	2004	
2N846B	160	3123	245	2004	
2N848					2N834
2N849	108	3452	86	2038	2N834
2N85	126	3006/160	52	2024	
2N850	108	3452	86	2038	2N834
2N851	108	3452	86	2038	2N834
2N852	108	3452	86	2038	2N834
2N858	159	3466	82	2032	2N2906
2N859	159	3466	82	2032	2N2906
2N86	126	3006/160	52	2024	
2N860	159	3466	82	2032	2N2906
2N861	159	3466	82	2032	2N2906
2N862	159	3466	82	2032	2N2906
2N863	159	3466	82	2032	2N2906
2N864	159	3466	82	2032	2N2906
2N864A	106	3466/159	21	2034	2N2906
2N865	106	3984	21	2034	2N2906
2N865A	106	3984	21	2034	2N2906
2N866	123A	3124/289	20	2051	2N2906
2N867	123A	3124/289	20	2051	2N2906
2N869	106	3984	21	2034	2N869A
2N869A	106	3984	21	2034	2N869A
2N87	126	3006/160	52	2024	
2N870	128		243	2030	2N3020
2N871	128		243	2030	2N3019
2N876	5400	3950			MCR102
2N877	5400	3950			MCR102
2N878	5401	3638/5402			MCR103
2N879	5402	3638			MCR104
2N88	126	3006/160	52	2024	
2N880	5403	3627/5404			MCR115
2N881	5404	3627			MCR120
2N882					MCR100-5
2N883					MCR100-6
2N884					MCR102
2N885					MCR102
2N886					MCR103
2N887					MCR104
2N888					MCR115
2N889					MCR120
2N89	126	3006/160	52	2024	
2N890					MCR100-5
2N891					MCR100-6
2N90	126	3006/160	52	2024	
2N902				2011	
2N903				2011	
2N904				2011	
2N905				2011	
2N907			63	2030	
2N908			18	2030	
2N909	123A		20	2051	
2N910	128		243	2030	2N910
2N911	128		243	2030	2N911

Industry Standard No.	ECG	SK	GE	RS 276-	MOTOR.
2N912	128		243	2030	
2N913	123A	3452/108	86	2038	
2N914	123A	3452/108	86	2038	2N914
2N914/46			17	2051	
2N914/51	108	3452	86	2038	
2N914A	123A	3452/108	86	2038	
2N915	123A	3444	20	2051	2N915
2N915A	123A		214	2038	
2N916	123A	3452/108	86	2038	2N916
2N916A	123A	3452/108	86	2038	
2N917	161	3019	39	2015	2N918
2N917/51			11	2015	
2N917A	161	3117	39	2015	
2N918	108	3117	86	2038	2N918
2N918/46			63	2030	
2N918/51			86	2038	
2N919	123A	3444	20	2051	
2N920	123A	3444	20	2051	2N834
2N921	123A	3444	20	2051	2N834
2N922	123A	3444	20	2051	
2N923	159	3466	82	2032	2N2906
2N924	159	3466	82	2032	2N2906
2N925	159	3466	82	2032	2N2906
2N926	159	3466	82	2032	2N2906
2N927	159	3466	82	2032	2N2906
2N928	159	3466	82	2032	2N2906
2N929	123A	3444	20	2051	2N930
2N929/46	123A		210	2051	
2N929A	123A	3444	20	2051	
2N930	123A	3444	20	2051	2N930
2N930/46	123A		210	2051	
2N930/TNT			17	2051	
2N930A	123A	3444	20	2051	
2N930A/46	123A		210	2051	
2N930B	123A		210	2051	
2N933	126		52	2024	
2N934	126		245	2004	
2N935	159	3466	82	2032	2N2907A
2N936	159	3466	82	2032	2N2907A
2N937	159	3466	82	2032	2N2907A
2N938	159	3466	82	2032	2N2907A
2N939	159	3466	82	2032	2N2907A
2N94	103	3011	59	2002	
2N940	159	3466	82	2032	2N2907A
2N941	159	3466	82	2032	2N2907A
2N942	159	3715	82	2032	
2N943	159	3466	82	2032	
2N944	159	3466	82	2032	
2N945	159	3466	82	2032	
2N946	159	3466	82	2032	2N2907A
2N947	123A	3444	20	2051	2N2907A
2N948					MCR102
2N949					MCR103
2N94A	101	3011	59	2002	
2N950					MCR104
2N951			20	2051	MCR120
2N955	101	3861	8	2002	
2N955A	101	3861	8	2002	
2N956	123A	3479	81	2051	2N956
2N957	123A	3444	20	2051	2N2501
2N958					2N3303
2N959					2N3303
2N96	102	3004	2	2007	
2N960	126		245	2004	
2N960/46	160		245	2004	
2N961	126		245	2004	
2N961/46			51	2004	
2N962	126		245	2004	
2N962/46			51	2004	
2N963	126		245	2004	
2N964	126		245	2004	
2N964/46	160		245	2004	
2N964A	126		245	2004	
2N965	126		245	2004	
2N966	126		245	2004	
2N967	126		245	2004	
2N968	160		245	2004	
2N969	126		245	2004	
2N97	103	3861/101	8	2002	
2N970	126		245	2004	
2N971	126		245	2004	
2N972	126		245	2004	
2N973	126		245	2004	
2N974	126		245	2004	
2N975	126		245	2004	
2N976	160		245	2004	
2N977	126	3123	245	2004	
2N978	106	3984	21	2034	2N2906
2N979	126		245	2004	
2N97A	101	3861	8	2002	
2N98	101	3861	8	2002	
2N980	126		245	2004	
2N981	128		243	2030	2N720A
2N982	126	3006/160	245	2004	
2N983	126	3006/160	245	2004	
2N984	126	3006/160	245	2004	
2N985	126		245	2004	
2N986	126		52	2024	2N2895
2N987	160	3006	245	2004	
2N988	123A	3452/108	86	2038	2N2221
2N989	123A	3452/108	86	2038	2N2221
2N98A	101	3861	8	2002	
2N99	101	3861	8	2002	
2N990	126	3007	245	2004	
2N991	126	3006/160	245	2004	
2N992	126	3006/160	245	2004	
2N993	126	3005	245	2004	
2N994	126	3008	245	2004	
2N995	106	3984	21	2034	2N3251
2N995A	106	3984	21	2034	
2N996	106	3984	21	2034	2N869A
2N997	172A	3156			2N720A
2N998	172A	3156			2N998
2N999	172A	3156			2N998
2NJ233B	312	3116	FET-2	2035	
2NJ50	126	3007	52	2024	
2NJ51	126	3007	52	2024	
2NJ52	126	3006/160	52	2024	
2NJ53	126	3006/160	52	2024	
2NJ59D			52	2024	
2NJ5A	100	3005	1	2007	
2NJ5D		3123	2	2007	
2NJ6	100	3005	1	2007	
2NJ8A	100	3005	1	2007	
2NJ9A	102	3004	2	2007	
2NJ9D	102	3004	2	2007	
2NL48			63	2030	
2NS121	102	3004	2	2007	
2NS31	102	3722	2	2007	
2NS32	102	3004	2	2007	
2NSM-1	125	3311	510,531	1114	
2NU/9WHT		3123	80		
2NU/9YEL		3123	80		
20A90	110MP			1123(2)	

Industry Standard No.	ECG	SK	GE	RS 276-	MOTOR.
20C72	158		53	2007	
02P1B	159	3466	82	2032	
02P1BC	189	3114/290	218	2026	
2P1M	5444	3634			
2P2150M	177		300	1122	
2P2M	5455	3597			
2PB187	102A	3004	53	2007	
2PD-358			1N34AS	1123	
2S-16E	116	3016	504A	1104	
2S001	123A	3124/289	20	2051	
2S002	123A	3124/289	20	2051	
2S003	123A	3124/289	20	2051	
2S004	123A	3124/289	20	2051	
2S005	123A	3124/289	20	2051	
2S006	108	3452	86	2038	
2S012			66	2048	
2S014	128	3020/123	243	2030	
2S017	128	3024	243	2030	
2S018	128	3024	243	2030	
2S019	128	3024	243	2030	
2S020	128	3024	243	2030	
2S021	129	3025	244	2027	
2S022	129	3025	244	2027	
2S023	129	3025	244	2027	
2S033	130	3027	14	2041	
2S034	130	3027	14	2041	
2S035	162	3438	35		
2S036	162	3438	35		
2S095A			20	2051	
2S10032	177	3100/519	300	1122	
2S101	123A	3444	20	2051	
2S102	123A	3444	20	2051	
2S103	123A	3024/128	243	2030	
2S104	123A	3024/128	243	2030	
2S109	126	3008	52	2024	
2S110	126	3007	52	2024	
2S111	100	3005	1	2007	
2S112	126	3008	52	2024	
2S1182D			20	2051	
2S12	100	3005	1	2007	
2S13	100	3005	1	2007	
2S131	123A	3444	20	2051	
2S134		3123	80		
2S14	102	3004	2	2007	
2S141	126	3006/160	52	2024	
2S142	126	3006/160	52	2024	
2S143	126	3006/160	52	2024	
2S144			51	2004	
2S145	126	3006/160	52	2024	
2S146	126	3006/160	52	2024	
2S148			51	2004	
2S15	102	3004	2	2007	
2S155	100	3005	1	2007	
2S159	100	3005	1	2007	
2S15A	102	3004	2	2007	
2S160	100	3005	1	2007	
2S163	102	3004	2	2007	
2S167	100	3005	1	2007	
2S1718/4454C	140A		ZD-10	562	
2S174	100	3005	1	2007	
2S175		3006	50	2004	
2S176	126	3006/160	52	2024	
2S1760			2	2007	
2S178	100	3005	1	2007	
2S179	102	3004	2	2007	
2S189	102A	3004	53	2007	
2S1K	116	3017B/117	504A	1104	
2S201			51	2004	
2S2114A	138A		ZD-7.5		
2S22	102	3004	2	2007	
2S24	102	3004	2	2007	
2S25	100	3005	1	2007	
2S26	121	3009	239	2006	
2S26A	121	3009	239	2006	
2S273	102A		53	2007	
2S30	100	3005	1	2007	
2S301	129	3025	244	2027	
2S3010	129	3025	244	2027	
2S302	129	3025	244	2027	
2S3020	159	3466	82	2032	
2S3020A		3114	82	2023	
2S3021	159	3466	82	2032	
2S3021A		3114	82	2023	
2S302A	129	3025	244	2027	
2S303	129	3025	244	2027	
2S3030	159	3466	82	2032	
2S3030A		3114	82	2023	
2S304	129	3025	244	2027	
2S3040	159	3466	82	2024	
2S3040A		3114	82	2023	
2S305		3025	221	2032	
2S306	159	3466	82	2032	
2S307	159	3715	82	2032	
2S307A		3715	82	2023	
2S31	100	3005	1	2007	
2S32	102	3004	2	2007	
2S321	159	3466	82	2032	
2S3210	159	3466	82	2032	
2S3210A		3114	82	2023	
2S321A		3114	82	2023	
2S322	159	3466	82	2032	
2S3220	159	3466	82	2032	
2S3220A		3114	82	2023	
2S3221		3466	82	2032	
2S3221A		3114	82	2023	
2S322A	159	3466	82	2032	
2S322AB		3114	82	2023	
2S323	159	3466	82	2032	
2S3230	159	3466	82	2032	
2S3230A		3114	82	2023	
2S323A		3114	82	2023	
2S324	159	3466	82	2032	
2S3240	159	3466	82	2032	
2S3240A		3114	82	2023	
2S324A		3114	82	2023	
2S326	159	3466	82	2032	
2S326A		3114	82	2023	
2S327	159	3466	82	2032	
2S327A		3114	89	2023	
2S32D				2004	
2S33	102	3004	2	2007	
2S3370			52	2024	
2S34	102	3004	2	2007	
2S35	126	3008	52	2024	
2S36	126	3008	52	2024	
2S363			20	2051	
2S37	102	3004	2	2007	
2S38	102	3004	2	2007	
2S39	102	3004	2	2007	
2S40	102	3004	2	2007	
2S41	104	3009	16	2006	
2S41A	104	3009	16	2006	
2S42	104	3009	16	2006	
2S43	102	3004	2	2007	

Industry Standard No.	ECG	SK	GE	RS 276-	MOTOR.
2S44	102	3004	2	2007	
2S45	100	3005	1	2007	
2S46	102	3003	2	2007	
2S47	102	3003	2	2007	
2S471-1			50	2004	
2S49	100	3005	1	2007	
2S494-YE		3114	89	2023	
2S501	123A	3444	20	2051	
2S502	123A	3444	20	2051	
2S503	123A	3444	20	2051	
2S51	100	3005	1	2007	
2S512	123A	3452/108	86	2038	
2S52	100	3005	1	2007	
2S53	100	3005	1	2007	
2S54	102	3004	2	2007	
2S56	102	3004	2	2007	
2S564R			21	2034	
2S57			51	2004	
2S58	126	3008	52	2024	
2S60	100	3005	1	2007	
00002S606	107	3039/316	11	2015	
2S644(S)			18	2030	
2S645			20	2051	
2S673C	159	3466	82	2032	
2S6856		3040	27		
2S701	123	3124/289	20	2051	
2S702	123	3124/289	20	2051	
2S703	123	3124/289	20	2051	
2S711	123A	3124/289	20	2051	
2S712	123A	3124/289	20	2051	
2S731	123A	3444	20	2051	
2S732	123A	3444	20	2051	
2S733	123A	3444	20	2051	
2S741	123	3024/128	243	2030	
2S741A	123A	3444	20	2051	
2S742	128	3024	243	2030	
2S742A	128	3024	243	2030	
2S743	154	3045/225	40	2012	
2S743A	154	3045/225	40	2012	
2S744	123	3024/128	243	2030	
2S744A	123A	3444	243	2051	
2S745	128	3024	243	2030	
2S745A	128	3045/225	243	2012	
2S746	154	3045/225	40	2012	
2S746A	154	3045/225	40	2012	
2S82	107		11	2015	
2S91	100	3005	1	2007	
2S92	100	3008	1	2007	
2S92A	100	3008	1	2007	
2S93	100	3008	1	2007	
2S93A	100	3008	1	2007	
2S95A	123A	3444	20	2051	
2S96	126	3006/160	52	2024	
2S97	126	3006/160	52	2024	
2S98	126	3006/160	52	2024	
2SA-4551		3006	51	2004	
2SA-4561		3006	51	2004	
2SA-NJ-101			22	2032	
2SA-NJ101	159	3466	82	2032	
2SA007H			53	2007	
2SA076F			51	2004	
2SA081C			53	2007	
2SA098R			51	2004	
2SA100	126	3004/102A	51	2024	
2SA1001					2N6438
2SA1002					2N6438
2SA1003					2N6438
2SA1007	88	3846/285			2N6231
2SA1007A	88	3846/285			
2SA1008	292				TIP32C
2SA100A	126	3004/102A	51	2005	
2SA100B	126	3004/102A	51	2024	
2SA100C	126	3004/102A	51	2024	
2SA100D		3004	51	2004	
2SA100E		3004	51	2004	
2SA100F		3004	51	2004	
2SA100G		3004	51	2004	
2SA100H		3004	51	2004	
2SA100J		3004	51	2004	
2SA100K		3004	51	2004	
2SA100M		3004	51	2004	
2SA1000R		3004	51	2004	
2SA100R		3004	51	2004	
2SA100X			51	2004	
2SA100Y		3004	51	2004	
2SA101	126	3007	52	2024	
2SA101-OR			51	2007	
2SA1010	332				TIP42C
2SA1010Y	332	3006/160	52	2024	
2SA1011	398				MJE15031
2SA1012	153$				TIP42A
2SA1015	290A	3114/290			
2SA1015-O	290A	3114/290	82		
2SA1015Y	290A	3114/290	82		
2SA1017	234	3867	221		
2SA1018	288		51	2004	
2SA1019	91	3867	221		
2SA101A	126	3007	52	2024	
2SA101AA	126	3007	52	2024	
2SA101AY	126	3007	52	2024	
2SA101B	126	3007	52	2024	
2SA101BA	126	3007	52	2024	
2SA101BB	126	3007	52	2024	
2SA101BC	126	3007	52	2024	
2SA101BX	126	3007	52	2024	
2SA101C	126	3007	52	2024	
2SA101CA	126	3007	52	2024	
2SA101CV	126	3007	52	2024	
2SA101CX	126	3007	52	2024	
2SA101D		3007	52	2005	
2SA101E	126	3007	52	2024	
2SA101F		3007	52	2005	
2SA101G		3007	52	2005	
2SA101H		3007	52	2005	
2SA101K		3007	52	2005	
2SA101L		3007	52	2005	
2SA101M		3007	52	2005	
2SA1010R		3007	52	2024	
2SA101QA	126		52	2024	
2SA101R		3007	52	2005	
2SA101V	126		52	2024	
2SA101X	126	3007	52	2024	
2SA101XBX		3007	52	2005	
2SA101Y	126	3007	52	2024	
2SA101YA		3007	52	2005	
2SA101Z	126	3007	52	2024	
2SA102	126	3007	52	2024	
2SA102(BA)		3007	52	2024	
2SA102-OR			51	2024	
2SA1020					TIP32
2SA1021	374	9042			
2SA1021-O	374	9042			
2SA1027R		3841	82		
2SA1028A			52	2024	
2SA1029	290A	3114/290	269	2032	
2SA1029B	290A	3114/290			
2SA1029C	290A	3114/290	269	2032	
2SA1029D	290A	3114/290			
2SA102A	126	3007	52	2024	
2SA102AA	126	3007	52	2024	
2SA102AB	126	3007	52	2024	
2SA102B	126	3007	52	2024	
2SA102BA	126	3007	52	2024	
2SA102BA-2		3007	52	2024	
2SA102BN	126	3007	52	2024	
2SA102C		3007	52	2024	
2SA102CA	126	3007	52	2024	
2SA102CA-1		3007	52	2024	
2SA102D		3007	52	2024	
2SA102E		3007	52	2024	
2SA102F		3007	52	2024	
2SA102G		3007	52	2024	
2SA102H		3007	52	2024	
2SA102K		3007	52	2024	
2SA102L		3007	52	2024	
2SA102M		3007	52	2024	
2SA1020R		3007	52	2024	
2SA102TV	126		52	2024	
2SA102TV-2		3007	52	2024	
2SA102X			52	2024	
2SA102Y		3007	52	2024	
2SA103	126	3005	52	2024	
2SA103(CA)			52	2004	
2SA1030	290A	3114/290			
2SA1030B	290A	3114/290			
2SA1030C	290A	3114/290			
2SA103A	126	3005	52	2024	
2SA103B	126	3005	52	2024	
2SA103C	126	3005	52	2024	
2SA103CA	126	3005	52	2024	
2SA103CAK	126	3005	52	2024	
2SA103CB			52	2004	
2SA103CG	126	3005	52	2024	
2SA103D		3005	52	2004	
2SA103DA	126	3005	52	2024	
2SA103E		3005	52	2004	
2SA103F		3005	52	2004	
2SA103G		3005	52	2004	
2SA103GA			52	2004	
2SA103K	126	3005	52	2024	
2SA103L		3005	52	2004	
2SA103M		3005	52	2004	
2SA1030R		3005	52	2004	
2SA103R		3005	52	2004	
2SA103X		3005	52	2004	
2SA103Y		3005	52	2004	
2SA104	126	3007	50	2024	
2SA1040	290A				2N6438
2SA1041					2N6438
2SA1042					2N6436
2SA1043					2N6438
2SA1044					2N6436
2SA1045					2N6052
2SA1046					2N6052
2SA104A		3007	50	2004	
2SA104B		3007	50	2004	
2SA104C		3007	50	2004	
2SA104D	126	3007	50	2024	
2SA104E		3007	50	2004	
2SA104F		3007	50	2004	
2SA104G		3007	50	2004	
2SA104H		3007	50	2004	
2SA104K		3007	50	2004	
2SA104L		3007	50	2004	
2SA104M		3007	50	2004	
2SA1040R		3007	50	2004	
2SA104P	126	3007	50	2024	
2SA104R		3007	50	2004	
2SA104X		3007	50	2004	
2SA104Y		3007	50	2004	
2SA105	126	3008	53	2004	
2SA105A		3008	53	2007	
2SA105B		3008	53	2007	
2SA105C		3008	53	2007	
2SA105D		3008	53	2007	
2SA105E		3008	53	2007	
2SA105G		3008	53	2007	
2SA105H		3008	53	2007	
2SA105K		3008	53	2007	
2SA105L		3008	53	2007	
2SA105M		3008	53	2007	
2SA1050R		3008	53	2007	
2SA105R		3008	53	2007	
2SA105X		3008	53	2007	
2SA105Y		3008	53	2007	
2SA106	126	3007	50	2004	
2SA1063	285				2N6228
2SA1064	285				2N6231
2SA1065	285				2N6231
2SA1067					2N6230
2SA1068					2N6231
2SA1069					TIP421B
2SA106A		3008	50	2004	
2SA106B		3007	50	2004	
2SA106C		3007	50	2004	
2SA106D		3007	50	2004	
2SA106E		3007	50	2004	
2SA106F		3007	50	2004	
2SA106G		3007	50	2004	
2SA106H		3007	50	2004	
2SA106K		3007	50	2004	
2SA106L		3007	50	2004	
2SA106M		3007	50	2004	
2SA1060R		3007	50	2004	
2SA106R		3007	50	2004	
2SA106X		3007	50	2004	
2SA106Y		3007	50	2004	
2SA107	126	3007	50	2004	
2SA107A		3007	50	2004	
2SA107B		3007	50	2004	
2SA107C		3007	50	2004	
2SA107D		3007	50	2004	
2SA107E		3007	50	2004	
2SA107F		3007	50	2004	
2SA107G		3007	50	2004	
2SA107H		3007	50	2004	
2SA107K		3007	50	2004	
2SA107L		3007	50	2004	
2SA107M		3007	50	2004	
2SA1070R		3007	50	2004	
2SA107R		3007	50	2004	
2SA107X		3007	50	2004	
2SA107Y		3007	50	2004	
2SA108	126	3007	50	2024	
2SA108A		3007	50	2004	
2SA108B		3007	50	2004	
2SA108C		3007	50	2004	
2SA108D		3007	50	2004	

Industry Standard No.	ECG	SK	GE	RS 276-	MOTOR.
2SA108E		3007	50	2004	
2SA108F		3007	50	2004	
2SA108G		3007	50	2004	
2SA108H		3007	50	2004	
2SA108K		3007	50	2004	
2SA108L		3007	50	2004	
2SA108M		3007	50	2004	
2SA108OR		3007	50	2004	
2SA108R		3007	50	2004	
2SA108X		3007	50	2004	
2SA108Y		3007	50	2004	
2SA109	126	3007	50	2024	
2SA109A		3007	50	2004	
2SA109B		3007	50	2004	
2SA109C		3007	50	2004	
2SA109D		3007	50	2004	
2SA109E		3007	50	2004	
2SA109F		3007	50	2004	
2SA109G		3007	50	2004	
2SA109K		3007	50	2004	
2SA109L		3007	50	2004	
2SA109M		3007	50	2004	
2SA109OR		3007	50	2004	
2SA109R		3007	50	2004	
2SA109X		3007	50	2004	
2SA109Y		3007	50	2004	
2SA110	126	3007	50	2024	
2SA110A		3007	50	2004	
2SA110B		3007	50	2004	
2SA110C		3007	50	2004	
2SA110D		3007	50	2004	
2SA110E		3007	50	2004	
2SA110F		3007	50	2004	
2SA110G		3007	50	2004	
2SA110K		3007	50	2004	
2SA110L		3007	50	2004	
2SA110M		3007	50	2004	
2SA110OR		3007	50	2004	
2SA110R		3007	50	2004	
2SA110X		3007	50	2004	
2SA110Y		3007	50	2004	
2SA111	126	3007	50	2024	
2SA1110	374#				MJE350
2SA1111					MJE15031
2SA1112					MJE15031
2SA111A		3007	50	2004	
2SA111B		3007	50	2004	
2SA111C		3007	50	2004	
2SA111D		3007	50	2004	
2SA111E		3007	50	2004	
2SA111F		3007	50	2004	
2SA111G		3007	50	2004	
2SA111K		3007	50	2004	
2SA111L		3007	50	2004	
2SA111M		3007	50	2004	
2SA111OR		3007	50	2004	
2SA111R		3007	50	2004	
2SA111X		3007	50	2004	
2SA111Y		3007	50	2004	
2SA112	126	3007	50	2024	
2SA112A		3007	50	2004	
2SA112B		3007	50	2004	
2SA112C		3007	50	2004	
2SA112D		3007	50	2004	
2SA112E		3007	50	2004	
2SA112F		3007	50	2004	
2SA112G		3007	50	2004	
2SA112GN		3007	50	2004	
2SA112H		3007	50	2004	
2SA112K		3007	50	2004	
2SA112L		3007	50	2004	
2SA112M		3007	50	2004	
2SA112OR		3007	50	2004	
2SA112R		3007	50	2004	
2SA112X		3007	50	2004	
2SA112Y		3007	50	2004	
2SA113	126	3007	50	2024	
2SA113A		3007	50	2004	
2SA113B		3007	50	2004	
2SA113C		3007	50	2004	
2SA113D		3007	50	2004	
2SA113E		3007	50	2004	
2SA113F		3007	50	2004	
2SA113G		3007	50	2004	
2SA113GN		3007	50	2004	
2SA113H		3007	50	2004	
2SA113J		3007	50	2004	
2SA113L		3007	50	2004	
2SA113M		3007	50	2004	
2SA113R		3007	50	2004	
2SA113X		3007	50	2004	
2SA113Y		3007	50	2004	
2SA114	126	3007	50	2005	
2SA114A		3007	50	2004	
2SA114B		3007	50	2004	
2SA114C		3007	50	2004	
2SA114D		3007	50	2004	
2SA114E		3007	50	2004	
2SA114F		3007	50	2004	
2SA114G		3007	50	2004	
2SA114H		3007	50	2004	
2SA114K		3007	50	2004	
2SA114L		3007	50	2004	
2SA114M		3007	50	2004	
2SA114OR		3007	50	2004	
2SA114R		3007	50	2004	
2SA114X		3007	50	2004	
2SA114Y		3007	50	2004	
2SA115	126	3007	50	2005	
2SA115A		3007	50	2004	
2SA115B		3007	50	2004	
2SA115C		3007	50	2004	
2SA115D		3007	50	2004	
2SA115E		3007	50	2004	
2SA115F		3007	50	2004	
2SA115G		3007	50	2004	
2SA115GN		3007	50	2004	
2SA115H		3007	50	2004	
2SA115J		3007	50	2004	
2SA115K		3007	50	2004	
2SA115L		3007	50	2004	
2SA115M		3007	50	2004	
2SA115OR		3007	50	2004	
2SA115R		3007	50	2004	
2SA115X		3007	50	2004	
2SA115Y		3007	50	2004	
2SA116	126	3006/160	51	2005	
2SA116A		3006	51	2004	
2SA116B		3006	51	2004	
2SA116C		3006	51	2004	
2SA116D		3006	51	2004	
2SA116E		3006	51	2004	
2SA116F		3006	51	2004	
2SA116G		3006	51	2004	
2SA116GN		3006	51	2004	
2SA116H		3006	51	2004	
2SA116J		3006	51	2004	
2SA116K		3006	51	2004	
2SA116L		3006	51	2004	
2SA116M		3006	51	2004	
2SA116OR		3006	51	2004	
2SA116R		3006	51	2004	
2SA116X		3006	51	2004	
2SA116Y		3006	51	2004	
2SA117	126	3006/160	51	2004	
2SA117A		3006	51	2004	
2SA117B		3006	51	2004	
2SA117C		3006	51	2004	
2SA117D		3006	51	2004	
2SA117E		3006	51	2004	
2SA117F		3006	51	2004	
2SA117G		3006	51	2004	
2SA117GN		3006	51	2004	
2SA117H		3006	51	2004	
2SA117J		3006	51	2004	
2SA117K		3006	51	2004	
2SA117L		3006	51	2004	
2SA117M		3006	51	2004	
2SA117OR		3006	51	2004	
2SA117R		3006	51	2004	
2SA117X		3006	51	2004	
2SA117Y		3006	51	2004	
2SA118	126	3006/160	51	2004	
2SA118A		3006	51	2004	
2SA118B		3006	51	2004	
2SA118C		3006	51	2004	
2SA118D		3006	51	2004	
2SA118E		3006	51	2004	
2SA118F		3006	51	2004	
2SA118G		3006	51	2004	
2SA118GN		3006	51	2004	
2SA118H		3006	51	2004	
2SA118J		3006	51	2004	
2SA118K		3006	51	2004	
2SA118L		3006	51	2004	
2SA118M		3006	51	2004	
2SA118OR		3006	51	2004	
2SA118R		3006	51	2004	
2SA118X		3006	51	2004	
2SA118Y		3006	51	2004	
2SA119	129			2032	
2SA12	102A	3003	53	2005	
2SA120	129	3003			
2SA121	126	3006/160	51	2024	
2SA121A		3006	51	2004	
2SA121B		3006	51	2004	
2SA121C		3006	51	2004	
2SA121D		3006	51	2004	
2SA121E		3006	51	2004	
2SA121F		3006	51	2004	
2SA121G		3006	51	2004	
2SA121GN		3006	51	2004	
2SA121H		3006	51	2004	
2SA121J		3006	51	2004	
2SA121K		3006	51	2004	
2SA121L		3006	51	2004	
2SA121M		3006	51	2004	
2SA121OR			51	2004	
2SA121R		3006	51	2004	
2SA121X		3006	51	2004	
2SA121Y		3006	51	2004	
2SA122	126	3006/160	52	2024	
2SA122A		3006	51	2004	
2SA122B		3006	51	2004	
2SA122C		3006	51	2004	
2SA122D		3006	51	2004	
2SA122E		3006	51	2004	
2SA122F		3006	51	2004	
2SA122G		3006	51	2004	
2SA122GN		3006	51	2004	
2SA122H		3006	51	2004	
2SA122K		3006	51	2004	
2SA122L		3006	51	2004	
2SA122M		3006	51	2004	
2SA122OR		3006	51	2004	
2SA122R		3006	51	2004	
2SA122X		3006	51	2004	
2SA122Y		3006	51	2004	
2SA123	126	3006/160	51	2024	
2SA123A		3006	51	2004	
2SA123B		3006	51	2004	
2SA123C		3006	51	2004	
2SA123D		3006	51	2004	
2SA123E		3006	51	2004	
2SA123F		3006	51	2004	
2SA123G		3006	51	2004	
2SA123GN		3006	51	2004	
2SA123H		3006	51	2004	
2SA123J		3006	51	2004	
2SA123K		3006	51	2004	
2SA123L		3006	51	2004	
2SA123M		3006	51	2004	
2SA123OR		3006	51	2004	
2SA123R		3006	51	2004	
2SA123X		3006	51	2004	
2SA123Y		3006	51	2004	
2SA124	126	3006/160	51	2024	
2SA124A		3006	51	2004	
2SA124B		3006	51	2004	
2SA124C		3006	51	2004	
2SA124D		3006	51	2004	
2SA124E		3006	51	2004	
2SA124F		3006	51	2004	
2SA124G		3006	51	2004	
2SA124GN		3006	51	2004	
2SA124H		3006	51	2004	
2SA124J		3006	51	2004	
2SA124K		3006	51	2004	
2SA124L		3006	51	2004	
2SA124M		3006	51	2004	
2SA124OR		3006	51	2004	
2SA124R		3006	51	2004	
2SA124X		3006	51	2004	
2SA124Y		3006	51	2004	
2SA125	126	3006/160	51	2024	
2SA125A		3006	51	2004	
2SA125B		3006	51	2004	
2SA125C		3006	51	2004	
2SA125D		3006	51	2004	
2SA125E		3006	51	2004	
2SA125F		3006	51	2004	
2SA125G		3006	51	2004	
2SA125GN		3006	51	2004	
2SA125H		3006	51	2004	
2SA125J		3006	51	2004	
2SA125K		3006	51	2004	
2SA125L		3006	51	2004	
2SA125M		3006	51	2004	

Industry Standard No.	ECG	SK	GE	RS 276-	MOTOR.
2SA125OR		3006	51	2004	
2SA125R		3006	51	2004	
2SA125X		3006	51	2004	
2SA125Y		3006	51	2004	
2SA126	126		245	2004	
2SA127			245	2004	
2SA128	158	3008	53	2007	
2SA128A		3008	53	2007	
2SA128B		3008	53	2007	
2SA128C		3008	53	2007	
2SA128D		3008	53	2007	
2SA128E		3008	53	2007	
2SA128F		3008	53	2007	
2SA128G		3008	53	2007	
2SA128GN		3008	53	2007	
2SA128H		3008	53	2007	
2SA128J		3008	53	2007	
2SA128K		3008	53	2007	
2SA128L		3008	53	2007	
2SA128M		3008	53	2007	
2SA128OR		3008	53	2007	
2SA128R		3008	53	2007	
2SA128X		3008	53		
2SA128Y		3008	53	2007	
2SA129	158	3008	53	2007	
2SA129A		3008	53	2007	
2SA129B		3008	53	2007	
2SA129C		3008	53	2007	
2SA129D		3008	53	2007	
2SA129E		3008	53	2007	
2SA129F		3008	53	2007	
2SA129G		3008	53		
2SA129GN		3008	53	2007	
2SA129H		3008	53	2007	
2SA129J		3008	53	2007	
2SA129K		3008	53	2007	
2SA129L		3008	53	2007	
2SA129M		3008	53	2007	
2SA129OR		3008	53	2007	
2SA129R		3008	53	2007	
2SA129X		3008	53	2007	
2SA129Y		3008	53	2007	
2SA12A	102A	3003	53	2005	
2SA12B	102A	3003	53	2005	
2SA12C	102A	3003	53	2005	
2SA12D	102A	3003	53	2005	
2SA12H	102A	3003	53	2005	
2SA12V	102A	3003	53	2005	
2SA13	102A	3005	51	2004	
2SA130	160	3005	51	2004	
2SA130A		3005	51	2004	
2SA130B		3005	51	2004	
2SA130C		3005	51	2004	
2SA130D		3005	51	2004	
2SA130E		3005	51	2004	
2SA130F		3005	51	2004	
2SA130G		3005	51	2004	
2SA130GN		3005	51	2004	
2SA130H		3005	51	2004	
2SA130J		3005	51	2004	
2SA130K		3005	51	2004	
2SA130L		3005	51	2004	
2SA130M		3005	51	2004	
2SA130OR			51	2004	
2SA130R		3005	51	2004	
2SA130X		3005	51	2004	
2SA130Y		3005	51	2004	
2SA131	126	3005	51	2024	
2SA131A		3005	51	2004	
2SA131B		3005	51	2004	
2SA131C		3005	51	2004	
2SA131D		3005	51	2004	
2SA131E		3005	51	2004	
2SA131F		3005	51	2004	
2SA131G		3005	51	2004	
2SA131GN		3005	51	2004	
2SA131H		3005	51	2004	
2SA131J		3005	51	2004	
2SA131K		3005	51	2004	
2SA131L		3005	51	2004	
2SA131M		3005	51	2004	
2SA131OR		3005	51	2004	
2SA131R		3005	51	2004	
2SA131X		3005	51	2004	
2SA131Y		3005	51	2004	
2SA132	126	3005	51	2024	
2SA132A		3005	51	2004	
2SA132B		3005	51	2004	
2SA132C		3005	51	2004	
2SA132D		3005	51	2004	
2SA132E		3005	51	2004	
2SA132F		3005	51	2004	
2SA132G		3005	51	2004	
2SA132H		3005	51	2004	
2SA132J		3005	51	2004	
2SA132K		3005	51	2004	
2SA132L		3005	51	2004	
2SA132M		3005	51	2004	
2SA132OR		3005	51	2004	
2SA132X		3005	51	2004	
2SA132Y		3005	51	2004	
2SA133	126	3006/160	51	2024	
2SA133A		3006	51	2004	
2SA133B		3006	51	2004	
2SA133C		3006	51	2004	
2SA133D		3006	51	2004	
2SA133E		3006	51	2004	
2SA133F		3006	51	2004	
2SA133G		3006	51	2004	
2SA133GN		3006	51	2004	
2SA133H		3006	51	2004	
2SA133J		3006	51	2004	
2SA133K		3006	51	2004	
2SA133L		3006	51	2004	
2SA133M		3006	51	2004	
2SA133OR		3006	51	2004	
2SA133R		3006	51	2004	
2SA133X		3006	51	2004	
2SA133Y		3006	51	2004	
2SA134	126	3006/160	51	2024	
2SA134A		3006	51	2004	
2SA134B		3006	51	2004	
2SA134C		3006	51	2004	
2SA134D		3006	51	2004	
2SA134E		3006	51	2004	
2SA134F		3006	51	2004	
2SA134G		3006	51	2004	
2SA134H		3006	51	2004	
2SA134J		3006	51	2004	
2SA134K		3006	51	2004	
2SA134L		3006	51	2004	
2SA134OR		3006	51	2004	
2SA134R		3006	51	2004	
2SA134X		3006	51	2004	
2SA134Y		3006	51	2004	
2SA135	160	3005	51	2004	
2SA135A		3005	51	2004	
2SA135B		3005	51	2004	
2SA135C		3005	51	2004	
2SA135D		3005	51	2004	
2SA135E		3005	51	2004	
2SA135F		3005	51	2004	
2SA135G		3005	51	2004	
2SA135GN		3005	51	2004	
2SA135H		3005	51	2004	
2SA135J		3005	51	2004	
2SA135K		3005	51	2004	
2SA135L		3005	51	2004	
2SA135M		3005	51	2004	
2SA135OR		3005	51	2004	
2SA135R		3005	51	2004	
2SA135X		3005	51	2004	
2SA135Y		3005	51	2004	
2SA136	126	3006/160	51	2024	
2SA136A		3006	51	2004	
2SA136B		3006	51	2004	
2SA136C		3006	51	2004	
2SA136D		3006	51	2004	
2SA136E		3006	51	2004	
2SA136F		3006	51	2004	
2SA136G		3006	51	2004	
2SA136GN		3006	51	2004	
2SA136H		3006	51	2004	
2SA136J		3006	51	2004	
2SA136K		3006	51	2004	
2SA136L		3006	51	2004	
2SA136M		3006	51	2004	
2SA136OR		3006	51	2004	
2SA136R		3006	51	2004	
2SA136X		3006	51	2004	
2SA136Y		3006	51	2004	
2SA137	126	3005	51	2024	
2SA137A		3005	51	2004	
2SA137B		3005	51	2004	
2SA137C		3005	51	2004	
2SA137D		3005	51	2004	
2SA137E		3005	51	2004	
2SA137F		3005	51	2004	
2SA137G		3005	51	2004	
2SA137GN		3005	51	2004	
2SA137H		3005	51	2004	
2SA137J		3005	51	2004	
2SA137K		3005	51	2004	
2SA137L		3005	51	2004	
2SA137M		3005	51	2004	
2SA137OR		3005	51	2004	
2SA137R		3005	51	2004	
2SA137X		3005	51	2004	
2SA137Y		3005	51	2004	
2SA138	102A		53	2005	
2SA139	126	3006/160	51	2024	
2SA139A		3006	51	2004	
2SA139B		3006	51	2004	
2SA139C		3006	51	2004	
2SA139D		3006	51	2004	
2SA139E		3006	51	2004	
2SA139F		3006	51	2004	
2SA139G		3006	51	2004	
2SA139GN		3006	51	2004	
2SA139J		3006	51	2004	
2SA139K		3006	51	2004	
2SA139L		3006	51	2004	
2SA139M		3006	51	2004	
2SA139OR		3006	51	2004	
2SA139R		3006	51	2004	
2SA139X		3006	51	2004	
2SA139Y		3006	51	2004	
2SA13A		3005	51	2004	
2SA13B		3005	51	2004	
2SA13C		3005	51	2004	
2SA13D		3005	51	2004	
2SA13G		3005	51	2004	
2SA13L		3005	51	2004	
2SA13M		3005	51	2004	
2SA13OR		3005	51	2004	
2SA13R		3005	51	2004	
2SA13X		3005	51	2004	
2SA13Y		3005	51	2004	
2SA14	102A	3005	51	2004	
2SA141	126	3007	51	2024	
2SA141A		3006	51	2004	
2SA141B	126	3006/160	51	2024	
2SA141C	126	3006/160	51	2024	
2SA141D		3006	51	2004	
2SA141E		3006	51	2004	
2SA141F		3006	51	2004	
2SA141G		3006	51	2004	
2SA141GN		3006	51	2004	
2SA141H		3006	51	2004	
2SA141K		3006	51	2004	
2SA141L		3006	51	2004	
2SA141M		3006	51	2004	
2SA141OR		3006	51	2004	
2SA141R		3006	51	2004	
2SA141X		3006	51	2004	
2SA141Y		3006	51	2004	
2SA142	126	3005	53	2024	
2SA142A	126	3005	53	2024	
2SA142B	126	3005	53	2024	
2SA142C	126	3005	53	2024	
2SA142D		3005	53	2024	
2SA142E		3005	53	2024	
2SA142F		3005	53	2024	
2SA142G		3005	53	2024	
2SA142GN		3005	53	2024	
2SA142H		3005	53	2024	
2SA142J		3005	53	2024	
2SA142K		3005	53	2024	
2SA142L		3005	53	2024	
2SA142M		3005	53	2024	
2SA142OR		3005	53	2024	
2SA142R		3005	53	2024	
2SA142X			53	2024	
2SA142Y		3005	53	2024	
2SA143	126	3006/160	51	2024	
2SA143A		3006	51	2004	
2SA143B		3006	51	2004	
2SA143C		3006	51	2004	
2SA143D		3006	51	2004	
2SA143E		3006	51	2004	
2SA143F		3006	51	2004	
2SA143G		3006	51	2004	
2SA143GN		3006	51	2004	
2SA143H		3006	51	2004	
2SA143J		3006	51	2004	
2SA143K		3006	51	2004	
2SA143L		3006	51	2004	
2SA143M		3006	51	2004	

Industry Standard No.	ECG	SK	GE	RS 276-	MOTOR.
2SA1430R		3006	51	2004	
2SA143R		3006	51	2004	
2SA143X		3006	51	2004	
2SA143Y		3006	51	2004	
2SA144	126	3005	51	2024	
2SA144A		3005	51	2004	
2SA144B		3005	51	2004	
2SA144C	126	3005	51	2024	
2SA144D		3005	51	2004	
2SA144E		3005	51	2004	
2SA144F		3005	51	2004	
2SA144G		3005	51	2004	
2SA144GN		3005	51	2004	
2SA144H		3005	51	2004	
2SA144J		3005	51	2004	
2SA144K		3005	51	2004	
2SA144L		3005	51	2004	
2SA144M		3005	51	2004	
2SA1440R		3005	51	2004	
2SA144R		3005	51	2004	
2SA144X		3005	51	2004	
2SA144Y		3005	51	2004	
2SA145	126	3005	51	2024	
2SA145A	126	3005	51	2024	
2SA145B		3005	51	2004	
2SA145C	126	3005	51	2024	
2SA145D		3005	51	2004	
2SA145E		3005	51	2004	
2SA145G		3005	51	2004	
2SA145GN		3005	51	2004	
2SA145K		3005	51	2004	
2SA145M		3005	51	2004	
2SA1450R		3005	51	2004	
2SA145R		3005	51	2004	
2SA145X		3005	51	2004	
2SA145Y		3005	51	2004	
2SA146	126	3005	51	2007	
2SA146A		3005	51	2004	
2SA146B		3005	51	2004	
2SA146C		3005	51	2004	
2SA146D		3005	51	2004	
2SA146E		3005	51	2004	
2SA146F		3005	51	2004	
2SA146G		3005	51	2004	
2SA146GN		3005	51	2004	
2SA146H		3005	51	2004	
2SA146J		3005	51	2004	
2SA146K		3005	51	2004	
2SA146L		3005	51	2004	
2SA146M		3005	51	2004	
2SA1460R		3005	51	2004	
2SA146X		3005	51	2004	
2SA146Y		3005	51	2004	
2SA147	126	3005	51	2007	
2SA147A		3005	51	2004	
2SA147B		3005	51	2004	
2SA147C		3005	51	2004	
2SA147D		3005	51	2004	
2SA147E		3005	51	2004	
2SA147F		3005	51	2004	
2SA147G		3005	51	2004	
2SA147H		3005	51	2004	
2SA147J		3005	51	2004	
2SA147K		3005	51	2004	
2SA147L		3005	51	2004	
2SA147M		3005	51	2004	
2SA1470R		3005	51	2004	
2SA147R		3005	51	2004	
2SA147X		3005	51	2004	
2SA147Y		3005	51	2004	
2SA148	126	3005	51	2007	
2SA148A		3005	51	2004	
2SA148B		3005	51	2004	
2SA148C		3005	51	2004	
2SA148D		3005	51	2004	
2SA148E		3005	51	2004	
2SA148F		3005	51	2004	
2SA148G		3005	51	2004	
2SA148GN		3005	51	2004	
2SA148H		3005	51	2004	
2SA148J		3005	51	2004	
2SA148K		3005	51	2004	
2SA148L		3005	51	2004	
2SA148M		3005	51	2004	
2SA1480R		3005	51	2004	
2SA148R		3005	51	2004	
2SA148X		3005	51	2004	
2SA148Y		3005	51	2004	
2SA149	126	3005	51	2007	
2SA149A		3005	51	2004	
2SA149B		3005	51	2004	
2SA149C		3005	51	2004	
2SA149D		3005	51	2004	
2SA149E		3005	51	2004	
2SA149F		3005	51	2004	
2SA149G		3005	51	2004	
2SA149GN		3005	51	2004	
2SA149H		3005	51	2004	
2SA149J		3005	51	2004	
2SA149K		3005	51	2004	
2SA149L		3005	51	2004	
2SA149M		3005	51	2004	
2SA1490R		3005	51	2004	
2SA149R		3005	51	2004	
2SA149X		3005	51	2004	
2SA149Y		3005	51	2004	
2SA14A		3005	51	2004	
2SA14B		3005	51	2004	
2SA14C		3005	51	2004	
2SA14D		3005	51	2004	
2SA14E		3005	51	2004	
2SA14G		3005	51	2004	
2SA14L		3005	51	2004	
2SA14M		3005	51	2004	
2SA140R		3005	51	2004	
2SA14R		3005	51	2004	
2SA14X		3005	51	2004	
2SA14Y		3005	51	2004	
2SA15	102A	3005	51	2005	
2SA15-6		3005	51	2004	
2SA150	126	3008	53	2024	
2SA150A		3008	53	2007	
2SA150B		3008	53	2007	
2SA150C		3008	53	2007	
2SA150D		3008	53	2007	
2SA150E		3008	53	2007	
2SA150F		3008	53	2007	
2SA150G		3008	53	2007	
2SA150GN		3008	53	2007	
2SA150H		3008	53	2007	
2SA150J		3008	53	2007	
2SA150K		3008	53	2007	
2SA150L		3008	53	2007	
2SA150M		3008	53	2007	
2SA1500R		3008	53	2007	
2SA150R		3008	53	2007	
2SA150X		3008	53	2007	
2SA150Y		3008	53	2007	
2SA151	126	3005	51	2004	
2SA151A		3005	51	2004	
2SA151B		3005	51	2004	
2SA151C		3005	51	2004	
2SA151D		3005	51	2004	
2SA151E		3005	51	2004	
2SA151F		3005	51	2004	
2SA151G		3005	51	2004	
2SA151GN		3005	51	2004	
2SA151H		3005	51	2004	
2SA151J		3005	51	2004	
2SA151K		3005	51	2004	
2SA151L		3005	51	2004	
2SA151M		3005	51	2004	
2SA1510R		3005	51	2004	
2SA151R		3005	51	2004	
2SA151X		3005	51	2004	
2SA151Y		3005	51	2004	
2SA152	126	3005	51	2004	
2SA152A		3005	51	2024	
2SA152B		3005	51	2024	
2SA152C		3005	51	2024	
2SA152D		3005	51	2024	
2SA152E		3005	51	2024	
2SA152F		3005	51	2024	
2SA152G		3005	51	2024	
2SA152GN		3005	51	2024	
2SA152H		3005	51	2024	
2SA152J		3005	51	2024	
2SA152K		3005	51	2024	
2SA152L		3005	51	2024	
2SA152M		3005	51	2024	
2SA1520R		3005	51	2024	
2SA152R		3005	51	2024	
2SA152X		3005	51	2024	
2SA152Y		3005	51	2024	
2SA153	126	3006/160	51	2024	
2SA153A		3006	51	2004	
2SA153B		3006	51	2004	
2SA153C		3006	51	2004	
2SA153D		3006	51	2004	
2SA153E		3006	51	2004	
2SA153F		3006	51	2004	
2SA153G		3006	51	2004	
2SA153GN		3006	51	2004	
2SA153H		3006	51	2004	
2SA153J		3006	51	2004	
2SA153K		3006	51	2004	
2SA153L		3006	51	2004	
2SA153M		3006	51	2004	
2SA1530R		3006	51	2004	
2SA153R		3006	51	2004	
2SA153X		3006	51	2004	
2SA153Y		3006	51	2004	
2SA154	126	3006/160	51	2024	
2SA154A		3006	51	2004	
2SA154B		3006	51	2004	
2SA154C		3006	51	2004	
2SA154D		3006	51	2004	
2SA154E		3006	51	2004	
2SA154F		3006	51	2004	
2SA154G		3006	51	2004	
2SA154GN		3006	51	2004	
2SA154H		3006	51	2004	
2SA154J		3006	51	2004	
2SA154K		3006	51	2004	
2SA154L		3006	51	2004	
2SA154M		3006	51	2004	
2SA1540R		3006	51	2004	
2SA154R		3006	51	2004	
2SA154X		3006	51	2004	
2SA154Y		3006	51	2004	
2SA155	126	3006/160	51	2024	
2SA155A		3006	51	2004	
2SA155B		3006	51	2004	
2SA155C		3006	51	2004	
2SA155D		3006	51	2004	
2SA155E		3006	51	2004	
2SA155F		3006	51	2004	
2SA155G		3006	51	2004	
2SA155GN		3006	51	2004	
2SA155H		3006	51	2004	
2SA155J		3006	51	2004	
2SA155K		3006	51	2004	
2SA155L		3006	51	2004	
2SA155M		3006	51	2004	
2SA1550R		3006	51	2004	
2SA155R		3006	51	2004	
2SA155X		3006	51	2004	
2SA155Y		3006	51	2004	
2SA156	126	3006/160	51	2024	
2SA156A		3006	51	2004	
2SA156B		3006	51	2004	
2SA156C		3006	51	2004	
2SA156D		3006	51	2004	
2SA156E		3006	51	2004	
2SA156F		3006	51	2004	
2SA156G		3006	51	2004	
2SA156GN		3006	51	2004	
2SA156H		3006	51	2004	
2SA156J		3006	51	2004	
2SA156K		3006	51	2004	
2SA156L		3006	51	2004	
2SA156M		3006	51	2004	
2SA1560R		3006	51	2004	
2SA156R		3006	51	2004	
2SA156X		3006	51	2004	
2SA156Y		3006	51	2004	
2SA157	126	3006/160	51	2024	
2SA157A		3006	51	2004	
2SA157B		3006	51	2004	
2SA157C		3006	51	2004	
2SA157D		3006	51	2004	
2SA157E		3006	51	2004	
2SA157F		3006	51	2004	
2SA157G		3006	51	2004	
2SA157GN		3006	51	2004	
2SA157H		3006	51	2004	
2SA157J		3006	51	2004	
2SA157K		3006	51	2004	
2SA157L		3006	51	2004	
2SA157M		3006	51	2004	
2SA1570R		3006	51	2004	
2SA157R		3006	51	2004	
2SA157X		3006	51	2004	
2SA157Y		3006	51	2004	
2SA159	126	3006/160	51	2024	
2SA159A		3006	51	2004	
2SA159B		3006	51	2004	
2SA159C		3006	51	2004	

Industry Standard No.	ECG	SK	GE	RS 276-	MOTOR.
2SA159D		3006	51	2004	
2SA159E		3006	51	2004	
2SA159F		3006	51	2004	
2SA159G		3006	51	2004	
2SA159GN		3006	51	2004	
2SA159H		3006	51	2004	
2SA159J		3006	51	2004	
2SA159K		3006	51	2004	
2SA159L		3006	51	2004	
2SA159M		3006	51	2004	
2SA159OR		3006	51	2004	
2SA159R		3006	51	2004	
2SA159X		3006	51	2004	
2SA159Y		3006	51	2004	
2SA15A		3005	51	2004	
2SA15B		3005	51	2004	
2SA15BK	102A	3005	51	2005	
2SA15BL	102A		51	2007	
2SA15BLU	102A		51	2005	
2SA15C		3005	51	2004	
2SA15D		3005	51	2004	
2SA15E		3005	51	2004	
2SA15F		3005	51	2004	
2SA15G		3005	51	2004	
2SA15H	102A	3005	51	2005	
2SA15K	102A	3005	51	2007	
2SA15L		3005	51	2004	
2SA15M		3005	51	2004	
2SA15OR		3005	51	2004	
2SA15R	102A	3005	51	2005	
2SA15RD			50	2004	
2SA15U	102A	3005	51	2007	
2SA15V	102A	3005	51	2005	
2SA15VR	102A		51	2007	
2SA15X			51	2004	
2SA15Y	102A	3005	51	2005	
2SA16	102A	3005	51	2005	
2SA160	126	3005	51	2007	
2SA160A		3005	51	2004	
2SA160B		3005	51	2004	
2SA160C		3005	51	2004	
2SA160D		3005	51	2004	
2SA160E		3005	51	2004	
2SA160F		3005	51	2004	
2SA160GN		3005	51	2004	
2SA160H		3005	51	2004	
2SA160J		3005	51	2004	
2SA160K		3005	51	2004	
2SA160L		3005	51	2004	
2SA160M		3005	51	2004	
2SA160OR		3005	51	2004	
2SA160R		3005	51	2004	
2SA160X		3005	51	2004	
2SA160Y		3005	51	2004	
2SA161	126	3006/160	51	2024	
2SA161A		3006	51	2004	
2SA161B		3006	51	2004	
2SA161C		3006	51	2004	
2SA161D		3006	51	2004	
2SA161E		3006	51	2004	
2SA161F		3006	51	2004	
2SA161G		3006	51	2004	
2SA161GN		3006	51	2004	
2SA161H		3006	51	2004	
2SA161J		3006	51	2004	
2SA161K		3006	51	2004	
2SA161L		3006	51	2004	
2SA161M		3006	51	2004	
2SA161OR		3006	51	2004	
2SA161R		3006	51	2004	
2SA161X		3006	51	2004	
2SA161Y		3006	51	2004	
2SA162	126	3006/160	51	2024	
2SA162A		3006	51	2004	
2SA162B		3006	51	2004	
2SA162C		3006	51	2004	
2SA162D		3006	51	2004	
2SA162E		3006	51	2004	
2SA162F		3006	51	2004	
2SA162G		3006	51	2004	
2SA162GN		3006	51	2004	
2SA162H		3006	51	2004	
2SA162J		3006	51	2004	
2SA162K		3006	51	2004	
2SA162L		3006	51	2004	
2SA162M		3006	51	2004	
2SA162OR		3006	51	2004	
2SA162R		3006	51	2004	
2SA162X		3006	51	2004	
2SA162Y		3006	51	2004	
2SA163	126	3006/160	51	2024	
2SA163A		3006	51	2004	
2SA163B		3006	51	2004	
2SA163C		3006	51	2004	
2SA163D		3006	51	2004	
2SA163E		3006	51	2004	
2SA163F		3006	51	2004	
2SA163G		3006	51	2004	
2SA163GN		3006	51	2004	
2SA163H		3006	51	2004	
2SA163J		3006	51	2004	
2SA163K		3006	51	2004	
2SA163L		3006	51	2004	
2SA163M		3006	51	2004	
2SA163OR		3006	51	2004	
2SA163R		3006	51	2004	
2SA163X		3006	51	2004	
2SA163Y		3006	51	2004	
2SA164	126	3006/160	51	2024	
2SA164A		3006	51	2004	
2SA164B		3006	51	2004	
2SA164C		3006	51	2004	
2SA164D		3006	51	2004	
2SA164E		3006	51	2004	
2SA164F		3006	51	2004	
2SA164G		3006	51	2004	
2SA164GN		3006	51	2004	
2SA164H		3006	51	2004	
2SA164J		3006	51	2004	
2SA164K		3006	51	2004	
2SA164L		3006	51	2004	
2SA164M		3006	51	2004	
2SA164OR		3006	51	2004	
2SA164R		3006	51	2004	
2SA164X		3006	51	2004	
2SA164Y		3006	51	2004	
2SA165	126	3006/160	51	2024	
2SA165A		3006	51	2004	
2SA165B		3006	51	2004	
2SA165C		3006	51	2004	
2SA165D		3006	51	2004	
2SA165E		3006	51	2004	
2SA165F		3006	51	2004	
2SA165G		3006	51	2004	
2SA165GN		3006	51	2004	
2SA165H		3006	51	2004	
2SA165J		3006	51	2004	
2SA165K		3006	51	2004	
2SA165L		3006	51	2004	
2SA165M		3006	51	2004	
2SA165OR		3006	51	2004	
2SA165R		3006	51	2004	
2SA165X		3006	51	2004	
2SA165Y		3006	51	2004	
2SA166	126	3006/160	51	2024	
2SA166A		3006	51	2004	
2SA166B		3006	51	2004	
2SA166C		3006	51	2004	
2SA166D		3006	51	2004	
2SA166E		3006	51	2004	
2SA166F		3006	51	2004	
2SA166G		3006	51	2004	
2SA166GN		3006	51	2004	
2SA166H		3006	51	2004	
2SA166J		3006	51	2004	
2SA166K		3006	51	2004	
2SA166L		3006	51	2004	
2SA166M		3006	51	2004	
2SA166OR		3006	51	2004	
2SA166R		3006	51	2004	
2SA166X		3006	51	2004	
2SA166Y		3006	51	2004	
2SA167	100	3005	1	2007	
2SA167A		3005	51	2004	
2SA167B		3005	51	2004	
2SA167C		3005	51	2004	
2SA167D		3005	51	2004	
2SA167E		3005	51	2004	
2SA167F		3005	51	2004	
2SA167G		3005	51	2004	
2SA167GN		3005	51	2004	
2SA167H		3005	51	2004	
2SA167J		3005	51	2004	
2SA167K		3005	51	2004	
2SA167L		3005	51	2004	
2SA167M		3005	51	2004	
2SA167OR		3005	51	2004	
2SA167R		3005	51	2004	
2SA167X		3005	51	2004	
2SA167Y		3005	51	2004	
2SA168	100	3005	51	2007	
2SA168A	100	3005	51	2007	
2SA168B		3005	51	2004	
2SA168C		3005	51	2004	
2SA168D		3005	51	2004	
2SA168E		3005	51	2004	
2SA168F		3005	51	2004	
2SA168G		3005	51	2004	
2SA168GN		3005	51	2004	
2SA168H		3005	51	2004	
2SA168J		3005	51	2004	
2SA168K		3005	51	2004	
2SA168L		3005	51	2004	
2SA168M		3005	51	2004	
2SA168OR		3005	51	2004	
2SA168R		3005	51	2004	
2SA168X		3005	51	2004	
2SA168Y		3005	51	2004	
2SA169	100	3005	1	2007	
2SA169A		3005	51	2004	
2SA169B		3005	51	2004	
2SA169C		3005	51	2004	
2SA169D		3005	51	2004	
2SA169E		3005	51	2004	
2SA169F		3005	51	2004	
2SA169G		3005	51	2004	
2SA169GN		3005	51	2004	
2SA169H		3005	51	2004	
2SA169J		3005	51	2004	
2SA169K		3005	51	2004	
2SA169L		3005	51	2004	
2SA169M		3005	51	2004	
2SA169OR		3005	51	2004	
2SA169R		3005	51	2004	
2SA169X		3005	51	2004	
2SA169Y		3005	51	2004	
2SA16A		3005	51	2004	
2SA16B		3005	51	2004	
2SA16C		3005	51	2004	
2SA16D		3005	51	2004	
2SA16E		3005	51	2004	
2SA16F		3005	51	2004	
2SA16L		3005	51	2004	
2SA16M		3005	51	2004	
2SA16OR		3005	51	2004	
2SA16R		3005	51	2004	
2SA16X		3005	51	2004	
2SA16Y		3005	51	2004	
2SA17	102A	3005	51	2005	
2SA170	100	3005	51	2007	
2SA170A		3005	51	2004	
2SA170B		3005	51	2004	
2SA170C		3005	51	2004	
2SA170D		3005	51	2004	
2SA170E		3005	51	2004	
2SA170F		3005	51	2004	
2SA170G		3005	51	2004	
2SA170GN		3005	51	2004	
2SA170H		3005	51	2004	
2SA170J		3005	51	2004	
2SA170K		3005	51	2004	
2SA170L		3005	51	2004	
2SA170M		3005	51	2004	
2SA170OR		3005	51	2004	
2SA170R		3005	51	2004	
2SA170X		3005	51	2004	
2SA170Y		3005	51	2004	
2SA171	100	3005	51	2007	
2SA171A		3005	51	2004	
2SA171B		3005	51	2004	
2SA171C		3005	51	2004	
2SA171D		3005	51	2004	
2SA171E		3005	51	2004	
2SA171F		3005	51	2004	
2SA171G		3005	51	2004	
2SA171GN		3005	51	2004	
2SA171H		3005	51	2004	
2SA171J		3005	51	2004	
2SA171K		3005	51	2004	
2SA171L		3005	51	2004	
2SA171M		3005	51	2004	
2SA171OR		3005	51	2004	
2SA171R		3005	51	2004	
2SA171X		3005	51	2004	
2SA171Y		3005	51	2004	
2SA172	100	3005	51	2007	
2SA172A	100	3005	51	2007	
2SA172B		3005	51	2004	

Industry Standard No.	ECG	SK	GE	RS 276-	MOTOR.
2SA172C		3005	51	2004	
2SA172D		3005	51	2004	
2SA172E		3005	51	2004	
2SA172F		3005	51	2004	
2SA172G		3005	51	2004	
2SA172GN		3005	51	2004	
2SA172H		3005	51	2004	
2SA172J		3005	51	2004	
2SA172K		3005	51	2004	
2SA172L		3005	51	2004	
2SA172M		3005	51	2004	
2SA172OR		3005	51	2004	
2SA172R		3005	51	2004	
2SA172X		3005	51	2004	
2SA172Y		3005	51	2004	
2SA173	102A	3005	51	2005	
2SA173A		3005	51	2004	
2SA173B	102A	3005	51	2005	
2SA173C		3005	51	2004	
2SA173D		3005	51	2004	
2SA173E		3005	51	2004	
2SA173F		3005	51	2004	
2SA173G		3005	51	2004	
2SA173GN		3005	51	2004	
2SA173H		3005	51	2004	
2SA173J		3005	51	2004	
2SA173K		3005	51	2004	
2SA173L		3005	51	2004	
2SA173M		3005	51	2004	
2SA173OR		3005	51	2004	
2SA173R		3005	51	2004	
2SA173X		3005	51	2004	
2SA173Y		3005	51	2004	
2SA174	102	3005	51	2007	
2SA174A		3005	51	2004	
2SA174B		3005	51	2004	
2SA174C		3005	51	2004	
2SA174D		3005	51	2004	
2SA174E		3005	51	2004	
2SA174F		3005	51	2004	
2SA174G		3005	51	2004	
2SA174GN		3005	51	2004	
2SA174H		3005	51	2004	
2SA174J		3005	51	2004	
2SA174K		3005	51	2004	
2SA174L		3005	51	2004	
2SA174M		3005	51	2004	
2SA174OR		3005	51	2004	
2SA174R		3005	51	2004	
2SA174X		3005	51	2004	
2SA174Y		3005	51	2004	
2SA175	126	3006/160	51	2024	
2SA175A		3006	51	2004	
2SA175B		3006	51	2004	
2SA175C		3006	51	2004	
2SA175D		3006	51	2004	
2SA175E		3006	51	2004	
2SA175F		3006	51	2004	
2SA175G		3006	51	2004	
2SA175GN		3006	51	2004	
2SA175H		3006	51	2004	
2SA175J		3006	51	2004	
2SA175K		3006	51	2004	
2SA175L		3006	51	2004	
2SA175M		3006	51	2004	
2SA175OR		3006	51	2004	
2SA175R		3006	51	2004	
2SA175X		3006	51	2004	
2SA175Y		3006	51	2004	
2SA176	126	3008	53	2024	
2SA176A		3008	53	2007	
2SA176B		3008	53	2007	
2SA176C		3008	53	2007	
2SA176D		3008	53	2007	
2SA176E		3008	53	2007	
2SA176F		3008	53	2007	
2SA176G		3008	53	2007	
2SA176GN		3008	53	2007	
2SA176H		3008	53	2007	
2SA176J		3008	53	2007	
2SA176K		3008	53	2007	
2SA176L		3008	53	2007	
2SA176M		3008	53	2007	
2SA176OR		3008	53	2007	
2SA176R		3008	53	2007	
2SA176X		3008	53	2007	
2SA176Y		3008	53	2007	
2SA178			51	2004	
2SA17A		3005	51	2004	
2SA17B		3005	51	2004	
2SA17C		3005	51	2004	
2SA17D		3005	51	2004	
2SA17E		3005	51	2004	
2SA17F		3005	51	2004	
2SA17G		3005	51	2004	
2SA17H	102A	3005	51	2005	
2SA17L		3005	51	2004	
2SA17OR		3005	51	2004	
2SA17R		3005	51	2004	
2SA17X		3005	51	2004	
2SA17Y		3005	51	2004	
2SA18	102A	3005	51	2005	
2SA180	126	3005	51	2004	
2SA180A		3005	51	2004	
2SA180B		3005	51	2004	
2SA180C		3005	51	2004	
2SA180D		3005	51	2004	
2SA180E		3005	51	2004	
2SA180F		3005	51	2004	
2SA180G		3005	51	2004	
2SA180GN		3005	51	2004	
2SA180H		3005	51	2004	
2SA180J		3005	51	2004	
2SA180K		3005	51	2004	
2SA180L		3005	51	2004	
2SA180M		3005	51	2004	
2SA180OR		3005	51	2004	
2SA180R		3005	51	2004	
2SA180X		3005	51	2004	
2SA180Y		3005	51	2004	
2SA181	126	3005	51	2007	
2SA181A		3005	51	2004	
2SA181B		3005	51	2004	
2SA181C		3005	51	2004	
2SA181D		3005	51	2004	
2SA181E		3005	51	2004	
2SA181F		3005	51	2004	
2SA181G		3005	51	2004	
2SA181GN		3005	51	2004	
2SA181H		3005	51	2004	
2SA181J		3005	51	2004	
2SA181K		3005	51	2004	
2SA181L		3005	51	2004	
2SA181M		3005	51	2004	
2SA181OR		3005	51	2004	
2SA181R		3005	51	2004	
2SA181X		3005	51	2004	
2SA181Y		3005	51	2004	
2SA182	100	3005	51	2007	
2SA182A		3005	51	2004	
2SA182B		3005	51	2004	
2SA182C		3005	51	2004	
2SA182D		3005	51	2004	
2SA182E		3005	51	2004	
2SA182F		3005	51	2004	
2SA182G		3005	51	2004	
2SA182GN		3005	51	2004	
2SA182H		3005	51	2004	
2SA182J		3005	51	2004	
2SA182K		3005	51	2004	
2SA182L		3005	51	2004	
2SA182M		3005	51	2004	
2SA182OR		3005	51	2004	
2SA182R		3005	51	2004	
2SA182X		3005	51	2004	
2SA182Y		3005	51	2004	
2SA183	126	3005	51	2005	
2SA183A		3005	51	2004	
2SA183B		3005	51	2004	
2SA183C		3005	51	2004	
2SA183D		3005	51	2004	
2SA183E		3005	51	2004	
2SA183F		3005	51	2004	
2SA183G		3005	51	2004	
2SA183GN		3005	51	2004	
2SA183H		3005	51	2004	
2SA183J		3005	51	2004	
2SA183K		3005	51	2004	
2SA183L		3005	51	2004	
2SA183M		3005	51	2004	
2SA183OR		3005	51	2004	
2SA183R		3005	51	2004	
2SA183X		3005	51	2004	
2SA183Y		3005	51	2004	
2SA184			51	2004	
2SA186			53	2007	
2SA187TV	102A		53	2007	
2SA188	126	3005	51	2024	
2SA188A		3005	51	2004	
2SA188B		3005	51	2004	
2SA188C		3005	51	2004	
2SA188D		3005	51	2004	
2SA188E		3005	51	2004	
2SA188F		3005	51	2004	
2SA188G		3005	51	2004	
2SA188GN		3005	51	2004	
2SA188H		3005	51	2004	
2SA188J		3005	51	2004	
2SA188K		3005	51	2004	
2SA188L		3005	51	2004	
2SA188M		3005	51	2004	
2SA188OR		3005	51	2004	
2SA188R		3005	51	2004	
2SA188X		3005	51	2004	
2SA188Y		3005	51	2004	
2SA189	126	3005	52	2024	
2SA189A		3005	51	2004	
2SA189B		3005	51	2004	
2SA189C		3005	51	2004	
2SA189D		3005	51	2004	
2SA189E		3005	51	2004	
2SA189F		3005	51	2004	
2SA189G		3005	51	2004	
2SA189GN		3005	51	2004	
2SA189H		3005	51	2004	
2SA189J		3005	51	2004	
2SA189K		3005	51	2004	
2SA189L		3005	51	2004	
2SA189M		3005	51	2004	
2SA189OR		3005	51	2004	
2SA189R		3005	51	2004	
2SA189X		3005	51	2004	
2SA189Y		3005	51	2004	
2SA18A		3005	51	2004	
2SA18B		3005	51	2004	
2SA18C		3005	51	2004	
2SA18D		3005	51	2004	
2SA18E		3005	51	2004	
2SA18F		3005	51	2004	
2SA18G		3005	51	2004	
2SA18H	102A		51	2005	
2SA18L		3005	51	2004	
2SA18M		3005	51	2004	
2SA18OR		3005	51	2004	
2SA18R		3005	51	2004	
2SA18X		3005	51	2004	
2SA18Y		3005	51	2004	
2SA19	126	3008	53	2024	
2SA190			51	2004	
2SA191			51	2004	
2SA192			51	2004	
2SA193			51	2004	
2SA194			51	2004	
2SA195			51	2004	
2SA196			51	2004	
2SA197	126	3006/160	51	2007	
2SA197A		3006	51	2004	
2SA197B		3006	51	2004	
2SA197C		3006	51	2004	
2SA197D		3006	51	2004	
2SA197E		3006	51	2004	
2SA197F		3006	51	2004	
2SA197G		3006	51	2004	
2SA197GN		3006	51	2004	
2SA197H		3006	51	2004	
2SA197J		3006	51	2004	
2SA197K		3006	51	2004	
2SA197L		3006	51	2004	
2SA197M		3006	51	2004	
2SA197OR		3006	51	2004	
2SA197R		3006	51	2004	
2SA197X		3006	51	2004	
2SA197Y		3006	51	2004	
2SA198	102A	3005	51	2007	
2SA198A		3005	51	2004	
2SA198B		3005	51	2004	
2SA198C		3005	51	2004	
2SA198D		3005	51	2004	
2SA198E		3005	51	2004	
2SA198F		3005	51	2004	
2SA198G		3005	51	2004	
2SA198GN		3005	51	2004	
2SA198H		3005	51	2004	
2SA198J		3005	51	2004	
2SA198K		3005	51	2004	
2SA198L		3005	51	2004	
2SA198M		3005	51	2004	
2SA198OR		3005	51	2004	

Industry Standard No.	ECG	SK	GE	RS 276-	MOTOR.
2SA198R		3005	51	2004	
2SA198X		3005	51	2004	
2SA198Y		3005	51	2004	
2SA199			51	2004	
2SA19A		3008	53	2007	
2SA19B		3008	53	2007	
2SA19C		3008	53	2007	
2SA19D		3008	53	2007	
2SA19E		3008	53	2007	
2SA19F		3008	53	2007	
2SA19G		3008	53	2007	
2SA19L		3008	53	2007	
2SA19M		3008	53	2007	
2SA19OR		3008	53	2007	
2SA19R		3008	53	2007	
2SA19X		3008	53	2007	
2SA19Y		3008	53	2007	
2SA20	126	3008	53	2024	
2SA200			53	2004	
2SA201	126	3005	51	2024	
2SA201-O	126	3005	51	2024	
2SA201-N			1	2024	
2SA201-O		3005	52	2024	
2SA201-OR			52	2024	
2SA201A	126	3008	51	2024	
2SA201B	126	3005	51	2024	
2SA201CL		3005	51	2024	
2SA201D		3005	51	2024	
2SA201E	126	3005	51	2024	
2SA201F		3005	51	2024	
2SA201G		3005	51	2024	
2SA201GN		3005	51	2024	
2SA201H		3005	51	2024	
2SA201J		3005	51	2024	
2SA201K		3005	51	2024	
2SA201L		3005	51	2024	
2SA201M		3005	51	2024	
2SA201N	126	3005	52	2024	
2SA201O			51	2024	
2SA201OR		3005	51	2024	
2SA201R		3005	51	2024	
2SA201TV	126	3005	51	2024	
2SA201TVO		3005	52	2024	
2SA201TVO	126		51	2024	
2SA201X		3005	51	2024	
2SA201Y		3005	51	2024	
2SA202	126	3005	53	2007	
2SA202-OR			52	2024	
2SA202A	126	3005	53	2024	
2SA202AP		3005	53	2024	
2SA202B	126	3005	53	2024	
2SA202C	126	3005	53	2024	
2SA202D	126	3005	53	2024	
2SA202D-4		3005	53	2024	
2SA202E		3005	53	2024	
2SA202F		3005	53	2024	
2SA202G		3005	53	2024	
2SA202GN		3005	53	2024	
2SA202H		3005	53	2024	
2SA202J		3005	53	2024	
2SA202K		3005	53	2024	
2SA202L		3005	53	2024	
2SA202M		3005	53	2024	
2SA202OR		3005	53	2024	
2SA202R		3005	53	2024	
2SA202X			53	2024	
2SA202Y		3005	53	2024	
2SA203	126	3007	52	2024	
2SA203A	126		52	2024	
2SA203AA	126	3005	53	2024	
2SA203B	126	3003	52	2024	
2SA203C			52	2024	
2SA203D			52	2024	
2SA203P	126	3124/289	52	2024	
2SA204	100	3004/102A	51	2007	
2SA204A		3004	51	2004	
2SA204B		3004	51	2004	
2SA204C		3004	51	2004	
2SA204D		3004	51	2004	
2SA204E		3004	51	2004	
2SA204F		3004	51	2004	
2SA204G		3004	51	2004	
2SA204GN		3004	51	2004	
2SA204H		3004	51	2004	
2SA204J		3004	51	2004	
2SA204K		3004	51	2004	
2SA204L		3004	51	2004	
2SA204M		3004	51	2004	
2SA204OR		3004	51	2004	
2SA204R		3004	51	2004	
2SA204X		3004	51	2004	
2SA204Y		3004	51	2004	
2SA205	100	3004/102A	51	2007	
2SA205A		3004	51	2004	
2SA205B		3004	51	2004	
2SA205C		3004	51	2004	
2SA205D		3004	51	2004	
2SA205E		3004	51	2004	
2SA205F		3004	51	2004	
2SA205G		3004	51	2004	
2SA205GN		3004	51	2004	
2SA205H		3004	51	2004	
2SA205J		3004	51	2004	
2SA205K		3004	51	2004	
2SA205L		3004	51	2004	
2SA205M		3004	51	2004	
2SA205OR		3004	51	2004	
2SA205R		3004	51	2004	
2SA205X		3004	51	2004	
2SA205Y		3004	51	2004	
2SA206	100	3004/102A	51	2007	
2SA206A		3004	51	2004	
2SA206B		3004	51	2004	
2SA206C		3004	51	2004	
2SA206D		3004	51	2004	
2SA206E		3004	51	2004	
2SA206F		3004	51	2004	
2SA206G		3004	51	2004	
2SA206GN		3004	51	2004	
2SA206H		3004	51	2004	
2SA206J		3004	51	2004	
2SA206K		3004	51	2004	
2SA206L		3004	51	2004	
2SA206M		3004	51	2004	
2SA206OR		3004	51	2004	
2SA206R		3004	51	2004	
2SA206X		3004	51	2004	
2SA206Y		3004	51	2004	
2SA207	100	3005	51	2007	
2SA207A		3005	51	2004	
2SA207B		3005	51	2004	
2SA207C		3005	51	2004	
2SA207D		3005	51	2004	
2SA207E		3005	51	2004	
2SA207F		3005	51	2004	
2SA207G		3005	51	2004	
2SA207GN		3005	51	2004	
2SA207H		3005	51	2004	
2SA207J		3005	51	2004	
2SA207K		3005	51	2004	
2SA207L		3005	51	2004	
2SA207M		3005	51	2004	
2SA207OR		3005	51	2004	
2SA207R		3005	51	2004	
2SA207X		3005	51	2004	
2SA207Y		3005	51	2004	
2SA208	176	3005	51	2007	
2SA208A		3005	51	2004	
2SA208B		3005	51	2004	
2SA208C		3005	51	2004	
2SA208D		3005	51	2004	
2SA208E		3005	51	2004	
2SA208F		3005	51	2004	
2SA208G		3005	51	2004	
2SA208GN		3005	51	2004	
2SA208H		3005	51	2004	
2SA208J		3005	51	2004	
2SA208K		3005	51	2004	
2SA208L		3005	51	2004	
2SA208M		3005	51	2004	
2SA208OR		3005	51	2004	
2SA208R		3005	51	2004	
2SA208X		3005	51	2004	
2SA208Y		3005	51	2004	
2SA209	176	3005	51	2007	
2SA209A		3005	51	2004	
2SA209B		3005	51	2004	
2SA209C		3005	51	2004	
2SA209D		3005	51	2004	
2SA209E		3005	51	2004	
2SA209F		3005	51	2004	
2SA209G		3005	51	2004	
2SA209GN		3005	51	2004	
2SA209H		3005	51	2004	
2SA209J		3005	51	2004	
2SA209K		3005	51	2004	
2SA209L		3005	51	2004	
2SA209M		3005	51	2004	
2SA209OR		3005	51	2004	
2SA209R		3005	51	2004	
2SA209X		3005	51	2004	
2SA209Y		3005	51	2004	
2SA20A		3008	53	2007	
2SA20B		3008	53	2007	
2SA20C		3008	53	2007	
2SA20D		3008	53	2007	
2SA20E		3008	53	2007	
2SA20F		3008	53	2007	
2SA20G		3008	53	2007	
2SA20L		3008	53	2007	
2SA20M		3008	53	2007	
2SA20OR		3008	53	2007	
2SA20R		3008	53	2007	
2SA20X		3008	53	2007	
2SA20Y		3008	53	2007	
2SA21	126	3008	53	2024	
2SA210	176	3005	51	2007	
2SA210A		3005	51	2004	
2SA210B		3005	51	2004	
2SA210C		3005	51	2004	
2SA210D		3005	51	2004	
2SA210E		3005	51	2004	
2SA210F		3005	51	2004	
2SA210G		3005	51	2004	
2SA210GN		3005	51	2004	
2SA210H		3005	51	2004	
2SA210J		3005	51	2004	
2SA210K		3005	51	2004	
2SA210L		3005	51	2004	
2SA210M		3005	51	2004	
2SA210OR		3005	51	2004	
2SA210R		3005	51	2004	
2SA210X		3005	51	2004	
2SA210Y		3005	51	2004	
2SA211	100	3005	51	2007	
2SA211A		3005	51	2004	
2SA211B		3005	51	2004	
2SA211C		3005	51	2004	
2SA211D		3005	51	2004	
2SA211E		3005	51	2004	
2SA211F		3005	51	2004	
2SA211G		3005	51	2004	
2SA211GN		3005	51	2004	
2SA211H		3005	51	2004	
2SA211J		3005	51	2004	
2SA211K		3005	51	2004	
2SA211L		3005	51	2004	
2SA211M		3005	51	2004	
2SA211OR		3005	51	2004	
2SA211R		3005	51	2004	
2SA211X		3005	51	2004	
2SA211Y		3005	51	2004	
2SA212	100	3005	51	2007	
2SA212A		3005	51	2004	
2SA212B		3005	51	2004	
2SA212C		3005	51	2004	
2SA212D		3005	51	2004	
2SA212E		3005	51	2004	
2SA212F		3005	51	2004	
2SA212G		3005	51	2004	
2SA212GN		3005	51	2004	
2SA212H		3005	51	2004	
2SA212J		3005	51	2004	
2SA212K		3005	51	2004	
2SA212L		3005	51	2004	
2SA212M		3005	51	2004	
2SA212OR		3005	51	2004	
2SA212R		3005	51	2004	
2SA212X		3005	51	2004	
2SA212Y		3005	51	2004	
2SA213	126	3006/160	52	2024	
2SA213A		3006	51	2004	
2SA213B		3006	51	2004	
2SA213C		3006	51	2004	
2SA213D		3006	51	2004	
2SA213E		3006	51	2004	
2SA213F		3006	51	2004	
2SA213G		3006	51	2004	
2SA213GN		3006	51	2004	
2SA213H		3006	51	2004	
2SA213J		3006	51	2004	
2SA213K		3006	51	2004	
2SA213L		3006	51	2004	
2SA213M		3006	51	2004	
2SA213OR		3006	51	2004	
2SA213R		3006	51	2004	
2SA213X		3006	51	2004	
2SA213Y		3006	51	2004	

Industry Standard No.	ECG	SK	GE	RS 276-	MOTOR.
2SA214	126	3006/160	51	2024	
2SA214A		3006	51	2004	
2SA214B		3006	51	2004	
2SA214C		3006	51	2004	
2SA214D		3006	51	2004	
2SA214E		3006	51	2004	
2SA214F		3006	51	2004	
2SA214G		3006	51	2004	
2SA214GN		3006	51	2004	
2SA214H		3006	51	2004	
2SA214J		3006	51	2004	
2SA214K		3006	51	2004	
2SA214L		3006	51	2004	
2SA214M		3006	51	2004	
2SA214OR		3006	51	2004	
2SA214R		3006	51	2004	
2SA214X		3006	51	2004	
2SA214Y		3006	51	2004	
2SA215	126	3006/160	51	2004	
2SA215A		3006	51	2004	
2SA215B		3006	51	2004	
2SA215C		3006	51	2004	
2SA215D		3006	51	2004	
2SA215E		3006	51	2004	
2SA215F		3006	51	2004	
2SA215G		3006	51	2004	
2SA215GN		3006	51	2004	
2SA215H		3006	51	2004	
2SA215J		3006	51	2004	
2SA215K		3006	51	2004	
2SA215L		3006	51	2004	
2SA215M		3006	51	2004	
2SA215OR		3006	51	2004	
2SA215R		3006	51	2004	
2SA215X		3006	51	2004	
2SA215Y		3006	51	2004	
2SA216	126	3006/160	51	2024	
2SA216A		3006	51	2004	
2SA216B		3006	51	2004	
2SA216C		3006	51	2004	
2SA216D		3006	51	2004	
2SA216E		3006	51	2004	
2SA216F		3006	51	2004	
2SA216G		3006	51	2004	
2SA216GN		3006	51	2004	
2SA216H		3006	51	2004	
2SA216J		3006	51	2004	
2SA216K		3006	51	2004	
2SA216L		3006	51	2004	
2SA216M		3006	51	2004	
2SA216OR		3006	51	2004	
2SA216R		3006	51	2004	
2SA216X		3006	51	2004	
2SA216Y		3006	51	2004	
2SA217	100	3005	1	2007	
2SA217A		3005	51	2004	
2SA217B		3005	51	2004	
2SA217C		3005	51	2004	
2SA217D		3005	51	2004	
2SA217E		3005	51	2004	
2SA217F		3005	51	2004	
2SA217G		3005	51	2004	
2SA217GN		3005	51	2004	
2SA217H		3005	51	2004	
2SA217J		3005	51	2004	
2SA217K		3005	51	2004	
2SA217L		3005	51	2004	
2SA217M		3005	51	2004	
2SA217OR		3005	51	2004	
2SA217R		3005	51	2004	
2SA217X		3005	51	2004	
2SA217Y		3005	51	2004	
2SA218	160	3008	53	2004	
2SA218A		3008	53	2004	
2SA218B		3008	53	2004	
2SA218C		3008	53	2004	
2SA218D		3008	53	2004	
2SA218E		3008	53	2004	
2SA218F		3008	53	2004	
2SA218G		3008	53	2004	
2SA218GN		3008	53	2004	
2SA218H		3008	53	2004	
2SA218J		3008	53	2004	
2SA218K		3008	53	2004	
2SA218L		3008	53	2004	
2SA218M		3008	53	2004	
2SA218OR		3008	53	2004	
2SA218R		3008	53	2004	
2SA218X		3008	53	2004	
2SA218Y		3008	53	2004	
2SA219	126	3006/160	51	2004	
2SA219A		3006	51	2004	
2SA219B		3006	51	2004	
2SA219C		3006	51	2004	
2SA219D		3006	51	2004	
2SA219E		3006	51	2004	
2SA219F		3006	51	2004	
2SA219G		3006	51	2004	
2SA219GN		3006	51	2004	
2SA219H		3006	51	2004	
2SA219J		3006	51	2004	
2SA219K		3006	51	2004	
2SA219L		3006	51	2004	
2SA219M		3006	51	2004	
2SA219OR		3006	51	2004	
2SA219R		3006	51	2004	
2SA219X		3006	51	2004	
2SA219Y		3006	51	2004	
2SA21A		3008	53	2007	
2SA21B		3008	53	2007	
2SA21C		3008	53	2007	
2SA21D		3008	53	2007	
2SA21E		3008	53	2007	
2SA21F		3008	53	2007	
2SA21G		3008	53	2007	
2SA21L		3008	53	2007	
2SA21M		3008	53	2007	
2SA21OR		3008	53	2007	
2SA21R		3008	53	2007	
2SA21X		3008	53	2007	
2SA21Y		3008	53	2007	
2SA22			51	2004	
2SA220	126	3008	53	2004	
2SA220A		3008	53	2007	
2SA220B		3008	53	2007	
2SA220C		3008	53	2007	
2SA220D		3008	53	2007	
2SA220E		3008	53	2007	
2SA220F		3008	53	2007	
2SA220G		3008	53	2007	
2SA220GN		3008	53	2007	
2SA220H		3008	53	2007	
2SA220J		3008	53	2007	
2SA220K		3008	53	2007	
2SA220L		3008	53	2007	
2SA220M		3008	53	2007	
2SA220OR		3008	53	2007	
2SA220R		3008	53	2007	
2SA220X		3008	53	2007	
2SA220Y		3008	53	2007	
2SA221	126	3007	53	2004	
2SA221-OR			53	2007	
2SA221A		3008	53	2007	
2SA221B		3008	53	2007	
2SA221C		3008	53	2007	
2SA221D		3008	53	2007	
2SA221E		3008	53	2007	
2SA221F		3008	53	2007	
2SA221G		3008	53	2007	
2SA221GN		3008	53	2007	
2SA221H		3008	53	2007	
2SA221J		3008	53	2007	
2SA221K		3008	53	2007	
2SA221L		3008	53	2007	
2SA221M		3008	53	2007	
2SA221OR		3008	53	2007	
2SA221R		3008	53	2007	
2SA221X		3008	53	2007	
2SA221Y		3008	53	2007	
2SA222	126	3007	53	2004	
2SA222A		3008	53	2007	
2SA222B		3008	53	2007	
2SA222C		3008	53	2007	
2SA222D		3008	53	2007	
2SA222E		3008	53	2007	
2SA222F		3008	53	2007	
2SA222G		3008	53	2007	
2SA222GN		3008	53	2007	
2SA222H		3008	53	2007	
2SA222J		3008	53	2007	
2SA222K		3008	53	2007	
2SA222L		3008	53	2007	
2SA222M		3008	53	2007	
2SA222OR		3008	53	2007	
2SA222R		3008	53	2007	
2SA222X		3008	53	2007	
2SA222Y		3008	53	2007	
2SA223	126	3006/160	51	2004	
2SA223A		3006	51	2004	
2SA223B		3006	51	2004	
2SA223C		3006	51	2004	
2SA223D		3006	51	2004	
2SA223E		3006	51	2004	
2SA223F		3006	51	2004	
2SA223G		3006	51	2004	
2SA223GN		3006	51	2004	
2SA223H		3006	51	2004	
2SA223J		3006	51	2004	
2SA223K		3006	51	2004	
2SA223L		3006	51	2004	
2SA223M		3006	51	2004	
2SA223OR		3006	51	2004	
2SA223R		3006	51	2004	
2SA223X		3006	51	2004	
2SA223Y		3006	51	2004	
2SA224	160	3008	53	2004	
2SA224A		3008	53	2007	
2SA224B		3008	53	2007	
2SA224C		3008	53	2007	
2SA224D		3008	53	2007	
2SA224E		3008	53	2007	
2SA224F		3008	53	2007	
2SA224G		3008	53	2007	
2SA224GN		3008	53	2007	
2SA224H		3008	53	2007	
2SA224J		3008	53	2007	
2SA224K		3008	53	2007	
2SA224L		3008	53	2007	
2SA224M		3008	53	2007	
2SA224OR		3008	53	2007	
2SA224R		3008	53	2007	
2SA224X		3008	53	2007	
2SA224Y		3008	53	2007	
2SA225	126	3006/160	51	2004	
2SA225A		3006	51	2004	
2SA225B		3006	51	2004	
2SA225C		3006	51	2004	
2SA225D		3006	51	2004	
2SA225E		3006	51	2004	
2SA225F		3006	51	2004	
2SA225G		3006	51	2004	
2SA225GN		3006	51	2004	
2SA225H		3006	51	2004	
2SA225J		3006	51	2004	
2SA225K		3006	51	2004	
2SA225L		3006	51	2004	
2SA225M		3006	51	2004	
2SA225OR		3006	51	2004	
2SA225R		3006	51	2004	
2SA225X		3006	51	2004	
2SA225Y		3006	51	2004	
2SA226	160		245	2004	
2SA227	160	3006	51	2004	
2SA227A		3006	51	2004	
2SA227B		3006	51	2004	
2SA227C		3006	51	2004	
2SA227D		3006	51	2004	
2SA227E		3006	51	2004	
2SA227F		3006	51	2004	
2SA227G		3006	51	2004	
2SA227GN		3006	51	2004	
2SA227H		3006	51	2004	
2SA227J		3006	51	2004	
2SA227K		3006	51	2004	
2SA227L		3006	51	2004	
2SA227M		3006	51	2004	
2SA227OR		3006	51	2004	
2SA227R		3006	51	2004	
2SA227X		3006	51	2004	
2SA227Y		3006	51	2004	
2SA228	160		245	2004	
2SA229	160	3006	51	2004	
2SA229A		3006	51	2004	
2SA229B		3006	51	2004	
2SA229C		3006	51	2004	
2SA229D		3006	51	2004	
2SA229E		3006	51	2004	
2SA229F		3006	51	2004	
2SA229G		3006	51	2004	
2SA229GN		3006	51	2004	
2SA229H		3006	51	2004	
2SA229J		3006	51	2004	
2SA229K		3006	51	2004	
2SA229L		3006	51	2004	
2SA229OR		3006	51	2004	
2SA229R		3006	51	2004	
2SA229X		3006	51	2004	
2SA229Y		3006	51	2004	

Industry Standard No.	ECG	SK	GE	RS 276-	MOTOR.
2SA23			51	2004	
2SA230	160	3006	51	2004	
2SA230A		3006	51	2004	
2SA230B		3006	51	2004	
2SA230C		3006	51	2004	
2SA230D		3006	51	2004	
2SA230E		3006	51	2004	
2SA230F		3006	51	2004	
2SA230G		3006	51	2004	
2SA230GN		3006	51	2004	
2SA230H		3006	51	2004	
2SA230J		3006	51	2004	
2SA230K		3006	51	2004	
2SA230L		3006	51	2004	
2SA230M		3006	51	2004	
2SA230OR			51	2004	
2SA230R		3006	51	2004	
2SA230X		3006	51	2004	
2SA230Y		3006	51	2004	
2SA231	176	3123	53	2007	
2SA232	176		53	2007	
2SA233	160	3007	50	2004	
2SA233A	160	3007	50	2004	
2SA233B	160	3007	50	2004	
2SA233C	160	3007	50	2004	
2SA233D		3007	50	2004	
2SA233E		3007	50	2004	
2SA233F		3007	50	2004	
2SA233G		3007	50	2004	
2SA233GN		3007	50	2004	
2SA233H		3007	50	2004	
2SA233J		3007	50	2004	
2SA233K		3007	50	2004	
2SA233L		3007	50	2004	
2SA233M		3007	50	2004	
2SA233OR		3007	50	2004	
2SA233R		3007	50	2004	
2SA233X		3007	50	2004	
2SA233Y		3007	50	2004	
2SA234	160	3007	51	2004	
2SA234A	160	3006	51	2004	
2SA234B	160	3006	51	2004	
2SA234C	160	3006	51	2004	
2SA234D		3006	51	2004	
2SA234E		3006	51	2004	
2SA234F		3006	51	2004	
2SA234G		3006	51	2004	
2SA234GN		3006	51	2004	
2SA234H		3006	51	2004	
2SA234J		3006	51	2004	
2SA234K		3006	51	2004	
2SA234L		3006	51	2004	
2SA234M		3006	51	2004	
2SA234OR		3006	51	2004	
2SA234R		3006	51	2004	
2SA234X		3006	51	2004	
2SA234Y		3006	51	2004	
2SA235	160	3006	51	2004	
2SA235A	160	3006	51	2004	
2SA235B	160	3006	51	2004	
2SA235C	160	3006	51	2004	
2SA235D		3006	51	2004	
2SA235E		3006	51	2004	
2SA235F		3006	51	2004	
2SA235G		3006	51	2004	
2SA235GN		3006	51	2004	
2SA235H		3006	50	2004	
2SA235K		3006	51	2004	
2SA235M		3006	51	2004	
2SA235OR		3006	51	2004	
2SA235R		3006	51	2004	
2SA235X		3006	51	2004	
2SA235Y		3006	51	2004	
2SA236	126	3007	50	2024	
2SA236A		3007	50	2004	
2SA236B		3007	50	2004	
2SA236C		3007	50	2004	
2SA236D		3007	50	2004	
2SA236E		3007	50	2004	
2SA236F		3007	50	2004	
2SA236G		3007	50	2004	
2SA236GN		3007	50	2004	
2SA236H		3007	50	2004	
2SA236J		3007	50	2004	
2SA236K		3007	50	2004	
2SA236L		3007	50	2004	
2SA236M		3007	50	2004	
2SA236OR		3007	50	2004	
2SA236R		3007	50	2004	
2SA236X		3007	50	2004	
2SA236Y		3007	50	2004	
2SA237	126	3008	53	2024	
2SA237A		3008	53	2007	
2SA237B		3008	53	2007	
2SA237C		3008	53	2007	
2SA237D		3008	53	2007	
2SA237E		3008	53	2007	
2SA237F		3008	53	2007	
2SA237G		3008	53	2007	
2SA237GN		3008	53	2007	
2SA237H		3008	53	2007	
2SA237J		3008	53	2007	
2SA237K		3008	53	2007	
2SA237L		3008	53	2007	
2SA237M		3008	53	2007	
2SA237OR		3008	53	2007	
2SA237R		3008	53	2007	
2SA237X		3008	53	2007	
2SA237Y		3008	53	2007	
2SA238	126$	3006/160	51	2024	
2SA238A		3006	51	2004	
2SA238B		3006	51	2004	
2SA238C		3006	51	2004	
2SA238D		3006	51	2004	
2SA238E		3006	51	2004	
2SA238F		3006	51	2004	
2SA238G		3006	51	2004	
2SA238GN		3006	51	2004	
2SA238H		3006	51	2004	
2SA238J		3006	51	2004	
2SA238K		3006	51	2004	
2SA238L		3006	51	2004	
2SA239	160	3006	51	2004	
2SA239A		3006	51	2004	
2SA239B		3006	51	2004	
2SA239C		3006	51	2004	
2SA239D		3006	51	2004	
2SA239E		3006	51	2004	
2SA239F		3006	51	2004	
2SA239G		3006	51	2004	
2SA239GN		3006	51	2004	
2SA239GREEN			51	2004	
2SA239H		3006	51	2004	
2SA239J		3006	51	2004	
2SA239K		3006	51	2004	
2SA239L		3006	51	2004	
2SA239M		3006	51	2004	
2SA239OR		3006	51	2004	
2SA239R		3006	51	2004	
2SA239RED			51	2004	
2SA239X		3006	51	2004	
2SA239Y		3006	51	2004	
2SA24			50	2004	
2SA240	160	3006	51	2004	
2SA240A	160	3006	51	2004	
2SA240B	160	3006	51	2004	
2SA240B2	160	3006	51	2004	
2SA240BL	160		51	2004	
2SA240C		3006	51	2004	
2SA240D		3006	51	2004	
2SA240E		3006	51	2004	
2SA240F		3006	51	2004	
2SA240G		3006	51	2004	
2SA240GN		3006	51	2004	
2SA240GREEN			50	2004	
2SA240H		3006	51	2004	
2SA240J		3006	51	2004	
2SA240K		3006	51	2004	
2SA240L		3006	51	2004	
2SA240M		3006	51	2004	
2SA240OR			51	2004	
2SA240R		3006	51	2004	
2SA240RED			51	2004	
2SA240X		3006	51	2004	
2SA240Y		3006	51	2004	
2SA241	160	3006	51	2004	
2SA241A		3006	51	2004	
2SA241B		3006	51	2004	
2SA241C		3006	51	2004	
2SA241D		3006	51	2004	
2SA241E		3006	51	2004	
2SA241F		3006	51	2004	
2SA241G		3006	51	2004	
2SA241GN		3006	51	2004	
2SA241H		3006	51	2004	
2SA241J		3006	51	2004	
2SA241K		3006	51	2004	
2SA241L		3006	51	2004	
2SA241M		3006	51	2004	
2SA241OR		3006	51	2004	
2SA241R		3006	51	2004	
2SA241X		3006	51	2004	
2SA241Y		3006	51	2004	
2SA242	160		245	2004	
2SA243	160		245	2004	
2SA243A				2005	
2SA244	160		245	2004	
2SA244A				2005	
2SA245	160		245	2004	
2SA246	126	3006/160	51	2024	
2SA246A		3006	51	2004	
2SA246B		3006	51	2004	
2SA246C		3006	51	2004	
2SA246D		3006	51	2004	
2SA246E		3006	51	2004	
2SA246F		3006	51	2004	
2SA246G		3006	51	2004	
2SA246GN		3006	51	2004	
2SA246H		3006	51	2004	
2SA246J		3006	51	2004	
2SA246K		3006	51	2004	
2SA246L		3006	51	2004	
2SA246M		3006	51	2004	
2SA246OR		3006	51	2004	
2SA246R		3006	51	2004	
2SA246V	126	3006/160	51	2024	
2SA246X		3006	51	2004	
2SA246Y		3006	51	2004	
2SA247	126	3006/160	51	2024	
2SA247A		3006	51	2004	
2SA247B		3006	51	2004	
2SA247C		3006	51	2004	
2SA247D		3006	51	2004	
2SA247E		3006	51	2004	
2SA247F		3006	51	2004	
2SA247G		3006	51	2004	
2SA247GN		3006	51	2004	
2SA247H		3006	51	2004	
2SA247J		3006	51	2004	
2SA247K		3006	51	2004	
2SA247L		3006	51	2004	
2SA247M		3006	51	2004	
2SA247OR		3006	51	2004	
2SA247R		3006	51	2004	
2SA247X		3006	51	2004	
2SA247Y		3006	51	2004	
2SA248	102A	3005	51	2007	
2SA248A		3005	51	2004	
2SA248B		3005	51	2004	
2SA248C		3005	51	2004	
2SA248D		3005	51	2004	
2SA248E		3005	51	2004	
2SA248F		3005	51	2004	
2SA248G		3005	51	2004	
2SA248GN		3005	51	2004	
2SA248H		3005	51	2004	
2SA248J		3005	51	2004	
2SA248K		3005	51	2004	
2SA248L		3005	51	2004	
2SA248M		3005	51	2004	
2SA248OR		3005	51	2004	
2SA248R		3005	51	2004	
2SA248X		3005	51	2004	
2SA248Y		3005	51	2004	
2SA249		3008		2004	
2SA25			50	2004	
2SA250	126$	3004/102A	53	2024	
2SA250A		3004	53	2007	
2SA250B		3004	53	2007	
2SA250C		3004	53	2007	
2SA250D		3004	53	2007	
2SA250E		3004	53	2007	
2SA250F		3004	53	2007	
2SA250G		3004	53	2007	
2SA250GN		3004	53	2007	
2SA250H		3004	53	2007	
2SA250J		3004	53	2007	
2SA250L		3004	53	2007	
2SA250M		3004	53	2007	
2SA250OR			53	2007	
2SA250X			53	2007	
2SA250Y		3004	53	2007	
2SA251	126	3006/160	51	2024	
2SA251A		3006	51	2004	
2SA251B		3006	51	2004	
2SA251C		3006	51	2004	
2SA251D		3006	51	2004	
2SA251E		3006	51	2004	

Industry Standard No.	ECG	SK	GE	RS 276-	MOTOR.
2SA251F		3006	51	2004	
2SA251G		3006	51	2004	
2SA251H		3006	51	2004	
2SA251J		3006	51	2004	
2SA251K		3006	51	2004	
2SA251L		3006	51	2004	
2SA251M		3006	51	2004	
2SA251OR		3006	51	2004	
2SA251R		3006	51	2004	
2SA251X		3006	51	2004	
2SA251Y		3006	51	2004	
2SA252	126	3006/160	51	2024	
2SA252A		3006	51	2004	
2SA252B		3006	51	2004	
2SA252C		3006	51	2004	
2SA252D		3006	51	2004	
2SA252E		3006	51	2004	
2SA252F		3006	51	2004	
2SA252G		3006	51	2004	
2SA252H		3006	51	2004	
2SA252J		3006	51	2004	
2SA252K		3006	51	2004	
2SA252L		3006	51	2004	
2SA252M		3006	51	2004	
2SA252OR		3006	51	2004	
2SA252R		3006	51	2004	
2SA252X		3006	51	2004	
2SA252Y		3006	51	2004	
2SA253	126	3006/160	51	2024	
2SA253A		3006	51	2004	
2SA253B		3006	51	2004	
2SA253C		3006	51	2004	
2SA253D		3006	51	2004	
2SA253E		3006	51	2004	
2SA253F		3006	51	2004	
2SA253G		3006	51	2004	
2SA253GN		3006	51	2004	
2SA253H		3006	51	2004	
2SA253J		3006	51	2004	
2SA253K		3006	51	2004	
2SA253L		3006	51	2004	
2SA253M		3006	51	2004	
2SA253OR		3006	51	2004	
2SA253R		3006	51	2004	
2SA253X		3006	51	2004	
2SA253Y		3006	51	2004	
2SA254	126	3005	51	2024	
2SA254A		3005	51	2004	
2SA254B		3005	51	2004	
2SA254C		3005	51	2004	
2SA254D		3005	51	2004	
2SA254E		3005	51	2004	
2SA254F		3005	51	2004	
2SA254G		3005	51	2004	
2SA254H		3005	51	2004	
2SA254J		3005	51	2004	
2SA254K		3005	51	2004	
2SA254L		3005	51	2004	
2SA254M		3005	51	2004	
2SA254OR		3005	51	2004	
2SA254R		3005	51	2004	
2SA254X		3005	51	2004	
2SA254Y		3005	51	2004	
2SA255	126	3005	52	2024	
2SA255A		3005	51	2004	
2SA255B		3005	51	2004	
2SA255C		3005	51	2004	
2SA255D		3005	51	2004	
2SA255E		3005	51	2004	
2SA255F		3005	51	2004	
2SA255G		3005	51	2004	
2SA255GN		3005	51	2004	
2SA255H		3005	51	2004	
2SA255J		3005	51	2004	
2SA255K		3005	51	2004	
2SA255L		3005	51	2004	
2SA255M		3005	51	2004	
2SA255OR		3005	51	2004	
2SA255R		3005	51	2004	
2SA255X		3005	51	2004	
2SA255Y		3005	51	2004	
2SA256	126	3008	53	2024	
2SA256A		3008	53	2007	
2SA256B		3008	53	2007	
2SA256C		3008	53	2007	
2SA256D		3008	53	2007	
2SA256E		3008	53	2007	
2SA256F		3008	53	2007	
2SA256G		3008	53	2007	
2SA256GN		3008	53	2007	
2SA256H		3008	53	2007	
2SA256J		3008	53	2007	
2SA256K		3008	53	2007	
2SA256L		3008	53	2007	
2SA256M		3008	53	2007	
2SA256OR		3008	53	2007	
2SA256R		3008	53	2007	
2SA256X		3008	53	2007	
2SA256Y		3008	53	2007	
2SA257		3007	50	2024	
2SA257A		3007	50	2004	
2SA257B		3007	50	2004	
2SA257C		3007	50	2004	
2SA257D		3007	50	2004	
2SA257E		3007	50	2004	
2SA257F		3007	50	2004	
2SA257G		3007	50	2004	
2SA257GN		3007	50	2004	
2SA257H		3007	50	2004	
2SA257J		3007	50	2004	
2SA257K		3007	50	2004	
2SA257L		3007	50	2004	
2SA257OR		3007	50	2004	
2SA257R		3007	50	2004	
2SA257X		3007	50	2004	
2SA257Y		3007	50	2004	
2SA258		3007	50	2024	
2SA258A		3007	50	2004	
2SA258B		3007	50	2004	
2SA258C		3007	50	2004	
2SA258D		3007	50	2004	
2SA258E		3007	50	2004	
2SA258F		3007	50	2004	
2SA258G		3007	50	2004	
2SA258GN		3007	50	2004	
2SA258H		3007	50	2004	
2SA258J		3007	50	2004	
2SA258K		3007	50	2004	
2SA258L		3007	50	2004	
2SA258M		3007	50	2004	
2SA258OR		3007	50	2004	
2SA258R		3007	50	2004	
2SA258X		3007	50	2004	
2SA258Y		3007	50	2004	

Industry Standard No.	ECG	SK	GE	RS 276-	MOTOR.
2SA259	126	3008	53	2024	
2SA259B		3008	53	2007	
2SA259C		3008	53	2007	
2SA259D		3008	53	2007	
2SA259E		3008	53	2007	
2SA259F		3008	53	2007	
2SA259G		3008	53	2007	
2SA259GN		3008	53	2007	
2SA259H		3008	53	2007	
2SA259J		3008	53	2007	
2SA259L		3008	53	2007	
2SA259M		3008	53	2007	
2SA259OR		3008	53	2007	
2SA259R		3008	53	2007	
2SA259X		3008	53	2007	
2SA259Y		3008	53	2007	
2SA26	100	3005	51	2007	
2SA260	160	3006	51	2024	
2SA260A		3006	51	2004	
2SA260B		3006	51	2004	
2SA260C		3006	51	2004	
2SA260D		3006	51	2004	
2SA260E		3006	51	2004	
2SA260F		3006	51	2004	
2SA260G		3006	51	2004	
2SA260GN		3006	51	2004	
2SA260H		3006	51	2004	
2SA260J		3006	51	2004	
2SA260K		3006	51	2004	
2SA260L		3006	51	2004	
2SA260M		3006	51	2004	
2SA260OR			51	2004	
2SA260R		3006	51	2004	
2SA260X		3006	51	2004	
2SA260Y		3006	51	2004	
2SA261	160	3006	51	2024	
2SA261A		3006	51	2004	
2SA261B		3006	51	2004	
2SA261C		3006	51	2004	
2SA261D		3006	51	2004	
2SA261E		3006	51	2004	
2SA261F		3006	51	2004	
2SA261G		3006	51	2004	
2SA261GN		3006	51	2004	
2SA261H		3006	51	2004	
2SA261K		3006	51	2004	
2SA261L		3006	51	2004	
2SA261M		3006	51	2004	
2SA261OR		3006	51	2004	
2SA261R		3006	51	2004	
2SA261X		3006	51	2004	
2SA261Y		3006	51	2004	
2SA262	160	3006	51	2024	
2SA262A		3006	51	2004	
2SA262B		3006	51	2004	
2SA262C		3006	51	2004	
2SA262D		3006	51	2004	
2SA262E		3006	51	2004	
2SA262F		3006	51	2004	
2SA262G		3006	51	2004	
2SA262GN		3006	51	2004	
2SA262H		3006	51	2004	
2SA262J		3006	51	2004	
2SA262K		3006	51	2004	
2SA262L		3006	51	2004	
2SA262M		3006	51	2004	
2SA262OR		3006	51	2004	
2SA262R		3006	51	2004	
2SA262X		3006	51	2004	
2SA262Y		3006	51	2004	
2SA263	160	3006	51	2024	
2SA263A		3006	51	2004	
2SA263B		3006	51	2004	
2SA263C		3006	51	2004	
2SA263D		3006	51	2004	
2SA263E		3006	51	2004	
2SA263F		3006	51	2004	
2SA263G		3006	51	2004	
2SA263GN		3006	51	2004	
2SA263H		3006	51	2004	
2SA263J		3006	51	2004	
2SA263K		3006	51	2004	
2SA263L		3006	51	2004	
2SA263M		3006	51	2004	
2SA263OR		3006	51	2004	
2SA263R		3006	51	2004	
2SA263X		3006	51	2004	
2SA263Y		3006	51	2004	
2SA264	160	3006	51	2024	
2SA264(1)			51	2004	
2SA264A		3006	51	2004	
2SA264B		3006	51	2004	
2SA264C		3006	51	2004	
2SA264D		3006	51	2004	
2SA264E		3006	51	2004	
2SA264F		3006	51	2004	
2SA264G		3006	51	2004	
2SA264GN		3006	51	2004	
2SA264H		3006	51	2004	
2SA264K		3006	51	2004	
2SA264L		3006	51	2004	
2SA264M		3006	51	2004	
2SA2640				2004	
2SA2640R		3006	51		
2SA264R		3006	51	2004	
2SA264X		3006	51	2004	
2SA264Y		3006	51	2004	
2SA265	160	3006	51	2024	
2SA265A		3006	51	2004	
2SA265B		3006	51	2004	
2SA265C		3006	51	2004	
2SA265D		3006	51	2004	
2SA265E		3006	51	2004	
2SA265F		3006	51	2004	
2SA265G		3006	51	2004	
2SA265GN		3006	51	2004	
2SA265H		3006	51	2004	
2SA265J		3006	51	2004	
2SA265K		3006	51	2004	
2SA265L		3006	51	2004	
2SA265M		3006	51	2004	
2SA265OR		3006	51	2004	
2SA265R		3006	51	2004	
2SA265X		3006	51	2004	
2SA265Y		3006	51	2004	
2SA266	126	3008	53	2024	
2SA266A		3008	53	2007	
2SA266B		3008	53	2007	
2SA266C		3008	53	2007	
2SA266D		3008	53	2007	
2SA266E		3008	53	2007	
2SA266F		3008	53	2007	
2SA266G		3008	53	2007	
2SA266GN		3008	53	2007	

Industry Standard No.	ECG	SK	GE	RS 276-	MOTOR.
2SA266GREEN			53	2007	
2SA266H		3008	53	2007	
2SA266J		3008	53	2007	
2SA266K		3008	53	2007	
2SA266L		3008	53	2007	
2SA266M		3008	53	2007	
2SA266OR		3008	53	2007	
2SA266R		3008	53	2007	
2SA266X		3008	53	2007	
2SA266Y		3008	53	2007	
2SA267	126	3006/160	51	2024	
2SA267A		3006	51	2004	
2SA267B		3006	51	2004	
2SA267C		3006	51	2004	
2SA267D		3006	51	2004	
2SA267E		3006	51	2004	
2SA267F		3006	51	2004	
2SA267G		3006	51	2004	
2SA267GN		3006	51	2004	
2SA267H		3006	51	2004	
2SA267J		3006	51	2004	
2SA267K		3006	51	2004	
2SA267L		3006	51	2004	
2SA267M		3006	51	2004	
2SA267OR		3006	51	2004	
2SA267R		3006	51	2004	
2SA267X		3006	51	2004	
2SA267Y		3006	51	2004	
2SA268	126	3006/160	51	2024	
2SA268A		3006	51	2004	
2SA268B		3006	51	2004	
2SA268C		3006	51	2004	
2SA268D		3006	51	2004	
2SA268E		3006	51	2004	
2SA268F		3006	51	2004	
2SA268G		3006	51	2004	
2SA268GN		3006	51	2004	
2SA268H		3006	51	2004	
2SA268J		3006	51	2004	
2SA268K		3006	51	2004	
2SA268L		3006	51	2004	
2SA268M		3006	51	2004	
2SA268OR		3006	51	2004	
2SA268R		3006	51	2004	
2SA268X		3006	51	2004	
2SA268Y		3006	51	2004	
2SA269	126	3008	53	2024	
2SA269A		3008	53	2007	
2SA269B		3008	53	2007	
2SA269C		3008	53	2007	
2SA269D		3008	53	2007	
2SA269E		3008	53	2007	
2SA269F		3008	53	2007	
2SA269GN		3008	53	2007	
2SA269H		3008	53	2007	
2SA269J		3008	53	2007	
2SA269K		3008	53	2007	
2SA269L		3008	53	2007	
2SA269M		3008	53	2007	
2SA269OR		3008	53	2007	
2SA269X		3008	53	2007	
2SA269Y		3008	53	2007	
2SA26A		3005	51	2004	
2SA26B		3005	51	2004	
2SA26C		3005	51	2004	
2SA26D		3005	51	2004	
2SA26E		3005	51	2004	
2SA26F		3005	51	2004	
2SA26G		3005	51	2004	
2SA26L		3005	51	2004	
2SA26M		3005	51	2004	
2SA26OR		3005	51	2004	
2SA26R		3005	51	2004	
2SA26X		3005	51	2004	
2SA26Y		3005	51	2004	
2SA27			51	2004	
2SA270	126	3007	50	2024	
2SA270A		3007	50	2004	
2SA270B		3007	50	2004	
2SA270C		3007	50	2004	
2SA270D		3007	50	2004	
2SA270E		3007	50	2004	
2SA270F		3007	50	2004	
2SA270G		3007	50	2004	
2SA270GN		3007	50	2004	
2SA270H		3007	50	2004	
2SA270J		3007	50	2004	
2SA270K		3007	50	2004	
2SA270L		3007	50	2004	
2SA270M		3007	50	2004	
2SA2700R			50	2004	
2SA270R		3007	50	2004	
2SA270X		3007	50	2004	
2SA270Y		3007	50	2004	
2SA271	126	3008	53	2024	
2SA271(2)			51	2007	
2SA271(3)			51	2007	
2SA271A		3008	53	2007	
2SA271B		3008	53	2007	
2SA271C		3008	53	2007	
2SA271D		3008	53	2007	
2SA271E		3008	53	2007	
2SA271F		3008	53	2007	
2SA271G		3008	53	2007	
2SA271GN		3008	53	2007	
2SA271H		3008	53	2007	
2SA271J		3008	53	2007	
2SA271K		3008	53	2007	
2SA271L		3008	53	2007	
2SA271M		3008	53	2007	
2SA271OR		3008	53	2007	
2SA271R		3008	53	2007	
2SA271X		3008	53	2007	
2SA271Y		3008	53	2007	
2SA272	126	3008	53	2024	
2SA272A		3000	53	2007	
2SA272B		3008	53	2007	
2SA272C		3008	53	2007	
2SA272D		3008	53	2007	
2SA272E		3008	53	2007	
2SA272F		3008	53	2007	
2SA272G		3008	53	2007	
2SA272GN		3008	53	2007	
2SA272H		3008	53	2007	
2SA272J		3008	53	2007	
2SA272K		3008	53	2007	
2SA272L		3008	53	2007	
2SA272M		3008	53	2007	
2SA272OR		3008	53	2007	
2SA272R		3008	53	2007	
2SA272X		3008	53	2007	
2SA272Y		3008	53	2007	
2SA273	126	3008	53	2024	
2SA273A		3008	53	2007	
2SA273B		3008	53	2007	
2SA273C		3008	53	2007	
2SA273D		3008	53	2007	
2SA273E		3008	53	2007	
2SA273F		3008	53	2007	
2SA273G		3008	53	2007	
2SA273GN		3008	53	2007	
2SA273H		3008	53	2007	
2SA273J		3008	53	2007	
2SA273K		3008	53	2007	
2SA273L		3008	53	2007	
2SA273M		3008	53	2007	
2SA273OR		3008	53	2007	
2SA273X		3008	53	2007	
2SA273Y		3008	53	2007	
2SA274	126	3008	53	2024	
2SA274A		3008	53	2007	
2SA274B		3008	53	2007	
2SA274C		3008	53	2007	
2SA274D		3008	53	2007	
2SA274E		3008	53	2007	
2SA274F		3008	53	2007	
2SA274G		3008	53	2007	
2SA274GN		3008	53	2007	
2SA274H		3008	53	2007	
2SA274J		3008	53	2007	
2SA274K		3008	53	2007	
2SA274L		3008	53	2007	
2SA274M		3008	53	2007	
2SA274OR		3008	53	2007	
2SA274R		3008	53	2007	
2SA274X		3008	53	2007	
2SA274Y		3008	53	2007	
2SA275	126	3008	53	2024	
2SA275A		3008	53	2007	
2SA275B		3008	53	2007	
2SA275C		3008	53	2007	
2SA275D		3008	53	2007	
2SA275E		3008	53	2007	
2SA275F		3008	53	2007	
2SA275G		3008	53	2007	
2SA275GN		3008	53	2007	
2SA275H		3008	53	2007	
2SA275J		3008	53	2007	
2SA275L		3008	53	2007	
2SA275M		3008	53	2007	
2SA275OR		3008	53	2007	
2SA275R		3008	53	2007	
2SA275X		3008	53	2007	
2SA275Y		3008	53	2007	
2SA276	160		245	2004	
2SA277	100	3005	51	2007	
2SA277A		3005	51	2004	
2SA277B		3005	51	2004	
2SA277C		3005	51	2004	
2SA277D		3005	51	2004	
2SA277E		3005	51	2004	
2SA277F		3005	51	2004	
2SA277G		3005	51	2004	
2SA277GN		3005	51	2004	
2SA277H		3005	51	2004	
2SA277J		3005	51	2004	
2SA277K		3005	51	2004	
2SA277L		3005	51	2004	
2SA277M		3005	51	2004	
2SA277OR		3005	51	2004	
2SA277X		3005	51	2004	
2SA277Y		3005	51	2004	
2SA278	100	3005	51	2007	
2SA278A		3005	51	2004	
2SA278B		3005	51	2004	
2SA278C		3005	51	2004	
2SA278D		3005	51	2004	
2SA278E		3005	51	2004	
2SA278F		3005	51	2004	
2SA278G		3005	51	2004	
2SA278GN		3005	51	2004	
2SA278J		3005	51	2004	
2SA278K		3005	51	2004	
2SA278L		3005	51	2004	
2SA278M		3005	51	2004	
2SA278OR		3005	51	2004	
2SA278R		3005	51	2004	
2SA278X		3005	51	2004	
2SA278Y		3005	51	2004	
2SA279	100	3005	51	2007	
2SA279A		3005	51	2004	
2SA279B		3005	51	2004	
2SA279C		3005	51	2004	
2SA279D		3005	51	2004	
2SA279E		3005	51	2004	
2SA279F		3005	51	2004	
2SA279G		3005	51	2004	
2SA279GN		3005	51	2004	
2SA279H		3005	51	2004	
2SA279J		3005	51	2004	
2SA279K		3005	51	2004	
2SA279L		3005	51	2004	
2SA279OR		3005	51	2004	
2SA279R		3005	51	2004	
2SA279X		3005	51	2004	
2SA279Y		3005	51	2004	
2SA28	126	3005	51	2024	
2SA280	100	3006/160	51	2024	
2SA280A		3006	51	2004	
2SA280B		3006	51	2004	
2SA280C		3006	51	2004	
2SA280D		3006	51	2005	
2SA280E		3006	51	2004	
2SA280F		3006	51	2004	
2SA280G		3006	51	2004	
2SA280GN		3006	51	2004	
2SA280H		3006	51	2004	
2SA280J		3006	51	2004	
2SA280K		3006	51	2004	
2SA280L		3006	51	2004	
2SA280M		3006	51	2004	
2SA280OR		3006	51	2004	
2SA280R		3006	51	2004	
2SA280X		3006	51	2004	
2SA280Y		3006	51	2004	
2SA281	100	3006/160	51	2024	
2SA281A		3006	51	2004	
2SA281B		3006	51	2004	
2SA281C		3006	51	2004	
2SA281D		3006	51	2004	
2SA281E		3006	51	2004	
2SA281F		3006	51	2004	
2SA281G		3006	51	2004	
2SA281GN		3006	51	2004	
2SA281H		3006	51	2004	
2SA281J		3006	51	2004	
2SA281K		3006	51	2004	
2SA281L		3006	51	2004	

Industry Standard No.	ECG	SK	GE	RS 276-	MOTOR.
2SA281M		3006	51	2004	
2SA281OR		3006	51	2004	
2SA281R		3006	51	2004	
2SA281X		3006	51	2004	
2SA281Y		3006	51	2004	
2SA282	102A	3005	51	2007	
2SA282A		3005	51	2004	
2SA282B		3005	51	2004	
2SA282C		3005	51	2004	
2SA282D		3005	51	2004	
2SA282E		3005	51	2004	
2SA282F		3005	51	2004	
2SA282G		3005	51	2004	
2SA282GN		3005	51	2004	
2SA282H		3005	51	2004	
2SA282J		3005	51	2004	
2SA282K		3005	51	2004	
2SA282L		3005	51	2004	
2SA282M		3005	51	2004	
2SA282OR		3005	51	2004	
2SA282R		3005	51	2004	
2SA282X		3005	51	2004	
2SA282Y		3005	51	2004	
2SA283	100	3005	51	2007	
2SA283A		3005	51	2004	
2SA283B		3005	51	2004	
2SA283C		3005	51	2004	
2SA283D		3005	51	2004	
2SA283E		3005	51	2004	
2SA283F		3005	51	2004	
2SA283G		3005	51	2004	
2SA283GN		3005	51	2004	
2SA283H		3005	51	2004	
2SA283J		3005	51	2004	
2SA283K		3005	51	2004	
2SA283L		3005	51	2004	
2SA283M		3005	51	2004	
2SA283OR		3005	51	2004	
2SA283R		3005	51	2004	
2SA283X		3005	51	2004	
2SA283Y		3005	51	2004	
2SA284	100	3005	51	2007	
2SA284A		3005	51	2004	
2SA284B		3005	51	2004	
2SA284C		3005	51	2004	
2SA284D		3005	51	2004	
2SA284E		3005	51	2004	
2SA284F		3005	51	2004	
2SA284G		3005	51	2004	
2SA284GN		3005	51	2004	
2SA284J		3005	51	2004	
2SA284K		3005	51	2004	
2SA284L		3005	51	2004	
2SA284M		3005	51		
2SA284OR		3005	51	2004	
2SA284R		3005	51	2004	
2SA284X		3005	51	2004	
2SA284Y		3005	51	2004	
2SA285	160	3008	53	2004	
2SA285B		3008	53	2007	
2SA285C		3008	53	2007	
2SA285E		3008	53	2007	
2SA285F		3008	53	2007	
2SA285G		3008	53	2007	
2SA285GN		3008	53	2007	
2SA285H		3008	53	2007	
2SA285J		3008	53	2007	
2SA285L		3008	53	2007	
2SA285M		3008	53	2007	
2SA285OR		3008	53	2007	
2SA285R		3008	53	2007	
2SA285X		3008	53	2007	
2SA285Y		3008	53	2007	
2SA286	160	3008	53	2004	
2SA286A		3008	53	2007	
2SA286B		3008	53	2007	
2SA286C		3008	53	2007	
2SA286D		3008	53	2007	
2SA286E		3008	53	2007	
2SA286F		3008	53	2007	
2SA286G		3008	53	2007	
2SA286GN		3008	53	2007	
2SA286H		3008	53	2007	
2SA286K		3008	53	2007	
2SA286M		3008	53	2007	
2SA286OR		3008	53	2007	
2SA286R		3008	53	2007	
2SA286X		3008	53	2007	
2SA286Y		3008	53	2007	
2SA287	160	3008	53	2004	
2SA287A		3008	53	2007	
2SA287B		3008	53	2007	
2SA287C		3008	53	2007	
2SA287D		3008	53	2007	
2SA287E		3008	53	2007	
2SA287F		3008	53	2007	
2SA287G		3008	53	2007	
2SA287GN		3008	53	2007	
2SA287H		3008	53	2007	
2SA287J		3008	53	2007	
2SA287K		3008	53	2007	
2SA287L		3008	53	2007	
2SA287M		3008	53	2007	
2SA287OR		3008	53	2007	
2SA287X		3008	53	2007	
2SA287Y		3008	53	2007	
2SA288	126	3008	53	2024	
2SA288A	126	3006/160	53	2024	
2SA288B		3008	53	2007	
2SA288C		3008	53	2007	
2SA288D		3008	53	2007	
2SA288E		3008	53	2007	
2SA288F		3008	53	2007	
2SA288G		3008	53	2007	
2SA288GN		3008	53	2007	
2SA288J		3008	53	2007	
2SA288K		3008	53	2007	
2SA288L		3008	53	2007	
2SA288M		3008	53	2007	
2SA288OR		3008	53	2007	
2SA288X		3008	53	2007	
2SA288Y		3008	53	2007	
2SA289	126	3006/160	51	2024	
2SA289A		3006	51	2004	
2SA289B		3006	51	2004	
2SA289C		3006	51	2004	
2SA289D		3006	51	2004	
2SA289E		3006	51	2004	
2SA289F		3006	51	2004	
2SA289G		3006	51	2004	
2SA289GN		3006	51	2004	
2SA289H		3006	51	2004	
2SA289J		3006	51	2004	
2SA289K		3006	51	2004	
2SA289L		3006	51	2004	
2SA289M		3006	51	2004	
2SA289OR		3006	51	2004	
2SA289R		3006	51	2004	
2SA289X		3006	51	2004	
2SA289Y		3006	51	2004	
2SA28A		3005	51	2004	
2SA28B		3005	51	2004	
2SA28C		3005	51	2004	
2SA28D		3005	51	2004	
2SA28E		3005	51	2004	
2SA28F		3005	51	2004	
2SA28G		3005	51	2004	
2SA28L		3005	51	2004	
2SA28M		3005	51	2004	
2SA28OR		3005	51	2004	
2SA28X		3005	51	2004	
2SA28Y		3005	51	2004	
2SA29	126	3008	53	2024	
2SA290	126	3006/160	52	2024	
2SA290A		3006	51	2004	
2SA290B		3006	51	2004	
2SA290C		3006	51	2004	
2SA290D		3006	51	2004	
2SA290E		3006	51	2004	
2SA290F		3006	51	2004	
2SA290G		3006	51	2004	
2SA290GN		3006	51	2004	
2SA290H		3006	51	2004	
2SA290J		3006	51	2004	
2SA290K		3006	51	2004	
2SA290L		3006	51	2004	
2SA290M		3006	51	2004	
2SA290OR			51	2004	
2SA290R		3006	51	2004	
2SA290X		3006	51	2004	
2SA290Y		3006	51	2004	
2SA291	126	3006/160	51	2024	
2SA291A		3006	51	2004	
2SA291B		3006	51	2004	
2SA291C		3006	51	2004	
2SA291D		3006	51	2004	
2SA291E		3006	51	2004	
2SA291F		3006	51	2004	
2SA291G		3006	51	2004	
2SA291GN		3006	51	2004	
2SA291H		3006	51	2004	
2SA291J		3006	51	2004	
2SA291K		3006	51	2004	
2SA291L		3006	51	2004	
2SA291M		3006	51	2004	
2SA291OR		3006	51	2004	
2SA291R		3006	51	2004	
2SA291X		3006	51	2004	
2SA291Y		3006	51	2004	
2SA292	126	3006/160	52	2024	
2SA292A		3006	51	2004	
2SA292B		3006	51	2004	
2SA292C		3006	51	2004	
2SA292D		3006	51	2004	
2SA292E		3006	51	2004	
2SA292F		3006	51	2004	
2SA292G		3006	51	2004	
2SA292GN		3006	51	2004	
2SA292H		3006	51	2004	
2SA292J		3006	51	2004	
2SA292K		3006	51	2004	
2SA292L		3006	51	2004	
2SA292M		3006	51	2004	
2SA292OR		3006	51	2004	
2SA292R		3006	51	2004	
2SA292X		3006	51	2004	
2SA292Y		3006	51	2004	
2SA293	126	3006/160	51	2024	
2SA293A		3006	51	2004	
2SA293B		3006	51	2004	
2SA293C		3006	51	2004	
2SA293D		3006	51	2004	
2SA293E		3006	51	2004	
2SA293G		3006	51	2004	
2SA293GN		3006	51	2004	
2SA293H		3006	51	2004	
2SA293J		3006	51	2004	
2SA293K		3006	51	2004	
2SA293L		3006	51	2004	
2SA293M		3006	51	2004	
2SA293OR		3006	51	2004	
2SA293R		3006	51	2004	
2SA293X		3006	51	2004	
2SA293Y		3006	51	2004	
2SA294	126	3007	51	2024	
2SA294A		3006	51	2004	
2SA294B		3006	51	2004	
2SA294C		3006	51	2004	
2SA294D		3006	51	2004	
2SA294E		3006	51	2004	
2SA294F		3006	51	2004	
2SA294GN		3006	51	2004	
2SA294H		3006	51	2004	
2SA294J		3006	51	2004	
2SA294K		3006	51	2004	
2SA294L		3006	51	2004	
2SA294M		3006	51	2004	
2SA294OR		3006	51	2004	
2SA294R		3006	51	2004	
2SA294X		3006	51	2004	
2SA294Y		3006	51	2004	
2SA295	126	3006/160	51	2024	
2SA295A		3006	51	2004	
2SA295B		3006	51	2004	
2SA295C		3006	51	2004	
2SA295D		3006	51	2004	
2SA295E		3006	51	2004	
2SA295F		3006	51	2004	
2SA295G		3006	51	2004	
2SA295GN		3006	51	2004	
2SA295H		3006	51	2004	
2SA295J		3006	51	2004	
2SA295K		3006	51	2004	
2SA295L		3006	51	2004	
2SA295OR		3006	51	2004	
2SA295R		3006	51	2004	
2SA295Y		3006	51	2004	
2SA296	126	3005	51	2004	
2SA296A		3005	51	2004	
2SA296B		3005	51	2004	
2SA296C		3005	51	2004	
2SA296D		3005	51	2004	
2SA296E		3005	51	2004	
2SA296F		3005	51	2004	
2SA296G		3005	51	2004	
2SA296GN		3005	51	2004	
2SA296H		3005	51	2004	
2SA296J		3005	51	2004	

Industry Standard No.	ECG	SK	GE	RS 276-	MOTOR.
2SA296K		3005	51	2004	
2SA296L		3005	51	2004	
2SA296M		3005	51	2004	
2SA296OR		3005	51	2004	
2SA296R		3005	51	2004	
2SA296X		3005	51	2004	
2SA296Y		3005	51	2004	
2SA297	126	3005	51	2004	
2SA297A		3005	51	2004	
2SA297B		3005	51	2004	
2SA297C		3005	51	2004	
2SA297D		3005	51	2004	
2SA297E		3005	51	2004	
2SA297F		3005	51	2004	
2SA297G		3005	51	2004	
2SA297H		3005	51	2004	
2SA297J		3005	51	2004	
2SA297K		3005	51	2004	
2SA297L		3005	51	2004	
2SA297M		3005	51	2004	
2SA297OR		3005	51	2004	
2SA297R		3005	51	2004	
2SA297X		3005	51	2004	
2SA297Y		3005	51	2004	
2SA298	160	3008	53	2004	
2SA298A		3008	53	2007	
2SA298B		3008	53	2007	
2SA298C		3008	53	2007	
2SA298D		3009	53	2007	
2SA298E		3009	53	2007	
2SA298F		3009	53	2007	
2SA298G		3008	53	2007	
2SA298GN		3009	53	2007	
2SA298H		3009	53	2007	
2SA298J		3009	53	2007	
2SA298K		3009	53	2007	
2SA298L		3009	53	2007	
2SA298M		3009	53	2007	
2SA298OR		3008	53	2007	
2SA298R		3009	53	2007	
2SA298X		3008	53	2007	
2SA298Y		3008	53	2007	
2SA299				2004	
2SA29A		3008	53	2007	
2SA29B		3008	53	2007	
2SA29C		3008	53	2007	
2SA29D		3008	53	2007	
2SA29E		3008	53	2007	
2SA29F		3008	53	2007	
2SA29G		3008	53	2007	
2SA29L		3008	53	2007	
2SA29M		3008	53	2007	
2SA29R		3008	53	2007	
2SA29X		3008	53	2007	
2SA29Y		3008	53	2007	
2SA30	102A	3005	51	2024	
2SA300				2004	
2SA301	126	3006/160	51	2024	
2SA301A		3006	51	2007	
2SA301B		3006	51	2007	
2SA301C		3006	51	2007	
2SA301D		3006	51	2007	
2SA301E		3006	51	2007	
2SA301F		3006	51	2007	
2SA301G		3006	51	2007	
2SA301GN		3006	51	2007	
2SA301H		3006	51	2007	
2SA301J		3006	51	2007	
2SA301K		3006	51	2007	
2SA301L		3006	51	2007	
2SA301M		3006	51	2007	
2SA301OR			51	2007	
2SA301R		3006	51	2007	
2SA301X		3006	51	2007	
2SA301Y		3006	51	2007	
2SA302	102A		53	2007	
2SA303	102A	3004	53	2007	
2SA304	100	3123	80	2005	
2SA304A		3123	80		
2SA304B		3123	80		
2SA304C		3123	80		
2SA304D		3123	80		
2SA304F		3123	80		
2SA304G		3123	80		
2SA304GN		3123	80		
2SA304J		3123	80		
2SA304K		3123	80		
2SA304L		3123	80		
2SA304M		3123	80		
2SA304OR		3123	80		
2SA304R		3123	80		
2SA304X		3123	80		
2SA304Y		3123	80		
2SA305	100	3005	51	2007	
2SA305-RED			52	2024	
2SA305-YELLOW			52	2024	
2SA305A		3005	51	2024	
2SA305B		3005	51	2024	
2SA305C		3005	51	2024	
2SA305D		3005	51	2024	
2SA305E		3005	51	2024	
2SA305F		3005	51	2024	
2SA305G		3005	51	2024	
2SA305GN		3005	51	2024	
2SA305H		3005	51	2024	
2SA305J		3005	51	2024	
2SA305K		3005	51	2024	
2SA305L		3005	51	2024	
2SA305M		3005	51	2024	
2SA305OR		3005	51	2024	
2SA305R		3005	51	2024	
2SA305X		3005	51	2024	
2SA305Y		3005	51	2024	
2SA306	160		245	2004	
2SA307	160	3008	53	2004	
2SA307A		3008	53	2004	
2SA307B		3008	53	2004	
2SA307C		3008	53	2004	
2SA307D		3008	53	2004	
2SA307E		3008	53	2004	
2SA307F		3008	53	2004	
2SA307G		3008	53	2004	
2SA307GN		3008	53	2004	
2SA307J		3008	53	2004	
2SA307K		3008	53	2004	
2SA307L		3008	53	2004	
2SA307M		3008	53	2004	
2SA307OR		3008	53	2004	
2SA307R		3008	53	2004	
2SA307X		3008	53	2004	
2SA307Y		3008	53	2004	
2SA308	126	3006/160	51	2024	
2SA308A		3006	51	2004	
2SA308B		3006	51	2004	
2SA308C		3006	51	2004	
2SA308D		3006	51	2004	
2SA308E		3006	51	2004	
2SA308F		3006	51	2004	
2SA308G		3006	51	2004	
2SA308GN		3006	51	2004	
2SA308H		3006	51	2004	
2SA308J		3006	51	2004	
2SA308K		3006	51	2004	
2SA308L		3006	51	2004	
2SA308M		3006	51	2004	
2SA308OR		3006	51	2004	
2SA308R		3006	51	2004	
2SA308X		3006	51	2004	
2SA308Y		3006	51	2004	
2SA309	126	3006/160	51	2024	
2SA309A		3006	51	2004	
2SA309B		3006	51	2004	
2SA309C		3006	51	2004	
2SA309D		3006	51	2004	
2SA309E		3006	51	2004	
2SA309F		3006	51	2004	
2SA309G		3006	51	2004	
2SA309GN		3006	51	2004	
2SA309H		3006	51	2004	
2SA309J		3006	51	2004	
2SA309K		3006	51	2004	
2SA309L		3006	51	2004	
2SA309M		3006	51	2004	
2SA309OR		3006	51	2004	
2SA309R		3006	51	2004	
2SA309X		3006	51	2004	
2SA309Y		3006	51	2004	
2SA30A		3005	51	2004	
2SA30B		3005	51	2004	
2SA30C		3005	51	2004	
2SA30D		3005	51	2004	
2SA30E		3005	51	2004	
2SA30F		3005	51	2004	
2SA30G		3005	51	2004	
2SA30L		3005	51	2004	
2SA30M		3005	51	2004	
2SA30OR			51	2004	
2SA30X		3005	51	2004	
2SA30Y		3005	51	2004	
2SA31	102A	3005	51	2007	
2SA310	126	3006/160	51	2024	
2SA310A		3006	51	2004	
2SA310B		3006	51	2004	
2SA310C		3006	51	2004	
2SA310D		3006	51	2004	
2SA310E		3006	51	2004	
2SA310F		3006	51	2004	
2SA310G		3006	51	2004	
2SA310GN		3006	51	2004	
2SA310H		3006	51	2004	
2SA310J		3006	51	2004	
2SA310K		3006	51	2004	
2SA310L		3006	51	2004	
2SA310M		3006	51	2004	
2SA310OR		3006	51	2004	
2SA310R		3006	51	2004	
2SA310X		3006	51	2004	
2SA310Y		3006	51	2004	
2SA311	102A	3008	53	2007	
2SA311A		3008	53	2007	
2SA311B		3008	53	2007	
2SA311C		3008	53	2007	
2SA311D		3008	53	2007	
2SA311E		3008	53	2007	
2SA311F		3008	53	2007	
2SA311G		3008	53	2007	
2SA311GN		3008	53	2007	
2SA311H		3008	53	2007	
2SA311J		3008	53	2007	
2SA311K		3008	53	2007	
2SA311L		3008	53	2007	
2SA311M		3008	53	2007	
2SA311OR		3008	53	2007	
2SA311R		3008	53	2007	
2SA311X		3008	53	2007	
2SA311Y		3008	53	2007	
2SA312	102A	3008	53	2007	
2SA312A		3008	53	2007	
2SA312B		3008	53	2007	
2SA312C		3008	53	2007	
2SA312D		3008	53	2007	
2SA312E		3008	53	2007	
2SA312F		3008	53	2007	
2SA312G		3008	53	2007	
2SA312GN		3008	53	2007	
2SA312H		3008	53	2007	
2SA312J		3008	53	2007	
2SA312K		3008	53	2007	
2SA312L		3008	53	2007	
2SA312M		3008	53	2007	
2SA312OR		3008	53	2007	
2SA312R		3008	53	2007	
2SA312X		3008	53	2007	
2SA312Y		3008	53	2007	
2SA313	126	3008	52	2024	
2SA313-BLUE				2005	
2SA313-GREEN				2005	
2SA313-RED			53	2005	
2SA313-YELLOW			53	2005	
2SA313A		3008	53	2007	
2SA313B		3008	53	2007	
2SA313C		3008		2007	
2SA313D		3008		2007	
2SA313E		3008		2007	
2SA313F		3008		2007	
2SA313G		3008		2007	
2SA313GN		3008		2007	
2SA313H		3008		2007	
2SA313J		3008		2007	
2SA313K		3008		2007	
2SA313L		3008		2007	
2SA313M		3008		2007	
2SA313OR		3008		2007	
2SA313R		3008		2007	
2SA313X		3008		2007	
2SA313Y		3008		2007	
2SA314	126	3008		2024	
2SA314-RED				2005	
2SA314-YELLOW				2005	
2SA314A		3008	53	2007	
2SA314B		3008	53	2007	
2SA314C		3008	53	2007	
2SA314D		3008	53	2007	
2SA314E		3008	53	2007	
2SA314F		3008	53	2007	
2SA314G		3008	53	2007	
2SA314GN		3008	53	2007	
2SA314H		3008	53	2007	

Industry Standard No.	ECG	SK	GE	RS 276-	MOTOR.
2SA314J		3008	53	2007	
2SA314K		3008	53	2007	
2SA314L		3008	53	2007	
2SA314M		3008	53	2007	
2SA314OR		3008	53	2007	
2SA314R		3008	53	2007	
2SA314X		3008	53	2007	
2SA314Y		3008	53	2007	
2SA315	126	3008	53	2024	
2SA315-RED			53	2005	
2SA315-YELLOW				2005	
2SA315A		3008	53	2007	
2SA315B		3008	53	2007	
2SA315C		3008	53	2007	
2SA315D		3008	53	2007	
2SA315E		3008	53	2007	
2SA315F		3008	53	2007	
2SA315G		3008	53	2007	
2SA315GN		3008	53	2007	
2SA315H		3008	53	2007	
2SA315J		3008	53	2007	
2SA315K		3008	53	2007	
2SA315L		3008	53	2007	
2SA315M		3008	53	2007	
2SA315OR		3008	53	2007	
2SA315R		3008	53	2007	
2SA315X		3008	53	2007	
2SA315Y		3008	53	2007	
2SA316	126	3008	53	2024	
2SA316-RED			53	2005	
2SA316-YELLOW				2005	
2SA316A		3008	53	2007	
2SA316B		3008	53	2007	
2SA316C		3008	53	2007	
2SA316D		3008	53	2007	
2SA316E		3008	53	2007	
2SA316F		3008	53	2007	
2SA316G		3008	53	2007	
2SA316GN		3008	53	2007	
2SA316H		3008	53	2007	
2SA316J		3008	53	2007	
2SA316K		3008	53	2007	
2SA316L		3008	53	2007	
2SA316M		3008	53	2007	
2SA316OR		3008	53	2007	
2SA316R		3008	53	2007	
2SA316X		3008	53	2007	
2SA316Y		3008	53	2007	
2SA31A		3005	51	2004	
2SA31B		3005	51	2004	
2SA31C		3005	51	2004	
2SA31D		3005	51	2004	
2SA31E		3005	51	2004	
2SA31F		3005	51	2004	
2SA31G		3005	51	2004	
2SA31L		3005	51	2004	
2SA31M		3005	51	2004	
2SA31OR		3005	51	2004	
2SA31X		3005	51	2004	
2SA31Y		3005	51	2004	
2SA32	102A	3005	51	2005	
2SA321	102A	3006/160	51	2004	
2SA321-1		3006	51	2004	
2SA321A		3006	51	2004	
2SA321B		3006	51	2004	
2SA321C		3006	51	2004	
2SA321D		3006	51	2004	
2SA321E		3006	51	2004	
2SA321F		3006	51	2004	
2SA321G		3006	51	2004	
2SA321GN		3006	51	2004	
2SA321H		3006	51	2004	
2SA321J		3006	51	2004	
2SA321K		3006	51	2004	
2SA321L		3006	51	2004	
2SA321M		3006	51	2004	
2SA321OR		3006	51	2004	
2SA321R		3006	51	2004	
2SA321X		3006	51	2004	
2SA321Y		3006	51	2004	
2SA322	102A	3008	53	2004	
2SA322A		3008	53	2007	
2SA322B		3008	53	2007	
2SA322C		3008	53	2007	
2SA322D		3008	53	2007	
2SA322E		3008	53	2007	
2SA322F		3008	53	2007	
2SA322G		3008	53	2007	
2SA322GN		3008	53	2007	
2SA322H		3008	53	2007	
2SA322K	100	3721	53	2007	
2SA322L		3008	53	2007	
2SA322M		3008	53	2007	
2SA322OR		3008	53	2007	
2SA322R		3008	53	2007	
2SA322X		3008	53	2007	
2SA322Y		3008	53	2007	
2SA323	160	3008	53	2004	
2SA323A		3008	53	2004	
2SA323B		3008	53	2004	
2SA323C		3008	53	2004	
2SA323D		3008	53	2004	
2SA323E		3008	53	2004	
2SA323F		3008	53	2004	
2SA323G		3008	53	2004	
2SA323GN		3008	53	2004	
2SA323J		3008	53	2004	
2SA323K		3008	53	2004	
2SA323L		3008	53	2004	
2SA323M		3008	53	2004	
2SA323R		3008	53	2004	
2SA323X		3008	53	2004	
2SA323Y		3008	53	2004	
2SA324	126	3006/160	51	2004	
2SA324A		3006	51	2004	
2SA324B		3006	51	2004	
2SA324C		3006	51	2004	
2SA324D		3006	51	2004	
2SA324E		3006	51	2004	
2SA324F	126	3006/160	51	2004	
2SA324G		3006	51	2004	
2SA324GN		3006	51	2004	
2SA324H		3006	51	2004	
2SA324K		3006	51	2004	
2SA324L		3006	51	2004	
2SA324M		3006	51	2004	
2SA324OR		3006	51	2004	
2SA324R		3006	51	2004	
2SA324X		3006	51	2004	
2SA324Y		3006	51	2004	
2SA325	126	3006/160	51	2024	
2SA325A		3006	51	2004	
2SA325B		3006	51	2004	
2SA325D		3006	51	2004	
2SA325E		3006	51	2004	
2SA325G		3006	51	2004	
2SA325GN		3006	51	2004	
2SA325H		3006	51	2004	
2SA325J		3006	51	2004	
2SA325K		3006	51	2004	
2SA325L		3006	51	2004	
2SA325M		3006	51	2004	
2SA325OR		3006	51	2004	
2SA325R		3006	51	2004	
2SA325X		3006	51	2004	
2SA325Y		3006	51	2004	
2SA326	126	3006/160	51	2024	
2SA326A		3006	51	2004	
2SA326B		3006	51	2004	
2SA326C		3006	51	2004	
2SA326D		3006	51	2004	
2SA326E		3006	51	2004	
2SA326F		3006	51	2004	
2SA326G		3006	51	2004	
2SA326GN		3006	51	2004	
2SA326H		3006	51	2004	
2SA326J		3006	51	2004	
2SA326K		3006	51	2004	
2SA326L		3006	51	2004	
2SA326M		3006	51	2004	
2SA326OR		3006	51	2004	
2SA326R		3006	51	2004	
2SA326X		3006	51	2004	
2SA326Y		3006	51	2004	
2SA327			245	2004	
2SA328				2004	
2SA329	126	3006/160	51	2024	
2SA329A	126	3008	51	2024	
2SA329B	126	3008	51	2024	
2SA329C		3006	51	2004	
2SA329D		3006	51	2004	
2SA329E		3006	51	2004	
2SA329F		3006	51	2004	
2SA329G		3006	51	2004	
2SA329GN		3006	51	2004	
2SA329H		3006	51	2004	
2SA329J		3006	51	2004	
2SA329K		3006	51	2004	
2SA329L		3006	51	2004	
2SA329M		3006	51	2004	
2SA329OR		3006	51	2004	
2SA329R		3006	51	2004	
2SA329X		3006	51	2004	
2SA329Y		3006	51	2004	
2SA32A		3005	51	2004	
2SA32B		3005	51	2004	
2SA32C		3005	51	2004	
2SA32D		3005	51		
2SA32E		3005	51	2004	
2SA32F		3005	51	2004	
2SA32G		3005	51	2004	
2SA32L		3005	51	2004	
2SA32M		3005	51	2004	
2SA32OR			51	2004	
2SA32X		3005	51	2004	
2SA32Y		3005	51	2004	
2SA33	102A		53	2007	
2SA330	126	3005	51	2007	
2SA330A		3005	51	2004	
2SA330B		3005	51	2004	
2SA330C		3005	51	2004	
2SA330D		3005	51	2004	
2SA330E		3005	51	2004	
2SA330F		3005	51	2004	
2SA330G		3005	51	2004	
2SA330GN		3005	51	2004	
2SA330H		3005	51	2004	
2SA330J		3005	51	2004	
2SA330K		3005	51	2004	
2SA330L		3005	51	2004	
2SA330M		3005	51	2004	
2SA330OR			51	2004	
2SA330X		3005	51	2004	
2SA330Y		3005	51	2004	
2SA331	126	3007	51	2024	
2SA331A		3006	51	2004	
2SA331B		3006	51	2004	
2SA331C		3006	51	2004	
2SA331D		3006	51	2004	
2SA331E		3006	51	2004	
2SA331F		3006	51	2004	
2SA331G		3006	51	2004	
2SA331GN		3006	51	2004	
2SA331H		3006	51	2004	
2SA331J		3006	51	2004	
2SA331K		3006	51	2004	
2SA331L		3006	51	2004	
2SA331M		3006	51	2004	
2SA331OR		3006	51	2004	
2SA331R		3006	51	2004	
2SA331X		3006	51	2004	
2SA331Y		3006	51	2004	
2SA332	160	3008	53	2007	
2SA332A		3008	53	2007	
2SA332B		3008	53	2007	
2SA332C		3008	53	2007	
2SA332D		3008	53	2007	
2SA332E		3008	53	2007	
2SA332F		3008	53	2007	
2SA332G		3008	53	2007	
2SA332GN		3008	53	2007	
2SA332H		3008	53	2007	
2SA332J		3008	53	2007	
2SA332K		3008	53	2007	
2SA332L		3008	53	2007	
2SA332M		3008	53	2007	
2SA332OR		3008	53	2007	
2SA332R		3008	53	2007	
2SA332X		3008	53	2007	
2SA332Y		3008	53	2007	
2SA333				2005	
2SA334				2005	
2SA335	126	3006/160	51	2024	
2SA335A		3006	51	2004	
2SA335B		3006	51	2004	
2SA335C		3006	51	2004	
2SA335D		3006	51	2004	
2SA335E		3006	51	2004	
2SA335F		3006	51	2004	
2SA335G		3006	51	2004	
2SA335GN		3006	51	2004	
2SA335H		3006	51	2004	
2SA335J		3006	51	2004	
2SA335K		3006	51	2004	
2SA335L		3006	51	2004	
2SA335M		3006	51	2004	
2SA335OR		3006	51	2004	
2SA335R		3006	51	2004	

Industry Standard No.	ECG	SK	GE	RS 276-	MOTOR.
2SA335X		3006	51	2004	
2SA335Y		3006	51	2004	
2SA336				2005	
2SA337	126	3006/160	51	2024	
2SA337A		3006	51	2004	
2SA337B		3006	51	2004	
2SA337C		3006	51	2004	
2SA337D		3006	51	2004	
2SA337E		3006	51	2004	
2SA337F		3006	51	2004	
2SA337G		3006	51	2004	
2SA337GN		3006	51	2004	
2SA337H		3006	51	2004	
2SA337J		3006	51	2004	
2SA337K		3006	51	2004	
2SA337L		3006	51	2004	
2SA337M		3006	51	2004	
2SA337OR		3006	51	2004	
2SA337R		3006	51	2004	
2SA337X		3006	51	2004	
2SA337Y		3006	51	2004	
2SA338	126	3008	53	2024	
2SA338A		3008	53	2007	
2SA338B		3008	53	2007	
2SA338C		3008	53	2007	
2SA338D		3008	53	2007	
2SA338E		3008	53	2007	
2SA338F		3008	53	2007	
2SA338G		3008	53	2007	
2SA338GN		3008	53	2007	
2SA338H		3008	53	2007	
2SA338J		3008	53	2007	
2SA338K		3008	53	2007	
2SA338L		3008	53	2007	
2SA338M		3008	53	2007	
2SA338OR		3008	53	2007	
2SA338R		3008	53	2007	
2SA338X		3008	53	2007	
2SA338Y		3008	53	2007	
2SA339	126	3008	53	2024	
2SA339A		3008	53	2004	
2SA339B		3008	53	2004	
2SA339C		3008	53	2004	
2SA339D		3008	53	2004	
2SA339E		3008	53	2004	
2SA339F		3008	53	2004	
2SA339G		3008	53	2004	
2SA339GN		3008	53	2004	
2SA339H		3008	53	2004	
2SA339J		3008	53	2004	
2SA339K		3008	53	2004	
2SA339L		3008	53	2004	
2SA339M		3008	53	2004	
2SA339OR		3008	53	2004	
2SA339R		3008	53	2004	
2SA339X		3008	53	2004	
2SA339Y		3008	53	2004	
2SA340	160		245	2004	
2SA341	160	3007	53	2004	
2SA341-OA	160		53	2004	
2SA341-OB	160	3007	53	2004	
2SA3410A		3006	51	2004	
2SA3410B		3007	51	2004	
2SA341A		3008	53	2007	
2SA341B		3008	53	2007	
2SA341C		3008	53	2007	
2SA341D		3008	53	2007	
2SA341E		3008	53	2007	
2SA341F		3008	53	2007	
2SA341G		3008	53	2007	
2SA341GN		3008	53	2007	
2SA341H		3008	53	2007	
2SA341J		3008	53	2007	
2SA341K		3008	53	2007	
2SA341L		3008	53	2007	
2SA341M		3008	53	2007	
2SA341OR		3008	53	2007	
2SA341R		3008	53	2007	
2SA341X		3008	53	2007	
2SA341Y		3008	53	2007	
2SA342	160	3008	51	2004	
2SA342A	160	3006	51	2004	
2SA342B		3006	51	2004	
2SA342C		3006	51	2004	
2SA342D		3006	51	2004	
2SA342E		3006	51	2004	
2SA342F		3006	51	2004	
2SA342G		3006	51	2004	
2SA342GN		3006	51	2004	
2SA342H		3006	51	2004	
2SA342J		3006	51	2004	
2SA342K		3006	51	2004	
2SA342L		3006	51	2004	
2SA342M		3006	51	2004	
2SA342OR		3006	51	2004	
2SA342R		3006	51	2004	
2SA342X		3006	51	2004	
2SA342Y		3006	51	2004	
2SA343	160	3006	51	2024	
2SA343A		3006	51	2004	
2SA343B		3006	51	2004	
2SA343C		3006	51	2004	
2SA343D		3006	51	2004	
2SA343E		3006	51	2004	
2SA343F		3006	51	2004	
2SA343G		3006	51	2004	
2SA343H		3006	51	2004	
2SA343J		3006	51	2004	
2SA343K		3006	51	2004	
2SA343L		3006	51	2004	
2SA343M		3006	51	2004	
2SA343OR		3006	51	2004	
2SA343R		3006	51	2004	
2SA343X		3006	51	2004	
2SA343Y		3006	51	2004	
2SA344	126	3006/160	51	2024	
2SA344A		3006	51	2004	
2SA344B		3006	51	2004	
2SA344C		3006	51	2004	
2SA344D		3006	51	2004	
2SA344E		3006	51	2004	
2SA344F		3006	51	2004	
2SA344G		3006	51	2004	
2SA344GN		3006	51	2004	
2SA344H		3006	51	2004	
2SA344J		3006	51	2004	
2SA344K		3006	51	2004	
2SA344L		3006	51	2004	
2SA344M		3006	51	2004	
2SA344OR		3006	51	2004	
2SA344R		3006	51	2004	
2SA344X		3006	51	2004	
2SA344Y		3006	51	2004	
2SA345	160		245	2004	
2SA346	160		245	2004	
2SA347	160		245	2004	
2SA348	160	3006	51	2004	
2SA348A		3006	51	2004	
2SA348B		3006	51	2004	
2SA348C		3006	51	2004	
2SA348D		3006	51	2004	
2SA348E		3006	51	2004	
2SA348F		3006	51	2004	
2SA348G		3006	51	2004	
2SA348GN		3006	51	2004	
2SA348H		3006	51	2004	
2SA348J		3006	51	2004	
2SA348K		3006	51	2004	
2SA348L		3006	51	2004	
2SA348M		3006	51	2004	
2SA348OR		3006	51	2004	
2SA348R		3006	51	2004	
2SA348X		3006	51	2004	
2SA348Y		3006	51	2004	
2SA349	160		245	2004	
2SA35	102A	3005	51	2024	
2SA350	126	3008	51	2024	
2SA350A	126	3008	51	2024	
2SA350AY			51	2004	
2SA350B		3008	51	2004	
2SA350BK		3008	51	2004	
2SA350C	126	3008	51	2024	
2SA350D		3008	51	2004	
2SA350E		3008	51	2004	
2SA350F		3008	51	2004	
2SA350G		3008	51	2004	
2SA350GN		3008	51	2004	
2SA350H	126	3008	51	2024	
2SA350J		3008	51	2004	
2SA350K		3008	51	2004	
2SA350L		3008	51	2004	
2SA350M		3008	51	2004	
2SA350OR		3008	51	2004	
2SA350R	126	3008	51	2024	
2SA350T	126	3008	51	2024	
2SA350TY	126	3008	51	2024	
2SA350X			51	2004	
2SA350Y	126	3008	53	2024	
2SA351	126	3007	50	2024	
2SA351A	126	3007	50	2024	
2SA351A-2		3007	50	2004	
2SA351B	126	3007	50	2024	
2SA351C		3007	50	2004	
2SA351D		3007	50	2004	
2SA351E		3007	50	2004	
2SA351F		3007	50	2004	
2SA351G		3007	50	2004	
2SA351GN		3007	50	2004	
2SA351GR	126		50	2024	
2SA351K		3007	50	2004	
2SA351L		3007	50	2004	
2SA351M		3007	50	2004	
2SA351OR		3007	50	2004	
2SA351R		3007	50	2004	
2SA351X		3007	50	2004	
2SA351Y		3007	50	2004	
2SA352	126	3008	51	2024	
2SA352A	126	3008	51	2024	
2SA352B	126	3008	51	2024	
2SA352C		3008	51	2004	
2SA352D		3008	51	2004	
2SA352E		3008	51	2004	
2SA352F		3008	51	2004	
2SA352G		3008	51	2004	
2SA352GN		3008	51	2004	
2SA352H		3008	51	2004	
2SA352J		3008	51	2004	
2SA352K		3008	51	2004	
2SA352L		3008	51	2004	
2SA352M		3008	51	2004	
2SA352OR		3008	51	2004	
2SA352R		3008	51	2004	
2SA352X			51	2004	
2SA352Y		3008	51	2004	
2SA353	126	3008	51	2024	
2SA353-AC	126		51	2024	
2SA353A	126	3008	51	2024	
2SA353AL		3008	51	2004	
2SA353B		3008	51	2004	
2SA353C	126	3008	51	2024	
2SA353CL		3008	51	2004	
2SA353D		3008	51	2004	
2SA353E		3008	51	2004	
2SA353F		3008	51	2004	
2SA353G		3008	51	2004	
2SA353GN		3008	51	2004	
2SA353H		3008	51	2004	
2SA353J		3008	51	2004	
2SA353K		3008	51	2004	
2SA353L		3008	51	2004	
2SA353M		3008	51	2004	
2SA353OR		3008	51	2004	
2SA353R		3008	51	2004	
2SA353X			51	2004	
2SA353Y		3008	51	2004	
2SA354	126	3007	51	2024	
2SA354-B			51	2004	
2SA354A	126	3007	51	2024	
2SA354B	126	3007	51	2024	
2SA354BK		3007	51	2004	
2SA354C		3007	51	2004	
2SA354D		3007	51	2004	
2SA354E		3007	51	2004	
2SA354F		3007	51	2004	
2SA354G		3007	51	2004	
2SA354GN		3007	51	2004	
2SA354H		3007	51	2004	
2SA354J		3007	51	2004	
2SA354K		3007	51	2004	
2SA354L		3007	51	2004	
2SA354M		3007	51	2004	
2SA354OR		3007	51	2004	
2SA354R		3007	51	2004	
2SA354X			51	2004	
2SA354Y		3007	51	2004	
2SA355	126	3006/160	51	2024	
2SA355A	126	3006/160	51	2024	
2SA355B		3006	51	2004	
2SA355C		3006	51	2004	
2SA355D		3006	51	2004	
2SA355E		3006	51	2004	
2SA355F		3006	51	2004	
2SA355G		3006	51	2004	
2SA355H		3006	51	2004	
2SA355J		3006	51	2004	
2SA355K		3006	51	2004	
2SA355L		3006	51	2004	
2SA355M		3006	51	2004	

Industry Standard No.	ECG	SK	GE	RS 276-	MOTOR.
2SA3550R		3006	51	2004	
2SA355R		3006	51	2004	
2SA355X		3006	51	2004	
2SA355Y		3006	51	2004	
2SA356	126	3008	53	2024	
2SA356A		3008	53	2004	
2SA356B		3008	53	2004	
2SA356C		3008	53	2004	
2SA356D		3008	53	2004	
2SA356E		3008	53	2004	
2SA356F		3008	53	2004	
2SA356G		3008	53	2004	
2SA356GN		3008	53	2004	
2SA356H		3008	53	2004	
2SA356J		3008	53	2004	
2SA356K		3008	53	2004	
2SA356L		3008	53	2004	
2SA356M		3008	53	2004	
2SA3560R		3008	53	2004	
2SA356R		3008	53	2004	
2SA357	126	3008	52	2024	
2SA357A		3008	53	2007	
2SA357B		3008	53	2007	
2SA357C		3008	53	2007	
2SA357D		3008	53	2007	
2SA357E		3008	53	2007	
2SA357F		3008	53	2007	
2SA357G		3008	53	2007	
2SA357GN		3008	53	2007	
2SA357H		3008	53	2007	
2SA357J		3008	53	2007	
2SA357K		3008	53	2007	
2SA357L		3008	53	2007	
2SA357M		3008	53	2007	
2SA3570R		3008	53	2007	
2SA357R		3008	53	2007	
2SA357X		3008	53	2007	
2SA357Y		3008	53	2007	
2SA358	126$	3007	51	2024	
2SA358-3		3006	51	2004	
2SA358A		3006	51	2004	
2SA358B		3006	51	2004	
2SA358C		3006	51	2004	
2SA358D		3006	51	2004	
2SA358E		3006	51	2004	
2SA358F		3006	51	2004	
2SA358G		3006	51	2004	
2SA358GN		3006	51	2004	
2SA358H		3006	51	2004	
2SA358J		3006	51	2004	
2SA358K		3006	51	2004	
2SA358L		3006	51	2004	
2SA358M		3006	51	2004	
2SA3580R		3006	51	2004	
2SA358R		3006	51	2004	
2SA358X		3006	51	2004	
2SA358Y		3006	51	2004	
2SA359	126	3006/160	51	2024	
2SA359A		3006	51	2007	
2SA359B		3006	51	2007	
2SA359C		3006	51	2007	
2SA359D		3006	51	2007	
2SA359E		3006	51	2007	
2SA359F		3006	51	2007	
2SA359G		3006	51	2007	
2SA359GN		3006	51	2007	
2SA359H		3006	51	2007	
2SA359J		3006	51	2007	
2SA359K		3006	51	2007	
2SA359L		3006	51	2007	
2SA359M		3006	51	2007	
2SA3590R		3006	51	2007	
2SA359R		3006	51	2007	
2SA359X		3006	51	2007	
2SA359Y		3006	51	2007	
2SA35A		3005	51	2004	
2SA35B		3005	51	2004	
2SA35C		3005	51	2004	
2SA35D		3005	51	2004	
2SA35E		3005	51	2004	
2SA35F		3005	51	2004	
2SA35G		3005	51	2004	
2SA35L		3005	51	2004	
2SA35M		3005	51	2004	
2SA350R			51	2004	
2SA35X		3005	51	2004	
2SA35Y		3005	51	2004	
2SA36	102A	3005	51	2007	
2SA360	160	3007	51	2004	
2SA360A		3006	51	2004	
2SA360B		3006	51	2004	
2SA360C		3006	51	2004	
2SA360E		3006	51	2004	
2SA360F		3006	51	2004	
2SA360G		3007	51	2004	
2SA360GN		3006	51	2004	
2SA360H		3007	51	2004	
2SA360J		3006	51	2004	
2SA360K		3007	51	2004	
2SA360L		3006	51	2004	
2SA360M		3006	51	2004	
2SA3600R		3006	51	2004	
2SA360R		3006	51	2004	
2SA360X		3006	51	2004	
2SA360Y		3006	51	2004	
2SA361	160	3007	50	2004	
2SA361A		3007	50	2004	
2SA361B		3007	50	2004	
2SA361C		3007	50	2004	
2SA361D		3007	50	2004	
2SA361E		3007	50	2004	
2SA361F		3007	50	2004	
2SA361G		3007	50	2004	
2SA361GN		3007	50	2004	
2SA361H		3007	50	2004	
2SA361J		3007	50	2004	
2SA361K		3007	50	2004	
2SA361L		3007	50	2004	
2SA361M		3007	50	2004	
2SA3610R		3007	50	2004	
2SA361R		3007	50	2004	
2SA361X		3007	50	2004	
2SA361Y		3007	50	2004	
2SA362	160$		245	2004	
2SA363	160$		245	2004	
2SA364	126	3006/160	51	2024	
2SA364A		3006	51	2004	
2SA364B		3006	51	2004	
2SA364C		3006	51	2004	
2SA364D		3006	51	2004	
2SA364E		3006	51	2004	
2SA364F		3006	51	2004	
2SA364G		3006	51	2004	
2SA364GN		3006	51	2004	

Industry Standard No.	ECG	SK	GE	RS 276-	MOTOR.
2SA364H		3006	51	2004	
2SA364J		3006	51	2004	
2SA364K		3006	51	2004	
2SA364L		3006	51	2004	
2SA364M		3006	51	2004	
2SA3640R		3006	51	2004	
2SA364R		3006	51	2004	
2SA364X		3006	51	2004	
2SA364Y		3006	51	2004	
2SA365	126	3006/160	51	2024	
2SA365A		3006	51	2004	
2SA365B		3006	51	2004	
2SA365C		3006	51	2004	
2SA365D		3006	51	2004	
2SA365E		3006	51	2004	
2SA365F		3006	51	2004	
2SA365G		3006	51	2004	
2SA365GN		3006	51	2004	
2SA365H		3006	51	2004	
2SA365J		3006	51	2004	
2SA365K		3006	51	2004	
2SA365L		3006	51	2004	
2SA365M		3006	51	2004	
2SA3650R		3006	51	2004	
2SA365R		3006	51	2004	
2SA365X		3006	51	2004	
2SA365Y		3006	51	2004	
2SA366	126	3006/160	51	2024	
2SA366A		3006	51	2024	
2SA366B		3006	51	2024	
2SA366C		3006	51	2024	
2SA366D		3006	51	2024	
2SA366E		3006	51	2024	
2SA366F		3006	51	2024	
2SA366G		3006	51	2024	
2SA366GN		3006	51	2024	
2SA366H		3006	51	2024	
2SA366J		3006	51	2024	
2SA366K		3006	51	2024	
2SA366L		3006	51	2024	
2SA366M		3006	51	2024	
2SA3660R		3006	51	2024	
2SA366R		3006	51	2024	
2SA366X		3006	51	2024	
2SA366Y		3006	51	2024	
2SA367	126	3007	51	2024	
2SA367A		3006	51	2004	
2SA367B		3006	51	2004	
2SA367C		3006	51	2004	
2SA367D		3006	51	2004	
2SA367E		3006	51	2004	
2SA367F		3006	51	2004	
2SA367G		3006	51	2004	
2SA367GN		3006	51	2004	
2SA367H		3006	51	2004	
2SA367J		3006	51	2004	
2SA367K		3006	51	2004	
2SA367L		3006	51	2004	
2SA367M		3006	51	2004	
2SA3670R		3006	51	2004	
2SA367R		3006	51	2004	
2SA367X		3006	51	2004	
2SA367Y		3006	51	2004	
2SA368	126	3007	50	2024	
2SA368A		3007	50	2004	
2SA368B		3007	50	2004	
2SA368C		3007	50	2004	
2SA368D		3007	50	2004	
2SA368E		3007	50	2004	
2SA368F		3007	50	2004	
2SA368G		3007	50	2004	
2SA368GN		3007	50	2004	
2SA368H		3007	50	2004	
2SA368J		3007	50	2004	
2SA368K		3007	50	2004	
2SA368L		3007	50	2004	
2SA368M		3007	50	2004	
2SA3680R		3007	50	2004	
2SA368R		3007	50	2004	
2SA368X		3007	50	2004	
2SA368Y		3007	50	2004	
2SA369	126	3008	53	2024	
2SA369A		3008	53	2004	
2SA369B		3008	53	2004	
2SA369C		3008	53	2004	
2SA369D		3008	53	2004	
2SA369E		3008	53	2004	
2SA369F		3008	53	2004	
2SA369G		3008	53	2004	
2SA369GN		3008	53	2004	
2SA369H		3008	53	2004	
2SA369J		3008	53	2004	
2SA369K		3008	53	2004	
2SA369L		3008	53	2004	
2SA369M		3008	53	2004	
2SA3690R		3008	53	2004	
2SA369R		3008	53	2004	
2SA369X		3008	53	2004	
2SA369Y		3008	53	2004	
2SA36A		3005	51	2004	
2SA36B		3005	51	2004	
2SA36C		3005	51	2004	
2SA36D		3005	51	2004	
2SA36E		3005	51	2004	
2SA36F		3005	51	2004	
2SA36G		3005	51	2004	
2SA36L		3005	51	2004	
2SA36M		3005	51	2004	
2SA360R		3005	51	2004	
2SA36R		3005	51	2004	
2SA36X		3005	51	2004	
2SA36Y		3005	51	2004	
2SA37	126	3005	51	2024	
2SA370			53	2007	
2SA371			53	2007	
2SA372	160$	3006/160	245	2004	
2SA373	102A$		245	2004	
2SA373A				2005	
2SA374	102A		53	2005	
2SA375	102A	3123	80	2004	
2SA375A		3123	80		
2SA375B		3123	80		
2SA375C		3123	80		
2SA375D		3123	80		
2SA375E		3123	80		
2SA375F		3123	80		
2SA375G		3123	80		
2SA375GN		3123	80		
2SA375H		3123	80		
2SA375J		3123	80		
2SA375K		3123	80		
2SA375L		3123	80		
2SA375M		3123	80		
2SA3750R		3123	80		

Industry Standard No.	ECG	SK	GE	RS 276-	MOTOR.
2SA375R		3123	80		
2SA375X		3123	80		
2SA375Y		3123	80		
2SA376	126	3006/160	51	2024	
2SA376A		3006	51	2004	
2SA376B		3006	51	2004	
2SA376C		3006	51	2004	
2SA376D		3006	51	2004	
2SA376E		3006	51	2004	
2SA376F		3006	51	2004	
2SA376GN		3006	51	2004	
2SA376H		3006	51	2004	
2SA376J		3006	51	2004	
2SA376K		3006	51	2004	
2SA376L		3006	51	2004	
2SA376OR		3006	51	2004	
2SA376R		3006	51	2004	
2SA376X		3006	51	2004	
2SA376Y		3006	51	2004	
2SA377	160	3006	51	2004	
2SA377A		3006	51	2004	
2SA377B		3006	51	2004	
2SA377C		3006	51	2004	
2SA377D		3006	51	2004	
2SA377E		3006	51	2004	
2SA377F		3006	51	2004	
2SA377G		3006	51	2004	
2SA377GN		3006	51	2004	
2SA377H		3006	51	2004	
2SA377J		3006	51	2004	
2SA377K		3006	51	2004	
2SA377L		3006	51	2004	
2SA377M		3006	51	2004	
2SA377OR		3006	51	2004	
2SA377R		3006	51	2004	
2SA377X		3006	51	2004	
2SA377Y		3006	51	2004	
2SA378	160	3006	245	2004	
2SA379	160	3006	245	2004	
2SA37A		3005	51	2004	
2SA37AC				2027	
2SA37B		3005	51	2004	
2SA37C		3005	51	2004	
2SA37D		3005	51	2004	
2SA37E		3005	51	2004	
2SA37F		3005	51	2004	
2SA37G		3005	51	2004	
2SA37L		3005	51	2004	
2SA37M		3005	51	2004	
2SA37OR		3005	51	2004	
2SA37R		3005	51	2004	
2SA37X		3005	51	2004	
2SA37Y		3005	51	2004	
2SA38	126	3005	51	2024	
2SA380	126	3008	50	2024	
2SA380A		3008	50	2004	
2SA380B		3008	50	2004	
2SA380C		3008	50	2004	
2SA380D		3008	50	2004	
2SA380E		3008	50	2004	
2SA380F		3008	50	2004	
2SA380G		3008	50	2004	
2SA380GN		3008	50	2004	
2SA380H		3008	50	2004	
2SA380J		3008	50	2004	
2SA380K		3008	50	2004	
2SA380L		3008	50	2004	
2SA380M		3008	50	2004	
2SA380OR		3008	50	2004	
2SA380R		3008	50	2004	
2SA380X			50	2004	
2SA380Y		3008	50	2004	
2SA381	126	3008	53	2024	
2SA381A		3008	53	2004	
2SA381B		3008	53	2004	
2SA381D		3008	53	2004	
2SA381E		3008	53	2004	
2SA381F		3008	53	2004	
2SA381G		3008	53	2004	
2SA381GN		3008	53	2004	
2SA381H		3008	53	2004	
2SA381K		3008	53	2004	
2SA381L		3008	53	2004	
2SA381M		3008	53	2004	
2SA381OR		3008	53	2004	
2SA381R		3008	53	2004	
2SA381X		3008	53	2004	
2SA381Y		3008	53	2004	
2SA382	126	3008	53	2024	
2SA382A		3008	53	2004	
2SA382B		3008	53	2004	
2SA382C		3008	53	2004	
2SA382D		3008	53	2004	
2SA382E		3008	53	2004	
2SA382F		3008	53	2004	
2SA382G		3008	53	2004	
2SA382GN		3008	53	2004	
2SA382H		3008	53	2004	
2SA382J		3008	53	2004	
2SA382K		3008	53	2004	
2SA382L		3008	53	2004	
2SA382M		3008	53	2004	
2SA382OR		3008	53	2004	
2SA382R		3008	53	2004	
2SA382X		3008	53	2004	
2SA382Y		3008	53	2004	
2SA383	126	3008	53	2024	
2SA383A		3008	53	2004	
2SA383B		3008	53	2004	
2SA383C		3008	53	2004	
2SA383D		3008	53	2004	
2SA383E		3008	53	2004	
2SA383F		3008	53	2004	
2SA383G		3008	53	2004	
2SA383GN		3008	53	2004	
2SA383H		3008	53	2004	
2SA383J		3008	53	2004	
2SA383K		3008	53	2004	
2SA383L		3008	53	2004	
2SA383M		3008	53	2004	
2SA383OR		3008	53	2004	
2SA383R		3008	53	2004	
2SA383X		3008	53	2004	
2SA383Y		3008	53	2004	
2SA384	126	3008	53	2024	
2SA384A		3008	53	2004	
2SA384B		3008	53	2004	
2SA384C		3008	53	2004	
2SA384D		3008	53	2004	
2SA384E		3008	53	2004	
2SA384F		3008	53	2004	
2SA384G		3008	53	2004	
2SA384GN		3008	53	2004	
2SA384H		3009	53	2004	
2SA384J		3009	53	2004	
2SA384K		3008	53	2004	
2SA384L		3008	53	2004	
2SA384M		3008	53	2004	
2SA384OR		3008	53	2004	
2SA384R		3008	53	2004	
2SA384X		3008	53	2004	
2SA384Y		3008	53	2004	
2SA385	126	3008	51	2024	
2SA385A	126	3008	51	2024	
2SA385B		3008	51	2004	
2SA385C		3008	51	2004	
2SA385D	126	3008	51	2024	
2SA385E		3008	51	2004	
2SA385F		3008	51	2004	
2SA385G		3008	51	2004	
2SA385GN		3008	51	2004	
2SA385H		3008	51	2004	
2SA385J		3008	51	2004	
2SA385K		3008	51	2004	
2SA385L		3008	51	2004	
2SA385M		3008	51	2004	
2SA385OR		3008	51	2004	
2SA385R		3008	51	2004	
2SA385X			51	2004	
2SA385Y		3008	51	2004	
2SA386				2005	
2SA387				2005	
2SA38A		3005	51	2004	
2SA38B		3005	51	2004	
2SA38C		3005	51	2004	
2SA38D		3005	51	2004	
2SA38E		3005	51	2004	
2SA38F		3005	51	2004	
2SA38G		3005	51	2004	
2SA38L		3005	51	2004	
2SA38M		3005	51	2004	
2SA38OR		3005	51	2004	
2SA38R		3005	51	2004	
2SA38X		3005	51	2004	
2SA38Y		3005	51	2004	
2SA39	126	3005	51	2024	
2SA391	100	3006/160	51	2024	
2SA391A		3006	51	2004	
2SA391B		3006	51	2004	
2SA391C		3006	51	2004	
2SA391D		3006	51	2004	
2SA391E		3006	51	2004	
2SA391F		3006	51	2004	
2SA391G		3006	51	2004	
2SA391GN		3006	51	2004	
2SA391H		3006	51	2004	
2SA391J		3006	51	2004	
2SA391K		3006	51	2004	
2SA391L		3006	51	2004	
2SA391M		3006	51	2004	
2SA391OR		3006	51	2004	
2SA391R		3006	51	2004	
2SA391X		3006	51	2004	
2SA391Y		3006	51	2004	
2SA392	100	3006/160	51	2024	
2SA392A		3006	51	2004	
2SA392B		3006	51	2004	
2SA392C		3006	51	2004	
2SA392D		3006	51	2004	
2SA392E		3006	51	2004	
2SA392F		3006	51	2004	
2SA392G		3006	51	2004	
2SA392GN		3006	51	2004	
2SA392H		3006	51	2004	
2SA392J		3006	51	2004	
2SA392K		3006	51	2004	
2SA392L		3006	51	2004	
2SA392M		3006	51	2004	
2SA392OR		3006	51	2004	
2SA392R		3006	51	2004	
2SA392X		3006	51	2004	
2SA392Y		3006	51	2004	
2SA393	100	3006/160	52	2024	
2SA393A	126	3006/160	52	2024	
2SA394	100	3006/160	51	2024	
2SA394A		3006	51	2004	
2SA394B		3006	51	2004	
2SA394C		3006	51	2004	
2SA394D		3006	51	2004	
2SA394E		3006	51	2004	
2SA394F		3006	51	2004	
2SA394G		3006	51	2004	
2SA394GN		3006	51	2004	
2SA394H		3006	51	2004	
2SA394J		3006	51	2004	
2SA394K		3006	51	2004	
2SA394L		3006	51	2004	
2SA394M		3006	51	2004	
2SA394OR		3006	51	2004	
2SA394R		3006	51	2004	
2SA394X		3006	51	2004	
2SA394Y		3006	51	2004	
2SA395	100	3006/160	51	2024	
2SA395A		3006	51	2004	
2SA395B		3006	51	2004	
2SA395C		3006	51	2004	
2SA395D		3006	51	2004	
2SA395E		3006	51	2004	
2SA395F		3006	51	2004	
2SA395G		3006	51	2004	
2SA395GN		3006	51	2004	
2SA395H		3006	51	2004	
2SA395J		3006	51	2004	
2SA395K		3006	51	2004	
2SA395L		3006	51	2004	
2SA395M		3006	51	2004	
2SA395OR		3006	51	2004	
2SA395R		3006	51	2004	
2SA395X		3006	51	2004	
2SA395Y		3006	51	2004	
2SA396	102	3722	2	2007	
2SA397	102	3722	2	2007	
2SA398	102	3006/160	51	2024	
2SA398A		3006	51	2004	
2SA398B		3006	51	2004	
2SA398C		3006	51	2004	
2SA398D		3006	51	2004	
2SA398E		3006	51	2004	
2SA398F		3006	51	2004	
2SA398G		3006	51	2004	
2SA398GN		3006	51	2004	
2SA398H		3006	51	2004	
2SA398J		3006	51	2004	
2SA398K		3006	51	2004	
2SA398L		3006	51	2004	
2SA398M		3006	51	2004	
2SA398OR		3006	51	2004	
2SA398X		3006	51	2004	

Industry Standard No.	ECG	SK	GE	RS 276-	MOTOR.
2SA398Y		3006	51	2004	
2SA399	102	3006/160	51	2024	
2SA399A		3006	51	2004	
2SA399B		3006	51	2004	
2SA399C		3006	51	2004	
2SA399D		3006	51	2004	
2SA399E		3006	51	2004	
2SA399F		3006	51	2004	
2SA399G		3006	51	2004	
2SA399GN		3006	51	2004	
2SA399H		3006	51	2004	
2SA399J		3006	51	2004	
2SA399K		3006	51	2004	
2SA399L		3006	51	2004	
2SA399M		3006	51	2004	
2SA399OR		3006	51	2004	
2SA399X		3006	51	2004	
2SA399Y		3006	51	2004	
2SA39A		3005	51	2004	
2SA39B		3005	51	2004	
2SA39C		3005	51	2004	
2SA39D		3005	51	2004	
2SA39E		3005	51	2004	
2SA39F		3005	51	2004	
2SA39G		3005	51	2004	
2SA39L		3005	51	2004	
2SA39M		3005	51	2004	
2SA39OR		3005	51	2004	
2SA39R		3005	51	2004	
2SA39X		3005	51	2004	
2SA39Y		3005	51	2004	
2SA40	102A	3005	51	2007	
2SA400	126	3008	53	2024	
2SA400A		3008	53	2007	
2SA400B		3008	53	2007	
2SA400C		3008	53	2007	
2SA400D		3008	53	2007	
2SA400E		3008	53	2007	
2SA400F		3008	53	2007	
2SA400G		3008	53	2007	
2SA400GN		3008	53	2007	
2SA400H		3008	53	2007	
2SA400J		3008	53	2007	
2SA400K		3008	53	2007	
2SA400L		3008	53	2007	
2SA400M		3008	53	2007	
2SA400OR			53	2007	
2SA400R		3008	53	2007	
2SA400X		3008	53	2007	
2SA400Y		3008	53	2007	
2SA401	126		245	2004	
2SA402	159		82	2032	
2SA403	160	3006	51	2004	
2SA403A		3006	51	2004	
2SA403B		3006	51	2004	
2SA403C		3006	51	2004	
2SA403D		3006	51	2004	
2SA403E		3006	51	2004	
2SA403F		3006	51	2004	
2SA403G		3006	51	2004	
2SA403GN		3006	51	2004	
2SA403H		3006	51	2004	
2SA403J		3006	51	2004	
2SA403K		3006	51	2004	
2SA403L		3006	51	2004	
2SA403M		3006	51	2004	
2SA403OR		3006	51	2004	
2SA403R		3006	51	2004	
2SA403X		3006	51	2004	
2SA403Y		3006	51	2004	
2SA404	160	3006	51	2004	
2SA404A		3006	51	2004	
2SA404B		3006	51	2004	
2SA404C		3006	51	2004	
2SA404D		3006	51	2004	
2SA404E		3006	51	2004	
2SA404F		3006	51	2004	
2SA404G		3006	51	2004	
2SA404GN		3006	51	2004	
2SA404H		3006	51	2004	
2SA404K		3006	51	2004	
2SA404L		3006	51	2004	
2SA404M		3006	51	2004	
2SA404OR		3006	51	2004	
2SA404R		3006	51	2004	
2SA404X		3006	51	2004	
2SA404Y		3006	51	2004	
2SA405	160		50	2004	
2SA405-0			50	2004	
2SA406	102A	3005	51	2005	
2SA406A		3005	51	2004	
2SA406B		3005	51	2004	
2SA406C		3005	51	2004	
2SA406D		3005	51	2004	
2SA406E		3005	51	2004	
2SA406F		3005	51	2004	
2SA406G		3005	51	2004	
2SA406GN		3005	51	2004	
2SA406H		3005	51	2004	
2SA406J		3005	51	2004	
2SA406K		3005	51	2004	
2SA406L		3005	51	2004	
2SA406M		3005	51	2004	
2SA406OR		3005	51	2004	
2SA406R		3005	51	2004	
2SA406X		3005	51	2004	
2SA406Y		3005	51	2004	
2SA407	102A	3005	51	2005	
2SA407A		3005	51	2004	
2SA407B		3005	51	2004	
2SA407C		3005	51	2004	
2SA407D		3005	51	2004	
2SA407E		3005	51	2004	
2SA407F		3005	51	2004	
2SA407G		3005	51	2004&	
2SA407GN		3005	51	2004	
2SA407H		3005	51	2004	
2SA407J		3005	51	2004	
2SA407K		3005	51	2004	
2SA407L		3005	51	2004	
2SA407M		3005	51	2004	
2SA407OR		3005	51	2004	
2SA407R		3005	51	2004	
2SA407X		3005	51	2004	
2SA407Y		3005	51	2004	
2SA408	126	3123	80	2004	
2SA408A		3123	80		
2SA408B		3123	80		
2SA408D		3123	80		
2SA408E		3123	80		
2SA408F		3123	80		
2SA408GN		3123	80		
2SA408H		3123	80		
2SA408J		3123	80		
2SA408K		3123	80		
2SA408L		3123	80		
2SA408M		3123	80		
2SA408OR		3123	80		
2SA408R		3123	80		
2SA408X		3123	80		
2SA408Y		3123	80		
2SA409	126	3123	80	2004	
2SA409A		3123	80		
2SA409B		3123	80		
2SA409C		3123	80		
2SA409D		3123	80		
2SA409E		3123	80		
2SA409F		3123	80		
2SA409G		3123	80		
2SA409GN		3123	80		
2SA409H		3123	80		
2SA409J		3123	80		
2SA409K		3123	80		
2SA409L		3123	80		
2SA409M		3123	80		
2SA409OR		3123	80		
2SA409R		3123	80		
2SA409X		3123	80		
2SA409Y		3123	80		
2SA40A		3005	51	2004	
2SA40B		3005	51	2004	
2SA40C		3005	51	2004	
2SA40D		3005	51	2004	
2SA40E		3005	51	2004	
2SA40F		3005	51	2004	
2SA40G		3005	51	2004	
2SA40L		3005	51	2004	
2SA40M		3005	51	2004	
2SA40OR			51	2004	
2SA40R		3005	51	2004	
2SA40X		3005	51	2004	
2SA40Y		3005	51	2004	
2SA41	160	3005	51	2004	
2SA410			245	2004	
2SA411			245	2004	
2SA412	126	3006/160	51	2024	
2SA412A		3006	51	2004	
2SA412B		3006	51	2004	
2SA412C		3006	51	2004	
2SA412D		3006	51	2004	
2SA412E		3006	51	2004	
2SA412F		3006	51	2004	
2SA412G		3006	51	2004	
2SA412GN		3006	51	2004	
2SA412H		3006	51	2004	
2SA412J		3006	51	2004	
2SA412K		3006	51	2004	
2SA412L		3006	51	2004	
2SA412M		3006	51	2004	
2SA412OR		3006	51	2004	
2SA412R		3006	51	2004	
2SA412X		3006	51	2004	
2SA412Y		3006	51	2004	
2SA413	126		245	2004	
2SA414	100	3005	51	2007	
2SA414A		3005	51	2004	
2SA414B		3005	51	2004	
2SA414C		3005	51	2004	
2SA414D		3005	51	2004	
2SA414E		3005	51	2004	
2SA414F		3005	51	2004	
2SA414G		3005	51	2004	
2SA414GN		3005	51	2004	
2SA414H		3005	51	2004	
2SA414J		3005	51	2004	
2SA414K		3005	51	2004	
2SA414L		3005	51	2004	
2SA414M		3005	51	2004	
2SA414OR		3005	51	2004	
2SA414R		3005	51	2004	
2SA414X		3005	51	2004	
2SA415	100	3005	51	2007	
2SA415A		3005	51	2004	
2SA415B		3005	51	2004	
2SA415C		3005	51	2004	
2SA415D		3005	51	2004	
2SA415E		3005	51	2004	
2SA415F		3005	51	2004	
2SA415G		3005	51	2004	
2SA415II		3005	51	2004	
2SA415J		3005	51	2004	
2SA415K		3005	51	2004	
2SA415L		3005	51	2004	
2SA415M		3005	51	2004	
2SA415OR		3005	51	2004	
2SA415R		3005	51	2004	
2SA415X		3005	51	2004	
2SA415Y		3005	51	2004	
2SA416	121	3014	3	2006	
2SA416A		3014	3	2006	
2SA416B		3014	3	2006	
2SA416C		3014	3	2006	
2SA416D		3014	3	2006	
2SA416E		3014	3	2006	
2SA416F		3014	3	2006	
2SA416G		3014	3	2006	
2SA416GN		3014	3	2006	
2SA416H		3014	3	2006	
2SA416J		3014	3	2006	
2SA416K		3014	3	2006	
2SA416L		3014	3	2006	
2SA416M		3014	3	2006	
2SA416OR		3014	3	2006	
2SA416R		3014	3	2006	
2SA416X		3014	3	2006	
2SA416Y		3014	3	2006	
2SA417	102$		245	2004	
2SA419	160		245	2004	
2SA41A		3005	51	2004	
2SA41B		3005	51	2004	
2SA41C		3005	51	2004	
2SA41D		3005	51	2004	
2SA41E		3005	51	2004	
2SA41F		3005	51	2004	
2SA41G		3005	51	2004	
2SA41L		3005	51	2004	
2SA41M		3005	51	2004	
2SA41OR		3005	51	2004	
2SA41R		3005	51	2004	
2SA41X		3005	51	2004	
2SA41Y		3005	51	2004	
2SA42	160	3123	245	2004	
2SA420	160	3006	51	2004	
2SA420A		3006	51	2004	
2SA420B		3006	51	2004	
2SA420C		3006	51	2004	
2SA420D		3006	51	2004	
2SA420E		3006	51	2004	

Industry Standard No.	ECG	SK	GE	RS 276-	MOTOR.
2SA420F		3006	51	2004	
2SA420G		3006	51	2004	
2SA420H		3006	51	2004	
2SA420J		3006	51	2004	
2SA420K		3006	51	2004	
2SA420L		3006	51	2004	
2SA420M		3006	51	2004	
2SA420OR			51	2004	
2SA420R		3006	51	2004	
2SA420X		3006	51	2004	
2SA420Y		3006	51	2004	
2SA421	160		245	2004	
2SA422	160		245	2004	
2SA423				2004	
2SA424				2004	
2SA425	160		245	2004	
2SA426	160		51	2004	
2SA426GN		3006	51	2004	
2SA427	126	3006/160	51	2024	
2SA427A		3006	51	2004	
2SA427B		3006	51	2004	
2SA427C		3006	51	2004	
2SA427D		3006	51	2004	
2SA427E		3006	51	2004	
2SA427F		3006	51	2004	
2SA427G		3006	51	2004	
2SA427GN		3006	51	2004	
2SA427H		3006	51	2004	
2SA427J		3006	51	2004	
2SA427K		3006	51	2004	
2SA427L		3006	51	2004	
2SA427M		3006	51	2004	
2SA427OR		3006	51	2004	
2SA427R		3006	51	2004	
2SA427X		3006	51	2004	
2SA427Y		3006	51	2004	
2SA428	126	3006/160	51	2024	
2SA428A		3006	51	2004	
2SA428B		3006	51	2004	
2SA428C		3006	51	2004	
2SA428D		3006	51	2004	
2SA428E		3006	51	2004	
2SA428F		3006	51	2004	
2SA428G		3006	51	2004	
2SA428GN		3006	51	2004	
2SA428H		3006	51	2004	
2SA428J		3006	51	2004	
2SA428K		3006	51	2004	
2SA428L		3006	51	2004	
2SA428M		3006	51	2004	
2SA428OR		3006	51	2004	
2SA428R		3006	51	2004	
2SA428X		3006	51	2004	
2SA428Y		3006	51	2004	
2SA429		3715	89		
2SA429-O		3715	89		
2SA429A		3715	89		
2SA429B		3715	89		
2SA429C		3715	89		
2SA429D		3715	89		
2SA429E		3715	89		
2SA429F		3715	89		
2SA429G		3715	89		
2SA429G-O		3715	89		
2SA429G-R		3715	89		
2SA429G-Y		3715	89		
2SA429GN		3715	89		
2SA429H		3715	89		
2SA429J		3715	89		
2SA429K		3715	89		
2SA429L		3715	89		
2SA429M		3715	89		
2SA429OR		3715	89		
2SA429R		3715	89		
2SA429X		3715	89		
2SA429Y		3715	89		
2SA43	126	3007	52	2024	
2SA430	160		245	2004	
2SA431	160		245	2004	
2SA431A	160		245	2004	
2SA432	160	3006	51	2004	
2SA432A	160	3006	51	2004	
2SA432B		3006	51	2004	
2SA432C		3006	51	2004	
2SA432D		3006	51	2004	
2SA432E		3006	51	2004	
2SA432F		3006	51	2004	
2SA432G		3006	51	2004	
2SA432GN		3006	51	2004	
2SA432H		3006	51	2004	
2SA432K		3006	51	2004	
2SA432L		3006	51	2004	
2SA432M		3006	51	2004	
2SA432OR		3006	51	2004	
2SA432R		3006	51	2004	
2SA432X		3006	51	2004	
2SA432Y		3006	51	2004	
2SA433	160	3006	51	2004	
2SA433A		3006	51	2004	
2SA433B		3006	51	2004	
2SA433C		3006	51	2004	
2SA433D		3006	51	2004	
2SA433E		3006	51	2004	
2SA433F		3006	51	2004	
2SA433G		3006	51	2004	
2SA433GN		3006	51	2004	
2SA433H		3006	51	2004	
2SA433K		3006	51	2004	
2SA433L		3006	51	2004	
2SA433M		3006	51	2004	
2SA433OR		3006	51	2004	
2SA433R		3006	51	2004	
2SA433X		3006	51	2004	
2SA433Y		3006	51	2004	
2SA434	160	3006	51	2004	
2SA434A		3006	51	2004	
2SA434B		3006	51	2004	
2SA434C		3006	51	2004	
2SA434D		3006	51	2004	
2SA434E		3006	51	2004	
2SA434F		3006	51	2004	
2SA434G		3006	51	2004	
2SA434GN		3006	51	2004	
2SA434H		3006	51	2004	
2SA434J		3006	51	2004	
2SA434K		3006	51	2004	
2SA434L		3006	51	2004	
2SA434M		3006	51	2004	
2SA434OR		3006	51	2004	
2SA434R		3006	51	2004	
2SA434X		3006	51	2004	
2SA434Y		3006	51	2004	
2SA435	160	3006	51	2004	
2SA435A	160	3006	51	2004	
2SA435B	160	3006	51	2004	
2SA435C		3006	51	2004	
2SA435D		3006	51	2004	
2SA435E		3006	51	2004	
2SA435F		3006	51	2004	
2SA435G		3006	51	2004	
2SA435GN		3006	51	2004	
2SA435H		3006	51	2004	
2SA435J		3006	51	2004	
2SA435L		3006	51	2004	
2SA435M		3006	51	2004	
2SA435OR		3006	51	2004	
2SA435R		3006	51	2004	
2SA435X			51	2004	
2SA435Y		3006	51	2004	
2SA436	126	3006/160	51	2024	
2SA436A		3006	51	2004	
2SA436B		3006	51	2004	
2SA436C		3006	51	2004	
2SA436D		3006	51	2004	
2SA436E		3006	51	2004	
2SA436F		3006	51	2004	
2SA436G		3006	51	2004	
2SA436GN		3006	51	2004	
2SA436H		3006	51	2004	
2SA436J		3006	51	2004	
2SA436K		3006	51	2004	
2SA437	126	3006/160	51	2024	
2SA437A		3006	51	2004	
2SA437B		3006	51	2004	
2SA437C		3006	51	2004	
2SA437D		3006	51	2004	
2SA437E		3006	51	2004	
2SA437F		3006	51	2004	
2SA437G		3006	51	2004	
2SA437GN		3006	51	2004	
2SA437H		3006	51	2004	
2SA437J		3006	51	2004	
2SA437K		3006	51	2004	
2SA437L		3006	51	2004	
2SA437M		3006	51	2004	
2SA437OR		3006	51	2004	
2SA437R		3006	51	2004	
2SA437X		3006	51	2004	
2SA437Y		3006	51	2004	
2SA438	126	3006/160	51	2024	
2SA438A		3006	51	2004	
2SA438B		3006	51	2004	
2SA438C		3006	51	2004	
2SA438D		3006	51	2004	
2SA438E		3006	51	2004	
2SA438F		3006	51	2004	
2SA438G		3006	51	2004	
2SA438GN		3006	51	2004	
2SA438H		3006	51	2004	
2SA438J		3006	51	2004	
2SA438K		3006	51	2004	
2SA438L		3006	51	2004	
2SA438M		3006	51	2004	
2SA438OR		3006	51	2004	
2SA438R		3006	51	2004	
2SA438X		3006	51	2004	
2SA438Y		3006	51	2004	
2SA44	126	3005	51	2007	
2SA440	160	3006	51	2004	
2SA440A	160	3007	51	2004	
2SA440AL		3006	51	2004	
2SA440B		3006	51	2004	
2SA440C		3006	51	2004	
2SA440D		3006	51	2004	
2SA440E		3006	51	2004	
2SA440F		3006	51	2004	
2SA440G		3006	51	2004	
2SA440GN		3006	51	2004	
2SA440H		3006	51	2004	
2SA440J		3006	51	2004	
2SA440K		3006	51	2004	
2SA440L		3006	51	2004	
2SA440M		3006	51	2004	
2SA440OR		3006	51	2004	
2SA440R		3006	51	2004	
2SA440X		3006	51	2004	
2SA440Y		3006	51	2004	
2SA446	102S	3006/160	51	2024	
2SA446A		3006	51	2007	
2SA446B		3006	51	2007	
2SA446C		3006	51	2007	
2SA446D		3006	51	2007	
2SA446E		3006	51	2007	
2SA446F		3006	51	2007	
2SA446G		3006	51	2007	
2SA446GN		3006	51	2007	
2SA446H		3006	5.	2007	
2SA446J		3006	51	2007	
2SA446K		3006	51	2007	
2SA446L		3006	51	2007	
2SA446M		3006	51	2007	
2SA446OR		3006	51	2007	
2SA446R		3006	51	2007	
2SA446X		3006	51	2007	
2SA446Y		3006	51	2007	
2SA447	126	3006/160	51	2024	
2SA447A		3006	51	2007	
2SA447B		3006	51	2007	
2SA447C		3006	51	2007	
2SA447D		3006	51	2007	
2SA447E		3006	51	2007	
2SA447F		3006	51	2007	
2SA447G		3006	51	2007	
2SA447GN		3006	51	2007	
2SA447H		3006	51	2007	
2SA447J		3006	51	2007	
2SA447K		3006	51	2007	
2SA447L		3006	51	2007	
2SA447M		3006	51	2007	
2SA447OR		3006	51	2007	
2SA447R		3006	51	2007	
2SA447X		3006	51	2007	
2SA447Y		3006	51	2007	
2SA448			245	2004	
2SA44A		3005	51	2004	
2SA44B		3005	51	2004	
2SA44C		3005	51	2004	
2SA44D		3005	51	2004	
2SA44E		3005	51	2004	
2SA44F		3005	51	2004	
2SA44G		3005	51	2004	
2SA44L		3005	51	2004	
2SA44M		3005	51	2004	
2SA44OR			51	2004	
2SA44R		3005	51	2004	
2SA44X		3005	51	2004	
2SA44Y		3005	51	2004	
2SA45	126	3006/160	52	2024	

Industry Standard No.	ECG	SK	GE	RS 276-	MOTOR.
2SA45-1	126	3006/160	52	2024	
2SA45-2	126	3006/160	52	2024	
2SA45-3	126	3006/160	52	2024	
2SA450	160$		245	2004	
2SA450H	160$		245	2004	
2SA451	160$		245	2004	
2SA451H	160$		245	2004	
2SA452	160$		245	2004	
2SA452H	160$		245	2004	
2SA453	126	3006/160	51	2024	
2SA453A		3006	51	2004	
2SA453B		3006	51	2004	
2SA453C		3006	51	2004	
2SA453D		3006	51	2004	
2SA453E		3006	51	2004	
2SA453F		3006	51	2004	
2SA453G		3006	51	2004	
2SA453GN		3006	51	2004	
2SA453H		3006	51	2004	
2SA453J		3006	51	2004	
2SA453K		3006	51	2004	
2SA453L		3006	51	2004	
2SA453M		3006	51	2004	
2SA453OR		3006	51	2004	
2SA453R		3006	51	2004	
2SA453X		3006	51	2004	
2SA453Y		3006	51	2004	
2SA454	126	3006/160	51	2024	
2SA454A		3006	51	2004	
2SA454B		3006	51	2004	
2SA454C		3006	51	2004	
2SA454D		3006	51	2004	
2SA454E		3006	51	2004	
2SA454F		3006	51	2004	
2SA454G		3006	51	2004	
2SA454GN		3006	51	2004	
2SA454H		3006	51	2004	
2SA454J		3006	51	2004	
2SA454K		3006	51	2004	
2SA454L		3006	51	2004	
2SA454OR		3006	51	2004	
2SA454R		3006	51	2004	
2SA454X		3006	51	2004	
2SA454Y		3006	51	2004	
2SA455	126	3006/160	51	2024	
2SA455A		3006	51	2004	
2SA455B		3006	51	2004	
2SA455C		3006	51	2004	
2SA455D		3006	51	2004	
2SA455E		3006	51	2004	
2SA455F		3006	51	2004	
2SA455G		3006	51	2004	
2SA455GN		3006	51	2004	
2SA455H		3006	51	2004	
2SA455J		3006	51	2004	
2SA455K		3006	51	2004	
2SA455L		3006	51	2004	
2SA455M		3006	51	2004	
2SA455OR		3006	51	2004	
2SA455R		3006	51	2004	
2SA455X		3006	51	2004	
2SA455Y		3006	51	2004	
2SA456	126	3006/160	51	2024	
2SA456A		3006	51	2004	
2SA456B		3006	51	2004	
2SA456C		3006	51	2004	
2SA456D		3006	51	2004	
2SA456E		3006	51	2004	
2SA456F		3006	51	2004	
2SA456G		3006	51	2004	
2SA456GN		3006	51	2004	
2SA456H		3006	51	2004	
2SA456J		3006	51	2004	
2SA456K		3006	51	2004	
2SA456L		3006	51	2004	
2SA456M		3006	51	2004	
2SA456OR		3006	51	2004	
2SA456R		3006	51	2004	
2SA456X		3006	51	2004	
2SA456Y		3006	51	2004	
2SA457	126	3006/160	51	2024	
2SA457A		3006	51	2004	
2SA457B		3006	51	2004	
2SA457C		3006	51	2004	
2SA457D		3006	51	2004	
2SA457E		3006	51	2004	
2SA457F		3006	51	2004	
2SA457G		3006	51	2004	
2SA457GN		3006	51	2004	
2SA457H		3006	51	2004	
2SA457J		3006	51	2004	
2SA457K		3006	51	2004	
2SA457L		3006	51	2004	
2SA457M		3006	51	2004	
2SA457OR		3006	51	2004	
2SA457R		3006	51	2004	
2SA457X		3006	51	2004	
2SA457Y		3006	51	2004	
2SA46				2005	
2SA460	160		245	2004	
2SA461	160		245	2004	
2SA462	160		245	2004	
2SA463	160		245	2004	
2SA464	160		245	2004	
2SA465		3114	21	2034	
2SA466	126	3006/160	51	2024	
2SA466-2	126	3006/160	51	2024	
2SA466-3	126	3006/160	51	2024	
2SA466A		3006	51	2004	
2SA466B		3006	51	2004	
2SA466BLK	126	3006/160	51	2024	
2SA466BLU	126	3006/160	51	2024	
2SA466C		3006	51	2004	
2SA466D		3006	51	2004	
2SA466E		3006	51	2004	
2SA466F		3006	51	2004	
2SA466G		3006	51	2004	
2SA466GN		3006	51	2004	
2SA466H		3006	51	2004	
2SA466J		3006	51	2004	
2SA466K		3006	51	2004	
2SA466L		3006	51	2004	
2SA466M		3006	51	2004	
2SA466OR		3006	51	2004	
2SA466R		3006	51	2004	
2SA466X		3006	51	2004	
2SA466Y		3006	51	2004	
2SA466YEL	126	3006/160	51	2024	
2SA467	290A	3466/159	89	2032	
2SA467-O	290A		89	2032	
2SA467-Y	290A		89	2032	
2SA467A		3114	89	2023	
2SA467B		3114	89	2023	
2SA467C		3114	89	2023	
2SA467D		3114	89	2023	
2SA467E		3114	89	2023	
2SA467F		3114	89	2023	
2SA467G	290A	3466/159	89	2032	
2SA467G-O	290A		82	2032	
2SA467G-O		3466	89	2023	
2SA467G-R	290A	3466/159	89	2032	
2SA467G-Y	290A	3466/159	89	2032	
2SA467GN		3114	89	2023	
2SA467H		3114	89	2023	
2SA467J		3114	89	2023	
2SA467K		3114	89	2023	
2SA467L		3114	89	2023	
2SA467M		3114	89	2023	
2SA467OR		3466	89	2023	
2SA467R		3114	89	2023	
2SA467X		3114	89	2023	
2SA467Y		3466	89	2023	
2SA468	126	3008	50	2024	
2SA468A		3008	50	2004	
2SA468B		3008	50	2004	
2SA468C		3008	50	2004	
2SA468D		3008	50	2004	
2SA468E		3008	50	2004	
2SA468F		3008	50	2004	
2SA468G		3008	50	2004	
2SA468GN		3008	50	2004	
2SA468H		3008	50	2004	
2SA468J		3008	50	2004	
2SA468K		3008	50	2004	
2SA468L		3008	50	2004	
2SA468M		3008	50	2004	
2SA468OR		3008	50	2004	
2SA468R		3008	50	2004	
2SA468X			50	2004	
2SA468Y		3008	50	2004	
2SA469	126	3007	50	2024	
2SA469A		3007	50	2004	
2SA469B		3007	50	2004	
2SA469C		3007	50	2004	
2SA469D		3007	50	2004	
2SA469E		3007	50	2004	
2SA469F		3007	50	2004	
2SA469G		3007	50	2004	
2SA469GN		3007	50	2004	
2SA469H		3007	50	2004	
2SA469J		3007	50	2004	
2SA469K		3007	50	2004	
2SA469L		3007	50	2004	
2SA469M		3007	50	2004	
2SA469OR		3007	50	2004	
2SA469R		3007	50	2004	
2SA469X			50	2004	
2SA469Y		3007	50	2004	
2SA47				2005	
2SA470	126	3007	51	2024	
2SA470A		3007	51	2004	
2SA470B		3007	51	2004	
2SA470C		3007	51	2004	
2SA470D		3007	51	2004	
2SA470E		3007	51	2004	
2SA470F		3007	51	2004	
2SA470G		3007	51	2004	
2SA470GN		3007	51	2004	
2SA470H		3007	51	2004	
2SA470J		3007	51	2004	
2SA470K		3007	51	2004	
2SA470L		3007	51	2004	
2SA470M		3007	51	2004	
2SA470OR		3007	51	2004	
2SA470R		3007	51	2004	
2SA470X			51	2004	
2SA470Y		3007	51	2004	
2SA471	126	3007	51	2024	
2SA471-1	126	3007	51	2024	
2SA471-2	126	3007	51	2024	
2SA471-3	126	3007	51	2024	
2SA471A		3007	51	2004	
2SA471B		3007	51	2004	
2SA471C		3007	51	2004	
2SA471D		3007	51	2004	
2SA471E		3007	51	2004	
2SA471F		3007	51	2004	
2SA471G		3007	51	2004	
2SA471GN		3007	51	2004	
2SA471H		3007	51	2004	
2SA471J		3007	51	2004	
2SA471K		3007	51	2004	
2SA471L		3007	51	2004	
2SA471M		3007	51	2004	
2SA471OR		3007	51	2004	
2SA471R		3007	51	2004	
2SA471X			51	2004	
2SA471Y		3007	51	2004	
2SA472	126	3007	50	2024	
2SA472-1			50	2004	
2SA472-2			50	2004	
2SA472-3			50	2004	
2SA472-4			50	2004	
2SA472-5			50	2004	
2SA472-6			50	2004	
2SA4728B			51	2004	
2SA472A	126	3007	50	2024	
2SA472B	126	3007	50	2024	
2SA472C	126	3007	50	2024	
2SA472D	126	3007	50	2024	
2SA472E	126	3007	50	2024	
2SA472F		3007	50	2004	
2SA472G		3007	50	2004	
2SA472GN		3007	50	2004	
2SA472H		3007	50	2004	
2SA472J		3007	50	2004	
2SA472K		3007	50	2004	
2SA472L		3007	50	2004	
2SA472M		3007	50	2004	
2SA472OR		3007	50	2004	
2SA472R		3007	50	2004	
2SA472X			50	2004	
2SA472Y		3007	50	2004	
2SA473	153	3083/197	69	2049	
2SA473(O)			69	2049	
2SA473(O)			69	2049	
2SA473-GR	153		69	2049	
2SA473-O	153		69	2049	
2SA473-R	153		69	2049	
2SA473-Y			69	2049	
2SA473B		3083	69	2049	
2SA473GR			69	2049	
2SA473R		3083	69	2049	
2SA473Y		3083	69	2049	
2SA474	126	3005	51	2024	
2SA474A		3005	51	2024	
2SA474B		3005	51	2024	
2SA474C		3005	51	2024	

Industry Standard No.	ECG	SK	GE	RS 276-	MOTOR.
2SA474D		3005	51	2024	
2SA474E		3005	51	2024	
2SA474F		3005	51	2024	
2SA474G		3005	51	2024	
2SA474GN		3005	51	2024	
2SA474H		3005	51	2024	
2SA474J		3005	51	2024	
2SA474K		3005	51	2024	
2SA474L		3005	51	2024	
2SA474M		3005	51	2024	
2SA474OR		3005	51	2024	
2SA474R		3005	51	2024	
2SA474X			51	2024	
2SA474Y		3005	51	2024	
2SA475		3007	50	2004	
2SA476	126	3008	53	2024	
2SA476A		3008	53	2024	
2SA476B		3008	53	2024	
2SA476C		3008	53	2024	
2SA476D		3008	53	2024	
2SA476E		3008	53	2024	
2SA476F		3008	53	2024	
2SA476G		3008	53	2024	
2SA476GN		3008	53	2024	
2SA476H		3008	53	2024	
2SA476J		3008	53	2024	
2SA476K		3008	53	2024	
2SA476L		3008	53	2024	
2SA476M		3008	53	2024	
2SA476OR		3008	53	2024	
2SA476R		3008	53	2024	
2SA476X		3008	53	2024	
2SA476Y		3008	53	2024	
2SA477	126	3006/160	51	2024	
2SA477A		3006	51	2024	
2SA477B		3006	51	2024	
2SA477C		3006	51	2024	
2SA477D		3006	51	2024	
2SA477E		3006	51	2024	
2SA477F		3006	51	2024	
2SA477G		3006	51	2024	
2SA477GN		3006	51	2024	
2SA477H		3006	51	2024	
2SA477J		3006	51	2024	
2SA477K		3006	51	2024	
2SA477L		3006	51	2024	
2SA477M		3006	51	2024	
2SA477OR		3006	51	2024	
2SA477R		3006	51	2024	
2SA477X		3006	51	2024	
2SA477Y		3006	51	2024	
2SA478	102A$	3006/160	51	2024	
2SA478A		3006	51	2004	
2SA478B		3006	51	2004	
2SA478C		3006	51	2004	
2SA478D		3006	51	2004	
2SA478E		3006	51	2004	
2SA478F		3006	51	2004	
2SA478G		3006	51	2004	
2SA478GN		3006	51	2004	
2SA478H		3006	51	2004	
2SA478J		3006	51	2004	
2SA478K		3006	51	2004	
2SA478L		3006	51	2004	
2SA478M		3006	51	2004	
2SA478OR		3006	51	2004	
2SA478R		3006	51	2004	
2SA478X		3006	51	2004	
2SA478Y		3006	51	2004	
2SA479	102A$	3006/160	51	2024	
2SA479A		3006	51	2004	
2SA479B		3006	51	2004	
2SA479C		3006	51	2004	
2SA479D		3006	51	2004	
2SA479E		3006	51	2004	
2SA479F		3006	51	2004	
2SA479G		3006	51	2004	
2SA479GN		3006	51	2004	
2SA479H		3006	51	2004	
2SA479J		3006	51	2004	
2SA479K		3006	51	2004	
2SA479L		3006	51	2004	
2SA479M		3006	51	2004	
2SA479OR		3006	51	2004	
2SA479R		3006	51	2004	
2SA479X		3006	51	2004	
2SA479Y		3006	51	2004	
2SA48			51	2004	
2SA480	159	3466	89	2032	
2SA480-OR			89	2032	
2SA480A		3118	89	2032	
2SA480B		3118	89	2032	
2SA480C		3118	89	2032	
2SA480D		3118	89	2032	
2SA480E		3118	89	2032	
2SA480F		3118	89	2032	
2SA480G		3118	89	2032	
2SA480GN		3118	89	2032	
2SA480H		3118	89	2032	
2SA480J		3118	89	2032	
2SA480K		3118	89	2032	
2SA480L		3118	89	2032	
2SA480M		3118	89	2032	
2SA480OR		3118	89	2032	
2SA480R		3118	89	2032	
2SA480X		3118	89	2032	
2SA480Y		3118	89	2032	
2SA482	159	3466	89	2032	
2SA482A		3114	89	2032	
2SA482B		3114	89	2032	
2SA482C		3114	89	2032	
2SA482D		3114	89	2032	
2SA482E		3114	89	2032	
2SA482F		3114	89	2032	
2SA482G		3114	89	2032	
2SA482GN		3114	89	2032	
2SA482H		3114	89	2032	
2SA482J		3114	89	2032	
2SA482K		3114	89	2032	
2SA482L		3114	89	2032	
2SA482M		3114	89	2032	
2SA482OR		3114	89	2032	
2SA482R		3114	89	2032	
2SA482X		3114	89	2032	
2SA482Y		3114	89	2032	
2SA483	218$	3085	234		2N6420
2SA484	323	3513			
2SA484R	323	3513			
2SA484Y	323	3513			
2SA485	323	3513			
2SA485R	323	3513			
2SA485Y	323	3513			
2SA486	323		69	2049	
2SA486-R			69	2049	
2SA486-RED			69	2049	
2SA486-Y			69	2049	
2SA486-YEL			69	2049	
2SA489	153	3083/197	69	2049	2N6126
2SA489-0			69	2049	
2SA489-O	153		69	2049	
2SA489-R	153		69	2049	
2SA489-Y	153		69	2049	
2SA489Y	153	3083/197			
2SA49	102A	3004	51	2024	
2SA490	153	3274	69	2049	2N6125
2SA490(POWER)	153		69	2049	
2SA490-0	153	3274	69	2049	
2SA490-O	292		69	2049	
2SA490-Y			69	2049	
2SA490A		3274	69	2049	
2SA490B		3274	69	2049	
2SA490C		3274	69	2049	
2SA490D		3274	69	2049	
2SA490E		3274	69	2049	
2SA490F		3274	69	2049	
2SA490G		3274	69	2049	
2SA490GN		3274	69	2049	
2SA490H		3274	69	2049	
2SA490J		3274	69	2049	
2SA490K		3274	69	2049	
2SA490L		3274	69	2049	
2SA490LBG1			69	2049	
2SA490M		3274	69	2049	
2SA490OR		3274	69	2049	
2SA490R		3274	69	2049	
2SA490X			69	2049	
2SA490Y		3274	69	2049	
2SA490YA		3274	69	2049	
2SA490YLBG1			69	2049	
2SA493	290A	3466/159	89	2032	
2SA493-0			89	2032	
2SA4930		3118	89	2023	
2SA493A		3118	89		
2SA493B		3118	89		
2SA493C		3118	89		
2SA493D		3118	89		
2SA493E		3118	89		
2SA493F		3118	89		
2SA493G		3118	89		
2SA493GR	234$	3466/159	89	2032	
2SA493H		3118	89		
2SA493J		3118	89		
2SA493K		3118	89		
2SA493L		3118	89		
2SA493M		3118	89		
2SA493OR		3466	89		
2SA493R		3118	89		
2SA493X		3118	89		
2SA493Y	290A	3466/159	89	2032	
2SA494	234	3466/159	65	2032	
2SA494(Y)			65	2050	
2SA494-GR	234	3466/159	65	2032	
2SA494-GR-1		3114	65	2050	
2SA494-O	234	3466/159	65	2032	
2SA494-OR		3114	65	2050	
2SA494-Y	234	3466/159	65	2032	
2SA4940		3114	89	2023	
2SA494A		3114	65	2050	
2SA494B		3114	65	2050	
2SA494C		3114	65	2050	
2SA494D		3114	65	2050	
2SA494E		3114	65	2050	
2SA494F		3114	65	2050	
2SA494G		3114	65	2050	
2SA494GN		3114	65	2050	
2SA494GR		3114	65	2050	
2SA494H		3114	65	2050	
2SA494J		3114	65	2050	
2SA494K		3114	65	2050	
2SA494L		3114	65	2050	
2SA494M		3114	65	2050	
2SA4940			65	2050	
2SA494OR		3114	65	2050	
2SA494R		3114	65	2050	
2SA494X		3114	65	2050	
2SA494Y		3114	65	2050	
2SA495	290A	3114/290	221	2032	
2SA495(O)			221	2032	
2SA495(R)			221	2032	
2SA495(Y)			221	2032	
2SA495-0	290A	3114/290	221	2032	
2SA495-1		3114	221	2032	
2SA495-G	290A	3114/290	221	2032	
2SA495-GN			221	2032	
2SA495-O	290A	3114/290	221	2032	
2SA495-OF			221	2032	
2SA495-OR		3114	221	2032	
2SA495-ORG			221	2032	
2SA495-ORG-G			221	2032	
2SA495-R	290A	3114/290	221	2032	
2SA495-RD		3114	221	2032	
2SA495-RED			221	2032	
2SA495-RED-G			221	2032	
2SA495-Y	290A	3114/290	221	2032	
2SA495-YEL			221	2032	
2SA495-YEL-G			221	2032	
2SA495-YL		3114	221	2032	
2SA4950		3466	89	2032	
2SA4950R		3118	89	2023	
2SA495A	290A	3114/290	221	2032	
2SA495B		3114	221	2032	
2SA495C		3114	221	2032	
2SA495D	290A	3114/290	221	2032	
2SA495E		3114	221	2032	
2SA495F		3114	221	2032	
2SA495G	290A$	3114/290	221	2032	
2SA495G-GR	290A$	3114/290	221	2032	
2SA495G-O	290A	3114/290	221	2032	
2SA495G-R	290A$	3114/290	221	2032	
2SA495G-Y	290A$	3114/290	221	2032	
2SA495GN			221	2032	
2SA495GR	290A		221	2032	
2SA495H	290A	3114/290	221	2032	
2SA495J	290A	3114/290	221	2032	
2SA495K		3114	221	2032	
2SA495L		3114	221	2032	
2SA495M		3114	221	2032	
2SA4950	290A		221	2032	
2SA4950F		3114	221	2032	
2SA4950R		3114	221	2032	
2SA495R	290A	3114/290	221	2032	
2SA495RD			221	2032	
2SA495RED			221	2032	
2SA495RED-G			221	2032	
2SA495W	290A	3114/290	221	2032	
2SA495W1		3114	221	2032	
2SA495WI			221	2032	
2SA495X	290A	3114/290	221	2032	

Industry Standard No.	ECG	SK	GE	RS 276-	MOTOR.
2SA495Y	290A	3114/290	221	2032	
2SA495YEL			221	2032	
2SA495YEL-G			221	2032	
2SA495YL			221	2032	
2SA496	374	9042	58	2025	2N4918
2SA496(O)			58	2025	
2SA496-O		9042	58	2025	
2SA496-O	374		58	2025	
2SA496-ORG			58	2025	
2SA496-R	374	9042	58	2025	
2SA496-RED			58	2025	
2SA496-Y	374	9042	58	2025	
2SA496-YEL			58	2025	
2SA4964		3191	58	2025	
2SA4960			58	2025	
2SA496ORG			58	2025	
2SA496R		9042	58	2025	
2SA496RED			58	2025	
2SA496Y	374	9042	58	2025	
2SA496YEL			58	2025	
2SA497	129	3025	244	2027	
2SA497-O			244	2027	
2SA497-ORG			244	2027	
2SA497-R			244	2027	
2SA497-RED			244	2027	
2SA497-Y			244	2027	
2SA497-YEL			244	2027	
2SA497R	129	3025	244	2027	
2SA497RED			244	2027	
2SA497Y		3025	244	2027	
2SA498	129	3025	244	2027	
2SA498Y	129	3025	244	2027	
2SA499	106$	3466/159	89	2032	
2SA499-O	106$	3466/159	89	2032	
2SA499-ORG			89	2032	
2SA499-R	106$	3466/159	89	2032	
2SA499-RED			89	2032	
2SA499-Y	106$	3466/159	89	2032	
2SA499-YEL			89	2032	
2SA499A		3118	89	2032	
2SA499B		3118	89	2032	
2SA499C		3118	89	2032	
2SA499D		3118	89	2032	
2SA499E		3118	89	2032	
2SA499F		3118	89	2032	
2SA499G		3118	89	2032	
2SA499GN		3118	89	2032	
2SA499H		3118	89	2032	
2SA499J		3118	89	2032	
2SA499K		3118	89	2032	
2SA499L		3118	89	2032	
2SA499M		3118	89	2032	
2SA4990			89	2032	
2SA499OR		3118	89	2032	
2SA499R		3118	89	2032	
2SA499X		3118	89	2032	
2SA499Y		3118	89	2032	
2SA49A		3004	51	2004	
2SA49B		3004	51	2004	
2SA49C		3004	51	2004	
2SA49D		3004	51	2004	
2SA49E		3004	51	2004	
2SA49F		3004	51	2004	
2SA49G		3004	51	2004	
2SA49L		3004	51	2004	
2SA49M		3004	51	2004	
2SA49OR		3004	51	2004	
2SA49R		3004	51	2004	
2SA49X			51	2004	
2SA49Y		3004	51	2004	
2SA50	102A	3005	51	2005	
2SA500	106$	3466/159	89	2032	
2SA500-O	106$		89	2032	
2SA500-ORG			89	2032	
2SA500-R	106$		89	2032	
2SA500-RED			89	2032	
2SA500-Y	106$	3466/159	89	2032	
2SA500-YEL			89	2032	
2SA500A		3118	89	2032	
2SA500B		3118	89	2032	
2SA500C		3118	89	2032	
2SA500D		3118	89	2032	
2SA500E		3118	89	2032	
2SA500F		3118	89	2032	
2SA500G		3118	89	2032	
2SA500GN		3118	89	2032	
2SA500H		3118	89	2032	
2SA500J		3118	89	2032	
2SA500K		3118	89	2032	
2SA500L		3118	89	2032	
2SA500M		3118	89	2032	
2SA500OR		3118	89	2032	
2SA500R		3118	89	2032	
2SA500X		3118	89	2032	
2SA500Y		3118	89	2032	
2SA501	129	3025	244	2027	
2SA502	290A	3466/159	89	2032	
2SA502-O			82	2032	
2SA502-OR				2032	
2SA502-R			82	2032	
2SA502-Y			82	2032	
2SA502A		3025	89	2032	
2SA502B		3025	89	2032	
2SA502C		3025	89	2032	
2SA502D		3025	89	2032	
2SA502E		3025	89	2032	
2SA502F		3025	89	2032	
2SA502G		3025	89	2032	
2SA502GN		3025	89	2032	
2SA502H		3025	89	2032	
2SA502J		3025	89	2032	
2SA502K		3025	89	2032	
2SA502L		3025	89	2032	
2SA502M		3025	89	2032	
2SA502OR		3025	89	2032	
2SA502R		3025	89	2032	
2SA502X		3025	89	2032	
2SA502Y		3025	89	2032	
2SA503	129	3025	48	2027	
2SA503-O	129	3025	48	2027	
2SA503-R	129	3025	48	2027	
2SA503-Y	129	3025	48	2027	
2SA503GR	129	3203/211	48	2026	
2SA504	129	3025	244	2027	
2SA504-O	129	3025	244	2027	
2SA504-R	129	3025	244	2027	
2SA504-Y	129	3025	244	2027	
2SA504GR	129	3203/211	244	2027	
2SA505	374	9042	58	2025	2N4919
2SA505-O	374	3191/185	58	2025	
2SA505-ORG			58	2025	
2SA505-R	374	9042	58	2025	
2SA505-RED			58	2025	
2SA505-Y	374	9042	58	2025	
2SA505-YEL			58	2025	
2SA5050			58	2025	
2SA505R			58	2025	
2SA505Y			58	2025	
2SA506	160		245	2004	
2SA507	160	3006	51	2004	
2SA507A		3006	51	2004	
2SA507B		3006	51	2004	
2SA507C		3006	51	2004	
2SA507D		3006	51	2004	
2SA507E		3006	51	2004	
2SA507F		3006	51	2004	
2SA507G		3006	51	2004	
2SA507GN		3006	51	2004	
2SA507H		3006	51	2004	
2SA507J		3006	51	2004	
2SA507K		3006	51	2004	
2SA507L		3006	51	2004	
2SA507OR		3006	51	2004	
2SA507R		3006	51	2004	
2SA507X		3006	51	2004	
2SA507Y		3006	51	2004	
2SA508	160		245	2004	
2SA509		3114	269	2032	
2SA509(A)			269	2032	
2SA509(O)			269	2032	
2SA509-O		3114	269	2032	
2SA509-O		3114	269	2032	
2SA509-OR		3114	269	2032	
2SA509-R		3114	269	2032	
2SA509-RD		3114	269	2032	
2SA509-Y		3114	269	2032	
2SA509-YE		3114	269	2032	
2SA509A		3114	269	2032	
2SA509B		3114	269	2032	
2SA509BL		3114	269	2032	
2SA509C		3114	269	2032	
2SA509D		3114	269	2032	
2SA509E		3114	269	2032	
2SA509F		3114	269	2032	
2SA509G		3114	269	2032	
2SA509G-O			269	2032	
2SA509G-Y			269	2032	
2SA509GN		3114	269	2032	
2SA509GR		3114	269	2032	
2SA509GR-1		3114	269	2032	
2SA509H		3114	269	2032	
2SA509J		3114	269	2032	
2SA509K		3114	269	2032	
2SA509L		3114	269	2032	
2SA509M		3114	269	2032	
2SA509OR		3114	269	2032	
2SA509Q			269	2032	
2SA509R		3114		2032	
2SA509RD			269	2032	
2SA509T		3114	269	2032	
2SA509V			269	2032	
2SA509X		3114	269	2032	
2SA509Y		3114	269	2032	
2SA509YE			269	2032	
2SA50A		3005	51	2004	
2SA50B		3005	51	2004	
2SA50C		3005	51	2004	
2SA50D		3005	51	2004	
2SA50E		3005	51	2004	
2SA50F		3005	51	2004	
2SA50G		3005	51	2004	
2SA50L		3005	51	2004	
2SA50M		3005	51	2004	
2SA50OR			51	2004	
2SA50R		3005	51	2004	
2SA50X		3005	51	2004	
2SA50Y		3005	51	2004	
2SA51	126	3005	51	2004	
2SA510	323	3466/159	89	2032	
2SA510-O	323	3466/159	89	2032	
2SA510-OR		3114	89	2023	
2SA510-R	323	3466/159	89	2032	
2SA510-RD		3025	89	2023	
2SA510A		3025	89	2023	
2SA510B		3025	89	2023	
2SA510C		3025	89	2023	
2SA510D		3025	89	2023	
2SA510E		3025	89	2023	
2SA510F		3025	89	2023	
2SA510G		3025	89	2023	
2SA510GN		3025	89	2023	
2SA510H		3025	89	2023	
2SA510J		3025	89	2023	
2SA510K		3025	89	2023	
2SA510L		3025	89	2023	
2SA510M		3025	89	2023	
2SA510OR			89	2023	
2SA510R		3025	89	2023	
2SA510X		3025	89	2023	
2SA510Y		3025	89	2023	
2SA511	323	3466/159	89	2032	
2SA511-G		3114	89	2023	
2SA511-O	323	3466/159	89	2032	
2SA511-OR		3114	89	2023	
2SA511-R	323	3466/159	89	2032	
2SA511-RD		3114	89	2023	
2SA511A		3114	89	2023	
2SA511B		3114	89	2023	
2SA511C		3114	89	2023	
2SA511D		3114	89	2023	
2SA511E		3114	89	2023	
2SA511F		3114	89	2023	
2SA511G		3114	89	2023	
2SA511GN		3114	89	2023	
2SA511H		3114	89	2023	
2SA511J		3114	89	2023	
2SA511K		3114	89	2023	
2SA511L		3114	89	2023	
2SA511M		3114	89	2023	
2SA511OR		3114	89	2023	
2SA511R		3114	89	2023	
2SA511X		3114	89	2023	
2SA511Y		3114	89	2023	
2SA512	323	3466/159	89	2032	
2SA512-O	323	3466/159	89	2032	
2SA512-OR		3114	89	2023	
2SA512-OR1		3114	89	2023	
2SA512-ORG			67	2023	
2SA512-R	323	3466/159	89	2032	
2SA512-RD		3025	89	2023	
2SA512-RED			67	2023	
2SA512A		3025	89	2023	
2SA512B		3025	89	2023	
2SA512C		3025	89	2023	
2SA512D		3025	89	2023	
2SA512E		3025	89	2023	
2SA512F		3025	89	2023	
2SA512G		3025	89	2023	

Industry Standard No.	ECG	SK	GE	RS 276-	MOTOR.
2SA512GN		3025	89	2023	
2SA512H		3025	89	2023	
2SA512J		3025	89	2023	
2SA512K		3025	89	2023	
2SA512L		3114	89	2023	
2SA512M		3025	89	2023	
2SA512OR			89	2023	
2SA512R		3025	89	2023	
2SA512X		3025	89	2023	
2SA512Y		3025	89	2023	
2SA513	323	3466/159	89	2032	
2SA513-O	323	3466/159	89	2032	
2SA513-OR		3114	89	2023	
2SA513-ORG			89	2023	
2SA513-R	323	3466/159	89	2032	
2SA513-RD		3114	89	2023	
2SA513-RED			89	2023	
2SA513A		3114	89	2023	
2SA513B		3114	89	2023	
2SA513C		3114	89	2023	
2SA513D		3114	89	2023	
2SA513E		3114	89	2023	
2SA513F		3114	89	2023	
2SA513G		3114	89	2023	
2SA513GN		3114	89	2023	
2SA513H		3114	89	2023	
2SA513J		3114	89	2023	
2SA513K		3114	89	2023	
2SA513L		3114	89	2023	
2SA513M		3114	89	2023	
2SA513OR		3114	89	2023	
2SA513R		3114	89	2023	
2SA513X		3114	89	2023	
2SA513Y		3114	89	2023	
2SA516	323	3025/129	244	2027	
2SA516A	323	3025/129	244	2027	
2SA517	126	3007	52	2024	
2SA518	126	3007	51	2024	
2SA518-G			51	2004	
2SA518A		3007	51	2004	
2SA518B		3007	51	2004	
2SA518C		3007	51	2004	
2SA518D		3007	51	2004	
2SA518E		3007	51	2004	
2SA518F		3007	51	2004	
2SA518G		3007	51	2004	
2SA518GN		3007	51	2004	
2SA518H		3007	51	2004	
2SA518J		3007	51	2004	
2SA518K		3007	51	2004	
2SA518L		3007	51	2004	
2SA518M		3007	51	2004	
2SA518OR		3007	51	2004	
2SA518R		3007	51	2004	
2SA518X			51	2004	
2SA518Y		3007	51	2004	
2SA51A		3005	51	2004	
2SA51B		3005	51	2004	
2SA51C		3005	51	2004	
2SA51D		3005	51	2004	
2SA51E		3005	51	2004	
2SA51F		3005	51	2004	
2SA51G		3005	51	2004	
2SA51L		3005	51	2004	
2SA51M		3005	51	2004	
2SA51OR		3005	51	2004	
2SA51R		3005	51	2004	
2SA51X		3005	51	2004	
2SA51Y		3005	51	2004	
2SA52	102A	3004	51	2024	
2SA522	159	3466	89	2024	
2SA522A	159	3466	89	2032	
2SA522AL		3114	89	2032	
2SA522B		3114	89	2032	
2SA522C		3114	89	2032	
2SA522D		3114	89	2032	
2SA522E		3114	89	2032	
2SA522F		3114	89	2032	
2SA522G		3114	89	2032	
2SA522GN		3114	89	2032	
2SA522H		3114	89	2032	
2SA522J		3114	89	2032	
2SA522K		3114	89	2032	
2SA522L		3114	89	2032	
2SA522M		3114	89	2032	
2SA522OR		3114	89	2032	
2SA522R		3114	89	2032	
2SA522X		3114	89	2032	
2SA522Y		3114	89	2032	
2SA524	106	3984		2024	
2SA525	160	3118	51	2004	
2SA525A	160	3006	51	2004	
2SA525B	160	3006	51	2004	
2SA525C		3006	51	2004	
2SA525D		3006	51	2004	
2SA525E		3006	51	2004	
2SA525F		3006	51	2004	
2SA525GN		3006	51	2004	
2SA525H		3006	51	2004	
2SA525J		3006	51	2004	
2SA525K		3006	51	2004	
2SA525L		3006	51	2004	
2SA525M		3006	51	2004	
2SA525OR		3006	51	2004	
2SA525R		3118	51	2004	
2SA525X		3006	51	2004	
2SA525Y		3118	51	2004	
2SA527		3025	244	2027	
2SA528		3025	244	2027	
2SA52A		3004	51	2004	
2SA52B		3004	51	2004	
2SA52C		3004	51	2004	
2SA52D		3004	51	2004	
2SA52E		3004	51	2004	
2SA52F		3004	51	2004	
2SA52G		3004	51	2004	
2SA52L		3004	51	2004	
2SA52M		3004	51	2004	
2SA52OR		3004	51	2004	
2SA52R		3004	51	2004	
2SA52X			51	2004	
2SA52Y		3004	51	2004	
2SA53	102A	3004	51	2024	
2SA530	159	3466	89	2032	
2SA530A		3118	89	2023	
2SA530B		3118	89	2023	
2SA530C		3118	89	2023	
2SA530D		3118	89	2023	
2SA530E		3118	89	2023	
2SA530F		3118	89	2023	
2SA530G		3118	89	2023	
2SA530GN		3118	89	2023	
2SA530GR		3118	89	2023	
2SA530H	159	3466	89	2032	

Industry Standard No.	ECG	SK	GE	RS 276-	MOTOR.
2SA530H1		3114	89	2023	
2SA530J		3118	89	2023	
2SA530K		3118	89	2023	
2SA530L		3118	89	2023	
2SA530M		3118	89	2023	
2SA530OR			89	2023	
2SA530R		3114	89	2023	
2SA530X		3118	89	2023	
2SA530Y		3118	89	2023	
2SA532	129	3025	244	2027	
2SA532A	129	3025	244	2027	
2SA532B	129	3025	244	2027	
2SA532C	129	3025	244	2027	
2SA532D	129	3025	244	2027	
2SA532E	129	3025	244	2027	
2SA532F	129	3025	244	2027	
2SA536			51	2004	
2SA537	129	3025	244	2027	
2SA537A	129	3025	244	2027	
2SA537AA	129		244	2027	
2SA537AB	129		244	2027	
2SA537AC	129		244		
2SA537AH	129	3025	244	2027	
2SA537B	129	3025	244	2027	
2SA537C	129	3025	244	2027	
2SA537H	129	3025	244	2027	
2SA538	102A		53	2007	
2SA538-G			21	2034	
2SA539	290A	3466/159	89	2032	
2SA539(K)			89	2032	
2SA539(L)			89	2032	
2SA539(M)			89	2032	
2SA539A		3114	89	2032	
2SA539B		3114	89	2032	
2SA539C		3114	89	2032	
2SA539D		3114	89	2032	
2SA539E		3114	89	2032	
2SA539F		3114	89	2032	
2SA539G		3114	89	2032	
2SA539GN		3114	89	2032	
2SA539H		3114	89	2032	
2SA539J		3114	89	2032	
2SA539K	290A	3466/159	89	2032	
2SA539L	290A	3466/159	89	2032	
2SA539M	290A	3466/159	89	2032	
2SA539OR		3114	89	2032	
2SA539R		3114	89	2032	
2SA539S	290A	3466/159	89	2032	
2SA539X		3114	89	2032	
2SA539Y		3114	89	2032	
2SA53A		3004	51	2004	
2SA53B		3004	51	2004	
2SA53C		3004	51	2004	
2SA53D		3004	51	2004	
2SA53E		3004	51	2004	
2SA53F		3004	51	2004	
2SA53G		3004	51	2004	
2SA53L		3004	51	2004	
2SA53M		3004	51	2004	
2SA53OR		3004	51	2004	
2SA53R		3004	51	2004	
2SA53X			51	2004	
2SA53Y		3004	51	2004	
2SA54	160	3005	51	2004	
2SA542		3466	89	2032	
2SA542A		3466	89	2032	
2SA542B		3466	89	2048	
2SA542C		3466	89	2032	
2SA542D		3466	89	2032	
2SA542E		3466	89	2032	
2SA542F		3466	89	2032	
2SA542G		3114	89	2032	
2SA542GN		3466	89	2032	
2SA542H		3466	89	2032	
2SA542J		3466	89	2032	
2SA542K		3466	89		
2SA542L		3466	89	2032	
2SA542M		3466	89	2032	
2SA542OR		3466	89	2032	
2SA542R		3466	89	2032	
2SA542X		3466	89	2032	
2SA542Y		3466	89	2032	
2SA543		3025	89	2050	
2SA544	129	3466/159	89	2032	
2SA544A		3114	89	2023	
2SA544B		3114	89	2023	
2SA544C		3114	89	2023	
2SA544D		3114	89	2023	
2SA544E		3114	89	2023	
2SA544F		3114	89	2023	
2SA544G		3114	89	2023	
2SA544GN		3114	89	2023	
2SA544H		3114	89	2023	
2SA544J		3114	89	2023	
2SA544K		3114	89	2023	
2SA544L		3114	89	2023	
2SA544M		3114	89	2023	
2SA544OR		3114	89	2023	
2SA544R		3114	89	2023	
2SA544X		3114	89	2023	
2SA544Y		3114	89	2023	
2SA545	193A	3138	67	2023	
2SA545(K)			67	2023	
2SA545(L)			67	2023	
2SA545A		3138	67	2023	
2SA545B		3138	67	2023	
2SA545C		3138	67	2023	
2SA545D		3138	67	2023	
2SA545E		3138	67	2023	
2SA545F		3138	67	2023	
2SA545G		3138	67	2023	
2SA545GN		3138	67	2023	
2SA545GRN	193A		67	2023	
2SA545H		3138	67	2023	
2SA545J		3138	67	2023	
2SA545K	193A	3138	67	2023	
2SA545KLM	193A	3138	67	2023	
2SA545L	193A	3138	67	2023	
2SA545LM	193A		67	2023	
2SA545M	193A	3138	67	2023	
2SA545OR		3138	67	2023	
2SA545R		3138	67	2023	
2SA545X		3138	67	2023	
2SA545Y		3138	67	2023	
2SA546	129	3025	244	2027	
2SA546A	129	3025	244	2027	
2SA546B	129	3025	244	2027	
2SA546D	129	3025			
2SA546E	129	3025	244	2027	
2SA546F	129	3025			
2SA546H	129	3025	244	2027	
2SA546J	129	3025			
2SA547	129/427		67	2023	
2SA547A	129/427		67	2023	

Industry Standard No.	ECG	SK	GE	RS 276-	MOTOR.
2SA548	159	3025/129	89	2027	
2SA548A	159	3025/129			
2SA548B	159	3025/129			
2SA548C	159	3025/129			
2SA548G		3025	89	2034	
2SA548GN		3025	89	2034	
2SA548OR		3025	89	2034	
2SA548R		3025	89	2034	
2SA548Y		3025	89	2034	
2SA549A	288		223		
2SA549AH	288		223		
2SA54A		3005	51	2004	
2SA54B		3005	51	2004	
2SA54C		3005	51	2004	
2SA54D		3005	51	2004	
2SA54E		3005	51	2004	
2SA54F		3005	51	2004	
2SA54G		3005	51	2004	
2SA54L		3005	51	2004	
2SA54M		3005	51	2004	
2SA54OR		3005	51	2004	
2SA54R		3005	51	2004	
2SA54X		3005	51	2004	
2SA54Y		3005	51	2004	
2SA55	126	3005	2	2007	
00002SA550	151A	3118	ZD-110	2024	
2SA550A	159	3118	82	2032	
2SA550A(Q)		3118	22	2032	
2SA550A(R)		3118	22	2032	
2SA550A(R,Q,S			22	2032	
2SA550A(S)			22	2032	
2SA550AB		3118	22	2027	
2SA550AQ	159	3118	22	2032	
2SA550AR	159		244	2027	
2SA550AS	159		244	2027	
2SA550B		3118	22	2027	
2SA550BC			244	2027	
2SA550BL		3118	22	2027	
2SA550C		3118	22	2027	
2SA550D		3118	22	2027	
2SA550P	159	3118	22	2032	
2SA550Q	159	3118	22	2032	
2SA550R	159	3118	22	2032	
2SA550S	159	3118	22	2032	
2SA550Y		3118	22	2027	
2SA551	129	3025	244	2027	
2SA551C	129	3025	244	2027	
2SA551D	129	3025	244	2027	
2SA551E	129	3025	244	2027	
2SA552	129	3025	244	2027	
2SA556	290A			2024	
2SA55A		3005	51	2004	
2SA55B		3005	51	2004	
2SA55C		3005	51	2004	
2SA55D		3005	51	2004	
2SA55E		3005	51	2004	
2SA55F		3005	51	2004	
2SA55G		3005	51	2004	
2SA55L		3005	51	2004	
2SA55M		3005	51	2004	
2SA55OR		3005	51	2004	
2SA55R		3005	51	2004	
2SA55X		3005	51	2004	
2SA55Y		3005	51	2004	
2SA56	126		245	2004	
2SA560	129	3025	22	2027	
2SA560A		3025	22	2032	
2SA561	159	3114/290	82	2032	
2SA561(O)			82	2032	
2SA561-O		3114	82	2032	
2SA561-GR			82	2032	
2SA561-GRN			82	2032	
2SA561-O	290A	3114/290	82	2032	
2SA561-OR		3114	82	2032	
2SA561-ORG			82	2032	
2SA561-R	290A	3114/290	82	2032	
2SA561-RD		3114	82	2032	
2SA561-RED			82	2032	
2SA561-Y	290A	3114/290	82	2032	
2SA561-YEL			82	2032	
2SA561-YL		3114	82	2032	
2SA561GR	290A$	3114/290	82	2032	
2SA561GRN		3114	82	2032	
2SA561R	290A	3114/290	82	2032	
2SA561RD			82	2032	
2SA561RED			82	2032	
2SA561Y	290A	3114/290	82	2032	
2SA561YEL			82	2032	
2SA561YL			82	2032	
2SA562	290A	3114/290	269	2032	
2SA562(O)			269	2032	
2SA562(Y)			269	2032	
2SA562-O	290A	3114/290	269	2032	
2SA562-GR			269	2032	
2SA562-GRN			269	2032	
2SA562-O	290A	3114/290	269	2032	
2SA562-OR			269	2032	
2SA562-ORG			269	2032	
2SA562-R		3114	269	2032	
2SA562-RD		3114	269	2032	
2SA562-RED			269	2032	
2SA562-Y	290A	3114/290	269	2032	
2SA562-YE		3114	269	2032	
2SA562-YEL			269	2032	
2SA5620		3114	89	2023	
2SA562D		3114	269	2032	
2SA562E		3114	269	2032	
2SA562G	290A	3114/290	269	2032	
2SA562GR	290A$	3114/290	269	2032	
2SA562GRN		3114	269	2032	
2SA562R	290A	3114/290	269	2032	
2SA562RD			269	2032	
2SA562RED			269	2032	
2SA562V			269	2032	
2SA562VO			269	2032	
2SA562Y	290A	3114/290	269	2032	
2SA562Y-TM	290A	3114/290			
2SA562YE			269	2032	
2SA562YEL			269	2032	
2SA564	290A	3247/234	65	2050	
2SA564(O)			65	2050	
2SA564(O)			65	2050	
2SA564(P)			65	2050	
2SA564(Q)			65	2050	
2SA564(R)			65	2050	
2SA564(S,T)			65	2050	
2SA564(T)			65	2050	
2SA564-O	290A	3247/234	65	2050	
2SA564-1			65	2050	
2SA564-O	234	3247	65	2050	
2SA564-OGD			65	2050	
2SA564-OR			65	2050	
2SA564-P			65	2050	
2SA564-Q	234		65	2050	

Industry Standard No.	ECG	SK	GE	RS 276-	MOTOR.
2SA564-R			65	2050	
2SA564A	290A	3247/234	65	2050	
2SA564A(P)			65	2050	
2SA564A(R)			65	2050	
2SA564A(S)			65	2050	
2SA564ABQ	290A	3247/234	65	2050	
2SA564ABQ-1		3247	65	2050	
2SA564AG			65	2050	
2SA564AK			65	2050	
2SA564AL			65	2050	
2SA564AO	290A	3247/234	65	2050	
2SA564AP	290A	3247/234	65	2050	
2SA564AQ	290A	3247/234	65	2050	
2SA564AQGD	290A	3247/234			
2SA564AR	290A	3247/234	65	2050	
2SA564AS	234	3247	65	2050	
2SA564AT	234	3247	65	2050	
2SA564B			65	2050	
2SA564C			65	2050	
2SA564D	290A	3247/234	65	2050	
2SA564E			65	2050	
2SA564F	290A	3247/234	65	2050	
2SA564FQ	290A	3247/234	65	2050	
2SA564FQ-1		3247	65	2050	
2SA564FR	290A	3247/234	65	2050	
2SA564FR-1		3247	65	2050	
2SA564G	290A	3247/234	65	2050	
2SA564GN			65	2050	
2SA564H	290A	3247/234	65	2050	
2SA564J	290A	3247/234	65	2050	
2SA564K			65	2050	
2SA564L			65	2050	
2SA564M			65	2050	
2SA564OR		3247	65	2050	
2SA564P	290A	3247/234	65	2050	
2SA564P.A			65	2050	
2SA564POR	290A	3247/234	65	2050	
2SA564Q	290A	3247/234	65	2050	
2SA564QGD	290A		65	2050	
2SA564QHD	290A	3247/234	65	2050	
2SA564QP			65	2050	
2SA564QR	290A	3247/234	65	2050	
2SA564R	290A	3247/234	65	2050	
2SA564S	234	3247	65	2050	
2SA564T	234	3247	65	2050	
2SA564X			65	2050	
2SA564XL			65	2050	
2SA564Y	290A	3247/234	65	2050	
2SA565	159	3466	89	2032	
2SA565(D,C)			89	2034	
2SA565A	159	3466	89	2032	
2SA565AB		3114	89	2034	
2SA565B	159	3466	89	2032	
2SA565BA		3114	89	2034	
2SA565C	159	3466	89	2032	
2SA565D	159	3466	89	2032	
2SA565E		3114	89	2034	
2SA565F		3114	89	2034	
2SA565G		3114	89	2034	
2SA565GN		3114	89	2034	
2SA565H		3114	89	2034	
2SA565J		3114	89	2034	
2SA565K	159	3466	82	2024	
2SA565L		3114	89	2034	
2SA565M		3114	89	2034	
2SA565OR		3114	89	2034	
2SA565R		3114	89	2034	
2SA565X		3114	89	2034	
2SA565Y		3114	89	2034	
2SA566	218	3085	234		2N6420
2SA566A	218	3085	234		
2SA566B	218	3085	234		
2SA566C	218$	3085	234		
2SA566H	218	3085			
2SA567	159	3466	89	2024	
2SA5670		3118	89	2023	
2SA567OR		3118	89	2023	
2SA567A	159	3114/290	89	2024	
2SA567B	159	3118	89	2024	
2SA567C	159	3114/290	89	2024	
2SA567D		3114	89	2024	
2SA567E		3114	89	2024	
2SA567F		3114	89	2024	
2SA567G		3114	89	2024	
2SA567GN		3114	89	2024	
2SA567GR		3118	89	2024	
2SA567H		3114	89	2024	
2SA567J		3114	89	2024	
2SA567K		3114	89	2024	
2SA567L		3114	89	2024	
2SA567M		3114	89	2024	
2SA567OR		3114	89	2024	
2SA567R		3118	89	2024	
2SA567X		3114	89	2024	
2SA567Y		3118	89	2024	
2SA568	290A	3118	89	2032	
2SA5680		3118	89	2023	
2SA568OR		3118	89	2023	
2SA568A		3118	89	2032	
2SA568B	290A	3118	89	2032	
2SA568C	290A	3118	89	2032	
2SA568D	290A	3118	89	2032	
2SA568E	290A	3118	89	2032	
2SA568F	290A	3118	89	2032	
2SA568G	290A	3118	89	2032	
2SA568GN	290A	3118	89	2032	
2SA568H	290A	3118	89	2032	
2SA568J	290A	3118	89	2032	
2SA568K	290A	3118	89	2032	
2SA568L	290A	3118	89	2032	
2SA568M	290A	3118	89	2032	
2SA568OR		3118	89	2032	
2SA568R	290A	3118	89	2032	
2SA568X	290A	3118	89	2032	
2SA568Y	290A	3118	89	2032	
2SA569	290A	3114/290	89	2032	
2SA5690		3118	89	2023	
2SA569OR		3118	89	2023	
2SA569A		3114	89	2023	
2SA569B		3114	89	2023	
2SA569C		3114	89	2023	
2SA569D		3114	89	2023	
2SA569E		3114	89	2023	
2SA569F		3114	89	2023	
2SA569G		3114	89	2023	
2SA569GN		3118	89	2023	
2SA569H		3114	89	2023	
2SA569J	290A	3466/159	89	2032	
2SA569K		3114	89	2023	
2SA569L		3114	89	2023	
2SA569M		3114	89	2023	
2SA569OR		3114	89	2023	
2SA569R		3118	89	2023	
2SA569X		3114	89	2023	

Industry Standard No.	ECG	SK	GE	RS 276-	MOTOR.
2SA569Y		3118	89	2023	
2SA56A			22	2024	
2SA57	160	3007	52	2024	
2SA570	290A	3466/159	89	2032	
2SA570A		3114	89	2032	
2SA570B		3114	89	2032	
2SA570C		3114	89	2032	
2SA570D		3114	89	2032	
2SA570E		3114	89	2032	
2SA570F		3114	89	2032	
2SA570G		3114	89	2032	
2SA570GN		3118	89	2032	
2SA570H		3114	89	2032	
2SA570J		3114	89	2032	
2SA570K		3114	89	2032	
2SA570L		3114	89	2032	
2SA570M		3114	89	2032	
2SA570OR		3114	89	2032	
2SA570R		3118	89	2032	
2SA570X		3114	89	2032	
2SA570Y		3118	89	2032	
2SA571	129$	3025/129	244	2027	
2SA572	234	3247	21	2050	
2SA572Y			21	2034	
2SA575	290A	3025/129			
2SA575K	290A	3025/129	67	2023	
2SA575L	290A	3025/129	67	2023	
2SA578		3247	65	2050	
2SA578A		3247	65	2050	
2SA578B		3247	65	2050	
2SA578C		3247	65	2050	
2SA579		3247	65	2050	
2SA579A		3247	65	2050	
2SA579B		3247	65	2050	
2SA579C		3247	65	2050	
2SA57A		3008	53	2007	
2SA57B		3008	53	2007	
2SA57C		3008	53	2007	
2SA57D		3008	53	2007	
2SA57E		3008	53	2007	
2SA57F		3008	53	2007	
2SA57G		3008	53	2007	
2SA57L		3008	53	2007	
2SA57M		3008	53	2007	
2SA57OR		3008	53	2007	
2SA57R		3008	53	2007	
2SA57X		3008	53	2007	
2SA57Y		3008	53	2007	
2SA58	160	3007	52	2024	
2SA581	129		51	2004	
2SA58A		3006	51	2004	
2SA58B		3006	51	2004	
2SA58C		3006	51	2004	
2SA58D		3006	51	2004	
2SA58E		3006	51	2004	
2SA58F		3006	51	2004	
2SA58G		3006	51	2004	
2SA58H		3006	51	2004	
2SA58J		3006	51	2004	
2SA58K		3006	51	2004	
2SA58L		3006	51	2004	
2SA58M		3006	51	2004	
2SA58OR		3006	51	2004	
2SA58R		3006	51	2004	
2SA58X		3006	51	2004	
2SA58Y		3006	51	2004	
2SA59	160	3007	52	2024	
2SA592Y	159	3466	82	2024	
2SA594	129	3025	244	2027	
2SA594-O	129		244	2027	
2SA594-R	129	3025	244	2027	
2SA594-Y	129	3025	244	2027	
2SA594N			21	2034	
2SA595C	129		244	2027	
2SA597	129$	3025/129	244	2027	
2SA599			67	2023	
2SA599(Y)			67	2023	
2SA599Y			67	2023	
2SA59A		3007	50	2004	
2SA59B		3007	50	2004	
2SA59C		3007	50	2004	
2SA59D		3007	50	2004	
2SA59E		3007	50	2004	
2SA59F		3007	50	2004	
2SA59G		3007	50	2004	
2SA59L		3007	50	2004	
2SA59M		3007	50	2004	
2SA59OR		3007	50	2004	
2SA59R		3007	50	2004	
2SA59X		3007	50	2004	
2SA59Y		3007	50	2004	
2SA60	126	3007	52	2024	
2SA603	159	3466	89	2032	
2SA603A		3118	89	2032	
2SA603B		3118	89	2032	
2SA603C		3118	89	2032	
2SA603D		3118	89	2032	
2SA603E		3118	89	2032	
2SA603F		3118	89	2032	
2SA603G		3118	89	2032	
2SA603GN		3118	89	2032	
2SA603H		3118	89	2032	
2SA603J		3118	89	2032	
2SA603K		3118	89	2032	
2SA603L		3118	89	2032	
2SA603M		3118	89	2032	
2SA603OR		3118	89	2032	
2SA603R		3118	89	2032	
2SA603X		3118	89	2032	
2SA603Y		3118	89	2032	
2SA604	129	3025	244	2027	
2SA605			244	2027	
2SA606	129	3025	244	2027	
2SA606S	129		244	2027	
2SA607	129/427	3025/129	89		
2SA607A	129/427	3025/129	89		
2SA607B	129/427	3025/129	89		
2SA607C	129/427	3025/129	89		
2SA607D	129/427	3025/129	89		
2SA607E		3025	89	2023	
2SA607F		3025	89	2023	
2SA607G		3025	89	2023	
2SA607GN		3025	89	2023	
2SA607H		3025	89	2023	
2SA607J		3025	89	2023	
2SA607K	129/427	3025/129	89	2023	
2SA607L	129/427	3025/129	89	2023	
2SA607M	129/427	3025/129	89	2023	
2SA607OR		3025	89	2023	
2SA607R		3025	89	2023	
2SA607S	129/427	3025/129			
2SA607X		3025	89	2023	
2SA607Y		3025	89	2023	
2SA608		3114	82	2032	

Industry Standard No.	ECG	SK	GE	RS 276-	MOTOR.
2SA608(C)			82	2032	
2SA608(D)			82	2032	
2SA608(F)			82	2032	
2SA608-O			82	2032	
2SA608-C			82	2032	
2SA608-D			82	2032	
2SA608-E			82	2032	
2SA608-F			82	2032	
2SA608-O		3114	82	2032	
2SA608-OR			82	2032	
2SA608A		3114	82	2032	
2SA608AF	159	3114/290			
2SA608B		3114	82	2032	
2SA608BL		3114	82	2032	
2SA608C		3114	82	2032	
2SA608D		3114	82	2032	
2SA608D(F)			82	2032	
2SA608E		3114	82	2032	
2SA608F		3114	82	2032	
2SA608G		3114	82	2032	
2SA608GN			82	2032	
2SA608H			82	2032	
2SA608J			82	2032	
2SA608K		3114	82	2032	
2SA608L		3114	82	2032	
2SA608M		3114	82	2032	
2SA608OR		3114	82	2032	
2SA608P		3114	82	2032	
2SA608R		3114	82	2032	
2SA608X			82	2032	
2SA608Y			82	2032	
00002SA609	159	3466	82	2032	
2SA609A		3466	82	2032	
2SA609B		3466	82	2032	
2SA609C		3466	82	2032	
2SA609D		3466	82	2032	
2SA609E		3466	82	2032	
2SA609F		3466	82	2032	
2SA609G		3466	82	2032	
2SA609GN			82	2023	
2SA609J			82	2023	
2SA609K			82	2023	
2SA609L			82	2023	
2SA609M			82	2023	
2SA609OR			82	2023	
2SA609R			82	2023	
2SA609Y			82	2023	
2SA60A		3006	51	2004	
2SA60B		3006	51	2004	
2SA60C		3006	51	2004	
2SA60D		3006	51	2004	
2SA60E		3006	51	2004	
2SA60F		3006	51	2004	
2SA60G		3006	51	2004	
2SA60H		3006	51	2004	
2SA60K		3006	51	2004	
2SA60L		3006	51	2004	
2SA60M		3006	51	2004	
2SA60OR			51	2004	
2SA60R		3006	51	2004	
2SA60X		3006	51	2004	
2SA60Y		3006	51	2004	
2SA61	160	3007	245	2004	
2SA610	290A	3466/159	82	2032	
2SA610B	290A	3466/159	82	2032	
2SA611	290A	3466/159	269	2032	
2SA611-4E	290A	3466/159	269	2032	
2SA6111			21	2034	
2SA612	129	3025	244	2027	
2SA613	218		234		2N4899
2SA614	218	3085	234		2N4900
2SA616	218	3085	234		2N3741
2SA616(1)	218		234		
2SA616(2)	218		234		
2SA617K	159	3466	82	2032	
2SA618K	159	3466	82	2032	
2SA61A		3005	51	2004	
2SA61C		3005	51	2004	
2SA61D		3005	51	2004	
2SA61E		3005	51	2004	
2SA61G		3005	51	2004	
2SA61K		3005	51	2004	
2SA61L		3005	51	2004	
2SA61OR		3005	51	2004	
2SA61R		3005	51	2004	
2SA61X		3005	51	2004	
2SA61Y		3005	51	2004	
2SA623	307	3193/187	274		D41E1
2SA623-O	307	3083/197	274		
2SA623A	307	3193/187	274		
2SA623B	307	3193/187	274		
2SA623C	307	3193/187	274		
2SA623D	307	3193/187	274		
2SA623G	307	3193/187	274		
2SA623R	307	3193/187	274		
2SA623Y	307	3193/187	274		
2SA624	307	3083/197	274		D41E5
2SA624A	307	3083/197	274		
2SA624B	307	3083/197	274		
2SA624C	307	3083/197	274		
2SA624D	307	3083/197	274		
2SA624E	307	3083/197	274		
2SA624G	307	3083/197	274		
2SA624GN	307	3083/197	274		
2SA624L	307	3083/197	274		
2SA624LG	307	3083/197	274		
2SA624R	307	3083/197	274		
2SA624Y	307	3083/197	274		
2SA626	88	3359/281	74	2043	2N6226
2SA626L	88	3359/281	74	2043	
2SA626M	88	3359/281			
2SA626N	88	3359/281			
2SA627	88	3359/281	74	2043	2N6226
2SA627L	88	3359/281			
2SA627M	88	3359/281			
2SA627N	88	3359/281			
2SA628		3114	82	2032	
2SA628(EF)			82	2032	
2SA628(F)			82	2032	
2SA628-O			82	2032	
2SA628-OR			82	2032	
2SA628A		3114	82	2032	
2SA628AA		3114	82	2032	
2SA628AD			82	2032	
2SA628AE			82	2032	
2SA628B		3118	82	2032	
2SA628C		3118	82	2032	
2SA628D		3114	82	2032	
2SA628E		3114	82	2032	
2SA628E,F			82	2032	
2SA628EF		3114	82	2032	
2SA628F		3114	82	2032	
2SA628G	234	3114/290	82	2032	
2SA628GN		3114	82	2032	

Industry Standard No.	ECG	SK	GE	RS 276-	MOTOR.
2SA628H		3114	82	2032	
2SA628J		3114	82	2032	
2SA628K		3114	82	2032	
2SA628L		3118	82	2032	
2SA628M		3114	82	2032	
2SA628OR		3114	82	2032	
2SA628R		3114	82	2032	
2SA628X		3114	82	2032	
2SA628Y		3114	82	2032	
2SA629	234	3114/290	89	2032	
2SA629A		3114	89	2032	
2SA629B		3114	89	2032	
2SA629C		3114	89	2032	
2SA629D		3114	89	2032	
2SA629E		3114	89	2032	
2SA629F		3114	89	2032	
2SA629G		3114	89	2032	
2SA629GN		3114	89	2032	
2SA629H		3114	89	2032	
2SA629J		3114	89	2032	
2SA629K		3114	89	2032	
2SA629L		3114	89	2032	
2SA629M		3114	89	2032	
2SA629OR		3114	89	2032	
2SA629R		3114	89	2032	
2SA629X		3114	89	2032	
2SA629Y		3114	89	2032	
2SA633			18		D41E1
2SA634	187A	9076	29		D41E5
2SA634A	187A	9076	29		
2SA634B	187A		29		
2SA634C	187A	9076	29		
2SA634D	187A	9076	29		
2SA634K	187A	9076	29		
2SA634L	187A	9076	29		
2SA634M	187A	9076	29		
2SA635					D41D7
2SA636	187	9076	248		2N6556
2SA636(4)K	187	3193	29	2025	
2SA636(4)L	187	3193	29	2025	
2SA636(L)			248	2025	
2SA636(M)			29	2025	
2SA6361			29	2025	
2SA6361,K			29	2025	
2SA636A	187	9076	248		
2SA636B	187	9076	248		
2SA636C	187	9076	248		
2SA636D	187	9076	248		
2SA636K	187	9076	248		
2SA636L	187	9076	248		
2SA636M	187	9076	248		
2SA636UK				2025	
2SA637	288	3715	221		
2SA638E,F			22	2032	
2SA6398		3114	223		
2SA64	102A	3008	53	2007	
2SA640	234	3118	82	2050	
2SA640(M)			82	2050	
2SA640A	234	3466/159	82	2050	
2SA640B	234	3466/159	82	2050	
2SA640C	234	3466/159	82	2050	
2SA640D	234	3466/159	82	2050	
2SA640E	234	3466/159	82	2050	
2SA640F	234	3114/290			
2SA640L	234	3466/159	82	2050	
2SA640M	234	3466/159	82	2050	
2SA640S	234	3466/159	82	2050	
2SA641	234	3841/294	69	2032	
2SA641A	234	3841/294	69	2032	
2SA641B	234	3841/294	69	2032	
2SA641BL		3841	69	2023	
2SA641C	234	3841/294	69	2032	
2SA641D	234	3841/294	69	2032	
2SA641G		3841	69	2032	
2SA641GR		3841	69	2032	
2SA641K		3841	69	2032	
2SA641L	234	3841/294	69	2032	
2SA641M	234	3841/294	69	2032	
2SA6410			69	2032	
2SA6410R		3841	69	2032	
2SA641R		3841	69	2032	
2SA641Y		3841	69	2032	
2SA642	290A	3114/290	82	2032	
2SA642A	290A	3114/290	82	2032	
2SA642B	290A	3114/290	82	2032	
2SA642C	290A	3114/290	82	2032	
2SA642D	290A	3114/290	82	2032	
2SA642E	290A	3114/290	82	2032	
2SA642F	290A	3114/290	82	2032	
2SA642G		3114	82	2032	
2SA642GN		3114	82	2032	
2SA642H		3114	82	2032	
2SA642J		3114	82	2032	
2SA642K		3114	82	2032	
2SA642L	290A	3114/290	82	2032	
2SA642M		3114	82	2032	
2SA642OR		3114	82	2032	
2SA642R	290A	3114/290	82	2032	
2SA642S	290A	3114/290	82	2032	
2SA642V	290A	3114/290	82	2032	
2SA642W	290A	3114/290	82	2032	
2SA642X		3114	82	2032	
2SA642Y	290A	3114/290	82	2032	
2SA643	193A	3138	67	2023	
2SA643(R)			67	2023	
2SA643(V,R)			67	2023	
2SA643(W)			67	2023	
2SA643A	193A	3138	67	2023	
2SA643B	193A	3138	67	2023	
2SA643C	193A	3138	67	2023	
2SA643D	193A	3138	67	2023	
2SA643E	193A	3138	67	2023	
2SA643F	193A	3138	67	2023	
2SA643R	193A	3138	67	2023	
2SA643V	193A	3138	67	2023	
2SA643W	193A		67	2023	
2SA645	211	9076/187A	253	2026	D41D10
2SA646	211	9076/187A	248		2N6556
2SA647		3930			2N6556
2SA648	281		74	2043	2N6230
2SA648A	281		74	2043	
2SA648B	281		74	2043	
2SA648C	281		74	2043	
2SA64A		3008	53	2007	
2SA64B		3008	53	2007	
2SA64C		3008	53	2007	
2SA64D		3008	53	2007	
2SA64E		3008	53	2007	
2SA64F		3008	53	2007	
2SA64G		3008	53	2007	
2SA64GN		3008	53	2007	
2SA64H		3008	53	2007	
2SA64J		3008	53	2007	
2SA64K		3008	53	2007	

Industry Standard No.	ECG	SK	GE	RS 276-	MOTOR.
2SA64L		3008	53	2007	
2SA64M		3008	53	2007	
2SA640R		3008	53	2007	
2SA64R		3008	53	2007	
2SA64X		3008	53	2007	
2SA64Y		3008	53	2007	
2SA65	102A	3005	51	2005	
2SA652					2N6420
2SA653					2N6420
2SA656	88		248		2N6228
2SA656A	88	3359/281			
2SA656L	88		248		
2SA656M	88		248		
2SA657	88	3173/219			2N6226
2SA657A	88	3173/219			
2SA658	88	3173/219	74	2043	2N6226
2SA658A	88	3173/219			
2SA659	290A	3114/290	89	2032	
2SA659(D)			89	2032	
2SA659(E)			89	2032	
2SA659A	290A	3114/290	89	2032	
2SA659B	290A	3114/290	89	2032	
2SA659C	290A	3114/290	82	2032	
2SA659D	290A	3114/290	89	2032	
2SA659E	290A	3114/290	89	2032	
2SA659F	290A	3114/290	89	2032	
2SA659G	290A	3114/290	89	2032	
2SA659L	290A	3114/290	89	2032	
2SA659P	290A	3114/290	89	2032	
2SA659R	290A	3114/290	89	2032	
2SA659Y	290A	3114/290	89	2032	
2SA65A		3005	51	2004	
2SA65B		3005	51	2004	
2SA65C		3005	51	2004	
2SA65D		3005	51	2004	
2SA65E		3005	51	2004	
2SA65F		3005	51	2004	
2SA65G		3005	51	2004	
2SA65K		3005	51	2004	
2SA65L		3005	51	2004	
2SA65M		3005	51	2004	
2SA650R		3005	51	2004	
2SA65R		3005	51	2004	
2SA65X		3005	51	2004	
2SA65Y		3005	51	2004	
2SA66	102A	3005	51	2005	
2SA661	290A	3450/298	272		
2SA661-O	290A	3450/298	272		
2SA661GR	290A$	3450/298	272		
2SA661R	290A	3450/298	272		
2SA661Y	290A	3450/298	272		
2SA663	219	3173	74	2043	2N6226
2SA663-R	219		51	2004	
2SA666	290A	3841/294	65	2050	
2SA6661QRS		3466	82	2032	
2SA6666QRS			65	2050	
2SA666A	290A	3841/294	65	2050	
2SA666B		3841	65	2050	
2SA666BL		3841	65	2050	
2SA666C		3841	65	2050	
2SA666D		3841	65	2050	
2SA666E		3841	65	2050	
2SA666H	290A	3841/294	65	2050	
2SA666HR	290A	3841/294	65	2050	
2SA666I			21	2034	
2SA666IQRS			21	2050	
2SA666Q	290A	3247/234	65	2050	
2SA666QRS	290A		65	2050	
2SA666R	290A	3247/234	65	2050	
2SA666S	234	3247	65	2050	
2SA666Y		3841	65	2050	
2SA66A		3005	51	2004	
2SA66B		3005	51	2004	
2SA66C		3005	51	2004	
2SA66D		3005	51	2004	
2SA66E		3005	51	2004	
2SA66F		3005	51	2004	
2SA66G		3005	51	2004	
2SA66K		3005	51	2004	
2SA66L		3005	51	2004	
2SA66M		3005	51	2004	
2SA660R		3005	51	2004	
2SA66R		3005	51	2004	
2SA66X		3005	51	2004	
2SA66Y		3005	51	2004	
2SA67	102A	3006/160	52	2024	
2SA670	153	3274	69	2049	2N6125
2SA670A	153	3274	69	2049	
2SA670B	153	3274	69	2049	
2SA670C	153	3274	69	2049	
2SA671	378	3274/153	69		2N6125
2SA671A	378	3274/153	69		
2SA671B	378	3274/153	69		
2SA671C	378	3274/153	69		
2SA671K	378	3274/153	69		
2SA671KA	378	3274/153	69		
2SA671KB	378	3274/153	69		
2SA671KC	378	3274/153	69		
2SA672	159	3114/290	82	2032	
2SA672(B)			82	2034	
2SA672A	159	3114/290	82	2032	
2SA672B	159	3114/290	82	2032	
2SA672C	159	3114/290	82	2032	
2SA673		3114	269	2032	
2SA673(B)	290A		269	2032	
2SA673(D)			269	2032	
2SA673-OR			269	2032	
2SA673A	290A	3114/290	269	2032	
2SA673A(C)			269	2032	
2SA673AA	290A		269	2032	
2SA673AB	290A	3114/290	269	2032	
2SA673AC	290A	3114/290	269	2032	
2SA673AD	290A	3114/290	269	2032	
2SA673AE	290A		269	2032	
2SA673AK	290A		269	2032	
2SA673AKA			269	2032	
2SA673AKB			269	2032	
2SA673AKC			269	2032	
2SA673AS	290A	3114/290	269	2032	
2SA673AS(C)			269	2032	
2SA673ASC	290A		269	2032	
2SA673B	290A	3114/290	269	2032	
2SA673C	290A	3114/290	269	2032	
2SA673C2	290A$		269	2032	
2SA673D	290A	3114/290	269	2032	
2SA673E		3114	269	2032	
2SA673F		3114	269	2032	
2SA673G		3114	269	2032	
2SA673GN		3114	269	2032	
2SA673H		3114	269	2032	
2SA673J		3114	269	2032	
2SA673K		3114	269	2032	
2SA673L		3114	269	2032	
2SA673M		3114	269	2032	

Industry Standard No.	ECG	SK	GE	RS 276-	MOTOR.
2SA6730R		3114	269	2032	
2SA673R			269	2032	
2SA673WT	290A$		269	2032	
2SA673X			269	2032	
2SA673Y			269	2032	
2SA675	290A	3466/159	82	2032	
2SA675A	290A	3466/159	82	2032	
2SA675B	290A	3466/159	82	2032	
2SA675C	290A	3466/159	82	2032	
2SA677	290A	3114/290	89	2024	
2SA677-O		3114	89	2032	
2SA677-5	290A	3114/290			
2SA677-6	290A	3114/290			
2SA677-O			82	2032	
2SA677-OR			82	2032	
2SA677A		3114	89	2032	
2SA677B		3114	89	2032	
2SA677C		3114	89	2032	
2SA677D		3114	89	2032	
2SA677E		3114	89	2032	
2SA677F		3114	89	2032	
2SA677G		3114	89	2032	
2SA677GN		3114	89	2032	
2SA677H		3114	89	2032	
2SA677HL	290A		82	2032	
2SA677J		3114	89	2032	
2SA677K		3114	89	2032	
2SA677L		3114	89	2032	
2SA677M		3114	89	2032	
2SA677OR		3114	89	2032	
2SA677R		3114	89	2032	
2SA677X		3114	89	2032	
2SA677Y		3114	89	2032	
2SA678	290A	3114/290	82	2032	
2SA678(C)	290A	3114/290	82	2032	
2SA678(SONY)			82	2032	
2SA678-5	290A	3114/290			
2SA678-6	290A	3114/290			
2SA678-O			82	2032	
2SA678-OR			82	2032	
2SA678A		3114	82	2032	
2SA678B		3114	82	2032	
2SA678C		3114	82	2032	
2SA678D		3114	82	2032	
2SA678E	290A	3114/290	82	2032	
2SA678F		3114	82	2032	
2SA678G		3114	82	2032	
2SA678GN		3114	82	2032	
2SA678H		3114	82	2032	
2SA678J		3114	82	2032	
2SA678K		3114	82	2032	
2SA678L		3114	82	2032	
2SA678M		3114	82	2032	
2SA678OR		3114	82	2032	
2SA678R		3114	82	2032	
2SA678X		3114	82	2032	
2SA678Y		3114	82	2032	
2SA679	281	3359	263	2043	MJ15016
2SA679-R			263	2043	
2SA679R	281		263	2043	
2SA679Y	281		263	2043	
2SA67A		3006	51	2004	
2SA67B		3006	51	2004	
2SA67C		3006	51	2004	
2SA67D		3006	51	2004	
2SA67E		3006	51	2004	
2SA67F		3006	51	2004	
2SA67G		3006	51	2004	
2SA67H		3006	51	2004	
2SA67K		3006	51	2004	
2SA67L		3006	51	2004	
2SA67OR		3006	51	2004	
2SA67R		3006	51	2004	
2SA67X		3006	51	2004	
2SA67Y		3006	51	2004	
2SA68				2005	
2SA680	281	3173/219	263	2043	2N5880
2SA680-R			263	2043	
2SA680R	281	3173/219	263	2043	
2SA680Y	281	3173/219	263	2043	
2SA681					MJE253
2SA682	374	3189A/183			MJE253
2SA683	294	3841	48		
2SA683P	294	3841	48		
2SA683Q	294	3841	48		
2SA683R	294	3841	48		
2SA683S	294	3841	48		
2SA684	294	3841	48		
2SA684Q	294	3841	48		
2SA684R	294	3841	48		
2SA685	290A	3867	82	2032	
2SA69	126	3006/160	52	2024	
2SA690D	159	3466	82	2032	
2SA693			269	2032	
2SA693C			269	2032	
2SA695		3450	272		
2SA695C		3450	272		
2SA695D		3450	272		
2SA696		3450	223		
2SA696C		3450	223		
2SA696D		3450	223		
2SA696E		3450	223		
2SA697		3450	48		
2SA697C		3450	48		
2SA698					MDS60
2SA699	187A	9076	248		D41E5
2SA699-O	187A		248		
2SA699-O	187A	9076	248		
2SA699A	187A	9076	248		
2SA699A-Q	187A		248		
2SA699AO	187A	9076	248		
2SA699AP	187A	9076	248		
2SA699AQ	187A	9076	248		
2SA699AR	187A	9076	248		
2SA699CF		9076	248		
2SA699F	187A	9076	248		
2SA699PF	187A		248		
2SA699Q	187A	9076	248		
2SA699R	187A	9076	248		
2SA69A		3006	51	2004	
2SA69B		3006	51	2004	
2SA69C		3006	51	2004	
2SA69D		3006	51	2004	
2SA69E		3006	51	2004	
2SA69F		3006	51	2004	
2SA69G		3006	51	2004	
2SA69H		3006	51	2004	
2SA69K		3006	51	2004	
2SA69L		3006	51	2004	
2SA69M		3006	51	2004	
2SA69OR		3006	51	2004	
2SA69R		3006	51	2004	
2SA69X		3006	51	2004	
2SA69Y		3006	51	2004	
2SA70	160	3008	51	2004	
2SA70-08			51	2004	
2SA70-OB		3008	50	2004	
2SA70-OB			51	2004	
2SA700	153	9076/187A	248		TIP30
2SA700(B)			29	2025	
2SA700B	153	9076/187A	248		
2SA700Y	153	9076/187A	248		
2SA701		3466	82	2032	
2SA701F		3466	82	2032	
2SA701FO			48	2034	
2SA701FJ		3466	82	2032	
2SA701FO	159	3466	82	2032	
2SA702		3247	221	2050	
2SA703	307	3203/211	274		D41E1
2SA703C	307		274		
2SA703D	307		274		
2SA703E	307		274		
2SA704	290A	3118	48	2024	
2SA704A		3118	89	2032	
2SA704B		3118	89	2032	
2SA704C		3118	89	2032	
2SA704D		3118	89	2032	
2SA704E		3118	89	2032	
2SA704F		3118	89	2032	
2SA704G		3118	89	2032	
2SA704GN		3118	89	2032	
2SA704H		3118	89	2032	
2SA704J		3118	89	2032	
2SA704K		3118	89	2032	
2SA704L		3118	89	2032	
2SA704M		3118	89	2032	
2SA704OR		3118	89	2032	
2SA704R		3118	89	2032	
2SA704X		3118	89	2032	
2SA704Y		3118	89	2032	
2SA705	290A	3841/294	48	2032	
2SA705A		3841	89	2032	
2SA705B		3841	89	2032	
2SA705C		3841	89	2032	
2SA705D		3841	89	2032	
2SA705E		3841	89	2032	
2SA705F		3841	89	2032	
2SA705G		3841	89	2032	
2SA705GN		3841	89	2032	
2SA705H		3841	89	2032	
2SA705J		3841	89	2032	
2SA705K		3841	89	2032	
2SA705L		3841	89	2032	
2SA705M		3841	89	2032	
2SA705OR		3841	89	2032	
2SA705R		3841	89	2032	
2SA705X			89	2032	
2SA705Y		3841	89	2032	
2SA706	189	3200	253	2026	MPSU55
2SA707	298	3450	67	2023	
2SA707V	298	3450	67	2023	
2SA708	129	3203/211	244	2027	
2SA708A	129	3203/211	244	2027	
2SA708B	129	3203/211	244	2027	
2SA708C	129	3203/211	244	2027	
2SA70A		3008	51	2004	
2SA70B		3008	51	2004	
2SA70C		3008	51	2004	
2SA70D		3008	51	2004	
2SA70E		3008	51	2004	
2SA70F	160	3008	51	2004	
2SA70G		3008	51	2004	
2SA70H		3008	51	2004	
2SA70K		3008	51	2004	
2SA70L	160	3008	51	2004	
2SA70MA	160	3008	51	2004	
2SA700A			51	2004	
2SA700R		3008	51	2004	
2SA70R		3008	51	2004	
2SA70X			51	2004	
2SA70Y		3008	51	2004	
2SA71	160	3008	51	2024	
2SA711		3984			2N5583
2SA714	281				2N6228
2SA715	185#	9042/374	58	2025	MJE170
2SA715A	185#		58	2025	
2SA715B	185#	9042/374	58	2025	
2SA715C	185#	9042/374	58	2025	
2SA715D	185#	9042/374	58	2025	
2SA715WB	185#	9076/187A	58	2025	
2SA715WBP	185#	9076/187A	248		
2SA715WT	185#	9042/374	58	2025	
2SA715WT(C,B)			67	2025	
2SA715WT-B	185#	9076/187A			
2SA715WT-C	185#	9076/187A			
2SA715WT-D	185#	9076/187A			
2SA715WTA	185#	9076/187A	58	2025	
2SA715WTB	185#	9042/374	58	2025	
2SA715WTC	185#	9042/374	58	2025	
2SA715WTD	185#	9076/187A	58	2025	
2SA717	129	3203/211	253	2026	
2SA718	159	3466	82	2032	
2SA719	290A	3114/290	269	2032	
2SA719(Q)			269	2032	
2SA719,R			269	2032	
2SA719K	290A		269	2032	
2SA719P	290A	3114/290	269	2032	
2SA719PQR	290A		269	2032	
2SA719Q	290A	3114/290	269	2032	
2SA719QR	290A		269	2032	
2SA719R	290A	3114/290	269	2032	
2SA719RS	290A	3114/290	269	2032	
2SA719S	290A	3114/290	269	2032	
2SA71A	160	3008	51	2024	
2SA71AB	160	3008	51	2004	
2SA71AC	160	3008	51	2024	
2SA71B	160	3008	51	2024	
2SA71BS	160	3008	51	2024	
2SA71C		3008	51	2004	
2SA71D	160	3008	51	2024	
2SA71E		3008	51	2004	
2SA71F		3008	51	2004	
2SA71G		3008	51	2004	
2SA71H		3008	51	2004	
2SA71K		3008	51	2004	
2SA71L		3008	51	2004	
2SA71M		3008	51	2004	
2SA71OR		3008	51	2004	
2SA71R		3008	51	2004	
2SA71X			51	2004	
2SA71Y	160	3008	51	2024	
2SA71YA		3008	51	2004	
2SA72	126	3006/160	52	2024	
2SA720	290A	3114/290	269		
2SA720O		3466	82	2032	
2SA720A	290A	3114/290	269		
2SA720B		3114	269		
2SA720C		3114	269		

Industry Standard No.	ECG	SK	GE	RS 276-	MOTOR.
2SA720D		3114	269		
2SA720L		3114	269		
2SA720P	290A	3114/290	269		
2SA720Q	290A	3114/290	269		
2SA720Q,R,S			22	2034	
2SA720R	290A	3114/290	269		
2SA720S	290A	3114/290	269		
2SA720Y		3114	269		
2SA721	234	3247	65	2050	
2SA721Q			65	2050	
2SA721R		3247	65	2050	
2SA721S	234	3247	65	2050	
2SA721T	234	3247	65	2050	
2SA721U	234	3247	65	2050	
2SA722	234	3247	65	2050	
2SA722S	234	3247	65	2050	
2SA722T	234	3247	65	2050	
2SA722U	234	3247	65	2050	
2SA723	290A	3114/290	82	2005	
2SA723A	290A	3114/290	82	2032	
2SA723B	290A	3114/290	82	2032	
2SA723C	290A	3114/290	82	2032	
2SA723D	290A	3114/290	82	2032	
2SA723E	290A	3114/290	82	2032	
2SA723F	290A	3114/290	82	2032	
2SA723R	290A	3114/290	82	2032	
2SA725		3247	65	2050	
2SA725F		3247	65	2050	
2SA725G		3247	65	2050	
2SA725H		3247	65	2050	
2SA726		3247	65	2050	
2SA726F		3247	65	2050	
2SA726G		3247	65	2050	
2SA726H		3247	65	2050	
2SA726Y			65	2050	
2SA728	290A		65	2050	
2SA728A	290A		67	2023	
2SA72BLU	126	3006/160	52	2024	
2SA72BLU-BLU	126	3006/160	52	2024	
2SA72BRN	126	3006/160	52	2024	
2SA72ORN	126	3006/160	52	2024	
2SA72WHT	126	3006/160	52	2024	
2SA73	160	3007	50	2004	
2SA730	193A	3138	82	2032	
2SA730P	193A	3138			
2SA730Q	193A	3138			
2SA730R	193A	3138			
2SA730S	193A	3138			
2SA731	193A	3466/159	82	2032	
2SA731R	193A	3138			
2SA733	290A	3114/290	48		
2SA733(P)	290A		48		
2SA733A	290A	3114/290	48		
2SA733AP	290A	3114/290	48		
2SA733AQ	290A	3114/290			
2SA733B	290A		48		
2SA733C	290A		48		
2SA733D	290A		48		
2SA733E	290A		48		
2SA733F	290A		48		
2SA733H	290A	3114/290	48		
2SA733I	290A	3114/290	48		
2SA733IQ	290A	3114/290	48		
2SA733P	290A	3114/290	48		
2SA733PQ	290A		48		
2SA733Q	290A	3114/290	48		
2SA733QP	290A	3114/290	48		
2SA733R	290A	3114/290	48		
2SA734	129		244	2027	
2SA735	159		82	2032	
2SA736	129	3118	244	2027	
2SA738	185#	9042/374	58	2025	MJE170
2SA738B	185#	9042/374	58	2025	
2SA738C	185#	9042/374	58	2025	
2SA738D	185#	9042/374	58		
2SA739					MJ6502
2SA73A		3007	50	2004	
2SA73B		3007	50	2004	
2SA73C		3007	50	2004	
2SA73D		3007	50	2004	
2SA73E		3007	50	2004	
2SA73F		3007	50	2004	
2SA73G		3007	50	2004	
2SA73H		3007	50	2004	
2SA73K		3007	50	2004	
2SA73L		3007	50	2004	
2SA73M		3007	50	2004	
2SA73OR		3007	50	2004	
2SA73R		3007	50	2004	
2SA73X		3007	50	2004	
2SA73Y		3007	50	2004	
2SA74	126	3007	51	2004	
2SA740	398	3930			
2SA740A	398	3930			
2SA741H	106$	3114/290	65	2050	
2SA742		3025	244	2027	
2SA742H		3025	244	2034	
2SA743	374#	9042/374	58	2025	
2SA743A	374#	9042/374	58	2025	
2SA743AA	374#		58	2025	
2SA743AB	374#	9042/374	58	2025	
2SA743AC	374#	9042/374	58	2025	
2SA743AD	374#		58	2025	
2SA743B	374#	9042/374	58	2025	
2SA743C	374#	9042/374	58	2025	
2SA743D	374#	9042/374	58	2025	
2SA744	219	3359/281	74	2043	
2SA745	281	3846/285			
2SA745A	281	3846/285			
2SA746	281	3846/285	74	2043	
2SA747	281	3846/285			
2SA747A	281	3846/285			
2SA748Q	153	3083/197	69	2049	
2SA748R			69	2049	
2SA749		3867	221		
2SA749A		3867	221		
2SA74A		3006	51	2004	
2SA74B		3006	51	2004	
2SA74C		3006	51	2004	
2SA74D		3006	51	2004	
2SA74E		3006	51	2004	
2SA74F		3006	51	2004	
2SA74G		3006	51	2004	
2SA74H		3006	51	2004	
2SA74K		3006	51	2004	
2SA74L		3006	51	2004	
2SA74M		3006	51	2004	
2SA74OR		3006	51	2004	
2SA74R		3006	51	2004	
2SA74X		3006	51	2004	
2SA74Y		3006	51	2004	
2SA75	126	3007	51	2024	
2SA751	383	3138/193A	67	2023	
2SA751(P)			67	2023	

Industry Standard No.	ECG	SK	GE	RS 276-	MOTOR.
2SA751P	383	3138/193A	67	2023	
2SA751Q	383	3138/193A	67	2023	
2SA751QR	383	3138/193A	67	2023	
2SA751R	383	3138/193A	67	2023	
2SA751S	383	3138/193A	67	2023	
2SA752	383		67	2023	
2SA752P	383		67	2023	
2SA752Q	383		67	2023	
2SA752R	383		67	2023	
2SA752S	383		67	2023	
2SA753	281	3359	263	2043	
2SA753A	281	3359	263	2043	
2SA753B	281	3359	263	2043	
2SA753C	281	3359	263	2043	
2SA754	153	3083/197	69	2049	
2SA754A	153	3083/197	69	2049	
2SA754B	153	3083/197	69	2049	
2SA754C	153	3083/197	69	2049	
2SA754D	153	3083/197	69	2049	
2SA755	153	3274	69	2049	2N6125
2SA755A	153	3274	69	2049	
2SA755B	153	3274	69	2049	
2SA755C	153	3274	69	2049	
2SA755D	153		69	2049	
2SA756	88	3173/219	263	2043	2N6226
2SA756A	88	3173/219	263	2043	
2SA756B	88	3173/219	263	2043	
2SA756C	88	3173/219	263	2043	
2SA757	88		263	2043	2N6227
2SA757A	88		263	2043	
2SA757B	88		263	2043	
2SA757C	88		263	2043	
2SA758	281		263	2043	2N6228
2SA758A	281		263	2043	
2SA758B	281		263	2043	
2SA758C	281		263	2043	
2SA75A		3005	51	2024	
2SA75B	126	3005	51	2024	
2SA75C		3005	51	2024	
2SA75D		3005	51	2024	
2SA75E		3005	51	2024	
2SA75F		3005	51	2024	
2SA75G		3005	51	2024	
2SA75H		3005	51	2024	
2SA75K		3005	51	2024	
2SA75L		3005	51	2024	
2SA75M		3005	51	2024	
2SA75OR		3005	51	2024	
2SA75R		3005	51	2024	
2SA75X		3005	51	2024	
2SA75Y		3005	51	2024	
2SA76	126	3006/160	51	2024	
2SA762	398$				2N6211
2SA763	234	3247	65	2050	
2SA763-W	234	3247	65	2050	
2SA763-WL-3	234	3247	65	2050	
2SA763-WL-4	234	3247	65	2050	
2SA763-WL-5	234	3247	65	2050	
2SA763-WL-6	234	3247	65	2050	
2SA763-WN	234	3247	65	2050	
2SA763-WN-3	234	3247	65	2050	
2SA763-WN-4	234	3247	65	2050	
2SA763-WN-5	234	3247	65	2050	
2SA763-WN-6	234	3247	65	2050	
2SA763-Y	234	3247	65	2050	
2SA763-YL	234	3247	65	2050	
2SA763-YL-3	234	3247	65	2050	
2SA763-YL-4	234	3247	65	2050	
2SA763-YL-5	234	3247	65	2050	
2SA763-YL-6	234	3247	65	2050	
2SA763-YN	234	3247	65	2050	
2SA763-YN-3	234	3247	65	2050	
2SA763-YN-4	234	3247	65	2050	
2SA763-YN-5	234	3247	65	2050	
2SA763-YN-6	234	3247	65	2050	
2SA764	197		234		2N6317
2SA765	197		234		2N6318
2SA766	398	3624	69	2049	2N6420
2SA766S	398	3624	69	2049	
2SA768	153	3274	250	2049	2N6125
2SA769	197	3083			2N6126
2SA76A		3006	51	2004	
2SA76B		3006	51	2004	
2SA76C		3006	51	2004	
2SA76D		3006	51	2004	
2SA76E		3006	51	2004	
2SA76F		3006	51	2004	
2SA76G		3006	51	2004	
2SA76H		3006	51	2004	
2SA76K		3006	51	2004	
2SA76L		3006	51	2004	
2SA76OR		3006	51	2004	
2SA76R		3006	51	2004	
2SA76X		3006	51	2004	
2SA76Y		3006	51	2004	
2SA77	126	3007	51	2024	
2SA770	197				2N6109
2SA771	332				2N6107
2SA772		3841	84		
2SA773	383	3841/294	48	2027	
2SA774	234			2024	
2SA775	292	3441			TIP30C
2SA775A	292	3441			
2SA775AA	292	3441			
2SA775AB	292	3441			
2SA775AC	292	3441			
2SA775B	292	3441			
2SA775C	292	3441	69		
2SA777	298	3450	272		
2SA777P	298	3450	272		
2SA777Q	298	3450	272		
2SA777R	298	3450	272		
2SA777S	298	3450	272		
2SA778	288	3715	223		
2SA778A	288$	3434/288	223		
2SA778AK	288	3434	223		
2SA778K	288	3715	223		
2SA779	187$	9076/187A	248		2N4918
2SA779K	187$	9076/187A	248		
2SA779KB	187$	9076/187A	248		
2SA779KC	187$	9076/187A	248		
2SA779KD	187$	9076/187A	248		
2SA77A	126	3006/160	51	2024	
2SA77B	126	3006/160	51	2024	
2SA77C	126	3006/160	51	2024	
2SA77D	126		51	2024	
2SA77E		3006	51	2004	
2SA77F		3006	51	2004	
2SA77G		3006	51	2004	
2SA77H		3006	51	2004	
2SA77K		3006	51	2004	
2SA77L		3006	51	2004	
2SA77M		3006	51	2004	
2SA77OR		3006	51	2004	

Industry Standard No.	ECG	SK	GE	RS 276-	MOTOR.
2SA77R		3006	51	2004	
2SA77X		3006	51	2004	
2SA77Y		3006	51	2004	
2SA78	102A	3006/160	52	2024	
2SA780		9076			2N4919
2SA780A		9076	D43C11		
2SA780AK	211$	9076/187A	248		
2SA780AKA	211$	9076/187A	248		
2SA780AKB	211$	9076/187A	248		
2SA780AKC	211$	9076/187A	248		
2SA780AKD	211$	9076/187A	248		
2SA786		3247	65	2032	
2SA786QL		3247	65		
2SA787			65	2050	
2SA789				2024	
2SA78B	102A		52	2024	
2SA78C	102A		52	2024	
2SA78D	102A		52	2024	
2SA79	102A	3123	245	2004	
2SA794	374#	9042/374			MJE253
2SA794A	374#	9042/374			
2SA794AP	374#	9042/374			
2SA794AQ	374#	9042/374			
2SA794AR	374#	9042/374			
2SA794AS	374#	9042/374			
2SA794P	374#	9042/374			
2SA794Q	374#	9042/374			
2SA794R	374#	9042/374			
2SA794S	374#	9042/374			
2SA795	374				MJE253
2SA701FJ			82	2032	
2SA80	126	3006/160	53	2024	
2SA800	395				2N4957
2SA807	88				2N3789
2SA808	88	3173/219			2N3790
2SA808A	88	3173/219			
2SA80A		3008	53	2007	
2SA80B		3008	53	2007	
2SA80D		3008	53	2007	
2SA80E		3008	53	2007	
2SA80F		3008	53	2007	
2SA80G		3008	53	2007	
2SA80H		3008	53	2007	
2SA80K		3008	53	2007	
2SA80L		3008	53	2007	
2SA80M		3008	53	2007	
2SA80OR		3008	53	2007	
2SA80R		3008	53	2007	
2SA80X		3008	53	2007	
2SA80Y		3008	53	2007	
2SA81	160	3008	245	2004	
2SA811			221	2032	
2SA812			221	2032	
2SA813			21	2034	
2SA814	292	3930			TIP30C
2SA815	292	3930			TIP30C
2SA816	398	3083/197			TIP30B
2SA817	298	3450	272		
2SA817A	383	3450/298			
2SA817AY		3450	272		
2SA818					MPSU60
2SA82	126	3008	53	2024	
2SA825	290A	3114/290	65	2050	
2SA825Q	290A		65	2050	
2SA825R	290A		65	2050	
2SA826	290A	3114/290	89	2032	
2SA826P	290A	3114/290	89	2032	
2SA826Q	290A	3114/290	89	2032	
2SA826R	290A		89	2032	
2SA826RY	290A	3114/290	89	2032	
2SA828A		3020	20	2032	
2SA82A		3008	53	2007	
2SA82B		3008	53	2007	
2SA82C		3008	53	2007	
2SA82D		3008	53	2007	
2SA82E		3008	53	2007	
2SA82F		3008	53	2007	
2SA82G		3008	53	2007	
2SA82H		3008	53	2007	
2SA82K		3008	53	2007	
2SA82M		3008	53	2007	
2SA82OR		3008	53	2007	
2SA82R		3008	53	2007	
2SA82X		3008	53	2007	
2SA82Y		3008	53	2007	
2SA83	126	3007	50	2024	
2SA835		3434	223	2026	MPSU60
2SA835H			223	2026	
2SA836	234	3114/290	244	2050	
2SA836C	234	3114/290			
2SA836D	234	3114/290	244	2050	
2SA836E	234	3114/290	244	2050	
2SA836F	234		244	2050	
2SA837	88		263	2043	2N6226
2SA838			334	2004	
2SA838B			338	2032	
2SA838C			334	2032	
2SA839	398				TIP32C
2SA83A		3007	50	2004	
2SA83B		3007	50	2004	
2SA83C		3008	50	2004	
2SA83D		3007	50	2004	
2SA83E		3007	50	2004	
2SA83F		3007	50	2004	
2SA83G		3007	50	2004	
2SA83H		3007	50	2004	
2SA83K		3007	50	2004	
2SA83L		3007	50	2004	
2SA83M		3007	50	2004	
2SA83OR		3007	50	2004	
2SA83R		3007	50	2004	
2SA83X		3007	50	2004	
2SA83Y		3007	50	2004	
2SA84	160	3007	50	2004	
2SA840	398$	3434/288	223		
2SA840JM		3715	223		
2SA841	234	3114/290	82	2032	
2SA841-GR	234		82	2032	
2SA842	234	3114/290			
2SA842-BL	234		65	2050	
2SA842-GR	234		65	2050	
2SA843	398				MJE15031
2SA844	234	3114/290	244	2027	
2SA844C	234	3114/290	244	2027	
2SA844D	234	3114/290	244	2027	
2SA844E	234	3114/290	244	2027	
2SA84A		3007	50	2004	
2SA84B		3007	50	2004	
2SA84C		3007	50	2004	
2SA84D		3007	50	2004	
2SA84E		3007	50	2004	
2SA84F		3007	50	2004	
2SA84G		3007	50	2004	
2SA84H		3007	50	2004	
2SA84K		3007	50	2004	
2SA84L		3007	50	2004	
2SA84M		3007	50	2004	
2SA84OR		3007	50	2004	
2SA84R		3007	50	2004	
2SA84X		3007	50	2004	
2SA84Y		3007	50	2004	
2SA85	126	3007	53	2024	
2SA854	290A	3247/234	65	2022	
2SA854A	290A		65		
2SA854Q	290A	3247/234	65	2022	
2SA85L		3007	53	2007	
2SA86		3123	245	2004	
2SA861		3200	29		MPSU51
2SA866		3191		2024	
2SA87	126	3006/160	51	2024	
2SA871				2024	
2SA872		3932	51		
2SA872A		3932	221		
2SA872AE		3932	221		
2SA872E		3932	221		
2SA872F		3932	221		
2SA877	88				2N5876
2SA878	88				2N6230
2SA879		3434	223		
2SA879P		3434	223		
2SA879Q		3434	223		
2SA87A		3006	51	2004	
2SA87B		3006	51	2004	
2SA87C		3006	51	2004	
2SA87D		3006	51	2004	
2SA87E		3006	51	2004	
2SA87F		3006	51	2004	
2SA87G		3006	51	2004	
2SA87H		3006	51	2004	
2SA87K		3006	51	2004	
2SA87L		3006	51	2004	
2SA87M		3006	51	2004	
2SA87OR		3006	51	2004	
2SA87R		3006	51	2004	
2SA87X		3006	51	2004	
2SA87Y		3006	51	2004	
2SA88	160		245	2004	
2SA880	234		52	2024	
2SA882	281				2N6231
2SA886	374#	9042/374	58	2025	
2SA886NQ	374#	9042/374			
2SA886NR	374#	9042/374			
2SA886P	374#	9042/374			
2SA886Q	374#	9042/374			
2SA886R	374#	9042/374			
2SA886V	374#	9042/374	58	2025	
2SA886VQ	374#	9042/374			
2SA886VR	374#		58	2025	
2SA887		9076			D41E7
2SA888	159		52	2024	
2SA889	159		52	2024	
2SA89	160		245	2004	
2SA890	159		52	2024	
2SA891	159		52	2024	
2SA893	91	3932	221		
2SA893A		3932	221		
2SA893AE	91	3932	221		
2SA893AF	91		221		
2SA897					MPSU55
2SA898	374				MJE350
2SA899	374	3930			MJE350
2SA90		3006	52	2024	
2SA900	374#				MJE210
2SA907	180				MJ15016
2SA908	285				MJ15002
2SA909	285$				MJ15023
2SA912	288	3715	221		
2SA912Q	288	3715	221		
2SA912R	288	3715	221		
2SA912S	288	3715	211		
2SA915		3867	221		
2SA916		3867	221		
2SA916K		3867	221		
2SA916L		3867	221		
2SA916M		3867	221		
2SA917	383		221		
2SA918				2032	
2SA92	126	3007	50	2024	
2SA921AS		3867	221		
2SA922	129				2N4918
2SA92A		3007	50	2004	
2SA92B		3007	50	2004	
2SA92C		3007	50	2004	
2SA92D		3007	50	2004	
2SA92E		3007	50	2004	
2SA92F		3007	50	2004	
2SA92G		3007	50	2004	
2SA92H		3007	50	2004	
2SA92K		3007	50	2004	
2SA92L		3007	50	2004	
2SA92M		3007	50	2004	
2SA92OR		3007	50	2004	
2SA92R		3007	50	2004	
2SA92X		3007	50	2004	
2SA92Y		3007	50	2004	
2SA93	126	3007	50	2024	
2SA939					MJE350
2SA93A		3007	50	2004	
2SA93B		3007	50	2004	
2SA93C		3007	50	2004	
2SA93E		3007	50	2004	
2SA93F		3007	50	2004	
2SA93G		3007	50	2004	
2SA93H		3007	50	2004	
2SA93K		3007	50	2004	
2SA93L		3007	50	2004	
2SA93M		3007	50	2004	
2SA93OR		3007	50	2004	
2SA93R		3007	50	2004	
2SA93X		3007	50	2004	
2SA93Y		3007	50	2004	
2SA94	126	3008	53	2024	
2SA940	398	3930			MJE15031
2SA945			53	2007	
2SA945-O	159		82	2032	
2SA945Y	159	3466	82	2032	
2SA949	383$	3867			MJE15031
2SA949Y		3867	221		
2SA94A		3008	53	2007	
2SA94C		3008	53	2007	
2SA94D		3008	53	2007	
2SA94E		3008	53	2007	
2SA94F		3008	53	2007	
2SA94G		3008	53	2007	
2SA94H		3008	53	2007	
2SA94K		3008	53	2007	
2SA94L		3008	53	2007	
2SA94M		3008	53	2007	

Industry Standard No.	ECG	SK	GE	RS 276-	MOTOR.
2SA940R		3008	53	2007	
2SA94R		3008	53	2007	
2SA94X		3008	53	2007	
2SA94Y		3008	53	2007	
2SA95				2005	
2SA950	290A	3841/294	244	2027	
2SA950-0	290A	3841/294	244	2027	
2SA950-0		3841	244	2027	
2SA950Y		3841	244	2027	
2SA952	290A	3114/290	272	2027	
2SA952L	290A		272		
2SA957		3930			MJE15031
2SA958		3930			MJE15031
2SA96				2005	
2SA962	189	3200	227	2026	MPSU55
2SA963	374				MJE171
2SA965	383$	3715			MJE15029
2SA966	294	3841			TIP32
2SA968	398$	3930			MJE15031
2SA968-0	398$	3930			
2SA968B	398$	3930			
2SA969	398	3624			MJ3238
2SA969Y	398	3624			
2SA97				2005	
2SA970	290A$	3932			
2SA971	285				2N6609
2SA98				2005	
2SA980	281				2N6229
2SA981	281				2N6230
2SA982	281				2N6231
2SA984	290A	3867			
2SA984E	290A	3867			
2SA99				2005	
2SAJ15GN		3005	51	2004	
2SANJ101	159	3466	82	2032	
2SAU/3H			51	2004	
2SAU03H		3006	51	2004	
2SAUUJ		3123	80		
2SB-2C	506		511	1114	
2SB-3783		3004	53	2007	
2SB-3812		3004	53	2007	
2SB-3813		3004	53	2007	
2SB-C731	116	3017B/117	504A	1104	
2SB-F1A	102A	3004	53	2007	
2SB-S131		3105	512		
2SB-S851	166	9075		1152	
2SB100	102	3004	53	2007	
2SB100A		3004	53	2007	
2SB100B		3004	53	2007	
2SB100C		3004	53	2007	
2SB100D		3004	53	2007	
2SB100E		3004	53	2007	
2SB100F		3004	53	2007	
2SB100G		3004	53	2007	
2SB100GN		3004	53	2007	
2SB100H		3004	53	2007	
2SB100J		3004	53	2007	
2SB100K		3004	53	2007	
2SB100L		3004	53	2007	
2SB100M		3004	53	2007	
2SB100OR		3004	53	2007	
2SB100R		3004	53	2007	
2SB100X		3004	53	2007	
2SB100Y		3004	53	2007	
2SB101	102	3004	53	2007	
2SB101A		3004	53	2007	
2SB101B		3004	53	2007	
2SB101C		3004	53	2007	
2SB101D		3004	53	2007	
2SB101E		3004	53	2007	
2SB101F		3004	53	2007	
2SB101G		3004	53	2007	
2SB101GN		3004	53	2007	
2SB101H		3004	53	2007	
2SB101J		3004	53	2007	
2SB101K		3004	53	2007	
2SB101L		3004	53	2007	
2SB101M		3004	53	2007	
2SB101OR		3004	53	2007	
2SB101R		3004	53	2007	
2SB101X		3004	53	2007	
2SB101Y		3004	53	2007	
2SB102		3004	53	2007	
2SB102A		3004	53	2007	
2SB102B		3004	53	2007	
2SB102C		3004	53	2007	
2SB102D		3004	53	2007	
2SB102E		3004	53	2007	
2SB102F		3004	53	2007	
2SB102G		3004	53	2007	
2SB102GN		3004	53	2007	
2SB102H		3004	53	2007	
2SB102J		3004	53	2007	
2SB102K		3004	53	2007	
2SB102L		3004	53	2007	
2SB102M		3004	53	2007	
2SB102OR		3004	53	2007	
2SB102R		3004	53	2007	
2SB102X		3004	53	2007	
2SB102Y		3004	53	2007	
2SB103	102	3004	53	2007	
2SB103A		3004	53	2007	
2SB103B		3004	53	2007	
2SB103C		3004	53	2007	
2SB103D		3004	53	2007	
2SB103E		3004	53	2007	
2SB103F		3004	53	2007	
2SB103G		3004	53	2007	
2SB103GN		3004	53	2007	
2SB103H		3004	53	2007	
2SB103J		3004	53	2007	
2SB103K		3004	53	2007	
2SB103L		3004	53	2007	
2SB103M		3004	53	2007	
2SB103OR		3004	53	2007	
2SB103R		3004	53	2007	
2SB103X		3004	53	2007	
2SB103Y		3004	53	2007	
2SB104		3004	53	2007	
2SB104A		3004	53	2007	
2SB104B		3004	53	2007	
2SB104C		3004	53	2007	
2SB104D		3004	53	2007	
2SB104E		3004	53	2007	
2SB104F		3004	53	2007	
2SB104G		3004	53	2007	
2SB104GN		3004	53	2007	
2SB104H		3004	53	2007	
2SB104J		3004	53	2007	
2SB104K		3004	53	2007	
2SB104L		3004	53	2007	
2SB104M		3004	53	2007	
2SB104OR		3004	53	2007	
2SB104R		3004	53	2007	
2SB104X		3004	53	2007	
2SB104Y		3004	53	2007	
2SB105	158	3123	80	2007	
2SB105A		3123	80	2007	
2SB105B		3123	80	2007	
2SB105C		3123	80	2007	
2SB105D		3123	80	2007	
2SB105E		3123	80	2007	
2SB105F		3123	80	2007	
2SB105G		3123	80	2007	
2SB105GN		3123	80	2007	
2SB105H		3123	80	2007	
2SB105K		3123	80	2007	
2SB105L		3123	80	2007	
2SB105M		3123	80	2007	
2SB105OR		3123	80	2007	
2SB105R		3123	80	2007	
2SB105X		3123	80	2007	
2SB105Y		3123	80	2007	
2SB106			53	2007	
2SB1060B		3041	55		
2SB107	121	3009	25	2006	
2SB107A	121	3009	25	2006	
2SB107B		3009	25		
2SB107C		3009	25		
2SB107D		3009	25		
2SB107E		3009	25		
2SB107F		3009	25		
2SB107G		3009	25		
2SB107GN		3009	25		
2SB107H		3009	25		
2SB107J		3009	25		
2SB107K		3009	25		
2SB107L		3009	25		
2SB107M		3009	25		
2SB107OR		3009	25		
2SB107R		3009	25		
2SB107X		3009	25		
2SB107Y		3009	25		
2SB108		3123	80	2007	
2SB108A		3123	80	2007	
2SB108B		3123	53	2007	
2SB108C		3123	80	2007	
2SB108D		3123	80	2007	
2SB108E		3123	80	2007	
2SB108F		3123	80	2007	
2SB108G		3123	80	2007	
2SB108GN		3123	80	2007	
2SB108H		3123	80	2007	
2SB108J		3123	80	2007	
2SB108K		3123	80	2007	
2SB108L		3123	80	2007	
2SB108M		3123	80	2007	
2SB108OR		3123	80	2007	
2SB108X		3123	80	2007	
2SB108Y		3123	80	2007	
2SB109			53	2007	
2SB110	102A	3003	53	2007	
2SB110A		3004	53	2007	
2SB110B		3004	53	2007	
2SB110C		3004	53	2007	
2SB110D		3004	53	2007	
2SB110E		3004	53	2007	
2SB110F		3004	53	2007	
2SB110G		3004	53	2007	
2SB110GN		3004	53	2007	
2SB110H		3004	53	2007	
2SB110J		3004	53	2007	
2SB110K		3004	53	2007	
2SB110L		3004	53	2007	
2SB110M		3004	53	2007	
2SB110OR		3004	53	2007	
2SB110R		3004	53	2007	
2SB110X		3004	53	2007	
2SB110Y		3004	53	2007	
2SB111	102A	3003	53	2007	
2SB111A		3004	53	2007	
2SB111B		3004	53	2007	
2SB111C		3004	53	2007	
2SB111D		3004	53	2007	
2SB111E		3004	53	2007	
2SB111F		3004	53	2007	
2SB111G		3004	53	2007	
2SB111GN		3004	53	2007	
2SB111H		3004	53	2007	
2SB111J		3004	53	2007	
2SB111K	102A	3004	53	2007	
2SB111L		3004	53	2007	
2SB111M		3004	53	2007	
2SB111OR		3004	53	2007	
2SB111R		3004	53	2007	
2SB111X		3004	53	2007	
2SB111Y		3004	53	2007	
2SB112	102A	3004	53	2007	
2SB112A		3004	53	2007	
2SB112B		3004	53	2007	
2SB112C		3004	53	2007	
2SB112D		3004	53	2007	
2SB112E		3004	53	2007	
2SB112F		3004	53	2007	
2SB112G		3004	53	2007	
2SB112H		3004	53	2007	
2SB112J		3004	53	2007	
2SB112K		3004	53	2007	
2SB112L		3004	53	2007	
2SB112M		3004	53	2007	
2SB112OR		3004	53	2007	
2SB112R		3004	53	2007	
2SB112X		3004	53	2007	
2SB112Y		3004	53	2007	
2SB113	102A	3004	53	2007	
2SB113A		3004	53	2007	
2SB113B		3004	53	2007	
2SB113C		3004	53	2007	
2SB113D		3004	53	2007	
2SB113E		3004	53	2007	
2SB113F		3004	53	2007	
2SB113G		3004	53	2007	
2SB113GN		3004	53	2007	
2SB113H		3004	53	2007	
2SB113J		3004	53	2007	
2SB113K		3004	53	2007	
2SB113L		3004	53	2007	
2SB113M		3004	53	2007	
2SB113OR		3004	53	2007	
2SB113R		3004	53	2007	
2SB113X		3004	53	2007	
2SB113Y		3004	53	2007	
2SB114	102A	3003	53	2007	
2SB114A		3004	53	2007	
2SB114B		3004	53	2007	
2SB114C		3004	53	2007	
2SB114D		3004	53	2007	
2SB114E		3004	53	2007	

Industry Standard No.	ECG	SK	GE	RS 276-	MOTOR.
2SB114F		3004	53	2007	
2SB114G		3004	53	2007	
2SB114GN		3004	53	2007	
2SB114H		3004	53	2007	
2SB114J		3004	53	2007	
2SB114K		3004	53	2007	
2SB114L		3004	53	2007	
2SB114M		3004	53	2007	
2SB114OR		3004	53	2007	
2SB114R		3004	53	2007	
2SB114X		3004	53	2007	
2SB114Y		3004	53	2007	
2SB115	102A	3003	53	2007	
2SB115A		3004	53	2007	
2SB115B		3004	53	2007	
2SB115C		3004	53	2007	
2SB115D		3004	53	2007	
2SB115E		3004	53	2007	
2SB115F		3004	53	2007	
2SB115G		3004	53	2007	
2SB115GN		3004	53	2007	
2SB115H		3004	53	2007	
2SB115J		3004	53	2007	
2SB115K		3004	53	2007	
2SB115L		3004	53	2007	
2SB115M		3004	53	2007	
2SB115OR		3004	53	2007	
2SB115R		3004	53	2007	
2SB115X		3004	53	2007	
2SB115Y		3004	53	2007	
2SB116	102A	3004	53	2007	
2SB116A		3004	53	2007	
2SB116B		3004	53	2007	
2SB116C		3004	53	2007	
2SB116D		3004	53	2007	
2SB116E		3004	53	2007	
2SB116F		3004	53	2007	
2SB116G		3004	53	2007	
2SB116GN		3004	53	2007	
2SB116H		3004	53	2007	
2SB116J		3004	53	2007	
2SB116K		3004	53	2007	
2SB116L		3004	53	2007	
2SB116M		3004	53	2007	
2SB116OR		3004	53	2007	
2SB116R		3004	53	2007	
2SB116X		3004	53	2007	
2SB116Y		3004	53	2007	
2SB117	102A	3004	53	2007	
2SB117A		3004	53	2007	
2SB117B		3004	53	2007	
2SB117C		3004	53	2007	
2SB117D		3004	53	2007	
2SB117E		3004	53	2007	
2SB117F		3004	53	2007	
2SB117G		3004	53	2007	
2SB117GN		3004	53	2007	
2SB117H		3004	53	2007	
2SB117J		3004	53	2007	
2SB117K	102A	3004	53	2007	
2SB117L		3004	53	2007	
2SB117M		3004	53	2007	
2SB117OR		3004	53	2007	
2SB117R		3004	53	2007	
2SB117X		3004	53	2007	
2SB117Y		3004	53	2007	
2SB118			53	2007	
2SB119	121	3009	25	2006	
2SB119A	121	3009	25	2006	
2SB119B		3009	25		
2SB119C		3009	25		
2SB119D		3009	25		
2SB119E		3009	25		
2SB119F		3009	25		
2SB119G		3009	25		
2SB119GN		3009	25		
2SB119H		3009	25		
2SB119J		3009	25		
2SB119K		3009	25		
2SB119L		3009	25		
2SB119M		3009	25		
2SB119OR		3009	25		
2SB119R		3009	25		
2SB119X		3009	25		
2SB119Y		3009	25		
2SB12				2005	
2SB120	102A	3004	53	2004	
2SB120A		3004	53	2004	
2SB120B		3004	53	2004	
2SB120C		3004	53	2004	
2SB120D		3004	53	2004	
2SB120E		3004	53	2004	
2SB120F		3004	53	2004	
2SB120G		3004	53	2004	
2SB120GN		3004	53	2004	
2SB120H		3004	53	2004	
2SB120J		3004	53	2004	
2SB120K		3004	53	2004	
2SB120L		3004	53	2004	
2SB120M		3004	53	2004	
2SB120OR		3004	53	2004	
2SB120R		3004	53	2004	
2SB120X		3004	53	2004	
2SB120Y		3004	53	2004	
2SB122	121	3009	25	2006	
2SB122A		3009	25		
2SB122B		3009	25		
2SB122C		3009	25		
2SB122D		3009	25		
2SB122E		3009	25		
2SB122F		3009	25		
2SB122G		3009	25		
2SB122GN		3009	25		
2SB122H		3009	25		
2SB122J		3009	25		
2SB122K		3009	25		
2SB122L		3009	25		
2SB122M		3009	25		
2SB122OR		3009	25		
2SB122R		3009	25		
2SB122X		3009	25		
2SB122Y		3009	25		
2SB123	104	3009	16	2006	
2SB123A	104	3009	16	2006	
2SB123B		3009	25		
2SB123C		3009	25		
2SB123D		3009	25		
2SB123E		3009	25		
2SB123F		3009	25		
2SB123G		3009	25		
2SB123GN		3009	25		
2SB123H		3009	25		
2SB123J		3009	25		
2SB123K		3009	25		
2SB123L		3009	25		
2SB123M		3009	25		
2SB123OR		3009	25		
2SB123R		3009	25		
2SB123X		3009	25		
2SB123Y		3009	25		
2SB124		3009	25	2006	
2SB124A		3009	25		
2SB124B		3009	25		
2SB124C		3009	25		
2SB124D		3009	25		
2SB124E		3009	25		
2SB124F		3009	25		
2SB124G		3009	25		
2SB124H		3009	25		
2SB124J		3009	25		
2SB124K		3009	25		
2SB124L		3009	25		
2SB124M		3009	25		
2SB124OR		3009	25		
2SB124R		3009	25		
2SB124X		3009	25		
2SB124Y		3009	25		
2SB126	121	3009	25	2006	
2SB126A	121	3009	25	2006	
2SB126B	121	3009	25		
2SB126C	121	3009	25		
2SB126D	121	3009	25		
2SB126E	121	3009	25		
2SB126F	121	3009	25	2006	
2SB126G	121	3009	25		
2SB126GN		3009	25		
2SB126H	121	3009	25		
2SB126J		3009	25		
2SB126K		3009	25		
2SB126L		3009	25		
2SB126M		3009	25		
2SB126OR		3009	25		
2SB126R		3009	25		
2SB126V	121	3009	25	2006	
2SB126X		3009	25		
2SB126Y		3009	25		
2SB127	121	3009	25	2006	
2SB127A	121	3009	25	2006	
2SB127B		3009	25		
2SB127C		3009	25		
2SB127D		3009	25		
2SB127E		3009	25		
2SB127F		3009	25		
2SB127G		3009	25		
2SB127GN		3009	25		
2SB127H		3009	25		
2SB127J		3009	25		
2SB127K		3009	25		
2SB127L		3009	25		
2SB127M	121	3009	25		
2SB127OR		3009	25		
2SB127R		3009	25		
2SB127X		3009	25		
2SB127Y		3009	25		
2SB128	127	3035	25	2006	
2SB128A	127	3009	25	2006	
2SB128B	127	3035	25		
2SB128C	127	3035	25		
2SB128D	127	3035	25		
2SB128E	127	3035	25		
2SB128F	127	3035	25		
2SB128G	127	3035	25		
2SB128GN		3035	25		
2SB128H	127	3035	25		
2SB128J		3035	25		
2SB128K		3035	25		
2SB128L		3035	25		
2SB128M		3035	25		
2SB128OR		3035	25		
2SB128R		3035	25		
2SB128V	127	3035	25	2006	
2SB128X		3035	25		
2SB128Y		3035	25		
2SB129	127	3009	25	2006	
2SB129A		3009	25		
2SB129B		3009	25		
2SB129C		3009	25		
2SB129D		3009	25		
2SB129E		3009	25		
2SB129F		3009	25		
2SB129G		3009	25		
2SB129GN		3009	25		
2SB129H		3009	25		
2SB129J		3009	25		
2SB129K		3009	25		
2SB129L		3009	25		
2SB129M		3009	25		
2SB129OR		3009	25		
2SB129R		3009	25		
2SB129X		3009	25		
2SB129Y		3009	25		
2SB13		3004	53	2007	
2SB130	131	3035	25	2006	
2SB130A	131		25	2006	
2SB130B		3035	25		
2SB130C		3035	25		
2SB130D		3035	25		
2SB130E		3035	25		
2SB130F		3035	25		
2SB130G		3035	25		
2SB130GN		3035	25		
2SB130H		3035	25		
2SB130J		3035	25		
2SB130K		3035	25		
2SB130L		3035	25		
2SB130M		3035	25		
2SB130OR		3035	25		
2SB130R		3035	25		
2SB130X		3035	25		
2SB130Y		3035	25		
2SB131	121	3009	25	2006	
2SB131A	121	3009	25	2006	
2SB131B		3009	25		
2SB131C		3009	25		
2SB131D		3009	25		
2SB131E		3009	25		
2SB131F		3009	25		
2SB131G		3009	25		
2SB131GN		3009	25		
2SB131H		3009	25		
2SB131J		3009	25		
2SB131K		3009	25		
2SB131L		3009	25		
2SB131M		3009	25		
2SB131OR		3009	25		
2SB131R		3009	25		
2SB131X		3009	25		
2SB131Y		3009	25		

Industry Standard No.	ECG	SK	GE	RS 276-	MOTOR.
2SB132	121	3009	25	2006	
2SB132A	121	3014	25	2006	
2SB132B		3009	25		
2SB132C		3009	25		
2SB132D		3009	25		
2SB132E		3009	25		
2SB132F		3009	25		
2SB132G		3009	25		
2SB132GN		3009	25		
2SB132H		3009	25		
2SB132J		3009	25		
2SB132L		3009	25		
2SB132M		3009	25		
2SB132OR		3009	25		
2SB132R		3009	25		
2SB132X		3009	25		
2SB132Y		3009	25		
2SB134	102A	3004	53	2007	
2SB134-D	102A		53	2007	
2SB134-E	102A		53	2007	
2SB134A		3004	53	2007	
2SB134B	102A	3004	53	2007	
2SB134C	102A	3004	53	2007	
2SB134D		3004	53	2007	
2SB134E	102A	3004	53	2007	
2SB134F		3004	53	2007	
2SB134G		3004	53	2007	
2SB134GN		3004	53	2007	
2SB134H		3004	53	2007	
2SB134J		3004	53	2007	
2SB134K		3004	53	2007	
2SB134L		3004	53	2007	
2SB134M		3004	53	2007	
2SB134OR		3004	53	2007	
2SB134R		3004	53	2007	
2SB134X		3004	53	2007	
2SB134Y		3004	53	2007	
2SB135	102A	3004	53	2007	
2SB135(C)			53	2007	
2SB135A	102A	3004	52	2007	
2SB135B	102A	3004	53	2007	
2SB135C	102A	3004	53	2007	
2SB135D	102A	3004	53	2007	
2SB135E	102A	3004	53	2007	
2SB135F	102A	3004	53	2007	
2SB135G		3004	53	2007	
2SB135GN		3004	53	2007	
2SB135H		3004	53	2007	
2SB135J		3004	53	2007	
2SB135K		3004	53	2007	
2SB135L		3004	53	2007	
2SB135M		3004	53	2007	
2SB135OR		3004	53	2007	
2SB135R		3004	53	2007	
2SB135X		3004	53	2007	
2SB135Y		3004	53	2007	
2SB136	102A	3004	53	2007	
2SB136(C)			53	2007	
2SB136-2	102A		53	2007	
2SB136-3	102A		53	2007	
2SB136A	102A	3004	53	2007	
2SB136B	102A	3004	53	2007	
2SB136C	102A	3004	53	2007	
2SB136D	102A	3004	53	2007	
2SB136E		3004	53	2007	
2SB136F		3004	53	2007	
2SB136G		3004	53	2007	
2SB136GN		3004	53	2007	
2SB136H		3004	53	2007	
2SB136J		3004	53	2007	
2SB136K		3004	53	2007	
2SB136L		3004	53	2007	
2SB136M		3004	53	2007	
2SB136OR		3004	53	2007	
2SB136R		3004	53	2007	
2SB136U	102A	3004	53	2007	
2SB136X			53	2007	
2SB137	121	3009	25	2006	
2SB137A		3009	25		
2SB137B		3009	25		
2SB137C		3009	25		
2SB137D		3009	25		
2SB137E		3009	25		
2SB137F		3009	25		
2SB137G		3009	25		
2SB137GN		3009	25		
2SB137H		3009	25		
2SB137K		3009	25		
2SB137L		3009	25		
2SB137OR		3009	25		
2SB137R		3009	25		
2SB137X		3009	25		
2SB137Y		3009	25		
2SB138	121	3009	25	2006	
2SB138A		3014	25		
2SB138B		3014	25		
2SB138C		3009	25		
2SB138D		3009	25		
2SB138E		3009	25		
2SB138G		3009	25		
2SB138GN		3009	25		
2SB138H		3009	25		
2SB138J		3009	25		
2SB138K		3009	25		
2SB138L		3009	25		
2SB138M		3009	25		
2SB138OR		3009	25		
2SB138R		3009	25		
2SB138X		3009	25		
2SB138Y		3009	25		
2SB14				2005	
2SB140	104	3009	25	2006	
2SB140A		3009	25		
2SB140B		3009	25		
2SB140C		3009	25		
2SB140D		3009	25		
2SB140E		3009	25		
2SB140F		3009	25		
2SB140G		3009	25		
2SB140GN		3009	25		
2SB140H		3009	25		
2SB140J		3009	25		
2SB140K		3009	25		
2SB140L		3009	25		
2SB140M		3009	25		
2SB140X		3009	25		
2SB140Y		3009	25		
2SB141	104	3009	16	2006	
2SB142	104	3009	25	2006	
2SB142A		3009	25		
2SB142B	104	3005	25	2006	
2SB142C	104	3005	25	2006	
2SB142D		3009	25		
2SB142E		3009	25		
2SB142F		3009	25		
2SB142G		3009	25		
2SB142GN		3009	25		
2SB142H		3009	25		
2SB142J		3009	25		
2SB142K		3009	25		
2SB142L		3009	25		
2SB142M		3009	25		
2SB142OR		3009	25		
2SB142R		3009	25		
2SB142X		3009	25		
2SB142Y		3009	25		
2SB143	104	3009	25	2006	
2SB143A		3009	25		
2SB143B		3009	25		
2SB143C		3009	25		
2SB143D		3009	25		
2SB143E		3009	25		
2SB143F		3009	25		
2SB143G		3009	25		
2SB143GN		3009	25		
2SB143H		3009	25		
2SB143J		3009	25		
2SB143K		3009	25		
2SB143L		3009	25		
2SB143M		3009	25		
2SB143OR		3009	25		
2SB143P	104	3009	25	2006	
2SB143X		3009	25		
2SB143Y		3009	25		
2SB144	104	3009	25	2006	
2SB144A		3009	25		
2SB144B		3009	25		
2SB144C		3009	25		
2SB144D		3009	25		
2SB144E		3009	25		
2SB144F		3009	25		
2SB144G		3009	25		
2SB144GN		3009	25		
2SB144H		3009	25		
2SB144J		3009	25		
2SB144K		3009	25		
2SB144L		3009	25		
2SB144M		3009	25		
2SB144OR		3009	25		
2SB144P	104	3009	25	2006	
2SB144R		3009	25		
2SB144X		3009	25		
2SB144Y		3009	25		
2SB145	104	3009	25	2006	
2SB145A		3009	25		
2SB145B		3009	25		
2SB145C		3009	25		
2SB145D		3009	25		
2SB145E		3009	25		
2SB145F		3009	25		
2SB145G		3009	25		
2SB145GN		3009	25		
2SB145H		3009	25		
2SB145J		3009	25		
2SB145K		3009	25		
2SB145L		3009	25		
2SB145M		3009	25		
2SB145OR		3009	25		
2SB145R		3009	25		
2SB145X		3009	25		
2SB145Y		3009	25		
2SB146	104	3009	25	2006	
2SB146A		3009	25		
2SB146B		3009	25		
2SB146C		3009	25		
2SB146D		3009	25		
2SB146E		3009	25		
2SB146F		3009	25		
2SB146G		3009	25		
2SB146GN		3009	25		
2SB146H		3009	25		
2SB146J		3009	25		
2SB146K		3009	25		
2SB146L		3009	25		
2SB146M		3009	25		
2SB146OR		3009	25		
2SB146R		3009	25		
2SB146X		3009	25		
2SB146Y		3009	25		
2SB147	104	3009	25	2006	
2SB147A		3009	25		
2SB147B		3009	25		
2SB147C		3009	25		
2SB147D		3009	25		
2SB147E		3009	25		
2SB147F		3009	25		
2SB147G		3009	25		
2SB147H		3009	25		
2SB147J		3009	25		
2SB147K		3009	25		
2SB147L		3009	25		
2SB147M		3009	25		
2SB147OR		3009	25		
2SB147X		3009	25		
2SB147Y		3009	25		
2SB149	104	3009	25	2006	
2SB149A		3009	25		
2SB149B		3009	25		
2SB149C		3009	25		
2SB149D		3009	25		
2SB149E		3009	25		
2SB149F		3009	25		
2SB149G		3009	25		
2SB149GN		3009	25		
2SB149H		3009	25		
2SB149J		3009	25		
2SB149K		3009	25		
2SB149L		3009	25		
2SB149M		3009	25		
2SB149OR		3009	25		
2SB149R		3009	25		
2SB149X		3009	25		
2SB149Y		3009	25		
2SB15			53	2007	
2SB150		3004	53	2007	
2SB150A		3004	53	2007	
2SB150B		3004	53	2007	
2SB150C		3004	53	2007	
2SB150D		3004	53	2007	
2SB150E		3004	53	2007	
2SB150F		3004	53	2007	
2SB150G		3004	53	2007	
2SB150GN		3004	53	2007	
2SB150H		3004	53	2007	
2SB150J		3004	53	2007	
2SB150K		3004	53	2007	
2SB150L		3004	53	2007	
2SB150M		3004	53	2007	

Industry Standard No.	ECG	SK	GE	RS 276-	MOTOR.
2SB1500R		3004	53	2007	
2SB150R		3004	53	2007	
2SB150X		3004	53	2007	
2SB150Y		3004	53	2007	
2SB151		3009	25	2006	
2SB151A		3009	25		
2SB151B		3009	25		
2SB151C		3009	25		
2SB151D		3009	25		
2SB151E		3009	25		
2SB151F		3009	25		
2SB151G		3009	25		
2SB151GN		3009	25		
2SB151H		3009	25		
2SB151J		3009	25		
2SB151K		3009	25		
2SB151L		3009	25		
2SB151M		3009	25		
2SB151OR		3009	25		
2SB151R		3009	25		
2SB151X		3009	25		
2SB151Y		3009	25		
2SB152	127	3009	16	2006	
2SB152A	127	3009	25		
2SB152B		3009	25		
2SB152C		3009	25		
2SB152D		3009	25		
2SB152E		3009	25		
2SB152F		3009	25		
2SB152G		3009	25		
2SB152GN		3009	25		
2SB152H		3009	25		
2SB152J		3009	25		
2SB152K		3009	25		
2SB152L		3009	25		
2SB152M		3009	25		
2SB152OR		3009	25		
2SB152R		3009	25		
2SB152X		3009	25		
2SB152Y		3009	25		
2SB153	102A	3004	53	2007	
2SB153A		3004	53	2007	
2SB153B		3004	53	2007	
2SB153C		3004	53	2007	
2SB153D		3004	53	2007	
2SB153E		3004	53	2007	
2SB153F		3004	53	2007	
2SB153G		3004	53	2007	
2SB153GN		3004	53	2007	
2SB153H		3004	53	2007	
2SB153J		3004	53	2007	
2SB153K		3004	53	2007	
2SB153L		3004	53	2007	
2SB153M		3004	53	2007	
2SB153OR		3004	53	2007	
2SB153R		3004	53	2007	
2SB153X		3004	53	2007	
2SB153Y		3004	53	2007	
2SB154	102A	3004	53	2007	
2SB154A		3004	53	2007	
2SB154B		3004	53	2007	
2SB154C		3004	53	2007	
2SB154D		3004	53	2007	
2SB154E		3004	53	2007	
2SB154F		3004	53	2007	
2SB154G		3004	53	2007	
2SB154GN		3004	53	2007	
2SB154H		3004	53	2007	
2SB154J		3004	53	2007	
2SB154K		3004	53	2007	
2SB154L		3004	53	2007	
2SB154M		3004	53	2007	
2SB154OR		3004	53	2007	
2SB154R		3004	53	2007	
2SB154X		3004	53	2007	
2SB154Y		3004	53	2007	
2SB155	102A	3003	53	2007	
2SB155A	102A	3004	53	2007	
2SB155B	102A	3004	53	2007	
2SB155C		3004	53	2007	
2SB155D		3004	53	2007	
2SB155E		3004	53	2007	
2SB155F		3004	53	2007	
2SB155G		3004	53	2007	
2SB155GN		3004	53	2007	
2SB155H		3004	53	2007	
2SB155J		3004	53	2007	
2SB155K		3004	53	2007	
2SB155L		3004	53	2007	
2SB155M		3004	53	2007	
2SB155OR		3004	53	2007	
2SB155R		3004	53	2007	
2SB155X		3004	53	2007	
2SB155Y		3004	53	2007	
2SB156	102A	3003	53	2007	
2SB156/7825B	102A	3004	53	2007	
2SB156A	102A	3003	53	2007	
2SB156AA	102A	3003	53	2007	
2SB156AB	102A	3003	53	2007	
2SB156AC	102A	3003	53	2007	
2SB156B	102A	3003	53	2007	
2SB156BC	102A	3003			
2SB156BK		3003	53	2007	
2SB156C	102A	3003	53	2007	
2SB156D	102A	3003	53	2007	
2SB156E		3003	53	2007	
2SB156F		3003	53	2007	
2SB156G		3003	53	2007	
2SB156GN		3003	53	2007	
2SB156H		3003	53	2007	
2SB156J		3003	53	2007	
2SB156K		3003	53	2007	
2SB156L		3003	53	2007	
2SB156M		3003	53	2007	
2SB156OR		3003	53	2007	
2SB156P	102A	3003	53	2007	
2SB156R		3003	53	2007	
2SB156X			53	2007	
2SB156Y		3003	53	2007	
2SB157	126	3004/102A	53	2005	
2SB157A		3004	53	2005	
2SB157B		3004	53	2005	
2SB157C		3004	53	2005	
2SB157D		3004	53	2005	
2SB157E		3004	53	2005	
2SB157F		3004	53	2005	
2SB157G		3004	53	2005	
2SB157GN		3004	53	2005	
2SB157H		3004	53	2005	
2SB157J		3004	53	2005	
2SB157K		3004	53	2005	
2SB157L		3004	53	2005	
2SB157M		3004	53	2005	
2SB157OR		3004	53	2005	
2SB157R		3004	53	2005	
2SB157X		3004	53	2005	
2SB157Y		3004	53	2005	
2SB158	126	3004/102A	53	2005	
2SB158A		3004	53	2004	
2SB158B		3004	53	2004	
2SB158C		3004	53	2004	
2SB158D		3004	53	2004	
2SB158E		3004	53	2004	
2SB158F		3004	53	2004	
2SB158G		3004	53	2004	
2SB158GN		3004	53	2004	
2SB158H		3004	53	2004	
2SB158J		3004	53	2004	
2SB158K		3004	53	2004	
2SB158L		3004	53	2004	
2SB158M		3004	53	2004	
2SB158OR		3004	53	2004	
2SB158R		3004	53	2004	
2SB158X		3004	53	2004	
2SB158Y		3004	53	2004	
2SB159	126	3004/102A	53	2005	
2SB159A		3004	53	2004	
2SB159B		3004	53	2004	
2SB159C		3004	53	2004	
2SB159D		3004	53	2004	
2SB159E		3004	53	2004	
2SB159F		3004	53	2004	
2SB159G		3004	53	2004	
2SB159GN		3004	53	2004	
2SB159H		3004	53	2004	
2SB159J		3004	53	2004	
2SB159K		3004	53	2004	
2SB159L		3004	53	2004	
2SB159M		3004	53	2004	
2SB159OR		3004	53	2004	
2SB159R		3004	53	2004	
2SB159X		3004	53	2004	
2SB159Y		3004	53	2004	
2SB160	126	3004/102A	53	2005	
2SB160A		3004	53	2004	
2SB160B		3004	53	2004	
2SB160C		3004	53	2004	
2SB160D		3004	53	2004	
2SB160E		3004	53	2004	
2SB160F		3004	53	2004	
2SB160G		3004	53	2004	
2SB160GN		3004	53	2004	
2SB160H		3004	53	2004	
2SB160J		3004	53	2004	
2SB160K		3004	53	2004	
2SB160L		3004	53	2004	
2SB160M		3004	53	2004	
2SB160OR			53	2004	
2SB160R		3004	53	2004	
2SB160X		3004	53	2004	
2SB160Y		3004	53	2004	
2SB161	102	3004	53	2007	
2SB161A		3004	53	2007	
2SB161B		3004	53	2007	
2SB161C		3004	53	2007	
2SB161D		3004	53	2007	
2SB161E		3004	53	2007	
2SB161F		3004	53	2007	
2SB161G		3004	53	2007	
2SB161GN		3004	53	2007	
2SB161H		3004	53	2007	
2SB161J		3004	53	2007	
2SB161K		3004	53	2007	
2SB161L		3004	53	2007	
2SB161M		3004	53	2007	
2SB161OR		3004	53	2007	
2SB161R		3004	53	2007	
2SB161X		3004	53	2007	
2SB161Y		3004	53	2007	
2SB162		3004	53	2007	
2SB162A		3004	53	2007	
2SB162B		3004	53	2007	
2SB162C		3004	53	2007	
2SB162D		3004	53	2007	
2SB162E		3004	53	2007	
2SB162F		3004	53	2007	
2SB162G		3004	53	2007	
2SB162GN		3004	53	2007	
2SB162H		3004	53	2007	
2SB162J		3004	53	2007	
2SB162K		3004	53	2007	
2SB162L		3004	53	2007	
2SB162M		3004	53	2007	
2SB162OR		3004	53	2007	
2SB162R		3004	53	2007	
2SB162X		3004	53	2007	
2SB162Y		3004	53	2007	
2SB163	102	3004	53	2007	
2SB163A		3004	53	2007	
2SB163B		3004	53	2007	
2SB163C		3004	53	2007	
2SB163D		3004	53	2007	
2SB163E		3004	53	2007	
2SB163F		3004	53	2007	
2SB163G		3004	53	2007	
2SB163GN		3004	53	2007	
2SB163H		3004	53	2007	
2SB163J		3004	53	2007	
2SB163K		3004	53	2007	
2SB163L		3004	53	2007	
2SB163M		3004	53	2007	
2SB163OR		3004	53	2007	
2SB163R		3004	53	2007	
2SB163X		3004	53	2007	
2SB163Y		3004	53	2007	
2SB164		3004	53	2007	
2SB164A		3004	53	2007	
2SB164B		3004	53	2007	
2SB164C		3004	53	2007	
2SB164D		3004	53	2007	
2SB164E		3004	53	2007	
2SB164F		3004	53	2007	
2SB164G		3004	53	2007	
2SB164GN		3004	53	2007	
2SB164H		3004	53	2007	
2SB164J		3004	53	2007	
2SB164L		3004	53	2007	
2SB164M		3004	53	2007	
2SB164OR		3004	53	2007	
2SB164R		3004	53	2007	
2SB164X		3004	53	2007	
2SB164Y		3004	53	2007	
2SB165	102	3004	53	2007	
2SB165A		3004	53	2007	
2SB165B		3004	53	2007	
2SB165C		3004	53	2007	
2SB165D		3004	53	2007	
2SB165E		3004	53	2007	

Industry Standard No.	ECG	SK	GE	RS 276-	MOTOR.
2SB165F		3004	53	2007	
2SB165G		3004	53	2007	
2SB165GN		3004	53	2007	
2SB165H		3004	53	2007	
2SB165J		3004	53	2007	
2SB165K		3004	53	2007	
2SB165M		3004	53	2007	
2SB165OR		3004	53	2007	
2SB165R		3004	53	2007	
2SB165X		3004	53	2007	
2SB165Y		3004	53	2007	
2SB166		3004	53	2007	
2SB166A		3004	53	2007	
2SB166B		3004	53	2007	
2SB166C		3004	53	2007	
2SB166D		3004	53	2007	
2SB166E		3004	53	2007	
2SB166F		3004	53	2007	
2SB166G		3004	53	2007	
2SB166GN		3004	53	2007	
2SB166H		3004	53	2007	
2SB166J		3004	53	2007	
2SB166K		3004	53	2007	
2SB166L		3004	53	2007	
2SB166M		3004	53	2007	
2SB166OR		3004	53	2007	
2SB166R		3004	53	2007	
2SB166X		3004	53	2007	
2SB166Y		3004	53	2007	
2SB167	158	3004/102A	53	2007	
2SB167A		3004	53	2007	
2SB167B		3004	53	2007	
2SB167BK		3004	53	2007	
2SB167C		3004	53	2007	
2SB167D		3004	53	2007	
2SB167E		3004	53	2007	
2SB167F		3004	53	2007	
2SB167G		3004	53	2007	
2SB167GN		3004	53	2007	
2SB167H		3004	53	2007	
2SB167J		3004	53	2007	
2SB167K		3004	53	2007	
2SB167L		3004	53	2007	
2SB167M		3004	53	2007	
2SB167OR		3004	53	2007	
2SB167R		3004	53	2007	
2SB167X		3004	53	2007	
2SB167Y		3004	53	2007	
2SB168	102A	3003	53	2007	
2SB168A		3004	53	2007	
2SB168B		3004	53	2007	
2SB168C		3004	53	2007	
2SB168D		3004	53	2007	
2SB168E		3004	53	2007	
2SB168F		3004	53	2007	
2SB168G		3004	53	2007	
2SB168GN		3004	53	2007	
2SB168H		3004	53	2007	
2SB168J		3004	53	2007	
2SB168K		3004	53	2007	
2SB168L		3004	53	2007	
2SB168M		3004	53	2007	
2SB168OR		3004	53	2007	
2SB168R		3004	53	2007	
2SB168X		3004	53	2007	
2SB168Y		3004	53	2007	
2SB169	102A	3003	53	2007	
2SB169A		3004	53	2007	
2SB169B		3004	53	2007	
2SB169C		3004	53	2007	
2SB169D		3004	53	2007	
2SB169E		3004	53	2007	
2SB169F		3004	53	2007	
2SB169G		3004	53	2007	
2SB169GN		3004	53	2007	
2SB169H		3004	53	2007	
2SB169J		3004	53	2007	
2SB169K		3004	53	2007	
2SB169L		3004	53	2007	
2SB169M		3004	53	2007	
2SB169OR		3004	53	2007	
2SB169R		3004	53	2007	
2SB169X		3004	53	2007	
2SB169Y		3004	53	2007	
2SB16A		3012	4		
2SB16B		3012	4		
2SB16C		3012	4		
2SB16D		3012	4		
2SB16E		3012	4		
2SB16F		3012	4		
2SB16G		3012	4		
2SB16GN		3012	4		
2SB16H		3012	4		
2SB16J		3012	4		
2SB16K		3012	4		
2SB16L		3012	4		
2SB16M		3012	4		
2SB16OR		3012	4		
2SB16R		3012	4		
2SB16X		3012	4		
2SB16Y		3012	4		
2SB170	102A	3004	53	2007	
2SB170A		3004	53	2007	
2SB170B		3004	53	2007	
2SB170C		3004	53	2007	
2SB170D		3004	53	2007	
2SB170E		3004	53	2007	
2SB170F		3004	53	2007	
2SB170G		3004	53	2007	
2SB170GN		3004	53	2007	
2SB170H		3004	53	2007	
2SB170J		3004	53	2007	
2SB170K		3004	53	2007	
2SB170L		3004	53	2007	
2SB170M		3004	53	2007	
2SB170OR			53	2007	
2SB170R		3004	53	2007	
2SB170X		3004	53	2007	
2SB170Y		3004	53	2007	
2SB171	102A	3004	53	2007	
2SB171A	102A	3004	53	2007	
2SB171B	102A	3004	53	2007	
2SB171C		3004	53	2007	
2SB171D		3004	53	2007	
2SB171E		3004	53	2007	
2SB171F		3004	53	2007	
2SB171G		3004	53	2007	
2SB171GN		3004	53	2007	
2SB171H		3004	53	2007	
2SB171J		3004	53	2007	
2SB171K		3004	53	2007	
2SB171L		3004	53	2007	
2SB171M		3004	53	2007	
2SB171OR		3004	53	2007	
2SB171X		3004	53	2007	
2SB171Y		3004	53	2007	
2SB172	102A	3003	53	2007	
2SB172A	102A	3003	53	2007	
2SB172A-1		3003	53	2007	
2SB172A-F		3003	53	2007	
2SB172AF	102A	3003	53	2007	
2SB172AL		3003	53	2007	
2SB172B	102A	3003	53	2007	
2SB172C	102A	3003	53	2007	
2SB172D	102A	3003	53	2007	
2SB172E	102A	3003	53	2007	
2SB172F	102A	3003	53	2007	
2SB172FN		3003	53	2007	
2SB172G		3003	53	2007	
2SB172GN		3003	53	2007	
2SB172H	102A	3003	53	2007	
2SB172J		3003	53	2007	
2SB172K		3003	53	2007	
2SB172L		3003	53	2007	
2SB172M		3003	53	2007	
2SB172OR		3003	53	2007	
2SB172P	102A		53	2007	
2SB172R	102A	3003	53	2007	
2SB172X		3003	53	2007	
2SB172Y		3003	53	2007	
2SB173	102A	3004	53	2007	
2SB173(C)			53	2007	
2SB173A	102A	3004	53	2007	
2SB173B	102A	3004	53	2007	
2SB173BL		3004	53	2007	
2SB173C	102A	3004	53	2007	
2SB173CL		3004	53	2007	
2SB173D		3004	53	2007	
2SB173E		3004	53	2007	
2SB173F		3004	53	2007	
2SB173G		3004	53	2007	
2SB173GN		3004	53	2007	
2SB173H		3004	53	2007	
2SB173J		3004	53	2007	
2SB173K		3004	53	2007	
2SB173L	102A	3004	53	2007	
2SB173M		3004	53	2007	
2SB173OR		3004	53	2007	
2SB173R		3004	53	2007	
2SB173X		3004	53	2007	
2SB173Y		3004	53	2007	
2SB174	102A	3004	53	2007	
2SB174A		3004	53	2007	
2SB174B		3004	53	2007	
2SB174C		3004	53	2007	
2SB174D		3004	53	2007	
2SB174E		3004	53	2007	
2SB174F		3004	53	2007	
2SB174G		3004	53	2007	
2SB174GN		3004	53	2007	
2SB174H		3004	53	2007	
2SB174K		3004	53	2007	
2SB174L		3004	53	2007	
2SB174M		3004	53	2007	
2SB174OR		3004	53	2007	
2SB174R		3004	53	2007	
2SB174X		3004	53	2007	
2SB174Y		3004	53	2007	
2SB175	102A	3004	53	2007	
2SB175(A)			53	2007	
2SB175(B)			53	2007	
2SB175(C)			53	2007	
2SB175A	102A	3004	53	2007	
2SB175B	102A	3004	53	2007	
2SB175B-1		3004	53	2007	
2SB175BL		3004	53	2007	
2SB175C	102A	3004	53	2007	
2SB175CL		3004	53	2007	
2SB175D		3004	53	2007	
2SB175E	102A	3004	53	2007	
2SB175F		3004	53	2007	
2SB175G		3004	53	2007	
2SB175GN		3004	53	2007	
2SB175H		3004	53	2007	
2SB175L		3004	53	2007	
2SB175M		3004	53	2007	
2SB175OR		3004	53	2007	
2SB175R		3004	53	2007	
2SB175X		3004	53	2007	
2SB175Y		3004	53	2007	
2SB176	102A	3003	53	2007	
2SB176-0	102A		53	2007	
2SB176-P	102A		53	2007	
2SB176-PR	102A		53	2007	
2SB176A		3003	53	2007	
2SB176B	102A	3003	53	2007	
2SB176C		3003	53	2007	
2SB176D		3003	53	2007	
2SB176E		3003	53	2007	
2SB176F		3003	53	2007	
2SB176G		3003	53	2007	
2SB176GN		3003	53	2007	
2SB176H		3003	53	2007	
2SB176J		3003	53	2007	
2SB176K		3003	53	2007	
2SB176L		3003	53	2007	
2SB176M	102A	3003	53	2007	
2SB176O		3003	52	2007	
2SB176OR		3003	53	2007	
2SB176P	102A	3003	53	2007	
2SB176PL		3003	53	2007	
2SB176PR		3003	52	2007	
2SB176PRC	102A	3003	53	2007	
2SB176R	102A	3003	53	2007	
2SB176R(1)		3003	53	2007	
2SB176RG		3003	53	2007	
2SB176X		3003	53	2007	
2SB176Y		3003	53	2007	
2SB177	102A	3004	53	2007	
2SB177A	102A	3004	53	2007	
2SB177B		3004	53	2007	
2SB177C		3004	53	2007	
2SB177D	102A	3004	53	2007	
2SB177E	102A	3004	53	2007	
2SB177F		3004	53	2007	
2SB177G		3004	53	2007	
2SB177GN		3004	53	2007	
2SB177H		3004	53	2007	
2SB177K		3004	53	2007	
2SB177L		3004	53	2007	
2SB177M	102A	3004	53	2007	
2SB177OR		3004	53	2007	
2SB177R	102A	3004	53	2007	
2SB177X		3004	53	2007	
2SB177Y		3004	53	2007	
2SB178	102A	3004	53	2007	
2SB178-0	102A		53	2007	
2SB178-0R			53	2007	

Industry Standard No.	ECG	SK	GE	RS 276-	MOTOR.
2SB178-S	102A		53	2007	
2SB1780		3004	53	2007	
2SB1780A		3004	53	2007	
2SB1780B		3004	53	2007	
2SB1780C		3004	53	2007	
2SB1780D		3004	53	2007	
2SB1780E		3004	53	2007	
2SB1780F		3004	53	2007	
2SB1780G		3004	53	2007	
2SB1780GN		3004	53	2007	
2SB1780H		3004	53	2007	
2SB1780J		3004	53	2007	
2SB1780K		3004	53	2007	
2SB1780L		3004	53	2007	
2SB1780M		3004	53	2007	
2SB17800R			53	2007	
2SB1780R		3004	53	2007	
2SB1780X		3004	53	2007	
2SB1780Y		3004	53	2007	
2SB178A	102A	3004	53	2007	
2SB178B		3004	53	2007	
2SB178BU				2007	
2SB178C	102A	3004	53	2007	
2SB178D	102A	3004	53	2007	
2SB178E		3004	53	2007	
2SB178F		3004	53	2007	
2SB178G		3004	53	2007	
2SB178GN		3004	53	2007	
2SB178H		3004	53	2007	
2SB178J		3004	53	2007	
2SB178K		3004	53	2007	
2SB178L		3004	53	2007	
2SB178M	102A	3004	53	2007	
2SB178N	102A	3004	53	2007	
2SB1780R		3004	53	2007	
2SB178R		3004	53	2007	
2SB178S	102A	3004	53	2007	
2SB178T	102A	3004	53	2007	
2SB178TC		3004	53	2007	
2SB178TS		3004	53	2007	
2SB178U	102A	3004	53		
2SB178V	102A	3004	53	2007	
2SB178X	102A		53	2007	
2SB178Y	102A	3004	53	2007	
2SB179			50	2004	
2SB17A		3012	4		
2SB17B		3012	4		
2SB17C		3012	4		
2SB17D		3012	4		
2SB17E		3012	4		
2SB17F		3012	4		
2SB17G		3012	4		
2SB17GN		3012	4		
2SB17H		3012	4		
2SB17J		3012	4		
2SB17K		3012	4		
2SB17L		3012	4		
2SB17M		3012	4		
2SB170R		3012	4		
2SB17R		3012	4		
2SB17X		3012	4		
2SB17Y		3012	4		
2SB180			53	2007	
2SB180A			53	2007	
2SB181			53	2007	
2SB181A			53	2007	
2SB182			53	2004	
2SB183	102A	3004	53	2007	
2SB183A		3004	53	2007	
2SB183B		3004	53	2007	
2SB183C		3004	53	2007	
2SB183D		3004	53	2007	
2SB183E		3004	53	2007	
2SB183F		3004	53	2007	
2SB183G		3004	53	2007	
2SB183GN		3004	53	2007	
2SB183H		3004	53	2007	
2SB183J		3004	53	2007	
2SB183K		3004	53	2007	
2SB183L		3004	53	2007	
2SB183M		3004	53	2007	
2SB1830R		3004	53	2007	
2SB183R		3004	53	2007	
2SB183X		3004	53	2007	
2SB183Y		3004	53	2007	
2SB184	102A	3004	53	2007	
2SB184A		3004	53	2007	
2SB184B		3004	53	2007	
2SB184C		3004	53	2007	
2SB184D		3004	53	2007	
2SB184E		3004	53	2007	
2SB184F		3004	53	2007	
2SB184G		3004	53	2007	
2SB184GN		3004	53	2007	
2SB184H		3004	53	2007	
2SB184J		3004	53	2007	
2SB184K		3004	53	2007	
2SB184L		3004	53	2007	
2SB184M		3004	53	2007	
2SB1840R		3004	53	2007	
2SB184R		3004	53	2007	
2SB184X		3004	53	2007	
2SB184Y		3004	53	2007	
2SB185	102A	3004	53	2007	
2SB185(0)	102A		53	2007	
2SB185(0)			53	2007	
2SB185(P)			53	2007	
00002SB185-0	102A	3004	53	2007	
2SB185-0			52	2024	
002SB18500	102A	3004	53	2007	
2SB1851		3004	53		
2SB185A		3004	53	2007	
002SB185AA	102A	3004	53	2007	
2SB185B			53	2007	
2SB185C			53	2007	
2SB185D			53	2007	
2SB185E			53	2007	
2SB185F	102A	3004	53	2007	
2SB185G			53	2007	
2SB185GN			53	2007	
2SB185H			53	2007	
2SB185I			53	2007	
2SB185J			53	2007	
2SB185L			53	2007	
2SB185M			53	2007	
2SB1850R			53	2007	
2SB185P	102A	3004	53	2007	
2SB185R			53	2007	
2SB185X			53	2007	
2SB185Y			53	2007	
00002SB186	102A	3004	53	2007	
2SB186(0)	102A		53	2007	
2SB186(0)			53	2007	
0002SB186-0	102A	3004	53	2007	
2SB186-1	102A		53	2007	
2SB186-7		3004	53	2007	
2SB186-K	102A		53	2007	
2SB186-0			53	2007	
2SB186-0R			53	2007	
0002SB1860		3004	53	2007	
2SB186A	102A	3004	53	2007	
2SB186AG	102A	3004	53	2007	
2SB186B	102A	3004	53	2007	
2SB186BY	102A	3004	53	2007	
2SB186C	102A	3004	53	2007	
2SB186D		3004	53	2007	
2SB186E		3004	53	2007	
2SB186F		3004	53	2007	
2SB186G	102A	3004	53	2007	
2SB186GN		3004	53	2007	
2SB186H	102A	3004	53	2007	
2SB186J		3004	53	2007	
2SB186K		3004	53	2007	
2SB186L	102A	3004	53	2007	
2SB186M		3004	53	2007	
2SB1860	102A		53	2007	
2SB1860R		3004	53	2007	
2SB186R		3004	53	2007	
2SB186X		3004	53	2007	
2SB186Y	102A	3004	53	2007	
00002SB187	102A	3004	53	2007	
2SB187(1)	102A		53	2007	
2SB187(K)			53	2007	
0002SB187(RED)	102A		53	2007	
2SB187-1			53	2007	
2SB187-0R			53	2007	
2SB187A		3004	53	2007	
2SB187AA	102A	3004	53	2007	
2SB187B	102A	3004	53	2007	
2SB187BK		3004	53	2007	
2SB187C	102A	3004	53	2007	
2SB187D	102A	3004	53	2007	
2SB187E		3004	53	2007	
2SB187F		3004	53	2007	
2SB187G	102A	3004	53	2007	
2SB187GN		3004	53	2007	
2SB187H	102A	3004	53	2007	
2SB187K	102A	3004	53	2007	
2SB187L		3004	53	2007	
2SB187M		3004	53	2007	
2SB1870R		3004	53	2007	
2SB187R	102A	3004	53	2007	
2SB187RED	102A		53	2007	
2SB187S	102A	3004	53	2007	
2SB187TV			53	2007	
2SB187X		3004	53	2007	
2SB187Y	102A	3004	53	2007	
2SB187YEL	102A		53	2007	
2SB188	102A	3004	53	2007	
2SB188A		3004	53	2007	
2SB188B		3004	53	2007	
2SB188C		3004	53	2007	
2SB188D		3004	53	2007	
2SB188E		3004	53	2007	
2SB188F		3004	53	2007	
2SB188G		3004	53	2007	
2SB188GN		3004	53	2007	
2SB188H		3004	53	2007	
2SB188J		3004	53	2007	
2SB188K		3004	53	2007	
2SB188L		3004	53	2007	
2SB188M		3004	53	2007	
2SB1880R		3004	53	2007	
2SB188R		3004	53	2007	
2SB188X		3004	53	2007	
2SB188Y		3004	53	2007	
2SB189		3004	53	2007	
2SB189A		3004	53	2007	
2SB189B		3004	53	2007	
2SB189C		3004	53	2007	
2SB189D		3004	53	2007	
2SB189E		3004	53	2007	
2SB189F		3004	53	2007	
2SB189G		3004	53	2007	
2SB189GN		3004	53	2007	
2SB189H		3004	53	2007	
2SB189J		3004	53	2007	
2SB189K		3004	53	2007	
2SB189L		3004	53	2007	
2SB189M		3004	53	2007	
2SB1890R		3004	53	2007	
2SB189R		3004	53	2007	
2SB189X		3004	53	2007	
2SB189Y		3004	53	2007	
2SB18A		3012	4		
2SB18B		3012	4		
2SB18C		3012	4		
2SB18D		3012	4		
2SB18E		3012	4		
2SB18F		3012	4		
2SB18G		3012	4		
2SB18GN		3012	4		
2SB18H		3012	4		
2SB18J		3012	4		
2SB18K		3012	4		
2SB18L		3012	4		
2SB18M		3012	4		
2SB180R		3012	4		
2SB18R		3012	4		
2SB18X		3012	4		
2SB18Y		3012	4		
2SB19		3009	25	2006	
2SB190			53	2007	
2SB192			53	2007	
2SB193			53	2007	
2SB194			53	2007	
2SB195			53	2007	
2SB196			53	2007	
2SB197			53	2007	
2SB198			53	2007	
2SB199	102	3004	53	2007	
2SB199A		3004	53	2007	
2SB199B		3004	53	2007	
2SB199C		3004	53	2007	
2SB199D		3004	53	2007	
2SB199E		3004	53	2007	
2SB199F		3004	53	2007	
2SB199G		3004	53	2007	
2SB199GN		3004	53	2007	
2SB199H		3004	53	2007	
2SB199J		3004	53	2007	
2SB199L		3004	53	2007	
2SB199M		3004	53	2007	
2SB1990R		3004	53	2007	
2SB199R		3004	53	2007	
2SB199X		3004	53	2007	
2SB199Y		3004	53	2007	
2SB19A		3009	25		

Industry Standard No.	ECG	SK	GE	RS 276-	MOTOR.
2SB19B		3009	25		
2SB19C		3009	25		
2SB19D		3009	25		
2SB19E		3009	25		
2SB19F		3009	25		
2SB19G		3009	25		
2SB19GN		3009	25		
2SB19H		3009	25		
2SB19J		3009	25		
2SB19K		3009	25		
2SB19L		3009	25		
2SB19M		3009	25		
2SB19OR		3009	25		
2SB19R		3009	25		
2SB19X		3009	25		
2SB19Y		3009	25		
2SB20		3009	25	2006	
2SB200	102A	3004	53	2007	
2SB200A	102A	3004	53	2007	
2SB200B		3004	53	2007	
2SB200C		3004	53	2007	
2SB200D		3004	53	2007	
2SB200E		3004	53	2007	
2SB200F		3004	53	2007	
2SB200G		3004	53	2007	
2SB200GN		3004	53	2007	
2SB200H		3004	53	2007	
2SB200J		3004	53	2007	
2SB200K		3004	53	2007	
2SB200L		3004	53	2007	
2SB200M		3004	53	2007	
2SB200OR		3004	53	2007	
2SB200R		3004	53	2007	
2SB200X		3004	53	2007	
2SB200Y		3004	53	2007	
2SB201	102A		53	2007	
2SB202	102A	3004	53	2007	
2SB202A		3004	53	2007	
2SB202B		3004	53	2007	
2SB202C		3004	53	2007	
2SB202D		3004	53	2007	
2SB202E		3004	53	2007	
2SB202F		3004	53	2007	
2SB202G		3004	53	2007	
2SB202GN		3004	53	2007	
2SB202H		3004	53	2007	
2SB202J		3004	53	2007	
2SB202L		3004	53	2007	
2SB202M		3004	53	2007	
2SB202OR		3004	53	2007	
2SB202R		3004	53	2007	
2SB202X		3004	53	2007	
2SB202Y		3004	53	2007	
2SB203A		3004	53	2007	
2SB203AA	179	3004/102A	53	2007	
2SB203B		3004	53	2007	
2SB203C		3004	53	2007	
2SB203D		3004	53	2007	
2SB203E		3004	53	2007	
2SB203F		3004	53	2007	
2SB203G		3004	53	2007	
2SB203GN		3004	53	2007	
2SB203H		3004	53	2007	
2SB203J		3004	53	2007	
2SB203K		3004	53	2007	
2SB203L		3004	53	2007	
2SB203M		3004	53	2007	
2SB203OR		3004	53	2007	
2SB203R		3004	53	2007	
2SB203X		3004	53	2007	
2SB203Y		3004	53	2007	
2SB204	179		76		
2SB205	179		76		
2SB206	179		76		
2SB20A		3009	25		
2SB20B		3009	25		
2SB20C		3009	25		
2SB20D		3009	25		
2SB20E		3009	25		
2SB20F		3009	25		
2SB20G		3009	25		
2SB20GN		3009	25		
2SB20H		3009	25		
2SB20J		3009	25		
2SB20K		3009	25		
2SB20L		3009	25		
2SB20M		3009	25		
2SB20OR		3009	25		
2SB20R		3009	25		
2SB20X		3009	25		
2SB20Y		3009	25		
2SB21		3009	16	2006	
2SB215	121	3009	25	2006	
2SB215A		3009	25		
2SB215B		3009	25		
2SB215C		3009	25		
2SB215D		3009	25		
2SB215E		3009	25		
2SB215F		3009	25		
2SB215G		3009	25		
2SB215GN		3009	25		
2SB215H		3009	25		
2SB215J		3009	25		
2SB215K		3009	25		
2SB215L		3009	25		
2SB215M		3009	25		
2SB215OR		3009	25		
2SB215R		3009	25		
2SB215X		3009	25		
2SB215Y		3035	25		
2SB216	121	3034	25	2006	
2SB216A	121	3034	25	2006	
2SB216B		3034	25		
2SB216C		3009	25		
2SB216D		3009	25		
2SB216E		3009	25		
2SB216F		3009	25		
2SB216G		3034	25		
2SB216GN		3034	25		
2SB216H		3034	25		
2SB216J		3034	25		
2SB216K		3009	25		
2SB216L		3009	25		
2SB216M		3009	25		
2SB216OR		3009	25		
2SB216R		3034	25		
2SB216X		3034	25		
2SB216Y		3009	25		
2SB217	121	3009	25	2006	
2SB217A	121	3009	25	2006	
2SB217B		3009	25		
2SB217C		3009	25		
2SB217D		3009	25		
2SB217E		3009	25		
2SB217F		3009	25		
2SB217G	121	3009	25	2006	
2SB217GN		3009	25		
2SB217H		3009	25		
2SB217K		3009	25		
2SB217L		3009	25		
2SB217M		3009	25		
2SB217OR		3009	25		
2SB217R		3009	25		
2SB217U	121	3009	25	2006	
2SB217X		3009	25		
2SB217Y		3009	25		
2SB218	102	3004	53	2007	
2SB218A		3004	53	2007	
2SB218B		3004	53	2007	
2SB218C		3004	53	2007	
2SB218D		3004	53	2007	
2SB218E		3004	53	2007	
2SB218F		3004	53	2007	
2SB218G		3004	53	2007	
2SB218GN		3004	53	2007	
2SB218H		3004	53	2007	
2SB218J		3004	53	2007	
2SB218K		3004	53	2007	
2SB218L		3004	53	2007	
2SB218M		3004	53	2007	
2SB218OR		3004	53	2007	
2SB218R		3004	53	2007	
2SB218X		3004	53	2007	
2SB218Y		3004	53	2007	
2SB219	102	3004	53	2007	
2SB219A		3004	53	2007	
2SB219B		3004	53	2007	
2SB219E		3004	53	2007	
2SB219F		3004	53	2007	
2SB219G		3004	53	2007	
2SB219H		3004	53	2007	
2SB219J		3004	53	2007	
2SB219K		3004	53	2007	
2SB219L		3004	53	2007	
2SB219R		3004	53	2007	
2SB219X		3004	53	2007	
2SB219Y		3004	53	2007	
2SB21A		3009	25		
2SB21B		3009	25		
2SB21C		3009	25		
2SB21D		3009	25		
2SB21E		3009	25		
2SB21F		3009	25		
2SB21G		3009	25		
2SB21GN		3009	25		
2SB21H		3009	25		
2SB21J		3009	25		
2SB21K		3009	25		
2SB21L		3009	25		
2SB21M		3009	25		
2SB21OR		3009	25		
2SB21R		3009	25		
2SB21X		3009	25		
2SB21Y		3009	25		
2SB22	102A	3004	53	2007	
2SB22-0	102A		53	2007	
2SB22/09-30100			53	2007	
2SB220	102	3004	53	2007	
2SB220A	102	3004	53	2007	
2SB220B		3004	53	2007	
2SB220C		3004	53	2007	
2SB220D		3004	53	2007	
2SB220E		3004	53	2007	
2SB220F		3004	53	2007	
2SB220GN		3004	53	2007	
2SB220H		3004	53	2007	
2SB220J		3004	53	2007	
2SB220K		3004	53	2007	
2SB220L		3004	53	2007	
2SB220M		3004	53	2007	
2SB220OR			53	2007	
2SB220R		3004	53	2007	
2SB220X		3004	53	2007	
2SB220Y		3004	53	2007	
2SB221	102	3004	53	2007	
2SB221A	102	3004	53	2007	
2SB221B		3004	53	2007	
2SB221C		3004	53	2007	
2SB221D		3004	53	2007	
2SB221E		3004	53	2007	
2SB221F		3004	53	2007	
2SB221G		3004	53	2007	
2SB221GN		3004	53	2007	
2SB221H		3004	53	2007	
2SB221J		3004	53	2007	
2SB221K		3004	53	2007	
2SB221L		3004	53	2007	
2SB221M		3004	53	2007	
2SB221OR		3004	53	2007	
2SB221R		3004	53	2007	
2SB221X		3004	53	2007	
2SB221Y		3004	53	2007	
2SB222	102	3004	53	2007	
2SB222A		3004	53	2007	
2SB222B		3004	53	2007	
2SB222C		3004	53	2007	
2SB222D		3004	53	2007	
2SB222E		3004	53	2007	
2SB222F		3004	53	2007	
2SB222G		3004	53	2007	
2SB222GN		3004	53	2007	
2SB222H		3004	53	2007	
2SB222J		3004	53	2007	
2SB222K		3004	53	2007	
2SB222L		3004	53	2007	
2SB222M		3004	53	2007	
2SB222OR		3004	53	2007	
2SB222R		3004	53	2007	
2SB222X		3004	53	2007	
2SB222Y		3004	53	2007	
2SB223	102	3004	53	2007	
2SB223A		3004	53	2007	
2SB223B		3004	53	2007	
2SB223C		3004	53	2007	
2SB223D		3004	53	2007	
2SB223E		3004	53	2007	
2SB223F		3004	53	2007	
2SB223G		3004	53	2007	
2SB223Y		3004	53	2007	
2SB224	102A	3004	53	2007	
2SB224A		3004	53	2007	
2SB224B		3004	53	2007	
2SB224C		3004	53	2007	
2SB224D		3004	53	2007	
2SB224E		3004	53	2007	
2SB224F		3004	53	2007	
2SB224G		3004	53	2007	
2SB224GN		3004	53	2007	

Industry Standard No.	ECG	SK	GE	RS 276-	MOTOR.
2SB224H		3004	53	2007	
2SB224K		3004	53	2007	
2SB224L		3004	53	2007	
2SB224M		3004	53	2007	
2SB224OR		3004	53	2007	
2SB224R		3004	53	2007	
2SB224X		3004	53	2007	
2SB224Y		3004	53	2007	
2SB225	102	3004	53	2007	
2SB225A		3004	53	2007	
2SB225B		3004	53	2007	
2SB225C		3004	53	2007	
2SB225D		3004	53	2007	
2SB225E		3004	53	2007	
2SB225F		3004	53	2007	
2SB225G		3004	53	2007	
2SB225GN		3004	53	2007	
2SB225H		3004	53	2007	
2SB225K		3004	53	2007	
2SB225L		3004	53	2007	
2SB225M		3004	53	2007	
2SB225OR		3004	53	2007	
2SB225R		3004	53	2007	
2SB225X		3004	53	2007	
2SB225Y		3004	53	2007	
2SB226	102A	3004	53	2007	
2SB226A		3004	53	2007	
2SB226B		3004	53	2007	
2SB226C		3004	53	2007	
2SB226D		3004	53	2007	
2SB226E		3004	53	2007	
2SB226F		3004	53	2007	
2SB226G		3004	53	2007	
2SB226GN		3004	53	2007	
2SB226J		3004	53	2007	
2SB226K		3004	53	2007	
2SB226L		3004	53	2007	
2SB226M		3004	53	2007	
2SB226OR		3004	53	2007	
2SB226R		3004	53	2007	
2SB226X		3004	53	2007	
2SB226Y		3004	53	2007	
2SB227	102A	3004	53	2007	
2SB227A		3004	53	2007	
2SB227B		3004	53	2007	
2SB227C		3004	53	2007	
2SB227D		3004	53	2007	
2SB227E		3004	53		
2SB227F		3004	53	2007	
2SB227G		3004	53	2007	
2SB227GN		3004	53	2007	
2SB227H		3004	53	2007	
2SB227J		3004	53	2007	
2SB227K		3004	53	2007	
2SB227L		3004	53	2007	
2SB227M		3004	53	2007	
2SB227OR		3004	53	2007	
2SB227R		3004	53	2007	
2SB227X		3004	53	2007	
2SB227Y		3004	53	2007	
2SB228	127	3009	25	2006	
2SB228A		3009	25		
2SB228B		3009	25		
2SB228C		3122	25		
2SB228D		3122	25		
2SB228E		3009	25		
2SB228F		3009	25		
2SB228G		3009	25		
2SB228GN		3009	25		
2SB228H		3009	25		
2SB228J		3009	25		
2SB228K		3009	25		
2SB228L		3009	25		
2SB228M		3009	25		
2SB228OR		3009	25		
2SB228R		3009	25		
2SB228X		3009	25		
2SB228Y		3009	25		
2SB229	127	3009	25	2006	
2SB229A		3009	25		
2SB229B		3009	25		
2SB229C		3009	25		
2SB229D		3009	25		
2SB229E		3009	25		
2SB229F		3009	25		
2SB229G		3009	25		
2SB229GN		3009	25		
2SB229J		3009	25		
2SB229K		3009	25		
2SB229L		3009	25		
2SB229M		3009	25		
2SB229OR		3009	25		
2SB229R		3009	25		
2SB229X		3009	25		
2SB229Y		3009	25		
2SB22A	102A		53	2007	
2SB22B	102A		53	2007	
2SB22C			53	2007	
2SB22D			53	2007	
2SB22E			53	2007	
2SB22F			53	2007	
2SB22G			53	2007	
2SB22GN			53	2007	
2SB22H	102A		53	2007	
2SB22I	102A			2007	
2SB22J			53	2007	
2SB22K			53	2007	
2SB22L			53	2007	
2SB22M			53	2007	
2SB22OR			53	2007	
2SB22P			53	2007	
2SB22R	102A		53	2007	
2SB22X			53	2007	
2SB22Y	102A		53	2007	
2SB23	102A	3004	53	2007	
2SB230	127	3009	25	2006	
2SB230A		3009	25		
2SB230B		3009	25		
2SB230C		3009	25		
2SB230D		3009	25		
2SB230E		3009	25		
2SB230F		3009	25		
2SB230G		3009	25		
2SB230GN		3009	25		
2SB230H		3009	25		
2SB230J		3009	25		
2SB230K		3009	25		
2SB230L		3009	25		
2SB230M		3009	25		
2SB230R		3009	25		
2SB230X		3009	25		
2SB230Y		3009	25		
2SB231	127	3035	25		
2SB231A		3035	25		
2SB231B		3035	25		
2SB231C		3035	25		
2SB231D		3035	25		
2SB231E		3035	25		
2SB231F		3035	25		
2SB231G		3035	25		
2SB231GN		3035	25		
2SB231H		3035	25		
2SB231J		3035	25		
2SB231K		3035	25		
2SB231L		3035	25		
2SB231M		3035	25		
2SB231OR		3035	25		
2SB231R		3035	25		
2SB231X		3035	25		
2SB231Y		3035	25		
2SB232	127	3035	25		
2SB232A		3035	25		
2SB232B		3035	25		
2SB232C		3035	25		
2SB232D		3035	25		
2SB232E		3035	25		
2SB232F		3035	25		
2SB232G		3035	25		
2SB232GN		3035	25		
2SB232H		3035	25		
2SB232K		3035	25		
2SB232L		3035	25		
2SB232M		3035	25		
2SB232OR		3035	25		
2SB232R		3035	25		
2SB232X		3035	25		
2SB232Y		3035	25		
2SB233	127	3764	25	2006	
2SB233A		3764	25		
2SB233B		3764	25		
2SB233C		3764	25		
2SB233D		3764	25		
2SB233E		3764	25		
2SB233F		3764	25		
2SB233GN		3764	25		
2SB233J		3004	25		
2SB233K		3764	25		
2SB233L		3764	25		
2SB233M		3764	25		
2SB233OR		3764	25		
2SB233R		3764	25		
2SB233Y		3764	25		
2SB234	127	3035	25		
2SB234A		3035	25		
2SB234B		3035	25		
2SB234C		3035	25		
2SB234D		3035	25		
2SB234E		3035	25		
2SB234F		3035	25		
2SB234G		3035	25		
2SB234GN		3035	25		
2SB234H		3035	25		
2SB234J		3035	25		
2SB234K		3035	25		
2SB234L		3035	25		
2SB234M		3035	25		
2SB234N	127	3035	25		
2SB234OR		3035	25		
2SB234R		3035	25		
2SB234X		3035	25		
2SB234Y		3035	25		
2SB235	213	3012/105	4		
2SB235A	213	3012/105	4		
2SB235B		3012	4		
2SB235C		3012	4		
2SB235D		3012	4		
2SB235E		3012	4		
2SB235F		3012	4		
2SB235G		3012	4		
2SB235GN		3012	4		
2SB235H		3012	4		
2SB235L		3012	4		
2SB235M		3012	4		
2SB235OR		3012	4		
2SB235R		3012	4		
2SB235X		3012	4		
2SB235Y		3012	4		
2SB236	105	3012	4		
2SB236A		3012	4		
2SB236B		3012	4		
2SB236C		3012	4		
2SB236D		3012	4		
2SB236E		3012	4		
2SB236F		3012	4		
2SB236G		3012	4		
2SB236GN		3012	4		
2SB236H		3012	4		
2SB236J		3012	4		
2SB236K		3012	4		
2SB236L		3012	4		
2SB236M		3012	4		
2SB236OR		3012	4		
2SB236R		3012	4		
2SB236X		3012	4		
2SB236Y		3012	4		
2SB237	105	3012	4		
2SB237-12A	105	3012	4		
2SB237-12B	105	3012	4		
2SB237A		3012	4		
2SB237B		3012	4		
2SB237C		3012	4		
2SB237D		3012	4		
2SB237E		3012	4		
2SB237F		3012	4		
2SB237G		3012	4		
2SB237GN		3012	4		
2SB237H		3012	4		
2SB237J		3012	4		
2SB237K		3012	4		
2SB237L		3012	4		
2SB237M		3012	4		
2SB237OR		3012	4		
2SB237R		3012	4		
2SB237X		3012	4		
2SB237Y		3012	4		
2SB238	158	3123	4	2007	
2SB238-12A	158	3012/105	4	2007	
2SB238-12B	158	3012/105	4	2007	
2SB238-12C	158	3012/105	4	2007	
2SB238A		3012	4		
2SB238B		3012	4		
2SB238C		3012	4		
2SB238D		3012	4		
2SB238E		3012	4		
2SB238F		3012	4		
2SB238G		3012	4		
2SB238GN		3012	4		
2SB238H		3012	4		

Industry Standard No.	ECG	SK	GE	RS 276-	MOTOR.
2SB238J		3012	4		
2SB238K		3012	4		
2SB238L		3012	4		
2SB238M		3012	4		
2SB238OR		3012	4		
2SB238R		3012	4		
2SB238X		3012	4		
2SB238Y		3012	4		
2SB239		3009	25	2006	
2SB239A		3009	25	2006	
2SB239B		3009	25		
2SB239C		3009	25		
2SB239D		3009	25		
2SB239E		3009	25		
2SB239F		3009	25		
2SB239G		3009	25		
2SB239GN		3009	25		
2SB239H		3009	25		
2SB239J		3009	25		
2SB239K		3009	25		
2SB239L		3009	25		
2SB239M		3009	25		
2SB239OR		3009	25		
2SB239R		3009	25		
2SB239X		3009	25		
2SB239Y		3009	25		
2SB23A		3004	53	2007	
2SB23B		3004	53	2007	
2SB23C		3004	53	2007	
2SB23D		3004	53	2007	
2SB23E		3004	53	2007	
2SB23F		3004	53	2007	
2SB23G		3004	53	2007	
2SB23GN		3004	53	2007	
2SB23H		3004	53	2007	
2SB23J		3004	53	2007	
2SB23K		3004	53	2007	
2SB23L		3004	53	2007	
2SB23M		3004	53	2007	
2SB23OR		3004	53	2007	
2SB23R		3004	53	2007	
2SB23X		3004	53	2007	
2SB23Y		3004	53	2007	
2SB24	102A	3004	53	2007	
2SB240		3009	25	2004	
2SB240A		3009	25	2004	
2SB240B		3009	25		
2SB240C		3009	25		
2SB240D		3009	25		
2SB240E		3009	25		
2SB240F		3009	25		
2SB240GN		3009	25		
2SB240H		3009	25		
2SB240J		3009	25		
2SB240K		3009	25		
2SB240L		3009	25		
2SB240M		3009	25		
2SB240OR		3009	25		
2SB240R		3009	25		
2SB240X		3009	25		
2SB240Y		3009	25		
2SB241		3004	53	2007	
2SB241A		3004	53	2007	
2SB241B		3004	53	2007	
2SB241C		3004	53	2007	
2SB241D		3004	53	2007	
2SB241E		3004	53	2007	
2SB241F		3004	53	2007	
2SB241G		3004	53	2007	
2SB241GN		3004	53	2007	
2SB241H		3004	53	2007	
2SB241J		3004	53	2007	
2SB241K		3004	53	2007	
2SB241L		3004	53	2007	
2SB241M		3004	53	2007	
2SB241OR		3004	53	2007	
2SB241R		3004	53	2007	
2SB241V		3035	53	2007	
2SB241X		3004	53	2007	
2SB241Y		3004	53	2007	
2SB242		3009	25	2007	
2SB242A		3009	25	2007	
2SB242B		3009	25	2007	
2SB242C		3009	25	2007	
2SB242D		3009	25	2007	
2SB242E		3009	25	2007	
2SB242F		3009	25	2007	
2SB242G		3009	25	2007	
2SB242GN		3009	25	2007	
2SB242H		3009	25	2007	
2SB242J		3009	25	2007	
2SB242K		3009	25	2007	
2SB242L		3009	25	2007	
2SB242M		3009	25	2007	
2SB242OR		3009	25	2007	
2SB242R		3009	25	2007	
2SB242X		3009	25	2007	
2SB242Y		3009	25	2007	
2SB243		3009	25		
2SB243A		3009	25		
2SB243B		3009	25		
2SB243C		3009	25		
2SB243D		3009	25		
2SB243E		3009	25		
2SB243F		3009	25		
2SB243G		3009	25		
2SB243GN		3009	25		
2SB243H		3009	25		
2SB243J		3009	25		
2SB243K		3009	25		
2SB243L		3009	25		
2SB243M		3009	25		
2SB243OR		3009	25		
2SB243R		3009	25		
2SB243X		3009	25		
2SB243Y		3009	25		
2SB244		3009	25		
2SB244A		3009	25		
2SB244B		3009	25		
2SB244C		3009	25		
2SB244D		3009	25		
2SB244E		3009	25		
2SB244F		3009	25		
2SB244G		3009	25		
2SB244GN		3009	25		
2SB244H		3009	25		
2SB244J		3009	25		
2SB244K		3009	25		
2SB244L		3009	25		
2SB244M		3009	25		
2SB244OR		3009	25		
2SB244R		3009	25		
2SB244X		3009	25		
2SB244Y		3009	25		
2SB245		3009	25		
2SB245A		3009	25		
2SB245B		3009	25		
2SB245C		3009	25		
2SB245D		3009	25		
2SB245E		3009	25		
2SB245F		3009	25		
2SB245G		3009	25		
2SB245H		3009	25		
2SB245J		3009	25		
2SB245L		3009	25		
2SB245M		3009	25		
2SB245OR		3009	25		
2SB245R		3009	25		
2SB245X		3009	25		
2SB245Y		3009	25		
2SB246	104	3009	25	2006	
2SB246A		3009	25		
2SB246B		3009	25		
2SB246C		3009	25		
2SB246D		3009	25		
2SB246E		3009	25		
2SB246F		3009	25		
2SB246G		3009	25		
2SB246GN		3009	25		
2SB246H		3009	25		
2SB246J		3009	25		
2SB246K		3009	25		
2SB246L		3009	25		
2SB246M		3009	25		
2SB246OR		3009	25		
2SB246R		3009	25		
2SB246X		3009	25		
2SB246Y		3009	25		
2SB247	104	3009	25	2006	
2SB247A		3009	25		
2SB247B		3009	25		
2SB247C		3009	25		
2SB247D		3009	25		
2SB247E		3009	25		
2SB247F		3009	25		
2SB247G		3009	25		
2SB247GN		3009	25		
2SB247H		3009	25		
2SB247J		3009	25		
2SB247K		3009	25		
2SB247L		3009	25		
2SB247M		3009	25		
2SB247OR		3009	25		
2SB247R		3009	25		
2SB247X		3009	25		
2SB247Y		3009	25		
2SB248	104	3009	25	2006	
2SB248A	104	3009	25	2006	
2SB248B		3009	25		
2SB248C		3009	25		
2SB248D		3009	25		
2SB248E		3009	25		
2SB248F		3009	25		
2SB248G		3009	25		
2SB248GN		3009	25		
2SB248H		3009	25		
2SB248J		3009	25		
2SB248K		3009	25		
2SB248L		3009	25		
2SB248M		3009	25		
2SB248OR		3009	25		
2SB248R		3009	25		
2SB248X		3009	25		
2SB248Y		3009	25		
2SB249	179	3009	25	2006	
2SB249A	179	3009	16	2006	
2SB249B		3009	25		
2SB249C		3009	25		
2SB249D		3009	25		
2SB249E		3009	25		
2SB249F		3009	25		
2SB249G		3009	25		
2SB249GN		3009	25		
2SB249H		3009	25		
2SB249J		3009	25		
2SB249K		3009	25		
2SB249L		3009	25		
2SB249M		3009	25		
2SB249OR		3009	25		
2SB249X		3009	25		
2SB249Y		3009	25		
2SB24A		3004	53	2007	
2SB24B		3004	53	2007	
2SB24C		3004	53	2007	
2SB24D		3004	53	2007	
2SB24E		3004	53	2007	
2SB24F		3004	53	2007	
2SB24G		3004	53	2007	
2SB24GN		3004	53	2007	
2SB24H		3004	53	2007	
2SB24J		3004	53	2007	
2SB24K		3004	53	2007	
2SB24L		3004	53	2007	
2SB24M		3004	53	2007	
2SB24OR		3004	53	2007	
2SB24R		3004	53	2007	
2SB24X		3004	53	2007	
2SB24Y		3004	53	2007	
2SB25	121	3009	25	2006	
2SB250	104	3009	25	2006	
2SB250A	104	3009	25	2006	
2SB250B		3009	25		
2SB250D		3009	25		
2SB250E		3009	25		
2SB250F		3009	25		
2SB250G		3009	25		
2SB250GN		3009	25		
2SB250H		3009	25		
2SB250K		3009	25		
2SB250L		3009	25		
2SB250M		3009	25		
2SB250X		3009	25		
2SB250Y		3009	25		
2SB251	127	3035	25		
2SB251A	127	3035	25		
2SB251B		3035	25		
2SB251C		3035	25		
2SB251D		3035	25		
2SB251E		3035	25		
2SB251F		3035	25		
2SB251G		3035	25		
2SB251GN		3035	25		
2SB251H		3035	25		
2SB251J		3035	25		
2SB251K		3035	25		
2SB251L		3035	25		
2SB251M		3035	25		
2SB251OR		3035	25		

Industry Standard No.	ECG	SK	GE	RS 276-	MOTOR.
2SB251R		3035	25		
2SB251X		3035	25		
2SB252	127	3009	25		
2SB252A	127	3035	25		
2SB252B		3009	25		
2SB252C		3009	25		
2SB252D		3009	25		
2SB252E		3009	25		
2SB252F		3009	25		
2SB252G		3009	25		
2SB252GN		3009	25		
2SB252H		3009	25		
2SB252J		3009	25		
2SB252L		3009	25		
2SB252M		3009	25		
2SB252OR		3009	25		
2SB252R		3009	25		
2SB252X		3009	25		
2SB252Y		3009	25		
2SB253	127	3035	25		
2SB253A	127	3009	25		
2SB253B		3035	25		
2SB253C		3035	25		
2SB253D		3035	25		
2SB253E		3035	25		
2SB253F		3035	25		
2SB253G		3035	25		
2SB253GN		3035	25		
2SB253H		3035	25		
2SB253K		3035	25		
2SB253L		3035	25		
2SB253M		3035	25		
2SB253OR		3035	25		
2SB253X		3035	25		
2SB253Y		3035	25		
2SB254	226	3009	25	2006	
2SB254A		3009	25		
2SB254B		3009	25		
2SB254C		3009	25		
2SB254D		3004	25		
2SB254E		3009	25		
2SB254F		3009	25		
2SB254G		3009	25		
2SB254GN		3009	25		
2SB254H		3009	25		
2SB254J		3009	25		
2SB254K		3009	25		
2SB254L		3009	25		
2SB254M		3009	25		
2SB254OR		3009	25		
2SB254R		3009	25		
2SB254X		3009	25		
2SB254Y		3009	25		
2SB255	226	3009	25	2006	
2SB255A		3009	25		
2SB255B		3009	25		
2SB255C		3009	25		
2SB255D		3009	25		
2SB255E		3009	25		
2SB255F		3009	25		
2SB255G		3009	25		
2SB255GN		3009	25		
2SB255H		3009	25		
2SB255J		3009	25		
2SB255K		3009	25		
2SB255L		3009	25		
2SB255M		3009	25		
2SB255OR		3009	25		
2SB255R		3009	25		
2SB255X		3009	25		
2SB255Y		3009	25		
2SB256	226	3052	30	2006	
2SB256A		3052	30		
2SB256B		3052	30		
2SB256C		3052	30		
2SB256D		3052	30		
2SB256E		3052	30		
2SB256F		3052	30		
2SB256G		3052	30		
2SB256GN		3052	30		
2SB256H		3052	30		
2SB256J		3052	30		
2SB256L		3052	30		
2SB256M		3052	30		
2SB256OR		3052	30		
2SB256R		3052	30		
2SB256X		3052	30		
2SB256Y		3052	30		
2SB257	102A	3004	53	2007	
2SB257A		3004	53	2007	
2SB257B		3004	53	2007	
2SB257C		3004	53	2007	
2SB257D		3004	53	2007	
2SB257E		3004	53	2007	
2SB257F		3004	53	2007	
2SB257G		3004	53	2007	
2SB257GN		3004	53	2007	
2SB257H		3004	53	2007	
2SB257K		3004	53	2007	
2SB257L		3004	53	2007	
2SB257M		3004	53	2007	
2SB257OR		3004	53	2007	
2SB257R		3004	53	2007	
2SB257X		3004	53	2007	
2SB257Y		3004	53	2007	
2SB258	2138	3012/105	4		
2SB258A		3012	4	2007	
2SB258B		3012	4	2007	
2SB258C		3012	4	2007	
2SB258D		3012	4	2007	
2SB258E		3012	4	2007	
2SB258F		3012	4	2007	
2SB258G		3012	4	2007	
2SB258GN		3012	4	2007	
2SB258H		3012	4	2007	
2SB258J		3012	4	2007	
2SB258K		3012	4	2007	
2SB258L		3012	4	2007	
2SB258M		3012	4	2007	
2SB258OR		3012	4	2007	
2SB258R		3012	4	2007	
2SB258X		3012	4	2007	
2SB258Y		3012	4	2007	
2SB259	213	3012/105	4		
2SB259A		3012	4		
2SB259B		3012	4		
2SB259C		3012	4		
2SB259D		3012	4		
2SB259E		3012	4		
2SB259F		3012	4		
2SB259GN		3012	4		
2SB259H		3012	4		
2SB259J		3012	4		
2SB259K		3012	4		
2SB259L		3012	4		
2SB259OR		3012	4		
2SB259R		3012	4		
2SB259X		3012	4		
2SB259Y		3012	4		
2SB25A		3009	25		
2SB25B	121	3006/160	25	2006	
2SB25C		3009	25		
2SB25D		3009	25		
2SB25E		3009	25		
2SB25F		3009	25		
2SB25G		3009	25		
2SB25GN		3009	25		
2SB25H		3009	25		
2SB25J		3009	25		
2SB25K		3009	25		
2SB25L		3009	25		
2SB25M		3009	25		
2SB25OR		3009	25		
2SB25R		3009	25		
2SB25X		3009	25		
2SB25Y		3009	25		
2SB26	104	3009	25	2006	
2SB260	213	3012/105	4		
2SB260A		3012	4		
2SB260B		3012	4		
2SB260C		3012	4		
2SB260D		3012	4		
2SB260E		3012	4		
2SB260F		3012	4		
2SB260G		3012	4		
2SB260GN		3012	4		
2SB260H		3012	4		
2SB260J		3012	4		
2SB260K		3012	4		
2SB260L		3012	4		
2SB260M		3012	4		
2SB260OR		3012	4		
2SB260R		3012	4		
2SB260X		3012	4		
2SB260Y		3012	4		
2SB261	102A	3004	53	2007	
2SB261A		3004	53	2007	
2SB261B		3004	53	2007	
2SB261C		3004	53	2007	
2SB261D		3004	53	2007	
2SB261E		3004	53	2007	
2SB261F		3004	53	2007	
2SB261G		3004	53	2007	
2SB261GN		3004	53	2007	
2SB261H		3004	53	2007	
2SB261J		3004	53	2007	
2SB261K		3004	53	2007	
2SB261L		3004	53	2007	
2SB261M		3004	53	2007	
2SB261OR		3004	53	2007	
2SB261R		3004	53	2007	
2SB261X		3004	53	2007	
2SB261Y		3004	53	2007	
2SB262	102A	3004	53	2007	
2SB262A		3004	53	2007	
2SB262B		3004	53	2007	
2SB262C		3004	53	2007	
2SB262D		3004	53	2007	
2SB262E		3004	53	2007	
2SB262F		3004	53	2007	
2SB262G		3004	53	2007	
2SB262GN		3004	53	2007	
2SB262H		3004	53	2007	
2SB262J		3004	53	2007	
2SB262K		3004	53	2007	
2SB262L		3004	53	2007	
2SB262M		3004	53	2007	
2SB262OR		3004	53	2007	
2SB262R		3004	53	2007	
2SB262X		3004	53	2007	
2SB262Y		3004	53	2007	
2SB263	102A	3004	53	2007	
2SB263A		3004	53	2007	
2SB263B		3004	53	2007	
2SB263C		3004	53	2007	
2SB263D		3004	53	2007	
2SB263E		3004	53	2007	
2SB263F		3004	53	2007	
2SB263G		3004	53	2007	
2SB263GN		3004	53	2007	
2SB263H		3004	53	2007	
2SB263J		3004	53	2007	
2SB263K		3004	53	2007	
2SB263L		3004	53	2007	
2SB263M		3004	53	2007	
2SB263OR		3004	53	2007	
2SB263R		3004	53	2007	
2SB263X		3004	53	2007	
2SB263Y		3004	53	2007	
2SB264	102A	3004	53	2007	
2SB264A		3004	53	2007	
2SB264B		3004	53	2007	
2SB264C		3004	53	2007	
2SB264D		3004	53	2007	
2SB264E		3004	53	2007	
2SB264F		3004	53	2007	
2SB264G		3004	53	2007	
2SB264GN		3004	53	2007	
2SB264H		3004	53	2007	
2SB264J		3004	53	2007	
2SB264K		3004	53	2007	
2SB264L		3004	53	2007	
2SB264M		3004	53	2007	
2SB264OR		3004	53	2007	
2SB264R		3004	53	2007	
2SB264X		3004	53	2007	
2SB264Y		3004	53	2007	
2SB265	102A	3004	53	2007	
2SB265A		3004	53	2007	
2SB265B		3004	53	2007	
2SB265C		3004	53	2007	
2SB265D		3004	53	2007	
2SB265E		3004	53	2007	
2SB265F		3004	53	2007	
2SB265G		3004	53	2007	
2SB265GN		3004	53	2007	
2SB265H		3004	53	2007	
2SB265J		3004	53	2007	
2SB265K		3004	53	2007	
2SB265L		3004	53	2007	
2SB265M		3004	53	2007	
2SB265OR		3004	53	2007	
2SB265X		3004	53	2007	
2SB265Y		3004	53	2007	
2SB266	102A	3004	53	2007	
2SB266A		3004	53	2007	
2SB266B		3004	53	2007	
2SB266C		3004	53	2007	

Industry Standard No.	ECG	SK	GE	RS 276-	MOTOR.
2SB266D		3004	53	2007	
2SB266E		3004	53	2007	
2SB266F		3004	53	2007	
2SB266G		3004	53	2007	
2SB266GN		3004	53	2007	
2SB266H		3004	53	2007	
2SB266J		3004	53	2007	
2SB266K		3004	53	2007	
2SB266L		3004	53	2007	
2SB266M		3004	53	2007	
2SB266OR		3004	53	2007	
2SB266P	102A	3004	53	2007	
2SB266Q	102A	3004	53	2007	
2SB266R		3004	53	2007	
2SB266X		3004	53	2007	
2SB266Y		3004	53	2007	
2SB267	102A	3004	53	2005	
2SB267A		3004	53	2007	
2SB267B		3004	53	2007	
2SB267C		3004	53	2007	
2SB267D		3004	53	2007	
2SB267E		3004	53	2007	
2SB267F		3004	53	2007	
2SB267G		3004	53	2007	
2SB267GN		3004	53	2007	
2SB267H		3004	53	2007	
2SB267J		3004	53	2007	
2SB267K		3004	53	2007	
2SB267L		3004	53	2007	
2SB267M		3004	53	2007	
2SB267OR		3004	53	2007	
2SB267R		3004	53	2007	
2SB267X		3004	53	2007	
2SB267Y		3004	53	2007	
2SB268	102A	3004	53	2007	
2SB268A		3004	53	2007	
2SB268B		3004	53	2007	
2SB268C		3004	53	2007	
2SB268D		3004	53	2007	
2SB268E		3004	53	2007	
2SB268F		3004	53	2007	
2SB268G		3004	53	2007	
2SB268GN		3004	53	2007	
2SB268H		3004	53	2007	
2SB268J		3004	53	2007	
2SB268K		3004	53	2007	
2SB268L		3004	53	2007	
2SB268M		3004	53	2007	
2SB268OR		3004	53	2007	
2SB268R		3004	53	2007	
2SB268X		3004	53	2007	
2SB268Y		3004	53	2007	
2SB269	102A	3004	53	2007	
2SB269A		3004	53	2007	
2SB269C		3004	53	2007	
2SB269D		3004	53	2007	
2SB269E		3004	53	2007	
2SB269F		3004	53	2007	
2SB269G		3004	53	2007	
2SB269GN		3004	53	2007	
2SB269J		3004	53	2007	
2SB269K		3004	53	2007	
2SB269L		3004	53	2007	
2SB269M		3004	53	2007	
2SB269OR		3004	53	2007	
2SB269R		3004	53	2007	
2SB269X		3004	53	2007	
2SB269Y		3004	53	2007	
2SB26A	104	3009	25	2006	
2SB26B		3009	25		
2SB26C		3009	25		
2SB26D		3009	25		
2SB26E		3009	25		
2SB26F		3009	25		
2SB26G		3009	25		
2SB26GN		3009	25		
2SB26H		3009	25		
2SB26J		3009	25		
2SB26K		3009	25		
2SB26L		3009	25		
2SB26M		3009	25		
2SB26OR		3009	25		
2SB26R		3009	25		
2SB26X		3009	25		
2SB26Y		3009	25		
2SB27	104	3009	16	2006	
2SB270	102A	3004	53	2007	
2SB270A	102A	3004	53	2007	
2SB270B	102A	3004	53	2007	
2SB270C	102A	3004	53	2007	
2SB270D	102A	3004	53	2007	
2SB270E	102A	3004	53	2007	
2SB270F		3004	53		
2SB270G		3004	53		
2SB270GN		3004	53		
2SB270H		3004	53		
2SB270J		3004	53		
2SB270K		3004	53		
2SB270L		3004	53		
2SB270M		3004	53		
2SB270OR		3004	53		
2SB270R		3004	53		
2SB270X		3004	53		
2SB270Y		3004	53		
2SB271	102A		53	2007	
2SB272	102A	3004	53	2007	
2SB272A		3004	53	2007	
2SB272B		3004	53	2007	
2SB272C		3004	53	2007	
2SB272D		3004	53	2007	
2SB272E		3004	53	2007	
2SB272F		3004	53	2007	
2SB272G		3004	53	2007	
2SB272GN		3004	53	2007	
2SB272J		3004	53	2007	
2SB272K		3004	53	2007	
2SB272L		3004	53	2007	
2SB272M		3004	53	2007	
2SB272OR		3004	53	2007	
2SB272R		3004	53	2007	
2SB272X		3004	53	2007	
2SB272Y		3004	53	2007	
2SB273	102A	3004	53	2007	
2SB273A		3004	53	2007	
2SB273B		3004	53	2007	
2SB273C		3004	53	2007	
2SB273D		3004	53	2007	
2SB273E		3004	53	2007	
2SB273F		3004	53	2007	
2SB273G		3004	53	2007	
2SB273GN		3004	53	2007	
2SB273H		3004	53	2007	
2SB273J		3004	53	2007	
2SB273K		3004	53	2007	
2SB273L		3004	53	2007	
2SB273M		3004	53	2007	
2SB273OR		3004	53	2007	
2SB273R		3004	53	2007	
2SB273X		3004	53	2007	
2SB273Y		3004	53	2007	
2SB274	127	3035	25		
2SB274A		3035	25	2007	
2SB274B		3035	25	2007	
2SB274C		3035	25	2007	
2SB274D		3035	25	2007	
2SB274E		3035	25	2007	
2SB274F		3035	25	2007	
2SB274G		3035	25	2007	
2SB274H		3035	25	2007	
2SB274J		3035	25	2007	
2SB274K		3035	25	2007	
2SB274L		3035	25	2007	
2SB274M		3035	25	2007	
2SB274OR		3035	25	2007	
2SB274R		3035	25	2007	
2SB274V			25	2007	
2SB274X		3035	25	2007	
2SB274Y		3035	25	2007	
2SB275	127	3035	25		
2SB275A		3035	25		
2SB275B		3035	25		
2SB275C		3035	25		
2SB275D		3035	25		
2SB275E		3035	25		
2SB275F		3035	25		
2SB275G		3035	25		
2SB275GN		3035	25		
2SB275H		3035	25		
2SB275J		3035	25		
2SB275K		3035	25		
2SB275L		3035	25		
2SB275M		3035	25		
2SB275OR		3035	25		
2SB275R		3035	25		
2SB275X		3035	25		
2SB275Y		3035	25		
2SB276	127	3035	25		
2SB276A		3035	25		
2SB276B		3035	25		
2SB276C		3035	25		
2SB276D		3035	25		
2SB276E		3035	25		
2SB276F		3035	25		
2SB276G		3035	25		
2SB276GN		3035	25		
2SB276H		3035	25		
2SB276J		3035	25		
2SB276K		3035	25		
2SB276L		3035	25		
2SB276M		3035	25		
2SB276OR		3035	25		
2SB276R		3035	25		
2SB276X		3035	25		
2SB276Y		3035	25		
2SB277				2004	
2SB27A		3009	25		
2SB27B		3009	25		
2SB27C		3009	25		
2SB27D		3009	25		
2SB27E		3009	25		
2SB27F		3009	25		
2SB27G		3009	25		
2SB27GN		3009	25		
2SB27H		3009	25		
2SB27J		3009	25		
2SB27K		3009	25		
2SB27L		3009	25		
2SB27M		3009	25		
2SB27OR		3009	25		
2SB27R		3009	25		
2SB27X		3009	25		
2SB27Y		3009	25		
2SB28	104	3009	25	2006	
2SB282	127	3009	25	2006	
2SB282A		3009	25		
2SB282B		3009	25		
2SB282C		3009	25		
2SB282D		3009	25		
2SB282E		3009	25		
2SB282G		3009	25		
2SB282GN		3009	25		
2SB282H		3009	25		
2SB282J		3009	25		
2SB282K		3009	25		
2SB282L		3009	25		
2SB282M		3009	25		
2SB282OR		3009	25		
2SB282R		3009	25		
2SB282X		3009	25		
2SB282Y		3009	25		
2SB283	121	3009	25	2006	
2SB283A		3009	25		
2SB283B		3009	25		
2SB283C		3009	25		
2SB283D		3009	25		
2SB283E		3009	25		
2SB283F		3009	25		
2SB283G		3009	25		
2SB283GN		3009	25		
2SB283H		3009	25		
2SB283J		3009	25		
2SB283K		3009	25		
2SB283L		3009	25		
2SB283M		3009	25		
2SB283OR		3009	25		
2SB283R		3009	25		
2SB283X		3009	25		
2SB283Y		3009	25		
2SB284	121	3009	25	2006	
2SB284A		3009	25		
2SB284B		3009	25		
2SB284C		3009	25		
2SB284D		3009	25		
2SB284E		3009	25		
2SB284F		3009	25		
2SB284G		3009	25		
2SB284H		3009	25		
2SB284J		3009	25		
2SB284K		3009	25		
2SB284L		3009	25		
2SB284M		3009	25		
2SB284OR		3009	25		
2SB284R		3009	25		
2SB284X		3009	25		
2SB284Y		3009	25		
2SB285	127	3009	25	2006	
2SB285A		3009	25		
2SB285B		3009	25		

Industry Standard No.	ECG	SK	GE	RS 276-	MOTOR.
2SB285C		3009	25		
2SB285D		3009	25		
2SB285E		3009	25		
2SB285F		3009	25		
2SB285G		3009	25		
2SB285GN		3009	25		
2SB285H		3009	25		
2SB285J		3009	25		
2SB285K		3009	25		
2SB285L		3009	25		
2SB285M		3009	25		
2SB285OR		3009	25		
2SB285R		3009	25		
2SB285X		3009	25		
2SB285Y		3009	25		
2SB28A		3009	25		
2SB28B		3009	25		
2SB28C		3009	25		
2SB28D		3009	25		
2SB28E		3009	25		
2SB28F		3009	25		
2SB28G		3009	25		
2SB28GN		3009	25		
2SB28H		3009	25		
2SB28J		3009	25		
2SB28K		3009	25		
2SB28L		3009	25		
2SB28M		3009	25		
2SB28OR		3009	25		
2SB28R		3009	25		
2SB28X		3009	25		
2SB28Y		3009	25		
2SB29	104	3009	16	2006	
2SB290	100	3004/102A	53	2007	
2SB290A		3004	53	2007	
2SB290B		3004	53	2007	
2SB290C		3004	53	2007	
2SB290D		3004	53	2007	
2SB290E		3004	53	2007	
2SB290F		3004	53	2007	
2SB290G		3004	53	2007	
2SB290GN		3004	53	2007	
2SB290H		3004	53	2007	
2SB290J		3004	53	2007	
2SB290K		3004	53	2007	
2SB290L		3004	53	2007	
2SB290M		3004	53	2007	
2SB290OR		3004	53	2007	
2SB290R		3004	53	2007	
2SB290X		3004	53	2007	
2SB290Y		3004	53	2007	
2SB291	100	3004/102A	53	2007	
2SB291-GREEN			53	2007	
2SB291-RED			53	2007	
2SB291-YELLOW			53	2007	
2SB291A		3004	53	2007	
2SB291B		3004	53	2007	
2SB291C		3004	53	2007	
2SB291D		3004	53	2007	
2SB291E		3004	53	2007	
2SB291F		3004	53	2007	
2SB291G		3004	53	2007	
2SB291GN		3004	53	2007	
2SB291H		3004	53	2007	
2SB291J		3004	53	2007	
2SB291K		3004	53	2007	
2SB291L		3004	53	2007	
2SB291M		3004	53	2007	
2SB291OR		3004	53	2007	
2SB291R		3004	53	2007	
2SB291X		3004	53	2007	
2SB291Y		3004	53	2007	
2SB292	100	3004/102A	53	2007	
2SB292-BLUE			53	2007	
2SB292-GREEN			53	2007	
2SB292-ORANGE			53	2007	
2SB292-RED			53	2007	
2SB292-YELLOW			53	2007	
2SB292A	100	3005	53	2007	
2SB292B		3004	53	2007	
2SB292C		3004	53	2007	
2SB292D		3004	53	2007	
2SB292E		3004	53	2007	
2SB292F		3004	53	2007	
2SB292G		3004	53	2007	
2SB292GN		3004	53	2007	
2SB292H		3004	53	2007	
2SB292J		3004	53	2007	
2SB292K		3004	53	2007	
2SB292L		3004	53	2007	
2SB292M		3004	53	2007	
2SB292OR		3004	53	2007	
2SB292R		3004	53	2007	
2SB292X		3004	53	2007	
2SB292Y		3004	53	2007	
2SB293	102A	3004	53	2007	
2SB293A		3004	53	2007	
2SB293B		3004	53	2007	
2SB293C		3004	53	2007	
2SB293D		3004	53	2007	
2SB293E		3004	53	2007	
2SB293F		3004	53	2007	
2SB293G		3004	53	2007	
2SB293GN		3004	53	2007	
2SB293H		3004	53	2007	
2SB293J		3004	53	2007	
2SB293K		3004	53	2007	
2SB293L		3004	53	2007	
2SB293M		3004	53	2007	
2SB293OR		3004	53	2007	
2SB293R		3004	53	2007	
2SB293X		3004	53	2007	
2SB293Y		3004	53	2007	
2SB294	102A	3123	80	2007	
2SB294A		3123	80	2007	
2SB294B		3123	80	2007	
2SB294C		3123	80	2007	
2SB294D		3123	80	2007	
2SB294E		3123	80	2007	
2SB294F		3123	80	2007	
2SB294G		3123	80	2007	
2SB294GN		3123	80	2007	
2SB294H		3123	80	2007	
2SB294J		3123	80	2007	
2SB294K		3123	80	2007	
2SB294L		3123	80	2007	
2SB294M		3123	80	2007	
2SB294OR		3123	80	2007	
2SB294R		3123	80	2007	
2SB294X		3123	80	2007	
2SB294Y		3123	80	2007	
2SB295	121$	3009	25	2006	
2SB295A		3009	25		
2SB295B		3009	25		
2SB295C		3009	25		
2SB295D		3009	25		
2SB295E		3009	25		
2SB295F		3009	25		
2SB295G		3009	25		
2SB295GN		3009	25		
2SB295H		3009	25		
2SB295J		3009	25		
2SB295K		3009	25		
2SB295L		3009	25		
2SB295M		3009	25		
2SB295OR		3009	25		
2SB295R		3009	25		
2SB295X		3009	25		
2SB295Y		3009	25		
2SB296	127	3035	25		
2SB296A		3035	25		
2SB296B		3035	25		
2SB296C		3035	25		
2SB296D		3035	25		
2SB296E		3035	25		
2SB296F		3035	25		
2SB296G		3035	25		
2SB296GN		3035	25		
2SB296H		3035	25		
2SB296J		3035	25		
2SB296K		3035	25		
2SB296L		3035	25		
2SB296M		3035	25		
2SB296OR		3035	25		
2SB296R		3035	25		
2SB296X		3035	25		
2SB296Y		3035	25		
2SB299	102A	3004	53	2007	
2SB299A		3004	53	2007	
2SB299B		3004	53	2007	
2SB299C		3004	53	2007	
2SB299D		3004	53	2007	
2SB299E		3004	53	2007	
2SB299F		3004	53	2007	
2SB299G		3004	53	2007	
2SB299GN		3004	53	2007	
2SB299H		3004	53	2007	
2SB299J		3004	53	2007	
2SB299K		3004	53	2007	
2SB299L		3004	53	2007	
2SB299M		3004	53	2007	
2SB299OR		3004	53	2007	
2SB299R		3004	53	2007	
2SB299X		3004	53	2007	
2SB299Y		3004	53	2007	
2SB29A		3009	25		
2SB29B		3009	25		
2SB29C		3009	25		
2SB29D		3009	25		
2SB29E		3009	25		
2SB29F		3009	25		
2SB29G		3009	25		
2SB29GN		3009	25		
2SB29H		3009	25		
2SB29J		3009	25		
2SB29K		3009	25		
2SB29L		3009	25		
2SB29M		3009	25		
2SB29OR		3009	25		
2SB29R		3009	25		
2SB29X		3009	25		
2SB29Y		3009	25		
2SB30	104	3009	25	2006	
2SB300	127	3035	25		
2SB300A		3035	25		
2SB300B		3035	25		
2SB300C		3035	25		
2SB300D		3035	25		
2SB300E		3035	25		
2SB300F		3035	25		
2SB300G		3035	25		
2SB300GN		3035	25		
2SB300H		3035	25		
2SB300J		3035	25		
2SB300K		3035	25		
2SB300L		3035	25		
2SB300M		3035	25		
2SB300R		3035	25		
2SB300X		3035	25		
2SB300Y		3035	25		
2SB301	127	3035	25		
2SB301A		3035	25		
2SB301B		3035	25		
2SB301C		3035	25		
2SB301D		3035	25		
2SB301E		3035	25		
2SB301F		3035	25		
2SB301G		3035	25		
2SB301GN		3035	25		
2SB301H		3035	25		
2SB301J		3035	25		
2SB301L		3035	25		
2SB301M		3035	25		
2SB301R		3035	25		
2SB301X		3035	25		
2SB301Y		3035	25		
2SB302	102A	3004	53	2007	
2SB302A		3004	53	2004	
2SB302B		3004	53	2004	
2SB302C		3004	53	2004	
2SB302D		3004	53	2004	
2SB302E		3004	53	2004	
2SB302F		3004	53	2004	
2SB302G		3004	53	2004	
2SB302J		3004	53	2004	
2SB302K		3004	53	2004	
2SB302L		3004	53	2004	
2SB302M		3004	53	2004	
2SB302OR		3004	53	2004	
2SB302R		3004	53	2004	
2SB302X		3004	53	2004	
2SB302Y		3004	53	2004	
00002SB303	102A	3004	53	2007	
2SB303(0)			53	2007	
0002SB303-0	102A	3004	53	2007	
0002SB3030		3004	53	2007	
2SB303A	102A	3004	53	2007	
2SB303B	102A	3004	53	2007	
2SB303BK		3004	53	2007	
2SB303C	102A	3004	53	2007	
2SB303D		3004	53	2007	
2SB303E		3004	53	2007	
2SB303F		3004	53	2007	
2SB303G		3004	53	2007	
2SB303GN		3004	53	2007	
2SB303H	102A	3004	53	2007	
2SB303J		3004	53	2007	
2SB303K	102A	3004	53	2007	

Industry Standard No.	ECG	SK	GE	RS 276-	MOTOR.
2SB303L		3004	53	2007	
2SB303M		3004	53	2007	
2SB303OR		3004	53	2007	
2SB303R		3004	53	2007	
2SB303X		3004	53	2007	
2SB303Y		3004	53	2007	
2SB304	158	3123	80	2007	
2SB304A	158	3123	80	2007	
2SB304B		3123	80		
2SB304C		3123	80		
2SB304D		3123	80		
2SB304E		3123	80		
2SB304F		3123	80		
2SB304G		3123	80		
2SB304GN		3123	80		
2SB304H		3123	80		
2SB304J		3123	80		
2SB304K		3123	80		
2SB304L		3123	80		
2SB304M		3123	80		
2SB304OR		3123	80		
2SB304R		3123	80		
2SB304X		3123	80		
2SB304Y		3123	80		
2SB309	127	3035	25		
2SB309A		3035	25		
2SB309B		3035	25		
2SB309C		3035	25		
2SB309D		3035	25		
2SB309E		3035	25		
2SB309F		3035	25		
2SB309G		3035	25		
2SB309GN		3035	25		
2SB309H		3035	25		
2SB309K		3035	25		
2SB309L		3035	25		
2SB309M		3035	25		
2SB309OR		3035	25		
2SB309R		3035	25		
2SB309X		3035	25		
2SB309Y		3035	25		
2SB30A		3009	25		
2SB30B		3009	25		
2SB30C		3009	25		
2SB30D		3009	25		
2SB30E		3009	25		
2SB30F		3009	25		
2SB30G		3009	25		
2SB30GN		3009	25		
2SB30H		3009	25		
2SB30J		3009	25		
2SB30K		3009	25		
2SB30L		3009	25		
2SB30M		3009	25		
2SB30OR		3009	25		
2SB30R		3009	25		
2SB30X		3009	25		
2SB30Y		3009	25		
2SB31	104	3009	25	2006	
2SB310	127	3035	25		
2SB310A		3035	25		
2SB310B		3035	25		
2SB310C		3035	25		
2SB310D		3035	25		
2SB310E		3035	25		
2SB310F		3035	25		
2SB310G		3035	25		
2SB310GN		3035	25		
2SB310H		3035	25		
2SB310J		3035	25		
2SB310K		3035	25		
2SB310L		3035	25		
2SB310M		3035	25		
2SB310R		3035	25		
2SB310X		3035	25		
2SB310Y		3035	25		
2SB311	127	3035	25		
2SB311A		3035	25		
2SB311B		3035	25		
2SB311C		3035	25		
2SB311D		3035	25		
2SB311E		3035	25		
2SB311F		3035	25		
2SB311G		3035	25		
2SB311GN		3035	25		
2SB311H		3035	25		
2SB311J		3035	25		
2SB311K		3035	25		
2SB311L		3035	25		
2SB311M		3035	25		
2SB311OR		3035	25		
2SB311R		3035	25		
2SB311X		3035	25		
2SB311Y		3035	25		
2SB312	127	3035	25		
2SB312A		3035	25		
2SB312B		3035	25		
2SB312C		3035	25		
2SB312D		3035	25		
2SB312E		3035	25		
2SB312F		3035	25		
2SB312G		3035	25		
2SB312GN		3035	25		
2SB312H		3035	25		
2SB312J		3035	25		
2SB312K		3035	25		
2SB312L		3035	25		
2SB312M		3035	25		
2SB312X		3035	25		
2SB312Y		3035	25		
2SB313	127	3035	25		
2SB313A		3035	25		
2SB313B		3035	25		
2SB313C		3035	25		
2SB313D		3035	25		
2SB313E		3035	25		
2SB313F		3035	25		
2SB313G		3035	25		
2SB313GN		3035	25		
2SB313H		3035	25		
2SB313J		3035	25		
2SB313K		3035	25		
2SB313L		3035	25		
2SB313M		3035	25		
2SB313OR		3035	25		
2SB313R		3035	25		
2SB313X		3035	25		
2SB313Y		3035	25		
2SB314		3004	53	2007	
2SB314A		3004	53	2007	
2SB314B		3004	53	2007	
2SB314C		3004	53	2007	
2SB314D		3004	53	2007	
2SB314E		3004	53	2007	
2SB314F		3004	53	2007	
2SB314GN		3004	53	2007	
2SB314H		3004	53	2007	
2SB314J		3004	53	2007	
2SB314K		3004	53	2007	
2SB314L		3004	53	2007	
2SB314M		3004	53	2007	
2SB314OR		3004	53	2007	
2SB314R		3004	53	2007	
2SB314X		3004	53	2007	
2SB314Y		3004	53	2007	
2SB315	102A	3004	53	2007	
2SB315A		3004	53	2007	
2SB315B		3004	53	2007	
2SB315C		3004	53	2007	
2SB315D		3004	53	2007	
2SB315E		3004	53	2007	
2SB315F		3004	53	2007	
2SB315G		3004	53	2007	
2SB315GN		3004	53	2007	
2SB315H		3004	53	2007	
2SB315J		3004	53	2007	
2SB315L		3004	53	2007	
2SB315M		3004	53	2007	
2SB315OR		3004	53	2007	
2SB315R		3004	53	2007	
2SB315X		3004	53	2007	
2SB315Y		3004	53	2007	
2SB316	102A	3004	53	2007	
2SB316A		3004	53	2007	
2SB316B		3004	53	2007	
2SB316C		3004	53	2007	
2SB316D		3004	53	2007	
2SB316E		3004	53	2007	
2SB316F		3004	53	2007	
2SB316G		3004	53	2007	
2SB316GN		3004	53	2007	
2SB316H		3004	53	2007	
2SB316OR		3004	53	2007	
2SB316R		3004	53	2007	
2SB316X		3004	53	2007	
2SB316Y		3004	53	2007	
2SB317	102A		53	2007	
2SB318	127	3014	3		
2SB318A		3014	3		
2SB318B		3014	3		
2SB318C		3014	3		
2SB318D		3014	3		
2SB318E		3014	3		
2SB318F		3014	3		
2SB318G		3014	3		
2SB318GN		3014	3		
2SB318H		3014	3		
2SB318J		3014	3		
2SB318K		3014	3		
2SB318L		3014	3		
2SB318M		3014	3		
2SB318OR		3014	3		
2SB318R		3014	3		
2SB318X		3014	3		
2SB318Y		3014	3		
2SB319	127	3035	25		
2SB319A		3035	25		
2SB319B		3035	25		
2SB319C		3035	25		
2SB319D		3035	25		
2SB319E		3035	25		
2SB319F		3035	25		
2SB319G		3035	25		
2SB319GN		3035	25		
2SB319H		3035	25		
2SB319J		3035	25		
2SB319K		3035	25		
2SB319L		3035	25		
2SB319M		3035	25		
2SB319OR		3035	25		
2SB319R		3035	25		
2SB319X		3035	25		
2SB319Y		3035	25		
2SB31A		3009	25		
2SB31B		3009	25		
2SB31C		3009	25		
2SB31D		3009	25		
2SB31E		3009	25		
2SB31F		3009	25		
2SB31G		3009	25		
2SB31GN		3009	25		
2SB31H		3009	25		
2SB31J		3009	25		
2SB31K		3009	25		
2SB31L		3009	25		
2SB31M		3009	25		
2SB31OR		3009	25		
2SB31R		3009	25		
2SB31X		3009	25		
2SB31Y		3009	25		
2SB32	102A	3004	53	2007	
2SB32-0	102A	3004	53	2007	
2SB32-1	102A	3004	53	2007	
2SB32-2	102A	3004	53	2007	
2SB32-4	102A		53	2007	
2SB320	127	3035	25		
2SB320A		3035	25		
2SB320B		3035	25		
2SB320C		3035	25		
2SB320D		3035	25		
2SB320E		3035	25		
2SB320F		3035	25		
2SB320G		3035	25		
2SB320GN		3035	25		
2SB320H		3035	25		
2SB320J		3035	25		
2SB320K		3035	25		
2SB320L		3035	25		
2SB320M		3035	25		
2SB320R		3035	25		
2SB320X		3035	25		
2SB320Y		3035	25		
2SB321	102A		53	2007	
2SB322	102A		53	2007	
2SB323	102A		53	2007	
2SB324	158	3004/102A	53	2007	
2SB324(E)		3004	53	2007	
2SB324(F)			53	2007	
2SB324(I)			53	2007	
2SB324(L)			53	2007	
2SB324(N)			53	2007	
2SB324,K			53	2007	
2SB324-OR			53	2007	
2SB324/4454C	158		53	2007	
2SB3240		3004	53	2007	
2SB3240A		3004	53	2007	
2SB3240B		3004	53	2007	
2SB3240C		3004	53	2007	

Industry Standard No.	ECG	SK	GE	RS 276-	MOTOR.
2SB3240D		3004	53	2007	
2SB3240E		3004	53	2007	
2SB3240F		3004	53	2007	
2SB3240G		3004	53	2007	
2SB3240GN		3004	53	2007	
2SB3240H		3004	53	2007	
2SB3240J		3004	53	2007	
2SB3240K		3004	53	2007	
2SB3240L		3004	53	2007	
2SB3240M		3004	53	2007	
2SB32400R		3004	53	2007	
2SB3240R		3004	53	2007	
2SB3240X		3004	53	2007	
2SB3240Y		3004	53	2007	
2SB3244H			53	2007	
2SB324A	158	3004/102A	53	2007	
2SB324B	158	3004/102A	53	2007	
2SB324C		3004	53	2007	
2SB324D	158	3004/102A	53	2007	
2SB324E	158	3004/102A	53	2007	
2SB324E-1	158		53	2007	
2SB324E-L			53	2007	
2SB324F	158	3004/102A	53	2007	
2SB324G	158	3004/102A	53	2007	
2SB324GN		3004	53	2007	
2SB324H	158	3004/102A	53	2007	
2SB324I	158	3004/102A	53	2007	
2SB324J	158	3004/102A	53	2007	
2SB324K	158	3004/102A	53	2007	
2SB324L	158	3004/102A	53	2007	
2SB324M		3004	53	2007	
2SB324N	158	3004/102A	53	2007	
2SB3240R		3004	53	2007	
2SB324P	158	3004/102A	53	2007	
2SB324R	158	3004/102A	53	2007	
2SB324S	158	3004/102A	53	2007	
2SB324V	158	3004/102A	53	2007	
2SB324X		3004	53	2007	
2SB324Y		3004	53	2007	
2SB326	102A	3004	53	2007	
2SB326A		3004	53	2007	
2SB326B		3004	53	2007	
2SB326C		3004	53	2007	
2SB326D		3004	53	2007	
2SB326E		3004	53	2007	
2SB326F		3004	53	2007	
2SB326G		3004	53	2007	
2SB326GN		3004	53	2007	
2SB326H		3004	53	2007	
2SB326J		3004	53	2007	
2SB326K		3004	53	2007	
2SB326L		3004	53	2007	
2SB326M		3004	53	2007	
2SB3260R		3004	53	2007	
2SB326R		3004	53	2007	
2SB326X		3004	53	2007	
2SB326Y		3004	53	2007	
2SB327	102A	3004	53	2007	
2SB327A		3004	53	2007	
2SB327B		3004	53	2007	
2SB327C		3004	53	2007	
2SB327D		3004	53	2007	
2SB327E		3004	53	2007	
2SB327F		3004	53	2007	
2SB327G		3004	53	2007	
2SB327GN		3004	53	2007	
2SB327H		3004	53	2007	
2SB327J		3004	53	2007	
2SB327K		3004	53	2007	
2SB327L		3004	53	2007	
2SB327M		3004	53	2007	
2SB3270R		3004	53	2007	
2SB327R		3004	53	2007	
2SB327X		3004	53	2007	
2SB327Y		3004	53	2007	
2SB328	102	3004	53	2007	
2SB328A		3004	53	2007	
2SB328B		3004	53	2007	
2SB328C		3004	53	2007	
2SB328D		3004	53	2007	
2SB328E		3004	53	2007	
2SB328F		3004	53	2007	
2SB328G		3004	53	2007	
2SB328GN		3004	53	2007	
2SB328H		3004	53	2007	
2SB328J		3004	53	2007	
2SB328K		3004	53	2007	
2SB328L		3004	53	2007	
2SB328M		3004	53	2007	
2SB3280R		3004	53	2007	
2SB328R		3004	53	2007	
2SB328X		3004	53	2007	
2SB328Y		3004	53	2007	
2SB329	102A	3004	53	2007	
2SB329A		3004	53	2007	
2SB329B		3004	53	2007	
2SB329C		3004	53	2007	
2SB329D		3004	53	2007	
2SB329E		3004	53	2007	
2SB329F		3004	53	2007	
2SB329G		3004	53	2007	
2SB329GN		3004	53	2007	
2SB329H		3004	53	2007	
2SB329J		3004	53	2007	
2SB329K	102A	3004	53	2007	
2SB329L		3004	53	2007	
2SB329M		3004	53	2007	
2SB3290R		3004	53	2007	
2SB329R		3004	53	2007	
2SB329X		3004	53	2007	
2SB329Y		3004	53	2007	
2SB32A		3004	53	2007	
2SB32B		3004	53	2007	
2SB32C		3004	53	2007	
2SB32D		3004	53	2007	
2SB32E		3004	53	2007	
2SB32F		3004	53	2007	
2SB32G		3004	53	2007	
2SB32GN		3004	53	2007	
2SB32H		3004	53	2007	
2SB32J		3004	53	2007	
2SB32K		3004	53	2007	
2SB32M		3004	53	2007	
2SB32N	102A		53	2007	
2SB320R		3004	53	2007	
2SB32R		3004	53	2007	
2SB32X		3004	53	2007	
2SB32Y		3004	53	2007	
2SB33	102A	3004	53	2007	
2SB33(5)		3004	2	2007	
2SB33-4	102A	3004	53	2007	
2SB33-5			53	2007	
2SB331	213	3012/105	4		
2SB331A		3012	4		
2SB331B		3012	4		
2SB331C		3012	4		
2SB331D		3012	4		
2SB331E		3012	4		
2SB331F		3012	4		
2SB331G		3012	4		
2SB331GN		3012	4		
2SB331H	213	3012/105	4		
2SB331HA	213		4		
2SB331HB	213		4		
2SB331HC	213		4		
2SB331J		3012	4		
2SB331K		3012	4		
2SB331L		3012	4		
2SB331M		3012	4		
2SB3310R		3012	4		
2SB331R		3012	4		
2SB331X		3012	4		
2SB331Y		3012	4		
2SB332	213	3012/105	4		
2SB332A		3012	4		
2SB332B		3012	4		
2SB332C		3012	4		
2SB332D		3012	4		
2SB332E		3012	4		
2SB332F		3012	4		
2SB332G		3012	4		
2SB332GN		3012	4		
2SB332H	213	3012/105	4		
2SB332HA	213		4		
2SB332HB	213		4		
2SB332HC	213		4		
2SB332J		3012	4		
2SB332K		3012	4		
2SB332L		3012	4		
2SB332M		3012	4		
2SB3320R		3012	4		
2SB332R		3012	4		
2SB332X		3012	4		
2SB332Y		3012	4		
2SB333	213	3012/105	4		
2SB333A		3012	4		
2SB333B		3012	4		
2SB333D		3012	4		
2SB333E		3012	4		
2SB333F		3012	4		
2SB333G		3012	4		
2SB333GN		3012	4		
2SB333H		3012	4		
2SB333J		3012	4		
2SB333K		3012	4		
2SB333L		3012	4		
2SB333M		3012	4		
2SB3330R		3012	4		
2SB333R		3012	4		
2SB333X		3012	4		
2SB333Y		3012	4		
2SB334		3012	4		
2SB334A		3012	4		
2SB334B		3012	4		
2SB334C		3012	4		
2SB334D		3012	4		
2SB334E		3012	4		
2SB334F		3012	4		
2SB334G		3012	4		
2SB334GN		3012	4		
2SB334H		3012	4		
2SB334J		3012	4		
2SB334K		3012	4		
2SB334L		3012	4		
2SB334M		3012	4		
2SB3340R		3012	4		
2SB334R		3012	4		
2SB334X		3012	4		
2SB334Y		3012	4		
2SB335	102	3722	2	2007	
2SB336	102A		53	2007	
2SB337	121	3009	25	2006	
2SB337-B		3009	25		
2SB337-OR			239	2006	
2SB337A	121	3009	25	2006	
2SB337B	121	3009	25	2006	
2SB337BK	121	3009	25	2006	
2SB337C		3009	25	2006	
2SB337D		3009	25	2006	
2SB337E		3009	25	2006	
2SB337F		3009	25	2006	
2SB337G		3009	25	2006	
2SB337GN		3009	25	2006	
2SB337H	121	3009	25	2006	
2SB337HA	121	3717	25	2006	
2SB337HB	121	3717	25	2006	
2SB337J		3009	25	2006	
2SB337K		3009	25		
2SB337L		3009	25		
2SB337LB		3009	25	2006	
2SB337M		3009	25	2006	
2SB3370R		3009	25		
2SB337R		3009	25	2006	
2SB337X		3009	25	2006	
2SB337Y		3009	25	2006	
2SB338	121	3009	25	2006	
2SB338A		3009	25		
2SB338B		3009	25		
2SB338C		3009	25		
2SB338D		3009	25		
2SB338E		3009	25		
2SB338F		3009	25		
2SB338G		3009	25		
2SB338GN		3009	25		
2SB338H	121	3009	25	2006	
2SB338HA	121	3717	25	2006	
2SB338HB	121	3717	25	2006	
2SB338J		3009	25		
2SB338K		3009	25		
2SB338L		3009	25		
2SB338M		3009	25		
2SB3380R		3009	25		
2SB338R		3009	25		
2SB338X		3009	25		
2SB338Y		3009	25		
2SB339	179	3034	76		
2SB339H	179	3034	76		
2SB33A		3004	53	2007	
2SB33B		3004	53	2007	
2SB33BK		3004	53	2007	
2SB33C	102A	3004	53	2007	
2SB33D	102A	3004	53	2007	
2SB33E	102A	3004	53	2007	
2SB33F	102A	3004	53	2007	
2SB33G		3004	53	2007	
2SB33GN		3004	53	2007	
2SB33H		3004	53	2007	
2SB33J		3004	53	2007	

Industry Standard No.	ECG	SK	GE	RS 276-	MOTOR.
2SB33K		3004	53	2007	
2SB33L		3004	53	2007	
2SB33M		3004	53	2007	
2SB330R		3004	53	2007	
2SB33R		3004	53	2007	
2SB33X		3004	53	2007	
2SB33Y		3004	53	2007	
2SB34	102	3003	53	2007	
2SB340	179	3034	25		
2SB340H	179	3034	25		
2SB341	127	3764	25		
2SB341H	127	3764	25		
2SB341S	127	3764	25		
2SB341V	127	3764	25		
2SB342	127	3035	25		
2SB342A		3035	25		
2SB342B		3035	25		
2SB342C		3035	25		
2SB342D		3035	25		
2SB342E		3035	25		
2SB342F		3035	25		
2SB342G		3035	25		
2SB342GN		3035	25		
2SB342H		3035	25		
2SB342J		3035	25		
2SB342K		3035	25		
2SB342L		3035	25		
2SB342M		3035	25		
2SB3420R		3035	25		
2SB342R		3035	25		
2SB342X		3035	25		
2SB342Y		3035	25		
2SB343	127	3035	25		
2SB343A		3035	25		
2SB343B		3035	25		
2SB343C		3035	25		
2SB343D		3035	25		
2SB343E		3035	25		
2SB343F		3035	25		
2SB343G		3035	25		
2SB343GN		3035	25		
2SB343H		3035	25		
2SB343J		3035	25		
2SB343K		3035	25		
2SB343L		3035	25		
2SB343M		3035	25		
2SB3430R		3035	25		
2SB343R		3035	25		
2SB343X		3035	25		
2SB343Y		3035	25		
2SB345	102A	3004	53	2007	
2SB345A		3004	53	2007	
2SB345B		3004	53	2007	
2SB345C		3004	53	2007	
2SB345D		3004	53	2007	
2SB345E		3004	53	2007	
2SB345F		3004	53	2007	
2SB345G		3004	53	2007	
2SB345GN		3004	53	2007	
2SB345H		3004	53	2007	
2SB345J		3004	53	2007	
2SB345K		3004	53	2007	
2SB345L		3004	53	2007	
2SB345M		3004	53	2007	
2SB3450R		3004	53	2007	
2SB345R		3004	53	2007	
2SB345X		3004	53	2007	
2SB345Y		3004	53	2007	
2SB346	102A	3004	53	2007	
2SB346(Q)			53	2007	
2SB346A		3004	53	2007	
2SB346B		3004	53	2007	
2SB346C		3004	53	2007	
2SB346D		3004	53	2007	
2SB346E		3004	53	2007	
2SB346F		3004	53	2007	
2SB346G		3004	53	2007	
2SB346GN		3004	53	2007	
2SB346H		3004	53	2007	
2SB346J		3004	53	2007	
2SB346K	102A	3004	53	2007	
2SB346L		3004	53	2007	
2SB346M		3004	53	2007	
2SB3460R		3004	53	2007	
2SB346Q	102A	3004	53	2007	
2SB346R		3004	53	2007	
2SB346X		3004	53	2007	
2SB346Y		3004	53	2007	
2SB347	102A	3004	53	2007	
2SB347A		3004	53	2007	
2SB347B		3004	53	2007	
2SB347C		3004	53	2007	
2SB347D		3004	53	2007	
2SB347E		3004	53	2007	
2SB347F		3004	53	2007	
2SB347G		3004	53	2007	
2SB347GN		3004	53	2007	
2SB347H		3004	53	2007	
2SB347J		3004	53	2007	
2SB347K		3004	53	2007	
2SB347L		3004	53	2007	
2SB347M		3004	53	2007	
2SB3470R		3004	53	2007	
2SB347R		3004	53	2007	
2SB347X		3004	53	2007	
2SB347Y		3004	53	2007	
2SB348	102A	3004	53	2007	
2SB348(Q)			52	2007	
2SB348A		3004	53	2007	
2SB348B		3004	53	2007	
2SB348C		3004	53	2007	
2SB348D		3004	53	2007	
2SB348E		3004	53	2007	
2SB348F		3004	53	2007	
2SB348G		3004	53	2007	
2SB348GN		3004	53	2007	
2SB348H		3004	53	2007	
2SB348J		3004	53	2007	
2SB348K		3004	53	2007	
2SB348L		3004	53	2007	
2SB348M		3004	53	2007	
2SB3480R		3004	53	2007	
2SB348Q	102A	3004	53	2007	
2SB348R	102A	3004	53	2007	
2SB348X			53	2007	
2SB348Y		3004	53	2007	
2SB349	102A		53	2007	
2SB34A		3004	53	2007	
2SB34B		3004	53	2007	
2SB34C		3004	53	2007	
2SB34D		3004	53	2007	
2SB34E		3004	53	2007	
2SB34F		3004	53	2007	
2SB34G		3004	53	2007	
2SB34GN		3004	53	2007	
2SB34H		3004	53	2007	
2SB34J		3004	53	2007	
2SB34K		3004	53	2007	
2SB34L		3004	53	2007	
2SB34M		3004	53	2007	
2SB34N	102	3722	53	2007	
2SB340R		3004	53	2007	
2SB34R		3004	53	2007	
2SB34X		3004	53	2007	
2SB34Y		3004	53	2007	
2SB35		3123	80	2007	
2SB350	102	3004	2	2007	
2SB350A		3004	53	2007	
2SB350B		3004	53	2007	
2SB350C		3004	53	2007	
2SB350D		3004	53	2007	
2SB350E		3004	53	2007	
2SB350F		3004	53	2007	
2SB350G		3004	53	2007	
2SB350GN		3004	53	2007	
2SB350H		3004	53	2007	
2SB350J		3004	53	2007	
2SB350K		3004	53	2007	
2SB350L		3004	53	2007	
2SB350M		3004	53	2007	
2SB3500R		3004	53	2007	
2SB350R		3004	53	2007	
2SB350X		3004	53	2007	
2SB350Y		3004	53	2007	
2SB351	105	3012	4		
2SB351A		3012	4		
2SB351B		3012	4		
2SB351C		3012	4		
2SB351D		3012	4		
2SB351E		3012	4		
2SB351F		3012	4		
2SB351G		3012	4		
2SB351GN		3012	4		
2SB351H		3012	4		
2SB351J		3012	4		
2SB351K		3012	4		
2SB351L		3012	4		
2SB351M		3012	4		
2SB3510R		3012	4		
2SB351R		3012	4		
2SB351X		3012	4		
2SB351Y		3012	4		
2SB352	213	3012/105	4		
2SB352A		3012	4		
2SB352B		3012	4		
2SB352C		3012	4		
2SB352D	213	3012/105	4		
2SB352E		3012	4		
2SB352F		3012	4		
2SB352G		3012	4		
2SB352GN		3012	4		
2SB352H		3012	4		
2SB352J		3012	4		
2SB352K		3012	4		
2SB352L		3012	4		
2SB352M		3012	4		
2SB3520R		3012	4		
2SB352R		3012	4		
2SB352X		3012	4		
2SB352Y		3012	4		
2SB353	213	3012/105	4		
2SB353A		3012	4		
2SB353B		3012	4		
2SB353C		3012	4		
2SB353D		3012	4		
2SB353E		3012	4		
2SB353F		3012	4		
2SB353G		3012	4		
2SB353GN		3012	4		
2SB353H		3012	4		
2SB353J		3012	4		
2SB353K		3012	4		
2SB353L		3012	4		
2SB353M		3012	4		
2SB3530R		3012	4		
2SB353R		3012	4		
2SB353X		3012	4		
2SB353Y		3012	4		
2SB354		3012	4		
2SB354A		3012	4		
2SB354B		3012	4		
2SB354C		3012	4		
2SB354D		3012	4		
2SB354E		3012	4		
2SB354F		3012	4		
2SB354G		3012	4		
2SB354GN		3012	4		
2SB354H		3012	4		
2SB354J		3012	4		
2SB354K		3012	4		
2SB354L		3012	4		
2SB354M		3012	4		
2SB3540R		3012	4		
2SB354R		3012	4		
2SB354X		3012	4		
2SB354Y		3012	4		
2SB355	104	3009	25	2006	
2SB355A		3009	25		
2SB355B		3009	25		
2SB355C		3009	25		
2SB355D		3009	25		
2SB355E		3009	25		
2SB355F		3009	25		
2SB355G		3009	25		
2SB355GN		3009	25		
2SB355H		3009	25		
2SB355J		3009	25		
2SB355K		3009	25		
2SB355L		3009	25		
2SB355M		3009	25		
2SB3550R		3009	25		
2SB355R		3009	25		
2SB355X		3009	25		
2SB355Y		3009	25		
2SB356	104	3009	25	2006	
2SB356A		3009	25		
2SB356B		3009	25		
2SB356C		3009	25		
2SB356D		3009	25		
2SB356E		3009	25		
2SB356F		3009	25		
2SB356G		3009	25		
2SB356GN		3009	25		
2SB356H		3009	25		
2SB356J		3009	25		
2SB356K		3009	25		
2SB356L		3009	25		
2SB356M		3009	25		

Industry Standard No.	ECG	SK	GE	RS 276-	MOTOR.
2SB3560R		3009	25		
2SB356R		3009	25		
2SB356X		3009	25		
2SB356Y		3009	25		
2SB357	127	3035	25		
2SB357A		3035	25		
2SB357B		3035	25		
2SB357C		3035	25		
2SB357D		3035	25		
2SB357E		3035	25		
2SB357F		3035	25		
2SB357G		3035	25		
2SB357GN		3035	25		
2SB357H		3035	25		
2SB357J		3035	25		
2SB357K		3035	25		
2SB357L		3035	25		
2SB357M		3035	25		
2SB3570R		3035	25		
2SB357R		3035	25		
2SB357X		3035	25		
2SB357Y		3035	25		
2SB358	127	3035	25		
2SB358A		3035	25		
2SB358B		3035	25		
2SB358C		3035	25		
2SB358D		3035	25		
2SB358E		3035	25		
2SB358F		3035	25		
2SB358G		3035	25		
2SB358GN		3035	25		
2SB358H		3035	25		
2SB358J		3035	25		
2SB358K		3035	25		
2SB358L		3035	25		
2SB358M		3035	25		
2SB3580R		3035	25		
2SB358R		3035	25		
2SB358X		3035	25		
2SB358Y		3035	25		
2SB359	127	3035	25		
2SB359A		3035	25		
2SB359B		3035	25		
2SB359C		3035	25		
2SB359D		3035	25		
2SB359E		3035	25		
2SB359F		3035	25		
2SB359G		3035	25		
2SB359GN		3035	25		
2SB359H		3035	25		
2SB359J		3035	25		
2SB359K		3035	25		
2SB359L		3035	25		
2SB359M		3035	25		
2SB3590R		3035	25		
2SB359R		3035	25		
2SB359Y		3035	25		
2SB35A		3123	80	2007	
2SB35B		3123	80	2007	
2SB35C		3123	80	2007	
2SB35D		3123	80	2007	
2SB35E		3123	80	2007	
2SB35F		3123	80	2007	
2SB35G		3123	80	2007	
2SB35GN		3123	80	2007	
2SB35H		3123	80	2007	
2SB35J		3123	80	2007	
2SB35K		3123	80	2007	
2SB35L		3123	80	2007	
2SB35M		3123	80	2007	
2SB350R		3123	80	2007	
2SB35R		3123	80	2007	
2SB35X		3123	80	2007	
2SB35Y		3123	80	2007	
2SB360	127	3035	25		
2SB360A		3035	25		
2SB360B		3035	25		
2SB360C		3035	25		
2SB360D		3035	25		
2SB360E		3035	25		
2SB360F		3035	25		
2SB360G		3035	25		
2SB360GN		3035	25		
2SB360H		3035	25		
2SB360J		3035	25		
2SB360K		3035	25		
2SB360L		3035	25		
2SB360M		3035	25		
2SB3600R		3035	25		
2SB360R		3035	25		
2SB360X		3035	25		
2SB360Y		3035	25		
2SB361	127	3035	25		
2SB361A		3035	25		
2SB361B		3035	25		
2SB361C		3035	25		
2SB361D		3035	25		
2SB361E		3035	25		
2SB361F		3035	25		
2SB361G		3035	25		
2SB361GN		3035	25		
2SB361H		3035	25		
2SB361J		3035	25		
2SB361K		3035	25		
2SB361L		3035	25		
2SB361M		3035	25		
2SB3610R		3035	25		
2SB361R		3035	25		
2SB361X		3035	25		
2SB361Y		3035	25		
2SB362	127	3014	3		
2SB362A		3014	3		
2SB362B		3014	3		
2SB362C		3014	3		
2SB362D		3014	3		
2SB362E		3014	3		
2SB362F		3014	3		
2SB362G		3014	3		
2SB362GN		3014	3		
2SB362H		3014	3		
2SB362J		3014	3		
2SB362K		3014	3		
2SB362L		3014	3		
2SB362M		3014	3		
2SB3620R		3014	3		
2SB362R		3014	3		
2SB362X		3014	3		
2SB362Y		3014	3		
2SB364	158	3004/102A	53	2007	
2SB364-OR			53	2007	
2SB364A		3004	53	2007	
2SB364B		3004	53	2007	
2SB364C		3004	53	2007	
2SB364D		3004	53	2007	
2SB364E		3004	53	2007	
2SB364F		3004	53	2007	
2SB364G		3004	53	2007	
2SB364GN		3004	53	2007	
2SB364H		3004	53	2007	
2SB364J		3004	53	2007	
2SB364K		3004	53	2007	
2SB364L		3004	53	2007	
2SB364M		3004	53	2007	
2SB3640R		3004	53		
2SB364R		3004	53	2007	
2SB364X			53	2007	
2SB364Y		3004	53	2007	
2SB365	158	3004/102A	53	2007	
2SB365A			53	2007	
2SB365B	158	3004/102A	53	2007	
2SB365C		3004	53	2007	
2SB365D		3004	53	2007	
2SB365E		3004	53	2007	
2SB365F		3004	53	2007	
2SB365G		3004	53	2007	
2SB365GN		3004	53	2007	
2SB365H		3004	53	2007	
2SB365J		3004	53	2007	
2SB365K		3004	53	2007	
2SB365L		3004	53	2007	
2SB365M		3004	53	2007	
2SB3650R		3004	53	2007	
2SB365R		3004	53	2007	
2SB365X			53	2007	
2SB365Y		3004	53	2007	
2SB366	127	3123	80	2007	
2SB366A		3123	80		
2SB366B		3123	80		
2SB366C		3123	80		
2SB366D		3123	80		
2SB366E		3123	80		
2SB366F		3123	80		
2SB366G		3123	80		
2SB366GN		3123	80		
2SB366H		3123	80		
2SB366J		3123	80		
2SB366K		3123	80		
2SB366L		3123	80		
2SB366M		3123	80		
2SB3660R		3123	80		
2SB366R		3123	80		
2SB366X		3123	80		
2SB366Y		3123	80		
2SB367	131	3198	30	2006	
2SB367(A)	131		30	2006	
2SB367(B)P			30	2006	
2SB367-4		3198	30	2006	
2SB367-5		3198	30	2006	
2SB367-OR			30	2006	
2SB367A	131	3198	30	2006	
2SB367AL		3198	30	2006	
2SB367B	131	3198	30	2006	
2SB367B-2			30	2006	
2SB367BL		3198	30	2006	
2SB367BP			30	2006	
2SB367C	131	3198	30	2006	
2SB367D		3198	30	2006	
2SB367E			30	2006	
2SB367F			30	2006	
2SB367G			30	2006	
2SB367H	131	3198	30	2006	
2SB367J		3198	30	2006	
2SB367K		3198	30	2006	
2SB367L			30	2006	
2SB367M		3198	30	2006	
2SB3670R		3198	30	2006	
2SB367P			30	2006	
2SB367R			30	2006	
2SB367X			30	2006	
2SB367Y			30	2006	
2SB368	1318	3052	30	2006	
2SB368-OR			30	2006	
2SB368A		3052	30	2006	
2SB368B	1318	3052	30	2006	
2SB368C		3052	30	2006	
2SB368D		3052	30	2006	
2SB368E		3052	30	2006	
2SB368F		3052	30	2006	
2SB368G		3052	30	2006	
2SB368GN		3052	30	2006	
2SB368H	1318	3052	30	2006	
2SB368J		3052	30	2006	
2SB368K		3052	30	2006	
2SB368L		3052	30	2006	
2SB368M		3052	30	2006	
2SB3680R		3052	30		
2SB368X		3052	30	2006	
2SB368Y		3052	30	2006	
2SB37	102A	3004	53	2007	
2SB370	102A	3004	53	2007	
2SB370-0			53	2007	
2SB370A	102A	3004	53	2007	
2SB370AA	102A		53	2007	
2SB370AB	102A		53	2007	
2SB370AC	102A		53	2007	
2SB370AHA	102A		53	2007	
2SB370AHB	102A		53	2007	
2SB370B	102A	3004	53	2007	
2SB370C	102A	3004	53	2007	
2SB370D	102A	3004	53	2007	
2SB370E		3004	53	2007	
2SB370F		3004	53	2007	
2SB370G		3004	53	2007	
2SB370GN		3004	53	2007	
2SB370H		3004	53	2007	
2SB370J		3004	53	2007	
2SB370K		3004	53	2007	
2SB370L		3004	53	2007	
2SB370M		3004	53	2007	
2SB3700R		3004	53	2007	
2SB370P	102A		53	2007	
2SB370PB	102A		53	2007	
2SB370R		3004	53	2007	
2SB370V	102A		53	2007	
2SB370X		3004	53	2007	
2SB370Y		3004	53	2007	
2SB371	102A	3004	53	2007	
2SB371A		3004	53	2007	
2SB371B		3004	53	2007	
2SB371C		3004	53	2007	
2SB371D	102A	3004	53	2007	
2SB371E		3004	53	2007	
2SB371F		3004	53	2007	
2SB371G		3004	53	2007	
2SB371GN		3004	53	2007	
2SB371H		3004	53	2007	
2SB371J		3004	53	2007	
2SB371K		3004	53	2007	

Industry Standard No.	ECG	SK	GE	RS 276-	MOTOR.
2SB371L		3004	53	2007	
2SB371M		3004	53	2007	
2SB371OR		3004	53	2007	
2SB371R		3004	53	2007	
2SB371X		3004	53	2007	
2SB371Y		3004	53	2007	
2SB372	176		53	2007	
2SB373	176	3009	25	2007	
2SB373A		3009	25		
2SB373B		3009	25		
2SB373C		3009	25		
2SB373D		3009	25		
2SB373E		3009	25		
2SB373F		3009	25		
2SB373G		3009	25		
2SB373GN		3009	25		
2SB373H		3009	25		
2SB373K		3009	25		
2SB373L		3009	25		
2SB373M		3009	25		
2SB373OR		3009	25		
2SB373R		3009	25		
2SB373X		3009	25		
2SB373Y		3009	25		
2SB375	127	3035	25		
2SB375-2B	127	3035	25		
2SB375-5B	127	3035	25		
2SB375A	127	3035	25		
2SB375A-2B	127	3035	25		
2SB375A-5B	127	3035	25		
2SB375A-NB	127	3035	25		
2SB375AL		3035	25		
2SB375B		3035	25		
2SB375C		3035	25		
2SB375D		3035	25		
2SB375E		3035	25		
2SB375F		3035	25		
2SB375G		3035	25		
2SB375GN		3035	25		
2SB375H		3035	25		
2SB375J		3035	25		
2SB375K		3035	25		
2SB375L		3035	25		
2SB375M		3035	25		
2SB375OR		3035	25		
2SB375R		3035	25		
2SB375TV	127	3035	25		
2SB375X		3035	25		
2SB375Y		3035	25		
2SB376	102A	3004	53	2007	
2SB376A		3004	53	2007	
2SB376B		3004	53	2007	
2SB376C		3004	53	2007	
2SB376D		3004	53	2007	
2SB376E		3004	53	2007	
2SB376F		3004	53	2007	
2SB376G	102A	3004	53	2007	
2SB376GN		3004	53	2007	
2SB376J		3004	53	2007	
2SB376K		3004	53	2007	
2SB376L		3004	53	2007	
2SB376M		3004	53	2007	
2SB376OR		3004	53	2007	
2SB376R		3004	53	2007	
2SB376X			53	2007	
2SB376Y		3004	53	2007	
2SB377	102	3004	53	2007	
2SB377B	102	3004	53	2007	
2SB378	102A	3004	53	2005	
2SB3783A		3004	53	2007	
2SB3783B		3004	53	2007	
2SB3783C		3004	53	2007	
2SB3783D		3004	53	2007	
2SB3783E		3004	53	2007	
2SB3783F		3004	53	2007	
2SB3783G		3004	53	2007	
2SB3783GN		3004	53	2007	
2SB3783H		3004	53	2007	
2SB3783J		3004	53	2007	
2SB3783K		3004	53	2007	
2SB3783L		3004	53	2007	
2SB3783M		3004	53	2007	
2SB3783OR		3004	53	2007	
2SB3783R		3004	53	2007	
2SB3783X		3004	53	2007	
2SB3783Y		3004	53	2007	
2SB378A	102A	3004	53	2007	
2SB378B		3004	53	2007	
2SB378C		3004	53	2007	
2SB378D		3004	53	2007	
2SB378E		3004	53	2007	
2SB378F		3004	53	2007	
2SB378GN		3004	53	2007	
2SB378J		3004	53	2007	
2SB378K		3004	53	2007	
2SB378L		3004	53	2007	
2SB378M		3004	53	2007	
2SB378OR		3004	53	2007	
2SB378R		3004	53	2007	
2SB378X			53	2007	
2SB378Y		3004	53	2007	
2SB379	102	3004	53	2007	
2SB379-2	102	3004	53	2007	
2SB379A	102	3004	53	2007	
2SB379B	102	3004	53	2007	
2SB379C		3004	53	2007	
2SB379D		3004	53	2007	
2SB379E		3004	53	2007	
2SB379F		3004	53	2007	
2SB379G		3004	53	2007	
2SB379GN		3004	53	2007	
2SB379H		3004	53	2007	
2SB379J		3004	53	2007	
2SB379K		3004	53	2007	
2SB379L		3004	53	2007	
2SB379M		3004	53	2007	
2SB379OR		3004	53	2007	
2SB379R		3004	53	2007	
2SB379Y		3004	53	2007	
2SB37A	102A	3004	53	2007	
2SB37B	102A	3004	53	2007	
2SB37C	102A	3004	53	2007	
2SB37D		3004	53	2007	
2SB37E	102A	3004	53	2007	
2SB37F	102A	3004	53	2007	
2SB37G		3004	53	2007	
2SB37GN		3004	53	2007	
2SB37H		3004	53	2007	
2SB37J		3004	53	2007	
2SB37K		3004	53	2007	
2SB37L		3004	53	2007	
2SB37M		3004	53	2007	
2SB37OR		3004	53	2007	
2SB37R		3004	53	2007	
2SB37X		3004	53	2007	
2SB37Y		3004	53	2007	
2SB38	102	3004	53	2007	
2SB380	102	3004	53	2007	
2SB380A	102	3004	53	2007	
2SB380B		3004	53	2007	
2SB380C		3004	53	2007	
2SB380D		3004	53	2007	
2SB380E		3004	53	2007	
2SB380F		3004	53	2007	
2SB380GN		3004	53	2007	
2SB380J		3004	53	2007	
2SB380K		3004	53	2007	
2SB380L		3004	53	2007	
2SB380OR		3004	53	2007	
2SB380R		3004	53	2007	
2SB380X			53	2007	
2SB380Y		3004	53	2007	
2SB381	102A	3004	53	2007	
2SB3812A		3004	53	2007	
2SB3812B		3004	53	2007	
2SB3812C		3004	53	2007	
2SB3812D		3004	53	2007	
2SB3812F		3004	53	2007	
2SB3812G		3004	53	2007	
2SB3812GN		3004	53	2007	
2SB3812H		3004	53	2007	
2SB3812J		3004	53	2007	
2SB3812K		3004	53	2007	
2SB3812L		3004	53	2007	
2SB3812M		3004	53	2007	
2SB3812OR		3004	53	2007	
2SB3812R		3004	53	2007	
2SB3812X		3004	53	2007	
2SB3812Y		3004	53	2007	
2SB3813A		3004	53	2007	
2SB3813B		3004	53	2007	
2SB3813C		3004	53	2007	
2SB3813D		3004	53	2007	
2SB3813E		3004	53	2007	
2SB3813F		3004	53	2007	
2SB3813G		3004	53	2007	
2SB3813GN		3004	53	2007	
2SB3813H		3004	53	2007	
2SB3813J		3004	53	2007	
2SB3813K		3004	53	2007	
2SB3813L		3004	53	2007	
2SB3813M		3004	53	2007	
2SB3813OR		3004	53	2007	
2SB3813R		3004	53	2007	
2SB3813X		3004	53	2007	
2SB3813Y		3004	53	2007	
2SB381A		3004	53	2007	
2SB381B		3004	53	2007	
2SB381C		3004	53	2007	
2SB381D		3004	53	2007	
2SB381E		3004	53	2007	
2SB381F		3004	53	2007	
2SB381G		3004	53	2007	
2SB381GN		3004	53	2007	
2SB381H		3004	53	2007	
2SB381J		3004	53	2007	
2SB381K		3004	53	2007	
2SB381L		3004	53	2007	
2SB381M		3004	53	2007	
2SB381OR		3004	53	2007	
2SB381R		3004	53	2007	
2SB381X		3004	53	2007	
2SB381Y		3004	53	2007	
2SB382	102A	3004	53	2007	
2SB382A		3004	53	2007	
2SB382B		3004	53	2007	
2SB382BK			53	2007	
2SB382BN			17	2007	
2SB382C		3004	53	2007	
2SB382D		3004	53	2007	
2SB382E		3004	53	2007	
2SB382F		3004	53	2007	
2SB382G		3004	53	2007	
2SB382GN		3004	53	2007	
2SB382H		3004	53	2007	
2SB382J		3004	53	2007	
2SB382K		3004	53	2007	
2SB382L		3004	53	2007	
2SB382OR		3004	53	2007	
2SB382R		3004	53	2007	
2SB382X		3004	53	2007	
2SB382Y		3004	53	2007	
2SB383	102A	3004	53	2007	
2SB383-1	102A	3004	53	2007	
2SB383-2	102A	3004	53	2007	
2SB383A		3004	53	2007	
2SB383B		3004	53	2007	
2SB383C		3004	53	2007	
2SB383D		3004	53	2007	
2SB383E		3004	53	2007	
2SB383F		3004	53	2007	
2SB383G		3004	53	2007	
2SB383GN		3004	53	2007	
2SB383H		3004	53	2007	
2SB383J		3004	53	2007	
2SB383K		3004	53	2007	
2SB383L		3004	53	2007	
2SB383M		3004	53	2007	
2SB383OR		3004	53	2007	
2SB383R		3004	53	2007	
2SB383X		3004	53	2007	
2SB383Y		3004	53	2007	
2SB384	102A	3004	53	2024	
2SB384A		3004	53	2007	
2SB384B		3004	53	2007	
2SB384C		3004	53	2007	
2SB384D		3004	53	2007	
2SB384E		3004	53	2007	
2SB384F		3004	53	2007	
2SB384G		3004	53	2007	
2SB384GN		3004	53	2007	
2SB384H		3004	53	2007	
2SB384J		3004	53	2007	
2SB384K		3004	53	2007	
2SB384L		3004	53	2007	
2SB384M		3004	53	2007	
2SB384OR		3004	53	2007	
2SB384R		3004	53	2007	
2SB384X		3004	53	2007	
2SB384Y		3004	53	2007	
2SB385	102A	3004	52	2024	
2SB385A		3004	53	2007	
2SB385B		3004	53	2007	
2SB385C		3004	53	2007	
2SB385D		3004	53	2007	
2SB385E		3004	53	2007	
2SB385F		3004	53	2007	
2SB385G		3004	53	2007	

Industry Standard No.	ECG	SK	GE	RS 276-	MOTOR.
2SB385GN		3004	53	2007	
2SB385H		3004	53	2007	
2SB385J		3004	53	2007	
2SB385K		3004	53	2007	
2SB385L		3004	53	2007	
2SB385M		3004	53	2007	
2SB385OR		3004	53	2007	
2SB385R		3004	53	2007	
2SB385X		3004	53	2007	
2SB385Y		3004	53	2007	
2SB386	102A	3123	80	2007	
2SB386A		3123	80		
2SB386B		3123	80		
2SB386C		3123	80		
2SB386D		3123	80		
2SB386E		3123	80		
2SB386F		3123	80		
2SB386G		3123	80		
2SB386GN		3123	80		
2SB386H		3123	80		
2SB386K		3123	80		
2SB386L		3123	80		
2SB386M		3123	80		
2SB386OR		3123	80		
2SB386R		3123	80		
2SB386X		3123	80		
2SB386Y		3123	80		
2SB387	102A	3004	53	2007	
2SB387A	102A	3004	53	2007	
2SB387B		3004	53	2007	
2SB387D		3004	53	2007	
2SB387E		3004	53	2007	
2SB387F		3004	53	2007	
2SB387G		3004	53	2007	
2SB387GN		3004	53	2007	
2SB387H		3004	53	2007	
2SB387J		3004	53	2007	
2SB387K		3004	53	2007	
2SB387L		3004	53	2007	
2SB387M		3004	53	2007	
2SB387OR		3004	53	2007	
2SB387R		3004	53	2007	
2SB387X		3004	53	2007	
2SB387Y		3004	53	2007	
2SB389	102A	3004	53	2005	
2SB389-0			66	2048	
2SB389A		3004	53	2007	
2SB389B		3004	53	2007	
2SB389BK		3004	53	2007	
2SB389C		3004	53	2007	
2SB389D		3004	53	2007	
2SB389E		3004	53	2007	
2SB389F		3004	53	2007	
2SB389G		3004	53	2007	
2SB389GN		3004	53	2007	
2SB389H		3004	53	2007	
2SB389J		3004	53	2007	
2SB389K		3004	53	2007	
2SB389L		3004	53	2007	
2SB389M		3004	53	2007	
2SB389OR		3004	53	2007	
2SB389R		3004	53	2007	
2SB389X		3004	53	2007	
2SB389Y		3004	53	2007	
2SB38A		3004	53	2007	
2SB38B		3004	53	2007	
2SB38C		3004	53	2007	
2SB38D		3004	53	2007	
2SB38E		3004	53	2007	
2SB38F		3004	53	2007	
2SB38G		3004	53	2007	
2SB38GN		3004	53	2007	
2SB38H		3004	53	2007	
2SB38J		3004	53	2007	
2SB38K		3004	53	2007	
2SB38L		3004	53	2007	
2SB38M		3004	53	2007	
2SB38OR		3004	53	2007	
2SB38R		3004	53	2007	
2SB38X		3004	53	2007	
2SB38Y		3004	53	2007	
2SB39	102A	3003	53	2007	
2SB390	127	3035	25		
2SB390A		3035	25		
2SB390B		3035	25		
2SB390C		3035	25		
2SB390D		3035	25		
2SB390E		3035	25		
2SB390F		3035	25		
2SB390G		3035	25		
2SB390GN		3035	25		
2SB390H		3035	25		
2SB390J		3035	25		
2SB390K		3035	25		
2SB390L		3035	25		
2SB390M		3035	25		
2SB390OR		3035	25		
2SB390R		3035	25		
2SB390X		3035	25		
2SB390Y		3035	25		
2SB391	121	3009	25	2006	
2SB391A		3009	25		
2SB391B		3009	25		
2SB391C		3009	25		
2SB391D		3009	25		
2SB391E		3009	25		
2SB391F		3009	25		
2SB391G		3009	25		
2SB391GN		3009	25		
2SB391H		3009	25		
2SB391J		3009	25		
2SB391K		3009	25		
2SB391L		3009	25		
2SB391M		3009	25		
2SB391OR		3009	25		
2SB391R		3009	25		
2SB391X		3009	25		
2SB391Y		3009	25		
2SB392	100	3004/102A	53	2007	
2SB392A		3004	53	2007	
2SB392B		3004	53	2007	
2SB392C		3004	53	2007	
2SB392D		3004	53	2007	
2SB392E		3004	53	2007	
2SB392F		3004	53	2007	
2SB392G		3004	53	2007	
2SB392GN		3004	53	2007	
2SB392H		3004	53	2007	
2SB392J		3004	53	2007	
2SB392K		3004	53	2007	
2SB392L		3004	53	2007	
2SB392M		3009	53	2007	
2SB392OR		3004	53	2007	
2SB392R		3004	53	2007	
2SB392X		3004	53	2007	
2SB392Y		3004	53	2007	
2SB393	100	3004/102A	53	2007	
2SB393A		3004	53	2007	
2SB393B		3004	53	2007	
2SB393C		3004	53	2007	
2SB393D		3004	53	2007	
2SB393E		3004	53	2007	
2SB393F		3004	53	2007	
2SB393G		3004	53	2007	
2SB393GN		3004	53	2007	
2SB393H		3004	53	2007	
2SB393J		3004	53	2007	
2SB393K		3004	53	2007	
2SB393L		3004	53	2007	
2SB393M		3004	53	2007	
2SB393OR		3004	53	2007	
2SB393R		3004	53	2007	
2SB393X		3004	53	2007	
2SB393Y		3004	53	2007	
2SB394	100	3004/102A	53	2007	
2SB394A		3004	53	2007	
2SB394B		3004	53	2007	
2SB394C		3004	53	2007	
2SB394D		3004	53	2007	
2SB394E		3004	53	2007	
2SB394F		3004	53	2007	
2SB394G		3004	53	2007	
2SB394GN		3004	53	2007	
2SB394H		3004	53	2007	
2SB394K		3004	53	2007	
2SB394L		3004	53	2007	
2SB394M		3004	53	2007	
2SB394OR		3004	53	2007	
2SB394R		3004	53	2007	
2SB394X		3004	53	2007	
2SB394Y		3004	53	2007	
2SB395	100	3004/102A	53	2007	
2SB395A		3004	53	2007	
2SB395B		3004	53	2007	
2SB395C		3004	53	2007	
2SB395D		3004	53	2007	
2SB395E		3004	53	2007	
2SB395F		3004	53	2007	
2SB395G		3004	53	2007	
2SB395GN		3004	53	2007	
2SB395H		3004	53	2007	
2SB395J		3004	53	2007	
2SB395K		3004	53	2007	
2SB395M		3004	53	2007	
2SB395OR		3004	53	2007	
2SB395R		3004	53	2007	
2SB395X		3004	53	2007	
2SB395Y		3004	53	2007	
2SB396	102	3004	53	2007	
2SB396A		3004	53	2007	
2SB396B		3004	53	2007	
2SB396C		3004	53	2007	
2SB396D		3004	53	2007	
2SB396E		3004	53	2007	
2SB396F		3004	53	2007	
2SB396G		3004	53	2007	
2SB396GN		3004	53	2007	
2SB396H		3004	53	2007	
2SB396J		3004	53	2007	
2SB396K		3004	53	2007	
2SB396L		3004	53	2007	
2SB396M		3004	53	2007	
2SB396OR		3004	53	2007	
2SB396R		3004	53	2007	
2SB396X		3004	53	2007	
2SB396Y		3004	53	2007	
2SB398			53	2007	
2SB39A		3004	53	2007	
2SB39B		3004	53	2007	
2SB39C		3004	53	2007	
2SB39D		3004	53	2007	
2SB39E		3004	53	2007	
2SB39F		3004	53	2007	
2SB39G		3004	53	2007	
2SB39GN		3004	53	2007	
2SB39H		3004	53	2007	
2SB39J		3004	53	2007	
2SB39K		3004	53	2007	
2SB39L		3004	53	2007	
2SB39M		3004	53	2007	
2SB39OR		3004	53	2007	
2SB39R		3004	53	2007	
2SB39X		3004	53	2007	
2SB39Y		3004	53	2007	
2SB40	102A	3004	53	2007	
2SB400	102A	3004	53	2007	
2SB400A	102A	3004	53	2007	
2SB400B	102A		53	2007	
2SB400BK		3004	53	2024	
2SB400BL	102A		53	2007	
2SB400C		3004	53	2024	
2SB400D		3004	53	2024	
2SB400E		3004	53	2024	
2SB400F		3004	53	2024	
2SB400G		3004	53	2024	
2SB400GN		3004	53	2024	
2SB400H		3004	53	2024	
2SB400J		3004	53	2024	
2SB400K	102A	3004	53	2007	
2SB400L		3004	53	2024	
2SB400M		3004	53	2024	
2SB400OR			53	2024	
2SB400R		3004	53	2024	
2SB400X		3004	53	2024	
2SB400Y		3004	53	2024	
2SB401	100	3004/102A	53	2007	
2SB401A		3004	53	2007	
2SB401B		3004	53	2007	
2SB401C		3004	53	2007	
2SB401D		3004	53	2007	
2SB401E		3004	53	2007	
2SB401F		3004	53	2007	
2SB401G		3004	53	2007	
2SB401GN		3004	53	2007	
2SB401H		3004	53	2007	
2SB401K		3004	53	2007	
2SB401L		3004	53	2007	
2SB401M		3004	53	2007	
2SB401OR		3004	53	2007	
2SB401R		3004	53	2007	
2SB401X		3004	53	2007	
2SB401Y		3004	53	2007	
2SB402	100	3004/102A	53	2007	
2SB402A		3004	53	2007	
2SB402B		3004	53	2007	
2SB402C		3004	53	2007	
2SB402D		3004	53	2007	
2SB402E		3004	53	2007	

Industry Standard No.	ECG	SK	GE	RS 276-	MOTOR.
2SB402F		3004	53	2007	
2SB402G		3004	53	2007	
2SB402GN		3004	53	2007	
2SB402H		3004	53	2007	
2SB402J		3004	53	2007	
2SB402K		3004	53	2007	
2SB402L		3004	53	2007	
2SB402M		3004	53	2007	
2SB4020R		3004	53	2007	
2SB402R		3004	53	2007	
2SB402X		3004	53	2007	
2SB402Y		3004	53	2007	
2SB403	100	3004/102A	53	2007	
2SB403A		3004	53	2007	
2SB403B		3004	53	2007	
2SB403C		3004	53	2007	
2SB403D		3004	53	2007	
2SB403E		3004	53	2007	
2SB403F		3004	53	2007	
2SB403G		3004	53	2007	
2SB403GN		3004	53	2007	
2SB403H		3004	53	2007	
2SB403J		3004	53	2007	
2SB403K		3004	53	2007	
2SB403L		3004	53	2007	
2SB403M		3004	53	2007	
2SB4030R		3004	53	2007	
2SB403R		3004	53	2007	
2SB403X		3004	53	2007	
2SB403Y		3004	53	2007	
00002SB405	158	3004/102A	53	2007	
2SB405(K)			53	2007	
2SB405-0			53	2007	
2SB405-1		3004	53	2007	
2SB405-2C	158	3004/102A	53	2007	
2SB405-3C	158	3004/102A	53	2007	
2SB405-4C	158	3004/102A	53	2007	
2SB405-0			53	2007	
2SB405-0R			53	2007	
2SB405-R		3004	53	2007	
2SB405A	158	3004/102A	53	2007	
2SB405AG		3004	53	2007	
2SB405B	158	3004/102A	53	2007	
2SB405BR	158		53	2007	
2SB405C	158	3004/102A	53	2007	
2SB405D	158	3004/102A	53	2007	
2SB405DK		3004	53	2007	
2SB405E	158	3004/102A	53	2007	
2SB405EK		3004	53	2007	
2SB405F		3004	53	2007	
2SB405G	158	3004/102A	53	2007	
2SB405GN		3004	53	2007	
2SB405H	158	3004/102A	53	2007	
2SB405J		3004	53	2007	
2SB405K	158	3004/102A	53	2007	
2SB405L		3004	53	2007	
2SB405M		3004	53	2007	
2SB4050R		3004	53	2007	
2SB405P	158		53	2007	
2SB405R	158	3004/102A	53	2007	
2SB405RE	158		53	2007	
2SB405X		3004	53	2007	
2SB405Y		3004	53	2007	
2SB407	121	3009	25	2006	
2SB407-0		3009	25		
2SB407-0	121		25	2006	
2SB407-0R			25	2006	
2SB407A		3009	25	2006	
2SB407B		3009	25	2006	
2SB407BK			25	2006	
2SB407C		3009	25	2006	
2SB407D		3009	25	2006	
2SB407E		3009	25	2006	
2SB407F		3009	25	2006	
2SB407G		3009	25	2006	
2SB407GN		3009	25	2006	
2SB407H		3009	25	2006	
2SB407J		3009	25	2006	
2SB407K		3009	25	2006	
2SB407M		3009	25	2006	
2SB4070R		3009	25		
2SB407R		3009	25	2006	
2SB407TV	121	3009	25	2006	
2SB407TV-2		3009	25	2006	
2SB407X			25	2006	
2SB407Y		3009	25	2006	
2SB408	102A	3004	53	2007	
2SB408A		3004	53	2007	
2SB408B		3004	53	2007	
2SB408C		3004	53	2007	
2SB408D		3004	53	2007	
2SB408E		3004	53	2007	
2SB408F		3004	53	2007	
2SB408G		3004	53	2007	
2SB408GN		3004	53	2007	
2SB408H		3004	53	2007	
2SB408J		3004	53	2007	
2SB408K		3004	53	2007	
2SB408L		3004	53	2007	
2SB408M		3004	53	2007	
2SB4080R		3004	53	2007	
2SB408R		3004	53	2007	
2SB408X		3004	53	2007	
2SB408Y		3004	53	2007	
2SB40A		3004	53	2007	
2SB40B		3004	53	2007	
2SB40C		3004	53	2007	
2SB40D		3004	53	2007	
2SB40E		3004	53	2007	
2SB40F		3004	53	2007	
2SB40G		3004	53	2007	
2SB40GN		3004	53	2007	
2SB40H		3004	53	2007	
2SB40J		3004	53	2007	
2SB40K		3004	53	2007	
2SB40L		3004	53	2007	
2SB40M		3004	53	2007	
2SB400R		3004	53	2007	
2SB40R		3004	53	2007	
2SB40X		3004	53	2007	
2SB40Y		3004	53	2007	
2SB41	104	3009	25	2006	
2SB410	127	3014	25		
2SB411	127	3014	25		
2SB411TV		3035	25		
2SB413	131	3014	3	2006	
2SB413A		3014	3	2006	
2SB413B		3014	3	2006	
2SB413C		3014	3	2006	
2SB413D		3014	3	2006	
2SB413E		3014	3	2006	
2SB413F		3014	3	2006	
2SB413G		3014	3	2006	
2SB413GN		3014	3	2006	
2SB413H		3014	3	2006	
2SB413J		3014	3	2006	
2SB413K		3014	3	2006	
2SB413L		3014	3	2006	
2SB413M		3014	3	2006	
2SB4130R		3014	3	2006	
2SB413R		3014	3	2006	
2SB413X		3014	3	2006	
2SB413Y		3014	3	2006	
2SB414	226	3014	3	2006	
2SB414B		3014	3	2006	
2SB414C		3014	3	2006	
2SB414D		3014	3	2006	
2SB414E		3014	3	2006	
2SB414F		3014	3	2006	
2SB414G		3014	3	2006	
2SB414GN		3014	3	2006	
2SB414H		3014	3	2006	
2SB414J		3014	3	2006	
2SB414K		3014	3	2006	
2SB414L		3014	3	2006	
2SB414M		3014	3	2006	
2SB4140R		3014	3	2006	
2SB414R		3014	3	2006	
2SB414X		3014	3	2006	
2SB414Y		3014	3	2006	
2SB415	158	3004/102A	53	2007	
2SB415-0R			53	2007	
2SB415A	158	3004/102A	53	2007	
2SB415B	158	3004/102A	53	2007	
2SB415C		3004	53	2007	
2SB415D		3004	53	2007	
2SB415E		3004	53	2007	
2SB415F		3004	53	2007	
2SB415G		3004	53	2007	
2SB415GN		3004	53	2007	
2SB415H		3004	53	2007	
2SB415J		3004	53	2007	
2SB415K		3004	53	2007	
2SB415L		3004	53	2007	
2SB415M		3004	53	2007	
2SB415MP	158(2)		53		
2SB4150R		3004	53		
2SB415R		3004	53	2007	
2SB415X			53	2007	
2SB415Y		3004	53	2007	
2SB416	100	3004/102A	53	2007	
2SB416A		3004	53	2007	
2SB416B		3004	53	2007	
2SB416C		3004	53	2007	
2SB416D		3004	53	2007	
2SB416E		3004	53	2007	
2SB416F		3004	53	2007	
2SB416G		3004	53	2007	
2SB416GN		3004	53	2007	
2SB416H		3004	53	2007	
2SB416J		3004	53	2007	
2SB416K		3004	53	2007	
2SB416L		3004	53	2007	
2SB416M		3004	53	2007	
2SB4160R		3004	53	2007	
2SB416R		3004	53	2007	
2SB416X		3004	53	2007	
2SB416Y		3004	53	2007	
2SB417	100	3004/102A	53	2007	
2SB417A		3004	53	2007	
2SB417B		3004	53	2007	
2SB417C		3004	53	2007	
2SB417D		3004	53	2007	
2SB417E		3004	53	2007	
2SB417F		3004	53	2007	
2SB417G		3004	53	2007	
2SB417GN		3004	53	2007	
2SB417H		3004	53	2007	
2SB417J		3004	53	2007	
2SB417K		3004	53	2007	
2SB417L		3004	53	2007	
2SB417M		3004	53	2007	
2SB4170R		3004	53	2007	
2SB417R		3004	53	2007	
2SB417X		3004	53	2007	
2SB417Y		3004	53	2007	
2SB41A		3009	25		
2SB41B		3009	25		
2SB41C		3009	25		
2SB41D		3009	25		
2SB41E		3009	25		
2SB41F		3009	25		
2SB41G		3009	25		
2SB41GN		3009	25		
2SB41H		3009	25		
2SB41J		3009	25		
2SB41K		3009	25		
2SB41L		3009	25		
2SB41M		3009	25		
2SB410R		3009	25		
2SB41R		3009	25		
2SB41X		3009	25		
2SB41Y		3009	25		
2SB42	121	3009	25	2006	
2SB421			53	2007	
2SB422	102	3004	53	2007	
2SB422A		3004	53	2007	
2SB422B		3004	53	2007	
2SB422C		3004	53	2007	
2SB422D		3004	53	2007	
2SB422E		3004	53	2007	
2SB422F		3004	53	2007	
2SB422G		3004	53	2007	
2SB422GN		3004	53	2007	
2SB422H		3004	53	2007	
2SB422J		3004	53	2007	
2SB422K		3004	53	2007	
2SB422L		3004	53	2007	
2SB422M		3004	53	2007	
2SB4220R		3004	53	2007	
2SB422R		3004	53	2007	
2SB422X		3004	53	2007	
2SB422Y		3004	53	2007	
2SB423	102	3004	53	2007	
2SB423A		3004	53	2007	
2SB423B		3004	53	2007	
2SB423C		3004	53	2007	
2SB423D		3004	53	2007	
2SB423E		3004	53	2007	
2SB423F		3004	53	2007	
2SB423G		3004	53	2007	
2SB423GN		3004	53	2007	
2SB423H		3004	53	2007	
2SB423J		3004	53	2007	
2SB423K		3004	53	2007	
2SB423L		3004	53	2007	
2SB423M		3004	53	2007	
2SB4230R		3004	53	2007	

Industry Standard No.	ECG	SK	GE	RS 276-	MOTOR.
2SB423R		3004	53	2007	
2SB423X		3004	53	2007	
2SB423Y		3004	53	2007	
2SB424	121	3014	25	2006	
2SB424A		3034	25		
2SB424B		3034	25		
2SB424C		3034	25		
2SB424D		3034	25		
2SB424E		3034	25		
2SB424F		3034	25		
2SB424G		3034	25		
2SB424GN		3034	25		
2SB424H		3034	25		
2SB424J		3034	25		
2SB424K		3034	25		
2SB424L		3034	25		
2SB424M		3034	25		
2SB424OR		3034	25		
2SB424R		3034	25		
2SB424X		3034	25		
2SB424Y		3034	25		
2SB425	121	3034	25		
2SB425A		3034	25		
2SB425B		3034	25		
2SB425C		3034	25		
2SB425D		3034	25		
2SB425E		3034	25		
2SB425F		3034	25		
2SB425G		3034	25		
2SB425H		3034	25		
2SB425J		3034	25		
2SB425K		3034	25		
2SB425L		3034	25		
2SB425M		3034	25		
2SB425R		3034	25		
2SB425X		3034	25		
2SB425Y	121	3034	25		
2SB426	121	3009	25	2006	
2SB426A		3009	25		
2SB426B		3009	25		
2SB426BL	121	3717	25	2006	
2SB426C		3009	25		
2SB426D		3009	25		
2SB426E		3009	25		
2SB426F		3009	25		
2SB426G		3009	25		
2SB426GN		3009	25		
2SB426J		3009	25		
2SB426K		3009	25		
2SB426L		3009	25		
2SB426M		3009	25		
2SB426OR		3009	25		
2SB426R	121	3009	25	2006	
2SB426X		3009	25		
2SB426Y	121	3009	25	2006	
2SB427	102A	3123	80	2007	
2SB427A		3123	80		
2SB427B		3123	80		
2SB427C		3123	80		
2SB427D		3123	80		
2SB427E		3123	80		
2SB427F		3123	80		
2SB427G		3123	80		
2SB427GN		3034	80		
2SB427H		3123	80		
2SB427J		3123	80		
2SB427K		3123	80		
2SB427L		3123	80		
2SB427M		3123	80		
2SB427OR		3034	80		
2SB427R		3123	80		
2SB427X		3123	80		
2SB427Y		3123	80		
2SB428	102A	3123	80	2007	
2SB428A		3123	80		
2SB428B		3123	80		
2SB428C		3123	80		
2SB428D		3123	80		
2SB428E		3123	80		
2SB428F		3123	80		
2SB428G		3123	80		
2SB428GN		3123	80		
2SB428H		3123	80		
2SB428J		3123	80		
2SB428K		3123	80		
2SB428L		3123	80		
2SB428M		3123	80		
2SB428OR		3123	80		
2SB428R		3123	80		
2SB428X		3123	80		
2SB428Y		3123	80		
2SB42A		3009	25		
2SB42B		3009	25		
2SB42C		3009	25		
2SB42D		3009	25		
2SB42E		3009	25		
2SB42F		3009	25		
2SB42G		3009	25		
2SB42GN		3009	25		
2SB42H		3009	25		
2SB42J		3009	25		
2SB42K		3009	25		
2SB42L		3009	25		
2SB42M		3009	25		
2SB42OR		3009	25		
2SB42R		3009	25		
2SB42X		3009	25		
2SB42Y		3009	25		
2SB43	102A	3003	53	2007	
2SB430	213		4		
2SB431	158		53	2007	
2SB432	127	3035	25		
2SB432A		3035	25		
2SB432B		3035	25		
2SB432C		3035	25		
2SB432D		3035	25		
2SB432E		3035	25		
2SB432F		3035	25		
2SB432G		3035	25		
2SB432GN		3035	25		
2SB432H		3035	25		
2SB432J		3035	25		
2SB432K		3035	25		
2SB432L		3035	25		
2SB432M		3035	25		
2SB432OR		3035	25		
2SB432R		3035	25		
2SB432X		3035	25		
2SB432Y		3035	25		
2SB433	213		4		
2SB434	153	3274	69	2049	
2SB434(0)			69	2049	
2SB434-0	153	3274	69	2049	
2SB434-R	153		69	2049	
2SB434-Y	153		69	2049	
2SB435	153	3274	53	2049	
2SB435-0	153		69	2049	
2SB435-O	153	3274	69	2049	
2SB435-R	153		69	2049	
2SB435-Y	153		69	2049	
2SB435R		3274	66	2049	
002SB435RY	153	3274	69	2049	
2SB435Y	153	3274	66	2049	
2SB436				2005	
2SB439	102A	3004	53	2007	
2SB439A	102A	3004	53	2007	
2SB439B		3004	53	2007	
2SB439C		3004	53	2007	
2SB439D		3004	53	2007	
2SB439E		3004	53	2007	
2SB439F		3004	53	2007	
2SB439G		3004	53	2007	
2SB439GN		3004	53	2007	
2SB439H		3004	53	2007	
2SB439J		3004	53	2007	
2SB439L		3004	53	2007	
2SB439M		3004	53	2007	
2SB439OR		3004	53	2007	
2SB439R		3004	53	2007	
2SB439X			53	2007	
2SB439Y		3004	53	2007	
2SB43A	102A	3004	53	2007	
2SB43B		3004	53	2024	
2SB43C		3004	53	2024	
2SB43D		3004	53	2024	
2SB43E		3004	53	2024	
2SB43F		3004	53	2024	
2SB43G		3004	53	2024	
2SB43GN		3004	53	2024	
2SB43H		3004	53	2024	
2SB43J		3004	53	2024	
2SB43K		3004	53	2024	
2SB43L		3004	53	2024	
2SB43M		3004	53	2024	
2SB43OR		3004	53	2024	
2SB43R		3004	53	2024	
2SB43X		3004	53	2024	
2SB43Y		3004	53	2024	
2SB44	102A	3004	53	2007	
2SB440	102A	3004	53	2007	
2SB440A		3004	53	2007	
2SB440B		3004	53	2007	
2SB440C		3004	53	2007	
2SB440E		3004	53	2007	
2SB440F		3004	53	2007	
2SB440G		3004	53	2007	
2SB440GN		3004	53	2007	
2SB440H		3004	53	2007	
2SB440K		3004	53	2007	
2SB440L		3004	53	2007	
2SB440M		3004	53	2007	
2SB440OR			53	2007	
2SB440R		3004	53	2007	
2SB440X		3004	53	2007	
2SB440Y		3004	53	2007	
2SB443	102A	3004	53	2007	
2SB443A	102A	3004	53	2007	
2SB443B	102A	3004	53	2007	
2SB443C		3004	53	2007	
2SB443D		3004	53	2007	
2SB443E		3004	53	2007	
2SB443F		3004	53	2007	
2SB443G		3004	53	2007	
2SB443GN		3004	53	2007	
2SB443H		3004	53	2007	
2SB443J		3004	53	2007	
2SB443K		3004	53	2007	
2SB443L		3004	53	2007	
2SB443M		3004	53	2007	
2SB443OR		3004	53	2007	
2SB443R		3004	53	2007	
2SB443X		3004	53	2007	
2SB443Y		3004	53	2007	
2SB444	126	3004/102A	53	2024	
2SB4440D		3004	53	2007	
2SB444A	126	3006/160	53	2024	
2SB444B	126	3006/160	53	2024	
2SB444C		3004	53	2007	
2SB444D		3004	53	2007	
2SB444E		3004	53	2007	
2SB444F		3004	53	2007	
2SB444G		3004	53	2007	
2SB444GN		3004	53	2007	
2SB444H		3004	53	2007	
2SB444K		3004	53	2007	
2SB444L		3004	53	2007	
2SB444M		3004	53	2007	
2SB444OR		3004	53	2007	
2SB444R		3004	53	2007	
2SB444X		3004	53	2007	
2SB444Y		3004	53	2007	
2SB445	226$	3014	3	2006	
2SB445A		3014	3	2006	
2SB445B		3014	3	2006	
2SB445C		3014	3	2006	
2SB445D		3014	3	2006	
2SB445E		3014	3	2006	
2SB445F		3014	3	2006	
2SB445G		3014	3	2006	
2SB445GN		3014	3	2006	
2SB445H		3014	3	2006	
2SB445J		3014	3	2006	
2SB445K		3014	3	2006	
2SB445L		3014	3	2006	
2SB445M		3014	3	2006	
2SB445OR		3014	3	2006	
2SB445R		3014	3	2006	
2SB445X		3014	3	2006	
2SB445Y		3014	3	2006	
2SB446	226$	3014	3	2006	
2SB446A		3014	3	2006	
2SB446B		3014	3	2006	
2SB446C		3014	3	2006	
2SB446D		3014	3	2006	
2SB446E		3014	3	2006	
2SB446F		3014	3	2006	
2SB446G		3014	3	2006	
2SB446GN		3014	3	2006	
2SB446H		3014	3	2006	
2SB446J		3014	3	2006	
2SB446K		3014	3	2006	
2SB446L		3014	3	2006	
2SB446M		3014	3	2006	
2SB446OR		3014	3	2006	
2SB446R		3014	3	2006	
2SB446X		3014	3	2006	
2SB446Y		3014	3	2006	
2SB447	127	3035	25		

Industry Standard No.	ECG	SK	GE	RS 276-	MOTOR.
2SB447A		3035	25		
2SB447B		3035	25		
2SB447C		3035	25		
2SB447D		3035	25		
2SB447E		3035	25		
2SB447F		3035	25		
2SB447G		3035	25		
2SB447GN		3035	25		
2SB447H		3035	25		
2SB447J		3035	25		
2SB447K		3035	25		
2SB447L		3035	25		
2SB447OR		3035	25		
2SB447R		3035	25		
2SB447X		3035	25		
2SB447Y		3035	25		
2SB448	131	3198	25	2006	
2SB448A		3198	25		
2SB448B		3198	25		
2SB448C		3198	25		
2SB448D		3198	25		
2SB448E		3198	25		
2SB448F		3198	25		
2SB448G		3198	25		
2SB448GN		3198	25		
2SB448H		3198	25		
2SB448K		3198	25		
2SB448L		3198	25		
2SB448M		3198	25		
2SB448OR		3198	25		
2SB448R		3198	25		
2SB448Y		3198	25		
2SB449	121	3009	25	2006	
2SB449A	121	3009	25		
2SB449B	121	3009	25		
2SB449C	121	3009	25		
2SB449D	121	3009	25		
2SB449E	121	3009	25		
2SB449F	121	3009	25	2006	
2SB449G		3009	25		
2SB449GN		3009	25		
2SB449H		3009	25		
2SB449J		3009	25		
2SB449K		3009	25		
2SB449L	121	3009	25		
2SB449M	121	3009	25		
2SB449OR		3009	25		
2SB449P	121	3717	25	2006	
2SB449PG		3009	25		
2SB449X		3009	25		
2SB449Y		3009	25		
2SB44A		3004	53	2024	
2SB44B		3004	53	2024	
2SB44C		3004	53	2024	
2SB44D		3004	53	2024	
2SB44E		3004	53	2024	
2SB44F		3004	53	2024	
2SB44G		3004	53	2024	
2SB44GN		3004	53	2024	
2SB44H		3004	53	2024	
2SB44J		3004	53	2024	
2SB44K		3004	53	2024	
2SB44L		3004	53	2024	
2SB44M		3004	53	2024	
2SB44OR		3004	53	2024	
2SB44R		3004	53	2024	
2SB44X		3004	53	2024	
2SB44Y		3004	53	2024	
2SB450	158		53	2007	
2SB450A	158		53	2007	
2SB451	158		53	2007	
2SB452	158	3004/102A	53	2007	
2SB452A	158	3004/102A	53	2007	
2SB453	158		53	2007	
2SB454			53	2007	
2SB455			53	2007	
2SB457	158	3004/102A	53	2007	
2SB457-C	158		53	2007	
2SB457A	158	3004/102A	53	2007	
2SB457AC	158	3004/102A	53	2007	
2SB457B		3004	53	2007	
2SB457C		3004	53	2007	
2SB457D		3004	53	2007	
2SB457E		3004	53	2007	
2SB457F		3004	53	2007	
2SB457G		3004	53	2007	
2SB457GN		3004	53	2007	
2SB457H		3004	53	2007	
2SB457J		3004	53	2007	
2SB457K		3004	53	2007	
2SB457L		3004	53	2007	
2SB457M		3004	53	2007	
2SB457OR		3004	53	2007	
2SB457R		3004	53	2007	
2SB457X		3004	53	2007	
2SB457Y		3004	53	2007	
2SB458	131	3052	30	2006	
2SB458A	131		30	2006	
2SB458B		3052	44	2006	
2SB458BC			30	2006	
2SB458BL			30	2006	
2SB458C		3052	30	2006	
2SB458D		3052	30		
2SB458E		3052	30		
2SB458F		3052	30		
2SB458G		3052	30		
2SB458GN		3052	30		
2SB458H		3052	30		
2SB458J		3052	30		
2SB458K		3052	30		
2SB458L		3052	30		
2SB458M		3052	30		
2SB458OR		3052	30		
2SB458R		3052	30		
2SB458X		3052	30		
2SB458Y		3052	30		
2SB459	102A	3004	53	2007	
2SB459-0	102A		53	2007	
2SB459A	102A	3004	53	2007	
2SB459B	102A	3004	53	2007	
2SB459C	102A	3004	53	2007	
2SB459C-2		3004	53	2007	
2SB459D	102A	3004	53	2007	
2SB459E		3004	53	2007	
2SB459F		3004	53	2007	
2SB459G		3004	53	2007	
2SB459GN		3004	53	2007	
2SB459H		3004	53	2007	
2SB459J		3004	53	2007	
2SB459K		3004	53	2007	
2SB459L		3004	53	2007	
2SB459M		3004	53	2007	
2SB459OR		3004	53	2007	
2SB459R		3004	53	2007	
2SB459X		3004	53	2007	
2SB459Y		3004	53	2007	
2SB46	102A	3003	53	2007	
2SB460	102A	3004	53	2015	
2SB460A	102A	3004	80	2007	
2SB460B	102A	3004	80	2007	
2SB461	176	3004/102A	80		
2SB461A		3004	80	2007	
2SB461B		3004	80	2007	
2SB461BL			53	2007	
2SB461C		3004	80	2007	
2SB461D		3004	80	2007	
2SB461E		3004	80	2007	
2SB461F		3004	80	2007	
2SB461G		3004	80	2007	
2SB461H		3004	80	2007	
2SB461J		3004	80	2007	
2SB461K		3004	80	2007	
2SB461L		3004	80	2007	
2SB461M		3004	80	2007	
2SB461OR		3004	80	2007	
2SB461R		3004	80	2007	
2SB461X		3004	80	2007	
2SB461Y		3004	80	2007	
2SB462	131$	3052	30	2006	
2SB462A		3052	30		
2SB462B		3052	30		
2SB462C		3052	30		
2SB462D		3052	30		
2SB462E		3052	30		
2SB462F		3052	30		
2SB462G		3052	30		
2SB462GN		3052	80		
2SB462H		3052	30		
2SB462K		3052	30		
2SB462L		3052	30		
2SB462M		3052	30		
2SB462OR		3052	30		
2SB462R		3052	30		
2SB462X		3052	30		
2SB462Y		3052	30		
2SB463	131	3052	49	2006	
2SB463(Y)	131		49		
2SB463-0			30	2006	
2SB463A		3052	49	2006	
2SB463B		3052	49	2006	
2SB463BL	131	3052	49	2006	
2SB463BLU			30	2006	
2SB463BLU-G			30	2006	
2SB463C		3052	49	2006	
2SB463D		3052	49	2006	
2SB463E	131	3052	44	2006	
2SB463F		3052	49	2006	
2SB463G		3052	49	2006	
2SB463G-BL			30	2006	
2SB463G-R			30	2006	
2SB463G-Y			30	2006	
2SB463GN		3052	49	2006	
2SB463H		3052	49	2006	
2SB463J		3052	49	2006	
2SB463K		3052	49	2006	
2SB463L		3052	49	2006	
2SB463M		3052	49	2006	
2SB463R	131	3052	49	2006	
2SB463RED			30	2006	
2SB463RED-G			30	2006	
2SB463X			49	2006	
2SB463XL		3052	49	2006	
2SB463Y	131	3052	44	2006	
2SB463YEL			30	2006	
2SB463YEL-G			30	2006	
2SB464	127	3035	25		
2SB464A		3035	25		
2SB464B		3035	25		
2SB464C		3035	25		
2SB464D		3035	25		
2SB464E		3035	25		
2SB464F		3035	25		
2SB464G		3035	25		
2SB464GN		3035	25		
2SB464H		3035	25		
2SB464J		3035	25		
2SB464K		3035	25		
2SB464M		3035	25		
2SB464OR		3035	25		
2SB464X		3035	25		
2SB464Y		3035	25		
2SB465	127	3035	25		
2SB465A		3035	25		
2SB465B		3035	25		
2SB465C		3035	25		
2SB465D		3035	25		
2SB465E		3035	25		
2SB465F		3035	25		
2SB465G		3035	25		
2SB465GN		3035	25		
2SB465H		3035	25		
2SB465J		3035	25		
2SB465K		3035	25		
2SB465L		3035	25		
2SB465M		3035	25		
2SB465R		3035	25		
2SB465X		3035	25		
2SB465Y		3035	25		
2SB466	226$	3009	25	2006	
2SB466A		3009	25		
2SB466B		3009	25		
2SB466C		3009	25		
2SB466D		3009	25		
2SB466E		3009	25		
2SB466F		3009	25		
2SB466G		3009	25		
2SB466GN		3009	25		
2SB466H		3009	25		
2SB466J		3009	25		
2SB466K		3009	25		
2SB466L		3009	25		
2SB466M		3009	25		
2SB466OR		3009	25		
2SB466R		3009	25		
2SB466X		3009	25		
2SB466Y		3009	25		
2SB467	226$	3009	25	2006	
2SB467A		3009	25		
2SB467C		3009	25		
2SB467D		3009	25		
2SB467E		3009	25		
2SB467F		3009	25		
2SB467G		3009	25		
2SB467GN		3009	25		
2SB467H		3009	25		
2SB467J		3009	25		
2SB467K		3009	25		
2SB467L		3009	25		

Industry Standard No.	ECG	SK	GE	RS 276-	MOTOR.
2SB467M		3009	25		
2SB467OR		3009	25		
2SB467X		3009	25		
2SB467Y		3009	25		
2SB468	127	3035	25		
2SB468A	127	3035	25		
2SB468B	127	3035	25		
2SB468B-5		3035	25		
2SB468C	127	3035	25		
2SB468D	127	3035	25		
2SB468E		3035	25		
2SB468F		3035	25		
2SB468G		3035	25		
2SB468GN		3035	25		
2SB468H		3035	25		
2SB468J		3035	25		
2SB468K		3035	25		
2SB468L		3035	25		
2SB468OR		3035	25		
2SB468X		3035	25		
2SB468Y		3035	25		
2SB46A		3004	53	2024	
2SB46B		3004	53	2024	
2SB46C		3004	53	2024	
2SB46D		3004	53	2024	
2SB46E		3004	53	2024	
2SB46F		3004	53	2024	
2SB46G		3004	53	2024	
2SB46GN		3004	53	2024	
2SB46H		3004	53	2024	
2SB46J		3004	53	2024	
2SB46K		3004	53	2024	
2SB46L		3004	53	2024	
2SB46M		3004	53	2024	
2SB46OR		3004	53	2024	
2SB46R		3004	53	2024	
2SB46X		3004	53	2024	
2SB46Y		3004	53	2024	
2SB47	102A	3003	53	2007	
2SB470	102A	3004	53	2007	
2SB470A		3004	53	2007	
2SB470B		3004	53	2007	
2SB470C		3004	53	2007	
2SB470D		3004	53	2007	
2SB470E		3004	53	2007	
2SB470F		3004	53	2007	
2SB470G		3004	53	2007	
2SB470GN		3004	53	2007	
2SB470H		3004	53	2007	
2SB470J		3004	53	2007	
2SB470K		3004	53	2007	
2SB470L		3004	53	2007	
2SB470M		3004	53	2007	
2SB470OR		3004	53	2007	
2SB470R		3004	53	2007	
2SB470X		3004	53	2007	
2SB470Y		3004	53	2007	
2SB471	121	3009	25	2006	
2SB471-2	121	3009	25	2006	
2SB471-0			16	2006	
2SB471A	121	3009	25	2006	
2SB471B	121	3009	25	2006	
2SB471C		3009	25		
2SB471D		3009	25	2006	
2SB471E		3009	25	2006	
2SB471F		3009	25	2006	
2SB471G		3009	25	2006	
2SB471GN		3009	25	2006	
2SB471H		3009	25	2006	
2SB471J		3009	25	2006	
2SB471K		3009	25	2006	
2SB471L		3009	25	2006	
2SB471M		3009	25	2006	
2SB471OR		3009	25		
2SB471R		3009	25	2006	
2SB471X		3009	25	2006	
2SB471Y		3009	25	2006	
2SB472	121$	3034	25	2006	
2SB472A	121$	3009	25	2006	
2SB472B	121$	3009	25	2006	
2SB472C		3034	25		
2SB472D		3034	25		
2SB472E		3034	25		
2SB472F		3034	25		
2SB472G		3034	25		
2SB472GN		3034	25		
2SB472H		3034	25		
2SB472J		3034	25		
2SB472K		3034	25		
2SB472M		3034	25		
2SB472OR		3034	25		
2SB472X		3034	25		
2SB472Y		3034	25		
2SB473	131	3198	30	2006	
2SB473(H)			30	2006(2)	
2SB473A	131	3198	30		
2SB473B	131	3198	30		
2SB473C	131	3198	30		
2SB473D	131	3198	30	2006	
2SB473E	131	3198	30		
2SB473F	131	3198	30	2006	
2SB473G	131	3198	30		
2SB473GN		3198	30		
2SB473H	131	3198	30	2006	
2SB473K		3198	30		
2SB473L		3198	30		
2SB473M		3198	30		
2SB473X		3198	30		
2SB473Y		3198	30		
2SB474	226	3082	49	2025	
2SB474-2	226	3082	49	2025	
2SB474-3	226	3082	49	2025	
2SB474-4	226		49	2025	
2SB474-6D	226		49	2025	
2SB474-OR			49	2025	
2SB474A			49	2025	
2SB474B			49	2025	
2SB474C			49	2025	
2SB474D			49	2025	
2SB474E			49	2025	
2SB474F			49	2025	
2SB474G			49	2025	
2SB474GN			49	2025	
2SB474H			49	2025	
2SB474J			49	2025	
2SB474K			49	2025	
2SB474L			49	2025	
2SB474M			49	2025	
2SB474MP	226		49	2025	
2SB474OR			49	2025	
2SB474R			49	2025	
2SB474S	226	3082	49	2025	
2SB474V10	226	3082	49	2025	
2SB474V4	226		49	2025	
2SB474X			49	2025	
2SB474Y	226	3082	49	2025	
2SB474YE1			49	2025	
2SB474YEL			49	2025	
2SB475	102A	3004	53	2007	
2SB475A	102A	3004	53	2007	
2SB475B	102A	3004	53	2007	
2SB475C	102A	3004	53	2007	
2SB475D	102A	3004	53	2007	
2SB475E	102A	3004	53	2007	
2SB475F	102A	3004	53	2007	
2SB475G	102A	3004	53	2007	
2SB475GN		3004	53	2007	
2SB475H		3004	53	2007	
2SB475J		3004	53	2007	
2SB475K		3004	53	2007	
2SB475L		3004	53	2007	
2SB475M		3004	53	2007	
2SB475OR		3004	53	2007	
2SB475P	102A	3004	53	2007	
2SB475PL		3004	53	2007	
2SB475Q	102A	3004	53	2007	
2SB475R		3004	53	2007	
2SB475X			53	2007	
2SB475Y		3004	53	2007	
2SB476	176		53	2007	
2SB477(C)			80	2007	
2SB477A		3123	80		
2SB477C		3123	80	2024	
2SB477D		3123	80		
2SB477E		3123	80		
2SB477F		3123	80		
2SB477G		3123	80		
2SB477GN		3123	80		
2SB477H		3123	80		
2SB477J		3123	80		
2SB477K		3123	80		
2SB477L		3123	80		
2SB477M		3123	80		
2SB477OR		3123	80		
2SB477R		3123	80		
2SB477X		3123	80		
2SB477Y		3123	80		
2SB47A		3004	53	2024	
2SB47B		3004	53	2024	
2SB47C		3004	53	2024	
2SB47D		3004	53	2024	
2SB47E		3004	53	2024	
2SB47F		3004	53	2024	
2SB47G		3004	53	2024	
2SB47GN		3004	53	2024	
2SB47H		3004	53	2024	
2SB47J		3004	53	2024	
2SB47K		3004	53	2024	
2SB47L		3004	53	2024	
2SB47M		3004	53	2024	
2SB47OR		3004	53	2024	
2SB47R		3004	53	2024	
2SB47X		3004	53	2024	
2SB47Y		3004	53	2024	
2SB48	102	3004	53	2007	
2SB481	131	3198	44(2)	2006(2)	
2SB481-OR			44	2006	
2SB481A	131		44	2006	
2SB481B	131		44	2006	
2SB481C	131		44	2006	
2SB481D	131		44	2006	
2SB481E	131		44	2006	
2SB481F	131		44	2006	
2SB481G	131		44	2006	
2SB481GN			44	2006	
2SB481H	131		44	2006	
2SB481J	131		44	2006	
2SB481K	131		44	2006	
2SB481L	131		44	2006	
2SB481M	131		44	2006	
2SB481OR			44	2006	
2SB481R			44	2006	
2SB481X	131		44	2006	
2SB482	102A	3004	53	2007	
2SB482A		3004	53	2007	
2SB482B		3004	53	2007	
2SB482C		3004	53	2007	
2SB482D		3004	53	2007	
2SB482E		3004	53	2007	
2SB482F		3004	53	2007	
2SB482G		3004	53	2007	
2SB482GN		3004	53	2007	
2SB482H		3004	53	2007	
2SB482J		3004	53	2007	
2SB482K		3004	53	2007	
2SB482L		3004	53	2007	
2SB482M		3004	53	2007	
2SB482V			53	2007	
2SB482X		3004	53	2007	
2SB482Y		3004	53	2007	
2SB483	179	3642	76		
2SB484	179	3642	76		
2SB485	179$	3642/179	76		
2SB486	102A	3004	53	2007	
2SB486A		3004	53	2007	
2SB486B		3004	53	2007	
2SB486C		3004	53	2007	
2SB486D		3004	53	2007	
2SB486E		3004	53	2007	
2SB486F		3004	53	2007	
2SB486G		3004	53	2007	
2SB486GN		3004	53	2007	
2SB486H		3004	53	2007	
2SB486J		3004	53	2007	
2SB486K		3004	53	2007	
2SB486L		3004	53	2007	
2SB486M		3004	53	2007	
2SB486OR		3004	53	2007	
2SB486R		3004	53	2007	
2SB486X		3004	53	2007	
2SB486Y		3004	53	2007	
2SB48A		3004	53	2024	
2SB48B		3004	53	2024	
2SB48C		3004	53	2024	
2SB48D		3004	53	2024	
2SB48E		3004	53	2024	
2SB48F		3004	53	2024	
2SB48G		3004	53	2024	
2SB48GN		3004	53	2024	
2SB48H		3004	53	2024	
2SB48J		3004	53	2024	
2SB48K		3004	53	2024	
2SB48L		3004	53	2024	
2SB48M		3004	53	2024	
2SB48OR		3004	53	2024	
2SB48R		3004	53	2024	
2SB48X		3004	53	2024	
2SB48Y		3004	53	2024	

Industry Standard No.	ECG	SK	GE	RS 276-	MOTOR.
2SB49	102	3004	53	2007	
2SB492	176	3845	80		
2SB492A			53	2007	
2SB492B	176	3845	53		
2SB492C			53	2007	
2SB492D			53	2007	
2SB492E			53	2007	
2SB492F			53	2007	
2SB492G			53	2007	
2SB492GN			53	2007	
2SB492H			53	2007	
2SB492J			53	2007	
2SB492K			53	2007	
2SB492L			53	2007	
2SB492M			53	2007	
2SB492OR			53	2007	
2SB492R			53	2007	
2SB492X			53	2007	
2SB492Y			53	2007	
2SB494	158		53	2007	
2SB495	158	3004/102A	53	2007	
2SB495A	158	3004/102A	53	2007	
2SB495B	158	3004/102A	53	2007	
2SB495C	158	3004/102A	53	2007	
2SB495D	158	3004/102A	53	2007	
2SB495E		3004	53	2007	
2SB495F		3004	53	2007	
2SB495G		3004	53	2007	
2SB495GN		3004	53	2007	
2SB495H		3004	53	2007	
2SB495J		3004	53	2007	
2SB495K		3004	53	2007	
2SB495L		3004	53	2007	
2SB495M		3004	53	2007	
2SB495OR		3004	53	2007	
2SB495R		3004	53	2007	
2SB495T	158	3004/102A	53	2007	
2SB495X		3004	53	2007	
2SB495Y		3004	53	2007	
2SB496	102A	3004	53	2007	
2SB496A		3004	53	2007	
2SB496B		3004	53	2007	
2SB496C		3004	53	2007	
2SB496D		3004	53	2007	
2SB496E		3004	53	2007	
2SB496F		3004	53	2007	
2SB496G		3004	53	2007	
2SB496GN		3004	53	2007	
2SB496H		3004	53	2007	
2SB496J		3004	53	2007	
2SB496K		3004	53	2007	
2SB496L		3004	53	2007	
2SB496M		3004	53	2007	
2SB496OR		3004	53	2007	
2SB496R		3004	53	2007	
2SB496X		3004	53	2007	
2SB496Y		3004	53	2007	
2SB497	102A	3004	53	2007	
2SB497A		3004	53	2007	
2SB497B		3004	53	2007	
2SB497C		3004	53	2007	
2SB497D		3004	53	2007	
2SB497E		3004	53	2007	
2SB497F		3004	53	2007	
2SB497G		3004	53	2007	
2SB497GN		3004	53	2007	
2SB497H		3004	53	2007	
2SB497J		3004	53	2007	
2SB497K		3004	53	2007	
2SB497L		3004	53	2007	
2SB497M		3004	53	2007	
2SB497OR		3004	53	2007	
2SB497R		3004	53	2007	
2SB497X		3004	53	2007	
2SB497Y		3004	53	2007	
2SB498	102A	3004	53	2007	
2SB498A		3004	53	2007	
2SB498B		3004	53	2007	
2SB498C		3004	53	2007	
2SB498D		3004	53	2007	
2SB498E		3004	53	2007	
2SB498F		3004	53	2007	
2SB498G		3004	53	2007	
2SB498GN		3004	53	2007	
2SB498H		3004	53	2007	
2SB498J		3004	53	2007	
2SB498K		3004	53	2007	
2SB498L		3004	53	2007	
2SB498M		3004	53	2007	
2SB498OR		3004	53	2007	
2SB498R		3004	53	2007	
2SB498X		3004	53	2007	
2SB498Y		3004	53	2007	
2SB49A		3004	53	2024	
2SB49B		3004	53	2024	
2SB49C		3004	53	2024	
2SB49D		3004	53	2024	
2SB49E		3004	53	2024	
2SB49F		3004	53	2024	
2SB49G		3004	53	2024	
2SB49GN		3004	53	2024	
2SB49H		3004	53	2024	
2SB49J		3004	53	2024	
2SB49K		3004	53	2024	
2SB49L		3004	53	2024	
2SB49M		3004	53	2024	
2SB49OR		3004	53	2024	
2SB49R		3004	53	2024	
2SB49X		3004	53	2024	
2SB49Y		3004	53	2024	
2SB50	102	3003	53	2007	
2SB502	197		250	2027	2N3741
2SB503	197		250	2027	2N3741
2SB503-R			52	2024	
2SB503A-R			52	2024	
2SB506	88	3846/285	263	2043	2N6228
2SB506A	88		263	2043	
2SB506B	88		263	2043	
2SB506C	88		263	2043	
2SB506D	88		263	2043	
2SB507	153	3083/197	69	2049	2N6125
2SB507E		3083	69	2049	
2SB508	153	3083/197	69	2049	
2SB509	153		69	2049	2N6126
2SB50A		3004	53	2024	
2SB50B		3004	53	2024	
2SB50C		3004	53	2024	
2SB50D		3004	53	2024	
2SB50E		3004	53	2024	
2SB50F		3004	53	2024	
2SB50G		3004	53	2024	
2SB50GN		3004	53	2024	
2SB50H		3004	53	2024	
2SB50J		3004	53	2024	
2SB50K		3004	53	2024	
2SB50L		3004	53	2024	
2SB50M		3004	53	2024	
2SB50OR		3004	53	2024	
2SB50R		3004	53	2024	
2SB50X		3004	53	2024	
2SB50Y		3004	53	2024	
2SB51	102	3003	53	2007	
2SB510	129		244	2027	
2SB510S	129	3203/211	244	2027	
2SB511	153	3274	69	2049	TIP32
2SB511C	153	3274	69	2049	
2SB511D	153	3274	69	2049	
2SB511E	153	3274	69	2049	
2SB512	153	3083/197	69	2049	
2SB512-O	153	3083/197			
2SB512A	197	3083	69	2049	
2SB512AP	197	3083			
2SB512AQ	197	3083			
2SB512P	153	3083/197	69	2049	
2SB512Q	153	3083/197			
2SB513	153	3083/197	69	2049	2N6126
2SB513A	197	3083	69	2049	
2SB513AP	197	3083			
2SB513AQ	197	3083			
2SB513P	153	3083/197	69	2049	
2SB513Q	153	3083/197	69	2049	
2SB513R	153		69	2049	
2SB514	153		69	2049	TIP32A
2SB515	153		69	2049	TIP32A
2SB516	102A		53	2007	
2SB516(C)			52	2024	
2SB516(D)			52	2024	
2SB516C	102A	3004	53	2007	
2SB516CD	102A	3004	53	2007	
2SB516CD(P)			53	2024	
2SB516D	102A	3004	53	2007	
2SB516P	102A	3004	53	2007	
2SB518			74	2043	2N6226
2SB519			74	2043	2N6227
2SB51A		3004	53	2007	
2SB51B		3004	53	2007	
2SB51C		3004	53	2007	
2SB51D		3004	53	2007	
2SB51E		3004	53	2007	
2SB51F		3004	53	2007	
2SB51G		3004	53	2007	
2SB51GN		3004	53	2007	
2SB51H		3004	53	2007	
2SB51J		3004	53	2007	
2SB51K		3004	53	2007	
2SB51L		3004	53	2007	
2SB51M		3004	53	2007	
2SB51OR		3004	53	2007	
2SB51R		3004	53	2007	
2SB51X		3004	53	2007	
2SB51Y		3004	53	2007	
2SB52	102	3003	53	2007	
2SB520	281				2N6228
2SB521					TIP42A
2SB522	197				TIP42A
2SB523	153		29	2052	2N5193
2SB524	153				2N5194
2SB525		3450	272		
2SB525C		3450	272		
2SB525D		3450	272		
2SB525E		3450	272		
2SB526	292	3189A/183			2N4920
2SB526C	292	3189A/183			
2SB526D	292	3189A/183			
2SB527	292				2N4920
2SB528	292	3441			2N4920
2SB528C	292	3441			
2SB528D	292	3441			
2SB529	153				2N5193
2SB52A		3004	53	2007	
2SB52B		3004	53	2007	
2SB52C		3004	53	2007	
2SB52D		3004	53	2007	
2SB52E		3004	53	2007	
2SB52F		3004	53	2007	
2SB52G		3004	53	2007	
2SB52GN		3004	53	2007	
2SB52H		3004	53	2007	
2SB52J		3004	53	2007	
2SB52K		3004	53	2007	
2SB52L		3004	53	2007	
2SB52M		3004	53	2007	
2SB52OR		3004	53	2007	
2SB52R		3004	53	2007	
2SB52X		3004	53	2007	
2SB52Y		3004	53	2007	
2SB53	102	3004	53	2007	
2SB530	281				2N6230
2SB531	281	3173/219			2N6226
2SB532	281	3846/285	266		2N6226
2SB532P	281	3846/285	266		
2SB532Q	281	3846/285	266		
2SB532R	281	3846/285	266		
2SB532S	281	3846/285	266		
2SB534	102A		53	2007	
2SB534(A)			53	2007	
2SB534A	102A		53	2007	
2SB535	158	3004/102A	53	2007	
2SB536	292	344U			2N6126
2SB536K	292	3441			
2SB536L	292	3441			
2SB536M	292	3441			
2SB537	292	3441	69	2049	2N6126
2SB537L	292	3441	69		
2SB537LM	292		69	2049	
2SB539	285	3846	266		2N6231
2SB539AQ	285	3846			
2SB539AR	285	3846			
2SB539AS	285	3846			
2SB539BQ	285	3846			
2SB539BR	285	3846			
2SB539BS	285	3846			
2SB539CQ	285	3846			
2SB539CR	285	3846			
2SB539CS	285	3846			
2SB539Q	285	3846			
2SB539R	285	3846			
2SB539S	285	3846			
2SB53A		3004	53	2007	
2SB53B		3004	53	2007	
2SB53C		3004	53	2007	
2SB53D		3004	53	2007	
2SB53E		3004	53	2007	
2SB53F		3004	53	2007	
2SB53G		3004	53	2007	
2SB53GN		3004	53	2007	
2SB53H		3004	53	2007	
2SB53J		3004	53	2007	

Industry Standard No.	ECG	SK	GE	RS 276-	MOTOR.
2SB53K		3004	53	2007	
2SB53L		3004	53	2007	
2SB53M		3004	53	2007	
2SB53OR		3004	53	2007	
2SB53R		3004	53	2007	
2SB53X		3004	53	2007	
2SB53Y		3004	53	2007	
2SB54	102A	3004	53	2007	
2SB541	88		263	2043	2N6230
2SB542		3114	221		
2SB542D		3114	221		
2SB544	294	3841	48		
2SB544D	294		48		
2SB544E	294	3841	48		
2SB544F	294	3841	48		
2SB544P1	193A	3841/294	48		
2SB544P1D	193A	3841/294	48		
2SB544P1E	193A	3841/294	48		
2SB544P1F	193A	3841/294	48		
2SB546	398	3930			MJE15031
2SB546AK	398	3930			
2SB546AL	398	3930			
2SB546E	398	3930			
2SB546EX	398	3930			
2SB546K	398	3930			
2SB546L	398	3930			
2SB546M	398	3930			
2SB547	398$	3930			MJE15031
2SB547A	398$	3930			
2SB547K	398$	3930			
2SB547L	398$	3930			
2SB547M	398$	3930			
2SB548	374	9042			2N4920
2SB548Q	374	9042			
2SB548R	374	9042			
2SB548S	374	9042			
2SB549	374	9042			2N4920
2SB549Q	374	9042			
2SB549R	374	9042			
2SB549S	374	9042			
2SB54A		3004	53	2024	
2SB54B	102A	3004	53	2007	
2SB54BA		3004	53	2024	
2SB54C		3004	53	2024	
2SB54D		3004	53	2024	
2SB54E	102A	3004	53	2007	
2SB54F	102A	3004	53	2007	
2SB54G		3004	53	2024	
2SB54GN		3004	53	2024	
2SB54H		3004	53	2024	
2SB54J		3004	53	2024	
2SB54K		3004	53	2024	
2SB54L		3004	53	2024	
2SB54L1		3004	53	2024	
2SB54M		3004	53	2024	
2SB54OR		3004	53	2024	
2SB54R		3004	53	2024	
2SB54X			53	2024	
2SB54Y	102A	3004	53	2007	
2SB55	102A	3004	53	2007	
2SB550	88		250	2027	
2SB552	285	3846			MJ15023
2SB554	285	3846	266		MJ15023
2SB554-O	285		266		
2SB554-R	285	3846	266		
2SB555	281	3359	263	2043	MJ15012
2SB555-O	281	3359	263	2043	
2SB555R		3359	263		
2SB556	281	3359			MJ15012
2SB556-R			52	2024	
2SB556R	281	3359			
2SB557	281	3359			2N6230
2SB557-O	281	3359			
2SB558	281	3173/219			2N6229
2SB558-O	281	3173/219			
2SB559	185				2N4918
2SB55A		3004	53	2024	
2SB55B		3004	53	2024	
2SB55C		3004	53	2024	
2SB55D		3004	53	2024	
2SB55E		3004	53	2024	
2SB55F		3004	53	2024	
2SB55G		3004	53	2024	
2SB55GN		3004	53	2024	
2SB55H		3004	53	2024	
2SB55J		3004	53	2024	
2SB55K		3004	53	2024	
2SB55L		3004	53	2024	
2SB55M		3004	53	2024	
2SB55OR		3004	53	2024	
2SB55R		3004	53	2024	
2SB55X		3004	53	2024	
2SB55Y		3004	53	2024	
2SB56	102A	3004	53	2007	
2SB560	383		53	2007	
2SB561		3450	244	2032	
2SB561B		3450	244	2032	
2SB561C		3450	244	2032	
2SB562	294	3841	48		
2SB562-C	294		48		
2SB562A	294	3841			
2SB562B	294	3841	48		
2SB562C	294	3841	48		
2SB562D	294	3841			
2SB564	294	3841	67		
2SB564K	294	3841			
2SB564L	294	3841	67		
2SB564M	294	3841			
2SB565		3004	53	2007	2N6125
2SB565A		3004	53	2007	
2SB565B		3004	53	2007	
2SB565C		3004	53	2007	
2SB565D		3004	53	2007	
2SB565E		3004	53	2007	
2SB565F		3004	53	2007	
2SB565G		3004	53	2007	
2SB565GN		3004	53	2007	
2SB565H		3004	53	2007	
2SB565J		3004	53	2007	
2SB565K		3004	53	2007	
2SB565L		3004	53	2007	
2SB565M		3004	53	2007	
2SB565OR		3004	53	2007	
2SB565R		3004	53	2007	
2SB565X		3004	53	2007	
2SB565Y		3004	53	2007	
2SB566	292	3083/197	250		2N6126
2SB566A	292	3083/197	250		
2SB566AC	292	3083/197			
2SB566AK	292	3083/197			
2SB566C	292	3083/197	250		
2SB566D	292	3083/197	250		
2SB566K	292	3083/197			
2SB567					MJE15031

Industry Standard No.	ECG	SK	GE	RS 276-	MOTOR.
2SB568	398	3930			MJE15031
2SB568B	398	3930			
2SB568C	398	3930			
2SB568D	398	3930			
2SB569	242				MJE3310
2SB56A	102A	3004	53	2007	
2SB56B	102A	3004	53	2007	
2SB56C	102A	3004	53	2007	
2SB56CK		3004	53	2024	
2SB56D		3004	53	2024	
2SB56E		3004	53	2024	
2SB56F		3004	53	2024	
2SB56G		3004	53	2024	
2SB56GN		3004	53	2024	
2SB56H		3004	53	2024	
2SB56J		3004	53	2024	
2SB56K		3004	53	2024	
2SB56L		3004	53	2024	
2SB56OR		3004	53	2024	
2SB56R		3004	53	2024	
2SB56X			53	2024	
2SB56Y		3004	53	2024	
2SB57	102A	3004	53	2007	
2SB570					MJE3311
2SB571					MJE3312
2SB572					2N5193
2SB573					2N5194
2SB574		3841			2N5195
2SB575					2N5193
2SB576					2N5194
2SB577					2N5195
2SB578					MJE2955
2SB579					2N5975
2SB57A		3004	53	2024	
2SB57B		3004	53	2024	
2SB57C		3004	53	2024	
2SB57D		3004	53	2024	
2SB57E		3004	53	2024	
2SB57F		3004	53	2024	
2SB57G		3004	53	2024	
2SB57GN		3004	53	2024	
2SB57H		3004	53	2024	
2SB57J		3004	53	2024	
2SB57K		3004	53	2024	
2SB57L		3004	53	2024	
2SB57M		3004	53	2024	
2SB57OR		3004	53	2024	
2SB57R		3004	53	2024	
2SB57X		3004	53	2024	
2SB57Y		3004	53	2024	
2SB58	158		53	2007	
2SB580					2N5976
2SB581					2N5976
2SB582					MJE6040
2SB583					MJE6041
2SB584					MJE6042
2SB585					2N6053
2SB586					2N6054
2SB587					2N6050
2SB588	281				2N6051
2SB589					2N6052
2SB59	102A	3004	53	2007	
2SB595	332				TIP42C
2SB596	197	3441/292	250	2027	2N6126
2SB596-0	197		250	2027	
2SB596-0		3441	250	2027	
2SB598		3841	244	2027	
2SB598F		3841	244	2027	
2SB59A		3004	53	2024	
2SB59B		3004	53	2024	
2SB59C		3004	53	2024	
2SB59D		3004	53	2024	
2SB59E		3004	53	2024	
2SB59F		3004	53	2024	
2SB59G		3004	53	2024	
2SB59GN		3004	53	2024	
2SB59J		3004	53	2024	
2SB59K		3004	53	2024	
2SB59L		3004	53	2024	
2SB59M		3004	53	2024	
2SB59OR		3004	53	2024	
2SB59R		3004	53	2024	
2SB59X		3004	53	2024	
2SB59Y		3004	53	2024	
2SB60	102A	3004	53	2007	
2SB600	88				MJ15012
2SB604	242				2N6126
2SB605	294	3841			
2SB60A	102A	3004	53	2007	
2SB60B		3004	53	2024	
2SB60C		3004	53	2024	
2SB60D		3004	53	2024	
2SB60E		3004	53	2024	
2SB60F		3004	53	2024	
2SB60G		3004	53	2024	
2SB60GN		3004	53	2024	
2SB60H		3004	53	2024	
2SB60J		3004	53	2024	
2SB60K		3004	53	2024	
2SB60L		3004	53	2024	
2SB60M		3004	53	2024	
2SB60OR		3004	53	2024	
2SB60R		3004	53	2024	
2SB60X		3004	53	2024	
2SB60Y		3004	53	2024	
2SB61	102A	3004	53	2007	
2SB616	381	3934			
2SB616Q	381	3934			
2SB616R	381	3934			
2SB616S	381	3934			
2SB617	381	3934			
2SB617A	381$	3934/381			
2SB617Q	381$	3934/381			
2SB617R	381$	3934/381			
2SB617S	381$	3934/381			
2SB618	381	3934			
2SB61A		3004	53	2007	
2SB61B		3004	53	2007	
2SB61C		3004	53	2007	
2SB61D		3004	53	2007	
2SB61E		3004	53	2007	
2SB61F		3004	53	2007	
2SB61G		3004	53	2007	
2SB61GN		3004	53	2007	
2SB61H		3004	53	2007	
2SB61J		3004	53	2007	
2SB61K		3004	53	2007	
2SB61L		3004	53	2007	
2SB61M		3004	53	2007	
2SB61OR		3004	53	2007	
2SB61R		3004	53	2007	
2SB61X		3004	53	2007	
2SB61Y		3004	53	2007	
2SB62	226	3009	25	2006	

Industry Standard No.	ECG	SK	GE	RS 276-	MOTOR.
2SB621	294	3841	48		
2SB621ANC	294	3841			
2SB621NC	294	3841			
2SB621R	294	3841	48		
2SB628	398$	3930			MJE15031
2SB628Q	398$	3930			
2SB628R	398$	3930			
2SB628S	398$	3930			
2SB62A		3009	25		
2SB62B		3009	25		
2SB62C		3009	25		
2SB62D		3009	25		
2SB62E		3009	25		
2SB62F		3009	25		
2SB62G		3009	25		
2SB62GN		3009	25		
2SB62H		3009	25		
2SB62J		3009	25		
2SB62K		3009	25		
2SB62L		3009	25		
2SB62M		3009	25		
2SB620R		3009	25		
2SB62R		3009	25		
2SB62X		3009	25		
2SB62Y		3009	25		
2SB63	131		44	2006	
2SB630		3930			MJE15031
2SB631	374				2N4920
2SB632	185			2025	2N4918
2SB632K	185	3191	58	2025	
2SB633	197				TIP42C
2SB63A		3034	25		
2SB63B		3034	25		
2SB63C		3034	25		
2SB63D		3034	25		
2SB63E		3034	25		
2SB63F		3034	25		
2SB63G		3034	25		
2SB63GN		3034	25		
2SB63H		3034	25		
2SB63K		3034	25		
2SB63L		3034	25		
2SB63M		3034	25		
2SB630R		3034	25		
2SB63R		3034	25		
2SB63X		3034	25		
2SB63Y		3034	25		
2SB64	127	3009	25	2006	
2SB641	290A	3912	269	2032	
2SB641Q	290A	3912			
2SB641R	290A	3912			
2SB641S	290A	3912	269	2032	
2SB642	290A	3912	221	2022	
2SB642P	290A	3912			
2SB642Q	290A	3912	221	2050	
2SB642R	234	3912	221	2050	
2SB643	290A	3912	269	2032	
2SB643Q	290A	3912	269	2032	
2SB643R	290A	3912	269	2032	
2SB643S	290A	3912	269	2032	
2SB644	290A	3912			
2SB644R	290A	3912			
2SB645	285$	3846/285			
2SB645E			20	2051	
2SB648	374#	9042/374			MJE350
2SB648A	374#	9042/374			
2SB648AB	374#	9042/374			
2SB648AC	374#	9042/374			
2SB648B	374#	9042/374			
2SB648C	374#	9042/374			
2SB648D	374#	9042/374			
2SB649	374#	9042/374			MJE350
2SB649A	374#	9042/374			
2SB649AB	374#	9042/374			
2SB649AC	374#	9042/374			
2SB649B	374#	9042/374			
2SB649C	374#	9042/374			
2SB649D	374#	9042/374			
2SB64A		3009	25		
2SB64B		3009	25		
2SB64C		3009	25		
2SB64D		3009	25		
2SB64E		3009	25		
2SB64F		3009	25		
2SB64G		3009	25		
2SB64H		3009	25		
2SB64J		3009	25		
2SB64K		3009	25		
2SB64L		3009	25		
2SB64M		3009	25		
2SB640R		3009	25		
2SB64R		3009	25		
2SB64X		3009	25		
2SB64Y		3009	25		
2SB65	102A	3004	53	2007	
2SB653	88	3359/281			2N6227
2SB653A	88	3359/281			
2SB653AB	88	3359/281			
2SB653AC	88	3359/281			
2SB654	88	3359/281			2N6227
2SB654A	88	3359/281			
2SB654AB	88	3359/281			
2SB654AC	88	3359/281			
2SB655	285	3846			MJ15002
2SB655A	285	3846			
2SB655AB	285	3846			
2SB655AC	285	3846			
2SB656	285	3846			MJ15002
2SB656A	285	3846			
2SB656AB	285	3846			
2SB656AC	285	3846			
2SB65A		3004	53	2007	
2SB65B		3004	53	2007	
2SB65C		3004	53	2007	
2SB65D		3004	53	2007	
2SB65F		3004	53	2007	
2SB65G		3004	53	2007	
2SB65GN		3004	53	2007	
2SB65H		3004	53	2007	
2SB65J		3004	53	2007	
2SB65K		3004	53	2007	
2SB65L		3004	53	2007	
2SB65M		3004	53	2007	
2SB650R		3004	53	2007	
2SB65R		3004	53	2007	
2SB65X		3004	53	2007	
2SB65Y		3004	53	2007	
2SB66	102A	3004	53	2007	
2SB668	262	3181A/264			TIP32A
2SB668A	262	3181A/264			
2SB669	262		29		TIP32B
2SB66A		3004	53	2007	
2SB66B		3004	53	2007	
2SB66C		3004	53	2007	
2SB66D		3004	53	2007	
2SB66E		3004	53	2007	
2SB66F		3004	53	2007	
2SB66G		3004	53	2007	
2SB66GN		3004	53	2007	
2SB66H	102A	3123	53	2007	
2SB66J		3004	53	2007	
2SB66K		3004	53	2007	
2SB66L		3004	53	2007	
2SB66M		3004	53	2007	
2SB660R		3004	53	2007	
2SB66R		3004	53	2007	
2SB66X		3004	53	2007	
2SB66Y		3004	53	2007	
2SB67		3004	53	2007	
2SB673	262				2N6042
2SB674	262				2N6041
2SB675	262				2N6040
2SB676	292				TIP127
2SB677	153				TIP125
2SB679					TIP117
2SB67A		3004	53	2007	
2SB67B		3004	53	2007	
2SB67C		3004	53	2007	
2SB67D		3004	53	2007	
2SB67E		3004	53	2007	
2SB67F		3004	53	2007	
2SB67G		3004	53	2007	
2SB67GN		3004	53	2007	
2SB67H		3004	53	2007	
2SB67J		3004	53	2007	
2SB67K		3004	53	2007	
2SB67L		3004	53	2007	
2SB67M		3004	53	2007	
2SB670R		3004	53	2007	
2SB67R		3004	53	2007	
2SB67X		3004	53	2007	
2SB67Y		3004	53	2007	
2SB68		3004	53	2007	
2SB681	285				MJ15002
2SB682	292	3189A/183			
2SB682B	292	3189A/183			
2SB682C	292	3189A/183			
2SB683	197		39	2015	
2SB684	294	3513	67		
2SB689	292				TIP42C
2SB68A		3004	53	2007	
2SB68B		3004	53	2007	
2SB68C		3004	53	2007	
2SB68D		3004	53	2007	
2SB68E		3004	53	2007	
2SB68F		3004	53	2007	
2SB68G		3004	53	2007	
2SB68GN		3004	53	2007	
2SB68H		3004	53	2007	
2SB68J		3004	53	2007	
2SB68K		3004	53	2007	
2SB68L		3004	53	2007	
2SB68M		3004	53	2007	
2SB680R		3004	53	2007	
2SB68R		3004	53	2007	
2SB68X		3004	53	2007	
2SB68Y		3004	53	2007	
2SB69	121	3009	25	2006	
2SB690	292				TIP42C
2SB691	391$				MJE4352
2SB692	391$				MJE4352
2SB693	252				2N6287
2SB694					MJ11015
2SB695					MJE4352
2SB696					2N6231
2SB697	281				MJ15002
2SB699		3189A	69	2049	
2SB699Q		3181A	69	2049	
2SB69A		3009	25		
2SB69B		3009	25		
2SB69C		3009	25		
2SB69D		3009	25		
2SB69E		3009	25		
2SB69F		3009	25		
2SB69G		3009	25		
2SB69GN		3009	25		
2SB69H		3009	25		
2SB69J		3009	25		
2SB69K		3009	25		
2SB69L		3009	25		
2SB69M		3009	25		
2SB690R		3009	25		
2SB69R		3009	25		
2SB69X		3009	25		
2SB69Y		3009	25		
2SB70			53	2007	
2SB707	197$				2N6107
2SB708	197				2N6107
2SB71	102A	3004	53	2007	
2SB711	262				2N6041
2SB712					2N6042
2SB713					MJE4352
2SB717					MJE350
2SB718					MJE350
2SB719	398				MJE15031
2SB71A		3004	53	2007	
2SB71B		3004	53	2007	
2SB71C		3004	53	2007	
2SB71D		3004	53	2007	
2SB71E		3004	53	2007	
2SB71F		3004	53	2007	
2SB71G		3004	53	2007	
2SB71GN		3004	53	2007	
2SB71H		3004	53	2007	
2SB71J		3004	53	2007	
2SB71K		3004	53	2007	
2SB71L		3004	53	2007	
2SB71M		3004	53	2007	
2SB710R		3004	53	2007	
2SB71R		3004	53	2007	
2SB71X		3004	53	2007	
2SB71Y		3004	53	2007	
2SB72	102	3004	53	2007	
2SB720					MJE15031
2SB722					MJ15002
2SB723					MJ15023
2SB724	153				TIP32A
2SB726	290A$	3450/298			
2SB726S	290A$	3450/298	82		
2SB727					MJE15029
2SB72A		3004	53	2007	
2SB72B		3004	53	2007	
2SB72C		3004	53	2007	
2SB72D		3004	53	2007	
2SB72E		3004	53	2007	
2SB72F		3004	53	2007	
2SB72G		3004	53	2007	
2SB72GN		3004	53	2007	

Industry Standard No.	ECG	SK	GE	RS 276-	MOTOR.
2SB72H		3004	53	2007	
2SB72J		3004	53	2007	
2SB72K		3004	53	2007	
2SB72L		3004	53	2007	
2SB72M		3004	53	2007	
2SB72OR		3004	53	2007	
2SB72R		3004	53	2007	
2SB72X		3004	53	2007	
2SB72Y		3004	53	2007	
2SB73	102A	3004	53	2007	
2SB73A	102A	3004	53	2007	
2SB73A-1		3004	53	2007	
2SB73B	102A	3004	53	2005	
2SB73C	102A	3004	53	2007	
2SB73D		3004	53	2007	
2SB73E		3004	53	2007	
2SB73F		3004	53	2007	
2SB73G		3004	53	2007	
2SB73GN		3004	53	2007	
2SB73GR	102A		53	2007	
2SB73H		3004	53	2007	
2SB73J		3004	53	2007	
2SB73K		3004	53	2007	
2SB73L		3004	53	2007	
2SB73M		3004	53	2007	
2SB73OR		3004	53	2007	
2SB73R		3004	53	2007	
2SB73S	102A		53	2007	
2SB73X		3004	53	2007	
2SB73Y		3004	53	2007	
2SB74	158	3004/102A	53	2004	
2SB741	383	3114/290	269	2032	
2SB743	185				MJE170
2SB744	185	3191			MJE172
2SB744A	185	3191			
2SB744Q	185	3191			
2SB744R	185	3191			
2SB745	234	3841/294			
2SB745A	234	3841/294			
2SB745S	234	3841/294			
2SB745T	234	3841/294			
2SB74A		3004	53	2007	
2SB74B		3004	53	2007	
2SB74C		3004	53	2007	
2SB74D		3004	53	2007	
2SB74E		3004	53	2007	
2SB74F		3004	53	2007	
2SB74G		3004	53	2007	
2SB74GN		3004	53	2007	
2SB74H		3004	53	2007	
2SB74J		3004	53	2007	
2SB74K		3004	53	2007	
2SB74L		3004	53	2007	
2SB74M		3004	53	2007	
2SB74OR		3004	53	2007	
2SB74R		3004	53	2007	
2SB74X		3004	53	2007	
2SB74Y		3004	53	2007	
2SB75	102A	3004	53	2007	
2SB750	262				TIP115
2SB751	262				MJE703T
2SB753	197				TIP42C
2SB754	391				2N6109
2SB75A	102A	3004	53	2007	
2SB75AH	102A	3004	53	2007	
2SB75B	102A	3004	53	2007	
2SB75C	102A	3004	53	2007	
2SB75C-4		3004	53	2024	
2SB75D		3004	53	2024	
2SB75E		3004	53	2024	
2SB75F	102A	3004	53	2007	
2SB75G		3004	53	2024	
2SB75GN		3004	53	2024	
2SB75H	102A	3004	53	2005	
2SB75J		3004	53	2024	
2SB75L		3004	53	2024	
2SB75LB	102A		53	2007	
2SB75M		3004	53	2024	
2SB75OR		3004	53	2024	
2SB75R		3004	53	2024	
2SB75X			53	2024	
2SB75Y		3004	53	2024	
2SB76	102A	3003	53	2007	
2SB76A		3004	53	2007	
2SB76B		3004	53	2007	
2SB76C		3004	53	2007	
2SB76D		3004	53	2007	
2SB76E		3004	53	2007	
2SB76F		3004	53	2007	
2SB76G		3004	53	2007	
2SB76GN		3004	53	2007	
2SB76H		3004	53	2007	
2SB76J		3004	53	2007	
2SB76K		3004	53	2007	
2SB76L		3004	53	2007	
2SB76M		3004	53	2007	
2SB76OR		3004	53	2007	
2SB76R		3004	53	2007	
2SB76X		3004	53	2007	
2SB76Y		3004	53	2007	
2SB77	102A	3004	53	2007	
2SB77(B)	102A	3004	53		
2SB77-C		3004	53		
2SB77-OR			53	2007	
2SB772					MJE170
2SB774	290A	3247/234	65	2050	
2SB774R		3247		2050	
2SB774S		3247	65	2050	
2SB77A	102A	3004	53	2007	
2SB77A/P		3004	52	2007	
2SB77AA	102A		53	2007	
2SB77AB	102A	3004	53	2007	
2SB77AC	102A		53	2007	
2SB77AD	102A		53	2007	
2SB77AH	102A	3004	53	2007	
2SB77AP	102A		53	2007	
2SB77B	102A	3004	53	2007	
2SB77B-11	102A	3004	53	2007	
2SB77C	102A	3004	53	2007	
2SB77D	102A	3004	53	2007	
2SB77E		3004	53	2007	
2SB77F		3004	53	2007	
2SB77G		3004	53	2007	
2SB77GN		3004	53	2007	
2SB77H	102A	3004	53	2007	
2SB77K		3004	53	2007	
2SB77L		3004	53	2007	
2SB77M		3004	53	2007	
2SB77OR		3004	53		
2SB77P			53	2007	
2SB77PD			53	2007	
2SB77R		3004	53	2007	
2SB77RED			53	2007	
2SB77V	102A	3004	53	2007	
2SB77VRED	102A		53	2007	
2SB77X		3004	53	2007	
2SB77Y		3004	53	2007	
2SB78	102A	3004	53	2007	
2SB78A		3004	53	2007	
2SB78B		3004	53	2007	
2SB78C		3004	53	2007	
2SB78D		3004	53	2007	
2SB78E		3004	53	2007	
2SB78F		3004	53	2007	
2SB78G		3004	53	2007	
2SB78GN		3004	53	2007	
2SB78H		3004	53	2007	
2SB78J		3004	53	2007	
2SB78K		3004	53	2007	
2SB78L		3004	53	2007	
2SB78M		3004	53	2007	
2SB78OR		3004	53	2007	
2SB78R		3004	53	2007	
2SB78X		3004	53	2007	
2SB78Y		3004	53	2007	
2SB79	102A	3004	53	2007	
2SB79A		3004	53	2007	
2SB79B		3004	53	2007	
2SB79C		3004	53	2007	
2SB79D		3004	53	2007	
2SB79E		3004	53	2007	
2SB79F		3004	53	2007	
2SB79G		3004	53	2007	
2SB79GN		3004	53	2007	
2SB79H		3004	53	2007	
2SB79J		3004	53	2007	
2SB79K		3004	53	2007	
2SB79L		3004	53	2007	
2SB79M		3004	53	2007	
2SB79OR		3004	53	2007	
2SB79R		3004	53	2007	
2SB79X		3004	53	2007	
2SB79Y		3004	53	2007	
2SB80	131	3009	25	2006	
2SB80A		3009	25		
2SB80B		3009	25		
2SB80C		3009	25		
2SB80D		3009	25		
2SB80E		3009	25		
2SB80F		3009	25		
2SB80G		3009	25		
2SB80GN		3009	25		
2SB80H		3009	25		
2SB80J		3009	25		
2SB80K		3009	25		
2SB80L		3009	25		
2SB80M		3009	25		
2SB80OR		3009	25		
2SB80R		3009	25		
2SB80X		3009	25		
2SB80Y		3009	25		
2SB81		3009	25	2006	
2SB81A		3009	25		
2SB81B		3009	25		
2SB81C		3009	25		
2SB81D		3009	25		
2SB81E		3009	25		
2SB81F		3009	25		
2SB81GN		3009	25		
2SB81H		3009	25		
2SB81J		3009	25		
2SB81K		3009	25		
2SB81L		3009	25		
2SB81M		3009	25		
2SB81OR		3009	25		
2SB81R		3009	25		
2SB81X		3009	25		
2SB81Y		3009	25		
2SB82		3009	25	2006	
2SB82A		3009	25		
2SB82B		3009	25		
2SB82C		3009	25		
2SB82D		3009	25		
2SB82E		3009	25		
2SB82F		3009	25		
2SB82G		3009	25		
2SB82GN		3009	25		
2SB82H		3009	25		
2SB82J		3009	25		
2SB82K		3009	25		
2SB82L		3009	25		
2SB82M		3009	25		
2SB82OR		3009	25		
2SB82R		3009	25		
2SB82X		3009	25		
2SB82Y		3009	25		
2SB83	104	3009	25	2006	
2SB83A		3009	25		
2SB83B		3009	25		
2SB83C		3009	25		
2SB83D		3009	25		
2SB83E		3009	25		
2SB83F		3009	25		
2SB83G		3009	25		
2SB83GN		3009	25		
2SB83H		3009	25		
2SB83J		3009	25		
2SB83K		3009	25		
2SB83L		3009	25		
2SB83M		3009	25		
2SB83OR		3009	25		
2SB83R		3009	25		
2SB83X		3009	25		
2SB83Y		3009	25		
2SB84	121	3014	3	2006	
2SB84A		3014	3	2006	
2SB84B		3014	3	2006	
2SB84C		3014	3	2006	
2SB84D		3014	3	2006	
2SB84E		3014	3	2006	
2SB84F		3014	3	2006	
2SB84G		3014	3	2006	
2SB84GN		3014	3	2006	
2SB84H		3014	3	2006	
2SB84J		3014	3	2006	
2SB84K		3014	3	2006	
2SB84L		3014	3	2006	
2SB84M		3014	3	2006	
2SB84OR		3014	3	2006	
2SB84R		3014	3	2006	
2SB84X		3014	3	2006	
2SB84Y		3014	3	2006	
2SB85	127	3004/102A	53	2007	
2SB85A		3004	53	2007	
2SB85B		3004	53	2007	
2SB85C		3004	53	2007	
2SB85D		3004	53	2007	
2SB85E		3004	53	2007	

Industry Standard No.	ECG	SK	GE	RS 276-	MOTOR.
2SB85F		3004	53	2007	
2SB85G		3004	53	2007	
2SB85GN		3004	53	2007	
2SB85H		3004	53	2007	
2SB85J		3004	53	2007	
2SB85K		3004	53	2007	
2SB85L		3004	53	2007	
2SB85M		3004	53	2007	
2SB85OR		3004	53	2007	
2SB85R		3004	53	2007	
2SB85X		3004	53	2007	
2SB85Y		3004	53	2007	
2SB87	127	3004/102A	53	2007	
2SB87A		3004	53	2007	
2SB87B		3004	53	2007	
2SB87C		3004	53	2007	
2SB87D		3004	53	2007	
2SB87E		3004	53	2007	
2SB87F		3004	53	2007	
2SB87G		3004	53	2007	
2SB87GN		3004	53	2007	
2SB87H		3004	53	2007	
2SB87J		3004	53	2007	
2SB87K		3004	53	2007	
2SB87L		3004	53	2007	
2SB87M		3004	53	2007	
2SB87OR		3004	53	2007	
2SB87R		3004	53	2007	
2SB87X		3004	53	2007	
2SB87Y		3004	53	2007	
2SB89	102A	3003	53	2007	
2SB89A	102A	3004	53	2007	
2SB89AH	102A	3123	53	2007	
2SB89C		3004	53	2007	
2SB89D		3004	53	2007	
2SB89E		3004	53	2007	
2SB89F		3004	53	2007	
2SB89G		3004	53	2007	
2SB89GN		3004	53	2007	
2SB89H	102A	3123	53	2007	
2SB89J		3004	53	2007	
2SB89K		3004	53	2007	
2SB89L		3004	53	2007	
2SB89M		3004	53	2007	
2SB89OR		3004	53	2007	
2SB89R		3004	53	2007	
2SB89X		3004	53	2007	
2SB89Y		3004	53	2007	
2SB90	102A	3004	53	2007	
2SB90A		3004	53	2024	
2SB90B		3004	53	2024	
2SB90C		3004	53	2024	
2SB90D		3004	53	2024	
2SB90E		3004	53	2024	
2SB90F		3004	53	2024	
2SB90G		3004	53	2024	
2SB90GN		3004	53	2024	
2SB90H		3004	53	2024	
2SB90J		3004	53	2024	
2SB90K		3004	53	2024	
2SB90L		3004	53	2024	
2SB90M		3004	53	2024	
2SB90OR		3004	53	2024	
2SB90R		3004	53	2024	
2SB90X		3004	53	2024	
2SB90Y		3004	53	2024	
2SB91	102A	3004	53	2007	
2SB91A		3004	53	2024	
2SB91B		3004	53	2024	
2SB91C		3004	53	2024	
2SB91D		3004	53	2024	
2SB91E		3004	53	2024	
2SB91F		3009	53	2024	
2SB91G		3004	53	2024	
2SB91GN		3004	53	2024	
2SB91H		3004	53	2024	
2SB91J		3004	53	2024	
2SB91K		3004	53	2024	
2SB91L		3004	53	2024	
2SB91M		3004	53	2024	
2SB91OR		3004	53	2024	
2SB91R		3004	53	2024	
2SB91X		3004	53	2024	
2SB91Y		3004	53	2024	
2SB92	102A	3003	53	2007	
2SB92A		3004	53	2007	
2SB92B		3004	53	2007	
2SB92C		3004	53	2007	
2SB92D		3004	53	2007	
2SB92E		3004	53	2007	
2SB92F		3004	53	2007	
2SB92G		3004	53	2007	
2SB92GN		3004	53	2007	
2SB92H		3004	53	2007	
2SB92J		3004	53	2007	
2SB92K		3004	53	2007	
2SB92L		3004	53	2007	
2SB92M		3004	53	2007	
2SB92X		3004	53	2007	
2SB92Y		3004	53	2007	
2SB93				2005	
2SB94	102A	3004	53	2007	
2SB94A		3004	53	2007	
2SB94B		3004	53	2007	
2SB94C		3004	53	2007	
2SB94D		3004	53	2007	
2SB94E		3004	53	2007	
2SB94F		3004	53	2007	
2SB94G		3004	53	2007	
2SB94GN		3004	53	2007	
2SB94H		3004	53	2007	
2SB94J		3004	53	2007	
2SB94K		3004	53	2007	
2SB94M		3004	53	2007	
2SB94OR		3004	53	2007	
2SB94R		3004	53	2007	
2SB94X		3004	53	2007	
2SB94Y		3004	53	2007	
2SB95	102A	3004	53	2007	
2SB95A		3004	53	2007	
2SB95B		3004	53	2007	
2SB95C		3004	53	2007	
2SB95D		3004	53	2007	
2SB95E		3004	53	2007	
2SB95F		3004	53	2007	
2SB95G		3004	53	2007	
2SB95GN		3004	53	2007	
2SB95H		3004	53	2007	
2SB95J		3004	53	2007	
2SB95L		3004	53	2007	
2SB95M		3004	53	2007	
2SB95OR		3004	53	2007	
2SB95R		3004	53	2007	
2SB95X		3004	53	2007	
2SB95Y		3004	53	2007	
2SB97	102A	3004	53	2007	
2SB97A		3004	53	2007	
2SB97B		3004	53	2007	
2SB97C		3004	53	2007	
2SB97D		3004	53	2007	
2SB97F		3004	53	2007	
2SB97G		3004	53	2007	
2SB97GN		3004	53	2007	
2SB97H		3004	53	2007	
2SB97K		3004	53	2007	
2SB97L		3004	53	2007	
2SB97M		3004	53	2007	
2SB97OR		3004	53	2007	
2SB97R		3004	53	2007	
2SB97X		3004	53	2007	
2SB97Y		3004	53	2007	
2SB98	102	3004	53	2007	
2SB98A		3004	53	2007	
2SB98B		3004	53	2007	
2SB98C		3004	53	2007	
2SB98D		3004	53	2007	
2SB98E		3004	53	2007	
2SB98F		3004	53	2007	
2SB98G		3004	53	2007	
2SB98GN		3004	53	2007	
2SB98H		3004	53	2007	
2SB98J		3004	53	2007	
2SB98K		3004	53	2007	
2SB98L		3004	53	2007	
2SB98M		3004	53	2007	
2SB98OR		3004	53	2007	
2SB98R		3004	53	2007	
2SB98X		3004	53	2007	
2SB98Y		3004	53	2007	
2SB99		3004	53	2007	
2SB99A		3004	53	2007	
2SB99B		3004	53	2007	
2SB99C		3004	53	2007	
2SB99D		3004	53	2007	
2SB99E		3004	53	2007	
2SB99F		3004	53	2007	
2SB99G		3004	53	2007	
2SB99GN		3004	53	2007	
2SB99H		3004	53	2007	
2SB99J		3004	53	2007	
2SB99K		3004	53	2007	
2SB99L		3004	53	2007	
2SB99M		3004	53	2007	
2SB99OR		3004	53	2007	
2SB99R		3004	53	2007	
2SB99X		3004	53	2007	
2SB99Y		3004	53	2007	
2SBF1	102A	3004	53	2007	
2SBF1A	102A	3004	53	2007	
2SBF2	102A	3004	53	2007	
2SBF2A		3004	53	2007	
2SBF5	121	3009	239	2006	
2SBM77		3004	53	2007	
2SBZ		3035	25		
2SC-1307			216	2053	
2SC-2C	506		511	1114	
2SC-313			11	2015	
2SC-4012		3124	47	2030	
2SC-4033		3018	61	2038	
2SC-923			62	2010	
2SC-F11		3018	61	2038	
2SC-F11A		3018	61	2038	
2SC-F11B		3018	61	2038	
2SC-F11C		3018	61	2038	
2SC-F11D		3018	61	2038	
2SC-F11E		3018	61	2038	
2SC-F11F		3018	61	2038	
2SC-F11G		3018	61	2038	
2SC-F11GN		3018	61	2038	
2SC-F11H		3018	61	2038	
2SC-F11J		3018	61	2038	
2SC-F11K		3018	61	2038	
2SC-F11L		3018	61	2038	
2SC-F11M		3018	61	2038	
2SC-F11OR		3018	61	2038	
2SC-F11R		3018	61	2038	
2SC-F11X		3018	61	2038	
2SC-F11Y		3018	61	2038	
2SC-F14		3018	61	2038	
2SC-F14A		3018	61	2038	
2SC-F14B		3018	61	2038	
2SC-F14C		3018	61	2038	
2SC-F14D		3018	61	2038	
2SC-F14E		3018	61	2038	
2SC-F14F		3018	61	2038	
2SC-F14G		3018	61	2038	
2SC-F14GN		3018	61	2038	
2SC-F14H		3018	61	2038	
2SC-F14J		3018	61	2038	
2SC-F14K		3018	61	2038	
2SC-F14L		3018	61	2038	
2SC-F14M		3018	61	2038	
2SC-F14OR		3018	61	2038	
2SC-F14R		3018	61	2038	
2SC-F14X		3018	61	2038	
2SC-F14Y		3018	61	2038	
2SC-F6			20	2051	
2SC-F8	237	3299	46		
2SC-NJ-100			62	2010	
2SC-NJ-107			18	2030	
2SC-NJ100	199		62	2010	
2SC-NJ107	128		243	2030	
2SC100		3444	88	2051	
2SC100-OY			88	2051	
2SC1000	199	3122	62	2010	
2SC1000(GR)			62	2010	
2SC1000-BL	199	3122	62	2010	
2SC1000-GR	199	3122	62	2010	
2SC1000-Y	199	3122	62	2010	
2SC1000A		3122	62	2010	
2SC1000B		3122	62	2010	
2SC1000BL	199	3122	62	2010	
2SC1000C		3122	62	2010	
2SC1000D		3122	62	2010	
2SC1000E		3122	62	2010	
2SC1000F		3122	62	2010	
2SC1000G		3122	62	2010	
2SC1000G-BL			62	2010	
2SC1000G-GR			62	2010	
2SC1000GN		3122	62	2010	
2SC1000GR	199	3122	62	2010	
2SC1000H		3122	62	2010	
2SC1000J		3122	62	2010	
2SC1000K		3122	62	2010	
2SC1000L		3122	62	2010	
2SC1000M		3122	62	2010	
2SC1000OR			62	2010	
2SC1000R			62	2010	

Industry Standard No.	ECG	SK	GE	RS 276-	MOTOR.
2SC1000SR	199	3122			
2SC1000X		3122	62	2010	
2SC1000Y	199	3122	62	2010	
2SC1001	361		261		
2SC1004	164	3133	37		BU204
2SC1004A	164	3710/238	37		BU205
2SC1004R		3133	37		
2SC1005	165	3115	38		BU207
2SC1005A	165	3115	38		
2SC1005B		3115	38		
2SC1005C		3115	38		
2SC1005D		3115	38		
2SC1005E		3115	38		
2SC1005F		3115	38		
2SC1005G		3115	38		
2SC1005GN		3115	38		
2SC1005H		3115	38		
2SC1005J		3115	38		
2SC1005K		3115	38		
2SC1005L		3115	38		
2SC1005M		3115	38		
2SC1005OR		3115	38		
2SC1005R		3115	38		
2SC1005X		3115	38		
2SC1005Y		3115	38		
2SC1006		3245	62	2010	
2SC1006A		3245	62	2010	
2SC1006B		3438	62	2010	
2SC1006C		3245	62	2010	
2SC1007	123A	3444	20	2051	
2SC1008	128		243	2030	
2SC1009			61	2038	
2SC100A		3122	88		
2SC100B		3122	88		
2SC100C		3122	88		
2SC100D		3122	88		
2SC100E		3122	88		
2SC100F		3122	88		
2SC100G		3122	88		
2SC100GN		3122	88		
2SC100J		3122	88		
2SC100L		3122	88		
2SC100M		3122	88		
2SC100R		3122	88		
2SC100X		3122	88		
2SC100Y		3122	88		
2SC101		3021	12	2005	2N5050
2SC1010		3122	88	2010	
2SC1010A		3122	88	2010	
2SC1010B		3122	88	2010	
2SC1010C		3122	88	2010	
2SC1010D		3122	88	2030	
2SC1010E		3122	88	2030	
2SC1010F		3122	88	2030	
2SC1010G		3122	88	2030	
2SC1010GN		3122	88	2030	
2SC1010H		3122	88	2030	
2SC1010J		3122	88	2030	
2SC1010K		3122	88	2030	
2SC1010L		3122	88	2030	
2SC1010M		3122	88	2030	
2SC10100R			88	2030	
2SC1010R		3122	88	2030	
2SC1010X		3122	88	2030	
2SC1010Y		3122	88	2030	
2SC1012	154	3044	40	2012	
2SC1012A	154	3045/225	40	2012	
2SC1012B		3045	40	2012	
2SC1012C		3045	40	2012	
2SC1012D		3045	40	2012	
2SC1012E	154	3045/225	40	2012	
2SC1012F		3045	40	2012	
2SC1012G		3045	40	2012	
2SC1012GN		3045	40	2012	
2SC1012H		3045	40	2012	
2SC1012J		3045	40	2012	
2SC1012K		3045	40	2012	
2SC1012L		3045	40	2012	
2SC1012M		3045	40	2012	
2SC1012OR			40	2012	
2SC1012R		3045	40	2012	
2SC1012X		3045	40	2012	
2SC1012Y		3045	40	2012	
2SC1013	300	3464	83		MDS26
2SC1013A		3464	83		
2SC1013B	300	3464	83		
2SC1013C	300	3464	83		
2SC1013D	300	3464	83		
2SC1013E		3464	83		
2SC1013F		3464	83		
2SC1013G		3464	83		
2SC1013GN		3464	83		
2SC1013H		3464	83		
2SC1013J		3464	83		
2SC1013K		3464	83		
2SC1013L		3464	83		
2SC1013LJ		3464	83		
2SC1013M		3464	83		
2SC1013OR		3464	83		
2SC1013PJ		3464	83		
2SC1013R		3464	83		
2SC1013X		3464	83		
2SC1013Y		3464	83		
2SC1014	300	3464	273		MDS27
2SC10148			28	2017	
2SC1014A		3464	273		
2SC1014B	300	3464	273		
2SC1014BY		3464	273		
2SC1014C	300	3464	273		
2SC1014CD	300	3464	273		
2SC1014D	300	3464	273		
2SC1014D1	300	3464	273		
2SC1014E	300	3464	273		
2SC1014F		3464	273		
2SC1014G		3464	273		
2SC1014GA		3464	273		
2SC1014GN		3464	273		
2SC1014H		3464	273		
2SC1014J		3464	273		
2SC1014K		3464	273		
2SC1014L		3464	273		
2SC1014LG		3464	273		
2SC1014LR		3464	273		
2SC1014M		3464	273		
2SC1014OR		3464	273		
2SC1014R		3464	273		
2SC1014W		3464	273		
2SC1014X		3464	273		
2SC1014Y		3464	273		
2SC1017	299	3298	236		
2SC1018	299	3298	236		
2SC1018A		3298	236		
2SC1018B	299	3298	236		
2SC1018C		3298	236		
2SC1018D		3298	236		
2SC1018E		3298	236		
2SC1018F		3298	236		
2SC1018G		3298	236		
2SC1018GN		3298	236		
2SC1018H		3298	236		
2SC1018J		3298	236		
2SC1018K		3298	236		
2SC1018L		3298	236		
2SC1018M		3298	236		
2SC1018R		3298	236		
2SC1018X		3298	236		
2SC1018Y		3298	236		
2SC1019C			66	2048	
2SC101A		3021	12	2005	
2SC101B		3021	12	2004	
2SC101C		3021	12	2004	
2SC101D		3021	12	2004	
2SC101E		3021	12	2004	
2SC101F		3021	12	2004	
2SC101G		3021	12	2004	
2SC101GN		3021	12	2004	
2SC101H		3021	12	2004	
2SC101J		3021	12	2004	
2SC101K		3021	12	2004	
2SC101L		3021	12	2004	
2SC101M		3021	12	2004	
2SC101OR		3021	12	2004	
2SC101R		3021	12	2004	
2SC101X		3021	12	2005	
2SC101XL		3006	12	2004	
2SC101Y		3021	12	2004	
2SC1023	107	3018	11	2051	
2SC1023(O)			61	2051	
2SC1023(O)			61	2051	
2SC1023-O	107	3039/316	61	2051	
2SC1023-Y	107	3039/316	61	2051	
2SC1023A		3018	61	2051	
2SC1023B		3018	61	2051	
2SC1023C		3018	61	2051	
2SC1023D		3018	61	2051	
2SC1023E		3018	61	2051	
2SC1023F		3018	61	2051	
2SC1023G	107	3018	11	2051	
2SC1023GN		3018	61	2051	
2SC1023H		3018	61	2051	
2SC1023J		3018	61	2051	
2SC1023K		3018	61	2051	
2SC1023L		3018	61	2051	
2SC1023M		3018	61	2051	
2SC1023OR		3018	61	2051	
2SC1023R		3018	61	2051	
2SC1023X		3018	61	2051	
2SC1023Y	107	3018	61	2051	
2SC1024	175	3026	246	2020	
2SC1024(D)			246	2020	
2SC1024(F)			246	2020	
2SC1024-D2	175		246	2020	
2SC1024-E		3026	246	2020	
2SC1024A		3026	246	2020	
2SC1024B	175	3026	246	2020	
2SC1024C	175	3026	246	2020	
2SC1024D	175	3026	246	2020	
2SC1024E	175	3026	246	2020	
2SC1024F	175	3026	246	2020	
2SC1024G		3026	246	2020	
2SC1024L		3026	246	2020	
2SC1024Y		3026	246	2020	
2SC1025	175	3626	246	2020	2N6233
2SC1025CTV	175	3626	246	2020	
2SC1025D	175	3626	246	2020	
2SC1025E	175	3626	246	2020	
2SC1025J	175		246	2020	
2SC1025MT	175	3626	246	2020	
2SC1026	107	3018	11	2015	
2SC1026(G)			61	2038	
2SC1026-O			61	2038	
2SC1026-R		3039	61	2038	
0002SC1026A	107	3018	11	2015	
0002SC1026B	107	3039/316	11	2038	
2SC1026BL		3039	61	2038	
0002SC1026C	107	3018	11	2015	
2SC1026D		3018	61	2038	
2SC1026E		3018	61	2038	
2SC1026F		3018	61	2038	
2SC1026G	107	3018	61	2015	
2SC1026GN		3018	61	2038	
2SC1026GR		3039	61	2038	
2SC1026H		3018	61	2038	
2SC1026J		3018	61	2038	
2SC1026K		3018	61	2038	
2SC1026L		3018	61	2038	
2SC1026M		3018	61	2038	
2SC1026OR		3018	61	2038	
2SC1026R			61	2038	
2SC1026X		3018	61	2038	
2SC1026Y	107	3039/316	61	2015	
2SC1027	163A	3439	36		
2SC103	123A	3444	86	2051	
2SC1030	87	3619	14	2041	2N5760
2SC1030-OR			14	2041	
2SC1030A	87	3619	14	2041	
2SC1030B	87	3619	14	2041	
2SC1030B2C			14	2041	
2SC1030C	87	3619	14	2041	
2SC1030D	87	3619	14	2041	
2SC1030E		3619	14	2041	
2SC1030F		3619	14	2041	
2SC1030G		3619	14	2041	
2SC1030H			14	2041	
2SC1030J		3619	14	2041	
2SC1030K		3619	14	2041	
2SC1030L		3619	14	2041	
2SC1030M		3619	14	2041	
2SC1030OR			14	2041	
2SC1030R			14	2041	
2SC1030X			14	2041	
2SC1030Y			14	2041	
2SC1031	286				2N3585
2SC1032	107	3018	11	2015	
2SC1032(Y)			61	2015	
0002SC1032A	107	3018	61	2015	
0002SC1032B	107	3018	61	2015	
2SC1032BL		3018	61	2015	
0002SC1032C	107	3018	61	2015	
2SC1032D		3018	61	2015	
2SC1032E		3018	61	2015	
2SC1032F		3018	61	2038	
2SC1032G	107	3018	61	2015	
2SC1032GN		3018	61	2015	
2SC1032H		3018	61	2015	
2SC1032J		3018	61	2015	
2SC1032K		3018	61		
2SC1032L		3018	61	2015	

Industry Standard No.	ECG	SK	GE	RS 276-	MOTOR.
2SC1032M		3018	61	2015	
2SC1032OR		3018	61	2015	
2SC1032R		3018	61	2015	
2SC1032X		3018	61	2015	
2SC1032Y	107	3018	61	2015	
2SC1033	287	3444/123A	18	2051	
2SC1033A	287	3444/123A	18	2051	
2SC1033B		3044	18	2051	
2SC1033C		3044	18	2051	
2SC1033D		3044	18	2051	
2SC1033E		3044	18	2051	
2SC1033F		3044	18	2051	
2SC1033G		3044	18	2051	
2SC1033GN		3044	18	2051	
2SC1033H		3044	18	2051	
2SC1033J		3044	18	2051	
2SC1033K		3044	18	2051	
2SC1033L		3044	18	2051	
2SC1033OR		3044	18	2051	
2SC1033R		3044	18	2051	
2SC1033X		3044	18	2051	
2SC1033Y		3044	18	2051	
2SC1034	277		260		BU204
2SC1035	161	3117	211	2015	
2SC1035A		3117	211	2016	
2SC1035B		3117	211	2016	
2SC1035C	161	3117	39	2015	
2SC1035D	161	3117	211	2015	
2SC1035E	161	3117	211	2015	
2SC1035F		3117	211	2016	
2SC1035GN		3117	211	2016	
2SC1035H		3117	211	2016	
2SC1035J		3117	211	2016	
2SC1035L		3117	211	2016	
2SC1035M		3117	211	2016	
2SC1035OR		3117	211	2016	
2SC1035R		3117	211	2016	
2SC1035X		3117	211	2016	
2SC1035Y		3117	211	2016	
2SC1036	161	3117	211	2015	
2SC1036A		3117	211	2016	
2SC1036B		3117	211	2016	
2SC1036C		3117	211	2016	
2SC1036D		3117	211	2016	
2SC1036E		3117	211	2016	
2SC1036F		3117	211	2016	
2SC1036G		3117	211	2016	
2SC1036GN		3117	211	2016	
2SC1036H		3117	211	2016	
2SC1036J		3117	211	2016	
2SC1036K		3117	211	2016	
2SC1036L		3117	211	2016	
2SC1036M		3117	211	2016	
2SC1036OR		3117	211	2016	
2SC1036R		3117	211	2016	
2SC1036X		3117	211	2016	
2SC1036Y		3117	211	2016	
2SC103A	123A	3444	86	2051	
2SC103B		3039	86	2010	
2SC103C		3039	86	2010	
2SC103D		3039	86	2010	
2SC103E		3039	86	2010	
2SC103F		3039	86	2010	
2SC103G		3039	86	2010	
2SC103GN		3039	86	2010	
2SC103H		3039	86	2010	
2SC103J		3039	86	2010	
2SC103K		3039	86	2010	
2SC103L		3039	86	2010	
2SC103M		3039	86	2010	
2SC103X		3039	86	2010	
2SC103Y		3039	86	2010	
2SC104	123A	3444	88	2051	
2SC1043	76$				MRF511
2SC1044	161	3716	39	2015	2N6304
2SC1045		3115	38		
2SC1045B	164	3115/165	38		
2SC1045C	164	3115/165	38		
2SC1045D	164	3115/165	38		
2SC1045E	164	3115/165	38		
2SC1045F		3115	38		
2SC1045G		3115	38		
2SC1045GN		3115	38		
2SC1045H		3115	38		
2SC1045J		3115	38		
2SC1045K		3115	38		
2SC1045L		3115	38		
2SC1045M		3115	38		
2SC1045OR		3115	38		
2SC1045R	164	3115/165	38		
2SC1045X		3115	38		
2SC1045Y		3115	38		
2SC1046	165	3115	38		BU207
2SC1046A		3115	38		
2SC1046B		3115	38		
2SC1046C		3115	38		
2SC1046D		3115	38		
2SC1046E		3115	38		
2SC1046F		3115	38		
2SC1046G		3115	38		
2SC1046GN		3115	38		
2SC1046H		3115	38		
2SC1046J		3115	38		
2SC1046K	165	3115	38		
2SC1046L		3115	38		
2SC1046M		3115	38		
2SC1046N	165	3115	38		
2SC1046OR		3115	38		
2SC1046R		3115	38		
2SC1046X		3115	38		
2SC1046Y		3115	38		
2SC1047	107	3132	60	2038	
2SC1047A		3132	60	2038	
2SC1047B	107	3132	60	2038	
2SC1047BC	107		60	2038	
2SC1047BCD	107	3132	60	2038	
2SC1047C	107	3132	60	2038	
2SC1047D	107	3132	60	2038	
2SC1047E	107		60	2038	
2SC1047F			60	2038	
2SC1047G			60	2038	
2SC1047GN			60	2038	
2SC1047GR			60	2038	
2SC1047H			60	2038	
2SC1047J			60	2038	
2SC1047K			60	2038	
2SC1047L			60	2038	
2SC1047M			60	2038	
2SC1047R		3132	60	2038	
2SC1047X			60	2038	
2SC1047Y			60	2038	
2SC1048	154	3040	40	2012	
2SC1048B	154	3040	40	2012	
2SC1048C	154	3040	40	2012	

Industry Standard No.	ECG	SK	GE	RS 276-	MOTOR.
2SC1048D	154	3040	40	2012	
2SC1048DC			40	2015	
2SC1048E	154	3040	40	2012	
2SC1048F	154	3040	40	2012	
2SC1048N	154		40	2012	
2SC104A	123A	3444	88	2051	
2SC104B		3122	88	2010	
2SC104C		3122	88	2010	
2SC104D		3122	88	2010	
2SC104E		3122	88	2010	
2SC104F		3122	88	2010	
2SC104G		3122	88	2010	
2SC104GN		3122	88	2010	
2SC104H		3122	88	2010	
2SC104J		3122	88	2010	
2SC104K		3122	88	2010	
2SC104L		3122	88	2010	
2SC104M		3122	88	2010	
2SC104OR		3122	88	2010	
2SC104R		3122	88	2010	
2SC104X		3122	88	2010	
2SC104Y		3122	88	2010	
2SC105	123A	3444	20	2051	
2SC1050	94	3559/162	35		MJ411
2SC1050C	94	3559/162			
2SC1050D	94	3559/162	35		
2SC1050E	94	3559/162	35		
2SC1050F	94	3559/162	35		
2SC1051	280	3559/162	262	2041	2N5760
2SC1051C		3559	262	2041	
2SC1051D	280	3559/162	262	2041	
2SC1051E	280	3559/162	262	2041	
2SC1051F	280	3559/162	262	2041	
2SC1051LC	280	3559/162	262	2041	
2SC1051LD	280	3559/162	262	2041	
2SC1051LE	280	3559/162	262	2041	
2SC1051LF	280	3559/162	262	2041	
2SC1054	161			2011	
2SC1055	384	3538			2N5430
2SC1055F	384	3538			
2SC1055H	384	3538	246		
2SC1056	154	3044	32	2012	
2SC1056A		3044	32		
2SC1056B		3044	32		
2SC1056C		3044	32		
2SC1056D		3044	32		
2SC1056E		3044	32		
2SC1056F		3044	32		
2SC1056G		3044	32		
2SC1056GN		3044	32		
2SC1056H		3044	32		
2SC1056J		3044	32		
2SC1056K		3044	32		
2SC1056L		3044	32		
2SC1056M		3044	32		
2SC1056OR		3044	32		
2SC1056R		3044	32		
2SC1056Y		3044	32		
2SC1059	124	3021	12		2N3739
2SC1059A		3021	12		
2SC1059B		3021	12		
2SC1059C		3021	12		
2SC1059D		3021	12		
2SC1059E		3021	12		
2SC1059F		3021	12		
2SC1059G		3021	12		
2SC1059GN		3021	12		
2SC1059H		3021	12		
2SC1059J		3021	12		
2SC1059K		3021	12		
2SC1059L		3021	12		
2SC1059M		3021	12		
2SC1059OR		3021	12		
2SC1059R		3021	12		
2SC1059Y		3021	12		
2SC106	282$	3048/329	46		
2SC1060	152	3893	66	2048	TIP31A
2SC1060(C,D)			66	2048	
2SC1060A	152	3893	66	2048	
2SC1060B	152	3893	66	2048	
2SC1060BL			66	2048	
2SC1060BM	152	3893	66	2048	
2SC1060BY			66	2048	
2SC1060C	152	3893	66	2048	
2SC1060D	152	3893	66	2048	
2SC1060E			66	2048	
2SC1060F			66	2048	
2SC1060G			66	2048	
2SC1060GN			66	2048	
2SC1060H			66	2048	
2SC1060J			66	2048	
2SC1060K			66	2048	
2SC1060L			66	2048	
2SC1060M			66	2048	
2SC1060R			66	2048	
2SC1060X			66	2048	
2SC1060Y			66	2048	
2SC1061	152	3054/196	66	2017	TIP31A
2SC1061(B)			66	2048	
2SC1061(C)			66	2048	
0002SC1061A	152	3893	66	2048	
2SC1061B	152	3054/196	66	2048	
2SC1061BM			66	2048	
2SC1061BT	152	3893	66	2048	
2SC1061C	152	3054/196	66	2048	
2SC1061D	152	3893	66	2048	
2SC1061KA	152		66	2048	
2SC1061KB	152		66	2048	
2SC1061KC	152		66	2048	
2SC1061T	152	3893	66	2048	
2SC1061T-B	152		66	2048	
2SC1061TB	152	3893	66	2048	
2SC1062	396		235	2012	
2SC1063			47	2030	
2SC106A		3048	46	2020	
2SC106B		3048	46	2020	
2SC106C		3048	46	2020	
2SC106G		3048	46	2020	
2SC106GN		3048	46	2020	
2SC106H		3048	46	2020	
2SC106J		3048	46	2020	
2SC106K		3048	46	2020	
2SC106L		3048	46	2020	
2SC106M		3048	46	2020	
2SC106OR		3048	46	2020	
2SC106R		3048	46	2020	
2SC106X		3048	46	2020	
2SC106Y		3048	46	2020	
2SC107			66	2048	
2SC1070		3018	39	2015	
2SC1071	123A	3444	20	2051	
2SC1072		3024	18	2030	
2SC1072A		3024	18	2030	
2SC1072B		3024	18	2030	

Industry Standard No.	ECG	SK	GE	RS 276-	MOTOR.
2SC1072C		3024	18	2030	
2SC1072D		3024	18	2030	
2SC1072E		3024	18	2030	
2SC1072F		3024	18	2030	
2SC1072G		3024	18	2030	
2SC1072GN		3024	18	2030	
2SC1072H		3024	18	2030	
2SC1072J		3024	18	2030	
2SC1072K		3024	18	2030	
2SC1072L		3024	18	2030	
2SC1072M		3024	18	2030	
2SC10720R		3024	18	2030	
2SC1072R		3024	18	2030	
2SC1072X		3024	18	2030	
2SC1072Y		3024	18	2030	
2SC1078	369	3131A	12		BU204
2SC1079	280	3297	262	2041	MJ15001
2SC1079R	280	3297	262	2041	
2SC1079Y	280	3079	262	2041	
2SC108	128	3024	18	2030	
2SC1080	280	3027/130	262	2041	MJ15001
2SC1080-R			18	2030	
2SC1080R	280	3027/130	262	2041	
2SC1080Y	280	3027/130	262	2041	
2SC1081					2N5646
2SC1085			246	2020	
2SC1086	389	3710/238	38		BU207
2SC1086A		3710	38		
2SC1086B		3710	38		
2SC1086C		3710	38		
2SC1086D		3710	38		
2SC1086E		3710	38		
2SC1086F		3710	38		
2SC1086G		3710	38		
2SC1086GN		3710	38		
2SC1086H		3710	38		
2SC1086J		3710	38		
2SC1086K		3710	38		
2SC1086L		3710	38		
2SC1086M	389	3710/238	38		
2SC10860R		3710	38		
2SC1086R		3710	38		
2SC1086X		3710	38		
2SC1086Y		3710	38		
2SC1088					MJE3439
2SC1089	375	3104A	27		MJE3439
2SC10898	375		27		
2SC1089B	375	3040	27		
2SC1089BD	375	3104A			
2SC1089C	375	3104A	27		
2SC1089D	375	3040	27		
2SC108A	128	3024	18	2030	
2SC108A-O	128		18	2030	
2SC108A-R	128		18	2030	
2SC108B		3024	18	2030	
2SC108C		3024	18	2030	
2SC108D		3024	18	2030	
2SC108E		3024	18	2030	
2SC108F		3024	18	2030	
2SC108G		3024	18	2030	
2SC108GN		3024	18	2030	
2SC108H		3024	18	2030	
2SC108J		3024	18	2030	
2SC108K		3024	18	2030	
2SC108L		3024	18	2030	
2SC108M		3024	18	2030	
2SC1080R		3024	18	2030	
2SC108R		3024	18	2030	
2SC108X		3024	18	2030	
2SC108Y		3024	18	2030	
2SC109	128	3024	18	2030	
2SC1090			40	2012	BFR90
2SC1090D			40	2012	
2SC1095	186A	3357	247	2052	
2SC1095(6)	186A	3357	247	2052	
2SC1095L	186A	3357	247	2052	
2SC1095M	186A	3357	247	2052	
2SC1096	186A	3248	28	2052	MDS26
2SC1096(M)	186A		28	2052	
2SC1096-3ZL	186A	3248			
2SC1096-3ZM	186A	3248	28	2052	
2SC1096-4ZL	186A	3248	28	2052	
2SC1096-OR			28	2052	
2SC10963ZM	186A	3202/210	247	2052	
2SC10964ZL	186A	3202/210	247	2052	
2SC1096A	186A	3248	28	2052	
2SC1096B	186A	3248	28	2052	
2SC1096C	186A	3248	28	2052	
2SC1096D	186A	3248	28	2052	
2SC1096E	186A	3248	28	2052	
2SC1096F			28	2052	
2SC1096G			28	2052	
2SC1096GN			28	2052	
2SC1096H			28	2052	
2SC1096J			28	2052	
2SC1096K	186A	3248	28	2052	
2SC1096L	186A	3248	28	2052	
2SC1096LM	186A		28	2052	
2SC1096M	186A	3248	28	2052	
2SC1096N	186A	3248	28	2052	
2SC10960R			247	2052	
2SC1096Q			28	2052	
2SC1096R			28	2052	
2SC1096W	186A	3248	28	2052	
2SC1096X			28	2052	
2SC1096Y			28	2052	
2SC1098	186	3357	28	2052	2N6552
2SC1098(4)K	186	3357	28	2052	
2SC1098(4)L	186	3357	28	2052	
2SC1098(L)			28	2052	
2SC1098(M)			28	2052	
2SC1098A	186	3357	28	2052	
2SC1098B	186		28	2052	
2SC1098C	186	3357	28	2052	
2SC1098D	186	3357	28	2052	
2SC1098K	186	3357			
2SC1098L	186	3357	28	2052	
2SC1098M	186	3357	28	2052	
2SC1099	165		36		BU207
2SC1099K	165		36		
2SC109A	128	3024	18	2030	
2SC109A-O	128		18	2030	
2SC109A-R	128		18	2030	
2SC109A-Y	128		18	2030	
2SC109B		3024	18	2030	
2SC109C		3024	18	2030	
2SC109D		3024	18	2030	
2SC109E		3024	18	2030	
2SC109F		3024	18	2030	
2SC109G		3024	18	2030	
2SC109G1			18	2030	
2SC109GN		3024	18	2030	
2SC109J		3024	18	2030	
2SC109K		3024	18	2030	

Industry Standard No.	ECG	SK	GE	RS 276-	MOTOR.
2SC109L		3024	18	2030	
2SC1090R		3024	18	2030	
2SC109R		3024	18	2030	
2SC109X		3024	18	2030	
2SC109Y		3024	18	2030	
2SC11	103A	3011	59	2002	
2SC110	123A	3444	88	2051	
2SC1100	389	3115/165	38		BU207
2SC1100A		3115	38		
2SC1100B		3115	38		
2SC1100C		3115	38		
2SC1100D		3115	38		
2SC1100E		3115	38		
2SC1100F		3115	38		
2SC1100G		3115	38		
2SC1100GN		3115	38		
2SC1100H		3115	38		
2SC1100J		3115	38		
2SC1100K		3115	38		
2SC1100L		3115	38		
2SC1100M		3115	38		
2SC1100X		3115	38		
2SC1100Y		3115	38		
2SC1101	164	3115/165	37		BU204
2SC1101A	164	3115/165	37		
2SC1101B	164	3115/165	37		
2SC1101C	164	3115/165	37		
2SC1101D	164	3115/165	37		
2SC1101E	164	3115/165	37		
2SC1101F	164	3115/165	37		
2SC1101G		3115	37		
2SC1101GN		3115	37		
2SC1101H		3115	37		
2SC1101J		3115	37		
2SC1101K		3115	37		
2SC1101L	164	3115/165	37		
2SC1101M		3115	38		
2SC11010R		3115	37		
2SC1101R		3115	37		
2SC1101X		3115	37		
2SC1101Y		3115	37		
2SC1102	124		12		2N3739
2SC1102(M)	124		12		
2SC1102A	124		12		
2SC1102B	124		12		
2SC1102C	124		12		
2SC1102K	124		12		
2SC1102L	124		12		
2SC1102M	124		12		
2SC1103	154	3044	40	2012	
2SC1103(A)	154		40	2012	
2SC1103A	154	3044	40	2012	
2SC1103B	154		40	2012	
2SC1103C	154		40	2012	
2SC1103L	154	3044	40	2012	
2SC1104	286	3079	12		2N3585
2SC1104A	286		12		
2SC1104B	286		12		
2SC1104C	286		12		
2SC1104L	286		12		
2SC1105	124	3021	12		2N3739
2SC1105A	124	3021	12		
2SC1105B	124	3021	12		
2SC1105C	124	3021	12		
2SC1105D		3021	12		
2SC1105E		3021	12		
2SC1105F		3021	12		
2SC1105G		3021	12		
2SC1105GN		3021	12		
2SC1105H		3021	12		
2SC1105J		3021	12		
2SC1105K	124	3021	12		
2SC1105L	124	3021	12		
2SC1105M	124	3021	12		
2SC11050R		3021	12		
2SC1105R		3021	12		
2SC1105X		3021	12		
2SC1105Y		3021	12		
2SC1106	94	3559/162	73		2N5840
2SC1106A	94	3559/162	73		
2SC1106B	94		73		
2SC1106C	94	3559/162	73		
2SC1106K	94	3559/162	73		
2SC1106L	94	3559/162	73		
2SC1106M	94	3559/162	73		
2SC1106P	94		73		
2SC1106Q	94	3559/162	73		
2SC1106R	94	3559/162			
2SC1107	291	3357/186A			2N6123
2SC1107O			28	2017	
2SC1107Q	291	3357/186A	247	2052	
2SC1107YG			247	2017	
2SC1108	291				2N6123
2SC1109	291				2N6123
2SC110A		3122	88	2030	
2SC110B		3122	88	2030	
2SC110C		3122	88	2030	
2SC110D		3122	88	2030	
2SC110E		3122	88	2030	
2SC110F		3122	88	2030	
2SC110G		3122	88	2030	
2SC110GN		3122	88	2030	
2SC110H		3122	88	2030	
2SC110J		3122	88	2030	
2SC110K		3122	88	2030	
2SC110L		3122	88	2030	
2SC110M		3122	88	2030	
2SC1100R			88	2030	
2SC110R		3122	88	2030	
2SC110X		3122	88	2030	
2SC110Y		3122	88	2030	
2SC111	123A	3444	46	2051	
2SC1110	87				2N6123
2SC1111	87	3297/280	262	2041	2N5634
2SC1112	87	3836/284			2N5634
2SC1113	384				MJ3247
2SC1114	94	3438	14		2N6542
2SC1115	280	3836/284	262	2041	2N5634
2SC1116	284	3836	40	2012	MJ15011
2SC1116-O	284	3836	40	2012	
2SC1116A	284	3836			
2SC1117	161	3039/316	86	2015	
2SC1117A		3039	86	2038	
2SC1117B		3039	86	2038	
2SC1117C		3039	86	2038	
2SC1117D		3039	86	2038	
2SC1117E		3039	86	2038	
2SC1117F		3039	86	2038	
2SC1117G		3039	86	2038	
2SC1117GN		3039	86	2038	
2SC1117H	161	3039/316	86	2038	
2SC1117J		3039	86	2038	
2SC1117K		3039	86	2038	
2SC1117L		3039	86	2038	

Industry Standard No.	ECG	SK	GE	RS 276-	MOTOR.
2SC1117M		3039	86	2038	
2SC1117OR		3039	86	2038	
2SC1117R		3039	86	2038	
2SC1117X		3039	86	2038	
2SC1117Y		3039	86	2038	
2SC1119					MRF902
2SC111A		3048	46	2030	
2SC111B		3048	46	2030	
2SC111C		3048	46	2030	
2SC111D	123A	3048/329	46	2030	
2SC111E		3048	46	2030	
2SC111F		3048	46	2030	
2SC111G		3048	46	2030	
2SC111GN		3048	46	2030	
2SC111H		3048	46	2030	
2SC111J		3048	46	2030	
2SC111K		3048	46	2030	
2SC111L		3048	46	2030	
2SC111M		3048	46	2030	
2SC111OR		3048	46	2030	
2SC111R		3048	46	2030	
2SC111X		3048	46	2030	
2SC111Y		3048	46	2030	
2SC112	128	3024	18	2030	
2SC1123	123AP	3018	61	2015	
2SC1123A		3018	61	2038	
2SC1123B		3018	61	2038	
2SC1123C		3018	61	2038	
2SC1123D		3018	61	2038	
2SC1123E		3018	61	2038	
2SC1123F		3018	61	2038	
2SC1123GN		3018	61	2038	
2SC1123H		3018	61	2038	
2SC1123J		3018	61	2038	
2SC1123K		3018	61	2038	
2SC1123L		3018	61	2038	
2SC1123M		3018	61	2038	
2SC1123OR		3018	61	2038	
2SC1123R		3018	61	2038	
2SC1123X		3018	61	2038	
2SC1123Y		3018	61	2038	
2SC1124	190	3865	217		MPSU04
2SC1124-OR			217	1122	
2SC1124E	190	3865	217		
2SC1124JA	190	3929			
2SC1124JD	190	3929			
2SC1125					MPSU10
2SC1126	229	3018	61	2015	
2SC1126A		3018	61	2038	
2SC1126B		3018	61	2038	
2SC1126E		3018	61	2038	
2SC1126F		3018	61	2038	
2SC1126GN		3018	61	2038	
2SC1126J		3018	61	2038	
2SC1126K		3018	61	2038	
2SC1126L		3018	61	2038	
2SC1126M		3018	61	2038	
2SC1126OR		3018	61	2038	
2SC1126R		3018	61	2038	
2SC1126X		3018	61	2038	
2SC1126Y		3018	61	2038	
2SC1127	171	3865	27		
2SC1127A		3865	27		
2SC1127B		3865	27		
2SC1127C		3865	27		
2SC1127D		3865	27		
2SC1127E	171	3865	27		
2SC1127F		3865	27		
2SC1127G		3865	27		
2SC1127GA		3865	27		
2SC1127GN		3865	27		
2SC1127H	171	3865	27		
2SC1127HE	171	3865			
2SC1127J		3865	27		
2SC1127JR	171	3865	27		
2SC1127K		3865	27		
2SC1127L		3865	27		
2SC1127M		3865	27		
2SC1127OR		3865	27		
2SC1127R		3865	27		
2SC1127Y		3865	27		
2SC1128	229	3246	61	2015	
2SC1128(3RD IF)	233			2009	
2SC1128(3RD-IF)			61	2015	
2SC1128(FINAL IF)	233			2009	
2SC1128(FINAL-IF)			61	2015	
2SC1128(M)	229	3246	61	2015	
2SC1128(S)	229	3246	61	2015	
2SC1128-0			61	2015	
2SC1128A		3246	61	2015	
2SC1128B		3246	61	2015	
2SC1128BL		3246	61	2015	
2SC1128C		3246	61	2015	
2SC1128D	229	3246	61	2015	
2SC1128G		3246	61	2015	
2SC1128H	229	3246	61	2015	
2SC1128M		3246	61	2015	
2SC1128R		3246	61	2015	
2SC1128S		3246	61	2015	
2SC1128S(3RDIF)	233		61	2015	
2SC1128Y		3246	61	2015	
2SC1129	229	3246	60	2015	
2SC1129(M)	229	3246	60	2015	
2SC1129(R)	229	3246	60	2015	
2SC1129-0		3246	60	2015	
2SC1129A		3246	60	2015	
2SC1129B		3246	60	2015	
2SC1129BL		3246	60	2015	
2SC1129C		3246	60	2015	
2SC1129G		3246	60	2015	
2SC1129M		3246	60	2015	
2SC1129R	229	3246	60	2015	
2SC1129Y		3246	60	2015	
2SC112A		3024	18	2030	
2SC112B		3024	18	2030	
2SC112C		3024	18	2030	
2SC112D		3024	18	2030	
2SC112E		3024	18	2030	
2SC112F		3024	18	2030	
2SC112G		3024	18	2030	
2SC112GN		3024	18	2030	
2SC112H		3024	18	2030	
2SC112J		3024	18	2030	
2SC112K		3024	18	2030	
2SC112L		3024	18	2030	
2SC112M		3024	18	2030	
2SC112OR		3024	18	2030	
2SC112R		3024	18	2030	
2SC112X		3024	18	2030	
2SC112Y		3024	18	2030	
2SC113	128	3024	18	2030	
2SC1130					2N6543
2SC1131					2N6542
2SC1132					BU207

Industry Standard No.	ECG	SK	GE	RS 276-	MOTOR.
2SC113A		3024	18	2030	
2SC113B		3024	18	2030	
2SC113C		3024	18	2030	
2SC113D		3024	18	2030	
2SC113E		3024	18	2030	
2SC113F		3024	18	2030	
2SC113G		3024	18	2030	
2SC113GN		3024	18	2030	
2SC113H		3024	18	2030	
2SC113J		3024	18	2030	
2SC113K		3024	18	2030	
2SC113L		3024	18	2030	
2SC113M		3024	18	2030	
2SC113OR		3024	18	2030	
2SC113R		3024	18	2030	
2SC113X		3024	18	2030	
2SC113Y		3024	18	2030	
2SC114	128	3024	18	2030	
2SC1140	386				2N6547
2SC1141	386				2N6546
2SC1142	283				MJ13015
2SC1143	283				MJ13014
2SC114A		3024	18	2030	
2SC114B		3024	18	2030	
2SC114C		3024	18	2030	
2SC114D		3024	18	2030	
2SC114E		3024	18	2030	
2SC114F		3024	18	2030	
2SC114G		3024	18	2030	
2SC114GN		3024	18	2030	
2SC114H		3024	18	2030	
2SC114J		3024	18	2030	
2SC114K		3024	18	2030	
2SC114L		3024	18	2030	
2SC114M		3024	18	2030	
2SC114OR		3024	18	2030	
2SC114R		3024	18	2030	
2SC114X		3024	18	2030	
2SC114Y		3024	18	2030	
2SC115	128	3024	18	2030	
2SC115-1			18	2030	
2SC115-2			18	2030	
2SC115-3			47	2030	
2SC115-43			47	2030	
2SC1151	164	3133	37		BU204
2SC1151A	164	3133	37		
2SC1152	162	3995	14		2N5840
2SC1152F	162	3995	14		
2SC1152G	162	3995	14		
2SC1153	389	3115/165	38		BU204
2SC1153A	389		38		
2SC1154	389	3439/163	36		MJ12003
2SC1155	190		217		D40D13
2SC1156	190	3104A	32		2N6543
2SC1156A		3104A	32		
2SC1156B		3104A	32		
2SC1156C		3104A	32		
2SC1156D		3104A	32		
2SC1156E		3104A	32		
2SC1156F		3104A	32		
2SC1156G		3104A	32		
2SC1156GN		3104A	32		
2SC1156H		3104A	32		
2SC1156K		3104A	32		
2SC1156L		3104A	32		
2SC1156M		3104A	32		
2SC1156OR		3104A	32		
2SC1156R		3104A	32		
2SC1156X		3104A	32		
2SC1156Y		3104A	32		
2SC1157	190	3104A	32		2N6553
2SC1157A		3104A	32		
2SC1157B		3104A	32		
2SC1157C		3104A	32		
2SC1157E		3104A	32		
2SC1157F		3104A	32		
2SC1157G		3104A	32		
2SC1157GN		3104A	32		
2SC1157H		3104A	32		
2SC1157J		3104A	32		
2SC1157K		3104A	32		
2SC1157L		3104A	32		
2SC1157M		3104A	32		
2SC1157OR		3104A	32		
2SC1157R		3104A	32		
2SC1157X		3104A	32		
2SC1157Y		3104A	32		
2SC1158	313$	3452/108	86	2038	
2SC1159	313$		11	2015	
2SC115A		3024	18	2030	
2SC115B		3024	18	2030	
2SC115C		3024	18	2030	
2SC115D		3024	18	2030	
2SC115E		3024	18	2030	
2SC115F		3024	18	2030	
2SC115G		3024	18	2030	
2SC115GN		3024	18	2030	
2SC115H		3024	18	2030	
2SC115J		3024	18	2030	
2SC115K		3024	18	2030	
2SC115L		3024	18	2030	
2SC115M		3024	18	2030	
2SC115OR		3024	18	2030	
2SC115R		3024	18	2030	
2SC115X		3024	18	2030	
2SC115Y		3024	18	2030	
2SC116	123	3047	46		
2SC1160	175		246	2020	2N3738
2SC1160K	175		246	2020	
2SC1160L	175		246	2020	
2SC1161	175	3021/124	246	2020	2N3738
2SC1162	184#	3190/184	57	2017	MJE180
2SC1162(C)			247	2052	
2SC1162(RF-PWR)			247	2052	
2SC1162-0	184#	3190/184			
2SC1162A	184#	9041/373	247	2052	
2SC1162B	184#	9041/373	247	2052	
2SC1162C	184#	9041/373	247	2052	
2SC1162C(RF-PWR)			247	2052	
2SC1162CP	184#	9041/373	247	2052	
2SC1162D	184#	9041/373	247	2052	
2SC1162MP	184#	3190/184	247	2052	
2SC1162WB	184#	9041/373	247	2052	
2SC1162WBP	184#		247	2052	
2SC1162WT	184#	9041/373	247	2052	
2SC1162WTA	184#	9041/373	247	2052	
2SC1162WTB	184#	9041/373	247	2052	
2SC1162WTC	184#	9041/373	247	2052	
2SC1162WTD	184#	9041/373	247	2052	
2SC1165	486	3195/311			
2SC1166	289A	3449/297	81	2030	
2SC1166-0	289A	3449/297	81	2030	
2SC1166-GR			81	2030	
2SC1166-0	289A	3449/297	81	2030	

Industry Standard No.	ECG	SK	GE	RS 276-	MOTOR.
2SC1166-R		3449	81	2030	
2SC1166-Y	289A		81	2030	
2SC1166D	289A	3449/297	81	2030	
2SC1166GR	289A	3449/297	81	2030	
2SC1166O	289A	3449/297	81	2030	
2SC1166R	289A		81	2030	
2SC1166X	289A	3449/297			
2SC1166Y	289A	3449/297	81	2030	
2SC1167	389	3710/238	38		BU204
2SC1168	124	3271	32		2N3739
2SC1168X	124	3271	32		
2SC116A		3047	46		
2SC116B		3047	46		
2SC116C		3047	46		
2SC116D		3047	46		
2SC116E		3047	46		
2SC116F		3047	46		
2SC116G		3047	46		
2SC116GN		3047	46		
2SC116H		3047	46		
2SC116J		3047	46		
2SC116K		3047	46		
2SC116L		3047	46		
2SC116M		3047	46		
2SC116OR		3047	46		
2SC116R		3047	46		
2SC116T	195A	3047	46		
2SC116Y		3047	46		
2SC117		3024	18	2030	
2SC1170	389	3115/165	38		BU207
2SC1170A	389	3710/238	38		BU208
2SC1170B	389	3710/238	38		
2SC1170B/FA-2	389	3710/238			
2SC1171	164	3133	38		BU204
2SC1172	238	3710	38		BU208
2SC1172A	238	3710	38		
2SC1172B	238	3710	38		
2SC1172C		3710	38		
2SC1172D		3710	38		
2SC1172E		3710	38		
2SC1172F		3710	38		
2SC1172G		3710	38		
2SC1172GN		3710	38		
2SC1172H		3710	38		
2SC1172J		3710	38		
2SC1172K		3710	38		
2SC1172L		3710	38		
2SC1172M		3710	38		
2SC1172R		3710	38		
2SC1172X		3710	38		
2SC1172Y		3710	38		
2SC1173	152		215		TIP31
2SC1173-O	152		215		
2SC1173-GR	152		215		
2SC1173-O	152	3054/196	215		
2SC1173-R	152		215		
2SC1173-Y	152	3054/196	215		
2SC1173A	152	3054/196	215		
2SC1173B		3054	215		
2SC1173C	152	3054/196	215		
2SC1173M	152	3054/196			
2SC1173R	152	3054/196	215		
2SC1173X	152		215		
2SC1173X(RF-PWR)			215	2053	
2SC1173XO	152	3197/235	215		
2SC1173XO	152		215		
2SC1173Y	152		215		
2SC1174	389	3710/238	38		MJ12003
2SC1174A		3710	38		
2SC1174B		3710	38		
2SC1174C		3710	38		
2SC1174D		3710	38		
2SC1174E		3710	38		
2SC1174F		3710	38		
2SC1174G		3710	38		
2SC1174GN		3710	38		
2SC1174H		3710	38		
2SC1174J		3710	38		
2SC1174K		3710	38		
2SC1174L		3710	38		
2SC1174M		3710	38		
2SC1174OR		3710	38		
2SC1174R		3710	38		
2SC1174X		3710	38		
2SC1174Y		3710	38		
2SC1175	289A	3122	210	2051	
2SC1175(D,E,F)			210	2051	
2SC1175C	289A	3122	210	2051	
2SC1175CTV			210	2051	
2SC1175D	289A	3122	210	2051	
2SC1175E	289A	3122	210	2051	
2SC1175E,F,D				2051	
2SC1175F	289A	3122	210	2051	
2SC117A		3024	18	2030	
2SC117B		3024	18	2030	
2SC117C		3024	18	2030	
2SC117D		3024	18	2030	
2SC117F		3024	18	2030	
2SC117G		3024	18	2030	
2SC117GN		3024	18	2030	
2SC117H		3024	18	2030	
2SC117J		3024	18	2030	
2SC117K		3024	18	2030	
2SC117L		3024	18	2030	
2SC117M		3024	18	2030	
2SC117OR		3024	18		
2SC117R		3024	18	2030	
2SC117X		3024	18	2030	
2SC117Y		3024	18	2030	
2SC118		3024	18	2030	
2SC1180	161		39	2015	
2SC1182			39	2038	
2SC1182B	161	3018	39	2015	
2SC1182C	161	3018	39	2015	
2SC1182D	161	3018	39	2015	
2SC1184	164		37		BU204
2SC1184A	164		37		
2SC1184B	164		37		
2SC1184C	164		37		
2SC1184D	164		37		
2SC1184E	164		37		
2SC1185	162	3438	38		2N5840
2SC1185A	162	3438	38		
2SC1185B	162	3438	38		
2SC1185C	162	3438	38		
2SC1185K	162	3438	38		
2SC1185L	162	3438	38		
2SC1185M	162	3438	38		
2SC1187	229	3246	61	2038	
2SC1189K	229	3857/231			
2SC1189L	229	3857/231	62	2038	
2SC1189M	229	3857/231			
2SC118A		3024	18	2030	
2SC118B		3024	18	2030	
2SC118C		3024	18	2030	
2SC118D		3024	18	2030	
2SC118E		3024	18	2030	
2SC118F		3024	18	2030	
2SC118G		3024	18	2030	
2SC118GN		3024	18	2030	
2SC118H		3024	18	2030	
2SC118J		3024	18	2030	
2SC118L		3024	18	2030	
2SC118M		3024	18	2030	
2SC118OR		3024	18	2030	
2SC118R		3024	18	2030	
2SC118X		3024	18	2030	
2SC118Y		3024	18	2030	
2SC119		3024	18	2030	
2SC1195	94	3836/284	35		2N5838
2SC1195FA-1	94		35		
2SC119A		3024	18	2030	
2SC119B		3024	18	2030	
2SC119C		3024	18	2030	
2SC119D		3024	18	2030	
2SC119E		3024	18	2030	
2SC119F		3024	18	2030	
2SC119G		3024	18	2030	
2SC119GN		3024	18	2030	
2SC119H		3024	18	2030	
2SC119J		3024	18	2030	
2SC119K		3024	18	2030	
2SC119L		3024	18	2030	
2SC119M		3024	18	2030	
2SC119OR		3024	18	2030	
2SC119R		3024	18	2030	
2SC119X		3024	18	2030	
2SC119Y		3024	18	2030	
2SC11A		3011	59		
2SC11B		3011	59		
2SC11C		3011	59		
2SC11D		3011	59		
2SC11E		3011	59		
2SC11F		3011	59		
2SC11G		3011	59		
2SC11GN		3011	59		
2SC11H		3011	59		
2SC11J		3011	59		
2SC11K		3011	59		
2SC11L		3011	59		
2SC11M		3011	59		
2SC11R		3011	59		
2SC11X		3011	59		
2SC11Y		3011	59		
2SC12	128	3047/128	18	2030	
2SC120	123A	3024/128	18	2030	
2SC1204	289A	3124/289	62	2010	
2SC1204B	289A	3124/289	62	2010	
2SC1204C	289A	3124/289	62	2010	
2SC1204D	199	3124/289	62	2010	
2SC1205	289A	3452/108	210	2038	
2SC1205A	289A	3452/108	210	2038	
2SC1205B	289A	3452/108	210	2038	
2SC1205C	289A	3452/108	210	2038	
2SC1206		3024	18	2030	
2SC1209		3124	63	2030	
2SC1209(C)			271	2030	
002SC1209C	297	3124/289	63	2030	
2SC1209D		3124	63	2030	
2SC1209E		3124	63	2030	
2SC120A		3024	18	2010	
2SC120B		3024	18	2010	
2SC120C		3024	18	2010	
2SC120D		3024	18	2010	
2SC120E		3024	18	2010	
2SC120F		3024	18	2010	
2SC120G		3024	18	2010	
2SC120GN		3024	18	2010	
2SC120H		3024	18	2010	
2SC120J		3024	18	2010	
2SC120K		3024	18	2010	
2SC120M		3024	18	2010	
2SC120OR		3024	18	2010	
2SC120Y		3024	18	2010	
2SC121	128	3024	18	2030	
2SC1210		3842	222		
2SC1210C		3842	222		
2SC1210D		3842	222		
2SC1210E		3842	222		
2SC1211		3020	47	2030	
2SC1211C			47	2030	
2SC1211D			47	2030	
2SC1211E		3020	47	2030	
2SC1212	373#	9041/373	55	2017	
2SC1212A	373#	9041/373	55	2017	
2SC1212AA	373#	9041/373	55	2017	
2SC1212AB	373#	9041/373	55	2017	
2SC1212ABWT	373#		55	2017	
2SC1212AC	373#	9041/373	55	2017	
2SC1212ACWT	373#		55	2017	
2SC1212AD	373#	9041/373	55	2017	
2SC1212AWT	373#	9041/373	55	2017	
2SC1212AWTA	373#		55	2017	
2SC1212AWTB	373#		55	2017	
2SC1212AWTC	373#		55	2017	
2SC1212AWTD	373#		55	2017	
2SC1212B	373#	9041/373	55	2017	
2SC1212C	373#	9041/373	55	2017	
2SC1212D	373#	9041/373	55	2017	
2SC1212WT	373#	9041/373	55	2017	
2SC1212WTA	373#	9041/373	55	2017	
2SC1212WTB	373#		55	2017	
2SC1212WTC	373#		55	2017	
2SC1212WTD	373#		55	2017	
2SC1213	289A	3122	268	2038	
2SC1213(B)			268	2038	
2SC1213-OR			268	2038	
2SC1213A	289A	3122	268	2038	
2SC1213A(C)			268	2038	
2SC1213AA	289A		268	2038	
2SC1213AB	289A	3122	268	2038	
2SC1213AC	289A	3122	268	2038	
2SC1213AD	289A		268	2038	
2SC1213AK	289A		268	2038	
2SC1213AKA	289A		268	2038	
2SC1213AKB	289A		268	2038	
2SC1213AKC	289A		268	2038	
2SC1213AKD	289A		268	2038	
2SC1213B	289A	3122	268	2038	
2SC1213BC	289A	3122	268	2038	
2SC1213C	289A	3122	268	2038	
2SC1213CD	289A		268	2038	
2SC1213D	289A	3122	268	2038	
2SC1213E		3122	268	2038	
2SC1213F		3122	268	2038	
2SC1213G		3122	268	2038	
2SC1213GN		3122	268	2038	
2SC1213H		3122	268	2038	

Industry Standard No.	ECG	SK	GE	RS 276-	MOTOR.
2SC1213J		3122	268	2038	
2SC1213K		3122	268	2038	
2SC1213L		3122	268	2038	
2SC1213M		3122	268	2038	
2SC1213OR		3122	268	2038	
2SC1213X			268	2038	
2SC1213Y		3122	268	2038	
2SC1214	289A	3124/289	47	2038	
2SC1214(B)			47	2038	
2SC1214A	289A	3124/289	47	2038	
2SC1214B	289A	3124/289	47	2038	
2SC1214C	289A	3124/289	47	2038	
2SC1214D	289A	3124/289	47	2038	
2SC1215	107	3293	61	2015	
2SC1215C		3293	61	2015	
2SC1215D		3293	61	2015	
2SC1215E		3293	61	2015	
2SC1215F		3293	61	2015	
2SC1215G		3293	61	2015	
2SC1215GN		3293	61	2015	
2SC1215H		3293	61	2015	
2SC1215J		3293	61	2015	
2SC1215K		3293	61	2015	
2SC1215L		3293	61	2015	
2SC1215M		3293	61	2015	
2SC1215OR		3293	61	2015	
2SC1215R	107	3293	61	2015	
2SC1215X			61	2015	
2SC1215Y		3293	61	2015	
2SC1216			61	2038	
2SC1217		3201	27		
2SC1218			243	2030	
2SC121A		3024	18	2010	
2SC121B		3024	18	2010	
2SC121C		3024	18	2010	
2SC121D		3024	18	2010	
2SC121E		3024	18	2010	
2SC121F		3024	18	2010	
2SC121G		3024	18	2010	
2SC121GN		3024	18	2010	
2SC121H		3024	18	2010	
2SC121J		3024	18	2010	
2SC121L		3024	18	2010	
2SC121M		3024	18	2010	
2SC121OR		3024	18	2010	
2SC121R		3024	18	2010	
2SC121X		3024	18	2010	
2SC121Y		3024	18	2010	
2SC122	128	3024	18	2030	
2SC1220	289A		18	2030	
2SC1220(E)			243	2030	
2SC1220-003	289A	3122	243	2030	
2SC1220A			18	2030	
2SC1220A(QPR)			243	2030	
2SC1220AP			18	2030	
2SC1220AQ			18	2030	
2SC1220AR			18	2030	
2SC1220E	289A	3024/128	243	2030	
2SC1220P			18	2030	
2SC1220Q			18	2030	
2SC1220R			18	2030	
2SC1222	199	3122	62	2010	
2SC1222A	199	3245	62	2010	
2SC1222B	199	3245	62	2010	
2SC1222C	199	3245	62	2010	
2SC1222D	199	3245	62	2010	
2SC1222E	199	3122	62	2010	
2SC1222U	199	3122	62	2010	
2SC1224					2N6591
2SC1226	186A	3357	215	2052	2N6548
2SC1226(A)			215	2052	
2SC1226(AP)			215	2052	
2SC1226(P)			215	2052	
2SC1226(R)			215	2052	
2SC1226-O		3357	215	2052	
2SC1226A	186A	3357	215	2052	
2SC1226A(P)			215	2052	
2SC1226A(Q)			215	2052	
2SC1226A(QPR)			215	2052	
2SC1226A(R)			215	2052	
2SC1226AC	186A	3357	215	2052	
2SC1226ACF			215	2052	
2SC1226AF			215	2052	
2SC1226AP	186A	3357	215	2052	
2SC1226AQ	186A	3357	215	2052	
2SC1226AR	186A	3357	215	2052	
2SC1226ARL	186A		215	2052	
2SC1226B	186A	3357	215	2052	
2SC1226BL			215	2052	
2SC1226C	186A	3357	215	2052	
2SC1226CF	186A	3357	215	2052	
2SC1226D	186A		215	2052	
2SC1226E			215	2052	
2SC1226F	186A	3357	215	2052	
2SC1226G			215	2052	
2SC1226H			215	2052	
2SC1226L	186A		215	2052	
2SC1226O	186A		215	2052	
2SC1226OR			215	2052	
2SC1226P	186A	3357	215	2052	
2SC1226Q	186A	3357	215	2052	
2SC1226QR	186A	3357	215	2052	
2SC1226R	186A	3357	215	2052	
2SC1226RL	186A		215	2052	
2SC1226RLP	186A		215	2052	
2SC1226RLQ	186A		215	2052	
2SC1226RLR	186A		215	2052	
2SC1226SC	186A	3357	215	2052	
2SC1226Y			215	2052	
2SC1227					MJ10006
2SC1228	385$				SDT13305
2SC1229			251		MJ10006
2SC122A		3024	18	2010	
2SC122B		3024	18	2010	
2SC122C		3024	18	2010	
2SC122D		3024	18	2010	
2SC122E		3024	18	2010	
2SC122F		3024	18	2010	
2SC122G		3024	18	2010	
2SC122GN		3024	18	2010	
2SC122H		3024	18	2010	
2SC122J		3024	18	2010	
2SC122K		3024	18	2010	
2SC122L		3024	18	2010	
2SC122M		3024	18	2010	
2SC122OR		3024	18	2010	
2SC122R		3024	18	2010	
2SC122X		3024	18	2010	
2SC122Y		3024	18	2010	
2SC123	128	3024	18	2030	
2SC1235	124		12		
2SC1235AL	124	3021	12		
2SC1235AM	124	3021	12		
2SC1237	235	3197	215		TIP31B

Industry Standard No.	ECG	SK	GE	RS 276-	MOTOR.
2SC1237E	235	3197	215		
2SC1239	282/427	3299/237	46		MRF8004
2SC123A		3024	18	2010	
2SC123B		3024	18	2010	
2SC123C		3024	18	2010	
2SC123D		3024	18	2010	
2SC123E		3024	18	2010	
2SC123F		3024	18	2010	
2SC123G		3024	18	2010	
2SC123GN		3024	18	2010	
2SC123H		3024	18	2010	
2SC123J		3024	18	2010	
2SC123K		3024	18	2010	
2SC123L		3024	18	2010	
2SC123M		3024	18	2010	
2SC123OR		3024	18	2010	
2SC123R		3024	18	2010	
2SC123X		3024	18	2010	
2SC123Y		3024	18	2010	
2SC124	128	3024	18	2030	
2SC1243	300	3464	273		D40K3
2SC1243-24	300	3464	273		
2SC1243C	300	3464	273		
2SC1243C1	300	3464	273		
2SC1243C2	300	3464	273		
2SC1243D	300	3464	273		
2SC1243D1	300	3464	273		
2SC1243D2	300	3464	273		
2SC1243E	300	3464	273		
2SC1244	123A	3444	20	2051	
2SC1247	289A	3842	210	2030	
2SC1247A	289A	3842			
2SC1247AF	289A	3842			
2SC124A		3024	18	2010	
2SC124B		3024	18	2010	
2SC124C		3024	18	2010	
2SC124D		3024	18	2010	
2SC124E		3024	18	2010	
2SC124F		3024	18	2010	
2SC124G		3024	18	2010	
2SC124GN		3024	18	2010	
2SC124H		3024	18	2010	
2SC124J		3024	18	2010	
2SC124K		3024	18	2010	
2SC124L		3024	18	2010	
2SC124M		3024	18	2010	
2SC124OR		3024	18	2010	
2SC124R		3024	18	2010	
2SC124X		3024	18	2010	
2SC124Y		3024	18	2010	
2SC125			245	2004	
2SC1251					MRF511
2SC1252	77				MRF517
2SC1253	278				2N5109
2SC1254	161	3716	39	2015	2N6304
2SC1256		3195			2N6255
2SC1257	349				2N5590
2SC1258	350				2N6081
2SC1259					2N6083
2SC1260			280		2N6304
2SC127	123A	3452/108	86	2038	
2SC1275	316				2N5031
2SC1276				2014	
2SC1278		3866	220		
2SC1278S		3866	220		
2SC1279	287	3244	220		
2SC1279S	287	3244	220		
2SC127A		3039	86	2038	
2SC127B		3039	86	2038	
2SC127C		3039	86	2038	
2SC127D		3039	86	2038	
2SC127E		3039	86	2038	
2SC127F		3039	86	2038	
2SC127G		3039	86	2038	
2SC127GN		3039	86	2038	
2SC127H		3039	86	2038	
2SC127J		3039	86	2038	
2SC127K		3039	86	2038	
2SC127L		3039	86	2038	
2SC127M		3039	86	2038	
2SC127OR		3039	86	2038	
2SC127R		3039	86	2038	
2SC127X		3039	86	2038	
2SC127Y		3039	86	2038	
2SC128	101	3862/103	59	2002	
2SC1280	172A	3156	64		
2SC1280A	172A	3156	64		
2SC1280AS	172A	3156	64		
2SC1280S	172A	3156	64		
2SC1285			212	2010	
2SC128A		3862	59	2002	
2SC128B		3862	59	2002	
2SC128C		3862	59	2002	
2SC128D		3862	59	2002	
2SC128E		3862	59	2002	
2SC128F		3862	59	2002	
2SC128G		3862	59	2002	
2SC128GN		3862	59	2002	
2SC128H		3862	59	2002	
2SC128J		3862	59	2002	
2SC128K		3862	59	2002	
2SC128L		3862	59	2002	
2SC128M		3862	59	2002	
2SC128OR		3862	59	2002	
2SC128R		3862	59	2002	
2SC128X		3862	59	2002	
2SC128Y		3862	59	2002	
2SC129	101	3861	59	2002	
2SC1292					2N5840
2SC1293	107	3132	61	2009	
2SC1293(3RD IF)				2009	
2SC1293(3RD-IF)			61	2009	
2SC1293(A)			61	2009	
2SC1293A	107	3132	61	2009	
2SC1293A(LAST IF)				2009	
2SC1293A(LAST-IF)			61	2009	
2SC1293B	107	3132	61	2009	
2SC1293B(3RD IF)				2009	
2SC1293B(3RD-IF)			61	2009	
2SC1293B(LAST IF)				2009	
2SC1293B(LAST-IF)			61	2009	
2SC1293C	107		61	2009	
2SC1293C(3RD IF)				2009	
2SC1293C(3RD-IF)			61	2009	
2SC1293D	107		61	2009	
2SC1295	165	3115	38		BU204
2SC1295-O	165		38		
2SC1295A	165		38		
2SC1296	165	3439/163A	36		
2SC1297					2N6082
2SC1298					2N6084
2SC129A		3861	59	2002	
2SC129B		3861	59	2002	
2SC129C		3861	59	2002	

Industry Standard No.	ECG	SK	GE	RS 276-	MOTOR.
2SC129D		3861	59	2002	
2SC129E		3861	59	2002	
2SC129F		3861	59	2002	
2SC129GN		3861	59	2002	
2SC129H		3861	59	2002	
2SC129J		3861	59	2002	
2SC129K		3861	59	2002	
2SC129L		3861	59	2002	
2SC129M		3861	59	2002	
2SC129OR		3861	59	2002	
2SC129R		3861	59	2002	
2SC129X		3861	59	2002	
2SC129Y		3861	59	2002	
2SC12A		3047	18	2030	
2SC12B		3047	18	2030	
2SC12C		3047	18	2030	
2SC12D		3047	18	2030	
2SC12E		3047	18	2030	
2SC12F		3047	18	2030	
2SC12G		3047	18	2030	
2SC12GN		3047	18	2030	
2SC12H		3047	18	2030	
2SC12J		3047	18	2030	
2SC12K		3047	18	2030	
2SC12L		3047	18	2030	
2SC12M		3047	18	2030	
2SC12OR		3047	18	2030	
2SC12R		3047	18	2030	
2SC12X		3047	18	2030	
2SC12Y		3047	18	2030	
2SC13	101	3861	59	2002	
2SC130	128	3024	18	2030	
2SC1303	346	3529	261		
2SC1304	286	3261/175	12		2N3739
2SC1304BK	286	3021/124	12		
2SC1306	235	3239/236	215		MRF340
2SC1306-1	235	3239/236			
2SC1306A		3239	215		
2SC1306I	235		215		
2SC1306K	235		215		
2SC1307	236	3239	216	2053	MRF485
2SC1307-1	236	3239	216	2053	
2SC1307K	236	3239			
2SC1308	238	3710	38		
2SC1308K	238	3710	38		
2SC1308N	238	3710	38		
2SC1309	165	3115	38		BU207
2SC130A		3024	18	2030	
2SC130B		3024	18	2030	
2SC130C		3024	18	2030	
2SC130D		3024	18	2030	
2SC130E		3024	18	2030	
2SC130F		3024	18	2030	
2SC130GN		3024	18	2030	
2SC130H		3024	18	2030	
2SC130J		3024	18	2030	
2SC130K		3024	18	2030	
2SC130L	128	3197/235	18	2030	
2SC130M		3024	18	2030	
2SC130OR		3024	18	2030	
2SC130R		3024	18	2030	
2SC130X		3024	18	2030	
2SC130Y		3024	18	2030	
2SC131	123A	3444	88	2051	
2SC1310	85		212	2010	
2SC1311	85		212	2010	
2SC1312		3899	62	2010	
2SC1312A		3899	62	2010	
2SC1312C		3899	62	2010	
2SC1312D		3899	62	2010	
2SC1312E		3899	62	2010	
2SC1312F		3899	62	2010	
2SC1312G		3899	62	2010	
2SC1312GN			62	2010	
2SC1312H		3899	62	2010	
2SC1312J			62	2010	
2SC1312K			62	2010	
2SC1312L			62	2010	
2SC1312M			62	2010	
2SC1312OR			62	2010	
2SC1312R		3899	62	2010	
2SC1312X			62	2010	
2SC1312Y		3899	62	2010	
2SC1312YF		3899	62	2010	
2SC1312YG		3899	62	2010	
2SC1312YH		3899	62	2010	
2SC1313		3250	62	2010	
2SC1313B		3250	62	2010	
2SC1313F		3250	62	2010	
2SC1313G		3250	62	2010	
2SC1313H		3250	62	2010	
2SC1313Y		3250	62	2010	
2SC1313YE		3250	62	2010	
2SC1313YG		3250	62	2010	
2SC1313YH		3250	62	2010	
2SC1315	278		324		
2SC1316	277		260		MJ4381
2SC1317	289A	3124/289	20	2051	
2SC1317(P)			210	2038	
2SC1317(R)			210	2038	
2SC1317(S)			210	2038	
2SC1317-OR			210	2038	
2SC1317A		3124	210	2038	
2SC1317B	289A	3124/289	210	2038	
2SC1317BC		3124	210	2038	
2SC1317C		3124	210	2038	
2SC1317D	289A	3124/289	210	2038	
2SC1317E		3124	210	2038	
2SC1317G		3124	210	2038	
2SC1317GR		3124	210	2038	
2SC1317L		3124	210	2038	
2SC1317OR		3124	210	2038	
2SC1317P	289A	3124/289	210	2038	
2SC1317Q	289A	3124/289	210	2038	
2SC1317R	289A	3124/289	210	2038	
2SC1317S	289A	3124/289	210	2038	
2SC1317T	289A		210	2038	
2SC1317V		3124	210	2038	
2SC1317Y		3124	210	2038	
2SC1318	289A	3124/289	210	2030	
2SC1318(P)			210	2030	
2SC1318(P,R)			210	2030	
2SC1318(Q)			210	2030	
2SC1318A	289A	3124/289	210	2030	
2SC1318B		3124	210	2030	
2SC1318C		3124	210	2030	
2SC1318E		3124	210	2030	
2SC1318F		3124	210	2030	
2SC1318G		3124	210	2030	
2SC1318GN		3124	210	2030	
2SC1318H		3124	210	2030	
2SC1318J		3124	210	2030	
2SC1318K		3124	210	2030	
2SC1318L		3124	210	2030	
2SC1318M		3124	210	2030	
2SC1318P	289A	3124/289	210	2030	
2SC1318PR			210	2030	
2SC1318Q	289A	3124/289	210	2030	
2SC1318QP			210	2030	
2SC1318QR	289A		210	2030	
2SC1318R	289A	3124/289	210	2030	
2SC1318S	289A	3124/289	210	2030	
2SC1318S,R			210	2030	
2SC1318X	289A	3124/289	210	2030	
2SC1318Y		3124	210	2030	
2SC1319				2013	
2SC131A		3122	88	2051	
2SC131B		3122	88	2051	
2SC131C		3122	88	2051	
2SC131D		3122	88	2051	
2SC131E		3122	88	2051	
2SC131F		3122	88	2051	
2SC131G		3122	88	2051	
2SC131GN		3122	88	2051	
2SC131H		3122	88	2051	
2SC131J		3122	88	2051	
2SC131L		3122	88	2051	
2SC131M		3122	88	2051	
2SC131OR		3122	88	2051	
2SC131R		3122	88	2051	
2SC131T		3122	18	2051	
2SC131Y		3122	88	2051	
2SC132	123A	3444	88	2051	
2SC1320(K)			61	2038	
2SC1320A		3018	61	2038	
2SC1320B		3018	61	2038	
2SC1320C		3018	61	2038	
2SC1320D		3018	61	2038	
2SC1320E		3018	61	2038	
2SC1320F		3018	61	2038	
2SC1320G		3018	61	2038	
2SC1320GN		3018	61	2038	
2SC1320H		3018	61	2038	
2SC1320J		3018	61	2038	
2SC1320K	107	3018	61	2015	
2SC1320L		3018	61	2038	
2SC1320M		3018	61	2038	
2SC1320OR			61	2038	
2SC1320R		3018	61	2038	
2SC1320X		3018	61	2038	
2SC1320Y		3018	61	2038	
2SC1321			39	2015	
2SC1322			262	2041	2N6250
2SC1324	278		20	2051	
2SC1324(C)			20	2051	
2SC1324C			20	2051	
2SC1325	238	3710	38		MJ12005
2SC1325A	238	3710	38		
2SC1325AK	238	3710	38		
2SC1325AL	238	3710	38		
2SC1327	199	3899	85	2010	
2SC1327FS	199	3899	85	2010	
2SC1327R	199	3899			
2SC1327S	199	3245	85	2010	
2SC1327T	199	3899	85	2010	
2SC1327TU	199	3899	85	2010	
2SC1327TV	199		85	2010	
2SC1327U	199	3899	85	2010	
2SC1328	199	3899	62	2010	
2SC1328(U)			62	2010	
2SC1328(U)(T)			62	2010	
2SC1328R	199	3899			
2SC1328S	199	3899			
2SC1328T	199	3899	62	2010	
2SC1328U	199	3899	62	2010	
2SC1329					MRF234
2SC132A		3122	88	2051	
2SC132B		3122	88	2051	
2SC132C		3122	88	2051	
2SC132D		3122	88	2051	
2SC132E		3122	88	2051	
2SC132F		3122	88	2051	
2SC132G		3122	88	2051	
2SC132GN		3122	88	2051	
2SC132H		3122	88	2051	
2SC132J		3122	88	2051	
2SC132K		3122	88	2051	
2SC132L		3122	88	2051	
2SC132M		3122	88	2051	
2SC132OR		3122	88	2051	
2SC132R		3122	88	2051	
2SC132X		3122	88	2051	
2SC132Y		3122	88	2051	
2SC133	123A	3444	88	2051	
2SC1330	192A	3137	63	2030	
2SC1330A	192A		63	2030	
2SC1330B	192A		63	2030	
2SC1330C	192A		63	2030	
2SC1330D	192A		63	2030	
2SC1330L	192A	3137	63	2030	
2SC1330R	192A	3137	63	2030	
2SC1331				2011	
2SC1335	199	3899	210	2010	
2SC1335(E)			210	2010	
2SC1335-OR			210	2010	
2SC1335A	199	3899	210	2010	
2SC1335B	199	3899	210	2010	
2SC1335C	199	3899	210	2010	
2SC1335D	199	3899	210	2010	
2SC1335E	199	3899	210	2010	
2SC1335F	199	3899	210	2010	
2SC1335G			210	2010	
2SC1335GN			210	2010	
2SC1335H			210	2010	
2SC1335J			210	2010	
2SC1335K			210	2010	
2SC1335L			210	2010	
2SC1335M			210	2010	
2SC1335OR			210	2010	
2SC1335R			210	2010	
2SC1335X			210	2010	
2SC1335Y			210	2010	
2SC1336		3357	86	2038	MRF902
2SC1336JK			86	2038	
2SC133A		3122	88	2051	
2SC133C		3122	88	2051	
2SC133D		3122	88	2051	
2SC133E		3122	88	2051	
2SC133F		3122	88	2051	
2SC133G		3122	88	2051	
2SC133GN		3122	88	2051	
2SC133H		3122	88	2051	
2SC133J		3122	88	2051	
2SC133K		3122	88	2051	
2SC133L		3122	88	2051	
2SC133M		3122	88	2051	
2SC133OR		3122	88	2051	
2SC133X		3122	88	2051	

Industry Standard No.	ECG	SK	GE	RS 276-	MOTOR.
2SC133Y		3122	88	2051	
2SC134	123A	3444	61	2051	
2SC1342	107	3124/289	61	2038	
2SC1342(A)			61	2038	
2SC1342(B)			61	2038	
2SC1342(C)			61	2038	
2SC1342-OR			61	2038	
2SC1342A	107	3124/289	61	2038	
2SC1342B	107	3124/289	61	2038	
2SC1342C	107	3124/289	61	2038	
2SC1342D		3124	61	2038	
2SC1342E		3124	61	2038	
2SC1342F		3124	61	2038	
2SC1342G		3124	61	2038	
2SC1342GN		3124	61	2038	
2SC1342H		3124	61	2038	
2SC1342J		3124	61	2038	
2SC1342K		3124	61	2038	
2SC1342L		3124	61	2038	
2SC1342M		3124	61	2038	
2SC13420R		3124	61	2038	
2SC1342R		3124	61	2038	
2SC1342X		3124	61	2038	
2SC1342Y		3124	61	2038	
2SC1343	280	3619	262	2041	MJ15011
2SC1343A	280	3619	262	2041	
2SC1343B	280	3619	262	2041	
2SC1343BL		3619	262	2041	
2SC1343C	280	3619	262	2041	
2SC1343D		3619	262	2041	
2SC1343E		3619	262	2041	
2SC1343F		3619	262	2041	
2SC1343G		3619	262	2041	
2SC1343G-R			262	2041	
2SC1343GN		3619		2041	
2SC1343H	280	3619	262	2041	
2SC1343HA	280	3619	262	2041	
2SC1343HB	280	3619	262	2041	
2SC1343J		3619	262	2041	
2SC1343K		3619	262	2041	
2SC1343L		3619	262	2041	
2SC1343M		3619	262	2041	
2SC13430		3619	262	2041	
2SC13430R		3619	262	2041	
2SC1343R		3619	262	2041	
2SC1343X		3619	262	2041	
2SC1343Y		3619	262	2041	
2SC1344	199	3899	62	2010	
2SC1344(E)			62	2010	
2SC1344C	199	3899	62	2010	
2SC1344D	199	3899	62	2010	
2SC1344E	199	3899	62	2010	
2SC1344F	199	3899	62	2010	
2SC1345	199	3899	212	2010	
2SC1345(E)			212	2010	
2SC1345C	199	3899	212	2010	
2SC1345D	199	3899	212	2010	
2SC1345E	199	3899	212	2010	
2SC1345F	199	3899	212	2010	
2SC1346	192A	3137	63	2030	
2SC1346(R)			63	2030	
2SC1346Q	192A•		63	2030	
2SC1346R	192A		63	2030	
2SC1346S	192A		63	2030	
2SC1347	192A	3137	88	2030	
2SC1347(Q)			88	2030	
2SC1347A		3137	88	2030	
2SC1347B		3137	88	2030	
2SC1347C		3137	88		
2SC1347D		3137	88		
2SC1347F		3137	88	2030	
2SC1347G		3137	88		
2SC1347L		3137	88	2030	
2SC1347Q	192A		88	2030	
2SC1347R	192A	3137	88	2030	
2SC1347RQ			88	2030	
2SC1347S	192A		88	2030	
2SC1347X		3137	88	2030	
2SC1347Y		3137	88	2030	
2SC1348	389	3439/163A	36		BU207
2SC134A		3018	61	2051	
2SC134B	123A	3444	61	2051	
2SC134C		3018	61	2051	
2SC134D		3018	61	2051	
2SC134E		3018	61	2051	
2SC134F		3018	61	2051	
2SC134G		3018	61	2051	
2SC134GN		3018	61	2051	
2SC134H		3018	61	2051	
2SC134J		3018	61	2051	
2SC134K		3018	61	2051	
2SC134L		3018	61	2051	
2SC134M		3018	61	2051	
2SC1340R		3018	61	2051	
2SC134R		3018	61	2051	
2SC134X		3018	61	2051	
2SC134Y		3018	61	2051	
2SC135	123A	3444	61	2051	
2SC1358	165	3710/238	38		BU208
2SC1358A	165	3710/238	38		
2SC1358K	165	3710/238	38		
2SC1358K1	165	3710/238	38		
2SC1358K2	165	3710/238	38		
2SC1358K3	165	3710/238	38		
2SC1358L	238	3710	38		
2SC1358M	238	3710			
2SC1358P	165	3710/238	38		
2SC1358Q	165	3710/238	38		
2SC1358R	165	3710/238	38		
2SC1359	85	3122	212	2038	
2SC1359(A)		3122	212	2038	
2SC1359(B)		3122	212	2038	
2SC1359(C)			212	2038	
2SC1359(C,B)		3122	212	2038	
2SC1359A	85	3122	212	2038	
2SC1359B	85	3122	212	2038	
2SC1359BC	85		212	2010	
2SC1359C	85	3122	212	2038	
2SC1359D	85	3122			
2SC1359Q	85		212	2010	
2SC135A		3018	61	2051	
2SC135B		3018	61	2051	
2SC135C		3018	61	2051	
2SC135D		3018	61	2051	
2SC135E		3018	61	2051	
2SC135F		3018	61	2051	
2SC135G		3018	61	2051	
2SC135GN		3018	61	2051	
2SC135H		3018	61	2051	
2SC135J		3018	61	2051	
2SC135L		3018	61	2051	
2SC135M		3018	61	2051	
2SC1350R		3018	61	2051	
2SC135R		3018	61	2051	

Industry Standard No.	ECG	SK	GE	RS 276-	MOTOR.
2SC135X		3018	61	2051	
2SC135Y		3018	61	2051	
2SC136	123A	3444	20	2051	
2SC1360		3132	62	2015	
2SC1361	289A	3124/289	20	2051	
2SC1362	289A	3124/289	62	2010	
2SC1363	289A	3124/289	20	2051	
2SC1364	289A	3124/289	210	2051	
2SC1364-6	289A	3124/289	210	2051	
2SC1364-OR			210	2051	
2SC1364A	289A	3124/289	210	2051	
2SC1364B		3124	88	2051	
2SC1364C		3124	210	2051	
2SC1364D		3124	210	2051	
2SC1364E		3124	210	2051	
2SC1364H		3124	210	2051	
2SC1364K		3124	210	2051	
2SC1364L		3124	210	2051	
2SC1364M		3124	210	2051	
2SC13640R		3124	210	2051	
2SC1364R		3124	210	2051	
2SC1364X		3124	210	2051	
2SC1364Y		3124	210	2051	
2SC1365	278$	3218/278			MRF517
2SC1366					MRF517
2SC1367	164	3115/165	37		BU204
2SC1367A	164	3131A/369	37		
2SC1368	184#	9041/373	57	2017	
2SC1368(B)			57	2017	
2SC1368B	184#	9041/373	57	2017	
2SC1368C	184#	9041/373	57	2017	
2SC1368D	184#	9041/373	57	2052	
2SC136D			20	2051	
2SC137	123A	3444	61	2051	
2SC1372	289A	3444/123A			
2SC1372Y	289A	3122	20	2051	
2SC1376H	123A$	3024/128			
2SC1377	236	3197/235	216	2053	
2SC137A		3018	61	2051	
2SC137B		3018	61	2051	
2SC137C		3018	61	2051	
2SC137D		3018	61	2051	
2SC137E		3018	61	2051	
2SC137F		3018	61	2051	
2SC137G		3018	61	2051	
2SC137GN		3018	61	2051	
2SC137H		3018	61	2051	
2SC137J		3018	61	2051	
2SC137K		3018	61	2051	
2SC137L		3018	61	2051	
2SC137M		3018	61	2051	
2SC1370R		3018	61	2051	
2SC137R		3018	61	2051	
2SC137X		3018	61	2051	
2SC137Y		3018	61	2051	
2SC138		3444	88	2051	
2SC1380	123A	3444	212	2010	
2SC1380-BL			212	2010	
2SC1380-GR			212	2010	
2SC1380A	123A	3444	212	2010	
2SC1380A-BL		3444	212	2010	
2SC1380A-GR		3444	212	2010	
2SC1380GR	123A	3444			
2SC1381					MJE182
2SC1382	373	3253/295	232		MJE182
2SC1382-O	373	3253/295			
2SC1382-O	373	3253/295	232		
2SC1382Y	373	3253/295	232		
2SC1383	293	3849	243	2030	
2SC1383(P,Q,R)			47	2030	
2SC1383(S)			47	2030	
2SC1383,RS		3849	47	2030	
2SC13838	128		243	2030	
2SC1383P	293	3849	47	2030	
2SC1383Q	293	3849	47	2030	
2SC1383R	293	3849	47	2030	
2SC1383RS	293	3849	47	2030	
2SC1383S	293	3849	47	2030	
2SC1383X	293	3849	47	2030	
2SC1384	293	3849	88	2030	
2SC1384(Q)			88	2030	
2SC1384-OR			88	2030	
2SC1384A		3849	88	2030	
2SC1384B		3849	88	2030	
2SC1384C		3849	88	2030	
2SC1384D		3849	88	2030	
2SC1384E		3849	88	2030	
2SC1384F		3849	88	2030	
2SC1384G		3849	88	2030	
2SC1384GN		3849	88	2030	
2SC1384H		3849	88	2030	
2SC1384J		3849	88	2030	
2SC1384K		3849	88	2030	
2SC1384L		3849	88	2030	
2SC1384M		3849	88	2030	
2SC13840R		3849	88	2030	
2SC1384P	293	3849			
2SC1384Q	293	3849	88	2030	
2SC1384Q,R			88	2030	
2SC1384R	293	3849	88	2030	
2SC1384S	293	3849	88	2030	
2SC1384X		3849	88	2030	
2SC1384Y		3849	88	2030	
2SC1385H	123A$	3024/128	47	2030	
2SC1386H		3024	63	2030	
2SC138A		3444	88	2051	
2SC138B		3122	88	2030	
2SC138C		3122	88	2030	
2SC138D		3122	88	2030	
2SC138E		3122	88	2030	
2SC138F		3122	88	2030	
2SC138G		3122	88	2030	
2SC138GN		3122	88	2030	
2SC138H		3122	88	2030	
2SC138J		3122	88	2030	
2SC138L		3122	88	2030	
2SC138M		3122	88	2030	
2SC1380R		3122	88	2030	
2SC138R		3122	88	2030	
2SC138S		3444	88	2051	
2SC138X		3122	88	2030	
2SC139		3444	88	2051	
2SC1390	123A	3444	20	2051	
2SC1390(L,Y)			20	2051	
2SC1390(V)			20	2051	
2SC1390(W)			20	2051	
2SC1390(X)			20	2051	
2SC1390(Y)			20	2051	
2SC13901	123A	3444	20	2051	
2SC1390I	123A	3018	20	2051	
2SC1390I(W)			20	2009	
2SC1390IW			20	2051	
2SC1390J	123A	3444	20	2051	
2SC1390J(X)			20	2009	

Industry Standard No.	ECG	SK	GE	RS 276-	MOTOR.
2SC1390JX			20	2051	
2SC1390K	123A	3444	20	2051	
2SC1390L	123A	3444	20	2051	
2SC1390V	123A	3444	20	2051	
2SC1390W	123A	3444	20	2051	
2SC1390WH	123A	3444	20	2051	
2SC1390WI	123A	3018	20	2051	
2SC1390WX	123A	3444	20	2051	
2SC1390X	123A	3444	20	2051	
2SC1390XJ	123A	3444	20	2051	
2SC1390XK	123A	3444	20	2051	
2SC1390Y	123A	3444	20	2051	
2SC1390YM	123A	3444	20	2051	
2SC1391	124	3021	12		2N3739
2SC1391VL	124	3021	12		
2SC1393	229	3246	62	2051	
2SC1393K	229	3246	62	2051	
2SC1393L	229	3246			
2SC1393M	229	3246	62	2051	
2SC1394	229	3246	17	2015	
2SC1394K	229	3246			
2SC1394L	229	3246			
2SC1394M	229	3246			
2SC1395		3293	11	2015	
2SC1398	152	3197/235	66	2048	
2SC1398P	152	3197/235	66	2048	
2SC1398Q	152	3197/235	66	2048	
2SC1399			62	2010	
2SC1399E		3024	62	2010	
2SC139A		3122	88	2030	
2SC139B		3122	88	2030	
2SC139C		3122	88	2030	
2SC139D		3122	88	2030	
2SC139E		3122	88	2030	
2SC139F		3122	88	2030	
2SC139G		3122	88	2030	
2SC139GN		3122	88	2030	
2SC139H		3122	88	2030	
2SC139J		3122	88	2030	
2SC139K		3122	88	2030	
2SC139M		3122	88	2030	
2SC139OR		3122	88	2030	
2SC139R		3122	88	2030	
2SC139X		3122	88	2030	
2SC139Y		3122	88	2030	
2SC13A		3861	59		
2SC13B		3861	59		
2SC13C		3861	59		
2SC13D		3861	59		
2SC13E		3861	59		
2SC13F		3861	59		
2SC13G		3861	59		
2SC13GN		3861	59		
2SC13H		3861	59		
2SC13J		3861	59		
2SC13K		3861	59		
2SC13L		3861	59		
2SC13M		3861	59		
2SC13OR		3861	59		
2SC13R		3861	59		
2SC13X		3861	59		
2SC13Y		3861	59		
2SC14	101	3861	59	2002	
2SC140	128	3024	18	2030	
2SC1402	280	3297	262	2041	2N5634
2SC1403	87	3836/284			2N5634
2SC1403A	87	3836/284			
2SC1406	192	3137	63	2030	
2SC1406(P)			63	2030	
2SC1406P			63	2030	
2SC1406Q		3137	63	2030	
2SC1407		3137	63	2030	
2SC1407(Q)			63	2030	
2SC14070	192		63	2030	
2SC1407B			63	2030	
2SC1407P			63	2030	
2SC1407Q			63	2030	
2SC1407R			63	2030	
2SC1407S			63	2030	
2SC1407X			63	2030	
2SC1409	375	3440/291	66		TIP47
2SC1409(B)			66	2048	
2SC1409A	375	3440/291	66		
2SC1409AA	375	3440/291			
2SC1409AB	375	3440/291	66		
2SC1409AC	375	3440/291			
2SC1409B	375	3440/291	66		
2SC1409C	375	3440/291	66		
2SC140A		3024	18	2030	
2SC140C		3024	18	2030	
2SC140D		3024	18	2030	
2SC140E		3024	18	2030	
2SC140F		3024	18	2030	
2SC140G		3024	18	2030	
2SC140GN		3024	18	2030	
2SC140H		3024	18	2030	
2SC140J		3024	18	2030	
2SC140K		3024	18	2030	
2SC140L		3024	18	2030	
2SC140M		3024	18	2030	
2SC140OR		3024	18	2030	
2SC140R		3024	18	2030	
2SC140X		3024	18	2030	
2SC140Y		3024	18	2030	
2SC141			210	2051	
2SC1410	375	3440/291			TIP47
2SC1410A	375	3440/291			
2SC1410AA	375	3440/291			
2SC1410AB	375	3440/291			
2SC1410AC	375	3440/291			
2SC1410B	375	3440/291	66		
2SC1410C	375	3440/291			
2SC1413	165	3710/238	38		BU207
2SC1413A	165	3710/238	38		
2SC1413B		3710	38		
2SC1413C		3710	38		
2SC1413D		3710	38		
2SC1413E		3710	38		
2SC1413F		3710	38		
2SC1413G		3710	38		
2SC1413GN		3710	38		
2SC1413K		3710	38		
2SC1413L		3710	38		
2SC1413R		3710	38		
2SC1413X		3710	38		
2SC1413Y		3710	38		
2SC1416	123A		62	2010	
2SC1416A		3244	62	2051	
2SC1416BL	123A		62	2051	
2SC1417	107	3293	61	2038	
2SC1417(V,G)			61	2015	
2SC1417(W)			61	2038	
2SC1417C	107		61	2038	
2SC1417D	107	3293	61	2038	
2SC1417D(U)	107		61	2038	
2SC1417DU	107		61	2038	
2SC1417F	107	3293	61	2038	
2SC1417G	107	3293	61	2038	
2SC1417H	107	3293	61	2038	
2SC1417U	107	3293	61	2038	
2SC1417V	107	3293	61	2038	
2SC1417VF		3018	61	2038	
2SC1417VW	107	3293	61	2015	
2SC1417W	107	3293	61	2038	
2SC1418	152	3054/196	66	2048	TIP31
2SC1418A	152	3054/196	66	2048	
2SC1418B	152	3054/196	66	2048	
2SC1418C	152	3054/196	66	2048	
2SC1418D	152	3054/196	66	2048	
2SC1419	152	3197/235	66	2048	TIP31
2SC1419A	152	3197/235	66	2048	
2SC1419B	152	3197/235	66	2048	
2SC1419C	152	3197/235	66	2048	
2SC1419D	152	3197/235	66	2048	
2SC142			210	2051	
2SC1424	316	3018	280		
2SC1428				2014	
2SC1429	152	3199/188	243	2018	MPSU01
2SC1429-1	152	3199/188	243	2018	
2SC1429-2	152	3199/188	243	2018	
2SC143			210	2051	
2SC1431	286				2N5050
2SC1433	283	3111			MJ411
2SC1434	386				2N6546
2SC1436	386				2N6249
2SC144			210	2051	
2SC1440	386				MJ15001
2SC1441	386				2N6249
2SC1444	196				2N5428
2SC1445	196				2N5430
2SC1446	376	3219	251		
2SC1446B	376	3219	251		
2SC1446C	376	3219	251		
2SC1446L	376	3219	251		
2SC1446LB	376	3219	251		
2SC1446P	376	3219	251		
2SC1446PQ	376	3219	251		
2SC1446Q	376	3219	251		
2SC1446R	376	3219	251		
2SC1447	396	3219	251		TIP47
2SC1447-O	396	3219	251		
2SC1447-O	396		251		
2SC1447A	396	3219			
2SC1447LB	396	3220/198	251		
2SC1447R	396	3219	251		
2SC1448	375	3929	32		TIP47
2SC1448A	375	3929	32		
2SC1448LB	375		32		
2SC1448P	375	3929	32		
2SC1448Q	375	3929	32		
2SC1448R	375	3929	32		
2SC1448S	375	3929	32		
2SC1449	184	3253/295	270	2017	MJE180
2SC1449(CB)	295		270	2017	
2SC1449CB			270	2017	
2SC1449L	184	3253/295			
2SC1449M	184	3253/295	270	2017	
2SC144A			210	2051	
2SC1450	286	3194	66	2048	2N3583
2SC14505			66	2048	
2SC1450S	286	3194	66	2048	
2SC1453			212	2010	
2SC1454	94	3560	35		MJ411
2SC1454-O	94		35		
2SC1456	124		32		2N3739
2SC1456L	124	3021	32		
2SC1456LM	124		32		
2SC1456M	124	3021	32		
2SC1463	385$				2N6543
2SC1466	369				2N3585
2SC1467	369$				MJ4381
2SC1468					SDT13304
2SC1469					SDT13305
2SC147	128	3048/329	18	2030	
2SC147-3	293	3048/329			
2SC147-4	293	3048/329			
2SC1472		3156	64		
2SC1472K	172A	3156	64		
2SC1473		3244	222		
2SC1473NC		3244	222		
2SC1473Q		3244	222		
2SC1473R		3244	222		
2SC1474		3849	83	2030	
2SC1474-3		3849	83	2030	
2SC1474-4		3849	83	2030	
2SC1474J		3849	83	2030	
2SC1474S		3849	83	2030	
2SC1475	315	3250	279		
2SC1475-1	315	3250	279		
2SC1475-3	315	3250	279		
2SC1475-4	315	3250	279		
2SC1475A	315	3250	279		
2SC1475D	315	3250	279		
2SC1475E	315	3250			
2SC1475HE	315	3250			
2SC1475K	315		279		
2SC1477	283		36		MJ10006
2SC147A		3048	18	2030	
2SC147B		3048	18	2030	
2SC147C		3048	18	2030	
2SC147D		3048	18	2030	
2SC147F		3048	18	2030	
2SC147G		3048	18	2030	
2SC147GN		3048	18	2030	
2SC147H		3048	18	2030	
2SC147J		3048	18	2030	
2SC147K		3048	18	2030	
2SC147L		3048	18	2030	
2SC147M		3048	18	2030	
2SC147R		3048	18	2030	
2SC147X		3048	18	2030	
2SC147Y		3048	18	2030	
2SC148	108	3452	61	2038	
2SC148A		3018	61	2038	
2SC148B		3018	61	2038	
2SC148D		3018	61	2038	
2SC148E		3018	61	2038	
2SC148F		3018	61	2038	
2SC148G		3018	61	2038	
2SC148GN		3018	61	2038	
2SC148H		3018	61	2038	
2SC148J		3018	61	2038	
2SC148K		3018	61	2038	
2SC148L		3018	61	2038	
2SC148M		3018	61	2038	
2SC148OR			61	2038	
2SC148R		3018	61	2038	
2SC148X		3018	61	2038	

Industry Standard No.	ECG	SK	GE	RS 276-	MOTOR.
2SC148Y		3018	61	2038	
2SC14A		3861	59		
2SC14B		3861	59		
2SC14C		3861	59		
2SC14D		3861	59		
2SC14E		3861	59		
2SC14F		3861	59		
2SC14G		3861	59		
2SC14GN		3861	59		
2SC14H		3861	59		
2SC14J		3861	59		
2SC14K		3861	59		
2SC14L		3861	59		
2SC14M		3861	59		
2SC140R		3861	59		
2SC14R		3861	59		
2SC14X		3861	59		
2SC14Y		3861	59		
2SC15	123A	3444	88	2051	
2SC15-0			88	2030	
2SC15-1	123A	3444	88	2051	
2SC15-2	123A	3444	88	2051	
2SC15-3	123A	3444	88	2051	
2SC150	123	3047	47	2030	
2SC150-0R			47	2030	
2SC1501	157	3747	232		MJE3439
2SC1501P	157	3747			
2SC1501Q	157	3747	232		
2SC1501R	157	3747	232		
2SC1504	162				MJ4381
2SC1505	198	3219	251		TIP48
2SC1505-1	198	3219			
2SC1505I	198		251		
2SC1505K	198	3219	251		
2SC1505L	198	3219	251		
2SC1505LA	198	3219	251		
2SC1505M	198	3219	251		
2SC1506	198	3219	251		TIP48
2SC1506K	198		251		
2SC1506L	198		251		
2SC1506M	198		251		
2SC1507	198	3219	251		TIP48
2SC1507A	198	3219	251		
2SC1507H	198	3219	251		
2SC1507J		3219	251		
2SC1507K	198	3219	251		
2SC1507L	198	3219	251		
2SC1507LK	198	3219			
2SC1507LM	198	3219	251		
2SC1507M	198	3219	251		
2SC1507R	198	3219			
2SC1509	297	3449	271	2030	
2SC1509P	297	3449	271	2030	
2SC1509Q	297	3449	271	2030	
2SC1509R	297	3449	271	2030	
2SC1509S	297	3449	271	2030	
2SC150A		3047	47	2030	
2SC150B		3047	47	2030	
2SC150C		3047	47	2030	
2SC150D		3047	47	2030	
2SC150E		3047	47	2030	
2SC150F		3047	47	2030	
2SC150G		3047	47	2030	
2SC150GN		3047	47	2030	
2SC150H		3047	47	2030	
2SC150J		3047	47	2030	
2SC150K		3047	47	2030	
2SC150L		3047	47	2030	
2SC150M		3047	47	2030	
2SC1500R		3047	47	2030	
2SC150R		3047	47	2030	
2SC150T	128	3047	47	2030	
2SC150X			47	2030	
2SC150Y		3047	47	2030	
2SC151	123	3265	88	2051	
2SC1514	376	3219	251		MJE3439
2SC1514B	376		251		
2SC1514BK	376	3219	251		
2SC1514BVC	376	3219	251		
2SC1514CVC	376	3219	251		
2SC1515		3244	222		
2SC1515K		3244	222		
2SC1516	186A				MJE3300
2SC1517					2N4922
2SC1517A	228A		D42C11		
2SC1518	293	3849	47	2030	
2SC1518R	293	3849			
2SC1518S	293	3849			
2SC1519	198	3219	251		2N6557
2SC151A		3265	88	2030	
2SC151B		3265	88	2030	
2SC151C	280	3265	88	2030	
2SC151D		3265	88	2030	
2SC151E		3265	88	2030	
2SC151G		3265	88	2030	
2SC151GN		3265	88	2030	
2SC151H	123	3265	88	2030	
2SC151J		3265	88	2030	
2SC151K		3265	88	2030	
2SC151L		3265	88	2030	
2SC151M		3265	88	2030	
2SC1510R		3265	88	2030	
2SC151R		3265	88	2030	
2SC151X			88	2030	
2SC151Y		3265	88	2030	
2SC152	123	3265	18	2030	
2SC1520	198	3220	251		2N6557
2SC1520-1	198		251		
2SC1520-1A	198	3220	251		
2SC1520-3A	198	3220	251		
2SC1520E	198	3220			
2SC1520EK	198	3220			
2SC1520EX	198	3220			
2SC1520I	198	3220	251		
2SC1520K	198	3220	251		
2SC1520K-1	198		251		
2SC1520KL	198	3220	251		
2SC1520L	198	3220	251		
2SC1520L-1	198		251		
2SC1520M	198	3220	251		
2SC1520M-1	198		251		
2SC1521	198$		251		2N6557
2SC1521K	198$	3103A/396	251		
2SC1521L	198$	3103A/396	251		
2SC1521LM	198$		251		
2SC1521M	198$	3103A/396			
2SC152A	123	3265	18	2030	
2SC152B	123	3265	18	2030	
2SC152C	123	3265	18	2030	
2SC152D		3265	18	2030	
2SC152E		3265	18	2030	
2SC152F		3265	18	2030	
2SC152G		3265	18	2030	
2SC152GN		3265	18	2030	

Industry Standard No.	ECG	SK	GE	RS 276-	MOTOR.
2SC152H	123	3265	18	2030	
2SC152J		3265	18	2030	
2SC152K		3265	18	2030	
2SC152L		3265	18	2030	
2SC152M		3265	18	2030	
2SC1520R		3265	18	2030	
2SC152R		3024	18	2030	
2SC152X			18	2030	
2SC152Y		3265	18	2030	
2SC1537	199	3245	63	2051	
2SC1537(S)			63	2051	
2SC1537-0		3122	63	2051	
2SC1537-0	199		63	2051	
2SC1537B	199	3122	63	2051	
2SC1537S	199	3122	63	2051	
2SC1538	199	3122	63	2010	
2SC1538A	199	3245	63	2010	
2SC1538S	199	3245	63	2010	
2SC1538S(A)			63	2010	
2SC1538SA	199	3245	63	2010	
2SC154	154	3040	235	2012	
2SC154(C)			235	2012	
2SC154-0R			235	2012	
2SC1542	199	3444/123A	212	2051	
2SC1546	172A		40	2012	
2SC1547	161	3716	280		
2SC154A		3040	235	2012	
2SC154B	154	3040	235	2012	
2SC154C	154	3040	235	2012	
2SC154D			235	2012	
2SC154E			235	2012	
2SC154F			235	2012	
2SC154G			235	2012	
2SC154GN			235	2012	
2SC154H	128	3040	235	2012	
2SC154J			235	2012	
2SC154K			235	2012	
2SC154L			235	2012	
2SC154M			235	2012	
2SC1540R			235	2012	
2SC154R			235	2012	
2SC154X			235	2012	
2SC154Y			235	2012	
2SC155	107	3452/108	86	2038	
2SC1550	154	3747/157	232		
2SC1556	195A	3765	219		
2SC155A		3039	86	2038	
2SC155B		3039	86	2038	
2SC155C		3039	86	2038	
2SC155D		3039	86	2038	
2SC155E		3039	86	2038	
2SC155F		3039	86	2038	
2SC155GN		3039	86	2038	
2SC155H		3039	86	2038	
2SC155J		3039	86	2038	
2SC155K		3039	86	2038	
2SC155L		3039	86	2038	
2SC155M		3039	86	2038	
2SC1550R		3039	86	2038	
2SC155R		3039	86	2038	
2SC155X		3039	86	2038	
2SC155Y		3039	86	2038	
2SC156	107	3452/108	86	2038	
2SC1566	157#	3747/157	232		
2SC1566F	228A	3747/157			
2SC1567	373#	9041/373	270		
2SC1567A	373#	9041/373	270		
2SC1567AP	373#	9041/373			
2SC1567AQ	373#	9041/373			
2SC1567AR	373#	9041/373			
2SC1567AS	373#	9041/373			
2SC1567P	373#	9041/373			
2SC1567Q	373#	9041/373			
2SC1567R	373#	9041/373			
2SC1567S	373#	9041/373			
2SC1568	184#	3190/184	57	2017	
2SC1568(R)			57	2017	
2SC1568R	184#	3054/196	57	2017	
2SC1569	376	3219	251		
2SC1569-0	376	3219	251		
2SC1569BK	376	3219	251		
2SC1569K	376	3219	251		
2SC1569LBO	376		251		
2SC1569LBQ	376	3219	251		
2SC1569LBR	376	3219	251		
2SC1569LBY	376	3219	251		
2SC1569Q	376		251		
2SC1569R	376	3219	251		
2SC1569Y	376	3219	251		
2SC156A		3039	86	2038	
2SC156B		3039	86	2038	
2SC156C		3039	86	2038	
2SC156D		3039	86	2038	
2SC156E		3039	86	2038	
2SC156F		3039	86	2038	
2SC156G		3039	86	2038	
2SC156GN		3039	86	2038	
2SC156H		3039	86	2038	
2SC156J		3039	86	2038	
2SC156K		3039	86	2038	
2SC156L		3039	86	2038	
2SC156M		3039	86	2038	
2SC1560R		3039	86	2038	
2SC156R		3039	86	2038	
2SC156X		3039	86	2038	
2SC156Y		3039	86	2038	
2SC157	123	3124/289	47	2051	
2SC1570	199	3124/289	212	2030	
2SC1570LH	199	3124/289	212	2030	
2SC1571	199	3124/289	212	2030	
2SC1571G	199	3124/289	212	2030	
2SC1571L	199		212	2030	
2SC1573	399	3433/287	222		
2SC1573P		3433	232		
2SC1573Q		3433	232		
2SC1576	385				SDT13304
2SC1577	386				SDT13305
2SC1578	386				MJ10014
2SC1579	386				MJ10013
2SC157A		3124	47	2030	
2SC157B		3124	47	2030	
2SC157C		3124	47	2030	
2SC157D		3124	47	2030	
2SC157E		3124	47	2030	
2SC157F		3124	47	2030	
2SC157G		3124	47	2030	
2SC157GN		3124	47	2030	
2SC157J		3124	47	2030	
2SC157K		3124	47	2030	
2SC157L		3124	47	2030	
2SC157M		3124	47	2030	
2SC1570R		3124	47	2030	
2SC157R		3124	47	2030	
2SC157Y		3124	47	2030	

Industry Standard No.	ECG	SK	GE	RS 276-	MOTOR.
2SC158	123	3124/289	47	2051	
2SC1580	386				MJ10014
2SC1584	284	3260			2N6249
2SC1585	388	3836/284			2N6249
2SC1585F	388	3836/284	11	2015	
2SC1585F,H			11	2015	
2SC1585H	388	3836/284	11	2015	
2SC1586	388		265	2047	2N6250
2SC1588	278		261		
2SC1589	474		236		MRF227
2SC158A		3124	47	2030	
2SC158B		3124	47	2030	
2SC158C		3124	47	2030	
2SC158D		3124	47	2030	
2SC158E		3124	47	2030	
2SC158F		3124	47	2030	
2SC158G		3124	47	2030	
2SC158GN		3124	47	2030	
2SC158H		3124	47	2030	
2SC158J		3124	47	2030	
2SC158K		3124	47	2030	
2SC158L		3124	47	2030	
2SC158M		3124	47	2030	
2SC1580R		3124	47	2030	
2SC158R		3124	47	2030	
2SC158X		3124	47	2030	
2SC158Y		3124	47	2030	
2SC159	123A	3444	61	2051	
2SC1590					MRF260
2SC1591					MRF262
2SC159A		3018	61	2038	
2SC159B		3018	61	2038	
2SC159C		3018	61	2038	
2SC159D		3018	61	2038	
2SC159E		3018	61	2038	
2SC159F		3018	61	2038	
2SC159GN		3018	61	2038	
2SC159H		3018	61	2038	
2SC159J		3018	61	2038	
2SC159K		3018	61	2038	
2SC159L		3018	61	2038	
2SC159M		3018	61	2038	
2SC1590R		3018	61	2038	
2SC159R		3018	61	2038	
2SC159X		3018	61	2038	
2SC159Y		3018	61	2038	
2SC15A		3122	88	2030	
2SC15B		3122	88	2030	
2SC15C		3122	88	2030	
2SC15D		3122	88	2030	
2SC15E		3122	88	2030	
2SC15F		3122	88	2030	
2SC15G		3122	88	2030	
2SC15GN		3122	88	2030	
2SC15H		3122	88	2030	
2SC15J		3122	88	2030	
2SC15K		3122	88	2030	
2SC15L		3122	88	2030	
2SC15M		3122	88	2030	
2SC150R		3122	88	2030	
2SC15R		3122	88	2030	
2SC15X		3122	88	2030	
2SC15Y		3122	88	2030	
2SC16	123A	3444	88	2051	
2SC160	123A	3444	61	2051	
2SC1600					MRF905
2SC1603	362				2N5944
2SC1604					MRF628
2SC1605A					MRF226
2SC1606					MRF327
2SC1609					2N6340
2SC160A		3018	61	2038	
2SC160B		3018	61	2038	
2SC160C		3018	61	2038	
2SC160D		3018	61	2038	
2SC160E		3018	61	2038	
2SC160F		3018	61	2038	
2SC160G		3018	61	2038	
2SC160GN		3018	61	2038	
2SC160H		3018	61	2038	
2SC160J		3018	61	2038	
2SC160K		3018	61	2038	
2SC160L		3018	61	2038	
2SC160M		3018	61	2038	
2SC1600R			61	2038	
2SC1600R				2048	
2SC160R			61	2038	
2SC160X		3018	61	2038	
2SC160Y		3018	61	2038	
2SC161					2N3447
2SC1610	327				2N6341
2SC1617	283	3439/163A	36		MJ411
2SC1618	87	3027/130	14	2041	2N5758
2SC1618B	87	3027/130	14	2041	
2SC1619	87	3027/130			2N5758
2SC162			81	2051	
2SC1621			214	2038	
2SC1622			39	2015	
2SC1623			212	2010	
2SC1624	291	3219			
2SC1624-O	291	3219			
2SC1624-Y	291	3219			
2SC1625	291	3219	251		
2SC1625-O	291		251		
2SC1625-O	291	3219			
2SC1625YLBGL1	291	3119/113	251		
2SC1626	375	3054/196			
2SC1626Y	375	3054/196			
2SC1627	289A	3449/297	271	2030	
2SC1627-O	289A	3449/297	243		
2SC1627-O			271	2030	
2SC1627-Y	289A	3449/297			
2SC1627AY	289A	3449/297	271		
2SC1627Y	289A	3024/128	271	2030	
2SC1628	190				MPSU04
2SC1629	86$	3563	75	2041	MJ1001
2SC1629A	86	3563	75	2041	
2SC1629AO		3563	75	2041	
2SC1629M	86	3563	75	2041	
2SC163	311	3047	18		
2SC1630	154				MPSU04
2SC1632	85	3124/289	62	2010	
2SC1633	85	3124/289	210	2010	
2SC1634	85	3124/289	20	2051	
2SC1636	108	3039/316	86	2038	
2SC1637	199			2033	
2SC1639	123A	3444	20	2051	
2SC163A		3024	18	2030	
2SC163B		3024	18	2030	
2SC163C		3024	18	2030	
2SC163D		3024	18	2030	
2SC163E		3024	18	2030	
2SC163F		3024	18	2030	
2SC163G		3024	18	2030	
2SC163GN		3024	18	2030	
2SC163H		3024	18	2030	
2SC163J		3024	18	2030	
2SC163K		3024	18	2030	
2SC163L		3024	18	2030	
2SC163M		3024	18	2030	
2SC1630R		3024	18	2030	
2SC163R		3024	18	2030	
2SC163X		3024	18	2030	
2SC163Y		3024	18	2030	
2SC164			45	2030	
2SC1641	123A	3122	20	2051	
2SC1641Q	123A	3122	20	2051	
2SC1641R	123A	3020/123	20	2051	
2SC1647	199	3245	62	2010	
2SC1647Q	199	3245	62	2010	
2SC1647RY	199	3245	62	2010	
2SC1648	199	3245	62	2010	
2SC1648E	199	3245	62	2010	
2SC1648EY	199	3245			
2SC1648S	199	3245	62	2010	
2SC1648SH	199	3245	62	2010	
2SC164A		3047	45	2030	
2SC164B		3047	45	2030	
2SC164C		3047	45	2030	
2SC164D		3047	45	2030	
2SC164E		3047	45	2030	
2SC164F		3047	45	2030	
2SC164G		3047	45	2030	
2SC164GN		3047	45	2030	
2SC164H		3047	45	2030	
2SC164J		3047	45	2030	
2SC164K		3047	45	2030	
2SC164L		3047	45	2030	
2SC1640R		3047	45	2030	
2SC164R		3122	45	2030	
2SC164X		3047	45	2030	
2SC164Y		3047	45	2030	
2SC165			88	2030	
2SC1658Q		3275	220		
2SC166	123A	3444	88	2051	
2SC1663	190	3865	226	2018	
2SC1663A	190	3865			
2SC1663H	190	3865	226	2018	
2SC1664		3913			2N6300
2SC1664A		3913	246	2020	
2SC1667	130	3836/284	265	2047	2N5758
2SC1667Q	130	3836/284			
2SC1667R	130	3836/284			
2SC1669	375				TIP47
2SC166A		3122	88	2015	
2SC166B		3122	88	2015	
2SC166C		3122	88	2015	
2SC166D		3122	88	2015	
2SC166E		3122	88	2015	
2SC166F		3122	88	2015	
2SC166G		3122	88	2015	
2SC166GN		3122	88	2015	
2SC166H		3122	88	2015	
2SC166J		3122	88	2015	
2SC166K		3122	88	2015	
2SC166L		3122	88	2015	
2SC166M		3122	88	2015	
2SC1660R		3122	88	2015	
2SC166R		3122	88	2015	
2SC166X		3122	88	2015	
2SC166Y		3122	88	2015	
2SC167	123A	3444	88	2051	
2SC1670	382$	3433/287	222		
2SC1670H	382$		222		
2SC1670J	382$	3433/287	222		
2SC1670JA	382$		222		
2SC1670JW	382$	3433/287	222		
2SC1672	327				2N6341
2SC1674	107	3132	61	2038	
2SC1674K	107	3132	61	2038	
2SC1674L	107	3132	61	2038	
2SC1674M	107	3132	61	2038	
2SC1675	85	3124/289	213	2051	
2SC1675B	85	3124/289			
2SC1675K	85	3124/289	213	2051	
2SC1675K2	85	3124/289			
2SC1675KZ	85	3124/289			
2SC1675L	85	3124/289	213	2051	
2SC1675M	85	3124/289	213	2051	
2SC1675Z	85	3124/289			
2SC1678	236	3197/235	322		
2SC1678B	236	3197/235			
2SC1678D	236	3197/235			
2SC1678E	236	3197/235	322		
2SC1678L	236	3197/235			
2SC1678T	236	3197/235			
2SC1679	236	3197/235	215		
2SC167A		3122	88	2015	
2SC167B		3122	88	2015	
2SC167C		3122	88	2015	
2SC167D		3122	88	2015	
2SC167E		3122	88	2015	
2SC167F		3122	88	2015	
2SC167G		3122	88	2015	
2SC167GN		3122	88	2015	
2SC167H		3122	88	2015	
2SC167K		3122	88	2015	
2SC167L		3122	88	2015	
2SC167M		3122	88	2015	
2SC1670R		3122	88	2015	
2SC167X		3122	88	2015	
2SC167Y		3122	88	2015	
2SC1681	199	3931	62	2010	
2SC1681-GR			62	2010	
2SC1681B	199	3931			
2SC1681BL	199	3931	62	2010	
2SC1681GR	199	3931	62	2010	
2SC1681V	199	3931	62	2010	
2SC1682	199	3124/289	62	2010	
2SC1682-BL	199		88	2030	
2SC1682-GR	199		88	2030	
2SC1682V	199	3124/289	62	2030	
2SC1683	375	3220/198	251		TIP47
2SC1683A	375		251		
2SC1683LA	375	3220/198	251		
2SC1683P	375	3220/198	251		
2SC1683Q	375		251		
2SC1683R	375	3220/198	251		
2SC1684	289A	3124/289	62	2010	
2SC1684BL	289A	3124/289	62	2010	
2SC1684P	289A	3124/289	62	2010	
2SC1684Q	289A	3124/289	62	2010	
2SC1684R	289A	3124/289	62	2010	
2SC1684S	199	3124/289	62	2010	
2SC1684T	199	3124/289	62	2010	
2SC1685	85	3124/289	62	2010	
2SC1685-O			62	2010	
2SC1685P	85	3124/289	62	2010	

Industry Standard No.	ECG	SK	GE	RS 276-	MOTOR.
2SC1685Q	85	3124/289	62	2010	
2SC1685R	85	3124/289	62	2010	
2SC1685S	85	3124/289	62	2010	
2SC1685T	199		62	2010	
2SC1686	229	3246	61	2038	
2SC1686B	229		61	2038	
2SC1686LL	229	3246			
2SC1687	229	3246	20	2009	
2SC1687C	229	3246			
2SC1687LL	229	3246			
2SC1688	229$	3246/229	81	2009	
2SC1689	360				2N5643
2SC16A	123A	3444	88	2051	
2SC16B		3122	88	2030	
2SC16C		3122	88	2030	
2SC16D		3122	88	2030	
2SC16E		3122	88	2030	
2SC16F		3122	88	2030	
2SC16G		3122	88	2030	
2SC16GN		3122	88	2030	
2SC16H		3122	88	2030	
2SC16J		3122	88	2030	
2SC16K		3122	88	2030	
2SC16L		3122	88	2030	
2SC16M		3122	88	2030	
2SC16OR		3122	88	2030	
2SC16R		3122	88	2030	
2SC16X		3122	88	2030	
2SC16Y		3122	88	2030	
2SC17	123A	3452/108	61	2038	
2SC170	123A	3444	86	2038	
2SC170A		3444	86	2038	
2SC170B		3444	86	2038	
2SC170C		3444	86	2038	
2SC170D		3444	86	2038	
2SC170E		3444	86	2038	
2SC170F		3444	86	2038	
2SC170G		3444	86	2038	
2SC170GN		3444	86	2038	
2SC170H		3444	86	2038	
2SC170J		3444	86	2038	
2SC170K		3444	86	2038	
2SC170L		3444	86	2038	
2SC170M		3444	86	2038	
2SC170OR			86	2038	
2SC170R	123A		86	2038	
2SC170X		3444	86	2038	
2SC170Y		3444	86	2038	
2SC171	123A	3019	86	2015	
2SC171A		3039	86	2038	
2SC171B		3039	86	2038	
2SC171C		3039	86	2038	
2SC171D		3039	86	2038	
2SC171E		3039	86	2038	
2SC171F		3039	86	2038	
2SC171G		3039	86	2038	
2SC171GN		3039	86	2038	
2SC171H		3039	86	2038	
2SC171J		3039	86	2038	
2SC171K		3039	86	2038	
2SC171L		3039	86	2038	
2SC171M		3039	86	2038	
2SC171OR		3039	86	2038	
2SC171R		3039	86	2038	
2SC171X		3039	86	2038	
2SC171Y		3039	86	2038	
2SC172	123A	3452/108	86	2038	
2SC1722	376$	3219	251		TIP48
2SC1722BK	376$		251		
2SC1722BKS	376$		251		
2SC1722S	376$	3219	251		
2SC1723	376$	3219	325		TIP48
2SC1723-02	376$		325		
2SC1723A	376$		325		
2SC1727	229	3246		2031	
2SC1728	302	3252	275		MPSU07
2SC1728-3	302	3252	275		
2SC1728D	302	3252	275		
2SC1728H	302		275		
2SC1729	344				MRF226
2SC172A	123A	3452/108	86	2038	
2SC172B		3039	86	2038	
2SC172C		3039	86	2038	
2SC172D		3039	86	2038	
2SC172E		3039	86	2038	
2SC172F		3039	86	2038	
2SC172G		3039	86	2038	
2SC172GN		3039	86	2038	
2SC172H		3039	86	2038	
2SC172J		3039	86	2038	
2SC172K		3039	86	2038	
2SC172L		3039	86	2038	
2SC172M		3039	86	2038	
2SC172OR		3039	86	2038	
2SC172R		3039	86	2038	
2SC172Y		3039	86	2038	
2SC173	101	3861	59	2002	
2SC1730	107	3293	17		
2SC1730L	107	3293	17		
2SC1739	123A	3122	20	2051	
2SC173A		3861	59		
2SC173B		3861	59		
2SC173C		3861	59		
2SC173D		3861	59		
2SC173E		3861	59		
2SC173F		3861	59		
2SC173G		3861	59		
2SC173GN		3861	59		
2SC173J		3861	59		
2SC173K		3861	59		
2SC173L		3861	59		
2SC173M		3861	59		
2SC173OR		3861	59		
2SC173R		3861	59		
2SC173X		3861	59		
2SC173Y		3861	59		
2SC174	123A	3019	86	2015	
2SC1740	85	3122	62	2051	
2SC1740L	85		62	2051	
2SC1740P	85	3122	62	2051	
2SC1740Q	85	3122	62	2051	
2SC1740QH		3122	62	2051	
2SC1740QJ		3122	62	2051	
2SC1740R	85	3122	62	2051	
2SC1740RH		3122	62	2051	
2SC1740S	85	3122	62	2051	
2SC1741	85	3124/289	47	2030	
2SC1741Q	85	3124/289			
2SC1749	376	3219	251		MJE340
2SC1749XD	376		251		
2SC174A	161	3019	86	2015	
2SC174B		3039	86	2038	
2SC174C		3039	86	2038	
2SC174D		3039	86	2038	
2SC174E		3039	86	2038	
2SC174F		3039	86	2038	
2SC174G		3039	86	2038	
2SC174GN		3039	86	2038	
2SC174H		3039	86	2038	
2SC174J		3039	86	2038	
2SC174K		3039	86	2038	
2SC174L		3039	86	2038	
2SC174M		3039	86	2038	
2SC174R		3039	86	2038	
2SC174X		3039	86	2038	
2SC174Y		3039	86	2038	
2SC175	101	3861	59	2002	
2SC1755	376	3219	251		MJE2360T
2SC1755A	376	3219	251		
2SC1755C	376	3219	251		
2SC1756	376	3219	251		MJE2360T
2SC1756-0	376	3219			
2SC1756A	376	3219	251		
2SC1756B	376	3219	251		
2SC1756C	376	3219	251		
2SC1756D	376	3219	251		
2SC1756E	376	3219	251		
2SC1756K	376	3220/198	251		
2SC1756M	376	3219	251		
2SC1756MC	376	3219	251		
2SC1756N	376	3219			
2SC1757	376	3219	251		MJE2360T
2SC1758	171	3201			
2SC175A		3861	59		
2SC175B	101	3861	59	2002	
2SC175BL		3861	59		
2SC175C		3861	59		
2SC175D		3861	59		
2SC175E		3861	59		
2SC175F		3861	59		
2SC175G		3861	59		
2SC175GN		3861	59		
2SC175H		3861	59		
2SC175J		3861	59		
2SC175K		3861	59		
2SC175L		3861	59		
2SC175M		3861	59		
2SC175OR		3861	59		
2SC175R		3861	59		
2SC175X		3861	59		
2SC175Y		3861	59		
2SC176	101	3861	59	2002	
2SC1760	306	3251	276		MPSU07
2SC1760-2	306	3251	276		
2SC1760-3	306	3251	276		
2SC1760H	306		276		
2SC1761		3251	276		MPSU01
2SC1766	85		62	2051	
2SC1766C	85	3122	62	2051	
2SC1768	86	3269			MJ3041
2SC176A		3861	59		
2SC176B		3861	59		
2SC176C		3861	59		
2SC176D		3861	59		
2SC176E		3861	59		
2SC176F		3861	59		
2SC176G		3861	59		
2SC176GN		3861	59		
2SC176H		3861	59		
2SC176J		3861	59		
2SC176K		3861	59		
2SC176L		3861	59		
2SC176OR		3861	59		
2SC176R		3861	75		
2SC176X		3861	59		
2SC176Y		3861	59		
2SC177	101	3861	59	2002	
2SC1775		3931	62	2051	
2SC1775E		3931	62	2051	
2SC1775F		3931	62	2051	
2SC1777	87	3027/130	14	2041	2N5882
2SC1778	229		214	2038	
2SC1779	229		278	2016	
2SC177A		3861	59		
2SC177B		3861	59		
2SC177C		3861	59		
2SC177D		3861	59		
2SC177E		3861	59		
2SC177F		3861	59		
2SC177G		3861	59		
2SC177GN		3861	59		
2SC177H		3861	59		
2SC177J		3861	59		
2SC177K		3861	59		
2SC177L		3861	59		
2SC177M		3861	59		
2SC177OR		3861	59		
2SC177R		3861	59		
2SC177X		3861	59		
2SC177Y		3861	59		
2SC178	101	3861	59	2002	
2SC1782	87				MJ15001
2SC1783	87				2N6249
2SC1784	284				MJ15001
2SC1785	388				2N6249
2SC1786	388				2N6250
2SC1787	199		85	2010	
2SC1788	289A	3124/289	47	2030	
2SC1788P	289A	3124/289			
2SC1788Q	289A	3124/289			
2SC1788R	289A	3124/289	47	2030	
2SC1789	107	3293	327	2038	
2SC178A		3861	59		
2SC178B		3861	59		
2SC178C		3861	59		
2SC178D		3861	59		
2SC178E		3861	59		
2SC178F		3861	59		
2SC178G		3861	59		
2SC178H		3861	59		
2SC178J		3861	59		
2SC178K		3861	59		
2SC178L		3861	59		
2SC178M		3861	59		
2SC178OR		3861	59		
2SC178R		3861	59		
2SC178X		3861	59		
2SC178Y		3861	59		
2SC179	103A	3011A	47	2002	
2SC1790	316	3716/161	278	2016	
2SC179A		3011A	47	2002	
2SC179B		3011A	47	2002	
2SC179C		3011A	47	2002	
2SC179D		3011A	47	2002	
2SC179E		3011A	47	2002	
2SC179F		3011A	47	2002	
2SC179G		3011A	47	2002	
2SC179GN		3011A	47	2002	

Industry Standard No.	ECG	SK	GE	RS 276-	MOTOR.
2SC179H		3011A	47	2002	
2SC179J		3011A	47	2002	
2SC179K		3011A	47	2002	
2SC179L		3011A	47	2002	
2SC179M		3011A	47	2002	
2SC179OR		3011A	47	2002	
2SC179R		3011A	47	2002	
2SC179X		3011A	47	2002	
2SC179Y		3011A	47	2002	
2SC17A	123A	3452/108	61	2038	
2SC17B		3018	61	2038	
2SC17C		3018	61	2038	
2SC17D		3018	61	2038	
2SC17E		3018	61	2038	
2SC17F		3018	61	2038	
2SC17G		3018	61	2038	
2SC17GN		3018	61	2038	
2SC17H		3018	61	2038	
2SC17J		3018	61	2038	
2SC17K		3018	61	2038	
2SC17L		3018	61	2038	
2SC17M		3018	61	2038	
2SC17OR		3018	61	2038	
2SC17R		3018	61	2038	
2SC17X		3018	61	2038	
2SC17Y		3018	61	2038	
2SC18	123A	3444	88	2051	
2SC180	103A	3835	59	2002-	
2SC1804					MRF321
2SC1805					MRF323
2SC1806					MRF5177A
2SC1807	311				BFY90
2SC1808					2N5944
2SC180A		3835	59	2002	
2SC180B		3835	59	2002	
2SC180C		3835	59	2002	
2SC180D		3835	59	2002	
2SC180E		3835	59	2002	
2SC180F		3835	59	2002	
2SC180G		3835	59	2002	
2SC180GN		3835	59	2002	
2SC180H		3835	59	2002	
2SC180J		3835	59	2002	
2SC180K		3835	59	2002	
2SC180L		3835	59	2002	
2SC180M		3835	59	2002	
2SC180OR			59	2002	
2SC180R		3835	59	2002	
2SC180X		3835	59	2002	
2SC180Y		3835	59	2002	
2SC181	103A	3835	47	2002	
2SC1811	399	3244	222		
2SC1811-1		3244	222		
2SC1811-2		3244	222		
2SC1811KC		3244	222		
2SC1815	85	3245/199	62	2051	
2SC1815-O		3245	62	2051	
2SC1815GR	85		62	2051	
2SC1815Y	85	3245/199	62	2051	
2SC1815YW	85		62	2051	
2SC1816	236$	3197/235	216	2053	
2SC1816E	236$	3239/236			
2SC1816HL		3197	216		
2SC1818	280				2N6340
2SC1819	376	3219	251		MJE2361T
2SC1819M	376	3219	251		
2SC1819RL	376		251		
2SC181A		3835	47	2002	
2SC181B		3835	47	2002	
2SC181C		3835	47	2002	
2SC181D		3835	47	2002	
2SC181E		3835	47	2002	
2SC181F		3835	47	2002	
2SC181G		3835	47	2002	
2SC181GN		3835	47	2002	
2SC181H		3835	47	2002	
2SC181J		3835	47	2002	
2SC181L		3835	47	2002	
2SC181M		3835	47	2002	
2SC181OR		3835	47	2002	
2SC181R		3835	47	2002	
2SC181X		3835	47	2002	
2SC181Y		3835	47	2002	
2SC182	313	3020/123	47	2016	
2SC182(Q)			47	2034	
2SC182(V)			47	2034	
2SC1826		3054	241	2020	TIP41B
2SC1826P	196	3054	241	2020	
2SC1826Q	196		241	2020	
2SC1826R	196		241	2020	
2SC1827	291	3188A/182			TIP41C
2SC1828	369	3131A	246		
2SC1829	94	3836/284	36		MJ3041
2SC182A		3124	47	2034	
2SC182B		3124	47	2034	
2SC182C		3124	47	2034	
2SC182D		3124	47	2034	
2SC182E		3124	47	2034	
2SC182F		3124	47	2034	
2SC182G		3124	47	2034	
2SC182GN		3124	47	2034	
2SC182H		3124	47	2034	
2SC182J		3124	47	2034	
2SC182K		3124	47	2034	
2SC182L		3124	47	2034	
2SC182M		3124	47	2034	
2SC182Q	313	3124/289	47	2016	
2SC182R		3124	47	2034	
2SC182V			47	2030	
2SC182X		3124	47	2034	
2SC182Y		3124	47	2034	
2SC183	313		20	2016	
2SC183(P)			20	2016	
2SC183(Q)(R)			20	2016	
2SC183(R)			20	2016	
2SC183-1			20	2016	
2SC183-OR			20	2016	
2SC1830					2N6578
2SC1831					2N6056
2SC1832	98				MJ10009
2SC183A			20	2016	
2SC183AP			20	2016	
2SC183B			20	2016	
2SC183BK			20	2016	
2SC183C			20	2016	
2SC183D			20	2016	
2SC183E	313		20	2016	
2SC183F			20	2016	
2SC183G			20	2016	
2SC183GN			20	2016	
2SC183H			20	2016	
2SC183J	313		20	2016	
2SC183K	313		20	2016	
2SC183L	313		20	2016	
2SC183M	313		20	2016	
2SC183OR			20	2016	
2SC183P	313		20	2016	
2SC183Q	313		20	2016	
2SC183R	313		20	2016	
2SC183S			20	2016	
2SC183W	313		20	2016	
2SC183X			20	2016	
2SC183Y			20	2016	
2SC184	313		61	2016	
2SC184(R)			61	2016	
2SC184-OR			61	2016	
2SC1840	199	3866			
2SC1844	199	3866			
2SC1846	295#	3253/295	336		MJE180
2SC1846B	295#	3253/295	336		
2SC1846P	295#	3253/295	336		
2SC1846Q	295#	3253/295	336		
2SC1846QRS	295#		336		
2SC1846R	295#	3253/295	336		
2SC1846S	295#	3253/295	336		
2SC1847	373#	9041/373	270		MJE181
2SC1847Q	373#	9041/373	270		
2SC1847R	373#	9041/373			
2SC1847VQ	373#	9041/373			
2SC1847VR	373#	9041/373			
2SC1848	186A	3239/236	28	2052	D40E7
2SC1848P	186A	3239/236			
2SC1848Q	186A	3239/236	28	2052	
2SC1848R	186A	3239/236	28	2052	
2SC1848V		3239	28	2052	
2SC184A			61	2016	
2SC184AP			61	2016	
2SC184B			61	2016	
2SC184BK			61	2016	
2SC184C			61	2016	
2SC184D			61	2016	
2SC184E			61	2016	
2SC184F			61	2016	
2SC184G			61	2016	
2SC184GN			61	2016	
2SC184H	313		61	2016	
2SC184J	313		61	2016	
2SC184K			61	2016	
2SC184L	313		61	2016	
2SC184M			61	2016	
2SC184OR			61	2016	
2SC184P			61	2016	
2SC184Q			61	2016	
2SC184R	313		61	2016	
2SC184X			61	2016	
2SC184Y			61	2016	
2SC185	313	3018	86	2016	
2SC1851	123AP		47	2030	
2SC1852	123AP		86	2038	
2SC1853	85		62	2051	
2SC1854	85	3122	61	2038	
2SC1854C	85	3122	61	2038	
2SC1854S	85	3122	61	2038	
2SC1855	229	3246	61	2038	
2SC1856	229	3132	86	2038	
2SC1856M	229$	3132			
2SC185A	313	3039/316	86	2016	
2SC185B		3039	86	2038	
2SC185C		3039	86	2038	
2SC185E		3039	86	2038	
2SC185F		3039	86	2038	
2SC185G		3039	86	2038	
2SC185GN		3039	86	2038	
2SC185H		3039	86	2038	
2SC185J	313	3039/316	86	2016	
2SC185K		3039	86	2038	
2SC185L		3039	86	2038	
2SC185M	313	3018	86	2016	
2SC185OR			86	2038	
2SC185Q	313	3018	86	2016	
2SC185R	313	3018	86	2016	
2SC185V	313	3018	86	2016	
2SC185X		3039	86	2038	
2SC185Y		3039	86	2038	
2SC186	107	3019	86	2015	
2SC1860	282	3104A			
2SC1860K	282	3104A			
2SC1863	384	3538			
2SC1864	384	3131A/369			
2SC1866	328	3946/385			2N5760
2SC1867	385	3946			
2SC1868	385	3946			SDT13304
2SC1869	328	3836/284			2N5634
2SC186A		3039	86	2038	
2SC186B		3039	86	2038	
2SC186C		3039	86	2038	
2SC186D		3039	86	2038	
2SC186E		3039	86	2038	
2SC186F		3039	86	2038	
2SC186G		3039	86	2038	
2SC186GN		3039	86	2038	
2SC186H		3039	86	2038	
2SC186J		3039	86	2038	
2SC186K		3039	86	2038	
2SC186L		3039	86	2038	
2SC186M		3039	86	2038	
2SC186OR		3039	86	2038	
2SC186R		3039	86	2038	
2SC186X		3039	86	2038	
2SC186Y		3039	86	2038	
2SC187	107	3019	86	2002	
2SC187(I)			86	2002	
2SC187-OR			86	2002	
2SC1870	388	3946/385			2N6546
2SC1871	385	3946			
2SC1871A	386	3946/385			
2SC1875	389	3710/238	38		MJ12003
2SC1875K	389	3710/238	38		
2SC1875L	389	3710/238	38		
2SC187A		3039	86	2002	
2SC187B		3039	86	2002	
2SC187C		3039	86	2002	
2SC187D		3039	86	2002	
2SC187E		3039	86	2002	
2SC187F		3039	86	2002	
2SC187G		3039	86	2002	
2SC187H		3039	86	2002	
2SC187I			86	2002	
2SC187J		3039	86	2002	
2SC187K		3039	86	2002	
2SC187L		3039	86	2002	
2SC187M		3039	86	2002	
2SC187OR		3039	86	2002	
2SC187R	107	3039/316	86	2002	
2SC187X		3039	86	2002	
2SC187Y	107	3039/316	86	2002	
2SC188	128	3024	18	2030	
2SC1880					TIP112

Industry Standard No.	ECG	SK	GE	RS 276-	MOTOR.
2SC1881					TIP110
2SC1883					TIP122
2SC1884					2N6301
2SC1885	399	3244	61		
2SC188A	128	3024	18	2030	
2SC188AB	128	3024	18	2030	
2SC188B		3024	18	2030	
2SC188C		3024	18	2030	
2SC188D		3024	18	2030	
2SC188E		3024	18	2030	
2SC188F		3024	18	2030	
2SC188G		3024	18	2030	
2SC188GN		3024	18	2030	
2SC188J		3024	18	2030	
2SC188K		3024	18	2030	
2SC188L		3024	18	2030	
2SC188M		3024	18	2030	
2SC188OR		3024	18	2030	
2SC188R		3024	18	2030	
2SC188X		3024	18	2030	
2SC188Y		3024	18	2030	
2SC189	128	3024	18	2030	
2SC1890	90	3931	220		
2SC1890A	90	3931	220		
2SC1890AD	90	3931	220		
2SC1890AE	90	3931	220		
2SC1890AF	90	3931			
2SC1890E	90	3931	220		
2SC1890F	90	3931	220		
2SC1891	389	3115/165	259		BU204
2SC1892	389	3710/238	259		BU205
2SC1893	389	3710/238	38		MJ12003
2SC1894	238	3111	259		BU208
2SC1894K	238		259		
2SC1895	238	3710	259		MJ12005
2SC1896	238	3710			MJ12005
2SC1898				2031	
2SC1899				2031	
2SC189A		3024	18	2030	
2SC189B		3024	18	2030	
2SC189C		3024	18	2030	
2SC189D	128	3024	18	2030	
2SC189E	128	3024	18	2030	
2SC189F	194	3024/128	18	2030	
2SC189G		3024	18	2030	
2SC189GN		3024	18	2030	
2SC189H		3024	18	2030	
2SC189J		3024	18	2030	
2SC189K		3024	18	2030	
2SC189L		3024	18	2030	
2SC189M		3024	18	2030	
2SC189OR		3024	18	2030	
2SC189R		3024	18	2030	
2SC189X		3024	18	2030	
2SC189Y		3024	18	2030	
2SC18A		3122	17	2030	
2SC18B		3122	88	2030	
2SC18C		3122	88	2030	
2SC18D		3122	88	2030	
2SC18E		3122	88	2030	
2SC18F		3122	88	2030	
2SC18G		3122	88	2030	
2SC18GN		3122	88	2030	
2SC18H		3122	88	2030	
2SC18J		3122	88	2030	
2SC18K		3122	88	2030	
2SC18L		3122	88	2030	
2SC18M		3122	88	2030	
2SC18OR		3122	88	2030	
2SC18R		3122	88	2030	
2SC18X		3122	88	2030	
2SC19	128	3444/123A	18	2030	
2SC190	128	3024	18	2030	
2SC1903	373		336		MJE341
2SC1904	373				MJE341
2SC1905	198	3219	251		MJE2361T
2SC1905H	198		251		
2SC1906	107	3293	86	2038	
2SC1907	107	3293			
2SC1908		3124	20	2038	
2SC1908E		3124	20	2038	
2SC1908H			20	2038	
2SC1909	235	3197	215		
2SC1909K	235	3197	215		
2SC1909R	235		215		
2SC190A		3024	18	2030	
2SC190B		3024	18	2030	
2SC190C		3024	18	2030	
2SC190D		3024	18	2030	
2SC190E		3024	18	2030	
2SC190F		3024	18	2030	
2SC190GN		3024	18	2030	
2SC190H		3024	18	2030	
2SC190J		3024	18	2030	
2SC190K		3024	18	2030	
2SC190L		3024	18	2030	
2SC190M		3024	18	2030	
2SC190OR			18	2030	
2SC190R		3024	18	2030	
2SC190X		3024	18	2030	
2SC190Y		3024	18	2030	
2SC191	123A	3444	88	2051	
2SC1919C			11	2015	
2SC191A		3122	88	2010	
2SC191B		3122	88	2010	
2SC191C		3122	88	2010	
2SC191D		3122	88	2010	
2SC191E		3122	88	2010	
2SC191F		3122	88	2010	
2SC191G		3122	88	2010	
2SC191GN		3122	88	2010	
2SC191H		3122	88	2010	
2SC191J		3122	88	2010	
2SC191K		3122	88	2010	
2SC191L		3122	88	2010	
2SC191M		3122	88	2010	
2SC191OR		3122	88	2010	
2SC191R		3122	88	2010	
2SC191X		3122	88	2010	
2SC191Y		3122	88	2010	
2SC192	123A	3444	88	2051	
2SC1921		3433	222		
2SC1922	389$	3710/238	38		MJ12003
2SC1923	229	3132	61	2038	
2SC1923-O	229	3132			
2SC1923A	229		61	2038	
2SC1923BN	229	3132	61	2038	
2SC1929	198	3219	249		TIP48
2SC1929Q	198	3219	249		
2SC1929R	198	3219	249		
2SC1929S	198	3219			
2SC192A		3122	88	2010	
2SC192B		3122	88	2010	
2SC192C		3122	88	2010	
2SC192D		3122	88	2010	
2SC192E		3122	88	2010	
2SC192F		3122	88	2010	
2SC192G		3122	88	2010	
2SC192GN		3122	88	2010	
2SC192H		3122	88	2010	
2SC192J		3122	88	2010	
2SC192K		3122	88	2010	
2SC192L		3122	88	2010	
2SC192M		3122	88	2010	
2SC192OR		3122	88	2010	
2SC192R		3122	88	2010	
2SC192X		3122	88	2010	
2SC192Y		3122	88	2010	
2SC193	123A	3444	88	2051	
2SC193A		3122	88	2010	
2SC193B		3122	88	2010	
2SC193C		3122	88	2010	
2SC193D		3122	88	2010	
2SC193E		3122	88	2010	
2SC193F		3122	88	2010	
2SC193G		3122	88	2010	
2SC193GN		3122	88	2010	
2SC193H		3122	88	2010	
2SC193J		3122	88	2010	
2SC193K		3122	88	2010	
2SC193L		3122	88	2010	
2SC193M		3122	88	2010	
2SC193OR		3122	88	2010	
2SC193R		3122	88	2010	
2SC193X		3122	88	2010	
2SC194	123A	3444	88	2051	
2SC1941		3866	220		
2SC1941K	399	3866	222		
2SC1941L	399	3866	222		
2SC1941M	399	3866	222		
2SC1942	389$	3710/238	259		MJ12003
2SC1942-07	165		259		
2SC1944	236	3239			
2SC1945					MRF342
2SC1946	344				MRF222
2SC1947	488				2N3924
2SC1949					MRF962
2SC194A		3122	88	2010	
2SC194B		3122	88	2010	
2SC194C		3122	88	2010	
2SC194D		3122	88	2010	
2SC194E		3122	88	2010	
2SC194F		3122	88	2010	
2SC194G		3122	88	2010	
2SC194GN		3122	88	2010	
2SC194H		3122	88	2010	
2SC194J		3122	88	2010	
2SC194K		3122	88	2010	
2SC194L		3122	88	2010	
2SC194M		3122	88	2010	
2SC194OR		3122	88	2010	
2SC194X		3122	88	2010	
2SC194Y		3122	88	2010	
2SC195	123A	3444	61	2051	
2SC1957	295	3197/235	270		
2SC1957B	295	3197/235			
2SC1957K	295		270		
2SC1959	85	3124/289			
2SC1959-O	85	3124/289			
2SC1959Y	85	3124/289	268	2051	
2SC195A		3018	61	2010	
2SC195B		3018	61	2010	
2SC195C		3018	61	2010	
2SC195D		3018	61	2010	
2SC195E		3018	61	2010	
2SC195F		3018	61	2010	
2SC195G		3018	61	2010	
2SC195GN		3018	61	2010	
2SC195H		3018	61	2010	
2SC195J		3018	61	2010	
2SC195K		3018	61	2010	
2SC195L		3018	61	2010	
2SC195M		3018	61	2010	
2SC195OR		3018	61	2010	
2SC195R		3018	61	2010	
2SC195X		3018	61	2010	
2SC195Y		3018	61	2010	
2SC196	123A	3444	61	2051	
2SC1962	191	3865	222		
2SC1962-O	191	3865			
2SC1962H	191	3865			
2SC1964	236	3197/235	215		
2SC1964D	236		215		
2SC1966					2N5944
2SC1967					2N5945
2SC1968					MRF641
2SC1969	236	3239	216	2053	MRF475
2SC1969B	236	3239	216	2053	
2SC1969BH	236		210	2053	
2SC1969H	236		216	2053	
2SC196A		3018	61	2010	
2SC196B		3018	61	2010	
2SC196C		3018	61	2010	
2SC196D		3018	61	2010	
2SC196E		3018	61	2010	
2SC196F		3018	61	2010	
2SC196G		3018	61	2010	
2SC196GN		3018	61	2010	
2SC196H		3018	61	2010	
2SC196J		3018	61	2010	
2SC196K		3018	61	2010	
2SC196L		3018	61	2010	
2SC196M		3018	61	2010	
2SC196OR		3018	61	2010	
2SC196R		3018	61	2010	
2SC196X		3018	61	2010	
2SC196Y		3018	61	2010	
2SC197	123A	3444	61	2051	
2SC1970	342$				MRF227
2SC1973	85	3849/293	285	2030	
2SC1974	236	3239	337		
2SC1975	235	3197	338		
2SC197A		3018	61	2010	
2SC197B		3018	61	2010	
2SC197C		3018	61	2010	
2SC197D		3018	61	2010	
2SC197E		3018	61	2010	
2SC197F		3018	61	2010	
2SC197G		3018	61	2010	
2SC197GN		3018	61	2010	
2SC197H		3018	61	2010	
2SC197J		3018	61	2010	
2SC197K		3018	61	2010	
2SC197L		3018	61	2010	
2SC197M		3018	61	2010	
2SC197OR		3018	61	2010	
2SC197R		3018	61	2010	
2SC197X		3018	61	2010	

Industry Standard No.	ECG	SK	GE	RS 276-	MOTOR.
2SC197Y		3018	61	2010	
2SC198			19	2041	
2SC1980	90	3866			
2SC1980S	90	3866			
2SC1980T	90	3866			
2SC1983		3041	241	2048	TIP111
2SC1984					TIP112
2SC1985	196				TIP41B
2SC1986	331				TIP41C
2SC1988					MRF904
2SC1989				2013	
2SC198H			19	2041	
2SC198S			19	2041	
2SC199	123A	3452/108	86	2038	
2SC1990			86	2038	
2SC1990B	107	3018	86	2038	
2SC1999				2014	
2SC199A		3039	86	2051	
2SC199B		3039	86	2051	
2SC199C		3039	86	2051	
2SC199D		3039	86	2051	
2SC199E		3039	86	2051	
2SC199F		3039	86	2051	
2SC199G		3039	86	2051	
2SC199GN		3039	86	2051	
2SC199H		3039	86	2051	
2SC199J		3039	86	2051	
2SC199K		3039	86	2051	
2SC199L		3039	86	2051	
2SC199M		3039	86	2051	
2SC1990R		3039	86	2051	
2SC199R		3039	86	2051	
2SC199X		3039	86	2051	
2SC199Y		3039	86	2051	
2SC19A		3024	18	2030	
2SC19B		3024	18	2030	
2SC19C		3024	18	2030	
2SC19D		3024	18	2030	
2SC19E		3024	18	2030	
2SC19F		3024	18	2030	
2SC19G		3024	18	2030	
2SC19GN		3024	18	2030	
2SC19H		3024	18	2030	
2SC19J		3024	18	2030	
2SC19K		3024	18	2030	
2SC19L		3024	18	2030	
2SC19M		3024	18	2030	
2SC190R		3024	18	2030	
2SC19R		3024	18	2030	
2SC19X		3024	18	2030	
2SC19Y		3024	18	2030	
2SC20	128	3024	18	2030	
2SC200	123A	3444	88	2051	
2SC2000	85	3124/289	210	2051	
2SC2000K	85	3124/289			
2SC2000L	85	3124/289	210	2051	
2SC2000M	85	3124/289			
2SC2001	85	3849/293	47	2038	
2SC2001L	85		47	2038	
2SC2002	85	3449/297			
2SC2002K	85	3449/297			
2SC2002L	85	3449/297			
2SC2002M	85	3449/297			
2SC2003	85	3449/297			
2SC2003K	85	3449/297			
2SC2003L	85	3449/297			
2SC2003M	85	3449/297			
2SC2009	229	3246	39	2015	
2SC200A		3122	88	2030	
2SC200B		3122	88	2030	
2SC200C		3122	88	2030	
2SC200D		3122	88	2030	
2SC200E		3122	88	2030	
2SC200F		3122	88	2030	
2SC200G		3122	88	2030	
2SC200GN		3122	88	2030	
2SC200H		3122	88	2030	
2SC200J		3122	88	2030	
2SC200K		3122	88	2030	
2SC200L		3122	88	2030	
2SC200M		3122	88	2030	
2SC2000R			88	2030	
2SC200R		3122	88	2030	
2SC200X		3122	88	2030	
2SC200Y		3122	88	2030	
2SC201	123A	3444	88	2051	
2SC2012	107	3124/289	20	2015	
2SC2017	386		36		
2SC2018	386		36		
2SC201A		3122	88	2030	
2SC201B		3122	88	2030	
2SC201C		3122	88	2030	
2SC201D		3122	88	2030	
2SC201E		3122	88	2030	
2SC201F		3122	88	2030	
2SC201G		3122	88	2030	
2SC201GN		3122	88	2030	
2SC201H		3122	88	2030	
2SC201J		3122	88	2030	
2SC201K		3122	88	2030	
2SC201L		3122	88	2030	
2SC201M		3122	88	2030	
2SC201R		3122	88	2030	
2SC201X		3122	88	2030	
2SC201Y		3122	88	2030	
2SC202	123A	3444	88	2051	
2SC2020	235	3239/236	329		
2SC2021	85	3449/297	62	2051	
2SC2021Q	85	3449/297	20	2051	
2SC2021R	85		20	2051	
2SC2021S	85	3449/297	20	2051	
2SC2024	295		331		2N4923
2SC2027	165				MJ12005
2SC2028	373	3253/295	270	2030	
2SC2028-2	373		270	2030	
2SC2028-B/20	373	3253/295	270		
2SC2028/2	373	3253/295	270	2030	
2SC2028B	373		270	2030	
2SC2028B/20	373		270	2030	
2SC2028B20			270	2030	
2SC2029	235	3197	333		
2SC2029-1	235	3197	333		
2SC2029-3	235		333		
2SC2029-B/10	235	3197	333		
2SC2029/1	235	3197			
2SC2029/3	235	3197	333		
2SC2029B	235	3197	333		
2SC2029B/10	235		333		
2SC2029C	235	3197	333		
2SC202A		3122	88	2030	
2SC202B		3122	88	2030	
2SC202C		3122	88	2030	
2SC202D		3122	88	2030	
2SC202E		3122	88	2030	

Industry Standard No.	ECG	SK	GE	RS 276-	MOTOR.
2SC202F		3122	88	2030	
2SC202G		3122	88	2030	
2SC202GN		3122	88	2030	
2SC202H		3122	88	2030	
2SC202J		3122	88	2030	
2SC202K		3122	88	2030	
2SC202L		3122	88	2030	
2SC202M		3122	88	2030	
2SC2020R		3122	88	2030	
2SC202R		3122	88	2030	
2SC202X		3122	88	2030	
2SC202Y		3122	88	2030	
002SC203		3005	88	2051	
2SC2034	282	3847	46		
2SC2036	373	3253/295	336		
002SC203A		3005	88	2030	
002SC203AA		3005	52	2024	
2SC203B		3122	88	2030	
2SC203C		3122	88	2030	
2SC203D		3122	88	2030	
2SC203E		3122	88	2030	
2SC203F		3122	88	2030	
2SC203G		3122	88	2030	
2SC203GN		3122	88	2030	
2SC203H		3122	88	2030	
2SC203J		3122	88	2030	
2SC203K		3122	88	2030	
2SC203L		3122	88	2030	
2SC203M		3122	88	2030	
2SC2030R		3122	88	2030	
2SC203R		3122	88	2030	
2SC203X		3122	88	2030	
2SC203Y		3122	88	2030	
2SC204	123A	3444	88	2051	
2SC2040	76$				MRF511
2SC2043	236	3239	332	2053	
2SC204A		3122	88	2030	
2SC204B		3122	88	2030	
2SC204C		3122	88	2030	
2SC204D		3122	88	2030	
2SC204E		3122	88	2030	
2SC204F	123A	3122	88	2030	
2SC204G		3122	88	2030	
2SC204GN		3122	88	2030	
2SC204H		3122	88	2030	
2SC204J		3122	88	2030	
2SC204K		3122	88	2030	
2SC204L		3122	88	2030	
2SC204M		3122	88	2030	
2SC2040R		3122	88	2030	
2SC204R		3122	88	2030	
2SC204X		3122	88	2030	
2SC204Y		3122	88	2030	
2SC205	123A	3444	88	2051	
2SC2057	229	3246	39	2016	
2SC2057-C			39	2016	
2SC2057C	229		39	2016	
2SC2057D	229	3246	39	2016	
2SC2057E	229		39	2016	
2SC2057F	229		39	2016	
2SC205A		3122	88	2030	
2SC205B		3122	88	2030	
2SC205C		3122	88	2030	
2SC205D		3122	88	2030	
2SC205F		3122	88	2030	
2SC205G		3122	88	2030	
2SC205GN		3122	88	2030	
2SC205H		3122	88	2030	
2SC205J		3122	88	2030	
2SC205K		3122	88	2030	
2SC205L		3122	88	2030	
2SC205M		3122	88	2030	
2SC2050R		3122	88	2030	
2SC205R		3122	88	2030	
2SC205X		3122	88	2030	
2SC205Y		3122	88	2030	
2SC206	161	3018	86	2038	
2SC206-OR			86	2038	
2SC2060	293	3024/128	47	2030	
2SC2060Q	293	3024/128	47	2030	
2SC2068	376	3201/171	251		D40N4
2SC2068FA-1	376		251		
2SC2068LB	376	3201/171	251		
2SC206A		3018	86	2038	
2SC206B		3018	86	2038	
2SC206C		3018	86	2038	
2SC206D		3018	86	2038	
2SC206E		3018	86	2038	
2SC206G		3018	86	2038	
2SC206GN		3018	86	2038	
2SC206H		3018	86	2038	
2SC206J		3018	86	2038	
2SC206K		3018	86	2038	
2SC206L		3018	86	2038	
2SC206M		3018	86	2038	
2SC2060R		3018	86	2038	
2SC206R		3018	86	2038	
2SC206RED			86	2038	
2SC206WHITE			86	2038	
2SC206X		3018	86	2038	
2SC206Y		3018	86	2038	
2SC2071					MJE3440
2SC2073	375	3929			TIP47
2SC2073D	375	3929			
2SC2073E	375	3929			
2SC2074	322	3252/302	275		
2SC2074C	322	3252/302	275		
2SC2074Y	322	3252/302	275		
2SC2075	236$	3197/235	215		
2SC2076	85	3122	210	2051	
2SC2076B	85		210	2051	
2SC2076C	85	3122	210	2051	
2SC2076CB		3122	210	2051	
2SC2076CD	85	3122	210	2051	
2SC2076D	85	3122	210	2051	
2SC2078	236	3197/235	215		
2SC2078B	236	3197/235			
2SC2078C	236	3197/235			
2SC2078D	236	3197/235			
2SC2078E	236	3197/235			
2SC2079		3253	270		
2SC208	161	3024/128	18	2030	
2SC2080					MJE180
2SC2081			47	2030	
2SC2085	376	3219	32		MJE2361T
2SC2085P	376		32		
2SC2085Q	376		32		
2SC2086		3842	47	2030	
2SC2088	199$	3866			
2SC208A		3024	18	2030	
2SC208B		3024	18	2030	
2SC208C		3024	18	2030	
2SC208D		3024	18	2030	
2SC208E		3024	18	2030	

Industry Standard No.	ECG	SK	GE	RS 276-	MOTOR.
2SC208F		3024	18	2030	
2SC208G		3024	18	2030	
2SC208GN		3024	18	2030	
2SC208H		3024	18	2030	
2SC208J		3024	18	2030	
2SC208K		3024	18	2030	
2SC208L		3024	18	2030	
2SC208M		3024	18	2030	
2SC208OR		3024	18	2030	
2SC208R		3024	18	2030	
2SC208X		3024	18	2030	
2SC208Y		3024	18	2030	
2SC2091	295#	3253/295	270		
2SC2092	235$	3197/235	337		
2SC2097					MRF247
2SC2098	236	3239			
2SC20A		3024	18	2030	
2SC20B		3024	18	2030	
2SC20C		3024	18	2030	
2SC20D		3024	18	2030	
2SC20E		3024	18	2030	
2SC20F		3024	18	2030	
2SC20G		3024	18	2030	
2SC20GN		3024	18	2030	
2SC20H		3024	18	2030	
2SC20J		3024	18	2030	
2SC20K		3024	18	2030	
2SC20L		3024	18	2030	
2SC20M		3024	18	2030	
2SC20OR			18	2030	
2SC20R		3024	18	2030	
2SC20X		3024	18	2030	
2SC20Y		3024	18	2030	
2SC21	130	3027	75	2041	
2SC210	128	3024	18	2030	
2SC2100					MRF412
2SC2109	123AP	3854			
2SC210A		3024	18	2030	
2SC210B		3024	18	2030	
2SC210C		3024	18	2030	
2SC210D		3024	18	2030	
2SC210E		3024	18	2030	
2SC210F		3024	18	2030	
2SC210G		3024	18	2030	
2SC210GN		3024	18	2030	
2SC210H		3024	18	2030	
2SC210J		3024	18	2030	
2SC210K		3024	18	2030	
2SC210L		3024	18	2030	
2SC210M		3024	18	2030	
2SC210OR			18	2030	
2SC210R		3024	18	2030	
2SC210X		3024	18	2030	
2SC210Y		3024	18	2030	
2SC211	128	3024	18	2030	
2SC2110			18	2030	
2SC211A		3024	18	2051	
2SC211B		3024	18	2051	
2SC211C		3024	18	2051	
2SC211D		3024	18	2051	
2SC211F		3024	18	2051	
2SC211G		3024	18	2051	
2SC211GN		3024	18	2051	
2SC211H		3024	18	2051	
2SC211J		3024	18	2051	
2SC211K		3024	18	2051	
2SC211L		3024	18	2051	
2SC211M		3024	18	2051	
2SC211OR		3024	18	2051	
2SC211R		3024	18	2051	
2SC211X		3024	18	2051	
2SC211Y		3024	18	2051	
2SC212	128	3275/194	18		
2SC2120	289A	3124/289	18	2030	
2SC2120-0		3124	243	2030	
2SC2120Y	289A	3124/289	243	2030	
2SC2121	283				MJ411
2SC2122	283	3467			MJ431
2SC2123		3115			MJ10014
2SC2126					MJ4380
2SC2127					2N6249
2SC2128					MJ10015
2SC212A		3024	18	2051	
2SC212B		3024	18	2051	
2SC212C		3024	18	2051	
2SC212D		3024	18	2051	
2SC212E		3024	18	2051	
2SC212F		3024	18	2051	
2SC212G		3024	18	2051	
2SC212GN		3024	18	2051	
2SC212H		3024	18	2051	
2SC212J		3024	18	2051	
2SC212K		3024	18	2051	
2SC212L		3024	18	2051	
2SC212M		3024	18	2051	
2SC212OR		3024	18	2051	
2SC212R		3024	18	2051	
2SC212X		3024	18	2051	
2SC212Y		3024	18	2051	
2SC213	282	3024/128	18	2030	
2SC2132	366$				MRF646
2SC2138					SDT13304
2SC2139	386				SDT13305
2SC213A		3024	18	2030	
2SC213B		3024	18	2030	
2SC213C		3024	18	2030	
2SC213D		3024	18	2030	
2SC213E		3024	18	2030	
2SC213F		3024	18	2030	
2SC213G		3024	18	2030	
2SC213GN		3024	18	2030	
2SC213H		3024	18	2030	
2SC213J		3024	18	2030	
2SC213K		3024	18	2030	
2SC213L		3024	18	2030	
2SC213M		3024	18	2030	
2SC213OR		3024	18	2030	
2SC213R		3024	18	2030	
2SC213X		3024	18	2030	
2SC213Y		3024	18	2030	
2SC214	282	3024/128	18	2030	
2SC2140					SDT13304
2SC2141	190	3929			
2SC2146	235	3239/236			
2SC2147					MJ10015
2SC2148					SDT13305
2SC214A		3024	18	2030	
2SC214B		3024	18	2030	
2SC214C		3024	18	2030	
2SC214D		3024	18	2030	
2SC214E		3024	18	2030	
2SC214F		3024	18	2030	
2SC214G		3024	18	2030	
2SC214GN		3024	18	2030	
2SC214H		3024	18	2030	
2SC214J		3024	18	2030	
2SC214K		3024	18	2030	
2SC214L		3024	18	2030	
2SC214M		3024	18	2030	
2SC214OR		3024	18	2030	
2SC214X		3024	18	2030	
2SC214Y		3024	18	2030	
2SC215	282	3024/128	18	2030	
2SC2151	386				MJ10014
2SC2159					MJ10015
2SC215A		3024	18	2030	
2SC215B		3024	18	2030	
2SC215C		3024	18	2030	
2SC215E		3024	18	2030	
2SC215F		3024	18	2030	
2SC215G		3024	18	2030	
2SC215GN		3024	18	2030	
2SC215H		3024	18	2030	
2SC215J		3024	18	2030	
2SC215K		3024	18	2030	
2SC215L		3024	18	2030	
2SC215M		3024	18	2030	
2SC215OR		3024	18	2030	
2SC215R		3024	18	2030	
2SC215X		3024	18	2030	
2SC215Y		3024	18	2030	
2SC216	128	3024	18	2030	
2SC2166	236	3197/235	215		
2SC2166D	236	3197/235			
2SC2167	375	3929			MJE15030
2SC2168		3929			MJE15030
2SC216A		3024	18	2030	
2SC216B		3024	18	2030	
2SC216C		3024	18	2030	
2SC216D		3024	18	2030	
2SC216E		3024	18	2030	
2SC216F		3024	18	2030	
2SC216G		3024	18	2030	
2SC216GN		3024	18	2030	
2SC216H		3024	18	2030	
2SC216J		3024	18	2030	
2SC216K		3024	18	2030	
2SC216L		3024	18	2030	
2SC216M		3024	18	2030	
2SC216OR		3024	18	2030	
2SC216R		3024	18	2030	
2SC216X		3024	18	2030	
2SC216Y		3024	18	2030	
2SC217	128	3024	18	2030	
2SC217A		3024	18	2030	
2SC217B		3024	18	2030	
2SC217C		3024	18	2030	
2SC217D		3024	18	2030	
2SC217E		3024	18	2030	
2SC217F		3024	18	2030	
2SC217G		3024	18	2030	
2SC217GN		3024	18	2030	
2SC217H		3024	18	2030	
2SC217J		3024	18	2030	
2SC217K		3024	18	2030	
2SC217L		3024	18	2030	
2SC217M		3024	18	2030	
2SC217OR		3024	18	2030	
2SC217R		3024	18	2030	
2SC217X		3024	18	2030	
2SC217Y		3024	18	2030	
2SC218	128	3444/123A	18	2051	
2SC2184	236	3239	329		
2SC2189					MJ15001
2SC218A	128	3444/123A	18	2051	
2SC218B		3124	18	2030	
2SC218C		3024	18	2030	
2SC218D		3024	18	2030	
2SC218E		3024	18	2030	
2SC218F		3024	18	2030	
2SC218G		3024	18	2030	
2SC218GN		3024	18	2030	
2SC218H		3024	18	2030	
2SC218J		3024	18	2030	
2SC218L		3024	18	2030	
2SC218M		3024	18	2030	
2SC218OR		3024	18	2030	
2SC218R		3024	18	2030	
2SC218X		3024	18	2030	
2SC218Y		3024	18	2030	
2SC2190					2N6545
2SC2191					2N6547
2SC2194	188		226	2018	
2SC2198		3913			2N6301
2SC2199		3563	75		MJ11018
2SC21A		3027	75	2041	
2SC21B		3027	75	2041	
2SC21C		3027	75	2041	
2SC21D		3027	75	2041	
2SC21E		3027	75	2041	
2SC21F		3027	75	2041	
2SC21G		3027	75	2041	
2SC21GN		3027	75	2041	
2SC21H		3027	75	2041	
2SC21J		3027	75	2041	
2SC21K		3027	75	2041	
2SC21L		3027	75	2041	
2SC21M		3027	75	2041	
2SC21OR		3027	75	2041	
2SC21R		3027	75	2041	
2SC21X		3027	75	2041	
2SC21Y		3027	75	2041	
2SC22		3024	18	2030	
2SC220	128	3024	18	2030	
2SC2204					MJ10016
2SC2207					MRF342
2SC2209		9041			MJE181
2SC220A		3024	18	2030	
2SC220B		3024	18	2030	
2SC220C		3024	18	2030	
2SC220D		3024	18	2030	
2SC220E		3024	18	2030	
2SC220F		3024	18	2030	
2SC220G		3024	18	2030	
2SC220GN		3024	18	2030	
2SC220H		3024	18	2030	
2SC220K		3024	18	2030	
2SC220L		3024	18	2030	
2SC220M		3024	18	2030	
2SC220OR			18	2030	
2SC220R			18	2030	
2SC220X		3024	18	2030	
2SC220Y		3024	18	2030	
2SC221	128	3024	18	2030	
2SC2210		3124		2013	
2SC2216	319P	3132			
2SC221A		3024	18	2030	
2SC221B		3024	18	2030	

Industry Standard No.	ECG	SK	GE	RS 276-	MOTOR.
2SC221C		3024	18	2030	
2SC221D		3024	18	2030	
2SC221E		3024	18	2030	
2SC221F		3024	18	2030	
2SC221G		3024	18	2030	
2SC221GN		3024	18	2030	
2SC221H		3024	18	2030	
2SC221J		3024	18	2030	
2SC221K		3024	18	2030	
2SC221L		3024	18	2030	
2SC221M		3024	18	2030	
2SC221OR		3024	18	2030	
2SC221R		3024	18	2030	
2SC221X		3024	18	2030	
2SC221Y		3024	18	2030	
2SC222	128	3024	18	2030	
2SC2220					MJ10016
2SC2222				2017	
2SC2228	399	3866	222		
2SC2228C	399	3866	222		
2SC2228D	399	3866	222		
2SC2228E		3866	222		
2SC2228F		3866	222		
2SC2228M	399	3866	222		
2SC2229	399	3244	220		TIP47
2SC2229-0		3244	220		
2SC2229M	399	3244	220		
2SC2229Y	399	3244	220		
2SC222A		3024	18	2030	
2SC222B		3024	18	2030	
2SC222C		3024	18	2030	
2SC222D		3024	18	2030	
2SC222E		3024	18	2030	
2SC222F		3024	18	2030	
2SC222G		3024	18	2030	
2SC222GN		3024	18	2030	
2SC222H		3024	18	2030	
2SC222J		3024	18	2030	
2SC222L		3024	18	2030	
2SC222M		3024	18	2030	
2SC222OR		3024	18	2030	
2SC222R		3024	18	2030	
2SC222X		3024	18	2030	
2SC222Y		3024	18	2030	
2SC223	282	3024/128	18	2030	
2SC2230	399	3866	222		TIP47
2SC2230A	399	3866			
2SC2230A-Y	399		222		
2SC2230AG	399	3866			
2SC2230AGR	399	3866	222		
2SC2230AY	399	3866	222		
2SC2230Y	399	3866	222		
2SC2231Y	198	3219	251		
2SC2233		3929			2N6497
2SC2235	382$	3275/194			TIP47
2SC2236	382	3849/293	270	2030	TIP31
2SC2236-0	382	3849/293	270	2030	
2SC2236-0			270	2030	
2SC2236Y	382	3849/293	270	2030	
2SC2238	375$	3929			TIP47
2SC2238A	375$	3929			
2SC2239		3194			2N5052
2SC223A		3024	18	2017	
2SC223B		3024	18	2017	
2SC223C		3024	18	2017	
2SC223D		3024	18	2017	
2SC223E		3024	18	2017	
2SC223F		3024	18	2017	
2SC223G		3024	18	2017	
2SC223GN		3024	18	2017	
2SC223H		3024	18	2017	
2SC223J		3024	18	2017	
2SC223L		3024	18	2017	
2SC223M		3024	18	2017	
2SC223OR		3024	18	2017	
2SC223R		3024	18	2017	
2SC223X		3024	18	2017	
2SC223Y		3024	18	2017	
2SC224	282	3024/128	18	2030	
2SC2240GR	199$		220		
2SC2242	376	3219	251		MJE2361T
2SC2242BK		3219	251		
2SC2243					2N6543
2SC2244					2N6545
2SC2245					SDT13304
2SC2246					2N6547
2SC2247	107				2N6543
2SC2248					2N6545
2SC2249					MJ10015
2SC224A		3024	18	2017	
2SC224B		3024	18	2017	
2SC224C		3024	18	2017	
2SC224D		3024	18	2017	
2SC224E		3024	18	2017	
2SC224F		3024	18	2017	
2SC224G		3024	18	2017	
2SC224GN		3024	18	2017	
2SC224H		3024	18	2017	
2SC224J		3024	18	2017	
2SC224K		3024	18	2017	
2SC224L		3024	18	2017	
2SC224M		3024	18	2017	
2SC224OR		3024	18	2017	
2SC224R		3024	18	2017	
2SC224X		3024	18	2017	
2SC224Y		3024	18	2017	
2SC225	128	3024	18	2030	
2SC2250					MJ10016
2SC2256	388	3836/284			2N6249
2SC2258	157#	3747/157	232		
2SC225A		3024	18	2017	
2SC225B		3024	18	2017	
2SC225C		3024	18	2017	
2SC225D		3024	18	2017	
2SC225E		3024	18	2017	
2SC225F		3024	18	2017	
2SC225G		3024	18	2017	
2SC225GN		3024	18	2017	
2SC225H		3024	18	2017	
2SC225J		3024	18	2017	
2SC225L		3024	18	2017	
2SC225M		3024	18	2017	
2SC225OR		3024	18	2017	
2SC225R		3024	18	2017	
2SC225X		3024	18	2017	
2SC225Y		3024	18	2017	
2SC226	128	3024	18	2030	
2SC2260	87				2N6249
2SC2261	87				2N6249
2SC2262	87				2N6249
2SC226A		3024	18	2030	
2SC226B		3024	18	2030	
2SC226C		3024	18	2030	
2SC226D		3024	18	2030	

Industry Standard No.	ECG	SK	GE	RS 276-	MOTOR.
2SC226E		3024	18	2030	
2SC226F		3024	18	2030	
2SC226G		3024	18	2030	
2SC226GN		3024	18	2030	
2SC226H		3024	18	2030	
2SC226J		3024	18	2030	
2SC226K		3024	18	2030	
2SC226L		3024	18	2030	
2SC226M		3024	18	2030	
2SC226OR		3024	18	2030	
2SC226R		3024	18	2030	
2SC226X		3024	18	2030	
2SC226Y		3024	18	2030	
2SC227	128	3024	18	2030	
2SC2270					2N5194
2SC2271	399	3433/287			
2SC2271C	399	3433/287			
2SC2271D	399	3433/287			
2SC2271E	399	3433/287			
2SC2271N	399	3433/287	224		
2SC2278	171$	3201/171	257		MJE3439
2SC227A		3024	18	2030	
2SC227B		3024	18	2030	
2SC227C		3024	18	2030	
2SC227D		3024	18	2030	
2SC227E		3024	18	2030	
2SC227F		3024	18	2030	
2SC227G		3024	18	2030	
2SC227GN		3024	18	2030	
2SC227H		3024	18	2030	
2SC227J		3024	18	2030	
2SC227K		3024	18	2030	
2SC227L		3024	18	2030	
2SC227M		3024	18	2030	
2SC227OR		3024	18	2030	
2SC227R		3024	18	2030	
2SC227X		3024	18	2030	
2SC227Y		3024	18	2030	
2SC228	128	3024	18	2030	
2SC228A		3024	18	2030	
2SC228B		3024	18	2030	
2SC228C		3024	18	2030	
2SC228D		3024	18	2030	
2SC228E		3024	18	2030	
2SC228F		3024	18	2030	
2SC228G		3024	18	2030	
2SC228GN		3024	18	2030	
2SC228H		3024	18	2030	
2SC228J		3024	18	2030	
2SC228K		3024	18	2030	
2SC228L		3024	18	2030	
2SC228M		3024	18	2030	
2SC228OR		3024	18	2030	
2SC228R		3024	18	2030	
2SC228X		3024	18	2030	
2SC228Y		3024	18	2030	
2SC229	282	3024/128	18	2030	
2SC2290	317				MRF421
2SC2292					SDT13305
2SC2293					SDT13305
2SC2298					MJE270
2SC229A		3024	18	2017	
2SC229B		3024	18	2017	
2SC229C		3024	18	2017	
2SC229D		3024	18	2017	
2SC229E		3024	18	2017	
2SC229F		3024	18	2017	
2SC229G		3024	18	2017	
2SC229GN		3024	18	2017	
2SC229H		3024	18	2017	
2SC229J		3024	18	2017	
2SC229K		3024	18	2017	
2SC229L		3024	18	2017	
2SC229M		3024	18	2017	
2SC229OR		3024	18	2017	
2SC229R		3024	18	2017	
2SC229X		3024	18	2017	
2SC229Y		3024	18	2017	
2SC22A		3024	18	2030	
2SC22B		3024	18	2030	
2SC22D		3024	18	2030	
2SC22E		3024	18	2030	
2SC22F		3024	18	2030	
2SC22G		3024	18	2030	
2SC22GN		3024	18	2030	
2SC22H		3024	18	2030	
2SC22J		3024	18	2030	
2SC22K		3024	18	2030	
2SC22L		3024	18	2030	
2SC22M		3024	18	2030	
2SC22OR		3024	18	2030	
2SC22R		3024	18	2030	
2SC22X		3024	18	2030	
2SC22Y		3024	18	2030	
2SC23	128/400	3024/128	18	2030	
2SC230	282	3444/123A	88	2051	
2SC2308	289A	3124/289		2014	
2SC2308B	289A	3124/289			
2SC2308C	289A	3124/289			
2SC2309	199			2014	
2SC230A		3122	88	2030	
2SC230B		3122	88	2030	
2SC230C		3122	88	2030	
2SC230D		3122	88	2030	
2SC230E		3122	88	2030	
2SC230F		3122	88	2030	
2SC230G		3122	88	2030	
2SC230GN		3122	88	2030	
2SC230H		3122	88	2030	
2SC230J		3122	88	2030	
2SC230K		3122	88	2030	
2SC230L		3122	88	2030	
2SC230M		3122	88	2030	
2SC230OR		3122	88	2030	
2SC230R		3122	88	2030	
2SC230X		3122	88	2030	
2SC230Y		3122	88	2030	
2SC231	282	3024/128	18	2030	
2SC2310	85			2014	
2SC2311					2N4922
2SC2312	152	3197/235			
2SC2314	295	3253	270		
2SC2314D	295	3253	270		
2SC2317		3054	66	2048	
2SC231A		3024	18	2030	
2SC231B		3024	18	2030	
2SC231C		3024	18	2030	
2SC231D		3024	18	2030	
2SC231E		3024	18	2030	
2SC231F		3024	18	2030	
2SC231G		3024	18	2030	
2SC231GN		3024	18	2030	
2SC231H		3024	18	2030	
2SC231J		3024	18	2030	

Industry Standard No.	ECG	SK	GE	RS 276-	MOTOR.
2SC231K		3024	18	2030	
2SC231L		3024	18	2030	
2SC231M		3024	18	2030	
2SC231OR		3024	18	2030	
2SC231R		3024	18	2030	
2SC231X		3024	18	2030	
2SC231Y		3024	18	2030	
2SC232	282	3024/128	18	2030	
2SC2321					2N5634
2SC2322					MJ15015
2SC2323					MJ15001
2SC2324					2N6038
2SC232A		3024	18	2030	
2SC232B		3024	18	2030	
2SC232C		3024	18	2030	
2SC232D		3024	18	2030	
2SC232E		3024	18	2030	
2SC232F		3024	18	2030	
2SC232G		3024	18	2030	
2SC232GN		3024	18	2030	
2SC232H		3024	18	2030	
2SC232J		3024	18	2030	
2SC232K		3024	18	2030	
2SC232L		3024	18	2030	
2SC232M		3024	18	2030	
2SC232OR		3024	18	2030	
2SC232R		3024	18	2030	
2SC232X		3024	18	2030	
2SC232Y		3024	18	2030	
2SC233	282	3024/128	18	2030	
2SC2331	375$				MJE13004
2SC2333	379				MJE13005
2SC2334	379				MJE15030
2SC2335	379				MJE13007
2SC2337	87	3836/284			2N5634
2SC2337A	87	3836/284			
2SC233A		3024	18	2030	
2SC233B		3024	18	2030	
2SC233C		3024	18	2030	
2SC233D		3024	18	2030	
2SC233F		3024	18	2030	
2SC233G		3024	18	2030	
2SC233GN		3024	18	2030	
2SC233H		3024	18	2030	
2SC233J		3024	18	2030	
2SC233L		3024	18	2030	
2SC233M		3024	18	2030	
2SC233OR		3024	18	2030	
2SC233R		3024	18	2030	
2SC233X		3024	18	2030	
2SC233Y		3024	18	2030	
2SC234	128	3024	66	2048	
2SC2344	375				TIP47
2SC234A		3024	66	2048	
2SC234B		3024	66	2048	
2SC234C		3024	66	2048	
2SC234D		3024	66	2048	
2SC234E		3024	66	2048	
2SC234F		3024	66	2048	
2SC234G		3024	66	2048	
2SC234GN		3024	66	2048	
2SC234H		3024	66	2048	
2SC234J		3024	66	2048	
2SC234K		3024	66	2048	
2SC234L		3024	66	2048	
2SC234M		3024	66	2048	
2SC234OR		3024	66	2048	
2SC234R		3024	66	2048	
2SC234X		3024	66	2048	
2SC234Y		3024	66	2048	
2SC235	293	3512	18	2030	
2SC235(0)	190		18		
2SC235-0	293		18	2030	
2SC2350			66	2048	
2SC2354	384	3131A/369	246		2N3739
2SC2356	386				SDT13305
2SC2357					MJ12010
2SC2358					MJ12010
2SC2359	379				MJ4381
2SC235A		3512	18	2030	
2SC235B		3512	18	2030	
2SC235C		3512	18	2030	
2SC235D		3512	18	2030	
2SC235E		3512	18	2030	
2SC235F		3512	18	2030	
2SC235G		3512	18	2030	
2SC235GN		3512	18	2030	
2SC235H		3512	18	2030	
2SC235J		3512	18	2030	
2SC235K		3512	18	2030	
2SC235L		3512	18	2030	
2SC235M		3512	18	2030	
2SC235OR		3512	18	2030	
2SC235R		3512	18	2030	
2SC235X		3512	18	2030	
2SC235Y	123AP	3512	18	2030	
2SC236	128	3024	18	2030	
2SC2366					MJ10016
2SC236A		3024	18	2030	
2SC236B		3024	18	2030	
2SC236C		3024	18	2030	
2SC236D		3024	18	2030	
2SC236E		3024	18	2030	
2SC236F		3024	18	2030	
2SC236G		3024	18	2030	
2SC236GN		3024	18	2030	
2SC236H		3024	18	2030	
2SC236J		3024	18	2030	
2SC236K		3024	18	2030	
2SC236L		3024	18	2030	
2SC236M		3024	18	2030	
2SC236OR		3024	18	2030	
2SC236R		3024	18	2030	
2SC236X		3024	18	2030	
2SC236Y		3024	18	2030	
2SC237	123A	3444	88	2051	
2SC2371		3244	222		MJE3439
2SC2373	379	3439/163A			MJE13006
2SC2373K	379	3440/291			
2SC2373L	379	3440/291			
2SC2373M	379	3440/291			
2SC2375	399	3866			
2SC2375E	399	3866			
2SC2375F	399	3866			
2SC237A		3122	88	2051	
2SC237B		3122	88	2051	
2SC237C		3122	88	2051	
2SC237D		3122	88	2051	
2SC237E		3122	88	2051	
2SC237F		3122	88	2051	
2SC237G		3122	88	2051	
2SC237GN		3122	88	2051	
2SC237H		3122	88	2051	
2SC237J		3122	88	2051	
2SC237K		3122	88	2051	
2SC237L		3122	88	2051	
2SC237M		3122	88	2051	
2SC237OR		3122	88	2051	
2SC237R		3122	88	2051	
2SC237X		3122	88	2051	
2SC237Y		3122	88	2051	
2SC238	123A	3024/128	18	2030	
2SC2388	386				2N6543
2SC238A		3024	18	2030	
2SC238B		3024	18	2030	
2SC238C		3024	18	2030	
2SC238D		3024	18	2030	
2SC238E		3024	18	2030	
2SC238F		3024	18	2030	
2SC238G		3024	18	2030	
2SC238GN		3024	18	2030	
2SC238H		3024	18	2030	
2SC238J		3024	18	2030	
2SC238K		3024	18	2030	
2SC238L		3024	18	2030	
2SC238M		3024	18	2030	
2SC238OR		3024	18	2030	
2SC238R		3024	18	2030	
2SC238X		3024	18	2030	
2SC238Y		3024	18	2030	
2SC239	123A	3444	88	2051	
2SC2393D		3197	216	2053	
2SC2394D			216	2053	
2SC2395					MRF433
2SC2397					MJE3055T
2SC239A		3122	88	2051	
2SC239B		3122	88	2051	
2SC239C		3122	88	2051	
2SC239D		3122	88	2051	
2SC239E		3122	88	2051	
2SC239F		3122	88	2051	
2SC239G		3122	88	2051'	
2SC239GN		3122	88	2051	
2SC239H		3122	88	2051	
2SC239J		3122	88	2051	
2SC239K		3122	88	2051	
2SC239L		3122	88	2051	
2SC239M		3122	88	2051	
2SC239OR		3122	88	2051	
2SC239R		3122	88	2051	
2SC239X		3122	88	2051	
2SC239Y		3122	88	2051	
2SC23A		3024	18	2030	
2SC23B		3024	18	2030	
2SC23D		3024	18	2030	
2SC23E		3024	18	2030	
2SC23F		3024	18	2030	
2SC23G		3024	18	2030	
2SC23GN		3024	18	2030	
2SC23H		3024	18	2030	
2SC23J		3024	18	2030	
2SC23K		3024	18	2030	
2SC23L		3024	18	2030	
2SC23M		3024	18	2030	
2SC23OR		3024	18	2030	
2SC23R		3024	18	2030	
2SC23X		3024	18	2030	
2SC23Y		3024	18	2030	
2SC24	128/400	3024/128	18	2030	
2SC240	130	3438	75		2N4347
2SC2402	284$				2N6546
2SC2403					MJ10015
2SC240A		3027	75	2041	
2SC240B		3027	75	2041	
2SC240C		3027	75	2041	
2SC240D		3027	75	2041	
2SC240E		3027	75	2041	
2SC240F		3027	75	2041	
2SC240G		3027	75	2041	
2SC240GN		3027	75	2041	
2SC240H		3027	75	2041	
2SC240J		3027	75	2041	
2SC240K		3027	75	2041	
2SC240L		3027	75	2041	
2SC240M		3027	75	2041	
2SC240OR		3027	75	2041	
2SC240R		3027	75	2041	
2SC240X		3027	75	2041	
2SC240Y		3027	75	2041	
2SC241	130	3438	35		2N3447
2SC241A		3027	75	2041	
2SC241B		3027	75	2041	
2SC241C		3027	75	2041	
2SC241D		3027	75	2041	
2SC241E		3027	75	2041	
2SC241F		3027	75	2041	
2SC241G		3027	75	2041	
2SC241GN		3027	75	2041	
2SC241H		3027	75	2041	
2SC241K		3027	75	2041	
2SC241L		3027	75	2041	
2SC241M		3027	75	2041	
2SC241OR		3027	75	2041	
2SC241R		3027	75	2041	
2SC241X		3027	75	2041	
2SC241Y		3027	75	2041	
2SC242	130	3438	75		2N4347
2SC2428					2N6249
2SC2429					2N6547
2SC242A		3027	75	2041	
2SC242B		3027	75	2041	
2SC242C		3027	75	2041	
2SC242D	130	3027	75	2041	
2SC242E		3027	75	2041	
2SC242F		3027	75	2041	
2SC242G		3027	75	2041	
2SC242GN		3027	75	2041	
2SC242H		3027	75	2041	
2SC242J		3027	75	2041	
2SC242K		3027	75	2041	
2SC242L		3027	75	2041	
2SC242M		3027	75	2041	
2SC242OR		3027	75	2041	
2SC242R		3027	75	2041	
2SC242X		3027	75	2041	
2SC242Y		3027	75	2041	
2SC243	280	3438	35		MJ410
2SC2430	328$				2N5633
2SC2431	328$				MJ15015
2SC2432	328$				2N5882
2SC2433	387$				MJ11016
2SC2434	387$				2N6327
2SC2435	97$				2N6059
2SC2436	98$				2N6059
2SC244	385	3027/130	75	2041	2N3447
2SC2442					MJ10016
2SC2443					MJ10016
2SC2448					SDT13305

Industry Standard No.	ECG	SK	GE	RS 276-	MOTOR.
2SC2449					SDT13305
2SC244A		3027	75	2041	
2SC244B		3027	75	2041	
2SC244C		3027	75	2041	
2SC244D		3027	75	2041	
2SC244E		3027	75	2041	
2SC244F		3027	75	2041	
2SC244G		3027	75	2041	
2SC244GN		3027	75	2041	
2SC244J		3027	75	2041	
2SC244K		3027	75	2041	
2SC244L		3027	75	2041	
2SC244M		3027	75	2041	
2SC244OR		3027	75	2041	
2SC244R		3027	75	2041	
2SC244X		3027	75	2041	
2SC244Y		3027	75	2041	
2SC245	385	3438	35		2N4347
2SC2450					SDT13305
2SC2451					SDT13305
2SC2452					SDT13305
2SC2453					SDT13305
2SC2456		3190	57	2017	
2SC246	385	3438	35		MJ410
2SC247	128	3024	18	2030	
2SC247A		3024	18	2030	
2SC247B		3024	18	2030	
2SC247C		3024	18	2030	
2SC247D		3024	18	2030	
2SC247E		3024	18	2030	
2SC247F		3024	18	2030	
2SC247G		3024	18	2030	
2SC247GN		3024	18	2030	
2SC247H		3024	18	2030	
2SC247J		3024	18	2030	
2SC247L		3024	18	2030	
2SC247M		3024	18	2030	
2SC247OR		3024	18	2030	
2SC247R		3024	18	2030	
2SC247X		3024	18	2030	
2SC247Y		3024	18	2030	
2SC248	123A	3024/128	18	2030	
2SC2481	373	9041			
2SC2482	399	3244	222		MJE2361T
2SC2487	284				2N5634
2SC2488	284				2N5634
2SC2489	284				2N5634
2SC248A		3024	18	2030	
2SC248B		3024	18	2030	
2SC248C		3024	18	2030	
2SC248D		3024	18	2030	
2SC248E		3024	18	2030	
2SC248F		3024	18	2030	
2SC248G		3024	18	2030	
2SC248J		3024	18	2030	
2SC248K		3024	18	2030	
2SC248L		3024	18	2030	
2SC248M		3024	18	2030	
2SC248OR		3024	18	2030	
2SC248R		3024	18	2030	
2SC248X		3024	18	2030	
2SC248Y		3024	18	2030	
2SC249	128	3024	18	2030	
2SC2492					2N5633
2SC2493					2N5634
2SC2497	373#	3253/295			
2SC2497Q	373#	3253/295	57	2017	
2SC2497R	373#	3253/295	57	2017	
2SC249A		3024	18	2051	
2SC249B		3024	18	2051	
2SC249C		3024	18	2051	
2SC249D		3024	18	2051	
2SC249E		3024	18	2051	
2SC249F		3024	18	2051	
2SC249G		3024	18	2051	
2SC249GN		3024	18	2051	
2SC249H		3024	18	2051	
2SC249J		3024	18	2051	
2SC249K		3024	18	2051	
2SC249L		3024	18	2051	
2SC249M		3024	18	2051	
2SC249OR		3024	18	2051	
2SC249R		3024	18	2051	
2SC249X		3024	18	2051	
2SC249Y		3024	18	2051	
2SC24A		3024	18	2030	
2SC24B		3024	18	2030	
2SC24C		3024	18	2030	
2SC24D		3024	18	2030	
2SC24E		3024	18	2030	
2SC24F		3024	18	2030	
2SC24G		3024	18	2030	
2SC24GN		3024	18	2030	
2SC24H		3024	18	2030	
2SC24J		3024	18	2030	
2SC24K		3024	18	2030	
2SC24L		3024	18	2030	
2SC24M		3024	18	2030	
2SC24OR		3024	18	2030	
2SC24R		3024	18	2030	
2SC24X		3024	18	2030	
2SC24Y		3024	18	2030	
2SC250	123A	3246/229	86	2015	
2SC2500					TIP31
2SC250A			86	2038	
2SC250B			86	2038	
2SC250C			86	2038	
2SC250D			86	2038	
2SC250E			86	2038	
2SC250F			86	2038	
2SC250G			86	2038	
2SC250GN			86	2038	
2SC250H			86	2038	
2SC250J			86	2038	
2SC250K			86	2038	
2SC250L			86	2038	
2SC250M			86	2038	
2SC250OR			86	2038	
2SC250R			86	2038	
2SC250X			86	2038	
2SC250Y			86	2038	
2SC251	316	3716/161	86	2015	
2SC2516					2N6497
2SC251A	316	3716/161	86	2015	
2SC251B			86	2038	
2SC251C			86	2038	
2SC251D			86	2038	
2SC251E			86	2038	
2SC251F			86	2038	
2SC251G			86	2038	
2SC251GN			86	2038	
2SC251H			86	2038	
2SC251J			86	2038	
2SC251K			86	2038	
2SC251L			86	2038	
2SC251M			86	2038	
2SC251OR			86	2038	
2SC251R			86	2038	
2SC251X			86	2038	
2SC251Y			86	2038	
2SC252	316	3716/161	86	2015	
2SC252A			86	2038	
2SC252B			86	2038	
2SC252C			86	2038	
2SC252D			86	2038	
2SC252E			86	2038	
2SC252F			86	2038	
2SC252G			86	2038	
2SC252GN			86	2038	
2SC252H			86	2038	
2SC252J			86	2038	
2SC252K			86	2038	
2SC252L			86	2038	
2SC252M			86	2038	
2SC252OR			86	2038	
2SC252R			86	2038	
2SC252X			86	2038	
2SC252Y			86	2038	
2SC253	316	3716/161	86	2015	
2SC2536					2N6499
2SC253A			86	2038	
2SC253B			86	2038	
2SC253C			86	2038	
2SC253D			86	2038	
2SC253E			86	2038	
2SC253F			86	2038	
2SC253G			86	2038	
2SC253GN			86	2038	
2SC253H			86	2038	
2SC253J			86	2038	
2SC253K			86	2038	
2SC253L			86	2038	
2SC253M			86	2038	
2SC253OR			86	2038	
2SC253R			86	2038	
2SC253X			86	2038	
2SC253Y			86	2038	
2SC254	128		246	2020	
2SC2541 (TO218)					SDT13305
2SC2562	377				TIP42A
2SC2568K		3190	57	2017	
2SC2568L		3190	57	2017	
2SC2568M		3190	57	2017	
2SC2569					2N5760
2SC2590	373#				MJE341
2SC26	123A	3444	20	2051	
2SC260			88	2030	
2SC2606		3201	27		
2SC261	195A		63	2030	
2SC2610	399	3244			
2SC262			88	2030	
2SC263		3452	86	2038	
2SC2637	228A	3220/198	251		
2SC263A		3039	86	2038	
2SC263B		3039	86	2038	
2SC263C		3039	86	2038	
2SC263D		3039	86	2038	
2SC263E		3039	86	2038	
2SC263F		3039	86	2038	
2SC263G		3039	86	2038	
2SC263GN		3039	86	2038	
2SC263H		3039	86	2038	
2SC263J		3039	86	2038	
2SC263K		3039	86	2038	
2SC263L		3039	86	2038	
2SC263M		3039	86	2038	
2SC263OR		3039	86	2038	
2SC263R		3039	86	2038	
2SC263X		3039	86	2038	
2SC263Y		3039	86	2038	
2SC264			210	2051	
2SC265			210	2051	
2SC266	313	3019	86	2015	
2SC266A		3039	86	2038	
2SC266B		3039	86	2038	
2SC266C		3039	86	2038	
2SC266D		3039	86	2038	
2SC266E		3039	86	2038	
2SC266F		3039	86	2038	
2SC266G		3039	86	2038	
2SC266GN		3039	86	2038	
2SC266H		3039	86	2038	
2SC266J		3039	86	2038	
2SC266K		3039	86	2038	
2SC266L		3039	86	2038	
2SC266M		3039	86	2038	
2SC266OR		3039	86	2038	
2SC266R		3039	86	2038	
2SC266X		3039	86	2038	
2SC266Y		3039	86	2038	
2SC267		3444	88	2051	
2SC267A		3444	88	2051	
2SC267B		3122	88	2030	
2SC267C		3122	88	2030	
2SC267D		3122	88	2030	
2SC267E		3122	88	2030	
2SC267F		3122	88	2030	
2SC267G		3122	88	2030	
2SC267GN		3122	88	2030	
2SC267H		3122	88	2030	
2SC267J		3122	88	2030	
2SC267K		3122	88	2030	
2SC267L		3122	88	2030	
2SC267M		3122	88	2030	
2SC267OR		3122	88	2030	
2SC267R		3122	88	2030	
2SC267X		3122	88	2030	
2SC267Y		3122	88	2030	
2SC268		3019	220	2030	
2SC268-OR			220	2030	
2SC268A		3024	220	2030	
2SC268B		3039	220	2030	
2SC268C		3039	220	2030	
2SC268D		3039	220	2030	
2SC268E		3039	220	2030	
2SC268F		3039	220	2030	
2SC268G		3039	220	2030	
2SC268GN		3039	220	2030	
2SC268H		3039	220	2030	
2SC268J		3039	220	2030	
2SC268K		3039	220	2030	
2SC268L		3039	220	2030	
2SC268M		3039	220	2030	
2SC268OR		3039	220	2030	
2SC268R		3039	220	2030	
2SC268X		3039	220	2030	
2SC268Y		3039	220	2030	
2SC269		3452	86	2038	

Industry Standard No.	ECG	SK	GE	RS 276-	MOTOR.
2SC269A		3039	86	2038	
2SC269B		3039	86	2038	
2SC269C		3039	86	2038	
2SC269D		3039	86	2038	
2SC269E		3039	86	2038	
2SC269F		3039	86	2038	
2SC269G		3039	86	2038	
2SC269GN		3039	86	2038	
2SC269H		3039	86	2038	
2SC269J		3039	86	2038	
2SC269K		3039	86	2038	
2SC269L		3039	86	2038	
2SC269M		3039	86	2038	
2SC269OR		3039	86	2038	
2SC269R		3039	86	2038	
2SC269X		3039	86	2038	
2SC269Y		3039	86	2038	
2SC26A		3122	88	2030	
2SC26B		3122	88	2030	
2SC26C		3122	88	2030	
2SC26D		3122	88	2030	
2SC26E		3122	88	2030	
2SC26F		3122	88	2030	
2SC26G		3122	88	2030	
2SC26GN		3122	88	2030	
2SC26H		3122	88	2030	
2SC26J		3122	88	2030	
2SC26K		3122	88	2030	
2SC26L		3122	88	2030	
2SC26M		3122	88	2030	
2SC26OR		3122	88	2030	
2SC26R		3122	88	2030	
2SC26X		3122	88	2030	
2SC26Y		3122	88	2030	
2SC27	123A	3024/128	18	2030	
2SC270	165		35		MJ411
2SC271		3452	86	2038	
2SC271A		3039	86	2038	
2SC271B		3039	86	2038	
2SC271C		3039	86	2038	
2SC271D		3039	86	2038	
2SC271E		3039	86	2038	
2SC271F		3039	86	2038	
2SC271G		3039	86	2038	
2SC271GN		3039	86	2038	
2SC271H		3039	86	2038	
2SC271J		3039	86	2038	
2SC271K		3039	86	2038	
2SC271L		3039	86	2038	
2SC271M		3039	86	2038	
2SC271OR		3039	86	2038	
2SC271R		3039	86	2038	
2SC271X		3039	86	2038	
2SC271Y		3039	86	2038	
2SC272		3452	86	2038	
2SC272A		3039	86	2038	
2SC272B		3039	86	2038	
2SC272C		3039	86	2038	
2SC272D		3039	86	2038	
2SC272E		3039	86	2038	
2SC272F		3039	86	2038	
2SC272G		3039	86	2038	
2SC272GN		3039	86	2038	
2SC272H		3039	86	2038	
2SC272J		3039	86	2038	
2SC272K		3039	86	2038	
2SC272L		3039	86	2038	
2SC272M		3039	86	2038	
2SC272OR		3039	86	2038	
2SC272R		3039	86	2038	
2SC272X		3039	86	2038	
2SC272Y		3039	86	2038	
2SC273	154		40	2012	
2SC277C	103A	3835	59	2002	
2SC278-OR			278	2016	
2SC27A		3024	18	2030	
2SC27B		3024	18	2030	
2SC27D		3024	18	2030	
2SC27E		3024	18	2030	
2SC27F		3024	18	2030	
2SC27G		3024	18	2030	
2SC27GN		3024	18	2030	
2SC27H		3024	18	2030	
2SC27J		3024	18	2030	
2SC27K		3024	18	2030	
2SC27L		3024	18	2030	
2SC27M		3024	18	2030	
2SC27OR		3024	18	2030	
2SC27R		3024	18	2030	
2SC27X		3024	18	2030	
2SC27Y		3024	18	2030	
2SC28	123	3124/289	47	2051	
2SC280A0			20	2051	
2SC281	123A	3444	212	2010	
2SC281(B)			212	2010	
2SC281-OR			212	2010	
2SC2818			20	2051	
2SC281A	123A	3444	212	2010	
2SC281B	123A	3444	212	2010	
2SC281BL		3124	212	2010	
2SC281C	123A	3444	212	2010	
2SC281C-EP	123A	3444	212	2010	
2SC281D	123A	3444	212	2010	
2SC281E		3124	212	2010	
2SC281EP	123A	3444	212	2010	
2SC281F		3124	212	2010	
2SC281G		3124	212	2010	
2SC281GN		3124	212	2010	
2SC281H	123A	3444	212	2010	
2SC281HA	123A	3444	212	2010	
2SC281HB	123A	3444	212	2010	
2SC281HC	123A	3444	212	2010	
2SC281J		3124	212	2010	
2SC281K		3124	212	2010	
2SC281L		3124	212	2010	
2SC281M		3124	212	2010	
2SC281OR		3124	212	2010	
2SC281R		3124	212	2010	
2SC281X		3124	212	2010	
2SC281Y		3124	212	2010	
2SC282	123A	3444	86	2051	
2SC282A		3039	86	2038	
2SC282B		3039	86	2038	
2SC282C		3039	86	2038	
2SC282D		3039	86	2038	
2SC282E		3039	86	2038	
2SC282F		3039	86	2038	
2SC282G		3039	86	2038	
2SC282GN		3039	86	2038	
2SC282H	123A	3444	86	2051	
2SC282HA	123A	3444	86	2051	
2SC282HB	123A	3444	86	2051	
2SC282HC	123A	3444	86	2051	
2SC282J		3039	86	2038	
2SC282K		3039	86	2038	
2SC282L		3039	86	2038	
2SC282M		3039	86	2038	
2SC282OR		3039	86	2038	
2SC282R		3039	86	2038	
2SC282X		3039	86	2038	
2SC282Y		3039	86	2038	
2SC283	123A	3444	88	2051	
2SC283A		3122	88	2010	
2SC283B		3122	88	2010	
2SC283C		3122	88	2010	
2SC283D		3122	88	2010	
2SC283E		3122	88	2010	
2SC283F		3122	88	2010	
2SC283G		3122	88	2010	
2SC283GN		3122	88	2010	
2SC283H		3122	88	2010	
2SC283J		3122	88	2010	
2SC283K		3122	88	2010	
2SC283L		3122	88	2010	
2SC283M		3122	88	2010	
2SC283OR		3122	88	2010	
2SC283R		3122	88	2010	
2SC283X		3122	88	2010	
2SC283Y		3122	88	2010	
2SC284	123A	3444	88	2051	
2SC284A		3122	88	2010	
2SC284B		3122	88	2010	
2SC284C		3122	88	2010	
2SC284D		3122	88	2010	
2SC284E		3122	88	2010	
2SC284F		3122	88	2010	
2SC284G		3122	88	2010	
2SC284GN		3122	88	2010	
2SC284H	123A	3444	88	2051	
2SC284HA	123A	3444	88	2051	
2SC284HB	123A	3444	88	2051	
2SC284HC	123A	3444	88	2051	
2SC284J		3122	88	2010	
2SC284K		3122	88	2010	
2SC284L		3122	88	2010	
2SC284M		3122	88	2010	
2SC284OR		3122	88	2010	
2SC284R		3122	88	2010	
2SC284X		3122	88	2010	
2SC284Y		3122	88	2010	
2SC285	311	3444/123A	88	2051	
2SC285A	311	3444/123A	88	2051	
2SC285B		3122	88	2030	
2SC285C		3122	88	2030	
2SC285D		3122	88	2030	
2SC285E		3122	88	2030	
2SC285F		3122	88	2030	
2SC285G		3122	88	2030	
2SC285GN		3122	88	2030	
2SC285H		3122	88	2030	
2SC285J		3122	88	2030	
2SC285K		3122	88	2030	
2SC285L		3122	88	2030	
2SC285M		3122	88	2030	
2SC285OR		3122	88	2030	
2SC285R		3122	88	2030	
2SC285X		3122	88	2030	
2SC285Y		3122	88	2030	
2SC286		3452	86	2038	
2SC286A		3039	86	2038	
2SC286B		3039	86	2038	
2SC286C		3039	86	2038	
2SC286D		3039	86	2038	
2SC286E		3039	86	2038	
2SC286F		3039	86	2038	
2SC286G		3039	86	2038	
2SC286GN		3039	86	2038	
2SC286H		3039	86	2038	
2SC286J		3039	86	2038	
2SC286K		3039	86	2038	
2SC286L		3039	86	2038	
2SC286M		3039	86	2038	
2SC286OR		3039	86	2038	
2SC286R		3039	86	2038	
2SC286X		3039	86	2038	
2SC286Y		3039	86	2038	
2SC287		3019	86	2016	
2SC287-OR			86	2016	
2SC287A		3018	86	2016	
2SC287B		3039	86	2016	
2SC287C		3039	86	2016	
2SC287D		3039	86	2016	
2SC287E		3039	86	2016	
2SC287F		3039	86	2016	
2SC287G		3039	86	2016	
2SC287GN		3039	86	2016	
2SC287H		3039	86	2016	
2SC287J		3039	86	2016	
2SC287K		3039	86	2016	
2SC287L		3039	86	2016	
2SC287M		3039	86	2016	
2SC287OR		3039	86	2016	
2SC287R		3039	86	2016	
2SC287X		3039	86	2016	
2SC287Y		3039	86	2016	
2SC288		3019	86	2016	
2SC2884			11	2015	
2SC288A		3444	86	2016	
2SC288A1			86	2030	
2SC288AB		3019	86	2030	
2SC288B		3039	86	2030	
2SC288C		3039	86	2030	
2SC288D		3039	86	2030	
2SC288E		3039	86	2030	
2SC288F		3039	86	2030	
2SC288G		3039	86	2030	
2SC288GN		3039	86	2030	
2SC288H		3039	86	2030	
2SC288J		3039	86	2030	
2SC288K		3039	86	2030	
2SC288L		3039	86	2030	
2SC288M		3039	86	2030	
2SC288OR		3039	86	2030	
2SC288R		3039	86	2030	
2SC288X		3039	86	2030	
2SC288Y		3039	86	2030	
2SC289		3452	86	2038	
2SC289A		3039	86	2038	
2SC289B		3039	86	2038	
2SC289C		3039	86	2038	
2SC289D		3039	86	2038	
2SC289E		3039	86	2038	
2SC289F		3039	86	2038	
2SC289G		3039	86	2038	
2SC289GN		3039	86	2038	
2SC289H		3039	86	2038	
2SC289J		3039	86	2038	
2SC289K		3039	86	2038	

Industry Standard No.	ECG	SK	GE	RS 276-	MOTOR.
2SC289L		3039	86	2038	
2SC289M		3039	86	2038	
2SC289OR		3039	86	2038	
2SC289R		3039	86	2038	
2SC289X		3039	86	2038	
2SC289Y		3039	86	2038	
2SC28A		3124	47	2010	
2SC28B		3124	47	2010	
2SC28C		3124	47	2010	
2SC28D		3124	47	2010	
2SC28E		3124	47	2010	
2SC28F		3124	47	2010	
2SC28G		3124	47	2010	
2SC28GN		3124	47	2010	
2SC28H		3124	47	2010	
2SC28J		3124	47	2010	
2SC28K		3124	47	2010	
2SC28L		3124	47	2010	
2SC28M		3124	47	2010	
2SC28OR		3124	47	2010	
2SC28R		3124	47	2010	
2SC28X		3124	47	2010	
2SC28Y		3124	47	2010	
2SC29	123	3019	86	2051	
2SC290		3765	46		
2SC291	128	3024	18	2030	
2SC291A		3024	18	2017	
2SC291B		3024	18	2017	
2SC291C		3024	18	2017	
2SC291D		3024	18	2017	
2SC291E		3024	18	2017	
2SC291F		3024	18	2017	
2SC291G		3024	18	2017	
2SC291GN		3024	18	2017	
2SC291H		3024	18	2017	
2SC291J		3024	18	2017	
2SC291K		3024	18	2017	
2SC291L		3024	18	2017	
2SC291M		3024	18	2017	
2SC291OR		3024	18	2017	
2SC291R		3024	18	2017	
2SC291X		3024	18	2017	
2SC291Y		3024	18	2017	
2SC292	282	3024/128	18	2030	
2SC292A		3024	18	2017	
2SC292B		3024	18	2017	
2SC292C		3024	18	2017	
2SC292D		3024	18	2017	
2SC292E		3024	18	2017	
2SC292F		3024	18	2017	
2SC292G		3024	18	2017	
2SC292GN		3024	18	2017	
2SC292H		3024	18	2017	
2SC292J		3024	18	2017	
2SC292K		3024	18	2017	
2SC292L		3024	18	2017	
2SC292M		3024	18	2017	
2SC292OR		3024	18	2017	
2SC292R		3024	18	2017	
2SC292X		3024	18	2017	
2SC292Y		3024	18	2017	
2SC293	282	3048/329	18		
2SC293A		3024	18	2030	
2SC293B		3024	18	2030	
2SC293C		3024	18	2030	
2SC293D		3024	18	2030	
2SC293E		3024	18	2030	
2SC293F		3024	18	2030	
2SC293G		3024	18	2030	
2SC293GN		3024	18	2030	
2SC293H		3024	18	2030	
2SC293J		3024	18	2030	
2SC293K		3024	18	2030	
2SC293L		3024	18	2030	
2SC293M		3024	18	2030	
2SC293OR		3024	18	2030	
2SC293R		3024	18	2030	
2SC293X		3024	18	2030	
2SC293Y		3024	18	2030	
2SC296	196	3452/108	86	2038	
2SC297			28	2017	
2SC298	282		248	2030	
2SC298-4	282	3024/128	243	2030	
2SC299	282/427	3124/289	47		
2SC299A		3124	47		
2SC299B		3124	47		
2SC299C		3124	47		
2SC299D		3124	47		
2SC299E		3124	47		
2SC299F		3124	47		
2SC299G		3124	47		
2SC299GN		3124	47		
2SC299H		3124	47		
2SC299J		3124	47		
2SC299K		3124	47		
2SC299L		3124	47		
2SC299M		3124	47		
2SC299OR		3124	47		
2SC299R		3124	47		
2SC299X		3124	47		
2SC299Y		3124	47		
2SC29A		3039	86	2038	
2SC29B		3039	86	2038	
2SC29C		3039	86	2038	
2SC29D		3039	86	2038	
2SC29E		3039	86	2038	
2SC29F		3039	86	2038	
2SC29G		3039	86	2038	
2SC29GN		3039	86	2038	
2SC29H		3039	86	2038	
2SC29J		3039	86	2038	
2SC29K		3039	86	2038	
2SC29L		3039	86	2038	
2SC29M		3039	86	2038	
2SC29OR		3039	86	2038	
2SC29R		3039	86	2038	
2SC29X		3039	86	2038	
2SC29Y		3039	86	2038	
2SC30	128	3024	17	2051	
2SC30-OR			17	2051	
2SC300	123A	3444	88	2051	
2SC300A		3444	88	2030	
2SC300B	123A	3444	88	2030	
2SC300C	123A	3444	88	2030	
2SC300D	123A	3444	88	2030	
2SC300E	123A	3444	88	2030	
2SC300F		3444	88	2030	
2SC300G		3444	88	2030	
2SC300GN		3444	88	2030	
2SC300H		3444	88	2030	
2SC300J		3444	88	2030	
2SC300K		3444	88	2030	
2SC300L		3444	88	2030	
2SC300M		3444	88	2030	
2SC300OR		3444	88	2030	
2SC300R		3444	88	2030	
2SC300X		3444	88	2030	
2SC300Y		3444	88	2030	
2SC301	123A	3444	88	2051	
2SC301A		3122	88	2030	
2SC301B	123A	3122	88	2030	
2SC301C	123A	3122	88	2030	
2SC301D	123A	3122	88	2030	
2SC301E	123A	3122	88	2030	
2SC301F		3122	88	2030	
2SC301G		3122	88	2030	
2SC301GN		3122	88	2030	
2SC301H		3122	88	2030	
2SC301J		3122	88	2030	
2SC301K		3122	88	2030	
2SC301L		3122	88	2030	
2SC301M		3122	88	2030	
2SC301OR		3122	88	2030	
2SC301R		3122	88	2030	
2SC301X		3122	88	2030	
2SC301Y		3122	88	2030	
2SC302	123A	3444	88	2051	
2SC302A		3122	88	2030	
2SC302B	123A	3122	88	2030	
2SC302C	123A	3122	88	2030	
2SC302D	123A	3122	88	2030	
2SC302E	123A	3122	88	2030	
2SC302F		3122	88	2030	
2SC302G		3122	88	2030	
2SC302GN		3122	88	2030	
2SC302H		3122	88	2030	
2SC302J		3122	88	2030	
2SC302K		3122	88	2030	
2SC302L		3122	88	2030	
2SC302M		3122	88	2030	
2SC302OR		3122	88	2030	
2SC302R		3122	88	2030	
2SC302X		3122	88	2030	
2SC302Y		3122	88	2030	
2SC303			63	2030	
2SC304			63	2030	
2SC305		3104A	32	2030	
2SC305A		3104A	32	2030	
2SC305B		3104A	32	2030	
2SC305C		3104A	32	2030	
2SC305D		3104A	32	2030	
2SC305E		3104A	32	2030	
2SC305F		3104A	32	2030	
2SC305G		3104A	32	2030	
2SC305GN		3104A	32	2030	
2SC305H		3104A	32	2030	
2SC305J		3104A	32	2030	
2SC305K		3104A	32	2030	
2SC305L		3104A	32	2030	
2SC305M		3104A	32	2030	
2SC305OR		3104A	32	2030	
2SC305R		3104A	32	2030	
2SC305X		3104A	32	2030	
2SC305Y		3104A	32	2030	
2SC306	128	3024	18	2030	
2SC306A		3024	18	2030	
2SC306B	128	3024	18	2030	
2SC306C	128	3024	18	2030	
2SC306D	128	3024	18	2030	
2SC306E	128	3024	18	2030	
2SC306F	128	3024	18	2030	
2SC306G		3024	18	2030	
2SC306GN		3024	18	2030	
2SC306H		3024	18	2030	
2SC306J		3024	18	2030	
2SC306K		3024	18	2030	
2SC306L		3024	18	2030	
2SC306M		3024	18	2030	
2SC306OR		3024	18	2030	
2SC306R		3024	18	2030	
2SC306X		3024	18	2030	
2SC306Y		3024	18	2030	
2SC307	128	3024	18	2030	
2SC307(CB-FINAL)			18	2030	
2SC307-OR			18	2030	
2SC307A		3024	18	2030	
2SC307B	128	3024	18	2030	
2SC307C	128	3024	18	2030	
2SC307D	128	3024	18	2030	
2SC307E	128	3024	18	2030	
2SC307F		3024	18	2030	
2SC307G		3024	18	2030	
2SC307GN		3024	18	2030	
2SC307H		3024	18	2030	
2SC307J		3024	18	2030	
2SC307K		3024	18	2030	
2SC307L		3024	18	2030	
2SC307M		3024	18	2030	
2SC307OR		3024	18	2030	
2SC307R		3024	18	2030	
2SC307T	195A	3048/329	18	2030	
2SC307X		3024	18	2030	
2SC307Y		3024	18	2030	
2SC308	128	3024	18	2030	
2SC308A		3024	18	2030	
2SC308B		3024	18	2030	
2SC308C		3024	18	2030	
2SC308D		3024	18	2030	
2SC308E		3024	18	2030	
2SC308F		3024	18	2030	
2SC308G		3024	18	2030	
2SC308GN		3024	18	2030	
2SC308H		3024	18	2030	
2SC308J		3024	18	2030	
2SC308K		3024	18	2030	
2SC308L		3024	18	2030	
2SC308M		3024	18	2030	
2SC308OR		3024	18	2030	
2SC308R		3024	18	2030	
2SC308X		3024	18	2030	
2SC308Y		3024	18	2030	
2SC309	128	3024	18	2030	
2SC309A		3024	18	2030	
2SC309B	128	3024	18	2030	
2SC309C	128	3024	18	2030	
2SC309D	128	3024	18	2030	
2SC309E	128	3024	18	2030	
2SC309F		3024	18	2030	
2SC309G		3024	18	2030	
2SC309GN		3024	18	2030	
2SC309H		3024	18	2030	
2SC309J		3024	18	2030	
2SC309K		3024	18	2030	
2SC309L		3024	18	2030	
2SC309M		3024	18	2030	
2SC309OR		3024	18	2030	
2SC309R		3024	18	2030	
2SC309X		3024	18	2030	

Industry Standard No.	ECG	SK	GE	RS 276-	MOTOR.
2SC309Y		3024	18	2030	
2SC30A		3024	17	2051	
2SC30B		3024	17	2051	
2SC30C		3024	17	2051	
2SC30D		3024	17	2051	
2SC30E		3024	17	2051	
2SC30F		3024	17	2051	
2SC30G		3024	17	2051	
2SC30GN		3024	17	2051	
2SC30H		3024	17	2051	
2SC30J		3024	17	2051	
2SC30L		3024	17	2051	
2SC30M		3024	17	2051	
2SC30OR			17	2051	
2SC30R		3024	17	2051	
2SC30X		3024	17	2051	
2SC30Y		3024	17	2051	
2SC31	128	3024	18	2030	
2SC310	282	3024/128	18	2030	
2SC310A		3024	18	2030	
2SC310B	282	3024/128	18	2030	
2SC310C	282	3024/128	18	2030	
2SC310D	282	3024/128	18	2030	
2SC310E	282	3024/128	18	2030	
2SC310F		3024	18	2030	
2SC310G		3024	18	2030	
2SC310GN		3024	18	2030	
2SC310H		3024	18	2030	
2SC310J		3024	18	2030	
2SC310K		3024	18	2030	
2SC310L		3024	18	2030	
2SC310M		3024	18	2030	
2SC310OR		3024	18	2030	
2SC310R		3024	18	2030	
2SC310X		3024	18	2030	
2SC310Y		3024	18	2030	
2SC311	195A	3048/329	45		
2SC311A		3047	45		
2SC311B		3047	45		
2SC311C		3047	45		
2SC311D		3047	45		
2SC311E		3047	45		
2SC311F		3047	45		
2SC311G		3047	45		
2SC311GN		3047	45		
2SC311H		3047	45		
2SC311J		3047	45		
2SC311K		3047	45		
2SC311L		3047	45		
2SC311M		3047	45		
2SC311OR		3047	45		
2SC311R		3047	45		
2SC311X		3047	45		
2SC311Y		3047	45		
2SC312	224	3049	46		
2SC312A		3048	46		
2SC312B		3048	46		
2SC312C		3048	46		
2SC312D		3048	46		
2SC312E		3048	46		
2SC312F		3048	46		
2SC312G		3048	46		
2SC312GN		3048	46		
2SC312H		3048	46		
2SC312J		3048	46		
2SC312K		3048	46		
2SC312L		3048	46		
2SC312M		3048	46		
2SC312OR		3048	46		
2SC312R		3048	46		
2SC312X		3048	46		
2SC312Y		3048	46		
2SC313	161	3019	11	2015	
2SC313-OR			11	2015	
2SC313A		3039	11	2015	
2SC313B		3039	11	2015	
2SC313C	161	3117	11	2015	
2SC313D		3039	11	2015	
2SC313E		3039	11	2015	
2SC313F		3039	11	2015	
2SC313G		3039	11	2015	
2SC313GN		3039	11	2015	
2SC313H	161	3117	11	2015	
2SC313J		3039	11	2015	
2SC313K		3039	11	2015	
2SC313L		3039	11	2015	
2SC313M		3039	11	2015	
2SC313OR		3039	11	2015	
2SC313R		3039	11	2015	
2SC313X		3039	11	2015	
2SC313Y		3039	11	2015	
2SC315	123A	3444	45	2051	
2SC315A		3047	45		
2SC315B		3047	45		
2SC315C		3047	45		
2SC315D		3047	45		
2SC315E		3047	45		
2SC315F		3047	45		
2SC315G		3047	45		
2SC315GN		3047	45		
2SC315H		3047	45		
2SC315J		3047	45		
2SC315K		3047	45		
2SC315L		3047	45		
2SC315M		3047	45		
2SC315OR		3047	45		
2SC315R		3047	45		
2SC315X		3047	45		
2SC315Y		3047	45		
2SC316	123A	3452/108	86	2038	
2SC316A		3039	86	2038	
2SC316B		3039	86	2038	
2SC316C		3039	86	2038	
2SC316D		3039	86	2038	
2SC316E		3039	86	2038	
2SC316F		3039	86	2038	
2SC316G		3039	86	2038	
2SC316GN		3039	86	2038	
2SC316H		3039	86	2038	
2SC316J		3039	86	2038	
2SC316K		3039	86	2038	
2SC316L		3039	86	2038	
2SC316M		3039	86	2038	
2SC316OR		3039	86	2038	
2SC316R		3039	86	2038	
2SC316X		3039	86	2038	
2SC316Y		3039	86	2038	
2SC317	123A	3444	88	2051	
2SC317A		3444	88	2030	
2SC317B		3444	88	2030	
2SC317C	123A	3444	88	2051	
2SC317D		3444	88	2030	
2SC317E		3444	88	2030	
2SC317F		3444	88	2030	
2SC317G		3444	88	2030	
2SC317GN		3444	88	2030	
2SC317H		3444	88	2030	
2SC317J		3444	88	2030	
2SC317K		3444	88	2030	
2SC317L		3444	88	2030	
2SC317M		3444	88	2030	
2SC317OR		3444	88	2030	
2SC317R		3444	88	2030	
2SC317X		3444	88	2030	
2SC317Y		3444	88	2030	
2SC318	123A	3444	88	2051	
2SC318A	123A	3444	88	2051	
2SC318AB		3124	88	2010	
2SC318B		3122	88	2010	
2SC318C		3122	88	2010	
2SC318D		3122	88	2010	
2SC318E		3122	88	2010	
2SC318F		3122	88	2010	
2SC318G		3122	88	2010	
2SC318GN		3122	88	2010	
2SC318H		3122	88	2010	
2SC318J		3122	88	2010	
2SC318K		3122	88	2010	
2SC318L		3122	88	2010	
2SC318M		3122	88	2010	
2SC318OR		3122	88	2010	
2SC318R		3122	88	2010	
2SC318X		3122	88	2010	
2SC318Y		3122	88	2010	
2SC319	311	3122	88	2051	2N4427
2SC319A		3122	88	2051	
2SC319B		3122	88	2051	
2SC319C		3122	88	2051	
2SC319D		3122	88	2051	
2SC319E		3122	88	2051	
2SC319F		3122	88	2051	
2SC319G		3122	88	2051	
2SC319GN		3122	88	2051	
2SC319H		3122	88	2051	
2SC319J		3122	88	2051	
2SC319K		3122	88	2051	
2SC319L		3122	88	2051	
2SC319M		3122	88	2051	
2SC319OR		3122	88	2051	
2SC319R		3122	88	2051	
2SC319X		3122	88	2051	
2SC319Y		3122	88	2051	
2SC31A		3024	18	2030	
2SC31B		3024	18	2030	
2SC31C		3024	18	2030	
2SC31D		3024	18	2030	
2SC31E		3024	18	2030	
2SC31F		3024	18	2030	
2SC31G		3024	18	2030	
2SC31GN		3024	18	2030	
2SC31H		3024	18	2030	
2SC31J		3024	18	2030	
2SC31K		3024	18	2030	
2SC31L		3024	18	2030	
2SC31M		3024	18	2030	
2SC31OR		3024	18	2030	
2SC31R		3024	18	2030	
2SC31X		3024	18	2030	
2SC31Y		3024	18	2030	
2SC32	128	3024	18	2030	
2SC320	472	3444/123A	88	2051	MRF607
2SC320A		3122	88	2051	
2SC320B		3122	88	2051	
2SC320C		3122	88	2051	
2SC320D		3122	88	2051	
2SC320E		3122	88	2051	
2SC320F		3122	88	2051	
2SC320G		3122	88	2051	
2SC320GN		3122	88	2051	
2SC320H		3122	88	2051	
2SC320J		3122	88	2051	
2SC320K		3122	88	2051	
2SC320L		3122	88	2051	
2SC320M		3122	88	2051	
2SC320OR		3122	88	2051	
2SC320R		3122	88	2051	
2SC320X		3122	88	2051	
2SC320Y		3122	88	2051	
2SC321	123A	3444	88	2051	
2SC321A		3122	88	2051	
2SC321B		3122	88	2051	
2SC321C		3122	88	2051	
2SC321D		3122	88	2051	
2SC321E		3122	88	2051	
2SC321F		3122	88	2051	
2SC321G		3122	88	2051	
2SC321GN		3122	88	2051	
2SC321H	123A	3444	88	2051	
2SC321HA	123A	3444	88	2051	
2SC321HB	123A	3444	88	2051	
2SC321HC	123A	3444	88	2051	
2SC321J		3122	88	2051	
2SC321K		3122	88	2051	
2SC321L		3122	88	2051	
2SC321M		3122	88	2051	
2SC321OR		3122	88	2051	
2SC321R		3122	88	2051	
2SC321X		3122	88	2051	
2SC321Y		3122	88	2051	
2SC323	123A	3444	88	2051	
2SC323A		3122	88	2010	
2SC323B		3122	88	2010	
2SC323C		3122	88	2010	
2SC323D		3122	88	2010	
2SC323E		3122	88	2010	
2SC323F		3122	88	2010	
2SC323G		3122	88	2010	
2SC323GN		3122	88	2010	
2SC323H		3122	88	2010	
2SC323J		3122	88	2010	
2SC323K		3122	88	2010	
2SC323L		3122	88	2010	
2SC323M		3122	88	2010	
2SC323OR		3122	88	2010	
2SC323R		3122	88	2010	
2SC323X		3122	88	2010	
2SC323Y		3122	88	2010	
2SC324	123A	3444	61	2051	
2SC324A	123A	3444	61	2051	
2SC324B		3018	61	2038	
2SC324C		3018	61	2038	
2SC324D		3018	61	2038	
2SC324E		3018	61	2038	
2SC324F		3018	61	2038	
2SC324G		3018	61	2038	
2SC324GN		3018	61	2038	
2SC324H	123A	3444	61	2051	
2SC324HA	123A	3444	61	2051	

Industry Standard No.	ECG	SK	GE	RS 276-	MOTOR.
2SC324J		3018	61	2038	
2SC324K		3018	61	2038	
2SC324L		3018	61	2038	
2SC324M		3018	61	2038	
2SC324OR		3018	61	2038	
2SC324R		3018	61	2038	
2SC324X		3018	61	2038	
2SC324Y		3018	61	2038	
2SC325	316	3893/152	66	2048	
2SC325A	316	3893/152	66	2048	
2SC325C	316	3893/152	66	2048	
2SC325E	316	3893/152	66	2048	
2SC328A	316	3444/123A	20	2051	
2SC329	316		11	2015	
2SC329B	316		11	2015	
2SC329C	316		11	2015	
2SC32A	128	3024	18	2030	
2SC32B		3024	18	2030	
2SC32C		3024	18	2030	
2SC32D		3024	18	2051	
2SC32E		3024	18	2030	
2SC32F		3024	18	2030	
2SC32G		3024	18	2030	
2SC32GN		3024	18	2030	
2SC32H		3024	18	2030	
2SC32J		3024	18	2030	
2SC32K		3024	18	2030	
2SC32L		3024	18	2030	
2SC32M		3024	18	2030	
2SC32OR		3024	18	2030	
2SC32R		3024	18	2030	
2SC32X		3024	18	2030	
2SC32Y		3024	18	2030	
2SC33	161	3122	88	2051	
2SC333H	123A$		261		
2SC333HA	123A$		261		
2SC333HB	123A$		261		
2SC333HC	123A$		261		
2SC335	123A$	3018	10	2010	
2SC337		3018	3	2006	
2SC337B	123A	3018	239	2006	
2SC33A		3122	88	2030	
2SC33B		3122	88	2030	
2SC33C		3122	88	2030	
2SC33D		3122	88	2030	
2SC33E		3122	88	2030	
2SC33F		3122	88	2030	
2SC33G		3122	88	2030	
2SC33GN		3122	88	2030	
2SC33H		3122	88	2030	
2SC33J		3122	88	2030	
2SC33K		3122	88	2030	
2SC33L		3122	88	2030	
2SC33M		3122	88	2030	
2SC33OR		3122	88	2030	
2SC33R		3122	88	2030	
2SC33X		3122	88	2030	
2SC33Y		3122	88	2030	
2SC34	103A	3835	59	2002	
2SC341V		3764	25		
2SC344			11	2015	
2SC344(Y)			11	2015	
2SC344Y			11	2015	
2SC348	123A	3117	211	2015	
2SC348A		3117	211	2016	
2SC348B		3117	211	2016	
2SC348C		3117	211	2016	
2SC348D		3117	211	2016	
2SC348E		3117	211	2016	
2SC348F		3117	211	2016	
2SC348G		3117	211	2016	
2SC348GN		3117	211	2016	
2SC348H		3117	211	2016	
2SC348J		3117	211	2016	
2SC348K		3117	211	2016	
2SC348L		3117	211	2016	
2SC348M		3117	211	2016	
2SC348OR		3117	211	2016	
2SC348X		3117	211	2016	
2SC348Y		3117	211	2016	
2SC349			20	2015	
2SC349R		3018	20	2015	
2SC34A		3835	59		
2SC34C		3835	59		
2SC34D		3835	59		
2SC34E		3835	59		
2SC34F		3835	59		
2SC34G		3835	59		
2SC34GN		3835	59		
2SC34H		3835	59		
2SC34J		3835	59		
2SC34K		3835	59		
2SC34L		3835	59		
2SC34M		3835	59		
2SC34OR		3835	59		
2SC34R		3835	59		
2SC34X		3835	59		
2SC34Y		3835	59		
2SC35	103A	3835	59	2002	
2SC350	123A	3444	88	2051	
2SC350A		3122	88	2010	
2SC350B		3122	88	2010	
2SC350C		3122	88	2010	
2SC350D		3122	88	2010	
2SC350E		3122	88	2010	
2SC350F		3122	88	2010	
2SC350G		3122	88	2010	
2SC350GN		3122	88	2010	
2SC350H	123A	3444	88	2051	
2SC350J		3122	88	2010	
2SC350K		3122	88	2010	
2SC350L		3122	88	2010	
2SC350M		3122	88	2010	
2SC350OR		3122	88	2010	
2SC350R		3122	88	2010	
2SC350X		3122	88	2010	
2SC350Y		3122	88	2010	
2SC351	107	3293	86	2015	
2SC351(FA)	107		86	2015	
2SC351A		3293	86	2038	
2SC351B		3293	86	2038	
2SC351C		3293	86	2038	
2SC351D		3293	86	2038	
2SC351E		3293	86	2038	
2SC351F		3293	86	2038	
2SC351FA1			86	2038	
2SC351G		3293	86	2038	
2SC351GN		3293	86	2038	
2SC351H		3293	86	2038	
2SC351J		3293	86	2038	
2SC351K		3293	86	2038	
2SC351L		3293	86	2038	
2SC351M		3293	86	2038	
2SC351OR		3293	86	2038	

Industry Standard No.	ECG	SK	GE	RS 276-	MOTOR.
2SC351R		3293	86	2038	
2SC351X			86	2038	
2SC351Y		3293	86	2038	
2SC352	123A	3024/128	47	2030	
2SC352-OR			47	2030	
2SC352A	123A	3024/128	47	2030	
2SC352AC			47	2030	
2SC352B		3024	47	2030	
2SC352C		3024	47	2030	
2SC352D		3024	47	2030	
2SC352E		3024	47	2030	
2SC352F		3024	47	2030	
2SC352G		3024	47	2030	
2SC352GN		3024	47	2030	
2SC352H		3024	47	2030	
2SC352J		3024	47	2030	
2SC352K		3024	47	2030	
2SC352L		3024	47	2030	
2SC352M		3024	47	2030	
2SC352OR		3024	47	2030	
2SC352R		3024	47	2030	
2SC352X		3024	47	2030	
2SC352Y		3024	47	2030	
2SC353	128	3024	18	2030	
2SC353A	128	3024	18	2030	
2SC353AC			18	2030	
2SC353B		3024	18	2030	
2SC353C		3024	18	2030	
2SC353D		3024	18	2030	
2SC353E		3024	18	2030	
2SC353F		3024	18	2030	
2SC353G		3024	18	2030	
2SC353GN		3024	18	2030	
2SC353H		3024	18	2030	
2SC353J		3024	18	2030	
2SC353K		3024	18	2030	
2SC353L		3024	18	2030	
2SC353M		3024	18	2030	
2SC353OR		3024	18	2030	
2SC353R		3024	18	2030	
2SC353X		3024	18	2030	
2SC353Y		3024	18	2030	
2SC354		3197	215		
2SC356		3444	86	2051	
2SC356A		3039	86	2051	
2SC356B		3039	86	2051	
2SC356C		3039	86	2051	
2SC356D		3039	86	2051	
2SC356E		3039	86	2051	
2SC356F		3039	86	2051	
2SC356G		3039	86	2051	
2SC356GN		3039	86	2051	
2SC356H		3039	86	2051	
2SC356J		3039	86	2051	
2SC356K		3039	86	2051	
2SC356L		3039	86	2051	
2SC356M		3039	86	2051	
2SC356OR		3039	86	2051	
2SC356R		3039	86	2051	
2SC356X		3039	86	2051	
2SC356Y		3039	86	2051	
2SC35A		3835	59		
2SC35B		3835	59		
2SC35C		3835	59		
2SC35E		3835	59		
2SC35F		3835	59		
2SC35G		3835	59		
2SC35GN		3835	59		
2SC35H		3835	59		
2SC35J		3835	59		
2SC35K		3835	59		
2SC35L		3835	59		
2SC35M		3835	59		
2SC35OR		3835	59		
2SC35R		3835	59		
2SC35X		3835	59		
2SC35Y		3835	59		
2SC36	103A	3010	59	2002	
2SC360	123A	3444	86	2051	
2SC360-OR			86	2051	
2SC360A		3039	86	2051	
2SC360B	123A	3039/316	86	2051	
2SC360C		3039	86	2051	
2SC360D	123A	3444	86	2051	
2SC360E		3039	86	2051	
2SC360F		3039	86	2051	
2SC360G		3039	86	2051	
2SC360GN		3039	86	2051	
2SC360H		3039	86	2051	
2SC360J		3039	86	2051	
2SC360K		3039	86	2051	
2SC360L		3039	86	2051	
2SC360M		3039	86	2051	
2SC360OR		3039	86	2051	
2SC360R		3039	86	2051	
2SC360X		3039	86	2051	
2SC360Y		3039	86	2051	
2SC361	289A	3018	61	2015	
2SC361A		3018	61	2038	
2SC361B		3018	61	2038	
2SC361C		3018	61	2038	
2SC361D		3018	61	2038	
2SC361E		3018	61	2038	
2SC361F		3018	61	2038	
2SC361G		3018	61	2038	
2SC361GN		3018	61	2038	
2SC361H		3018	61	2038	
2SC361J		3018	61	2038	
2SC361K		3018	61	2038	
2SC361L		3018	61	2038	
2SC361M		3018	61	2038	
2SC361OR		3018	61	2038	
2SC361R		3018	61	2038	
2SC361X		3018	61	2038	
2SC361Y		3018	61	2038	
2SC362	289A	3124/289	47	2015	
2SC362A		3124	47	2038	
2SC362B		3124	47	2038	
2SC362C		3124	47	2038	
2SC362D		3124	47	2038	
2SC362E		3124	47	2038	
2SC362F		3124	47	2038	
2SC362G		3124	47	2038	
2SC362GN		3124	47	2038	
2SC362H		3124	47	2038	
2SC362J		3124	47	2038	
2SC362K		3124	47	2038	
2SC362L		3124	47	2038	
2SC362M		3124	47	2038	
2SC362OR		3124	47	2038	
2SC362R		3124	47	2038	
2SC362X		3124	47	2038	
2SC362Y		3124	47	2038	
2SC363	85	3444/123A	86	2051	

Industry Standard No.	ECG	SK	GE	RS 276-	MOTOR.
2SC363-OR			86	2051	
2SC363A		3039	86	2051	
2SC363B		3039	86	2051	
2SC363C		3039	86	2051	
2SC363D		3039	86	2051	
2SC363E		3039	86	2051	
2SC363F		3039	86	2051	
2SC363G		3039	86	2051	
2SC363GN		3039	86	2051	
2SC363H		3039	86	2051	
2SC363J		3039	86	2051	
2SC363K		3039	86	2051	
2SC363L		3039	86	2051	
2SC363M		3039	86	2051	
2SC363OR		3039	86	2051	
2SC363R		3039	86	2051	
2SC363X		3039	86	2051	
2SC363Y		3039	86	2051	
2SC366	85	3444/123A	88	2051	
2SC366-ORG-G		3244	88		
2SC366-RED-G		3244	88		
2SC366A		3122	88		
2SC366B		3122	88		
2SC366C		3122	88		
2SC366D		3122	88		
2SC366E		3122	88		
2SC366F		3122	88		
2SC366G	85	3122	88		
2SC366GN		3122	88		
2SC366H		3122	88		
2SC366J		3122	88		
2SC366K		3122	88		
2SC366L		3122	88		
2SC366M		3122	88		
2SC366OR		3122	88		
2SC366R	85	3122	88		
2SC366X		3122	88		
2SC366Y	85	3122	88		
2SC367	85	3444/123A	47	2051	
2SC367-ORG-G		3244	47		
2SC367-RED-G		3244	47		
2SC367-YEL-G		3244	47		
2SC367A		3124	47		
2SC367B		3124	47		
2SC367C		3124	47		
2SC367D		3124	47		
2SC367E		3124	47		
2SC367F		3124	47		
2SC367G	85	3124/289	47		
2SC367GN		3124	47		
2SC367H		3124	47		
2SC367J		3124	47		
2SC367K		3124	47		
2SC367L		3124	47		
2SC367M		3124	47		
2SC367OR		3124	47		
2SC367R	85	3124/289	47		
2SC367X		3124	47		
2SC367Y	85	3124/289	47		
2SC368	199	3122	88	2010	
2SC368-BL			212	2010	
2SC368-GR			212	2010	
2SC368A		3122	88	2010	
2SC368B		3122	88	2010	
2SC368BL	199		88	2010	
2SC368C		3122	88	2010	
2SC368D		3122	88	2010	
2SC368E		3122	88	2010	
2SC368F		3122	88	2010	
2SC368G		3122	88	2010	
2SC368GN		3122	88	2010	
2SC368GR	199		88	2010	
2SC368H		3122	88	2010	
2SC368J		3122	88	2010	
2SC368K		3122	88	2010	
2SC368L		3122	88	2010	
2SC368M		3122	88	2010	
2SC368OR		3122	88	2010	
2SC368R		3122	88	2010	
2SC368V	199		88	2010	
2SC368X		3122	88	2010	
2SC368Y		3122	88	2010	
2SC369	199	3250/315	47	2051	
2SC369-BL			62	2010	
2SC369-BLU-G		3250	47	2010	
2SC369-GR			62	2010	
2SC369-GRN-G		3250	47	2010	
2SC369A		3250	47	2010	
2SC369B		3250	47	2010	
2SC369BL	199	3250/315	47	2051	
2SC369C		3250	47	2010	
2SC369D		3250	47	2010	
2SC369E		3250	47	2010	
2SC369F		3250	47	2010	
2SC369G	199	3250/315	47	2051	
2SC369G-BL	199		20	2051	
2SC369G-GR	199		20	2051	
2SC369G-V	199		47	2010	
2SC369G/BL		3250	47	2010	
2SC369G/GR		3250	47	2010	
2SC369GBL	199		47	2051	
2SC369GGR	199		47	2051	
2SC369GN			47	2010	
2SC369GR	199	3250/315	47	2051	
2SC369H		3250	47	2010	
2SC369J		3250	47	2010	
2SC369K		3250	47	2010	
2SC369L		3250	47	2010	
2SC369M		3250	47	2010	
2SC369OR		3250	47	2010	
2SC369R		3250	47	2010	
2SC369V	199	3250/315	20	2051	
2SC369X			47	2010	
2SC369Y		3250	47	2010	
2SC36B		3010	59		
2SC36C		3010	59		
2SC36D		3010	59		
2SC36E		3010	59		
2SC36F		3010	59		
2SC36G		3010	59		
2SC36GN		3010	59		
2SC36H		3010	59		
2SC36J		3010	59		
2SC36K		3010	59		
2SC36L		3010	59		
2SC36M		3010	59		
2SC36OR		3010	59		
2SC36R		3010	59		
2SC36X		3010	59		
2SC36Y		3010	59		
2SC37	123A	3444	86	2051	
2SC370	289A	3452/108	61	2038	
2SC370-0		3018	61	2038	
2SC370-G		3245	61*	2038	

Industry Standard No.	ECG	SK	GE	RS 276-	MOTOR.
2SC370-0			61	2038	
2SC370-T		3245	61	2038	
2SC370A		3018	61	2038	
2SC370B		3018	61	2038	
2SC370C		3018	61	2038	
2SC370D		3018	61	2038	
2SC370E		3018	61	2038	
2SC370F	289A	3452/108	61	2038	
2SC370G	289A	3452/108	61	2038	
2SC370GN		3018	61	2038	
2SC370H	289A	3452/108	61	2038	
2SC370J	289A	3452/108	61	2038	
2SC370K	289A	3452/108	61	2038	
2SC370L		3018	61	2038	
2SC370M		3018	61	2038	
2SC370OR		3018	61	2038	
2SC370R		3018	61	2038	
2SC370X		3018	61	2038	
2SC370Y		3018	61	2038	
2SC371	289A	3122	61	2051	
2SC371(O)	289A	3122	61	2051	
2SC371(O)			61	2051	
2SC371-O	289A	3122	61	2051	
2SC371-0	289A	3122	61	2051	
2SC371-OR			61	2051	
2SC371-ORG-G		3122	61	2051	
2SC371-R	289A	3122	61	2051	
2SC371-R-1	289A	3122	61	2051	
2SC371-RED-G		3122	61	2051	
2SC371-T		3122	61	2051	
2SC3710		3018	20	2051	
2SC371A		3122	61	2051	
2SC371B	289A	3122	61	2051	
2SC371C		3122	61	2051	
2SC371D		3122	61	2051	
2SC371E		3122	61	2051	
2SC371F		3122	61	2051	
2SC371G	289A	3122	61	2051	
2SC371G-O			61	2051	
2SC371G-R			61	2051	
2SC371GN		3122	61	2051	
2SC371H		3122	61	2051	
2SC371J		3122	61	2051	
2SC371K		3122	61	2051	
2SC371L		3122	61	2051	
2SC371M		3122	61	2051	
2SC3710	289A	3122	61	2051	
2SC371OR		3122	61	2051	
2SC371R	289A	3122	61	2051	
2SC371R-1			61	2051	
2SC371RED-G			61	2051	
2SC371T		3122	61	2051	
2SC371X		3122	61	2051	
2SC371Y		3122	61	2051	
2SC372	85	3245/199	61	2009	
2SC372(3RD-IF)			61	2009	
2SC372(H)			61	2009	
2SC372(O)	85	3245/199	61	2009	
2SC372(Y)			61	2009	
2SC372-O		3245	61	2009	
2SC372-1	85	3245/199	61	2009	
2SC372-2	85	3245/199	61	2009	
2SC372-0	85	3245/199	61	2009	
2SC372-OR			61	2009	
2SC372-ORG			61	2009	
2SC372-ORG-G			61	2009	
2SC372-R	85	3245/199	61	2009	
2SC372-WORR				2009	
2SC372-Y	85	3245/199	61	2009	
2SC372-YEL-G			61	2009	
2SC372-Z	85	3245/199	61	2009	
2SC372/4454C	85	3245/199	61	2009	
2SC3720		3122	61	2038	
2SC3720A		3018	61	2038	
2SC3720B		3018	61	2038	
2SC3720C		3018	61	2038	
2SC3720D		3018	61	2038	
2SC3720F		3018	61	2038	
2SC3720G		3018	61	2038	
2SC3720GN		3018	61	2038	
2SC3720H		3018	61	2038	
2SC3720J		3018	61	2038	
2SC3720K		3018	61	2038	
2SC3720L		3018	61	2038	
2SC3720OR			61	2038	
2SC3720R			61	2038	
2SC3720X		3018	61	2038	
2SC3720Y		3018	61	2038	
2SC3721		3018	62	2010	
2SC3721GR		3245	62	2010	
2SC3724		3444	61	2051	
2SC372A		3245	61	2009	
2SC372AR			61	2009	
2SC372B		3245	61	2009	
2SC372BL	85	3245/199	61	2009	
2SC372C		3245	61	2009	
2SC372D		3245	61	2009	
2SC372E		3245	61	2009	
2SC372F		3245	61	2009	
2SC372G	85	3245/199	61	2009	
2SC372G-O			61	2009	
2SC372G-Y			61	2009	
2SC372GN		3245	61	2009	
2SC372GR	85	3245/199	61	2009	
2SC372H	85	3245/199	61	2009	
2SC372J		3245	61	2009	
2SC372K		3245	61	2009	
2SC372L		3245	61	2009	
2SC372M		3245	61	2009	
2SC3720		3245	61	2051	
2SC372OR		3245	61	2009	
2SC372R		3245	61	2009	
2SC372X	85	3245/199	61	2009	
2SC372Y	85	3245/199	61	2009	
2SC372Y(3RD IF)	233			2009	
2SC372Y1		3245	61		
2SC372YEL			61	2009	
2SC372YEL-G			61	2009	
2SC372Z			61	2009	
2SC373	85	3245/199	20	2051	
2SC373(GR)			212	2010	
2SC373-14		3245	212	2010	
2SC373-G			212	2010	
2SC373-0			212	2010	
2SC373-OR			212	2010	
2SC373A		3245	212	2010	
2SC373AL		3245	212	2010	
2SC373B		3245	212	2010	
2SC373BL	85	3245/199	212	2010	
2SC373C		3245	212	2010	
2SC373D		3245	212	2010	
2SC373E		3245	212	2010	
2SC373F		3245	212	2010	
2SC373G	85	3245/199	212	2010	

Industry Standard No.	ECG	SK	GE	RS 276-	MOTOR.
2SC373GN		3245	212	2010	
2SC373GR	85	3245/199	212	2010	
2SC373H		3245	212	2010	
2SC373J		3245	212	2010	
2SC373K		3245	212	2010	
2SC373L		3245	212	2010	
2SC373M		3245	212	2010	
2SC373OR		3245	212	2010	
2SC373R		3245	212	2010	
2SC373W	85	3245/199	62	2010	
2SC373X		3245	212	2010	
2SC373Y		3245	212	2010	
2SC374	199	3444/123A	212	2010	
2SC374(BL)			212	2010	
2SC374(V)			212	2010	
2SC374-BL	199		212	2010	
2SC374-OR			212	2010	
2SC374-V	199		212	2010	
2SC374A		3444	212	2010	
2SC374B		3444	212	2010	
2SC374BL		3444	212		
2SC374BLK		3444	212	2010	
2SC374C		3444	212	2010	
2SC374D		3444	212	2010	
2SC374E		3444	212	2010	
2SC374F		3444	212	2010	
2SC374G		3444	212	2010	
2SC374GN		3444	212	2010	
2SC374H		3444	212	2010	
2SC374J		3444	212	2010	
2SC374JA	199		212	2010	
2SC374K		3444	212	2010	
2SC374L		3444	212	2010	
2SC374M		3444	212	2010	
2SC374OR		3444	212	2010	
2SC374R		3444	212	2010	
2SC374V		3444	212	2010	
2SC374X		3444	212	2010	
2SC374Y		3444	212	2010	
2SC375	107	3132	86		
2SC375-0	107		86		
2SC375-0		3132	86	2015	
2SC375-Y	107	3132	86		
2SC375A		3132	86	2038	
2SC375B		3132	86	2038	
2SC375C		3132	86	2038	
2SC375D		3132	86	2038	
2SC375E		3132	86	2038	
2SC375F		3132	86	2038	
2SC375G		3132	86	2038	
2SC375GN		3132	86	2038	
2SC375H		3132	86	2038	
2SC375J		3132	86	2038	
2SC375K		3132	86	2038	
2SC375L		3132	86	2038	
2SC375M		3132	86	2038	
2SC375OR		3132	86	2038	
2SC375R		3132	86	2038	
2SC375X		3132	86	2038	
2SC376	85	3024/128	18	2030	
2SC376A		3024	18	2010	
2SC376B		3024	18	2010	
2SC376C		3024	18	2010	
2SC376D		3024	18	2010	
2SC376E		3024	18	2010	
2SC376F		3024	18	2010	
2SC376G		3024	18	2010	
2SC376GN		3024	18	2010	
2SC376H		3024	18	2010	
2SC376J		3024	18	2010	
2SC376K		3024	18	2010	
2SC376L		3024	18	2010	
2SC376M		3024	18	2010	
2SC376OR		3024	18	2010	
2SC376R		3024	18	2010	
2SC376X		3024	18	2010	
2SC376Y		3024	18	2010	
2SC377	85	3854/123AP	17	2051	
2SC377-BN			17	2051	
2SC377-BRN			17	2051	
2SC377-0			17	2051	
2SC377-OR			17	2051	
2SC377-ORG			17	2051	
2SC377-R			17	2051	
2SC377-RED			17	2051	
2SC377A	85	3854/123AP	17	2051	
2SC377B		3854	17	2051	
2SC377BN			17	2051	
2SC377BRN			17	2051	
2SC377C		3854	17	2051	
2SC377D		3854	17	2051	
2SC377E		3854	17	2051	
2SC377F		3854	17	2051	
2SC377G		3854	17	2051	
2SC377GN		3854	17	2051	
2SC377H		3854	17	2051	
2SC377J		3854	17	2051	
2SC377K		3854	17	2051	
2SC377L		3854	17	2051	
2SC377M		3854	17	2051	
2SC3770			17	2051	
2SC377OR		3854	17	2051	
2SC377R		3854	17	2051	
2SC377RED			17	2051	
2SC377X		3854	17	2051	
2SC377Y		3854	17	2051	
2SC378	289A	3444/123A	61	2051	
2SC378-0	289A		212	2010	
2SC378-ORG			61	2010	
2SC378-R	289A		212	2010	
2SC378-RED			61	2010	
2SC378-Y	289A		212	2010	
2SC378-YEL		3246	61	2010	
2SC378A		3018	61	2010	
2SC378B		3018	61	2010	
2SC378C		3018	61	2010	
2SC378D		3018	61	2010	
2SC378E		3018	61	2010	
2SC378F		3018	61	2010	
2SC378G		3018	61	2010	
2SC378GN		3018	61	2010	
2SC378H		3018	61	2010	
2SC378J		3018	61	2010	
2SC378K		3018	61	2010	
2SC378L		3018	61	2010	
2SC378M		3018	61	2010	
2SC378OR		3018	61	2010	
2SC378R		3018	61	2010	
2SC378X		3018	61	2010	
2SC378Y		3018	61	2010	
2SC379	85	3444/123A	88	2051	
2SC379A		3122	88	2010	
2SC379B		3122	88	2010	
2SC379C		3122	88	2010	
2SC379D		3122	88	2010	
2SC379E		3122	88	2010	
2SC379F		3122	88	2010	
2SC379G		3122	88	2010	
2SC379GN		3122	88	2010	
2SC379H		3122	88	2010	
2SC379J		3122	88	2010	
2SC379K		3122	88	2010	
2SC379L		3122	88	2010	
2SC379M		3122	88	2010	
2SC379OR		3122	88	2010	
2SC379R		3122	88	2010	
2SC379X		3122	88	2010	
2SC379Y		3122	88	2010	
2SC37A		3039	86	2038	
2SC37B		3039	86	2038	
2SC37C		3039	86	2038	
2SC37D		3039	86	2038	
2SC37E		3039	86	2038	
2SC37F		3039	86	2038	
2SC37G		3039	86	2038	
2SC37GN		3039	86	2038	
2SC37H		3039	86	2038	
2SC37J		3039	86	2038	
2SC37K		3039	86	2038	
2SC37L		3039	86	2038	
2SC37M		3039	86	2038	
2SC37OR		3039	86	2038	
2SC37R		3039	86	2038	
2SC37X		3039	86	2038	
2SC37Y		3039	86	2038	
2SC38	123A	3444	45	2051	
2SC380	85	3245/199	61	2015	
2SC380(R)			61	2015	
2SC380-0	85		61	2015	
2SC380-0/4454C			61	2015	
2SC380-BRN			61	2015	
2SC380-0	85	3245/199	61	2015	
2SC380-0/4454C			61	2015	
2SC380-OR			61	2015	
2SC380-ORG			61	2015	
2SC380-R		3245	61	2015	
2SC380-RED			61	2015	
2SC380-Y	85		61	2015	
2SC380-YEL			61	2015	
2SC380/4454C			61	2015	
2SC3800		3018	61	2038	
2SC3800A		3018	61	2038	
2SC3800B		3018	61	2038	
2SC3800C			61	2038	
2SC3800D		3018	61	2038	
2SC3800E		3018	61	2038	
2SC3800F		3018	61	2038	
2SC3800G		3018	61	2038	
2SC3800GN		3018	61	2038	
2SC3800H		3018	61	2038	
2SC3800J		3018	61	2038	
2SC3800K		3018	61	2038	
2SC3800L		3018	61	2038	
2SC3800M		3018	61	2038	
2SC38000R			61	2038	
2SC3800R			61	2038	
2SC3800X		3018	61	2038	
2SC3800Y		3018	61	2038	
2SC380A	85	3245/199	61	2015	
2SC380A(D)			61	2015	
2SC380A(O)	85		61	2015	
2SC380A(R)			61	2013	
2SC380A(Y)			61	2015	
2SC380A+0)			61	2015	
2SC380A-0		3245	61	2015	
2SC380A-0(TV)	85		61	2015	
2SC380A-0	85		61	2015	
2SC380A-0(TV)	85		61	2015	
2SC380A-R	85	3245/199	61	2015	
2SC380A-R(TV)	85		61	2015	
2SC380A-Y			61	2015	
2SC380AO	85		61	2015	
2SC380AR		3245	61	2015	
2SC380ATV	85	3245/199	61	2015	
2SC380AY	85	3245/199	61	2015	
2SC380B		3245	61	2015	
2SC380B-Y		3245	61	2015	
2SC380BY			61	2015	
2SC380C		3245	61	2015	
2SC380C-Y		3245	61	2015	
2SC380CY			61	2015	
2SC380D	85	3245/199	61	2015	
2SC380D-Y		3245	61	2015	
2SC380DY			61	2015	
2SC380E		3245	61	2015	
2SC380E-Y		3245	61	2015	
2SC380EY			61	2015	
2SC380F		3245	61	2015	
2SC380F-Y		3245	61	2015	
2SC380FY			61	2015	
2SC380G		3245	61	2015	
2SC380GN		3245	61	2015	
2SC380H		3245	61	2015	
2SC380J		3245	61	2015	
2SC380K		3245	61	2015	
2SC380L		3245	61	2015	
2SC380M		3245	61	2015	
2SC3800		3245	61	2015	
2SC3800R		3245	61	2015	
2SC380R	85	3245/199	61	2015	
2SC380R/4454C			61	2015	
2SC380RED			61	2015	
2SC380X		3245	61	2015	
2SC380Y	85	3245/199	61	2015	
2SC380YEL			61	2015	
2SC381	107	3122	61	2015	
2SC381(BN)			61	2015	
2SC381-0	107	3122	61	2015	
2SC381-BN			61	2015	
2SC381-BRN			61	2015	
2SC381-0	107	3122	61	2015	
2SC381-OR			61	2015	
2SC381-ORG			61	2015	
2SC381-R	107		61	2015	
2SC381-RED			61	2015	
2SC381A		3122	61	2015	
2SC381B		3122	61	2015	
2SC381BN	107	3122	61	2015	
2SC381BN-1		3122	61	2015	
2SC381BRN			61	2015	
2SC381C		3122	61	2015	
2SC381D		3122	61	2015	
2SC381E		3122	61	2015	
2SC381F		3122	61	2015	
2SC381G		3122	61	2015	
2SC381GN		3122	61	2015	
2SC381H		3122	61	2015	
2SC381J		3122	61	2015	

Industry Standard No.	ECG	SK	GE	RS 276-	MOTOR.
2SC381K		3122	61	2015	
2SC381L		3122	61	2015	
2SC381M		3122	61	2015	
2SC381O			61	2015	
2SC381OR		3122	61	2015	
2SC381R	107	3122	61	20156	
2SC381RED			61	2015	
2SC381RL		3122	61	2015	
2SC381X		3122	61	2015	
2SC381Y		3122	61	2015	
2SC382		3132	86	2016	
2SC382(BL)			86	2016	
2SC382(BN)			86	2016	
2SC382(R)			86	2016	
2SC382-BK			86	2016	
2SC382-BK(1)			86	2016	
2SC382-BK(2)			86	2016	
2SC382-G			86	2016	
2SC382-GR			86	2016	
2SC382-GY			86	2016	
2SC382-OR			86	2016	
2SC382-R			86	2016	
2SC382-V		3132	86	2016	
2SC382A		3132	86	2016	
2SC382B		3132	86	2016	
2SC382BK		3132	86	2016	
2SC382BK1			86	2016	
2SC382BK2			86	2016	
2SC382BL		3132	86	2016	
2SC382BN		3132	86	2016	
2SC382BR		3132	86	2016	
2SC382C		3132	86	2016	
2SC382D		3132	86	2016	
2SC382E		3132	86	2016	
2SC382F		3132	86	2016	
2SC382G		3132	86	2016	
2SC382GN		3132	86	2016	
2SC382GR		3132	86	2016	
2SC382GY			86	2016	
2SC382H		3132	86	2016	
2SC382J		3132	86	2016	
2SC382K		3132	86	2016	
2SC382L		3132	86	2016	
2SC382M		3132	86	2016	
2SC382OR		3132	86	2016	
2SC382R		3132	86	2016	
2SC382V		3132	86	2016	
2SC382W		3132	86	2016	
2SC382W,R			86	2016	
2SC382X		3132	86	2016	
2SC382Y		3132	86	2016	
2SC383	85	3124/289	213	2010	
2SC383(3RD IF)				2009	
2SC383(3RD-IF)			213	2010	
2SC383(FINAL IF)				2009	
2SC383(FINAL-IF)			213	2010	
2SC383(T)			213	2010	
2SC383-OR			213	2010	
2SC383A		3124	213	2010	
2SC383B		3124	213	2010	
2SC383C		3124	213	2010	
2SC383D		3124	213	2010	
2SC383E		3124	213	2010	
2SC383F	85	3124/289	213	2010	
2SC383G	85	3124/289	213	2016	
2SC383GN		3124	61	2010	
2SC383H		3124	213	2010	
2SC383J		3124	213	2010	
2SC383K		3124	61	2010	
2SC383L		3124	213	2010	
2SC383M		3124	213	2010	
2SC383OR		3124	213	2010	
2SC383R		3124	213	2010	
2SC383T	85		213	2010	
2SC383T(LAST IF)				2009	
2SC383T(LAST-IF)			213	2010	
2SC383W	85	3124/289	213	2010	
2SC383X		3124	213	2010	
2SC383Y	85	3124/289	213	2010	
2SC384	107	3132	61	2038	
2SC384(O)			61	2038	
2SC384(Y)			61	2038	
2SC384-O		3132	61	2038	
2SC384-O	107		61	2038	
2SC384A		3132	61	2038	
2SC384B		3132	61	2038	
2SC384C		3132	61	2038	
2SC384D		3132	61	2038	
2SC384E		3132	61	2038	
2SC384F		3132	61	2038	
2SC384G		3132	61	2038	
2SC384GN		3132	61	2038	
2SC384H		3132	61	2038	
2SC384J		3132	61	2038	
2SC384K		3132	61	2038	
2SC384L		3132	61	2038	
2SC384M		3132	61	2038	
2SC384OR		3132	61	2038	
2SC384R		3132	61	2038	
2SC384X		3132	61	2038	
2SC384Y	107	3132	61	2038	
2SC385	107	3293	86	2015	
2SC3854			11	2015	
2SC385A	107	3293	86	2015	
2SC385B		3293	86	2038	
2SC385C		3293	86	2038	
2SC385D		3293	86	2038	
2SC385E		3293	86	2038	
2SC385F		3293	86	2038	
2SC385G		3293	86	2038	
2SC385GN		3293	86	2038	
2SC385H		3293	86	2038	
2SC385J		3293	86	2038	
2SC385K		3293	86	2038	
2SC385L		3293	86	2038	
2SC385M		3293	86	2038	
2SC385OR		3293	86	2038	
2SC385R		3293	86	2038	
2SC385X			86	2038	
2SC385Y		3293	86	2038	
2SC386	107	3452/108	86	2038	
2SC386-O			86	2038	
2SC386A		3039	86	2038	
2SC386A-O(TV)	107	3452/108	86	2038	
2SC386AO			86	2038	
2SC386B		3039	86	2038	
2SC386C		3039	86	2038	
2SC386D		3039	86	2038	
2SC386E		3039	86	2038	
2SC386F		3039	86	2038	
2SC386G		3039	86	2038	
2SC386GN		3039	86	2038	
2SC386H		3039	86	2038	
2SC386J		3039	86	2038	
2SC386K		3039	86	2038	
2SC386L		3039	86	2038	
2SC386M		3039	86	2038	
2SC386OR		3039	86	2038	
2SC386R		3039	86	2038	
2SC386X		3039	86	2038	
2SC386Y		3039	86	2038	
2SC387	107	3293	214		
2SC387A	107	3293	214		
2SC387A(FA-3)	107		214		
2SC387AG	107		214		
2SC387B		3293	214		
2SC387C		3293	214		
2SC387D		3293	214		
2SC387E		3293	214		
2SC387F		3293	214		
2SC387G	107	3293	214		
2SC387GN		3293	214		
2SC387H		3293	214		
2SC387J		3293	214		
2SC387K		3293	214		
2SC387L		3293	214		
2SC387M		3293	214		
2SC387OR		3293	214		
2SC387R		3293	214		
2SC387X		3293	214		
2SC387Y		3293	214		
2SC388	85	3132	61	2009	
2SC388-OR			61	2009	
2SC388A	85	3132	61	2009	
2SC388A(3RD IF)				2009	
2SC388A(3RD-IF)			61	2009	
2SC388ATV	85	3132	61	2009	
2SC388B		3132	61	2009	
2SC388C		3132	61	2009	
2SC388D		3132	61	2009	
2SC388E		3132	61	2009	
2SC388F		3132	61	2009	
2SC388G		3132	61	2009	
2SC388GN		3132	61	2009	
2SC388H		3132	61	2009	
2SC388J		3132	61	2009	
2SC388K		3132	61	2009	
2SC388L		3132	61	2009	
2SC388M		3132	61	2009	
2SC388OR		3132	61	2009	
2SC388R		3132	61	2009	
2SC388X		3132	61	2009	
2SC388Y		3132	61	2009	
2SC389	161	3019	211	2015	
2SC389-O	161	3019	211	2015	
2SC389-O			61	2038	
2SC389-OR			61	2038	
2SC389A		3019	211	2038	
2SC389B		3019	211	2038	
2SC389C		3019	211	2038	
2SC389D		3019	211	2038	
2SC389E		3019	211	2038	
2SC389F		3019	211	2038	
2SC389G		3019	211	2038	
2SC389GN		3019	211	2038	
2SC389H		3019	211	2038	
2SC389J		3019	211	2038	
2SC389K		3019	211	2038	
2SC389L		3019	211	2038	
2SC389M		3019	211	2038	
2SC389OR		3019	211		
2SC389R	161	3019	211	2015	
2SC389X			211	2038	
2SC389Y		3019	211	2038	
2SC38A		3047	45		
2SC38B		3047	45		
2SC38C		3047	45		
2SC38D		3047	45		
2SC38E		3047	45		
2SC38F		3047	45		
2SC38G		3047	45		
2SC38H		3047	45		
2SC38J		3047	45		
2SC38K		3047	45		
2SC38L		3047	45		
2SC38M		3047	45		
2SC38OR		3047	45		
2SC38R		3047	45		
2SC38X		3047	45		
2SC38Y		3047	45		
2SC39	108	3452	86	2038	
2SC390	316	3039	211	2015	
2SC390A		3039	211	2016	
2SC390B		3117	211	2016	
2SC390C		3117	211	2016	
2SC390D		3117	211	2016	
2SC390E		3117	211	2016	
2SC390F		3117	211	2016	
2SC390G		3117	211	2016	
2SC390GN		3117	211	2016	
2SC390H		3117	211	2016	
2SC390J		3117	211	2016	
2SC390K		3117	211	2016	
2SC390L		3117	211	2016	
2SC390M		3117	211	2016	
2SC390OR		3117	211	2016	
2SC390R		3117	211	2016	
2SC390X		3117	211	2016	
2SC390Y		3117	211	2016	
2SC391		3039	86	2016	
2SC391A		3039	86	2016	
2SC391B		3039	86	2016	
2SC391C		3039	86	2016	
2SC391D		3039	86	2016	
2SC391E		3039	86	2016	
2SC391F		3039	86	2016	
2SC391G		3039	86	2016	
2SC391GN		3039	86	2016	
2SC391H		3039	86	2016	
2SC391J		3039	86	2016	
2SC391K		3039	86	2016	
2SC391L		3039	86	2016	
2SC391M		3039	86	2016	
2SC391OR		3039	86	2016	
2SC391R		3039	86	2016	
2SC391X		3039	86	2016	
2SC391Y		3039	86	2016	
2SC392	316	3452/108	86	2038	
2SC392A		3039	86	2016	
2SC392B		3039	86	2016	
2SC392C		3039	86	2016	
2SC392D		3039	86	2016	
2SC392E		3039	86	2016	
2SC392F		3039	86	2016	
2SC392G		3039	86	2016	
2SC392GN		3039	86	2016	
2SC392H		3039	86	2016	
2SC392J		3039	86	2016	

Industry Standard No.	ECG	SK	GE	RS 276-	MOTOR.
2SC392K		3039	86	2016	
2SC392L		3039	86	2016	
2SC392M		3039	86	2016	
2SC392OR		3039	86	2016	
2SC392R		3039	86	2016	
2SC392X		3039	86	2016	
2SC392Y		3039	86	2016	
2SC394	289A	3124/289	61	2038	
2SC394(O)		3124	61	2038	
2SC394(0)		3124	61	2038	
2SC394-O	289A	3124/289	61	2038	
2SC394-GR			61	2038	
2SC394-GRN			61	2038	
2SC394-0	289A	3124/289	61	2038	
2SC394-OR			61	2038	
2SC394-ORG			61	2038	
2SC394-R			61	2038	
2SC394-RED			61	2038	
2SC394-Y		3124	61	2038	
2SC394-YEL			61	2038	
2SC3940		3018	61	2038	
2SC3940A		3018	61	2038	
2SC3940B		3018	61	2038	
2SC3940C		3018	61	2038	
2SC3940D		3018	61	2038	
2SC3940E		3018	61	2038	
2SC3940F		3018	61	2038	
2SC3940G		3018	61	2038	
2SC3940GN		3018	61	2038	
2SC3940H		3018	61	2038	
2SC3940J		3018	61	2038	
2SC3940K		3018	61	2038	
2SC3940L		3018	61	2038	
2SC3940M		3018	61	2038	
2SC39400R			61	2038	
2SC3940R		3018	61	2038	
2SC3940X		3018	61	2038	
2SC3940Y		3018	61	2038	
2SC394A		3124	61	2038	
2SC394AP		3124	61	2038	
2SC394B		3124	61	2038	
2SC394C		3124	61	2038	
2SC394D		3124	61	2038	
2SC394E		3124	61	2038	
2SC394F		3124	61	2038	
2SC394G		3124	61	2038	
2SC394GN		3124	61	2038	
2SC394GR	289A	3124/289	61	2038	
2SC394GRN			61	2038	
2SC394H		3124	61	2038	
2SC394J		3124	61	2038	
2SC394K		3124	61	2038	
2SC394L		3124	61	2038	
2SC394M		3124	61	2038	
2SC394O		3124	61	2038	
2SC394OR		3124	61	2038	
2SC394R	289A	3124/289	61	2038	
2SC394RED			61	2038	
2SC394W	289A	3124/289	61	2038	
2SC394X	289A	3124/289	61	2038	
2SC394Y	289A	3124/289	61	2038	
2SC394YEL			61	2038	
2SC395	123A	3444	88	2051	
2SC395A	123A	3444	88	2051	
2SC395A-O			210	2051	
2SC395A-ORG			88	2051	
2SC395A-R			210	2051	
2SC395A-RED			88	2051	
2SC395A-Y			210	2051	
2SC395A-YEL			88	2051	
2SC395B		3122	88	2051	
2SC395C		3122	88	2051	
2SC395D		3122	88	2051	
2SC395E		3122	88	2051	
2SC395F		3122	88	2051	
2SC395G		3122	88	2051	
2SC395GN		3122	88	2051	
2SC395H		3122	88	2051	
2SC395J		3122	88	2051	
2SC395K		3122	88	2051	
2SC395L		3122	88	2051	
2SC395M		3122	88	2051	
2SC395OR		3122	88	2051	
2SC395R	123A	3444	88	2051	
2SC395X		3122	88	2051	
2SC395Y		3122	88	2051	
2SC396	311	3444/123A	88	2051	
2SC396A		3122	88	2030	
2SC396B		3122	88	2030	
2SC396C		3122	88	2030	
2SC396D		3122	88	2030	
2SC396E		3122	88	2030	
2SC396F		3122	88	2030	
2SC396G		3122	88	2030	
2SC396GN		3122	88	2030	
2SC396GR			88	2030	
2SC396H		3122	88	2030	
2SC396J		3122	88	2030	
2SC396K		3122	88	2030	
2SC396L		3122	88	2030	
2SC396M		3122	88	2030	
2SC396OR		3122	88	2030	
2SC396R		3122	88	2030	
2SC396X		3122	88	2030	
2SC396Y		3122	88	2030	
2SC397	161	3452/108	86	2038	
2SC397A		3039	86	2038	
2SC397B		3039	86	2038	
2SC397C		3039	86	2038	
2SC397D		3039	86	2038	
2SC397E		3039	86	2038	
2SC397F		3039	86	2038	
2SC397G		3039	86	2038	
2SC397GN		3039	86	2038	
2SC397H		3039	86	2038	
2SC397J		3039	86	2038	
2SC397K		3039	86	2038	
2SC397L		3039	86	2038	
2SC397M		3039	86	2038	
2SC397OR		3039	86	2038	
2SC397R		3039	86	2038	
2SC397X		3039	86	2038	
2SC397Y		3039	86	2038	
2SC398	319P	3018	61	2015	
2SC398(PA-1)	319P	3018	61	2038	
2SC398A		3018	61	2038	
2SC398B		3018	61	2038	
2SC398C		3018	61	2038	
2SC398D		3018	61	2038	
2SC398E		3018	61	2038	
2SC398F		3018	61	2038	
2SC398FA1			61	2038	
2SC398G		3018	61	2038	
2SC398GN		3018	61	2038	
2SC398H		3018	61	2038	
2SC398J		3018	61	2038	
2SC398K		3018	61	2038	
2SC398L		3018	61	2038	
2SC398M		3018	61	2038	
2SC398OR		3018	61	2038	
2SC398R		3018	61	2038	
2SC398X		3018	61	2038	
2SC398Y		3018	61	2038	
2SC399	319P	3018	61	2015	
2SC399A		3018	61	2038	
2SC399B		3018	61	2038	
2SC399C		3018	61	2038	
2SC399D		3018	61	2038	
2SC399E		3018	61	2038	
2SC399F		3018	61	2038	
2SC399FA1			11	2038	
2SC399G		3018	61	2038	
2SC399GN		3018	61	2038	
2SC399H		3018	61	2038	
2SC399J		3018	61	2038	
2SC399K		3018	61	2038	
2SC399L		3018	61	2038	
2SC399M		3018	61	2038	
2SC399OR		3018	61	2038	
2SC399R		3018	61	2038	
2SC399X		3018	61	2038	
2SC399Y		3018	61	2038	
2SC39A	108	3452	86	2038	
2SC39B		3039	86	2038	
2SC39C		3039	86	2038	
2SC39D		3039	86	2038	
2SC39E		3039	86	2038	
2SC39F		3039	86	2038	
2SC39G		3039	86	2038	
2SC39GN		3039	86	2038	
2SC39H		3039	86	2038	
2SC39J		3039	86	2038	
2SC39K		3039	86	2038	
2SC39L		3039	86	2038	
2SC39M		3039	86	2038	
2SC39OR		3039	86	2038	
2SC39R		3039	86	2038	
2SC39X		3039	86	2038	
2SC39Y		3039	86	2038	
2SC3T2Y	123A	3444	20	2051	
2SC40	108	3452	86	2038	
2SC400	123A	3444	88	2051	
2SC400-O	123A		88	2051	
2SC400-GR	123A		88	2010	
2SC400-0		3444	212	2010	
2SC400-R	123A	3444	88	2051	
2SC400-Y	123A	3444	88	2051	
2SC400A		3122	88	2010	
2SC400B		3122	88	2010	
2SC400C		3122	88	2010	
2SC400D		3122	88	2010	
2SC400E		3122	88	2010	
2SC400F		3122	88	2010	
2SC400G		3122	88	2010	
2SC400GN		3122	88	2010	
2SC400H		3122	88	2010	
2SC400J		3122	88	2010	
2SC400K		3122	88	2010	
2SC400L		3122	88	2010	
2SC400M		3122	88	2010	
2SC4000R		3122	88	2010	
2SC400R		3122	88	2010	
2SC400X		3122	88	2010	
2SC400Y		3122	88	2010	
2SC401	289A	3124/289	61	2051	
2SC4012A		3124	47	2030	
2SC4012B		3124	47	2030	
2SC4012C		3124	47	2030	
2SC4012D		3124	47	2030	
2SC4012E		3124	47	2030	
2SC4012F		3124	47	2030	
2SC4012G		3124	47	2030	
2SC4012GN		3124	47	2030	
2SC4012H		3124	47	2030	
2SC4012J		3124	47	2030	
2SC4012L		3124	47	2030	
2SC4012M		3124	47	2030	
2SC4012X		3124	47	2030	
2SC4012Y		3124	47	2030	
2SC401A	289A	3124/289	61	2038	
2SC401B		3124	61	2038	
2SC401C		3124	61	2038	
2SC401D		3124	61	2038	
2SC401E		3124	61	2038	
2SC401F		3124	61	2038	
2SC401G		3124	61	2038	
2SC401GN		3124	61	2038	
2SC401H		3124	61	2038	
2SC401J		3124	61	2038	
2SC401K		3124	61	2038	
2SC401L		3124	61	2038	
2SC401M		3124	61	2038	
2SC4010R		3124	61	2038	
2SC401R		3124	61	2038	
2SC401X			61	2038	
2SC401Y		3124	61	2038	
2SC402	289A	3444/123A	47	2051	
2SC402A	289A	3444/123A	47	2051	
2SC402B		3124	47	2010	
2SC402C		3124	47	2010	
2SC402D		3124	47	2010	
2SC402E		3124	47	2010	
2SC402F		3124	47	2010	
2SC402G		3124	47	2010	
2SC402GN		3124	47	2010	
2SC402H		3124	47	2010	
2SC402J		3124	47	2010	
2SC402K		3124	47	2010	
2SC402L		3124	47	2010	
2SC402M		3124	47	2010	
2SC4020R		3124	47	2010	
2SC402R		3124	47	2010	
2SC402X		3124	47	2010	
2SC402Y		3124	47	2010	
2SC403	289A	3124/289	62	2038	
2SC403(C)	289A	3124/289	62	2038	
2SC403(SONY)			62	2038	
2SC403,A,B				2038	
2SC403-OR			62	2038	
2SC4033A		3018	61	2038	
2SC4033B		3018	61	2038	
2SC4033C		3018	61	2038	
2SC4033D		3018	61	2038	
2SC4033E		3018	61	2038	
2SC4033F		3018	61	2038	
2SC4033G		3018	61	2038	
2SC4033GN		3018	61	2038	
2SC4033H		3018	61	2038	

Industry Standard No.	ECG	SK	GE	RS 276-	MOTOR.
2SC4033J		3018	61	2038	
2SC4033K		3018	61	2038	
2SC4033L		3018	61	2038	
2SC4033M		3018	61	2038	
2SC4033OR		3018	61	2038	
2SC4033R		3018	61	2038	
2SC4033X		3018	61	2038	
2SC4033Y		3018	61	2038	
2SC403A	289A	3124/289	62	2038	
2SC403AL		3124	62	2038	
2SC403B		3124	62	2038	
2SC403B(SONY)			62	2038	
2SC403C	289A	3124/289	62	2038	
2SC403C(SONY)			62	2038	
2SC403C,A				2038	
2SC403C,B,A				2038	
2SC403CG		3124	62	2038	
2SC403D		3124	62	2038	
2SC403E		3124	62	2038	
2SC403F		3124	62	2038	
2SC403G		3124	62	2038	
2SC403GN		3124	62	2038	
2SC403H		3124	62	2038	
2SC403J		3124	62	2038	
2SC403K		3124	62	2038	
2SC403L		3124	62	2038	
2SC403M		3124	62	2038	
2SC403OR		3124	62	2038	
2SC403R		3124	62	2038	
2SC403X		3124	62	2038	
2SC403Y		3124	62	2038	
2SC404	289A	3444/123A	61	2051	
2SC404A		3018	61	2010	
2SC404B		3018	61	2010	
2SC404C		3018	61	2010	
2SC404D		3018	61	2010	
2SC404E		3018	61	2010	
2SC404F		3018	61	2010	
2SC404G		3018	61	2010	
2SC404GN		3018	61	2010	
2SC404H		3018	61	2010	
2SC404J		3018	61	2010	
2SC404K		3018	61	2010	
2SC404L		3018	61	2010	
2SC404M		3018	61	2010	
2SC404OR		3018	61	2010	
2SC404R		3018	61	2010	
2SC404X		3018	61	2010	
2SC404Y		3018	61	2010	
2SC405	123A	3452/108	86	2038	
2SC405A		3039	86	2051	
2SC405B		3039	86	2051	
2SC405C		3039	86	2051	
2SC405D		3039	86	2051	
2SC405E		3039	86	2051	
2SC405F		3039	86	2051	
2SC405G		3039	86	2051	
2SC405GN		3039	86	2051	
2SC405H		3039	86	2051	
2SC405J		3039	86	2051	
2SC405K		3039	86	2051	
2SC405L		3039	86	2051	
2SC405M		3039	86	2051	
2SC405OR		3039	86	2051	
2SC405R		3039	86	2051	
2SC405X		3039	86	2051	
2SC405Y		3039	86	2051	
2SC406	123A	3011A	86	2038	
2SC406A		3011A	86	2051	
2SC406B		3011A	86	2051	
2SC406C		3011A	86	2051	
2SC406D		3011A	86	2051	
2SC406E		3011A	86	2051	
2SC406F		3011A	86	2051	
2SC406G		3011A	86	2051	
2SC406GN		3011A	86	2051	
2SC406H		3011A	86	2051	
2SC406J		3011A	86	2051	
2SC406K		3011A	86	2051	
2SC406L		3011A	86	2051	
2SC406M		3011A	86	2051	
2SC406OR		3011A	86	2051	
2SC406R		3011A	86	2051	
2SC406X		3011A	86	2051	
2SC406Y		3011A	86	2051	
2SC407	385	3438	35		MJ15011
2SC408	385	3439/163A	36		MJ15011
2SC409	385	3439/163A	36		2N6249
2SC40A		3039	86	2038	
2SC40B		3039	86	2038	
2SC40C		3039	86	2038	
2SC40D		3039	86	2038	
2SC40E		3039	86	2038	
2SC40F		3039	86	2038	
2SC40G		3039	86	2038	
2SC40GN		3039	86	2038	
2SC40H		3039	86	2038	
2SC40J		3039	86	2038	
2SC40K		3039	86	2038	
2SC40L		3039	86	2038	
2SC40OR			86	2038	
2SC40R		3039	86	2038	
2SC40X		3039	86	2038	
2SC40Y		3039	86	2038	
2SC41	163A	3439	59		MJ410
2SC410	385	3439/163A	36		2N6249
2SC411	385	3439/163A	36		2N6546
2SC4116		3026	66	2048	
2SC4116A		3026	66	2048	
2SC4116B		3026	66	2048	
2SC4116C		3026	66	2048	
2SC4116D		3026	66	2048	
2SC4116E		3026	66	2048	
2SC4116F		3026	66	2048	
2SC4116G		3026	66	2048	
2SC4116GN		3026	66	2048	
2SC4116H		3026	66	2048	
2SC4116J		3026	66	2048	
2SC4116K		3026	66	2048	
2SC4116L		3026	66	2048	
2SC4116M		3026	66	2048	
2SC4116OR		3026	66	2048	
2SC4116R		3026	66	2048	
2SC4116X		3026	66	2048	
2SC4116Y		3026	66	2048	
2SC412	385	3439/163A	36		2N6546
2SC41B		3439	59		
2SC41C		3439	59		
2SC41F		3439	59		
2SC41GN		3439	59		
2SC41H		3439	59		
2SC41J		3439	59		
2SC41M		3439	59		
2SC41OR		3439	59		

Industry Standard No.	ECG	SK	GE	RS 276-	MOTOR.
2SC41R		3439	59		
2SC41TV	163A	3439	59		
2SC41X		3439	59		
2SC41Y		3439	59		
2SC42	163A	3079	35		MJ410
2SC423	123A	3444	20	2051	
2SC423-O			20	2051	
2SC423A		3024	20	2051	
2SC423B	123A	3444	20	2051	
2SC423C	123A	3444	20	2051	
2SC423D	123A	3444	20	2051	
2SC423E	123A	3444	20	2051	
2SC423F	123A	3444	20	2051	
2SC423G		3024	18	2051	
2SC423GN		3024	20	2051	
2SC423H		3024	20	2051	
2SC423J		3024	20	2051	
2SC423K		3024	20	2051	
2SC423L		3024	20	2051	
2SC423M		3024	20	2051	
2SC423OR		3024	20	2051	
2SC423R		3024	20	2051	
2SC423X		3024	20	2051	
2SC423Y		3024	20	2051	
2SC424	123A	3444	86	2051	
2SC424D	123A	3444	86	2051	
2SC425	123A	3444	18	2051	
2SC425A		3024	18	2030	
2SC425B	123A	3444	18	2051	
2SC425C	123A	3444	18	2051	
2SC425D	123A	3444	18	2051	
2SC425E	123A	3444	18	2051	
2SC425F	123A	3444	18	2051	
2SC425G		3024	18	2030	
2SC425GN		3024	18	2030	
2SC425H		3024	18	2030	
2SC425K		3024	18	2030	
2SC425L		3024	18	2030	
2SC425M		3024	18	2030	
2SC425OR		3024	18	2030	
2SC425R		3024	18	2030	
2SC425X		3024	18	2030	
2SC425Y		3024	18	2030	
2SC426			17	2051	
2SC427			17	2051	
2SC428			17	2051	
2SC429	313	3452/108	61	2038	
2SC429A		3018	61	2038	
2SC429B		3018	61	2038	
2SC429C		3018	61	2038	
2SC429D		3018	61	2038	
2SC429E		3018	61	2038	
2SC429F		3018	61	2038	
2SC429G		3018	61	2038	
2SC429GN		3018	61	2038	
2SC429H		3018	61	2038	
2SC429J	313	3452/108	61	2038	
2SC429K		3018	61	2038	
2SC429L		3018	61	2038	
2SC429M		3018	61	2038	
2SC429OR		3018	61	2038	
2SC429R		3018	61	2038	
2SC429X	313	3452/108	61	2038	
2SC429Y		3018	61	2038	
2SC42A	163A	3079	35		MJ410
2SC42B		3079	35		
2SC42C		3079	35		
2SC42E		3079	35		
2SC42F		3079	35		
2SC42G		3079	35		
2SC42GN		3079	35		
2SC42H		3079	35		
2SC42J		3079	35		
2SC42K		3079	35		
2SC42L		3079	35		
2SC42M		3079	35		
2SC42OR		3079	35		
2SC42R		3079	35		
2SC42X		3079	35		
2SC42Y		3079	35		
2SC43	163A	3438	35		2N4347
2SC430	313	3452/108	61	2038	
2SC430A		3018	61	2038	
2SC430B		3018	61	2038	
2SC430C		3018	61	2038	
2SC430D		3018	61	2038	
2SC430E		3018	61	2038	
2SC430F		3018	61	2038	
2SC430G		3018	61	2038	
2SC430GN		3018	61	2038	
2SC430H	313	3452/108	61	2038	
2SC430J		3018	61	2038	
2SC430K		3018	61	2038	
2SC430L		3018	61	2038	
2SC430M		3018	61	2038	
2SC430OR		3018	61	2038	
2SC430R		3018	61	2038	
2SC430W	313	3452/108	61	2038	
2SC430X		3018	61	2038	
2SC430Y		3018	61	2038	
2SC431					2N6341
2SC432					2N6341
2SC433		3444	20	2051	MJ15022
2SC434					MJ15022
2SC435					MJ10000
2SC436					MJ10000
2SC44	163A	3438	35		2N4347
2SC440	311		20	2051	
2SC441	311	3124/289	47	2051	
2SC441A		3124	47	2030	
2SC441B		3124	47	2030	
2SC441C		3124	47	2030	
2SC441D		3124	47	2030	
2SC441E		3124	47	2030	
2SC441F		3124	47	2030	
2SC441G		3124	47	2030	
2SC441GN		3124	47	2030	
2SC441H		3124	47	2030	
2SC441J		3124	47	2030	
2SC441K		3124	47	2030	
2SC441L		3124	47	2030	
2SC441M		3124	47	2030	
2SC441OR		3124	47	2030	
2SC441R		3124	47	2030	
2SC441X		3124	47	2030	
2SC441Y		3124	47	2030	
2SC442	311	3124/289	47	2051	
2SC442A		3124	47	2030	
2SC442B		3124	47	2030	
2SC442C		3124	47	2030	
2SC442D		3124	47	2030	
2SC442E		3124	47	2030	
2SC442F		3124	47	2030	
2SC442G		3124	47	2030	

Industry Standard No.	ECG	SK	GE	RS 276-	MOTOR.
2SC442GN		3124	47	2030	
2SC442H		3124	47	2030	
2SC442J		3124	47	2030	
2SC442K		3124	47	2030	
2SC442L		3124	47	2030	
2SC442M		3124	47	2030	
2SC442OR		3124	47	2030	
2SC442R		3124	47	2030	
2SC442X		3124	47	2030	
2SC442Y		3124	47	2030	
2SC443	128	3024	18	2030	
2SC443A		3024	18	2030	
2SC443B		3024	18	2030	
2SC443C		3024	18	2030	
2SC443D		3024	18	2030	
2SC443E		3024	18	2030	
2SC443F		3024	18	2030	
2SC443G		3024	18	2030	
2SC443GN		3024	18	2030	
2SC443H		3024	18	2030	
2SC443J		3024	18	2030	
2SC443K		3024	18	2030	
2SC443L		3024	18	2030	
2SC443M		3024	18	2030	
2SC443OR		3024	18	2030	
2SC443R		3024	18	2030	
2SC443X		3024	18	2030	
2SC443Y		3024	18	2030	
2SC444			32	2030	
2SC445		3104A	63		
2SC445A		3104A	32		
2SC445B		3104A	32		
2SC445C		3104A	32		
2SC445D		3104A	32		
2SC445E		3104A	32		
2SC445F		3104A	32		
2SC445G		3104A	32		
2SC445GN		3104A	32		
2SC445H		3104A	32		
2SC445J		3104A	32		
2SC445K		3104A	32		
2SC445L		3104A	32		
2SC445M		3104A	32		
2SC445OR		3104A	32		
2SC445R		3104A	32		
2SC445X		3104A	32		
2SC445Y		3104A	32		
2SC45	123A	3444	20	2051	
2SC454	85	3124/289	210	2051	
2SC454(A)	85	3124/289	210	2051	
2SC454(B)		3124	210	2051	
2SC454-3		3124	210	2051	
2SC454-5		3124	210	2051	
2SC454-OR			210	2051	
2SC454A	85	3124/289	210	2051	
2SC454B	85	3124/289	210	2051	
2SC454B-6		3124	210	2051	
2SC454BL		3124	210	2051	
2SC454C	85	3124/289	210	2051	
2SC454D	85	3124/289	210	2051	
2SC454E		3124	210	2051	
2SC454F		3124	210	2051	
2SC454G		3124	210	2051	
2SC454GN		3124	210	2051	
2SC454H		3124	210	2051	
2SC454J		3124	210	2051	
2SC454K		3124	210	2051	
2SC454L	85	3124/289	210	2051	
2SC454LA	85	3124/289	210	2051	
2SC454M		3124	210	2051	
2SC454OR		3124	210	2051	
2SC454R		3124	210	2051	
2SC454X		3124	210	2051	
2SC454Y		3124	210	2051	
2SC455	289A	3018	20	2038	
2SC455-OR			20	2038	
2SC455A		3018	20	2038	
2SC455B		3018	20	2038	
2SC455C		3018	20	2038	
2SC455D		3018	20	2038	
2SC455E		3018	20	2038	
2SC455F		3018	20	2038	
2SC455G		3018	20	2038	
2SC455GN		3018	20	2038	
2SC455H		3018	20	2038	
2SC455J		3018	20	2038	
2SC455K		3018	20	2038	
2SC455L		3018	20	2038	
2SC455M		3018	20	2038	
2SC455OR		3018	20	2038	
2SC455R		3018	20	2038	
2SC455X		3018	61	2038	
2SC455Y		3018	20	2038	
2SC456	195A	3048/329	46		
2SC456-O	195A	3048/329	46		
2SC456A	195A	3048/329	46		
2SC456B		3048	46	2030	
2SC456C		3048	46	2030	
2SC456D	195A	3024/128	46		
2SC456E		3048	46	2030	
2SC456F		3048	46	2030	
2SC456G		3048	46	2030	
2SC456GN		3048	46	2030	
2SC456H		3048	46	2030	
2SC456J		3048	46	2030	
2SC456K		3048	46	2030	
2SC456L		3048	46	2030	
2SC456M		3048	46	2030	
2SC456OR		3048	46	2030	
2SC456R		3048	46	2030	
2SC456X		3048	46	2030	
2SC456Y		3048	46	2030	
2SC458	85	3124/289	210	2038	
2SC458(C)	85	3124/289	210	2038	
2SC458(C,D)			210	2038	
2SC458(D)			210	2038	
2SC458(LG)			210	2038	
2SC458-4		3124	210	2038	
2SC458-5		3124	210	2038	
2SC458-O			210	2038	
2SC458-OR			210	2038	
2SC458A	85	3124/289	210	2038	
2SC458AD	85	3124/289	210	2038	
2SC458AK			210	2038	
2SC458B	85	3124/289	20	2051	
2SC458B-D		3124	210	2038	
2SC458BC	85	3124/289	210	2038	
2SC458BD			210	2038	
2SC458BK		3124	210	2038	
2SC458BL	85	3124/289	210	2038	
2SC458BLG		3137	210	2038	
2SC458BM			210	2038	
2SC458C	85	3124/289	20	2051	
2SC458C/L6			268	2038	
2SC458CL			210	2038	
2SC458CLG	85	3124/289	210	2038	
2SC458CM	85	3124/289	210	2038	
2SC458D	85	3124/289	210	2038	
2SC458E			210	2038	
2SC458F			210	2038	
2SC458G	85	3124/289	210	2038	
2SC458GLB	85	3124/289	268	2038	
2SC458GN			210	2038	
2SC458H			210	2038	
2SC458J			210	2038	
2SC458K	85	3124/289	210	2038	
2SC458KA	85		210	2038	
2SC458KB	85		210	2038	
2SC458KC	85		210	2038	
2SC458KD	85		210	2038	
2SC458L	85	3124/289	210	2038	
2SC458L(C)			210	2038	
2SC458L6			210	2038	
2SC458LB	85	3124/289	210	2038	
2SC458LC	85	3124/289	210	2038	
2SC458LD			210	2038	
2SC458LG	85	3124/289	210	2038	
2SC458LG(B)	85	3124/289	210	2038	
2SC458LG(C)			210	2038	
2SC458LG(D)			210	2038	
2SC458LGA	85		210	2038	
2SC458LGB	85	3124/289	210	2038	
2SC458LGBM	85		210	2038	
2SC458LGC	85	3124/289	268	2051	
2SC458LGC-6		3124	210	2038	
2SC458LGD	85	3124/289	210	2038	
2SC458LGO	85	3124/289	210	2038	
2SC458LGR		3124	210	2038	
2SC458LGS	85		210	2038	
2SC458M	85	3124/289	210	2038	
2SC458OR			210	2038	
2SC458P	85	3124/289	210	2038	
2SC458R		3124	210	2038	
2SC458RGS	85		210	2038	
2SC458TOK			210	2038	
2SC458V	85		210	2038	
2SC458VC	85		210	2038	
2SC458X			210	2038	
2SC458Y			210	2038	
2SC459	85	3122	61	2030	
2SC459A		3122	61	2051	
2SC459B	85		61	2030	
2SC459C	85	3122	61	2051	
2SC459D	85	3122	61	2051	
2SC459E		3122	61	2051	
2SC459F		3122	61	2051	
2SC459G		3122	61	2051	
2SC459GN		3122	61	2051	
2SC459H		3122	61	2051	
2SC459J		3122	61	2051	
2SC459K		3122	61	2051	
2SC459L		3122	61	2051	
2SC459M		3122	61	2051	
2SC459OR		3122	61	2051	
2SC459R		3122	61	2051	
2SC459X		3122	61	2051	
2SC459Y		3122	61	2051	
2SC46	128	3024	18	2030	
00002SC460	107	3122	11	2051	
2SC460(A)	85	3122	61	2051	
2SC460(B)	85	3122	61	2051	
2SC460-5		3122	61	2051	
2SC460-B		3122	61	2051	
2SC460-C		3122	61	2051	
2SC460-OR			61	2051	
2SC460A	85	3122	86	2038	
2SC460AB	85	3122			
2SC460B	85	3122	86	2051	
00002SC460C	108	3122	86	2038	
2SC460D	85	3122	61	2051	
2SC460E		3122	61	2051	
2SC460F	85	3122	61	2051	
2SC460G	85	3122	61	2051	
2SC460GB	85		61	2051	
2SC460GN		3122	61	2051	
2SC460H	85	3122	61	2051	
2SC460J		3122	61	2051	
2SC460K	85	3122	61	2051	
2SC460L	85	3122	210	2051	
2SC460M		3122	61	2051	
2SC460OR		3122	61	2051	
2SC460R		3122	61	2051	
2SC460X		3122	61	2051	
2SC460Y		3122	61	2051	
00002SC461	108	3122	86	2038	
2SC461(BF)			61	2038	
2SC461-BF			61	2038	
2SC461-A		3122	61	2038	
2SC461-B		3122	61	2038	
2SC461A	85	3122	61	2038	
2SC461AL		3122	61	2038	
2SC461B	85	3122	61	2038	
2SC461BF			61	2038	
2SC461BK		3122	61	2038	
2SC461BL		3122	61	2038	
2SC461C	85	3122	61	2038	
2SC461E	85		61	2038	
2SC461EP		3122	61	2038	
2SC461L	85	3122	61	2038	
2SC462			20	2051	
2SC463	161	3019	86	2016	
2SC463A		3039	86	2015	
2SC463B		3039	86	2015	
2SC463C		3039	86	2015	
2SC463D		3039	86	2015	
2SC463E		3039	86	2015	
2SC463F		3039	86	2015	
2SC463G		3039	86	2015	
2SC463GN		3039	86	2015	
2SC463H	161	3117	86	2016	
2SC463J		3039	86	2015	
2SC463K		3039	86	2015	
2SC463L		3039	86	2015	
2SC463M		3039	86	2015	
2SC463OR		3039	86	2015	
2SC463R		3039	86	2015	
2SC463X		3039	86	2015	
2SC463Y		3039	86	2015	
2SC464	316	3018	61	2015	
2SC464A		3018	61	2038	
2SC464B		3018	61	2038	
2SC464C	316	3137/192A	61	2015	
2SC464D		3018	61	2038	
2SC464E		3018	61	2038	
2SC464F		3018	61	2038	
2SC464G		3018	61	2038	
2SC464GN		3018	61	2038	
2SC464H		3018	61	2038	

Industry Standard No.	ECG	SK	GE	RS 276-	MOTOR.
2SC464J		3018	61	2038	
2SC464K		3018	61	2038	
2SC464L		3018	61	2038	
2SC464M		3018	61	2038	
2SC464OR		3018	61	2038	
2SC464R		3018	61	2038	
2SC464X		3018	61	2038	
2SC464Y		3018	61	2038	
2SC465	316	3018	61	2015	
2SC465A		3018	61	2038	
2SC465B		3018	61	2038	
2SC465C		3018	61	2038	
2SC465D		3018	61	2038	
2SC465E		3018	61	2038	
2SC465F		3018	61	2038	
2SC465G		3018	61	2038	
2SC465GN		3018	61	2038	
2SC465H		3018	61	2038	
2SC465J		3018	61	2038	
2SC465K		3018	61	2038	
2SC465L		3018	61	2038	
2SC465M		3018	61	2038	
2SC465OR		3018	61	2038	
2SC465R		3018	61	2038	
2SC465X		3018	61	2038	
2SC465Y		3018	61	2038	
2SC466	316	3018	61	2015	
2SC466A		3018	61	2038	
2SC466B		3018	61	2038	
2SC466C		3018	61	2038	
2SC466D		3018	61	2038	
2SC466E		3018	61	2038	
2SC466F		3018	61	2038	
2SC466G		3018	61	2038	
2SC466GN		3018	61	2038	
2SC466H	316	3117	61	2015	
2SC466J		3018	61	2038	
2SC466K		3018	61	2038	
2SC466L		3018	61	2038	
2SC466M		3018	61	2038	
2SC466OR		3018	61	2038	
2SC466R		3018	61	2038	
2SC466X		3018	61	2038	
2SC466Y		3018	61	2038	
2SC467	108		16	2006	
2SC468	123A	3444	20	2051	
2SC468(LGR)	123A	3137/192A	20		
2SC468A	123A	3444	20	2051	
2SC468B	123A	3137/192A	220		
2SC469	313	3018	61	2016	
2SC469-OR			61	2016	
2SC469A	313	3018	61	2016	
2SC469B		3018	61	2016	
2SC469C		3018	61	2016	
2SC469D		3018	61	2016	
2SC469E		3018	61	2016	
2SC469F	313	3452/108	61	2016	
2SC469G		3018	61	2016	
2SC469GN		3018	61	2016	
2SC469H		3018	61	2016	
2SC469J		3018	61	2016	
2SC469K	313	3018	61	2016	
2SC469L		3018	61	2016	
2SC469M		3018	61	2016	
2SC469OR		3018	61	2016	
2SC469Q	313	3018	61	2016	
2SC469R	313	3018	61	2016	
2SC469X		3018	61	2016	
2SC469Y		3018	61	2016	
2SC46A		3024	18	2030	
2SC46B		3024	18	2030	
2SC46C		3024	18	2030	
2SC46E		3024	18	2030	
2SC46F		3024	18	2030	
2SC46G		3024	18	2030	
2SC46GN		3024	18	2030	
2SC46H		3024	18	2030	
2SC46J		3024	18	2030	
2SC46K		3024	18	2030	
2SC46L		3024	18	2030	
2SC46M		3024	18	2030	
2SC46OR		3024	18	2030	
2SC46X		3024	18	2030	
2SC46Y		3024	18	2030	
2SC47	128	3444/123A	88	2051	
2SC470	154	3045/225	32	2012	
2SC470-3			32	2012	
2SC470-4			32	2012	
2SC470-5			32	2012	
2SC470-6			32	2012	
2SC470A		3045	32	2012	
2SC470B		3045	32	2012	
2SC470C		3045	32	2012	
2SC470D		3045	32	2012	
2SC470E		3045	32	2012	
2SC470F		3045	32	2012	
2SC470G		3045	32	2012	
2SC470GN		3045	32	2012	
2SC470H		3045	32	2012	
2SC470J		3045	32	2012	
2SC470K		3045	32	2012	
2SC470L		3045	32	2012	
2SC470M		3045	32	2012	
2SC470OR		3045	32	2012	
2SC470R		3045	32	2012	
2SC470X		3045	32	2012	
2SC470Y		3045	32	2012	
2SC471			47	2030	
2SC472		3018	61	2038	
2SC472A		3018	61	2038	
2SC472B		3018	61	2038	
2SC472C		3018	61	2038	
2SC472D		3018	61	2038	
2SC472E		3018	61	2038	
2SC472F		3018	61	2038	
2SC472G		3018	61	2038	
2SC472GN		3018	61	2038	
2SC472H		3018	61	2038	
2SC472J		3018	61	2038	
2SC472K		3018	61	2038	
2SC472L		3018	61	2038	
2SC472M		3018	61	2038	
2SC472OR		3018	61	2038	
2SC472R		3018	61	2038	
2SC472X		3018	61	2038	
2SC472Y	123A	3018	61	2012	
2SC474	123A	3444	20	2051	
2SC475	313	3444/123A	47	2051	
2SC475A		3124	47	2010	
2SC475B		3124	47	2010	
2SC475C		3124	47	2010	
2SC475D		3124	47	2010	
2SC475E		3124	47	2010	
2SC475F		3124	47	2010	
2SC475G		3124	47	2010	
2SC475GN		3124	47	2010	
2SC475H		3124	47	2010	
2SC475J		3124	47	2010	
2SC475K	313	3444/123A	47	2051	
2SC475L		3124	47	2010	
2SC475M		3124	47	2010	
2SC475OR		3124	47	2010	
2SC475R		3124	47	2010	
2SC475X		3124	47	2010	
2SC475Y		3124	47	2010	
2SC476	313	3444/123A	88	2051	
2SC476A		3122	88	2010	
2SC476B		3122	88	2010	
2SC476C		3122	88	2010	
2SC476D		3122	88	2010	
2SC476E		3122	88	2010	
2SC476F		3122	88	2010	
2SC476G		3122	88	2010	
2SC476GN		3122	88	2010	
2SC476H		3122	88	2010	
2SC476J		3122	88	2010	
2SC476K		3122	88	2010	
2SC476L		3122	88	2010	
2SC476M		3122	88	2010	
2SC476OR		3122	88	2010	
2SC476R		3122	88	2010	
2SC476X		3122	88	2010	
2SC476Y		3122	88	2010	
2SC477	161	3452/108	86	2038	
2SC477A		3039	86	2038	
2SC477B		3039	86	2038	
2SC477C		3039	86	2038	
2SC477D		3039	86	2038	
2SC477E		3039	86	2038	
2SC477F		3039	86	2038	
2SC477G		3039	86	2038	
2SC477GN		3039	86	2038	
2SC477H		3039	86	2038	
2SC477J		3039	86	2038	
2SC477K		3039	86	2038	
2SC477L		3039	86	2038	
2SC477M		3039	86	2038	
2SC477OR		3039	86	2038	
2SC477R		3039	86	2038	
2SC477X		3039	86	2038	
2SC477Y		3039	86	2038	
2SC478	123A	3018	210		
2SC478(D)	123A	3048/329	210		
2SC478-4	123A	3047	210		
2SC478A		3047	210		
2SC478B		3047	210		
2SC478C		3047	210		
2SC478D	123A	3047	210		
2SC478E		3047	210		
2SC478F		3047	210		
2SC478G		3047	210		
2SC478GN		3047	210		
2SC478H		3047	210		
2SC478J		3047	210		
2SC478K		3047	210		
2SC478L		3047	210		
2SC478M		3047	210		
2SC478OR		3047	210		
2SC478R		3047	210		
2SC478X		3047	210		
2SC478Y		3047	210		
2SC479	128	3024	18	2030	
2SC479A		3024	18	2030	
2SC479B		3024	18	2030	
2SC479C		3024	18	2030	
2SC479D		3024	18	2030	
2SC479E		3024	18	2030	
2SC479F		3024	18	2030	
2SC479G		3024	18	2030	
2SC479GN		3024	18	2030	
2SC479H	128	3024	18	2030	
2SC479J		3024	18	2030	
2SC479K		3024	18	2030	
2SC479L		3024	18	2030	
2SC479M		3024	18	2030	
2SC479OR		3024	18	2030	
2SC479R		3024	18	2030	
2SC479X		3024	18	2030	
2SC479Y		3024	18	2030	
2SC47A		3122	88	2030	
2SC47B		3122	88	2030	
2SC47C		3122	88	2030	
2SC47D		3122	88	2030	
2SC47E		3122	88	2030	
2SC47G		3122	88	2030	
2SC47GN		3122	88	2030	
2SC47H		3122	88	2030	
2SC47J		3122	88	2030	
2SC47K		3122	88	2030	
2SC47L		3122	88	2030	
2SC47M		3122	88	2030	
2SC47OR		3122	88	2030	
2SC47R		3122	88	2030	
2SC47X		3122	88	2030	
2SC47Y		3122	88	2030	
2SC48	128	3124/289	47	2051	
2SC481	195A	3765	219		
2SC481A		3765	219		
2SC481B		3765	219		
2SC481C		3765	219		
2SC481D		3765	219		
2SC481E		3765	219		
2SC481F		3765	219		
2SC481G		3765	219		
2SC481GN		3765	219		
2SC481H		3765	219		
2SC481J		3765	219		
2SC481K		3765	219		
2SC481L		3765	219		
2SC481M		3765	219		
2SC481OR		3765	219		
2SC481R		3765	219		
2SC481X	195A	3765	219		
2SC481Y		3765	219		
2SC482	195A	3512	219		
2SC482-O	195A		219		
2SC482-GR	195A	3512	219		
2SC482-Y	195A		210		
2SC482A		3512	219		
2SC482B		3512	219		
2SC482C		3512	219		
2SC482D		3512	219		
2SC482E		3512	219		
2SC482F		3512	219		
2SC482G		3512	219		
2SC482GN		3512	219		
2SC482GR	195A	3512	219		
2SC482GRY		3512	219		

Industry Standard No.	ECG	SK	GE	RS 276-	MOTOR.
2SC482H		3512	219		
2SC482J		3512	219		
2SC482K		3512	219		
2SC482L		3512	219		
2SC482M		3512	210		
2SC482OR		3512	219		
2SC482R		3512	219		
2SC482X	195A	3512	219		
2SC482Y	195A	3512	219		
2SC483					2N3583
2SC484	324	3024/128	32		
2SC484A		3024	32		
2SC484B		3024	32		
2SC484BL	324	3748/282	32		
2SC484C		3024	32		
2SC484D		3024	32		
2SC484E		3024	32		
2SC484F		3024	32		
2SC484G		3024	32		
2SC484GN		3024	32		
2SC484H		3024	32		
2SC484J		3024	32		
2SC484K		3024	32		
2SC484L		3024	32		
2SC484M		3024	32		
2SC484OR		3024	32		
2SC484R	324	3024/128	32		
2SC484X		3024	32		
2SC484Y	324	3024/128	32		
2SC485	324	3748/282	18		
2SC485A		3748	18		
2SC485B		3748	18		
2SC485BL	324	3748/282	18		
2SC485C	324	3748/282	18		
2SC485D		3748	18		
2SC485E		3748	18		
2SC485F		3748	18		
2SC485G		3748	18		
2SC485GN		3748	18		
2SC485H		3748	18		
2SC485J		3748	18		
2SC485K		3748	18		
2SC485L		3748	18		
2SC485M		3748	18		
2SC485OR		3748	18		
2SC485R		3748	18		
2SC485X		3748	18		
2SC485Y	324	3748/282	18		
2SC486	324	3748/282	18		
2SC486-R			18	2030	
2SC486-RED			18	2030	
2SC486-Y			18	2030	
2SC486-YEL			18	2030	
2SC486A		3748	243	2030	
2SC486B		3748	243	2030	
2SC486BL	324		18		
2SC486C		3748	243	2030	
2SC486D		3748	243	2030	
2SC486E		3748	243	2030	
2SC486F		3748	243	2030	
2SC486G		3748	243	2030	
2SC486GN		3748	243	2030	
2SC486H		3748	243	2030	
2SC486J		3748	243	2030	
2SC486K		3748	243	2030	
2SC486L		3748	243	2030	
2SC486M		3748	243	2030	
2SC486OR		3748	243	2030	
2SC486R		3748	243	2030	
2SC486X		3748	243	2030	
2SC486Y	324	3748/282	243		
2SC487	175	3131A/369	246	2020	2N3583
2SC488	175	3131A/369	246	2020	2N6233
2SC489	175	3026	66	2020	2N3441
2SC489-R			66	2048	
2SC489-Y			66	2048	
2SC489A		3026	66	2048	
2SC489B		3026	66	2048	
2SC489C		3026	66	2048	
2SC489D		3026	66	2048	
2SC489E		3026	66	2048	
2SC489F		3026	66	2048	
2SC489G		3026	66	2048	
2SC489GN		3026	66	2048	
2SC489H		3026	66	2048	
2SC489J		3026	66	2048	
2SC489K		3026	66	2048	
2SC489L		3026	66	2048	
2SC489M		3026	66	2048	
2SC489OR		3026	66	2048	
2SC489R		3026	66	2048	
2SC489X		3026	66	2048	
2SC489Y	175	3026	66	2020	
2SC48A		3124	47	2030	
2SC48B		3124	47	2030	
2SC48C	128	3444/123A	47	2051	
2SC48D		3124	47	2030	
2SC48E		3124	47	2030	
2SC48F		3124	47	2030	
2SC48G		3124	47	2030	
2SC48GN		3124	47	2030	
2SC48H		3124	47	2030	
2SC48J		3124	47	2030	
2SC48K		3124	47	2030	
2SC48L		3124	47	2030	
2SC48M		3124	47	2030	
2SC48OR		3124	47	2030	
2SC48R		3124	47	2030	
2SC48X		3124	47	2030	
2SC48Y		3124	47	2030	
2SC49	128	3024	18	2030	
2SC490	175	3271	216	2020	2N3766
2SC490-BLU			216	2053	
2SC490-R			216	2053	
2SC490-RED		3271	216	2053	
2SC490-Y			216	2053	
2SC490-YEL		3271	216	2053	
2SC491	175	3271	66	2020	2N5050
2SC491-BL			66	2048	
2SC491-BLU			66	2048	
2SC491-R			66	2048	
2SC491-RED			66	2048	
2SC491-Y			66	2048	
2SC491-YEL			66	2048	
2SC491A		3271	66	2048	
2SC491B		3271	66	2048	
2SC491BL	175	3271	66	2020	
2SC491C		3271	66	2048	
2SC491D		3271	66	2048	
2SC491E		3271	66	2048	
2SC491F		3271	66	2048	
2SC491G		3271	66	2048	
2SC491GN		3271	66	2048	
2SC491H		3271	66	2048	
2SC491J		3271	66	2048	
2SC491K		3271	66	2048	
2SC491L		3271	66	2048	
2SC491M		3271	66	2048	
2SC491OR		3271	66	2048	
2SC491R	175	3271	66	2020	
2SC491X		3271	66	2048	
2SC491Y	175	3271	66	2020	
2SC492	280	3123	35		2N4347
2SC493	87	3027/130	75	2041	2N4347
2SC493-BL	87		75	2041	
2SC493-R	87		75	2041	
2SC493-Y	87		75	2041	
2SC493A		3027	75	2041	
2SC493B		3027	75	2041	
2SC493C		3027	75	2041	
2SC493D		3027	75	2041	
2SC493E		3027	75	2041	
2SC493F		3027	75	2041	
2SC493G		3027	75	2041	
2SC493GN		3027	75	2041	
2SC493H		3027	75	2041	
2SC493J		3027	75	2041	
2SC493K		3027	75	2041	
2SC493L		3027	75	2041	
2SC493M		3027	75	2041	
2SC493OR		3027	75	2041	
2SC493R		3027	75	2041	
2SC493X		3027	75	2041	
2SC493Y		3027	75	2041	
2SC494		3122	255	2041	2N3447
2SC494-BL			255	2041	
2SC494-R			255	2041	
2SC494-Y			255	2041	
2SC494A		3029	255	2041(2)	
2SC494B		3029	255	2041(2)	
2SC494BL		3027	255	2041(2)	
2SC494C		3029	255	2041(2)	
2SC494D		3029	255	2041(2)	
2SC494E		3029	255	2041(2)	
2SC494F		3029	255	2041(2)	
2SC494G		3029	255	2041(2)	
2SC494GN		3029	255	2041(2)	
2SC494H		3029	255	2041(2)	
2SC494J		3029	255	2041(2)	
2SC494K		3029	255	2041(2)	
2SC494L		3029	255	2041(2)	
2SC494M		3029	255	2041(2)	
2SC494OR		3029	255	2041(2)	
2SC494R		3029	255	2041(2)	
2SC494X		3029	255	2041(2)	
2SC494Y		3029	255	2041(2)	
2SC495	295	3253	270		2N4923
2SC495-O	295		270		
2SC495-O	295	3253	270		
2SC495-R	295	9041/373	270		
2SC495-Y	295	9041/373	270		
2SC495A		9041	270		
2SC495B		9041	270		
2SC495C		9041	270		
2SC495D		9041	270		
2SC495T	295	9041/373	270		
2SC495Y	295		270		
2SC496	295	9041/373	57		2N4921
2SC496(0)	295		57		
2SC496-O	295	9041/373	57		
2SC496-O	295		57		
2SC496-R	295	9041/373	57		
2SC496-Y	295	9041/373	57		
2SC496A		9041	57		
2SC496B		9041	57		
2SC496C		9041	57		
2SC496Y	295		57		
2SC497	128	3024	18	2030	
2SC497-O	128		18	2030	
2SC497-O			18	2030	
2SC497-OR			18	2030	
2SC497-ORG			18	2030	
2SC497-R	128		18	2030	
2SC497-RED			18	2030	
2SC497-Y	128		18	2030	
2SC497A		3024	18	2030	
2SC497B		3024	18	2030	
2SC497C		3024	18	2030	
2SC497D		3024	18	2030	
2SC497E		3024	18	2030	
2SC497F		3024	18	2030	
2SC497G		3024	18	2030	
2SC497GN		3024	18	2030	
2SC497H		3024	18	2030	
2SC497J		3024	18	2030	
2SC497K		3024	18	2030	
2SC497L		3024	18	2030	
2SC497M		3024	18	2030	
2SC497OR		3024	18	2030	
2SC497R		3024	18	2030	
2SC497RED			18	2030	
2SC497X		3024	18	2030	
2SC497Y		3024	18	2030	
2SC498	128	3024	47	2030	
2SC498-O	128		47	2030	
2SC498-O			47	2030	
2SC498-OR			47	2030	
2SC498-ORG			47	2030	
2SC498-R	128		47	2030	
2SC498-RED			47	2030	
2SC498-Y	128		47	2030	
2SC498-YEL			47	2030	
2SC498A		3024	47	2030	
2SC498B		3024	47	2030	
2SC498C		3024	47	2030	
2SC498D		3024	47	2030	
2SC498E		3024	47	2030	
2SC498F		3024	47	2030	
2SC498G		3024	47	2030	
2SC498GN		3024	47	2030	
2SC498H		3024	47	2030	
2SC498J		3024	47	2030	
2SC498K		3024	47	2030	
2SC498L		3024	47	2030	
2SC498M		3024	47	2030	
2SC498OR		3024	47	2030	
2SC498R		3024	47	2030	
2SC498RED			47	2030	
2SC498X		3024	47	2030	
2SC498Y		3024	47	2030	
2SC498YEL			47	2030	
2SC499	382	3244	220		
2SC499-R		3244	220		
2SC499-R(FA-1)	194	3244	220		
2SC499-RED		3244	220		
2SC499-RY	194	3244	220		
2SC499-Y(FA-1)	194	3244	220		
2SC499-YEL		3244	220		

Industry Standard No.	ECG	SK	GE	RS 276-	MOTOR.
2SC499A		3244	27		
2SC499B		3244	220		
2SC499C		3244	220		
2SC499D		3244	220		
2SC499E		3244	220		
2SC499F		3244	220		
2SC499G		3244	220		
2SC499GN		3244	220		
2SC499H		3244	220		
2SC499J		3244	220		
2SC499K		3244	220		
2SC499L		3244	220		
2SC499M		3244	220		
2SC499OR		3244	27		
2SC499R	194	3244	220		
2SC499X		3244	220		
2SC499Y	194	3244	220		
2SC49A		3024	18	2030	
2SC49B		3024	18	2030	
2SC49C		3024	18	2030	
2SC49D		3024	18	2030	
2SC49E		3024	18	2030	
2SC49F		3024	18	2030	
2SC49G		3024	18	2030	
2SC49GN		3024	18	2030	
2SC49H		3024	18	2030	
2SC49J		3024	18	2030	
2SC49K		3024	18	2030	
2SC49L		3024	18	2030	
2SC49M		3024	18	2030	
2SC49OR		3024	18	2030	
2SC49X		3024	18	2030	
2SC49Y	128	3027/130	18	2030	
2SC50	101	3011	47	2002	
2SC500	154	3040	27	2012	
2SC500A		3040	27		
2SC500B		3040	27		
2SC500C		3040	27		
2SC500D		3040	27		
2SC500E		3040	27		
2SC500F		3040	27		
2SC500G		3040	27		
2SC500GN		3040	27		
2SC500H		3040	27		
2SC500J		3040	27		
2SC500K		3040	27		
2SC500L		3040	27		
2SC500M		3040	27		
2SC500OR		3040	27		
2SC500R	154	3040	27	2012	
2SC500X		3040	27		
2SC500Y	154	3045/225	27	2012	
2SC501	128	3024	18	2030	
2SC501-O			18	2030	
2SC501-ORG			18	2030	
2SC501-R			18	2030	
2SC501-RED			18	2030	
2SC501-Y			18	2030	
2SC501-YEL			18	2030	
2SC501A		3024	18	2030	
2SC501B		3024	18	2030	
2SC501C		3024	18	2030	
2SC501D		3024	18	2030	
2SC501E		3024	18	2030	
2SC501F		3024	18	2030	
2SC501G		3024	18	2030	
2SC501GN		3024	18	2030	
2SC501H		3024	18	2030	
2SC501J		3024	18	2030	
2SC501K		3024	18	2030	
2SC501L		3024	18	2030	
2SC501M		3024	18	2030	
2SC501OR		3024	18	2030	
2SC501R		3024	18	2030	
2SC501X		3024	18	2030	
2SC501Y		3024	18	2030	
2SC502	195A	3048/329	18		
2SC502A		3024	18	2030	
2SC502B		3024	18	2030	
2SC502C		3024	18	2030	
2SC502D		3024	18	2030	
2SC502E		3024	18	2030	
2SC502F		3024	18	2030	
2SC502G		3024	18	2030	
2SC502GN		3024	18	2030	
2SC502H		3024	18	2030	
2SC502J		3024	18	2030	
2SC502K		3024	18	2030	
2SC502L		3024	18	2030	
2SC502M		3024	18	2030	
2SC502OR		3024	18	2030	
2SC502R		3024	18	2030	
2SC502X		3024	18	2030	
2SC502Y		3024	18	2030	
2SC503	128	3024	18	2030	
2SC503-GR			18	2030	
2SC503-O	128	3024	18	2030	
2SC503-Y	128	3024	18	2030	
2SC503A		3024	18	2030	
2SC503B		3024	18	2030	
2SC503C		3024	18	2030	
2SC503D		3024	18	2030	
2SC503E		3024	18	2030	
2SC503F		3024	18	2030	
2SC503G		3024	18	2030	
2SC503GN		3024	18	2030	
2SC503GR	128	3024	18	2030	
2SC503H		3024	18	2030	
2SC503J		3024	18	2030	
2SC503K		3024	18	2030	
2SC503L		3024	18	2030	
2SC503M		3024	18	2030	
2SC503OR		3024	18	2030	
2SC503R		3024	18	2030	
2SC503X		3024	18	2030	
2SC504	128	3024	18	2030	
2SC504-GR			18	2030	
2SC504-O	128	3024	18	2030	
2SC504-Y	128	3024	18	2030	
2SC504A		3024	18	2030	
2SC504B		3024	18	2030	
2SC504C		3024	18	2030	
2SC504D		3024	18	2030	
2SC504E		3024	18	2030	
2SC504F		3024	18	2030	
2SC504G		3024	18	2030	
2SC504GN		3024	18	2030	
2SC504GR	128	3024	18	2030	
2SC504H		3024	18	2030	
2SC504J		3024	18	2030	
2SC504K		3024	18	2030	
2SC504L		3024	18	2030	
2SC504M		3024	18	2030	
2SC504OR		3024	18	2030	

Industry Standard No.	ECG	SK	GE	RS 276-	MOTOR.
2SC504R		3024	18	2030	
2SC504X		3024	18	2030	
2SC505	154	3044	32	2012	
2SC505-O	154		32	2012	
2SC505-R	154		32	2012	
2SC505A		3044	32		
2SC505B		3044	32		
2SC505C		3044	32		
2SC505D		3044	32		
2SC505E		3044	32		
2SC505F		3044	32		
2SC505G		3044	32		
2SC505GN		3044	32		
2SC505H		3044	32		
2SC505J		3044	32		
2SC505K		3044	32		
2SC505L		3044	32		
2SC505M		3044	32		
2SC505O		3044	32		
2SC505OR		3044	32		
2SC505R		3044	32		
2SC505X		3044	32		
2SC505Y		3044	32		
2SC506	154	3044	32	2012	
2SC506-O	154	3045/225	32	2012	
2SC506-R	154	3044	32	2012	
2SC506A		3044	32		
2SC506B		3044	32		
2SC506C		3044	32		
2SC506D		3044	32		
2SC506E		3044	32		
2SC506F		3044	32		
2SC506G		3044	32		
2SC506GN		3044	32		
2SC506H		3044	32		
2SC506J		3044	32		
2SC506K		3044	32		
2SC506L		3044	32		
2SC506M		3044	32		
2SC506OR		3045	32		
2SC506X		3044	32		
2SC506Y		3044	32		
2SC507	154	3044	32	2012	
2SC507-O	154		32	2012	
2SC507-R	154		32	2012	
2SC507-Y	154		32	2012	
2SC507A		3044	32		
2SC507B		3044	32		
2SC507C		3044	32		
2SC507D		3044	32		
2SC507E		3044	32		
2SC507F		3044	32		
2SC507G		3044	32		
2SC507GN		3044	32		
2SC507H		3044	32		
2SC507J		3044	32		
2SC507K		3044	32		
2SC507L		3044	32		
2SC507M		3044	32		
2SC507OR		3044	32		
2SC507R		3044	32		
2SC507X		3044	32		
2SC507Y		3044	32		
2SC508	384	3261/175	246	2020	2N6233
2SC509	289A	3124/289	81	2038	
2SC509(O)	289A		81	2038	
2SC509(O)			81	2038	
2SC509(Y)			81	2038	
2SC509-O	289A		81	2038	
2SC509-Y			81	2038	
2SC509G	289A	3124/289	81	2038	
2SC509G-O			81	2038	
2SC509G-Y			81	2038	
2SC509Y	289A	3124/289	81	2038	
2SC50A	101	3011	47	2002	
2SC50B		3011	47	2030	
2SC50C		3011	47	2030	
2SC50E		3011	47	2030	
2SC50F		3011	47	2030	
2SC50G		3011	47	2030	
2SC50GN		3011	47	2030	
2SC50H		3011	47	2030	
2SC50J		3011	47	2030	
2SC50K		3011	47	2030	
2SC50L		3011	47	2030	
2SC50M		3011	47	2030	
2SC50OR		3011	47	2030	
2SC50R		3011	47	2030	
2SC50X		3011	47	2030	
2SC50Y		3011	47	2030	
2SC51	128	3024	18	2030	
2SC510	324	3024/128	32		
2SC5100		3104A	32		
2SC5100A		3104A	32		
2SC5100B		3104A	32		
2SC5100C		3104A	32		
2SC5100D		3104A	32		
2SC5100E		3104A	32		
2SC5100F		3104A	32		
2SC5100G		3104A	32		
2SC5100GN		3104A	32		
2SC5100H		3104A	32		
2SC5100J		3104A	32		
2SC5100K		3104A	32		
2SC5100L		3104A	32		
2SC5100M		3104A	32		
2SC5100R		3104A	32		
2SC5100Y		3104A	32		
2SC510A		3024	32		
2SC510B		3024	32		
2SC510C		3024	32		
2SC510D		3024	32		
2SC510E		3024	32		
2SC510F		3024	32		
2SC510G		3024	32		
2SC510GN		3024	32		
2SC510H		3024	32		
2SC510J		3024	32		
2SC510K		3024	32		
2SC510L		3024	32		
2SC510M		3024	32		
2SC510OR		3024	32		
2SC510R		3024	32		
2SC510X		3024	32		
2SC510Y		3024	32		
2SC511	324	3104A	32		
2SC5110		3104A	32		
2SC5110A		3104A	32		
2SC5110B		3104A	32		
2SC5110C		3104A	32		
2SC5110D		3104A	32		
2SC5110E		3104A	32		
2SC5110F		3104A	32		
2SC5110G		3104A	32		

Industry Standard No.	ECG	SK	GE	RS 276-	MOTOR.
2SC5110GN		3104A	32		
2SC5110H		3104A	32		
2SC5110J		3104A	32		
2SC5110K		3104A	32		
2SC5110L		3104A	32		
2SC5110M		3104A	32		
2SC5110R		3104A	32		
2SC5110X		3104A	32		
2SC5110Y		3104A	32		
2SC511A		3104A	32		
2SC511B		3104A	32		
2SC511C		3104A	32		
2SC511D		3104A	32		
2SC511E		3104A	32		
2SC511F		3104A	32		
2SC511G		3104A	32		
2SC511GN		3104A	32		
2SC511H		3104A	32		
2SC511J		3104A	32		
2SC511K		3104A	32		
2SC511L		3104A	32		
2SC511M		3104A	32		
2SC511OR		3104A	32		
2SC511R		3104A	32		
2SC511X		3104A	32		
2SC511Y		3104A	32		
2SC512	324	3024/128	18	2030	
2SC512-O		3024	18	2030	
2SC512-O	324		18	2030	
2SC512-ORG			18	2030	
2SC512-R	324		18	2030	
2SC512-RED			18	2030	
2SC512A		3024	18	2030	
2SC512B		3024	18	2030	
2SC512C		3024	18	2030	
2SC512D		3024	18	2030	
2SC512E		3024	18	2030	
2SC512F		3024	18	2030	
2SC512G		3024	18	2030	
2SC512GN		3024	18	2030	
2SC512H		3024	18	2030	
2SC512J		3024	18	2030	
2SC512K		3024	18	2030	
2SC512L		3025	18	2030	
2SC512M		3024	18	2030	
2SC5120		3104A	18	2030	
2SC5120R		3024	18	2030	
2SC512R		3024	D42C7		
2SC512X		3024	18	2030	
2SC512Y		3024	18	2030	
2SC513	324	3024/128	18	2030	
2SC513-O			18	2030	
2SC513-O	324	3024/128	18	2030	
2SC513-ORG			18	2030	
2SC513-R			18	2030	
2SC513-RED			18	2030	
2SC513A		3024	18	2030	
2SC513B		3024	18	2030	
2SC513C		3024	18	2030	
2SC513D		3024	18	2030	
2SC513E		3024	18	2030	
2SC513F		3024	18	2030	
2SC513G		3024	18	2030	
2SC513GN		3024	18	2030	
2SC513H		3024	18	2030	
2SC513J		3024	18	2030	
2SC513K		3024	18	2030	
2SC513L		3024	18	2030	
2SC513M		3024	18	2030	
2SC5130		3104A	18	2030	
2SC5130R		3024	18	2030	
2SC513R	324	3104A	18	2030	
2SC513X		3024	18	2030	
2SC513Y		3024	18	2030	
2SC514	124	3021	12		
2SC514A		3021	12		
2SC514B		3021	12		
2SC514C		3021	12		
2SC514D		3021	12		
2SC514E		3021	12		
2SC514F		3021	12		
2SC514G		3021	12		
2SC514GN		3021	12		
2SC514H		3021	12		
2SC514J		3021	12		
2SC514K		3021	12		
2SC514L		3021	12		
2SC514M		3021	12		
2SC514OR		3021	12		
2SC514R		3021	12		
2SC514X		3021	12		
2SC514Y		3021	12		
2SC515	124	3021	12		2N3739
2SC515A	124	3021	246		
2SC515A(BK)	124		12		
2SC515AM	124	3021	12		
2SC515AX	124	3021	12		
2SC515B		3021	12		
2SC515BK	124	3021	12		
2SC515C		3021	12		
2SC515D		3021	12		
2SC515E		3021	12		
2SC515J		3021	12		
2SC515M		3021	12		
2SC515X		3021	12		
2SC515Y	124	3021	12		
2SC516	128	3024	18	2030	
2SC516A	128	3024	18	2030	
2SC516B		3024	18	2030	
2SC516C		3024	18	2030	
2SC516D		3024	18	2030	
2SC516E		3024	18	2030	
2SC516F		3024	18	2030	
2SC516G		3024	18	2030	
2SC516GN		3024	18	2030	
2SC516H		3024	18	2030	
2SC516J		3024	18	2030	
2SC516K		3024	18	2030	
2SC516L		3024	18	2030	
2SC516M		3024	18	2030	
2SC5160R		3024	18	2030	
2SC516R		3024	18	2030	
2SC516X		3024	18	2030	
2SC516Y		3024	18	2030	
2SC517	237	3049/224	46		
2SC517A		3049	46		
2SC517B		3049	46		
2SC517C	237	3049/224	46		
2SC517D		3049	46		
2SC517E		3049	46		
2SC517F		3049	46		
2SC517G		3049	46		
2SC517GN		3049	46		
2SC517H		3049	46		
2SC517J		3049	46		
2SC517K		3049	46		
2SC517L		3049	46		
2SC517M		3049	46		
2SC5170R		3049	46		
2SC517R		3049	46		
2SC517X		3049	46		
2SC517Y		3049	46		
2SC518			38		2N3448
2SC518A					2N3448
2SC519	87	3438	35		2N5759
2SC519A	87	3438	35		2N5760
2SC51A		3024	18	2030	
2SC51B		3024	18	2030	
2SC51C		3024	18	2030	
2SC51D		3024	18	2030	
2SC51E		3024	18	2030	
2SC51F		3024	18	2030	
2SC51G		3024	18	2030	
2SC51GN		3024	18	2030	
2SC51H		3024	18	2030	
2SC51J		3024	18	2030	
2SC51K		3024	18	2030	
2SC51L		3024	18	2030	
2SC51M		3024	18	2030	
2SC510R		3024	18	2030	
2SC51R		3024	18	2030	
2SC51X		3024	18	2030	
2SC51Y		3024	18	2030	
2SC52	123A	3444	88	2051	
2SC520	87	3510	14	2041	2N3448
2SC520A	87	3510	14	2041	2N3448
2SC521	87	3027/130	14	2041	2N3447
2SC521A	87	3027/130	14	2041	2N3448
2SC522	282/427	3045/225	256		
2SC522-O	282/427		256		
2SC522-R	282/427		256		
2SC523	282/427		256		
2SC523-O	282/427		256		
2SC523-R	225		256		
2SC524	282/427	3045/225	256		
2SC524-O	225		256		
2SC524-R	225		256		
2SC525	237	3045/225	256		
2SC525-O	237		256		
2SC525-R	237		256		
2SC526	154	3040	32		
2SC526A		3040	32		
2SC526B		3040	32		
2SC526C		3040	32		
2SC526D		3040	32		
2SC526E		3040	32		
2SC526F		3040	32		
2SC526G		3040	32		
2SC526GN		3040	32		
2SC526H		3040	32		
2SC526J		3040	32		
2SC526K		3040	32		
2SC526L		3040	32		
2SC526M		3040	32		
2SC5260R		3040	32		
2SC526R		3040	32		
2SC526Y		3040	32		
2SC527			39	2015	
2SC528		3245	212	2010	
2SC529	85	3444/123A	20	2051	
2SC529A	85	3444/123A	61	2051	
2SC529B		3018	61	2038	
2SC529C		3018	61	2038	
2SC529D		3018	61	2038	
2SC529E		3018	61	2038	
2SC529F		3018	61	2038	
2SC529G		3018	61	2038	
2SC529GN		3018	61	2038	
2SC529H		3018	61	2038	
2SC529J		3018	61	2038	
2SC529K		3018	61	2038	
2SC529L		3018	61	2038	
2SC529M		3018	61	2038	
2SC5290R		3018	61	2038	
2SC529R		3018	61	2038	
2SC529X		3018	61	2038	
2SC529Y		3018	61	2038	
2SC52A		3122	88	2030	
2SC52B		3122	88	2030	
2SC52C		3122	88	2030	
2SC52D		3122	88	2030	
2SC52E		3122	88	2030	
2SC52F		3122	88	2030	
2SC52G		3122	88	2030	
2SC52GN		3122	88	2030	
2SC52H		3122	88	2030	
2SC52J		3122	88	2030	
2SC52K		3122	88	2030	
2SC52L		3122	88	2030	
2SC52M		3122	88	2030	
2SC520R		3122	88	2030	
2SC52X		3122	88	2030	
2SC52Y		3122	88	2030	
2SC53	123A	3444	88	2051	
2SC531		3122	20	2051	
0002SC531F	123A	3122	20	2051	
00002SC535	107	3293	11	2038	
2SC535(B)	107	3293	86	2038	
2SC535-OR			86	2038	
2SC5359		3452	86	2038	
2SC535A	107	3293	86	2038	
2SC535ABC			86	2038	
2SC535AL		3293	86	2038	
2SC535B	107	3293	86	2038	
2SC535C	107	3293	86	2038	
2SC535D		3293	86	2038	
2SC535E		3293	86	2038	
2SC535F		3293	86	2038	
2SC535G	107	3293	86	2038	
2SC535GN		3293	86	2038	
2SC535H		3293	86	2038	
2SC535J		3293	86	2038	
2SC535K		3293	86	2038	
2SC535L		3293	86	2038	
2SC535M		3293	86	2038	
2SC5350R		3293	86	2038	
2SC535R		3293	86	2038	
2SC535X		3293	86	2038	
2SC535Y		3293	86	2038	
2SC536	199	3122	62	2010	
2SC536(C)			212	2010	
2SC536(D)			212	2010	
2SC536(E)			212	2010	
2SC536-D			212	2010	
2SC536-E			212	2010	
2SC536-F			212	2010	
2SC536-G			212	2010	
2SC536-OR			212	2010	

Industry Standard No.	ECG	SK	GE	RS 276-	MOTOR.
2SC536A		3122	212	2010	
2SC536A(3RD IF)	233			2009	
2SC536A(3RD-IF)			212	2010	
2SC536AG		3122	212	2010	
2SC536B		3122	212	2010	
2SC536C		3122	212	2010	
2SC536D		3122	212	2010	
2SC536DK		3122	212	2010	
2SC536E		3122	212	2010	
2SC536ED			212	2010	
2SC536EH		3122	212	2010	
2SC536EJ		3122	212	2010	
2SC536EN		3122	212	2010	
2SC536EP		3122	212	2010	
2SC536ER		3122	212	2010	
2SC536ET		3122	212	2010	
2SC536EZ			212	2010	
2SC536F		3122	212	2010	
2SC536F1		3122	212	2010	
2SC536F2		3122	212	2010	
2SC536FC		3122	212	2010	
2SC536FP			212	2010	
2SC536FS			212	2010	
2SC536FS6		3122	212	2010	
2SC536FZ		3122	62	2010	
2SC536G		3122	212	2010	
2SC536G-1			212	2010	
2SC536G1			212	2010	
2SC536G2			212	2010	
2SC536GF			212	2010	
2SC536GJ			212	2010	
2SC536GK		3122	212	2010	
2SC536GL		3122	212	2010	
2SC536GM			212	2010	
2SC536GN			212	2010	
2SC536GP		3122	212	2010	
2SC536GT		3122	212	2010	
2SC536GV		3122	212	2010	
2SC536GY		3122	212	2010	
2SC536GZ		3122	212	2010	
2SC536H		3122	212	2010	
2SC536J			212	2010	
2SC536K		3122	212	2010	
2SC536L			212	2010	
2SC536M			212	2010	
2SC536NP			212	2010	
2SC536OR			212	2010	
2SC536Q			212	2010	
2SC536R			212	2010	
2SC536W		3122	212	2010	
2SC536X		3122	212	2010	
2SC536XL			212	2010	
2SC536Y			212	2010	
2SC537	289A	3122	20	2051	
2SC537(F)	289A	3122	212	2009	
2SC537(G)	199	3122	212	2009	
2SC537-01	289A	3122	212	2009	
2SC537-EV		3122	212	2009	
2SC5370		3124	47	2030	
2SC5370A		3124	47	2030	
2SC5370B		3124	47	2030	
2SC5370C		3124	47	2030	
2SC5370D		3124	47	2030	
2SC5370E		3124	47	2030	
2SC5370F		3124	47	2030	
2SC5370G		3124	47	2030	
2SC5370GN		3124	47	2030	
2SC5370H		3124	47	2030	
2SC5370J		3124	47	2030	
2SC5370K		3124	47	2030	
2SC5370L		3124	47	2030	
2SC5370M		3124	47	2030	
2SC53700R			47	2030	
2SC5370R		3124	47	2030	
2SC5370X		3124	47	2030	
2SC5370Y		3124	47	2030	
2SC537ALC			212	2009	
2SC537B			212	2009	
2SC537BK		3122	212	2009	
2SC537C	289A	3122	212	2009	
2SC537C7			212	2009	
2SC537D	289A	3122	212	2009	
2SC537D1			212	2009	
2SC537D2	289A	3122	212	2009	
2SC537E	289A	3122	212	2009	
2SC537EF	289A	3122	212	2009	
2SC537EH	289A	3122	212	2009	
2SC537EJ	289A	3122	212	2009	
2SC537EK	289A	3122	212	2009	
2SC537EV			212	2009	
2SC537F	289A	3124/289	20	2009	
2SC537F-C7	199	3122	212	2009	
2SC537F1	289A	3122	212	2009	
2SC537F2	289A	3122	212	2009	
2SC537FC	289A	3122	212	2009	
2SC537FC7			212	2009	
2SC537FJ		3122	212	2009	
2SC537FK		3122	212	2009	
2SC537FV	199	3122	212	2009	
2SC537G	199	3122	212	2009	
2SC537G1		3122	212	2009	
2SC537G2	199	3122	212	2009	
2SC537GF	199	3122	212	2009	
2SC537GFL		3122	212	2009	
2SC537GI	199	3122	212	2009	
2SC537H	199	3122	212	2009	
2SC537HT	199	3122	212	2009	
2SC537W	199	3122	212	2009	
2SC537WF		3122	212	2009	
2SC538	123A	3444	212	2051	
2SC538(P)			212	2051	
2SC538(R)			212	2051	
2SC538-Q			212	2051	
2SC538A	123A	3444	212	2051	
2SC538A(Q)			212	2051	
2SC538A-P			212	2051	
2SC538A-R			212	2051	
2SC538AQ	123A	3444	212	2051	
2SC538AR			212	2051	
2SC538AS			212	2051	
2SC538K			212	2051	
2SC538P	123A	3444	212	2051	
2SC538Q	123A	3444	121	2051	
2SC538R	123A	3444	212	2051	
2SC538S	123A	3444	212	2051	
2SC538T	123A	3444	212	2051	
2SC539	123A	3444	20	2051	
2SC539(L)(K)			20	2010	
2SC539(R)			20	2010	
2SC539K	123A	3444	20	2051	
2SC539L	123A	3444	20	2051	
2SC539R	123A	3444	20	2051	
2SC539S	123A	3444	20	2051	
2SC539T		3444	20	2010	
2SC53A		3122	88	2030	
2SC53B		3122	88	2030	
2SC53C		3122	88	2030	
2SC53D		3122	88	2030	
2SC53E		3122	88	2030	
2SC53F		3122	88	2030	
2SC53G		3122	88	2030	
2SC53GN		3122	88	2030	
2SC53H		3122	88	2030	
2SC53J		3122	88	2030	
2SC53L		3122	88	2030	
2SC530R		3122	88	2030	
2SC53R		3122	88	2030	
2SC53X		3122	88	2030	
2SC53Y		3122	88	2030	
2SC54	123A	3444	88	2051	
2SC540	313	3444/123A	88	2051	
2SC540A		3122	88	2010	
2SC540B		3122	88	2010	
2SC540C		3122	88	2010	
2SC540D		3122	88	2010	
2SC540E		3122	88	2010	
2SC540F		3122	88	2010	
2SC540G		3122	88	2010	
2SC540GN		3122	88	2010	
2SC540H		3122	88	2010	
2SC540J		3122	88	2010	
2SC540K		3122	88	2010	
2SC540L		3122	88	2010	
2SC540M		3122	88	2010	
2SC5400R		3122	88	2010	
2SC540R		3122	88	2010	
2SC540X		3122	88	2010	
2SC540Y		3122	88	2010	
2SC541			28	2017	
2SC542			28	2017	
2SC543		3018	61	2048	
2SC54300R			61	2038	
2SC543A		3018	61	2038	
2SC543B		3018	61	2038	
2SC543C		3018	61	2038	
2SC543D		3018	61	2038	
2SC543E		3018	61	2038	
2SC543F		3018	61	2038	
2SC543G		3018	61	2038	
2SC543GN		3018	61	2038	
2SC543H		3018	61	2038	
2SC543J		3018	61	2038	
2SC543K		3018	61	2038	
2SC543L		3018	61	2038	
2SC543M		3018	61	2038	
2SC5430R		3018	61	2038	
2SC543R		3018	61	2038	
2SC543X		3018	61	2038	
2SC543Y		3018	61	2038	
2SC544	107	3452/108	61	2038	
2SC544A		3018	61	2038	
2SC544AG		3018	61	2038	
2SC544B		3018	61	2038	
2SC544C	107	3452/108	61	2038	
2SC544D	107	3452/108	61	2038	
2SC544D(VHF)			61	2038	
2SC544E	107	3452/108	61	2038	
2SC544F		3018	61	2038	
2SC544G		3018	61	2038	
2SC544GN		3018	61	2038	
2SC544H		3018	61	2038	
2SC544J		3018	61	2038	
2SC544K		3018	61	2038	
2SC544L		3018	61	2038	
2SC544M		3018	61	2038	
2SC5440R		3018	61	2038	
2SC544R		3018	61	2038	
2SC544X		3018	61	2038	
2SC544Y		3018	61	2038	
2SC545	107	3018	11	2015	
2SC545A	107	3018	11	2015	
2SC545B	107	3018	11	2015	
2SC545C	107	3018	11	2015	
2SC545D	107	3018	11	2015	
2SC545E	107	3018	11	2015	
2SC546			39	2015	
2SC546K				2015	
2SC547	3118		28	2017	
2SC548		3218	28	2017	
2SC549			28	2025	
2SC54A		3122	88	2030	
2SC54B		3122	88	2030	
2SC54C		3122	88	2030	
2SC54D		3122	88	2030	
2SC54E		3122	88	2030	
2SC54F		3122	88	2030	
2SC54G		3122	88	2030	
2SC54GN		3122	88	2030	
2SC54H		3122	88	2030	
2SC54J		3122	88	2030	
2SC54K		3122	88	2030	
2SC54L		3122	88	2030	
2SC54M		3122	88	2030	
2SC540R		3122	88	2030	
2SC54X		3122	88	2030	
2SC54Y		3122	88	2030	
2SC55	123A	3444	61	2051	
2SC550			28	2017	
2SC551			66	2048	
2SC552			66	2048	
2SC553			66	2048	
2SC554			20	2048	
2SC555	311	3195			
2SC556			61	2038	
2SC558	389	3439/163A	73		MJ3029
2SC559			47	2030	
2SC55A		3018	61	2038	
2SC55B		3018	61	2038	
2SC55C		3018	61	2038	
2SC55D		3018	61	2038	
2SC55E		3018	61	2038	
2SC55F		3018	61	2038	
2SC55G		3018	61	2038	
2SC55GN		3018	61	2038	
2SC55H		3018	61	2038	
2SC55J		3018	61	2038	
2SC55K		3018	61	2038	
2SC55L		3018	61	2038	
2SC55M		3018	61	2038	
2SC550R		3018	61	2038	
2SC55R		3018	61	2038	
2SC55X		3018	61	2038	
2SC55Y		3018	61	2038	
2SC56	107	3018	47	2015	
2SC560	128	3104A	32	2030	
2SC560A		3104A	32		
2SC560B		3104A	32		
2SC560C		3104A	32		

Industry Standard No.	ECG	SK	GE	RS 276-	MOTOR.
2SC560D		3104A	32		
2SC560E		3104A	32		
2SC560F		3104A	32		
2SC560G		3104A	32		
2SC560GN		3104A	32		
2SC560H		3104A	32		
2SC560J		3104A	32		
2SC560K		3104A	32		
2SC560L		3104A	32		
2SC560M		3104A	32		
2SC560OR		3104A	32		
2SC560R		3104A	32		
2SC560X		3104A	32		
2SC560Y		3104A	32		
2SC561	108	3452	11	2038	
2SC561A		3018	11	2038	
2SC561B		3018	11	2038	
2SC561C		3018	11	2038	
2SC561D		3018	11	2038	
2SC561E		3018	11	2038	
2SC561F		3018	11	2038	
2SC561G		3018	11	2038	
2SC561GN		3018	11	2038	
2SC561H		3018	11	2038	
2SC561J		3018	11	2038	
2SC561K		3018	11	2038	
2SC561L		3018	11	2038	
2SC561M		3018	11	2038	
2SC561OR		3018	11	2038	
2SC561R		3018	11	2038	
2SC561X		3018	11	2038	
2SC561Y		3018	11	2038	
2SC562	161	3018	61	2015	
2SC562-O	161	3018	61	2015	
2SC562-OR			61	2015	
2SC562A		3018	61	2015	
2SC562B		3018	61	2015	
2SC562C		3018	61	2015	
2SC562D		3018	61	2015	
2SC562E		3018	61	2015	
2SC562F		3018	61	2015	
2SC562G		3018	61	2015	
2SC562GN		3018	61	2015	
2SC562H		3018	61	2015	
2SC562J		3018	61	2015	
2SC562K		3018	61	2015	
2SC562L		3018	61	2015	
2SC562M		3018	61	2015	
2SC562OR		3018	61	2015	
2SC562R		3018	61	2015	
2SC562X		3018	61	2015	
2SC562Y	161	3018	61	2015	
2SC563	161	3018	61	2015	
2SC563(3RDIF)	161		61	2015	
2SC563-P			61	2015	
2SC563-G			61	2015	
2SC563-OR			61	2015	
2SC563A	161	3018	61	2015	
2SC563A(3RDIF)			61	2015	
2SC563B		3018	61	2015	
2SC563C		3018	61	2015	
2SC563D		3018	61	2015	
2SC563E		3018	61	2015	
2SC563F		3018	61	2015	
2SC563G		3018	61	2015	
2SC563GN		3018	61	2015	
2SC563H		3018	61	2015	
2SC563J		3018	61	2015	
2SC563K		3018	61	2015	
2SC563L		3018	61	2015	
2SC563M		3018	61	2015	
2SC563OR		3018	61	2015	
2SC563R		3018	61	2015	
2SC563X		3018	61	2015	
2SC563Y		3018	61	2015	
2SC564	123A		18	2030	
2SC564(Q)			18	2030	
2SC564(Q)(R)			18	2030	
2SC564A	123A	3024/128	18	2030	
2SC564B		3024	18	2030	
2SC564C		3024	18	2030	
2SC564D		3024	18	2030	
2SC564E		3024	18	2030	
2SC564F		3024	18	2030	
2SC564G		3024	18	2030	
2SC564GN		3024	18	2030	
2SC564H		3024	18	2030	
2SC564J		3024	18	2030	
2SC564K		3024	18	2030	
2SC564L		3024	18	2030	
2SC564M		3024	18	2030	
2SC564OR		3024	18	2030	
2SC564P	123A	3024/128	18	2030	
2SC564PL		3024	18	2030	
2SC564Q	123A	3024/128	18	2030	
2SC564QC		3024	18	2030	
2SC564R	123A	3024/128	18	2030	
2SC564S	123A		18	2030	
2SC564T	123A	3114/290	243	2030	
2SC564X		3024	18	2030	
2SC564Y		3024	18	2030	
2SC566	3168	3529	88	2051	
2SC566A		3529	88	2030	
2SC566B		3529	88	2030	
2SC566C		3529	88	2030	
2SC566D		3529	88	2030	
2SC566E		3529	88	2030	
2SC566F		3529	88	2030	
2SC566G		3529	88	2030	
2SC566GN		3529	88	2030	
2SC566H		3529	88	2030	
2SC566J		3529	88	2030	
2SC566K		3529	88	2030	
2SC566L		3529	88	2030	
2SC566M		3529	88	2030	
2SC566OR		3529	88	2030	
2SC566R		3529	88	2030	
2SC566X		3529	88	2030	
2SC566Y		3529	88	2030	
2SC567	316	3039	86	2015	2N5031
2SC567A		3039	86	2038	
2SC567B		3039	86	2038	
2SC567C		3039	86	2038	
2SC567D		3039	86	2038	
2SC567E		3039	86	2038	
2SC567F		3039	86	2038	
2SC567G		3039	86	2038	
2SC567GN		3039	86	2038	
2SC567H		3039	86	2038	
2SC567J		3039	86	2038	
2SC567K		3039	86	2038	
2SC567L		3039	86	2038	
2SC567M		3039	86	2038	
2SC567OR		3039	86	2038	
2SC567R		3039	86	2038	
2SC567X		3039	86	2038	
2SC567Y		3039	86	2038	
2SC568	316	3716/161	86	2038	2N5031
2SC568A			86	2038	
2SC568B			86	2038	
2SC568C			86	2038	
2SC568D			86	2038	
2SC568E			86	2038	
2SC568F			86	2038	
2SC568G			86	2038	
2SC568GN			86	2038	
2SC568H			86	2038	
2SC568J			86	2038	
2SC568K			86	2038	
2SC568L			86	2038	
2SC568M			86	2038	
2SC568OR			86	2038	
2SC568R			86	2038	
2SC568X			86	2038	
2SC568Y			86	2038	
2SC56A		3124	47	2030	
2SC56B		3124	47	2030	
2SC56C		3124	47	2030	
2SC56D		3124	47	2030	
2SC56E		3124	47	2030	
2SC56F		3124	47	2030	
2SC56G		3124	47	2030	
2SC56GN		3124	47	2030	
2SC56H		3124	47	2030	
2SC56J		3124	47	2030	
2SC56K		3124	47	2030	
2SC56L		3124	47	2030	
2SC56M		3124	47	2030	
2SC56OR		3124	47	2030	
2SC56R		3124	47	2030	
2SC56X		3124	47	2030	
2SC56Y		3124	47	2030	
2SC57			63	2030	
2SC571	346	3765/195A	46		2N4427
2SC572	475		28	2017	2N3926
2SC573	476		66	2048	2N3927
2SC58	154	3040	27	2012	
2SC580	128	3765/195A	32		
2SC580A		3047	32		
2SC580B		3047	32		
2SC580C		3047	32		
2SC580D		3047	32		
2SC580E		3047	32		
2SC580F		3047	32		
2SC580G		3047	32		
2SC580GN		3047	32		
2SC580H		3047	32		
2SC580J		3047	32		
2SC580K		3047	32		
2SC580L		3047	32		
2SC580M		3047	32		
2SC580OR		3047	32		
2SC580R		3047	32		
2SC580T		3047	32		
2SC580X		3047	32		
2SC580Y		3047	32		
2SC581			39	2015	
2SC582	124	3021	12		2N3739
2SC582A	124	3021	12		
2SC582B	124	3021	12		
2SC582BC	124	3021	12		
2SC582BX	124		12		
2SC582BY	124		12		
2SC582C	124	3021	12		
2SC582D		3021	12		
2SC582E		3021	12		
2SC582EA	124		12		
2SC582F		3021	12		
2SC582G		3021	12		
2SC582GN		3021	12		
2SC582H		3021	12		
2SC582J		3021	12		
2SC582K		3021	12		
2SC582L		3021	12		
2SC582M		3021	12		
2SC582OR		3021	12		
2SC582R		3021	12		
2SC582X		3021	12		
2SC582Y		3021	12		
2SC583	316		261		
2SC583G	316		261		
2SC585			66	2048	2N3632
2SC586	280	3438	35		MJ410
2SC587	123A	3444	88	2051	
2SC587A	123A	3444	88	2051	
2SC587B	123A	3122	88	2010	
2SC587C	123A	3122	88	2010	
2SC587D		3122	88	2010	
2SC587E		3122	88	2010	
2SC587F		3122	88	2010	
2SC587G		3122	88	2010	
2SC587GN		3122	88	2010	
2SC587H		3122	88	2010	
2SC587J		3122	88	2010	
2SC587K		3122	88	2010	
2SC587L		3122	88	2010	
2SC587M		3122	88	2010	
2SC587OR		3122	88	2010	
2SC587R		3122	88	2010	
2SC587X		3122	88	2010	
2SC587Y		3122	88	2010	
2SC588	123A	3444	88	2051	
2SC588A		3122	88	2030	
2SC588B		3122	88	2030	
2SC588C		3122	88	2030	
2SC588D		3122	88	2030	
2SC588E		3122	88	2030	
2SC588F		3122	88	2030	
2SC588G		3122	88	2030	
2SC588GN		3122	88	2030	
2SC588H		3122	88	2030	
2SC588J		3122	88	2030	
2SC588K		3122	88	2030	
2SC588L		3122	88	2030	
2SC588M		3122	88	2030	
2SC588OR		3122	88	2030	
2SC588R		3122	88	2030	
2SC588X		3122	88	2030	
2SC588Y		3122	88	2030	
2SC589	154	3044	32	2012	
2SC589A		3044	32		
2SC589B		3044	32		
2SC589C		3044	32		
2SC589D		3044	32		
2SC589E		3044	32		
2SC589F		3044	32		
2SC589G		3044	32		
2SC589GN		3044	32		

Industry Standard No.	ECG	SK	GE	RS 276-	MOTOR.
2SC589H		3044	32		
2SC589J		3044	32		
2SC589K		3044	32		
2SC589L		3044	32		
2SC589M		3044	32		
2SC5890R		3044	32		
2SC589R		3044	32		
2SC589X		3044	32		
2SC589Y		3044	32		
2SC58A	154	3040	27	2012	
2SC58AC			20	2012	
2SC58B		3040	27	2012	
2SC58D		3040	27	2012	
2SC58E		3040	27	2012	
2SC58F		3040	27	2012	
2SC58G		3040	27	2012	
2SC58GN		3040	27	2012	
2SC58H		3040	27	2012	
2SC58J		3040	27	2012	
2SC58K		3040	27	2012	
2SC58L		3040	27	2012	
2SC58M		3040	27	2012	
2SC580R		3040	27	2012	
2SC58R		3040	27	2012	
2SC58X		3040	27	2012	
2SC58Y		3040	27	2012	
2SC59	128	3024	18	2030	
2SC590	154	3024/128	18	2038	
2SC590A		3024	18	2030	
2SC590B		3024	18	2030	
2SC590C		3024	18	2030	
2SC590D		3024	18	2030	
2SC590E		3024	18	2030	
2SC590F		3024	18	2030	
2SC590G		3024	18	2030	
2SC590GN		3024	18	2030	
2SC590H		3024	18	2030	
2SC590J		3024	18	2030	
2SC590K		3024	18	2030	
2SC590L		3024	18	2030	
2SC590M		3024	18	2030	
2SC5900R		3024	18	2030	
2SC590X		3024	18	2030	
2SC590Y	289A	3024/128	18	2038	
2SC591		3104A	32		
2SC591A		3104A	32		
2SC591B		3104A	32		
2SC591C		3104A	32		
2SC591D		3104A	32		
2SC591E		3104A	32		
2SC591F		3104A	32		
2SC591G		3104A	32		
2SC591GN		3104A	32		
2SC591H		3104A	32		
2SC591J		3104A	32		
2SC591K		3104A	32		
2SC591L		3104A	32		
2SC591M		3104A	32		
2SC5910R		3104A	32		
2SC591R		3104A	32		
2SC591X		3104A	32		
2SC591Y		3104A	32		
2SC592			28	2017	
2SC593	123A	3444	20	2051	
2SC594	123A	3024/128	18	2030	
2SC594-0			47	2030	
2SC594-R			47	2030	
2SC594-Y			47	2030	
2SC594A		3024	18	2030	
2SC594B		3024	18	2030	
2SC594C		3024	18	2030	
2SC594D		3024	18	2030	
2SC594E		3024	18	2030	
2SC594F		3024	18	2030	
2SC594G		3024	18	2030	
2SC594GN		3024	18	2030	
2SC594H		3024	18	2030	
2SC594J		3024	18	2030	
2SC594K		3024	18	2030	
2SC594L		3024	18	2030	
2SC594M		3024	18	2030	
2SC5940R		3024	18	2030	
2SC594R		3024	18	2030	
2SC594X		3024	18	2030	
2SC594Y		3024	18	2030	
2SC595	123A	3444	18	2051	
2SC595A		3122	88	2030	
2SC595B		3122	88	2030	
2SC595C		3122	88	2030	
2SC595D		3122	88	2030	
2SC595E		3122	88	2030	
2SC595F		3122	88	2030	
2SC595G		3122	88	2030	
2SC595GN		3122	88	2030	
2SC595H		3122	88	2030	
2SC595J		3122	88	2030	
2SC595K		3122	88	2030	
2SC595L		3122	88	2030	
2SC595M		3122	88	2030	
2SC5950R		3122	88	2030	
2SC595R		3122	88	2030	
2SC595X		3122	88	2030	
2SC595Y		3122	88	2030	
2SC596	123A	3444	88	2051	
2SC596A		3122	88	2030	
2SC596B		3122	88	2030	
2SC596C		3122	88	2030	
2SC596D		3122	88	2030	
2SC596E		3122	88	2030	
2SC596F		3122	88	2030	
2SC596G		3122	88	2030	
2SC596GN		3122	88	2030	
2SC596H		3122	88	2030	
2SC596J		3122	88	2030	
2SC596K		3122	88	2030	
2SC596L		3122	88	2030	
2SC596M		3122	88	2030	
2SC5960R		3122	88	2030	
2SC596R		3122	88	2030	
2SC596X		3122	88	2030	
2SC596Y		3122	88	2030	
2SC597		3197	215		2N3553
2SC598			28	2017	2N3926
2SC599			66	2048	
2SC59A		3024	18	2030	
2SC59B		3024	18	2030	
2SC59C		3024	18	2030	
2SC59D		3024	18	2030	
2SC59E		3024	18	2030	
2SC59F		3024	18	2030	
2SC59G		3024	18	2030	
2SC59GN		3024	18	2030	
2SC59H		3024	18	2030	
2SC59J		3024	18	2030	
2SC59K		3024	18	2030	
2SC59L		3024	18	2030	
2SC59M		3024	18	2030	
2SC590R		3024	18	2030	
2SC59R		3024	18	2030	
2SC59X		3024	18	2030	
2SC59Y		3024	18	2030	
2SC6	195A	3765	46		
2SC60	103A	3011	59	2002	
2SC600			66	2048	2N3927
2SC601	311		277		
2SC602	161		60	2015	
2SC604			11	2015	
2SC605	313	3018	278	2016	
2SC605(L)			278	2016	
2SC605(Q)	313		278	2016	
2SC605-OR			278	2016	
2SC605A		3018	278	2016	
2SC605B		3018	278	2016	
2SC605C		3018	278	2016	
2SC605D		3018	278	2016	
2SC605E		3018	278	2016	
2SC605F		3018	278	2016	
2SC605G		3018	278	2016	
2SC605GN		3018	278	2016	
2SC605H		3018	278	2016	
2SC605J		3018	278	2016	
2SC605K	313	3018	278	2016	
2SC605L	313	3018	278	2016	
2SC605M	313	3018	278	2016	
2SC6050R		3018	278	2016	
2SC605Q	313	3018	61	2016	
2SC605R		3018	278	2016	
2SC605TW		3039	278	2016	
2SC605X		3018	278	2016	
2SC605Y		3018	278	2016	
2SC606	313	3452/108	86	2038	
2SC606(VHF)			278	2016	
2SC606A	313	3018	278	2016	
2SC606B		3018	278	2016	
2SC606C		3018	278	2016	
2SC606D		3018	278	2016	
2SC606E		3018	278	2016	
2SC606F		3018	278	2016	
2SC606G		3018	278	2016	
2SC606GN		3018	278	2016	
2SC606H		3018	278	2016	
2SC606J		3018	278	2016	
2SC606K		3018	278	2016	
2SC606L		3018	278	2016	
2SC606M		3018	278	2016	
2SC606N			278	2016	
2SC6060R		3018	278	2016	
2SC606R		3018	278	2016	
2SC606X		3018	278	2016	
2SC606Y		3018	278	2016	
2SC608	237	3765/195A	18		
2SC608A		3765	18		
2SC608AA		3765	18		
2SC608B		3765	18		
2SC608C		3765	18		
2SC608D		3765	18		
2SC608E	237	3765/195A	18		
2SC608F		3765	18		
2SC608G		3765	18		
2SC608GN		3765	18		
2SC608H		3765	18		
2SC608J		3765	18		
2SC608K		3765	18		
2SC608L		3765	18		
2SC608M		3765	18		
2SC6080R		3765	18		
2SC608R		3765	18		
2SC608T	237	3765/195A	18		
2SC608X		3765	18		
2SC608Y		3765	18		
2SC609	237	3765/195A	46		
2SC609A		3765	46		
2SC609B		3765	46		
2SC609C		3765	46		
2SC609D		3765	18		
2SC609E		3765	46		
2SC609F	237	3765/195A	46		
2SC609G		3765	46		
2SC609GN		3765	46		
2SC609H		3765	46		
2SC609J		3765	46		
2SC609K		3765	46		
2SC609L		3765	46		
2SC609M		3765	46		
2SC6090R		3765	46		
2SC609R		3765	46		
2SC609T	237	3765/195A	46		
2SC609X		3765	46		
2SC609Y	237	3765/195A	46		
2SC60A		3011	59		
2SC60B		3011	59		
2SC60C		3011	59		
2SC60D		3011	59		
2SC60E		3011	59		
2SC60F		3011	59		
2SC60G		3011	59		
2SC60GN		3011	59		
2SC60H		3011	59		
2SC60J		3011	59		
2SC60K		3011	59		
2SC60L		3011	59		
2SC60M		3011	59		
2SC60R		3011	59		
2SC60X		3011	59		
2SC60Y		3011	59		
2SC61	128	3024	18	2030	
2SC610		3024	18	2032	
2SC610A		3024	18	2030	
2SC610B		3024	18	2030	
2SC610C		3024	18	2030	
2SC610D		3024	18	2030	
2SC610E		3024	18	2030	
2SC610F		3024	18	2030	
2SC610G		3024	18	2030	
2SC610GN		3024	18	2030	
2SC610H		3024	18	2030	
2SC610J		3024	18	2030	
2SC610K		3024	18	2030	
2SC610L		3024	18	2030	
2SC610M		3024	18	2030	
2SC6100R		3024	18	2030	
2SC610R		3024	18	2030	
2SC610X		3024	18	2030	
2SC610Y		3024	18	2030	
2SC611	316	3452/108	86	2038	
2SC611A		3039	86	2038	
2SC611B		3039	86	2038	
2SC611C		3039	86	2038	

Industry Standard No.	ECG	SK	GE	RS 276-	MOTOR.
2SC611D		3039	86	2038	
2SC611E		3039	86	2038	
2SC611F		3039	86	2038	
2SC611G		3039	86	2038	
2SC611GN		3039	86	2038	
2SC611H		3039	86	2038	
2SC611J		3039	86	2038	
2SC611K		3039	86	2038	
2SC611L		3039	86	2038	
2SC611M		3039	86	2038	
2SC611OR		3039	86	2038	
2SC611R		3039	86	2038	
2SC611X		3039	86	2038	
2SC611Y		3039	86	2038	
2SC612	316	3452/108	86	2038	
2SC612A		3039	86	2038	
2SC612B		3039	86	2038	
2SC612C		3039	86	2038	
2SC612D		3039	86	2038	
2SC612E		3039	86	2038	
2SC612F		3039	86	2038	
2SC612G		3039	86	2038	
2SC612GN		3039	86	2038	
2SC612H		3039	86	2038	
2SC612J		3039	86	2038	
2SC612K		3039	86	2038	
2SC612L		3039	86	2038	
2SC612M		3039	86	2038	
2SC612OR		3039	86	2038	
2SC612R		3039	86	2038	
2SC612X		3039	86	2038	
2SC612Y		3039	86	2038	
2SC613		3452	86	2038	
2SC613A		3039	86	2038	
2SC613B		3039	86	2038	
2SC613C		3039	86	2038	
2SC613D		3039	86	2038	
2SC613E		3039	86	2038	
2SC613F		3039	86	2038	
2SC613G		3039	86	2038	
2SC613GN		3039	86	2038	
2SC613H		3039	86	2038	
2SC613J		3039	86	2038	
2SC613K		3039	86	2038	
2SC613L		3039	86	2038	
2SC613M		3039	86	2038	
2SC613OR		3039	86	2038	
2SC613R		3039	86	2038	
2SC613X		3039	86	2038	
2SC613Y		3039	86	2038	
2SC614	195A	3048/329	18		
2SC614A		3024	18	2030	
2SC614B		3024	18	2030	
2SC614C	195A	3024/128	18		
2SC614D	195A	3047	18		
2SC614E	195A	3048/329	18		
2SC614F	195A	3024/128	18		
2SC614G	195A	3024/128	18		
2SC614GN		3024	18	2030	
2SC614H		3024	18	2030	
2SC614J		3024	18	2030	
2SC614K		3024	18	2030	
2SC614L		3024	18	2030	
2SC614M		3024	18	2030	
2SC614OR		3024	18	2030	
2SC614R		3024	18	2030	
2SC614X		3024	18	2030	
2SC614Y		3024	18	2030	
2SC615	195A	3024/128	46		
2SC615A	195A	3024/128	46		
2SC615B	195A	3024/128	46		
2SC615C	195A	3024/128	46		
2SC615D	195A	3024/128	46		
2SC615E	195A	3024/128	46		
2SC615F	195A	3048/329	46		
2SC615G	195A	3024/128	46		
2SC615GN		3024	46		
2SC615H		3024	46		
2SC615J		3024	46		
2SC615K		3024	46		
2SC615L		3024	46		
2SC615M		3024	46		
2SC615OR		3024	46		
2SC615R		3024	46		
2SC615X		3024	46		
2SC615Y		3024	46		
2SC618	161			2011	
2SC618A	161			2011	
2SC619		3444	88	2051	
2SC619(B)			88	2051	
2SC619(C)			88	2051	
2SC619A		3122	88	2051	
2SC619B		3444	88	2051	
2SC619C		3444	88	2051	
2SC619D		3444	88	2051	
2SC619E		3122	88	2051	
2SC619F		3122	88	2051	
2SC619G		3122	88	2051	
2SC619GN		3122	88	2051	
2SC619H		3122	88	2051	
2SC619J		3122	88	2051	
2SC619K		3122	88	2051	
2SC619L		3122	88	2051	
2SC619M		3122	88	2051	
2SC619OR		3122	88	2051	
2SC619R		3122	88	2051	
2SC619X		3122	88	2051	
2SC619Y		3122	88	2051	
2SC61A		3024	18	2030	
2SC61B		3024	18	2030	
2SC61C		3024	18	2030	
2SC61D		3024	18	2030	
2SC61E		3024	18	2030	
2SC61F		3024	18	2030	
2SC61G		3024	18	2030	
2SC61GN		3024	18	2030	
2SC61H		3024	18	2030	
2SC61J		3024	18	2030	
2SC61K		3024	18	2030	
2SC61L		3024	18	2030	
2SC61M		3024	18	2030	
2SC61OR		3024	18	2030	
2SC61R		3024	18	2030	
2SC61Y		3024	18	2030	
2SC62	123A	3444	88	2051	
2SC620		3124	20	2051	
2SC620(C)			20	2051	
2SC620(D)			20	2051	
2SC620-OR			20	2051	
2SC620A		3124	20	2051	
2SC620B		3124	20	2051	
2SC620C		3124	20	2051	
2SC620CD		3124	20	2051	
2SC620D		3124	20	2051	
2SC620DE		3124	20	2051	
2SC620E		3124	20	2051	
2SC620F		3124	20	2051	
2SC620G		3124	20	2051	
2SC620GN		3124	20	2051	
2SC620H		3124	20	2051	
2SC620J		3124	20	2051	
2SC620K		3124	20	2051	
2SC620L		3124	20	2051	
2SC620M		3124	20	2051	
2SC620OR		3124	20	2051	
2SC620R		3124	20	2051	
2SC620X		3124	20	2051	
2SC620Y		3124	20	2051	
2SC621	85	3444/123A	88	2051	
2SC621A		3122	88	2030	
2SC621B		3122	88	2030	
2SC621C		3122	88	2030	
2SC621D		3122	88	2030	
2SC621E		3122	88	2030	
2SC621F		3122	88	2030	
2SC621G		3122	88	2030	
2SC621GN		3122	88	2030	
2SC621H		3122	88	2030	
2SC621J		3122	88	2030	
2SC621K		3122	88	2030	
2SC621L		3122	88	2030	
2SC621M		3122	88	2030	
2SC621OR		3122	88	2030	
2SC621R		3122	88	2030	
2SC621X		3122	88	2030	
2SC621Y		3122	88	2030	
2SC622	123A	3444	88	2051	
2SC622A		3122	88	2030	
2SC622B		3122	88	2030	
2SC622C		3122	88	2030	
2SC622D		3122	88	2030	
2SC622E		3122	88	2030	
2SC622F		3122	88	2030	
2SC622G		3122	88	2030	
2SC622GN		3122	88	2030	
2SC622H		3122	88	2030	
2SC622J		3122	88	2030	
2SC622K		3122	88	2030	
2SC622L		3122	88	2030	
2SC622M		3122	88	2030	
2SC622OR		3122	88	2030	
2SC622R		3122	88	2030	
2SC622X		3122	88	2030	
2SC622Y		3122	88	2030	
2SC626	123A	3444	20	2051	
2SC627	154		40	2012	
2SC628	311		243	2030	MRF225
2SC628E			17	2051	
2SC628F			17	2051	
2SC629	107	3018	61	2015	
2SC629-31		3039	61	2038	
2SC629-41		3039	61	2038	
2SC629A		3018	61	2038	
2SC629B		3018	61	2038	
2SC629C		3018	61	2038	
2SC629D		3018	61	2038	
2SC629E		3018	61	2038	
2SC629F		3018	61	2038	
2SC629G		3018	61	2038	
2SC629GN		3018	61	2038	
2SC629H		3018	61	2038	
2SC629J		3018	61	2038	
2SC629K		3018	61	2038	
2SC629L		3018	61	2038	
2SC629M		3018	61	2038	
2SC629OR		3018	61	2038	
2SC629R		3018	61	2038	
2SC629X		3018	61	2038	
2SC629Y		3018	61	2038	
2SC62A		3122	88	2030	
2SC62B		3122	88	2030	
2SC62C		3122	88	2030	
2SC62D		3122	88	2030	
2SC62E		3122	88	2030	
2SC62F		3122	88	2030	
2SC62G		3122	88	2030	
2SC62GN		3122	88	2030	
2SC62H		3122	88	2030	
2SC62J		3122	88	2030	
2SC62L		3122	88	2030	
2SC62M		3122	88	2030	
2SC62OR		3122	88	2030	
2SC62R		3122	88	2030	
2SC62X		3122	88	2030	
2SC62Y		3122	88	2030	
2SC63	108	3452	86	2038	
2SC631	85	3124/289	20	2051	
2SC631A	85	3124/289	20	2051	
2SC631B		3124	20	2030	
2SC631C		3124	20	2030	
2SC631D		3124	20	2030	
2SC631E		3124	20	2030	
2SC631F		3124	20	2030	
2SC631G		3124	20	2030	
2SC631GN		3124	20	2030	
2SC631H		3124	20	2030	
2SC631J		3124	20	2030	
2SC631K		3124	20	2030	
2SC631L		3124	20	2030	
2SC631M		3124	20	2030	
2SC631OR		3124	20	2030	
2SC631R		3124	20	2030	
2SC631X		3124	20	2030	
2SC631Y		3124	20	2030	
2SC632	85	3124/289	47	2051	
2SC632(1)		3124	47	2051	
2SC632-OR			47	2051	
2SC632A	85	3124/289	47	2051	
2SC632B		3124	47	2051	
2SC632C		3124	47	2051	
2SC632D		3124	47	2051	
2SC632E		3124	47	2051	
2SC632F		3124	47	2051	
2SC632G		3124	47	2051	
2SC632GN		3124	47	2051	
2SC632H		3124	47	2051	
2SC632J		3124	47	2051	
2SC632K		3124	47	2051	
2SC632L		3124	47	2051	
2SC632M		3124	47	2051	
2SC632OR		3124	47	2051	
2SC632R		3124	47	2051	
2SC632X		3124	47	2051	
2SC632Y		3124	47	2051	
2SC633	85	3124/289	47	2051	
2SC633-7	85	3124/289	47	2051	
2SC633-OR			47	2051	
2SC633A	85	3124/289	47	2051	

Industry Standard No.	ECG	SK	GE	RS 276-	MOTOR.
2SC633B		3124	47	2051	
2SC633C		3124	47	2051	
2SC633D		3124	47	2051	
2SC633E		3124	47	2051	
2SC633F		3124	47	2051	
2SC633G	85	3124/289	47	2051	
2SC633GN		3124	47	2051	
2SC633H	85	3124/289	47	2051	
2SC633J		3124	47	2051	
2SC633K		3124	47	2051	
2SC633L		3124	47	2051	
2SC633M		3124	47	2051	
2SC633OR		3124	47	2051	
2SC633R		3124	47	2051	
2SC633X		3124	47	2051	
2SC633Y		3124	47	2051	
2SC634	85	3124/289	220	2051	
2SC634(2)		3124	220	2051	
2SC634-0			220	2051	
2SC634-OR			220	2051	
2SC634A	85	3124/289	220	2051	
2SC634AK		3124	220	2051	
2SC634AL		3124	220	2051	
2SC634AXL		3124	220	2051	
2SC634B		3124	220	2051	
2SC634C		3124	220	2051	
2SC634D		3124	220	2051	
2SC634E		3124	220	2051	
2SC634F		3124	220	2051	
2SC634G		3124	220	2051	
2SC634GN		3124	220	2051	
2SC634H		3124	220	2051	
2SC634J		3124	220	2051	
2SC634K		3124	220	2051	
2SC634L		3124	220	2051	
2SC634M		3124	220	2051	
2SC634OR		3124	220	2051	
2SC634R		3124	220	2051	
2SC634X		3124	220	2051	
2SC634Y		3124	220	2051	
2SC635			28	2051	2N3632
2SC635A			28	2051	
2SC636		3083	66	2048	2N3632
2SC637	475		28	2017	2N3926
2SC638	476		66	2048	2N3927
2SC638C		3018	66	2048	
2SC63A		3039	86	2038	
2SC63B		3039	86	2038	
2SC63D		3039	86	2038	
2SC63E		3039	86	2038	
2SC63F		3039	86	2038	
2SC63G		3039	86	2038	
2SC63GN		3039	86	2038	
2SC63H		3039	86	2038	
2SC63J		3039	86	2038	
2SC63K		3039	86	2038	
2SC63L		3039	86	2038	
2SC63M		3039	86	2038	
2SC63OR		3039	86	2038	
2SC63R		3039	86	2038	
2SC63X		3039	86	2038	
2SC63Y		3039	86	2038	
2SC64	128	3045/225	32	2012	
2SC640	313	3444/123A	88	2051	
2SC640A		3122	88	2010	
2SC640B	313	3444/123A	20	2051	
2SC640C		3122	88	2010	
2SC640D		3122	88	2010	
2SC640E		3122	88	2010	
2SC640F		3122	88	2010	
2SC640G		3122	88	2010	
2SC640GN		3122	88	2010	
2SC640J		3122	88	2010	
2SC640K		3122	88	2010	
2SC640L		3122	88	2010	
2SC640M		3122	88	2010	
2SC640R		3122	88	2010	
2SC640X		3122	88	2010	
2SC641	85	3452/108	86	2038	
2SC641A		3039	86	2051	
2SC641B	85	3452/108	86	2038	
2SC641C	85	3039/316	86	2051	
2SC641D		3039	86	2051	
2SC641E		3039	86	2051	
2SC641F		3039	86	2051	
2SC641G		3039	86	2051	
2SC641GN		3039	86	2051	
2SC641H		3039	86	2051	
2SC641J		3039	86	2051	
2SC641K	85	3039/316	86	2051	
2SC641L		3039	86	2051	
2SC641M		3039	86	2051	
2SC641OR		3039	86	2051	
2SC641R		3039	86	2050	
2SC641X		3039	86	2051	
2SC641Y		3039	86	2051	
2SC642	164	3133	37		BU204
2SC642A	164	3710/238	37		
2SC643	389	3710/238	259		BU204
2SC643A	389	3710/238	259		
2SC643B		3710	259		
2SC643C		3710	259		
2SC643D		3710	259		
2SC643E		3710	259		
2SC643F		3710	259		
2SC643G		3710	259		
2SC643GN		3710	259		
2SC643H		3710	259		
2SC643J		3710	259		
2SC643K		3710	259		
2SC643L		3710	259		
2SC643M		3710	259		
2SC644	199	3245	62	2010	
2SC644(F)			62	2010	
2SC644(H)			62	2010	
2SC644(R)			62	2010	
2SC644(R,S)			62	2010	
2SC644(S)			62	2010	
2SC644-OR			62	2010	
2SC644A			62	2010	
2SC644B			62	2010	
2SC644C	199	3245	62	2010	
2SC644D			62	2010	
2SC644E			62	2010	
2SC644F	199	3245	62	2010	
2SC644F(H)(S)			62	2010	
2SC644FH			62	2010	
2SC644FHS			62	2010	
2SC644FR	199	3245	62	2010	
2SC644FS	199	3245	62	2010	
2SC644G		3245	62	2010	
2SC644GN			62	2010	
2SC644H	199	3245	62	2010	
2SC644H(S)			62	2010	
2SC644HR	199	3245	62	2010	
2SC644HS	199	3245	62	2010	
2SC644J			62	2010	
2SC644K			62	2010	
2SC644L			62	2010	
2SC644M			62	2010	
2SC644OR			62	2010	
2SC644P	199	3245	62	2010	
2SC644PJ	199	3245	62	2010	
2SC644Q	199	3245	62	2010	
2SC644R	199	3245	62	2010	
2SC644RF	199	3245			
2SC644RST	199		62	2010	
2SC644S	199	3245	62	2010	
0002SC644S,R,Q			62	2010	
2SC644T	199	3245	62	2010	
2SC644X			62	2010	
2SC644Y			62	2010	
2SC645		3444	61	2015	
2SC645-OR			61	2015	
2SC645A		3444	61	2015	
2SC645B		3444	61	2015	
2SC645B-1		3018	61	2015	
2SC645C		3018	61	2015	
2SC645D		3018	61	2015	
2SC645E		3018	61	2015	
2SC645F		3018	61	2015	
2SC645G		3018	61	2015	
2SC645GN		3018	61	2015	
2SC645GR		3018	61	2015	
2SC645H		3018	61	2015	
2SC645J		3018	61	2015	
2SC645K		3018	61	2015	
2SC645L		3018	61	2015	
2SC645M		3018	61	2015	
2SC645N		3018	61	2015	
2SC645OR		3018	61	2015	
2SC645R		3018	61	2015	
2SC645V			61	2015	
2SC645X		3018	61	2015	
2SC645Y		3018	61	2015	
2SC646	328	3563	255	2041	2N3447
2SC647	328	3561	241	2041	2N3448
2SC647P	328	3561			
2SC647Q	328	3561	241	2041	
2SC647R	328	3561	241	2041	
2SC648	123A	3124/289	62	2010	
2SC648A		3124	86	2038	
2SC648B		3124	86	2038	
2SC648C		3124	86	2038	
2SC648D		3124	86	2038	
2SC648E		3124	86	2038	
2SC648F		3124	86	2038	
2SC648G		3124	86	2038	
2SC648H	123A	3124/289	62	2010	
2SC648J		3124	86	2038	
2SC648K		3124	86	2038	
2SC648L		3124	86	2038	
2SC648M		3124	86	2038	
2SC648OR		3124	86	2038	
2SC648R		3124	86	2038	
2SC648X		3124	86	2038	
2SC648Y		3124	86	2038	
2SC649	123A	3039/316	86	2038	
2SC649A	123A	3039/316	86	2038	
2SC649B	123A	3039/316	86	2038	
2SC649C		3039	86	2038	
2SC649D		3039	86	2038	
2SC649E		3039	86	2038	
2SC649F		3039	86	2038	
2SC649G		3039	86	2038	
2SC649GN		3039	86	2038	
2SC649H		3039	86	2038	
2SC649J		3039	86	2038	
2SC649K		3039	86	2038	
2SC649L		3039	86	2038	
2SC649M		3039	86	2038	
2SC649OR		3039	86	2038	
2SC649R		3039	86	2038	
2SC649X		3039	86	2038	
2SC649Y		3039	86	2038	
2SC64A		3045	32		
2SC64B		3045	32		
2SC64C		3045	32		
2SC64D		3045	32		
2SC64E		3045	32		
2SC64F		3045	32		
2SC64G		3045	32		
2SC64GN		3045	32		
2SC64H		3045	32		
2SC64K		3045	32		
2SC64L		3045	32		
2SC64M		3045	32		
2SC64OR		3045	32		
2SC64R		3045	32		
2SC64X		3045	32		
2SC64Y		3045	32		
2SC64Y-RST			62	2010	
2SC65	154	3024/128	40	2012	
2SC65-0			40	2012	
2SC65-OR			40	2012	
2SC650	123A	3122	62	2010	
2SC650-OR			62	2010	
2SC650-Y			62	2010	
2SC650A	123A	3124/289	62	2010	
2SC650B	123A	3122	62	2010	
2SC650C		3124	62	2010	
2SC650D		3124	62	2010	
2SC650E		3124	62	2010	
2SC650F		3124	60	2010	
2SC650G		3124	62	2010	
2SC650GN		3124	62	2010	
2SC650H		3124	62	2010	
2SC650J		3124	62	2010	
2SC650K		3124	62	2010	
2SC650L		3124	62	2010	
2SC650M		3124	62	2010	
2SC650OR			62	2010	
2SC650R		3124	62	2010	
2SC650X		3124	62	2010	
2SC650Y		3124	62	2010	
2SC651	311		277		2N4428
2SC652	311		277		2N5943
2SC653	316	3716/161	39	2015	
2SC654	311	3444/123A	20	2051	
2SC654A		3122	88	2030	
2SC654B		3122	88	2030	
2SC654C		3122	88	2030	
2SC654D		3122	88	2030	
2SC654E		3122	88	2030	
2SC654F		3122	88	2030	
2SC654G		3122	88	2030	
2SC654GN		3122	88	2030	
2SC654H		3122	88	2030	

Industry Standard No.	ECG	SK	GE	RS 276-	MOTOR.
2SC654J		3122	88	2030	
2SC654K		3122	88	2030	
2SC654L		3122	88	2030	
2SC654M		3122	88	2030	
2SC6540R		3122	88	2030	
2SC654R		3122	88	2030	
2SC654X		3122	88	2030	
2SC654Y		3122	88	2030	
2SC655		3444	20	2051	
2SC655A		3122	88	2010	
2SC655B		3122	88	2010	
2SC655C		3122	88	2010	
2SC655D		3122	88	2010	
2SC655E		3122	88	2010	
2SC655F		3122	88	2010	
2SC655G		3122	88	2010	
2SC655GN		3122	88	2010	
2SC655H		3122	88	2010	
2SC655J		3122	88	2010	
2SC655K		3122	88	2010	
2SC655L		3122	88	2010	
2SC655M		3122	88	2010	
2SC6550R		3122	88	2010	
2SC655R		3122	88	2010	
2SC655X		3122	88	2010	
2SC655Y		3122	88	2010	
2SC656		3039	11	2015	
2SC656A		3039	86	2038	
2SC656B		3039	86	2038	
2SC656C		3039	86	2038	
2SC656D		3039	86	2038	
2SC656E		3039	86	2038	
2SC656F		3039	86	2038	
2SC656G		3039	86	2038	
2SC656GN		3039	86	2038	
2SC656H		3039	86	2038	
2SC656J		3039	86	2038	
2SC656K		3039	86	2038	
2SC656L		3039	86	2038	
2SC656M		3039	86	2038	
2SC6560R		3039	86	2038	
2SC656R		3039	86	2038	
2SC656X		3039	86	2038	
2SC656Y		3039	86	2038	
2SC657	107	3018	11	2015	
2SC657A		3018	61	2038	
2SC657B		3018	61	2038	
2SC657C		3018	61	2038	
2SC657D		3018	61	2038	
2SC657E		3018	61	2038	
2SC657F		3018	61	2038	
2SC657G		3018	61	2038	
2SC657GN		3018	61	2038	
2SC657H		3018	61	2038	
2SC657J		3018	61	2038	
2SC657K		3018	61	2038	
2SC657L		3018	61	2038	
2SC657M		3018	61	2038	
2SC6570R		3018	61	2038	
2SC657R		3018	61	2038	
2SC657X		3018	61	2038	
2SC657Y		3018	61	2038	
2SC658	107	3452/108	86	2038	
2SC658A	107	3452/108	86	2038	
2SC658B		3039	86	2038	
2SC658C		3039	86	2038	
2SC658D		3039	86	2038	
2SC658E		3039	86	2038	
2SC658F		3039	86	2038	
2SC658G		3039	86	2038	
2SC658GN		3039	86	2038	
2SC658H		3039	86	2038	
2SC658J		3039	86	2038	
2SC658K		3039	86	2038	
2SC658L		3039	86	2038	
2SC658M		3039	86	2038	
2SC6580R		3039	86	2038	
2SC658R		3039	86	2038	
2SC658X		3039	86	2038	
2SC658Y		3039	86	2038	
2SC659	107	3452/108	86	2038	
2SC659A		3039	86	2038	
2SC659B		3039	86	2038	
2SC659C		3039	86	2038	
2SC659D		3039	86	2038	
2SC659E		3039	86	2038	
2SC659F		3039	86	2038	
2SC659G		3039	86	2038	
2SC659GN		3039	86	2038	
2SC659H		3039	86	2038	
2SC659J		3039	86	2038	
2SC659K		3039	86	2038	
2SC659L		3039	86	2038	
2SC659M		3039	86	2038	
2SC6590R		3039	86	2038	
2SC659R		3039	86	2038	
2SC659X		3039	86	2038	
2SC659Y		3039	86	2038	
2SC65A		3044	40	2012	
2SC65B	154	3044	40	2012	
2SC65C		3044	40	2012	
2SC65D		3044	40	2012	
2SC65E		3044	40	2012	
2SC65F		3044	40	2012	
2SC65G		3044	40	2012	
2SC65GN		3044	40	2012	
2SC65H		3044	40	2012	
2SC65K		3044	40	2012	
2SC65L		3044	40	2012	
2SC65M		3044	40	2012	
2SC65N	154		40	2012	
2SC650R		3044	40	2012	
2SC65R		3044	40	2012	
2SC65X		3044	40	2012	
2SC65Y	154	3044	40	2012	
2SC65Y(B)			40	2012	
2SC65YA	154	3044	40	2012	
2SC65YB	154		40	2012	
2SC65YTV	154	3045/225	40	2012	
2SC65YTV1		3045	40	2012	
2SC66	154	3045/225	32	2012	
2SC660				2011	
2SC661				2011	
2SC662	108	3452	86	2038	
2SC662A		3018	61	2038	
2SC662B		3039	86	2038	
2SC662C		3039	86	2038	
2SC662D		3039	86	2038	
2SC662E		3039	86	2038	
2SC662F		3039	86	2038	
2SC662G		3039	86	2038	
2SC662GN		3039	86	2038	
2SC662H		3039	86	2038	
2SC662J		3039	86	2038	
2SC662K		3039	86	2038	
2SC662L		3039	86	2038	
2SC662M		3039	86	2038	
2SC6620R		3039	86	2038	
2SC662R		3039	86	2038	
2SC662X		3039	86	2038	
2SC662Y		3039	86	2038	
2SC663	161	3117	39	2015	
2SC663A		3117	211	2016	
2SC663B		3117	211	2016	
2SC663C		3117	211	2016	
2SC663D		3117	211	2016	
2SC663E		3117	211	2016	
2SC663F		3117	211	2016	
2SC663G		3117	211	2016	
2SC663GN		3117	211	2016	
2SC663H		3117	211	2016	
2SC663J		3117	211	2016	
2SC663K		3117	211	2016	
2SC663L		3117	211	2016	
2SC663M		3117	211	2016	
2SC6630R		3117	211	2016	
2SC663R		3117	211	2016	
2SC663X		3117	211	2016	
2SC663Y		3117	211	2016	
2SC664	130	3027	75		2N5758
2SC664A		3027	75		
2SC664B	130	3027	75		
2SC664C	130	3027	75		
2SC664D		3027	75		
2SC664E		3027	75		
2SC664F		3027	75		
2SC664L		3027	75		
2SC6640R		3027	75		
2SC664R		3027	75		
2SC664Y		3027	75		
2SC665	87	3297/280	262	2041	2N5760
2SC665H	87	3297/280	262	2041	
2SC665HA	87	3297/280	262	2041	
2SC665HB	87	3111	262	2041	
2SC667			39	2015	
2SC668	107	3293	86	2038	
2SC668(C)			211	2038	
2SC668(D)			211	2038	
2SC668-0	107		211	2038	
2SC668-0R			211	2038	
2SC668A	107	3293	211	2038	
2SC668B	107	3293	211	2038	
2SC668B1	107		211	2038	
2SC668BC2	107		211	2038	
2SC668C	107	3293	211	2038	
2SC668C1	107	3293	211	2038	
2SC668C2			211	2038	
2SC668CD	107	3293	211	2038	
0002SC668D	108	3452	11	2015	
2SC668D0	107	3293	211	2038	
2SC668D1	107	3293	211	2038	
2SC668DE	107		211	2038	
2SC668D0	107		211	2038	
2SC668DV	107	3293	211	2038	
2SC668DX	107	3293	211	2038	
2SC668DZ	107	3293	211	2038	
2SC668E	107	3293	211	2038	
2SC668E1	107	3293	211	2038	
2SC668E2	107	3293	211	2038	
2SC668EP	107	3293	211	2038	
2SC668EV	107	3293	211	2038	
2SC668EX	107		211	2038	
2SC668F	107	3293	211	2038	
2SC668G		3293	211	2038	
2SC668GN		3293	211	2038	
2SC668H		3293	211	2038	
2SC668K		3293	211	2038	
2SC668L		3293	211	2038	
2SC668M		3293	211	2038	
2SC6680R		3293	211	2038	
2SC668R		3293	211	2038	
2SC668SP	107	3293			
2SC668X		3293	211	2038	
2SC668Y		3293	211	2038	
2SC66A		3045	32		
2SC66B		3045	32		
2SC66C		3045	32		
2SC66D		3045	32		
2SC66E		3045	32		
2SC66F		3045	32		
2SC66G		3045	32		
2SC66GN		3045	32		
2SC66H		3045	32		
2SC66J		3045	32		
2SC66K		3045	32		
2SC66L		3045	32		
2SC66M		3045	32		
2SC660R		3045	32		
2SC66R		3045	32		
2SC66X		3045	32		
2SC66Y		3045	32		
2SC67		3444	88	2051	
2SC670	124		12		
2SC673			67	2015	
2SC673(B)			67	2015	
2SC673B		3114	67	2015	
2SC673C		3715	67	2015	
2SC673C2			67	2015	
2SC673D			67	2015	
2SC674	107	3018	39	2015	
2SC674(D)			39	2015	
2SC674(F)			39	2015	
2SC674(G)			39	2015	
2SC674-B			39	2015	
2SC674-F			39	2015	
2SC674B	107	3117	39	2015	
2SC674C	107	3117	39	2015	
2SC674CK	107	3117	39	2015	
2SC674CL	107	3117	39	2015	
2SC674CV	107	3716/161	39	2015	
2SC674CZ	107	3716/161	39	2015	
2SC674D	107	3117	39	2015	
2SC674E	107	3132	39	2015	
2SC674F	107	3117	39	2015	
2SC674G	107	3716/161	39	2015	
2SC674V	107	3018	39	2015	
2SC675	87	3467/283	36		2N6306
2SC676					2N6306
2SC677	87		22	2032	2N6306
2SC678		3268			2N6306
2SC679					2N3585
2SC67A		3122	88	2030	
2SC67B		3122	88	2030	
2SC67C		3122	88	2030	
2SC67D		3122	88	2030	
2SC67E		3122	88	2030	
2SC67F		3122	88	2030	
2SC67G		3122	88	2030	

Industry Standard No.	ECG	SK	GE	RS 276-	MOTOR.
2SC67GN		3122	88	2030	
2SC67H		3122	88	2030	
2SC67J		3122	88	2030	
2SC67K		3122	88	2030	
2SC67L		3122	88	2030	
2SC67M		3122	88	2030	
2SC67OR		3122	88	2030	
2SC67R		3122	88	2030	
2SC67X		3122	88	2030	
2SC67Y		3122	88	2030	
2SC68		3444	88	2051	
2SC680	286	3194	66	2020	2N5052
2SC680(A)			66	2020	
2SC680A	286	3194	66	2020	
2SC680B		3194	66	2020	
2SC680C		3194	66	2020	
2SC680G		3194	66	2020	
2SC680GN		3194	66	2020	
2SC680H		3194	66	2020	
2SC680J		3194	66	2020	
2SC680K		3194	66	2020	
2SC680L		3194	66	2020	
2SC680M		3194	66	2020	
2SC680R	286	3194	66	2020	
2SC680X		3194	66	2020	
2SC681	283	3439/163A	36		MJ15011
2SC681A	283	3439/163A	36		
2SC681ARD	283	3439/163A			
2SC681AYL	283	3439/163A	36		
2SC681B	283	3439/163A	36		
2SC681YL	283	3439/163A	36		
2SC682	161	3117	61	2015	
2SC682(B)			61	2015	
2SC682-OR			61	2015	
2SC682A	161	3117	61	2015	
2SC682B	161	3117	61	2015	
2SC682C		3117	61	2015	
2SC682D		3117	61	2015	
2SC682E	161	3117	61	2015	
2SC682F		3117	61	2015	
2SC682G		3117	61	2015	
2SC682GN		3117	61	2015	
2SC682H		3117	61	2015	
2SC682J		3117	61	2015	
2SC682K		3117	61	2015	
2SC682L		3117	61	2015	
2SC682M		3117	61	2015	
2SC682OR		3117	61	2015	
2SC682R		3117	61	2015	
2SC682X		3117	61	2015	
2SC682Y		3117	61	2015	
2SC683	161	3018	61	2038	
2SC683(B)			61	2038	
2SC683-OR			61	2038	
2SC683A	161	3018	61	2038	
2SC683B	161	3018	61	2038	
2SC683C		3018	61	2038	
2SC683D		3018	61	2038	
2SC683E		3018	61	2038	
2SC683F		3018	61	2038	
2SC683G		3018	61	2038	
2SC683GN		3018	61	2038	
2SC683H		3018	61	2038	
2SC683J		3018	61	2038	
2SC683K		3018	61	2038	
2SC683L		3018	61	2038	
2SC683M		3018	61	2038	
2SC683OR		3018	61	2038	
2SC683R	161	3018	61	2038	
2SC683S			61	2038	
2SC683V	161	3018	61	2038	
2SC683X		3018	61	2038	
2SC683Y		3018	61	2038	
2SC684	107	3293	86	2038	
2SC684-OR			11	2015	
2SC684A	107	3293	86	2038	
2SC684B	107	3293	86	2038	
2SC684BK	107	3293	86	2038	
2SC684C		3293	86	2015	
2SC684D		3293	86	2015	
2SC684E		3293	86	2015	
2SC684F	107	3293	86	2038	
2SC684G		3293	86	2015	
2SC684GN		3293	86	2015	
2SC684H		3293	86	2015	
2SC684J		3293	86	2015	
2SC684K		3293	86	2015	
2SC684L		3293	86	2015	
2SC684M		3293	86	2015	
2SC684OR		3293	86		
2SC684R	107	3293	86	2015	
2SC684X		3293	86	2015	
2SC684Y		3293	86	2015	
2SC685	124	3021	12		2N3739
2SC685(O)	124		12		
2SC685-O		3021	12		
2SC685A	124	3021	12		
2SC685B	124	3021	12		
2SC685BK	124	3021	12		
2SC685GN		3021	12		
2SC685GU	124	3021	12		
2SC685H	124		12		
2SC685P	124	3021	12		
2SC685S	124	3021	12		
2SC685Y	124	3021	12		
2SC686	154	3045/225	32	2012	
2SC686A		3045	32	2012	
2SC686B		3045	32	2012	
2SC686C		3045	32	2012	
2SC686D		3045	32	2012	
2SC686E		3045	32	2012	
2SC686F		3045	32	2012	
2SC686G		3045	32	2012	
2SC686GN		3045	32	2012	
2SC686H		3045	32	2012	
2SC686J		3045	32	2012	
2SC686K		3045	32	2012	
2SC686L		3045	32	2012	
2SC686M		3045	32	2012	
2SC686OR		3045	32	2012	
2SC686R		3045	32	2012	
2SC686X		3045	32	2012	
2SC686Y		3045	32	2012	
2SC687	280	3079	35		MJ410
2SC687A		3079	35		
2SC687B		3079	35		
2SC687C		3079	35		
2SC687D		3079	35		
2SC687E		3079	35		
2SC687F		3079	35		
2SC687G		3079	35		
2SC687GN		3079	35		
2SC687H		3079	35		
2SC687J		3079	35		
2SC687K		3079	35		
2SC687L		3079	35		
2SC687M		3079	35		
2SC687OR		3079	35		
2SC687R		3079	35		
2SC687X		3079	35		
2SC687Y		3079	35		
2SC688		3018	61	2038	
2SC688A		3018	61	2038	
2SC688B		3018	61	2038	
2SC688C		3018	61	2038	
2SC688D		3018	61	2038	
2SC688E		3018	61	2038	
2SC688F		3018	61	2038	
2SC688G		3018	61	2038	
2SC688GN		3018	61	2038	
2SC688H		3018	61	2038	
2SC688J		3018	61	2038	
2SC688K		3018	61	2038	
2SC688L		3018	61	2038	
2SC688M		3018	61	2038	
2SC688OR		3018	61	2038	
2SC688R		3018	61	2038	
2SC688X		3018	61	2038	
2SC688Y		3018	61	2038	
2SC689		3122	88	2051	
2SC689A		3122	88	2010	
2SC689B		3122	88	2010	
2SC689C		3122	88	2010	
2SC689D		3122	88	2010	
2SC689E		3122	88	2010	
2SC689G		3122	88	2010	
2SC689GN		3122	88	2010	
2SC689H		3122	88	2051	
2SC689J		3122	88	2010	
2SC689K		3122	88	2010	
2SC689L		3122	88	2010	
2SC689M		3122	88	2010	
2SC689OR		3122	88	2010	
2SC689R		3122	88	2010	
2SC689X		3122	88	2010	
2SC689Y		3122	88	2010	
2SC68A		3122	88	2030	
2SC68B		3122	88	2030	
2SC68C		3122	88	2030	
2SC68D		3122	88	2030	
2SC68E		3122	88	2030	
2SC68F		3122	88	2030	
2SC68G		3122	88	2030	
2SC68GN		3122	88	2030	
2SC68H		3122	88	2030	
2SC68J		3122	88	2030	
2SC68K		3122	88	2030	
2SC68L		3122	88	2030	
2SC68M		3122	88	2030	
2SC68OR		3122	88	2030	
2SC68R		3122	88	2030	
2SC68X		3122	88	2030	
2SC68Y		3122	88	2030	
2SC69	128	3024	18	2030	
2SC690	3598		66	2048	
2SC691			28	2017	
2SC692			28	2017	
2SC693	199	3124/289	62	2010	
2SC693-OR			62	2010	
2SC693A		3124	62	2010	
2SC693B		3124	62	2010	
2SC693C		3124	62	2010	
2SC693D		3124	62	2010	
2SC693E	199	3124/289	62	2010	
2SC693EB	199	3124/289	62	2010	
2SC693ET	199	3124/289	62	2010	
2SC693F	199	3124/289	62	2010	
2SC693FC	199	3124/289	62	2010	
2SC693FL	199	3124/289	62	2010	
2SC693FP			62	2010	
2SC693FU	199	3124/289	62	2010	
2SC693G	199	3124/289	62	2010	
2SC693GL	199	3124/289	62	2010	
2SC693GN		3124	62	2010	
2SC693GS	199	3124/289	62	2010	
2SC693GU	199	3124/289	62	2010	
2SC693GZ	199	3124/289	62	2010	
2SC693H	199	3124/289	62	2010	
2SC693J		3124	62	2010	
2SC693K		3124	62	2010	
2SC693L		3124	62	2010	
2SC693M		3124	62	2010	
2SC693NP			62	2010	
2SC693OR		3124	62	2010	
2SC693R		3124	62	2010	
2SC693U			62	2010	
2SC693X		3124	62	2010	
2SC693Y		3124	62	2010	
2SC694	199	3124/289	47	2010	
2SC694A		3124	47	2016	
2SC694B		3124	47	2016	
2SC694C		3124	47	2016	
2SC694D		3124	47	2016	
2SC694E	199	3124/289	47	2010	
2SC694F	199	3122	47	2010	
2SC694G	199	3122	47	2010	
2SC694GN		3124	47	2016	
2SC694H		3124	47	2016	
2SC694J		3124	47	2016	
2SC694K		3124	47	2016	
2SC694L		3124	47	2016	
2SC694M		3124	47	2016	
2SC694OR		3124	47	2016	
2SC694R		3124	47	2016	
2SC694X		3124	47	2016	
2SC694Y		3124	47	2016	
2SC694Z	199	3039/316	47	2010	
2SC695	313		11	2015	
2SC696	128	3048/329	46		
2SC696(D)	128	3048/329	46		
2SC696-4		3047	46		
2SC696A	128	3124/289	46		
2SC696AB	128		46		
2SC696AD	128		46		
2SC696AE	128		46		
2SC696AF	128		46		
2SC696AH	128		46		
2SC696AI	128		46		
2SC696B	128	3024	46		
2SC696BL		3024	46		
2SC696C		3024	46		
2SC696D	128	3048/329	46		
2SC696E	128	3048/329	46		
2SC696F	128	3048/329	46		
2SC696G	128	3047	46		
2SC696GN		3047	46		
2SC696H	128	3024	46		
2SC696I	128	3048/329	46		

Industry Standard No.	ECG	SK	GE	RS 276-	MOTOR.
2SC696J		3047	46		
2SC696K		3047	46		
2SC696L		3047	46		
2SC696M		3047	46		
2SC696OR		3047	46		
2SC696R		3024	46		
2SC696X		3047	46		
2SC696Y		3047	46		
2SC697	282/427	3299/237	46		
2SC697A	282/427	3299/237	46		
2SC697AB	282/427	3299/237			
2SC697AD	282/427	3299/237			
2SC697AE	282/427	3299/237			
2SC697AF	282/427	3299/237			
2SC697AH	282/427	3299/237			
2SC697AI	282/427	3299/237			
2SC697AJ	282/427	3299/237			
2SC697B	282/427	3299/237	46		
2SC697D	282/427	3299/237	46		
2SC697E	282/427	3299/237	46		
2SC697F	282/427	3299/237	46		
2SC697H	282/427	3299/237	46		
2SC697I	282/427	3299/237			
2SC697J	282/427	3299/237	46		
2SC699			63	2030	
2SC69A		3024	18	2030	
2SC69B		3024	18	2030	
2SC69C		3024	18	2030	
2SC69D		3024	18	2030	
2SC69E		3024	18	2030	
2SC69F		3024	18	2030	
2SC69G		3024	18	2030	
2SC69GN		3024	18	2030	
2SC69H		3024	18	2030	
2SC69J		3024	18	2030	
2SC69K		3024	18	2030	
2SC69L		3024	18	2030	
2SC69M		3024	18	2030	
2SC69OR		3024	18	2030	
2SC69R		3024	18	2030	
2SC69X		3024	18	2030	
2SC69Y		3024	18	2030	
2SC7	130		14	2041	
2SC70	154	3018	32	2012	
2SC700			63	2030	
2SC701		3124	47		
2SC701A		3124	47	2030	
2SC701B		3122	47	2030	
2SC701C		3124	47	2030	
2SC701D		3124	47	2030	
2SC701E		3124	47	2030	
2SC701F		3124	47	2030	
2SC701G		3124	47	2030	
2SC701GN		3124	47	2030	
2SC701H		3124	47	2030	
2SC701J		3124	47	2030	
2SC701K		3124	47	2030	
2SC701L		3124	47	2030	
2SC701M		3124	47	2030	
2SC701OR		3124	47	2030	
2SC701R		3124	47	2030	
2SC701X		3124	47	2030	
2SC701Y		3124	47	2030	
2SC702		3124	47	2051	
2SC702A		3124	47	2030	
2SC702B		3124	47	2030	
2SC702C		3124	47	2030	
2SC702D		3124	47	2030	
2SC702E		3124	47	2030	
2SC702F		3124	47	2030	
2SC702G		3124	47	2030	
2SC702GN		3124	47	2030	
2SC702H		3124	47	2030	
2SC702J		3124	47	2030	
2SC702K		3124	47	2030	
2SC702L		3124	47	2030	
2SC702M		3124	47	2030	
2SC702OR			47	2030	
2SC702R		3124	47	2030	
2SC702X		3124	47	2030	
2SC702Y		3124	47	2030	
2SC703			66	2048	
2SC704			66	2048	
2SC705	107	3039/316	86	2015	
2SC705A		3039	86	2038	
2SC705B	107	3039/316	86	2015	
2SC705C	107	3039/316	86	2015	
2SC705D	107	3039/316	86	2015	
2SC705E	107	3039/316	86	2015	
2SC705F	107	3039/316	86	2015	
2SC705G		3039	86	2038	
2SC705GN		3039	86	2038	
2SC705J		3039	86	2038	
2SC705K		3039	86	2038	
2SC705L		3039	86	2038	
2SC705M		3039	86	2038	
2SC705OR		3039	86	2038	
2SC705R		3039	86	2038	
2SC705TV	107		86	2015	
2SC705TV(3RD IF)	233			2009	
2SC705TV(3RD-IF)			86	2051	
2SC705TVV			86	2038	
2SC705TW			86	2038	
2SC705X		3039	86	2038	
2SC705Y		3039	86	2038	
2SC706				2011	
2SC707	316	3039	11	2015	
2SC707A		3039	86	2038	
2SC707B		3039	86	2038	
2SC707C		3039	86	2038	
2SC707D		3039	86	2038	
2SC707E		3039	86	2038	
2SC707F		3039	86	2038	
2SC707G		3039	86	2038	
2SC707GN		3039	86	2038	
2SC707H	316	3039	86	2015	
2SC707K		3039	86	2038	
2SC707L		3039	86	2038	
2SC707M		3039	86	2038	
2SC707OR		3039	86	2038	
2SC707R		3039	86	2038	
2SC707X		3039	86	2038	
2SC707Y		3039	86	2038	
2SC708	128	3024	18	2030	
2SC708(A)			18	2030	
2SC708(B)			18	2030	
2SC708(C)			18	2030	
2SC708-OR			18	2030	
2SC708A	128	3024	18	2030	
2SC708AA	128		18	2030	
2SC708AB	128	3024	18	2030	
2SC708AC	128	3024	18	2030	
2SC708AH	128	3024	18	2030	
2SC708AHA	128		18	2030	
2SC708AHB	128		18	2030	
2SC708AHC	128		18	2030	
2SC708B	128	3024	18	2030	
2SC708C	128	3024	18	2030	
2SC708D		3024	18	2030	
2SC708E		3024	18	2030	
2SC708F		3024	18	2030	
2SC708G		3024	18	2030	
2SC708GN		3024	18	2030	
2SC708H		3024	18	2030	
2SC708HA			18	2030	
2SC708HB			18	2030	
2SC708L		3024	18	2030	
2SC708M		3024	18	2030	
2SC708OR		3024	18	2030	
2SC708R		3024	18	2030	
2SC708X		3024	18	2030	
2SC708Y		3024	18	2030	
2SC709	85	3122	88	2051	
2SC709(B)(C)			88	2051	
2SC709(C)			88	2051	
2SC709A		3122	88	2051	
2SC709B	85	3122	88	2051	
2SC709C	85	3122	88	2051	
2SC709CD	85	3122	88	2051	
2SC709D	85	3122	88	2051	
2SC709E		3122	88	2051	
2SC709F		3122	88	2051	
2SC709G		3122	88	2051	
2SC709GN		3122	88	2051	
2SC709H		3122	88	2051	
2SC709J		3122	88	2051	
2SC709L		3122	88	2051	
2SC709M		3122	88	2051	
2SC709OR		3122	88	2051	
2SC709X		3122	88	2051	
2SC709Y		3122	88	2051	
2SC70A		3044	32		
2SC70B		3044	32		
2SC70C		3044	32		
2SC70D		3044	32		
2SC70E		3044	32		
2SC70F		3044	32		
2SC70G		3044	32		
2SC70GN		3044	32		
2SC70H		3044	32		
2SC70J		3044	32		
2SC70K		3044	32		
2SC70L		3044	32		
2SC70M		3044	32		
2SC70OR		3044	32		
2SC70R		3044	32		
2SC70X		3044	32		
2SC70Y		3044	32		
2SC71	101	3861	59	2002	
2SC710		3356	211	2009	
2SC710(B)		3356	211	2009	
2SC710(C)		3356	211	2009	
2SC710(D)		3356	211	2009	
2SC710-1		3356	211	2009	
2SC710-2		3356	211	2009	
2SC710-4		3356	211	2009	
2SC710-OR		3356	211	2009	
2SC7108B		3122	20	2051	
2SC710AL		3356	211	2009	
0002SC710B	123A	3018	20	2051	
2SC710B2			211	2009	
2SC710BC		3356	211	2009	
0002SC710C	123A	3018	20	2051	
2SC710D		3356	211	2009	
2SC710DE			211	2009	
2SC710E		3356	211	2009	
2SC710F		3356	211	2009	
2SC710G		3356	211	2009	
2SC710GN		3356	211	2009	
2SC710H		3356	211	2009	
2SC710K		3356	211	2009	
2SC710L		3356	211	2009	
2SC710M		3356	211	2009	
2SC710OR			211	2009	
2SC710R			211	2009	
2SC710X		3356	211	2009	
2SC710XL		3356	211	2009	
2SC710Y		3356	211	2009	
2SC711	85	3899	62	2010	
2SC711(D)			62	2010	
2SC711(E)	85	3899	62	2010	
2SC711(F)			62	2010	
2SC711,A,F,G			62	2010	
2SC711-OR			62	2010	
2SC711A		3899	62	2010	
2SC711A(E)			62	2010	
2SC711AD	85	3899			
2SC711AE	85	3899	62	2010	
2SC711AF	85	3899	62	2010	
2SC711AG		3899	62	2010	
2SC711AN			62	2010	
2SC711B		3899	62	2010	
2SC711C		3899	62	2010	
2SC711D	85	3899	62	2010	
2SC711E	85	3899	62	2010	
2SC711F	85	3899	62	2010	
2SC711FG		3899	62	2010	
2SC711G		3899	62	2010	
2SC711GN			62	2010	
2SC711H	85	3899	62	2010	
2SC711J			62	2010	
2SC711L			62	2010	
2SC711M			62	2010	
2SC711OR			62	2010	
2SC711R			62	2010	
2SC711X			62	2010	
2SC711Y			62	2010	
2SC712		3124	47	2051	
2SC712(D)			47	2051	
2SC712-CD		3124	47	2051	
2SC712A		3124	47	2051	
2SC712B		3124	47	2051	
2SC712C		3124	47	2051	
2SC712CD		3124	47	2051	
2SC712D		3124	47	2051	
2SC712DC			47	2051	
2SC712E		3124	47	2051	
2SC712F		3124	47	2051	
2SC712G		3124	47	2051	
2SC712GN		3124	47	2051	
2SC712H		3124	47	2051	
2SC712J		3124	47	2051	
2SC712K		3124	47	2051	
2SC712L		3124	47	2051	
2SC712M		3124	47	2051	
2SC712OR		3124	47	2051	
2SC712R		3124	47	2051	
2SC712W		3124	47	2051	

Industry Standard No.	ECG	SK	GE	RS 276-	MOTOR.
2SC712X			47	2051	
2SC712Y		3124	47	2051	
2SC713		3122	88	2051	
2SC713A		3122	88	2051	
2SC713B		3122	88	2051	
2SC713C		3122	88	2051	
2SC713D		3122	88	2051	
2SC713E		3122	88	2051	
2SC713F		3122	88	2051	
2SC713G		3122	88	2051	
2SC713GN		3122	88	2051	
2SC713H		3122	88	2051	
2SC713J		3122	88	2051	
2SC713K		3122	88	2051	
2SC713L		3122	88	2051	
2SC713M		3122	88	2051	
2SC713OR		3122	88	2051	
2SC713R		3122	88	2051	
2SC713X		3122	88	2051	
2SC713Y		3122	88	2051	
2SC714		3122	88	2051	
2SC714A		3122	88	2051	
2SC714B		3122	88	2051	
2SC714C		3122	88	2051	
2SC714D		3122	88	2051	
2SC714E		3122	88	2051	
2SC714F		3122	88	2051	
2SC714G		3122	88	2051	
2SC714GN		3122	88	2051	
2SC714H		3122	88	2051	
2SC714J		3122	88	2051	
2SC714K		3122	88	2051	
2SC714L		3122	88	2051	
2SC714M		3122	88	2051	
2SC714OR		3122	88	2051	
2SC714R		3122	88	2051	
2SC714X		3122	88	2051	
2SC714Y		3122	88	2051	
2SC715	85	3124/289	47	2009	
2SC715-OR			20	2009	
2SC715A	85	3124/289	47	2009	
2SC715B	85	3124/289	47	2009	
2SC715C	85	3124/289	47	2009	
2SC715D	85	3124/289	47	2009	
2SC715E	85	3124/289	47	2009	
2SC715EJ	85	3124/289	47	2009	
2SC715EV	85	3124/289	47	2009	
2SC715F	85	3124/289	47	2009	
2SC715G		3124	47	2009	
2SC715GN			47	2009	
2SC715H			47	2009	
2SC715J		3124	47	2009	
2SC715K		3124	47	2009	
2SC715L		3124	47	2009	
2SC715M		3124	47	2009	
2SC715R		3124	47	2009	
2SC715X			47	2009	
2SC715XL	85	3124/289	47	2009	
2SC715Y		3124	47	2009	
2SC716	289A	3122	86	2051	
2SC716A		3122	86	2030	
2SC716B	289A	3122	86	2051	
2SC716C	289A	3122	86	2051	
2SC716D	289A	3122	86	2051	
2SC716E	289A	3122	86	2051	
2SC716F	289A	3122	86	2051	
2SC716G		3122	86	2051	
2SC716GN			86	2030	
2SC716H			86	2030	
2SC716J		3122	86	2030	
2SC716K		3122	86	2030	
2SC716L		3122	86	2030	
2SC716M		3122	86	2030	
2SC716OR			86	2030	
2SC716R		3122	86	2030	
2SC716X			86	2030	
2SC716Y		3122	86	2030	
2SC717	107	3132	17	2015	
2SC717(3RD-IF)			17	2015	
2SC717(LAST-IF)			17	2015	
2SC717B	107	3132	17	2015	
2SC717BK	107	3132	17	2015	
2SC717BLK	107	3132	17	2015	
2SC717C	107	3132	17	2015	
2SC717E	107	3132	17	2015	
2SC717F		3132	17	2015	
2SC717G			17	2015	
2SC717GN			17	2015	
2SC717H		3132	17	2015	
2SC717K		3132	17	2015	
2SC717L		3132	17	2015	
2SC717M		3132	17	2015	
2SC717X			17	2015	
2SC719		3124	21	2034	
2SC719Q			21	2034	
2SC71A		3861	59		
2SC71B		3861	59		
2SC71C		3861	59		
2SC71D		3861	59		
2SC71E		3861	59		
2SC71F		3861	59		
2SC71G		3861	59		
2SC71GN		3861	59		
2SC71H		3861	59		
2SC71J		3861	59		
2SC71M		3861	59		
2SC71OR		3861	59		
2SC71R		3861	59		
2SC71X		3861	59		
2SC72	101	3861	59	2002	
2SC721			20	2051	
2SC722	107	3849/293	61	2038	
2SC723	107		20	2051	
2SC723BL			20	2051	
2SC725	289A	3444/123A	20	2051	
2SC725-O	289A		20	2051	
2SC727	287	3024/128	18	2012	
2SC727A	287	3024/128	18	2030	
2SC727B	287	3024/128	18	2030	
2SC727C		3024	18	2030	
2SC727D		3024	18	2030	
2SC727E		3024	18	2030	
2SC727F		3024	18	2030	
2SC727G		3024	18	2030	
2SC727GN		3024	18	2030	
2SC727H		3024	18	2030	
2SC727J		3024	18	2030	
2SC727K		3024	18	2030	
2SC727L		3024	18	2030	
2SC727M		3024	18	2030	
2SC727R		3024	18	2030	
2SC727X		3024	18	2030	
2SC727Y		3024	18	2030	
2SC728	287	3045/225	32	2012	

Industry Standard No.	ECG	SK	GE	RS 276-	MOTOR.
2SC728A	287	3045/225	32	2051	
2SC728B	287	3045/225	32	2051	
2SC728C		3045	32	2051	
2SC728D		3045	32	2051	
2SC728E		3045	32	2051	
2SC728F		3045	32	2051	
2SC728G		3045	32	2051	
2SC728GN		3045	32	2051	
2SC728H		3045	32	2051	
2SC728J		3045	32	2051	
2SC728K		3045	32	2051	
2SC728L		3045	32	2051	
2SC728M		3045	32	2051	
2SC728OR		3045	32	2051	
2SC728R		3045	32	2051	
2SC728X		3045	32	2051	
2SC728Y		3045	32	2051	
2SC72A		3861	59		
2SC72B		3861	59		
2SC72C		3861	59		
2SC72D		3861	59		
2SC72E		3861	59		
2SC72F		3861	59		
2SC72G		3861	59		
2SC72GN		3861	59		
2SC72H		3861	59		
2SC72J		3861	59		
2SC72K		3861	59		
2SC72L		3861	59		
2SC72M		3861	59		
2SC72OR		3861	59		
2SC72R		3861	59		
2SC72X		3861	59		
2SC72Y		3861	59		
2SC73	101	3861	59	2002	
2SC730	346	3218/278	219		
2SC731	311	3195	45	2030	
2SC731R		3195	45	2030	
2SC732	199	3245	62	2010	
2SC732(BL)			62	2010	
2SC732-B			62	2010	
2SC732-BL			62	2010	
2SC732-BLU			62	2010	
2SC732-G			62	2010	
2SC732-GR			62	2010	
2SC732-GRN			62	2010	
2SC732-OR			62	2010	
2SC732-V			62	2010	
2SC732-V10			62	2010	
2SC732-VIO			62	2010	
2SC732A		3245	62	2010	
2SC732B	199	3245	62	2010	
2SC732BL	199	3245	62	2010	
2SC732BL-1		3245	62	2010	
2SC732BLU			62	2010	
2SC732C		3245	62	2010	
2SC732D		3245	62	2010	
2SC732E		3245	62	2010	
2SC732F		3245	62	2010	
2SC732G		3245	62	2010	
2SC732GN		3245	62	2010	
2SC732GR	199	3245	62	2010	
2SC732GR/4454C	199	3245	62	2010	
2SC732GRB			62	2010	
2SC732GRN			62	2010	
2SC732H		3245	62	2010	
2SC732J		3245	62	2010	
2SC732L		3245	62	2010	
2SC732M		3245	62	2010	
2SC732OR		3245	62	2010	
2SC732R		3245	62	2010	
2SC732S	199	3245	62	2010	
2SC732V	199	3245	62	2010	
2SC732V10			62	2010	
2SC732VIO			62	2010	
2SC732X		3245	62	2010	
2SC732Y	199	3245	62	2010	
2SC733	199	3245	62	2010	
2SC733(GR)	199		62	2010	
2SC733-O	199	3245	62	2010	
2SC733-B			62	2010	
2SC733-BL			62	2010	
2SC733-BLU			62	2010	
2SC733-G			62	2010	
2SC733-GR			62	2010	
2SC733-GRN			62	2010	
2SC733-O	199	3245	62	2010	
2SC733-OR			62	2010	
2SC733-ORG		3245	62	2010	
2SC733-Y	199		62	2010	
2SC733-YEL		3245	62	2010	
2SC7335-BL	199	3444/123A	20	2051	
2SC733A		3245	62	2010	
2SC733B	199	3245	62	2010	
2SC733BL	199	3245	62	2010	
2SC733BLK		3245	62	2010	
2SC733BLU			62	2010	
2SC733C		3245	62	2010	
2SC733D		3245	62	2010	
2SC733E		3245	62	2010	
2SC733ER			62	2010	
2SC733F		3245	62	2010	
2SC733G		3245	62	2010	
2SC733GN		3245	62	2010	
2SC733GR	199	3245	62	2010	
2SC733GRN			62	2010	
2SC733H		3245	62	2010	
2SC733J		3245	62	2010	
2SC733K		3245	62	2010	
2SC733L		3245	62	2010	
2SC733M		3245	62	2010	
2SC733O			62	2010	
2SC733OR		3245	62	2010	
2SC733Q	199	3245	62	2010	
2SC733R	199	3245	62	2010	
2SC733S			62	2010	
2SC733S-BL		3245	62	2010	
2SC733V	199	3245	62	2010	
2SC733X		3245	62	2010	
2SC733Y	199	3245	62	2010	
2SC733YEL			62	2010	
2SC734	289A	3245/199	212	2030	
2SC734(O)			212	2030	
2SC734(R)		3245	212	2030	
2SC734(Y)			212	2030	
2SC734-O	289A	3245/199	212	2030	
2SC734-G			212	2030	
2SC734-GR			212	2030	
2SC734-GRN		3245	212	2030	
2SC734-O	289A	3245/199	212	2030	
2SC734-OR			212	2030	
2SC734-ORG		3245	212	2030	
2SC734-OY			212	2030	
2SC734-R	289A	3245/199	212	2030	

Industry Standard No.	ECG	SK	GE	RS 276-	MOTOR.
2SC734-RED		3245	212	2030	
2SC734-Y	289A		212	2030	
2SC734-YEL		3245	212	2030	
2SC734A		3245	212	2030	
2SC734B		3245	212	2030	
2SC734C		3245	212	2030	
2SC734D		3245	212	2030	
2SC734E		3245	212	2030	
2SC734F		3245	212	2030	
2SC734G		3245	212	2030	
2SC734GN		3245	212	2030	
2SC734GR	289A	3245/199	212	2030	
2SC734GRN			212	2030	
2SC734H		3245	212	2030	
2SC734J		3245	212	2030	
2SC734K		3245	212	2030	
2SC734K/GR		3245	212	2030	
2SC734L		3245	212	2030	
2SC734M		3245	212	2030	
2SC734O	289A	3245/199	212	2030	
2SC734OR		3245	212	2030	
2SC734R		3245	212	2030	
2SC734RED			212	2030	
2SC734X		3245	212	2030	
2SC734Y	289A	3245/199	212	2030	
2SC734YEL			212	2030	
2SC735	289A	3122	20	2051	
2SC735(O)			210	2038	
2SC735(FA-3)	289A		210	2038	
2SC735(O)	289A	3122	210	2038	
2SC735(Y)			210	2038	
2SC735-O	289A	3122	210	2038	
2SC735-GRN		3122	210	2038	
2SC735-O	289A	3122	210	2038	
2SC735-OR			210	2038	
2SC735-ORG		3122	210	2038	
2SC735-ORN			210	2038	
2SC735-OY			20	2038	
002SC735-OY				2051	
2SC735-R			210	2038	
2SC735-RED		3122	210	2038	
2SC735-Y	289A	3122	210	2038	
2SC735-YEL		3122	210	2038	
2SC735/4454C	289A		210	2038	
002SC7350Y	289A		20	2038	
2SC7354	289A	3444/123A	20	2051	
2SC735A		3122	210	2038	
2SC735B	289A	3122	210	2038	
2SC735C		3122	210	2038	
2SC735D		3122	210	2038	
2SC735E		3122	210	2038	
2SC735F	289A	3122	210	2038	
2SC735FA3			210	2038	
2SC735G	289A	3122	210	2038	
2SC735GN		3122	210	2038	
2SC735GR	289A	3122	210	2038	
2SC735GRN			210	2038	
2SC735H	289A	3122	210	2038	
2SC735J	289A	3122	210	2038	
2SC735K	289A	3122	210	2038	
2SC735L	289A	3122	210	2038	
2SC735M		3122	210	2038	
2SC7350		3122	210	2038	
2SC7350R		3122	210	2038	
2SC7350RN	289A		210	2038	
002SC7350Y			20	2051	
2SC735R	289A	3122	210	2038	
2SC735RED			210	2038	
2SC735X		3122	210	2038	
2SC735Y	289A	3122	210	2038	
2SC735Y/4454C	289A		210	2038	
2SC735YEL			210	2038	
2SC736	87	3027/130	75	2041	2N4347
2SC736A		3027	75	2041	
2SC736B		3027	75	2041	
2SC736C		3027	75	2041	
2SC736D		3027	75	2041	
2SC736E		3027	75	2041	
2SC736F		3027	75	2041	
2SC736G		3027	75	2041	
2SC736GN		3027	75	2041	
2SC736H		3027	75	2041	
2SC736J		3027	75	2041	
2SC736K		3027	75	2041	
2SC736L		3027	75	2041	
2SC736M		3027	75	2041	
2SC7360R		3027	75	2041	
2SC736R		3018	75	2041	
2SC736X		3027	75	2041	
2SC736Y		3027	75	2041	
2SC737		3444	20	2051	
2SC737Y		3444	20	2051	
2SC738		3122	86	2038	
2SC738A	107	3122	86	2038	
2SC738B		3122	86	2038	
2SC738C		3122	86	2038	
2SC738D		3122	86	2038	
2SC738E		3122	86	2038	
2SC738F		3018	86	2038	
2SC738G			86	2038	
2SC738GN			86	2038	
2SC738H			86	2038	
2SC738J			86	2038	
2SC738K			86	2038	
2SC738L			86	2038	
2SC738M			86	2038	
2SC7380R			86	2038	
2SC738R			86	2038	
2SC738X			86	2038	
2SC738Y			86	2038	
2SC739		3018	61	2015	
2SC739A	107	3018	61	2038	
2SC739B		3018	61	2038	
2SC739C		3122	61	2015	
2SC739D		3018	61	2038	
2SC739E		3018	61	2038	
2SC739F		3018	61	2038	
2SC739G		3018	61	2038	
2SC739GN		3018	61	2038	
2SC739H		3018	61	2038	
2SC739K		3018	61	2038	
2SC739L		3018	61	2038	
2SC739M		3018	61	2038	
2SC7390R		3018	61	2038	
2SC739R		3018	61	2038	
2SC739Y		3018	61	2038	
2SC73A		3861	59		
2SC73B		3861	59		
2SC73C		3861	59		
2SC73D		3861	59		
2SC73E		3861	59		
2SC73F		3861	59		
2SC73G		3861	59		
2SC73GN		3861	59		
2SC73H		3861	59		
2SC73J		3861	59		
2SC73K		3861	59		
2SC73L		3861	59		
2SC73M		3861	59		
2SC730R		3861	59		
2SC73R		3861	59		
2SC73X		3861	59		
2SC73Y		3861	59		
2SC74	123	3452/108	61	2038	
2SC74-O			61	2038	
2SC74-R			61	2038	
2SC740	108	3039/316	86	2038	
2SC740A		3039	86	2038	
2SC740B		3039	86	2038	
2SC740C		3039	86	2038	
2SC740D		3039	86	2038	
2SC740E		3039	86	2038	
2SC740F		3039	86	2038	
2SC740G		3039	86	2038	
2SC740GN		3039	86	2038	
2SC740H		3039	86	2038	
2SC740J		3039	86	2038	
2SC740K		3039	86	2038	
2SC740L		3039	86	2038	
2SC740M		3039	86	2038	
2SC7400R			86	2038	
2SC740R			86	2038	
2SC740X		3039	86	2038	
2SC740Y		3039	86	2038	
2SC741	486	3122	88	2051	
2SC741B		3122	88	2030	
2SC741C		3122	88	2030	
2SC741D		3122	88	2030	
2SC741E		3122	88	2030	
2SC741F		3122	88	2030	
2SC741GN		3122	88	2030	
2SC741H		3122	88	2030	
2SC741J		3122	88	2030	
2SC741K		3122	88	2030	
2SC741L		3122	88	2030	
2SC741M		3122	88	2030	
2SC7410R		3122	88	2030	
2SC741R		3122	88	2030	
2SC741X		3122	88	2030	
2SC741Y		3122	88	2030	
2SC743A		3045	40	2012	
2SC744			17	2030	
2SC744A		3024	17	2030	
2SC746A		3045	32	2012	
2SC746B		3045	32		
2SC746C		3045	32		
2SC746D		3045	32		
2SC746E		3045	32		
2SC746F		3045	32		
2SC746GN		3045	32		
2SC746H		3045	32		
2SC746J		3045	32		
2SC746K		3045	32		
2SC746L		3045	32		
2SC746M		3045	32		
2SC7460R		3045	32		
2SC746R		3045	32		
2SC746X		3045	32		
2SC746Y		3045	32		
2SC748		3039	86	2038	
2SC748A		3039	86	2038	
2SC748C		3039	86	2038	
2SC748D		3039	86	2038	
2SC748E		3039	86	2038	
2SC748F		3039	86	2038	
2SC748G		3039	86	2038	
2SC748GN		3039	86	2038	
2SC748H		3039	86	2038	
2SC748J		3039	86	2038	
2SC748K		3039	86	2038	
2SC748L		3039	86	2038	
2SC748M		3039	86	2038	
2SC7480R		3039	86	2038	
2SC748R		3039	86	2038	
2SC748X		3039	86	2038	
2SC748Y		3039	86	2038	
2SC74A		3018	61	2038	
2SC74B		3018	61	2038	
2SC74C		3018	61	2038	
2SC74D		3018	61	2038	
2SC74E		3018	61	2038	
2SC74F		3018	61	2038	
2SC74G		3018	61	2038	
2SC74GN		3018	61	2038	
2SC74H		3018	61	2038	
2SC74J		3018	61	2038	
2SC74K		3018	61	2038	
2SC74L		3018	61	2038	
2SC74M		3018	61	2038	
2SC74R		3018	61	2038	
2SC74X		3018	61	2038	
2SC74Y		3018	61	2038	
2SC75	101	3861	59	2002	
2SC752	85	3122	88	2051	
2SC752-ORG-G			88	2051	
2SC752-RED-G			88	2051	
2SC752-YEL-G			88	2051	
2SC752A		3122	88	2051	
2SC752B		3122	88	2051	
2SC752C		3122	88	2051	
2SC752D		3122	88	2051	
2SC752E		3122	88	2051	
2SC752F		3122	88	2051	
2SC752G	85	3122	88	2051	
2SC752G-O	85		210	2051	
2SC752G-R	85		210	2051	
2SC752G-Y	85		210	2051	
2SC752GA		3122	88	2051	
2SC752GN		3122	88	2051	
2SC752H		3122	88	2051	
2SC752J		3122	88	2051	
2SC752K		3122	88	2051	
2SC752L		3122	88	2051	
2SC752M		3122	88	2051	
2SC7520R		3122	88	2051	
2SC752R	85	3122	88	2051	
2SC752X		3122	88	2051	
2SC752Y	85	3122	88	2051	
2SC756	282	3748	264		
2SC756-1	282	3748	264		
2SC756-1-1	282	3748	264		
2SC756-1-2	282	3748	264		
2SC756-1-3	282	3748	264		
2SC756-1-4	282	3748	264		
2SC756-2	282	3748	264		
2SC756-2-1	282	3748	264		
2SC756-2-2	282	3748	264		
2SC756-2-3	282	3748	264		

Industry Standard No.	ECG	SK	GE	RS 276-	MOTOR.
2SC756-2-4	282	3748	264		
2SC756-2-5	282	3748	264		
2SC756-3	282	3748	264		
2SC756-3-1	282	3748	264		
2SC756-3-2	282	3748	264		
2SC756-3-3	282	3748	264		
2SC756-3-4	282	3748	264		
2SC756-4	282	3748	264		
2SC756-4-1	282	3748	264		
2SC756-4-2	282	3748	264		
2SC756-4-3	282	3748	264		
2SC756-4-4	282	3748	264		
2SC756-5	282	3748			
2SC756A	282	3765/195A	264		
2SC756A-1	282	3765/195A	264		
2SC756A-2	282	3765/195A	264		
2SC756A-3	282	3765/195A	264		
2SC756A-4	282	3765/195A	264		
2SC756B		3765	264		
2SC756C	282	3765/195A	264		
2SC756D	282	3765/195A	264		
2SC756E	282	3765/195A	264		
2SC756F		3765	264		
2SC756G		3765	264		
2SC756GN		3765	264		
2SC756H	282	3765/195A	264		
2SC756J		3765	264		
2SC756K		3765	264		
2SC756L		3765	264		
2SC756M		3765	264		
2SC756OR		3765	264		
2SC756R		3765	264		
2SC756X		3765	264		
2SC756Y		3765	264		
2SC757	313	3444/123A			
2SC758					2N6307
2SC758OR		3018	61	2038	
2SC759					2N6306
2SC75A		3861	59		
2SC75B		3861	59		
2SC75B-1		3861	59		
2SC75C		3861	59		
2SC75E		3861	59		
2SC75F		3861	59		
2SC75G		3861	59		
2SC75GN		3861	59		
2SC75H		3861	59		
2SC75J		3861	59		
2SC75K		3861	59		
2SC75L		3861	59		
2SC75M		3861	59		
2SC75OR		3861	59		
2SC75R		3861	59		
2SC75X		3861	59		
2SC75Y		3861	59		
2SC76	101	3861	59	2002	
2SC760					2N6306
2SC761	161	3039/316	39	2015	
2SC761(Y)			39	2015	
2SC761A		3039	39	2015	
2SC761B		3039	39	2015	
2SC761C		3039	39	2015	
2SC761D		3039	39	2015	
2SC761E		3039	39	2015	
2SC761F		3039	39	2015	
2SC761G		3039	39	2015	
2SC761GN		3039	39	2015	
2SC761H		3039	39	2015	
2SC761J		3039	39	2015	
2SC761K		3039	39	2015	
2SC761L		3039	39	2015	
2SC761M		3039	39	2015	
2SC761OR		3039	39	2015	
2SC761R		3039	39	2015	
2SC761X		3039	39	2015	
2SC761Y	161	3039/316	39	2015	
2SC761Z	161	3039/316	39	2015	
2SC762	161	3039/316	214	2015	
2SC762B		3039	214	2015	
2SC762C		3039	214	2015	
2SC762D		3039	214	2015	
2SC762E		3039	214	2015	
2SC762F		3039	214	2015	
2SC762G		3039	214	2015	
2SC762GN		3039	214	2015	
2SC762H		3039	214	2015	
2SC762J		3039	214	2015	
2SC762K		3039	214	2015	
2SC762L		3039	214	2015	
2SC762M		3039	214	2015	
2SC762R		3039	214	2015	
2SC762X		3039	214	2015	
2SC762Y		3039	214	2015	
2SC763		3122	61	2038	
2SC763(C)	107	3122	61	2038	
2SC763-OR			61	2038	
2SC763A	107	3122	61	2038	
2SC763B		3122	61	2038	
2SC763C		3122	61	2038	
2SC763CD			61	2038	
2SC763D		3122	61	2038	
2SC763E		3122	61	2038	
2SC763F		3122	61	2038	
2SC763G		3122	61	2038	
2SC763GN		3122	61	2038	
2SC763H		3122	61	2038	
2SC763J		3122	61	2038	
2SC763K		3122	61	2038	
2SC763L		3122	61	2038	
2SC763M		3122	61	2038	
2SC763OR		3122	61	2038	
2SC763X		3122	61	2038	
2SC763Y		3122	61	2038	
2SC765		3027	75	2041	
2SC765A		3027	75	2041	
2SC765B		3027	75	2041	
2SC765C		3027	75	2041	
2SC765D		3027	75	2041	
2SC765E		3027	75	2041	
2SC765F		3027	75	2041	
2SC765G		3027	75	2041	
2SC765GN		3027	75	2041	
2SC765H		3027	75	2041	
2SC765J		3027	75	2041	
2SC765K		3027	75	2041	
2SC765L		3027	75	2041	
2SC765M		3027	75	2041	
2SC765OR		3027	75	2041	
2SC765R		3027	75	2041	
2SC765X		3027	75	2041	
2SC765Y		3027	75	2041	
2SC766			75	2041	
2SC767			75	2041	
2SC768	328	3027/130	75	2041	2N3055

Industry Standard No.	ECG	SK	GE	RS 276-	MOTOR.
2SC768A		3027	75	2041	
2SC768B		3027	75	2041	
2SC768C		3027	75	2041	
2SC768D		3027	75	2041	
2SC768E		3027	75	2041	
2SC768F		3027	75	2041	
2SC768G		3027	75	2041	
2SC768GN		3027	75	2041	
2SC768H		3027	75	2041	
2SC768J		3027	75	2041	
2SC768K		3027	75	2041	
2SC768L		3027	75	2041	
2SC768M		3027	75	2041	
2SC768OR		3027	75	2041	
2SC768R		3027	75	2041	
2SC768X		3027	75	2041	
2SC768Y		3027	75	2041	
2SC769					2N5633
2SC76A		3861	59		
2SC76B		3861	59		
2SC76C		3861	59		
2SC76D		3861	59		
2SC76E		3861	59		
2SC76F		3861	59		
2SC76G		3861	59		
2SC76GN		3861	59		
2SC76H		3861	59		
2SC76J		3861	59		
2SC76K		3861	59		
2SC76L		3861	59		
2SC76M		3861	59		
2SC76OR		3861	59		
2SC76R		3861	59		
2SC76X		3861	59		
2SC76Y		3861	59		
2SC77	101	3861	59	2002	
2SC770					MJ15011
2SC771		3018	61	2038	MJ15011
2SC771A		3018	61	2038	
2SC771B		3018	61	2038	
2SC771BX		3018	61	2038	
2SC771C		3018	61	2038	
2SC771D		3018	61	2038	
2SC771E		3018	61	2038	
2SC771F		3018	61	2038	
2SC771G		3018	61	2038	
2SC771GN		3018	61	2038	
2SC771H		3018	61	2038	
2SC771J		3018	61	2038	
2SC771K		3018	61	2038	
2SC771L		3018	61	2038	
2SC771M		3018	61	2038	
2SC771OR		3018	61	2038	
2SC771R		3018	61	2038	
2SC771X		3018	61	2038	
2SC771Y		3018	61	2038	
00002SC772	107	3132	11	2015	
2SC772-OR			61	2038	
2SC772A		3132	61	2038	
2SC772B		3132	61	2038	
2SC772BG		3132	61	2038	
2SC772BH		3132	61	2038	
2SC772BV			61	2038	
2SC772BX		3132	61	2038	
2SC772BY		3132	61	2038	
0002SC772C	107	3132	11	2015	
2SC772C1		3132	61	2038	
2SC772C2		3132	61	2038	
2SC772CA			61	2015	
2SC772CK		3132	61	2038	
2SC772CL		3132	61	2038	
2SC772CS		3132	61	2038	
2SC772CU		3132	61	2038	
2SC772CV		3132	61	2038	
2SC772CX		3132	61	2038	
2SC772D		3132	61	2038	
2SC772DJ		3132	61	2038	
2SC772DU		3132	61	2038	
2SC772DV			61	2038	
2SC772DX		3132	61	2038	
2SC772DY		3132	61	2038	
2SC772E		3132	61	2038	
2SC772F		3132	61	2038	
2SC772G		3132	61	2038	
2SC772GN		3132	61	2038	
2SC772H		3132	61	2038	
2SC772J		3132	61	2038	
2SC772K		3132	61	2038	
2SC772KB		3132	61	2038	
2SC772KC			61	2038	
2SC772KD		3132	61	2038	
2SC772KD1		3132	61	2038	
2SC772KD2		3132	61	2038	
2SC772L		3132	61	2038	
2SC772M		3132	61	2038	
2SC772OR		3132	61	2038	
2SC772R		3132	61	2038	
2SC772RB-D		3132	61	2038	
2SC772RD		3132	61	2038	
2SC772RS-D			61	2038	
2SC772X		3132	61	2038	
2SC772Y		3132	61	2038	
2SC773		3132	210	2051	
2SC773(E)			210	2051	
2SC773A		3132	210	2051	
2SC773B		3132	210	2051	
2SC773C		3132	210	2051	
2SC773D		3132	210	2051	
2SC773E		3132	210	2051	
2SC773F		3132	210	2051	
2SC773G			210	2051	
2SC773GN			210	2051	
2SC773H			210	2051	
2SC773J		3132	210	2051	
2SC773K		3132	210	2051	
2SC773L		3132	210	2051	
2SC773M		3132	210	2051	
2SC773OR			210	2051	
2SC773R		3132	210	2051	
2SC773X			210	2051	
2SC773Y		3132	210	2051	
2SC774	195A	3047	219		
2SC774B		3047	219		
2SC774C		3047	219		
2SC774D		3047	219		
2SC774E		3047	219		
2SC774F		3047	219		
2SC774G		3047	219		
2SC774GN		3047	219		
2SC774H		3047	219		
2SC774J		3047	219		
2SC774K		3047	219		
2SC774L		3047	219		
2SC774M		3047	219		

Industry Standard No.	ECG	SK	GE	RS 276-	MOTOR.
2SC7740R		3047	219		
2SC774R		3047	219		
2SC774X		3047	219		
2SC774Y		3047	219		
2SC775	195A	3047	219		
2SC775A		3047	219		
2SC775B		3047	219		
2SC775C		3047	219		
2SC775D		3047	219		
2SC775E		3047	219		
2SC775F		3047	219		
2SC775G		3047	219		
2SC775GN		3047	219		
2SC775H		3047	219		
2SC775J		3047	219		
2SC775K		3047	219		
2SC775L		3047	219		
2SC775M		3047	219		
2SC7750R		3047	219		
2SC775R		3047	219		
2SC775X		3047	219		
2SC775Y		3047	219		
2SC776	195A	3024/128	219		
2SC776(Y)	195A	3024/128	219		
2SC776A		3024	219		
2SC776B		3024	219		
2SC776C		3024	219		
2SC776D		3024	219		
2SC776E		3024	219		
2SC776F		3024	219		
2SC776G		3024	219		
2SC776GN		3024	219		
2SC776H		3024	219		
2SC776J		3024	219		
2SC776K		3024	219		
2SC776L		3024	219		
2SC776M		3024	219		
2SC7760R		3024	219		
2SC776R		3024	219		
2SC776X		3024	219		
2SC776Y	195A	3024/128	219		
2SC777	237	3299	46		
2SC777A		3299	46		
2SC777AP		3299	46		
2SC777B		3299	46		
2SC777C		3299	46		
2SC777D		3299	46		
2SC777E		3299	46		
2SC777F		3299	46		
2SC777G		3299	46		
2SC777GN		3299	46		
2SC777H		3299	46		
2SC777J		3299	46		
2SC777K		3299	46		
2SC777L		3299	46		
2SC777M		3299	46		
2SC7770R		3299	46		
2SC777R		3299	46		
2SC777X		3299	46		
2SC777Y		3299	46		
2SC778	237	3299	46		
2SC778A		3299	46		
2SC778B	237	3299	46		
2SC778C		3299	46		
2SC778D	237	3299	46		
2SC778E		3299	46		
2SC778F		3299	46		
2SC778G		3299	46		
2SC778GN		3299	46		
2SC778H		3299	46		
2SC778J		3299	46		
2SC778K		3299	46		
2SC778L		3299	46		
2SC778M		3299	46		
2SC7780R		3299	46		
2SC778R		3299	46		
2SC778X		3299	46		
2SC778Y		3299	46		
2SC779	286	3194	267		2N3739
2SC779-O	286		267		
2SC779-R	286		267		
2SC779-Y	286		267		
2SC779R	286	3194	267		
2SC779Y	286		267		
2SC77A		3861	59		
2SC77B		3861	59	2002	
2SC77C	101	3861	59	2002	
2SC77D		3861	59		
2SC77F		3861	59		
2SC77G		3861	59		
2SC77GN		3861	59		
2SC77H		3861	59		
2SC77J		3861	59		
2SC77K		3861	59		
2SC77L		3861	59		
2SC77M		3861	59		
2SC770R		3861	59		
2SC77R		3861	59		
2SC77X		3861	59		
2SC77Y		3861	59		
2SC77Z		3861	59		
2SC78	101	3861	59	2002	
2SC780	194	3045/225	32		
2SC780-RED-G		3244	32		
2SC780-YEL-G		3244	32		
2SC780A		3045	32		
2SC780A/G		3045	32		
2SC780AG	194	3433/287	32		
2SC780AG-O	194		32		
2SC780AG-O	194	3433/287	32		
2SC780AG-R	194	3433/287	32		
2SC780AG-Y	194	3433/287	32		
2SC780B		3045	32		
2SC780C		3045	32		
2SC780D		3045	32		
2SC780E		3045	32		
2SC780F		3045	32		
2SC780G	194	3045/225	32		
2SC780GA		3045	32		
2SC780GN		3045	32		
2SC780H		3045	32		
2SC780J		3045	32		
2SC780K		3045	32		
2SC780L		3045	32		
2SC780M		3045	32		
2SC780R		3045	32		
2SC780X		3045	32		
2SC780Y		3045	32		
2SC781	195A	3765	219		
2SC781A		3765	219		
2SC781AK		3765	219		
2SC781B		3765	219		
2SC781C		3765	219		
2SC781D		3765	219		
2SC781E		3765	219		
2SC781F		3765	219		
2SC781G		3765	219		
2SC781H		3765	219		
2SC781J		3765	219		
2SC781K		3765	219		
2SC781M		3765	219		
2SC781R		3765	219		
2SC781X		3765	219		
2SC781Y		3765	219		
2SC782	286	3131A/369	12		2N3739
2SC782A		3131A	12		
2SC782B		3131A	12		
2SC782C		3131A	12		
2SC782D		3131A	12		
2SC782E		3131A	12		
2SC782F		3131A	12		
2SC782G		3131A	12		
2SC782GN		3131A	12		
2SC782H		3131A	12		
2SC782J		3131A	12		
2SC782K		3131A	12		
2SC782L		3131A	12		
2SC7820R		3131A	12		
2SC782R		3131A	12		
2SC782X		3131A	12		
2SC782Y		3131A	12		
2SC783	286	3194	12		2N3738
2SC783Y	286	3194			
2SC784		3246	60	2038	
2SC784(BN)		3246	60	2038	
2SC784-O		3246	60	2038	
2SC784-6		3246	60	2038	
2SC784-B			60	2038	
2SC784-BN		3246	60	2038	
2SC784-BRN			60	2038	
2SC784-O		3246	60	2038	
2SC784-OR			60	2038	
2SC784-ORG			60	2038	
2SC784-R		3246	60	2038	
2SC784-RED			60	2038	
2SC784-Y			60	2038	
2SC7840	229	3018			
2SC784A		3246	60	2038	
2SC784B		3246	60	2038	
2SC784BN		3246	60	2038	
2SC784BN-1		3246	60	2038	
2SC784BRN			60	2038	
2SC784C		3246	60	2038	
2SC784D		3246	60	2038	
2SC784E		3246	60	2038	
2SC784F		3246	60	2038	
2SC784G		3246	60	2038	
2SC784GN		3246	60	2038	
2SC784H		3246	60	2038	
2SC784J		3246	60	2038	
2SC784K		3246	60	2038	
2SC784L		3246	60	2038	
2SC784M		3246	60	2038	
2SC7840		3246	60	2038	
2SC7840R		3246	60	2038	
2SC784Q			60	2038	
2SC784R	229	3246	60	2038	
2SC784R/4454C			60	2038	
2SC784RA		3246	60	2038	
2SC784RED			60	2038	
2SC784X		3246	60	2038	
2SC784Y		3246	60	2038	
2SC785		3246	60	2038	
2SC785(E)(D)			60	2038	
2SC785(O)			60	2038	
2SC785-O			60	2038	
2SC785-B			60	2038	
2SC785-BN			60	2038	
2SC785-BRN			60	2038	
2SC785-O		3246	60	2038	
2SC785-ORG			60	2038	
2SC785-R			60	2038	
2SC785-RED			60	2038	
2SC785-Y			60	2038	
2SC785-YEL			60	2038	
2SC785A		3246	60	2038	
2SC785B		3246	60	2038	
2SC785BL		3246	60	2038	
2SC785BN		3246	60	2038	
2SC785BR		3246	60	2038	
2SC785BRN			60	2038	
2SC785C		3246	60	2038	
2SC785D		3246	60	2038	
2SC785E		3246	60	2038	
2SC785F		3246	60	2038	
2SC785G		3246	60	2038	
2SC785GN		3246	60	2038	
2SC785GR		3246	60	2038	
2SC785H		3246	60	2038	
2SC785K		3246	60	2038	
2SC785L		3246	60	2038	
2SC785M		3246	60	2038	
2SC7850			60	2038	
2SC785R		3246	60	2038	
2SC785RA		3246	60	2038	
2SC785RED			60	2038	
2SC785V		3246	60	2038	
2SC785X		3246	60	2038	
2SC785Y		3246	60	2038	
2SC785YEL			60	2038	
2SC786	161	3018	86	2015	
2SC786-O	161	3018			
2SC786-O	161	3018			
2SC786A		3018	86	2015	
2SC786B		3018	86	2015	
2SC786C		3018	86	2015	
2SC786D		3018	86	2015	
2SC786E		3018	86	2015	
2SC786F		3018	86	2015	
2SC786G		3018	86	2015	
2SC786GN		3018	86	2015	
2SC786H		3018	86	2015	
2SC786J		3018	86	2015	
2SC786K		3018	86	2015	
2SC786L		3018	86	2015	
2SC786M		3018	86	2015	
2SC7860R		3018	86	2015	
2SC786R	161	3018	86	2015	
2SC786X		3018	86	2015	
2SC786Y		3018	86	2015	
2SC787	161	3039/316	214	2015	
2SC787A	161	3039/316	214	2015	
2SC787B		3039	214	2015	
2SC787C		3039	214	2015	
2SC787D		3039	214	2015	
2SC787E		3039	214	2015	
2SC787F		3039	214	2015	
2SC787G		3039	214	2015	

Industry Standard No.	ECG	SK	GE	RS 276-	MOTOR.
2SC787GN		3039	214	2015	
2SC787H		3039	214	2015	
2SC787K		3039	214	2015	
2SC787L		3039	214	2015	
2SC787M		3039	214	2015	
2SC787OR		3039	214	2015	
2SC787R		3039	214	2015	
2SC787X		3039	214	2015	
2SC787Y		3039	214	2015	
2SC788	154	3044	32	2012	
2SC788B		3044	32	2012	
2SC788C		3044	32	2012	
2SC788D		3044	32	2012	
2SC788E		3044	32	2012	
2SC788F		3044	32	2012	
2SC788G		3044	32	2012	
2SC788GN		3044	32	2012	
2SC788J		3044	32	2012	
2SC788K		3044	32	2012	
2SC788L		3044	32	2012	
2SC788M		3044	32	2012	
2SC788OR		3044	32	2012	
2SC788R		3044	32	2012	
2SC788X		3044	32	2012	
2SC788Y		3044	32	2012	
2SC789	152	3054/196	66	2048	2N6123
2SC789-O	152	3054/196	66	2048	
2SC789-O	152		66	2048	
2SC789-R	152		66	2048	
2SC789-Y	152		66	2048	
2SC789A		3054	66	2048	
2SC789B		3054	66	2048	
2SC789C		3054	66	2048	
2SC789D		3054	66	2048	
2SC789E		3054	66	2048	
2SC789F		3054	66	2048	
2SC789G		3054	66	2048	
2SC789GN		3054	66	2048	
2SC789H		3054	66	2048	
2SC789J		3054	66	2048	
2SC789K		3054	66	2048	
2SC789L		3054	66	2048	
2SC789M		3054	66	2048	
2SC789OR		3054	66	2048	
2SC789R	152	3054/196	66	2048	
2SC789X		3054	66	2048	
2SC789Y	152	3054/196	66	2048	
2SC78A		3861	59		
2SC78B		3861	59		
2SC78C		3861	59		
2SC78D		3861	59		
2SC78E		3861	59		
2SC78F		3861	59		
2SC78G		3861	59		
2SC78GN		3861	59		
2SC78H		3861	59		
2SC78J		3861	59		
2SC78K		3861	59		
2SC78L		3861	59		
2SC78M		3861	59		
2SC78OR		3861	59		
2SC78R		3861	59		
2SC78X		3861	59		
2SC78Y		3861	59		
2SC79	108	3452	86	2038	
2SC790	152	3054/196	66	2048	TIP31A
2SC790(0)			66	2048	
2SC790-O	152	3440/291			
2SC790-O	152		66	2048	
2SC790Y	152	3054/196	66	2048	
2SC791	175	3562	12	2020	2N5050
2SC791-OR			12	2020	
2SC791A		3562	12	2020	
2SC791B		3562	12	2020	
2SC791C		3562	12	2020	
2SC791D		3562	12	2020	
2SC791E		3562	12	2020	
2SC791F		3562	12	2020	
2SC791FA1			12	2020	
2SC791G		3562	12	2020	
2SC791GN		3562	12	2020	
2SC791H		3562	12	2020	
2SC791J		3562	12	2020	
2SC791K		3562	12	2020	
2SC791M		3562	12	2020	
2SC791OR		3562	12	2020	
2SC791R		3562	12	2020	
2SC791X		3562	12	2020	
2SC791Y		3562	12	2020	
2SC792		3438	35		2N5840
2SC793	130	3027	14	2041	2N5758
2SC793-BLU			14	2041	
2SC793-R			14	2041	
2SC793-RED			14	2041	
2SC793-YEL			14	2041	
2SC793A		3027	14	2041	
2SC793B		3027	14	2041	
2SC793BL	130		14	2041	
2SC793C		3027	14	2041	
2SC793E		3027	14	2041	
2SC793F		3027	14	2041	
2SC793G		3027	14	2041	
2SC793GN		3027	14	2041	
2SC793H		3027	14	2041	
2SC793J		3027	14	2041	
2SC793K		3027	14	2041	
2SC793L		3027	14	2041	
2SC793M		3027	14	2041	
2SC793OR		3027	14	2041	
2SC793R	130	3027	14	2041	
2SC793X			14	2041	
2SC793Y	130	3027	14	2041	
2SC794		3027	75	2041	2N5758
2SC794A		3027	75	2041	
2SC794B		3027	75	2041	
2SC794C		3027	75	2041	
2SC794D		3027	75	2041	
2SC794E		3027	75	2041	
2SC794F		3027	75	2041	
2SC794G		3027	75	2041	
2SC794GN		3027	75	2041	
2SC794H		3027	75	2041	
2SC794J		3027	75	2041	
2SC794K		3027	75	2041	
2SC794L		3027	75	2041	
2SC794OR		3027	75	2041	
2SC794R	130	3027	75	2041	
2SC794RA		3027	75	2041	
2SC794X		3027	75	2041	
2SC794Y		3027	75	2041	
2SC795	124	3021	12		2N3739
2SC795A	124	3021	12		
2SC795B		3021	12		
2SC795C		3021	12		
2SC795D		3021	12		
2SC795F		3021	12		
2SC795G		3021	12		
2SC795GN		3021	12		
2SC795H		3021	12		
2SC795J		3021	12		
2SC795L		3021	12		
2SC795M		3021	12		
2SC795OR		3021	12		
2SC795R		3021	12		
2SC795X		3021	12		
2SC795Y		3021	12		
2SC796	123A	3444	88	2051	
2SC796A		3122	88	2030	
2SC796B		3122	88	2030	
2SC796C		3122	88	2030	
2SC796D		3122	88	2030	
2SC796E		3122	88	2030	
2SC796F		3122	88	2030	
2SC796G		3122	88	2030	
2SC796GN		3122	88	2030	
2SC796H		3122	88	2030	
2SC796J		3122	88	2030	
2SC796K		3122	88	2030	
2SC796L		3122	88	2030	
2SC796M		3122	88	2030	
2SC796OR		3122	88	2030	
2SC796R		3122	88	2030	
2SC796X		3122	88	2030	
2SC796Y		3122	88	2030	
2SC797	128	3024	18	2030	
2SC797A		3024	18	2051	
2SC797B		3024	18	2051	
2SC797C		3024	18	2051	
2SC797D		3024	18	2051	
2SC797E		3024	18	2051	
2SC797F		3024	18	2051	
2SC797G		3024	18	2051	
2SC797GN		3024	18	2051	
2SC797H		3024	18	2051	
2SC797J		3024	18	2051	
2SC797K		3024	18	2051	
2SC797L		3024	18	2051	
2SC797M		3024	18	2051	
2SC797OR		3024	18	2051	
2SC797R		3024	18	2051	
2SC797X		3024	18	2051	
2SC797Y		3024	18	2051	
2SC798	195A	3024/128	18	2030	
2SC798A		3024	18	2030	
2SC798B		3024	18	2030	
2SC798C		3024	18	2030	
2SC798D		3024	18	2030	
2SC798E		3024	18	2030	
2SC798G		3024	18	2030	
2SC798GN		3024	18	2030	
2SC798H		3024	18	2030	
2SC798J		3024	18	2030	
2SC798K		3024	18	2030	
2SC798L		3024	18	2030	
2SC798M		3024	18	2030	
2SC798OR		3024	18	2030	
2SC798R		3024	18	2030	
2SC798X		3024	18	2030	
2SC798Y		3024	18	2030	
2SC799	237	3299	46		
2SC799-4		3299	46		
2SC799A		3299	46		
2SC799AP		3299	46		
2SC799B		3299	46		
2SC799C		3299	46		
2SC799D		3299	46		
2SC799E		3299	46		
2SC799F		3299	46		
2SC799G		3299	46		
2SC799GN		3299	46		
2SC799H		3299	46		
2SC799J		3299	46		
2SC799K	237	3299	46		
2SC799L		3299	46		
2SC799M		3299	46		
2SC799OR		3299	46		
2SC799R		3299	46		
2SC799X		3299	46		
2SC799Y		3299	46		
2SC79A		3039	86	2038	
2SC79B		3039	86	2038	
2SC79C		3039	86	2038	
2SC79D		3039	86	2038	
2SC79E		3039	86	2038	
2SC79F		3039	86	2038	
2SC79G		3039	86	2038	
2SC79GN		3039	86	2038	
2SC79H		3039	86	2038	
2SC79J		3039	86	2038	
2SC79K		3039	86	2038	
2SC79L		3039	86	2038	
2SC79M		3039	86	2038	
2SC79OR		3039	86	2038	
2SC79R		3039	86	2038	
2SC79X		3039	86	2038	
2SC79Y		3039	86	2038	
2SC80	161	3452/108	86	2038	
2SC800		3039	86	2015	
2SC800A		3039	86		
2SC800B		3039	86		
2SC800C		3039	86		
2SC800D		3039	86		
2SC800E		3039	86		
2SC800F		3039	86		
2SC800G		3039	86		
2SC800GN		3039	86		
2SC800H		3039	86		
2SC800J		3039	86		
2SC800K		3039	86		
2SC800L		3039	86		
2SC800M		3039	86		
2SC800OR		3039	86		
2SC800R		3039	86		
2SC800X		3039	86		
2SC800Y		3039	86		
2SC802	128		63	2030	
2SC803	195A	3024/128	18	2030	
2SC803A		3024	18		
2SC803B		3024	18		
2SC803C		3024	18		
2SC803D		3024	18		
2SC803E		3024	18		
2SC803F		3024	18		
2SC803G		3024	18		
2SC803H		3024	18		
2SC803J		3024	18		
2SC803K		3024	18		
2SC803L		3024	18		

Industry Standard No.	ECG	SK	GE	RS 276-	MOTOR.
2SC803M		3024	18		
2SC803OR		3024	18		
2SC803R		3024	18		
2SC803X		3024	18		
2SC803Y		3024	18		
2SC804			214	2038	
2SC804H	313		241	2020	
2SC805	154	3040	27	2012	
2SC805A		3040	27	2012	
2SC805B		3040	27	2012	
2SC805C		3040	27	2012	
2SC805D		3040	27	2012	
2SC805E		3040	27	2012	
2SC805F		3040	27	2012	
2SC805G		3040	27	2012	
2SC805GN		3040	27	2012	
2SC805H		3040	27	2012	
2SC805J		3040	27	2012	
2SC805L		3040	27	2012	
2SC805OR		3040	27	2012	
2SC805R		3040	27	2012	
2SC805X		3040	27	2012	
2SC805Y		3040	27	2012	
2SC806	283	3439/163A	38		MJ431
2SC806A	283	3439/163A	38		
2SC806B		3439	38		
2SC806C		3439	38		
2SC806D		3439	38		
2SC806E		3439	38		
2SC806F		3439	38		
2SC806G		3439	38		
2SC806J		3439	38		
2SC806L		3439	38		
2SC806M		3439	38		
2SC806OR		3439	38		
2SC806R		3439	38		
2SC806X		3439	38		
2SC806Y		3439	38		
2SC807	385	3439/163A	36		MJ413
2SC807A	385	3439/163A	36		
2SC807AK		3079	36		
2SC808	388				MJ411
2SC80A		3039	86	2038	
2SC80B		3039	86	2038	
2SC80C		3039	86	2038	
2SC80D		3039	86	2038	
2SC80E		3039	86	2038	
2SC80F		3039	86	2038	
2SC80G		3039	86	2038	
2SC80GN		3039	86	2038	
2SC80H		3039	86	2038	
2SC80J		3039	86	2038	
2SC80K		3039	86	2038	
2SC80L		3039	86	2038	
2SC80M		3039	86	2038	
2SC80OR		3039	86	2038	
2SC80R		3039	86	2038	
2SC80X		3039	86	2038	
2SC80Y		3039	86	2038	
2SC81		3124	81	2030	
2SC811	161			2011	
2SC812				2014	
2SC812A		3049	46		
2SC812B		3049	46		
2SC812C		3049	46		
2SC812D		3049	46		
2SC812E		3049	46		
2SC812F		3049	46		
2SC812G		3049	46		
2SC812GN		3049	46		
2SC812H		3049	46		
2SC812J		3049	46		
2SC812K		3049	46		
2SC812L		3049	46		
2SC812M		3049	46		
2SC812OR		3049	46		
2SC812R		3049	46		
2SC812Y		3049	46		
2SC814	192A	3124/289	47	2030	
2SC8146			47	2030	
2SC814A		3124	47	2051	
2SC814B		3124	47	2051	
2SC814C		3124	47	2051	
2SC814D		3124	47	2051	
2SC814E		3124	47	2051	
2SC814F		3124	47	2051	
2SC814G		3124	47	2051	
2SC814GN		3124	47	2051	
2SC814H		3124	47	2051	
2SC814J		3124	47	2051	
2SC814K		3124	47	2051	
2SC814M		3124	47	2051	
2SC814OR		3124	47	2051	
2SC814R		3124	47	2051	
2SC814X		3124	47	2051	
2SC814Y		3124	47	2051	
2SC815	289A	3124/289	210	2030	
2SC815(M)	289A	3124/289	210	2030	
2SC815-1		3124	210	2030	
2SC815A	289A	3124/289	210	2030	
2SC815B	289A	3124/289	210	2030	
2SC815BK		3124	210	2030	
2SC815C	289A	3124/289	210	2030	
2SC815D		3124	210	2030	
2SC815E		3124	210	2030	
2SC815F	289A	3124/289	210	2030	
2SC815G		3124	210	2030	
2SC815GN		3124	210	2030	
2SC815H		3124	210	2030	
2SC815J		3124	210	2030	
2SC815K	289A	3124/289	210	2030	
2SC815K,L		3124	210	2030	
2SC815L	289A	3124/289	210	2030	
2SC815LJ		3124	210	2030	
2SC815M	289A	3124/289	210	2030	
2SC815OR		3124	210	2030	
2SC815R		3124	210	2030	
2SC815S	289A	3124/289	210	2051	
2SC815SA	289A	3124/289	210	2030	
2SC815SC	289A	3124/289	210	2030	
2SC815X		3124	210	2030	
2SC815Y		3124	210	2030	
2SC816	128	3024	18	2030	
2SC816A		3024	18	2030	
2SC816B		3024	18	2030	
2SC816C		3024	18	2030	
2SC816D		3024	18	2030	
2SC816E		3024	18	2030	
2SC816F		3024	18	2030	
2SC816G		3024	18	2030	
2SC816GN		3024	18	2030	
2SC816H		3024	18	2030	
2SC816K	128	3024	18	2030	
2SC816L		3024	18	2030	
2SC816M		3024	18	2030	
2SC816OR		3024	18	2030	
2SC816R		3024	18	2030	
2SC816X		3024	18	2030	
2SC816Y		3024	18	2030	
2SC818	154	3045/225	32	2012	
2SC818A		3045	32		
2SC818B		3045	32		
2SC818C		3045	32		
2SC818D		3045	32		
2SC818E		3045	32		
2SC818F		3045	32		
2SC818G		3045	32		
2SC818GN		3045	32		
2SC818H		3045	32		
2SC818J		3045	32		
2SC818K		3045	32		
2SC818L		3045	32		
2SC818M		3045	32		
2SC818OR		3045	32		
2SC818R		3045	32		
2SC818X		3045	32		
2SC818Y		3045	32		
2SC81A		3124	47	2030	
2SC81B		3124	47	2030	
2SC81C		3124	47	2030	
2SC81D		3124	47	2030	
2SC81E		3124	47	2030	
2SC81F		3124	47	2030	
2SC81G		3124	47	2030	
2SC81GN		3124	47	2030	
2SC81H		3124	47	2030	
2SC81J		3124	47	2030	
2SC81K		3124	47	2030	
2SC81L		3124	47	2030	
2SC81M		3124	47	2030	
2SC81OR		3124	47	2030	
2SC81R		3124	47	2030	
2SC821	346		261		2N4427
2SC822	346		261		MRF607
2SC823			261		2N5943
2SC824	278$		261		2N5943
2SC825	286	3021/124	12		2N3585
2SC825A	286	3021/124	12		
2SC825B	286	3021/124	12		
2SC825C	286	3021/124	12		
2SC825D		3021	12		
2SC825E		3021	12		
2SC825F		3021	12		
2SC825G		3021	12		
2SC825GN		3021	12		
2SC825H		3021	12		
2SC825J		3021	12		
2SC825K		3021	12		
2SC825L		3021	12		
2SC825M		3021	12		
2SC825OR		3021	12		
2SC825R		3021	12		
2SC825X		3021	12		
2SC825Y		3021	12		
2SC826	128	3024	18	2030	
2SC826A		3024	18	2030	
2SC826B		3024	18	2030	
2SC826C		3024	18	2030	
2SC826D		3024	18	2030	
2SC826E		3024	18	2030	
2SC826F		3024	18	2030	
2SC826G		3024	18	2030	
2SC826H		3024	18	2030	
2SC826J		3024	18	2030	
2SC826K		3024	18	2030	
2SC826L		3024	18	2030	
2SC826M		3024	18	2030	
2SC826OR		3024	18	2030	
2SC826R		3024	18	2030	
2SC826X		3024	18	2030	
2SC826Y		3024	18	2030	
2SC827	128	3024	18	2030	
2SC827A	128	3024	18	2030	
2SC827B	128	3024	18	2030	
2SC827C	128	3024	18	2030	
2SC827D		3024	18	2030	
2SC827E		3024	18	2030	
2SC827F		3024	18	2030	
2SC827G		3024	18	2030	
2SC827GN		3024	18	2030	
2SC827H		3024	18	2030	
2SC827J		3024	18	2030	
2SC827K		3024	18	2030	
2SC827M		3024	18	2030	
2SC827OR		3024	18	2030	
2SC827R		3024	18	2030	
2SC827X		3024	18	2030	
2SC827Y		3024	18	2030	
2SC828	85	3122	20	2051	
2SC828(H)			61	2010	
2SC828(N)			61	2010	
2SC828(O)			61	2010	
2SC828(P)			61	2010	
2SC828(P)(Q)			61	2010	
2SC828(Q)			61	2010	
2SC828(R)			61	2010	
2SC828(R)(S)			61	2010	
2SC828(R,Q,P)			61	2010	
2SC828(R,S,T)			61	2010	
2SC828(S)			61	2010	
2SC828(T)			61	2010	
2SC828-0	85		61	2010	
2SC828-0P	85		61	2010	
2SC828-0R			61	2010	
2SC828A	85	3866	61	2010	
2SC828A(P)			61	2010	
2SC828A(Q)			61	2010	
2SC828A(R)			61	2010	
2SC828A0		3866	61	2010	
2SC828AP	85	3866	61	2010	
2SC828AQ	85	3866	61	2010	
2SC828AR	85	3866	61	2010	
2SC828AS	85	3866	61	2010	
2SC828AT	85	3866			
2SC828B	85		61	2010	
2SC828C	85		61	2010	
2SC828D			61	2010	
2SC828E	85		61	2010	
2SC828F	85		61	2010	
2SC828FR	85	3866	61	2010	
2SC828G			61	2010	
2SC828GN			61	2010	
2SC828H	85	3018	20	2051	
2SC828HR	85	3866	61	2010	
2SC828K	85		61	2010	
2SC828L			61	2010	
2SC828LR	85	3866	61	2010	
2SC828LS	85	3866	61	2010	

Industry Standard No.	ECG	SK	GE	RS 276-	MOTOR.
2SC828M			61	2010	
2SC828N	85		61	2010	
2SC828OR			61	2010	
2SC828P	85	3866	61	2010	
2SC828PQ	85	3866	61	2010	
2SC828Q	85	3866	20	2051	
2SC828Q-6		3866	61	2010	
2SC828QRS	85		61	2010	
2SC828R	85	3866	61	2010	
2SC828R-1		3866	61	2010	
2SC828RA			61	2010	
2SC828RH			61	2010	
2SC828RS	85		61	2010	
2SC828RST	85		61	2010	
2SC828S	85	3866	61	2010	
2SC828T	85	3866	61	2010	
2SC828W	85		61	2010	
2SC828X			61	2010	
2SC828Y	85		61	2010	
2SC828YL			61	2010	
00002SC829	107	3122	11	2015	
2SC829(Y)			20	2051	
2SC829-OR			20	2051	
2SC829/4454C				2051	
2SC829O			20	2051	
2SC829A	85	3122	20	2051	
2SC829AK		3122	20	2051	
2SC829B	85	3122	20	2051	
2SC829B/4454C	85	3122	20	2051	
2SC829BC	85	3122	20	2051	
2SC829BJ		3122	20	2051	
2SC829BK		3122	20	2051	
2SC829BY	85	3122	20	2051	
2SC829C	85	3122	20	2051	
2SC829CL		3122	20	2051	
2SC829D	85	3122	20	2051	
2SC829E		3122	20	2051	
2SC829F	85		20	2051	
2SC829G		3122	20	2051	
2SC829GN		3122	20	2051	
2SC829H		3122	20	2051	
2SC829K		3122	20	2051	
2SC829L		3122	20	2051	
2SC829M		3122	20	2051	
2SC829OR		3122	20	2051	
2SC829R	85	3122	20	2051	
2SC829X	85	3122	20	2051	
2SC829Y	85	3122	20	2051	
2SC82BN			11	2015	
2SC82R			11	2015	
2SC83			11	2015	
2SC830	175	3562	66	2020	
2SC830A	175	3562	66	2020	
2SC830B	175	3562	66	2020	
2SC830BKB	175	3026			
2SC830C	175	3562	66	2020	
2SC830D		3562	66	2020	
2SC830E		3562	66	2020	
2SC830F		3562	66	2020	
2SC830G		3562	66	2020	
2SC830GN		3562	66	2020	
2SC830H		3562	66	2020	
2SC830J		3562	66	2020	
2SC830K		3562	66	2020	
2SC830L		3562	66	2020	
2SC830M		3562	66	2020	
2SC830OR			66	2020	
2SC830R		3562	66	2020	
2SC830X			66	2020	
2SC830Y		3562	66	2020	
2SC831			69	2049	2N3927
2SC833	384		20	2051	2N6235
2SC833BL	384	3124/289	62	2051	
2SC834L			10	2051	
2SC835	107	3122	11	2015	
2SC836M	107	3444/123A	20	2051	
2SC837	107	3018	61	2015	
2SC837(K)			61	2038	
2SC837(KL)			61	2038	
2SC837(L)			61	2038	
2SC837A		3018	61	2038	
2SC837B		3018	61	2038	
2SC837C		3018	61	2038	
2SC837D		3018	61	2038	
2SC837E		3018	61	2038	
2SC837F	107	3018	61	2015	
2SC837G		3018	61	2038	
2SC837GN		3018	61	2038	
2SC837H	107	3018	61	2015	
2SC837J		3018	61	2038	
2SC837K	107	3018	61	2015	
2SC837KL			61	2038	
2SC837L	107	3018	61	2015	
2SC837M		3018	61	2038	
2SC837OR		3018	61	2038	
2SC837R		3018	61	2038	
2SC837WF	107		61	2015	
2SC837X		3018	61	2038	
2SC837Y		3018	61	2038	
2SC838	289A	3122	20	2051	
2SC838(A)			20	2051	
2SC838(E)			20	2051	
2SC838(F)			20	2051	
2SC838(H)	289A	3122	20	2051	
2SC838(J)	289A	3122	20	2051	
2SC838(K)	289A	3122	20	2051	
2SC838(L)	289A	3122	20	2051	
2SC838(M)	289A	3122	20	2051	
2SC838(O)			20	2051	
2SC838-2		3122	20	2051	
2SC838-O			20	2051	
2SC838O	289A	3452/108	86	2038	
2SC838A	289A	3122	20	2051	
2SC838B	289A	3122	20	2051	
2SC838BL			20	2051	
2SC838C	289A	3122	20	2051	
2SC838D	289A	3122	20	2051	
2SC838E	289A	3122	20	2051	
2SC838F	289A	3122	20	2051	
2SC838H	289A	3122	20	2051	
2SC838HF			20	2051	
2SC838J	289A	3122	20	2051	
2SC838K	289A	3122	20	2051	
2SC838L	289A	3122	20	2051	
2SC838M	289A	3122	20	2051	
2SC838S	289A	3122	20	2051	
2SC839	85	3124/289	61	2030	
2SC839(F)			61	2030	
2SC839(H)	85	3124/289	61	2030	
2SC839(J)	85	3124/289	61	2030	
2SC839(JI)			61	2030	
2SC839(L)	85	3124/289	61	2030	
2SC839(M)	85	3124/289	61	2030	
2SC839-E			61	2030	
2SC839-F			61	2030	
2SC839A	85	3124/289	61	2030	
2SC839B	85	3124/289	61	2030	
2SC839C	85	3124/289	61	2030	
2SC839D	85	3124/289	61	2030	
2SC839E	85	3124/289	61	2030	
2SC839F	85	3124/289	61	2030	
2SC839G		3124	61	2030	
2SC839GN		3124	61	2030	
2SC839H	85	3124/289	61	2030	
2SC839J	85	3124/289	61	2030	
2SC839JH			61	2030	
2SC839JI	85	3124/289	61	2030	
2SC839K	85	3124/289	61	2030	
2SC839L	85	3124/289	61	2030	
2SC839M	85	3124/289	61	2030	
2SC839N	85	3124/289	61	2030	
2SC839OR		3124	61	2030	
2SC839R	85	3124/289	61	2030	
2SC839S	85	3124/289	61	2030	
2SC839X		3124	61	2030	
2SC839Y	85	3124/289	61	2030	
2SC840	175	3538	246	2020	2N5050
2SC840(P)			246	2020	
2SC840A	175	3538	246	2020	2N5051
2SC840AC	175		246	2020	
2SC840B		3538	246	2020	
2SC840C		3538	246	2020	
2SC840D		3538	246	2020	
2SC840E		3538	246	2020	
2SC840F		3538	246	2020	
2SC840G		3538	246	2020	
2SC840GN		3538	246	2020	
2SC840H	175	3538	246	2020	
2SC840HP	175		246	2020	
2SC840J		3538	246	2020	
2SC840K		3538	246	2020	
2SC840L		3538	246	2020	
2SC840M		3538	246	2020	
2SC840OR			246	2020	
2SC840P	175	3538	246	2020	
2SC840PQ	175		246	2020	
2SC840Q			246	2020	
2SC840R		3538	241	2020	
2SC840X		3538	246	2020	
2SC840Y		3538	246	2020	
2SC844	311	3444/123A	244	2027	
2SC847	123A	3444	88	2051	
2SC847A	123A	3122	88	2030	
2SC847B	123A	3122	88	2030	
2SC847C	123A	3122	88	2030	
2SC847D		3122	88	2030	
2SC847E		3122	88	2030	
2SC847F		3122	88	2030	
2SC847G		3122	88	2030	
2SC847GN		3122	88	2030	
2SC847H		3122	88	2030	
2SC847J		3122	88	2030	
2SC847K		3122	88	2030	
2SC847L		3122	88	2030	
2SC847M		3122	88	2030	
2SC847OR		3122	88	2030	
2SC847R		3122	88	2030	
2SC847X		3122	88	2030	
2SC848	123A	3444	88	2051	
2SC848A	123A	3122	88	2030	
2SC848B	123A	3122	88	2030	
2SC848C	123A	3122	88	2030	
2SC848D		3122	88	2030	
2SC848E		3122	88	2030	
2SC848F		3122	88	2030	
2SC848G		3122	88	2030	
2SC848GN		3122	88	2030	
2SC848H		3122	88	2030	
2SC848J		3122	88	2030	
2SC848K		3122	88	2030	
2SC848L		3122	88	2030	
2SC848M		3122	88	2030	
2SC848OR		3122	88	2030	
2SC848R		3122	88	2030	
2SC848X		3122	88	2030	
2SC848Y		3122	88	2030	
2SC849	123A	3444	88	2051	
2SC849A	123A	3122	88	2030	
2SC849B	123A	3122	88	2030	
2SC849C	123A	3122	88	2030	
2SC849D		3122	88	2030	
2SC849E		3122	88	2030	
2SC849F		3122	88	2030	
2SC849G		3122	88	2030	
2SC849GN		3122	88	2030	
2SC849H		3122	88	2030	
2SC849J		3122	88	2030	
2SC849K		3122	88	2030	
2SC849L		3122	88	2030	
2SC849M		3122	88	2030	
2SC849OR		3122	88	2030	
2SC849R		3122	88	2030	
2SC849X		3122	88	2030	
2SC849Y		3122	88	2030	
2SC850	123A	3444	88	2051	
2SC850A	123A	3122	88	2030	
2SC850B	123A	3122	88	2030	
2SC850C	123A	3122	88	2030	
2SC850D		3122	88	2030	
2SC850E		3122	88	2030	
2SC850F		3122	88	2030	
2SC850G		3122	88	2030	
2SC850GN		3122	88	2030	
2SC850H		3122	88	2030	
2SC850J		3122	88	2030	
2SC850K		3122	88	2030	
2SC850L		3122	88	2030	
2SC850M		3122	88	2030	
2SC850OR			88	2030	
2SC850R		3122	88	2030	
2SC850X		3122	88	2030	
2SC850Y		3122	88	2030	
2SC851		3027	75	2041	
2SC851A		3027	75	2041	
2SC851B		3027	75	2041	
2SC851C		3027	75	2041	
2SC851D		3027	75	2041	
2SC851E		3027	75	2041	
2SC851F		3027	75	2041	
2SC851G		3027	75	2041	
2SC851GN		3027	75	2041	
2SC851H		3027	75	2041	
2SC851K		3027	75	2041	
2SC851L		3027	75	2041	
2SC851M		3027	75	2041	
2SC851OR		3027	75	2041	
2SC851R		3027	75	2041	
2SC851X		3027	75	2041	

Industry Standard No.	ECG	SK	GE	RS 276-	MOTOR.
2SC851Y		3027	75	2041	
2SC852	311		39	2015	2N5943
2SC852A			39	2015	
2SC853	192A	3122	88	2030	
2SC853-OR			88	2030	
2SC853A	192A	3122	88	2030	
2SC853B	192A	3122	88	2030	
2SC853C	192A	3122	88	2030	
2SC853D		3122	88	2030	
2SC853E		3122	88	2030	
2SC853F		3122	88	2030	
2SC853G		3122	88	2030	
2SC853GN		3122	88	2030	
2SC853H		3122	88	2030	
2SC853J		3122	88	2030	
2SC853K	192A	3122	88	2030	
2SC853KLM	192A	3020/123	88	2030	
2SC853L	192A	3122	88	2030	
2SC853M		3122	88	2030	
2SC853OR		3122	88	2030	
2SC853R		3122	88	2030	
2SC853X		3122	88	2030	
2SC853Y		3122	88	2030	
2SC856	154	3040	220	2012	
2SC856-02	154	3040	220	2012	
2SC856-OR			220	2012	
2SC856A		3040	220	2012	
2SC856B		3040	220	2012	
2SC856C	154	3040	220	2012	
2SC856D		3040	220	2012	
2SC856E		3040	220	2012	
2SC856F		3040	220	2012	
2SC856G		3040	220	2012	
2SC856GN		3040	220	2012	
2SC856J		3040	220	2012	
2SC856K		3040	220	2012	
2SC856L		3040	220	2012	
2SC856M		3040	220	2012	
2SC856OR		3040	220	2012	
2SC856R		3040	220	2012	
2SC856X		3040	220	2012	
2SC856Y		3040	220	2012	
2SC857	154	3040	40	2012	
2SC857H	154	3040	40	2012	
2SC857K		3040	40	2012	
00002SC858	199	3245	62	2010	
2SC858A		3124	47	2051	
2SC858B		3124	47	2051	
2SC858C		3124	47	2051	
2SC858D		3124	47	2051	
2SC858E	199	3245	47	2051	
2SC858F	199	3245	47	2051	
2SC858FG	199	3245	47	2051	
2SC858G	199	3245	47	2051	
2SC858GA		3124	47	2051	
2SC858GN		3124	47	2051	
2SC858H		3124	47	2051	
2SC858J		3124	47	2051	
2SC858K		3124	47	2051	
2SC858L		3124	47	2051	
2SC858M		3124	47	2051	
2SC858OR		3124	47	2051	
2SC858R		3124	47	2051	
2SC858X		3124	47	2051	
2SC858Y		3124	47	2051	
2SC859	199	3124/289	47	2010	
2SC859A		3124	47	2010	
2SC859B		3124	47	2010	
2SC859D		3124	47	2010	
2SC859E	199	3124/289	47	2010	
2SC859F	199	3124/289	47	2010	
2SC859FG	199	3124/289	47	2010	
2SC859G	199	3124/289	47	2010	
2SC859GK	199	3124/289	47	2010	
2SC859GL		3124	47	2010	
2SC859GM		3124	47	2010	
2SC859GN		3124	47	2010	
2SC859H		3124	47	2010	
2SC859J		3124	47	2010	
2SC859K		3124	47	2010	
2SC859L		3124	47	2010	
2SC859M		3124	47	2010	
2SC859OR		3124	47	2010	
2SC859R		3124	47	2051	
2SC859X		3124	47	2010	
2SC859Y		3124	47	2010	
2SC860	161	3117	211	2015	
2SC860A		3117	211	2038	
2SC860B		3117	211	2038	
2SC860C	161	3117	211	2015	
2SC860D	161	3117	211	2015	
2SC860E	161	3117	211	2015	
2SC860F		3117	211	2038	
2SC860G		3117	211	2038	
2SC860GN		3117	211	2038	
2SC860H		3117	211	2038	
2SC860J		3117	211	2038	
2SC860K		3117	211	2038	
2SC860L		3117	211	2038	
2SC860M		3117	211	2038	
2SC860OR		3246	211	2038	
2SC860R		3246	211	2038	
2SC860X		3117	211	2038	
2SC860Y		3117	211	2038	
2SC861					MJ3029
2SC862					MJ3030
2SC863	161	3117	211	2015	
2SC863A		3117	211	2016	
2SC863B		3117	211	2016	
2SC863C		3117	211	2016	
2SC863D	161	3117	211	2016	
2SC863E		3117	211	2016	
2SC863F		3117	211	2016	
2SC863G		3117	211	2016	
2SC863GN		3117	211	2016	
2SC863H		3117	211	2016	
2SC863J		3117	211	2016	
2SC863K		3117	211	2016	
2SC863L		3117	211	2016	
2SC863M		3117	211	2016	
2SC863OR		3117	211	2016	
2SC863R		3117	211	2016	
2SC863X		3117	211	2016	
2SC863Y		3117	211	2016	
2SC864	161	3132	46	2015	
2SC866	237	3299	46		
2SC866A		3299	46		
2SC866B		3299	46		
2SC866C		3299	46		
2SC866D		3299	46		
2SC866E		3299	46		
2SC866F		3299	46		
2SC866G		3299	46		
2SC866GN		3299	46		
2SC866H		3299	46		
2SC866J		3299	46		
2SC866K		3299	46		
2SC866L		3299	46		
2SC866M		3299	46		
2SC866OR		3299	46		
2SC866R		3299	46		
2SC866X		3299	46		
2SC866Y		3299	46		
2SC867	277		32		2N3739
2SC867A	277		32		
2SC867B	277		32		
2SC868		3045	32	2012	
2SC868A		3045	32		
2SC868B		3045	32		
2SC868C		3045	32		
2SC868D		3045	32		
2SC868E		3045	32		
2SC868F		3045	32		
2SC868G		3045	32		
2SC868GN		3045	32		
2SC868H		3045	32		
2SC868J		3045	32		
2SC868K		3045	32		
2SC868L		3045	32		
2SC868M		3045	32		
2SC868OR		3045	32		
2SC868R		3045	32		
2SC868X		3045	32		
2SC868Y		3045	32		
2SC869		3045	32	2012	
2SC869A		3045	32		
2SC869B		3045	32		
2SC869C		3045	32		
2SC869D		3045	32		
2SC869E		3045	32		
2SC869F		3045	32		
2SC869G		3045	32		
2SC869GN		3045	32		
2SC869J		3045	32		
2SC869K		3045	32		
2SC869L		3045	32		
2SC869M		3045	32		
2SC869OR		3045	32		
2SC869R		3045	32		
2SC869X		3045	32		
2SC869Y		3045	32		
2SC87	123A	3444	88	2051	
2SC870	85	3124/289	20	2051	
00002SC870A	123A	3124/289	20	2051	
00002SC870B	123A	3124/289	20	2051	
2SC870BL	85	3124/289	47	2051	
2SC870C	85	3124/289	20	2051	
2SC870D	85	3124/289	47	2051	
2SC870E	85	3124/289	47	2051	
2SC870F	85	3124/289	47	2051	
2SC870FL			47	2030	
2SC870G			47	2051	
2SC870GN			47	2051	
2SC870H		3124	47	2051	
2SC870J			47	2051	
2SC870K			47	2051	
2SC870L			47	2051	
2SC870M			47	2051	
2SC870OR			47	2051	
2SC870R			47	2030	
2SC870X			47	2051	
2SC870Y			47	2051	
2SC871	85	3124/289	62	2051	
2SC871-G			62	2051	
2SC871A		3124	62	2051	
2SC871AM		3124	62	2051	
2SC871B		3124	62	2051	
2SC871BL	85	3124/289	62	2051	
2SC871C		3124	62	2051	
2SC871D	85	3124/289	62	2051	
2SC871E	85	3124/289	62	2051	
2SC871F	85	3124/289	62	2051	
2SC871G	85	3124/289	62	2051	
2SC871GN		3124	62	2051	
2SC871H		3124	62	2051	
2SC871J		3124	62	2051	
2SC871K		3124	62	2051	
2SC871L		3124	62	2051	
2SC871M		3124	62	2051	
2SC871OR		3124	62	2051	
2SC871R		3124	62	2051	
2SC871X		3124	62	2051	
2SC871Y		3124	62	2051	
2SC875	128	3024	243	2030	
2SC875(D)			243	2030	
2SC875(E)			243	2030	
2SC875(F)			243	2030	
2SC875-1	128	3024	243	2030	
2SC875-1C	128	3024	243	2030	
2SC875-1D	128	3024	243	2030	
2SC875-1E	128	3024	243	2030	
2SC875-1F	128		243	2030	
2SC875-2	128	3024	243	2030	
2SC875-2C	128	3024	243	2030	
2SC875-2D	128	3024	243	2030	
2SC875-2E	128	3024	243	2030	
2SC875-2F	128		243	2030	
2SC875-3	128	3024	243	2030	
2SC875-3C	128	3024	243	2030	
2SC875-3D	128	3024	243	2030	
2SC875-3E	128	3024	243	2030	
2SC875-3F	128		243	2030	
2SC875B		3024	243	2030	
2SC875BR	128		243	2030	
2SC875C	128	3024	243	2030	
2SC875D	128	3024	243	2030	
2SC875DL		3024	243	2030	
2SC875E	128	3024	243	2030	
2SC875EL		3024	243	2030	
2SC875F	128	3024	243	2030	
2SC875G		3024	243	2030	
2SC875GN		3024	243	2030	
2SC875J		3024	243	2030	
2SC875K		3024	243	2030	
2SC875L		3024	243	2030	
2SC875M		3024	243	2030	
2SC875OR		3024	243	2030	
2SC875X		3024	243	2030	
2SC875Y		3024	243	2030	
2SC876	128	3024	18	2030	
2SC876(F)			18	2030	
2SC876A		3024	18	2030	
2SC876B		3024	18	2030	
2SC876C	128	3024	18	2030	
2SC876D	128	3024	18	2030	
2SC876E	128	3024	18	2030	
2SC876F	128	3024	18	2030	
2SC876G		3024	18	2030	

Industry Standard No.	ECG	SK	GE	RS 276-	MOTOR.
2SC876GN		3024	18	2030	
2SC876H		3024	18	2030	
2SC876J		3024	18	2030	
2SC876K		3024	18	2030	
2SC876L		3024	18	2030	
2SC876M		3024	18	2030	
2SC876OR		3024	18	2030	
2SC876R		3024	18	2030	
2SC876TV	128	3044/154	18	2030	
2SC876TV(D)			18	2030	
2SC876TV(E)			18	2030	
2SC876TVD	128		18	2030	
2SC876TVE	128	3024	18	2030	
2SC876TVEF	128		18	2030	
2SC876TVF		3024	18	2030	
2SC876X		3024	18	2030	
2SC876Y		3024	18	2030	
2SC879			11	2015	
2SC87A		3122	88	2030	
2SC87B		3122	88	2030	
2SC87C		3122	88	2030	
2SC87D		3122	88	2030	
2SC87E		3122	88	2030	
2SC87F		3122	88	2030	
2SC87G		3122	88	2030	
2SC87GN		3122	88	2030	
2SC87H		3122	88	2030	
2SC87J		3122	88	2030	
2SC87K		3122	88	2030	
2SC87L		3122	88	2030	
2SC87M		3122	88	2030	
2SC87OR		3122	88	2030	
2SC87R		3122	88	2030	
2SC87X		3122	88	2030	
2SC87Y		3122	88	2030	
2SC88	128	3045/225	32	2012	
2SC881	192A	3137	63	2030	
2SC881A	192A	3137	63	2030	
2SC881B	192A	3137	63	2030	
2SC881C	192A	3137	63	2030	
2SC881D	192A	3137	63	2030	
2SC881E		3137	63	2030	
2SC881F		3137	63	2030	
2SC881G		3137	63	2030	
2SC881GN		3137	63	2030	
2SC881H		3137	63	2030	
2SC881K	192A	3137	63	2030	
2SC881L	192A	3137	63	2030	
2SC881M	192A	3137	63	2030	
2SC881OR		3137	63	2030	
2SC881R		3137	63	2030	
2SC881X		3137	63	2030	
2SC881Y		3137	63	2030	
2SC884					2N5050
2SC885					2N6307
2SC886					2N6306
2SC887					MJ410
2SC888					MJ410
2SC889	130	3438	35		MJ410
2SC88A	128	3045/225	32	2012	
2SC88B		3045	32		
2SC88C		3045	32		
2SC88D		3045	32		
2SC88E		3045	32		
2SC88F		3045	32		
2SC88G		3045	32		
2SC88GN		3045	32		
2SC88H		3045	32		
2SC88J		3045	32		
2SC88K		3045	32		
2SC88L		3045	32		
2SC88M		3045	32		
2SC88OR		3045	32		
2SC88R		3045	32		
2SC88X		3045	32		
2SC88Y		3045	32		
2SC89	101	3861	59	2002	
2SC890	487		277		MRF515
2SC891	363$		28	2017	2N5645
2SC892	364$		66	2048	2N5646
2SC893	282/427		28	2017	
2SC894	289A	3444/123A	88	2051	
2SC894A		3122	88	2051	
2SC894B		3122	88	2051	
2SC894C		3122	88	2051	
2SC894D		3122	88	2051	
2SC894E		3122	88	2051	
2SC894F		3122	88	2051	
2SC894G		3122	88	2051	
2SC894GN		3122	88	2051	
2SC894H		3122	88	2051	
2SC894J		3122	88	2051	
2SC894K		3122	88	2051	
2SC894L		3122	88	2051	
2SC894M		3122	88	2051	
2SC894OR		3122	88	2051	
2SC894R		3122	88	2051	
2SC894X		3122	88	2051	
2SC894Y		3122	88	2051	
2SC895	162		246	2020	2N3441
2SC896	123A	3444	88	2051	
2SC896A		3122	88		
2SC896B		3122	88		
2SC896C		3122	88		
2SC896D		3122	88		
2SC896E		3122	88		
2SC896F		3122	88		
2SC896G		3122	88		
2SC896GN		3122	88		
2SC896J		3122	88		
2SC896K		3122	88		
2SC896L		3122	88		
2SC896M		3122	88		
2SC896OR		3122	88		
2SC896R		3122	88		
2SC896X		3122	88		
2SC896Y		3122	88		
2SC897	280	3535/181	262	2041	2N5760
2SC897A	280	3535/181	262	2041	
2SC897B	280	3535/181	262	2041	
2SC897C	280	3297	262	2041	
2SC898	280	3297	61	2041	2N5760
2SC898A	280	3535/181	61	2041	
2SC898B	280	3535/181	61	2041	
2SC898C	280	3297	61	2041	
2SC898D		3535	61		
2SC898E		3535	61		
2SC898F		3535	61		
2SC898G		3535	61		
2SC898GN		3535	61		
2SC898H		3535	61		
2SC898J		3535	61		
2SC898K		3535	61		
2SC898L		3535	61		
2SC898M		3535	61		
2SC898OR		3535	61		
2SC898X		3535	61		
2SC898Y		3535	61		
2SC899	123A	3444	61	2051	
2SC899A		3122	61	2051	
2SC899B		3122	61	2051	
2SC899C		3122	61	2051	
2SC899D		3122	61	2051	
2SC899E		3122	88	2038	
2SC899F		3122	61	2051	
2SC899G		3122	61	2051	
2SC899GN		3122	61	2051	
2SC899H		3122	61	2051	
2SC899J		3122	61	2051	
2SC899K	123A	3444	61	2051	
2SC899L		3122	61	2051	
2SC899M		3122	61	2051	
2SC899OR		3122	61	2051	
2SC899R		3122	61	2051	
2SC899X		3122	61	2051	
2SC899Y		3122	61	2051	
2SC89A		3861	59		
2SC89B		3861	59		
2SC89C		3861	59		
2SC89D		3861	59		
2SC89E		3861	59		
2SC89F		3861	59		
2SC89G		3861	59		
2SC89GN		3861	59		
2SC89H		3861	59		
2SC89J		3861	59		
2SC89K		3861	59		
2SC89L		3861	59		
2SC89M		3861	59		
2SC89OR		3861	59		
2SC89R		3861	59		
2SC89X		3861	59		
2SC89Y		3861	59		
2SC9	236		216	2053	
2SC90	101	3861	59	2002	
2SC900	199	3899	62	2010	
2SC900(E)			62	2010	
2SC900(E)(L)			62	2010	
2SC900(F)	199	3899	62	2010	
2SC900(L)	199	3250/315	62	2010	
2SC900(U)			62	2010	
2SC900-OR			62	2010	
2SC900A	199	3899	62	2010	
2SC900AF	199	3899			
2SC900B	199	3899	62	2010	
2SC900C	199	3899	62	2010	
2SC900D	199	3899	62	2010	
2SC900E	199	3899	62	2010	
2SC900EF	199	3899			
2SC900EU	199	3899			
2SC900F	199	3899	62	2010	
2SC900G			62	2010	
2SC900J			62	2010	
2SC900K			62	2010	
2SC900L	199	3899	62	2010	
2SC900M	199	3899	62	2010	
2SC900OR			62	2010	
2SC900R			62	2010	
2SC900S			62	2010	
2SC900SA			62	2010	
2SC900SB			62	2010	
2SC900SC			62	2010	
2SC900SD			62	2010	
2SC900U	199	3899	62	2010	
2SC900U/E	199	3899			
2SC900UE	199	3899			
2SC900V	199	3899			
2SC900VE	199	3899	62	2010	
2SC900X			62	2010	
2SC900Y			62	2010	
2SC901	163A	3836/284	35		2N6306
2SC9011E			20	2051	
2SC9011F			20	2051	
2SC9011H			20	2051	
2SC9012-HG			21	2034	
2SC9012-HH			21	2034	
2SC901A	163A	3836/284	35		2N6306
2SC901B		3836	35		
2SC901C		3836	35		
2SC901F		3836	35		
2SC901J		3836	35		
2SC901K		3836	35		
2SC901_		3836	35		
2SC901M		3836	35		
2SC901R		3836	35		
2SC901Y		3836	35		
2SC902	284				2N5634
2SC903	289A	3244	220		
2SC904	289A	3124/289	220		
2SC905		3244	220		
2SC906	123A	3444	20	2051	
2SC906(F)			20	2032	
2SC906F	123A	3444	20	2051	
2SC907	123A	3122	62	2010	
2SC907A	123A	3122	62	2010	
2SC907AC	123A	3245/199	62	2010	
2SC907AD	123A	3245/199	62	2010	
2SC907AH	123A	3122	62	2010	
2SC907B		3122	62	2030	
2SC907C	123A	3122	62	2010	
2SC907D	123A	3122	62	2010	
2SC907E		3122	88	2030	
2SC907F		3122	62	2030	
2SC907G		3122	62	2030	
2SC907GN		3122	62	2030	
2SC907H	123A	3122	62	2010	
2SC907HA	123A	3122	62	2010	
2SC907J		3122	62	2030	
2SC907K		3122	62	2030	
2SC907L		3122	62	2030	
2SC907M		3122	62	2030	
2SC907OR		3122	62	2030	
2SC907R		3122	62	2030	
2SC907X		3122	62	2030	
2SC907Y		3122	62	2030	
2SC909			28	2017	
2SC90A		3861	59		
2SC90B		3861	59		
2SC90C		3861	59		
2SC90D		3861	59		
2SC90E		3861	59		
2SC90F		3861	59		
2SC90G		3861	59		
2SC90GN		3861	59		
2SC90H		3861	59		
2SC90J		3861	59		
2SC90K		3861	59		
2SC90L		3861	59		

Industry Standard No.	ECG	SK	GE	RS 276-	MOTOR.
2SC90M		3861	59		
2SC90R		3861	59		
2SC90X		3861	59		
2SC90Y		3861	59		
2SC91	101	3861	59	2002	
2SC911			28	2017	
2SC912	85		86	2038	
2SC912M			212	2010	
2SC913	123A$	3444/123A	88	2051	
2SC913A		3122	88		
2SC913B		3122	88		
2SC913C		3122	88		
2SC913D		3122	88		
2SC913E		3122	88		
2SC913F		3122	88		
2SC913G		3122	88		
2SC913GN		3122	88		
2SC913H		3122	88		
2SC913J		3122	88		
2SC913K		3122	88		
2SC913L		3122	88		
2SC913M		3122	88		
2SC913OR		3122	88		
2SC913R		3122	88		
2SC913X		3122	88		
2SC913Y		3122	88		
2SC914	123A$	3244	220		
2SC915	123A$	3244	220		
2SC916			66	2048	
2SC917	161	3444/123A	88	2038	
2SC917(K)			88	2038	
2SC917A		3444	88	2038	
2SC917B		3444	88	2038	
2SC917C		3444	88	2038	
2SC917D		3444	88	2038	
2SC917E		3444	88	2038	
2SC917F		3444	88	2038	
2SC917G		3444	88	2038	
2SC917GN		3444	88	2038	
2SC917H		3444	88	2038	
2SC917J		3444	88	2038	
2SC917K	161	3444/123A	88	2038	
2SC917L		3444	88	2038	
2SC917M		3444	88	2038	
2SC917OR		3444	88	2038	
2SC917R		3444	88	2038	
2SC917X			88	2038	
2SC917Y		3444	88	2038	
2SC918	161	3018	61	2015	
2SC918A	161	3018	61	2038	
2SC918AL		3039	61	2038	
2SC918B		3018	61	2038	
2SC918C		3018	61	2038	
2SC918D		3018	61	2038	
2SC918E		3018	61	2038	
2SC918F		3018	61	2038	
2SC918G		3018	61	2038	
2SC918GN		3018	61	2038	
2SC918H		3018	61	2038	
2SC918J		3018	61	2038	
2SC918K		3018	61	2038	
2SC918L		3018	61	2038	
2SC918LF		3039	61	2038	
2SC918M		3018	61	2038	
2SC918OR		3018	61	2038	
2SC918R		3018	61	2038	
2SC918X		3018	61	2038	
2SC918XL		3039	86	2038	
2SC918Y		3018	61	2038	
2SC91A		3861	59		
2SC91B		3861	59		
2SC91C		3861	59		
2SC91D		3861	59		
2SC91E		3861	59		
2SC91F		3861	59		
2SC91G		3861	59		
2SC91GN		3861	59		
2SC91H		3861	59		
2SC91J		3861	59		
2SC91K		3861	59		
2SC91L		3861	59		
2SC91M		3861	59		
2SC91OR		3861	59		
2SC91R		3861	59		
2SC91X		3861	59		
2SC91Y		3861	59		
2SC92	282		66	2048	
2SC920		3018	61	2038	
2SC920-OQ		3018	61	2038	
2SC920-OR			61	2038	
2SC920A		3018	61	2038	
2SC920B		3018	61	2038	
2SC920C		3018	61	2038	
2SC920CL		3018	61	2038	
2SC920D		3018	61	2038	
2SC920E		3018	61	2038	
2SC920F		3018	61	2038	
2SC920G		3018	61	2038	
2SC920GN		3018	61	2038	
2SC920H		3018	61	2038	
2SC920L		3018	61	2038	
2SC920M		3018	61	2038	
2SC920OR		3018	61	2038	
2SC920Q		3122	61	2038	
2SC920R		3122	61	2038	
2SC920X		3018	61	2038	
2SC920Y		3018	61	2038	
2SC921		3018	61	2015	
2SC921(L)			61	2015	
2SC921(VHF)			61	2015	
2SC921A		3018	61	2015	
2SC921B		3018	61	2015	
2SC921C		3018	61	2015	
2SC921C1			61	2015	
2SC921CL		3018	61	2038	
2SC921D		3018	61	2015	
2SC921E		3018	61	2015	
2SC921F		3018	61	2015	
2SC921G		3018	61	2015	
2SC921GN		3018	61	2015	
2SC921H		3018	61	2015	
2SC921J		3018	61	2015	
2SC921K		3018	61	2015	
2SC921L		3018	61	2015	
2SC921M		3018	61	2015	
2SC921OR		3018	61	2015	
2SC921R		3018	61	2015	
2SC921W		3018	61	2015	
2SC921X		3018	61	2015	
2SC921Y		3018	61	2015	
2SC922	107	3018	17	2051	
2SC922A	107		17	2051	
2SC922B	107		17	2051	
2SC922C	107		17	2051	

Industry Standard No.	ECG	SK	GE	RS 276-	MOTOR.
2SC922K	107		17	2051	
2SC922L	107	3018	17	2051	
2SC922M	107	3018	17	2051	
2SC923	199	3124/289	62	2010	
2SC923(E)(F)			62	2010	
2SC923A	199	3124/289	62	2010	
2SC923B	199	3124/289	62	2010	
2SC923C	199	3124/289	62	2010	
2SC923D	199	3124/289	62	2010	
2SC923E	199	3124/289	62	2010	
2SC923F	199	3124/289	62	2010	
2SC923G		3124	62	2010	
2SC923GN		3124	62	2010	
2SC923H		3124	62	2010	
2SC923K		3124	62	2010	
2SC923L		3124	62	2010	
2SC923M		3124	62	2010	
2SC923OR		3124	62	2010	
2SC923R		3124	62	2010	
2SC923X		3124	62	2010	
2SC923Y		3124	62	2010	
2SC924	107	3452/108	61	2038	
2SC924A		3124	61	2038	
2SC924B		3124	61	2038	
2SC924C		3124	61	2038	
2SC924D		3124	61	2038	
2SC924E	107	3452/108	61	2038	
2SC924F	107	3452/108	61	2038	
2SC924G		3124	61	2038	
2SC924GN		3124	47	2038	
2SC924H		3124	61	2038	
2SC924J		3124	61	2038	
2SC924K		3124	61	2038	
2SC924L		3124	61	2038	
2SC924M	107	3452/108	61	2038	
2SC924OR		3124	61	2038	
2SC924R		3124	61	2038	
2SC924X		3124	61	2038	
2SC924Y		3124	61	2038	
2SC925	85	3444/123A	47	2051	
2SC925(M)			47	2030	
2SC925A		3124	47	2030	
2SC925B		3124	47	2030	
2SC925C		3124	47	2030	
2SC925D		3124	47	2030	
2SC925E		3124	47	2030	
2SC925F		3124	47	2030	
2SC925G		3124	47	2030	
2SC925GN		3124	47	2030	
2SC925H		3124	47	2030	
2SC925J		3124	47	2030	
2SC925K		3124	47	2030	
2SC925L		3124	47	2030	
2SC925M		3124	47	2030	
2SC925OR		3124	47	2030	
2SC925R		3124	47	2030	
2SC925X		3124	47	2030	
2SC925Y		3124	47	2030	
2SC926		3244	220	2012	
2SC926(A)			220	2012	
2SC926-OR			220	2012	
2SC926A		3244	220	2012	
2SC926B		3244	220	2012	
2SC926C		3244	220	2012	
2SC926D		3244	220	2012	
2SC926E		3244	220	2012	
2SC926F		3244	220	2012	
2SC926G		3244	220	2012	
2SC926GN		3244	220	2012	
2SC926H		3244	220	2012	
2SC926J		3244	220	2012	
2SC926K		3244	220	2012	
2SC926L		3244	220	2012	
2SC926M		3244	220	2012	
2SC926OR		3244	32	2012	
2SC926R		3244	220	2012	
2SC926X		3244	220	2012	
2SC926Y		3244	220	2012	
2SC927	161	3246/229	39	2015	
2SC927(D)			39	2015	
2SC927(E)			39	2015	
2SC927A	161	3246/229	39	2016	
2SC927B	161	3246/229	39	2015	
2SC927C	161	3246/229	39	2016	
2SC927C(E)			39	2015	
2SC927C(K)			39	2015	
2SC927CJ	161	3246/229	39	2015	
2SC927CK			39	2015	
2SC927CT	161	3246/229	39	2015	
2SC927CU	161	3246/229	39	2015	
2SC927CW	161		39	2015	
2SC927D	161	3246/229	39	2015	
2SC927DD	161	3716			
2SC927E	161	3246/229	39	2015	
2SC927E,Z			39	2015	
2SC927F		3246	39	2015	
2SC927G		3246	39	2015	
2SC927GN		3246	39	2015	
2SC927H		3246	39	2015	
2SC927J		3246	39	2015	
2SC927K		3246	39	2015	
2SC927L		3246	39	2015	
2SC927M		3246	39	2015	
2SC927OR		3246	39	2015	
2SC927R		3246	39	2015	
2SC927X		3246	39	2015	
2SC927XL		3246	39	2015	
2SC927Y		3246	39	2015	
2SC927Z			39	2015	
2SC928	161	3117	211	2015	
2SC928A		3117	211	2016	
2SC928B	161	3117	211	2015	
2SC928C	161	3117	211	2015	
2SC928D	161	3117	211	2015	
2SC928E	161	3117	211	2015	
2SC928F		3117	211	2016	
2SC928G		3117	211	2016	
2SC928GN		3117	211	2016	
2SC928H		3117	211	2016	
2SC928J		3117	211	2016	
2SC928K		3117	211	2016	
2SC928L		3117	211	2016	
2SC928M		3117	211	2016	
2SC928OR		3117	211	2016	
2SC928R		3117	211	2016	
2SC928X		3117	211	2016	
2SC928Y		3117	211	2016	
2SC929	107	3132	62	2010	
2SC929(O)			86	2038	
2SC929(E)			86	2038	
2SC929-O	107		86	2038	
2SC929-0			86	2038	
2SC929A		3132	61	2038	
2SC929B	107	3132	86	2038	

Industry Standard No.	ECG	SK	GE	RS 276-	MOTOR.
2SC929C	107	3132	86	2038	
2SC929C1	107	3132	86	2038	
2SC929D	107	3132	86	2038	
2SC929D1	107	3132	86	2038	
2SC929DE	107	3132	86	2038	
2SC929DP	107	3132	86	2038	
2SC929DU	107	3132	86	2038	
2SC929DV	107	3132	86	2038	
2SC929E	107	3132	86	2038	
2SC929ED	107	3132	86	2038	
2SC929EZ			86	2038	
2SC929F	107	3132	86	2038	
2SC929FK	107	3132	86	2038	
2SC929G		3132	86	2038	
2SC929GN		3132	86	2038	
2SC929H	107	3132	86	2038	
2SC929J		3132	86	2038	
2SC929K		3132	86	2038	
2SC929L		3132	86	2038	
2SC929M		3132	86	2038	
2SC929NP	107		86	2038	
2SC929OR		3132	86	2038	
2SC929R		3132	86	2038	
2SC929X		3132	86	2038	
2SC929Y		3132	86		
2SC93	282		66	2048	
2SC930	107	3356	62	2010	
2SC930(D)			60	2038	
2SC930(E)			60	2038	
2SC930-OR			60	2038	
2SC930A		3356	60	2038	
2SC930B	107	3356	60	2038	
2SC930BB	107	3356	60	2038	
2SC930BK	107	3356	60	2038	
2SC930BV	107	3356	60	2038	
2SC930C	107	3356	60	2038	
2SC930CK	107	3356	60	2038	
2SC930CL	107	3356	60	2038	
2SC930CS	107	3356	60	2038	
2SC930D	107	3245/199	62	2010	
2SC930DB		3356	60	2038	
2SC930DC		3356	60	2038	
2SC930DE	107		60	2038	
2SC930DH	107	3356	60	2038	
2SC930DK			60	2038	
2SC930DS	107	3356	60	2038	
2SC930DT	107	3356	60	2038	
2SC930DT-2	107	3356	60	2038	
2SC930DX	107	3356	60	2038	
2SC930DZ	107	3356	60	2038	
2SC930E	107	3245/199	62	2010	
2SC930EP	107	3356	60	2038	
2SC930ET	107	3356	60	2038	
2SC930EV	107	3356	60	2038	
2SC930EX	107	3356	60	2038	
2SC930F	107	3356	60	2038	
2SC930G		3356	60	2038	
2SC930GN		3356	60	2038	
2SC930H		3356	60	2038	
2SC930J	107	3356	60	2038	
2SC930K		3356	60	2038	
2SC930L	107	3356	60	2038	
2SC930M		3356	60	2038	
2SC930NP	107		60	2038	
2SC930OR		3356	60	2038	
2SC930R		3356	60	2038	
2SC930X		3356	60	2038	
2SC930Y		3356	60	2038	
2SC931		3357	247	2052	MJE205
2SC931C		3357	247	2052	
2SC931D		3357	247	2052	
2SC931E		3357	247	2052	
2SC932		3192	28	2048	2N5977
2SC932(E)			28	2017	
2SC932A		3192	28	2017	
2SC932B		3192	28	2017	
2SC932BK		3192	28	2017	
2SC932C		3192	28	2017	
2SC932D		3192	28	2017	
2SC932E		3192	28	2048	
2SC932F		3192	28	2017	
2SC932G		3192	28	2017	
2SC932GN		3192	28	2017	
2SC932H		3192	28	2017	
2SC932J		3192	28	2017	
2SC932K		3192	28	2017	
2SC932L		3192	28	2017	
2SC932M		3192	28	2017	
2SC932OR		3054	28	2017	
2SC932R		3192	28	2017	
2SC932X		3192	28	2017	
2SC932Y		3192	28	2017	
2SC933	289A	3122	210	2051	
2SC933(D)			210	2051	
2SC933(F)			210	2051	
2SC933(G)			210	2051	
2SC933A		3122	210	2051	
2SC933B		3122	210	2051	
2SC933BB	289A	3122	210	2051	
2SC933C	289A	3122	210	2051	
2SC933D	289A	3122	210	2051	
2SC933D(F)			210	2051	
2SC933E	289A	3122	210	2051	
2SC933E(F)			210	2051	
2SC933F	289A	3122	210	2051	
2SC933FP	289A	3122	210	2051	
2SC933FPC	289A	3122	210	2051	
2SC933FPD	289A	3122	210	2051	
2SC933FPE	289A	3122	210	2051	
2SC933FPF	289A	3122	210	2051	
2SC933FPG		3122	210	2051	
2SC933G		3122	210	2051	
2SC933GN		3122	210	2051	
2SC933H		3122	210	2051	
2SC933J		3122	210	2051	
2SC933K		3122	210	2051	
2SC933L		3122	210	2051	
2SC933M		3122	210	2051	
2SC933OR		3122	210	2051	
2SC933R		3122	210	2051	
2SC933X		3122	210	2051	
2SC933Y		3122	210	2051	
2SC934	123A	3444	18	2051	
2SC934-O			20	2051	
2SC934A		3024	18	2051	
2SC934B		3024	18	2051	
2SC934C	123A	3444	18	2051	
2SC934D	123A	3444	18	2051	
2SC934E	123A	3444	18	2051	
2SC934F	123A	3444	18	2051	
2SC934G	123A	3444	18	2051	
2SC934GN		3024	18	2051	
2SC934H		3024	18	2051	
2SC934J		3024	18	2051	

Industry Standard No.	ECG	SK	GE	RS 276-	MOTOR.
2SC934K		3024	18	2051	
2SC934L		3024	18	2051	
2SC934M		3024	18	2051	
2SC934OR		3024	18	2051	
2SC934P	123A	3444	18	2051	
2SC934Q			20	2051	
2SC934R		3024	18	2051	
2SC934X		3024	18	2051	
2SC934Y		3024	18	2051	
2SC935	162	3560	38		2N5840
2SC935A		3560	38		
2SC935B		3560	38		
2SC935C		3560	38		
2SC935D		3560	38		
2SC935E		3560	38		
2SC935F		3560	38		
2SC935G		3560	38		
2SC935GN		3560	38		
2SC935H		3560	38		
2SC935J		3560	38		
2SC935K		3560	38		
2SC935M		3560	38		
2SC935OR		3560	38		
2SC935R		3560	38		
2SC935X		3560	38		
2SC935Y		3560	38		
2SC936	164	3133	38		BU204
2SC936A	164	3133	38		
2SC936B		3133	38		
2SC936BK	164	3133	38		
2SC936C		3133	38		
2SC936D		3133	38		
2SC936E		3133	38		
2SC936F		3133	38		
2SC936G		3133	38		
2SC936GN		3133	38		
2SC936H		3133	38		
2SC936J		3133	38		
2SC936K		3133	38		
2SC936L		3133	38		
2SC936M		3133	38		
2SC936OR		3133	38		
2SC936R		3133	38		
2SC936X		3133	38		
2SC936Y		3133	38		
2SC937	389	3710/238	38		BU204
2SC937(BK)	389		38		
2SC937(YL)	389		38		
2SC937-01	389		38		
2SC937A	389	3710/238	38		
2SC937B	389	3710/238	38		
2SC937BK		3710	38		
2SC937C		3710	38		
2SC937D		3710	38		
2SC937E		3710	38		
2SC937F		3710	38		
2SC937G		3710	38		
2SC937GN		3710	38		
2SC937H		3710	38		
2SC937J		3710	38		
2SC937K		3710	38		
2SC937L		3710	38		
2SC937M		3710	38		
2SC937OR		3710	38		
2SC937R		3710	38		
2SC937X		3710	38		
2SC937Y		3710	38		
2SC937YL	389	3710/238	38		
2SC938	289A	3444/123A	20	2051	
2SC938-O			20	2030	
2SC938A	289A	3444/123A	20	2051	
2SC938B	289A	3444/123A	20	2051	
2SC938C	289A	3444/123A	20	2051	
2SC938D		3124	20	2030	
2SC938E		3124	20	2030	
2SC938F		3124	20	2030	
2SC938G		3124	20	2030	
2SC938GN		3124	20	2030	
2SC938H		3124	20	2030	
2SC938J		3124	20	2030	
2SC938K		3124	20	2030	
2SC938L		3124	20	2030	
2SC938M		3124	20	2030	
2SC938OR		3124	20	2030	
2SC938R		3124	20	2030	
2SC938X		3124	20	2030	
2SC938Y		3124	20	2030	
2SC939	328	3619	36		MJ15001
2SC939D	328	3619	36		
2SC939L	328	3619	36		
2SC94	282		66	2048	
2SC940	283	3439/163A	73		2N6249
2SC940L	283	3439/163A	73		
2SC940M	283	3439/163A	73		
2SC940P	283	3439/163A			
2SC940Q	283	3439/163A	73		
2SC940R	283	3439/163A			
2SC941	289A	3124/289	210	2051	
2SC941(O),(R)			210	2051	
2SC941-O	289A	3124/289			
2SC941-O	289A	3124/289	210	2051	
2SC941-OY			210	2051	
2SC941-R	289A	3124/289	210	2051	
2SC941-Y	289A	3124/289	210	2051	
2SC941K			210	2051	
2SC941R	289A	3124/289	210	2051	
2SC941TM	289A		210		
2SC941TM-O	289A		210		
2SC941Y	289A	3124/289	210	2051	
2SC943	123A	3444	88	2051	
2SC943A	123A	3444	88	2051	
2SC943B	123A	3444	88	2051	
2SC943C	123A	3444	88	2051	
2SC943D		3122	88	2051	
2SC943E		3122	88	2051	
2SC943F		3122	88	2051	
2SC943G		3122	88	2051	
2SC943GN		3122	88	2051	
2SC943H		3122	88	2051	
2SC943J		3122	88	2051	
2SC943K		3122	88	2051	
2SC943L		3122	88	2051	
2SC943M		3122	88	2051	
2SC943OR		3122	88	2051	
2SC943R		3122	88	2051	
2SC943X		3122	88	2051	
2SC943Y		3122	88	2051	
2SC944	85	3124/289	20	2051	
2SC944K	85	3124/289	20	2051	
2SC944S	85	3124/289			
00002SC945	289A	3124/289	62	2010	
2SC945(K)			212	2010	
2SC945(L)			212	2010	
2SC945(P)			212	2010	

Industry Standard No.	ECG	SK	GE	RS 276-	MOTOR.
2SC945(Q)			212	2010	
2SC945(R)	85	3124/289	212	2010	
2SC945(TK)			212	2010	
2SC945(TK,P)			212	2010	
2SC945(TP)			212	2010	
2SC945(TQ)			212	2010	
2SC945(TQ,Q)			212	2010	
2SC945-O			212	2010	
2SC945-O	85		212	2010	
2SC945-OR			212	2010	
2SC945-R	85		212	2010	
2SC945A	85	3124/289	212	2010	
2SC945A/D	85		212	2010	
2SC945AK	85	3124/289	212	2010	
2SC945AP	85	3124/289	212	2010	
2SC945AQ	85	3124/289	212	2010	
2SC945AR	85	3124/289	212	2010	
2SC945B	85	3124/289	212	2010	
2SC945C	85	3124/289	212	2010	
2SC945CK	85	3124/289	212	2010	
2SC945D	85	3124/289	212	2010	
2SC945E	85	3124/289	212	2010	
2SC945F	85	3124/289	212	2010	
2SC945G	85	3124/289	212	2010	
2SC945GN		3124	212	2010	
2SC945H		3124	212	2010	
2SC945J		3124	212	2010	
2SC945K	85	3124/289	212	2010	
2SC945L	85	3124/289	212	2010	
2SC945LP	85	3124/289	212	2010	
2SC945LPQ	85		212	2010	
2SC945LQ	85	3124/289	212	2010	
2SC945M	85	3124/289	212	2010	
2SC945O	85	3124/289	212	2010	
2SC945OR		3124	212	2010	
2SC945P	85	3124/289	212	2010	
2SC945PJ		3124	212	2010	
2SC945PO	85		212	2010	
2SC945PQ	85		212	2010	
2SC945Q	85	3124/289	62	2010	
2SC945QL	85	3124/289	212	2010	
2SC945QP	85	3124/289	212	2010	
2SC945QR	85		212	2010	
2SC945R	85	3124/289	212	2010	
2SC945RA		3124	212	2010	
2SC945S	85	3124/289	212	2010	
2SC945T	85	3124/289	212	2010	
2SC945TK	85	3124/289	212	2010	
2SC945TP	85	3124/289	212	2010	
2SC945TQ	85	3124/289	212	2010	
2SC945TR	85	3124/289	212	2010	
2SC945U	85	3124/289			
2SC945X	85	3124/289	212	2010	
2SC945Y	85	3124/289	212	2010	
2SC947	161	3039/316	214	2015	
2SC947A		3039	214	2015	
2SC947B		3039	214	2015	
2SC947C		3039	214	2015	
2SC947D		3039	214	2015	
2SC947E		3039	214	2015	
2SC947F		3039	214	2015	
2SC947G		3039	214	2015	
2SC947GN		3039	214	2015	
2SC947H		3039	214	2038	
2SC947J		3039	214	2015	
2SC947K		3039	214	2015	
2SC947L		3039	214	2015	
2SC947M		3039	214	2015	
2SC947OR		3039	214	2015	
2SC947R		3039	214	2015	
2SC947X		3039	214	2015	
2SC947Y		3039	214	2015	
2SC948	161	3019	214	2015	
2SC948A		3039	214	2015	
2SC948B		3039	214	2015	
2SC948D		3039	214	2015	
2SC948E	161	3039/316	214	2015	
2SC948F		3039	214	2015	
2SC948G		3039	214	2015	
2SC948GN		3039	214	2015	
2SC948H		3039	214	2015	
2SC948J		3039	214	2015	
2SC948K		3039	214	2015	
2SC948L		3039	214	2015	
2SC948M		3039	214	2015	
2SC948OR		3039	214	2015	
2SC948R		3039	214	2015	
2SC948X		3039	214	2015	
2SC948Y		3039	214	2015	
2SC95	128	3045/225	32	2012	
2SC957	319P	3039/316	86	2015	
2SC957A		3039	86	2038	
2SC957AL		3039	86	2038	
2SC957B		3039	86	2038	
2SC957C		3039	86	2038	
2SC957D		3039	86	2038	
2SC957E		3039	86	2038	
2SC957F		3039	86	2038	
2SC957G		3039	86	2038	
2SC957GN		3039	86	2038	
2SC957H		3039	86	2038	
2SC957J		3039	86	2038	
2SC957K		3039	86	2038	
2SC957L		3039	86	2038	
2SC957M		3039	86	2038	
2SC957OR		3039	86	2038	
2SC957R		3039	86	2038	
2SC957X		3039	86	2038	
2SC957XL		3039	86	2038	
2SC957Y		3039	86	2038	
2SC959	128	3104A	243	2030	
2SC959A	128	3104A	243	2030	
2SC959B	128	3104A	243	2030	
2SC959C	128	3104A	243	2030	
2SC959D	128	3104A	243	2030	
2SC959E		3104A	243		
2SC959F		3104A	243		
2SC959G		3104A	243		
2SC959GN		3104A	243		
2SC959H		3104A	243		
2SC959J		3104A	243		
2SC959K		3104A	243		
2SC959L		3104A	243		
2SC959M	128	3104A	243	2030	
2SC959OR		3104A	243		
2SC959R		3104A	243		
2SC959S	128		243	2030	
2SC959SA	128		243	2030	
2SC959SB	128		243	2030	
2SC959SC	128		243	2030	
2SC959SD	128		243	2030	
2SC959X		3104A	243		
2SC959Y		3104A	243		
2SC95A		3045	32		
2SC95B		3045	32		
2SC95C		3045	32		
2SC95D		3045	32		
2SC95E		3045	32		
2SC95F		3045	32		
2SC95G		3045	32		
2SC95GN		3045	32		
2SC95H		3045	32		
2SC95J		3045	32		
2SC95K		3045	32		
2SC95L		3045	32		
2SC95M		3045	32		
2SC95OR		3045	32		
2SC95R		3045	32		
2SC95X		3045	32		
2SC95Y		3045	32		
2SC960	128/427	3024/128	256		
2SC960A	128/427	3024/128	256		
2SC960B	128/427	3024/128	256		
2SC960C	128/427		256		
2SC960D	128/427	3024/128	256		
2SC960K	128/427	3024/128			
2SC960L	128/427	3024/128			
2SC960M	128/427	3024/128			
2SC960S	128/427	3024/128	256		
2SC960SA	128/427		256		
2SC960SB	128/427		256		
2SC960SC	128/427		256		
2SC960SD	128/400		256		
2SC961					2N5759
2SC962					2N5758
2SC963	123A			2014	
2SC964	123A			2014	
2SC965	123A			2014	
2SC966	123A	3444	88	2051	
2SC966A		3122	88	2030	
2SC966B		3122	88	2030	
2SC966C		3122	88	2030	
2SC966D		3122	88	2030	
2SC966E		3122	88	2030	
2SC966F		3122	88	2030	
2SC966G		3122	88	2030	
2SC966GN		3122	88	2030	
2SC966H		3122	88	2030	
2SC966J		3122	88	2030	
2SC966K		3122	88	2030	
2SC966L		3122	88	2030	
2SC966M		3122	88	2030	
2SC966OR		3122	88	2030	
2SC966R		3122	88	2030	
2SC966X		3122	88	2030	
2SC966Y		3122	88	2030	
2SC967	123A	3444	88	2051	
2SC967A		3122	88	2030	
2SC967B		3122	88	2030	
2SC967C		3122	88	2030	
2SC967D		3122	88	2030	
2SC967E		3122	88	2030	
2SC967F		3122	88	2030	
2SC967G		3122	88	2030	
2SC967GN		3122	88	2030	
2SC967H		3122	88	2030	
2SC967J		3122	88	2030	
2SC967K		3122	88	2030	
2SC967L		3122	88	2030	
2SC967M		3122	88	2030	
2SC967OR		3122	88	2030	
2SC967R		3122	88	2030	
2SC967X		3122	88	2030	
2SC967Y		3122	88	2030	
2SC968	123A	3124/289	20	2051	
2SC968A			47	2051	
2SC968B			47	2051	
2SC968C			47	2051	
2SC968D			47	2051	
2SC968E			47	2051	
2SC968F			47	2051	
2SC968G			47	2051	
2SC968GN			47	2051	
2SC968H			47	2051	
2SC968J			47	2051	
2SC968K			47	2051	
2SC968L			47	2051	
2SC968M			47	2051	
2SC968OR			47	2051	
0002SC968P	128	3124/289	243	2030	
2SC968R			47	2051	
2SC968X			47	2051	
2SC968Y			47	2030	
2SC97		3024	18	2030	
2SC971		3018	61	2030	
2SC971A		3018	61		
2SC971B		3018	61		
2SC971BK		3124	61		
2SC971C		3018	61		
2SC971D		3018	61		
2SC971E		3018	61		
2SC971F		3018	61		
2SC971G		3018	61		
2SC971GN		3018	61		
2SC971H		3018	61		
2SC971J		3018	61		
2SC971K		3018	61		
2SC971L		3018	61		
2SC971M		3018	61		
2SC971OR		3018	61		
2SC971R		3018	61		
2SC971X		3018	61		
2SC971Y		3018	61		
2SC972	128	3024	18	2030	
2SC972A		3024	18	2030	
2SC972B		3024	18	2030	
2SC972C	128	3024	18	2030	
2SC972D	128	3024	18	2030	
2SC972E	128	3024	18	2030	
2SC972F		3024	18	2030	
2SC972G		3024	18	2030	
2SC972GN		3024	18	2030	
2SC972H		3024	18	2030	
2SC972J		3024	18	2030	
2SC972K		3024	18	2030	
2SC972L		3024	18	2030	
2SC972M		3024	18	2030	
2SC972OR		3024	18	2030	
2SC972R		3024	18	2030	
2SC972X		3024	18	2030	
2SC972Y		3024	18	2030	
2SC976TV			18	2030	
2SC979	123A	3122			
2SC97A		3024	18	2030	
2SC97B		3024	18	2030	
2SC97C		3024	18	2030	
2SC97D		3024	18	2030	
2SC97E		3024	18	2030	

Industry Standard No.	ECG	SK	GE	RS 276-	MOTOR.
2SC97F		3024	18	2030	
2SC97G		3024	18	2030	
2SC97GN		3024	18	2030	
2SC97H		3024	18	2030	
2SC97J		3024	18	2030	
2SC97K		3024	18	2030	
2SC97L		3024	18	2030	
2SC97M		3024	18	2030	
2SC970R		3024	18	2030	
2SC97R		3024	18	2030	
2SC97X		3024	18	2030	
2SC97Y		3024	18	2030	
2SC98	123A	3452/108	86	2038	
2SC980	85	3122			
2SC980G	85	3122			
2SC9800	85	3122			
2SC980Y	85	3122			
2SC981					2N5430
2SC982	172A	3156	64		
2SC983		3244	27		
2SC983-0		3244	27		
2SC983-0		3244	27		
2SC983-R		3244	27		
2SC983-Y		3244	27		
2SC983A		3244	27		
2SC983C		3244	27		
2SC983R		3244	27		
2SC983S		3244	27		
2SC983Y		3244	27		
2SC984	128	3444/123A	18	2051	
2SC984A	128	3444/123A	18	2051	
2SC984B	128	3444/123A	18	2051	
2SC984C	128	3444/123A	18	2051	
2SC984D		3444	18	2051	
2SC984E		3444	18	2051	
2SC984F		3444	18	2051	
2SC984G		3444	18	2051	
2SC984GN		3444	18	2051	
2SC984H	128	3444/123A	18	2051	
2SC984J		3444	18	2051	
2SC984L		3444	18	2051	
2SC984M		3444	18	2051	
2SC9840R		3444	18	2051	
2SC984R		3444	18	2051	
2SC984X			18	2051	
2SC984Y		3444	18	2051	
2SC985			86	2038	
2SC985A			86	2010	
2SC987			60	2015	
2SC987A			60	2015	
2SC988			39	2015	2N6304
2SC988A			39	2015	MRF904
2SC988B			39	2015	
2SC98A		3039	86	2051	
2SC98B		3039	86	2051	
2SC98C		3039	86	2051	
2SC98D		3039	86	2051	
2SC98E		3039	86	2051	
2SC98F		3039	86	2051	
2SC98G		3039	86	2051	
2SC98GN		3039	86	2051	
2SC98H		3039	86	2051	
2SC98J		3039	86	2051	
2SC98K		3122	88	2051	
2SC98L		3039	86	2051	
2SC98M		3039	86	2051	
2SC980R		3039	86	2051	
2SC98R		3039	86	2051	
2SC98X		3039	86	2051	
2SC98Y		3039	86	2051	
2SC99	123A	3452/108	86	2038	
2SC990			66	2048	2N5646
2SC991	311	3444/123A	88	2051	
2SC991A		3122	88	2030	
2SC991B		3122	88	2030	
2SC991C		3122	88	2030	
2SC991D		3122	88	2030	
2SC991E		3122	88	2030	
2SC991F		3122	88	2030	
2SC991G		3122	88	2030	
2SC991GN		3122	88	2030	
2SC991H		3122	88	2030	
2SC991J		3122	88	2030	
2SC991K		3122	88	2030	
2SC991L		3122	88	2030	
2SC991M		3122	88	2030	
2SC9910R		3122	88	2030	
2SC991R		3122	88	2030	
2SC991X		3122	88	2030	
2SC991Y		3122	88	2030	
2SC992	311	3444/123A	88	2051	
2SC992A		3122	88	2030	
2SC992B		3122	88	2030	
2SC992C		3122	88	2030	
2SC992D		3122	88	2030	
2SC992E		3122	88	2030	
2SC992F		3122	88	2030	
2SC992G		3122	88	2030	
2SC992GN		3122	88	2030	
2SC992H		3122	88	2030	
2SC992J		3122	88	2030	
2SC992K		3122	88	2030	
2SC992L		3122	88	2030	
2SC992M		3122	88	2030	
2SC9920R		3122	88	2030	
2SC992R		3122	88	2030	
2SC992X		3122	88	2030	
2SC992Y		3122	88	2030	
2SC993	123A		243	2030	
2SC993D	128		243	2030	
2SC993E			17	2051	
2SC994	486	3195/311	17	2051	
2SC995	154	3044	32	2012	
2SC995A		3044	32		
2SC995B		3044	32		
2SC995C		3044	32		
2SC995E		3044	32		
2SC995F		3044	32		
2SC995G		3044	32		
2SC995GN		3044	32		
2SC995H		3044	32		
2SC995J		3044	32		
2SC995K		3044	32		
2SC995L		3044	32		
2SC995M		3044	32		
2SC9950R		3044	32		
2SC995R		3044	32		
2SC995X		3044	32		
2SC995Y		3044	32		
2SC996	154/427	3045/225	256		
2SC997	161	3018	61	2015	
2SC997A		3018	61	2038	
2SC997B		3018	61	2038	
2SC997C		3018	61	2038	
2SC997D		3018	61	2038	
2SC997E		3018	61	2038	
2SC997F		3018	61	2038	
2SC997G		3018	61	2038	
2SC997GN		3018	61	2038	
2SC997H		3018	61	2038	
2SC997J		3018	61	2038	
2SC997K		3018	61	2038	
2SC997L		3018	61	2038	
2SC997M		3018	61	2038	
2SC9970R		3018	61	2038	
2SC997R		3018	61	2038	
2SC997X		3018	61	2038	
2SC997Y		3018	61	2038	
2SC998	487	3195/311			
2SC999	389	3115/165	38		BU205
2SC999A	389	3115/165	38		
2SC999B		3115	38		
2SC999C		3115	38		
2SC999D		3115	38		
2SC999E		3115	38		
2SC999F		3115	38		
2SC999G		3115	38		
2SC999GN		3115	38		
2SC999H		3115	38		
2SC999J		3115	38		
2SC999K		3115	38		
2SC999L		3115	38		
2SC999M		3115	38		
2SC9990R		3115	38		
2SC999R		3115	38		
2SC999X		3115	38		
2SC999Y		3115	38		
2SC99A		3039	86	2051	
2SC99B		3039	86	2051	
2SC99C		3039	86	2051	
2SC99D		3039	86	2051	
2SC99E		3039	86	2051	
2SC99F		3039	86	2051	
2SC99G		3039	86	2051	
2SC99GN		3039	86	2051	
2SC99H		3039	86	2051	
2SC99J		3039	86	2051	
2SC99K		3039	86	2051	
2SC99L		3039	86	2051	
2SC99M		3039	86	2051	
2SC990R		3039	86	2051	
2SC99R		3039	86	2051	
2SC99X		3039	86	2051	
2SC99Y		3039	86	2051	
2SCD78			66	2048	
2SCF-2	123A	3444	20	2051	
2SCF-8		3048	46		
2SCF1	108	3452	61	2038	
2SCF11	313	3122	278	2016	
2SCF12A		3049	46		
2SCF12B		3049	46		
2SCF12C		3049	46		
2SCF12D		3049	46		
2SCF12E		3049	46		
2SCF12F		3049	46		
2SCF12G		3049	46		
2SCF12GN		3049	46		
2SCF12H		3049	46		
2SCF12J		3049	46		
2SCF12K		3049	46		
2SCF12L		3049	46		
2SCF12M		3049	46		
2SCF120R		3049	46		
2SCF12R		3049	46		
2SCF12X		3049	46		
2SCF12Y		3049	46		
2SCF14	233		210	2009	
2SCF1A		3018	61	2038	
2SCF1B		3018	61	2038	
2SCF1C		3018	61	2038	
2SCF1D		3018	61	2038	
2SCF1E		3018	61	2038	
2SCF1F		3018	61	2038	
2SCF1G		3018	61	2038	
2SCF1GN		3018	61	2038	
2SCF1H		3018	61	2038	
2SCF1J		3018	61	2038	
2SCF1K		3018	61	2038	
2SCF1L		3018	61	2038	
2SCF1M		3018	61	2038	
2SCF10R		3018	61	2038	
2SCF1R		3018	61	2038	
2SCF1X		3018	61	2038	
2SCF1Y		3018	61	2038	
2SCF2	123A	3444	61	2051	
2SCF2A		3018	61	2038	
2SCF2B		3018	61	2038	
2SCF2C		3018	61	2038	
2SCF2D		3018	61	2038	
2SCF2E		3018	61	2038	
2SCF2G		3018	61	2038	
2SCF2GN		3018	61	2038	
2SCF2H		3018	61	2038	
2SCF2J		3018	61	2038	
2SCF2K		3018	61	2038	
2SCF2L		3018	61	2038	
2SCF2M		3018	61	2038	
2SCF2R		3018	61	2038	
2SCF2X		3018	61	2038	
2SCF2Y		3018	61	2038	
2SCF3		3048	46		
2SCF3A		3048	46		
2SCF3B		3048	46		
2SCF3C		3048	46		
2SCF3D		3048	46		
2SCF3E		3048	46		
2SCF3F		3048	46		
2SCF3G		3048	46		
2SCF3GN		3048	46		
2SCF3H		3048	46		
2SCF3J		3048	46		
2SCF3K		3048	46		
2SCF3L		3048	46		
2SCF3M		3048	46		
2SCF30R		3048	46		
2SCF3R		3048	46		
2SCF3X		3048	46		
2SCF3Y		3048	46		
2SCF5	123A	3122	45	2051	
2SCF5A		3047	45	2051	
2SCF5B		3047	45	2051	
2SCF5C		3047	45	2051	
2SCF5D		3047	45	2051	
2SCF5E		3047	45	2051	
2SCF5F		3047	45	2051	
2SCF5G		3047	45	2051	
2SCF5GN		3047	45	2051	
2SCF5H		3047	45	2051	

Industry Standard No.	ECG	SK	GE	RS 276-	MOTOR.
2SCF5J		3047	45	2051	
2SCF5K		3047	45	2051	
2SCF5L		3047	45	2051	
2SCF5M		3047	45	2051	
2SCF5OR		3047	45	2051	
2SCF5R		3046	45	2051	
2SCF5X		3047	45	2051	
2SCF5Y		3047	45	2051	
2SCF6	195A	3047	18		
2SCF6A		3047	18	2030	
2SCF6B		3047	18	2030	
2SCF6C		3047	18	2030	
2SCF6D		3047	18	2030	
2SCF6E		3047	18	2030	
2SCF6F		3047	18	2030	
2SCF6G		3047	18	2030	
2SCF6GN		3047	18	2030	
2SCF6H		3047	18	2030	
2SCF6J		3047	18	2030	
2SCF6K		3047	18	2030	
2SCF6L		3047	18	2030	
2SCF6M		3047	18	2030	
2SCF6OR		3047	18	2030	
2SCF6R		3047	18	2030	
2SCF6X		3047	18	2030	
2SCF6Y		3047	18	2030	
2SCF8	237	3049/224	46		
2SCF812		3049	46		
2SCF8A		3048	46		
2SCF8B		3048	46		
2SCF8C		3048	46		
2SCF8D		3048	46		
2SCF8E		3048	46		
2SCF8F		3048	46		
2SCF8G		3048	46		
2SCF8GN		3048	46		
2SCF8H		3048	46		
2SCF8J		3048	46		
2SCF8K		3048	46		
2SCF8L		3048	46		
2SCF8OR		3048	46		
2SCF8R		3048	46		
2SCF8X		3048	46		
2SCF8Y		3048	46		
2SCI090B			40	2012	
2SCM39J		3018	61	2038	
2SCM39X		3018	61	2038	
2SCM93D		3036	14	2041	
2SCM95K		3021	12		
2SCM98F		3024	18	2030	
2SCNJ100	199	3122	62	2010	
2SCNJ107		3020	18	2030	
2SCS183A		3124	47	2030	
2SCS183B		3124	47	2030	
2SCS183C		3124	47	2030	
2SCS183D		3124	47	2030	
2SCS183E	123	3124/289	20	2051	
2SCS183F		3124	47	2030	
2SCS183G		3124	47	2030	
2SCS183GN		3124	47	2030	
2SCS183H		3124	47	2030	
2SCS183J		3124	47	2030	
2SCS183K		3124	47	2030	
2SCS183L		3124	47	2030	
2SCS183M		3124	47	2030	
2SCS183OR		3124	47	2030	
2SCS183R		3124	47	2030	
2SCS183Y		3124	47	2030	
2SCS184		3124	47	2030	
2SCS184A		3124	47	2030	
2SCS184B		3124	47	2030	
2SCS184C		3124	47	2030	
2SCS184D		3124	47	2030	
2SCS184E	123	3124/289	47	2051	
2SCS184F		3124	47	2030	
2SCS184G		3124	47	2030	
2SCS184GN		3124	47	2030	
2SCS184H		3124	47	2030	
2SCS184J	108	3452	47	2038	
2SCS184K		3124	47	2030	
2SCS184L		3124	47	2030	
2SCS184M		3124	47	2030	
2SCS184OR		3124	47	2030	
2SCS184R		3124	47	2030	
2SCS184X		3124	47	2030	
2SCS184Y		3124	47	2030	
2SCS429		3018	61	2038	
2SCS429A		3018	61	2038	
2SCS429B		3018	61	2038	
2SCS429C		3018	61	2038	
2SCS429D		3018	61	2038	
2SCS429E		3018	61	2038	
2SCS429F		3018	61	2038	
2SCS429G		3018	61	2038	
2SCS429GN		3018	61	2038	
2SCS429H		3018	61	2038	
2SCS429J	108	3452	61	2038	
2SCS429K		3018	61	2038	
2SCS429L		3018	61	2038	
2SCS429M		3018	61	2038	
2SCS429OR		3018	61	2038	
2SCS429R		3018	61	2038	
2SCS429X		3018	61	2038	
2SCS429Y		3018	61	2038	
2SCS430		3018	61	2038	
2SCS430A		3018	61	2038	
2SCS430B		3018	61	2038	
2SCS430C		3018	61	2038	
2SCS430D		3018	61	2038	
2SCS430E		3018	61	2038	
2SCS430F		3018	61	2038	
2SCS430GN		3018	61	2038	
2SCS430H	108	3452	61	2038	
2SCS430J		3018	61	2038	
2SCS430K		3018	61	2038	
2SCS430L		3018	61	2038	
2SCS430M		3018	61	2038	
2SCS430R		3018	61	2038	
2SCS430X		3018	61	2038	
2SCS430Y		3018	61	2038	
2SCS461		3018	61	2038	
2SCS461A		3018	61	2038	
2SCS461B		3018	61	2038	
2SCS461C		3018	61	2038	
2SCS461D		3018	61	2038	
2SCS461E		3018	61	2038	
2SCS461F	108	3018	61	2038	
2SCS461G		3018	61	2038	
2SCS461GN		3018	61	2038	
2SCS461H		3018	61	2038	
2SCS461J		3018	61	2038	
2SCS461K		3018	61	2038	
2SCS461L		3018	61	2038	
2SCS461M		3018	61	2038	
2SCS4610R		3018	61	2038	
2SCS461R		3018	61	2038	
2SCS461X		3018	61	2038	
2SCS461Y		3018	61	2038	
2SCS469		3018	61	2038	
2SCS469A		3018	61	2038	
2SCS469B		3018	61	2038	
2SCS469C		3018	61	2038	
2SCS469D		3018	61	2038	
2SCS469E		3018	61	2038	
2SCS469F	108	3018	61	2038	
2SCS469G		3018	61	2038	
2SCS469GN		3018	61	2038	
2SCS469H		3018	61	2038	
2SCS469J		3018	61	2038	
2SCS469K		3018	61	2038	
2SCS469L		3018	61	2038	
2SCS469M		3018	61	2038	
2SCS469R		3018	61	2038	
2SCS469X		3018	61	2038	
2SCS469Y		3018	61	2038	
2SCU64M		3047	45		
2SD-1283		3010	59		
2SD-258		3026	8		
2SD-F1	101	3010	59	2002	
2SD-F1A	102A	3010	59	2007	
2SD-F1B		3010	59		
2SD-F1C		3010	59		
2SD-F1D		3010	59		
2SD-F1E		3010	59		
2SD-F1F		3010	59		
2SD-F1G		3010	59		
2SD-F1GN		3010	59		
2SD-F1H		3010	59		
2SD-F1J		3010	59		
2SD-F1K		3010	59		
2SD-F1L		3010	59		
2SD-F1M		3010	59		
2SD-F1R		3010	59		
2SD-F1X		3010	59		
2SD-F1Y		3010	59		
2SD041					2N3716
2SD100	103A	3835	59	2002	
2SD100A	103A	3835	59	2002	
2SD100B		3835	59		
2SD100C		3835	59		
2SD100D		3835	59		
2SD100E		3835	59		
2SD100F		3835	59		
2SD100G		3835	59		
2SD100GN		3835	59		
2SD100H		3835	59		
2SD100J		3835	59		
2SD100K		3835	59		
2SD100L		3835	59		
2SD100M		3835	59		
2SD100R		3835	59		
2SD100X		3835	59		
2SD100Y		3835	59		
2SD101	101	3862/103	8	2002	
2SD102	175	3538	47	2020	2N3583
2SD102-O	175	3538	47	2020	
2SD102-R	175	3538	47	2020	
2SD102-Y	175	3538	47	2020	
2SD102A		3538	47	2030	
2SD102B		3538	47	2030	
2SD102C		3538	47	2030	
2SD102D		3538	47	2030	
2SD102E		3538	47	2030	
2SD102F		3538	47	2030	
2SD102G		3538	47	2030	
2SD102GN		3538	47	2030	
2SD102H		3538	47	2030	
2SD102J		3538	47	2030	
2SD102K		3538	47	2030	
2SD102L		3538	47	2030	
2SD102M		3538	47	2030	
2SD102OR		3538	47	2030	
2SD102R		3538	47	2030	
2SD102X		3538	47	2030	
2SD102Y		3538	47	2030	
2SD103	175	3026	246	2020	2N5050
2SD103-O	175	3026	246	2020	
2SD103-R	175	3131A/369	246	2020	
2SD103-Y	175	3026	246	2020	
2SD104	103A	3835	59	2002	
2SD104A		3835	59		
2SD104B		3835	59		
2SD104C		3835	59		
2SD104D		3835	59		
2SD104E		3835	59		
2SD104F		3835	59		
2SD104G		3835	59		
2SD104GN		3835	59		
2SD104H		3835	59		
2SD104K		3835	59		
2SD104L		3835	59		
2SD104M		3835	59		
2SD104OR		3835	59		
2SD104R		3835	59		
2SD104X		3835	59		
2SD104Y		3835	59		
2SD105	103A	3835	59	2002	
2SD105A		3835	59		
2SD105B		3835	59		
2SD105C		3835	59		
2SD105D		3835	59		
2SD105E		3835	59		
2SD105F		3835	59		
2SD105G		3835	59		
2SD105GN		3835	59		
2SD105H		3835	59		
2SD105J		3835	59		
2SD105K		3835	59		
2SD105L		3835	59		
2SD105M		3835	59		
2SD105OR		3835	59		
2SD105R		3835	59		
2SD105X		3835	59		
2SD105Y		3835	59		
2SD107					2N6056
2SD108					2N6056
2SD11	103	3010	59	2002	
2SD110	280	3438	35		2N5634
2SD110-O		3438	35		
2SD110-0	280		35		
2SD110-R	280	3438	35		
2SD110-Y	280	3438	35		
2SD111	280	3438	35		2N5632
2SD111-0	280		35		
2SD111-ORG			35	2041	
2SD111-R	280	3438	35		
2SD111-RED			35	2041	
2SD111-Y	280	3438	35		

Industry Standard No.	ECG	SK	GE	RS 276-	MOTOR.
2SD111-YEL			35	2041	
2SD113	181	3535	75	2041	MJ802
2SD113-O	181		75	2041	
2SD113-ORG			75	2041	
2SD113-R	181		75	2041	
2SD113-RED			75	2041	
2SD113-Y	181		75	2041	
2SD113-YEL			75	2041	
2SD114	181	3535	75	2041	2N5686
2SD114-O	181		75	2041	
2SD114-ORG			75	2041	
2SD114-R	181		75	2041	
2SD114-RED			75	2041	
2SD114-Y	181		75	2041	
2SD114-YEL			75	2041	
2SD116	130				2N5758
2SD117	328				2N5760
2SD1173			247	2052	
2SD118	280	3036	262	2041	2N5760
2SD118-B			262	2041	
2SD118-BLU			262	2041	
2SD118-R			262	2041	
2SD118-RED			262	2041	
2SD118-Y			262	2041	
2SD118-YEL			262	2041	
2SD118A		3036	262	2041	
2SD118BL	280	3036	262	2041	
2SD118C		3036	262	2041	
2SD118D		3036	262	2041	
2SD118R	280	3036	262	2041	
2SD118Y	280	3036	262	2041	
2SD119	280	3036	262	2041	2N5758
2SD119-BL			262	2041	
2SD119-BLU			262	2041	
2SD119-R			262	2041	
2SD119-RED			262	2041	
2SD119-Y			262	2041	
2SD119-YEL			262	2041	
2SD119A		3036	262	2041	
2SD119B		3036	262	2041	
2SD119BL	280	3036	262	2041	
2SD119C		3036	262	2041	
2SD119D		3036	262	2041	
2SD119R	280	3036	262	2041	
2SD119Y	280	3036	262	2041	
2SD11A		3010	59		
2SD11B		3010	59		
2SD11C		3010	59		
2SD11D		3010	59		
2SD11E		3010	59		
2SD11F		3010	59		
2SD11G		3010	59		
2SD11GN		3010	59		
2SD11H		3010	59		
2SD11J		3010	59		
2SD11K		3010	59		
2SD11L		3010	59		
2SD11M		3010	59		
2SD11OR		3010	59		
2SD11R		3010	59		
2SD11X		3010	59		
2SD11Y		3010	59		
2SD12	130	3027	75	2041	2N5758
2SD120	324	3024/128	264		
2SD120A	324	3024/128	264		
2SD120B	324	3024/128	264		
2SD120C	324	3024/128	264		
2SD120D		3024	264		
2SD120E		3024	264		
2SD120F		3024	264		
2SD120G		3024	264		
2SD120GN		3024	264		
2SD120H	324	3024/128	264		
2SD120HA	324	3202/210	264		
2SD120HB	324	3202/210	264		
2SD120HC	324	3202/210	264		
2SD120J		3024	264		
2SD120K		3024	264		
2SD120L		3024	264		
2SD120M		3024	264		
2SD120R		3024	264		
2SD120X		3024	264		
2SD120Y		3024	264		
2SD121	324	3024/128	18		
2SD121A	324	3024/128	18		
2SD121B	324	3024/128	18		
2SD121C		3024	18	2030	
2SD121D		3024	18	2030	
2SD121E		3024	18	2030	
2SD121F		3024	18	2030	
2SD121G		3024	18	2030	
2SD121GN		3024	18	2030	
2SD121H	324	3024/128	18		
2SD121HA	324	3202/210	18		
2SD121HB	324	3202/210	18		
2SD121J		3024	18	2030	
2SD121K		3024	18	2030	
2SD121L		3024	18	2030	
2SD121M		3024	18	2030	
2SD121OR		3024	18	2030	
2SD121R		3024	18	2030	
2SD121X		3024	18	2030	
2SD121Y		3024	18	2030	
2SD124	87	3027/130	75	2041	2N5758
2SD124A	87	3297/280	75	2041	2N5758
2SD124AH	87	3297/280	75	2041	
2SD124AHA	87	3297/280	75	2041	
2SD124AHB	87	3297/280	75	2041	
2SD124B		3027	75	2041	
2SD124C		3027	75	2041	
2SD124E		3027	75	2041	
2SD124F		3027	75	2041	
2SD124G		3027	75	2041	
2SD124GN		3027	75	2041	
2SD124H		3027	75	2041	
2SD124J		3027	75	2041	
2SD124K		3027	75	2041	
2SD124L		3027	75	2041	
2SD124M		3027	75	2041	
2SD124OR		3027	75	2041	
2SD124R		3027	75	2041	
2SD124X		3027	75	2041	
2SD124Y		3027	75	2041	
2SD125	87	3297/280	75	2041	2N5758
2SD125A	87	3297/280	75	2041	2N5758
2SD125AH	87	3297/280	75	2041	
2SD125AHA	87	3297/280	75	2041	
2SD125AHB		3297	75	2041	
2SD125B		3027	75	2041	
2SD125C		3027	75	2041	
2SD125E		3027	75	2041	
2SD125F		3027	75	2041	
2SD125G		3027	75	2041	
2SD125GN		3027	75	2041	

Industry Standard No.	ECG	SK	GE	RS 276-	MOTOR.
2SD125H		3027	75	2041	
2SD125J		3027	75	2041	
2SD125K		3027	75	2041	
2SD125L		3027	75	2041	
2SD125M		3027	75	2041	
2SD125OR		3027	75	2041	
2SD125R		3027	75	2041	
2SD125X		3027	75	2041	
2SD125Y		3027	75	2041	
2SD126	87	3297/280	262	2041	2N5760
2SD126A	87	3297/280	262	2041	
2SD126AH	87	3297/280	262	2041	
2SD126AHA	87	3297/280	262	2041	
2SD126AHB	87	3297/280	262	2041	
2SD126H	87	3297/280	262	2041	
2SD126HA	87	3297/280	262	2041	
2SD126HB	87	3297/280	262	2041	
2SD127	103A	3835	59	2002	
2SD127A	103A	3835	59	2002	
2SD127B		3835	59		
2SD127C		3835	59		
2SD127D		3835	59		
2SD127E		3835	59		
2SD127F		3835	59		
2SD127G		3835	59		
2SD127GN		3835	59		
2SD127H		3835	59		
2SD127J		3835	59		
2SD127K		3835	59		
2SD127L		3835	59		
2SD127M		3835	59		
2SD127OR		3835	59		
2SD127R		3835	59		
2SD127X		3835	59		
2SD127Y		3835	59		
2SD128	103A	3835	59	2002	
2SD128A	103A	3835	59	2002	
2SD128B		3835	59		
2SD128C		3835	59		
2SD128D		3835	59		
2SD128E		3835	59		
2SD128F		3835	59		
2SD128G		3835	59		
2SD128GN		3835	59		
2SD128H		3835	59		
2SD128J		3835	59		
2SD128K		3835	59		
2SD128L		3835	59		
2SD128M		3835	59		
2SD128OR		3835	59		
2SD128R		3835	59		
2SD128X		3835	59		
2SD128Y		3835	59		
2SD129	175	3026	246	2020	2N3767
2SD129-BL	175		246	2020	
2SD129-R	175		246	2020	
2SD129-Y	175		246	2020	
2SD12A		3027	75		
2SD12B		3027	75		
2SD12C		3027	75		
2SD12E		3027	75		
2SD12F		3027	75		
2SD12G		3027	75		
2SD12GN		3027	75		
2SD12H		3027	75		
2SD12J		3027	75		
2SD12K		3027	75		
2SD12L		3027	75		
2SD12M		3027	75		
2SD12OR		3027	75		
2SD12R		3027	75		
2SD12X		3027	75		
2SD12Y		3027	75		
2SD13		3012	4		
2SD130	175	3562	233	2020	2N3766
2SD130(BL)			233	2020	
2SD130(Y)			233	2020	
2SD130-BLU			233	2020	
2SD130-R	175		233	2020	
2SD130-RED			233	2020	
2SD130-Y	175		233	2020	
2SD130-YEL		3562	233	2020	
2SD130A		3562	233	2020	
2SD130B		3562	233	2020	
2SD130BL	175	3562	233	2020	
2SD130C		3562	233	2020	
2SD130D		3562	233	2020	
2SD130E		3562	233	2020	
2SD130F		3562	233	2020	
2SD130G		3562	233	2020	
2SD130GN		3562	233	2020	
2SD130H		3562	233	2020	
2SD130J		3562	233	2020	
2SD130K		3562	233	2020	
2SD130OR			233	2020	
2SD130R			233	2020	
2SD130X			233	2020	
2SD130Y	175	3562	233	2020	
2SD131					2N5758
2SD1318			243	2030	
2SD132			75	2041	2N6338
2SD134	123	3124/289	47	2051	
2SD1347		3124	47	2030	
2SD134A		3124	47	2030	
2SD134C		3124	47	2030	
2SD134D		3124	47	2030	
2SD134F		3124	47	2030	
2SD134G		3124	47	2030	
2SD134GN		3124	47	2030	
2SD134H		3124	47	2030	
2SD134J		3124	47	2030	
2SD134K		3124	47	2030	
2SD134L		3124	47	2030	
2SD134M		3124	47	2030	
2SD134OR		3124	47	2030	
2SD134R		3124	47	2030	
2SD134X		3124	47	2030	
2SD134Y		3124	47	2030	
2SD136		3021	12		
2SD136D		3021	12	2030	
2SD136E		3021	12	2030	
2SD136F		3021	12	2030	
2SD136G		3021	12	2030	
2SD136GN		3021	12	2030	
2SD136J		3021	12	2030	
2SD136K		3021	12	2030	
2SD136L		3021	12	2030	
2SD136M		3021	12	2030	
2SD136OR		3021	12	2030	
2SD136R		3021	12	2030	
2SD136X		3021	12	2030	
2SD136Y		3021	12	2030	
2SD137		3021	12		
2SD137A		3021	12		

Industry Standard No.	ECG	SK	GE	RS 276-	MOTOR.
2SD137B		3021	12		
2SD137C		3021	12		
2SD137D		3021	12		
2SD137F		3021	12		
2SD137G		3021	12		
2SD137H		3021	12		
2SD137J		3021	12		
2SD137K		3021	12		
2SD137L		3021	12		
2SD137M		3021	12		
2SD137OR		3021	12		
2SD137R		3021	12		
2SD137X		3021	12		
2SD137Y		3021	12		
2SD138					2N3738
2SD139					2N3739
2SD141	152	3893	75	2048	2N3766
2SD141A		3893	75	2048	
2SD141B		3893	75	2048	
2SD141C		3893	75	2048	
2SD141E		3893	75	2048	
2SD141F			75	2048	
2SD141G			75	2048	
2SD141GN			75	2048	
2SD141H		3893	75	2048	
2SD141H01	152	3893	75	2048	
2SD141H9Z	152	3893	75	2048	
2SD141J		3893	75	2048	
2SD141K			75	2048	
2SD141L			75	2048	
2SD141M			75	2048	
2SD141OR			75	2048	
2SD141R			75	2048	
2SD141X			75	2048	
2SD141Y			75	2048	
2SD142	175	3562	18	2020	2N3766
2SD142A		3562	18	2020	
2SD142B		3562	18	2020	
2SD142C		3562	18	2020	
2SD142D		3562	18	2020	
2SD142E		3562	18	2020	
2SD142F		3562	18	2020	
2SD142G		3562	18	2020	
2SD142GN		3562	18	2020	
2SD142H		3562	18	2020	
2SD142J		3562	18	2020	
2SD142K		3562	18	2020	
2SD142L		3562	18	2020	
2SD142M	175	3562	18	2020	
2SD142OR		3562	18	2020	
2SD142R		3562	18	2020	
2SD142X		3024	18	2020	
2SD142Y		3562	18	2020	
2SD143	175	3026	246	2020	2N3767
2SD144	175	3131A/369	246	2020	2N3767
2SD145	175	3131A/369	246	2020	
2SD146	226	3026	75	2041	2N4912
2SD146A		3027	75	2041	
2SD146B		3027	75	2041	
2SD146C		3027	75	2041	
2SD146D		3027	75	2041	
2SD146E		3027	75	2041	
2SD146F		3027	75	2041	
2SD146G		3027	75	2041	
2SD146GN		3027	75	2041	
2SD146H		3027	75	2041	
2SD146J		3027	75	2041	
2SD146K		3027	75	2041	
2SD146L		3027	75	2041	
2SD146OR		3027	75	2041	
2SD146R		3027	75	2041	
2SD146UK		3026	75	2041	
2SD146VK			75	2041	
2SD146X		3027	75	2041	
2SD146Y		3027	75	2041	
2SD147		3027	75	2041	2N4912
2SD147A		3027	75	2041	
2SD147B		3027	75	2041	
2SD147C		3027	75	2041	
2SD147D		3027	75	2041	
2SD147E		3027	75	2041	
2SD147F		3027	75	2041	
2SD147GN		3027	75	2041	
2SD147H		3027	75	2041	
2SD147J		3027	75	2041	
2SD147K		3027	75	2041	
2SD147L		3027	75	2041	
2SD147M		3027	75	2041	
2SD147OR		3027	75	2041	
2SD147R		3027	75	2041	
2SD147X		3027	75	2041	
2SD147Y		3027	75	2041	
2SD148					2N4912
2SD15	87	3027/130	75	2041	2N5758
2SD150	175	3562	12	2020	2N3583
2SD150A		3562	12		
2SD150B		3562	12		
2SD150C		3562	12		
2SD150E		3562	12		
2SD150F		3562	12		
2SD150G		3562	12		
2SD150GN		3562	12		
2SD150H		3562	12		
2SD150J		3562	12		
2SD150K		3562	12		
2SD150L		3562	12		
2SD150M		3562	12		
2SD150Y		3562	12		
2SD151	328	3036	14	2041	2N5632
2SD151A		3036	14	2041	
2SD151B		3036	14	2041	
2SD151C		3036	14	2041	
2SD151E		3036	14	2041	
2SD151F		3036	14	2041	
2SD151G		3036	14	2041	
2SD151GN		3036	14	2041	
2SD151H		3036	14	2041	
2SD151J		3036	14	2041	
2SD151K		3036	14	2041	
2SD151L		3036	14	2041	
2SD151M		3036	14	2041	
2SD151OR		3036	14	2041	
2SD151R		3036	14	2041	
2SD151Y		3036	14	2041	
2SD152	375	3026	66		2N3583
2SD152(L)			66	2020	
2SD152(M)			66	2020	
2SD152A		3026	66	2020	
2SD152B		3026	66	2020	
2SD152C		3026	66	2020	
2SD152D		3026	66	2020	
2SD152E		3026	66	2020	
2SD152F		3026	66	2020	
2SD152G		3026	66	2020	

Industry Standard No.	ECG	SK	GE	RS 276-	MOTOR.
2SD152GN		3026	66	2020	
2SD152H		3026	66	2020	
2SD152J		3026	66	2020	
2SD152K		3026	66	2020	
2SD152L		3026	66	2020	
2SD152M		3026	66	2020	
2SD152OR		3026	66	2020	
2SD152R		3026	66	2020	
2SD152X		3026	66	2020	
2SD152Y		3026	66	2020	
2SD154	152	3054/196	66	2048	2N3767
2SD154A		3054	12		
2SD154B		3054	12		
2SD154C		3054	12		
2SD154GN		3054	12		
2SD154H		3054	12		
2SD154J		3054	12		
2SD154K		3054	12		
2SD154L		3054	12		
2SD154M		3054	12		
2SD154R		3054	12		
2SD154Y		3054	12		
2SD155	175	3026	246	2020	2N3767
2SD155-H		3026	246		
2SD155H	175		246	2020	
2SD155K	175		246	2020	
2SD155L	175	3026	246	2020	
2SD156	124	3021	12		2N3738
2SD156B	124	3021	12		
2SD156C	124	3021	12		
2SD156D		3021	12		
2SD156E		3021	12		
2SD156F	124	3021	12		
2SD156G		3021	12		
2SD156GN		3021	12		
2SD156H		3021	12		
2SD156K		3021	12		
2SD156L		3021	12		
2SD156M		3021	12		
2SD156OR		3021	12		
2SD156R		3021	12		
2SD156X		3021	12		
2SD156Y		3021	12		
2SD157	124	3261/175	12		2N3739
2SD157A	124	3261/175	12		
2SD157B	124	3261/175	12		
2SD157C	124	3261/175	12		
2SD157D		3261	12		
2SD157G		3261	12		
2SD157GN		3261	12		
2SD157H		3261	12		
2SD157J		3261	12		
2SD157K		3261	12		
2SD157L		3261	12		
2SD157OR		3261	12		
2SD157R		3261	12		
2SD157Y		3261	12		
2SD158	286	3021/124	12		2N3738
2SD158A	286	3021/124	12		
2SD158B	286	3021/124	12		
2SD158C	286	3021/124	12		
2SD158D		3021	12		
2SD158E		3021	12		
2SD158F	286	3021/124	12		
2SD158GN		3021	12		
2SD158H		3021	12		
2SD158J		3021	12		
2SD158K		3021	12		
2SD158L		3021	12		
2SD158M		3021	12		
2SD158OR		3021	12		
2SD158R		3021	12		
2SD158X		3021	12		
2SD158Y		3021	12		
2SD159	286	3021/124	12		2N3739
2SD159A	286	3021/124	12		
2SD159B	286	3021/124	12		
2SD159C	286	3021/124	12		
2SD159D		3021	12		
2SD159E		3021	12		
2SD159F	286	3021/124	12		
2SD159G		3021	12		
2SD159GN		3021	12		
2SD159H		3021	12		
2SD159J		3021	12		
2SD159K		3021	12		
2SD159L		3021	12		
2SD159M		3021	12		
2SD159OR		3021	12		
2SD159R		3021	12		
2SD159X		3021	12		
2SD159Y		3021	12		
2SD15A		3027	75	2041	
2SD15B		3027	75	2041	
2SD15C		3027	75	2041	
2SD15D		3027	75	2041	
2SD15E		3027	75	2041	
2SD15F		3027	75	2041	
2SD15G		3027	75	2041	
2SD15GN		3027	75	2041	
2SD15H		3027	75	2041	
2SD15J		3027	75	2041	
2SD15K		3027	75	2041	
2SD15L		3027	75	2041	
2SD15M		3027	75	2041	
2SD15OR		3027	75	2041	
2SD15R		3027	75	2041	
2SD15X		3027	75	2041	
2SD15Y		3027	75	2041	
2SD16	87	3027/130	75	2041	2N5758
2SD161	328	3079	59	2002	2N5633
2SD161A		3079	59		
2SD161B		3079	59		
2SD161C		3079	59		
2SD161D		3079	59		
2SD161E		3079	59		
2SD161F		3079	59		
2SD161G		3079	59		
2SD161GN		3079	59		
2SD161H		3079	59		
2SD161J		3079	59		
2SD161K		3079	59		
2SD161L		3079	59		
2SD161M		3079	59		
2SD161OR		3079	59		
2SD161R		3079	59		
2SD161X		3079	59		
2SD161Y		3079	59		
2SD162	103A	3011A	59	2002	
2SD162A		3011A	59		
2SD162B		3011A	59		
2SD162C		3011A	59		
2SD162D		3011A	59		
2SD162E		3011A	59		

Industry Standard No.	ECG	SK	GE	RS 276-	MOTOR.
2SD162F		3011A	59		
2SD162G		3011A	59		
2SD162GN		3011A	59		
2SD162H		3011A	59		
2SD162J		3011A	59		
2SD162K		3011A	59		
2SD162L		3011A	59		
2SD162M		3011A	59		
2SD162OR		3011A	59		
2SD162R		3011A	59		
2SD162X		3011A	59		
2SD162Y		3011A	59		
2SD163	130	3027	75	2041	2N3715
2SD163A		3027	75	2041	
2SD163B		3027	75	2041	
2SD163C		3027	75	2041	
2SD163D		3027	75	2041	
2SD163E		3027	75	2041	
2SD163F		3027	75	2041	
2SD163G		3027	75	2041	
2SD163GN		3027	75	2041	
2SD163H		3027	75	2041	
2SD163J		3027	75	2041	
2SD163K		3027	75	2041	
2SD163L		3027	75	2041	
2SD163M		3027	75	2041	
2SD163OR		3027	75	2041	
2SD163R		3027	75	2041	
2SD163X		3027	75	2041	
2SD163Y		3027	75	2041	
2SD164	130	3027	75	2041	2N5632
2SD164A		3027	75	2041	
2SD164B		3027	75	2041	
2SD164C		3027	75	2041	
2SD164D		3027	75	2041	
2SD164E		3027	75	2041	
2SD164F		3027	75	2041	
2SD164G		3027	75	2041	
2SD164GN		3027	75	2041	
2SD164H		3027	75	2041	
2SD164J		3027	75	2041	
2SD164K		3027	75	2041	
2SD164L		3027	75	2041	
2SD164M		3027	75	2041	
2SD164OR		3027	75	2041	
2SD164R		3027	75	2041	
2SD164X		3027	75	2041	
2SD164Y		3027	75	2041	
2SD165	385	3438	35		2N5634
2SD166	385	3438	35		MJ15011
2SD167	103A	3011A	59	2002	
2SD167A		3011A	59		
2SD167B		3011A	59		
2SD167C		3011A	59		
2SD167E		3011A	59		
2SD167F		3011A	59		
2SD167G		3011A	59		
2SD167GN		3011A	59		
2SD167H		3011A	59		
2SD167J		3011A	59		
2SD167K		3011A	59		
2SD167L		3011A	59		
2SD167M		3011A	59		
2SD167OR		3011A	59		
2SD167R		3011A	59		
2SD167X		3011A	59		
2SD167Y		3011A	59		
2SD168					2N6385
2SD16A		3027	75	2041	
2SD16B		3027	75	2041	
2SD16C		3027	75	2041	
2SD16D		3027	75	2041	
2SD16E		3027	75	2041	
2SD16F		3027	75	2041	
2SD16G		3027	75	2041	
2SD16GN		3027	75	2041	
2SD16H		3027	75	2041	
2SD16J		3027	75	2041	
2SD16K		3027	75	2041	
2SD16L		3027	75	2041	
2SD16M		3027	75	2041	
2SD16OR		3027	75	2041	
2SD16R		3027	75	2041	
2SD16X		3027	75	2041	
2SD16Y		3027	75	2041	
2SD17	87	3027/130	75	2041	2N5760
2SD170	103A	3835	59	2002	
2SD170A	103A		59	2002	
2SD170A/PB		3835	59		
2SD170AA	103A		59	2002	
2SD170AB	103A		59	2002	
2SD170AC	103A		59	2002	
2SD170B	103A		59	2002	
2SD170BC	103A	3010	59	2002	
2SD170C	103A	3835	59	2002	
2SD170PB	103A		59	2002	
2SD171	283				2N6543
2SD172	130	3027	75	2041	2N5877
2SD172A	130	3027	75	2041	
2SD172B	130	3027	75	2041	
2SD172C	130	3027	75	2041	
2SD172D		3027	75	2041	
2SD172E		3027	75	2041	
2SD172F		3027	75	2041	
2SD172G		3027	75	2041	
2SD172GN		3027	75	2041	
2SD172H		3027	75	2041	
2SD172J		3027	75	2041	
2SD172K		3027	75	2041	
2SD172L		3027	75	2041	
2SD172M		3024	18	2041	
2SD172OR		3027	75	2041	
2SD172R		3027	75	2041	
2SD172X		3027	75	2041	
2SD172Y		3027	75	2041	
2SD173	130	3027	75	2041	2N5632
2SD173A	130	3027	75	2041	
2SD173B	130	3027	75	2041	
2SD173C	130	3027	75	2041	
2SD173D		3027	75	2041	
2SD173E		3027	75	2041	
2SD173F		3027	75	2041	
2SD173G		3027	75	2041	
2SD173GN		3027	75	2041	
2SD173H		3027	75	2041	
2SD173J		3027	75	2041	
2SD173K		3027	75	2041	
2SD173M		3027	75	2041	
2SD173OR		3027	75	2041	
2SD173X		3027	75	2041	
2SD173Y		3027	75	2041	
2SD174	130	3027	14	2041	2N5877
2SD174A		3027	75		
2SD174B		3027	75		
2SD174C		3027	75		
2SD174D		3027	75		
2SD174E		3027	75		
2SD174F	130	3027	75		
2SD174G		3027	75		
2SD174GN		3027	75		
2SD174H		3027	75		
2SD174J		3027	75		
2SD174K		3027	75		
2SD174L		3027	75		
2SD174M		3027	75		
2SD174R		3027	75		
2SD174X		3027	75		
2SD174Y		3027	75		
2SD175	130	3027	75	2041	2N5632
2SD175A		3027	75	2041	
2SD175B	130	3027	75	2041	
2SD175C		3027	75		
2SD175D		3027	75	2041	
2SD175E		3027	75	2041	
2SD175F	130	3027	75	2041	
2SD175G		3027	75	2041	
2SD175GN		3027	75	2041	
2SD175H		3027	75	2041	
2SD175J		3027	75	2041	
2SD175K		3027	75	2041	
2SD175L		3027	75	2041	
2SD175M	130	3027	75	2041	
2SD175OR		3027	75	2041	
2SD175R		3027	75	2041	
2SD175X		3027	75	2041	
2SD175Y		3027	75	2041	
2SD176	130	3027	75	2041	2N5632
2SD176A		3027	75	2041	
2SD176B		3027	75	2041	
2SD176C		3027	75	2041	
2SD176D		3027	75	2041	
2SD176E		3027	75	2041	
2SD176F		3027	75	2041	
2SD176G		3027	75	2041	
2SD176GN		3027	75	2041	
2SD176H		3027	75	2041	
2SD176J		3027	75	2041	
2SD176K		3027	75	2041	
2SD176L		3027	75	2041	
2SD176M		3027	75	2041	
2SD176OR		3027	75	2041	
2SD176X		3027	75	2041	
2SD176Y		3027	75	2041	
2SD177	328	3438	35		2N5634
2SD178	103A	3835	47	2002	
2SD178A	103A	3835	47	2002	
2SD178B		3835	47	2002	
2SD178C		3835	47	2002	
2SD178D		3835	47	2002	
2SD178E		3835	47	2002	
2SD178F		3835	47	2002	
2SD178G		3835	47	2002	
2SD178GN		3835	47	2002	
2SD178H		3835	47	2002	
2SD178J		3835	47	2002	
2SD178K		3835	47	2002	
2SD178L		3835	47	2002	
2SD178M		3835	47	2002	
2SD178Q	103A		47	2002	
2SD178T	103A		59	2002	
2SD178X		3835	47	2002	
2SD178Y		3835	47	2002	
2SD17A		3027	75	2041	
2SD17C		3027	75	2041	
2SD17D		3027	75	2041	
2SD17E		3027	75	2041	
2SD17F		3027	75	2041	
2SD17GN		3027	75	2041	
2SD17H		3027	75	2041	
2SD17J		3027	75	2041	
2SD17K		3027	75	2041	
2SD17L		3027	75	2041	
2SD17M		3027	75	2041	
2SD17R		3027	75	2041	
2SD17X		3027	75	2041	
2SD17Y		3027	75	2041	
2SD18	87	3438	35		MJ15011
2SD18(RECTIFIER)	116(2)		504A(2)	1104	
2SD180	87	3297/280	75	2041	2N5758
2SD180A	87	3297/280	75	2041	
2SD180B	87	3297/280	75	2041	
2SD180C	87	3297/280	75	2041	
2SD180D	87	3297/280	75	2041	
2SD180F		3297	75	2041	
2SD180G		3297	75		
2SD180GN		3297	75	2041	
2SD180H		3297	75	2041	
2SD180J		3297	75	2041	
2SD180K	87	3297/280	75	2041	
2SD180L	87	3297/280	75	2041	
2SD180M	87	3297/280	75	2041	
2SD180N	87	3297/280			
2SD180OR			75	2041	
2SD180R		3297	75	2041	
2SD180X		3297	75	2041	
2SD180Y		3297	75	2041	
2SD181	385	3438	35		MJ15001
2SD182		3024	18	2030	
2SD182A		3024	18	2030	
2SD182B		3024	18	2030	
2SD182C		3024	18	2030	
2SD182D		3024	18	2030	
2SD182E		3024	18	2030	
2SD182F		3024	18	2030	
2SD182G		3024	18	2030	
2SD182GN		3024	18	2030	
2SD182H		3024	18	2030	
2SD182J		3024	18	2030	
2SD182K		3024	18	2030	
2SD182L		3024	18	2030	
2SD182OR		3024	18	2030	
2SD182R		3024	18	2030	
2SD182X		3024	18	2030	
2SD182Y		3024	18	2030	
2SD183		3024	18	2030	
2SD183B		3024	18	2030	
2SD183E		3024	18	2030	
2SD183GN		3024	18	2030	
2SD183K		3024	18	2030	
2SD183L		3024	18	2030	
2SD183OR		3024	18	2030	
2SD183R		3024	18	2030	
2SD184		3530	18	2048	
2SD184A		3530	18	2048	
2SD184B		3530	18	2048	
2SD184C		3530	18	2048	
2SD184D		3530	18	2048	
2SD184E		3530	18	2048	

Industry Standard No.	ECG	SK	GE	RS 276-	MOTOR.
2SD184F			18	2048	
2SD184H			18	2048	
2SD184J		3530	18	2048	
2SD184K		3530	18	2048	
2SD184L		3530	18	2048	
2SD184M		3530	18	2048	
2SD1840R		3530	18	2048	
2SD184Y		3530	18	2048	
2SD185			66	2048	
2SD186	103A	3835	59	2002	
2SD186A	103A	3835	59	2002	
2SD186B	103A	3835	59	2002	
2SD186C		3835	59		
2SD186D		3835	59		
2SD186E		3835	59		
2SD186F		3835	59		
2SD186G		3835	59		
2SD186GN		3835	59		
2SD186H		3835	59		
2SD186J		3835	59		
2SD186K		3835	59		
2SD186L		3835	59		
2SD186M		3835	59		
2SD1860R		3835	59		
2SD186R		3835	59		
2SD186X		3835	59		
2SD186Y		3835	59		
2SD187	103A	3835	59	2002	
2SD187-OR			59	2002	
2SD187A	103A	3835	59	2002	
2SD187B		3835	59	2002	
2SD187C		3835	59	2002	
2SD187D		3835	59	2002	
2SD187E		3835	59	2002	
2SD187F		3835	59	2002	
2SD187G		3835	59	2002	
2SD187GN		3835	59	2002	
2SD187H		3835	59	2002	
2SD187J		3835	59	2002	
2SD187K		3835	59	2002	
2SD187L		3835	59	2002	
2SD187M		3835	59	2002	
2SD1870R		3835	59	2002	
2SD187R	103A	3835	59	2002	
2SD187X		3835	59	2002	
2SD187Y	103A	3835	59	2002	
2SD188	87	3297/280	75	2041	2N5758
2SD188A	87	3297/280	75	2041	
2SD188B	87	3297/280	75	2041	
2SD188C	87	3297/280	75	2041	
2SD188D		3297	75	2041	
2SD188E		3297	75	2041	
2SD188F		3297	75	2041	
2SD188G		3297	75	2041'	
2SD188GN		3297	75	2041	
2SD188H		3297	75	2041	
2SD188J		3297	75	2041	
2SD188L	87	3297/280	75	2041	
2SD188M	87	3297/280	75	2041	
2SD1880R		3297	75	2041	
2SD188R		3297	75	2041	
2SD188X		3297	75	2041	
2SD188Y		3297	75	2041	
2SD189	280	3438	75		2N5758
2SD189A	280	3438	75		2N5758
2SD189B		3027	75	2041	
2SD189C		3027	75	2041	
2SD189D		3027	75	2041	
2SD189E		3027	75	2041	
2SD189F		3027	75	2041	
2SD189G		3027	75	2041	
2SD189GN		3027	75	2041	
2SD189H		3027	75	2041	
2SD189J		3027	75	2041	
2SD189K		3027	75	2041	
2SD189L		3027	75	2041	
2SD189M		3027	75	2041	
2SD1890R		3027	75	2041	
2SD189R	280	3027/130	75	2041	
2SD189X		3027	75	2041	
2SD189Y		3027	75	2041	
2SD19	101	3862/103	59	2002	
2SD190	124	3021	12		
2SD190B		3021	12		
2SD190C		3021	12		
2SD190D		3021	12		
2SD190E		3021	12		
2SD190F		3021	12		
2SD190GN		3021	12		
2SD190H		3021	12		
2SD190J		3021	12		
2SD190K		3021	12		
2SD190L		3021	12		
2SD190M		3021	12		
2SD1900R		3021	12		
2SD190R		3021	12		
2SD190X		3021	12		
2SD190Y		3021	12		
2SD191	103A	3835	59	2002	
2SD192	103A	3835	59	2002	
2SD193	103A		59	2002	
2SD194	103A	3835	59	2002	
2SD195	103A	3011A	75	2002	
2SD195A	103A	3011A	59	2002	
2SD195B		3011A	59		
2SD195C		3011A	59		
2SD195D		3011A	59		
2SD195E		3011A	59		
2SD195F		3011A	59		
2SD195G		3011A	59		
2SD195GN		3011A	59		
2SD195H		3011A	59		
2SD195J		3011A	59		
2SD195K		3011A	59		
2SD195L		3011A	59		
2SD195M		3011A	59		
2SD1950R		3011A	59		
2SD195R		3011A	59		
2SD195X		3011A	59		
2SD195Y		3011A	59		
2SD198	94	3559/162	35		2N5840
2SD198A	94	3559/162	35		
2SD198AP	94	3559/162	35		
2SD198AQ	94	3559/162	35		
2SD198AR	94	3559/162	35		
2SD198B		3559	35		
2SD198C		3559	35		
2SD198D		3559	35		
2SD198F		3559	35		
2SD198G		3559	35		
2SD198H	94	3559/162	35		
2SD198HQ	94	3559/162	35		
2SD198HR	94	3559/162	35		
2SD198J		3559	35		
2SD198K		3559	35		
2SD1980R		3559	35		
2SD198P		3559	35		
2SD198Q	94	3559/162	35		
2SD198R	94	3559/162	35		
2SD198S	94	3559/162	35		
2SD198V	94	3559/162	35		
2SD198X		3559	35		
2SD198Y		3559	35		
2SD199	164	3439/163A	35		BU204
2SD19A		3862	59		
2SD19B		3862	59		
2SD19C		3862	59		
2SD19E		3862	59		
2SD19F		3862	59		
2SD19G		3862	59		
2SD19GN		3862	59		
2SD19H		3862	59		
2SD19J		3862	59		
2SD19K		3862	59		
2SD19M		3862	59		
2SD19Y		3862	59		
2SD20	101	3862/103	59	2002	
2SD200	389	3115/165	38		BU204
2SD200A	389	3115/165	259		
2SD200B		3115	38		
2SD200C		3115	38		
2SD200D		3115	38		
2SD200E		3115	38		
2SD200F		3115	38		
2SD200G		3115	38		
2SD200GN		3115	38		
2SD200H		3115	38		
2SD200J		3115	38		
2SD200K		3115	38		
2SD200L		3115	38		
2SD200M		3115	38		
2SD2000R		3115	38		
2SD200X		3115	38		
2SD200Y		3115	38		
2SD201	87	3027/130	75	2041	2N5758
2SD201(O)	87		75	2041	
2SD201-O			75	2041	
2SD201A		3027	75	2041	
2SD201B		3027	75	2041	
2SD201C		3027	75	2041	
2SD201F		3027	75	2041	
2SD201G		3027	75	2041	
2SD201GN		3027	75	2041	
2SD201H		3027	75	2041	
2SD201J		3027	75	2041	
2SD201K		3027	75	2041	
2SD201L		3027	75	2041	
2SD201M	87	3027/130	75	2041	
2SD201M(O)			75	2041	
2SD201M(O,Y)			75	2041	
2SD201M(Y)			75	2041	
2SD201MO			75	2041	
2SD201MY	87		75	2041	
2SD2010	87		75	2041	
2SD201Q	87		75	2041	
2SD201X		3027	75	2041	
2SD201Y	87	3027/130	75	2041	
2SD202	87	3438	35		2N5759
2SD203	87	3438	35		2N5760
2SD204	128	3024	18	2030	
2SD204(L)			18	2030	
2SD204A		3024	18	2030	
2SD204B		3024	18	2030	
2SD204BL		3024	18	2030	
2SD204C		3024	18	2030	
2SD204D		3024	18	2030	
2SD204E		3024	18	2030	
2SD204F		3024	18	2030	
2SD204G		3024	18	2030	
2SD204GA		3024	18	2030	
2SD204GN		3024	18	2030	
2SD204H		3024	18	2030	
2SD204J		3024	18	2030	
2SD204K		3024	18	2030	
2SD204L	128	3024	18	2030	
2SD204M		3024	18	2030	
2SD204R		3024	18	2030	
2SD204Y		3024	18	2030	
2SD205	128/427	3024/128	18		
2SD205(L)			18	2030	
2SD205(M)			18	2030	
2SD205A		3024	18	2030	
2SD205B		3024	18	2030	
2SD205C		3024	18	2030	
2SD205D		3024	18	2030	
2SD205E		3024	18	2030	
2SD205F		3024	18	2030	
2SD205G		3024	18	2030	
2SD205GN		3024	18	2030	
2SD205H		3024	18	2030	
2SD205J		3024	18	2030	
2SD205L		3024	18	2030	
2SD205M		3024	18	2030	
2SD2050R		3024	18	2030	
2SD205R		3024	18	2030	
2SD205X		3024	18	2030	
2SD205Y		3024	18	2030	
2SD206		6629			2N5877
2SD207					2N5632
2SD208	284	6631/5531			2N5634
2SD20A		3862	59		
2SD20B		3862	59		
2SD20C		3010	59		
2SD20D		3862	59		
2SD20E		3862	59		
2SD20F		3862	59		
2SD20G		3862	59		
2SD20GN		3862	59		
2SD20H		3862	59		
2SD20J		3862	59		
2SD20K		3862	59		
2SD20L		3862	59		
2SD20M		3862	59		
2SD20R		3862	59		
2SD20X		3862	59		
2SD20Y		3862	59		
2SD21	101	3862/103	59	2002	
2SD211	130	3027	75	2041	2N5877
2SD211-OR			75	2041	
2SD211A		3027	75	2041	
2SD211B		3027	75	2041	
2SD211C		3027	75	2041	
2SD211D		3027	75	2041	
2SD211E		3027	75	2041	
2SD211F		3027	75	2041	
2SD211G		3027	75	2041	
2SD211GN		3027	75	2041	
2SD211H		3027	75	2041	

Industry Standard No.	ECG	SK	GE	RS 276-	MOTOR.
2SD211J		3027	75	2041	
2SD211K		3027	75	2041	
2SD211L		3027	75	2041	
2SD211M	130	3027	75	2041	
2SD211OR		3027	75	2041	
2SD211R		3027	75	2041	
2SD211X		3027	75	2041	
2SD211Y	130	3027	75	2041	
2SD212	130	3027	75	2041	2N5632
2SD212A		3027	75	2041	
2SD212B		3027	75	2041	
2SD212C		3027	75	2041	
2SD212D		3027	75	2041	
2SD212E		3027	75	2041	
2SD212F		3027	75	2041	
2SD212G		3027	75	2041	
2SD212GN		3027	75	2041	
2SD212H		3027	75	2041	
2SD212J		3027	75	2041	
2SD212K		3027	75	2041	
2SD212L		3027	75	2041	
2SD212M		3027	75	2041	
2SD212OR		3027	75	2041	
2SD212R		3027	75	2041	
2SD212X		3027	75	2041	
2SD212Y		3027	75	2041	
2SD213	280	3836/284	35		2N5633
2SD214	280	3836/284			2N5634
2SD215	128	6629/5529	8	2002	
2SD216	128	6629/5529			
2SD217	87	3297/280	35		2N5633
2SD217E	87	3297/280			
2SD217K	87	3297/280			
2SD217L	87	3297/280			
2SD217M	87	3297/280			
2SD218	87	3619	35		2N5634
2SD218K	87	3619			
2SD218L	87	3619			
2SD218M	87	3619			
2SD219	128	3024	18	2030	
2SD219A		3024	18	2030	
2SD219B		3024	18	2030	
2SD219C		3024	18	2030	
2SD219D		3024	18	2030	
2SD219E		3024	18	2030	
2SD219F	128/427	3024/128	18	2030	
2SD219G		3024	18	2030	
2SD219GN		3024	18	2030	
2SD219H		3024	18	2030	
2SD219J		3024	18	2030	
2SD219K		3024	18	2030	
2SD219L		3024	18	2030	
2SD219M		3024	18	2030	
2SD219OR		3024	18	2030	
2SD219R		3024	18	2030	
2SD219X		3024	18	2030	
2SD219Y		3024	18	2030	
2SD21A		3862	59		
2SD21B		3862	59		
2SD21C		3862	59		
2SD21D		3862	59		
2SD21E		3862	59		
2SD21F		3862	59		
2SD21G		3862	59		
2SD21GN		3862	59		
2SD21H		3862	59		
2SD21J		3862	59		
2SD21K		3862	59		
2SD21L		3862	59		
2SD21M		3862	59		
2SD21OR		3862	59		
2SD21R		3862	59		
2SD21X		3862	59		
2SD21Y		3862	59		
2SD22	101	3862/103	59	2002	
2SD220	128	3024	18	2030	
2SD220A		3024	18	2030	
2SD220B		3024	18	2030	
2SD220C		3024	18	2030	
2SD220D		3024	18	2030	
2SD220E		3024	18	2030	
2SD220F	128/427	3024/128	18	2030	
2SD220G		3024	18	2030	
2SD220GN		3024	18	2030	
2SD220H		3024	18	2030	
2SD220J		3024	18	2030	
2SD220K		3024	18	2030	
2SD220L		3024	18	2030	
2SD220M		3024	18	2030	
2SD220OR		3024	18	2030	
2SD220R			18	2030	
2SD220X		3024	18	2030	
2SD220Y		3024	18	2030	
2SD221	128	3024	243	2030	
2SD222	324/427	3024/128	18	2030	
2SD222A		3024	18	2030	
2SD222B		3024	18	2030	
2SD222C		3024	18	2030	
2SD222D		3024	18	2030	
2SD222E		3024	18	2030	
2SD222F		3024	18	2030	
2SD222G		3024	18	2030	
2SD222GN		3024	18	2030	
2SD222H		3024	18	2030	
2SD222J		3024	18	2030	
2SD222K		3024	18	2030	
2SD222L		3024	18	2030	
2SD222M		3024	18	2030	
2SD222OR		3024	18	2030	
2SD222R		3024	18	2030	
2SD222X		3024	18	2030	
2SD222Y		3024	18	2030	
2SD223	324/427	3049/224	32		
2SD223A		3045	32		
2SD223B		3045	32		
2SD223C		3045	32		
2SD223D		3045	32		
2SD223E		3045	32		
2SD223F		3045	32		
2SD223G		3045	32		
2SD223GN		3045	32		
2SD223H		3045	32		
2SD223J		3045	32		
2SD223K		3045	32		
2SD223L		3045	32		
2SD223M		3045	32		
2SD223OR		3045	32		
2SD223R		3045	32		
2SD223X		3045	32		
2SD223Y		3045	32		
2SD226	175	3562	66	2020	2N3766
2SD226-0	175		66	2020	
2SD226-0			66	2048	
2SD226A	175	3562	66	2020	
2SD226A(0)			66	2048	
2SD226AP	175	3562	66	2020	
2SD226B	175	3562	66	2020	
2SD226BP	175	3562	66	2020	
2SD226C		3562	66	2048	
2SD226E		3562	66	2048	
2SD226F		3562	66	2048	
2SD226G		3562	66	2048	
2SD226GN		3562	66	2048	
2SD226H		3562	66	2048	
2SD226J		3562	66	2048	
2SD226K		3562	66	2048	
2SD226L		3562	66	2048	
2SD226M		3562	66	2048	
2SD2260	175	3562	66	2020	
2SD2260R		3562	66	2048	
2SD226P	175	3562	66	2020	
2SD226Q	175	3562	66	2020	
2SD226R	175	3562	66	2048	
2SD226X		3562	66	2048	
2SD226Y		3562	66	2048	
2SD227	85	3124/289	62	2030	
2SD227(PANASONIC)			62	2030	
2SD227(R)			62	2030	
2SD227-175			62	2030	
2SD227-OR			62	2030	
2SD227A	85	3124/289	62	2030	
2SD227B	85	3124/289	62	2030	
2SD227C	85	3124/289	62	2030	
2SD227D	85	3124/289	62	2009	
2SD227E	85	3124/289	62	2030	
2SD227F	85	3124/289	62	2030	
2SD227G		3124	62	2030	
2SD227GN		3124	62	2030	
2SD227H		3124	62	2030	
2SD227J		3124	62	2030	
2SD227K		3124	62	2030	
2SD227L	85	3124/289	62	2030	
2SD227LP		3124	62	2030	
2SD227M		3124	62	2030	
2SD227OR		3124	62	2030	
2SD227R	85	3124/289	62	2030	
2SD227S	85	3124/289	62	2030	
2SD227V	85	3124/289	62	2030	
2SD227W	85	3124/289	62	2030	
2SD227X	85	3124/289	62	2030	
2SD227Y		3124	62	2030	
2SD228	289A	3122	88	2030	
2SD228A	289A	3122	88	2051	
2SD228B	289A	3122	88	2051	
2SD228C	289A	3122	88	2051	
2SD228D	289A	3122	88	2051	
2SD228E	289A	3122	88	2051	
2SD228F		3122	88	2051	
2SD228G		3122	88	2051	
2SD228GN		3122	88	2051	
2SD228H		3122	88	2051	
2SD228J		3122	88	2051	
2SD228K		3122	88	2051	
2SD228L		3122	88	2051	
2SD228M		3122	88	2051	
2SD228OR		3122	88	2051	
2SD228R		3122	88	2051	
2SD228X		3122	88	2051	
2SD228Y		3122	88	2051	
2SD22A		3862	59		
2SD22B		3862	59		
2SD22C		3862	59		
2SD22D		3862	59		
2SD22E		3862	59		
2SD22F		3862	59		
2SD22G		3862	59		
2SD22GN		3862	59		
2SD22H		3862	59		
2SD22J		3862	59		
2SD22K		3862	59		
2SD22L		3862	59		
2SD22M		3862	59		
2SD22OR		3862	59		
2SD22R		3862	59		
2SD22X		3862	59		
2SD22Y		3862	59		
2SD23	101	3862/103	59	2002	
2SD231	181				2N5302
2SD232	181				2N6275
2SD234	152	3054/196	66	2048	TIP31A
2SD234-O	152	3054/196	66	2048	
2SD234-0	152	3054/196	66	2048	
2SD234-ORG			66	2048	
2SD234-R	152		66	2048	
2SD234-RED			66	2048	
2SD234-Y	152		66	2048	
2SD2340			66	2048	
2SD234A		3054	66	2048	
2SD234B		3054	66	2048	
2SD234C		3054	66	2048	
2SD234D		3054	66	2048	
2SD234E		3054	66	2048	
2SD234F		3054	66	2048	
2SD234G		3054	66	2048	
2SD234GA		3054	66	2048	
2SD234GN		3054	66	2048	
2SD234GR		3054	66	2048	
2SD234H		3054	66	2048	
2SD234J		3054	66	2048	
2SD234K		3054	66	2048	
2SD234L		3054	66	2048	
2SD234M		3054	66	2048	
2SD234N		3054	66	2048	
2SD234OR		3054	66	2048	
2SD234R	152	3054/196	66	2048	
2SD234X		3054	66	2048	
2SD234Y		3054	66	2048	
2SD235	152	3054/196	66	2048	TIP31A
2SD235(0)			66	2048	
2SD235(Y)			66	2048	
2SD235-0	152	3054/196	66	2048	
2SD235-0	152	3054/196	66	2048	
2SD235-OR			66	2048	
2SD235-ORG			66	2048	
2SD235-R	152		66	2048	
2SD235-RED			66	2048	
2SD235-Y	152		66	2048	
2SD2350		3054	66	2048	
2SD235A		3054	66	2048	
2SD235B		3054	66	2048	
2SD235C		3054	66	2048	
2SD235D	152	3054/196	66	2048	
2SD235E		3054	66	2048	
2SD235F		3054	66	2048	
2SD235G	152	3054/196	66	2048	
2SD235GN		3054	66	2048	
2SD235GR	152	3054/196	66	2048	
2SD235H		3054	66	2048	

Industry Standard No.	ECG	SK	GE	RS 276-	MOTOR.
2SD235J		3054	66	2048	
2SD235L		3054	66	2048	
2SD235LBY	152		66	2048	
2SD235M		3054	66	2048	
2SD235OR		3054	66	2048	
2SD235R	152	3054/196	66	2048	
2SD235RED			66	2048	
002SD235RY	152	3054/196	66	2048	
2SD235X		3054	66	2048	
2SD235Y	152	3054/196	66	2048	
2SD236	175	3026	23	2020	2N4912
2SD236(O2Y)			23	2020	
2SD236A		3026	23	2020	
2SD236B		3026	23	2020	
2SD236C		3026	23	2020	
2SD236D		3026	23	2020	
2SD236E		3026	23	2020	
2SD236F		3026	23	2020	
2SD236G		3026	23	2020	
2SD236GN		3026	23	2020	
2SD236H		3026	23	2020	
2SD236J		3026	23	2020	
2SD236K		3026	23	2020	
2SD236L		3026	23	2020	
2SD236M		3026	23	2020	
2SD236R		3026	23	2020	
2SD236X		3026	23	2020	
2SD236Y		3026	23	2020	
2SD237	175				2N4912
2SD238	175	3562	246	2020	2N3583
2SD238F	175	3562	246	2020	
2SD23A		3862	59		
2SD23B		3862	59		
2SD23C		3862	59		
2SD23D		3862	59		
2SD23E		3862	59		
2SD23F		3862	59		
2SD23G		3862	59		
2SD23GN		3862	59		
2SD23H		3862	59		
2SD23J		3862	59		
2SD23K		3862	59		
2SD23L		3862	59		
2SD23M		3862	59		
2SD23OR		3862	59		
2SD23R		3862	59		
2SD23X		3862	59		
2SD23Y		3862	59		
2SD24	124	3021	12		2N3739
2SD241	152				2N3766
2SD241H			14	2041	
2SD242	196				2N3767
2SD243	291				MJ3247
2SD244					MJ3248
2SD246	389	3115/165	38		BU208
2SD246A		3115	38		
2SD246B		3115	38		
2SD246C		3115	38		
2SD246D		3115	38		
2SD246E		3115	38		
2SD246F		3115	38		
2SD246G		3115	38		
2SD246GN		3115	38		
2SD246H		3115	38		
2SD246J		3115	38		
2SD246K		3115	38		
2SD246L		3115	38		
2SD246M		3115	38		
2SD246OR		3115	38		
2SD246R		3115	38		
2SD246X		3115	38		
2SD246Y		3115	38		
2SD247					2N5758
2SD249	181				2N5302
2SD24A		3021	12		
2SD24B	124	3021	12		
2SD24C	124	3021	12		
2SD24CK	124	3021	12		
2SD24D	124	3021	12		
2SD24E	124	3021	12		
2SD24F	124	3021	12		
2SD24G		3021	12		
2SD24GN		3021	12		
2SD24H		3021	12		
2SD24J		3021	12		
2SD24K	124	3021	12		
2SD24KC	124	3021	12		
2SD24KD	124	3021	12		
2SD24KE	124		12		
2SD24L		3021	12		
2SD24M		3021	12		
2SD24OR		3021	12		
2SD24R		3021	12		
2SD24X		3021	12		
2SD24Y	124	3021	12		
2SD24Y-K	124		12		
2SD24YB	124	3021	12		
2SD24YC		3021	12		
2SD24YD	124	3021	12		
2SD24YE	124	3021	12		
2SD24YF	124		12		
2SD24YK	124	3021	12		
2SD24YL	124		12		
2SD24YLC	124	3021	12		
2SD24YLD	124	3021	12		
2SD24YLE	124	3021	12		
2SD24YM	124	3021	12		
2SD24YS	124	3021			
2SD25	103A	3010	59	2002	
2SD250	181				2N6328
2SD251	286				2N5052
2SD254	175	3026	246	2020	2N3767
2SD255	175	3026	246	2020	2N3767
2SD256	152	3239/236	216	2053	2N3766
2SD257	196		246	2020	2N3767
2SD258	291		12		MJ3247
2SD259	291	3021/124	12		MJ3248
2SD25A		3010	59		
2SD25B		3010	59		
2SD25C		3010	59		
2SD25D		3010	59		
2SD25E		3010	59		
2SD25G		3010	59		
2SD25GN		3010	59		
2SD25H		3010	59		
2SD25J		3010	59		
2SD25K		3010	59		
2SD25L		3010	59		
2SD25M		3010	59		
2SD25OR		3010	59		
2SD25R		3010	59		
2SD25X		3010	59		
2SD25Y		3010	59		
2SD26	87	3027/130	75	2041	2N5758
2SD260		3559			2N5758
00002SD261	192	3137	63	2030	
2SD261(L)			63	2030	
2SD261(O)			63	2030	
2SD261(Q)			63	2030	
2SD261(R)	192A		63	2030	
2SD261(U)	192A		63	2030	
2SD261(V)			63	2030	
2SD261-O			63	2030	
2SD261-O		3137	63	2030	
2SD261A	192A	3137	63	2030	
2SD261B	192A	3137	63	2030	
2SD261C	192A	3137	63	2030	
2SD261D	192A	3137	63	2030	
2SD261E	192A	3137	63	2030	
2SD261F	192A	3137	63	2030	
2SD261G		3137	63	2030	
2SD261GN		3137	63	2030	
2SD261H		3137	63	2030	
2SD261J		3137	63	2030	
2SD261K		3137	63	2030	
2SD261L	192A	3137	63	2030	
2SD261M		3137	63	2030	
2SD261O	192A		63	2030	
2SD261OR		3137	63	2030	
2SD261P	192A	3137	63	2030	
2SD261Q	192A	3137	63	2030	
2SD261R	192A	3137	63	2030	
2SD261S	192A	3137	63	2030	
2SD261U		3137	63	2030	
2SD261V	192A	3137	63	2030	
2SD261W	192A	3137	63	2030	
2SD261X		3137	63	2030	
2SD261Y		3137	63	2030	
2SD262		3835			2N6546
2SD265					2N6545
2SD266					2N6545
2SD26A	87	3027/130	75	2041	2N5758
2SD26B	87	3027/130	75	2041	2N5758
2SD26C	87	3027/130	75	2041	2N5760
2SD26D		3027	75	2041	
2SD26E		3027	75	2041	
2SD26F		3027	75	2041	
2SD26G		3027	75	2041	
2SD26GN		3027	75	2041	
2SD26H		3027	75	2041	
2SD26J		3027	75	2041	
2SD26K		3027	75	2041	
2SD26L		3027	75	2041	
2SD26M		3027	75	2041	
2SD26OR		3027	75	2041	
2SD26R		3027	75	2041	
2SD26X		3027	75	2041	
2SD26Y		3027	75	2041	
2SD271					MJ4401
2SD272					MJ4401
2SD273					2N6545
2SD274					2N6545
2SD28	175	3026	66	2020	2N3767
2SD283	162	3438	35		MJ3247
2SD284	384	3438	35		MJ3247
2SD285	162	3438	35		MJ3247
2SD286					MJ15011
2SD287	284	3836	265	2047	MJ15011
2SD287A	284	3836			
2SD287AQ	284	3836			
2SD287AR	284	3836			
2SD287AS	284	3836			
2SD287B	284	3836			
2SD287BR	284	3836			
2SD287BS	284	3836			
2SD287C	284	3836			
2SD287CQ	284	3836			
2SD287CR	284	3836			
2SD287CS	284	3836			
2SD288	196	3197/235	66	2048	TIP31B
2SD288A	196		66	2048	
2SD288B	196		66	2048	
2SD288C	196		66	2048	
2SD288K	196	3197/235	66	2048	
2SD288L	196	3197/235	66	2048	
2SD288M	196	3197/235			
2SD289	196	3197/235	66	2048	TIP31B
2SD289A	196		66	2048	
2SD289B	196		66	2048	
2SD289C	196		66	2048	
2SD289K	196	3197/235			
2SD289L	196	3197/235			
2SD289M	196	3197/235			
2SD28A		3026	66	2048	
2SD28B		3026	66	2048	
2SD28C		3026	66	2048	
2SD28D		3026	66	2048	
2SD28E		3026	66	2048	
2SD28F		3026	66	2048	
2SD28G		3026	66	2048	
2SD28GN		3026	66	2048	
2SD28H		3026	66	2048	
2SD28J		3026	66	2048	
2SD28K		3026	66	2048	
2SD28L		3026	66	2048	
2SD28M		3026	66	2048	
2SD28OR		3026	66	2048	
2SD28R		3026	66	2048	
2SD28X		3026	66	2048	
2SD28Y		3026	66	2048	
2SD29	175	3026	66	2020	2N3767
2SD290	175		246	2020	2N5428
2SD290L	175		246	2020	
2SD291	175	3026	246	2020	2N3767
2SD291(R)	175		246	2020	
2SD291-O		3026	246	2020	
2SD291A	175	3026			
2SD291B		3026	246	2020	
2SD291BL		3026	246	2020	
2SD291C		3026	246	2020	
2SD291D		3026	246	2020	
2SD291E		3026	246	2020	
2SD291F		3026	246		
2SD291G		3026	246	2020	
2SD291GA			246	2020	
2SD291GN			246	2020	
2SD291H		3026	246	2020	
2SD291J		3026	246	2020	
2SD291K		3026	246	2020	
2SD291L		3026	246	2020	
2SD291M		3026	246	2020	
2SD291OR		3026	246	2020	
2SD291R			246	2020	
2SD291X			246	2020	
2SD291Y			246	2020	
2SD292	175	3026	66	2020	2N3767
2SD292-O		3026	66	2048	
2SD292A		3026	66	2048	

Industry Standard No.	ECG	SK	GE	RS 276-	MOTOR.
2SD292B		3026	66	2048	
2SD292BL		3026	66	2048	
2SD292C		3026	66	2048	
2SD292D		3026	66	2048	
2SD292E		3026	66	2048	
2SD292F		3026	66	2048	
2SD292G		3026	66	2048	
2SD292GA			66	2048	
2SD292GN			66	2048	
2SD292H		3026	66	2048	
2SD292J		3026	66	2048	
2SD292K		3026	66	2048	
2SD292L		3026	66	2048	
2SD292M		3026	66	2048	
2SD292OR		3026	66	2048	
2SD292R			66	2048	
2SD292X			66	2048	
2SD292Y			66	2048	
2SD293					2N6547
2SD294					2N6547
2SD295					MJ13335
2SD296					MJ13335
2SD297	175		246	2020	MJ3248
2SD299	389	3710/238	259		MJ12004
2SD299A		3710	259		
2SD299B	389	3710/238	259		
2SD299C		3710	259		
2SD299E		3710	259		
2SD299F		3710	259		
2SD299G		3710	259		
2SD299GN		3710	259		
2SD299H		3710	259		
2SD299J		3710	259		
2SD299K		3710	259		
2SD299M		3710	259		
2SD299OR		3710	259		
2SD299R		3710	259		
2SD299SL	389		259		
2SD299V	389	3710/238	259		
2SD299X		3710	259		
2SD299Y		3710	259		
2SD29A		3026	66	2048	
2SD29B		3026	66	2048	
2SD29E		3026	66	2048	
2SD29F		3026	66	2048	
2SD29GN		3026	66	2048	
2SD29J		3026	66	2048	
2SD29K		3026	66	2048	
2SD29L		3026	66	2048	
2SD29M		3026	66	2048	
2SD29OR		3026	66	2048	
2SD29R		3026	66	2048	
2SD29X		3026	66	2048	
2SD29Y		3026	66	2048	
2SD30	103A	3835	59	2002	
2SD30-O	103A		59	2002	
2SD30-OR			59	2002	
2SD30-N	103A	3835	59	2002	
2SD30-0		3835	59	2002	
2SD30-OR			59	2002	
2SD300	389	3710/238	38		MJ12004
2SD300A		3710	38		
2SD300B	389	3710/238	38		
2SD300C		3710	38		
2SD300D		3710	38		
2SD300G		3710	38		
2SD300H		3710	38		
2SD300J		3710	38		
2SD301	99				2N6385
2SD30A		3835	59	2002	
2SD30B		3835	59	2002	
2SD30C		3835	59	2002	
2SD30D		3835	59	2002	
2SD30E		3835	59	2002	
2SD30F		3835	59	2002	
2SD30G	103A	3835	59	2002	
2SD30GN		3835	59	2002	
2SD30H		3835	59	2002	
2SD30J		3835	59	2002	
2SD30K		3835	59	2002	
2SD30L		3835	59	2002	
2SD30M		3835	59	2002	
2SD30N			59	2002	
2SD30OR			59	2002	
2SD30P		3835	59	2002	
2SD30R		3835	59	2002	
2SD30X		3835	59	2002	
2SD30Y		3835	59	2002	
2SD31	103A	3835	59	2002	
2SD310					2N6547
2SD311					2N6547
2SD312	164	3079	35		2N6543
2SD312A		3079	35		
2SD312B		3079	35		
2SD312C		3079	35		
2SD312D		3079	35		
2SD312E		3079	35		
2SD312F		3079	35		
2SD312G		3079	35		
2SD312GN		3079	35		
2SD312H		3079	35		
2SD312J		3079	35		
2SD312K		3079	35		
2SD312L		3079	35		
2SD312M		3079	35		
2SD312X		3079	35		
2SD312Y		3079	35		
2SD313	152	3620	241	2048	TIP31A
2SD313(D,E)			241	2048	
2SD313(DE)			241	2048	
2SD313A		3620	241	2048	
2SD313B		3620	241	2048	
2SD313C	152	3620	241	2048	
2SD313D	152	3620	241	2048	
2SD313DE			241	2048	
2SD313E	152	3620	241	2048	
2SD313F		3620	241	2048	
2SD313G		3620	241	2048	
2SD313GN		3620	241	2048	
2SD313H		3620	241	2048	
2SD313K	152		241	2048	
2SD313L		3620	241	2048	
2SD313M	152	3620	241	2048	
2SD313N		3620	241	2048	
2SD313R		3620	241	2048	
2SD313Y		3620	241	2048	
2SD314	152	3054/196	28	2048	TIP31A
2SD314A		3054	28	2048	
2SD314B		3054	28	2048	
2SD314C	152	3054/196	28	2048	
2SD314D	152	3054/196	28	2048	
2SD314E	152	3054/196	28	2020	
2SD314F		3054	28	2048	
2SD314GN		3054	28	2048	
2SD314H		3054	28	2048	
2SD314L		3054	28	2048	
2SD314M		3054	28	2048	
2SD314N		3054	28	2048	
2SD314R		3054	28	2048	
2SD314Y		3054	28	2048	
2SD315	152	3026	233	2020	2N3766
2SD315C	152		233	2020	
2SD315D	152	3026	233	2020	
2SD315E		3026	233	2020	
2SD316	280	3836/284			2N3716
2SD317	152	3054/196	66	2048	TIP31A
2SD317A	196	3054	66	2048	
2SD317A(F)			66	2017	
2SD317A(F)(P)			66	2017	
2SD317AF			66	2017	
2SD317AP	196		66	2017	
2SD317F	152	3054/196	66	2048	
2SD317P	152	3054/196	66	2048	
2SD318	152		66	2048	TIP31A
2SD318(O)	152		66	2048	
2SD318-O			66	2017	
2SD318A	396		66	2048	
2SD318B	152		66	2048	
2SD318P	152		66	2048	
2SD318Q	152	3054/196	66	2017	
2SD319	284		75	2041	2N5633
2SD31B		3835	59		
2SD31C		3835	59		
2SD31D	103A	3835	59	2002	
2SD31E		3835	59		
2SD31F		3835	59		
2SD31G		3835	59		
2SD31GN		3835	59		
2SD31H		3835	59		
2SD31J		3835	59		
2SD31K		3835	59		
2SD31L		3835	59		
2SD31M		3835	59		
2SD31OR		3835	59		
2SD31R		3835	59		
2SD31X		3835	59		
2SD31Y		3835	59		
2SD32	103A	3835	59	2002	
2SD320	162	3438	35		2N5840
2SD321	283	3467	36		2N6306
2SD322	280	3297	262	2041	MJ4247
2SD322A	280	3297	262	2041	
2SD322B	280	3297	262	2041	
2SD322C	280	3297	262	2041	
2SD323	280	3297	262	2041	MJ4248
2SD323A	280	3297	262	2041	
2SD323B	280	3297	262	2041	
2SD323C	280	3297	262	2041	
2SD324	124		12		2N3739
2SD325	152	3197/235	28	2017	TIP31
2SD325-OR			28	2017	
2SD325A		3197	28	2017	
2SD325B		3197	28	2017	
2SD325C	152	3197/235	28	2017	
2SD325D	152	3197/235	28	2017	
2SD325E	152	3197/235	28	2017	
2SD325F		3197	28	2017	
2SD325G		3197	28	2017	
2SD325GN		3197	28	2017	
2SD325H		3197	28	2017	
2SD325J		3197	28	2017	
2SD325K		3197	28	2017	
2SD325L		3197	28	2017	
2SD325M		3197	28	2017	
2SD325OR		3197	28	2017	
2SD325R		3197	28	2017	
2SD325X		3197	28	2017	
2SD325Y		3197	28	2017	
2SD326	124	3021	12		2N3739
2SD326A		3021	12		
2SD326B		3021	12		
2SD326C		3021	12		
2SD326D		3021	12		
2SD326E		3021	12		
2SD326L		3021	12		
2SD326M		3021	12		
2SD326OR		3021	12		
2SD326X		3021	12		
2SD326Y		3021	12		
2SD327	85	3444/123A	20	2030	
2SD327A	85	3444/123A	20	2030	
2SD327B	85	3444/123A	20	2030	
2SD327C	85	3444/123A	20	2030	
2SD327D	85	3444/123A	20	2030	
2SD327E	85	3444/123A	20	2030	
2SD327F	85	3444/123A	20	2030	
2SD327L		3122	20	2030	
2SD327OR		3122	20	2030	
2SD327R		3122	20	2030	
2SD327Y	85	3122	20	2030	
2SD328S	128		252	2018	
2SD32A		3835	59		
2SD32B		3835	59		
2SD32C		3835	59		
2SD32D		3835	59		
2SD32E		3835	59		
2SD32F		3835	59		
2SD32G		3835	59		
2SD32GN		3835	59		
2SD32H		3835	59		
2SD32J		3835	59		
2SD32K		3835	59		
2SD32L		3835	59		
2SD32M		3835	59		
2SD32OR		3835	59		
2SD32R		3835	59		
2SD32X		3835	59		
2SD32Y		3835	59		
2SD33	103A	3011A	59	2002	
2SD330	152	3239/236	66	2048	TIP31A
2SD330D	152		66	2048	
2SD330E	152	3239/236	66	2048	
2SD331					TIP31A
2SD334	280	3438	35		2N5759
2SD334A	280	3438	35		
2SD334R	280	3438	35		
2SD335			75	2041	2N5758
2SD336	297	3122	63	2030	
2SD336R	297		63	2030	
2SD336Y	297		63	2030	
2SD338			75	2041	2N5758
2SD339	280		75	2041	2N5758
2SD33A		3011A	59	2015	
2SD33B		3011A	59	2015	
2SD33C	103A	3011A	59	2011	
2SD33D		3011A	59	2015	
2SD33E		3011A	59	2015	
2SD33F		3011A	59	2015	

Industry Standard No.	ECG	SK	GE	RS 276-	MOTOR.
2SD33G		3011A	59	2015	
2SD33GN		3011A	59	2015	
2SD33H		3011A	59	2015	
2SD33J		3011A	59	2015	
2SD33K		3011A	59	2015	
2SD33L		3011A	59	2015	
2SD33OR		3011A	59	2015	
2SD33X		3011A	59	2015	
2SD33Y		3011A	59	2015	
2SD34	103A	3835	47	2002	
2SD340	280				MJ15015
2SD341	130		74	2043	MJ15015
2SD341H	130		74	2043	
2SD342					TIP31B
2SD343	152	3054/196	66	2048	TIP31B
2SD343A		3054	28	2017	
2SD343B		3054	28	2017	
2SD343C		3054	28	2017	
2SD343D		3054	28	2017	
2SD343H		3054	28	2017	
2SD343J		3054	28	2017	
2SD343K		3054	28	2017	
2SD344					TIP31B
2SD345					TIP31B
2SD346					TIP41A
2SD347					TIP41A
2SD348	238	3710	259		MJ12005
2SD34A		3835	47		
2SD34B		3835	47		
2SD34C		3835	47		
2SD34D		3835	47		
2SD34E		3835	47		
2SD34F		3835	47		
2SD34G		3835	47		
2SD34GN		3835	47		
2SD34H		3835	47		
2SD34J		3835	47		
2SD34K		3835	47		
2SD34L		3835	47		
2SD34M		3835	47		
2SD34OR		3835	47		
2SD34R		3835	47		
2SD34X		3835	47		
2SD34Y		3835	47		
2SD35	103A	3835	59	2002	
2SD350	165	3710/238	259		MJ12004
2SD350A	165	3710/238	259		
2SD350Q	165	3710/238	259		
2SD350T	165	3710/238	259		
2SD351	283		36		2N6545
2SD352	103A	3010	59	2002	
2SD352A	103A	3010			
2SD352B	103A	3010			
2SD352C	103A	3010			
2SD352D	103A	3010	59	2002	
2SD352E	103A	3010	59	2002	
2SD352F	103A	3010	59	2002	
2SD352S	103A	3010			
2SD353	162	3133/164	37		2N5838
2SD355		3849	271	2030	
2SD355C			271	2030	
2SD355D		3849	271	2030	
2SD355E		3849	271	2030	
2SD356	291	3188A/182			2N4923
2SD356C	291	3188A/182			
2SD356D	291	3188A/182			
2SD357	291				2N4923
2SD358	291	3440			2N4923
2SD358C	291	3440			
2SD358D	291	3440			
2SD359	152	3239/236	247	2052	2N5190
2SD359C	152	3239/236	247	2052	
2SD359C2	152		247	2052	
2SD359D	152	3239/236	247	2052	
2SD359D1	152		247	2052	
2SD359D2	152		247	2052	
2SD359E		3239	247	2052	
2SD36	103A	3835	59	2002	
2SD360	152	3893	66	2048	2N5190
2SD3601		3197	215		
2SD360C	152	3893	66	2048	
2SD360D	152	3893	66	2048	
2SD360E		3893	66	2048	
2SD361	152				2N5191
2SD363					MJ10015
2SD364					MJ10016
2SD365	152	3054/196	66	2048	TIP31A
2SD365-O	152	3054/196	66	2048	
2SD365A	196	3054	66	2048	
2SD365AP	196	3188A/182			
2SD365AQ	196	3188A/182			
2SD365B	152	3054/196	66	2048	
2SD365H	152	3054/196	66	2048	
2SD365P	152	3054/196	66	2048	
2SD365Q	152	3054/196	66	2048	
2SD366	152	3054/196	66	2048	TIP31A
2SD366-O	152	3054/196	66	2048	
2SD366AP	196	3188A/182			
2SD366AQ	196	3188A/182			
2SD366P	152	3054/196	66	2048	
2SD366Q	152	3054/196	66	2048	
2SD367	103A	3835	59	2002	
2SD367A	103A	3835	59	2002	
2SD367B	103A	3835	59	2002	
2SD367C	103A	3835	59	2002	
2SD367D	103A	3835	59	2002	
2SD367E	103A	3835	59	2002	
2SD367F	103A	3835	59	2002	
2SD367H		3835	59	2030	
2SD367J		3835	59	2030	
2SD367K		3835	59	2030	
2SD367L		3835	59	2030	
2SD367M		3835	59	2030	
2SD367OR		3835	59	2030	
2SD367P	103A	3835	59	2002	
2SD367R		3835	59	2030	
2SD367X		3835	59	2030	
2SD367Y		3835	59	2030	
2SD368	238		59		MJ12005
2SD369	284				2N3716
2SD36A		3835	59		
2SD36B		3835	59		
2SD36C		3835	59		
2SD36D		3835	59		
2SD36E		3835	59		
2SD36F		3835	59		
2SD36G		3835	59		
2SD36GN		3835	59		
2SD36H		3835	59		
2SD36J		3835	59		
2SD36K		3835	59		
2SD36L		3835	59		
2SD36M		3835	59		
2SD36OR		3835	59		
2SD36R		3835	59		
2SD36X		3835	59		
2SD36Y		3835	59		
2SD37	103A	3010	59	2002	
2SD371	87	3027/130			2N5758
2SD372					MJ10015
2SD373					MJ10015
2SD374					MJ10016
2SD375	328	3895			MJ13330
2SD376	385				MJ13331
2SD377	385				MJ13334
2SD379	280	3836/284	265	2047	2N5758
2SD379P	280	3836/284	265	2047	
2SD379Q	280	3836/284	265	2047	
2SD379R	280	3836/284	265	2047	
2SD379S	280	3836/284	265	2047	
2SD37A	103A	3010	59	2002	
2SD37B	103A	3010	59	2002	
2SD37C	103A	3010	59	2002	
2SD37D		3010	59		
2SD37E		3010	59		
2SD37F		3010	59		
2SD37G		3010	59		
2SD37GN		3010	59		
2SD37H		3010	59		
2SD37K		3010	59		
2SD37L		3010	59		
2SD37M		3010	59		
2SD37OR		3010	59		
2SD37R		3010	59		
2SD37X		3010	59		
2SD37Y		3010	59		
2SD38	103A	3835	47	2002	
2SD380	165	3710/238	259		MJ12005
2SD380A	165	3710/238			
2SD381	291	3440			MJE15030
2SD381K	291	3440			
2SD381L	291	3440			
2SD381M	291	3440			
2SD381N	291	3440			
2SD382	291	3440	66		MJE15030
2SD382L	291	3440			
2SD382LM	291		66		
2SD382M	291	3440			
2SD382N	291	3440			
2SD383	283	3467	36		MJ411
2SD384					2N6301
2SD385					2N6301
2SD386	375	3929			MJE13004
2SD386A	375	3929	251		
2SD386AC	375	3929			
2SD386AD	375	3929			
2SD386AE	375	3929			
2SD386C	375	3929			
2SD386D	375	3929	251		
2SD386E	375	3929			
2SD386F	375	3929			
2SD386Y	375	3929	251		
2SD387	375	3929			MJE13004
2SD387AC	375	3929			
2SD387AD	375	3929			
2SD387AE	375	3929			
2SD387AF	375	3929			
2SD387C	375	3929			
2SD387D	375	3929			
2SD387E	375	3929			
2SD387F	375	3929			
2SD388	87	3269			MJ4247
2SD389	152	3054/196	66	2048	TIP31A
2SD389(O)			66	2048	
2SD389(LP)	152		66	2048	
2SD389(O)	152		66	2048	
2SD389(O,LP,P)				2048	
2SD389(P)	152		66	2048	
2SD389-O	152		66	2048	
2SD389-OP	152		66	2048	
2SD389A	196	3054	66	2048	
2SD389AF	196	3054	66	2048	
2SD389AFO	152	3054/196	66	2048	
2SD389AFO	196		66	2048	
2SD389AFP	196	3054	66	2048	
2SD389AP	196	3054	66	2048	
2SD389AQ		3054	66	2048	
2SD389B	152	3054/196	66	2048	
2SD389BL	152		66	2048	
2SD389BLB	152	3054/196	66	2048	
2SD389BLB-O	152	3054/196			
2SD389BLB-O	152		66	2048	
2SD389BLB-P	152	3054/196	66	2048	
2SD389BP	152	3054/196	66	2048	
2SD389L			66	2048	
2SD389LB	152		66	2048	
2SD389LBP			66	2048	
2SD389LP	152	3054/196	66	2048	
2SD389P	152	3054/196	66	2048	
2SD389Q	152	3054/196	66	2048	
2SD38A		3835	47		
2SD38B		3835	47		
2SD38C		3835	47		
2SD38D		3835	47		
2SD38E		3835	47		
2SD38F		3835	47		
2SD38G		3835	47		
2SD38GN		3835	47		
2SD38H		3835	47		
2SD38J		3835	47		
2SD38K		3835	47		
2SD38L		3835	47		
2SD38M		3835	47		
2SD38OR		3835	47		
2SD38R		3835	47		
2SD38X		3835	47		
2SD38Y		3835	47	2002	
2SD390	152	3054/196	66	2048	TIP31A
2SD390(O)	152		66	2048	
2SD390A	196	3054			
2SD390P	152	3054/196	66	2048	
2SD390Q	152	3054/196	28	2017	
2SD392		3244	220		
2SD393					SDT13305
2SD394					SDT13305
2SD395					SDT13305
2SD396					2N6547
2SD400	382	3137/192A	47	2030	
2SD400D	382	3849/293	47	2030	
2SD400E	382	3849/293	47	2030	
2SD400F	382	3849/293	47	2030	
2SD400P1		3849	47	2030	
2SD400P1D		3849	47	2030	
2SD400P1E		3849	47	2030	
2SD400P1F		3849	47	2030	
2SD400P2			47	2030	
2SD401	375	3929			TIP47
2SD401A	375	3929			

Industry Standard No.	ECG	SK	GE	RS 276-	MOTOR.
2SD401AK	375	3929			
2SD401AL	375	3929			
2SD401E	375	3929			
2SD401EK	375	3929			
2SD401EX	375	3929			
2SD401K	375	3929			
2SD401L	375	3929			
2SD401M	375	3929			
2SD402	375	3929			TIP47
2SD402A	375	3929			
2SD402AK	375	3929			
2SD402AL	375	3929			
2SD402AM	375	3929			
2SD404					TIP120
2SD41	130	3027	75	2041	
2SD414	373	9041			MJE341
2SD414Q	373	9041			
2SD414R	373	9041			
2SD414S	373	9041			
2SD415	373	9041			MJE341
2SD415Q	373	9041			
2SD415R	373	9041			
2SD415S	373	9041			
2SD416	238		259		MJ12005
2SD417	87				2N6306
2SD418	165		38		MJ12005
2SD41A		3027	75	2041	
2SD41B		3027	75	2041	
2SD41C		3027	75	2041	
2SD41D		3027	75	2041	
2SD41E		3027	75	2041	
2SD41F		3027	75	2041	
2SD41G		3027	75	2041	
2SD41GN		3027	75	2041	
2SD41H		3027	75	2041	
2SD41J		3027	75	2041	
2SD41K		3027	75	2041	
2SD41L		3027	75	2041	
2SD41M		3027	75	2041	
2SD41OR		3027	75	2041	
2SD41R		3027	75	2041	
2SD41X		3027	75	2041	
2SD41Y		3027	75	2041	
2SD422	384				MJ4380
2SD423	384				MJ4380
2SD424	284	3836	265	2047	MJ15001
2SD424-O	284		265	2047	
2SD424-R	284		265	2047	
2SD425	280	3297	262	2041	2N5634
2SD425-R			262	2041	
2SD4250	280		262	2041	
2SD426	280	3535/181			2N5633
2SD426-R			262	2041	
2SD4260	280	3535/181			
2SD426R	280	3535/181			
2SD427	280	3297			2N5759
2SD427R	280	3297			
2SD428	87	3027/130	14	2041	2N5758
2SD429					2N6547
2SD43	103A	3011A	59	2002	
2SD430					2N5759
2SD431					2N5633
2SD432					2N5634
2SD433					MJ15011
2SD434					MJ13330
2SD435					MJ13332
2SD436					MJ13333
2SD437					SDT13305
2SD438	382	3250/315	47	2030	
2SD438E	382	3250/315	47	2030	
2SD43A	103A	3011A	59	2002	
2SD43B		3011A	59		
2SD43C		3011A	59		
2SD43D		3011A	59		
2SD43E		3011A	59		
2SD43F		3011A	59		
2SD43G		3011A	59		
2SD43GN		3011A	59		
2SD43H		3011A	59		
2SD43J		3011A	59		
2SD43K		3011A	59		
2SD43L		3011A	59		
2SD43M		3011A	59		
2SD43OR		3011A	59		
2SD43R		3011A	59		
2SD43X		3011A	59		
2SD43Y		3011A	59		
2SD44	101	3011A	59	2002	
2SD44A		3011A	59		
2SD44B		3011A	59		
2SD44C		3011A	59		
2SD44D		3011A	59		
2SD44E		3011A	59		
2SD44F		3011A	59		
2SD44G		3011A	59		
2SD44GN		3011A	59		
2SD44H		3011A	59		
2SD44J		3011A	59		
2SD44K		3011A	59		
2SD44L		3011A	59		
2SD44M		3011A	59		
2SD44R		3011A	59		
2SD44X		3011A	59		
2SD44Y		3011A	59		
2SD45	162	3438	35		2N5760
2SD457					MJ10015
2SD458	283	3439/163A	36		SDT13305
2SD459	263				TIP121
2SD46	162	3438	35		2N5760
2SD460					TIP122
2SD461					MJ411
2SD463	243				2N6056
2SD464					2N6056
2SD467	85	3449/297	243	2030	
2SD467-C	85		243		
2SD467A	85	3449/297			
2SD467B	85	3449/297	243	2030	
2SD467C	85	3449/297	243	2030	
2SD468	293	3849	47	2030	
2SD468A	293	3849	47	2030	
2SD468AC		3849	47	2030	
2SD468B	293	3849	47	2030	
2SD468C	293	3849	47	2030	
2SD468D	293	3849	47	2030	
2SD468E		3849	47	2030	
2SD468F		3849	47	2030	
2SD468G		3849	47	2030	
2SD468GN		3849	47	2030	
2SD468L		3849	47	2030	
2SD468LN		3849	47	2030	
2SD468Y		3849	47	2030	
2SD47	328	3438	35		2N5758
2SD470	321	3844	260		
2SD470A	321	3844	260	2020	
2SD470B	321	3844	260		
2SD471	293	3849	268	2038	
2SD471K	293	3849	268	2038	
2SD471L	293	3849	268	2038	
2SD471M	293	3849	268	2038	
2SD471S	293	3849			
2SD475					2N6122
2SD476	196	3054	241	2020	2N6123
2SD476A	196	3054	241	2020	
2SD476B	196	3054	241	2020	
2SD476C	196	3054	241	2020	
2SD476D	196	3054	241	2020	
2SD476YL	196		241	2020	
2SD477	152	3054/196	66	2048	
2SD478	375	3929			TIP47
2SD478B	375	3929			
2SD478C	375	3929			
2SD478D	375	3929			
2SD479					2N6037
2SD48			66	2048	
2SD480					2N6038
2SD481					2N6039
2SD482					2N5655
2SD483					2N5656
2SD484					2N5657
2SD485					2N5190
2SD486					2N5191
2SD487					2N5192
2SD488					2N4921
2SD489					2N4922
2SD49	291	3026	66	2020	2N5050
2SD490					2N4923
2SD491					MJE3055
2SD492					2N3055
2SD493					2N5977
2SD494					2N5978
2SD495					2N8979
2SD496					MJE6043
2SD497					MJE6044
2SD498					MJE6045
2SD499					MJE3055
2SD49A		3026	66	2048	
2SD49B		3026	66	2048	
2SD49C		3026	66	2048	
2SD49D		3026	66	2048	
2SD49E		3026	66	2048	
2SD49F		3026	66	2048	
2SD49G		3026	66	2048	
2SD49GN		3026	66	2048	
2SD49J		3026	66	2048	
2SD49K		3026	66	2048	
2SD49L		3026	66	2048	
2SD49M		3026	66	2048	
2SD490R		3026	66	2048	
2SD49R		3026	66	2048	
2SD49X		3026	66	2048	
2SD49Y		3026	66	2048	
2SD50	87	3027/130	75	2041	2N5758
2SD500					MJE3055
2SD501					2N5991
2SD502					2N6055
2SD503					2N6056
2SD504					2N6057
2SD505					2N6058
2SD506					2N6059
2SD50A		3027	75	2041	
2SD50B		3027	75	2041	
2SD50C		3027	75	2041	
2SD50D		3027	75	2041	
2SD50E		3027	75	2041	
2SD50F		3027	75	2041	
2SD50G		3027	75	2041	
2SD50GN		3027	75	2041	
2SD50H		3027	75	2041	
2SD50J		3027	75	2041	
2SD50K		3027	75	2041	
2SD50L		3027	75	2041	
2SD50M		3027	75	2041	
2SD500R			75	2041	
2SD50R		3027	75	2041	
2SD50X		3027	75	2041	
2SD50Y		3027	75	2041	
2SD51	87	3027/130	75	2041	2N5758
2SD517	389	3710/238			MJ12003
2SD518					MJ4380
2SD519	384				MJ13015
2SD51A	87	3027/130	75	2041	
2SD51B		3027	75	2041	
2SD51C		3027	75	2041	
2SD51D		3027	75	2041	
2SD51E		3027	75	2041	
2SD51F		3027	75	2041	
2SD51G		3027	75	2041	
2SD51GN		3027	75	2041	
2SD51H		3027	75	2041	
2SD51J		3027	75	2041	
2SD51K		3027	75	2041	
2SD51L		3027	75	2041	
2SD51M		3027	75	2041	
2SD51OR		3027	75	2041	
2SD51R		3027	75	2041	
2SD51X		3027	75	2041	
2SD51Y		3027	75	2041	
2SD52	87				2N5758
2SD522					2N5632
2SD523	243				2N6055
2SD524	249				2N6056
2SD525	331	3440/291	66	2048	TIP41C
2SD525-O	331		66	2048	
2SD525-O	331	3440/291	66	2048	
2SD525R	331	3440/291			
2SD525Y	331	3440/291	66	2048	
2SD526	241	3054/196	241	2020	TIP41B
2SD526-O	241		241		
2SD526-O	241		241	2020	
2SD53	87	3027/130	75	2041	2N5759
2SD531		3440			TIP41C
2SD533	283		35		MJ423
2SD538	386				SDT13305
2SD539					MJ13015
2SD53A	87	3027/130	75	2041	
2SD53B		3027	75	2041	
2SD53C		3027	75	2041	
2SD53E		3027	75	2041	
2SD53F		3027	75	2041	
2SD53G		3027	75	2041	
2SD53GN		3027	75	2041	
2SD53H		3027	75	2041	
2SD53J		3027	75	2041	
2SD53K		3027	75	2041	
2SD53L		3027	75	2041	
2SD53M		3027	75	2041	
2SD530R		3027	75	2041	
2SD53R		3027	75	2041	

Industry Standard No.	ECG	SK	GE	RS 276-	MOTOR.
2SD53X		3027	75	2041	
2SD53Y		3027	75	2041	
2SD544					TIP41C
2SD545	293	3849	47	2030	
2SD545E	293	3849			
2SD545F	293	3849	47	2030	
2SD545G			47	2030	
2SD546	369	3131A	12	2020	
2SD55	181		75	2041	2N6328
2SD551	284	3836			
2SD551-0	284	3836			
2SD552	388				2N6250
2SD553	377				TIP41B
2SD554					2N3584
2SD555	87				MJ13015
2SD558					MPSU07
2SD55A	181		75	2041	
2SD56	277	3026	66	2020	2N3738
2SD562	108	3452	86	2038	
2SD562A		3039	86	2053	
2SD562B		3039	86	2053	
2SD562C		3039	86	2053	
2SD562D		3039	86	2053	
2SD562E		3039	86	2053	
2SD562K		3039	86	2053	
2SD562L		3039	86	2053	
2SD562OR		3039	86	2053	
2SD562R		3039	86	2053	
2SD562X		3039	86	2053	
2SD562Y		3039	86	2053	
2SD56A		3026	66	2048	
2SD56B		3026	66	2048	
2SD56C		3026	66	2048	
2SD56D		3026	66	2048	
2SD56E		3026	66	2048	
2SD56F		3026	66	2048	
2SD56G		3026	66	2048	
2SD56GN		3026	66	2048	
2SD56H		3026	66	2048	
2SD56J		3026	66	2048	
2SD56K		3026	66	2048	
2SD56L		3026	66	2048	
2SD56M		3026	66	2048	
2SD56OR		3026	66	2048	
2SD56R		3026	66	2048	
2SD56X		3026	66	2048	
2SD56Y		3026	66	2048	
2SD57	175	3131A/369	246	2020	2N3766
2SD570	196	3054	241	2020	2N6123
2SD571	293	3124/289			
2SD571K	293	3124/289	279		
2SD571L	293	3124/289			
2SD571M	293	3124/289			
2SD572					MJ10013
2SD573					MJ10014
2SD574					MJ11016
2SD575	389	3111	259		
2SD575L	389		259		
2SD577	165	3115	38		MJ12004
2SD58	175	3131A/369	246	2020	2N3766
2SD582	284	3182/243	265	2041	
2SD582A	284	3182/243	265	2041	
2SD586	380	3933		2020	
2SD586Q	380	3933			
2SD586R	380	3933		2020	
2SD586S	380	3933			
2SD587	380	3933			
2SD587A	380	3933			
2SD587AQ	380	3933			
2SD587AR	380	3933			
2SD587AS	380	3933			
2SD587Q	380	3933			
2SD587R	380	3933			
2SD587S	380	3933			
2SD588	380	3933			
2SD588Q	380	3933			
2SD588R	380	3933			
2SD588S	380	3933			
2SD589					MJ12005
2SD59	162	3438	35		
2SD591	199		62	2010	
2SD591R	199	3124/289	62	2010	
2SD592	85	3849/293	47	2051	
2SD592R	85	3849/293	47	2051	
2SD597	87				2N5758
2SD598	87				2N5759
2SD599	199	3245	62	2010	
2SD60	162	3438	35		2N5760
2SD600	373				2N4923
2SD604		3560	73		MJ3041
2SD605					MJ3042
2SD606					MJ10014
2SD608	375	3929			TIP47
2SD608Q	375	3929			
2SD608R	375	3929			
2SD608S	375	3929			
2SD61	103A	3010	59	2002	
2SD610	286				TIP47
2SD612	184	3253/295	57	2017	MJE520
2SD612E		3253	57	2017	
2SD612K	184	3253/295	57	2017	
2SD613	377$				TIP41C
2SD617	247$	3948/247			
2SD61A		3010	59		
2SD61B		3010	59		
2SD61C		3010	59		
2SD61D		3010	59		
2SD61E		3010	59		
2SD61F		3010	59		
2SD61G		3010	59		
2SD61GN		3010	59		
2SD61H		3010	59		
2SD61J		3010	59		
2SD61K		3010	59		
2SD61L		3010	59		
2SD61M		3010	59		
2SD61OR		3010	59		
2SD61R		3010	59		
2SD61X		3010	59		
2SD61Y		3010	59		
2SD62	103A	3010	59	2002	
2SD622	384				MJ4381
2SD625	228A	3103A/396	257		
2SD626					MJ10012
2SD627	389	3710/238	259		MJ12004
2SD628	247				2N6059
2SD629	247				2N6059
2SD62A		3010	59		
2SD62B		3010	59		
2SD62C		3010	59		
2SD62D		3010	59		
2SD62E		3010	59		
2SD62G		3010	59		
2SD62GN		3010	59		
2SD62H		3010	59		
2SD62J		3010	59		
2SD62K		3010	59		
2SD62L		3010	59		
2SD62M		3010	59		
2SD62R		3010	59		
2SD62X		3010	59		
2SD62Y		3010	59		
2SD63	103A	3011A	59	2002	
2SD630					2N5302
2SD631					2N5302
2SD632	94	3559/162	73		2N5840
2SD632P	94	3559/162	73		
2SD632Q	94	3559/162	73		
2SD632R	94		73		
2SD633	261				TIP122
2SD634	261				TIP121
2SD635	261				TIP120
2SD636		3911	62	2009	
2SD636-0		3122	62	2009	
2SD636-0			62	2009	
2SD636P		3911	62	2051	
2SD636Q		3911	62	2051	
2SD636R		3911	62	2051	
2SD637		3911	62		
2SD637Q		3911	62		
2SD637S		3911	62		
2SD637T		3124	62		
2SD639		3911	210	2051	
2SD63A		3011A	59		
2SD63B		3011A	59		
2SD63C		3011A	59		
2SD63E		3011A	59		
2SD63F		3011A	59		
2SD63G		3011A	59		
2SD63GN		3011A	59		
2SD63H		3011A	59		
2SD63J		3011A	59		
2SD63K		3011A	59		
2SD63L		3010	59		
2SD63M		3011A	59		
2SD63OR		3011A	59		
2SD63R		3011A	59		
2SD63X		3011A	59		
2SD63Y		3011A	59		
2SD64	103A	3011A	59	2002	
2SD640	283	3439/163A			2N6545
2SD641	386	3995			
2SD642					MJ10016
2SD643					MJ10015
2SD644					MJ10016
2SD645					MJ10016
2SD646					MJ10016
2SD649	165	3115	259		MJ12004
2SD649Q	165		259		
2SD64A		3011A	59		
2SD64B		3011A	59		
2SD64C		3011A	59		
2SD64D		3011A	59		
2SD64E		3011A	59		
2SD64F		3011A	59		
2SD64G		3011A	59		
2SD64GN		3011A	59		
2SD64H		3011A	59		
2SD64J		3011A	59		
2SD64K		3011A	59		
2SD64M		3011A	59		
2SD64OR		3011A	59		
2SD64R		3011A	59		
2SD64X		3011A	59		
2SD64Y		3011A	59		
2SD65	103A	3011A	59	2002	
2SD65-1	103A	3011A	59	2002	
2SD650					MJ3042
2SD656	175	3538	246	2020	
2SD65A		3011A	59	2002	
2SD65B		3011A	59	2002	
2SD65C		3011A	59	2002	
2SD65D		3011A	59	2002	
2SD65E		3011A	59	2002	
2SD65F		3011A	59	2002	
2SD65G		3011A	59	2002	
2SD65GN		3011A	59	2002	
2SD65H		3011A	59	2002	
2SD65J		3011A	59	2002	
2SD65K		3011A	59	2002	
2SD65L		3011A	59	2002	
2SD65M		3011A	59	2002	
2SD65OR		3011A	59	2002	
2SD65R		3011A	59	2002	
2SD65X		3011A	59	2002	
2SD65Y		3011A	59	2002	
2SD66	103A	3011A	59	2002	
2SD663					MJ3042
2SD665	284$	3836/284			2N6249
2SD666	382	3275/194	220		
2SD666AC	382		220		
2SD666C	382	3275/194	220		
2SD668	373#	9041/373			MJE344
2SD668A	373#	9041/373			
2SD668AB	373#	9041/373			
2SD668AC	373#	9041/373			
2SD668B	373#	9041/373			
2SD668C	373#	9041/373			
2SD668D	373#	9041/373			
2SD669	373#	3219			MJE344
2SD669A	373#	3219			
2SD669AB	373#	3219			
2SD669AC	373#	3219			
2SD669AD	373#	3219			
2SD669B	373#	3219			
2SD669C	373#	3219			
2SD669D	373#	3219			
2SD66A		3011A	59		
2SD66B		3011A	59		
2SD66C		3011A	59		
2SD66D		3011A	59		
2SD66E		3011A	59		
2SD66F		3011A	59		
2SD66G		3011A	59		
2SD66GN		3011A	59		
2SD66H		3011A	59		
2SD66J		3011A	59		
2SD66K		3011A	59		
2SD66L		3011A	59		
2SD66M		3011A	59		
2SD66OR		3011A	59		
2SD66R		3011A	59		
2SD66X		3011A	59		
2SD66Y		3011A	59		
2SD67	280	3438	35		2N5759
2SD670	249				2N6578
2SD672	162				2N5840
2SD673	87	3297/280			2N5759

Industry Standard No.	ECG	SK	GE	RS 276-	MOTOR.
2SD673A	87	3297/280			
2SD673AB	87	3297/280			
2SD673AC	87	3297/280			
2SD673C	87	3114/290			
2SD674	87	3297/280			2N5759
2SD674A	87	3297/280			
2SD674AB	87	3297/280			
2SD674AC	87	3297/280			
2SD675	284	3836			2N5760
2SD675A	284	3836			
2SD675AA	284	3836			
2SD675AB	284	3836			
2SD675AC	284	3836			
2SD676	284	3836			2N5760
2SD676A	284	3836			
2SD676AA	284	3836/281			
2SD676AB	284	3836			
2SD676AC	284	3836			
2SD677					2N6543
2SD678	261	3180/263			TIP110
2SD678A	261	3180/263			
2SD679	261	3180/263			TIP111
2SD679A	261	3180/263			
2SD67B	280	3438	35		
2SD67C	280	3438	35		
2SD67D	280	3438	35		
2SD67E	280	3438	35		
2SD68	130	3027	255	2041	2N5758
2SD686	261				TIP122
2SD687	261				MJE800T
2SD689	291				TIP112
2SD68A		3027	75	2041	
2SD68B		3027	75	2041	
2SD68C		3027	75	2041	
2SD68D	130	3027	75	2041	
2SD68E		3027	75	2041	
2SD68F		3027	75	2041	
2SD68G		3027	75	2041	
2SD68GN		3027	75	2041	
2SD68H		3027	75	2041	
2SD68J		3027	75	2041	
2SD68L		3027	75	2041	
2SD68M		3027	75	2041	
2SD680R		3027	75	2041	
2SD68R		3027	75	2041	
2SD68X		3027	75	2041	
2SD68Y		3027	75	2041	
2SD69	130	3111	14	2041	2N5760
2SD690	384				2N5428
2SD691	261				2N6301
2SD692	243	3182			2N6056
2SD693					MJ10012
2SD694					MJ10015
2SD695					MJ10015
2SD696					MJ10015
2SD70	175	3027/130	75	2020	2N3766
2SD702					MJ10015
2SD703					MJ10016
2SD704	377	3620			
2SD705					MJ10012
2SD706					MJ10013
2SD707					MJ10013
2SD708					MJ10013
2SD709					MJ3041
2SD70A		3027	75	2041	
2SD70B		3027	75	2041	
2SD70C		3027	75	2041	
2SD70D		3027	75	2041	
2SD70E		3027	75	2041	
2SD70F		3027	75	2041	
2SD70G		3027	75	2041	
2SD70GN		3027	75	2041	
2SD70H		3027	75	2041	
2SD70J		3027	75	2041	
2SD70K		3027	75	2041	
2SD70L		3027	75	2041	
2SD70M		3027	75	2041	
2SD700R			75	2041	
2SD70R		3027	75	2041	
2SD70X		3027	75	2041	
2SD70Y		3027	75	2041	
2SD71	175		59	2020	2N5050
2SD710					MJ10004
2SD712	379	3188A/182			
2SD712D	291	3188A/182			
2SD716	390				TIP41C
2SD717	390				D44H10
2SD718					MJE15028
2SD71L	175		59	2038	
2SD72	103A	3010	59	2002	
2SD72-2C	103A	3010	59	2002	
2SD72-3C	103A	3010	59	2002	
2SD72-4C	103A	3010	59	2002	
2SD72-6		3010	59	2002	
2SD72-0R			59	2002	
2SD720D			59	2002	
2SD720E			59	2002	
2SD720PJ		3010	59		
2SD721					2N6045
2SD722					2N6045
2SD723	291				TIP31C
2SD724	375$	3929			MJE13004
2SD725	165	3111	259		MJ12005
2SD725-06	238		259		
2SD726	291				TIP31C
2SD727	392$				2N4347
2SD728	390				2N5760
2SD729	251				2N6284
2SD72A	103A	3010	59	2002	
2SD72B	103A	3010	59	2002	
2SD72BR	103A		59	2002	
2SD72C	103A	3010	59	2002	
2SD72D		3010	59	2002	
2SD72E		3010	59	2002	
2SD72EJ		3010	59		
2SD72F		3010	59	2002	
2SD72G			59	2002	
2SD72GA		3010	59	2002	
2SD72H		3010	59	2002	
2SD72J		3010	59	2002	
2SD72K	103A	3010	59	2002	
2SD72L		3010	59	2002	
2SD72M		3010	59	2002	
2SD720R		3010	59		
2SD72P	103A		59	2002	
2SD72R		3010	59	2002	
2SD72RE	103A		59	2002	
2SD72X		3010	59	2002	
2SD72Y		3010	59	2002	
2SD73	87	3836/284	262	2041	2N5758
2SD731					2N6306
2SD732	87				2N6306
2SD733	280				MJ15001
2SD735	388$		8	2002	
2SD73A	87	3836/284	262	2041	
2SD73B	87	3836/284	262	2041	
2SD73C	87	3836/284	262	2041	
2SD73E	87	3836/284	262	2041	
2SD74	87	3836/284	262	2041	2N5760
2SD748	87	3559/162			2N5838
2SD748A	87	3559/162			
2SD749	283				2N6543
2SD74A	87	3836/284	262	2041	
2SD74B	87	3836/284	262	2041	
2SD74C	87	3836/284	262	2041	
2SD74D	87	3836/284	262	2041	
2SD74E	87	3836/284	262	2041	
2SD75	103A	3011A	59	2002	
2SD751					MJ423
2SD752					MJ15001
2SD753		3836			2N6249
2SD755	90	3866			
2SD756	90	3866			
2SD756A	90$	3866			
2SD756E	90	3866			
2SD757					MJE3440
2SD758		3929	251		MJE3440
2SD758C		3929	251		
2SD759	375				TIP47
2SD75A	103A	3011A	59	2002	
2SD75AH	103A	3835	59	2002	
2SD75B	103A	3011A	59	2002	
2SD75C	103A	3011A	59	2002	
2SD75D		3011A	59		
2SD75E		3011A	59		
2SD75F		3011A	59		
2SD75G		3011A	59		
2SD75GN		3011A	59		
2SD75H	103A	3011A	59	2002	
2SD75J		3011A	59		
2SD75K		3011A	59		
2SD75L		3011A	59		
2SD75M		3011A	59		
2SD750R		3011A	59		
2SD760					TIP47
2SD761	375$				TIP47
2SD762	152	3239/236			TIP31A
2SD764	389	3710/238	38		MJ12002
2SD765	389	3710/238	38		MJ12003
2SD766	369	3131A	246		2N3739
2SD766P	369	3131A			
2SD766Q	369	3131A	246		
2SD766R	369	3131A	246		
2SD768					2N6045
2SD77	103A	3011	59	2002	
2SD77-A		3011	59	2002	
2SD773K		3849		2030	
2SD77A	103A	3011	59	2002	
2SD77AH	103A		59	2002	
2SD77B	103A	3011	59	2002	
2SD77C	103A	3011	59	2002	
2SD77D	103A	3011	59	2002	
2SD77E		3011	59	2002	
2SD77F		3011	59	2002	
2SD77G		3011	59	2002	
2SD77GN		3011	59	2002	
2SD77H	103A	3011	59	2002	
2SD77J		3011	59	2002	
2SD77L		3011	59	2002	
2SD77M		3011	59	2002	
2SD770R		3011	59	2002	
2SD77P	103A	3011	59	2002	
2SD77R		3011	59	2002	
2SD77X		3011	59	2002	
2SD77Y		3011	59	2002	
2SD78	282	3748	264		
2SD789	315	3250			
2SD78A	282		264		
2SD78B	282	3748	264		
2SD78C	282		264		
2SD78D	282		264		
2SD79	282/427		256		
2SD792	165	3710/238			
2SD792S		3710	259		
2SD792T		3710	259		
2SD793					MJE180
2SD794		3190			MJE182
2SD797	387				MJE802
2SD79A	282/427		256		
2SD79B	282/427		256		
2SD79C	282/427		256		
2SD79D	282/427		256		
2SD80	87	3027/130	75	2041	2N5758
2SD800					2N5840
2SD801					2N6545
2SD802					2N6545
2SD803	247	3948			2N6059
2SD804	152	3893			
2SD804HP		3893		2048	
2SD805					MJ10016
2SD80A		3027	75	2041	
2SD80B		3027	75	2041	
2SD80C		3027	75	2041	
2SD80D		3027	75	2041	
2SD80E		3027	75	2041	
2SD80F		3027	75	2041	
2SD80G		3027	75	2041	
2SD80GN		3027	75	2041	
2SD80H		3027	75	2041	
2SD80J		3027	75	2041	
2SD80K		3027	75	2041	
2SD80L		3027	75	2041	
2SD800R			75	2041	
2SD80R		3027	75	2041	
2SD80X		3027	75	2041	
2SD80Y		3027	75	2041	
2SD81	87	3027/130	75	2041	2N5758
2SD811					MJ12010
2SD81A		3027	75	2041	
2SD81B		3027	75	2041	
2SD81C		3027	75	2041	
2SD81E		3027	75	2041	
2SD81F		3027	75	2041	
2SD81G		3027	75	2041	
2SD81GN		3027	75	2041	
2SD81H		3027	75	2041	
2SD81J		3027	75	2041	
2SD81K		3027	75	2041	
2SD81L		3027	75	2041	
2SD81M		3027	75	2041	
2SD810R		3027	75	2041	
2SD81R		3027	75	2041	
2SD81X		3027	75	2041	
2SD81Y		3027	75	2041	
2SD82	87	3027/130	75	2041	2N5758
2SD82-0R			75	2041	
2SD823	379				MJE15030
2SD829	107		11	2015	

Industry Standard No.	ECG	SK	GE	RS 276-	MOTOR.
2SD82A		3027	75	2041	
2SD82B		3027	75	2041	
2SD82C		3027	75	2041	
2SD82D		3027	75	2041	
2SD82E		3027	75	2041	
2SD82F		3027	75	2041	
2SD82G		3027	75	2041	
2SD82GN		3027	75	2041	
2SD82H		3027	75	2041	
2SD82J		3027	75	2041	
2SD82K		3027	75	2041	
2SD82L		3027	75	2041	
2SD82M		3027	75	2041	
2SD82OR		3027	75	2041	
2SD82R		3027	75	2041	
2SD82X		3027	75	2041	
2SD82Y		3027	75	2041	
2SD83	87	3438	35		2N5760
2SD836	261				TIP110
2SD837	261				TIP120
2SD839					MJE800T
2SD84	87	3438	35		MJ15011
2SD840					MJE802T
2SD843	331				MJE15028
2SD844	390				2N6290
2SD867	280				2N5633
2SD870	89	3710/238			
2SD872	379				2N6499
2SD873	388				2N3773
2SD876		3054	66	2048	
2SD877	175				2N3441
2SD878	328				2N3055
2SD88	87	3438	35		2N5758
2SD880	152	3440/291			TIP31A
2SD880-O	152	3440/291			
2SD882					MJE180
2SD88A	87	3438	35		
2SD90	152	3026	18	2048	2N3766
2SD903	89	3710/238			MJ12005
2SD90A		3026	18	2048	
2SD90B		3026	18	2048	
2SD90C		3026	18	2048	
2SD90D		3026	18	2048	
2SD90E		3026	18	2048	
2SD90F		3026	18	2048	
2SD90G			18	2048	
2SD90GN		3026	18	2048	
2SD90H			18	2048	
2SD90J		3026	18	2048	
2SD90K		3026	18	2048	
2SD90L		3026	18	2048	
2SD90M		3026	18	2048	
2SD90OR			18	2048	
2SD90X			18	2048	
2SD90Y		3026	18	2048	
2SD91	152	3026	66	2048	2N3766
2SD91A		3026	66	2048	
2SD91B		3026	66	2048	
2SD91C		3026	66	2048	
2SD91D		3026	66	2048	
2SD91E		3026	66	2048	
2SD91F	152	3026	66	2048	
2SD91G			66	2048	
2SD91GN		3026	66	2048	
2SD91H			66	2048	
2SD91J		3026	66	2048	
2SD91L		3026	66	2048	
2SD91M		3026	66	2048	
2SD91OR		3026	66	2048	
2SD91R		3026	66	2048	
2SD91X			66	2048	
2SD91Y		3026	66	2048	
2SD92	291	3562	246	2020	2N3583
2SD92D	291	3562	246	2020	
2SD93		3268	72		2N5051
2SD94	375		12		2N5052
2SD950	389				MJ12004
2SD951	389				MJ12004
2SD952	389				MJ12004
2SD953	165				MJ12005
2SD96	103A	3010	59	2002	
2SD96A		3010	59		
2SD96B		3010	59		
2SD96C		3010	59		
2SD96D		3010	59		
2SD96E		3010	59		
2SD96F		3010	59		
2SD96G		3010	59		
2SD96GN		3010	59		
2SD96H		3010	59		
2SD96J		3010	59		
2SD96K		3010	59		
2SD96L		3010	59		
2SD96OR		3010	59		
2SD96R		3010	59		
2SD96X		3010	59		
2SD96Y		3010	59		
2SDF1	101	3861	59	2002	
2SDF1A		3011	59	2002	
2SDF2A		3027	75	2041	
2SDF2B		3027	75	2041	
2SDF2C		3027	75	2041	
2SDF2D		3027	75	2041	
2SDF2E		3027	75	2041	
2SDF2F		3027	75	2041	
2SDF2G		3027	75	2041	
2SDF2GN		3027	75	2041	
2SDF2H		3027	75	2041	
2SDF2J		3027	75	2041	
2SDF2K		3027	75	2041	
2SDF2L		3027	75	2041	
2SDF2M		3027	75	2041	
2SDF2OR		3027	75	2041	
2SDF2R		3027	75	2041	
2SDF2X		3027	75	2041	
2SDF2Y		3027	75	2041	
2SDL2F		3010	59		
2SDU36C		3021	12		
2SDU37E		3021	12		
2SDU47X		3027	75	2041	
2SDU57M		3021	12		
2SDU78OR		3124	47	2030	
2SDU84GN		3024	18	2030	
2SDU9OR		3010	59		
2SE4002	199	3011	62	2010	
2SE4002-1		3011	59		
2SE629	108	3452	86	2038	
2SF.T212			16	2006	
2SF100	5483	3942			
2SF101	5401	3638/5402		1067	
2SF102	5402	3638		1067	
2SF102A	5402	3638		1067	
2SF103	5403	3627/5404			
2SF104	5404	3627			
2SF105	5405	3951			
2SF106	5405	3951		1020	
2SF107	5405	3951			
2SF108	5405	3951			
2SF1089	5483	3942			
2SF1090	5484	3943/5485			
2SF1091	5485	3943			
2SF1092	5486	3944/5487			
2SF1099	5454	6754			
2SF110	5406	3952			
2SF1100	5455	3597			
2SF1110	5401	3638/5402			
2SF1111(GAK)	5402	3638			
2SF1112	5404	3627			
2SF1113(GAK)	5405	3951			
2SF1114(GAK)	5405	3951			
2SF1169	5402	3638			
2SF1170	5404	3627			
2SF1171	5405	3951			
2SF1172	5405	3951			
2SF1177	5402	3638			
2SF1178	5404	3627			
2SF1180	5531	6631			
2SF1188	230	3042	700		
2SF1188A	230	3042	700		
2SF1188B	230	3042	700		
2SF1188C	230	3042	700		
2SF1188D	230	3042	700		
2SF1188E	230	3042	700		
2SF1188F	230	3042	700		
2SF1188G	230	3042	700		
2SF1188H	230	3042	700		
2SF1188K	230	3042	700		
2SF1188L	230	3042	700		
2SF1188M	230	3042	700		
2SF1188N	230	3042	700		
2SF1188R	230	3042	700		
2SF1188Y	230	3042	700		
2SF1189	231	3857	700		
2SF1189A	231	3857	700		
2SF1189C	231	3857	700		
2SF1189D	231	3857	700		
2SF1189E	231	3857	700		
2SF1189F	231	3857	700		
2SF1189G	231	3857	700		
2SF1189H	231	3857	700		
2SF1189K	231	3857	700		
2SF1189L	231	3857	700		
2SF1189M	231	3857	700		
2SF1189N		3857	700		
2SF1189R	231	3857	700		
2SF1189Y	231	3857	700		
2SF1271	5400	3950			
2SF131	5540	6615/5515			
2SF132	5541	6642/5542			
2SF133	5542	6642			
2SF135	5543		2N685		
2SF138	5546		2N689		
2SF139	5528	6629/5529	2N689		
2SF14	5483	3942			
2SF140	5500	6642/5542			
2SF1422	230	3042			
2SF149	5529	6629			
2SF16	5484	3943/5485			
2SF17	5485	3943			
2SF18	5485	3943			
2SF19	5486	3944/5487			
2SF200	5486	3944/5487			
2SF224	5483	3942			
2SF226	5484	3943/5485			
2SF227	5485	3943			
2SF228	5486	3944/5487			
2SF229	5403	3627/5404			
2SF230	5520	6621/5521			
2SF231	5521	6621			
2SF232	5522	6622			
2SF233	5523	6624/5524			
2SF234	5524	6624			
2SF235	5525	6627/5527			
2SF236	5526	6627/5527			
2SF238	5540	6615/5515			
2SF239	5541	6642/5542			
2SF248	314	3898			
2SF248A	314	3898			
2SF248T				2048	
2SF253	5483	3942			
2SF254	5483	3942			
2SF255	5484	3943/5485			
2SF256	5484	3943/5485			
2SF257	5485	3943			
2SF258	5485	3943			
2SF259	5486	3944/5487			
2SF263	5483	3942			
2SF264	5483	3942			
2SF265	5484	3943/5485			
2SF266	5484	3943/5485			
2SF267	5485	3943			
2SF268	5485	3943			
2SF269	5486	3944/5487			
2SF280	5520	6621/5521			
2SF281	5521	6621			
2SF282	5522	6622			
2SF283	5523	6624/5524			
2SF284	5524	6624			
2SF285	5525	6627/5527			
2SF286	5526	6627/5527			
2SF287	5527	6627			
2SF288	5527	6627			
2SF289	5528	6629/5529			
2SF290	5540	6642/5542			
2SF291	5541	6642/5542			
2SF292	5542	6642			
2SF31	5541	6642/5542			
2SF31A	5541	6642/5542			
2SF32	5542	6642			
2SF324A	5483	3942			
2SF325A	5483	3942			
2SF326A	5484	3943/5485			
2SF327A	5484	3943/5485			
2SF329A	5486	3944/5487			
2SF32A	5542	6642			
2SF330A	5487	3943/5485			
2SF357				1067	
2SF380	5491	6791			
2SF381	5491	6791			
2SF382	5492	6792			
2SF382A	5492	6792			
2SF383	5494	6794			
2SF383A	5494	3794			
2SF384	5494	3794			
2SF384A	5494	6794			
2SF385	5496	6796			
2SF385A	5496	6796			
2SF386	5496	6796			

Industry Standard No.	ECG	SK	GE	RS 276-	MOTOR.
2SF40	5540	6642/5542			
2SF40A	5540	6642/5542			
2SF41F			FET-2	2035	
2SF431	5528	6629/5529			
2SF432	5529	6629			
2SF434	5531	6631			
2SF439	5531	6631			
2SF521	5401	3638/5402			
2SF522	5402	3638		1067	
2SF523	5404	3627			
2SF524	5405	3951			
2SF525	5405	3951			
2SF526	5406	3952			
2SF527	5406	3952			
2SF546	5483	3942			
2SF547	5486	3944/5487			
2SF55	5520	6621/5521			
2SF56	5521	6621			
2SF567	5492	6792			
2SF57	5522	6622			
2SF572	5404	3627			
2SF58	5523	6624/5524			
2SF59	5524	6624			
2SF60	5525	6627/5527			
2SF609	5514	3613			
2SF61	5526	6627/5527			
2SF610	5515	6615			
2SF611	5515	6615			
2SF62	5527	6627			
2SF623	5517	3615			
2SF624	5518	3653			
2SF625	5518	3653			
2SF656	5402	3638			
2SF657	5402	3638			
2SF658	5404	3627			
2SF659(GAK)	5405	3951			
2SF664S	5402	3638			
2SF667	5483	3942			
2SF668	5485	3943			
2SF674S	5402	3638			
2SF676S	5402	3638			
2SF71	5520	6621/5521			
2SF72	5521	6621			
2SF73	5522	6622			
2SF74	5523	6624/5524			
2SF75	5524	6624			
2SF753	5526	6627/5527			
2SF754	5527	6627			
2SF755	5528	6629/5529			
2SF756	5529	6629			
2SF76	5526	6627/5527			
2SF77	5527	6627	C230D		
2SF8	282	3748			
2SF84	5541	6642/5542			
2SF84A	5541	6642/5542			
2SF85	5542	6642			
2SF85A	5542	6642			
2SF941	5454	6754		1067	
2SF942	5455	3597		1067	
2SF944(GAK)	5402	3638			
2SF945(GAK)	5404	3627			
2SF946(KAG)	5402	3638			
2SF947(KAG)	5404	3627			
2SF948(KAG)	5405	3951			
2SF949(KAG)	5405	3951			
2SF98	5523	6624/5524			
2SF99	5525	6627/5527			
2SFT212	121	3009	239	2006	
2SG536GN	123A	3124/289	20	2051	
2SH11	6401			2029	
2SH12	6400A			2029	
2SH18	6401			2029	
2SH18K	6401			2029	
2SH18L	6401			2029	
2SH18M	6401			2029	
2SH18N	6401			2029	
2SH19	6401			2029	
2SH19K	6401			2029	
2SH19L	6401			2029	
2SH19M	6401			2029	
2SH19N	6401			2029	
2SH20	6401			2029	
2SH203	100	3005	1	2007	
2SH21	6400A		2N2160	2029	
2SH22	6409			2029	
2SH643	193		67	2023	
2SH678			21	2034	
2SJ11	312	3112	FET-1	2028	
2SJ12	312	3112	FET-1	2028	
2SJ2A	116	3017B/117	504A	1104	
2SJ4A	116	3017B/117	504A	1104	
2SJ60A	116	3017B/117	504A	1104	
2SJ8A	125	3033	510,531	1114	
2SK1033B	312	3116	FET-2	2035	
2SK104	312	3834/132	FET-2	2035	
2SK104H	312	3834/132	FET-2	2035	
2SK106			FET-1	2028	
2SK11		3112	FET-1	2035	
2SK11-0			FET-1	2035	
2SK11-R		3112	FET-1	2035	
2SK11-Y		3112	FET-1	2035	
2SK12		3112	FET-1	2035	
2SK12-GR			FET-1	2035	
2SK12-0			FET-1	2035	
2SK12-R		3112	FET-1	2035	
2SK12-Y		3112	FET-1	2035	
2SK13	312	3448	FET-1	2035	2N5457
2SK15-GR			FET-1	2035	
2SK15-0			FET-1	2035	
2SK15-R		3112	FET-1	2035	
2SK15-Y		3112	FET-1	2035	
2SK16	312		277		
2SK16H	312		277		
2SK16HB	312		277		
2SK16HC	312		277		
2SK17	312	3448	FET-1	2035	
2SK17(0)			FET-1	2028	
2SK17-0	312	3448	FET-1	2035	
2SK17-OR		3112	FET-1	2028	
2SK170		3448	FET-1	2035	
2SK170R		3448	FET-1	2035	
2SK17A	312	3448	FET-1	2035	
2SK17B	312	3448	FET-1	2035	
2SK17BL	312	3448	FET-1	2035	
2SK17GR	312	3977/456	FET-1	2035	2N5457
2SK170					2N5457
2SK17R	312	3448	FET-1	2035	2N5457
2SK17Y	312	3448	FET-1	2035	2N5484
2SK19	459	3834/132	FET-2	2035	
2SK19(BL)			FET-2	2035	
2SK19(GR)	459	3834/132	FET-2	2035	
2SK19-14		3834	FET-2	2035	
2SK19-BL			FET-2	2035	
2SK19-GR	459		FET-2	2035	

Industry Standard No.	ECG	SK	GE	RS 276-	MOTOR.
2SK19-Y		3834	FET-2	2035	
2SK19A		3834	FET-2	2035	
2SK19B		3834	FET-2	2035	
2SK19BB		3834	FET-2	2035	
2SK19BL	459	3834/132	FET-2	2035	2N5484
2SK19FET			FET-2	2035	
2SK19G					2N5484
2SK19GB	459		FET-2	2035	
2SK19GC	459	3834/132	FET-2	2035	
2SK19GE	459	3834/132	FET-2	2035	
2SK19GR	459	3834/132	FET-2	2035	
2SK19H	459	3834/132	FET-2	2035	
2SK19K	459		FET-2	2035	
2SK19V	459		FET-2	2035	
2SK19Y	459	3834/132	FET-2	2035	2N5486
2SK22Y	312	3116	FET-1	2035	
2SK23	312	3448	FET-2	2035	2N5486
2SK23A	312	3448	FET-1	2035	
2SK23A540	312		FET-2	2035	
2SK24	312	3112	FET-2	2028	
2SK24C	312			2028	
2SK24D	312			2028	
2SK24DR	312	3112	FET-1	2028	
2SK24E	312	3112	FET-1	2028	
2SK24F	312			2028	
2SK24G	312			2028	
2SK25	312	3448	FET-1	2035	
2SK25C	312	3448	FET-1	2035	
2SK25D	312	3448	FET-1	2035	
2SK25E	312	3448	FET-1	2035	
2SK25ET	312	3448	FET-1	2035	
2SK25F	312	3448	FET-1	2035	
2SK25G	312	3448	FET-1	2035	
2SK2SET			FET-1	2028	
2SK30		3112	FET-1	2035	
2SK30(0)			FET-1	2035	
2SK30-0		3112	FET-1	2035	
2SK30-0		3112	FET-1	2035	
2SK30-R		3112	FET-1	2035	
2SK30-Y		3112	FET-1	2035	
2SK304	312	3448	FET-1	2035	
2SK30A		3112	FET-1	2035	
2SK30A(D)			FET-1	2035	
2SK30A-Y		3112	FET-1	2035	
2SK30AD		3112	FET-1	2035	
2SK30AGR		3112	FET-1	2035	
2SK30AO			FET-1	2035	
2SK30AY		3112	FET-1	2035	
2SK30D		3112	FET-1	2035	
2SK30GR		3112	FET-1	2035	
2SK30R			FET-1	2035	
2SK30Y		3112	FET-1	2035	
2SK31	312	3448	FET-1	2035	
2SK31(C)			FET-2	2035	
2SK31C	312	3448	FET-1	2035	
2SK32					2N5486
2SK32B	312	3834/132	FET-1	2035	
2SK33	312	3834/132	FET-2	2035	
2SK33(E)	312	3834/132	FET-2	2035	
2SK33D	312		FET-2	2035	
2SK33E	312	3834/132	FET-2	2035	
2SK33F	312	3834/132	FET-2	2035	
2SK33GR	312	3834/132	FET-2	2035	
2SK33H	312	3834/132	FET-2	2035	
2SK34	312	3448	FET-2	2035	
2SK34(E)	312	3448	FET-1		
2SK34A		3448	FET-1	2035	
2SK34B	312	3448	FET-2	2035	
2SK34C	312	3448	FET-2	2035	
2SK34D	312	3448	FET-2	2035	
2SK34E	312	3448	FET-2	2035	
2SK34GR	312	3448			
2SK35		3112	FET-1	2028	2N5485
2SK35-0		3112	FET-1	2028	
2SK35-1		3112	FET-1	2028	
2SK35-2		3112	FET-1	2028	
2SK35A		3112	FET-1	2028	
2SK35BL		3112	FET-1	2028	
2SK35C		3112	FET-1	2028	
2SK35GN		3112	FET-1	2028	
2SK35R			FET-1	2028	
2SK35Y		3112	FET-1	2028	
2SK37	312	3448	FET-1	2035	
2SK37(K)			FET-2	2035	
2SK37H	312	3448	FET-1	2035	
2SK37K	312	3448	FET-1	2035	
2SK37L	312	3448	FET-1	2035	
2SK39		3834	FET-4	2036	
2SK39B			FET-4	2036	
2SK39P			FET-4	2036	
2SK39Q			FET-1	2028	
2SK40	459	3112	FET-2	2028	
2SK40(C)			FET-1	2028	
2SK40-3	459			2028	
2SK40A	459	3112		2028	
2SK40B	459	3112		2028	
2SK40C	459	3112	FET-1	2028	
2SK40D	459	3112		2028	
2SK41	451	3834/132	FET-2	2035	
2SK41C	451	3834/132			
2SK41D	451	3834/132	FET-2		
2SK41E	451	3834/132	FET-2	2035	
2SK41E1		3834	FET-2		
2SK41E2	451		FET-2	2035	
2SK41F	451	3834/132	FET-2	2035	
2SK42	312	3834/132	FET-1	2035	
2SK42-CM1	312	3834/132	FET-1	2035	
2SK42-CMI			FET-2	2035	
2SK42CM1	312	3834/132	FET-1	2035	
2SK43	312	3112	FET-1	2028	
2SK43-OR			FET-1	2028	
2SK43A		3112	FET-1	2028	
2SK43B		3112	FET-1	2028	
2SK43C		3112	FET-1	2028	
2SK43D		3112	FET-1	2028	
2SK43E		3112	FET-1	2028	
2SK43F		3112	FET-1	2028	
2SK43GN		3112	FET-1	2028	
2SK43H		3112	FET-1	2028	
2SK43J		3112	FET-1	2028	
2SK43K		3112	FET-1	2028	
2SK43L		3112	FET-1	2028	
2SK43M		3112	FET-1	2028	
2SK43OR		3112	FET-1	2028	
2SK43R		3112	FET-1	2028	
2SK43X		3112	FET-1	2028	
2SK43Y		3112	FET-1	2028	
2SK44	457		FET-1	2028	
2SK44(D)			FET-1	2028	
2SK44C	457	3112	FET-1	2028	
2SK44D	457	3112	FET-1	2028	
2SK45B		3990		2036	
2SK47	457	3834/132	FET-1	2035	
2SK47M	457	3834/132	FET-1	2035	

Industry Standard No.	ECG	SK	GE	RS 276-	MOTOR.
2SK49	312	3834/132	FET-2	2035	
2SK491			FET-1	2028	
2SK49E	312	3834/132			
2SK49E2	312		FET-2	2035	
2SK49F	312	3834/132	FET-2	2035	
2SK49H	312	3834/132	FET-2	2035	
2SK49H1	312	3834/132	FET-2	2035	
2SK49H2	312		FET-2	2035	
2SK49HK	312	3834/132	FET-2	2035	
2SK49I	312	3834/132	FET-2	2035	
2SK49M	312	3834/132	FET-2	2035	
2SK54	312	3116	FET-2	2035	
2SK54J	312	3834/132			
2SK55	312	3834/132	FET-2	2035	
2SK55C	312	3834/132	FET-2		
2SK55D	312	3834/132	FET-2	2035	
2SK55DE	312	3834/132	FET-2	2035	
2SK55E	312	3834/132	FET-2	2035	
2SK55R	312	3834/132	FET-2	2035	
2SK61	451	3834/132	FET-2	2035	
2SK61GR	451		FET-2	2035	
2SK61Y	451		FET-2	2035	
2SK68		3448	FET-1	2028	
2SK68-L		3448	FET-1	2028	
2SK68A		3448	FET-1	2028	
2SK68AL		3448	FET-1	2028	
2SK68AM		3448	FET-1	2028	
2SK68L		3448	FET-1	2028	
2SK68M		3448	FET-1	2028	
2SK68Q		3448	FET-1	2028	
2SK68Y		3448	FET-1	2028	
2SK83	451	3116	FET-2	2035	
2SK84	459	3116	FET-2	2035	
2SM100	5685	3521/5686			
2SM106	5685	3521/5686			
2SM116	5685	3521/5686			
2SM117	5686	3521			
2SM610B	159	3025/129	82	2032	
2S012			66	2048	
2S033			19	2041	
2S034			19	2041	
2S035			19	2041	
2S036			19	2041	
2SQ371	108	3452	86	2038	
2SR1K	116	3311	504A	1104	
2SR24			FET-1	2028	
2SR677			67	2023	
2SR68AM			FET-2	2035	
2SV341V	127	3035	25		
2T11	158	3004/102A	53	2007	
2T12	158	3004/102A	53	2007	
2T13	158	3004/102A	53	2007	
2T14	158	3004/102A	53	2007	
2T14A	126		52	2024	
2T15	158	3004/102A	53	2007	
2T15X3			11	2015	
2T16	158	3004/102A	53	2007	
2T17	158	3004/102A	53	2007	
2T172	123A	3444	20	2051	
2T20	160	3004/102A	245	2004	
2T2001	102	3004	2	2007	
2T201	160	3006	245	2004	
2T202	123A	3444	20	2051	
2T203	160	3006	245	2004	
2T204	160	3007	245	2004	
2T204A	160	3007	245	2004	
2T205	160	3007	245	2004	
2T205A	160	3007	245	2004	
2T21	158	3004/102A	53	2007	
2T2102			63	2030	
2T22	158	3004/102A	53	2007	
2T23	158	3004/102A	53	2007	
2T230	102	3004	2	2007	
2T231	102	3004	2	2007	
2T24	158	3004/102A	53	2007	
2T25	158		53	2007	
2T26	158	3004/102A	53	2007	
2T2708	123A	3444	20	2051	
2T2785	123A	3444	20	2051	
2T2857	123A	3444	20	2051	
2T2P			2	2007	
2T3	102A	3004	53	2007	
2T3011	121	3009	239	2006	
2T3021	121	3009	239	2006	
2T3022	121	3009	239	2006	
2T3030	121	3009	239	2006	
2T3031	121	3009	239	2006	
2T3032	121	3009	239	2006	
2T3033	121	3009	239	2006	
2T3041	121	3009	239	2006	
2T3042	121	3009	239	2006	
2T3043	121	3009	239	2006	
2T311	158	3004/102A	53	2007	
2T312	158	3004/102A	53	2007	
2T313	158	3004/102A	53	2007	
2T314	158	3004/102A	53	2007	
2T315	158	3004/102A	53	2007	
2T321	158	3123	53	2005	
2T322	158	3004/102A	53	2005	
2T323	158	3004/102A	53	2007	
2T324	158	3004/102A	53	2007	
2T383	158	3123	53	2007	
2T40	123A	3444	20	2051	
2T402	123	3124/289	20	2051	
2T403	123A	3444	20	2051	
2T404	123A	3444	20	2051	
2T41	123A	3444	20	2051	
2T42	123A	3444	20	2051	
2T43	123A	3444	20	2051	
2T44	123A	3444	20	2051	
2T501	116	3017B/117	504A	1104	
2T502	116	3016	504A	1104	
2T503	116	3016	504A	1104	
2T504	116	3016	504A	1104	
2T505	116	3017B/117	504A	1104	
2T506	116	3017B/117	504A	1104	
2T507	125	3032A	510,531	1114	
2T508	125	3032A	510,531	1114	
2T509	125	3033	510,531	1114	
2T51	103	3862	8	2002	
2T510	125	3033	510,531	1114	
2T513	101	3861	8	2002	
2T52	101	3861	8	2002	
2T520	101	3861	8	2002	
2T521	101	3861	8	2002	
2T522	103	3862	8	2002	
2T523	103	3862	8	2002	
2T524	101	3861	8	2002	
2T53	101	3861	8	2002	
2T54	101	3011	8	2002	
2T55	101	3011	8	2002	
2T551	101	3011	8	2002	
2T552	103	3862	8	2002	
2T56	101	3011	8	2002	
2T57	101	3011	8	2002	
2T58	101	3011	8	2002	
2T61	103	3862	8	2002	
2T62	103	3862	8	2002	
2T63	103	3862	8	2002	
2T64	103	3862	8	2002	
2T64R	103	3862	8	2002	
2T65	103	3862	8	2002	
2T650	101	3861	8	2002	
2T65R	103	3862	8	2002	
2T66	103	3862	8	2002	
2T66R	103	3862	8	2002	
2T67	101	3011	8	2002	
2T681	103	3862	8	2002	
2T682	103	3862	8	2002	
2T69	103	3862	8	2002	
2T71	101	3861	8	2002	
2T72	101	3861	8	2002	
2T73	101	3861	8	2002	
2T73R	101	3861	8	2002	
2T74	101	3861	8	2002	
2T75	101	3861	8	2002	
2T75R	101	3861	8	2002	
2T76	101	3861	8	2002	
2T76R	101	3861	8	2002	
2T77	101	3861	8	2002	
2T77R	101	3861	8	2002	
2T78	101	3011	8	2002	
2T78R	101	3011	8	2002	
2T82	101	3861	8	2002	
2T83	101		8	2002	
2T84	103	3862	8	2002	
2T85	103	3862	8	2002	
2T85A	103	3862	8	2002	
2T86	103	3862	8	2002	
2T89	103	3862	8	2002	
2T918	123A	3444	20	2051	
2T919	107	3039/316	11	2015	
2TN15	100	3005	1	2007	
2TN32	100	3005	1	2007	
2TN45A	102	3004	2	2007	
2TN48	100	3005	1	2007	
2TN49	100	3005	1	2007	
2TN52	100	3005	1	2007	
2TN53	100	3005	1	2007	
2TN56	102	3004	2	2007	
2TN95	102	3004	2	2007	
2TN95A	102	3004	2	2007	
2V362	102	3004	2	2007	
2V363	102	3004	2	2007	
2V464	100	3005	1	2007	
2V465	100	3005	1	2007	
2V466	100	3005	1	2007	
2V467	100	3005	1	2007	
2V482	100	3005	1	2007	
2V483	100	3005	1	2007	
2V484	100	3005	1	2007	
2V485	160		245	2004	
2V486	100	3005	1	2007	
2V559	160		245	2004	
2V560	160		245	2004	
2V561	160		245	2004	
2V562	160		245	2004	
2V563	160		245	2004	
2V631	100	3005	1	2007	
2V632	100	3005	1	2007	
2V633	100	3005	1	2007	
2VR10	5125A	3391			
2VR100	5156A	3422			
2VR100A	5156A	3422			
2VR100B	5156A	3422			
2VR10A	5125A	3391			
2VR10B	5125A	3391			
2VR11	5126A	3392			
2VR110	5157A	3423			
2VR110A	5157A	3423			
2VR110B	5157A	3423			
2VR11A	5126A	3392	5ZD-11		
2VR11B	5126A	3392			
2VR12	5127A	3393			
2VR120	5158A	3424			
2VR120A	5158A	3424			
2VR120B	5158A	3424			
2VR12A	5127A	3393	5ZD-12		
2VR12B	5127A	3393			
2VR13	5128A	3394			
2VR130	5159A	3425			
2VR130A	5159A	3425			
2VR130B	5159A	3425			
2VR13A	5128A	3394	5ZD-13		
2VR13B	5128A	3394			
2VR14	5129A	3395			
2VR14A	5129A	3395	5ZD-14		
2VR14B	5129A	3395			
2VR15	5130A	3396			
2VR150	5161A	3427			
2VR150A	5161A	3427			
2VR150B	5161A	3427			
2VR15A	5130A	3396			
2VR15B	5130A	3396			
2VR16	5131A	3397			
2VR160	5162A	3428			
2VR160A	5162A	3428			
2VR160B	5162A	3428			
2VR16A	5131A	3397			
2VR16B	5131A	3397			
2VR18	5133A	3399			
2VR180	5164A	3430			
2VR180A	5164A	3430			
2VR180B	5164A	3430			
2VR18A	5133A	3399			
2VR18B	5133A	3399			
2VR20	5135A	3401			
2VR200	5166A	3432			1M200Z810
2VR200A	5166A	3432			
2VR200B	5166A	3432			
2VR20A	5135A	3401			
2VR20B	5135A	3401			
2VR22	5136A	3402			
2VR22A	5136A	3402			
2VR22B	5136A	3402			
2VR24	5137A	3403			
2VR24A	5137A	3403			
2VR24B	5137A	3403			
2VR27	5139A	3405			
2VR27A	5139A	3405			
2VR27B	5139A	3405			
2VR28	5140A	3406	5ZD-28		
2VR28A	5140A	3406	5ZD-28		
2VR28B	5140A	3406	5ZD-28		
2VR30	5141A	3407			
2VR30A	5141A	3407			
2VR30B	5141A	3407			
2VR33	5142A	3408			

Industry Standard No.	ECG	SK	GE	RS 276-	MOTOR.
2VR33A	5142A	3408			
2VR33B	5142A	3408			
2VR36	5143A	3409			
2VR36A	5143A	3409			
2VR36B	5143A	3409			
2VR39	5144A	3410			
2VR39A	5144A	3410			
2VR39B	5144A	3410			
2VR43	5145A	3411			
2VR43A	5145A	3411			
2VR43B	5145A	3411			
2VR47	5146A	3412			
2VR47A	5146A	3412			
2VR47B	5146A	3412			
2VR5.6	5117A	3383			
2VR5.6A	5117A	3383			
2VR5.6B	5117A	3383			
2VR51	5147A	3413			
2VR51A	5147A	3413			
2VR51B	5147A	3413			
2VR56	5148A	3414			
2VR56A	5148A	3414			
2VR56B	5148A	3414			
2VR6	5118A	3384			
2VR6.2	5118A	3385/5119A			1M6.2ZS10
2VR6.2A	5118A	3385/5119A			
2VR6.2B	5119A	3385			
2VR6.8	5120A	3386			
2VR6.8A	5120A	3386			
2VR6.8B	5120A	3386			
2VR62	5150A	3416			
2VR62A	5150A	3416			
2VR62B	5150A	3416			
2VR67	5151A	3417			
2VR67A	5151A	3417			
2VR67B	5151A	3417			
2VR68	5151A	3417			
2VR68A	5151A	3417			
2VR68B	5151A	3417			
2VR6A	5118A	3384			
2VR6B	5118A	3384			
2VR7.5	5121A	3387			
2VR7.5A	5121A	3387			
2VR7.5B	5121A	3387			
2VR75	5152A	3418			
2VR75B	5152A	3418			
2VR8.2	5122A	3388			
2VR8.2A	5122A	3388			
2VR8.2B	5122A	3388	5ZD-8.2		
2VR8.5	5123A	3389			
2VR8.5A	5123A	3389			
2VR8.5B	5123A	3389	5ZD-8.7		
2VR80	5153A	3419			
2VR80A	5153A	3419			
2VR80B	5153A	3419			
2VR9.1	5124A	3390			
2VR9.1A	5124A	3390			
2VR9.1B	5124A	3390			
2VR90	5155A	3421			
2VR90A	5155A	3421			
2VR90B	5155A	3421			
2VR91	5155A	3421			
2VR91A	5155A	3421			
2VR91B	5155A	3421			
2W005					MDA200
2W02					MDA202
2W04					MDA204
2W06					MDA206
2W08					MDA208
2W10					MDA210
2W3A	116	3017B/117	504A	1104	
2W4A	116	3016	504A	1104	
2W5A	116	3017B/117	504A	1104	
2W6A	125	3032A	510,531	1114	
2W7A	125	3032A	510,531	1114	
2W9A	125	3033	510,531	1114	
2WMT1	5944		5048		
2WMT10	6002	7202	5108		
2WMT6	5952	3501/5994			
2WMT8	5998		5108		
2X9A116	116	3016	504A	1104	
2XAA111	109			1123	
2XAA112	109	3087		1123	
2XAA113	109	3087		1123	
02Z-10A	140A	3061	ZD-10	562	
02Z-6.2A	137A			561	
02Z-7.5A		3059	ZD-7.5		
02Z-8.2A		3136	ZD-8.2	562	
02Z-9.1A	139A	3060	ZD-9.1	562	
02Z102A	139A	3060	ZD-9.1	562	
02Z10A	5019A	3061/140A	ZD-10	562	
02Z10A-U	140A		ZD-10	562	
02Z11A	5074A	3786/5020A	ZD-11	563	
02Z12A	142A	3062	ZD-12	563	
02Z12GR	5021A	3062/142A	ZD-12	563	
02Z13A	143A	3750	ZD-13	563	
02Z15A	145A	3063	ZD-15	564	
02Z16A	5075A	3751	ZD-16	564	
02Z18A	5077A	3752	ZD-18		
02Z24A	5081A	3797/5031A	ZD-24		
02Z5.6	136A	3057		561	
02Z5.6A	136A	3057	ZD-5.6	561	
02Z6.2	137A	3058	ZD-6.2	561	
02Z6.2A	5013A	3779	ZD-6.2	561	
02Z6.2W	137A	3779/5013A		561	
02Z6.8A	5071A	3058/137A	ZD-6.8	561	
02Z62A	137A	3058	ZD-6.2	561	
02Z7.5A	138A	3059	ZD-7.5		
02Z8.2A	5072A	3136	ZD-8.2	562	
02Z9.1	5018A	3060/139A			
02Z9.1A	139A	3060	ZD-9.1	562	
003	110MP	3709			
003(MCINTOSH)	724		IC-86		
003-00	229	3246	61		
003-001	116	3017B/117	504A	1104	
03-0018-0	116	3031A	504A	1104	
03-0018-6	124	3031A	12		
03-0020-0	102	3004	2	2007	
03-0021-0	109	3087		1123	
03-0022	102	3004	2	2007	
003-002200	113A	3119/113	6GC1		
03-0023-0	102	3004	2	2007	
3-0033	123A	3444	20	2051	
003-004200	110MP	3709	1N34AS	1123	
003-005400	109	3087		1123	
03-0063-04	177		300	1122	
003-006700	109	3091		1123	
003-007500	110MP	3091	1N60	1123	
003-009000	109	3090		1123	
003-009100	145A	3063	ZD-15	564	
003-009200	142A	3062	ZD-12	563	
003-009400	116	3017B/117	504A	1104	
003-009600		3091		1123	
003-009700	142A	3062	ZD-12	563	

Industry Standard No.	ECG	SK	GE	RS 276-	MOTOR.
003-009900	116		504A	1104	
003-01	229	3018	61	2038	
003-010000	142A	3062	ZD-12	563	
03-034-042	519		514	1122	
3-041	129		244	2027	
3-1477	5804	9007		1144	
03-156B	102A	3003	53	2007	
03-1585/G	123A	3444	20	2051	
03-160	109	3088		1123	
3-19	124	3021	12		
3-20	127	3764	25		
3-215	129		244	2027	
3-233	197		250	2027	
03-3016	116	3311	504A	1104	
3-30173	181		75	2041	
03-460C	108	3452	86	2038	
03-461B	108	3452	86	2038	
03-535A	108	3452	86	2038	
03-57-001	126	3008	52	2024	
03-57-002	126	3007	52	2024	
03-57-003	160		245	2004	
03-57-101	126	3008	52	2024	
03-57-102	160		245	2004	
03-57-200	160	3006	245	2004	
03-57-201	126	3008	52	2024	
03-57-202	160		245	2004	
03-57-301	102	3003	2	2007	
03-57-302	102A	3004	53	2007	
03-57-304	102	3004	2	2007	
03-57-501	121	3009	239	2006	
3-7	129		244	2027	
03-931051	109	3087	1N34AS	1123	
03-931601	116	3017B/117	504A	1104	
03-931609	116	3016	504A	1104	
03-931641	177	3100/519	300	1122	
03-931642	177	3100/519	300	1122	
03-931645	177	3100/519	300	1122	
03-931771	109	3087	1N34AS	1123	
03-931971	116	3016	504A	1104	
03-933935	5086A	3338	ZD-39		
03-933943	142A	3062	ZD-12	563	
03-936011	506	3130	511	1114	
003-H03	102	3004	2		
003-H03				2007	
3.58MC	358		41		
3/4Z12D10		3062	ZD-12	563	
3/4Z12D5		3062	ZD-12	563	
3A-200			510	1114	
03A02	293	3124/289	47	2030	
03A03	123A	3018	11	2015	
03A04	290A	3118	21	2034	
03A05	123A	3444	20	2051	MR501
03A06	294	3466/159	82	2032	
03A09	312	3448	FET-2	2035	
3A1	156		512		MR501
03A10	222	3065	FET-4	2036	
3A100	156		512		MR501
3A1000					MR510
03A11		3444	20	2051	
03A12	123AP	3444/123A	20	2051	
3A15	5800	9003		1142	MR501
3A1510	125	3033	510,531	1114	
3A152	116	3016	504A	1104	
3A154	116	3017B/117	504A	1104	
3A156	116	3017B/117	504A	1104	
3A158	125	3033	510,531	1114	
3A2					MR502
3A200	116	3016	504A	1104	MR502
3A2510	125	3033	510,531	1114	
3A252	116	3016	504A	1104	
3A254	116	3031A	504A	1104	
3A256	116	3017B/117	504A	1104	
3A258	125	3032A	510,531	1114	
3A30	5800	9003		1142	MR501
3A300	5803	9006		1144	MR504
3A4					MR504
3A400		3051			MR504
3A50					MR501
3A500	5805	9008			MR506
3A6					MR506
3A600		3051			MR506
3A8					MR508
3A800		3051			MR508
3A90	109			1123	
3AF1					MR501
3AF10					MR510
3AF2	156		512		MR502
3AF3					MR504
3AF4	5803	9006		1144	MR504
3AF6	5806	3848			MR506
3AF8	5808	9009			MR508
3AFR1					MR851
3AFR2					MR852
3AFR3					MR854
3AFR4					MR854
3AFR6					MR856
3AS1	116	3016	504A	1104	
3AS2	116	3016	504A	1104	
3B15	102	3722	2	2007	
3B15-1	102	3722	2	2007	
3BB-20801			504A	1104	
3BB-20B01	166	9075		1152	
3BF1	156				MR501
3BF10	156				MR510
3BF2	156				MR502
3BF3					MR504
3BF4	156				MR504
3BF6	156				MR506
3BF8	156				MR508
3BFR1					MR851
3BFR2					MR852
3BFR3					MR854
3BFR4					MR854
3BFR6					MR856
3BS1	116	3016	504A	1104	
3BS2	116	3016	504A	1104	
3BZ61	156	3051	512	1143	
3C05	5830		5004		
3C05R	5831		5005		
3C1	5982	3609/5986	5096		
3C10	6002	7202	5108		
3C100	5848	7048	5012		
3C100R	5849	7049	5013		
3C10R	5835		5005		
3C15	5834		5004		
3C15R	5835		5005		
3C2	5986	3609	5096		
3C20	5834		5004		
3C20R	5835		5005		
3C30	5836		5004		
3C30R	5837		5005		
3C38(TRANSISTOR)	108		86	2038	
3C4	5990	3608	5100		
3C40	5838		5004		

Industry Standard No.	ECG	SK	GE	RS 276-	MOTOR.
3C40R	5839		5005		
3C50	5840	3584/5862	5008		
3C50R	5841	3517/5883	5009		
306	5994	3501	5104		
3060	5842	7042	5008		
306030	5400	3950			
3C60R	5843	3517/5883	5009		
3C6100	5454	6754		1067	
3C70	5846	3516	5012		
3C70R	5847		5013		
308	5998		5108		
3080	5846	3516	5012		
3080R	5847		5013		
3090	5848	7048	5012		
3C90R	5849	7049	5013		
3CC13	5834	3500/5882			
3CD13	5835	3517/5883			
3CFS10					1N4007
3CFS15					MR1-1600
3CS1	116	3016	504A	1104	
3CS2	116	3016	504A	1104	
3D-702	5081A		ZD-24		
3DS1	116	3016	504A	1104	
3DS2	116	3016	504A	1104	
3DS3	156	3761/514	512		
3DZ61		3051		1143	
3E-1	123A		20	2051	
3E-2	123A	3444	20	2051	
3E-27	102A		53	2007	
3E-28	158		53	2007	
3E-29	158		53	2007	
3E-3	123A	3444	20	2051	
3E-4	109			1123	
3E-64	116		504A	1104	
3E-65	116		504A	1104	
3E05	5800	9003		1142	MR501
3E1					MR501
3E10	5809	9010			MR510
3E2	5802			1143	MR502
3E4					MR504
3E6	5806	3848			MR506
3E8	5808	9009			MR508
3ES1	116	3016	504A	1104	
3ES2	116	3016	504A	1104	
3F10	5832		5004		MR1121
3F100	5848		5012		MR1130
3F100(SCR)	5402	3638			
3F100D	5848		5012		
3F10D	5832		5004		
3F10D-C	5832		5004		
3F15	5834		5004		
3F150(SCR)	5403	3627/5404			
3F20	5834		5004		MR1122
3F200(SCR)	5404	3627			
3F20D	5834		5004		
3F20D-C	5834	3051/156	5004		
3F30	5836		5004		MR1124
3F30(SCR)	5400	3950			
3F30D	5836		5004		
3F40	5834				MR1124
3F40D	5838		5004		
3F40D-C	5838		5004		
3F5	5830		5004		
3F50	5840	3500/5882	5008		MR1126
3F50D	5840	3584/5862	5008		
3F60	5842		5008		MR1126
3F60(SCR)	5401	3638/5402			
3F60D	5842		5008		
3F80	5846	3516	5012		MR1128
3F80D	5846		5012		
3FC13	5838	3500/5882			
3FD13	5839	3517/5883			
3FR10	5835	3517/5883	5005		
3FR100	5849	7049	5013		
3FR15	5835		5005		
3FR20	5835	3517/5883	5005		
3FR30	5837	3517/5883	5005		
3FR40	5839	3517/5883	5005		
3FR5	5831	3517/5883	5005		
3FR50	5841	3517/5883	5009		
3FR60	5843	3517/5883	5009		
3FR80	5847		5013		
3FS1	116	3016	504A	1104	
3FS2	116	3016	504A	1104	
3G1510	125	3033	510,531	1114	
3G152	116	3016	504A	1104	
3G154	116	3017B/117	504A	1104	
3G156	116	3017B/117	504A	1104	
3G158	125	3033	510,531	1114	
3G2	457	3834/132	FET-1	2035	
3G2510	125	3033	510,531	1114	
3G252	116	3016	504A	1104	
3G254	116	3031A	504A	1104	
3G256	116	3017B/117	504A	1104	
3G258	125	3033	510,531	1114	
3G8	116	3016	504A	1104	
3GA	116	3031A	504A	1104	
3GC12	5838	3051/156	5004		
3GS1	116	3016	504A	1104	
3GS2	116	3017B/117	504A	1104	
3GZ61		3051		1144	
003H03				2007	
3HS1	116	3017B/117	504A	1104	
3HS2	116	3017B/117	504A	1104	
3J100	5402	3638			
3J15	5400	3950			
3J30	5400	3950			
3J60	5401	3638/5402			
3JC12	5842	7042	5008		
3K15	5400	3950			
3L03					MR850
3L05					MR850
3L1030	5400	3950			
3L2015	5400	3950			
3L4-2001			1N60	1123	
3L4-2001-1	109	3088		1123	
3L4-2001-1A	109	3088		1123	
3L4-2001-3	177	3100/519	300	1122	
3L4-2001-4	177	3100/519	300	1122	
3L4-2003-1	109	3088	1N60	1123	
3L4-2003-3	109	3100/519	300	1122	
3L4-2003-4	109	3088	1N60	1123	
3L4-3001-1	177	3100/519	300	1122	
3L4-3001-5	116	3031A	504A	1104	
3L4-3001-7	177	3100/519	300	1122	
3L4-3001-8	116	3031A	504A	1104	
3L4-3002-01	519		514	1122	
3L4-3002-10	177	3100/519	300	1122	
3L4-3002-13			504A	1104	
3L4-3002-25	177	3100/519	300	1122	
3L4-3002-31	177	3100/519	300	1122	
3L4-3002-32	177	3100/519	300	1122	
3L4-3002-7	177	3088	300	1122	
3L4-3503-1		3126	90		
3L4-3503-5	614	3126	90		
3L4-3503-6	614	3327	90		
3L4-3503-7	614	3327			
3L4-3505-1	5072A		ZD-8.2	562	
3L4-3505-2	5071A	3059/138A	ZD-6.8	561	
3L4-3505-3	138A	3059	ZD-7.5		
3L4-3505-4	5074A		ZD-11.0	563	
3L4-3506-12	5071A	3334	ZD-6.8	561	
3L4-3506-2	5071A	3058/137A	5ZD-15		
3L4-3506-21	136A	3057	ZD-5.6	561	
3L4-3506-29	145A	3063	ZD-15	564	
3L4-3506-3	138A		ZD-7.5		
3L4-3506-31	136A	3056/135A	ZD-5.1		
3L4-3506-40	5075A	3751	ZD-16	564	
3L4-3506-7	5072A	3136	ZD8.2	562	
3L4-5007-3			20	2051	
3L4-6001-01	130	3027	14	2041	
3L4-6004	175		246	2020	
3L4-6005-1	152	3893	66	2048	
3L4-6005-2	184	3190	57	2017	
3L4-6005-3	184	3190	57	2017	
3L4-6005-5	152	3893	66	2048	
3L4-6005-55	152	3893	66	2048	
3L4-6006-1	124	3021	12		
3L4-6007-02	123A	3444	20	2051	
3L4-6007-03	123A	3444	20	2051	
3L4-6007-04	123A	3444	20	2051	
3L4-6007-08	123A	3444	20	2051	
3L4-6007-09	123A	3444	20	2051	
3L4-6007-1	123A	3444	20	2051	
3L4-6007-10	161	3018	39	2015	
3L4-6007-11	161	3018	39	2015	
3L4-6007-12	161	3018	39	2015	
3L4-6007-13	161	3018	39	2015	
3L4-6007-14	161	3018	39	2015	
3L4-6007-15	199		62	2010	
3L4-6007-16	199		62	2010	
3L4-6007-17	107		11	2015	
3L4-6007-19	161	3018	39	2015	
3L4-6007-2	123A	3444	20	2051	
3L4-6007-20	161	3018	39	2015	
3L4-6007-21	107	3018	11	2015	
3L4-6007-22	161	3018	39	2015	
3L4-6007-23	107	3018	11	2015	
3L4-6007-3	123A	3444	20	2051	
3L4-6007-34	159		82	2032	
3L4-6007-35	107	3018	11	2015	
3L4-6007-37	123A	3444	20	2051	
3L4-6007-38	123A	3444	20	2051	
3L4-6007-4	128	3024	243	2030	
3L4-6007-41	229	3024/128	63	2030	
3L4-6007-51	108	3452	86	2038	
3L4-6010-03	123A	3444	20	2051	
3L4-6010-3		3122	61	2051	
3L4-6010-4	129	3025	244	2027	
3L4-6010-6	123A	3444	20	2051	
3L4-6010-8	129	3025	67	2032	
3L4-6011-02	129		244	2027	
3L4-6011-1	196	3054	241	2020	
3L4-6011-11	153	3084	69	2049	
3L4-6011-12	153		69	2049	
3L4-6011-14	153	3084	69	2049	
3L4-6011-2	211	3084	253	2026	
3L4-6011-3	211	3084	253	2026	
3L4-6011-52	211	3084	253	2026	
3L4-6011-53	211	3084	253	2026	
3L4-6011-9	153	3084	69	2049	
3L4-6012-02	152	3893	66	2048	
3L4-6012-06	196		241	2020	
3L4-6012-2	152	3893	66	2048	
3L4-6012-3	152	3893	66	2048	
3L4-6012-4	182	3041	55		
3L4-6012-5	196	3041	241	2020	
3L4-6012-55	196	3041	241	2020	
3L4-6012-56	196	3054	241	2020	
3L4-6012-58	196	3041	241	2020	
3L4-6012-6	196	3041	241	2020	
3L4-6012-8	196	3041	241	2020	
3L4-6013-02	153		69	2049	
3L4-6013-15	153	3084	69	2049	
3L4-6013-2	153	3084	59	2049	
3L4-6013-3	153	3084	69	2049	
3L4-6013-4	183	3084	56	2027	
3L4-6013-5	153	3084	69	2049	
3L4-6013-55	153	3084	69	2049	
3L4-6013-56	153	3083/197	69	2049	
3L4-6013-58	153	3084	69	2049	
3L4-6013-6	197	3084	250	2027	
3L4-6013-8	197	3084	250	2027	
3L4-6015-01	123A	3444	20	2051	
3L4-6017-01	159	3466	82	2032	
3L4-6020-01	238	3710	259		
3L4-6021-01	247	3948			
3L4-6503-1	222	3050/221	FET-4	2036	
3L4-6503-2	222	3050/221	FET-4	2036	
3L4-7004-1	6402	3628			
3L4-9002-01	786	3140	IC-227		
3L4-9002-1	786	3140	IC-227		
3L4-9004-01	743		IC-214		
3L4-9004-1	743		IC-214		
3L4-9004-3	743		IC-214		
3L4-9004-4	743	3172	IC-214		
3L4-9004-51	743		IC-214		
3L4-9006-51	726	3129			
3L4-9007-01	737	3375	IC-15		
3L4-9007-1	737	3375	IC-16		
3L4-9007-51	737	3375	IC-16		
3L4-9020	1232	3852			
3L4-9020-1	1288	3852/1232		703	
3L46007-1			20	2051	
3L46007-2			20	2051	
3L49007-51	737	3375	IC-16		
3LA-6007-4			63	2030	
3LC12	5846	3516	5012		
3MA	118	3066	CR-1		
3MC	160	3007	245	2004	
3MS10	116	3016	504A	1104	
3MS20	116	3016	504A	1104	
3MS30	116	3016	504A	1104	
3MS40	116	3016	504A	1104	
3MS5	116	3016	504A	1104	
3MS50	116	3016	504A	1104	
3N112	106	3984	21	2034	
3N113	106	3984	21	2034	
3N114			221	2032	
3N115			221	2032	
3N116			221	2032	
3N117			221	2032	
3N118			221	2032	
3N119			221	2032	
3N120			11	2015	
3N121			11	2015	
3N123			221	2050	
3N124					3N124

Industry Standard No.	ECG	SK	GE	RS 276-	MOTOR.
3N125					3N125
3N126					3N126
3N127			60	2015	
3N128	220	3990	FET-3	2036	3N128
3N129			221	2024	
3N130			221	2032	
3N131			221	2032	
3N132			221	2032	
3N133			221	2032	
3N134			221	2024	
3N135			221	2032	
3N136			221	2032	
3N138		3531			3N128
3N139		3531			3N128
3N140	221	3065/222		2036	3N140
3N141	221	3065/222			3N140
3N142	220	3990		2036	3N128
3N143	220	3990		2036	3N128
3N145					2N4352
3N146					2N4352
3N149					2N4352
3N150					2N4352
3N152	220	3990		2036	3N128
3N153		3990			3N128
3N154		3990			3N128
3N155					3N155
3N155A					3N155A
3N156					3N156
3N157					3N157
3N157A					3N157A
3N158					3N158
3N158A					3N158A
3N159	454	3991			3N140
3N161					2N4352
3N163					2N4352
3N164					2N4352
3N169					3N169
3N170					3N170
3N171					3N171
3N172					2N4352
3N173					2N4352
3N174					2N4352
3N175	465				2N4351
3N176	465				2N4351
3N177	465				2N4351
3N178					3N158
3N179					3N158
3N180					3N158
3N184					3N157
3N185					3N157
3N186					3N155
3N187	222	3065			3N201
3N200	454	3991	FET-4	2036	3N209
3N201	454	3991	FET-4	2036	3N201
3N201A	222	3991/454	FET-4	2036	
3N202	454	3991		2036	3N202
3N203	454	3991	FET-4	2036	3N202
3N204	454	3991			3N201
3N205	454	3991			3N202
3N206	454	3991			3N203
3N209					3N209
3N21		3123	52	2024	
3N211	454$	3187			3N211
3N212		3187			3N212
3N213	222	3187	FET-4	2036	3N213
3N22	101	3011	8	2002	
3N23	101	3011	8	2002	
3N23A	101	3011	8	2002	
3N23B	101	3011	8	2002	
3N23C	101	3011	8	2002	
3N25/501			51	2004	
3N29	101	3011	8	2002	
3N30	101	3011	8	2002	
3N31	101	3011	8	2002	
3N34	160		245	2004	
3N35	123A	3444	20	2051	
3N35A	160		245	2004	
3N36	101	3011	8	2002	
3N37	101	3011	8	2002	
3N49		3012	4		
3N50	105	3012	4		
3N51	105	3012	4		
3N52	105	3012	4		
3N71	107	3039/316	11	2015	
3N72	107	3039/316	11	2015	
3N73	107	3039/316	11	2015	
3N74		3245	212	2010	
3N75		3245	212	2010	
3N76		3245	212	2010	
3N77		3245	212	2010	
3N78		3245	212	2010	
3N79		3245	212	2010	
3N87			11	2015	
3N88			11	2015	
3N90		3114	82	2032	
3N91		3114	82	2032	
3N92		3114	82	2032	
3N93		3114	82	2032	
3N94		3114	82	2032	
3N95		3114	82	2032	
3N98					MFE3004
3N99					MFE3004
3NC12	5848	7048	5012		
3NV15	525	3925			
3R10	5125A	3391			
3R100	5156A	3422			
3R100A	5156A	3422			
3R100B	5156A	3422			
3R10A	5125A	3391			
3R10B	5125A	3391			
3R11	5126A	3392			
3R110	5157A	3423			
3R110A	5157A	3423			
3R110B	5157A	3423			
3R11A	5126A	3392			
3R11B	5126A	3392			
3R12	5127A	3393			
3R120	5158A	3424			
3R120A	5158A	3424			
3R120B	5158A	3424			
3R12A	5127A	3393			
3R12B	5127A	3393			
3R13	5128A	3394			
3R130	5159A	3425			
3R130A	5159A	3425			
3R130B	5159A	3425			
3R13A	5128A	3394	5ZD-13		
3R13B	5128A	3394			
3R140	5160A	3426			
3R140A	5160A	3426			
3R140B	5160A	3426			
3R15	5130A	3396			
3R150	5161A	3427			
3R150A	5161A	3427			
3R150B	5161A	3427			
3R15A	5130A	3396			
3R15B	5130A	3396			
3R16	5131A	3397			
3R160	5162A	3428			
3R160A	5162A	3428			
3R160B	5162A	3428	35		
3R16A	5131A	3397	5ZD-16		
3R16B	5131A	3397			
3R18	5133A	3399			
3R180	5164A	3430			
3R180A	5164A	3430			
3R180B	5164A	3430			
3R18A	5133A	3399			
3R18B	5133A	3399			
3R20	5135A	3401			
3R200	5166A	3432			
3R200A	5166A	3432			
3R200B	5166A	3432			
3R20A	5135A	3401			
3R20B	5135A	3401			
3R22	5136A	3402			
3R22A	5136A	3402			
3R22B	5136A	3402			
3R24	5137A	3403			
3R24A	5137A	3403			
3R24B	5137A	3403			
3R27	5139A	3405			
3R27A	5139A	3405			
3R27B	5139A	3405			
3R30	5141A	3407			
3R30A	5141A	3407			
3R30B	5141A	3407			
3R33	5142A	3408			
3R33A	5142A	3408			
3R33B	5142A	3408			
3R35	5143A	3409			
3R35A	5143A	3409			
3R35B	5143A	3409			
3R36	5143A	3409			
3R39A	5144A	3410			
3R39B	5144A	3410			
3R43	5145A	3411			
3R43A	5145A	3411			
3R43B	5145A	3411			
3R47	5146A	3412			
3R47A	5146A	3412			
3R47B	5146A	3412			
3R51	5147A	3413			
3R51A	5147A	3413			
3R51B	5147A	3413			
3R56	5148A	3414			
3R56A	5148A	3414			
3R56B	5148A	3414			
3R62	5150A	3416			
3R62A	5150A	3416			
3R62B	5150A	3416			
3R68	5151A	3417			
3R68A	5151A	3417			
3R68B	5151A	3417			
3R7.5	5121A	3387			
3R7.5A	5121A	3387			
3R7.5B	5121A	3387			
3R75	5152A	3418			
3R75A	5152A	3418			
3R75B	5152A	3418			
3R8.2	5122A	3388			
3R8.2A	5122A	3388			
3R8.2B	5122A	3388			
3R82	5153A	3419			
3R82A	5153A	3419			
3R82B	5153A	3419			
3R9.1	5124A	3390			
3R9.1A	5124A	3390			
3R9.1B	5124A	3390			
3R91	5155A	3421			
3R91A	5155A	3421			
3R91B	5155A	3421			
3S004	123A	3019	20	2051	
3S05E	5800	9003		1142	
3S105					MR501
3S11					MR501
3S12					MR502
3S14					MR504
3S16					MR506
3S1E	5801	9004		1142	
3S2E	5802	9005		1143	
3S30B	312	3448	FET-1	2035	
3S3E	5803	9006		1144	
3S4E	5804			1144	
3S5E	5805	9008			
3S6E	5806	3848			
3SA324	160		245	2004	
3SB-B732	116		504A	1104	
3SB347	102A		53	2007	
3SB629	116	3016	504A	1104	
3SCBR05					MDA970A1
3SCBR1					MDA970A2
3SCBR2					MDA970A3
3SD313	196		241	2020	
3SD313C	196		241	2020	
3SF1					MR851
3SF2					MR852
3SF4					MR854
3SK-30B			FET-3	2036	
3SK-35			FET-4	2036	
3SK-39			FET-4	2036	
3SK14		3990		2036	
3SK20				2036	
3SK20H		3112		2036	2N5457
3SK20HW				2036	
3SK20HY				2036	
3SK21			FET-4	2036	
3SK21H		3112	FET-4	2036	2N5457
3SK22		3116	FET-3	2036	2N5485
3SK22-Y		3116		2036	
3SK22GR		3116	FET-3	2036	
3SK22Y		3116	FET-3	2036	
3SK23	451	3116	FET-1	2035	2N5485
3SK28		3116			2N5485
3SK29		3990		2036	
3SK30		3065	FET-1	2035	
3SK30A		3065	FET-1	2035	
3SK30B		3065	FET-3	2035	
3SK30C		3065	FET-1	2035	
3SK32			FET-4	2036	
3SK32(B)			FET-4	2036	
3SK32A				2036	
3SK32B		3050	FET-4	2036	
3SK32B-6		3050		2036	
3SK32C				2036	
3SK32D				2036	
3SK32E		3050		2036	
3SK32E-4		3050		2036	

Industry Standard No.	ECG	SK	GE	RS 276-	MOTOR.
3SK33	220	3990		2036	
3SK34	222	3116	FET-1	2028	
3SK34C			FET-1	2028	
3SK35	221	3050	FET-4	2036	
3SK35-BL	221		FET-4	2036	
3SK35-GR	221		FET-4	2036	
3SK35-Y	221		FET-4	2036	
3SK35BL		3050	FET-4	2036	
3SK35G	221	3050	FET-4	2036	
3SK37	222	3050/221	FET-4	2036	
3SK39	222	3065	FET-4	2036	
3SK390	221	3065/222			
3SK39E	222		FET-4	2036	
3SK39P	222		FET-4	2036	
3SK39Q	222	3065	FET-4	2036	
3SK39R	222	3065	FET-4	2036	
3SK3E			FET-4	2036	
3SK40	454	3050/221	FET-4	2036	
3SK40I	454		FET-4	2036	
3SK40L	454	3050/221			
3SK40M	454	3050/221	FET-4	2036	
3SK40ML	454	3050/221			
3SK41	221	3834/132	FET-4	2036	
3SK41(L)	221		FET-4	2036	
3SK411			FET-4	2036	
3SK41C	221		FET-4	2036	
3SK41L	221	3834/132	FET-4	2036	
3SK41M	221	3834/132	FET-4	2036	
3SK44	222	3448	FET-4	2036	
3SK44W	222	3448			
3SK45	221	3065/222	FET-4	2036	
3SK45-B	221	3065/222			
3SK45-B-09	221	3065/222	FET-4	2036	
3SK45B	221	3065/222	FET-4	2036	
3SK45B09	221	3065/222	FET-4	2036	
3SK49	222	3050/221	FET-4	2036	
3SK49E2			FET-4	2036	
3SK49Q	222	3065	FET-4	2036	
3SK59	222	3050/221	FET-4	2036	
3SK59BL	222	3050/221	FET-4	2036	
3SK59GR	222	3050/221	FET-4	2036	
3SK59Y	222	3050/221			
3SM0					MR510
3SM2					MR502
3SM4	5804	9007		1144	MR504
3SM6					MR506
3SM8					MR508
3T201	101	3861	8	2002	
3T202	101	3861	8	2002	
3T203	101	3861	8	2002	
3T501	116	3016	504A	1104	
3T502	116	3017B/117	504A	1104	
3T503	116	3016	504A	1104	
3T504	116	3016	504A	1104	
3T505	116	3017B/117	504A	1104	
3T506	116	3017B/117	504A	1104	
3T507	125	3032A	510,531	1114	
3T508	125	3032A	510,531	1114	
3T509	125	3033	510,531	1114	
3T510	125	3033	510,531	1114	
3TE120	130	3036	14	2041	
3TE140	130	3027	14	2041	
3TE150			63	2030	
3TE160			63	2030	
3TE230	130	3027	14	2041	
3TE240	130	3027	14	2041	
3TE250			63	2030	
3TE260			63	2030	
3TX003	130	3027	14	2041	
3TX004	130	3027	14	2041	
3TX620	362				2N5944
3TX621	362				2N5944
3TX622	363				2N5945
3TX820	362				2N5944
3TX822	363				2N5945
3TZ10	5125A	3391			
3TZ100	5156A	3422			
3TZ100A	5156A	3422			
3TZ100B	5156A	3422			
3TZ100C	5156A	3422			
3TZ100D	5156A	3422			
3TZ10A	5125A	3391			
3TZ10B	5125A	3391			
3TZ10C	5125A	3391			
3TZ10D	5125A	3391	5ZD-10		
3TZ11	5126A	3392			
3TZ110	5157A	3423			
3TZ110A	5157A	3423			
3TZ110B	5157A	3423			
3TZ110C	5157A	3423			
3TZ110D	5157A	3423			
3TZ11A	5126A	3392			
3TZ11B	5126A	3392			
3TZ11C	5126A	3392			
3TZ11D	5126A		5ZD-11		
3TZ12	5127A	3393			
3TZ120	5158A	3424			
3TZ120A	5158A	3424			
3TZ120B	5158A	3424			
3TZ120C	5158A	3424			
3TZ120D	5158A	3424			
3TZ12A	5127A	3393	5ZD-12		
3TZ12B	5127A	3393			
3TZ12C	5127A	3393			
3TZ12D	5127A	3393			
3TZ13	5128A	3394			
3TZ130	5159A	3425			
3TZ130A	5159A	3425			
3TZ130B	5159A	3425			
3TZ130C	5159A	3425			
3TZ130D	5159A	3425			
3TZ13A	5128A	3394			
3TZ13B	5128A	3394			
3TZ13C	5128A	3394			
3TZ13D	5128A	3394	5ZD-13		
3TZ14	5129A	3395			
3TZ140	5160A	3426			
3TZ140A	5160A	3426			
3TZ140B	5160A	3426			
3TZ140C	5160A	3426			
3TZ140D	5160A	3426			
3TZ14A	5129A	3395			
3TZ14B	5129A	3395			
3TZ14C	5129A	3395			
3TZ14D	5129A		5ZD-14		
3TZ15	5130A	3396			
3TZ150	5161A	3427			
3TZ150A	5161A	3427			
3TZ150B	5161A	3427			
3TZ150C	5161A	3427			
3TZ150D	5161A	3427			
3TZ15A	5130A	3396			
3TZ15B	5130A	3396			
3TZ15C	5130A	3396			
3TZ15D	5130A	3396			
3TZ16	5131A	3397			
3TZ160	5162A	3428			
3TZ160A	5162A	3428			
3TZ160B	5162A	3428			
3TZ160C	5162A	3428			
3TZ160D	5162A	3428			
3TZ16A	5131A	3397			
3TZ16B	5131A	3397			
3TZ16C	5131A	3397			
3TZ16D	5131A		5ZD-16		
3TZ17	5132A	3398	5ZD-17		
3TZ17A	5132A	3398	5ZD-17		
3TZ17B	5132A	3398	5ZD-17		
3TZ17C	5132A	3398	5ZD-17		
3TZ17D	5132A	3398	5ZD-17		
3TZ18	5133A	3399			
3TZ180	5164A	3430			
3TZ180A	5164A	3430			
3TZ180B	5164A	3430			
3TZ180C	5164A	3430			
3TZ180D	5164A	3430			
3TZ18A	5133A	3399			
3TZ18B	5133A	3399			
3TZ18C	5133A	3399			
3TZ18D	5133A	3399			
3TZ19	5134A	3400	5ZD-19		
3TZ19A	5134A	3400			
3TZ19B	5134A	3400			
3TZ19C	5134A	3400			
3TZ19D	5134A	3400			
3TZ20	5135A	3401			
3TZ200	5166A	3432			
3TZ200A	5166A	3432			
3TZ200B	5166A	3432			
3TZ200C	5166A	3432			
3TZ200D	5166A	3432			
3TZ20A	5135A	3401			
3TZ20B	5135A	3401			
3TZ20C	5135A	340105135A			
3TZ20D	5135A	3401			
3TZ22	5136A	3402			
3TZ22A	5136A	3402			
3TZ22B	5136A	3402			
3TZ22C	5136A	3402			
3TZ22D	5136A	3402			
3TZ24	5137A	3403			
3TZ24A	5137A	3403			
3TZ24B	5137A	3403			
3TZ24C	5137A	3403			
3TZ24D	5137A	3403			
3TZ27	5139A	3405			
3TZ27A	5139A	3405			
3TZ27B	5139A	3405			
3TZ27C	5139A	3405			
3TZ27D	5139A	3405			
3TZ3.6	5112A	3378			
3TZ3.6A	5112A	3378			
3TZ3.6B	5112A	3378			
3TZ3.6C	5112A	3378			
3TZ3.6D	5112A	3378			
3TZ3.9	5113A	3379			
3TZ3.9A	5113A	3379			
3TZ3.9B	5113A	3379			
3TZ3.9C	5113A	3379			
3TZ3.9D	5113A	3379			
3TZ30	5141A	3407			
3TZ30A	5141A	3407			
3TZ30B	5141A	3407			
3TZ30C	5141A	3407			
3TZ30D	5141A	3407			
3TZ33	5142A	3408			
3TZ33A	5142A	3408			
3TZ33B	5142A	3408			
3TZ33C	5142A	3408			
3TZ33D	5142A	3408			
3TZ36	5143A	3409			
3TZ36A	5143A	3409			
3TZ36B	5143A	3409			
3TZ36C	5143A	3409			
3TZ36D	5143A	3409			
3TZ39	5144A	3410			
3TZ39A	5144A	3410			
3TZ39B	5144A	3410			
3TZ39C	5144A	3410			
3TZ4.3	5114A	3380			
3TZ4.3A	5114A	3380			
3TZ4.3B	5114A	3380			
3TZ4.3C	5114A	3380			
3TZ4.3D	5114A	3380			
3TZ4.7A	5115A	3381			
3TZ4.7B	5115A	3381			
3TZ4.7C	5115A	3381			
3TZ4.7D	5115A	3381			
3TZ43	5145A	3411			
3TZ43A	5145A	3411			
3TZ43B	5145A	3411			
3TZ43C	5145A	3411			
3TZ43D	5145A	3411			
3TZ47	5146A	3412			
3TZ47A	5146A	3412			
3TZ47B	5146A	3412			
3TZ47C	5146A	3412			
3TZ47D	5146A	3412			
3TZ5.1	5116A	3382			
3TZ5.1A	5116A	3382			
3TZ5.1B	5116A	3382			
3TZ5.1C	5116A	3382			
3TZ5.1D	5116A	3382			
3TZ5.6	5117A	3383			
3TZ5.6A	5117A	3383			
3TZ5.6B	5117A	3383			
3TZ5.6C	5117A	3383			
3TZ5.6D	5117A	3383			
3TZ50	5147A	3413			
3TZ50A	5147A	3413			
3TZ50B	5147A	3413			
3TZ50C	5147A	3413			
3TZ50D	5147A	3413			
3TZ51	5147A	3413			
3TZ51A	5147A	3413			
3TZ51B	5147A	3413			
3TZ51C	5147A	3413			
3TZ51D	5147A	3413			
3TZ56	5148A	3414			
3TZ56A	5148A	3414			
3TZ56B	5148A	3414			
3TZ56C	5148A	3414			
3TZ56D	5148A	3414			
3TZ6.2C	5119A	3385			
3TZ6.2D	5119A	3385			
3TZ6.8	5120A	3386			
3TZ6.8A	5120A	3386			
3TZ6.8B	5120A	3386	5ZD-6.8		

Industry Standard No.	ECG	SK	GE	RS 276-	MOTOR.
3TZ6.8C	5120A	3386			
3TZ6.8D	5120A	3386			
3TZ62	5150A	3416			
3TZ62A	5150A	3416			
3TZ62B	5150A	3416			
3TZ62C	5150A	3416			
3TZ62D	5150A	3416			
3TZ68	5151A	3417			
3TZ68A	5151A	3417			
3TZ68B	5151A	3417			
3TZ68C	5151A	3417			
3TZ68D	5151A	3417			
3TZ7.5	5121A	3387			
3TZ7.5A	5121A	3387			
3TZ7.5B	5121A	3387			
3TZ7.5C	5121A	3387			
3TZ7.5D	5121A	3387	5ZD-7.5		
3TZ75	5152A	3418			
3TZ75A	5152A	3418			
3TZ75B	5152A	3418			
3TZ75C	5152A	3418			
3TZ75D	5152A	3418			
3TZ8.2	5122A	3388			
3TZ8.2A	5122A	3388			
3TZ8.2B	5122A	3388	5ZD-8.2		
3TZ8.2C	5122A	3388			
3TZ8.2D	5122A	3388			
3TZ82	5153A	3419			
3TZ82A	5153A	3419			
3TZ82B	5153A	3419			
3TZ82C	5153A	3419			
3TZ82D	5153A	3419			
3TZ9.1	5124A	3390			
3TZ9.1A	5124A	3390			
3TZ9.1B	5124A	3390			
3TZ9.1C	5124A	3390			
3TZ9.1D	5124A	3390	5ZD-9.1		
3TZ91	5155A	3421			
3TZ91A	5155A	3421			
3TZ91B	5155A	3421			
3TZ91C	5155A	3421			
3VR10	5125A	3391			
3VR100	5156A	3422			
3VR100A	5156A	3422			
3VR100B	5156A	3422			
3VR10A	5125A	3391			
3VR10B	5125A	3391	5ZD-10		
3VR11	5126A	3392			
3VR110	5157A	3423			
3VR110A	5157A	3423			
3VR110B	5157A	3423			
3VR11A	5126A	3392			
3VR11B	5126A	3392	5ZD-11		
3VR12	5127A	3393			
3VR120	5158A	3424			
3VR120A	5158A	3424			
3VR120B	5158A	3424			
3VR12A	5127A	3393			
3VR12B	5127A	3393			
3VR13	5128A	3394			
3VR130	5159A	3425			
3VR130A	5159A	3425			
3VR130B	5159A	3425			
3VR13A	5128A	3394			
3VR13B	5128A	3394	5ZD-13		
3VR15	5130A	3396			
3VR150	5161A	3427			
3VR150A	5161A	3427			
3VR150B	5161A	3427			
3VR15A	5130A	3396			
3VR15B	5130A	3396			
3VR16	5131A	3397			
3VR160	5162A	3428			
3VR160A	5162A	3428			
3VR160B	5162A	3428			
3VR16A	5131A	3397			
3VR16B	5131A	3397	5ZD-16		
3VR18	5133A	3399			
3VR180	5164A	3430			
3VR180A	5164A	3430			
3VR180B	5164A	3430			
3VR18A	5133A	3399			
3VR18B	5133A	3399			
3VR20	5135A	3401			
3VR200	5166A	3432			
3VR200A	5166A	3432			
3VR200B	5166A	3432			
3VR20A	5135A	3401			
3VR20B	5135A	3401			
3VR22	5136A	3402			
3VR22A	5136A	3402			
3VR22B	5136A	3402			
3VR24	5137A	3403			
3VR24A	5137A	3403			
3VR24B	5137A	3403			
3VR27	5139A	3405			
3VR27A	5139A	3405			
3VR27B	5139A	3405			
3VR30	5141A	3407			
3VR30A	5141A	3407			
3VR30B	5141A	3407			
3VR33	5142A	3408			
3VR33A	5142A	3408			
3VR33B	5142A	3408			
3VR36	5143A	3409			
3VR36A	5143A	3409			
3VR36B	5143A	3409			
3VR39	5144A	3410			
3VR39A	5144A	3410			
3VR39B	5144A	3410			
3VR43	5145A	3411			
3VR43A	5145A	3411			
3VR43B	5145A	3411			
3VR47	5146A	3412			
3VR47A	5146A	3412			
3VR47B	5146A	3412			
3VR5.6	5117A	3383			
3VR5.6A	5117A	3383			
3VR5.6B	5117A	3383			
3VR51	5147A	3413			
3VR51A	5147A	3413			
3VR51B	5147A	3413			
3VR56	5148A	3414			
3VR56A	5148A	3414			
3VR56B	5148A	3414			
3VR6	5118A	3384			
3VR6.2	5119A	3385			
3VR6.2A	5119A	3385			
3VR6.2B	5119A	3385			
3VR6.8	5120A	3386			
3VR6.8A	5120A	3386			
3VR6.8B	5120A	3386	5ZD-6.8		
3VR62	5150A	3416			
3VR62A	5150A	3416			

Industry Standard No.	ECG	SK	GE	RS 276-	MOTOR.
3VR62B	5150A	3416			
3VR67	5151A	3417			
3VR67A	5151A	3417			
3VR67B	5151A	3417			
3VR68	5151A	3417			
3VR68A	5151A	3417			
3VR68B	5151A	3417			
3VR6A	5118A	3384			
3VR6B	5118A	3384			
3VR7.5	5121A	3387			
3VR7.5B	5121A	3387	5ZD-7.5		
3VR75	5152A	3418			
3VR75A	5152A	3418			
3VR75B	5152A	3418			
3VR8.2	5122A	3388			
3VR8.2A	5122A	3388			
3VR8.2B	5122A	3388			
3VR8.5	5123A	3389			
3VR8.5A	5123A	3389			
3VR8.5B	5123A	3389	5ZD-8.7		
3VR82	5153A	3419			
3VR82A	5153A	3419			
3VR82B	5153A	3419			
3VR9.1	5124A	3390			
3VR9.1A	5124A	3390			
3VR9.1B	5124A	3390	5ZD-9.1		
3VR90	5155A	3421			
3VR90A	5155A	3421			
3VR90B	5155A	3421			
3VR91	5155A	3421			
3VR91A	5155A	3421			
3VR91B	5155A	3421			
3X11/1			504A	1104	
3Z12	5127A	3393			
3Z12A	5127A	3393			
3Z12B	5127A	3393			
3Z15	5130A	3396			
3Z15A	5130A	3396			
3Z15B	5130A	3396			
3Z18	5133A	3399			
3Z18A	5133A	3399			
3Z18B	5133A	3399			
3Z21	5135A	3401			
3Z21A	5135A	3401			
3Z21B	5135A	3401			
3Z24	5137A	3403			
3Z24A	5137A	3403			
3Z24B	5137A	3403			
3Z27	5139A	3405			
3Z27A	5139A	3405			
3Z27B	5139A	3405			
3Z30	5141A	3407			
3Z30A	5141A	3407			
3Z30B	5141A	3407			
3Z33	5142A	3408			
3Z33A	5142A	3408			
3Z33B	5142A	3408			
3Z36	5143A	3409			
3Z36A	5143A	3409			
3Z36B	5143A	3409			
3Z6.8	5120A	3386			
3Z6.8A	5120A	3386			
3Z6.8B	5120A	3386			
3Z9.0	5124A	3390			
3Z9.0A	5124A	3390			
3Z9.0B	5124A	3390			
004	1098	3162/725			
004(MCINTOSH)	725		IC-19		
004-00	123A	3444	20	2051	
004-00(LAST IF)	233			2009	
004-00(LAST-IF)			210	2051	
04-000653	519		514	1122	
04-000655-1	519			1122	
04-00072-01	103A	3010	59	2002	
004-001	179	3642	76		
04-00156-03	158	3004/102A	53	2007	
004-002000	116	3017B/117	504A	1104	
004-002700	116	3016	504A	1104	
004-002800	116	3017B/117	504A	1104	
004-002900	113A		6GC1		
004-003000	116	3017B/117	CR-3	1104	
004-003100	119	3109	CR-2		
004-003200	118	3066	CR-1		
004-003300	116	3016	504A	1104	
004-003400	116	3016	504A	1104	
0004-003500	116	3016	511	1114	
004-003600	116	3016	504A	1104	
004-003700	120	3110	CR-3		
4-003721-00	175		246	2020	
004-003900	116	3313	504A	1104	
004-004000	116	3016	504A	1104	
004-004100	116	3016	504A	1104	
04-00460-03	123A	3444	20	2051	
04-00461-02	107	3018	11	2015	
4-005	725		IC-19		
04-00535-02	107	3018	11	2015	
04-00535-06	108	3018	86	2038	
4-006	725		IC-19		
4-007	703A	3157	IC-12		
4-008	703A	3157	IC-12		
4-009	720	9014	IC-7		
004-00900			300	1122	
004-009200			1N60	1123	
04-01585-06	123A	3444	20	2051	
04-01585-07	123A	3444	20	2051	
04-01585-08	107	3018	11	2015	
04-02090-02	123A	3444	20	2051	
4-0294	130		14	2041	
004-03100	119	3109	CR-2		
004-03200	118	3066	CR-1		
004-03300	116	3017B/117	504A	1104	
004-03500	116	3017B/117	504A	1104	
004-03600	116	3017B/117	504A	1104	
004-03700	116	3017B/117	504A	1104	
4-0485	108	3452	86	2038	
4-0498	128		243	2030	
4-0563	130		14	2041	
04-07150-01	187A	9076	248		
4-082-664-0001	704	3023	IC-205		
004-1	189	3083/197	218	2026	
4-1001	128		243	2030	
04-11620-01	186A	3357	247	2052	
4-12-1A7-1	123A	3124/289	20	2051	
4-142	131	3052	44	2006	
4-142-1(SEARS)	176		80		
4-14A17-1	121	3009	239	2006	
4-1520-00150	1027	3153			
4-1540-11510	1237	3707			
4-1544	199	3124/289	62	2010	
4-1545	123A	3444	20	2051	
4-1546	176	3123	80		
04-15850-06	107	3018	11	2015	
4-1723	5067A		ZD-3.9		
4-1724	177	3100/519	300	1122	

Industry Standard No.	ECG	SK	GE	RS 276-	MOTOR.
4-1726	177	3100/519	300	1122	
4-1790	123A	3444	20	2051	
4-1791	199	3245	62	2010	
4-1792	192	3137	63	2030	
4-1807	116	3311	504A	1104	
4-18341		3126	90		
4-1848	131MP	3840	44(2)	2006(2)	
004-2	152	3126	90		
4-2020	109	3087		1123	
4-2020-03-700	506		511	1114	
4-2020-03173	116	3311	504A	1104	
4-2020-03200	116	3032A	504A	1104	
4-2020-03500	109	3087		1123	
4-2020-03571			1N34AS	1123	
4-2020-03600	109	3087		1123	
4-2020-03700	506	3043	511	1114	
4-2020-03800	506	3043	511	1114	
4-2020-03900	125	3032A	510,531	1114	
4-2020-0500	113A	3098/150A	6GC1		
4-2020-05000	113A	3119/113	6GC1		
4-2020-05200	116	3032A	504A	1104	
4-2020-05400	113A	3119/113	6GC1		
4-2020-05600	110MP	3087	1N34AS	1123(2)	
4-2020-05800	177	3100/519	300	1122	
4-2020-06100	177	3100/519	300	1122	
4-2020-06200	177	3100/519	300	1122	
4-2020-06300	119	3109	CR-2		
4-2020-06400	177	3100/519	300	1122	
4-2020-06500	506	3130	511#(3)	1114	
4-2020-06600	5022A	3093	ZD-13	563	
4-2020-06700	506	3032A	511	1114	
4-2020-06800			504A	1104	
4-2020-07300	116	3016	504A	1104	
4-2020-07500	137A	3058	ZD-6.2	561	
4-2020-07600	116	3031A	504A	1104	
4-2020-07601		3016	504A	1104	
4-2020-07700	116	3016	504A	1104	
4-2020-07800	506	3130	511	1114	
4-2020-07801	506	3130	511	1114	
4-2020-07802			511	1114	
4-2020-07900		3119	6GC1		
4-2020-08000	506	3033	511	1114	
4-2020-08001		3016	300	1122	
4-2020-08200			300	1122	
4-2020-08500	116	3032A	504A	1104	
4-2020-08600	110MP	3088	1N60	1123(2)	
4-2020-08900			504A	1104	
4-2020-09200	5077A	3752	ZD-18		
4-2020-09400	119	3109	CR-2		
4-2020-10100	177	3100/519	300	1122	
4-2020-10500			504A	1104	
4-2020-11300	5071A	3334	ZD-6.8	561	
4-2020-11500	5074A		ZD-11.0	563	
4-2020-11900	5077A	3752			
4-2020-12000	5077A	3752	ZD-18		
4-2020-12300	5074A	3092/141A	ZD-11.0	563	
4-2020-12400	142A	3062	ZD-12	563	
4-2020-12700	145A		ZD-15	564	
4-2020-13300	5085A	3337	ZD-36		
4-2020-13600	519	3017B/117	504A	1104	
4-2020-14000	151A	3099	ZD-110		
4-2020-14100	5096A	3346	ZD-100		
4-2020-14400	506	3130	511	1114	
4-2020-14500	116	3017B/117	504A	1104	
4-2020-14600	506	3130	511	1114	
4-2020-15100	506	3130	511	1114	
4-2020-15600	177	3100/519	300	1122	
4-2020-16200	177		300	1122	
4-2020-20900	506	3843			
4-2020-22900	525	3925			
4-2020-23000	506	3843			
4-2020-23200	506	3843			
4-2020-23300	525	3925			
4-2020-8000	125		510,531	1114	
4-2020-8700	125		510,531	1114	
4-2020035000	109			1123	
4-202003571	109			1123	
4-202012000	5077A		ZD-18		
4-2021-04170	166	9075	510#(4)	1152	
4-2021-04470	125		509	1114	
4-2021-04570	506	3043	511	1114	
4-2021-04770	125		510,531	1114	
4-2021-04870	506	3043	511	1114	
4-2021-04970	116	3017B/117	504A	1104	
4-2021-05000	178MP	3100/519	300(2)	1122(2)	
4-2021-05070	127	3113/516			
4-2021-05170	112		1N82A		
4-2021-05470	138A	3059	ZD-7.5		
4-2021-05870	109	3087		1123	
4-2021-06970	506	3043	511	1114	
4-2021-07370	503	3068			
4-2021-07470	177	3100/519	300	1122	
4-2021-07570	505	3067/502	CR-7		
4-2021-07670	177	3100/519	300	1122	
4-2021-07970	506	3843	511	1114	
4-2021-08070	505	3108	CR-7		
4-2021-08270			504A	1104	
4-2021-08570		3336	ZD-22		
4-2021-09070			ZD-11	563	
4-2021-09370	116	3017B/117	504A	1104	
4-2021-10270		3311	504A	1104	
4-2021-10470	506	3843	504A	1104	
4-2021-10870	5074A		ZD-11.0	563	
4-2021-14970	506	3843			
4-2021-15770	506	3843			
4-2021-7370	503		CR-5		
4-202104170	116		504A	1104	
4-202104570	506	3843	511	1114	
4-202104770	158		53	2007	
4-202104870	506	3843	511	1114	
4-202104970	177		300	1122	
4-202105470	138A		ZD-7.5		
4-202A16	109	3088		1123	
4-202R101	116	3031A	504A	1104	
4-2040-08000			300	1122	
4-2060-02300	1004	3102/710			
4-2060-02400	1004	3102/710			
4-2060-02600	1080	3284			
4-2060-02900	1080	3284			
4-2060-03900	1080	3284			
4-2060-04000	712	3072	IC-2		
4-2060-04200	1094		IC-157		
4-2060-04600	712	3072	IC-2		
4-2060-04800	1122		IC-40		
4-2060-04900	1122		IC-40		
4-2060-05200	1159	3290	IC-72		
4-2060-07200	1094		IC-157		
4-2060-07500	1158	3289	IC-71		
4-2060-09100	1183	3475/1050			
4-2060-09200	738	3167			
4-2060-09700	738	3167			
4-2060-09800	1178	3480			
4-2061-04970	712		IC-2		
4-2061-05370	1109	3711			
4-2061-05470	712		IC-2		
4-2073	160	3008	245	2004	
4-245	181		75	2041	
4-265	175		246	2020	
4-274	128		243	2030	
4-279	126	3008	52	2024	
4-280	126	3008	52	2024	
4-282	109	3088		1123	
4-283	185		58	2025	
4-288	128		243	2030	
4-30203845	129	3025	244	2027	
4-3022861	108	3452	86	2038	
4-3023190	181	3036	75	2041	
4-3023212	123A	3444	20	2051	
4-3023221	123A	3444	20	2051	
4-3023222	129	3025	244	2027	
4-3023223	128	3024	243	2030	
4-3023843	130	3027	14	2041	
4-3023844	128	3024	243	2030	
4-3025763	108	3452	86	2038	
4-3025764	108	3452	86	2038	
4-3025765	108	3452	86	2038	
4-3025766	123A	3444	20	2051	
4-3025767	108	3452	86	2038	
4-3033	116	3100/519	504A	1104	
4-3034	177	3100/519	300	1122	
4-3036	1029		IC-162		
4-324	312			2028	
4-3540012	116	3016	504A	1104	
04-38190-01	312	3116	FET-1	2035	
4-397	128	3024	243	2030	
4-398	128	3024	243	2030	
4-399	108	3452	86	2038	
4-400	108	3452	86	2038	
4-432	160	3006	245	2004	
4-433	108	3452	86	2038	
4-434	108	3452	86	2038	
4-435	104	3009	16	2006	
4-436		3057	ZD-5.6	561	
4-436(SEARS)	136A		ZD-5.6	561	
4-437		3126	90		
4-438	124		12		
4-44-0012-PT2	160		245	2004	
04-440032-002	159	3466	82	2032	
04-440032-008	159	3466	82	2032	
4-443	108	3452	86	2038	
4-458	181		75	2041	
4-46	199	3122	62	2010	
04-46000-02	229	3018	20	2038	
4-464	152	3893	66	2048	
4-47		3122	62	2010	
4-47(SEARS)	123A		20	2051	
4-48	226MP	3052	49(2)	2025(2)	
4-48(SEARS)	226MP	3086	49(2)	2025(2)	
4-490	181		75	2041	
4-4A-1A7-1	121	3009	239	2006	
4-50			504A	1104	
4-50(SEARS)	116		504A	1104	
4-5145	123A	3444	20	2051	
04-57-303	102	3004	2	2007	
4-65-1A7-1	160	3006	245	2004	
4-65-2A7-1	103A	3010	59	2002	
4-65-4A7-1	101	3011	8	2002	
4-65A17-1	160	3007	245	2004	
4-65B17-1	160	3008	245	2004	
4-65C17-1	121	3009	239	2006	
4-66-1A7-1	102A	3003	53	2007	
4-66-2A7-1	102A	3004	53	2007	
4-66-3A7-1	102A	3004	53	2007	
04-67000-01	159	3138/193A	21	2032	
4-684120-3	108	3452	86	2038	
4-685285-3	108	3452	86	2038	
4-686105-3	125	3031A	510,531	1114	
4-686106-3	125	3032A	510,531	1114	
4-686107-3	108	3452	86	2038	
4-686108-3	108	3452	86	2038	
4-686112-3	108	3452	86	2038	
4-686114-3	108	3452	86	2038	
4-686116-3	117	3017B	504A	1104	
4-686118-3	108	3452	86	2038	
4-686119-3	108	3452	86	2038	
4-686120-3	108	3452	86	2038	
4-686126-3	108	3452	86	2038	
4-686127-3	108	3452	86	2038	
4-686130-3	175	3026	246	2020	
4-686131-3	108	3452	86	2038	
4-686132-3	123A	3444	20	2051	
4-686139-3	125	3032A	510,531	1114	
4-686140-3	108	3452	86	2038	
4-686143-3	123A	3444	20	2051	
4-686144-3	123A	3444	20	2051	
4-686145-3	154	3040	40	2012	
4-686147-3	117	3017B	504A	1104	
4-686148-3	117	3017B	504A	1104	
4-686149-3	116	3016	504A	1104	
4-686150-3	117	3017B	504A	1104	
4-686151-3	117	3017B	504A	1104	
4-686163-3	102A	3004	53	2007	
4-686169-3	108	3452	86	2038	
4-686170-3	129	3025	244	2027	
4-686171-3	108	3452	86	2038	
4-686172-3	108	3452	86	2038	
4-686173-3	123A	3444	20	2051	
4-686177-3	117	3017B	504A	1104	
4-686179-3	125	3032A	510,531	1114	
4-686182-3	128	3024	243	2030	
4-686183-3	123A	3444	20	2051	
4-686184-3	125	3311	510,531	1114	
4-686186-3	125	3311	510,531	1114	
4-686189-3	125	3031A	510,531	1114	
4-686195-3	102A	3004	53	2007	
4-686196-3	102A	3004	53	2007	
4-686199-3	125	3031A	510,531	1114	
4-686201-3	125	3311	510,531	1114	
4-686207-3	108	3452	86	2038	
4-686208-3	108	3452	86	2038	
4-686209-3	108	3452	86	2038	
4-686212-3	125	3031A	510,531	1114	
4-686213-3	121	3009	239	2006	
4-686224-3	108	3452	86	2038	
4-686226-3	175	3026	246	2020	
4-686227-3	502	3067	CR-4		
4-686228-3	108	3452	86	2038	
4-686229-3	129	3025	244	2027	
4-686230-3	129	3025	244	2027	
4-686231-3	123A	3444	20	2051	
4-686232-3	154	3040	40	2012	
4-686234-3	196	3041	241	2020	
4-686235-3	129	3025	244	2027	
4-686238-3	129	3025	244	2027	
4-686244-3	108	3452	86	2038	
4-686251-3	108	3452	86	2038	
4-686252-3	191	3044/154	249		
4-686256-1	126	3008	52	2024	

Industry Standard No.	ECG	SK	GE	RS 276-	MOTOR.
4-686256-2	126	3008	52	2024	
4-686256-3	126	3008	52	2024	
4-686257-3	123A	3444	20	2051	
4-68681-2	102A	3004	53	2007	
4-68681-3	102A	3004	53	2007	
4-68682-3	123A	3444	20	2051	
4-68687-3	117	3017B	504A	1104	
4-68689-3	125	3031A	510,531	1114	
4-68695-3	108	3452	86	2038	
4-68697-3	125	3032A	510,531	1114	
4-6B-3A7-1	121MP	3015	239(2)	2006(2)	
4-74-3A7-1	100	3005	1	2007	
4-77A17-1	105	3012	4		
4-77B17-1	121MP	3013	239(2)	2006(2)	
4-77C17-1	121	3014	239	2006	
004-8000	121	3009	239	2006	
04-8054-3	116	3175	504A	1104	
04-8054-4	116		504A	1104	
04-8054-7	116	3175	504A	1104	
4-8134842	191	3044/154	249		
4-850	107	3018	11	2015	
4-851	123A	3124/289	20	2051	
4-852	109	3088		1123	
4-853	109	3088		1123	
4-854	109	3088		1123	
4-855	109	3088		1123	
4-856	5072A		ZD-8.2	562	
4-857	109	3088		1123	
4-882				1123	
4-88A17-1	121	3717	239	2006	
4-88B17-1	160	3006	245	2004	
4-88C17-1	100	3005	1	2007	
4-8P-2A7-1	102A	3004	53	2007	
4-92-1A7-1	105	3012	4		
4-9L-4A7-1	102A	3003	53	2007	
4-JBD1	108		86	2038	
4A-1	121	3009	239	2006	
04A-1-12-7	121	3009	239	2006	
4A-1-70	121	3009	239	2006	
4A-1-70-12	121	3009	239	2006	
4A-1-70-12-7	121	3009	239	2006	
4A-1-A-7B	121	3717	239	2006	
4A-10	121	3009	239	2006	
4A-11	121	3009	239	2006	
4A-12	121	3009	239	2006	
4A-13	121	3009	239	2006	
4A-14	121	3009	239	2006	
4A-15	121	3009	239	2006	
4A-16	121	3009	239	2006	
4A-17	121	3009	239	2006	
4A-18	121	3009	239	2006	
4A-19	121	3009	239	2006	
4A-1A	121	3009	239	2006	
4A-1A0	121	3009		2006	
4A-1A0R	121	3009		2006	
4A-1A1	121	3009	239	2006	
4A-1A19	121	3009	239	2006	
4A-1A2	121	3009	239	2006	
4A-1A21	121	3009	239	2006	
4A-1A3	121	3009	239	2006	
4A-1A3P	121	3009	239	2006	
4A-1A4	121	3009	239	2006	
4A-1A4-7	121	3009	239	2006	
4A-1A5	121	3009	239	2006	
4A-1A5L	121	3009	239	2006	
4A-1A6	121	3009	239	2006	
4A-1A6-4	121	3009	239	2006	
4A-1A7	121	3009	239	2006	
4A-1A7-1	121	3009	239	2006	
4A-1A8	121	3009	239	2006	
4A-1A82	121	3009	239	2006	
4A-1A9	121	3009	239	2006	
4A-1A9G	121	3009	239	2006	
4A-1AO			239	2006	
4A-1AOR		3717	239	2006	
4A009	276	3296			
04A1	121	3717	239	2006	
04A1-12	121	3717	239	2006	
4A1122	5994	9024/5094A	5104		
4A132	5944		5048		
4A162	5994	3501	5104		
4A2122	5994	9024/5094A	5104		
4A232	5994	3501	5104		
4A262	5994	3501	5104		
4A8-1A5	160		245	2004	
4A8-1A7-1	160		245	2004	
4AJ4DX520	116	3016	504A	1104	
4AJ4DX52D	116	3017B/117	504A	1104	
4C1	5982	3609/5986	5096		
4C10	6002	7202	5108		
4C2	5986	3609	5096		
4C28	128	3444/123A	20	2030	
4C29	128	3444/123A	20	2030	
4C30	128	3444/123A	20	2030	
4C31	128	3444/123A	20	2030	
4C4	5990	3608	5100		
4C43			81	2051	
4C6	5994	3501	5104		
4C8	5998		5108		
4D20	123A	3444	20	2051	
4D21	123A	3444	20	2051	
4D22	123A	3444	20	2051	
4D24	123A	3444	20	2051	
4D25	123A	3444	20	2051	
4D26	123A	3444	20	2051	
4D4	116	3031A	504A	1104	1N4004
4D6	116	3017B/117	504A	1104	1N4005
4D8	116	3017B/117	504A	1104	
4FB10					1N4934
4FB20					1N4935
4FB30					1N4936
4FB40					1N4936
4FB5					1N4933
4FC					1N4934
4FC10					1N4934
4FC20					1N4935
4FC30					IN4936
4FC40					1N4936
4FC5					1N4933
4G1122	5994	3501	5104		
4G132	5994	3501	5104		
4G162	5994	3501	5104		
4G2	451	3116			2N5485
4G2122	5994	3501	5104		
4G232	5994	3501	5104		
4G262	5994	3501	5104		
4G8	125	3031A	510,531	1114	
4GA	116	3031A	504A	1104	
4I29200602				1123(2)	
4I39104002	116		504A	1104	
4J24X539			ZD-12	563	
4J56			52	2024	
4JA10DX3	116	3016	504A	1104	
4JA10DX32	116	3016	504A	1104	
4JA10EX3	116	3016	504A	1104	
4JA11BX4	5874		5032		
4JA16MR700M	125	3017B/117	510,531	1114	
4JA211A	116	3016	504A	1104	
4JA27DR700	5878		5036		
4JA2FX355	116	3016	504A	1104	
4JA2X355	116	3016	504A	1104	
4JA38DR700	5998		5108		
4JA4DR700	116	3017B/117	504A	1104	
4JA4DX520	116	3016	504A	1104	
4JA6MR700	116	3017B/117	504A	1104	
4JA70MR700	6156	7356			
4JB2C6	112	3089	1N82A		
4JBC12	112	3089	1N82A		
4JD1A17	100	3005	1	2007	
4JD1A73	102	3004	2	2007	
4JD3B1	101	3861	8	2002	
4JD5E29	6400A			2029	
4JX11C2848	128		243	2030	
4JX16A567	123A	3444	20	2051	
4JX16A569			8	2002	
4JX16A667	123A	3444	20	2051	
4JX16A667/G			17	2051	
4JX16A667/O		3122	17	2051	
4JX16A667/R		3122	17	2051	
4JX16A667/Y		3122	17	2051	
4JX16A6670		3444	20	2051	
4JX16A667G	123A	3444	20	2051	
4JX16A6670	123A			2051	
4JX16A667R	123A	3444	20	2051	
4JX16A667Y	123A	3444	20	2051	
4JX16A668	123A	3444	20	2051	
4JX16A668/G		3122	17	2051	
4JX16A668/O		3122	17	2051	
4JX16A668/Y		3122	17	2051	
4JX16A6680		3444	20	2051	
4JX16A668G	123A	3444	20	2051	
4JX16A6680	123A			2051	
4JX16A668Y	123A	3444	20	2051	
4JX16A669	123A	3444	20	2051	
4JX16A669G	123A	3444	20	2051	
4JX16A669Y	123A	3444	20	2051	
4JX16A670	123A	3444	20	2051	
4JX16A670G	123A	3444	20	2051	
4JX16B670/B		3122	17	2051	
4JX16B670/R		3122	17	2051	
4JX16B670/Y		3122	17	2051	
4JX16B670B	123A		20	2051	
4JX16B670G	123A	3444	20	2051	
4JX16B670R	123A	3444	20	2051	
4JX16B670Y	123A	3444	20	2051	
4JX16E3860	123A	3444	20	2051	
4JX16E3890	123A	3444	20	2051	
4JX16E3960	123A	3444	20	2051	
4JX1A520	102	3004	2	2007	
4JX1A520B	102	3004	2	2007	
4JX1A520C	102	3004	2	2007	
4JX1A520D	100	3005	1	2007	
4JX1A520E	100	3005	1	2007	
4JX1A813	160	3011	245	2004	
4JX1C1224	102	3722	2	2007	
4JX1C707	160		245	2004	
4JX1C850	102	3004	2	2007	
4JX1C850A	100	3005	1	2007	
4JX1D925	102A		53	2007	
4JX1E596	123A	3444	20	2051	
4JX1E821	100	3004/102A	1	2007	
4JX1E850	103	3862	8	2002	
4JX24X539	142A		ZD-12	563	
4JX2816	103	3862	8	2002	
4JX2825	103	3862	8	2002	
4JX29A529		3466	82	2032	
4JX29A826	159	3466	82	2032	
4JX29A829	159			2032	
4JX2A60	100	3005	1	2007	
4JX2A601	103	3862	8	2002	
4JX2A616	103A	3835	59	2002	
4JX2A801	101	3861	8	2002	
4JX2A816	103	3862	8	2002	
4JX2A822	103	3862	8	2002	
4JX5E670	6400A			2029	
4JX7A972	123A	3444	20	2051	
4JX8D404	121	3009	239	2006	
4JX8P404	121	3009	239	2006	
4JX8P409	121	3009	239	2006	
4JZ4X539	142A		ZD-12	563	
4JZ4XL12	142A		ZD-12	563	
4K100	5402	3638			
4K30	5400	3950			
4K60	5401	3638/5402			
4L100	5402	3638			
4L30	5400	3950			
4N25	3040		4N25		4N25
4N26	3040				4N26
4N27	3040		4N27		4N27
4N28	3040		4N28		4N28
4N29			4N29		4N29
4N30					4N30
4N31					4N31
4N32			4N32		4N32
4N33			4N33		4N33
4N35	3041		4N35		4N35
4N36	3041		4N36		4N36
4N37	3041		4N37		4N37
4N38					4N38
4N39			4N39		4N39
4N40			4N40		4N40
4N45					4N32
4N46					4N32
4RCM30	5484	3943/5485			
04S05		3298	236		
04S06	235	3197	337		
4SD46-2	109			1123	
4T501	116	3016	504A	1104	
4T502	116	3016	504A	1104	
4T503	116	3016	504A	1104	
4T504	116	3016	504A	1104	
4T505	116	3017B/117	504A	1104	
4T506	116	3017B/117	504A	1104	
4T507	125	3032A	510,531	1114	
4T508	125	3032A	510,531	1114	
4T509	125	3033	510,531	1114	
4T510	125	3033	510,531	1114	
4T6.2	137A		ZD-6.2	561	
4T6.2A	137A		ZD-6.2	561	
4T6.2B	137A		ZD-6.2	561	
4W01-13	142A	3062	ZD-12	563	
4ZL	186A	3202/210	247	2052	
005(MCINTOSH)	724		IC-86		
05-00000-00	109	3087		1123	
05-000104		3312		1104	
05-00060-00	109	3088		1123	
05-00060-01	109	3088		1123	
05-00085-00		3126	90		

Industry Standard No.	ECG	SK	GE	RS 276-	MOTOR.
05-001100	116	3311			
05-00160-01	109	3088		1123	
05-002139	611	3324			
005-010400	522	3303			
005-010800	522	3303			
005-010900	522	3303			
005-011000	522	3303			
005-011200	522	3303			
005-012200	534	3305			
005-013500	532	3300/517			
005-013600	531	3301			
005-02	123A	3444	20	2051	
05-02160-01	177	3100/519	300	1122	
05-02638-01		3126	90		
05-03016-01	116	3311	504A	1104	
05-040004	552	3998/506			
05-040006	552	3998/506			
05-04001-02	116	3311	504A	1104	
05-04800-02	5072A	3136	ZD-8.2	562	
05-060110	5074A	3139	ZD-11	563	
05-090092	139A	3060			
05-110046		3843	511	1114	
05-110106	5071A	3334	ZD-6.8		
05-110107		3334	ZD-6.8	561	
05-110108	138A	3059	ZD-7.5		
05-110442	177	9091		1122	
05-111011		3169	ZD-11A	563	
05-112404	506	3843	511	1114	
05-112406		3080	5	1114	
05-123103	552	3998/506			
05-141025			504A	1104	
05-150046	605	3864			
05-170034	109	3087	1N34AS	1123	
05-170060	109	3088	1N60	1123	
05-174002	116	3312			
05-174004		3312		1104	
05-180034	109	3087	1N34AS	1123	
05-180053	177	9091		1122	
05-180188	109	3087	1N34AS	1123	
05-180553	612	3325			
05-180953	519	3175	514	1122	
05-18155			300	1122	
05-181555	177	3175	300	1122	
05-181658	614	3327			
05-182076	519	3100	514	1122	
05-182236	611	3325/612			
05-182688	614	3327	90		
05-190061	125	3313/116	504A	1104	
05-200310	614	3325/612	ECG614		
5-3/830136.T			510,531	1114	
5-30082.1003	117	3017B			
5-30082.3	125	3032A	510,531	1114	
5-30082.4	125	3032A	510,531	1114	
5-30086.1	116	3016	504A	1104	
5-30088.1	116	3016	504A	1104	
5-30088.1002	125	3032A			
5-30088.1003	125	3032A			
5-30088.1004	125	3032A			
5-30088.2	117	3017B	504A	1104	
5-30088.3	125	3032A	510,531	1114	
5-30094.1	116	3016	504A	1104	
5-30095.1	116	3016	504A	1104	
5-30098.1	116	3016	504A	1104	
5-30099.1	116	3016	504A	1104	
5-30099.3	116	3016	504A	1104	
5-30099.4	116	3016	504A	1104	
5-30106.1	116	3016	504A	1104	
5-30106.1001	125	3033			
5-30109.1	116	3016	504A	1104	
5-30111.1	117	3017B	504A	1104	
5-30111.1001	125	3031A			
5-30113.1	116	3016	504A	1104	
5-30119.1	503	3068	CR-5		
5-30120.1	116	3016	504A	1104	
5-30122.1	117	3017B	504A	1104	
5-30132.1	118	3066	CR-1		
5-30136.T	125	3032A			
05-320301	610		300	1122	
05-328518	600	3327/614			
05-330150	177	9091	300	1122	
05-330161		3100	514	1122	
05-331091	5018A	3781/5015A			
05-340301	610	3323			
05-370151	503	3068	CR-5		
05-429602	116	3311			
05-472209	614	3327	90		
05-480086	5017A	3749/5073A			
05-480204	5067A	3331	ZD-3.9		
05-480205	5010A	3056/135A	ZD-5.0		
05-480306	5070A	9021	ZD-6.0	561	
05-490005	138A	3059			
05-490090		3088	1N60	1123	
05-490091		3087	1N34A	1123	
05-490095		3087	1N34A	1123	
05-540001	116	3311	504A	1104	
05-540082	5072A	3136	ZD-8.2	562	
05-540091	139A	3060			
05-540094	139A	3060			
05-540112		3311	504A	1104	
05-610046	109	3087	1N34AS	1123	
05-69669-01	5071A	3334	ZD6.8	561	
5-70004503	123A	3020/123	20	2051	
5-70005452	123A	3124/289	20	2051	
5-70005503	123A	3124/289	20	2051	
5-7000901504	123A	3124/289	20	2051	
05-740012	552	3998/506			
05-750010	116	3313	504A	1104	
05-780251	614	3327	90		
5-8	123A	3444	20	2051	
05-800015		3843	511	1114	
05-860002	552	3998/506	511	1114	
05-931601	116	3313	504A	1104	
05-931609	552	3312	504A	1104	
05-931642	177	3175	300	1122	
05-931645	177	3175	300	1122	
05-931771	109	3087	1N34AS	1123	
05-931971	116	3312	504A	1104	
05-932510	109	3088	1N60	1123	
05-933935		3338	ZD-39		
05-933943		3062	ZD-12	563	
05-933945	5081A	3151	ZD-24		
05-933948		3145	ZD-17		
05-933949	5005A	3771			
05-933950	5075A	3751	ZD-16		
05-935201	552	3312	504A	1104	
05-936010	506	3843	511	1114	
05-936470	177	3175	300	1122	
05-950177	5077A	3752	ZD-18		
05-990094	139A	3060	ZD-9.1	562	
5-S30082.1003			504A	1104	
5-S30088.1002			510,531	1114	
5-S30088.1003			510,531	1114	
5-S30088.1004			510,531	1114	
5-S30106.1001			510,531	1114	

Industry Standard No.	ECG	SK	GE	RS 276-	MOTOR.
5-S30111.1001			510,531	1114	
5A					1N4004
5A-D	116	3017B/117	504A	1104	
05A01		3100	300	1122	
05A03	109	3087	1N34AS	1123	
05A05		3100	300	1122	
05A06	519	3100	514	1122	
05A07	116	3311	504A	1104	
05A08	614	3327			
5A1	116	3016	504A	1104	1N4002
5A10	125	3033	510,531	1114	1N4007
5A10C	125		510,531	1114	
5A10D	125	3081	510,531	1114	
5A10D-C	125	3033	510,531	1114	
5A15	5400	3950			
5A2	116	3016	504A	1104	1N4003
5A3	116	3016	504A	1104	1N4004
5A4	116	3016	504A	1104	1N4004
5A4D	116	3016	504A	1104	
5A4D-C	116	3016	504A	1104	
5A5	116	3017B/117	504A	1104	1N4005
5A5D	116	3017B/117	504A	1104	
5A6	116	3017B/117	504A	1104	1N4005
5A6D	116	3017B/117	504A	1104	
5A6D-C	116	3017B/117	504A	1104	
5A8	125	3032A	510,531	1114	1N4006
5A8D	125	3032A	509	1114	
5A8DC	125		510,531	1114	
5AA5-1	105	3012	4		
5AD10	156	3033	512		
5AD10C	156	3033	512		
5AD10D	156	3033	512		
5AD8	156	3033	512		
5AD8DC	156	3033	512		
5B-1	116(4)	3311	504A	1104	
5B-15H	116		504A	1104	
5B-15H(SHARP)	506		511	1114	
5B-2	116		504A	1104	
5B-2-H5W	116		504A	1104	
5B-2-H5W(SHARP)	506		511	1114	
5B1	116(4)	3311			
05B1Z	712	3072	IC-2		
5B2	156	3051	512		
05B2Z	712	3072	IC-2		
5B3	116	3017B/117	504A	1104	
5B4	169	3678		1104	
5C004	960	3591			
5C010	788	3147			
5D-5A-L	5068A	3332	ZD-4.3		
5D1	116	3016	504A	1104	
5D10	125		510,531	1114	
5D2	116		504A	1104	
5D46	109	3087			
5D8	125		510,531	1114	
5E1	116	3016	504A	1104	
5E2	116	3016	504A	1104	
5E3			504A	1104	
5E4	116	3016	504A	1104	
5E4850/56-0002	128		243	2030	
5E5	116	3017B/117	504A	1104	
5E6	125	3017B/117	504A	1104	
5E8	125	3033	510,531	1114	
5F1	5830		5044		
5F15	5944	3501/5994	5104		
5F20	5944	3501/5994	5104		
5F5	5850		5016		
5G-D		3311	504A	1104	
5G2	451	3116			2N5485
5G8	125	3033	510,531	1114	
5GA	116	3017B/117	504A	1104	
5GB	116	3100/519	300	1104	
5GD	116	3016	504A	1104	
5GF	125	3031A	510,531	1114	
5GFH	116	3016	504A	1104	
5GJ	125	3032A	510,531	1114	
5GJ/FR1N		3016	504A	1104	
5GJFR1N	116		504A	1104	
5GL	116	3017B/117	504A	1104	
5H	116	3311	504A	1104	
5H3P	5850		5016		
5H3PN	5851		5017		
5H4D1	116	3016	504A	1104	
5H750M	116	3016	504A	1104	
5J-F1	116	3016	504A	1104	
5J3P	5870		5032		
5K10	125		510,531	1114	
5L15	525	3925			
5L1626	7447	7447			
5M100	5910		5076		
5M40	5900	3608/5990	5068		
5M50	5902		5072		
5M60	5904	7104	5072		
5M80	5908		5076		
5MA10	125	3033	510,531	1114	
5MA2	116	3016	504A	1104	
5MA4	116	3017B/117	504A	1104	
5MA5	116	3017B/117		1104	
5MA6	116	3017B/117	504A	1104	
5MA8	125	3032A	510,531	1114	
5MAK			504A	1104	
5MF1	116	3016	504A	1104	
5MS10	116	3016	504A	1104	
5MS20	116	3016	504A	1104	
5MS30	116	3016	504A	1104	
5MS40	116	3016	504A	1104	
5MS5	116	3016	504A	1104	
5MS50	116	3016	504A	1104	
5N1	116	3016	504A	1104	
5N2	5874		5032		
5N3	5874		5032		
50				1123	
5P05M	5461	3685			
5P1M	5462	3686			
5P2	5874		5032		
5P2M	5463	3572			
5P3	5874		5032		
5P4M	5465	3687			
5Q3	5981	3698/5987	5097		
5RC15	5483	3942			
5RC20	5483	3942			
5RC25	5484	3943/5485			
5RC30	5484	3943/5485			
5RC40	5485	3943			
5RC40A	5485	3943			
5RC50	5486	3944/5487			
5RC50A	5486	3944/5487			
5RC50B	5486	3944/5487			
5RC60	5487	3944			
5RC60A	5487	3944			
5RCL15	5483	3942			
5RCL20	5483	3942			
5SCBR05					MDA2500
5SCBR1					MDA2501
5SCBR2					MDA2502

Industry Standard No.	ECG	SK	GE	RS 276-	MOTOR.
5SCBR4					MDA2504
5SCBR6					MDA2506
5TB2	5483	3942			
5TB3	5484	3943/5485			
5TB4	5485	3943			
5TB5	5486	3944/5487			
5TB6	5487	3944			
05V-50	116	3311	504A	1104	
05Z	138A	3059			
05Z-10	140A	3061	ZD-10	562	
05Z-6.2L	137A		ZD-6.2	561	
05Z-9.1U	139A	3060			
05Z10	140A		ZD-10	562	
05Z11	5074A			563	
05Z12	142A			563	
05Z13	143A	3750	ZD-13	563	
05Z15	145A			564	
05Z15U	5024A	3063/145A			
05Z16	5075A	3751	ZD-16	564	
05Z18	5077A	3752			
05Z18L	5027A	3793			
05Z5.1L	5010A	3056/135A	ZD-5.1		
05Z5.6				561	
5Z5338					1N5338A
5Z5364					1N5364A
05Z6.2	137A			561	
05Z6.2L	5013A	3779			
05Z6.8	5071A			561	
05Z7.5	138A		ZD-7.5		
05Z7.5-UNI		3059	ZD-7.5		
05Z8.2	5072A	3136	ZD-8.2	562	
05Z9.1	139A			562	
05Z9.1L	139A	3060	ZD-9.1	562	
5ZS10	5125A	3391			
5ZS100	5156A	3422			
5ZS100A	5156A	3422			
5ZS100B	5156A	3422			
5ZS10A	5125A	3391	5ZD-10		
5ZS10B	5125A	3391			
5ZS11	5126A	3392			
5ZS11A	5126A	3392			
5ZS11B	5126A	3392	5ZD-11		
5ZS12	5127A	3393			
5ZS12A	5127A	3393			
5ZS12B	5127A	3393	5ZD-12		
5ZS13	5128A	3394			
5ZS13A	5128A	3394	5ZD-13		
5ZS13B	5128A	3394			
5ZS14	5129A	3395			
5ZS14A	5129A	3395	5ZD-14		
5ZS14B	5129A	3395			
5ZS15	5130A	3396			
5ZS15A	5130A	3396			
5ZS15B	5130A	3396			
5ZS16	5131A	3397	5ZD-16		
5ZS16A	5131A	3397	5ZD-16		
5ZS16B	5131A	3397	5ZD-16		
5ZS17	5132A	3398	5ZD-17		
5ZS17A	5132A	3398	5ZD-17		
5ZS17B	5132A	3398	5ZD-17		
5ZS18	5133A	3399			
5ZS18A	5133A	3399			
5ZS18B	5133A	3399			
5ZS19	5134A	3400	5ZD-19		
5ZS19A	5134A	3400	5ZD-19		
5ZS19B	5134A	3400	5ZD-19		
5ZS20	5135A	3401			
5ZS20A	5135A	3401			
5ZS20B	5135A	3401			
5ZS22	5136A	3402			
5ZS22A	5136A	3402			
5ZS22B	5136A	3402			
5ZS24	5137A	3403			
5ZS24A	5137A	3403			
5ZS24B	5137A	3403			
5ZS25	5138A	3404	5ZD-25		
5ZS25A	5138A	3404	5ZD-25		
5ZS25B	5138A	3404	5ZD-25		
5ZS27	5139A	3405			
5ZS27A	5139A	3405			
5ZS27B	5139A	3405			
5ZS28	5140A	3406	5ZD-28		
5ZS28A	5140A	3406	5ZD-28		
5ZS28B	5140A	3406	5ZD-28		
5ZS3.3	5111A	3377			
5ZS3.3A	5111A	3377			
5ZS3.3B	5111A	3377			
5ZS3.6A	5112A	3378			
5ZS3.6B	5112A	3378			
5ZS3.9	5113A	3379			
5ZS3.9A	5113A	3379			
5ZS30	5141A	3407			
5ZS30A	5141A	3407			
5ZS30B	5141A	3407			
5ZS33	5142A	3408			
5ZS33A	5142A	3408			
5ZS33B	5142A	3408			
5ZS36	5143A	3409			
5ZS36A	5143A	3409			
5ZS36B	5143A	3409			
5ZS39	5144A	3410			
5ZS39A	5144A	3410			
5ZS39B	5144A	3410			
5ZS4.3	5114A	3380			
5ZS4.3A	5114A	3380			
5ZS4.3B	5114A	3380			
5ZS43	5145A	3411			
5ZS43A	5145A	3411			
5ZS43B	5145A	3411			
5ZS47	5146A	3412			
5ZS47A	5146A	3412			
5ZS47B	5146A	3412			
5ZS5.1	5116A	3382			
5ZS5.1A	5116A	3382			
5ZS5.1B	5116A	3382			
5ZS5.6	5117A	3383			
5ZS5.6A	5117A	3383			
5ZS5.6B	5117A	3383			
5ZS51	5147A	3413			
5ZS51A	5147A	3413			
5ZS51B	5147A	3413			
5ZS6.0A	5118A	3384			
5ZS6.0B	5118A	3384			
5ZS6.2	5119A	3385			
5ZS6.2A	5119A	3385			
5ZS6.2B	5119A	3385			
5ZS60	5043A	3809	ZD-60		
5ZS60A	5043A	3809	ZD-60		
5ZS60B	5149A	3809/5043A	ZD-60		
5ZS68	5151A	3417			
5ZS68A	5151A	3417			
5ZS68B	5151A	3417			
5ZS7.5	5121A	3387			
5ZS7.5A	5121A	3387	5ZD-7.5		

Industry Standard No.	ECG	SK	GE	RS 276-	MOTOR.
5ZS7.5B	5121A	3387	5ZD-7.5		
5ZS75	5152A	3418			
5ZS75A	5152A	3418			
5ZS75B	5152A	3418			
5ZS8.2	5122A	3388			
5ZS8.2A	5122A	3388			
5ZS8.2B	5122A	3388			
5ZS8.7	5123A	3389			
5ZS8.7A	5123A	3389	5ZD-8.7		
5ZS8.7B	5183A	3389/5123A			
5ZS82	5153A	3419			
5ZS82A	5153A	3419			
5ZS82B	5153A	3419			
5ZS87	5153A	3420/5154A			
5ZS87A	5153A	3420/5154A			
5ZS87B	5154A	3420			
5ZS9.1	5124A	3390			
5ZS9.1A	5124A	3390	5ZD-9.1		
5ZS9.1B	5124A	3390			
5ZS91	5155A	3421			
5ZS91A	5155A	3421			
5ZS91B	5155A	3421			
006-0000004	519		514	1122	
006-0000134	123A	3444	20	2051	
006-0000135	159	3466	82	2032	
6-0000139	128		243	2030	
006-0000146	7400	7400	7400	1801	
006-0000147	7410	7410	7410	1807	
6-0000155A	102	3004	2	2007	
6-0000158	121		239	2006	
006-0000162	74107	74107			
6-000105	102	3722	2	2007	
6-000140	175		246	2020	
006-0004270-001		3590	IC-250		
006-0004443	312			2028	
6-0004799	5635	3533			
006-0004956	129		244	2027	
006-0005191	129		244	2027	
6-0005193	152	3893	66	2048	
006-00055182	234		65	2050	
6-000555-2	175		246	2020	
6-007-3	708		IC-10		
006-02	159	3466	20	2032	
6-04	123A	3335/5079A	20	2051	
6-0451	123A	3335/5079A	20	2051	
6-0452	123A	3335/5079A	20	2051	
6-04GRN	123A	3335/5079A	20	2051	
6-04ORN	123A			2051	
6-04ORNN			20	2051	
6-04S1	123A	3335/5079A	20	2051	
6-04S2	123A	3335/5079A	20	2051	
6-05	123A	3335/5079A	20	2051	
6-05F	123A	3335/5079A	20	2051	
6-05YEL	123A	3335/5079A	20	2051	
6-11	123A	3335/5079A	20	2051	
6-1260039	160	3008	245	2004	
6-1260039A	160	3008	245	2004	
6-13	102A	3004	53	2007	
6-137	130	3510	14	2041	
6-138	234	3511	65	2050	
6-138(PWR)	130		14	2041	
6-139	128		243	2030	
6-140	175		246	2020	
6-155	102	3722	2	2007	
6-158	121	3717	239	2006	
6-19	123A	3335/5079A	20	2051	
6-24	175		246	2020	
6-2708	123A	3122	62	2038	
6-30	123A	3335/5079A	20	2051	
6-31	159	3466	82	2032	
6-31A	159	3466	82	2032	
6-38	159	3466	82	2032	
6-38A	159	3466	82	2032	
6-4	123A	3335/5079A	20	2051	
6-4799	5635	3533			
6-49	234	3118	65	2050	
6-490001	175		246	2020	
6-50	126	3006/160	52	2024	
6-5193	152	3893	66	2048	
6-53	102A	3004	53	2007	
6-53/63			2	2007	
6-5363	102A		53	2007	
6-53A	102A	3004	53	2007	
6-53F	102A		53	2007	
6-5552	175		246	2020	
6-59010	116	3016	504A	1104	
6-60	102A	3008	53	2007	
6-60-P			2	2007	
6-60A	160	3008	245	2004	
6-60B	102A	3008	53	2007	
6-60C	160	3008	245	2004	
6-60D	102A	3008	53	2007	
6-60E	160	3008	245	2004	
6-60F	160	3008	245	2004	
6-60P	102A	3008	53	2007	
6-60T	160		245	2004	
6-60X	160	3123	245	2004	
6-61	102A	3008	53	2007	
6-61-P			2	2007	
6-61A	160	3008	245	2004	
6-61B	102A	3008	53	2007	
6-61D	102A	3008	53	2007	
6-61E	160	3008	245	2004	
6-61F	160	3008	245	2004	
6-61P	102A	3008	53	2007	
6-61T	160	3008	245	2004	
6-61X	160	3123	245	2004	
6-62	102A	3008	53	2007	
6-62-P			2	2007	
6-62A	102A	3008	53	2007	
6-62B	102A	3008	53	2007	
6-62C	160	3008	245	2004	
6-62D	102A	3008	245	2005	
6-62E	160	3008	245	2004	
6-62F	126	3008	52	2024	
6-62P	126	3008	52	2024	
6-62X	126	3006/160	52	2024	
6-63	102A	3004	53	2007	
6-63A	102A	3004	53	2007	
6-63T	102A		53	2007	
006-6400902	519		514	1122	
006-6450032	123A	3122	20	2051	
6-6450036	123A	3444	20	2051	
6-6490001	175		246	2020	
6-6490004	130		14	2041	
6-65	102A		53	2007	
6-65T	102A		53	2007	
6-66	102A		53	2007	
6-66T	102A		53	2007	
6-67	102A		53	2007	
6-67T	102A		53	2007	
6-69	160	3006	245	2004	
6-69X	160	3006	245	2004	
6-70	126	3006/160	52	2024	

Industry Standard No.	ECG	SK	GE	RS 276-	MOTOR.
6-71	126	3006/160	52	2024	
6-72	126	3006/160	52	2024	
6-84F	160		245	2004	
6-856	128		243	2030	
6-85F	160		245	2004	
6-87	102A		53	2007	
6-88	102A	3009	53	2007	
6-88(AUTOMATIC)	131		44	2006	
6-89	126	3006/160	52	2024	
6-89X	101	3861	8	2002	
6-90	123A	3444	20	2051	
6-9029-15D	123A	3444	20	2051	
6-9029-15E	123A	3444	20	2051	
6-9029-20J	129		244	2027	
6-93	123A	3444	20	2051	
6.2SR1	137A	3058	ZD-6.2	561	
6.2SR1A	137A	3058	ZD-6.2	561	
6.2SR2	137A	3058	ZD-6.2	561	
6.2SR2A	137A	3058	ZD-6.2	561	
6.2SR3	137A	3058	ZD-6.2	561	
6.2SR3A	137A	3058	ZD-6.2	561	
6.2SR4	137A	3058	ZD-6.2	561	
6.2SR4A	137A	3058	ZD-6.2	561	
6A100	5852		5016		
6A1000	5868	7068	5028		MR1130
6A10227	123A	3444	20	2051	
6A10228	128	3122	243	2030	
6A10229	131		44	2006	
6A10422	123A	3444	20	2051	
6A10423	123A	3444	20	2051	
6A10520	123A	3444	20	2051	
6A10622	158		53	2007	
6A10624	102A	3123	53	2007	
6A10851	123A	3444	20	2051	
6A10855	123A	3444	20	2051	
6A10F					SPECIAL
6A11180	123A	3444	20	2051	
6A11223	161	3338/5086A	39	2015	
6A11301	158	3008	53	2007	
6A11665	102A	3123	53	2007	
6A11668	102A	3123	53	2007	
6A12515	158		53	2007	
6A12516	102A	3123	53	2007	
6A12517	158		53	2007	
6A12677	108	3452	86	2038	
6A12678	100	3721	1	2007	
6A12679	108	3452	86	2038	
6A12680	126		52	2024	
6A12681	123A	3444	20	2051	
6A12682	123A	3444	20	2051	
6A12683	123A	3444	20	2051	
6A12684	102A		53	2007	
6A12685	102A		53	2007	
6A12725	123A	3444	20	2051	
6A12788	123A	3444	20	2051	
6A12789	123A	3444	20	2051	
6A12889	160		245	2004	
6A12988	154	3024/128	40	2012	
6A12989	102A	3004	53	2007	
6A12990	102A	3004	53	2007	
6A12992	103A	3010	59	2002	
6A12993	103	3862	8	2002	
6A15	5850		5016		
6A16399	123A	3444	20	2051	
6A200	5854		5016		
6A30	5850		5016		
6A300	5856		5020		
6A400	5858		5020		
6A50	5850		5016		
6A600	5862		5024		
6A6F					MR1366
6A700	5866		5028		MR1128
6A8-1A5L	160		245	2004	
6A800	5866		5028		MR1128
6A8F					SPECIAL
6A900	5868	7068	5028		MR1130
6AL1					MR751
6AL2					MR752
6AL3					MR754
6AL4					MR754
6AL6					MR756
6ALR1					MR821
6ALR2					MR822
6ALR3					MR824
6ALR4					MR824
6ALR6					MR826
6B-3	121MP	3015	239(2)	2006(2)	
06B-3-12	121MP	3015	239(2)	2006(2)	
06B-3-12-7	121MP	3015	239(2)	2006(2)	
6B-3-70	121MP	3015	239(2)	2006(2)	
6B-3-70-12	121MP	3015	239(2)	2006(2)	
6B-30	121MP	3015	239(2)	2006(2)	
6B-31	121MP	3015	239(2)	2006(2)	
6B-32	121MP	3015	239(2)	2006(2)	
6B-33	121MP	3015	239(2)	2006(2)	
6B-34	121MP	3015	239(2)	2006(2)	
6B-35	121MP	3015	239(2)	2006(2)	
6B-36	121MP	3015	239(2)	2006(2)	
6B-37	121MP	3015	239(2)	2006(2)	
6B-38	121MP	3015	239(2)	2006(2)	
6B-39	121MP	3015	239(2)	2006(2)	
6B-3A	121MP	3015	239(2)	2006(2)	
6B-3A0	121MP	3015	239(2)	2006(2)	
6B-3A1	121MP	3015	239(2)	2006(2)	
6B-3A19	121MP	3015	239(2)	2006(2)	
6B-3A2	121MP	3015	239(2)	2006(2)	
6B-3A21	121MP	3015	239(2)	2006(2)	
6B-3A3	121MP	3015	239(2)	2006(2)	
6B-3A3P	121MP	3015	239(2)	2006(2)	
6B-3A4	121MP	3015	239(2)	2006(2)	
6B-3A4-7	121MP	3015	239(2)	2006(2)	
6B-3A4-7B	121MP	3718	239(2)	2006(2)	
6B-3A5	121MP	3015	239(2)	2006(2)	
6B-3A5L	121MP	3015	239(2)	2006(2)	
6B-3A6	121MP	3015	239(2)	2006(2)	
6B-3A7	121MP	3015	239(2)	2006(2)	
6B-3A7-1	121MP	3015	239(2)	2006(2)	
6B-3A8	121MP	3015	239(2)	2006(2)	
6B-3A82	121MP	3015	239(2)	2006(2)	
6B-3A9	121MP	3015	239(2)	2006(2)	
6B-3A9G	121MP	3015	239(2)	2006(2)	
6B-3A0R	121MP		239(2)	2006(2)	
06B1M	722	3161	IC-9		
06B1Z	722	3161	IC-9		
006B2M	722	3161	IC-9		
06B2Z	722	3161	IC-9		
6C05	5850		5016		
6C05R	5851		5017		
6C10	5852		5016		
6C100	5868	7068	5028		
6C100R	5869	7069	5029		
6C10R	5853		5017		
6C15	5854		5016		
6C15R	5855		5017		
6C20	5854		5016		
6C20R	5855		5017		
6C30	5856		5020		
6C30R	5857		5021		
6C40	5858		5020		
6C40R	5859		5021		
6C50	5860		5024		
6C50R	5861		5025		
6C60	5862		5024		
6C60R	5863		5025		
6C70	5866		5028		
6C70R	5867		5029		
6C80	5866		5028		
6C80R	5867		5029		
6C90	5868	7068	5028		
6C90R	5869	7069	5029		
6CC11	5944		5048		
6CC13	5854	3500/5882			
6CD13	5855	3517/5883			
6D0000105	102	3722	2	2007	
6D122	102A	3004	53	2007	
6D122R	102A	3004	53	2007	
6D122T	102A		53	2007	
6D122TC	102A		53	2007	
6D122TH	102A		53	2007	
6D122U	102A		53	2007	
6D122V	102A	3123	53	2007	
6D122W	102A		53	2007	
6D122Y	102A		53	2007	
6DC11	5994	3501	5104		
6DM		3136	ZD-8.2	562	
6E4850/56-0001	123A	3444	20	2051	
6E4850/56-0002	128		243	2030	
6F10	5852		5016		
6F10-D	5852		5016		
6F100	5868	7068	5028		
6F100-D	5868	7068	5028		
6F100A	5868	7068	5028		
6F100A,B					MR1130
6F100B	5868	7068	5028		
6F10A	5852		5016		
6F10B	5852		5016		
6F15	5854		5016		
6F15A	5854		5016		
6F15B	5854		5016		
6F20	5854		5016		
6F20A	5854		5016		
6F20B	5854		5016		
6F30	5856		5020		
6F30-D	5856		5020		
6F30A	5856		5020		
6F30B	5856		5020		
6F40	5858		5020		
6F40-D	5858		5020		
6F40A	5858		5020		
6F40B	5858		5020		
6F5	5850		5016		
6F5-D	5850		5016		
6F50	5860	3500/5882	5024		
6F50-D	5860		5024		
6F50A	5860		5024		
6F50B	5860		5024		
6F5A	5850		5016		
6F5B	5850		5016		
6F60	5862	3500/5882	5024		
6F60-D	5862		5024		
6F60A	5862		5024		
6F60B	5862		5024		
6F70	5866		5028		
6F70A	5866		5028		
6F70A,B					MR1128
6F70B	5866		5028		
6F80	5866		5028		
6F80-D	5866		5028		
6F80A	5866		5028		
6F80A,B					MR1128
6F80B	5866		5028		
6F90	5868	7068	5028		
6F90A	5868	7068	5028		
6F90A,B					MR1130
6F90B	5868	7068	5028		
6FC13	5858	3500/5882			
6FD13	5859	3517/5883			
6FL10					1N3880
6FL20					1N3881
6FL30					1N3883
6FL40					1N3883
6FL5					1N3879
6FL50					MR1366
6FL60					MR1366
6FR10	5853	3517/5883	5017		
6FR10-D	5853	3517/5883	5017		
6FR100	5869	7069	5029		
6FR100A	5869	7069	5029		
6FR100B	5869	7069	5029		
6FR10A	5853	3517/5883	5017		
6FR10B	5853	3517/5883	5017		
6FR15	5855	3517/5883	5017		
6FR15A	5855	3517/5883	5017		
6FR15B	5855	3517/5883	5017		
6FR20	5855	3517/5883	5017		
6FR20-D	5855	3517/5883	5017		
6FR20A	5855	3517/5883	5017		
6FR20B	5855	3517/5883	5017		
6FR30	5857	3517/5883	5021		
6FR30A	5857	3517/5883	5021		
6FR30B	5857	3517/5883	5021		
6FR40	5859	3517/5883	5021		
6FR40-D	5859	3517/5883	5021		
6FR40A	5859	3517/5883	5021		
6FR40B	5859	3517/5883	5021		
6FR5	5851	3517/5883	5017		
6FR5-D	5851	3517/5883	5017		
6FR50	5861		5025		
6FR50A	5861		5025		
6FR50B	5861		5024		
6FR58	5851		5017		
6FR5A	5851	3517/5883	5017		
6FR60	5863		5025		
6FR60-D	5863		5025		
6FR60A	5863		5025		
6FR60B	5863		5025		
6FR70	5867		5029		
6FR70A	5867		5029		
6FR70B	5867		5029		
6FR80	5867		5029		
6FR80-D	5867		5029		
6FR80A	5867		5029		
6FR80B	5867		5029		
6FR90	5869	7069	5029		
6FR90A	5869	7069	5029		
6FR90B	5869	7069	5029		
6FT10					1N3880
6FT20					1N3881

Industry Standard No.	ECG	SK	GE	RS 276-	MOTOR.
6FT30					1N3883
6FT40					1N3883
6FT5					1N3879
6FT50					MR1366
6FT60					MR1366
6FV10					1N3880
6FV20					1N3881
6FV30					1N3883
6FV40					1N3883
6FV5					1N3879
6FV50					MR1366
6FV60					MR1366
6G4	113A	3119/113	6GC1		
6G8	116	3017B/117	504A	1104	
6GA1750	116		504A	1104	
6GA175D	116	3016	504A	1104	
6GC1	113A	3119/113	6GC1		
6GC12	5858	3500/5882	5020		
6GC1BY1	113A	3119/113	6GC1		
6GD1	114	3120	6GD1	1104	
6GX1	115	3121	6GX1		
6GX1BY1	115	3119/113	6GX1		
6HB050R					SPECIAL
6HB100R					SPECIAL
6HB200R					SPECIAL
6HB400R					SPECIAL
6HB600R					SPECIAL
6I0148-1	128		243	2030	
6JC12	5862		5024		
6L122	102A	3123	53	2007	
6LC12	5866		5028		
6M4	116	3017B/117	504A	1104	
6M404-1	116	3017B/117	504A	1104	
6M404-2	116	3017B/117	504A	1104	
6M404-3	116	3017B/117	504A	1104	
6M404-4	116	3017B/117	504A	1104	
6M404-5	116	3017B/117	504A	1104	
6M404-6	116	3017B/117	504A	1104	
6M404-7	116	3017B/117	504A	1104	
6MC	160	3007	245	2004	
6NC12	5868	7068	5028		
06P1C	121	3717	239	2006	
6PPL10	5312	3985			
6PPL20	5313	3986			
6PPL40	5314	3987			
6PPL5	5312	3985			
6PPL60	5315	3988			
6PPL80	5316	3989			
6R522PC7BAD1	116		504A	1104	
6RS18PH10BNB1	118		CR-1		
6RS18PH110BEB1	118	3066	CR-1		
6RS18PH110BHB1	118	3066	CR-1		
6RS18PH110BMB1	118	3066	CR-1		
6RS18PH110BNB1		3066	CR-1		
6RS36PH13BJJ1	119		CR-2		
6RS36PH13BJK1	119	3109	CR-2		
6RS36PH13BKJ1	119		CR-2		
6RS36PH13BLJ1	119	3109	CR-2		
6RS51GX12	120	3110	CR-3		
6RS56PHL3BKJ1	119	3110/120	CR-2		
6RS6PH13BCJ1	119	3109	CR-2		
6RS6PH13BJJ1	119	3109	CR-2		
6RS6PH13BKJ1	119	3109	CR-2		
6RS6PH13BLJ1	119	3109	CR-2		
6RS6PH13BMJ1	119	3109	CR-2		
6RS7PH130BCB1	118	3066	CR-1		
6RS7PH136AB1	119		CR-2		
6RS7PH30BCB1	118	3066	CR-1		
6RW62HY	116	3016	504A	1104	
6SB10	5312	3985			
6SB100	5312	3985			
6SB20	5313	3986			
6SB200	5313	3986			
6SB40	5314	3987			
6SB400	5314	3987			
6SB5	5312	3985			
6SB50	5312	3985			
6SB60	5315	3988			
6SB600	5315	3988			
6SB80	5316	3989			
6SB800	5316	3989			
6V-200	136A	3057	ZD-5.6	561	
6X97047A01	123A	3444	20	2051	
6X97047A02	102A		53	2007	
6X97174A01	177	3100/519	300	1122	
6X97174XA08	177	3100/519	300	1122	
007-00	171	3201	27		
7-0002	116	3017B/117	504A	1104	
7-0002-00	128		243	2030	
7-0003	166	9075	BR-600	1152	
7-0004	116	3017B/117	504A	1104	
7-0005	109	3087		1123	
7-0006	109	3087		1123	
7-0008	116	3017B/117	504A	1104	
7-0009	5220A	3017B/117			
7-0011-00	128		243	2030	
7-0012-00	129		244	2027	
7-0013	177	3100/519	300	1122	
7-0014	159	3466	82	2032	
7-0015	123A	3444	20	2051	
007-0030	175		246	2020	
7-0030-00	175		246	2020	
007-0040-00	130		14	2041	
007-0051	128		243	2030	
7-0051-00	128	3024	243	2030	
007-0074	175	3026	246	2020	
007-0112	152	3893	66	2048	
7-0112-00	152	3893	66	2048	
007-0112-03	152	3893	66	2048	
7-0112-04	152	3893	66	2048	
7-0112-05	152	3893	66	2048	
7-0115			255	2041	
7-0115-000	390		255	2041	
7-0197	390		255	2041	
7-0197-00	390	3509/5695	255	2041	
007-0214-00	312			2028	
007-0214-01	312			2028	
07-07113	290A	3114/290	269	2032	
07-07119	102A	3004	53	2007	
07-07124	123A	3122	20	2051	
07-07125	123A	3444	20	2051	
07-07129	229	3018	61	2038	
07-07139	123A	3444	20	2051	
07-07141	237	3299	46		
07-07156	123A	3444	20	2051	
07-07158	312	3448	FET-1	2035	
07-07159	312	3112	FET-1	2028	
07-07161	724		IC-86		
07-07163	289A	3138/193A	268	2038	
07-07164	237	3299	46		
07-07165	186A	3357	247	2052	
07-07166	199	3018	20	2010	
07-07167	103A	3010	59	2002	
7-1(SARKES)	123A	3444	20	2051	

Industry Standard No.	ECG	SK	GE	RS 276-	MOTOR.
7-1(STANDEL)	121	3717	239	2006	
7-10	161	3039/316	39	2015	
7-10(SARKES)	161	3716	39	2015	
07-1075-01	102A	3004	53	2007	
07-1075-02	102A		53	2007	
7-11(SARKES)	161	3716	39	2015	
7-11(STANDEL)	125(4)		510,531	1114	
007-112-04	152	3893	66	2048	
07-1156-03	102A		53	2007	
7-117-02	221	3065/222		2036	
7-12		3027	14	2041	
7-12(STANDEL)	130		14	2041	
7-13		3027	14	2041	
7-13(STANDEL)	130		14	2041	
07-1458-85	123A	3124/289	20	2051	
7-14A	129	3025	244	2027	
7-15(SARKES)	123A	3444	20	2051	
7-15(STANDEL)	125(4)		510,531	1114	
7-16	107		11	2015	
7-16(SARKES)	123A	3444	20	2051	
007-1668601	910		IC-251		
007-1668602	910D		IC-252		
007-166870I	911		IC-253		
007-16687C1	911		IC-253		
007-1669602	941M		IC-265	007	
007-1669901	923D		IC-260	1740	
007-1695001	7400	7400	7400	1801	
007-1695101	7420	7420	7400	1809	
007-1695301	7404	7404	7404	1802	
007-1695701	7440	7440			
007-1695901	7410	7410	7410	1807	
007-1696001	7485	7485	7485	1826	
007-1696101	7414	7414			
007-1696201	7402	7402	7402	1811	
007-1696301	7437	7437			
007-1696801	74145	74145	74145	1828	
007-1696901	7406	7406	7406	1821	
007-1697801	7441	7441	7441	1804	
007-1698301	74192	74192	74192	1831	
007-1698401	74193	74193	74193	1820	
007-1699301	7408	7408	7408	1822	
007-1699801	7474	7474	7474	1818	
7-17	123A	3444	20	2051	
7-17(SARKES)	123A		20	2051	
7-18(SARKES)	123A	3444	20	2051	
7-19(SARKES)	123A	3444	20	2051	
7-19(STANDEL)	124		12		
7-2		3013	3	2006	
7-2(SARKES)	123A	3444	20	2051	
7-2(STANDEL)	104MP	3720	16(2)	2006(2)	
7-20			11	2015	
7-20(SARKES)	123A	3444	20	2051	
07-2012-04	102A	3006/160	53	2007	
7-21(SARKES)	161	3716	39	2015	
7-22			11	2015	
7-22(SARKES)	161	3716	39	2015	
7-23		3018	11	2015	
7-23(SARKES)	161	3093	39	2015	
7-24(SARKES)	161	3093	39	2015	
7-25(SARKES)	161	3093	39	2015	
007-25005-01	519		514	1122	
007-25013-01	519		514	1122	
007-25016-01	519		514	1122	
7-2580504	5494	6794			
7-26	107	3039/316	11	2015	
7-28		3039	39	2015	
7-28(STANDEL)	218		234		
7-29		3126	90		
7-29(STANDEL)	175		246	2020	
7-3(SARKES)	123A	3444	20	2051	
7-3(STANDEL)	129		244	2027	
07-3012-04	102A	3006/160	53	2007	
07-3015-05	102A	3008	53	2007	
07-3080-06	160	3006	245	2004	
07-3350-57	126	3006/160	52	2024	
7-3401A	126		52	2024	
7-36			11	2015	
7-39	107		11	2015	
7-4	108	3452	86	2038	
7-4(SARKES)	123A	3444	20	2051	
7-4(STANDEL)	129		244	2027	
7-40	222	3050/221	FET-4	2036	
07-4233-19	160	3006	245	2004	
07-4235-13	160	3006	245	2004	
07-4235-73	160	3006	245	2004	
7-43			11	2015	
7-44	108	3452	86	2038	
7-45			11	2015	
7-466201				2041	
7-5	123A	3444	20	2051	
7-5(SARKES)	123A	3444	20	2051	
07-5085-36		3126	90		
07-5134-14	109	3087		1123	
07-5134-14A	109	3087		1123	
07-5134-14B	109	3087		1123	
07-5134-14C	109	3087		1123	
07-5160-15	110MP	3089/112	1N60	1123(2)	
07-5160-15A	112	3089	1N82A		
07-5160-15B	112	3089	1N82A		
07-5160-15C	112	3089	1N82A		
07-5331-86	137A	3057/136A	ZD-6.2	561	
07-5331-86A	136A	3057	ZD-5.6	561	
07-5331-86B	136A	3057	ZD-5.6	561	
07-5331-86C	136A	3057	ZD-5.6	561	
07-5331-86D	136A	3057	ZD-5.6	561	
7-59-001/3477		3088	1N60	1123	
7-59-0013477	109			1123	
7-59-005/3477			BR-600	1152	
7-59-010/3477			2	2007	
7-59-0103477	158		53	2007	
7-59-019/3477		3018	20	2051	
7-59-0193477	108	3452	86	2038	
7-59-020/3477		3018	20	2051	
7-59-0203477	108	3452	86	2038	
7-59-021/3477		3018	20	2051	
7-59-0213477	108	3452	86	2038	
7-59-022/3477			17	2051	
7-59-0223477	108	3452	86	2038	
7-59-023/3477		3018	17	2051	
7-59-0233477	108	3452	86	2038	
7-59-024/3477		3124	10	2051	
7-59-0243477	123A	3444	20	2051	
7-59-029/3477		3004	1	2007	
7-59-0293477	102A		53	2007	
7-59-060/3477		3004	2	2007	
7-59-0603477	102A		53	2007	
7-59-068	123A	3444	20	2051	
7-6	108	3452	86	2038	
7-6(SARKES)	123A	3444	20	2051	
7-6006-00	5801	9004		1142	
07-6015-16	102A	3004	53	2007	
007-6016-00	519		514	1122	
007-6060-00	519		514	1122	
7-7	108	3452	86	2038	

Industry Standard No.	ECG	SK	GE	RS 276-	MOTOR.
7-7(SARKES)	123A	3444	20	2051	
7-73004-02	176	3845	80		
7-73004-03	176	3845	80		
7-73004-04	176	3845	80		
7-73004-1	176	3845	80		
7-7340102	126		52	2024	
007-74004-01	159	3466	82	2032	
007-74008-01	159	3466	82	2032	
007-7450301	130	3027	14	2041	
007-74655-02	123A	3122	20	2051	
007-74655-06	123A	3122	20	2051	
007-74659-01	128		243	2030	
007-74659-04	128		243	2030	
007-74659-06	128		243	2030	
007-74661-01	123A	3122	20	2051	
007-7466101	123A	3122	20	2051	
7-7466201	181		75	2041	
7-8	108	3018	86	2038	
7-8(SARKES)	123A	3444	20	2051	
7-8(STANDEL)	168			1173	
7-9	161	3039/316	39	2015	
7-9(SARKES)	161	3716	39	2015	
7.Z17B	5076A	9022			
7A1011			18	2030	
7A1011(GE)	128		243	2030	
7A1011(SHERWOOD)	128		243	2030	
7A30	107	3039/316	11	2015	
7A30(GE)	128		243	2030	
7A30(SHERWOOD)	123A	3444	20	2051	
7A31	107	3039/316	11	2015	
7A31(GE)	128		243	2030	
7A31(SHERWOOD)	123A	3444	20	2051	
7A32	107	3039/316	11	2015	
7A32(GE)	128		243	2030	
7A32(SHERWOOD)	128		243	2030	
7A35			63	2030	
7A35(GE)	128		243	2030	
7A995			18	2030	
7A995(GE)	128		243	2030	
7A995(SHERWOOD)	128		243	2030	
7B-3B1	169	3107/5307	BR-600		
7B1	124	3021	12		
7B13	124	3021	12		
07B1Z	708	3135/709	IC-10		
7B2	124	3021	12		
07B27	708		IC-10		
07B2B	708	3135/709	IC-10		
07B2Z	788	3829	IC-229		
07B3B	708	3135/709	IC-10		
07B3C	708	3135/709	IC-10		
07B3D	708	3135/709	IC-10		
07B3M	708	3135/709	IC-10		
07B3Z	708	3135/709	IC-10		
7C1	124	3021	12		
7C13			63	2030	
7C2	124	3021	12		
7C3	124	3021	12		
7D	116	3017B/117	504A	1104	
7D1	124	3021	12		
7D13			63	2030	
7D2	124	3021	12		
7D210	125	3033	510,531	1114	
7D210A	125	3033	510,531	1114	
7D3	124	3021	12		
7E1	124	3021	12		
7E13			63	2030	
7E2	124	3021	12		
7E3	124	3021	12		
7F13			63	2030	
7G1	124	3021	12		
7G13			63	2030	
7G2	124	3021	12		
7G3	124	3021	12		
7G4	124	3021	12		
7K705M	113A	3119/113	6GC1		
7L6-0105	175		246	2020	
7L6-0444-1(NPN)			241	2020	
7L6-0444-1(PNP)			250	2027	
7L6-0495-14			504A	1104	
7L6-0531-1(NPN)			241	2020	
7L6-0531-1(PNP)			250	2027	
7L6-0531-19(NPN)			28	2017	
7L6-0531-19(PNP)			29	2025	
7MA60	116	3017B/117	504A	1104	
7TB1	5491	6791			
7TB2	5483	3942			
7TB3	5484	3943/5485			
7TB4	5507	3943/5485			
7TB5	5528	6629/5529			
7TB6	5496	6796			
7TB8	5531	6631			
7VM705M	113A	3119/113	6GC1		
7X070	138A		ZD-7.5		
08	139A	3060	ZD-9.1	562	
8-0001300	125	3031A	510,531	1114	
8-0001400	125	3031A	510,531	1114	
8-0024-1	108	3452	86	2038	
8-0024-2	108	3452	86	2038	
8-0024-3	123	3124/289	20	2051	
8-00243	123A	3444	20	2051	
08-0040			504A	1104	
8-0050100	123A	3444	20	2051	
8-0050300	103A	3010	59	2002	
8-0050400	160	3006	245	2004	
8-0050500	160	3008	245	2004	
8-0050600	103A	3010	59	2002	
8-0050700	121	3009	239	2006	
8-0051500	123A	3444	20	2051	
8-0051600	129	3025	244	2027	
8-005202	123A	3444	20	2051	
8-0052102	123A	3444	20	2051	
8-0052302	123A	3444	20	2051	
8-0052402	130	3027	14	2041	
8-0052600	123A	3444	20	2051	
8-0052700	129	3025	244	2027	
8-0052800	103A	3010	59	2002	
8-0053001	123A	3444	20	2051	
8-0053300	128	3024	243	2030	
8-0053400	123A	3444	20	2051	
8-0053600	108	3452	86	2038	
8-0053702	130MP	3029	15	2041(2)	
8-0060	100	3006/160	1	2007	
8-0062	102	3004	2	2007	
8-0104900	160	3008	245	2004	
8-0105200	160	3008	245	2004	
8-0105300	160	3008	245	2004	
8-0205400	102A	3004	53	2007	
8-0205600	102A	3004	53	2007	
8-0222631U	102A		53	2007	
8-0236400	102A	3004	53	2007	
8-0236430	102A	3004	53	2007	
008-024-00	116	3017B/117	504A	1104	
8-0243900	102A	3004	53	2007	
8-0318250	123A	3444	20	2051	
8-0337390	123A	3444	20	2051	
8-0338030	108	3452	86	2038	
8-0338040	108	3452	86	2038	
8-0339430	108	3452	86	2038	
8-0339440	108	3452	86	2038	
8-0383840	108	3452	86	2038	
8-0383930	108	3452	86	2038	
8-0383940	123A	3444	20	2051	
8-0389910	123A	3444	20	2051	
8-0389930	123A	3444	20	2051	
8-0414120	130	3027	14	2041	
8-0414130	130	3027	14	2041	
8-0421980	123A	3444	20	2051	
08-08111	109	3088		1123	
08-08112	110MP	3088	1N60	1123	
08-08117	177	3100/519	300	1122	
08-08119	177	3100/519	300	1122	
08-08120	145A		ZD-15	564	
08-08122	116	3017B/117	504A	1104	
08-08125	139A	3060	ZD-9.1	562	
08-0821	116	3016	504A	1104	
8-1	175		246	2020	
8-1(BENDIX)	175		246	2020	
8-1074	152	3893	66	2048	
8-1075	390		255	2041	
8-22	116	3016	504A	1104	
8-2409501	123A	3444	20	2051	
8-2410300	129	3025	244	2027	
8-25	116	3016	504A	1104	
8-28-045	601	3463			
08-302152			61	2038	
8-38	116	3016	504A	1104	
8-4(BENDIX)	123A	3444	20	2051	
8-40-027	316	3039			
8-619-030-007	102A	3004	53	2007	
8-619-030-008	102A	3004	53	2007	
8-619-030-009	102A	3004	53	2007	
8-619-030-011	109	3087		1123	
8-619-030-012	116	3017B/117	504A	1104	
8-619-030-014	102A	3004	53	2007	
8-619-030-015	131	3198	44	2006	
8-619-030-016	102A		53	2007	
8-619-030-017	102A		53	2007	
8-639-001-095	116	3017B/117	504A	1104	
8-697-020-567	102A	3004	53	2007	
8-697-020-568	102A	3004	53	2007	
8-697-020-569	102A	3004	53	2007	
8-697-020-570	123A	3444	20	2051	
8-697-020-571	109	3087		1123	
8-710-222-21	116	3016	504A	1104	
8-719-026-11	109	3087	1N34AS	1123	
8-719-112-24	5021A	3787		563	
8-719-113-25	5022A	3788			
8-719-122-00	605	3864			
8-719-143-07	5008A	3774			
8-719-156-23	5011A	3777			
8-719-156-27	5011A	3777			
8-719-168-07	5014A			561	
8-719-182-25	5016A	3782			
8-719-200-02	552	3311	504A	1104	
8-719-205-10	116	3016	504A	1104	
8-719-305-15	525	3998/506	511		
8-719-320-11	552	3998/506	511	1114	
8-719-320-31	552	3998/506	511	1114	
8-719-422-21	109	3088	1N60	1123	
8-719-713-93	612	3325			
8-719-768-71	612	3325			
8-719-815-55	177	3175	300	1122	
8-719-900-63	116	3311	504A	1104	
8-719-900-93	552	3998/506	511		
8-719-901-02	116	3311	504A	1104	
8-719-901-13	116	3313		1104	
8-719-901-19	552	3998/506	511	1114	
8-719-901-24	5457	3598			
8-719-901-92	116	3318	530		
8-719-901-93	552	3081/125	530		
8-719-903-09		3998	511	1114	
8-719-906-15	506		511	1114	
8-719-906-24	5081A	3151	ZD-24		
8-719-908-03	116	3311	504A	1104	
8-719-911-54	156	3051	512		
8-719-912-54	156	3051	512		
8-719-923-76	519		514	1122	
8-719-930-12	142A			563	
8-719-937-10	140A	3061	ZD-10		
8-719-941-13	156	3051			
8-721-323-00	158		53	2007	
8-722-923-00	107		11	2015	
8-723-302-00	312	3834/152		2028	
8-723-650	103	3010	8	2002	
8-724-034-00	107		11	2015	
8-724-733-30	123A	3444	20	2051	
8-726-357-10	123A	3124/289	47	2030	
8-729-118-76	165	3710/238	38		
8-729-133-53	379	3893/152			
8-729-213-01	287	3433	220		
8-729-213-12	287		220		
8-729-304-92	374	9042			
8-729-306-92	373	9041			
8-729-309-06	199	3244	220		
8-729-309-36		3865	221		
8-729-311-42	94		14		
8-729-316-12	375	3893/152	66	2048	
8-729-322-78	171	3201	257		
8-729-331-53	375	3219			
8-729-341-34	165	3710/238	38		
8-729-345-42	94	3115/165	35		
8-729-372-31	198	3219	325		
8-729-372-52	238	3710	259		
8-729-375-01	289A	3124/289			
8-729-447-53	158	3035	53	2007	
8-729-468-43	294	3841	67		
8-729-612-77	159		82		
8-729-663-47	123AP	3124/289	210	2051	
8-729-665-47	315	3124/289	62		
8-729-671-14	107	3356	211		
8-729-803-04	107	3356	60	2038	
8-750-105-11	1096	3709/110MP			
8-759-101-60	1045	3072/712	IC-2		
8-759-110-31	1245	3878			
8-759-113-53	1246	3879			
8-759-140-01	4001B	4001			
8-759-157-40	615	3468/1185	ZD-33		
8-759-157-60	1185		IC-139		
8-759-240-11	4011B	4011			
8-759-240-13	4013B	4013			
8-759-240-27	4027B	4027			
8-759-424-00	712		IC-2		
8-759-425-00	712		IC-2		
8-759-600-95	1308	3833			
8-759-651-34	1096	3709/110MP			
8-759-651-35	1004	3365			
8-759-812-01	1003	3288	IC-43		
8-759-904-69	4069	4069			

Industry Standard No.	ECG	SK	GE	RS 276-	MOTOR.
8-760-335-10	293	3849	83	2030	
8-760-343-10	124	3021	32	2030	
8-760-413-10	315	3250	279		
8-760-514-10	294	3841	84		
8-760-523-10	294	3841			
8-762-020-00	288	3434	223		
8-763-113-00	306	3251	276		
8-765-170-01	191	3865	222		
8-765-422-00	312	3834/132			
8-765-500-00	291	3440			
8-765-510-00	292	3441			
8-81250108	199	3038	62	2010	
8-81250109	123A	3444	20	2051	
8-902-0706-071	123A	3444	20	2051	
8-905-013-752	116	3017B/117	504A	1104	
8-905-013-759	116	3017B/117	504A	1104	
8-905-013-760	116	3017B/117	504A	1104	
8-905-014-008	118		CR-1		
8-905-014-017	123A	3444	20	2051	
8-905-198-001	116	3017B/117	504A	1104	
8-905-198-004	116	3017B/117	504A	1104	
8-905-198-005	116	3017B/117	504A	1104	
8-905-198-007	116	3017B/117	504A	1104	
8-905-198-008	116	3017B/117	504A	1104	
8-905-198-010	116	3017B/117	504A	1104	
8-905-198-034	116	3017B/117	504A	1104	
8-905-305-004	110MP	3709			
8-905-305-007	109	3087		1123	
8-905-305-020	109	3087		1123	
8-905-305-023	110MP	3709			
8-905-305-055	109	3087		1123	
8-905-305-318	109	3087		1123	
8-905-305-327	109			1123	
8-905-305-330	109	3087		1123	
8-905-305-336	109	3087		1123	
8-905-305-338	109	3087		1123	
8-905-305-339	109	3087		1123	
8-905-305-342	109	3087		1123	
8-905-305-348	109	3087		1123	
8-905-305-400	116	3033	504A	1104	
8-905-305-405	109	3087		1123	
8-905-305-555	109	3087		1123	
8-905-305-561	109	3087		1123	
8-905-305-580	109	3087		1123	
8-905-305-635	109	3087		1123	
8-905-313-007	110MP	3709			
8-905-313-008	110MP	3709			
8-905-313-010	109	3087		1123	
8-905-313-011	109	3087		1123	
8-905-313-018	110MP	3709			
8-905-313-100	109	3087		1123	
8-905-313-101	109	3087		1123	
8-905-313-120	109	3087		1123	
8-905-405-002	116	3017B/117	504A	1104	
8-905-405-026	116	3017B/117	504A	1104	
8-905-405-069	116	3017B/117	504A	1104	
8-905-405-077	109	3087		1123	
8-905-405-098	177	3100/519	300	1122	
8-905-405-105	125	3033	510,531	1114	
8-905-405-134	116	3017B/117	504A	1104	
8-905-405-146	116	3017B/117	504A	1104	
8-905-405-160	125	3033	510,531	1114	
8-905-405-170	125	3033	510,531	1114	
8-905-405-206	116		504A	1104	
8-905-405-838	109	3087		1123	
8-905-406-020	177	3175	300	1122	
8-905-413-092	116	3017B/117	504A	1104	
8-905-421-109	136A	3057	ZD-5.6	561	
8-905-421-118	139A	3060	ZD-9.1	562	
8-905-421-128	145A	3063	ZD-15	564	
8-905-421-215	5069A		ZD-4.7		
8-905-421-228	139A	3060	ZD-9.1	562	
8-905-421-234	142A	3062	ZD-12	563	
8-905-421-239	145A	3063	ZD-15	564	
8-905-421-300	177	3100/519	300	1122	
8-905-421-315	150A	3098	ZD-82		
8-905-421-319	142A	3062	ZD-12	563	
8-905-421-715	145A	3063	ZD-15	564	
8-905-605-016	102A	3004	53	2007	
8-905-605-030	102A	3004	53	2007	
8-905-605-032	102A	3004	53	2007	
8-905-605-050	102A	3004	53	2007	
8-905-605-051	102A	3004	53	2007	
8-905-605-075	102A	3004	53	2007	
8-905-605-090	102A	3004	53	2007	
8-905-605-091	102A	3004	53	2007	
8-905-605-105	103A	3010	59	2002	
8-905-605-108	103A	3010	59	2002	
8-905-605-109	103A	3010	59	2002	
8-905-605-110	103A	3010	59	2002	
8-905-605-111	103A	3010	59	2002	
8-905-605-112	103A	3010	59	2002	
8-905-605-113	103A	3010	59	2002	
8-905-605-120	158	3004/102A	53	2007	
8-905-605-123	158	3004/102A	53	2007	
8-905-605-124	158	3004/102A	53	2007	
8-905-605-125	158	3004/102A	53	2007	
8-905-605-126	158	3004/102A	53	2007	
8-905-605-127	158	3004/102A	53	2007	
8-905-605-128	158	3004/102A	53	2007	
8-905-605-129	158	3004/102A	53	2007	
8-905-605-230	102A	3004	53	2007	
8-905-605-232	102A	3004	53	2007	
8-905-605-234	102A	3004	53	2007	
8-905-605-250	158	3004/102A	53	2007	
8-905-605-255	158	3004/102A	53	2007	
8-905-605-260	158	3004/102A	53	2007	
8-905-605-264	158	3004/102A	53	2007	
8-905-605-266	158	3004/102A	53	2007	
8-905-605-268	158	3004/102A	53	2007	
8-905-605-269	158	3004/102A	53	2007	
8-905-605-292	102A	3004	53	2007	
8-905-605-305	102A	3004	53	2007	
8-905-605-320	160	3006	245	2004	
8-905-605-365	103A	3010	59	2002	
8-905-605-384	103A	3010	59	2002	
8-905-605-390	103A	3010	59	2002	
8-905-605-607	131		44	2006	
8-905-605-624	121	3009	239	2006	
8-905-605-635	121	3009	239	2006	
8-905-605-636	121	3009	239	2006	
8-905-605-637	121	3009	239	2006	
8-905-605-644	161	3132	39	2015	
8-905-605-650	131		44	2006	
8-905-605-775	127	3035	25		
8-905-605-908	127	3035	25		
8-905-606-001	160	3006	245	2004	
8-905-606-003	160	3006	245	2004	
8-905-606-007	160	3006	245	2004	
8-905-606-008	160	3006	245	2004	
8-905-606-010	160	3006	245	2004	
8-905-606-016	160	3006	245	2004	
8-905-606-051	160	3006	245	2004	
8-905-606-075	160	3006	245	2004	

Industry Standard No.	ECG	SK	GE	RS 276-	MOTOR.
8-905-606-077	160	3006	245	2004	
8-905-606-090	160	3006	245	2004	
8-905-606-105	160	3006	245	2004	
8-905-606-106	160	3006	245	2004	
8-905-606-120	160	3006	245	2004	
8-905-606-142	160	3006	245	2004	
8-905-606-152	160	3006	245	2004	
8-905-606-153	160		245	2004	
8-905-606-154	160	3006	245	2004	
8-905-606-155	160	3006	245	2004	
8-905-606-158	160	3006	245	2004	
8-905-606-165	160	3006	245	2004	
8-905-606-168	160	3006	245	2004	
8-905-606-180	160	3006	245	2004	
8-905-606-211	160	3006	245	2004	
8-905-606-225	160	3006	245	2004	
8-905-606-241	160	3006	245	2004	
8-905-606-255	160	3006	245	2004	
8-905-606-256	160	3006	245	2004	
8-905-606-349	160	3006	245	2004	
8-905-606-350	160		245	2004	
8-905-606-351	160	3006	245	2004	
8-905-606-352	160	3006	245	2004	
8-905-606-360	160	3006	245	2004	
8-905-606-375	160	3006	245	2004	
8-905-606-390	160	3006	245	2004	
8-905-606-391	160	3006	245	2004	
8-905-606-392	160	3006	245	2004	
8-905-606-405	160	3006	245	2004	
8-905-606-419	160	3006	245	2004	
8-905-606-420	160	3006	245	2004	
8-905-606-423	160	3006	245	2004	
8-905-606-720	121	3009	239	2006	
8-905-606-750	102A	3004	53	2007	
8-905-606-800	102A	3004	53	2007	
8-905-606-815	102A	3004	53	2007	
8-905-606-817	102A	3004	53	2007	
8-905-606-885	102A	3004	53	2007	
8-905-613-010	102A	3004	53	2007	
8-905-613-015	103A	3010	59	2002	
8-905-613-062	103A	3010	59	2002	
8-905-613-070	158		53	2007	
8-905-613-071	158	3004/102A	53	2007	
8-905-613-131	158	3004/102A	53	2007	
8-905-613-132	158	3004/102A	53	2007	
8-905-613-133	158	3004/102A	53	2007	
8-905-613-160	158	3004/102A	53	2007	
8-905-613-210	121	3717	239	2006	
8-905-613-215	121	3009	239	2006	
8-905-613-232	127	3035	25		
8-905-613-240	131		44	2006	
8-905-613-241	131		44	2006	
8-905-613-242	131		44	2006	
8-905-613-245	131		44	2006	
8-905-613-250	121	3009	239	2006	
8-905-613-265	131	3717/121	44	2006	
8-905-613-266	131	3198	44	2006	
8-905-613-277	131	3717/121	44	2006	
8-905-613-282	131	3717/121	44	2006	
8-905-613-283	131	3717/121	44	2006	
8-905-613-284	131	3717/121	44	2006	
8-905-613-295	127	3035	25		
8-905-613-555	131		44	2006	
8-905-613-640	102A	3004	53	2007	
8-905-613-710	102A	3004	53	2007	
8-905-613-955	102A	3004	53	2007	
8-905-615-156	102A	3004	53	2007	
8-905-705-112	123A	3444	20	2051	
8-905-705-403	123A	3444	20	2051	
8-905-705-405	123A	3444	20	2051	
8-905-705-410	128	3024	243	2030	
8-905-706-010	171	3103A/396	27		
8-905-706-044	161	3132	39	2015	
8-905-706-055	161	3132	39	2015	
8-905-706-060	161	3132	39	2015	
8-905-706-067	171	3103A/396	27		
8-905-706-068	171	3103A/396	27		
8-905-706-070	161	3132	39	2015	
8-905-706-071	161	3132	39	2015	
8-905-706-075	161	3132	39	2015	
8-905-706-080	161	3132	39	2015	
8-905-706-101	161	3132	39	2015	
8-905-706-104	123A	3444	20	2051	
8-905-706-110	161	3132	39	2015	
8-905-706-112			39	2015	
8-905-706-201	123A	3444	20	2051	
8-905-706-202	123A	3444	20	2051	
8-905-706-203	123A	3444	20	2051	
8-905-706-206	123A	3444	20	2051	
8-905-706-208	123A	3444	20	2051	
8-905-706-211	123A	3444	20	2051	
8-905-706-215	123A	3444	20	2051	
8-905-706-235	123A	3444	20	2051	
8-905-706-236	123A	3444	20	2051	
8-905-706-238	123A	3444	20	2051	
8-905-706-239	123A	3444	20	2051	
8-905-706-240	123A	3444	20	2051	
8-905-706-242	123A	3444	20	2051	
8-905-706-244	123A	3444	20	2051	
8-905-706-245	123A	3444	20	2051	
8-905-706-246	123A	3444	20	2051	
8-905-706-247	159	3466	82	2032	
8-905-706-250	123A	3444	20	2051	
8-905-706-251	159	3466	82	2032	
8-905-706-253	159	3466	82	2032	
8-905-706-254	159	3466	82	2032	
8-905-706-255	159	3466	82	2032	
8-905-706-256	159	3466	82	2032	
8-905-706-257	123A	3444	20	2051	
8-905-706-260	123A	3444	20	2051	
8-905-706-263	123A	3444	20	2051	
8-905-706-280	159	3466	82	2032	
8-905-706-286	159	3466	82	2032	
8-905-706-287	159	3466	82	2032	
8-905-706-288	159	3466	82	2032	
8-905-706-289	159	3466	82	2032	
8-905-706-290	159	3466	82	2032	
8-905-706-336	123A	3444	20	2051	
8-905-706-545	129	3025	.244	2027	
8-905-706-555	130	3027	14	2041	
8-905-706-556	130		14	2041	
8-905-706-557	130	3027	14	2041	
8-905-706-606	123A	3444	20	2051	
8-905-706-730	161	3122	39	2015	
8-905-706-790	160	3006	245	2004	
8-905-706-801	184	3054/196	57	2017	
8-905-706-901	312	3112		2028	
8-905-707-254	123A	3444	20	2051	
8-905-707-265	123A	3444	20	2051	
8-905-707-313	123A	3444	20	2051	
8-905-713-058	159	3466	82	2032	
8-905-713-101	130	3027	14	2041	
8-905-713-110	184	3054/196	57	2017	
8-905-713-556	130	3027	14	2041	

Industry Standard No.	ECG	SK	GE	RS 276-	MOTOR.
8-905-713-810	129	3025	244	2027	
8-906-706-112	161	3132			
8-9V	5072A	3136	ZD-8.2	562	
8.75VZENER			ZD-9.1	562	
8A01	109	3087		1123	
8A1002	172A	3156	64		
8A1003	172A	3156	64		
8A10521	131	3052	44	2006	
8A10625	131	3198	44	2006	
8A11083	131	3052	44	2006	
8A11667	116		504A	1104	
8A11721	131	3052	44	2006	
8A12359	131		44	2006	
8A12789	123A	3444	20	2051	
8A12991	104	3009	16	2006	
8A13164	131	3052	44	2006	
8A13718	158	3004/102A	53	2007	
08A159-007	177	3100/519	300	1122	
08A161-001	HIDIV-1	3868/DIV-1			
08A161-002	HIDIV-1	3868/DIV-1			
08A161-003	HIDIV-1	3868/DIV-1			
08A165-001	177		300	1122	
08A8300-2	123A		20	2051	
8AN10	5945		5104		
8AN100	5910	7110	5076		
8AN20	5945	3501/5994	5104		
8AN30	5878		5036		
8AN40	5900	7100	5068		
08B1M	725	3162	IC-19		
08B2D	725		IC-19		
08B2M	725	3162	IC-19		
8C145-01	218		234		
8C200	159	3466	82	2032	
8C201	159	3466	82	2032	
8C202	159	3466	82	2032	
8C203	159	3466	82	2032	
8C204	159	3466	82	2032	
8C205	159	3466	82	2032	
8C206	159	3466	82	2032	
8C207	159	3466	82	2032	
8C430	159	3466	82	2032	
8C430K	159	3466	82	2032	
8C440	159	3466	82	2032	
8C440K	159	3466	82	2032	
8C443	159	3466	82	2032	
8C443K	159	3466	82	2032	
8C445	159	3466	82	2032	
8C445K	159	3466	82	2032	
8C449	159	3466	82	2032	
8C449K	159	3466	82	2032	
8C450	159	3466	82	2032	
8C460	159	3466	82	2032	
8C460K	159	3466	82	2032	
8C463	159	3466	82	2032	
8C463K	159	3466	82	2032	
8C465	159	3466	82	2032	
8C465K	159	3466	82	2032	
8C466	159	3466	82	2032	
8C466K	159	3466	82	2032	
8C467	159	3466	82	2032	
8C467K	159	3466	82	2032	
8C468	159	3466	82	2032	
8C468K	159	3466	82	2032	
8C469	159	3466	82	2032	
8C469K	159	3466	82	2032	
8C470	159	3466	82	2032	
8C470K	159	3466	82	2032	
8C700	159	3466	82	2032	
8C700A	159	3466	82	2032	
8C700B	159	3466	82	2032	
8C702	159	3466	82	2032	
8C702A	159	3466	82	2032	
8C702B	159	3466	82	2032	
8C704	159	3466	82	2032	
8C740	159	3466	82	2032	
8C7400	159	3466	82	2032	
8C740G	159	3466	82	2032	
8C740M	159	3466	82	2032	
8C742	159	3466	82	2032	
8C7420	159	3466	82	2032	
8C742G	159	3466	82	2032	
8C742M	159	3466	82	2032	
8C91	139A	3060	ZD-9.1	562	
8CG15	116	3016	504A	1104	
8CG15RE	116	3016	504A	1104	
8D	160	3006	245	2004	
8D10	125	3033	510,531	1114	
8D4	116	3031A	504A	1104	1N4004
8D6	116	3017B/117	504A	1104	1N4005
8D8	125	3033	510,531	1114	
8E	160	3006	245	2004	
8E(AUTOMATIC)	126		52	2024	
8F	160	3006	245	2004	
8F(AUTOMATIC)	126		52	2024	
8G7	125	3032A	510,531	1114	
8GA	125	3032A	510,531	1114	
8H303	121	3014	239	2006	
8L	160		245	2004	
8L201	121	3009	239	2006	
8L201B	121	3009	239	2006	
8L201C	121	3009	239	2006	
8L201R	121	3009	239	2006	
8L201V	121	3009	239	2006	
8L301V	104	3009	16	2006	
8L404	121	3009	239	2006	
8M-26102	358		42		
8MOB10					MRF840
8MOB15					MRF842
8MOB2					MRF870
8MOB30					MRF844
8MOB45					MRF846
8MOB5					MRF840
8P	160		245	2004	
08P-12-12	102A		53	2007	
08P-2	102A	3004	53	2007	
08P-2-12-7	102A	3004	53	2007	
8P-2-70	102A	3004	53	2007	
8P-2-70-12	102A	3004	53	2007	
8P-2-70-12-7	102A	3004	53	2007	
8P-20	102A	3004	53	2007	
8P-21	102A	3004	53	2007	
8P-22	102A	3004	53	2007	
8P-23	102A	3004	53	2007	
8P-24	102A	3004	53	2007	
8P-25	102A	3004	53	2007	
8P-26	102A	3004	53	2007	
8P-27	102A	3004	53	2007	
8P-28	102A	3004	53	2007	
8P-29	102A	3004	53	2007	
8P-2A	102A	3004	53	2007	
8P-2A0	102A	3004		2007	
8P-2A0R	102A	3004		2007	
8P-2A1	102A	3004	53	2007	
8P-2A19	102A	3004	53	2007	
8P-2A2	102A	3004	53	2007	
8P-2A21	102A	3004	53	2007	
8P-2A3	102A	3004	53	2007	
8P-2A3P	102A	3004	53	2007	
8P-2A4	102A	3004	53	2007	
8P-2A4-7	102A	3004	53	2007	
8P-2A4-7B	102A		53	2007	
8P-2A5	102A	3004	53	2007	
8P-2A5L	102A	3004	53	2007	
8P-2A6	102A	3004	53	2007	
8P-2A6-2	102A	3004	53	2007	
8P-2A7	102A	3004	53	2007	
8P-2A7-1	102A	3004	53	2007	
8P-2A8	102A	3004	53	2007	
8P-2A82	102A	3004	53	2007	
8P-2A9	102A	3004	53	2007	
8P-2A9G	102A	3004	53	2007	
8P-2A0			53	2007	
8P-2A0R			53	2007	
8P-404			16	2006	
8P-404R			239	2006	
8P-505		3052	30	2006	
8P111	312			2028	
8P1555	179	3642	76		
8P202	131	3198	44	2006	
8P345	152	3893	66	2048	
8P40	104	3009	16	2006	
8P404	104	3009	16	2006	
8P404B	121	3009	239	2006	
8P404F	104	3009	16	2006	
8P404M	121	3009	239	2006	
8P404M-1		3009	3	2006	
8P404N	121	3009	239	2006	
8P404ORN	104	3009	16	2006	
8P404R	121	3009	239	2006	
8P404T	131	3009	44	2006	
8P404V	121	3009	239	2006	
8P415C	121	3009	239	2006	
8P416C	121	3009	239	2006	
8P445	153		69	2049	
8P505	131	3052	44	2006	
8P50S	131		44	2006	
8P73BLU	152	3893		2048	
8P73GRN	152	3893	66	2048	
8P73YEL	152	3893	66	2048	
8P70BLU			66	2048	
8P880	105		4		
8P880B	105		4		
8P9253	151A	3084	ZD-110		
8PC60	121	3009	239	2006	
8PS60	121	3009	239	2006	
8Q-000003-11	234		65	2050	
8Q-3-01			62	2010	
8Q-3-02			21	2034	
8Q-3-04	103A	3835	53	2002	
8Q-3-10	199		62	2010	
8Q-3-11	159	3466	82	2032	
8Q-3-12			10	2051	
8Q-3-13	128	3024	243	2030	
8Q-3-14	159	3466	82	2032	
8Q-3-23	155	3839	43		
8Q-7-01			504A	1104	
8Q-7-02			504A	1104	
8Q-7-03			504A	1104	
009-00	165	3111	38		
009-000	165		38		
09-002012	123A	3122	20	2051	
9-003	107		11	2015	
09-004	718	3159	IC-8		
09-005	703A		IC-12		
9-006	123A		IC-12	2051	
09-007	750	3280	IC-219		
09-010	746	3234	IC-217		
09-011	720	9014	IC-7		
09-017	722	3161	IC-9		
09-018	788	3829	IC-229		
09-033006	103A	3010	59	2002	
09-3-2123		3202	247	2052	
09-300002	126	3008	52	2024	
09-300005	158	3005	53	2007	
09-300006	126	3006/160	52	2024	
09-300007	126	3004/102A	52	2024	
09-300011	126	3005	2	2024	
09-300012	126	3007	52	2024	
09-300015	126	3007	52	2024	
09-300016	126	3007	52	2024	
09-300017	102A	3005	53	2007	
09-300021	160	3006	245	2004	
09-300024	160	3006	245	2004	
09-300026	234			2050	
09-300027	126	3006/160	52	2024	
09-300028	160	3006	245	2004	
09-300029	126	3006/160	52	2024	
09-300036	234	3114/290	65	2050	
09-300037	159	3466	82	2032	
09-300037A	121	3114/290	239	2006	
09-300043	129	3114/290	239	2027	
09-300059	159	3466	82	2032	
09-300061	159	3466	82	2032	
09-300062	159	3466	82	2032	
09-300063	159	3466	82	2032	
09-300064	290A	3114/290	269	2032	
09-300067	187	3193	29	2025	
09-300068	187	3193	29	2025	
09-300070	290A	3114/290	21	2032	
09-300071	294		48		
09-300072	193		67	2023	
09-300073	187A	9076	29		
09-300074	159	3466	67	2032	
09-300076	294	3114/290	48		
09-300077	159	3466	82	2032	
09-300078	126	3083/197	52	2024	
09-300079	126	3006/160	52	2024	
09-300080	290A	3114/290	269	2032	
09-300081	290A	3114/290	269	2032	
09-300084			50	2004	
09-300090	153	3083/197	69	2049	
09-30011			52	2024	
09-30012			50	2004	
09-3002006	229	3018	61	2038	
09-300307	159	3466	82	2032	
09-30063	159	3466	82	2032	
09-301001	102A	3004	53	2007	
09-301002	102A	3004	53	2007	
09-301002-6	102A	3004	53	2007	
09-301003	102A	3004	53	2007	
09-301004	102A	3004	53	2007	
09-301005	102A	3004	53	2007	
09-301006	102A	3004	53	2007	
09-301007	102A	3004	53	2007	
09-301008	158	3004/102A	53	2007	
09-301008-18	102A	3004	53	2007	
09-301009	102A	3004	53	2007	
09-301010	104	3009	16	2006	

Industry Standard No.	ECG	SK	GE	RS 276-	MOTOR.
09-301012			53	2007	
09-301014	158	3004/102A	53	2007	
09-301015	176	3123	80		
09-301016	102A	3004	53	2007	
09-301019	158	3004/102A	53	2007	
09-301020	102A	3004	53	2007	
09-301022	158	3004/102A	53	2007	
09-301023	102A	3004	53	2007	
09-301024	131		44	2006	
09-301025	102A	3004	53	2007	
09-301025-6	102A	3004	53	2007	
09-301026	102A	3004	53	2007	
09-301027	158	3004/102A	53	2007	
09-301030	226MP	3086	49(2)	2025(2)	
09-301031	176	3845	80		
09-301032	102A	3004	53	2007	
09-301034	131		44	2006	
09-301036	102A	3004	53	2007	
09-301039			60	2015	
09-301048	102A	3004	53	2007	
09-30104B	102A	3004	53	2007	
09-301052	121	3009	239	2006	
09-301054	158	3004/102A	53	2007	
09-301056	102A	3004	53	2007	
09-301066	158	3004/102A	53	2007	
09-301071	127	3024/128	25		
09-301072	102A	3004	53	2007	
09-301073	127	3035	25		
09-301074	102A	3004	53	2007	
09-301075	131	3052	44	2006	
09-301077	153		69	2049	
09-301079	127	3035	25		
09-30126	102A	3004	53	2007	
09-302002	229	3018	61	2038	
09-302003	229	3122	61	2038	
09-302004	233	3122	210	2009	
09-302005	108	3452	86	2038	
09-302006	229	3018	61	2038	
09-302007	123A	3444	20	2051	
09-302009	108	3452	86	2038	
09-302010	108	3452	86	2038	
09-302012	123A	3444	20	2051	
09-302014	107	3124/289	11	2015	
09-302015	195A	3047	46		
09-302016	233	3018	210	2009	
09-302017	108	3452	86	2038	
09-302019		3040	20	2051	
09-302020	289A	3038	268	2038	
09-302030	195A	3048/329	46		
09-302032	107	3039/316	11	2015	
09-302033	123A	3444	20	2051	
09-302034	123A	3444	20	2051	
09-302035	195A	3048/329	46		
09-302036	229	3039/316	61	2038	
09-302037	229	3018	61	2038	
09-302037(IC)	1006	3358			
09-302038	199	3022/1188	62	2010	
09-302039	123A	3444	20	2051	
09-302040	289A	3124/289	268	2038	
09-302041			20	2051	
09-302044	107	3124/289	11	2015	
09-302045	123A	3444	20	2051	
09-302045-12	123A	3444	20	2051	
09-302046	300	3464	273		
09-302050	237	3299	46		
09-302051	237	3299	46		
09-302053	199	3122	212	2010	
09-302054	123A	3444	20	2051	
09-302055	195A	3047	46		
09-302056	237	3299	46		
09-302058	123A	3444	20	2051	
09-302060	108	3452	86	2038	
09-302061	316	3039	11		
09-302062	123A	3444	20	2051	
09-302063	107	3018	11	2015	
09-302068	195A	3047	46		
09-302072	107	3117	11	2015	
09-302073	107	3018	11	2015	
09-302074	123A	3444	20	2051	
09-302074(SHARP)			62	2010	
09-302075	192	3122	63	2030	
09-302078	123A	3444	20	2051	
09-302079	107	3018	11	2015	
09-302080	186A	3357	28	2052	
09-302081	195A	3047	46		
09-302082	195A	3048/329	46		
09-302083	152	3054/196	66	2048	
09-302085	199	3124/289	62	2010	
09-302086	199	3018	62	2010	
09-302090	128	3122	243	2030	
09-302090(DIODE)	177			1122	
09-302092	107	3122	11	2015	
09-302093	199	3122	62	2010	
09-302095	229	3018	61	2038	
09-302097	199	3245	62	2010	
09-302099	154	3040	40	2012	
09-302101	123A	3444	20	2051	
09-302102	299	3054/196	236		
09-302103			11	2015	
09-302106	123A	3444	20	2051	
09-302107	199	3124/289	62	2010	
09-302113	186	3192	28	2017	
09-302114	107	3018	11	2015	
09-302115	108	3452	86	2038	
09-302116	192	3137	63	2030	
09-302117	237	3299	46		
09-302118	123A	3444	20	2051	
09-302119	235	3024/128	215		
09-302121	186A	3357	247	2052	
09-302122	130	3027	14	2041	
09-302123	186A	3357	28	2052	
09-302124	123A	3444	20	2051	
09-302125	199	3124/289	62	2010	
09-302126	186A	3357	247	2052	
09-302127	199	3122	62	2010	
09-302128	161	3018	61	2015	
09-302129	107	3132	11	2015	
09-302130	194	3045/225	220		
09-302131	123A	3444	20	2051	
09-302132	152	3054/196	66	2048	
09-302135	299	3047	236		
09-302136	295	3190/184	270		
09-302138	107	3018	11	2015	
09-302139	199	3122	62	2010	
09-302140	123A	3444	20	2051	
09-302141	108	3452	86	2038	
09-302142	107	3018	11	2015	
09-302143	107	3018	11	2015	
09-302144			11	2015	
09-302145			11	2015	
09-302146	124	3021	12		
09-302148	123A	3444	20	2051	
09-302149	108	3452	86	2038	
09-302150	287		40		
09-302151	107	3018	11	2015	
09-302152	107	3117	11	2015	
09-302153	123A	3444	20	2051	
09-302155	293	3137/192A	47	2030	
09-302156	124	3021	12		
09-302157	164	3133	37		
09-302158	163A	3439	36		
09-302159	162	3438	35		
09-302160	124	3021	12		
09-302161	193	3025/129	67	2023	
09-302162	108	3452	86	2038	
09-302164	152	3054/196	66	2048	
09-302165	123A	3444	20	2051	
09-302166	282	3748	264		
09-302169	237	3299	46		
09-302170	187	3193	29	2025	
09-302171	128	3024	243	2030	
09-302172	123A	3444	20	2051	
09-302173	289A	3122	243	2038	
09-302174			60	2015	
09-302175	123A	3444	20	2051	
09-302176	163A	3111	36		
09-302177	163A	3439	36		
09-302185	124	3021	12		
09-302186	198	3104A	251		
09-302187	165	3115	38		
09-302188	286	3021/124	267		
09-302189	123A	3444	20	2051	
09-302190	108	3452	86	2038	
09-302191		3018	17	2051	
09-302192	236	3197/235	216	2053	
09-302193	236	3197/235	216	2053	
09-302194	199	3124/289	62	2010	
09-302199			60	2015	
09-302200	233		210	2009	
09-302201	161	3018	39	2015	
09-302202	315	3250	279		
09-302203	107		11	2015	
09-302204	123A		11	2051	
09-302206	229	3018	61	2038	
09-302207	229	3122	61	2038	
09-302212	295	3048/329	270		
09-302215	123A	3444	20	2051	
09-302216	107	3018	11	2015	
09-302218	175	3131A/369	246	2020	
09-302219	302	3252	275		
09-302220	107	3018			
09-302222	297	3122	210	2030	
09-302224	229	3018	61	2038	
09-302225	229	3124/289	213	2038	
09-302226	192	3020/123	63	2030	
09-302227	123A			2051	
09-302236	152	3054/196	66	2048	
09-302237	198		32		
09-302238	238	3710	259		
09-302240	161			2015	
09-302241	108	3018	61	2038	
09-302242	229			2038	
09-302243	171		27		
09-302244		3122	210	2051	
09-302245	238	3710	259		
09-303005	175		246	2020	
09-303006	103A	3835	59	2002	
09-303012	103A	3835	8	2002	
09-303013	103A	3010	59	2002	
09-303018	152	3054/196	66	2048	
09-303019	192	3137	63	2030	
09-303021	196	3054	241	2020	
09-303022	152	3054/196	66	2048	
09-303023	103A	3835	59	2002	
09-303025	123A	3444	20	2051	
09-303028	124	3021	12		
09-303029	198	3054/196	251		
09-303030	103A	3010	59	2002	
09-303031	152	3893		2048	
09-303032	152	3054/196	66	2048	
09-303033	152	3054/196	66	2048	
09-303042			61	2038	
09-30313	102A	3004	53	2007	
09-30318	152	3054/196	66	2048	
09-30319	192	3137	63	2030	
09-304011	160	3006	245	2004	
09-304012	159	3466	82	2032	
09-304017	312	3448	FET-1	2035	
09-304019			11	2015	
09-304042	229	3018	20	2038	
09-304043	229	3018	61	2038	
09-304044	123A	3444	20	2051	
09-304045	123A	3444	20	2051	
09-304046	171	3104A	27		
09-304047	159	3466	82	2032	
09-304048		3018	20	2051	
09-304049	159	3466	82	2032	
09-304050	159	3466	82	2032	
09-304051	159	3466	82	2032	
09-304052	191	3104A	249		
09-304055	182	3104A	55		
09-304056	165	3115	38		
09-304057	164	3111	37		
09-304058	123A	3444	20	2051	
09-304140	152	3893	66	2048	
09-305006	161	3117	39	2015	
09-305007			11	2015	
09-30501	199		62	2010	
09-305011	161	3018	39	2015	
09-305014	312	3448	FET-2	2035	
09-305021	312	3112	FET-1	2028	
09-305023	312	3112	FET-1	2028	
09-305024	159	3114/290	82	2032	
09-305031	312	3112	FET-1	2028	
09-305032	312	3448	FET-1	2035	
09-305033	107	3018	11	2015	
09-305034	123A	3444	20	2051	
09-305036			11	2015	
09-305040	222	3050/221	FET-4	2036	
09-305041	107	3018	11	2015	
09-305048	199	3124/289	62	2010	
09-305049	300	3464	273		
09-305050	229	3018	61	2038	
09-305051	229	3018	61	2038	
09-305052	199	3124/289	62	2010	
09-305058	234	3025/129	65	2050	
09-305062	123A	3444	20	2051	
09-305063	123A	3444	20	2051	
09-305064	123A	3444	20	2051	
09-305065	123A	3444	20	2051	
09-305066	123A	3444	20	2051	
09-305067	123A	3444	20	2051	
09-305068	123A	3444	20	2051	
09-305069	108	3452	86	2038	
09-305070	108	3452	86	2038	
09-305071	108	3452	86	2038	
09-305072	108	3452	86	2038	
09-305073	159	3466	82	2032	

Industry Standard No.	ECG	SK	GE	RS 276-	MOTOR.
09-305074	108	3452	86	2038	
09-305075	129	3025	244	2027	
09-305076	128	3024	243	2030	
09-305077	123A	3444	20	2051	
09-305091	237	3299	46		
09-305092	297	3122	271	2030	
09-305093	229	3018	61	2038	
09-305094	229	3018	61	2038	
09-305095	235	3054/196	215		
09-305096	229		61	2038	
09-305123	199	3124/289	62	2010	
09-305124		3048	279		
09-305126	199	3122	62	2010	
09-305131	102A			2007	
09-305132	161	3039/316	39	2015	
09-305133	312	3112	FET-1	2028	
09-305134	129	3083/197	244	2027	
09-305135	312	3116	FET-1	2035	
09-305136	235	3197	215		
09-305137	236	3197/235	216	2053	
09-305138	390	3027/130	255	2041	
09-305139	123A	3444	20	2051	
09-305140	152	3054/196	66	2048	
09-305148	123A	3122	10	2051	
09-305149	159	3138/193A	21	2032	
09-305152	123A	3122	10	2051	
09-306-083	116	3017B/117	504A	1104	
09-306002	109	3087		1123	
09-306008		3126	90		
09-306009	109	3087		1123	
09-306010	109	3087		1123	
09-3060111		3100	300	1122	
09-306012	109	3087		1123	
09-306014		3126	90		
09-306018		3126	90		
09-306019	110MP	3709	1N60	1123(2)	
09-306020	109	3087		1123	
09-306023		3126	90		
09-306024	109	3087	ZD-15	1123	
09-306024(ZENER)	145A			564	
09-306028		3110	511		
09-306030	506	3843	511	1114	
09-306031	506	3125	511	1114	
09-306033	116	3017B/117	504A	1104	
09-306034	116	3311	504A	1104	
09-306036	109	3088		1123	
09-306037	109	3088		1123	
09-306039		3126	1N60	1123	
09-306040	109	3087		1123	
09-306042		3017B	504A	1104	
09-306042(DIO)			509	1114	
09-306042(RECT)		3017B	504A	1104	
09-306042(ZENER)	5072A		ZD-8.2	562	
09-306046	116	3311	504A	1104	
09-306047	109			1123	
09-306049	109	3087		1123	
09-306050	116	3032A	504A	1104	
09-306051	109	3087		1123	
09-306052	5014A	3780	ZD-6.8	561	
09-306053	156	3051	512		
09-306054	116	3017B/117	504A	1104	
09-306055	137A	3058	ZD-6.2	561	
09-306057	110MP	3709	1N60	1123(2)	
09-306058			1N34AS	1123	
09-306059	116	3311	504A	1104	
09-306060	177	3100/519	300	1122	
09-306061	109	3088		1123	
09-306062	177	3100/519	300	1122	
09-306063	116	3032A	504A	1104	
09-306064	109			1123	
09-306073	5070A	9021	ZD-6.2	561	
09-306077	112	3089	1N82A		
09-306083		3032A	504A	1104	
09-306088	116	3032A	504A	1104	
09-306089	112	3089	1N82A		
09-306091	109	3087		1123	
09-306093	109	3087		1123	
09-306100	116	3031A	504A	1104	
09-306101	113A	3119/113	6GC1		
09-306102			509	1114	
09-306103	116	3017B/117	504A	1104	
09-306104	116	3110/120	504A	1104	
09-306106	139A		ZD-9.1	562	
09-306107	109	3087		1123	
09-306108	109	3087		1123	
09-306109	136A	3057	ZD-5.6	561	
09-306110	177	3100/519	300	1122	
09-306110(ZENER)	145A			564	
09-306111	177	3100/519	300	1122	
09-306112	116	3100/519	504A	1104	
09-306113	177	3100/519	300	1122	
09-306114	116	3017B/117	504A	1104	
09-306115	116	3031A	504A	1104	
09-306119	116	3311	504A	1104	
09-306124	139A	3060	ZD-9.1	562	
09-306125	116	3017B/117	504A	1104	
09-306126		3126	90		
09-306127	143A	3750	ZD-13	563	
09-306129	177	3017B/117	300	1122	
09-306134	177	3100/519	300	1122	
09-306135	519	3100	514	1122	
09-306138	116		504A	1104	
09-306141	506	3843	511	1114	
09-306144	506	3843	511	1114	
09-306145	177	3100/519	300	1122	
09-306148	177	3130	300	1122	
09-306149	116	3017B/117	504A	1123	
09-306151	177	3100/519	300	1122	
09-306154	177	3100/519	300	1122	
09-306157	116	3031A	504A	1104	
09-306158	139A	3060	ZD-9.1	562	
09-306159	177	3100/519	300	1122	
09-306160	116	3017B/117	504A	1104	
09-306161	177	3100/519	300	1122	
09-306162	116		504A	1104	
09-306163	177	3100/519	300		
09-3061633				1122	
09-306165	139A	3060	ZD-9.1	562	
09-306167		3126	90		
09-306168	178MP	3100/519	300(2)	1122(2)	
09-306169	116		504A	1104	
09-306170	177	3175	300	1122	
09-306171	177	3175	300	1122	
09-306172	116	3031A	504A	1104	
09-306173	5022A	3788	ZD-13	563	
09-306176	116	3032A	504A	1104	
09-306177	116	3311	504A	1104	
09-306178	177		300	1122	
09-306179	142A	3062	ZD-12	563	
09-306180	139A	3060	ZD-9.1	562	
09-306181	138A	3059	ZD-7.5		
09-306183	137A	3058	ZD-6.2	561	
09-306191	5010A	3776	ZD-5.1		
09-306192	116	3032A	504A	1104	

Industry Standard No.	ECG	SK	GE	RS 276-	MOTOR.
09-306193	113A	3119/113	6GC1		
09-306194	139A	3060	ZD-9.1	562	
09-306195	177	3100/519	300	1122	
09-306196	5072A	3136	ZD-8.2	562	
09-306197	138A	3059	ZD-7.5		
09-306198	177	3100/519	300	1122	
09-306199	177	3100/519	300	1122	
09-306200		3100	1N34AS	1123	
09-306201		3126	90		
09-306202	177	3100/519	300	1122	
09-306205	116	3017B/117	504A	1104	
09-306206	177	3100/519	300	1122	
09-306208	5070A	9021	ZD-6.2	561	
09-306209	112	3089	1N82A		
09-306210	112	3126	1N82A		
09-306211	177	3100/519	300	1122	
09-306211(RECT)			300	1122	
09-306211(ZENER)	5015A		ZD-7.5		
09-306212	5015A	3100/519	ZD-7.5		
09-306212(RECT)			511	1114	
09-306212(ZENER)	5015A		ZD-7.5		
09-306213	116	3311	504A	1104	
09-306214	116	3311	504A	1104	
09-306215	5071A	3334	ZD-6.8	561	
09-306216	112	3089	1N82A		
09-306219	177	3088	300	1122	
09-306220	177	3100/519	300	1122	
09-306221	177	3100/519	300	1122	
09-306222	109			1123	
09-306223	177	3100/519	300	1122	
09-306224	116		504A	1104	
09-306225			509	1114	
09-306226	125		510,531	1114	
09-306227	156		512		
09-306228	140A		ZD-10	562	
09-306229	110MP	3709	1N60	1123(2)	
09-306230		3126	90		
09-306231	177	3100/519	300	1122	
09-306232	139A	3060	ZD-9.1	562	
09-306233	177	3017B/117	300	1122	
09-306235	140A		ZD-10	562	
09-306236	177	3100/519	300	1122	
09-306237	506	3843	511	1114	
09-306238	5072A	3136	ZD-8.2	562	
09-306239	139A	3060	ZD-9.1	562	
09-306241	139A	3060	ZD-9.1	562	
09-306242	5072A	3059/138A	ZD-8.2	562	
09-306243	5075A	3751	ZD-16	564	
09-306244	519	3100	514	1122	
09-306245	116	3017B/117	504A	1104	
09-306247	139A	3060	ZD-9.1	562	
09-306248	177	3100/519	300	1122	
09-306249	116	3311	504A	1104	
09-306250	116	3311	504A	1104	
09-306251		3126	90		
09-306253		3126	90		
09-306254	116	3311	504A	1104	
09-306255	116	3311	504A	1104	
09-306257			300	1122	
09-306258			504A	1104	
09-306259			504A	1104	
09-306260	506	3843	511	1114	
09-306263	116	3031A	504A	1104	
09-306264	116	3031A	504A	1104	
09-306265		3126	90		
09-306266	177	3100/519	300	1122	
09-306268	5075A	3751	ZD-16	564	
09-306270			1N34AS	1123	
09-306274	125	3017B/117	510,531	1114	
09-306275	139A		ZD-9.1	562	
09-306276	177	3100/519	300	1122	
09-306277		3059	ZD-8.2	562	
09-306278	144A	3094	ZD-14	564	
09-306283	177	3100/519	300	1122	
09-306285	116	3031A	504A	1104	
09-306286		3058	ZD-6.8	561	
09-306287	139A	3060	ZD-9.1	562	
09-306288	177	3100/519	300	1122	
09-306289	5068A	3333/5069A	ZD-4.7		
09-306290	109	3088		1123	
09-306291	177		300	1122	
09-306300	116		504A	1104	
09-306302			504A	1104	
09-306303	116		504A	1104	
09-306309	177	3100/519	300	1122	
09-306310	505	3068/503	CR-7		
09-306311	125	3016	510,531	1114	
09-306312	116	3311	504A	1104	
09-306313	177	3100/519	300	1122	
09-306314	5072A		ZD-8.2	562	
09-306315	116	3017B/117	504A	1104	
09-306323	116	3031A	504A	1104	
09-306324	5075A	3751	ZD-16	564	
09-306325	5071A	3334	ZD-6.8	561	
09-306326	177	3100/519	300	1122	
09-306327	139A	3060	ZD-9.1	562	
09-306330	109	3088			
09-306331	109	3088		1123	
09-306332	139A	3060	ZD-9.1	562	
09-306333	116	3017B/117	504A	1104	
09-306334	109			1123	
09-306335	110MP	3709	1N60	1123(2)	
09-3063353				1104	
09-306336	109	3088		1123	
09-306339	109			1123	
09-306341	116		504A	1104	
09-306349	109	3088		1123	
09-306350	116	3311	504A	1104	
09-306351	139A	3060	ZD-9.1	562	
09-306352		3126	90		
09-306353	116	3017B/117	504A		
09-306354	5072A	3059/138A	ZD-8.2	562	
09-306355	5077A	3752	ZD-17		
09-306356	5072A	3059/138A	ZD-8.2	562	
09-306359	611	3126	90		
09-30636	110MP	3709		1123(2)	
09-306365	116	3311	504A	1104	
09-306366		3017B	504A	1104	
09-306367	145A	3063	ZD-15	564	
09-306368	177	3100/519	300	1122	
09-306370	109	3088		1123	
09-306373	177	3100/519	300	1122	
09-306374		3126	90		
09-306375	139A	3060	ZD-9.1	562	
09-306376	116	3088	504A	1104	
09-306377	137A	3054/196	ZD-6.2	561	
09-306378	5072A	3136	ZD-8.2	562	
09-306379	156	3051	512		
09-306380	5067A	3331	ZD-3.9		
09-306381	5072A	3136	ZD-8.2	562	
09-306382	139A	3060	ZD-9.1	562	
09-306383	5070A	9021	ZD-6.2	561	
09-306384	116		504A	1104	
09-306389	116	3311	504A	1104	

Industry Standard No.	ECG	SK	GE	RS 276-	MOTOR.
09-306390	177	3100/519	300	1122	
09-306391	142A	3062	ZD-12	563	
09-306392	506	3843	511	1114	
09-306393	504		CR-6		
09-306394	116	3311	504A	1104	
09-306401	139A			562	
09-306417	116		509	1104	
09-306418	506	3843	511	1114	
09-306419	5071A		ZD-6.8	561	
09-306420		3108	CR-6		
09-306421	116	3311	504A	1104	
09-306422	116	3311	504A	1104	
09-306423	506	3843	504A	1114	
09-306424	125	3051/156	510	1114	
09-306425	506	3843	511	1114	
09-306426	177	3100/519	300	1122	
09-306427	116	3017B/117	504A	1104	
09-306428	5074A	3139	ZD-11	563	
09-306429	5074A	3139	ZD-11	563	
09-306430	503	3060/139A	CR-5		
09-306431		3017B	504A	1104	
09-306432	116	3017B/117	504A	1104	
09-306433	116	3017B/117	504A	1104	
09-307039	177	3175	300	1122	
09-307043	116		504A	1104	
09-307045	177	3100/519	300	1122	
09-307055	177	3100/519	300	1122	
09-307075	177	3100/519	300	1122	
09-307080	177	3100/519	300	1122	
09-307081	177	3100/519	300	1122	
09-307082	138A	3100/519	ZD-7.5		
09-307083	177	3017B/117	300	1122	
09-307084	116	9075/166	504A	1104	
09-307085	177	3100/519	300	1122	
09-307088	5071A	3334	ZD-6.8	561	
09-307089	177	3100/519	300	1122	
09-308002	724	3525	IC-86		
09-308003	724	3525	IC-86		
09-308004	703A	3157	IC-12		
09-308008	1003		IC-43		
09-308009	1046	3471	IC-118		
09-308010	726	3129			
09-308011	1054	3457	IC-45		
09-308013	703A	3157	IC-12		
09-308017	704	3023	IC-205		
09-308019	703A	3157	IC-12		
09-308022	7400		7400	1801	
09-308025	1076		IC-125		
09-308026	1091*		IC-126		
09-308027	1047		IC-116		
09-308028	1077		IC-124		
09-308029	1086		IC-142		
09-308030	1089*		IC-120		
09-308031	1048		IC-122		
09-308033	1045	3072/712	IC-1045		
09-308034	1006	3358	IC-38		
09-308036	1103	3281	IC-94		
09-308037	1029		IC-162		
09-308038	1142	3485	IC-128		
09-308041	1100	3223	IC-92		
09-308043	1103	3281	IC-94		
09-308045	715	3076	IC-6		
09-308046	715	3076	IC-6		
09-308047	713	3077/790	IC-5		
09-308048	713	3077/790	IC-5		
09-308050	1002	3481	IC-69		
09-308052	1082	3461	IC-140		
09-308053	1075A	3877			
09-308058	1005	3723	IC-42		
09-308059	1100	3223	IC-92		
09-308061	1097	3446			
09-308062	1135	3876	IC-314		
09-308063	1003	3288	IC-43		
09-308064	1006	3358	IC-38		
09-308066	1106		IC-90		
09-308067	1107	3526/947	IC-276		
09-308069	1047		IC-116		
09-308070	1108		IC-87		
09-308071	720	9014	IC-128		
09-308072			11	2015	
09-308076	1052	3249	IC-135		
09-308078	804	3455	IC-27		
09-308079	714	3075	IC-4		
09-308080	1155	3231			
09-308084	1127	3243/1160	IC-197		
09-308089	1004	3102/710	IC-149		
09-308090	749	3168	IC-97		
09-308094	1243	3731	IC-113		
09-308095	1087		IC-103		
09-308096	1092	3472	IC-130		
09-308098	712	3072			
09-308099	791	3149	IC-231		
09-308100	712	3072	IC-2		
09-308102	1049	3470			
09-30864	1006		IC-38		
09-309006	123A	3018	39	2051	
09-309007	108	3452	86	2038	
09-309012	123A	3444	20	2051	
09-309013	108	3452	86	2038	
09-309023	123A	3444	20	2051	
09-309024	108	3452	86	2038	
09-309027	108	3452	86	2038	
09-309028	108	3452	86	2038	
09-309029	172A	3156	64		
09-309030	193	3025/129	67	2023	
09-309031	192	3024/128	63	2030	
09-309032	108	3452	86	2038	
09-309038	159	3466	82	2032	
09-309042	159	3466	82	2032	
09-309049	123A	3444	20	2051	
09-309050	123A	3444	20	2051	
09-309059	199	3245	62	2010	
09-309060		3444	20	2051	
09-309061	289A		268	2038	
09-309062	195A	3048/329	46		
09-309063	297	3137/192A	271	2030	
09-309064	123A	3444	20	2051	
09-309065	229	3018	61	2038	
09-309069	107	3018	11	2015	
09-309070	199	3124/289	62	2010	
09-309071	186A	3357	28	2017	
09-309072	107	3122	11	2015	
09-309073	108	3452	86	2038	
09-309074	312	3448	FET-1	2035	
09-309075	103A	3010	59	2002	
09-309076	123A	3444	20	2051	
09-309672	107		11	2015	
09-32124	108	3452	86	2038	
9-5108	160	3006	245	2004	
9-511			245	2004	
9-5110	160	3006	245	2004	
9-5111	160	3006		2004	
9-5112	101	3861	8	2002	
9-5113	101	3861	8	2002	

Industry Standard No.	ECG	SK	GE	RS 276-	MOTOR.
9-5114	101	3861	8	2002	
9-511410100	102A	3004	53	2007	
9-511410200	102A	3004	53	2007	
9-511410900	102A	3004	53	2007	
9-511413500	102A	3004	53	2007	
9-51141400	121	3717	239	2006	
9-511511500	125	3032A	510,531	1114	
9-5116	160	3006	245	2004	
9-5117	160	3006	245	2004	
9-5118	160	3006	245	2004	
9-5119	160	3006	245	2004	
9-5120	160	3006	245	2004	
9-5120A	100	3005	1	2007	
9-5121	160	3006	245	2004	
9-5122	160	3006	245	2004	
9-5123	160	3006	245	2004	
9-5124	160	3006	245	2004	
9-5125	108	3452	86	2038	
9-5126	108	3452	86	2038	
9-5127	108	3452	86	2038	
9-5128	108	3452	86	2038	
9-5129	108	3452	86	2038	
9-5130	108	3452	86	2038	
9-5131	108	3452	86	2038	
9-5201	102A	3004	53	2007	
9-5202	103A	3835	59	2002	
9-5203	102A	3835/103A	53	2007	
9-5204	102A	3835/103A	53	2007	
9-5208	102A	3835/103A	53	2007	
9-5209	102A	3835/103A	53	2007	
9-5212	102A	3835/103A	53	2007	
9-5213	102A	3835/103A	53	2007	
9-5214	102A	3835/103A	53	2007	
9-5216	123A	3122	20	2051	
9-5217	102A	3004	53	2007	
9-5218	102A	3004	53	2007	
9-5220	128	3024	243	2030	
9-5221	123A	3444	20	2051	
9-5222-1	102A	3835/103A	53	2007	
9-5222-2	103A	3010	59	2002	
9-5223	108	3452	86	2038	
9-5224-1	102A	3004	53	2007	
9-5224-2	103A	3835	59	2002	
9-5225	123A	3444	20	2051	
9-5226-003	129	3025	244	2027	
9-5226-004	128	3024	243	2030	
9-5226-1	129	3025	244	2027	
9-5226-2	123A	3444	20	2051	
9-5226-3	129	3025	244	2027	
9-5226-4	128	3024	243	2030	
9-5227	123A	3444	20	2051	
9-5250	121	3009	239	2006	
9-5250-1	121MP	3013	239(2)	2006(2)	
9-5251	121	3009	239	2006	
9-5252	124	3021	12		
9-5252-1	124	3021	12		
9-5252-2	124	3021	12		
9-5252-3	124	3021	12		
9-5252-4	124	3021	12		
9-5257	121MP	3013	239(2)	2006(2)	
9-5296	123A	3444	20	2051	
9-905-606-001			51	2004	
9-9101	126	3006/160	52	2024	
9-9102	126	3006/160	52	2024	
9-9103	126	3006/160	52	2024	
9-9104	102A	3004	53	2007	
9-9105	160	3006	245	2004	
9-9106	160	3006	245	2004	
9-9107	160	3006	245	2004	
9-9108	160	3006	245	2004	
9-9109-1	123A	3444	20	2051	
9-9109-2	123A	3444	20	2051	
9-9120	160	3006	245	2004	
9-9121	160	3006	245	2004	
9-9201	102A	3004	53	2007	
9-9202	102A	3004	53	2007	
9-9203	102A	3004	53	2007	
9.S011			2	2007	
9.S037			20	2051	
09A02	746	3234			
09A04	74LS193	74LS193			
09A05	4020B	4020			
09A07	787	3146			
09A08	1239	3708/1169			
9A1	358		42		
9A8-1A64	160		245	2004	
9ACW			FET-1	2028	
9D1			300	1122	
9D11	177	3100/519	300	1122	
9D12	109	3087		1123	
9D13	116	3017B/117	300	1104	
9D14	177	3100/519	300	1122	
9D141003-12	139A	3060	ZD-9.1	562	
9D15	139A	3060	ZD-9.1	562	
9D16	109	3087		1123	
9DI	177			1122	
9DI1		3100	300	1122	
9DI2	109	3088	1N60	1123	
9DI3	177	3100/519	300	1122	
9DI5		3060	ZD-9.1	562	
9DT1100310	109	3087	86	1123	
9GR2	123A	3444	20	2051	
9H00DC	74H00	74H00			
9H00PC	74H00	74H00			
9H04DC	74H04	74H04			
9H04PC	74H04	74H04			
9L-4	102A	3003	53	2007	
09L-4-12	102A	3003	53	2007	
09L-4-12-7	102A	3003	53	2007	
9L-4-70	102A	3003	53	2007	
9L-4-70-12	102A	3003	53	2007	
9L-4-70-12-7	102A	3003	53	2007	
9L-40	102A	3003	53	2007	
9L-41	102A	3003	53	2007	
9L-42	102A	3003	53	2007	
9L-43	102A	3003	53	2007	
9L-44	102A	3003	53	2007	
9L-45	102A	3003	53	2007	
9L-46	102A	3003	53	2007	
9L-47	102A	3003	53	2007	
9L-48	102A	3003	53	2007	
9L-49	102A	3003	53	2007	
9L-4A	102A	3003	53	2007	
9L-4A0	102A	3003	53	2007	
9L-4A0R	102A	3003	53	2007	
9L-4A1	102A	3003	53	2007	
9L-4A19	102A	3003	53	2007	
9L-4A2	102A	3003	53	2007	
9L-4A21	102A	3003	53	2007	
9L-4A3	102A	3003	53	2007	
9L-4A3P	102A	3003	53	2007	
9L-4A4	102A	3003	53	2007	
9L-4A4-7	102A	3003	53	2007	
9L-4A4-7B	102A		53	2007	

Industry Standard No.	ECG	SK	GE	RS 276-	MOTOR.
9L-4A5	102A	3003	53	2007	
9L-4A5L	102A	3003	53	2007	
9L-4A6	102A	3003	53	2007	
9L-4A6-1	102A	3003	53	2007	
9L-4A7	102A	3003	53	2007	
9L-4A7-1	102A	3003	53	2007	
9L-4A8	102A	3003	53	2007	
9L-4A82	102A	3003	53	2007	
9L-4A9	102A	3003	53	2007	
9L-4A90	102A	3003	53	2007	
9LR2	113A	3119/113	6GC1		
9LR2-1	113A	3119/113	6GC1		
9LR2-2	114	3120	6GD1	1104	
9LR2-24	113A	3119/113	6GC1		
9LR2-3	113A	3119/113	6GC1		
9LR2-4	115	3121	6GX1		
9LR2-S	113A	3119/113	6GC1		
9LR2-S1	113A	3119/113	6GC1		
9LR21	113A	3119/113	6GC1		
9LS00	74LS00	74LS00			
9LS109	74LS109A	74LS109			
9LS163	74LS163A	74LS163			
9LS195	74LS195A	74LS195			
9LS253	74LS253	74LS253			
9LS74	74LS74A	74LS74			
9LS93	74LS93	74LS93			
9N00	7400	7400	7400	1801	
9N00DC	7400	7400		1801	
9N00PC	7400	7400		1801	
9N01DC	7401	7401			
9N01PC	7401	7401			
9N02		7402	7402	1811	
9N02DC	7402	7402		1811	
9N02PC	7402	7402		1811	
9N04		7404	7404	1802	
9N04DC	7404	7404		1802	
9N04PC	7404	7404		1802	
9N05DC	7405	7405			
9N05PC	7405	7405			
9N06		7406	7406	1821	
9N06DC	7406	7406		1821	
9N06PC	7406	7406		1821	
9N08		7408	7408	1822	
9N08DC	7408	7408		1822	
9N08PC	7408	7408		1822	
09N1	123A	3444	20	2051	
9N10		7410	7410	1807	
9N107DC	74107	74107			
9N107PC	74107	74107			
9N10DC	7410	7410		1807	
9N10PC	7410	7410		1807	
9N123		74123	74123	1817	
9N123DC	74123	74123		1817	
9N123PC	74123	74123		1817	
9N13		7413	7413	1815	
9N13DC	7413	7413		1815	
9N13PC	7413	7413		1815	
9N14DC	7414	7414			
9N14PC	7414	7414			
9N16DC	7416	7416			
9N16PC	7416	7416			
9N20		7420	7420	1809	
9N20DC		7420		1809	
9N20PC	7420	7420		1809	
9N27		7427	7427	1823	
9N27DC	7427	7427		1823	
9N27PC	7427	7427		1823	
9N30DC	7430	7430			
9N30PC	7430	7430			
9N32		7432	7432	1824	
9N32DC	7432	7432		1824	
9N32PC	7432	7432		1824	
9N37DC	7437	7437			
9N37PC	7437	7437			
9N38DC	7438	7438			
9N40DC	7440	7440			
9N40PC	7440	7440			
9N51		7451	7451	1825	
9N51DC	7451	7451		1825	
9N51PC	7451	7451		1825	
9N54DC	7454	7454			
9N54PC	7454	7454			
9N73		7473	7473	1803	
9N73DC	7473	7473		1803	
9N73PC	7473	7473		1803	
9N74		7474	7474	1818	
9N74DC	7474	7474		1818	
9N74PC	7474	7474		1818	
9N76		7476	7476	1813	
9N76DC	7476	7476		1813	
9N76PC	7476	7476		1813	
9N86		7486	7486	1827	
9N86DC	7486	7486		1827	
9N86PC	7486	7486		1827	
90C69-1			504A	1104	
9RE1	116		504A	1104	
9S00DC	74S00	74S00			
9S00PC	74S00	74S00			
9S011	102A		53	2007	
9S037	123A		20	2051	
9S04DC	74S04	74S04			
9S04PC	74S04	74S04			
9TR1	107	3122	11	2015	
9TR10	123A	3122	20	2051	
9TR11	195A	3047	46		
9TR11001-01	107		11	2015	
9TR2	123A	3444	20	2051	
9TR21001-02	123A	3444	20	2051	
9TR3	199	3122	62	2010	
9TR31001-03	123A		20	2051	
9TR4	195A	3047	46		
9TR5	237	3299	46		
9TR6	195A	3047	46		
9TR61001-07	128		243	2030	
9TR7	123A	3444	20	2051	
9TR8	152	3893	66	2048	
9TR9	289A	3047	268	2038	
9TR91001-09	130		14	2041	
9TRZ1001-02	123A	3444	20	2051	
010	173BP	3999	305		
10-0009			28	2052	
10-001	1254	3880	IC-172		
10-002	229	3122	213	2038	
10-003	199	3124/289	212	2010	
10-004	312	3448	FET-2	2035	
10-005	312	3963/1249			
10-006	295	3253	270		
10-007	235		333		
10-008	289A	3122	81	2038	
10-009	186A	3197/235	28		
10-010	177	3100/519	300	1122	
10-012	116	3311	504A	1104	
10-013	139A	3060	ZD-9.1	562	
10-014	229	3065/222			

Industry Standard No.	ECG	SK	GE	RS 276-	MOTOR.
10-015	614		90		
10-016	136A	3057	ZD-5.6	561	
10-017	138A	3124/289			
10-018	139A	3849/293			
10-080009		3452	86	2038	
10-080010	123A	3444	20	2051	
10-085001	109	3087		1123	
10-085004	110MP		1N60	1123	
10-085005	177	3087	300	1122	
10-085006	116	3311	504A	1104	
10-085009	116	3311	504A	1104	
10-085010	116	3311	504A	1104	
10-085013		3126	90		
10-085014		3088	1N60	1123	
10-085018		3088	1N60	1123	
10-085025	109	3087		1123	
10-085026	116	3017B/117	504A	1104	
10-085027	135A	3056	ZD-5.1		
10-085030		3311	504A	1104	
10-1		3117	39	2015	
10-10112	177	3100/519	300	1122	
10-102005	125	3031A	510,531	1114	
10-12	116	3016	504A	1104	
10-13-002-003	130		14	2041	
10-13-002-004	130		14	2041	
10-13-002-3	130		14	2041	
10-13-002-4	130		14	2041	
10-13002-003	130	3027	14	2041	
10-13002-004	130	3027	14	2041	
10-13030-004	181		75	2041	
10-13030-005	181		75	2041	
10-13159-002	175		246	2020	
10-13I59-002	175		246	2020	
10-2	161	3039/316	39	2015	
10-26-123-313	152	3893	66	2048	
10-2SA49	126	3018	52	2024	
10-2SB54	102A	3004	53	2007	
10-2SB56	102A	3004	53	2007	
10-2SC380	107	3018	11	2015	
10-2SC80	108	3452	86	2038	
10-2SC94	108	3452	86	2038	
10-3159-002	175		246	2020	
10-374101	130		14	2041	
10-42	116	3016	504A	1104	
10-47674-01	722	3161	IC-9		
010-6742	173BP	3999	305		
010-6744	173BP	3999	305		
010-694		3044	27		
010-694(AMPEX)	154		40	2012	
10-7	116	3016	504A	1104	
10-D1	116		504A	1104	
10-I3002-004	130			2041	
10A	160		245	2004	
10A590B	116	3016	504A	1104	
10AG10	125	3033	510,531	1114	
10AG2	116	3016	504A	1104	
10AG4	116	3017B/117	504A	1104	
10AG6	116	3017B/117	504A	1104	
10AG8	125	3032A	510,531	1114	
10AL2	116	3016	504A	1104	
10AL6	116	3017B/117	504A	1104	
10AS	116	3016	504A	1104	
10AT10	125	3033	510,531	1114	
10AT2	116	3016	504A	1104	
10AT4	116	3017B/117	504A	1104	
10AT6	116	3017B/117	504A	1104	
10AT8	125	3033	510,531	1114	
10B	160	3016	245	2004	MR1121
10B-2	116	3311	504A	1104	
10B-2-B1W			504A	1104	
10B-2-N1W	116(2)		504A	1104	
10B-4	120	3031A	CR-3		
10B-4-C4	120	3110	CR-3		
10B-Y	116	3110/120	504A	1104	
10B1	116	3017B/117	504A	1104	1N4002
10B10	125	3051/156	512		1N4007
10B1051	108	3452	86	2038	
10B1055	108	3452	86	2038	
10B2	116	9005/5802	504A	1143	1N4003
10B2-B1W	166	9075		1152	
10B3	116	9006/5803	510	1144	1N4004
10B4	116	9007/5804	CR-3	1144	1N4004
10B5	116	9008/5805	504A		1N4005
10B551	108	3452	86	2038	
10B551-2			39	2015	
10B551-3			39	2015	
10B553	108	3452	86	2038	
10B553-2			39	2015	
10B553-3			39	2015	
10B555	108	3452	86	2038	
10B555-2			214	2038	
10B555-3			214	2038	
10B556	108	3452	86	2038	
10B556-2			214	2038	
10B556-3			214	2038	
10B6	116	3848/5806	504A		1N4005
10B62	140A		ZD-10	562	
10B8	125	3051/156	510,531	1114	1N4006
10BR					1N3880
10C	116	3031A	504A	1104	
10C05	116	3017B/117	504A	1104	
10C1	116	3016	504A	1104	
10C10	125	3033	510,531	1114	
10C2	116	3016	504A	1104	
10C3	116	3031A	504A	1104	
10C4	116	3031A	504A	1104	
10C4D	116	3031A	504A	1104	
10C5	116	3017B/117	504A	1104	
10C573	108	3452	86	2038	
10C573-2			39	2015	
10C573-3			39	2015	
10C574	108	3452	86	2038	
10C574-2			39	2015	
10C574-3			39	2015	
10C6	116	3017B/117	504A	1104	
10C8	125	3032A	510,531	1114	
10D	116	3311	504A	1104	
10D-02	116		504A	1104	
10D-05	116	3311	504A	1104	
10D-06	116	3311	504A	1104	
10D-1	116		504A	1104	
10D-2	552		511	1114	
10D-2(CROWN)	506		511	1114	
10D-2B	116	3017B/117	511	1114	
10D-2B(-4)	116	3017B/117	504A	1104	
10D-2B(4)			504A	1104	
10D-2B-4	506	3017B/117	504A	1104	
10D-4	116		509	1114	
10D-5		3031A	511	1114	
10D-6	116	3313	504A	1104	
10D-7K	125	3032A	509	1114	
10D-V	116	3031A	504A	1104	
10D0.5			504A	1104	
10D05	116	3311	504A	1104	

Industry Standard No.	ECG	SK	GE	RS 276-	MOTOR.
10D08	5072A	3136	ZD-8.2	562	
10D1	116	3311	504A	1104	1N5392
10D10	125	3033	510,531	1114	1N5399
10D2	116	9000/552	504A	1102	1N5393
10D2(CROWN)	506		511	1114	
10D2(DAMPER)	506		511	1114	
10D2L	116	3031A	511	1114	
10D3	116	3016	504A	1104	1N5394
10D3G	116	3311	504A	1104	
10D4	116	3312	504A	1104	1N5395
10D4C	116		504A	1104	
10D4D	116		504A	1104	
10D4E	116		504A	1104	
10D4L	116	3130	511	1114	
10D5	116	3017B/117	504A	1104	1N5396
10D5A	506	3843	511	1114	
10D5B		3311	504A	1104	
10D5C	116		504A	1104	
10D5D	506	3843	511	1114	
10D5E	116	3311	504A	1104	
10D5F	116	3017B/117	504A	1104	
10D6	116	3017B/117	504A	1104	1N5397
10D6D	116		504A	1104	
10D6E	116	3017B/117	504A	1104	
10D7	125	3032A	510,531	1114	
10D7F	125		509	1114	
10D8	552	3032A	510,531	1114	1N5398
10DB	116(4)		504A(4)	1104	
10DB1	166	9075		1152	
10DB1P	5304				MDA101A
10DB2				1172	
10DB2A	167	3647		1172	
10DB2P	5304				MDA102A
10DB3P					MDA104A
10DB4				1173	
10DB4A	168	3648		1173	
10DB4P	5304				MDA104A
10DB6A	168	3678/169		1173	
10DB6A-C	168	3678/169		1173	
10DB6P	5305	3676			MDA920A7
10DB8P	5306	3677			
10DBF	116		504A	1104	
10DC	116	3031A	504A(2)	1104	
10DC-1	116(2)		504A(2)	1104	
10DC-1(SANYO)	135A		ZD-5.1		
10DC-1N	113A		6GC1		
10DC-1R	116		504A	1104	
10DC-2	156(2)	3032A	512(2)	1104	
10DC-2B	116(2)		504A(2)	1104	
10DC-2C	116	3311	504A	1104	
10DC-2F	116(2)		504A(2)	1104	
10DC-2J	116	3311	504A	1104	
10DC-4	116	3106/5304	504A	1104	
10DC-4R	116	3106/5304	504A	1104	
10DC0.5			504A(2)	1104	
10DC05(RED)	116		504A	1104	
10DC05N	113A	3031A	6GC1		
10DC05R	116	3031A	504A	1104	
10DC05R(BOOST)	115		6GX1		
10DC0B		3311	504A	1104	
10DC0H		3311	504A	1104	
0000010DC1	116	9001/113A	504A	1104	
10DC1(BLACK)			504A	1104	
10DC1(RED)			504A	1104	
10DC1BLACK				1104	
10DC1N	125	9001/113A			
10DC1R	125	9002	504A	1104	
10DC2			504A	1104	
10DC2F	116	3311	504A	1104	
10DC4	116		504A	1104	
10DC4R	116		504A	1104	
10DC5		3031A	504A	1104	
10DC8	125		510,531	1114	
10DC8R			509	1114	
10DC05R	116		504A	1104	
10DC0B	116		504A	1104	
10DC0H	116		504A	1104	
10DG	116		504A	1104	
10DI		3080	509	1114	
10D0.5	116		504A	1104	
10DRV	116	3311	504A	1104	
10DV	116		504A	1104	
10DX2	116		504A	1104	
10DZ	116	3311	504A	1104	
10E-1	116	3311	504A	1104	
10E-2	116	3311	504A	1104	
10E-4D	116	3311	504A	1104	
10E-7L		3311	504A	1104	
10E1	116	3311	504A	1104	
10E1LF	116		504A	1104	
10E2	552	3311	504A	1104	
10E6	116		504A	1104	
10EB1	140A	3061	ZD-10	562	
10F5	5852		5016		
10G1051	108	3452	86	2038	
10G1052	108	3452	86	2038	
10G4	125	3033	510,531	1114	
10GA	125	3033	510,531	1114	
10H	116	3016	504A	1104	
10H1051	108	3452	86	2038	
10H1053	108	3452	86	2038	
10H3	5874		5032		
10H3P	5852	3017B/117	5016		MR1121
10H3PN	5853		5017		
10H551	108	3452	86	2038	
10H553	108	3452	86	2038	
10HB050					MDA2500
10HB100					MDA2501
10HB200					MDA2502
10HB400					MDA2504
10HB600					MDA2506
10HR3P	5818				1N3880
10I10	125	3311		1114	
10J2	116	3311	504A	1104	
10J2F	116	3311	504A	1104	
10J3P	5872		5032		
10JH3	6050		5128		
10K	116	3016	504A	1104	
10K-1	116		504A	1104	
10K80	505	3108	CR-7		
10L15	525	3925			
10LZ3.3D5					10M3.3AZ5
10LZ7.5D5					10M7.5AZ5
10M	116	3016	504A	1104	
10M10	5982	3609/5986	5096		
10M100	6002	7202	5108		
10M20	5986	3609	5096		
10M30	5988	3608/5990	5100		
10M40	5990	3608	5100		
10M50	5992	3501/5994	5104		
10M60	5994	3501	5104		
10M80	5998		5108		
10N1	116	3016	504A	1104	
10N2	5874		5032		
10N3	5874		5032		
100-202			1N60	1123	
10P1	159	3466	82	2032	
10P1A	159	3466	82	2032	
10P1S	5522	6622			
10P2	5874		5032		
10P2S	5524	6624			
10P3	5874		5032		
10P3S	5526	6627/5527			
10P4S	5527	6627			
10P5S	5528	6629/5529			
10P6S	5529	6629			
10P8S	5531	6631			
10Q3	5944		5048		
10R10B	125	3033	510,531	1114	
10R11B	506	3843	511	1114	
10R12B	506	3843	511	1114	
10R13B	506	3843	511	1114	
10R14B	506	3843	511	1114	
10R15B	525	3925			
10R1B	116	3017B/117	504A	1104	
10R2B	116	3016	504A	1104	
10R3B	116	3017B/117	504A	1104	
10R4B	116	3017B/117	504A	1104	
10R5B	116	3017B/117	504A	1104	
10R6B	116	3017B/117	504A	1104	
10R7B	125	3032A	510,531	1114	
10R8B	125	3032A	510,531	1114	
10R9B	125	3033	510,531	1114	
10RC10	5522	6622	MR-3		
10RC10A	5522		MR-3		
10RC15	5523	6624/5524	MR-3		
10RC15A	5523	6624/5524	MR-3		
10RC2	5520	6621/5521	MR-3		
10RC20	5524	6624	MR-3		
10RC20A	5524	6624	MR-3		
10RC25	5525	6627/5527	MR-3		
10RC25A	5544		MR-3		
10RC2A	5520	6621/5521	MR-3		
10RC30	5526	6627/5527	MR-3		
10RC30A	5526	6627/5527	MR-3		
10RC40	5527	6627	MR-3		
10RC40A	5527	6627	MR-3		
10RC5	5521	6621	MR-3		
10RC50	5528	6629/5529			
10RC50A	5528	6629/5529			
10RC5A	5521	6621	MR-3		
10RC60	5529	6629			
10RC60A	5529	6629			
10RC80	5531	6631			
10RC80A	5531	6631			
10T200	5232A	3311	504A	1104	
10TB5	5528	6629/5529			
10TB6	5529	6629			
10TB8	5531	6631			
10V	140A		ZD-10	562	
10VJ	140A	3061	ZD-10	562	
10Z12	5188A		ZD-12		
011-00	191	3104A	249		
11-0399	121	3009	239	2006	
11-0400	121	3009	239	2006	
11-0422	289A	3444/123A	20	2051	
11-0423	172A	3156	64		
11-0429	116	3031A	504A	1104	
11-0430	177	3175	300	1122	
11-0769	116	3031A	504A	1104	
11-0770	185	3191	58	2025	
11-0771	116	3031A	504A	1104	
11-0772	184	3054/196	57	2017	
11-0773	185	3083/197	58	2025	
11-0774	289A	3444/123A	20	2051	
11-0775	172A	3156	64		
11-0778	289A	3444/123A	20	2051	
11-0781	177	3175	300	1122	
11-085001	109	3087		1123	
11-085003	116	3311	504A	1104	
11-085004	109	3087		1123	
11-085005	109	3087		1123	
11-085007	109	3087		1123	
11-085008	109	3091		1123	
11-085010	123A	3444	20	2051	
11-085012	177		300	1122	
11-085013	116	3016	504A	1104	
11-085014		3091		1123	
11-085015	109	3087		1123	
11-085022	109	3087		1123	
11-085024	116	3031A	504A	1104	
11-102-001	166	9075	504A	1152	
11-102001	116	3311	504A	1104	
11-102003	166	9075	504A	1152	
11-108002	166	9075		1152	
11-11911-1	181		75	2041	
11-120007	116	3311	504A	1104	
11-1500	5842	3025/129			
11-1592	116		504A	1104	
11-27070	199		62	2010	
11-691501	234	3247	65	2050	
11-691502	234	3247	65	2050	
11-691504	159	3466	82	2032	
11-9-156438	519		514	1122	
011-H01	100	3005	1	2007	
11/1	177(3)	3311	504A	1104	
11/1&12/1			504A	1104	
11/1+12/1			504A	1104	
11/10	506	3843	511	1114	
11/15	116		504A	1104	
11/15(BOOST)	119		CR-2		
11/15-92	119		CR-2		
11/1592			504A	1104	
11/20		3109	CR-2		
11B1052	108	3452	86	2038	
11B1055	108	3452	86	2038	
11B551	108	3452	86	2038	
11B551-2		3245	212	2010	
11B551-3		3245	212	2010	
11B552	108	3452	86	2038	
11B552-2		3245	212	2010	
11B552-3		3245	212	2010	
11B554	108	3452	86	2038	
11B554-2		3245	212	2010	
11B554-3		3245	212	2010	
11B555	108	3452	86	2038	
11B555-2		3245	212	2010	
11B555-3		3245	212	2010	
11B556			63	2030	
11B556-2		3244	220		
11B556-3		3244	220		
11B560			63	2030	
11B560-2		3244	220		
11B560-3		3244	220		
11C1051	108	3452	86	2038	
11C1053	108	3452	86	2038	
11C1057	108	3452	86	2038	
11C10B1	124	3021	12		

Industry Standard No.	ECG	SK	GE	RS 276-	MOTOR.
11C11B1	124	3021	12		
11C1536	128	3024	243	2030	
11C1B1	124	3021	12		
11C211B20			63	2030	
11C3B1	124	3021	12		
11C3B3	124	3021	12		
11C44	974	3965			
11C55	974	3965			
11C551	108	3452	86	2038	
11C551-2		3245	212	2010	
11C551-3		3245	212	2010	
11C553	108	3452	86	2038	
11C553-2		3245	212	2010	
11C553-3		3245	212	2010	
11C557	108	3452	86	2038	
11C557-2			39	2015	
11C557-3			39	2015	
11C5B1	124	3021	12		
11C702			47	2030	
11C7B1	124	3021	12		
011H01			1	2007	
11J2	116	3017B/117	504A	1104	
11J2F	116	3017B/117	504A	1104	
11R05S	5870		5032		
11R10S	5890	7090	5044		
11R1S	5872		5032		
11R2S	5874		5032		
11R3S	5876		5036		
11R4S	5878		5036		
11R5S	5880		5040		
11R6S	5882		5040		
11R8S	5886		5044		
11RC10	5522	6622			
11RC20	5524	6624			
11RC30	5526	6627/5527			
11RC40	5527	6627			
11RC50	5528	6629/5529			
11RC60	5529	6629			
11RC80	5531	6631			
11RT1			21	2034	
11T4	5408	6753/5453		1067	
11T4S	5453	6753		1067	
012-00	182	3104A	55		
012-0121-001	177		300	1122	
12-085005	109	3087		1123	
12-085006	109	3091		1123	
12-085009	109	3126		1123	
12-085029	109	3087		1123	
12-085031	116	3017B/117	504A	1104	
12-085034	109	3087		1123	
12-085035	109	3087		1123	
12-085038	109	3087		1123	
12-085040	116	3031A	504A	1104	
12-085041	110MP	3709	1N60	1123(2)	
12-087003	109	3087		1123	
12-087004	177	3175	300	1122	
12-1	154	3444/123A	40	2012	
12-1-100	100	3005	1	2007	
12-1-102	100	3005	1	2007	
12-1-103	100	3005	1	2007	
12-1-104	100	3005	1	2007	
12-1-105	100	3005	1	2007	
12-1-106	102A	3004	53	2007	
12-1-107	102A	3004	53	2007	
012-1-12	123A	3444	20	2051	
012-1-12-7	123A	3444	20	2051	
12-1-120	102A	3004	53	2007	
12-1-128	100	3005	1	2007	
12-1-135	160	3006	245	2004	
12-1-137	160	3006	245	2004	
12-1-138	160	3008	245	2004	
12-1-139	160	3006	245	2004	
12-1-148	102A	3004	53	2007	
12-1-150	160	3007	245	2004	
12-1-157	160	3006	245	2004	
12-1-161	100	3005	1	2007	
12-1-162	100	3005	1	2007	
12-1-164	102A	3004	53	2007	
12-1-179	100	3005	1	2007	
12-1-180	100	3005	1	2007	
12-1-184	102A	3004	53	2007	
12-1-186	100	3005	1	2007	
12-1-189	126	3008	52	2024	
12-1-190	160	3007	245	2004	
12-1-191	102A	3004	53	2007	
12-1-226	102A		53	2007	
12-1-227	102A	3004	53	2007	
12-1-228	160	3006	245	2004	
12-1-229	160	3006	245	2004	
12-1-230	160	3006	245	2004	
12-1-231	160	3006	245	2004	
12-1-232	102A	3004	53	2007	
12-1-233	160	3006	245	2004	
12-1-234	100	3005	1	2007	
12-1-235	100	3005	1	2007	
12-1-236	100	3005	1	2007	
12-1-240	100	3005	1	2007	
12-1-241	100	3005	1	2007	
12-1-242	126	3008	52	2024	
12-1-243	126	3008	52	2024	
12-1-244	126	3008	52	2024	
12-1-246	102A	3004	53	2007	
12-1-254	100	3005	1	2007	
12-1-256	160	3007	245	2004	
12-1-257	126	3008	52	2024	
12-1-258	160	3007	245	2004	
12-1-259	160	3007	245	2004	
12-1-260	160	3007	245	2004	
12-1-266	102A	3004	53	2007	
12-1-267	102A	3004	53	2007	
12-1-270	121	3009	239	2006	
12-1-271	121	3009	239	2006	
12-1-272	102A	3004	53	2007	
12-1-273	100	3005	1	2007	
12-1-274	102A	3003	53	2007	
12-1-275	100	3005	1	2007	
12-1-276	123A	3444	20	2051	
12-1-277	123A	3444	20	2051	
12-1-278	123A	3444	20	2051	
12-1-279	123A	3444	20	2051	
12-1-289	100	3005	1	2007	
12-1-70	123A	3444	20	2051	
12-1-70-12	123A	3444	20	2051	
12-1-70-12-7	123A	3444	20	2051	
12-1-73	100	3005	1	2007	
12-1-74	100	3005	1	2007	
12-1-75	100	3005	1	2007	
12-1-76	100	3005	1	2007	
12-1-78	100	3005	1	2007	
12-1-83	100	3005	1	2007	
12-1-91	100	3005	1	2007	
12-1-92	100	3005	1	2007	
12-1-93	100	3005	1	2007	
12-1-95	102A	3004	53	2007	

Industry Standard No.	ECG	SK	GE	RS 276-	MOTOR.
12-1-96	102A	3004	53	2007	
12-10	123A	3444	20	2051	
12-100001	116	3017B/117	504A	1104	
12-100003	116	3311	504A	1104	
12-100008	116	3031A	504A	1104	
12-100027	129		244	2027	
12-100047	130		14	2041	
12-101001	123A	3444	20	2051	
012-1020-005	177	3100/519	300	1122	
12-102001	116	3031A	504A	1104	
012-1021-001		3100	300	1122	
012-1022-002	114	3120	6GD1	1104	
012-1023-007	136A	3057	ZD-5.6	561	
012-1024-001	177	3100/519	300	1122	
012-1025-002	156	3051	512		
012-103002	152	3893	66	2048	
12-11	123A	3444	20	2051	
12-12	123A	3444	20	2051	
12-13	123A	3444	20	2051	
12-14	123A	3444	20	2051	
12-15	123A	3444	20	2051	
12-16	123A	3444	20	2051	
12-17	123A	3444	20	2051	
12-18	123A	3444	20	2051	
12-19	123A	3444	20	2051	
12-1A	123A	3444	20	2051	
12-1A0	123A	3444		2051	
12-1A0R	123A	3444		2051	
12-1A1	123A	3444	20	2051	
12-1A19	123A	3444	20	2051	
12-1A2	123A	3444	20	2051	
12-1A21	123A	3444	20	2051	
12-1A3	123A	3444	20	2051	
12-1A3P	123A	3444	20	2051	
12-1A4	123A	3444	20	2051	
12-1A4-7	123A	3444	20	2051	
12-1A4-7B	123A	3444	20	2051	
12-1A5	123A	3444	20	2051	
12-1A5L	123A	3444	20	2051	
12-1A6	123A	3444	20	2051	
12-1A6A	123A	3444	20	2051	
12-1A7	123A	3444	20	2051	
12-1A7-1	123A	3444	20	2051	
12-1A8	123A	3444	20	2051	
12-1A82	123A	3444	20	2051	
12-1A9	123A	3444	20	2051	
12-1A9G	123A	3444	20	2051	
12-1A0			20	2051	
12-1A0R			20	2051	
12-2	225	3045	256		
12-200009	177		300	1122	
12-21-050	234		65	2050	
12-23163-3	107	3039/316	61	2038	
12-4	123A	3444	20	2051	
12-680701-05	5547	3505			
0012-911	116		504A	1104	
12-CAM	159		82	2032	
012-H02	100	3005	1	2007	
12-PG 01	234			2050	
12-PG-01			65	2050	
12/1		3311	6GC1		
12/100	113A		6GC1		
12/1N	115	3121	6GX1		
12/1N10	178MP	3100/519	300(2)	1122(2)	
12/3	113A		6GC1		
12/3-04	113A	3119/113	6GC1		
12A1000	5890	7090	5044		MR1130
12A1027-P5	177		300	1122	
12A10F					SPECIAL
12A6240	100	3005	1	2007	
12A6F					MR1376
12A700	5886		5044		MR1128
12A7239P1	102	3722	2	2007	
12A800	5886		5044		MR1128
12A8F					SPECIAL
12A900	5890	7090	5044		MR1130
12A9244-1	160		245	2004	
12A9244-P2	160		245	2004	
12A9275	102	3004	2	2007	
12A9275-1	102	3004	2	2007	
12AA2	101	3861	8	2002	
12B-2	116	3311	504A	1104	
12B-2-BIP-M			510	1114	
12B-2B1P-M	116		504A	1104	
12C05	5870		5032		
12C05R	5871		5033		
12C10	5872		5032		
12C100	5890	7090	5044		
12C100R	5891	7091	5045		
12C10R	5873		5033		
12C15	5874		5032		
12C15R	5875		5033		
12C2	116	3016	504A	1104	
12C20	5874		5032		
12C20R	5875		5033		
12C2P-114	166	9075	504A	1152	
12C30	5876		5036		
12C30R	5877		5037		
12C40	5876		5036		
12C40R	5879		5037		
12C50	5880		5040		
12C50R	5881		5041		
12C5Y	142A	3062	ZD-12	563	
12C60	5882		5040		
12C60R	5883		5041		
12C70	5886		5044		
12C80	5886		5044		
12C80R	5887		5045		
12C90	5890	7090	5044		
12C90R	5891	7091	5045		
12CC12	5944	3500/5882			
12CD12	5945	3517/5883			
12CF11	5887		5045		
12CLN	123A	3124/289	20	2051	
12CTQ030					MBR1530CT
12CTQ035					MBR1535CT
12CTQ040					MBR1545CT
12CTQ045					MBR1545CT
12CZP-114	116(4)		504A	1104	
012E	123A	3444	20	2051	
12F10	5872		5032		
12F100	5890	7090	5044		
12F100A	5890	7090	5044		
12F100B	5890	7090	5044		MR1130
12F10A	5872		5032		
12F10B	5872		5032		
12F15	5874		5032		
12F15A	5874		5032		
12F15B	5874		5032		
12F20	5874		5032		
12F20A	5874		5032		
12F20B	5874		5032		
12F30	5876		5036		
12F30A	5876		5036		

Industry Standard No.	ECG	SK	GE	RS 276-	MOTOR.
12F30B	5876		5036		
12F40	5878		5036		
12F40A	5878		5036		
12F40B	5878		5036		
12F5	5870		5032		
12F50	5880	3500/5882	5040		
12F50A	5880		5040		
12F50B	5880		5040		
12F5A	5870		5032		
12F5B	5870		5032		
12F60	5882	3500	5040		
12F60A	5882		5040		
12F60B	5882		5040		
12F70B	5886		5044		
12F80	5886		5044		
12F80A	5886		5044		
12F80B	5886		5044		MR1128
12F90	5890	7090	5044		
12F90A	5890	7090	5044		
12F90B	5890	7090	5044		
12FC12	5948	3500/5882			
12FD12	5949	3517/5883			
12FL10,502					1N3890
12FL20,502					1N3891
12FL30,502					1N3893
12FL40,502					1N3893
12FL5,502					1N3889
12FL50,502					MR1376
12FL60,502					MR1376
12FR10	5873	3517/5883	5033		
12FR100	5891	7091	5045		
12FR100A	5891	7091	5045		
12FR100B	5891	7091	5045		
12FR10A	5873		5033		
12FR10B	5875	3517/5883	5033		
12FR15	5875	3517/5883	5033		
12FR15A	5875	3517/5883	5033		
12FR15B	5875	3517/5883	5033		
12FR20	5875	3517/5883	5033		
12FR20A	5875	3517/5883	5033		
12FR20B	5875	3517/5883	5033		
12FR30	5877	3517/5883	5037		
12FR30A	5877	3517/5883	5037		
12FR30B	5877	3517/5883	5037		
12FR40	5879	3517/5883	5037		
12FR40A	5879	3517/5883	5037		
12FR40B	5879	3517/5883	5037		
12FR5	5871	3517/5883	5033		
12FR50	5881	3517/5883	5041		
12FR50A	5881	3517/5883	5041		
12FR50B	5881	3517/5883	5041		
12FR5A	5871	3517/5883	5033		
12FR5B	5871	3517/5883	5033		
12FR60	5883	3517	5041		
12FR60A	5883	3517	5041		
12FR60B	5883	3517	5041		
12FR70B	5887		5045		
12FR80	5887		5045		
12FR80A	5887		5045		
12FR80B	5887		5045		
12FR90	5891	7091	5045		
12FR90A	5891	7091	5045		
12FR90B	5891	7091	5045		
12FT10					1N3890
12FT20					1N3891
12FT30					1N3893
12FT40					1N3893
12FT5					1N3889
12FT50					MR1376
12FT60					MR1376
12FV10					1N3890
12FV20					1N3891
12FV30					1N3893
12FV40					1N3893
12FV5					1N3889
12FV50					MR1376
12FV60					MR1376
12G4	506	3843	511	1114	
12GC11	5948	3500/5882	5036		
012H01			1	2007	
12J2	116	3016	504A	1104	
12J2F	116	3017B/117	504A	1104	
12JC11	5952	3500/5882	5040		
12LC11	5998	3500/5882	5044		
12LF11	5887		5045		
12M2	121	3009	239	2006	
12MC	160	3007	245	2004	
12MZ	160		245	2004	
12NC11	6002	7090/5890	5044		
12NF11	5891	7091	5045		
12P2	177	3311	300	1122	
12RCM10	5522	6622			
12RCM5	5521	6621			
12V	142A		ZD-12	563	
12X047	123A	3018	20	2051	
12Z6F	134A	3055	ZD-3.6		
13-0002	140A	3786/5020A	ZD-10	562	
13-0003	519	3100	514	1122	
13-0003(PACE)	519		514	1122	
13-0004	109	3091	1N295	1123	
13-0006	159	3025/129	82	2032	
13-0006A	159	3114/290	82	2032	
13-0009	229	3018	61	2038	
13-0010	229	3018	61	2038	
13-0014	175	3026	246	2020	
13-0015	116	3017B/117	504A	1104	
13-0017	5085A	3337	ZD-36		
13-0020	229	3018	61	2038	
13-0021	123A	3444	20	2051	
13-0022	123A	3444	61	2051	
13-0024	123A	3444	20	2051	
13-0026666-001	128		243	2030	
13-0028	195A	3765	46		
13-0029		3098	ZD-9.1	562	
13-0032	130	3027	14	2041	
13-0035	195A	3765	46		
13-004	109	3087		1123	
13-0040	229	3018	61	2038	
13-0041	123A	3444	20	2051	
13-0043	159	3025/129	82	2032	
13-0043A	159	3114/290	82	2032	
13-0044	159	3025/129	82	2032	
13-0044A	159	3114/290	82	2032	
13-0048	123A	3444	20	2051	
13-0049	184	3190	57	2017	
13-0050	125	3017B/117	510,531	1114	
13-0058	123A	3444	20	2051	
13-006	159	3025/129	82	2032	
13-0061	159	3114/290	82	2032	
13-0061A	159	3114/290	82	2032	
13-0062	229	3122	61	2038	
13-0062-1	229	3444/123A	20	2051	
13-0063	229	3018	61	2038	
13-0064	311	3047	20		

Industry Standard No.	ECG	SK	GE	RS 276-	MOTOR.
13-0065	229	3122	61	2038	
13-0078	5130A		5ZD-15		
13-0079	329	3048	46		
13-0079A	329	3048	219		
13-0091	816	3242			
13-0095	316	3039			
13-0097	182	3188A	55		
13-0099	5800	9003			
13-0104	74145	74145			
13-0106	5092A	3343	ZD-68		
13-0108	1115	3184			
13-0117	614	3327			
13-0118	222	3050/221	FET-4	2036	
13-0123	923D	3165			
13-0161	981	3724			
13-0164	5069A	3775/5009A	ZD-4.7		
13-0165	222		FET-4	2036	
13-0178	293		268	2038	
13-0321-10	123A	3444	20	2051	
13-0321-11	123A	3444	20	2051	
13-0321-12	123A	3444	20	2051	
13-0321-14	108	3452	86	2038	
13-0321-15	108	3452	86	2038	
13-0321-16	108	3452	86	2038	
13-0321-17	108	3452	86	2038	
13-0321-21	108	3452	86	2038	
13-0321-5	123A	3444	20	2051	
13-0321-6	123A	3444	20	2051	
13-0321-7	123A	3444	20	2051	
13-0321-8	123A	3444	20	2051	
13-0321-81	123A	3444	20	2051	
13-0321-9	123A	3444	20	2051	
13-076001	724		IC-86		
13-076002	703A	3157	IC-12		
13-085002		3088	300	1122	
13-085012	109	3087		1123	
13-085015	177	3100/519	300	1122	
13-085022	177	3100/519	300	1122	
13-085023	177		300	1122	
13-085024	125	3031A	510,531	1114	
13-085026	177	3100/519	300	1122	
13-085027	125	3031A	510,531	1114	
13-085028	139A	3060	ZD-9.1	562	
13-085029	110MP	3709	1N34AS	1123(2)	
13-085039	116	3016	504A	1104	
13-085042	139A	3060	ZD-9.1	562	
13-087005	102A	3004	53	1122	
13-087027	116	3031A	504A	1104	
13-1-6	703A	3157	IC-12		
13-10-4	HIDIV-2	3869/DIV-2	FR-9		
13-10-6	703A	3157	IC-12		
13-100000	804	3455	IC-27		
13-1000000	804	3455	IC-27		
13-10102-1	116	3016	504A	1104	
13-102001			504A	1104	
13-1023-1	77		261		
13-1032-5	108	3452	86	2038	
13-10320-14	161	3338/5086A	39	2015	
13-10321-1	108	3452	86	2038	
13-10321-10	108	3452	86	2038	
13-10321-11	108	3452	86	2038	
13-10321-12	108	3452	86	2038	
13-10321-14	108	3452	86	2038	
13-10321-15	108	3452	86	2038	
13-10321-16	108	3452	86	2038	
13-10321-17	108	3452	86	2038	
13-10321-2	108	3452	86	2038	
13-10321-20	108	3452	86	2038	
13-10321-21	108	3452	86	2038	
13-10321-24		3018	1N82A		
13-10321-26	108	3452	86	2038	
13-10321-29	107	3018	11	2015	
13-10321-3	116	3016	504A	1104	
13-10321-30	108	3452	86	2038	
13-10321-31	107	3018	11	2015	
13-10321-32	107	3018	11	2015	
13-10321-34	555		300	1122	
13-10321-35	229		11	2015	
13-10321-36	229		11	2015	
13-10321-37	222	3065	FET-4	2036	
13-10321-41	229	3452/108	86	2038	
13-10321-42	112		11	2015	
13-10321-43	108	3452	86	2038	
13-10321-46	229	3018	61	2038	
13-10321-47	161	3338/5086A	39	2015	
13-10321-5	108	3452	86	2038	
13-10321-50	316	3338/5086A	39	2015	
13-10321-51	108	3452	86	2038	
13-10321-53		3050	FET-4	2036	
13-10321-54	177	3100/519	300	1122	
13-10321-55	177	3100/519	300	1122	
13-10321-59			17	2051	
13-10321-6	108	3452	86	2038	
13-10321-62	229		61	2038	
13-10321-65			11	2015	
13-10321-66	229		11	2015	
13-10321-67	108		11	2015	
13-10321-7	108	3452	86	2038	
13-10321-70	454		17	2051	
13-10321-71	229		11	2015	
13-10321-72			11	2015	
13-10321-75	319P		11	2015	
13-10321-76	161	3716	17	2051	
13-10321-77	108	3452	86	2038	
13-10321-78	161	3716	39	2015	
13-10321-79	108	3716/161	39	2015	
13-10321-8	108	3452	86	2038	
13-10321-9	108	3452	86	2038	
13-1032176	161	3716	39	2015	
13-105698-1	128	3024	243	2030	
13-11-6	703A	3157	IC-12		
13-11-6(SEARS)	703A		IC-12		
13-12-4	HIDIV-2	3869/DIV-2	FR-9		
13-12001-0	109	3088		1123	
13-12002-0	109			1123	
13-12003-0	109	3088	1N60	1123	
13-13021-15			11	2015	
13-13532-1	159		82	2032	
13-14065-77			20	2051	
13-14085-1	108	3452	86	2038	
13-14085-10	102A	3004	53	2007	
13-14085-11	158	3004/102A	53	2007	
13-14085-12	158	3004/102A	53	2007	
13-14085-121		3041	66	2048	
13-14085-122		3124	20	2051	
13-14085-125	199	3122			
13-14085-126	184	3190	55		
13-14085-13	159	3025/129	82	2032	
13-14085-14	172A	3156	64		
13-14085-15	123A	3444	20	2051	
13-14085-15A	128	3024	243	2030	
13-14085-16	107	3018	11	2015	
13-14085-17	107	3018	11	2015	
13-14085-18	102A	3004	53	2007	

Industry Standard No.	ECG	SK	GE	RS 276-	MOTOR.
13-14085-2	108	3452	86	2038	
13-14085-23	102A	3004	53	2007	
13-14085-24	108	3452	86	2038	
13-14085-25	158	3004/102A	53	2007	
13-14085-26	229	3018	11	2015	
13-14085-27	108	3452	86	2038	
13-14085-28	160	3006	245	2004	
13-14085-29	124	3021	12		
13-14085-3	107	3018	11	2015	
13-14085-30	160	3006	245	2004	
13-14085-31	126	3006/160	52	2024	
13-14085-32	126	3006/160	52	2024	
13-14085-33	126	3006/160	52	2024	
13-14085-34	123A	3444	20	2051	
13-14085-35	102A	3004	53	2007	
13-14085-4	107	3018	11	2015	
13-14085-41	199	3124/289	20	2051	
13-14085-49	199	3124/289	62	2010	
13-14085-50	123A	3444	20	2051	
13-14085-54	289A	3124/289	17	2051	
13-14085-6	123A	3444	20	2051	
13-14085-60	102A	3004	53	2007	
13-14085-7	123A	3444	20	2051	
13-14085-71	102A	3004	53	2007	
13-14085-72	123A		20	2051	
13-14085-74	108	3452	86	2038	
13-14085-75	108	3452	86	2038	
13-14085-76	108	3452	86	2038	
13-14085-77	108	3452	86	2038	
13-14085-83	123A	3444	20	2051	
13-14085-84	123A	3444	20	2051	
13-14085-85	123A	3444	20	2051	
13-14085-86	187	3193	29	2025	
13-14085-87	159		82	2032	
13-14085-88	152	3893	28	2017	
13-14085-89	123A	3444	20	2051	
13-14085-9	102A	3004	53	2007	
13-14085-91	128	3024	243	2030	
13-14085-92	129	3025	244	2027	
13-14085-93	126	3054/196	52	2024	
13-14085-94			18	2030	
13-14085-95	199	3122	62	2010	
13-14085-96	199	3122	62	2010	
13-14085-97	199	3122	62	2010	
13-14094-1	109	3087		1123	
13-14094-11	109	3088		1123	
13-14094-12	116	3031A	504A	1104	
13-14094-13		3126	90		
13-14094-14		3088	1N60	1123	
13-14094-15	109	3087		1123	
13-14094-16	116	3016	504A	1104	
13-14094-17	116	3031A	504A	1104	
13-14094-2	109	3087		1123	
13-14094-24	116	3311	504A	1104	
13-14094-3	109			1123	
13-14094-33	177		300	1122	
13-14094-36	601	3463			
13-14094-38	116		504A	1104	
13-14094-39	116	3311	504A	1104	
13-14094-42	116	3311	504A	1104	
13-14094-5	109	3087		1123	
13-14094-54	116	3311	504A	1104	
13-14094-6	125	3311	510,531	1114	
13-14094-8	110MP	3709	1N60	1123(2)	
13-14094-9	109			1123	
13-14097-7	177		300	1122	
13-14261-1	116	3311	504A	1104	
13-14261-3	116	3311	504A	1104	
13-14278-1	614		90		
13-14278-2	614	3126	90		
13-14279-1	103A	3835	53	2002	
13-14604-1	121MP	3015	239(2)	2006(2)	
13-14604-1A	121MP	3013	239(2)	2006(2)	
13-14604-1B	121MP	3013	239(2)	2006(2)	
13-14604-1C	121MP	3013	239(2)	2006(2)	
13-14604-1D	121MP	3013	239(2)	2006(2)	
13-14604-1E	121MP	3013	239(2)	2006(2)	
13-14605-1	128	3024	243	2030	
13-14606-1	123A	3444	20	2051	
13-14627-1	116	3016	504A	1104	
13-14627-4	116	3031A	504A	1104	
13-14735	121	3009	239	2006	
13-14735-1	121	3014	239	2006	
13-14735A	121	3009	239	2006	
13-14777-1	121MP	3014	239(2)	2006(2)	
13-14777-1A	121MP	3013	239(2)	2006(2)	
13-14777-1B	121MP	3013	239(2)	2006(2)	
13-14777-1C	121MP	3013	239(2)	2006(2)	
13-14777-1D	121MP	3013	239(2)	2006(2)	
13-14778-1	121MP	3015	239(2)	2006(2)	
13-14778-1A	121MP	3013	239(2)	2006(2)	
13-14778-1B	121MP	3013	239(2)	2006(2)	
13-14778-1C	121MP	3013	239(2)	2006(2)	
13-14778-1D	121MP	3013	239(2)	2006(2)	
13-14858-1			504A	1104	
13-14879-1	145A	3063	ZD-15	564	
13-14879-2	142A	3062	ZD-12	563	
13-14879-3	145A	3063	ZD-15	564	
13-14879-4	138A	3059	ZD-7.5		
13-14879-5		3063	ZD-16	564	
13-14879-6	5079A	3335	ZD-20		
13-14879-7	5079A	3335	ZD-20		
13-14886-1	126	3006/160	52	2024	
13-14887-1	126	3006/160	52	2024	
13-14888-3	102	3004	2	2007	
13-14889-1	126	3006/160	52	2024	
13-14890-1	109	3088		1123	
13-15465-1	178MP	3031A	300(2)	1122(2)	
13-15804-1	123A	3124/289	20	2051	
13-15805-1	102	3004	2	2007	
13-15806-1	121	3034	239	2006	
13-15808-1	123AP	3018	39	2051	
13-15808-2	161	3124/289	39	2015	
13-15809-1	154	3045/225	40	2012	
13-15810-1	108	3452	86	2038	
13-15833-1	128	3024	243	2030	
13-15835-1	161	3018	39	2015	
13-15836-1	102	3004	2	2007	
13-15840-1	123A	3444	20	2051	
13-15840-2	123A	3444	20	2051	
13-15841-1	108	3452	86	2038	
13-15842-1	297	3444/123A	20	2009	
13-15865-1	123A	3444	20	2051	
13-16104-8	116	3016	504A	1104	
13-16104-9	116	3032A	504A	1104	
13-16105-3	119	3109	CR-2		
13-16106-1	118	3066	CR-1		
13-16106-3	503		CR-6		
13-16106-4	504		CR-6		
13-16106-5	504		CR-6		
13-16106-6	504		CR-6		
13-16106-7	504		CR-6		
13-16219-7	113A		6GC1		
13-16235-8	109	3087		1123	
13-16247-3	116	3016	504A	1104	
13-16570-1	159	3114/290	82	2032	
13-16570-1A	159	3114/290	82	2032	
13-16570-2	159	3114/290	82	2032	
13-16570-2A	159	3114/290	82	2032	
13-16592-1	127	3012/105	25		
13-16607-1	127	3035	25		
13-16608-1	127	3034	25		
13-16744-1	108	3452	86	2038	
13-16769-1	123A	3444	20	2051	
13-17-6	123A*	3122	20		
13-17-6(SEAR)			20	2051	
13-17-6(SEARS)	123A			2051	
13-17174-1	125	3032A	504A	1104	
13-17174-2	125	3032A	504A	1104	
13-17174-3	125	3100/519	504A	1104	
13-17174-4	156	3051	512		
13-17174-5	125	3032A	511	1114	
13-17204-1	177	3087	300	1122	
13-17546-2			300	1122	
13-17557-1	116	3016	504A	1104	
13-17569-1	177	3017B/117	300	1122	
13-17569-2	177		300	1122	
13-17595-2			300	1122	
13-17596-1	120	3110	504A	1104	
13-17596-10	519	3100	514	1122	
13-17596-2	519	3100	300	1122	
13-17596-2(RECT)			504A	1104	
13-17596-3	519	3100	300	1122	
13-17596-4	519	3100	300	1122	
13-17596-5	519	3100	300	1122	
13-17596-6	519	3100			
13-17596-7	519	3100	300	1122	
13-17596-8	519	3100	300	1122	
13-17596-9	519	3100	300	1122	
13-17607-1	127	3035	25		
13-17607B	127	3035	25		
13-17608-1	127	3034	25		
13-17608-2	127	3034	25		
13-17608B	127	3034	25		
13-17608C	127	3034	25		
13-17609-1	127	3034	25		
13-17825-1	116	3016	504A	1104	
13-17917-1	552	9000			
13-17917-2	552	9000			
13-17918-1	130	3027	14	2041	
13-18-4	HIDIV-2	3869/DIV-2	FR-9		
13-18032-1	102A		53	2007	
13-18033-1	102A		53	2007	
13-18034	121	3009	239	2006	
13-18034-1	121	3009	239	2006	
13-18034A	121	3009	239	2006	
13-18087-1	123A	3444	20	2051	
13-18087-2	123A	3444	20	2051	
13-18158-1	123A	3444	20	2051	
13-18198-1	179	3015	239(2)	2006(2)	
13-18282	124	3021	12		
13-18282-1	124	3021	12		
13-18304-1	102A	3004	53	2007	
13-18359	124	3021	12		
13-18359-1	124	3021	12		
13-18359-3	124	3021	12		
13-18359A	124	3021	12		
13-1836-1	123A	3335/5079A	20	2051	
13-18363	123	3124/289	20	2051	
13-18363-1	123A	3444	20	2051	
13-18363-1A	123	3124/289	20	2051	
13-18364-1	123A	3444	20	2051	
13-18365-1	199	3124/289	62	2010	
13-18458-1	116	3016	504A	1104	
13-1847-1	196		241	2020	
13-18481-1	5802	9005	504A	1143	
13-18481-2	5800	9005/5802		1142	
13-18481-3	5802	9005		1143	
13-18642-1	179	3642	76		
13-18642-2	179	3642	76		
13-18642-20	179(2)		76(2)		
13-18642-21	179(2)		76(2)		
13-18642-2D	179	3642	76		
13-18642-2E	179	3642	76		
13-18642-2F	179	3642	76		
13-18642-3	179	3642	76		
13-18642-3A	179	3642	76		
13-18642-3B	179	3642	76		
13-18654-1	103A	3010	59	2002	
13-18671-1	102A		53	2007	
13-18671-1A	102A		53	2007	
13-18671-1B	102A		53	2007	
13-18671-1C	102A		53	2007	
13-18924-1			MR-5	1067	
13-18924-2	5401	3638/5402			
13-18924-3	5400	3950			
13-18924-4	5400	3950			
13-18924-5	5400	3638/5402			
13-18924-6	5403	3627/5404			
13-18924-8	5400	3638/5402			
13-18927-1	123A	3124/289	20	2051	
13-18927-1A	128	3124/289	243	2030	
13-18944-1	100	3004/102A	1	2007	
13-18944-2	100	3005	1	2007	
13-18946-1	160	3006	245	2004	
13-18946-2	160		245	2004	
13-18947-1	160	3006	245	2004	
13-18948-1	160	3006	245	2004	
13-18948-2	160	3006	245	2004	
13-18949-1	108	3452	86	2038	
13-18950-1	108	3452	86	2038	
13-18950-2			51	2004	
13-18951-1	126	3006/160	52	2024	
13-18951-2	126	3006/160	52	2024	
13-19-4	HIDIV-2	3869/DIV-2	FR-9		
13-19776-1	159		82	2032	
13-20-4	HIDIV-2	3869/DIV-2	FR-9		
13-21-4	HIDIV-2		FR-9		
13-21606-1	127	3034	25		
13-22017-0	177	3100/519	300	1122	
13-22154-500	610	3323	90		
13-22156-0		3126	90		
13-22319-0	5074A		ZD-11.0	563	
13-22452-0	116		504A	1104	
13-22463-1	116	3311	504A	1104	
13-22581	123A	3122	20	2051	
13-22581-1	123A	3124/289	20	2051	
13-22582-1	159	3025/129	82	2032	
13-22582-1A	159	3114/290	82	2032	
13-22606-0	177	3100/519	514	1122	
13-22609-0	177	3100/519	300	1122	
13-22690-1	312	3448	FET-1	2035	
13-22692-1	312	3448	FET-1	2035	
13-22692-2	312	3448	FET-1	2035	
13-22739-1	121MP	3009	239(2)	2006(2)	
13-22741	121	3009	239	2006	
13-22741-1	121MP	3009	239(2)	2006(2)	
13-22741-2	121MP	3013	239(2)	2006(2)	

Industry Standard No.	ECG	SK	GE	RS 276-	MOTOR.
13-23001-2	229		61	2038	
13-23002-2	229		11	2038	
13-23013-2	229	3018	60	2015	
13-23160-2	233	3018	210	2009	
13-23160-3	319P	3018	283	2016	
13-23160-4	123A	3444	20	2051	
13-23160-5	229	3311	60	2038	
13-23163-2	107	3039/316	61	2038	
13-23309-5	123A	3018	20	2051	
13-23323-4	289A		20	2051	
13-23323-6	108	3020/123	20	2051	
13-23324-6	123A	3124/289	20	2051	
13-23325-5	159	3114/290	21	2032	
13-23326-6	299	3298	236		
13-23327-4	192	3124/289	62	2010	
13-23338-3	199	3124/289	62	2010	
13-23338-4	199		62	2010	
13-23339-2	199		62	2010	
13-23339-3	199	3124/289	60	2010	
13-23505-2	297	3137/192A	81		
13-23506-2	298	3434/288	223		
13-23507-0	184		57	2017	
13-23508-0	185		58	2025	
13-23510-4	287	3193/187	62	2010	
13-23517-2	152	3893			
13-23543-1	124	3021	12		
13-23543-2		3021	12		
13-23594-1	130	3027	14	2041	
13-23785-1	102A	3004	53	2007	
13-23822	319P	3452/108	86	2038	
13-23822-1	229	3854/123AP	39	2015	
13-23824			20	2051	
13-23824-1	229	3246	283	2015	
13-23824-1(3RDIF)	233		210	2009	
13-23824-2	319P	3039/316	39	2015	
13-23824-3	233	3246/229	20	2009	
13-23825-1	154	3124/289	40	2012	
13-23826-1	159	3466	82	2032	
13-23826-1A	159	3466	82	2032	
13-23826-2	159	3466	82	2032	
13-23826-2A	159	3466	82	2032	
13-23826-3	159	3466	82	2032	
13-23826-3A	159	3466	82	2032	
13-2384-2	161	3338/5086A	39	2015	
13-23840-1	128	3024	243	2030	
13-23892-5			63	2030	
13-23916-1	123A	3444	20	2051	
13-23917-1	177	3087	300	1122	
13-25226-1			20	2051	
13-25343-1		3021	12		
13-26-6	718	3159	IC-8		
13-26009-1	161	3039/316	39	2015	
13-26377-1	131		44	2006	
13-26377-2	155	3839	43		
13-26386-1	159	3466	82	2032	
13-26386-1A	159	3466	82	2032	
13-26386-2	159	3466	82	2032	
13-26386-2A	159	3466	82	2032	
13-26386-3	159	3466	82	2032	
13-26386-4	159	3138/193A	21	2034	
13-26576-1	161	3039/316	39	2015	
13-26576-2	161	3039/316	39	2015	
13-26577-1	161	3018	39	2015	
13-26577-2	161	3117	39	2015	
13-26577-3	161	3039/316	39	2015	
13-26614-1	156	3032A	511		
13-26666	128	3024	243	2030	
13-26666-1	128	3024	243	2030	
13-27-6	711	3070	IC-207		
13-27-6(SEARS)	711		IC-207		
13-27050-1	103	3010	59	2002	
13-27404-1	123AP	3444/123A	20	2051	
13-27404-2	123A	3444	20	2051	
13-27432-1	194	3024/128	243	2030	
13-27432-2	123AP	3854	243	2051	
13-27433-1	123A	3444	20	2051	
13-27443-1	128	3024	243	2030	
13-27596-5			300	1122	
13-27974-1	157	3201/171	232		
13-27974-2		3747	232		
13-27974-3		3747	232		
13-27974-4		3747	232		
13-28-6	706	3101	IC-43		
13-28-6(SEARS)	706		IC-43		
13-28222-1	183	3189A	56	2027	
13-28222-2	182	3188A	55		
13-28222-3	182	3054/196	55		
13-28222-4	182	3188A	55		
13-28336-1	185	3191	58	2025	
13-28336-2	184	3054/196	57	2017	
13-28336-20		3191/185	57,58		
13-283361-1	197	3083	250	2027	
13-28386-1	129		244	2027	
13-28391-1	159	3466	82	2032	
13-28391-1A	159	3466	82	2032	
13-28391-2	159	3466	82	2032	
13-28391-2A	159	3466	82	2032	
13-28392-1		3024	243	2030	
13-28392-2		3137	243	2030	
13-28392-3		3024	243	2030	
13-28393-1	159	3124/289	244	2027	
13-28393-1A	159	3025/129	244	2027	
13-28393-2	159	3025/129	244	2027	
13-28393-2A	159	3025/129	244	2027	
13-28393-3	159		67	2023	
13-28394-1	128	3024	243	2030	
13-28394-2	128	3024	243	2030	
13-28394-3	128	3024	243	2030	
13-28394-4	129	3025	244	2027	
13-28394-4A	129	3025	244	2027	
13-28394-5	129	3025	244	2027	
13-28394-5A	129	3025	244	2027	
13-28394-6	129	3025	244	2027	
13-28396-1	390		255	2041	
13-28396-I			255	2041	
13-28432-1	194	3045/225	220		
13-28432-2	154		40	2012	
13-28469-2	152	3054/196	66	2048	
13-28471-1		3024	243	2030	
13-28471-1(METAL)	128		243	2030	
13-28532-1	182	3036	55		
13-28583-1	222	3050/221	FET-4	2036	
13-28584	108	3452	86	2038	
13-28584-1	108	3452	86	2038	
13-28654-1	312	3448	FET-1	2035	
13-28654-2	312	3448	FET-1	2035	
13-28654-3	452	3448	FET-1	2035	
13-28654-4	452	3112	FET-1	2028	
13-28654-5		3116	FET-2	2028	
13-29-6	712	3072	IC-148		
13-29-6(SEARS)	712		IC-2		
13-29033-1	123AP	3854	20	2051	
13-29033-2	123AP	3854	62	2051	
13-29033-3	123AP	3854	62	2010	
13-29033-4	199	3854/123AP	62	2010	
13-29033-5	123AP	3854	62	2010	
13-29033-6	123AP	3854	20	2038	
13-29165-1	156	3031A	512		
13-29165-2	156	3051	512		
13-29392-2	108	3020/123	20	2051	
13-29432-1	123A	3444	20	2051	
13-29437-1	191	3044/154	249		
13-29656-1	118		CR-1		
13-29663-1	506	3843	511	1114	
13-29687-2	177	3100/519	300	1122	
13-29775-1	172A	3156	64		
13-29775-2	172A/410		64		
13-29775-3	172A/410		64		
13-29776-1	159	3466	82	2032	
13-29776-1A	159	3466	82	2032	
13-29776-2	159	3466	82	2032	
13-29776-3	159	3466	82	2032	
13-29867-1	177	3090/109	300	1122	
13-29867-2	177	9091	300	1122	
13-29867-2(SUPP)			504A	1104	
13-29867-2(SW)			300	1122	
13-29947-1	123AP	3018	20	2015	
13-29974-1			39	2015	
13-29974-4	157	3104A	232		
13-2I606-1	127	3764	25		
13-2P64	152	3054/196	66	2048	
13-2P64-1	152	3893	66	2048	
13-3-6	726	3022/1188			
13-30-6	780	3141	IC-222		
13-3015121-1	238	3710			
13-3015124-1	397	3528			
13-30281	109			1123	
13-31013-1	123	3018	20	2051	
13-31013-1(SYLV.)	159		82	2032	
13-31013-1/2	159	3466	82	2032	
13-31013-2			22	2032	
13-31013-3	113A		6GC1		
13-31013-4	108	3452	86	2038	
13-31013-5	161	3018	39	2015	
13-31013-6	116	3017B/117	504A	1104	
13-31014-1	116	3017B/117	504A	1104	
13-31014-2	113A	3119/113	6GC1		
13-31014-3	116	3017B/117	504A	1104	
13-31014-4	112	3089	1N82A		
13-31014-6	552	3017B/117	504A	1104	
13-32362-1	123AP	3018	62	2010	
13-32364-1	159	3466	82	2032	
13-32366-1	229	3452/108	86	2038	
13-32366-2	108	3452	86	2038	
13-32630-1	123AP		243	2030	
13-32630-2	192		63	2030	
13-32630-3	123AP	3024/128	243	2030	
13-32630-4	192		63	2030	
13-32631		3025	67	2023	
13-32631-1	159		244	2027	
13-32631-2	193		67	2023	
13-32631-3	159	3025/129	244	2027	
13-32631-4	193		67	2023	
13-32632-1		3054	226,252		
13-32634-1	186	3192	28	2017	
13-32635-1	187	3193	29	2025	
13-32636-1	152	3054/196	66	2048	
13-32638-1	184	3054/196	57	2017	
13-32640-1	152	3054/196	66	2048	
13-32642-1	184	3054/196	57	2017	
13-331182-1	162	3079	35		
13-33172-1	558	3998/506	511	1114	
13-33172-2	558	3998/506	511	1114	
13-33173-1	220	3990	FET-2	2035	
13-33174-1	171	3201	27		
13-33174-2	171	3104A	27		
13-33175-1	172A	3156	64		
13-33175-2	172A	3241/232	64		
13-33176-1	171	3104A	27		
13-33177-1	613	3126	90		
13-33178-1	193	3025/129	67	2023	
13-33179-1	5099A		ZD-140		
13-33179-2	145A	3063	ZD-15	564	
13-33179-4	140A	3094/144A	ZD-10	562	
13-33179-5	5079A	3335	ZD-20		
13-33179-6	142A		ZD-12	563	
13-33179-7	5097A		ZD-120		
13-33179-8	5082A	3753	ZD-25		
13-33180-1	188	3199	226	2018	
13-33181-1	389	3710/238	38		
13-33181-2	389	3710/238	38		
13-33181-3	389	3710/238	38		
13-33182-1	162	3438	35		
13-33182-2	162	3438	35		
13-33182-8	5099A		ZD-140		
13-33183-1	5421	3503		1067	
13-33184-1	6402	3628	FET-2		
13-33185-1	210	3202	252	2018	
13-33186-1	147A	3095	ZD-33		
13-33187-1	5047A	3098/150A	ZD-82		
13-33187-10	5055A	3821	ZD-150		
13-33187-11	5021A	3750/143A	ZD-13	563	
13-33187-12	135A	3056	ZD-5.6	561	
13-33187-13	5052A	3347/5097A	ZD-120		
13-33187-14	5006A	3055/134A	ZD-3.6		
13-33187-15	5032A	3753/5082A	ZD-25		
13-33187-18	5024A	3063/145A	ZD-15	564	
13-33187-19	5008A	3774	ZD-4.3		
13-33187-2	5050A	3346/5096A	ZD-100		
13-33187-26	5015A	3781	ZD-7.5	561	
13-33187-3	5029A	3401/5135A	ZD-20		
13-33187-4	5031A		ZD-24		
13-33187-5	5048A	9024/5094A	ZD-87		
13-33187-6	5019A	3061/140A	ZD-10	562	
13-33187-7	5019A	3785	ZD-10	562	
13-33187-8		3819	ZD-130		
13-33187-9		3796	ZD-22		
13-33188-1	182	3188A	55		
13-33188-2	130	3027	14	2041	
13-33189-1	507	3043	511		
13-33190-1		3648		1173	
13-33350-1	123A	3444	20	2051	
13-33376-1	116	3032A	504A	1104	
13-33595-1	123A	3444	20	2051	
13-33595-2	123A	3444	20	2051	
13-33595-3	123A	3444	20	2051	
13-33742-1		3054	28	2017	
13-33742-1PLASTIC	188		226	2018	
013-339	116	3017B/117	504A	1104	
13-33925-1	184	3054/196	57	2017	
13-33959-1	5400	3950			
13-34001-1		3156	64		
13-34002-1	196	3054	241	2020	
13-34002-2	196		241	2020	
13-34002-3	241	3188A/182	57	2020	
13-34002-4	196		241	2020	
13-34002-5	241	3188A/182		2020	
13-34003-1	192A	3199/188	226	2018	

Industry Standard No.	ECG	SK	GE	RS 276-	MOTOR.
13-34003-1(METAL)	128		244	2027	
13-34004-1	189	3200	218	2026	
13-34004-1(METAL)	129		244	2027	
13-34004-1A		3025	58	2025	
13-34045-1	108	3452	86	2038	
13-34045-2	108	3452	86	2038	
13-34046-1	186	3192	28	2017	
13-34046-2	186	3192	28	2017	
13-34046-3	186	3192	28	2017	
13-34046-4	186	3192	28	2017	
13-34046-5	291	3192/186	28	2017	
13-34047-1	187	3193	29	2025	
13-34047-2	187	3193	29	2025	
13-34047-3	187	3083/197	29	2025	
13-34047-4	292	3193/187	29	2025	
13-34056-1	177	3100/519	300	1122	
13-34057-1	116	3017B/117	504A	1104	
13-34089-2	157	3103A/396	232		
13-34089-4	157	3103A/396	232		
13-34367-1	159	3025/129	82	2032	
13-34367-3	159	3118	244	2032	
13-34367-30			65	2050	
13-34368-1	5802	9005		1143	
13-34369-1	106	3025/129	21	2034	
13-34371-1	128	3024	243	2030	
13-34372-1	190	3199/188	226	2018	
13-34372-2	186		28	2017	
13-34373-1	189	3200	218	2026	
13-34373-2	187	3193	29	2025	
13-34374-1	181		75	2041	
13-34375-1	312	3448	FET-1	2035	
13-34375-2	312	3116	FET-2	2035	
13-34378-1	312	3448	FET-1	2035	
13-34378-2	312	3116	FET-2	2035	
13-34378-3	312	3116	FET-2	2035	
13-34381			20	2051	
13-34381-1	199	3122	62	2010	
13-34381-2	199	3122	20	2051	
13-34616-1			241	2020	
13-34617-1			250	2027	
13-34684-1	181		75	2041	
13-34838-1	196	3054	241	2020	
13-34839-1	197	3083	250	2027	
13-34901-1	5814	3051/156	510	1114	
13-34940-1	159		21	2034	
13-35059-1	712	3072	IC-15		
13-35089-1	228A	3104A	257		
13-35089-2	228A	3103A/396	257		
13-35089-3	228A	3103A/396	257		
13-35089-4	228A	3103A/396	257		
13-35226-1	123AP	3122	20	2051	
13-35257-1	157	3747	232		
13-35257-2	157	3747	232		
13-35257-3		3747	232		
13-35257-4		3747	232		
13-35324-1	265	3860	64		
13-35472-1	198		251		
13-35550	108	3452	86	2038	
13-35550-1	194	3122	62	2010	
13-35621-1	109	3088		1123	
13-35792-1	102A	3004	53	2007	
13-35807-1	194	3275	220		
13-35807-2	123A	3122	20	2051	
13-36314-1	241	3188A/182	57	2020	
13-36386-1	159	3466	82	2032	
13-36440-1	284	3621	265	2047	
13-36441-1		3846	266		
13-36442-1	331	3054/196			
13-36443-1	332	3083/197	250	2027	
13-36444-3		3190	57	2017	
13-36445-3		3191	58	2025	
13-36508-1	188		226	2018	
13-36509-1	189		218	2026	
13-37526-1	188	3199	226	2018	
13-37527-1	189	3200	218	2026	
13-37708-1	284	3836	38		
13-37868-1	552	3843	511	1114	
13-37869-1	287	3433	249		
13-37870-1	165	3111	38		
13-37870-2	165	3111	38		
13-37900-1	220	3990	FET-2	2036	
13-37905-1	147A		ZD-33		
13-37933-2	504	3108/505	CR-6		
13-38-6	906		IC-246		
13-39003-0	5400	3950			
13-39004-1	152	3054/196	241	2020	
13-39004-2	152	3893	66	2048	
13-39046-3	186	3192	28	2017	
13-39047-3	187	3193	29	2025	
13-39072-1	551	9000/552	511	1114	
13-39072-2	551	9000/552	511	1114	
13-39072-7	551	9000/552			
13-39073-1	506	3016	504A	1104	
13-39074-1	291	3192/186	28	2017	
13-39098-1		3192	28	2017	
13-39099-1	152	3054/196	66	2048	
13-39100-1	153	3083/197	69	2049	
13-39114-1	123AP	3024/128	63	2051	
13-39114-2	123AP	3024/128	222	2051	
13-39114-3	192	3854/123AP	63	2030	
13-39115-1	288	3466/159	67	2023	
13-39115-2	193	3434/288	223		
13-39115-3	159	3138/193A	67	2023	
13-39174-1	191		249		
13-39607-4				1000	
13-39678-1	5604	3666/5605		1000	
13-39678-3	5604			1000	
13-39678-4	5604			1000	
13-39696-3	152	3893			
13-39819-1	153	3083/197	69	2049	
13-39819-2	153		69	2049	
13-39851-1	194	3275	220		
13-39860-1	116	3311	504A	1104	
13-39863-1		3095	ZD-33		
13-39867-2			511	1114	
13-39884-1	152	3054/196	66	2048	
13-39884-2	152		66	2048	
13-39884-3			66	2048	
13-39970-1	159	3466	82	2032	
13-40-6	715	3076	IC-6		
13-40083-1	159	3118	21	2034	
13-40083-2	159	3984/106	21	2034	
13-40312-1	199		85	2010	
13-40340-1		3188A		2020	
13-40342-1	327		265	2047	
13-40344-1	373	3190/184	57	2017	
13-40345-1	374	3191/185	58	2025	
13-40346-1	181		74	2043	
13-40347-1	180		75	2043	
13-4085-121	199		62	2010	
13-4085-122	199		62	2010	
13-4085-41	199		62	2010	
13-41-6	790	3454	IC-18		
13-41122-1	125	3311	504A	1104	
13-41122-2	552	3998/506	504A	1104	
13-41122-4	125	3311		1104	
13-41123-2	125	3311	509	1114	
13-41123-4	125	3017B/117	504A	1104	
13-41628-2	291	3893/152	66	2048	
13-41628-3	186	3054/196	66	2048	
13-41628-5	291	3192/186	28	2052	
13-41629-4	292	3274/153	69	2025	
13-41738-1	284	3027/130	14	2041	
13-42-6	714	3075	IC-4		
13-43005-1	291	3054/196	241		
13-43112-1	312	3834/132	FET-2	2035	
13-43250-1	177	3863/600	300	1122	
13-43382-1	503	3068	CR-5		
13-43463-1	238	3710	259		
13-43463-2	165	3710/238	259		
13-43633-1	283	3111	38		
13-43634-1	159	3446/1097	82	2032	
13-43635-1	196	3054	241	2048	
13-43766-1	116	3311	504A	1104	
13-43773-1	123AP	3854	62	2051	
13-43777-1	552	9000	504A		
13-43777-2	552	9000	504A		
13-43790-1	186	3192	241	2017	
13-43791-1	187	3193	250	2025	
13-43956-1	515	3081/125	510	1114	
13-44290	312			2028	
13-44291	312			2028	
13-45016-1	389	3710/238	38		
13-45018-1	228A	3232/191	27		
13-45321-1	171	3201			
13-45539-1	238	3710			
13-4800869145	129		244	2027	
13-4P63	152	3054/196	66	2048	
13-4P63-1	152	3893	66	2048	
13-4P63-I	152			2048	
13-5018-6	941M		IC-265	007	
13-5020-6	923D		IC-260	1740	
13-50385-1	374	3191/185			
13-50484-1	100	3005	1	2007	
13-50486-1	100	3005	1	2007	
13-50631-1	100	3005	1	2007	
13-50944-1	100	3005	1	2007	
13-52-6	722	3161	IC-9		
13-52-G	722		IC-9		
13-53-6	909	3590	IC-249		
13-547604	519		514	1122	
13-55009-1	127	3035	25		
13-55009-2	127	3035	25		
13-55010-1	503	3068	CR-5		
13-55018-4	128	3024	243	2030	
13-55020-1	108	3452	86	2038	
13-55029-1	116	3031A	504A	1104	
13-55030-1	506	3843	511	1114	
13-55031-1	506	3843	511	1114	
13-55031-2	506	3843	511	1114	
13-55031-3	506	3843	511	1114	
13-55046-1	109	3087		1123	
13-55061-1	123A	3444	20	2051	
13-55061-1(AGC)			18	2030	
13-55061-1(SOUND)			20	2051	
13-55061-2	123A	3444	20	2051	
13-55061-2(SYL)			20	2051	
13-55061-2(WARDS)			18	2030	
13-55062-1	154	3045/225	40	2012	
13-55063-1	108	3452	86	2038	
13-55064-1	128	3024	243	2030	
13-55064-1(SYL)			23	2020	
13-55064-1(WARDS)			63	2030	
13-55065-1	108	3452	86	2038	
13-55066-1	123A	3444	20	2051	
13-55066-2	123A	3444	20	2051	
13-55066-2(SYL)			18	2030	
13-55066-2(WARDS)			18	2030	
13-55067-1	123A	3444	20	2051	
13-55068-1	123A	3444	20	2051	
13-55069-1	159	3466	82	2032	
13-55069-1A	159	3466	82	2032	
13-55078-1	506	3843	511	1114	
13-55166-1	109	3088		1123	
13-55166-1/2439-2	109			1123	
13-55166-1/3464	109			1123	
13-55332-1	142A	3062	ZD-12	563	
13-55333-1	178MP	3089/112	300(2)	1122	
13-56-6	738	3167	IC-29		
13-57-6	739	3235	IC-30		
13-59-6	804	3455	IC-27		
13-60-6	722	3161	IC-9		
13-61-6	804	3455	IC-27		
13-64-6	712	3072	IC-2		
13-67-6	739	3235	IC-30		
13-67539-1	166	9075	504A	1152	
13-67539-1/3464	166	9075		1152	
13-67544-1	136A	3057	ZD-5.6	561	
13-67544-1/3464	136A		ZD-5.6	561	
13-67583-5	108	3452	86	2038	
13-67583-6	123A	3444	20	2051	
13-67583-6/3464	107		11	2015	
13-67585-4	123A	3444	20	2051	
13-67585-4/2439-2			20	2051	
13-67585-4/3464	107		11	2015	
13-67585-5	123A	3444	20	2051	
13-67585-5/2439-2	108		86	2038	
13-67585-5/3464	123A	3444	20	2051	
13-67585-6/2439-2	108		86	2038	
13-67585-7			10	2051	
13-67585-7/2439-2	123A		20	2051	
13-67586-3	123A	3444	20	2051	
13-67586-3/3464	123A		20	2051	
13-67590-1	177	3100/519	300	1122	
13-67590-1/2439-2	109			1123	
13-67590-1/3464	177		300	1122	
13-67599-3	102A	3004	53	2007	
13-67599-3/2439-2	102A		53	2007	
13-67599-3/3464	158		53	2007	
13-67599-9		3004	52	2024	
13-67600-3		3126	90		
13-67600-5		3126	90		
13-68617-1	123A	3444	20	2051	
013-768	5084A	3755	ZD-30		
13-85943-1	113A	3119/113	6GC1		
13-85943-2	113A	3119/113	6GC1		
13-85943-3	113A	3119/113	6GC1		
13-85962-1	109	3862/103	1N295	1123	
13-86416-1	103	3862	8	2002	
13-86420-1	101	3861	8	2002	
13-87433-1	101	3861	8	2002	
13-87539-4			504A	1104	
13-88302	117		504A	1104	
13-9-4	HIDIV-2	3869/DIV-2	FR-9		
13-9-6	703A	3157	IC-12		
0013-911	116		504A	1104	
13-94096-2	102A	3004	53	2007	
13BKJ1		3109	CR-2		

Industry Standard No.	ECG	SK	GE	RS 276-	MOTOR.
13BKJ1(GE)	119		CR-2		
13D4	116	3017B/117	504A	1104	
13DD02F	140A	3061	ZD-10	562	
13J2	116	3016	504A	1104	
13J2F	116	3017B/117	504A	1104	
13P1	116	3016	504A	1104	
13P2	177	3311	300	1122	
13P2(RECTIFIER)	125		510,531	1114	
13RC10	5542	6615/5515	MR-3		
13RC15	5543		MR-3		
13RC2	5540	6615/5515	MR-3		
13RC20	5543		MR-3		
13RC25	5544		MR-3		
13RC30	5544		MR-3		
13RC40	5545		MR-3		
13RC5	5541	6615/5515	MR-3		
13S	502	3067	CR-4		
13Z-1005	109	3087		1123	
13Z6AF	5067A	3331	ZD-3.9		
13Z6F	5067A	3331	ZD-3.9		
14 601 28	123A	3444			
14 601 29	123A	3444			
14 806 12	123A	3444		2051	
14 807 23	123A	3444			
14 809 23	123A	3444			
14 854 12	123A	3444			
14 862 32	123A	3444			
14 865 12	123A	3444			
14 866 32	123A	3444			
14-0072-1		3017B	504A	1104	
14-0072-1(PHILCO)	116		504A	1104	
14-0072-2		3017B	504A	1104	
14-0072-2(PHILCO)			504A	1104	
14-0072-3		3017B	504A	1104	
14-0072-3(PHILCO)	116		504A	1104	
14-0086-1	129		244	2027	
14-0104-1	152	3893	66	2048	
14-0104-2	153		69	2049	
14-0104-3	129		244	2027	
14-0104-4	128		243	2030	
14-0104-5	172A	3156	64		
14-0104-6	5066A	3330	ZD-3.3		
14-0104-7	123A	3444	20	2051	
14-0110-1			60	2015	
14-0110-2			60	2015	
14-1	123A	3444	20	2051	
14-10	109			1123	
14-1007-00	705A		IC-3		
14-2	123A	3444	20	2051	
14-2000-01	172A	3156	64		
14-2000-02	172A	3156	64		
14-2000-03	172A	3156	64		
14-2000-04	172A	3156	64		
14-2000-05	172A	3156	64		
14-2000-23	172A	3156	64		
14-2002-01	312	3112		2028	
14-2007-00	705A	3134	IC-3		
14-2007-00B	707		IC-206		
14-2007-01	707		IC-3		
14-2007-02	705A	3134	IC-3		
14-2007-03	707		IC-206		
14-2007-04	705A	3134	IC-3		
14-2008-01	708	3135/709	IC-10		
14-2009-01	814		IC-243		
14-2010-01	713	3077/790	IC-5		
14-2010-02	713	3156/172A	IC-5		
14-2011-01	779A	3240/779-1			
14-2012-01	801	3160	IC-35		
14-2014-01	815	3255	IC-244		
14-2052-23	172A		64		
14-2053-23	172A	3156	64		
14-2054-01	731	3170	IC-13		
14-2056-01	OBS-NLA		IC-277		
14-3	123A	3444	20	2051	
14-32430	108	3452	86	2038	
14-40325A	130	3027	14	2041	
14-40363A	130	3027	14	2041	
14-40369A	130	3027	14	2041	
14-40421A	130	3027	14	2041	
14-40464A	130	3027	14	2041	
14-40465A	130	3027	14	2041	
14-40466A	130	3027	14	2041	
14-40471	130	3027	14	2041	
14-40934-1	130	3027	14	2041	
14-501-01	113A	3119/113	6GC1		
14-501-02	113A	3119/113	6GC1		
14-502-01			1N295	1123	
14-503-01	114	3120	6GD1	1104	
14-503-02	114	3120	6GD1		
14-503-03	114	3120	6GD1	1104	
14-503-04	114	3120	6GD1		
14-503-08	114	3120	6GD1	1104	
14-504-01	109	3087		1123	
14-504-04	114	3120	6GD1	1104	
14-507-01	177		300	1122	
14-509-01	142A	3062	ZD-12	563	
14-509-02	135A	3056	ZD-5.1		
14-510-01	109	3087		1123	
14-511-01	109	3087		1123	
14-512-01	109	3087		1123	
14-513-01	109	3087		1123	
14-514-01	109	3087		1123	
14-514-01/51	109			1123	
14-514-02	177	3100/519	300	1122	
14-514-03	506	3843	511	1114	
14-514-03/53	506	3843	511	1114	
14-514-05	109	3088		1123	
14-514-05/55	109			1123	
14-514-06	109	3087		1123	
14-514-07	506	3843	511	1114	
14-514-07/57	506	3843	511	1114	
14-514-08	109	3087		1123	
14-514-08/58	109			1123	
14-514-09	109	3087		1123	
14-514-10	109	3087		1123	
14-514-10/60	109			1123	
14-514-11	110MP	3088	1N60	1123	
14-514-12	519	3087	514	1122	
14-514-13	110MP	3709011OMP		1123(2)	
14-514-13/63	110MP	3709		1123(2)	
14-514-14	506	3843	511	1114	
14-514-15	177(2)	3100/519	300(2)	1122	
14-514-17	177	3100/519	300	1122	
14-514-17/67	506	3843	511	1114	
14-514-18	506	3843	511	1114	
14-514-18/68	506	3843	511	1114	
14-514-19	177		300	1122	
14-514-20	178MP	3100/519	300(2)	1122(2)	
14-514-20/70	506	3843	511	1114	
14-514-21	109	3087		1123	
14-514-22	109	3087		1123	
14-514-22/72	109			1123	
14-514-23	506	3843	511	1114	
14-514-23/73	506	3843	511	1114	
14-514-55	109	3088		1123	
14-514-61	109	3088		1123	
14-514-62	519	3100	514	1122	
14-514-64	519	3100	514	1122	
14-514-65	178MP	3100/519	300(2)	1122(2)	
14-514-66		3843	511	1114	
14-514-70	177	3100/519	300	1122	
14-514-72	109	3091		1123	
14-514-74		3843	511	1114	
14-514-75			504A	1104	
14-515-01	142A	3062	ZD-12	563	
14-515-02	135A	3056	ZD-5.1		
14-515-03	142A	3062	ZD-12	563	
14-515-04	138A	3059	ZD-7.5		
14-515-05	5077A	3752	ZD-18		
14-515-06/66	109			1123	
14-515-07	145A	3063	ZD-15	564	
14-515-09	145A		ZD-15	564	
14-515-11	145A		ZD-15	564	
14-515-12			ZD-9.1	562	
14-515-13	138A	3059	ZD-7.5		
14-515-15	135A	3124/289	ZD-5.1		
14-515-16			ZD-9.1	562	
14-515-18	141A	3092	ZD-11.5		
14-515-19			ZD-9.1	562	
14-515-23	5024A	3790	ZD-15	564	
14-515-24			ZD-9.1	562	
14-515-25	142A		ZD-12	563	
14-515-26	138A		ZD-7.5		
14-515-27	5071A	3334	ZD-6.8	561	
14-515-28		3753	ZD-25		
14-515-29	151A		ZD-110		
14-515-30	5071A	3334	ZD-6.8	561	
14-515-73	145A	3063	ZD-15	564	
014-556	105	3459	4		
14-557-10	102A		53	2007	
14-564-08			53	2007	
14-566-08			52	2024	
14-568-04			52	2024	
14-569-09	160		245	2004	
14-572-10			10	2051	
14-573-10	121	3009	239	2006	
14-574-10	121	3009	239	2006	
14-575-10	123A	3122	20	2051	
14-576-10			53	2007	
14-577-10	102A	3004	53	2007	
14-578-10	121	3009	239	2006	
14-579-10	121	3717	239	2006	
14-580-01	160		245	2004	
14-581-01	160		245	2004	
14-582-01	126	3006/160	52	2024	
14-583-01	123A	3444	20	2051	
14-584-01	102A		53	2007	
14-585-01	160		245	2004	
14-586-01	104	3009	16	2006	
14-587-01	160		245	2004	
14-588-01	160		245	2004	
14-589-01	121	3009	239	2006	
14-590-01	121	3009	239	2006	
14-591-01	160		245	2004	
14-593-01	6400A			2029	
14-593-03	6400A			2029	
14-600-01	160	3006	245	2004	
14-600-02	160		245	2004	
14-600-04	160	3006	245	2004	
14-600-10		3006	51	2004	
14-600-11	160		245	2004	
14-600-13	160		245	2004	
14-600-16	160	3006	245	2004	
14-600-19	160	3008	245	2004	
14-600-20	160		245	2004	
14-600-22	160		245	2004	
14-601-01	121	3009	239	2006	
14-601-02	179	3642	76		
14-601-03	121	3009	239	2006	
14-601-04	121	3009	239	2006	
14-601-05	121	3009	239	2006	
14-601-06	121	3009	239	2006	
14-601-07	121	3009	239	2006	
14-601-08	104	3009	16	2006	
14-601-09	121	3009	239	2006	
14-601-10	130	3027	14	2041	
14-601-11	121	3009	239	2006	
14-601-12	130	3027	14	2041	
14-601-13	130	3027	14	2041	
14-601-14	162	3438	35		
14-601-15	130	3027	14	2041	
14-601-15A	130	3027	14	2041	
14-601-16	130	3027	14	2041	
14-601-16A	130	3027	14	2041	
14-601-17	184	3027/130	57	2017	
14-601-18	390		255	2041	
14-601-20	390		255	2041	
14-601-23	390	3027/130	255	2041	
14-601-24	390		255	2041	
14-601-26	181	3079	75	2041	
14-601-27	165	3115	38		
14-601-28	123A	3444	20	2051	
14-601-29			20	2051	
14-602-01	123A	3444	20	2051	
14-602-02	123A	3444	20	2051	
14-602-03	123A	3444	20	2051	
14-602-04	102A		53	2007	
14-602-05	102A	3004	53	2007	
14-602-05A	102A	3004	53	2007	
14-602-06	158		53	2007	
14-602-07	158		53	2007	
14-602-08	158		53	2007	
14-602-09	158		53	2007	
14-602-10	102A		53	2007	
14-602-11	159	3114/290	59	2032	
14-602-11A	159	3114/290	59	2032	
14-602-12	123A	3444	20	2051	
14-602-13	123A	3444	20	2051	
14-602-14	123A	3444	20	2051	
14-602-15	102A		53	2007	
14-602-16	123A	3444	20	2051	
14-602-17	123A	3444	20	2051	
14-602-18	128	3124/289	243	2030	
14-602-19	154	3045/225	40	2012	
14-602-20	159	3466	82	2032	
14-602-20A	159	3466	82	2032	
14-602-21	103A	3835	59	2002	
14-602-22	123A	3444	20	2051	
14-602-23	123A	3444	20	2051	
14-602-24	128	3024	243	2030	
14-602-25		3122	62	2010	
14-602-26	123A	3444	20	2051	
14-602-28	129	3025	244	2027	
14-602-28A	129	3025	244	2027	
14-602-29	128	3124/289	243	2030	
14-602-30	128	3124/289	243	2030	
14-602-31	161	3018	39	2015	
14-602-32	159	3466	82	2032	

Industry Standard No.	ECG	SK	GE	RS 276-	MOTOR.
14-602-32A	159	3466	82	2032	
14-602-34	161	3018	39	2015	
14-602-35	123A	3444	20	2051	
14-602-36	154	3045/225	40	2012	
14-602-37	154	3045/225	40	2012	
14-602-41	108	3452	86	2038	
14-602-42	159	3466	82	2032	
14-602-42A	159	3466	82	2032	
14-602-43	128	3024	243	2030	
14-602-44	159	3466	82	2032	
14-602-44A	159	3466	82	2032	
14-602-45		3122	60	2015	
14-602-46	199	3018	62	2010	
14-602-46A	199	3124/289	62	2010	
14-602-47	159	3466	82	2032	
14-602-47A	159	3466	82	2032	
14-602-48	123A	3444	20	2051	
14-602-49	128	3024	243	2030	
14-602-50	123A	3444	20	2051	
14-602-51	158		53	2007	
14-602-52	103A	3835	59	2002	
14-602-54	193	3114/290	67	2023	
14-602-54A	159	3466	82	2032	
14-602-55	192	3024/128	63	2030	
14-602-55A	123A	3444	20	2051	
14-602-56	159	3466	82	2032	
14-602-56A	159	3466	82	2032	
14-602-58	159	3466	82	2032	
14-602-580	159	3466	82	2032	
14-602-58A	159	3466	82	2032	
14-602-59	128	3024	243	2030	
14-602-60	129	3025	244	2027	
14-602-600	159	3466	82	2032	
14-602-61	123A	3444	20	2051	
14-602-62	123A	3444	20	2051	
14-602-63	199	3039/316	62	2010	
14-602-64	234		65	2050	
14-602-65	128		243	2030	
14-602-66	129		244	2027	
14-602-67	175		246	2020	
14-602-68	159	3466	82	2032	
14-602-69	123A	3444	20	2051	
14-602-70	192		63	2030	
14-602-71	193		67	2023	
14-602-72	128		243	2030	
14-602-73	129		244	2027	
14-602-74	192	3024/128	63	2030	
14-602-75	193	3025/129	67	2023	
14-602-76	198	3220	251		
14-602-77	107	3018	11	2015	
14-602-77B	107	3018	11	2015	
14-602-78	123A	3444	20	2051	
14-602-79	129	3025	244	2027	
14-602-79A	129	3025	244	2027	
14-602-80	123A	3444	20	2051	
14-602-81	123A	3444	20	2051	
14-602-85	129		244	2027	
14-602-87	123A	3444	20	2051	
14-602-88	159	3466	82	2032	
14-602-89	123A	3444	20	2051	
14-602-90	159	3466	82	2032	
14-603-01	128	3024	243	2030	
14-603-02	161	3018	39	2015	
14-603-02A	128	3018	243	2030	
14-603-03	123A	3444	20	2051	
14-603-04	123A	3444	20	2051	
14-603-05	107	3039/316	11	2015	
14-603-05-2	107	3018	11	2015	
14-603-06	107	3039/316	11	2015	
14-603-07	128	3024	243	2030	
14-603-08	161	3716	39	2015	
14-603-09	161	3716	39	2015	
14-603-10	123A	3444	20	2051	
14-603-11	123A	3444	20	2051	
14-603-12	108	3452	86	2038	
14-603-13	161	3716	39	2015	
14-604-02		3009	43,44		
14-604-03(PNP)			30	2006	
14-604-07	121		239	2006	
14-604-08	121	3009	239	2006	
14-607-29	129		244	2027	
14-607-29A	129	3025	244	2027	
14-608-01	184	3054/196	57	2017	
14-608-02	184	3054/196	57	2017	
14-608-02A	196	3054	241	2020	
14-608-03	184	3190	57	2017	
14-608-04	184	3190	57	2017	
14-609-00	152	3893	66	2048	
14-609-01	191	3232	249		
14-609-01A	191	3044/154	249		
14-609-02	171	3044/154	27		
14-609-02A	191	3044/154	249		
14-609-03	152	3054/196	66	2048	
14-609-03A	196	3054	241	2020	
14-609-04	152	3054/196	66	2048	
14-609-05	191	3044/154	249		
14-609-06	152	3893	66	2048	
14-609-08	152	3893	66	2048	
14-609-09	196		241	2020	
14-609-49A	108	3452	86	2038	
014-611	159	3466	82	2032	
14-651-12	123A	3444	20	2051	
014-652	159	3466	82	2032	
14-652-12	161		39	2015	
014-652C	159	3466	82	2032	
14-653-21	161	3117	39	2015	
14-654-21	128	3132	243	2030	
14-655-13	123A	3444	20	2051	
14-656-21	123A	3444	20	2051	
14-659-12	123A	3444	20	2051	
14-660-12	123A	3444	20	2051	
14-661-21	161	3122	39	2015	
014-680	123A	3444	20	2051	
014-686	123A	3444	20	2051	
014-698	123A	3444	20	2051	
14-700-01	312		FET-1	2035	
14-700-02	312		FET-2	2028	
14-700-03	312	3448	FET-1	2035	
14-700-04	312	3448	FET-1	2035	
14-700-05	312	3448	FET-1	2035	
14-700-06	312	3448	FET-1	2035	
14-710-21	312	3448	FET-1	2035	
14-713-31	312	3448	FET-1	2035	
14-713-32	312	3448	FET-1	2035	
14-714-13	312	3448	FET-1	2035	
014-754	234		65	2050	
014-772	159	3466	82	2032	
014-784	123A	3444	20	2051	
14-800-32	123A	3444	20	2051	
14-801-12			20	2051	
14-801-23	199	3122	62	2010	
14-802-12	123A	3444	20	2051	
14-803 12	159			2032	
14-803-12	159	3466	82	2032	
14-803-32	234	3247	65	2050	
14-804-12	159	3466	82	2032	
14-805-12	123A	3444	20	2051	
14-806-12	123A	3444	20	2051	
14-806-23	123A	3444	20	2051	
14-807-12	106	3984	21	2034	
14-807-23			20	2051	
14-808-12	159	3466	82	2032	
14-809-23	123A	3444	20	2051	
14-809-32	123A	3444	20	2051	
14-811-12	194		220		
14-850-12	161	3716	39	2015	
14-851-12	161	3716	39	2015	
14-851-32	123A	3444	20	2051	
14-853-23	123A	3444	20	2051	
14-854-12	123A	3444	20	2051	
14-855-32	159	3466	82	2032	
14-856-23	159	3466	82	2032	
14-857-12	159	3466	82	2032	
14-857-32			244	2027	
14-857-79	159	3466	82	2032	
14-858-12	123A	3444	20	2051	
14-861-12	106	3984	21	2034	
014-862	199	3122	62	2010	
14-862-23	123A	3444	20	2051	
14-862-32	123A	3444	20	2051	
14-863-23	159	3466	82	2032	
14-864-12	123A	3444	20	2051	
14-864-23	159	3466	82	2032	
14-865-12	123A	3444	20	2051	
14-866-32	123A	3444	20	2051	
14-867-32	159	3466	82	2032	
14-900-12	191	3232	249		
14-901-12	171	3104A	27		
14-902-23	152		66	2048	
14-903-23	152	3054/196	66	2048	
14-904-12	191	3104A	249		
14-905-23	152	3054/196	66	2048	
14-906-13	184	3190	57	2017	
14-907-13	184	3190	57	2017	
14-908-23	152	3054/196	66	2048	
14-909-23		3054	66	2048	
14-910-13	184	3190	57	2017	
0014-911	177		300	1122	
14-911-13	184	3190	57	2017	
14/514-12/62	506		511	1114	
14/514-14/64	506		511	1114	
14/514-19/69	506		511	1114	
14A	121	3009	239	2006	
014A-12	121	3009	239	2006	
014A-12-7	121	3009	239	2006	
14A0	121	3009		2006	
14A1	121	3009	239	2006	
14A1-A82	121	3717	239	2006	
14A10	121	3009	239	2006	
14A10R	121	3009	239	2006	
14A11	121	3009	239	2006	
14A12	121	3009	239	2006	
14A13	121	3009	239	2006	
14A13P	121	3009	239	2006	
14A14	121	3009	239	2006	
14A14-7	121	3009	239	2006	
14A14-7B	121	3717	239	2006	
14A15	121	3009	239	2006	
14A15L	121	3009	239	2006	
14A16	121	3009	239	2006	
14A16-5	121	3009	239	2006	
14A17	121	3009	239	2006	
14A17-1	121	3009	239	2006	
14A18	121	3009	239	2006	
14A19	121	3009	239	2006	
14A19G	121	3009	239	2006	
14A2	121	3009	239	2006	
14A3	121	3009	239	2006	
14A4	121	3009	239	2006	
14A5	121	3009	239	2006	
14A6	121	3009	239	2006	
14A7	121	3009	239	2006	
14A8	121	3009	239	2006	
14A8-1	160	3007	245	2004	
14A8-1-12	160	3007	245	2004	
14A9	121	3009	239	2006	
14AO			239	2006	
14B348-003	523	3306			
14B348-1	523	3306	528		
14B348-2	523	3306	528		
14B348-3	523	3306	528		
14B348-4	523	3306	528		
14B348-5	523	3306	528		
14I4-180	175		246	2020	
14J2	116	3016	504A	1104	
14J2F	116	3017B/117	504A	1104	
14LN033	1060		IC-50		
14LN034	1181	3706	IC-61		
14LQ007	1003	3288	IC-43		
14MW69	102A	3004	53	2007	
14P2	116	3311	504A	1104	
14Z6AF	5068A	3332	ZD-4.3		
14Z6F	5068A	3332	ZD-4.3		
015	108	3452	86	2038	
015(LAMBDA)	912		IC-172		
15-0009-05	196		241	2020	
015-002	116	3016	504A	1104	
015-006	116	3016	504A	1104	
15-008-1	106	3984	21	2034	
15-01742	159	3466	82	2032	
15-01913-00	159	3466	82	2032	
15-01999	123A	3444	20	2051	
15-02155	128		243	2030	
15-02757-00	199		62	2010	
15-02762-00	159	3466	82	2032	
15-02762-1	159	3466	82	2032	
15-02762-2	159	3466	82	2032	
15-02979	159	3466	82	2032	
15-03014-00	123A	3444	20	2051	
15-03051-00	107		11	2015	
15-03068	130		14	2041	
15-03099	159	3466	82	2032	
15-03100	123A	3444	20	2051	
15-033-0				1104	
15-03409-0	159	3466	82	2032	
15-03409-02	159	3466	82	2032	
15-03409-1	159	3466	82	2032	
15-05302	123A	3444	20	2051	
15-05369	123A	3444	20	2051	
15-05415	218		234		
15-05650	123A	3444	20	2051	
15-082019	123A	3444	20	2051	
15-085002	109	3087		1123	
15-085003	109	3087		1123	
15-085004	113A		6GC1		
15-085005	177	3100/519	300	1122	
15-085006	116		504A	1104	
15-085007	116		504A	1104	

Industry Standard No.	ECG	SK	GE	RS 276-	MOTOR.
15-085008	177	3125	300	1122	
15-085009	109	3087		1123	
15-085015			504A	1104	
15-085016			504A	1104	
15-085018	112	3089	1N82A		
15-085027	109			1123	
15-085032	109			1123	
15-085033	125		510,531	1114	
15-085037	109	3089/112		1123	
15-085038	113A	3119/113	6GC1		
15-085039	113A	3119/113	6GC1		
15-085040	116		504A	1104	
15-085041	125	3311	510,531	1114	
15-085043	116	3031A	504A	1104	
15-085047	113A	3119/113	6GC1		
15-085061	109			1123	
15-088002	159	3114/290	82	2032	
15-088002A	159	3114/290	82	2032	
15-088003	107	3019	11	2015	
15-088004	108	3452	86	2038	
15-08800U	108	3452	86	2038	
15-09090-01	159	3466	82	2032	
15-09338	123A	3444	20	2051	
15-09587	199		62	2010	
15-09650	128		243	2030	
15-09980	123A	3444	20	2051	
15-1	123A	3444	20	2051	
15-100001	116	3017B/117	504A	1104	
15-100002	116	3032A	504A	1104	
15-100003	178MP	3311	300(2)	1122(2)	
15-100004	116	3031A	504A	1104	
15-10062-0	128		243	2030	
15-103022			504A	1104	
15-108001	178MP	3119/113	300(2)	1122(2)	
15-108002	113A	3119/113	6GC1		
15-108003	116	3017B/117	504A	1104	
15-108004	116	3017B/117	504A	1104	
15-108005	116	3032A	504A	1104	
15-108006	116	3032A	504A	1104	
15-108007	120	3110	CR-3		
15-108008	177	3125	300	1122	
15-108009	113A	3119/113	6GC1		
15-108010	116	3017B/117	504A	1104	
15-108011	116	3031A	504A	1104	
15-108012	118	3066	CR-1		
15-108013	506	3843	511	1114	
15-108014	120		CR-3		
15-108015	116(3)		504A(3)	1104	
15-108016	116	3017B/117	504A	1104	
15-108017	118	3066	CR-1		
15-108020	116	3032A	504A	1104	
15-108021	116	3032A	504A	1104	
15-108022	116	3032A	504A	1104	
15-108023		3119	6GC1		
15-108024		3017B	504A	1104	
15-108025	177	3100/519	300	1122	
15-108026	506	3843	511	1114	
15-108027			511	1114	
15-108028		3081	CR-2		
15-108029			511	1114	
15-108030	502		CR-4		
15-108031			504A	1104	
15-108032			504A	1104	
15-108034			504A	1104	
15-108035	506	3843	511	1114	
15-108036	116	3017B/117	504A	1104	
15-108037	116	3311	504A	1104	
15-108038	120	3110	CR-3		
15-108040	506	3843	511	1114	
15-108041	506	3843	511	1114	
15-108042	177	3100/519	300	1122	
15-108043	506	3843	511	1114	
15-108044	506	3843	511	1114	
15-108045	502	3067	CR-4		
15-108046	125	3081	510,531	1114	
15-108047	506	3843	511	1114	
15-108048	506	3843	511	1114	
15-108049	116	3031A	504A	1104	
15-108050	116	3017B/117	504A	1104	
15-123060	539	3309	538		
15-123065	152	3054/196	66	2048	
15-123100	283	3467	73		
15-123101	177	3100/519	300	1122	
15-123102	506	3843	511	1114	
15-123103	506	3843	511	1114	
15-123104	503	3068	CR-5		
15-123105	116	3311	504A	1104	
15-123106	5015A	3059/138A	ZD-7.5		
15-123230	165	3710/238	38		
15-123231	503	3068	CR-5		
15-123242	506	3843	509	1114	
15-123243	125	3313/116	504A	1104	
15-123297	506	3998			
15-123298	503	3068			
15-123300	116	3998/506			
15-123302	186A	3357			
15-123303	187A	9076			
15-14471-1	1085	3476	IC-104		
15-14504-1	794	3974			
15-166N	107		11	2015	
15-190C19	5100A		ZD-150		
15-2	123A	3444	20	2051	
15-20A-90M	109			1123	
15-20A70	110MP	3709		1123(2)	
15-20A90	110MP	3709		1123(2)	
15-2210921				2004	
15-2221011	102A	3004	53	2007	
15-22210111	102A	3004	53	2007	
15-22210131	160	3006	245	2004	
15-22210300	160	3006	245	2004	
15-22210921	160	3007	245	2004	
15-22211021	160	3006	245	2004	
15-22211200	102A	3004	53	2007	
15-22211328	102A	3004	53	2007	
15-22211921	160	3006	245	2004	
15-22214400	160	3006	245	2004	
15-22214411	160	3006	245	2004	
15-22214435	160	3006	245	2004	
15-22214821	160	3006	245	2004	
15-22214831	160	3006	245	2004	
15-22216500	102A	3004	53	2007	
15-22216600	160	3007	245	2004	
15-22217400	160	3006	245	2004	
15-22223720	123A	3444	20	2051	
15-26587	703A		IC-12		
15-26587-1	703A	3157	IC-12		
15-3	159	3466	82	2032	
15-30	159	3466	82	2032	
15-3015129-1	797	3158			
15-3015130-1	740A	3328			
15-3015131-1	1232	3852			
15-31015-1	1022	3481/1002	IC-102		
15-33201	712	3072	IC-2		
15-33201-1	712	3072	IC-2		
15-33201-2	712	3072	IC-2		
15-34005-1	727	3071	IC-210		
15-34048-1	708	3135/709	IC-10		
15-34048-2	708	3135/709	IC-10		
15-34049	722	3161	IC-9		
15-34049-1	722	3161	IC-9		
15-34049-3	722	3161	IC-9		
15-34202-1	727	3071	IC-210		
15-34379-1	720	9014	IC-7		
15-34408-2	708		IC-10		
15-34452	709	3135	IC-11		
15-34452-1	709	3135	IC-11		
15-34502-1	778A		IC-220		
15-34502-2	778A		IC-220		
15-34503	722	3161	IC-9		
15-34503-1	722	3161	IC-9		
15-34503-2	722	3161	IC-9		
15-34906-1	799		IC-34		
15-35059-1	712	3072	IC-148		
15-35059-2	712	3072	IC-148		
15-36446-1	743		IC-214		
15-36995-1	744	3171	IC-215	2022	
15-37534-1	794	3974			
15-37534-2	794	3974			
15-37702-1	714	3075	IC-4		
15-37703-1	715	3076	IC-6		
15-37704-1	790	3454	IC-18		
15-37833-1	909D	3590	IC-250		
15-37833-2	909D	3590	IC-250		
15-3793	797	3158			
15-39060-1	747	3279	IC-218		
15-39061-1	795	3237	IC-232		
15-39075-1	738	3167	IC-29		
15-39098-1			28	2017	
15-39209-1	791	3149	IC-231		
15-39600-1	794	3974			
15-39600-2	794	3974			
15-4	159	3466	82	2032	
15-40	159	3466	82	2032	
15-40183-1	801	3160	IC-35		
15-41545-1	794	3974			
15-41627-2	821	3882			
15-41764	822	3919			
15-41764-1	822	3919			
15-41856-1	795		IC-232		
15-43098-1	819	3928			
15-43251-1	747		IC-218		
15-43251-2	747		IC-218		
15-43312-1	820	3927			
15-43312-2	820	3927			
15-43703-1	738	3167	IC-29		
15-43705-1	4011B	4011			
15-43706-1	4093B	4093			
15-45141-1	4011B	4011			
15-45184-1	4024B	4024	4024		
15-45185-1	4016B		4016		
15-45186-1	987	3643			
15-45300-1	712	3072	IC-148		
15-5	159	3466	82	2032	
15-50	159	3466	82	2032	
15-505753-2	911		IC-253		
15-505753-3	911D		IC-254		
15-71420-1	722	3159/718	IC-9		
15-71420-1	722	3161	IC-9		
15-875-075-001	159	3466	82	2032	
15-875-075-003	123A	3444	20	2051	
0015-911	116		504A	1104	
15B1				1143	
15B10	5809	9010			
15B2	116		504A	1104	
15B4				1144	
15B6	5806	3848			
15B8	5808	9009			
15BD11		3313	504A	1104	
15C05	5800	9003		1142	
15C1	5801	9004		1142	
15C10	5809	9010			
15C2	5802	9005		1143	
15C2D	5802	9005		1143	
15C4	5804	9007		1144	
15C6	5806	3051/156			
15C8	5808	9009			
15E10	5809	9010			
15E6	5806	3848			
15E7	5808	9009			
15E8	5808	9009			
15F10	5809	9010			
15F6	5806	3848			
15F8	5808	9009			
15H3P	5854		5016		
15J2	116	3017B/117	504A	1104	
15J2F	116	3017B/117	504A	1104	
15N1	116	3016	504A	1104	
15N2(POWER)	5874		5032		
15N3(POWER)	5874		5032		
15P2	177	3311	300	1122	
15P2(POWER)	5874		5032		
15P3(POWER)	5874		5032		
15Q3	5944		5048		
15S05	5800	9003	504A	1142	
15S1	5801	9004	504A	1142	
15S2	5802	9005	504A	1143	
15S4	5804	9007	504A	1144	
15S6	5806	3848	504A	1104	
15Z3	159	3466	82	2032	
15Z6AF	5069A	3333	ZD-4.7		
15Z6F	5069A	3333	ZD-4.7		
016	5637	3939/5636			
16-147191229	123A	3124/289	20	2051	
16-17		3116	FET-2	2035	
16-171191368	123A	3124/289	20	2051	
16-19(SYMPHONIC)	129		244	2027	
16-2	116	3016	504A	1104	
16-20(SYMPHONIC)			243	2030	
16-207190405	102A	3004	53	2007	
16-21426	161	3018	39	2015	
16-501190016	125	3031A	510,531	1114	
16-736	108	3452	86	2038	
16A1	103	3862	8	2002	
16A1(FLEETWOOD)		3444	20	2051	
16A1938	123A	3444	20	2051	
16A2	103	3862	8	2002	
16A2(FLEETWOOD)		3444	20	2051	
16A545-7	123A	3444	20	2051	
16A667-GRN		3245	212	2010	
16A667-ORG		3245	212	2010	
16A667-RED		3245	212	2010	
16A667-YEL		3245	212	2010	
16A668-GRN		3245	212	2010	
16A668-ORG		3245	212	2010	
16A668-YEL		3245	212	2010	
16A669-GRN		3245	212	2010	
16A669-YEL		3245	212	2010	
16A787	102	3722	2	2007	

Industry Standard No.	ECG	SK	GE	RS 276-	MOTOR.
16B-3A82	121MP	3718	239(2)	2006(2)	
16B1		3244	220		
016B12	123A	3122	20	2051	
16B2			210	2051	
16B670-GRN		3156	64	2014	
16B670-RED		3244	220		
16B670-YEL		3244	220		
016B810	123A	3122	20	2051	
016B812	123A	3122	20	2051	
16C-4	116(4)	3311	504A	1104	
16C-4P	166	9075		1152	
16C025B	5520	6621/5521			
16C025C	5520	6621/5521			
16C050B	5521	6621			
16C050C	5521	6621			
16C10C	5522	6622			
16C15B	5523	6624/5524			
16C15C	5523	6624/5524			
16C20B	5524	6624			
16C20C	5524	6624			
16C25B	5525	6627/5527			
16C25C	5525	6627/5527			
16C30B	5526	6627/5527			
16C30C	5526	6627/5527			
16C35B	5527	6627			
16C35C	5527	6627			
16C4	116	3016	504A	1104	
16C40B	5527	6627			
16C40C	5527	6627			
16C45B	5528	6629/5529			
16C4B1P	166	9075	BR-600	1152	
16C50B	5528	6629/5529			
16C50C	5528	6629/5529			
16C60B	5529	6629			
16C60C	5529	6629			
16C70B	5530	6632			
16C70C	5530	6632			
16C80B	5531	6631			
16C80C	5531	6631			
16E1330		3122	20	2051	
16E1330(GE)	123A		20	2051	
16F10	5894		5064		MR2001S
16F100	5910	7110	5076		MR2010S
16F15	5896		5064		MR2002S
16F20	5896	7096	5064		MR2002S
16F30	5898		5068		MR2004S
16F40	5900	7100	5068		MR2004S
16F5	5892		5064		MR2000S
16F50	5902		5072		MR2006S
16F60	5904	7104	5072		MR2006S
16F80	5928		5076		MR2008S
16G2	123A	3444	20	2051	
16G27	312			2028	
16G28	177		300	1122	
16G28A	177		300	1122	
16G5	116		504A	1104	
16GN	107	3019	11	2015	
16J1	108	3452	86	2038	
16J2	108	3452	86	2038	
16J2(DIODE)	116		504A	1104	
16J2F	116	3017B/117	504A	1104	
16J3			20	2051	
16K1	108	3452	86	2038	
16K2	108	3452	86	2038	
16K3	108	3452	86	2038	
16L2	108	3452	86	2038	
16L22	108	3452	86	2038	
16L23	108	3452	86	2038	
16L24		3245	212	2010	
16L25		3245	212	2010	
16L3	108	3452	86	2038	
16L4			212	2010	
16L42	123A	3444	20	2051	
16L43	123A	3444	20	2051	
16L44	123A	3444	20	2051	
16L45		3245	212	2010	
16L5	108	3452	86	2038	
16L62	123A	3444	20	2051	
16L63	123A	3444	20	2051	
16L64	107	3019	11	2015	
16L65		3245	212	2010	
16P2	178MP	3311	300(2)	1122	
16P2881	172A	3156	64		
16P3367	172A	3156	64		
16RC10A	5542	6615/5515	MR-3		
16RC15A	5543		MR-3		
16RC20A	5543		MR-3		
16RC25A	5544		MR-3		
16RC2A	5540	6615/5515	MR-3		
16RC30A	5544		MR-3		
16RC40A	5545		MR-3		
16RC5A	5541	6615/5515	MR-3		
16U1		3122	20	2051	
16U1(HEATH KIT)	123A			2051	
16U1(HEATH-KIT)			20	2051	
16X1	123A	3122	20	2051	
16X10	116		504A	1104	
16X2	123A	3444	20	2051	
16X39	177	3100/519	300	1122	
16Z4	142A	3062	ZD-12	563	
16Z6	135A	3056	ZD-5.1		
16Z6A	135A	3056	ZD-5.1		
16Z6AF	135A		ZD-5.1		
16Z6F	135A		ZD-5.1		
17-1	151A	3099	ZD-110		
17-10	116	3016	504A	1104	
17-12096-1	909	3590	IC-249		
17-12097-1	910		IC-251		
17-410	116	3311	504A	1104	
17-443	128	3024	243	2030	
17-451	123A	3444	20	2051	
17-457	123A	3444	20	2051	
17-458	129	3024/128	244	2027	
17-459	159	3466	82	2032	
17-459A	159	3466	82	2032	
17-50(FISHER)	130		14	2041	
17-RLB	234		65	2050	
17A4422-1	121	3009	239	2006	
017E824	123A	3444	20		
017EB24				2051	
17P2	177	3311	300	1122	
17S08	6358	6558			
17S08R	6359	6559			
17S10	6358	6558			
17S10R	6359	6559			
17TB1	5522	6622			
17TB2	5524	6624			
17TB3	5526	6627/5527			
17TB4	5527	6627			
17TB5	5528	6629/5529			
17TB6	5529	6629			
17TB8	5531	6631			
17Z6	136A	3057	ZD-5.6	561	
17Z6A	136A	3057	ZD-5.6	561	
17Z6AF	136A		ZD-5.6	561	
17Z6F	136A		ZD-5.6	561	
0018		3020	10	2051	
018-00001	159	3466	82	2032	
018-00002	159	3466	82	2032	
018-00003	123A	3444	20	2051	
018-00004	187	3193	29	2025	
018-00005	186	3192	28	2017	
018-00006	116	3031A	504A	1104	
018-00007	116	3031A	504A	1104	
018-00008	116		504A	1104	
018-00009	116	3031A	504A	1104	
018-0009				1104	
18-085001	116	3017B/117	504A	1104	
18-085002	142A	3062	ZD-12	563	
18-148A	123A	3444	20	2051	
18-177-1	175		246	2020	
18-198-2	124		12		
18-22-17	116		504A	1104	
18-3539-1	128		243	2030	
18-605	912		IC-172		
18-605-RO	912		IC-172		
18-606	724		IC-86		
18/T2C			14	2041	
18A8-1	160	3007	245	2004	
18A8-1-12	160	3007	245	2004	
18A8-1-127	160	3007	245	2004	
18A8-1L	160	3007	245	2004	
18A8-1L8	160	3007	245	2004	
18AA-1-82	121	3717	239	2006	
018B	107	3039/316	11	2015	
18DB10A	170	3649			MDA920A10
18DB10A-C	170	3649			
18DB2A	167	3647		1172	MDA920A4
18DB2A-C	167			1172	
18DB4A	168	3648		1173	MDA920A6
18DB4A-C	168	3648		1173	
18DB6A	169	3107/5307			MDA920A7
18DB8A	170	3107/5307			MDA920A8
18FA10					1N4934
18FA20					1N4935
18FA30					1N4936
18FA40					1N4936
18FA5					1N4933
18FB10					1N4934
18FB20					1N4935
18FB30					1N4936
18FB40					1N4936
18FB5					1N4933
18FC10					1N4934
18FC20					1N4935
18FC30					1N4936
18FC40					1N4936
18FC5					1N4933
18J2	125	3032A	510,531	1114	
18J2F	125	3032A	510,531	1114	
18P-2A82	102A		53	2007	
18P2	116	3311	504A	1104	
18RC10	5542	6615/5515	MR-3		
18RC2	5540	6615/5515	MR-3		
18RC20	5543		MR-3		
18RC25	5544		MR-3		
18RC30	5544		MR-3		
18RC40	5545		MR-3		
18RC5	5541	6615/5515	MR-3		
18Z6	137A	3058	ZD-6.2	561	
18Z6A	137A	3058	ZD-6.2	561	
18Z6AF	137A		ZD-6.2	561	
18Z6F	137A		ZD-6.2	561	
19-00-3485	175	3026	246	2020	
019-00009	123A	3444	20	2051	
019-00010	123A	3444	20	2051	
019-001918	109	3087		1123	
019-001980	109	3087		1123	
019-002691	5066A	3330	ZD-3.3		
019-002718	109	3087		1123	
019-002935	116	3016	504A	1104	
019-002964	177	3175	300	1122	
019-00301980	109	3087	86	1123	
019-003315	160	3007	245	2004	
019-003317	103	3862	8	2002	
019-003318	103	3862	8	2002	
019-003319	103	3862	8	2002	
019-003324	100	3005	1	2007	
019-003342	100	3005	1	2007	
019-003343	100	3005	1	2007	
019-003349	128	3444/123A	243	2051	
019-003411	140A	3061	ZD-10	562	
019-003415	158	3004/102A	53	2007	
019-003416	158	3004/102A	53	2007	
019-003420	116	3017B/117	504A	1104	
019-003485	175	3026	246	2020	
019-003637	128		243	2030	
019-003675-196	123A	3444	20	2051	
019-003675-203	123A	3444	20	2051	
019-003675-205	199	3245	62	2010	
019-003675-207	123A	3444	20	2051	
019-003675-231	159	3466	82	2032	
019-003675-232	159	3466	82	2032	
019-003675-234	159	3466	82	2032	
019-003675-246	123A	3444	20	2051	
019-003675-257	159		82	2032	
019-003676-334	519		514	1122	
019-003691	195A	3048/329	46		
019-003692	195A	3024/128	243		
019-003777	126	3007	52	2024	
019-003778	160	3007	245	2004	
019-003870-013	116	3016	504A	1104	
019-003870-020	116	3016	504A	1104	
019-003928	5012A	3778	ZD-6.0	561	
019-003929	108	3018	86	2038	
019-003931	159	3466	82	2032	
019-003932	123A	3444	20	2051	
019-003934	123A	3444	20	2051	
019-003935	224	3049	14	2041	
019-004094	199	3245	62	2010	
019-004111	123A	3444	20	2051	
019-004428-002	123A	3444	20	2051	
019-004558	159	3466	82	2032	
019-005006	123A	3444	20	2051	
019-005010	159	3466	82	2032	
019-005021	123A	3444	20	2051	
019-005043	109	3087		1123	
019-005045	5806	3848			
019-005157	161	3716	39	2015	
019-005179	159	3466	82	2032	
19-020-001	126		52	2024	
19-020-002	126		52	2024	
19-020-003	102A		53	2007	
19-020-005	126		52	2024	
19-020-007	102A		53	2007	
19-020-015	102	3722	2	2007	
19-020-019	213		254		
19-020-030	104	3719			

Industry Standard No.	ECG	SK	GE	RS 276-	MOTOR.
19-020-031	160	3006	245	2004	
19-020-032	160	3006	245	2004	
19-020-033	100	3005	1	2007	
19-020-034	102A	3004	53	2007	
19-020-035	102A	3004	53	2007	
19-020-036	102A	3004	53	2007	
19-020-037	108	3452	86	2038	
19-020-038	128	3048/329	243	2030	
19-020-043	123A	3444	20	2051	
19-020-043A	123A	3444	20	2051	
19-020-044	161	3452/108	86	2038	
19-020-045	175	3021/124	246	2020	
19-020-046	154		40	2012	
19-020-048	161	3452/108	86	2038	
19-020-050	128/401	3893/152	243	2030	
19-020-052	108	3452	86	2038	
19-020-056	152		66	2048	
19-020-058	123A	3444	20	2051	
19-020-066	152	3893	66	2048	
19-020-067	123A	3444	20	2051	
19-020-070	229	3018	61	2038	
19-020-071	107	3122	11	2015	
19-020-072	311	3039/316	277		
19-020-073	123A	3444	20	2051	
19-020-074	123A	3444	20	2051	
19-020-075	128	3024	243	2030	
19-020-076	195A	3048/329	46		
19-020-077	195A	3048/329	46		
19-020-078	237	3299	46		
19-020-079	703A	3157	IC-12		
19-020-100	129		244	2027	
19-020-101	186	3192	28	2017	
19-020-102	187	3193	29	2025	
19-020-114	290A	3466/159	82	2032	
19-020-115	312	3448	FET-1	2035	
19-020-44	108	3452	86	2038	
19-020071	123A	3444	20	2051	
19-040-002	116	3016	504A	1104	
19-040-003	116	3311	504A	1104	
19-040-004	116	3016	504A	1104	
19-076001	1003	3288	IC-43		
19-080-001	177	3100/519	300	1122	
19-080-002	116	3017B/117	504A	1104	
19-080-008	177	3100/519	300	1122	
19-080-009	109	3087		1123	
19-080-014	5010A	3776	ZD-5.1		
19-085005	109	3087		1123	
19-085010	116	3017B/117	504A	1104	
19-085016		3126	90		
19-085017	138A		ZD-7.5		
19-085018	109	3088		1123	
19-085022	116	3031A	504A	1104	
19-090-007	5063A	3837			
19-090-008	5074A		ZD-11.5		
19-090-008A	140A		ZD-10	562	
19-090-014	5010A	3776	ZD-5.1		
19-090-015	134A		ZD-3.6		
19-09234	941		IC-263	010	
19-09973-0	7401	7401			
19-09981-0	923		IC-259		
19-1	159	3466	82	2032	
19-10	159	3466	82	2032	
19-100001	125	3033	510,531	1114	
19-10298-00	941M		IC-265	007	
19-10415-00	923D		IC-260	1740	
19-130-001	781	3169	IC-223		
19-130-001(DIODE)	177		300	1122	
19-130-004	74145	74145		1828	
19-130-005	7490	7490		1808	
19-15840	128		243	2030	
19-19420	108	3452	86	2038	
19-2	159	3466	82	2032	
19-2-02616	123A	3444	20	2051	
19-20	159	3466	82	2032	
19-3	159	3466	82	2032	
19-30	159	3466	82	2032	
019-301980	109	3087		1123	
19-3349	128	3047	243	2030	
19-3415	158	3004/102A	53	2007	
19-3416	158	3004/102A	53	2007	
19-3692	128	3048/329	243	2030	
19-3934-643	128	3024	243	2030	
19-3935-641	190	3104A	217		
19-C	119		CR-2		
19A115024-P4	116	3017B/117	504A	1104	
19A115024-P6	125	3033	510,531	1114	
19A115056-P1	127	3035	25		
19A115061-P1	123A	3444	20	2051	
19A115061-P2	123A	3444	20	2051	
19A115077-P1	102	3004	2	2007	
19A115077-P2	102	3004	2	2007	
19A115086-P1	109	3087		1123	
19A115087-P1	126	3006/160	52	2024	
19A115089	5874		5032		
19A115094-P1	105	3012	4		
19A115098	126	3006/160	52	2024	
19A115098-P1	126	3006/160	52	2024	
19A115099-P1	126	3006/160	52	2024	
19A115100-P1	116	3017B/117	504A	1104	
19A115101	121	3009	239	2006	
19A115101-P1	121	3009	239	2006	
19A115102-P1	123A	3444	20	2051	
19A115103-P1	101	3861	8	2002	
19A115108-P1	123A	3444	20	2051	
19A115108-P2	123A	3444	20	2051	
19A115123-2	123A	3444	20	2051	
19A115123-P1	123A	3444	20	2051	
19A115123-P2	123A	3444	20	2051	
19A115129-2	103	3862	8	2002	
19A115129-P1	103	3862	8	2002	
19A115140-P1	160	3006	245	2004	
19A115140-P2	160	3006	245	2004	
19A115142-P1	123A	3444	20	2051	
19A115142-P2	123A	3444	20	2051	
19A115145-P3	116	3017B/117	504A	1104	
19A115145-P4	116	3017B/117	504A	1104	
19A115157-1	123A	3444	20	2051	
19A115167-2	123A	3444	20	2051	
19A115178-P1	159	3466	82	2032	
19A115178-P2	159	3466	82	2032	
19A115180-2	129	3025	244	2027	
19A115184-P1	121	3717	239	2006	
19A115192-P1	160	3006	245	2004	
19A115192-P2	160	3006	245	2004	
19A115200-P1	152	3054/196	66	2048	
19A115201-P1	101	3861	8	2002	
19A115201-P2	101	3861	8	2002	
19A115208	100	3721	1	2007	
19A115208-P1	100	3721	1	2007	
19A115208-P2	100	3721	1	2007	
19A115238-2	128		243	2030	
19A115238-P1	128		243	2030	
19A115245-P1	123A	3444	20	2051	
19A115245-P2	123A	3444	20	2051	

Industry Standard No.	ECG	SK	GE	RS 276-	MOTOR.
19A115249-1	108	3452	86	2038	
19A115250	519		514	1122	
19A115253-P1	123A	3444	20	2051	
19A115253-P2	123A	3444	20	2051	
19A115267P1	121	3009	239	2006	
19A115268	121	3717	239	2006	
19A115281-P1	102	3004	2	2007	
19A115300	128		243	2030	
19A115300-1	128	3024	243	2030	
19A115300-2	128	3024	243	2030	
19A115300-P1	186	3192	28	2017	
19A115300-P2	186	3192	28	2017	
19A115300-P3	186	3192	28	2017	
19A115301-P1	100	3005	1	2007	
19A115301-P2	100	3005	1	2007	
19A115304-2	128	3024	243	2030	
19A115315-P1	123A	3444	20	2051	
19A115315-P2	123A	3444	20	2051	
19A115322	112	3089	1N82A		
19A115322-P1	112	3089	1N82A		
19A115328-1	123A	3444	20	2051	
19A115330	123A	3444	20	2051	
19A115341P1	121	3009	239	2006	
19A115342-1	108	3452	86	2038	
19A115342-2	161	3019	39	2015	
19A115342-P1	123A	3444	20	2051	
19A115342-P2	123A	3444	20	2051	
19A115359-P1	123A	3444	20	2051	
19A115359-P2	123A	3444	20	2051	
19A115361-P1	121	3009	239	2006	
19A115362-P1	123A	3444	20	2051	
19A115362-P2	123A	3444	20	2051	
19A115371-1	177		300	1122	
19A115376	121	3717	239	2006	
19A115385-P1	121	3009	239	2006	
19A115410-P1	123A	3444	20	2051	
19A115410-P2	123A	3444	20	2051	
19A115440-1	108	3452	86	2038	
19A115440-2	108	3452	86	2038	
19A115441-1	108	3452	86	2038	
19A115458-P1	159	3466	82	2032	
19A115458-P2	159	3466	82	2032	
19A115460-P1	5545		MR-3		
19A115487-P1	105	3012	4		
19A115527	175	3026	246	2020	
19A115527-1	175		246	2020	
19A115527-P1	175	3538	246	2020	
19A115528-P1			ZD-6.2	561	
19A115528-P3	139A	3060	ZD-9.1	562	
19A11552B-P1	137A		ZD-6.2	561	
19A115531-P1	127	3025/129	25		
19A115540-P1	105	3012	4		
19A115546-P1	101	3861	8	2002	
19A115546-P2	101	3861	8	2002	
19A115548-P1			1	2007	
19A115552-P1			17	2051	
19A115552-P2			17	2051	
19A115552P1	123A	3444	20	2051	
19A115552P2	123A	3444	20	2051	
19A115553-P1			51	2004	
19A115554-P1			51	2004	
19A115554-P2			51	2004	
19A115556-P1			1	2007	
19A115561	121	3009	239	2006	
19A115561-1	121	3717	239	2006	
19A115562P2	159		82	2032	
19A115567-P1			51	2004	
19A115567-P2			51	2004	
19A115569-P1			504A	1104	
19A115569-P2	116	3017B/117	504A	1104	
19A115591P1	123A	3444	20	2051	
19A115591P2	123A	3444	20	2051	
19A115623-P1	124	3021	12		
19A115623-P2	124	3021	12		
19A115628-P1	160	3006	245	2004	
19A115628-P2	160		245	2004	
19A115635-1	160	3006	245	2004	
19A115635-P1			51	2004	
19A115636-P1	160	3006	245	2004	
19A115653-P1	159	3466	82	2032	
19A115653-P2	159	3466	82	2032	
19A115654-P1	159	3466	82	2032	
19A115654-P2	159	3466	82	2032	
19A115661P1	519		514	1122	
19A115665-P1	160	3006	245	2004	
19A115665-P2	160	3006	245	2004	
19A115666-1	108	3452	86	2038	
19A115673-P1	101	3861	8	2002	
19A115673-P2	101	3861	8	2002	
19A115674-P1	102A		53	2007	
19A115674-P2	102A	3004	53	2007	
19A115683-P1	5545		MR-3		
19A115688-P1	159	3466	82	2032	
19A115688-P2	159	3466	82	2032	
19A115706-1	159	3466	82	2032	
19A115706-2	159	3466	82	2032	
19A115706-P1	159	3466	82	2032	
19A115706-P2	159	3466	82	2032	
19A115720-1	123A	3444	20	2051	
19A115720-2	123A	3444	20	2051	
19A115728-1	123A	3444	20	2051	
19A115728-2	123A	3444	20	2051	
19A115747-P1	5545		MR-3		
19A115768-1	159	3466	82	2032	
19A115768-2	159	3466	82	2032	
19A115768-3	159	3466	82	2032	
19A115768-P1	159	3466	82	2032	
19A115768-P2	159	3466	82	2032	
19A115779P1	159	3466	82	2032	
19A115783-1	124		12		
19A115786	123A	3444	20	2051	
19A115786A	123A	3444	20	2051	
19A115810P1	311		277		
19A115817P1	124		12		
19A115818	181		75	2041	
19A115852P1	159	3466	82	2032	
19A115889-P1	128		243	2030	
19A115889-P2	128		243	2030	
19A115889-P3	128		243	2030	
19A115910P1	123A	3444	20	2051	
19A115925-1	108	3452	86	2038	
19A115944P1	123A	3444	20	2051	
19A115944P2	123A	3444	20	2051	
19A115976P1	129		244	2027	
19A116081	177		300	1122	
19A116118-1	152	3893	66	2048	
19A116118-2	152	3893	66	2048	
19A116118-I	152		66	2048	
19A116118P1	152	3893	66	2048	
19A116118P2	152	3893	66	2048	
19A116180-11	7454	7454			
19A116180-18	7486	7486		1827	
19A116180-24	7490	7490		1808	
19A116180-27	7492	7492		1819	

Industry Standard No.	ECG	SK	GE	RS 276-	MOTOR.
19A116180-29	7495	7495			
19A116180-7	7440	7440			
19A116180P1	7400	7400		1801	
19A116180P11	7454	7454			
19A116180P15	7473	7473		1803	
19A116180P16	7474	7474		1818	
19A116180P18	7486	7486		1827	
19A116180P2	7401	7401			
19A116180P20	7404	7404		1802	
19A116180P3	7402			1811	
19A116180P4	7410	7410		1807	
19A116180P5	7420	7420		1809	
19A116180P7	7440	7440			
19A116223P1	159	3466	82	2032	
19A116297P3	941		IC-263	010	
19A116297P3-10	941		IC-263	010	
19A116297P3-9	941		IC-263	010	
19A116375	153		69	2049	
19A116375P1	153		69	2049	
19A116408-1	159	3466	82	2032	
19A116445	709	3135	IC-11		
19A11644P1	708	3135/709	IC-10		
19A116549P1	909	3590	IC-249		
19A116623	912		IC-172		
19A116631P1	123A	3444	20	2051	
19A116753P1	181		75	2041	
19A116755P1	123A	3444	20	2051	
19A116761P1	130		14	2041	
19A116774-P1	123A	3444	20	2051	
19A116796-1	704	3023	IC-205		
19A116797-1	710	3102	IC-89		
19A116841P1	923		IC-259		
19A116865	123A	3444	20	2051	
19A123160-1	108	3452	86	2038	
19A123160-2	108	3452	86	2038	
19A126265-1	160		245	2004	
19A126265-2	160		245	2004	
19A126813	130	3027	14	2041	
19A126813A	130	3027	14	2041	
19A126826-P2	181		75	2041	
19A129207P1	123A	3444	20	2051	
19A13-4339	1115	3184			
19AI15527-1	175			2020	
19AR11	116	3017B/117	504A	1104	
19AR12	116	3017B/117	504A	1104	
19AR13-1	160	3006	245	2004	
19AR13-2	160	3006	245	2004	
19AR13-3	160	3006	245	2004	
19AR13-4	160	3006	245	2004	
19AR14-1	102A	3004	53	2007	
19AR14-2	102A	3004	53	2007	
19AR16-1	102A	3004	53	2007	
19AR16-2	102A	3004	53	2007	
19AR17	116	3017B/117	504A	1104	
19AR18	160	3006	245	2004	
19AR19-1	102A	3004	53	2007	
19AR19-2	102A	3004	53	2007	
19AR2	116		504A	1104	
19AR20	123A	3444	20	2051	
19AR21	106(4)	3118	21(4)		
19AR24	160	3006	245	2004	
19AR25	102A	3004	53	2007	
19AR26	102A	3004	53	2007	
19AR27	102A	3004	53	2007	
19AR29	113A	3119/113	6GC1		
19AR29-1	113A	3119/113	6GC1		
19AR3	109			1123	
19AR30	118	3066	CR-1		
19AR31	121	3009	239	2006	
19AR32	102A	3004	53	2007	
19AR34	116	3017B/117	504A	1104	
19AR35	124	3021	12		
19AR36	123A	34440123A	20	2051	
19AR4	114		6GD1	1104	
19AR4-1	114	3120	6GD1		
19AR5	116	3017B/117	504A	1104	
19AR6-1	103A	3010	59	2002	
19AR6-2	103A	3010	59	2002	
19AR6-3	103A	3010	59	2002	
19AR7-1	102A	3004	53	2007	
19AR7-2	102A	3004	53	2007	
19B200011-P1	5874	3500/5882	5032		
19B200011-P2	5874	3500/5882	5032		
19B200011-P3	5874	3500/5882	5032		
19B200011-P4	5874	3500/5832	5032		
19B200011-P5	116	3017B/117	504A	1104	
19B2000129-P1	100	3005	1	2007	
19B2000130-P1	160	3006	245	2004	
19B2000130-P2	160	3006	245	2004	
19B2000132-P1	102A	3004	53	2007	
19B2000132-P2	102A	3004	53	2007	
19B2000132-P3	102A	3004	53	2007	
19B2000132-P4	102A	3004	53	2007	
19B200054-P1	102A	3004	53	2007	
19B200061-P1	102A		53	2007	
19B200061-P2	102A	3004	53	2007	
19B200061-P3	102A	3004	53	2007	
19B200061-P4	102A	3004	53	2007	
19B200063-P1	102A	3004	53	2007	
19B200065-P1	101	3861	8	2002	
19B200065-P2	101	3861	8	2002	
19B200129-P1			1	2007	
19B200130-P1			51	2004	
19B200130-P2			51	2004	
19B200132-P1			2	2007	
19B200132-P2			2	2007	
19B200132-P3			2	2007	
19B200132-P4			2	2007	
19B200210-P1	102A	3004	53	2007	
19B200210-P2	102A	3004	53	2007	
19B200210-P3	102A	3004	53	2007	
19B200248-P1	5455	3597	MR-5	1067	
19B200248-P2	5455		MR-5	1067	
19B200248-P3	5455	3597	MR-5	1067	
19B200249-P1	177	3175	300	1122	
19B200249-P2	177	3175	300	1122	
19B200249-P3			300	1122	
19B200379-P1	139A	3060	ZD-9.1	562	
19C300073-P1	102	3004	2	2007	
19C300073-P2	102	3004	2	2007	
19C300073-P3	102	3004	2	2007	
19C300073-P4	102	3004	2	2007	
19C300073-P5	102	3004	2	2007	
19C300073-P6	102	3004	2	2007	
19C300074-P2	102	3004	2	2007	
19C300076-P1	116	3017B/117	504A	1104	
19C300076-P2	116	3017B/117	504A	1104	
19C300076-P3	116	3017B/117	504A	1104	
19C300076-P4	116	3017B/117	504A	1104	
19C300076-P5	116	3017B/117	504A	1104	
19C300076-P6	116	3017B/117	504A	1104	
19C300113-P1	121	3009	239	2006	
19C300114-P1	123A	3444	20	2051	
19C300114-P2	123A	3444	20	2051	
19C300114P1	123A	3444	20	2051	
19C300114P2	123A	3444	20	2051	
19C300114P3	123A	3444	20	2051	
19C300115-1	128		243	2030	
19C300115-P1	128	3024	243	2030	
19C300128-P1	102	3004	2	2007	
19C300128-P2	102	3004	2	2007	
19C300128-P3	102	3004	2	2007	
19C300128-P4	102		2	2007	
19C300128-P5	102	3722	2	2007	
19C300128-P6	102	3004	2	2007	
19C300128-P7	102	3004	2	2007	
19C300128-P8	102	3004	2	2007	
19C300138-P4	102	3004	2	2007	
19C300138-P8	102	3722	2	2007	
19C3001414P2				2051	
19C300216-P1	160	3006	245	2004	
19C300216-P2	160	3004/102A	245	2004	
19C300LL5-1				2030	
19L-4A82	102A		53	2007	
19P1	177	3175	300	1122	
19P2	116	3311	504A	1104	
19Q820	519		514	1122	
19QC17	107		11	2015	
19QC19	234		65	2050	
19Z6AF	5071A	3334	ZD-6.8	561	
19Z6F	5071A	3334	ZD-6.8	561	
020-00011	110MP	3709		1123(2)	
020-00012	110MP	3709		1123(2)	
020-00023	126	3008	52	2024	
020-00024	107	3018	11	2015	
020-00025	107	3018	11	2015	
020-00026	108	3452	86	2038	
020-00027	108	3452	86	2038	
020-00028	107	3018	11	2015	
020-00030	109	3088		1123	
020-00031		3126	90		
20-00229-001	108	3452	86	2038	
20-00444-001	108	3452	86	2038	
20-1	108	3452	86	2038	
020-1110-004		3114	21	2034	
020-1110-004C	159	3466	82	2032	
020-1110-005	234	3114/290	65	2050	
020-1110-006		3006	51	2004	
020-1110-008	128	3024	243	2030	
020-1110-009		3025	54	2027	
020-1110-010		3122	62	2010	
020-1110-011		3122	20	2051	
020-1110-012	199	3122	62	2010	
020-1110-013	199	3122	62	2010	
020-1110-014	129	3122	244	2027	
020-1110-016	312	3112	FET-1	2028	
020-1110-017		3122	20	2051	
020-1110-018		3112	FET-1	2028	
020-1110-021	312	3112	FET-1	2028	
020-1110-022	312			2028	
020-1110-025			53	2007	
020-1110-027	234	3114/290	65	2050	
020-1110-038	194	3275	220		
20-1111-002	130	3027	14	2041	
20-1111-003	130	3027	35	2041	
020-1111-004		3083	69	2049	
020-1111-005		3054	66	2048	
020-1111-007			14	2041	
020-1111-008	130	3027	14	2041	
020-1111-009	219		74	2043	
020-1111-017	129	3025	244	2027	
020-1111-018	128	3024	243	2030	
020-1111-019	130		14	2041	
020-1111-038	128		243	2030	
020-1111-080	219		74	2043	
020-1112-001	123A	3444	20	2051	
020-1112-002		3116	FET-2	2035	
020-1112-003	161	3716	39	2015	
020-1112-004	107	3122	11	2015	
020-1112-005	312	3112	FET-1	2028	
020-1112-006	312	3448	FET-1	2035	
020-1112-007	312			2028	
020-1112-008	312	3116	FET-2	2028	
020-1112-009	312			2028	
020-1114-003	725	3162	IC-19		
020-1114-005	725	3162	IC-19		
020-1114-006	718	3159	IC-8		
020-1114-007	703A		IC-12		
020-1114-008	703A		IC-12		
020-1114-009	720	9014	IC-7		
020-1114-013	725		IC-19		
020-1114-015	941M		IC-265	007	
020-114-007	703A	3157	IC-12		
020-114-008	703A	3157	IC-12		
020-114-009	720	9014	IC-7		
20-16-3			1N34AS	1123	
20-161002	177		300	1122	
20-1680-143	116	3017B/117	504A	1104	
20-1680-174	128	3024	243	2030	
20-1680-175	109	3087		1123	
20-1680-189	158	3004/102A	53	2007	
20-22-08	116	3311	504A	1104	
20-JLM	159		53	2032	
20A-70	109			1123	
20A-90	109			1123	
20A-90H	109			1123	
20A-90M	109			1123	
20A0007	102A	3004	53	2007	
20A0009	102A	3004	53	2007	
20A0015	102A	3004	53	2007	
20A0017	121	3009	239	2006	
20A0041	121	3009	239	2006	
20A0042	121	3009	239	2006	
20A0053	123A	3444	20	2051	
20A0054	116	3017B/117	504A	1104	
20A0055	128		243	2030	
20A0059	189	3083/197	218	2026	
20A0060	189	3083/197	218	2026	
20A0073	123A	3444	20	2051	
20A0074	121	3009	239	2006	
20A0075	129	3025	244	2027	
20A0076	128	3024	243	2030	
20A1	5801	9004		1142	1N4002
20A10	5809	9010			1N4007
20A10849	123A	3444	20	2051	
20A10F					SPECIAL
20A11	177	3100/519	300	1122	
20A13	177	3175	300	1122	
20A2	5802			1143	1N4003
20A3	5803	9006		1144	1N4004
20A4	5804	9007		1144	1N4004
20A5	5805	9008			1N4005
20A6	5806	3848			1N4005
20A6F					MR1386
20A70	109	3087		1123	
20A79	109	3087		1123	
20A8	5808	9009			1N4006
20A8F					SPECIAL

Industry Standard No.	ECG	SK	GE	RS 276-	MOTOR.
20A90	109	3088	1N60	1123	
20A90LP	109		1N60	1123	
20A90M	109	3088		1123	
20A90MLP	110MP	3088	1N60	1123	
20A90P	110MP	3088			
20A90Z	110MP	3088		1123(2)	
20A9M	109	3087		1123	
20AS	116	3016	504A	1104	
20B					MR1122
20B1M	703A		IC-12		
20B2M	703A	3157	IC-12		
20B3AH	703A	3157	IC-12		
20B409	135A	3056	ZD-5.1		
20B410	135A	3056	ZD-5.1		
20BR					1N3881
20BS	116	3016	504A	1104	
20C	116	3017B/117	504A	1104	
20C1	5801	9004		1142	
20C10	5808	9009			
20C2	5802	9005		1143	
20C3	5803	9006		1144	
20C4	5804	9007		1144	
20C5	5805	9008			
20C6	5806	3848-5806			
20C71	102A	3123	53	2007	
20C72	102A	3004	53	2007	
20C9	5808	9009			
20CTQ030					MBR2530CT
20CTQ035					MBR2535CT
20CTQ040					MBR2545CT
20CTQ045					MBR2545CT
20D1					1N5392
20D10					1N5399
20D2					1N5393
20D3					1N5394
20D4					1N5395
20D5					1N5396
20D6					1N5397
20D8					1N5398
20E-54-5411	5695	3509			
20E1114	358		42		
20F1	5834	3051/156			
20F10	5914				MR2001S
20F20	5916				MR2002S
20F30	5918				MR2004S
20F40	5920				MR2004S
20FQ020					MBR3520
20FQ030					MBR3535
20FQ035					MBR3535
20FQ040					MBR3545
20FQ045					MBR3545
20H	116	3016	504A	1104	
20H3	5874	3500/5882	5032		
20H3P	5854	3017B/117	5016		MR1122
20H3PN	5855	3517/5883	5017		
20HA3	5994	3501	5104		
20HR3P	5818				1N3881
20J2	5874	3500/5882	5032		
20J3P	5874	3500/5882	5032		
20JH3	6026	7226	5128		
20K	116	3016	504A	1104	
20M	116	3016	504A	1104	
20M10	6022		5128		
20M100	6044	7244	5140		
20M20	6026	7226	5128		
20M30	6030		5132		
20M40	6034	7234	5132		
20M50	6038		5136		
20M60	6040	7240	5136		
20M80	6042		5140		
20MC	160	3007	245	2004	
20N1	116	3016	504A	1104	
20N2	5874	3500/5882	5032		
20N3	5874	3500/5882	5032		
20P1S	5542	6642			
20P2	5874	3500/5882	5032		
20P3	5874	3517/5883	5032		
20Q3	5897		5065		
20R2	5980	3610	5096		
20R3	312	3608/5990	5100		
20S3	5990	3608	5100		
20S4	5990	3608	5100		
20S5	5834	3051/156	5004		
20TB1	5542	6642			
20V-HG	102A	3123	53	2007	
20Z6AF	138A	3059	ZD-7.5		
20Z6F	138A	3059	ZD-7.5		
021	128		243	2030	
21-0101	192		63	2030	
021-0121-00	123A	3444	20	2051	
021-0137-00	128		243	2030	
021-0224-00	129		244	2027	
21-1	123A	3122	20	2051	
21-1L	123A	3124/289	20	2051	
21-2			11	2015	
21-28	121	3009	239	2006	
21-32	126	3008	52	2024	
21-33	126	3008	52	2024	
21-34	102	3004	2	2007	
21-35	124	3021	12		
21-36	102	3003	2	2007	
21-37	102	3004	2	2007	
21-4			11	2015	
21-6			11	2015	
21-606-0001	177		300	1122	
21-606-0001-00H	177		300	1122	
21-608-4148-006	519		514	1122	
21-609-3595-009	177		300	1122	
21-7			11	2015	
21-810	116	3016	504A	1104	
21-810-2	116	3016	504A	1104	
21-B1	718	3159	IC-8		
21/3	114	3120	6GC1	1104	
21/3.92	114	3120	6GD1		
21A001-00			504A	1104	
21A001-000	116(3)	3311	504A(3)	1104	
21A002	113A	3119/113	6GC1		
21A002-000	113A	3119/113	6GC1		
21A004-000	506	3843	511	1114	
21A005	120	3110	CR-3		
21A005-000	102A	3004	53	2007	
21A006-000	116	3110/120	504A	1104	
21A007-000	116	3031A	504A	1104	
21A008-000	116		504A	1104	
21A008-001	116	3016	504A	1104	
21A008-002		3016	511	1114	
21A008-003	116	3016	504A	1104	
21A008-008	177	3100/519	300	1122	
21A008-016	120	3110	CR-3		
21A009	177	3100/519	300	1122	
21A009-000	109	3091		1123	
21A009-002	110MP	3091		1123	
21A009-008	177	3100/519	300	1122	
21A009-009	177	3100/519	300	1122	

Industry Standard No.	ECG	SK	GE	RS 276-	MOTOR.
21A015-001	102A	3004	53	2007	
21A015-002	155		43		
21A015-003	131	3052	44	2006	
21A015-004	108	3452	86	2038	
21A015-005	103A	3010	59	2002	
21A015-006	102A	3004	53	2007	
21A015-008	159	3466	82	2032	
21A015-008A	159	3466	82	2032	
21A015-009	159	3466	82	2032	
21A015-009A	159	3466	82	2032	
21A015-011	159	3466	82	2032	
21A015-011A	159	3466	82	2032	
21A015-012	159	3466	82	2032	
21A015-012A	159	3466	82	2032	
21A015-013	123A	3444	20	2051	
21A015-014	108	3452	86	2038	
21A015-016	108		86	2038	
21A015-018	128	3024	243	2030	
21A015-019	128	3024	243	2030	
21A015-020	123A	3444	20	2051	
21A015-021	155	3839	43		
21A015-022	131		44	2006	
21A015-025	159	3466	82	2032	
21A015-026	128	3024	243	2030	
21A015-027	123A	3444	20	2051	
21A020-001	116	3016	504A	1104	
21A020-005	109	3091		1123	
21A020-006	116	3017B/117	504A	1104	
21A037-001	5021A	3787	ZD-12	563	
21A037-003	142A	3032A	ZD-12	563	
21A037-006	5100A		ZD-150		
21A037-008	142A	3062	ZD-12	563	
21A037-009	5079A	3335	ZD-20		
21A037-012	5072A		ZD-8.2	562	
21A037-016	5096A	3346	ZD-100		
21A037-017	5076A	9022	ZD-16	564	
21A037-018	143A	3750	ZD-13	563	
21A037-020	5074A	3092/141A	ZD-11	563	
21A037-021	5098A		ZD-130		
21A038-000	102A	3004	53	2007	
21A039-000	102A	3004	53	2007	
21A040-000	102A		53	2007	
21A040-001	109			1123	
21A040-003	108	3452	86	2038	
21A040-004	108	3452	86	2038	
21A040-005	102A	3004	53	2007	
21A040-007	108	3452	86	2038	
21A040-010	108	3452	86	2038	
21A040-014	102A		53	2007	
21A040-015	5527	3834/132	FET-1	2035	
21A040-016	108	3452	86	2038	
21A040-017	108	3452	86	2038	
21A040-019	108	3452	86	2038	
21A040-020	123A	3124/289	20	2051	
21A040-021	102A	3004	53	2007	
21A040-022	102A	3004	53	2007	
21A040-023	107		11	2015	
21A040-024	107		11	2015	
21A040-025	107		11	2015	
21A040-031	126	3008	52	2024	
21A040-032	123A	3444	20	2051	
21A040-033	123A	3444	20	2051	
21A040-033A	123A	3444	20	2051	
21A040-034	123A	3444	20	2051	
21A040-035	176	3123	80		
21A040-036	158	3004/102A	53	2007	
21A040-037	123A	3444	20	2051	
21A040-045	229	3018	61	2038	
21A040-046	233	3018	210	2009	
21A040-047	233	3018	210	2009	
21A040-049	129	3114/290	244	2027	
21A040-050	129	3114/290	244	2027	
21A040-051	128	3122	243	2030	
21A040-051(DIO)			ZD-9.1	562	
21A040-051(XSTR)			63	2030	
21A040-052	152	3054/196	66	2048	
21A040-053	229	3018	61	2038	
21A040-054	107	3018	11	2015	
21A040-055	107	3018	11	2015	
21A040-056	123A	3444	20	2051	
21A040-057	102A	3004	53	2007	
21A040-058	102A	3004	53	2007	
21A040-059	159	3114/290	82	2032	
21A040-060	158	3004/102A	53	2007	
21A040-061	158	3010	53	2007	
21A040-063	107	3018	11	2015	
21A040-064	199		62	2010	
21A040-065	199		62	2010	
21A040-066	199	3124/289	62	2010	
21A040-067	199	3124/289	62	2010	
21A040-068	1075A	3877			
21A040-077	123A	3444	20	2051	
21A040-078	123A	3444	20	2051	
21A040-079	102A	3004	53	2007	
21A040-081	158		53	2007	
21A040-082	199	3020/123	62	2010	
21A040-083	199		62	2010	
21A040-091	229	3018	61	2038	
21A040-092	123A	3444	20	2051	
21A040-36	102A		53	2007	
21A040-37	123A	3444	20	2051	
21A040-44	116	3311	504A	1104	
21A040-54	229		61	2038	
21A045-000	160		245	2004	
21A048-000	160	3006	245	2004	
21A049-000	160	3006	245	2004	
21A050-000	160	3006	245	2004	
21A050-001	160	3006	245	2004	
21A050-004	108	3452	86	2038	
21A051-000	100	3721	1	2007	
21A053-000	102A	3004	53	2007	
21A054-000	102A	3004	53	2007	
21A055-000	102A	3004	53	2007	
21A062-000	106	3984	21	2034	
21A063-000			2	2007	
21A064-000	121	3014	239	2006	
21A074-000	102A	3004	53	2007	
21A079-000	116	3017B/117	504A	1104	
21A097-000	121	3009	239	2006	
21A101-001	706	3101	IC-43		
21A101-001(RECT)			504A	1104	
21A101-002	711	3070	IC-207		
21A101-004	1047		IC-116		
21A101-005	1046	3471	IC-118		
21A101-006	1089*		IC-120		
21A101-007	1048		IC-122		
21A101-008	1077		IC-124		
21A101-009	1076		IC-125		
21A101-010	1091*		IC-126		
21A101-011	1084		IC-127		
21A101-012	1086		IC-142		
21A101-013	OBS-NLA		IC-166		
21A101-014	749		IC-97		
21A101-015	747	3279	IC-218		

Industry Standard No.	ECG	SK	GE	RS 276-	MOTOR.
21A101-016	715	3076	IC-6		
21A101-017	714	3075	IC-4		
21A101-018	1050	3475	IC-123		
21A101-1	706	3101	IC-43		
21A101-2	711	3070	IC-207		
21A102-001	116	3017B/117	504A	1104	
21A102-002	170	3017B/117	504A	1104	
21A103-005		3126	90		
21A103-006	110MP	3088	1N60	1123(2)	
21A103-007	116	3311	504A	1104	
21A103-010	109	3087		1123	
21A103-011	109			1123	
21A103-012	125	3016	510,531	1114	
21A103-013	125	3033	510,531	1114	
21A103-015	125	3033	510,531	1114	
21A103-016	110MP	3088	1N60	1123	
21A103-017	109	3087		1123	
21A103-018	116	3031A	504A	1104	
21A103-019	109	3087		1123	
21A103-021	116	3311	504A	1104	
21A103-022	109	3087		1123	
21A103-044	116	3311	504A	1104	
21A103-045	5074A	3092/141A	ZD-11.0	563	
21A103-046	109	3087		1123	
21A103-047		3126	90		
21A103-048	110MP	3088	1N60	1123(2)	
21A103-049	138A	3059	ZD-7.5		
21A103-050	116	3031A	504A	1104	
21A103-052	109	3088		1123	
21A103-055	109	3088		1123	
21A103-058	116	3311	504A	1104	
21A103-060		3126	90		
21A103-064	139A	3060	ZD-9.1	562	
21A103-065	177	3100/519	300	1122	
21A103-069		3126	90		
21A103-070	116	3311	504A	1104	
21A103-104	125	3311	510,531	1114	
21A103-064			ZD-9.1	562	
21A105-001	108	3452	86	2038	
21A105-004	128	3024	243	2030	
21A105-006	128	3024	243	2030	
21A108-001	177	3100/519	300	1122	
21A108-002	139A	3060	ZD-9.1	562	
21A108-003	519	3100	514	1122	
21A108-004	177	3100/519	300	1122	
21A109-001	109	3088		1123	
21A109-002	109	3088		1123	
21A109-003	110MP			1123(2)	
21A109-022	109	3088		1123	
21A110-001	116	3017B/117	504A	1104	
21A110-002	116	3017B/117	504A	1104	
21A110-003	116	3116	504A	1104	
21A110-004	125	3100/519	510,531	1114	
21A110-005	506	3843	511	1114	
21A110-006	116	3125	511	1114	
21A110-007	116	3017B/117	504A	1104	
21A110-008	506	3843	511	1114	
21A110-009	506	3843	511	1114	
21A110-010	116		504A	1104	
21A110-012	125	3311	504A	1104	
21A110-013	506	3125	511	1114	
21A110-014	506	3843	511	1114	
21A110-071	525	3925	511	1114	
21A110-072	125	3081	510	1114	
21A111-001	177	3100/519	300	1122	
21A111-002	601	3463	300	1122	
21A112-001	159	3466	82	2032	
21A112-002	129	3114/290	244	2027	
21A112-003	159	3114/290	82	2032	
21A112-004	193	3138	67	2023	
21A112-006	506	3039/316	39	1114	
21A112-007	107	3018	11	2015	
21A112-010	161	3117	39	2015	
21A112-013	123A	3444	20	2051	
21A112-015	123A	3444	20	2051	
21A112-017	123A	3444	20	2051	
21A112-018	123A	3444	20	2051	
21A112-019	199	3122	61	2010	
21A112-020	123A	3444	20	2051	
21A112-023	164	3133	37		
21A112-025	124	3021	12		
21A112-029	124	3021	12		
21A112-031	162	3438	35		
21A112-033	124	3079	12		
21A112-036	165	3115	259		
21A112-045	199	3124/289	62	2010	
21A112-046	199	3124/289	212	2010	
21A112-047	159	3466	82	2032	
21A112-048	187	3193	29	2025	
21A112-049	186	3192	28	2017	
21A112-050	123A	3444	20	2051	
21A112-058	199	3124/289	212	2010	
21A112-062	123A	3444	20	2010	
21A112-063	123A	3444	61	2010	
21A112-065	159	3466	82	2032	
21A112-070	198	3103A/396	251		
21A112-071	198	3220	251		
21A112-074	194	3275	220		
21A112-075	159	3466	82	2032	
21A112-084	229	3018	61	2038	
21A112-085	123A	3444	20	2051	
21A112-086	108	3452	86	2038	
21A112-087	108	3452	86	2038	
21A112-088	123A	3444	20	2051	
21A112-089	123A	3444	20	2051	
21A112-090	123A	3444	20	2051	
21A112-091	123A	3444	20	2051	
21A112-092	123A	3444	20	2051	
21A112-093	159	3466	82	2032	
21A112-094	153	3083/197	69	2049	
21A112-095	152	3054/196	66	2048	
21A112-096	124		12		
21A112-098	198	3220	251		
21A112-099	198	3103A/396	251		
21A112-100	159	3466	82	2032	
21A112-101	123A	3444	20	2038	
21A112-102	123A	3466/159	221	2032	
21A112-103	165	3115	38		
21A112-104	123A	3444	20	2051	
21A112-107	162	3438	35		
21A112-124	283	3027/130	14	2041	
21A112-125	292	3441			
21A112095			66	2048	
21A113-002	312	3448	FET-1	2035	
21A118-008	198	322	251		
21A118-029	152	3893	66	2048	
21A118-031	289A		268	2038	
21A118-032	193	3114/290	67	2023	
21A118-049	152	3197/235	216	2053	
21A118-063	291	3440			
21A118-124	283	3438	14		
21A119-005	109	3087	1N34AS	1123	
21A119-008	143A		ZD-13	563	
21A119-030	125	3311	504A	1104	

Industry Standard No.	ECG	SK	GE	RS 276-	MOTOR.
21A119-040	605	3864			
21A119-041	136A	3057	ZD-5.6	561	
21A119-068	506	3843	511	1114	
21A119-073	5072A		ZD-8.7	562	
21A120-001	749	3168	IC-97		
21A120-002	1109	3279/747	IC-218		
21A120-008	712	3072	IC-2		
21A2	113A	3119/113	6GC1		
21A3	118	3066	CR-1		
21A4	119	3109	CR-2		
21A404-066	199	3245	62	2010	
21A500	120	3110	CR-3		
21A500-000	120	3110	CR-3		
21A6	116	3016	504A	1104	
21A7	116	3016	504A	1104	
21B-14	116	3016	504A	1104	
21B-17	116		504A	1104	
21B1M	718	3159	IC-8		
21B1Z	718	3159	IC-8		
21B2Z	718	3159	IC-8		
21FQ030					MBR3535
21FQ035					MBR3535
21FQ040					MBR3545
21FQ045					MBR3545
21K60	109	3087		1123	
21M006	126	3006/160	52	2024	
21M007	100	3006/160	1	2007	
21M018		3126	90		
21M020			22	2032	
21M022	159	3466	82	2032	
21M025	187	3193	29	2025	
21M026	187	3193	29	2025	
21M027	193	3138	67	2023	
21M028	187	3193	29	2025	
21M084	123A	3444	20	2051	
21M085	123A	3444	20	2051	
21M086	123A	3444	20	2051	
21M087	199	3245	62	2010	
21M091	199	3018	62	2010	
21M093	229	3018	11	2015	
21M094	107	3018	11	2015	
21M095	199	3018	62	2010	
21M096	199	3245	62	2010	
21M099	229	3018	11	2015	
21M122	123A	3444	20	2051	
21M123	123A	3444	20	2051	
21M124	199	3124/289	62	2010	
21M125	123A	3444	20	2051	
21M137	199	3124/289	62	2010	
21M138	199	3124/289	65	2050	
21M139	123A	3444	20	2051	
21M140	107	3018	11	2015	
21M146	123A	3444	20	2051	
21M149	123A	3444	20	2051	
21M150	123A	3444	20	2051	
21M151	107	3018	11	2015	
21M152	107	3018	11	2015	
21M153	107	3018	11	2015	
21M154	107	3018	11	2015	
21M160	199	3122	62	2010	
21M161	199	3018	62	2010	
21M170	199	3124/289	62	2010	
21M174	199	3124/289	62	2010	
21M178	107	3018	11	2015	
21M179	107	3018	11	2015	
21M180	186	3192	28	2017	
21M181	186	3192	28	2017	
21M182	107	3020/123	11	2015	
21M183	186	3192	28	2017	
21M184	186	3192	28	2017	
21M185	128	3024	243	2030	
21M186	123A	3444	20	2051	
21M188	107	3018	11	2015	
21M192	192	3137	63	2030	
21M193	192	3137	63	2030	
21M196	312	3834/132	FET-1	2035	
21M200	123A	3444	20	2051	
21M205	123A	3444	20	2051	
21M214	139A	3060	ZD-9.1	562	
21M224	312	3834/132	FET-1	2035	
21M228		3088	18	2030	
21M248	116	3311	504A	1104	
21M283	116	3016	504A	1104	
21M286	186	3192	28	2017	
21M288	109	3088		1123	
21M288(DIODE)	109			1123	
21M289	110MP	3087	1N34AS	1123(2)	
21M302	116	3016	504A	1104	
21M307	139A	3061/140A	ZD-9.1	562	
21M312	116	3017B/117	504A	1104	
21M315	116	3311	504A	1104	
21M316		3126	90		
21M317	116	3311	504A	1104	
21M323	109			1123	
21M325	110MP	3087	IN34AS	1123(2)	
21M330	177	3100/519	300	1122	
21M345	187	3193	29	2025	
21M355	159	3466	82	2032	
21M366	123A	3444	20	2051	
21M367	186	3192	28	2017	
21M369	186	3192	28	2017	
21M37	199	3124/289			
21M386	116	3100/519	504A	1104	
21M386(PWR)	116		504A	1104	
21M387	128	3018	243	2030	
21M395	187	3193	29	2025	
21M408	199	3124/289	62	2010	
21M412	312	3448	FET-1	2035	
21M415	177	3100/519	300	1122	
21M416	116	3311	504A	1104	
21M417	116	3311	504A	1104	
21M419	116	3017B/117	504A	1104	
21M432	109	3088		1123	
21M432(REG)	143A		ZD-13	563	
21M433	116	3031A	504A	1104	
21M434	116	3311	504A	1104	
21M435	116	3311	504A	1104	
21M436	116	3311	504A	1104	
21M437	116	3017B/117	504A	1104	
21M443	187	3193	29	2025	
21M446	199	3124/289	62	2010	
21M448	128	3024	243	2030	
21M455	128	3122	243	2030	
21M459	193	3138	67	2023	
21M465	185	3084	58	2025	
21M466	184	3054/196	57	2017	
21M469	116	3031A	504A	1104	
21M476	108	3452	86	2038	
21M481	108	3452	86	2038	
21M485	1142	3485	IC-128		
21M487	116	3311	504A	1104	
21M488	123A	3444	20	2051	
21M492		3126	90		
21M493	140A	3061	ZD-10	562	

Industry Standard No.	ECG	SK	GE	RS 276-	MOTOR.
21M502		3126	90		
21M506	799	3238	IC-34		
21M519	116	3311	504A	1104	
21M520	123A	3444	20	2051	
21M526		3126	90		
21M532	1006	3358	IC-38		
21M534	312	3448	FET-1	2035	
21M541	297	3122	271	2030	
21M545	116	3311	504A	1104	
21M550	199	3124/289	62	2010	
21M556	186	3192	28	2017	
21M562	177	3100/519	300	1122	
21M562(DIODE)	177		300	1122	
21M563	123A	3444	20	2051	
21M568	110MP	3088	1N60	1123(2)	
21M577	229	3018	61	2038	
21M578	123A	3444	20	2051	
21M579	123A	3444	20	2051	
21M581	159	3466	82	2032	
21M582	1124		720		
21M583		3126	90		
21M584	139A	3060	ZD-9.1	562	
21M585	139A	3060	ZD-9.1	562	
21M586	178MP	3100/519	300(2)	1122(2)	
21M588	722	3161	IC-9		
21M589	812		IC-242		
21M590	116		504A	1104	
21M594	109	3087		1123	
21M599	722		IC-9		
21M60	123A	3444	20	2051	
21M600	722		IC-9		
21M603	199	3124/289	62	2010	
21M604	234		65	2050	
21M605	123A	3444	20	2051	
21M606	192	3137	63	2030	
21M623	199	3122			
21M028	187			2025	
21MW132			2	2007	
21R05S	5980	3610			
21R10S	6002	7202	5108		
21R1S	5982	3609/5986	5096		
21R2S	5986	3609	5096		
21R3S	5988	3608/5990	5100		
21R4S	5990	3608	5100		
21R5S	5992	3501/5994	5104		
21R6S	5994	3501	5104		
21R8S	5998		5108		
21RC10	5522	6622			
21RC20	5524	6624			
21RC30	5526	6627/5527			
21RC40	5527	6627			
21RC50	5528	6629/5529			
21RC60	5529	6629			
21RC80	5531	6631			
21Z6AF	5072A	3136	ZD-8.2	562	
21Z6F	5072A	3136	ZD-8.2	562	
022	177		300	1122	
22-001001	123A	3122	20	2051	
22-001002	108	3452	86	2038	
22-001003	108	3452	86	2038	
22-001004	108	3452	86	2038	
22-001005	108	3452	86	2038	
22-001006	123A	3444	20	2051	
22-001007	123A	3444	20	2051	
22-001008	175	3538	246	2020	
22-001009	175	3538	246	2020	
22-001010	159	3466	82	2032	
22-002001	131	3052	44	2006	
22-002006	102	3123	2	2007	
22-002007	102A	3004	53	2007	
22-002008	131	3052	44	2006	
22-002009	131	3052	44	2006	
22-004003	109	3091		1123	
22-004004	125	3033	510,531	1114	
022-006500	123	3020	20	2051	
022-009600			20	2051	
022-0163-00	519		514	1122	
22-1	107	3124/289	IC-12	2015	
22-1-005		3088	1N60	1123	
22-1-044		3175	300	1122	
22-1-075		3311	504A	1104	
22-1-129		3088	1N60	1123	
22-1-131		3060	ZD-9.1		
22-1-132		9021	ZD-6.0		
22-1-138			504A	1104	
22-1-5			1N60	1123	
22-1-70			300	1122	
022-1110-005C	159		82	2032	
022-2823-001	156	3051	512		
022-2823-002	139A	3060	ZD-9.1	562	
022-2823-003	109	3088	1N60	1123	
022-2823-004	177	3100/519	300	1122	
022-2823-005	519	3100	300	1122	
022-2823-006	109	3087	1N34AS	1123	
022-2823-007	109	3088	1N60	1123	
022-2823-008	109	3087	1N34AS	1123	
022-2823-010	177	3175			
022-2823-011	116	3017B/117	504A	1104	
022-2823-501	612	3325	103-176		
022-2823-503	614	3327			
022-2844-001	1169	3708			
022-2844-002	123AP	3708/1169	IC-12		
022-2844-501	988	3973			
022-2876-002	103A	3835	59	2002	
022-2876-003	123A	3245/199	61	2038	
022-2876-004	159	3114/290	221		
022-2876-005	312	3112	FET-1	2028	
022-2876-006	312	3834/132	FET-2	2035	
022-2876-007	123AP	3356	211	2016	
022-2876-008	222	3050/221	FET-4	2036	
022-2876-009	312	3116	FET-2	2035	
022-2876-010	295	3253			
022-2876-011	235	3239/236	215		
022-2876-012	236	3197/235	216	2053	
022-2876-013	221	3065/222			
022-2876-014	152	3893			
022-2876-501	159	3114/290			
022-2876-502	312	3112			
022-2876-503	312	3834/132			
022-2876-504	312	3834/132			
022-2876-505	312	3448			
022-2876-506	236	3239			
022-3504-040	102	3004	2	2007	
022-3504-060	102	3004	2	2007	
022-3505-910	102	3004	2	2007	
022-3511-770	160	3007	245	2004	
022-3511-780	160	3007	245	2004	
022-3511-790	160	3007	245	2004	
022-3511780				2004	
022-3511790				2004	
022-3516-380	126	3007	52	2024	
022-3640-050	121	3009	239	2006	
022-3640-080	108	3452	86	2038	
022-3640-081	195A	3048/329	46		
022-3640-082	195A	3048/329	46		
022-3640-253	158	3004/102A	53	2007	
022-3901-001	109	3087		1123	
022-3902-001	109	3087		1123	
022-3905-001	116	3017B/117	504A	1104	
022-5311-770	126	3006/160	52	2024	
022-5311-780	126	3006/160	52	2024	
022-5311-798			52	2024	
022-5511-790	126	3006/160			
022.3504-040	102		2	2007	
022.3504-060	102		2	2007	
022.3505-910	102		2	2007	
022.3511-770	160		245	2004	
022.3511-780	160		245	2004	
022.3511-790	160		245	2004	
022.3516-380	126		52	2024	
022.3640-050	121		239	2006	
022.3640-080	108	3452	86	2038	
022.3640-081	195A		46		
022.3640-082	195A		46		
022.3901-001	109			1123	
022.3902-001	109			1123	
022.3905-001	116		504A	1104	
022A	703A	3157	IC-12		
22A001-17	605	3864	504A	1104	
22B1B	739	3235	IC-30		
022D	116	3031A	504A	1104	
22E89	176	3845	80		
22N1215				2024	
22N1319				2007	
22R22	5986	3609	5096		
22RC10	5542	6642	MR-3		
22RC15	5543		MR-3		
22RC2	5540	6642/5542	MR-3		
22RC20	5543		MR-3		
22RC25	5544		MR-3		
22RC30	5544		MR-3		
22RC40	5545		MR-3		
22RC5	5541	6642/5542	MR-3		
22Z6AF	139A	3060	ZD-9.1	562	
22Z6F	139A	3060	ZD-9.1	562	
23	123A	3444	20	2020	
23-0003	5802	3016	504A	1104	
23-0004	116	3017B/117	504A	1104	
23-0010	116	3500/5882	504A	1104	
23-0017	116	3017B/117	504A	1104	
23-0018	116	3017B/117	510,531	1114	
23-001R03A10	519	3100	514	1122	
23-001R03AA10	519	3100	514	1122	
23-1	159	3466	82	2032	
23-10	159	3466	82	2032	
23-2	159	3466	82	2032	
23-20	159	3466	82	2032	
23-3	159	3466	82	2032	
23-30	159	3466	82	2032	
23-5006	102	3004			
23-5007	102	3004			
23-5009	179	3009	239	2006	
23-5014	102A	3004	53	2007	
23-5017	102A	3004	53	2007	
23-5020	289A	3444/123A	20	2051	
23-5021	289A	3444/123A	20	2051	
23-5022	289A	3444/123A	20	2051	
23-5023	289A	3444/123A	20	2051	
23-5024	289A	3444/123A	20	2051	
23-5025	289A	3444/123A	20	2051	
23-5026	289A	3444/123A	20	2051	
23-5027	289A	3444/123A	20	2051	
23-5029	289A	3444/123A	20	2051	
23-5031	175	3026	246	2020	
23-5033	123A	3444	20	2051	
23-5034	176	3845	80		
23-5035	130	3027	14	2041	
23-5037	128/427	3024/128	18		
23-5038	130	3027	14	2041	
23-5039	128	3024	243	2030	
23-5041	130	3027	14	2041	
23-5042	121	3014	239	2006	
23-5044	172A	3156	64		
23-5045	129/427	3466/159	82	2032	
23-5051	225	3045			
23-5052	123	3444	20	2051	
23-6001-16			53	2007	
23-6001-17			53	2007	
23-6001-20			53	2007	
23-6001-21			53	2007	
23-6001-23			53	2007	
23-LLB	159	3466	82	2032	
23-PT274-120	161	3132	39	2015	
23-PT274-121	123A	3444	20	2051	
23-PT274-122	159	3466	82	2032	
23-PT274-123	161	3132	39	2015	
23-PT274-125	177	3170/731	300	1122	
23-PT275-121	161	3039/316	39	2015	
23-PT275-122	108	3452	86	2038	
23-PT275-124	112	3089	1N82A		
23-PT283-122	161	3132	39	2015	
23-PT283-124	161	3132	39	2015	
23-PT283-125	177	3170/731	300	1122	
23-PT284-122	160	3006	245	2004	
23-PT284-123	160	3006	245	2004	
23B-210-025	121	3009	239	2006	
23B-210-230-2		3034	25		
23B-210067-001	358A		41		
23B-210067-002	358A		41		
23B114044	123A	3444	20	2051	
23B114053	128	3024	243	2030	
23B114054	128	3024	243	2030	
23B210679-1	118	3066	CR-1		
23BSC101	116	3016	504A	1104	
23C025B	5540	6642/5542			
23C025C	5540	6642/5542			
23C050B	5541	6642/5542			
23C050C	5541	6642/5542			
23C10B	5542	6642			
23C10C	5542	6642			
23C50B	5546		C137E		
23C60B	5547		C137M		
23D		3156	64		
23E001-1	123A	3444	20	2051	
23J2	177	3311	300	1122	
23TB1	5542	6642			
23Z6AF	140A	3061	ZD-10	562	
23Z6F	140A	3061	ZD-10	562	
024	218	3083/197	234		
24(SHARP)	5082A		ZD-25		
24-0003714-1	123A	3444	20	2051	
24-000451	123A	3444	20	2051	
24-000452	128		243	2030	
24-000457	123A	3444	20	2051	
24-000653-1	123A	3444	20	2051	
24-001326	199	3245	62	2010	
24-001327-1	123A	3444	20	2051	
24-001354	199	3245	62	2010	

Industry Standard No.	ECG	SK	GE	RS 276-	MOTOR.
24-002	123A	3444	20	2051	
24-016			11	2015	
24-016-001			21	2034	
24-016-005			11	2015	
24-198	116	3016	504A	1104	
24-28201	177	3100/519	300	1122	
24-3564	108	3452	86	2038	
24-602-25	123A	3444	20	2051	
24-AWH	159	3466	82	2032	
24-DP1			504A	1104	
24A	123A	3444	20	2051	
24A1	108	3452	86	2038	
24B	123A	3444	20	2051	
24B1	123A	3444	20	2051	
24B1AH	731	3170	IC-13		
24B1B	731	3170	IC-13		
24B1Z	731	3170	IC-13		
24DP1	116	3016	504A	1104	
24E-001	177		300	1122	
24E-002	112	3089	1N82A		
24E-002-C	112		1N82A		
24E-006	177	3175	300	1122	
24E-022	112	3089	1N82A		
24J2	177	3311	300	1122	
24M125	199	3024/128	62	2010	
24MW 656	107			2015	
24MW1022	199	3018	62	2010	
24MW1023	123A	3444	20	2051	
24MW1024	123A	3444	20	2051	
24MW1025	199	3124/289	62	2010	
24MW1028	1003	3288	IC-43		
24MW1029	109	3087		1123	
24MW1030	109	3087		1123	
24MW1031	159	3466	82	2032	
24MW1038	108	3452	86	2038	
24MW1040	158	3004/102A	53	2007	
24MW1043	109	3088		1123	
24MW1049	159	3466	82	2032	
24MW1051	109	3088		1123	
24MW1052		3126	90		
24MW1057	107	3018	11	2015	
24MW1058	107	3018	11	2015	
24MW1059	123A	3444	20	2051	
24MW1060	199	3122	62	2010	
24MW1061	159	3466	82	2032	
24MW1062(NPN)			63	2030	
24MW1062(PNP)			67	2023	
24MW1063		3126	90		
24MW1065	140A	3061	ZD-10	562	
24MW1066	116	3311	504A	1104	
24MW1067	109	3126		1123	
24MW1068	123A	3444	20	2051	
24MW1069	123A	3444	20	2051	
24MW107	102	3004	2	2007	
24MW1071	116	3017B/117	504A	1104	
24MW1081	107	3018	11	2015	
24MW1082	108	3452	86	2038	
24MW1083	102A	3004	53	2007	
24MW1084	102A	3004	53	2007	
24MW1089	123A	3444	20	2051	
24MW1092	109	3311		1123	
24MW1096	123A	3444	20	2051	
24MW11	100	3005	1	2007	
24MW1106	107	3018	11	2015	
24MW1107	5069A	3333	ZD-4.7		
24MW1108	116	3031A	504A	1104	
24MW1109	177	3100/519	300	1122	
24MW111	102A	3008	53	2007	
24MW1110	1032		IC-164		
24MW1112		3126	90		
24MW1113	116	3031A	504A	1104	
24MW1115	102A	3004	53	2007	
24MW1116		3010	54		
24MW1116(NPN)			59	2002	
24MW1116(PNP)			53	2007	
24MW1118	156	3051	512		
24MW1120	123A	3444	20	2051	
24MW1122	222	3050/221		2036	
24MW1123	116	3311	504A	1104	
24MW1124	113A	3311	6GC1		
24MW1125	140A	3061	ZD-10	562	
24MW1141	123A	3444	20	2051	
24MW1143	184	3041	57	2017	
24MW1144	116	3311	504A	1104	
24MW1146	116	3311	504A	1104	
24MW1147	123A	3444	20	2051	
24MW115	102	3004	2	2007	
24MW1152	192	3024/128	63	2030	
24MW116	102	3004	2	2007	
24MW1161	128	3122	243	2030	
24MW1162	116	3311	504A	1104	
24MW119	123A	3444	20	2051	
24MW122	110MP	3709		1123(2)	
24MW130	103	3862	8	2002	
24MW132	102	3004	2	2007	
24MW15	102	3004	2	2007	
24MW152	126		52	2024	
24MW157	160	3008	245	2004	
24MW16	102	3004	2	2007	
24MW175	117	3017B	504A	1104	
24MW178	102A	3004	53	2007	
24MW179	102A	3004	53	2007	
24MW185	102A	3004	53	2007	
24MW187	102A		53	2007	
24MW192	166	9075		1152	
24MW196	117	3017B	504A	1104	
24MW197	116	3017B/117	504A	1104	
24MW199	109	3087		1123	
24MW205	126	3008	52	2024	
24MW207	138A	3059	ZD-7.5		
24MW208	116(2)			1104	
24MW227	116(2)	3031A		1104	
24MW241	116(2)	3311		1104	
24MW243	109	3087		1123	
24MW244	110MP	3709		1123(2)	
24MW246	125	3031A	510,531	1114	
24MW256	158	3004/102A	53	2007	
24MW263	102	3004	2	2007	
24MW267	116	3311	504A	1104	
24MW268	116	3311	504A	1104	
24MW269	116	3031A	504A	1104	
24MW27	102A	3006/160	53	2007	
24MW271	126	3006/160	52	2024	
24MW28	102A	3004	53	2007	
24MW287	107	3018	11	2015	
24MW29	102	3004	2	2007	
24MW303	126	3006/160	52	2024	
24MW331	139A	3060	ZD-9.1	562	
24MW333	123A	3444	20	2051	
24MW34	102A	3008	53	2007	
24MW351	160		245	2004	
24MW352	126	3008	52	2024	
24MW353	126	3008	52	2024	
24MW361	107	3018	11	2015	
24MW368	126	3007	52	2024	
24MW370	102A	3004	53	2007	
24MW372	123A	3124/289	20	2051	
24MW384	102A	3004	53	2007	
24MW43	102	3004	2	2007	
24MW44	160	3006	245	2004	
24MW441	102A	3008	53	2007	
24MW454	123A	3444	20	2051	
24MW458	123A	3444	20	2051	
24MW460	123A	3444	20	2051	
24MW461	123A	3444	20	2051	
24MW535	107	3018	11	2015	
24MW55	160	3006	245	2004	
24MW59	160	3006	245	2004	
24MW593	107	3018	11	2015	
24MW594	107	3039/316	11	2015	
24MW595	107	3018	11	2015	
24MW596	107	3039/316	11	2015	
24MW597	107	3039/316	11	2015	
24MW598	102A	3004	53	2007	
24MW599	102A	3004	53	2007	
24MW60	102	3004	2	2007	
24MW600	102A		53	2007	
24MW601	102A	3004	53	2007	
24MW602	125	3031A	510,531	1114	
24MW603	109			1123	
24MW605	125	3031A	510,531	1114	
24MW607	115		6GX1		
24MW608	102A		53	2007	
24MW609	123A	3444	20	2051	
24MW61	126	3007	52	2024	
24MW613	102A		53	2007	
24MW614	102A	3004	53	2007	
24MW615	102A	3004	53	2007	
24MW618	131MP	3840	44#(2)	2006(2)	
24MW619	116	3311	504A	1104	
24MW652		3116	FET-2	2035	
24MW653	107	3018	11	2015	
24MW654	108	3452	86	2038	
24MW655	123A	3444	20	2051	
24MW656	107	3018	11	2015	
24MW657	108	3452	86	2038	
24MW658	123A	3444	20	2051	
24MW659	123A	3444	20	2051	
24MW660	128	3122	243	2030	
24MW661	159	3466	82	2032	
24MW662	152	3893	66	2048	
24MW663	128	3024	243	2030	
24MW664		3126	90		
24MW665	110MP	3088	1N60	1123(2)	
24MW667	177	3100/519	300	1122	
24MW669	116	9075/166	504A	1104	
24MW670	5073A	3749	ZD-9.1	562	
24MW671	116	3017B/117	504A	1104	
24MW673	107	3018	11	2015	
24MW674	128	3024	243	2030	
24MW675	108	3452	86	2038	
24MW676	123A	3444	20	2051	
24MW677	128	3122	243	2030	
24MW69	102	3004	2	2007	
24MW70	102	3004	2	2007	
24MW700	107	3018	11	2015	
24MW714	128	3024	243	2030	
24MW721	116	3017B/117	504A	1104	
24MW723	312	3834/132	FET-1	2035	
24MW724	108	3452	86	2038	
24MW725	108	3452	86	2038	
24MW726		3126	90		
24MW727	129	3025	244	2027	
24MW734		3126	90		
24MW736	312	3834/132	FET-1	2035	
24MW737	107	3018	11	2015	
24MW738	107	3018	11	2015	
24MW739	108	3452	86	2038	
24MW74	126	3006/160	52	2024	
24MW740	123A	3444	20	2051	
24MW741	158	3004/102A	53	2007	
24MW742	5071A	3334	ZD-6.8	561	
24MW743	140A	3061	ZD-10	562	
24MW744	177	3100/519	300	1122	
24MW76		3126	90		
24MW760	123A	3444	20	2051	
24MW768	116	3311	504A	1104	
24MW77	100	3005	1	2007	
24MW771	109			1123	
24MW772	116		504A	1104	
24MW773	123A	3444	20	2051	
24MW774	123A	3444	20	2051	
24MW775	123A	3444	20	2051	
24MW776	123A	3444	20	2051	
24MW777	102A	3004	53	2007	
24MW778	152	3054/196	66	2048	
24MW779	125	3031A	510,531	1114	
24MW7796			20	2051	
24MW78	102	3004	2	2007	
24MW780	102A	3004	53	2007	
24MW781	102A	3004	53	2007	
24MW782	102A	3004		2007	
24MW783	158	3004/102A	53	2007	
24MW785	177(2)	3088	300#(2)	1122	
24MW789	102A	3004	53	2007	
24MW790	123A	3444	20	2051	
24MW793	107	3018	11	2015	
24MW795	123A	3444	20	2051	
24MW796	123A	3444		2051	
24MW797	123A	3444	20	2051	
24MW799	102A	3123	53	2007	
24MW801	123A	3444	20	2051	
24MW805	107	3039/316	11	2015	
24MW807	123A	3444	20	2051	
24MW808	123A	3444	20	2051	
24MW809	123A	3444	20	2051	
24MW812	107	3018	11	2015	
24MW813	107	3018	11	2015	
24MW814	107	3018	11	2015	
24MW815	107	3018	11	2015	
24MW816	126	3006/160	52	2024	
24MW817	123A	3444	20	2051	
24MW818	123A	3444	20	2051	
24MW819	158	3004/102A	53	2007	
24MW820	110MP	3088	1N60	1123(2)	
24MW823	123A	3444	20	2051	
24MW824	102A	3004	53	2007	
24MW825	177	3100/519	300	1122	
24MW826	199	3124/289	62	2010	
24MW827	108	3452	86	2038	
24MW828	176	3845			
24MW829	125	3031A	510,531	1114	
24MW83	102	3004	2	2007	
24MW84	102	3004	2	2007	
24MW851	116	3311	504A	1104	
24MW852	108	3452	86	2038	
24MW853	102A	3004	53	2007	
24MW854	123A	3444	20	2051	

Industry Standard No.	ECG	SK	GE	RS 276-	MOTOR.
24MW855	123A	3444	20	2051	
24MW856	102A	3004	53	2007	
24MW857	102A	3004	53	2007	
24MW858	177	3100/519	300	1122	
24MW860	109	3087		1123	
24MW861	177	3100/519	300	1122	
24MW862	116	3031A	504A	1104	
24MW863	107	3018	11	2015	
24MW864	125	3031A	510,531	1114	
24MW865	108	3452	86	2038	
24MW867	116	3311	504A	1104	
24MW87	109	3087		1123	
24MW871	116	3031A	504A	1104	
24MW874	123A	3444	20	2051	
24MW892	102A	3004	53	2007	
24MW893	102A	3004	53	2007	
24MW894	177	3100/519	300#(2)	1122	
24MW899	123A	3444	20	2051	
24MW924	125	3031A	510,531	1114	
24MW950		3126	90		
24MW953	107	3018	11	2015	
24MW954	123A	3444	20	2051	
24MW955		3126	90		
24MW956	177	3100/519	300	1122	
24MW957	161	3018	39	2015	
24MW958	161	3039/316	39	2015	
24MW961	123A	3444	20	2051	
24MW964	199	3124/289	62	2010	
24MW965	199	3124/289	62	2010	
24MW967	109	3087		1123	
24MW973	158	3004/102A	53	2007	
24MW974	117	3031A	504A	1104	
24MW975	117	3031A	504A	1104	
24MW976	159	3466	82	2032	
24MW977	152	3054/196	66	2048	
24MW978	184	3036	57	2017	
24MW988	123A	3444	20	2051	
24MW989	312	3834/132	FET-1	2035	
24MW990	199	3018	62	2010	
24MW991	126	3008	52	2024	
24MW992	123A	3444	20	2051	
24MW994	131	3052	44	2006	
24MW995	116	3311	504A	1104	
24MW997	1024	3152	720		
24MW998	139A	3060	ZD-9.1	562	
24R			17	2051	
24R2	5990	3608	5100		
24T-002	108	3452	86	2038	
24T-009	199		62	2010	
24T-011-001	161	3716	39	2015	
24T-011-003	107		11	2015	
24T-011-008	108	3452	86	2038	
24T-011-011	159	3466	82	2032	
24T-011-013	229		61	2038	
24T-011-015			11	2015	
24T-013-003	161	3039/316	39	2015	
24T-013-005	108	3452	86	2038	
24T-015-013			20	2051	
24T-016	107	3019	11	2015	
24T-016-001	108	3452	86	2038	
24T-016-005	108	3452	86	2038	
24T-016-010	107	3019	11	2015	
24T-016-013	108	3452	86	2038	
24T-016-015	108	3452	86	2038	
24T-016-016	108	3452	86	2038	
24T-016-024			11	2015	
24T-016-0B			11	2015	
24T-026-001	312			2028	
24T002			11	2015	
24T011-008	108		86	2038	
24T011-012	161	3716	39	2015	
24T013003	107	3039/316	11	2015	
24T013005	107	3039/316	11	2015	
24T016	107		11	2015	
24T016001	107	3039/316	11	2015	
24T016005	107	3039/316	11	2015	
24T021	161	3132	39	2015	
24TB1	5522	6622			
24TB2	5524	6624			
24TB3	5526	6627/5527			
24TB4	5527	6627			
24TB5	5528	6629/5529			
24TB6	5529	6629			
24TB8	5531	6631			
24Z6AF	5074A		ZD-11.0	563	
24Z6F	5074A		ZD-11.0	563	
025	130	3027	14	2041	
25-000453	159	3466	82	2032	
25-000456-1	234		65	2050	
25-000462	159	3466	82	2032	
25-001328	106	3984	21	2034	
25-0060-4	123A	3444	20	2051	
025-009600			11	2015	
025-1	730	3143			
025-100003	108	3452	86	2038	
025-100004	108	3452	86	2038	
025-100008	112	3089	1N82A		
025-100009	108	3452	86	2038	
025-100011		3126	90		
025-100012	161	3018	39	2015	
025-100013	108	3452	86	2038	
025-100014	108	3452	86	2038	
025-100015	128	3452/108	86	2038	
025-100016	116	3311	504A	1104	
025-100017	199	3020/123	62	2010	
025-100018	123A	3444	20	2051	
025-100024			504A	1104	
025-100026	107	3019	11	2015	
025-100027	109	3088		1123	
025-100028	116		504A	1104	
025-100029	116	3031A	504A	1104	
025-100030	123A	3444	20	2051	
025-100031	158	3004/102A	53	2007	
025-100035	116	3311	504A	1104	
025-100036	161	3018	39	2015	
025-100037	161	3018	39	2015	
025-100038	161	3018	39	2015	
025-100039		3126	90		
025-100040	123A	3444	20	2051	
025-10029	125	3033	510,531	1114	
025-10030	123A	3444	20	2051	
025-10031	102A	3004	53	2007	
25-3			1N60	1123	
25-37833-2	909D		IC-250		
25-5	116	3017B/117	21	1104	
25-7	125(4)	3033	510,531	1114	
25-MEF	159	3466	82	2032	
25A	108	3452	86	2038	
25A1	108	3444/123A	20	2051	
25A1262-005	108	3452	86	2038	
25A1273-001	123A	3444	20	2051	
25A1281-001	108	3452	86	2038	
25A2	123A	3444	20	2051	
25A473Y			69	2049	

Industry Standard No.	ECG	SK	GE	RS 276-	MOTOR.
25A561Y			22	2032	
25AM624	108	3452	86	2038	
25B	108	3444/123A	86	2051	
25B-1	108	3018	86	2038	
025B-YEL	123A	3444	20	2051	
25B1	108	3452	86	2038	
25B1C	730	3143			
25B1T	730	3143			
25B2	229	3018	61	2038	
25B21			20	2051	
25B2T	730	3143			
25B378	102A	3004	53	2007	
25C05	5980	3610	5096		
25C05R	5981	3698/5987	5097		
25C10	5982	3609/5986	5096		
25C100	6002	7202	5108		
25C100R	6003	7203	5109		
25C10R	5983	3698/5987	5097		
25C15	5986	3609	5096		
25C15R	5987	3698	5097		
25C20	5986	3609	5096		
25C206	108	3452	86	2038	
25C30	5988	3608/5990	5100		
25C30R	5991	3518/5995	5101		
25C40	5990	3608	5100		
25C40R	5991	3518/5995	5101		
25C50	5992	3501/5994	5104		
25C50R	5993	3518/5995	5105		
25C60	5994	3501	5104		
25C60R	5995	3518	5105		
25C70R	5999		5109		
25C80	5998		5108		
25C80R	5999		5109		
25C90	6002	7202	5108		
25C90R	6003	7203	5109		
25CS58LGBM	123A	3124/289	20	2051	
25D10	125	3033	510,531	1114	
25F-B1F	167	3647	510	1172	
25FQ010					1N5829
25FQ015					1N5829
25FQ020					1N5829
25FQ025					1N5830
25FQ030					1N5830
25G10	6022		5128		
25G100	6044	7244	5140		
25G10R	6023		5129		
25G20	6026	7226	5128		
25G20R	6027	7227	5129		
25G40	6034		5132		
25G40R	6035		5133		
25G5	6020	7220	5128		
25G5R	6021		5129		
25G60	6040	7240	5136		
25G80	6042		5140		
25GC12	6034	3608/5990	5100		
25H10	6050		5128		
25H100	6072	7272	5140		
25H100A	6072	7272	5140		
25H10A	6050		5128		
25H15	6054	7254	5128		
25H15A	6054	7254	5128		
25H2	123A		20	2051	
25H20A	6054	7254	5128		
25H25	6058		5132		
25H25A	6058		5132		
25H30	6058		5132		
25H30A	6058		5132		
25H40	6060	7260	5132		
25H40A	6060		5132		
25H5	6050		5128		
25H50	6064	7264	5136		
25H50A	6064	7264	5136		
25H5A	6050		5128		
25H60	6064	7264	5136		
25H60A	6064	7264	5136		
25H70	6068		5140		
25H70A	6068		5140		
25H80	6068		5140		
25H80A	6068		5140		
25H90	6072	7272	5140		
25H90A	6072	7272	5140		
25HB10	6022		5128		
25HB15	6026	7226	5128		
25HB20	6026	7226			
25HB25	6030		5132		
25HB30	6030		5132		
25HB35	6034	7234	5132		
25HB40	6034	7234	5132		
25HB5	6020	7220	5128		
25HB50	6038		5136		
25HB60	6040	7240	5136		
25J2	177	3311	300	1122	
25JC12	6040	3501/5994	5104		
25K10	125	3033	510,531	1114	
25LC12	6042		5108		
25N15	145A	3063	ZD-15	564	
25N27	146A	3064	ZD-27		
25NC12	6044	7202/6002	5108		
25PW10					1N3492
25PW20					1N3493
25PW30					1N3494
25PW40					1N3495
25PW5					1N3491
25PW50					MR328
25PW60					MR328
25R	108	3452	86	2038	
25T-002	107	3039/316	11	2015	
25T1	102	3004	2	2007	
25TB1	5542	6642			
25Z6	142A	3062	ZD-12	563	
25Z6A	142A	3062	ZD-12	563	
25Z6AF	142A		ZD-12	563	
25Z6F	142A		ZD-12	563	
026-1000-20			16	2006	
026-100003	121	3009	239	2006	
026-100004	121MP	3013	239(2)	2006(2)	
026-100005	102	3004	2	2007	
026-100012	102	3004	2		
026-100013	128	3024	243	2030	
026-100017	123	3020	20	2051	
026-100018	102	3004	2	2007	
026-100019	129	3025	244	2027	
026-100020	121MP	3013	239(2)	2006(2)	
026-100026	123A	3444	20	2051	
026-100028	121	3009	239	2006	
026-10012				2007	
26-16162-1	358		42		
26-4701508	125	3032A	510,531	1114	
26/P19503			18	2030	
26D00505	116		504A	1104	
26J2	177	3311	300	1122	
26MW613	102A	3004	53	2007	
26P33108	164		37		
26R2	5994	3501	5104		

Industry Standard No.	ECG	SK	GE	RS 276-	MOTOR.
26R2S	5994	3501	5104		
26T1	102	3004	2	2007	
26TB1	5522	6622			
26TB2	5524	6624			
26TB3	5526	6627/5527			
26TB4	5527	6627			
26TB5	5528	6629/5529			
26TB6	5529	6629			
26TB8	5531	6631			
26Z6AF	143A	3750	ZD-13	563	
26Z6F	143A	3750	ZD-13	563	
027-000296	116	3017B/117	504A	1104	
027-000306	116	3016	504A	1104	
027-000312	116	3016	504A	1104	
27-226	113A	3119/113	6GC1		
027-300226	113A	3119/113	6GC1		
27-C226	113A	3119/113	6GC1		
27A10489-101-11	199	3245	62	2010	
27A10533	159	3466	82	2032	
27C226		3119	6GC1		
27D127	198	3103A/396	251		
27J2	177	3311	300	1122	
27P1	109	3087		1123	
27T401	100	3005	1	2007	
27T402	100	3005	1	2007	
27T403	102	3004	2	2007	
27T404	102	3004	2	2007	
27T405	102	3004	2	2007	
27T406	121	3009	239	2006	
27T407	105	3012	4		
27T408	101	3011	8	2002	
27T409	123	3124/289	20	2051	
27T410	103	3010	8	2002	
27T411	123	3124/289	20	2051	
27T412	126	3006/160	52	2024	
27Z6	144A	3094	ZD-14	564	
27Z6A	144A	3094	ZD-14	564	
27Z6AF	144A		ZD-14	564	
27Z6F	144A		ZD-14	564	
028	124		512		
28-1-01			504A	1104	
28-1-02			504A	1104	
28-13-01	116		504A	1104	
28-14-01	116		504A	1104	
28-15-01	116		504A	1104	
28-15-02	116		504A	1104	
28-18-01	116(4)			1104	
28-19-01	116	3017B/117	504A	1104	
28-20-01	116		504A	1104	
28-20-02	116		504A	1104	
28-21-01	116	3016	504A	1104	
28-22-01	116	3016	504A	1104	
28-22-02	116	3017B/117	504A	1104	
28-22-03	116	3017B/117	504A	1104	
28-22-04	116	3016	504A	1104	
28-22-05	116		504A	1104	
28-22-06	116		504A	1104	
28-22-07	116	3016	504A	1104	
28-22-10	116	3016	504A	1104	
28-22-11	156	3016	512		
28-22-12	116	3017B/117	504A	1104	
28-22-13	506	3843	511	1114	
28-22-14	116	3017B/117	504A	1104	
28-22-15	116	3016	504A	1104	
28-22-16	156	3017B/117	512		
28-22-17	116	3017B/117	504A	1104	
28-22-18	156		512		
28-22-19	119		CR-2		
28-22-20	156		512		
28-22-21	116	3311	504A	1104	
28-22-22	506	3843	511	1114	
28-23-01	118	3066	CR-1		
28-24-01	119		CR-2		
28-25-01	116	3016	504A	1104	
28-254566-1	116	3016	504A	1104	
28-26-01	120	3016	CR-3		
28-29-01	116	3017B/117	504A	1104	
028-300-226	110MP	3119/113			
28-31-00	513	3443	513		
28-31-01	505	3108	CR-7		
28-31-02		3443	513		
28-32-0X	500A	3304	527		
28-35-01	522	3303	523		
28-6-01	116	3016	504A	1104	
28-65-01	116	3017B/117	504A	1104	
28-7-01	116		504A	1104	
28-819-172	107		11	2015	
28A477			52	2024	
28J2	116	3311	504A	1104	
28M018		3126	90		
28P1	109	3087		1123	
28R2	5998		5108		
28R25	5998		5108		
28Z6	145A	3063	ZD-15	564	
28Z6A	145A	3063	ZD-15	564	
28Z6AF	145A		ZD-15	564	
28Z6F	145A		ZD-15	564	
29-505	109	3087		1123	
29-HCL	159	3466	82	2032	
29A4	129	3025	244	2027	
29B1	790	3454	IC-230		
29B17	790	3454	IC-230		
29B1B	790	3454	IC-18		
29B1Z	790	3454	IC-230		
29P1	134A	3055	ZD-3.6		
29S	138A		ZD-7.5		
29V0038H03	102	3004	2	2007	
29V008M01	100	3006/160	1	2007	
29V011H01	100	3006/160	1	2007	
29V012H01	100	3006/160	1	2007	
29V069C02	163A	3439	36		
29V12H01	102A	3004	53	2007	
030	156	3051	512		
30-004-001	179	3642	76		
30-005072	390		255	2041	
030-007-0	129	3114/290	244	1114	
030-034-0	103A	3011	59	2002	
30-090	184		57	2017	
30-1,007	152	3893			
30-8054-7		3311	504A	1104	
30-8057-13	178MP		300#(2)	1122(2)	
30A10F					SPECIAL
30A6F					MR1396
30A8F					SPECIAL
30AS	116	3016	504A	1104	
30B					MR1123
30B5	116	3016	504A	1104	
30BR					1N3882
30BS	116	3017B/117	504A	1104	
30C	116	3017B/117	504A	1104	1N4004
30CTQ030					SPECIAL
30CTQ035					SPECIAL
30CTQ040					SPECIAL
30CTQ045					SPECIAL
30D1	156		512	1143	
30D2	125	3051/156	510	1114	
30DB10T					MDA3510
30DB2T					MDA970A3
30DB8T					MDA3508
30F1	5836		5008		
30FQ030					MBR3535
30FQ045					MBR3545
30FQ30A					SPECIAL
30FQ35A					SPECIAL
30FQ40A					SPECIAL
30FQ45A					SPECIAL
30H	116	3017B/117	504A	1104	
30H3	5874		5032		
30H3P	5856	3031A	5020		MR1123
30H3PN	5857		5021		
30HR3P	5820				1N3882
30J2	5874	3500/5882	5032		
30J3P	5876		5036		
30JH3	6058		5132		
30K	116	3016	504A	1104	
30M	116	3017B/117	504A	1104	
30N2	5874		5032		
30N3	5874		5032		
30P1	109	3091		1123	
30P2	5874		5032		
30P3	5874		5032		
30QHC030					SPECIAL
30QHC045					SPECIAL
30R1	5801	9004		1142	
30R10	156	3051	512		
30R25	6002	7202	5108		
30R3	5804	9007		1144	
30R6	5806	3848			
30R8	156	3051	512		
30S05	5800	9003		1142	
30S1	5801	9004	512	1142	MR501
30S10	5809	9010	512		MR510
30S2	5805	9008	512	1143	MR502
30S3	5803	9006	512	1144	MR504
30S4	5804		512	1144	MR504
30S5	5805	9008	512		MR506
30S6	5806	3848	512		MR506
30S8	5808	9009	512		MR508
30V-H6	102A		53	2007	
30V-HG	102A	3123	53	2007	
031	156	3006/160	512		
31-0001	160	3006	245	2004	
31-0002	160	3006	245	2004	
31-0003	160	3006	245	2004	
31-0004	160	3008	245	2004	
31-0005	126	3004/102A	52	2024	
31-0006	102A	3008	53	2007	
31-0007	123A	3444	20	2051	
31-0008	102A	3004	53	2007	
31-0009	123A	3444	20	2051	
31-0010	175	3026	246	2020	
31-0012	199	3124/289	62	2010	
31-0013	199	3124/289	62	2010	
31-0015	160	3008	245	2004	
31-0016	160	3008	245	2004	
31-0017	158	3003	53	2007	
31-0018	102A	3004	53	2007	
31-002-0	123AP	3245/199	61	2051	
31-0025	102A	3004	53	2007	
31-0026	102A	3004	53	2007	
31-0033	102A		53	2007	
31-0035	158	3004/102A	53	2007	
31-0039	109	3087		1123	
31-0041	126	3006/160	52	2024	
31-0042	126	3006/160	52	2024	
31-0048	107	3018	11	2015	
31-0049	107	3018	11	2015	
31-0050	107	3018	11	2015	
31-0051	108	3452	86	2038	
31-0052	199	3018	62	2010	
31-0053	102A	3004	53	2007	
31-0054	107	3018	11	2015	
31-0055	193	3114/290	67	2023	
31-0065	126	3008	52	2024	
31-0066	186	3192	28	2017	
31-0068	123A	3444	20	2051	
31-0069	123A	3444	20	2051	
31-0070	102A	3004	53	2007	
31-0075	158		53	2007	
31-0080	123A	3444	20	2051	
31-0081	123A	3444	20	2051	
31-0082	123A	3444	20	2051	
31-0083	128	3024	243	2030	
31-0084	123A	3444	20	2051	
31-0085	123A	3444	20	2051	
31-0097	108	3452	86	2038	
31-0098	107	3018	11	2015	
31-0099	199	3124/289	62	2010	
31-0100	199	3124/289	62	2010	
31-0101	192	3054/196	63	2030	
31-0102	193	3083/197	67	2023	
31-0103	108	3452	86	2038	
31-0104	123A	3444	20	2051	
31-0105	158	3004/102A	53	2007	
31-0106	123A	3444	20	2051	
31-0107	102A	3004	53	2007	
31-0108	390	3006/160	52	2024	
31-011	177		300	1122	
31-0115	123A	3444	20	2051	
31-0116	123A	3444	20	2051	
31-0123	126	3006/160	52	2024	
31-0124	126	3006/160	52	2024	
31-0132	126	3006/160	52	2024	
31-0134	126	3008	52	2024	
31-0135	126	3008	52	2024	
31-0139	126	3008	52	2024	
31-0141.	160	3006	245	2004	
31-0148	102A		53	2007	
31-0150	160	3006	245	2004	
31-0153	102A	3004	53	2007	
31-0161	160	3004/102A	245	2004	
31-0163	160		245	2004	
31-0165	160		245	2004	
31-0166	160		245	2004	
31-0168	160	3006	245	2004	
31-0170	160	3008	245	2004	
31-0171	160		245	2004	
31-0172	102A	3004	53	2007	
31-0175	127	3035	25		
31-0177	123A	3444	20	2051	
31-0178	126	3008	52	2024	
31-0180	160	3006	245	2004	
31-0181	160		245	2004	
31-0182	158	3004/102A	53	2007	
31-0183	158		53	2007	
31-0184	126	3006/160	52	2024	
31-0187	123	3124/289	20	2051	
31-0188	102A	3004	53	2007	

Industry Standard No.	ECG	SK	GE	RS 276-	MOTOR.
31-0189	102A		53	2007	
31-0190	126	3008	52	2024	
31-0191	126	3006/160	52	2024	
31-0192	121	3009	239	2006	
31-0196	121	3035	239	2006	
31-020	HIDIV-1		FR-8		
31-0205	102A		53	2007	
31-0206	107	3039/316	11	2015	
31-021	HIDIV-2		FR-9		
31-0217	126	3008	52	2024	
31-0228	126	3006/160	52	2024	
31-0229	158	3004/102A	53	2007	
31-0230	123A		20	2051	
31-0239	123A	3124/289	20	2051	
31-0240	121	3009	239	2006	
31-0241	160	3006	245	2004	
31-0241-1	160	3006	245	2004	
31-0242	108	3452	86	2038	
31-0243	108	3452	86	2038	
31-0246	123A		20	2051	
31-0247	102	3004	2	2007	
31-0248				2006(2)	
31-025	102	3004	2	2007	
31-025-0	123AP	3444/123A	211	2051	
31-0253	100	3005	1	2007	
31-027-0	199	3245	62		
31-058	123A	3444	20	2051	
031-058-0	300	3464	273		
31-069-0	235	3239/236	215		
31-091-3	236	3239	210	2053	
31-093	177		300	1122	
031-098-0	235		357		
31-1	123A	3444	20	2051	
31-10			11	2015	
31-1012	941M		IC-265	007	
31-16	123A	3444	20	2051	
31-194	125	3032A	510,531	1114	
31-195	116	3017B/117	504A	1104	
31-21004900	160	3006	245	2004	
31-21007744	160	3006	245	2004	
31-21024033	160	3006	245	2004	
31-21024044	160	3006	245	2004	
31-21047111	160	3006	245	2004	
31-2104733	160	3006	245	2004	
31-21050611	160	3006	245	2004	
31-21050622	160	3006	245	2004	
31-22005400	102A	3004	53	2007	
031A	123A	3444	20	2051	
31R2	5986	3609	5096		
31Z6	5005A	3771	ZD-3.3		
31Z6A	5005A	3771	ZD-3.3		
032	156	3051	512		
32-0000	109	3087		1123	
32-0001	109	3087		1123	
32-0002	109	3087		1123	
32-0003	109	3087		1123	
32-0004	109	3087		1123	
32-0005	139A	3060	ZD-9.1	562	
32-0007	110MP	3709	1N60	1123(2)	
32-0008	109	3087		1123	
32-0013	109	3088		1123	
32-0022	177	3100/519	300	1122	
32-0023	109	3087		1123	
32-0025	5072A		ZD-8.2	562	
32-0026	116	3017B/117	504A	1104	
32-0029	109	3087		1123	
32-0036	109	3087		1123	
32-0037	116	3311	504A	1104	
32-0038	116	3311	504A	1104	
32-0039	110MP			1123	
32-0042	116	3031A	504A	1104	
32-0043		3126	90		
32-0044		3126	90		
32-0045	116	3016	504A	1104	
32-0046	116	3311	504A	1104	
32-0047	116	3031A	504A	1104	
32-0048	117	3031A	504A	1104	
32-0049	140A	3061	ZD-10	562	
32-0050	116	3311	504A	1104	
32-0057	177	3100/519	300	1122	
32-0059	116	3031A	504A	1104	
32-0060	116	3017B/117	504A	1104	
32-0061	116	3032A	504A	1104	
32-0062	113A	3119/113	6GC1		
32-0063	177	3100/519	300	1122	
32-107261-1	910		IC-251		
32-12066-10	160		245	2004	
32-13843-2	123A	3444	20	2051	
32-16591	104MP	3720	16(2)	2006(2)	
32-16599	121MP	3015	239(2)	2006(2)	
32-18537	109	3087		1123	
32-18539	110MP	3709		1123(2)	
32-20738	123A	3444	20	2051	
32-20739	159	3025/129	82	2032	
32-23555-1	704	3022/1188	IC-205		
32-23555-2	704	3022/1188	IC-205		
32-23555-3	704	3022/1188	IC-205		
32-23555-4	704	3023	IC-205		
32-29778-1	500A	3304			
32-29778-2	500A	3304	527		
32-29778-3	500A	3303/522	529		
32-29778-4	500A	3304			
32-33057-1	500A	3304			
32-33057-2	522	3303	517		
32-33057-3	500A	3304	527		
32-33057-4	500A	3304	527		
32-33057-5	500A	3304	527		
32-33057-6	500A	3304			
32-33094-1	522	3303	517		
32-33094-2	522	3900/536A	522		
32-33094-3	522	3303	517		
32-33094-4	522	3303	517		
32-33094-5	522	3900/536A	522		
32-33094-6	522	3303	523		
32-35894-1	500A	3304	527		
32-35894-2	522	3303	523		
32-35894-3	522	3303	523		
32-35894-4	522	3303	523		
32-35894-5	500A	3304	527		
32-35894-6	522	3303	523		
32-35894-7	522	3303	523		
32-39091-1	523	3306	528		
32-39091-2	523	3306	528		
32-39091-3	526A	3306/523	521		
32-39091-4	523	3306	528		
32-39091-5	523	3306	528		
32-39091-6	523	3306	528		
32-39091-7	526A	3306/523	521		
32-39091-8	526A	3306/523	521		
32-39091-9	523	3306	528		
32-39704-1	522	3303	523		
32-39704-2	522	3303	523		
32-43737-1	556	3307/529	529		
32-805483-1	910		IC-251		

Industry Standard No.	ECG	SK	GE	RS 276-	MOTOR.
32-805483-2	910		IC-251		
32-807071-1	911		IC-253		
32-807072-1	909	3590	IC-249		
32E64	129	3025	244	2027	
32R2	5986	3609	5096		
32Z6	134A	3055	ZD-3.6		
32Z6A	134A	3055	ZD-3.6		
33-0002	116	3311	504A	1104	
33-0006	116	3016	504A	1104	
33-0023	116	3017B/117	504A	1104	
33-00234-B	175		246	2020	
33-0024	116	3311	504A	1104	
33-00243	124		12		
33-0025	138A	3059	ZD-7.5		
33-0026	116	3031A	504A	1104	
33-0029	116	3017B/117	504A	1104	
33-0030	116	3017B/117	504A	1104	
33-0031	116(3)	3017B/117	504A(3)	1104	
33-0036			511	1114	
33-0039	105		4		
33-00706A	123A	3444	20	2051	
33-00742	129		244	2027	
033-014-0	222	3065	FET-4	2036	
33-016	159	3466	82	2032	
33-048	128		243	2030	
33-050	175		246	2020	
33-052	130	3510	14	2041	
33-070	123A	3444	20	2051	
33-0706	123A	3444	20	2051	
33-071	123A	3444	20	2051	
33-084		3444	20	2051	
33-086	159	3466	82	2032	
33-090	184	3054/196	57	2017	
33-096	185	3083/197	58	2025	
33-1000-00	102	3004	2	2007	
33-1001-00	102	3004	2	2007	
33-1002-00	131MP	3840	44(2)	2006(2)	
33-1004-00	121	3009	239	2006	
33-1009-01	102A	3004	53	2007	
33-1019-00	102A	3004	53	2007	
33-1020-00	102A	3004	53	2007	
33-1021-00	102A	3123	53	2007	
33-108			255	2041	
33-1390-1	HIDIV-12	3868/DIV-1	FR-8		
33-1390-2	HIDIV-12	3868/DIV-1	FR-8		
33-1390-3	HIDIV-12	3868/DIV-1	FR-8		
033-3			18	2030	
033-31(SYLVANIA)	123A		20	2051	
33-4-3	358		42		
33-4-3A	358		42		
33-7-3	358		42		
33-8-3	358		42		
033A	199		62	2010	
33G59019	113A	3119/113	6GC1		
33G59024	116	3017B/117	504A	1104	
33G59113	119	3109	CR-2		
33G59121	116	3017B/117	504A	1104	
33G59122	116	3031A	504A	1104	
33H50	108	3452	86	2038	
33K59	312	3116	FET-2	2035	
33R	5082A	3753	ZD-25		
33R-09	139A	3060	ZD-9.1	562	
33Z6	134A	3055	ZD-3.6		
33Z6A	134A	3055	ZD-3.6		
034-001-0	109		1N34AS	1123	
34-0012	358		42		
34-0037-102	177		300	1122	
34-028-0	109		1N60	1123	
34-029-0	614	3126			
034-032-0	519	3100	514	1122	
34-1000	130	3027	14	2041	
34-1000A	130	3027	14	2041	
34-1001	219		74	2043	
34-1002	152	3893	66	2048	
34-1003	153	3083/197	69	2049	
34-1006	157	3103A/396	232		
34-1007	234	3247	65	2050	
34-1008	234	3247	65	2050	
34-1009	199	3245	62	2010	
34-1010	123A	3444	20	2051	
34-1011	106	3984	21	2034	
34-1013	159	3466	82	2032	
34-1017	194		220		
34-1019	199	3245	62	2010	
34-1022	159	3466	82	2032	
34-1026	175		246	2020	
34-1027	218		234		
34-1028	130	3027	14	2041	
34-1028A	130	3027	14	2041	
34-1029	219	3173	74	2043	
34-1041	726	3129			
34-11016		3067	CR-4		
34-119	160	3006	245	2004	
34-143-12	159	3466	82	2032	
34-194	941M		IC-265	007	
34-2001-1	109			1123	
34-220	160	3006	245	2004	
34-221	160	3006	245	2004	
34-298	160	3006	245	2004	
34-3015-28	159	3466	82	2032	
34-3015-46	107		11	2015	
34-3015-47	107		11	2015	
34-3015-49	107		11	2015	
34-3022-7			1N60	1123	
34-34-6015-43	123A	3444	20	2051	
34-6	160	3006	245	2004	
34-6000-10	160	3006	245	2004	
34-6000-11	160	3006	245	2004	
34-6000-12	160	3006	245	2004	
34-6000-13	160	3006	245	2004	
34-6000-14	160	3006	245	2004	
34-6000-15	102A	3004	53	2007	
34-6000-16	160	3004/102A	245	2004	
34-6000-17	160	3006	245	2004	
34-6000-18	160	3005	245	2004	
34-6000-19	160	3005	245	2004	
34-6000-20	160	3006	245	2004	
34-6000-25	160	3006	245	2004	
34-6000-26	160	3006	245	2004	
34-6000-27	102A		53	2007	
34-6000-28	102A	3004	53	2007	
34-6000-29	102A	3004	53	2007	
34-6000-3	160	3005	245	2004	
34-6000-30	102A		53	2007	
34-6000-31	102A	3004	53	2007	
34-6000-32	102A	3004	53	2007	
34-6000-33	102A	3004	53	2007	
34-6000-34	102A	3004	53	2007	
34-6000-4	102A	3004	53	2007	
34-6000-5	102A	3004	53	2007	
34-6000-58	160	3006	245	2004	
34-6000-59	160	3006	245	2004	
34-6000-6	102A	3004	53	2007	
34-6000-60	160	3006	245	2004	

Industry Standard No.	ECG	SK	GE	RS 276-	MOTOR.
34-6000-61	160	3006	245	2004	
34-6000-62	160	3006	245	2004	
34-6000-63	160	3006	245	2004	
34-6000-64	123A	3444	20	2051	
34-6000-65	160	3122	245	2004	
34-6000-66	160		245	2004	
34-6000-67	160	3006	245	2004	
34-6000-68	160	3006	245	2004	
34-6000-69	123A	3444	20	2051	
34-6000-7	102A	3004	53	2007	
34-6000-70	123A	3444	20	2051	
34-6000-71	123A	3444	20	2051	
34-6000-72	233	3132	210	2009	
34-6000-76	160	3006	245	2004	
34-6000-77	160	3006	245	2004	
34-6000-78	160	3006	245	2004	
34-6000-79	160	3006	245	2004	
34-6000-8	102A	3004	53	2007	
34-6000-80	160	3006	245	2004	
34-6000-81	160	3006	245	2004	
34-6000-82	160	3006	245	2004	
34-6000-83	102A	3008	53	2007	
34-6000-84	102A	3008	53	2007	
34-6000-85	102A	3008	53	2007	
34-6000-9	160	3006	245	2004	
34-60001-63			20	2051	
34-6001-1	104	3009	16	2006	
34-6001-10	102A	3004	53	2007	
34-6001-11	102A	3004	53	2007	
34-6001-12	102A	3018	53	2007	
34-6001-13	102A	3004	53	2007	
34-6001-14	102A	3004	53	2007	
34-6001-15	159	3466	82	2032	
34-6001-16	102A	3004	53	2007	
34-6001-17	102A	3004	53	2007	
34-6001-18	102A	3004	53	2007	
34-6001-19	102A	3004	53	2007	
34-6001-20	102A	3004	53	2007	
34-6001-21	102A	3004	53	2007	
34-6001-22	102A	3004	53	2007	
34-6001-23	102A	3004	53	2007	
34-6001-26	102A	3004	53	2007	
34-6001-28	158(2)	3004/102A	53(2)		
34-6001-29	102A	3004	53	2007	
34-6001-3	108	3452	86	2038	
34-6001-30	102A	3004	53	2007	
34-6001-31	102A	3004	53	2007	
34-6001-33	102A	3004	53	2007	
34-6001-34	123A	3122	20	2051	
34-6001-41	102A		53	2007	
34-6001-42	102A	3004	53	2007	
34-6001-43	102A	3005	53	2007	
34-6001-44	102A	3005	53	2007	
34-6001-47	102A	3004	53	2007	
34-6001-48	123A	3444	20	2051	
34-6001-49	123A	3444	20	2051	
34-6001-5	123A	3444	20	2051	
34-6001-50	128	3124/289	243	2030	
34-6001-51	128	3024	243	2030	
34-6001-51MR-12	128		243	2030	
34-6001-52	123A	3444	20	2051	
34-6001-53	123A	3444	20	2051	
34-6001-54	123A	3444	20	2051	
34-6001-55	123A	3444	20	2051	
34-6001-56	123A	3444	20	2051	
34-6001-57	123A	3444	20	2051	
34-6001-58	123A	3444	20	2051	
34-6001-6	108	3452	86	2038	
34-6001-60	123A	3444	20	2051	
34-6001-61	123A	3444	20	2051	
34-6001-62	123A	3444	20	2051	
34-6001-63	123A	3444	20	2051	
34-6001-64	191	3018	249		
34-6001-65	154	3040	40	2012	
34-6001-66	158	3004/102A	53	2007	
34-6001-69	123A	3444	20	2051	
34-6001-7	102A	3004	53	2007	
34-6001-70	123A	3444	20	2051	
34-6001-71	123		20	2051	
34-6001-72	102A	3004	53	2007	
34-6001-73	123A	3444	20	2051	
34-6001-74	123A	3444	20	2051	
34-6001-76	102A	3004	53	2007	
34-6001-77	123A	3444	20	2051	
34-6001-78	128	3124/289	243	2030	
34-6001-79	121	3018	239	2006	
34-6001-8	102A	3004	53	2007	
34-6001-80	128	3124/289	243	2030	
34-6001-82	128	3045/225	243	2030	
34-6001-83	128	3024	243	2030	
34-6001-84	103A	3835	59	2002	
34-6001-85	128	3124/289	243	2030	
34-6001-86	129	3025	244	2027	
34-6001-9	102A	3004	53	2007	
34-6002-1	175		246	2020	
34-6002-10	121	3009	239	2006	
34-6002-11	121	3009	239	2006	
34-6002-13	121	3009	239	2006	
34-6002-14	121	3009	239	2006	
34-6002-17	104	3009	16	2006	
34-6002-18	104	3009	16	2006	
34-6002-18A	104	3009	16	2006	
34-6002-19	121	3009	239	2006	
34-6002-2	121	3009	239	2006	
34-6002-20	104	3009	16	2006	
34-6002-21	124	3021	12		
34-6002-22	104	3009	16	2006	
34-6002-22A	104	3009	16	2006	
34-6002-23			74	2043	
34-6002-24	127	3764	25		
34-6002-26	124	3021	12		
34-6002-27			19	2041	
34-6002-28	175	3131A/369	246	2020	
34-6002-29	175	3131A/369	246	2020	
34-6002-3	121	3009	239	2006	
34-6002-30	281	3035	25		
34-6002-31	127	3035	25		
34-6002-32	130	3027	14	2041	
34-6002-32A	130	3027	14	2041	
34-6002-33	179	3642	76		
34-6002-34	121	3009	239	2006	
34-6002-35	181		75	2041	
34-6002-36	180		74	2043	
34-6002-37	181		75	2041	
34-6002-38	180		74	2043	
34-6002-4	121	3009	239	2006	
34-6002-41	186	3192	28	2017	
34-6002-42	187	3193	29	2025	
34-6002-43	196	3026	241	2020	
34-6002-46	124	3021	12		
34-6002-49	127	3035	25		
34-6002-5	121	3009	239	2006	
34-6002-50	152	3054/196	66	2048	
34-6002-52	152	3054/196	66	2048	
34-6002-54			20	2051	
34-6002-55			20	2051	
34-6002-56	152	3054/196	66	2048	
34-6002-57	153	3083/197	69	2049	
34-6002-58	163A	3439	36		
34-6002-59	165	3111	38		
34-6002-6	121	3009	239	2006	
34-6002-61	162	3438	35		
34-6002-62	124	3021	12		
34-6002-63	165	3111	38		
34-6002-64	165	3111	38		
34-6002-7	121	3009	239	2006	
34-6002-8	121	3009	239	2006	
34-6002-9	121	3009	239	2006	
34-6005-1	160	3124/289	245	2004	
34-6005-2	175	3131A/369	246	2020	
34-6005-3	175	3131A/369	246	2020	
34-6006-1			66	2048	
34-6007-1	123A	3444	20	2051	
34-6007-10			60	2015	
34-6007-11			60	2015	
34-6007-12			60	2015	
34-6007-13			60	2015	
34-6007-14			62	2010	
34-6007-2	123A	3444	20	2051	
34-6007-3	123A	3444	20	2051	
34-6007-4			63	2030	
34-6007-5			10	2051	
34-6007-6			20	2051	
34-6007-7			20	2051	
34-6007-8			63	2030	
34-6007-9			10	2051	
34-6008	102A	3004	53	2007	
34-6009	102A	3004	53	2007	
34-6015-1	123A	3444	20	2051	
34-6015-10	123A	3444	20	2051	
34-6015-11	123A	3444	20	2051	
34-6015-12	123AP	3444/123A	20	2051	
34-6015-13	123A	3444	20	2051	
34-6015-14	123A	3444	20	2051	
34-6015-15	319P	3117	283	2016	
34-6015-16	319P	3018	283	2016	
34-6015-17	319P	3018	283	2016	
34-6015-18	161	3018	39	2015	
34-6015-19	319P	3018	39	2015	
34-6015-2	123A	3444	20	2051	
34-6015-20	161	3018	39	2015	
34-6015-21	123A	3444	20	2051	
34-6015-22	161	3018	39	2015	
34-6015-23	128	3024	243	2030	
34-6015-24	128	3124/289	243	2030	
34-6015-25	128	3018	243	2030	
34-6015-26	159	3466	82	2032	
34-6015-27	229	3117	61	2038	
34-6015-28	191	3024/128	249		
34-6015-29	319P	3117	283	2016	
34-6015-3	123A	3444	20	2051	
34-6015-30	128	3024	243	2030	
34-6015-31	161	3117	39	2015	
34-6015-32	160		245	2004	
34-6015-33	160		245	2004	
34-6015-34	160	3005	245	2004	
34-6015-35	160		245	2004	
34-6015-36	160		245	2004	
34-6015-37	233	3018	210	2009	
34-6015-38	233	3039/316	210	2009	
34-6015-39	160		245	2004	
34-6015-4	123A	3444	20	2051	
34-6015-40	160		245	2004	
34-6015-41	123A	3444	20	2051	
34-6015-42	159	3466	82	2032	
34-6015-42A	123A	3444	20	2051	
34-6015-43		3124	10	2051	
34-6015-43A	123A	3444	20	2051	
34-6015-44	123A	3444	20	2051	
34-6015-44A	102A	3004	53	2007	
34-6015-46	229	3444/123A	20	2051	
34-6015-47	229	3018	61	2038	
34-6015-48	229	3018	61	2038	
34-6015-49	108	3452	86	2038	
34-6015-5	123A	3444	20	2051	
34-6015-50	107	3018	11	2015	
34-6015-51	128	3018	243	2030	
34-6015-52	229	3018	61	2038	
34-6015-54	123A	3444	20	2051	
34-6015-59	154		40	2012	
34-6015-6	123A	3444	20	2051	
34-6015-60	123A	3444	20	2051	
34-6015-61	233		210	2009	
34-6015-62	233	3117	210	2009	
34-6015-63	123A	3444	20	2051	
34-6015-64	171	3054/196	27		
34-6015-7	123A	3444	20	2051	
34-6015-8	107	3018	11	2015	
34-6015-80	123A	3444	20	2051	
34-6015-9	123A	3018	20	2051	
34-6016-11	102A	3004	53	2007	
34-6016-12	129		244	2027	
34-6016-13			29	2025	
34-6016-14	123A	3444	20	2051	
34-6016-15	159	3466	82	2032	
34-6016-15A	159	3466	82	2032	
34-6016-16	123A	3444	20	2051	
34-6016-17	108	3452	86	2038	
34-6016-18	123A	3444	20	2051	
34-6016-19	123A	3444	20	2051	
34-6016-2	123A	3444	20	2051	
34-6016-22	128	3024	243	2030	
34-6016-23	129	3025	244	2027	
34-6016-23A	129	3025	244	2027	
34-6016-24	123A	3444	20	2051	
34-6016-25	123A	3444	20	2051	
34-6016-26	123A	3444	20	2051	
34-6016-27	128	3024	243	2030	
34-6016-28	160	3008	245	2004	
34-6016-29	160	3008	245	2004	
34-6016-3	123A	3444	20	2051	
34-6016-30	128	3024	243	2030	
34-6016-31	210	3054/196	252	2018	
34-6016-32	159	3466	82	2032	
34-6016-32A	129	3025	244	2027	
34-6016-33	128	3024	243	2030	
34-6016-4	123A	3444	20	2051	
34-6016-41	194	3275	220		
34-6016-44			63	2030	
34-6016-45			66	2048	
34-6016-46			69	2049	
34-6016-47	159	3466	82	2032	
34-6016-49	123A	3444	20	2051	
34-6016-49A	123A	3444	20	2051	
34-6016-50	102A	3003	53	2007	
34-6016-51	192		63	2030	
34-6016-53	210	3054/196	252	2018	
34-6016-54	211	3083/197	253	2026	

Industry Standard No.	ECG	SK	GE	RS 276-	MOTOR.
34-6016-56	194	3275	220		
34-6016-59			67	2023	
34-6016-59(NPN)	192		63	2030	
34-6016-59(PNP)	193		67	2023	
34-6016-6			10	2051	
34-6016-60	159	3466	82	2032	
34-6016-63	123A	3444	20	2051	
34-6016-64	234	3114/290	65	2050	
34-6016-65	187	3193	29	2025	
34-6016-7	123A	3444	20	2051	
34-6016-8	123A	3444	20	2051	
34-6017-3	172A	3156	64		
34-601703	172A	3156	64		
34-6018-2	312	3116	FET-2	2028	
34-6075-46			20	2051	
34-8001-43	126	3006/160	52	2024	
34-8002-1	109	3087	86	1123	
34-8002-2	109	3087		1123	
34-8002-3	109	3087		1123	
34-8002-4	109	3087		1123	
34-8002-5	109	3087		1123	
34-8002-6	109	3087		1123	
34-8002-7	109	3087		1123	
34-8003	116	3017B/117	504A	1104	
34-8022			1N60	1123	
34-8022-1	109			1123	
34-8022-2	109			1123	
34-8022-3	109			1123	
34-8022-4	109			1123	
34-8022-5	109			1123	
34-8022-6	109	3088		1123	
34-8022-6(PHILCO)	110MP			1123(2)	
34-8022-7	109	3087		1123	
34-8022-77	109			1123	
34-8026-1			504A	1104	
34-8026-2			504A	1104	
34-8026-3			504A	1104	
34-8026-4			504A	1104	
34-8027	112		1N82A		
34-8028	358		42		
34-8034	113A		6GC1		
34-8034-1	116	3017B/117	504A	1104	
34-8034-2	116	3017B/117	504A	1104	
34-8034-3	116	3017B/117	504A	1104	
34-8034-4	116	3017B/117	42	1104	
34-8034-7	113A	3119/113	6GC1		
34-8036-1	116	3017B/117	504A	1104	
34-8036-2	116	3017B/117	504A	1104	
34-8036-3	116	3017B/117	504A	1104	
34-8036-4	116	3017B/117	504A	1104	
34-8037	113A	3119/113	6GC1		
34-8037-1	113A	3119/113	6GC1		
34-8037-2	113A	3119/113	6GC1		
34-8037-3	113A	3119/113	6GC1		
34-8037-4	113A	3119/113	6GC1		
34-8040-2	116	3017B/117	504A	1104	
34-8042-1	116	3017B/117	504A	1104	
34-8042-2	116	3017B/117	504A	1104	
34-8042-3	116		504A	1104	
34-8043-4	358		42		
34-8047-1	178MP		300(2)	1122(2)	
34-8047-2			504A	1104	
34-8048-1	116	3017B/117	504A	1104	
34-8048-2	116	3016	504A	1104	
34-8048-3			504A	1104	
34-8048-4			504A	1104	
34-8048-5	116	3017B/117	504A	1104	
34-8050-1			504A	1104	
34-8050-10	125	3033	510,531	1114	
34-8050-14	116	3017B/117	504A	1104	
34-8050-2	116	3017B/117	504A	1104	
34-8050-5			504A	1104	
34-8050-6			504A	1104	
34-8050-7	116	3017B/117	504A	1104	
34-8050-8			504A	1104	
34-8050-9			504A	1104	
34-8051	116	3017B/117	504A	1104	
34-8053-2	118	3066	CR-1		
34-8053-3	118	3066	CR-1		
34-8053-4	118	3066	CR-1		
34-8053-7	118		CR-1		
34-8054-1	116	3016	504A	1104	
34-8054-10	116	3051/156	504A	1104	
34-8054-11	116	3017B/117	504A	1104	
34-8054-12	116	3017B/117	504A	1104	
34-8054-13	116		504A	1104	
34-8054-14	116	3032A	504A	1104	
34-8054-15	116	3031A	504A	1104	
34-8054-16	507	3315	504A	1104	
34-8054-17	515	9098	504A	1104	
34-8054-18	116	3017B/117	504A	1104	
34-8054-2	116	3016	504A	1104	
34-8054-23	116	3016	504A	1104	
34-8054-24	506	3843	511	1114	
34-8054-25	506	3843	511	1114	
34-8054-27	116	3016	504A	1104	
34-8054-3			504A	1104	
34-8054-4	116		504A	1104	
34-8054-5	116	3016	504A	1104	
34-8054-6	116	3057/136A	504A	1104	
34-8054-6(PHILCO)	136A		ZD-5.6	561	
34-8054-7	116	3311	504A	1104	
34-8054-8	506	3843	511	1114	
34-8054-9		3016	504A	1104	
34-8055-2	116(3)	3016	CR-3	1104	
34-8055-3	116	3016	504A	1104	
34-8056-1	119	3109	CR-2		
34-8057-1		3126	90		
34-8057-10	5071A	3334	ZD-6.8	561	
34-8057-11	116	3081/125	504A	1104	
34-8057-12	145A	3063	ZD-15	564	
34-8057-13	177(2)	3100/519	300(2)	1122	
34-8057-14	142A	3062	ZD-12	563	
34-8057-15	119	3109	CR-2		
34-8057-16	138A	3059	ZD-7.5		
34-8057-18	116	3311	504A	1104	
34-8057-19		3126	90		
34-8057-22	506	3843	511	1114	
34-8057-23	109	3087		1123	
34-8057-24	506	3843	511	1114	
34-8057-25	109	3087		1123	
34-8057-26	109	3087		1123	
34-8057-27	138A	3059	ZD-7.5		
34-8057-28	116	3087	504A	1104	
34-8057-29	109	3088		1123	
34-8057-3	110MP	3088	1N60	1123(2)	
34-8057-30	109	3087		1123	
34-8057-31	5080A	3336	ZD-22		
34-8057-32	136A	3057	ZD-5.6	561	
34-8057-33	142A	3062	ZD-12	563	
34-8057-34	178MP		300(2)	1122(2)	
34-8057-37	135A	3056	ZD-5.1		
34-8057-38	147A		ZD-33		
34-8057-39		3126	511	1114	
34-8057-4	5076A	9022	ZD-17		
34-8057-40	177		300	1122	
34-8057-41	142A		ZD-12	563	
34-8057-42	5014A	3780	ZD-6.8	561	
34-8057-43	5027A	3793	ZD-18		
34-8057-44	5029A	3795	ZD-20		
34-8057-45	116		504A	1104	
34-8057-49	5014A	3780	ZD-6.8	561	
34-8057-5	177		300	1122	
34-8057-52	109			1123	
34-8057-53	142A	3062	ZD-12	563	
34-8057-55	5007A	3773	ZD-3.9		
34-8057-56		3092	ZD-11	563	
34-8057-6	177	3087	300	1122	
34-8057-7	135A		ZD-5.1		
34-8057-8	178MP	3031A	300(2)	1122(2)	
34-8057-9	138A	3060/139A	ZB-7.5	562	
34-8057-9(PHILCO)	139A		ZD-9.1	562	
34-8058	120	3110	CR-3		
34-8058-1	120	3110	CR-3	1103	
34-8058-2	120	3110			
34-8058-7	120	3110	CR-3		
34-8059-1	230	3042			
34-8059-2	231	3857			
34-8061-1	502	3067	CR-4		
34-8062-1	505	3108	CR-7		
34-8062-4	504	3108/505	CR-6		
34-9037-1	113A	3119/113	6GC1		
034A	194	3275	220		
34E31	108	3452	86	2038	
34E3L				2038	
34H31	108	3452	86	2038	
34M14Z	144A	3094	ZD-14	564	
34M14Z10	144A	3094	ZD-14	564	
34M14Z5	144A	3094	ZD-14	564	
34P1AA	159	3466	82	2032	
34P4	177	3100/519	300	1122	
34R2	5990	3608	5100		
34Z140	144A	3094	ZD-14	564	
34Z14D10	144A	3094	ZD-14	564	
34Z14D5	144A	3094	ZD-14	564	
035	156	3051	512		
35(RCA)	159	3466	82	2032	
35-003-001	116		504A	1104	
35-010-04	519		514	1122	
035-013-0	611	3324			
35-020-21	128		243	2030	
35-1000	125	3033	510,531	1114	
35-1003	125	3032A	510,531	1114	
35-1004	116	3017B/117	504A	1104	
35-1005	116	3017B/117	504A	1104	
35-1006	5892		5064		
35-1008	5802	9005		1143	
35-1014	177	3175	300	1122	
35-1029	125	3033	510,531	1114	
35-210631-01	5635	3533			
35-210631-02	5635	3533			
35-39306001	123A	3444	20	2051	
35-39306002	123A	3444	20	2051	
35-39306003	123A		20	2051	
35-ALD	159	3466	82	2032	
35B611	167	3647	504A	1172	
35BL611	167	3647	504A	1172	
35C05	5980	3610	5096		
35C05R	5981	3698/5987	5097		
35C10	5982	3609/5986	5096		
35C100	6002	7202	5108		
35C100R	6003	7203	5109		
35C10R	5983	3698/5987	5097		
35C15	5986	3609	5096		
35C15R	5987		5097		
35C20	5986	3609	5096		
35C20R	5987	3698	5097		
35C30	5988	3608/5990	5100		
35C30R	5991	3518/5995	5101		
35C40	5990	3608	5100		
35C40R	5991	3518/5995	5101		
35C50	5992		5104		
35C50R	5993		5105		
35C60	5994		5104		
35C60R	5995		5105		
35C70	5998		5108		
35C70R	5999		5109		
35C80	5998		5108		
35C80R	5999		5109		
35C90	6002	7202	5108		
35C90R	6003	7203	5109		
35H10	5986	3609	5096		
35H10R	5987	3698	5097		
35H20	5986	3609	5096		
35H30	5990	3608	5100		
35H40	5990	3608	5100		
35H5	5980	3610	5096		
35HR20	5987	3698	5097		
35HR30	5991	3518/5995	5101		
35HR40	5991	3518/5995	5101		
35MB10A					MDA3510
35MB1A					MDA3501
35MB2A					MDA3502
35MB4A					MDA3504
35MB5A					MDA3500
35MB6A					MDA3506
35MB8A					MDA3508
35P1	102	3004	2	2007	
35P2	102	3004	2	2007	
35P2C	102	3004	2	2007	
35P4	177		300	1122	
35T1	102	3004	2	2007	
35Z6	5009A	3775	ZD-4.7		
35Z6A	5009A	3775	ZD-4.7		
36(SEARS)	160		245	2004	
0036-001	159	3466	82	2032	
36-0041	1142	3485	IC-128		
36-0083	1140	3473	IC-138		
36-1	107	3114/290	11	2015	
36-6015-46	107		11	2015	
36-6016-59	193		67	2023	
36-6343	142A	3062	ZD-12	563	
36D-32	177	3100/519	300	1122	
36E004-1	109	3088		1123	
36J003-1	102A	3004	53	2007	
36P1	102	3004	2	2007	
36P1C	102	3004	2	2007	
36P1F	102	3004	2	2007	
36P2F	102	3004	2	2007	
36P3	102	3004	2	2007	
36P3A	102	3004	2	2007	
36P3C	102	3004	2	2007	
36P4	102	3004	2	2007	
36P4C	102	3004	2	2007	
36P5	102	3004	2	2007	
36P5C	102	3004	2	2007	
36P6C	102A		53	2007	
36P7	102A	3004	53	2007	

Industry Standard No.	ECG	SK	GE	RS 276-	MOTOR.
36P7C	102A	3004	53	2007	
36P7T	102A	3004	53	2007	
36P8	102A	3004	53	2007	
36P8C	102A	3004	53	2007	
36R2	5994	3501	5104		
36R28	5994	3501	5104		
36T1	102	3118	2	2007	
36Z6	135A	3056	ZD-5.1		
36Z6A	135A	30560135A	ZD-5.1		
037	123A	3444	20	2051	
37-19201	234		65	2050	
37-193MP	199		62	2010	
37-210	5697	3522			
37-21401	199		62	2010	
37-8-01	358		42		
37-8-1	358		42		
37T1	126	3008	52	2024	
37Z6	136A	3057	ZD-5.6	561	
37Z6A	136A	3057	ZD-5.6	561	
38	116	3122	504A	2030	
38-11016	502	3067	CR-4		
38-8057-6	519		514	1122	
38A4148-000	501B	3069/501A	520		
38A4148-001	501B	3069/501A	520		
38A64C	138A	3059	ZD-7.5		
38P1	121	3009	239	2006	
38P1C	121	3009	239	2006	
38R2	5998		5108		
38R28	5998		5108		
38T1	102	3004	2	2007	
38Z6	137A	3058	ZD-6.2	561	
38Z6A	137A	3058	ZD-6.2	561	
039	129	3025	244	2027	
39(SHARP)	142A		ZD-12	563	
39-02	116		504A	1104	
039-033-0	988	3973			
39-033-2	988	3973			
39-047-1	829	3891			
39-059-0	1214	3736			
39-060-0	1260	3744			
39-061-0	1192	3445			
39-077-0	834	3569			
39-13			ZD-12	563	
39A69-2	177	3100/519	300	1122	
39A9	102	3004	2	2007	
39P1	121	3009	239	2006	
39P1C	121	3009	239	2006	
39PC1			16	2006	
39R-26	146A		ZD-27		
39T1	102	3004	2	2007	
39Z6	5014A	3780	ZD-6.8	561	
39Z6A	5014A	3780	ZD-6.8	561	
40-0068-2	128		243	2030	
40-035-0	4011B	4011	4011		
40-0502	117	3017B	504A	1104	
40-065-19-016	7440	7440			
40-065-19-027	7475	7475		1806	
40-065-19-029	7486	7486		1827	
40-065-19-030	7495	7495			
40-09297	519		514	1122	
40-09437	234	3247	65	2050	
40-09952	234	3247	65	2050	
40-102430	124		12		
40-11253	234	3247	65	2050	
40-601	102	3004	2	2007	
40A100	5982	3609/5986	5096		1N1184A
40A150	5986	3609	5096		
40A200	5986	3609	5096		1N1186A
40A300	5988	3608/5990	5100		
40A400	5990	3608	5100		1N1188A
40A50	5980	3610	5096		1N1183A
40A500	5992	3501/5994	5104		
40A600	5994	3501	5104		1N1190A
40AS	116	3016	504A	1104	
40B					MR1124
40B5	116	3016	504A	1104	
40BR					1N3883
40BS	116	3017B/117	504A	1104	
40C	116	3031A	504A	1104	1N4004
40C2PWBV1SP	106	3118	21	2034	
40D1					1N5401
40D1547	160	3007	245	2004	
40D2					1N5402
40D4					1N5404
40D6					1N5406
40D6665A03	116	3016	504A	1104	
40D8					1N5407
40H	116	3016	504A	1104	
40H3	5874	3500/5882	5032		
40H3P	5858	3017B/117	5020		MR1124
40H3PN	5859	3517/5883	5021		
40HP10	5982	3609/5986	5096		1N1184A
40HP100	6002	7202	5108		
40HP15	5986	3609	5096		1N1186A
40HP20	5986	3609	5096		1N1186A
40HP30	5988	3608/5990	5100		1N1188A
40HP40	5990	3608	5100		1N1188A
40HP5	5980	3610	5096		1N1183A
40HP50	5992	3501/5994	5104		1N1190A
40HP60	5994	3501	5104		1N1190A
40HP70	5998		5108		
40HP80	5998		5108		
40HP90	6002	7202	5108		
40HFR10	5983	3698/5987	5097		
40HFR100	6003	7203	5109		
40HFR15	5987	3698	5097		
40HFR20	5987	3698	5097		
40HFR30	5989	3518/5995	5101		
40HFR40	5991	3518/5995	5101		
40HFR5	5981	3698/5987	5097		
40HFR50	5993	3518/5995	5105		
40HFR60	5995	3518	5105		
40HFR70	5999		5109		
40HFR80	5999		5109		
40HFR90	6003	7203	5109		
40HR3P	5820				1N3883
40J2	116	3016	504A	1104	
40J3P	5878	3500/5882	5036		
40JH3	6060		5132		
40JH3R	6061	7261	5133		
40JZ	116	3017B/117	504A	1104	
40K	116	3016	504A	1104	
40KR	116	3016	504A	1104	
40M	116	3016	504A	1104	
40N1	116	3016	504A	1104	
40NJ	116	3017B/117	504A	1104	
40P1	102	3004	2	2007	
40P2	102	3004	2	2007	
40P2(POWER)	5874		5032		
40P3	5874	3500/5882	5032		
40Q3	5920	7100/5900	5068		
40Q4	5900	7100	5068		
40R28	6002	7202	5108		
40R3	5990	3608	5100		
4083	5990	3608	5100		
4085	5838		5004		
40Y3P	116	3016	504A	1104	
041	123A	3444	20	2051	
41-001	120	3110	CR-3		
41-0318	130	3027	14	2041	
41-0318A	130	3027	14	2041	
41-032	116(4)	3017B/117	504A(4)	1104	
41-0499	123A	3444	20	2051	
41-0500	129	3025	244	2027	
41-0500A	129	3025	244	2027	
41-0606	198		251		
41-0609	198	3220	251		
41-0905	142A	3062	ZD-12	563	
41-0906	5066A	3330	ZD-3.3		
41-0909	175	3538	246	2020	
41-J2	116		504A	1104	
041A	222	3050/221		2036	
41B581014	128	3024	243	2030	
41B581144-P001	175		246	2020	
41C-407	1003	3288	IC-43		
41C-409	1110	3229			
41E3-1			1N295	1123	
41J2		3016	504A	1104	
41M-19			11	2015	
41N	161	3018	39	2015	
41N1	108	3452	86	2038	
41N2	108	3452	86	2038	
41N2A	108	3452	86	2038	
41N2AA	108	3452	86	2038	
41N2B	108	3452	86	2038	
41N2M	161	3132	86	2038	
41N3	108	3452	86	2038	
41Z6	5016A	3782	ZD-8.2	562	
41Z6A	5016A	3782	ZD-8.2	562	
042	123A	3444	20	2051	
042(I.C.)	710		IC-89		
42-0094-399			52	2024	
42-051	125	3032A	510,531	1114	
42-14027	116	3016	504A	1104	
42-16599	121	3009	239	2006	
42-17143	102	3004	2	2007	
42-17443	116	3016	504A	1104	
42-17443A	116	3016	504A	1104	
42-17444	123	3124/289	20	2051	
42-18109	176	3845	80		
42-18111	123	3124/289	20	2051	
42-18310	124	3021	12		
42-19642	128	3024	243	2030	
42-19643	129	3025	244	2027	
42-19644	123	3124/289	20	2051	
42-19645	125		504A	1114	
42-19670	123	3124/289	20	2051	
42-19671	102A	3004	53	2007	
42-19681	110MP	3709		1123(2)	
42-19682	160	3005	245	2004	
42-19683	108	3452	86	2038	
42-19792	160	3006	245	2004	
42-19840	123A	3444	20	2051	
42-19862	103A	3010	59	2002	
42-19862A	158	3004/102A	53	2007	
42-19863	102A	3004	53	2007	
42-19863A	102A	3004	53	2007	
42-19864	102A	3004	53	2007	
42-19864A	102A	3004	53	2007	
42-19865	116	3016	504A	1104	
42-19917	143A	3750	ZD-13	563	
42-2	1111		1N82A		
42-20222	102A	3004	53	2007	
42-20738	123	3124/289	20	2051	
42-20739	129	3025	244	2027	
42-20960	175	3026	246	2020	
42-20961	130	3027	14	2041	
42-20961A	130	3027	14	2041	
42-21232	129	3025	244	2027	
42-21233	128	3024	243	2030	
42-21234	123A	3444	20	2051	
42-21362	110MP	3709	504A	1123(2)	
42-21400	116		504A	1104	
42-21401	161	3018	39	2015	
42-21402	161	3018	39	2015	
42-21403	126	3008	52	2024	
42-21404	103A	3835	59	2002	
42-21405	158	3025/129	53	2007	
42-21406	102A	3004	53	2007	
42-21407	123	3124/289	20	2051	
42-21408	116	3031A	504A	1104	
42-21443	121MP	3015	239(2)	2006(2)	
42-21866	116	3016	504A	1104	
42-22008	159	3466	82	2032	
42-22008A	159	3466	82	2032	
42-22009	172A	3156	64		
42-22154			63	2030	
42-22158	123A	3444	20	2051	
42-22532	107	3018	11	2015	
42-22533	123A	3444	20	2051	
42-22534	102A	3004	53	2007	
42-22535	158	3004/102A	53	2007	
42-22535Q			2	2007	
42-22536		3126	90		
42-22537	109	3087		1123	
42-22538	177	3100/519	300	1122	
42-22539		3087	1N34AS	1123	
42-22539(DIODE)	109			1123	
42-22540	119	3016	CR-2		
42-22755	109	3087		1123	
42-22778	160	3006	245	2004	
42-22779	160	3006	245	2004	
42-22780	160	3006	245	2004	
42-22781	160	3006	245	2004	
42-22784	160	3006	245	2004	
42-22785	108	3018	86	2038	
42-22786	123A	3444	20	2051	
42-22787	123A	3444	20	2051	
42-22809	123A	3124/289	20	2051	
42-22810	159	3025/129	86	2032	
42-22810A	159	3114/290	86	2032	
42-22811	123A	3124/289	20	2051	
42-22812	123A	3124/289	20	2051	
42-22834	121MP	3009	239(2)	2006(2)	
42-22835	116	3311	504A	1104	
42-22847	123A	3124/289	20	2051	
42-23348	123A	3124/289	20	2051	
42-23349	123A	3124/289	20	2051	
42-23350	116	3031A	504A	1104	
42-23350A	116	3016	504A	1104	
42-23459	152	3054/196	66	2048	
42-23541	159	3466	82	2032	
42-23541A	159	3466	82	2032	
42-23542	123A		20	2051	
42-23622	102A		53	2007	
42-23960	107		11	2015	
42-23960P	107	3018	11	2015	
42-23961	107		11	2015	

Industry Standard No.	ECG	SK	GE	RS 276-	MOTOR.
42-23961P	107		11	2015	
42-23962	107		11	2015	
42-23962P	107	3018	11	2015	
42-23963	107		11	2015	
42-23963P	107	3018	11	2015	
42-23964	123A	3124/289	20	2051	
42-23964P	123A	3124/289	20	2051	
42-23965	126		52	2024	
42-23965P	160	3006	245	2004	
42-23966	123A	3124/289	20	2051	
42-23966P	123A	3124/289	20	2051	
42-23967	158		53	2007	
42-23967P	158	3004/102A	53	2007	
42-23968	121	3717	239	2006	
42-23968P	121	3009	239	2006	
42-23969	109	3088		1123	
42-23970		3126	90		
42-23972	110MP	3709	1N60	1123(2)	
42-23975	116	3311	504A	1104	
42-23977	5809	9010			
42-24263	172A	3156	64		
42-24387	177	3175	300	1122	
42-27202	116	3311	504A	1104	
42-27277	199	3124/289	62	2010	
42-27278	116	3311	504A	1104	
42-27372		3018	60	2015	
42-27373		3018	60	2015	
42-27374		3124	20	2051	
42-27375		3124	20	2051	
42-27376		3004	2	2007	
42-27378		3087	1N34AS	1123	
42-27379		3126	90		
42-27380		3087	1N34AS	1123	
42-27381		3087	1N34AS	1123	
42-27463		3311	504A	1104	
42-27529	229		60	2038	
42-27530	229		61	2038	
42-27533		3122	210	2010	
42-27534	199	3122	212	2010	
42-27535	199	3122	62	2010	
42-27536	159	3114/290	221	2032	
42-27537	289A	3122	210	2038	
42-27538	153	3191/185	69	2049	
42-27539	152	3190/184	66	2048	
42-27541	142A	3062	ZD-12	563	
42-27542	601	3463			
42-27543	109	3088	1N60	1123	
42-27544		3088	1N60	1123(2)	
42-28056	123A	3124/289	20	2051	
42-28057	158	3004/102A	53	2007	
42-28058	116	3311	504A	1104	
42-28199	109	3088		1123	
42-28200		3126	90		
42-28201	177		300	1122	
42-28202	116	3031A	504A	1104	
42-28203	108	3452	86	2038	
42-28204	108	3452	86	2038	
42-28205	123A	3018	20	2051	
42-28206	108	3452	86	2038	
42-28207	123A	3444	20	2051	
42-28208	159	3466	82	2032	
42-28210	123A		20	2051	
42-28211	159	3466	82	2032	
42-28212	184	3054/196	57	2017	
42-28213	185	3083/197	58	2025	
42-30092	199		62	2010	
42-7			504A	1104	
42-9029-31B	199	3245	62	2010	
42-9029-31L	199	3245	62	2010	
42-9029-31M	123A	3444	20	2051	
42-9029-31P	199	3245	62	2010	
42-9029-31Q	199	3245	62	2010	
42-9029-31R	123A	3444	20	2051	
42-9029-31V	123A		20	2051	
42-9029-40C	123A	3444	20	2051	
42-9029-40F	234		65	2050	
42-9029-40L	123A	3444	20	2051	
42-9029-40T	128		243	2030	
42-9029-40U	129		244	2027	
42-9029-40W	234	3247	65	2050	
42-9029-40X	159	3466	82	2032	
42-9029-40Y	123A	3444	20	2051	
42-9029-60A	123A	3444	20	2051	
42-9029-60B	234	3247	65	2050	
42-9029-60C	123A	3444	20	2051	
42-9029-60D	199	3245	62	2010	
42-9029-60Q	159	3466	82	2032	
42-9029-60W	129		244	2027	
42-9029-70C	123A	3444	20	2051	
42-9029-70D	159	3466	82	2032	
42-9029-70E	159	3466	82	2032	
42-9029-70F	123A	3444	20	2051	
42-9029-70G	129		244	2027	
42-9029-70J	128		243	2030	
42-9029-70K	128		243	2030	
42-9029-70P	123A	3444	20	2051	
42-9029-70Q	234	3247	65	2050	
42-Z6A	139A	3060	ZD-9.1	562	
042A	710	3102	IC-89		
42A11	116	3016	504A	1104	
42A14	109	3087		1123	
42A23	116	3016	504A	1104	
42B16	116	3016	504A	1104	
42B2	116	3016	504A	1104	
42J2	116	3017B/117	504A	1104	
42R2	5874		5032		
42X210	103	3862	8	2002	
42X230	102A	3004	53	2007	
42X233	102	3004	2	2007	
42X244	116	3016	504A	1104	
42X244B	116	3016	504A	1104	
42X245	116	3016	504A	1104	
42X245B	116	3016	504A	1104	
42X25	116	3016	504A	1104	
42X308		3004	2	2007	
42X309	102	3004	2	2007	
42X310	103	3862	8	2002	
42X311	102	3004	2	2007	
42X32	116	3016	504A	1104	
42Z6	139A	3060	ZD-9.1	562	
0043-004793	722		IC-9		
43-0203845	129	3025	244	2027	
43-022861	108	3452	86	2038	
43-023190	181	3036	75	2041	
43-023212	123A	3020/123	20	2051	
43-023221	123A	3124/289	20	2051	
43-023222	129	3025	244	2027	
43-023223	128	3024	243	2030	
43-023843	130	3027	14	2041	
43-023844	128	3024	243	2030	
43-025763	108	3452	86	2038	
43-025764	108	3018	86	2038	
43-025765	108	3452	86	2038	
43-025766	123A	3124/289	20	2051	
43-025767	108	3452	86	2038	
43-025834	131	3052	44	2006	
43-540012	116	3016	504A	1104	
43-VD-09	128		243	2030	
43A111449	160	3005	245	2004	
43A113534	177		300	1122	
43A114346-1	519		514	1122	
43A114346-P1	519		514	1122	
43A114832-3	519		514	1122	
43A126932	128		243	2030	
43A128340-1	123A		20	2051	
43A128340-2	123A		20	2051	
43A128340-3	123A	3122	20	2051	
43A128340-4	123A		20	2051	
43A128342-1	123A	3122	20	2051	
43A128342-2	123A	3122	20	2051	
43A128342-3	123A		20	2051	
43A128342-4	123A		20	2051	
43A128342-5	123A		20	2051	
43A128342-6	123A		20	2051	
43A128342-7	123A		20	2051	
43A144188-5	5675	3520/5677			
43A145291-1	159	3466	82	2032	
43A145291-2	159	3466	82	2032	
43A162445P1	199		62	2010	
43A162455-1	128		243	2030	
43A165137P1	130		14	2041	
43A165137P3	130		14	2041	
43A165137P4	130		14	2041	
43A167207P1	159	3466	82	2032	
43A167207P2	159	3466	82	2032	
43A167229-01	177		300	1122	
43A167851	123A		20	2051	
43A167885-P1	181		75	2041	
43A167885-P2	181	3036	75	2041	
43A167885-P3	181		75	2041	
43A167886P4	218		234		
43A168016P1	123A		20	2051	
43A168064-1	159	3466	82	2032	
43A168064P1	159	3466	82	2032	
43A168440	912		IC-172		
43A168440-1	912		IC-172		
43A168481	941D		IC-264		
43A168481P1	941D		IC-264		
43A175989P1	519		514	1122	
43A176002	159	3466	82	2032	
43A180002-P1	123A		20	2051	
43A180002-P2	123A		20	2051	
43A212040-1	909	3590	IC-249		
43A212040-2	909D	3590	IC-250		
43A212042-P2	910D		IC-252		
43A212042P1	910		IC-251		
43A212042P2	910D		IC-252		
43A212067	130		14	2041	
43A212090P1	199		62	2010	
43A223006P1	74H00	74H00			
43A223007	7401	7401			
43A223009	7402	7402		1811	
43A223015	7430	7430			
43A223025	7473	7473		1803	
43A223026P1	7474	7474		1818	
43A223028	7476	7476		1813	
43A223029P1	7442	7442			
43A223030	7495	7495			
43A223034P1	7493A	7493			
43A223046-1	923		IC-259		
43A223060-1	128		243	2030	
43B140883-1	123A	3444	20	2051	
43B168450-1	159	3466	82	2032	
43B168495-1	159	3466	82	2032	
43B168566-P1	159	3466	82	2032	
43B168610	123A	3444	20	2051	
43B168613-1	199	3245	62	2010	
43C168567	123A	3444	20	2051	
43C216408P1	74H00	74H00			
43C216410P1	74H04	74H04			
43C216447	74193	74193		1820	
43C216447P1	74193	74193		1820	
43N3	123	3124/289	20	2051	
43N6	123	3124/289	20	2051	
43P1	100	3005	1	2007	
43P2	102	3004	2	2007	
43P3	100	3005	1	2007	
43P4	102	3004	2	2007	
43P4C	102	3004	2	2007	
43P6	102	3004	2	2007	
43P6A	102	3004	2	2007	
43P6C	100	3004/102A	1	2007	
43P7	102	3004	2	2007	
43P7A	102	3004	2	2007	
43P7C	102	3004	2	2007	
43X16A567	123		20	2051	
43Z6	140A	3061	ZD-10	562	
43Z6A	140A	3061	ZD-10	562	
44-13	126	3006/160	52	2024	
44-2020-07800	506	3843			
44-44886G01	289A		268	2038	
44-530	116	3031A	504A	1104	
044-9667-02	123A	3444	20	2051	
44A-1A5	121	3717	239	2006	
44A-417014-001	181		75	2041	
44A319819-1	128		243	2030	
44A332168-001	941		IC-263	010	
44A332168-002	941		IC-263	010	
44A332169-002	941		IC-263	010	
44A333413	923		IC-259		
44A333463-001	123A	3444	20	2051	
44A333464	159	3466	82	2032	
44A333464-1	159	3466	82	2032	
44A335854-001	912		IC-172		
44A353980-002	175	3131A/369	246	2020	
44A354637-001	175	3026	246	2020	
44A355565-001	130		14	2041	
44A355565001	130		14	2041	
44A358624-003	129		244	2027	
44A359497-001	128		243	2030	
44A359497-002	128		243	2030	
44A390202-001	909		IC-249		
44A390202-003	909		IC-249		
44A390205	716		IC-208		
44A390243-001	175		246	2020	
44A390244-001	130	3027	14	2041	
44A390247	123A	3444	20	2051	
44A390248-001	159	3466	82	2032	
44A390249	123A	3444	20	2051	
44A390251	123A		20	2051	
44A390251-001	123A		20	2051	
44A390256-001	159	3466	82	2032	
44A390261	159	3466	82	2032	
44A390264-001	128		243	2030	
44A391505	129		244	2027	
44A391593-001	181	3036	75	2041	
44A393605	941D		IC-264		
44A393611	923D		IC-260	1740	

Industry Standard No.	ECG	SK	GE	RS 276-	MOTOR.
44A393611-001	923D		IC-260	1740	
44A395909-1	181		75	2041	
44A395986-001	152	3893	66	2048	
44A395992-001	128		243	2030	
44A395992-002	128		243	2030	
44A395994-001	123A	3444	20	2051	
44A397905	159	3466	82	2032	
44A417031-001	159	3466	82	2032	
44A417032-001	152	3893	66	2048	
44A417033-001	130		14	2041	
44A417034	192		63	2030	
44A417063-001	128		243	2030	
44A417716	130		14	2041	
44A417716-001	130	3027	14	2041	
44A417756-1	197		250	2027	
44A417779-001	941M			007	
44A41779-001			IC-265	007	
44A418041-001	159	3466	82	2032	
44A4417714	181		75	2041	
44A4417714-001	181		75	2041	
44B1	738	3167	IC-29		
44B1B	738	3167	IC-29		
44B1Z	738	3167	IC-29		
44B238203-1	159	3466	82	2032	
44B238208-001	175		246	2020	
44B238208-002	175		246	2020	
44B238208002	175		246	2020	
44B238246	159	3466	82	2032	
44B311097	123A	3444	20	2051	
44L			20	2051	
44P1	160	3008	245	2004	
44R2	5858		5020		
44R2R	5859		5021		
44T-100-119	1104	3225			
44T-100-120	1155	3231	IC-179	705	
44T-300-100	964	3630			
44T-300-102	1155		IC-179	705	
44T-300-103		3122	62	2051	
44T-300-104	229	3122	213	2038	
44T-300-105	312	3448	FET-2	2035	
44T-300-106	199	3124/289	212	2010	
44T-300-107	295	3253	270		
44T-300-108	295	3197/235	270		
44T-300-109	235	3197	215		
44T-300-110		3246	61	2038	
44T-300-111	199	3124/289	62	2010	
44T-300-112	199	3124/289	212	2010	
44T-300-113	312	3448	FET-2	2035	
44T-300-91	614		90		
44T-300-92	139A	3056/135A	ZD-5.6	561	
44T-300-93	140A	3061	ZD-10	562	
44T-300-94	138A	3059	ZD-7.5		
44T-300-95		3311	504A	1104	
44T-300-96	177	3100/519	300	1122	
44T-300-97	109		1N60	1123	
44T-300-99	1104		IC-91		
045	221	3050		2026	
045-1		3039	11	2015	
045-1(SYLVANIA)	108	3452	86	2038	
045-2		3039	11	2015	
045-2(SYLVANIA)	108	3452	86	2038	
45A2FX355	116	3016	504A	1104	
45AN4AA	123A	3444	20	2051	
45B17	804	3455	IC-27		
45B1AH	804	3455	IC-27		
45B1D	804	3455	IC-27		
45B1Z	804	3455	IC-27		
45N1	103	3010	8	2002	
45N2	103	3010	8	2002	
45N2A	103A	3835	59	2002	
45N2M	123A	3444	20	2051	
45N3	123A	3444	20	2051	
45N4	123A	3444	20	2051	
45N4M	123A	3444	20	2051	
45NP	108	3452	86	2038	
45X1A502C	102	3004	2	2007	
45X1A520C	102A	3123	53	2007	
45X2	100	3005	1	2007	
046-0134	177	3175	300	1122	
46-06311-3	107		11	2015	
046-07037	5873	3517/5883	5065		
046-0909	177	3017B/117	504A	1104	
046-1			28	2017	
046-1(SYLVANIA)	186		28	2017	
46-119-3	161	3018	39	2015	
46-13101-3	1080	3284	IC-98		
46-13103-3	1094		IC-157		
46-13105-3	1122		IC-40		
46-13121-3	712		IC-114		
46-13124-3	738	3167	IC-29		
46-13125-3	1183	3480/1178			
46-13131-3	1009	3499	IC-77		
46-13133-3	1200	3714			
46-13145-3	712	3072	IC-148		
46-13172-3		3920	IC-293		
46-13173-3		3477	IC-103		
46-1340-3	711	3070	IC-207		
46-1343-3	1003	3288	IC-43		
46-1346-3	711	3070	IC-207		
46-1347-3	710	3102	IC-89		
46-1348-3	710	3102	IC-89		
46-1352-3	707		IC-206		
46-1356-3	710	3102	IC-89		
46-1357-3	1004	3365	IC-149		
46-1361-3	712	3072	IC-2		
46-136279P1	909	3590	IC-249		
46-136284-P1	923		IC-259		
46-136284-P2	923D		IC-260	1740	
46-1365-3	749	3168	IC-97		
46-1366-3	748	3236			
46-1369-3	1004	3365	IC-149		
46-1370-3	1109	3711	IC-99		
46-1382-3	1108		IC-87		
46-1384-3	712		IC-2		
46-1393-3	1130	3478	IC-111		
46-1394-3	1131	3286	IC-109		
46-1395-3	1134	3489	IC-106		
46-1396-3	1128	3488	IC-105		
46-1397-3	1132	3287	IC-110		
46-1398-3	1133	3490	IC-107		
46-156741P2	941D		IC-264		
46-16261	116		504A	1104	
46-16261-3	116	3311	504A	1104	
46-163-3	102A	3004	53	2007	
46-203200P39	949		IC-25		
46-29-6	712		IC-148		
46-34-3	116		504A	1104	
046-40209	5872	3500/5882	5064		
46-41763C01	5072A		ZD-8.2	562	
46-5002-1	703A		IC-12		
46-5002-10	706	3288/1003	IC-43		
46-5002-11	790	3454	IC-230		
46-5002-12	714	3075	IC-4		
46-5002-13	715	3076	IC-6		
46-5002-14	739	3235	IC-30		
46-5002-15	712	3072	IC-2		
46-5002-16	780	3141	IC-222		
46-5002-17	746	3234	IC-217		
46-5002-18	747	3279	IC-218		
46-5002-2	703A		IC-12		
46-5002-20	923D	3165	IC-260	1740	
46-5002-21	782		IC-224		
46-5002-23	740A	3328	IC-31		
46-5002-26	712		IC-2		
46-5002-27	779A		IC-13		
46-5002-28	749	3168	IC-97		
46-5002-3	717		IC-209		
46-5002-30	799	3238			
46-5002-33	731	3170	IC-13		
46-5002-4	703A	3157	IC-12		
46-5002-5	782		IC-224		
46-5002-6	711	3070	IC-207		
46-5002-7	703A	3157	IC-12		
46-5002-8	748	3236			
46-5002-8B	748	3236			
46-5002-9	718	3159	IC-12		
46-500230	799	3238	IC-34		
46-61249-3	116		504A	1104	
46-61267-3	177	3100/519	300	1122	
46-61307-3	177	3100/519	300	1122	
46-6661-2	117		504A	1104	
46-67120A13	116	9075/166	504A	1104	
46-6829			11	2015	
46-80309-3	177	3100/519	300	1122	
46-81187-3	171	3201	27		
46-8257-3	123A	3122	20	2051	
46-836380-3	177		300	1122	
46-840-3	102A		53	2007	
46-84120-3	108	3452	86	2038	
46-85285-3	108	3452	86	2038	
46-86-3	116	3016	504A	1104	
46-8601-3	116	3031A	504A	1104	
46-861-3	112	3089	1N82A		
46-8610-3	102A	3004	53	2007	
46-86101-3	107	3039/316	11	2015	
46-86102-3	126		52	2024	
46-86105-3	119	3031A	CR-2		
46-86106-3	506	3843	511	1114	
46-86107-3	161	3018	39	2015	
46-86108-3	161	3018	39	2015	
46-86109-3	107	3039/316	11	2015	
46-8611	102A	3004	53	2007	
46-8611-3	102A	3004	53	2007	
46-8611-4	116		504A	1104	
46-86110-3	107	3039/316	11	2015	
46-86112-2	161	3117	39	2015	
46-86112-3	319P	3018	283	2016	
46-86113-3	161	3039/316	39	2015	
46-86114-3	161	3018	39	2015	
46-861148-3	116	3017B/117	504A	1104	
46-861149-3	506	3843	511	1114	
46-86115-3	103A	3835	59	2002	
46-86116-3	117	3017B	504A	1104	
46-86117-3	161	3018	39	2015	
46-861179-3	506	3843	511	1114	
46-86118-3	161	3018	39	2015	
46-861187-3	177	3100/519	300	1122	
46-86119-3	161	3018	39	2015	
46-8612-3	160		245	2004	
46-86120-3	107	3018	11	2015	
46-86121-3	123A	3444	20	2051	
46-86122-3	123A	3444	20	2051	
46-86123-3	160	3005	245	2004	
46-86125-3	131	3198	44	2006	
46-86126-3	107	3018	11	2015	
46-86127-3	107	3018	11	2015	
46-8613	112	3089	1N82A		
46-8613-3	126	3008	52	2024	
46-8613-3(SYNC)			50	2004	
46-86130-3	286	3026	246	2020	
46-86131-3	233	3018	210	2009	
46-86132-3	289A	3444/123A	210	2038	
46-86133-3	107	3039/316	11	2015	
46-86134-3	112	3089	1N82A		
46-86135-3	131		44	2006	
46-86136-3	121	3009	239	2006	
46-86138-3	119		CR-2		
46-86139-3	116	3032A	504A	1104	
46-8614-3	101	3861	8	2002	
46-86140-3	108	3452	86	2038	
46-86141-3	125		510,531	1114	
46-86143-3	123A	3444	20	2051	
46-86144-3	123A	3444	20	2051	
46-86145-3	123A	3444	20	2051	
46-86146-3	125	3031A	510,531	1114	
46-86147-3	125	3032A	510,531	1114	
46-86148	125		510,531	1114	
46-86148-3	116	3051/156	504A	1104	
46-86149-3	116	3016	504A	1104	
46-8615-3	121	3009	239	2006	
46-86150-3	117	3017B	504A	1104	
46-86151-3	117	3017B	504A	1104	
46-86152-3	123A	3444	20	2051	
46-8616-3	110MP	3087	25		
46-86163-3	102A	3124/289	53	2007	
46-86165-3	102A		53	2007	
46-86166-3			2	2007	
46-86168-3	177	3087	300	1122	
46-86169-3	123A	3444	20	2051	
46-8617-3	121	3009	239	2006	
46-86170-3	159	3114/290	82	2032	
46-86171-3	123A	3444	20	2051	
46-86172-3	107	3018	11	2015	
46-86173-3	154	3124/289	40	2012	
46-86177-3	116	3017B/117	504A	1104	
46-86178-3	178MP	3100/519	300(2)	1122(2)	
46-86179-3	506	3843	511	1114	
46-86180-3	118	3066	CR-1		
46-86182-3	154	3024/128	40	2012	
46-86183-3	154	3020/123	40	2012	
46-86184	177	3311	300	1122	
46-86184-3	177	3100/519	300	1122	
46-86185-3	115		6GX1		
46-86186-3	177	3100/519	300	1122	
46-86187-3	177	3311	300	1122	
46-86189-3	175	3131A/369	246	2020	
46-8619	116(3)		504A(3)	1104	
46-8619-3	109	3088		1123	
46-86192-3	123A	3444	20	2051	
46-86194-3	5073A	3062/142A	ZD-9.1	562	
46-86195-3	102A	3004	53	2007	
46-86196-3	158	3004/102A	53	2007	
46-86198-3	176	3845	80		
46-86199-3	116	3031A	504A	1104	
46-862-3	160			2004	
46-8620-3	506	3843	511	1114	
46-86200-3	110MP	3087	1N34AS	1123	
46-86201-3			504A	1104	

Industry Standard No.	ECG	SK	GE	RS 276-	MOTOR.
46-86201-3(POWER)	116		504A	1104	
46-86207-3	107	3018	11	2015	
46-86208-3	108	3452	86	2038	
46-86209-3	107	3018	11	2015	
46-8621-3	166	3647/167		1152	
46-86210-3	154	3045/225	40	2012	
46-86211-3	103A	3010	59	2002	
46-86212-3	116	3031A	504A	1104	
46-86213-3	121	3009	239	2006	
46-86214-3	109			1123	
46-8622-3	127	3113/516			
46-86220-3	113A	3119/113	6GC1		
46-86224-3	161	3018	39	2015	
46-86225-3	138A	3059	ZD-7.5		
46-86226-3	175	3131A/369	246	2020	
46-86227-3	502	3067	CR-4		
46-86228-3	123A	3444	20	2051	
46-86229-3	159	3466	82	2032	
46-86229-3A	159	3466	82	2032	
46-8623-3	109			1123	
46-86230-3	159	3466	82	2032	
46-86231-3	123A	3444	20	2051	
46-86232-3	154	3040	40	2012	
46-86233-3	128	3024	243	2030	
46-86234-3	184	3054/196	57	2017	
46-86235-3	185	3083/197	58	2025	
46-86237-3	502		CR-4		
46-86238-3	159	3466	82	2032	
46-86238-3A	159	3466	82	2032	
46-86239-3	107	3018	11	2015	
46-8624-3	123A	3444	20	2051	
46-86240-3	229	3018	61	2038	
46-86244-3	108	3452	86	2038	
46-86247-2	123A	3444	20	2051	
46-86247-3	123A	3444	20	2051	
46-86249-3		3126	90		
46-8625	112	3089	1N82A		
46-8625-1	126	3008	52	2024	
46-8625-2	126	3008	52	2024	
46-8625-3	126	3008	52	2024	
46-8625-4	126	3008	52	2024	
46-8625-5	126	3008	52	2024	
46-8625-6	126	3008	52	2024	
46-86250-3	177	3100/519	300	1122	
46-86251-3	107	3018	11	2015	
46-86252-3	289A	3018	20	2051	
46-86253-3	109	3087		1123	
46-86254-3	178MP	3100/519	300(2)	1122(2)	
46-86256-1	126	3008	52	2024	
46-86256-2	126	3008	52	2024	
46-86256-3	102A	3008	53	2007	
46-86257-3	123A	3444	20	2051	
46-86261-3	116	3311	504A	1104	
46-86262-3	108	3122	86	2038	
46-86264-3	116(2)	3452/108	504A(2)	1104	
46-86265-3	107	3018	11	2015	
46-86266-3	109	3087		1123	
46-86267-3	116	3017B/117	504A	1104	
46-86268-3	123A	3444	20	2051	
46-86269-3	108	3452	86	2038	
46-86270-3	116(3)	3311	504A(3)	1104	
46-86271-3	116	3017B/117	504A	1104	
46-86274-3	123A	3444	20	2051	
46-86279-3	502	3067	CR-4		
46-86280-3	506	3843	511	1114	
46-86281-3	506	3843	511	1114	
46-86282-3	506	3843	511	1114	
46-86283-3	159	3466	82	2032	
46-86284-3	139A	3125	ZD-9.1	562	
46-86285-3	161	3018	39	2015	
46-86286-3	112		1N82A		
46-86288-3	112		1N82A		
46-86289-3	614	3327	90	1122	
46-8629	107	3019	11	2015	
46-86290-3	116(3)		504A(3)	1104	
46-86291-3	502	3067	CR-4		
46-86292-3	504	3108/505	CR-6		
46-86293-3	159	3466	82	2032	
46-86294-3	505	3108	CR-6		
46-86295-3	107	3018	11	2015	
46-86296-3	143A	3750	ZD-13	563	
46-86299-3	107	3018	11	2015	
46-8630			1N60	1123	
46-86300-3	160	3006	245	2004	
46-86301-3	107	3018	11	2015	
46-86302-3	107	3018	11	2015	
46-86303-3	116(2)	3031A	504A(2)	1104	
46-86304-3	114		6GD1	1104	
46-86305-3		3126	90		
46-86307-3	116	3032A	504A	1104	
46-86308-3	506	3843	511	1114	
46-86309-3	519	3100	514	1122	
46-8631-3	102A		53	2007	
46-86310-3	123A	3444	20	2051	
46-86311-3			11	2015	
46-86312-3	113A		6GC1		
46-86313-3	503	3068	CR-5		
46-86314-3	108	3451	86	2038	
46-86314-3A	108	3452	86	2038	
46-86315-3			18	2030	
46-86316-3	312	3834/132	FET-1	2035	
46-86317-3	152	3893	66	2048	
46-86318-3	154		40	2012	
46-86319-3	157	3103A/396	232		
46-86320-3	506	3843	511	1114	
46-86321-3	116	3032A	504A	1104	
46-86322-3	116	3017B/117	504A	1104	
46-86323-3	140A	3061	ZD-10	562	
46-86324-3	506	3843	511	1114	
46-86326-3	116	3032A	504A	1104	
46-86327-3	171	3104A	27		
46-86328-3	506	3843	511	1114	
46-86330-3	501B	3069/501A	520		
46-86331-3	506	3843	511	1114	
46-86332-3	113A	3119/113	6GC1		
46-86334-3	504	3108/505	CR-6		
46-86335-3	152	3893	66	2048	
46-86336-3	113A	3112	6GC1		
46-86337-3			504A	1104	
46-86338-3	177	3100/519	300	1122	
46-86339-3	116	3016	504A	1104	
46-8634-3	121	3009	239	2006	
46-86341-3	144A	3094	ZD-14	564	
46-86342-3			21	2034	
46-86343-3	177		300	1122	
46-86344-3	171	3044/154	27		
46-86345-3	287	3433	27		
46-86346-3	171	3044/154	27		
46-86347-3	196	3054	241	2020	
46-86348-3	152	3054/196	66	2048	
46-86349-3	175	3026	246	2020	
46-86349-9	175		246	2020	
46-86350-3	175	3026	246	2020	
46-86351-3	116	3032A	504A	1104	
46-86352-3	107	3018	11	2015	
46-86353-3	199	3018	11	2015	
46-86354-3	107	3018	11	2015	
46-86355-2			504A	1104	
46-86355-3	116	3032A	504A	1104	
46-86357-3	108	3018	86	2038	
46-86358-3	116	3017B/117	504A	1104	
46-8636-3	126	3005	52	2024	
46-86360-3	186	3192	28	2017	
46-86364-3	116	3017B/117	504A	1104	
46-86365-3	116	3031A	504	1104	
46-86371-3	128	3024	243	2030	
46-86373-3	102A	3004	53	2007	
46-86374-3	152	3054/196	66	2048	
46-86375-3	123A	3444	20	2051	
46-86376-3	123A	3452/108	86	2038	
46-86377-3	159	3466	82	2032	
46-86378-3	123A	3444	20	2051	
46-86379-3	142A	3062	ZD-12	563	
46-8638			FET-2	2035	
46-8638-3	121	3009	239	2006	
46-86380-3	178MP	3100/519	300(2)	1122(2)	
46-86381	128	3020/123	243	2030	
46-86381-3	128	3024	243	2030	
46-86382-3	506	3843	511	1114	
46-86383-3	116	3031A	504A	1104	
46-86384-3	124	3021	12		
46-86386-3	124	3021	12		
46-86387-3	124		12		
46-86388-3	130	3027	14	2041	
46-86389-3	165	3115	38		
46-8638P-3			18	2030	
46-86390-3	164	3111	37		
46-86391-3	506	3843	511	1114	
46-86392-3	5077A	3793/5027A	ZD-18		
46-86393-2	506	3843	511	1114	
46-86393-3	506	3843	511	1114	
46-86394-3	137A	3058	ZD-6.2	561	
46-86394-3(10)			ZD-6.2	561	
46-86395-3	506	3843	511	1114	
46-86396-3	221	3050	FET-4	2036	
46-86397-3	108	3452	86	2038	
46-86398-3			63	2030	
46-86399-3	159	3466	82	2032	
46-864-3	108	3452	86	2038	
46-86400-3	152	3054/196	66	2048	
46-86401-3	144A	3094	ZD-14	564	
46-86402-3	116	3016	504A	1104	
46-86403-3	129	3025	244	2027	
46-86404-3	123A	3444	20	2051	
46-86405-3	128	3020/123	243	2030	
46-86406-3	129	3114/290	244	2027	
46-86407-3	123A	3444	20	2051	
46-86408-3	123A	3444	20	2051	
46-86409-3	123A	3444	20	2051	
46-86411-3	185	3083/197	58	2025	
46-86412-3	159	3466	82	2032	
46-86415-3	165	3111	38		
46-86416-3	116	3016	504A	1104	
46-86419-3	123A	3444	20	2051	
46-8642		3126	90		
46-86420-3	116	3031A	504A	1104	
46-86421-3	139A	3060	ZD-9.1	562	
46-86422-3	177		300	1122	
46-86424-3	159	3466	82	2032	
46-86425-3	129	3025	244	2027	
46-86426-3	128	3024	243	2030	
46-86427-3	128	3024	243	2030	
46-86428-3	177	3175	300	1122	
46-86429-3	234	3025/129	65	2050	
46-8643	112	3089	1N82A		
46-8643-3	112	3089	1N82A		
46-86430-3	102A	3004	53	2007	
46-86431-3	177	3100/519	300	1122	
46-86432-3	5096A		ZD-100		
46-86433-3	116	3311	504A	1104	
46-86434-3	199	3018	62	2010	
46-86435-3	107	3018	11	2015	
46-86436-3	177	3087	300	1122	
46-86437-3	116	3311	504A	1104	
46-86438-3	116	3311	504A	1104	
46-86439-3	124	3021	12		
46-8644-3	109	3087		1123	
46-86440-3	506	3843	511	1114	
46-86442-3	5072A	3136	ZD-8.2	562	
46-86444-3	513		513		
46-86446-3	124		12		
46-86446-9	124		12		
46-86447-3	5083A	3754	ZD-28		
46-86451-3	504	3108/505	CR-6		
46-86454-3	504	3108/505	CR-6		
46-86455-3	162	3438	35		
46-86456-3	138A	3059	ZD-7.5		
46-86457-3	5013A	3779	ZD-6.2	561	
46-86458-3	5074A	3092/141A	ZD-11.0	563	
46-86459-3	196	3054	241	2020	
46-8646	222	3050/221	FET-4	2036	
46-8646-3	109	3087		1123	
46-86460-3	5070A	9021	ZD-6.2	561	
46-86461-3	238	3710	259		
46-86462-3	5074A	3092/141A	ZD-11.0	563	
46-86463-3	116	3031A	504A	1104	
46-86464-3	145A		ZD-15	564	
46-86466-3	198	3104A	251		
46-86467-3	5096A	3346	ZD-100		
46-86469-3	198	3220	251		
46-8647	107	3018	11	2015	
46-8647-3	116	3017B/117	504A	1104	
46-86474-3	147A		ZD-33		
46-86475-3	162	3438	35		
46-86476-3	175	3626	246	2020	
46-86477-3	165	3115	38		
46-8648			FET-2	2035	
46-86481-3	177	3100/519	300	1122	
46-86482-3	198	3220	251		
46-86483-3	5072A	3136	ZD-8.2	1122	
46-86483-3(ZENER)	5072A		ZD-8.2	562	
46-86484-3	109	3087	1N34A	1123	
46-86485-3	123A	3122	243	2030	
46-86486-3	165	3111	38		
46-86487-3	5074A	3092/141A	ZD-11.0	563	
46-86488-3	125	3016	504A	1104	
46-86489-3	506	3843	511	1114	
46-8649			FET-2	2035	
46-86492-3	238	3710	38		
46-86494			510	1114	
46-86494-3	525	3081/125	530	1114	
46-86496-3	175	3026	246	2020	
46-86497-3	116	3016	504A	1104	
46-86499-3	125	3032A			
46-865-3	160		245	2004	
46-8650			11	2015	
46-86500-3	291	3054/196	32		
46-86501-3	5013A	3779		561	

Industry Standard No.	ECG	SK	GE	RS 276-	MOTOR.
46-86503-3	116	3311	504A	1104	
46-86504-3	5074A	3092/141A	ZD-11	563	
46-86505-3	376	3190/184	251		
46-86506-3	198	3104A	251		
46-86507-3	552	3998/506	511	1114	
46-86508-3	198	3220	32		
46-86509-3	198	3104A	32		
46-8651			17	2051	
46-86511-3	506	3843	511	1114	
46-86512-3	229	3246	81	2051	
46-86513-3	289A	3138/193A	212	2010	
46-86514-3	294	3466/159	269	2032	
46-86515-3	153		69	2049	
46-86516-3	152	3054/196	66	2048	
46-86517-3	129	3114/290	244	2027	
46-86518-3	127	3035	25		
46-86519-3	506	3843	504A	1114	
46-8652			11	2015	
46-86520-3	503	3068	CR-5		
46-86521-3	138A		ZD-7.5		
46-86522-3	162	3559	35		
46-86525-1	506	3843	511	1114	
46-86525-3	506	3843	511	1114	
46-86526-3	198	3220			
46-86529-3	5081A	3151	ZD-24		
46-8653			11	2015	
46-86530-3	5143A		ZD-36		
46-86531-3	233	3132	61	2009	
46-86533-3	375	3440/291	241		
46-86534-3	292	3441			
46-86535-3	199		213	2010	
46-86536-3	199	3124/289	212	2010	
46-86538-3	319P	3018	39	2015	
46-8654			11	2015	
46-86540-3	297	3122	210	2051	
46-86543-3	199	3124/289	62	2010	
46-86546-3	290A	3114/290	269	2032	
46-86547-3	504	3108/505	CR-6		
46-86550-3	293	3849	47	2030	
46-86551-3	290A	3182/243	265	2047	
46-86552-3	198	3219			
46-86553-3	125	3080	509	1114	
46-86554-3	5143A	3409			
46-86556-3	198	3220	251		
46-86557-3	238	3710	38		
46-86558-3	156	3051	512		
46-86559-3	5012A	9021/5070A		561	
46-8656			20	2051	
46-86560-3	525	3925	530		
46-86561-3	94		73		
46-86562-3	125	3032A	509	1114	
46-86563-3	165		38		
46-86565-3	129		244	2027	
46-86566-3	5071A		ZD-6.8	561	
46-86567-3	198	3201/171	251		
46-86568-3			246	2020	
46-86569-3	5465	3687			
46-86570-3			ZD-12	563	
46-86571-3	6402	3628			
46-86572-3	289A	3244	27		
46-86574-3	129	3025	244	2027	
46-86578-3	290A		269	2032	
46-86581-3		3083	69	2049	
46-86583-3	188	3199	226	2018	
46-86584-3	189	3200	227	2026	
46-86589-3	289A	3024/128	243	2030	
46-86590-3	94		35		
46-86591-3	287	3024/128	221	2030	
46-866-3	160		245	2004	
46-8660-3	102A	3123	53	2007	
46-8661-3	116	3016	504A	1104	
46-86627-3	5025A	3791			
46-86628-3	287	3866			
46-86630-3	287	3244	230		
46-8664-3	102A	3004	53	2007	
46-86646-3	375	3929			
46-86647-3	287	3433	224		
46-86648-3		3866	222		
46-8665-3	102A	3004	53	2007	
46-8666-3	102A	3004	53	2007	
46-86678-3	287	3095/147A	222		
46-86679-3	290A	3114/290			
46-8668-3	102A		53	2007	
46-867-3	154	3045/225	40	2012	
46-8671-3	103A	3010	59	2002	
46-8672-3	161	3117	39	2015	
46-8676-3	177	3100/519	300	1122	
46-8677-2	108		86	2038	
46-8677-3	316	3019	11	2015	
46-8679-1	102A	3004	53	2007	
46-8679-2	102A	3004	53	2007	
46-8679-3	102A	3004	53	2007	
46-868-3	160		245	2004	
46-8680-1	102A	3004	53	2007	
46-8680-2	102A	3004	53	2007	
46-8680-3	102A	3004	53	2007	
46-86807-3	117	3017B	504A	1104	
46-8681-1	102A	3004	53	2007	
46-8681-2	102A	3004	53	2007	
46-8681-3	158	3004/102A	53	2007	
46-8682-2	123A	3444	20	2051	
46-8682-3	199	3124/289	62	2010	
46-8687-3	116	3311	504A	1104	
46-8688-3	112	3087	1N82A		
46-8689-3	116	3311	504A	1104	
46-869-3	102A		53	2007	
46-8694-3	115		6GX1		
46-8695-3	123A	3444	20	2051	
46-8697-3	116	3032A	504A	1104	
46-88676-3				2020	
46-96434-3	107		11	2015	
46AR1	116	3017B/117	504A	1104	
46AR10	116	3031A	504A	1104	
46AR11	116	3017B/117	504A	1104	
46AR12	116	3017B/117	504A	1104	
46AR13	125	3032A	510,531	1114	
46AR15	125	3033	510,531	1114	
46AR16	125	3033	510,531	1114	
46AR18	125	3033	510,531	1114	
46AR2	116	3311	504A	1104	
46AR21	116	3017B/117	504A	1104	
46AR27	116	3017B/117	504A	1104	
46AR28	116	3017B/117	504A	1104	
46AR29	116	3017B/117	504A	1104	
46AR3	116	3311	504A	1104	
46AR35	116	3017B/117	504A	1104	
46AR4	116	3017B/117	504A	1104	
46AR5	116	3017B/117	504A	1104	
46AR50	125	3032A	510,531	1114	
46AR52	125	3033	510,531	1114	
46AR59	125	3033	510,531	1114	
46AR6	116	3016	504A	1104	
46AR7	116	3017B/117	504A	1104	
46AR8	116	3016	504A	1104	

Industry Standard No.	ECG	SK	GE	RS 276-	MOTOR.
46AR9	116	3017B/117	504A	1104	
46AX1	116	3017B/117	504A	1104	
46AX10	116	3016	504A	1104	
46AX11	116	3017B/117	504A	1104	
46AX12	125	3033	510,531	1114	
46AX13	125	3032A	510,531	1114	
46AX14	125	3033	510,531	1114	
46AX16	116	3017B/117	504A	1104	
46AX17	116	3016	504A	1104	
46AX19	116	3016	504A	1104	
46AX2	116	3311	504A	1104	
46AX21	116	3017B/117	504A	1104	
46AX3	116	3311	504A	1104	
46AX30	116	3016	504A	1104	
46AX34	116	3017B/117	504A	1104	
46AX4	116	3017B/117	504A	1104	
46AX5	116	3016	504A	1104	
46AX52	116	3017B/117	504A	1104	
46AX54	116	3017B/117	504A	1104	
46AX55	125	3032A	510,531	1114	
46AX56	125	3032A	510,531	1114	
46AX59	125	3033	510,531	1114	
46AX7	116	3016	504A	1104	
46AX70	125	3033	510,531	1114	
46AX8	116	3017B/117	504A	1104	
46AX82	125	3033	510,531	1114	
46AX84	125		510,531	1114	
46AX85	125	3033	510,531	1114	
46AX9	116	3017B/117	504A	1104	
46B-3A5	121MP	3718	239#(2)	2006(2)	
46BD1	116	3017B/117	504A	1104	
46BD101	125	3033	510,531	1114	
46BD11	116	3016	504A	1104	
46BD12	116	3017B/117	504A	1104	
46BD14	116	3017B/117	504A	1104	
46BD19	116	3017B/117	504A	1104	
46BD2	116	3017B/117	504A	1104	
46BD25	116	3017B/117	504A	1104	
46BD27	125	3033	510,531	1114	
46BD30	125	3033	510,531	1114	
46BD32	125	3033	510,531	1114	
46BD33	125	3033	510,531	1114	
46BD34	125	3033	510,531	1114	
46BD38	125	3033	510,531	1114	
46BD39	125	3033	510,531	1114	
46BD5	116	3017B/117	504A	1104	
46BD52	125	3033	510,531	1114	
46BD8	116	3017B/117	504A	1104	
46BD9	116	3017B/117	504A	1104	
46BR10	116	3017B/117	504A	1104	
46BR11	116	3017B/117	504A	1104	
46BR15	116	3017B/117	504A	1104	
46BR17	116	3017B/117	504A	1104	
46BR18	125	3033	510,531	1114	
46BR21	125	3033	510,531	1114	
46BR27	125	3032A	510,531	1114	
46BR5	116	3017B/117	504A	1104	
46BR62	125	3033	510,531	1114	
46BR63	125	3033	510,531	1114	
46BR64	125	3033	510,531	1114	
46BR68	125	3033	510,531	1114	
46BR7	116	3016	504A	1104	
46BR9	116	3031A	504A	1104	
46BX2	116	3311	504A	1104	
46BX3	116	3311	504A	1104	
46R2	5882	3500	5040		
46R2R	5863		5025		
46R2S	5862	3500/5882	5024		
047	221	3050		2036	
047(DIODE)	177		300	1122	
047-1			1N82A	2025	
047-1(SYLVANIA)	187		29	2025	
47-2	112	3019	1N82A		
47-2(BRADFORD)	108		86	2038	
47-2(XSTR)			11	2015	
47-4	112	3089	1N82A		
47C23-3			21	2034	
47P1	121	3009	239	2006	
47Z102536-P1	177		300	1122	
048	220			2036	
048(ZENER)	5075A	3751	ZD-16	564	
48-01	116	3016	504A	1104	
48-01-003	473		243	2030	
48-01-004	108	3452	86	2038	
48-01-005	123A		20	2051	
48-01-007	128	3024	243	2030	
48-01-010	108	3452	86	2038	
48-01-017	107		11	2015	
48-01-027	161	3716	39	2015	
48-01-031	107		11	2015	
48-01-049	199		62	2010	
48-03-002	159	3466	82	2032	
48-03-04046103	128		243	2030	
48-03-04093403	130		14	2041	
48-03-041840-2	130		14	2041	
48-03-05013702	128		243	2030	
48-03-10111102	129		244	2027	
48-03-10744702	192		63	2030	
48-03-10744802	129		244	2027	
48-03005A01	177	3100/519	300	1122	
48-03005A03	109	3087	1N34AS	1123	
48-03005A05	177	3100/519	300	1122	
48-03005A06	519	3100	514	1122	
48-03005A07	177	3311	504A	1104	
48-03005A08	177	3327/614			
48-03073A06	5072A	3136	ZD-8.2	562	
48-03073A08	142A	3339/5087A	ZD-43		
48-03073A09	139A	3060	ZD-9.1	562	
48-03073A10	5000A	3766			
48-05-001	116		504A	1104	
48-05-011	177		300	1122	
48-06-001	109			1123	
48-10001-A01	116	3017B/117	504A	1104	
48-10001-A03		3017B	504A	1104	
48-10001-A030-1	116	3017B/117	504A	1104	
48-1005	116	3016	504A	1104	
48-10062A01	116	3311	504A	1104	
48-10062A01A	116		504A	1104	
48-10062A02	116	3031A	504A	1104	
48-10062A04	116	3017B/117	504A	1104	
48-10062A05	116	3017B/117	504A	1104	
48-10062A05A	116	3017B/117	504A	1104	
48-10073A01	102A		53	2007	
48-10073A02	102	3004	2	2007	
48-10074A01	102A		53	2007	
48-10074A02	102	3004	2	2007	
48-10074A03	102A		53	2007	
48-10075A01	121MP	3013	239(2)	2006(2)	
48-10075A02	121MP	3013	239(2)	2006(2)	
48-10075A03	121MP	3013	239(2)	2006(2)	
48-10075A04	121MP	3013	239(2)	2006(2)	
48-10075A05	121MP	3013	239(2)	2006(2)	
48-10075A06	121MP	3013	239(2)	2006(2)	
48-10075A07	121MP	3013	239(2)	2006(2)	

Industry Standard No.	ECG	SK	GE	RS 276-	MOTOR.
48-10075A08	121MP	3013	239(2)	2006(2)	
48-10079A01	160		245	2004	
48-10079A02	160		245	2004	
48-10103A01	121MP	3013	239(2)	2006(2)	
48-10103A02	121MP	3013	239(2)	2006(2)	
48-10103A03	121MP	3013	239(2)	2006(2)	
48-10103A04	121MP	3013	239(2)	2006(2)	
48-10103A05	121MP	3013	239(2)	2006(2)	
48-10103A06		3013	16	2006	
48-10103A07	121MP	3013	239(2)	2006(2)	
48-10103A08	121MP	3013	239(2)	2006(2)	
48-10103A09	121MP	3013	239(2)	2006(2)	
48-10103A10	121MP	3013	239(2)	2006(2)	
48-10103A11		3013	16	2006	
48-103083	105		4		
48-10577A01	177		300	1122	
48-10577A04	125		510,531	1114	
48-10641D62	137A		ZD-6.2	561	
48-11-005	140A		ZD-10	562	
48-11-010	5041A	3807	ZD-51		
48-110	113A		6GC1		
48-115107				1114	
48-12091A	181		75	2041	
48-123173			20	2051	
48-123522	126	3008	52	2024	
48-123536	126	3008	52	2024	
48-123802	123A	3444	20	2051	
48-123803	123A	3444	20	2051	
48-124158	102A	3123	53	2007	
48-124159	102A	3123	53	2007	
48-124175	102A	3123	53	2007	
48-124204	121	3009	239	2006	
48-124206	121	3717			
48-124216	101	3861	8	2002	
48-124217	101	3861	8	2002	
48-124218	101	3861	8	2002	
48-124219	102	3004	2	2007	
48-124220	101	3861	8	2002	
48-124221	101	3861	8	2002	
48-124246	121	3009	239	2006	
48-124247	121	3009	239	2006	
48-124255	126	3006/160	52	2024	
48-124256	126	3006/160	52	2024	
48-124258	158		53	2007	
48-124259	158		53	2007	
48-124275	158		53	2007	
48-124276	158	3123	53	2007	
48-124279	158	3123	53	2007	
48-124285	121	3009	239	2006	
48-124286	102	3004	2	2007	
48-124296	160		245	2004	
48-124297	102	3004	2	2007	
48-124300	158	3123	53	2007	
48-124302	121	3009	239	2006	
48-124303	102	3004	2	2007	
48-124304	102	3004	2	2007	
48-124305	126	3006/160	52	2024	
48-124306	102	3004	2	2007	
48-124307	100	3005	1	2007	
48-124308	100	3005	1	2007	
48-124309	102	3004	2	2007	
48-124310	126	3006/160	52	2024	
48-124311	126	3006/160	52	2024	
48-124312	126	30060160	52	2024	
48-124314	100	3005	1	2007	
48-124315	100	3005	1	2007	
48-124316	126	3008	52	2024	
48-124318	102	3004	2	2007	
48-124319	102	3004	2	2007	
48-124322	158	3123	53	2007	
48-124327	100	3005	1	2007	
48-124328	100	3005	1	2007	
48-124329	105		4		
48-124332	121	3009	239	2006	
48-124343	100	3005	1	2007	
48-124344	100	3005	1	2007	
48-124345	100	3005	1	2007	
48-124346	126	3006/160	52	2024	
48-124347	126	3008	52	2024	
48-124348	126	3008	52	2024	
48-124349	126	3006/160	52	2024	
48-124350	126	3006/160	52	2024	
48-124351	126	3008	52	2024	
48-124352	126	3008	52	2024	
48-124353	102	3004	2	2007	
48-124354	102	3004	2	2007	
48-124355	102	3004	2	2007	
48-124356	121	3009	239	2006	
48-124357	100	3005	1	2007	
48-124358	100	3005	1	2007	
48-124359	100	3005	1	2007	
48-124360	126	3006/160	52	2024	
48-124363	160		245	2004	
48-124364	126	3006/160	52	2024	
48-124365	126	3006/160	52	2024	
48-124366	126	3006/160	52	2024	
48-124367	126	3006/160	52	2024	
48-124368	160		245	2004	
48-124370	100	3005	1	2007	
48-124371	100	3005	1	2007	
48-124373	102A	3123	53	2007	
48-124377	126	3008	52	2024	
48-124378	100	3005	1	2007	
48-124379	100	3005	1	2007	
48-124380	100	3005	1	2007	
48-124388	126		52	2024	
48-124389	100	3721	1	2007	
48-124398	100	3005	1	2007	
48-12443	102A	3123	53	2007	
48-124443	100	3005	1	2007	
48-124444	100		1	2007	
48-124445	100	3005	1	2007	
48-124446	100	3005	1	2007	
48-124804	108	3452	86	2038	
48-124805	108	3452	86	2038	
48-124808	108	3452	86	2038	
48-125204	121	3009	239	2006	
48-125208	121	3009	239	2006	
48-125219	121	3717			
48-125228	126	3006/160	52	2024	
48-125229	100	3005	1	2007	
48-125230	100	3005	1	2007	
48-125231	100	3005	1	2007	
48-125232	100	3005	1	2007	
48-125233	101	3861	8	2002	
48-125234	101	3861	8	2002	
48-125235	101	3861	8	2002	
48-125236	101	3861	8	2002	
48-125237	100	3005	1	2007	
48-125238	100	3005	1	2007	
48-125239	100	3005	1	2007	
48-125240	100	3005	1	2007	
48-125242	100	3005	1	2007	
48-125252	105		4		

Industry Standard No.	ECG	SK	GE	RS 276-	MOTOR.
48-125266	121	3717			
48-125267	121	3009	239	2006	
48-125271	158	3123	53	2007	
48-125276	158	3123	53	2007	
48-125278	126	3006/160	52	2024	
48-125282	102A		53	2007	
48-125285	102A		53	2007	
48-125286	105		4		
48-125288	121	3014	239	2006	
48-125294	102A		53	2007	
48-125296	100	3005	1	2007	
48-125298	121	3717			
48-125299	105		4		
48-125332	121	3009	239	2006	
48-127021	5071A	3334	ZD-6.8	561	
48-128093	126	3008	52	2024	
48-128094	158	3008	53	2007	
48-128095	126	3008	52	2024	
48-128096	126	3006/160	52	2024	
48-128219	160		245	2004	
48-128239	101	3861	8	2002	
48-128303	100	3005	1	2007	
48-129934	121	3009	239	2006	
48-129935	121	3009	239	2006	
48-129936	121	3009	239	2006	
48-129937	121	3009	239	2006	
48-13001A01	124		12		
48-13309X3	152	3893	66	2048	
48-134101	126	3006/160	52	2024	
48-134145			28	2017	
48-134173	123A	3444	20	2051	
48-134190A1G	161	3018	39	2015	
48-134302	121	3009	239	2006	
48-134372	126	3008	52	2024	
48-134387	109			1123	
48-134404	126	3006/160	52	2024	
48-134405	126	3006/160	52	2024	
48-134406	126	3006/160	52	2024	
48-134407	158	3123	53	2007	
48-134408	102A	3123	53	2007	
48-134411	160		245	2004	
48-134412	160		245	2004	
48-134413	160		245	2004	
48-134414	126	3006/160	52	2024	
48-134415	100	3005	1	2007	
48-134416	100	3005	1	2007	
48-134417	100	3005	1	2007	
48-134418	100	3005	1	2007	
48-134419	100	3005	1	2007	
48-134420	100	3005	1	2007	
48-134421	100	3005	1	2007	
48-134422	100	3005	1	2007	
48-134423	100	3005	1	2007	
48-134424	100	3005	1	2007	
48-134425	100	3005	1	2007	
48-134426	100	3005	1	2007	
48-134427	100	3005	1	2007	
48-134428	100	3005	1	2007	
48-134429	100	3005	1	2007	
48-134430	105		4		
48-134431	176	3845	80		
48-134432	100	3005	1	2007	
48-134433	100	3005	1	2007	
48-134434	126	3006/160	52	2024	
48-134439	160		245	2004	
48-134443	100	3005	1	2007	
48-134444	100	3005	1	2007	
48-134445	100	3005	1	2007	
48-134446	100	3005	1	2007	
48-134447	121	3009	239	2006	
48-134448	121	3009	239	2006	
48-134449	121	3009	239	2006	
48-134450	100	3005	1	2007	
48-134454	126	3006/160	52	2024	
48-134456	126	3006/160	52	2024	
48-134457	126	3006/160	52	2024	
48-134458	100	3005	1	2007	
48-134459	127	3764	25		
48-134462	100	3005	1	2007	
48-134463	121	3009	239	2006	
48-134464	123A	3444	20	2051	
48-134465	123A	3444	20	2051	
48-134466	100	3005	1	2007	
48-134467	5896		5064		
48-134468	100	3005	1	2007	
48-134469	100	3005	1	2007	
48-134470	100	3005	1	2007	
48-134471	100	3005	1	2007	
48-134472	100	3005	1	2007	
48-134473	100	3005	1	2007	
48-134474	100	3005	1	2007	
48-134475	100	3005	1	2007	
48-134476	100	3005	1	2007	
48-134477	100	3005	1	2007	
48-134478	129	3025	244	2027	
48-134478A	129	3025	244	2027	
48-134479	126	3006/160	52	2024	
48-134480	126	3006/160	52	2024	
48-134481	126	3006/160	52	2024	
48-134482	158	3123	53	2007	
48-134483	158	3123	53	2007	
48-134484	160		245	2004	
48-134485	160		245	2004	
48-134486	160		245	2004	
48-134487	121	3009	239	2006	
48-134488	121	3009	239	2006	
48-134493	121	3009	239	2006	
48-134494	100	3005	1	2007	
48-134495	100	3005	1	2007	
48-134496	100	3005	1	2007	
48-134499	100	3005	1	2007	
48-134500	100	3005	1	2007	
48-134501	100	3005	1	2007	
48-134504	160		245	2004	
48-134506	160		245	2004	
48-134507	160	3005	245	2004	
48-134508	126	3006/160	52	2024	
48-134509	100	3005	1	2007	
48-134510	100	3005	1	2007	
48-134512	100	3005	1	2007	
48-134514	126	3006/160	52	2024	
48-134519	121	3009	239	2006	
48-134520	101	3861	8	2002	
48-134521	126		52	2024	
48-134522	126		52	2024	
48-134524	160		245	2004	
48-134525	159	3466	82	2032	
48-134525A	159	3466	82	2032	
48-134526	160		245	2004	
48-134535	100	3005	1	2007	
48-134536	126	3006/160	52	2024	
48-134537	109	3008		1123	
48-134538	100	3005	1	2007	
48-134539	100	3005	1	2007	

Industry Standard No.	ECG	SK	GE	RS 276-	MOTOR.
48-134540	100	3005	1	2007	
48-134541	100	3005	1	2007	
48-134542	100	3005	1	2007	
48-134543	100	3005	1	2007	
48-134544	100	3005	1	2007	
48-134545	126	3006/160	52	2024	
48-134547	126	3006/160	52	2024	
48-134552	144A		14	564	
48-134553	100	3005	1	2007	
48-134554	100	3005	1	2007	
48-134555	100	3005	1	2007	
48-134556	100	3005	1	2007	
48-134557	100	3005	1	2007	
48-134558	100	3005	1	2007	
48-134559	100	3005	1	2007	
48-134560	121	3009	239	2006	
48-134561	126	3006/160	52	2024	
48-134562	100	3005	1	2007	
48-134563	100	3005	1	2007	
48-134564	100	3005	1	2007	
48-134565	100	3005	1	2007	
48-134567	100	3005	1	2007	
48-134570	121	3009	239	2006	
48-134571	121	3717			
48-134572	102A		53	2007	
48-134573	102	3004	2	2007	
48-134574	121	3009	239	2006	
48-134575	121	3009	239	2006	
48-134576	126		52	2024	
48-134577	126		52	2024	
48-134578	126		52	2024	
48-134579	160		245	2004	
48-134582	121	3717	239	2006	
48-134583	121	3717	239	2006	
48-134584	176	3845	80		
48-134585	176	3845	80		
48-134587	109	3087		1123	
48-134588	109	3087		1123	
48-134591	126	3006/160	52	2024	
48-134592	121	3009	239	2006	
48-134600	126	3006/160	52	2024	
48-134601	126	3006/160	52	2024	
48-134602	126	3006/160	52	2024	
48-134603	100	3005	1	2007	
48-134604	100	3005	1	2007	
48-134605	126	3008	52	2024	
48-134606	121	3717	239	2006	
48-134610	100	3005	1	2007	
48-134611	121	3009	239	2006	
48-134612	121	3009	239	2006	
48-134613	121	3009	239	2006	
48-134620	121	3717			
48-134621	102A	3004	53	2007	
48-134622	105	3012	4		
48-134623	127	3764	25		
48-134625	100	3005	1	2007	
48-134626	100	3005	1	2007	
48-134630	195A	3765			
48-134631	100	3005	1	2007	
48-134632	102A	3004	53	2007	
48-134633	158	3123	53	2007	
48-134634	121	3009	239	2006	
48-134635	126	3006/160	52	2024	
48-134636	100	3005	1	2007	
48-134637	100	3005	1	2007	
48-134638	121	3009	239	2006	
48-134639	121	3009	239	2006	
48-134640	179	3642	76		
48-134641	100	3005	1	2007	
48-134643	148A	3096	ZD-55		
48-134644	121	3009	239	2006	
48-134645	121	3009	239	2006	
48-134646	121	3009	239	2006	
48-134647	121	3009	239	2006	
48-134648	154	3045/225	40	2012	
48-134649	121	3009	239	2006	
48-134651	121	3014	239	2006	
48-134652	127	3764	25		
48-134653	139A	3060	ZD-9.1	562	
48-134654	123A	3444	20	2051	
48-134655	100	3005	1	2007	
48-134656	100	3005	1	2007	
48-134657	100	3005	1	2007	
48-134659	144A	3094	ZD-14	564	
48-13466			8	2002	
48-134661	121	3717			
48-134663	147A	3095	ZD-33		
48-134664	123A	3444	20	2051	
48-134665	123A	3444	20	2051	
48-134666	123A	3444	20	2051	
48-134667	123A	3444	20	2051	
48-134668	123A	3444	20	2051	
48-134669	123A	3444	20	2051	
48-134670	121	3009	239	2006	
48-134671	156		512		
48-134672	121	3009	239	2006	
48-134673	123A	3444	20	2051	
48-134674	123A	3444	20	2051	
48-134675	123A	3444	20	2051	
48-134676	160		245	2004	
48-134677	160		245	2004	
48-134678	160		245	2004	
48-134679	160		245	2004	
48-134680	126	3008	52	2024	
48-134681	126	3006/160	52	2024	
48-134682	126	3006/160	52	2024	
48-134683	126	3006/160	52	2024	
48-134684	126	3008	52	2024	
48-134689	128	3024	243	2030	
48-134690	123A	3444	20	2051	
48-134691	123A	3444	20	2051	
48-134692	179	3642	76		
48-134693	160		245	2004	
48-134694	160		245	2004	
48-134695	179	3642	76		
48-134696	121	3009	239	2006	
48-134697	126	3006/160	52	2024	
48-134698	135A	3056	ZD-5.1		
48-134699	150A	3098	ZD-82		
48-13470	108	3452	86	2038	
48-134700	101	3861	8	2002	
48-134701	130	3027	14	2041	
48-134701A	130	3027	14	2041	
48-134702	159	3466	82	2032	
48-134702A	159	3466	82	2032	
48-134703	123A	3444	20	2051	
48-134704	150A	3098	ZD-82		
48-134705	123A	3444	20	2051	
48-134706	108	3452	86	2038	
48-134709	108	3452	86	2038	
48-134711	126	3006/160	52	2024	
48-134713	108	3452	86	2038	
48-134714	123A	3444	20	2051	
48-134715	130	3027	14	2041	
48-134715A	130	3027	14	2041	
48-134717	108	3452	86	2038	
48-134718	123A	3444	20	2051	
48-134719	108	3452	86	2038	
48-134720	123A	3444	20	2051	
48-134721	123A	3444	20	2051	
48-134722	121	3009	239	2006	
48-134723	121	3009	239	2006	
48-134724	108	3452	86	2038	
48-134725	108	3452	86	2038	
48-134726	123A	3444	20	2051	
48-134727	104	3009	16	2006	
48-134728	151A	3099	ZD-110		
48-134729	179	3642	76		
48-134730	121	3009	239	2006	
48-134731	104	3009	16	2006	
48-134732	123A	3444	20	2051	
48-134733	123A	3444	20	2051	
48-134733A	123A	3444	20	2051	
48-134734	123A	3444	20	2051	
48-134734A	123A	3444	20	2051	
48-134737	123A	3444	20	2051	
48-134738	121	3009	239	2006	
48-134739	123A	3444	20	2051	
48-134739A	124		12		
48-134740	179	3642	76		
48-134741	179	3642	76		
48-134742	179	3642	76		
48-134743	179	3642	76		
48-134744	121	3717	239	2006	
48-134745	159	3466	82	2032	
48-134745A	159	3466	82	2032	
48-134746	121	3009	239	2006	
48-134747	104MP	3720	16(2)	2006(2)	
48-134748	179	3642	76		
48-134749	179	3642	76		
48-134750	121	3009	239	2006	
48-134751	121	3009	239	2006	
48-134752	179	3642	76		
48-134753	179	3642	76		
48-134756	161	3117	39	2015	
48-134757	121	3009	239	2006	
48-134758	121	3009	239	2006	
48-134759	121	3009	239	2006	
48-134760	179	3642	76		
48-134761	179	3642	76		
48-134763	121	3009	239	2006	
48-134764	121	3009	239	2006	
48-134765	123A	3444	20	2051	
48-134766	121	3009	239	2006	
48-134767	121	3009	239	2006	
48-134768	123A	3444	20	2051	
48-134769	116	3017B/117	504A	1104	
48-134772	108	3452	86	2038	
48-134773	108	3452	86	2038	
48-134774	108	3452	86	2038	
48-134775	123A	3444	20	2051	
48-134776	123A	3444	20	2051	
48-134777	108	3452	86	2038	
48-134779	108	3452	86	2038	
48-134780	108	3452	86	2038	
48-134781	177	3100/519	300	1122	
48-134782	123A	3444	20	2051	
48-134783	108	3452	86	2038	
48-134784	108	3452	86	2038	
48-134785	123A	3444	20	2051	
48-134786	108	3452	86	2038	
48-134787	108	3039/316	86	2038	
48-134788	179	3642	76		
48-134789	107	3039/316	11	2015	
48-134790	116		504A	1104	
48-134791	123A	3444	20	2051	
48-134792	6400A			2029	
48-134795	126	3006/160	52	2024	
48-134796	126	3008	52	2024	
48-134797	126	3006/160	52	2024	
48-134798	126	3006/160	52	2024	
48-134800	108	3452	86	2038	
48-134801	123A	3444	20	2051	
48-134802		3122	20	2051	
48-134803		3122	20	2051	
48-134804	123A	3444	20	2051	
48-134805	229	3018	61	2038	
48-134806	108	3452	86	2038	
48-134807	123A	3444	20	2051	
48-134808	123A	3444	20	2051	
48-134809	123A	3444	20	2051	
48-13481	123A	3444	20	2051	
48-134810	199	3124/289	62	2010	
48-134811	123A	3444	20	2051	
48-134813	199	3245	62	2010	
48-134814	108	3452	86	2038	
48-134815	159	3466	82	2032	
48-134815A	159	3466	82	2032	
48-134816	177	3100/519	300	1122	
48-134817	123A	3444	20	2051	
48-134818	108	3452	86	2038	
48-134819	154	3044	40	2012	
48-134820	108	3452	86	2038	
48-134821	108	3452	86	2038	
48-134822	123A	3444	20	2051	
48-134823	199	3124/289	62	2010	
48-134824	123A	3444	20	2051	
48-134825	229	3018	61	2038	
48-134826	108	3452	86	2038	
48-134827	108	3452	86	2038	
48-134828	108	3452	86	2038	
48-134829	159	3466	82	2032	
48-134829A	159	3466	82	2032	
48-134830	159	3114/290	223		
48-134830A	234	3114/290	65	2050	
48-134831	234	3114/290	65	2050	
48-134831A	234	3114/290	65	2050	
48-134832	234	3114/290	65	2050	
48-134832A	234	3114/290	65	2050	
48-134833	159	3466	82	2032	
48-134833A	159	3466	82	2032	
48-134837	107	3039/316	11	2015	
48-134838	128	3024	243	2030	
48-134839	123A	3444	20	2051	
48-134840	123A	3444	20	2051	
48-134841	123A	3444	20	2051	
48-134842	123A	3444	20	2051	
48-134843	154	3044	40	2012	
48-134844	123A	3444	20	2051	
48-134845	107	3039/316	11	2015	
48-134846	287	3433	222		
48-134847	123A	3444	20	2051	
48-134848	123A	3444	20	2051	
48-134850	141A	3092	ZD-11.5		
48-134851	141A	3092	ZD-11.5		
48-134852	123A	3444	20	2051	
48-134853	154	3045/225	40	2012	
48-134854	123A	3444	20	2051	

Industry Standard No.	ECG	SK	GE	RS 276-	MOTOR.
48-134855	108	3452	86	2038	
48-134856	176	3845	80		
48-134857	229	3018	61	2038	
48-134858	147A	3095	ZD-33		
48-134859	126	3008	52	2024	
48-134860	126	3006/160	52	2024	
48-134861	126	3008	52	2024	
48-134862	126	3008	52	2024	
48-134865	159	3466	82	2032	
48-134865A	159	3466	82	2032	
48-134866	159	3466	82	2032	
48-134866A	159	3466	82	2032	
48-134867	159	3466	82	2032	
48-134867A	159	3466	82	2032	
48-134868	159	3466	82	2032	
48-134868A	159	3466	82	2032	
48-134869	159	3466	82	2032	
48-134869A	159	3466	82	2032	
48-134870	159	3466	82	2032	
48-134870A	159	3466	82	2032	
48-134871	159	3466	82	2032	
48-134871A	159	3466	82	2032	
48-134872	124		12		
48-134879	108	3452	86	2038	
48-134880	126	3008	52	2024	
48-134882	130		14	2041	
48-134884	130		14	2041	
48-134885	124		12		
48-134888	121	3014	239	2006	
48-134889	123A	3444	20	2051	
48-134891	108	3452	86	2038	
48-134892	108	3452	86	2038	
48-134893	108	3452	86	2038	
48-134894	123A	3444	20	2051	
48-134895	123A	3444	20	2051	
48-134896	123A	3444	20	2051	
48-134897	123A	3444	20	2051	
48-134898	154	3045/225	40	2012	
48-134899	123A	3444	20	2051	
48-134900	162	3438	35		
48-134901	163A	3439	36		
48-134902	108	3452	86	2038	
48-134903	123A	3444	20	2051	
48-134904	107	3039/316	11	2015	
48-134904A1G	107	3132	11	2015	
48-134904F	107		11	2015	
48-134905	123A	3444	20	2051	
48-134906	123A	3444	20	2051	
48-134907	104	3009	16	2006	
48-134908	108	3452	86	2038	
48-134909	159	3466	82	2032	
48-134909A	159	3466	82	2032	
48-134910	159	3466	82	2032	
48-134910A	159	3466	82	2032	
48-134910F	159	3466	82	2032	
48-134911	159	3466	82	2032	
48-134912	5072A	3122	ZD-8.2	562	
48-134913	159	3466	82	2032	
48-134913A	159	3466	82	2032	
48-134914	159	3466	82	2032	
48-134914A	159	3466	82	2032	
48-134915	159	3466	82	2032	
48-134915A	159	3466	82	2032	
48-134916	178MP		300(2)	1122(2)	
48-134917	178MP		300(2)	1122(2)	
48-134918	123A	3122	20	2051	
48-134919	154	3044	40	2012	
48-134920	124	3021	12		
48-134921	506	3843	511	1114	
48-134922	107	3039/316	11	2015	
48-134923	107	3039/316	11	2015	
48-134924	107	3039/316	11	2015	
48-134925	107	3039/316	11	2015	
48-134926	107	3039/316	11	2015	
48-134927	154	3045/225	40	2012	
48-134928	123A	3444	20	2051	
48-134929	123A	3444	20	2051	
48-134930	121	3009	239	2006	
48-134931	101	3861	8	2002	
48-134932	233	3039/316	210	2009	
48-134932(3RD LF)				2009	
48-134932(3RD-IF)			210	2051	
48-134933	123A	3444	20	2051	
48-134933E	123A	3444	20	2051	
48-134934	127	3764	25		
48-134935	123A	3444	20	2051	
48-134936	175	3131A/369	246	2020	
48-134937	108	3452	86	2038	
48-134938	121	3014	239	2006	
48-134939	506	3843	511	1114	
48-134940	159	3466	82	2032	
48-134940A	159	3466	82	2032	
48-134941	128	3024	243	2030	
48-134942	123A	3444	20	2051	
48-134943	159	3466	82	2032	
48-134943A	159	3466	82	2032	
48-134944	312	3112	FET-1	2028	
48-134945	107	3039/316	11	2015	
48-134946	108	3452	86	2038	
48-134947	121	3009	239	2006	
48-134948	108	3452	86	2038	
48-134949	108	3452	86	2038	
48-134950	108	3452	86	2038	
48-134951	129	3025	244	2027	
48-134951A	129	3025	244	2027	
48-134952	123A	3444	20	2051	
48-134953	128	3024	243	2030	
48-134954	110MP	3709		1123(2)	
48-134955	149A	3097	ZD-62		
48-134956	100	3005	1	2007	
48-134957	139A	3060	ZD-9.1	562	
48-134958	116		504A	1104	
48-134959	116	3311	504A	1104	
48-134960	108	3452	86	2038	
48-134961	107	3039/316	11	2015	
48-134962	107	3039/316	11	2015	
48-134963	107	3039/316	11	2015	
48-134964	107	3039/316	11	2015	
48-134965	107	3039/316	11	2015	
48-134966	107	3039/316	11	2015	
48-134967	159	3466	82	2032	
48-134967A	159	3466	82	2032	
48-134969	130		14	2041	
48-134970	123A	3444	20	2051	
48-134971	148A	3096	ZD-55		
48-134972	124	3021	12		
48-134973	159	3466	82	2032	
48-134973A	159	3466	82	2032	
48-134974	121	3009	239	2006	
48-134975	159	3466	82	2032	
48-134975A	159	3466	82	2032	
48-134977	121	3009	239	2006	
48-134978	506	3843	511	1114	
48-134979	108	3452	86	2038	

Industry Standard No.	ECG	SK	GE	RS 276-	MOTOR.
48-134980	123A	3444	20	2051	
48-134981	107	3039/316	11	2015	
48-134982	128	3024	243	2030	
48-134983	108	3452	86	2038	
48-134985	108	3452	86	2038	
48-134987	185	3191	58	2025	
48-134988	123A	3444	20	2051	
48-134989	159	3466	82	2032	
48-134989A	159	3466	82	2032	
48-134990	116	3311	504A	1104	
48-134991	135A	3056	ZD-5.1		
48-134992	123A	3444	20	2051	
48-134993	136A	3057	ZD-5.6	561	
48-134994	123A	3444	20	2051	
48-134995	163A	3439	36		
48-134996	123A	3444	20	2051	
48-134997	199	3122	62	2010	
48-134998	157	3103A/396	232		
48-136665	123A	3444	20	2051	
48-137000	143A	3750	ZD-13	563	
48-137001	127	3764	25		
48-137002	154	3045/225	40	2012	
48-137003	123A	3444	20	2051	
48-137004	108	3452	86	2038	
48-137005	128	3024	243	2030	
48-137006	108	3452	86	2038	
48-137007	123A	3444	20	2051	
48-137008	130	3027	14	2041	
48-137008A	130	3027	14	2041	
48-137010	123A	3444	20	2051	
48-137011	190		217		
48-137013	123A	3444	20	2051	
48-137014	123A	3444	20	2051	
48-137015	199	3122	62	2010	
48-137017	144A	3094	ZD-14	564	
48-137019	123A	3444	20	2051	
48-137020	159	3466	82	2032	
48-137020A	159	3466	82	2032	
48-137021	159		ZD-82	2032	
48-137022	123A	3444	20	2051	
48-137023	312			2028	
48-137024	5400	3950			
48-137025	104	3719	16	2006	
48-137026	104	3719	16	2006	
48-137027	130	3027	14	2041	
48-137027A	130	3027	14	2041	
48-137029	116	3017B/117	504A	1104	
48-137030	241	3092/141A	57	2020	
48-137031	121	3009	239	2006	
48-137032	159	3466	82	2032	
48-137032A	159	3466	82	2032	
48-137033	108	3452	86	2038	
48-137034	146A	3064	ZD-27		
48-137035	154	3045/225	40	2012	
48-137036	130	3027	14	2041	
48-137036A	130	3027	14	2041	
48-137037	184	3190	57	2017	
48-137039	176	3845	80		
48-137040	107		11	2015	
48-137041	128	3024	243	2030	
48-137043	123A	3444	20	2051	
48-137044	123A	3444	20	2051	
48-137045	159	3466	82	2032	
48-137045A	159	3466	82	2032	
48-137046	159	3466	82	2032	
48-137047	123A	3444	20	2051	
48-137048	144A	3094	ZD-14	564	
48-137049	180		74	2043	
48-137053	130	3027	14	2041	
48-137053A	130	3027	14	2041	
48-137055	108	3452	86	2038	
48-137056	123A	3444	20	2051	
48-137057	123A	3444	20	2051	
48-137058	6400A			2029	
48-137059	107		11	2015	
48-137061	159	3466	82	2032	
48-137062	5033A	3799	ZD-27		
48-137065	6407	3523			
48-137066	106	3984	21	2034	
48-137067	159	3466	82	2032	
48-137067A	159	3466	82	2032	
48-137068	159	3466	82	2032	
48-137068A	159	3466	82	2032	
48-137069	159	3466	82	2032	
48-137069A	159	3466	82	2032	
48-13707	220	3990	FET-2	2036	
48-137070	220	3990	FET-3	2036	
48-137071	107	3039/316	11	2015	
48-137072	123A	3444	20	2051	
48-137073	123A	3444	20	2051	
48-137074	116	3017B/117	504A	1104	
48-137075	107	3039/316	11	2015	
48-137076	107	3039/316	11	2015	
48-137077	107	3039/316	11	2015	
48-137078		3009	16	2006	
48-137079	130	3027	14	2041	
48-137079A	130	3027	14	2041	
48-137080	197	3041	250	2027	
48-137081	502	3067	CR-4		
48-137082	503	3068	CR-5		
48-137083	123A	3444	20	2051	
48-137088	128	3024	243	2030	
48-137089	123A	3444	20	2051	
48-137090	159	3466	82	2032	
48-137090A	159	3466	82	2032	
48-137091	188		226	2018	
48-137092	188		226	2018	
48-137093	190		217		
48-137095	184	3190	57	2017	
48-137096	123A	3444	20	2051	
48-137098	5800	9003		1142	
48-137101	123A	3444	20	2051	
48-137102	104	3009	16	2006	
48-137104	108	3452	86	2038	
48-137105	108	3452	86	2038	
48-137106	123A	3444	20	2051	
48-137107	123A	3444	20	2051	
48-137108	123A	3444	20	2051	
48-137109	123A	3444	20	2051	
48-137110	123A	3444	20	2051	
48-137111	123A	3444	20	2051	
48-137112	506	3843	511	1114	
48-137113	191		249		
48-137114	505		CR-7		
48-137115	123A	3444	20	2051	
48-137116	181		75	2041	
48-137118	121	3009	239	2006	
48-137119	121	3009	239	2006	
48-137120	121	3009	239	2006	
48-137121	179	3642	76		
48-137122	121	3009	239	2006	
48-137123	121	3009	239	2006	
48-137124	121	3009	239	2006	
48-137125	179	3642	76		

Industry Standard No.	ECG	SK	GE	RS 276-	MOTOR.
48-137126	108	3452	86	2038	
48-137127	159	3466	82	2032	
48-137127A	159	3466	82	2032	
48-137128	184		57	2017	
48-137130	139A	3066/118	ZD-9.1	562	
48-137132	146A		ZD-27		
48-137133	507	3311			
48-137134	163A	3439	36		
48-137136	108	3452	86	2038	
48-137137	123A	3444	20	2051	
48-137138	123A	3444	20	2051	
48-137139	123A	3444	20	2051	
48-137140	108	3452	86	2038	
48-137142	5444	3634			
48-137143	116	3017B/117	504A	1104	
48-137144	108	3452	86	2038	
48-137145	184	3054/196	57	2017	
48-137146	184	3054/196	57	2017	
48-137147	184	3054/196	57	2017	
48-137148	184	3054/196	57	2017	
48-137149	188	3199	226	2018	
48-137153	185	3083/197	58	2025	
48-137154	185	3083/197	58	2025	
48-137155	185	3083/197	58	2025	
48-137156	185	3083/197	58	2025	
48-137157	185	3083/197	58	2025	
48-137158	108	3452	86	2038	
48-137160	189		218	2026	
48-137164	139A	3066/118	ZD-9.1	562	
48-137165	6400A			2029	
48-137166	108	3452	86	2038	
48-137167	178MP		300(2)	1122(2)	
48-137168	189		218	2026	
48-137169	188		226	2018	
48-137170	137A	3058	ZD-6.2	561	
48-137171	123AP	3854	20	2051	
48-137171D	123A	3444	20	2051	
48-137172	123A	3444	20	2051	
48-137173	159	3466	82	2032	
48-137173A	159	3466	82	2032	
48-137174	123A	3444	20	2051	
48-137175	130	3027	14	2041	
48-137175A	130	3027	14	2041	
48-137176	159	3466	82	2032	
48-137176A	159	3466	82	2032	
48-137177	142A	3062	ZD-12	563	
48-137178	127	3764	ZD-25		
48-137179	163A	3439	36		
48-137180	130	3027	14	2041	
48-137180A	130	3027	14	2041	
48-137183	5882		5040		
48-137184	5484	3943/5485			
48-137185	197		250	2027	
48-137188	141A	3092	ZD-11.5		
48-137190	108	3452	86	2038	
48-137191	108	3452	86	2038	
48-137192	123A	3444	20	2051	
48-137193	158		53	2007	
48-137194	108	3452	86	2038	
48-137195	159	3466	82	2032	
48-137196	108	3452	86	2038	
48-137197	161	3039/316	39	2015	
48-137198	116	3017B/117	504A	1104	
48-137199	102	3005	2	2007	
48-137200	184	3190	57	2017	
48-137202	188		226	2018	
48-137203	163A	3439	36		
48-137205	125	3016	510,531	1114	
48-137206	123A	3444	20	2051	
48-137207	124	3021	12		
48-137208	116	3017B/117	504A	1104	
48-137209	143A	3750	ZD-13	563	
48-137210	137A		ZD-6.2	561	
48-137211	184		57	2017	
48-137212	125	3311	510,531	1114	
48-137213	104	3009	16	2006	
48-137214	131		44	2006	
48-137215	104	3009	16	2006	
48-137216	104	3009	16	2006	
48-137217	104	3009	16	2006	
48-137218	104	3009	16	2006	
48-137219	104	3009	16	2006	
48-137220	104	3009	16	2006	
48-137234	127	3764	25		
48-137235	127	3764	25		
48-137238	194	3275	220		
48-137239	191	3232	249		
48-137240	189		218	2026	
48-137251	130	3027	14	2041	
48-137256	185	3041	58	2025	
48-137257	123A	3444	20	2051	
48-137258	185		58	2025	
48-137259	197	3083	250	2027	
48-137260	123A	3444	20	2051	
48-137265	123A	3444	20	2051	
48-137266	147A		ZD-33		
48-137267	131	3052	44	2006	
48-137268	131	3052	44	2006	
48-137269	131	3052	44	2006	
48-137270	131	3052	44	2006	
48-137271	131	3052	44	2006	
48-137272	142A	3062	ZD-12	563	
48-137277	184	3190	57	2017	
48-137278	194	3275	220		
48-137279	5014A	3780	ZD-6.8	561	
48-137280	5400	3950			
48-137281	5404	3627			
48-137282	6400A			2029	
48-137290	125	3033	510,531	1114	
48-137291	116	3016	504A	1104	
48-137295	5404	3627			
48-137297	141A		ZD-11.5		
48-137298	144A		ZD-14	564	
48-137299	109	3087		1123	
48-137300	199		62	2010	
48-137301	125		510,531	1114	
48-137302	125		510,531	1114	
48-137303	185		58	2025	
48-137304	185		58	2025	
48-137306	5030A	3796	ZD-22		
48-137307	128	3024	243	2030	
48-137308	131	3052	44	2006	
48-137309	196		241	2020	
48-137310	197		250	2027	
48-137311	241	3188A/182	57	2020	
48-137312	153	3083/197	69	2049	
48-137313	219		74	2043	
48-137314	189		218	2026	
48-137315	128		243	2030	
48-137316	125	3016	510,531	1114	
48-137318	159	3466	82	2032	
48-137319	188	3199	226	2018	
48-137320	189	3200	218	2026	
48-137321	193		67	2023	
48-137322	5011A	3777	ZD-5.6	561	
48-137323	184	3190	57	2017	
48-137324	159	3466	82	2032	
48-137325	199	3245	62	2010	
48-137326	162	3438	35		
48-137327	5804	9007		1144	
48-137328	5016A	3782	ZD-8.2	562	
48-137329	104	3719	16	2006	
48-137330	5024A	3790	ZD-15	564	
48-137331	242	3189A/183	58	2027	
48-137332	194	3275	220		
48-137333	130		14	2041	
48-137334	141A		ZD-11.5		
48-137336	123A	3444	20	2051	
48-137337	5010A	3776	ZD-5.1		
48-137338	5486	3944/5487			
48-137339	108	3452	86	2038	
48-137340	5804	9007		1144	
48-137341	165		38		
48-137342	127	3764	25		
48-137342-P47	127	3764	25		
48-137343	312	3834/132	FET-1	2035	
48-137344	130		14	2041	
48-137347	552	3031A	510,531	1114	
48-137348	552	3031A	510,531	1114	
48-137349	5442	3634/5444			
48-137350	123A	3444	20	2051	
48-137351	229	3018	61	2038	
48-137352	108	3452	86	2038	
48-137353	123A	3444	20	2051	
48-137354	123A	3444	20	2051	
48-137355	108	3452	86	2038	
48-137364	154	3044	40	2012	
48-137365	144A		ZD-14	564	
48-137366	159	3466	82	2032	
48-137367	127	3764	25		
48-137368	130		14	2041	
48-137369	241	3188A/182	57	2020	
48-137370	242	3189A/183	58	2027	
48-137371	108	3452	86	2038	
48-137372	108	3452	86	2038	
48-137373	123A	3444	20	2051	
48-137374	123A	3444	20	2051	
48-137375	108	3452	86	2038	
48-137376	108	3452	86	2038	
48-137377	123A	3444	20	2051	
48-137378	123A	3444	20	2051	
48-137379	159	3466	82	2032	
48-137380	159	3466	82	2032	
48-137381	159	3466	82	2032	
48-137382	159	3466	82	2032	
48-137383	159	3466	82	2032	
48-137384	123A	3444	20	2051	
48-137385	177	3100/519	300	1122	
48-137386	194	3275	220		
48-137387	137A		ZD-6.2	561	
48-137388	108	3452	86	2038	
48-137390	199	3245	62	2010	
48-137391	159	3466	82	2032	
48-137392	172A	3156	64		
48-137393	5019A	3785	ZD-10	562	
48-137394	5024A	3790	ZD-15	564	
48-137395	5404	3627			
48-137396	152	3893	66	2048	
48-137397	513	3443	513		
48-137398	123A	3444	20	2051	
48-137399	123A	3444	20	2051	
48-137400	107		11	2015	
48-137415	154	3044	40	2012	
48-137421	5072A	3136	ZD-8.2	562	
48-137437	152	3054/196	66	2048	
48-137472	183	3083/197	56	2027	
48-137473	184	3054/196	57	2017	
48-137476	191	3232	249		
48-137483	108	3452	86	2038	
48-137487	614	3126	90		
48-137488	222	3050/221	FET-4	2036	
48-137491	161	3018	39	2015	
48-137495	109	3087		1123	
48-137497	109			1123	
48-137498	123A	3444	20	2051	
48-137500	123A	3444	20	2051	
48-137501	185	3189A/183	69	2049	
48-137502	159	3466	82	2032	
48-137504	159	3466	82	2032	
48-137505	188	3024/128	226	2018	
48-137506	152	3893	66	2048	
48-137507	153		69	2049	
48-137509	123A	3444	20	2051	
48-137514	177	3100/519	300	1122	
48-137524	283	3467	36		
48-137526	241	3188A/182	57	2020	
48-137527	242	3189A/183	58	2027	
48-137528	286	3194	267		
48-137530	123A	3444	20	2051	
48-137531	5444	3634			
48-137533	506	3843	511	1114	
48-137535	124		12		
48-137539	165	3439/163A	36		
48-137540	153	3080	69	2049	
48-137543	123A	3444	20	2051	
48-137546	506	3843	511	1114	
48-137548	164		37		
48-137549	184	3054/196	57	2017	
48-137550	185	3083/197	58	2025	
48-137551	506	3843	511	1114	
48-137552	186	3192			
48-137553	187	3193			
48-137561	185	3189A/183			
48-137562	242	3189A/183	58	2027	
48-137563	143A	3750			
48-137566	153		69	2049	
48-137567	222	3050/221	FET-4	2036	
48-137573	177	3100/519	300	1122	
48-137577	5073A	3749	ZD-9.1	562	
48-137610	189	3084	218	2026	
48-137612	107	3018	17	2051	
48-137621	5073A	3749			
48-137801	502	3067	CR-4		
48-137855	123A	3444	20	2051	
48-137978	121	3717	239	2006	
48-137988	128	3024	243	2030	
48-137998	123A	3444	20	2051	
48-15	130	3027	14	2041	
48-155001	165		38		
48-155002	502	3067			
48-155006	154		40	2012	
48-155035	290A	3466/159	269	2032	
48-155039	109	3087	1N34A	1123	
48-155041	552	3311	504A	1104	
48-155042	186A	3357	215	2052	
48-155044	175	3131A/369			
48-155045	290A	3247/234	65	2050	
48-155046	297	3124/289	210	2030	

Industry Standard No.	ECG	SK	GE	RS 276-	MOTOR.
48-155047	519	3100	514	1122	
48-155048	5074A	3139	ZD-11		
48-155050	5404	3627			
48-155051	287	3433	222		
48-155052	605	3864			
48-155056	552	3843			
48-155058	238	3710	259		
48-155059	198	3219	251		
48-155060	519	3100	300	1122	
48-155061	109	3088	1N60	1123	
48-155062	152	3893			
48-155063	116	3312	504A	1104	
48-155065	290A	3114/290	65	2050	
48-155070	198	3747/157	232		
48-155071	199		62		
48-155073	199		62	2010	
48-155074	186A	3357	215	2052	
48-155077	519	3175	300	1122	
48-155078	110MP	3709			
48-155080	5070A	9021			
48-155081	5074A	3139	ZD-11	563	
48-155083	125	3080	509	1114	
48-155087	229	3018	61	2038	
48-155088	123A	3246/229	81	2051	
48-155092	5404	3627			
48-155093	289A	3124/289			
48-155095	290A	3114/290	269	2032	
48-155097	186	3192	28	2052	
48-155098	187	3193			
48-155099	552	3311	504A	1104	
48-155107	552	3843	510	1114	
48-155108	506	3843	509	1114	
48-155110	198	3219	251		
48-155113	5404	3627			
48-155114	109	3088	1N60	1123	
48-155116	187A	9076			
48-155119	289A	3124/289	210	2038	
48-155121	293	3849			
48-155122	294	3841			
48-155125	506	3843	511	1114	
48-155126	116	3311	504A	1104	
48-155128	552	3843			
48-155130	198	3219	251		
48-155131	605	3864			
48-155136	116	3313	504A	1104	
48-155139	506	3843			
48-155140	86	3710/238	75	2041	
48-155146	5013A	3058/137A	ZD-6.2	561	
48-155148	601	3864/605			
48-155152	525	3925	530		
48-155153	198	3219	251		
48-155154	199	3124/289	62	2010	
48-155156	290A	3114/290	269	2032	
48-155157	86	3948/247			
48-155158	5092A	3341/5089A	ZD-51		
48-155159	177	3100/519	300	1122	
48-155178	552	3311	504A	1104	
48-155182	5077A	3752	ZD-18		
48-155184	5013A	3779			
48-155189	165	3710/238	259		
48-155193	116	3311	504A	1104	
48-155198		3080	509	1114	
48-155210				563	
48-155213	198	3219	251		
48-155221		3244	222		
48-155222	287	3244			
48-155223	552	3998/506	511	1114	
48-155224	165	3710/238	259		
48-155225	5070A	9021	ZD-6.0		
48-155235	116	3312	504A	1104	
48-155236	116	3312	504A	1104	
48-171-A06	102A		53	2007	
48-17162A06	102A		53	2007	
48-17162A10	158		53	2007	
48-17162A13	124		12		
48-17162A17	102A		53	2007	
48-17162A22	102A		53	2007	
48-17271A03	102A		53	2007	
48-191807	116		504A	1104	
48-191A01	116	3017B/117	504A	1104	
48-191A01-9	116		504A	1104	
48-191A02	116	3031A	504A	1104	
48-191A03	116	3017B/117	504A	1104	
48-191A04	116	3311	504A	1104	
48-191A05	506	3125	511	1114	
48-191A05A	506	3843	511	1114	
48-191A06		3032A	504A	1104	
48-191A07	116	3032A	504A	1104	
48-191A07A	116		504A	1104	
48-191A08	116	3031A	504A	1104	
48-191A09			504A	1104	
48-191A11	116		504A	1104	
48-21598B01	102A	3024/128	53	2007	
48-217241	181		75	2041	
48-21785J61	311		277		
48-22310K51	311		277		
48-232796	181		75	2041	
48-3003A02	293	3124/289	47	2030	
48-3003A03	107	3018	11	2015	
48-3003A04	106	3118	21	2034	
48-3003A05	123A	3444	20	2051	
48-3003A06	159	3466	82	2032	
48-3003A09	312	3448	FET-2	2035	
48-3003A10	222	3065	FET-4	2036	
48-3003A11	123A	3444	20	2051	
48-3003A12	123A	3444	20	2051	
48-32000	142A		ZD-12	563	
48-34816	177		300	1122	
48-355002	123A	3444	20	2051	
48-355004	199	3245	62	2010	
48-355005	198	3220	251		
48-355006	234	3247	65	2050	
48-355007	159	3466	82	2032	
48-355008	110MP	3709		1123(2)	
48-355009	109			1123	
48-355012	287	3433	222		
48-355013	506	3843	511	1114	
48-355014	178MP		300(2)	1122(2)	
48-355016	116		504A	1104	
48-355023	116		504A	1104	
48-355025	116		504A	1104	
48-355029	5404	3627			
48-355035	177		300	1122	
48-355036	177		300	1122	
48-355037	297		271	2030	
48-355038	198	3220	251		
48-355039	186A	3357	247	2052	
48-355040	152	3893	66	2048	
48-355042	175		246	2020	
48-355043	238	3710	259		
48-355044	198	3220	251		
48-355045	601	3463	504A	1104	
48-355046	605	3864	504A	1104	
48-355047	504	3108/505	CR-6		

Industry Standard No.	ECG	SK	GE	RS 276-	MOTOR.
48-355048	552	3032A	511	1114	
48-355049	116		504A	1104	
48-355050	5082A	3753	ZD-25		
48-355052	123A	3444	20	2051	
48-355053	229	3444/123A	20	2038	
48-355054	171	3747/157	232		
48-355055	193		67	2023	
48-355056	321	3844			
48-355057	290A		269	2032	
48-355058	288	3434	223		
48-355059	152	3893	66	2048	
48-355062	5074A		ZD-11.0	563	
48-35P1	102	3004	2	2007	
48-36P1	102	3004	2	2007	
48-36P3	102	3004	2	2007	
48-37312	242	3189A/183	58	2027	
48-39P1	121	3009	239	2006	
48-39P3	102	3004	2	2007	
48-40004S05		3298	236		
48-40004S06	235	3197	337		
48-40118B01	159	3466	82	2032	
48-40118B01A	159	3466	82	2032	
48-40170-G01	199	3245		2010	
48-401700-G01			62	2010	
48-40170G01	199	3122	20	2051	
48-40171G01	123A	3444	20	2051	
48-40172G01	131		44	2006	
48-40212	128		243	2030	
48-40235G01	116	3016	504A	1104	
48-40235G02	506	3843	504A	1104	
48-40246-G01	199		62	2010	
48-40246G01	199	3122	20	2051	
48-40246G02	123A	3444	20	2051	
48-40247G01		3039	11	2015	
48-40247G02	199	3245	62	2010	
48-40382J01	186	3192	28	2017	
48-40383J01	187	3193	29	2025	
48-40458A04	136A		ZD-5.6	561	
48-40458A06	5072A	3136	ZD-8.2	562	
48-40458A064	5072A		ZD-8.2	562	
48-40516G01		3126	90		
48-40516G02		3126	90		
48-40606J01	123A	3444	20	2051	
48-40606J02	199	3245	62	2010	
48-40607J01	199	3245	62	2010	
48-40734P01	295	3253	270		
48-40738P01	177	3100/519	300	1122	
48-40739P01	116	3311	504A	1104	
48-40764P01	235	3197	322		
48-41266G01	506	3843	511	1114	
48-41508A01		3311	504A	1104	
48-41508A02	116		504A	1104	
48-41763G01	139A	3136/5072A	ZD-8.2	562	
48-41763G02	5072A	3059/138A	ZD-8.2	562	
48-41763G03	139A	3059/138A	ZD-8.2	562	
48-4176301	5072A	3136	ZD-8.2	562	
48-41768G01	109	3088	1N60	1123	
48-41784J03	184	3190	57	2017	
48-41784J04	186A	3357	247	2052	
48-41785J03	185	3191	58	2025	
48-41785J04	187A	9076	248		
48-41815J02	107	3018	11	2015	
48-41816J01	107	3018	11	2015	
48-41816J02	107	3018	212	2010	
48-41816J03	229	3018			
48-41873J02	139A	3060	ZD-9.1	562	
48-41873J03	140A	3061	ZD-10	562	
48-41884J03	184		57	2017	
48-42098B01	159	3466	82	2032	
48-42098B01A	159	3466	82	2032	
48-42383A01		3126	90		
48-42884P01	295	3253	270		
48-42885P01	235	3197	337		
48-42899J01	177		300	1122	
48-43238G01	159	3466	82	2032	
48-43240G01	152	3893	66	2053	
48-43241G01	153		69	2049	
48-43265G01	116	3100/519	504A	1104	
48-43351A01	102A		53	2007	
48-43351A02	107	3039/316	11	2015	
48-43351A03	108	3452	86	2038	
48-43351A04	108	3452	86	2038	
48-43351A05	108	3452	86	2038	
48-43354A81	123A	3444	20	2051	
48-43354A82	123A	3444	20	2051	
48-43354A83	158		53	2007	
48-43467J01	312		FET-1	2035	
48-43992J01	108	3018	61	2038	
48-43P3	102	3004	2	2007	
48-43P4	102	3004	2	2007	
48-44080J01	601	3463			
48-44080J05	139A	3060	ZD-9.1	562	
48-44125A07	125		510,531	1114	
48-44883G01	152		66	2048	
48-44884G01	153		69	2049	
48-44885G01	199	3122	62	2010	
48-44885G02	123A	3444	20	2051	
48-44886G01	289A	3122	268	2038	
48-44887G0	116		504A	1104	
48-45323G01	612	3325	90		
48-45N2	199	3245	62	2010	
48-46669H01	177		300	1122	
48-4937	233		210	2009	
48-56P1	100	3721	1	2007	
48-57120A01	116	3017B/117	504A	1104	
48-57B2	121	3009	239	2006	
48-57B42	121	3009	239	2006	
48-60022A13	123A	3444	20	2051	
48-60022A14	175	3131A/369	246	2020	
48-60022A97	109	3087		1123	
48-60022A98	116	3311	504A	1104	
48-60077A06	109	3087		1123	
48-60154A01	110MP	3709		1123(2)	
48-61074B01	109	3087		1123	
48-61767B01	109	3087		1123	
48-62334A01	110MP			1123	
48-62334A02	109	3087		1123	
48-63005A66	123A	3444	20	2051	
48-63005A72	123A	3444	20	2051	
48-63006A56	109	3087		1123	
48-63026A45	124	3021	12		
48-63026A46	108	3452	86	2038	
48-63026A47	123A	3444	20	2051	
48-63026A48	123A	3444	20	2051	
48-63029A16	160		245	2004	
48-63029A17	158		53	2007	
48-63029A18	102	3004	2	2007	
48-63029A19	102	3004	2	2007	
48-63029A20	109	3087		1123	
48-63029A60	160		245	2004	
48-63029A90	126	3006/160	52	2024	
48-63029A91	102	3004	2	2007	
48-63029A92	100	3005	1	2007	
48-63029A93	102	3004	2	2007	

Industry Standard No.	ECG	SK	GE	RS 276-	MOTOR.
48-63029A94	102	3004	2	2007	
48-63044A05	102A		53	2007	
48-63075A72	160		245	2004	
48-63075A73	160		245	2004	
48-63075A74	126	3006/160	52	2024	
48-63075A75	160		245	2004	
48-63075A76	100	3005	1	2007	
48-63075A78	109			1123	
48-63076A52	123A	3444	20	2051	
48-63076A81	107	3039/316	11	2015	
48-63076A82	107	3039/316	11	2015	
48-63076A83	123A	3444	20	2051	
48-63077A03	100	3005	1	2007	
48-63077A10	123A	3444	20	2051	
48-63077A11	109			1123	
48-63077A29	108	3452	86	2038	
48-63077A30	123A	3444	20	2051	
48-63077A31	123A	3444	20	2051	
48-63077A32	109	3087		1123	
48-63078A52	107	3039/316	11	2015	
48-63078A54	107		11	2015	
48-63078A59	100	3005	1	2007	
48-63078A60	100	3005	1	2007	
48-63078A61	100	3005	1	2007	
48-63078A62	158	3123	53	2007	
48-63078A63	160		245	2004	
48-63078A64	100	3005	1	2007	
48-63078A65	126	3006/160	52	2024	
48-63078A66	158		53	2007	
48-63078A68	158	3123	53	2007	
48-63078A69	158		53	2007	
48-63078A70	123A	3444	20	2051	
48-63078A71	123A	3444	20	2051	
48-63079A97	123A	3444	20	2051	
48-63081A82	160		245	2004	
48-63082A15	100	3005	1	2007	
48-63082A16	158		53	2007	
48-63082A24	126	3006/160	52	2024	
48-63082A25	123A	3444	20	2051	
48-63082A26	123A	3444	20	2051	
48-63082A27	123A	3444	20	2051	
48-63082A45	123A	3444	20	2051	
48-63082A71	123A	3444	20	2051	
48-63084A03	102A	3123	53	2007	
48-63084A04	102A	3123	53	2007	
48-63084A05	102A	3123	53	2007	
48-63084A06	109	3087		1123	
48-63086A16	116		504A	1104	
48-63086A19	158		53	2007	
48-63590A01	109	3091		1123	
48-64169	116		504A	1104	
48-644587	109	3087		1123	
48-644676	160	3008	245	2004	
48-644677	160	3008	245	2004	
48-644678	102A	3123	53	2007	
48-644679	158	3123	53	2007	
48-644681	109	3087		1123	
48-645867	160	3008	245	2004	
48-646954	116	3017B/117	504A	1104	
48-647311	109	3087		1123	
48-647313	109	3087		1123	
48-647713			1N295	1123	
48-647829	116	3017B/117	504A	1104	
48-64978A10	121MP	3013	239(2)	2006(2)	
48-64978A11	121MP	3013	239	2006(2)	
48-64978A24	121MP	3013	239(2)	2006(2)	
48-64978A27	160		245	2004	
48-64978A28	160		245	2004	
48-64978A29	160		245	2004	
48-64978A39	222	3050/221		2036	
48-64978A40	159	3466	82	2032	
48-64978A40A	159	3466	82	2032	
48-64978A41	159	3466	82	2032	
48-64978A41A	159	3466	82	2032	
48-65108A23	102A(2)		53(2)		
48-65108A62	102A(2)		53(2)		
48-65112A65	108	3039/316	86	2038	
48-65112A67	108	3039/316	86	2038	
48-65112A68	161	3716	39	2015	
48-65112A73	112	3089	1N82A		
48-65113A84	112	3089	1N82A		
48-65113A88	108	3452	86	2038	
48-65118A64	108	3452	86	2038	
48-65123A67	108	3452	86	2038	
48-65123A94	123A	3444	20	2051	
48-65123A95	108	3452	86	2038	
48-65132A79	160		245	2004	
48-65144A72	108	3452	86	2038	
48-65145A74	116		504A	1104	
48-65146A61	161	3117	39	2015	
48-65146A62	161	3117	39	2015	
48-65146A63	161	3122	39	2015	
48-65147A72	123A	3444	20	2051	
48-65173A78	161	3117	39	2015	
48-65174A24	161	3716	39	2015	
48-65177A77	129	3025	244	2027	
48-65177A77A	129	3025	244	2027	
48-65831A02	115		6GX1		
48-65837A02	109	3087		1123	
48-65937A02	109	3087		1123	
48-660370A05	116		504A	1104	
48-66037A03	116	3017B/117	504A	1104	
48-66037A04	116	3017B/117	504A	1104	
48-66037A05	116	3017B/117	504A	1104	
48-66037A08	116	3017B/117	504A	1104	
48-66037A10	116	3017B/117	504A	1104	
48-66037A12	116	3017B/117	504A	1104	
48-66544A88	358		42		
48-66629A02	116	3311	504A	1104	
48-66629A03	116	3017B/117	504A	1104	
48-66629A05	116	3311	504A	1104	
48-66629A06	116		504A	1104	
48-66653A001	116(3)		504A(3)	1104	
48-66653A002	116(3)		504A(3)	1104	
48-66653A003	118	3066	CR-1		
48-66653A005	118	3066	CR-1		
48-66653A02	120	3110	CR-3		
48-66653APT001	120		CR-3		
48-66653APT003	118		CR-1		
48-66653APT015	118	3066	CR-1		
48-66653APTOU1	120		CR-3		
48-66653APT1003	118	3066	CR-1		
48-66654A02	116	3017B/117	504A	1104	
48-66865A01	358		42		
48-67020A11	109	3087		1123	
48-6712	113A		6GC1		
48-67120A01	116	3016	504A	1104	
48-67120A02	177(2)	3016	300(2)	1122	
48-67120A03	177	3016	300	1122	
48-67120A04	125	3016	510,531	1114	
48-67120A05	125	3016	510,531	1114	
48-67120A06	116	3017B/117	504A	1104	
48-67120A0607			504A	1104	
48-67120A07	116	3017B/117	504A	1104	
48-67120A08	119		CR-2		
48-67120A09	109	3016		1123	
48-67120A10	125	3031A	510,531	1114	
48-67120A11	177	3100/519	300	1122	
48-67120A13	177	3175	300	1122	
48-67120AB	177		300	1122	
48-6712A02	178MP		300(2)	1122(2)	
48-6712A11	177		300	1122	
48-6720A02	177	3175	300	1122	
48-674297U	112	3089	1N82A		
48-674970	112	3089	1N82A		
48-67926A01	116	3017B/117	504A	1104	
48-68688A79	116		504A	1104	
48-68688A79B	116		504A	1104	
48-69394A01	123A		20	2051	
48-69723A01	513	3443	513		
48-69723A02	513	3443	513		
48-69723AD2-185	513	3443	513		
48-69723B02	513	3443	513		
48-711052	109	3087		1123	
48-7322230	358		42		
48-732230	358		42		
48-733746	116	3017B/117	504A	1104	
48-739300	109	3087		1123	
48-741255	178MP		300(2)	1122(2)	
48-741280	109	3087		1123	
48-741656	178MP		300(2)	1122(2)	
48-741724	178MP		300(2)	1122(2)	
48-741752	114		6GD1	1104	
48-742698	114		6GD1		
48-742970	112	3089	1N82A		
48-746831	116	3017B/117	504A	1104	
48-751656	113A		6GC1		
48-751724	115		6GX1		
48-752497	116	3017B/117	504A	1104	
48-754153	114		6GD1	1104	
48-77	358		42		
48-80313A46	507	3315			
48-82095C01	116		504A	1104	
48-82095C02	116		504A	1104	
48-82095C03	116		504A	1104	
48-82095C64	116		504A	1104	
48-82139G01	109			1123	
48-82139G02	109			1123	
48-82178A01	109			1123	
48-82178A02	109			1123	
48-82178A03	109			1123	
48-82178A04	109			1123	
48-82178A05	109			1123	
48-82178A06	109			1123	
48-82178A07	109			1123	
48-82178A08	109			1123	
48-82178A09	109			1123	
48-82178A10	109			1123	
48-82178A11	109			1123	
48-82178A12	109			1123	
48-82178A13	109			1123	
48-82256C01	5013A	3797/5031A	ZD-6.2	561	
48-82256C02	5014A	3780	ZD-6.8	561	
48-82256C03		3775	ZD-4.7		
48-82256C05	5045A	3343/5092A	ZD-68		
48-82256C06	5074A		ZD-11.5		
48-82256C07	5009A	3775	ZD-4.7		
48-82256C08	141A	3782/5016A	ZD-8.2	562	
48-82256C09	5016A	3782	ZD-8.2	562	
48-82256C10	144A		ZD-14	564	
48-82256C11	5019A	3785	ZD-10	562	
48-82256C12	5011A	3777	ZD-5.6	561	
48-82256C13	5023A	3789	ZD-14	564	
48-82256C14	5024A	3790	ZD-15	564	
48-82256C15	5010A		ZD-5.1		
48-82256C16	5016A	3782	ZD-8.2	562	
48-82256C17	5021A	3787	ZD-12	563	
48-82256C18	5018A		ZD-9.1	562	
48-82256C19	5014A	3780	ZD-6.8	561	
48-82256C20	5033A	3799	ZD-27		
48-82256C22	139A		ZD-9.1	562	
48-82256C23	5014A	3780	ZD-6.8	561	
48-82256C24	5027A	3793	ZD-18		
48-82256C25	5021A	3787	ZD-12	563	
48-82256C26	5005A	3771	ZD-3.3		
48-82256C27	151A		ZD-110		
48-82256C28	140A		ZD-10	562	
48-82256C29	5090A	3342	ZD-56		
48-82256C30	5096A	3346	ZD-100		
48-82256C31	5031A		ZD-24		
48-82256C32	5019A	3785	ZD-10	562	
48-82256C34	5020A	3786	ZD-11	563	
48-82256C35	135A		ZD-5.1		
48-82256C36	5083A	3754	ZD-28		
48-82256C37	5071A	3334	ZD-6.8	561	
48-82256C38	5018A		ZD-9.1	562	
48-82256C39	5079A	3335	ZD-20		
48-82256C40	140A		ZD-10	562	
48-82256C42	5082A	3753	ZD-25		
48-82256C43	139A		ZD-9.1	562	
48-82256C44	5015A	3781	ZD-7.5		
48-82256C45	5017A	3783	ZD-9.1	562	
48-82256C46	5067A	3331	ZD-3.9		
48-82256C47	5014A	3780	ZD-6.8	561	
48-82256C48	5022A	3788	ZD-13	563	
48-82256C49	5031A	3797	ZD-24		
48-82256C50	5022A	3788	ZD-13	563	
48-82256C51	135A		ZD-5.1		
48-82256C53	5077A	3752	ZD-18		
48-82256C54	142A		ZD-12	563	
48-82256C56	139A		ZD-9.1	562	
48-82256C57	5079A	3335	ZD-20		
48-82256C58	5086A	3338	ZD-39		
48-82256C59	145A		ZD-15	564	
48-82292A01	109			1123	
48-82292A02	109			1123	
48-82292A03	109			1123	
48-82292A04	109			1123	
48-82292A05	109			1123	
48-82292B03				1123	
48-82392B01	177		300	1122	
48-82392B02	177		300	1122	
48-82392B03	177		300	1122	
48-82392B04	177		300	1122	
48-82392B05	177		300	1122	
48-82392B06	177		300	1122	
48-82392B07	177		300	1122	
48-82392B08	177		300	1122	
48-82392B09	177		300	1122	
48-82392B10	177		300	1122	
48-82392B11	177		300	1122	
48-82392B12	177		300	1122	
48-82392B13	177		300	1122	
48-82392B14	177		300	1122	
48-82392B15	177		300	1122	
48-8240C06	116		504A	1104	
48-8240C07	519		514	1122	
48-8240C08	116		504A	1104	

Industry Standard No.	ECG	SK	GE	RS 276-	MOTOR.
48-8240009	177		300	1122	
48-8240010	506	3843	511	1114	
48-8240011	519		514	1122	
48-8240012	506	3843	511	1114	
48-8240013	177		300	1122	
48-8240014	519		514	1122	
48-8240015	177		300	1122	
48-8240016	519		514	1122	
48-8240017	519		514	1122	
48-8240018	519		514	1122	
48-82420001	177		300	1122	
48-82420002	177		300	1122	
48-82420003	519		514	1122	
48-82420004	519		514	1122	
48-82420005	177		300	1122	
48-82466H	125		510,531	1114	
48-82466H01	116		504A	1104	
48-82466H02	116		504A	1104	
48-82466H03	116		504A	1104	
48-82466H04	116		504A	1104	
48-82466H06	116		504A	1104	
48-82466H07	116		504A	1104	
48-82466H12	506	3843	511	1114	
48-82466H13	506	3843	511	1114	
48-82466H14	506	3843	511	1114	
48-82466H15	506	3843	511	1114	
48-82466H16	506	3843	511	1114	
48-82466H17		3843	511	1114	
48-82466H18	506	3843	511	1114	
48-82466H19	506	3843	511	1114	
48-82466H21	506	3843	511	1114	
48-82466H25	506	3843	511	1114	
48-82466H27		3843	511	1114	
48-82466H28	506	3843	511	1114	
48-82732C04	5944		5048		
48-82965F	5635	3533			
48-83461E01	5075A	3751	ZD-16	564	
48-83461E02	5022A	3788	ZD-13	563	
48-83461E03	5066A	3330	ZD-3.3		
48-83461E04	5093A	3344	ZD-75		
48-83461E05		9021	ZD-6.2	561	
48-83461E06	5096A	3346	ZD-100		
48-83461E07	5075A	3751	ZD-16	564	
48-83461E08	147A		ZD-33		
48-83461E09	5090A	3342	ZD-56		
48-83461E10	5010A	3776	ZD-5.1		
48-83461E11	5082A	3753	ZD-25		
48-83461E12	146A		ZD-27		
48-83461E13		3332	ZD-4.3		
48-83461E14	5096A	3346	ZD-100		
48-83461E15	5018A	3784	ZD-9.1	562	
48-83461E16	5095A	3345	ZD-91		
48-83461E17	5001A	3767			
48-83461E18	5027A	3793	ZD-18		
48-83461E19	5067A	3331	ZD-3.9		
48-83461E20	146A		ZD-27		
48-83461E21	5129A	3395	5ZD-14		
48-83461E22	5029A	3795	ZD-20		
48-83461E23	510		SS-3DB3		
48-83461E24	5028A	3794	ZD-19		
48-83461E25	5014A	3780	ZD-6.8	561	
48-83461E26	5031A	3797	ZD-24		
48-83461E27	5012A	3778	ZD-6.0	561	
48-83461E28	5141A	3407			
48-83461E30	148A		ZD-55		
48-83461E31	5022A	3788	ZD-13	563	
48-83461E32	5016A		ZD-8.2	562	
48-83461E33	5027A	3793	ZD-18		
48-83461E34	5117A	3383			
48-83461E35	5137A	3403	5ZD-24		
48-83461E36	137A		ZD-6.2	561	
48-83461E37	5077A	3752	ZD-18		
48-83461E38	136A		ZD-5.6	561	
48-83461E39	5135A	3401			
48-83461E40	5010A	3776	ZD-5.1		
48-83461E41	135A		ZD-5.1		
48-83461E42	5071A		ZD-6.8	561	
48-83461E43		3772	ZD-3.6		
48-83461E44	5115A	3381			
48-83741C01	5465	3687			
48-83750G01	123A		20	2051	
48-83750G02	121	3717			
48-8375D02	5455	3597		1067	
48-83875D01	5454	6754		1067	
48-83875D04	5431			1067	
48-83875D05	5422			1067	
48-83875D06	5455			1067	
48-859248	128	3024	243	2030	
48-859428			18	2030	
48-8613	112		1N82A		
48-8613-3	126		52	2024	
48-86148	116	3031A	504A	1104	
48-86168-3	109	3087		1123	
48-8619-3	109	3087		1123	
48-86200-3	109	3087		1123	
48-86289-3	177	3088	300	1122	
48-863030	109			1123	
48-86308-3	506	3843	511	1114	
48-86343-3	109	3088		1123	
48-86376-3	123A	3444	20	2051	
48-86429-3	234		65	2050	
48-86475-3	162	3438	35		
48-867716	109			1123	
48-869001	102A		53	2007	
48-86904GF	160		245	2004	
48-869052	487	3765/195A			
48-869061	195A	3765			
48-869083	121	3717			
48-869087B	104	3719	16	2006	
48-869090	127	3764	25		
48-869091	121	3717			
48-869092	103	3862	8	2002	
48-869093	103	3862	8	2002	
48-869099B	104	3719	16	2006	
48-869104	121	3717			
48-869125	195A	3765			
48-869138	128	3024	243	2030	
48-869141	121	3717	239	2006	
48-869142	121	3717	239	2006	
48-869145	129		244	2027	
48-869147	195A	3765			
48-869148	102	3722	2	2007	
48-869170	128	3024	243	2030	
48-869174	195A	3765			
48-869177	176	3845	80		
48-869182	121	3717	239	2006	
48-869184	128		243	2030	
48-869186	195A	3765			
48-869187	195A	3765			
48-869197	199		62	2010	
48-869198	102A		53	2007	
48-869202	121	3717	239	2006	
48-869205	179	3642	76		
48-869206	6400A			2029	

Industry Standard No.	ECG	SK	GE	RS 276-	MOTOR.
48-869209	195A	3765	243	2030	
48-869221	128		243	2030	
48-869225	175		246	2020	
48-869226-0	123A	3444	20	2051	
48-869228	128	3024	243	2030	
48-869237	121	3717	239	2006	
48-869241	121	3717	239	2006	
48-869244	130		14	2041	
48-869248	123A	3444	20	2051	
48-869249	102	3004	2	2007	
48-869250	102	3722	2	2007	
48-869251	105		4		
48-869253	102	3004	2	2007	
48-869254	103	3862	8	2002	
48-869255	121	3717	239	2006	
48-869256	6401			2029	
48-869257	129		244	2027	
48-869259	130		14	2041	
48-869260	195A	3765			
48-869263	128		243	2030	
48-869264	6400A			2029	
48-869265	198	3220	251		
48-869266	161	3716	39	2015	
48-869269	199	3245	62	2010	
48-869271	195A	3765			
48-869273C	163A	3439	36		
48-869274	175		246	2020	
48-869278	130		14	2041	
48-869279C	162	3438	35		
48-869282	102	3004	2	2007	
48-869283	103	3862	8	2002	
48-869285	195A	3765			
48-869286	198	3220	251		
48-869287	124		12		
48-869293	199	3245	62	2010	
48-869294	121	3717			
48-869301	175		246	2020	
48-869302	130		14	2041	
48-869306	314	3898			
48-869308	129		244	2027	
48-869309	175		246	2020	
48-869312	123A	3444	20	2051	
48-869316	476		246	2020	
48-869320	198	3220	251		
48-869321	130		14	2041	
48-869325	123A	3444	20	2051	
48-869329	123A	3444	20	2051	
48-869334	159	3466	82	2032	
48-869337	162	3438	35		
48-869338	199	3245	62	2010	
48-869342	121	3717	239	2006	
48-869355	195A	3765			
48-869380	128		243	2030	
48-869384	199	3245	62	2010	
48-869389	289A		268	2038	
48-869393	175		246	2020	
48-869400	129		244	2027	
48-869408	162	3438	35		
48-869409	199	3245	62	2010	
48-869412	234	3247	65	2050	
48-869413	159	3466	82	2032	
48-869416	199	3245	62	2010	
48-869420	121	3717			
48-869426	129	3025	244	2027	
48-869427	179	3642	76		
48-869430	195A	3765			
48-869431	195A	3765			
48-869432	129		244	2027	
48-869435	129		244	2027	
48-869436	121	3717	239	2006	
48-869444	123A	3444	20	2051	
48-869447	199	3245	62	2010	
48-869450	161	3716	39	2015	
48-869458	121	3717			
48-869464	128	3024	243	2030	
48-869465	198	3220	251		
48-869467	234	3247	65	2050	
48-869474	199	3245	62	2010	
48-869475	102	3004	2	2007	
48-869475A	102	3004	2	2007	
48-869476	103	3862	8	2002	
48-869476A	103	3862	8	2002	
48-869480	181		75	2041	
48-869481	161	3716	39	2015	
48-869486	199	3245	62	2010	
48-869491	195A	3765	243	2030	
48-869497	199	3245	62	2010	
48-869515	130		14	2041	
48-869517	121	3717			
48-869519	195A	3765	243	2030	
48-869520	129		244	2027	
48-869521	192		63	2030	
48-869525	123A	3444	20	2051	
48-869526	159	3466	82	2032	
48-869536	225		256		
48-869547	199	3245	62	2010	
48-869550	121	3717	239	2006	
48-869562	128		243	2030	
48-869563	123A	3444	20	2051	
48-869568	123A	3444	20	2051	
48-869570	123A	3444	20	2051	
48-869571	159	3466	82	2032	
48-869576	152	3893	66	2048	
48-869582	185		58	2025	
48-869591	346		243	2030	
48-869594	199	3245	62	2010	
48-869599	128		243	2030	
48-869610	476		246	2020	
48-869618	184		57	2017	
48-869628	181		75	2041	
48-869631	346		243	2030	
48-869639	181		75	2041	
48-869640	188		226	2018	
48-869641	189		218	2026	
48-869649	159	3466	82	2032	
48-869660	390		255	2041	
48-869661	152	3893	66	2048	
48-869676	196		241	2020	
48-869677	197		250	2027	
48-869681	129		244	2027	
48-869701	197		250	2027	
48-869703-3	128		243	2030	
48-869767	123A	3444	20	2051	
48-90066A01	113A		6GC1		
48-90068A01	120	3110	20		
48-90158A01	506	3843	511	1114	
48-90165A01	159	3466	82	2032	
48-90165A01A	159	3466	82	2032	
48-90172A01	123A	3444	20	2051	
48-90210A01	109	3087		1123	
48-90222A08	109			1123	
48-90229A01	116	3017B/117	504A	1104	
48-90232A01	161	3117	39	2015	
48-90232A03	108	3452	86	2038	

Industry Standard No.	ECG	SK	GE	RS 276-	MOTOR.
48-90232A04	108	3452	86	2038	
48-90232A05	123A	3444	20	2051	
48-90232A06	129	3025	244	2027	
48-90232A06A	129	3025	244	2027	
48-90232A07	124	3021	12		
48-90232A08	128	3024	243	2030	
48-90232A09	129	3025	244	2027	
48-90232A09A	129	3025	244	2027	
48-90232A10	108	3452	86	2038	
48-90232A11	123A	3444	20	2051	
48-90232A12	129	3122	244	2027	
48-90232A12A	129	3025	244	2027	
48-90232A13	123A	3444	20	2051	
48-90232A14	312	3834/132	FET-1	2035	
48-90232A15	129	3025	244	2027	
48-90232A15A	129	3025	244	2027	
48-90232A16	124		12		
48-90232A17	161	3117	39	2015	
48-90232A18	161	3117	39	2015	
48-90232A19	108	3452	86	2038	
48-90233A01	109	3087		1123	
48-90233A04	116		504A	1104	
48-90233A05	112		1N82A		
48-90233A06	109	3087		1123	
48-90233A07	134A		ZD-3.6		
48-90233A08	109			1123	
48-90234A01	116	3017B/117	504A	1104	
48-90234A02	117		504A	1104	
48-90234A03	120		CR-3		
48-90234A05	503	3068			
48-90234A11	175		246	2020	
48-90234A12	506	3843	511	1114	
48-90234A13	289A		210	2038	
48-90234A14	290A		269	2032	
48-90234A36	373	3190/184	57	2017	
48-90234A38	234	3247	65	2050	
48-90234A39	312	3834/132	FET-2		
48-90234A58	605	3864			
48-90234A64		3245	62		
48-90234A66	112	3089	1N82A		
48-90234A99		3247	65	2050	
48-90235A01	113A	3119/113	6GC1		
48-90343A06	199	3444/123A	61	2010	
48-90343A52	165	3115	259		
48-90343A53	506	3843	511	1114	
48-90343A54	552	3843	511	1114	
48-90343A55	5010A	3056/135A	ZD-5.1		
48-90343A56	175	3026	246	2020	
48-90343A57	287		222		
48-90343A58	288		223		
48-90343A59	234	3114/290	65	2050	
48-90343A61	198	3220	32		
48-90343A62	125	3081	510	1114	
48-90343A64	116	3312	504A	1104	
48-90343A66	109	3088	1N60	2015	
48-90343A68	290A	3114/290	269	2032	
48-90343A73	199	3122	61		
48-90343A75	229	3132	60		
48-90343A76		3357	215	2052	
48-90343A77	289A	3124/289	210		
48-90343A79	199	3245	85		
48-90343A80	229	3122	212		
48-90343A83	289A	3124/289	62		
48-90343A85	373	3253/295	270		
48-90343A88	601	3463			
48-90343A89	605	3864			
48-90343A91	116	3312	504A	1104	
48-90343A92	116	3312	504A	1104	
48-90343A93	116	3311	504A	1104	
48-90420A01		3911	62		
48-90420A02	187A	3191/185	29		
48-90420A03	519	3175	300	1122	
48-90420A04		3175	300	1122	
48-90420A05	294	3841	48		
48-90420A06	153	3083/197	69	2049	
48-90420A73		3433	222		
48-90420A79	116	3312	504A	1104	
48-90420A80	116	3312	504A	1104	
48-90420A92	376	3747/157	232		
48-90420A97	552	3313/116	504A	1104	
48-90420A98	552	3843	511	1114	
48-97046A02	102A		53	2007	
48-97046A03	102A	3123	53	2007	
48-97046A04	108	3452	86	2038	
48-97046A05	107	3039/316	11	2015	
48-97046A06	107	3039/316	11	2015	
48-97046A07	107	3039/316	11	2015	
48-97046A08	126	3006/160	52	2024	
48-97046A09	126	3006/160	52	2024	
48-97046A10	102A	3123	53	2007	
48-97046A14	5069A	3330/5066A	ZD-4.7		
48-97046A15	131	3198	44	2006	
48-97046A16	126		52	2024	
48-97046A17	107		11	2015	
48-97046A18	108	3452	86	2038	
48-97046A18		3198	44	2006	
48-97046A20	107		11	2015	
48-97046A21	107		11	2015	
48-97046A22	123A	3444	20	2051	
48-97046A23	123A	3444	20	2051	
48-97046A24	123A	3444	20	2051	
48-97046A25	161	3716	39	2015	
48-97046A26	159	3466	82	2032	
48-97046A27	159	3466	82	2032	
48-97046A28	123A	3444	20	2051	
48-97046A29	128		243	2030	
48-97046A30	152	3893	66	2048	
48-97046A31	131	3052	44	2006	
48-97046A32	102A		53	2007	
48-97046A33	102A		53	2007	
48-97046A34	102A	3123	53	2007	
48-97046A36	129		244	2027	
48-97046A37	158		53	2007	
48-97046A38	152	3893	66	2048	
48-97046A39	129		244	2027	
48-97046A40	128		243	2030	
48-97046A42	123A	3444	20	2051	
48-97046A43	123A	3444	20	2051	
48-97046A45	161	3716	39	2015	
48-97046A46	123A	3444	20	2051	
48-97046A47	312	3834/132	FET-1	2035	
48-97046A48	312	3834/132	FET-1	2035	
48-97046A50	123A	3444	20	2051	
48-97046A51	108	3452	86	2038	
48-97046A52	123A	3444	20	2051	
48-97046A53	102	3722	2	2007	
48-97046A54	102A		53	2007	
48-97046A55	102A		53	2007	
48-97046A56	102A		53	2007	
48-97046A57	102A		53	2007	
48-97048A01	177	3100/519	300	1122	
48-97048A02	109	3087		1123	
48-97048A04	116	3017B/117	504A	1104	
48-97048A05	110MP	3709		1123(2)	
48-97048A06	109			1123	
48-97048A07	116		504A	1104	
48-97048A08	142A		ZD-12	563	
48-97048A10	116		504A	1104	
48-97048A16	142A		ZD-12	563	
48-97048A17	177		300	1122	
48-97048A18	116	3311	504A	1104	
48-97048A19	109			1123	
48-97048A20	125		510,531	1114	
48-971-A95	123A	3444	20	2051	
48-97127A01	116		504A	1104	
48-97127A012	123A	3444	20	2051	
48-97127A013	123A	3444	20	2051	
48-97127A015	106	3984	86	2034	
48-97127A018	123A	3444	20	2051	
48-97127A019	128		243	2030	
48-97127A02	108	3039/316	86	2038	
48-97127A03	108	3039/316	86	2038	
48-97127A04	128	3024	243	2030	
48-97127A05	158		53	2007	
48-97127A06	107	3039/316	11	2015	
48-97127A09	158		53	2007	
48-97127A12	123A	3444	20	2051	
48-97127A13	123A	3444	20	2051	
48-97127A15	106	3984	21	2034	
48-97127A18	123A	3444	20	2051	
48-97127A19	123A	3444	20	2051	
48-97127A20	102A		53	2007	
48-97127A22	102A		53	2007	
48-97127A23	158		53	2007	
48-97127A24	123A	3444	20	2051	
48-97127A29	123A	3444	20	2051	
48-97127A30	102A		53	2007	
48-97127A31	102A		53	2007	
48-97127A32	102	3722	2	2007	
48-97127A33	123A	3444	20	2051	
48-97162A01	108	3039/316	86	2038	
48-97162A02	108	3039/316	86	2038	
48-97162A03	126	3008	52	2024	
48-97162A04	123A	3444	20	2051	
48-97162A05	123A	3444	20	2051	
48-97162A06	158		53	2007	
48-97162A07	158		53	2007	
48-97162A08	158		53	2007	
48-97162A09	123A	3444	20	2051	
48-97162A11	102A		53	2007	
48-97162A12	123A	3444	20	2051	
48-97162A15	123A	3444	20	2051	
48-97162A16	102A		53	2007	
48-97162A18	102A	3123	53	2007	
48-97162A19	102A	3123	53	2007	
48-97162A20	102A		53	2007	
48-97162A21	123A		20	2051	
48-97162A23	123A		20	2051	
48-97162A24	102A		53	2007	
48-97162A25	102A		53	2007	
48-97162A26	107		11	2015	
48-97162A28	161		39	2015	
48-97162A30	161	3716	39	2015	
48-97162A31	161	3716	39	2015	
48-97162A32	161	3716	39	2015	
48-97162A33	123A	3444	20	2051	
48-97162A34	158		53	2007	
48-97162A35	184		57	2017	
48-97162A69	184		57	2017	
48-97168A01	109	3091		1123	
48-97168A02	116	3311	504A	1104	
48-97168A03	110MP	3091		1123(2)	
48-97168A04	109			1123	
48-97168A06	116		504A	1104	
48-97168A07	109	3091		1123	
48-97168A09	109			1123	
48-97168A10	116		504A	1104	
48-97168A11	116	3311	504A	1104	
48-97168A13	109			1123	
48-97168A14	116		504A	1104	
48-97172A01	116		5048	1104	
48-97177A01	312	3834/132	FET-1	2035	
48-97177A02	161	3117	39	2015	
48-97177A03	161	3117	39	2015	
48-97177A04	108	3452	86	2038	
48-97177A06	312	3834/132	FET-1	2035	
48-97177A07	108	3452	86	2038	
48-97177A08	108	3452	86	2038	
48-97177A09	123A	3444	20	2051	
48-97177A10	188	3199	226	2018	
48-97177A11	189	3200	218	2026	
48-97177A12	123A	3444	20	2051	
48-97177A13	123A	3444	20	2051	
48-97177A14	159	3466	82	2032	
48-97177A14A	159	3466	82	2032	
48-97177A15	109	3087		1123	
48-97177H01	312			2028	
48-971A04	108	3452	86	2038	
48-971A05	123A	3444	20	2051	
48-971A13	124		12		
48-971A203	160		245	2004	
48-97221A01	102A		53	2007	
48-97221A02	102A		53	2007	
48-97221A03	102A	3123	53	2007	
48-97221A04	102A		53	2007	
48-97221A05	102A		53	2007	
48-97222A01	109			1123	
48-97222A02	116		504A	1104	
48-97238A01	126	3006/160	52	2024	
48-97238A02	126	3006/160	52	2024	
48-97238A03	126	3008	52	2024	
48-97238A04	123A	3122	20	2051	
48-97238A05	102A		53	2007	
48-97238A06	131		44	2006	
48-97238A07	102A		53	2007	
48-97239A01	109			1123	
48-97270A01	116		504A	1104	
48-97270A02	109			1123	
48-97271A01	102A	3123	53	2007	
48-97271A02	102A	3123	53	2007	
48-97271A03	102A	3123	53	2007	
48-97271A04	102A	3006/160	53	2007	
48-97271A05	160	3008	245	2004	
48-97271A06	160	3006	245	2004	
48-97271A2	102A		53	2007	
48-97271A3	102A		53	2007	
48-97271A4	102A		53	2007	
48-97271A5	102A		53	2007	
48-97271A6	102A		53	2007	
48-97305A02	136A		ZD-5.6	561	
48-97305A03	166	9075		1152	
48-97305A05	136A		ZD-5.6	561	
48-97762A02	108	3452	86	2038	
48-97768A06	125		510,531	1114	
48-K869575	108	3452	86	2038	
48-P02597A	123A	3444	20	2051	
48-V34816	177	3100/519	300	1122	
48A124315		3004	2	2007	

Industry Standard No.	ECG	SK	GE	RS 276-	MOTOR.
48A124327	102	3722	2	2007	
48A40458A06			ZD-8.2	562	
48A41508A01	116	3016	504A	1104	
48A41508A02		3016	504A	1104	
48A42383A01	614	3126	90		
48B407624P1	123A	3444	20	2051	
48B41226G01	506	3843	511	1114	
48B41266G01			511	1114	
48B41768G01	109	3088		1123	
48B43265G01	116	3311	504A	1104	
48B62334A01			1N60	1123	
48B63494A01		3126	90		
48B66629A01			1N60	1123	
48B66629A02			1N34AS	1123	
48B66629A03		3016	504A	1104	
48B66629A05			504A	1104	
48B67020A11			1N34AS	1123	
48C-40235-602	116		504A	1104	
48C125233			8	2002	
48C125235			8	2002	
48C125236			8	2002	
48C125237		3004	52	2024	
48C134587		3091	1N295	1123	
48C134816		3100	300	1122	
48C40235C01			504A	1104	
48C40235G01	116	3311	504A	1104	
48C40235G02	506	3843	511	1114	
48C40235G1			504A	1104	
48C40524A02			300	1122	
48C42428A01			1N60	1123	
48C61074B01	109	3087		1123	
48C61767B01			1N60	1123	
48C65832A02			1N60	1123	
48C65837A01			1N60	1123	
48C65837A02	109	3088		1123	
48C66037A03		3016	504A	1104	
48C66037A04		3016	504A	1104	
48C66037A05		3016	504A	1104	
48C66037A10		3016	504A	1104	
48C66037A12		3016	504A	1104	
48C66629A02			504A	1104	
48C66653A02		3110	504A	1104	
48C67120A02			300	1122	
48C67120A05			504A	1104	
48C67926A01		3016	504A	1104	
48C751656		3119	6GC1		
48D63590A01			1N60	1123	
48D66037A03		3016	504A	1104	
48D66037A04		3016	504A	1104	
48D66037A05		3016	504A	1104	
48D66037A08		3016	504A	1104	
48D66653A02		3110	CR-3		
48D66653APT.005		3066	CR-1		
48D66653APT001		3110	CR-3		
48D66653APT003		3066	CR-1		
48D66653APT015		3066	CR-1		
48D66654A02		3016	504A	1104	
48D67	177		300	1122	
48D67120A01		3016	504A	1104	
48D67120A02	116	3017B/117	6GC1	1104	
48D67120A05			504A	1104	
48D67120A06		3016	504A	1104	
48D67120A07		3016	504A	1104	
48D67120A08		3016	CR-2		
48D67120A11	177	3124/289		1123	
48D67120A11(DIO)			1N34AS	1123	
48D67120A11(XSTR)			20	2051	
48D67120A13	177	3100/519	300	1122	
48D67120A13(C)			300	1122	
48D67120H02			504A	1104	
48D69723A01	513	3443	513		
48D69723A02		3443	513		
48D69723A02-185	513	3443	513		
48D82420C01	177		300	1122	
48D82420C02	177		300	1122	
48D82420C03	519		514	1122	
48D82420C04	519		514	1122	
48D82420C05	177		300	1122	
48D8375D02	5455	3597		1067	
48D83875D01	5454	6754		1067	
48D83875D04	5431			1067	
48D83875D05	5422			1067	
48D83875D06	5455			1067	
48G10346A01		3100	1N60	1123	
48G10346A02		3087	300	1122	
48G13001A01	124	3538	12		
48K125230		3009	16	2006	
48K134450			2	2007	
48K134458		3004	2	2007	
48K134482	102	3722	2	2007	
48K134483	102	3722	2	2007	
48K134494		3008	51	2004	
48K134495		3008	51	2004	
48K134496		3008	51	2004	
48K134583	121	3717	239	2006	
48K134584	121	3717	239	2006	
48K134587			1N60	1123	
48K134601		3008	51	2004	
48K134796		3008	51	2004	
48K134798		3008	51	2004	
48K35P1		3004	52	2024	
48K36P1		3004	52	2024	
48K36P3		3004	52	2024	
48K39P1		3009	16	2006	
48K43P3		3004	52	2024	
48K43P4		3004	52	2024	
48K45N2			8	2002	
48K544539			1N60	1123	
48K56P1		3123	2	2007	
48K57B2		3009	16	2006	
48K57B42			16	2006	
48K640675			1N60	1123	
48K64169		3016	504A	1104	
48K644681	109	3088		1123	
48K646954		3016	504A	1104	
48K647311			1N60	1123	
48K647713			1N60	1123	
48K647769			1N60	1123	
48K647829		3016	504A	1104	
48K741255		3119	6GC1		
48K741752		3120	6GD1		
48K742698		3120	6GD1		
48K746831		3016	504A	1104	
48K751724		3121	6GX1		
48K752497		3016	504A	1104	
48K863030	109			1123	
48K867716	109			1123	
48K869001		3005	1	2007	
48K869228	128	3024	243	2030	
48K869269	199	3245	62	2010	
48K869293	199	3245	62	2010	
48K869309	175		246	2020	
48K869342	121	3717	239	2006	
48K869409	199	3245	62	2010	
48K869447	199	3245	62	2010	
48K869474	199	3245	62	2010	
48K869486	199	3245	62	2010	
48K869515	130		14	2041	
48M355002		3018	20	2051	
48M355004			61	2038	
48M355005		3103A	32		
48M355006		3114	48		
48M355007		3114	21	2034	
48M355008		3087	1N34AS	1123	
48M355009		3087	1N34AS	1123	
48M355012		3044	40	2012	
48M355013		3130	511	1114	
48M355014		3100	300	1122	
48M355016		3017B	504A	1104	
48M355021		3080	509	1114	
48M355023		3017B	504A	1104	
48M355025		3016	504A	1104	
48M355035		3100	300	1122	
48M355037		3024	63	2030	
48M355039		3054	28	2017	
48M355040		3054	66	2048	
48M355042		3538	246	2020	
48M355043		3710	259		
48M355045		3311	504A	1104	
48M355046	601	3463	504A	1104	
48M355048			504A	1104	
48M355049			504A	1104	
48M355052		3018	20	2051	
48M355053		3018	20	2051	
48M355055			29	2025	
48M355056		3131A	32		
48M355059		3054	66	2048	
48M355062		3139	ZD-11	563	
48NSP1035	121	3717	239	2006	
48P-2A5	102A		53	2007	
48P1	102A	3004	53	2007	
48P217241	181	3511	75	2041	
48P232796	181	3036	75	2041	
48P60022A97			1N60	1123	
48P60022A98		3126	90		
48P60077A06			1N34AS	1123	
48P63005A72		3124	20	2051	
48P63006A56			1N60	1123	
48P63076A81		3018	17	2051	
48P63076A82		3018	17	2051	
48P63077A03			2	2007	
48P63077A11			1N60	1123	
48P63077A31		3018	20	2051	
48P63077A32			1N34AS	1123	
48P63077A52			1N60	1123	
48P63078A52		3018	17	2051	
48P63078A62			2	2007	
48P63078A69			11	2015	
48P63078A70		3018	17	2051	
48P63078A71	123A	3444	20	2051	
48P63078A86(NPN)			8	2002	
48P63078A86(PNP)			2	2007	
48P63079A97		3018	20	2051	
48P63082A24			17	2051	
48P63082A25		3018	10	2051	
48P63082A26			20	2051	
48P63082A27		3018	18	2030	
48P63082A45	108	3452	86	2038	
48P63082A71	108	3452	86	2038	
48P63086A18			10	2051	
48P64978A27			245	2004	
48P64978A28			245	2004	
48P64978A29			245	2004	
48P65112A65		3019	17	2051	
48P65112A73	112	3089	1N82A		
48P65113A88		3019	17	2051	
48P65118A64			17	2051	
48P65123A67		3019	11	2015	
48P65123A95		3019	11	2015	
48P65144A72		3019	11	2015	
48P65146A61	107	3018	11	2015	
48P65146A62		3018	11	2015	
48P65146A63	108	3452	86	2038	
48P65173A78	108	3452	86	2038	
48P65174A24	161	3019	39	2015	
48P65193A56			11	2015	
48P65194A92			11	2015	
48Q134722			3	2006	
48R10001-A01		3016	504A	1104	
48R10001-A03			504A	1104	
48R10001-A030-1			504A	1104	
48R100620A02			504A	1104	
48R100620A04			504A	1104	
48R100620A05			504A	1104	
48R10062A01			504A	1104	
48R10062A02		3016	504A	1104	
48R10073A02			52	2024	
48R10074A02			52	2024	
48R134407		3004	2	2007	
48R134545	160		245	2004	
48R134573		3004	2	2007	
48R134582	121	3717	239	2006	
48R134587			1N60	1123	
48R134606	121	3717	239	2006	
48R134621	102A	3004	53	2007	
48R134632	102A	3004	53	2007	
48R134665		3010	8	2002	
48R134666			18	2030	
48R134671		3016	504A	1104	
48R134722		3009	16	2006	
48R2	5866		5028		
48R25	5866		5028		
48R660370A05			504A	1104	
48R859428		3024	18	2030	
48R869090	127	3764	25		
48R869092	103	3862	8	2002	
48R869093	103	3862	8	2002	
48R869138		3024	18	2030	
48R869141	121	3717	239	2006	
48R869142	121	3717	239	2006	
48R869145	129		244	2027	
48R869148		3004	2	2007	
48R869170	128	3024	243	2030	
48R869197	199		62	2010	
48R869202	121	3717	239	2006	
48R869205	121	3717	239	2006	
48R869206	6400A			2029	
48R869209	195A	3765	243	2030	
48R869225	175		246	2020	
48R869241	121	3717	239	2006	
48R869244	130		14	2041	
48R869248		3124	20	2051	
48R869249		3004	2	2007	
48R869250	102	3722	2	2007	
48R869251	105		4		
48R869253		3004	2	2007	
48R869254		3010	8	2002	
48R869255	121	3717	239	2006	

Industry Standard No.	ECG	SK	GE	RS 276-	MOTOR.
48R869256	6401			2029	
48R869257	129		244	2027	
48R869264	6400A			2029	
48R869282		3004	2	2007	
48R869283		3010	8	2002	
48R869301	175		246	2020	
48R869302	130		14	2041	
48R869306	314	3898			
48R869312	123A	3444	20	2051	
48R869321	130		14	2041	
48R869325	123A	3444	20	2051	
48R869329	123A	3444	20	2051	
48R869334	159	3466	82	2032	
48R869338	199	3245	62	2010	
48R869384	199	3245	62	2010	
48R869389	289A		268	2038	
48R869412	234		65	2050	
48R869413	159	3466	82	2032	
48R869416	199	3245	62	2010	
48R869426		3025	21	2034	
48R869432	129		244	2027	
48R869436	121	3717	239	2006	
48R869444	123A	3444	20	2051	
48R869464		3024	18	2030	
48R869467	234		65	2050	
48R869475		3004	2	2007	
48R869475A		3004	2	2007	
48R869476		3010	8	2002	
48R869476A		3010	8	2002	
48R869491	195A	3765	243	2030	
48R869519	195A	3765	243	2030	
48R869520	129		244	2027	
48R869521	192		63	2030	
48R869525	123A	3444	20	2051	
48R869526	159	3466	82	2032	
48R869547	199	3245	62	2010	
48R869550	121	3717	239	2006	
48R869562	128		243	2030	
48R869563	123A	3444	20	2051	
48R869568	123A	3444	20	2051	
48R869570	123A	3444	20	2051	
48R869571	159	3466	82	2032	
48R869582	185		58	2025	
48R869594	199	3245	62	2010	
48R869599	311		277		
48R869618	184		57	2017	
48R869640	188		226	2018	
48R869641	189	3245/199	218	2026	
48R869649	159	3466	82	2032	
48R869676	196	3054	241	2020	
48R869767	123A	3444	20	2051	
48R869773	316	3039			
48R869844	316	3039			
48R969681	129		244	2027	
48S10062A01		3031A	504A	1104	
48S10062A02			504A	1104	
48S10062A05		3016	504A	1104	
48S10062A05A		3016	504A	1104	
48S10346A02	109	3087		1123	
48S10577A01	177	3100/519	300	1122	
48S10577A02	177		300	1122	
48S10577A11	177	3100/519	300	1122	
48S10577A13		3100	300	1122	
48S10641D62	137A	3058	ZD-6.2	561	
48S13270	131	3052	44	2006	
48S134404			2	2007	
48S134405		3124	51	2004	
48S134406			51	2004	
48S134407		3004	51	2004	
48S134408		3004	2	2007	
48S134458			52	2024	
48S134587		3087	1N34AS	1123	
48S134666		3124	20	2051	
48S134695		3009	3	2006	
48S134718		3124	20	2051	
48S134719		3124	20	2051	
48S134720		3124	20	2051	
48S134721			10	2051	
48S134732		3124	10	2051	
48S134733		3124	20	2051	
48S134733A		3124	20	2051	
48S134734		3124	10	2051	
48S134734A		3124	20	2051	
48S134736		3016	504A	1104	
48S134737		3124	20	2051	
48S134739		3021	12		
48S134739A		3021	12		
48S134746		3009	3	2006	
48S134747		3009	16	2006	
48S134751			16	2006	
48S134756		3019	11	2015	
48S134758			239	2006	
48S134759		3009	16	2006	
48S134760			16	2006	
48S134761		3009	16	2006	
48S134765		3024	10	2051	
48S134766		3009	3	2006	
48S134767		3009	3	2006	
48S134768		3124	18	2030	
48S134773		3018	20	2051	
48S134774		3018	20	2051	
48S134775		3018	20	2051	
48S134776		3018	20	2051	
48S134783		3018	17	2051	
48S134784		3018	17	2051	
48S134785		3018	17	2051	
48S134789			20	2051	
48S134790			504A	1104	
48S134797		3008	17	2051	
48S134804		3018	17	2051	
48S134805		3018	17	2051	
48S134807		3018	17	2051	
48S134809		3124	20	2051	
48S134810		3124	20	2051	
48S134811		3124	20	2051	
48S134814			82	2032	
48S134815	159	3466	82	2032	
48S134816	177	3100/519	300	1122	
48S134820			17	2051	
48S134821			21	2034	
48S134823		3124	20	2051	
48S134825		3018	17	2051	
48S134826		3018	17	2051	
48S134827		3018	17	2051	
48S134830			22	2032	
48S134831		3118	22	2032	
48S134832		3124	22	2032	
48S134837		3018	17	2051	
48S134838		3124	20	2051	
48S134840		3018	20	2051	
48S134841		3020	20	2051	
48S134842		3124	20	2051	
48S134843		3124	27		
48S134844		3018	20	2051	

Industry Standard No.	ECG	SK	GE	RS 276-	MOTOR.
48S134845		3018	20	2051	
48S134846		3124	20	2051	
48S134851			ZD-9.1	562	
48S134853		3124	18	2030	
48S134854		3124	10	2051	
48S134855		3018	20	2051	
48S134857		3018	20	2051	
48S134858		3095	ZD-33		
48S134860		3005	1	2007	
48S134861		3005	1	2007	
48S134862		3005	1	2007	
48S134872		3021	12		
48S134879		3018	17	2051	
48S134888			3	2006	
48S134889			10	2051	
48S134894		3018	20	2051	
48S134898		3124	20	2051	
48S134899		3024	20	2051	
48S134900		3079	36		
48S134901		3027	36		
48S134902	108	3452	86	2038	
48S134903	123A	3444	20	2051	
48S134904		3018	17	2051	
48S134905		3124	10	2051	
48S134906		3124	10	2051	
48S134908			10	2051	
48S134909		3004	21	2034	
48S134910		3025	21	2034	
48S134912		3124	ZD-8.2	562	
48S134913			82	2032	
48S134915			82	2032	
48S134916	113A	3119/113	6GC1		
48S134917	178MP	3119/113	300(2)	1122(2)	
48S134918		3020	20	2051	
48S134919	154	3124/289	40	2012	
48S134920		3021	12		
48S134921	116	3017B/117	504A	1104	
48S134922		3018	17	2051	
48S134923			17	2051	
48S134924			20	2051	
48S134925		3018	20	2051	
48S134926		3018	20	2051	
48S134927		3044	20	2051	
48S134932	233	3018	39	2015	
48S134933	123A	3444	20	2051	
48S134934		3035	25		
48S134935		3124	18	2030	
48S134936	196	3026	241	2020	
48S134937		3018	17	2051	
48S134938		3034	25		
48S134939	116	3032A	504A	1104	
48S134941		3024	20	2051	
48S134942		3024	18	2030	
48S134943		3025	22	2032	
48S134944	312	3834/132	FET-1	2035	
48S134945(A2C			20	2051	
48S134946	108	3452	86	2038	
48S134947		3009	16	2006	
48S134948		3018	20	2051	
48S134949		3018	20	2051	
48S134950		3018	20	2051	
48S134952		3124	20	2051	
48S134953		3024	18	2030	
48S134956		3006	51	2004	
48S134957			ZD-9.1	562	
48S134958		3311	504A	1104	
48S134959		3031A	504A	1104	
48S134960	108	3452	86	2038	
48S134961		3018	20	2051	
48S134962		3018	20	2051	
48S134963		3018	20	2051	
48S134964		3018	20	2051	
48S134970	108	3452	86	2038	
48S134972		3021	12		
48S134974		3009	16	2006	
48S134978		3017B	CR-2		
48S134979	108	3452	86	2038	
48S134981	161	3018	39	2015	
48S134988	192	3024/128	63	2030	
48S134989	193	3025/129	67	2023	
48S134990		3311	504A	1104	
48S134992		3024	20	2051	
48S134995		3111	36		
48S134997	123A	3444	20	2051	
48S137000	142A	3062	ZD-12	563	
48S137001		3035	25		
48S137002		3040	27		
48S137003		3018	20	2051	
48S137006	108	3452	86	2038	
48S137007			63	2030	
48S137014		3124	18	2030	
48S137015	199	3124/289	62	2010	
48S137017		3094	ZD-15	564	
48S137021	5072A		ZD-8.2	562	
48S137021(10)			ZD-8.2	562	
48S137022			18	2030	
48S137029			504A	1104	
48S137031		3009	3	2006	
48S137032	159	3025/129	82	2032	
48S137033			17	2051	
48S137040			504A	1104	
48S137041	128	3024	243	2030	
48S137044			20	2051	
48S137045		3025	21	2034	
48S137047			18	2030	
48S137055			17	2051	
48S137056			17	2051	
48S137057			17	2051	
48S137070		3116	FET-2	2035	
48S137074			504A	1104	
48S137081		3067	CR-4		
48S137082	503	3068	CR-5		
48S137093		3104A	28	2017	
48S137101			1N34AS	1123	
48S137106			18	2030	
48S137107	123A	3444	20	2051	
48S137108		3122	20	2051	
48S137109		3122	20	2051	
48S137110	123A	3018	17	2051	
48S137111		3124	20	2051	
48S137113		3044	27		
48S137114	505	3108	CR-7		
48S137115	123A	3444	20	2051	
48S137127	159	3025/129	82	2032	
48S137133	177	3031A	300	1122	
48S137145			28	2017	
48S137158		3018	11	2015	
48S137160		3200	84	2026	
48S137164		3121	ZD-9.1	562	
48S137167	178MP	3100/519	300(2)	1122(2)	
48S137168	189	3200	218	2026	
48S137169	188	3199	226	2018	
48S137170		3058	ZD-5.6	561	
48S137171	123A	3444	20	2051	

Industry Standard No.	ECG	SK	GE	RS 276-	MOTOR.
48S137171(D)			20	2051	
48S137172	123A	3444	20	2051	
48S137173	159	3025/129	82	2032	
48S137174	123A	3122	18	2030	
48S137190		3018	20	2051	
48S137191		3018	20	2051	
48S137192		3018	20	2051	
48S137203		3111	36		
48S137206		3124	18	2030	
48S137207		3021	12		
48S137208			504A	1104	
48S137260			20	2051	
48S137266	147A	3095	ZD-33		
48S137270		3052	30	2006	
48S137272	147A	3095	ZD-33		
48S137299	109	3087		1123	
48S137300	123A	3444	20	2051	
48S137308			30	2025	
48S137309	152	3054/196	66	2048	
48S137310	153	3083/197	69	2049	
48S137311	152	3054/196	66	2048	
48S137312	153	3083/197	69	2049	
48S137314		3118	22	2032	
48S137315	123A	3444	20	2051	
48S137321		3118	22	2032	
48S137323	196	3054	241	2020	
48S137330	145A	3063	ZD-15	564	
48S137331			29	2025	
48S137341	165	3115	38		
48S137342		3035	25		
48S137343	312	3834/132	FET-2	2035	
48S137344	130	3133/164	14	2041	
48S137347	552	3032A	504A	1104	
48S137348	552	3032A	504A	1104	
48S137350		3018	17	2051	
48S137351		3018	20	2051	
48S137364	191	3044/154	249		
48S137364(H)			40	2012	
48S137369	152	3054/196	66	2048	
48S137370	153	3083/197	69	2049	
48S137386	194	3018	220		
48S137387		3058	ZD-6.2	561	
48S137397	513	3443	513		
48S137415		3103A	27		
48S137415(I)			40	2012	
48S137442	5080A	3336	ZD-22		
48S137472			29	2025	
48S137473			28	2017	
48S137476	123A	3232/191	20	2051	
48S137495			1N34AS	1123	
48S137498	123A	3444	20	2051	
48S137512	247		20	2051	
48S137524	163A	3439	36		
48S137528	124	3131A/369	12		
48S137530	123A	3444	20	2051	
48S137533	506	3843	511	1114	
48S137535	124	3538	12		
48S137539	165	3439/163A	36		
48S137543	123A	3018	20	2051	
48S137546	506	3843	511	1114	
48S137548	162	3559	35		
48S137551	506	3843	511	1114	
48S137572	188	3199	83	2018	
48S137855			20	2051	
48S155001	165	3115	38		
48S155002	502	3067	CR-4		
48S155005	175	3538	246	2020	
48S155006	154	3045/225	40	2012	
48S155013	152	3054/196	66	2048	
48S155014			69	2049	
48S155034	198	3220	251		
48S155035	290A	3114/290	269	2032	
48S155037	116	3016	504A	1104	
48S155039	109	3087	1N34AS	1123	
48S155040	116	3016	504A	1104	
48S155041	116	3016	504A	1104	
48S155042	186A	3357	215	2052	
48S155043	198		32		
48S155044	175	3538	12	2020	
48S155045	234	3114/290	65	2050	
48S155046	297	3122	210	2030	
48S155047	177	3100/519	300	1122	
48S155048	5074A	3139	ZD-11	563	
48S155050	5404	3627			
48S155051	287		232		
48S155052	605	3864			
48S155053	130	3027	75	2041	
48S155054	116	3313	ZD-10	562	
48S155054(ZENER)			ZD-10	562	
48S155056	552	3843	511	1114	
48S155058	238	3710	259		
48S155059	198	3220	251		
48S155060	178MP	3100/519	514	1122	
48S155061	109	3130	1N60	1123	
48S155062	152	3054/196	66	2048	
48S155063	116	3016	504A	1104	
48S155066	153	3083/197	69	2049	
48S155067	130	3027	75	2041	
48S155068	5633	3533/5635			
48S155069	198		251		
48S155070	198	3104A	32		
48S155072	198	3103A/396	32		
48S155073	199		62	2010	
48S155074	186A	3357	247	2052	
48S155076	165	3115	38		
48S155077	177	3100/519	300	1122	
48S155078	109	3087	1N34AS	1123	
48S155079	116	3016	504A	1104	
48S155080	5070A	9021	ZD-6.0	561	
48S155081	5074A	3139	ZD-11	563	
48S155083	125	3080	509	1114	
48S155084	5074A	3139	ZD-11	563	
48S155085		3080	509	1114	
48S155087	229	3018	20	2051	
48S155088	123A	3137/192A	81	2051	
48S155089	199	3444/123A	61	2010	
48S155090	238	3115/165	259		
48S155092	5404	3627			
48S155094	194	3137/192A	81	2051	
48S155095	194	3114/290	21	2034	
48S155096	198	3103A/396	32		
48S155099	116	3016	504A	1104	
48S155100	116	3311	504A	1104	
48S155103	142A	3062	ZD-12	563	
48S155104	145A	3063	ZD-15	564	
48S155105	162	3079	35		
48S155106	116	3311	504A	1104	
48S155107	506	3843	509	1114	
48S155108	506	3843	509	1114	
48S155110	198	3220	251		
48S155116	187A	9076	29	2025	
48S155117	297	3122	210	2051	
48S155118	283	3467	73		
48S155121	293	3849	47	2030	

Industry Standard No.	ECG	SK	GE	RS 276-	MOTOR.
48S155122	294	3025/129	48		
48S155123	192	3137	88	2030	
48S155126	116	3311	504A	1104	
48S155127	5069A		ZD-4.7		
48S155128	506	3843			
48S155140	238	3710	259		
48S191A02	116	3016	504A	1104	
48S191A04		3311	504A	1104	
48S191A04(A)			504A	1104	
48S191A05	506	3843	504A	1104	
48S191A05A	506	3843	511	1114	
48S191A06		3016	504A	1104	
48S191A07	116	3032A	504A	1104	
48S191A08	506	3843	504A	1104	
48S32000			ZD-12	563	
48S40170G01	199	3124/289	62	2010	
48S40171G01		3124	18	2030	
48S40172G01		3052	30	2006	
48S40235G02	116	3311	504A	1104	
48S40241G01			62	2010	
48S40246G01	199	3444/123A	20	2051	
48S40247G01	199	3444/123A	62	2010	
48S40247G02		3444	20	2051	
48S40382J01	186	3192	28	2017	
48S40383J01	187	3193	29	2025	
48S40606G01			20	2051	
48S40606G02			62	2010	
48S40606J02	199	3122	62	2010	
48S40607G01			62	2010	
48S40607J01	199	3122	62	2010	
48S40662G04			66	2048	
48S41508A01			504A	1104	
48S43240G01	152	3054/196	66	2048	
48S43241G01	152	3893	66	2048	
48S43991J01	108		86	2038	
48S43992J01	108		86	2038	
48S44657G01	1022		IC-102		
48S44883G01	152	3893	66	2048	
48S44884G01	153		69	2049	
48S44885G01	123A	3444	20	2051	
48S44885G02	123A	3444	20	2051	
48S44886G01	289A	3122	243	2030	
48S44887G01			300	1122	
48S45323G01	612	3325			
48S50	218		234		
48S623334A01			1N60	1123	
48S65123A67			11	2015	
48S67120A13		3100	300	1122	
48SP134804			10	2051	
48SP134826			11	2015	
48SP134837			11	2015	
48SP134855			10	2051	
48SP134857			11	2015	
48SP134894			10	2051	
48SP134897			10	2051	
48SP134903			10	2051	
48SP134904			11	2015	
48SP134905			10	2051	
48SP134906			10	2051	
48SP134933			10	2051	
48SP134937			11	2015	
48T66544A88	558		42		
48X134902		3019	11	2015	
48X134970			20	2051	
48X644681			1N60	1123	
48X90232A01			17	2051	
48X90232A02			17	2051	
48X90232A03		3018	17	2051	
48X90232A04			17	2051	
48X90232A05		3124	18	2030	
48X90232A06			21	2034	
48X90232A07		3021	18	2030	
48X90232A08			17	2051	
48X90232A09			21	2034	
48X90232A10		3018	17	2051	
48X90232A11			17	2051	
48X90232A12			21	2034	
48X90232A13			20	2051	
48X90232A14			FET-1	2028	
48X90232A15			21	2034	
48X90232A17		3018	11	2015	
48X90232A18		3018	11	2015	
48X90232A19		3018	11	2015	
48X90232A20			11	2015	
48X90233A01			1N60	1123	
48X90233A02			1N34AS	1123	
48X90233A04			504A	1104	
48X90233A06			1N60	1123	
48X90233A07		3062	ZD-12	563	
48X90234A01		3031A	504A	1104	
48X90234A02		3017B	504A	1104	
48X90234A03			504A	1104	
48X90234A05	505	3068	CR-5		
48X97046A15			30	2006	
48X97046A16		3006	51	2004	
48X97046A17		3018	17	2051	
48X97046A18		3018	20	2051	
48X97046A19		3018	20	2051	
48X97046A20		3018	20	2051	
48X97046A21		3018	20	2051	
48X97046A22		3124	20	2051	
48X97046A23		3124	20	2051	
48X97046A24			18	2030	
48X97046A25			17	2051	
48X97046A31			30	2006	
48X97046A34			2	2007	
48X97046A36			67	2023	
48X97046A48	158	3004/102A	53	2007	
48X97046A50			20	2051	
48X97046A51	108	3452	86	2038	
48X97046A52	128	3124/289	243	2030	
48X97046A53	102A	3004	53	2007	
48X97046A54	102A	3004	53	2007	
48X97046A55	102A	3004	53	2007	
48X97046A60	123A	3444	20	2051	
48X97046A61	123A	3444	20	2051	
48X97046A62	123A	3444	20	2051	
48X97048A02			1N34AS	1123	
48X97048A04			504A	1104	
48X97048A05		3088	1N60	1123	
48X97048A06	109	3088		1123	
48X97048A07		3311	504A	1104	
48X97048A08		3062	ZD-12	563	
48X97048A10	116	3311	504A	1104	
48X97048A18			17	2051	
48X97048A19	109	3087		1123	
48X97127A01			504A	1104	
48X97162A01	108	3452	86	2038	
48X97162A02	108	3452	86	2038	
48X97162A04	108	3452	86	2038	
48X97162A05	123A	3444	20	2051	
48X97162A06	102A	3004	53	2007	
48X97162A07	102A	3004	53	2007	
48X97162A09	108	3452	86	2038	

Industry Standard No.	ECG	SK	GE	RS 276-	MOTOR.
48X97162A10	108	3452	86	2038	
48X97162A21	123A	3444	20	2051	
48X97162A36	158	3004/102A	53	2007	
48X97162A37	158	3004/102A	53	2007	
48X97168A01	109	3088		1123	
48X97168A02		3126	90		
48X97168A03	109	3088		1123	
48X97168A04	109	3087		1123	
48X97168A06		3031A	504A	1104	
48X97168A07			1N60	1123	
48X97168A10		3311	504A	1104	
48X97168A11		3126	90		
48X97168A16		3311	504A	1104	
48X97172A01	116	3031A	504A	1104	
48X97177A03		3018	20	2051	
48X97177A10			83	2018	
48X97177A11			84	2026	
48X97177A12		3124	18	2030	
48X97177A13		3018	20	2051	
48X97177A14		3025	21	2034	
48X97177A15			1N60	1123	
48X97177H01	312		FET-1	2028	
48X97222A01			1N60	1123	
48X97222A02		3311	504A	1104	
48X97238A01	126	3006/160	52	2024	
48X97238A02	126	3006/160	52	2024	
48X97238A03	126	3006/160	52	2024	
48X97238A04	123A	3444	20	2051	
48X97238A05	102	3004	2	2007	
48X97238A06	131MP	3840	44(2)	2006(2)	
48X97239A01	109	3087		1123	
48X97271A01			504A	1104	
48X97271A02			1N60	1123	
48X97305A02			ZD-5.6	561	
049	5081A		ZD-24		
49-1	108	3452	86	2038	
49-1042	116	3311	504A	1104	
49-3112	116	3016	504A	1104	
49-61963	5444	3634			
49-62139	128		243	2030	
49A0000	7475	7475		1806	
49A0002-000	7473	7473		1803	
49A0005-000	7410	7410		1807	
49A0006-000	7420			1809	
49A0012-000	7474	7474		1818	
49A0103-001	923		IC-259		
49L-4A5	102A		53	2007	
49P1C	121	3009	239	2006	
49S1	177	3100/519	300	1122	
49X90232A05			20	2051	
050-0011-00			504A	1104	
50-30307-07	294		48		
50-40101-04	108		86	2038	
50-40101-05	108		86	2038	
50-40102-04	123A		211	2051	
50-40102-05	229		211	2016	
50-40105-08	123A	3024/128	81	2051	
50-40106-09	159		21	2034	
50-40201-08	123A		20	2051	
50-40201-09	123A		20	2051	
50-40201-10	123A		20	2051	
50-40204-10	159	3118	21	2034	
50-40205-09	129		82	2032	
50-40306-07	293		47	2030	
50-81402-04	123A	3444			
50-81502-03	229	3854/123AP			
50-81502-04	229	3854/123AP			
50A	116	3016	504A	1104	
50A102	126	3008	52	2024	
50A103	129	3006/160	244	2027	
50A103K	126	3006/160	52	2024	
50A52	126	3008	52	2024	
50AS	116	3017B/117	504A	1104	
50B173-C	102A		53	2007	
50B173-S	102A	3004	53	2007	
50B175A	102A	3004	53	2007	
50B175B	102A	3004	53	2007	
50B175C	102A		53	2007	
50B324	158	3004/102A	53	2007	
50B364	158	3004/102A	53	2007	
50B415	158	3004/102A	53	2007	
50B423	102A	3004	53	2007	
50B5	116	3016	504A	1104	
50B54	102A	3004	53	2007	
50BU75-C	102A	3004	53	2007	
50C	116	3017B/117	504A	1104	
50C05	6020	7220	5128		
50C10	6022		5128		
50C100	6044	7244	5140		
50C100R	6045	7245	5141		
50C1047	108	3452	86	2038	
50C10R	6023		5129		
50C20	6026	7226	5128		
50C20R	6027	7227	5129		
50C30	6030		5132		
50C30R	6031		5133		
50C371	123A	3124/289	20	2051	
50C372	123A	3124/289	20	2051	
50C373	123A	3124/289	20	2051	
50C374	123A	3444	20	2051	
50C380-0	107	3018	11	2015	
50C380-OR				2015	
50C380-OR	107		11	2015	
50C394-0	107	3018	11	2015	
50C394-R	107	3018	11	2015	
50C40	6034	7234	5132		
50C401	130	3027	14	2041	
50C40R	6035		5133		
50C50	6038		5136		
50C50R	6039		5137		
50C538	123A	3444	20	2051	
50C60	6040	7240	5136		
50C60R	6041		5137		
50C644	123A	3444	20	2051	
50C70	6042		5140		
50C70R	6043		5142		
50C784	108	3452	86	2038	
50C784-R	107	3018	11	2015	
50C80	6042		5140		
50C80R	6043		5142		
50C828	123A	3444	20	2051	
50C829	108	3452	86	2038	
50C829B	108	3452	86	2038	
50C829C	108	3452	86	2038	
50C838	123A	3444	20	2051	
50C90	6044	7244	5140		
50C90R	6045	7245	5141		
50CJ139	123A	3444	20	2051	
50D10	125	3033	510,531	1114	
50D2	116	3016	504A	1104	
50D4	116	3016	504A	1104	
50D8	125	3032A	510,531	1114	
50E05	116	3016	504A	1104	
50E1	116	3016	504A	1104	
50E10	125	3033	510,531	1114	
50E2	116	3016	504A	1104	
50E3	116	3016	504A	1104	
50E4	116	3016	504A	1104	
50E5	116	3017B/117	504A	1104	
50E6	116	3017B/117	504A	1104	
50E7	125	3032A	510,531	1114	
50E8	125	3032A	510,531	1114	
50F1	5840	3584/5862	5008		
50F5	5860		5024		
50FB3L					MDA970A4
50FB4L					MDA970A5
50H3P	5860	3500/5882	5024		MR1125
50H3PN	5861		5025		
50HQ020					MBR6020
50HQ030					MBR6035
50HQ035					MBR6035
50HQ040					MBR6045
50HQ045					MBR6045
50J1	5805	9008			
50J2P	5880		5040		
50J3P	5880		5040		
50JH3	6064	7264	5136		
50M	116	3017B/117	504A	1104	
50M33Z5	5142A		5ZD-33		
50P2	126	3006/160	52	2024	
50P3	126	3006/160	52	2024	
50R2S	5868	7068	5028		
50S5	5840		5008		
50T200	5296A				50M200ZS10
50T6.8	5247A				50M6.8ZS10
50Z11	141A	3092	ZD-11.5		
50Z110	151A	3099	ZD-110		
50Z110A	151A	3099	ZD-110		
50Z110B	151A	3099	ZD-110		
50Z110C	151A	3099	ZD-110		
50Z11A	141A	3092	ZD-11.5		
50Z11B	141A	3092	ZD-11.5		
50Z11C	141A	3092	ZD-11.5		
50Z12	142A	3062	ZD-12	563	
50Z12A	142A	3062	ZD-12	563	
50Z12B	142A	3062	ZD-12	563	
50Z12C	142A	3062	ZD-12	563	
50Z14	144A	3094	ZD-14	564	
50Z14A	144A	3094	ZD-14	564	
50Z14B	144A	3094	ZD-14	564	
50Z14C	144A	3094	ZD-14	564	
50Z15	145A	3063	ZD-15	564	
50Z15A	145A	3063	ZD-15	564	
50Z15B	145A	3063	ZD-15	564	
50Z15C	145A	3063	ZD-15	564	
50Z27	146A	3064	ZD-27		
50Z27A	146A	3064	ZD-27		
50Z27B	146A	3064	ZD-27		
50Z27C	146A	3064	ZD-27		
50Z33	147A	3095	ZD-33		
50Z33A	147A	3095	ZD-33		
50Z33B	147A	3095	ZD-33		
50Z33C	147A	3095	ZD-33		
50Z56	149A	3097	ZD-62		
50Z56A	149A	3097	ZD-62		
50Z56B	149A	3097	ZD-62		
50Z56C	149A	3096/148A	ZD-62		
50Z62	149A	3097	ZD-62		
50Z62A	149A	3097	ZD-62		
50Z62B	149A	3097	ZD-62		
50Z62C	149A	3097	ZD-62		
50Z82	150A	3098	ZD-82		
50Z82A	150A	3098	ZD-82		
50Z82B	150A	3098	ZD-82		
50Z82C	150A	3098	ZD-82		
50Z9.1	139A	3066/118	ZD-9.1	562	
50Z9.1A	139A	3066/118	ZD-9.1	562	
50Z9.1B	139A	3066/118	ZD-9.1	562	
50Z9.1C	139A	3060	ZD-9.1	562	
51	108	3452	86	2038	
051-0003	177	3016	300	1122	
051-0006	116	3017B/117	504A	1104	
051-0010-00	1023		IC-272		
051-0010-01	1023		IC-272		
051-0010-02	1023		IC-272		
051-0010-03	1023		IC-272		
051-0010-04	1023		IC-272		
051-0010-05	1023		IC-272		
051-0011-00	1103	3281	IC-94		
051-0011-00-04	1103	3281	IC-94		
051-0011-00-05	1107	3526/947	IC-276		
051-0011-01	1103	3281	IC-94		
051-0011-02	1103	3281	IC-94		
051-0011-03	1103	3281	IC-94		
051-0011-04	1103	3281	IC-94		
051-0012-00	720	9014	IC-7		
051-0016-00			504A	1104	
051-0017-00	722		IC-9		
051-0020	177	3175	300	1122	
051-0020-00	1087	3477	IC-103		
051-0020-02	1087	3477	IC-103		
051-0020-02/03	1087	3477			
051-0020-03	1087	3477			
051-0021-00	746	3234	IC-217		
051-0022-00	709	3135	IC-11		
051-0024	139A	3060	ZD-9.1	562	
051-0035-01	1087	3477	IC-171		
051-0035-02	1087	3477	IC-171		
051-0035-03	1087	3477			
051-0035-04	1087	3477	IC-171		
051-0036-00	1037	3371	IC-170		
051-0036-01	1037	3364	IC-170		
051-0036-0102	1037	3371			
051-0036-02	1037	3364	IC-170		
051-0036-02/03	1037	3371			
051-0036-03	1037	3364	IC-170		
051-0038-00	1005	3723	IC-42		
051-0046	123A	3444	IC-20	2051	
051-0047	123A	3444	IC-20	2051	
051-0049	107	3020/123	11	2015	
051-0050-00	801	3160	IC-155		
051-0050-01	801	3160	IC-155		
051-0055-02	1155	3231	IC-179	705	
051-0055-0203	1155	3231			
051-0055-03	1155	3231	IC-179	705	
051-0062	160	3007	245	2004	
051-0063	160	3007	245	2004	
051-0068-02	1003		IC-43		
051-0079	126	3008	52	2024	
051-0086-00	1226	3763			
051-0088-00	1217	3700	IC-76		
051-0100-00	1167	3732			
051-0107	159	3466	82	2032	
051-0151	175	3026	246	2020	
051-0155	123A	3444	20	2051	
051-0156	195A	3765	46		
051-0157	195A	3765	46		
51-02006-12	5072A		ZD-8.2	562	

Industry Standard No.	ECG	SK	GE	RS 276-	MOTOR.
51-02007-12	5018A	3784	ZD-9.1	562	
51-03007-06	116	3031A	504A	1104	
51-03009A02	746	3234			
51-03009A04	74LS193	74LS193			
51-03009A05	4020B	4020			
51-03009A07	787	3146			
51-03009A08	1239	3708/1169			
51-04001-01	109	3088	1N60	1123	
51-04488D03	727	3071	IC-210		
51-06002-00	610	3323	90		
51-08001-11	519	3100	514	1122	
51-10276A01	704	3023	IC-205		
51-10302A01	703A	3157	IC-12		
51-10382A	718	3159	IC-8		
51-10382A01	722	3161	IC-9		
51-10393A01	748	3236			
51-10408A01	710	3102	IC-89		
51-10422A	718	3159	IC-8		
51-10422A01	718	3159	IC-8		
51-10422A02	718	3159	IC-8		
51-10425A01	798	3216	IC-234		
51-10432A01	717		IC-209		
51-10437A01	718	3159	IC-8		
51-10534A04	755		IC-212		
51-10541A01	917	3675/971	IC-258		
51-10559A01	722	3161	IC-9		
51-10566A01	722	3161	IC-9		
51-10566A02	720	9014	IC-7		
51-1056FA01	722	3161	IC-9		
51-1056GA01	722		IC-9		
51-10592A01	722	3161	IC-9		
51-10594A01	723	3144	IC-15		
51-10594C01	723		IC-15		
51-1059A01	723		IC-15		
51-10611A11	7400			1801	
51-10611A12	7404	7404		1802	
51-10611A16	7475	7475		1806	
51-10617A01	718	3159	IC-8		
51-10619A01	723		IC-15		
51-10631A01	708		IC-10		
51-10637A01	781	3169	IC-223		
51-10650A01	5944	3465/778A	IC-220		
51-10655A01	798		IC-234		
51-10655A05	749		IC-97		
51-10655A17	4001B		4001	2401	
51-10655A19	4013B		4013	2413	
51-10655A21	4016B		4016		
51-10655B01	798	3216	IC-234		
51-10655B13	712	3072			
51-10655C05	749		IC-97		
51-10658A01	788	3147	IC-229		
51-10658A02	788	3829	IC-229		
51-10658A03	788	3829	IC-229		
51-10672A	736		IC-17		
51-10672A01	736		IC-17		
51-10711A01	743		IC-214		
51-10715A01	941M		IC-265	007	
51-110276A01	704	3023	IC-205		
51-13-14	177		300	1122	
51-13747A03	529	3306/523			
51-13753A01	1196	3725			
51-13753A02	1164	3727			
51-13753A04	1162	3072/712	IC-148		
51-13753A06	1173	3729			
51-13753A08	1168	3728			
51-13753A10	1164	3727	IC-301		
51-13753A11	712	3072	IC-148		
51-13753A18	1173	3729	IC-302		
51-13753A19	797	3158	IC-233		
51-13753A20	1176	3210			
51-13753A40	1231	3832			
51-21043-06	610	3323			
51-25789H	909	3590	IC-249		
51-33801A01	712		IC-2		
51-4	157	3747	232		
51-40464P01	801	3160	IC-35		
51-42211P01	1170	3745	IC-65		
51-42908501	801	3160			
51-42908J01	801	3160	IC-35		
51-43684B	735		IC-212		
51-43684B01	755		IC-212		
51-44789J01	1234		IC-181		
51-44789J02	1234		IC-181		
51-44837J04	1239	3078/789	IC-64		
51-47-20	128		243	2030	
51-47-21	159	3466	82	2032	
51-47-23	123A	3444	20	2051	
51-47-24	123A	3444	20	2051	
51-47-28	519		514	1122	
51-47-34	234		65	2050	
51-70177A01	798	3216	IC-234		
51-70177A02	748	3236			
51-70177A03	738	3167	IC-29		
51-70177A05	749	3168	IC-97		
51-70177B02	748	3236			
51-84320A16	912		IC-172		
51-84320A32	916	3550	IC-257		
51-90305A04	1186	3171/744	IC-175		
51-90305A20	1123	3743	IC-281		
51-90305A21	1263	3920	IC-293		
51-90305A61	712	3072	IC-148		
51-90433A10	4069	4050	4050		
51-90433A11	4050B	4050	4050		
51-90433A13	4016B	4016	4016		
51-90433A30	958	3699			
51A180-4	161	3117	39	2015	
51A524004-01	941D		IC-264		
51C43684B	735		IC-212		
51D0177A02	748	3236			
51D170	102	3004	2	2007	
51D176	102A		53	2007	
51D188	100		1	2007	
51D189	100	3721	1	2007	
51D70177A01	798	3216	IC-234		
51D70177A02	748	3736/1214			
51D70177A03	738	3167	IC-29		
51D70177B02	748	3236			
51G10679A03	738	3167	IC-29		
51G10679A13	748	3236	IC-115		
51GX14	120	3110	CR-3		
51GX3	120	3110	CR-3		
51HQ045					MBR6045
51IN60P	109	3087		1123	
51IS34S	109	3087		1123	
51LN60P				1123	
51LS34S				1123	
51M33801A01	1162	3072/712	IC-2		
51M70177A01	798	3216	IC-234		
51M70177A02	748	3236			
51M70177A03	738	3167	IC-29		
51M70177A05	749	3168	IC-97		
51M70177B01	798	3216	IC-234		
51M70177B02	748	3236			
51N3M	180	3018	74	2043	
51P2	126	3006/160	52	2024	
51P4	126	3006/160	52	2024	
51R04488D03	727	3071	IC-210		
51R40000S13	1232	3852			
51R84320A02	724		IC-86		
51R84320A03	912		IC-172		
51R84320A05	912		IC-172		
51R84320A16	912		IC-172		
51R84320A32	916	3550	IC-257		
51S10072A01	736		IC-17		
51S10276A01	704	3023	IC-205		
51S10302A01	703A	3157	IC-12		
51S10382A	718	3159	IC-8		
51S10382A01	718	3159	IC-8		
51S10393A01	748	3236			
51S10408A01	710	3102	IC-89		
51S10422A	718	3159	IC-8		
51S10422A01	718	3159	IC-8		
51S10425A01	798	3216	IC-234		
51S10432A01	717		IC-209		
51S10437A01	718	3159	IC-8		
51S10536A01	912		IC-172		
51S10541A01	917		IC-258		
51S10553A01	917		IC-258		
51S10559A	722		IC-9		
51S10559A01	722	3161	IC-9		
51S10566A01	722	3161	IC-9		
51S10566A02	722		IC-9		
51S1056FA01	722	3161	IC-9		
51S1056GA01	722	3161	IC-9		
51S10592A01	722	3161	IC-9		
51S10594A01	723	3144	IC-15		
51S10611A11	7400	7400		1801	
51S10611A12	7404	7404		1802	
51S10611A16	7475	7475		1806	
51S10617A01	718	3159	IC-8		
51S10619A01	723		IC-15		
51S10631A01	708		IC-10		
51S10637A01	781	3169	IC-223		
51S10638A01	789	3078			
51S10650A01	778A		IC-220		
51S10655A01	798	3216	IC-234		
51S10655A01-3	798	3216	IC-234		
51S10655A03	738	3167	IC-29		
51S10655A03A	738	3167	IC-29		
51S10655A05	749	3168	IC-97		
51S10655A17	4001B	4001	4001		
51S10655A18	4011B	4011	4011		
51S10655A19	4013B		4013		
51S10655B01	798	3216	IC-234		
51S10655B03	738	3167	IC-29		
51S10655B13	712	3072			
51S10655C05	749		IC-97		
51S10658A01	788	3829	IC-229		
51S10672A	736		IC-17		
51S10672A01	736		IC-17		
51S10711A01	743		IC-214		
51S10715A01	941M		IC-265	007	
51S110276A01	704		IC-205		
51S13753A01	1196	3725			
51S13753A02	1164	3727	IC-301		
51S13753A03	1161		IC-300		
51S13753A06	1173	3729	IC-302		
51S13753A08	1168	3728	IC-66		
51S13753A09	712	3072	IC-148		
51S13753A10	1164	3727	IC-301		
51S13753A11	712	3072	IC-2		
51S1834	109	3087		1123	
51S40464P01	801	3160			
51S44789J01	1234		IC-181		
51S44789J02	1234	3487	IC-181		
51S70177A01	798		IC-234		
052	172A	3156	64		
52-010-106-0	294	3841			
52-010-109-0	288	3118	82	2032	
52-010-151-0	282	3748	264		
52-020-108-0	123A	3444	20	2051	
52-020-173-0	299	3197/235	236		
52-025-004-0	390	3027/130	255	2041	
52-025-006-0	152	3893			
52-040-009-0	1003		IC-43		
52-040-010-0	1006		IC-38		
52-040-015-0	1135	3876			
52-050-021-0	109	3088		1123	
52-051-017-0		3087	300	1123	
52-052-004-0	110MP	3709			
52-053-005-0	147A	3095	ZD-33		
52-053-013-0	139A	3060	ZD-9.1	562	
52-1			504A	1104	
52-4	157	3103A/396	232		
052A	172A	3156	64		
52A011	116	3017B/117	504A	1104	
52A011-1	125	3311	510,531	1114	
52A4	177		300	1122	
52BBIA	116		504A	1104	
52BBLA				1104	
52C04	102A	3004	53	2007	
52D189	158	3123	53	2007	
52DS-18	116		504A	1104	
52HQ030					MBR6035
52HQ035					MBR6035
52HQ040					MBR6045
52HQ045					MBR6045
52SD-1		3311	504A	1104	
52SD1	116		504A	1104	
053	722		IC-9		
53-0051-2	116	3016	504A	1104	
53-0082-1003	117		504A	1104	
53-0082-3	125		510,531	1114	
53-0086-1	116		504A	1104	
53-0088-1	116		504A	1104	
53-0088-1002	125		510,531	1114	
53-0088-1003	125		510,531	1114	
53-0088-1004	125		510,531	1114	
53-0088-2	117		504A	1104	
53-0088-3	125		510,531	1114	
53-0094-1	116		504A	1104	
53-0095-1	116		504A	1104	
53-0098-1	116		504A	1104	
53-0099-1	116		504A	1104	
53-0099-3	116		504A	1104	
53-0099-4	116		504A	1104	
53-0106-1	116		504A	1104	
53-0106-1001	125		510,531	1114	
53-0109-1	116		504A	1104	
53-0111-1	117		504A	1104	
53-0111-1001	125		510,531	1114	
53-0113-1	116		504A	1104	
53-0119-1	503		CR-5		
53-0120-1	116		504A	1104	
53-0122-1	117		504A	1104	
53-0132-1	118		CR-1		
53-0136T	125		510,531	1114	
053-1	722	3161	IC-9		

Industry Standard No.	ECG	SK	GE	RS 276-	MOTOR.
53-1086	116	3080	504A	1104	
53-1110	123A	3444	20	2051	
53-1173	172A	3156	64		
53-1362	186	3192	28	2017	
53-1516	159	3466	82	2032	
53-1517	5455	3597		1067	
53-1519	109			1123	
53-1967	186	3192	28	2017	
53A001-1	109	3087		1123	
53A001-10	142A		ZD-12	563	
53A001-12	116	3311	504A	1104	
53A001-2	110MP	3709	1N60	1123(2)	
53A001-3	116	3311	504A	1104	
53A001-32	177		300	1122	
53A001-33	144A		ZD-14	564	
53A001-34	125	3051/156	510,531	1114	
53A001-35	116	3311	504A	1104	
53A001-36	177		300	1122	
53A001-4	125	3081	510,531	1114	
53A001-5		3334	ZD-6.8	561	
53A001-6		3126	90		
53A001-9	116	3311	504A	1104	
53A0016		3126	96		
53A002-1	116		504	1104	
53A006-1	109	3087		1123	
53A008-1	109			1123	
53A0081	109			1123	
53A009-1	113A		6GC1		
53A010-1	116	3032A	504A	1104	
53A011-1		3311	504A	1104	
53A014-1	506	3843	511	1114	
53A015-1	506	3843	511	1114	
53A017-1	506	3043	511	1114	
53A018-1	178MP	3100/519	300(2)	1122	
53A019-1	177		300	1122	
53A020-1	177	3100/519	300	1122	
53A022-1			504A	1104	
53A022-10	502	3067	CR-4		
53A022-2	109	3088		1123	
53A022-3	178MP	3100/519	300(2)	1122(2)	
53A022-4	116	3016	504A	1104	
53A022-5	506	3031A	511	1114	
53A022-6	139A	3060	ZD-9.1	562	
53A022-7	139A	3060	ZD-9.1	562	
53A030-1	177	3100/519	300	1122	
53B001-1	109	3087		1123	
53B001-2	109	3087		1123	
53B001-3	109	3126		1123	
53B001-5	116(2)	3311	504A(2)	1104	
53B001-6	116		504A	1104	
53B001-7	177	3100/519	300	1122	
53B001-8		3126	90		
53B001-9		3100	300	1122	
53B003-1	116	3031A	504A	1104	
53B003-2	116(2)	3311	504A(2)	1104	
53B004-1	109	3087		1123	
53B005-2	109	3087		1123	
53B006-1			504A	1104	
53B007-1	109	3119/113		1123	
53B008-1	112		1N82A		
53B010-1	113A	3088	6GC1		
53B010-2	113A	3119/113	6GC1		
53B010-3	116	3031A	504A	1104	
53B010-4	116	3017B/117	504A	1104	
53B010-5	156	3032A	512		
53B010-6	177		300	1122	
53B010-7	110MP		1N34AS	1123(2)	
53B010-8	506	3125	511	1114	
53B010-9	507		504A	1104	
53B011-1	116(3)		504A(3)	1104	
53B011-2	116	3017B/117	504A	1104	
53B011-3	118	3066	CR-1		
53B012-1	165	3031A	38		
53B013-1	177		300	1122	
53B014-1	177		300	1122	
53B015-1	138A	3059	ZD-7.5		
53B018-1	166	9075	504A	1152	
53B019-1	502		CR-4		
53B019-2	116	3017B/117	504A	1104	
53B019-3	113A		6GC1		
53B020-2	116	3311	504A	1104	
53B020-3		3126	90		
53C001			1N60	1123	
53C001-1	109	3087		1123	
53C001-3	177	3100/519			
53C001-7	125(2)	9001/113A			
53C001-8	125(2)	9002			
53C002-1	169	3678	BR-600		
53C003-1	116	3311	504A	1104	
53C005-3	116	3031A	504A	1104	
53C006-1	109	3087		1123	
53C006-2	109	3087		1123	
53C006-52	109			1123	
53C007-1	116	3311	504A	1104	
53C008-1	178MP		300(2)	1122(2)	
53C009-1	116	3311	504A	1104	
53C009-2	109	3087		1123	
53C011-4	502		CR-4		
53C012-1	116		504A	1104	
53C013-1	134A		ZD-3.6		
53C014-1	116		504A	1104	
53C015-1	116		504A	1104	
53C016-1	116		504A	1104	
53C017-1	116		504A	1104	
53C020-1	109	3087		1123	
53C021-1	169	3678	BR-600		
53C022-1	116		504A	1104	
53CD01-1	109	3087		1123	
53D001-4	177	3100/519	300	1122	
53D001-7	116	3311	504A	1104	
53D001-8	177	3100/519	300	1122	
53D002-1	109			1123	
53D002-2	109			1123	
53D003-1	116		504A	1104	
53D003-2	116	3017B/117	504A	1104	
53E001-1	109			1123	
53E003-1	109	3087		1123	
53E003-2	177	3100/519	300	1122	
53E006-1	109	3087		1123	
53E009-1		3126	90		
53E010-1	116	3311	504A	1104	
53E011-1	166	9075	BR-600	1152	
53E011-2	137A	3058	ZD-6.2	561	
53F001-1	116	3311	504A	1104	
53F002-1	116	3311	504A	1104	
53H001-1	116		504A	1104	
53H001-2	109			1123	
53J001-1	156	3032A	512		
53J002-1	156	3032A	512		
53J002-2	117	3017B	504A	1104	
53J003-1	116		504A	1104	
53J003-2	177		300	1122	
53J004-1	137A	3058	ZD-6.2	561	
53J004-2			ZD-13	563	

Industry Standard No.	ECG	SK	GE	RS 276-	MOTOR.
53J004-3	138A	3062/142A	ZD-7.5		
53K001-11	116		504A	1104	
53K001-14	139A	3060	ZD-9.1	562	
53K001-15	614	3327	90		
53K001-18	139A		ZD-9.1	562	
53K001-2	109			1123	
53K001-4	5070A	9021	ZD-6.2	561	
53K001-5	109	3088		1123	
53K001-6	116	3311	504A	1104	
53K001-6(5,7)			504A	1104	
53K001-7	116	3311	504A	1104	
53K001-7(6,8)			504A	1104	
53K001-8	116	3311	504A	1104	
53K001-9	117	3031A	504A	1104	
53L001-1			1N34AS	1123	
53L001-10	109		1N60	1123	
53L001-14	506	3081/125	510	1114	
53L001-15	177	3100/519	514	1122	
53L001-5	506	3843	504A	1114	
53N001-1			1N34AS	1123	
53N001-2	116	3031A	504A	1104	
53N001-3	116	3311	504A	1104	
53N001-4	5080A	3336	ZD-22		
53N001-5	177	3100/519	300	1122	
53N001-6		3126	90		
53N001-7	109	3088		1123	
53N001-9	5081A		ZD-24		
53N002-2	116	3311	504A	1104	
53N003-1	109	3088		1123	
53N003-2	109	3088		1123	
53N003-3	110MP	3709	1N60	1123(2)	
53N004-1		3126	90		
53N004-10	110MP	3709	1N60	1123(2)	
53N004-11	109	3088		1123	
53N004-12	177	3100/519	300	1122	
53N004-13		3126	90		
53N004-14	109	3087		1123	
53N004-5	109	3087		1123	
53N004-6	109	3088		1123	
53N004-7	116	3031A		1104	
53N004-8	116	3031A		1104	
53N004-9	116	3031A	504A	1104	
53N49	105	3012	4		
53P151	123	3124/289	20	2051	
53P153	121	3014	239	2006	
53P157	102	3004	2	2007	
53P158	123	3124/289	20	2051	
53P159	123	3124/289	20	2051	
53P161	123	3124/289	20	2051	
53P162	123	3124/289	20	2051	
53P163	123	3124/289	20	2051	
53P165	123	3124/289	20	2051	
53P166	129	3124/289	4	2027	
53P169	128	3024	243	2030	
53P170	129	3025	244	2027	
53SC-15		3126	90		
53T001-1	109	3087		1123	
53T001-2		3126	90		
53T001-3	116	3311	504A	1104	
53T001-4	109	3088		1123	
53T001-5	109	3088		1123	
53T001-6	109	3087		1123	
53T001-7	177	3100/519	300	1122	
53U001-1	109			1123	
53U001-2	177		300	1122	
53V001-2	109	3088		1123	
53W001-2	614	3327			
53Y001-1	116	3311	504A	1104	
054	123A	3452/108	86	2051	
54-1	123A	3444	20	2051	
54A	123A	3444	20	2051	
54B	123A	3444	20	2051	
54BLK	108	3452	86	2038	
54BLU	123A	3444	20	2051	
54BRN	108	3452	86	2038	
54C	123A	3444	20	2051	
54D	123A	3444	20	2051	
54E	287	3433	222		
54F	123A	3444	20	2051	
54GRN	123A	3444	20	2051	
54LS00DM					SN54LS00J
54LS00FM					SN54LS00W
54LS02DM					SN54LS02J
54LS02FM					SN54LS02W
54LS03DM					SN54LS03J
54LS03FM					SN54LS03W
54LS04DM					SN54LS04J
54LS04FM					SN54LS04W
54LS05DM					SN54LS05J
54LS05FM					SN54LS05W
54LS08DM					SN54LS08J
54LS08FM					SN54LS08W
54LS09DM					SN54LS09J
54LS09FM					SN54LS09W
54LS10DM					SN54LS10J
54LS10FM					SN54LS10W
54LS11DM					SN54LS11J
54LS11FM					SN54LS11W
54LS132DM					SN54LS132J
54LS132FM					SN54LS132W
54LS133DM					SN54LS133J
54LS133FM					SN54LS133W
54LS136DM					SN54LS136J
54LS136FM					SN54LS136W
54LS138DM					SN54LS138J
54LS138FM					SN54LS138W
54LS139DM					SN54LS139J
54LS139FM					SN54LS139W
54LS13DM					SN54LS13J
54LS13FM					SN54LS13W
54LS14DM					SN54LS14J
54LS14FM					SN54LS14W
54LS151DM					SN54LS151J
54LS151FM					SN54LS151W
54LS153DM					SN54LS153J
54LS153FM					SN54LS153W
54LS155DM					SN54LS155J
54LS155FM					SN54LS155W
54LS156DM					SN54LS156J
54LS156FM					SN54LS156W
54LS157DM					SN54LS157J
54LS157FM					SN54LS157W
54LS158DM					SN54LS158J
54LS158FM					SN54LS158W
54LS15DM					SN54LS15J
54LS15FM					SN54LS15W
54LS168DM					SN54LS168J
54LS168FM					SN54LS168W
54LS169DM					SN54LS169J
54LS169FM					SN54LS169W
54LS170DM					SN54LS170J
54LS170FM					SN54LS170W
54LS173DM					SN54LS173J
54LS173FM					SN54LS173W

Industry Standard No.	ECG	SK	GE	RS 276-	MOTOR.
54LS174DM					SN54LS174J
54LS174FM					SN54LS174W
54LS175DM					SN54LS175J
54LS175FM					SN54LS175W
54LS181DM					SN54LS181J
54LS181FM					SN54LS181W
54LS190DM					SN54LS190J
54LS190FM					SN54LS190W
54LS191DM					SN54LS191J
54LS191FM					SN54LS191W
54LS192DM					SN54LS192J
54LS192FM					SN54LS192W
54LS193DM					SN54LS193J
54LS193FM					SN54LS193W
54LS196DM					SN54LS196J
54LS196FM					SN54LS196W
54LS197DM					SN54LS197J
54LS197FM					SN54LS197W
54LS20DM					SN54LS20J
54LS20FM					SN54LS20W
54LS21DM					SN54LS21J
54LS21FM					SN54LS21W
54LS22DM					SN54LS22J
54LS22FM					SN54LS22W
54LS251DM					SN54LS251J
54LS251FM					SN54LS251W
54LS253DM					SN54LS253J
54LS253FM					SN54LS253W
54LS259DM					SN54LS259J
54LS259FM					SN54LS259W
54LS260DM					SN54LS260J
54LS260FM					SN54LS260W
54LS266DM					SN54LS266J
54LS266FM					SN54LS266W
54LS26DM					SN54LS26J
54LS26FM					SN54LS26W
54LS279DM					SN54LS279J
54LS279FM					SN54LS279W
54LS27DM					SN54LS27J
54LS27FM					SN54LS27W
54LS28DM					SN54LS28J
54LS28FM					SN54LS28W
54LS293DM					SN54LS293J
54LS293FM					SN54LS293W
54LS298DM					SN54LS298J
54LS298FM					SN54LS298W
54LS299DM					SN54LS299J
54LS299FM					SN54LS299W
54LS30DM					SN54LS30J
54LS30FM					SN54LS30W
54LS32DM					SN54LS32J
54LS32FM					SN54LS32W
54LS33DM					SN54LS33J
54LS33FM					SN54LS33W
54LS37DM					SN54LS37J
54LS37FM					SN54LS37W
54LS38DM					SN54LS38J
54LS38FM					SN54LS38W
54LS390DM					SN54LS390J
54LS390FM					SN54LS390W
54LS393DM					SN54LS393J
54LS393FM					SN54LS393W
54LS395DM					SN54LS395J
54LS395FM					SN54LS395W
54LS40DM					SN54LS40J
54LS40FM					SN54LS40W
54LS42DM					SN54LS42J
54LS42FM					SN54LS42W
54LS490DM					SN54LS490J
54LS490FM					SN54LS490W
54LS51DM					SN54LS51J
54LS51FM					SN54LS51W
54LS54DM					SN54LS54J
54LS54FM					SN54LS54W
54LS55DM					SN54LS55J
54LS55FM					SN54LS55W
54LS670DM					SN54LS670J
54LS670FM					SN54LS670W
54LS74DM					SN54LS74AJ
54LS74FM					SN54LS74AW
54LS85DM					SN54LS85J
54LS85FM					SN54LS85W
54LS86DM					SN54LS86J
54LS86FM					SN54LS86W
54LS90DM					SN54LS90J
54LS90FM					SN54LS90W
54LS92DM					SN54LS92J
54LS92FM					SN54LS92W
54LS93DM					SN54LS93J
54LS93FM					SN54LS93W
54LS95BDM					SN54LS95BJ
54LS95BFM					SN54LS95BW
540RN	108	3452	86	2038	
54RED	108	3452	86	2038	
54V001-2	109			1123	
54WHT	123A	3444	20	2051	
54YEL	123A	3444	20	2051	
055	123A	3025/129	20	2051	
55-001	506	3843	511	1114	
55-1			1N34AS	1123	
55-1016	102A	3004	53	2007	
55-1026	123A	3444	20	2051	
55-1027	103A	3835	59	2002	
55-1029	158	3004/102A	53	2007	
55-1031	158	3004/102A	53	2007	
55-1032	103A	3835	59	2002	
55-1034	123A	3444	20	2051	
55-1082	123A	3444	20	2051	
55-1083	159	3466	82	2032	
55-1083A	159	3466	82	2032	
55-1084	128	3024	243	2030	
55-1085	159	3466	82	2032	
55-1085A	159	3466	82	2032	
55-152579	159	3466	82	2032	
55-509276-2	218		234		
55-509276-4	218		234		
55-641	172A	3156	64		
55-642	128	3024	243	2030	
55-643	129	3025	244	2027	
55P2	126	3006/160	52	2024	
55P3	126	3006/160	52	2024	
55Z4	140A	3060/139A	ZD-10	562	
056	159	3466	82	2032	
56-1	109	3087		1123	
56-10	109	3087		1123	
56-11	109	3087		1123	
56-13	148A		55		
56-15	506	3032A	511	1114	
56-16	135A	3056	ZD-5.1		
56-18	151A	3099	ZD-110		
56-19	139A	3060	ZD-9.1	562	
56-2	109	3087		1123	
56-20	109	3087		1123	
56-21	112		1N82A		

Industry Standard No.	ECG	SK	GE	RS 276-	MOTOR.
56-234	108	3452	86	2038	
56-24	177	3100/519	300	1122	
56-25	145A	3063	ZD-15	564	
56-26	109	3087		1123	
56-27	177	3100/519	300	1122	
56-28	177	3100/519	300	1122	
56-29	151A	3099	ZD-110		
56-3	109	3087		1123	
56-31	138A	3059	ZD-7.5		
56-32	143A	3750	ZD-13	563	
56-33	117		504A	1104	
56-35	123A	3444	20	2051	
56-36	5075A	3751	ZD-16	564	
56-4	109	3087		1123	
56-44	135A	3056	ZD-5.1		
56-45	5079A	3335	ZD-20		
56-46	139A	3060	ZD-9.1	562	
56-47	146A	3064	ZD-27		
56-48	151A	3099	ZD-110		
56-4826	229	3018	212	2010	
56-4827	199	3124/289	62	2010	
56-4829	199	3122	61	2038	
56-4830	293	3849	47	2030	
56-4831	294	3841	67	2023	
56-4832	177	3100/519	300	1122	
56-4833	1073		IC-63		
56-4834	801	3160	IC-35		
56-4835	614	3126	90		
56-4836	601	3463			
56-4837	142A	3062	ZD-12	563	
56-4839	116		504A	1104	
56-4885	5072A		ZD-8.2	562	
56-4886	109	3087	1N34AS	1123	
56-5	177	3016	300	1122	
56-50	5066A	3330	ZD-3.3		
56-51	142A	3062	ZD-12	563	
56-52	116	3017B/117	504A	1104	
56-53	141A	3092	ZD-11.5		
56-54	5072A		ZD-8.2	562	
56-55	5086A	3338	ZD-39		
56-5561	143A	3864/605			
56-56	519	3100	300	1122	
56-57	142A	3062	ZD-12	563	
56-58	5071A	3334	ZD-6.8	561	
56-59	135A	3056	ZD-5.1		
56-6	5071A	3334	ZD-6.8	561	
56-62	139A	3060	ZD-9.1	562	
56-63	5071A	3334	ZD-6.8	561	
56-66	5086A	3338	ZD-39		
56-67	140A	3061	ZD-10	562	
56-68	5092A	3343	ZD-68		
56-7			1N34AS	1123	
56-70	134A	3055	ZD-3.6		
56-71	5071A	3334	ZD-6.8	561	
56-72	5087A	3338/5086A	ZD-39		
56-73	177		300	1122	
56-76	5084A	3755	ZD-30		
56-78	117		504A	1104	
56-79	5084A	3755	ZD-30		
56-8	109	3087		1123	
56-8086	107	3018	11	2015	
56-8086A	107	3018	11	2015	
56-8086B	107	3018	11	2015	
56-8086C	107	3018	11	2015	
56-8087	107	3018	11	2015	
56-8087B	107	3018	11	2015	
56-8087C	107	3018	11	2015	
56-8088	107	3018	11	2015	
56-8088A	107	3018	11	2015	
56-8088C	107	3018	11	2015	
56-8089	123A	3444	20	2051	
56-8089A	123A	3444	20	2051	
56-8089C	123A	3444	20	2051	
56-8090	123A	3444	20	2051	
56-8090A	123A	3444	20	2051	
56-8090C	123A	3444	20	2051	
56-8091	102A	3004	53	2007	
56-8091A	102A	3004	53	2007	
56-8091B	102A	3004	53	2007	
56-8091C	102A	3004	53	2007	
56-8091D	102A	3004	53	2007	
56-8092	102A	3004	53	2007	
56-8092A	102A	3004	53	2007	
56-8092B	102A	3004	53	2007	
56-8093	109	3088		1123	
56-8094		3126	54		
56-8095	110MP	3709	1N60	1123(2)	
56-8096	138A	3059	ZD-7.5		
56-8097	116	3311	504A	1104	
56-8098	159	3466	82	2032	
56-8098A	159	3466	82	2032	
56-8098B	159	3466	82	2032	
56-8098C	159	3466	82	2032	
56-8099	158		53	2007	
56-8100	103A	3010	59	2002	
56-8100A	103A	3010	59	2002	
56-8100B	103A	3010	59	2002	
56-8100C	103A	3010	59	2002	
56-8100D	103A	3010	59	2002	
56-8101	158	3087	53	2007	
56-8196	199	3124/289	62	2010	
56-8197	199	3124/289	62	2010	
56-8198	116	3311	504A	1104	
56-8199	116	3311	504A	1104	
56-86412-3	184		57	2017	
56-89			1N34AS	1123	
56-93-2867	177		300	1122	
56A1-1	703A	3157	IC-12		
56A17-1	791	3149	IC-231		
56A20-1	783	3215	IC-225		
56A22-1	123A	3444	20	2051	
56A23-1	815	3255	IC-244		
56A3-1	712	3072	IC-148		
56A4-1	714	3075	IC-4		
56A42-1	1116	3969	IC-279		
56A5-1	715	3076	IC-6		
56A55	830	9030			
56A55-1	830	9030			
56A6-1	790	3454	IC-230		
56A65-1	1231	3832			
56A7-1	147A	3045/225	ZD-33		
56A9-1	748	3236			
56B22-1	123A	3444	20	2051	
56C1-1	703A	3157	IC-12		
56C3-1	712	3072	IC-2		
56C4-1	714	3075	IC-4		
56C5-1	790	3076/715	IC-18		
56C6-1	713	3454/790	IC-5		
56C7-1	147A	3095	ZD-33		
56C9-1	748	3236			
56D1-1	703A	3157	IC-12		
56D17-1	791	3149	IC-231		
56D20-1	783	3215	IC-21		
56D23-1	815	3255			

Industry Standard No.	ECG	SK	GE	RS 276-	MOTOR.
56D3-1	712	3072	IC-2		
56D4-1	714	3075	IC-4		
56D5-1	715	3076	IC-6		
56D55	830	9030			
56D6-1	790	3454	IC-230		
56D8-1	814	3072/712	IC-243		
56D9-1	748	3236			
56L101	714	3075	IC-4		
56L102	715	3076	IC-6		
56L103	790	3454	IC-230		
56P1	160		245	2004	
56P2	160	3008	245	2004	
56P3	160	3004/102A	245	2004	
56P4	160	3008	245	2004	
56P4P	160		245	2004	
56Z6	5116A	3382			
56Z6A	5116A	3382			
057	123A	3444	20	2051	
57-0004503	123A	3444	20	2051	
57-0005452	123A	3444	20	2051	
57-0005503	123A	3444	20	2051	
57-0006	116	3016	504A	1104	
57-000901504	123A	3444	20	2051	
57-00901504				2051	
57-01491-B	123A	3444	20	2051	
57-01491-C	128		243	2030	
57-01494C	128		243	2030	
57-1	116	3017B/117	504A	1104	
57-12	116	3017B/117	504A	1104	
57-13	116	3017B/117	504A	1104	
57-15	116	3017B/117	504A	1104	
57-17	116	3017B/117	504A	1104	
57-18	5980	3610	5096		
57-2	116	3017B/117	504A	1104	
57-20	116	3017B/117	504A	1104	
57-21	116	3017B/117	504A	1104	
57-22	116	3017B/117	504A	1104	
57-23	116	3017B/117	504A	1104	
57-24	116	3017B/117	504A	1104	
57-25	117		504A	1104	
57-26	117		504A	1104	
57-27	116	3016	504A	1104	
57-28	116	3017B/117	504A	1104	
57-29	117		504A	1104	
57-31	116	3017B/117	504A	1104	
57-32	113A		6GC1		
57-33	116	3017B/117	504A	1104	
57-36		3066	CR-1		
57-38	116(3)		504A(3)	1104	
57-39	5524	6624			
57-4015-27	102A		53	2007	
57-4015-28	158		53	2007	
57-4018-60	130		14	2041	
57-42	156	3016	512		
57-42A	156	3016	512		
57-43	117		504A	1104	
57-46	116	3017B/117	504A	1104	
57-49	117		504A	1104	
57-52	173BP	3999	305		
57-55	5402	3638			
57-56	173BP	3999	305		
57-58	116	3314	504A	1104	
57-59	116	3314	504A	1104	
57-6	116	3017B/117	504A	1104	
57-60	116		504A	1104	
57-61	231	3857			
57-62	230	3857/231			
57-65	116		504A	1104	
57-83	532	3301/531	525		
57-90	530	3308	540		
57-98	530	3308	540		
57A1-1	109			1123	
57A1-10	160		245	2004	
57A1-100	160		245	2004	
57A1-101	160		245	2004	
57A1-102	160		245	2004	
57A1-103	160		245	2004	
57A1-104	102A		53	2007	
57A1-105	102A		53	2007	
57A1-106	102A		53	2007	
57A1-107	160		245	2004	
57A1-108	160		245	2004	
57A1-109	160		245	2004	
57A1-11	102	3722	2	2007	
57A1-110	160		245	2004	
57A1-111	102A		53	2007	
57A1-112	160		245	2004	
57A1-113	160		245	2004	
57A1-114	160		245	2004	
57A1-115	160		245	2004	
57A1-116	102A		53	2007	
57A1-117	102A		53	2007	
57A1-118	102A		53	2007	
57A1-119	121	3717	239	2006	
57A1-12	160		245	2004	
57A1-120	160		245	2004	
57A1-121	102A		53	2007	
57A1-122	154		40	2012	
57A1-123	123A	3444	20	2051	
57A1-124	123A	3444	20	2051	
57A1-13	160		245	2004	
57A1-14	102A		53	2007	
57A1-15	160		245	2004	
57A1-16	160		245	2004	
57A1-17	158		53	2007	
57A1-18	158		53	2007	
57A1-19	158		53	2007	
57A1-2	109			1123	
57A1-20	158		53	2007	
57A1-21	158		53	2007	
57A1-22	160		245	2004	
57A1-23	102A		53	2007	
57A1-24	160		245	2004	
57A1-25	160		245	2004	
57A1-26	102A		53	2007	
57A1-27	102A		53	2007	
57A1-28	102A		53	2007	
57A1-3	103A	3835	59	2002	
57A1-30	160		245	2004	
57A1-31	160		245	2004	
57A1-32	160		245	2004	
57A1-33	160		245	2004	
57A1-34	102A		53	2007	
57A1-35	160		245	2004	
57A1-36	160		245	2004	
57A1-37	160		245	2004	
57A1-38	158		53	2007	
57A1-4	103A	3835	59	2002	
57A1-40	102A		53	2007	
57A1-41	160		245	2004	
57A1-42	158		53	2007	
57A1-43	102A		53	2007	
57A1-44	102A		53	2007	
57A1-45	160		245	2004	
57A1-46	160		245	2004	
57A1-47	160		245	2004	
57A1-48	160		245	2004	
57A1-49	160		245	2004	
57A1-5	103A	3835	59	2002	
57A1-50	160		245	2004	
57A1-51	123A	3444	20	2051	
57A1-52	106	3984	21	2034	
57A1-53	102A		53	2007	
57A1-54	109			1123	
57A1-56	158		53	2007	
57A1-57	160		245	2004	
57A1-58	102A		53	2007	
57A1-59	102A		53	2007	
57A1-6	103A	3835	59	2002	
57A1-60	158		53	2007	
57A1-61	160		245	2004	
57A1-62	109			1123	
57A1-63	117		504A	1104	
57A1-64	109			1123	
57A1-65	112		1N82A		
57A1-66	102A		53	2007	
57A1-67	160		245	2004	
57A1-68	102A		53	2007	
57A1-69	160		245	2004	
57A1-7	158		53	2007	
57A1-70	102A		53	2007	
57A1-71	102A		53	2007	
57A1-72	160		245	2004	
57A1-73	160		245	2004	
57A1-74	160		245	2004	
57A1-75	123A	3444	20	2051	
57A1-76	159		82	2032	
57A1-76A	159	3466	82	2032	
57A1-77	158		63	2007	
57A1-78	103A	3835	59	2002	
57A1-79	102A		53	2007	
57A1-8	102A		53	2007	
57A1-80	160		245	2004	
57A1-81	160		245	2004	
57A1-82	102A		53	2007	
57A1-83	102A		53	2007	
57A1-84	160		245	2004	
57A1-85	160		245	2004	
57A1-86	160		245	2004	
57A1-87	160		245	2004	
57A1-89	160		245	2004	
57A1-9	160		245	2004	
57A1-90	102A		53	2007	
57A1-91	102A		53	2007	
57A1-92	102A		53	2007	
57A1-93	102A		53	2007	
57A1-94	160		245	2004	
57A1-95	102A		53	2007	
57A1-96	160		245	2004	
57A1-97	102A		53	2007	
57A1-98	160		245	2004	
57A1-99	160		245	2004	
57A10-1	108	3452	86	2038	
57A10-2	108	3452	86	2038	
57A100-11	131		44	2006	
57A100-7	102A		53	2007	
57A101-4	108	3452	86	2038	
57A102-4	161	3716	39	2015	
57A103-4	161	3716	39	2015	
57A104-1	154	3040	40	2012	
57A104-2	154	3040	40	2012	
57A104-3	154	3040	40	2012	
57A104-4	154	3040	40	2012	
57A104-5	154	3040	40	2012	
57A104-6	154	3040	40	2012	
57A104-7	154	3040	40	2012	
57A104-8	154	3040	40	2012	
57A104-8-6	128	3024	243	2030	
57A105-12	123A	3444	20	2051	
57A106-12	159	3466	82	2032	
57A107-1	108	3452	86	2038	
57A107-2	108	3452	86	2038	
57A107-3	108	3452	86	2038	
57A107-4	108	3452	86	2038	
57A107-5	108	3452	86	2038	
57A107-6	108	3452	86	2038	
57A107-8	108	3452	86	2038	
57A108-1	106	3984	21	2034	
57A108-2	106	3984	21	2034	
57A108-3	106	3984	21	2034	
57A108-4	106	3984	21	2034	
57A108-5	106	3984	21	2034	
57A108-6	106	3984	21	2034	
57A108-6-8	129	3025	244	2027	
57A108-6A	159	3984/106	82	2032	
57A108-7	106	3984	21	2034	
57A108-8	106	3984	21	2034	
57A109-9	128		243	2030	
57A10A-8-6	108	3452	86	2038	
57A11-1	123A	3444	20	2051	
57A110-9	129		244	2027	
57A112-9	129		244	2027	
57A112-9A	129		244	2027	
57A1127	102A		53	2007	
57A113-9	128		243	2030	
57A1130	160		245	2004	
57A1131	160		245	2004	
57A1132	160		245	2004	
57A114-9	129		244	2027	
57A114-9A	129		244	2027	
57A1143	102A		53	2007	
57A115-9	129		244	2027	
57A115-9A	129		244	2027	
57A116-9	129		244	2027	
57A116-9A	129		244	2027	
57A117-9	123A	3444	20	2051	
57A118-12	123A	3444	20	2051	
57A1186	160		245	2004	
57A119-12	123A	3444	20	2051	
57A119-2	161	3716	39	2015	
57A12-1	154		40	2012	
57A12-2	154		40	2012	
57A12-3	175		246	2020	
57A12-4	128		243	2030	
57A12-5	175		246	2020	
57A120-12	123A	3444	20	2051	
57A121-9	123A	3444	20	2051	
57A122-9	129		244	2027	
57A122-9A	159	3466	82	2032	
57A123-10	175		246	2020	
57A124-10	104	3719	16	2006	
57A125-9	123A	3444	20	2051	
57A126	102	3722	2	2007	
57A126-1			62	2010	
57A126-12	107	3018	11	2015	
57A127	157	3747	232		
57A128-9	128		243	2030	
57A129-9	128		243	2030	

Industry Standard No.	ECG	SK	GE	RS 276-	MOTOR.
57A130	126		52	2024	
57A130-9	129		244	2027	
57A130-9A	159	3466	82	2032	
57A131	126		52	2024	
57A131-10	128		243	2030	
57A132	126		52	2024	
57A132-10	129		244	2027	
57A132-29	1142	3485	IC-128		
57A132-9	160		245	2004	
57A133-12	159	3466	82	2032	
57A134-12	108	3452	86	2038	
57A135-12	123A	3444	20	2051	
57A136-1	154		40	2012	
57A136-10	154		40	2012	
57A136-11	154		40	2012	
57A136-12	123A	3444	20	2051	
57A136-2	154		40	2012	
57A136-3	154		40	2012	
57A136-4	154		40	2012	
57A136-5	154		40	2012	
57A136-6	154		40	2012	
57A136-7	154		40	2012	
57A136-8	154		40	2012	
57A136-9	154		40	2012	
57A137-12	159	3466	82	2032	
57A137-12A	159	3466	82	2032	
57A138-4	319P	3132	86	2038	
57A138-4(3RD LF)				2009	
57A138-4(3RD-IF)			210	2051	
57A138-4(LAST LF)				2009	
57A138-4(LAST-IF)			210	2051	
57A138-4-6	319P	3452/108	86	2038	
57A139-1	161	3117	39	2015	
57A139-2	161	3117	39	2015	
57A139-3	161	3117	39	2015	
57A139-4	319P	3117	283	2016	
57A139-4-6	108	3452	86	2038	
57A14-1	128		243	2030	
57A14-2	128		243	2030	
57A14-3	128		243	2030	
57A140-12	123A	3444	20	2051	
57A141-1	108	3452	86	2038	
57A141-2	108	3452	86	2038	
57A141-3	108	3452	86	2038	
57A141-4	161	3117	39	2015	
57A142-1	108	3452	86	2038	
57A142-2	199	3018	62	2010	
57A142-3	108	3452	86	2038	
57A142-4	233	3132	210	2009	
57A142-4(3RDIF)	233		210	2051	
57A142-4(3RDLF)				2009	
57A142-7	161	3716	39	2015	
57A143	102	3722	2	2007	
57A143-1	108	3452	86	2038	
57A143-10	108	3452	86	2038	
57A143-11	108	3452	86	2038	
57A143-12	199	3018	62	2010	
57A143-2	108	3452	86	2038	
57A143-3	108	3452	86	2038	
57A143-4	108	3452	86	2038	
57A143-5	108	3452	86	2038	
57A143-6	108		86	2038	
57A143-7	108	3452	86	2038	
57A143-8	108	3452	86	2038	
57A143-9	108	3452	86	2038	
57A144-12	199	3018	62	2010	
57A145-12	159	3466	82	2032	
57A145-12A	159	3466	82	2032	
57A146-12	108	3452	86	2038	
57A147-12	159	3466	82	2032	
57A147-12A	159	3466	82	2032	
57A148-1	102A		53	2007	
57A148-10	129		244	2027	
57A148-11	129		244	2027	
57A148-12	159	3466	82	2032	
57A148-12A	129	3114/290	244	2027	
57A148-2	129		244	2027	
57A148-3	129		244	2027	
57A148-4	129		244	2027	
57A148-5	129		244	2027	
57A148-6	129		244	2027	
57A148-7	129		244	2027	
57A148-8	129		244	2027	
57A148-9	129		244	2027	
57A149-12	312	3112	FET-1	2035	
57A15-1	123A	3444	20	2051	
57A15-2	123A	3444	20	2051	
57A15-3	123A	3444	20	2051	
57A15-4	123A	3444	20	2051	
57A15-5	159	3466	82	2032	
57A15-50	159	3466	82	2032	
57A150-12	312	3116	FET-1	2035	
57A151-6	108	3452	86	2038	
57A152-1	123A	3444	20	2051	
57A152-10	123A	3444	20	2051	
57A152-11	123A	3444	20	2051	
57A152-12	108	3452	86	2038	
57A152-2	123A	3444	20	2051	
57A152-3	123A	3444	20	2051	
57A152-4	123A	3444	20	2051	
57A152-5	123A	3444	20	2051	
57A152-6	123A	3444	20	2051	
57A152-7	123A	3444	20	2051	
57A152-8	123A	3444	20	2051	
57A152-9	123A	3444	20	2051	
57A153-1	123A	3444	20	2051	
57A153-2	123A	3444	20	2051	
57A153-3	123A	3444	20	2051	
57A153-4	123A	3444	20	2051	
57A153-5	123A	3444	20	2051	
57A153-6	123A	3444	20	2051	
57A153-7	123A	3444	20	2051	
57A153-8	123A	3444	20	2051	
57A153-9	123A	3444	20	2051	
57A155-10	175		246	2020	
57A156	128		243	2030	
57A156-9	123A	3444	20	2051	
57A157	159	3466	82	2032	
57A157-9	159	3466	82	2032	
57A157-90	159	3466	82	2032	
57A157-9A	159	3466	82	2032	
57A158-1	124		12		
57A158-10	124	3021	12		
57A158-2	124		12		
57A158-3	124		12		
57A158-5	124		12		
57A158-6	124		12		
57A158-7	124		12		
57A158-8	124		12		
57A158-9	124		12		
57A159-12	159	3466	82	2032	
57A159-12A	159	3466	82	2032	
57A16	126		52	2024	
57A16-1	123A	3444	20	2051	
57A160-1	108	3452	86	2038	
57A160-2	108	3452	86	2038	
57A160-3	108	3452	86	2038	
57A160-4	108	3452	86	2038	
57A160-8	154	3132	40	2012	
57A164-4	161	3018	39	2015	
57A166-12	123A	3444	20	2051	
57A167-9	192	3024/128	63	2030	
57A168	126		52	2024	
57A168-9	193	3025/129	67	2023	
57A169	102	3722	2	2007	
57A170	102	3722	2	2007	
57A172-8	190	3103A/396	27		
57A174-8	159	3466	82	2032	
57A175-12	159	3466	82	2032	
57A177-12	161	3018	39	2015	
57A178-12	159	3466	82	2032	
57A179-4	161	3246/229	39	2015	
57A180	160		245	2004	
57A180-4		3246	39	2009	
57A180-4(3RDIF)	233		210	2009	
57A180-4(3RDLF)				2009	
57A181-12	123A	3444	20	2051	
57A182-10			20	2051	
57A182-12	123AP	3444/123A	20	2051	
57A184	160		245	2004	
57A184-12	123A	3444	20	2051	
57A185-12	159	3466	82	2032	
57A186	126		52	2024	
57A186-11	165	3115	38		
57A186-12	165	3115	38		
57A187	160		245	2004	
57A187-12	291	3054/196	66		
57A188	160		245	2004	
57A188-12	292	3083/197	69		
57A189	102	3722	2	2007	
57A189-8	159	3466	82	2032	
57A19	159	3466	82	2032	
57A19-1	159	3466	82	2032	
57A19-10	159	3466	82	2032	
57A19-1A	159	3466	82	2032	
57A19-2	159	3466	82	2032	
57A19-20	159	3466	82	2032	
57A19-3	159	3466	82	2032	
57A19-30	159	3466	82	2032	
57A191-12	123A	3444	20	2051	
57A192-10	124	3021	12		
57A193-11	171	3201	27		
57A193-12			20	2051	
57A194-11	154	3244	40	2012	
57A194-11(PULSE)	154		40	2012	
57A194-12	5404	3627			
57A194-1L			40	2012	
57A195-10	124	3131A/369	66	2048	
57A196-10	94	3027/130	14	2041	
57A197-12	159	3466	82	2032	
57A198-11	165	3111	38		
57A199-11	165		38		
57A199-4	123A	3444	20	2051	
57A2-1	126		52	2024	
57A2-101	123A	3444	20	2051	
57A2-102	123A	3444	20	2051	
57A2-103	123A	3444	20	2051	
57A2-104	160		245	2004	
57A2-105	160		245	2004	
57A2-113	123A	3444	20	2051	
57A2-116	123A	3444	20	2051	
57A2-126	123A	3444	20	2051	
57A2-149	160		245	2004	
57A2-15	102A	3004	53	2007	
57A2-153	123A	3444	20	2051	
57A2-157	160		245	2004	
57A2-158	160		245	2004	
57A2-159	160		245	2004	
57A2-19	126	3008	52	2024	
57A2-192	123A	3444	20	2051	
57A2-22	160		245	2004	
57A2-23	102A		53	2007	
57A2-24	102A	3004	53	2007	
57A2-25	158		53	2007	
57A2-26	160		245	2004	
57A2-27	123A	3444	20	2051	
57A2-28	123A	3444	20	2051	
57A2-29	102A		53	2007	
57A2-30	160		245	2004	
57A2-31	160		245	2004	
57A2-32	102A		53	2007	
57A2-33	102A		53	2007	
57A2-34	102A		53	2007	
57A2-35	160		245	2004	
57A2-36	102A		53	2007	
57A2-37	160		245	2004	
57A2-38	128		243	2030	
57A2-39	102A		53	2007	
57A2-4		3004	53	2007	
57A2-40	160		245	2004	
57A2-41	160		245	2004	
57A2-42	160		245	2004	
57A2-43	102A		53	2007	
57A2-44	102A		53	2007	
57A2-45	102A		53	2007	
57A2-46	102A		53	2007	
57A2-47	175		246	2020	
57A2-48	160		245	2004	
57A2-50	160		245	2004	
57A2-51	160		245	2004	
57A2-52	102A		53	2007	
57A2-58	175		246	2020	
57A2-59	123A	3444	20	2051	
57A2-60	102A		53	2007	
57A2-61	158		53	2007	
57A2-62	123A	3444	20	2051	
57A2-63	123A	3444	20	2051	
57A2-64	123A	3444	20	2051	
57A2-65	160		245	2004	
57A2-66	160		245	2004	
57A2-67	160		245	2004	
57A2-68	160		245	2004	
57A2-70	159	3466	82	2032	
57A2-70A	159	3466	82	2032	
57A2-71	159	3466	82	2032	
57A2-71A	159	3466	82	2032	
57A2-72	102A		53	2007	
57A2-73	123A	3444	20	2051	
57A2-75	160		245	2004	
57A2-77	160		245	2004	
57A2-78	102A		53	2007	
57A2-80	160		245	2004	
57A2-83	102A		53	2007	
57A2-84	175		246	2020	
57A2-85	123A	3444	20	2051	
57A2-87	123A	3444	20	2051	
57A2-88	102A		53	2007	
57A2-89	160		245	2004	

Industry Standard No.	ECG	SK	GE	RS 276-	MOTOR.
57A2-90	160		245	2004	
57A2-93	160		245	2004	
57A2-97	123A	3444	20	2051	
57A2.4	102A			2007	
57A20-1	108	3452	86	2038	
57A200-12	123A	3444	20	2051	
57A201-13	123A	3444	20	2051	
57A201-14	159	3466	82	2032	
57A202-13	123A	3444	20	2051	
57A203-14	123A	3444	20	2051	
57A204-14	123A	3444	20	2051	
57A205-14	379	3054/196	69	2049	
57A206-14	153	3083/197	69	2049	
57A207-8	154	3040	40	2012	
57A208-8	171	3201	27		
57A21	107		11	2015	
57A21-1	108	3452	86	2038	
57A21-10	108	3452	86	2038	
57A21-11			11	2015	
57A21-12	107	3018	11	2015	
57A21-13	107	3018	11	2015	
57A21-14	107	3018	11	2015	
57A21-15	108	3452	86	2038	
57A21-16	161	3039/316	39	2015	
57A21-17	161	3018	39	2015	
57A21-18	161	3019	39	2015	
57A21-2	108	3452	11	2038	
57A21-21			11	2015	
57A21-26			11	2015	
57A21-3	108	3452	86	2038	
57A21-4	108	3452	86	2038	
57A21-45	107	3018	11	2015	
57A21-5	108	3452	86	2038	
57A21-6	108	3452	86	2038	
57A21-7	108	3452	86	2038	
57A21-8	123A	3444	20	2051	
57A21-9	108	3452	86	2038	
57A211-8	190	3103A/396	27		
57A213-11	165	3115	38		
57A214-12	188	3054/196	226	2018	
57A214-2			57	2017	
57A2141-14	128		243	2030	
57A215-12	159		82	2032	
57A216-12	159	3466	82	2032	
57A219-14	289A	3024/128	268	2038	
57A22-1	179	3642	76		
57A22-2	179	3642	76		
57A220-14	290A	3025/129	269	2032	
57A23	129		244	2027	
57A23-1	129		244	2027	
57A23-2	129		244	2027	
57A23-3	129		244	2027	
57A235-12	159	3466	82	2032	
57A236-11	171	3040	27		
57A24	161	3716	39	2015	
57A24-1	123A	3444	20	2051	
57A24-2	123A	3444	20	2051	
57A24-3	123A	3444	20	2051	
57A24-4	123A	3444	20	2051	
57A240-14	290A	3025/129	269	2032	
57A241-14	289A	3024/128	268	2038	
57A243-10	5404	3627			
57A244-14	196	3054	241	2020	
57A245-14	197	3083	250	2027	
57A249-4	161	3018	39	2015	
57A250-14	152	3054/196	66	2048	
57A251-14	197	3083	69	2049	
57A252-1	123A	3444	20	2051	
57A253-14	123A	3444	20	2051	
57A256-10		3027	14	2041	
57A258-8	159	3466	82	2032	
57A259-10	5404	3627			
57A261-10	154	3040	40	2012	
57A263-11	238	3710	259		
57A265-4	123A	3122	20	2051	
57A267-4	222	3050/221	FET-4	2036	
57A268-9	123A	3122	20	2051	
57A27-1	123A	3444	20	2051	
57A27-2	108	3452	86	2038	
57A273-14	263	3935/270			
57A277-14	153	3083/197	69	2049	
57A278-14	152	3054/196	66	2048	
57A279-14	152	3054/196	66	2048	
57A28	704		IC-205		
57A280-14	289A	3024/128	268	2038	
57A281-14	290A	3114/290	269	2032	
57A282-12	123A	3444	20	2051	
57A283-11		3122	40	2012	
57A285-10	5645	3659/56006			
57A286-10	152	3054/196	66	2048	
57A29	706	3288/1003	IC-43		
57A29-1	706	3288/1003	IC-43		
57A29-2	706	3101	IC-43		
57A295-8	171	3232/191	27		
57A3-10	104MP	3720	16(2)	2006(2)	
57A3-11	104MP	3720	16(2)	2006(2)	
57A3-12	121MP	3718	239(2)	2006(2)	
57A3-4	102A		53	2007	
57A3-5	102A		53	2007	
57A3-6	102A		53	2007	
57A3-7	104MP	3720	16(2)	2006(2)	
57A3-8	104MP	3720	16(2)	2006(2)	
57A3-9	104MP	3720	16(2)	2006(2)	
57A305-12	159	3466	82		
57A31-1	312	3834/132	FET-1	2035	
57A31-2			FET-2	2035	
57A31-3			FET-2	2035	
57A31-4	312	3112	FET-1	2028	
57A310-14	292	3441			
57A311-14	291	3440			
57A312-11	171	3201	27		
57A313-11	165	3710/238			
57A32-1	724	3525	IC-86		
57A32-10	722	3161	IC-9		
57A32-11	1075A	3877			
57A32-16	722		IC-9		
57A32-19	1103	3281	IC-94		
57A32-2	1142	3288/1003	IC-128		
57A32-21	799		IC-34		
57A32-22	720	9014	IC-7		
57A32-25	1124		IC-119		
57A32-27	1124		IC-119		
57A32-28	1142		IC-7		
57A32-29	1142	3485	IC-128		
57A32-3	1039		IC-159		
57A32-32	801	3160	IC-35		
57A4-1	121	3717	239	2006	
57A4-2	121		239	2006	
57A4-4	121	3717	239	2006	
57A403-12	159	3466			
57A404-12	297	3449			
57A405-12	298	3450			
57A5	103	3862	8	2002	
57A5-1	126		52	2024	
57A5-10	126		52	2024	
57A5-2	126		52	2024	
57A5-3	102A		53	2007	
57A5-4	126		52	2024	
57A5-5	102A		53	2007	
57A5-6	108	3452	86	2038	
57A5-7	108	3452	86	2038	
57A5-8	108	3452	86	2038	
57A5-9	126		52	2024	
57A6(NPN)	199		62	2010	
57A6(PNP)	102	3722	22	2007	
57A6-1	102	3722	2	2007	
57A6-10	124		12		
57A6-11	123A	3444	20	2051	
57A6-12	121	3717	239	2006	
57A6-14	124		12		
57A6-15	129		244	2027	
57A6-16	126		52	2024	
57A6-17	123		20	2051	
57A6-19(DIODE)	120		CR-3		
57A6-2	121	3717	239	2006	
57A6-20	103A	3835	59	2002	
57A6-21	103A	3835	59	2002	
57A6-22	102	3722	2	2007	
57A6-23	121	3717	239	2006	
57A6-24	124		12		
57A6-25	102A		53	2007	
57A6-26	129		244	2027	
57A6-26A	129		244	2027	
57A6-27	123		20	2051	
57A6-29	123		20	2051	
57A6-3	121	3717	239	2006	
57A6-30	123		20	2051	
57A6-31	129		244	2027	
57A6-31A	129		244	2027	
57A6-32	123		20	2051	
57A6-33	129		244	2027	
57A6-33A	129		244	2027	
57A6-4	123A	3444	20	2051	
57A6-5	103A	3835	59	2002	
57A6-6	103	3862	8	2002	
57A6-6A	100	3721	1	2007	
57A6-6B	100	3721	1	2007	
57A6-6C	100	3721	1	2007	
57A6-7	123		20	2051	
57A6-8	121	3717	239	2006	
57A6-9	123A	3444	20	2051	
57A68	102A		53	2007	
57A7-1	108	3452	86	2038	
57A7-10	123A	3444	20	2051	
57A7-15	123A	3444	20	2051	
57A7-17	123A	3444	20	2051	
57A7-18	123A	3444	20	2051	
57A7-2	108	3452	86	2038	
57A7-20	123A	3444	20	2051	
57A7-3	108	3452	86	2038	
57A7-4	108	3452	86	2038	
57A7-5	108	3452	86	2038	
57A7-6	108	3452	86	2038	
57A7-7	108	3452	86	2038	
57A7-8	128		243	2030	
57A7-9	123A	3444	20	2051	
57A9-1	160		245	2004	
57A9-2	121	3717	239	2006	
57B-102-4			20	2051	
57B100-11	131		44	2006	
57B100-7	102A		53	2007	
57B101-4	108	3452	86	2038	
57B102-4	161	3018	39	2015	
57B103-4	161	3018	39	2015	
57B104-8	154	3045/225	39	2015	
57B105-12	123A	3444	20	2051	
57B106-12	159	3466	82	2032	
57B107-8	123A	3444	20	2051	
57B108-6	159	3466	82	2032	
57B108-6A	159	3466	82	2032	
57B109-9	128	3024	243	2030	
57B110-9	129	3025	244	2027	
57B112-9	129	3025	244	2027	
57B112-9A	129	3025	244	2027	
57B1127	102A		53	2007	
57B113-9	128	3024	243	2030	
57B1130	160		245	2004	
57B1131	160		245	2004	
57B114-9	129	3025	244	2027	
57B114-9A	129	3025	244	2027	
57B1143	102A		53	2007	
57B115-9	129	3025	244	2027	
57B115-9A	129	3025	244	2027	
57B116-9	129	3025	244	2027	
57B116-9A	129	3025		2027	
57B117-9	123A	3444	20	2051	
57B118-12	123A	3444	20	2051	
57B1186	160		245	2004	
57B119-12	123A	3444	20	2051	
57B119-2	161	3716	39	2015	
57B12-4			18	2030	
57B120-12	123A	3444	20	2051	
57B121-9	123A	3466/159	20	2051	
57B122-9	159	3466	82	2032	
57B122-9A	159	3466	82	2032	
57B123-10	175	3131A/369	246	2020	
57B124-10	104	3009	16	2006	
57B125-9	123A	3444	20	2051	
57B126-12	123A	3444	20	2051	
57B128-9	128	3024	243	2030	
57B129-9	123A	3444	20	2051	
57B130-9	159	3466	82	2032	
57B130-9A	159	3466	82	2032	
57B131-10	152	3893	66	2048	
57B132-10	153	3025/129	69	2049	
57B133-12	159	3466	82	2032	
57B134-12	108	3452	86	2038	
57B135-12	123A	3118	20	2051	
57B136-1	154	3044	40	2012	
57B136-10	154	3044	40	2012	
57B136-11	154	3044	40	2012	
57B136-12	123A	3444	20	2051	
57B136-2	154	3044	40	2012	
57B136-3	154	3044	40	2012	
57B136-4	154	3044	40	2012	
57B136-5	154	3044	40	2012	
57B136-6	154	3044		2012	
57B136-7	154	3044	40	2012	
57B136-8	154	3044	40	2012	
57B136-9	154	3044	40	2012	
57B137-12	159	3466	82	2032	
57B137-12A	159	3466	82	2032	
57B138-4	319P	3444/123A	20	2051	
57B138-4(LAST LF)				2009	
57B138-4(LAST-IF)			210	2051	
57B139-4	319P	3039/316	283	2016	
57B140-12	123A	3444	20	2051	
57B141-1	108	3452	86	2038	

Industry Standard No.	ECG	SK	GE	RS 276-	MOTOR.
57B141-2	108	3452	86	2038	
57B141-3	108	3452	86	2038	
57B141-4	161	3018	39	2015	
57B142-1	108	3452	86	2038	
57B142-2	108	3452	86	2038	
57B142-3	108	3452	86	2038	
57B142-4	233	3018	210	2009	
57B143-1	108	3452	86	2038	
57B143-10	108	3452	86	2038	
57B143-11	108	3452	86	2038	
57B143-12	123A	3444	20	2051	
57B143-2	108	3452	86	2038	
57B143-3	108	3452	86	2038	
57B143-4	108	3452	86	2038	
57B143-5	108	3452	86	2038	
57B143-6	108	3452	86	2038	
57B143-7	108	3452	86	2038	
57B143-8	108	3452	86	2038	
57B143-9	108	3452	86	2038	
57B144-12	123A	3444	20	2051	
57B145-12	159	3466	82	2032	
57B145-12A	159	3466	82	2032	
57B146-12	123A	3444	20	2051	
57B147-12	159	3466	82	2032	
57B147-12A	159	3466	82	2032	
57B148-1	129	3025	244	2027	
57B148-10	129	3025	244	2027	
57B148-11	129	3025	244	2027	
57B148-12	129	3025	244	2027	
57B148-12A	129	3025	244	2027	
57B148-2	129	3025	244	2027	
57B148-3	129	3025	244	2027	
57B148-4	129	3025	244	2027	
57B148-5	129	3025	244	2027	
57B148-6	129	3025	244	2027	
57B148-7	129	3025	244	2027	
57B148-8	129	3025	244	2027	
57B148-9	129	3025	244	2027	
57B149-12	312	3116	FET-1	2035	
57B150-12	312	3116	FET-1	2035	
57B151-6	108	3452	86	2038	
57B152-1	123A	3444	20	2051	
57B152-10	123A	3444	20	2051	
57B152-11	123A	3444	20	2051	
57B152-12	108	3452	86	2038	
57B152-2	123A	3444	20	2051	
57B152-3	123A	3444	20	2051	
57B152-4	123A	3444	20	2051	
57B152-5	123A	3444	20	2051	
57B152-6	123A	3444	20	2051	
57B152-7	123A	3444	20	2051	
57B152-8	123A	3444	20	2051	
57B152-9	123A	3444	20	2051	
57B153-1	123A	3444	20	2051	
57B153-2	123A	3444	20	2051	
57B153-3	123A	3444	20	2051	
57B153-4	123A	3444	20	2051	
57B153-5	123A	3444	20	2051	
57B153-6	123A	3444	20	2051	
57B153-7	123A	3444	20	2051	
57B153-8	123A	3444	20	2051	
57B153-9	123A	3444	20	2051	
57B155-10	175	3131A/369	246	2020	
57B156-9	123A	3444	20	2051	
57B157-9	159	3466	82	2032	
57B157-9A	159	3466	82	2032	
57B158-1	124	3021	12		
57B158-10	175	3021/124	246	2020	
57B158-2	124	3021	12		
57B158-3	124	3021	12		
57B158-4	124	3021	12		
57B158-5	124	3021	12		
57B158-6	124	3021	12		
57B158-7	124	3021	12		
57B158-8	124	3021	12		
57B158-9	124	3021	12		
57B159-12	159	3466	82	2032	
57B159-12A	159	3466	82	2032	
57B160-1	108	3452	86	2038	
57B160-2	108	3452	86	2038	
57B160-3	108	3452	86	2038	
57B160-4	108	3452	86	2038	
57B160-5	108	3452	86	2038	
57B160-6	108	3452	86	2038	
57B160-7	108	3452	86	2038	
57B160-8	108	3452	86	2038	
57B162-12			63	2030	
57B163-12	129	3025	244	2027	
57B163-12A	129	3025	244	2027	
57B165-11	128		243	2030	
57B166-12	108	3452	86	2038	
57B166-9A			244	2027	
57B167-9	128	3024	243	2030	
57B168	102A		53	2007	
57B168-9	129	3025	67	2027	
57B168-99			244	2027	
57B169	102A		53	2007	
57B169-12	312	3834/132	FET-1	2035	
57B170	102A		53	2007	
57B170-9	128	3024	243	2030	
57B171-9	129	3025	244	2027	
57B172-8	190		217		
57B175-12	159	3466	82	2032	
57B175-9	130	3025/129	14	2041	
57B175-9A	130	3027	14	2041	
57B178-12	159	3466	82	2032	
57B180	160		245	2004	
57B181-12	194	3275	220		
57B182-12	123A	3444	20	2051	
57B184	160		245	2004	
57B184-12	123A	3444	20	2051	
57B185-12	159	3466	82	2032	
57B186			51	2004	
57B186-11	238	3710	38		
57B187	160		245	2004	
57B188	160		245	2004	
57B188-12	292	3441	29		
57B189-8	159	3466	82	2032	
57B191-12	123A	3444	20	2051	
57B192-10	124		12		
57B193-11	171	3232/191	27		
57B194-11	123A	3444	20	2051	
57B195-10	124	3021	12		
57B196-10	94		14	2041	
57B197-12	159	3114/290	82	2032	
57B198-11	165	3115	38		
57B199-11	165	3115	38		
57B2-1	160	3452/108	245	2004	
57B2-10	158		53	2007	
57B2-101	123A	3444	20	2051	
57B2-102	123A	3444	20	2051	
57B2-103	123A	3444	20	2051	
57B2-104	160		245	2004	
57B2-105	160		245	2004	

Industry Standard No.	ECG	SK	GE	RS 276-	MOTOR.
57B2-11	160	3006	245	2004	
57B2-113	123A	3444	20	2051	
57B2-116	123A	3444	20	2051	
57B2-12	102A	3008	53	2007	
57B2-126	123A	3444	20	2051	
57B2-13	160	3006	245	2004	
57B2-14	160	3006	245	2004	
57B2-149	160		245	2004	
57B2-15	158	3004/102A	53	2007	
57B2-153	123A	3444	20	2051	
57B2-157	160		245	2004	
57B2-158	160		245	2004	
57B2-159	160		245	2004	
57B2-16	158		53	2007	
57B2-17	160	3008	245	2004	
57B2-18	160	3008	245	2004	
57B2-19	160	3008	245	2004	
57B2-192	123A	3444	20	2051	
57B2-2	160		245	2004	
57B2-20	160		245	2004	
57B2-21	158		53	2007	
57B2-22	160	3006	245	2004	
57B2-23	102A	3005	53	2007	
57B2-24	102A	3123	53	2007	
57B2-25	158	3004/102A	53	2007	
57B2-26	160		245	2004	
57B2-27	123A	3444	20	2051	
57B2-28	123A	3444	20	2051	
57B2-29	102A		53	2007	
57B2-3	158	3004/102A	53	2007	
57B2-30	160		245	2004	
57B2-31	160		245	2004	
57B2-32	102A		53	2007	
57B2-33	102A(2)		53(2)		
57B2-34	102A		53	2007	
57B2-35	160		245	2004	
57B2-36	102A		53	2007	
57B2-37	160		245	2004	
57B2-38	128		243	2030	
57B2-39	102A		53	2007	
57B2-4	158		53	2007	
57B2-40	160		245	2004	
57B2-41	160		245	2004	
57B2-42	160		245	2004	
57B2-43	102A		53	2007	
57B2-44	102A		53	2007	
57B2-45	102A		53	2007	
57B2-46	102A(2)		53(2)		
57B2-47	175	3131A/369	246	2020	
57B2-48	160		245	2004	
57B2-5	103	3862	8	2002	
57B2-50	160		245	2004	
57B2-51	160		245	2004	
57B2-52	102A		53	2007	
57B2-57		3024	18	2030	
57B2-58	175	3131A/369	246	2020	
57B2-59	123A	3444	20	2051	
57B2-6	102A	3008	53	2007	
57B2-60	102A		53	2007	
57B2-61	158(2)		53(2)		
57B2-62	123A	3444	20	2051	
57B2-63	123A	3444	20	2051	
57B2-64	123A	3444	20	2051	
57B2-65	160		245	2004	
57B2-66	160		245	2004	
57B2-67	160		245	2004	
57B2-68	160		245	2004	
57B2-7	158	3004/102A	53	2007	
57B2-70	159	3466	82	2032	
57B2-70A	159	3466	82	2032	
57B2-71	159	3466	82	2032	
57B2-71A	159	3466	82	2032	
57B2-72	102A		53	2007	
57B2-73	123A	3444	20	2051	
57B2-75	160		245	2004	
57B2-77	160		245	2004	
57B2-78	102A		53	2007	
57B2-79		3122	20	2051	
57B2-8	103A	3010	59	2002	
57B2-80	160		245	2004	
57B2-83	102A		53	2007	
57B2-84	175	3131A/369	246	2020	
57B2-85	123A	3444	20	2051	
57B2-87	123A	3444	20	2051	
57B2-88	102A		53	2007	
57B2-89	160		245	2004	
57B2-9	160		245	2004	
57B2-90	160		245	2004	
57B2-93	160		245	2004	
57B2-97	123A	3444	20	2051	
57B200-12	123A	3444	20	2051	
57B201-14	159	3114/290	82	2032	
57B202-13	123A	3444	20	2051	
57B203-14	199	3245	62	2010	
57B205-14	292	3054/196	66	2048	
57B206-14	153	3083/197	69	2049	
57B207-8	154	3044	40	2012	
57B208-8	171		27		
57B21	108	3452	86	2038	
57B21-1	108		86	2038	
57B21-12	108	3452	86	2038	
57B21-13	108	3452	86	2038	
57B21-14	108	3452	86	2038	
57B21-15	108	3452	86	2038	
57B21-16	108	3452	86	2038	
57B21-17	161	3039/316	39	2015	
57B21-18	108	3452	86	2038	
57B21-2	108	3452	86	2038	
57B21-3	108	3452	86	2038	
57B21-4	108	3452	86	2038	
57B21-5	108	3452	86	2038	
57B21-6	161	3018	39	2015	
57B21-7	161	3018	39	2015	
57B21-9	107	3018	11	2015	
57B211-8	190	3104A	217		
57B213-11	165	3027/130	38		
57B216-12	159	3466	82	2032	
57B219-14	289A		268	2038	
57B220-14	290A		269	2032	
57B235-12	159	3466	82	2032	
57B236-11	171	3103A/396	27		
57B240-14	290A		269	2032	
57B241-14	289A		268	2038	
57B243-10	5404	3627			
57B244-14	196	3054	241	2020	
57B245-14	197		250	2027	
57B249-4	161	3132	39	2015	
57B250-14	152	3054/196	66	2048	
57B251-14	197	3083	69	2049	
57B253-14	123A	3444	20	2051	
57B256-10	94		14	2041	
57B258-8	159	3466	82	2032	
57B263-11	238	3710	259		
57B27-2			20	2051	

Industry Standard No.	ECG	SK	GE	RS 276-	MOTOR.
57B273-14	263	3935/270			
57B274-14	262	3936/271			
57B277-14	153		69	2049	
57B278-14	152		66	2048	
57B279-14	152		66	2048	
57B280-14	289A		268	2038	
57B281-14	290A		269	2032	
57B282-12	123A		20	2051	
57B283-11	154		40	2012	
57B285-10	5645	3659/56006			
57B286-10			66	2048	
57B3-1		3122	20,82		
57B3-10	104MP	3720	16(2)	2006(2)	
57B3-11	104MP	3720	16(2)	2006(2)	
57B3-12	121MP	3718	239(2)	2006(2)	
57B3-13		3122	20,82		
57B3-2		3122	20,82		
57B3-4	102A		53	2007	
57B3-5	102A		53	2007	
57B3-6	102A		53	2007	
57B3-7	104MP	3720	16(2)	2006(2)	
57B3-8	104MP	3720	16(2)	2006(2)	
57B3-9	104MP	3720	16(2)	2006(2)	
57B30-12			11	2015	
57B4-1	121	3009	239	2006	
57B4-2	121	3009	239	2006	
57B4-4	121	3009	239	2006	
057B474H	161	3132	39	2015	
57B5-6			62	2010	
57B5-7			62	2010	
57B5-8			62	2010	
57B6	102	3004	2	2007	
57B6-11			20	2051	
57B6-12			16	2006	
57B6-19			20	2051	
57B6-4			20	2051	
57B6-9			20	2051	
57C10-1	108	3452	86	2038	
57C10-2	108	3452	86	2038	
57C104-8			39	2015	
57C105-12			20	2051	
57C109-9	128	3024	243	2030	
57C11-1	123A	3444	20	2051	
57C110-9	129	3025	244	2027	
57C12-1	154	3045/225	40	2012	
57C12-2	154	3044	40	2012	
57C12-3	175		246	2020	
57C12-4	128	3024	243	2030	
57C12-5	175	3131A/369	246	2020	
57C121-9	123A	3444	20	2051	
57C122-9	129	3025	244	2027	
57C14-1	128	3024	243	2030	
57C14-2	128	3024	243	2030	
57C14-3	128	3024	243	2030	
57C142-4	161	3117	39	2015	
57C148-12	129	3025	244	2027	
57C148-12A	129	3025	244	2027	
57C15-1	123A	3444	20	2051	
57C15-2	123A	3444	20	2051	
57C15-3	123A	3444	20	2051	
57C15-4	123A	3444	20	2051	
57C15-5	159	3466	82	2032	
57C15-50	159	3466	82	2032	
57C156-9	123A	3444	20	2051	
57C157-9	159	3466	82	2032	
57C157-90	159	3466	82	2032	
57C16	126	3006/160	52	2024	
57C16-1	123A	3444	20	2051	
57C164-4	161	3716	39	2015	
57C19-1	159	3466	82	2032	
57C19-1A	159	3466	82	2032	
57C20-1	108	3452	86	2038	
57C21	107		11	2015	
57C21-5	107		11	2015	
57C22-1	179	3642	76		
57C22-2	179	3642	76		
57C23	129		244	2027	
57C23-1	129	3025	244	2027	
57C23-2	129	3025	244	2027	
57C23-3	129	3025	244	2027	
57C24-1	123A	3444	20	2051	
57C24-2	123A	3444	20	2051	
57C24-3	123A	3444	20	2051	
57C24-4	123A	3444	20	2051	
57C27-1	123A	3444	20	2051	
57C27-2	108	3998/506	86	2038	
57C28	704	3023	IC-205		
57C29	706	3101	IC-43		
57C29-1	706	3101	IC-43		
57C29-2	706	3101	IC-43		
57C5	103	3862	8	2002	
57C5-1	126	3006/160	52	2024	
57C5-10	126	3007	52	2024	
57C5-2		3006	51	2004	
57C5-3	102A	3006/160	53	2007	
57C5-4	126	3006/160	52	2024	
57C5-5	102A	3008	53	2007	
57C5-6	108	3452	86	2038	
57C5-7	108	3452	86	2038	
57C5-8	108	3452	86	2038	
57C5-9	126	3007	52	2024	
57C6	199		62	2010	
57C6-1	102	3004	2	2007	
57C6-10	124	3021	12		
57C6-11	123A	3444	20	2051	
57C6-12	121	3009	239	2006	
57C6-14	124	3021	12		
57C6-15	129	3025	244	2027	
57C6-16	126	3006/160	52	2024	
57C6-17	123	3124/289	20	2051	
57C6-18		3122	20,82		
57C6-19	123	3124/289	20	2051	
57C6-2	121	3009	239	2006	
57C6-20	103A	3835	59	2002	
57C6-21	103A	3010	59	2002	
57C6-22	102	3004	2	2007	
57C6-23	121	3009	239	2006	
57C6-24	124	3021	12		
57C6-25	102A	3004	53	2007	
57C6-26	129	3025	244	2027	
57C6-26A	129	3025	244	2027	
57C6-27	123	3124/289	20	2051	
57C6-28		3122	20,82		
57C6-29	123	3124/289	20	2051	
57C6-3	121	3009	239	2006	
57C6-30	123	3124/289	20	2051	
57C6-31	129	3025	244	2027	
57C6-31A	129	3025	244	2027	
57C6-32	123	3124/289	20	2051	
57C6-33	129	3025	244	2027	
57C6-33A	129	3025	244	2027	
57C6-4	123A	3444	20	2051	
57C6-5	103A	3010	59	2002	
57C6-6	103	3862	8	2002	
57C6-6A	100	3005	1	2007	
57C6-6B	100	3005	1	2007	
57C6-6C	100	3005	1	2007	
57C6-7	123	3124/289	20	2051	
57C6-8	121	3009	239	2006	
57C6-9	123A	3444	20	2051	
57C68	102A		53	2007	
57C7-1	108	3452	86	2038	
57C7-10	123A	3444	20	2051	
57C7-15	123A	3444	20	2051	
57C7-17	123A	3444	20	2051	
57C7-18	123A	3444	20	2051	
57C7-2	108	3452	86	2038	
57C7-20	123A	3444	20	2051	
57C7-3	108	3452	86	2038	
57C7-4	108	3452	86	2038	
57C7-5	108	3452	86	2038	
57C7-6	108	3452	86	2038	
57C7-7	108	3452	86	2038	
57C7-8	128	3452/108	243	2030	
57C7-9	123A	3444	20	2051	
57C9-2	121	3009	239	2006	
57D1-1	109	3087		1123	
57D1-10	160		245	2004	
57D1-100	160		245	2004	
57D1-101	160		245	2004	
57D1-102	160		245	2004	
57D1-103	160		245	2004	
57D1-104	102A		53	2007	
57D1-105	102A	3004	53	2007	
57D1-106	102A		53	2007	
57D1-107	160	3006	245	2004	
57D1-108	160		245	2004	
57D1-109	160		245	2004	
57D1-11	102	3004	2	2007	
57D1-110	160		245	2004	
57D1-111	102A	3004	53	2007	
57D1-112	160		245	2004	
57D1-113	160		245	2004	
57D1-114	160	3006	245	2004	
57D1-115	160		245	2004	
57D1-116	102A		53	2007	
57D1-117	102A		53	2007	
57D1-118	102A		53	2007	
57D1-119	121	3009	239	2006	
57D1-12	160		245	2004	
57D1-120	160	3004/102A	245	2004	
57D1-121	102A	3004	53	2007	
57D1-122	154	3045/225	40	2012	
57D1-123	123A	3444	20	2051	
57D1-124	123A	3444	20	2051	
57D1-13	160		245	2004	
57D1-14	102A		53	2007	
57D1-15	160		245	2004	
57D1-16	160		245	2004	
57D1-17	158		53	2007	
57D1-18	158		53	2007	
57D1-19	158		53	2007	
57D1-2	109	3087		1123	
57D1-20	158		53	2007	
57D1-21	158		53	2007	
57D1-22	160		245	2004	
57D1-23	102A		53	2007	
57D1-24	160		245	2004	
57D1-25	160		245	2004	
57D1-26	102A	3003	53	2007	
57D1-27	102A	3003	53	2007	
57D1-28	102A		53	2007	
57D1-3	103A	3835	59	2002	
57D1-30	160		245	2004	
57D1-31	160		245	2004	
57D1-32	160		245	2004	
57D1-33	160		245	2004	
57D1-34	102A		53	2007	
57D1-35	160		245	2004	
57D1-36	160		245	2004	
57D1-37	160		245	2004	
57D1-38	158		53	2007	
57D1-39			51	2004	
57D1-4	103A	3835	59	2002	
57D1-40	102A		53	2007	
57D1-41	160		245	2004	
57D1-42	158	3003	53	2007	
57D1-43	102A	3004	53	2007	
57D1-44	102A	3004	53	2007	
57D1-45	160		245	2004	
57D1-46	160		245	2004	
57D1-47	160		245	2004	
57D1-48	160		245	2004	
57D1-49	160		245	2004	
57D1-5	103A	3835	59	2002	
57D1-50	160		245	2004	
57D1-51	123A	3444	20	2051	
57D1-52	106	3118	21	2034	
57D1-53	102A		53	2007	
57D1-54	109	3087		1123	
57D1-55		3122	20,82		
57D1-56	158	3004/102A	53	2007	
57D1-57	160		245	2004	
57D1-58	102A		53	2007	
57D1-59	102A		53	2007	
57D1-6	103A	3835	59	2002	
57D1-60	158		53	2007	
57D1-61	160		245	2004	
57D1-62	109	3087		1123	
57D1-63	117		504A	1104	
57D1-64	109			1123	
57D1-65	112	3089	1N82A		
57D1-66	102A		53	2007	
57D1-67	160		245	2004	
57D1-68	102A		53	2007	
57D1-69	160		245	2004	
57D1-7	158		53	2007	
57D1-70	102A	3004	53	2007	
57D1-71	102A		53	2007	
57D1-72	160		245	2004	
57D1-73	160		245	2004	
57D1-74	160		245	2004	
57D1-75	123A	3444	20	2051	
57D1-76	159	3466	82	2032	
57D1-76A	159	3466	82	2032	
57D1-77	158		53	2007	
57D1-78	103A	3835	59	2002	
57D1-79	102A		53	2007	
57D1-8	102A		53	2007	
57D1-80	160	3005	245	2004	
57D1-81	160	3008	245	2004	
57D1-82	102A		53	2007	
57D1-83	102A		53	2007	
57D1-84	160	3005	245	2004	
57D1-85	160	3008	245	2004	
57D1-86	160	3008	245	2004	
57D1-87	160	3008	245	2004	
57D1-88		3007	51	2004	

Industry Standard No.	ECG	SK	GE	RS 276-	MOTOR.
57D1-89	160	3007	245	2004	
57D1-9	160		245	2004	
57D1-90	102A		53	2007	
57D1-91	102A		53	2007	
57D1-92	102A		53	2007	
57D1-93	102A		53	2007	
57D1-94	160		245	2004	
57D1-95	102A		53	2007	
57D1-96	160	3006	245	2004	
57D1-97	102A	3008	53	2007	
57D1-98	160		245	2004	
57D1-99	160		245	2004	
57D107-8	108	3452	86	2038	
57D1127	102	3004	2	2007	
57D1130	160		245	2004	
57D1131	160		245	2004	
57D1132	160		245	2004	
57D1143	102	3004	2	2007	
57D1186	160		245	2004	
57D126	102	3004	2	2007	
57D127	157	3004/102A	232		
57D130	126	3007	52	2024	
57D131	126	3007	52	2024	
57D132	126	3007	52	2024	
57D132-9	160		245	2004	
57D136-12	123A	3444	20	2051	
57D14-1	123A	3444	20	2051	
57D14-2	123A	3444	20	2051	
57D14-3	123A	3444	20	2051	
57D143	102	3004	2	2007	
57D156	102	3004	2	2007	
57D168	126	3006/160	52	2024	
57D169	102	3004	2	2007	
57D170	102	3004	2	2007	
57D180	100	3005	1	2007	
57D184	100	3005	1	2007	
57D186	126	3007	52	2024	
57D187	126	3006/160	52	2024	
57D188		3004	52	2024	
57D189	102	3004	2	2007	
57D19	159	3466	82	2032	
57D19-1	159	3466	82	2032	
57D19-10	159	3466	82	2032	
57D19-2	159	3466	82	2032	
57D19-20	159	3466	82	2032	
57D19-3	159	3466	82	2032	
57D19-30	159	3466	82	2032	
57D24	161	3716	39	2015	
57D24-1	108	3452	86	2038	
57D24-2	108	3452	86	2038	
57D24-3	108	3452	86	2038	
57D29-2	706		IC-43		
57D3-6	102	3004	2	2007	
57D4-1	121	3009	239	2006	
57D4-2	121	3009	239	2006	
57D5-1	126	3006/160	52	2024	
57D5-2	126	3006/160	52	2024	
57D5-4	126	3006/160	52	2024	
57D6-10	124	3021	12		
57D6-12	121	3009	239	2006	
57D6-19	123	3124/289	20	2051	
57D6-4	123A	3444	20	2051	
57D68	102	3004	2	2007	
57D9-1	160	3009	245	2004	
57D9-2	121	3009	239	2006	
57DG-23	121	3014	239	2006	
57DG-32	121	3717	239	2006	
57L1-1	160		245	2004	
57L1-10	160		245	2004	
57L1-11	160		245	2004	
57L1-12	160		245	2004	
57L1-2	160		245	2004	
57L1-3	160		245	2004	
57L1-4	160		245	2004	
57L1-5			53	2007	
57L1-6	102A		53	2007	
57L1-7	158(2)		53(2)		
57L1-8	102A		53	2007	
57L1-9	160		245	2004	
57L103-13	717		IC-209		
57L105-12	128		243	2030	
57L106-9	312			2028	
57L2-1	128		243	2030	
57L2-2	123A	3444	20	2051	
57L3-1	123A	3444	20	2051	
57L3-4	123A	3444	20	2051	
57L5-1	104	3009	16	2006	
57M1-1	160		245	2004	
57M1-10	160		245	2004	
57M1-11	160		245	2004	
57M1-12	160		245	2004	
57M1-13	129	3025	244	2027	
57M1-13A	129	3025	244	2027	
57M1-14	123A	3444	20	2051	
57M1-15	123A	3444	20	2051	
57M1-16	103A	3835	59	2002	
57M1-17	160		245	2004	
57M1-18	103A	3835	59	2002	
57M1-19	123A	3444	20	2051	
57M1-2	160		245	2004	
57M1-20	123A	3444	20	2051	
57M1-21	129	3025	244	2027	
57M1-21A	129	3025	244	2027	
57M1-22	129	3025	244	2027	
57M1-23	123A	3444	20	2051	
57M1-24	123A	3444	20	2051	
57M1-25	129	3025	244	2027	
57M1-25A	129	3025	244	2027	
57M1-26	123A	3444	20	2051	
57M1-27	123A	3444	20	2051	
57M1-28	123A	3444	20	2051	
57M1-29	123A	3444	20	2051	
57M1-3	160		245	2004	
57M1-30	123A	3444	20	2051	
57M1-31	123A	3444	20	2051	
57M1-32	123A	3444	20	2051	
57M1-33	128	3024	243	2030	
57M1-34	128	3024	243	2030	
57M1-35	158		53	2007	
57M1-4	160		245	2004	
57M1-5	102A		53	2007	
57M1-6	102A		53	2007	
57M1-7	158(2)		53(2)		
57M1-8	158		53	2007	
57M1-9	160		245	2004	
57M2-1	103A	3835	59	2002	
57M2-10	129	3025	244	2027	
57M2-10A	129	3025	244	2027	
57M2-11	128	3024	243	2030	
57M2-14	128	3024	243	2030	
57M2-15	129	3025	244	2027	
57M2-15A	129	3025	244	2027	
57M2-16	154	3045/225	40	2012	
57M2-17	154	3045/225	40	2012	
57M2-18	128	3024	243	2030	
57M2-2	103A	3835	59	2002	
57M2-3	158		53	2007	
57M2-4	158		53	2007	
57M2-6	103A	3835	59	2002	
57M2-7	128	3024	243	2030	
57M2-8	158		53	2007	
57M2-9	103A	3835	59	2002	
57M3-1	130	3027	14	2041	
57M3-10N	155	3839	43		
57M3-10P	131		44	2006	
57M3-11	155	3839	43		
57M3-12	131		44	2006	
57M3-1A	130	3027	14	2041	
57M3-2	130		14	2041	
57M3-3	128	3024	243	2030	
57M3-4	130	3027	14	2041	
57M3-4A	130	3027	14	2041	
57M3-5	175		246	2020	
57M3-6	130	3027	14	2041	
57M3-6A	130	3027	14	2041	
57M3-7	104	3009	16	2006	
57M3-8	104	3009	16	2006	
57M3-9N	155	3839	43		
57M3-9P	131	3198	44	2006	
57P1	102	3004	2	2007	
57S-06	5070A	9021		561	
57X11	505	3108	CR-7		
57X14			504A	1104	
58-000516	724		IC-86		
58-1		3124	61	2038	
58-1(TRUETONE)	123A	3444	20	2051	
58B2-14	160	3006	245	2004	
58T	457	3116			
059	124	3021	12		
59E001-1	177	3100/519	300	1122	
59P2C	102A		53	2007	
59Z6	5120A	3386			
59Z6A	5120A	3386	5ZD-6.8		
60-211040	181		75	2041	
60-3			504A	1104	
60B					MR1126
60BR					MR1366
60C	116	3016	504A	1104	1N4005
60CR					IN4937
60D	116	3017B/117	504A	1104	
60DE10	125	3033	510,531	1114	
60F1	5842	7042	5008		
60F5	5862		5024		
60FB1L					MDA801
60FB2L			BR-206		MDA802
60FB3L					MDA804
60FB4L					MDA804
60FB5L					MDA806
60FB6L					MDA806
60H	116	3017B/117	504A	1104	
60H3P	5862	3017B/117	5024		MR1126
60H3PN	5863		5025		
60HF10	6022		5128		1N1184A
60HF100	6044		5141		
60HF15	6026	7226	5128		
60HF20	6026	7226	5128		1N1186A
60HF25	6030		5132		
60HF30	6030		5132		1N1187A
60HF35	6034	7234	5132		
60HF40	6034	7234	5132		1N1188A
60HF45	6038		5136		
60HF5	6020	7220	5128		
60HF50	6038		5136		1N1189A
60HF60	6040	7240	5136		1N1190A
60HR3P	5822				MR1366
60J1	5806	3848			
60J2	116	3017B/117	504A	1104	
60J2P	5882		5040		
60J3P	5882		5040		
60JH3	6064	7264	5136		
60L20	525	3925			
60LA	156	3051	512		
60M	116	3017B/117	504A	1104	
60P19009	162	3438	35		
60P22204	191	3232	249		
60S5	5842	7042	5008		
60T4	5400	3950			
61-1053-1	196		241	2020	
61-1130	102A		53	2007	
61-1131	102A		53	2007	
61-1215	102A		53	2007	
61-1320	116	3017B/117	504A	1104	
61-1400	123A	3444	20	2051	
61-1401	123A	3444	20	2051	
61-1402	123A	3444	20	2051	
61-1403	123A	3444	20	2051	
61-1404	123A	3444	20	2051	
61-1763	123A	3444	20	2051	
61-1764	128		243	2030	
61-1765	116		504A	1104	
61-1906	131	3198	44	2006	
61-1907	102A		53	2007	
61-1934	102A		53	2007	
61-1935	102A		53	2007	
61-259	109			1123	
61-260039	126	3008	52	2024	
61-260039A	126	3008	52	2024	
61-309-458	128		243	2030	
61-309449	130	3511	14	2041	
61-3096-90	197		250	2027	
61-309686	199	3245	62	2010	
61-309687	128		243	2030	
61-309688	129		244	2027	
61-309689	152	3054/196	66	2048	
61-309690	153	3083/197	69	2049	
61-59395	109	3087		1123	
61-607	102A		53	2007	
61-608	102A		53	2007	
61-654	102A		53	2007	
61-655	102A		53	2007	
61-656	102A		53	2007	
61-746	123A	3444	20	2051	
61-747	128	3024	243	2030	
61-751	123A	3444	20	2051	
61-754	123A	3444	20	2051	
61-755	123A	3444	20	2051	
61-756	117		504A	1104	
61-7728	116	3016	504A	1104	
61-782	121	3009	239	2006	
61-813	128	3024	243	2030	
61-814	123A	3444	20	2051	
61-815	123A	3444	20	2051	
61-820	117		504A	1104	
61-8968	119	3109	CR-2		
61-8969	118	3066	CR-1		
61-8969M		3066	CR-1		
61-926	117		504A	1104	
61-928	102A		53	2007	

Industry Standard No.	ECG	SK	GE	RS 276-	MOTOR.
61-929	102A		53	2007	
61A001-10	1082	3461	IC-140		
61A001-11	1142	3485	IC-128		
61A001-12	1052	3249	IC-135		
61A023-2	1045		IC-2		
61A030-6	712	3072	IC-2		
61A030-9	1083		IC-275		
61B0015-1	102A	3123	53	2007	
61B002-1	160		245	2004	
61B003-1	126	3006/160	52	2024	
61B004-1	102A	3004	53	2007	
61B005-1	102A	3004	53	2007	
61B006-1	102A	3004	53	2007	
61B007-1	108		86	2038	
61B007-2	108	3039/316	86	2038	
61B009-1	102A	3004	53	2007	
61B015-1	102A	3004	53	2007	
61B016-1	102A	3123	53	2007	
61B017-1	102A	3004	53	2007	
61B018-1	102A	3123	53	2007	
61B019-1	102A	3004	53	2007	
61B020-1	102A	3004	53	2007	
61B021-1	102A	3004	53	2007	
61B022-2	102A	3004	53	2007	
61B022-3	102A	3004	53	2007	
61B023-1	102A	3004	53	2007	
61B026-1	102A	3005	53	2007	
61B027-1	158	3004/102A	53	2007	
61B042-9	1003		IC-43		
61B1C	783	3215			
61B2Z	783	3215	IC-225		
61B45-14	102A		53	2007	
61C001-1	123	3124/289	20	2051	
61C001-2		3743	IC-281		
61C001-3T	712	3072			
61C001-4	1263	3920	IC-293		
61C002-1	102	3004	2	2007	
61C003-1	102	3004	2	2007	
61C004-1	128	3024	243	2030	
61C005-1	121	3009	239	2006	
61C66-1	HIDIV-1	3868/DIV-1			
61C66-10	HIDIV-1	3868/DIV-1			
61C66-2	HIDIV-1	3868/DIV-1			
61C66-3	HIDIV-1	3868/DIV-1			
61C66-4	HIDIV-1	3868/DIV-1			
61C66-5	HIDIV-1	3868/DIV-1			
61C66-6	HIDIV-1	3868/DIV-1			
61C66-7	HIDIV-1	3868/DIV-1			
61C66-8	HIDIV-1	3868/DIV-1			
61C66-9	HIDIV-1	3868/DIV-1			
61E01	5075A	3751	ZD-16	564	
61E02	5022A	3788	ZD-13	563	
61E03	5066A	3330	ZD-3.3		
61E04	5093A	3344	ZD-75		
61E05	5070A		ZD-6.2	561	
61E06	5096A	3346	ZD-100		
61E07	5075A	3751	ZD-16	564	
61E08	147A		ZD-33		
61E09	5090A	3342	ZD-56		
61E10	5010A	3776	ZD-5.1		
61E11	5082A	3753	ZD-25		
61E12	146A		ZD-27		
61E13	5068A	3332	ZD-4.3		
61E14	5096A	3346	ZD-100		
61E15	5018A	3784	ZD-9.1	562	
61E16	5095A	3345	ZD-91		
61E17	5001A	3767			
61E18	5027A	3793	ZD-18		
61E19	5067A	3331	ZD-3.9		
61E20	146A		ZD-27		
61E21	5129A	3395	5ZD-14		
61E22	5029A	3795	ZD-20		
61E23	5010A	3776	ZD-5.1		
61E24	5028A	3794	ZD-19		
61E25	5014A	3780	ZD-6.8	561	
61E26	5031A	3797	ZD-24		
61E27	5012A	3778	ZD-6.0	561	
61E28	5141A	3407	5ZD-30		
61E30	148A		ZD-55		
61E31	5022A	3788	ZD-13	563	
61E32	5016A	3782	ZD-8.2	562	
61E33	5027A	3793	ZD-18		
61E34	5117A	3383			
61E35	5137A	3403	5ZD-24		
61E36	137A		ZD-6.2	561	
61E37	5077A	3752	ZD-18		
61E38	136A		ZD-5.6	561	
61E39	5135A	3401			
61E40	5010A	3776	ZD-5.1		
61E41	135A		ZD-5.1		
61E42	5071A	3334	ZD-6.8	561	
61E43	5006A	3772	ZD-3.6		
61E44	5115A	3381			
61J001-1	123A	3444	20	2051	
61J002-1	123A	3444	20	2051	
61J003-1	123A	3444	20	2051	
61J004-1	102A	3004	53	2007	
61J2	116	3017B/117	504A	1104	
61K001-10	801	3160	IC-35		
61K001-12	1011		IC-79		
61K001-13	1185	3468	IC-139		
61K001-9	1011		IC-79		
61L001-3	1213	3704	IC-150		
61P1	126	3006/160	52	2024	
61P10	126	3006/160	52	2024	
61P1D	160		245	2004	
61T4	5401	3638/5402			
61Z6	5122A	3388			
61Z6A	5183A	3388/5122A	5ZD-8.2		
000062	129	3025	20	2027	
62-10234	109	3087		1123	
62-10655	109	3087		1123	
62-112524			1N60	1123	
62-113998			504A	1104	
62-114267			11	2015	
62-118825			509	1114	
62-12034	109	3087		1123	
62-125528			300	1122	
62-126321			300	1122	
62-126856			504A	1104	
62-128343			1	2007	
62-129604			11	2015	
62-130045			300	1122	
62-130046			300	1122	
62-130047			ZD-9.1	562	
62-130139			21	2034	
62-130761			ZD-15	564	
62-130762			ZD-15	564	
62-132497			21	2034	
62-13258	102A		53	2007	
62-13259	103A	3835	59	2002	
62-13261	116		504A	1104	
62-13477	116	3017B/117	504A	1104	
62-13494	102A		53	2007	

Industry Standard No.	ECG	SK	GE	RS 276-	MOTOR.
62-15318	109	3087		1123	
62-15483	116	3017B/117	504A	1104	
62-16013	113A		6GC1		
62-16711	116	3017B/117	504A	1104	
62-16712	113A	3119/113	6GC1		
62-16769	109	3087		1123	
62-16841	109	3087		1123	
62-16905	123A	3444	20	2051	
62-16918	102A		53	2007	
62-16919	175	3131A/369	246	2020	
62-17232	112	3089	1N82A		
62-17390	160		245	2004	
62-17391	160		245	2004	
62-17550	123A	3444	20	2051	
62-18135	116	3017B/117	504A	1104	
62-18337	113A	3119/113	6GC1		
62-18415	100	3005	1	2007	
62-18416	100	3005	1	2007	
62-18417	100	3005	1	2007	
62-18418	160		245	2004	
62-18419	160		245	2004	
62-18420	102A		53	2007	
62-18421	102A		53	2007	
62-18422	160		245	2004	
62-18423	100	3005	1	2007	
62-18424	100	3005	1	2007	
62-18425	123A	3444	20	2051	
62-18426	154		40	2012	
62-18427	104	3009	16	2006	
62-18428	121	3717	239	2006	
62-18429	179	3642	76		
62-18430	102A		53	2007	
62-18431	116	3017B/117	504A	1104	
62-18434	116	3017B/117	504A	1104	
62-18435	116	3017B/117	504A	1104	
62-18436	116	3017B/117	504A	1104	
62-18438	116	3017B/117	504A	1104	
62-18641	123A	3444	20	2051	
62-18642	123A	3444	20	2051	
62-18643	123A	3444	20	2051	
62-18782	177	3100/519	300	63-8555	
62-18828	123A	3444	20	2051	
62-19115	116	3017B/117	504A	1104	
62-19260	112	3089	1N82A		
62-19280	123A	3444	20	2051	
62-19452	159	3466	82	2032	
62-19516	123A	3444	20	2051	
62-19548	123A	3444	20	2051	
62-19581	107	3039/316	11	2015	
62-19620	112		1N82A		
62-19734	113A	3119/113	6GC1		
62-19749	116	3017B/117	504A	1104	
62-19814	116	3017B/117	504A	1104	
62-19837	123A	3444	20	2051	
62-19838	123A	3444	20	2051	
62-19846	109	3087		1123	
62-20154	159	3466	82	2032	
62-20154A	159	3466	82	2032	
62-20155	123A	3444	20	2051	
62-20223	177	3100/519	300	1122	
62-20240	123A	3444	20	2051	
62-20241	123A	3444	20	2051	
62-20242	123A	3444	20	2051	
62-20243	123A	3444	20	2051	
62-20244	159	3466	82	2032	
62-20244A	159	3466	82	2032	
62-20319	177	3100/519	300	1122	
62-20360	123A	3444	20	2051	
62-20437	177	3100/519	300	1122	
62-20565	119		CR-2		
62-20597	177	3100/519	300	1122	
62-20643	145A	3063	ZD-15	564	
62-21369	116		504A	1104	
62-21496	123A	3444	20	2051	
62-21552	177		300	1122	
62-21573	5103A	3353	ZD-180		
62-21574	5077A	3752	ZD-18		
62-21586	711		IC-207		
62-21587	712		IC-2		
62-21683			21	2034	
62-22038	123A	3444	20	2051	
62-22039	123A	3444	20	2051	
62-22250	123A	3444	20	2051	
62-22251	123A	3444	20	2051	
62-22524			21	2034	
62-22529	106	3984	21	2034	
62-26597	177	3100/519	300	1122	
62-26851	160		245	2004	
62-3597-1			21	2034	
62-3597-2			21	2034	
62-7567	123A	3444	20	2051	
62-8555			20	2051	
62-8781	160		245	2004	
62A01	116	3017B/117	504A	1104	
62A02	116	3031A	504A	1104	
62A04	116	3017B/117	504A	1104	
62A05	116	3017B/117	504A	1104	
62A11868	123A	3444	20	2051	
62A11871	159	3466	82	2032	
62B046-1	152	3054/196	66	2048	
62B046-2	152	3054/196	66	2048	
62B046-3	152	3893	66	2048	
62B046-4	152	3054/196	66	2048	
62J2	116	3016	504A	1104	
62R2R	5944		5048		
62T4	5402	3638			
62Z6	5124A	3390			
62Z6A	5124A	3390	5ZD-9.1		
63-10035	160		245	2004	
63-10036	160		245	2004	
63-10037	102A		53	2007	
63-10038	102A		53	2007	
63-10062	175		246	2020	
63-10064	116	3017B/117	504A	1104	
63-10145	160		245	2004	
63-10146	160		245	2004	
63-10147	102A		53	2007	
63-10148	160		245	2004	
63-10149	160		245	2004	
63-10150	160		245	2004	
63-10151	102A		53	2007	
63-10152	102A		53	2007	
63-10153	102A		53	2007	
63-10154	102A		53	2007	
63-10156	102A		53	2007	
63-10158	102A		53	2007	
63-10159	102A		53	2007	
63-10188	123A	3122	20	2051	
63-10195	160		245	2004	
63-10196	160		245	2004	
63-10200	100	3005	1	2007	
63-10375	160		245	2004	
63-10376	160		245	2004	
63-10377	123A	3444	20	2051	

Industry Standard No.	ECG	SK	GE	RS 276-	MOTOR.
63-10378	121	3009	239	2006	
63-10383	103A	3835	59	2002	
63-10384	158		53	2007	
63-10408	102A		53	2007	
63-10708	123A	3444	20	2051	
63-10709	116	3017B/117	504A	1104	
63-10725	123A	3444	20	2051	
63-10732	123A	3444	20	2051	
63-10733	123A	3444	20	2051	
63-10734	123A	3444	20	2051	
63-10735	123A	3444	20	2051	
63-10736	123A	3444	20	2051	
63-10737	123A	3444	20	2051	
63-10739	109	3087		1123	
63-10860	123A	3444	20	2051	
63-11025	123A	3444	20	2051	
63-11055	160		245	2004	
63-11073	158		53	2007	
63-11074	109	3087		1123	
63-11143	123A	3444	20	2051	
63-11144	102A		53	2007	
63-11147	5071A		ZD-6.8	561	
63-11148	167	3647	504A	1172	
63-11215	117		504A	1104	
63-11289	123A	3444	20	2051	
63-11290	175		246	2020	
63-11291	116	3017B/117	504A	1104	
63-11468	123A	3444	20	2051	
63-11469	123A	3444	20	2051	
63-11470	123A	3444	20	2051	
63-11471	123A	3444	20	2051	
63-11472	123A	3444	20	2051	
63-11474	102A		53	2007	
63-11496	160		245	2004	
63-11497	158		53	2007	
63-11582	160		245	2004	
63-11584	160		245	2004	
63-11585	100	3721	1	2007	
63-11586	102A		53	2007	
63-11659	5071A		ZD-6.8	561	
63-11660	123A	3444	20	2051	
63-11661	102A		53	2007	
63-11757	123A	3444	20	2051	
63-11758	123A	3444	20	2051	
63-11759	123A	3444	20	2051	
63-11762	125	3033	510,531	1114	
63-11825		3122	20	2051	
63-11831	123A	3444	20	2051	
63-11832	123A	3444	20	2051	
63-11833	123A	3444	20	2051	
63-11878	175	3131A/369	246	2020	
63-11879	109	3087		1123	
63-11881	116	3017B/117	504A	1104	
63-11916	123A	3444	20	2051	
63-11934	123A	3444	20	2051	
63-11935	123A	3444	20	2051	
63-11936	128		243	2030	
63-11937	123A	3444	20	2051	
63-11938	175	3131A/369	246	2020	
63-11957	116	3017B/117	504A	1104	
63-11989	128	3024	243	2030	
63-11990	185		58	2025	
63-11991	130	3027	14	2041	
63-11991A	130	3027	14	2041	
63-12003	123A	3444	20	2051	
63-12004	123A	3444	20	2051	
63-12062	123A	3444	20	2051	
63-12077	117		504A	1104	
63-12110	5071A	3334	ZD-6.8	561	
63-12154	159	3466	82	2032	
63-12154A	159	3466	82	2032	
63-12156	159	3466	82	2032	
63-12156A	159	3466	82	2032	
63-12157	159	3466	82	2032	
63-12157A	159	3466	82	2032	
63-12158	109	3087		1123	
63-12272	123A	3444	20	2051	
63-12273	162	3438	35		
63-12287	167	3647	504A	1172	
63-12316	102A		53	2007	
63-12317	102A		53	2007	
63-12366	125		510,531	1114	
63-12605	123A	3444	20	2051	
63-12607	109	3087		1123	
63-12608	123A	3444	20	2051	
63-12609	123A	3444	20	2051	
63-12610	160		245	2004	
63-12641	123A	3444	20	2051	
63-12642	123A	3444	20	2051	
63-12645	109	3087		1123	
63-12669	158		53	2007	
63-12670	158		53	2007	
63-12696	123A	3444	20	2051	
63-12697	123A	3444	20	2051	
63-12698	158		53	2007	
63-12706	123A	3444	20	2051	
63-12707	123A	3444	20	2051	
63-12750	123A	3444	20	2051	
63-12751	123A	3444	20	2051	
63-12752	123A	3444	20	2051	
63-12753	123A	3444	20	2051	
63-12754	109	3087		1123	
63-12755	109	3087		1123	
63-12756	109	3087		1123	
63-12757	109	3087		1123	
63-12874	123A	3444	20	2051	
63-12875	123A	3444	20	2051	
63-12876	102A		53	2007	
63-12877	123A	3444	20	2051	
63-12878	123A	3444	20	2051	
63-12879	123A	3444	20	2051	
63-12880	102A		53	2007	
63-12881	102A		53	2007	
63-12933	123A	3444	20	2051	
63-12940	123A	3444	20	2051	
63-12941	123A	3444	20	2051	
63-12942	123A	3444	20	2051	
63-12943	123A	3444	20	2051	
63-12944		3122	20	2051	
63-12945	158		53	2007	
63-12946	123A	3444	20	2051	
63-12947	158		53	2007	
63-12948	123A	3444	20	2051	
63-12949	123A	3444	20	2051	
63-12950	123A	3444	20	2051	
63-12951	123A	3444	20	2051	
63-12952	123A	3444	20	2051	
63-12953	123A	3444	20	2051	
63-12954	175		246	2020	
63-12989	128	3024	243	2030	
63-12990	128	3024	243	2030	
63-13025	160		245	2004	
63-13080	109	3087		1123	
63-13214	128	3024	243	2030	

Industry Standard No.	ECG	SK	GE	RS 276-	MOTOR.
63-13215	128	3024	243	2030	
63-13216	162	3438	35		
63-13322	159	3466	82	2032	
63-13322A	159	3466	82	2032	
63-13323	158		53	2007	
63-13419	123A	3444	20	2051	
63-13438	123A	3444	20	2051	
63-13440	123A	3444	20	2051	
63-13441	123A	3444	20	2051	
63-13839	160		245	2004	
63-13840	102A		53	2007	
63-13842	109			1123	
63-13864	123A	3444	20	2051	
63-13899	160		245	2004	
63-13903	116		504A	1104	
63-13919	116		504A	1104	
63-13926	312	3448	FET-1	2035	
63-13927	123A	3444	20	2051	
63-13954	162		35		
63-14032	123A	3444	20	2051	
63-14051	123A	3444	20	2051	
63-14052	123A	3444	20	2051	
63-14057	123A	3444	20	2051	
63-14135	727		IC-210		
63-14195	116		504A	1104	
63-15345	718	3159	IC-8		
63-15483	116	3017B/117	504A	1104	
63-16918	102A		53	2007	
63-17390	160		245	2004	
63-18135	117		504A	1104	
63-18416	100	3005	1	2007	
63-18418	160		245	2004	
63-18419	160		245	2004	
63-18420	102A		53	2007	
63-18421	102A		53	2007	
63-18423	160		245	2004	
63-18424	160		245	2004	
63-18426		3045	27		
63-18427	104	3009	16	2006	
63-18430	102A		53	2007	
63-18643	123A	3444	20	2051	
63-19173	173BP	3999	305		
63-19280	123A	3444	20	2051	
63-19282	123A	3444	20	2051	
63-22724	109	3087	1N295	1123	
63-23041	102A		53	2007	
63-250128-2	112	3089	1N82A		
63-25179	102A		53	2007	
63-25180	102A		53	2007	
63-25181	102A		53	2007	
63-25182	102A		53	2007	
63-25261	158		53	2007	
63-25281	102A		53	2007	
63-25282	102A		53	2007	
63-25720	102A		53	2007	
63-25726	160		245	2004	
63-25727	102A		53	2007	
63-25728	102A		53	2007	
63-25729	102A		53	2007	
63-25933	109	3087		1123	
63-25942	102A		53	2007	
63-25944	102A		53	2007	
63-25946	100	3005	1	2007	
63-26382	109	3087		1123	
63-26597	116	3017B/117	504A	1104	
63-26849	100	3005	1	2007	
63-26850	160		245	2004	
63-26851	102A		53	2007	
63-27278	100	3005	1	2007	
63-27279	100	3005	1	2007	
63-27280	100	3005	1	2007	
63-27281	102A		53	2007	
63-27366	160		245	2004	
63-27367	100	3005	1	2007	
63-27483	117		504A	1104	
63-27500	160		245	2004	
63-27622	116	3017B/117	504A	1104	
63-28250	109	3087		1123	
63-28348	160		245	2004	
63-28358	160		245	2004	
63-28390	102A		53	2007	
63-28399	102A		53	2007	
63-28426	154		40	2012	
63-28888	109	3087		1123	
63-29383	117		504A	1104	
63-29451	121	3009	239	2006	
63-29459	121	3009	239	2006	
63-29461	123A	3444	20	2051	
63-29661	160		245	2004	
63-29662	100	3005	1	2007	
63-29663	100	3005	1	2007	
63-29664	100	3005	1	2007	
63-29665	102A		53	2007	
63-29666	102A		53	2007	
63-29819	160		245	2004	
63-29820	160		245	2004	
63-29821	160		245	2004	
63-29862	160		245	2004	
63-29863	100	3005	1	2007	
63-3954	160		245	2004	
63-5058	5105A	3355	ZD-200		
63-7246	158		53	2007	
63-7247	102A		53	2007	
63-7248	103A		59	2002	
63-7396	102A		53	2007	
63-7397	102A		53	2007	
63-7398	102A		53	2007	
63-7399	102A		53	2007	
63-7420	102A		53	2007	
63-7421	123A	3444	20	2051	
63-7433	116	3017B/117	504A	1104	
63-7538	160		245	2004	
63-7541	160		245	2004	
63-7547	100	3005	1	2007	
63-7548	160		245	2004	
63-7549	103A	3835	59	2002	
63-7564	102A		53	2007	
63-7565	103A	3835	59	2002	
63-7567	123A	3444	20	2051	
63-7579	160		245	2004	
63-7580	160		245	2004	
63-7581	160		245	2004	
63-7582	160		245	2004	
63-7596	102A		53	2007	
63-7660	160		245	2004	
63-7670	123A	3444	20	2051	
63-7871	102A		53	2007	
63-7872	102A		53	2007	
63-7873	102A		53	2007	
63-8119	160		245	2004	
63-8120	158		53	2007	
63-8376	160		245	2004	
63-8377	160		245	2004	
63-8378	160		245	2004	

Industry Standard No.	ECG	SK	GE	RS 276-	MOTOR.
63-8379	160		245	2004	
63-8380	102A		53	2007	
63-8381	109	3087		1123	
63-8473	103A	3835	59	2002	
63-8512	175		246	2020	
63-8555	123A	3444	20	2051	
63-8590	121	3717	239	2006	
63-8677	HIDIV-4	3871/DIV-4			
63-8678	HIDIV-4	3871/DIV-4			
63-8679	HIDIV-4	3871/DIV-4			
63-8680	HIDIV-4	3871/DIV-4			
63-8685	116	3017B/117	504A	1104	
63-8699	160		245	2004	
63-8700	160		245	2004	
63-8701	123A	3444	20	2051	
63-8702	123A	3444	20	2051	
63-8703	102A		53	2007	
63-8704	102A		53	2007	
63-8705	103A	3835	59	2002	
63-8706	121	3009	239	2006	
63-8707	130	3027	14	2041	
63-8707A	130	3027	14	2041	
63-8819	117		504A	1104	
63-8824	116	3017B/117	504A	1104	
63-8825	5072A		ZD-8.2	562	
63-8945	175		246	2020	
63-8954	160		245	2004	
63-8955	109	3087		1123	
63-8970	HIDIV-4	3871/DIV-4			
63-9072	160		245	2004	
63-9337	123A	3444	20	2051	
63-9338	123A	3444	20	2051	
63-9339	123A	3444	20	2051	
63-9340	103A	3835	59	2002	
63-9341	123A	3444	20	2051	
63-9516	123A	3444	20	2051	
63-9517	160		245	2004	
63-9518	123A	3444	20	2051	
63-9519	102A		53	2007	
63-9520	102A		53	2007	
63-9521	102A		53	2007	
63-9523	109	3087		1123	
63-9659	102A		53	2007	
63-9664	160		245	2004	
63-9665	160		245	2004	
63-9783	5072A		ZD-8.2	562	
63-9787	117		504A	1104	
63-9829	123A	3444	20	2051	
63-9830	123A	3444	20	2051	
63-9831	123A	3444	20	2051	
63-9832	123A	3444	20	2051	
63-9833	123A	3444	20	2051	
63-9847	123A	3444	20	2051	
63-9876	160		245	2004	
63-9877	160		245	2004	
63-9893	HIDIV-1	3870/DIV-3	FR-8		
63-9894	HIDIV-1	3870/DIV-3	FR-8		
63-9895	HIDIV-1	3870/DIV-3	FR-8		
63-9896	HIDIV-1	3870/DIV-3	FR-8		
63-9897	HIDIV-3	3870/DIV-3	FR-8		
63-9898	HIDIV-3	3870/DIV-3	FR-8		
63-9898-02	HIDIV-3	3870/DIV-3	FR-8		
63-9941	160		245	2004	
63-9942	142A	3062	ZD-12	563	
63J2	116	3016	504A	1104	
63N1	163A	3439	36		
63N50	105	3012	4		
63P3	102	3004	2	2007	
64-1			1N34AS	1123	
64-8054-6			504A	1104	
64-J2	116		504A	1104	
64J2		3016	504A	1104	
64N1	162	3438	35		
64R2	5878		5036		
64T1	100	3005	1	2007	
065	175	3016	504A	2020	
65(DIODE)	116		504A	1104	
65(TRANSISTOR)	107		11	2015	
065-001	312	3448	FET-1	2035	
065-002	312	3448	FET-1	2035	
065-004	123A	3444	20	2051	
065-006	199	3122	62	2010	
065-007	186A	3202/210	247	2052	
065-008	192	3137	63	2030	
065-012	177	3100/519	300	1122	
065-013	109	3087		1123	
065-014	110MP	3709	1N60	1123(2)	
065-015	116	3017B/117	504A	1104	
065-016	156	3051	512		
65-080001	226MP		49(2)	2025(2)	
65-085002	109			1123	
65-085003	109			1123	
65-085004	142A	3062	ZD-12	563	
65-085010	109	3087		1123	
65-085012	109	3087		1123	
65-085013	116	3031A	504A	1104	
65-1	123A	3006/160	20	2051	
065-1-12	160	3006	245	2004	
065-1-12-7	160	3006	245	2004	
65-1-70	160	3006	245	2004	
65-1-70-12	160	3006	245	2004	
65-1-70-12-7	160	3006	245	2004	
65-10	160	3006	245	2004	
65-11	160	3006	245	2004	
65-12	160	3006	245	2004	
65-13	160	3006	245	2004	
65-14	160	3088	245	2004	
65-15	160	3006	245	2004	
65-16	160	3006	245	2004	
65-17	160	3006	245	2004	
65-18	160	3006	245	2004	
65-19	160	3006	245	2004	
65-1A	160	3006	245	2004	
65-1A0	160	3006		2004	
65-1A0R	160	3006		2004	
65-1A1	160	3006	245	2004	
65-1A19	160	3006	245	2004	
65-1A2	160	3006	245	2004	
65-1A21	160	3006	245	2004	
65-1A3	160	3006	245	2004	
65-1A3P	160	3006	245	2004	
65-1A4	160	3006	245	2004	
65-1A4-7	160	3006	245	2004	
65-1A4-7B	160		245	2004	
65-1A5	160	3006	245	2004	
65-1A5L	160	3006	245	2004	
65-1A6	160	3006	245	2004	
65-1A6-5	160	3006	245	2004	
65-1A7	160	3006	245	2004	
65-1A7-1	160	3006	245	2004	
65-1A8	160	3006	245	2004	
65-1A82	160	3006	245	2004	
65-1A9	160	3006	245	2004	
65-1A9G	160	3006	245	2004	
65-1A0			245	2004	
65-1A0R			245	2004	
065-2	103A	3010	59	2002	
065-2-12	103A	3010	59	2002	
65-2-70	103A	3010	59	2002	
65-2-70-12	103A	3010	59	2002	
65-2-70-12-7	103A	3010	59	2002	
65-20	103A	3010	59	2002	
65-21	103A	3010	59	2002	
065-210	236	3197/235	216	2053	
65-22	103A	3010	59	2002	
65-23	103A	3010	59	2002	
65-24	103A	3010	59	2002	
65-25	103A	3010	59	2002	
65-26	103A	3010	59	2002	
65-27	103A	3010	59	2002	
65-28	103A	3010	59	2002	
65-29	103A	3010	59	2002	
65-2A	103A	3010	59	2002	
65-2A0	103A	3010	59	2002	
65-2A0R	103A	3010	59	2002	
65-2A1	103A	3010	59	2002	
65-2A19	103A	3010	59	2002	
65-2A2	103A	3010	59	2002	
65-2A21	103A	3010	59	2002	
65-2A3	103A	3010	59	2002	
65-2A3P	103A	3010	59	2002	
65-2A4	103A	3010	59	2002	
65-2A4-7	103A	3010	59	2002	
65-2A4-7B	103A	3835	59	2002	
65-2A5	103A	3010	59	2002	
65-2A5L	103A	3010	59	2002	
65-2A6	103A	3010	59	2002	
65-2A6-1	103A	3010	59	2002	
65-2A7	103A	3010	59	2002	
65-2A7-1	103A	3010	59	2002	
65-2A8	103A	3010	59	2002	
65-2A82	103A	3010	59	2002	
65-2A9	103A	3010	59	2002	
65-2A9G	103A	3010	59	2002	
65-4	101	3861	8	2002	
65-4-70	101	3861	8	2002	
65-4-70-12	101	3861	8	2002	
65-4-70-12-7	101	3861	8	2002	
65-40	101	3861	8	2002	
65-41	101	3861	8	2002	
65-42	101	3861	8	2002	
65-43	101	3861	8	2002	
65-44	101	3861	8	2002	
65-45	101	3861	8	2002	
65-46	101	3861	8	2002	
65-47	101	3861	8	2002	
65-48	101	3861	8	2002	
65-49	101	3861	8	2002	
65-4A	101	3861	8	2002	
65-4A0	101	3011	8	2002	
65-4A0R	101	3011	8	2002	
65-4A1	101	3861	8	2002	
65-4A19	101	3861	8	2002	
65-4A2	101	3861	8	2002	
65-4A21	101	3861	8	2002	
65-4A3	101	3861	8	2002	
65-4A3P	101	3861	8	2002	
65-4A4	101	3861	8	2002	
65-4A4-7	101	3861	8	2002	
65-4A4-7B	101	3861	8	2002	
65-4A5	101	3861	8	2002	
65-4A5L	101	3861	8	2002	
65-4A6	101	3861	8	2002	
65-4A6-2	101	3011	8	2002	
65-4A7	101	3861	8	2002	
65-4A7-1	101	3861		2002	
65-4A8	101	3861	8	2002	
65-4A82	101	3861	8	2002	
65-4A9	101	3861	8	2002	
65-4A9G	101	3861	8	2002	
65-744238	114		6GD1	1104	
65-80001		3086	49	2025	
65-K11305-0001			514	1122	
65-L11305-0001			514	1122	
65-L11324-0001			514	1122	
65-P11305-0001	519		514	1122	
65-P11308-0001	177		300	1122	
65-P11311-0001	177		300	1122	
65-P11324-0001	519		514	1122	
65-P11325-0001	519		514	1122	
65A	159	3007	245	2032	
065A-12	160	3007	245	2004	
065A-12-7	160	3007	245	2004	
65A-70	160	3007	245	2004	
65A-70-12	160	3007	245	2004	
65A-70-12-7	160	3007	245	2004	
65A0	160	3007	245	2004	
65A1	160	3007	245	2004	
65A10	160	3007	245	2004	
65A10R	160	3007	245	2004	
65A11573	199		62	2010	
65A119	160	3007	245	2004	
65A12	160	3007	245	2004	
65A121	160	3007	245	2004	
65A13	160	3007	245	2004	
65A13P	160	3007	245	2004	
65A14	160	3007	245	2004	
65A14-7	160	3007	245	2004	
65A14-7B	160		245	2004	
65A15	160	3007	245	2004	
65A15L	160	3007	245	2004	
65A16	160	3007	245	2004	
65A16-3	160	3007	245	2004	
65A17	160	3007	245	2004	
65A17-1	160	3007	245	2004	
65A18	160	3007	245	2004	
65A182	160	3007	245	2004	
65A19	160	3007	245	2004	
65A19G	160	3007	245	2004	
65A2	160	3007	245	2004	
65A3	160	3007	245	2004	
65A4	160	3007	245	2004	
65A5	160	3007	245	2004	
65A6	160	3007	245	2004	
65A7	160	3007	245	2004	
65A8	160	3007	245	2004	
65A9	160	3007	245	2004	
65B	159	3008	82	2024	
065B-12	126	3008	52	2024	
065B-12-7	126	3008	52	2024	
65B-70	126	3008	52	2024	
65B-70-12	126	3008	52	2024	
65B-70-12-7	126	3008	52	2024	
65B0		3008	52		
65B1	126	3114/290	52	2024	
65B10	126	3008	52	2024	
65B10R	126	3008	52	2024	
65B11	126	3008	52	2024	

Industry Standard No.	ECG	SK	GE	RS 276-	MOTOR.
65B119	126	3008	52	2024	
65B12	126	3008	52	2024	
65B121	126	3008	52	2024	
65B13	126	3008	52	2024	
65B13P	126	3008	52	2024	
65B14	126	3008	52	2024	
65B14-7	126	3008	52	2024	
65B14-7B	126		52	2024	
65B15	126	3008	52	2024	
65B15L	126	3008	52	2024	
65B16	126	3008	52	2024	
65B16-2	126	3008	52	2024	
65B17	126	3008	52	2024	
65B17-1	126	3008	52	2024	
65B18	126	3008	52	2024	
65B182	126	3008	52	2024	
65B19	126	3008	52	2024	
65B19G	126	3008	52	2024	
65B3	126	3008	52	2024	
65B4	126	3008	52	2024	
65B5	126	3008	52	2024	
65B6	126	3008	52	2024	
65B7	126	3008	52	2024	
65B8	126	3008	52	2024	
65B9	126	3008	52	2024	
65B0	126			2024	
65C	159	3009	82	2006	
065C-12	121	3009	239	2006	
065C-12-7	121	3009	239	2006	
65C-70	121	3009	239	2006	
65C-70-12	121	3009	239	2006	
65C-70-12-7	121	3009	239	2006	
65C0	121	3009	239	2006	
65C01			51	2004	
65C02			51	2004	
65C03			51	2004	
65C1	159	3466	82	2032	
65C10	121	3009	239	2006	
65C10R	121	3009	239	2006	
65C11	121	3009	239	2006	
65C119	121	3009	239	2006	
65C12	121	3009	239	2006	
65C121	121	3009	239	2006	
65C13	121	3009	239	2006	
65C13P	121	3009	239	2006	
65C14	121	3009	239	2006	
65C14-7	121	3009	239	2006	
65C14-7B	121	3717	239	2006	
65C15	121	3009	239	2006	
65C15L	121	3009	239	2006	
65C16	121	3009	239	2006	
65C16-4	121	3009	239	2006	
65C17	121	3009	239	2006	
65C17-1	121	3009	239	2006	
65C18	121	3009	239	2006	
65C182	121	3009	239	2006	
65C19	121	3009	239	2006	
65C19G	121	3009	239	2006	
65C2	121	3009	239	2006	
65C3	121	3009	239	2006	
65C4	121	3009	239	2006	
65C5	121	3009	239	2006	
65C6	121	3009	239	2006	
65C7	121	3009	239	2006	
65C8	121	3009	239	2006	
65C9	121	3009	239	2006	
65D	159	3466	82	2032	
65D1	159	3466	82	2032	
65E	159	3466	82	2032	
65E1	159	3466	82	2032	
65F	159	3466	82	2032	
65F1	159	3466	82	2032	
65J2	116		504A	1104	
65P117	116	3016	504A	1104	
65P124	116		504A	1104	
65P124-1	116		504A	1104	
65P124-2	116	3016	504A	1104	
65P153	116	3016	504A	1104	
65P155	116	3016	504A	1104	
65P206	116	3016	504A	1104	
65P284	116	3016	504A	1104	
65P297	116	3016	504A	1104	
65T1	102A	3123	53	2007	
066	128		243	2030	
66-&11I20				2032	
66-1	102A	3003	53	2007	
066-1-12	102A	3003	53	2007	
066-1-12-7	102A	3003	53	2007	
66-1-70	102A	3003	53	2007	
66-1-70-12	102A	3003	53	2007	
66-1-70-12-7	102A	3003	53	2007	
66-10	102A	3003	53	2007	
66-11	102A	3003	53	2007	
66-12	102A	3003	53	2007	
66-127119	123A	3444	20	2051	
66-13	102A	3003	53	2007	
66-14	102A	3003	53	2007	
66-15	102A	3003	53	2007	
66-16	102A	3003	53	2007	
66-17	102A	3003	53	2007	
66-18	102A	3003	53	2007	
66-19	102A	3003	53	2007	
66-1A	102A	3003	53	2007	
66-1A0	102A	3003	53	2007	
66-1A0R	102A	3003	53	2007	
66-1A1	102A	3003	53	2007	
66-1A19	102A	3003		2007	
66-1A2	102A	3003	53	2007	
66-1A21	102A	3003	53	2007	
66-1A3	102A	3003	53	2007	
66-1A3P	102A	3003	53	2007	
66-1A4	102A	3003	53	2007	
66-1A4-7	102A	3003	53	2007	
66-1A4-7B	102A		53	2007	
66-1A5	102A	3003	53	2007	
66-1A5L	102A	3003	53	2007	
66-1A6	102A	3003	53	2007	
66-1A6-3	102A	3003	53	2007	
66-1A7	102A	3003	53	2007	
66-1A7-1	102A	3003	53	2007	
66-1A8	102A	3003	53	2007	
66-1A82	102A	3003	53	2007	
66-1A9	102A	3003	53	2007	
66-1A9G	102A	3003	53	2007	
66-1AA19			53	2007	
66-2	102A	3004	53	2007	
066-2-12	102A	3004	53	2007	
066-2-12-7	102A	3004	53	2007	
66-2-70	102A	3004	53	2007	
66-2-70-12	102A	3004	53	2007	
66-2-70-12-7	102A	3004	53	2007	
66-20	102A	3004	53	2007	
66-21	102A	3004	53	2007	
66-22	102A	3004	53	2007	

Industry Standard No.	ECG	SK	GE	RS 276-	MOTOR.
66-2246	116	3031A	504A	1104	
66-23	102A	3004	53	2007	
66-24	102A	3004	53	2007	
66-25	102A	3004	53	2007	
66-26	102A	3004	53	2007	
66-27	102A	3004	53	2007	
66-28	102A	3004	53	2007	
66-29	102A	3004	53	2007	
66-2A	102A	3004	53	2007	
66-2A0	102A	3004	53	2007	
66-2A0R	102A	3004	53	2007	
66-2A1	102A	3004	53	2007	
66-2A19	102A	3004	53	2007	
66-2A2	102A	3004	53	2007	
66-2A21	102A	3004	53		
66-2A3	102A	3004	53	2007	
66-2A3P	102A	3004	53	2007	
66-2A4	102A	3004	53		
66-2A4-7	102A	3004	53	2007	
66-2A4-7B	102A		53	2007	
66-2A5	102A	3004	53	2007	
66-2A5L	102A	3004	53	2007	
66-2A6	102A	3004	53	2007	
66-2A6-4	102A	3004	53	2007	
66-2A7	102A	3004	53	2007	
66-2A7-1	102A	3004	53	2007	
66-2A8	102A	3004	53	2007	
66-2A82	102A	3004	53	2007	
66-2A9	102A	3004	53	2007	
66-2A9G	102A	3004	53	2007	
66-3	102A	3004	53	2007	
066-3-12	102A	3004	53	2007	
066-3-12-7	102A	3004	53	2007	
66-3-70	102A	3004	53	2007	
66-3-70-12	102A	3004	53	2007	
66-3-70-12-7	102A	3004	53	2007	
66-30	102A	3004	53	2007	
66-31	102A		53	2007	
66-32	102A		53	2007	
66-33	102A	3004	53	2007	
66-34	102A	3004	53	2007	
66-35	102A	3004	53	2007	
66-36	102A	3004	53	2007	
66-37	102A	3004	53	2007	
66-38	102A	3004	53	2007	
66-3A	102A	3004	53	2007	
66-3A0	102A	3004	53	2007	
66-3A0R	102A	3004	53	2007	
66-3A1	102A	3004	53	2007	
66-3A19	102A	3004	53	2007	
66-3A2	102A	3004	53	2007	
66-3A21	102A	3004	53	2007	
66-3A3	102A	3004	53	2007	
66-3A3P	102A	3004	53	2007	
66-3A4	102A	3004	53	2007	
66-3A4-7	102A	3004	53	2007	
66-3A4-7B	102A		53	2007	
66-3A5	102A	3004	53	2007	
66-3A5L	102A	3004	53	2007	
66-3A6	102A	3004	53	2007	
66-3A6C	102A	3004	53	2007	
66-3A7	102A	3004	53	2007	
66-3A7-1	102A	3004	53	2007	
66-3A8	102A	3004	53	2007	
66-3A82	102A	3004	53	2007	
66-3A9	102A	3004	53	2007	
66-3A9G	102A	3004	53	2007	
66-6023	102A	3004	53	2007	
66-6023-00	102A	3004	53	2007	
66-6024-00	102A	3004	53	2007	
66-6025-00	102A	3004	53	2007	
66-6026-00	102A	3004	53	2007	
66-6027-00	102A	3004	53	2007	
66-6028-00	102A	3004	53	2007	
66-6030-00	116	3016	504A	1104	
66-6031-00	116	3016	504A	1104	
66-6033	102A	3004	53	2007	
66-8504	125		510,531	1114	
66-F29-1	123A	3444	20	2051	
66-P11112-0001	199	3245	62	2010	
66-P11120	159		82	2032	
66-P11139	129		244	2027	
66-P11141	129		244	2027	
66A00008A	123A	3444	20	2051	
66A00010A	123A	3444	20	2051	
66A10298	128		243	2030	
66A10310	234		65	2050	
66A10319	519		514	1122	
66B-3A5L	121MP	3718	239(2)	2006(2)	
66C6-1	HIDIV-1	3868/DIV-1	FR-8		
66C6-10	HIDIV-1	3868/DIV-1	FR-8		
66C6-2	HIDIV-1	3868/DIV-1	FR-8		
66C6-3	HIDIV-1	3868/DIV-1	FR-8		
66C6-4	HIDIV-1	3868/DIV-1	FR-8		
66C6-5	HIDIV-1	3868/DIV-1	FR-8		
66C6-6	HIDIV-1	3868/DIV-1	FR-8		
66C6-7	HIDIV-1	3868/DIV-1	FR-8		
66C6-8	HIDIV-1	3868/DIV-1	FR-8		
66C6-9	HIDIV-1	3868/DIV-1	FR-8		
66F-001			504A	1104	
66F-001-1			504A	1104	
66F-010			66	2048	
66F-054-1		3304	517		
66F-054-2		3304	517		
66F-054-3	500A	3304	527		
66F-054-4		3304	527		
66F-084-1			FET-2	2035	
66F001	116		504A	1104	
66F001-1	116		504A	1104	
66F015	704	3023	IC-205		
66F016-1	177		300	1122	
66F016-2	177		300	1122	
66F017	178MP		300(2)	1122(2)	
66F018	177		300	1122	
66F020-1	157	3747	232		
66F020-2	157	3747	232		
66F021-1	161	3716	39	2015	
66F022-1	161	3716	39	2015	
66F023-1	159		82	2032	
66F024-1	159		82	2032	
66F025-1	128		243	2030	
66F026-1	199	3245	62	2010	
66F027-1	123A	3444	20	2051	
66F028-1	123A	3444	20	2051	
66F029-1	123A	3444	20	2051	
66F039-1	172A	3156	64		
66F039-2	172A	3156	64		
66F041-1	159		82	2032	
66F042-1	161	3716	39	2015	
66F054-3	500A		527		
66F054-4	500A		527		
66F057-1	123A	3444	20	2051	
66F057-2	123A	3444	20	2051	
66F058-2	188	3199	226	2018	

Industry Standard No.	ECG	SK	GE	RS 276-	MOTOR.
66F069-1	188		226	2018	
66F069-2	189		218	2026	
66F074-1	171	3201	27		
66F074-2	171	3201	27		
66F074-3	171	3201			
66F074-4	171	3201	27		
66F077-1	712	3072	IC-2		
66F112-1	522	3303	523		
66F112-2	522	3303	523		
66F1271	790		IC-18		
66F133-1	198		251		
66F1551	738		IC-29		
66F159-1	523	3306	528		
66F159-2	523	3306	528		
66F159-3	523	3306	528		
66F181-1	522	3303	523		
66J2	116	3017B/117	504A	1104	
66M	123A	3444	20	2051	
66R2	5882	3500	5040		
66R2S	5883	3500/5882	5041		
66S-159-1	523	3306	528		
66S00000A	128		243	2030	
66X0003	358		42		
66X0003-001	358		42		
66X0003-1	358		42		
66X0007-104	108	3452	86	2038	
66X0020-000	109	3087		1123	
66X0020-001	109	3087		1123	
66X0023-001	116	3016	504A	1104	
66X0023-002	116	3016	504A	1104	
66X0023-003	116	3031A	504A	1104	
66X0023-004	116	3016	504A	1104	
66X0023-005	116	3016	504A	1104	
66X0023-006	116		504A	1104	
66X0023-007	116	3016	504A	1104	
66X0023-008	116	3016	504A	1104	
66X0023-009	125	3081	510	1114	
66X0023-1	116	3016	504A	1104	
66X0024-000	113A	3119/113	6GC1		
66X0025-000	113A	3119/113	6GC1		
66X0025-000-001	113A	3119/113	6GC1		
66X0025-001	113A	3119/113	6GC1		
66X0028-001	116	3017B/117	504A	1104	
66X0028-008	125		510,531	1114	
66X0033-000	116	3016	CR-3	1104	
66X0033-001	116	3016	CR-3	1104	
66X0035-001	118	3066	CR-1		
66X0036-001	506	3109/119	CR-2	1114	
66X0036-002	119	3125	511	1114	
66X0037-001	116	3016	504A	1104	
66X0038-001	506	3031A	511	1114	
66X0039-001	109	3087		1123	
66X0040-003	137A	3058	ZD-6.2	561	
66X0040-004	5084A	3755	ZD-30		
66X0040-005	140A	3060/139A	ZD-10	562	
66X0040-007	5081A	3151	ZD-24		
66X0040-008	5100A		ZD-150		
66X0040-009	145A	3063	ZD-15	564	
66X0040-011	5068A		ZD-4.3		
66X0040-012	149A	3097	ZD-62		
66X0041-001	120	3110	CR-3		
66X0043-001	177	3087	300	1122	
66X0044-001	177	3311	300	1122	
66X0044-100	177		300	1122	
66X0045-001	501B	3069/501A	520		
66X0046-001	519	3100	300	1122	
66X0047-001	109	3087	1N34AS	1123	
66X0047-901	109	3087		1123	
66X0048-001	116	3311	300	1122	
66X0049-001	194	3088	220		
66X0049-002	109	3088		1123	
66X0049-100	110MP	3088	1N60	1123	
66X0050-001	614	3126	90		
66X0051-001	109	3088		1123	
66X0053-001	515	3314			
66X0054-001	515	9098	300	1122	
66X0054-002	515	3314			
66X0055-001	506	3016	511	1114	
66X0056-001		3100	300	1122	
66X0060-001	522	3303	523		
66X0060-002	522	3303	523		
66X0062-001	177	3100/519	300	1122	
66X041-001			504A	1104	
66X053-001	116	3016	504A	1104	
66X19			504A	1104	
66X20	109			1123	
66X21	113A	3016	6GC1		
66X21S	113A	3119/113	6GC1		
66X23	116	3016	504A	1104	
66X24	116	3016	504A	1104	
66X25	113A	3016	6GC1		
66X25-0	113A	3119/113	6GC1		
66X26	116	3016	504A	1104	
66X29			300	1122	
66X36-1	506	3130	511	1114	
66X41-1	120	3110	CR-3		
66XZ18	113A	3119/113	6GC1		
66Z6	5128A	3394			
66Z6A	5128A	3394	5ZD-13		
67-1000-00	116	3032A	504A	1104	
67-1003-00	116	3017B/117	504A	1104	
67A8926	234		65	2050	
67A9060	199	3245	62	2010	
67J2	125	3032A	510,531	1114	
67J2A	125	3032A	510,531	1114	
67P1	121	3009	25	1114	
67P1C	121	3717	25	1114	
67P2	121	3009	25	1114	
67P2C	121	3717	25	1114	
67P3	121	3009	25	1114	
67P3C	121	3717	25	1114	
67Z6	5129A	3395			
67Z6A	5129A	3395	5ZD-14		
68 A 8318-P1	159			2032	
68-110-02	159	3466	82	2032	
68-20102-2701	724		IC-86		
68A7252	177		300	1122	
68A7355P1	234		65	2050	
68A7366-1	199	3245	62	2010	
68A7368	128		243	2030	
68A7370-1	159	3466	82	2032	
68A7370-P3	159	3466	82	2032	
68A7380-1	123A	3444	20	2051	
68A7380-2	128		243	2030	
68A7382-P1	159	3466	82	2032	
68A7672P1	941		IC-263	010	
68A7702P1	194	3275	220		
68A7715P1	123A	3444	20	2051	
68A7734P1	159	3466	82	2032	
68A7860-P1	923		IC-259		
68A8225P001	5626	3633			
68A8318-P1		3466	82	2032	
68A8319001	130		14	2041	
68A8321	123A		20	2051	
68A9025	7400	7400		1801	
68A9026	74H00	74H00			
68A9027	7402	7402		1811	
68A9028	7404	7404		1802	
68A9030	7410	7410		1807	
68A9032	7406	7406		1821	
68A9033	7420	7420		1809	
68A9035	7430	7430			
68A9036	7437	7437			
68A9037	7438	7438			
68A9041	7475	7475		1806	
68A9042	7476	7476		1813	
68A9048	74151	74151			
68A9049	74153	74153			
68P-2A5L	102A		53	2007	
68P1	121	3009	239	2006	
68P1B	121	3009	239	2006	
68R2	5886		5044		
68R2S	5886		5044		
68X0003	358		42		
68X0003-001	358		42		
68X0040-004	5084A	3755			
68X0040-005	140A		ZD-10	562	
68Z6	5130A	3396			
68Z6A	5130A	3396			
069	123A	3444	20	2051	
69-001	230	3042			
69-1810		3018	17	2051	
69-1811		3018	17	2051	
69-1812		3018	17	2051	
69-1813		3018	17	2051	
69-1814		3020	18	2030	
69-1815		3114	21	2034	
69-1816		3122	62	2010	
69-1817		3114	21	2034	
69-1818		3054	57	2017	
69-1819		3083	58	2025	
69-1820		3088	1N60	1123	
69-1821		3126	90		
69-1822		3100	300	1122	
69-1823		3311	504A	1104	
69-2246	116	3311	504A	1104	
69-2401	804	3455	IC-27		
69-2403	804	3455	IC-27		
69-2922	109	3088	1N60	1123	
69-3116	804	3455	IC-27		
69A49728-P1	177		300	1122	
69AJ110	218		234		
69B1Z	715	3076	IC-6		
69B2Z	715	3076	IC-6		
69L-4A5L	102A		53	2007	
69N1	108	3452	86	2038	
69SP112	129		244	2027	
070	130	3027	14	2041	
070-001	160	3006	245	2004	
070-004	125	3033	510,531	1114	
070-005	125	3033	510,531	1114	
070-006	125	3033	510,531	1114	
070-007	125	3033	510,531	1114	
070-008	125	3033	510,531	1114	
070-009	125	3033	510,531	1114	
070-010	125	3033	510,531	1114	
070-011	140A	3061	ZD-10	562	
070-013	125	3033	510,531	1114	
070-014	125	3033	510,531	1114	
070-015	125	3033	510,531	1114	
070-016	125	3033	510,531	1114	
070-017	125	3033	510,531	1114	
070-019	116	3017B/117	504A	1104	
070-020	100	3005	1	2007	
070-021	145A	3063	ZD-15	564	
070-022	177	3175	300	1122	
070-024	140A	3061	ZD-10	562	
070-025	5093A	3344	ZD-75		
070-027	5809	9010			
070-028	125	3033	510,531	1114	
070-030	125	3033	510,531	1114	
070-031	156	3051	512		
070-032	125	3033	510,531	1114	
070-033	125	3033	510,531	1114	
070-035	156	3051	512		
070-036	5809	9010			
070-041	156	3051	512		
070-043	5809	9010			
070-047	177	3175	300	1122	
070-048	5075A	3751	ZD-16	564	
070-049	5081A	3151	ZD-24		
70-270050	125	3031A	510,531	1114	
70-943-083-002	519		514	1122	
70-943-083-003	519		514	1122	
70-943-722-001	123A	3444	20	2051	
70-943-754-002	123A	3444	20	2051	
70-943-762-001	123A	3444	20	2051	
70-943-772-002	123A	3444	20	2051	
70-943-773-001	234	3247	65	2050	
70.00.730	126	3006/160	50	2004	
70.01.704	123A	3122	62	2010	
70B1Z	714		IC-4		
70F40	116	3016	504A	1104	
70H10	6050		5128		
70H100	6072	7272	5140		
70H100A	6072	7272	5140		
70H100AR	6073	7273			
70H100R	6073	7273			
70H10A	6050		5128		
70H10AR	6050		5129		
70H10R	6050		5129		
70H15	6054	7254	5128		
70H15A	6054	7254	5128		
70H20	6054	7254	5128		
70H20A	6054	7254	5128		
70H25	6058		5132		
70H25A	6058		5132		
70H25AR	6059		5133		
70H25R	6059		5133		
70H30	6058		5132		
70H30A	6058		5132		
70H30AR	6059		5133		
70H30R	6059		5133		
70H40	6060	7260	5132		
70H40A	6060	7260	5132		
70H40AR	6061	7261			
70H40R	6061	7261			
70H5	6048		5128		
70H50	6064	7264	5136		
70H50A	6064	7264	5136		
70H5A	6048		5128		
70H5AR	6049		5129		
70H5R	6049		5129		
70H60	6064	7264	5136		
70H60A	6064		5136		
70H70	6068		5140		
70H70A	6068		5140		
70H70AR	6069		5140		

Industry Standard No.	ECG	SK	GE	RS 276-	MOTOR.
70H70R	6069		5140		
70H80	6068		5140		
70H80A	6068		5140		
70H80AR	6069		5140		
70H80R	6069		5140		
70H90	6072	7272	5140		
70H90A	6072	7272	5140		
70H90AR	6073	7273			
70H90R	6073	7273			
70N1	123A	3124/289	20	2051	
70N1M		3124	62	2010	
70N2	123	3124/289	20	2051	
70N3	107		11	2015	
70N4	123A	3444	20	2051	
70R2S	5890	7090	5044		
70S40	116	3016	504A	1104	
70T40	116	3016	504A	1104	
70U10	6354	6554			
70U100	6358	6558			
70U100R	6359	6559			
70U10R	6355	6555			
70U15A	6354	6554			
70U15AR	6355	6555			
70U20	6354	6554			
70U20R	6355	6555			
70U25A	6354	6554			
70U25AR	6355	6555			
70U40	6354	6554			
70U40R	6355	6555			
70U60	6356	6556			
70U70A	6358	6558			
70U70AR	6359	6559			
70U80	6358	6558			
70U80R	6359	6559			
70U90A	6358	6558			
70U90AR	6359	6559			
70UW15	6354	6554			
70UW15R	6355	6555			
70UW25	6354	6554			
70UW25R	6355	6555			
70UW35	6354	6554			
70UW45	6356	6556			
70UW65	6358	6558			
70UW65R	6359	6559			
71-126268	123A	3444	20	2051	
71-13/51/60	519		514	1122	
71-70177A02	748	3236			
71-70177A03	738	3167	IC-29		
71-70177A05	749	3168	IC-97		
71-70177B02	743		IC-214		
71D70177A02	748	3236			
71D70177A03	738	3167	IC-29		
71M70177A02	748	3236			
71M70177A03	738	3167	IC-29		
71N1	124	3021	12		
71N1B	123A	3444	20	2051	
71N1T	124	3021	12		
71N2	124	3021	12		
71N2T	124	3021	12		
072	184	3102/710	57	2017	
072(I.C.)	710		IC-89		
72-028-9-002	306		276		
72-11	116		504A	1104	
72-15	116	3311	504A	1104	
72-18	177	3100/519	300	1122	
72-9			511	1114	
72B1Z	739		IC-30		
72N1	108	3452	86	2038	
72N1B	128	3024	243	2030	
72N2	108	3452	86	2038	
72N2B	123A	3444	20	2051	
72Z	116	3016	504A	1104	
73-15	145A	3063	ZD-15	564	
73-17	5075A	3751	ZD-16	564	
73-31	136A	3057	ZD-5.6	561	
73A01	176	3123	80		
73A02	176	3123	80		
73A03			52	2024	
73A60-11	506	3843	511	1114	
73A64-1	178MP		300(2)	1122(2)	
73B-140-003-5	123A	3444	20	2051	
73B-140-005-1	159	3466	82	2032	
73B-140005-4	159		82	2032	
73B140-004	127	3764	25		
73B140385-001	128		243	2030	
73B140585-21	194	3275	220		
73B140585-22	194	3275	220		
73B140585-23	194	3275	220		
73B140585-24	194	3275	220		
73B140585-25	194	3275	220		
73B140585-26	194	3275	220		
73B140585-27	194	3275	220		
73B14077-1	HIDIV-4	3871/DIV-4			
73B14077-2	HIDIV-4	3871/DIV-4			
73B141789-1	815	3255			
73C18028-11	152	3893	66	2048	
73C18028-12				2048	
73C180475	712	3072	IC-2		
73C180475-1	712	3072	IC-2		
73C180475-4	712	3072	IC-2		
73C180475-7	712	3072	IC-2		
73C180475-8	712	3072	IC-2		
73C180475-9	712	3072	IC-2		
73C180475004	712	3072	IC-2		
73C180476-5	712	3108/505	IC-2		
73C180497-4	154		40	2012	
73C180499-5	154		40	2012	
73C180499-6	154		40	2012	
73C180828-12	152	3893	66		
73C180829-11	152	3893	66	2048	
73C180829-12	152	3054/196	66	2048	
73C180830-11	153		69	2049	
73C180830-12	153	3083/197	69	2049	
73C180831-1	159	3466	82	2032	
73C180831-2	159	3466	82	2032	
73C180831-3	234		65	2050	
73C180837-1	713	3077/790	IC-5		
73C180837-2	713	3077/790	IC-5		
73C180837-3	713	3077/790	IC-5		
73C180837P2	713	3077/790	IC-5		
73C180838-1	749	3168	IC-97		
73C180838-2	749		IC-97		
73C180838-3	749	3168	IC-97		
73C180843-1	783	3215	IC-225		
73C180843-2	783	3215	IC-225		
73C180843-3	783	3215	IC-225		
73C180843-4	783	3215	IC-225		
73C181254	522	3303			
73C181260-1	HIDIV-4	3871/DIV-4			
73C181260-11	HIDIV-4	3871/DIV-4			
73C181260-12	HIDIV-4	3871/DIV-4			
73C181260-13	HIDIV-4	3871/DIV-4			
73C181260-14	HIDIV-4	3871/DIV-4			
73C181260-15	HIDIV-4	3871/DIV-4			
73C181260-16	HIDIV-4	3871/DIV-4			
73C181260-17	HIDIV-4	3871/DIV-4			
73C181260-18	HIDIV-4	3871/DIV-4			
73C181260-2	HIDIV-4	3871/DIV-4			
730181260-3	HIDIV-4	3871/DIV-4			
73C181260-4	HIDIV-4	3871/DIV-4			
73C181260-5	HIDIV-4	3871/DIV-4			
73C181260-6	HIDIV-4	3871/DIV-4			
73C181260-7	HIDIV-4	3871/DIV-4			
73C181260-8	HIDIV-4	3871/DIV-4			
73C181260-9	HIDIV-4	3871/DIV-4			
73C182051	522		517		
73C182051-1	522	3303			
73C182051-2	522	3303			
73C182051-3	522	3303			
73C182077	728	3073	IC-22		
73C182080-33	154		40	2012	
73C182081-31	123A	3444	20	2051	
73C182082-31	159		82	2032	
73C182088-31	154		40	2012	
73C182186	712		IC-2		
73C182186-1	712	3072	IC-2		
73C182186-2	712	3072	IC-2		
73C182186-3	712	3072	IC-2		
73C182186-4	712	3072	IC-2		
73N1	161	3018	39	2015	
73N1B	123A	3444	20	2051	
73N51	105	3012	4		
74	108	3039/316	86	2038	
74-01-772	199	3245	62	2010	
074-1(PHILCO)	186		28	2017	
74-3	100	3005	1	2007	
74-3-70	100	3005	1	2007	
74-3-70-12	100	3005	1	2007	
74-3-70-12-7	100	3005	1	2007	
74-30	100	3005	1	2007	
74-31	100	3005	1	2007	
74-32	100	3005	1	2007	
74-33	100	3005	1	2007	
74-34	100	3005	1	2007	
74-35	100	3005	1	2007	
74-36	100	3005	1	2007	
74-37	100	3005	1	2007	
74-38	100	3005	1	2007	
74-39	100	3005	1	2007	
74-3A	100	3005	1	2007	
74-3A0	100	3005	1	2007	
74-3A0R	100	3721	1	2007	
74-3A1	100	3005	1	2007	
74-3A19	100	3005	1	2007	
74-3A2	100	3005	1	2007	
74-3A21	100	3005	1	2007	
74-3A3	100	3005	1	2007	
74-3A3P	100	3005	1	2007	
74-3A4	100	3005	1	2007	
74-3A4-7	100	3005	1	2007	
74-3A4-7B	100		1	2007	
74-3A5	100	3005	1	2007	
74-3A5L	100	3005	1	2007	
74-3A6	100	3005	1	2007	
74-3A6-3	100	3005	1	2007	
74-3A7	100	3005	1	2007	
74-3A7-1	100	3005	1	2007	
74-3A8	100	3005	1	2007	
74-3A82	100	3005	1	2007	
74-3A9	100	3005	1	2007	
74-3A9G	100	3005	1	2007	
74A01	176	3123	80		
74A02	176	3123	80		
74A03	176	3123	80		
74H00PC	74H00	74H00			
74H04PC	74H04	74H04			
74LS00	74LS00	74LS00			
74LS00DC					SN74LS00J
74LS00N	74LS00	74LS00		1900	
74LS00PC	74LS00				SN74LS00N
74LS02	74LS02	74LS02			
74LS02DC					SN74LS02J
74LS02PC	74LS02				SN74LS02N
74LS03	74LS03	74LS03			
74LS03DC					SN74LS03J
74LS03PC	74LS03				SN74LS03N
74LS04	74LS04	74LS04			
74LS04DC					SN74LS04J
74LS04PC	74LS04	74LS04			SN74LS04N
74LS05	74LS05	74LS05			
74LS05DC					SN74LS05J
74LS05PC	74LS05				SN74LS05N
74LS08	74LS08	74LS08			
74LS08DC					SN74LS08J
74LS08PC	74LS08				SN74LS08N
74LS09DC					SN74LS09J
74LS09PC					SN74LS09N
74LS10	74LS10	74LS10			
74LS107	74LS107	74LS107			
74LS107A	74LS107	74LS107			
74LS109	74LS109A	74LS109			
74LS10DC					SN74LS10J
74LS10PC	74LS10	74LS10			SN74LS10N
74LS11	74LS11	74LS11			
74LS11DC					SN74LS11J
74LS11PC	74LS11				SN74LS11N
74LS123	74LS123	74LS123			
74LS132DC	74LS132				SN74LS132J
74LS132PC	74LS132				SN74LS132N
74LS133DC					SN74LS133J
74LS133PC					SN74LS133N
74LS136	74LS136	74LS136			
74LS136DC					SN74LS136J
74LS136PC	74LS136				SN74LS136N
74LS138	74LS138	74LS138			
74LS138DC					SN74LS138J
74LS138PC	74LS138	74LS138			SN74LS138N
74LS139DC	74LS139				SN74LS139J
74LS139PC	74LS139				SN74LS139N
74LS13DC	74LS13				SN74LS13J
74LS13PC	74LS13				SN74LS13N
74LS14	74LS14	74LS14			
74LS14DC					SN74LS14J
74LS14PC	74LS14				SN74LS14N
74LS151	74LS151	74LS151			
74LS151DC					SN74LS151J
74LS151PC	74LS151				SN74LS151N
74LS153	74LS153	74LS153			
74LS153DC					SN74LS153J
74LS153PC	74LS153				SN74LS153N
74LS155DC	74LS155				SN74LS155J
74LS155PC	74LS155				SN74LS155N
74LS156DC					SN74LS156J
74LS156PC					SN74LS156N
74LS157	74LS157	74LS157			
74LS157DC					SN74LS157J
74LS157PC	74LS157				SN74LS157N
74LS158DC					SN74LS158J

Industry Standard No.	ECG	SK	GE	RS 276-	MOTOR.
74LS158PC					SN74LS158N
74LS15DC	74LS15				SN74LS15J
74LS15PC	74LS15				SN74LS15N
74LS161A	74LS161A	74LS161			
74LS163	74LS163A	74LS163			
74LS163A	74LS163A	74LS163			
74LS164	74LS164	74LS164			
74LS168DC					SN74LS168J
74LS168PC					SN74LS168N
74LS169DC					SN74LS169J
74LS169PC					SN74LS169N
74LS170DC	74LS170				SN74LS170J
74LS170PC	74LS170				SN74LS170N
74LS173DC					SN74LS173J
74LS173PC					SN74LS173N
74LS174	74LS174	74LS174			
74LS174DC					SN74LS174J
74LS174PC	74LS174				SN74LS174N
74LS175	74LS175	74LS175			
74LS175DC					SN74LS175J
74LS175PC	74LS175				SN74LS175N
74LS181DC					SN74LS181J
74LS181PC					SN74LS181N
74LS190DC	74LS190				SN74LS190J
74LS190PC	74LS190				SN74LS190N
74LS191	74LS191	74LS191			
74LS191DC					SN74LS191J
74LS191PC	74LS191				SN74LS191N
74LS192DC	74LS192				SN74LS192J
74LS192PC	74LS192				SN74LS192N
74LS193	74LS193	74LS193			
74LS193DC					SN74LS193J
74LS193N	74LS193	74LS193			
74LS193PC	74LS193				SN74LS193N
74LS195	74LS195A	74LS195			
74LS196DC					SN74LS196J
74LS196PC					SN74LS196N
74LS197DC					SN74LS197J
74LS197PC					SN74LS197N
74LS20	74LS20	74LS20			
74LS20DC					SN74LS20J
74LS20PC	74LS20	74LS20			SN74LS20N
74LS21DC	74LS21				SN74LS21J
74LS21PC	74LS21				SN74LS21N
74LS221	74LS221	74LS221			
74LS22DC	74LS22				SN74LS22J
74LS22PC	74LS22				SN74LS22N
74LS241	74LS241	74LS240			
74LS251DC					SN74LS251J
74LS251PC					SN74LS251N
74LS253	74LS253	74LS253			
74LS253DC					SN74LS253J
74LS253PC	74LS253				SN74LS253N
74LS257	74LS257	74LS257			
74LS258	74LS258	74LS258			
74LS259	74LS259	74LS259			
74LS259DC					SN74LS259J
74LS259PC	74LS259				SN74LS259N
74LS260DC					SN74LS260J
74LS260PC					SN74LS260N
74LS266	74LS266	74LS266			
74LS266DC					SN74LS266J
74LS266PC	74LS266				SN74LS266N
74LS26DC	74LS26				SN74LS26J
74LS26PC	74LS26				SN74LS26N
74LS27	74LS27	74LS27			
74LS273	74LS273	74LS273			
74LS279DC	74LS279				SN74LS279J
74LS279PC	74LS279				SN74LS279N
74LS27DC					SN74LS27J
74LS27PC	74LS27				SN74LS27N
74LS280	74LS280	74LS280			
74LS28DC					SN74LS28J
74LS28PC					SN74LS28N
74LS293DC					SN74LS293J
74LS293PC					SN74LS293N
74LS298	74LS298	74LS298			
74LS298DC					SN74LS298J
74LS298PC	74LS298				SN74LS298N
74LS299DC					SN74LS299J
74LS299PC					SN74LS299N
74LS30	74LS30	74LS30			
74LS30DC					SN74LS30J
74LS30PC	74LS30				SN74LS30N
74LS32	74LS32	74LS32			
74LS32DC					SN74LS32J
74LS32PC	74LS32				SN74LS32N
74LS33DC					SN74LS33J
74LS33PC					SN74LS33N
74LS367	74LS367	74LS367			
74LS377	74LS377	74LS377			
74LS37DC	74LS37				SN74LS37J
74LS37PC	74LS37				SN74LS37N
74LS38	74LS38	74LS38			
74LS38DC					SN74LS38J
74LS38PC	74LS38				SN74LS38N
74LS390DC					SN74LS390J
74LS390PC					SN74LS390N
74LS393DC					SN74LS393J
74LS393PC					SN74LS393N
74LS395DC					SN74LS395J
74LS395PC					SN74LS395N
74LS40DC	74LS40				SN74LS40J
74LS40PC	74LS40				SN74LS40N
74LS42	74LS42	74LS42			
74LS42DC					SN74LS42J
74LS42PC	74LS42				SN74LS42N
74LS490DC					SN74LS490J
74LS490PC					SN74LS490N
74LS51	74LS51	74LS51			
74LS51DC					SN74LS51J
74LS51PC	74LS51				SN74LS51N
74LS54DC	74LS54				SN74LS54J
74LS54PC	74LS54				SN74LS54N
74LS55DC	74LS55				SN74LS55J
74LS55PC	74LS55				SN74LS55N
74LS670DC	74LS670				SN74LS670J
74LS670PC					SN74LS670N
74LS73	74LS73	74LS73			
74LS74	74LS74A	74LS74			
74LS74DC					SN74LS74AJ
74LS74PC	74LS74A				SN74LS74AN
74LS75	74LS75	74LS75			
74LS83	74LS83A	74LS83			
74LS83A	74LS83A	74LS83			
74LS85	74LS85	74LS85			
74LS85DC					SN74LS85J
74LS85PC					SN74LS85N
74LS86	74LS86	74LS86			
74LS86DC					SN74LS86J
74LS86PC	74LS86				SN74LS86N
74LS90DC	74LS90				SN74LS90J
74LS90PC	74LS90				SN74LS90N
74LS92DC					SN74LS92J

Industry Standard No.	ECG	SK	GE	RS 276-	MOTOR.
74LS92PC					SN74LS92N
74LS93	74LS93	74LS93		1925	
74LS93DC					SN74LS93J
74LS93PC	74LS93				SN74LS93N
74LS95BDC					SN74LS95BJ
74LS95BPC					SN74LS95BN
74N1	123A	3122	20	2051	
74P1	159	3025/129	82	2032	
74P1M	129		244	2027	
74Q1262	160	3004/102A	245	2004	
74Q22881			31MP	2025(2)	
74S00PC	74S00	74S00			
74S04PC	74S04	74S04			
74S573					MCM7643
75	116	3016	504A	1104	
075-045037	7474			1818	
75-461	102A		53	2007	
75B1Z	801	3160	IC-35		
75D01	5454	6754		1067	
75D02	5455			1067	
75D04	5431			1067	
75D05	5422			1067	
75D06	5455			1067	
75D1	116	3016	504A	1104	
75D10	125	3033	510,531	1114	
75D2	116	3016	504A	1104	
75D8	125	3032A	510,531	1114	
75E1	116	3016	504A	1104	
75E10	125	3033	510,531	1114	
75E2	116	3016	504A	1104	
75E3	116	3016	504A	1104	
75E4	116	3017B/117	504A	1104	
75E5	116	3016	504A	1104	
75E6	116	3017B/117	504A	1104	
75E7	125	3032A	510,531	1114	
75E8	125	3032A	510,531	1114	
75F05	116		504A	1104	
75HQ030					MBR7530
75HQ035					MBR7535
75HQ040					MBR7540
75HQ045					MBR7545
75KBP005					MDA100A
75KBP02					MDA102A
75KBP04					MDA104A
75KBP06		3678			MDA106A
75KBP08					MDA108A
75KBP10					MDA110A
75N1	154	3124/289	40	2012	
75N12		3062	ZD-12	563	
75N27		3064	ZD-27		
75N5		3063	ZD-15	564	
75N5.6			ZD-5.6	561	
75N5AA	123A	3444	20	2051	
75N6.2			ZD-6.2	561	
75R10B	125	3033	510,531	1114	
75R11B	506	3843	511	1114	
75R12B	506	3843	511	1114	
75R13B	506	3843	511	1114	
75R14B	506	3843	511	1114	
75R1B	116	3017B/117	504A	1104	
75R2B	116	3017B/117	504A	1104	
75R3B	116	3017B/117	504A	1104	
75R4B	116	3031A	504A	1104	
75R5B	116	3017B/117	504A	1104	
75R6B	116	3017B/117	504A	1104	
75R7B	125	3032A	510,531	1114	
75R8B	125	3032A	510,531	1114	
75R9B	125	3033	510,531	1114	
75W-005	116(4)	3020/123	504A(4)	1104	
75W005	116(4)	3020/123			
75W005M					MDA100A
75W02M					MDA102A
75W04M					MDA104A
75W06M		3676			MDA106A
75W08M		3677			MDA108A
75W10M					MDA110A
75Z11	141A	3092	ZD-11.5		
75Z110	151A	3099	ZD-110		
75Z110A	151A	3099	ZD-110		
75Z110B	151A	3099	ZD-110		
75Z110C	151A	3099	ZD-110		
75Z11A	141A	3092	ZD-11.5		
75Z11B	141A	3092	ZD-11.5		
75Z11C	141A	3092	ZD-11.5		
75Z12	142A	3062	ZD-12	563	
75Z12A	142A	3062	ZD-12	563	
75Z12B	142A	3062	ZD-12	563	
75Z12C	142A	3062	ZD-12	563	
75Z13	143A	3750	ZD-13	563	
75Z13A	143A	3750	ZD-13	563	
75Z13B	143A	3750	ZD-13	563	
75Z13C	143A	3750	ZD-13	563	
75Z14	144A	3094	ZD-14	564	
75Z14A	144A	3094	ZD-14	564	
75Z14B	144A	3094	ZD-14	564	
75Z14C	144A	3094	ZD-14	564	
75Z15	145A	3063	ZD-15	564	
75Z15A	145A	3063	ZD-15	564	
75Z15B	145A	3063	ZD-15	564	
75Z15C	145A	3063	ZD-15	564	
75Z27	146A	3064	ZD-27		
75Z27A	146A	3064	ZD-27		
75Z27B	146A	3064	ZD-27		
75Z27C	146A	3064	ZD-27		
75Z33	147A	3095	ZD-33		
75Z33A	147A	3095	ZD-33		
75Z33B	147A	3095	ZD-33		
75Z33C	147A	3095	ZD-33		
75Z56	148A	3096	ZD-55		
75Z56A	148A	3096	ZD-55		
75Z56B	148A	3096	ZD-55		
75Z56C	148A	3096	ZD-55		
75Z62	149A	3097	ZD-62		
75Z62A	149A	3097	ZD-62		
75Z62B	149A	3097	ZD-62		
75Z62C	149A	3097	ZD-62		
75Z82	150A	3098	ZD-82		
75Z82A	150A	3098	ZD-82		
75Z82B	150A	3098	ZD-82		
75Z82C	150A	3098	ZD-82		
75Z9.1	139A	3060	ZD-9.1	562	
75Z9.1A	139A	3066/118	ZD-9.1	562	
75Z9.1B	139A	3060	ZD-9.1	562	
75Z9.1C	139A	3066/118	ZD-9.1	562	
76	126	3006/160	52	2024	
76-0105	175		246	2020	
76-042-9-006	289A		268	2038	
76-1		3118	18	2030	
76-11770	104	3009	16	2006	
76-12965-26	109			1123	
76-13570-39	108	3452	86	2038	
76-13570-59	108	3452	86	2038	
76-13570-65	112	3089	1N82A		
76-13848-23	112	3089	1N82A		

Industry Standard No.	ECG	SK	GE	RS 276-	MOTOR.
76-13866-17	108	3452	86	2038	
76-13866-18	108	3452	86	2038	
76-13866-19	108	3452	86	2038	
76-13866-19(VHF)			11	2015	
76-13866-20	108	3452	86	2038	
76-13866-59	108	3452	86	2038	
76-13866-62	108	3452	86	2038	
76-14090-1	128	3024	243	2030	
76-14196-1	110MP	3087	1N34AS	1123(2)	
76-14327-1	500A		527		
76-14327-2	500A	3304	527		
76-14327-3	500A	3304	527		
76-14327-4	500A	3304	527		
76-14327-5	500A	3304	527		
76-14327-6	500A	3304	527		
76-14327-7	500A	3304	527		
76-14327-8	500A	3304	527		
76-I(SYLVANIA)			82	2032	
76-L(SYLVANIA)				2032	
76B1Z	742		IC-213		
76G1450J	116		504A	1104	
76N1	123A	3444	20	2051	
76N1(REMOTE)			20	2051	
76N1(VID)			62	2010	
76N1B	128	3024	243	2030	
76N1M	123A	3444	20	2051	
76N2	123A	3444	20	2051	
76N2369-000	123A	3444	20	2051	
76N2369-001	123A	3444	20	2051	
76N2B	128	3024	243	2030	
76N3B	128	3024	243	2030	
76S1030-000	128		243	2030	
77	100	3005	1	2007	
77-001	230	3042			
77-270877-2	121	3717	239	2006	
77-270878-2	121	3717	239	2006	
77-270993-1	117		504A	1104	
77-271025-1	102A		53	2007	
77-271026-1	102A		53	2007	
77-271027-1	102A		53	2007	
77-271029-1	160		245	2004	
77-271029-2	160		245	2004	
77-271031-1	109	3087		1123	
77-271032-1	109	3087		1123	
77-271036-1	102A		53	2007	
77-271037-1	102A		53	2007	
77-271038-1	160	3984/106	245	2004	
77-271039-1	102A		53	2007	
77-271166-2			50	2004	
77-271166-3	160		245	2004	
77-271374-1	116	3017B/117	504A	1104	
77-271453-1	123A	3444	20	2051	
77-271490	128	3024	245	2030	
77-271490-1	128	3024	245	2030	
77-271491-1	121	3009	239	2006	
77-27166-2	160		245	2004	
77-271798-1	184		57	2017	
77-271798-2	199		62	2010	
77-271798-3	184		57	2017	
77-271818-1	129		244	2027	
77-271819-1	123A	3444	20	2051	
77-271967-1	123A	3444	20	2051	
77-27198-3	199		62	2010	
77-272913-1	152	3893	66	2048	
77-272914-1	153		69	2049	
77-272999-1	128		243	2030	
77-273001-2	123A	3444	20	2051	
77-273001-3	160		245	2004	
77-273004-1	158		53	2007	
77-273715-1	152	3893	66	2048	
77-273716-1	153		69	2049	
77-273738-1	152	3893	66	2048	
77-273739-1	153		69	2049	
077A	105	3012	4		
077A-12	105	3012	4		
077A-12-7	105	3012	4		
77A-70	105	3012	4		
77A-70-12	105	3012	4		
77A-70-12-7	105	3012	4		
77A0	105	3012	4		
77A01	177	3100/519	300	1122	
77A02	177	3100/519	300	1122	
77A1	105	3012	4		
77A10	105	3012	4		
77A10R	105	3012	4		
77A11	105	3100/519	300	1122	
77A119	105	3012	4		
77A12	105	3012	4		
77A121	105	3012	4		
77A13	105	3100/519	300	1122	
77A13P	105	3012	4		
77A14	105	3012	4		
77A14-7	105	3012	4		
77A14-7B	105		4		
77A15	105	3012	4		
77A15L	105	3012	4		
77A16	105	3012	4		
77A16-1	105	3012	4		
77A17	105	3012	4		
77A17-1	105	3012	4		
77A18	105	3012	4		
77A182	105	3012	4		
77A19	105	3012	4		
77A19G	105	3012	4		
77A2	105	3012	4		
77A3	105	3012	4		
77A4	105	3012	4		
77A5	105	3012	4		
77A6	105	3012	4		
77A7	105	3012	4		
77A8	105	3012	4		
77A9	105	3012	4		
077B	121MP	3013	239(2)	2006(2)	
077B-12	121MP	3013	239(2)	2006(2)	
077B-12-7	121MP	3013	239(2)	2006(2)	
77B-70	121MP	3013	239(2)	2006(2)	
77B-70-12	121MP	3013	239(2)	2006(2)	
77B-70-12-7	121MP	3013	239(2)	2006(2)	
77B0	121MP	3013	239(2)	2006(2)	
77B1	121MP	3013	239(2)	2006(2)	
77B10	121MP	3013	239(2)	2006(2)	
77B10R	121MP	3013	239(2)	2006(2)	
77B11	121MP	3013	239(2)	2006(2)	
77B119	121MP	3013	239(2)	2006(2)	
77B12		3013	239(2)	2006(2)	
77B12(IC)	788	3829			
77B121	121MP	3013	239(2)	2006(2)	
77B13	121MP	3013	239(2)	2006(2)	
77B13P	121MP	3013	239(2)	2006(2)	
77B14	121MP	3013	239(2)	2006(2)	
77B14-7	121MP	3013	239(2)	2006(2)	
77B14-7B	121MP	3718	239(2)	2006(2)	
77B15	121MP	3013	239(2)	2006(2)	
77B15L	121MP	3013	239(2)	2006(2)	
77B16	121MP	3013	239(2)	2006(2)	
77B16-2	121MP	3013	239(2)	2006(2)	
77B17	121MP	3013	239(2)	2006(2)	
77B17-1	121MP	3013	239(2)	2006(2)	
77B18	121MP	3013	239(2)	2006(2)	
77B182	121MP	3013	239(2)	2006(2)	
77B19	121MP	3013	239(2)	2006(2)	
77B19G	121MP	3013	239(2)	2006(2)	
77B2	121MP	3013	239(2)	2006(2)	
77B3	121MP	3013	239(2)	2006(2)	
77B4	121MP	3013	239(2)	2006(2)	
77B5	121MP	3013	239(2)	2006(2)	
77B6	121MP	3013	239(2)	2006(2)	
77B7	121MP	3013	239(2)	2006(2)	
77B8	121MP	3013	239(2)	2006(2)	
77B9	121MP	3013	239(2)	2006(2)	
77C	121	3014	239	2006	
077C-12	121	3014	239	2006	
077C-12-7	121	3014	239	2006	
77C-70	121	3014	239	2006	
77C-70-12	121	3014	239	2006	
77C-70-12-7	121	3014	239	2006	
77C0	121	3014	239	2006	
77C1	121	3014	239	2006	
77C10	121	3014	239	2006	
77C10R	121	3014	239	2006	
77C11	121	3014	239	2006	
77C119	121	3014	239	2006	
77C12	121	3014	239	2006	
77C121	121	3014	239	2006	
77C13	121	3014	239	2006	
77C13P	121	3014	239	2006	
77C14	121	3014	239	2006	
77C14-7	121	3014	239	2006	
77C14-7B	121	3717	239	2006	
77C15	121	3014	239	2006	
77C15L	121	3014	239	2006	
77C16	121	3014	239	2006	
77C16-3	121	3014	239	2006	
77C17	121	3014	239	2006	
77C17-1	121	3014	239	2006	
77C18	121	3014	239	2006	
77C182	121	3014	239	2006	
77C19	121	3014	239	2006	
77C19G	121	3014	239	2006	
77C2	121	3014	239	2006	
77C3	121	3014	239	2006	
77C4	121	3014	239	2006	
77C5	121	3014	239	2006	
77C6	121	3014	239	2006	
77C7	121	3014	239	2006	
77C710891-2	941		IC-263	010	
77C8	121	3014	239	2006	
77C9	121	3014	239	2006	
77IU/O					2N4352
77N1	123A		20	2051	
77N2	123A	3444	20	2051	
77N2B	128	3024	243	2030	
77N3	123		20	2051	
77N4	123A	3444	20	2051	
77N5	123A	3444	20	2051	
77N6	123A	3444	20	2051	
77S1017	519		514	1122	
78-001	231	3857			
078-0016	116	3016	504A	1104	
078-1696	116	3016	504A	1104	
078-2400	116	3016	504A	1104	
78-254566-1	116	3016	504A	1104	
78-254566-4	116	3016	504A	1104	
78-271030-1	117		504A	1104	
78-271143-1	116	3016	504A	1104	
78-271199-1	109	3087		1123	
78-271228-1	109	3087		1123	
78-271383-1	142A		ZD-12	563	
78-271383-2	146A		ZD-27		
78-271383-3	145A		ZD-15	564	
78-271383-4	145A		ZD-15	564	
78-272160-1	116		504A	1104	
78-272212-1	121		239	2006	
78-273002-1	109			1123	
78-273008			504A	1104	
78-273008-1	116		504A	1104	
78-273085	116		504A	1104	
078-5001	116		504A	1104	
78-5009	121	3009	239	2006	
78A200010P4	7400	7400		1801	
78BLK	100	3005	1	2007	
78C01	108	3452	86	2038	
78C02	108	3452	86	2038	
78E67-2	199		62	2010	
78GRN	100	3005	1	2007	
78L05	977	3462	VR-100		
78L05-AV	977	3462			
78L05A	977	3462			
78L05AVP	977	3462			
78L05J	977	3462			
78L05LP	977	3462			
78L06	988	3831			
78L06C	988	3831			
78L08	981		VR-106		
78L08A	981	3724			
78L08AC	981		VR-106		
78L08AWC	981	3724			
78L08LP	981	3724			
78L62	988	3973			
78L62AWC	988	3973			
78L62WV	988	3973			
78L82AC	981	3724			
78N1	123A	3444	20	2051	
78N2B	123A	3444	20	2051	
78RED	100	3005	1	2007	
78YEL	100	3005	1	2007	
079	185	3191	58	2025	
79E104-2	519		514	1122	
79F015	109	3087		1123	
79F114-1	128	3024	243	2030	
79F114-2A	129	3025	244	2027	
79F114-3	128	3024	243	2030	
79F114-4	129	3025	244	2027	
79F114-4A	129	3025	244	2027	
79F150	358		42		
79F150-2	358		42		
79F151	118		CR-1		
79F153	120		CR-3		
79P1	102A	3004	53	2007	
080	184	3190	57	2017	
80-001300	125	3031A	510,531	1114	
80-001400	125	3031A	510,531	1114	
80-050100	123A	3444	20	2051	
80-050300	103A	3010	59	2002	
80-050400	160	3006	245	2004	
80-050500	160	3008	245	2004	
80-050600	103A	3010	59	2002	
80-050700	121	3009	239	2006	

Industry Standard No.	ECG	SK	GE	RS 276-	MOTOR.
80-051500	123A	3444	20	2051	
80-051600	129	3025	244	2027	
80-052102	123A	3444	20	2051	
80-052202	123A	3444	20	2051	
80-052302	123A	3444	20	2051	
80-052402	130	3027	14	2041	
80-052600	123A	3444	20	2051	
80-052700	129	3025	244	2027	
80-052800	103A	3010	59	2002	
80-053001	123A	3444	20	2051	
80-053300	128	3024	243	2030	
80-053400	123A	3444	20	2051	
80-053600	108	3452	86	2038	
80-053702	130MP	3029	15	2041(2)	
80-104900	126	3008	52	2024	
80-105200	126	3008	52	2024	
80-105300	126	3008	52	2024	
80-205400	102A	3004	53	2007	
80-205600	102A	3004	53	2007	
80-2226314	102A	3004	53	2007	
80-236400	102A	3004	53	2007	
80-236430	102A	3004	53	2007	
80-243900	102A	3004	53	2007	
80-308-2	123A	3444	20	2051	
80-318250	123A	3444	20	2051	
80-337390	123A	3444	20	2051	
80-338030	108	3452	86	2038	
80-338040	108	3452	86	2038	
80-339430	108	3452	86	2038	
80-339440	108	3452	86	2038	
80-383840	108	3452	86	2038	
80-383930	108	3452	86	2038	
80-383940	123A	3444	20	2051	
80-389910	123A	3444	20	2051	
80-389930	123A	3444	20	2051	
80-414120	130	3027	14	2041	
80-414130	130	3027	14	2041	
80-421980	123A	3444	20	2051	
80-60-1	109	3087		1123	
80A5	125		510,531	1114	
80AS	125	3032A	510,531	1114	
80B					MR1128
80C		3033			1N4006
80H	125	3032A	510,531	1114	
80H3	5866		5028		
80H3P		3032A			MR1128
80L20	525	3925			
80P1	102A	3004	53	2007	
80P2	129		244	2027	
80P2B	129		244	2027	
80P3	129		244	2027	
80P3B	129		244	2027	
081	184	3190	IC-18	2017	
081(I.C.)	790		IC-230		
081(IC)	790	3454			
081-1	790	3454	IC-18		
81-23860400-3	188	3054/196	226	2018	
81-23860400A	188	3054/196	226	2018	
81-23860400B	188	3054/196	226	2018	
81-27123100-3	116	3311	504A	1104	
81-27123150-8	109	3087		1123	
81-27123300-3	116	3311	504A	1104	
81-27123307-3	116		504A	1104	
81-27125140-7	123A	3444	20	2051	
81-27125140-7A	123A	3444	20	2051	
81-27125140-7B	123A	3444	20	2051	
81-27125160-5	123A	3444	20	2051	
81-27125160-5A	123A	3444	20	2051	
81-27125160-5B	123A	3444	20	2051	
81-27125270-2	123A	3444	20	2051	
81-27125270-2A	123A	3444	20	2051	
81-27125270-2B	123A	3444	20	2051	
81-27125300-7	123A	3444	20	2051	
81-27125530-9	188	3122	226	2018	
81-27125530-9A	188	3122	226	2018	
81-27125530-9B	188	3122	226	2018	
81-27126100-0	188	3199	226	2018	
81-27126130-7	121	3014	239	2006	
81-27126130-7A	121	3014	239	2006	
81-27126130-7B	121	3014	239	2006	
81-46123001-3	109	3088		1123	
81-46123002-1		3126	90		
81-46123004-5	116		504A	1104	
81-46123004-7	116	3031A	504A	1104	
81-46123005	116	3031A	504A	1104	
81-46123006-2	109	3088		1123	
81-46123010-4	5444	3634			
81-46123011-2	116	3311	504A	1104	
81-46123012-2	116	3311	504A	1104	
81-46123013-8	109	3087		1123	
81-46123014-6	116	3311	504A	1104	
81-46123015-3	109	3088		1123	
81-46123018-7	116	3311	504A	1104	
81-46123022-9	116	3311	504A	1104	
81-46123023-7	177	3100/519	300	1122	
81-46123034-4	116(2)	3311	504A(2)	1104	
81-46123035-1	116(2)	3311	504A(2)	1104	
81-46123038-5	177	3100/519	300	1122	
81-46123043-5	116		504A	1104	
81-46123044-3	5072A		ZD-8.2	562	
81-46125001-1	126	3007	52	2024	
81-46125002-9	102A	3004	53	2007	
81-46125003-7	102A	3004	53	2007	
81-46125004-5	102A	3004	53	2007	
81-46125005-2	158	3004/102A	53	2007	
81-46125006-0	108	3452	86	2038	
81-46125007-8	107	3018	11	2015	
81-46125009-4	102A	3004	53	2007	
81-46125010-2	102A	3004	53	2007	
81-46125011-0	102A	3004	53	2007	
81-46125012-8	107	3018	11	2015	
81-46125013-6	107	3018	11	2015	
81-46125016-9	123A	3444	20	2051	
81-46125018-5	158	3004/102A	53	2007	
81-46125019-3	123A	3444	20	2051	
81-46125026-8	123A	3444	20	2051	
81-46125027-6	123A	3444	20	2051	
81-46125028-4	121	3004/102A	239	2006	
81-46125029-2	102A	3004	53	2007	
81-46125030-0	107	3018	11	2015	
81-46125032-6	107	3018	11	2015	
81-46125033-4	107	3018	11	2015	
81-46125053-2	199		62	2010	
81-46125063-1	199		62	2010	
81-46125065-6	312			2028	
81-46125071-4	159	3466	82	2032	
81-46128001-8	1097	3446			
81-46128002-6	1110	3229			
81P3	129		244	2027	
81T2	128	3024	243	2030	
082	712	3072	IC-2		
082-1	712	3072	IC-2		
82-4	116		504A	1104	
82-409501	123A	3444	20	2051	
82-410300	129	3025	244	2027	
082.115.015	159	3466	82	2032	
082A	712		IC-2		
082A(PACKARDBELL)	712		IC-2		
82M432B2	915			017	
82M667B2	925		IC-261		
82S110					MCM93415
82S111					MCM93425
82S137					MCM7643
82S141					MCM7641
82S181					MCM7681
82S185					MCM7685
83	129		244	2027	
83-1056	121	3009	239	2006	
83-2			511	1114	
83-43-3	118	3016	CR-1		
83-829	116	3017B/117	504A	1104	
83-880	116	3017B/117	504A	1104	
83A30-1			504A	1104	
83B38-1	109			1123	
83C30-1			504A	1104	
83N52	105	3012	4		
83P1	159	3466	82	2032	
83P1A	159	3466	82	2032	
83P1B	159	3466	82	2032	
83P1BC	159	3466	82	2032	
83P1M	159	3466	82	2032	
83P1MC	159	3466	82	2032	
83P2	159	3466	82	2032	
83P2A	159	3466	82	2032	
83P2AA	159	3466	82	2032	
83P2AA1	159	3466	82	2032	
83P2B	129	3025	244	2027	
83P2M	159	3466	82	2032	
83P2M1	159	3466	82	2032	
83P2N	159	3466	82	2032	
83P3	159	3466	82	2032	
83P3A	159	3466	82	2032	
83P3AA	159	3466	82	2032	
83P3AA1	159	3466	82	2032	
83P3B	159	3466	82	2032	
83P3B1	159	3466	82	2032	
83P3M	159	3466	82	2032	
83P3M1	159	3466	82	2032	
83P4	159	3466	82	2032	
83PS	159		82	2032	
84	121	3009	239	2006	
84-44-3	118		CR-1		
84A	121	3009	239	2006	
084A-12	121	3009	239	2006	
084A-12-7	121	3009	239	2006	
84A-70	121	3009	239	2006	
84A-70-12	121	3009	239	2006	
84A-70-12-7	121	3009	239	2006	
84A0	121	3009	239	2006	
84A1	121	3009	239	2006	
84A10	121	3009	239	2006	
84A10R	121	3009	239	2006	
84A12	121	3009	239	2006	
84A121	121	3009	239	2006	
84A13	121	3009	239	2006	
84A13P	121	3009	239	2006	
84A14	121	3009	239	2006	
84A14-7	121	3009	239	2006	
84A14-7B	121	3717	239	2006	
84A15	121	3009	239	2006	
84A15L	121	3009	239	2006	
84A16	121	3009	239	2006	
84A16B	121	3009	239	2006	
84A17	121	3009	239	2006	
84A17-1	121	3009	239	2006	
84A18	121	3009	239	2006	
84A182	121	3009	239	2006	
84A19	121	3009	239	2006	
84A19G	121	3009	239	2006	
84A2	121	3009	239	2006	
84A20	177	3031A	300	1122	
84A3	121	3009	239	2006	
84A4	121	3009	239	2006	
84A5	121	3009	239	2006	
84A6	121	3009	239	2006	
84A7	121	3009	239	2006	
84A8	121	3009	239	2006	
84A9	121	3009	239	2006	
84AA1	121	3009	239	2006	
84AA19	121	3009	239	2006	
84B	121	3009	239	2006	
84C01	157	3747	232		
84G01	157	3747	232		
085	130	3027	14	2041	
85-370-2 BLU	121	3717		2006	
85-370-2(BLU)			239	2006	
85-5	116	3016	504A	1104	
85-5056-2			20	2051	
85-5058-2			239	2006	
85A500161	5626	3633			
85C04	123A	3444	20	2051	
85M02704	910		IC-251		
0086	148A	3096	16	2006	
86 A 86A327	123A			2051	
86-0001		3126	90		
86-0002	109	3088		1123	
86-0005	139A	3060	ZD-9.1	562	
86-0006	116	3031A	504A	1104	
86-0007	115	3121	6GX1		
86-0007-004	123A	3444	20	2051	
86-0008	109	3088	1N60	1123	
86-001	109			1123	
86-0010	116		504A	1104	
86-0012	116		504A	1104	
86-0012-001			20	2051	
86-0013	116		504A	1104	
86-0016-01	116	3016	504A	1104	
86-0022-001	123A	3444	20	2051	
86-0029-001	123A	3444	20	2051	
86-0031-001	123A	3444	20	2051	
86-0033-007	131MP	3840	44(2)	2006(2)	
86-0036-001	159		82	2032	
086-005132-02	123A		20	2051	
86-005135-2	199	3466/159	82	2010	
86-0509		3126	90		
86-0510	5071A		ZD-6.8	561	
86-0511	116	3311	504A	1104	
86-0513	109	3100/519	1N60	1123	
86-0514		3126	90		
86-0515	177		300	1122	
86-0516	116	3016	504A	1104	
86-0518	611	3364			
86-0521	142A	3062	ZD-12	563	
86-0522	177	3100/519	514	1122	
86-1-3	116	3016	504A	1104	
86-10-1	109	3087	1N34AS	1123	
86-10-2	101	3861	8	2002	
86-100-2	160	3006	245	2004	

Industry Standard No.	ECG	SK	GE	RS 276-	MOTOR.
86-100002	229	3018	61	2038	
86-100003		3018	20	2051	
86-100004	229	3018	61	2038	
86-100005	123A	3444	20	2051	
86-100006	229	3018	20	2038	
86-100007	229	3018	61	2038	
86-100008	123A	3444	20	2051	
86-100009	159	3114/290	21	2032	
86-100010	293	3137/192A	81	2030	
86-100011	294	3138/193A	48		
86-100012		3126	90		
86-100013	177	3100/519	300	1122	
86-100014	177	3100/519	300	1122	
86-10003	229			2038	
86-10006			61	2038	
86-10009		3466	82	2032	
86-10000			61	2038	
86-10010			47	2030	
86-10014			300	1122	
86-101-2	160	3008	245	2004	
86-102-2	160	3006	245	2004	
86-102-3	5401	3638/5402			
86-103-2	102A		53	2007	
86-103-3	5448	3636			
86-103-9	5448	3636			
86-104-3	552	3175	511	1114	
86-105-3	506	3175	511	1114	
86-106-3	522	3304/500A	523		
86-107-2	160		245	2004	
86-108-2	126	3008	52	2024	
86-109-1	5069A	3332/5068A	ZD-4.3		
86-109-2	126	3008	52	2024	
86-11-1	5080A	3336	ZD-22		
86-11-2	101	3861	8	2002	
86-110-1	5071A	3334	ZD-6.8	561	
86-110-2	123A	3444	20	2051	
86-110-3	5402	3638	MR-5		
86-111-1	5080A	3336	ZD-22		
86-111-2	126	3008	52	2024	
86-111-3	116	3311	504A	1104	
86-112-2	160	3006	245	2004	
86-113-1	5081A	3151	ZD-24		
86-113-3	5414	3954		1067	
86-114-1	5072A	3136	ZD-8.2	562	
86-114-2	102A	3004	53	2007	
86-114-3	513	3443	513		
86-115-2	102A	3008	53	2007	
86-116-1		3126	90		
86-116-2	126	3008	52	2024	
86-116-3	552	3843	511	1114	
86-117-1	5082A	3753	ZD-25		
86-117-2	160	3006	245	2004	
86-117-3	506	3130	511	1114	
86-118-1	140A	3061	ZD-10	562	
86-118-3	506	3130	511	1114	
86-119-1	5085A	3337	ZD-36		
86-119-2	123A	3006/160	20	2051	
86-12-1	112	3089	1N82A		
86-12-2	101	3861	8	2002	
86-120-2	121	3009	239	2006	
86-121-2	411/128		45		
86-123-2	123A	3124/289	20	2051	
86-125-1	109	3087		1123	
86-126-2	102A	3004	53	2007	
86-127-2	121	3009	239	2006	
86-127-3	529	3307	529		
86-128-2	102A	3004	53	2007	
86-128-3	125	3051/156	510	1114	
86-129-2	102A	3004	53	2007	
86-13-2	103A	3010	59	2002	
86-130-2	102A	3004	53	2007	
86-131-2	102A	3008	53	2007	
86-132-2	102A	3006/160	53	2007	
86-133-2	102A	3005	53	2007	
86-135-1	5072A	3059/138A	ZD-8.2	562	
86-135-2	160	3006	245	2004	
86-136-2	160	3006	245	2004	
86-138-2	108	3452	86	2038	
86-138-3			11	2015	
86-139-1	5079A	3335	ZD-20		
86-139-2	123A	3444	20	2051	
86-139-3	116		504A	1104	
86-1392	123A	3444	20	2051	
86-14-1	110MP		1N60	1123	
86-14-2	103A	3010	59	2002	
86-141-2	121	3009	239	2006	
86-142-2	121	3009	239	2006	
86-143-2	123A	3444	20	2051	
86-144-2	123A	3444	20	2051	
86-145-1	138A		ZD-7.5		
86-146-1	109	3087	1N34AS	1123	
86-146-2	121	3009	239	2006	
86-146-3			504A	1104	
86-147-1	116	3100/519	300	1122	
86-147-2	121	3009	239	2006	
86-147-3	116	3017B/117	504A	1104	
86-148-1	5133A		ZD-18		
86-149-1	5088A		ZD-47		
86-149-2	160	3006	245	2004	
86-15-1	110MP	3709	1N34AS	1123(2)	
86-150-2	160	3006	245	2004	
86-150-3	5096A		ZD-100		
86-151-2	160	3007	245	2004	
86-152-2	102A		53	2007	
86-155-2	123A	3444	20	2051	
86-156-2	102A	3004	53	2007	
86-156-2A	102	3004	2	2007	
86-157-2		3124	20	2051	
86-158-2	123A	3444	20	2051	
86-159-2	102A	3004	53	2007	
86-16-2	102A	3004	53	2007	
86-160-2	129/411		53		
86-161-2	128	3124/289	243	2030	
86-162-2	160	3006	245	2004	
86-163-2	160	3006	245	2004	
86-164-2	160	3006	245	2004	
86-165-2	157	3747	232		
86-166-2	123A	3444	20	2051	
86-169-2	102A	3004	53	2007	
86-170-2	128	3124/289	243	2030	
86-171-2	123A	3444	20	2051	
86-172-2	102A		53	2007	
86-173-2	121	3009	239	2006	
86-173-9	121	3009	239	2006	
86-175-2	123A	3444	20	2051	
86-176-2	102A		53	2007	
86-177-2	157	3021/124	232		
86-178-2	159	3466	82	2032	
86-178-20	159	3466	82	2032	
86-179-2	160	3006	245	2004	
86-18-1	113A	3119/113	6GC1		
86-18-1A	113A	3119/113	6GC1		
86-18-2	160	3007	245	2004	
86-180-2	160	3006	245	2004	
86-181-2	160	3006	245	2004	
86-182-2	123A	3444	20	2051	
86-183-2	159	3466	82	2032	
86-183-20	159	3466	82	2032	
86-185-2	319P	3018	283	2016	
86-186-2	161	3117	39	2015	
86-188-2	123A	3444	20	2051	
86-189-2	123A	3444	20	2051	
86-19-2	121	3009	239	2006	
86-190-2	123A	3444	20	2051	
86-191-2	123A	3444	20	2051	
86-192-2	123A	3444	20	2051	
86-193-2	123A	3444	20	2051	
86-194-2	123A	3444	20	2051	
86-195-2	123A	3444	20	2051	
86-196-2	123A	3444	20	2051	
86-197-2	123A	3444	20	2051	
86-198-2	123A	3444	20	2051	
86-199-2	123A	3444	20	2051	
86-20-1	109	3091		1123	
86-20-2	160	3008	245		
86-201-2	123A	3444	20	2051	
86-202-2	123A	3444	20	2051	
86-204-2	161	3018	39	2015	
86-205-2	161	3117	39	2015	
86-207-2	128	3024	243	2030	
86-208-2	128	3024	243	2030	
86-21-1	116	3016	504A	1104	
86-21-2	102A	3005	53	2007	
86-210-2	128	3024	243	2030	
86-211-2	128	3024	243	2030	
86-212-2	194		220		
86-213-2	154	3045/225	40	2012	
86-214-2	154	3045/225	40	2012	
86-215-2	154	3045/225	40	2012	
86-216-2	159	3466	82	2032	
86-217-2	159	3466	82	2032	
86-217-20	159	3466	82	2032	
86-218-2	159	3466	82	2032	
86-218-20	159	3466	82	2032	
86-219-2	159	3466	82	2032	
86-22-1	109	3088		1123	
86-22-2	102A	3005	53	2007	
86-22-3	116	3016	504A	1104	
86-221-2	163A	3439	36		
86-222-2	163A	3439	36		
86-224-2	163A	3439	36		
86-225-2	163A	3439	36		
86-227-2	124	3021	12		
86-228-2	157	3103A/396	232		
86-23-1	507	3311	504A	1104	
86-23-2	102A	3003	53	2007	
86-23-3	116	3016	504A	1104	
86-230-2	121	3009	239	2006	
86-231-2	121	3009	239	2006	
86-232-2	121	3009	239	2006	
86-233-2	159	3466	82	2032	
86-234-2	128	3024	243	2030	
86-235-2	121	3009	239	2006	
86-236-2	124	3021	12		
86-237-2	123A	3444	20	2051	
86-238-2	123A	3444	20	2051	
86-24-2	103A	3010	59	2002	
86-243-2	108	3452	86	2038	
86-244-2	108	3452	86	2038	
86-245-2	108	3452	86	2038	
86-246-2	159	3466	82	2032	
86-246-20	159	3466	82	2032	
86-247-2	123A	3444	20	2051	
86-248-2	121	3009	239	2006	
86-249-2	102A		53	2007	
86-249-9		3021	12		
86-25-1			ZD-12	563	
86-25-2	103A	3010	59	2002	
86-250-2	123A	3444	20	2051	
86-251-2	159	3466	82	2032	
86-251-20	159	3466	82	2032	
86-253-2	160	3006	245	2004	
86-254-2	160	3006	245	2004	
86-255-2	123A	3444	20	2051	
86-256-2	123A	3444	20	2051	
86-257-2	157	3021/124	232		
86-259-2	124	3021	12		
86-26-1		3016	504A	1104	
86-26-2	101	3861	8	2002	
86-260-2	124	3021	12		
86-261-2	124	3021	12		
86-262-0	165		38		
86-262-2	161	3117	39	2015	
86-262-9		3115	38		
86-263-2	161	3117	39	2015	
86-264-2	123A	3444	20	2051	
86-265-2	123A	3444	20	2051	
86-266-2	128	3024	243	2030	
86-267-2	128	3024	243	2030	
86-27-1	177	3031A	300	1122	
86-27-2	102A	3005	53	2007	
86-271-2	175		246	2020	
86-272-2	175		246	2020	
86-273-2	128	3024	243	2030	
86-275-2	124	3021	12		
86-276-2	159	3466	82	2032	
86-276-20	159	3466	82	2032	
86-277-2	123A	3444	20	2051	
86-278-2	160	3006	245	2004	
86-279-2	160	3006	245	2004	
86-28-2	102A	3005	53	2007	
86-28-3	116	3016	504A	1104	
86-280-2	126	3006/160	52	2024	
86-281-2	126	3008	52	2024	
86-282-2	126	3008	52	2024	
86-283-2	102A	3008	53	2007	
86-284-2	188	3024/128	226	2018	
86-286-2	159	3466	82	2032	
86-286-20	159	3466	82	2032	
86-287-2	157	3103A/396	232		
86-289-2	161	3117	39	2015	
86-29-2	102A	3003	53	2007	
86-290-2	161	3117	39	2015	
86-291-2	128	3124/289	243	2030	
86-291-9	123A	3444	20	2051	
86-292-2	127	3764	25		
86-292-9	127	3764			
86-293-2	123A	3444	20	2051	
86-294-2	159	3466	82	2032	
86-294-20	159	3466	82	2032	
86-295-2	102A	3008	53	2007	
86-296-2	160	3006	245	2004	
86-297-2	102A		53	2007	
86-298-2	159	3114/290	82	2032	
86-298-20	159	3114/290	82	2032	
86-3-1	114	3120	6GD1	1104	
86-3-3	116	3017B/117	504A	1104	
86-30-1			504A	1104	

Industry Standard No.	ECG	SK	GE	RS 276-	MOTOR.
86-30-2	102A	3004	53	2007	
86-30-3	116	3016	504A	1104	
86-300-2	102A		53	2007	
86-301-2	103A	3835	59	2002	
86-303-2	102A	3004	53	2007	
86-304-2	102A	3004	53	2007	
86-305-2	102A	3004	53	2007	
86-306-2		3122	18	2030	
86-308-2	123A	3124/289	20	2051	
86-309-2	123A	3444	20	2051	
86-31-1	142A	3062	ZD-12	563	
86-31-2	101	3861	8	2002	
86-310-2	123A	3444	20	2051	
86-311-2	126	3008	52	2024	
86-312-2	160	3008	245	2004	
86-313-2	121	3009	239	2006	
86-316-2	154	3045/225	40	2012	
86-317-2	121	3009	239	2006	
86-319-2	121	3009	239	2006	
86-32-1	116	3031A	504A	1104	
86-32-2	102A	3005	53	2007	
86-320-2	160		245	2004	
86-321-2	160		245	2004	
86-322-2	160		245	2004	
86-323-2	123A	3444	20	2051	
86-324-2	123A	3444	20	2051	
86-327-2	123A	3444	20	2051	
86-328-2	123A	3444	20	2051	
86-329-2	129	3025	244	2027	
86-33-2	102A	3005	53	2007	
86-330-2	128	3024	243	2030	
86-334-2	129	3025	244	2027	
86-336-2	128	3024	243	2030	
86-339-2	123A	3444	20	2051	
86-339-9	123A	3444	20	2051	
86-34-3	116	3016	504A	1104	
86-340-2	159	3466	82	2032	
86-340-20	159	3466	82	2032	
86-342-2	123A	3444	20	2051	
86-344-2	184	3054/196	57	2017	
86-345-2	185	3025/129	58	2025	
86-347-2	160		245	2004	
86-348-2	160	3008	245	2004	
86-35-1	139A	3066/118	ZD-9.1	562	
86-35-2	103A	3010	59	2002	
86-35-3	116	3017B/117	504A	1104	
86-353-2	121	3009	239	2006	
86-354-2	121	3009	239	2006	
86-359-2	123A	3444	20	2051	
86-36-1			504A	1104	
86-36-2	160	3007	245	2004	
86-362-2	123A	3444	20	2051	
86-363-2	160	3006	245	2004	
86-365-2	123A	3444	20	2051	
86-366-2	160		245	2004	
86-367-2	160	3006	245	2004	
86-368-2	160	3006	245	2004	
86-37-1	139A	3060	ZD-9.1	562	
86-37-2	160	3007	245	2004	
86-37-3	120	3110	CR-3		
86-370-2	179	3642	76	2006	
86-370-2 GRN	121	3717			
86-370-2 ORN	121	3717			
86-370-2 VIO	121	3717			
86-370-2(GRN)			239	2006	
86-370-2(ORN)			239	2006	
86-370-2(VIO)			239	2006	
86-370-2YEL	121	3642/179	239	2006	
86-373-2	160	3006	245	2004	
86-374-2	160	3006	245	2004	
86-376-2	160	3006	245	2004	
86-379-2	123A	3444	20	2051	
86-38-1			504A	1104	
86-38-2	160	3007	245	2004	
86-381-2	161	3716	39	2015	
86-386-2	108	3452	86	2038	
86-389-2	123	3018	20	2051	
86-39-2	102A	3003	53	2007	
86-390-2	123A	3444	20	2051	
86-391-2	123A	3444	20	2051	
86-392-2	102A		53	2007	
86-393-2	128	3024	243	2030	
86-396-2	185	3025/129	58	2025	
86-399-1	123A	3444	20	2051	
86-399-2	123A	3444	20	2051	
86-399-9	123A	3444	20	2051	
86-4-1	116	3016	504A	1104	
86-4-2	101	3861	8	2002	
86-40-3	116	3016	504A	1104	
86-400-2	199	3275/194	62	2010	
86-403-2	123A	3444	20	2051	
86-406-2	159	3466	82	2032	
86-407-2	159	3466	82	2032	
86-41-1	178MP	3100/519	300(2)	1122(2)	
86-41-1(SEARS)	177		300	1122	
86-412-2	175	3538	246	2020	
86-416-2	108	3452	86	2038	
86-417-2	108	3452	86	2038	
86-419-2	102A	3004	53	2007	
86-42-1	123A		20	2051	
86-42-3	116	3016	504A	1104	
86-420-2	123A	3444	20	2051	
86-421-2	102A	3004	53	2007	
86-422-2	289A	3849/293	268	2038	
86-422-3	188	3199	226	2018	
86-423-2	294	3114/290	269	2032	
86-423-3	189	3200	218	2026	
86-428-2		3024		2030	
86-428-9	128	3024	243	2030	
86-431-2	129	3025	244	2027	
86-431-9	129	3025	244	2027	
86-44-2	101	3861	8	2002	
86-44-3	118	3066	CR-1		
86-440-2	128	3024	243	2030	
86-441-2	128	3024	243	2030	
86-442-2	161	3018	39	2015	
86-444-2	199	3020/123	62	2010	
86-445-2	123A	3444	20	2051	
86-449-2	126	3006/160	52	2024	
86-449-9	160	3006	245	2004	
86-45-1	109	3091		1123	
86-45-2	102A	3003	53	2007	
86-45-3	119	3109	CR-2		
86-452-2	128	3024	243	2030	
86-457-2	123A	3444	20	2051	
86-458-2	123A	3444	20	2051	
86-459-2	159	3466	82	2032	
86-46-2	102A	3005	53	2007	
86-46-3	116	3016	504A	1104	
86-460-2	123A	3444	20	2051	
86-461-2	123A	3444	20	2051	
86-462-2	123A	3444	20	2051	
86-463-2	128	3024	243	2030	
86-463-3		3024	18	2030	
86-463-3(SEARS)	128		243	2030	
86-464-2	222	3050/221	FET-4	2036	
86-465-2		3018	20	2051	
86-467-2	108	3452	86	2038	
86-47-2	102A	3005	53	2007	
86-47-3		3110	CR-3		
86-472-2	123A	3444	20	2051	
86-475-2	159	3466	82	2032	
86-476-2	102A	3004	53	2007	
86-477-2	312	3112	FET-1	2028	
86-48-1	110MP	3709		1123(2)	
86-48-2	102A	3005	53	2007	
86-480-9	131		3	2006	
86-481-1	123A	3444	20	2051	
86-481-2	123A	3444	20	2051	
86-482-2	159	3466	82	2032	
86-483-2	123A	3444	20	2051	
86-483-3	123A	3444	20	2051	
86-484-2	123A	3444	20	2051	
86-485-2	123A	3444	20	2051	
86-486-2	123A	3444	20	2051	
86-487-2	124	3021	12		
86-487-3	124	3021	12		
86-488-2	108	3452	86	2038	
86-49-1	109			1123	
86-49-2	102A	3003	53	2007	
86-49-3	116	3016	504A	1104	
86-490-2	108	3452	86	2038	
86-491-2	108	3452	86	2038	
86-493-2	123A	3444	20	2051	
86-494-2	123A	3444	20	2051	
86-495-2	123A	3444	20	2051	
86-496-2	123A	3444	20	2051	
86-497-2	102A	3004	53	2007	
86-5-2	103A	3835	59	2002	
86-50-2	102A	3004	53	2007	
86-50-3	116	3016	504A	1104	
86-500-2	312			2028	
86-5000-2	102A	3004	53	2007	
86-5000-3	116	3017B/117	504A	1104	
86-5001-2	102A	3123	53	2007	
86-5001-3	116	3017B/117	504A	1104	
86-5002-3	116	3017B/117	504A	1104	
86-5003-2	103A	3011	59	2002	
86-5003-3	116	3017B/117	504A	1104	
86-5004-2	103A	3011	59	2002	
86-5005-2	103A	3835	59	2002	
86-5006-2	102A	3004	52	2007	
86-5006-3	116	3017B/117	504A	1104	
86-5007-2	103A	3011	59	2002	
86-5007-3	109	3087		1123	
86-5008-2	103A	3011	59	2002	
86-5009-3	116	3016	504A	1104	
86-501-2	159	3466	82	2032	
86-5010-3	177	3016	300	1122	
86-5011-2	103A	3011	59	2002	
86-5011-3	177	3016	300	1122	
86-5012-2	103A	3011	59	2002	
86-5012-3	116	3017B/117	504A	1104	
86-5013-2	103A	3011	59	2002	
86-5015-2	103A	3011	59	2002	
86-5015-3	177	3016	300	1122	
86-5016-2	103A	3011	59	2002	
86-5017-2	103A	3011	59	2002	
86-5018-2	123A	3011	20	2051	
86-502-2	123A	3444	20	2051	
86-5024-3	116	3016	504A	1104	
86-5026-2	103A	3011	59	2002	
86-5027-2	102A		53	2007	
86-5027-3	177	3016	300	1122	
86-5028-3	5802	9005	510	1143	
86-5029-2	103A	3011	59	2002	
86-5029-3	116	3016	504A	1104	
86-5032-3	5802	9005	510	1143	
86-5034-2	103A	3835	59	2002	
86-5037-3	177		300	1122	
86-5039-2	121	3014	239	2006	
86-5040-2	123A	3444	20	2051	
86-5041-2	123A	3444	20	2051	
86-5042-2	102A	3123	52	2007	
86-5043-2	121	3009	239	2006	
86-5044-2	123A	3444	59	2051	
86-5045-2	123A	3444	20	2051	
86-5046-2	123A	3444	20	2051	
86-5047-2	103A	3011	59	2002	
86-5048-2	103A	3011	59	2002	
86-5049-2	123A	3444	20	2051	
86-5050-2	123A	3444	20	2051	
86-5051-2	123A	3444	20	2051	
86-5052-2	158		53	2007	
86-5055-2	123A	3444	20	2051	
86-5056-2	123A	3444	17	2051	
86-5057-2	121	3717	239	2006	
86-5058-2	121	3009	3	2006	
86-506-2	184	3188A/182	57	2017	
86-5060-2	103A	3011	5	2002	
86-5061-2	103A	3011	59	2002	
86-5062-2	103A	3011	5	2002	
86-5063-2	102A	3123	53	2007	
86-5064-2	129	3025	244	2027	
86-5064-2A	129	3025	244	2027	
86-5065-2	123	3124/289	20	2051	
86-5067-2	102A	3004	53	2007	
86-507-2	185	3083/197	58	2025	
86-5070-2	128		243	2030	
86-5073-2	128	3024	243	2030	
86-5074-2	128		243	2030	
86-5075-2	192		63	2030	
86-5079-2	159(2)	3114/290	82(2)		
86-5080-2	103A	3011	59	2002	
86-5081-2	123A	3444	20	2051	
86-5082-2	129	3025	244	2027	
86-5082-2A	129	3025	244	2027	
86-5083-2	121	3717	239	2006	
86-5084-2	130	3027	14	2041	
86-5084-2A	130	3027	14	2041	
86-5085-2	175	3131A/369	246	2020	
86-5086-2	103A	3835	59	2002	
86-5087-2	103A	3011	59	2002	
86-5088-2	121	3009	239	2006	
86-5089-2	121	3717	239	2006	
86-509-2	102A	3005	53	2007	
86-5090-2	121	3717	239	2006	
86-5091-2	102A	3123	53	2007	
86-5093-2	128	3024	243	2030	
86-5095-2	312	3112	FET-1	2028	
86-5096-2	312	3112	FET-1	2028	
86-5097-2	123A	3444	20	2051	
86-5099-2	123A(2)	3124/289	20(2)		
86-51-2			504A	1104	
86-51-3	116	3016	504A	1104	
86-510-2	128	3024	243	2030	
86-5100-2	184	3190	57	2017	
86-5101-2	130	3027	14	2041	

Industry Standard No.	ECG	SK	GE	RS 276-	MOTOR.
86-5101-2A	130	3027	14	2041	
86-5102-2	152	3893	66	2048	
86-5103-2	123A	3444	20	2051	
86-5104-2	193	3114/290	67	2023	
86-5105-2	172A	3156	64		
86-5106-2	175		246	2020	
86-5107-2	152	3893	66	2048	
86-5108-2	152	3893	66	2048	
86-5109-2	411/128	3044/154	28		
86-511-9	294		61	2038	
86-5110-2	123A	3444	20	2051	
86-5111-2	123A	3444	20	2051	
86-5112-2	130		14	2041	
86-5112-2(THOMAS)	130		20	2041	
86-5113-2	121	3717	239	2006	
86-5114-2	123A	3444	20	2051	
86-5117-2	123A	3444	20	2051	
86-512-2	105	3012	4		
86-5122-2	312	3112	FET-1	2028	
86-5125-2	121	3717	239	2006	
86-513-2	233	3065/222	210	2009	
86-514-2	123A	3444	20	2051	
86-515-2	194	3018	220		
86-515-2(SEARS)	123A	3444	20	2051	
86-52-1	116	3311	504A	1104	
86-520-2	123A	3444	20	2051	
86-521-2	194	3124/289	220		
86-525-2	107	3039/316	11	2015	
86-526-2	199	3018	62	2010	
86-527-2	159	3466	82	2032	
86-528-2	159	3466	82	2032	
86-529-2	152	3054/196	66	2048	
86-530-2	153	3083/197	69	2049	
86-533-2	159	3466	82	2032	
86-534-2	123A	3444	20	2051	
86-5342-2			53	2007	
86-536-2			10	2051	
86-537-2			10	2051	
86-538-2			10	2051	
86-539-2	123A	3444	20	2051	
86-54-2	102A		53	2007	
86-54-3	116	3016	504A	1104	
86-540-2	221	3050	FET-4	2036	
86-541-2	172A	3156	64		
86-543-2	128		243	2030	
86-544-2	152	3893	66	2048	
86-547-2	159	3466	82	2032	
86-548-2	123A	3444	20	2051	
86-549-2	172A	3156	64		
86-55-3	120	3110	CR-3		
86-550-2	194	3122	220		
86-551-2	123A	3444	20	2051	
86-552-2	159	3466	82	2032	
86-554-2	123A	3444	20	2051	
86-555-2	290A	3466/159	82	2032	
86-556-2	171	3104A	27		
86-557-2	194	3275	220		
86-559-2	123A	3444	20	2051	
86-56-3	120	3110	CR-3	1104	
86-56-3C	120	3110	CR-3		
86-56-3F	120	3110	CR-3		
86-560-2	123A	3444	20	2051	
86-561-2	123A	3444	20	2051	
86-563-2	165	3115	38		
86-563-9	165	3115	38		
86-564-2	283	3467-283	20	2051	
86-564-3	165	3115	38		
86-564-9	165	3115	38		
86-565-2	123A	3444	20	2051	
86-566-2	194	3275	220		
86-567-2	128	3020/123	243	2030	
86-568-2	196	3054	241	2020	
86-57-3	116	3016	504A	1104	
86-570-2	159	3114/290	223		
86-572-2	6402	3628			
86-572-3	6402	3628			
86-573-2	123A	3444	20	2051	
86-574-2		3100	6GC1	1122	
86-574-2(DIODE)	178MP		300(2)	1122(2)	
86-58-1	139A	3060	ZD-9.1	562	
86-58-2	123A	3010	20	2051	
86-58-3	116	3016	504A	1104	
86-59-1	116		504A	1104	
86-59-2	102A	3005	53	2007	
86-59-3	116	3016	504A	1104	
86-5911-3	116		504A	1104	
86-593-2	229	3018	61	2038	
86-593-9	229	3018	60	2038	
86-594-2	229	3018	61	2038	
86-5943-2	121	3009	239	2006	
86-595-2	123A	3444	20	2051	
86-596-2	107	3018	11	2015	
86-597-2	107	3018	11	2015	
86-598-2	123A	3444	20	2051	
86-599-2	123A	3444	20	2051	
86-6-2	103A	3010	59	2002	
86-60-1	109			1123	
86-60-2	102A	3005	53	2007	
86-60-3	116	3017B/117	504A	1104	
86-600-2	159	3466	82	2032	
86-601-2	192	3054/196	63	2030	
86-602-2	193	3083/197	67	2023	
86-604-2			66	2048	
86-605-2	222	3050/221	FET-4	2036	
86-606-2	222	3050/221	FET-4	2036	
86-607-2	199	3122	283	2016	
86-608-2	234	3118	65	2050	
86-609-2	297	3122	271	2030	
86-609-9	297	3137/192A	271	2030	
86-61-1	142A	3093	ZD-12	563	
86-61-2	102A	3003	53	2007	
86-610-2	298	3450	272		
86-610-9	290A	3114/290	269	2032	
86-611-2	194	3275	220		
86-612-2	154	3044	40	2012	
86-613-2	291	3054/196	28		
86-614-2	292	3441	29		
86-615-2	194	3275	220		
86-616-2	129		244	2027	
86-619-2	229	3018	61	2038	
86-62-1	177	3100/519	300	1122	
86-62-2	121	3009	239	2006	
86-62-3	116	3016	504A	1104	
86-620-2	229	3018	61	2038	
86-621-2	229	3018	61	2038	
86-622-2	159	3466	82	2032	
86-624-2	124	3104A	12		
86-624-9	124	3021	12		
86-625-2	222	3050/221	FET-4	2036	
86-626-2	238	3710	38		
86-626-9	165	3111	38		
86-628-2	171	3104A	27		
86-628-9	171	3104A	27		
86-629-2	154	3044	40	2012	

Industry Standard No.	ECG	SK	GE	RS 276-	MOTOR.
86-63-2	121	3009	239	2006	
86-63-3	116	3016	504A	1104	
86-630-2	194	3124/289	220		
86-631-2	171	3104A	27		
86-632-2	222	3050/221	FET-4	2036	
86-633-2	165	3111	38		
86-633-9	238	3710	259		
86-64-1	109	3087		1123	
86-646-2	123A	3444	20	2051	
86-648-2	198	3054/196	251		
86-649-2	194	3124/289	220		
86-65-1	552	3062/142A	ZD-12	563	
86-65-3	116	3017B/117	504A	1104	
86-65-4	177		300	1122	
86-65-9	116	3031A	504A	1104	
86-650-2	128	3122	243	2030	
86-651-2	287	3433	222		
86-655-2	123A	3444	20	2051	
86-659-2	189	3200	84	2026	
86-66-1	5079A	3335	ZD-20		
86-660-2	188	3199	83	2018	
86-661-2	123A	3122	20	2051	
86-663-2	152	3054/196	66	2048	
86-664-2	171	3232/191	27		
86-665-2	130	3027	14	2041	
86-665-9		3027	14	2041	
86-668-2	6401		2N2160	2029	
86-669-2	159	3114/290	22	2032	
86-67-0	116		504A	1104	
86-67-1	143A	3750	ZD-13	563	
86-67-2	125	3032A	510,531	1114	
86-67-3	552	3100/519	504A	1104	
86-67-8		3017B	504A	1104	
86-67-9	116	3032A	504A	1104	
86-671-2	263	3180			
86-672-2	171	3104A	32		
86-673-2	210	3202	252	2018	
86-674-2	128	3512	45		
86-675-2	128	3122	20	2051	
86-68-3	166	9075	BR-600	1152	
86-68-3A	166	9075		1152	
86-7-1	116	3016	504A	1104	
86-70-1	177	3100/519	300	1122	
86-702-2	123A	3444	20	2051	
86-72-2	102A	3003	53	2007	
86-72-3	125	3031A	510,531	1114	
86-73-1	5079A	3335	ZD-20		
86-73-2	102A	3005	53	2007	
86-74-1	177	3087	300	1122	
86-74-2	102A	3003	53	2007	
86-74-9	177	3100/519	300	1122	
86-75-2	102A	3005	53	2007	
86-75-3		3032A	504A	1104	
86-75-3(SEARS)	116		504A	1104	
86-76-1	177	3100/519	300	1122	
86-76-2	103A	3010	59	2002	
86-77-1	177	3100/519	300	1122	
86-77-2	102A	3004	53	2007	
86-78-1	177	3100/519	300	1122	
86-78-2	102A	3005	53	2007	
86-78-3			504A	1104	
86-79-2	102A	3005	53	2007	
86-8-2	121	3009	239	2006	
86-80-1	116	3311	504A	1104	
86-80-2	102A	3004	53	2007	
86-80-3	116	3311	504A	1104	
86-81-2	103A	3011	59	2002	
86-82-2	102A	3004	53	2007	
86-83-2	102A	3004	53	2007	
86-84-1	116	3126	504A	1104	
86-84-2	102A	3004	53	2007	
86-85-1	142A	3062	ZD-12	563	
86-85-3	116	3032A	504A	1104	
86-86-2	160	3008	245	2004	
86-86560-3	506	3925/525			
86-87-2	160	3008	245	2004	
86-87-3		3126	90		
86-88-1	5079A	3335	ZD-20		
86-88-2	160	3006	245	2004	
86-88-3	109	3088		1123	
86-89-1	116	3016	504A	1104	
86-89-2	160	3006	245	2004	
86-89-3	116	3311	504A	1104	
86-9-1	113A	3119/113	6GC1		
86-9-3	116	3016	504A	1104	
86-90-2	160	3006	245	2004	
86-91-1	5081A	3151	ZD-24		
86-91-2	160	3006	245	2004	
86-92-1			ZD-12	563	
86-92-2	160	3006	245	2004	
86-93-1	5075A	3751	ZD-16	564	
86-93-2	102A	3004	53	2007	
86-94-1	147A	3095	ZD-33		
86-95-1	5137A	3403	ZD-24		
86-95-2	102A	3123	53	2007	
86-96-1	5166A	3432			
86-97-1	113A	3119/113	6GC1		
86-98-2	102A	3004	53	2007	
86-99-2	160	3008	245	2004	
86A318	102A	3004	53	2007	
86A327	199	3245	62	2010	
86A332	130		14	2041	
86A334	123A	3444	20	2051	
86A335	159	3466	82	2032	
86A336	123A	3444	20	2051	
86A338	175		246	2020	
86A339	218		234		
86A350	123A	3444	20	2051	
86A86A327			20	2051	
86C04	185		58	2025	
86P1AA	159		82	2032	
86X-55-1	728		IC-22		
86X-56-1	729		IC-23		
86X00011-001	102	3722	2	2007	
86X0006-001	123A	3444	20	2051	
86X0007-001	123A	3444	20	2051	
86X0007-004	108	3452	10	2038	
86X0007-104	123A	3444	20	2051	
86X0007-204	108	3124/289	20	2038	
86X0008-001	123A	3444	10	2051	
86X0009-001	121	3009	16	2006	
86X0011-001	126	3008	2	2024	
86X0012-001	123	3124/289	10	2051	
86X0013-001	126	3006/160	52	2024	
86X0014-001	100	3006/160	1	2007	
86X0015-001	121	3009	239	2006	
86X0016-001	159	3466	82	2032	
86X0016-001A	159	3466	82	2032	
86X0017-001	102A	3004	53	2007	
86X0018-001	102A	3004	53	2007	
86X0019-001		3018	11	2015	
86X0022-001	123A	3124/289	20	2051	
86X0024-001	704	3023	IC-205		
86X0025-001	123	3124/289	20	2051	

Industry Standard No.	ECG	SK	GE	RS 276-	MOTOR.
86X0025-001(TIK)			10	2051	
86X0027-001	704	3023	IC-205		
86X0028-001	124	3021	12		
86X0029-001	123A	3444	20	2051	
86X0030-001	121MP	3014	239	2006(2)	
86X0030-100	121MP	3013	239	2006(2)	
86X0031-001	123A	3444	20	2051	
86X0031-002	123A	3444	20	2051	
86X0031-003	123A	3444	20	2051	
86X0032-001	123A	3444	20	2051	
86X0033-001	131		44	2006	
86X0034-001	233	3124/289	210	2009	
86X0035-001	123A	3444	20	2051	
86X0036-001		3124	22	2032	
86X0036-001A	159	3466	82	2032	
86X0037-001		3862/103	54		
86X0037-002	102A	3004	53	2007	
86X0037-100		3010	54		
86X0038-001	108	3452	86	2038	
86X0040-00	123A	3444	20	2051	
86X0040-001	123A	3444	20	2051	
86X0041-001	159	3466	82	2032	
86X0041-001A	159	3466	82	2032	
86X0042-001	242	3191/185	58	2027	
86X0042-002	241	3190/184	57	2020	
86X0043-001	108	3452	86	2038	
86X0044-001	159	3466	82	2032	
86X0044-001A	159	3466	82	2032	
86X0045-001	123A	3444	20	2051	
86X0046-001	159	3466	82	2032	
86X0047-001	159	3466	82	2032	
86X0048-001	123A	3444	20	2051	
86X0049-001	194	3275	220		
86X0050-001	123A	3444	20	2051	
86X0051-001	123A	3444	20	2051	
86X0052-001	319P	3117	39	2015	
86X0053-001	712	3072	IC-2		
86X0054-001	123A	3444	20	2051	
86X0055-001	728	3073	IC-22		
86X0056-001	729	3074	IC-23		
86X0058-001	123A	3444	20	2051	
86X0058-002	123A	3444	20	2051	
86X0058-003	123A	3444	20	2051	
86X0059-001	184	3054/196	57	2017	
86X0059-002	182	3054/196	55		
86X006-001	123A	3444	20	2051	
86X0060-001	107	3018	11	2015	
86X0061-001	107	3018	11	2015	
86X0062-001	107	3124/289	11	2015	
86X0063-001	123A	3444	20	2051	
86X0064-001	799	3238	IC-34		
86X0065-001	171	3201	27		
86X0066-001	159	3114/290	82	2032	
86X0066-003	159	3114/290	21	2034	
86X0067-001	310	3856			
86X0068-001	308	3855			
86X0069-001	230	3042			
86X007-004	123A	3444	20	2051	
86X007-034	123A	3444	20	2051	
86X0070-001	6402	3628			
86X0071-001	194	3024/128	243	2030	
86X0072-001	159	3025/129	82	2032	
86X0073-001	188	3054/196	226	2018	
86X0073-002	157	3199/188	226	2018	
86X0074-001	331		66	2017	
86X0075-001	332		58	2025	
86X0076-001	228A	3220/198	257		
86X0077-001	230	3042			
86X0077-002	231	3042/230			
86X0078-001	231	3857			
86X0079-001	123A	3444	20	2051	
86X0080-001	242	3189A/183	58	2027	
86X0080-002	241	3188A/182	57	2020	
86X0081-001	5457	3598	MR-5	1067	
86X0081-002	5457	3597/5455	MR-5		
86X0083-001	198	3104A			
86X0084-001	815	3255	IC-244		
86X0089-001	172A	3156	64		
86X0090-001	123A	3124/289	20	2051	
86X037-001	103		8	2002	
86X2	179	3009	239	2006	
86X3	100	3004/102A	53	2007	
86X34-1	123A	3444	20	2051	
86X46	159	3466	82	2032	
86X47	159	3466	82	2032	
86X483-2			18	2030	
86X53-1	712	3072	IC-2		
86X55-1	728	3073	IC-22		
86X56-1	729	3074	IC-23		
86X56-2	729	3074	IC-23		
86X6	123A	3444	20	2051	
86X6-1	123A	3444	20	2051	
86X6-4-518	123A	3444	20	2051	
86X6029-001	108	3452	86	2038	
86X7-2	123A	3444	20	2051	
86X7-3	123A	3444	20	2051	
86X7-4	123A	3444	20	2051	
86X7-6	107	3124/289	11	2015	
86X7-6013	108	3452	86	2038	
86X8-1	123A	3444	20	2051	
86X8-2	123A	3444	20	2051	
86X8-3	123A	3124/289	20	2051	
86X8-4	123A	3444	20	2051	
0087	146A	3719/104	ZD-27	2006	
87-0001	312	3834/132	FET-1	2035	
87-0002	229	3018	61	2038	
87-0002-1	229	3132	61	2038	
87-0003	229	3039/316	61	2038	
87-0004	1006	3358	IC-38		
87-0005	199	3124/289	62	2010	
87-0006	192	3124/289	63	2030	
87-0009	199	3124/289	62	2010	
87-0013	199	3124/289	62	2010	
87-0014	123A	3444	20	2051	
87-0015	126	3008	52	2024	
87-0016	126	3005	52	2024	
87-0017	126	3008	52	2024	
87-0018	102A	3004	53	2007	
87-0019	102A	3004	53	2007	
87-0020	102A	3004	53	2007	
87-0021	102A	3004	53	2007	
87-0023-7			61	2038	
87-0023-T	199	3122	61		
87-0027	108	3452	86	2038	
87-0028			67	2023	
87-0029	193		67	2023	
87-0203-1	289A	3137/192A	268	2038	
87-0212-1	199	3124/289	62	2010	
87-0217	1185	3468	IC-139		
87-0218-U	199		85	2010	
87-0227	199	3124/289	62	2010	
87-0228	291	3440			
87-0229	292	3441			
87-0230	199	3245	62	2010	
87-0230-1	199	3245	62	2010	
87-0231	199	3245	62	2010	
87-0233	1242	3483	IC-291		
87-0234	801	3160	IC-155		
87-0235	229	3038	212	2010	
87-0235-C	229	3038	212		
87-0235A	229		N-212	2038	
87-0235B	199	3038	212	2010	
87-0235C	229		212	2038	
87-0236-Q	293	3849	47		
87-0236-R	293	3849	47		
87-0236-S	293	3849	47		
87-0236Q			47	2030	
87-0236R	293		47	2030	
87-0236S			47	2030	
87-0237-R	294	3841	48		
87-0237-S	294	3841	48		
87-0238	199	3124/289	62	2010	
87-0239-A	287	3124/289	222		
87-0240	1223	3493			
87-0246	1223	3493	IC-295		
87-10-0	109	3087	1N34AS	1123	
87-10-1	110MP	3087	1N34AS	1123	
87-104-3	506	3130	511	1114	
87-190XX-001	177		300	1122	
87-218-U	199	3124/289	85		
87-423-2	290A	3114/290	269	2032	
87-5-3	120		CR-3		
87-56-3		3110	CR-3		
87-593-2	229		60	2015	
87-67-3	116	3017B/117	504A	1104	
87B02	188		226	2018	
87M42899B39	735		IC-212		
87S228					MCM7681
87S296					MCM7641
88-0550	1115	3184	IC-278		
88-125	116	3016	504A	1104	
88-1250108	199	3038	62	2010	
88-1250109	123A	3444	20	2051	
88-18920	1058	3459	IC-49		
088-2	125		510,531	1114	
88-20372	788	3829	IC-49		
88-20404	801	3160			
88-3	116		504A	1104	
88-77-1	177	3100/519	300	1122	
88-831	116	3016	504A	1104	
88-832	116	3016	504A	1104	
88-833	116	3016	504A	1104	
88-9132	OBS-NLA		IC-277		
88-9132P	OBS-NLA		IC-277		
88-9302	744	3171	IC-215	2022	
88-9302R	744		IC-215	2022	
88-9302RS			IC-24	2022	
88-9302S	744		IC-215	2022	
88-9304	804	3455	IC-27		
88-9574	737	3375	IC-16		
88-9575		3163	IC-211		
88-9779	736		IC-17		
88-9779P	736		IC-17		
88-9841R	789	3078			
88-9842P	723	3144	IC-15		
88-9842R	723	3144	IC-15		
88-9842S	723	3144	IC-15		
88A7522	949		IC-25		
88A752271	949		IC-25		
088B	160	3006	245	2004	
088B-12	160	3006	245	2004	
088B-12-7	160	3006	245	2004	
88B-70	160	3006	245	2004	
88B-70-12	160	3006	245	2004	
88B-70-12-7	160	3006	245	2004	
88B0	160	3006	245	2004	
88B1	160	3006	245	2004	
88B10	160	3006	245	2004	
88B10R	160	3006	245	2004	
88B11	160	3006	245	2004	
88B119	160	3006	245	2004	
88B12	160	3006	245	2004	
88B121	160	3006	245	2004	
88B13	160	3006	245	2004	
88B13P	160	3006	245	2004	
88B14	160	3006	245	2004	
88B14-7	160	3006	245	2004	
88B14-7B	160		245	2004	
88B15	160	3006	245	2004	
88B15L	160	3006	245	2004	
88B16	160	3006	245	2004	
88B16E	160	3006	245	2004	
88B17	160	3006		2004	
88B17-1	160	3006	245	2004	
88B18	160	3006	245	2004	
88B182	160	3006	245	2004	
88B19	160	3006	245	2004	
88B19G	160	3006	245	2004	
88B2	160	3006	245	2004	
88B3	160	3006	245	2004	
88B4	160	3006	245	2004	
88B5	160	3006	245	2004	
88B6	160	3006	245	2004	
88B7	160	3006	245	2004	
88B725B	709		IC-11		
88B8	160	3006	245	2004	
88B9	160	3006	245	2004	
88C	100	3005	1	2007	
088C-12	100	3005	1	2007	
088C-12-7	100	3005	1	2007	
88C-70	100	3005	1	2007	
88C-70-12	100	3721	1	2007	
88C-70-12-7	100	3005	1	2007	
88C0	100	3005	1	2007	
88C1	100	3005	1	2007	
88C10	100	3005	1	2007	
88C10R	100	3005	1	2007	
88C11	100	3005	1	2007	
88C119	100	3005	1	2007	
88C12	100	3005	1	2007	
88C121	100	3005	1	2007	
88C13	100	3005	1	2007	
88C13P	100	3005	1	2007	
88C14	100	3005	1	2007	
88C14-7	100	3005	1	2007	
88C14-7B	100	3721	1	2007	
88C15	100	3005	1	2007	
88C15L	100	3005	1	2007	
88C16	100	3005	1	2007	
88C16D	100	3005	1	2007	
88C17	100	3005	1	2007	
88C17-1	100	3005	1	2007	
88C18	100	3005	1	2007	
88C182	100	3005	1	2007	
88C19	100	3005	1	2007	
88C19G	100	3005	1	2007	
88C2	100	3005	1	2007	
88C3	100	3005	1	2007	

Industry Standard No.	ECG	SK	GE	RS 276-	MOTOR.
88C4	100	3005	1	2007	
88C5	100	3005	1	2007	
88C6	100	3005	1	2007	
88C7	100	3005	1	2007	
88C8	100	3005	1	2007	
88C9	100	3005	1	2007	
88CC-70-12	100	3005	1	2007	
089-2	228A	3103A/396	257		
089-214	108	3452	86	2038	
089-215	108	3452	86	2038	
089-216	108	3452	86	2038	
089-220	126	3006/160	52	2024	
089-222	102A	3003	53	2007	
089-223	123A	3444	20	2051	
089-226	123A	3444	20	2051	
089-231	158	3004/102A	53	2007	
089-233	103	3862	8	2002	
089-235		3126	90		
089-236	109	3088		1123	
089-241	177	3100/519	300	1122	
089-248	109	3088		1123	
089-252	117	3031A	504A	1104	
089-293	109	3088		1123	
089-3	228A	3103A/396	257		
089-4(SYLVANIA)	157	3747	232		
89WJ75/46N			1N34AS	1123	
90(SHARP)	138A		ZD-7.5		
90-110	237	3299	46		
90-111	186A	3357	247	2052	
90-112	315	3047	279		
90-140	199	3124/289	212	2010	
90-175	237	3299	46		
90-176	186A	3357	247	2052	
90-177	315	3048/329	279		
90-178	222	3050/221	FET-4	2036	
90-179	312	3116	FET-2	2035	
90-180	229	3018	20	2051	
90-181	199	3122	18	2030	
90-2213-00-18	123A	3444	20	2051	
90-30	123A	3444	20	2051	
90-31	289A	3024/128	243	2030	
90-32	123A	3124/289	20	2051	
90-33	199	3124/289	62	2010	
90-35	1102	3224	IC-93		
90-36	1100	3223	IC-92		
90-37	1104		51	2004	
90-37(IC)	1104	3225	IC-91		
90-37(TRANSISTOR)	160		245	2004	
90-38	186A	3192/186	28	2017	
90-39	7474			1818	
90-45	161	3018	39	2015	
90-450	235	3197	215		
90-451	186A	3357	247	2052	
90-452	123A	3018	61	2038	
90-453	123A	3018	61	2038	
90-454	315	3048/329	279		
90-455	233	3122	210	2051	
90-457	123A	3018	61	2038	
90-458	199	3124/289	210	2051	
90-459	199	3124/289	20	2051	
90-46	313	3019	39	2015	
90-47	313	3444/123A	20	2051	
90-48	123A	3444	20	2051	
90-49	316	3452/108	86	2038	
90-50	312	3834/132	FET-1	2035	
90-54	160		245	2004	
90-55	312	3448	FET-1	2035	
90-56		3124	53	2007	
90-57	123A	3444	20	2051	
90-58	102A	3050/221	53	2007	
90-59	160		245	2004	
90-60		3018	53	2007	
90-600	152	3054/196	66	2048	
90-601	123A	3018	61	2038	
90-602	123A	3018	211	2016	
90-603	199	3122	20	2051	
90-604	229	3018	60	2015	
90-605	199	3124/289	20	2051	
90-606	312	3116	FET-2	2035	
90-607	312	3116	FET-2	2035	
90-608	312	3112	FET-2	2035	
90-609	235	3197	215		
90-61	123A	3444	20	2051	
90-610	236	3197/235	216	2053	
90-612	123A	3018	61	2038	
90-613	312	3116	FET-2	2035	
90-614	199	3124/289	62	2010	
90-615	1052	3249			
90-62	312	3448	FET-1	2035	
90-65	123A	3444	20	2051	
90-66	289A	3024/128	243	2030	
90-67	7474			1818	
90-69	123A	3444	20	2051	
90-70	199	3124/289	62	2010	
90-71	123A	3124/289	20	2051	
90-72	1102	3224	IC-93		
90-73	1100	3223	IC-92		
90-74	1104	3225	IC-91		
90-75	186A	3357	28	2017	
090A64-1	109	3087		1123	
90T2	107	3039/316	11	2015	
91-3	109			1123	
91-4	160		245	2004	
91-46	109			1123	
91A	108	3452	86	2038	
91A01	116	3017B/117	504A	1104	
91A018	116		504A	1104	
91A02	116	3017B/117	504A	1104	
91A03	116	3017B/117	504A	1104	
91A04	125	3311	510,531	1114	
91A05	116	3444/123A	504A	1104	
91A06	116	3016	504A	1104	
91A08	506	3843	504A	1104	
91A11	116	3311	504A	1104	
91AJ150	175		246	2020	
91B	108	3452	86	2038	
91BGRN	108	3452	86	2038	
91C	123A	3444	20	2051	
91D	123A	3444	20	2051	
91E	123A	3444	20	2051	
91F	108	3452	86	2038	
91N1	110MP	3018	11	1123(2)	
91N1B	161	3716	39	2015	
91T6		3245	212	2010	
092-1	105	3012		1123	
092-1-12	105	3012	4		
92-1-70	105	3012			
92-1-70-12	105	3012	4		
92-1-70-12-7	105	3012	4		
92-10	105	3012	4		
92-1001	109			1123	
92-11	105	3012	4		
92-11-1				1104	
92-12	105	3012	4		

Industry Standard No.	ECG	SK	GE	RS 276-	MOTOR.
92-13	105	3012	4		
92-14	105	3012	4		
92-15	105	3012	4		
92-16	105	3012	4		
92-17	105	3012	4		
92-18	105	3012	4		
92-19	105	3012	4		
92-1A	105	3012	4		
92-1A0	105	3012	4		
92-1A0R	105	3012	4		
92-1A1	105	3012	4		
92-1A19	105	3012	4		
92-1A2	105	3012	4		
92-1A21	105	3012	4		
92-1A3	105	3012	4		
92-1A3P	105	3012	4		
92-1A4	105	3012	4		
92-1A4-7	105	3012	4		
92-1A4-7B	105		4		
92-1A5	105	3012	4		
92-1A5L	105	3012	4		
92-1A6	105	3012	4		
92-1A6-1	105	3012	4		
92-1A7	105	3012	4		
92-1A7-1	105	3012	4		
92-1A8	105	3012	4		
92-1A82	105	3012	4		
92-1A9	105	3012	4		
92-1A90	105	3012	4		
92-30942	123A		20	2051	
92-64-1				1122(2)	
92A11-1	116		504A	1104	
92A64-1	178MP		300(2)	1122	
92B12-2	116		504A	1104	
92B1C	723		IC-15		
92BIC	723		IC-15		
92L102-2			504A	1104	
92N1	108	3452	86	2038	
92N1B	108	3452	86	2038	
92T6		3245	212	2010	
93-13	177		300	1122	
93-188	5693	3652			
93-302	116		504A	1104	
93-3151	505		CR-7		
93-SE-124	130		14	2041	
93.20.709	167	3106/5304	510	1114	
93.20.714	167	3106/5304	510	1114	
93.24.401	109	3087	1N34AS	1123	
93.24.601	109	3087	1N34AS	1123	
93.24.604	109	3087	1N34AS	1123	
93A1-20			504A	1104	
93A1-21			504A	1104	
93A10-1			504A	1104	
93A102-2	143A	3750	ZD-13	563	
93A104-1	116(4)	3311	BR-600	1172	
93A105-1	109	3087	1N34AS	1123	
93A11			504A	1104	
93A110-1	109	3087	1N34AS	1123	
93A12-1	116	3016	504A	1104	
93A12-3			504A	1104	
93A120			504A	1104	
93A15-1	117		504A	1104	
93A1D-1	116		504A	1104	
93A2	117		504A	1104	
93A25-1	109	3088	1N60	1123	
93A25-2	110MP	3088	1N34AS	1123(2)	
93A25-3	109	3088	1N34AS	1123	
93A27-1	109	3087		1123	
93A27-2			504A	1104	
93A27-3			300	1122	
93A27-8	177	3087	300	1122	
93A30-1			504A	1104	
93A30-3			504A	1104	
93A31-1	177		300	1122	
93A33-1	109	3087		1123	
93A38-1	109	3087		1123	
93A39-1	5081A	3750/143A	ZD-24		
93A39-11	5071A	3334	ZD-6.8	561	
93A39-12	142A	3062	ZD-12	563	
93A39-13	5074A	3092/141A	ZD-11.0	563	
93A39-14	5082A	3753	ZD-25		
93A39-15	123A	3444	ZD-20	2051	
93A39-17			11	2015	
93A39-19	136A	3057	ZD-5.6	561	
93A39-2			ZD-5.6	561	
93A39-24	145A	3063	ZD-15	564	
93A39-25	149A	3097	ZD-62		
93A39-26			ZD-5.6	561	
93A39-28	5098A		ZD-130		
93A39-3			ZD-9.1	562	
93A39-30			ZD-10	562	
93A39-31	143A	3093	ZD-13	563	
93A39-34	5067A	3331	ZD-3.9		
93A39-35	5086A	3338	ZD-39		
93A39-40	147A	3095	ZD-33		
93A39-43	142A	3062	ZD-12	563	
93A39-44	5074A	3139	ZD-11	563	
93A39-45	5081A	3151	ZD-24		
93A39-48	5076A		ZD-17		
93A39-49	5005A	3330/5066A	ZD-3.3		
93A39-5	5079A	3335	ZD-20		
93A39-50	5075A	3751	ZD-16	564	
93A39-6			ZD-10	562	
93A39-7	5079A	3335	ZD-20		
93A39-8	5081A		ZD-24		
93A3912	177		300	1122	
93A3D-2	116	3017B/117	504A	1104	
93A4-2	116	3017B/117	504A	1104	
93A40-1	506	3130	511	1114	
93A41-2	110MP	3088		1123	
93A41-5	109	3126		1123	
93A42-1			504A	1104	
93A42-2			504A	1104	
93A42-7	116	3031A	504A	1104	
93A42-8			504A	1104	
93A43-1	112	3089	1N82A		
93A43-2	112	3089	1N82A		
93A45-1			504A	1104	
93A45-2			504A	1104	
93A47			504A	1104	
93A48-1			300	1122	
93A48-2			504A	1104	
93A5-10	113A	3119/113	6GC1		
93A5-2	113A	3120/114	6GC1		
93A5-9	113A	3119/113	6GC1		
93A51-3	116		504A	1104	
93A52-1	116	3312	504A	1104	
93A52-2			300	1122	
93A52-7		3130	511	1114	
93A53-2	120	3110	CR-3		
93A53-3	116		504A	1104	
93A55-1	5074A	3092/141A	ZD-11.0	563	
93A56-1			504A	1104	
93A57-1	118	3066	CR-1		

Industry Standard No.	ECG	SK	GE	RS 276-	MOTOR.
93A58-1			504A	1104	
93A59-1	112	3089	1N82A		
93A6-1	116	3017B/117	504A	1104	
93A6-2	116	3017B/117	504A	1104	
93A60-1	506	3130	511	1114	
93A60-10	506	3843	511	1114	
93A60-11	506	3130	511	1114	
93A60-14	116	3313	504A	1104	
93A60-2	116	3031A	504A	1104	
93A60-3	506	3031A	511	1114	
93A60-5	177	3100/519	300	1122	
93A60-6	177	3100/519	300	1122	
93A60-7	552	3311	504A	1104	
93A60-8	116	3031A	504A	1104	
93A60-80			504A	1104	
93A60-9	116	3312	504A	1104	
93A63-1		3126	90		
93A64-1	177	3087	1N34AS	1122	
93A64-2	177	3175	300	1122	
93A64-2(APC)	178MP		300(2)	1122(2)	
93A64-3	177	3100/519	300	1122	
93A64-5	177	3175	300	1122	
93A64-7	177	3175	300	1122	
93A67-1	116	3031A	504A	1104	
93A69-1	506	3032A	511	1114	
93A69-2	177	3100/519	300	1122	
93A71-1	506	3033	511	1114	
93A75-1	120	3110	CR-3		
93A76-1		3126	90		
93A77-1	109	3089/112	1N34AS	1123	
93A78-1	116	3032A	504A	1104	
93A79-6	125	3032A	510,531	1114	
93A8-1	109	3088		1123	
93A80-1	142A	3136/5072A	ZD-12	563	
93A83-1	110MP	3088	1N60	1123(2)	
93A9	102	3004	2	2007	
93A9-1	102A		53	2007	
93A9-2	102A		53	2007	
93A9-3	100	3005	1	2007	
93A9-4	100	3005	1	2007	
93A91-1	522	3303	517		
93A91-2	522	3303	517		
93A91-3	522	3303	527		
93A91-4	522	3303	517		
93A91-5	522	3303			
93A91-6	522	3303			
93A93-1	504	3108/505	CR-6		
93A93-2	504	3108/505	CR-6		
93A96-1	500A	3304	527		
93A96-2	500A	3304			
93A96-3	500A	3304	527		
93A97-1	116	3312	504A	1104	
93A97-2	116	3016	504A	1104	
93A97-3	116	3311	504A	1104	
93A97-37	116		504A	1104	
93A99-2	538	3310			
93A99-3	538	3310			
93A99-4	538	3310			
93A99-5	539	3309			
93A99-6	539	3309			
93A99-7	539	3309	538		
93A99-8	539	3309	538		
93B1-1	116	3017B/117	504A	1104	
93B1-10	116	3017B/117	504A	1104	
93B1-11	116	3017B/117	504A	1104	
93B1-12	116	3017B/117	504A	1104	
93B1-13	116	3017B/117	504A	1104	
93B1-14	116	3017B/117	504A	1104	
93B1-15	116	3017B/117	504A	1104	
93B1-16	116	3017B/117	504A	1104	
93B1-17	116	3017B/117	504A	1104	
93B1-18	116	3016	504A	1104	
93B1-2	116	3017B/117	504A	1104	
93B1-20	116(3)	3016	504A(3)	1104	
93B1-21	116(3)	3016	504A(3)	1104	
93B1-3	116	3017B/117	504A	1104	
93B1-4	116	3017B/117	504A	1104	
93B1-5	116	3017B/117	504A	1104	
93B1-6	116	3016	504A	1104	
93B1-7	116	3017B/117	504A	1104	
93B1-8	116	3017B/117	504A	1104	
93B1-9	116	3017B/117	504A	1104	
93B12-1	116	3016	504A	1104	
93B12-2	116	3016	504A	1104	
93B12-3	116	3016	504A	1104	
93B122			504A	1104	
93B123			504A	1104	
93B19-1	120	3110	CR-3		
93B1C	787		IC-228		
93B2-1	116(3)	3017B/117	504A(3)	1104	
93B20-1	117		504A	1104	
93B20-2	117		504A	1104	
93B20-3	117		504A	1104	
93B22-3	358		42		
93B24-2	116	3016	504A	1104	
93B24-3	116	3016	504A	1104	
93B25-1	109			1123	
93B25-2	109			1123	
93B25-3	109	3087		1123	
93B27-1	109	3087		1123	
93B27-2	116	3016	504A	1104	
93B27-3	110MP	3016	1N60	1123(2)	
93B27-4			300	1122	
93B27-5			300	1122	
93B27-8	177		300	1122	
93B3-3	358		42		
93B3-4	358		42		
93B30-1	116	3017B/117	504A	1104	
93B30-3	116	3016	504A	1104	
93B38-1	109	3087		1123	
93B38-5	109	3087		1123	
93B39-1			ZD-9.1	562	
93B39-12	142A	3062	ZD-12	563	
93B39-13	142A	3062	ZD-12	563	
93B39-3			ZD-9.1	562	
93B39-31			ZD-13	563	
93B39-5		3335	ZD-20		
93B39-6	140A		ZD-10	562	
93B39-7		3335	ZD-20		
93B41	109			1123	
93B41-1	109	3087		1123	
93B41-12	116	3017B/117	504A	1104	
93B41-14	116	3017B/117	504A	1104	
93B41-2		3091		1123	
93B41-20	117		504A	1104	
93B41-29		3087	1N295	1123	
93B41-3	109	3087		1123	
93B41-4	116	3016	504A	1104	
93B41-6	116	3017B/117	504A	1104	
93B41-8	116	3017B/117	504A	1104	
93B42-1	116	3017B/117	504A	1104	
93B42-10	116	3017B/117	504A	1104	
93B42-11	116	3017B/117	504A	1104	
93B42-12	116	3017B/117	504A	1104	
93B42-2	116	3016	504A	1104	
93B42-3	116	3017B/117	504A	1104	
93B42-4	116	3017B/117	504A	1104	
93B42-5			504A	1104	
93B42-6	116	3017B/117	504A	1104	
93B42-7	116	3311	504A	1104	
93B42-8	116	3017B/117	504A	1104	
93B42-9	116	3017B/117	504A	1104	
93B44-1	112		1N82A		
93B45-1	116	3016	504A	1104	
93B45-2	116	3017B/117	504A	1104	
93B45-3	125		510,531	1114	
93B46-1			1N60	1123	
93B47-1	116	3016	504A	1104	
93B48-1	177	3100/519	300	1122	
93B48-2	177	3016	300	1122	
93B48-3	177	3100/519	300	1122	
93B48-4	177	3100/519	300	1122	
93B5-1	113A	3119/113	6GC1		
93B5-10	113A	3119/113	6GC1		
93B5-2	113A		6GC1		
93B5-3	113A	3119/113	6GC1		
93B5-3-6	113A	3119/113	6GC1		
93B5-3-8	113A	3119/113	6GC1		
93B5-3-9	113A	3119/113	6GC1		
93B5-4	113A	3120/114	6GC1		
93B5-5	113A	3119/113	6GC1		
93B5-6	113A	3119/113	6GC1		
93B5-7	113A		6GC1		
93B5-8	113A	3119/113	6GC1		
93B5-9	113A	3119/113	6GC1		
93B51-3	116	3017B/117	504A	1104	
93B52-1	506	3843	511	1114	
93B52-2	506	3843	511	1114	
93B53-1	116(3)	3110/120	504A(3)		
93B53-2	120	3110	CR-3		
93B57-1	118	3066	CR-1		
93B58-1	506	3016	511	1114	
93B59-1	112		1N82A		
93B60-10			511	1114	
93B60-11			511	1114	
93B60-3	506	3843	511	1114	
93B60-8			300	1122	
93B64-1	177	3100/519	300	1122	
93B64-2	177		511	1114	
93B65-1	125	3033	510,531	1114	
93B65-2			509	1114	
93B67-1			511	1114	
93B71-1			509	1114	
93B77-1			1N34AS	1123	
93B8-1	109	3087		1123	
93B91-1	522	3303	527		
93B91-2	522	3303			
93B91-3	522	3303			
93B91-4	522	3303			
93B91-5	522	3303			
93B91-6	522	3303			
93B96-1	500A	3304	517		
93B96-2	500A	3304			
93B96-3	500A	3304			
93B97-1			504A	1104	
93C1-20	116	3016	504A	1104	
93C1-2021	120	3110	CR-3		
93C1-21	116	3016	504A	1104	
93C103-4			504A	1104	
93C118-1	125	3033	510,531	1114	
93C118-2	125	3033	510,531	1114	
93C12-1			504A	1104	
93C12-2			504A	1104	
93C12-3			504A	1104	
93C16-2	116	3016	504A	1104	
93C18-1	125		510,531	1114	
93C18-2	125		510,531	1114	
93C19-1	116	3110/120	504A	1104	
93C2-6		3087	1N295	1123	
93C21-1			1N295	1123	
93C21-2			1N295	1123	
93C21-3			1N295	1123	
93C21-4			1N295	1123	
93C21-5			1N295	1123	
93C21-6			1N295	1123	
93C21-7			1N295	1123	
93C21-8			1N295	1123	
93C218	109	3087		1123	
93C22-1	358		42		
93C22-3	358		42		
93C24-1			504A	1104	
93C24-2	116	3016	504A	1104	
93C24-3			504A	1104	
93C24-4			504A	1104	
93C25-2	110MP		1N60	1123(2)	
93C25-3	109	3091		1123	
93C25-4			504A	1104	
93C26-1			504A	1104	
93C26-2			504A	1104	
93C26-3			504A	1104	
93C26-4			504A	1104	
93C26-5			504A	1104	
93C26-7		3119	6GC1		
93C26-8			504A	1104	
93C26-9		3119	6GC1		
93C267	114		6GD1	1104	
93C27-1		3087	1N34AS	1123	
93C27-2		3100	300	1122	
93C27-3		3100	300	1122	
93C27-4			504A	1104	
93C27-5			1N34AS	1123	
93C27-6			1N34AS	1123	
93C27-7			300	1122	
93C28-4			504A	1104	
93C30-1	116	3016	504A	1104	
93C30-3	116	3016	504A	1104	
93C39-1	134A	3055	ZD-3.6		
93C39-10	138A	3059	ZD-7.5		
93C39-11	123		20	2051	
93C39-12	142A	3062	ZD-12	563	
93C39-13	142A	3062	ZD-12	563	
93C39-19			ZD-5.6	561	
93C39-2	142A	3062	ZD-12	563	
93C39-24			ZD-15	564	
93C39-3	139A	3060	ZD-9.1	562	
93C39-6	142A	3062	ZD-12	563	
93C39-7		3335	ZD-20		
93C39-8	5081A		ZD-24		
93C40	119	3109	CR-2		
93C40-1	119	3109	CR-2		
93C42-2	116	3016	504A	1104	
93C42-7	116	3311	504A	1104	
93C5-10	113A	3119/113	6GC1		
93C5-2	114		6GD1		
93C5-3	114		6GD1	1104	
93C5-4	114		6GD1		
93C5-5	113A	3119/113	6GC1		
93C5-6	113A	3119/113	6GC1		

Industry Standard No.	ECG	SK	GE	RS 276-	MOTOR.
93C5-7	113A	3119/113	6GC1		
93C5-8	113A	3119/113	6GC1		
93C5-9	113A	3119/113	6GC1		
93C51-3	114		6GD1	1104	
93C52-1	506	3016	511	1114	
93C53-2	120	3016	CR-3		
93C55-1	109			1123	
93C59	113A	3119/113	6GC1		
93C6-7	113A		6GC1		
93C6-9	113A		6GC1		
93C60-10			511	1114	
93C60-3	506	3843	511	1114	
93C60-5	177	3175	300	1122	
93C60-6	177		300	1122	
93C60-7	552		504A	1104	
93C60-9		3175	504A	1104	
93C64-1	519	3087	514	1122	
93C64-2	519		514	1122	
93C64-3	519		514	1122	
93C69-1	506	3843	511	1114	
93C7-1	109			1123	
93C7-2	177		300	1122	
93C7-3	177		300	1122	
93C77-1			1N34AS	1123	
93C8-1	109	3091		1123	
93C89-1			511	1114	
93D112-57	112	3089	1N82A		
93D60-5	519		514	1122	
93D91-1	522	3303	523		
93D91-2	522	3303	523		
93D91-3	522	3900/536A	523		
93D91-4	522	3303	523		
93D91-5	522	3303	523		
93D91-6	522	3303	523		
93D96-1	500A	3304	527		
93D96-2	500A	3304	527		
93D96-3	500A	3304	527		
93D99-2	538	3310	537		
93D99-3	538	3310	537		
93D99-4	538	3310	537		
93D99-5	539	3309	538		
93D99-6	539	3309	538		
93D99-7	539	3309	538		
93D99-8	539	3309	538		
93ERH-1X1128	712	3072			
93ERH-DX0155//	552	3311		1104	
93ERH-DX0156//	116			1104	
93EVHD1N4148//	519	3100		1122	
93K2-1	113A	3119/113	6GC1		
93L00	74LS195A	74LS195			
93L101-2	116	3017B/117	504A	1104	
93L102-2	116	3017B/117	504A	1104	
93L103-4	116	3017B/117		1104	
93L104-5	156		512		
93L107-2	116	3017B/117	504A	1104	
93L3-1	117		504A	1104	
93L3-2	117		504A	1104	
93L5-1	116	3017B/117	504A	1104	
93L5-2	117		504A	1104	
93L5-3	156		512		
93L5-4	116	3017B/117	504A	1104	
93L5-5	156		512		
93L5-6	116	3017B/117	504A	1104	
93L5-7	116	3017B/117	504A	1104	
93L5-8	156		512		
93M8-1	117		504A	1104	
93N9-1		3303	517		
93P1AA	159	3114/290	82	2032	
93SC165			14	2041	
93SC165133	130		14	2041	
93SC165133A	130		14	2041	
93SE165	181	3036	75	2041	
93T6			212	2010	
094	714	3075	IC-4		
094-007	519	3100	514	1122	
094-010	116	3017B/117	504A	1104	
094-011	125	3080	510,531	1114	
094-012	506	3843	511	1114	
094-013	121	3009	239	2006	
094-014	109	3088		1123	
094-1	714	3075	IC-4		
94-1066-1	116	3016	504A	1104	
94-42-9	109	3088		1123	
094A	714	3075	IC-4		
94A-1A6-4	121	3717	239	2006	
94A80-1			ZD-12	563	
94B1C	724		IC-86		
94N1	123A	3444	20	2051	
94N1B	123A	3444	20	2051	
94N1R	123A	3444	20	2051	
94N1V	128	3024	243	2030	
94N2	123A	3444	20	2051	
94N2P	155	3839	43		
94T1	100	3005	1	2007	
095	715	3076	IC-6		
095-1	715	3076	IC-6		
95-108	160	3006	245	2004	
95-110	160	3006	245	2004	
95-111	126	3006/160	52	2024	
95-112	101	3861	8	2002	
95-113	101	3861	8	2002	
95-114	101	3861	8	2002	
95-11410100	102A	3004	53	2007	
95-11410200	102A	3004	53	2007	
95-11410900	102A	3004	53	2007	
95-11413500	102A	3004	53	2007	
95-11414000	102	3009		2007	
95-114140000			2	2007	
95-11511500	125	3032A	510,531	1114	
95-116	160	3006	245	2004	
95-117	160	3006	245	2004	
95-118	160	3006	245	2004	
95-119	160	3006	245	2004	
95-120	160	3006	245	2004	
95-120A	100	3005	1	2007	
95-121	160	3006	245	2004	
95-122	160	3006	245	2004	
95-123	160	3006	245	2004	
95-124	160	3006	245	2004	
95-125	108	3452	86	2038	
95-126	108	3452	86	2038	
95-127	108	3452	86	2038	
95-128	108	3452	86	2038	
95-129	108	3452	86	2038	
95-130	108	3452	86	2038	
95-131	108	3452	86	2038	
95-201	102A	3004	53	2007	
95-202	103A	3010	59	2002	
95-203	102A	3004	53	2007	
95-204	102A	3004	53	2007	
95-208	102A	3004	53	2007	
95-209	102A	3004	53	2007	
95-212	102A	3004	53	2007	
95-213	102A	3004	53	2007	
95-214	102A	3004	53	2007	
95-216	123A	3122	20	2051	
95-217	102A	3004	53	2007	
95-218	102A	3004	53	2007	
95-220	128	3024	243	2030	
95-221	123A	3122	20	2051	
95-222-1	102A	3004	53	2007	
95-222-2	103A	3010	59	2002	
95-223	108	3452	86	2038	
95-224-1	102A	3004	53	2007	
95-224-2	103A	3010	59	2002	
95-225	123A	3122	20	2051	
95-226-003	129	3025	244	2027	
95-226-004	128	3024	243	2030	
95-226-1	129	3025	244	2027	
95-226-2	123A	3444	20	2051	
95-226-3	129	3025	244	2027	
95-226-4	128	3024	243	2030	
95-227	123A	3444	20	2051	
95-250	121	3009	239	2006	
95-250-1	121MP	3013	239(2)	2006(2)	
95-251	121	3009	239	2006	
95-252	124	3021	12		
95-252-1	124	3021	12		
95-252-2	124	3021	12		
95-252-3	124	3021	12		
95-252-4	124	3021	12		
95-257	121MP	3013	239(2)	2006(2)	
95-296	123A	3122	20	2051	
095A	715	3076	IC-6		
096	713		IC-5		
96-0008	109	3088		1123	
96-056-234	108	3452	86	2038	
096-1	713		IC-5		
96-138-2	108	3452	86	2038	
96-5005-01	171	3103A/396	27		
96-5007-01	109	3087		1123	
96-5022-01	116	3016	504A	1104	
96-5022-1	116	3016	504A	1104	
96-5023-01	116	3016	504A	1104	
96-5026-01	121	3009	239	2006	
96-5032-01	102A	3004	53	2007	
96-5033-01	102A	3004	53	2007	
96-5033-02	102A	3004	53	2007	
96-5033-03	102A	3004	53	2007	
96-5033-04	102A	3004	53	2007	
96-5045-01	121	3009	239	2006	
96-5046-01	116	3017B/117	504A	1104	
96-5059-01	109	3087		1123	
96-5062-01	160	3006	245	2004	
96-5064-01	121	3009	239	2006	
96-5076-01	158	3004/102A	53	2007	
96-5080-02	123A	3444	20	2051	
96-5081-01	121	3009	239	2006	
96-5082-01	116	3017B/117	504A	1104	
96-5085-01		3004	53	2007	
96-5085-02	102A	3004	53	2007	
96-5086-02	121	3009	239	2006	
96-5087-01	109			1123	
96-5088-01	116	3017B/117	504A	1104	
96-5091-01	142A	3062	ZD-12	563	
96-5094-01(TRANS)			52	2024	
96-5095-01	160	3006	245	2004	
96-5096-01	116		504A	1104	
96-5098-01	102A	3004	53	2007	
96-5099-01	160	3004/102A	245	2004	
96-5100-01	121	3009	239	2006	
96-5100-03	121	3009	239	2006	
96-5101-01	102A	3004	53	2007	
96-5102-01	102A	3004	53	2007	
96-5103-01	116	3017B/117	504A	1104	
96-5106-01	5802	9005		1143	
96-5107-01	128	3024	243	2030	
96-5107-02	128	3024	243	2030	
96-5109-01	116	3017B/117	504A	1104	
96-5109-02	116	3017B/117	504A	1104	
96-5110-02	145A	3063	ZD-15	564	
96-5110-03	145A	3063	ZD-15	564	
96-5112-01	156		512		
96-5113-01	116	3017B/117	504A	1104	
96-5115-01	123A	3444	20	2051	
96-5115-02	123A	3444	20	2051	
96-5115-03	123A	3444	20	2051	
96-5115-04	123A	3444	20	2051	
96-5115-05	123A	3444	20	2051	
96-5116-01	140A	3061	ZD-10	562	
96-5117-01	130	3016	14	2041	
96-5117-91	130	3027/U30	14	2041	
96-5118-01	177	3063/145A	300	1122	
96-5119-01	116(2)	3017B/117	504A(2)	1104	
96-5120-01	169	3107/5307			
96-5121-01	116(2)	3017B/117	504A(2)	1104	
96-5124-02	138A	3059	ZD-7.5		
96-5125-01	121	3009	239	2006	
96-5131-01	161	3132	39	2015	
96-5132-01	124	3021	12		
96-5133-01-02	142A	3062	ZD-12	563	
96-5133-02	142A		ZD-12	563	
96-5135-01	124	3021	12		
96-5138-01	160	3006	245	2004	
96-5139-01	160	3006	245	2004	
96-5140-01	160	3006	245	2004	
96-5141-01	160		245	2004	
96-5143-01	121	3009	239	2006	
96-5143-02	121	3009	239	2006	
96-51430-02	121	3009	239	2006	
96-5148-01	121	3009	239	2006	
96-5149-01	5802	9005		1143	
96-5152-01	123A	3444	20	2051	
96-5152-03	123A	3444	20	2051	
96-5153-01	123A	3444	20	2051	
96-5153-03	123A	3444	20	2051	
96-5155-01	121	3009	239	2006	
96-5161-01	192	3122	63	2030	
96-5162-03	130	3027	14	2041	
96-5162-04	130	3027	14	2041	
96-5163-01	161	3132	39	2015	
96-5164-02	130	3027	14	2041	
96-5164-03	130	3027	14	2041	
96-5165-01	129	3024/128	244	2027	
96-5166-01	116(2)	3017B/117	504A(2)	1104	
96-5170-01	128	3024	243	2030	
96-5174-01	161	3132	39	2015	
96-5175-01	161	3132	39	2015	
96-5176-01	129	3025	244	2027	
96-5177-01	123A	3444	20	2051	
96-5178-01	125	3033	510,531	1114	
96-5180-01	128	3024	243	2030	
96-5180-02	128	3024	243	2030	
96-5184-01	5802	9005		1143	
96-5187-01	123A	3444	20	2051	
96-5190-01	175	3124/289	246	2020	
96-5191-01	175	3538	246	2020	
96-5192-01	121	3009	239	2006	

Industry Standard No.	ECG	SK	GE	RS 276-	MOTOR.
96-5193-01	156	3051	512		
96-5194-01	156	3051	512		
96-5195-01	5809	9010			
96-5196-01	116	3017B/117	504A	1104	
96-5198-01	161	3132	39	2015	
96-5199-01	161	3132	39	2015	
96-5201-01	130	3027	14	2041	
96-5203-01	128	3024	243	2030	
96-5204-01	128	3024	243	2030	
96-5205-01	103A	3010	59	2002	
96-5207-01	130	3027	14	2041	
96-5208-01	128		243	2030	
96-5209-01	129	3157/703A	244	2027	
96-5213-01	123A	3444	20	2051	
96-5214-01	5400	3950			
96-5215-01	159	3466	82	2032	
96-5219-01	171	3103A/396	27		
96-5220-01(NPN)	123A	3444	20	2051	
96-5220-01(PNP)	129		244	2027	
96-5221-01	123A	3444	20	2051	
96-5225-01	152	3054/196	66	2048	
96-5228-01	123A	3444	20	2051	
96-5229-01	123A	3444	20	2051	
96-5230-01	129		244	2027	
96-5231-01	128	3035	243	2030	
96-5232-01	196	3054	241	2020	
96-5232-02	196	3054	66	2048	
96-5232-03	196		241	2020	
96-5235-01	161	3132	39	2015	
96-5236-01	161	3132	39	2015	
96-5237-01	123A	3444	20	2051	
96-5238-01	703A	3157	IC-12		
96-5238-02	703A	3157	IC-12		
96-5241-01	5809	9010			
96-5244-01	128	3024	243	2030	
96-5245-01	196		241	2020	
96-5246-01	5986	3609	5096		
96-5248-01	5071A	3334	ZD-6.8	561	
96-5248-04	142A	3064/146A	ZD-12	563	
96-5248-06	145A		ZD-15	564	
96-5248-09	5067A	3331	ZD-3.9		
96-5248-11	5067A	3331	ZD-3.9		
96-5248-12	5085A	3337	ZD-36		
96-5249-01	5127A	3393			
96-5249-02	5120A	3386	5ZD-6.8		
96-5249-03	5133A	3399			
96-5250-01	5127A	3393			
96-5250-02	5139A	3405	5ZD-27		
96-5250-03	5141A	3407			
96-5250-06	5130A	3396			
96-5250-07	5125A	3391			
96-5252-01	128	3024	243	2030	
96-5254-01	177	3175	300	1122	
96-5255-01	123A	3444	20	2051	
96-5256-01	128	3024	243	2030	
96-5257-01	123A	3444	20	2051	
96-5258-01	234	3025/129	65	2050	
96-5259-01	161	3132	39	2015	
96-5260-01	161	3132	39	2015	
96-5262-01	189		218	2026	
96-5263-01	188		226	2018	
96-5267-01	196	3021/124	241	2020	
96-5269-01	6401			2029	
96-5281-01	123A	3444	20	2051	
96-5282-01	159	3466	82	2032	
96-5283-01	159		82	2032	
96-5284-01	162	3438	35		
96-5285-01	130		246	2020	
96-5290-01	123A	3444	20	2051	
96-5303-01	153		69	2049	
96-5310-01	280	3297	262	2041	
96-5314-01	123A	3444	20	2051	
96-5315-01	181	3036	75	2041	
96-5316-01	197		250	2027	
96-5320-01	189		218	2026	
96-5333-01	116		504A	1104	
96-5334-01	161	3716	39	2015	
96-5334-04	5133A	3399			
96-5334-05	5135A	3401			
96-5344-01	5125A	3391	5ZD-10		
96-5344-03	5130A	3396			
96-5345-01	125		510,531	1114	
96-5346-01	199	3245	62	2010	
96-5348-01	152	3893	66	2048	
96-5349-01	153		69	2049	
96-5356-01	153		69	2049	
96-5357-01	152	3893	66	2048	
96-5363-01	109			1123	
96-5364-01	128		243	2030	
96-5365-01	159	3466	82	2032	
96-5370-01	284	3836	265	2047	
96-5371-01	285	3846	266		
96-5374-01	804	3455	IC-27		
96-5376-01	725		IC-19		
96-5378-01	121	3717	239	2006	
096A	713	3077/790	IC-5		
96B-3A65	121MP	3013	239(2)	2006(2)	
96N(AIRLINE)	108		86	2038	
96N1	124	3021	12		
96N927	108	3452	86	2038	
96N932	108	3452	86	2038	
96NPT	108	3452	86	2038	
96XZ0778-44N			1N34AS	1123	
96XZ077844N	109			1123	
96XZ0868/15X			1N34AS	1123	
96XZ6050/25N	161	3716	39	2015	
96XZ6051-28N	126	3006/160	52	2024	
96XZ6051-35N	126	3006/160	52	2024	
96XZ6051-36N	126	3006/160	52	2024	
96XZ6052-52N			20	2051	
96XZ6052/52N	123A	3444	20	2051	
96XZ6053-09N	102	3004	2	2007	
96XZ6053-10N	126	3006/160	52	2024	
96XZ6053-11N			20	2051	
96XZ6053-24N	126	3006/160	52	2024	
96XZ6053-27N	102	3004	2	2007	
96XZ6053-51N			53	2007	
96XZ6053/11N	123A	3444	20	2051	
96XZ6053/24N	158		53	2007	
96XZ6053/27N	102A		53	2007	
96XZ6053/35N	123A	3444	20	2051	
96XZ6053/36N	123A	3444	20	2051	
96XZ6053/38N	128	3024	243	2030	
96XZ6053/51N	102A	3004	53	2007	
96XZ6054/45X	131MP	3840	44(2)	2006(2)	
96XZ778/21N	109	3087		1123	
96XZ778/27N	113A	3119/113	6GC1		
96XZ778/44N	109	3087		1123	
96XZ801/06N	121	3009	239	2006	
96XZ801/10N	121	3009	239	2006	
96XZ801/14N	123A	3444	20	2051	
96XZ801/34X	121	3009	239	2006	
96XZ801/37N	160	3006	245	2004	
96XZ801/50N	102A	3004	53	2007	

Industry Standard No.	ECG	SK	GE	RS 276-	MOTOR.
097	780	3141	IC-222		
097-1	783	3215	IC-225		
097A	780	3141	IC-222		
97A83	104	3009	16	2006	
97N2	101	3861	8	2002	
97N2U	103A	3010	59	2002	
97P1	102	3004	2	2007	
97P1U	158	3004/102A	53	2007	
98-1	102A	3123	53	2007	
98-2	102A	3123	53	2007	
98-24320-2	175		246	2020	
98-3	102A	3123	53	2007	
98-301	116	3017B/117	504A	1104	
98-302	117		504A	1104	
98-4	102A	3123	53	2007	
98-5	102A	3123	53	2007	
98-6	102A	3123	53	2007	
98-7	102A	3123	53	2007	
98A12518	116	3017B/117	504A	1104	
098G1219			510,531	1114	
098GI219	125	3051/156	510		
98P-2A6-2	102A		53	2007	
98P1	159	3114/290	82	2032	
98P10	159	3114/290	82	2032	
98P1P	131	3198	44	2006	
98T2			210	2051	
099-1&58PHILCO)				2020	
099-1(PHILCO)	152		66	2048	
099-1(SYL)	152		66	2048	
99-101	126		52	2024	
99-102	126		52	2024	
99-103	126		52	2024	
99-104	102A		53	2007	
99-105	160		245	2004	
99-106	160		245	2004	
99-107	160		245	2004	
99-108	160		245	2004	
99-109-1	123A	3444	20	2051	
99-109-2	123A	3444	20	2051	
99-120	160		245	2004	
99-121	160		245	2004	
99-201	102A		53	2007	
99-202	102A		53	2007	
99-203	102A		53	2007	
99-PWR			16	2006	
99-S-075	196		241	2020	
99AT6	102	3004	2	2007	
99B5	102	3004	2	2007	
99BA6	100	3005	1	2007	
99BE6	100	3005	1	2007	
99K7	101	3861	8	2002	
99L-4A6-1	102A		53	2007	
99L6	103	3862	8	2002	
99L6(SHARP)	128		243	2030	
99P1	159	3025/129	82	2032	
99P10	159	3466	82	2032	
99P117	129		244	2027	
99P1M	159	3025/129	244	2032	
99P2	129	3025	244	2027	
99P2B		3025		2027	
99P3	102A		53	2007	
99P3AA	102A		53	2007	
99P3C	129		244	2027	
99P5	159	3466	82	2032	
99PIM	129	3025			
99PLM				2027	
99S001	121	3009	239	2006	
99S002	102A	3004	53	2007	
99S003	100	3004/102A	1	2007	
99S004	158	3004/102A	53	2007	
99S004A	102A	3004	53	2007	
99S005	102A	3004	53	2007	
99S006	160	3006	245	2004	
99S007	160	3006	245	2004	
99S010	102A	3004	53	2007	
99S010A	102A	3004	53	2007	
99S011	102A	3004	53	2007	
99S011A	102A	3004	53	2007	
99S012	123A	3124/289	20	2051	
99S012A	123A	3124/289	20	2051	
99S012E	123A	3444	20	2051	
99S013	121MP	3014	239	2006(2)	
99S013A	121MP	3014	239	2006(2)	
99S014	121	3009	239	2006	
99S014A	121	3009	239	2006	
99S015	121	3009	239	2006	
99S016	108	3452	86	2038	
99S016-1	108	3452	86	2038	
99S017	108	3452	86	2038	
99S018	108	3452	86	2038	
99S018A	108	3452	86	2038	
99S019	108	3452	86	2038	
99S019A	108	3452	86	2038	
99S019B	108	3452	86	2038	
99S020	123	3124/289	20	2051	
99S022	703A	3157	IC-12		
99S022-1	703A	3157	IC-12		
99S025	123A	3444	20	2051	
99S025A	123A	3444	20	2051	
99S031	161	3018	39	2015	
99S032	161	3018	39	2015	
99S032(3RD LF)				2009	
99S032(3RD-IF)			210	2051	
99S033	199	3040	62	2010	
99S033A	123A	3444	20	2051	
99S034	194	3275	220		
99S035	123A	3444	20	2051	
99S036	123A	3444	20	2051	
99S036(TELEDYNE)			18	2030	
99S037	108	3452	86	2038	
99S038	123A	3444	20	2051	
99S039	159	3466	82	2032	
99S039A	159	3466	82	2032	
99S040	194	3275	220		
99S041	222	3050/221	FET-4	2036	
99S042	710	3102	IC-89		
99S044	161	3132	39	2015	
99S045	221	3116	FET-3	2036	
99S045A	222	3050/221		2036	
99S046	222	3050/221	FET-4	2036	
99S047	175	3131A/369	246	2020	
99S053	722	3161	IC-9		
99S053-1	722	3161	IC-9		
99S055	229	3018	61	2038	
99S056	229	3018	61	2038	
99S057	179	3642	76		
99S060-1	194	3275	220		
99S061-1	194	3275	220		
99S062-1	159		82	2032	
99S063-1	228A		257		
99S067-1	161	3018	39	2015	
99S07	101	3861	8	2002	
99S070-1	194	3044/154	220		
99S072	710	3102	IC-89		

Industry Standard No.	ECG	SK	GE	RS 276-	MOTOR.
99S073	129	3114/290	244	2027	
99S074	128	3024	243	2030	
99S075	196	3893/152	241	2020	
99S077-1	194	3044/154	220		
99S079	165		38		
99S079-1	165	3111	38		
99S079-2	165		38		
99S081	790	3454	IC-18		
99S081-1	790	3454	IC-18		
99S082	712		IC-2		
99S082-1	712	3072	IC-2		
99S083-1	196	3054	241	2020	
99S084-1	159	3466	82	2032	
99S085	123A	3444	20	2051	
99S087-1	154		40	2012	
99S090-1	108	3452	86	2038	
99S091-1	188	3024/128	226	2018	
99S092-1	189	3025/129	218	2026	
99S094	714		IC-4		
99S094-1	714	3075	IC-4		
99S095	715		IC-6		
99S095-1	715	3076	IC-6		
99S096	713		IC-5		
99S096-1	713	3077/790	IC-5		
99S097	783	3215	IC-225		
99S097-1	783	3215	IC-225		
99S099	197	3083	250	2027	
99S099-1	153		69	2049	
99S100	196	3054	241	2020	
99S100-1	152	3893	66	2048	
99S101-1	154	3045/225	40	2012	
99S102-1	190	3103A/396	217		
99S103-1	181	3111	14	2041	
99S103-2	181	3111	75	2041	
99S103-3	181	3111	14	2041	
99S105-1	184	3104A	57	2017	
99SA7	103A	3011	59	2002	
99SK5	101	3861	8	2002	
99SK7	101	3861	8	2002	
99SQ7	101	3861	8	2002	
0100	116	3311	504A	1104	
100-001-01/2228-3	117		504A	1104	
100-003-40/2228-3	109			1123	
100-003-40/228-3				1123	
100-00310-09	177	3100/519	300	1122	
100-00340-00	109	3087		1123	
100-0051	109	3088		1123	
100-007-10/2228/3	177		300	1122	
100-007-13/2228-3	109			1123	
100-007-13/228-3				1123	
100-00910-07	109	3088	1N60	1123	
100-00914-10	110MP	3088	1N60	1123	
100-011-20	177	3100/519	300	1122	
100-011-50/2228-3	177		300	1122	
100-01110-01	177	3126	300	1122	
100-01120-09	177	3100/519	300	1122	
100-0124	109	3088		1123	
100-0125	109	3087		1123	
100-0495-15	159	3114/290	221		
100-0673-04	290A	3114/290	269	2032	
100-1(PHILCO)	153		69	2049	
100-10	136A	3057	ZD-5.6	561	
100-10121-05	116	3311	504A	1104	
100-10132-00	116	3311	504A	1104	
100-11	177	3100/519	300	1122	
100-12	109	3088		1123	
100-120	177	3100/519	300	1122	
100-125	177	3100/519	300	1122	
100-13	125	3311	504A	1104	
100-130	177	3100/519	300	1122	
100-132-00			504A	1104	
100-135	136A	3057	ZD-5.6	561	
100-136	109	3088		1123	
100-137	177	3100/519	300	1122	
100-138	116	3311	504A	1104	
100-139	139A	3060	ZD-9.1	562	
100-14	139A	3060	ZD-9.1	562	
100-15	116		504A	1104	
100-160	109	3087		1123	
100-161	177	3100/519	300	1122	
100-162	116	3017B/117	504A	1104	
100-180	109	3088	1N60	1123	
100-181	109	3088	1N60	1123	
100-184	177	3100/519	300	1122	
100-198	106(4)	3017B/117	21(4)		
100-215	109	3088	1N60	1123	
100-216	177	3100/519	300	1122	
100-217	116	3311	504A	1104	
100-218	5071A		ZD-6.8	561	
100-219	5072A	3136	ZD-8.2	562	
100-22363-08		3126	90		
100-286	143A	3750	ZD-13	563	
100-4107	107		11	2015	
100-435	177	3100/519	300	1122	
100-436	109	3088	1N60	1123	
100-437	5071A		ZD-6.8	561	
100-438	116	3311	504A	1104	
100-4790	159		82	2032	
100-4846-001	128		243	2030	
100-520	116	3311	504A	1104	
100-521	5072A		ZD-8.2	562	
100-522	139A		ZD-9.1	562	
100-523	139A		ZD-9.1	562	
100-525	5801	9004	ZD-24		
100-526		3126	90		
100-527	116	3311	504A	1104	
100-5338	128		243	2030	
100-5338-001	128		243	2030	
100-5765-001	175		246	2020	
100A	116	3016	504A	1104	
100B					MR1130
100B63	102	3004	2	2007	
100C		3125	509	1114	1N4007
100C(ADMIRAL)	506		511	1114	
100C-4R	116		504A	1104	
100D10	125	3033	510,531	1114	
100H3P					MR1130
100JB01L					MDA2501
100JB02L					MDA2502
100JB04L					MDA2504
100JB05L					MDA2500
100JB06L					MDA2506
100K10	125		510,531	1114	
100L20	525	3925			
100N1	161	3018	39	2015	
100N1AS	161		39	2015	
100N1P	161	3018	39	2015	
100N1P(SOUND)			20	2051	
100N1P(VID)			39	2015	
100N3	319P	3039/316	283	2016	
100N3P	161	3039/316	39	2015	
100PB05P					MDA2500
100PB1P					MDA2501
100PB2P					MDA2502

Industry Standard No.	ECG	SK	GE	RS 276-	MOTOR.
100PB3P					MDA2504
100PB4P					MDA2504
100PB6P					MDA2506
100R10B	125	3033	510,531	1114	
100R11B	506	3843	511	1114	
100R12B	506	3843	511	1114	
100R13B	506	3843	511	1114	
100R14B	506	3843	511	1114	
100R1B	116	3017B/117	504A	1104	
100R20B	525	3925			
100R2B	116	3016	504A	1104	
100R3B	116	3031A	504A	1104	
100R4B	116	3031A	504A	1104	
100R5B	116	3017B/117	504A	1104	
100R6B	116	3017B/117	504A	1104	
100R7B	125	3032A	510,531	1114	
100R8B	125	3032A	510,531	1114	
100R9B	125	3033	510,531	1114	
100T2	130	3027	14	2041	
100T2A	130	3027	14	2041	
100W1	108	3452	86	2038	
100X2	130	3027	14	2041	
100X6	130	3027	14	2041	
100X6A	130	3027	14	2041	
0000101	126	3008	300	1122	
0101-0034	126	3006/160	52	2024	
0101-0060A	108	3452	86	2038	
0101-0222	126	3006/160	52	2024	
0101-0439	193	3025/129	67	2023	
0101-0448	187	3193	29	2025	
0101-0448A	187	3193	29	2025	
0101-0466	187	3193	29	2025	
0101-0491	123A	3444	20	2051	
0101-0531	108	3452	86	2038	
0101-0540	123A	3444	20	2051	
101-1(ADMIRAL)	161	3716	39	2015	
101-12	100	3005	1	2007	
101-15	102A	3123	53	2007	
101-2(ADMIRAL)	108	3452	86	2038	
101-3(ADMIRAL)	108	3452	86	2038	
101-4(ADMIRAL)	108	3452	86	2038	
0101-439	129	3025	244	2027	
101-6742(RCA)	173BP		305		
101-6744(RCA)	173BP		305		
101A	160	3123	245	2004	
101B	160	3123	245	2004	
101B6	116	3016	504A	1104	
101M	160	3123	245	2004	
101P1	159	3466	82	2032	
101P10	159	3466	82	2032	
0102	116	3008	504A	2024	
102-02	109	3087		1123	
0102-0371	102A	3004	53	2007	
102-0373-00	199	3122	212	2010	
102-0394-25	229	3018	61	2038	
102-0454-02	289A	3018	210	2051	
102-0460-02	233	3018	61	2038	
102-0461-02	229	3018	61	2038	
102-0495-00	295		270		
102-0495-20	295		270		
102-0535-02	229	3018	86	2038	
102-0732-28	199	3122	62	2010	
102-0735-25	289A	3122	210	2038	
102-0828-17	199	3122	61	2038	
102-0945-16	199	3124/289	212	2010	
102-0945-17	199	3124/289	212	2010	
102-0945-38	199	3124/289	212	2010	
102-0945-39	199	3124/289	212	2010	
102-1047-03	229	3018	60	2015	
102-1061-01	152	3054/196	66	2017	
102-1166-25	297	3122	81	2051	
102-1317-18	289A		210	2051	
102-1335-04	199		210	2051	
102-1342-02	229	3018	61	2038	
102-1384-17	293	3849	88	2030	
102-1675-11	229	3124/289	213	2038	
102-1675-12	229	3124/289	61	2038	
102-1678-00	235	3197	322		
102-207	109	3087		1123	
102-339	177	3100/519	300	1122	
102-4	161	3117	39	2015	
102-412	519	3100	514	1122	
102-842	106	3984			
102B6	116	3311	504A	1104	
102D	116	3016	504A	1104	
102P1	159	3466	82	2032	
102P10	159	3466	82	2032	
0000103	102A	3004	53	2007	
0103-0014	123A	3444	20	2051	
0103-0014(R,S)			18	2030	
0103-0014/4460	123A	3444	20	2051	
0103-0014R	128	3024	243	2030	
0103-0014S	128	3024	243	2030	
0103-0014T			18	2030	
0103-0014U			18	2030	
0103-0051	128	3024	243	2030	
0103-0060	108	3452	86	2038	
0103-0060(B)			17	2051	
0103-00608			20	2051	
0103-0060A		3122	17	2051	
0103-0060B	108	3452	86	2038	
0103-0088	123A	3444	20	2051	
0103-0088/4460	199	3124/289		2010	
0103-0088H	123A	3444	20	2051	
0103-0088R	123A	3444	20	2051	
0103-0088S	123A	3444	20	2051	
0103-0191	108	3452	86	2038	
103-0227-18	123A	3122	62	2051	
103-0235-85	152	3054/196	66	2048	
0103-0389	108	3452	86	2038	
0103-0419	186	3192	28	2017	
0103-0419A	186	3192	28	2017	
0103-0473	123A	3444	20	2051	
0103-0482	123A	3444	20	2051	
0103-0491	123A	3444	20	2051	
0103-0491/4460	123A	3444	20	2051	
0103-0492	199	3124/289	62	2010	
0103-0503	128	3024	243	2030	
0103-0503S	128	3024	243	2030	
0103-0504	123A	3444	20	2051	
0103-0521	108	3452	86	2038	
0103-0521(B)			17	2051	
0103-0521B	108	3452	86	2038	
0103-0531	108	3452	86	2038	
0103-0531/4460	107	3122	11	2015	
0103-0540	123A	3444	20	2051	
0103-0568	107	3018	11	2015	
0103-0568/4460	107	3018	11	2015	
0103-0607	128	3024	243	2030	
0103-0616	128	3024	243	2030	
103-101	178MP		300(2)	1122(2)	
103-102	110MP	3709	1N60	1123(2)	
103-104		3089	90		
103-105	5081A	3151	ZD-24		

Industry Standard No.	ECG	SK	GE	RS 276-	MOTOR.
103-105-01	5081A	3151	ZD-24		
103-112	506	3843	511	1114	
103-114	109	3087		1123	
103-131	177	9091	300	1122	
103-136			ZD-15	564	
103-140	5071A	3334	ZD-6.8	561	
103-140A	5071A			561	
103-141	519	3311	514	1122	
103-142	177	9091	300	1122	
103-142(DET)	178MP		300(2)	1122	
103-142-01	177	3175	300	1122	
103-144	5137A	3404/5138A	5ZD-24		
103-144-01	5137A	3151/5081A	ZD-24		
103-145	177	3311	300	1122	
103-145-01	116	3100/519	300	1122	
103-146	616	3324/611	90		
103-158	142A	3062	ZD-12	563	
103-159	177	3100/519	300	1122	
103-160	506	3843	511	1114	
103-176	616	3319	90		
103-178	177		300	1122	
103-178-01	177		300	1122	
103-185			504A	1102	
103-189	614	3126	90		
103-19	109	3091		1123	
103-191	116	3031A	504A	1104	
103-192	178MP	3088	300(2)	1122(2)	
103-193	506	3843	511	1114	
103-194	5121A	3781/5015A	ZD-7.5		
103-196	506	3843	511	1114	
103-20	113A	3119/113	6GC1		
103-202	109	3089/112	1N82A		
103-202-GE			1N82A	1123	
103-203	116	3032A	504A	1104	
103-206	5162A	3428	ZD-150		
103-208	5098A	3820/5054A	ZD-130		
103-212	5080A	3797/5031A	ZD-22		
103-212-76		3313	504A		
103-215	513	3443	513		
103-216	116	3032A	504A	1104	
103-22	109	3087		1123	
103-222	177	3100/519	300	1122	
103-228	116	3032A	504A	1104	
103-23	110MP	3090/109		1123	
103-23-01	109	3090		1123	
103-23-1		3090		1123	
103-231	5075A	3751	ZD-16	564	
103-236	147A	3095	ZD-33		
103-237	615	3095/147A	ZD-33		
103-239	503	3068	CR-5		
103-239-01	503	3068	CR-5		
103-239-02	504	3108/505	CR-6		
103-240	177	3100/519	300	1122	
103-244	506	3843	511	1114	
103-245	116	3016	504A	1104	
103-246	5074A	3092/141A	ZD-11.0	563	
103-247	506	3843	511	1114	
103-248	5081A	3064/146A	ZD-24		
103-252	5077A	3752	ZD-18		
103-254	116	3313	504A	1104	
103-254-01	116	3313	504A	1104	
103-256	5128A		ZD-13	563	
103-258	505	3108	CR-7		
103-261	116	3031A	504A	1104	
103-261-0	116	3031A	504A	1104	
103-261-01	116	3311	504A	1104	
103-261-02	525	3100/519	300	1122	
103-261-04	552	3311	504A	1102	
103-263	506	3843	511	1114	
103-270	5086A	3338	ZD-39		
103-271	109	3709/110MP	1N34AS	1123(2)	
103-272	139A	3060	ZD-9.1	562	
103-275-02	503		CR-5		
103-276	5081A	3151			
103-278	5081A	3151	ZD-24		
103-279-01	142A	3062		563	
103-279-09	5009A	3775			
103-279-11				561	
103-279-12				561	
103-279-13				561	
103-279-13A	5013A	3779			
103-279-14			ZD-6.8	561	
103-279-14A	5014A	3780		561	
103-279-16			ZD-8.2	562	
103-279-16A	5016A	3782			
103-279-17				562	
103-279-18	139A	3784/5018A		562	
103-279-19				562	
103-279-20	5074A	3786/5020A	ZD-11	563	
103-279-21	142A	3787/5021A	ZD-12	563	
103-279-21A	5021A	3787		563	
103-279-22				563	
103-279-23				564	
103-279-24				564	
103-279-25				564	
103-279-28A	5078A	9023	ZD-19		
103-279-37	5085A	3803/5037A	ZD-36		
103-279-37A	5085A		ZD-36		
103-28	358		42		
103-284	552	3032A	509	1103	
103-284A	552	3318	510		
103-287	525	3998/506	511	1114	
103-289	5081A	3151	ZD-24		
103-29	116	3016	504A	1104	
103-292	5081A		ZD-24		
103-293-02	504	3108/505	CR-6		
103-295-02		9091	300	1122	
103-298-05A	552	3318	504A		
103-3			1N60	1123	
103-301-09	5071A		ZD-6.8		
103-301-17A	143A	3750	ZD-13		
103-305	525	3925	530		
103-308A	5012A			563	
103-31	109	3087		1123	
103-312	515	9098	511		
103-315-03A	116		504A	1104	
103-316-04	558	9098/515			
103-32	113A	3119/113	6GC1		
103-34	109	3087		1123	
0103-389	108	3452	86	2038	
103-39	614	3126	90		
103-4	108	3452	86	2038	
103-42	177	3100/519	300	1122	
103-43	113A	3119/113	6GC1		
103-44	109	3091		1123	
103-47	614		90		
103-47-01	614	3327	90		
103-49	112	3089	1N82A		
103-51	177	3032A	6GC1	1122	
0103-512M	186	3192	28	2017	
103-59				1104	
103-60	112	3089	1N82A		
103-61	112	3089	1N82A		
103-65	112	3089	1N82A		
103-71	358		42		
103-73	109	3087		1123	
103-74	109	3088		1123	
103-76	116	3311	504A	1104	
103-79	109	3087		1123	
103-82	116	3017B/117	504A	1104	
103-84				564	
103-87	109	3087		1123	
103-89	358		42		
103-90	110MP	3709	504A	1104	
0103-93	199	3124/289	62	2010	
0103-94	123A	3444	20	2051	
0103-95	158		53	2007	
0103-9531/4460	108	3452	86	2038	
103-96	552	3750/143A	ZD-12	563	
103-Z9000		3151	ZD-24		
103-Z9001		3090		1123	
103-Z9003		3062	ZD-12	563	
103-Z9004		3095	ZD-33		
103-Z9006		3056	ZD-5.0		
103-Z9007		3057	ZD-5.6	561	
103-Z9008		9021	ZD-6.2	561	
103-Z9010		3061	ZD-10	562	
103-Z9012		3094	ZD-14	564	
103-Z9013		3751	ZD-15	564	
103B6	116	3311	504A	1104	
103G3125	116	3016	504A	1104	
103G3125A	116	3016	504A	1104	
103P(AIRLINE)	159			2032	
103P935	159	3466	82	2032	
103P935A	159	3466	82	2032	
103PA	159	3466	82	2032	
103PNAIRLINEE			82	2032	
0000104	102A	3004	40	2007	
0104-0013	128	3024	243	2030	
104-1		3126	90		
104-17	126	3006/160	52	2024	
104-17(RCA)	159			2032	
104-170	159	3466	82	2032	
104-17NRCAE		3466	82	2032	
104-19	126	3006/160	52	2024	
104-21	126	3006/160	52	2024	
104-8	154	3045/225	40	2012	
104B6	116	3311	504A	1104	
104H01	128	3024	243	2030	
104T2	130	3027	14	2041	
104T2A	130	3027	14	2041	
105	159	3466	82	2032	
105(ADMIRAL)	123A	3444		2051	
105(JULIETTERECT)				1104	
105-001-04		3122	20	2051	
105-001-05		3018	10	2051	
105-001-07	233	3122	20	2051	
105-001-08	108	3452	86	2038	
105-00106-00	108	3452	86	2038	
105-00107-09	107	3018	11	2015	
105-00108-07	108	3452	86	2038	
0105-0012	312	3116	FET-1	2035	
105-003-06	123A	3444	20	2051	
105-003-09	123A	3444	20	2051	
105-005-04/2228-3	108		86	2038	
105-005-12	108	3452	86	2038	
105-006-08	123A	3444	20	2051	
105-008-04/2228-3	108		86	2038	
105-009-21/2228-3	108		86	2038	
105-02	109	3087		1123	
105-02004-09	108	3452	86	2038	
105-02005-07	107	3018	11	2015	
105-02006-05	107	3018	11	2015	
105-02008-01	107	3018	11	2015	
105-03	109	3091		1123	
105-060-09	123A	3444	20	2051	
105-06004-00	108	3452	86	2038	
105-06007-05	123A	3444	20	2051	
105-08243-05	123A	3444	20	2051	
105-085-33	123A	3122	20	2051	
105-085-54	123A	3444	20	2051	
105-12	123A	3444	20	2051	
105-24191-04	108	3452	86	2038	
105-28196-07	123A	3122	20	2051	
105-904-85		3018	17	2051	
105-904-86		3018	17	2051	
105-904-87		3018	17	2051	
105-931-91/2228-3	123A		20	2051	
105-941-97/2228-3	108		86	2038	
105B6	116	3311	504A	1104	
105NADMIRALE			20	2051	
106	116		504A	1104	
106-001	229	3018	61	2038	
106-002	229	3018	61	2038	
106-003	199	3122	62	2010	
106-004	186A	3357	247	2052	
106-005	295	3253	270		
106-006	235	3197	215		
106-007	519	3100	514	1122	
106-008	109	3087		1123	
106-009	177	3100/519	300	1122	
106-010	139A	3060	ZD-9.1	562	
106-011	116	3017B/117	504A	1104	
106-013	156	3051	512		
106-1	116		504A	1104	
106-111	116		504A	1104	
106-12	159	3466	82	2032	
106-120	159	3466	82	2032	
106-351	229	3018	61	2038	
106A	5454			1067	
106B	5455			1067	
106B6	116	3311	504A	1104	
106C	5456			1020	
106F	5461	3685			
106KBO	129		244	2027	
106KBA	129	3025	244	2027	
106M	312	3448	FET-2	2035	
106P1	121	3009	239	2006	
106P1AG	121	3009	239	2006	
106P1T	121	3717	239	2006	
106Q	5461	3685			
106RED	159		82	2032	
106Y	5461	3685			
107-0021-00	6400A		2N2160	2029	
107-3088			210	2051	
107-8	123A	3122	20	2051	
107A	160		245	2004	
107A(SCR)	5454			1067	
107B	160		245	2004	
107B(SCR)	5455			1067	
107B6	116	3311	504A	1104	
107BRN	123A	3122	20	2051	
107C	5456			1020	
107F	5461	3685			
107M	160	3116	245	2004	
107N1	128	3024	243	2030	
107N2	123A	3124/289	20	2051	
107Q	5461	3685			

Industry Standard No.	ECG	SK	GE	RS 276-	MOTOR.
107Y	5461	3685			
108	116	3016	504A	1104	
108(PARFISA)	109			1123	
108(FARTISA)				1123	
108-002			61	2038	
108-0049-08	312	3116	FET-2	2035	
108-0068-12	312	3112	FET-1	2028	
108-1	160	3006	245	2004	
108-2	160	3006	245	2004	
108-3	160	3006	245	2004	
108-4	160	3006	245	2004	
108-6	159	3466	82	2032	
108-60	159	3466	82	2032	
108-74			1N60	1123	
108A	5454			1067	
108A4	116	3016	504A	1104	
108B	5455			1067	
108B4	116		504A	1104	
108B6	116	3031A	504A	1104	
108C	5456			1020	
108E-E2	116	3032A	504A	1104	
108F	5453			1067	
108GRN	159		82	2032	
108Q	5452	6752		1067	
108T2	327	3945			
108Y	5452	6752		1067	
109	123A		20	2051	
109-036500	109			1123	
109-1		3122	20	2051	
109-1(RCA)	123A		20	2051	
109-192			1N60	1123	
109A949CE	5693	3652			
109A949CM-P1	5624	3632			
109B6	116	3031A	504A	1104	
109T2	327	3945			
0110	116	3466/159	504A	1104	
110(BOOST)	119		CR-2		
0110-0011	116	3121/115	504A	1104	
0110-0141	116	3311	504A	1104	
0110-0141/4460	116	3311	504A	1104	
110-01563-00	102A	3004	53	2007	
0110-0209	116	3016	504A	1104	
110-629	116	3016	504A	1104	
110-635	116	3016	504A	1104	
110-636	125	3016	510,531	1114	
110-672	116	3016	504A	1104	
110-684	116	3016	504A	1104	
110-763	109			1123	
110B6	125	3031A	510,531	1114	
110BH1	118	3066	CR-1		
110P1	159	3466	82	2032	
110P1AA	159	3466	82	2032	
110P1M	159	3466	82	2032	
110P2			21	2034	
0111	116	3311	504A	1104	
111-1	116	3016	504A	1104	
111-4-2020-04600	712		IC-2		
111-4-2020-06100	177		300	1122	
111-4-2020-06200			511	1114	
111-4-2020-0800	116	3017B/117	504A	1104	
111-4-2020-08000	116		504A	1104	
111-4-2020-08600	109			1123	
111-4-2020-11300	5071A		ZD-6.8	561	
111-4-2020-11500	5074A		ZD-11	563	
111-4-2020-13300	5085A		ZD-36		
111-4-2020-14400	506		511	1114	
111-4-2020-14500			504A	1104	
111-4-2020-14600	116		504A	1104	
111-4-2020-15100	506		511	1114	
111-4-2020-15600	177		300	1122	
111-4-2060-04000	712		IC-2		
111-4-2060-04600	712		IC-2		
111-4-2060-05200	1159		IC-72		
111-4-2060-2600	1080	3284			
111-6910			50	2004	
111-6935	126	3006/160	52	2024	
111-731			FET-2	2035	
111N2C	130		14	2041	
111N4	130	3027	14	2041	
111N4A	130	3027	14	2041	
111N4B	130		14	2041	
111N4C	130		14	2041	
111N6	130		14	2041	
111N6C	152	3893	66	2048	
111N8C	196		241	2020	
111P3B	218		234		
111P5	219		74	2043	
111P5C	104	3719	16	2006	
111P7C	104	3719	16	2006	
111T2	154	3045/225	40	2012	
111Z4	411	3092/141A			
0112	116	3031A	504A	1104	
112-000088	123A		20	2051	
112-000172	159	3466	82	2032	
112-000185	159	3466	82	2032	
112-000187	159	3466	82	2032	
112-000267	160	3007	245	2004	
112-001	100	3005	1	2007	
112-0011 A				2051	
0112-0019	109	3088		1123	
112-002	126	3006/160	52	2024	
0112-0026	109	3087		1123	
0112-0028	109	3088		1123	
0112-0028-6438		3088	1N60	1123	
0112-0028/4460	109	3088		1123	
112-003	102	3004	2	2007	
0112-0037	109	3087		1123	
112-004	102	3004	2	2007	
0112-0046	109	3087		1123	
0112-0073	109	3088		1123	
0112-0082	109	3088		1123	
112-011 A	123A	3444			
112-011A			20	2051	
112-034923	102A	3004	53	2007	
112-1	312	3025/129			
112-10	159	3466	82	2032	
112-1A82	123A	3444	20	2051	
112-2	129	3025	244	2027	
112-200525	102A	3004	53	2007	
112-202147	121	3014	239	2006	
112-203053	128	3024	252	2018	
112-203055	130	3027	14	2041	
112-203391	199	3124/289	62	2010	
112-2A	129	3025	244	2027	
112-361	128	3024	243	2030	
112-362	224		46		
112-363	130	3027	14	2041	
112-500-0-50	177	3100/519	300	1122	
112-500-0-501	177	3016	300	1122	
112-520	108	3452	86	2038	
112-521	108	3452	86	2038	
112-522	108	3452	86	2038	
112-523	123A	3444	20	2051	
112-524	121	3009	239	2006	
112-525	128	3024	243	2030	
112-526	224		46		
112-527	224		46		
112-601-0-102	116	3016	504A	1104	
112-7	159	3466	82	2032	
112-7292955	179	3642	76		
112-8	159	3466	82	2032	
112-826	116	3017B/117	504A	1104	
0113-0027		3126	90		
113-039	116	3017B/117	504A	1104	
113-118	123	3124/289	20	2051	
113-321	116		504A	1104	
113-392	116	3017B/117	504A	1104	
113-398	108	3452	86	2038	
113-938	108	3452	86	2038	
113-998	117	3017B	504A	1104	
113A7739	116	3016	504A	1104	
113N1AG	172A	3156	64		
113N2	172A	3156	64		
0114-0017	139A	3060	ZD-9.1	562	
0114-0026/4460	5014A	3780	ZD-6.8	561	
0114-0090	139A	3060	ZD-9.1	562	
114-013	116	3017B/117	504A	1104	
0114-0260	5071A		ZD-6.8	561	
114-1(PHILCO)	192		63	2030	
114-118	108	3452	86	2038	
114-267	108	3452	86	2038	
114-4-2020-14500	116		504A	1104	
114-42020-14500	116		504A	1104	
114A-1-82	121	3717	239	2006	
114N2P	184	3041	57	2017	
114N4U	103A	3835	59	2002	
114P1P	185	3084	58	2025	
114P3U	158/410		53	2007	
115-039	116	3017B/117	504A	1104	
115-063	121	3009	239	2006	
115-1	123	3124/289	20	2051	
115-13		3018	20	2051	
115-225	123A	3444	20	2051	
115-227	160	3006	245	2004	
115-228	160	3006	245	2004	
115-229	160	3006	245	2004	
115-268	121	3009	239	2006	
115-269	121	3009	239	2006	
115-275	126	3006/160	52	2024	
115-281	121MP	3013	239(2)	2006(2)	
115-282	121MP	3013	239(2)	2006(2)	
115-283	121MP	3013	239(2)	2006(2)	
115-284	121MP	3013	239(2)	2006(2)	
115-4	123	3124/289	20	2051	
115-559	116	3017B/117	504A	1104	
115-599	116	3017B/117	504A	1104	
115-867	117	3017B	504A	1104	
115-875	123A	3444	20	2051	
115C-8	912		IC-172		
115C5	784	3524	IC-236		
115C7	786	3140	IC-227		
115Z4	145A	3063	ZD-15	564	
116,666	519		514	1122	
116-052	116	3017B/117	504A	1104	
116-068	127	3764	25		
116-072	126	3006/160	52	2024	
116-073	108	3452	86	2038	
116-074	123A	3444	20	2051	
116-075	124	3021	12		
116-078	123A	3444	20	2051	
116-079	108	3452	86	2038	
116-080	108	3452	86	2038	
116-082	108	3452	86	2038	
116-083	108	3452	86	2038	
116-085	123A	3444	20	2051	
116-086	127	3035	25		
116-087	127	3035	25		
116-088	127	3035	25		
116-089	127	3035	25		
116-091	102A	3004	53	2007	
116-092	123A	3444	20	2051	
116-1	124	3021	12		
116-198	108	3452	86	2038	
116-199	108	3452	86	2038	
116-200	108	3452	86	2038	
116-201	102A	3004	53	2007	
116-202	160	3006	245	2004	
116-203	102A	3004	53	2007	
116-206	102A	3004	53	2007	
116-207	160	3007	245	2004	
116-208	160	3007	245	2004	
116-209	160	3007	245	2004	
116-588	123A	3444	20	2051	
116-683	160	3006	245	2004	
116-684	160	3006	245	2004	
116-685	102A	3004	53	2007	
116-686	102A	3004	53	2007	
116-687	103A	3010	59	2002	
116-756	126	3006/160	52	2024	
116-757	102A	3004	53	2007	
116-875	123A	3444	20	2051	
116-997	102A	3004	53	2007	
116C3475	181	3036	75	2041	
0117-02	221			2036	
117-1	124	3021	12		
117-134-11		3311	504A	1104	
117-2			53	2007	
117-2(HEATHKIT)	158(4)		53(4)		
117-A40	504	3108/505	CR-6		
117-A40-7350	504		CR-6		
118-02900	519		514	1122	
118-02902	519		514	1122	
118-030	177		300	1122	
118-1	108	3452	86	2038	
118-2	108	3452	86	2038	
118-3	108	3452	86	2038	
118-4	108	3452	86	2038	
119-0016	105	3012	4		
119-0054	123A	3444	20	2051	
119-0055	159	3466	82	2032	
119-0056	123A	3444	20	2051	
119-0068	175	3026	246	2020	
119-0075	181	3036	75	2041	
119-0077	128	3024	243	2030	
119-6511	116	3017B/117	504A	1104	
120-00-19	102	3004	2	2007	
120-000190	160		245	2004	
120-001-300	116		504A	1104	
120-001190	160	3006	245	2004	
120-001192	100	3005	1	2007	
120-001195	102	3004	2	2007	
120-001300	116	3016	504A	1104	
120-001301	110MP	3709		1123(2)	
120-001795	102	3722	2	2007	
120-001798	195A	3765	46		
120-00190	126	3006/160	52	2024	
120-00195	102A		504A	1104	
120-002	912		IC-172		

Industry Standard No.	ECG	SK	GE	RS 276-	MOTOR.
120-002012	102A		504A	1104	
120-002013	102A	3004	504A	1104	
120-002014	102A	3004	504A	1104	
120-00213	126	3006/160	52	2024	
120-002213	160	3007	245	2004	
120-002214	160	3006	245	2004	
120-002216	160	3007	245	2004	
120-002513	126	3006/160	52	2024	
120-002515	126	3006/160	52	2024	
120-002518	126	3006/160	52	2024	
120-002520	158	3006/160	53	2007	
120-002521	102A	3004	53	2007	
120-002656	126	3006/160	52	2024	
120-002748	102A	3004	53	2007	
120-003147	177	3100/519	300	1122	
120-003148	116	3031A	504A	1104	
120-003149	125		510,531	1114	
120-003150	131	3052	44	2006	
120-003151	195A	3765	46		
120-004048	126	3006/160	52	2024	
120-004061	125	3033	510,531	1114	
120-004480	123A	3444	20	2051	
120-004482	123A	3444	20	2051	
120-004483	123A	3444	20	2051	
120-004492	126	3006/160	52	2024	
120-004493	158	3004/102A	53	2007	
120-004494	158	3004/102A	53	2007	
120-004495	158	3004/102A	53	2007	
120-004496	108	3452	86	2038	
120-004497	108	3452	86	2038	
120-004498	109	3087		1123	
120-004499	109	3087		1123	
120-004503	116	3017B/117	504A	1104	
120-004629	178MP		300(2)	1122(2)	
120-004722	160		245	2004	
120-004723	108	3452	86	2038	
120-004724	108	3452	86	2038	
120-004725	108	3452	86	2038	
120-004727	102A	3004	53	2007	
120-004728	158	3123	53	2007	
120-004729	102A	3004	53	2007	
120-004730	109	3088		1123	
120-004877	177	3100/519	300	1122	
120-004878	116	3017B/117	504A	1104	
120-004879	139A	3060	ZD-9.1	562	
120-004880	123A	3444	20	2051	
120-004881	108	3452	86	2038	
120-004882	123A	3444	20	2051	
120-004883	123A	3444	20	2051	
120-004884	191	3047	249		
120-004885	224	3049	46		
120-004886	224	3049	46		
120-004887	131	3009	44	2006	
120-004888	101	3861	8	2002	
120-005291	108	3452	86	2038	
120-005292	108	3452	86	2038	
120-005293	108	3452	86	2038	
120-005294	108	3452	86	2038	
120-005295	108	3452	86	2038	
120-005296	108	3452	86	2038	
120-005297	108	3452	86	2038	
120-005298	108	3452	86	2038	
120-005299	109	3087		1123	
120-005300		3126	90		
120-006604	159	3466	82	2032	
120-01193	102A		53	2007	
120-02213	160		245	2004	
120-1	123A	3444	20	2051	
120-190	102A		53	2007	
120-2	123A	3444	20	2051	
120-3	123A	3444	20	2051	
120-7	123A	3444	20	2051	
120-8	123A	3444	20	2051	
120-8A	123A	3444	20	2051	
120A02	177	3100/519	300	1122	
120A11	177	3100/519	300	1122	
120A13	177	3100/519	300	1122	
120BLU	123A	3444	20	2051	
120C					MR1-1200
120P1	159	3466	82	2032	
120P1M	159	3466	82	2032	
121	222	3444/123A	39	2051	
0121(AIRLINE)	185		58	2025	
121(SEARS)	123A	3444	20	2051	
121-0041	137A	3058	ZD-6.2	561	
121-1	159	3466	82	2032	
121-10	102A	3123	53	2005	
121-100	101	3861	8	2002	
121-1003	238	3710	259		
121-1004	123AP	3124/289		2014	
121-1005	159	3138/193A	82	2024	
121-1006	152	3202/210	66	2048	
121-1007	129	3025	48	2026	
121-1008	291	3440	233		
121-1009	292	3441	234		
121-101	160		245	2004	
121-1010				2038	
121-1012	262	3897			
121-1013	261	3896			
121-1014	190	3250/315	249		
121-1015	165	3710/238			
121-1016	288	3114/290	269	2032	
121-1017	287	3124/289	268		
121-1019	159	3466	22	2024	
121-102	160	3005	245	2004	
121-1020	188	3449/297	47	2018	
121-1021	189	3450/298	48	2026	
121-1024	222	3050/221	FET-4	2036	
121-1028	286	3131A/369	12		
121-1028-01	379	3219	325		
121-1029	165	3710/238	259		
121-103	160	3005	245	2004	
121-1030	222	3065	FET-4	2036	
121-1032	102	3004	2	2007	
121-1033	238	3004/102A	2	2007	
121-1034	102	3004	2	2007	
121-1035	297	3449	271	2007	
121-1036	298	3450	272	2007	
121-1037	171	3201	27		
121-104	160	3005	245	2004	
121-1040	123AP	3444/123A	20	2051	
121-1043	159		82		
121-1049		3054	66		
121-105	160	3005	245	2004	
121-106	102	3004	2	2007	
121-107	102	3004	2	2007	
121-11	102A	3123	53	2007	
121-1124	121	3009	239	2006	
121-113	108	3006/160	61	2038	
121-1134	121	3014	239	2006	
121-119	160		245	2004	
121-12	102A	3123	53	2007	
121-120	158	3004/102A	53	2007	
121-128	100	3005	1	2007	

Industry Standard No.	ECG	SK	GE	RS 276-	MOTOR.
121-132	160	3006	245	2004	
121-1330	100	3005	1	2007	
121-134	160		245	2004	
121-135	160	3006	245	2004	
121-1350	100	3005	1	2007	
121-136	160		245	2004	
121-1360	100	3005	1	2007	
121-137	160	3006	245	2004	
121-138	160	3008	245	2004	
121-139	160	3006	245	2004	
121-1390	100	3005	1	2007	
121-14		3123	1	2007	
121-14(COLUMBIA)	128		243	2030	
121-14(ZENITH)	102A		504A	1104	
121-1400	100	3005	1	2007	
121-1410	101	3011	8	2002	
121-145	100	3008	1	2007	
121-146	100	3008	1	2007	
121-147	100	3005	1	2007	
121-148	102A	3003	53	2007	
121-15	101	3861	8	2002	
121-150	126	3008	52	2024	
121-151	102A		53	2007	
121-152	102A		53	2007	
121-153	126	3008	52	2024	
121-154	126	3008	52	2024	
121-157	160	3006	245	2004	
121-16	101	3861	8	2002	
121-160	100	3005	1	2007	
121-161	126	3005	52	2024	
121-162	126	3005	52	2024	
121-163	102A	3003	53	2005	
121-164	102A	3004	53	2005	
121-167			53	2007	
121-168	102A	3004			
121-169	103A	3010			
121-17	101	3861	8	2002	
121-170	102A	3008	82	2032	
121-171	121	3717	239	2006	
121-172	103A	3010			
121-178	102A	3004			
121-179	160	3005	245	2004	
121-18	124		12		
121-18(ZENITH)	102A		53		
121-180	160	3005	245	2004	
121-181	160		245	2004	
121-184	102A	3004	53	2007	
121-185	160		245	2004	
121-186	126	3005	52	2024	
121-186D	102A	3004			
121-187	126	3006/160	52	2024	
121-189	126	3008	52	2024	
121-189B	102A	3008			
121-19	102A	3004	53	2007	
121-190	102A	3004	53	2007	
121-190E	102A	3004			
121-191	102A	3004	53	2007	
121-192	102A	3003	53	2007	
121-193	102A	3004	53	2007	
121-195B	123A		20	2051	
121-1RED	159		82	2032	
121-200			2	2007	
121-205	100	3005	1	2007	
121-206	100	3005	1	2007	
121-207	100	3005	1	2007	
121-208	100	3005	1	2007	
121-209	100	3005	1	2007	
121-21	101	3861	8	2002	
121-210	100	3005	1	2007	
121-211	100	3005	1	2007	
121-212	100	3005	1	2007	
121-213	100	3005	1	2007	
121-219	100	3005	1	2007	
121-22	101	3861	8	2002	
121-220	100	3005	1	2007	
121-221	100	3005	1	2007	
121-222	100	3005	1	2007	
121-225	100	3004/102A	1	2007	
121-226	102A	3004	53	2007	
121-227	102A	3004	53	2007	
121-228	126	3006/160	52	2024	
121-229	126	3006/160	52	2024	
121-230	126	3006/160	52	2024	
121-231	126	3006/160	52	2024	
121-232	126	3004/102A	52	2024	
121-233	126	3006/160	52	2024	
121-234	100	3005	1	2007	
121-235	100	3005	1	2007	
121-236	100	3005	1	2007	
121-237	103A	3835	59	2002	
121-238	103A	3835	59	2002	
121-239	102A		53	2007	
121-24	101	3861	8	2002	
121-240	126	3005	52	2024	
121-240X	102A		20	2007	
121-241	154	3044	40	2012	
121-241(ZENITH)	126		52		
121-242	160	3008	245	2004	
121-243	160	3008	245	2004	
121-244	160	3008	245	2004	
121-245	102A		53	2007	
121-246	102A	3004	53	2007	
121-247	103A	3835	59	2002	
121-248	103A	3835	59	2002	
121-25	101	3861	8	2002	
121-254	100	3005	1	2007	
121-256	126	3007	52	2024	
121-257	126	3008	52	2024	
121-258	126	3007	52	2024	
121-259	126	3007	52	2024	
121-26	101	3861	8	2002	
121-260	126	3007	52	2024	
121-261	126	3006/160	52	2024	
121-262	126	3006/160	52	2024	
121-263	126	3006/160	52	2024	
121-266	102	3004	2	2007	
121-267	102	3004	2	2007	
121-268	160		245	2004	
121-269	160		245	2004	
121-27	102	3004	2	2007	
121-270	121	3009	239	2006	
121-271	121	3009	239	2006	
121-272	102A	3004	53	2007	
121-273	102A	3005	53	2007	
121-274	102A	3003	53	2007	
121-275	102A	3005	53	2007	
121-276	123	3124/289	20	2051	
121-277	123	3124/289	20	2051	
121-278	123	3124/289	20	2051	
121-279	411/128	3124/289	20		
121-283	161	3132	39	2015	
121-284	160		245	2004	
121-286			10	2051	
121-287			2	2007	

Industry Standard No.	ECG	SK	GE	RS 276-	MOTOR.
121-288			10	2051	
121-289	5081A	3151	52	2024	
121-290	126	3007	52	2024	
121-291	102A	3004	53	2007	
121-292	126	3006/160	52	2024	
121-293	126	3004/102A	52	2024	
121-294	126	3006/160	52	2024	
121-295	126	3006/160	52	2024	
121-296	126	3006/160	52	2024	
121-297	126	3006/160	52	2024	
121-298	126	3006/160	52	2024	
121-299	126	3006/160	52	2024	
121-300	102A	3123	53	2007	
121-301	102A	3123	53	2007	
121-302	101	3861	8	2002	
121-303	108	3452	86	2038	
121-304	160	3006	245	2004	
121-305	102A	3004	53	2007	
121-306	102A	3004	53	2007	
121-307	102A	3004	53	2007	
121-308	121	3009	239	2006	
121-309	102A	3004	53	2007	
121-31	109	3087	1N34AS	1123	
121-310	102A	3005	53	2007	
121-311	102A	3004	53	2007	
121-311B	102A		53	2007	
121-311C				2007	
121-311D				2007	
121-311E				2007	
121-311F				2007	
121-312	160	3008	245	2004	
121-313	160	3008	245	2004	
121-314	102A	3004	53	2007	
121-315	124	3021	12		
121-316	108	3452	86	2038	
121-317	108	3452	86	2038	
121-318	108	3452	86	2038	
121-318L	108	3452	86	2038	
121-319	102A	3004	53	2005	
121-320	102A	3004	53	2005	
121-321	108	3452	86	2038	
121-327	102A	3004	53	2007	
121-328	102A	3004	53	2007	
121-329	160		245	2004	
121-33	102A		53	2007	
121-330	160		245	2004	
121-331	160		245	2004	
121-332	160		245	2004	
121-333	160	3005	245	2004	
121-334	160	3005	245	2004	
121-335	160	3005	245	2004	
121-336	160		245	2004	
121-34	102A	3004	53	2007	
121-345	108	3452	86	2038	
121-347	102A	3004	53	2007	
121-348	102A	3004	53	2007	
121-349	160	3006	245	2004	
121-350	160	3006	245	2004	
121-351	160	3006	245	2004	
121-352	160	3006	245	2004	
121-353	160	3006	245	2004	
121-354	100	3721	1	2007	
121-356	160	3006	245	2004	
121-357	160		245	2004	
121-358	160	3006	245	2004	
121-359	160	3006	245	2004	
121-360	126	3006/160	52	2024	
121-361	154	3045/225	40	2012	
121-362	158		53	2007	
121-363	121	3014	239	2006	
121-364	123A	3444	20	2051	
121-365	123A	3444	20	2051	
121-366	123A	3444	20	2051	
121-367	123A	3444	20	2051	
121-368	158	3123	53	2007	
121-369	123A	3444	20	2051	
121-370	127	3035	25		
121-371	121	3009	239	2006	
121-372	102	3004	2	2007	
121-373	158	3004/102A	53	2007	
121-373B				2007	
121-373C				2007	
121-373D				2007	
121-373E				2007	
121-373F				2007	
121-373G				2007	
121-373H				2007	
121-373I				2007	
121-373J				2007	
121-373K				2007	
121-373L				2007	
121-373M				2007	
121-373N				2007	
121-3730				2007	
121-373P				2007	
121-373Q				2007	
121-374	102A	3004	53	2007	
121-375	102A	3004	53	2007	
121-377	161	3018	39	2015	
121-378	161	3018	39	2015	
121-379	161	3018	39	2015	
121-380	161	3018	39	2015	
121-381	126	3008	52	2024	
121-382	121	3717	239	2006	
121-383	160	3132	39	2015	
121-384	160	3006	245	2004	
121-385	160	3006	245	2004	
121-388	158	3004/102A	53	2007	
121-389	121	3009	239	2006	
121-395	102A	3004	53	2007	
121-396	102A	3004	53	2007	
121-397	100	3005	1	2007	
121-398	121	3009	239	2006	
121-399	102A	3004	53	2007	
121-400	158	3004/102A	53	2007	
121-401	158	3004/102A	53	2007	
121-403	158	3004/102A	53	2007	
121-404	123A	3444	20	2051	
121-406	179	3642	76		
121-408	102A	3004	53	2007	
121-409	102A	3004	53	2007	
121-410			2	2007	
121-411	160	3006	245	2004	
121-412	160	3006	245	2004	
121-413	160	3006	245	2004	
121-414	160	3006	245	2004	
121-415	160	3006	245	2004	
121-415B	160		245	2004	
121-416			53	2007	
121-417	159	3114/290	82	2032	
121-418	179(2)	3009	76(2)		
121-419	179	3642	76		
121-420	102A		53	2007	
121-421	158		53	2007	
121-422	123	3124/289	20	2051	
121-423	123A	3444	20	2051	
121-425	158	3004/102A	53	2007	
121-425A				2007	
121-425B				2007	
121-425C				2007	
121-425D				2007	
121-425E				2007	
121-425F				2007	
121-425G				2007	
121-425H				2007	
121-425I				2007	
121-425J				2007	
121-425K				2007	
121-425L				2007	
121-425M				2007	
121-425N				2007	
121-4250				2007	
121-425P				2007	
121-425Q				2007	
121-426	160	3006	245	2004	
121-427	160	3006	245	2004	
121-428	160	3006	245	2004	
121-429	160	3006	245	2004	
121-43	102A		53	2007	
121-430	123AP	3444/123A	20	2051	
121-430B	123AP	3444/123A	20	2051	
121-430CL	123AP	3444/123A	20	2051	
121-431	128	3124/289	243	2030	
121-432	160	3006	245	2004	
121-433	123AP	3444/123A	20	2051	
121-433CL	123AP	3444/123A	20	2051	
121-434	123AP	3124/289	268	2038	
121-434H	123AP	3444/123A	20	2038	
121-435	123AP	3444/123A	20	2051	
121-436	124	3021	12		
121-437	102A	3004	53	2007	
121-44	126	3007	52	2024	
121-441	159	3466	82	2032	
121-442	123AP	3444/123A	20	2051	
121-444	159	3466	82	2032	
121-445	154	3040	40	2012	
121-446	159	3466	82	2032	
121-447	123AP	3444/123A	20	2051	
121-448	123A	3444	20	2051	
121-449	162	3559	35		
121-45	126	3005	52	2024	
121-450	123A	3444	20	2051	
121-451	124	3021	12		
121-452	163A	3439	38		
121-453	108	3018	39	2015	
121-46	102A	3004	53	2007	
121-460	161	3117	39	2015	
121-461	161	3117	39	2015	
121-462	161	3018	39	2015	
121-47	102A	3004	53	2007	
121-470	161	3018	39	2015	
121-471	161	3018	39	2015	
121-472	108	3452	86	2038	
121-473	154	3044	40	2012	
121-48	160	3008	245	2004	
121-480	107	3039/316	11	2015	
121-481	108	3452	86	2038	
121-482	108	3452	86	2038	
121-483	108	3452	86	2038	
121-49	160	3007	245	2004	
121-490	102A	3004	53	2007	
121-491	126	3004/102A	52	2024	
121-492	126	3004/102A	52	2024	
121-493	126	3008	52	2024	
121-494	126	3004/102A	52	2024	
121-495	159	3114/290	82	2032	
121-496	159	3466	82	2032	
121-497	159	3466	82	2032	
121-497WHT	159		82	2032	
121-498	229	3452/108	86	2038	
121-499	123AP	3444/123A	20	2051	
121-499-01	123AP		20	2051	
121-50	101	3861	8	2002	
121-500	161	3018	39	2015	
121-501	161	3716		2015	
121-502	161	3018	39	2015	
121-503	229	3246	60	2016	
121-504	229	3117	61	2016	
121-505	229	3018	61	2016	
121-505(3LEADS)			60	2015	
121-505501	123A		20	2051	
121-506	229	3018	61	2016	
121-5065	123A	3444	20	2051	
121-507		3246	60	2016	
121-508	229	3246	60	2016	
121-509	229	3246	60	2015	
121-51	101	3861	8	2002	
121-510	161	3018	39	2015	
121-52	102A	3004	53	2007	
121-520	108	3452	86	2038	
121-521	161	3018	39	2015	
121-521(MOTOROLA)			17	2051	
121-522	233	3246/229	210	2009	
121-523	161	3018	39	2015	
121-524	233	3246/229	210	2009	
121-524A	229		210	2009	
121-526	229	3018	210	2009	
121-52A	102A	3018			
121-53	100	3005	1	2007	
121-538	160	3006	245	2004	
121-538B	160		245	2004	
121-539	160	3006	245	2004	
121-54	126	3005	52	2024	
121-540	160	3006	245	2004	
121-540B	160		245	2004	
121-541	160		245	2004	
121-541B	160		245	2004	
121-542	160		245	2004	
121-542B	160		245	2004	
121-543	102A	3004	53	2007	
121-544	102A	3004	53	2007	
121-546	229	3452/108	86	2038	
121-546B	108	3452	86	2038	
121-547	108	3452	86	2038	
121-551	108	3452	86	2038	
121-552	160	3123	245	2004	
121-553	160	3123	245	2004	
121-554	126	3006/160	52	2024	
121-555	126	3006/160	52	2024	
121-557	103A	3835	59	2002	
121-558	103A	3835	59	2002	
121-560	108	3452	86	2038	
121-580	161	3018	39	2015	
121-581	123A	3124/289	20	2051	
121-582	124	3021	12		
121-583			39	2015	
121-584	121	3009			
121-585	316	3039	11	2015	

Industry Standard No.	ECG	SK	GE	RS 276-	MOTOR.
121-585B	316	3039	11	2015	
121-587	123AP	3444/123A	20	2051	
121-588	175	3026	246	2020	
121-59	103A	3835	59	2002	
121-6	103A	3011	59	2002	
121-60	103A	3011	59	2002	
121-600	123AP	3018	20	2051	
121-600(ZENITH)	123AP		40	2012	
121-601	160		245	2004	
121-602	159	3466	82	2032	
121-603	129	3466/159	22	2032	
121-608	159	3114/290	82	2032	
121-61	102A	3003	53	2007	
121-610	123A	3124/289	20	2051	
121-612	108	3293/107	11	2015	
121-612-16	108	3452	86	2038	
121-613	108	3452	86	2038	
121-613-16	108	3452	86	2038	
121-614	108	3452	86	2038	
121-614-9	108	3452	86	2038	
121-615	106	3984	21	2034	
121-616	108	3452	86	2038	
121-62	126	3005	52	2024	
121-629	123A	3444	20	2051	
121-63	160	3005	245	2004	
121-630	108	3452	86	2038	
121-632	102A	3004	53	2007	
121-633	102A	3004	53	2007	
121-634	102A		53	2007	
121-635	102A		53	2007	
121-636	102A	3123	53	2007	
121-636B				2007	
121-636C				2007	
121-636D				2007	
121-637	108	3452	86	2038	
121-638	108	3452	86	2038	
121-638B	108	3452	86	2038	
121-639	123AP	3444/123A	20	2051	
121-639CL	128	3024	243	2030	
121-64	102A	3003	53	2007	
121-640	102A	3004	53	2007	
121-641	103A	3010	59	2002	
121-642				2038	
121-643				2038	
121-644				2038	
121-645				2015	
121-646				2051	
121-647				2051	
121-648				2051	
121-649				2051	
121-65	126	3005	52	2024	
121-650				2002	
121-651				2015	
121-652				2051	
121-653				2015	
121-654				2032	
121-655				2038	
121-656				2038	
121-657				2051	
121-658				2038	
121-659			61	2038	
121-66	126	3005	52	2024	
121-660	123A	3444	20	2051	
121-661				2032	
121-662	123A	3444	20	2051	
121-663				2007	
121-664				2030	
121-665				2041	
121-667				2032	
121-668				2051	
121-67	126	3005	52	2024	
121-671	123AP	3444/123A	20	2051	
121-672				2051	
121-673				2002	
121-674				2007	
121-675	123AP	3444/123A	20	2051	
121-676	128	3024	243	2030	
121-677	123AP	3444/123A	20	2051	
121-678	123A	3444	20	2051	
121-678GREEN			20	2051	
121-678YELLOW				2051	
121-679	159	3466	82	2032	
121-679GREEN				2032	
121-679YELLOW				2032	
121-68	102A	3004	53	2007	
121-680			82	2032	
121-681			82	2032	
121-682			82	2032	
121-683			82	2032	
121-684			61	2038	
121-687	316	3039	11	2015	
121-69	102A	3004	53	2007	
121-692	161	3117	39	2015	
121-695	123AP	3444/123A	20	2051	
121-697	160	3006	245	2004	
121-698	160	3006	245	2004	
121-699	159	3466	82	2032	
121-699-02	159		82	2032	
121-7	103A	3861/101	59	2002	
121-70	101	3861	8	2002	
121-701	123A	3444	20	2051	
121-702			17	2051	
121-703	192	3137	47	2030	
121-704	161	3117	39	2015	
121-705			51	2004	
121-706	123A	3444	20	2051	
121-707	185	3025/129	58	2025	
121-708	184	3024/128	57	2017	
121-709	185	3025/129	58	2025	
121-71	101	3861	8	2002	
121-710	184	3024/128	57	2017	
121-711	123A	3444	20	2051	
121-712	157	3103A/396	232		
121-713	124	3021	12	2009	
121-714	106	3006/160	245	2004	
121-716			221	2050	
121-719	152	3054/196	66	2048	
121-72	102A	3005	53	2007	
121-722	128	3024	243	2030	
121-723	107	3018	11	2015	
121-725			82	2032	
121-726	130	3027	14	2041	
121-726A	130	3027	14	2041	
121-73	160	3005	245	2004	
121-730	123A	3444	20	2051	
121-731	312	3448	FET-1	2035	
121-732	161	3018	39	2015	
121-733			61	2038	
121-734	102	3004	2	2007	
121-735	108	3018	86	2038	
121-735B	107		11	2015	
121-737	123A	3444	20	2051	
121-737CL	128	3024	243	2030	
121-739			11	2015	
121-74	160	3005	245	2004	
121-742	108	3452	86	2038	
121-743	154	3040	40	2012	
121-743-01	191	3232	249		
121-743-01(T05)			40	2012	
121-743-031	196	3054			
121-744	123AP	3040	40	2012	
121-745	123AP	3444/123A	20	2051	
121-746	129	3466/159	82	2032	
121-748	123AP	3444/123A	20	2051	
121-75	160	3005	245	2004	
121-751	123A	3444	20	2051	
121-752	172A	3156	64		
121-753	108	3452	86	2038	
121-754	229	3117	61	2038	
121-755	190	3232/191	217		
121-756	312	3448	FET-1	2035	
121-758	164	3115/165	37		
121-758X	164	3115/165	37		
121-759	165	3115	38		
121-759X	165	3115	38		
121-76	160	3005	245	2004	
121-760	161	3117	39	2015	
121-761	161	3117	39	2015	
121-762	101	3861	8	2002	
121-762CL	103A	3010	59	2002	
121-764	123A	3444	20	2051	
121-765	129	3025	244	2027	
121-765-01	193	3114/290	269	2032	
121-766	128	3024	243	2030	
121-766-01	192	3124/289	268		
121-767	123AP	3124/289	62	2010	
121-767CL	123A	3444	20	2051	
121-768	123AP	3020/123	20	2051	
121-768CL	123A	3444	20	2051	
121-770	152	3893	66	2048	
121-770CL	196	3054	241	2020	
121-770X	152(2)		66(2)		
121-772			66	2048	
121-772CL	196	3054	241	2020	
121-772X	184(2)	3190/184	57(2)		
121-773	123A	3444	20	2051	
121-773CL	128	3024	243	2030	
121-774	159	3114/290	82	2032	
121-774CL	129	3025	244	2027	
121-775	161	3232/191	39	2015	
121-776	154	3103A/396	40	2012	
121-777	154	3466/159	229	2032	
121-777-01	154	3201/171	229	2012	
121-779	161	3018	11	2015	
121-78	160	3005	245	2004	
121-782		3050		2036	
121-783	222	3050/221	FET-4	2036	
121-784	222	3050/221	FET-4	2036	
121-785	222	3050/221	FET-4	2036	
121-786	222	3050/221	FET-4	2036	
121-787	222	3050/221	FET-4	2036	
121-787-01			17	2051	
121-79	160	3123	245	2004	
121-792	154	3040	40	2012	
121-793	121	3717	239	2006	
121-795	128	3024			
121-8	103A	3835	59	2002	
121-80	100	3005	1	2007	
121-801				2032	
121-802				2015	
121-803	153	3083/197	69	2049	
121-804	152	3054/196	66	2048	
121-805				2017	
121-806				2048	
121-807				2015	
121-808	152	3054/196	57	2017	
121-809				2048	
121-81	100	3005	1	2007	
121-812				2051	
121-813	128	3024			
121-819	108	3452	86	2038	
121-82	100	3005	1	2007	
121-821	164	3133	37		
121-822	171	3054/196	27		
121-823	161	3132	39	2015	
121-824	161	3132	39	2015	
121-825	123AP	3452/108	86	2038	
121-826	222	3065	FET-4	2036	
121-827	108	3452	86	2038	
121-829	162	3079	35		
121-83	100	3005	1	2007	
121-830	100	3009	1	2007	
121-831	165	3115	259		
121-834	108	3452	86	2038	
121-835	108	3452	86	2038	
121-836	123A	3444	20	2051	
121-837	123A	3444	20	2051	
121-838	159	3114/290	82	2032	
121-84	100	3005	1	2007	
121-840			86	2038	
121-841	108	3452	86	2038	
121-843	154	3024/128	217		
121-843A			63	2030	
121-844	128	3024	243	2030	
121-845	129	3114/290	244	2027	
121-846	108	3452	86	2038	
121-847			39	2015	
121-848	108	3452	86	2038	
121-849	107	3018	11	2015	
121-85	100	3005	1	2007	
121-850	123AP	3444/123A	86	2051	
121-851	108	3452	86	2038	
121-853	196	3054	241	2020	
121-853X			28	2017	
121-854			66	2048	
121-855	229	3452/108	86	2038	
121-856	123A	3018	20	2051	
121-857	108	3452	86	2038	
121-858	312	3448	FET-1	2035	
121-86	100	3005	1	2007	
121-860	312	3112		2028	
121-861	159	3466	82	2032	
121-862	123AP	3020/123	20	2051	
121-863	123AP	3444/123A	20	2051	
121-865	159	3466	82	2032	
121-868	171	3201	27		
121-868-01	171	3201	27	2012	
121-868-02	171	3201	27		
121-869	108	3452	86	2038	
121-87	100	3005	1	2007	
121-872			86	2038	
121-873	153	3083/197	69	2049	
121-874	152	3054/196	66	2048	
121-875	159	3114/290	82	2032	
121-876			62	2010	
121-877	123AP	3444/123A	20	2051	
121-878	128	3124/289	243	2030	
121-879	129	3114/290	244	2027	

Industry Standard No.	ECG	SK	GE	RS 276-	MOTOR.
121-88	100	3005	1	2007	
121-880	196	3054	241	2020	
121-881	123AP	3444/123A	20	2051	
121-883	229	3018	39	2015	
121-884	108	3452	86	2038	
121-885	229	3018	11	2015	
121-886	153	3083/197	69	2049	
121-887	152	3054/196	66	2048	
121-888	123AP	3444/123A	20	2051	
121-889	123A	3444	20	2051	
121-89	100	3005	1	2007	
121-895	123AP	3854	20	2015	
121-895A	123AP	3854	86	2038	
121-898	108	3452	86		
121-899	108	3452	86	2038	
121-9	102A	3123	2	2005	
121-90	100	3005	1	2007	
121-900	108	3452	86	2038	
121-907			61	2038	
121-909			86	2038	
121-91	100	3005	1	2007	
121-910			86	2038	
121-911	396	3103A	232		
121-911X	396	3103A			
121-913			20	2051	
121-914			20	2051	
121-915			20	2051	
121-916			20	2051	
121-92	100	3005	1	2007	
121-921-01	152	3893			
121-924	229	3117	39	2015	
121-925	108	3452	86	2038	
121-926	153	3084	69	2049	
121-927	152	3054/196	66	2048	
121-927-01	152	3054/196			
121-928			62	2010	
121-929			61	2038	
121-93	100	3005	1	2007	
121-930			61	2038	
121-931	123AP	3444/123A	20	2051	
121-932	108	3452	66	2048	
121-933	159	3715	267		
121-94	100	3005	1	2007	
121-943			61	2038	
121-944			20	2051	
121-945			61	2038	
121-946			11	2015	
121-95	102	3004	2	2007	
121-950	229	3444/123A	86	2051	
121-951	229	3018	210	2009	
121-952	129	3025	67	2032	
121-953	222	3050/221	FET-4	2038	
121-954	229	3246	61	2038	
121-96	102	3004	2	2007	
121-966	152	3054/196	66	2048	
121-966-01	152	3054/196	66	2048	
121-967	196	3054		2020	
121-968	108	3018	86		
121-969	292	3441	250		
121-969-02	153	3441/292		2049	
121-970	291	3440	241		
121-970-02	152	3440/291	241		
121-972	123AP	3444/123A	20	2051	
121-972-01	123AP	3444/123A		2051	
121-973	159	3466	21	2027	
121-974	108	3018	11	2015	
121-975	123AP	3444/123A	20	2051	
121-976	196	3054	66	2048	
121-976-01	196	3054			
121-977	197	3084	69	2049	
121-977-01	197	3083			
121-978	159	3466	82	2032	
121-980	189	3200	218	2026	
121-980-01	189	3200	218	2026	
121-982	123AP	3444/123A	20	2051	
121-983	229		11	2015	
121-984	229		11	2015	
121-985	165	3115	38		
121-986	159	3466	82	2032	
121-987	291	3440	241		
121-987-02	196	3440/291	241	2020	
121-988	292	3441	250		
121-988-02	197	3441/292	250	2027	
121-989	171	3104A	27		
121-990	171	3104A	27		
121-992	291	3054/196	66		
121-992(V.REG)			29	2025	
121-992-01	291		66	2048	
121-993	286	3194	246		
121-994	292	3083/197	69		
121-995			11	2015	
121-996	163A	3439	28		
121-997	153	3083/197	69	2049	
121-999				2016	
121-Z9000		3444	20	2051	
121-Z9000A		3444	20	2051	
121-Z9001		3710	259		
121-Z9002		3190	57	2017	
121-Z9003			221	2032	
121-Z9004			53	2007	
121-Z9005		3841	244	2027	
121-Z9006		3009	76	2006	
121-Z9013				2005	
121-Z9031				2002	
121G1241			511	1114	
121G3019	123A	3444	20	2051	
121G3020	123A	3444	20	2051	
121G3051	311		277		
121G3051D	311		277		
121G3115	311	3024/128	277		
121G3123	311		277		
121GI241	506	3130	511	1114	
121GL241				1114	
121J688-1	107		11	2015	
121J688-2	107		11	2015	
121S2	6402	3628			
121U2AT	6402	3628	2N2160		
121U3AC	6402	3628			
0122(AIRLINE)	184		57	2017	
122-1	123A	3444	20	2051	
122-1028	105	3012	4		
122-1028A	105	3012	4		
122-1625	121	3009	239	2006	
122-1648	126		52	2024	
122-1962	103	3862	8	2002	
122-2	123A	3444	20	2051	
122-229	100	3005	1	2007	
122-6	123A	3444	20	2051	
122-7	199	3018	62	2010	
122-7(RCA)	199		62	2010	
122-80	116	3017B/117	504A	1104	
122-A484	108	3452	86	2038	
122GRN	159		82	2032	
122N2	283		18	2030	

Industry Standard No.	ECG	SK	GE	RS 276-	MOTOR.
122YEL	159		82	2032	
0123	185	3041	58	2025	
0123(WARDS)	185		58	2025	
123-001	724		IC-86		
123-002	220	3990	FET-3	2036	
123-003	312	3112	FET-1	2028	
123-004	123A	3444	20	2051	
123-005	199	3124/289	62	2010	
123-006	159	3466	82	2032	
123-007	199	3122	62	2010	
123-008	236	3197/235	216	2053	
123-009	235	3197	215		
123-010	123A	3444	20	2051	
123-011	186A	3357	28	2017	
123-011A	186A	3357	247	2017	
123-012	289A	3138/193A	268	2038	
123-013	109	3088		1123	
123-015	109	3087		1123	
123-016	177	3100/519	300	1122	
123-017	172A	3100/519	64		
123-017(DIODE)	519	3100	514	1122	
123-018	506	3843	511	1114	
123-019	5072A	3136	ZD-8.2	562	
123-020	139A	3060	ZD-9.1	562	
123-021	116	3017B/117	504A	1104	
123-022	177	3100/519	300	1122	
123-024	519		514	1122	
123-025	177	3100/519	300	1122	
123B-001		3027	255	2041	
123B-002	186A	3357	28	2017	
123B-003	177	3100/519	300	1122	
123B-004	5072A	3136	ZD-8.2	562	
123B-005	137A	3058	ZD-6.2	561	
123B-006	5075A	3751	ZD-16	564	
123N1	164	3133	37		
123S-437	128		243	2030	
123S425	128		243	2030	
0124	184	3024/128	57	2017	
0124(HOFFMAN)	184		57	2017	
0124(KNIGHT)	159		82	2032	
0124(WARDS)	184		57	2017	
124-0028	125	3032A	510,531	1114	
124-0165	5994	3501	5104		
124-0178	116	3031A	504A	1104	
124-1	121	3009	239	2006	
124-N16	101	3861	8	2002	
0124A	159	3114/290	82	2032	
124J490	116	3016	504A	1104	
124N1	101	3861	8	2002	
124N16	123A	3444	20	2051	
0125	123A	3444	20	2051	
125-1			21	2034	
125-121	128		243	2030	
125-4(RCA)		3764	25		
125-402		3717	239	2006	
125-403		3717	239	2006	
125-410	175		246	2020	
125-425C			53	2007	
125-655			61	2038	
125-A000-F01	5883		5041		
125-B415	152		66	2048	
125A134	102A		53	2007	
125A137	128	3024	243	2030	
125A137A	128		243	2030	
125AS251	175		246	2020	
125B132	123A	3444	20	2051	
125B133	159	3466	82	2032	
125B139	128		243	2030	
125B410	175	3026	246	2020	
125C211	123A		20	2051	
12503	784	3524	IC-236		
12503RB	784	3524	IC-236		
125P1	159	3466	82	2032	
125P116	159	3466	82	2032	
125P1M	159	3466	82	2032	
125PI	159		82	2032	
125PL				2032	
0126	123A	3444	20	2051	
0126(WARDS)	189	3200	218	2026	
126-12	123A	3444	20	2051	
126-4	116	3016	504A	1104	
126-40	909	3590	IC-249		
126-7		3017B	504A	1104	
126-7(ARVIN)	116		504A	1104	
126N1	103A	3835	59	2002	
126N2	103A	3835	59	2002	
126P1	158	3025/129	53	2007	
127	123A	3444	20	2051	
127-115	199	3245	62	2010	
127-7	101	3011	8	2002	
127A905F01	5995	3518	5105		
128	108	3444/123A	86	2038	
128-853			28	2017	
128-9050	130		14	2041	
128A157F01	5995	3518	5105		
128A157F02	5995	3518	5105		
128C213H01	910		IC-251		
128N2	128	3018	243	2030	
128N4	108	3452	86	2038	
128WHT	123A	3444	20	2051	
129-10	121	3014	239	2006	
129-11	100	3006/160	1	2007	
129-13	121	3009	239	2006	
129-14	123	3124/289	20	2051	
129-15	123	3124/289	20	2051	
129-16	123A	3444	20	2051	
129-17	102	3004	2	2007	
129-18	102	3004	2	2007	
129-20	159	3466	82	2032	
129-21	123A	3444	20	2051	
129-23	175	3026	246	2020	
129-27	161	3018	39	2015	
129-30	103A	3011	59	2002	
129-31	102A		53	2007	
129-32	102A	3004	53	2007	
129-33	152	3893	66	2048	
129-33(PILOT)	123A	3444	20	2051	
129-34	159	3466	82	2032	
129-34(PILOT)	159		82	2032	
129-4	176	3845	80		
129-5	121	3009	239	2006	
129-6	104	3009	16	2006	
129-7	104	3009	16	2006	
129-8	102A	3004	53	2007	
129-8-1	102A	3004	53	2007	
129-8-1A	102A	3004	53	2007	
129-8-2	102A	3004	53	2007	
129-9	121	3009	239	2006	
129BRN	123A	3444	20	2051	
129N1	161	3716	39	2015	
129WHT	123A	3122	20	2051	
130	107	3024/128	11	2030	
130-013	5695	3509			
130-104	121	3014	239	2006	

Industry Standard No.	ECG	SK	GE	RS 276-	MOTOR.
130-105	5081A	3151	ZD-24		
130-112	107		11	2015	
130-138	108	3452	86	2038	
130-144	77	3024/128	277		
130-146	130		14	2041	
130-149	159	3466	82	2032	
130-150	77	3024/128	277		
130-152	316	3117	39	2015	
130-172	77	3024/128	277		
130-174	77	3024/128	277		
130-185	161	3039/316	39	2015	
130-191-00	128		243	2030	
130-200	77	3039/316			
130-220	316	3039			
130-240	161	3132	39	2015	
130-245	195A	3024/128	46		
130-30189	177	3031A	300	1122	
130-30192	116		504A	1104	
130-30256	125	3031A	510,531	1114	
130-30261	166	9075		1152	
130-30265	177	3100/519	300	1122	
130-30266	177		300	1122	
130-30274	177	3100/519	300	1122	
130-30281	109	3087		1123	
130-30301	109	3087		1123	
130-30313	116		504A	1104	
130-30702	177	3100/519	300	1122	
130-338-00	116	3017B/117	504A	1104	
130-398-99	116	3017B/117	504A	1104	
130-40089	101	3861	8	2002	
130-40095	102A	3004	53	2007	
130-40096	103A	3010	59	2002	
130-40214	123A	3444	20	2051	
130-40215	123A	3444	20	2051	
130-40216	199	3245	62	2010	
130-40229	109	3088		1123	
130-40236	102A	3004	53	2007	
130-40294	123A	3444	20	2051	
130-40304	108	3452	86	2038	
130-40311	123A	3444	20	2051	
130-40312	123A	3444	20	2051	
130-40313	123A	3444	20	2051	
130-40314	103A	3835	59	2002	
130-40315	159	3466	82	2032	
130-40317	123A	3444	20	2051	
130-40318	123A	3444	20	2051	
130-40347	103A	3010	59	2002	
130-40349		3052	43,44		
130-40352	102A	3004	53	2007	
130-40357	123A	3444	20	2051	
130-40362	108	3452	86	2038	
130-40421	108	3452	86	2038	
130-40429	159	3466	82	2032	
130-40456	158	3123	53	2007	
130-40459	108	3452	86	2038	
130-40883	199	3122	62	2010	
130-40896	123A	3444	20	2051	
130-40901	199	3122	62	2010	
130-40922	123A	3444	20	2051	
130-V10			11	2015	
130-VLO				2015	
130A17268	116	3311	504A	1104	
130ORN	107	3039/316		2015	
130YE	116	3311	504A	1104	
0131	123A	3444	20	2051	
131(ARVIN)	108		86	2038	
131(SEARS)	108		86	2038	
0131-000100	102A	3004	53	2007	
0131-000101	102A	3004	53	2007	
0131-000102	102A	3004	53	2007	
0131-000192	121	3009	239	2006	
0131-000335	159	3466	82	2032	
0131-000336	121	3009	239	2006	
0131-000337	121	3009	239	2006	
0131-000418	160	3006	245	2004	
0131-000419	160	3006	245	2004	
0131-000473	123A	3444	20	2051	
0131-000498	160	3006	245	2004	
0131-000561	128	3024	243	2030	
131-000562	121	3717	239	2006	
0131-000563	102A	3004	53	2007	
0131-000704	123A	3444	20	2051	
0131-000802	160	3006	245	2004	
0131-000859	160	3006	245	2004	
0131-000862	160	3006	245	2004	
0131-000863	160	3006	245	2004	
131-001-007	130		14	2041	
131-001007	130		14	2041	
0131-001050	158		53	2007	
0131-001056	158	3004/102A	53	2007	
0131-001182	160	3006	245	2004	
0131-001314	160	3006	245	2004	
0131-001328	159	3466	82	2032	
0131-001329	159	3466	82	2032	
0131-001332	160	3006	245	2004	
0131-001417	123A	3444	20	2051	
0131-001418	123A	3444	20	2051	
0131-001419	102A	3004	53	2007	
0131-001420	159	3025/129	82	2032	
0131-001421	123A	3444	20	2051	
0131-001422	123A	3444	20	2051	
0131-001423	123A	3444	20	2051	
0131-001424	123A	3444	20	2051	
0131-001425	121	3009	239	2006	
0131-001426	102A	3004	53	2007	
0131-001427	129	3025	244	2027	
0131-001428	129	3025	244	2027	
0131-001429	128	3024	243	2030	
0131-001430	128	3024	243	2030	
0131-001433	160	3006	245	2004	
0131-001434	160	3006	245	2004	
0131-001435	160	3006	245	2004	
0131-001436	160	3006	245	2004	
0131-001438	106	3118	21	2034	
0131-001439	159	3466	82	2032	
0131-001464	123A	3444	20	2051	
0131-001597	130	3027	14	2041	
0131-001697	160	3006	245	2004	
0131-001864	123A	3444	20	2051	
0131-002049	153	3083/197	69	2049	
0131-002068	130	3027	14	2041	
0131-0026	116	3311	504A	1104	
0131-002656	102A	3004	53	2007	
0131-003029	160	3006	245	2004	
0131-0035	116		504A	1104	
0131-004323	123A	3444	20	2051	
0131-004367	130		14	2041	
0131-0044	116	3311	504A	1104	
0131-004746	159		82	2032	
0131-004792	199	3245	62	2010	
131-005-353-1	197		250	2027	
131-005-807	128		243	2030	
131-005-808	129		244	2027	
0131-0053	116	3017B/117	504A	1104	

Industry Standard No.	ECG	SK	GE	RS 276-	MOTOR.
0131-005347	199	3245	62	2010	
0131-005348	199	3245	62	2010	
0131-005349	199	3245	62	2010	
0131-005350	199	3245	62	2010	
0131-005351	159	3466	82	2032	
0131-005352	152	3893	66	2048	
0131-005353	153	3084	69	2049	
131-005353-1	197		250	2027	
131-00561				2030	
131-005807	128	3024	243	2030	
131-005808	129	3025	244	2027	
0131-026	116	3017B/117	504A	1104	
131-043-67	130		14	2041	
131-04367	130		14	2041	
131-045-60	128		243	2030	
131-045-61	129		244	2027	
0131-04560	175		243	2020	
131-04561	129		244	2027	
0131-053	116	3017B/117	504A	1104	
131-2			28	2017	
0131-4328	159	3466	82	2032	
131A	113A		504A	1104	
131A246P01	5995	3518	5105		
131A246P02	5995	3518	5105		
131A8471	941		IC-263	010	
131N2	191	3054/196	249		
131N2(MAGNAVOX)			226	2018	
131N2G	152	3893	66	2048	
0132	130	3027	14	2041	
132-001	102A	3004	53	2007	
132-002	123A	3444	20	2051	
132-003	171	3201	27		
132-004	123A	3444	20	2051	
132-005	123A	3444	20	2051	
132-007	129	3025	244	2027	
132-008	161	3132	39	2015	
132-009	161	3132	39	2015	
132-010	102A	3004	53	2007	
132-011	123A	3444	20	2051	
132-014	171	3103A/396	27		
132-015	108	3452	86	2038	
132-017	123A	3444	20	2051	
132-018	123A	3444	20	2051	
132-019	160	3006	245	2004	
132-020	160	3006	245	2004	
132-021	123A	3444	20	2051	
132-021B	128	3024	243	2030	
132-022	128	3024	243	2030	
132-023	123A	3444	20	2051	
132-024	185	3083/197	58	2025	
132-025	162	3438	35		
132-026	123A	3444	20	2051	
132-027	160	3006	245	2004	
132-028	124	3021	12		
132-029	234	3025/129	65	2050	
132-030	123A	3444	20	2051	
132-031	234	3118	65	2050	
132-032	129	3025	244	2027	
132-033	124	3021	12		
132-038	128	3024	243	2030	
132-039	129	3025	244	2027	
132-041	123A	3444	20	2051	
132-042	123A	3444	20	2051	
132-045	221	3050		2036	
132-046	184	3054/196	57	2017	
132-047	221	3050		2036	
132-048	220	3990		2036	
132-049	312	3448	FET-1	2035	
132-050	123A	3444	20	2051	
132-051	123A	3444	20	2051	
132-052A	172A	3156	64		
132-054	123A	3444	20	2051	
132-055	123A	3444	20	2051	
132-056	159	3466	82	2032	
132-057	123A	3444	20	2051	
132-059	124	3021	12		
132-062	123A	3444	20	2051	
132-063	123A	3444	20	2051	
132-065	175	3538	246	2020	
132-066	128	3024	243	2030	
132-069	123A	3444	20	2051	
132-070	130	3027	14	2041	
132-072	184	3054/196	57	2017	
132-074	159	3466	82	2032	
132-075	123A	3444	20	2051	
132-076	161	3132	39	2015	
132-077	123A	3444	20	2051	
132-078	157	3103A/396	232		
132-079	185	3083/197	58	2025	
132-080	184	3054/196	57	2017	
132-081	184	3054/196	57	2017	
132-082	161	3132	39	2015	
132-085	130	3027	14	2041	
132-087	161	3132	39	2015	
132-090	102A	3004	53	2007	
132-185	161	3132	39	2015	
132-2	151A		ZD-110		
132-3	152	3054/196	66	2048	
132-4	152	3893	66	2048	
132-5	196	3054	241	2020	
132-501	123A	3444	20	2051	
132-502	123A	3444	20	2051	
132-503	123A	3444	20	2051	
132-504	123A	3444	20	2051	
132-515	124	3021	12		
132-516	124	3021	12		
132-521	124	3021	12		
132-522	124	3021	12		
132-523	124	3021	12		
132-524	124	3021	12		
132-525	124	3021	12		
132-526	124	3021	12		
132-539	123A	3444	20	2051	
132-540	123A	3444	20	2051	
132-541	130	3027	14	2041	
132-542	130MP	3029	15	2041(2)	
132-90	102A	3004	53	2007	
132N1	128	3024	243	2030	
0133	159	3466	82	2032	
133-001	726	3129			
133-001B	726	3129			
133-002	710	3102	IC-89		
133-003	724	3525	IC-86		
133-004	725	3162	IC-19		
133-005	724	3525	IC-86		
133-1	188	3054/196	226	2018	
133-3	152	3054/196	66	2048	
133P80029	7492	7492			
133P80057	909	3590	IC-249		
133P80067	941D		IC-264		
133P80104	941M		IC-265	007	
0134	123A	3444	20	2051	
134-1	189	3083/197	218	2026	
134-1P		3025	21	2034	

Industry Standard No.	ECG	SK	GE	RS 276-	MOTOR.
134B1037	717		IC-209		
134B1038-13			62	2010	
134B1038-21			10	2051	
134B1038-22			62	2010	
134B1038-4			62	2010	
134B1038-8			62	2010	
134B1040-7			63	2030	
134P1	234	3025/129	65	2050	
134P1(REMOTE)			67	2023	
134P1(VID)			21	2034	
134P1A	159	3466	82	2032	
134P1AA	129	3118	244	2027	
134P1M	159	3466	82	2032	
134P2	129	3025	244	2027	
134P4	159	3466	82	2032	
134P4AA	159	3466	82	2032	
134P4M	159	3466	82	2032	
134P6			21	2034	
0135-1	5801	3027/130		1142	
135-3	116	3311	504A	1104	
135C44322-542	123A	3444	20	2051	
135N1	154	3044	40	2012	
135N1M	154		40	2012	
135ORN	123A	3122	20	2051	
135P4	117	3311	504A	1104	
136-000100	358		42		
136-000200	358		42		
136-000300	358		42		
136-12			18	2030	
136P1	106	3984	21	2034	
136RED	123A	3122	20	2051	
0137	219	3173	21	2043	
137(ADMIRAL)	159		82	2032	
137-003	5697	3522			
137-12			21	2034	
137-684	125	3033	510,531	1114	
137-718	116	3017B/117	504A	1104	
137-737	5802	9005		1143	
137-759	145A	3063	ZD-15	564	
137-824	109	3087		1123	
137-828	125	3033	510,531	1114	
138-4	123A	3444	20	2051	
139-4	108	3452	86	2038	
139N1	229	3018	61	2038	
139N1D	229		61	2038	
139N2	161	3018	39	2015	
140-0007	123A	3444	20	2051	
140-013	116	3311	504A	1104	
0140-5	199	3124/289	62	2010	
0140-6	123A		20	2051	
0140-7	186	3192	28	2017	
0140-8	116	3311	504A	1104	
0140-9	116	3311	504A	1104	
140C					MR1-1400
140N1			14	2041	
140N1C	130		14	2041	
140N2	130	3027	14	2041	
141 402	161	3716		2015	
141-003	109	3087		1123	
141-078-0001	177		300	1122	
141-4			17	2051	
141-402			39	2015	
141-430		3124	20	2051	
0142	108	3452	86	2038	
142-001	123A	3444	20	2051	
142-002	199	3122	62	2010	
142-002(RECT.)	125		510,531	1114	
142-002(REGENCY)	199		62	2010	
142-003	199	3122	62	2010	
142-004	152	3054/196	66	2048	
142-005	289A	3122	268	2038	
142-006	186A	3357	247	2052	
142-006(DIODE)	177		300	1122	
142-006(REGENCY)	186A		247	2052	
142-007	282	3048/329	264		
142-008	152	3054/196	66	2048	
142-009	177	3100/519	300	1122	
142-010	177	3100/519	300	1122	
142-011	109	3087		1123	
142-012	139A	3060	ZD-9.1	562	
142-013	116	3017B/117	504A	1104	
142-014	116	3017B/117	504A	1104	
142-015	5075A	3751	ZD-16	564	
142-016	5071A	3334	ZD-6.8	561	
142-3	123A	3444	20	2051	
142-4	123A	3444	20	2051	
142N1	289A	3018	268	2038	
142N1P	161	3018	39	2015	
142N3	123AP	3444/123A	20	2051	
142N3P		3124	20	2051	
142N3T	123A	3444	20	2051	
142N4	123A	3444	20	2051	
142N5	123A	3444	20	2051	
142N6	108	3452	86	2038	
143			30	2006	
143-1	912		IC-172		
143-12			17	2051	
144-1	519	3100	514	1122	
144-12			20	2051	
144-3	177		300	1122	
144-4	154	3044	40	2012	
144A-1	121	3009	239	2006	
144A-1-12	121	3009	239	2006	
144A-1-12-3				2006	
144A-1-12-8	121	3009	239	2006	
144N1	171/403	3013	27		
144N10	171/403	3103A/396	28		
144N2	171	3201	27		
144N4				2012	
145-100	140A	3061	ZD-10	562	
145-118	139A	3060	ZD-9.1	562	
145-12			21	2034	
145-2		3201	27		
145-2(SYL.)	171		27		
145-470	5088A	3340	ZD-47		
145-T1B	126	3006/160	52	2024	
145A9254	116	3016	504A	1104	
145A9786	116	3016	504A	1104	
145N1	161	3018	39	2015	
145N1(LAST LF)				2009	
145N1(LAST-IF)			210	2051	
145N1P	161	3018	39	2015	
145NIP			20	2051	
145T1B			51	2004	
146-12			20	2051	
146-T1	121	3009	239	2006	
146B-3	121MP	3015	239(2)	2006(2)	
146B-3-12	121MP	3015	239(2)	2006(2)	
146B-3-12-8	121MP	3015	239(2)	2006(2)	
146D-1			300	1122	
146D1	178MP	3311	300(2)	1122(2)	
146D1B112	178MP		300(2)	1122(2)	
146N3	123A	3444	20	2051	
146N5	123A	3444	20	2051	

Industry Standard No.	ECG	SK	GE	RS 276-	MOTOR.
146T1		3014	16	2006	
147-7009-01	159	3466	82	2032	
147-7016-01	199	3245	62	2010	
147-7031-01	123AP	3444/123A		2030	
147-T1	121	3009	239	2006	
147N1	293	3122	47	2030	
147P2	294	3114/290	48		
147T1		3014	16	2006	
148-1			18	2030	
148-12			21	2034	
148-134622	105	3012	4		
148-3	506	3843	511	1114	
148N1	128	3124/289	243	2030	
148N2	123A	3444	20	2051	
148N212	123A	3444	20	2051	
148N3	128	3122	243	2030	
148P-2	102A	3004	53	2007	
148P-2-12	102A	3004	53	2007	
148P-2-12-8	102A	3004	53	2007	
149-12			FET-1	2028	
149-142-G1	177	3175	300	1122	
149L-4	102A	3003	53	2007	
149L-4-12	102A	3003	53	2007	
149L-4-12-8	102A	3003	53	2007	
149N1	184	3054/196	57	2017	
149N2	184	3190	57	2017	
149N2002D	152	3893	66	2048	
149N2004	152	3054/196	66	2048	
149N2B	241	3188A/182	57	2020	
149N2D	152	3893	66	2048	
149N4	184	3054/196	57	2017	
149N4B	241	3188A/182	57	2020	
149P1	185	3083/197	58	2025	
149P1B	242	3189A/183	58	2027	
149P1D	152	3893	66	2048	
149P2001D	153		69	2049	
149P2003	153	3083/197	69	2049	
149P3	185	3191	58	2025	
149P4B	242	3189A/183	58	2027	
150	116	3016	504A	1104	
150-001-005	109			1123	
150-001-9-005	109	3088		1123	
150-001-9-007	109	3087		1123	
150-002-9-001	109	3091		1123	
150-004-9-001	109	3087		1123	
150-005-9-001	109	3087		1123	
150-006-9-001	109	3088		1123	
150-012-9-001	506	3130	511	1123	
150-013-9-001	109	3088	1N60	1123	
150-014-9-001	109	3088	1N60	1123	
150-015-9-001	110MP	3709	1N60	1123	
150-018-9-001			504A	1104	
150-022-9-001			504A	1104	
150-030-9-002	178MP	3100/519	300(2)	1122	
150-040-9-002	177	3100/519	300	1122	
150-066-9-001	519	3100	300	1122	
150-1	177	3100/519	300	1122	
150-12			FET-1	2028	
150-1N	161	3716	39	2015	
150D			ZD-15	564	
150K100A	6158	7358			
150K80A	6158	7358			
150N1	161	3018	39	2015	
150N2	123A	3018	20	2051	
150N3	161	3018	39	2015	
150R10B	156	3051	512		
150R1B	156		512		
150R2B	5802	9005		1143	
150R3B	5803	9006		1144	
150R4B	5804	9007		1144	
150R5B	5806	3848			
150R6B	5806	3848			
150R7B	5808	9009			
150R8B	156	3051	512		
150R9B	156	3051	512		
151-001-9-001	177	3100/519	300	1122	
151-002-9-001	177	3100/519	300	1122	
151-0040	101	3861	8	2002	
151-0040-00	101	3861	8	2002	
151-006-9-001	147A		ZD-33		
151-0087-00	159	3466	82	2032	
151-0096-00	128		243	2030	
151-0103-00	123A	3444	20	2051	
151-011-9-001	177	3100/519	300	1122	
151-011-9-011	116	3016	504A	1104	
151-012-9-001	139A	3843	ZD-9.1	562	
151-0124-00	159	3466	82	2032	
151-0127-00	123A	3444	20	2051	
151-013-9-001	178MP		300(2)	1122(2)	
151-0136	128		243	2030	
151-0136-00	128		243	2030	
151-0136-02	128		243	2030	
151-0138	108	3452	86	2038	
151-014-9-001	507	3017B/117			
151-014-9-1	177		300	1122	
151-0140	130		14	2041	
151-0140-00	130		14	2041	
151-0141	175		246	2020	
151-0141-00	175		246	2020	
151-0148	175		246	2020	
151-0149	175		246	2020	
151-0149-1A	175		246	2020	
151-015-9-001	519	3100	514	1122	
151-018-9-001	116		504A	1104	
151-0188-00	159	3466	82	2032	
151-0190-00	123A	3444	20	2051	
151-0208	129		244	2027	
151-0208-00-AA	129		244	2027	
151-021-001			300	1122	
151-021-9-001	177	3100/519	300	1122	
151-0210	124		12		
151-0210-00	124		12		
151-0211	128		243	2030	
151-0211-00	128		243	2030	
151-0211-01	128		243	2030	
151-0217	175		246	2020	
151-0217-1	175		246	2020	
151-0219-00	234		65	2050	
151-022-9-001	177	3017B/117	300	1122	
151-022-9-002	177		300	1122	
151-0221-00	159	3466	82	2032	
151-0221-02	159	3466	82	2032	
151-0223-00	123A	3444	20	2051	
151-0224-00	123A	3444	20	2051	
151-023-9-001	116	3017B/117	504A	1104	
151-0238	101	3861	8	2002	
151-024-9-001	177	3100/519	300	1122	
151-0241	124		12		
151-0241-00	124		12		
151-025-9-001	117		504A	1104	
151-0250-00	194		220		
151-0251	124		12		
151-0259-00	107		11	2015	
151-0274-00	194		220		

Industry Standard No.	ECG	SK	GE	RS 276-	MOTOR.
151-0275	181		75	2041	
151-0275-00	181		75	2041	
151-029-9-001	177	3100/519	300	1122	
151-029-9-002	177		300	1122	
151-029-9-003	116	9075/166	504A	1104	
151-030-9-001	177	3100/519	300	1122	
151-030-9-002	178MP	3100/519	300(2)	1122(2)	
151-030-9-003	137A	3058	ZD-6.2	561	
151-030-9-004	614	3126	90		
151-030-9-005	116	3311	504A	1104	
151-030-9-006	177	3100/519	300	1122	
151-030-9-009	177		300	1122	
151-0302-00	123A	3444	20	2051	
151-031-9-001	137A	3060/139A	ZD-6.2	561	
151-031-9-003	137A		ZD-6.2	561	
151-031-9-006		3060	ZD-9.1	562	
151-0315	124		12		
151-0315-00	124		12		
151-0316	124		12		
151-032-004			300	1122	
151-032-9-001	177	3100/519	300	1122	
151-032-9-002	177	3100/519	300	1122	
151-032-9-004	177		300	1122	
151-0325-00	159	3466	82	2032	
151-0336-00	130		14	2041	
151-0336-00-A	130		14	2041	
151-0337-00	130		14	2041	
151-0337-00-A	130		14	2041	
151-034-9-001	177	3100/519	300	1122	
151-0341-00	199	3245	62	2010	
151-0341-00-A	199		62	2010	
151-0342-00	234		65	2050	
151-0347-00	194		220		
151-035-0-003	177		300	1122	
151-035-9-001	519	3100	300	1122	
151-035-9-003	177	3100/519	300	1122	
151-035-9-004	519	3100	300	1122	
151-039-9-003		3126	90		
151-040-7-003	116	3311	504A	1104	
151-040-9-001	177	3100/519	300	1122	
151-040-9-002	177	3100/519	300	1122	
151-040-9-003	116	3017B/117	504A	1104	
151-0407-00	194		220		
151-0410-00	234		65	2050	
151-0417-00	106	3984	21	2034	
151-042-9-001	177	3100/519	300	1122	
151-0424-00	123A	3444	20	2051	
151-0427-00	107		11	2015	
151-045	126		52	2024	
151-045-9-001	116	3017B/117	504A	1104	
151-045-9-002	177	3100/519	300	1122	
151-045-9-003	506	3130	511	1114	
151-0453-00	234		65	2050	
151-0456-00	199	3245	62	2010	
151-0458-00	159	3466	82	2032	
151-0459-00	159	3466	82	2032	
151-046-9-001	116	3017B/117	504A	1104	
151-0471-00	107		11	2015	
151-049-9-001	177	3100/519	300	1122	
151-049-9-002	177		300	1122	
151-049-9-007	601	3463			
151-051-9-001	519	3100	514	1122	
151-059-9-001	519	3100	300	1122	
151-060-9-001	519		300	1122	
151-062-9-001	611	3126	90		
151-064-9-001	519	3100	300	1122	
151-064-9-002	519	3100	300	1122	
151-066-9-001	177	3100/519	300	1122	
151-067-9-001	177	3100/519	514	1122	
151-069-9-001	519	3100	300	1122	
151-070-9-001	613	3327/614			
151-072-9-001	610	3100/519	300	1122	
151-096-1C	128		243	2030	
151-0I49	175		246	2020	
151-1	116	3311	504A	1104	
151-1002	126		52	2024	
151-148-00-BC	175		246	2020	
151-150	128		243	2030	
151-150-1B	128		243	2030	
151-211	128		243	2030	
151-267-9-001	177		300	1122	
151-32-9-004			300	1122	
151-6			17	2051	
151M11	108	3452	86	2038	
151N1	108	3452	86	2038	
151N11	108	3452	86	2038	
151N116	108	3452	86	2038	
151N2	123A	3444	20	2051	
151N4	123A	3444	20	2051	
151N5	123A	3444	20	2051	
152-0047	5804	9007		1144	
152-0047-00	5804	9007		1144	
152-006-9-001	147A	3095	ZD-33		
152-0061-00	177		300	1122	
152-008-7-001		3126	90		
152-008-9-001	139A	3060	ZD-9.1	562	
152-012-9-001	139A	3060	ZD-9.1	562	
152-019-9-001	137A		ZD-6.2	561	
152-021-9-001	141A		ZD-11.5		
152-0242-00	177		300	1122	
152-029-9-002	147A	3095	ZD-33		
152-0333-00	519		514	1122	
152-042-9-001	5072A	3136	ZD-8.2	562	
152-042-9-002	5071A	3058/137A	ZD-6.8	561	
152-047	116		504A	1104	
152-047-9-001	137A	3058	ZD-6.2	561	
152-047-9-004	137A		ZD-6.2	561	
152-051-9-001	139A	3060	ZD-9.1	562	
152-051-9-002			ZD-6.2	561	
152-052-9-002	137A	3058	ZD-6.2	561	
152-054-9-001	5072A	3136	ZD-8.2	562	
152-057-9-001	5075A	3751	ZD-16	564	
152-079-9-001	135A		ZD-5.6		
152-079-9-002	137A		ZD-6.2	561	
152-079-9-003	137A		ZD-6.2	561	
152-079-9-004	138A		ZD-7.5		
152-082-9-001	139A	3060	ZD-9.1	562	
152-087-9-001	139A	3060			
152-12			17	2051	
152-141-1	519		514	1122	
152-221011	102A	3004	53	2007	
152CB	5626	3633			
152N-2			75	2041	
152N2	130		14	2041	
152N2C	181	3036	75	2041	
152P1B	218		234		
152P1C	180		74	2043	
152SC711(D)			ZD-6.8	561	
153-004-9-001		3126	90		
153-008-9-001	614	3126	90		
153-9			18	2030	
153N1	196	3054	241	2020	
153N1C	152	3893	66	2048	
153N2C	152	3893	66	2048	

Industry Standard No.	ECG	SK	GE	RS 276-	MOTOR.
153N3	152	3893			
153N4C	152	3893	66	2048	
153N5C	152	3041	66	2048	
153N6	152	3893	66	2048	
154-000-9-001	614	3327			
154-001-9-001		3126	90		
154-004-9-001	613	3327/614			
154A3675-105	102	3004	2	2007	
154A3676	126	3008	52	2024	
154A3676-205	102	3004	2	2007	
154A3677	126	3008	52	2024	
154A3679	102	3004	2	2007	
154A3679-5110	126	3008	52	2024	
154A3680	121	3009	239	2006	
154A3992	116	3016	504A	1104	
154A5941	123A	3444	20	2051	
154A5943	124	3021	12		
154A5943-1	124	3021	12		
154A5944-410			10	2051	
154A5944-413			62	2010	
154A5945-519			62	2010	
154A5946	123A	3444	20	2051	
154A5946-622			18	2030	
154A5946-624			20	2051	
154A5946-667	129		244	2027	
154A5947-7732	106	3984	21	2034	
154A8681	102	3004	2	2007	
154T1	160	3007	245	2004	
154T1A	160		245	2004	
154T1B	160		245	2004	
155N1	196	3041	241	2020	
155T1	160		245	2004	
155U		3112			2N4416
156	123A	3444	20	2051	
156-0013-00-A	910		IC-251		
156-0015-00	909	3590	IC-249		
156-0048-00	912		IC-172		
156-0049-00	941		IC-263	010	
156-0053-00	923		IC-259		
156-0067-00	941M		IC-265	007	
156-0067-06	941M		IC-265	007	
156-0071-00	923D		IC-260	1740	
156-0148-00	7404			1802	
156-0151-00	915			017	
156-0176-00	309K	3629			
156-0197	912		IC-172		
156-0197-00	912		IC-172		
156-0197-3306	912		IC-172		
156-032	130		14	2041	
156-043	130	3027	14	2041	
156-043A	130	3027	14	2041	
156-053	130		14	2041	
156-063	130		14	2041	
156-083			75	2041	
156-084			75	2041	
156-104			75	2041	
156-83			75	2041	
156T1	160		245	2004	
156WHT	123A	3444	20	2051	
157	159	3025/129	82	2032	
157-009-9-001	5072A		ZD-8.2	562	
157-100	140A	3061	ZD-10	562	
157N3	188	3054/196	226	2018	
157P4	189	3083/197	218	2026	
157T1	160	3006	245	2004	
157T1A			51	2004	
157YEL	159		82	2032	
158(SEARS)	123A	3444	20	2051	
158-045-0027	128		243	2030	
158-10	124	3021	12		
158P1M	129	3025	244	2027	
158P2	159	3466	82	2032	
158P2M	159	3466	82	2032	
159-12			21	2034	
159T1	160		245	2004	
160	142A	3062	ZD-12	563	
160-8			17	2051	
160P08908	165		38		
160T1	160		245	2004	
161-001I	289A	3122	20	2051	
161-011J	289A	3122	20	2051	
161-012H	290A		82		
161-012J	159	3138/193A	82	2032	
161-014H	289A	3244	222		
161-015G	288	3434	223		
161-016H	289A	3122	20	2051	
161-016K	123A	3122	20	2051	
161-249-1001	949		IC-25		
161-2NC	130		14	2041	
161-5	196		241	2020	
161-INC	130		14	2041	
161-LNC				2041	
161N1C	130		14	2041	
161N2	130MP	3029	15	2041(2)	
161N4	130	3027	14	2041	
161N4C	130	3036	14	2041	
161T1	160		245	2004	
161T2	108	3452	86	2038	
162-005A	116	3311	504A	1104	
162-005B	116	3312			
162-005D	116	3311			
162J2	116	3016	504A	1104	
162N2	184	3190	57	2017	
162P1	185	3191	58	2025	
162T1	160	3006	245	2004	
162T2	108	3452	86	2038	
163J2	116	3017B/117	504A	1104	
164J2	116	3031A	504A	1104	
165	113A	3119/113	6GC1		
165-1A82	160		245	2004	
165-2A82	103A		59	2002	
165-432-2-48-1	519		514	1122	
165-4A82	101	3861	8	2002	
165A-182	160		245	2004	
165A4378P2	177		300	1122	
165A4383	123A	3444	20	2051	
165B-182	160		245	2004	
165C-182	121	3717	239	2006	
165J2	116	3017B/117	504A	1104	
165N1			20	2051	
166	115	3121	6GX1		
166-1A82	102A		53	2007	
166-2A82	102A		53	2007	
166-3A82	102A		53	2007	
166J2	116	3017B/117	504A	1104	
167-008A	1004	3365			
167-009A	1196	3725			
167-9	192	3024/128	63	2030	
167J2	125	3032A	510,531	1114	
167M1				2051	
167N1	123A	3444	20		
167N2	123A	3444	20	2051	
168-002A	5455	3597			
168-003A	5404	3627			

Industry Standard No.	ECG	SK	GE	RS 276-	MOTOR.
168-107-001	177		300	1122	
168-9	193	3025/129	67	2023	
168J2	156		512		
168N1	123A	3444	20	2051	
168P1	121MP	3013	239(2)	2006(2)	
169-257	124	3021	12		
169-284	124	3021	12		
170(RCA)	161	3716	39	2015	
170-1	177	3100/519	300	1122	
170-9	128		243	2030	
171-001-9-001	121	3009	239	2006	
171-003-9-001	124	3021	12		
171-015-9-001	121	3009	239	2006	
171-016-9-001	131	3052	44	2006	
171-1		3100	300	1122	
171-9	129		244	2027	
172-001	224	3049	46		
172-001-9-001	224	3049	46		
172-003-9-001	130	3027	14	2041	
172-003-9-001A	130	3027	14	2041	
172-006-9-001	186	3192	28	2017	
172-007-9-001	237	3299	46		
172-008-9-001	237	3299	46		
172-009-9-001	237	3299	46		
172-010-9-001	152	3054/196	66	2048	
172-011-9-001	186A	3357	247	2052	
172-013-9-001	236	3197/235	216	2053	
172-014-9-001	152	3054/196	66	2048	
172-014-9-002	282	3049/224	264		
172-014-9-003	152	3054/196	66	2048	
172-014-9-007	152	3054/196	66	2048	
172-024-9-002	235	3197	215		
172-024-9-003	236	3197/235	216	2053	
172-024-9-004	152	3202/210	66	2048	
172-024-9-005	130	3036	14	2041	
172-028-9-001	235	3197	215		
172-029-9-001	236	3197/235	216	2053	
172-031-9-003	152	3054/196	66	2048	
172-038-9-001	295		270		
172-038-9-002	235	3239/236	215		
172-038-9-003	186A	3357	247	2052	
172-044-9-001			66	2048	
173-1		3112	FET-1	2028	
173-1(SYLVANIA)	312	3448	FET-1	2035	
173-15	5075A	3751	ZD-16	564	
173A-4490-5	128		243	2030	
173A-4490-7	152	3054/196	66	2048	
173A-4491-5	152	3054/196	66	2048	
173A04490-1	108	3452	86	2038	
173A04490-2	108	3452	86	2038	
173A3936	121	3009	239	2006	
173A3963	121	3009	239	2006	
173A3970	102	3722	2	2007	
173A3981	116	3016	504A	1104	
173A3981-1	116	3016	504A	1104	
173A4057	123A	3444	20	2051	
173A419-2	121	3009	239	2006	
173A4348	102A		53	2007	
173A4349	102A		53	2007	
173A4389-1	102A		53	2007	
173A4390	102A		53	2007	
173A4391	128		243	2030	
173A4393	116	3016	504A	1104	
173A4394-1			300	1122	
173A4399	123A	3444	20	2051	
173A4416	123A	3444	20	2051	
173A4419	121	3009	239	2006	
173A4419-1	121	3009	239	2006	
173A4419-10	121	3717	239	2006	
173A4419-2	121	3009	239	2006	
173A4419-3	121	3009	239	2006	
173A4419-4	121	3717	239	2006	
173A4419-5	121	3717	239	2006	
173A4419-6	121	3717	239	2006	
173A4419-7	121	3717	239	2006	
173A4419-8	121	3009	239	2006	
173A4419-9	121	3009	239	2006	
173A4420	121	3009	239	2006	
173A4420-1	121	3009	239	2006	
173A4420-5	121	3009	239	2006	
173A4421-1	121	3009	239	2006	
173A4422-1	121	3009	239	2006	
173A4424	120	3004/102A	CR-3		
173A4436	121	3009	239	2006	
173A4469	121	3009	239	2006	
173A4470-11	123A	3444	20	2051	
173A4470-13	123A	3444	20	2051	
173A4470-32	123A	3444	20	2051	
173A4473-5			62	2010	
173A4483-1	159	3466	82	2032	
173A4483-2	159	3466	82	2032	
173A4489-2	128		243	2030	
173A4490-2			28	2017	
173A4490-5	128	3024	243	2030	
173A4490-7	152	3893	66	2048	
173A4491-2	130	3027	14	2041	
173A4491-2A	130	3027	14	2041	
173A4491-4	130	3027	14	2041	
173A4491-5	152	3893	66	2048	
173A4491-7	130	3027	14	2041	
173A4491-8	152	3893	66	2048	
174-002-9-001	103A	3010	59	2002	
174-1	116	3031A	504A	1104	
174-1(PHILCO)	190		217		
174-2	116	3031A	504A	1104	
174-2(SYL.)	171		27		
174-20989-22	130		14	2041	
174-20989-23	124		12		
174-25566-01	129	3025	244	2027	
174-25566-21	128		243	2030	
174-25566-48	108	3019			
174-25566-50	128		243	2030	
174-25566-62	129		244	2027	
174-25566-63	128		243	2030	
174-25566-76	128		243	2030	
174-3	116	3031A	504A	1104	
174-3A82	100	3721	1	2007	
175-006-9-001	102A	3004	53	2007	
175-007-9-001	102A	3004	53	2007	
175-008-9-001	102A	3004	53	2007	
175-2(PHILCO)	172A		64		
176-003	108	3452	86	2038	
176-003-9-001	108	3452	86	2038	
176-004	128	3024	243	2030	
176-004-9-001	108	3452	86	2038	
176-005	108	3452	86	2038	
176-005-9-001	108	3452	86	2038	
176-006	108	3452	86	2038	
176-006-9-001	108	3452	86	2038	
176-006-9-002	199	3245	62	2010	
176-007	108	3452	86	2038	
176-007-9-001	108	3452	86	2038	
176-008-9-001	123A	3444	20	2051	
176-014-9-001	123A	3444	20	2051	

Industry Standard No.	ECG	SK	GE	RS 276-	MOTOR.
176-016-9-001	289A	3122	268	2038	
176-017-9-001	123A	3444	20	2051	
176-018-9-001	195A	3047	46		
176-024-9-001	123A	3444	20	2051	
176-024-9-002	235	3197	215		
176-024-9-003	236	3197/235	216	2053	
176-024-9-004	186	3192	28	2017	
176-025-9-001	199	3122	62	2010	
176-025-9-002	107	3039/316	11	2015	
176-026-9-001	229	3018	61	2038	
176-029-9-001	289A	3122	20	2038	
176-029-9-002	195A	3047	46		
176-029-9-003	195A	3047	46		
176-029-9-004	295	3253	270		
176-031-9-001	289A	3138/193A	268	2038	
176-031-9-002	199	3124/289	62	2010	
176-037-9-001	229	3122	61	2038	
176-037-9-002		3122	272		
176-037-9-003	199	3122	62	2010	
176-037-9-004	199	3122	62	2010	
176-039-9-001	229		61	2038	
176-040-9-001	130	3027	14	2041	
176-042-9-001	229	3018	61	2038	
176-042-9-002	123A	3444	20	2051	
176-042-9-003	199	3018	61	2010	
176-042-9-004	199	3124/289	62	2010	
176-042-9-005	186A	3357	247	2017	
176-042-9-006	289A	3124/289	268	2038	
176-042-9-007	152	3054/196	66	2048	
176-043-9-002	289A		210	2051	
176-044-9-001	236	3239	216	2053	
176-044-9-002	316	3039	280		
176-047-9-001	289A	3124/289	268	2038	
176-047-9-002	123A	3444	20	2051	
176-047-9-003	123A	3122	20	2051	
176-048-9-002	289A	3124/289	210	2051	
176-049-9-002	289A	3027/130	210	2051	
176-054-9-001	123A	3444	20	2051	
176-055-9-004	152	3054/196	66	2048	
176-056-9-001	229	3018	60	2015	
176-056-9-003	229	3018	60	2015	
176-056-9-005	297	3122	81	2030	
176-060-9-001	229	3039/316	60	2015	
176-060-9-002	229	3018	20	2051	
176-060-9-003	199	3124/289	212	2010	
176-060-9-004	295	3122	210	2051	
176-062-9-001	199	3124/289	212	2010	
176-065-9-001	229	3124/289	213	2038	
176-072-9-005	233	3018	61	2038	
176-073-9-001	107	3018	17	2051	
176-073-9-002	236	3197/235	216	2053	
176-073-9-003	315		335		
176-073-9-012	236	3197/235	216	2053	
176-074-9-001	229	3018	61	2038	
176-074-9-002	295		270		
176-074-9-003	235	3197	215		
176-074-9-004	128	3020/123	88	2030	
176-075-9-001	229		60	2015	
176-075-9-003	229	3122	20		
176-075-9-006	295		336		
176-075-9-008	306		336		
176-087-9-001	295		215		
177-001	129	3025	244	2027	
177-001-9-001	129	3025	244	2027	
177-006-9-001	159	3466	82	2032	
177-006-9-002	290A	3114/290	269	2032	
177-007-9-001	234	3114/290	65	2050	
177-012-9-001	159	3466	82	2032	
177-018-9-001	294	3025/129	65	2050	
177-019-9-003	159	3114/290	221		
177-020-9-001	294	3025/129	65		
177-023-9-001	153	3083/197	58	2049	
177-025-9-001	290A		269	2032	
177-025-9-002	129	3114/290	21	2034	
177A-1-82	105		4		
177A01	798	3216	IC-234		
177A01A	798		IC-234		
177A02	748	3236			
177A03	738	3167	IC-29		
177A05	749	3168	IC-97		
177B-1-82	121MP	3718	239(2)	2006(2)	
177B01	798	3216	IC-234		
177B02	748	3236			
177C-1-82	121MP	3718	239(2)	2006(2)	
178-1(PHILCO)	193		67	2023	
179-4	140A	3061	ZD-10	562	
179-46447-01	910		IC-251		
179-46447-03	909D	3590	IC-250		
179-46447-07	923		IC-259		
179-46447-08	941		IC-263	010	
179-46447-16	911		IC-253		
179-46447-17	940		IC-262		
179-46447-21	915			017	
180-1	116	3311	504A	1104	
180N1	233	3132	210	2009	
180N1P		3132	61	2038	
180T2	130	3027	140	2041	
180T2A	130	3027	14	2041	
180T2B	130	3027	14	2041	
180T2C	130	3027	19	2041	
181-000100	760	3157/703A	IC-12		
181-000200	718	3159	IC-8		
181-003-9-001	6408	9084A			
181N1		3117	60	2015	
181N1D		3117	39	2015	
181N2		3117	60	2015	
181N2D		3117	39	2015	
181T2A	130	3027	14	2041	
181T2B	280	3027/130	14	2041	
181T2C	130	3027	14	2041	
182-009-9-001	312	3448	FET-1	2035	
182-014-9-002	312	3112	FET-1	2028	
182-014-9-003	312	3112	FET-1	2028	
182-015-9-001	312	3448	FET-1	2035	
182-021-9-001	312	3448	FET-1	2035	
182-029-9-001	312	3116	FET-2	2035	
182-038-9-001	222	3050/221	FET-4	2036	
182-039-9-001	312	3116	FET-2	2035	
182-044-9-001	312	3116	FET-2	2035	
182-044-9-002	312	3116	FET-2	2035	
182-045-9-001	312	3116	FET-1	2035	
182-046-9-001	312	3112	FET-2	2028	
182-056-9-001	312		FET-2	2035	
182-138-9-001	222	3050/221	FET-4	2036	
182B2003JDP1	128		243	2030	
182T2A	284	3836			
182T2B	284	3836			
182T2C	284	3836			
183-1(SYLVANIA)				1067	
183P1	129	3025	244	2027	
184A-1	121	3009	239	2006	
184A-1-12	121	3009	239	2006	
184A-1-12-7	121		239	2006	
184A-1L	121	3009	239	2006	

Industry Standard No.	ECG	SK	GE	RS 276-	MOTOR.
184A-1L8	121	3009	239	2006	
184R	313		278	2016	
184T2C	162	3438	35		
185-001	186A	3357	28	2017	
185-002	229	3018	61	2038	
185-003	123A		20	2051	
185-004	123A		20	2051	
185-005	199	3122	62	2010	
185-006	289A	3138/193A	268	2038	
185-007	152	3054/196	66	2048	
185-008	282	3049/224	264		
185-009	123A	3444	20	2051	
185-010	199	3124/289	62	2010	
185-011	177	3100/519	300	1122	
185-012	177	3100/519	300	1122	
185-013	109	3088		1123	
185-014	116	3017B/117	504A	1104	
185-015	139A	3060	ZD-9.1	562	
185-6			504A	1104	
185-6(RCA)	116		504A	1104	
185-6(RECT)			504A	1104	
185-736	175		246	2020	
186-001	229	3018	61	2038	
186-002	123A	3444	20	2051	
186-003	199	3122	62	2010	
186-004	199	3122	62	2010	
186-005	199	3122	62	2010	
186-006	289A	3138/193A	268	2038	
186-007	123A	3444	20	2051	
186-008	186A	3357	247	2017	
186-009	186A	3357	28	2017	
186-010	282	3748	264		
186-011	177	3100/519	300	1122	
186-011(DIODE)	177		300	1122	
186-012	177	3100/519	300	1122	
186-013	109	3087		1123	
186-014	116	3017B/117	504A	1104	
186-015	177	3100/519	300	1122	
186-016	139A	3060	ZD-9.1	562	
186-4-127			8	2002	
186B-3	121MP	3015	239(2)	2006(2)	
186B-3-12	121MP	3015	239(2)	2006(2)	
186B-3-127	121MP	3015	239(2)	2006(2)	
186B-3L	121MP	3015	239(2)	2006(2)	
186B-3L8	121MP	3015	239(2)	2006(2)	
186N1	161	3132	39	2015	
186N1(LAST LF)				2009	
186N1(LAST-IF)			210	2051	
187-1	150A	3098	ZD-82		
187-2	151A	3099	ZD-110		
187-6	140A	3061	ZD-10	562	
187-7	5081A		ZD-24		
188-5	5675	3178A			
188-68-35	120		CR-3		
188-70-48	120	3110	CR-3		
188-826	121	3717	239	2006	
188B-1-82	160		245	2004	
188C-1-82	100		1	2007	
188P-2	102A	3004	53	2007	
188P-2-12	102A	3004	53	2007	
188P-2-127	102A	3004	53	2007	
188P-2L	102A	3004	53	2007	
188P-2L8	102A	3004	53	2007	
189	108	3452	86	2038	
189L-4	102A	3003	53	2007	
189L-4-12	102A	3003	53	2007	
189L-4-127	102A	3003	53	2007	
189L-4L	102A	3003	53	2007	
189L-4L8	102A	3003	53	2007	
189N1	238	3710	259		
189N1G	238	3710	259		
190N1	190	3232/191	217		
190N1C	184	3054/196	57	2017	
190N3	184	3190	57	2017	
190N3C	184	3054/196	57	2017	
190V039H18	118		CR-1		
192-1A82	105		4		
195	199	3245	62	2010	
195N1	152	3054/196	66	2048	
195N1C		3054	66	2048	
195N1D	152	3893	66	2048	
195N3	152		66	2048	
195P2	153	3083/197	69	2049	
195P2C	197	3083	250	2027	
195P4	153		69	2049	
196-654	116	3016	504A	1104	
199-POWER	121	3717	239	2006	
0200	5944	3016	504A	1104	
200-007	108	3452	86	2038	
200-010	108	3452	86	2038	
200-011	128	3452/108	243	2030	
200-015	108	3452	86	2038	
200-016	123	3124/289	20	2051	
200-018	175	3026	246	2020	
200-052	159	3466	82	2032	
200-053	312	3116	FET-1	2035	
200-055	108	3452	86	2038	
200-056		3452	86	2038	
200-057	123A	3444	20	2051	
200-058	123A	3444	20	2051	
200-064	312	3834/132	FET-1	2035	
200-076	152	3893	66	2048	
200-12	128		243	2030	
200-6582	156	3032A	512		
200-6582-22	117	3017B	504A	1104	
200-846	123A	3444	20	2051	
200-862	123A	3444	20	2051	
200-863	123A	3444	20	2051	
200A	103	3862	8	2002	
200AGEL3731	127	3764			
200R10B	156	3051	512		
200R1B	156	3051	512		
200R3B	5803	9006		1144	
200R5B	5805	3848/5806			
200R6B	5806	3848			
200R7B	156	3051	512		
200R8B	156	3051	512		
200R9B	156	3051	512		
200RB4	5804			1144	
200X2100-022	1004	3365	IC-149		
200X2110-269	712	3072	IC-148		
200X2120-012		3168	IC-97		
200X2120-033		3284	IC-98		
200X2501-708	1196	3725			
200X3110-607	162	3559	73		
200X3151-432	376	3219	251		
200X3172-208	198	3219	251		
200X3174-006	123AP	3122	62	2051	
200X3174-014	123AP	3122	62	2051	
200X3174-021	123AP	3122	62	2051	
200X3190-604	108	3293/107	86		
200X3192-101	287	3433	222		
200X3206-800	198	3219	251		
200X3222-907	287	3244	222		
200X3223-025	287	3244	222		
200X3224-007	287	3244	222		
200X4082-614	159	3114/290	89		
200X4085-415	159	3247/234	89	2050	
200X4094-001	292	3441			
200X4101-500	159	3114/290	82		
200X4547-806	375	3219			
200X8000-026	109	3088	1N60	1123	
200X8010-102	605	3864			
200X8010-165	177	9091	300	1122	
200X8130-171	506	3998	511	1114	
200X8220-531	145A	3063	ZD-15	564	
200X8220-878	5018A	3784		562	
200X9120-224		3311	504A	1104	
0201	5944	3466/159	82	2032	
201-0723A	5073A	3749			
201-15	102A		53	2007	
201-25-4343-12	108		86	2038	
201-254323-12	108	3452	86	2038	
201-254323-13	108	3452	86	2038	
201-254343-12	108	3452	86	2038	
201-254343-13	107		11	2015	
201-254343-22	161		39	2015	
201-254343-26	161		39	2015	
201-254343-28	107		11	2015	
201-254343-30	107		11	2015	
201-254343-33	107		11	2015	
201-254343-34			60	2015	
201-254343-49			86	2038	
201-283818-1	222			2036	
201-283818-2	222			2036	
201-283818-3	222			2036	
201A	160	3123	245	2004	
201A0723	5072A	3059/138A	ZD-8.2	562	
201B	160	3123	245	2004	
201S		6648	2N692		
201U			2N681		2N4416
201X2000-118	109	3088	1N60	1123	
201X2010-144	519	3100	514	1122	
201X2010-159	519	3100	514	1122	
201X2100-126	552	3318	511		
201X2100-164	525	3925	530		
201X2120-009	552	3998/506	511	1114	
201X2220-118		3784		562	
201X3120-255	116	3311	504A	1104	
201X3130-109	125	3080	509	1114	
202			53	2007	
202-1	189	3083/197	218	2026	
202-5-2300-01710	116		504A	1104	
202-5-9531-01010	177		300	1122	
202A	103	3862	8	2002	
202N1	189	3083/197	218	2026	
202P	5530	6648/5548			
202P1	189		218	2026	
202P2	189		218	2026	
202S	5531	6631			
203	116	3033A	504A	1104	
203-1	222	3050/221		2036	
203-4	222			2036	
203A6137			ZD-6.2	561	
203B		6627	2N1847A		
203P	5530	6648/5548			
203S	5531	3112			
203X3189-408	238	3111	259		
204-1	177		300	1122	
0205	722	3161	IC-9		
0205(DYNACO)	722		IC-9		
205-142-G1	178MP		300(2)	1122(2)	
205A2-210	102A	3004	53	2007	
206-709-001	909		IC-249		
207A	160		245	2004	
207A1	160	3006	245	2004	
207A10	123A	3444	20	2051	
207A14	234	3247	65	2050	
207A16	175		246	2020	
207A16A	175		246	2020	
207A17	199	3245	62	2010	
207A20	104	3719	16	2006	
207A20A	104	3719	16	2006	
207A25	107		11	2015	
207A27	161	3716	39	2015	
207A29	123A	3444	20	2051	
207A3	102A		53	2007	
207A30	152	3893	66	2048	
207A31	123A	3444	20	2051	
207A33	152	3893	66	2048	
207A35	123A	3444	20	2051	
207A7	102A		53	2007	
207A9	108	3452	86	2038	
207B	160		245	2004	
207M	160		245	2004	
207V073C04	159	3466	82	2032	
209-1	159	3466	82	2032	
209-30	152	3893	66	2048	
209-31	109	3088		1123	
209-32	177	3100/519	300	1122	
209-846	123A	3444	20	2051	
209-862	123A	3444	20	2051	
209-863	123A	3444	20	2051	
209P1	159	3466	82	2032	
0210	5944		5048		
210ATTF3638	159		82	2032	
210BWTF4121	106	3984	21	2034	
0211	5944		5048		
211-40140-18	130	3027	14	2041	
211-58	5804	9007		1144	
211A6380-1	130		14	2041	
211A6380-3	152	3054/196	66	2048	
211A6381-1	175	3893/152	246	2020	
211A6381-2	152	3054/196	66	2048	
211A6381-I	175			2020	
211A6382-2	152	3893	66	2048	
211AESU3055	130		14	2041	
211ATPE3391	159	3466	82	2032	
211AVPF3415	123A	3444	20	2051	
211AVTE4275	123A	3444	20	2051	
21124	141A	3092	ZD-11.5		
212-00102	500A	3304	527		
212-00104	500A	3304	527		
212-013	500A		518		
212-102	521	3304/500A	539		
212-103	521	3304/500A	518		
212-104	521	3304/500A			
212-105	521	3304/500A			
212-106	521	3304/500A			
212-108	501B	3069/501A	520		
212-109	501B	3069/501A	520		
212-110	501B	3069/501A	520		
212-128	521	3304/500A	539		
212-129	521	3304/500A	539		
212-130	521	3304/500A	539		
212-130X	521	3304/500A			
212-131	521	3304/500A	517		
212-132	521	3304/500A	517		

Industry Standard No.	ECG	SK	GE	RS 276-	MOTOR.
212-133	521	3304/500A	539		
212-134	521	3304/500A			
212-135	521	3304/500A	539		
212-136	521	3304/500A	539		
212-137	521	3304/500A			
212-138	521	3304/500A			
212-139	521	3304/500A	527		
212-139-01	521	3304/500A	527		
212-139-02	521	3304/500A	527		
212-139-03	522	3303	523		
212-140	521	3304/500A	539		
212-140-01	521	3304/500A			
212-141	523	3306	528		
212-141-01	523	3306	528		
212-141-02	526A	3306/523	521		
212-141-03	526A	3306/523	521		
212-141-04	526A	3306/523	521		
212-141-05	526A	3306/523			
212-142	522	3303	523	1122	
212-142-01	522		523		
212-143	530	3308	540		
212-145	528	3306/523	528		
212-145-01	528	3306/523	528		
212-145-02	528	3306/523	528		
212-146	528	3306/523	521		
212-146-01	528	3306/523			
212-146-02	528	3306/523			
212-18	116	3016	504A	1104	
212-192	116		504A	1104	
212-21	116	3017B/117	504A	1104	
212-22			504A	1104	
212-23	116	3016	504A	1104	
212-25	116	3016	504A	1104	
212-254	116	3311	504A	1104	
212-27	116	3016	504A	1104	
212-33	116	3016	504A	1104	
212-35	116	3016	504A	1104	
212-36	116	3016	504A	1104	
212-37	116	3016	504A	1104	
212-38			504A	1104	
212-39	116	3016	504A	1104	
212-40	116	3016	504A	1104	
212-41	116	3016	504A	1104	
212-42	116	3016	504A	1104	
212-46	120	3016	CR-3		
212-47	116	3016	504A	1104	
212-48	116	9086/518	CR-1		
212-49	116	3017B/117	504A	1104	
212-50	116	3016	504A	1104	
212-505			20	2051	
212-507			20	2051	
212-51	116	3016	504A	1104	
212-57				1104	
212-58	116	3016	504A	1104	
212-59			504A	1104	
212-60			504A	1104	
212-61	116	3031A	504A	1104	
212-62	116	3016	504A	1104	
212-63	120	3110	CR-3		
212-64	116	3017B/117	504A	1104	
212-65	116	3016	504A	1104	
212-66	502	3067	CR-4		
212-67	502	3067	CR-4		
212-68	506	3843	CR-2		
212-695	123A	3444	20	2051	
212-699	159	3466	82	2032	
212-7		3311	504A	1104	
212-70	116	3017B/117	504A	1104	
212-71	116	3311	504A	1104	
212-72	116(4)		300	1122	
212-741-005	5484	3943/5485			
212-75			504A	1104	
212-76	116(4)	3313/116	504A	1104	
212-76-02	116	3313	504A	1104	
212-77	116	3016	504A	1104	
212-79	116	3016	504A	1104	
212-80	506	3032A	511	1114	
212-85	118	3066	CR-1		
212-85B	118	3066	CR-1		
212-92	116	3017B/117	504A	1104	
212-94	116	3031A	504A	1104	
212-94B	116	3311	504A	1104	
212-95	504	3032A	CR-6		
212-96	519	3100		1103	
212-997	197	3083			
212-Z9000		3051	512		
212-Z9001		3647	BR-600		
212-Z9539	539	3309			
212-Z9549	549	3901/554			
212-Z9554	554	3901			
215-37567	199	3245	62	2010	
215-51	116	3016	504A	1104	
215-58	116	3016	504A	1104	
215-76(GE)	116		504A	1104	
215-76GE		3017B	504A	1104	
215Z4	145A	3063	ZD-15	564	
216-001-001	152	3893	66	2048	
217(RCA)	161		39	2015	
217-1	108	3452	86	2038	
217-76-02			504A	1104	
218-22	123A	3444	20	2051	
218-23	123A	3444	20	2051	
218-24	123A	3444	20	2051	
218-25	123A	3444	20	2051	
218-26	128	3028	243	2030	
218-A-464-PM	5695	3509			
218A4164P1M	5695	3509			
220	175		246	2020	
220-001001	123A	3444	20	2051	
220-001002	123A	3444	20	2051	
220-001011	108	3452	86	2038	
220-001012	108	3452	86	2038	
220-002001	131	3052	44	2006	
220-003001	116	3016	504A	1104	
220-008001	312	3834/132	FET-1	2035	
0221	5944	3124/289	5048		
221(SEARS)	123A		20	2051	
221-0048	712		IC-2		
221-104	700	9028			
221-105	701	9026			
221-106	702	9025			
221-107	744	3171		2022	
221-108	788	3147	IC-229		
221-141	815	3255	IC-244		
221-30	909	3590	IC-249		
221-31	703A	3157	IC-12		
221-32	703A	3157	IC-12		
221-34	708	3135/709	IC-10		
221-36	705A	3134	IC-3		
221-37	705A	3134	IC-3		
221-37A	705A	3134			
221-39	705A	3134	IC-3		
221-39A	705A	3134			
221-41	707		IC-3		

Industry Standard No.	ECG	SK	GE	RS 276-	MOTOR.
221-42	714	3075	IC-4		
221-42-1	714		IC-4		
221-43	715	3076	IC-6		
221-43-1	715		IC-6		
221-45	731	3170	IC-13		
221-45-01	731	3170	IC-13		
221-46	713	3077/790	IC-5		
221-46-1	713		IC-5		
221-48	712	3072	IC-148		
221-48-01	712	3072			
221-49	739	3235	IC-30		
221-51	713	3077/790	IC-5		
221-52	713	3077/790	IC-5		
221-62	790	3454			
221-62-1	790	3454	IC-230		
221-64(RECT)				1114	
221-65	722	3161	IC-9		
221-69	791	3149	IC-231		
221-79	722	3161	IC-9		
221-79-01	722	3161	IC-9		
221-86		3206	IC-244		
221-86-01	815	3206			
221-87	714	3205/982	IC-4		
221-87-01	982	3205			
221-89	736	3163	IC-17		
221-90	723	3144	IC-15		
221-91	801	3160	IC-35		
221-91-01	801	3160	IC-35		
221-93	976	3596			
221-96	818	3207			
221-96-01	818	3207			
221-97	843	3208			
221-98	1231	3832			
221-Z9008		3074	IC-23		
221-Z9010		3167	IC-29		
221-Z9013		3168	IC-97		
221-Z9020		3165	IC-260	1740	
221-Z9021		3288	IC-43		
221-Z9022		3365	IC-14[illegible]		
221-Z9027		3160	IC-35		
222-1	5995	3518	5105		
222-2	5995	3518	5105		
223	108	3452	86	2038	
223P1	159	3466	82	2032	
224HACA0723	923D		IC-260	1740	
224N1	199	3444/123A			
225A6946-P000	7400	7400		1801	
225A6946-P004	7404	7404		1802	
225A6946-P010	7410	7410		1807	
225A6946-P020	7420	7420		1809	
225A6946-P093	7493A	7493			
226-1	123A	3122			
226-1(SYLVANIA)	123A	3444	20	2051	
226-3	129	3025	244	2027	
226-4	128	3024	243	2030	
226N1	123A	3444			
227-200001	113A	3119/113	6GC1		
227-2000D1	113A		6GC1		
228N3	302	3252			
229-0-180-33			11	2015	
229-0-180-34			11	2015	
229-0014			1N295	1123	
229-0026			51	2004	
229-0027			53	2007	
229-0028			53	2007	
229-0029			53	2007	
229-0030			53	2007	
229-0035			1N295	1123	
229-0037			1N295	1123	
229-0038			51	2004	
229-0041			53	2007	
229-0049			1N295	1123	
229-0050-13	123A	3444	20	2051	
229-0050-14	123A	3444	20	2051	
229-0050-15	123A	3444	20	2051	
229-0055			53	2007	
229-0056			53	2007	
229-0057			1N295	1123	
229-0062			53	2007	
229-0077			51	2004	
229-0079			51	2004	
229-0080			53	2007	
229-0082			51	2004	
229-0083			51	2004	
229-0085			53	2007	
229-0086			50	2004	
229-0087			51	2004	
229-0088			53	2007	
229-0089			51	2004	
229-0090			51	2004	
229-0091			53	2007	
229-0092			53	2007	
229-0093			1N34AS	1123	
229-0095			50	2004	
229-0097			53	2007	
229-0098			50	2004	
229-0099			51	2004	
229-0100			53	2007	
229-0102			504A	1104	
229-0105			1N34AS	1123	
229-0106			51	2004	
229-0107			1N295	1123	
229-0108			1N295	1123	
229-0110			51	2004	
229-0111			50	2004	
229-0112			50	2004	
229-0116			16	2006	
229-0117			504A	1104	
229-0119			504A	1104	
229-0120			504A	1104	
229-0121			51	2004	
229-0123			51	2004	
229-0124			53	2007	
229-0125			53	2007	
229-0129			50	2004	
229-0130			53	2007	
229-0131			51	2004	
229-0132			51	2004	
229-0133			53	2007	
229-0135			504A	1104	
229-0136			51	2004	
229-0137			53	2007	
229-0138			53	2007	
229-0139			53	2007	
229-0140			53	2007	
229-0141			1N295	1123	
229-0142			53	2007	
229-0143			53	2007	
229-0144			20	2051	
229-0145			51	2004	
229-0146			53	2007	
229-0147			504A	1104	
229-0149			20	2051	
229-0150			20	2051	

Industry Standard No.	ECG	SK	GE	RS 276-	MOTOR.
229-0151			20	2051	
229-0151-3	108	3452	86	2038	
229-0152			20	2051	
229-0154			20	2051	
229-0162			504A	1104	
229-0180-119	161	3018	39	2015	
229-0180-123	123A	3444	20	2051	
229-0180-124	108	3452	86	2038	
229-0180-149	108	3452	86	2038	
229-0180-32	161	3018	39	2015	
229-0180-33	161	3018	39	2015	
229-0180-34	108	3452	86	2038	
229-0182-65	109			1123	
229-0185-2	108	3452	86	2038	
229-0185-3	108	3452	86	2038	
229-0190-29	108	3452	86	2038	
229-0190-30	161	3716	39	2015	
229-0190-31	107	3018	11	2015	
229-0190-90	123A	3444	20	2051	
229-0191-29	107	3018	11	2015	
229-0191-30	107	3018	11	2015	
229-0192-18	222	3050/221	FET-4	2036	
229-0192-19	108	3452	86	2038	
229-0192-20	312	3834/132	FET-1	2035	
229-0204-23	108	3452	86	2038	
229-0204-4	108	3452	86	2038	
229-0204-6	161	3117	39	2015	
229-0204-6(VHF)			39	2015	
229-0210-14	108	3452	86	2038	
229-0210-19	107		11	2015	
229-0214-40	108	3452	86	2038	
229-0220-14	112		1N82A		
229-0220-19	108	3452	86	2038	
229-0220-9	108	3452	86	2038	
229-0240-20	112	3089	1N82A		
229-0240-25	161	3019	39	2015	
229-0248-45	108	3452	86	2038	
229-0250-10	108	3452	86	2038	
229-0250-11	112		1N82A		
229-0260-18	107	3018	11	2015	
229-1054-5	116	3017B/117	504A	1104	
229-1054-82	116	3017B/117	504A	1104	
229-1054-85			504A	1104	
229-1054-9	116		504A	1104	
229-1200-29	112		1N82A		
229-1200-36	123A	3122	20	2051	
229-1301-19	177		300	1122	
229-1301-20	177		300	1122	
229-1301-21	5400	3950			
229-1301-22	161	3716	39	2015	
229-1301-23	199		62	2010	
229-1301-24	123A		20	2051	
229-1301-25	506	3843	511	1114	
229-1301-26	190		217		
229-1301-27	172A	3156	64		
229-1301-28	193		67	2023	
229-1301-29	5097A	3347	ZD-120		
229-1301-30	5079A	3335	ZD-20		
229-1301-31	5097A		ZD-120		
229-1301-34	196	3192/186	28	2017	
229-1301-35	152	3893	66	2048	
229-1301-36	153		69	2049	
229-1301-37	192		63	2030	
229-1301-38	193		67	2023	
229-1301-39	794	3974			
229-1301-41	790	3454	IC-230		
229-1301-42	747	3279	IC-218		
229-1301-43	795		IC-232		
229-1301-44	738	3167	IC-29		
229-1301-63	165		38		
229-1301-64	130		14	2041	
229-1513-46			18	2030	
229-180-32			39	2015	
229-485-2			11	2015	
229-5100-15U	108	3452	86	2038	
229-5100-15V	108	3452	86	2038	
229-5100-224	108	3452	86	2038	
229-5100-225	108	3452	86	2038	
229-5100-226	108	3452	86	2038	
229-5100-227	126	3008	52	2024	
229-5100-228	108	3452	86	2038	
229-5100-231	109	3087		1123	
229-5100-232	116	3017B/117	504A	1104	
229-5100-233	116	3031A	504A	1104	
229-5100-234	116(3)	3017B/117	504A(3)	1104	
229-5100-235	506	3017B/117	511	1114	
229-5100-31V	161	3018	11	2015	
229-5100-32	161	3716	39	2015	
229-5100-32V	161	3018	39	2015	
229-5100-33V	108	3452	86	2038	
229-6011			18	2030	
230-0006	109	3088		1123	
230-0014	177	3100/519	300	1122	
230-0023	139A	3060	ZD-9.1	562	
230-006			1N60	1123	
231-000-001	102A		53	2007	
231-0000-01	158	3004/102A	53	2007	
231-0004	130	3027	14	2041	
231-0004-01	199	3444/123A	20	2051	
231-0004-03	199	3642/179	20	2051	
231-0006-03	179	3009	76		
231-0006B			3	2006	
231-0008	175	3026	12		
231-0009	102	3004	53	2007	
231-0011	121	3014	239	2006	
231-0013	128	3024	243	2030	
231-0015	121	3009	239	2006	
231-0025	123AP	3122			
231-0026	159	3114/290			
231-0027	159	3114/290			
231-0028	123AP	3024/128			
231-0029	159	3025/129			
231-0032	129	3114/290			
231-006B	121	3717	239	2006	
232-0001	116		504A	1104	
232-0006	116		504A	1104	
232-0006-02	177		514	1122	
232-0007	5873	3517/5883	5065		
232-0008	5983	3698/5987	5097		
232-0009-31	142A		ZD-12	563	
232-0014	5874	3517/5883	5065		
232-1006	116	3017B/117	512		
232-1009	116		504A	1104	
232-1011	116	3016	504A	1104	
232N1	123AP	3444/123A	20	2051	
232N2	108	3452	20	2051	
233(SEARS)	128		243	2030	
0234	116	3311	504A	1104	
00234-B	175		246	2020	
235	100	3005	1	2007	
236-0005	7400	7400		1801	
236-0007	7404	7404		1802	
236-0008	7405	7405			
236-0009	7473	7473		1803	
239A7920	196		241	2020	
239A7921-1	197		250	2027	
241-15A	160		245	2004	
241ABNG4148	519		514	1122	
241B	128		243	2030	
242-997	6401			2029	
0243	5838		5004		
0243-001	175		246	2020	
0244	116	3311	504A	1104	
0245	5842	3584/5862	5008		
246P1	159	3118	21	2034	
247-016-013	108	3452	86	2038	
247-255	116		504A	1104	
247-256	103A	3835	59	2002	
247-257	123A	3444	20	2051	
247-621	116	3017B/117	504A	1104	
247-623	102A		53	2007	
247-624	121	3717	239	2006	
247-625	128		243	2030	
247-626	128		243	2030	
247-629	123A	3444	20	2051	
247AS-C0450-001	923		IC-259		
247AS-C1249-001	123A	3444	20	2051	
0248	156	3051	512		
248-38104-1	128	3024	243	2030	
249-1L	229	3246	61	2038	
249N1	229	3246	61	2038	
250-0359	152	3893	66	2048	
250-0373	199	3245	62	2010	
250-0380	229	3018	61	2038	
250-0700			62	2010	
250-0711	199	3245	62	2010	
250-0712	123A	3444	20	2051	
250-1213	289A	3122	268	2038	
250-1312	199	3122	62	2010	
250JB01L					MDA2501
250JB02L					MDA2502
250JB04L					MDA2504
250JB05L	5322	3679			MDA210
250JB06L					MDA2506
250JB1L	5322	3679			
250JB2L	5322	3680			
250JB4L	5324	3681	BR-425		
250JB6L	5326	3682			
250MDA4R					MDA1204
250MDA6R					MDA1206
250R10B	5809	9010			
250R1B	5801	9004		1142	
250R2B	5802	9005		1143	
250R3B	5803	9006		1144	
250R4B	5804	9007		1144	
250R5B	5805	9008			
250R6B	5806	3051/156			
250R7B	5808	9009			
250R8B	5808	9009			
250R9B	5809	9010			
251M1	103	3862	8	2002	
252-113002-001	911		IC-253		
258-1	188		226	2018	
258-2	189		218	2026	
260-10-006	312	3834/132	FET-1	2035	
260-10-010	195A	3048/329	46		
260-10-011	177	3100/519	300	1122	
260-10-016	159	3466	82	2032	
260-10-020	123A	3444	20	2051	
260-10-021	289A	3138/193A	268	2038	
260-10-023	199	3122	62	2010	
260-10-024	152	3054/196	66	2048	
260-10-025	109	3087		1123	
260-10-026	229	3018	61	2038	
260-10-027	290A		269	2032	
260-10-031		3126	90		
260-10-032	177		300	1122	
260-10-033	116	3017B/117	504A	1104	
260-10-035	1102	3224	IC-93		
260-10-036	1100	3223	IC-92		
260-10-039	290A	3114/290	269	2032	
260-10-040	229		61	2038	
260-10-041	297		271	2030	
260-10-042	199	3122	62	2010	
260-10-043	236	3197/235	216	2053	
260-10-044	5072A	3136	ZD-8.2	562	
260-10-046	137A	3058	ZD-6.2	561	
260-10-047	177	3100/519	300	1122	
260-10-048	109	3087		1123	
260-10-049	177	3100/519	300	1122	
260-10-051	229	3018	61	2038	
260-10-052	154	3122	40	2012	
260-10-053	236	3197/235	216	2053	
260-10-054	152	3054/196	66	2048	
260-10-055	236	3197/235	216	2053	
260-10-056	316	3122	280		
260-10-057	519	3100	514	1122	
260-10-20	123A	3122	20	2051	
260-10-46	137A		ZD-6.2	561	
260-61-011	177	3100/519	300	1122	
260-61-047	177	3100/519	300	1122	
260-61-067	177	3100/519	300	1122	
260D00401	102A	3004	53	2007	
260D00402	102A		53	2007	
260D00403	102A	3004	53	2007	
260D00404	102A		53	2007	
260D00507	109	3057/136A		1123	
260D02501	102A	3004	53	2007	
260D02601	102A	3005	53	2007	
260D02701	158	3004/102A	53	2007	
260D0404		3004	52	2024	
260D04501		3004	52	2024	
260D04701		3004	53	2007	
260D05701	108	3452	86	2038	
260D05704	107	3039/316	11	2015	
260D05707	108	3452	86	2038	
260D05709		3018	17	2051	
260D07201	154		40	2012	
260D07412	123A	3444	20	2051	
260D07901	128		243	2030	
260D08013	107	3018	11	2015	
260D08201			20	2051	
260D08214	210	3041	252	2018	
260D08514	102A	3004	53	2007	
260D08601	128		243	2030	
260D08701	128		243	2030	
260D08801	123A	3444	20	2051	
260D08912	158	3004/102A	53	2007	
260D09001	123A	3444	20	2051	
260D09301	130		14	2041	
260D09314	123A	3444	20	2051	
260D09413	102A	3004	53	2007	
260D09612	128	3024	243	2030	
260D106A1	123A	3444	20	2051	
260D13701	199	3124/289	62	2010	
260D13702	123A	3444	20	2051	
260D13704	102A	3004	53	2007	
260D15901	199	3122	212	2010	

Industry Standard No.	ECG	SK	GE	RS 276-	MOTOR.
260D15902	289A	3122	210	2038	
260D7201			40	2012	
260P01209	121	3009	239	2006	
260P02903	123A	3444	17	2051	
260P029033			20	2051	
260P02903A	123A	3444	20	2051	
260P02908	123A	3444	20	2051	
260P03001	158	3004/102A	53	2007	
260P03201	161	3039/316	39	2015	
260P03201A	161	3039/316	39	2015	
260P04001	123A	3444	20	2051	
260P04002	123A	3444	20	2051	
260P04003	123A	3444	20	2051	
260P04004	123A	3444	20	2051	
260P04502	123A	3444	20	2051	
260P04503	123A	3444	20	2051	
260P04505	123A	3444	20	2051	
260P05402	107	3018	11	2015	
260P05402A	107	3018	11	2015	
260P05801	108	3452	86	2038	
260P05901	161	3039/316	39	2015	
260P05901A	161	3039/316	39	2015	
260P06901	108	3452	86	2038	
260P06902	108	3452	86	2038	
260P06903	108	3452	86	2038	
260P06904	123A	3444	20	2051	
260P07001	123A	3444	20	2051	
260P07002	123A	3444	20	2051	
260P07004	108	3452	86	2038	
260P07301	123A	3444	20	2051	
260P07502	131	3052	44	2006	
260P07601	158	3004/102A	53	2007	
260P0770		3124	10	2051	
260P07701	123A	3444	20	2051	
260P07702	123A	3444	20	2051	
260P07703	123A	3444	47	2051	
260P07704	123A	3444	20	2051	
260P07705	123A	3444	20	2051	
260P07707	123A	3444	20	2051	
260P07901	108	3452	86	2038	
260P08001	108	3452	86	2038	
260P08201	159	3466	82	2032	
260P08401	108	3452	86	2038	
260P08601	175		246	2020	
260P08801	123A	3444	20	2051	
260P08801A	123A	3444	20	2051	
260P08901	165	3115	38		
260P08908	283	3111	73		
260P09201	161	3019	39	2015	
260P09402	124	3021	12		
260P09501	164	3133	37		
260P09508	164		37		
260P09701	165	3111	38		
260P09902	123A	3444	20	2051	
260P10003	128	3024	243	2030	
260P10003A	128	3024	243	2030	
260P10005	128	3024	212	2030	
260P1003			18	2030	
260P1030			40	2012	
260P10301	154	3040	40	2012	
260P10403	108	3452	86	2038	
260P10501	108	3452	86	2038	
260P10502	108	3452	86	2038	
260P1060	108	3452	86	2038	
260P10601	161	3019	39	2015	
260P10602	108	3452	86	2038	
260P10801	154	3045/225	40	2012	
260P11101	108	3452	86	2038	
260P11101A	108	3452	86	2038	
260P11102	233	3132	210	2009	
260P11209	165	3111	38		
260P11302	123A	3444	20	2051	
260P11303	123A	3444	20	2051	
260P11304	123A	3444	20	2051	
260P11305	123A	3444	20	2051	
260P11403	159	3466	82	2032	
260P11502	123A	3444	20	2051	
260P11503	123A	3444	20	2051	
260P11504	123A	3444	20	2051	
260P11505	123A	3444	20	2051	
260P12001	123A	3444	20	2051	
260P12002	123A	3444	20	2051	
260P12401	128	3024	243	2030	
260P12481	128		243	2030	
260P12701	152	3893	66	2048	
260P1300	126	3005	52	2024	
260P13001	126	3006/160	52	2024	
260P13604	128		243		
260P13701			20	2051	
260P13702			20	2051	
260P13704	129		244	2027	
260P14101	123A	3444	210	2051	
260P14102	123A	3444	20	2051	
260P14103	123A	3444	20	2051	
260P14104			18	2030	
260P14105	123A	3444	20	2051	
260P141103	123A	3444	20	2051	
260P14202	196	3054	241	2020	
260P14407	375	3219			
260P15100	124	3021	12		
260P15108	124	3021	12		
260P15201	159	3466	82	2032	
260P15202	159	3466	82	2032	
260P15203	159	3466	82	2032	
260P16009	175	3026	246	2020	
260P1601	161	3018	39	2015	
260P1610			39	2015	
260P16101	161	3018	39	2015	
260P16202	124	3021	12		
260P16208	124	3021	12		
260P16301	108	3452	86	2038	
260P16302	108	3452	86	2038	
260P16304			243	2030	
260P16502	159	3466	82	2032	
260P16503	290A	3466/159	82	2032	
260P16504	159	3114/290	82	2032	
260P16603	290A	3466/159	269	2032	
260P16802	298	3138/193A	48		
260P17002	128	3024	243	2030	
260P17101	123AP	3444/123A	211	2051	
260P17102	123AP	3444/123A	20	2051	
260P17103	123AP	3444/123A	20	2051	
260P17104	123A	3444	20	2051	
260P17105	123A	3444	211	2051	
260P17106	123A	3444	20	2051	
260P17201	108	3452	86	2038	
260P17501	123A	3444	62	2010	
260P17502	199	3124/289	62	2010	
260P17503	123A	3444	20	2051	
260P1760	107		11	2015	
260P17601	229	3018	61	2038	
260P17602	108	3452	86	2038	
260P17603	108	3452	86	2038	
260P17701	199	3124/289	62	2010	

Industry Standard No.	ECG	SK	GE	RS 276-	MOTOR.
260P17702	161	3132	39	2015	
260P17704	199	3124/289	62	2010	
260P19009	162	3079	38		
260P19101	123A	3444	20	2051	
260P19103	123A	3444	20	2051	
260P19108	164	3133	37		
260P19208	165	3439/163A	38		
260P19501	123A	3444	20	2051	
260P19503	199	3124/289	62	2010	
260P19909	162	3079	38		
260P20101	175	3021/124	246	2020	
260P2100	102A	3005	53	2007	
260P21001	100	3005	1	2007	
260P21002	100	3005	1	2007	
260P21102	186	3192	28	2017	
260P21106	210	3054/196	252	2018	
260P21208	124	3021	12		
260P21308	186	3192	28	2017	
260P21608	165	3079	38		
260P21901	162	3438	35		
260P21908	162	3438	73		
260P22001	312	3834/132	FET-1	2035	
260P22002	312	3834/132	FET-1	2035	
260P22003	312	3834/132	FET-2	2035	
260P22101	154	3040	40	2012	
260P22203	154	3104A	27		
260P22204	191	3104A	249		
260P22801	186A	3357	28	2052	
260P24008	130	3027	14	2041	
260P24108	165		38		
260P24308	164	3133	37		
260P24408	163A	3439	36		
260P24803	175	3021/124	246	2020	
260P24901	161	3117	39	2015	
260P26201	288	3114/290	223		
260P26301	187A	3203/211	29	2025	
260P28107	289A	3122	268	2038	
260P28401	152	3893	66	2048	
260P28701	186	3192	28	2017	
260P31303	289A	3020/123	47	2030	
260P31402	297	3137/192A	63	2030	
260P33108	165	3115	38		
260P33408	164	3111	37		
260P34008	292	3083/197	69		
260P34202	196		241	2020	
260P34602	376		325		
260P35101	376	3220/198	32		
260P35301	287	3244	222		
260P35401	376		32		
260P36001	290A	3118	221	2032	
260P36501	108		61	2038	
260P37108	291	3054/196	66		
260P39008	165	3115	38		
260P4002	123A	3444	20	2051	
260P70403	108	3018	86	2038	
260P70501	108	3018	86	2038	
260P70502	108	3452	86	2038	
260Z00109	108	3452	86	2038	
260Z00209	108	3452	86	2038	
260Z00309	108	3452	86	2038	
260Z00401	192	3047	63	2030	
260Z00402	123A	3444	20	2051	
260Z00703	158	3004/102A	53	2007	
260Z01201	158	3004/102A	53	2007	
261RAX	192		63	2030	
264-701508	125	3032A	510,531	1114	
264D00101	177	3100/519	300	1122	
264D00209	177	3100/519	300	1122	
264D00505	116	3311	504A	1104	
264D00507	5070A	9021	ZD-6.2	561	
264D00612	109	3088		1123	
264D00701	109	3088		1123	
264D00801	110MP	3088	1N60	1123(2)	
264D00901	109	3088		1123	
264D01001	109	3087		1123	
264D01112	116	3311	504A	1104	
264D0701	110MP			1123(2)	
264P00401	109	3087		1123	
264P00501	113A	3119/113	6GC1		
264P00502	113A		6GC1		
264P00506	113A	3119/113	6GC1		
264P00601	116	3031A	504A	1104	
264P00602	125	3311	510,531	1114	
264P0080	109			1123	
264P00801	109	3087	1N34AS	1123	
264P01011	116		504A	1104	
264P01012	116		504A	1104	
264P01301	109	3088	1N34AS	1123	
264P01305	109	3088	1N34AS	1123	
264P01306	109		1N60	1123	
264P01350			1N60	1123	
264P01508	119	3032A	CR-2		
264P01701	506	3017B/117	511	1114	
264P02001	116	3017B/117	504A	1104	
264P02104	118		CR-1		
264P02301	506	3017B/117	511	1114	
264P02402	116	3017B/117	504A	1104	
264P02501	135A	3056	ZD-5.1		
264P02502	5072A	3136			
264P02503		3784		562	
264P03001	116	3032A	504A	1104	
264P03301	5070A	9021	ZD-6.2	561	
264P03302	5074A	3062/142A	ZD-11.0	563	
264P03303	142A	3062	ZD-12	563	
264P03401	112	3089	1N82A		
264P03601	506	3125	511	1114	
264P03603	506	3043	511	1114	
264P03604	558	3130	511	1114	
264P03605	506	3033	511	1114	
264P03606		3017B	511	1114	
264P03607	116	3016	504A	1104	
264P03701			504A	1104	
264P03703	156	3100/519	512		
264P03802	109	3087		1123	
264P04003	145A	3063	ZD-15	564	
264P04005	136A	3058/137A	ZD-5.6	561	
264P04206	504		CR-6	1104	
264P04301	116	3032A	504A	1104	
264P04303	125	3033	510,531	1114	
264P04402	116	3311	504A	1104	
264P04501	519	3100	514	1122	
264P04502	519	3100	514	1122	
264P04507	519	3100	514	1122	
264P04701	506	3043	511	1114	
264P04702	116	3032A	504A	1104	
264P04703	116	3032A	504A	1104	
264P04705	506	3130	511	1114	
264P04801	116	3031A	504A	1104	
264P04901	116	3032A	504A	1104	
264P05001	116	3311	504A	1104	
264P05002	116	3311	504A	1104	
264P05502	611	3325/612			
264P05902		3126	90		
264P06301	102A	3004	53	2007	

Industry Standard No.	ECG	SK	GE	RS 276-	MOTOR.
264P06601	116	3032A	504A	1104	
264P06603	506	3130	511	1114	
264P06606	116		504A	1104	
264P07501	613	3327/614			
264P08002	116	3311	504A	1104	
264P08801	116	3311	504A	1104	
264P08901	502	3067	CR-4		
264P08902	503	3068	CR-5		
264P09001	125	3016	510,531	1114	
264P09101	506	3130	511	1114	
264P09301	142A	3062	ZD-12	563	
264P09501	506	3130	511	1114	
264P09703	140A		ZD-10	562	
264P09707	5067A	3331	ZD-3.9		
264P09801	116		504A	1104	
264P10102	552	3016	504A	1104	
264P10103	116	3016	504A	1104	
264P10105	116	3311	504A	1104	
264P10201	552	3125	511	1114	
264P10308	145A		ZD-15	564	
264P10502	142A	3139/5074A	ZD-11	563	
264P10502(ZENER)	5074A		ZD-11.0	563	
264P10906	5075A	3751	ZD-16	564	
264P11003	138A		ZD-6.8	561	
264P11007	5085A		ZD-36		
264P11009	147A	3095	ZD-33		
264P13002	506	3311	511	1114	
264P14002	503		CR-5		
264P14701	125	3081	510	1114	
264P17401	5071A	3334	ZD-6.8	561	
264P17402	143A	3093	ZD-13	563	
264Z00103		3126	90		
264Z00201	116	3311	504A	1104	
264Z00701	109	3088		1123	
265D00702	177	3100/519	300	1122	
265P03301	601	3463	504A	1104	
265Z00101	177	3100/519	300	1122	
266P001-01	704	3023	IC-205		
266P001-02	704		IC-205		
266P00101	704	3023	IC-205		
266P00101(XSTR)			20	2051	
266P00102	704	3023	IC-205		
266P00301	1096	3703			
266P00801	1096	3703			
266P01002	1096	3709/110MP			
266P10101	749	3168	IC-97		
266P10103	749	3168	IC-97		
266P10201	747	3279	IC-218		
266P10202	747	3279	IC-218		
266P30102	712	3072	IC-148		
266P30103	712		IC-2		
266P30109	712	3072	IC-2		
266P30201	1075A	3877			
266P30601	1029		IC-162		
266P30706	1103	3281	IC-94		
266P60304	1183	3475/1050			
266P6034	1183	3475/1050			
266P60502	738	3167	IC-29		
266P71301	719		IC-28		
269M01201	519	3175	300	1122	
269P19009	162	3438	35		
269V004-H01	113A	3119/113	6GC1		
270-950-030	123A	3444	20	2051	
270-950-037-02	128	3024	243	2030	
274	729	3074	IC-23		
276-007			IC-263	010	
276-010		3553	IC-263	010	
276-1050				1050	
276-1079				1067	
276-1101				1011	
276-1102				1102	
276-1103				1103	
276-1104			504A	1104	
276-1114			510	1114	
276-1122			1N60	1123	
276-1141				1141	
276-1142				1143	
276-1143				1143	
276-1144				1144	
276-1151				1151	
276-1152				1152	
276-1801		7400	7400		
276-1803		7473	7473		
276-1808		7490	7490		
276-1811		7402	7402		
276-1813		7476	7476		
276-1818		7474	7474		
276-1822		7408	7408		
276-1828		74145	74145	1828	
276-2003				2005	
276-2003/RS2003				2005	
276-2009			210	2051	
276-2009/RS2009			210	2051	
276-2014			210	2051	
276-2014/RS2014			210	2051	
276-2017			66	2048	
276-2017/RS2017			66	2048	
276-2018			252	2018	
276-2018/RS2018			252	2018	
276-2025			69	2049	
276-2025/RS2025			69	2049	
276-2026			253	2026	
276-2026/RS2026			253	2026	
276-2028			FET-1	2028	
276-2030			20	2051	
276-2030/RS2030			20	2051	
276-2033			20	2051	
276-2033/RS2033			20	2051	
276-2035			FET-1	2028	
276-2417		4017	4017		
276-561			ZD-6.2	561	
276-562				562	
276-563			ZD-12	563	
276-564			ZD-15	564	
0276.8A		3334	ZD-6.8	561	
276A01	704	3023	IC-205		
280-0001	1100		IC-92		
280-0002	801	3160	IC-35		
280-07	5547	3656/5564			
281	123A	3444	20	2051	
284HC	123A	3444	20	2051	
288(SEARS)	116		504A	1104	
290-1003	177		300	1122	
290V02H69		3452	86	2038	
290V034C01	125	3032A	510,531	1114	
291-04	116	3016	504A	1104	
291-20	116	3017B/117	504A	1104	
292-10	116		504A	1104	
294	159	3114/290	82	2032	
294-42-9	109	3087		1123	
295L001H01	116	3017B/117	504A	1104	
295L001M01	116	3017B/117	504A	1104	
295L001M02	116	3017B/117	504A	1104	
295L002H01			504A	1104	
295L002M03	117		504A	1104	
295L003M01	116	3017B/117	504A	1104	
295V002H01	116	3017B/117	504A	1104	
295V003H01	118	3066	CR-1		
295V003M01	118	3066	CR-1		
295V005H02	116	3017B/117	504A	1104	
295V005H03	109	3087		1123	
295V006H01	116	3017B/117	504A	1104	
295V006H02	116	3017B/117	504A	1104	
295V006H03	116	3016	504A	1104	
295V006H05	116	3016	504A	1104	
295V006H06	117		504A	1104	
295V006H07	116	3016	504A	1104	
295V006H08	117		504A	1104	
295V006H09	116	3017B/117	504A	1104	
295V007H02	116	3017B/117	504A	1104	
295V008H01	116	3017B/117	504A	1104	
295V012H01	116	3017B/117	504A	1104	
295V012H02	116	3017B/117	504A	1104	
295V012H03	116	3017B/117	504A	1104	
295V012H06	116	3017B/117	504A	1104	
295V014H01	116	3017B/117	504A	1104	
295V014H04	116	3016	504A	1104	
295V014H7			504A	1104	
295V015H02	116	3017B/117	504A	1104	
295V016H01	116	3017B/117	504A	1104	
295V017H01	116	3016	504A	1104	
295V020B01	116	3016	504A	1104	
295V020H01	116	3016	504A	1104	
295V023H01	116	3016	504A	1104	
295V027C01	116	3016	504A	1104	
295V027C01-1	116	3016	504A	1104	
295V028C01	116	3017B/117	504A	1104	
295V028C02	116	3016	504A	1104	
295V028C03	116	3017B/117	504A	1104	
295V028C04	125	3017B/117	510,531	1114	
295V029C01	116	3016	504A	1104	
295V029C02	116	3017B/117	504A	1104	
295V031B	119	3109	CR-2		
295V031B02	120	3110	CR-3		
295V031C02	120	3110	CR-3		
295V033B01	118	3066	CR-1		
295V034C01	119	3109	CR-2		
295V035C01	116	3017B/117	504A	1104	
295V041H04	121	3009	239	2006	
296	172A	3893/152	64		
296(REGENCY)	152		66	2048	
296(SEARS)	172A		64		
296-18-9	102A		53	2007	
296-19-9	102A		53	2007	
296-42-9	109	3087	300	1122	
296-46-9	160		245	2004	
296-50-9	123A	3444	20	2051	
296-51-9	123A	3444	20	2051	
296-55-9	199	3245	62	2010	
296-56-9	123A	3444	20	2051	
296-58-9	128	3048/329	243	2030	
296-59-9	123A	3444	20	2051	
296-60-9	102A	3004	53	2007	
296-61-9	131	3052	44	2006	
296-62-9	102A	3004	53	2007	
296-64-9-1	158	3004/102A	53	2007	
296-77-9	123A	3444	20	2051	
296-81-9	128	3047	243	2030	
296-86	229	3122	61	2038	
296-98-9	123A	3122	212	2010	
296L002B01	519	3087	514	1122	
296L003B01	142A	3062	ZD-12	563	
296V001H01	358		42		
296V001M01	358		42		
296V002H01	109	3087		1123	
296V002H02	109	3087		1123	
296V002H05	109	3087		1123	
296V002H06	109	3087		1123	
296V002H07	109	3087		1123	
296V002H08	109	3016		1123	
296V002M01	109	3087		1123	
296V004H01	113A	3119/113	6GC1		
296V006H02	109	3087		1123	
296V006H03	116	3016	504A	1104	
296V006H07			504A	1104	
296V007H02	109	3087		1123	
296V011H03	143A		ZD-13	563	
296V012H01	109	3087		1123	
296V015B01	109	3087		1123	
296V015H01	109	3087		1123	
296V017H01	142A	3062	ZD-12	563	
296V018B01	142A	3062	ZD-12	563	
296V019B01			ZD-12	563	
296V019B04	141A	3061/140A	ZD-11.5		
296V020B01	177	3087	300	1122	
296V020B02	116		504A	1104	
296V024B01		3087	1N34AS	1123	
296V024B02	177	3087	300	1122	
296V034C01	119		CR-2		
297C011H01			51	2004	
297L001H01	103A	3835	59	2002	
297L001H02	103A	3835	59	2002	
297L001H03	102A(2)		53(2)		
297L001M01	103A	3835	59	2002	
297L001M02	102A	3835/103A	53	2007	
297L001M03	102A(2)		53(2)		
297L002H01	102A		53	2007	
297L005H01	102A		53	2007	
297L006H01	123A	3444	20	2051	
297L006H02	123A	3444	20	2051	
297L007C	199	3245	62	2010	
297L007C02	123A	3444	20	2051	
297L007C03	199	3245	62	2010	
297L007H01	123A	3444	20	2051	
297L007H02	123A	3444	20	2051	
297L007H03	123A(2)	3122	20(2)		
297L007H03/C03			20	2051	
297L008C02	128(2)		243(2)		
297L008H01	128(2)		243(2)		
297L010C	234		65	2050	
297L010C01	234		65	2050	
297L011C01	107		11	2015	
297L012C-01	106	3984	21	2034	
297L012C01	159	3466	82	2032	
297L013B01	123A	3444	20	2051	
297L013B02	159	3466	82	2032	
297L015C01	199	3245	62	2010	
297V002H03	103A	3835	59	2002	
297V002H04	103A	3835	59	2002	
297V002H05	103A	3835	59	2002	
297V002M04	103A	3011	59	2002	
297V002M05	103A	3011	59	2002	
297V003			2	2007	
297V003H02	102A(2)		53(2)		
297V003H03	102	3004	2	2007	
297V003H06	102A(2)	3004/102A	53(2)		
297V003H08	102A(2)		53(2)		
297V003H09	102A	3004	53	2007	

Industry Standard No.	ECG	SK	GE	RS 276-	MOTOR.
297V003M01	102	3004	2	2007	
297V003M07	102	3004	2	2007	
297V004H01	102A		53	2007	
297V004H010			2	2007	
297V004H03	102A	3004	53	2007	
297V004H04	102A		53	2007	
297V004H06	102A	3004	53	2007	
297V004H08	102A		53	2007	
297V004H09	102A		53	2007	
297V004H10	102	3004	2	2007	
297V004H11	102A		53	2007	
297V004H14	102	3004	2	2007	
297V004H15	102	3004	2	2007	
297V004H16	102	3004	2	2007	
297V004M01	102	3004	2	2007	
297V005H01	102A		53	2007	
297V008H01	100	3005	1	2007	
297V008M01	160		245	2004	
297V010H01	102A(2)		53(2)		
297V010M01	102A(2)		53(2)		
297V011			1	2007	
297V011H01	126	3008	52	2024	
297V011H02	100	3005	1	2007	
297V012			1	2007	
297V012H01	126	3005	52	2024	
297V012H02	100	3005	1	2007	
297V012H03	100	3005	1	2007	
297V012H04	160		245	2004	
297V012H05	100	3005	1	2007	
297V012H06	100	3005	1	2007	
297V012H07	160		245	2004	
297V012H08	100	3008	1	2007	
297V012H09	100	3005	1	2007	
297V012H10	126	3008	52	2024	
297V012H11	160	3008	245	2004	
297V012H12	160		245	2004	
297V012H13	160		245	2004	
297V012H14	102	3008	2	2007	
297V012H15	126	3008	52	2024	
297V017H01	100	3005	1	2007	
297V017H02	100	3005	1	2007	
297V018H01	102A		53	2007	
297V019B01	100	3005	1	2007	
297V019B04	141A	3092	ZD-11.5		
297V019H01	100	3005	1	2007	
297V020H01	160		245	2004	
297V020H02	100	3005	1	2007	
297V020M01	100	3005	1	2007	
297V021H01	100	3005	1	2007	
297V021H02	100	3005	1	2007	
297V021H03	100	3005	1	2007	
297V022H01	100	3005	1	2007	
297V024H01	160		245	2004	
297V024H03	160	3006	245	2004	
297V025H02	102	3004	2	2007	
297V025H03	102A(2)		53(2)		
297V025H04	102	3004	2	2007	
297V025H05	102	3004	2	2007	
297V025H15	102	3004	2	2007	
297V026H01	126	3008	52	2024	
297V026H03	100	3005	1	2007	
297V027C01	116	3017B/117	504A	1104	
297V027H01	102	3004	2	2007	
297V032H01	102	3004	2	2007	
297V033H01	102	3004	2	2007	
297V034H01	126	3008	52	2024	
297V035H01	126	3008	52	2024	
297V036H01	126	3008	52	2024	
297V036H02	126	3007	52	2024	
297V037B02	102A	3004	53	2007	
297V037H01	102	3004	2	2007	
297V037H02	102	3004	2	2007	
297V038H01	100	3005	1	2007	
297V038H02	160	3005	245	2004	
297V038H03	160	3005	245	2004	
297V038H04	160	3004/102A	245	2004	
297V038H05	100	3005	1	2007	
297V038H06	126	3008	52	2024	
297V038H07	102	3005	2	2007	
297V038H09	100	3005	1	2007	
297V038H10	126	3008	52	2024	
297V038H11	126	3008	52	2024	
297V038H12	126	3008	52	2024	
297V040H01	102	3004	2	2007	
297V040H08	102	3004	2	2007	
297V040H09			53	2007	
297V040H10	102	3004	2	2007	
297V040H11	102	3004	2	2007	
297V040H12	102	3004	2	2007	
297V040H13	102	3004	2	2007	
297V040H15	121	3009	239	2006	
297V040H16	102	3004	2	2007	
297V041H01	121MP	3009	239(2)	2006(2)	
297V041H02	121	3009	239	2006	
297V041H03	121	3009	239	2006	
297V041H04	121	3009	239	2006	
297V041H05	104	3014	16	2006	
297V041H06	121	3014	239	2006	
297V041H07	104	3009	16	2006	
297V041H15			16	2006	
297V042C01	102	3004	2	2007	
297V042C02	102	3004	2	2007	
297V042C03	102	3004	2	2007	
297V042C04	102	3004	2	2007	
297V042H01	102	3004	2	2007	
297V042H02	102	3004	2	2007	
297V042H03	126	3006/160	52	2024	
297V042H04	126	3006/160	52	2024	
297V043H01	102	3004	2	2007	
297V043H02	100	3005	1	2007	
297V044H01	100	3005	1	2007	
297V045H01	126	3006/160	52	2024	
297V045H02	126	3006/160	52	2024	
297V049H01	123	3124/289	20	2051	
297V049H03	123	3124/289	20	2051	
297V049H04	123	3124/289	20	2051	
297V049H05	123A	3444	20	2051	
297V049H06	128	3024	243	2030	
297V050C02	102A	3004	53	2007	
297V050H01	120	3004/102A	CR-3		
297V050H02	120	3004/102A	CR-3		
297V050H03	102	3004	2	2007	
297V051C03	102	3004	2	2007	
297V051C04	102	3004	2	2007	
297V051H01	102	3004	2	2007	
297V051H02	102	3004	2	2007	
297V051H03	102	3004	2	2007	
297V051H04	102A	3004	53	2007	
297V052C01	102	3004	2	2007	
297V052H01	102A	3004	53	2007	
297V052H02	102	3004	2	2007	
297V052H04	102A	3004	53	2007	
297V053C01	102	3004	2	2007	
297V053H01	102A	3004	53	2007	
297V053H02	102A		53	2007	
297V054C01	100	3005	1	2007	
297V054C02	100	3005	1	2007	
297V054H01	126	3005	52	2024	
297V054H02	126	3005	52	2024	
297V055C01	100	3005	1	2007	
297V055H01	126	3005	52	2024	
297V057H01	158(2)	3123	53(2)		
297V057H02	158	3123	53	2007	
297V059H01	123	3020	20	2051	
297V059H02	123	3124/289	20	2051	
297V059H03	123	3124/289	20	2051	
297V060H01	124	3021	12		
297V060H02	124	3021	12		
297V060H03	124	3021	12		
297V061C01	130	3020/123	14	2041	
297V061C01A	130	3027	14	2041	
297V061C02	130	3124/289	14	2041	
297V061C02A	130	3027	14	2041	
297V061C03	123	3124/289	20	2051	
297V061C04	123	3124/289	20	2051	
297V061C05	128	3024	243	2030	
297V061C06	123	3124/289	20	2051	
297V061C07	123A	3124/289	20		
297V061H01	123A	3444	20	2051	
297V061H02	123A	3444	20	2051	
297V061H03	123A	3444	20	2051	
297V062C01	121	3009	239	2006	
297V062C05	121	3009	239	2006	
297V062C06	128	3024	243	2030	
297V063C01	126	3006/160	52	2024	
297V063H01			51	2004	
297V064B01		3006	50	2004	
297V064H01	160		245	2004	
297V065C01	126	3008	52	2024	
297V065C02	126	3008	52	2024	
297V065C03	126	3008	52	2024	
297V065H01	100	3005	1	2007	
297V065H02	100	3005	1	2007	
297V065H03	100	3005	1	2007	
297V070C01	126	3006/160	52	2024	
297V070H49	108	3452	86	2038	
297V071C03	124	3021	12		
297V071H03	124	3021	12		
297V072C01	108	3452	86	2038	
297V072C03	108	3452	86	2038	
297V072C04	108	3452	86	2038	
297V072C05	128	3024	243	2030	
297V072C06	123A	3444	20	2051	
297V073C01	129	3025	244	2027	
297V073C02	129	3025	244	2027	
297V073C03	159	3466	82	2032	
297V073C04	159	3466	82	2032	
297V074C01	161	3117	39	2015	
297V074C02	123	3124/289	20	2051	
297V074C03	123	3124/289	20	2051	
297V074C04	123	3124/289	20	2051	
297V074C06	123A	3444	20	2051	
297V074C07	123A	3444	20	2051	
297V074C08	123A	3444	20	2051	
297V074C09	108	3452	86	2038	
297V074C10	128	3124/289	243	2030	
297V074C11	129	3025	244	2027	
297V074C12	128	3024	243	2030	
297V076B01	102	3004	2	2007	
297V076C01	102	3004	2	2007	
297V077C01	126	3006/160	52	2024	
297V078C01	108	3452	86	2038	
297V078C02	108	3452	86	2038	
297V080C01	129	3025	244	2027	
297V081C01	102A	3004	53	2007	
297V082B01	128	3024	243	2030	
297V082B02	129	3025	244	2027	
297V082B03	129	3025	244	2027	
297V083C01	159	3466	82	2032	
297V083C02	123A	3444	20	2051	
297V083C03	129	3025	244	2027	
297V083C04	128	3024	243	2030	
297V084C01	157	3747	232		
297V085C01	123A	3444	20	2051	
297V085C02	123A	3444	20	2051	
297V085C03	123A	3444	20	2051	
297V085C04	123A	3444	20	2051	
297V086C01	159	3466	82	2032	
297V086C02	123A	3444	20	2051	
297V086C03	123A	3444	20	2051	
297V086C04	185		58	2025	
297V087B02	188		226	2018	
299POWER		3012	4		
0300	116	3016	504A	1104	
300-0003-002	177		300	1122	
300-0003-003	177		300	1122	
300C043	128		243	2030	
300E	6154		5132		
300R1B	5801	9004		1142	
300R2B	5802	9005		1143	
300R3B	5803	9006		1144	
300R5B	5805	9008			
300R7B	5808	9009			
300R8B	5808	9009			
300U100A	6358	6558			
300U100AR	6359	6559			
300U10A	6354	6554			
300U10AR	6355	6555			
300U15A	6354	6554			
300U15AR	6355	6555			
300U20A	6354	6554			
300U20AR	6355	6555			
300U25A	6354	6554			
300U25AR	6355	6555			
300U30A	6354	6554			
300U30AR	6355	6555			
300U40A	6354	6554			
300U40AR	6355	6555			
300U50A	6356	6556			
300U5A	6354	6554			
300U5AR	6355	6555			
300U60A	6356	6556			
300U70A	6358	6558			
300U70AR	6359	6559			
300U80A	6358	6558			
300U80AR	6359	6559			
300U90A	6358	6558			
300U90AR	6359	6559			
301	102	3004	2	1104	
301-1	120	3110	CR-3		
0301-3055-00	130		14	2041	
301-576-14	740A	3328	IC-31		
301-576-3	708		IC-10		
301-576-4	7400	7400		1801	
301U100	6358	6560			
301U100R	6359	6561			
301U80	6358	6560			
301U80R	6359	6561			

Industry Standard No.	ECG	SK	GE	RS 276-	MOTOR.
302	102	3016	2	2007	
302-679-1	299	3298	236		
302-680	299	3298	236		
302A	5986	3610/5980	5096		
302B	5982	3609/5986	5096		
302C	5986	3609	5096		
302D	5986	3609	5096		
302E	5988	3608/5990	5100		
302F	5988	3608/5990	5100		
302G	5990	3608	5100		
302H	5990	3608	5100		
302K	5992	3501/5994	5104		
302M	5994	3501	5104		
302P	5998		5108		
302S	5998		5108		
302V	6002	7202	5108		
302Z	6002	7202	5108		
303-1		3466	48		
303-2	129	3025	244	2027	
303A	5986	3610/5980	5096		
303B	5982	3609/5986	5096		
303C	5986	3609	5096		
303D	5986	3609	5096		
303E	5988	3608/5990	5100		
303F	5988	3608/5990	5100		
303G	5990	3608	5100		
303H	5990	3608	5100		
303K	5992	3501/5994	5104		
303M	5994	3501	5104		
303P	5998		5108		
303S	5998		5108		
303V	6002	7202	5108		
303Z	6002	7202	5108		
0304	116	3017B/117	504A	1104	
304A	5874		5032		
304B	5872	3016	5032	1104	
304C	5878		5032		
304D	5878		5032		
304F	5876		5036		
304H	5878		5036		
304K	5880		5040		
304M	5882		5040		
304P	5886		5044		
304S	5886		5044		
304V	5890	7090	5044		
304Z	5890	7090	5044		
305-047-9-002	724		IC-86		
305A	5830	3051/156	5004		
305B	5832	3051/156	5004		
305C	5834	3051/156	5004		
305D	5834	3051/156	5004		
305F	5836		5004		
305H	5838		5004		
305K	5840	3584/5862	5008		
305M	5842	7042	5008		
305P	5846	3516	5012		
305S	5846	3516	5012		
305V	5848	7048	5012		
305Z	5848	7048	5012		
306-1	123A	3024/128	20	2051	
0307	125	3033	510,531	1114	
307-001-9-001	703A	3157	IC-12		
307-005-9-001	1003	3288	IC-43		
307-007-9-001	1100	3223	IC-92		
307-007-9-002	1102	3224	IC-93		
307-008-9-001	724	3525	IC-86		
307-009-9-002	1102	3224	IC-93		
307-020-9-001	1100	3223	IC-92		
307-029-1-001	1100	3223	IC-92		
307-047-9-002	724		IC-86		
307-095-9-003	977	3462			
307-107-9-001	1235	3637			
307-107-9-003		3712	IC-179	705	
307-107-9-005	1197	3733			
307-107-9-006	1207	3713			
307-112-9-001	1194	3484			
307-112-9-002	977	3462			
307-112-9-003	977	3462			
307-112-9-005	973	3233			
307-112-9-007	1082	3461	IC-140		
307-113-9-001	4011B		4011	2411	
307-113-9-003	977	3462			
307-115-9-001	1082	3461			
307-131-9-003	1260	3744			
307-131-9-006	1226	3763			
307-133-9-004	1192	3445			
307-143-9-002	1249	3963			
307-143-9-003	1278	3726			
307-152-9-012	4011B		4011	2411	
307A	116	3016	504A	1104	
307B	116	3016	504A	1104	
307C	116	3016	504A	1104	
307D	116	3016	504A	1104	
307F	116	3016	504A	1104	
307H	116	3016	504A	1104	
307K	116	3016	504A	1104	
307M	116	3017B/117	504A	1104	
309(CATALINA)	159		82	2032	
309-324-613	118		CR-1		
309-324-616	109			1123	
309-327-601	138A		ZD-7.5		
309-327-608	138A		ZD-7.5		
309-327-802	138A		ZD-7.5		
309-327-803	125		510,531	1114	
309-327-910	116		504A	1104	
309-327-916	109			1123	
309-327-926	123A	3444	20	2051	
309-327-927	116		504A	1104	
309-327-931	102A		53	2007	
309-327-932	116		504A	1104	
310	116	3016	504A	1104	
310-068	126	3006/160	52	2024	
310-123	126	3006/160	52	2024	
310-124	126	3006/160	52	2024	
310-139	126	3006/160	52	2024	
310-187	123	3124/289	20	2051	
310-188	102	3004	2	2007	
310-189	100	3005	1	2007	
310-190	126	3008	52	2024	
310-191	126	3006/160	52	2024	
310-192	121	3009	239	2006	
310-4	116	3016	504A	1104	
310-68	126	3006/160	52	2024	
310-8028-001	5527	6627			
0311	116	3016	504A	1104	
311-0126-001	519		514	1122	
311-0139-001	177		300	1122	
311D589-P2	159	3466	82	2032	
311D883-P01	177		300	1122	
311D916P01	199	3245	62	2010	
31124	141A	3092	ZD-11.5		
0312	116	3016	504A	1104	
0314	116	3017B/117	504A	1104	
314-6005-1	184		57	2017	
314-6006	175	3026	246	2020	
314-6007-1	123A	3444	20	2051	
314-6007-2	123A	3444	20	2051	
314-6007-3	123A	3444	20	2051	
314-6010-3			61	2051	
315			504A	1104	
316-0227-001	910		IC-251		
0317	125	3033	510,531	11144	
317-0083-001	159		82	2032	
317-0139-001	129		244	2027	
317-8504-001	123A	3444	20	2051	
318-2	297	3137/192A	271	2030	
319C	123A	3444	20	2051	
0320	116	3016	504A	1104	
320A	116	3016	504A	1104	
320B	116	3016	504A	1104	
320C	116	3016	504A	1104	
320D	116	3016	504A	1104	
320F	116	3016	504A	1104	
320H	116	3016	504A	1104	
320K	116	3016	504A	1104	
320M	116	3017B/117	504A	1104	
320P	125	3033	510,531	1114	
320S	125	3032A	510,531	1114	
320Z	125	3033	510,531	1114	
0321	116	3016	504A	1104	
321-264	154		40	2012	
0322	116	3016	504A	1104	
322(CATALINA)	123A		20	2051	
322-0147	116		504A	1104	
322-1	175	3538	246	2020	
322T1	102A	3123	53	2007	
323T1	102A	3123	53	2007	
324	102A	3017B/117	53	1104	
324-0011	113A		6GC1		
324-0012	358		42		
324-0014	109	3087		1123	
324-0015	120	3110	CR-3		
324-0016	160		245	2004	
324-0026	160		245	2004	
324-0027	126	3008	52	2024	
324-0028	126	3008	52	2024	
324-0029	102A	3004	53	2007	
324-0030	158		53	2007	
324-0035	109	3087		1123	
324-0037	109	3087		1123	
324-0038	126	3008	52	2024	
324-0041	102	3004	2	2007	
324-0049	109	3087		1123	
324-0055	102	3004	2	2007	
324-0056	158	3004/102A	53	2007	
324-0057	109	3087		1123	
324-0062	158		53	2007	
324-0074	102A	3004	53	2007	
324-0077	160		245	2004	
324-0079	126	3006/160	52	2024	
324-0080	158		53	2007	
324-0082	160		245	2004	
324-0083	160	3006	245	2004	
324-0085	158	3004/102A	53	2007	
324-0086	160	3006	245	2004	
324-0087	160	3006	245	2004	
324-0088	102	3004	2	2007	
324-0089	102A	3008	53	2007	
324-0090	100	3008	1	2007	
324-0091	102A	3004	53	2007	
324-0092	102A	3004	53	2007	
324-0093	102A	3004	53	2007	
324-0093(PHILCO)				1123	
324-0095	160	3006	245	2004	
324-0097	158		53	2007	
324-0098	126	3006/160	52	2024	
324-0099	126	3006/160	52	2024	
324-0100	102A		53	2007	
324-0102	117		504A	1104	
324-0105	109	3087		1123	
324-0106	126	3006/160	52	2024	
324-0107	110MP	3709	1N60	1123(2)	
324-0107-01	109			1123	
324-0108	109	3091		1123	
324-0110	160		245	2004	
324-0111	160		245	2004	
324-0112	160		245	2004	
324-0114	112		1N82A		
324-0115	179	3642	76		
324-0116	127	3034	25		
324-0117	116	3016	504A	1104	
324-0118	5896	7096	5064		
324-0119	116	3016	504A	1104	
324-012			50	2004	
324-0120	116	3016	504A	1104	
324-0121	126	3006/160	52	2024	
324-0122	103A	3835	59	2002	
324-0123	160		245	2004	
324-0124	158		53	2007	
324-0125	158		53	2007	
324-0126	131		44	2006	
324-0128	127	3035	25		
324-0129	126	3006/160	52	2024	
324-0130	126	3007	52	2024	
324-0131	160	3008	245	2004	
324-0132	160		245	2004	
324-0133	102	3004	2	2007	
324-0134	103	3862	8	2002	
324-0135	116	3016	504A	1104	
324-0136	126	3006/160	52	2024	
324-0137	126	3007	52	2024	
324-0138	126	3006/160	52	2024	
324-0139	102	3722	2	2007	
324-0140	102	3722	2	2007	
324-0141	109	3087		1123	
324-0142	102	3004	2	2007	
324-0143	102	3004	2	2007	
324-0144	102	3004	2	2007	
324-0145	126	3007	52	2024	
324-0146	102	3004	2	2007	
324-0147	116	3016	504A	1104	
324-0149	108	3452	86	2038	
324-0150	108	3452	86	2038	
324-0151	123	3124/289	20	2051	
324-0152	123	3124/289	20	2051	
324-0154	123	3124/289	20	2051	
324-0155			50	2004	
324-0160	109	3087		1123	
324-0162	116	3016	504A	1104	
324-0187	160	3006	245	2004	
324-019	128	3024	243	2030	
324-1	108	3452	86	2038	
324-132	160	3006	245	2004	
324-144	102	3004	2	2007	
324-6005-5	123A	3444	20	2051	
324-6011	128	3024	243	2030	
324-6011(NPN)			18	2030	
324-6011(PNP)			2	2007	

Industry Standard No.	ECG	SK	GE	RS 276-	MOTOR.
324-6013	128	3024	243	2030	
324T1	102A	3123	53	2007	
324T2	102A		53	2007	
325-0025-327	109	3087		1123	
325-0025-329	102A	3004	53	2007	
325-0025-330	102A	3004	53	2007	
325-0025-331	102A	3004	53	2007	
325-0028-79	102A	3004	53	2007	
325-0028-80	102A	3004	53	2007	
325-0028-81	102A	3004	53	2007	
325-0028-82			52	2024	
325-0028-83	102A	3004	53	2007	
325-0028-84	108	3452	86	2038	
325-0028-85	160	3006	245	2004	
325-0028-86	109	3087		1123	
325-0028-87	109	3087		1123	
325-0028-89	116	3017B/117	504A	1104	
325-0030-315	102A	3004	53	2007	
325-0030-317	102A	3004	53	2007	
325-0030-318	102A	3004	53	2007	
325-0030-319	102A	3004	53	2007	
325-0031-303	123A	3444	20	2051	
325-0031-304	123	3124/289	20	2051	
325-0031-305	123	3124/289	20	2051	
325-0031-306	158	3004/102A	53	2007	
325-0031-310	123A	3444	20	2051	
325-0031-335	109	3087		1123	
325-0031-338	116	3311	504A	1104	
325-0036-536	102A	3006/160	53	2007	
325-0036-562	109	3087		1123	
325-0036-564	126	3008	52	2024	
325-0036-565	126	3006/160	52	2024	
325-0042-311	166	9075		1152	
325-0042-351	123A	3444	20	2051	
325-0047-516	102A	3004	53	2007	
325-0047-517	116	3016	504A	1104	
325-0054-310	102A	3004	53	2007	
325-0054-311	102A	3004	53	2007	
325-0054-312	116	3016	504A	1104	
325-0076-306	123A	3444	20	2051	
325-0076-307	123A	3444	20	2051	
325-0076-308	123A	3444	20	2051	
325-0076-315	116	3311	504A	1104	
325-0081-100	123A	3444	20	2051	
325-0081-101	123A	3444	20	2051	
325-0081-102	158	3004/102A	53	2007	
325-0081-109	167	3647	510	1172	
325-0081-110	116	3311	504A	1104	
325-0135-B	116	3016	504A	1104	
325-0141-23	116	3016	504A	1104	
325-0500-12	199	3124/289	62	2010	
325-0500-13	199	3124/289	62	2010	
325-0574-30	123A	3444	20	2051	
325-0574-31	123A	3444	20	2051	
325-0670	102A	3004	53	2007	
325-0670-1	102A	3004	53	2007	
325-0670-16	116		504A	1104	
325-0670-7	102A		53	2007	
325-0670A	102A		53	2007	
325-1370-18	102A	3004	53	2007	
325-1370-19	123A	3444	20	2051	
325-1370-20	123A	3444	20	2051	
325-1375-10	126	3008	52	2024	
325-1375-11	126	3008	52	2024	
325-1375-12	126	3008	52	2024	
325-1376-53	102A	3008	53	2007	
325-1376-54	126	3008	52	2024	
325-1376-55	126	3004/102A	52	2024	
325-1376-56	102A	3004	53	2007	
325-1376-57	102A	3004	53	2007	
325-1376-58	102A	3004	53	2007	
325-1376-60	109	3087		1123	
325-1378-18	107	3018	11	2015	
325-1378-19	107	3018	11	2015	
325-1378-20	102A	3004	53	2007	
325-1378-21	158	3004/102A	53	2007	
325-1378-22	102A	3004	53	2007	
325-1441-10	116	3016	504A	1104	
325-1441-11	116	3016	504A	1104	
325-1442-8	102	3004	2	2007	
325-1442-9	124	3021	12		
325-1446-26	123	3124/289	20	2051	
325-1446-27	123	3124/289	20	2051	
325-1446-28	123	3124/289	20	2051	
325-1446-29	116	3016	504A	1104	
325-1513-29	123	3124/289	20	2051	
325-1513-30	123	3124/289	20	2051	
325-1513-46	128	3024	243	2030	
325-1771-15	123A	3444	20	2051	
325-1771-16	107	3124/289	11	2015	
325-4610-100	125	3031A	510,531	1114	
325T1	102A	3123	53	2007	
326T1	102A	3123	53	2007	
0327	125	3033	510,531	1114	
330-1304-8	108	3452	86	2038	
331-1	196	3041	241	2020	
332-2911	130		14	2041	
332-2912	181		75	2041	
332-3562			255	2041	
332-4009	177		300	1122	
333-1			21	2034	
334-377	162	3438	35		
335-1	165	3115	38		
335A	5981	3698/5987	5097		
335B	5983	3698/5987	5097		
335C	5987	3698	5097		
335D	5987	3698	5097		
335F	5989	3518/5995	5101		
335H	5991	3518/5995	5101		
335K	5993	3518/5995	5105		
335M	5995	3518	5105		
335P	5999		5109		
335S	5999		5109		
335V	6003	7203	5109		
335Z	6003	7203	5109		
336A	5981	3698/5987	5097		
336B	5983	3698/5987			
336C	5987	3698	5097		
336D	5987	3698	5097		
336F	5989	3518/5995	5101		
336H	5991	3518/5995	5101		
336K	5993	3518/5995	5105		
336M	5995	3518	5105		
336S	5999		5109		
336V	6003	7203	5109		
336Z	6003	7203	5109		
337A	5871		5033		
337B	5873		5033		
337C	5879		5033		
337D	5879		5033		
337F	5877		5037		
337H	5879		5037		
337K	5881		5041		
337M	5883		5041		

Industry Standard No.	ECG	SK	GE	RS 276-	MOTOR.
337P	5887		5045		
337S	5887		5045		
337V	5891	7091	5045		
337Z	5891	7091	5045		
339-529-001	109			1123	
339-529-002	140A		ZD-10	562	
341A	5850		5016		
341B	5852		5016		
341C	5854		5016		
341D	5854		5016		
341F	5856		5020		
341H	5856		5020		
341K	5860		5024		
341M	5862		5024		
341P	5866		5028		
341S	5866		5028		
341V	5868	7068	5028		
341Z	5868	7068	5028		
342-1	128	3024	18	2030	
344-1	128		243	2030	
344-6000-2	123A	3444	20	2051	
344-6000-3	108	3452	86	2038	
344-6000-3A	108	3452	86	2038	
344-6000-4	123A	3444	20	2051	
344-6000-5	123A	3444	20	2051	
344-6000-5A	123A	3444	20	2051	
344-6001-1	128	3124/289	243	2030	
344-6001-2	128	3124/289	243	2030	
344-6002-3	123A	3444	20	2051	
344-6005-1	123A	3444	20	2051	
344-6005-2	123A	3444	20	2051	
344-6005-5	123A	3444	20	2051	
344-6005-6	177	3100/519	300	1122	
344-6006-6	109			1123	
344-6011-1	128	3124/289	243	2030	
344-6011-2	128	3024	243	2030	
344-6012-1	129	3024/128	244	2027	
344-6012-3	129		244	2027	
344-6013-1B	128	3024	243	2030	
344-6013-4	128	3124/289	243	2030	
344-6014-1B	129		244	2027	
344-6015-10	108	3452	86	2038	
344-6015-11	108	3452	86	2038	
344-6015-7	108	3452	86	2038	
344-6015-7A	108	3452	86	2038	
344-6015-8	107	3018	11	2015	
344-6015-9	107	3018	11	2015	
344-6017-1	159	3466	82	2032	
344-6017-2	123A	3444	20	2051	
344-6017-3	123A	3444	20	2051	
344-6017-4	128		243	2030	
344-6017-5	123A	3444	20	2051	
344-6017-6	108	3452	86	2038	
345-2	128	3024	18	2030	
346B	5853		5017		
346C	5855		5017		
346D	5855		5017		
346F	5857		5021		
346H	5857		5021		
346K	5861		5025		
346M	5863		5025		
346P	5867		5029		
346S	5867		5029		
346V	5869	7069	5029		
346Z	5869	7069	5029		
349-0002-001	519		514	1122	
349-0002-002	519		514	1122	
349-1	128	3024	243	2030	
349-2	128	3024	243	2030	
349-212-003	909	3590	IC-249		
350	100	3005	1	2007	
351-029-020				010	
351-1008-010	716		IC-208		
351-1011-022	724	3525	IC-86		
351-1011-032	724	3525	IC-86		
351-1017-010	703A		IC-12		
351-1027-010	716		IC-208		
351-1029-020	941		IC-263	010	
351-1035	923		IC-259		
351-1035-020	923		IC-259		
351-1041-010	912		IC-172		
351-1042-020	910		IC-251		
351-3031	177		300	1122	
351-7140-010	909	3590	IC-249		
351-7189-010	910		IC-251		
352		3004	2	2007	
352(TRANSISTOR)	102	3722			
352-0092-020	128		243	2030	
352-0195-000	123A	3444	20	2051	
352-0197-000	123A	3444		2051	
352-0197-001			20	2051	
352-0206-001	123A		20	2051	
352-0219-000	159	3466	82	2032	
352-0316-00	123A	3444	20	2051	
352-0318-00	123A	3444	20	2051	
352-0318-001	123A	3444	20	2051	
352-0319-000	123A	3444	20	2051	
352-0322-010	123A	3444	20	2051	
352-0349-000	123A	3444	20	2051	
352-0364-000	128		243	2030	
352-0364-010	128		243	2030	
352-0365-000	123A	3444	20	2051	
352-0400-00	123A	3444	20	2051	
352-0400-010	123A	3444	20	2051	
352-0400-030	123A	3444	20	2051	
352-0403-010	154		40	2012	
352-043-010	128	3024	243	2030	
352-0433-00	123A	3444	20	2051	
352-0477-00	123A	3444	20	2051	
352-0479-010	128	3024	243	2030	
352-0506-000	123A	3444	20	2051	
352-0519-00	123A	3444	20	2051	
352-0546-00	123A	3444	20	2051	
352-0549-000	199		62	2010	
352-0551-010	159	3466	82	2032	
352-0551-021	159	3466	82	2032	
352-0569-00	123A	3444	20	2051	
352-0569-010	123A	3444	20	2051	
352-0569-020	123A	3444	20	2051	
352-0579-00	123A	3444	20	2051	
352-0579-010	123A	3444	20	2051	
352-0579-020	123A	3444	20	2051	
352-0581-011	175		246	2020	
352-0581-020	175	3026	246	2020	
352-0581-021	175		246	2020	
352-0581-030	175	3026	246	2020	
352-0581-031	175		246	2020	
352-0583-011	130		14	2041	
352-0596-010	123A	3444	20	2051	
352-0596-020	123A	3444	20	2051	
352-0596-030	123A	3444	20	2051	
352-0606-011	175		246	2020	
352-0610-030	159	3466	82	2032	
352-0610-040	159	3466	82	2032	

Industry Standard No.	ECG	SK	GE	RS 276-	MOTOR.
352-0629-010	123A	3444	20	2051	
352-0630	107		11	2015	
352-0630-010	107		11	2015	
352-0636-010	159	3466	82	2032	
352-0636-020	159	3466	82	2032	
352-0638	199		62	2010	
352-0653-010	107		11	2015	
352-0653-020	107		11	2015	
352-0658-010	161	3716	39	2015	
352-0658-020	161	3716	39	2015	
352-0658-030	161	3716	39	2015	
352-0658-040	161	3716	39	2015	
352-0658-050	161	3716	39	2015	
352-0661-010	123A	3444	20	2051	
352-0661-020	123A	3444	20	2051	
352-0667-010	123A	3444	20	2051	
352-0675-010	123A	3444	20	2051	
352-0675-020	123A	3444	20	2051	
352-0675-030	123A	3444	20	2051	
352-0675-040	123A	3444	20	2051	
352-0675-050	123A	3444	20	2051	
352-0677-010	130		14	2041	
352-0677-011	130		14	2041	
352-0677-020	130		14	2041	
352-0677-021	130		14	2041	
352-0677-030	130		14	2041	
352-0677-031	130		14	2041	
352-0677-040				2041	
352-0677-041	130		14	2041	
352-0677-051	130		14	2041	
352-0677-40	130		14	2041	
352-0680-010	123A	3444	20	2051	
352-0680-020	123A	3444	20	2051	
352-0711-021	124		12		
352-0713-030	123A	3444	20	2051	
352-0749-010	390		255	2041	
352-0754-020	159	3466	82	2032	
352-0766-010	128		243	2030	
352-0773-010	234	3247	65	2050	
352-0773-020	234	3247	65	2050	
352-0773-030	234	3247	65	2050	
352-0778-010	159	3466	82	2032	
352-0783-020	128		243	2030	
352-0809	123A	3444	20	2051	
352-0816-010	128	3024	243	2030	
352-0848-020	159	3466	82	2032	
352-0950-010	106	3984	21	2034	
352-0950-020	106	3984	21	2034	
352-0959-010	159	3466	82	2032	
352-0959-020	159	3466	82	2032	
352-0959-030	159	3466	82	2032	
352-5	171		27		
352-7500-010	123A	3444	20	2051	
352-7500-450	123A	3444	20	2051	
352-7500-861	124		12		
352-8000-010	123A	3444	20	2051	
352-8000-020	123A	3444	20	2051	
352-8000-030	123A	3444	20	2051	
352-8000-040	123A	3444	20	2051	
352-9014-00	128		243	2030	
352-9036-00	123A	3444	20	2051	
352-9079-00	123A	3444	20	2051	
352-9094-00	194		220		
352-9103-000	123A	3444	20	2051	
353		3004	2	2007	
353(I.C.)	736		IC-17		
353(TRANSISTOR)	102	3722			
353-2575-00	177		300	1122	
353-2655-000	177		300	1122	
353-3024-000	519		514	1122	
353-3083-000	177		300	1122	
353-3273-00	177		300	1122	
353-3289-000	519		514	1122	
353-3338-000	519		514	1122	
353-3339-000	177		300	1122	
353-3627-010	519		514	1122	
353-3627-020	519		514	1122	
353-3663-010	519		514	1122	
353-3687-010	519		514	1122	
353-3687-020	519		514	1122	
353-9001-001	126	3008	52	2024	
353-9001-002	126	3007	52	2024	
353-9001-003	126	3008	52	2024	
353-9002-002	126	3008	52	2024	
353-9008-001	129	3025	244	2027	
353-9012-001	102A		53	2007	
353-9201-001	121MP	3009	239(2)	2006(2)	
353-9203-001	152	3893	66	2048	
353-9301-001	128	3024	243	2030	
353-9301-002	160	3006	245	2004	
353-9301-004	129		244	2027	
353-9304-001	159	3466	82	2032	
353-9304-004	129	3025	244	2027	
353-9306-001	123A	3444	20	2051	
353-9306-002	123A	3444	20	2051	
353-9306-003	123A	3444	20	2051	
353-9306-004	123A	3444	20	2051	
353-9306-005	123A	3444	20	2051	
353-9306-006	199	3245	62	2010	
353-9306-007	199	3245	62	2010	
353-9310-001	123A	3444	20	2051	
353-9312-001	102A	3004	53	2007	
353-9314-001	123A	3444	20	2051	
353-9315-001	123A	3444	20	2051	
353-9317-001	129	3114/290	244	2027	
353-9318-001	199	3122	62	2010	
353-9318-002	199	3122	62	2010	
353-9319-001	123A	3444	20	2051	
353-9319-002	123A	3444	20	2051	
353-9502-001	152	3054/196	66	2048	
354(CHRYSLER)	723		IC-15		
354-3052	102	3722	2	2007	
354-3127-1	123A	3444	20	2051	
354-9001-001	110MP	3088	504A	1104	
354-9101-002	110MP	3088	504A	1123(2)	
354-9101-006	116	3017B/117	504A	1104	
354-9102-001	116	3017B/117	504A	1104	
354-9110-001	116	3311	504A	1104	
355D6	161	3716	39	2015	
355D7	222			2036	
355D8	161	3716	39	2015	
355D9	123A	3444	20	2051	
358-1	152	3893			
359A	116	3311	504A	1104	
359B	116	3017B/117	504A	1104	
359C	116	3017B/117	504A	1104	
359D	116	3017B/117	504A	1104	
359F	116	3031A	504A	1104	
359H	116	3031A	504A	1104	
359K	116	3017B/117	504A	1104	
359M	116	3017B/117	504A	1104	
359P	125	3032A	510,531	1114	
359S	125	3032A	510,531	1114	
359V	125	3033	510,531	1114	

Industry Standard No.	ECG	SK	GE	RS 276-	MOTOR.
359Z	125	3033	510,531	1114	
360-1		3024	18	2030	
360-1(RCA)	128		243	2030	
360-32	177	3100/519	300	1122	
361-1	186	3192	28	2017	
362-3	171		27		
362A10	192	3512	63	2030	
363A					MR850
363B					MR851
363D					MR852
363F					MR854
363H					MR854
363K					MR856
363M					MR856
364-1	159	3118	21	2034	
364-1(SYLVANIA)	159		82	2032	
364-10048	312	3112		2028	
364-6004	184	3041	57	2017	
364B14	5496	6796			
365-1	123A	3444	20	2051	
365T1	100	3005	1	2007	
366-1		3018	20	2051	
366-1(SYLVANIA)	108		86	2038	
366-2		3122	20	2051	
366-2(SYLVANIA)	108		86	2038	
366A	5850		5016		
366B	5852	3016	5016	1104	
366C	5854		5016		
366D	5854		5016		
366F	5856		5020		
366H	5858		5020		
366K	5860		5024		
366M	5862		5024		
366P	5991		5109		
366RA	5851		5017		
366RB	5853		5017		
366RC	5855		5017		
366RD	5855		5017		
366RF	5857		5021		
366RK	5861		5025		
366RM	5863		5025		
366RP	5867		5029		
366RS	5867		5029		
366RV	5869	7069	5029		
366RZ	5869	7069	5029		
366S	5866		5028		
366V	5868	7068	5028		
366Z	5868	7068	5028		
367(SEARS)	160		245	2004	
367A	5874		5032		
367B	5872		5033		
367C	5878		5032		
367D	5878		5032		
367F	5876		5036		
367H	5878		5036		
367K	5880		5040		
367M	5882		5040		
367P	5886		5044		
367RA	5875		5033		
367RB	5873		5033		
367RC	5879		5033		
367RD	5879		5033		
367RF	5877		5037		
367RH	5879		5037		
367RK	5881		5041		
367RM	5883		5041		
367RP	5887		5045		
367RS	5887		5045		
367RV	5891	7091	5045		
367RZ	5891	7091	5045		
367S	5886		5044		
367V	5890	7090	5044		
367Z	5890	7090	5044		
368A	5892		5064		
368B	5894		5064		
368C	5896	7096	5064		
368D	5896	7096	5064		
368F	5898		5068		
368H	5900	7100	5068		
368K	5902		5072		
368M	5904	7104	5072		
368P	5908		5076		
368RA	5893		5065		
368RB	5895		5065		
368RC	5897		5065		
368RD	5897		5065		
368RF	5899		5069		
368RH	5921		5069		
368RK	5903		5073		
368RM	5925		5073		
368RP	5909		5077		
368RS	5909		5077		
368RV	5911	7111	5077		
368RZ	5911	7111	5077		
368S	5908		5076		
368V	5910	7110	5076		
368Z	5910	7110	5076		
369-2	116	3016	504A	1104	
369-3	116	3016	504A	1104	
370-1	165	3115	38		
370-116764	923		IC-259		
371A	5980	3610	5096		
371B	5982	3609/5986	5096		
371C	5986	3609	5096		
371D	5986	3609	5096		
371F	5988	3608/5990	5100		
371H	5990	3608	5100		
371K	5992	3501/5994	5104		
371M	5994	3501	5104		
371P	5998		5108		
371RA	5981		5097		
371RB	5983	3698/5987	5097		
371RC	5987	3698	5097		
371RD	5987	3698	5097		
371RF	5989	3518/5995	5101		
371RH	5991	3518/5995	5101		
371RK	5993	3518/5995	5105		
371RM	5995	3518	5105		
371RP	5999		5109		
371RS	5999		5109		
371RV	6003	7203	5109		
371RZ	6003	7203	5109		
371S	5998		5108		
371V	6002	7202	5108		
371Z	6002	7202	5108		
375-1005	123A		20	2051	
376-0099	7441	7441		1804	
378-44	130	3027	14	2041	
378-44A	130	3027	14	2041	
380-0057-000	124		12		
380-0171-000	129	3025	244	2027	
380-1000	519		514	1122	
380-1001	519		514	1122	
380H61	116	3016	504A	1104	

Industry Standard No.	ECG	SK	GE	RS 276-	MOTOR.
380K62	116	3017B/117	504A	1104	
380M63	116	3017B/117	504A	1104	
383				2005	
384A	116	3017B/117	504A	1104	
384B	116	3017B/117	504A	1104	
384C	116	3017B/117	504A	1104	
384D	116	3031A	504A	1104	
384F	116	3017B/117	504A	1104	
384H	116	3017B/117	504A	1104	
384K	125	3033	510,531	1114	
384M	125	3033	510,531	1114	
384P	125	3033	510,531	1114	
384S	125	3033	510,531	1114	
384V	125	3033	510,531	1114	
384Z	125		510,531	1114	
385-1	128		243	2030	
385A	116	3017B/117	504A	1104	
385B	116	3017B/117	504A	1104	
385C	116	3016	504A	1104	
385D	116	3017B/117	504A	1104	
385F	116	3031A	504A	1104	
385H	116	3031A	504A	1104	
385K	116	3017B/117	504A	1104	
385KW	116	3017B/117	504A	1104	
385M	116	3017B/117	504A	1104	
385P	125	3032A	510,531	1114	
385S	125	3032A	510,531	1114	
385Z	125		510,531	1114	
386-1	159	3466	82	2032	
386-1(SYLVANIA)	159		82	2032	
386-1102-P1	123A	3444	20	2051	
386-1102-P2	123A	3444	20	2051	
386-1102-P3	123A	3444	20	2051	
386-1AY	116	3017B/117	504A	1104	
386-1CY	116	3017B/117	504A	1104	
386-1FY	116	3017B/117	504A	1104	
386-40	130		14	2041	
386-7118P1	108	3452	86	2038	
386-7178-P001	199		62	2010	
386-7178P1	123A	3444	20	2051	
386-7181P2	128	3047	243	2030	
386-7182-P001	195A		46		
386-7182P1	195A	3048/329	46		
386-7183P1	130	3027	14	2041	
386-7183P1A	130	3027	14	2041	
386-7184P1	129	3025	244	2027	
386-7185P1	123A	3444	20	2051	
386-7188P1	108	3452	86	2038	
386-7243-P001	161	3716	39	2015	
386-7254-P202	129		244	2027	
386-7270-P2	181		75	2041	
386-7316-P1	128		243		
386-7316-P2	128		243	2030	
386-7316-PU				2030	
386AK	116	3017B/117	504A	1104	
386AW	116	3031A	504A	1104	
386AX	116	3017B/117	504A	1104	
386AY	116	3017B/117	504A	1104	
386BW	116	3016	504A	1104	
386BY	116	3017B/117	504A	1104	
386CW	116	3017B/117	504A	1104	
386CX	116	3016	504A	1104	
386CY	116	3016	504A	1104	
386DW	116	3016	504A	1104	
386DY	116	3016	504A	1104	
386FW	116	3031A	504A	1104	
386FX	116	3031A	504A	1104	
386FY	116	3017B/117	504A	1104	
386K	116	3017B/117	504A	1104	
386KX	116	3017B/117	504A	1104	
386KY	116	3017B/117	504A	1104	
386MW	116	3017B/117	504A	1104	
386MY	116	3017B/117	504A	1104	
388A	507				1N4933
388B	506	3843	511	1114	1N4934
388C					1N4935
388D					1N4935
388F					1N4936
388H					1N4936
388K					1N4937
388M					1N4937
392-1		3024	63	2030	
392-1(SYLVANIA)			243	2030	
393-1		3025	67	2023	
393-1(SYLVANIA)	129		244	2027	
394-1571-1	519		514	1122	
394-1592-1	177		300	1122	
394-1602-1	177		300	1122	
394-3003-1	123A	3444	20	2051	
394-3003-3	123A	3444	20	2051	
394-3003-7	123A	3444	20	2051	
394-3003-9	123A	3444	20	2051	
394-3005-2	128		243	2030	
394-3074-2	102	3004	2	2007	
394-3074-5	102	3004	2	2007	
394-3097-1	102	3004	2	2007	
394-3097-2	102	3004	2	2007	
394-3102-1	101	3011A	8	2002	
394-3127-1	128	3024	243	2030	
394-3127-2	128	3024	243	2030	
394-3127-3	128	3024	243	2030	
394-3127-4	130		14	2041	
394-3135	135A	3027/130	ZD-5.1		
394-3135A	130	3027	14	2041	
394-3137-1	198		251		
394-3141-1	128		243	2030	
394-3145	159	3466	82	2032	
396-7178P1	108	3452	86	2038	
398-13222-1	923D		IC-260	1740	
398-13223-1	7400	7400		1801	
398-13224-1	7404	7404		1802	
398-13225-1	7405	7405			
398-13226-1	7401	7401			
398-13227	941M		IC-265	007	
398-13632-1	7404	7404		1802	
398B	5801	9004		1142	
399A	5800	9003		1142	
399B	5802	9005		1143	
399C	5802	9005		1143	
399D	5802	9005		1143	
0400	116	3311	504A	1104	
400(QUASAR)	506		511	1114	
400-1362-101	128		243	2030	
400-1362-102	128		243	2030	
400-1362-201	128		243	2030	
400-1371-101	123A	3444	20	2051	
400-1417-101	519		514	1122	
400-1569-101	199	3245	62	2010	
400-1596	177		300	1122	
400-2023-101	123A	3444	20	2051	
400-2023-201	123A	3444	20	2051	
400A	109	3087		1123	
400C-11958	116		504A	1104	
400D	109	3087		1123	

Industry Standard No.	ECG	SK	GE	RS 276-	MOTOR.
0401	116	3311	504A	1104	
0402	116	3031A	504A	1104	
402-004-02	519		514	1122	
402A	5980	3610	5096		
402B	5982	3609/5986	5096		
402C	5986	3609	5096		
402D	5986	3609	5096		
402F	5988	3608/5990	5100		
402H	5990	3608	5100		
402K	5992	3501/5994	5104		
402M	5994	3501	5104		
402P	5998		5108		
402S	5998		5108		
402V	6002	7202	5108		
402Z	6002	7202	5108		
403-009/07	130		14	2041	
403-1	109	3087		1123	
403A	5980	3610	5096		
403B	5986	3609	5096		
403C	5986	3609	5096		
403D	5986	3609	5096		
403F	5988	3608/5990	5100		
403H	5990	3608	5100		
403K	5992	3501/5994	5104		
403M	5994	3501	5104		
403P	5998		5108		
403S	5998		5108		
403V	6002	7202	5108		
403Z	6002	7202	5108		
404-2		3122	11	2015	
404-2(SYLVANIA)	123A		20	2051	
404A	5874		5032		
404B	5872	3500/5882	5032		
404B(NCR)	129		244	2027	
404C	5878	3500/5882	5032		
404D	5878	3500/5882	5032		
404F	5876	3500/5882	5036		
404H	5878	3500/5882	5036		
404K	5880	3500/5882	5040		
404M	5882	3500	5040		
404P	5886		5044		
404S	5886		5044		
404V	5890	7090	5044		
404Z	5890	7090	5044		
405A	5940	3607	5048		
405B	5944	3501/5994	5048		
405C	5944	3501/5994	5048		
405D	5944	3501/5994	5048		
407A	5850	3500/5882	5016		
407B	5852	3500/5882	5016		
407C	5854	3500/5882	5016		
407D	5854	3500/5882	5016		
407F	5856	3500/5882	5020		
407H	5858	3500/5882	5020		
407K	5860	3500/5882	5024		
407M	5862	3500/5882	5024		
407P	5866		5028		
407RA	5851	3517/5883	5017		
407RB	5853	3517/5883	5017		
407RC	5855	3517/5883	5017		
407RD	5855	3517/5883	5017		
407RF	5857	3517/5883	5021		
407RH	5859	3517/5883	5021		
407RK	5861	3517/5883	5025		
407RM	5863	3517/5883	5025		
407RP	5867		5029		
407RS	5867		5029		
407RV	5869	7069	5029		
407RZ	5869	7069	5029		
407S	5866		5028		
407V	5868	7068	5028		
407Z	5868	7068	5028		
408A	5874		5032		
408B	5872	3500/5882	5032		
408C	5878	3500/5882	5032		
408D	5878	3500/5882	5032		
408F	5876	3500/5882	5036		
408H	5878	3500/5882	5036		
408K	5880	3500/5882	5040		
408M	5882	3500	5040		
408P	5886		5044		
408RA	5875	3517/5883	5033		
408RB	5873	3517/5883	5033		
408RC	5875	3517/5883	5033		
408RD	5879	3517/5883	5033		
408RF	5877	3517/5883	5037		
408RH	5877	3517/5883	5037		
408RK	5881	3517/5883	5041		
408RM	5883	3517	5041		
408RP	5887		5045		
408RS	5887		5045		
408RV	5891		5045		
408RZ	5891		5045		
408S	5886		5044		
408V	5890	7090	5044		
408Z	5890	7090	5044		
409A	5892		5064		
409B	5894		5064		
409C	5896	7096	5064		
409D	5896	7096	5064		
409F	5898		5069		
409H	5900	7100	5068		
409K	5902		5072		
409M	5904	7104	5072		
409P	5908		5076		
409RA	5893		5065		
409RB	5895		5065		
409RC	5897		5065		
409RD	5897		5065		
409RF	5899		5069		
409RH	5901		5069		
409RK	5903		5073		
409RM	5925		5073		
409RP	5909		5077		
409RS	5909		5077		
409RV	5911	7111	5077		
409RZ	5911	7111	5077		
409S	5908		5076		
409V	5910	7110	5076		
409Z	5910	7110	5076		
410	283	3311	504A	1104	
410-012-0150	102A	3004	53	2007	
410-013-0240	102A	3004	53	2007	
0411	116	3311	504A	1104	
411-237	128	3024	243	2030	
412	100	3031A	1	2007	
412(ZENER)	5072A		ZD-8.2	562	
412-1A5	123A		20	2051	
412Z4	141A	3092	ZD-11.5		
413	283	3190/184	66	2048	
413(CONCERT HALL)	177			1122	
413(CONCERT-HALL)			300	1122	
413(E.F.JOHNSON)	241		57	2020	
413(TRUETONE)	177		300	1122	

Industry Standard No.	ECG	SK	GE	RS 276-	MOTOR.
414			20	2051	
414A-15	121	3717	239	2006	
00415	159	3466	82	2032	
416A	6050		5128		
416B	6050		5128		
416C	6054		5128		
416D	6054		5128		
416F	6058		5132		
416H	6060		5132		
416K	6064		5136		
416M	6064		5136		
416P	6068		5140		
416RA	6050		5129		
416RB	6050		5129		
416RC	6055		5129		
416RD	6055		5129		
416RF	6061		5133		
416RH	6061		5133		
416RK	6065		5137		
416RM	6065		5137		
416RP	6069		5140		
416RS	6069		5140		
416RV	6073		5141		
416RZ	6073		5141		
416S	6068		5140		
416V	6072		5140		
416Z	6072		5140		
417-100	128		243	2030	
417-101	162	3438	35		
417-102	106	3984	21	2034	
417-103	102A		53	2007	
417-104	175		246	2020	
417-105	123A	3444	20	2051	
417-106	123A	3444	20	2051	
417-107		3122	20	2051	
417-108	123A	3444	20	2051	
417-108-13163	199	3245	62	2010	
417-109	123AP	3024/128	243	2030	
417-109-13163	123A	3444	20	2051	
417-11	160		245	2004	
417-110	123A	3444	20	2051	
417-110-13163	123A	3444	20	2051	
417-111	129		244	2027	
417-112	127	3764	25		
417-113	179	3642	76		
417-114	128	3024	243	2030	
417-114-13163	123A	3444	20	2051	
417-115	154	3045/225	40	2012	
417-115-13173	128		243	2030	
417-116	159	3466	82	2032	
417-116-13165	159	3466	82	2032	
417-118	123A	3444	20	2051	
417-119	704	3023	IC-205		
417-12	160		245	2004	
417-120	179	3642	76		
417-121	103A	3835	59	2002	
417-122	102A		53	2007	
417-124	108	3452	86	2038	
417-125	108	3452	86	2038	
417-125-12903	107		11	2015	
417-126	123AP	3444/123A	20	2051	
417-126-12903	199	3245	62	2010	
417-127	123	3124/289	20	2051	
417-128	128	3048/329	243	2030	
417-129	123A	3444	20	2051	
417-13	160		245	2004	
417-132	159	3466	82	2032	
417-133	128	3024	243	2030	
417-134	123A	3444	20	2051	
417-134-13271	123A	3444	20	2051	
417-135	123A	3444	20	2051	
417-136	128	3024	243	2030	
417-137	128	3024	243	2030	
417-138	129	3025	244	2027	
417-139	130	3036	14	2041	
417-139-13286	181		75	2041	
417-139A	130	3027	14	2041	
417-14	160		245	2004	
417-140	312		FET-1	2028	
417-141	121	3009	239	2006	
417-142	179	3642	76		
417-143	160		245	2004	
417-144	184		57	2017	
417-145	185	3191	58	2025	
417-146	158		53	2007	
417-147	158		53	2007	
417-148	158		53	2007	
417-149	158		53	2007	
417-150	158		53	2007	
417-151	158		53	2007	
417-152	158		53	2007	
417-153	159	3466	82	2032	
417-153-1341			65	2050	
417-153-13431	234	3247		2050	
417-154	107	3039/316	11	2015	
417-155	128	3024	243	2030	
417-155-13163	123A	3444	20	2051	
417-158	162	3438	35		
417-159	157	3103A/396	232		
417-16	160		245	2004	
417-160	131		44	2006	
417-161	172A	3156	64		
417-162	130		14	2041	
417-167			FET-1	2028	
417-168	159	3466	82	2032	
417-169	312	3112	FET-1	2028	
417-17	102A		53	2007	
417-170	129	3025	244	2027	
417-171	123AP	3444/123A	20	2051	
417-171-13163	123A	3444	20	2051	
417-172	123A	3444	20	2051	
417-172-13271	123A	3444	20	2051	
417-175	196		241	2020	
417-175-12993	152	3893	66	2048	
417-176	159	3466	82	2032	
417-177	105	3012	4		
417-178	128		243	2030	
417-18	158		53	2007	
417-180	128		243	2030	
417-181	129		244	2027	
417-182	159	3466	82	2032	
417-183	6400A			2029	
417-184	159	3466	82	2032	
417-185	123A	3444	20	2051	
417-187	6400A			2029	
417-19	107	3039/316		2015	
417-190	107	3039/316	11	2015	
417-192	123A	3444	20	2051	
417-193	128		243	2030	
417-194	312	3112	FET-1	2028	
417-195	157	3103A/396	232		
417-196	159	3466	82	2032	
417-196-13262	234	3247	65	2050	
417-197	123A	3444	20	2051	
417-199	175	3131A/369	246	2020	
417-2	160		245	2004	
417-20	121	3009	239	2006	
417-200	159	3466	82	2032	
417-201	159	3466	82	2032	
417-202	711	3070	IC-207		
417-203	196		241	2020	
417-204	162	3438	35		
417-205	107	3039/316	11	2015	
417-206	220	3990		2036	
417-207	220	3990		2036	
417-21	158		53	2007	
417-211	312	3834/132	FET-1	2035	
417-212	130	3027	14	2041	
417-212A	130	3027	14	2041	
417-213	123A	3444	20	2051	
417-214	181		75	2041	
417-214-13286	181		75	2041	
417-215	130		14	2041	
417-215-13286	130		14	2041	
417-215A	130	3027	14	2041	
417-216	121	3009	239	2006	
417-217	123A	3444	20	2051	
417-218			20	2051	
417-219			20	2051	
417-22	160		245	2004	
417-221			253	2026	
417-222	172A	3156	64		
417-223			FET-4	2036	
417-224	128	3024	243	2030	
417-225	185	3191	58	2025	
417-226-1	123A	3444	20	2051	
417-226-13163	199	3245	62	2010	
417-227	157	3103A/396	232		
417-228	123A	3444	20	2051	
417-229	123A	3444	20	2051	
417-23	160		245	2004	
417-231	312	3112	FET-1	2028	
417-232	198	3220	251		
417-233	128	3024	243	2030	
417-233-13163	123A	3444	20	2051	
417-234	129	3025	244	2027	
417-234-13165	159	3466	82	2032	
417-235	159	3466	82	2032	
417-235-13262	159	3466	82	2032	
417-237	128	3024	243	2030	
417-237-13163	128		243	2030	
417-239	163A	3439	36		
417-240		3065	FET-4	2036	
417-241	326	3746	FET-1	2028	
417-242-8181	159	3466	82	2032	
417-243	161	3716	39	2015	
417-244	123A	3444	20	2051	
417-244-12903	199		62	2010	
417-245	171	3201	27		
417-246	312			2028	
417-247	128		243	2030	
417-248	163A	3439	36		
417-25	160		245	2004	
417-250	128	3045/225	243	2030	
417-252	312			2028	
417-253	312			2028	
417-254	181		75	2041	
417-255	129		244	2027	
417-257	128		243	2030	
417-258	161	3716	39	2015	
417-26	160		245	2004	
417-260	129		244	2027	
417-260-50127	159	3466	82	2032	
417-262	161	3716	39	2015	
417-27	160		245	2004	
417-272	265	3860			
417-273	130	3027	14	2041	
417-273-13286	130		14	2041	
417-28	158		53	2007	
417-282			14	2041	
417-283			20	2051	
417-283-13271	199	3220/198	62	2010	
417-284			21	2034	
417-286		3079	262	2041	
417-289	197	3083	250	2027	
417-29	179	3642	76		
417-294	287	3044/154	222		
417-298	331		241	2020	
417-29BLK	179		76		
417-29GRN	179		76		
417-29WHT	179		76		
417-30	121	3717	239	2006	
417-31	160		245	2004	
417-32	121	3009	239	2006	
417-33	160		245	2004	
417-35	160		245	2004	
417-36	160		245	2004	
417-37	160		245	2004	
417-38	160		245	2004	
417-39	160		245	2004	
417-4-00226	196		241	2020	
417-40	102A		53	2007	
417-41	102A		53	2007	
417-42	179	3642	76		
417-43	129	3025	244	2027	
417-44	121	3009	239	2006	
417-45	121	3009	239	2006	
417-46	121	3009	239	2006	
417-47	102A		53	2007	
417-48	102A	3004	53	2007	
417-49	128	3024	243	2030	
417-5	102A		53	2007	
417-50	160		245	2004	
417-51	158		53	2007	
417-52	102A		53	2007	
417-53	160		245	2004	
417-54	160		245	2004	
417-56	160		245	2004	
417-57	160		245	2004	
417-58	160		245	2004	
417-59	128	3024	243	2030	
417-6	158		53	2007	
417-60	160	3014	245	2004	
417-62	104	3009	16	2006	
417-66	160		245	2004	
417-67	123A	3444	20	2051	
417-68	160	3008	245	2004	
417-69	123A	3444	20	2051	
417-7	123A		20	2051	
417-70	160		245	2004	
417-71	160		245	2004	
417-72	160		245	2004	
417-73	102A		53	2007	
417-74	102A		53	2007	
417-75	102A		53	2007	
417-76	160		245	2004	
417-77	123A	3444	20	2051	
417-78	102A		53	2007	

Industry Standard No.	ECG	SK	GE	RS 276-	MOTOR.
417-79	160	3123	245	2004	
417-801	123AP		20	2051	
417-801-12903	123A	3444	20	2051	
417-81	6400A			2029	
417-811	194	3024/128			
417-821-13163	128		243	2030	
417-822-13262	129		244	2027	
417-83	108	3452	86	2038	
417-84			86	2038	
417-85			86	2038	
417-852			66	2048	
417-863			FET-4	2036	
417-87	128	3024	243	2030	
417-88	128	3024	243	2030	
417-89	128	3024	243	2030	
417-90	121	3009	239	2006	
417-91	123A	3444	20	2051	
417-92	123A	3444	20	2051	
417-93	123A	3444	20	2051	
417-93-12903	123A	3444	20	2051	
417-94	123A	3444	20	2051	
417-99	121	3009	239	2006	
417A	5986	3610/5980	5096		1N248B
417B	5982	3609/5986	5096		1N249B
417C	5986	3609	5096		1N250B
417D	5986	3609	5096		1N250B
417F	5988	3608/5990	5100		1N1196
417H	5990	3608	5100		1N1196
417K	5992	3501/5994	5104		1N1198
417M	5994	3501	5104		1N1198
417P	5998		5108		
417RA	5987	3698	5097		
417RB	5983	3698/5987	5097		
417RC	5987	3698	5097		
417RD	5987	3698	5097		
417RF	5989	3518/5995	5101		
417RH	5991	3518/5995	5101		
417RK	5993	3518/5995	5105		
417RM	5995	3518	5105		
417RP	5999		5109		
417RS	5999		5109		
417RV	6003	7203	5109		
417RZ	6003	7203	5109		
417S	5998		5108		
417V	6002		5108		
417Z	6002		5108		
0418	121MP	3013	239(2)	2006(2)	
418A	5986		5096		1N1183
418B	5982		5096		1N1184
418C	5986		MR-2		1N1186
418D	5986		5096		1N1186
418F	5988		5100		1N1188
418H	5990		5100		1N1188
418K	5992		5104		1N1190
418M	5994		5104		1N1190
418P	5998		5108		
418RA	5987		5097		
418RB	5983		5097		
418RC	5987		5097		
418RD	5987		5097		
418RF	5989		5101		
418RH	5991		5101		
418RK	5993		5105		
418RM	5995		5105		
418RP	5999		5109		
418RS	5999		5109		
418RV	6003		5109		
418RZ	6003		5109		
418S	5998		5108		
418V	6002	7202	5108		
418Z	6002	7202	5108		
419A	5986	3610/5980	5096		1N1183A
419B	5982	3609/5986	5096		1N1184A
419C	5986	3609	5096		1N1186A
419D	5986	3609	5096		1N1186A
419F	5988	3608/5990	5100		1N1188A
419H	5990	3608	5100		1N1188A
419K	5992	3501/5994	5104		1N1190A
419M	5994	3501	5104		1N1190A
419P	5998		5108		
419RA	5987	3698	5097		
419RB	5983	3698/5987	5097		
419RC	5987	3698	5097		
419RD	5987	3698	5097		
419RF	5989	3518/5995	5101		
419RH	5991	3518/5995	5101		
419RK	5993	3518/5995	5105		
419RM	5995	3518	5105		
419RP	5999		5109		
419RS	5999		5109		
419RV	6003	7203	5109		
419S	5998		5108		
419V	6002	7202	5108		
420-2003-173	125	3032A	510,531	1114	
420-2005-000	113A	3119/113	6GC1		
420-2104-570	125	3032A	510,531	1114	
420-2106-270	505	3108	CR-7		
420T1	102A		53	2007	
421-10	102A		53	2007	
421-11	102A	3004	53	2007	
421-11B	102A	3123	53	2007	
421-12	102A	3004	53	2007	
421-12B	102A	3123	53	2007	
421-13	102A	3004	53	2007	
421-13B	102A		53	2007	
421-13C	102	3722	2	2007	
421-14	102A	3004	53	2007	
421-14B	102A	3123	53	2007	
421-15	102A	3004	53	2007	
421-15B	102A		53	2007	
421-16	126	3008	52	2024	
421-17	126	3008	52	2024	
421-18	124	3021	12		
421-19	102	3004	2	2007	
421-20	126	3006/160	52	2024	
421-20B	126	3006/160	52	2024	
421-21	126	3006/160	52	2024	
421-21B	126	3006/160	52	2024	
421-22	126	3006/160	52	2024	
421-22B	126	3006/160	52	2024	
421-24	121	3009	239	2006	
421-25	121	3009	239	2006	
421-26	126	3008	52	2024	
421-4027	116	3016	504A	1104	
421-6	160	3008	245	2004	
421-6599	121	3009	239	2006	
421-6B	160		245	2004	
421-7	160	3008	245	2004	
421-7143	102A	3004	53	2007	
421-7443	116	3016	504A	1104	
421-7443A	116	3016	504A	1104	
421-7444	123A	3444	20	2051	
421-7B	160		245	2004	
421-8	160	3008	245	2004	
421-8109	102A	3004	53	2007	
421-8111	123A	3444	20	2051	
421-8310	124	3021	12		
421-8B	160		245	2004	
421-9	102A		53	2007	
421-9644	123A	3444	20	2051	
421-9670	123A	3444	20	2051	
421-9671	102A	3004	53	2007	
421-9682	100	3005	1	2007	
421-9683	108	3452	86	2038	
421-9792	160	3006	245	2004	
421-9840	123A	3444	20	2051	
421-9862	103A	3010	59	2002	
421-9862A	102A	3004	53	2007	
421-9863	102A	3004	53	2007	
421-9863A	102A	3004	53	2007	
421-9864	102A	3004	53	2007	
421-9864A	102A	3004	53	2007	
421-9865	116	3016	504A	1104	
421T1	102A		53	2007	
422-0222	102A	3004	53	2007	
422-0738	123A	3444	20	2051	
422-0739	129	3025	244	2027	
422-0960	175	3026	246	2020	
422-0961	130	3027	14	2041	
422-1232	129	3025	244	2027	
422-1233	128	3024	243	2030	
422-1234	123A	3444	20	2051	
422-1362	116	3016	504A	1104	
422-1401	108	3452	86	2038	
422-1402	108	3452	86	2038	
422-1403	126	3008	52	2024	
422-1404	128	3024	243	2030	
422-1405	129	3025	244	2027	
422-1406	102A	3004	53	2007	
422-1407	123A	3444	20	2051	
422-1408	125	3031A	510,531	1114	
422-1443	121MP	3015	239(2)	2006(2)	
422-1866	116	3016	504A	1104	
422-2008	129	3025	244	2027	
422-2158	128	3024	243	2030	
422-2532	108	3452	86	2038	
422-2533	123A	3444	20	2051	
422-2534	123A	3444	20	2051	
422-2535	102A	3004	53	2007	
422-2540	116	3016	504A	1104	
422-2778	160	3006	245	2004	
422-2779	160	3006	245	2004	
422-2780	160	3006	245	2004	
423AB1AB1					MDA201
423AB1AF1					MDA2501
423BB1AB1					MDA202
423BB1AF1					MDA2502
423DB1AB1					MDA204
423DB1AF1					MDA2504
423FB1AB1					MDA200
423FB1AF1					MDA2500
423MB1AB1					MDA206
423MB1AF1					MDA2506
423NB1AB1					MDA208
423PB1AB1					MDA210
424-9001	175		246	2020	
429-0092-1	160	3006	245	2004	
429-0092-2	102A	3004	53	2007	
429-0092-3	102A	3004	53	2007	
429-0092-56	109	3087		1123	
429-0093-69	126	3006/160	52	2024	
429-0093-71	125	3126	510,531	1114	
429-0094-39	126	3008	51	2024	
429-0910-50	160		245	2004	
429-0910-51	102A	3123	53	2007	
429-0910-52	102A	3123	53	2007	
429-0910-53	110MP	3709		1123(2)	
429-0910-54	109	3087		1123	
429-0958-41	166	9075	504A	1152	
429-0958-42	123A	3444	20	2051	
429-0958-43	137A	3058	ZD-6.2	561	
429-0981-12	107	3039/316	11	2015	
429-0985-12	107	3018	11	2015	
429-0986-12	108	3452	86	2038	
429-0989-68	116	3311	504A	1104	
429-10001-0A	5870		5032		
429-10002-0A	5874		5032		
429-10036-0A	519		514	1122	
429-10036-0B	519		514	1122	
429-10054-0A	519		514	1122	
429-20004-0B	177		300	1122	
430		3124	20	2051	
430(ZENITH)	123A	3444	20	2051	
430-10034	123A	3444	20	2051	
430-10034-06	123A	3444	20	2051	
430-10047-0C	128		243	2030	
430-10053-0	123A	3444	20	2051	
430-10053-0A	123A	3444	20	2051	
430-1044-0A	123A	3444	20	2051	
430-20013-0B	159	3466	82	2032	
430-20018-0A	159	3466	82	2032	
430-20021	159	3466	82	2032	
430-20023-0A	159	3466	82	2032	
430-20026-0	159	3466	82	2032	
430-203845	129	3025	244	2027	
430-22861	108	3452	86	2038	
430-23190	181	3036	75	2041	
430-23212	123A	3444	20	2051	
430-23221	123A	3444	20	2051	
430-23222	129	3025	244	2027	
430-23223	128	3024	243	2030	
430-23843	130	3027	14	2041	
430-23844	128	3024	243	2030	
430-25762	312	3834/132	FET-1	2035	
430-25763	108	3452	86	2038	
430-25764	108	3452	86	2038	
430-25765	108	3452	86	2038	
430-25766	123A	3444	20	2051	
430-25767	108	3452	86	2038	
430-25834	131	3052	44	2006	
430-31	116	3311	504A	1104	
430-85	158		53	2007	
430-86	123A	3444	20	2051	
430-87	123A	3444	20	2051	
430CL	123A	3444	20	2051	
431-26551A	722	3161	IC-9		
432-1		3054	18	2030	
433		3124	20	2051	
433(ZENITH)	123A		20	2051	
433-1		3122	18	2051	
433-1(SYLVANIA)			20	2051	
433CL	123A	3124/289	20	2051	
433M852	129	3025	244	2027	
434	199		62	2010	
435-21026-0A	7400	7400		1801	
435-21027-0A	7402	7402		1811	
435-21028-0A	7404	7404		1802	
435-21029-0A	7408	7408		1822	

Industry Standard No.	ECG	SK	GE	RS 276-	MOTOR.
435-21030-0A	7410	7410		1807	
435-21033-0A	7420	7420		1809	
435-21034-0A	7451	7451		1825	
435-21035-0A	7486	7486		1827	
435-23006-0A	7473	7473		1803	
435-23007-0A	7474	7474		1818	
435-40012	116	3016	504A	1104	
435A	5981	3698/5987	5097		
435B	5983	3698/5987	5097		
435C	5987	3698	5097		
435D	5987	3698	5097		
435F	5989	3518/5995	5101		
435H	5991	3518/5995	5101		
435K	5993	3518/5995	5105		
435M	5995	3518	5105		
435P	6069		5140		
435S	6069		5140		
435V	6073		5140		
435Z	6073		5140		
436-10010-0A	7492	7492		1819	
436-403-001	123A		20	2051	
436-404-002	159	3466	82	2032	
436A	5981	3698/5987	5097		
436B	5987	3698	5097		
436C	5987	3698	5097		
436D	5987	3698	5097		
436F	5991	3518/5995	5101		
436H	5991	3518/5995	5101		
436K	5995	3518	5105		
436M	5995	3518	5105		
436P	5999		5109		
436S	5999		5109		
436V	6003	7203	5109		
436Z	6003	7203	5109		
437A	5871		5033		
437B	5873		5044		
437C	5879		5033		
437D	5879		5033		
437F	5877		5037		
437H	5879		5037		
437K	5881		5041		
437M	5883		5041		
437P	5886		5044		
437S	5886		5044		
437V	5890	7090	5044		
441A	5850		5016		
441B	5852		5016		
441C	5854		5016		
441D	5854		5016		
441F	5856		5020		
441H	5858		5020		
441K	5860		5024		
441M	5862		5024		
441P	5866		5028		
441S	5866		5028		
441V	5868	7068	5028		
441Z	5868	7068	5028		
442-10	705A	3134	IC-3		
442-11	748	3102/710			
442-16	722	3161	IC-9		
442-20	703A	3157	IC-12		
442-20-14299	703A		IC-12		
442-21	778A		IC-220		
442-28	708	3135/709	IC-10		
442-30	309K	3629			
442-30-2897	309K	3629			
442-33	713	3077/790	IC-5		
442-39	975	3641			
442-4	785	3254	IC-226		
442-41	725	3162	IC-19		
442-46	801	3160	IC-35		
442-5	780	3141	IC-222		
442-55	795		IC-232		
442-56	747	3279	IC-218		
442-59	783	3215	IC-225		
442-620	815	3255			
442-7	909D	3590	IC-250		
442-8	703A	3157	IC-12		
442-9	720	9014	IC-7		
443-1	7400	7400		1801	
443-12	7410			1807	
443-13	7475	7475		1806	
443-16	7476	7476		1813	
443-162	74193	74193		1820	
443-18	7404	7404		1802	
443-2	7420	7420		1809	
443-35	7441	7441			
443-36	7447	7447		1805	
443-44	7413	7413		1815	
443-44-2854	7413	7413		1815	
443-45	7408	7408		1822	
443-46	7402	7402		1811	
443-5	7473	7473		1803	
443-6	7474	7474		1818	
443-628	74196	74196		1833	
443-65	7427	7427		1823	
443-66	74192			1831	
443-7	7490	7490		1808	
443-7-16088	7490	7490		1808	
443-71	74H00	74H00			
443-77	7438	7438			
443-87	74145	74145		1828	
444(SEARS)	123A		20	2051	
444-0012-PT2	160		245	2004	
444-012-P1	160		245	2004	
445-0023	130		14	2041	
445-0023-P1	130		14	2041	
445-0023-P3	130		14	2041	
445-0023-P4	130		14	2041	
445-0034-1	181		75	2041	
445-0300P3	5626	3633			
446A	5851		5017		
446B	5854		5017		
446C	5855		5017		
446D	5855		5017		
446F	5859		5021		
446F(W.E.)	5809	9010			
446H	5859		5021		
446K	5861		5025		
446M	5863		5025		
446P	5866		5028		
446S	5866		5028		
446V	5868	7068	5028		
446Z	5868	7068	5028		
447			20	2051	
447(ZENITH)	123A	3444	20	2051	
448A662	128	3024	243	2030	
449	184	3041	57	2017	
0450	116	3311	504A	1104	
450-1167-1	159		82	2032	
450-1167-2	123A	3444	20	2051	
450-1261	123A	3444	20	2051	
451D	5944		5048		
454-A2534-1	116	3016	504A	1104	

Industry Standard No.	ECG	SK	GE	RS 276-	MOTOR.
454-A2534-10	116		504A	1104	
454A104	128	3024	243	2030	
454A25	116		504A	1104	
455-1	126	3006/160	52	2024	
0460	116	3311	504A	1104	
460-1009	177		300	1122	
460-1013	116		504A	1104	
461-1006	159	3466	82	2032	
461-1014	129		244	2027	
461-1048	129		244	2027	
461-1055-01	159	3466	82	2032	
461-2001	234	3247	65	2050	
462-0119	123A	3444	20	2051	
462-1000	123A	3444	20	2051	
462-1007	128		243	2030	
462-1009-01	123A	3444	20	2051	
462-1016	128		243	2030	
462-1019	128		243	2030	
462-1038-01	199	3245	62	2010	
462-1059	198	3220	251		
462-1061	123A	3444	20	2051	
462-1063	123A	3444	20	2051	
462-1066-01	199	3245	62	2010	
462-2002	123A	3444	20	2051	
462-2004	199	3245	62	2010	
462-I007	128		243	2030	
464-100-19	109	3087		1123	
464-103-19	109	3087		1123	
464-106-19	109	3087		1123	
464-110-19	110MP		1N34AS	1123	
464-111-19	109	3087		1123	
464-113-19	109	3087		1123	
464-119-19	110MP		1N34AS	1123	
464-280-19	117		504A	1104	
464-285-15	116		504A	1104	
464-285-19	117		504A	1104	
464-311-19	140A		ZD-10	562	
464.062.15			ZD-6.2	561	
465-005-19	102A		53	2007	
465-032-19	160		245	2004	
465-036-19	102A		53	2007	
465-042-19	160		245	2004	
465-045-19	160		245	2004	
465-049-19	160		245	2004	
465-061-19	160		245	2004	
465-067-19	158		53	2007	
465-072-19	102A		53	2007	
465-073-19	102A		53	2007	
465-075-19	102A		53	2007	
465-080-19	102A		53	2007	
465-082-19	102A		53	2007	
465-086-19	160		245	2004	
465-106-19	123A	3122	20	2051	
465-108-19	158		53	2007	
465-115-19	102A		53	2007	
465-132-19	102A		53	2007	
465-137-19	131		44	2006	
465-146-19	160		245	2004	
465-163-19	102A		53	2007	
465-165-19	102A		53	2007	
465-166-19	131MP	3840	44#(2)	2006(2)	
465-181-19	107	3039/316	11	2015	
465-191-15	102A		53	2007	
465-199-19			ZD-9.1	562	
465-1A5	160		245	2004	
465-206-19	131		44	2006	
465-223-19	160		245	2004	
465-2A5	103A		59	2002	
465-4A5	101	3861	8	2002	
465A-15	160		245	2004	
465B-15	160		245	2004	
465C-15	121	3717	239	2006	
466-1A5	102A		53	2007	
466-2A5	102A		53	2007	
466-3A5	102A		53	2007	
469-199-19	5078A	9023	ZD-19		
469-646-3	128		243	2030	
471-010	116	3016	504A	1104	
471-1(PLASTIC)	188		226	2018	
472-0309-001	128		243	2030	
472-0445-001	128		243	2030	
472-0491-001	123A	3444	20	2051	
472-0946-001	130	3027	14	2041	
472-0946-002	130		14	2041	
472-1198-001	123A	3444	20	2051	
473A31	121	3009	239	2006	
473B5	102A		53	2007	
473B6-2	102	3004	2	2007	
473B6-2A	102	3004	2	2007	
473B6-4	102	3004	2	2007	
473B6-5	102	3004	2	2007	
473B6-7	102	3004	2	2007	
474-004	116	3017B/117	504A	1104	
474-025	116	3016	504A	1104	
474-3A5	100	3721	1	2007	
474A410AXP001	124		12		
474A410BEP2	152	3893	66	2048	
474A410BW-2	152	3893	66	2048	
475-018	5802	9005		1143	
477-0376-001	909	3590	IC-249		
477-0376-002	909D		IC-250		
477-0404-002	910		IC-251		
477-0412-004	7473	7473		1803	
477-0542-001	941		IC-263	010	
477A15	105		4		
477B15	121MP	3718	239(2)	2006(2)	
477C15	121MP	3718	239(2)	2006(2)	
478C	5874	3501/5994	5104		
479-0547-001	519		514	1122	
479-0663-005	177		300	1122	
479-1006-002	5529	6629			
479-1012-001	519		514	1122	
479-1013-001	519		514	1122	
479-1013-002	519		514	1122	
479-1055-001	519		514	1122	
479-1066	5529	6629			
479-1163-001	519		514	1122	
479-1229-001	519		514	1122	
479-1248-001	519		514	1122	
480-9(SEARS)	131		44	2006	
481-201-A	128		243	2030	
481-201-B	128		243	2030	
481-34842	191	3044/154	249		
483-3141	123A		20	2051	
484A15	121	3717	239	2006	
486-1551	123A		20	2051	
486-40235-G02	116		504A	1104	
488-2(SEARS)	108		86	2038	
488B15	160		245	2004	
488C15	100	3721	1	2007	
490-2(SEARS)	108		86	2038	
491-2(SEARS)	108		86	2038	
491A948	128	3024	243	2030	
492-1A5	105		4		

Industry Standard No.	ECG	SK	GE	RS 276-	MOTOR.
499			20	2051	
499(CHRYSLER)	152		66	2048	
499(ZENITH)	123A		20	2051	
499-1	108	3452	86	2038	
0500	177	3175	300	1122	
500B10	125		510,531	1114	MR1-1200
500B15					MR1-1600
500RS20	525	3925			
500S10	125		510,531	1114	
500S20	525	3925			
501-068	282	3049/224	264		
501-363-2	5802	9005		1143	
501ES001M	108	3452	86	2038	
501T1	160		245	2004	
502			20	2051	
503-T10192	156	3051	512		
503-T21271	109	3088		1123	
503-T21472	109	3087		1123	
503-T21651		3126	90		
503T1	160		245	2004	
504-067	1232	3852			
504-080-000	HIDIV-1		FR-8		
504T1	160		245	2004	
505ES105	176	3123	80		
505T1	160		245	2004	
506T1	160		245	2004	
507T1	160		245	2004	
508ES020P			21	2034	
508ES021P			21	2034	
508T1	160		245	2004	
509			20	2051	
509(SEARS)	123A		20	2051	
509C801G02					MDA970A2
509C901G01					MDA970A1
509C901G03					MDA970A3
509C901G04					MDA970A3
509ES0258	176		80		
509ES025P		3123	2	2007	
509R	290A		269	2032	
509Y	290A		269	2032	
510A90	109	3088		1123	
510ED46	109	3088		1123	
510ES030M			63	2030	
510ES031M			63	2030	
510IN60	109	3087		1123	
510IS34	109	3087		1123	
510LN60				1123	
510LS34				1123	
511-3	5531	6631			
511-515	123	3124/289	20	2051	
511-519	123	3124/289	20	2051	
511-898	116		504A	1104	
511ES035P			63	2030	
511ES036P			63	2030	
511S345			1N60	1123	
512ES040P	176	3123	80		
512RED	123A	3122	20	2051	
513-891	114		6GD1	1104	
513ES045P	176	3123	80		
514-033338	289A		20	2051	
514-042791	177		514	1122	
514-044910	159		65	2050	
514-047830	196	3893/152	66	2048	
514-054214	130		14	2041	
515	123	3124/289	20	2051	
515-074			17	2051	
515-10408A-01	710	3102	IC-89		
515-10408A01	710		IC-89		
515-299	116		504A	1104	
515-521	108	3452	86	2038	
515ES045M			21	2034	
515ES046M			21	2034	
515ORN	123A	3444	20	2051	
516	123	3124/289	20	2051	
0516-3101-250			278	2016	
0516-3101-406			278	2016	
516ES047M			21	2034	
516ES048M			21	2034	
517-0021	116	3016	504A	1104	
517-0025	116	3016	504A	1104	
517-0031	116	3016	504A	1104	
517-0033	116	3016	504A	1104	
517-518	123	3124/289	20	2051	
518-499	116	3017B/117	504A	1104	
519-1			20	2051	
519-1(RCA)	123A	3444	20	2051	
519ES067M			17	2051	
519ES068M			17	2051	
520-301	128		243	2030	
520ES070M			21	2034	
520T1	100	3005	1	2007	
521-094	116	3017B/117	504A	1104	
521-145	109	3087		1123	
521ES071M			21	2034	
521T1	100	3005	1	2007	
522			20	2051	
522(ZENITH)	107		11	2015	
522-726	116	3017B/117	504A	1104	
522-893	113A		6GC1		
522-958	170	3649	504A	1104	
522ES075M			63	2030	
522ES076M			63	2030	
523-0001-001	116		504A	1104	
523-0001-002	116	3017B/117	504A	1104	
523-0001-003	116		504A	1104	
523-0001-004	116		504A	1104	
523-0001-005	116		504A	1104	
523-0001-006	116		504A	1104	
523-0001-007	125		510,531	1114	
523-0001-008	125		510,531	1114	
523-0001-009	125		510,531	1114	
523-0001-010	125		510,531	1114	
523-0006-002	519	3100	514	1122	
523-0007-001	519	3100	514	1122	
523-0009-041	610	3323			
523-0009-049	614	3327	90		
523-0013-002	116	3017B/117	504A	1104	
523-0013-201	519	3032A	514	1122	
523-0017-001	156	3051	512		
523-006-002	519		300	1122	
523-0501-002	116	3311	504A	1104	
523-0501-003	116	3017B/117	504A	1104	
523-1000-001	116	3017B/117	504A	1104	
523-1000-067	109	3087		1123	
523-1000-294	109			1123	
523-1000-295	109	3087		1123	
523-1000-326	604	3087		1123	
523-1000-881	116	3017B/117	300	1122	
523-1000-882	116	3017B/117	504A	1104	
523-1000-883	519	3017B/117	514	1122	
523-1000882	116	3017B/117	504A	1104	
523-1002-326	109	3087		1123	
523-1500-002	116	3017B/117	504A	1104	
523-1500-067	109	3087	1N34AS	1123	

Industry Standard No.	ECG	SK	GE	RS 276-	MOTOR.
523-1500-803			514	1122	
523-1500-881	116	3311	1N34AS	1123	
523-1500-883	519	3100	514	1122	
523-2001-100	140A	3061	ZD-10	562	
523-2003-0-1	109			1123	
523-2003-001	109	3087		1123	
523-2003-100	140A	3061	ZD-10	562	
523-2003-101	5096A		ZD-100		
523-2003-110	5074A		ZD-11.0	563	
523-2003-120	142A		ZD-12	563	
523-2003-130	143A	3750	ZD-13	563	
523-2003-140	144A		ZD-14	564	
523-2003-150	145A	3063	ZD-15	564	
523-2003-160	5075A	3751	ZD-16	564	
523-2003-170	5076A	9022	ZD-17		
523-2003-180	5077A	3752	ZD-18		
523-2003-190	5078A	9023	ZD-19		
523-2003-200	5079A	3335	ZD-20		
523-2003-220	5080A	3336	ZD-22		
523-2003-240	5081A		ZD-24		
523-2003-250	5082A	3753	ZD-25		
523-2003-270	146A		ZD-27		
523-2003-300	5084A	3755	ZD-30		
523-2003-330	147A		ZD-33		
523-2003-339	5066A	3330	ZD-3.3		
523-2003-360	5085A	3337	ZD-36		
523-2003-369	134A		ZD-3.6		
523-2003-390	5086A	3338	ZD-39		
523-2003-399	5067A	3331	ZD-3.9		
523-2003-430	5087A	3339	ZD-43		
523-2003-439	5068A	3332	ZD-4.3		
523-2003-450	5088A	3340	ZD-47		
523-2003-470	5088A	3340	ZD-47		
523-2003-479	5069A	3333	ZD-4.7		
523-2003-500	5089A	3341	ZD-51		
523-2003-510	5089A	3341	ZD-51		
523-2003-519	135A	3056	ZD-5.0		
523-2003-520	5089A	3341	ZD-51		
523-2003-560	5090A	3342	ZD-56		
523-2003-569	136A		ZD-5.6	561	
523-2003-620	149A		ZD-62		
523-2003-629	137A		ZD-6.2	561	
523-2003-680	5092A	3343	ZD-68		
523-2003-689	5071A	3334	ZD-6.8	561	
523-2003-750	5093A	3344	ZD-75		
523-2003-759	138A		ZD-7.5		
523-2003-820	150A		ZD-82		
523-2003-829	5072A		ZD-8.2	562	
523-2003-910	5095A	3345	ZD-91		
523-2003-919	139A		ZD-9.1	562	
523-2004-100	5125A	3391	5ZD-10		
523-2005-100	140A	3061	ZD-10	562	
523-2005-279	5063A	3837			
523-2005-369	134A		ZD-3.6		
523-2005-759	138A		ZD-7.5		
523-2005-829	5072A		ZD-8.2	562	
523-2005-919	139A		ZD-9.1	562	
523-2503-100	140A	3061	ZD-10		
523-2503-150	145A	3063	ZD-15	564	
523-2503-519		3056	ZD-5.1		
523-2503-689	5071A	3334	ZD-6.8	561	
523-2509-100	140A	3391/5125A	5ZD-10		
523-4001-001	156(4)	3051/156	512(4)		
523AB1AF1					MDA2501
523BB1BF1					MDA2502
523DB1AF1					MDA2504
523ES077M			63	2030	
523ES078M			63	2030	
523FB1AF1					MDA2500
523MB1AF1					MDA2506
524-457	109	3087		1123	
524WHT	123A	3444	20	2051	
525-212	125		510,531	1114	
525-24	116		504A	1104	
525-26	125(4)		510,531	1114	
525-498	116	3017B/117	504A	1104	
525-877	109	3087		1123	
525BF	5483	3942			
526-376	116	3017B/117	504A	1104	
527-062	113A		6GC1		
527-1			11	2015	
527-798	117		504A	1104	
528-325	117		504A	1104	
530-073-31		3057	ZD-5.6	561	
530-082-1003	117		504A	1104	
530-082-3	125		510,531	1114	
530-082-4	125		510,531	1114	
530-086-1	116		504A	1104	
530-088-1	116		504A	1104	
530-088-1002	125		510,531	1114	
530-088-1003	125		510,531	1114	
530-088-1004	125		510,531	1114	
530-088-2	117		504A	1104	
530-088-3	125		510,531	1114	
530-094-1	116		504A	1104	
530-095-1	116		504A	1104	
530-098-1	116		504A	1104	
530-099-1	116		504A	1104	
530-099-3	116		504A	1104	
530-099-4	116		504A	1104	
530-106-1	116		504A	1104	
530-106-1001	125		510,531	1114	
530-109-1	116		504A	1104	
530-111-1	116	3016	504A	1104	
530-111-1001	125		504A	1114	
530-113-1	116		504A	1104	
530-119-1	503		CR-5		
530-120-1	116		504A	1104	
530-122-1	117		504A	1104	
530-132-1	118		CR-1		
530-136-T	125		510,531	1114	
532-341	116	3016	504A	1104	
532-341A	116	3016	504A	1104	
535A	107		11	2015	
536-1			39	2015	
536-1(RCA)	161	3716	39	2015	
536-2		3122	39	2015	
536-2(RCA)	123A	3444	20	2051	
536D	123A	3444	20	2051	
536D9	123A	3444	20	2051	
536F	123A	3444	20	2051	
536F(JVC)	199	3245	62	2010	
536FS	123A	3444	20	2051	
536FU	123A	3444	20	2051	
536G(WARDS)	123A	3444	20	2051	
536GT	199	3245	62	2010	
536GU(WARDS)	128		243	2030	
536J2F	116	3311	504A	1104	
537D	123A	3444	20	2051	
537E	123A	3444	20	2051	
537FS	108	3452	86	2038	
537FV	123A	3444	20	2051	
537FY	123A	3444	20	2051	
537J2F	116	3017B/117	504A	1104	

Industry Standard No.	ECG	SK	GE	RS 276-	MOTOR.
538J2F	116	3017B/117	504A	1104	
539J2F	116	3031A	504A	1104	
540-008	5804	3016	504A	1104	
540-010	116	3031A	510,531	1114	
540-013	144A		ZD-14	564	
540-014	125	3016	504A	1104	
540-015	5801	9005/5802	510	1143	
540-028	177		514	1122	
540-028-00	519		514	1122	
540-033-00	519		514	1122	
540J2F	116	3031A	504A	1104	
542-1033	128		243	2030	
542-1034	130		14	2041	
544-2002-008	912	3543	IC-172		
544-2003-002	748	3236			
544-2003-005	968	3593			
544-2004-001	OBS-NLA		IC-277		
544-2006-001	1115	3184	IC-278		
544-2006-011	1232	3852		703	
544-2009-555		3564	IC-269		
544-2020-002	987	3643			
544-3001-103	4001B		4001	2401	
544-3001-140	4049	4049			
546	108	3452	86	2038	
000546-1	123A	3122	20	2051	
547J2F	125	3033	510,531	1114	
549-1	106	3984	21	2034	
549-2	159	3466	82	2032	
550-026-00	123A	3444	20	2051	
550-027-00	159	3466	82	2032	
551	108	3452	86	2038	
551-008-00	941M		IC-265	007	
555	955M		IC-269	1723	
555-3			18	2030	
555-3(RCA)	128		243	2030	
555-4	128	3024	243	2030	
556J	978	3689			
556L	978	3689			
0557-010	116	3311	504A	1104	
559-1	159	3466	82	2032	
559-1516-001	128	3026	243	2030	
560-2	123A	3444	20	2051	
565-072	128		243	2030	
565-073				2048	
565-074	123A	3444	20	2051	
565-1	128	3024	243	2030	
566H	1103	3281			
566H(IC)	1052	3249			
566H-L	1052	3249			
566H-M	1052	3249			
566H-N	1052	3249			
566H2(IC)	1052	3249			
567-0003-011	123A	3122	20	2051	
570-004503	123A	3444	20	2051	
570-005503	123A	3444	20	2051	
570-1	159	3466	82	2032	
571-844	130	3027	14	2041	
571-844A	130	3027	14	2041	
572-0040-051	121	3009	239	2006	
572-1			20	2051	
572-683	123A	3444	20	2051	
573-469	123A	3444	20	2051	
573-472	108	3452	86	2038	
573-474	108	3452	86	2038	
573-474A	108	3452	86	2038	
573-475	108	3452	86	2038	
573-479	123A	3444	20	2051	
573-480	123A	3444	20	2051	
573-481	123A	3444	20	2051	
573-491	108	3452	86	2038	
573-494	108	3452	86	2038	
573-495	108	3452	86	2038	
573-507	108	3452	86	2038	
573-509	108	3452	86	2038	
573-515	124	3021	12		
573-518	160	3006	245	2004	
573-529	102A	3004	53	2007	
573-532	128	3024	243	2030	
574	159	3466	82	2032	
574-1			21	2034	
574-844	162	3438	35		
575-0004-035	190		217		
0575-005	109	3087		1123	
575-028	116	3016	504A	1104	
575-042	116	3016	504A	1104	
575-048	116	3016	504A	1104	
575-050	116	3016	504A	1104	
575C2	1140	3473	IC-138		
576-0001-002		3466	82	2032	
576-0001-003	102A	3004	53	2007	
576-0001-004	123	3024/128	20	2051	
576-0001-005	123	3124/289	20	2051	
576-0001-006	108	3452	86	2038	
576-0001-008	123A	3444	20	2051	
576-0001-009	102A	3004	53	2007	
576-0001-012	123A	3444	20	2051	
576-0001-013	159	3466	82	2032	
576-0001-014	102A	3004	53	2007	
576-0001-018	123A	3444	20	2051	
576-0002-001	241	3188A/182	57	2020	
576-0002-002	104	3009	16	2006	
576-0002-003	130		14	2041	
576-0002-004	102A	3004	53	2007	
576-0002-005	121	3717	239	2006	
576-0002-006	123A	3444	20	2051	
576-0002-008	159	3466	82	2032	
576-0002-009	105		4		
576-0002-010	105		4		
576-0002-011	184	3190	57	2017	
576-0002-012	102A	3004	53	2007	
576-0002-013	103A	3010	59	2002	
576-0002-026	241	3188A/182	57	2020	
576-0002-029	241	3188A/182	57	2020	
576-0003-001	108	3452	86	2038	
576-0003-002	108	3452	86	2038	
576-0003-003	108	3452	86	2038	
576-0003-004	108	3452	86	2038	
576-0003-005	108	3452	86	2038	
576-0003-006	108	3452	86	2038	
576-0003-007	108	3452	86	2038	
576-0003-008	126	3006/160	52	2024	
576-0003-009	126	3007	245	2024	
576-0003-010	160	3008	245	2004	
576-0003-011	123A	3444	20	2051	
576-0003-012	106	3118	21	2034	
576-0003-013	126	3007	52	2024	
576-0003-014	126	3007	52	2024	
576-0003-015	126	3007	52	2024	
576-0003-017	159	3466	82	2032	
576-0003-018	108	3452	86	2038	
576-0003-019	159	3466	82	2032	
576-0003-020	108	3452	86	2038	
576-0003-021	108	3452	86	2038	
576-0003-022	199		62	2010	
576-0003-023	161	3039/316	39	2015	
576-0003-024	126	3007	52	2024	
576-0003-025	160	3007	245	2004	
576-0003-026	161	3117	39	2015	
576-0003-027	108	3452	86	2038	
576-0003-028	108	3452	86	2038	
576-0003-029	107	3018	11	2015	
576-0003-12(NPN)	123A		20	2051	
576-0003-12(PNP)	159		82	2032	
576-0003-224	222	3065	FET-4	2036	
576-0004-001	329	3048	40	2012	
576-0004-004	329	3047	FET-2	2035	
576-0004-005	329	3048	46		
576-0004-006	329	3046	86		
576-0004-007	329	3048	46		
576-0004-008	311	3047	277		
576-0004-009	329	3048	46		
576-0004-010	123A	3122	20	2051	
576-0004-011	329	3048	46		
576-0004-012	329	3048	46		
576-0004-013	195A	3048/329	243	2030	
576-0004-02	128		243	2030	
576-0004-035	190	3046	217		
576-0004-104	329	3298/299	236		
576-0004-105	235	3197	215		
576-0005-001	6400A			2029	
576-0005-004	128		243	2030	
576-0006-003	312	3112	FET-1	2028	
576-0006-011	108	3452	86	2038	
576-0006-221	221	3050		2036	
576-0006-222	222	3050/221	FET-4	2036	
576-0006-227	222	3065	FET-4	2036	
576-0007-001	172A	3156	64		
576-001	102A	3004	53	2007	
576-001-013			21	2034	
576-002-001	186	3192	28	2017	
576-002-004			2	2007	
576-003-009	160	3006	50	2004	
576-0036-212	195A	3765	20	2051	
576-0036-213	195A	3048/329	46		
576-0036-847	123	3124/289	20	2051	
576-0036-913	195A	3765	46		
576-0036-916	123A	3444	20	2051	
576-0036-917	123A	3444	20	2051	
576-0036-918	108	3452	86	2038	
576-0036-919	108	3452	86	2038	
576-0036-920	123A	3444	20	2051	
576-0036-921	123A	3444	20	2051	
576-0040-051	121	3009	239	2006	
576-0040-251	130		14	2041	
576-0040-253	102A		53	2007	
576-0040-254	121	3717	239	2006	
576-005	102A	3004	53	2007	
576-1	233	3132	210	2009	
576-1(RCA)	161	3716	39	2015	
576-2000-278	105		4		
576-2000-990	126	3007	52	2024	
576-2000-993	126	3007	52	2024	
576-2001-970	105		4		
576-3	723		IC-15		
577B196	5635	3533			
577H(IC)	1082	3461			
577R819H01	128	3024	243	2030	
579R925H01	910		IC-251		
580-029	116	3016	504A	1104	
580R304H01	159	3466	82	2032	
586-024	909	3590	IC-249		
586-2	123A	3444	20	2051	
586-437	1171	3565			
586-617	74154	74154			
588-40-201	1100	3223	IC-92		
588-40-202	1142	3485	IC-128		
588-40-203	1087	3477	IC-103		
588U	312	3834/132	FET-1	2035	
590-591731	194	3275	220		
590-591811	194	3275	220		
590-593031	123A	3444	20	2051	
592A2	1195	3469			
593D742-1	123A		20	2051	
595-1	123A	3122	20	2051	
595-1(SYLVANIA)	123A	3444	20	2051	
595-2	123A	3444	20	2051	
595-2(SYLVANIA)	123A	3444	20	2051	
596-2	177		300	1122	
596-5	177	3100/519	300	1122	
597-1		3118	21	2034	
597-1(RCA)	159		82	2032	
599P3430	199	3245	62	2010	
600	116	3311	504A	1104	
600-104-308	158	3004/102A	53	2007	
600-188-1-13	123A	3444	20	2051	
600-188-1-20	123A	3444	20	2051	
600-188-1-21	128		243	2030	
600-188-1-22	194		220		
600-188-1-23	123A	3444	20	2051	
600-207-801	199	3124/289	62	2010	
600-224-605	199	3122	62	2010	
600-229-201	289A	3122	210	2051	
600-301-801	123A	3444	20	2051	
600A39	724		IC-86		
600C	177		300	1122	
600X0091-086	123A	3444	20	2051	
600X0092-086	108	3452	86	2038	
600X0093-086	161	3117	39	2015	
600X0094-086	161	3117	39	2015	
600X0095-086	159	3466	82	2032	
600X0096-066	109	3088		1123	
600X0097-066	109	3088		1123	
600X0099-066	116	3017B/117	504A	1104	
600X0100-066	177		300	1122	
600X0101-066	142A	3093	ZD-12	563	
600X0141-000			11	2015	
600X0143-000			11	2015	
600X0175-000			11	2015	
600X0195-000			11	2015	
601	116	3311	504A	1104	
601-0100792	128		243	2030	
601-0100793	123A	3444	20	2051	
601-0100810	234		65	2050	
601-0100865	7413	7413		1815	
601-040	160	3006	245	2004	
601-054	126	3008	52	2024	
601-065	103A	3010	59	2002	
601-1	123A	3444	20	2051	
601-1(DIODE)	177		300	1122	
601-1(RCA)	123A	3444	20	2051	
601-1(TRANSISTOR)	123A		20	2051	
601-113	108	3452	86	2038	
601-2	123A	3444	20	2051	
601C	177		300	1122	
601X0048-066	116	3032A	504A	1104	
601X0049-066	506	3125	511	1114	
601X0149-086	123A	3444	20	2051	
601X0150-066	109	3088		1123	

Industry Standard No.	ECG	SK	GE	RS 276-	MOTOR.
601X0151-066	109	3088		1123	
601X0152-066	116	3032A	504A	1104	
601X0224-066	116	3100/519	504A	1104	
601X0225-066	178MP	3100/519	300(2)	1122(2)	
601X0226-066	143A	3750	ZD-13	563	
601X0226-0666			ZD-12	563	
601X0227-066	116	3032A	504A	1104	
601X0375-006	506	3043	511	1114	
601X0402-038		3750	ZD-13	563	
601X0417-086	159	3466	82	2032	
0602	5878		5036		
602-032	121	3009	239	2006	
602-040	102A	3004	53	2007	
602-075	160	3006	245	2004	
602-113	108	3452	86	2038	
602-56	159	3466	82	2032	
602-60	159	3466	82	2032	
602-61	108	3452	86	2038	
602X0008-002	199	3124/289	62	2010	
602X0018-000	123A	3444	20	2051	
602X0018-002	199	3124/289	62	2010	
602X0019-000	116	3311	504A	1104	
603-020	160	3006	245	2004	
603-030	160	3006	245	2004	
603-040	160	3006	245	2004	
603-113	108	3452	86	2038	
604	123A	3444	20	2051	
604(SEARS)	123A	3444	20	2051	
604-030	160	3006	245	2004	
604-080	160	3006	245	2004	
604-113	108	3452	86	2038	
604B	116	3016	504A	1104	
604C	177		300	1122	
605	123	3124/289	20	2051	
605-030	102A	3004	53	2007	
605-113	108	3452	86	2038	
0606	5890		5044		
606-020	102A	3004	53	2007	
606-113	116	3016	504A	1104	
606-6003-101	519		514	1122	
606-6003-102	519		514	1122	
606-6021-101	177		300	1122	
606-9601-101	123A	3444	20	2051	
606-9602-101	123A	3444	20	2051	
607-030	123A	3444	20	2051	
607-113	116	3016	504A	1104	
608	116	3311	504A	1104	
608-030	116	3016	504A	1104	
608-101	116	3016	504A	1104	
608-112	127	3034	25		
608-113	116	3016	504A	1104	
608-2	128	3024	243	2030	
608-3			18	2030	
609-020	160	3006	245	2004	
609-030	116	3016	504A	1104	
609-112	123A	3444	20	2051	
609-113	116	3016	504A	1104	
610	116	3311	504A	1104	
610-001-103	116	3311	504A	1104	
610-017-706	177	3100/519	300	1122	
610-035	102A	3004	53	2007	
610-035-1	102A	3004	53	2007	
610-036	102A	3004	53	2007	
610-036-1	102A	3004	53	2007	
610-036-2	102A	3004	53	2007	
610-036-3	102A	3004	53	2007	
610-036-4	102A	3004	53	2007	
610-036-5	102A	3004	53	2007	
610-036-7	102A	3004	53	2007	
610-036-8	102A	3004	53	2007	
610-039	121	3009	239	2006	
610-039-1	121	3009	239	2006	
610-040	102A	3004	53	2007	
610-040-1	102A	3004	53	2007	
610-040-2	102A	3004	53	2007	
610-041	108	3452	86	2038	
610-041-1	108	3452	86	2038	
610-041-2	108	3452	86	2038	
610-041-3	108	3452	86	2038	
610-042	108	3452	86	2038	
610-042-1	108	3452	86	2038	
610-043	102A	3004	53	2007	
610-043-1	102A		53	2007	
610-043-2	102A		53	2007	
610-043-3	102A		53	2007	
610-043-4	102A		53	2007	
610-043-6	102A		53	2007	
610-043-7	102A		53	2007	
610-045	108	3452	86	2038	
610-045-1	108	3452	86	2038	
610-045-2	108	3452	86	2038	
610-045-3	123A	3444	20	2051	
610-045-4	123A	3444	20	2051	
610-046-7	102A		53	2007	
610-050	160	3006	245	2004	
610-050-1	160		245	2004	
610-050-2	160		245	2004	
610-050-3	160		245	2004	
610-051	160	3006	245	2004	
610-051-1	160		245	2004	
610-051-2	160		245	2004	
610-051-4	160		245	2004	
610-052	126	3008	52	2024	
610-052-1	126		52	2024	
610-053	160	3006	245	2004	
610-053-1	160		245	2004	
610-053-2	160		245	2004	
610-055	160	3006	245	2004	
610-055-1	160		245	2004	
610-055-2	160		245	2004	
610-055-3	160		245	2004	
610-056	126	3008	52	2024	
610-056-1	126		52	2024	
610-056-2	126		52	2024	
610-056-3	126		52	2024	
610-056-4	126		52	2024	
610-067	121	3009	239	2006	
610-067-1	121	3717	239	2006	
610-067-2	121	3717	239	2006	
610-067-3	121	3717	239	2006	
610-068	121	3009	239	2006	
610-068-1	121	3717	239	2006	
610-069	108	3452	86	2038	
610-069-1	108	3452	86	2038	
610-070	123A	3444	20	2051	
610-070-1	123A	3444	20	2051	
610-070-2	123A	3444	20	2051	
610-070-3	123A	3444	20	2051	
610-071	124	3021	12		
610-071-1	124		12		
610-072	108	3452	86	2038	
610-072-1	108	3452	86	2038	
610-072-2	108	3452	86	2038	
610-073	108	3452	86	2038	
610-073-1	108	3452	86	2038	
610-074	126	3008	52	2024	
610-074-1	126		52	2024	
610-076	123A	3444	20	2051	
610-076-1	123A	3444	20	2051	
610-076-2	123A	3444	20	2051	
610-077	123A	3444	20	2051	
610-077-1	123A	3444	20	2051	
610-077-2	123A	3444	20	2051	
610-077-3	123A	3444	20	2051	
610-077-4	123A	3444	20	2051	
610-077-5	123A	3444	20	2051	
610-078	123A	3444	20	2051	
610-078-1	123A	3444	20	2051	
610-079	102A	3004	53	2007	
610-079-1	102A		53	2007	
610-080	102A	3004	53	2007	
610-080-1	102A		53	2007	
610-083	129	3025	244	2027	
610-083-1	129		244	2027	
610-083-2	129		244	2027	
610-083-3	129		244	2027	
610C	116	3016	504A	1104	
612	116	3016	504A	1104	
612-1			28	2017	
612-16		3018	20	2051	
612-16(ZENITH)	108		86	2038	
612-16A	108	3452	86	2038	
612-1A5L	123A	3444	20	2051	
612-60039	126	3008	52	2024	
612-60039A	126	3008	52	2024	
612C	177		300	1122	
613		3018	20	2051	
613(ZENITH)	108		86	2038	
613-4	154		40	2012	
613-72	108	3452	86	2038	
614	116	3311	504A	1104	
614(ZENITH)	108		86	2038	
614-1	128		243	2030	
614-118	177		300	1122	
614-12	108	3018	86	2038	
614-2		3122	20	2051	
614-3	123A	3444	20	2051	
614A-1-5L	121	3717	239	2006	
614A1-5L				2006	
614C	177		300	1122	
614X1	102A	3008	53	2007	
614X10	102A	3004	53	2007	
614X2	102A	3008	53	2007	
614X3	102A	3008	53	2007	
614X4	102A	3008	53	2007	
614X5	102A	3004	53	2007	
614X6	102A	3004	53	2007	
614X7	102A	3004	53	2007	
614X8	103A	3010	59	2002	
614X9	102A	3004	53	2007	
615-1		3024	11	2015	
616	116	3311	504A	1104	
616-1			21	2034	
616C	177		300	1122	
617-10	123A	3444	20	2051	
617-117	152	3054/196	66	2048	
617-15	109	3087		1123	
617-156	109	3088		1123	
617-161	123A	3444	20	2051	
617-162	116		504A	1104	
617-163	116		504A	1104	
617-17	109	3087		1123	
617-29	123A	3444	20	2051	
617-46	116	3031A	504A	1104	
617-50	102A	3004	53	2007	
617-52	102A	3004	53	2007	
617-53	116	3031A	504A	1104	
617-54	160	3006	245	2004	
617-55	160		245	2004	
617-56	126	3006/160	52	2024	
617-57	160	3006	245	2004	
617-58	160		245	2004	
617-62	116(2)	3311	504A(2)	1104	
617-63	123A	3444	20	2051	
617-64	123A	3444	20	2051	
617-65	161	3716	39	2015	
617-66	116(2)		504A(2)	1104	
617-67	123A	3444	20	2051	
617-68	123A	3444	20	2051	
617-69	158		53	2007	
617-70	102A	3123	53	2007	
617-71	123A	3444	20	2051	
617-87	123	3122	20	2051	
618	116	3311	504A	1104	
618C	177		300	1122	
620	116	3311	504A	1104	
620-1	159	3466	82	2032	
620-56			21	2034	
620C	177		300	1122	
622	116	3311	504A	1104	
622-1	128	3024	243	2030	
622-1(RCA)	128		243	2030	
622-2	128		243	2030	
622C	177		300	1122	
623	307		274		
623(RCA)	152		66	2048	
623-0	307		274		
623-1	186	3192	28	2017	
623-2003-100	140A		ZD-10	562	
623A	307		274		
623B	307		274		
623C	307		274		
623D	307		274		
623SN	6356	6556			
624	116	3031A	504A	1104	
624(RCA)	153		69	2049	
624-0005	113A		6GC1		
624-0006	113A		6GC1		
624-0007	113A		6GC1		
624-0009	116	3016	504A	1104	
624-0010	116	3016	504A	1104	
624-0011			1N34AS	1123	
624-1	185	3191	58	2025	
625-1		3024	20	2051	
625-1(RCA)	128		243	2030	
626	123A	3444	20	2051	
626(RECT.)	125		510,531	1114	
626-1	123A	3444	20	2051	
627-1	159	3466	82	2032	
628-3	188		226	2018	
629-3	189		218	2026	
629A02	116	3017B/117	504A	1104	
630-002	109	3087		1123	
630-052	116	3017B/117	504A	1104	
630-076	123A	3444	20	2051	
630-077	140A	3061	ZD-10	562	
630-079	109	3087		1123	
630-086	148A		ZD-55		

Industry Standard No.	ECG	SK	GE	RS 276-	MOTOR.
631-1	129	3025	244	2027	
631-3(SYLVANIA)	129		244	2027	
632-1(SYLVANIA)	188		226	2018	
632-3	196	3054	241	2020	
634-1		3018	11	2015	
635	159	3466	82	2032	
635-1		3050	FET-4	2036	
635-1(RCA)	221			2036	
637(RCA)	162		35		
637-1	162	3079	35		
637-1(RCA)	162		35		
638	123A	3444	20	2051	
638-1	128	3085	243	2030	
638-1(RCA)	218		234		
638H	175		246	2020	
638HJ	175	3026	246	2020	
639		3024	20	2051	
639(ZENITH)	123A	3444	20	2051	
639CL	128	3024	243	2030	
640-1(SYLVANIA)	152		66	2048	
642-028	109	3087		1123	
642-068	113A		6GC1		
642-102	109	3087		1123	
642-116	160		245	2004	
642-117	102A		53	2007	
642-119	109	3087		1123	
642-126	177	3100/519	300	1122	
642-132	109	3087		1123	
642-147	160		245	2004	
642-150	102A		53	2007	
642-151	158		53	2007	
642-152	131		44	2006	
642-173	160		245	2004	
642-174	123A	3444	20	2051	
642-176	121	3009	239	2006	
642-186	138A		ZD-7.5		
642-199	109	3087		1123	
642-202	160		245	2004	
642-206	104	3719	16	2006	
642-207	160		245	2004	
642-216	166	9075	504A	1152	
642-217	131		44	2006	
642-219	116	3017B/117	504A	1104	
642-221	109	3088		1123	
642-229	107	3018	11	2015	
642-230	107		11	2015	
642-236	5071A		ZD-6.8	561	
642-242	123A	3444	20	2051	
642-246	123A	3444	20	2051	
642-254	107	3039/316	11	2015	
642-255	145A	3063	ZD-15	564	
642-260	107	3039/316	11	2015	
642-261	184	3190	57	2017	
642-264	104	3719	16	2006	
642-266	185	3191	58	2025	
642-268	107	3124/289	11	2015	
642-269	107		11	2015	
642-270	107		11	2015	
642-271	121MP	3013	239(2)	2006(2)	
642-272	121	3009	239	2006	
642-274	107		11	2015	
642-275	177	3100/519	300	1122	
642-277	103A	3835	59	2002	
642-281	177		300	1122	
642-304	117		504A	1104	
642-306	128		243	2030	
642-316	104	3719	16	2006	
642-319	123A	3444	20	2051	
642AS4076-101	107			2015	
642AS407G-101			11	2015	
643RIX	193		67	2023	
643RLX				2023	
644-1	185	3191	58	2025	
647-1	165		38		
0648R	5071A	3334	ZD-6.8	561	
649-1	165		38		
650	109	3087		1123	
650-105	102	3004	2	2007	
650-106	102	3004	2	2007	
650-107	102	3004	2	2007	
650-108	102	3003	2	2007	
650-109	103	3862	8	2002	
650-110	116	3017B/117	504A	1104	
650C0	5006A	3772	ZD-3.6		
650C1	5007A	3773	ZD-3.9		
650C2	5007A	3773	ZD-3.9		
650C5	5008A	3774	ZD-4.3		
650C6	5008A	3774	ZD-4.3		
651	5009A	3775	ZD-4.7		
651C0	5009A	3775	ZD-4.7		
651C1	5009A	3775	ZD-4.7		
651C2	5009A	3775	ZD-4.7		
651C3	5009A	3775	ZD-4.7		
651C4	5010A	3776	ZD-5.1		
651C5	5010A	3776	ZD-5.1		
651C6	5010A	3776	ZD-5.1		
651C7	5010A	3776	ZD-5.1		
651C8	5010A	3776	ZD-5.1		
651C9	5011A	3777	ZD-5.6	561	
652	5012A	3778	ZD-6.0	561	
652C0	136A	3057	ZD-5.6	561	
652C1	5011A	3057/136A	ZD-5.6	561	
652C2	5011A		ZD-5.6	561	
652C3	5012A	3778	ZD-6.0	561	
652C4	5012A	3778	ZD-6.0	561	
652C5	5012A	3778	ZD-6.0	561	
652C6	5012A	3778	ZD-6.0	561	
652C7	137A	3058	ZD-6.2	561	
652C8	5013A	3779	ZD-6.2	561	
000653	123A	3122	20	2051	
653-202	108	3452	86	2038	
653C1	5014A	3780	ZD-6.8	561	
653C2	5014A	3780	ZD-6.8	561	
653C3	5014A	3780	ZD-6.8	561	
653C4	5014A	3780	ZD-6.8	561	
653C6	5015A	3781	ZD-7.5		
653C7	5015A	3781	ZD-7.5		
653C9	5016A	3782	ZD-8.2	562	
654-1			FET-2	2035	
654-1(SYLVANIA)	312	3448	FET-1	2035	
654C9	5018A	3784	ZD-9.1	562	
655C9	5019A	3785	ZD-10	562	
656-136	102A		53	2007	
656-137	102A		53	2007	
656-138	158		53	2007	
656-139	102A		53	2007	
656-141	116		504A	1104	
656-142	109	3087		1123	
656-2	182	3104A	55		
656-4	154	3044	40	2012	
657-31	161	3716	39	2015	
660-125	128	3024	243	2030	
660-126	123A	3444	20	2051	
660-127	108	3452	86	2038	
660-128	128	3047	243	2030	
660-131	123A	3444	20	2051	
660-134	123	3124/289	20	2051	
660-145	128	3024	243	2030	
660-220	123A	3444	20	2051	
660-221	123A	3444	20	2051	
660-222	123A	3444	20	2051	
660-224	158	3004/102A	53	2007	
660-225	123A	3444	20	2051	
660-227	102A	3004	53	2007	
660-228	103A	3010	59	2002	
660-230	116	3031A	504A	1104	
660B	102	3004	2	2007	
662A21				2007	
662A4				2007	
665-1A5L	160		245	2004	
665-2A5L	103A		59	2002	
665-4A5L	101	3861	8	2002	
665A-1-5L	160		245	2004	
665B-1-5L	126		52	2024	
665C-1-5L	121	3717	239	2006	
666-1		3114	21	2034	
666-1(RCA)	128		243	2030	
666-1-A-5L	102A		53	2007	
666-2-A-5L	102A		53	2007	
666-3A-5L	102A		53	2007	
668CS	108	3452	86	2038	
669	159	3466	82	2032	
669-1(RCA)	165		38		
669A459H01	910D		IC-252		
669A469H01	911D		IC-254		
673-1					MDA920A3
673-1S					MDA101A
673-2					MDA920A4
673-2S					MDA102A
673-3					MDA920A5
673-3S					MDA104A
673-4					MDA920A6
673-4S					MDA104A
673-5					MDA920A7
673-5S					MDA106A
673-6					MDA920A7
673-6S					MDA106A
674-3A5L	100	3721	1	2007	
675-153	158	3004/102A	53	2007	
675-154	102A	3004	53	2007	
675-155	102A	3004	53	2007	
675-156	102A	3004	53	2007	
675-158	109			1123	
675-206	131MP	3840	44(2)	2006(2)	
676-1	233	3246/229	210	2009	
677A-1-5L	105		4		
677B-1-5L	121MP		239(2)	2006(2)	
677C-1-5L	121MP	3718	239(2)	2006(2)	
679-1	773				MDA2501
679-2	5340				MDA2502
679-3	5342				MDA2504
679-4	5342				MDA2504
679-5	5342				MDA2506
679-6	5342				MDA2506
680-1	753	3018	39	2015	MDA2501
680-1(RCA)	161	3716	39	2015	
680-1(TRANSISTOR)	161		39	2015	
680-2	5340				MDA2502
680-3					MDA2504
680-4					MDA2504
680-5					MDA2506
680-6					MDA2506
684-652423-1	116		504A	1104	
684A-1-5L	121	3717	239	2006	
0684R			ZD-6.8	561	
686-0012	128	3024	243	2030	
686-0112	128	3024	243	2030	
686-0130	128	3024	243	2030	
686-0165	128		243	2030	
686-0210	175	3054/196	246	2020	
686-0210-0	175	3026	246	2020	
686-0243	130	3027	14	2041	
686-0243-0	130	3027	14	2041	
686-0325-0	159	3466	82	2032	
686-143	130	3027	14	2041	
686-143A	130	3027	14	2041	
686-229-0	128		243	2030	
686-257-0	199	3245	62	2010	
686-2700	159	3466	82	2032	
687-1	191	3232	249		
688B-1-5L	160		245	2004	
688C-1-5L	100	3721	1	2007	
690C092H88			504A	1104	
690L-021H25	152	3893	66	2048	
690L-021H26	153		69	2049	
690L270H02	121	3009	239	2006	
690L297H01	160		245	2004	
690L297H02	121	3009	239	2006	
690V0103H27	123A	3444	20	2051	
690V010H40	161	3018	39	2015	
690V010H41	108	3452	86	2038	
690V010H42	126	3005	52	2024	
690V028H28	108	3452	86	2038	
690V028H48	108	3452	86	2038	
690V028H69	108	3452	86	2038	
690V028H89	108	3452	86	2038	
690V02H69	108	3452	86	2038	
690V031H33	116	3016	504A	1104	
690V034H29	126	3008	52	2024	
690V034H30	102A	3004	53	2007	
690V034H31	102A	3004	53	2007	
690V034H32	109	3087		1123	
690V034H39	102	3004	2	2007	
690V034H74	109	3087		1123	
690V037H91		3087		1123	
690V037H92	120	3110	CR-3		
690V038H22	156	3016	512		
690V038H23	113A		6GC1		
690V039H17	119	3109	CR-2		
690V039H18		3066	CR-1		
690V039H22	156		512		
690V039H52	116	3016	504A	1104	
690V039H54	120	3110	CR-3		
690V040H56		3006	51	2004	
690V040H57	126	3006/160	52	2024	
690V040H58	126	3006/160	52	2024	
690V040H59	126	3006/160	52	2024	
690V040H60	126	3006/160	52	2024	
690V040H61	102	3004	2	2007	
690V040H62	102	3004	2	2007	
690V040H63	109			1123	
690V041H08	116	3016	504A	1104	
690V043H62	158	3004/102A	53	2007	
690V043H63	158	3004/102A	53	2007	
690V047H54	126	3008	52	2024	
690V047H55	126	3008	52	2024	
690V047H56	100	3005	1	2007	
690V047H57	100	3005	1	2007	

Industry Standard No.	ECG	SK	GE	RS 276-	MOTOR.
690V047H58	102	3004	2	2007	
690V047H59	102A	3004	53	2007	
690V047H60	102	3004	2	2007	
690V047H61	109	3087		1123	
690V047H97	123A	3444	20	2051	
690V049H81	108	3452	86	2038	
690V052H23	102	3003	2	2007	
690V052H24	102	3004	2	2007	
690V052H50	109	3087		1123	
690V052H63	160		245	2004	
690V052H68	109	3087		1123	
690V053H57	116	3031A	504A	1104	
690V054H20	102A		53	2007	
690V054H21	102	3004	2	2007	
690V056H27	160	3007	245	2004	
690V056H29	160	3007	245	2004	
690V056H30	160	3007	245	2004	
690V056H31	160	3008	245	2004	
690V056H32	160	3008	245	2004	
690V056H33	102A	3004	53	2007	
690V056H34	102A	3004	53	2007	
690V056H89	126	3008	52	2024	
690V056H90	102	3004	2	2007	
690V057H25	160	3006	245	2004	
690V057H27	102	3004	2	2007	
690V057H28	102	3004	2	2007	
690V057H59	126	3006/160	52	2024	
690V057H62	126	3006/160	52	2024	
690V059H20	102A	3004	53	2007	
690V059H21	158	3004/102A	53	2007	
690V059H52	102A	3004	53	2007	
690V059H55	102A	3004	53	2007	
690V059H63	109	3088		1123	
690V060H58	108	3452	86	2038	
690V060H59	108	3452	86	2038	
690V061H98	102	3004	2	2007	
690V061H99	102	3004	2	2007	
690V062H47	102A		53	2007	
690V063H14	126	3008	52	2024	
690V063H15	126	3008	52	2024	
690V063H16	102	3004	2	2007	
690V063H17	102	3004	2	2007	
690V063H50	102A	3004	53	2007	
690V063H51	102	3004	2	2007	
690V066H44	126	3006/160	52	2024	
690V066H45	126	3006/160	52	2024	
690V066H46	102	3004	2	2007	
690V066H47	102	3004	2	2007	
690V066H48	109	3087		1123	
690V066H49	126	3006/160	52	2024	
690V066H89	160	3007	245	2004	
690V067H09	109	3087		1123	
690V067H35	103A	3835	59	2002	
690V068H29	126	3008	52	2024	
690V068H30	102	3004	2	2007	
690V068H31	102	3004	2	2007	
690V068H32	109			1123	
690V069H39	116	3016	504A	1104	
690V070H49	108	3452	86	2038	
690V070H98	108	3452	86	2038	
690V073H59	126	3006/160	52	2024	
690V073H60	109	3087		1123	
690V073H85	126	3008	52	2024	
690V075H62	128	3024	243	2030	
690V075H68	108	3452	86	2038	
690V077H34	126	3006/160	52	2024	
690V077H35	126	3008	52	2024	
690V077H36	126	3006/160	52	2024	
690V077H37	102	3004	2	2007	
690V077H73	102A		53	2007	
690V080H36	161	3018	39	2015	
690V080H37	126	3006/160	52	2024	
690V080H38	154	3045/225	40	2012	
690V080H39	102	3006/160	2	2007	
690V080H40	126	3006/160	52	2024	
690V080H41	123	3124/289	20	2051	
690V080H42	127	3034	25		
690V080H43	127	3034	25		
690V080H44	102	3004	2	2007	
690V080H45	163A	3111	36		
690V080H47	116	3017B/117	504A	1104	
690V080H49	506	3125	511	1114	
690V080H50	506	3032A	511	1114	
690V080H51	506	3843	511	1114	
690V080H52	116	3017B/117	504A	1104	
690V080H53	167	3647	504A	1172	
690V080H54	156	3032A	512		
690V080H91	116	3017B/117	504A	1104	
690V081H07	108	3452	86	2038	
690V081H08	160		245	2004	
690V081H09	121	3009	239	2006	
690V081H40	166	9075		1152	
690V081H91	506	3125	511	1114	
690V081H92	145A	3063	ZD-15	564	
690V081H96	103A	3835	59	2002	
690V081H97	104	3009	16	2006	
690V082H40			504A	1104	
690V082H47	158	3004/102A	53	2007	
690V083H89	109	3087		1123	
690V084H60	126	3004/102A	52	2024	
690V084H61	102A	3004	53	2007	
690V084H62	123A	3444	20	2051	
690V084H63	102A	3004	53	2007	
690V084H94	108	3452	86	2038	
690V084H95	108	3452	86	2038	
690V084H96	108	3452	86	2038	
690V085	158	3004/102A	53	2007	
690V085H42	126	3006/160	52	2024	
690V085H44	158	3004/102A	53	2007	
690V086H39	102	3004	2	2007	
690V086H51	123A	3444	20	2051	
690V086H52	108	3452	86	2038	
690V086H86	159	3466	82	2032	
690V086H87	108	3452	86	2038	
690V086H88	123A	3444	20	2051	
690V086H89	129	3025	244	2027	
690V086H90	128	3024	243	2030	
690V086H91	116	3017B/117	504A	1104	
690V086H92	116	3031A	504A	1104	
690V086H94	161	3039/316	39	2015	
690V086H95	161	3039/316	39	2015	
690V086H96	108	3452	86	2038	
690V088H20	177	3100/519	300	1122	
690V088H44	108	3452	86	2038	
690V088H45	108	3452	86	2038	
690V088H46	107	3018	11	2015	
690V088H47	107	3018	11	2015	
690V088H48	108	3452	86	2038	
690V088H49	107	3018	11	2015	
690V088H50	123A	3444	20	2051	
690V088H51	123A	3444	20	2051	
690V088H52	102A	3004	53	2007	
690V089H46	107	3018	11	2015	
690V089H86	107	3018	11	2015	
690V089H89	123A	3444	20	2051	
690V089H90	102A	3004	53	2007	
690V089H91	116		504A	1104	
690V08H36			10	2051	
690V092H52	123A	3444	20	2051	
690V092H54	123A	3444	20	2051	
690V092H81	123A	3444	20	2051	
690V092H84	123A	3444	20	2051	
690V092H85	109	3088		1123	
690V092H88	117		504A	1104	
690V092H96	123A	3444	20	2051	
690V092H97	123A	3444	20	2051	
690V094H17	130	3021/124	14	2041	
690V094H18	102A	3004	53	2007	
690V094H19	102A	3004	53	2007	
690V094H20	102A	3004	53	2007	
690V094H21	123A	3444	20	2051	
690V094H35	116	3311	504A	1104	
690V097H59	102A	3004	53	2007	
690V097H62	123A	3444	20	2051	
690V098H48	123A	3444	20	2051	
690V098H49	123A	3444	20	2051	
690V098H50	123A	3444	20	2051	
690V098H51	131		44	2006	
690V098H52	109	3087		1123	
690V098H53	116	3017B/117	504A	1104	
690V099H59	158	3004/102A	53	2007	
690V099H79	123A	3444	20	2051	
690V102H39	103A	3010	59	2002	
690V102H40	177		300	1122	
690V102H71	123A	3444	20	2051	
690V102H72		3126	90		
690V102H96	126	3005	52	2024	
690V103H23	108	3452	86	2038	
690V103H24	108	3452	86	2038	
690V103H25	108	3452	86	2038	
690V103H26	108	3452	86	2038	
690V103H27	108	3452	86	2038	
690V103H28	107	3018	11	2015	
690V103H29		3020	20	2051	
690V103H30	128		243	2030	
690V103H31	123A	3444	20	2051	
690V103H32		3020	20	2051	
690V103H33	123A	3444	20	2051	
690V103H53	142A		ZD-12	563	
690V103H54	109	3088		1123	
690V104H53	102A	3004	53	2007	
690V104H54	102A	3004	53	2007	
690V105H-21			51	2004	
690V105H-24			52	2024	
690V105H-25			20	2051	
690V105H19			20	2051	
690V105H21			51	2004	
690V105H22			20	2051	
690V105H24			2	2007	
690V105H25			20	2051	
690V105H26			51	2004	
690V105H27			17	2051	
690V105H28			17	2051	
690V105H29			17	2051	
690V105H31			504A	1104	
690V105H32	139A	3060	ZD-9.1	562	
690V109H44	116	3031A	504A	1104	
690V109H46	107	3018	11	2015	
690V109H72		3126	90		
690V110H30	108	3452	86	2038	
690V110H31	108	3452	86	2038	
690V110H32	108	3452	86	2038	
690V110H33	108	3452	86	2038	
690V110H34	128	3124/289	243	2030	
690V110H36	128	3124/289	243	2030	
690V110H55	129	3025	244	2027	
690V110H88		3088	1N60	1123	
690V110H89		3054	28	2017	
690V111H63		3126	90		
690V114H29	108	3452	86	2038	
690V114H30	123A	3444	20	2051	
690V114H31	108	3452	86	2038	
690V114H33	123A	3444	20	2051	
690V114H36	116	3311	504A	1104	
690V116H19	108	3452	86	2038	
690V116H20	107	3018	11	2015	
690V116H21	123A	3444	20	2051	
690V116H22	312		FET-1	2028	
690V116H23	159	3466	82	2032	
690V116H24	128		243	2030	
690V116H25	185	3191	58	2025	
690V116H26	184		57	2017	
690V116H40	142A		ZD-12	563	
690V116H41	116		504A	1104	
690V118H57	177	3100/519	300	1122	
690V118H58	116	3100/519	504A	1104	
690V118H59	108	3452	86	2038	
690V118H60	159	3466	82	2032	
690V118H61	159	3466	82	2032	
690V118H62	102A		53	2007	
690V119H13		3126	90		
690V119H14	116	3088	504A	1104	
690V119H15	177	3100/519	300	1122	
690V119H54	158	3004/102A	53	2007	
690V119H94	126	3008	52	2024	
690V119H95	160	3006	245	2004	
690V119H96	160	3006	245	2004	
690V119H97	160	3008	245	2004	
690V120H89		3018	20	2051	
690V38H25	358		42		
690V68H32	109	3087		1123	
690V73H60	109	3087		1123	
690V037H91	109			1123	
691C844	128		243	2030	
693EP	123A	3444	20	2051	
693FS	123A	3444	20	2051	
693G	123A	3444	20	2051	
693GT	123A	3444	20	2051	
693GV	199	3124/289	62	2010	
694D	123A	3444	20	2051	
694E	123A	3444	20	2051	
698V102H39			8	2002	
699	123A	3444	20	2051	
0700	116	3016	504A	1104	
700-04	312		FET-1	2028	
700-110	128	3192/186	28	2017	
700-113	130	3027	14	2041	
700-133	159	3466	82	2032	
700-134	123AP	3444/123A	20	2051	
700-135	123AP		243	2030	
700-136	129		244	2027	
700-137	116	3016	504A	1104	
700-154	85	3444/123A	20	2051	
700-155	199	3124/289	62	2010	
700-156	123AP	3245/199	62	2010	
700-157	289A	3019			
700-159	139A		ZD-9.1	562	
700-244	384	3026			

Industry Standard No.	ECG	SK	GE	RS 276-	MOTOR.
700-321	177		ZD-3.3		
700-325	85	3444/123A	20	2051	
700-848A	908		IC-248		
700A-858-318	199		62	2010	
700A-858-328	199	3245	62	2010	
700A858-285	116	3311	504A	1104	
700A858-286	116	3311	504A	1104	
700A858-318	123A	3444	20	2051	
700A858-319	199	3124/289	62	2010	
700A858-322	177		300	1122	
700A858-328	123A	3444	20	2051	
0701	116	3018	504A	1104	
0702	116	3017B/117	504A	1104	
702-0002	107		11	2015	
702-810	113A		6GC1		
0703	108	3452	86	2038	
703 056 (4)	123A			2051	
703-056(4)			20	2051	
703-1	128		243	2030	
703-2	128	3024	243	2030	
703B	128	3024	243	2030	
703HC	703A		IC-12		
0704	116	3122	504A	2051	
705-600007	5633	3938			
706BC	OBS-NLA		IC-277		
706BPC	OBS-NLA		IC-277		
707	241	3188A/182	57	2020	
707W00048	5651	3506/5642			
707W00073	5401	3638/5402			
707W00084	6402	3628			
707W00086	5009A	3775			
707W00087	5014A	3780			
707W00091	157	3747			
707W00093	152	3893			
707W00104	5004A	3770			
707W00107	5007A	3773			
707W00109	5009A	3775			
707W00110	5010A	3776			
707W00114	5014A	3780			
707W00116	5016A	3782			
707W00119	5019A	3785			
707W00177	5077A	3752			
707W00205	5011A	3777			
707W00218	172A	3156			
707W00227	5453	6753			
707W00243	5454	6754			
707W00244	5454	6754			
707W00274	5801	9004			
707W00322	5454	6754			
707W00345	5402	3638			
707W00405	5077A	3752			
709BE					MC1709G
709BH					MC1709F
709CE					MC1709CG
709CH					MC1709CF
709CJ	909D				MC1709CP2
709DC	909D	3590	IC-250		
709HC	909	3551	IC-249		
709PC	909D	3590	IC-250		
710	116	3311	504A	1104	
710BE					MC1710G
710CE	910				MC1710CG
710DC	910D		IC-252		
710HC	910	3553	IC-251		
710PC	910D		IC-252		
711-001	390		255	2041	
711BE					MC1711G
711BN					MC1711L
711CE	911				MC1711CG
711CJ	911D				MC1711CP
711DC	911D		IC-254		
711HC	911		IC-253		
711PC	911D		IC-254		
715EN	123A	3018	20	2051	
715FB	107	3039/316	11	2015	
715HC	915			017	
716HC	716		IC-208		
720-10	130MP	3029			
720-18	159	3466			
720-19	175	3538			
720-21	312	3112			
720-23	284	3621			
720-27	188	3199			
720-2N1302	101	3011			
720-2N1305	102	3721/100			
720-2N1371	102	3004			
720-2N1637	126	3008			
720-2N2147	121	3014			
720-2N2895	128	3024			
720-2N2923	199	3444/123A			
720-2N3440	396	3044/154			
720-2N3645	129	3025			
720-2N3904	123AP	3444/123A			
720-2N3906	159	3466			
720-2N4221	102A	3112			
720-30	199	3444/123A			
720-31	130	3027			
720-32	152	3054/196			
720-35	159	3466			
720-35019	102	3004	2	2007	
720-35019A	102A	3004	53	2007	
720-37	128	3024			
720-46	289A	3444/123A			
720-47	123A	3444			
720-48	159	3466			
720-49	199	3250/315			
720-50	159	3466			
720-51	123A	3444			
720-52	292	3441			
720-61	186	3083/197			
720-62	187	3274/153			
720-64	129	3025			
720-69	172A	3156			
720-71	183	3189A			
720-74	157	3747			
720DC	744		IC-215	2022	
720PC	744		IC-215	2022	
720SDC	744		IC-215	2022	
720SPC	744		IC-215	2022	
723BE					MC1723G
723CE					MC1723CG
723CJ					MC1723CL
723DC	923D	3165	IC-260	1740	
723HC	923	3164	IC-259		
723HM	923D	3164/923			
723PC	923D	3165	IC-260	1740	
725EHC	925		IC-261		
725HC	925		IC-261		
0727-50	116	3016	504A	1104	
729-3	128	3024	243	2030	
729DC	720	9014	IC-7		
729PC	720	9014	IC-7		
0731	128		243	2030	
732DC	718	3159	IC-8		

Industry Standard No.	ECG	SK	GE	RS 276-	MOTOR.
732PC	718	3159	IC-8		
733-0100-699	187		29	2025	
733-0			62	2010	
733W00021	941M		IC-265	007	
733W00022	74141	74141			
733W00035	941D		IC-264		
733W00039	7490	7490		1808	
733W00059	309K	3629			
733W00133	7493A	7493			
734EU	312	3834/132	FET-1	2035	2N4416
734U		3112			2N4416
737		3024	20	2051	
737(ZENITH)	123A	3444	20	2051	
739		3162	IC-19		
739DC	725	3162	IC-19		
739H01	123A	3444	20	2051	
739PC	725	3162	IC-19		
740-2001-306	1029		IC-162		
740-2001-307	722	3161	IC-9		
740-2002-111	708	3135/709	IC-10		
740-2007-120	1087	3477	IC-103		
740-5853-300	1005	3723	IC-42		
740-5903-301	1006	3358	IC-38		
740-8120-160	1005	3723	IC-42		
740-8160-190	1003	3288	IC-43		
740-9000-017	1047		IC-116		
740-9000-544	1142	3485			
740-9000-554	1142	3485	IC-128		
740-9000-566	1052	3249	IC-135		
740-9003-301	1006	3358	IC-38		
740-9007-046	1108		IC-87		
740-9007-205	1155	3231	IC-179	705	
740-9011-322	1037		IC-170		
740-9016-105	720	9014	IC-7		
740-9017-092	1023		IC-272		
740-9037-120	1087	3477	IC-103		
740-9037-204	1153	3282			
740-9607-205	1155	3231	IC-179	705	
740HC	940		IC-262		
741BE					MC1741G
741BH					MC1741F
741BN					MC1741L
741CE					MC1741CG
741DC	941D	3552/941M	IC-264		
741HC	941	3526/947	IC-263	010	
741PC	941D		IC-264		
741TC	941M	3552	IC-265	007	
742	109	3087		2015	
742-1	188		226	2018	
742C2030-020	130		14	2041	
743			40	2012	
744			18	2030	
746HC	707	3134/705A			
746PC	790	3454	IC-230		
747BE					MC1747G
747BN					MC1747L
747CE	947				MC1747CG
747DC	947D	3556			
747HC	947	3526	IC-268		
747PC	947D	3556			
748			20	2051	
748(ZENITH)	123A	3444	20	2051	
748BE	1171	3565			MC1748G
748CE	1171	3565			MC1748CG
748CV	975	3641			
748HC	1171	3565			
749	949	3166	IC-25		
749(BENDIX)	949		IC-25		
749-8160-190	1003		IC-43		
749DHC	949	3166	IC-25		
750-045	175		246	2020	
750-137	102	3004	2	2007	
750-138	102	3004	2	2007	
750-139	102	3004	2	2007	
750-140	102	3004	2	2007	
750-141	116	3031A	504A	1104	
750-35019	102A		53	2007	
750A858-285	116	3311	504A	1104	
750A858-319		3124	10	2051	
750A858-328	199	3124/289	62	2010	
750A858-448	152	3893	66	2048	
750C838-123	123A	3444	20	2051	
750C838-124	123A	3444	20	2051	
750C838-125	123A	3444	20	2051	
750C871-1	HIDIV-1	3868/DIV-1	FR-8		
750C871-2	HIDIV-1	3868/DIV-1	FR-8		
750C871-3	HIDIV-1	3868/DIV-1	FR-8		
750C871-4	HIDIV-1	3868/DIV-1	FR-8		
750C871-5	HIDIV-1	3868/DIV-1	FR-8		
750C871-6	HIDIV-1	3868/DIV-1	FR-8		
750C871-7	HIDIV-1	3868/DIV-1	FR-8		
750C871-8	HIDIV-1	3868/DIV-1	FR-8		
750D858-211	125		510,531	1114	
750D858-212	123A	3444	20	2051	
750D858-213	108	3452	86	2038	
750M63-104	160		245	2004	
750M63-105	158		53	2007	
750M63-115	102A		53	2007	
750M63-116	160		245	2004	
750M63-117	160		245	2004	
750M63-119	123A	3444	20	2051	
750M63-120	123A	3444	20	2051	
750M63-146	123A	3444	20	2051	
750M63-147	158		53	2007	
750M63-148	158		53	2007	
750M63-149	116		504A	1104	
751-2001-212	116		504A	1104	
751-6300-001	116		504A	1104	
751-9001-124	116	3311	504A	1104	
753-0100-699	187	3193	29	2025	
753-0101-047	108	3452	86	2038	
753-0101-226	186	3192	28	2017	
753-0160-699			29	2025	
753-1303-801	229	3246	61	2038	
753-1372-100	123A	3122	61	2038	
753-1644-100	199	3124/289	62	2010	
753-1828-001	123A	3444	20	2051	
753-2000-003	123A	3444	20	2051	
753-2000-004	123A	3444	20	2051	
753-2000-006	175		246	2020	
753-2000-007	229	3018	61	2038	
753-2000-008	123A	3444	20	2051	
753-2000-009	123A	3444	20	2051	
753-2000-011	123A	3444	20	2051	
753-2000-100	199	3122	62	2010	
753-2000-101	159	3466	82	2032	
753-2000-107	128	3020/123	243	2030	
753-2000-460	107	3018	11	2015	
753-2000-463	131MP	3840	44(2)	2006(2)	
753-2000-535	229	3016	61	2038	
753-2000-710	108	3452	86	2038	
753-2000-711	123A	3444	62	2051	
753-2000-735	123A	3444	20	2051	
753-2000-870	123A	3444	20	2051	

Industry Standard No.	ECG	SK	GE	RS 276-	MOTOR.
753-2000-871	123A	3444	20	2051	
753-2001-173	152	3054/196	66	2048	
753-2100-001	123A	3444	20	2051	
753-2100-002	131		44	2006	
753-2100-008	123A	3444	20	2051	
753-3000-535	229		61	2038	
753-4000-010	199	3124/289	62	2010	
753-4000-011	123A	3444	20	2051	
753-4000-024	312	3112		2028	
753-4000-025	312	3834/132	FET-1	2035	
753-4000-101	123A	3444	20	2051	
753-4000-537	123A	3444	20	2051	
753-4000-668	107	3018	11	2015	
753-4000-929	107	3018	11	2015	
753-4001-474	131		44	2006	
753-4001-931	184	3041	57	2017	
753-4001-932	152	3054/196	66	2048	
753-4004-248	159	3466	82	2032	
753-5751-359	107	3246/229	11	2015	
753-5751359	229		61	2038	
753-5851-359	107		11	2015	
753-6000-002	199	3124/289	62	2010	
753-6000-019	312	3834/132	FET-1	2035	
753-8400-230	199	3124/289	62	2010	
753-8500-380	199	3124/289	62	2010	
753-8510-470	128	3024	243	2030	
753-9000-019	312	3834/132	FET-1	2035	
753-9000-096	186	3192	28	2017	
753-9000-839	199	3018	62	2010	
753-9000-922	107	3122	11	2015	
753-9001-674	229	3018	61	2038	
753-9001-675	229	3124/289	213	2038	
753-9010-021	6400A		2N2160	2029	
753-9010-235	152	3893	66	2048	
753-9020-784	229	3018	60	2015	
753-9050-785	229	3018	60	2015	
753TC	736	3163	IC-17		
754-0102-139	611	3126	90		
754-1003-030	110MP	3088	1N34AS	1123	
754-2000-001	177	3032A	300	1122	
754-2000-002	116	3017B/117	504A	1104	
754-2000-005	116	3311	504A	1104	
754-2000-009	109	3088		1123	
754-2000-011	177	3100/519	300	1122	
754-2009-150	177	3100/519	300	1122	
754-2509-150	177	3100/519	300	1122	
754-2720-021	177	3100/519	300	1122	
754-4000-088	109	3087		1123	
754-4000-188	109			1123	
754-4000-410	109			1123	
754-4000-553	612	3325	90		
754-5000-021	116	3311	504A	1104	
754-5700-282			ZD-8.2	562	
754-5710-219	5072A	3136	ZD-8.2	562	
754-5750-282	5072A	3136	ZD-8.2	562	
754-5750-283	177	3100/519	300	1122	
754-5750-284	116	3311	504A	1104	
754-5850-284	5072A		ZD-8.2	562	
754-5853-300	1005	3723	IC-42		
754-5900-090	109	3088		1123	
754-8130-140		3126	90		
754-900-124	116		504A	1104	
754-9000-082	5072A	3136	ZD-8.2	562	
754-9000-351		3126	90		
754-9000-460	109	3088		1123	
754-9000-473	519	3100	300	1122	
754-9000-953	116	3311	504A	1104	
754-9001-124	116		504A	1104	
754-9002-687	614	3126	90		
754-9030-009	5073A	3749	ZD-9.1	562	
754-9052-473	519	3100	300	1122	
754-9053-090	139A		ZD-9.1	562	
755-422494	159		82	2032	
755-845049	519		514	1122	
756	116	3017B/117	504A	1104	
758DC	743		IC-214		
758PC	743		IC-214		
762-105-00	129		244	2027	
762-110	234		65	2050	
762-120	129		244	2027	
762-2	231	3857			
763-1	128		243	2030	
766-100999	390		255	2041	
767		3124	20	2051	
767(ZENITH)	123A	3444	20	2051	
767CL	123A	3444	20	2051	
767DC	722	3161	IC-9		
767PC	722	3161	IC-9		
0770	185	3083/197	58	2025	
770-045	152	3893	66	2048	
0772	184	3054/196	57	2017	
772-101-00	199	3245	62	2010	
772-110	123A	3444	20	2051	
772-120-00	128		243	2030	
772-121-00	128		243	2030	
772A	107	3018	11	2015	
772B	107	3018	11	2015	
772B1	107	3039/316	11	2015	
772BJ	107	3018	11	2015	
772BL	107	3018	11	2015	
772BM	107	3018	11	2015	
772BN	107	3018	11	2015	
772BY	107	3018	11	2015	
772C	107	3018	11	2015	
772CC	107		11	2015	
772D	107	3018	11	2015	
772D1	107	3039/316	11	2015	
772DC	107		11	2015	
772DG	107		11	2015	
772E	107	3018	11	2015	
772EH	107	3039/316	11	2015	
772F	107	3018	11	2015	
772FE	107	3039/316	11	2015	
772G	107	3018	11	2015	
0773	185	3041	20	2025	
773(ZENITH)	123A	3444	20	2051	
773CL	128	3024	243	2030	
773RED	108	3452	86	2038	
774		3025	21	2034	
774(ZENITH)	159		82	2032	
774CL	129	3025	244	2027	
774ORN	108	3452	86	2038	
775-1(SYLVANIA)	172A		64		
775BRN	108	3452	86	2038	
0776-0160	123A		20	2051	
0776-0195	159		82	2032	
776-1	159	3025/129			
776-1(SYLVANIA)	159		82	2032	
776-151	123A	3444	20	2051	
776-183	123A	3444	20	2051	
776-2(PHILCO)	123A	3444	20	2051	
776GRN	123A	3444	20	2051	
776Y	224		46		
779BLU	108	3452	86	2038	

Industry Standard No.	ECG	SK	GE	RS 276-	MOTOR.
780DC	714	3075	IC-4		
780PC	714	3075	IC-4		
780WHT	123A	3444	20	2051	
781DC	715	3076	IC-6		
781PC	715	3076	IC-6		
783RED	108	3452	86	2038	
784ORN	108	3452	86	2038	
785YEL	123A	3444	20	2051	
786	108	3452	86	2038	
787BLU	108	3452	86	2038	
787PC	797	3158	IC-233		
791	123A	3444	20	2051	
792-238	116	3031A	504A	1104	
792-286	102	3004	2	2007	
792-287	102	3004	2	2007	
792-288	102	3004	2	2007	
792-289	102	3004	2	2007	
792-290	102	3004	2	2007	
792-292	109	3087		1123	
800-000-025	500A	3304	527		
800-0004-P090	724		IC-86		
800-001-031-1	159	3466	82	2032	
800-001-034	123A	3444	20	2051	
800-001-106-1	128		243	2030	
800-002-00	110MP	3090/109	1N34AS	1123	
800-003-00	109	3087		1123	
800-004-00	142A	3062	ZD-12	563	
800-005-00	109	3087		1123	
800-006-00	116	3016	504A	1104	
800-007-00	113A		6GC1		
800-008-00	120		CR-3		
800-010-00	156		512		
800-011-00	119	3109	CR-2		
800-012-00	118	3066	CR-1		
800-013-00	116		504A	1104	
800-01300-504			504A	1104	
800-014-00	116	3031A	504A	1104	
800-014-01	116		504A	1104	
800-01400		3031A	504A	1104	
800-016-00	177	3031A	300	1122	
800-017-00	116(4)		504A(4)	1104	
800-018-00	156	3311	512		
800-018-02	156		512		
800-020-00	109	3087		1123	
800-021-00	177		300	1122	
800-022-00	109	3087		1123	
800-023-00	142A	3062	ZD-12	563	
800-024-00	116	3311	504A	1104	
800-028-00	5075A	3751	ZD-16	564	
800-029-30	506	3125	511	1114	
800-030-00	5071A	3334	ZD-6.8	561	
800-032-00	116	3017B/117	504A	1104	
800-034-00	5120A		5ZD-6.8		
800-035-00	142A	3062	ZD-12	563	
800-036-00	177	3100/519	300	1122	
800-037-00	156		512		
800-038-00	116	3017B/117	504A	1104	
800-039-00	109	3087		1123	
800-040-00	177	3100/519	300	1122	
800-041-00	116	3017B/117	504A	1104	
800-042-00	177	3100/519	300	1122	
800-043-00	156		512		
800-045-00	5414	3954			
800-046-00	116	3017B/117	504A	1104	
800-050-00	156		512		
800-051-00	156		512		
800-101-101-1	123A	3444	20	2051	
800-101-101-2	128		243	2030	
800-101-102-1	123A	3444	20	2051	
800-101-108-1	159	3466	82	2032	
800-101-114-1	129		244	2027	
800-102-001	519		514	1122	
800-102-101-1	519		514	1122	
800-122	124	3021	12		
800-158	124	3021	12		
800-172	124	3021	12		
800-180	124		12		
800-181	124		12		
800-196	104MP	3720	16(2)	2006(2)	
800-203	124	3021	12		
800-204	124	3021	12		
800-205	158		53	2007	
800-250-102	123	3124/289	20	2051	
800-253	121MP	3013	239(2)	2006(2)	
800-256		3103A	12		
800-256(METAL)	124		12		
800-256(PLASTIC)	157		232		
800-282	506	3843	511	1114	
800-284	129	3025	244	2027	
800-289		3054/196	57,58		
800-294		3024/128	243,244		
800-310(CP)(NPN)			59	2002	
800-310(CP)(PNP)			53	2002	
800-321	124	3021	12		
800-329	104	3719	16	2006	
800-401	124		12		
800-501-00	123A	3444	20	2051	
800-501-01	123A	3444	20	2051	
800-501-02	123A	3444	20	2051	
800-501-03	123A	3444	20	2051	
800-501-04	123A	3444	20	2051	
800-501-11	123A	3444	20	2051	
800-501-22	123A	3444	20	2051	
800-50100	123A	3124/289	20	2051	
800-502-00	102A	3004	53	2007	
800-503-00		3010	54		
800-50300	103A	3010	59	2002	
800-504-00	160	3006	245	2004	
800-50400	160	3006	245	2004	
800-505-00	102A	3005	53	2007	
800-50500	126	3008	52	2024	
800-506-00	103A	3010	59	2002	
800-50600	103A	3010	59	2002	
800-507-00	121	3009	239	2006	
800-50700	121	3009	239	2006	
800-508-00	123A	3444	20	2051	
800-509-00	123A	3444	20	2051	
800-510-00	130	3027	14	2041	
800-510-01	130	3027	14	2041	
800-511-00	129	3025	244	2027	
800-512-00	128	3024	243	2030	
800-513-00	128	3024	243	2030	
800-514-00	123A	3444	20	2051	
800-515-00	128	3024	243	2030	
800-51500	123A	3444	20	2051	
800-516-00	129	3025	244	2027	
800-51600	129	3025	244	2027	
800-517-00	109	3087		1123	
800-518-00	121	3009	239	2006	
800-521-01	123A	3444	20	2051	
800-521-02	128	3024	243	2030	
800-521-03	128	3024	243	2030	
800-52102	123A	3444	20	2051	
800-522-01	123A	3124/289	20	2051	

Industry Standard No.	ECG	SK	GE	RS 276-	MOTOR.
800-522-02	123A	3444	20	2051	
800-522-03	128	3024	243	2030	
800-522-04	123A	3444	20	2051	
800-52202	123A	3444	20	2051	
800-523-01	159	3466	82	2032	
800-523-02	159	3466	82	2032	
800-52302	123A	3444	20	2051	
800-524-02	130	3027	14	2041	
800-524-02A	130	3027	14	2041	
800-524-03	130	3027	14	2041	
800-524-03A	130	3027	14	2041	
800-524-04A	130		14	2041	
800-52402	130	3027	14	2041	
800-525-03	129	3124/289	244	2027	
800-525-04	159	3466	82	2032	
800-525-04A	130	3027	14	2041	
800-526-00	123A	3444	20	2051	
800-52600	123A	3444	20	2051	
800-527-00	159	3466	82	2032	
800-52700	129	3025	244	2027	
800-528	103A	3010	59	2002	
800-528-00	103A	3010	59	2002	
800-52800	103A	3010	59	2002	
800-529-00	123A	3444	20	2051	
800-530-00	123A	3444	20	2051	
800-530-01	123A	3444	20	2051	
800-53001	123	3124/289	20	2051	
800-533-00	152	3054/196	66	2048	
800-533-01	184	3190	57	2017	
800-53300	128	3024	243	2030	
800-534-00	123A	3444	20	2051	
800-534-00(XSTR)			10	2051	
800-534-01	123A	3444	20	2051	
800-53400	123A	3444	20	2051	
800-535-00	312	3834/132	FET-1	2035	
800-535-01	312	3834/132	FET-1	2035	
800-536-00	108	3452	86	2038	
800-53600	108	3018	86	2038	
800-537-01	103A	3835	59	2002	
800-537-02	130MP	3027/130	15	2041(2)	
800-537-03	158	3027/130	53	2007	
800-53702	130MP	3029	15	2041(2)	
800-538-00	123A	3444	20	2051	
800-544-00	123A	3444	20	2051	
800-544-10	123A	3444	20	2051	
800-544-20	123A	3444	20	2051	
800-544-30	123A	3444	20	2051	
800-546-00	152	3054/196	66	2048	
800-547-00	159		69	2032	
800-548-00	123A	3444	20	2051	
800-550-00	162	3438	20	2051	
800-550-10	163A	3439	36		
800-552-00	191		249		
800-557-00	161	3018	39	2015	
800-616	HIDIV-3	3870/DIV-3	FR-10		
800-669B			243	2030	
800-669B(NPN)			47	2030	
800-764	165	3710/238			
800-791	521	3304/500A			
801-04900	126	3008	52	2024	
801-05200	126	3008	52	2024	
801-05300	126	3008	52	2024	
801B	123A	3444	20	2051	
802-05400	102A	3004	53	2007	
802-05600	102A	3004	53	2007	
802-22631U	102A	3004	53	2007	
802-36400	102A	3004	53	2007	
802-36430	102A	3004	53	2007	
802-43900	102A	3004	53	2007	
803-18250	123A	3444	20	2051	
803-37390	123A	3444	20	2051	
803-38030	108	3452	86	2038	
803-38040	108	3452	86	2038	
803-39430	108	3452	86	2038	
803-39440	108	3452	86	2038	
803-83840	108	3452	86	2038	
803-83930	108	3452	86	2038	
803-83940	123A	3444	20	2051	
803-89910	123A	3444	20	2051	
803-89930	123A	3444	20	2051	
804	123A	3444	20	2051	
804-14120	130	3027	14	2041	
804-14130	130	3027	14	2041	
804-21980	123A	3444	20	2051	
804A765-2	5809	9010			
805-1060	109	3087		1123	
805-10600	109			1123	
807-1(PHILCO)	194		220		
808-206	116	3311	504A	1104	
808-304	102A	3004	53	2007	
808-305	102A	3004	53	2007	
808-306	102A	3004	53	2007	
808-307	102A	3004	53	2007	
808-308	102A	3004	53	2007	
808-309	103A	3010	59	2002	
808-310	102A	3004	53	2007	
808-311	102A	3004	53	2007	
808-312	109	3087		1123	
809BE					MC1776G
809CE					MC1776CG
810	788	3147	IC-229		
815-1810	102A		53	2007	
815-181D	102	3004	2	2007	
817-275	175		246	2020	
818WHT	123A	3444	20	2051	
819-1	128		243	2030	
822	123A	3444	20	2051	
822-1	116	3017B/117	504A	1104	
822-1(SYLVANIA)	161	3716	39	2015	
822-2	116	3017B/117	504A	1104	
822-3	116	3017B/117	504A	1104	
822-4	116	3016	504A	1104	
822-5	116	3016	504A	1104	
822A	123A	3444	20	2051	
822ABLU	123A	3444	20	2051	
822B	123A	3444	20	2051	
823AE					MC1723G
823B	123A	3444	20	2051	
823WHT	123A	3444	20	2051	
824-09501	123A	3444	20	2051	
824-1		3018	18	2030	
824-1(SYLVANIA)	161	3716	39	2015	
824-10300	129	3025	244	2027	
826-1	159	3466	82	2032	
827BRN	123A	3444	20	2051	
828GRN	123A	3444	20	2051	
828S	123A	3444	20	2051	
829	159	3466	82	2032	
829A	159	3114/290	223		
829B	159	3466	82	2032	
829C	159	3466	82	2032	
829D	159	3466	82	2032	
829DE				2032	
829E	159	3466	82	2032	
829F	159	3466	82	2032	
830	102A	3123	53	2007	
0831	234	3114/290	65	2050	
833	159	3466	82	2032	
834-250-011	152	3893	66	2048	
834-6066	123A	3444	20	2051	
840-0300-577	1082	3461			
847BLK	123A	3444	20	2051	
851-0372-130	116	3031A	504A	1104	
852-92-43	185	3191			
853-0300-632	123A	3444	20	2051	
853-0300-634	187	3193	29	2025	
853-0300-643	193	3025/129	67	2023	
853-0300-644	199	3124/289	62	2010	
853-0300-900	123A	3444	20	2051	
853-0300-923	123A	3444	20	2051	
853-0301-096	186A	3357	28	2017	
853-0301-317	289A	3122	268	2038	
853-0373-110	123A	3444	20	2051	
854-0372-020	116	3031A	504A	1104	
858	199	3245	62	2010	
858GS	123A	3444	20	2051	
859GK	199	3245	62	2010	
860-022-01	181		75	2041	
863-254B	177	3100/519	300	1122	
863-567B	177	3100/519	300	1122	
863-776B	177	3100/519	300	1122	
866-20-2				2004	
866-6	130	3054/196	14	2041	
866-6(BENDIX)	130		14	2041	
880-101-00	703A	3157	IC-12		
880-102-00	726	3129			
880-207-000	116	3017B/117	504A	1104	
880-250-001				2006	
880-250-010				2006	
880-250-011				2017	
880-250-102	123A	3444	20	2051	
880-250-107	129	3025	244	2027	
880-250-108	123A	3444	20	2051	
880-250-109	123A	3444	20	2051	
880-250108				2010	
881-250-102	123A	3444	20	2051	
881-250-107	129	3025	244	2027	
881-250-108	123A	3444	20	2051	
881-250-109	123A	3444	20	2051	
881-250108	199	3038	62	2010	
884-250-001	121	3009	239	2006	
884-250-010	121	3009	239	2006	
884-250-011	184	3024/128	57	2017	
901-000-6-51	102A	3004	53	2007	
901-5553-7	612		90		
902-000-2-04	199	3020/123	210	2051	
902-000-8-04	289A	3124/289	210	2051	
902-001-7-18	229	3018	60		
902-002-3-06	199	3124/289	62	2010	
902-003-0-012			213	2038	
902-003-0-12	229	3124/289	213		
902-003-3-17	199	3124/289	212	2010	
902-003-6-006			62	2010	
902-003-6-06	199	3122	62		
903			39	2015	
903-00390	109	3087	1N34AS	1123	
903-00391	610	3323			
903-00393	177	9091	300	1122	
903-00394	116	3311	504A	1104	
903-100B	177	3100/519	300	1122	
903-101B		3126	90		
903-103B	109	3088		1123	
903-104B	116	3311	504A	1104	
903-105B	116	3311	504A	1104	
903-108	109		1N34AS	1123	
903-108B	109	3088		1123	
903-10B	109			1123	
903-110B	611	3126	90		
903-112B	177	3100/519	300	1122	
903-113B	109	3088		1123	
903-114B	109	3087		1123	
903-115B	109			1123	
903-116B	177	3100/519	300	1122	
903-117B	116		504A	1104	
903-118B	177	3100/519	300	1122	
903-119B	139A	3060	ZD-9.1	562	
903-120B	136A	3057	ZD-5.6	561	
903-121B	177	3100/519	300	1122	
903-123B		3126	90		
903-12B	109	3087		1123	
903-136B	177	3100/519			
903-137B	614	3126			
903-13B	117	3088	504A	1104	
903-141B	614	3126	90		
903-14B	116	3311	504A	1104	
903-156B	116	3311	504A	1104	
903-15B	116	3311	504A	1104	
903-164B	116	3311	504A	1104	
903-166B	5071A	3334	ZD-6.8	561	
903-167B	110MP	3088	1N60	1123(2)	
903-168B	109	3088	1N60	1123	
903-169B	116	3031A	504A	1104	
903-16B	109	3088		1123	
903-17	156	3032A	512		
903-1718	614	3327			
903-171B	614	3088	90		
903-177B	177	3100/519	300	1122	
903-178	177	3100/519	300	1122	
903-179	139A	3060	ZD-9.1	562	
903-17A	156	3032A	512		
903-17B	156	3311	512		
903-17C	156	3032A	512		
903-180	5071A	3334	ZD-6.8	561	
903-18B	109	3087		1123	
903-197	116		504A	1104	
903-19B		3126	90		
903-1B		3126	90		
903-20	177		300	1122	
903-208	5304		504A		
903-212	109		1N34AS	1123	
903-23	109	3087		1123	
903-23A	109	3087		1123	
903-23B	109	3087		1123	
903-23C	109	3087		1123	
903-23D	109	3087		1123	
903-23E	109	3087		1123	
903-25B	177		300	1122	
903-26B	134A		ZD-3.6		
903-27	109	3088	1N60	1123	
903-27B	109	3088		1123	
903-28B	116	3311	504A	1104	
903-29B	109	3088		1123	
903-3	123A	3444	20	2051	
903-303	116	3311	504A	1104	
903-30B	109	3087		1123	
903-311	177	3175	300	1122	
903-330	116		504A	1104	
903-332	177	3175	300	1122	

Industry Standard No.	ECG	SK	GE	RS 276-	MOTOR.
903-333	136A	3057	ZD-5.6	561	
903-334	156	3051	512		
903-335	5071A	3334	ZD-6.8	561	
903-337	140A	3061	ZD-10	562	
903-34	109	3088		1123	
903-34A	109	3088		1123	
903-34B	109	3088		1123	
903-34C	109	3088		1123	
903-34D	109	3088		1123	
903-34E	109	3088		1123	
903-35B			90	1114	
903-36	125	3032A	510,531	1114	
903-36A	125	3032A	510,531	1114	
903-36B	125	3032A	510,531	1114	
903-36C	125	3032A	510,531	1114	
903-36D	125	3032A	510,531	1114	
903-36F	125	3032A	510,531	1114	
903-37B	109	3088		1123	
903-3G	123A	3444	20	2051	
903-41B	177	3100/519	300	1122	
903-42B	177	3100/519	300	1122	
903-43B	109	3088		1123	
903-440B		3126	90		
903-45B	109	3087		1123	
903-47	612	3325	90		
903-47B	612	3126	90		
903-48B	177	3100/519	300	1122	
903-49B	116	3311	504A	1104	
903-51B	109	3088		1123	
903-52B	116		504A	1104	
903-54B	109	3088		1123	
903-58B	177	3100/519	300	1122	
903-65B	109	3087		1123	
903-67B	614	3126	90		
903-68B	506	3017B/117	511	1114	
903-69B	116	3311	504A	1104	
903-71B		3126	90		
903-72B	177	3100/519	300	1122	
903-73B	5022A	3788	ZD-13	563	
903-79B	116	3311	504A	1104	
903-80B		3126	90		
903-82B	177	3100/519	300	1122	
903-83B	109	3087		1123	
903-84B	177	3100/519	300	1122	
903-85B		3126	90		
903-8B	110MP	3088	1N60	1123(2)	
903-92B	109	3088		1123	
903-95B	177		300	1122	
903-96B	614	3126	90		
903-97B	142A	3062	ZD-12	563	
903-99B	614	3126	90		
903-9B	109	3088		1123	
903Y00212	117	3031A	504A	1104	
903Y002149	123A	3444	20	2051	
903Y002150	199	3124/289	62	2010	
903Y002151	187	3193	29	2025	
903Y002152	186	3192	28	2017	
903Y00228	139A	3060	ZD-9.1	562	
904			39	2015	
904-95	123A	3444	20	2051	
904-95A	123A	3444	20	2051	
904-95B	199	3124/289	62	2010	
904-96B	123A	3444	20	2051	
904-97B	109			1123	
904A			39	2015	
905			39	2015	
905-00160	743	3172			
905-102	7490	7490	7490	1808	
905-105	1004	3365	IC-149		
905-106	1308	3833			
905-125	4001B	4001		2401	
905-126	4011B	4011	4011	2411	
905-27	804		IC-27		
905-28	804		IC-27		
905-30B	1003	3288	IC-43		
905-38B	722	3161	IC-9		
905-39B	1075A	3877			
905-46B	722	3161	IC-9		
905-5	1024		720		
905-60	1003	3288			
905-61	1115	3184			
908D1	109			1123	
909(RCA)	107		11	2015	
909-27125-140	123A	3122	20	2051	
909-27125-160	107	3039/316	11	2015	
910			39	2015	
910X1	158		53	2007	
910X10	158		53	2007	
910X2	158		53	2007	
910X3	102A		53	2007	
910X4	102A		53	2007	
910X5	102A		53	2007	
910X6	102A		53	2007	
910X7	158		53	2007	
910X8	158		53	2007	
910X9	158		53	2007	
911-000-3-02	1055	3494			
912-1A6A	123A	3444	20	2051	
914-000-2-00	177	3100/519	300	2035	
914-000-4-00	109	3088	1N60	1123	
914-000-6-00	177	3100/519	300	1122	
914-000-7-00	612	3325	90		
914-001-1-00	177	3100/519	300	1122	
914-001-7-00	110MP	3088	1N60	1123	
914A-1-6-5	121	3717	239	2006	
914F298-1	128	3024	243	2030	
916-31001-1	102A	3004	53	2007	
916-31001-1B	102A		53	2005	
916-31001-7B	102A		53	2007	
916-31003-5B	102A		53	2007	
916-31007-5	102A		53	2007	
916-31007-5B	102A		53	2007	
916-31012-6	102A	3004	53	2007	
916-31012-6B	102A		53	2007	
916-31019-3	126	3008	52	2024	
916-31019-3B	126	3008	52	2024	
916-31024-3	123A	3444	20	2051	
916-31024-3B	107		11	2015	
916-31024-5	107	3018	11	2015	
916-31024-5B	108	3452	86	2038	
916-31025-4	161	3018	39	2015	
916-31025-4B	161	3117	39	2015	
916-31025-5	123A	3444	20	2051	
916-31025-5B	107	3039/316	11	2015	
916-31026-8B	123A	3444	20	2051	
916-31026-9B	102A		53	2005	
916-32000-7	109	3087		1123	
916-32003-2	109	3091		1123	
916-32006-2	109			1123	
916-33003-2	116(4)	3031A	504A(4)	1104	
917-1201-0	1003	3288	IC-43		
919-00-1445	136A		ZD-5.6	561	
919-00-1445-002	142A	3062	ZD-12	563	
919-00-1445-1	138A		ZD-7.5		

Industry Standard No.	ECG	SK	GE	RS 276-	MOTOR.
919-00-1445-2	142A		ZD-12	563	
919-00-1449-1	156		512		
919-00-2440-1	116	3017B/117	504A	1104	
919-00-2440-2	116	3017B/117	504A	1104	
919-00-3309		3335	ZD-20		
919-00-3309-1	5079A	3335	ZD-20		
919-00-3309-2	5127A	3393	5ZD-12		
919-00-3309-3	5131A	3397			
919-00-4326	177	3100/519	300	1122	
919-00-4799	177	3100/519	300	1122	
919-00-5045	117		504A	1104	
919-00-7109	117		504A	1104	
919-00-7394	177		300	1122	
919-00-7766	116	3017B/117	504A	1104	
919-00-7776	116		504A	1104	
919-00-9929	177	3100/519	300	1122	
919-001172	177	3016	504A	1104	
919-001172-1	177	3016	504A	1104	
919-001445-2	142A	3062	ZD-12	563	
919-001449	5802	3051/156	512		
919-001449-1	5802	3051/156	512		
919-001449-3	5802	3051/156	512		
919-002240-1	5802	3051/156	512		
919-002440	156	3051	512		
919-002440-1	156	3051	512		
919-002440-2	5804	3051/156	512		
919-003309	5079A	3335	ZD-20		
919-003309-1	5077A	3752	ZD-18		
919-003309-2	142A	3062			
919-004326	177	3016	504A	1104	
919-004799	116	3016	504A	1104	
919-005045	5802	9005		1143	
919-007109	116	3016	504A	1104	
919-007109RA	116		504A	1104	
919-007394	116	3016	504A	1104	
919-007776	5801	3016	504A	1104	
919-007776RA	117		504A	1104	
919-008862	5802	3051/156	512		
919-009459RA	116		504A	1104	
919-01-0459	116	3017B/117	504A	1104	
919-01-0623	116	3017B/117	504A	1104	
919-01-0829	117		504A	1104	
919-01-0829-1	117		504A	1104	
919-01-0867	109	3087		1123	
919-01-0873	177	3100/519	300	1122	
919-01-1172-1	177		300	1122	
919-01-1211	116	3017B/117	504A	1104	
919-01-1212	116	3017B/117	504A	1104	
919-01-1213	139A	3060	ZD-9.1	562	
919-01-1214	5072A		ZD-8.2	562	
919-01-1215	177	3100/519	300	1122	
919-01-1307	177	3100/519	300	1122	
919-01-1339	117		504A	1104	
919-01-1340	5127A	3393			
919-01-3035	139A		ZD-9.1	562	
919-01-3058	135A	3056	ZD-5.1		
919-01-3072	177	3100/519	300	1122	
919-010454	519(2)	3016			
919-010454-1	519(3)	3016			
919-010454-2	519(4)	3016			
919-010459	5802	3051/156	512		
919-010623	5802	3051/156	512		
919-010829	5801	3051/156	512		
919-010829-1	5802	3051/156	512		
919-010867	109	3087		1123	
919-010873	177		504A	1104	
919-010873-050	519		514	1122	
919-011211	5801	3051/156	512		
919-011212	116	3016	504A	1104	
919-011215	177		504A	1104	
919-011307	177		504A	1104	
919-011339	156	3051	512		
919-011340	5186A		ZD-12	563	
919-012414	116	3017B/117			
919-013036	5801	3051/156	512		
919-013044	116	3124/289	504A	1104	
919-013058	142A		ZD-12	563	
919-013059	177	3062/142A	300	1122	
919-013060	177		504A	1104	
919-013061	156	3051	512		
919-013067	177		300	1122	
919-013072	116		504A	1104	
919-013079	156	3051	512		
919-013081	177		504A	1104	
919-013082	177		300	1122	
919-013677	116	3016			
919-013732	142A	3062			
919-015618-1	5079A	3335	ZD-20		
919-017406-010	5135A	3391/5125A	5ZD-10		
919-017406-011	5136A	3402			
919-017406-013	5130A	3396			
919-017406-056	5133A	3399			
919-017406-057	5125A	3391			
919-72-2	5642	3506			
921-01122		3444	20	2051	
921-01123		3444	20	2051	
921-01124		3854	20	2051	
921-01125		3138	48		
921-01127	123AP	3444/123A	20	2051	
921-01129	172A	3156	64		
921-01130		3138	48		
921-01131		3190	57	2017	
921-1009	152	3054/196	66	2048	
921-100B	102A	3004	53	2007	
921-1010	315	3250	279		
921-1011	130	3027	75	2041	
921-1013	289A	3018	62	2038	
921-1014	229	3716/161	61	2015	
921-1016	290A	3114/290	82	2032	
921-1017	289A	3124/289	210	2051	
921-1019		3112	FET-1	2028	
921-102	107	3018	11	2015	
921-1020	190	3199/188	217		
921-1021	211	3200/189	253	2026	
921-1022	289A	3124/289	62	2010	
921-102A	107	3018	11	2015	
921-102B	107	3018	11	2015	
921-103B	106	3984	21	2034	
921-105B	158	3004/102A	53	2007	
921-106B	199	3452/108	86	2038	
921-109B	123A	3444	20	2051	
921-10B	160	3008	245	2004	
921-110B	129		244	2027	
921-111B	199	3122	62	2010	
921-112B	106	3984	21	2034	
921-114B	199	3018	62	2010	
921-115B	199	3018	62	2010	
921-116B	199	3018	62	2010	
921-117B	123A	3444	20	2051	
921-118B	158	3004/102A	53	2007	
921-119B	108	3452	86	2038	
921-11B	160	3008	245	2004	
921-120B	123A	3444	20	2051	
921-123B	123A	3444	20	2051	

Industry Standard No.	ECG	SK	GE	RS 276-	MOTOR.
921-124B	123A	3444	20	2051	
921-125B	123A	3444	20	2051	
921-126B	312	3834/132	FET-1		
921-127B	199	3444/123A	20	2051	
921-128B	123A	3444	20	2051	
921-129B	108	3452	86	2038	
921-12B	160	3004/102A	245	2004	
921-133	199	3245	212	2010	
921-133B	199	3122	212	2010	
921-13B	160		245	2004	
921-140B	158		53	2007	
921-141B	107	3018	11	2015	
921-142B	107	3018	11	2015	
921-143B	107	3018	11	2015	
921-145B	107	3018	11	2015	
921-147B	102A	3004	53	2007	
921-148B	102A		53	2007	
921-14B	102A	3004	53	2007	
921-150B	102A	3004	53	2007	
921-1528	107		11	2015	
921-152B	199	3039/316	11	2015	
921-153B	102A	3004	53	2007	
921-154B	199	3122	62	2010	
921-155B	199	3444/123A	20	2051	
921-156B	124	3021	12		
921-157B	220	3990	FET-3	2036	
921-158B	108	3452	86	2038	
921-159B	123A	3444	20	2051	
921-15B	160		245	2004	
921-160B	159	3466	82	2032	
921-161B	123A	3444	20	2051	
921-163B	186	3192	28	2017	
921-16B	160		245	2004	
921-170B	108	3452	86	2038	
921-171B	108	3452	86	2038	
921-172B	108	3452	86	2038	
921-173B	108	3452	86	2038	
921-174B	108	3452	86	2038	
921-176B	107	3018	11	2015	
921-177B	107	3018	11	2015	
921-17B	160		245	2004	
921-181B	229	3018	61	2038	
921-182B	129	3114/290	244	2027	
921-188B	128	3024	243	2030	
921-189B	123A	3444	20	2051	
921-191B	123A	3444	20	2051	
921-195B	123A	3444	20	2051	
921-196	199	3245	62	2010	
921-196B	199	3444/123A	62	2010	
921-197B	159	3466	82	2032	
921-198BX		3054/196	57,58		
921-1A	160	3006	245	2004	
921-1B	160	3006	245	2004	
921-20	108	3452	86	2038	
921-200B	123A	3444	20	2051	
921-202B	199	3039/316	62	2010	
921-203B	312	3834/132	FET-1	2035	
921-204B	107	3039/316	11	2015	
921-205B	199	3124/289	62	2010	
921-206	199	3245	62	2010	
921-206B	199	3444/123A	62	2010	
921-207	199	3124/289	62		
921-207B	199	3124/289	62	2010	
921-208B	199	3124/289	62	2010	
921-209B	199	3124/289	212	2010	
921-20A	108	3452	86	2038	
921-20B	108	3452	86	2038	
921-20BK	123A	3444	20	2051	
921-21	108	3452	86	2038	
921-210B	107	3018	11	2015	
921-211B	107	3018	11	2015	
921-212B	108	3452	86	2038	
921-213B	107	3018	11	2015	
921-214B	123A	3444	20	2051	
921-215B	123A	3444	20	2051	
921-216B	102A		53	2007	
921-217B	102A	3004	53	2007	
921-21A	108	3452	86	2038	
921-21B	108	3452	86	2038	
921-21BK	108	3452	86	2038	
921-22	108	3452	86	2038	
921-222B	102A	3004	53	2007	
921-223B	102A	3004	53	2007	
921-224B	158	3004/102A	53	2007	
921-225B	123A	3444	20	2051	
921-226B	107	3018	11	2015	
921-227B	158		53	2007	
921-228B	123A	3444	20	2051	
921-229B	123A	3444	20	2051	
921-22A	108	3452	86	2038	
921-22B	108	3452	86	2038	
921-22BG	123A	3444	20	2051	
921-23	108	3452	86	2038	
921-230B	192	3024/128	63	2030	
921-231B	312	3448	FET-1	2035	
921-232B	107	3018	11	2015	
921-233B	107	3018	11	2015	
921-234B	123A	3444	20	2051	
921-235B	107	3018	11	2015	
921-236B	128	3024	243	2030	
921-237B	123A	3444	20	2051	
921-238B	102A	3004	53	2007	
921-239B	199	3245	62	2010	
921-23A	108	3452	86	2038	
921-23B	108	3452	86	2038	
921-23BK	123A	3444	20	2051	
921-240B	199	3245	62	2010	
921-241B	128		18		
921-242B	158		53	2007	
921-243B	158	3004/102A	53	2007	
921-244B	158		53	2007	
921-24B	102A		53	2007	
921-252B	123A	3444	20	2051	
921-254B	159	3466	82	2032	
921-255B	123A	3444	20	2051	
921-256X	158	3004/102A	53	2007	
921-257B	199	3039/316	11	2015	
921-258B	199	3039/316	11	2015	
921-25B	160		245	2004	
921-26	123A	3444	20	2051	
921-264B	108	3018	61	2038	
921-265	108	3018	61	2038	
921-265B	108	3018	61	2038	
921-266	108		20	2038	
921-266B	108	3452	86	2038	
921-267	108		20	2038	
921-267B	108	3452	86	2038	
921-268B	123A	3444	20	2051	
921-269B	123A	3444	20	2051	
921-26A	123A	3444	20	2051	
921-26B	160	3124/289	245	2004	
921-27	102A	3004	53	2007	
921-270B	129	3025	244	2027	
921-272B	123A	3444	20	2051	
921-273B	102A	3004	53	2007	
921-274B	102A	3004	53	2007	
921-275B	289A	3444/123A	20	2038	
921-275R	123A	3444	20	2051	
921-276B	289A	3444/123A	20	2051	
921-27A	102A	3004	53	2007	
921-27B	123A	3444	20	2051	
921-28	123A	3444	20	2051	
921-281B	199	3124/289	62	2010	
921-282B	158	3004/102A	53	2007	
921-28A	123A	3444	20	2051	
921-28B	123A	3444	20	2051	
921-28BLU	123A	3444	20	2051	
921-291B	123AP	3444/123A	20	2051	
921-292	159		82		
921-292B	294	3466/159	82	2032	
921-296B	159	3466	82	2032	
921-29B	159	3466	82	2032	
921-2A	160	3006	245	2004	
921-2B	160	3006	245	2004	
921-30	108	3452	86	2038	
921-301	229		20	2051	
921-301B	229	3018	61	2038	
921-303B	123AP	3444/123A	20	2051	
921-304B	123AP	3444/123A	20	2051	
921-305B	123A	3444	20	2051	
921-306B	123AP	3444/123A	20	2051	
921-307	123AP		20	2051	
921-307B	123AP	3444/123A	20	2051	
921-308	159		21	2032	
921-308B	159	3114/290	82	2032	
921-309	123AP		18	2051	
921-309B	123AP	3444/123A	20	2051	
921-30A	108	3452	86	2038	
921-30B	108	3452	86	2038	
921-31	108	3452	86	2038	
921-312B	229	3018	61	2038	
921-313B	108	3452	86	2038	
921-314B	123A	3444	20	2051	
921-315B	129	3025	244	2027	
921-318B	102A	3004	53	2007	
921-319B	102A	3004	53	2007	
921-31A	108	3452	86	2038	
921-31B	108	3452	86	2038	
921-32	108	3452	86	2038	
921-325B	108	3452	86	2038	
921-326B	107	3444/123A	20	2051	
921-327B	158	3004/102A	53	2007	
921-32A	108	3452	86	2038	
921-32B	108	3452	86	2038	
921-33	108	3452	86	2038	
921-332B	159	3466	82	2032	
921-333B	106	3984	21	2034	
921-333P	106	3984	21	2034	
921-334B	108	3452	86	2038	
921-335B	108	3452	86	2038	
921-336B	108	3452	86	2038	
921-337B	128	3024	243	2030	
921-338B	108	3452	86	2038	
921-339B	123AP	3444/123A	20	2051	
921-33B	108	3452	86	2038	
921-34	108	3452	86	2038	
921-342B	177	3100/519	300	1122	
921-345B	123A	3444	20	2051	
921-348B	159	3466	82	2032	
921-349	123AP		61	2038	
921-349B	229	3018	61	2038	
921-34A	108	3452	86	2038	
921-34B	108	3452	86	2038	
921-35	102A	3004	53	2007	
921-350B	229	3018	61	2038	
921-351	123AP	3018	20	2051	
921-351B	123AP	3444/123A	20	2051	
921-352B	123AP	3444/123A	20	2051	
921-353	123AP	3444/123A	20	2051	
921-353B	123AP	3444/123A	20	2051	
921-354B	123AP	3444/123A	20	2051	
921-3558	123AP	3018			
921-355B	123AP	3444/123A	20	2051	
921-357B	299		236		
921-35A	102A	3004	53	2007	
921-35B	102A	3004	53	2007	
921-360B	123AP	3444/123A	20	2051	
921-369	123A	3444	20	2051	
921-36B	102A	3004	53	2007	
921-379	229	3018	61	2038	
921-37B	102A	3123	53	2007	
921-38	102A	3004	53	2007	
921-382	268	3180/263			
921-38A	102A	3004	53	2007	
921-38B	102A	3004	53	2007	
921-39	102A	3004	53	2007	
921-39A	102A	3004	53	2007	
921-39B	102A	3004	53	2007	
921-3A	160	3006	245	2004	
921-3B	160	3006	52	2024	
921-40	102A	3004	53	2007	
921-403	172A	3156	64		
921-404	232	3245/199	258		
921-405	159	3466	82	2032	
921-407	123AP		62	2010	
921-408	123AP	3245/199	62	2010	
921-40A	102A	3004	53	2007	
921-40B	102A	3004	53	2007	
921-410			47	2030	
921-41B	102A		53	2007	
921-428	108		20	2038	
921-429	291	3440			
921-430	292	3441			
921-431	152	3054/196			
921-432	153	3083/197			
921-433	153	3083/197			
921-434	152	3054/196			
921-43B	123A	3444	20	2051	
921-43E			20	2051	
921-449	123AP		20	2051	
921-44B	158		53	2007	
921-45	102A	3004	53	2007	
921-450	123AP		20	2051	
921-45A	102A	3004	53	2007	
921-45B	102A	3004	53	2007	
921-46	123A	3444	20	2051	
921-462	123AP		20	2051	
921-463	123AP		20	2051	
921-464	123AP		20	2051	
921-46A	123A	3444	20	2051	
921-46B	103A	3835	59	2002	
921-46BK	123A	3444	20	2051	
921-47	123A	3444	20	2051	
921-470	123AP		20	2051	
921-47A	123A	3444	20	2051	
921-47B	129	3025	244	2027	
921-47BL	123A	3444	20	2051	
921-48B	123A	3444	20		

Industry Standard No.	ECG	SK	GE	RS 276-	MOTOR.
921-49B	123A	3444	20	2051	
921-4A	160	3006	245	2004	
921-4B	126	3006/160	52	2024	
921-50B	123A	3444	20	2051	
921-51B	102A	3004	53	2007	
921-52B	102A	3004	53	2007	
921-53B	158		53	2007	
921-54B	103A	3835	59	2002	
921-55B	161	3117	39	2015	
921-56B	161	3018	39	2015	
921-57B	161	3018	39	2015	
921-58B	161	3018	39	2015	
921-59B	107	3018	11	2015	
921-5A	103A	3010	59	2002	
921-5B	103A	3010	59	2002	
921-60B	128	3024	243	2030	
921-61B	124	3021	12		
921-62	107	3018	11	2015	
921-62A	107	3018	11	2015	
921-62B	107	3018	11	2015	
921-63	107	3018	11	2015	
921-63A	107	3018	11	2015	
921-63B	107	3018	11	2015	
921-64	107	3018	11	2015	
921-64A	107	3018	11	2015	
921-64B	107	3018	11	2015	
921-64C	107	3018	11	2015	
921-65	126	3005	52	2024	
921-65A	126	3005	52	2024	
921-65B	126	3005	52	2024	
921-66	126	3005	52	2024	
921-66A	126	3005	52	2024	
921-66B	126	3005	52	2024	
921-67	102A	3004	53	2007	
921-67A	102A	3004	53	2007	
921-67B	102A	3004	53	2007	
921-68	102A	3004	53	2007	
921-68A	102A	3004	53	2007	
921-68B	102A	3004	53	2007	
921-69	158	3004/102A	53	2007	
921-69A	158	3004/102A	53	2007	
921-69B	158	3004/102A	53	2007	
921-6A	103A	3010	59	2002	
921-6B	103A	3010	59	2002	
921-7	123A	3124/289	20	2051	
921-70	159	3466	82	2032	
921-70A	159	3466	82	2032	
921-70B	159	3466	82	2032	
921-71	103A	3835	59	2002	
921-71A	103A	3835	59	2002	
921-71B	103A	3835	59	2002	
921-72B	108	3452	86	2038	
921-73B	123A	3444	20	2051	
921-77B	123A	3444	20	2051	
921-7B	103A	3835	59	2002	
921-8	123A	3444	20	2051	
921-84	107	3018	11	2015	
921-84A	107	3018	11	2015	
921-84B	107	3018	11	2015	
921-85	107	3018	11	2015	
921-85A	107	3018	11	2015	
921-85B	107	3018	11	2015	
921-86	107	3018	11	2015	
921-86A	107	3018	11	2015	
921-86B	107	3018	11	2015	
921-88	124	3021	12		
921-88A	124	3021	12		
921-88B	124	3021	12		
921-92B	199	3018	62	2010	
921-93B	123A	3444	20	2051	
921-97B	108	3452	86	2038	
921-98B	108	3452	86	2038	
921-99B	123A	3444	20	2051	
921-9B	160	3008	245	2004	
924-16598	102A		53	2007	
924-17945	121	3009	239	2006	
924-2209	172A	3156	64		
926C193-1	123A	3444	20	2051	
926C193-2	123A	3444	20	2051	
926C193-P1M4165	123A	3444	20	2051	
926C206P1	519		514	1122	
929(O)		3245	62	2010	
929C	199	3018	62	2010	
929CA	199	3124/289	62	2010	
929CU	199	3124/289	62	2010	
929D(JVC)	199	3245	62	2010	
929DX	199	3039/316	62	2010	
930(OV)				2010	
930(OV)	199	3245	62	2010	
930B	199	3018	62	2010	
930C	199	3039/316	62	2010	
930C(WARDS)	199		62	2010	
930C613-P1	909	3590	IC-249		
930D	199	3018	62	2010	
930D(WARDS)	199		62	2010	
930DC	9930				MC830L
930DM					MC930L
930DU	199	3245	62	2010	
930DX	199	3018	62	2010	
930DZ	107		11	2015	
930E	199	3018	62	2010	
930E(JVC)	199		62	2010	
930E(WARDS)	199		62	2010	
930EX	199	3018	62	2010	
930FC					MC930F
930FM					MC930F
930HC					MC830G
930HM					MC930G
930PC	9930				MC830P
930X1	123A	3444	20	2051	
930X10	102A	3004	53	2007	
930X2	123A	3444	20	2051	
930X3	123A	3018	20	2051	
930X4	108	3452	86	2038	
930X5	108	3452	86	2038	
930X6	126	3005	52	2024	
930X7	102A	3004	53	2007	
930X8	102A	3004	53	2007	
930X9	102A	3004	53	2007	
931	177		300	1122	
932DC	9932				MC832L
932DM					MC932L
932FC					MC932F
932FM					MC932F
932HC					MC832G
932HM					MC932G
932PC	9932				MC832P
933DC	9933				MC833L
933DM					MC933L
933FC					MC933F
933FM					MC933F
933HC					MC833G
933HM					MC933G
933PC	9933				MC833P

Industry Standard No.	ECG	SK	GE	RS 276-	MOTOR.
935-1	123A	3444	20	2051	
935DC	9935				MC840L
935DM					MC940L
935FC					MC940F
935FM					MC940F
935PC	9935				MC840P
936 NPN	184	3190		2017	
936(NPN)			57	2017	
936-10	116	3016	504A	1104	
936-20	116	3031A	504A	1104	
936DC	9936				MC836L
936DM					MC936L
936FC					MC936F
936FM					MC936F
936PC	9936				MC836P
936PNP	185		58	2025	
937DC	9937				MC837L
937DM					MC937L
937FC					MC937F
937FM					MC937F
937PC	9937				MC837P
938-3	283	3467	36		
939-12	177		300	1122	
941-026-0001	177		300	1122	
941T1	176	3123	80		
942H2	1195	3469			
943-086	519		514	1122	
943-087	519		514	1122	
943-087-1	519		514	1122	
943-105-001	519		514	1122	
943-721-001	159	3466	82	2032	
943-728-001	128		243	2030	
943-742-002	199	3245	62	2010	
944DC	9944				MC844L
944DM					MC944L
944FC					MC944F
944FM					MC944F
944HC					MC844G
944HM					MC944G
944PC	9944				MC844P
945	6400A			2029	
945DC	9945				MC845L
945DM					MC944L
945FC					MC945F
945FM					MC945F
945HC					MC845G
945HM					MC945G
945PC	9945				MC845P
946DC	9946				MC846L
946DM					MC946L
946FC					MC946F
946FM					MC946F
946HC					MC846G
946HM					MC946G
946PC	9946				MC846P
947-1	123AP	3018	39	2015	
947-1(SYLVANIA)	108		86	2038	
948DC	9948				MC848L
948DM					MC948L
948FC					MC948F
948FM					MC948F
948HC					MC848G
948HM					MC948G
948PC	9948				MC848P
949DC	9949				MC849L
949DM					MC949L
949FC					MC949F
949FM					MC949F
949HC					MC849G
949HM					MC949G
949PC	9949				MC849P
950DC	9950				MC850L
950DM					MC950L
950FC					MC950F
950FM					MC950F
950HC					MC850G
950HM					MC950G
950PC	9950				MC850P
951			47	2030	
951-1		3004	2	2007	
951-1(SYLVANIA)	126		52	2024	
951DC	9951				MC851L
951DM					MC951L
951FC					MC951F
951FM					MC951F
951HC					MC851G
951HM					MC951G
951PC	9951				MC851P
952			18	2030	
953			18	2030	
955-1	160	3006	245	2004	
955-2	160	3006	245	2004	
955-3	160	3006	245	2004	
958-023	159	3466	82	2032	
961DC	9961				MC861L
961DM					MC961L
961FC					MC961F
961FM					MC961F
961HC					MC861G
961HM					MC961G
961PC	9961				MC861P
962DC	9962				MC862L
962DM					MC962L
962FC					MC962F
962FM					MC962F
962HC					MC862G
962HM					MC962G
962PC	9962				MC862P
963DC	9963				MC863L
963DM					MC963L
963FC					MC963F
963FM					MC963F
963HC					MC863G
963HM					MC963G
963PC	9963				MC863P
964-16598	126	3006/160	52	2024	
964-16599	104	3013	16	2006	
964-17142	102	3004	2	2007	
964-17443	116	3016	504A	1104	
964-17444	123A	3444	20	2051	
964-174443	116		504A	1104	
964-17887	104	3009	16	2006	
964-17945	121	3717	239	2006	
964-19862		3010	54		
964-19862A	102A		53	2007	
964-19863	102A	3004	53	2007	
964-19864	102A	3004	53	2007	
964-19865	116	3016	504A	1104	
964-20738	123A	3444	20	2051	
964-20739	129	3025	244	2027	
964-2073B	123A	3122	20	2051	
964-21866	116	3031A	504A	1104	
964-22008	129	3241/232	258		
964-22009	172A	3156	64		

Industry Standard No.	ECG	SK	GE	RS 276-	MOTOR.
964-2209	172A		64		
964-22158	128	3024	63	2030	
964-24387	128		243	2030	
964-24584	123A	3444	20	2051	
964-25046	128		243	2030	
964-27986	210	3122	252	2018	
965-1A6-5	160		245	2004	
965-2A6-1	103A		59	2002	
965-4A6-2	101	3861	8	2002	
965A16-3	160		245	2004	
965B16-2	126		52	2024	
965C-16-4	121	3717	239	2006	
965T1	102	3004	2	2007	
966-1A6-3	102A		53	2007	
966-2A6-4	102A		53	2007	
966-3A6C	102A		53	2007	
971-B6	282	3748			
972-659B-0	159	3114/290	221	2027	
972D7	116	3017B/117	504A	1104	
972X1	160	3006	245	2004	
972X10	102A	3004	53	2007	
972X11	102A	3004	53	2007	
972X12	102A	3004	53	2007	
972X2	160	3006	245	2004	
972X3	160	3006	245	2004	
972X4	160	3006	245	2004	
972X5	160	3006	245	2004	
972X6	126	3008	52	2024	
972X7	126	3008	52	2024	
972X8	126	3008	52	2024	
972X9	102A	3004	53	2007	
974-1(SYLVANIA)	157	3716/161	39	2015	
974-2		3103A	27		
974-2(SYLVANIA)	157	3747	232		
974-3	157	3747	232		
974-3A6-3	157	3721/100	1	2007	
974-4		3104A	27		
974-4(SYLVANIA)	157	3747	232		
976-0036-921	177		300	1122	
977-1	177	3100/519	300	1122	
977-10B	116	3017B/117	504A	1104	
977-11B	116	3017B/117	504A	1104	
977-13B	116	3311	504A	1104	
977-14B	113A	3119/113	6GC1		
977-18B	116	3016	504A	1104	
977-19	116	3311	504A	1104	
977-19B	116	3016	504A	1104	
977-20B	116	3311	504A	1104	
977-21B	166	3311	504A(4)	1104	
977-22B	116	3311	504A	1104	
977-23B	116	3311	504A	1104	
977-24B	166	9075	BR-600	1152	
977-25B	116	3311	504A	1104	
977-27B	116	3311	504A	1104	
977-28B	116	3311	504A	1104	
977-2B	116	3031A	504A	1104	
977-33	521	3304/500A			
977-35	523	3304/500A			
977-35-02	526A	3306/523			
977-35-03	526A	3306/523			
977-36	526A	3306/523			
977-37	526A	3306/523			
977-38	522	3303			
977-3B	125	3031A	510,531	1114	
977-40	530	3308			
977-41	528	3306/523			
977-42	528	3306/523			
977-4B	166	9075	BR-600	1152	
977-64197	197	3245/199	250	2027	
977-6B	166	9075	BR-600	1152	
977-8B	116		504A	1104	
977-Z9522	522	3303			
977-Z9529	529	3307			
977-Z9530	530	3308			
977-Z9531	531	3301			
977-Z9532	532	3303/522			
977-Z9533	533	3302			
977-Z9534	534	3305			
977-Z9535	535	3307/529			
977A1-6-1	105		4		
977B1-6-2	121MP	3718	239(2)	2006(2)	
977C1-6-3	121MP	3718	239(2)	2006(2)	
978-1923	129		244	2027	
984A-1-6B	121	3717	239	2006	
987T1	102	3004	2	2007	
988T1	102	3004	2	2007	
989T1	102	3004	2	2007	
990T1	102	3004	2	2007	
991-00-1172	177	3100/519	300	1122	
991-00-1172-1	177	3100/519	300	1122	
991-00-1219	123A	3444	20	2051	
991-00-1221	102A		53	2007	
991-00-1222	102A		53	2007	
991-00-1449	117		504A	1104	
991-00-1449-1	117		504A	1104	
991-00-2232	123A		20	2051	
991-00-2248	123A		20	2051	
991-00-2298	123A		20	2051	
991-00-2356	123A		20	2051	
991-00-2356/K	123A	3444	20	2051	
991-00-2440	117		504A	1104	
991-00-2440-1	117		504A	1104	
991-00-2440-2	117		504A	1104	
991-00-2873	123A	3444	20	2051	
991-00-2888	128	3024	243	2030	
991-00-3144	123A	3444	20	2051	
991-00-3304	123A	3444	20	2051	
991-00-7394			504A	1104	
991-00-7776	116	3017B/117	504A	1104	
991-00-8393	123A	3444	20	2051	
991-00-8393A	123A	3444	20	2051	
991-00-8393M	123A	3444	20	2051	
991-00-8394	123A	3444	20	2051	
991-00-8394A	123A	3444	20	2051	
991-00-8394AH	123A	3444	20	2051	
991-00-8395	123A	3444	20	2051	
991-002232	289A	3019	17	2051	
991-002298	199	3124/289	20	2051	
991-002356	199	3124/289	20	2051	
991-002873	199	3019	17	2051	
991-002888	194	3045/225			
991-003144	192A	3124/289			
991-003304	289A	3018	17	2051	
991-008393	289A	3124/289	10	2051	
991-008394	289A	3124/289			
991-008394A	289A	3124/289			
991-01-0098	159	3466	82	2032	
991-01-0099	121	3717	239	2006	
991-01-0461	172A	3156	64		
991-01-0462	159	3466	82	2032	
991-01-1216	121	3009	239	2006	
991-01-1217	158		53	2007	
991-01-1219	123A	3444	20	2051	
991-01-1220	123A	3444	20	2051	
991-01-1221	102A		53	2007	
991-01-1222	102A		53	2007	
991-01-1223	102A		53	2007	
991-01-1224	102A		53	2007	
991-01-1225	159	3466	82	2032	
991-01-1305	128	3024	243	2030	
991-01-1306	123A	3444	20	2051	
991-01-1312	123A	3444	20	2051	
991-01-1314	128		243	2030	
991-01-1315	129		244	2027	
991-01-1316	107	3039/316	11	2015	
991-01-1317	130		14	2041	
991-01-1318	123A	3444	20	2051	
991-01-1319	159	3466	82	2032	
991-01-1705	123A	3444	20	2051	
991-01-1706	312	3112	FET-1	2028	
991-01-2328	159	3466	82	2032	
991-01-2686	129		244	2027	
991-01-3044	123A	3444	20	2051	
991-01-3055	312	3112	FET-1	2028	
991-01-3056	123A	3444	20	2051	
991-01-3057	123A	3444	20	2051	
991-01-3058	159	3466	82	2032	
991-01-3063	130	3027	14	2041	
991-01-3068	123A	3444	53	2051	
991-01-3170	196		241	2020	
991-01-3543	172A	3156	64		
991-01-3544	123A	3444	20	2051	
991-01-3599	159	3466	82	2032	
991-01-3683	123A	3444	20	2051	
991-01-3740	123A	3444	20	2051	
991-01-5000	197		250	2027	
991-01-5001	196		241	2020	
991-01-5062	197		250	2027	
991-01-5063	196		241	2020	
991-010098	159		22	2032	
991-010462	289A	3444/123A	20	2051	
991-011217	158	3004/102A	53	2007	
991-011219	123A	3444	20	2051	
991-011220	123A	3444	20	2051	
991-011221	102A	3123	245	2004	
991-011222	102A	3004	53	2007	
991-011223	102A	3004	53	2007	
991-011224	102A	3123			
991-011225	159	3466	82	2032	
991-011305	128	3124/289	243	2030	
991-011306	123A	3444	20	2051	
991-011312	123A	3444	20	2051	
991-011313	123A	3122	63	2030	
991-011314	128	3122	243	2030	
991-011315	129	3114/290			
991-011316	123A	3124/289			
991-011318	123A	3444	20	2051	
991-011319	159	3114/290	67	2023	
991-011576	312	3118	FET-1	2028	
991-011705	289A	3019			
991-011706	312	3118	FET-1	2028	
991-012328	159	3114/290	65	2050	
991-012637	102A	3004			
991-012686	123AP	3466/159	82	2032	
991-013044	289A		17	2051	
991-013056	289A	3444/123A	20	2051	
991-013057	289A	3019	62	2010	
991-013058	159	3025/129	21	2034	
991-013063	152		14	2041	
991-013068	289A	3019	20	2051	
991-013316	289A	3019			
991-013543	172A	3134/705A	64		
991-013544	199	3024/128	17	2051	
991-013683	289A	3018			
991-015587	123A	3444	20	2051	
991-015614	290A	3466/159	82	2032	
991-015615	289A	3444/123A	20	2051	
991-015663	172A	3156	64		
991-016274	289A	3444/123A	20	2051	
991-016614	289A	3124/289			
991-016727	172A	3134/705A			
991-016788	289A	3124/289			
991-017456	312	3118			
991-018047	123AP	3124/289			
991-2P	129		244	2027	
991-3N	128		243	2030	
991T1	102	3004	2	2007	
992-00-1192	104	3009	16	2006	
992-00-2271	130		14	2041	
992-00-2298	123A	3444	20	2051	
992-00-3139	130	3027	14	2041	
992-00-3139A	130	3027	14	2041	
992-00-3144	123A	3444	20	2051	
992-00-3172	124	3021	12		
992-00-4091	130		14	2041	
992-00-4092	130		14	2041	
992-00-8870	121	3009	239	2006	
992-00-8890	179	3642	76		
992-00-8890L	179	3642	76		
992-001192	121	3009	3	2006	
992-002271	130	3027	14	2041	
992-00271	130		14	2041	
992-00271A	130	3027	14	2041	
992-003139	130	3027	14	2041	
992-003172	384	3026	246	2020	
992-004091	130	3027	14	2041	
992-004092	384	3026	14	2041	
992-008-890	121MP	3014	239(2)	2006(2)	
992-008870	179		3	2006	
992-008890	121	3014	3	2006	
992-01-1216	121	3009	239	2006	
992-01-1218	121	3009	239	2006	
992-01-1317	130		14	2041	
992-01-3684	154		40	2012	
992-01-3705	175		246	2020	
992-01-3738	123A	3444	20	2051	
992-010404	128	3024			
992-011216	121		25		
992-011218	121	3009	16	2006	
992-011317	152		14	2041	
992-013684	128	3024			
992-013705	384	3026			
992-013738	199	3124/289			
992-017169	130	3027	14	2041	
992-020054	243	3182			
992-02271	130		14	2041	
992-08890	121	3717	239	2006	
992-10	116		504A	1104	
992-1A6-1	105		4		
992-531-01	166	9075	504A	1152	
992T1	102A	3004	53	2007	
995-01-6130	153		69	2049	
995-01-6131	152	3893	66	2048	
998-0061114	123A	3444	20	2051	
998-0200816	123A	3444	20	2051	
999-4601	128		243	2030	
1000-12	139A		ZD-9.1	562	
1000-129	116	3311	504A	1104	

Industry Standard No.	ECG	SK	GE	RS 276-	MOTOR.
1000-130	177		300	1122	
1000-131	109			1123	
1000-132	139A		ZD-9.1	562	
1000-133	144A		ZD-14	564	
1000-135	107	3018	11	2015	
1000-136	123A	3444	20	2051	
1000-137	123A	3444	20	2051	
1000-138	152	3893	66	2048	
1000-139	195A	3047	46		
1000-140	299	3298	217		
1000-141	195A	3048/329	46		
1000-142	152	3893	66	2048	
1000-17	109	3088		1123	
1000-25	703A	3157	IC-12		
1001	102A	3004	53	2007	
1001(GE)	712		IC-2		
1001(JULIETTE)	123A	3444	20	2051	
1001-0036/4460	720	9014	IC-7		
1001-0091/4460	1140	3473	IC-138		
1001-01	128	3047	243	2030	
1001-02	123A	3444	20	2051	
1001-02(COURIER)	107		11	2015	
1001-03	123A	3444	20	2051	
1001-04	123A	3444	20	2051	
1001-05	123A	3444	20	2051	
1001-06	123A	3444	20	2051	
1001-07	195A	3048/329	46		
1001-08	237	3299	46		
1001-09	130		14	2041	
1001-10	109	3088		1123	
1001-11	116	3031A	504A	1104	
1001-12	139A	3060	ZD-9.1	562	
1001-5	5642		62		
1001-6	5642		62		
1001-7663	102	3722	2	2007	
1002	234	3004/102A	65	2050	
1002(SQUELCH)	102A	3004	53	2007	
1002-01	312	3112	FET-2	2028	
1002-02	229	3018	61	2038	
1002-0219	116	3016	504A	1104	
1002-02A	108	3452	86	2038	
1002-03	123A	3444	20	2051	
1002-04	123A	3444	20	2051	
1002-04-1	108	3452	86	2038	
1002-05	102A	3004	53	2007	
1002-08	109	3087		1123	
1002-09	109	3087		1123	
1002-17	109	3087		1123	
1002-17(DIODE)	109			1123	
1002-25	726	3129			
1002-4404	142A	3062	ZD-12	563	
1002-6219	116(4)	3311	504A	1104	
1002-68	229	3018	61	2038	
1002A(JULIETTE)	107		11	2015	
1002A-1	129		244	2027	
1002A-2	128		243	2030	
1003	102A	3004	53	2007	
1003(E.F.JOHNSON)			53	2007	
1003(JULIETTE)	126		52	2024	
1003-01	107	3018	11	2015	
1003-0567	1196	3725			
1003-11	177	3311	300	1122	
1003-13	116		504A	1104	
1003-2084	5070A	9021			
1003-6754	199	3124/289	62	2010	
1004	123	3024/128	20	2051	
1004(2SC537)	123A	3444	20	2051	
1004(DIODE)	109			1123	
1004(G.E.)	161	3716	39	2015	
1004(JULIETTE)	158@108		86	2038	
1004-01	158	3004/102A	53	2007	
1004-02	195A	3047	46		
1004-03	123A	3444	20	2051	
1004-0780	199	3124/289	85	2010	
1004-0798	230	3042			
1004-17	108	3452	86	2038	
1004P	129		244	2027	
1005	123A	3444	20	1123	
1005(2SC537)	123A	3444	20	2051	
1005(JULIETTE)	102A		53	2007	
1005-02	195A	3048/329	46		
1005-03	123A	3444	20	2051	
1005-17	102A	3004	53	2007	
1005-20	116		504A	1104	
1005-3	123A		20	2051	
1005M19	159	3466	82	2032	
1006		3124	61	2038	
1006(DIODE)	109			1123	
1006(G.E.)	161	3716	39	2015	
1006(JULIETTE)			20	2051	
1006(JULIETTE)				2007	
1006-17	125		510,531	1114	
1006-1737	145A	3063	ZD-15	564	
1006-24	116		504A	1104	
1006-48	229	3122	20	2051	
1006-5977	177	3126	90		
1006-5985	140A	3061	ZD-10	562	
1006-78	160		245	2004	
1006-9292	109	3088	1N60	1123	
1006-93	102A		53	2007	
1007		3311	504A	1123	
1007(2SB405)	158		53	2007	
1007(JULIETTE)	126		52	2024	
1007-0951	116	3311	504A	1104	
1007-1124	110MP	3709	1N34AS	1123	
1007-17	102A	3004	53	2007	
1007-17(DIODE)	109			1123	
1007-3054	229		61	2038	
1007-3062	229		61	2038	
1007-3088	229	3124/289	210		
1007-3153	289A	3122	210	2051	
1007-3344	172A		64		
1008	128	3024	20	2030	
1008(E.F.JOHNSON)			53	2007	
1008(JULIETTE)	131		44	2006	
1008(POWER)	131		44	2006	
1008-02	123A	3444	20	2051	
1008-17	104	3719	16	2006	
1009	102A	3004	53	2007	
1009(E.F.JOHNSON)			53	2007	
1009(G.E.)	161	3716	39	2015	
1009(SEARS)	102A	3004	53	2007	
1009-01	222	3116	FET-2	2035	
1009-02	123A	3444	20	2051	
1009-02-16	123A	3444	20	2051	
1009-03	195A	3047	46		
1009-03-17	128	3047	243	2030	
1009-04	237	3299	46		
1009-04-17	195A	3048/329	46		
1009-05	300	3464	273		
1009-05A	196	3041	241	2020	
1009-06	138A	3059	ZD-7.5		
1009-07	177	3100/519	300	1122	
1009-08	177	3100/519	300	1122	

Industry Standard No.	ECG	SK	GE	RS 276-	MOTOR.
1009-09	116	3017B/117	504A	1104	
1009-127	312	3448	FET-1	2035	
1009-17	123A	3444	20	2051	
1009-2			20	2051	
1010	140A	3100/519	ZD-10	562	
1010(GE)	233		210	2009	
1010(JULIETTE)	116		504A	1104	
1010-14	195A	3047	46		
1010-143	177	3100/519	300	1122	
1010-145	109			1123	
1010-17	191	3047	249		
1010-17(R.F.)	195A		46		
1010-7738	159	3114/290	22	2032	
1010-78	160		245	2004	
1010-7928	123A	3017B/117	20	2051	
1010-7936	196	3054	66	2048	
1010-7951		3044	18	2030	
1010-7993	128	3024	18	2030	
1010-8025	165	3115	38		
1010-8041	128	3024	18	2030	
1010-8066	233	3018	20	2051	
1010-8082	123A	3124/289	17	2051	
1010-8090	123A	3124/289	20	2051	
1010-8116	506	3843	511	1114	
1010-8132	5080A		ZD-22		
1010-8173	110MP	3709	1N34AS	1123	
1010-87	160		245	2004	
1010-89	126		52	2024	
1010-9486	116		504A	1104	
1010-9494	116	3311	504A	1104	
1010-9932	738	3167	IC-29		
1010-9940	712	3072	IC-2		
1010-9965	783	3215	IC-21		
1010-9973	713	3077/790	IC-5		
1010A	519	3100	300	1122	
1011	109	3017B/117		1123	
1011(VO-6C)	116		504A	1104	
1011-01	195A	3048/329	46		
1011-02	5071A	3334	ZD-6.8	561	
1011-0302	178MP	3100/519	300	1122	
1011-11	191		249		
1011-11(FOCUS)	118		CR-1		
1011-11(R.F.)	108	3452	86	2038	
1011M57P01	129		244	2027	
1011M62P01	129		244	2027	
1012		3288	504A	1104	
1012(G.E.)	159		82	2032	
1012(GE)	159		82	2032	
1012(I.C.)	1003		IC-43		
1012-17	109	3087		1123	
1012GE			21	2034	
1013	109	3118		1123	
1013(E.F.JOHNSON)	159		82	2032	
1013(GE)	159		82	2032	
1013-15	229	3018	61	2038	
1013-16	102A	3004	52	2024	
1014		3004	53	1104	
1014(E.F.JOHNSON)			53	2007	
1014-25	703A		IC-12		
1015	145A	3063	ZD-15	564	
1016	234	3114/290	65	2050	
1016(GE)	159		82	2032	
1016-17	116	3017B/117	504A	1104	
1016-77	109	3088		1123	
1016-78	116		504A	1104	
1016-79	116	3032A	504A	1104	
1016-80	703A		IC-12		
1016-81	152	3893	66	2048	
1016-83	108	3452	86	2038	
1016-84	108	3452	86	2038	
1016-85	313	3018	278	2016	
1018	154	3024/128	40	2012	
1018-25	1056	3458	IC-48		
1018-3259	177	3100/519	300	1122	
1018-6963	116	3016	504A	1104	
1018-9884	177	3100/519	300	1122	
1019-1385	506	3125	511	1114	
1019-3852	123A	3444	20	2051	
1019-6699	109	3088	1N60	1123	
1019-74	102A		53	2007	
1020	116	3017B/117	504A	1104	
1020-17	123A	3444	20	2051	
1020-25	1075A	3877			
1021			FET-2	2035	
1021-17	102	3004	2	2007	
1021-25	1100	3223	IC-92		
1022-5548	116	3311	504A	1104	
1023-17	102	3004	2	2007	
1023G	123A	3444	20	2051	
1023G(GE)	123A	3444	20	2051	
1024	116	3311	504A	1104	
1024-17	104	3009	16	2006	
1024G		3018	20	2051	
1024G(GE)	123A	3444	20	2051	
1025G		3018	20	2051	
1025G(GE)	199	3444/123A	20	2051	
1026(GE)	229		61	2038	
1026G	123A	3444	20	2051	
1026G(GE)	123A	3444	20	2051	
1027	146A	3064	ZD-27		
1027(G.E.)	108		86	2038	
1027G	128	3044/154	243	2030	
1027G(GE)	128		243	2030	
1028			17	2051	
1028(G.E.)	123A		20	2051	
1028G	123A	3444	20	2051	
1028G(GE)	123A	3444	20	2051	
1029			17	2051	
1029(G.E.)	123A		20	2051	
1029G	123A	3444	20	2051	
1029G(GE)	123A	3444	20	2051	
1030	159		82	2032	
1030(G.E.)	165		38		
1030-17	109	3087		1123	
1030-21	199	3124/289	62	2010	
1030-25	7473	7473		1803	
1031-17	156		512		
1031-25	7401	7401			
1032	102	3004	2	2007	
1033	142A	3095/147A	ZD-12	563	
1033-0911	177	3100/519	300	1122	
1033-0983	177	3100/519	300	1122	
1033-0991	177	3100/519	300	1122	
1033-1	126	3008	52	2024	
1033-1270	5404	3627			
1033-17	116	3017B/117	504A	1104	
1033-1916	109	3087	1N34AS	1123	
1033-2	126	3008	52	2024	
1033-3	126	3008	52	2024	
1033-4	126	3008	52	2024	
1033-5	102	3004	2	2007	
1033-6	103	3862	8	2002	
1033-7	102	3004	2	2007	
1033-8	116	3017B/117	504A	1104	

Industry Standard No.	ECG	SK	GE	RS 276-	MOTOR.
1034	158	3031A	53	2007	
1034-17	123A	3444	20	2051	
1034-43	101	3861	8	2002	
1035	158		53	2007	
1035-80	199	3122	62	2010	
1036	158	3004/102A	53	2007	
1038(G.E.)	128		243	2030	
1038-1	123A	3444	20	2051	
1038-1-10	123A	3444	20	2051	
1038-10	123A	3444	20	2051	
1038-15	123A	3444	20	2051	
1038-15CL	123A	3444	20	2051	
1038-1697	116	3311	504A	1104	
1038-1788	506	3130	511	1114	
1038-18	123A	3444	20	2051	
1038-1804	1046		IC-118		
1038-1853	198		32		
1038-18CL	123A	3444	20	2051	
1038-21	123A	3444	20	2051	
1038-23	123A	3444	20	2051	
1038-23CL	123A	3444	20	2051	
1038-24	123A		20	2051	
1038-6	123A	3444	20	2051	
1038-6CL	123A	3444	20	2051	
1038-8	123A	3444	20	2051	
1038-9922	234	3247	65	2050	
1039-0060	290A	3114/290	269	2032	
1039-01	123A	3444	20	2051	
1039-0433	186A	3197/235	215	2052	
1039-0441	289A	3024/128	210	2038	
1039-0458	297	3124/289	210	2030	
1039-0482	107	3039/316	11	2015	
1039-0516	198		32		
1039-0961	199	3866	61	2038	
1039-1290	238	3710	259		
1040	116	3031A	504A	1104	
1040-01	123A	3124/289	62	2010	
1040-02	175	3026	246	2020	
1040-03	123A	3444	20	2051	
1040-04	195A	3047	46		
1040-05	237	3299	46		
1040-07	177	3100/519	300	1122	
1040-08	109	3088		1123	
1040-09	140A	3061	ZD-10	562	
1040-09(ZENER)	140A		ZD-10	562	
1040-10	116	3100/519	504A	1104	
1040-11	128		243	2030	
1040-155	123A	3444	20	2051	
1040-2	123A	3444	20	2051	
1040-59	160		245	2004	
1040-7	128	3024	243	2030	
1040-80	103A	3835	59	2002	
1040-81	116		504A	1104	
1040-9068	159	3114/290	221		
1040-9332	116	3100/519	300	1122	
1040-9373	5081A	3151	ZD-24		
1041-109	116	3017B/117	504A(2)	1104	
1041-63	116	3017B/117	504A	1104	
1041-64	116	3017B/117	504A	1104	
1041-65	109	3088		1123	
1041-66	177	3100/519	300	1122	
1041-67	138A	3059	ZD-7.5		
1041-70	312	3448	FET-1	2035	
1041-71	229	3122	61	2038	
1041-72	123A	3444	20	2051	
1041-73	123A	3444	20	2051	
1041-74	152	3054/196	66	2048	
1041-75	123A	3444	20	2051	
1041-76	190	3047	217		
1041-77	195A	3049/224	46		
1042-01	222	3116	FET-3	2036	
1042-02	312	3026	FET-2	2028	
1042-03	199	3122	62	2010	
1042-04	123A	3444	20	2051	
1042-05	289A	3138/193A	268	2038	
1042-06	159	3466	82	2032	
1042-07	123A	3444	20	2051	
1042-08	235	3197	215		
1042-09	236	3197/235	216	2053	
1042-10	226MP	3086	49(2)	2025(2)	
1042-11	724	3525	IC-86		
1042-12	109	3087		1123	
1042-13	109	3087	1N60	1123	
1042-14	178MP	3100/519	300(2)	1122(2)	
1042-15	519	3100	514	1122	
1042-16	177	3100/519	300	1122	
1042-17	116	3051/156	504A	1104	
1042-18	139A	3060	ZD-9.1	562	
1042-19	5072A	3136	ZD-8.2	562	
1042-23	177	3100/519	300	1122	
1042-3	109	3088		1123	
1042-7	123A	3122	20	2051	
1042-7938	749	3171/744			
1042-8	235	3197	215		
1043-0049	177		300	1122	
1043-07	123A	3444	212	2051	
1043-10	177	3100/519	300	1122	
1043-1229	289A	3124/289	210	2051	
1043-1260	289A		20	2051	
1043-1278	129		82	2032	
1043-1286	152	3054/196	66	2048	
1043-1294	153	3083/197	69	2049	
1043-1328	184	3054/196	217	2018	
1043-1344	722	3161	IC-9		
1043-1534	142A	3016	504A	1104	
1043-1583	605	3327/614			
1043-7309	124	3026	12		
1043-7358	198	3054/196	66	2048	
1043-7374	159	3114/290	21	2034	
1043-7382	116	3311	504A	1104	
1044	116	3031A	504A	1104	
1044-0295	159	3114/290	22	2032	
1044-6888	298	3450	272		
1044-7035	984	3207/818			
1044-7043	1174	3186			
1044-7049	1174	3186			
1044-8983	177	3100/519	300	1122	
1044-9544	128	3137/192A	243	2030	
1045-0518	116	3313	504A	1104	
1045-0534	116	3311	504A	1104	
1045-2951	123A	3122	20	2051	
1045-3082	128	3122	243	2030	
1045-7802	177	3100/519	300	1122	
1045-7828	152	3054/196	66	2048	
1045-7836	153	3083/197	69	2049	
1045-7844	130	3027	14	2041	
1045-7851	130	3027	14	2041	
1047-25	1052	3249			
1048-6421	177	3100/519	300	1122	
1048-9839	5072A		ZD-8.2	562	
1048-9870	109	3088	1N60	1123	
1048-9888	110MP	3709	1N60	1123	
1048-9904	229	3018	20	2051	
1048-9912	123A	3020/123	20	2051	

Industry Standard No.	ECG	SK	GE	RS 276-	MOTOR.
1048-9920	199	3124/289	62	2010	
1048-9938	614	3126	90		
1048-9946	116	3311	504A	1104	
1048-9987	177	3100/519	300	1122	
1048-9995	116	3311	504A	1104	
1049-0035	294		48		
1049-0060	229	3018	61	2038	
1049-0092	199	3124/289	62	2010	
1049-0100	199	3122	61	2038	
1049-0167	293	3849	47	2030	
1049-1744	123A	3122	61	2038	
1049-3435	116	3311	504A	1104	
1050	116	3017B/117	504A	1104	
1050(GE)	506		511	1114	
1050-21	199	3124/289	62	2010	
1050-64	116		504A	1104	
1050B		3114	21	2034	
1051	506	3130	511	1114	
1051(GE)	506		511	1114	
1052-17	102A	3004	53	2007	
1052-6390	1161		IC-300		
1052-6408	1164	3727	IC-301		
1052-6416	1173	3729	IC-302		
1054-8295	513		513		
01057-1	184	3054/196	57	2017	
1057-17	102A		53	2007	
1057-2071	199	3137/192A	62		
1057-9140			ZD-11	563	
1059-2848	506	3843	504A	1114	
1059-7961	116	3313	504A	1104	
1059-9140	5074A	3139	ZD-11		
1060	116	3017B/117	504A	1104	
1060-17	102A	3004	53	2007	
1060-17(DRIVER)			53	2007	
1060-17(PRE-AMP)			2	2007	
1060-6564	294	3138/193A	48		
1060-9428	287		232		
1061-6274	238	3710	38		
1061-6282	162	3559	35		
1061-6290	506	3843	511	1114	
1061-8312	159	3114/290	221		
1061-8320	290A	3138/193A	61	2038	
1061-8338	196	3054	241	2020	
1061-8346	198	3104A	32		
1061-8353	198	3104A	32		
1061-8361	125	3032A	509	1114	
1061-8379	116	3311	504A	1104	
1061-8387	506	3843	504A	1104	
1061-8668	198	3104A	32		
1061-8807	290A	3114/290	67	2032	
1061-8908	198	3104A	32		
1061-8916	116	3311	504A	1104	
1061-8924	116	3311	504A	1104	
1061-9068	159	3114/290	221		
1061-9153	1128	3488	IC-105		
1061-9161	1134	3489	IC-106		
1061-9526	1200	3714			
1061-9666	1130	3478	IC-111		
1061-9856	1133	3490	IC-107		
1062-0615	108		67	2017	
1062-6018	159		82	2032	
1062-6414	294	3138/193A	48		
1062-7511	165	3115	38		
1062GE	177	3100/519	300	1122	
1063-3553	109	3087	1N34AS	1123	
1063-3592	1073	3496			
1063-4145	5071A	3334	ZD-6.8	561	
1063-4806	310	3856			
1063-4814	308	3855			
1063-4939	1175	3212			
1063-5019	797	3158	IC-233		
1063-5142	171	3104A	27		
1063-5209	147A		ZD-33		
1063-5381	123A	3018	20	2051	
1063-5423	159	3114/290	82	2032	
1063-5431	159	3114/290	82	2032	
1063-5449	159	3114/290	82	2032	
1063-5704	198		32		
1063-6454	5071A	3334	ZD-6.8	561	
1063-6926	159	3114/290	82	2032	
1063-79	116		504A	1104	
1063-7916	171	3104A	27		
1063-8369	162	3559	35		
1063-8435	234	3247	65	2050	
1063-8591	109	3088	1N60	1123	
1063-8963	289A	3122	210	2051	
1063-8971	125	3051/156	510	1114	
1064(G.E.)	712		IC-2		
1064(GE)	712		IC-2		
1064-4417	175		246	2020	
1064-6032	130	3027	14	2041	
1065(G.E.)	748	3236			
1065-2055	1173	3729	IC-302		
1065-3225			66	2048	
1065-3525		3054	66		
1065-4861	7400	7400		1801	
1065-6775	140A	3061	ZD-10	562	
1065-9928	116	3016	504A	1104	
1065-9936	198		251		
1065-9944	175	3538	12	2020	
1067	154	3024/128	40	2012	
1067(GE)	154		40	2012	
1068-17	237	3299	46		
1068-17A	224	3049	46		
1069		3054	66	2048	
1069(GE)	152		66	2048	
1069-7032		3027	75	2041	
1070	125	3032A	510,531	1114	
1070-0623	125	3032A	509	1114	
1070-0631	198	3220	251		
1071		3083	69	2049	
1071(GE)	153		69	2049	
1071-3642	130		14	2041	
1071-4913	289A	3024/128	210	2038	
1072	193	3138	67	2023	
1072G		3138	67	2023	
1072K			67	2023	
1072K(GE)	193		67	2023	
1073	163A	3111	36		
1074	163A	3439	259		
1074(GE)	163A		36		
1074-03	199	3122	62	2010	
1074-115	199	3122	62	2010	
1074-116	186A	3357	28	2017	
1074-117	390	3027/130	255	2041	
1074-118	519	3100	514	1122	
1074-119	116	3017B/117	504A	1104	
1074-120	139A	3060	ZD-9.1	562	
1074-121	5075A	9022/5076A	ZD-17		
1074-122	5072A	3059/138A	ZD-8.2	562	
1074-123	5013A	3058/137A	ZD-6.2	561	
1074-124	109			1123	
1074-24	109	3087		1123	
1074T	238	3710	259		

Industry Standard No.	ECG	SK	GE	RS 276-	MOTOR.
1075	233	3039/316	210	2009	
1075(GE)	749		IC-97		
1075BLUE(GE)	749		IC-97		
1076(GE)	749		IC-97		
1076-0999	294	3138/193A	48		
1076-1377	161	3132	61	2038	
1076-1484	177	3100/519	300	1122	
1076-1559	154	3044	40	2012	
1076-1674	125	3080	509	1114	
1076BLUE	749		IC-97		
1077		3077	IC-5		
1077(GE)	713		IC-5		
1077-0261	116		504A	1104	
1077-07	123A	3018	20	2051	
1077-2283	294	3841			
1077-2325	109	3087	1N34AS	1123	
1077-2341	177	3100/519	300	1122	
1077-2366	116	3311	504A	1104	
1077-2382	1003	3288	IC-43		
1077-2390	1006	3358	IC-38		
1077-2408	1052	3249	IC-135		
1077-2760	109	3087	1N34AS	1123	
1077-3836	116	3311	504A	1104	
1077-3844	1218	3740			
1077-9296	109		1N60	1123	
1079		3141	IC-21		
1079(GE)	783		IC-225		
1079-01	177	3100/519	300	1122	
1079-85	159	3466	82	2032	
1079-89	177	3100/519	300	1122	
1079GREEN	783		IC-225		
1080	125	3032A	510,531	2030	
1080-01	229	3122	11	2015	
1080-03	123A	3122	62	2010	
1080-05	282	3748	264		
1080-06	186A	3357	247	2017	
1080-07	289A	3122	210	2038	
1080-08	116	3017B/117	504A	1104	
1080-10	139A	3060	ZD-9.1	562	
1080-130	152	3054/196	66	2048	
1080-20	123A	3444	61	2015	
1080-21	199	3124/289	62	2010	
1080-5364	130	3027	14	2041	
1080-6396	123A	3444	20	2051	
1080-7584			243	2030	
1080G	129	3114/290	244	2027	
1080G(GE)	193		67	2023	
1081-3087		3018	271	2030	
1081-3186	147A	3095	ZD-33		
1081-3285	5511	3502			
1081-3293	5401	3638/5402			
1081-3301	123A	3018	20	2051	
1081-3319	123A	3444	20	2051	
1081-3343	238	3710	259		
1081-3350	190	3199/188	217		
1081-3368	152	3054/196	66	2048	
1081-3475	123A	3444	20	2051	
1081-3541	1175	3212			
1081-4000	159	3466	82	2032	
1081-4010	159	3466	82	2032	
1081-4739	903	3540			
1081-7104	390	3027/130	255	2041	
1081-9464	123A	3124/289	20	2051	
1081K94-6	941		IC-263	010	
1081K94-7	909	3590	IC-249		
1081K94-9	941		IC-263	010	
1084	125	3032A	523	1114	
1084-9784	159	3466	82	2032	
1086		3303	523		
1087			ZD-15	2030	
1087-01	299	3054/196	236		
1087-02	123A	3122	20	2051	
1087-2380			243	2030	
1089 6199	159			2032	
1089-6199		3466	82	2032	
1089-9307	912		IC-172		
1090	125	3033	510,531	1114	
1091	5080A	3336	ZD-22		
1091(GE)	5080A		ZD-22		
1092		3079	75	2041	
1092(GE)	181		75	2041	
1092-16	177	3100/519	300	1122	
1095-01	312	3116	FET-2	2035	
1095J2F	116	3017B/117	504A	1104	
1096-11	195A	3048/329	46		
1096-12	199	3124/289	212	2010	
1096J2F	116	3017B/117	504A	1104	
1097	783	3215	IC-225		
1097(GE)	186	3192	28	2017	
1097-85	294	3138/193A	48		
1098-14	152	3893	66	2048	
1098-15	195A	3048/329	46		
1099-0950	128	3024	243	2030	
1100			29	2025	
1100(GE)	187		29	2025	
1100-75	195A	3765	46		
1100-9446	107	3293	61		
1100-9453	107	8293	61		
1100-9461	123AP	3444/123A	20	2051	
1100-9479	123AP	3854	20	2051	
1100-9487	519		514	1122	
1101	5063A	3837			
1101-8181	124		12		
1101-9072	5526	6627/5527			
1102	134A	3055	ZD-3.6		
1102(ZENER)	5005A	3771	ZD-3.3		
1102-17	158	3004/102A	53	2007	
1102-17A	102A	3004	53	2007	
1102-63	101	3861	8	2002	
1103	134A	3055	ZD-3.6		
1103-88	116		504A	1104	
1104	5009A	3033	ZD-4.7		
1104-94	102A		53	2007	
1104-95	103A	3835	59	2002	
1104-96	116		504A	1104	
1105	5011A	3777	ZD-5.6	561	
1105-15	121	3717	239	2006	
1106	5014A	3780	ZD-6.8	561	
1106-29	116		504A	1104	
1106-36	116		504A	1104	
1106-97	108	3452	86	2038	
1106-99	128		243	2030	
1107	5015A	3781	ZD-7.5		
1108	5016A	3782	ZD-8.2	562	
1108-73	116		504A	1104	
1109	5018A	3784	ZD-9.1	562	
1109(GE)	196		241	2020	
1110	5019A	3785	ZD-10	562	
1110-86	116		504A	1104	
1111	5020A	3786	ZD-11	563	
1111(GE)	152		66	2048	
1111-17	160		245	2004	
1111-18	160		245	2004	
1111-504-08000	HIDIV-1		FR-8		

Industry Standard No.	ECG	SK	GE	RS 276-	MOTOR.
1111P	1002	3481	IC-69		
1112	5021A	3787	ZD-12	563	
1112-78	128		243	2030	
1112-79	195A	3765	46		
1112-8	159	3466	82	2032	
1113	5023A	3789	ZD-14	564	
1113(GE)	153		69	2049	
1113-03	123A	3444	20	2051	
1113-13	160		245	2004	
1113-2875	102	3722	2	2007	
1115	165	3063/145A	38		
1115(GE)	165		36		
1115-16	116		504A	1104	
1116	5026A	3054/196	ZD-17		
1116(GE)	186	3192	28	2017	
1116-42	116		504A	1104	
1116-6527	130		14	2041	
1116-6535	128		243	2030	
1117(MC)			20	2051	
1117-76	116		504A	1104	
1118	5027A	3793	ZD-18		
1118-17	109	3017B/117		1123	
1118-20	116		504A	1104	
1119		3104A	27		
1119(GE)	171		27		
1119-17	109	3087		1123	
1119-4628	181		75	2041	
1119-54	160		245	2004	
1119-55	160		245	2004	
1119-56	160	3006	245	2004	
1119-57	102A		53	2007	
1119-58	103A		59	2002	
1119-59	102A		53	2007	
1119-8132	123A	3444	20	2051	
1120	5029A	3795	ZD-20		
1120-17	177		300	1122	
1120-18	116		504A	1104	
1121-17	116	3017B/117	504A	1104	
1122	5030A	3796			
01122-0073	109	3087		1123	
1122-96	126		52	2024	
1123(JULIETTE)	116		504A	1104	
1123-3335	175	3026	246	2020	
1123-55	108	3452	86	2038	
1123-56	108	3452	86	2038	
1123-57	108	3452	86	2038	
1123-58	108	3452	86	2038	
1123-59	108	3452	86	2038	
1123-60	123A	3444	20	2051	
1124	121	3009	239	2006	
1124A	121	3009	239	2006	
1124B	121	3009	239	2006	
1124C	104	3009	16	2006	
1125		3054	28	2017	
1125(GE)	186	3192	28	2017	
1125-2582	159	3466	82		
1125-2608	5014A			561	
1127	146A	3064	ZD-27		
1127-2077	910		IC-251		
1128-17	102A		53	2007	
1129		3072	IC-2		
1129(GE)	712		IC-2		
1132-2(RCA)	184		57	2017	
1133	147A	3095	ZD-33		
1133-14	785	3254	IC-226		
1136(GE)	783		IC-225		
1136GREEN	783		IC-225		
1137		3077	IC-5		
1137(GE)	713		IC-5		
1138	116	3016	504A	1104	
1138(GE)	728		IC-22		
1138RED(GE)	728		IC-22		
1139-17	116	3016	504A	1104	
1140-17	102A	3123	53	2007	
1145	100	3005	1	2007	
1145-17	102A		53	2007	
1146	100	3005	1	2007	
1147		3114	21	2034	
1147(GE)	159		82	2032	
1147-08	960		VR-102		
1147-09	4011B		4011	2411	
1148-17	158	3004/102A	53	2007	
1151(GE)	749		IC-97		
1151BLUE(GE)	749		IC-97		
1152	234		65	2050	
1152(GE)	749		IC-97		
1152BLUE(GE)	749		IC-97		
1153(GE)	783		IC-225		
1153GREEN	783		IC-225		
1154		3777	IC-21		
1154(GE)	783		IC-225		
1154GREEN	783		IC-225		
1156	1194	3484	513		
1156(GE)	513		513		
1156H	1194	3484			
1157	128	3024	243	2030	
1164(GE)	712		IC-2		
1164B	6026		5128		
1164D	6034		5132		
1164F	6040		5136		
1164H	6042		5140		
1164K	6044		5140		
1165	712	3072	IC-2		
1166(GE)	124		12		
1166-7821	102	3722	2	2007	
1169		3104A	40	2012	
1173(GE)	712		IC-2		
1174(GE)	749		IC-97		
1178		3072	IC-2		
1178(GE)	712		IC-2		
1178-3453	102	3004	2	2007	
1179(GE)	712		IC-2		
1179-0334	5529	6629			
1180-0182	102	3722	2	2007	
1183-17	195A	3048/329	46		
1184G	129	3025	244	2027	
1184G(GE)	193		67	2023	
1186		3118	22	2032	
1186(GE)	159		82	2032	
1186-1		3126	90		
1186-3		3126	90		
1187	128	3044/154	243	2030	
1187(GE)	192		63	2030	
1190	165	3115	38		
1192	102	3004	2	2007	
1192(GE)	712		IC-2		
1192(XSTR)			2	2007	
1193(GE)	712		IC-2		
1202		3124	20	2051	
1203	5240A	3124/289			
1203(GE)	123A	3444	20	2051	
1203-169	123A	3444	20	2051	
1204	5242A	3024/128			
1204(GE)	128		243	2030	

Industry Standard No.	ECG	SK	GE	RS 276-	MOTOR.
1205	116	3031A			
1205(GE)	123A	3444	20	2051	
1206	5247A	3024/128			
1206(GE)	128		243	2030	
1206-17	110MP	3088	1N34AS	1123	
1207	5248A	3018	20		
1207(GE)	128		243	2030	
1207-17	109	3088		1123	
1208	5249A	3018			
1208(GE)	123A		20	2051	
1210-17	313	3122	278	2016	
1210-17(MIXER)			10	2051	
1210-17(OSC)			20	2051	
1210-17B	123A	3444	20	2051	
1212	5254A	3100/519	300	1122	
1212(ROSS)			300	1122	
1212-4	235	3197	215		
1214	159	3114/290	82	2032	
1217SK333	311		277		
1227-17	123A	3444	20	2051	
1228-17	123A	3444	20	2051	
1229H	108	3452	86	2038	
1236-3750	159	3466	82	2032	
1236-3776	123A	3444	20	2051	
1239 5752	175			2020	
1239-5752			246	2020	
1241-719	126	3008	52	2024	
1241A	102	3004	2	2007	
1248	102A	3004	53	2007	
1251-1-1	222	3050/221	FET-4	2036	
1254		3114	21	2034	
1254(GE)	159		82	2032	
1263A	116	3016	504A	1104	
1264		3077	IC-18		
1264(GE)	713		IC-5		
1269	184		57	2017	
1272	123A	3444	20	2051	
1277-17	102A	3123	53	2007	
1278	749		IC-97		
1278(GE)	749		IC-97		
1282		3024	28	2017	
1284	108	3452	86	2038	
1285		3025	29	2025	
1285(GE)	187		29	2025	
1289			504A	1104	
1294		3114	21	2034	
1294(GE)	159		82	2032	
1300-1	128		243	2030	
1300A	128		243	2030	
1300RN			11	2015	
1301-1	160		245	2004	
1301-2	160		245	2004	
1302	5111A	3377			
1303	5113A	3379			
1305	5117A	3383			
1306	5120A	3386			
1307	5121A	3387			
1307H	722		IC-9		
1308	5122A	3388	5ZD-8.2		
1309	5124A	3390			
1310	5125A	3391			
1311	5126A	3392	5ZD-11		
1312	5127A	3393	5ZD-12		
1313	5129A		5ZD-14		
1314	159	3466	82	2032	
1314(GE)	159		82	2032	
1315	123A	3444	20	2051	
1316	123A	3444	20	2051	
1316-17	102A	3004	53	2007	
1317-17	102A		53	2007	
1320	102	3004	2	2007	
1321-4051	5995	3518	5105		
1321-7724	102	3722	2	2007	
1321-7732	102	3722	2	2007	
1326PC	739	3235	IC-30		
1329	102	3004	2	2007	
1330	102	3004	2	2007	
01339	116		504A	1104	
1339(GE)	181		75	2041	
1340	100	3005	1	2007	
1341	234	3118	65	2050	
1344-3767	101	3861	8	2002	
1344-7321	102	3722	2	2007	
1345(GE)	186		28	2017	
1347	739	3235	IC-30		
1347-17	102A	3004	53	2007	
1348	739	3235	IC-30		
1348A14H01	7476	7476		1813	
1349-17	103A	3010	59	2002	
1350	100	3005	1	2007	
1351	738	3167	IC-29		
1352	738	3167	IC-29		
1358GE	186		28	2017	
1360	102	3004	2	2007	
1362-17	102A	3004	53	2007	
1362-17A	102A	3004	53	2007	
1364-17	102A	3004	53	2007	
1368C/D	123A	3124/289	20	2051	
1369-6	815	3255			
1371	513	3443	513		
1371-17	103A	3010	59	2002	
1373-17	233	3018	210	2009	
1373-17A	107	3018	11	2015	
1373-17AL	108	3452	86	2038	
1374-17	289A	3124/289	268	2038	
1374-17A	289A	3124/289	268	2038	
1374-17AC	123A	3444	20	2051	
1376-001	5697	3522			
1376-002	5697	3522			
1380C	1196	3725			
1384(GE)	196		241	2020	
1385(GE)	197		250	2027	
1390	100	3005	1	2007	
1396	101	3861	8	2002	
1400	100	3005	1	2007	
1402B	123A		210	2051	
1402C	123A		210	2051	
1402E	123A	3018	20	2051	
1405-17	116		504A	1104	
1410	100	3005	1	2007	
1410-102	177		514	1122	
1410-102(ROGERS)	117		504A	1104	
1410-167	116	3017B/117	504A	1104	
1410-169	519		514	1122	
1410-169(ROGERS)	177		300	1122	
1410-171	109	3087		1123	
1411-130	5071A		ZD-7.5		
1411-135	142A		ZD-12	563	
1411-136	140A		ZD-10	562	
1411-137	137A	3334/5071A	ZD-6.8	561	
1412-1	123A	3444	20	2051	
1412-1-12	123A	3444	20	2051	
1412-1-12-8	123A	3444	20	2051	
1412-170	116	3016	504A	1104	
1412-177	5452	3954/5414			
1412-182	5802	9005	510	1143	
1413-159	213	3012/105	4		
1413-160	102	3004	53	2007	
1413-168	179	3009	239	2006	
1413-172	179	3009	239	2006	
1413-175	102	3004	53	2007	
1413-178	121	3014	239	2006	
1414-157	128	3024	243	2030	
1414-158	159	3114/290	32		
1414-158(ROGERS)	159		82	2032	
1414-173	123AP	3025/129	243	2030	
1414-174	289A	3444/123A	20	2051	
1414-176	159	3466	82	2032	
1414-179	251	3027/130	14	2041	
1414-180	243	3026	246	2020	
1414-183	289A	3444/123A	20	2051	
1414-184	128	3044/154	243	2030	
1414-185	129	3025	244	2027	
1414-186	128	3024	243	2030	
1414-187	129	3025	244	2027	
1414-188	181	3036	14	2041	
1414-189	128	3024	243	2030	
1414A	121	3009	239	2006	
1414A-12	121	3009	239	2006	
1414A-12-8	121	3009	239	2006	
1415	123A	3444	20	2051	
1417-177	105	3012	4		
1420-1-1	107	3018	11	2015	
1420-2-2	161	3019	39	2015	
001422	123A	3444	20	2051	
1424	123A	3444	20	2051	
1428	129		244	2027	
1431 8349	123A	3444		2051	
1431-7184	181	3036	75	2041	
1431-8349			20	2051	
1434	291	3054/196			
1436-17	102A		53	2007	
1449	160		245	2004	
1455-7-4	107	3018	11	2015	
1458CE					MC1458CG
1458CP1	778A	3465	IC-220		
1458P1	778A	3465	IC-220		
1459	102	3004	2	2007	
1461	738	3167			
1462	74123		74123	1817	
1463	123A	3444	20	2051	
1465	123A	3444	20	2051	
1465-1	160	3006	245	2004	
1465-1-12	160	3006	245	2004	
1465-1-12-8	160	3006	245	2004	
1465-2	103A	3010	59	2002	
1465-2-12	103A	3010	59	2002	
1465-2-12-8	103A	3010	59	2002	
1465-4	101	3861	8	2002	
1465-4-12	101	3861	8	2002	
1465-4-12-8	101	3861	8	2002	
1465A	160	3007	245	2004	
1465A-12	160	3007	245	2004	
1465A-12-8	160	3007	245	2004	
1465B	126	3008	52	2024	
1465B-12	126	3008	52	2024	
1465B-12-8	126	3008	52	2024	
1465C	121	3009	239	2006	
1465C-12	121	3009	239	2006	
1465C-12-8	121	3009	239	2006	
1466-1	102A	3003	53	2007	
1466-1-12	102A	3003	53	2007	
1466-1-12-8	102A	3003	53	2007	
1466-2	102A	3004	53	2007	
1466-2-12	102A	3004	53	2007	
1466-2-12-8	102A	3004	53	2007	
1466-3	102A	3004	53	2007	
1466-3-12	102A	3004	53	2007	
1466-3-12-8	102A	3004	53	2007	
1471-4406	910		IC-251		
1471-4414	726	3022/1188			
1471-4729	5995	3518	5105		
1471-4778	123A	3444	20	2051	
1471-4802	130		14	2041	
1472 8349	199			2010	
1472-8349		3245	62	2010	
1473-4255	128		243	2030	
1474-3	100	3005	1	2007	
1474-3-12	100	3005	1	2007	
1474-3-12-8	100	3005	1	2007	
1474-3736	911		IC-253		
1476 1118	128			2030	
1476-1118			243	2030	
1476-17	102A	3004	53	2007	
1476-17-6	102A	3004	53	2007	
1477-3352	906	3548	IC-246		
1477-3436	904	3542			
1477A	105	3012	4		
1477A-12	105	3012	4		
1477A-12-8	105	3012	4		
1477B	121MP	3013	239(2)	2006(2)	
1477B-12	121MP	3013	239(2)	2006(2)	
1477B-12-8	121MP	3013	239(2)	2006(2)	
1477C	121MP	3014	239(2)	2006(2)	
1477C-12	121MP	3014	239(2)	2006(2)	
1477C-12-8	121MP	3014	239(2)	2006(2)	
1479 7963	199			2010	
1479 7989	123A	3444			
1479-0240	74H00	74H00			
1479-0273	941		IC-263	010	
1479-7963		3245	62	2010	
1479-7971	74H04	74H04			
1479-7989			20	2051	
1479-8011	923		IC-259		
1479-8029	159	3466	82	2032	
1482	123A	3444	20	2051	
1482-17	195A	3047	46		
1483	128	3122	243	2030	
1484A	121	3009	239	2006	
1484A-12	121	3009	239	2006	
1484A-12-8	121	3009	239	2006	
1487-17	195A	3765	46		
1488(SEARS)	116		504A	1104	
1488B	160	3006	245	2004	
1488B-12	160	3006	245	2004	
1488B-12-8	160	3006	245	2004	
1488C	100	3005	1	2007	
1488C-12	100	3005	1	2007	
1488C-12-8	100	3005	1	2007	
1489-17	109	3087		1123	
1492-1	105	3012	4		
1492-1-12	105	3012	4		
1492-1-12-8	105	3012	4		
1493-17	123A	3444	20	2051	
1501	229	3018	61	2038	
1502B	229	3018	39	2038	
1502D	229	3018	61	2038	
1505-9850	124		12		

Industry Standard No.	ECG	SK	GE	RS 276-	MOTOR.
1507-7183	175		246	2020	
1510-2718	102	3722	2	2007	
1512	109	3090		1123	
1515		3010	20	2051	
1515(NPN)	199	3245	62	2010	
1515(PNP)	101		8	2002	
1521B	218		234		
1524	123	3124/289	20	2051	
1526	160		245	2004	
1534-8931	175	3026	246	2020	
1540	123A	3444	20	2051	
1542	74145	74145	74145	1828	
1543-00	519		514	1122	
1548-17	175	3131A/369	246	2020	
1550	109	3087		1123	
1550-17	143A	3750	ZD-13	563	
1553-17	159	3466	82	2032	
1559-17	121	3009	239	2006	
1559-17A	121	3009	239	2006	
1561-0404		3027	77		
1561-0408	181		75	2041	
1561-0410	181		75	2041	
1561-0604			14	2041	
1561-0608	181		75	2041	
1561-0610	181		75	2041	
1561-0615	181		75	2041	
1561-0803			75	2041	
1561-0804			75	2041	
1561-0805			75	2041	
1561-0808	181	3036	75	2041	
1561-0810	181	3036	75	2041	
1561-0815	181		75	2041	
1561-1004			75	2041	
1561-1005			75	2041	
1561-1010	181	3036	75	2041	
1561-1015	181		75	2041	
1561-17	121	3717	239	2006	
1561A603			14	2041	
1561A608			75	2041	
1561A615			75	2041	
1562	116	3311	504A	1104	
1567	123A	3444	20	2051	
1567-0	123A	3444	20	2051	
1567-2	123A	3444	20	2051	
1572	1196	3725			
1573-00	128		243	2030	
1573-01	128		243	2030	
1573PQ	287	3433	222		
1574-01	234		65	2050	
1582	159	3466	82	2032	
1582-0408			75	2041	
1582-0410			75	2041	
1582-0415			75	2041	
1582-0508			75	2041	
1582-0510			75	2041	
1582-0608			75	2041	
1582-0610			75	2041	
1582-0615			75	2041	
1582-0803			75	2041	
1582-0804			75	2041	
1582-0805			75	2041	
1582-0808			75	2041	
1582-0810			75	2041	
1582-0815			75	2041	
1582-1003			75	2041	
1582-1004			75	2041	
1582-1005			75	2041	
1582-1008			75	2041	
1582-1010			75	2041	
1582-1015			75	2041	
1585H	107		11	2015	
1607A80	7406	7406		1821	
1611-17	116	3017B/117	504A	1104	
1612SK24E	312	3112	FET-1	2028	
1615C	113A	3119/113	6GC1		
1616C	114	3120	6GD1		
1617C	115	3121	6GX1		
1618	152	3197/235			
1627 1843	177			1122	
1627-1843			300	1122	
1634-17	313	3122	278	2016	
1634-17-14A	108	3452	86	2038	
1673-0475	128		243	2030	
1674-17	116	3017B/117	504A	1104	
1676-1991	128		243	2030	
1677-1149	923D		IC-260	1740	
1678	384	3197/235	215		
1679 7391	159			2032	
1679-7391		3466	82	2032	
1687-17	107	3122	11	2015	
1699-17	177	3100/519	300	1122	
1702M	293		210	2051	
1703-8662	519		514	1122	
1705-4834	123A	3444	20	2051	
1705-5351	128		243	2030	
1710	289A	3122	210	2038	
1711	123A	3444	20	2051	
1711-17	123A	3444	20	2051	
1711MC	123A	3444	20	2051	
1712-17	123A	3444	20	2051	
1713-17	116	3311	504A	1104	
1714-0402	175	3026	246	2020	
1714-0405		3026	66	2048	
1714-0602	175	3026	246	2020	
1714-0605	175	3026	246	2020	
1714-0802	175	3026	246	2020	
1714-0805	175	3026	246	2020	
1714-1002	175	3026	246	2020	
1714-1005	175	3026	246	2020	
1721-0725	218		234		
1723-0405	327	3036	75	2041	
1723-0410	327	3036	75	2041	
1723-0605	327	3036	75	2041	
1723-0610	327	3036	75	2041	
1723-0805	327	3036	75	2041	
1723-0810	327	3036	75	2041	
1723-1005	327	3036	75	2041	
1723-1010	327	3036	75	2041	
1723-1205	327	3036	75	2041	
1723-1210	327	3036	75	2041	
1723-1405	327	3036	75	2041	
1723-1410	327	3036	75	2041	
1723-1605	327	3036	75	2041	
1723-1610	327	3036	75	2041	
1723-17	123A	3444	20	2051	
1723-1805	181	3036	75	2041	
1723-1810	181	3036	75	2041	
1729-7284	923		IC-259		
1741-0051	7400	7400	7400	1801	
1741-0069	74H00	74H00			
1741-0085	7401	7401	IC-194		
1741-0119	7402	7402	7402	1811	
1741-0143	7404	7404	7404	1802	
1741-0150	74H04	74H04			
1741-0176	7405	7405			
1741-0200	7408	7408	7408	1822	
1741-0234	7410	7410	7410	1807	
1741-0325	7420	7420	7420	1809	
1741-0416	7430	7430			
1741-0440	7437	7437			
1741-0473	7440	7440			
1741-0564	7451	7451	7451	1825	
1741-0622	7454	7454			
1741-0747	7475	7475	7475	1806	
1741-0804	7486	7486	7486	1827	
1741-0895	7493A	7493			
1741-0952	7495	7495			
1741-1018	74107	74107			
1741-1042	74150		74150	1829	
1741-1075	74151	74151			
1741-1190	7441	7441	7441	1804	
1741-1224	7442	7442			
1741-1257	74121	74121			
1743-0610	327	3945			
1743-0630	327	3945			
1743-0820	327	3945			
1743-1010	327	3945			
1743-1030	327	3945			
1743-1220	327	3945			
1743-1410	327	3945			
1743-1420	327	3945			
1743-1430	327	3945			
1743-1620	327	3945			
1750-103	177		300	1122	
1751-17	199	3122	62	2010	
1751G036	123A	3444	20	2051	
1756-17	177	3100/519	300	1122	
1761-17	108	3452	86	2038	
1763-0415	181	3036	75	2041	
1763-0420	181	3036	75	2041	
1763-0425	181	3036	75	2041	
1763-0615	181	3036	75	2041	
1763-0620	181	3036	75	2041	
1763-0625	181	3036	75	2041	
1763-0815	181	3036	75	2041	
1763-0825	181	3036	75	2041	
1763-1015	181	3036	75	2041	
1763-1020	181	3036	75	2041	
1763-1025	181	3036	75	2041	
1763-1215	181	3036	75	2041	
1763-1220	181	3036	75	2041	
1763-1225	181	3036	75	2041	
1763-1415	181	3036	75	2041	
1763-1420	181	3036	75	2041	
1763-1425	181	3036	75	2041	
1766	142A		ZD-12	563	
1777-17	102A	3004	53	2007	
1778-17	109	3087		1123	
1789-17	128		243	2030	
1792-17	108	3452	86	2038	
1794-17	139A	3059/138A	ZD-9.1	562	
1794-17(10)			ZD-9.1	562	
1799-17	123A	3444	20	2051	
1800-17	123A	3444	20	2051	
1800PC	9800				MC1800P
1801	161	3132	39	2015	
1801PC	9801				MC1801P
1802L	294		48		
1802M	294		48		
1802PC	9802				MC1802P
1803PC	9803				MC1803P
1804-17	154	3024/128	40	2012	
1804PC	9804				MC1804P
1805	7401	7401			
1805PC	9805				MC1805P
1806	7404	7404		1802	
1806-17	130	3027	14	2041	
1806-17A	130	3027	14	2041	
1806PC	9806				MC1806P
1807	7442	7442			
1807PC	9807				MC1807P
1808	7490	7490		1808	
1808PC	9808				MC1808P
1808T10				1152	
1809			300	1122	
1809PC	9809				MC1809P
1810PC					MC1810P
1811-110-10111	112		1N82A		
1811-110-10118	119		CR-2		
1811-110-10191	118		CR-1		
1811PC	9811				MC1811P
1812-1	123A	3444	20	2051	
1812-1-12	123A	3444	20	2051	
1812-1-127	123A	3444	20	2051	
1812-1L	123A	3444	20	2051	
1812-1L8	123A	3444	20	2051	
1812PC	9812				MC1812P
1813PC	9813				MC1813P
1814A	121	3009	239	2006	
1814A-12	121	3009	239	2006	
1814A-127	121	3009	239	2006	
1814AL	121	3009	239	2006	
1814AL-8	121	3717	239	2006	
1814PC	9814				MC1814P
1818	177		300	1122	
1820-0054	7400	7400		1801	
1820-0055	7490	7490		1808	
1820-0058	909	3590	IC-249		
1820-0063	7451	7451		1825	
1820-0068	7410	7410		1807	
1820-0069	7420	7420		1809	
1820-0070	7430	7430			
1820-0075	7473	7473		1803	
1820-0077	7474	7474		1818	
1820-0099	7493A	7493			
1820-0125	911		IC-253		
1820-0174	7404	7404		1802	
1820-0216-1	941M		IC-265	007	
1820-0217-1	941M		IC-265	007	
1820-0240	906		IC-246		
1820-0248	909	3590	IC-249		
1820-0261	74121	74121			
1820-0301	7475	7475		1806	
1820-0306	724	3525	IC-86		
1820-0328	7402	7402		1811	
1820-0398	910D		IC-252		
1820-0430	309K	3629			
1820-0476	915			017	
1820-0491	74145	74145			
1820-0707	74141	74141			
1820-0870	7408	7408		1822	
1820-0894	7404	7404		1802	
1820-0995	7447	7447			
1820-1037	7446	7446			
1820-1038	7448	7448			
1820-1064	74164	74164			
1820-1111	7440	7440			
1820-1172	7416	7416			

Industry Standard No.	ECG	SK	GE	RS 276-	MOTOR.
1821-0001	912		IC-172		
1821-0002	912		IC-172		
1826-0009	925		IC-261		
1826-0010	923		IC-259		
1826-0044	725		IC-19		
1826-0055	911D		IC-254		
1826-0070	941D		IC-264		
1826-0075	914		IC-256		
1826-0166	910		IC-251		
1827-17	102A		504A	1104	
1835-17	199	3122	62	2010	
1840-17	131	3052	44	2006	
1841-17	123A	3444	20	2051	
1843-17	161	3716	39	2015	
1844-17	159	3466	82	2032	
1845-17	161	3132	39	2015	
1846-17	177	3100/519	300	1122	
1848-17	116		504A	1104	
1849	116	3031A	504A	1104	
1849R	116(2)	3031A	504A(2)	1104	
1850	116(2)	3031A	504A	1104	
1850-0040	102	3722	2	2007	
1850-0040-1	102	3722	2	2007	
1850-0060	102	3722	2	2007	
1850-0062	102	3722	2	2007	
1850-0062-1	102	3722	2	2007	
1850-0101	102	3722	2	2007	
1850-0184	102	3722	2	2007	
1850-0184-1	102	3722	2	2007	
1850-17	159	3466	82	2032	
1850R	116(2)	3031A	504A	1104	
1851-17	117	3100/519	504A	1104	
1852-17	108	3039/316	86	2038	
1853-0001-1	159	3466	82	2032	
1853-0041	129		244	2027	
1853-0045	129		244	2027	
1853-0069	106	3984	21	2034	
1853-0081	159	3466	82	2032	
1853-0089		3466	82	2032	
1853-0215	129		244	2027	
1854-0003	123A	3444	20	2051	
1854-0005	123A	3444	20	2051	
1854-0022-1	128		243	2030	
1854-0033	123A	3444	20	2051	
1854-0060			62	2010	
1854-0090	128		243	2030	
1854-0090-1	128		243	2030	
1854-0231		3716	39	2015	
1854-0245	181	3036	75	2041	
1854-0265	175		246	2020	
1854-0274	128		243	2030	
1854-0332	128		243	2030	
1854-0353	123A	3444	20	2051	
1854-0387	199		62	2010	
1854-0417	107		11	2015	
1854-0432	123A	3444	20	2051	
1854-0438-1	124		12		
1854-0490-1	181		75	2041	
1854-0498	128		243	2030	
1854-0563	130	3027	14	2041	
1854-17	116	3311	504A	1104	
1854-SRK1-1	199	3245	62	2010	
1855-17	116	3311	504A	1104	
1858		3861	8	2002	
1858-0004	906		IC-246		
1858-0019	917		IC-258		
1858-0023	916	3550	IC-257		
1858A	139A	3060	ZD-9.1	562	
1859-14	121	3009	239	2006	
1859-16	121	3009	239	2006	
1859-17	312	3116	FET-1	2035	
1859R	116		504A	1104	
1865-1	160	3006	245	2004	
1865-1-12	160	3008	245	2004	
1865-1-127	160	3006	245	2004	
1865-1L	160	3006	245	2004	
1865-2	103A	3010	59	2002	
1865-2-12	103A	3010	59	2002	
1865-2L	103A	3010	59	2002	
1865-2L8	103A	3010	59	2002	
1865-4	101	3861	8	2002	
1865-4-12	101	3861	8	2002	
1865-4-127	101	3861		2002	
1865-4L	101	3861	8	2002	
1865-4L8	101	3861	8	2002	
1865A	160	3007	245	2004	
1865A-12	160	3007	245	2004	
1865A-127	160	3007	245	2004	
1865AL8	160	3007	245	2004	
1865B	126	3008	52	2024	
1865B-12	126	3008	52	2024	
1865B-127	126	3008	52	2024	
1865BL	126	3008	52	2024	
1865BL8	126	3008	52	2024	
1865C	121	3009	239	2006	
1865C-12	121	3009	239	2006	
1865C-127	121	3009	239	2006	
1865CL	121	3009	239	2006	
1865CL8	121	3009	239	2006	
1866-1	102A	3003	53	2007	
1866-1-12	102A	3003	53	2007	
1866-1-127	102A	3003	53	2007	
1866-17	123A	3024/128	20	2051	
1866-1L	102A	3003	53	2007	
1866-1L8	102A	3003	53	2007	
1866-2	102A	3004	53	2007	
1866-2-12	102A	3004	53	2007	
1866-2-127	102A	3004	53	2007	
1866-2L	102A	3004	53	2007	
1866-2L8	102A	3004	53	2007	
1866-3	102A	3004	53	2007	
1866-3-12	102A	3004	53	2007	
1866-3-127	102A	3004	53	2007	
1866-3L	102A	3004	53	2007	
1866-3L8	102A	3004	53	2007	
1867-17	159	3466	82	2032	
1872-3	506	3843	511	1114	
1872-6	177		300	1122	
1874-3	100	3005	1	2007	
1874-3-12	100	3005	1	2007	
1874-3-127	100	3005	1	2007	
1874-3L	100	3005	1	2007	
1874-3L8	100	3005	1	2007	
1877A	105	3012	4		
1877A-12	105	3012	4		
1877A-127	105	3012	4		
1877AL	105	3012	4		
1877AL8	105	3012	4		
1877B	121MP	3013	239(2)	2006(2)	
1877B-12	121MP	3013	239(2)	2006(2)	
1877B-127	121MP	3013	239(2)	2006(2)	
1877BL	121MP	3013	239(2)	2006(2)	
1877BL8	121MP	3013	239(2)	2006(2)	
1877C	121MP	3014	239(2)	2006(2)	
1877C-12	121MP	3014	239(2)	2006(2)	
1877C-127	121MP	3014	239(2)	2006(2)	
1877CL	121MP	3014	239(2)	2006(2)	
1877CL8	121MP	3014	239(2)	2006(2)	
1879-17	123A	3047	20	2051	
1879-17A	128	3047	243	2030	
1880-17	108	3452	86	2038	
1881-17	108	3452	86	2038	
1882-17	123A	3444	20	2051	
1883-17	123A	3444	20	2051	
1884-0088	5511	3502/5513			
1884-17	123A	3444	20	2051	
1884A	121	3009	239	2006	
1884A-12	121	3009	239	2006	
1884A-127	121	3009	239	2006	
1884AL	121	3009	239	2006	
1884AL8	121	3009	239	2006	
1885-17	154	3311	40	2012	
1885-17(RECT.)	116		504A	1104	
1888B	160	3006	245	2004	
1888B-12	160	3009	245	2004	
1888B-127	160	3006	245	2004	
1888BL	160	3006	245	2004	
1888BL8	160	3006	245	2004	
1888C	100	3005	1	2007	
1888C-12	100	3005	1	2007	
1888C-127	100	3005	1	2007	
1888CL	100	3005	1	2007	
1888CL8	100	3005	1	2007	
1889-17	129	3025	244	2027	
1890-17	108	3452	86	2038	
1892-1	105	3012	4		
1892-1-12	105	3012	4		
1892-1-127	105	3012	4		
1892-1L	105	3012	4		
1892-1L8	105	3012	4		
1893-17	123A	3444	20	2051	
1901-0028	5804	9007		1144	
1901-0036	5804	9007		1144	
1901-0045	5802	9005		1143	
1901-0388	5804	9007		1144	
1901-0389	5804	9007		1144	
1901-1011	907		IC-247		
1906-17	103A	3010	59	2002	
1907-6553	941D		IC-264		
1915-17	123A	3444	20	2051	
1916-17	195A	3047	46		
1917-17	102A		53	2007	
1919-17	102A	3004	53	2007	
1919-17A	102A	3004	53	2007	
1923-17	108	3452	86	2038	
1923-17-1	108	3452	86	2038	
1925-17	229	3122	61	2038	
1929-17	123A	3444	20	2051	
1931-17	107	3018	11	2015	
1931-17A	108	3452	86	2038	
1932-17	123A	3444	20	2051	
1933-17	128	3024	243	2030	
1934-17	312	3448	FET-1	2035	
1935-17	234	3025/129	65	2050	
1936-17	175	3026	246	2020	
1937-17	103A		59	1122	
1937-17(DIODE)	177		300	1122	
1940-17	159	3466	82	2032	
1941-17	116	3017B/117	504A	1104	
1945-17	131	3052	44	2006	
1946-17	102A		53	2007	
1947-17	110MP	3087	1N60	1123(2)	
1949-17	116	3017B/117	504A	1104	
1950-17	116	3031A	504A	1104	
1951-17	126	3006/160	52	2024	
1952-17	161	3117	39	2015	
1954-17	102A	3004	53	2007	
1955-17	131		44	2006	
1956-17	109	3087		1123	
1958-17	103A	3010	59	2002	
1960-17	102A	3004	53	2007	
1961-17	123A	3444	20	2051	
1966-17	123A	3444	20	2051	
1968-17	152	3054/196	66	2048	
1969-17	152	3054/196	66	2048	
1970-16	116	3311	504A	1104	
1970-17	116	3017B/117	504A	1104	
1971-17	125		510,531	1114	
1972-17	128		243	2030	
1973-17	102A		53	2007	
1974-17	102A		53	2007	
1977-17	109	3088		1123	
1979-17			300	1122	
1979-808-10	159	3466	82	2032	
1980-17	109	3088		1123	
1981-17	177		300	1122	
1982-17	116	3031A	504A	1104	
1983-17	108	3452	86	2038	
1984-17	123A	3444	20	2051	
1998-17	107	3018	11	2015	
1999	123A	3444	20	2051	
1999-17	107	3018	11	2015	
2000-004	977	3462			
2000-005	1194	3484			
2000-006	1195	3469			
2000-007	1082		IC-140		
2000-008	977	3462			
2000-009	973D	3233/973			
2000-017	1192	3445			
2000-019	1082	3461			
2000-021	1198	3963/1249			
2000-024	1195	3469			
2000-025	1194	3484			
2000-031	1249	3963			
2000-032	1278	3726			
2000-101	312	3116	FET-2	2035	
2000-102	222	3050/221	FET-4	2036	
2000-103	222	3050/221	FET-4	2036	
2000-104	312	3112	FET-1	2028	
2000-105	312	3116	FET-2	2035	
2000-107	312		FET-2	2035	
2000-111		3834	FET-2		
2000-125		3175	300	1122	
2000-201	290A	3114/290	269	2032	
2000-202	129	3114/290	21	2034	
2000-203	289A	3124/289	210	2051	
2000-204	233	3018	61	2038	
2000-205	229	3018	61	2038	
2000-206	123AP	3122	210	2051	
2000-207	295		270		
2000-208	235	3197	215		
2000-209	128	3020/123	88	2030	
2000-210	123A	3122	210	2051	
2000-211	236	3856/310	215		
2000-212	236	3197/235	216	2053	
2000-213	229	3124/289	213	2038	
2000-214	316	3018	17	2051	
2000-215	315		335		

Industry Standard No.	ECG	SK	GE	RS 276-	MOTOR.
2000-216	152	3054/196	66	2048	
2000-218	234	3025/129	48	2032	
2000-219	152	3893			
2000-220	152	3893			
2000-221	130	3027	14	2041	
2000-245	199	3245	62		
2000-258	85		212		
2000-265		3114	82		
2000-266		3245	62		
2000-267		3122	61		
2000-268	85	3356	211		
2000-269		3849	271		
2000-270	295	3253	270		
2000-271	235	3197	333		
2000-275	85		213		
2000-276	293		47	2030	
2000-301	109	3088	1N60	1123	
2000-302	519	3100	300	1122	
2000-303	519	3100	300	1122	
2000-304	116	3311	504A	1104	
2000-305	5018A	3060/139A	ZD-9.1	562	
2000-308	139A	3060	ZD-9.1	562	
2000-309	116	3126	90		
2000-312	5075A	3751	ZD-16	564	
2000-317	519	3100	300	1122	
2000-318	109	3100/519	300	1123	
2000-320	116	3311	504A	1104	
2000-321	135A		ZD-5.1		
2000-322	137A		ZD-6.2	561	
2000-323	177		90		
2000-324	138A		ZD-6.2	561	
2000-325	138A	3059	ZD-7.5		
2000-326	5075A	3751	ZD-16	564	
2000-327	139A	3126	ZD-9.1	562	
2000-328	137A		ZD-6.2	561	
2000-329	137A		ZD-6.2	561	
2000-332	519	3100	514	1122	
2000-339	601	3463			
2000-341	610	3323			
2000-342	605	3864			
2000-344	613	3327/614			
2000-345	177		514	1122	
2000-348	156	3051	512		
2001		3054	66	2048	
2001(E.F.JOHNSON)			66	2048	
2001-17	123A	3444	20	2051	
2002	121	3009	239	2006	
2002(E.F.JOHNSON)			239	2006	
2002-2	717		IC-209		
2003	130		5065	2041	
2003-17	123A	3444	20	2051	
2003-5	968	3593			
2003JDP1	128		243	2030	
2004			53	2007	
2004(TRANSISTOR)	102A		53	2007	
2004-01	107		11	2015	
2004-02	229		61	2038	
2004-03	123A	3444	20	2051	
2004-04	123A	3444	62	2051	
2004-05	123A		81	2051	
2004-06	159		82	2032	
2004-14	123A	3444	20	2051	
2004-67	177		300	1122	
2005	121		239	2006	
2005-1	712	3072	IC-2		
2005-2	712	3072	IC-2		
2005-2981	519		514	1122	
2006	123A	3444	20	2051	
2006-1	722	3161	IC-9		
2006-17	5022A	3788	ZD-13	563	
2006-2	722	3161	IC-9		
2007-01	121	3009	239	2006	
2007-1	708		IC-10		
2007-2	708	3135/709	IC-10		
2007-3	708	3135/709	IC-10		
2008	175		246	2020	
2008-1	725	3144/723	IC-19		
2008-17	123A	3444	20	2051	
2008-2	725	3144/723	IC-19		
2009	105	3012	4		
2010	105	3012	4		
2010-01	177	3017B/117	300	1122	
2010-02	116	3017B/117	504A	1104	
2010-03	177	3100/519	300	1122	
2010-17		3112	FET-1	2028	
2010-5409	912	3543	IC-172		
2011	184	3054/196	57	2017	
2012	102	3004	2	2007	
2012(E.F.JOHNSON)			2	2007	
2012-5472	923		IC-259		
2013(E.F.JOHNSON)	103A		59	2002	
2014-6684	941		IC-263	010	
2015	160	3006	245	2004	
2015-00	126	3006/160	52	2024	
2015-1	130	3027	14	2041	
2015-1A	130	3027	14	2041	
2015-2	130	3027	14	2041	
2015-2A	130	3027	14	2041	
2015-3	130	3027	14	2041	
2015-3A	130	3027	14	2041	
2015-4	130		14	2041	
2015-5	130		14	2041	
2015-6	130		14	2041	
2015-7	130		14	2041	
2017-107	159	3466	82	2032	
2017-108	186A	3192/186	28	2017	
2017-109	152	3893	66	2048	
2017-110	154	3024/128	40	2012	
2017-111	601	3463	504A	1104	
2017-113	5071A	3059/138A	ZD-7.5		
2017-114	116	3017B/117	504A	1104	
2017-115	123A	3444	20	2051	
2018-01	123A	3444	20	2051	
2019-45	519	3100	514	1122	
2020	160	3006	245	2004	
2020(RECT)	5896		5064		
2020-00	126	3006/160	52	2024	
2020-01	159	3114/290	82	2035	
2020-02	161	3716	39	2015	
2020-03	316		280		
2020-04	312		FET-1	2028	
2020-05	199	3245	62	2010	
2020-06	123A		81	2051	
2020-07	159		82	2032	
2020-1	703A	3157	IC-12		
2020-2	703A	3157	IC-12		
2020-3	703A	3157	IC-12		
2020R	5897		5065		
2021	160	3006	245	2004	
2021-00	126	3006/160	52	2024	
2021-05	116	3312	504A	1104	
2021-1	718	3159	IC-8		
2021-17	5022A	3788	ZD-13	563	
2021-2	718	3159	IC-8		
2022-01		3114	21	2034	
2022-03	229	3018	61	2038	
2022-04	235	3197	215		
2022-05	123A	3122	210	2051	
2022-06	222	3050/221	FET-4	2036	
2022-07	109	3100/519	1N60	1123	
2022-08	116	3017B/117	504A	1104	
2022-244	123A	3018	20	2051	
2023-41	139A		ZD-9.1	562	
2024-1	731	3170	IC-13		
2025-1	730	3143			
2025-2	730	3143			
2026	123A	3444	20	2051	
2026-00	123A	3444	20	2051	
2026-4	241	3188A/182	57	2020	
2026-5	241	3188A/182	57	2020	
2027	116	3048/329	220	2051	
2027(E.F.JOHNSON)	123A		220	2051	
2027(R.F.)	195A		46		
2027-00	195A	3765	220	2051	
2027R.F.			220	2051	
2028	108	3452	86	2038	
2028-00	108	3452	86	2038	
2029-1	790	3454	IC-230		
2029-2	790	3454	IC-230		
2030-1	715	3076	IC-6		
2031-1	714	3075	IC-4		
2031-17	116	3017B/117	504A	1104	
2032-33	108	3452	86	2038	
2032-34	108	3452	86	2038	
2032-35	294	3841	269	2032	
2032-36	312	3448	FET-1	2035	
2032-37			20	2051	
2032-40			81	2051	
2035-5100-53660	199	3124/289	212	2010	
2035-5100-69372	199	3124/289	62	2010	
2036-58	153		69	2049	
2036-59	152	3893	66	2048	
2036-68	909	3590	IC-249		
2036-72	909	3590	IC-249		
2036-93	909	3590	IC-249		
2039-2	101	3861	8	2002	
2040-17	140A	3061	ZD-10	562	
2041-01	229	3122	212	2038	
2041-02	287	3122	62	2010	
2041-03	186A	3357	247	2052	
2041-04	235	3197	215		
2041-05	177	3100/519	300	1122	
2041-06	139A	3060	ZD-9.1	562	
2042-17	102A	3004	53	2007	
2043-17	159	3466	82	2032	
2044-1	738	3167	IC-29		
2044-17	123	3124/289	20	2051	
2045-1	804	3455	IC-27		
2045-17	116	3311	504A	1104	
2047A2-288	102A		53	2007	
2048-17	229	3018	62	2010	
2049-32	614		90		
2052	116	3311	504A	1104	
2055-118	1193	3701			
2055-32	152	3893			
2056-04	1082	3461	IC-140		
2056-05	973	3233			
2056-06	1160		IC-196		
2056-10	159	3841/294			
2056-75	312		FET-2	2035	
02057-1	185	3083/197	58	2025	
2057A-120	108	3452	86	2038	
2057A-429	108	3452	86	2038	
2057A-430			21	2034	
2057A10-64	123A	3444	20	2051	
2057A100			14	2041	
2057A100-10			52	2024	
2057A100-14		3010	54		
2057A100-16	130MP	3027/130	15	2041(2)	
2057A100-17	128	3024	243	2030	
2057A100-18		3010	54		
2057A100-21	102A		53	2007	
2057A100-23	158	3004/102A	53	2007	
2057A100-24	102A	3004	53	2007	
2057A100-26	130MP	3036	15	2041(2)	
2057A100-30		3011	54		
2057A100-30(NPN)			59	2002	
2057A100-30(PNP)			53	2007	
2057A100-34	158	3004/102A	53	2007	
2057A100-35		3010	54		
2057A100-35(NPN)			59	2002	
2057A100-35(PNP)			53	2007	
2057A100-4		3004/102A	54		
2057A100-40(NPN)			63	2030	
2057A100-40(PNP)			67	2023	
2057A100-41	158	3004/102A	53	2007	
2057A100-44	158	3004/102A	53	2007	
2057A100-45(NPN)			28	2017	
2057A100-45(PNP)			29	2025	
2057A100-47		3190	57,58		
2057A100-47(NPN)			57	2017	
2057A100-47(PNP)			58	2025	
2057A100-48	158	3004/102A	53	2007	
2057A100-49	184	3054/196	57	2017	
2057A100-50		3202/210	28,29		
2057A100-51	159	3466	82	2032	
2057A100-53	107	3124/289	11	2015	
2057A100-55	102A		53	2007	
2057A100-58(NPN)			57	2017	
2057A100-58(PNP)			58	2025	
2057A100-62(NPN)			47	2030	
2057A100-66		3083/197	57,58		
2057A100-66(NPN)			57	2017	
2057A100-66(PNP)			58	2025	
2057A100-67	297	3137/192A	271	2030	
2057A100-8	102A	3004	53	2007	
2057A100-9	102A		53	2007	
2057A2-103	123A	3444	20	2051	
2057A2-109	107	3018	11	2015	
2057A2-110	107		11	2015	
2057A2-113	123A	3444	20	2051	
2057A2-116	107	3018	11	2015	
2057A2-117	107	3018	11	2015	
2057A2-117(OSC)			20	2051	
2057A2-119	108	3452	86	2038	
2057A2-120	107	3018	11	2015	
2057A2-121	123A	3444	20	2051	
2057A2-122	123A	3124/289	20	2051	
2057A2-127	108	3452	86	2038	
2057A2-128	107	3018	11	2015	
2057A2-131	123A	3444	20	2051	
2057A2-143			17	2051	
2057A2-145	123A	3444	20	2051	
2057A2-146	123A	3444	20	2051	
2057A2-147	102A	3123	53	2007	
2057A2-148	102A	3004	53	2007	
2057A2-149		3006	50	2004	
2057A2-150	129	3025	244	2027	

Industry Standard No.	ECG	SK	GE	RS 276-	MOTOR.
2057A2-151	128	3024	243	2030	
2057A2-152	123A	3444	20	2051	
2057A2-153	123A	3444	20	2051	
2057A2-154	123A	3444	20	2051	
2057A2-155	123A	3444	20	2051	
2057A2-156	128	3124/289	243	2030	
2057A2-157	161	3018	39	2015	
2057A2-158	161	3018	39	2015	
2057A2-159	160	3018	245	2004	
2057A2-163	108	3452	86	2038	
2057A2-165	102A	3004	53	2007	
2057A2-166	160	3006	245	2004	
2057A2-167	103A	3010	59	2002	
2057A2-179	108	3452	86	2038	
2057A2-180	161	3716	39	2015	
2057A2-181	161	3716	39	2015	
2057A2-182	159	3466	82	2032	
2057A2-183	159	3466	82	2032	
2057A2-184	123A	3444	20	2051	
2057A2-185	161	3716	39	2015	
2057A2-187	161	3716	39	2015	
2057A2-192	161	3124/289	39	2015	
2057A2-193	161	3716	39	2015	
2057A2-195	161	3117	39	2015	
2057A2-196	161	3716	39	2015	
2057A2-197	161	3716	39	2015	
2057A2-198	159	3466	82	2032	
2057A2-199	128		243	2030	
2057A2-200	106	3984	21	2034	
2057A2-201	108	3452	86	2038	
2057A2-202	161	3716	39	2015	
2057A2-203	106	3984	21	2034	
2057A2-204	161	3716	39	2015	
2057A2-205	160		245	2004	
2057A2-206	102	3004	2	2007	
2057A2-207	161	3018	39	2015	
2057A2-208	123A	3444	20	2051	
2057A2-209	123A	3444	20	2051	
2057A2-210	102A		53	2007	
2057A2-211	131		44	2006	
2057A2-212	199	3124/289	62	2010	
2057A2-215	123A	3444	20	2051	
2057A2-216	107	3018	11	2015	
2057A2-217	107	3018	11	2015	
2057A2-218	107	3018	11	2015	
2057A2-219	107	3038	11	2015	
2057A2-220	107	3018	11	2015	
2057A2-221	107	3018	11	2015	
2057A2-222	123A	3444	20	2051	
2057A2-223	124	3021	12		
2057A2-224	108	3452	86	2038	
2057A2-225	123A	3444	20	2051	
2057A2-226	123A	3444	20	2051	
2057A2-227			20	2051	
2057A2-228			10	2051	
2057A2-229			21	2034	
2057A2-230	128	3024	243	2030	
2057A2-231	160	3006	245	2004	
2057A2-232	160	3006	245	2004	
2057A2-234			18	2030	
2057A2-237	107	3018	11	2015	
2057A2-241	102A	3004	53	2007	
2057A2-249	199	3122	62	2010	
2057A2-251	161	3124/289	39	2015	
2057A2-252	160	3006	245	2004	
2057A2-257	199	3018	62	2010	
2057A2-258	161	3039/316	39	2015	
2057A2-259	107	3039/316	11	2015	
2057A2-260	199	3122	62	2010	
2057A2-261	154	3044	40	2012	
2057A2-262	199	3124/289	62	2010	
2057A2-263	126	3005	52	2024	
2057A2-264	123A	3444	20	2051	
2057A2-265	127	3764	25		
2057A2-27			20	2051	
2057A2-272	199	3039/316	62	2010	
2057A2-273	199	3018	62	2010	
2057A2-274	199	3018	62	2010	
2057A2-275	199	3018	62	2010	
2057A2-276	123A	3444	18	2051	
2057A2-277	59	3466/159	82	2032	
2057A2-278	123A	3444	20	2051	
2057A2-279	123A	3444	20	2051	
2057A2-28			2	2007	
2057A2-280	123A	3444	20	2051	
2057A2-281	123A	3444	20	2051	
2057A2-284	128	3124/289	243	2030	
2057A2-285	123A	3124/289	20	2051	
2057A2-288	102A	3004	53	2007	
2057A2-289	123A	3122	20	2051	
2057A2-290	199	3124/289	62	2010	
2057A2-294	123A	3444	20	2051	
2057A2-295	128	3024	243	2030	
2057A2-296	123A	3444	20	2051	
2057A2-297	123A	3444	20	2051	
2057A2-298	159	3466	82	2032	
2057A2-300	123A	3444	20	2051	
2057A2-301			66	2048	
2057A2-302	104	3009	16	2006	
2057A2-303	123A	3444	20	2051	
2057A2-304	107	3018	11	2015	
2057A2-305	107	3018	11	2015	
2057A2-306	123A	3444	20	2051	
2057A2-307	159	3466	82	2032	
2057A2-309	108	3452	86	2038	
2057A2-310	108	3452	86	2038	
2057A2-311	108	3452	86	2038	
2057A2-313	108	3452	86	2038	
2057A2-314	108	3452	86	2038	
2057A2-316	123A	3444	20	2051	
2057A2-317	158	3004/102A	53	2007	
2057A2-319			10	2051	
2057A2-322	107	3018	11	2015	
2057A2-323	107	3018	11	2015	
2057A2-324	123A	3444	20	2051	
2057A2-325	107	3018	11	2015	
2057A2-326	107	3018	11	2015	
2057A2-329	102A	3004	53	2007	
2057A2-331	107	3018	11	2015	
2057A2-332	123A	3444	20	2051	
2057A2-333	199	3124/289	62	2010	
2057A2-334	199	3124/289	62	2010	
2057A2-341	123A	3444	20	2051	
2057A2-342	108	3452	86	2038	
2057A2-343	159	3466	82	2032	
2057A2-352	199	3124/289	62	2010	
2057A2-353	159	3466	82	2032	
2057A2-356	107	3018	11	2015	
2057A2-359	159	3466	82	2032	
2057A2-37	126	3006/160	52	2024	
2057A2-370	199	3122	62	2010	
2057A2-373	199	3245	62	2010	
2057A2-374	123A	3444	20	2051	
2057A2-385	199	3245	62	2010	
2057A2-386	108	3452	86	2038	
2057A2-387	123A	3444	20	2051	
2057A2-390	123A	3444	20	2051	
2057A2-391	199	3122	62	2010	
2057A2-392	108	3452	86	2038	
2057A2-393	108	3452	86	2038	
2057A2-394	108	3452	86	2038	
2057A2-395	107	3018	11	2015	
2057A2-396	123A	3444	20	2051	
2057A2-397	159	3466	82	2032	
2057A2-398	123A	3444	20	2051	
2057A2-399	123A	3444	20	2051	
2057A2-400	159	3466	82	2032	
2057A2-401	123A	3444	20	2051	
2057A2-402	108	3452	86	2038	
2057A2-403	159	3466	82	2032	
2057A2-404	199	3024/128	62	2010	
2057A2-405	199	3122	62	2010	
2057A2-406	159	3466	82	2032	
2057A2-412	123A	3444	20	2051	
2057A2-427	289A		268	2038	
2057A2-428	199	3124/289	62	2010	
2057A2-430	159	3466	82	2032	
2057A2-432	229	3018	61	2038	
2057A2-433	123A	3444	20	2051	
2057A2-434	123A	3444	20	2051	
2057A2-436	199	3245	62	2010	
2057A2-445	312	3834/132	FET-1	2035	
2057A2-446	298	3450	272		
2057A2-448	108	3452	86	2038	
2057A2-449	123A	3444	20	2051	
2057A2-452	123A	3444	20	2051	
2057A2-454	199	3018	62	2010	
2057A2-457	159	3466	82	2032	
2057A2-463	123A	3444	20	2051	
2057A2-464	123A	3444	20	2051	
2057A2-465	108	3452	86	2038	
2057A2-466	107	3018	11	2015	
2057A2-468	128	3044/154	243	2030	
2057A2-475	199	3018	62	2010	
2057A2-477	229	3018	61	2038	
2057A2-478	229	3018	61	2038	
2057A2-479	123A	3444	20	2051	
2057A2-480		3114	67	2023	
2057A2-483	229		61	2038	
2057A2-484	293	3849	47	2030	
2057A2-486			63	2030	
2057A2-487	123A	3122	20	2051	
2057A2-489	159	3466	82	2032	
2057A2-49			50	2004	
2057A2-501	107	3018	11	2015	
2057A2-502	199	3018	62	2010	
2057A2-503	107	3018	11	2015	
2057A2-504	108	3452	86	2038	
2057A2-505	108	3452	86	2038	
2057A2-507	229	3122	61	2038	
2057A2-508	229	3122	61	2038	
2057A2-509	108	3452	86	2038	
2057A2-510	123A	3122	20	2051	
2057A2-511	123A	3444	20	2051	
2057A2-518	199	3122	62	2010	
2057A2-524	297		271	2030	
2057A2-526	229	3132	61	2038	
2057A2-527	229	3018	61	2038	
2057A2-529	234	3114/290	65	2050	
2057A2-530		3122	18	2030	
2057A2-539	107	3018	11	2015	
2057A2-540	107	3018	11	2015	
2057A2-541	229		61	2038	
2057A2-542	199	3245	62	2010	
2057A2-543	199	3245	62	2010	
2057A2-558	229	3018	20	2051	
2057A2-559	123A	3018	20	2051	
2057A2-560	199	3124/289	62	2010	
2057A2-561	159	3114/290	21	2034	
2057A2-60	126	3005	52	2024	
2057A2-61	158	3004/102A	53	2007	
2057A2-62	123A	3444	20	2051	
2057A2-64	123A	3444	20	2051	
2057A2-65	126	3006/160	52	2024	
2057A2-66	126	3006/160	52	2024	
2057A2-80	126	3006/160	52	2024	
2057A2-81	124	3021	12		
2057A2-84	124	3021	12		
2057A2-87	229	3124/289	61	2038	
2057A2-988	102A	3004	53	2007	
2057A32-26	1142	3485	IC-128		
2057A32-33	801	3160	IC-35		
2057A42-477	229		61	2038	
2057B-113	108	3018	86	2038	
2057B-59	199	3245	62	2010	
2057B-84	124		12		
2057B-85	123A	3444	20	2051	
2057B100-1	102A		53	2007	
2057B100-11	131		44	2006	
2057B100-12	108	3452	86	2038	
2057B100-13	102A		53	2007	
2057B100-16	130MP	3027/130	15	2041(2)	
2057B100-17	128	3024	243	2030	
2057B100-4	103A	3835	59	2002	
2057B100-6	102A		53	2007	
2057B100-7		3004	2	2007	
2057B100-8	102A		53	2007	
2057B100-9	102A		53	2007	
2057B101-4	123A	3444	20	2051	
2057B102-4	123A	3444	20	2051	
2057B103-4	123A	3444	20	2051	
2057B104-8	154	3045/225	40	2012	
2057B106-12	159	3466	82	2032	
2057B107-8	128	3024	243	2030	
2057B108-6	159	3466	82	2032	
2057B109-9	128	3024	243	2030	
2057B110-9	129	3025	244	2027	
2057B111-9		3024/128	243,244		
2057B112-9	129	3025	244	2027	
2057B113-9	128	3024	243	2030	
2057B114-9	129	3025	244	2027	
2057B115-9	129	3025	244	2027	
2057B116-9	129	3025	244	2027	
2057B117-9	123A	3444	20	2051	
2057B118-12	123A	3444	20	2051	
2057B119-2	123A	3122	20	2051	
2057B120-12	123A	3122	20	2051	
2057B121-9	129	3025	244	2027	
2057B122-9	129	3025	244	2027	
2057B123-10	124	3021	12		
2057B124-10	121	3009	239	2006	
2057B125-9	123A	3122	20	2051	
2057B126-12	128	3024	243	2030	
2057B129-9	128	3024	243	2030	
2057B141-4	123A	3444	20	2051	
2057B142-4	123A	3444	20	2051	
2057B143-12	123A	3444	20	2051	
2057B144-12	128	3024	243	2030	

Industry Standard No.	ECG	SK	GE	RS 276-	MOTOR.
2057B145-12	129	3025	244	2027	
2057B146-12	123A	3444	20	2051	
2057B147-12	159	3466	82	2032	
2057B149-12	312	3112		2028	
2057B151-6	123A	3444	20	2051	
2057B152-12	123A	3444	20	2051	
2057B153-9	128	3024	243	2030	
2057B155-10	175		246	2020	
2057B156-9	128	3024	243	2030	
2057B157-9	162	3438	35		
2057B158-10	175		246	2020	
2057B159-12	159	3466	82	2032	
2057B163-12	129	3025	244	2027	
2057B168	102A		53	2007	
2057B169	102A		53	2007	
2057B175-9	162	3438	35		
2057B186	160		245	2004	
2057B2-101	107	3018	11	2015	
2057B2-102	107	3018	11	2015	
2057B2-103	108	3452	86	2038	
2057B2-104	126	3008	52	2024	
2057B2-105	126	3008	52	2024	
2057B2-107	102A		53	2007	
2057B2-108	108	3452	86	2038	
2057B2-109	108	3452	86	2038	
2057B2-110	108	3452	86	2038	
2057B2-111	108	3452	86	2038	
2057B2-112	108	3452	86	2038	
2057B2-113	123A	3444	20	2051	
2057B2-114	107		11	2015	
2057B2-115	107	3039/316	11	2015	
2057B2-116	107	3018	11	2015	
2057B2-117	107	3018	11	2015	
2057B2-118	126		52	2024	
2057B2-119	108	3452	86	2038	
2057B2-120	108	3452	86	2038	
2057B2-121	123A	3444	20	2051	
2057B2-122	123A	3444	20	2051	
2057B2-123	123A	3444	20	2051	
2057B2-124	102A	3005	53	2007	
2057B2-125	107	3018	11	2015	
2057B2-127	108	3452	86	2038	
2057B2-128	108	3452	86	2038	
2057B2-129	102A		53	2007	
2057B2-130	123A	3444	20	2051	
2057B2-133	121	3009	239	2006	
2057B2-134	127	3034	25		
2057B2-135	102A	3004	53	2007	
2057B2-136	127	3035	25		
2057B2-137	102A		53	2007	
2057B2-138	161	3018	39	2015	
2057B2-139	161	3018	39	2015	
2057B2-14	108	3452	86	2038	
2057B2-140	154	3044	40	2012	
2057B2-141	126	3006/160	52	2024	
2057B2-142	102A	3004	53	2007	
2057B2-143	107		11	2015	
2057B2-149	160	3006	245	2004	
2057B2-150	153	3025/129	69	2049	
2057B2-151	152	3893	66	2048	
2057B2-152	123A	3444	20	2051	
2057B2-153	123A	3444	20	2051	
2057B2-154	123A	3444	20	2051	
2057B2-155	123A	3444	20	2051	
2057B2-157	160	3006	245	2004	
2057B2-158	160	3006	245	2004	
2057B2-159	160	3006	245	2004	
2057B2-160	108	3452	86	2038	
2057B2-161	108	3452	86	2038	
2057B2-162	108	3452	86	2038	
2057B2-192	161	3124/289	39	2015	
2057B2-206	102A		53	2007	
2057B2-23	102A	3003	53	2007	
2057B2-27			10	2051	
2057B2-28	102A	3124/289	53	2007	
2057B2-29	102A	3004	53	2007	
2057B2-32	102	3004	2	2007	
2057B2-34	102	3004	2	2007	
2057B2-35	126	3006/160	52	2024	
2057B2-37	126	3008	52	2024	
2057B2-38	123A	3444	20	2051	
2057B2-4	102	3004	2	2007	
2057B2-41	126	3008	52	2024	
2057B2-42	126	3008	52	2024	
2057B2-43	102	3004	2	2007	
2057B2-44	102	3004	2	2007	
2057B2-45	102	3004	2	2007	
2057B2-46	103	3862	8	2002	
2057B2-47	124	3021	12		
2057B2-48	126	3008	52	2024	
2057B2-49	102	3004	2	2007	
2057B2-50	126	3008	52	2024	
2057B2-51	126	3008	52	2024	
2057B2-52	102A	3004	53	2007	
2057B2-57	102	3124/289	2	2007	
2057B2-58	124	3021	12		
2057B2-59	123	3124/289	20	2051	
2057B2-60	126	3005	52	2024	
2057B2-61	158	3004/102A	53	2007	
2057B2-62	123A	3444	20	2051	
2057B2-63	123	3124/289	20	2051	
2057B2-64	108	3452	86	2038	
2057B2-65	126	3006/160	52	2024	
2057B2-66	126	3006/160	52	2024	
2057B2-67	126	3006/160	52	2024	
2057B2-68	126	3006/160	52	2024	
2057B2-69	123A	3444	20	2051	
2057B2-70	126	3006/160	52	2024	
2057B2-71	126	3006/160	52	2024	
2057B2-72	102	3004	2	2007	
2057B2-73	123	3124/289	20	2051	
2057B2-77	126	3006/160	52	2024	
2057B2-78	102	3004	2	2007	
2057B2-79	126	3008	52	2024	
2057B2-80	126	3008	52	2024	
2057B2-81	126	3006/160	52	2024	
2057B2-83	102	3004	2	2007	
2057B2-84	124	3021	12		
2057B2-85	108	3452	86	2038	
2057B2-86	102	3004	2	2007	
2057B2-87	108	3452	86	2038	
2057B2-88	126	3006/160	52	2024	
2057B2-89	160	3006	245	2004	
2057B2-90	160	3006	245	2004	
2057B2-93	160	3006	245	2004	
2057B2-94	102A		53	2007	
2057B2-97	123A	3444	20	2051	
2057B2-99	102A		53	2007	
2057B206	102A		53	2007	
2057B2A2-118	160	3006	245	2004	
2057B45-14	102A		53	2007	
2057S2-276			20	2051	
2058	126		52	2024	
2058-02	312		FET-2	2035	

Industry Standard No.	ECG	SK	GE	RS 276-	MOTOR.
2060-024	116	3016	504A	1104	
2061-1	783	3215	IC-225		
2061-45	614	3327	90		
2061A45-38	177	3100/519	300	1122	
2061A45-47	102	3722	2	2007	
2061A45-72	116		504A	1104	
2061A45-93	177		300	1122	
2061B45-14	158		53	2007	
2061B45-35	116	3031A	504A	1104	
2062-17	237	3299	46		
2062-17A	237	3299	46		
2063-17	199	3122	62	2010	
2063-17-12	123A	3444	20	2051	
2064		3122	20	2051	
2064(CROWN)	123A	3444	20	2051	
2064-17	5072A	3136	ZD-8.2	562	
2065-03	123A	3124/289	210	2051	
2065-04	229	3124/289	213	2038	
2065-06	235	3197	337		
2065-07	295	3122	210	2051	
2065-08	519		514	1122	
2065-0824-10110	1183	3475/1050			
2065-17	177	3100/519	300	1122	
2065-52	977	3462			
2065-54	199	3124/289	212	2010	
2065-55	222	3112	FET-4	2036	
2066-17	177	3100/519	300	1122	
2067-1	715		IC-6		
2069-1	715		IC-6		
2071	181	3036	75	2041	
2074-17	312	3116	FET-1	2035	
2075-1	801	3160	IC-35		
2076	177	3100/519	300	1122	
2076-1	742		IC-213		
2077-07			1N60	1123	
2077-1	788	3829	IC-229		
2079-40	221		FET-4	2036	
2079-42	5072A		ZD-8.1	562	
2079-92	1166	3827			
2079-93	139A		ZD-9.1	562	
2081-17	186A	3357	28	2017	
2081-6	153		69	2049	
2082-6	152	3054/196	66	2048	
2084-17	125	3017B/117	510,531	1114	
2085-17	184	3054/196	57	2017	
2087-46	139A		ZD-9.1	562	
2090		3045	27		
2090(CROWN)	154		40	2012	
2090A43-1	112	3089	1N82A		
2090A53-2	120	3110	CR-3		
2091-02	1081A	3243/1160			
2091-49	1006	3358	IC-38		
2091-50	1005	3723	IC-42		
2093A12-1	116	3016	504A	1104	
2093A2-289	108	3452	86	2038	
2093A25-3	109	3087		1123	
2093A33-1	109	3087		1123	
2093A38-1	109	3087		1123	
2093A38-10	109	3087		1123	
2093A38-11	177		300	1122	
2093A38-13	116	3031A	504A	1104	
2093A38-14	109			1123	
2093A38-15	109			1123	
2093A38-21	109	3088		1123	
2093A38-22	110MP		1N60	1123	
2093A38-23	102	3004	2	2007	
2093A38-24	128	3024	243	2030	
2093A38-27	109	3087		1123	
2093A38-28	116	3311	504A	1104	
2093A38-29		3126	90		
2093A38-30	109	3088		1123	
2093A38-31	109	3088		1123	
2093A38-32	109	3088		1123	
2093A38-33	109	3087		1123	
2093A38-34	109			1123	
2093A38-35	110MP	3709	1N60	1123(2)	
2093A38-37	177	3100/519	300	1122	
2093A38-4		3311	504A	1104	
2093A38-40	109	3087		1123	
2093A38-5	109	3088		1123	
2093A38-8		3126	90		
2093A39-12	142A	3062	ZD-12	563	
2093A3D-2	117	3017B	504A	1104	
2093A3D-20	131		44	2006	
2093A4-2	117	3017B	504A	1104	
2093A41	109	3087		1123	
2093A41-101		3126	90		
2093A41-102	177		300	1122	
2093A41-103	116	3311	504A	1104	
2093A41-104	116	3031A	504A	1104	
2093A41-105	177	3100/519	300	1122	
2093A41-108	116	3311	504A	1104	
2093A41-110	116	3311	504A	1104	
2093A41-112	167	3647	BR-206	1172	
2093A41-113	143A	3750	ZD-13	563	
2093A41-115	116		504A	1104	
2093A41-116	116	3100/519	504A	1104	
2093A41-117		3126	90		
2093A41-125	139A	3060			
2093A41-126	116	3311	504A	1104	
2093A41-129	177		300	1122	
2093A41-130		3126	90		
2093A41-131	116	3311	504A	1104	
2093A41-139	116	3311	504A	1104	
2093A41-14	109	3088		1123	
2093A41-141	109			1123	
2093A41-148	109	3087		1123	
2093A41-150	150A	3087	ZD-82		
2093A41-151	116	3311	504A	1104	
2093A41-152	116		504A	1104	
2093A41-153	5070A	9021	ZD-6.2	561	
2093A41-154	109	3087		1123	
2093A41-155	177		300	1122	
2093A41-156		3311	300	1122	
2093A41-157		3126	90		
2093A41-158	177	3100/519	300	1122	
2093A41-159	139A	3060	ZD-9.1	562	
2093A41-16	139A	3060	ZD-9.1	562	
2093A41-161		3126	90		
2093A41-163	139A		ZD-9.1	562	
2093A41-164	177		300	1122	
2093A41-165	177	3100/519	300	1122	
2093A41-166		3126	90		
2093A41-167	109	3088		1123	
2093A41-169	109	3088		1123	
2093A41-171	177	3100/519	300	1122	
2093A41-172	177	3100/519	300	1122	
2093A41-173	116	3311	504A	1104	
2093A41-180		3311	504A	1104	
2093A41-181	109			1123	
2093A41-182	116		504A	1104	
2093A41-185	116	3311	504A	1104	
2093A41-186	139A	3060	ZD-9.1	562	
2093A41-187	109	3087		1123	

Industry Standard No.	ECG	SK	GE	RS 276-	MOTOR.
2093A41-189	116	3311	504A	1104	
2093A41-194	135A		ZD-5.1		
2093A41-196	109			1123	
2093A41-197	116	3311	504A	1104	
2093A41-2	109	3087		1123	
2093A41-200		3126	90		
2093A41-23		3126	90		
2093A41-24	139A	3060	ZD-9.1	562	
2093A41-24A	139A	3060	ZD-9.1	562	
2093A41-27		3126	90		
2093A41-28			510	1114	
2093A41-29	110MP	3088	1N60	1123(2)	
2093A41-38	109	3088		1123	
2093A41-40	102	3722	2	2007	
2093A41-41	102	3722	2	2007	
2093A41-42	125	3031A	510,531	1114	
2093A41-43	125	3311	510,531	1114	
2093A41-45	116		504A	1104	
2093A41-49	116	3031A	504A	1104	
2093A41-5		3126	90		
2093A41-50	109	3091		1123	
2093A41-51	116	3031A	504A	1104	
2093A41-53	138A	3059	ZD-7.5		
2093A41-54	116	3031A	504A	1104	
2093A41-55	116		504A	1104	
2093A41-56	142A		ZD-12	563	
2093A41-57	116		504A	1104	
2093A41-58	115		6GX1		
2093A41-59	109	3088		1123	
2093A41-6	125	3311	510,531	1114	
2093A41-60			ZD-12	563	
2093A41-61			504A	1104	
2093A41-62	116	3311	504A	1104	
2093A41-63	135A		ZD-5.1		
2093A41-65	125	3032A	510,531	1114	
2093A41-66	116	3031A	504A	1104	
2093A41-67	506	3130	511	1114	
2093A41-68	109			1123	
2093A41-69	506	3843	511	1114	
2093A41-70	5071A	3334	ZD-6.8	561	
2093A41-72	506	3130	511	1114	
2093A41-73	506	3130	511	1114	
2093A41-74	503		CR-5		
2093A41-75	116	3311	504A	1104	
2093A41-76	177	3100/519	300	1122	
2093A41-77	125	3032A	510,531	1114	
2093A41-78	116		504A	1104	
2093A41-79		3126	90		
2093A41-8			504A	1104	
2093A41-81	116	3311	504A	1104	
2093A41-82	5022A	3788	ZD-13	563	
2093A41-84			504A	1104	
2093A41-85		3335	ZD-20		
2093A41-86	116	3031A	504A	1104	
2093A41-87	158		53	2007	
2093A41-88	116	3017B/117	504A	1104	
2093A41-89	116	3016	504A	1104	
2093A41-90	135A	3056	ZD-5.1		
2093A41-91		3126	90		
2093A41-92	109	3088		1123	
2093A41-95		3126	90		
2093A41-96	506	3100/519	511	1114	
2093A42-7	116	3031A	504A	1104	
2093A43-2	112	3089	1N82A		
2093A5-10	113A	3119/113	6GC1		
2093A5-2	114	3120	6GD1	1104	
2093A52-1	125	3032A	510,531	1114	
2093A57-1	118	3066	CR-1		
2093A59-1	112	3089	1N82A		
2093A6-1	117	3017B	504A	1104	
2093A6-2	117	3017B	504A	1104	
2093A69-1	125	3032A	510,531	1114	
2093A71-1	125	3033	510,531	1114	
2093A75-1	120	3110	CR-3		
2093A77-1	109	3087		1123	
2093A78-1	125	3032A	510,531	1114	
2093A79-6	125	3032A	510,531	1114	
2093A8-1	109	3088		1123	
2093A80-1	142A	3062	ZD-12	563	
2093A9-3	100	3005	1	2007	
2093A9-4	100	3005	1	2007	
2093B11-21	110MP		1N60	1123	
2093B38-14	116	3016	504A	1104	
2093B38-4	177		300	1122	
2093B38-9	112		1N82A		
2093B4-6	116	3016	504A	1104	
2093B41-10	116	3016	504A	1104	
2093B41-11	109	3087		1123	
2093B41-12	116	3016	504A	1104	
2093B41-14			504A	1104	
2093B41-18	116	3017B/117	504A	1104	
2093B41-20	116	3017B/117	504A	1104	
2093B41-21	116	3017B/117	504A	1104	
2093B41-22	116	3017B/117	504A	1104	
2093B41-23		3126	90		
2093B41-24	139A		ZD-9.1	562	
2093B41-25			504A	1104	
2093B41-28	116	3031A	504A	1104	
2093B41-29	110MP		1N60	1123	
2093B41-32			504A	1104	
2093B41-34	116		504A	1104	
2093B41-35	116		504A	1104	
2093B41-37	116		504A	1104	
2093B41-6	116	3016	504A	1104	
2093B41-62			504A	1104	
2093B41-8	116	3311	504A	1104	
2093B41-9	116	3016	504A	1104	
2101	199	3245	62	2010	
2101-03	177		300	1122	
2102	116		504A	1104	
2102-010	109	3087		1123	
2102-014	116	3017B/117	504A	1104	
2102-017	116	3017B/117	504A	1104	
2102-025	109	3087		1123	
2102-028	109	3087		1123	
2102-029	178MP	3100/519	300	1122	
2102-031	614		90		
2102-032	139A		ZD-9.1	562	
2102-074	116	3016	504A	1104	
2105		3126	90		
2106-119	102A	3004	53	2007	
2106-120	160	3006	245	2004	
2106-121	102A	3004	53	2007	
2106-122	102A	3004	53	2007	
2106-123	102A	3004	53	2007	
2106-124	109	3087		1123	
2106-17	5070A	9021	ZD-6.2	561	
2110N-132	161	3122	39	2015	
2110N-133	161	3122	39	2015	
2110N-134	175	3026	246	2020	
2110N-41	177	3100/519	300	1122	
2110N-42	116	3031A	504A	1104	
2112-17	102A	3004	53	2007	
2113	113A	3119/113	6GC1		
2114					MCM2114
2114-0	154	3044	40	2012	
2114-17	158		53	2007	
2115A					MCM2115A
2115AL					MCM21L15A
2116-17	116	3311	504A	1104	
2117					MCM4116
2118					MCM4517
2119-101-090	1196	3725			
2120-17	5075A	3751	ZD-16	564	
2121-17	199	3124/289	62	2010	
2122-17	109			1123	
2122-3	172A	3156	64		
2125A					MCM2125A
2125AL					MCM21L25A
2127-17	229	3018	60	2015	
2132	186	3192	28	2017	
2132(GE)	123A	3444	20	2051	
2132-17	199	3124/289	62	2010	
2134-17	313	3122	278	2016	
2135E	735		IC-212		
2136	737	3375	IC-16		
2136D	737	3375	IC-16		
2136P	737	3375	IC-16		
2136PC	737	3375	IC-16		
2147H					MCM2147H
2148					MCM2148
2148-17	116	3311	504A	1104	
2150-17	139A	3060	ZD-9.1	562	
2151-17	109	3088		1123	
2152	116	3311	504A	1104	
2158-1541	123A	3444	20	2051	
2158-1558	159	3466	82	2032	
2160	6400A			2029	
2160-17	186A	3357	28	2017	
2163-17	175	3026	246	2020	
2167-17	237	3299	46		
2168-17	186A	3202/210	247	2017	
2180-151	108	3452	86	2038	
2180-152	108	3452	86	2038	
2180-153	123A	3444	20	2051	
2180-154	123A	3444	20	2051	
2180-155	175	3026	246	2020	
2180-41	178MP	3087	300(2)	1122	
2181-17	290A	3122	269	2032	
2182-17	116	3017B/117	504A	1104	
2195-17	199	3124/289	62	2010	
2196-17	109	3088		1123	
2197-17	107	3018	11	2015	
2198-17	116		504A	1104	
2199-17	116	3311	504A	1104	
2200-17	116	3311	504A	1104	
2202-17	186A	3357	28	2017	
2203	5072A	3059/138A	ZD-8.2	562	
2203-17	5072A	3059/138A	ZD-8.2	562	
2204-17	123	3124/289	20	2051	
2205		3126	90		
2205-17		3126	90		
2206	116	3311	504A	1104	
2206-17	116	3311	504A	1104	
2207-17	199	3124/289	62	2010	
2208-17	128	3122	243	2030	
2209-17	177	3100/519	300	1122	
2210-17	152	3893	66	2048	
2211-17	153		69	2049	
2211B	116	3031A	504A	1104	
2212-17	199	3018	62	2010	
2213-17	107	3018	11	2015	
2214-17	229	3122	62	2038	
2215-17	160	3006	245	2004	
2218-17		3126	90		
2220-17	159	3466	82	2032	
2224-17	229	3018	61	2038	
2225-17	229	3039/316	61	2038	
2226-17	229	3018	61	2038	
2231-17	178MP	3100/519	300(2)	1122(2)	
2232-17	109	3088		1123	
2233-17	116	3311	504A	1104	
2243	121	3014	239	2006	
2244-1	5991	3193/187	29	2025	
2245-17	186	3192	28	2017	
2246-17	192	3122	63	2030	
2252		3031A	504A	1104	
2261-17	294	3841			
2263	123A	3444	20	2051	
2269	159	3466	82	2032	
2270	123A	3444	20	2051	
2270-5	128		243	2030	
2271	199	3245	62	2010	
2272	159	3466	82	2032	
2275-17	123A	3444	20	2051	
2277P	804	3455	IC-27		
2279-13	109	3087		1123	
2280-13	109	3087		1123	
2281-13	109	3087		1123	
2282-13	109	3087		1123	
2282-17	109	3088		1123	
2282-17(RAT.DET)	110MP			1123(2)	
2283-13	109	3087		1123	
2283-17	116	3311	504A	1104	
2284-13	109	3087		1123	
2284-17	107	3018	11	2015	
2285-17	116	3100/519	504A	1104	
2289-13	178MP		300(2)	1122(2)	
2290-13	109	3087		1123	
2290-17	123A	3444	20	2051	
2291-17	107	3018	11	2015	
2295	175	3538	246	2020	
2300.036.096	159	3466	82	2032	
2302	116	3031A	504A	1104	
2302-17	233	3122	210	2009	
2304	177	3175	300	1122	
2309	139A	3066/118	ZD-9.1	562	
2312-17	184	3190	57	2017	
2315-046	116	3017B/117	504A	1104	
2316E					MCM68A316E
2319-17	116		504A	1104	
2320-17	123A	3444	20	2051	
2321-17	123A	3444	20	2051	
2322-17	315	3048/329	279		
2323-17	136A	3058/137A	ZD-5.6	561	
2328	177	3100/519	300	1122	
2328-17	177	3100/519	300	1122	
2330-021	138A	3059	ZD-7.5		
2330-191	125	3033	510,531	1114	
2330-201	125	3311	510,531	1114	
2330-252	125	3311	510,531	1114	
2332					MCM68A332
2334-17	186A	3357	247	2017	
2335-17	312	3448	FET-1	2035	
2336-17	312	3116	FET-2	2035	
2337-17	123A	3444	20	2051	
2338-17	199	3124/289	62	2010	
2339-17	5072A	3059/138A	ZD-8.2	562	

Industry Standard No.	ECG	SK	GE	RS 276-	MOTOR.
2340-17	315	3048/329	279		
2341-17	5071A	3334	ZD-6.8	561	
2343-17	5072A	3136	ZD-8.2	562	
2347-17	104	3009	16	2006	
2359-17	222	3050/221	FET-4	2036	
2361-17	289A	3444/123A	20	2051	
2362-1		3122	20	2051	
2362-1(SYLVANIA)	108		86	2038	
2362-17	136A	3057	ZD-5.6	561	
02375-A	175		246	2020	
2381-17	159	3118	82	2032	
2382-17	152	3054/196	66	2048	
2396-17	145A	3063	ZD-15	564	
2399-17	235	3197	215		
2400-17	177	3017B/117	300	1122	
2400-23	116	3016	504A	1104	
2400-27	116	3016	504A	1104	
2402	116	3031A	504A	1104	
2402-17		3126	90		
2402-453	102A	3004	53	2007	
2402-454	102A	3004	53	2007	
2402-455	102A	3123	53	2007	
2402-456	131	3052	44	2006	
2402-457	102A	3004	53	2007	
2402-459	109	3087		1123	
2402-461	116	3311	504A	1104	
2402-462	116	3311	504A	1104	
2402-463	116	3311	504A	1104	
2403	177	3170/731	300	1122	
2405	177	3175	300	1122	
2405-453	102A	3004	53	2007	
2405-454	102A	3123	53	2007	
2405-455	102A	3123	53	2007	
2405-456	102A	3123	53	2007	
2405-457	102A	3123	53	2007	
2405-458	109	3087		1123	
2405-459	116	3311	504A	1104	
2405-462	116	3016	504A	1104	
2405-463	116	3016	504A	1104	
2408-17	152	3054/196	247	2052	
2408-326	102A	3004	53	2007	
2408-328	102A	3123	53	2007	
2408-329	158	3004/102A	53	2007	
2408-330	109	3087		1123	
2408-331	116(2)			1104	
2409-17	519	3100	514	1122	
2410-17	116	3017B/117	504A	1104	
2411	116		504A	1104	
2411-17		3017B	509	1114	
2417	124	3021	12		
2418-17	237	3299	46		
2427		3039	11	2015	
2427(RCA)	108		86	2038	
2432-1(RCA)	710		IC-89		
2434(RCA)	710		IC-89		
2434-1(RCA)	710		IC-89		
2436-17	5075A	3751			
2437-17	236	3197/235	216	2053	
2438-17	109	3087			
2443		3124	11	2015	
2443(RCA)	123A	3444	20	2051	
2444(RCA)	124		12		
2445	108	3452	86	2038	
2445(RCA)	711		IC-207		
2445-1(RCA)	711		IC-207		
2446		3122	20	2051	
2446(RCA)	123A	3444	20	2051	
2446-1(RCA)	121	3717	239	2006	
2447		3124	20	2051	
2447(RCA)	123A	3444	20	2051	
2448		3025	11	2015	
2448(RCA)	159		82	2032	
2448-17	123A	3018	211	2016	
2449-17	199	3122	20	2051	
2450		3018	20	2051	
2450(RCA)	108		86	2038	
2450-17	312	3112	FET-2	2035	
2451-17	5072A		ZD-8.2	562	
2452-17	139A		ZD-9.1	562	
2454-17	5801	9004	510	1114	
2460-13			504A	1104	
2460-14(RCA)	156		512		
2460-17	5075A	3751			
2472-5632	123A		20	2051	
2473		3018	17	2051	
2473(RCA)	108		86	2038	
2473-2109	7408			1822	
2474			40	2012	
2474(RCA)	154		40	2012	
2475		3124	11	2015	
2475(RCA)	123A	3444	20	2051	
2476		3018	17	2051	
2476(RCA)	108		86	2038	
2477		3018	17	2051	
2477(RCA)	108		86	2038	
2477-173	519		514	1122	
2478	126	3008	52	2024	
2478A	126	3008	52	2024	
2478B	126	3008	52	2024	
2482		3025	52	2024	
2482(RCA)	129		244	2027	
2487B	160	3008	245	2004	
2488	126	3008	52	2024	
2488A	126	3008	52	2024	
2489	126	3008	52	2024	
2489A	160	3008	245	2004	
2490	102A	3004	53	2007	
2490A	102A	3004	53	2007	
2491	124	3021	12		
2491A	124	3021	12		
2491B	124	3021	12		
2494		3035	25		
2494(RCA)	127		25		
2495	128	3124/289	11	2015	
2495(RCA)	123A	3444	20	2051	
2495-012	126	3008	52	2024	
2495-013	126	3008	52	2024	
2495-014	102A	3004	53	2007	
2495-078	160	3006	245	2004	
2495-079	160	3006	245	2004	
2495-080	102A	3004	53	2007	
2495-082	160	3006	245	2004	
2495-166-1	108	3452	86	2038	
2495-166-2	123A	3444	20	2051	
2495-166-4	108	3452	86	2038	
2495-166-8	108	3452	86	2038	
2495-166-9	108	3452	86	2038	
2495-200	126	3008	52	2024	
2495-376	160	3006	245	2004	
2495-377	160	3006	245	2004	
2495-378	160	3006	245	2004	
2495-388	102A	3004	53	2007	
2495-488-1	160		245	2004	
2495-488-2	160		245	2004	
2495-489	125	3031A	510,531	1114	
2495-520	108	3018	86	2038	
2495-521	108	3018	86	2038	
2495-522-1	108	3452	86	2038	
2495-522-4	123A	3444	20	2051	
2495-523-1	108	3452	86	2038	
2495-529	123A	3444	20	2051	
2495-567-2	102A		53	2007	
2495-567-3	102A		53	2007	
2495-586-2	102A		53	2007	
2496		3034	25		
2496(RCA)	127	3764	25		
2496-125-2	123A	3444	20	2051	
2497-473	102A	3004	53	2007	
2497-496	102A	3004	53	2007	
2498-163	128	3024	243	2030	
2498-507-2	108	3452	86	2038	
2498-507-3	108	3452	86	2038	
2498-508-2	108	3452	86	2038	
2498-508-3	108	3452	86	2038	
2498-513	125	3031A	510,531	1114	
2498-903-2	108	3452	86	2038	
2498-903-3	108	3452	86	2038	
2500		3035	25		
2500(RCA)	127	3764	25		
2502	129	3025	244	2027	
2502(RCA)	102A		53	2007	
2510-101		3018	61	2038	
2510-102		3018	61	2038	
2510-103	199	3245	62	2010	
2510-104	123A	3444	20	2051	
2510-105	186	3192	28	2017	
2510-31	109	3088		1123	
2510-32	116	3311	504A	1104	
2516	712	3072	IC-2		
2516(RCA)	712		IC-2		
2516-1	712	3072	IC-2		
2516-1(RCA)	712		IC-2		
2519-17	152	3893			
2520-17	152	3893			
2522	116	3311	504A	1104	
2523	116	3311	504A	1104	
2546		3124	11	2015	
2546(RCA)	123A	3444	20	2051	
2548A0157	124		12		
2549	219		74	2043	
2554-3	730	3143			
2559		3073	IC-22		
2559(RCA)	728		243		
2559-1		3073	IC-22		
2559-1(RCA)	728		IC-22		
2560	729	3074	IC-23		
2560(RCA)	729		IC-23		
2560-1		3074	IC-23		
2560-1(RCA)	729		IC-23		
2577	121	3014	239	2006	
2580	729	3074			
2584	123	3124/289	20	2051	
2603-180	102A	3123	53	2007	
2603-181	102A		53	2007	
2603-182	102A	3123	53	2007	
2603-183	126	3123	52	2024	
2603-184	123A	3444	20	2051	
2603-186	109	3087		1123	
2606-286	102A	3004	53	2007	
2606-287	102A	3004	53	2007	
2606-288	176	3845	80		
2606-291	102A	3004	53	2007	
2606-292	126	3008	52	2024	
2606-294	108	3452	86	2038	
2606-295	160	3006	245	2004	
2606-296	109	3088		1123	
2606-299	116	3311	504A	1104	
2606-303	5069A	3330/5066A	ZD-4.7		
2607					MCM68A308
2612	102A	3004	53	2007	
2614					MCM21L14
2616					MCM68A316E
2620	154	3045/225	40	2012	
2626-1	912		IC-172		
2632					MCM68A332
2633(RCA)	108		86	2038	
2634		3018	11	2015	
2634(RCA)	108		86	2038	
2634-1	108	3452	86	2038	
2636	108	3452	86	2038	
2664					MCM68A364
2665-1	723		IC-15		
2665-2	723	3144	IC-15		
2665-3	723		IC-15		
2667			61	2038	
2690					MCM4116
2700	160	3006	245	2004	
2703-384	102A		53	2007	
2703-385	102A	3123	53	2007	
2703-386	102A	3123	53	2007	
2703-387	176	3845	80		
2703-388	158		53	2007	
2703-389	109	3087		1123	
2703-390	116(2)			1104	
2704			1N34AS	1123	
2704-384	102A	3004	53	2007	
2704-385	102A	3004	53	2007	
2704-386	102A	3004	53	2007	
2704-387	158	3004/102A	53	2007	
2704-388	109	3087		1123	
2704-389	116	3311	504A	1104	
2716					MCM2716
2762	116	3016	504A	1104	
2763	116	3016	504A	1104	
2766	116	3016	504A	1104	
2774	135A	3056	ZD-5.1		
2777	116	3016	504A	1104	
2779	116	3016	504A	1104	
2780	121	3009	239	2006	
2780(AIRLINE)	121MP	3718	239#(2)	2006(2)	
2780-3	121MP	3009	239#(2)	2006(2)	
2780-4	121	3009	239	2006	
2780-5	121	3009	239	2006	
2781	102	3004	2	2007	
2784	116	3016	504A	1104	
2786	116	3016	504A	1104	
2787	123A	3444	20	2051	
2789	109	3087		1123	
2791	126	3006/160	52	2024	
2794	116	3051/156	504A	1104	
2795A	139A	3066/118	ZD-9.1	562	
2796	289A	3087	ZD-9.1	1123	
2797	289A	3006/160	245	2004	
2798	159	3466	82	2032	
2799	290A	3006/160	52	2024	
2802	125	3032A	510,531	1114	
2802-1	781	3169	IC-223		
2802-1(BENDIX)	781		IC-223		

Industry Standard No.	ECG	SK	GE	RS 276-	MOTOR.
2802-2	781	3169	IC-223		
2842-056	175		246	2020	
2842-875	181	3036	75	2041	
2853-1	186	3192	28	2017	
2853-2	152		66	2048	
2853-3	186	3192	28	2017	
2854	123A	3444	20	2051	
2854-1	186	3192	28	2017	
2854-2	152	3893	66	2048	
2854-3	186	3192	28	2017	
2855-1	186	3192	28	2017	
2855-2	152	3893	66	2048	
2855-3	186	3192	28	2017	
2856-1	186	3192	28	2017	
2856-2	152	3893	66	2048	
2856-3	186	3192	28	2017	
2900-007	108	3452	86	2038	
2901-010	121	3009	239	2006	
2904-003			17	2051	
2904-008	121	3009	239	2006	
2904-014	121	3009	239	2006	
2904-016	126	3007	52	2024	
2904-029	126	3007	52	2024	
2904-030	224	3049	46		
2904-032	128	3024	243	2030	
2904-033	108	3122	86	2038	
2904-034	123	3124/289	20	2051	
2904-035	123	3124/289	20	2051	
2904-037	195A	3048/329	46		
2904-038	159	3025/129	82	2032	
2904-038H05	100	3721	1	2007	
2904-045	123A	3114/290	212	2010	
2904-053	229	3018	61	2038	
2904-054	123A	3444	212	2010	
2904-057	184	3178A	57	2017	
2904-058	188	3199	226	2018	
2904-059	329	3048	219		
2904-061		3197	215		
2906-005	746	3234	IC-217		
2906-006	1192	3445			
2925	123A	3122	20	2051	
2957A100-25			23	2020	
3001		3452	86	2038	
3001A	116	3017B/117	504A	1104	
3001H	5531	6631			
3002	108	3036	86	1123	
3003		3018	11	2015	
3003(SEARS)	107		11	2015	
3004(SEARS)	107		11	2015	
3004-856	102A	3004	53	2007	
3005(SEARS)	123A	3444	20	2051	
3005-861	123A	3444	20	2051	
3006(SEARS)	123A	3444	20	2051	
003007	116	3313		1104	
3007(SEARS)	123A	3444	20	2051	
3008	160	3175	245	1122	
3008(CB)	126	3007	52	2024	
3008(E.F.JOHNSON)	160		245	2004	
03008-1	123A	3444	20	2051	
3009	160	3007	52	2024	
3009(E.F.JOHNSON)	160		245	2004	
3009(MIXER)	126	3007	52	2024	
3009(SEARS)	102A		53	2007	
3010	126	3008	52	2024	
3010(E.F.JOHNSON)	126		52	2024	
3010(IF)	126	3007	52	2024	
3010(SEARS)	102A		53	2007	
3011	123A	3444	20	2051	
3011(E.F.JOHNSON)	123A		20		
3012	106	3007	21	2034	
3012(EFJOHNSON)			21	2034	
3012(NPN)	108		86	2038	
3012(PNP)	159		82	2032	
3012(SEARS)	176		80		
3013(SEARS)	176		80		
3014(SEARS)	102A		53	2007	
003015		3311	504A	1104	
003016	109	3090	1N60	1123	
3016B	601	3463			
3017	159	3100/519	82	2032	
3017(E.F.JOHNSON)	159		82	2032	
3018	108	3124/289	86	2038	
3019	159	3466	82	2032	
3019(EFJOHNSON)	126		52	2024	
3020	108	3452	86	2038	
3021	108	3452	86	2038	
3022	128		243	2030	
3023	161	3311	39	1104	
3024	160	3007	245	2004	
3024(E.F.JOHNSON)	160		245	2004	
3025	160	3007	245	2004	
3026	123A	3444	20	2051	
3027	161	3039/316	39	2015	
3028	108	3452	86	2038	
3029	107	3018	11	2015	
3034		3024	18	2030	
3034(RCA)	128		243	2030	
3054	175	3131A/369	246	2020	
3055-1	130		14	2041	
3055-3	130		14	2041	
3064T	780	3141	IC-222		
3065D	712		IC-2		
3065E	712		IC-2		
3066D	728		IC-22		
3066E	728		IC-22		
3067D	729		IC-23		
3067E	729		IC-23		
3069		3017B	504A	1104	
3069(ARVIN)	116		504A	1104	
3074A20	116	3016	504A	1104	
3074A21	116	3016	504A	1104	
3075D	723		IC-15		
3075E	723		IC-15		
3075PC	723		IC-15		
3076HC	781	3169	IC-223		
3076PC	781	3169	IC-223		
003102	139A			562	
3105	5980	3610	5096		
3107-108-40401	530	3308	540		
3107-108-40404	530		540		
3107-108-40501	523	3306	528		
3107-204-9000	123A	3444	20	2051	
3107-204-90010	123A	3444	20	2051	
3107-204-90020	123A	3444	20	2051	
3107-204-90070	121	3717	239	2006	
3107-204-90080	108	3452	86	2038	
3107-204-90100	107		11	2015	
3107-204-90140	104	3719	16	2006	
3107-204-90150	199	3245	62	2010	
3107-204-90180	210	3202	252	2018	
3107-204-90182	152	3893	66	2048	
3107-204-90190	104	3719	16	2006	
3110	5982	3609/5986	5096		
3110R	5983	3698/5987	5097		
3111	123A	3060/139A	20	2051	
3112	102A	3003	53	2007	
3113	123A	3056/135A	20	2051	
003114		3100	ZD-12	563	
3115	5986	3609	5096		
3115R	5987	3698	5097		
3120	5986	3609	5096		
3120R	5987	3698	5097		
3125	5988	3608/5990	5100		
3125R	5989	3518/5995	5101		
3130	5988	3608/5990	5100		
3130-3193-501	7445	7445			
3130-3193-507	74145	74145			
3130-3193-512	787	3146	IC-228		
3130-3248-801	1115	3184	IC-278		
3130R	5989	3518/5995	5101		
3135	5990	3608	5100		
3140	5990	3608	5100		
3140R	5991	3518/5995	5101		
3145	5992	3501/5994	5104		
3146-977	130	3027	14	2041	
3150	5992	3501/5994	5104		
3150R	5993	3518/5995	5105		
3152-159	175	3026	246	2020	
3152-170	130	3027	14	2041	
3160	5994	3501	5104		
3170	5998		5108		
3180	5998		5108		
3202-51-01	128	3024	243	2030	
3202-5H01	128	3024	243	2030	
3222	129		244	2027	
3223	128		243	2030	
3225	6030		5132		
3225R	6031		5133		
3227-E	108	3452	86	2038	
3232					MC3232AL
3235	5990	3608	5100		
3242					MC3242AL
3245	5992	3501/5994	5104		
3270	5998		5108		
3295-001	124		12		
3301(SEARS)	107		11	2015	
003307		3100	514	1122	
3322-6	109	3088		1123	
3367	172A	3156	64		
3370	108	3452	86	2038	
3373	181		75	2041	
3374-3	358		42		
3377	116		504A	1104	
3391	199	3124/289	62	2010	
3391(SEARS)	199		62	2010	
3391A	199	3124/289	62	2010	
3391A(SEARS)	199		62	2010	
3425	100	3005	1	2007	
3434	100	3005	1	2007	
3435	100	3005	1	2007	
003449	229			2038	
3456	176	3845	80		
3458	160	3005	245	2004	
003460	199	3008		2010	
003461	199	3004/102A		2010	
003473	236	3003			
3476		3117		2012	
3476(RCA)	161	3716	39	2015	
3499	130		14	2041	
3500	100	3087	1	2007	
003501	1003	3288	18	2030	
3502(RCA)	704		IC-205		
3502-1(RCA)	704		IC-205		
3502-2(RCA)	704		IC-205		
3503		3021	12		
3503(RCA)	124		12		
3504		3123	2	2007	
3504(RCA)	160		245	2004	
3505	109	3087		1123	
3505(AIRLINE)	178MP		300(2)	1122(2)	
3505(RCA)	123A	3444	20	2051	
3506	143A	3122	ZD-13	563	
3506(RCA)	123A	3444	20	2051	
3507		3124	18	2030	
3507(RCA)	102A			2007	
3507(SEARS)	107		11	2015	
3507(WARDS)	123A	3444	20	2051	
3508	128	3018	243	2030	MCM68A308
3508(RCA)	123A	3444	20	2051	
3508(SEARS)	161	3716	39	2015	
3508(WARDS)	108		86	2038	
3509	108	3452	86	2038	
3509(SEARS)	123A	3444	20	2051	
3509(WARDS)	123A	3444	20	2051	
3510		3018	17	2051	
3510(RCA)	123A	3444	20	2051	
3510(SEARS)	108		86	2038	
3510(WARDS)	108		86	2038	
3511	108	3452	86	2038	
3511(SEARS)	312	3448	FET-1	2035	
3511(WARDS)	312	3448	FET-1	2035	
3512	121	3009	239	2006	
3512(RCA)	312	3448	FET-1	2035	
3512(SEARS)	312	3448	FET-1	2035	
3512(WARDS)	312	3448	FET-1	2035	
3513	159	3025/129	82	2032	
3513(RCA)	123A	3444	20	2051	
3513(SEARS)	159		82	2032	
3513(WARDS)	159		82	2032	
3514	121	3124/289	239	2006	
3514(RCA)	127	3764	25		
3514(SEARS)	107		11	2015	
3514(WARDS)	123A	3444	20	2051	
003515	801	3160	16	2006	
3515(RCA)	121	3717	239	2006	
003516	1217	3700	239	2006	
3516(RCA)	161	3716	39	2015	
3516(WARDS)	108		8	2038	
3517	102A	3004	53	2007	
3518		3004	11	2015	
3518(RCA)	161	3716	39	2015	
3519		3122	10	2051	
3519(RCA)	123A	3444	20	2051	
3519-1		3122	20	2051	
3519-1(RCA)	199		62	2010	
3519-2		3038	60	2015	
3520		3021	ZD-12	563	
3520(DIO)			ZD-12	563	
3520(RCA)	124		12		
3520(WARDS)	142A		ZD-12	563	
3520-1	124	3021	12		
3521			18	2030	
3521(SEARS)	123A	3444	20	2051	
3521-1			11	2015	
003522	1052	3249	21	2034	
3522(SEARS)	159		82	2032	
3523		3025	63	2030	
3523(RCA)	129		244	2027	

Industry Standard No.	ECG	SK	GE	RS 276-	MOTOR.
3523(SEARS)	123A	3444	20	2051	
3524		3018	67	2023	
3524(RCA)	108		86	2038	
3524(SEARS)	159		82	2032	
3524-1	108	3452	86	2038	
3524-1(RCA)	108		86	2038	
3524-2		3018	11	2015	
3524-2(RCA)	108		86	2038	
3525		3122	29	2025	
3525(RCA)	123A	3444	20	2051	
3525(SEARS)	185		58	2025	
003526	1170	3745	28	2017	
3526(RCA)	123A	3444	20	2051	
3526(SEARS)	184		57	2017	
3527		3018	20	2051	
3527(RCA)	108		86	2038	
3527-1		3018	11	2015	
3528-1	785	3254	IC-226		
3528-1(RCA)	785		IC-226		
3529	116	3081/125	504A	1104	
3529A	156	3081/125	512		
3529B	156	3081/125	512		
3529C	156	3081/125	512		
3530		3019	11	2015	
3530(RCA)	108		86	2038	
3530-1			11	2015	
3530-2		3019	11	2015	
3531-021-000	7413	7413		1815	
3531-030-000	788	3829	IC-229		
3531-031-000	720	9014	IC-7		
3532		3124	40	2012	
3532(RCA)	123A	3444	20	2051	
3532-1	128	3024	243	2030	
3533	722	3025/129	IC-9		
3533(RCA)	129		244	2027	
3533-1	129	3025	244	2027	
3534			51	2004	
3534(RCA)	160		245	2004	
3535	319P	3246/229	17	2051	
3535(RCA)	319P	3716/161	39	2015	
3535-110-50008	804	3455	IC-27		
3535-110-50009	804	3455	IC-27		
003536	1194	3484	17	2051	
3536(RCA)	123AP	3444/123A	20	2051	
3536-1	123AP	3444/123A	20	2051	
3536-2	123AP		39	2015	
3537		3132	17	2051	
3537(RCA)	108		86	2038	
3538		3122	17	2051	
3538(RCA)	123A	3444	20	2051	
3538-7	230	3042			
3538-8	231	3857			
3539		3039	18	2030	
3539(RCA)	108		86	2038	
3539-307-001	108	3452	86	2038	
3539-307-002	108	3452	86	2038	
3540		3005	21	2034	
3540(RCA)	159		82	2032	
3540-1		3114	22	2032	
3540-1(GE)	159		82	2032	
3541		3122	18	2030	
3541(RCA)	123A	3444	20	2051	
003542	7403	3003			
3543		3122	17	2051	
3543(RCA)	123A	3444	20	2051	
3544	102A	3122	2	2007	
3544(RCA)	128		243	2030	
3544-1	123A	3444	20	2051	
3545		3045	230	2012	
3545(RCA)	154		40	2012	
3546		3004	17	2051	
3546(RCA)	123A	3444	20	2051	
3546-1(RCA)	123A	3444	20	2051	
3546-2(RCA)	123A	3444	20	2051	
3547		3045	27		
3547(RCA)	225		256		
3548		3122	20	2051	
3548(RCA)	123A	3444	20	2051	
3549		3114	21	2034	
3549(RCA)	159		82	2032	
3549-1		3118	21	2034	
3549-1(RCA)	159		82	2032	
3549-2		3466	21	2034	
3549-2(RCA)	159		82	2032	
3550		3004	53	2007	
3551	126	3122	52	2024	
3551(RCA)	123A	3444	20	2051	
3551A	126	3007	52	2024	
3551A(BLU)			52	2024	
3551A(RCA)	123A	3444	20	2051	
3551A-BLU	126		52	2024	
3551A-GRN	126		52	2024	
3552	225	3045	256		
3552(RCA)	154		40	2012	
3552-1			40	2012	
3552-1(RCA)	191		249		
3553	225	3045	256		
3553(RCA)	154		40	2012	
3553-3	128		243	2030	
3554		3122	18	2030	
3554(RCA)	123A	3444	20	2051	
3555		3124	18	2030	
3555(RCA)	123A	3444	20	2051	
3555-1(RCA)	128		243	2030	
3555-3	128	3024	243	2030	
3556		3124	18	2030	
3556-1	123A	3444	20	2051	
3558		3122	20	2051	
3558(RCA)	123A	3444	20	2051	
3559		3114	21	2034	
3559(RCA)	159		82	2032	
3559-1	159	3466	82	2032	
3560		3122	20	2051	
3560(RCA)	123A	3444	20	2051	
3560-1			18	2030	
3560-1(RCA)	107		11	2015	
3560-2	128	3122	243	2030	
3560-2(RCA)	123A	3444	20	2051	
3561	123A	3444	20	2051	
3561(RCA)	123A	3444	20	2051	
3561-1		3122	18	2030	
3561-1(GE)	128		243	2030	
3561-1(RCA)	123A	3444	20	2051	
3562		3114	21	2034	
3562(RCA)	159		82	2032	
3563		3114	21	2034	
3563(RCA)	159		82	2032	
3564		3004	52	2024	
3564(RCA)	162		35		
3565	124	3024/128	12		
3565(RCA)	128		243	2030	
3565-1		3122	17	2051	
3566	124	3021	12		
3566(RCA)	128		243	2030	

Industry Standard No.	ECG	SK	GE	RS 276-	MOTOR.
3567		3054	12		
3567(RCA)	124		12		
3567-2		3041	66	2048	
3567-2(RCA)	157	3747	232		
3568		3018	39	2015	
3568(RCA)	108		86	2038	
3568(WARDS)	108		86	2038	
3569	161	3122	39	2015	
3569(RCA)	123A	3444	20	2051	
3569-1			20	2051	
3570	159	3466	82	2032	
3570(RCA)	159		82	2032	
3570-1	159	3466	82	2032	
3570P	159	3466	82	2032	
03571	109	3088	18	1123	
3571(RCA)	123A	3444	20	2051	
3571-1	161	3018	39	2015	
3571R	123A	3444	20	2051	
3572		3124	11	2015	
3572(RCA)	123A	3444	20	2051	
3572-3	108	3452	86	2038	
3574	159	3466	82	2032	
3574(RCA)	159		82	2032	
3574-1	129	3114/290	244	2027	
3574-1(RCA)	159		82	2032	
3576	233	3132	210	2009	
3576(RCA)	108		86	2038	
3577		3009	20	2051	
3577(RCA)	123A	3444	20	2051	
3577-1	123A	3444	20	2051	
3578-1	102A		53	2007	
3579	161	3117	39	2015	
3579(RCA)	161	3716	39	2015	
3581		3114	21	2034	
3581(RCA)	159		82	2032	
3582	225	3045	256		
3582(RCA)	154		40	2012	
3583	5804			1144	
3583(RCA)	127	3764	25		
3584(RCA)	124		12		
3585-6	231	3857			
3585-7	230	3042			
3585-8	231	3857			
3586		3004	52	2024	
3586(RCA)	123A	3444	20	2051	
3586-2			60	2015	
3588	123A	3444	20	2051	
3588(RCA)	222			2036	
3588-2		3065	FET-4	2036	
3589	123A	3444	20	2051	
3590	129	3025	244	2027	
3591	128	3024	243	2030	
3592(RCA)	129		244	2027	
3597	159	3466	82	2032	
3597-1	159	3466	82	2032	
3597-1(RCA)	159		82	2032	
3597-2	159	3466	82	2032	
3598(RCA)	108		86	2038	
3598-2		3004		2027	
3600	102	3004	2	2007	
3601	128	3444/123A	243	2030	
3601(RCA)	123A	3444	20	2051	
3601-1	123A	3444	20	2051	
3603		3005	50	2004	
3603(RCA)	108		86	2038	
3603-1			20	2051	
3604		3007	64		
3604(RCA)	108		86	2038	
3604-3			20	2051	
3607	101	3861	8	2002	
3608(RCA)	128		243	2030	
3608-1			18	2030	
3608-1(RCA)	190		217		
3608-2		3024	18	2030	
3608-2(RCA)	128		243	2030	
3609	101	3861	8	2002	
3610			11	2015	
3610(RCA)	108		86	2038	
3611			28	2017	
3612			28	2017	
3612(RCA)	152		66	2048	
3613	191	3007	249		
3613(RCA)	171		27		
3613-2		3104A	27		
3613-2(GE)	191	3232	249		
3613-3	191	3104A	249		
3613-3(RCA)	191		249		
3614-1	123A	3444	20	2051	
3614-3	123A	3444	20	2051	
3615		3024	18	2030	
3615(RCA)	128		243	2030	
3615-1			18	2030	
3616		3025	21	2034	
3616-1		3025	22	2032	
3616-1(RCA)	129		244	2027	
3617		3019	50	2004	
3618	221	3050	FET-4	2036	
3618(RCA)	108		86	2038	
3618-1	121	3717	239	2006	
3620		3114	21	2034	
3620(RCA)	159		82	2032	
3620-1	159	3466	82	2032	
3621(RCA)	152		66	2048	
3622(RCA)	128		243	2030	
3622-1	128		243	2030	
3622-2		3024	18	2030	
3622-2(RCA)	128		243	2030	
3624A					MCM7641
3625			20	2051	MCM7643
3625(RCA)	123A	3444	20	2051	
3625-1	128		243	2030	
3626	123A	3444	20	2051	
3627	159	3466	82	2032	
3627(RCA)	159		82	2032	
3627-1	159	3466	82	2032	
3628		3011			MCM7681
3628-2(RCA)	171		27		
3628-3	188	3199	226	2018	
3629-3	189	3200	218	2026	
3631	159	3466	82	2032	
3631(RCA)	152		66	2048	
3631-1	123A	3444	20	2051	
3631-1(RCA)	152		66	2048	
3632-1	196	3054	241	2020	
3632-2(RCA)	186		28	2017	
3632-3	196	3054	241	2020	
3633-1	710	3104A	IC-89		
3633-1(GE)	171		27		
3634		3111	35		
3634-1	163A	3439	36		
3634.0011	123A	3444	20	2051	
3634.2011	159	3466	82	2032	
3635		3050	FET-4	2036	
3635(RCA)	222			2036	

Industry Standard No.	ECG	SK	GE	RS 276-	MOTOR.
3635-1	222		FET-4	2036	
3635-2			FET-4	2036	
3636	124		12		
3637		3004	52	2024	
3637(RCA)	162		35		
3637-1		3079	19	2041	
3638			23	2020	
3638-1	218	3085	234		
3640(RCA)	196		241	2020	
3646-2(RCA)	108		86	2038	
3647		3710	35		
3647-1	165	3085	38		
3648(RCA)	127	3764	25		
3649		3115	36		
3649-1	165	3115	38		
3651			21	2034	
3652		3018	50	2004	
3652-2	229	3018	61	2038	
3656-1	182	3188A	55		
3656-2	182	3188A	55		
3656-4	182	3188A	55		
3657-1	229	3018	61	2038	
3657-2	229	3018	61	2038	
3662-2	165		38		
3665		3008	19	2041	
3665-2	130	3027	14	2041	
3666		3008	21	2034	
3666(RCA)	128		243	2030	
3669	165		38		
3676	233	3004/102A	210	2009	
3676(RCA)	233		210	2009	
3676-1		3246	60	2015	
3677(RCA)	711		IC-207		
3677-1	711	3141/780	IC-222		
3677-1(RCA)	780		IC-222		
3677-2	780	3141	IC-222		
3677-2(RCA)	780		IC-222		
3677-3	780	3070/711	IC-222		
3677-3(RCA)	780		IC-222		
3679	191	3008	249		
3679-1	191		249		
3680	161	3117	39	2015	
3680-1	161	3716	39	2015	
3681			66	2048	
3681-1	904	3054/196			
3682		3191	69	2049	
3682(RCA)	185		58	2025	
3683			19	2041	
3683-1	130	3027	14	2041	
3683-2	130	3027			
3686		3072	52	2024	
3686(RCA)	102A		53	2007	
3686-1	712	3072	IC-2		
3687	154	3044	40	2012	
3687-1	171		27		
3688-1	5404	3627			
3693(ARVINE)	108		86	2038	
3700-153	116	3016	504A	1104	
3701	137A	3057/136A	ZD-6.2	561	
3702	137A	3057/136A	ZD-6.2	561	
3703	137A	3057/136A	ZD-6.2	561	
3704	137A	3057/136A	ZD-6.2	561	
3706	123A	3444	20	2051	
3714H1	130	3027	14	2041	
3714H1A	130	3027	14	2041	
3721	175		246	2020	
3742	188		226	2018	
3746	100	3005	1	2007	
3746-00	129		244	2027	
3746-01	129		244	2027	
3750	160	3005	245	2004	
3755-1	5014A	3780	ZD-6.8	561	
3755-2	5014A	3780	ZD-6.8	561	
3771	181		75	2041	
3772-1	181	3511	75	2041	
3772-2	181	3511	75	2041	
3819		3116	FET-2	2035	
3819(RCA)	312	3448	FET-1	2035	
3843	152	3016	66	2048	
3843(SEARS)	196		241	2020	
3851	126	3008	52	2024	
3852	100	3004/102A	1	2007	
3867	123A	3444	20	2051	
3868	195A	3765	46		
3878	154		40	2012	
3881	161	3039/316	39	2015	
3907	102	3004	2	2007	
3907(2N404A)	126		52	2024	
3907/2N404A			51	2004	
3961	126	3006/160	52	2024	
3961(G.E.)	160		245	2004	
3970	100	3004/102A	1	2007	
3970(G.E.)	102A		53	2007	
3970CL	102A	3004	53	2007	
3980-002	175		246	2020	
3991-303-112	785	3254	IC-226		
3999	123A	3444	20	2051	
4000	4000		4000		
4001	4001B	3047	40	2012	
4001(IC)	1001		4001	2401	
4001(MAGNAVOX)	177		300	1122	
4001-151	125		510,531	1114	
4001-222	160		245	2004	
4001-223	160		245	2004	
4001-224	102A		53	2007	
4001-225	102A		53	2007	
4001-226	102A		53	2007	
4001-228	190		217		
4001-230	109	3087		1123	
4001B	154		277		
4002	4002B	3124/289	46		
4002(PACE)	123A	3122	20	2051	
4002(PENNCREST)	123A	3444	20	2051	
4002(SEARS)	109			1123	
4003(SEARS)	109			1123	
4003E	123A	3444	20	2051	
4004	195A	3765	219		
4004(PENNCREST)	102A		53	2007	
4004(SEARS)	195A	3765	46	1123	
4005		3048	46		
4005(CB)	195A	3048/329	46		
4005(SEARS)	109			1123	
4006	4006B	3046	86	2004	
4006(E.F.JOHNSON)	128		243	2030	
4006(SEARS)	109			1123	
4007	195A	3048/329	46		
4007(SEARS)	109			1123	
4008		3047	277		
4008(E.F.JOHNSON)	128		243	2030	
4008(SEARS)	109			1123	
4008B	4008B		277		
4009	195A	3004/102A	46		
4009(PENNCREST)	158		53	2007	
4009(SEARS)	109			1123	
4010	123A	3122	20	2051	
4010(E.F.JOHNSON)	108		86	2038	
4010(PENNCREST)	109			1123	
4010(SEARS)	109			1123	
4011	4011B	3048/329	46		
4011(E.F.JOHNSON)	195A		46		
4011(IC)	4011B		4011	2411	
4011(PENNCREST)	177		300	1122	
4011(SEARS)	109			1123	
4011-PC	4011B			2411	
4011PC	4011B		4011	2411	
4012			18	2030	
4012(E.F.JOHNSON)	195A		46		
4012(IC)			4012	2412	
4012(SEARS)	109			1123	
4013	159	3049/224	82		
4013(E.F.JOHNSON)	195A		46		
4013(IC)			4013	2413	
4013(RECTIFIER)	116		504A	1104	
4014	116		504A	1104	
4014-000-10160	102A		53	2007	
4014-200-30110	177	3100/519	300	1122	
4016	4016B		4016		
4017			4017	2417	
4020	785	3254	IC-226		
4020(CMOS)			4020	2420	
4021	123A	3444	20	2051	
4021(IC)			4021	2421	
4022		3444	20	2051	
4023			4023	2423	
4025	4025B		4025		
4027			4027	2427	MCM4027
4032P	1011		IC-79		
4032P-8M4	1011		IC-79		
4035(RCA)	190		217		
4039-00	199	3245	62	2010	
4039-01	199	3245	62	2010	
4041-000-10150	116		504A	1104	
4041-000-10160	102A	3004	53	2007	
4041-000-1018	109			1123	
4041-000-20120	102A	3004	53	2007	
4041-000-30180	102A	3004	53	2007	
4041-000-40270	107		11	2015	
4041-000-40300	107		11	2015	
4041-000-60170	107		11	2015	
4041-000-60200	107		11	2015	
4041-000-80100	126		52	2024	
4041-200-10140100				1123	
4041-200-10180	109	3088		1123	
4041-200-40100	109	3087		1123	
4046(SEARS)	123A	3444	20	2051	
4049	4049		4049	2449	
4049(IC)			4049	2449	
04049B	116		504A	1104	
4050	755		4050	2450	
4051			4051	2451	
4051-300-10150	116	3031A	504A	1104	
4053	4053B	3247/234	65	2050	
4055		4055	4055		
4057	123A	3444	20	2051	
4066		3444	20	2051	
4080-187-0507	128		243	2030	
4080-838-0001	175		246	2020	
4080-838-2	175		246	2020	
4080-838-3	175		246	2020	
4080-866-0006	196		241	2020	
4080-866-1	175		246	2020	
4080-866-2	175		246	2020	
4080-879-0001	196		241	2020	
4082-501-0001	121	3717	239	2006	
4082-748-0002	136A		ZD-5.6	561	
4085	123A	3444	20	2051	
4086	159	3466	82	2032	
4087	159	3466	82	2032	
4099A	189		218	2026	
4101-685	116	3016	504A	1104	
4117-1	949		IC-25		
4117-2	949		IC-25		
4126		3466	82		
4126(E.F.JOHNSON)	159		82	2032	
4149	5292A	3501/5994			
4150-01	123A		20	2051	
4151-01	159	3466	82	2032	
4164P1M	5695	3509			
4167		3018	17	2051	
4167(AIRLINE)	161	3716	39	2015	
4167(PENNCREST)	161	3716	39	2015	
4167(SEARS)	108		86	2038	
4168		3018	17	2051	
4168(PENNCREST)	161	3716	39	2015	
4168(SEARS)	108		86	2038	
4168(WARDS)	161	3716	39	2015	
4169		3018	17	2051	
4169(PENNCREST)	161	3093	39	2015	
4169(SEARS)	108		86	2038	
4169(WARDS)	161	3716	39	2015	
04170	166	9075		1152	
0004201	102A	3004	53	2007	
0004202	102A	3004	53	2007	
0004203	131	3198	44	2006	
4216	130		14	2041	
4218	192	3190/184	63	2030	
4219	129		244	2027	
4241	5282A	3014			
4247	104	3009	16	2006	
4306	128		243	2030	
4309(AIRLINE)	123A	3444	20	2051	
4310 (AIRLINE)	159			2032	
4310(AIRLINE)			82	2032	
4312(RCA)	124		12		
4313	102	3004	2	2007	
4315	102	3004	2	2007	
4322-542	123A		20	2051	
4331	121	3009	239	2006	
4339	1115	3184			
4347	104	3009	16	2006	
4348	102	3004	2	2007	
4349	102	3004	2	2007	
4351.0012			1N34AS	1123	
4351.0013			1N34AS	1123	
4351.0031			1N34AS	1123	
4354	109	3087		1123	
4354.0012			504A	1104	
4360D				2048	
4363	126	3008	52	2024	
4363BLU	126		52	2024	
4363GRN	126		52	2024	
4363ORN	126		52	2024	
4363WHT	126		52	2024	
4364	126	3006/160	52	2024	
4365	126	3008	52	2024	
4366	126	3005	52	2024	
4367	126	3005	52	2024	
4367-001	129	3025	244	2027	

Industry Standard No.	ECG	SK	GE	RS 276-	MOTOR.
4368	160		245	2004	
4398	102	3123	2	2007	
4403	116	3313		1104	
4419-4	180		74	2043	
4437	5276A	3146/787			
4437-1	787	3146	IC-228		
4437-1(RCA)	787		IC-228		
4437-2	787	3146	IC-228		
4437-2(RCA)	787		IC-228		
4437-3	787	3146	IC-228		
4437-3(RCA)	787		IC-228		
4438	5278A	3147			
4438-1	788	3147	IC-229		
4438-1(RCA)	788	3829	IC-229		
4438-2	788	3147	IC-229		
4438-2(RCA)	788	3829	IC-229		
4438-3	788	3147	IC-229		
4438-3(RCA)	788	3829	IC-229		
4439-1	789	3078			
4439-2	789	3078			
4442	159	3466	82	2032	
4442-3	188		226	2018	
4442-366	196		241	2020	
4450	102	3004	2	2007	
4451	100	3004/102A	1	2007	
4454	126	3006/160	52	2024	
4456	126	3006/160		2024	
4457	126	3006/160	52	2024	
4459	127	3764	25		
4460	720	9014			
4460-2	789	9014/720			
4460-3	789	9014/720			
4462	102	3004	2	2007	
4463	121	3009	239	2006	
4464	123	3124/289	20	2051	
4465	123	3124/289	20	2051	
4466ORN	102	3004	2	2007	
4467	176	3123	80		
4468BRN	102	3004	2	2007	
4469RED	102	3004	2	2007	
4470	123	3124/289	20	2051	
4470-31	123		20	2051	
4470-32	123		20	2051	
4470-33	123		20	2051	
4470-ORN	176		80		
4470M-32	123	3124/289	20	2051	
4471ORN	102	3004	2	2007	
4471YEL	102	3004	2	2007	
4472GRN	102	3004	2	2007	
4473	102A	3123	53	2007	
4473-1	108	3452	86	2038	
4473-11	108	3452	86	2038	
4473-12	123A	3444	20	2051	
4473-2	108	3452	86	2038	
4473-3	108	3452	86	2038	
4473-4	123A	3444	20	2051	
4473-5	123A	3444	20	2051	
4473-5X	123A	3444	20	2051	
4473-6	108	3452	86	2038	
4473-7	108	3452	86	2038	
4473-8	108	3452	86	2038	
4473-9	123A	3444	20	2051	
4473-M-12	128	3024	243	2030	
4473-M-3			20	2051	
4473-M3	128	3024	243	2030	
4473-N	128	3024	243	2030	
4474YEL	102	3004	2	2007	
4475GRN	102	3004	2	2007	
4476BLU	102	3004	2	2007	
4477PUR	102	3004	2	2007	
4477V10	102A	3123	53	2007	
4478	129	3025	244	2027	
4483	128	3024	243	2030	
4484	100	3005	1	2007	
4484-1	106	3984	21	2034	
4484-2	106	3984	21	2034	
4485	100	3984/106	1	2007	
4485-1	106	3984	21	2034	
4486	100	3005	1	2007	
4490	128		243	2030	
4490-1	108	3452	86	2038	
4490-7	152	3893	66	2048	
4491-4	130		14	2041	
4491-5	152	3893	66	2048	
4491-6	184	3041	57	2017	
4491-7	130		14	2041	
4491-8	152	3893	66	2048	
4491-9	184	3041	57	2017	
4501	126	3006/160	52	2024	
4509	126	3006/160	52	2024	
4510		3004	2	2007	
4511	4511B	4511			
4511B	4511B	4511			
4518			4518	2490	
4545	126	3006/160	52	2024	
4545BLU	126		52	2024	
4545WHT	126		52	2024	
4552	144A	3094	ZD-14	564	
4553BLU	102	3004	2	2007	
4553BRN	102	3004	2	2007	
4553GRN	102	3004	2	2007	
4553ORN	102	3004	2	2007	
4553RED	102	3004	2	2007	
4553V10	102A	3123	53	2007	
4553YEL	102	3004	2	2007	
4562	102	3004	2	2007	
4563	102	3004	2	2007	
4564	102	3004	2	2007	
4565	102	3004	2	2007	
004567	116	3016	2	1104	
4570	121	3717	239	2006	
4573	105	3012	4		
4582BRN	121	3009	239	2006	
4583RED	121	3009	239	2006	
4584GRN	121	3009	239	2006	
4586	126		52	2024	
4587	108	3452	86	2038	
4589	126	3006/160	52	2024	
4590	106	3984	21	2034	
4594	123		20	2051	
4595	126	3006/160	52	2024	
4596	102	3004	2	2007	
4597	105	3012	4		
4597GRN	105	3012	4		
4597RED	105		4		
4603	126	3006/160	52	2024	
4604	126	3006/160	52	2024	
4605	126	3006/160	52	2024	
4605RED	126		52	2024	
4607	102	3004	2	2007	
4608	121	3009	239	2006	
4619RED	121	3009	239	2006	
4620GRN	121	3009	239	2006	
4621	126	3006/160	52	2024	

Industry Standard No.	ECG	SK	GE	RS 276-	MOTOR.
4622	105	3012	4		
4623	127	3764	25		
4624	123		20	2051	
4627	102	3004	2	2007	
4630	123		20	2051	
4632	126	3008	52	2024	
4640	179	3642	76		
4640P	179	3642	76		
4648	154	3045/225	40	2012	
4649	121	3009	239	2006	
4652	127	3764	25		
4653	139A	3060	ZD-9.1	562	
4663	147A	3095	ZD-33		
4677	126	3006/160	52	2024	
4684-120-3	108	3018	86	2038	
4685-285-3	108	3452	86	2038	
4686-105-3	125	3031A	510,531	1114	
4686-106-3	125	3032A	510,531	1114	
4686-107-3	108	3452	86	2038	
4686-108-3	108	3452	86	2038	
4686-112-3	108	3452	86	2038	
4686-114-3	108	3452	86	2038	
4686-116-3	117	3017B	504A	1104	
4686-118-3	108	3452	86	2038	
4686-119-3	108	3452	86	2038	
4686-120-3	108	3452	86	2038	
4686-126-3	108	3452	86	2038	
4686-127-3	108	3452	86	2038	
4686-130-3	175	3026	246	2020	
4686-131-3	108	3452	86	2038	
4686-132-3	123A	3444	20	2051	
4686-139-3	125	3032A	510,531	1114	
4686-140-3	108	3452	86	2038	
4686-143-3	123A	3444	20	2051	
4686-144-3	123A	3444	20	2051	
4686-145-3	194	3040	220		
4686-147-3	117	3017B	504A	1104	
4686-148-3	117	3017B	504A	1104	
4686-149-3	116	3016	504A	1104	
4686-150-3	117	3017B	504A	1104	
4686-151-3	117	3017B	504A	1104	
4686-163-3	102A	3004	53	2007	
4686-169-3	108	3452	86	2038	
4686-170-3	129	3025	244	2027	
4686-171-3	108	3452	86	2038	
4686-172-3	108	3452	86	2038	
4686-173-3	123A	3444	20	2051	
4686-177-3	117	3017B	504A	1104	
4686-179-3	125	3032A	510,531	1114	
4686-182-3	128	3024	243	2030	
4686-183-3	123A	3444	20	2051	
4686-184-3	125	3311	510,531	1114	
4686-186-3	125	3311	510,531	1114	
4686-189-3	125	3031A	510,531	1114	
4686-195-3	102A	3004	53	2007	
4686-196-3	102A	3004	53	2007	
4686-199-3	125	3031A	510,531	1114	
4686-201-3	125	3311	510,531	1114	
4686-207-3	108	3452	86	2038	
4686-208-3	108	3452	86	2038	
4686-209-3	108	3452	86	2038	
4686-210-3	225	3045	256		
4686-212-3	125	3031A	510,531	1114	
4686-213-3	121	3009	239	2006	
4686-224-3	108	3452	86	2038	
4686-226-3	175	3026	246	2020	
4686-227-3	502	3067	CR-4		
4686-228-3	108	3452	86	2038	
4686-229-3	129	3025	244	2027	
4686-230-3	129	3025	244	2027	
4686-231-3	123A	3444	20	2051	
4686-232-3	154	3040	40	2012	
4686-234-3	196	3041	241	2020	
4686-235-3	129	3025	244	2027	
4686-238-3	129	3025	244	2027	
4686-244-3	108	3452	86	2038	
4686-251-3	108	3452	86	2038	
4686-252-3	191	3044/154	249		
4686-256-1	126	3008	52	2024	
4686-256-2	126	3008	52	2024	
4686-256-3	126	3008	52	2024	
4686-257-3	123A	3444	20	2051	
4686-81-2	102A	3004	53	2007	
4686-81-3	102A	3004	53	2007	
4686-82-3	123A	3444	20	2051	
4686-87-3	117	3017B	504A	1104	
4686-89-3	125	3031A	510,531	1114	
4686-95-3	108	3452	86	2038	
4686-97-3	125	3032A	510,531	1114	
4689	128	3024	243	2030	
4699	150A	3098	ZD-82		
4700	101	3861	8	2002	
4701	130	3027	14	2041	
4702	179	3642	76		
4704	150A	3098	ZD-82		
4705	123		20	2051	
4706	108	3452	86	2038	
4709	108	3452	86	2038	
4714	123	3124/289	20	2051	
4715	130	3027	14	2041	
4715A	130	3027	14	2041	
4722	121	3717	239	2006	
4722BLU	121	3009	239	2006	
4722GRN	121	3009	239	2006	
4722ORN	121	3009	239	2006	
4722PUR	121	3009	239	2006	
4722RED	121	3009	239	2006	
4722YEL	121	3009	239	2006	
4727	121	3009	239	2006	
4728	151A	3099	ZD-110		
4729	179	3642	76		
4730	121	3009	239	2006	
4732	123A	3444	20	2051	
4733	123A	3444	20	2051	
4734	123A	3444	20	2051	
4737	123A	3444	20	2051	
4745	196	3114/290	241	2020	
004746	159	3114/290	11	2032	
4756	161	3039/316	39	2015	
004763	177	3175	300	1122	
4765	123A	3444	20	2051	
4768	123A	3444	20	2051	
04770	109	3088		1123	
4778-8(RCA)	156		512		
4781	197	3083	250	2027	
4789	107	3039/316	11	2015	
4792	6400A	3293/107	11	2015	
4793(KLH)	722		IC-9		
4794(KLH)	781	3169	IC-223		
4800-200	102A		53	2007	
4800-220	102A		53	2007	
4800-221	102A		53	2007	
4800-222	102A		53	2007	
4800-223	131MP	3840	44(2)	2006(2)	

Industry Standard No.	ECG	SK	GE	RS 276-	MOTOR.
4800-224	507		504A	1104	
4801-0000-001	159		21	2034	
4801-0000-003	123A	3018	62	2010	
4801-0000-010	123A		62	2010	
4801-0000-016	123A	3018	211	2016	
4801-0000-035	108	3018	17	2051	
4801-0000-060	159	3138/193A	21	2034	
4801-0000-095	161	3716			
4801-00154	177	3100/519	300	1122	
4801-00628	109	3087		1123	
4801-00629	109	3087		1123	
4801-00801	139A		ZD-9.1	562	
4801-1100-0011			17	2051	
4801-1100-011	121	3009	239	2006	
4801-1284-200	106	3984			
4802-0000-002	184		57	2017	
4802-00002	108	3452	86	2038	
4802-00003	123A	3444	20	2051	
4802-00004	159	3466	82	2032	
4802-00005	130	3027	14	2041	
4802-00005A	130	3027	14	2041	
4802-00006	123A	3444	20	2051	
4802-00007	195A	3048/329	46		
4802-00008	237	3299	46		
4802-00009	123A	3444	20	2051	
4802-00010	312	3834/132	FET-1	2035	
4802-00012	123A	3444	20	2051	
4802-00014	289A	3138/193A	268	2038	
4802-00015	123A	3444	20	2051	
4802-00016	300	3464	273		
4802-00017	195A	3048/329	46		
4802-00019	237	3299	46		
4802-2000-012	116	3017B/117	504A	1104	
4802-2000-019	116	3017B/117	504A	1104	
4802-3268-000	299	3298	236		
4802-3274-200	182	3188A	55		
4804-3267-901	299	3298	236		
4805-1241-200	519	3100	514	1122	
4806-0000-004	116	3311	504A	1104	
4806-0000-005	156		512		
4808-0000-007	135A		ZD-5.1		
4808-0000-009	5072A		ZD-8.2	562	
4808-0000-037	5090A		ZD-56		
4809-0000-001	610	3323	90		
4811-0000-015	312		FET-2	2035	
4811-0000-025	312	3116	FET-2	2035	
4815	159	3466	82	2032	
4819	154	3044	40	2012	
4820	108	3452	86	2038	
4820-0201	519		514	1122	
4821	108	3452	86	2038	
4822-130-30132	5071A	3334	ZD-6.8	561	
4822-130-30192	116	3017B/117	504A	1104	
4822-130-30193	136A		ZD-5.6	561	
4822-130-30256	116	3017B/117	504A	1104	
4822-130-30259	116	3017B/117	504A	1104	
4822-130-30261	167	3647	504A	1172	
4822-130-30264	135A	3056	ZD-5.1		
4822-130-30281	109	3087		1123	
4822-130-30284	5069A	3330/5066A	ZD-4.7		
4822-130-30287	138A	3059	ZD-7.5		
4822-130-30301	109			1123	
4822-130-30311	109	3087		1123	
4822-130-30312	109	3087		1123	
4822-130-30414	166	9075		1152	
4822-130-40095	158		53	2007	
4822-130-40096	103A	3835	59	2002	
4822-130-40132	130		14	2041	
4822-130-40182	177		300	1122	
4822-130-40184	123A	3444	20	2051	
4822-130-40212	155	3839	43		
4822-130-40213	131		44	2006	
4822-130-40214	107		11	2015	
4822-130-40215	107		11	2015	
4822-130-40216	107		11	2015	
4822-130-40229	109			1123	
4822-130-40233	104	3719	16	2006	
4822-130-40235	102A		53	2007	
4822-130-40236	102A		53	2007	
4822-130-40252	160		245	2004	
4822-130-40255	160		245	2004	
4822-130-40304	161	3716	39	2015	
4822-130-40311	107		11	2015	
4822-130-40312	107		11	2015	
4822-130-40313	107		11	2015	
4822-130-40314	103A	3024/128	59	2002	
4822-130-40315	159	3466	82	2032	
4822-130-40317	107		11	2015	
4822-130-40318	107		11	2015	
4822-130-40333			20	2051	
4822-130-40343	123A	3444	20	2051	
4822-130-40348		3114	21	2034	
4822-130-40354	123A	3444	20	2051	
4822-130-40356	128	3024	243	2030	
4822-130-40361	123A	3444	20	2051	
4822-130-40369	106	3984	21	2034	
4822-130-40441	160		245	2004	
4822-130-40454	123A	3444	20	2051	
4822-130-40456	158		53	2007	
4822-130-40477	106	3984	21	2034	
4822-130-40508	106	3984	21	2034	
4822-130-40537	184	3190	57	2017	
4822-130-40614	106	3984	21	2034	
4822-130-50221	116		504A	1104	
4822-130-50228	168	3648		1173	
4823-0018	175		246	2020	
4823-0031-01	234		65	2050	
4824-0014	199	3245	62	2010	
4824-0014-02	199	3245	62	2010	
4824-28-1	311		277		
4824-33	128		243	2030	
4825	229	3018	61	2038	
4826	108	3452	86	2038	
4828-4	109	3087		1123	
4837	108	3452	86	2038	
4838	154	3045/225	40	2012	
4839	123A	3444	20	2051	
4840	123A	3444	20	2051	
4841	123A	3444	20	2051	
4842	123A	3444	20	2051	
4843	154	3045/225	40	2012	
4844	159	3466	82	2032	
4845	108	3452	86	2038	
4846-1	949		IC-25		
4846-2	949		IC-25		
4850	141A	3092	ZD-11.5		
4851	107	3092/141A	11	2015	
4851(ZENER)	141A		ZD-11.5		
4852	123A	3444	20	2051	
4853	154	3044	40	2012	
4854	123A	3444	20	2051	
4855	108	3452	86	2038	
4856-0101	123A	3444	20	2051	
4856-0106	159	3466	82	2032	
4856-0107	123A	3444	20	2051	
4856-0109	123A	3444	20	2051	
4856-0110	123A	3444	20	2051	
4857	229	3018	61	2038	
4858	147A	3095	ZD-33		
4860-1-3	311		277		
4872	124	3021	12		
4882	130	3027	14	2041	
4885	5667A	3632/5624			
004887	142A	3062	ZD-12	563	
4888A	121	3009	239	2006	
4888B	121	3009	239	2006	
4907-976	102A	3004	53	2007	
4927			20	2051	
04970	125	3032A	510,531	1114	
5001	6400A			2029	
5001-002	123A	3444	211	2051	
5001-004	299	3298	236		
5001-010	300	3464	273		
5001-014	123A	3444	20	2051	
5001-020	123A	3444	20	2051	
5001-021	289A	3124/289	268	2038	
5001-032	229	3018	61	2038	
5001-037	229	3122	20	2051	
5001-038	199	3124/289	62	2010	
5001-043	199	3124/289	62	2010	
5001-044	299	3054/196	236		
5001-046	222	3116	FET-3	2036	
5001-047	312	3116	FET-3	2028	
5001-048	159	3466	82	2032	
5001-049	226MP	3086	49(2)	2025(2)	
5001-050	235	3197	215		
5001-053	152	3054/196	66	2048	
5001-064	186A	3357	28	2017	
5001-066	159	3466	82	2032	
5001-068	282	3748	264		
5001-069	123A	3444	20	2051	
5001-070	289A	3138/193A	268	2038	
5001-071	236	3197/235	216	2053	
5001-072	123A	3444	20	2051	
5001-074	199	3122	212	2010	
5001-075	186A	3357	247	2017	
5001-080	109	3088		1123	
5001-083	177	3100/519	300	1122	
5001-089	125	3032A	509	1114	
5001-107	177	3100/519	300	1122	
5001-117	116	3017B/117	504A	1104	
5001-120	519	3100	514	1122	
5001-125	139A	3060	ZD-9.1	562	
5001-128	519	3100	514	1122	
5001-129	116	3017B/117	504A	1104	
5001-130	5072A	3136	ZD-8.2	562	
5001-131	137A	3058	ZD-6.2	561	
5001-134	109	3088		1123	
5001-135	116		504A	1104	
5001-141	109	3088	1N60	1123	
5001-143	5075A	3751	ZD-16	564	
5001-144	177	3100/519	300	1122	
5001-145	177	3100/519	300	1122	
5001-146	519	3100	514	1122	
5001-152	139A	3060	ZD-9.1	562	
5001-156	177	3100/519	300	1122	
5001-160	136A	3057	ZD-5.6	561	
5001-161	109	3088	1N60	1123	
5001-162	177	3100/519	300	1122	
5001-163	116	3311	504A	1104	
5001-164	177	3100/519	300	1122	
5001-196	614	3126	90		
5001-197	135A	3056	IC-5.1		
5001-505	199	3122	62	2010	
5001-506	289A	3138/193A	268	2038	
5001-508	130	3036	14	2041	
5001-509	159	3466	82	2032	
5001-510	108	3452	86	2038	
5001-511	199	3124/289	62	2010	
5001-512	103A	3010	59	2002	
5001-513	295		270		
5001-514	235	3197	322		
5001-517	306		276		
5001-539	297	3122	210	2051	
5001-540	290A	3114/290	21	2034	
5001-541	229	3038	212	2010	
5001-542	123A	3018	20	2051	
5001-543	229	3018	60	2015	
5001-544	229	3038	212	2010	
5001-545	199	3122	61	2038	
5002-001	724		IC-86		
5002-007	1100	3709/110MP	IC-92		
5002-030	1167	3732			
5002-031	1155	3231	IC-179	705	
5036-1	390		255	2041	
5036-2	390		255	2041	
5051-300-10150	116	3311	504A	1104	
5052	102	3722	2	2007	
5059-0236	159		82	2032	
5065	123A		20	2051	
5082-2301					MBD301
5082-2302					MBD301
5082-2303					MBD201
5082-2305					MBD301
5082-2350					MBD102
5082-2400					MBD102
5082-2520					MBD102
5082-2565					MBD102
5082-2787					MBD502
5082-2800					MBD701/702
5082-2810					MBD201
5082-2811					MBD201
5082-2817					MBD101
5082-2824					MBD502
5082-2835					MBD101/102
5082-2900					MBD201
5082-3042					MPN3401
5082-3043					MPN3401
5082-3168					MPN3402
5082-3188					MPN3401
5082-4203					MRD500
5082-4204					MRD500
5082-4207					MRD500
5082-4220					MRD500
5082-4350					MOC1006
5082-4351					MOC1005
5082-4352					MOC1005
5082-4370					4N32
5082-4371					4N32
5084-600-4521-0	519		514	1122	
5084-600-452I-0	519		514	1122	
5085	160		245	2004	
5093	108	3452	86	2038	
5096	312	3448	FET-1	2035	
5101N60	110MP	3709	1N60	1123(2)	
5101S34	109			1123	
5113T	704	3023	IC-205		
5118	219		74	2043	

Industry Standard No.	ECG	SK	GE	RS 276-	MOTOR.
5120A90	109	3087		1123	
5122	142A	3062	ZD-12	563	
5124	145A	3063	ZD-15	564	
5130	146A	3064	ZD-27		
5132	147A	3095	ZD-33		
5139	149A	3097	ZD-62		
5142	150A	3098	ZD-82		
5145	151A	3099	ZD-110		
5158	797	3158	IC-233		
5158(RCA)	728		IC-22		
5158-1	797	3073/728	IC-233		
5158-1(RCA)	797		IC-233		
5158-2	797	3158			
5159-1	797	3158			
5203N1	178MP	3031A	300(2)	1122(2)	
5203RNI	115	3121	6GX1		
5205NI	116	3311	504A	1104	
5205NL				1104	
05206-00	199		62	2010	
5210DC-1	5801	9004	504A	1142	
5213-4000	5882		5040		
5217			300	1122	
5222	142A	3062	ZD-12	563	
5224		3063	ZD-15	564	
5224(ZENER)	145A		ZD-15	564	
5226-1	159	3466	82	2032	
5226-2	123A	3444	20	2051	
5230	146A	3064	ZD-27		
5232	147A	3095	ZD-33		
5239	149A	3097	ZD-62		
5242	150A	3098	ZD-82		
5245	151A	3099	ZD-110		
5253	104	3717/121	16	2006	
5258	128		243	2030	
5259	196		241	2020	
5300-2	5626	3633			
5300-3	5626	3633			
5300-82-1003	117		504A	1104	
5300-82-3	125		510,531	1114	
5300-82-4	125		510,531	1114	
5300-86-1	116		504A	1104	
5300-88-1	116		504A	1104	
5300-88-1002	125		510,531	1114	
5300-88-1003	125		510,531	1114	
5300-88-1004	125		510,531	1114	
5300-88-2	117		504A	1104	
5300-88-3	125		510,531	1114	
5300-94-1	116		504A	1104	
5300-95-1	116		504A	1104	
5300-98-1	116		504A	1104	
5300-99-1	116		504A	1104	
5300-99-3	116		504A	1104	
5300-99-4	116		504A	1104	
5301-06-1	116		504A	1104	
5301-06-1001	125		510,531	1114	
5301-09-1	116		504A	1104	
5301-11-1	117		504A	1104	
5301-11-1001	125		510,531	1114	
5301-13-1	116		504A	1104	
5301-19-1	503		CR-5		
5301-20-1	116		504A	1104	
5301-22-1	117		504A	1104	
5301-32-1	118		CR-1		
5301-36-T	125		510,531	1114	
5313-461B	108	3452	86	2038	
5320-003	175	3026	246	2020	
5324	5130A	3396			
5325	5026A	3792	ZD-17		
5326	5133A	3399			
5327	5135A	3401			
5328	5136A	3402			
5329	5138A	3404	5ZD-25		
5330	5139A	3405			
5331	5141A	3407			
5332	5142A	3408	5ZD-33		
5333	5143A	3409			
5334	5144A	3410			
5335	5145A	3411			
5336	5146A	3412			
5337	5147A	3413			
5340	5151A	3417			
5341	5152A	3418			
5342	5153A	3419			
5343	5155A	3421			
5344	5156A	3422			
5345(ZENER)	5157A	3423			
5345-1	1174	3186			
5346	5158A	3424			
5348	5161A	3427			
5350	5164A	3430			
5351	5166A	3432			
5361-1N60P	110MP	3088	1N60	1123(2)	
5380-21	116	3311	504A	1104	
5380-71	123A	3444	20	2051	
5380-72	123A	3444	20	2051	
5380-73	128	3027/130	243	2030	
5380-73(POWER)	175		246	2020	
5403-MS		3060	ZD-9.1	562	
5416	116	3016	504A	1104	
5459	312			2028	
5464	102A		53	2007	
05470	138A	3059	ZD-7.5		
5489(KLH)	718		IC-8		
5504	181		75	2041	
5505	181		75	2041	
5506	5002A	3768			
5507	5004A	3770			
5508	5005A	3771	ZD-3.3		
5509	5006A	3772	ZD3.6		
5510	5007A	3773	ZD-3.9		
5511	5008A	3774	ZD-4.3		
5512	5009A	3775	ZD-4.7		
5513	5010A	3776	ZD-5.1		
5514	5011A	3057/136A	ZD-5.6	561	
5515	5013A	3779	ZD-6.2	561	
5516	5014A	3780	ZD-6.8	561	
5517	5015A	3781	ZD-7.5		
5518	5016A	3782	ZD-8.2	562	
5519	5018A	3784	ZD-9.1	562	
5520	5019A	3785	ZD-10	562	
5521	5020A	3786	ZD-11	563	
5522	5021A	3787	ZD-12	563	
5522-8	5802	9005		1143	
5523	5022A	3788	ZD-13	563	
5524	5185A	3790/5024A	ZD-15	564	
5525	5025A	3791		564	
5526	5027A	3793	ZD-18		
5527	5029A	3795	ZD-20		
5528	5030A	3796	ZD-22		
5529	5031A	3797	ZD-24		
5530	5033A	3799	ZD-27		
5531	5035A	3801	ZD-30		
5532	5036A	3802	ZD-33		
5533	5037A	3803	ZD-36		
5534	5038A	3804	ZD-39		
5535	5039A	3805	ZD-43		
5536	5040A	3806	ZD-47		
5537	5041A	3807	ZD-51		
5538	5042A	3808	ZD-56		
5539	5044A		ZD-62		
5540	5045A	3811	ZD-68		
5541	5046A	3812	ZD-75		
5553	128		243	2030	
5565-001	130		14	2041	
005575	181		75	2041	
5601-MS	116	3311	504A	1104	
5601-NI	116		504A	1104	
5601-NL				1104	
5608-1	814		IC-243		
5609-1	748	3236			
5611-628	159	3466	82	2032	
5611-628(F)			21	2034	
5611-628F	159	3466	82	2032	
5611-673	159	3466	82	2032	
5611-6730			21	2034	
5611-673D	159	3466	82	2032	
5611-695	193	3025/129	67	2023	
5611-695C	193	3025/129	67	2023	
5612-370	102A	3004	53	2007	
5612-370C	102A		53	2007	
5612-75(C)			1	2007	
5612-75C	102A	3004	53	2007	
5612-77C	102A	3004	53	2007	
5613-1209	192	3024/128	63	2030	
5613-1209C	192	3024/128	63	2030	
5613-1213D	128	3024	243	2030	
5613-1327(T)		3124	85		
5613-1327T	199		85	2010	
5613-1335	123A	3444	20	2051	
5613-1335(E)			85	2010	
5613-1335D	123A	3444	20	2051	
5613-1342	229	3018	11	2015	
5613-1342B	229	3018			
5613-1342C	229	3018	11	2015	
5613-1359(B)		3038	212		
5613-1359B	229		212	2010	
5613-1684(T)		3122	62		
5613-1684T	199		62	2010	
5613-1788(R)		3020	88		
5613-1788R	128		88	2030	
5613-1846Q	306		336		
5613-2086	293	3842			
5613-458	289A	3444/123A	20	2051	
5613-4581C	123A	3444	20	2051	
5613-458B	289A	3444/123A	20	2051	
5613-458C	289A	3444/123A	20	2051	
5613-458D	289A	3444/123A	20	2051	
5613-458LGC	289A	3444/123A	20	2051	
5613-460	107	3018	11	2015	
5613-460A	107	3018	11	2015	
5613-460B	107	3018	11	2015	
5613-460C	107	3018	11	2015	
5613-461	229	3018	11	2015	
5613-461(B)			20	2051	
5613-461C	229	3018	11	2015	
5613-46B	108	3452	86	2038	
5613-535	107	3018	11	2015	
5613-535(B)			20	2051	
5613-535A	107	3018	11	2015	
5613-535B	107	3018	11	2015	
5613-535C	107	3018	11	2015	
5613-558C	123A	3444	20	2051	
5613-710B	123AP	3444/123A	211	2051	
5613-711	123A	3444	20	2051	
5613-711(E)			10	2051	
5613-711E	123A	3444	62	2010	
5613-828(S)		3122	61		
5613-828S	199		61	2038	
5613-870	123A	3444	20	2051	
5613-870(F)			20	2051	
5613-870F	123A	3444	20	2051	
5613-871	199	3245	62	2010	
5613-871(F)			62	2010	
5613-871F	199	3122	62	2010	
5614-359C		3041	247	2052	
5614-360D		3197	337		
5614-77C	103A	3010	59	2002	
5615-1	1176	3210			
5615-2	1176	3210			
5617-2	1175	3212			
5631-1N34A	109			1123	
5631-1N60	109	3088		1123	
5631-1N60P	109	3088		1123	
5631-20A90	110MP	3709	1N60	1123	
5631-MA150	177	3100/519	300	1122	
5632-51B01-02	116	3311			
5632-HZ11A	140A	3061	ZD-10	562	
5632-S1B01-02	116		504A	1104	
5632-W03B	116	3311	504A	1104	
5632-W06A	116	3031A	504A	1104	
5632-W06B	116	3031A	504A	1104	
5633-1S85		3126	90		
5633-MV201	611	3325/612			
5635-HZ11A	140A		ZD-10	562	
5635-ZB1-10	140A		ZD-10	562	
5641-MV11	116	3031A	504A	1104	
5652-AN217	1060		IC-50		
5652-AN217BB	1060		IC-50		
5652-HA1156	801	3160	IC-35		
5652-HA1325	OBS-NLA	3373/1238			
5661-06B3	5400	3950			
5680	129		244	2027	
5701	159	3466	82	2032	
5710	229		211	2038	
5721	123A	3444	20	2051	
5722-3000			51	2004	
5723P	801	3160	IC-35		
5724P	801	3160	IC-35		
5757	781	3169	IC-223		
5766-25	102	3004	2	2007	
5815	723	3144	IC-15		
5847			66	2048	
5847(RCA)	184		57	2017	
5861	125		510,531	1114	
5862			60	2015	
5909-001	181		75	2041	
5923	5801	3051/156	504A	1104	
6001	1115	3184	IC-278		
6003	312	3112	FET-1	2028	
6007-3	708		IC-10		
06008	1027	3007	245	2004	
6008(E.F.JOHNSON)	160		245	2004	
6013	312	3448	FET-2	2035	
6019	177		300	1122	
6099-2	196	3054	241	2020	
6100	158		53	2007	
6100-35	158	3004/102A	53	2007	
06102	234	3100/519	300(2)	1122(2)	

Industry Standard No.	ECG	SK	GE	RS 276-	MOTOR.
06115	196		ZD-8.2	562	
6115(G.E.)	5072A		ZD-8.2	562	
6129-P1	519		514	1122	
6136	123A	3444	20	2051	
6151		3024	18	2030	
6151(RCA)	128		243	2030	
6154	126	3006/160	52	2024	
6155	126	3006/160	52	2024	
6158	108	3452	86	2038	
6158-3	108	3452	86	2038	
6162	126	3006/160	52	2024	
6171-17		3843	511	1114	
6171-18	506	3843	511	1114	
6171-28	116	3175	504A	1104	
6181-1	154	3045/225	40	2012	
6185-3	108	3452	86	2038	
6201	159	3466	82	2032	
6221	221			2036	
6227	222	3065	FET-4	2036	
06246-00	123A	3444	20	2051	
6284	128		243	2030	
6285	129		244	2027	
6313	160		245	2004	
6341					MCM7641
6343-1	123A	3444	20	2051	
6351		3050	FET-4	2036	MCM7643
6367	123A	3444	20	2051	
6367-1	123A	3444	20	2051	
6377-1(SYLVANIA)	131		44	2006	
6377-2(SYLVANIA)	155		43		
6380-1	130		14	2041	
6380-3	152	3054/196	66	2048	
6381					MCM7681
6381-1	175		246	2020	
6381-2	152	3054/196	66	2048	
6382-2	152	3893	66	2048	
6432-3	139A	3060	ZD-9.1	562	
6440	102	3004	2	2007	
6445	100	3004/102A	1	2007	
6452	102	3004	2	2007	
06501	138A				MCM5101
6507(AIRLINE)	108		86	2038	
06508	5073A				MCM6508
6514	123A		61	2051	
6517		3044	20	2051	
6518					MCM6518
6523-34			504A	1104	
6551	722	3161	IC-9		
6605J					MC3443P
6605L					MC3443P
6629	125	3017B/117	510,531	1114	
6629-A05	125	3017B/117	510,531	1114	
6651-486	128	3024	243	2030	
6762			82	2032	
6818	105	3012	4		
6854K90-062	159	3466	82	2032	
6854K90-074	123A	3444	20	2051	
6855K90	106	3984	21	2034	
6954K90-074	123A	3444	20	2051	
6990	160	3006	245	2004	
7001		3122	20	2051	
7001(DARLINGTON)	172A		64		
7001(E.F.JOHNSON)			20	2051	
7002			20	2051	
7005G(LOWREY)	123A	3444	20	2051	
7006			17	2051	
7014			20	2051	
7015			20	2051	
7049	1115	3184			
7112	177	3018	300	1122	
7112(TRANSISTOR)	229		61	2038	
7113	123A	3444	20	2051	
7115		3452	86	2038	
7115(STELMA)	708		IC-10		
7116	229	3018	61	2038	
7117		3018	20	2051	
7117(GE)	108		86	2038	
7118	108	3452	86	2038	
7120A(G.E.)	712		IC-2		
7122	108	3452	86	2038	
7122-5	123A	3444	20	2051	
7123	108	3452	86	2038	
7124	108	3452	86	2038	
7125	108	3452	86	2038	
7126	108	3452	86	2038	
7127	108	3452	86	2038	
7128	108	3452	86	2038	
7129	123A	3444	20	2051	
7130A		3072	IC-2		
7131	108	3452	86	2038	
7132	108	3452	86	2038	
7133	108	3452	86	2038	
7134	108	3452	86	2038	
7149A	712		IC-2		
7171	123A	3444	20	2051	
7171(GE)	123A		20	2051	
7172	123A	3444	20	2051	
7172-54	105	3012	4		
7173	108	3452	86	2038	
7174	108	3452	86	2038	
7175	108	3452	86	2038	
7176	123A	3444	20	2051	
7177	108	3452	86	2038	
7178	108	3452	86	2038	
7204A	804	3455	IC-27		
7204A(SEARS)	804		IC-27		
7206-0	120		CR-3		
7209A	120	3110	CR-3		
7211-9	116	3016	504A	1104	
7212-3	116	3016	504A	1104	
7212-3A	116	3016	504A	1104	
7213	152	3031A	66	2048	
7213(DIODE)	116		504A	1104	
7213-0	117	3017B	504A	1104	
7213-7	116	3016	504A	1104	
7214	108	3452	86	2038	
7214(LOWREY)	130		14	2041	
7214-6	125	3031A	510,531	1114	
7214A	130	3027	14	2041	
7214S			504A	1104	
7215	108	3452	86	2038	
7215-0	102A	3004	53	2007	
7215-1	125	3032A	510,531	1114	
7215-2	125	3031A	510,531	1114	
7216	108	3452	86	2038	
7217	108	3452	86	2038	
7218	108	3452	86	2038	
7219	108	3452	86	2038	
7219-3	131	3052	44	2006	
7220	108	3452	86	2038	
7221	108	3452	86	2038	
7221A	712	3072	IC-2		
7232	108	3452	86	2038	
7233	229	3018	61	2038	
7233S	229	3018	17	2051	
7234	108	3452	86	2038	
7235	108	3452	86	2038	
7236	229	3018	61	2038	
7236S	229	3018	60	2015	
7237	108	3452	86	2038	
7238	108	3452	86	2038	
7239	102	3722	2	2007	
7252	152	3054/196	66	2048	
7253		3045	27		
7253(LOWREY)	154		40	2012	
7261	229	3018	61	2038	
7262	108	3452	86	2038	
7264	229	3018	61	2038	
7301-1	129		244	2027	
7302			20	2051	
7303			21	2034	
7303-1	159	3466	82	2032	
7303-2	159	3466	82	2032	
7306	123A	3444	20	2051	
7306-1	123A	3444	20	2051	
7306-3	123A	3444			
7306-4	123A	3444	20	2051	
7306-5	123A	3444	20	2051	
7306-7	123A	3444			
7310(G.E.)	124		12		
7311	124	3021	12		
7311(GE)	124		12		
7312	175	3026	246	2020	
7313		3077	IC-5		
7313(IC)	713		IC-5		
7313(TRANSISTOR)	124		82	2032	
7314	116		504A	1104	
7316	188		226	2018	
7316-1		3199	28	2017	
7317	124	3021	12		
7317-1			29	2025	
7318	123A	3444	20	2051	
7318-1	297	3854/123AP	20	2051	
7318-2	123A	3444	20	2051	
7320-1	130		14	2041	
7321	194		220		
7321-1			20	2051	
7322	175		246	2020	
7322-1		3538	12		
7325-1	128	3854/123AP	243	2030	
7330-4	159	3466			
7334	128		243	2030	
7335	165		38		
7339-1	159	3466			
7340	106	3984	21	2034	
7340-2	123A	3444	20	2051	
7341	505		CR-7		
7342-1			504A	1104	
7344	177		300	1122	
7344-1	128		243	2030	
7345-2	128		243	2030	
7346-1	123A	3444			
7347-1	159	3466			
7349	128		243	2030	
7351	185	3083/197	58	2025	
7356-2	130		14	2041	
7358-1	152		66	2048	
7359-1	185		58	2025	
7362-1	198	3201/171	27		
7363-1	159	3466	82	2032	
7364-1	196	3054	241	2020	
7364-6053P1	195A	3048/329	46		
7370	165		38		
7372-1	222			2036	
7381-2	128		243	2030	
7398-1	190	3232/191			
7398-6117P1	123A	3444	20	2051	
7398-6118P1	123A	3444	20	2051	
7398-6119P	123A	3122	20	2051	
7398-6119P1	123A	3444	20	2051	
7398-6120P	300	3192/186	273		
7398-6120P1	300	3192/186	273		
7399-4	152	3440/291			
7400	7400	7400	7400	1801	
7400-6A	7400	7400	7400	1801	
7400-9A	7400	7400	7400	1801	
7400/9N00	7400	7400	7400	1801	
7400A	7400		7400	1801	
7400PC	7400	7400	7400	1801	
7401-1	159	3466			
7401-6A	7401	7401	IC-194		
7401-9A	7401	7401	IC-194		
7401PC	7401		IC-194		
7402	708		7402	1811	
7402-1	159	3466			
7402-6A	7402	7402	7402	1811	
7402-9A	7402	7402	7402	1811	
7402PC	7402	7402	7402	1811	
7403	505		CR-7		
7404		7404	7404	1802	
7404-6A	7404	7404	7404	1802	
7404-9A	7404	7404	7404	1802	
7404A	7404		7404	1802	
7404PC	7404	7404	7404	1802	
7405-6A	7405	7405			
7405-9A	7405	7405			
7405PC	7405	7405			
7406	7406		7406	1821	
7406PC	7406	7406	7406	1821	
7408			7408	1822	
7408-6A	7408	7408	7408	1822	
7408-9A	7408	7408	7408	1822	
7408A	7408		7408	1822	
7408N	7408	7408	7408		
7408PC	7408	7408	7408	1822	
7410	7410		7410	1807	
7410-6A	7410	7410	7410	1807	
7410-9A	7410	7410	7410	1807	
7410DC				1807	
7410N	7410	7410	7410	1807	
7410PC	7410	7410	7410	1807	
7413	184	3054/196	7413	2017	
7413(IC)			7413	1815	
7413PC	7413	7413	7413	1815	
7414	152		66	2048	
7414PC	7414	7414			
7416PC	7416	7416			
7419	708		IC-10		
7420	153	3083/197	7420	2049	
7420(IC)	7420		7420	1809	
7420-6A	7420	7420	7420	1809	
7420-9A	7420	7420	7420	1809	
7420DC				1809	
7420PC	7420	7420	7420	1809	
7423	708	3054/196	66	2048	
7425	108	3452	86	2038	
7426	108	3452	86	2038	
7427	108	3452	7427	2038	

Industry Standard No.	ECG	SK	GE	RS 276-	MOTOR.
7427(IC)			7427	1823	
7427DC				1823	
7427PC	7427	7427	7427	1823	
7428		3452	86	2038	
7429	123A	3122	20	2051	
7430	123A	3122	20	2051	
7430-6A	7430	7430			
7430-9A	7430	7430			
7431	123A	3122	20	2051	
7432	123A	3122	7432	2051	
7432(IC)			7432	1824	
7432DC				1824	
7432PC	7432	7432	7432	1824	
7433		3122	20	2051	
7437(GE)	198		251		
7437PC	7437	7437			
7438PC	7438	7438			
7440	708		IC-10		
7440-6A	7440	7440			
7440-9A	7440	7440			
7440PC	7440	7440			
7441			7441	1804	
7441-6A	7441	7441	7441	1804	
7441-9A	7441	7441	7441	1804	
7441DC	7441	7441	7441	1804	
7441PC	7441	7441	7441	1804	
7442DC	7442	7442			
7442PC	7442	7442			
7445DC	7445	7445			
7445PC	7445	7445			
7446ADC	7446	7446			
7446APC	7446	7446			
7446PC	7446	7446			
7447	7447		7447	1805	
7447BDC	7447	7447	7447	1805	
7447BPC	7447	7447	7447	1805	
7447DC	7447	7447	7447	1805	
7447PC	7447	7447	7447	1805	
7448			7448	1816	
7448DC	7448	7448	7448	1816	
7448PC	7448	7448	7448	1816	
7451			7451	1825	
7451-6A	7451	7451	7451	1825	
7451-9A	7451	7451	7451	1825	
7451DC				1825	
7451PC	7451	7451	7451	1825	
7454-6A	7454	7454			
7454-9A	7454	7454			
7454PC	7454	7454			
7473			7473	1803	
7473-6A	7473	7473	7473	1803	
7473-9A	7473	7473	7473	1803	
7473PC	7473	7473	7473	1803	
7474	7474		7474	1818	
7474-6A	7474	7474	7474	1818	
7474-9A	7474	7474	7474	1818	
7474/9N74	7474		7474	1818	
7474PC	7474	7474	7474	1818	
7475			7475	1806	
7475-6A	7475	7475	7475	1806	
7475-9A	7475	7475	7475	1806	
7475DC	7475	7475	7475	1806	
7475PC	7475	7475	7475	1806	
7476	7476	7476	7476	1813	
7476-6A	7476	7476	7476	1813	
7476-9A	7476	7476	7476	1813	
7476PC	7476	7476	7476	1813	
7485			7485	1826	
7485DC	7485	7485	7485	1826	
7485PC	7485	7485	7485	1826	
7486			7486	1827	
7486DC				1827	
7486PC	7486	7486	7486	1827	
7490	7490		7490	1808	
7490-6A	7490	7490	7490	1808	
7490-9A	7490	7490	7490	1808	
7490DC	7490	7490	7490	1808	
7490PC	7490	7490	7490	1808	
7492		7492	7492	1819	
7492-6A	7492	7492	7492	1819	
7492-9A	7492	7492	7492	1819	
7492DC	7492	7492	7492	1819	
7492PC	7492	7492	7492	1819	
7493	7493A	7493			
7493/9393	7493A	7493			
7493DC	7493A	7493			
7495DC	7495	7495			
7495PC	7495	7495			
7501	289A	3444/123A	20	2051	
7502	199	3444/123A	20	2051	
7503	159	3466	82	2032	
7505	289A	3444/123A	20	2051	
7506	123A	3444	20	2051	
7507	128		243	2030	
7508	129		244	2027	
7509	199	3444/123A	20	2051	
7510	384	3033	510,531	1114	
7510(TRANSISTOR)	175		246	2020	
7511	289A	3019			
7513	128	3192/186	28	2017	
7514	130	3027	14	2041	
7515	123A	3444	20	2051	
7516	123A	3444	20	2051	
7517	123A	3444	20	2051	
7518	123A	3444	20	2051	
7519	289A	3444/123A	20	2051	
7550	177	3330/5066A	ZD-3.3		
7553	188	3199	226	2018	
7554	139A		ZD-9.1	562	
7568	116	3016	504A	1104	
7574	138A	3059	ZD-7.5		
7585	123A	3444	20	2051	
7586	123A	3444	20	2051	
7586(GE)	123A	3444	20	2051	
7587	123A	3444	20	2051	
7587(GE)	123A	3444	20	2051	
7588	123A	3444	20	2051	
7588(GE)	123A	3444	20	2051	
7589	123A	3444	20	2051	
7590	123A	3444	20	2051	
7590(GE)	123A	3444	20	2051	
7591(GE)	123A	3444	20	2051	
7593-2	108	3452	86	2038	
7623	159	3004/102A			
7626	159	3138/193A	21	2034	
7637	123A	3444	20	2051	
7638	294	3018	48		
7639	294	3018	20	2051	
7641	123A	3444	20	2051	
7642	108	3452	86	2038	
7675	123A	3444	20	2051	
7676	123A	3444	20	2051	
7701	125		510,531	1114	
7702-1A	125	3033	510,531	1114	
7704-1	125	3033	510,531	1114	
7706-1	125	3033	510,531	1114	
7707	5065A	3838			
7708	5111A	3377			
7708-1	125	3033	510,531	1114	
7709	5112A	3378			
7710	5112A	3378			
7710-1		3033	509	1114	
7711	5114A	3380			
7711-1	125	3033	510,531	1114	
7712	5115A	3381			
7712-1	125	3033	510,531	1114	
7713	135A		ZD-5.1		
7713-1	125	3033	510,531	1114	
7714	5117A	3383			
7715	5119A	3385			
7717	5121A	3387	5ZD-7.5		
7718	5122A	3388	5ZD-8.2		
7719	5124A	3390	5ZD-9.1		
7720	5125A	3391	5ZD-10		
7721	5126A		5ZD-11		
7722		3126	90		
7724	5130A	3396	5ZD-15		
7725	5131A	3397	5ZD-16		
7726	5133A	3399	5ZD-18		
7727	5135A	3401			
7728	5136A	3402	5ZD-22		
7729	5137A	3403	5ZD-24		
7730	5139A	3405	5ZD-27		
7731	5141A	3407	5ZD-30		
7732	5142A	3408	5ZD-33		
7733	5143A	3298/299			
7734	5144A	3410			
7735	5145A	3411			
7736	5146A	3412			
7737	5147A	3413			
7738	5148A	3414			
7739	5150A	3416			
7741	5152A	3418			
7805KC	309K	3629			
7805UC	960	3591			
7808	964	3630			
7808UC	964	3630			
7810	108		86	2038	
7811	108		86	2038	
7812	108		86	1771	
7813	108		86	2038	
7814	108		86	2038	
7815	108		86	2038	
7816	123A		20	2051	
7817	123A		20	2051	
7818	123A		20	2051	
7824UC	972	3670			
7885-1	181		75	2041	
7885-2	181		75	2041	
7885-3	181		75	2041	
7909	128		243	2030	
7920-1	196		241	2020	
7921	5802	9005		1143	
7921-1	197		250	2027	
7962	116	3311	504A	1104	
7991	199	3124/289	62	2010	
7992	123A	3444	20	2051	
7993	158	3004/102A	53	2007	
8000-00001-068	102A	3004	53	2007	
8000-00003-033	123A	3444	20	2051	
8000-00003-034	233	3122	210	2009	
8000-00003-035	289A	3138/193A	268	2038	
8000-00003-036	289A	3138/193A	268	2038	
8000-00003-037	290A	3114/290	269	2032	
8000-00003-038	102A	3004	53	2007	
8000-00003-039	102A	3004	53	2007	
8000-00003-040	131	3052	44	2006	
8000-00003-041	299	3047	236		
8000-00003-042	299	3047	236		
8000-00003-043	299	3048/329	236		
8000-00003-044	116	3087		1104	
8000-00003-045	109	3087		1123	
8000-00003-046	177	3100/519	300	1122	
8000-00004-003			211	2016	
8000-00004-044	116		504A	1104	
8000-00004-045	109	3087		1123	
8000-00004-060	109	3087		1123	
8000-00004-061	177	3100/519	300	1122	
8000-00004-062	177	3100/519	300	1122	
8000-00004-063	109	3088		1123	
8000-00004-064	116	3100/519	504A	1104	
8000-00004-065	139A	3060	ZD-9.1	562	
8000-00004-066	177	3100/519	300	1122	
8000-00004-067	177	3017B/117	300	1122	
8000-00004-068	116	3017B/117	504A	1104	
8000-00004-079	123A	3444	20	2051	
8000-00004-080	312	3112	FET-2	2028	
8000-00004-081	312	3834/132	FET-1	2035	
8000-00004-082	123A	3444	20	2051	
8000-00004-083	237	3299	46		
8000-00004-084	237	3299	46		
8000-00004-085	289A	3122	268	2038	
8000-00004-086	103A	3010	59	2002	
8000-00004-087	300	3464	273		
8000-00004-088	102A	3004	53	2007	
8000-00004-089	199	3466/159	82	2032	
8000-00004-090	724		IC-86		
8000-00004-184	177	3100/519	300	1122	
8000-00004-185	186	3192	28	2017	
8000-00004-239	137A	3058	ZD-6.2	561	
8000-00004-241	130	3122	14	2041	
8000-00004-242	289A	3122	268	2038	
8000-00004-243	123A	3444	20	2051	
8000-00004-248		3126	90		
8000-00004-298	229	3018	61	2038	
8000-00004-299			61	2038	
8000-00004-300			61	2038	
8000-00004-301			212	2010	
8000-00004-305	703A		IC-12		
8000-00004-307	1075A	3877			
8000-00004-85	123A	3444	20	2051	
8000-00004-P079	123A	3122	20	2038	
8000-00004-P080	312	3834/132	FET-1	2035	
8000-00004-P081	312	3834/132	FET-1	2035	
8000-00004-P082	128	3122	20	2030	
8000-00004-P083	237	3299	46		
8000-00004-P084	237	3299	46		
8000-00004-P085	289A	3138/193A	268	2038	
8000-00004-P086	289A	3444/123A	20	2051	
8000-00004-P088	102A	3004	53	2007	
8000-00004-P089	159	3025/129	82	2032	
8000-00004-P090	724		IC-86		
8000-00004-P185	184	3192/186	57	2017	
8000-00005-001	312	3834/132	FET-1	2035	
8000-00005-002	123A	3444	20	2051	
8000-00005-003	199	3018	62	2010	
8000-00005-004	199	3122	62	2010	
8000-00005-005	199	3124/289	62	2010	

Industry Standard No.	ECG	SK	GE	RS 276-	MOTOR.
8000-00005-007	123A	3444	20	2051	
8000-00005-008	175	3026	246	2020	
8000-00005-009	195A	3047	46		
8000-00005-010	195A	3047	46		
8000-00005-011	237	3299	46		
8000-00005-012	175		246	2020	
8000-00005-014	1104	3225	IC-91		
8000-00005-015	109	3088		1123	
8000-00005-016	109	3087		1123	
8000-00005-017	109	3087		1123	
8000-00005-018	109	3088		1123	
8000-00005-019	138A	3059	ZD-7.5		
8000-00005-020	137A	3058	ZD-6.2	561	
8000-00005-021	139A	3060	ZD-9.1	562	
8000-00005-022	116	3031A.	504A	1104	
8000-00005-023	109	3088		1123	
8000-00005-055	123A	3444	62	2051	
8000-00005-152	116	3031A	504A	1104	
8000-00006-001	195A	3047	46		
8000-00006-002	237	3299	46		
8000-00006-003	123A	3444	20	2051	
8000-00006-004	159	3466	82	2032	
8000-00006-005	107	3018	11	2015	
8000-00006-006	130	3027	14	2041	
8000-00006-007	109	3088	1N60	1123	
8000-00006-008	177	3100/519	300	1122	
8000-00006-009	139A	3060	ZD-9.1	562	
8000-00006-010	147A	3095	ZD-33		
8000-00006-146	137A	3057/136A	ZD-6.2	561	
8000-00006-147	125	3081	510,531	1114	
8000-00006-190	130	3027	44	2006	
8000-00006-201	116		504A	1104	
8000-00006-230	237	3299	264		
8000-00006-230(3)	282		264		
8000-00006-231	116	3017B/117	504A	1104	
8000-00006-232	139A	3060	ZD-9.1	562	
8000-00006-280	123A	3124/289	20	2051	
8000-00006-281	519	3100	300	1122	
8000-00009-089	199	3245	62	2010	
8000-00009-174	123A	3444	20	2051	
8000-00009-177	229	3018	61	2038	
8000-00009-178	312	3834/132	FET-1	2035	
8000-00009-280	123A	3122	210		
8000-0001-017	312	3112			
8000-00010-017	312	3112	FET-1	2028	
8000-00010-109	177	3100/519	300	1122	
8000-00011-004	123A	3444	20	2051	
8000-00011-041	116	3017B/117	504A	1104	
8000-00011-042	177	3088	514	1122	
8000-00011-043	139A	3060	ZD-9.1	562	
8000-00011-044	116	3017B/117	504A	1104	
8000-00011-045	601		300	1122	
8000-00011-046	109	3100/519		1123	
8000-00011-047	123AP	3444/123A	20	2051	
8000-00011-048	123A	3444	20	2051	
8000-00011-049	289A	3124/289	268	2038	
8000-00011-050	152	3054/196	66	2048	
8000-00011-051	195A	3049/224	46		
8000-00011-052	237	3299	46		
8000-00011-053	222	3116	FET-4	2036	
8000-00011-054	312	3834/132	FET-1	2035	
8000-00011-055	312	3834/132	FET-1	2035	
8000-00011-060	109	3087		1123	
8000-00011-064			ZD-9.1	562	
8000-00011-086	103A	3835	59	2002	
8000-00011-103	138A	3059	ZD-7.5		
8000-00011-104	116	3017B/117	504A	1104	
8000-00011-105	156	3051	512		
8000-00011-151	195A	3765	46		
8000-00011-166	614	3327			
8000-00012-038	112	3089	1N82A		
8000-00012-039	123A	3122	20	2051	
8000-00012-040	192	3137	63	2030	
8000-00012-041	703A	3087	IC-12		
8000-00016-127	1108		IC-92		
8000-00028-037	229	3018	61	2038	
8000-00028-038	289A	3122	268	2038	
8000-00028-039	236	3054/196	216	2053	
8000-00028-040		3839	43		
8000-00028-041	152	3054/196	66	2048	
8000-00028-042	7404	7404		1802	
8000-00028-044	74145	74145			
8000-00028-045	177	3100/519	300	1122	
8000-00028-047	140A	3061	ZD-10	562	
8000-00028-048		3311	6GX1		
8000-00028-206	123A	3122	61	2038	
8000-00029-006	123A	3444	20	2051	
8000-00029-007	123A	3444	20	2051	
8000-00030-007	123A	3444	20	2051	
8000-00030-008	315	3048/329	279		
8000-00030-009	103A	3835	59	2002	
8000-00030-010	116	3017B/117	504A	1104	
8000-00032-025	123A	3444	20	2051	
8000-00032-026	161	3716	39	2015	
8000-00032-027	103A	3010	59	2002	
8000-00032-028	236	3197/235	216	2053	
8000-00032-030	1058	3459	IC-49		
8000-00032-031	724		IC-86		
8000-00035-001	116	3017B/117	504A	1104	
8000-00035-002	116	3017B/117	504A	1104	
8000-00035-003	229	3018	61	2038	
8000-00038-001	235	3197	270		
8000-00038-002	974	3965			
8000-00038-003	74161	74161			
8000-00038-004	7400	7400		1801	
8000-00038-006	7493A	7493			
8000-00038-007	7474	7474		1818	
8000-00038-008	177	3100/519	300	1122	
8000-00038-009	109	3087		1123	
8000-00038-010	611	3324	90		
8000-0004-004			82	2032	
8000-0004-042	109	3087		1123	
8000-0004-044			504A	1104	
8000-0004-063	109	3087		1123	
8000-0004-066	177	3100/519	300	1122	
8000-0004-086	103A		59	2002	
8000-0004-P-061			504A	1104	
8000-0004-P060				1123	
8000-0004-P061			300	1122	
8000-0004-P062		3031A	300	1122	
8000-0004-P063			1N60	1123	
8000-0004-P064			300	1122	
8000-0004-P065			ZD-9.1	562	
8000-0004-P066			300	1122	
8000-0004-P067			504A	1104	
8000-0004-P068			504A	1104	
8000-0004-P079		3444	20	2051	
8000-0004-P080		3448	FET-1	2028	
8000-0004-P081		3448	FET-1	2028	
8000-0004-P082		3444	20	2051	
8000-0004-P083	195A	3048/329	46		
8000-0004-P084	195A	3048/329	243		
8000-0004-P085		3020	20	2051	
8000-0004-P086		3124	59	2051	

Industry Standard No.	ECG	SK	GE	RS 276-	MOTOR.
8000-0004-P088		3004	53	2007	
8000-0004-P089		3466	82	2032	
8000-0004-P090	724	3525	IC-86		
8000-0004-P185	186	3192	28	2017	
8000-0004-P060	109			1123	
8000-0004-P061	177			1122	
8000-0004-P062	177			1122	
8000-0004-P063	109			1123	
8000-0004-P064	177			1122	
8000-0004-P065	139A			562	
8000-0004-P066	177			1122	
8000-0004-P067	116			1104	
8000-0004-P068	116			1104	
8000-0004-P079	123A	3444		2051	
8000-0004-P080	312			2028	
8000-0004-P081	312			2035	
8000-0004-P082	123A	3444		2051	
8000-0004-P083	128			2030	
8000-0004-P084	128			2030	
8000-0004-P085	123A	3444		2051	
8000-0004-P086	103A			2002	
8000-0004-P088	102A			2007	
8000-0004-P089	159			2032	
8000-00041-015	109	3087	1N34AS	1123	
8000-00041-016	109	3088	1N60	1123	
8000-00041-018	139A	3060	ZD-9.1	562	
8000-00041-019	177	3100/519	300	1122	
8000-00041-040	229	3039/316	60	2015	
8000-00041-041	199	3124/289	212	2010	
8000-00041-042	199	3124/289	212	2010	
8000-00041-043	186A	3357	28	2017	
8000-00041-044	235	3239/236	215		
8000-00041-046	229	3018	213	2038	
8000-00042-007	519	3100	514	1122	
8000-00042-013	222	3050/221	FET-4	2036	
8000-00042-014	6402	3628			
8000-00042-015	5401	3638/5402			
8000-00043-019	229	3018	60	2015	
8000-00043-020	158	3004/102A	53	2007	
8000-00043-021	139A	3060	ZD-9.1	562	
8000-00043-064	302	3252	275	2012	
8000-00043-065	235	3197	216	2053	
8000-00043-067	135A		ZD-5.1		
8000-00043-068	139A	3060	ZD-9.1	562	
8000-00045-004	7447	7447			
8000-00049-010	116	3031A	504A	1104	
8000-00049-020	614		90		
8000-00049-021	5010A	3056/135A	ZD-5.1		
8000-00049-027	1166	3827			
8000-00049-053	229	3124/289	213	2038	
8000-00049-054	229	3124/289	213	2038	
8000-00049-055	123A	3122	62	2051	
8000-00049-056	159	3118	82	2032	
8000-00049-057	123A	3122	62	2051	
8000-00049-059	293		47	2030	
8000-00049-060	235	3197	215		
8000-00049-061	293	3024/128	47	2030	
8000-00049-062	312	3116	FET-2	2035	
8000-0005-001	312	3834/132	FET-1	2035	
8000-0005-002	107		11	2015	
8000-0005-003	107		11	2015	
8000-0005-004			212	2010	
8000-0005-007	123A	3444	20	2051	
8000-0005-008	130		14	2041	
8000-0005-009	123A	3444	20	2051	
8000-0005-010	128		243	2030	
8000-0005-011	128		243	2030	
8000-0005-015	109			1123	
8000-0005-016	109			1123	
8000-0005-017	109			1123	
8000-0005-018	109			1123	
8000-0005-019	138A		ZD-7.5		
8000-0005-020	137A		ZD-6.2	561	
8000-0005-021	139A		ZD-9.1	562	
8000-0005-022	116		504A	1104	
8000-0005-023	109			1123	
8000-00053-004	746		IC-217		
8000-00054-002	981	3724			
8000-00057-011	139A		ZD-9.1	562	
8000-00058-003			ZD-6.2	561	
8000-00058-004	140A	3061	ZD-10	562	
8000-00058-008	1260	3744			
8000-00058-009	1192	3445	IC-181		
8000-0009-089	123A		212	2010	
8000-0028-042				1802	
8000-004-P061	125	3031A	510,531	1114	
8000-004-P063				1123	
8000-004-P064	125	3031A	510,531	1114	
8000-004-P067	125	3031A	510,531	1114	
8000-004-P068	125	3031A	510,531	1114	
8000-004-P089	129	3025	244	2027	
8000-0058-006	746	3234	IC-217		
8003-114	123A	3444	20	2051	
8003-115	116	3017B/117	504A	1104	
8005(PENNCREST)	181		75	2041	
8005-3	704	3023	IC-205		
8007-0	704	3022/1188	IC-205		
8007-1	704	3022/1188	IC-205		
8007-3	704	3023	IC-205		
8007-4	704	3023	IC-205		
8008-1	704	3022/1188	IC-205		
8008-3	704	3023	IC-205		
8009-0	704	3023	IC-205		
8009-4	704	3023	IC-205		
8010-171	1045		IC-2		
8010-173	312	3834/132	FET-1	2035	
8010-174	161	3018	39	2015	
8010-175	161	3018	39	2015	
8010-176	123A	3444	20	2051	
8010-52	177	3100/519	300	1122	
8010-53	109	3088		1123	
8020-202	177	3175	300	1122	
8020-203	116	3017B/117	504A	1104	
8020-204	199	3245	62	2010	
8020-205	123A	3444	20	2051	
8020-206	152	3893	66	2048	
08050	116	3017B/117	504A	1104	
8070-4	102A	3004	53	2007	
8071-4	102A	3004	53	2007	
8072-4	102A	3004	53	2007	
8073-4	102A	3004	53	2007	
8074-4	123A	3444	20	2051	
8075-4	123A	3444	20	2051	
8098	749		IC-97		
8102-206		3311	504A	1104	
8102-207		3124	20	2051	
8102-208		3124	62	2010	
8102-209		3024	20	2051	
8102-210			31MP	2006(2)	
8200-202	199	3124/289	62	2010	
8200-203	199	3124/289	62	2010	
8200-204	226	3082	49	2025	
8210-1203	123A	3444	20	2051	
8216	8216				MC8T26AL

Industry Standard No.	ECG	SK	GE	RS 276-	MOTOR.
8225P1	5626	3635			
8226					MC8T28L
8281	123A		20	2051	
8281-1	123A	3444	20	2051	
8300-8	116	3016	504A	1104	
8301	159	3466	82	2032	
8302	123A	3444	20	2051	
8303	129	3025	244	2027	
8304	128	3024	243	2030	
8319-001	130		14	2041	
8394	199	3245	62	2010	
8400-1	129	3025	244	2027	
8400-1A	129	3025	244	2027	
8400-1B	129	3025	244	2027	
8405	159	3466	82	2032	
8440-121	312	3112	FET-1		
8440-122	123A	3018	17		
8440-123	123A	3122	20		
8440-124	123A	3122	210		
8440-126	199	3124/289	212		
8440P1	912		IC-172		
8471		3054	28	2017	
8471(SYLVANIA)	188		226	2018	
8500-201	102A		53	2007	
8500-202	102A	3123	53	2007	
8500-203	102A	3123	53	2007	
8500-204	131		44	2006	
8500-206	116	3016	504A	1104	
8503	107		11	2015	
8504	123A	3444	20	2051	
8509	123A	3444	20	2051	
8517	293	3849	47	2030	
8530	123A	7490			
8540	159	3466	82	2032	
8554-9	128	3024	243	2030	
8564-1	310	3856			
8564-5	310	3856			
8564-6	308	3855			
8564-7	310	3856			
8564-8	308	3855			
8600	123A	3444	20	2051	
8601	159	3466	82	2032	
8602	123A	3444	20	2051	
8606	108	3452	86	2038	
8607	108	3452	86	2038	
8609	108	3452	86	2038	
8611	108	3452	86	2038	
8614 007 0	123A	3444		2051	
8620	175		246	2020	
8624-003	129		244	2027	
8710-161	312	3834/132	FET-1	2035	
8710-162	108	3452	86	2038	
8710-163	123A	3444	20	2051	
8710-164	123A	3444	20	2051	
8710-165	123A	3444	20	2051	
8710-166	123A	3444	20	2051	
8710-167	123A	3444	20	2051	
8710-168	123A	3444	20	2051	
8710-169	159	3466	82	2032	
8710-170	186	3192	28	2017	
8710-171	1100	3223	IC-92		
8710-172	1142	3485	IC-128		
8710-51		3126	90		
8710-52	177	3100/519	300	1122	
8710-53	109	3088		1123	
8710-54	139A	3060	ZD-9.1	562	
8710-55	116	3311	504A	1104	
8800-201	116	3031A	504A	1104	
8800-202	123A	3444	20	2051	
8800-203	123A	3444	20	2051	
8800-204	123A	3444	20	2051	
8800-205	152	3893	66	2048	
8800-206	116	3031A	504A	1104	
8840-121			FET-2	2035	
8840-122			17	2051	
8840-123			61	2038	
8840-124			210	2051	
8840-126			212	2010	
8840-161	312	3448	FET-1	2035	
8840-162	107	3018	11	2015	
8840-163	123A	3444	20	2051	
8840-164	123A	3444	20	2051	
8840-165	123A	3444	20	2051	
8840-166	123A	3444	20	2051	
8840-167	123A	3444	20	2051	
8840-168	123A	3444	20	2051	
8840-169	186	3192	28	2017	
8840-170	1100	3223	IC-92		
8840-171	720	9014	IC-7		
8840-52		3126	90		
8840-53	177	3100/519	300	1122	
8840-54	109	3088		1123	
8840-55	139A	3060	ZD-9.1	562	
8840-56	116	3311	504A	1104	
8864-1	116	3016	504A	1104	
8867	159	3466	82	2032	
8868-6	128	3024	243	2030	
8868-7	199	3038	62	2010	
8868-8	123A	3444	20	2051	
8880-3	128	3024	243	2030	
8883-2	121	3009	239	2006	
8883-4	128	3024	243	2030	
8886-2	123A	3444	20	2051	
8888-00005-013	295	3253	270		
8910-141	312	3834/132	FET-1	2035	
8910-142	229	3018	61	2038	
8910-143	123A	3444	20	2051	
8910-144	199	3245	62	2010	
8910-145	123A	3444	20	2051	
8910-146	186A	3357	247	2052	
8910-147	1100	3223	IC-92		
8910-148	1142	3485	IC-128		
8910-51		3126	90		
8910-52	177	3100/519	300	1122	
8910-53	110MP	3088	1N60	1123(2)	
8910-54	5072A		ZD8.2	562	
8910-55	116		504A	1104	
8999-115	104	3009	16	2006	
8999-201	116	3004/102A	504A	1104	
8999-202	102A	3004	53	2007	
8999-203	158	3004/102A	53	2007	
8999-205	116	3031A	504A	1104	
09004	718	3159	IC-8		
09005	703A	3157	IC-12		
9005-0	121	3009	239	2006	
09006	703A	3157	IC-12		
09007	750	3280	IC-219		
09010	746	3234	IC-217		
09011	720	3444/123A	20	2051	
9011(G)			20	2051	
9011E	229	3452/108	86	2038	
9011F	123AP	3018	61	2038	
9011F(TUNER)			86	2038	
9011G	123AP	3018	61	2038	

Industry Standard No.	ECG	SK	GE	RS 276-	MOTOR.
9011H	229	3018	61	2038	
9012HE	102A	3004	53	2007	
9012HF	102A	3004	53	2007	
9012HG		3114	21	2034	
9013G	128		20	2051	
9013H	123A	3444	20	2051	
9013HF	123A	3444	20	2051	
9013HG	123A	3444	20	2051	
9013HH	123A	3444	20	2051	
9014	123AP	3444/123A	20	2051	
9014(D)			10	2051	
9014B	123AP	3444/123A	20	2051	
9014C	123AP	3444/123A	20	2051	
9014D	123A	3444	20	2051	
9015	159		48		
9015C	159	3466	82	2032	
9016	229	7404	86	2038	
9016(F)			20	2051	
9016(G)			20	2051	
9016D	229	3452/108	86	2038	
9016E	229	3452/108	86	2038	
9016F	229	3452/108	86	2038	
9016G	108	3018	61	2038	
09017	722	7405	IC-9		
09018	788	3452/108	86	2038	
9018D	229	3452/108	86	2038	
9018E	229		61	2038	
9018F	108	3018	61	2038	
9018G	108	3018	61	2038	
9033	123A	3444	20	2051	
9033(SYLVANIA)	123A	3444	20	2051	
9033-1	123A	3444	20	2051	
9033-2	199	3245	62	2010	
9033-3	199	3245	62	2010	
9033-4	199	3245	62	2010	
9033-5	199	3245	62	2010	
9033BROWN	123A	3444	20	2051	
9033G		3018	18	2030	
9033G(SYLVANIA)	123A	3444	20	2051	
9033GREEN	199	3245	62	2010	
90330	123AP	3245/199			
90330RANGE	199		62	2010	
9033Q	199		62	2010	
9033RED	199	3245	62	2010	
9033WHITE	199	3245	62	2010	
9050-1	910D		IC-252		
9093DC	9093				MC853L
9093DM					MC953L
9093FC					MC853F
9093FM					MC953F
9093PC	9093				MC853P
9094DC	9094				MC856L
9094DM					MC956L
9094FC					MC856F
9094FM					MC964F
9094PC	9094				MC856P
9097DC	9097				MC855L
9097DM					MC955L
9097FC					MC855F
9097FM					MC955F
9097PC	9097				MC855P
9099DC	9099				MC852L
9099DM					MC952L
9099FC					MC852F
9099FM					MC952F
9099PC	9099				MC852P
9100-1	116	3016	504A	1104	
9109DC	9109				MC689TL
9109DM					MC689TL
9110DC	9110				MC690L
9110DM					MC690TL
9112DC	9112				MC691L
9112DM					MC691TL
9128-1503-001	177		300	1122	
9135DC					MC835L
9135FM					MC935F
9135PC	9135				MC835P
9144-60	787	3146	IC-228		
9144-61	788	3829	IC-229		
9157DC					MC857L
9157FM					MC957F
9157PC	9157				MC857P
9158DC					MC858L
9158PC	9158				MC858P
9203-8		3033A	509	1114	
9279	128		243	2030	
9300	108	3452	86	2038	
9300-5	116	3016	504A	1104	
9300-6	116	3016	504A	1104	
9300-7	116	3016	504A	1104	
9300A	108	3452	86	2038	
9300B	108	3452	86	2038	
9300Z	108	3452	86	2038	
9301-1	116	3016	504A	1104	
9302-2	117	3017B	504A	1104	
9302-2A	117	3017B	504A	1104	
9302-3	116	3016	504A	1104	
9307	7448	7448			
9311	74154	74154			
9314	108	3452	86	2038	
9315DC	7441	7441		1804	
9315PC	7441	7441		1804	
9316	74161	74161			
9317B	7446	7446			
9317C	7446	7446			
9330-006-11112	116		504A	1104	
9330-011-70112	102A		53	2007	
9330-092-90112	137A		ZD-6.2	561	
9330-228-60112	177		300	1122	
9330-229-20112	116		504A	1104	
9330-229-60112	107		11	2015	
9330-229-70112	107		11	2015	
9330-688-30112	199	3245	62	2010	
9330-767-60112	159	3466	82	2032	
9330-908-10112	159	3466	82	2032	
9345DC	7445	7445			
9345PC	7445	7445			
9348-3	116	3016	504A	1104	
9352DC	7442	7442			
9352PC	7442	7442			
9357A	7446	7446			
9357ADC	7446	7446			
9357APC	7446	7446			
9357B	7447	7447	7447	1805	
9357BDC	7447	7447	7447	1805	
9357BPC	7447	7447	7447	1805	
9358	7448	7448		1816	
9358DC	7448	7448	7448	1816	
9358PC	7448	7448	7448	1816	
9360DC	74192	74192	74192	1831	
9360PC	74192	74192	74192	1831	
9366DC	74193	74193	74193	1820	
9366PC	74193	74193	74193	1820	
9367-1	128	3024	243	2030	

Industry Standard No.	ECG	SK	GE	RS 276-	MOTOR.
9375DC	7475	7475	7475	1806	
9375PC	7475	7475	7475	1806	
9385DC	7485		7485	1826	
9385PC	7485	7485	7485	1826	
9390	102	3004	2	2007	
9390DC	7490	7490	7490	1808	
9390PC	7490	7490	7490	1808	
9391	102	3004	2	2007	
9392DC	7492	7492	7492	1819	
9392PC	7492	7492	7492	1819	
9393	7493A	7493			
9393DC	7493A	7493			
9393PC	7493A	7493			
9395DC	7495	7495			
9395PC	7495	7495			
9400-8	102A	3004	53	2007	
9400-9	102A	3004	53	2007	
9401-7	102A	3004	53	2007	
9403-2	121	3009	239	2006	
9403-3	160	3007	245	2004	
9403-6	160	3007	245	2004	
9403-7	102A	3004	53	2007	
9403-8	126	3008	52	2024	
9403-9	102A	3004	53	2007	
9404-0	121	3009	239	2006	
9404-2	128	3047	243	2030	
9404-3	195A	3048/329	46		
9404-9	175	3026	246	2020	
9405-0	128	3024	243	2030	
9405-1	128	3024	243	2030	
9405-2	129	3025	244	2027	
9409-4	175	3026	246	2020	
9410A	123A	3018	20	2051	
9426B	108	3018	86	2038	
9426C	108	3018	86	2038	
9440	177	3100/519	300	1122	
09500	199	3245	62	2010	
09501	199	3245	62	2010	
9502	199	3122	62	2010	
09502-8	123A	3444	20	2051	
9510-1	160	3007	245	2004	
9510-2	160	3007	245	2004	
9510-3	126	3008	52	2024	
9510-7	160	3006	245	2004	
9513	108	3452	86	2038	
9521	722		IC-9		
9564	102A		53	2007	
9582	185		58	2025	
9600	123A	3444	20	2051	
9600-5	123A	3444	20	2051	
9600C	108	3452	86	2038	
9600F	108	3452	86	2038	
9600G	108	3452	86	2038	
9600H	108	3452	86	2038	
9601	108	3452	86	2038	
9601-12	108	3452	86	2038	
9604F	108	3452	86	2038	
9606	5014A	3780	ZD-6.8	561	
9607	5015A	3781	ZD-7.5		
9614DC					MC75S110L
9614DM					MC75S110L
9615DC	9615				MC75108L
9615DM	9615$				MC55108L
9615FM					MC55108L
9616CDC					MC1488L
9616DM					MC1488L
9616EDC					MC1488L
9617DC					MC1489AL
9617K	128	3124/289	243	2030	
9618	229	3122	211	2038	
9620DC					MC75S110L
9620DM					MC75S110L
9621DC					MC75108L
9621DM					MC55108L
9623		3018	20	2051	
9623F	108	3452	86	2038	
9623G	108	3452	86	2038	
9623H	108	3452	86	2038	
9624DC					MMH0026CL
9624DM					MMH0026CL
9625DC					MMH0026CL
9625DM					MMH0026CL
9625F	108	3452	86	2038	
9625H	108	3452	86	2038	
9627CDC					MC1489AL
9627DM					MC1489AL
9630C	108	3452	86	2038	
9636AT					MC3488AP
9637T					MC3486P
9638T					MC3487P
9640D					MC3443P
9640DC					MC3440AP
9640J					MC3443P
9640NC					MC3440AP
9644	177		300	1122	
9646	177	3100/519	300	1122	
9650-001	116	3016	504A	1104	
9652H	159	3466	82	2032	
9665DC	2011				MC1411L
9665PC	2011	3975			MC1411P
9666DC	2012				MC1412L
9666PC	2012	9092			MC1412P
9667DC	2013				MC1413L
9667PC	2013	9093			MC1413P
9668DC	2014				MC1416L
9668PC	2014	9094			MC1416P
9692-1	723	3144	IC-15		
9693-1	787	3146	IC-228		
9694-1	724	3525	IC-86		
9696H	128	3122	243	2030	
09800	159	3466	82	2032	
09800-12	129	3025	244	2027	
9803	177		300	1122	
9842FF	723		IC-15		
9861B-43	109	3087		1123	
9920-3-6	125		510,531	1114	
9920-4	102A		53	2007	
9920-5	102A		53	2007	
9920-6-1	123A		20	2051	
9920-6-2	123A	3444	20	2051	
9920-7-2	123A	3444	20	2051	
9921-7	102A		53	2007	
9921-8	102A		53	2007	
9925-0	121		239	2006	
9925-2	124	3717/121	12		
9925-2-1	124		12		
9925-2-2	124		12		
9970	5004A	3770			
9971	5006A	3772	ZD-3.6		
9972	5008A	3774	ZD-4.3		
9973	5010A	3776	ZD-5.1		
10003	181	3036	75	2041	
10010	5870		5032		
10020	5074A		ZD-11.0	563	
10031	113A	3119/113	6GC1		
10032	102A		53	2007	
10036	102A	3004	53	2007	
10036-001	129	3025	244	2027	
10037	102A	3004	53	2007	
10038	102A	3123	53	2007	
10039	102A	3123	53	2007	
10100	125		510,531	1114	MC10500
10101					MC10501
10102					MC10502
10103					MC10503
10104					MC10504
10105					MC10505
10106					MC10506
10107					MC10507
10109					MC10509
10110			510,531	1114	MC10110
10111					MC10111
10113					MC10513
10114					MC10514
10115					MC10515
10116					MC10516
10117					MC10517
10118					MC10518
10119					MC10519
10121					MC10521
10123					MC10123
10124					MC10524
10125					MC10525
10129					MC10129
10130					MC10530
10131					MC10531
10132					MC10132
10133					MC10533
10134					MC10134
10135					MC10535
10136					MC10536
10137					MC10537
10139					MCM10139
10141					MC10541
10144					MCM10544
10145					MCM10545
10149					MCM10549
10158					MC10558
10159					MC10559
10160					MC10560
10161					MC10561
10162					MC10562
10164					MC10564
10165					MC10565
10170					MC10570
10171					MC10571
10172					MC10572
10173					MC10173
10174					MC10574
10175					MC10575
10176					MC10576
10178					MC10578
10179					MC10579
10180	116		504A	1104	MC10580
10181	109	3088		1123	MC10581
10186					MC10586
10188					MC10188
10189					MC10189
10190					MC10590
10191					MC10591
10210					MC10610
10211					MC10611
10212					MC10612
10216					MC10616
10226/2	123A	3444	20	2051	
10231					MC10631
10300	159	3466	82	2032	
10300-12	129	3025	244	2027	
10302-01	7430	7430			
10302-02	7420	7420		1809	
10302-03	7410	7410		1807	
10302-04	7400	7400		1801	
10302-05	7451	7451		1825	
10302-06	7405	7405			
10386			504A	1104	
10416-009	123A	3444	20	2051	
10416-010	128		243	2030	
10508		3114	21	2034	
010562				2032	
10650A01	778A		IC-220		
10655A01	798	3216	IC-234		
10655A01-3	798	3216	IC-234		
10655A03	738	3167	IC-29		
10655A03A	738	3167	IC-29		
10655A05	749		IC-97		
10655A13	748	3236			
10655B01	798	3216			
10655B02	748	3101/706			
10655B03	738	3167	IC-29		
10655B13		3072	IC-148		
10655C05	749		IC-97		
10658A01	788	3829	IC-229		
10895	519		514	1122	
10909	116	3311	504A	1104	
11000	234		65	2050	
011119	110MP	3709		1123(2)	
0011193	5072A		ZD-8.2	562	
11200-1	7410	7410		1807	
11202-1	7404	7404		1802	
11203-1	7493A	7493			
11204-1	74193	74193		1820	
11205-1	7420	7420		1809	
11207-1	7402	7402		1811	
11208-1	7430	7430			
11213-1	7474	7474		1818	
11214-1	7440	7440			
11216-1	7400	7400		1801	
11233-2	7454	7454			
11236-1	128		243	2030	
11236-2	224		46		
11236-3	130		14	2041	
11252	109	3087		1123	
11252-0	108	3452	86	2038	
11252-1	108	3452	86	2038	
11252-2	108	3452	86	2038	
11252-3	123A	3444	20	2051	
11252-4	121	3717	239	2006	
11252-5	128		243	2030	
11252-6	224		46		
11252-7	224		46		
11273-1	74121	74121			
11282-6	116		504A	1104	
11303-9	116		504A	1104	
11305-0001	519		514	1122	
11332-1	116		504A(3)	1104	
11339-2	116		504A	1104	
11339-8	108	3452	86	2038	
11352-78	519		514	1122	

Industry Standard No.	ECG	SK	GE	RS 276-	MOTOR.
11386-9	505	3108	CR-7		
11393-8	108	3452	86	2038	
11399-8	117		504A	1104	
11401-3	116		504A	1104	
11426-7	108	3452	86	2038	
11503-9	116		504A	1104	
11506-3	121	3717	239	2006	
11522-5	123A	3444	20	2051	
11522-7	160		245	2004	
11522-8	160		245	2004	
11522-9	160		245	2004	
11526-8	121	3717	239	2006	
11526-9	121	3717	239	2006	
11527-5	126		52	2024	
11528-1	121MP	3718	239(2)	2006(2)	
11528-2	121MP	3718	239(2)	2006(2)	
11528-3	121MP	3718	239(2)	2006(2)	
11528-4	121MP	3718	239(2)	2006(2)	
11555-9	116		504A	1104	
11559-9	116		504A	1104	
11586-7	117		504A	1104	
11587-5	123A	3444	20	2051	
11605-2	116		504A	1104	
11606-8	127	3764	25		
11607-2	126		52	2024	
11607-3	108	3452	86	2038	
11607-4	123A	3444	20	2051	
11607-5	124		12		
11607-8	123A	3444	20	2051	
11607-9	108	3452	86	2038	
11608-0	108	3452	86	2038	
11608-2	108	3452	86	2038	
11608-3	108	3452	86	2038	
11608-5	123A	3444	20	2051	
11608-6	127	3764	25		
11608-7	127	3764	25		
11608-8	127	3764	25		
11608-9	127	3764	25		
11609-1	102A		53	2007	
11609-2	123A	3444	20	2051	
11619-8	108	3452	86	2038	
11619-9	108	3452	86	2038	
11620-0	108	3452	86	2038	
11620-1	108	3452	86	2038	
11620-2	160		245	2004	
11620-3	102A		53	2007	
11620-6	102A		53	2007	
11620-7	160		245	2004	
11620-8	160		245	2004	
11620-9	160		245	2004	
11658-8	123A	3444	20	2051	
11668-3	160		245	2004	
11668-4	160		245	2004	
11668-5	102A		53	2007	
11668-6	102A		53	2007	
11668-7	103A	3835	59	2002	
11675-6	126		52	2024	
11675-7	102A		53	2007	
11687-5	123A	3444	20	2051	
11699-7	102A		53	2007	
11746	116	3016	504A	1104	
011950	519		514	1122	
011956	519		514	1122	
12011	177		300	1122	
012013-1	123A	3444	20		
012015	128		243	2030	
12020-02	152	3893	66	2048	
12044-0021	175	3026	246	2020	
12047-0023	192		63	2030	
12048-0011	129		244	2027	
0012060	109	3087		1123	
012085	175		246	2020	
012099-1	129		244	2027	
12101	177	3100/519	300	1122	
12110-0	100	3005	1	2007	
12110-3	100	3005	1	2007	
12110-4	100	3005	1	2007	
12110-5	100	3005	1	2007	
12110-6	102A	3004	53	2007	
12110-7	102A	3004	53	2007	
12112-0	102A	3004	53	2007	
12112-8	100	3005	1	2007	
12112-C	123A	3444	20	2051	
12112-D	123A	3444	20	2051	
12112-E	123A	3444	20	2051	
12112-F	123A	3444	20	2051	
12112C			20	2051	
12112D			20	2051	
12112E			20	2051	
12112F			20	2051	
12113-5	160	3006	245	2004	
12113-7	160	3006	245	2004	
12113-8	160	3008	245	2004	
12113-9	160	3006	245	2004	
12114-8	102A	3004	53	2007	
12114L					MCM21L14
12115-0	160	3007	245	2004	
12115-7	160	3006	245	2004	
12116-1	100	3005	1	2007	
12116-2	100	3005	1	2007	
12116-4	102A	3004	53	2007	
12117-9	100	3005	1	2007	
12118-0	100	3005	1	2007	
12118-4	102A	3004	53	2007	
12118-6	100	3005	1	2007	
12118-9	126	3008	52	2024	
12119-0	160	3007	245	2004	
12119-1	102A	3004	53	2007	
12119-2	102A	3004	53	2007	
12122-5	102A	3004	53	2007	
12122-6	102A	3004	53	2007	
12122-7	102A	3004	53	2007	
12122-8	160	3006	245	2004	
12122-9	160	3006	245	2004	
12123-0	160	3006	245	2004	
12123-1	160	3006	245	2004	
12123-2	102A	3004	53	2007	
12123-3	160	3006	245	2004	
12123-4	100	3005	1	2007	
12123-5	100	3005	1	2007	
12123-6	100	3005	1	2007	
12124-0	100	3005	1	2007	
12124-1	100	3005	1	2007	
12124-2	126	3008	52	2024	
12124-3	126	3008	52	2024	
12124-4	126	3008	52	2024	
12124-6	102A	3004	53	2007	
12125-4	100	3005	1	2007	
12125-6	160	3007	245	2004	
12125-7	126	3008	52	2024	
12125-8	160	3007	245	2004	
12125-9	160	3007	245	2004	
12126-0	160	3007	245	2004	
12126-6	102A	3004	53	2007	

Industry Standard No.	ECG	SK	GE	RS 276-	MOTOR.
12126-7	102A	3004	53	2007	
12127-0	121	3009	239	2006	
12127-1	121	3009	239	2006	
12127-2	102A	3004	53	2007	
12127-3	100	3005	1	2007	
12127-4	102A	3003	53	2007	
12127-5	100	3005	1	2007	
12127-6	123A	3444	20	2051	
12127-7	123A	3444	20	2051	
12127-8	123A	3444	20	2051	
12127-9	123A	3444	20	2051	
12128-9	100	3005	1	2007	
12147					MCM2147
12163	104	3009	16	2006	
12173	100	3005	1	2007	
12174	100	3005	1	2007	
12175	100	3005	1	2007	
12176	100	3005	1	2007	
12178	104	3005	16	2006	
12180	126		52	2024	
12183	100	3005	1	2007	
12191	100	3005	1	2007	
12192	100	3005	1	2007	
12193	100	3005	1	2007	
12195	102A	3004	53	2007	
12196	102A	3004	53	2007	
12255-235	519		514	1122	
12429		3066	CR-1		
12429(HEATHKIT)	118		CR-1		
12454	233		210	2009	
12536	130	3027	14	2041	
12536A	130	3027	14	2041	
12537	219		74	2043	
12538	175		246	2020	
12539	184	3190	57	2017	
12546	154	3045/225	40	2012	
12550	156	3051	512		
12593	123A	3444	20	2051	
12594	159	3114/290	82	2032	
12602	116	3016	504A	1104	
12685	506	3843	511	1114	
12720	116	3016	504A	1104	
12736	116	3016	504A	1104	
12746	116	3017B/117	504A	1104	
12768	116	3016	504A	1104	
12786	116	3016	504A	1104	
12788	116	3016	504A	1104	
12803	116	3016	504A	1104	
12808	109	3087		1123	
12837	116	3016	504A	1104	
12844	116	3031A	504A	1104	
12850	109	3087		1123	
12871	113A	3119/113	6GC1		
12888	159	3466	82	2032	
13002-3	130		14	2041	
13002-4	130		14	2041	
13030-4	181		75	2041	
13150L				1122	
13159-2	175		246	2020	
13162	159	3466	82	2032	
13213E4252	128		243	2030	
13217E-1513	311		277		
13217E1513	311		277		
13298	130		14	2041	
13299	211	3203	253	2026	
013339	117	3016	504A	1104	
013339(RECTIFIER)	125		510,531	1114	
13517	912		IC-172		
13782	116	3016	504A	1104	
014002-1	177		300	1122	
014007-1	519		514	1122	
014007-2	519		514	1122	
014024	519		514	1122	
14027	116	3016	504A	1104	
14126-1	116		504A	1104	
14207	947		IC-268		
14303	123A	3444	20	2051	
014382	121	3009	239	2006	
014558	154	3044	40	2012	
14573	105	3012	4		
14588-17	5801	3051/156			
14588-18	5801	3051/156			
14588-9	116	3016	504A	1104	
14692	128		243	2030	
14995	128		243	2030	
14996-1	128		243	2030	
14996-2	128		243	2030	
15009	102A		53	2007	
15024	121	3009	239	2006	
15027	121	3009	239	2006	
15027(DIODE)	109			1123	
015040/7	7404	7404		1802	
0015107	506	3125	511	1114	
15122	177	3100/519	300	1122	
15354-14	100	3004/102A			
15354-2	102	3004			
15354-3	104	3009	239	2006	
15354-6	101	3011			
15486	128		243	2030	
15809-1	123	3124/289	20	2051	
15810-1		3124	20	2051	
15820-1	123		20	2051	
15835-1	123	3020	20	2051	
15840-1	123A	3444	20	2051	
15841-1	123A	3444	20	2051	
15927	104	3009	16	2006	
16001	130	3027	14	2041	
16029	196	3054	241	2020	
16039	196	3054	241	2020	
16065	128	3218/278	243	2030	
16073	5643	3519			
16082	128	3024	243	2030	
16083	130		14	2041	
16088	7490	3027/130		1808	
16111	5635	3533			
16113	152	3054/196	66	2048	
16114	175	3027/130	246	2020	
16115	196	3054	241	2020	
16162	5626	3633			
16163	196	3054	241	2020	
16164	152	3054/196	66	2048	
16165	196	3083/197	241	2020	
16166	152	3893	66	2048	
16167	197	3083	250	2027	
16169	197	3083	250	2027	
16170	5635	3533			
16175	197	3083	250	2027	
16176	130	3027	14	2041	
16181	152	3054/196	66	2048	
16182	152	3054/196	66	2048	
16190	108	3039/316	86	2038	
16191	128		243	2030	
16194	108	3452	86	2038	
16201	130	3027	14	2041	

Industry Standard No.	ECG	SK	GE	RS 276-	MOTOR.
16207	152	3054/196	66	2048	
16208	5635	3533			
16230	130	3027	14	2041	
16232	124	3021	12		
16234	130	3027	14	2041	
16235	130		14	2041	
16237	199	3245	62	2010	
16238	5675	3520/5677			
16239	129	3103A/396	244	2027	
16240	130	3027	14	2041	
16241	152	3054/196	66	2048	
16254	128		243	2030	
16259	124		12		
16261	130		14	2041	
16266	130		14	2041	
16267	181	3036	75	2041	
16277	196		241	2020	
16279	197		250	2027	
16287	130		14	2041	
16289	5635	3533			
16291	5635	3533			
16292	130	3027	14	2041	
16299	130		14	2041	
16305	152	3893	66	2048	
16306	197		250	2027	
16319	130		14	2041	
16320	130		14	2041	
16334	152	3893	66	2048	
16335	152	3893	66	2048	
16336	152	3893	66	2048	
16338	130		14	2041	
16341	196		241	2020	
16342	197		250	2027	
16598	102	3260	2	2007	
16599	121	3009	239	2006	
16681	116	3031A	504A	1104	
16958	102	3004	2	2007	
16959	121	3009	239	2006	
17002	116	3031A	504A	1104	
17045	159	3442	82	2032	
17047-1	102	3004	2	2007	
17142	102	3004	2	2007	
17143	102	3004	2	2007	
17144	123	3124/289	20	2051	
17412-5	123A	3444	20	2051	
17443	116	3016	504A	1104	
17444	123	3124/289	20	2051	
17607-1	127	3764	25		
17887	121	3009	239	2006	
17945	121	3009	239	2006	
17973A	147A	3095	ZD-33		
018069	129	3025	244	2027	
018077	128	3024	243	2030	
18109	102	3004	2	2007	
18310	124	3021	12		
18410-141	312	3448	FET-1	2035	
18410-142	107		11	2015	
18410-143	123A	3444	20	2051	
18410-144	199	3122	62	2010	
18410-145	123A	3124/289	210	2051	
18410-146	123A	3444	20	2051	
18410-147	186A	3357	247	2052	
18410-148	788	3829	IC-229		
18410-41		3126	90		
18410-42	109	3088		1123	
18410-43	5072A		ZD-8.2	562	
18493	126	3006/160	52	2024	
18509	123A	3444	20	2051	
18529	100	3004/102A	1	2007	
18530	102	3004	2	2007	
18540	160	3005	245	2004	
18541	160	3006	245	2004	
18555	123A	3444	20	2051	
18600-151	107	3444/123A	11	2015	
18600-152	123A		20	2051	
18600-153	123A	3444	20	2051	
18600-154	1100	3223	IC-92		
18600-155	720	9014	IC-7		
18600-156	1058	3459	IC-49		
18600-51		3126	90		
18600-52	177	3100/519	300	1122	
18600-53	109	3088		1123	
18600-54	5072A	3059/138A	ZD-8.2	562	
18601	102	3004	2	2007	
18611	102	3004	2	2007	
18731	102	3004	2	2007	
19042	116	3017B/117	504A	1104	
19278	124	3021	12		
19420	108	3452	86	2038	
19500-253	199	3245	62	2010	
19645	123A	3444	20	2051	
19680	159	3466	82	2032	
19865	116	3017B/117	504A	1104	
20011	159	3466	82	2032	
20103	199	3245	62	2010	
020156	100	3123	53	2007	
20294	241	3188A/182	57	2020	
20295	241	3188A/182		2048	
020425-3	196		241	2020	
020426-3	197		250	2027	
20738	123	3020	20	2051	
20739	129	3025	244	2027	
20810-21	116		504A	1104	
20810-22	116		504A	1104	
20810-91	199	3124/289	62	2010	
20810-92	199	3124/289	62	2010	
20810-93	128	3024	243	2030	
20810-94	186	3192	28	2017	
20989	124		12		
21008-002	506	3043	511	1114	
021154	116	3016	504A	1104	
21201	128		243	2030	
21221	128		243	2030	
21280	181		75	2041	
21290	128		243	2030	
21606-1	127	3764	25		
21676A	128		243	2030	
22008	159	3466	82	2032	
22009	172A	3156	64		
22158	123A	3444	20	2051	
22164-000	519		514	1122	
0022481	131		44	2006	
22595-000	159	3466	82	2032	
22605-005	159	3466	82	2032	
22635-002	123A	3444	20	2051	
22635-003	123A	3444	20	2051	
22810-173	123A	3444	20	2051	
22810-174	199	3245	62	2010	
22881	226MP	3086	49(2)	2025(2)	
22939	198	3044/154	251		
23114-046	123A	3444	20	2051	
23114-050	129	3025	244	2027	
23114-051	129	3025	244	2027	
23114-052	154	3040	40	2012	
23114-053	123A	3444	20	2051	
23114-054	123A	3444	20	2051	
23114-056	108	3452	86	2038	
23114-057	108	3452	86	2038	
23114-060	108	3452	86	2038	
23114-061	102A	3004	53	2007	
23114-070	130	3027	14	2041	
23114-078	108	3452	86	2038	
23114-082	123A	3444	20	2051	
23114-095	123A	3444	20	2051	
23114-097	127	3034	25		
23114-104	108	3452	86	2038	
23115-022	118	3066	CR-1		
23115-042	125	3032A	510,531	1114	
23115-057	123A	3444	20	2051	
23115-058	123A	3444	20	2051	
23115-072	125	3032A	510,531	1114	
23115-085	156	3051	512		
23125-037	108	3452	86	2038	
23311-006	121	3009	239	2006	
23316	123A	3444	20	2051	
23606	312	3116	FET-1	2035	
0023645	107	3039/316	11	2051	
23648	128		243	2030	
23754	152	3893	66	2020	
023762	130	3027	14	2041	
23785			52	2024	
23785(SYLVANIA)	102A		53	2007	
23785-1			52	2024	
23785-1(SYLVANIA)	102A		53	2007	
23826			22	2032	
23826(SYLVANIA)	159		82	2032	
0023828	123A	3444	20	2051	
0023829	107	3039/316	11	2015	
23879	5529	6629			
24002			11	2015	
24198	126	3017B/117	52	2024	
24198(RECT.)	116		504A	1104	
24451	711	3070	IC-207		
24560A	124		12		
24785	102A		53	2007	
25011(HONEYWELL)	123A		20	2051	
025026	116	3017B/117	504A	1104	
025056	116	3016	504A	1104	
025072	125		510,531	1114	
25114-101	102A	3004	53	2007	
25114-102	102A	3004	53	2007	
25114-103	102A	3004	53	2007	
25114-104	102A	3004	53	2007	
25114-116	123A	3444	20	2051	
25114-121	108	3452	86	2038	
25114-130	123A	3444	20	2051	
25114-143	124	3021	12		
25114-161	123A	3444	20	2051	
25115-115	116	3311	504A	1104	
25201-001	109	3087		1123	
25202-002	116	3311	504A	1104	
25211-200	138A		ZD-7.5		
25260-61-067			300	1122	
25566-01	126		52	2024	
25566-21	128		243	2030	
25566-50	128		243	2030	
25566-62	129		244	2027	
25566-63	128		243	2030	
25566-76	128		243	2030	
25642-020	126		52	2024	
25642-030	126		52	2024	
25642-031	126		52	2024	
25642-040	126		52	2024	
25642-041	126		52	2024	
25642-110	126		52	2024	
25642-115	126		52	2024	
25642-120	126		52	2024	
25651-020	102A		53	2007	
25651-021	102A		53	2007	
25651-033	102A		53	2007	
25655-055	102A		53	2007	
25655-056	102A		53	2007	
25657-050	102A		53	2007	
25658-120	131	3198	44	2006	
25658-121	131	3198	44	2006	
25661-020	121	3009	239	2006	
25661-022	104	3719	16	2006	
25671-020	161	3039/316	39	2015	
25671-021	161	3039/316	39	2015	
25671-023	161	3039/316	39	2015	
25672-016	154		40	2012	
25762-010	225	3045	256		
25762-012	225	3045	256		
25810-161	123A	3444	20	2051	
25810-162	123A	3444	20	2051	
25810-163	123A	3444	20	2051	
25810-164	1100	3223	IC-92		
25810-165	720	9014	IC-7		
25810-166	1087	3477	IC-103		
25810-51	116	3311	504A	1104	
25810-52	177	3100/519	300	1122	
25810-53	109	3088		1123	
25810-54	5072A	3136	ZD-8.2	562	
25810-55	109			1123	
25840-161	123A	3444	20	2051	
25840-162	199	3124/289	62	2010	
25840-163	192	3122	63	2030	
25840-165	1142	3485	IC-128		
25840-166	1087	3477	IC-103		
25840-167	1081A*		IC-7		
25840-51	116	3311	504A	1104	
25840-52	116	3100/519	504A	1104	
25840-53	109	3088		1123	
25840-54	5072A	3136	ZD-8.2	562	
25840-55	109	3087		1123	
26235-1	116	3017B/117	504A	1104	
026237	123AP	3019	20	2051	
26587	703A	3157	IC-12		
26587-1	703A	3157	IC-12		
26666-1	128		243	2030	
26810-151	229		61	2038	
26810-152	159	3466	82	2032	
26810-153	199	3245	62	2010	
26810-154	123A	3444	20	2051	
26810-155	289A	3020/123	268	2038	
26810-157	801	3160	IC-35		
26810-158	1052	3249	IC-135		
26810-159	1155	3231	IC-179	705	
26810-51	109	3088		1123	
26810-52	140A	3061	ZD-10	562	
27113-100	116	3311	504A	1104	
27123-050	116	3016	504A	1104	
27123-070	116	3016	504A	1104	
27123-100	116	3016	504A	1104	
27123-120	116	3016	504A	1104	
27125-080	123A	3444	20	2051	
27125-090	123A		20	2051	
27125-110	128		243	2030	

Industry Standard No.	ECG	SK	GE	RS 276-	MOTOR.
27125-120	102A		53	2007	
27125-140	123A	3444	20	2051	
27125-150	102A	3004	53	2007	
27125-160	123A	3444	20	2051	
27125-170	102A	3004	53	2007	
27125-270	123A	3444	20	2051	
27125-300	123A	3444	20	2051	
27125-310	103A	3010	59	2002	
27125-330	102A	3004	53	2007	
27125-340	102A	3004	53	2007	
27125-350	102A	3004	53	2007	
27125-360	100	3005	1	2007	
27125-370	123A	3444	20	2051	
27125-380	128	3024	243	2030	
27125-460	199	3038	62	2010	
27125-470	199	3038	62	2010	
27125-480	102A	3004	53	2007	
27125-490	103A	3862/103	59	2002	
27125-500	123A	3444	20	2051	
27125-530	123A	3444	20	2051	
27125-540	102A	3004	53	2007	
27125-550	102A	3444/123A	53	2007	
27126-060	131	3052	44	2006	
27126-090	121	3717	239	2006	
27126-100	130	3510	14	2041	
27127-550	123A	3124/289	20	2051	
27840-161	123A	3444	20	2051	
27840-162	123A	3444	20	2051	
27840-163	1087		IC-103		
27840-41	116	3311	504A	1104	
27840-42	109	3088		1123	
27840-43	177	3100/519	300	1122	
27910-12150	123		20	2051	
27910-12153	102	3722	2	2007	
27913-14200	5626	3633			
28222-3(SYLVANIA)	182		55		
28287	109	3087		1123	
28396	390		255	2041	
28474	130	3027	14	2041	
28810-171	312	3448	FET-1	2035	
28810-172	107	3018	11	2015	
28810-173		3018	20	2051	
28810-174		3124	20	2051	
28810-175	1082	3461	IC-140		
28810-176	1142	3485	IC-128		
28810-177		3249	IC-135		
28810-61	116	3311	504A	1104	
28810-62		3126	90		
28810-63	177	3100/519	300	1122	
28810-64	110MP	3088		1123(2)	
28810-65	5072A	3136	ZD-8.2	562	
28977	128		243	2030	
29076-005	161	3716	39	2015	
29076-006	161	3716	39	2015	
29076-023	159	3466	82	2032	
29625					MCM7641
29631					MCM7681
29651					MCM7685
29810-171	312	3834/132			
29810-178	186A	3357			
29810-179	788	3829	IC-229		
29810-180	801	3160	IC-35		
030010	123A	3122	20	2051	
030010-1	123A	3122	20	2051	
030011	128	3024	243	2030	
030011-1	128	3024	243	2030	
030011-2	128	3024	243	2030	
30201	158	3004/102A	53	2007	
30202	158	3123	53	2007	
30203	121	3009	239	2006	
30204	102	3004	2	2007	
30206	102	3004	2	2007	
30207	102	3004	2	2007	
30208	100	3004/102A	1	2007	
30208-1	102A	3004	53	2007	
30208-2	102	3004	2	2007	
30210	123	3124/289	20	2051	
30211	121	3009	239	2006	
30213	126	3008	52	2024	
30214	126	3008	52	2024	
30215	126	3007	52	2024	
30215(RCA)	121	3717	239	2006	
30216	102	3007	2	2007	
30216(RCA)	121	3717	239	2006	
30217	126	3008	52	2024	
30218	102	3008	2	2007	
30218(RCA)	160		245	2004	
30219	123	3124/289	20	2051	
30221	126	3006/160	52	2024	
30222	126	3006/160	52	2024	
30223	126	3006/160	52	2024	
30224	123	3124/289	20	2051	
30226	123	3124/289	20	2051	
30227	123A	3444	20	2051	
30228	123A	3444	20	2051	
30229	123A	3444	20	2051	
30230	126	3006/160	52	2024	
30231	100	3007	1	2007	
30234	124	3021	12		
30235	123A	3444	20	2051	
30236	175		246	2020	
30238	160	3006	245	2004	
30239		3006	51	2004	
30240	160	3006	245	2004	
30241	123A	3444	20	2051	
30242	123A	3444	20	2051	
30243	123A	3444	20	2051	
30244	102	3004	2	2007	
30245	124	3021	12		
30246	121	3009	239	2006	
30246A	121	3009	239	2006	
30247	126		52	2024	
30248	123	3124/289	20	2051	
30253	123A	3444	20	2051	
30254	175	3124/289	246	2030	
30256	175	3026	246	2020	
30257	124	3021	12		
30259	123	3124/289	20	2051	
30263	102A		53	2007	
30267		3004	54		
30268	123A	3444	20	2051	
30269	123A	3444	20	2051	
30270	159	3114/290	82	2032	
30271	185	3083/197	58	2025	
30272	184	3054/196	57	2017	
30273	160		245	2004	
30274	126		52	2024	
30276	130	3027	14	2041	
30278	129	3025	244	2027	
30289	123A	3444	20	2051	
30290	159	3466	82	2032	
30291(NPN)			18	2030	
30291(PNP)			21	2034	
30292	108	3452	86	2038	
30293	102A	3006/160	53	2007	
30294	152	3893	66	2048	
30302	104	3009	16	2006	
30302(RCA)	102A		53	2007	
030512-1	123A	3444	20	2051	
030512-2	123A	3444	20	2051	
030515	123A	3444	20	2051	
030515-4	123A	3444	20	2051	
030527	123A	3444	20	2051	
030531-1	105	3122	4		
030536	123A	3444	20	2051	
030536-1			20	2051	
030537	123A	3444	20	2051	
030537-1	123A	3444	20	2051	
030537-2	123A	3444	20	2051	
030538	123A	3444	20	2051	
030539-1	130		14	2041	
030542	123A	3444	20	2051	
030542-1	123A	3444	20	2051	
030543	123A	3444	20	2051	
030543-1	123A	3444	20	2051	
030543-2	123A	3444	20	2051	
030548				2051	
030812-1	103A	3010	59	2002	
030828				2010	
030930				2038	
31001	123A	3444	20	2051	
31003	123A	3444	20	2051	
31004-1	153		69	2049	
31005	159	3466	82	2032	
31006	129		244	2027	
31009	123A	3444	20	2051	
31015	234		65	2050	
31032-0	129		244	2027	
031033	109	3087		1123	
031034	116	3017B/117	504A	1104	
031040	109	3087		1123	
031450	5802	9005	512	1143	
33188-2	130		14	2041	
33201-1	712	3072	IC-2		
33201-2	712	3072	IC-2		
33324		3126	90		
33509-1	159	3466	82	2032	
33563	123A	3124/289	20	2015	
033571	129		244	2027	
033589	193A	3466/159	82	2032	
33989-2069	121	3717	239	2006	
34001	4001B	3006/160			
34002	4002B	3006/160			
34005-1	727	3071	IC-210		
34007	4007	4007			
34011	4011B	4011			
34020PC	4020B	4020			
34021PC	4021B	4021			
34022	121	3009	239	2006	
34027	4027B	3010			
34043	4043B	3005			
34044	128		243	2030	
34048	126		52	2024	
34048-1	708		IC-10		
34049-1	722	3161	IC-9		
34052PC	4052B	4052			
34055PC	4055B	4055			
34098		3004	52	2024	
34099		3005	52	2024	
34100		3008	52	2024	
34118		3008	245	2004	
34119	100	3005	1	2007	
34174	116	3016	504A	1104	
34201	156		512		
34202-1	727	3071	IC-210		
34208	130		14	2041	
34219	100	3005	1	2007	
34220	100	3005	1	2007	
34221	100	3005	1	2007	
34262	102A		53	2007	
34298	121	3009	239	2006	
34315	121	3009	239	2006	
34342	160	3008	245	2004	
34379-1	720	9014	IC-7		
34389	160	3008	245	2004	
34394	116	3016	504A	1104	
34405	5802	9005		1143	
34423	160		245	2004	
34424	506	3125	511	1114	
34425	121	3009	239	2006	
34452-1	709	3135	IC-11		
34459	127	3014			
34493P	102	3004	2	2007	
34502-1	778A	3465	IC-220		
34503-1	722	3161	IC-9		
34518PC	4518B	4518			
34526	121	3717	239	2006	
34553	160	3004/102A	245	2004	
34588A	128		243	2030	
34675	160		245	2004	
34715	104	3009	16	2006	
34871	102	3722	2	2007	
34894		3016	504A	1104	
34923	158		53	2007	
34942	160	3004/102A	245	2004	
34966	176	3845	80		
35001	130	3024/128	14	2041	
35002	175	3007	246	2020	
35004	108	3452	86	2038	
35044	121	3009	239	2006	
35045	102	3004	2	2007	
35059-1	712	3072	IC-2		
35070	160	3008	245	2004	
35084	104	3009	16	2006	
35086	102	3722	2	2007	
35144	121	3014	239	2006	
35168	126	3008	52	2024	
35169	160	3008	245	2004	
35170	160	3008	245	2004	
35201	104	3009	16	2006	
35210	128		243	2030	
35212	128	3122	243	2030	
35218	176	3123	80		
35219	142A	3004/102A	ZD-12	563	
35231	121	3009	239	2006	
35242			10	2051	
35259			11	2015	
35260	104	3009	16	2006	
35287	116	3016	504A	1104	
35287A	116	3016	504A	1104	
35289	116	3017B/117	504A	1104	
35303	128		243	2030	
35306	116	3311	504A	1104	
35333	125	3311	510,531	1114	
35349	121	3009	239	2006	
35383	128		243	2030	
35405	175	3004/102A	246	2020	
35449	108	3452	86	2038	

Industry Standard No.	ECG	SK	GE	RS 276-	MOTOR.
35452-2	158		53	2007	
35454	102	3004	2	2007	
35454-1	158		53	2007	
35454-2	158		53	2007	
35454-3	158		53	2007	
35500	116	3016	504A	1104	
35590	102A	3004	53	2007	
35604	5804	9007		1144	
35628	102A	3004	53	2007	
35677	102A	3008	53	2007	
35678	102A	3008	53	2007	
35728	121	3009	239	2006	
35792	102A	3004	53	2007	
35815	160	3008	245	2004	
35816	102A	3008	53	2007	
35817	102A	3008	53	2007	
35818	126	3008	52	2024	
35819	158	3004/102A	53	2007	
35820	102	3004	2	2007	
35820-1	158	3004/102A	53	2007	
35820-2	158	3004/102A	53	2007	
35820-3	158	3004/102A	53	2007	
35821B					MRF902B
35821E					MRF902
35822B					MRF902B
35822E					MRF902
35824	102A	3008	53	2007	
35824A					MRF904
35825B					MRF902B
35825E					MRF902
35860			504A	1104	
35885A	121	3014	239	2006	
35885B	121	3014	239	2006	
35888	128		243	2030	
35950	102	3722	2	2007	
35951	121	3009	239	2006	
35952	102	3004	2	2007	
35953	102	3004	2	2007	
35954	102	3004	2	2007	
35955	102	3004	2	2007	
036001	703A	3157	IC-12		
36091R	5883		5041		
36145	128		243	2030	
36147	5802	9005		1143	
36201	116	3017B/117	504A	1104	
36203	121	3019	239	2006	
36212	108	3452	86	2038	
36212V1	108	3452	46	2038	
36212V1(RF PWR)	195A	3765			
36213	188	3048/329	226	2018	
36213A2	195A	3048/329	46		
36274	175		246	2020	
36303	121	3009	239	2006	
36304	121	3009	239	2006	
36304-4	121	3009	239	2006	
36312	121	3009	239	2006	
36320	175		246	2020	
36340	176	3845	80		
36344		3021	12		
36359-4	121MP	3718	239(2)	2006(2)	
36370-05490	175		246	2020	
36387	128		243	2030	
36395	121	3009	239	2006	
36466	128		243	2030	
36477	121	3009	239	2006	
36503		3031A	504A	1104	
36503(AIRCASTLE)	116		504A	1104	
36508	109	3530		1123	
36534	102	3027/130	2	2007	
36535	117	3027/130	504A	1104	
36537	117	3034	504A	1104	
36539	142A	3062	ZD-12	563	
36545	130		14	2041	
36549	116	3016	504A	1104	
36554	116	3016	504A	1104	
36555	116	3009	504A	1104	
36557	102	3004	2	2007	
36558	102	3004	2	2007	
36559	126	3006/160	52	2024	
36560	126	3006/160	52	2024	
36563	126	3006/160	52	2024	
36564	116	3017B/117	504A	1104	
36577	129	3025	244	2027	
36578	108	3452	86	2038	
36579	128	3024	243	2030	
36580	123A	3444	20	2051	
36581	108	3452	86	2038	
36582	312	3448	FET-1	2035	
36591	5802	9005		1143	
36634	124	3021	12		
36662	311		277		
36673	176	3845	80		
36680	311		277		
36680A	311		277		
36682	128		243	2030	
36687	121	3717	239	2006	
36695	176	3845	80		
36748	128		243	2030	
36800-2	121	3717	239	2006	
36800-3	121	3717	239	2006	
36800-4	121	3717	239	2006	
36800-5	121		239	2006	
36800-6	121	3717	239	2006	
36800-7	121	3717	239	2006	
36803	311	3024/128	277		
36816	100	3005	1	2007	
36846	130		14	2041	
36847	108	3452	86	2038	
36848	311	3024/128	277		
36849	311	3024/128	277		
36850	311	3024/128	277		
36855	130		14	2041	
36892	130	3027	14	2041	
36896	104	3009	16	2006	
36910	121	3009	239	2006	
36913	195A	3195/311	46		
36917	123A	3444	20	2051	
36918	108	3452	86	2038	
36919	108	3452	86	2038	
36920	123A	3444	20	2051	
36921	123A	3444	20	2051	
36946	130		14	2041	
36953	130		14	2041	
36971	121	3717	239	2006	
36997	311	3024/128	277		
37077	175		246	2020	
037085	130	3036	14	2041	
37126-1	116	3016	504A	1104	
37267	130		14	2041	
37269	129		244	2027	
37278	160	3004/102A	245	2004	
37279	103A	3010	59	2002	
37280	128		243	2030	
37287	128		243	2030	

Industry Standard No.	ECG	SK	GE	RS 276-	MOTOR.
37334	161	3510	39	2015	
37383	108	3452	86	2038	
37384	108	3452	86	2038	
37393	128		243	2030	
37431	311	3024/128	277		
37432	311	3024/128	277		
37433	311	3024/128	277		
37445	128		243	2030	
37464	128	3024	243	2030	
37475	130	3510	14	2041	
37476	181	3511	75	2041	
37478	129	3025			
37484	175		246	2020	
37486	159		82	2032	
37510-161	199	3122	62	2010	
37510-162	123A	3444	20	2051	
37510-163	123A	3444	20	2051	
37510-164	1100	3223	IC-92		
37510-165	1052	3249			
37510-166	1052	3249	IC-135		
37510-52	177	3100/519	300	1122	
37510-53	110MP	3088	1N60	1123(2)	
37510-54	5072A	3059/138A	ZD-8.2	562	
37549	102A	3004	53	2007	
37550	102A	3004	53	2007	
37551	102A	3004	53	2007	
37552	103A	3010	59	2002	
37563	130		14	2041	
37584	124	3021	12		
37585	123	3124/289	20	2051	
37599	175		246	2020	
37628	311		277		
37649	128	3024	243	2030	
37663	130		14	2041	
37664	129		244	2027	
37677	102	3722	2	2007	
37680	116	3017B/117	504A	1104	
37682	5547	3505			
37694	128	3024	243	2030	
37694A	108	3452	86	2038	
37694B	108	3452	86	2038	
37702-1	714	3075	IC-4		
37703-1	715	3076	IC-6		
37704-1	790	3454	IC-230		
37725	154		40	2012	
37730	124	3021	12		
37740	129		244	2027	
37741	128		243	2030	
37763	175		246	2020	
37764	129		244	2027	
37767	128		243	2030	
37793	129		244	2027	
37800	128	3024	243	2030	
37806	128		243	2030	
37833	100	3721	1	2007	
37833-2	909D	3590	IC-250		
37840	128		243	2030	
37847	128		243	2030	
37884	123A	3444	20	2051	
37888	130	3027	14	2041	
37894	311	3024/128	277		
37899	128		243	2030	
37900	175		246	2020	
37913	175		246	2020	
37918	129		244	2027	
37932	278		261		
37966	129		244	2027	
37967	181	3036	75	2041	
37974	181		75	2041	
37975	128		243	2030	
37982	128		243	2030	
37986-3563	107		11	2015	
37986-4040	107		11	2015	
37986-4046	107		11	2015	
37987	116	3017B/117	504A	1104	
37992	5529	6629			
38045	128		243	2030	
38049	181		75	2041	
38052	5802	9005		1143	
38057	102A		53	2007	
38058	128		243	2030	
38063	311	3024/128	277		
38074	5804	9007		1144	
38091	158		53	2007	
38093	126		52	2024	
38094	121MP	3014	239(2)	2006(2)	
38095	159		82	2032	
38112	124		12		
38120	154		40	2012	
38121	154		40	2012	
38122	195A	3765	46		
38137	130	3510	14	2041	
38138	130	3510	14	2041	
38166	130	3510	14	2041	
38174	116	3017B/117	504A	1104	
38175	103	3862	8	2002	
38176	102A	3004	53	2007	
38177	102	3004	2	2007	
38178	123	3124/289	20	2051	
38182	128		243	2030	
38190	311	3024/128	277		
38199	101	3011A	8	2002	
38200	101	3011A	8	2002	
38207	108	3452	86	2038	
38208	108	3452	86	2038	
38209	100	3005	1	2007	
38246	161	3452/108	86	2038	
38246A	108	3452	86	2038	
38265-00000	784	3524	IC-236		
38265-00010	784	3524	IC-236		
38265-00020	784	3524	IC-236		
38265-00030	784	3524	IC-236		
38267	181		75	2041	
38268	181	3036	75	2041	
38269	158		53	2007	
38270	128	3024	243	2030	
38271	128	3025/129	243	2030	
38272	130	3027	14	2041	
38279	311		277		
38281	278	3024/128	261		
38283	123A	3444	20	2051	
38332	5626	3633			
38334	128		243	2030	
38335			255	2041	
38354	128		243	2030	
38361	128		243	2030	
38378	220			2036	
38385	5806	3848			
38388	129		244	2027	
38397	181	3036	75	2041	
38398	128		243	2030	
38424	128		243	2030	
38432	128		243	2030	
38443	175		246	2020	

Industry Standard No.	ECG	SK	GE	RS 276-	MOTOR.
38446-00000	724	3525	IC-86		
38446-00010	724	3525	IC-86		
38446-00020	724	3525	IC-86		
38446-00030	724		IC-86		
38448	186	3192	28	2017	
38458	129		244	2027	
38468	128		243	2030	
38473	181	3036	75	2041	
38474	130	3027	14	2041	
38475	128		243	2030	
38476	128		243	2030	
38478	199	3245	62	2010	
38491	181	3036	75	2041	
38494	130	3027	14	2041	
38495	128		243	2030	
38496	129		244	2027	
38497	128		243	2030	
38510-161	312	3834/132	FET-1	2035	
38510-162	229	3018	61	2038	
38510-163	123A	3444	20	2051	
38510-164	199	3122	62	2010	
38510-165	192		63	2030	
38510-166	123A	3444	20	2051	
38510-167	123A	3444	20	2051	
38510-168	186A	3357	247	2052	
38510-169	788	3829	IC-229		
38510-170	801	3160	IC-35		
38510-171	1087	3193/187			
38510-330	177	3100/519	300	1122	
38510-331	5070A	9021	ZD-6.2	561	
38510-350	199	3124/289	62	2010	
38510-51		3126	90		
38510-52		3088	1N60	1123	
38510-53	140A	3061	ZD-10	562	
38510-54	177		300	1122	
38511	108	3452	86	2038	
38511A	108	3452	86	2038	
38513-50360C			255	2041	
38533	126		52	2024	
38551	128		243	2030	
38563	222			2036	
38588	128		243	2030	
38608	5547	3505			
38626	130		14	2041	
38654	129		244	2027	
38659	128		243	2030	
38680	160		245	2004	
38681	160		245	2004	
38685	102A		53	2007	
38716	128	3024	243	2030	
38725	128		243	2030	
38731(KALOF)	130		14	2041	
38733	152	3893	66	2048	
38734	129		244	2027	
38735	128		243	2030	
38736	128		243	2030	
38737	129		244	2027	
38785	108	3452	86	2038	
38786	108	3452	86	2038	
38787	161	3132	39	2015	
38788	123A	3444	20	2051	
38789	195A	3048/329	46		
38804	390		255	2041	
38837	128		243	2030	
38840	5626	3633			
38869	128		243	2030	
38870	129		244	2027	
38885	5527	6627			
38890	5626	3633			
38897	130	3027	14	2041	
38916	128		243	2030	
38920	108	3452	86	2038	
38921	108	3452	86	2038	
38927	5626	3633			
38965	181	3036	75	2041	
38970	5529	6629			
38971	311		277		
38976	5527	6627			
38986	311		277		
38996	154		40	2012	
39005	5547	3505			
39034	123A		20	2051	
39039	5626	3633			
39042	177		300	1122	
39043	311		277		
39053	160	3004/102A	245	2004	
39060-1	747	3279	IC-218		
39075-1	738	3167	IC-29		
39096	123A	3444	20	2051	
39097	128		243	2030	
39114	129		244	2027	
39119	5527	6627			
39123	124		12		
39127	130		14	2041	
39140	181		75	2041	
39148	130		14	2041	
39149	124		12		
39196	181		75	2041	
39213	130		14	2041	
39231	128		243	2030	
39238	128		243	2030	
39248	128		243	2030	
39250	129		244	2027	
39251	130		14	2041	
39252	128		243	2030	
39255	128		243	2030	
39285	175		246	2020	
39299	5527	6627			
39302	152	3893	66	2048	
39311	128		243	2030	
39329	128		243	2030	
39331	108	3452	86	2038	
39343			255	2041	
39369	130		14	2041	
39377	124		12		
39393	228A	3103A/396			
39414	130		14	2041	
39440	129		244	2027	
39443	128		243	2030	
39455	181	3036	75	2041	
39458	184	3054/196	57	2017	
39462	128		243	2030	
39465	181		75	2041	
39466	181	3036	75	2041	
39477	124		12		
39485	128		243	2030	
39486	128		243	2030	
39492	130		14	2041	
39510	157	3747	232		
39542	5642	3507			
39558	5695	3509			
39561	128		243	2030	
39581			255	2041	
39587	128		243	2030	
39616	181	3036	75	2041	
39617	128		243	2030	
39618	129		244	2027	
39619	129		244	2027	
39635	181	3036	75	2041	
39669	5626	3633			
39705	128	9038/346	243	2030	
39713	128		243	2030	
39715	124		12		
39730	108	3452	86	2038	
39731	108	3452	86	2038	
39741	192		63	2030	
39750	152	3893	66	2048	
39750(ORRTRONICS)	184		57	2017	
39751	181	3036	75	2041	
39767	152	3893	66	2048	
39789	108	3452	86	2038	
39789(POWER)	152		66	2048	
39803A			255	2041	
39804	5802	9005		1143	
39805	5626	3633			
39811	5642	3507			
39819			255	2041	
39824	152	3054/196	66	2048	
39830	5547	3505			
39835	128		243	2030	
39840	5626	3633			
39842	128	3103A/396	243	2030	
39853	129		244	2027	
39863	128		243	2030	
39864	128		243	2030	
39865	129		244	2027	
39868	128		243	2030	
39876	192		63	2030	
39893	121	3717	239	2006	
39901-0001	105	3102/710	4		
39902	5444	3634			
39919	192		63	2030	
39920	128		243	2030	
39921	130	3027	14	2041	
39922			255	2041	
39923	194		220		
39940	128		243	2030	
39948	152	3893	66	2048	
39952	5695	3509			
39954	130		14	2041	
39981	152	3893	66	2048	
040001	107	3020/123	11	2015	
40004	160	3008	245	2004	
40005	160		245	2004	
40006	160	3008	245		
40022	104	3009	16	2006	
40024	116	3017B/117	504A	1104	
40024VM	116	3311	504A	1104	
40034	102A		53	2007	
40034-1	102A	3004	53	2007	
40034-2	102A	3004	53	2007	
40034-3	102A	3004	53	2007	
40034VM	102A	3004	53	2007	
40035	102A		53	2007	
40035-1	102A	3004	53	2007	
40035-2	102A	3004	53	2007	
40035-3	102A	3004	53	2007	
40036	102A		53	2007	
40036-1	102A	3004	53	2007	
40036-2	102A	3004	53	2007	
40036-3	102A	3004	53	2007	
40037	103A	3835	59	2002	
40037-1	103A	3010	59	2002	
40037-2	103A	3010	59	2002	
40037-3	103A	3010	59	2002	
40037VM	103A	3010	59	2002	
40038	102A		53	2007	
40038-1	102A	3004	53	2007	
40038-2	102A	3004	53	2007	
40038-3	102A	3004	53	2007	
40038VM	102A	3010	53	2007	
040048			49	2025	
40050	104	3009	16	2006	
40051	121	3009	239	2006	
40051-2	121	3717	239	2006	
40053	128		243	2030	
40080	329	3046	243	2030	MRF8003
40081	329	3048	46		MRF8003
40081V1	329	3048	46		
40082	329	3048	46		MRF8004
40082A2	329	3048	46		
40084	123A	3024/128	20	2051	
40108	5870	3500/5882	5032		
40108R	5871	3517/5883	5033		
40109	5872	3500/5882	5032		
40109R	5873	3517/5883	5033		
40110	5874	3500/5882	MR-1		
40110R	5879	3517/5883			
40111	5876	3500/5882	5036		
40111R	5877	3517/5883	5037		
40112	5878	3500/5882	5036		
40112R	5879	3517/5883	5037		
40113	5880	3500/5882	5040		
40113R	5881	3517/5883	5041		
40114R	5883	3517	5041		
40115	5886	3500/5882	5044		MR1128
40115R	5887		5045		
40143	116		504A	1104	
40151	130		14	2041	
40208	5980	3610	5096		1N248B
40208R	5981	3698/5987	5097		
40209		3501	MR-2		1N249B
40209R	5983	3698/5987	5097		
40210	5986	3609	5096		1N250B
40210R	5987	3698	5097		
40211	5988	3608/5990	5100		1N1196
40211R	5989	3518/5995	5101		
40212	5990	3608	5100		1N1196
40212R	5991	3518/5995	5101		
40213	5992	3501/5994	5104		1N1198
40213R	5993	3518/5995	5105		
40214	5994	3501	5104		1N1198
40214R	5995	3518	5105		
40217	123A	3444	20	2051	
40218	123A	3444	20	2051	
40219	123A	3444	20	2051	
40220	123A	3444	20	2051	
40221	123A	3444	20	2051	
40222	123A	3444	20	2051	
40231	161	3854/123AP	39	2015	
40232	161	3444/123A	20	2051	
40233	161	3124/289	39	2015	
40234	161	3018	39	2015	
40235	161	3018	39	2015	
40236	161	3018	39	2015	
40237	161	3018	39	2015	
40238	161	3117	39	2015	
40239	161	3117	39	2015	

Industry Standard No.	ECG	SK	GE	RS 276-	MOTOR.
40240	161	3117	39	2015	MRF501
40242	161	3122	39	2015	
40243	161	3122	39	2015	
40244	161	3018	86	2015	
40245	161	3018	39	2015	
40246	161	3716	39	2015	
40250	384	3026	246	2020	2N4231A
40250V1	404/175	3026	28		
40251	130	3027	14	2041	2N6569
40253	102A	3004	53	2007	
40254	104	3009	16	2006	
40256	191	3232	249		
40259	161	3500/5882	39	2015	
40260	161		39	2015	
40261	126		52	2024	
40262	126	3008	52	2024	
40263	102A	3004	52	2024	
40263(RCA)	160		245	2004	
40264	124	3021	12		
40264V1	124		12		
40265	116	3131A/369	504A	1104	
40266	5801	9004		1142	MR501
40267	5802	9005		1143	MR502
40268	126		245	2004	
40269	100	3005	1	2007	
40279			28	2017	2N3375
40280	346		20	2051	2N4427
40280R	5995		5105		
40281	475		28	2017	2N3920
40282	476		246	2020	2N3927
40283	123A	3444	20	2051	
40290	473		20	2051	2N3553
40291			D40E7		2N3632
40292			66	2048	
40294	161	3039/316	86	2038	
40295	316	3039	86	2038	
40296	316	3716/161	39	2015	
40305			28	2017	
40306			28	2017	
40307			66	2048	
40309	128	3024	243	2030	
40309V1	128	3024	243	2030	
40309V2	128	3024	243	2030	
40310	175	3026	246	2020	2N4231A
40310V1	175	3026	246	2020	
40311	128	3024	226	2018	
40311A	188	3199	226	2018	
40311B	188	3199	226	2018	
40312	175	3026	246	2020	2N4232A
40312V1	175	3026	246	2020	
40313		3021	12		2N4240
40314	128	3024	243	2030	
40314V1	188	3024/128	226	2018	
40314V2	188	3024/128	226	2018	
40315	128	3024	243	2030	
40315V1		3024	226	2018	
40315V2	188	3024/128	226	2018	
40316	175	3026	246	2020	2N4231A
40317	128	3024	243	2030	
40317L			63	2030	
40317S			63	2030	
40317V1	188	3024/128	226	2018	
40317V2	188	3024/128	226	2018	
40318		3021	12		2N4240
40319	129	3025	244	2027	
40319L			67	2023	
40319S	129		67	2023	
40319V1	188	3025/129	226	2018	
40319V2	188	3025/129	226	2018	
40320	128	3024	243	2030	
40320L			63	2030	
40320S			63	2030	
40320V1	188	3024/128	226	2018	
40320V2	188	3024/128	226	2018	
40321	396	3104A	251		
40321V1	396	3044/154	226	2018	
40321V2	396	3044/154	226	2018	
40322	286	3021/124	12		2N4240
40323	128	3024	243	2030	
40323V1	188	3024/128	226	2018	
40323V2	188	3024/128	226	2018	
40324	175	3026	246	2020	2N4231A
40325	130	3027	14	2041	2N6569
40326	128	3024	243	2030	
40326L			63	2030	
40326S			63	2030	
40326V2	188	3024/128	226	2018	
40327	397	3104A	243	2030	
40327V2	188	3045/225	226	2018	
40328	124	3021	12		2N4240
40329	102A	3004	53	2007	
40340					2N5071
40341					2N3950
40346	396	3104A	251		
40346V1	191	3537	249		
40346V2	396/427	3045/225	256		
40347	324		243	2030	
40347V1		3536	28	2017	
40347V2	225		256		
40348V1		3536	D40E7		
40348V2	225		256		
40349V2	225		256		
40350	161		243	2030	
40351	161	3452/108	86	2038	
40352	161	3452/108	86	2038	
40354	154	3040	40	2012	
40355	154	3040	40	2012	
40359	102A	3004	2	2007	
40360	128	3024	243	2030	
40360V1	188	3024/128	226	2018	
40360V2	188	3024/128	226	2018	
40361	128	3024	243	2030	
40361V2	188	3024/128	226	2018	
40361V3	188	3199	226	2018	
40362	129	3025	244	2027	
40362V1	189	3025/129	218	2026	
40362V2	189	3025/129	218	2026	
40363	130	3027	14	2041	2N5877
40364	175	3562	246	2020	2N4233A
40366	128	3024	243	2030	
40367	324	3044/154	243	2030	
40368		3530	66	2048	
40369	130	3079	14	2041	2N5877
40372	175	3026	246	2020	2N3054
40373	175		246	2020	2N3441
40374	124	3021	12		2N3583
40375		3562			2N5428
40378		3503	MR-4		
40379	5527	3503	MR-4		
40383	116		504A	1104	
40385	396		243	2030	
40385V1	188	3199	226	2018	
40385V2	188	3199	226	2018	
40389		3045	28	2017	
40390	191	3044/154	249		
40391	129/401		48		
40392	128	3045/225	256		
40394		3025	48		
40395	102A	3004	53	2007	
40396			8	2002	
40396/P			53	2007	
40396N	103A		59	2002	
40396P	102A		53	2007	
40397	123A	3444	20	2051	
40398	123A	3444	20	2051	
40399	123A	3444	20	2051	
40400	123A	3444	20	2051	
40403	102	3005	1	2007	
40404	278	3444/123A	20	2051	
40405	123A	3444	20	2051	
40406	129	3025	244	2027	
40406L			67	2023	
40406S	129		67	2023	
40407	128	3024	244	2027	
40407L			63	2030	
40407S			63	2030	
40408	129	3024/128	217		
40409	128	3024	243	2030	
40410	129	3025	244	2027	
40411	181	3036	75	2041	MJ802
40412	396	3104A	251		
40412V1		3537	32		
40412V2	396/427	3045/225	256		
40413	108	3039/316	86	2038	
40414	108	3039/316	86	2038	
40421	121	3014	14	2041	
40422	124	3021	12		
40423	124	3021	12		
40424	175	3021/124	246	2020	
40425	124	3021	12		
40426	124	3021	12		
40427	124	3021	12		
40429	56006	3683/5511			
40430	5512	3684			
40432	123A		20	2051	
40437V1	188	3199	226	2018	
40437V2	188	3199	226	2018	
40438V1	190		217		
40438V2	190		217		
40439	191	3035	249		
40439V1	191	3232	249		
40439V2	191	3232	249		
40440	127	3035	25		
40442	127	3113/516			
40444	327		75	2041	
40446	224	3049	46		MRF8004
40450			63	2030	
40451			63	2030	
40452			63	2030	
40453			63	2030	
40454			63	2030	
40455			63	2030	
40456			63	2030	
40456(RCA)	123A	3444	20	2051	
40457	128	3024	243	2030	
40458	128	3122	46	2030	
40459	154	3040	40	2012	
40461	312		FET-1	2028	
40461-2	128		243	2030	
40462	104	3009	16	2006	
40464	130	3027	14	2041	
40465	130	3027	14	2041	
40466	130		14	2041	
40467					3N128
40467A	220	3990		2036	3N128
40468	312			2028	3N128
40468A	220	3990		2036	3N128
40469	311	3039/316	86	2038	
40470	108	3039/316	86	2038	
40471	311		14	2041	
40472	161	3452/108	86	2038	
40473	123A	3444	20	2051	
40474	123A	3444	20	2051	
40475	161	3452/108	86	2038	
40476	161		20	2051	
40477	123A	3444	20	2051	
40478	161	3018	86	2038	
40479	161	3018	86	2038	
40480	108	3452	86	2038	
40481	161	3452/108	86	2038	
40482	108	3452	86	2038	
40487	160		245	2004	
40488	126		52	2024	
40489	160		245	2004	
40490	102A	3004	53	2007	
40491	124	3021	12		
40500	123A	3444	20	2051	
40501	128	3024	243	2030	
40502	5624	3507			
40503	5626	3633			
40504	5444	3502/5513			
40505	5446	3502/5513			
40506	5448	3502/5513			
40507	5444	3634			
40508	5446	3503			
40509	5624	3632			
40510	5626	3633			
40511	5624	3632			
40512	5626	3633			
40513			255	2041	MJE3055T
40514			255	2041	MJE3055T
40517	123	3039/316	20	2051	
40518	123		20	2051	
40519	123A		20	2051	
40525	5650	3583/5641			
40526	5651	3506/5642			
40527	5652	3506/5642			
40528	5650	3506/5642			
40529	5651	3506/5642			
40530	5652	3506/5642			
40537	129	3025	244	2027	
40537L	129		244	2027	
40537S	129		244	2027	
40538	129	3025	244	2027	
40538L	129		244	2027	
40538S	129		244	2027	
40539	128	3024	243	2030	
40539L	128		243	2030	
40539S	128		243	2030	
40542			255	2041	2N5978
40543			255	2041	2N5978
40544	225	3045	256		
40546	124	3021	12		
40547	124	3021	12		
40553	230	3042			
40554	230	3042			
40555	230	3042			
40559					3N128

Industry Standard No.	ECG	SK	GE	RS 276-	MOTOR.
40559A	220	3990		2036	3N128
40576	56006	3507			
40577	123A	3444	20	2051	
40578	128	3195/311	243	2030	
40581	329	3048	46		MRF8004
40582	224	3049	46		MRF8004
40583	6408	3523/6407			
40594	188	3512	226	2018	
40595	323	3513	218	2026	
40595VX	129		244	2027	
40600	221	3050	FET-4	2036	3N201
40601	221	3050	FET-4	2036	3N202
40602	221	3050		2036	3N203
40603	221	3050		2036	3N211
40604	221	3050		2036	3N213
40605			28	2017	
40608	278		261		2N5943
40611	128	3024	243	2030	
40612	121	3014	239	2006	
40613	152	3054/196	66	2048	TIP31
40616	128	3024	243	2030	
40618	152	3054/196	66	2048	TIP31
40621	152	3054/196	66	2048	TIP31
40622	152	3054/196	66	2048	TIP31
40623	121MP	3009	239(2)	2006(2)	
40624	196	3054	241	2020	TIP41A
40625	128/401	3024/128			
40626	121	3014	239	2006	
40627	196	3054	241	2020	TIP41A
40628	128/401	3024/128			
40629	152	3054/196	66	2048	TIP31
40630	152	3054/196	66	2048	TIP31
40631	152	3054/196	66	2048	TIP31A
40632	196	3054	241	2020	TIP31A
40633		3027	255	2041	
40634	129	3025	244	2027	
40635	128	3024	243	2030	
40636	284	3027/130	265	2047	2N5878
40637	123A	3444	20	2051	
40637A			47	2030	2N4072
40638	5624	3632			
40639	5626	3633			
40640	230	3042			
40641	231	3857			
40642	515	9098			MR817
40643	515	9098			MR817
40644	515	9098			MR817
40648	124	3261/175			
40650	171		27		
40654		3577	C122B		
40655		3503	C122D		
40662	5693	3652			
40663	5695	3509			
40664	175		246	2020	
40665			66	2048	2N3375
40666			28	2017	2N3632
40668	5633	3938			
40669	5635	3533			
40672	5697	3522			
40673	222	3050/221	FET-4	2036	
40680		3655	C228A2		
40681		3655	C228B2		
40682		3656	C228D2		
40683		3657	C228M2		
40689	5697	3662/56024			
40690	5697	3663/56026			
40691	5651	3506/5642			
40692	5652	3506/5642			
40702	5693	3652			
40703	5695	3509			
40704	5697	3522			
40707	5693	3652			
40708	5695	3509			
40710	5697	3522			
40713	5673	3508/5675			
40714	5675	3508			
40716	5675	3507			
40719	5673	3508/5675			
40720	5675	3508			
40721	5633	3938			
40722	5635	3633/5626			
40737	5514	3504/5508	C222A		
40738	5514	3504/5508	C222B		
40739	5515	3504/5508	C222D		
40740	5516	3504/5508	C222M		
40741	5491	6791			
40742	5492	6792			
40743	5494	6794			
40744	5496	6796			
40745		3504	C220A2		
40746		3504	C220B2		
40747		3504	C220D2		
40748		3504	C220M2		
40749	5514	3613	C232A		
40750	5514	3613			
40751	5515	6615			
40752	5516	6616			
40763	102A		53	2007	
40766	5650	3506/5642			
40777	5673	3508/5675			
40778	5675	3508			
40781	5673	3508/5675			
40782(I.C.)	804		IC-27		
40785	5673	3508/5675			
40786	5675	3508			
40789	5683	3509/5695			
40790	5697	3522			
40793	5693	3652			
40794	5695	3509			
40796	5677	3520			
40798	5677	3520			
40808	125		510,531	1114	
40809	125		510,531	1114	
40816(RCA)	196		241	2020	
40819	454	3065/222		2036	3N201
40820	454	3065/222	FET-4	2036	3N201
40821	454	3065/222	FET-4	2036	3N201
40822	454	3065/222		2036	3N201
40823	222	3065	FET-4	2036	3N201
40828	102A		53	2007	
40829		3625			2N6316
40830					2N6315
40831					2N6315
40833		3503	C122M		
40841	221	3065/222	FET-4	2036	3N203
40842	5636	3939			
40850	124	3559/162	12		2N4240
40852	283	3467	36		2N6543
40852(VM)	102A		53	2007	
40853	283	3467	36		2N6546
40853(VM)	158		53	2007	
40854		3559	36		2N6546
40867	5462	3686	C122A		
40868	5463	3572	C122B		
40869	5465	3687	C122D		
40871		3440			TIP41C
40872		3441			TIP42C
40873					TIP41B
40874					TIP41B
40875					TIP41C
40876					TIP41A
40880	230	3042			
40885	228A	3104A	257		MPSU10
40886	228A	3103A/396	257		MPSU10
40887	228A	3103A/396	257		2N6559
40888	230	3042			
40889	231	3857			
40890	515	9098			
40891	515	9098			
40892	515	3314			
40893					2N5946
40894	316	3039	86	2038	2N5179
40895	316	3039	86	2038	2N5179
40896	316	3039	86	2038	2N5179
40897	316	3039	39	2015	2N5179
40900	5632	3937			
40901	5633	3938			
40902	5635	3533			
40910	175	3026	246	2020	2N4231A
40911		3026			2N4233A
40912		3021			2N5050
40913		3538			2N5051
40915					2N5031
40934					MRF616
40934-1	130	3027	14	2041	
40936					2N5070
40938		3505	C137N		
40940					MRF5175
40941					MRF313
40953			45		MRF207
40954		3176			MRF208
40955	345	3177/351			MRF238
40956		3501			MR870
40957		3501			MR871
40958		3501			MR872
40959		3501			MR874
40960		3501			MR876
40964	486	3765/195A			MRF515
40965	486	3765/195A			MRF515
40967	362				2N5944
40968	364				2N5946
40970	366				MRF644
40971	367				MRF646
40972	472		45		MRF607
40973	350	3176			2N6081
40974	351	3177			2N6082
40975			45		2N3553
40976			45		2N3553
40977	359				2N5642
41001-4	116	3031A	504A	1104	
41008					MRF628
41008A	362				2N5944
41009					MRF616
41009A	362				2N5944
41010	364				2N5946
41012		3561			2N5038
41013		3561			2N6339
41014	5633	3938			
41015	5635	3533			
41020-618	125	3031A	510,531	1114	
41023-224	125	3031A	510,531	1114	
41023-225	125	3031A	510,531	1114	
41024					2N5108
41025					MRF321
41026					MRF323
41027					MRF321
41028					MRF323
41038					MRF905
41051	123A	3444	20	2051	
41052	129	3025	244	2027	
41053	128	3024	243	2030	
41173	519		514	1122	
41175	199	3245	62	2010	
41176	123A	3444	20	2051	
41177	159		82	2032	
41178	184	3190	57	2017	
41179	185	3191	58	2025	
41180	909D	3590	IC-250		
041200-30110			300	1122	
41342	185	3041	58	2025	
41344	184	3041	57	2017	
41440	159	3466	82	2032	
41500	152	3054/196	241	2020	TIP29
41501	197	3084	250	2027	TIP30
41502	128		243	2030	
41503	129		244	2027	
41504	152	3893	66	2048	TIP31
41505	228A	3103A/396	257		MPSU10
41506	162	3560	35		2N6543
41570	102A		53	2007	
41616	116	3016	504A	1122	
41689	108	3452	86	2038	
41694	108	3452	86	2038	
41850	116	3017B/117			
42020	109	3087		1123	
42020-737	140A	3061	ZD-10	562	
42021	125	3032A	510,531	1114	
42025-850	142A	3062	ZD-12	563.	
42065	175		246	2020	
42221	116	3031A	504A	1104	
42302	126		52	2024	
42304	158		53	2007	
42305	102A		53	2007	
42311	126		52	2024	
42321	176	3845	80		
42322	102A		53	2007	
42323	124		12		
42324	102	3722	2	2007	
42342	175	3026	246	2020	
42396	312	3050/221	FET-1	2028	
42464	123A	3444	20	2051	
42942	152	3054/196	66	2048	
043001	108	3452	86	2038	
43021-017	123A	3444	20	2051	
43021-198	128	3024	243	2030	
43022-860	108	3452	86	2038	
43044	123A	3444	20	2051	
43045	123A	3444	20	2051	
43046	121	3717	239	2006	
43054	199	3444/123A	20	2051	
43055	199	3444/123A	20	2051	
43060	130	3027	14	2041	
43062	177		300	1122	
43072	5079A	3335	ZD-20		
43074	179	3009	239	2006	
43082	128		243	2030	
43088	128		243	2030	
43089	129		244	2027	

Industry Standard No.	ECG	SK	GE	RS 276-	MOTOR.
43090	706		IC-43		
43095	130		14	2041	
43104	284	3260	265	2047	2N5631
43107	129	3025	244	2027	
43114	130	3027	14	2041	
43115	192	3024/128	63	2030	
43116	129	3025	244	2027	
43117	128	3024	243	2030	
43118	182	3188A	55		
43119	182	3188A	55		
43120	182	3188A	55		
43122	128		243	2030	
43127	159	3466	82	2032	
43139	123A	3444	20	2051	
43163	152	3054/196	66	2048	
43165	128		243	2030	
43168	130	3027	14	2041	
43200	7400			1801	
43201	7404			1802	
43202	7408			1822	
43205	7473			1803	
43296	312	3834/132	FET-1	2035	
43959	109	3087		1123	
43992-2	101	3861	8	2002	
44001	125	3311	510,531	1114	
44002	519	3311	514	1122	
44002(RCA)	116		504A	1104	
44003	125	3311	510,531	1114	
44004	125	3312	510,531	1114	
44005	125	3313/116	510,531	1114	
44006	125	3032A	510,531	1114	
44007	125	3051/156	510,531	1114	
44208	152	3893	66	2048	
44209	153		69	2049	
44465-3	116	3016	504A	1104	
44465-4	116	3016	504A	1104	
44465-5	116	3016	504A	1104	
44465-6	116	3016	504A	1104	
44616-1	102	3003	2	2007	
44699	152	3054/196	66	2048	
44763	188	3024/128	226	2018	
44764	189		218	2026	
44766	312			2028	
44933	515	3315/507			
44934	515	3315/507			
44935	515	3316			
44936	515	3317	504A	1104	
44937	116	3318	504A	1104	
44938	116	9098/515	504A	1104	
44967-2	102	3004	2	2007	
45122	159	3466	82	2032	
45184	123A	3444	20	2051	
45190		3054	D44H4		
45191		3054	D44H7		
45192		3054	D44H10		
45193		3083	D45H4		
45194		3083	D45H7		
45195		3083	D45H10		
45337-A	199	3245	62	2010	
45337-C	159	3466	82	2032	
45354	128		243	2030	
45380	718	3159	IC-8		
45381	722	3161	IC-9		
45381(VOFM)	722		IC-9		
45385	747	3279	IC-218		
45386	750	3280	IC-219		
45387	804	3455	IC-27		
45390	743		IC-214		
45391	788	3829	IC-229		
45393	804		IC-27		
45395	804	3455	IC-27		
45411	6408	3523/6407			
45495-2	101	3861	8	2002	
45810-161	312	3116	FET-2	2035	
45810-162	123A	3018	62	2010	
45810-163	123A	3018	61	2038	
45810-164	123A	3122	210	2051	
45810-165	192	3137	63	2030	
45810-166	199	3124/289	212	2010	
45810-167	788	3829	IC-229		
45810-168	801	3160	IC-35		
45810-169	1087		IC-103		
45810-170	1155	3231	IC-179	705	
45810-51	116	3311	504A	1104	
45810-52		3126	90		
45810-53	109	3088	1N60	1123	
45810-54	140A	3061	ZD-10	562	
46140-2	116	3016	504A	1104	
46287-4	110MP	3087	1N34AS	1123	
46490-2	101	3861	8	2002	
46590-2	103	3010	8	2002	
46591-2	103	3010	8	2002	
46592-2	103	3010	8	2002	
46593-2	103	3010	8	2002	
46631-2	101	3861	8	2002	
46774-1	101	3861	8	2002	
46775-2	101	3861	8	2002	
46776-2	102	3003	2	2007	
46914	5802	9005		1143	
47126	116	3016	504A	1104	
47126-003	116	3031A	504A	1104	
47126-1	116	3016	504A	1104	
47126-10	156	3032A	512		
47126-12	116		504A	1104	
47126-1A	116	3016	504A	1104	
47126-2	116	3016	504A	1104	
47126-3	116	3311	504A	1104	
47126-3A	116	3016	504A	1104	
47126-4	116	3016	504A	1104	
47126-4A	116	3016	504A	1104	
47126-6	116	3016	504A	1104	
47126-7	116	3311	504A	1104	
47126-8	156	3032A	512		
47126-9	116	3032A	504A	1104	
47127-3	116		504A	1104	
47394-2	102	3004	2	2007	
47645-2	103	3010	8	2002	
47737-2	102	3003	2	2007	
48004-07	123A	3444	20	2051	
48004-08	123A	3444	20	2051	
48009	312		FET-2	2035	
48009(1)	312		FET-2	2035	
48106	169	3678	504A	1104	
48287	109	3087		1123	
48287-4	110MP	3087	1N34AS	1123	
48385-2	103	3862	8	2002	
48751-028			21	2034	
48937-2	126	3008	52	2024	
48939-2	126	3008	52	2024	
49058-2	103	3862	8	2002	
49092	128		243	2030	
49138-2	101	3861	8	2002	
49139-2	102	3003	2	2007	
49341	102A	3004	53	2007	

Industry Standard No.	ECG	SK	GE	RS 276-	MOTOR.
49751-163			30	2006	
49939-2	160		245	2004	
50009	804	3455	IC-27		
50137-2	128	3024	243	2030	
50200-12	128		243	2030	
50200-1B	152	3893	66	2048	
50200-24	152	3893	66	2048	
50200-8	196	3534	241	2020	
50200-9	152	3054/196	66	2048	
50201-4	153		69	2049	
50202-1	123A	3444	20	2051	
50202-12	128		243	2030	
50202-13	123A		20	2051	
50202-14	123A		20	2051	
50202-2	123A		20	2051	
50202-23	123A		20	2051	
50202-24	128		243	2030	
50202-9	194		220		
50203-12	159	3466	82	2032	
50203-8	159	3466	82	2032	
50212-19	177		300	1122	
50212-28	177		300	1122	
50212-30	177		300	1122	
50213-5	519		514	1122	
50280-3	175		246	2020	
50308-0100	152	3893	66	2048	
50447-4	121	3009	239	2006	
50477-4	121	3014	239	2006	
50505-01	109	3087		1123	
50745-01	116	3311	245	1104	
50745-02	116	3031A	245	1104	
50745-03	156	3031A	512		
50745-04	156	3031A	512		
50745-05	156	3031A	512		
50745-06	116	3016	504A	1104	
50745-07	116	3016	504A	1104	
50745-08	116	3031A	504A	1104	
50957-03	108	3452	86	2038	
51194	128	3024	243	2030	
51194-01	128	3024	243	2030	
51194-02	128	3024	243	2030	
51194-03	128	3024	243	2030	
51213	123A	3444	20	2051	
51213-01	123A	3444	20	2051	
51213-02	123A	3124/289	20	2051	
51213-03	123A	3444	20	2051	
51213-2	123A	3444	20	2051	
51300	175		246	2020	
51428-01	123A	3444	20	2051	
51429	199	3245	62	2010	
51429-02	123A	3444	20	2051	
51429-03	123A	3444	20	2051	
51429-3	123A	3444	20	2051	
51441	123A	3444	20	2051	
51441-01	123A	3444	20	2051	
51441-02	123A	3444	20	2051	
51441-03	123A	3444	20	2051	
51442	123A	3124/289	20	2051	
51442-01	123A	3124/289	20	2051	
51442-02	123A	3124/289	20	2051	
51442-03	123A	3124/289	20	2051	
51545	123A	3444	20	2051	
51547	123A	3444	20	2051	
51650	121	3014	239	2006	
52215-00	130		14	2041	
52329	130	3027	14	2041	
52360	130	3027	14	2041	
53016R	177	3100/519	300	1122	
53088-4	125	3032A	510,531	1114	
53092-1	109	3087	1N34AS	1123	
53093-1	113A	3119/113	6GC1		
53099-1	116	3017B/117	504A	1104	
53099-3	116	3017B/117	504A	1104	
53200-22	123A	3444	20	2051	
53200-23	123A	3444	20	2051	
53200-51	123A	3444	20	2051	
53200-74	199		62	2010	
53201-01	102A		53	2007	
53201-11	102A		53	2007	
53201-51	224		46		
53203-72	128		243	2030	
53300-31	125		510,531	1114	
53300-41	116		504A	1104	
53301-01	125		510,531	1114	
53400-01	123A	3444	20	2051	
053492	123A	3122	20	2051	
53704D	230	3042	700		
54806P1	5547		C137M		
55001	7400	7400	7400	1801	
55002	7402	7402	7402	1811	
55003	7404	7404	7404	1802	
55004	7405	7405			
55005	7410	7410		1807	
55006	7420	7420	7410	1807	
55007	7430	7430	7420	1809	
55008	7442	7442			
55009	7447	7447			
55011	7474	7474	7474		
55012	7476	7476	7476	1813	
55024	7416	7416			
55027	7413	7413	7413	1815	
55029	7454	7454			
55032	7401	7401	IC-194		
55034	7440	7440			
55036	7406	7406	7406	1821	
55107ADM					MC55107L
55107BDM					MC55107L
55108ADM					MC55108L
55108BDM					MC55108L
55110DM					MC75S110L
55121DM					MC8T13L
55122DM					MC8T14L
55166	109	3088	1N34AS	1123	
55170-1	229	3018	61	2038	
55206				1104	
55207				1114	
55207DM					MC55107L
55208DM					MC55108L
055210	117	3016	504A	1104	
055210H	116	3311	504A	1104	
055228	125	3032A	510,531	1114	
055228H	125	3032A	510,531	1114	
55325DM					MC55325L
55325FM					MC55325L
55606	123A	3444	20	2051	
55810-161	161	3124/289	39	2015	
55810-162		3018		2051	
55810-163	161	3124/289	39	2015	
55810-164	161	3018	39	2015	
55810-166	175	3027/130	246	2020	
55810-167	703A	3157	IC-12		
55810-51	109	3087		1123	
55810-52	109	3087		1123	
55974-1	910		IC-251		

Industry Standard No.	ECG	SK	GE	RS 276-	MOTOR.
55986-1	909	3590	IC-249		
57000-5452	108	3452	86	2038	
57000-5503	123		20	2051	
057001	109	3087		1123	
57001-01	128		243	2030	
057001H	109	3087		1123	
057005	109			1123	
057005H	109			1123	
57009	234	3247	65	2050	
057040	104	3009	16	2006	
57174	5642	3507			
57202	5626	3633			
057293	177		300	1122	
057500			1N34AS	1123	
057524		3126	90		
58215-01	123A		20	2051	
58810-160	312	3834/132	FET-1	2035	
58810-161	107		11	2015	
58810-162	123A	3444	20	2051	
58810-163	123A	3444	20	2051	
58810-164	199	3245	62	2010	
58810-165	123A	3444	20	2051	
58810-166	123A	3444	20	2051	
58810-167	199	3245	62	2010	
58810-168	123A	3444	20	2051	
58810-169	186	3192	28	2017	
58810-170	1100	3223	IC-92		
58810-171	1142	3485	IC-128		
58810-172	1087	3477	IC-103		
58810-80	116		504A	1104	
58810-82	177		300	1122	
58810-83	109			1123	
58810-85	116		504A	1104	
58810-86	140A	3061	ZD-10	562	
58840-111	116	3311	504A	1104	
58840-112		3126	90		
58840-113	177	3100/519	300	1122	
58840-114	110MP	3709	1N60	1123(2)	
58840-116	116	3031A	504A	1104	
58840-117	140A	3061	ZD-10	562	
58840-191			FET-2	2035	
58840-192	107	3018	11	2015	
58840-193	123A	3444	20	2051	
58840-194	123A	3444	20	2051	
58840-195	123A	3444	20	2051	
58840-196	123A	3444	20	2051	
58840-197	123A	3444	20	2051	
58840-198	123A	3444	20	2051	
58840-199	123A	3444	20	2051	
58840-200	186	3192	28	2017	
58840-201	1100	3223	IC-92		
58840-202	720	9014	IC-7		
58840-203	1087	3477	IC-103		
059395	109	3087	1N34AS	1123	
59395(RCA)	109			1123	
59471-1	116	3016			
59557-48	102A		53	2007	
59557-49	116		504A	1104	
59625-1	102A	3466/159	82	2032	
59625-10	159	3466	82	2032	
59625-11	102A	3466/159	82	2032	
59625-12	102A	3466/159	82	2032	
59625-2	102A	3466/159	82	2032	
59625-3	102A	3466/159	82	2032	
59625-4	102A	3466/159	82	2032	
59625-5	102A	3466/159	82	2032	
59625-6	102A	3466/159	82	2032	
59625-7	102A	3466/159	82	2032	
59625-8	102A	3466/159	82	2032	
59625-9	102A	3466/159	82	2032	
59840	177	3016	300	1122	
59840-1	519	3087		1123	
59844-1	116	3017B/117	504A	1104	
59844-2	116	3033	510,531	1114	
59844-3	116	3016	504A	1104	
59987-1	102A		53	2007	
59988-1	128		243	2030	
59989-1	129		244	2027	
59990-1	121	3717	239	2006	
59991-1	142A		ZD-12	563	
60024			255	2041	
60031	128		243	2030	
60034	5695	3509			
60041	130		14	2041	
60045	5642	3507			
60046	130		14	2041	
60047	130		14	2041	
60048	108	3452	86	2038	
60059	5642	3519/5643			
60076	181		75	2041	
60085	130	3027	14	2041	
60091	128		243	2030	
60106	128		243	2030	
60108	5547	3505			
60115	181		75	2041	
60127	130		14	2041	
60130	130	3027	14	2041	
60132		3054	28	2017	
60133		3054	66	2048	
60136	5626	3633			
60142	181		75	2041	
60154	129	3027/130	244	2027	
60172	192		63	2030	
60175	152	3893	66	2048	
60182	5642	3507			
60187	181		75	2041	
60194	128		243	2030	
60201	175		246	2020	
60202	5626	3633			
60205	130		14	2041	
60213(TRANSISTOR)	130		14	2041	
60215-1	109			1123	
60216	152	3893	66	2048	
60219	175		246	2020	
60228	128		243	2030	
60234	181		75	2041	
60237	196		241	2020	
60243	130		14	2041	
60245	5626	3507			
60246	5626	3507			
60256	5483	3942			
60259	5642	3507			
60270	5651	3507			
60294	128		243	2030	
60304	5528	6629/5529			
60306	5624	3506/5642			
60308	124		12		
60314	108	3533/5635	86	2038	
60317	5642	3507			
60319	5642	3509/5695			
60322	5693	3652			
60323	5642	3509/5695			
60335	128		243	2030	
60337	221			2036	
60339	181		75	2041	
60341	5652	3509/5695			
60350	181	3036	75	2041	
60353	724		IC-86		
60360	5635	3533			
60373	515	9098			
60380	181		75	2041	
60395	123A	3444	20	2051	
60407	186	3192	28	2017	
60408	152	3893	66	2048	
60410	5633	3938			
60413	186	3192	28	2017	
60415	5635	3533			
60416	196	3054	241	2020	
60417	128		243	2030	
60423	128		243	2030	
60424	5675	3520/5677			
60428	128		243	2030	
60451	5626	3633			
60457	188		226	2018	
60458	189		218	2026	
60465	130		14	2041	
60466	5633	3938			
60478	5643	3519			
60496	5675	3520/5677			
60533	5675	3520/5677			
60550	5643	3519			
60559	124		12		
60572	5633	3938			
60579	5626	3507			
60594	5693	3652			
60597	128		243	2030	
60613	5626	3507			
60615	5643	3519			
60632	175		246	2020	
60650	5635	3533			
60659	128		243	2030	
60663	5626	3633			
60667	5675	3520/5677			
60673	5642	3507			
60677	128		243	2030	
60678	196		241	2020	
60679	152	3893	66	2048	
60680	128		243	2030	
60682	128		243	2030	
60684	154		40	2012	
60685	124		12		
60694	5626	3507			
60697	128		243	2030	
60700	128		243	2030	
60701	129		244	2027	
60703	128		243	2030	
60710	130		14	2041	
60719	152	3893	66	2048	
60719-1	159	3466	82	2032	
60720	128		243	2030	
60732	5626	3633			
60733	5626	3633			
60741	5642	3507			
60761	124		12		
60770	104	3719	16	2006	
60784	124		12		
60793	220	3990		2036	
60806	5642	3507			
60810	181		75	2041	
60816	5635	3533			
60817	5635	3533			
60833	5635	3533			
60835	152	3893	66	2048	
60837	130		14	2041	
60838	152	3893	66	2048	
60879	5643	3519			
60885	181	3036	75	2041	
60886	152	3893	66	2048	
60916	5635	3533			
60944	130		14	2041	
60947	129		244	2027	
60966	152		66	2048	
60968	5675	3893/152			
60973	130		14	2041	
60977	196		241	2020	
60978	124		12		
60987	152	3893	66	2048	
60991	181		75	2041	
60994	128		243	2030	
60998	5643	3519			
61003-4	158	3004/102A	53	2007	
61007	390		255	2041	
61008-8	102A	3004	53	2007	
61008-8-1	102A		53	2007	
61008-8-2	102A		53	2007	
61009-1	108	3018	86	2038	
61009-1-1	108	3452	86	2038	
61009-1-2	108	3452	86	2038	
61009-2	108	3018	86	2038	
61009-2-1	108	3452	86	2038	
61009-4	123A	3444	20	2051	
61009-4-1	123A	3444	20	2051	
61009-6	108	3452	86	2038	
61009-6-1	108	3452	86	2038	
61009-9	129	3025	244	2027	
61009-9-1			244	2027	
61009-9-2	129		244	2027	
61009-9-3	102A		53	2007	
61010-0	108	3452	86	2038	
61010-0-1	108	3452	86	2038	
61010-2-1	126		52	2024	
61010-6	121	3009	239	2006	
61010-6-1	121	3717	239	2006	
61010-7-1	108	3452	86	2038	
61010-7-2	123A	3444	20	2051	
61011-0	129	3025	244	2027	
61011-0-1	129		244	2027	
61011-2	116	3016	504A	1104	
61011-3-2	123A	3444	20	2051	
61012	130		14	2041	
61012-4-1	101		8	2002	
61012-5-1	160		245	2004	
61013-2-1	123A	3444	20	2051	
61013-4-1	129		244	2027	
61013-9-1	108	3452	86	2038	
61015-0-1	108	3452	86	2038	
61019	130		14	2041	
61026	311	3195	277		
61035	196		241	2020	
61049	123A	3444	20	2051	
61086-1	233		210	2009	
61088	5801	3848/5806		1142	
61093	5642	3507			
61096	5642	3507			
61102	152	3893	66	2048	
61102-0	126	3008	52	2024	
61104	5643	3519			
61122	5529	6629			

Industry Standard No.	ECG	SK	GE	RS 276-	MOTOR.
61133	108	3452	86	2038	
61134	5643	3519			
61157	5642	3507			
61158	5642	3507			
61163	551	3857/231			
61173	181		75	2041	
61189	5675	3520/5677			
61190	5642	3506			
61193	152	3893	66	2048	
61209	128		243	2030	
61218	127	3764	25		
61219	128		243	2030	
61224	5642	3507			
61234	130		14	2041	
61239	187	3193	29	2025	
61242	186	3192	28	2017	
61244	129		244	2027	
61252	196		241	2020	
61260	5496	6796			
61273	311	3195	277		
61275	128		243	2030	
61282	311		277		
61285	152	3893	66	2048	
61286	152	3893	66	2048	
61291	5642	3507			
61295	5675	3520/5677			
61359	128		243	2030	
061366	289A	3444/123A	17	2051	
61367	130		14	2041	
61369/4367	130		14	2041	
61370/4560	128		243	2030	
61371/4561	129		244	2027	
61376	5635	3533			
61384	5695	3509			
61399	5626	3507			
61408	5529	6629			
61418	152	3893	66	2048	
61436	156	3032A	512		
61443	1003	3288	IC-43		
61451	181		75	2041	
61452	5635	3533			
61456	181	3036	75	2041	
61532	5675	3520/5677			
61534	196		241	2020	
61538	128	3103A/396	243	2030	
61558	161	3716	39	2015	
61562	128		243	2030	
61566	722	3161	IC-9		
61596	5675	3520/5677			
61612	5697	3522			
61613	5697	3522			
61636	152	3893	66	2048	
61652	5697	3522			
61661	108	3452	86	2038	
61663	108	3452	86	2038	
61664	5642	3507			
61666	129		244	2027	
61667	128		243	2030	
61733	128		243	2030	
61755	161	3018	39	2015	
61762	5633	3938			
61770	5642	3507			
61772	152	3893	66	2048	
61774	129		244	2027	
61807	5802	9005		1143	
61819	5547	3505			
61828	128	3024	243	2030	
61831	5635	3533			
61832	5697	3522			
61841	128		243	2030	
61865			69	2049	
61866			66	2048	
61868	130	3036	14	2041	
61872	5635	3533			
61875	152	3893	66	2048	
61917	128		243	2030	
61926			241	2020	
61932	5675	3520/5677			
61937	129		244	2027	
61957	5529	6629			
61958	152	3893	66	2048	
61965	124		12		
61981	152	3893	66	2048	
61997	196		241	2020	
62004	130		14	2041	
62005-1	130		14	2041	
62013	124		12		
62019	128		243	2030	
62032	102A		53	2007	
62075	5529	6629			
62119	124		12		
62140	125		510,531	1114	
62142	175		246	2020	
62143	130		14	2041	
62144	196		241	2020	
62156	152	3893	66	2048	
62177	121	3717	239	2006	
62185	128		243	2030	
62191	5635	3533			
62192	128		243	2030	
62193	5697	3522			
62203	192		63	2030	
62204	129		244	2027	
62207	5642	3507			
62211	5635	3533			
62212	5675	3520/5677			
62221			57	2017	
62229	124		12		
62232	5642	3507			
62238	5635	3533			
62243	128		243	2030	
62274	5642	3507			
62277	197	3083	250	2027	
62279	197		250	2027	
62282	5529	6629			
62287	181	3036	75	2041	
62334	5675	3520/5677			
62342	5635	3533			
62382	130		14	2041	
62398	128		243	2030	
62404	128		243	2030	
62446	128		243	2030	
62449	108	3452	86	2038	
62452	128		243	2030	
62457	5642	3507			
62511	196		241	2020	
62512	197		250	2027	
62540	128		243	2030	
62547	238	3710	259		
62571	152	3893	62	2048	
62584	129	3513	244	2027	
62612	128		243	2030	
62660	175		246	2020	
62665	124		12		
62676			66	2048	
62681	152	3893	66	2048	
62689	124		12		
62708	129		244	2027	
62746	5643	3519			
62759	197		250	2027	
62763	175		246	2020	
62792	130		14	2041	
62950	186	3192	28	2017	
63234-1	909		IC-249		
63282	159	3466	82	2032	
63841					MCM7685
63900-229	128		243	2030	
64071-1	102	3004	2	2007	
64074			218	2026	
64075			69	2049	
64076			66	2048	
64090	230	3042			
64197			69	2049	
65804-62	160		245	2004	
65804-63	102A		53	2007	
66007-4	128	3024	243	2030	
66008-2	102A	3004	53	2007	
66009-5	121	3014	239	2006	
66010-3	121	3009	239	2006	
66600		3126	90		
66682-42	116	3016	504A	1104	
67001	128		243	2030	
67003	128		243	2030	
67055	177	3100/519	300	1122	
67085-0	121MP	3013	239(2)	2006(2)	
67085-0-1	121MP	3718	239(2)	2006(2)	
67193-82	100	3721	1	2007	
67193-85	102	3004	2	2007	
67401	173BP	3710/238			
67544	136A	3710/238	ZD-5.6	561	
67586	123A	3260	20	2051	
67590	109	3088		1123	
67599	158	3004/102A	53	2007	
67802	108	3452	86	2038	
68177	116	3016	504A	1104	
68177A	116	3016	504A	1104	
68504-62	160	3006	245	2004	
68504-63	102A	3004	53	2007	
68504-76	110MP			1123	
68504-77	109	3087		1123	
68504-78	116	3016	504A	1104	
68617	123A	3444	20	2051	
68645-03	116		504A	1104	
68895-13	102	3004	2	2007	
68994-1	941		IC-263	010	
69107-42	126	3006/160	52	2024	
69107-43	160	3006	245	2004	
69107-44	126	3006/160	52	2024	
69107-45	102	3004	2	2007	
69111	116	3016	504A	1104	
69213-78	116	3016	504A	1104	
70008-0	130	3027	14	2041	
70008-3	130	3027	14	2041	
70019-1	175	3026	246	2020	
70019-5	175	3026	246	2020	
70023-0-00	123A	3444	20	2051	
70023-1-00	123A	3444	20	2051	
70055	177		300	1122	
70064-7	116	3016	504A	1104	
70066-3	116	3016	504A	1104	
70066-4	116	3016	504A	1104	
70087-31	128		243	2030	
70158-9-00	129		244	2027	
70167-8-00	108	3452	86	2038	
70177A01	798	3216	IC-234		
70177A02	748	3236			
70177A03	738	3167	IC-29		
70177A05	749	3168	IC-97		
70177B02	748	3236			
70205-8A	125	3032A	510,531	1114	
70231	108	3452	86	2038	
70260-11	108	3452	86	2038	
70260-12	108	3452	86	2038	
70260-13	108	3452	86	2038	
70260-14	123A	3444	20	2051	
70260-15	123A	3444	20	2051	
70260-16	123A	3444	20	2051	
70260-18	130		14	2041	
70260-19	129		244	2027	
70260-20	123A	3444	20	2051	
70260-29	129		244	2027	
70270-05	125		510,531	1114	
70270-23	116		504A	1104	
70270-38	125		510,531	1114	
70372-2	116	3016	300	1122	
70398-1	101	3444/123A	20	2051	
70399-1	6400A			2029	
70399-2	6401			2029	
70432-16	116		504A	1104	
70434	121	3009	239	2006	
70511	123A	3444	20	2051	
70581-1	119		300	1122	
71006-28	116		504A	1104	
71006-30	116		504A	1104	
000071090	131	3198	44	2006	
71119-1	177		300	1122	
71119-2	519	3087		1123	
71119-3	519	3016			
000071120	187	3193	29	2025	
000071130	187A	3193/187	248		
000071131	187A	9076	248		
000071150	159	3466	82	2032	
000071151	159	3466	82	2032	
71193-2	102A	3004	53	2007	
71226-1	289A	3444/123A	20	2051	
71226-10	123A	3444	20	2051	
71226-115	289A	3019			
71226-15	289A	3444/123A	20	2051	
71226-2	289A	3444/123A	20	2051	
71226-3	289A	3444/123A	20	2051	
71226-4	289A	3444/123A	20	2051	
71226-5	289A	3122	20	2051	
71226-6	289A	3444/123A	20	2051	
71266-4	123A		20	2051	
71411-1	141A	3092	ZD-11.5		
71411-2	142A	3062	ZD-12	563	
71411-3	5077A	3752	ZD-18		
71411-4	5079A	3335	ZD-20		
71411-5	5081A		ZD-24		
71411-6	146A	3064			
71412-4	289A	3444/123A	20	2051	
71412-5	289A	3444/123A	20	2051	
71447-1	192A	3024/128	243	2030	
71447-2	192A	3444/123A	20	2051	
71447-3	192A	3024/128	243	2030	
71448	121	3717	239	2006	
71448-1	121	3009	239	2006	
71448-2	121	3717	239	2006	

Industry Standard No.	ECG	SK	GE	RS 276-	MOTOR.
71448-3	121	3717	239	2006	
71448-4	121	3717	239	2006	
71448-5	121	3717	239	2006	
71448-6	121	3717	239	2006	
71448-7	121	3717	239	2006	
71449	5802	9005		1143	
71449-1	5801	9005/5802	510	1143	
71467-1	177	3087	300	1122	
71488-4	121	3009	239	2006	
71488-5	121	3009	239	2006	
71488-6	121	3009	239	2006	
71588-1	116	3016	504A	1104	
71588-2	116	3016	504A	1104	
71588-5	116	3016	504A	1104	
71588-6	116	3016	504A	1104	
71667-1	177		300	1122	
71677-1	519(2)	3016			
71686-4	457	3112	FET-1	2028	
71686-5	459	3116	FET-1	2028	
71686-6	459	3116	FET-1	2035	
71687	159		67	2023	
71687-1	159	3114/290	67	2023	
71687-101	159		67	2023	
71748-1	312		FET-1	2028	
71778	109			1123	
71779	116	3311	504A	1104	
71780	5009A	3775	ZD-4.7		
71783	169	3678	BR-600		
71794	156	3311	512		
071818	129		244	2027	
71818-1	159	3466	82	2032	
71819-1	123A	3444	20	2051	
71961-1	128	3024			
71962-1	129	3025			
71963-1	199	3444/123A	20	2051	
71964-1	130	3027			
71965-1	177	3016			
72003	176	3123	80		
72004	176	3123	80		
72006	176	3123	80		
72013	109	3088		1123	
000072020	116	3031A	42	1104	
72025			504A	1104	
72026			504A	1104	
72036	115		6GX1		
72041	116	3016	504A	1104	
000072050	116	3311	504A	1104	
72053	113A	3119/113	6GC1		
72054-1	172A	3156	64		
72058	116	3017B/117	504A	1104	
72058A	116	3016	504A	1104	
72059	358		42		
72060	109	3016	CR-3	1123	
72079-1	116	3016			
72079-2	125	3016			
72079-3	125	3031A			
72079-4	125	3016			
72080	109	3087		1123	
72089			1N34AS	1123	
000072090	109	3088		1123	
72097	116	3016	504A	1104	
72101	116	3016	504A	1104	
72103	116	3016	504A	1104	
72104			1N34AS	1123	
72106		3126	90		
72109	116	3110/120	504A	1104	
72110	118	3066	CR-1		
72111	119	3109	CR-2		
72113	116	3016	504A	1104	
72113A	116	3016	504A	1104	
72114	123A	3444	20	2051	
72115	123A	3444	20	2051	
72116	123A	3444	20	2051	
72117	102A	3004	53	2007	
72119	116	3016	504A	1104	
72123	116	3016	504A	1104	
72123A	116	3016	504A	1104	
72128	110MP	3091	1N34AS	1123(2)	
72128A	109	3088		1123	
72129	109	3091		1123	
72129A	109	3087		1123	
000072130	177	3175	504A	1122	
72130-1	116	3017B/117	504A	1104	
72135	138A	3059	ZD-7.5		
72135-C	138A		ZD-7.5		
72137	506	3130	511	1114	
72145	116	3017B/117	504A	1104	
72145A	116(5)		504A(5)		
72146	177	3100/519	300	1122	
72147	177	3100/519	300	1122	
72148	113A	3119/113	6GC1		
000072150	139A	3004/102A	53	562	
72151	123A	3444	20	2051	
72151(RECTIFIER)	156		512		
72152	178MP	3100/519	300(2)	1122(2)	
72152(TELEDYNE)			300	1122	
72158	116	3100/519	504A	1104	
000072160	109	3088		1123	
72161-1	506	3130	511	1114	
72162-1	522	3303	523		
72162-2	500A	3303/522	523		
72163-1	178MP	3100/519	300(2)	1122(2)	
72165-1	5097A		ZD-120		
72165-3	5068A	3332	ZD-4.3		
72165-5	5099A	3349	ZD-140		
72165-6	5101A	3351	ZD-160		
72167-2	5801	3051/156			
72168-2	5081A	3151	ZD-24		
72168-3	137A	3058	ZD-6.2	561	
72171-1	116		504A	1104	
72172-1	125	3032A	510,531	1114	
72174-1	166	9075	BR-600	1152	
72176	116	3311	504A	1104	
72180	523	3136/5072A	528	562	
72185	7476	7476		1813	
72190	158	3060/139A	53	562	
72191	103A	3835	59	2002	
72193	131		44	2006	
72197	177		300	1122	
72204	123A	3444	20	2051	
72206	123A	3444	20	2051	
72207	123A	3444	20	2051	
72784-21	100	3005	1	2007	
72784-22	102A	3004	53	2007	
72784-23	102A	3004	53	2007	
72797-80	126		52	2024	
72797-81	160		245	2004	
72799-41	102A		53	2007	
72813-10	102A		53	2007	
72847-51	102A		53	2007	
72856-63	121	3009	239	2006	
72874-52	123A	3444	20	2051	
72879-39	126		52	2024	
72879-40	126		52	2024	
72923-08	160		245	2004	
72941-33	102A		53	2007	
72949-10	108	3452	86	2038	
72951-95	108	3452	86	2038	
72951-96	108	3452	86	2038	
72963-14	123A	3444	20	2051	
72979-80	108	3452	86	2038	
000073070	199	3124/289	62	2010	
000073080	199	3245	62	2010	
000073090	123A	3124/289	20	2051	
000073100	123A	3444	20	2051	
73100-9	102A	3004	53	2007	
000073110	128	3024	243	2030	
000073120	123A	3124/289	20	2051	
000073130	123A	3124/289	20	2051	
000073140	229	3246	61	2038	
000073230	123A	3018	20	2051	
000073231	123A	3018	20	2051	
000073280	186	3192	28	2017	
000073290	123A	3124/289	20	2051	
000073300	192	3137	63	2030	
000073301	192	3137	63	2030	
000073302	192	3137	63	2030	
000073303	192	3137	63	2030	
000073305	192		63	2030	
000073310	123A	3124/289	20	2051	
000073320	186	3192	28	2017	
000073332	123A	3018	20	2051	
000073333	123A	3444	20	2051	
000073350	199	3245	62	2010	
000073351	199	3124/289	62	2010	
000073360	199	3245	62	2010	
000073361	199	3124/289	62	2010	
000073370		3124	210	2051	
000073373	199	3245	62	2010	
000073374	199	3245	62	2010	
000073380	186	3192	28	2017	
000073381	186A	3357	247	2052	
000073390	123A	3124/289	20	2051	
000073391	123A	3124/289	20	2051	
74004-1	724	3525	IC-86		
74004-2	724	3525	IC-86		
000074010	1006	3358	IC-38		
000074020	1045	3102/710	IC-2		
000074030	1142	3485	IC-128		
74107PC	74107	74107			
74121PC	74121	74121			
74123	74123	74123	74123	1817	
74123PC	74123	74123		1817	
74141DC	74141	74141			
74141PC	74141	74141			
74145	74145		74145	1828	
74145DC	74145	74145	74145	1828	
74145PC	74145	74145	74145	1828	
74150			74150	1829	
74150DC	74150	74150		1829	
74150PC	74150	74150		1829	
74151DC	74151	74151			
74151PC	74151	74151			
74153DC	74153	74153			
74153PC	74153	74153			
74154			74154	1834	
74154DC	74154	74154		1834	
74154PC	74154	74154		1834	
74163N	74163	74163			
74164DC	74164	74164			
74164PC	74164	74164			
74192			74192	1831	
74192DC	74192	74192		1831	
74192PC	74192	74192		1831	
74193	74193		74193	1820	
74193DC	74193	74193		1820	
74193PC	74193	74193		1820	
74196			74196	1833	
74196DC	74196	74196		1833	
74196PC	74196	74196		1833	
74200-8	116	3016	504A	1104	
74200-9	116	3016	504A	1104	
74651-02	123A	3444	20	2051	
74662	181		75	2041	
075005	109			1123	
75107ADC					MC75107L
75107APC					MC75107P
75107BDC					MC75107L
75107BPC					MC75107P
75108ADC					MC75108L
75108APC					MC75108P
75108BDC					MC75108L
75108BPC					MC75108P
75110DC					MC758110L
75110PC					MC758110P
75121DC					MC8T13L
75121PC					MC8T13P
75122DC					MC8T14L
75122PC					MC8T14P
75123DC					MC8T23L
75123PC					MC8T23P
75124DC					MC8T24L
75124PC					MC8T24P
75145-3	128		243	2030	
75207DC					MC75107L
75207PC					MC75107P
75208DC					MC75108L
75208PC					MC75108P
75230-9	116	3016	504A	1104	
75325DC					MC75325L
75325PC					MC75325P
75491ADC					MC75491P
75491APC	75491B				MC75491P
75491DC					MC75491P
75491PC	75491B				MC75491P
75492ADC					MC75492P
75492APC	75492B				MC75492P
75492DC					MC75492P
75492PC	75492B				MC75492P
75561-1	159		ZD-82	2032	
75561-16	123A	3444	20	2051	
75561-18	234	3247	65	2050	
75561-2	159	3466	82	2032	
75561-20	128		243	2030	
75561-21	234	3247	65	2050	
75561-28	123A	3444	20	2051	
75561-3	123A	3444	20	2051	
75561-31	159	3466	82	2032	
75561-32	128		243	2030	
75561-33	123A	3444	20	2051	
75568-3	107		11	2015	
75596-1	234	3247	65	2050	
75596-2	234	3247	65	2050	
75596-3	234	3247	65	2050	
75596-4	234	3247	65	2050	
75613-1	199		62	2010	
75613-2	199		62	2010	
75614-1	123A	3444	20	2051	

Industry Standard No.	ECG	SK	GE	RS 276-	MOTOR.
75616-6	108	3452	86	2038	
75617-1	159	3466	82	2032	
75617-2	159	3466	82	2032	
75700-03-01	121	3717	239	2006	
75700-04	123A	3444	20	2051	
75700-04-01	123A	3444	20	2051	
75700-05	123A	3444	20	2051	
75700-05-01	123A	3444	20	2051	
75700-05-02	123A	3444	20	2051	
75700-05-03	123A	3444	20	2051	
75700-08	123A	3444	20	2051	
75700-08-02	123A	3444	20	2051	
75700-09-01	123A	3444	20	2051	
75700-09-21	123A	3444	20	2051	
75700-13-01	129		244	2027	
75700-14-02	116		504A	1104	
75700-14-05	125		510,531	1114	
75700-22-01	175		246	2020	
75700-24-02	125		510,531	1114	
75702-15-21	116		504A	1104	
75702-15-24	117		504A	1104	
75702-15-34	125		510,531	1114	
75803-1	153		69	2049	
75803-2	153		69	2049	
75803-3	153		69	2049	
75810-17	108	3452	86	2038	
75960CH	102	3004	2	2007	
76236	123A	3444	20	2051	
76251	390		255	2041	
76600P	746	3234	IC-217		
76675	109			1123	
76675A	109			1123	
76675B	109			1123	
76797-2	923		IC-259		
77001	230	3042			
77052-3	102A	3004	53	2007	
77052-4	102A	3004	53	2007	
77053-2	102A	3004	53	2007	
77068-3170756	124		12		
77190-7	116	3016	504A	1104	
77190-8	125	3031A	510,531	1114	
77271-3	116	3311	504A	1104	
77271-4	125	3031A	510,531	1114	
77271-8	160	3006	245	2004	
77272-0	102A	3004	53	2007	
77272-1	102A	3004	53	2007	
77272-5	102A	3004	53	2007	
77272-7	102A	3004	53	2007	
77272-9	102A	3004	53	2007	
77273-2	102A	3004	53	2007	
77273-3	102A	3004	53	2007	
77273-6	102A	3004	53	2007	
77273-7	102A	3004	53	2007	
77489	112		1N82A		
77561-27	159	3466	82	2032	
78001	231	3857			
78331	159	3466	82	2032	
78399	130		14	2041	
78524-39-01	125		510,531	1114	
78527-75-01	102A		53	2007	
78527-76-01	102A		53	2007	
78527-78-01	102A		53	2007	
78527-79-01	102A		53	2007	
78894	116		504A	1104	
78896	358		42		
78972	112		1N82A		
79408	139A	3060	ZD-9.1	562	
79855	108	3452	86	2038	
79856	108	3452	86	2038	
79922	175	3131A/369	246	2020	
79949	148A	3096	ZD-55		
79985	109			1123	
79992	152	3893	66	2048	
080001	126	3008	52	2024	
080003	102A	3004	53	2007	
080004	102A	3004	53	2007	
080006	108	3452	86	2038	
080021	108	3452	86	2038	
080022	108	3452	86	2038	
080023	108	3452	86	2038	
080026	160	3006	245	2004	
080027	160	3006	245	2004	
080028	160	3006	245	2004	
080040	116	3016	504A	1104	
080041	108	3452	86	2038	
080042	108	3452	86	2038	
080043	102A	3004	53	2007	
080047	102A	3004	53	2007	
080048	121	3009	239	2006	
080050	116	3009	239	2006	
080052	102A	3004	53	2007	
80053	704	3022/1188	IC-205		
080059	108	3452	86	2038	
080060	108	3452	86	2038	
080061	160	3006	245	2004	
80070	704	3022/1188	IC-205		
80071	704	3006/160	IC-205	2004	
080072	126	3008	52	2024	
80073	704	3023	IC-205	2007	
80074	704	3022/1188	IC-205		
80081	726	3022/1188			
80083	704	3022/1188	IC-205		
80090	704	3022/1188	IC-205		
80091	76				MRF511
80094	704	3022/1188	IC-205		
80099					MRF525
80114	706	3006/160	IC-43	2004	
80131	241	3188A/182	57	2020	
80167	76				MRF511
080206	126	3008	52	2024	
080224	126	3007	52	2024	
080225	126	3006/160	52	2024	
080228	126	3007	52	2024	
80231	76				MRF511
080236	126	3008	52	2024	
080244	160	3006	245	2004	
080245	160	3006	245	2004	
80249-910787	163A	3439	36		
080253	126	3008	52	2024	
080258	160	3007	245	2004	
080266	160	3006	245	2004	
080267	160	3006	245	2004	
080269	160	3006	245	2004	
080274	126	3005	52	2024	
080275	126	3008	52	2024	
080276	126	3008	52	2024	
080277	126	3006/160	52	2024	
80287	706	3288/1003	IC-43		
80416C	121	3014	239	2006	
80540	123A	3444	20	2051	
80544	123A	3444	20	2051	
80545	123A	3444	20	2051	
80710	712	3072	IC-2		
80755	781	3169	IC-223		
80757	727		IC-210		
80807	130		14	2041	
80813VM	123A	3444	20	2051	
80814VM	123A	3444	20	2051	
80815VM	123A	3444	20	2051	
80816VM	123A	3444	20	2051	
80817VM	103A	3010	59	2002	
80818VM	102A	3004	53	2007	
80827	706	3288/1003	IC-43		
80902-1	128		243	2030	
80904-1	128		243	2030	
081001	102A	3004	53	2007	
081018	102A	3004	53	2007	
081019	102A	3004	53	2007	
081026	158	3004/102A	53	2007	
081027	158	3004/102A	53	2007	
081029	158	3004/102A	53	2007	
081038	102A	3004	53	2007	
081042	131MP	3840	239(2)	2006(2)	
081046	102A	3004	53	2007	
081047	102A	3004	53	2007	
081048	102A	3004	53	2007	
081049	102A	3004	53	2007	
081050	102A	3004	53	2007	
081056	102A	3004	53	2007	
081059	158	3004/102A	53	2007	
81170-6	123A	3444	20	2051	
81177	731	3170	IC-13		
81336-1	727	3071	IC-210		
81336-2	727	3071	IC-210		
81404-4A	102A	3004	53	2007	
81410-145	199		62	2010	
81500-3	102A	3004	53	2007	
81501-5	102A	3004	53	2007	
81502-0	100	3005	1	2007	
81502-0A	100	3005	1	2007	
81502-0B	100	3005	1	2007	
81502-1	100	3005	1	2007	
81502-1A	100	3005	1	2007	
81502-1B	100	3005	1	2007	
81502-2	102A	3004	53	2007	
81502-2A	102A	3004	53	2007	
81502-2B	102A	3004	53	2007	
81502-3B	102A	3004	53	2007	
81502-4	102A	3004	53	2007	
81502-4A	102A	3004	53	2007	
81502-4B	102A	3004	53	2007	
81502-5	102A		53	2007	
81502-5A	100	3005	1	2007	
81502-5B	100	3005	1	2007	
81502-6	101	3861	8	2002	
81502-6A	101	3861	8	2002	
81502-6B	101	3861	8	2002	
81502-6C	101	3861	8	2002	
81502-6D	101	3861	8	2002	
81502-7	100	3005	1	2007	
81502-7A	100	3005	1	2007	
81502-7B	100	3005	1	2007	
81502-7C	100	3005	1	2007	
81502-8	100	3005	1	2007	
81502-8A	100	3005	1	2007	
81502-8B	100	3005	1	2007	
81502-8C	100	3005	1	2007	
81502-9	102A	3004	53	2007	
81502-9A	102A	3004	53	2007	
81502-9B	102A	3004	53	2007	
81502-9C	102A	3004	53	2007	
81503-0	102A	3004	53	2007	
81503-0A	102A	3004	53	2007	
81503-0B	102A	3004	53	2007	
81503-1	102A	3004	53	2007	
81503-1A	102A	3004	53	2007	
81503-1B	102A	3004	53	2007	
81503-3	102A		53	2007	
81503-3A	102A		53	2007	
81503-4	102A	3004	53	2007	
81503-4A	102A	3004	53	2007	
81503-4B	102A	3004	53	2007	
81503-4C	102A	3004	53	2007	
81503-6	100	3005	1	2007	
81503-6A	100	3005	1	2007	
81503-6B	100	3005	1	2007	
81503-6C	100	3005	1	2007	
81503-7	100	3005	1	2007	
81503-7A	100	3005	1	2007	
81503-7B	100	3005	1	2007	
81503-7C	100	3005	1	2007	
81503-8	100	3005	1	2007	
81503-8B	102A	3004	53	2007	
81503-8C	102A	3004	53	2007	
81504-1	100	3005	1	2007	
81504-1A	100	3005	1	2007	
81504-1B	100	3005	1	2007	
81504-1C	100	3005	1	2007	
81504-3	100	3005	1	2007	
81504-3A	100	3005	1	2007	
81504-3B	100	3005	1	2007	
81504-3C	100	3005	1	2007	
81505-8	102A	3004	53	2007	
81505-8A	102A	3004	53	2007	
81505-8B	102A	3004	53	2007	
81505-8C	102A	3004	53	2007	
81505-8X	102A	3004	53	2007	
81506-4	126	3008	52	2024	
81506-4A	126	3008	52	2024	
81506-4B	126	3008	52	2024	
81506-4C	126	3008	52	2024	
81506-5	100	3005	1	2007	
81506-5A	100	3005	1	2007	
81506-5B	100	3005	1	2007	
81506-5C	100	3005	1	2007	
81506-6	100	3005	1	2007	
81506-6A	100	3005	1	2007	
81506-6B	100	3005	1	2007	
81506-6C	100	3005	1	2007	
81506-7	126	3009	52	2024	
81506-7A	126	3008	52	2024	
81506-7B	126	3008	52	2024	
81506-7C	126	3008	52	2024	
81506-8	126	3008	52	2024	
81506-8A	126	3008	52	2024	
81506-8B	126	3008	52	2024	
81506-8C	126	3008	52	2024	
81506-9	102A	3004	53	2007	
81506-9A	102A	3004	53	2007	
81506-9B	102A	3004	53	2007	
81506-9C	102A	3004	53	2007	
81507-0	102A	3004	53	2007	
81507-0A	102A	3004	53	2007	
81507-0B	102A	3004	53	2007	
81507-0C	102A	3004	53	2007	
81507-0D	102A	3004	53	2007	
81507-4	102A	3004	53	2007	
81507-5	103A	3010	53	2002	

Industry Standard No.	ECG	SK	GE	RS 276-	MOTOR.
81507-6	103A	3010	59	2002	
81510-3	100	3005	1	2007	
81510-4	102A	3004	53	2007	
81510-5	100	3005	1	2007	
81511-4	102A	3004	53	2007	
81511-5	100	3005	1	2007	
81511-6	100	3005	1	2007	
81511-7	100	3005	1	2007	
81511-8	102A	3004	53	2007	
81512-0	102A	3004	53	2007	
81512-0A	102A	3004	53	2007	
81512-0B	102A	3004	53	2007	
81512-0C	102A	3004	53	2007	
81512-0D	102A	3004	53	2007	
81512-0E	102A	3004	53	2007	
81513-3	123A	3444	20	2051	
81513-6	102A	3004	53	2007	
81513-7	121	3009	239	2006	
81513-8	116	3016	504A	1104	
81513-9	102A	3003	53	2007	
81514-2	116	3016	504A	1104	
81515-8	102A	3004	53	2007	
81516-0	102A	3004	53	2007	
81516-0A	102A	3004	53	2007	
81516-0B	102A	3004	53	2007	
81516-0C	102A	3004	53	2007	
81516-0D	102A	3004	53	2007	
81516-0E	102A	3004	53	2007	
81516-0F	102A	3004		2007	
81516-0G	102A	3004	53	2007	
81516-0H	102A	3004	53	2007	
81516-0I	102A	3004	53	2007	
81516-0J	102A	3004	53	2007	
082006	123A	3444	20	2051	
082019	123A	3444	20	2051	
082020	233	3018	210	2009	
082022	195A	3048/329	46		
082025	237	3299	46		
082028	289A	3138/193A	268	2038	
082029	237	3299	46		
082033	195A	3048/329	46		
82716	108	3452	86	2038	
82965F	5635	3533			
83008	116	3016	504A	1104	
83272	159	3466	82	2032	
83741C01	5465	3687			
84001	129	3025	244	2032	
84001A	129	3025	244	2027	
84001B	129	3025	244	2027	
084001C	121	3114/290	239	2006	
84011	4011B		4011	2411	
84011U	4011B		4011	2411	
84380-1	911		IC-253		
085002	109	3087		1123	
085003	177	3100/519	300	1122	
085004	109	3087		1123	
085005	110MP	3709	1N34AS	1123(2)	
085006	109	3087		1123	
085016	109	3087		1123	
085026	110MP	3709	1N34AS	1123(2)	
85549	123A	3444	20	2051	
86001	109			1123	
86257	129		244	2027	
86287	123A	3444	20	2051	
86313	786	3140	IC-227		
86452	102	3004	2	2007	
86458	124		12		
86458-00	124		12		
86812	123	3124/289	20	2051	
86822	123	3124/289	20	2051	
86832	102	3004	2	2007	
86842	102	3004	2	2007	
087003	102A	3004	53	2007	
87532	128		243	2030	
87756	177	3100/519	300	1122	
87757	123A	3444	20	2051	
87758	159	3466	82	2032	
87759	159	3466	82	2032	
88060-141	312	3448	FET-1	2035	
88060-142	229	3018	61	2038	
88060-143	123A	3444	20	2051	
88060-144	199	3124/289	62	2010	
88060-145	128	3124/289	243	2030	
88060-146	186	3192	28	2017	
88060-147	1100	3223	IC-92		
88060-51		3126	90		
88060-52	177	3100/519	300	1122	
88060-53	110MP	3088	1N60	1123(2)	
88060-54	5072A	3136	ZD-8.2	562	
88060-55	177	3100/519	300	1122	
88510-171	312	3834/132	FET-1	2035	
88510-172	107	3018	11	2015	
88510-173	123A	3444	20	2051	
88510-174	199	3124/289	62	2010	
88510-175	123A	3444	20	2051	
88510-176	186	3192	28	2017	
88510-177	1100	3223	IC-92		
88510-178	1142	3485	IC-128		
88510-51		3126	90		
88510-52	177	3100/519	300	1122	
88510-53	110MP	3088	1N60	1123(2)	
88510-54	5072A		ZD-8.2	562	
88510-55	177		300	1122	
88641	116	3311	504A	1104	
88686	123	3024/128	20	2051	
88687	123A	3444	20	2051	
88688	123A	3444	20	2051	
88796-1-1	5529	6629			
88801-3-1	175		246	2020	
88803	128	3024	243	2030	
88803-2-1	128		243	2030	
88803-3-1	128		243	2030	
88832	121	3009	239	2006	
88834	128	3024	243	2030	
88862	123A	3444	20	2051	
89999-205			504A	1104	
90050	121	3009	239	2006	
90203-6	116	3032A	504A	1104	
90209-172	123A	3444	20	2051	
90209-182	123A	3444	20	2051	
90209-246	234		65	2050	
90326-001	123A	3444	20	2051	
90330-001	159	3466	82	2032	
90429	123A	3444	20	2051	
90431	194		220		
90432	159	3466	82	2032	
90448	194		220		
90934-35	199	3245	62	2010	
91001	116	3016	504A	1104	
91021	101	3861	8	2002	
91271	128		243	2030	
91272	128		243	2030	
91273	128		243	2030	
91274	130		14	2041	
91371	198	3220	251		
91411	128		243	2030	
91605	123A	3444	20	2051	
92138-01	941		IC-263	010	
93005	116	3016	504A	1104	
93006	116	3016	504A	1104	
93007	116	3016	504A	1104	
93011	116	3016	504A	1104	
93012	116	3016	504A	1104	
93018	5072A		ZD-8.2	562	
93022	116	3017B/117	504A	1104	
93022A	116	3017B/117	504A	1104	
93023	116	3016	504A	1104	
93027	117	3017B	504A	1104	
93028	116	3017B/117	504A	1104	
93030	140A	3061	ZD-10	562	
93141DC	74141	74141			
93141PC	74141	74141			
93145DC	74145	74145		1828	
93145PC	74145	74145		1828	
93150DC	74150	74150		1829	
93150PC	74150	74150		1829	
93151DC	74151	74151			
93151PC	74151	74151			
93153DC	74153	74153			
93153PC	74153	74153			
93154DC	74154	74154		1834	
93154PC	74154	74154		1834	
93164DC	74164	74164			
93164PC	74164	74164			
93196DC	74196	74196		1833	
93196PC	74196	74196		1833	
93415					MCM93415
93422					MCM93422
93425					MCM93425
93448					MCM7641
93451					MCM7681
93453					MCM7643
94000	102A		53	2007	
94001	102A		53	2007	
94002	102A		53	2007	
94003	158		53	2007	
94004	104	3009	16	2006	
94005	158		53	2007	
94006	102A		53	2007	
94007	160		245	2004	
94008	102A	3004	53	2007	
94009	158	3004/102A	53	2007	
94010	179	3642	76		
094013	121	3717	245	2004	
94014	102A		53	2007	
94015	102A		53	2007	
94016	102A		53	2007	
94017	158	3004/102A	53	2007	
94018	158		53	2007	
94019	158		53	2007	
94020	158		53	2007	
94021	158		53	2007	
94022	158		53	2007	
94023	103A	3835	59	2002	
94024	121	3009	239	2006	
94025	121	3009	239	2006	
94026	121	3009	239	2006	
94027	123A	3444	20	2051	
94028	160		245	2004	
94029	103A	3835	59	2002	
94030	102A		53	2007	
94032	121	3009	239	2006	
94033	160	3007	245	2004	
94034	121	3009	239	2006	
94035	160		245	2004	
94036	160	3007	245	2004	
94037	159	3466	82	2032	
94037(EICO)	102A		53	2007	
94038	126	3008	52	2024	
94038(EICO)	102A		53	2007	
94039	102A	3004	53	2007	
94040	121	3009	239	2006	
94041	128	3024	243	2030	
94042	128	3047	243	2030	
94043	195A	3048/329	46		
94044	108	3452	86	2038	
94047	123A	3444	20	2051	
94048	123A	3444	20	2051	
94049	175	3026	246	2020	
94050	128	3024	243	2030	
94051	128	3024	243	2030	
94051(EICO)	154		40	2012	
94052	128	3025/129	243	2030	
94062	192	3512	63	2030	
94063	129	3025	244	2027	
94064	129	3025	244	2027	
94065	130	3027	14	2041	
94065A	130	3027	14	2041	
94066	128	3024	243	2030	
94067	129		244	2027	
94068	129	3513	244	2027	
94070	128	3024	243	2030	
94094	175	3026	246	2020	
94094(EICO)	130		14	2041	
94152	7408	7408		1822	
94330	910		IC-251		
94835-145-00	128		243	2030	
95000	109	3087		1123	
95001	109			1123	
95002		3087		1123	
95003	177	3100/519	300	1122	
95004	109	3087		1123	
95007	109	3087		1123	
95008	109			1123	
95012	177		300	1122	
95014	109	3087		1123	
95015	116	3017B/117	504A	1104	
95015(EICO)	177		300	1122	
95016	117		504A	1104	
95017	109	3087		1123	
95018	109	3087		1123	
95101	160	3007	245	2004	
95102	160	3008	245	2004	
95103	160	3005	245	2004	
95107	160	3006	245	2004	
95108	160	3006	245	2004	
95109				2005	
95110	160	3006	245	2004	
95111	160	3008	245	2004	
95112	101	3861	8	2002	
95113	101	3861	8	2002	
95114	101	3861	8	2002	
95115	101	3861	8	2002	
95116	126	3006/160	52	2024	
95117	101	3861	8	2002	
95118	126	3006/160	52	2024	
95119	126	3006/160	52	2024	
95120	126	3006/160	52	2024	

Industry Standard No.	ECG	SK	GE	RS 276-	MOTOR.
95120A	100	3005	1	2007	
95121	160	3006	245	2004	
95122	126	3006/160	52	2024	
95123	126	3006/160	52	2024	
95124		3006	51	2004	
95125	108	3452	86	2038	
95126	108	3452	86	2038	
95127	108	3452	86	2038	
95128	108	3452	86	2038	
95129	108	3452	86	2038	
95130	108	3452	86	2038	
95131	108	3452	86	2038	
95132	222	3050/221	FET-4	2036	
95133	312	3112	FET-1	2028	
95170-1	229	3018	20	2038	
95170-2	229	3452/108	20	2038	
95170-2(DIO)			1N60	1123	
95170-2(XSTR)			20	2038	
95171-1	289A	3124/289	86	2038	
95171-2	289A	3124/289	20	2051	
95171-3	289A	3124/289	268	2038	
95171-4	289A	3124/289	11	2038	
95172-2	102A		53	2007	
95173-1	102A		53	2007	
95201	102	3004	2	2007	
95202	103	3862	8	2002	
95203	102	3003	2	2007	
95204	102	3004	2	2007	
95208	102	3004	2	2007	
95209	102	3004	2	2007	
95211	101	3861	8	2002	
95212	102	3004	2	2007	
95213	102	3004	2	2007	
95214	102	3004	2	2007	
95216	123	3124/289	20	2051	
95216RED	123		20	2051	
95216YEL	123		20	2051	
95217	102	3004	2	2007	
95218	102	3004	2	2007	
95219	102	3004	2	2007	
95220	128	3024	243	2030	
95221	289A	3122	20	2038	
95222-1	102	3004	2	2007	
95222-2	103	3862	8	2002	
95223	123A	3444	20	2051	
95224-1	102A	3004	53	2007	
95224-2	103A	3010	8	2002	
95224-3	102A	3004	53	2007	
95224-4	103A	3010	59	2002	
95225	123	3124/289	20	2051	
95226-003	193		67	2023	
95226-004	192		63	2030	
95226-1	128	3025/129	243	2030	
95226-2	123	3124/289	20	2051	
95226-3	288	3025/129	223		
95226-4	287	3433	222		
95227	159	3466	82	2032	
95227-1	159	3466	82	2032	
95228	312	3112		2028	
95229	106	3984	21	2034	
95231	312	3112	FET-1	2028	
95232	159	3466	82	2032	
95233	128		243	2030	
95235	5400	3950			
95239-1	129	3114/290	82	2027	
95240-1	159	3466	82	2032	
95241-1	294	3841	48		
95241-3	294	3024/128	243	2030	
95242-1	108	3020/123	20	2038	
95250	121	3009	239	2006	
95250-1	121	3013	239	2006	
95251	121	3009	239	2006	
95252	124	3021	12		
95252-1	124	3021	12		
95252-2	124	3021	12		
95252-3	124	3021	12		
95252-4	124	3021	12		
95253		3009	3	2006	
95255-000	102	3722	2	2007	
95257	104MP	3720	16(2)	2006(2)	
95258-1	188	3054/196	226	2018	
95258-2	189	3083/197	218	2026	
95261-1	152	3893	66	2048	
95261-2	153	3083/197			
95262-1	184	3190	57	2017	
95262-2	185	3191	58	2025	
95263-1	184	3054/196	57	2017	
95263-2	185	3083/197	58	2025	
95285-1	188		226	2018	
95286	743		IC-214		
95291	722	3161	IC-9		
95292	722		IC-9		
95293	804	3455	IC-27		
95294	722	3161	IC-9		
95296	172A	3156	64		
95298	704	3023	IC-205		
95332-1	177		300	1122	
96147-001	5675	3508			
96148-001	5675	3520/5677			
96457-1	123A	3444	20	2051	
96458-1	159	3466	82	2032	
96481	128		243	2030	
97202-222	5077A	3752	ZD-18		
97251-3	119	3109	CR-2		
97680	106	3984	21	2034	
98484-001	181	3036	75	2041	
98496-001	5675	3508			
98497-001	5675	3520/5677			
99023-3	116		504A	1104	
99101	126	3008	52	2024	
99102	126	3008	52	2024	
99103	160	3008	245	2004	
99104	160	3004/102A	245	2004	
99105	160	3006	245	2004	
99106	160	3006	245	2004	
99107	160	3006	245	2004	
99108	160	3006	245	2004	
99109-1	123	3124/289	20	2051	
99109-2	123	3124/289	20	2051	
99120	126	3006/160	52	2024	
99121	126	3007	52	2024	
99201	102A		53	2005	
99201-110	142A	3777/5011A	ZD-12	563	
99201-110(RCA)	113A		6GC1		
99201-208	137A	3775/5009A	ZD-6.2	561	
99201-210	136A	3777/5011A	ZD-5.6	561	
99201-211	137A	3779/5013A	ZD-6.2	561	
99201-212	137A	3780/5014A	ZD-6.2	561	
99201-216	140A	3785/5019A	ZD-10	562	
99201-219	142A	3788/5022A	ZD-12	563	
99201-221	145A	3791/5025A	ZD-15	564	
99201-228	147A	3802/5036A			
99201-242	5097A	3818/5052A	ZD-120		
99201-312	5071A	3780/5014A	ZD-6.8	561	

Industry Standard No.	ECG	SK	GE	RS 276-	MOTOR.
99201-316	140A	3785/5019A	Z-10	562	
99201-319	5022A	3788	ZD-13	563	
99201-325	5081A	3151	ZD-24		
99201-331		3805	ZD-43		
99202	102A		53	2007	
0099202-128	147A	3095	ZD-33		
99202-22		3752	ZD-15	564	
99202-220	143A	3063/145A	ZD-13	563	
99202-221	5075A	3751	ZD-16	564	
99202-222	5077A	3752	ZD-18		
99202-226	146A	3064	ZD-27		
99202-228	147A	3095	ZD-33		
99203	102A		53	2007	
99203-005	116	3313	504A	1104	
99203-006	116		504A	1104	
0099203-007	116	3313	504A	1104	
99203-3	177	3033A	300	1122	
99203-5	116	3033A	504A	1104	
99203-5(DIODE)	177		300	1122	
99203-6	116	3033A	504A	1104	
99203-6(RCA)	116		504A	1104	
99203-7	125	3033A	510,531	1114	
99203-8	125	3033A	510,531	1114	
99203-9	116	3033A	504A	1104	
99204	102A	3004	53	2007	
99205	102A	3004	53	2007	
99206-1	123A	3444	20	2051	
99206-2	123A	3444	20	2051	
99207-2	123A	3444	20	2051	
99210-204	134A		ZD-3.6		
99210-210	136A		ZD-5.6	561	
99217	102	3004	2	2007	
99218	102	3004	2	2007	
99240-269	199	3245	62	2010	
99240-292	219		74	2043	
99250	121	3009	239	2006	
99250-1			3	2006	
99252	124	3021	12		
99252-1	124	3021	12		
99252-2		3021	12		
99252-2(METAL)	124		12		
99252-2(PLASTIC)	157		232		
99252-3	124	3021	12		
99252-4	152	3893	66	2048	
100017	109	3088	1N60	1123	
100092	123A	3444	20	2051	
100093	123A	3444	20	2051	
100292	199	3245	62	2010	
100412	116	3017B/117	504A	1104	
100449	358		42		
100471	113A	3119/113	6GC1		
100520	116	3016	504A	1104	
100581	113A	3119/113	6GC1		
100617	116	3031A	504A	1104	
100624	116	3016	504A	1104	
100678	126	3007	52	2024	
100693	102A	3004	53	2007	
100768	5521	6621			
100802			20	2051	
100828-003	177		300	1122	
100844	109	3087		1123	
101078	160	3007	245	2004	
101087	160	3007	245	2004	
101089	126	3008	52	2024	
101119	123A	3444	20	2051	
101154	177	3100/519	300	1122	
101185	123A	3444	20	2051	
101403	116	3016	504A	1104	
101434	108	3018	86	2038	
101435	128	3047	243	2030	
101436	195A	3048/329	46		
101497	159	3466	82	2032	
101568	181		75	2041	
101678	101	3861	8	2002	
101973	102A	3004	53	2007	
101974	102A	3004	53	2007	
102001	159	3466	82	2032	
102002	123A	3444	20	2051	
102005	7490	7490		1808	
102009	177		300	1122	
102209	128		243	2030	
102249	358		42		
102260	159	3466	82	2032	
102263	159	3466	82	2032	
102989	109	3088		1123	
103318	116	3016	504A	1104	
103443	101	3861	8	2002	
103514	519		514	1122	
103521	123A	3444	20	2051	
103562	102A	3004	53	2007	
103872	113A	3119/113	6GC1		
104009	126	3007	52	2024	
104059	126	3007	52	2024	
104080	103A	3010	59	2002	
104081	116	3016	504A	1104	
104152	109	3087		1123	
104213	116	3016	504A	1104	
104273	116	3017B/117	504A	1104	
104325	116	3311	504A	1104	
104389	123A	3444	20	2051	
104444	102A	3004	53	2007	
104615	116	3311	504A	1104	
104719	128		243	2030	
104762	519		514	1122	
104825	1179	3737			
105064	120	3110	CR-3		
105180	128		243	2030	
105325	1155	3231			
105330	358		42		
105432	123A		20	2051	
105468	106	3984	21	2034	
105517	109	3087		1123	
105604	358	3110/120	42		
106379	117	3016	504A	1104	
106525	1167	3732			
106625	1166	3827			
106719	1110	3229			
107225	1192	3445			
107268	115	3121	6GX1		
107274	102	3004	2	2007	
107474	113A	3119/113	6GC1		
107540	116	3017B/117	504A	1104	
107625	1234		IC-181		
107628	115	3121	6GX1		
107729	112	3089	1N82A		
109308	113A		6GC1		
109328	113A	3119/113	6GC1		
109474	113A	3175	6GC1		
110043	116	3016	504A	1104	
110075	195A		46		
110263	158	3011	53	2007	
110351	146A	3016	ZD-27		
110388	116	3016	504A	1104	
110494	158	3004/102A	53	2007	

Industry Standard No.	ECG	SK	GE	RS 276-	MOTOR.
110495	103A	3010	59	2002	
110496	116	3016	504A	1104	
110515	104	3009	16	2006	
110610	109	3087		1123	
110629	116	3016	504A	1104	
110636	116	3016	504A	1104	
110667	519		514	1122	
110669	128	3024	243	2030	
110697	123A	3444	20	2051	
110699	128	3047	243	2030	
110873	116	3016	504A	1104	
110957	102A		53	2007	
110958	103A	3835	59	2002	
110959	102A		53	2007	
111001	102	3722	2	2007	
111011	102	3722	2	2007	
111012	102	3722	2	2007	
111013	102	3722	2	2007	
111086	116	3016	504A	1104	
111117	160	3007	245	2004	
111118	160	3007	245	2004	
111193-001	311	3024/128	277		
111207	109	3087		1123	
111278	128	3024	243	2030	
111279	195A	3048/329	46		
111303	123A	3444	20	2051	
111313	160	3007	245	2004	
111489	116	3313			
111516	116	3016	504A	1104	
111605	109	3087		1123	
111642	116	3016	504A	1104	
111776	116	3016	504A	1104	
111820	116	3016	504A	1104	
111943	108	3452	86	2038	
111945	129		244	2027	
111954	160	3006	245	2004	
111955	160	3006	245	2004	
111956	160	3006	245	2004	
111957	102A	3004	53	2007	
111958	103A	3010	59	2002	
111959	102A	3004	53	2007	
112001	160		245	2004	
112002	160		245	2004	
112011	160		245	2004	
112017	116	3016	504A	1104	
112018	116	3016	504A	1104	
112032	160		245	2004	
112041	126		52	2024	
112071	102	3722	2	2007	
112296	160	3008	245	2004	
112297	102A	3004	53	2007	
112329	116	3017B/117	504A	1104	
112329(AMPHENOL)	177		300	1122	
112330	109			1123	
112355	108	3452	86	2038	
112356	123A	3444	20	2051	
112357	123A	3444	20	2051	
112358	123A	3444	20	2051	
112359	123A	3444	20	2051	
112360	128	3024	243	2030	
112361	128	3024	243	2030	
112362	224	3049	46		
112520	123A	3444	20	2051	
112521	123A	3444	20	2051	
112522	123A	3444	20	2051	
112523	123A	3444	20	2051	
112524	110MP	3087	1N34AS	1123	
112525	188	3024/128	226	2018	
112526	109	3088		1123	
112527	224	3049	46		
112528	156	3016	512		
112529	109	3087		1123	
112529(AMPHENOL)	177		300	1122	
112530	139A	3060	ZD-9.1	562	
112531	116	3017B/117	504A	1104	
112826	116	3016	504A	1104	
0112945	166	9075		1152	
0112945(ELGIN)	166			1152	
113039	116	3016	504A	1104	
113182	159	3466	82	2032	
113321	116(3)	3110/120	CR-3		
113348	123A	3444	20	2051	
113391	119	3109	CR-1		
113392	116	3016	504A	1104	
113397	177	3066/118	CR-1		
113398	116	3019	504A	1104	
113438	123A	3444	20	2051	
113524	123A	3444	20	2051	
113875	130		14	2041	
113876	181		75	2041	
113938	108	3452	86	2038	
113942	128		243	2030	
113998	116	3843	511	1114	
114013	116(5)	3016	504A(5)	1104	
114143-1	108	3452	86	2038	
114267	222	3019	FET-4	2036	
114401	5529	6629			
114504	176	3845	80		
114504A03	176	3845	80		
114525	108	3452	86	2038	
115039	116	3016	504A	1104	
115060-102	519		514	1122	
115063	121MP	3009	239(2)	2006(2)	
115099	614	3126	90		
115101	109	3087		1123	
115123	138A	3059	ZD-7.5		
115167	123A	3444	20	2051	
115225	123A	3444	20	2051	
115227	160	3006	245	2004	
115228	160	3006	245	2004	
115229	160	3006	245	2004	
115268	121	3009	239	2006	
115269	121	3009	239	2006	
115270-101	106	3984	21	2034	
115275	126	3008	52	2024	
115281	121	3014	239	2006	
115282	121	3014	239	2006	
115283	121	3014	239	2006	
115284	121	3014	239	2006	
115300-1	128		243	2030	
115304	195A	3765	46		
115440	108	3452	86	2038	
115504	160	3004/102A	245	2004	
115517-001	159	3466	82	2032	
115524	125		510,531	1114	
115527-1	175		246	2020	
115529	125		510,531	1114	
115559	116	3016	504A	1104	
115559A	116	3017B/117	504A	1104	
115599	116	3016	504A	1104	
115720	123A	3444	20	2051	
115728	123A	3444	20	2051	
115783	124		12		
115792	153		69	2049	
115810P2	128		243	2030	
115817	124		12		
115867	507	3017B/117	504A	1104	
115875	123A	3444	20	2051	
115910	108	3452	86	2038	
115925	108	3452	86	2038	
116021			300	1122	
116021(SEARS)	177		300	1122	
116048	109	3087		1123	
116052	116	3016	504A	1104	
116054	116	3017B/117	504A	1104	
116068	127	3035	25		
116072	126	3008	52	2024	
116073	108	3452	86	2038	
116074	123A	3444	20	2051	
116075	124	3021	12		
116076	123A	3444	20	2051	
116077	123A	3444	20	2051	
116078	123A	3444	20	2051	
116078(RCA)	159		82	2032	
116079	108	3452	86	2038	
116080	108	3452	86	2038	
116081	154	3040	40	2012	
116082	108	3452	86	2038	
116083	108	3452	86	2038	
116084	100	3025/129	1	2007	
116084(RCA)	129		244	2027	
116085	123A	3444	20	2051	
116086	127	3035	25		
116087	127	3124/289	25		
116088	127	3044/154	25		
116089	127	3034	25		
116090		3025	22	2032	
116091	102A	3025/129	53	2007	
116092	123A	3444	20	2051	
116093	121	3009	239	2006	
116118	152	3893	66	2048	
116118-2	152	3893	66	2048	
116119	107	3452/108	86	2038	
116148	123		20	2051	
116198	108	3452	86	2038	
116199	108	3018	86	2038	
116200	108	3452	86	2038	
116201	102A	3004	53	2007	
116202	160	3006	245	2004	
116203	102A	3003	53	2007	
116204	103A	3835	59	2002	
116205	102A	3004	53	2007	
116206	102A	3003	53	2007	
116207	160	3006	245	2004	
116208	160	3006	245	2004	
116209	160	3006	245	2004	
116273	109			1123	
116279	124		12		
116284	129	3025	244	2027	
116286	102A	3004	53	2007	
116314	112	3089	1N82A		
116588	123A	3124/289	20	2051	
116623	912		IC-172		
116628	102A		53	2007	
116683	160	3006	245	2004	
116684	160	3006	245	2004	
116685	102A	3004	53	2007	
116686	102A	3004	53	2007	
116687	103A	3835	59	2002	
116756	126	3006/160	52	2024	
116757	102A	3004	53	2007	
116796P1	704		IC-205		
116875	123A	3444	20	2051	
116988	158	3004/102A	53	2007	
116996	102A	3123	53	2007	
116997	102A	3004	53	2007	
116998	102A	3004	53	2007	
117145	116	3016	504A	1104	
117145A	116	3016	504A	1104	
117208	102A	3004	53	2007	
117209	102A	3004	53	2007	
117210	102A	3004	53	2007	
117616	100	3005	1	2007	
117617	100	3005	1	2007	
117618	160	3007	245	2004	
117658	160	3007	245	2004	
117659	109	3088		1123	
117724	126	3008	52	2024	
117725	160	3006	245	2004	
117726	160	3006	245	2004	
117727	102A	3004	53	2007	
117728	102A	3004	53	2007	
117729	612	3126	90		
117730	109	3709/110MP	1N60	1123(2)	
117760	109	3088		1123	
117823	108	3452	86	2038	
117824	126	3008	52	2024	
117866	160	3006	245	2004	
117867	158	3004/102A	53	2007	
118200	123A	3444	20	2051	
118244	116(4)	3016	504A(4)	1104	
118279	124	3021	12		
118284	159	3466	82	2032	
118335	519		514	1122	
118361	704	3023	IC-205		
118527-04	177		300	1122	
118686	124	3021	12		
118713	123A	3444	20	2051	
118822	107	3018	86	2038	
118825	116	3016	504A	1104	
118873	116	3031A	504A	1104	
119013	160	3006	245	2004	
119119(PAIR)	110MP	3709			
119199	109	3088		1123	
119228-001	159	3466	82	2032	
119232-001	123A	3444	20	2051	
119258-001	123A	3444	20	2051	
119264-001	519		514	1122	
119298-001	129		244	2027	
119412	107		11	2015	
119414	108	3018	86	2038	
119507	177		300	1122	
119526	126	3008	52	2024	
119554	161	3018	86	2038	
119555	161	3018	86	2038	
119556	161	3018	86	2038	
119557	161	3018	86	2038	
119594	140A	3061	ZD-10	562	
119596	177	3311	300	1122	
119597	177	9091	300	1122	
119609	704	3023	IC-205		
119635	123AP	3018	20	2051	
119636	123A	3444	20	2051	
119650	124	3538	12		
119661	614		90		
119662	112	3089	1N82A		
119697	177		300	1122	
119721	121	3122	239	2006	

Industry Standard No.	ECG	SK	GE	RS 276-	MOTOR.
119722	127	3173/219	25		
119723	127	3034	25		
119724	123A	3014	20	2051	
119725	123A	3444	20	2051	
119726	123A	3444	20	2051	
119727	158		53	2007	
119728	128		243	2030	
119730	159	3466	82	2032	
119822	128	3024	243	2030	
119823	161	3122	39	2015	
119824	161	3117	39	2015	
119825	161	3117	39	2015	
119919	109	3091		1123	
119956	177		300	1122	
119982	123A	3444	20	2051	
119983	129	3025	244	2027	
120068	519		514	1122	
120073	123A	3275/194	20	2051	
120074	123A	3444	20	2051	
120075	102A	3004	53	2007	
120085	123A	3444	20	2051	
120143	102A	3004	53	2007	
120144	102A	3004	53	2007	
120231	116(4)		504A(4)	1104	
120471	125	3032A	510,531	1114	
120481	123A	3444	20	2051	
120482	123A	3444	20	2051	
120483	123A	3444	20	2051	
120503	116		504A	1104	
120504	142A	3062	ZD-12	563	
120544	116	3016	504A	1104	
120545	102A	3005	53	2007	
120546	102A	3004	53	2007	
120617	109	3088	1N60	1123	
120818	173BP	3999	305		
120909-24.4	102A	3004	53	2007	
121151	102A		53	2007	
121152	102A		53	2007	
121153	100	3005	1	2007	
121154	100	3005	1	2007	
121180	5802	9005	512	1143	
121243	121	3009	239	2006	
121244	179	3034	25		
121467	159	3466	82	2032	
121468	116	3311	504A	1104	
121655	123A	3444	20	2051	
121658	123A	3444	20	2051	
121659	159	3114/290	82	2032	
121660	123A	3444	20	2051	
121661		3444/123A	243,244		
121661(RCA)	123A		20	2051	
121662	123A	3444	20	2051	
121663	123A	3444	20	2051	
121664	123	3124/289	20	2051	
121680	116		504A	1104	
122061	101	3861	8	2002	
122074	123A	3444	20	2051	
122111	101	3861	8	2002	
122112	101	3861	8	2002	
122129	116	3016	504A	1104	
122166	109	3087		1123	
122199	704		IC-205		
122243	102A	3004	53	2007	
122244	102A	3004	53	2007	
122517	108	3452	86	2038	
122518	108	3452	86	2038	
122519	199	3444/123A	20	2051	
122519(IFAMP)			10	2051	
122519(SW)			10	2051	
122520	109	3124/289		1123	
122664	123A	3444	20	2051	
122665	123A	3444	20	2051	
122725	126	3006/160	52	2024	
122788	116		504A	1104	
122792	121	3009	239	2006	
122901	102A	3004	53	2007	
122902	107	3018	11	2015	
122904	107	3018	61	2038	
123004	116		504A	1104	
123139	123AP	3024/128	243	2030	
123160	108	3452	86	2038	
123243	128		243	2030	
123244	126	3004/102A	52	2024	
123274	289A	3444/123A	20	2051	
123275	124	3021	12		
123275-14	124	3021	12		
123276	614	3126	90		
123296	116		504A	1104	
123375	124	3021	12		
123379	102A	3004	53	2007	
123429	108	3452	86	2038	
123430	108	3452	86	2038	
123431	108	3452	86	2038	
123511	126	3006/160	52	2024	
123702	116(4)	3031A	504A(4)	1104	
123703	128		243	2030	
123726	614	3126	90		
123791	102A	3004	53	2007	
123792	121	3009	239	2006	
123804		3311	504A	1104	
123804-1	116		504A	1104	
123805	102A	3100/519	53	2007	
123805(DIODE)	177		300	1122	
123806	102A	3004	53	2007	
123807	123A	3444	20	2051	
123808	131MP	3052	44,	2006	
123809	102A		53	2007	
123813	177		300	1122	
123877	102A	3123	53	2007	
123940	128	3466/159	82	2032	
123941	123AP	3124/289	20	2051	
123944	129		40	2012	
123971	159	3466	82	2032	
123991	159	3466	82	2032	
124024	107	3018	11	2015	
124047	159	3114/290	82	2032	
124097	126	3004/102A	52	2024	
124098	116	3031A	504A	1104	
124263	108	3452	86	2038	
124390	107	3122			
124412	108	3452	86	2038	
124511	130		14	2041	
124557	123A	3444	20	2051	
124616	129	3025	244	2027	
124623	107	3018	11	2015	
124624	107	3018	11	2015	
124625	160	3006	245	2004	
124626	102A	3004	53	2007	
124634	106	3984	21	2034	
124753	123AP	3444/123A	20	2051	
124754	123AP	3854	210	2009	
124755	159	3466	82	2032	
124756	123AP	3444/123A	20	2051	
124757	319P	3246/229	283	2016	
124759	123A	3444	20	2051	
124812	116	3016	504A	1104	
125105	116	3033A	504A	1104	
125126	142A	3062	ZD-12	563	
125127	116	3311	504A	1104	
125135	123A	3444	20	2051	
125137	108	3452	86	2038	
125138	108	3039/316	86	2038	
125139	123A	3444	20	2051	
125140	123A	3444	20	2051	
125141	123AP	3018	20	2051	
125142	129	3025	82	2032	
125143	123A	3122	20	2051	
125144	229	3246	39	2015	
125261	112		1N82A		
125263	108	3452	86	2038	
125264	108	3452	86	2038	
125329	107	3018	11	2015	
125330	126	3007	52	2024	
125389	107	3444/123A	20	2051	
125389-14	123A	3444	20	2051	
125390	107	3444/123A	20	2051	
125390(RCA)	107		11	2015	
125392	108	3452	86	2038	
125394	107	3444/123A	20	2051	
125397	177	3100/519	300	1122	
125399	612	3126	90		
125458	116	3031A	504A	1104	
125471	116	3017B/117	504A	1104	
125474	107	3444/123A	20	2051	
125475	107	3444/123A	20	2051	
125475-14	108	3452	86	2038	
125488			511	1114	
125499	5074A		ZD-11.0	563	
125519	289A	3444/123A	20	2051	
125528	177	3175	300	1122	
125529	506	3031A	504A	1104	
125549	116	3031A	504A	1104	
125588	177	3100/519	300	1122	
125703	121	3014	239	2006	
125707	129		21	2034	
125761	121	3014	239	2006	
125787	166	9075		1152	
125790	160	3008	245	2004	
125835	116	3017B/117	504A	1104	
125844	506	3843	511	1114	
125848	506	3843	511	1114	
125856			504A	1104	
125884			300	1122	
125944	108	3452	86	2038	
125964	116	3100/519	504A	1104	
125972	160	3008	245	2004	
125992		3332	ZD-7.5		
125993	116	3105	504A	1104	
125994	107	3452/108	86	2038	
125994(RCA)			11	2015	
125994-14	108	3452	86	2038	
125995	107	3452/108	86	2038	
126023	108	3452	86	2038	
126024	108	3452	86	2038	
126025	161	3018	39	2015	
126093	102A	3004	53	2007	
126093-1	102A	3004	53	2007	
126093-2	102A	3004	53	2007	
126093-3	102A		53	2007	
126093-4	102A	3004	53	2007	
126131	116	3017B/117	504A	1104	
126138	124	3021	12		
126148	116	3017B/117	504A	1104	
126149	612	3126	90		
126150	123A	3444	20	2051	
126156	123A	3444	20	2051	
126176	116	3311	504A	1104	
126177	109	3087		1123	
126184	160		245	2004	
126185	126	3006/160	52	2024	
126186	160		245	2004	
126187	102A	3004	53	2007	
126188	124	3021	12		
126276	102A	3004	53	2007	
126320	116	3313	504A	1104	
126321	177	3175	300	1122	
126331	123A	3444	20	2051	
126334	123AP	3444/123A	20	2051	
126413	166	9075	504A	1152	
126524	159	3466	82	2032	
126525	123A	3444	20	2051	
126526	123A	3444	20	2051	
126527	506	3843	511	1114	
126582			ZD-15	564	
126604	711	3070	IC-207		
126609	711	3070	IC-207		
126670	108	3452	86	2038	
126697	102A		53	2007	
126698	108	3452	86	2038	
126699	123A	3122	20	2051	
126700	159	3114/290	82	2032	
126701	127	3444/123A	20	2051	
126702	123A	3444	20	2051	
126703	128	3104A	243	2030	
126704	123A	3018	20	2051	
126705	154	3045/225	40	2012	
126706	123A	3444	20	2051	
126707	159	3466	82	2032	
126708	123A	3122	20	2051	
126709	154	3044	40	2012	
126710	154	3044	40	2012	
126711	123A	3444	20	2051	
126712	123AP	3444/123A	20	2051	
126713	123A	3444	20	2051	
126714	123A	3444	20	2051	
126715	159	3466	82	2032	
126716	123AP	3124/289	20	2051	
126717	123AP	3122	20	2051	
126718	159	3114/290	82	2032	
126719	159	3114/290	82	2032	
126720	123AP	3122	243	2030	
126721	128	3024	243	2030	
126722	124	3045/225	12		
126724	129	3025	244	2027	
126725	128	3024	243	2030	
126726	124	3045/225	12		
126772	124		12		
126826(DIODE)	506		511	1114	
126836(REGULATOR)	5074A		ZD-11.0	563	
126849	167	3647	504A	1172	
126851	5071A	3780/5014A	ZD-6.8	561	
126852	145A	3063	ZD-15	564	
126853	5103A		ZD-180		
126855	116	3081/125	504A	1104	
126856	506	3031A	511	1114	
126857	515	9098	511		
126858	515	9098	511		
126859	515	9098	511		

Industry Standard No.	ECG	SK	GE	RS 276-	MOTOR.
126860	118	3066	CR-1		
126861	116	3066/118	504A	1104	
126862	119	3109	ZD-15	564	
126863	5074A	3061/140A	20	2051	
126863(REGULATOR)	5097A		ZD-120		
126864	147A		ZD-33		
126871	710	3102	IC-89		
126885	125	3032A	510,531	1114	
126898	230	3857/231	700		
126899	230	3042	700		
126900	162	3438	35		
126945	102A	3004	53	2007	
127017	109	3090		1123	
127102	116	3017B/117	504A	1104	
127112	102A		53	2007	
127114	102A	3004	53	2007	
127166	785	3254	IC-226		
127176	116	3031A	504A	1104	
127177	5074A	3059/138A	ZD-7.5		
127214	312	3448	FET-1	2035	
127263	123A	3122	20	2051	
127297	102A		53	2007	
127303	102A		53	2007	
127354	123AP	3444/123A	20	2051	
127355	199	3444/123A	20	2051	
127376	128	3024	243	2030	
127379	125	3031A	510,531	1114	
127382	145A	3063	ZD-15	564	
127393			20	2051	
127396	116	3016	504A	1104	
127397	158	3004/102A	53	2007	
127398	123A	3124/289			
127399	123A	3124/289			
127474	177		300	1122	
127529	108	3444/123A	20	2051	
127532	112	3089		1123	
127589	102A	3004	53	2007	
127590	176	3845	80		
127693	108	3452	86	2038	
127695	116	3311	504A	1104	
127712	154	3045/225	40	2012	
127784	109			1123	
127792	108	3452	86	2038	
127793	108	3452	86	2038	
127794	108	3452	86	2038	
127798	152		66	2048	
127828	179	3642	76		
127835	156	3311	512		
127845	128	3024	243	2030	
127899	123AP	3444/123A	20	2051	
127962	102A	3004	53	2007	
127978	184	3054/196	57	2017	
127980	222	3050/221	FET-4	2036	
127992	116	3032A	512		
127993	177	3031A	300	1122	
128056	152	3054/196	66	2048	
128057	153	3083/197	69	2049	
128256	116		504A	1104	
128343	102A	3004	52	2024	
128474	177	3175	300	1122	
128474(RCA)	178MP		6GC1		
128938	126	3006/160	52	2024	
128940	102A	3024/128	53	2007	
129028	109			1123	
129029	116	3122	504A	1104	
129029(DIO)			504A	1104	
129029(XSTR)			10	2051	
129049		3122	20	2051	
129050	108	3018	61	2038	
129051	128	3024	243	2030	
129095	553	3322	300	1122	
129144	108	3452	86	2038	
129145	229	3039/316	17	2051	
129146	289A	3039/316	20	2051	
129147	123A	3444	20	2051	
129157	109	3088		1123	
129158	109	3088		1123	
129203	504	3068/503	CR-6		
129213	116	3031A	504A	1104	
129241	169	3678	BR-600		
129286	102A	3004	53	2007	
129334	116	3311		1104	
129347		3006	50	2004	
129348	125	3311	510,531	1114	
129348(DIO)			1N34AS	1123	
129348(XSTR)			2	2007	
129359	612	3126	90		
129375	177	3100/519	300	1122	
129389	160	3008	245	2004	
129392	107	3018	11	2015	
129392-14	108	3452	86	2038	
129393	107	3018	17	2051	
129393-14	108	3452	86	2038	
129394	107	3452/108	86	2038	
129394-14	108	3452	86	2038	
129424	451	3116	FET-4	2036	
129425	199	3122	17	2051	
129474	109	3088		1123	
129475	601	3100/519	300	1122	
129494		3087	504A	1104	
129507			53	2007	
129508			53	2007	
129509	199	3039/316	11	2015	
129510	199	3039/316	17	2051	
129511	199	3039/316	17	2051	
129512	199	3039/316	17	2051	
129513	199	3039/316	17	2051	
129556	112	3089	1N82A		
129571	107	3452/108	86	2038	
129572		3039	20	2051	
129573	107	3039/316	11	2015	
129574	107	3452/108	86	2038	
129604	313	3019	11	2015	
129682-001	519		514	1122	
129697	159	3114/290	21	2034	
129698		3018	20	2051	
129699	159	3114/290	82	2032	
129708-101	519		514	1122	
129759	116	3017B/117	504A	1104	
129802	102A	3004	53	2007	
129821	904		IC-289		
129871	710	3102	IC-89		
129897	123AP	3452/108	86	2038	
129899	123AP	3444/123A	20	2051	
129903	134A	3771/5005A	ZD-3.6		
129904		3784	ZD-9.1	562	
129904(RCA)	5073A	3784/5018A	ZD-9.1	562	
129938	137A	3779/5013A	ZD-6.2	561	
129940	143A	3063/145A	ZD-13	563	
129946	145A	3063	ZD-15	564	
129949	123A	3444	20	2051	
129979	107	3452/108	86	2038	
129980	222	3050/221	FET-4	2036	
130013			20	2051	
130014	130	3027	14	2041	
130026	536A	3900			
130040	172A	3156	64		
130044	140A	3785/5019A	ZD-10	562	
130045	177	3311	300	1122	
130046	178MP	3100/519	300	1122	
130047	140A	3785/5019A	ZD-10	562	
130052		3031A	504A	1104	
130096	148A	3824/5058A	ZD-55		
130110	125		510,531	1114	
130122	712	3072	IC-2		
130130	780	3141	IC-222		
130132	5077A	3752	ZD-18		
130139	159	3118	82	2032	
130172	128		243	2030	
130174	128		243	2030	
130200-00	102A	3004	53	2007	
130200-02	102A	3004	53	2007	
130215	159	3114/290	21	2034	
130221	909	3590	IC-249		
130253	234		65	2050	
130278	108	3018	86	2038	
130328	142A	3062	ZD-12	563	
130338-01	116	3016	504A	1104	
130380	142A	3062	ZD-12	563	
130389-00	116	3016	504A	1104	
130400-95	102A	3004	53	2007	
130400-96	103A	3010	59	2002	
130402-36	102A	3004	53	2007	
130403-04	108	3018	86	2038	
130403-13	123A	3444	20	2051	
130403-17	123A	3444	20	2051	
130403-18	123A	3444	20	2051	
130403-47	103A	3010	59	2002	
130403-52	102A	3004	53	2007	
130403-62	108	3452	86	2038	
130404-21	108	3452	86	2038	
130404-29	129	3025	244	2027	
130404-59	108	3452	86	2038	
130474	188		226	2018	
130536	123A	3466/159	82	2032	
130537	123A	3444	20	2051	
130537-1	289A	3018			
130537-2	289A	3018			
130537-3	289A	3018			
130537-4	289A	3018			
130537-5	289A	3018			
130607	116	3016	504A	1104	
130751	712	3072	IC-148		
130761	5075A	3751	ZD-15	564	
130762	145A	3142	ZD-15	564	
130793	107		11	2015	
131005	5527	6627			
131006	5527	6627			
131050			504A	1104	
131075	152	3054/196	66	2048	
131095-2	175		246	2020	
131139	171	3104A	249		
131140	123AP	3104A	249		
131148	506	3843	511	1114	
131161	196	3054	241	2020	
131214	112	3089	1N82A		
131221	108	3452	86	2038	
131239	196	3054	241	2020	
131240	123A	3444	20	2051	
131241	159	3466	82	2032	
131242	159	3025/129	82	2032	
131242-12	129	3025	244	2027	
131243	123AP	3024/128	20	2051	
131243-12	123A	3444	20	2051	
131245	116	3016	504A	1104	
131257	181	3054/196	75	2041	
131262	106	3984	21	2034	
131311	123AP	3444/123A	20	2051	
131318	506	3051/156	511	1114	
131346	230	3042	700		
131347	231	3857	700		
131454	123AP	3506/5642			
131475	506	3130	512		
131476	506	9098/515	504A	1104	
131501	116		300	1104	
131501(DIODE)	177		300	1122	
131502	177	3032A	300	1122	
131502(RCA)			504A	1104	
131543	161	3132	39	2015	
131544	108	3132	39	2015	
131545	108	3018	39	2015	
131647	159	3118	82	2032	
131648	108	3452	86	2038	
131710	234		65	2050	
131721	5801	3051/156			
131815	156	3051	512		
131815(RCA)	112		1N82A		
131844	108	3452	86	2038	
131848	29	3452/108	86	2038	
131848(RCA)	152		66	2048	
131849	152	3054/196	66	2048	
131950	116	3032A	504A	1104	
132148	506	3313/116	511	1114	
132149	116	3033A	504A	1104	
132175	128	3024	243	2030	
132176	159	3466	22	2032	
132285	159	3466	82	2032	
132313	730	3143			
132314	728	3073	IC-22		
132315	729	3074	IC-23		
132325	230	3042	700		
132326	230	3857/231	700		
132327	128	3024	243	2030	
132328	128	3024	243	2030	
132329	123AP	3444/123A	20	2051	
132416	142A		ZD-12	563	
132418	506	3843	511	1114	
132445	188	3054/196	226	2018	
132446	188	3054/196	226	2018	
132447	189	3083/197	218	2026	
132448	189	3083/197	250	2027	
132478	159	3466			
132488	189	3200	218	2026	
132492	184		ZD-62		
132495	184	3041	57	2017	
132498	159	3466	82	2032	
132499	152	3054/196	57	2017	
132500	128	3512	243	2030	
132501	506	3843	511	1114	
132509	506	3843	511	1114	
132547	506	3843	511	1114	
132548	506	3843	511	1114	
132549	506	3843	511	1114	
132553	519		514	1122	
132558	213		254		
132571	185	3083/197	58	2025	
132573	152	3054/196	66	2048	
132574	503	3083/197	250	2027	

Industry Standard No.	ECG	SK	GE	RS 276-	MOTOR.
132616	5071A	3780/5014A	ZD-6.8	561	
132634	536A	3900			
132642	289A	3444/123A	20	2051	
132643	289A	3444/123A	20	2051	
132645	177		300	1122	
132650	6410			2029	
132697	152	3893	66	2048	
132776	184	3027/130	14	2041	
132814	116	3313	504A	1104	
132815	156	3051	512		
132823	123AP	3854	243	2051	
132824	123AP	3095/147A	ZD-33		
132830	159	3025/129	82	2032	
132842	147A		ZD-33		
132865	136A	3777/5011A	ZD-12	563	
132912	109			1123	
132915			1N34AS	1123	
132966	503	3068	CR-5		
133026	503	3068			
133171	319P	3117	39	2015	
133176			63	2030	
133177	128		243	2030	
133178	123AP	3444/123A	20	2051	
133182	159	3466	82	2032	
133218	123AP	3038	62	2010	
133249	123AP	3854	20	2051	
133253	159	3466	82	2032	
133265	171	3104A	27		
133266	116	3051/156	512		
133275	123AP	3854	243	2030	
133390	177	3175	300	1122	
133390(PWR. RECT)	116			1104	
133390(PWR.-RECT)			504A	1104	
133543	5085A	3311	504A	1104	
133550	181	3036	75	2041	
133573	186	3192	28	2017	
133576	128		243	2030	
133600	730	3143			
133615	506	3843	511	1114	
133616	506	3843	511	1114	
133684	130		14	2041	
133685	130		14	2041	
133690	123AP	3854	20	2051	
133743	123A	3444	20	2051	
133823	175		246	2020	
133923	181		75	2041	
133925	181		75	2041	
133950	116	3017B/117	504A	1104	
134074	112	3089	1N82A		
134142	108	3018	86	2038	
134143	123AP	3444/123A	20	2051	
134144	108	3018	86	2038	
134144-1			11	2015	
134155	128		243	2030	
134180	109	3087	1N34AS	1123	
134263	108	3452	86	2038	
134264	112	3089	1N82A		
134277	116	3031A			
134279	185		58	2025	
134280	184	3190	57	2017	
134281	180		74	2043	
134282	181		75	2041	
134335	5022A	3788	ZD-13	563	
134417	229	3019	86	2038	
134419	108	3452	86	2038	
134442	172A	3452/108	86	2038	
134444	145A	3063	ZD-15	564	
134450	222	3050/221	FET-4	2036	
134587			1N34AS	1123	
134771	196	3054	241	2020	
134772	171	3104A	27		
134773	163A		36		
134774	188	3054/196	12		
134777	505	3108	CR-7		
134857	229		61	2038	
134989	128	3124/289	243	2030	
134993	136A	3777/5011A	ZD-5.6	561	
135029	5642	3506			
135244	503		CR-5		
135281	116	3311	504A	1104	
135284	116	3031A	504A	1104	
135286	159	3466	82	2032	
135320	173BP	3999	305		
135324	221	3050	FET-4	2036	
135341	506	3109/119	511	1114	
135347	106	3984	21	2034	
135351	218	3085	234		
135352	162	3438	35		
135380	506	3175	511	1114	
135386	515	3175	504A	1104	
135406	5097A	3818/5052A	ZD-120		
135571	177	3031A	300	1122	
135691	532	3300/517	517		
135716	171	3104A	27		
135734	506	3843	511	1114	
135735	196	3054	241	2020	
135739	196	3192/186	28	2017	
135744	127	3035	25		
135872	177	3087		1123	
135932	173BP	3999	305		
135963	221	3050	FET-4	2036	
136066	154		40	2012	
136145	787	3146	IC-228		
136146	788	3829	IC-229		
136147	789	3078			
136162	177	3100/519	300	1122	
136163	177	3100/519	300	1122	
136164	614	3126	90		
136165	123AP	3452/108	86	2038	
136168	229	3018	61	2038	
136239	123AP	3452/108	86	2038	
136240	229	3018	61	2038	
136281	234		65	2050	
136282	199	3245	62	2010	
136423	129	3025	244	2027	
136424	128	3024	243	2030	
136430	123AP	3018	62	2010	
136555	725		IC-19		
136605	125	3843	510,531	1114	
136606	506	3843	511	1114	
136634	142A	3788/5022A	ZD-13	563	
136635	116	3080	504A	1104	
136648	152	3054/196	66	2048	
136688	177	3100/519	300	1122	
136696	128	3024	243	2030	
136721	147A	3095	ZD-33		
136766	160	3984/106	21	2034	
137028	177		300	1122	
137028-001	177		300	1122	
137031	532	3301/531	519		
137057-001	519		514	1122	
137065	189	3054/196	218	2026	
137066	188	3199	226	2018	
137075	506	3843	511	1114	
137093	102A	3004	53	2007	
137127	229	3452/108	86	2038	
137155	159	3114/290	244	2027	
137241	128	3024	243	2030	
137245	780	3141	IC-222		
137338	229	3246	39	2009	
137338(3RD IF)	233			2009	
137338(3RD-IF)			210	2009	
137339	123AP	3444/123A	20	2051	
137340	159	3466	272		
137340(DIODE)	115		6GX1		
137352	163A	3439	36		
137369	184	3054/196	57	2017	
137383	123A	3117	39	2015	
137388	161	3132	39	2015	
137527	152	3893	66	2048	
137606	552	9000	511	1114	
137607	165	3710/238	38		
137614	123A	3444	20	2051	
137643		3300	519		
137646	532	3300/517	525		
137647	5069A		ZD-5.6	561	
137648	128	3232/191	243	2030	
137652	116	3313	504A	1104	
137655	5081A	3797/5031A	ZD-24		
137693	532	3300/517	519		
137718	163A	3115/165	38		
137780	616	3320			
137855	503		CR-5		
137875	123AP	3122	20	2051	
137875(I.C.)	941M		IC-265	007	
137875(U.C.)			IC-265	007	
137876	5642	3519/5643	1N60	1123	
138001	199	3245	62	2010	
138019	532	3300/517	519		
138019-001	199	3245	62	2010	
138035-001	128		243	2030	
138049-001	234	3247	65	2050	
138049-004	234	3247	65	2050	
138107	149A	3097	ZD-62		
138121	189		218	2026	
138172	116	9000/552	504A	1104	
138173	116	3175	511	1104	
138174	146A	3064	ZD-27		
138175	116	3313	504A	1122	
138191	123AP	3444/123A	20	2051	
138192	152	3054/196	66	2048	
138193	185	3191	69	2025	
138194	130	3027	14	2041	
138195	284	3027/130	14	2041	
138196	177	3175	300	1122	
138311	7400	7400		1801	
138312	7401	7401			
138313	7402	7402		1811	
138314	7404	7404		1802	
138315	7408	7408		1822	
138318	7420			1809	
138320	74193	74193		1820	
138376	159	3466	82	2032	
138378	123A	3444	20	2051	
138379	152	3893	66	2048	
138380	789	3078			
138381	7432	7432		1824	
138403	7473	7473		1803	
138421	532	3301/531	525		
138429	5404	3627			
138681	949	3166	IC-25		
138699	797		IC-233		
138699(RCA)	797	3158	IC-233		
138736	525	3109/119	CR-2	1114	
138752	531	3300/517	524		
138763	123AP	3024/128	243	2030	
138789-1	123A	3444	20	2051	
138789-2	123A	3444	20	2051	
138789-3	123A	3444	20	2051	
138789-4	199	3245	62	2010	
138789-5	199	3245	62	2010	
138907	532	3304/500A	525		
138946	222	3018	52	2024	
138974	137A	3775/5009A	ZD6.2	561	
139001	531	3300/517	524		
139017	191	3232	249		
139029	519		514	1122	
139044	190	3044/154	217		
139064	185	3083/197			
139065	177	3100/519	300(2)	1122(2)	
139069		3341	69	2049	
139266	188	3199	226	2018	
139267	189		218	2026	
139268	123AP	3275/194	271	2030	
139269	194	3275	220		
139270	175	3538	246	2020	
139295	165	3535/181	38		
139328	177	3031A	300	1122	
139362	123AP	3444/123A	20	2051	
139455	159	3466	82	2032	
139569			20	2051	
139605	116		504A	1104	
139605(AFC)	178MP		300(2)	1122(2)	
139618	172A	3156	64		
139634	109	3088		1123	
139696	128		243	2030	
139706	177	9091	300	1122	
140259	154	3201/171	27		
140290	159	3466	82	2032	
140371	159	3466	82	2032	
140372	159	3466	82	2032	
140501	128	3024	243	2030	
140503	506	3843	511	1114	
140506	182	3188A	55		
140612	181		75	2041	
140622	123A	3444	20	2051	
140623	159	3466	82	2032	
140624	194	3275	220		
140625	197	3083	250	2027	
140626	196	3054	241	2020	
140693	116(2)		504A(2)	1104	
140694	116(2)		504A(2)	1104	
140715	505		CR-7		
140763	230	3042	700		
140764	231	3857	700		
140858-12	123A	3444	20	2051	
140971	552	3998/506	511	1114	
140972	552	3998/506	511	1114	
140973	145A	3063	ZD-15	564	
140976	165	3115	38		
140977	165	3115	38		
140979	196	3054	241	2020	
140995	503	3068	CR-5		
141003			47	2030	
141008	128	3444/123A	271	2030	
141018	159	3466	82	2032	
141019	128		243	2030	
141020	196	3041	241	2020	

Industry Standard No.	ECG	SK	GE	RS 276-	MOTOR.
141134	984	3185			
141187	5077A	3752	ZD-18		
141227	159		82	2032	
141254	533	3302			
141255	310	3856			
141256	308	3855			
141270	1175	3212			
141279	797	3158	IC-233		
141280	1176	3210			
141295	171	3201	27		
141302	147A	3095	ZD-33		
141330	123AP	3444/123A	20	2051	
141331	123AP	3854	20	2051	
141332	222	3065	FET-4	2036	
141335	128		243	2030	
141343	159	3466	82	2032	
141344	159		82	2032	
141345	159	3466	82	2032	
141355	128		243	2030	
141367	165		38		
141370-1	165		38		
141402	161	3716	39	2015	
141421	159		82	2032	
141429	142A	3787/5021A	ZD-12	563	
141464	533	3302			
141489	125	3033A	509	1114	
141551	146A	3768/5002A			
141558	123A		20	2051	
141711	159	3118	82	2032	
141738P63-1	159	3466	82	2032	
141746	142A	3788/5022A			
141747	140A	3785/5019A			
141767	128	3024	243	2030	
141783	128	3024	18	2030	
141849	116		504A	1104	
141872-6	116	3031A	504A	1104	
141873	142A	3062			
141967	5087A		ZD-43		
142010-02	713		IC-5		
142251	772A	3201/171	27		
142341	985	3214			
142348	194		220		
142349	923D		IC-260	1740	
142421	123AP	3444/123A			
142569	552	3175	511	1114	
142648	778A	3465	IC-220		
142670	147A	3095	ZD-33		
142671	171	3201	27		
142681	5511	3683			
142683	123AP	3444/123A	20	2051	
142684	123AP		20	2051	
142686	123AP	3854	20	2051	
142689	165	3710/238	259		
142690	190	3232/191	217		
142691	152	3054/196	66	2048	
142711	123AP	3854	20	2051	
142718	1175	3212			
142719	986	3918			
142728	123AP	3302/533			
142738	5081A	3151	ZD-24		
142814	532	3301/531			
142838	159	3466	82	2032	
142839	159	3466	82	2032	
142903	985	3214			
143033	1175	3212			
143041		3433	271	017	
143042	128	3024	243	2030	
143162	112	3089	1N82A		
143227	159	3466			
143316	123AP	3854	271	2051	
143561	986	3918			
143594	525	3925	511	1114	
143595	109	3087	504A	1104	
143777	5080A	3336	ZD-22		
143790	5463	3502/5513			
143791	159	3466	82		
143792	123AP	3854	20	2051	
143793	123AP		243	2030	
143794	123AP	3854	20	2051	
143795	123AP		20	2051	
143796	123AP	3275/194	271	2030	
143797	128	3275/194	271	2030	
143798	128		45	2030	
143802	159	3466	82	2032	
143803	159	3466	82	2032	
143804	123AP	3854	20	2051	
143805	123AP		20	2051	
143806	159	3466	82	2032	
143807	159	3466	82	2032	
143822	986	3918			
143837	177	9091	300	1122	
143847	5081A		ZD-24		
143913	531	3302/533			
143963	159		82	2032	
143964	190		217		
144011	1123	3743	IC-281		
144012	1263	3920	IC-293		
144017	1268	3921			
144019	1223	3493			
144026	712	3072	IC-148		
144027	1168	3728			
144030	234	3247	65	2050	
144031	290A	3912	269	2032	
144033	374	9042	58	2025	
144034	199	3245	62		
144035	199	3250/315	61		
144037	229	3293/107			
144039	289A	3911	210		
144040	199	3250/315	85		
144043	199	3124/289			
144044	199	3124/289	62		
144045	373	9041			
144047	289A	3124/289			
144050	125	3311	510	1114	
144051	125	3081	510	1114	
144052	116	3311	504A	1104	
144056	5013A	3334/5071A			
144057	605	3864			
144076	153	3083/197	69	2049	
144178	941M		IC-265	007	
144182	295	3849/293			
144581	519	3100	300	1122	
144582	229	3911	212		
144585	109	3088	1N60	1123	
144691	519	3100	514	1122	
144761	290A	3114/290			
144856	184	3253/295	57	2017	
144857	187A	3189A/183			
144858	234	3247	65	2050	
144859	312	3834/132	FET-2		
144860	116	3311	504A	1104	
144861	137A	3779/5013A			
144863			65	2050	
144921	958	3699			
144922	5067A	3331			
144923	615		ZD-33		
144968	4011B	4011			
144969	4016B	4016			
144970	4016B	4050			
144971	4069	4069			
144973	199	3450/298			
144974	177	3100/519			
145113	615		ZD-33		
145134-526	121	3717	239	2006	
145172	189		218	2026	
145173	123AP		20	2051	
145208	537	3302/533	536		
145258	128	3122	243	2030	
145395	123AP		86	2038	
145398	123AP		20	2051	
145410	159		82	2032	
145513	5401	3638/5402			
145648	238	3710	259		
145671	165	3710/238	38		
145718	533	3302			
145776	159	3466	272		
145838	199	3124/289	62		
145839	116	3311	504A	1104	
145840	294	3841	48		
145966	234	3247			
146081	153	3274	69	2049	
146136	552	3998/506	511		
146137	552	3998/506	511		
146138	147A	3095	ZD-33		
146139	152	3893	66	2048	
146141	123AP	3444/123A	20	2051	
146142	123AP	3466/159	20	2051	
146143	190	3232/191	217		
146144-2	123A	3444	20	·2051	
146151	985	3214			
146152	986	3918			
146153-1	123A	3444	20	2051	
146260	5018A			562	
146286-01	124	3021	12		
146316	525	3925	530		
146359	159	3466			
146466-1	130		14	2041	
146484	123AP	3854	271		
146512	199	3124/289	62		
146568	234	3912			
146569	187A	3189A/183	248	2025	
146570	289A	3124/289	210		
146571	519	3100	300	1122	
146572	199		62		
146575	109		1N34AS	1123	
146576	519	3100	514	1122	
146596	373	9041	270		
146823	238	3710	259		
146826	171	3201	27		
146896	712	3072			
146898	374	3191/185			
146899	234	3912			
146901	234	3841/294			
146902	234	3247		2050	
146904	289A	3911			
146905	312	3834/132			
146906	125	9002			
146907	125	9001/113A			
146908	5011A	3777			
146912	519	3100			
147015	125	3033A	509	1114	
147112-7	129	3025	244	2027	
147115	123A	3444	20	2051	
147115-5	123A	3444	20	2051	
147115-6	123A	3444	20	2051	
147115-7	199	3245	62	2010	
147115-8	199	3245	62	2010	
147115-9	161	3716	39	2015	
147122P7	199	3245	62	2010	
147187-2-14	125		510,531	1114	
147245-0-1	108	3452	86	2038	
147350-2-1	704		IC-205		
147351-4-1	127	3764	25		
147351-5-1	121	3717	239	2006	
147352-0-1	124		12		
147353-0-1	161	3716	39	2015	
147355	128		243	2030	
147356-9-1	108	3452	86	2038	
147357-0-1	129		244	2027	
147357-1-1	123A	3444	20	2051	
147357-2-1	108	3452	86	2038	
147357-4-1	129		244	2027	
147357-7-1	123A	3444	20	2051	
147357-9-1	108	3452	86	2038	
147359-0-1	129		244	2027	
147360-1			10	2051	
147363-1	123A	3444	20	2051	
147477-7-1	140A		ZD-10	562	
147477-8-10	125		510,531	1114	
147477-8-2	125		510,531	1114	
147477-8-3	125		510,531	1114	
147477-8-7	125		510,531	1114	
147477-8-8	125		510,531	1114	
147478-8-2	117		504A	1104	
147513	123A	3444	20	2051	
147549-1	159	3466	82	2032	
147549-2	159	3466	82	2032	
147555-1	123A	3444	20	2051	
147617-11	506	3043	511	1114	
147617-12	506	3043	511	1114	
147624-1	197	3083	250	2027	
147663	159	3466	82		
147664	123AP	3444/123A	20	2051	
147665	123AP	3466/159	20	2051	
147676-1	233		210	2009	
147704-6-4	142A		ZD-12	563	
147922-2	112	3089	1N82A		
147993	125	3033A			
148751-147	108	3452	86	2038	
150045	128		243	2030	
150046A	128		243	2030	
150060	175		246	2020	
150070	181		75	2041	
150095	128		243	2030	
150117	123A	3122	61	2038	
150580	941M		IC-265	007	
150580-2285	941M		IC-265	007	
150622	912		IC-172		
150714-1	199	3245	62	2010	
150730	128		243	2030	
150741	123A	3444	20	2051	
150742	159	3466	82	2032	
150753	159	3466	82	2032	
150758	159	3466	82	2032	
150762	159	3466	82	2032	
150763	123A	3444	20	2051	
150768	123A	3444	20	2051	
150771	159	3466	82	2032	

Industry Standard No.	ECG	SK	GE	RS 276-	MOTOR.
150787	128		243	2030	
150796	128		243	2030	
150834	519		514	1122	
150865	106	3984	21	2034	
151476			511	1114	
152473K			300	1122	
153107	152	3893	66	2048	
153270	941		IC-263	010	
155103	923		IC-259		
156931	108	3452	86	2038	
157004	129		244	2027	
157008	123A		20	2051	
157506	912		IC-172		
157564	915			017	
157575	309K	3629			
157644	923		IC-259		
157681	912		IC-172		
157800	923D		IC-260	1740	
157800-2760	923D		IC-260	1740	
160196	106	3984	21	2034	
160201	909		IC-249		
161001	110MP	3091		1123	
161006	109	3087		1123	
161015	110MP	3709	1N60	1123(2)	
161016	109	3087		1123	
161021	177	3016	300	1122	
161022	143A	3750	ZD-13	563	
161027	112		1N82A		
161029	116	3016	504A	1104	
161030	116	3017B/117	504A	1104	
161031	116	3017B/117	504A	1104	
161032	506	3843	511	1114	
161033	506	3109/119	511	1114	
161034	112		1N82A		
161037	116	3016	504A	1104	
161038	109	3087		1123	
161199	137A	3057/136A	ZD-6.2	561	
161705	102A	3004	53	2007	
161918-28	123A	3444	20	2051	
161919-29	128		243	2030	
162002-033	121	3009	239	2006	
162002-039	109			1123	
162002-040	126	3008	52	2024	
162002-041	126	3008	52	2024	
162002-042	126	3008	52	2024	
162002-062	121	3014	239	2006	
162002-062A	121	3014	239	2006	
162002-071	181		75	2041	
162002-081	128		243	2030	
162002-082	128		243	2030	
162002-085	199	3245	62	2010	
162002-090	108	3452	86	2038	
162002-095	121	3009	239	2006	
162002-101	130		14	2041	
162002-39	109	3087		1123	
162002-40	126	3008	52	2024	
162002-41	126	3008	52	2024	
162002-71	130		14	2041	
162201	358		42		
165358	142A	3062	ZD-12	563	
165392	107	3018	11	2015	
165572	109	3088		1123	
165597			504A	1104	
165667	126	3006/160	52	2024	
165668	123A	3444	20	2051	
165735	128	3024	243	2030	
165736	128	3024	243	2030	
165737	234	3024/128	243	2030	
165738	175	3026	246	2020	
165739	116	3017B/117	504A	1104	
165740	144A	3094	ZD-14	564	
165827	123A	3444	20	2051	
165828	123A	3444	20	2051	
165931	107	3018	11	2015	
165932	107	3018	11	2015	
165976	102A	3004	53	2007	
165995	108	3452	86	2038	
166040			504A	1104	
166272	108	3452	86	2038	
166273	109	3087		1123	
166400	158	3004/102A	53	2007	
166593	116	3031A	504A	1104	
166726	116	3017B/117	504A	1104	
166881	116		504A	1104	
166882	102A	3004	53	2007	
166883	102A	3004	53	2007	
166906	107	3018	11	2015	
166907	1002	3481			
166908	126	3008	52	2024	
166909	126	3008	52	2024	
166917	123AP	3444/123A	243	2030	
166918	184	3036	57	2017	
166919	185	3083/197	58	2025	
166920	116	3311	504A	1104	
166921	139A	3059/138A	ZD-7.5		
166922	116	3031A	504A	1104	
166985	143A	3750	ZD-13	563	
166997	158		53	2007	
167034	116	3032A	504A	1104	
167263	107	3444/123A	20	2051	
167285	131	3052	44	2006	
167413	116	3031A	504A	1104	
167540	289A	3444/123A	20	2051	
167541	199	3444/123A	20	2051	
167542	295	3036	57	2017	
167543	116	3311	504A	1104	
167544	116	3311	504A	1104	
167562	109			1123	
167569	123A	3024/128	243	2030	
167572	110MP	3709	1N60	1123(2)	
167679	102A		53	2007	
167680	158	3004/102A	53	2007	
167688	123A	3444	20	2051	
167690	159	3466	82	2032	
167691	152	3893	66	2048	
167956	199	3444/123A	20	2051	
167957	199	3444/123A	20	2051	
167958	175	3026	246	2020	
167998	102A	3004	53	2007	
167999	102A	3004	53	2007	
168339	116	3031A	504A	1104	
168373	116	3311			
168405	199	3444/123A	20	2051	
168567	107	3018	11	2015	
168651	199	3124/289	20	2051	
168652	611	3126	90		
168653	137A	3058	ZD-6.2	561	
168657	108	3452	86	2038	
168658	108	3452	86	2038	
168659	108	3452	86	2038	
168660	128	3122	243	2030	
168692	605		504A	1104	
168706	177	3100/519	300	1122	
168716	199	3245	62	2010	
168906	158	3004/102A	53	2007	
168907	102A	3004	53	2007	
168910	109	3088	1N60	1123	
168953	103A	3010	59	2002	
168954	102A	3004	53	2007	
168983	102A		53	2007	
168984	158	3004/102A	53	2007	
169113	116		504A	1104	
169114	177	3100/519	300	1122	
169115	109			1123	
169116	116	3031A	504A	1104	
169117	177	3100/519	300	1122	
169175	158	3004/102A	53	2007	
169194	107	3018	11	2015	
169195	108	3452	86	2038	
169196	107	3018	11	2015	
169197	123A	3444	20	2051	
169198		3126	90		
169199	136A	3057	ZD-5.6	561	
169216	124		12		
169359	102A	3004	53	2007	
169360	102A	3004	53	2007	
169361	102A	3004	53	2007	
169362	109	3088		1123	
169363	116	3032A	504A	1104	
169501	109	3087		1123	
169505	107	3124/289	11	2015	
169558	177	3100/519	300	1122	
169565	116	3031A	504A	1104	
169570			504A	1104	
169574	107	3444/123A	20	2051	
169590	116	3311	504A	1104	
169678	199	3124/289			
169679	123A	3444	20	2051	
169680	123A	3444	20	2051	
169765	116	3031A	504A	1104	
169771	289A	3444/123A	20	2051	
169773	102A	3004	53	2007	
170128	234	3466/159	82	2032	
170132-1(NPN)			59	2002	
170132-1(PNP)			53	2007	
170133	116	3031A	504A	1104	
170294	107	3444/123A	20	2051	
170297	116	3311	504A	1104	
0170301	177	3100/519	300	1122	
170307-1	121	3009	239	2006	
170308	289A	3444/123A	20	2051	
170370	177	3126	300	1122	
170373	109	3087		1123	
170376	121	3009	239	2006	
170376-1	121	3009	239	2006	
170388	107	3018	61	2038	
170407-1	121	3014	239	2006	
170479-1	121	3009	239	2006	
170666-1	121MP	3009	239(2)	2006(2)	
170668-1	121MP	3013	239(2)	2006(2)	
170733-1	178MP	3100/519	300(2)	1122(2)	
170750	116	3311	504A	1104	
170753-1	161	3018	39	2015	
170756-1	161	3018	39	2015	
170783-3	103A	3010	59	2002	
170794	107	3018	61	2038	
170827-1		3052	44,43		
170850-1	121MP	3013	239(2)	2006(2)	
170856	116	3017B/117	504A	1104	
170857	177	3100/519	300	1122	
170890-1(NPN)			75	2041	
170890-1(PNP)			74	2043	
170891-1	181		75	2041	
170906	161	3019	39	2015	
170906-1	161	3019	39	2015	
170956	722		IC-9		
170964	1030	3372	IC-163		
170965	116	3058/137A	504A	1104	
170967	289A	3124/289	504A	1104	
170967(RECTIFIER)	116		504A	1104	
170967-1	123A	3444	20	2051	
170968-1	123A	3444	20	2051	
170970-1	116	3031A	504A	1104	
171003	127	3034	25		
171003(SEARS)	108		86	2038	
171003(TOSHIBA)	123A	3444	20	2051	
171004	127	3035	25		
171004(SEARS)	121	3717	239	2006	
171005	126	3010	52	2024	
171005(TOSHIBA)	103A		59	2002	
171009	107	3019	11	2015	
171009(SEARS)	108		86	2038	
171009(TOSHIBA)	123A	3444	20	2051	
171015(TOSHIBA)	124		12		
171016	160	3006	245	2004	
171016(SEARS)	102A		53	2007	
171017	102A	3004	53	2007	
171018	102A		53	2007	
171019	116	3017B/117	504A	1104	
171026		3124	10	2051	
171026(SEARS)	102A		53	2007	
171026(TOSHIBA)	123A	3444	20	2051	
171027	123A	3444	20	2051	
171028	108	3452	86	2038	
171029		3018	11	2015	
171029(SEARS)	108		86	2038	
171029(TOSHIBA)	107		11	2015	
171030	161	3018	39	2015	
171030(SEARS)	108		86	2038	
171030(TOSHIBA)	123A	3444	20	2051	
171031	161	3018	39	2015	
171031(SEARS)	108		86	2038	
171031(TOSHIBA)	107		11	2015	
171032	108	3452	86	2038	
171033	108	3452	86	2038	
171033-1	241	3188A/182	57	2020	
171033-1(PNP)			56	2027	
171033-2(NPN)			57	2017	
171033-2(PNP)		3191	58	2025	
171034	107	3452/108	86	2038	
171038	128	3018	243	2030	
171039		3006	50	2004	
171039(SEARS)	160		245	2004	
171039(TOSHIBA)	160		245	2004	
171040		3124	10	2051	
171040(SEARS)	123A	3444	20	2051	
171040(TOSHIBA)	123A	3444	20	2051	
171044		3018	11	2015	
171044(SEARS)	128		243	2030	
171044(TOSHIBA)	123A	3444	20	2051	
171045	108	3452	86	2038	
171046	123A	3444	20	2051	
171047(TOSHIBA)	162		35		
171048	108	3452	86	2038	
171049	102A	3004	53	2007	
171052	108	3452	86	2038	
171053-1	180	3036	74	2043	
171054	108	3452	86	2038	

Industry Standard No.	ECG	SK	GE	RS 276-	MOTOR.
171090-1	108	3452	86	2038	
171092-1	145A	3093	ZD-15	564	
171139-1	108	3452	86	2038	
171140-1	108	3452	86	2038	
171141-1	108	3452	86	2038	
171144-1			11	2015	
171149-016	177		300	1122	
171149-017			300	1122	
171149-024	116		504A	1104	
171162-004	199	3038	62	2010	
171162-005	123A	3444	20	2051	
171162-006	123A	3444	20	2051	
171162-008	123A	3444	20	2051	
171162-009	123A	3444	20	2051	
171162-025	176	3845	80		
171162-026	192	3018	63	2030	
171162-027	108	3452	86	2038	
171162-039		3126	90		
171162-042	109	3087		1123	
171162-072	158	3004/102A	53	2007	
171162-073	158	3004/102A	53	2007	
171162-074	102A	3004	53	2007	
171162-075	102A	3004	53	2007	
171162-076	102A	3004	53	2007	
171162-080	102A	3004	53	2007	
171162-081	158	3004/102A	53	2007	
171162-082	121	3014	239	2006	
171162-083	131	3052	44	2006	
171162-086	121	3014	239	2006	
171162-089	131		44	2006	
171162-090	131MP	3840	44(2)	2006(2)	
171162-095	199	3122	62	2010	
171162-100	199	3245	62	2010	
171162-108	176	3004/102A	80		
171162-113	199	3024/128	20	2051	
171162-118	161	3117	39	2015	
171162-119	123A	3444	20	2051	
171162-120	102A		53	2007	
171162-121	102A		53	2007	
171162-124	154		40	2012	
171162-125	154		40	2012	
171162-126	154		40	2012	
171162-128	108	3452	86	2038	
171162-129	108	3452	86	2038	
171162-130	108	3452	86	2038	
171162-131	108	3452	86	2038	
171162-132	123A	3444	20	2051	
171162-143	123A	3444	20	2051	
171162-161	123A	3444	20	2051	
171162-162	123A	3444	20	2051	
171162-163	128	3124/289	243	2030	
171162-164	210	3041	252	2018	
171162-169	160	3008	245	2004	
171162-172	138A	3059	ZD-7.5		
171162-180	199	3124/289	62	2010	
171162-186	229	3018	61	2038	
171162-187	229		61	2038	
171162-188	123A	3024/128	18	2030	
171162-190	123A	3444	20	2051	
171162-191	123A	3444	20	2051	
171162-193	129		244	2027	
171162-195	211	3084	253	2026	
171162-196	177	3100/519	300	1122	
171162-197	177	3060/139A	300	1122	
171162-202	123A	3444	20	2051	
171162-204	199	3245	62	2010	
171162-235	210		252	2018	
171162-247	199	3245	62	2010	
171162-252	116	3311	504A	1104	
171162-265	152	3893	66	2048	
171162-269	109	3087		1123	
171162-270	110MP	3709	1N60	1123(2)	
171162-271	177	3100/519	300	1122	
171162-272	139A		ZD-9.1	562	
171162-278	229	3018	61	2038	
171162-279	229	3018	61	2038	
171162-280	229		61	2038	
171162-285	199	3122	62	2010	
171162-286	123A	3444	20	2051	
171162-287	300	3464	273		
171162-288	199	3245	62	2010	
171162-291	152	3893	66	2048	
171162-292	139A		ZD-9.1	562	
171162-U90			20	2051	
171174-1	130MP	3029	15	2041(2)	
171174-2	300MP	3054/196	273MP		
171174-3	196		241	2020	
171175-1(NPN)		3190	57	2017	
171175-1(PNP)		3191	58	2025	
171179-027	1045	3101/706	IC-2		
171179-036	1003	3101/706	IC-43		
171179-045	1087	3477	IC-103		
171179-051	1006	3358	IC-38		
171179-070	1135	3876			
171206-1	108	3452	86	2038	
171206-10			11	2015	
171206-11			11	2015	
171206-13			11	2015	
171206-2	108	3452	86	2038	
171206-4	108	3452	86	2038	
171206-5	108	3452	86	2038	
171206-6	222	3050/221	FET-4	2036	
171206-7			39	2015	
171206-8			61	2038	
171206-9			11	2015	
171207-1	108	3452	86	2038	
171207-2	108	3452	86	2038	
171207-3			11	2015	
171207-4			FET-2	2035	
171217-1	131	3198	44	2006	
171217-3			61	2038	
171217-4			300	1122	
171305	116(2)	3311	504A(2)	1104	
171318	313		11	2015	
171319	313		11	2015	
171320			11	2015	
171338-1			11	2015	
171338-2			11	2015	
171338-3			11	2015	
171416	116	3311	504A	1104	
171522	102A	3004	53	2007	
171553	199	3018	62	2010	
171554	199	3444/123A	20	2051	
171555	234	3466/159	82	2032	
171556	298	3450	272		
171557	297	3124/289	271	2030	
171558	199	3444/123A	20	2051	
171559	199	3444/123A	20	2051	
171560	140A	3061	ZD-10	562	
171561	116	3031A	504A	1104	
171657	116	3311	504A	1104	
171676	199	3124/289	62	2010	
171677	199	3444/123A	20	2051	
171678	199	3124/289	62	2010	

Industry Standard No.	ECG	SK	GE	RS 276-	MOTOR.
171814	611	3126	90		
171840	177	3100/519	300	1122	
171841	177	3100/519	300	1122	
171842	136A	3057	ZD-5.6	561	
171843	177	3100/519	300	1122	
171894		3126	90		
171915	107	3452/108	86	2038	
171916	102A	3004	53	2007	
171917	102A	3004	53	2007	
171982	1072		IC-59		
171983	107	3018	11	2015	
171984	614	3126	90		
172165	177		300	1122	
172201	177	3100/519	300	1122	
172252	1003	3101/706	IC-43		
172253	177	3100/519	300	1122	
172254		3126	90		
172272	1006	3358	IC-38		
172336	159	3025/129	244	2027	
172463	152	3893	66	2048	
172547	177	3100/519	300	1122	
172551	116	3311	504A	1104	
172643	152	3893	66	2048	
172721	117	3031A	504A	1104	
172722	614	3100/519	300	1122	
172761	123AP	3018	39	2015	
172762	193		48		
172763	192	3849/293	47	2030	
172816	158	3004/102A	53	2007	
175006-181	160	3006	245	2004	
175006-182	160	3006	245	2004	
175006-183	160	3006	245	2004	
175006-184	160	3006	245	2004	
175006-185	160	3006	245	2004	
175006-186	102A	3004	53	2007	
175006-187	108	3452	86	2038	
175007-275	123A	3444	20	2051	
175007-276	123A	3444	20	2051	
175007-277	128	3026	243	2030	
175027-021	123A	3444	20	2051	
175027-022	104	3719	16	2006	
175043-023	121	3009	239	2006	
175043-058	123A	3444	20	2051	
175043-059	123A	3444	20	2051	
175043-060	123A	3444	20	2051	
175043-062	108	3452	86	2038	
175043-063	108	3452	86	2038	
175043-064	108	3452	86	2038	
175043-065	131	3052	44	2006	
175043-066	109			1123	
175043-068	109			1123	
175043-069	5072A		ZD-8.2	562	
175043-081	131MP	3840	44(2)	2006(2)	
175043-100	108	3452	86	2038	
175043-107	108	3452	86	2038	
175043-81	121	3009	239	2006	
177105	102	3722	2	2007	
180010-001	912		IC-172		
181003-7	108	3452	86	2038	
181003-8	108	3452	86	2038	
181003-9	108	3452	86	2038	
181012	199	3245	62	2010	
181015	159	3466	82	2032	
181023	123A	3444	20	2051	
181030	159	3466	82	2032	
181034	159	3466	82	2032	
181038	219		74	2043	
181073	177		300	1122	
181214	123A	3444	20	2051	
181503-6	108	3452	86	2038	
181503-7	108	3452	86	2038	
181503-9	108	3452	86	2038	
181504-1	108	3452	86	2038	
181504-2	123A	3444	20	2051	
181504-7	108	3452	86	2038	
181506-7	108	3452	86	2038	
181515-4	128		243	2030	
181515-6	128		243	2030	
181515-7	128		243	2030	
181515-9	128		243	2030	
181619	519	3466/159	82	1122	
181681	177		300	1122	
183013	718	3159	IC-8		
183014	722	3161	IC-9		
183015	107	3018	11	2015	
183016	107	3018	11	2015	
183017	123A	3444	20	2051	
183018	107	3018	11	2015	
183019	107	3018	11	2015	
183030	123A	3444	20	2051	
183031	123A	3444	20	2051	
183032	159	3466	82	2032	
183033	177	3100/519	300	1122	
183034	192	3054/196	63	2030	
183035	193	3083/197	67	2023	
183044	804	3455	IC-27		
183532	941		IC-263	010	
184798	519		514	1122	
185022	506	3843	511	1114	
185236	123AP	3018	20	2051	
185335	519		514	1122	
185411	177		300	1122	
185704	116		504A	1104	
186342A	128	3024	243	2030	
187217	159	3466	82	2032	
187218	123A	3444	20	2051	
188056	519		514	1122	
188165	130		14	2041	
188180	159	3466	82	2032	
188226	129		244	2027	
188400-34	5547	3505			
188660-01	941		IC-263	010	
190404	140A	3061	ZD-10	562	
190425	121	3009	239	2006	
190425A	121	3009	239	2006	
190426	123A	3444	20	2051	
190427	160		245	2004	
190428	123A	3444	20	2051	
190429	159	3466	82	2032	
190714	161	3716	39	2015	
190715	123A	3444	20	2051	
190716	109	3087		1123	
193207	941		IC-263	010	
193717	177	3100/519	300	1122	
194086-3	176	3845	80		
194243	128		243	2030	
194474-8	121	3717	239	2006	
194917	506	3843	504A	1104	
195601-6	102A		53	2007	
195617	109	3087	1N34AS	1123	
195648-6	116		504A	1104	
196023-1	123A	3444	20	2051	
196023-2	123A	3444	20	2051	
196058-4	121	3717	239	2006	

Industry Standard No.	ECG	SK	GE	RS 276-	MOTOR.
196064-3	102A		53	2007	
196148-0	121	3717	239	2006	
196183-5	121	3717	239	2006	
196183-7	102A		53	2007	
196184-3	116		504A	1104	
196259-4	116		504A	1104	
196501-7	121	3717	239	2006	
196607-9	121	3717	239	2006	
196779-9	128		243	2030	
196779-9-1	128		243	2030	
196780-1	128		243	2030	
197464	519		514	1122	
198003-1	123A	3444	20	2051	
198003-2	123A	3444	20	2051	
198005-1	195A	3765	46		
198007-3	123A	3444	20	2051	
198010-1	126	3008	52	2024	
198013-P1	123A	3444	20	2051	
198014-1	128		243	2030	
198020-1	128		243	2030	
198020-2	128		243	2030	
198020-3	128		243	2030	
198023-1	123A	3444	20	2051	
198023-2	128		243	2030	
198023-3	123A	3444	20	2051	
198023-4	123A	3444	20	2051	
198023-5	123A	3444	20	2051	
198024	159	3466	82	2032	
198030	123A	3444	20	2051	
198030-2	123A	3444	20	2051	
198030-3	123A	3444	20	2051	
198030-4	123A	3444	20	2051	
198030-6	123A	3444	20	2051	
198030-7	123A	3444	20	2051	
198031-1	123A	3444	20	2051	
198031-2	123A	3444	20	2051	
198034-1	130	3027	14	2041	
198034-2	130	3510	14	2041	
198034-3	130	3027	14	2041	
198034-4	130	3027	14	2041	
198034-5	130		14	2041	
198035-1	128		243	2030	
198035-3	128		243	2030	
198036-1	159	3466	82	2032	
198038-1	175		246	2020	
198038-3	175	3026	246	2020	
198038-4	175	3026	246	2020	
198039-0507	130	3027	14	2041	
198039-1	175		246	2020	
198039-3	175		246	2020	
198039-501	175		246	2020	
198039-503	175		246	2020	
198039-506	130		14	2041	
198039-507	130		14	2041	
198039-6	130	3027	14	2041	
198039-7	130	3027	14	2041	
198042-2	123A	3444	20	2051	
198042-3	123A	3444	20	2051	
198045-4	128		243	2030	
198047-1	128		243	2030	
198047-2	128		243	2030	
198047-3	128		243	2030	
198047-5	128		243	2030	
198047-6	128		243	2030	
198048-1	128		243	2030	
198048-2	128		243	2030	
198049-1	175		246	2020	
198050	159	3466	82	2032	
198051-1	123A	3444	20	2051	
198051-2	123A	3444	20	2051	
198051-3	123A	3444	20	2051	
198051-4	123A	3444	20	2051	
198063-1	153		69	2049	
198064-1	181	3511	75	2041	
198065-1	129		244	2027	
198065-3	129		244	2027	
198067-1	123A	3444	20	2051	
198072-1	128		243	2030	
198074-1	129		244	2027	
198075-1	198	3220	251		
198077-1	128		243	2030	
198078-1	129		244	2027	
198079-1	181		75	2041	
198079-2	130		14	2041	
198400	909		IC-249		
198410-1	909	3590	IC-249		
198581-1	123A	3444	20	2051	
198581-2	123A	3444	20	2051	
198581-3	123A	3444	20	2051	
198763-1	5870		5032		
198763-10	5877		5037		
198763-11	5878		5036		
198763-12	5879		5037		
198763-13	5880		5040		
198763-14	5881		5041		
198763-15	5882		5040		
198763-16	5883		5041		
198763-2	5871		5033		
198763-3	5874		5032		
198763-4	5875		5033		
198763-5	5874		5032		
198763-6	5875		5033		
198763-7	5874		5032		
198763-8	5875		5033		
198763-9	5876		5036		
198764-13	5994	3501	5104		
198764-5	5986	3609	5096		
198764-6	5987	3698	5097		
198765-1	5980	3610	5096		
198765-10	5991	3518/5995	5101		
198765-11	5992	3501/5994	5104		
198765-12	5993	3518/5995	5105		
198765-13	5994	3501	5104		
198765-14	5995	3518	5105		
198765-2	5981	3698/5987	5097		
198765-3	5982	3609/5986	5096		
198765-4	5983	3698/5987	5097		
198765-5	5986	3609	5096		
198765-6	5987		5097		
198765-7	5988	3608/5990	5100		
198765-8	5989	3518/5995	5101		
198765-9	5990	3608	5100		
198766-1	5529	6629			
198766-2	5529	6629			
198766-3	5529	6629			
198766-5	5529	6629			
198766-7	5529	6629			
198766-8	5529	6629			
198773-1	519		514	1122	
198775-1	519		514	1122	
198776-1	519		514	1122	
198779-1	519		514	1122	
198785	519		514	1122	
198785-1	519		514	1122	
198785-2	519		514	1122	
198794-1	101	3861	8	2002	
198799-4	519		514	1122	
198800-1	5626	3633			
198809-1	177		300	1122	
198810-1	519		514	1122	
198813-1	519		514	1122	
198818	5626	3507			
199596	177	3100/519	300	1122	
199807	194		220		
199919	167	3647	504A	1172	
199985	116(3)	3110/120	CR-3		
200028-7-28	102A		53	2007	
200062-5-31	102A		53	2007	
200062-5-32	102A		53	2007	
200062-5-33	102A		53	2007	
200062-5-34	102A		53	2007	
200062-5-36	116		504A	1104	
200062-6-32	116		504A	1104	
200064-6-103	108	3452	86	2038	
200064-6-104	160		245	2004	
200064-6-105	108	3452	86	2038	
200064-6-106	160		245	2004	
200064-6-107	123A	3444	20	2051	
200064-6-108	102A		53	2007	
200064-6-109	127	3764	25		
200064-6-110	127	3764	25		
200064-6-111	102A		53	2007	
200064-6-115	116		504A	1104	
200064-6-119	116		504A	1104	
200064-6-120	116		504A	1104	
200067	159	3466	82	2032	
200076	123A	3444	20	2051	
200200	199	3245	62	2010	
200200-700	199	3245	62	2010	
200220	159	3466	82	2032	
200251-5377	123A	3444	20	2051	
200252	128		243	2030	
200259-700	218		234		
200433	159	3466	82	2032	
200648-26	109			1123	
200781-702	124		12		
201034	124		12		
0201201	1003		IC-43		
202315	109	3087		1123	
202315(THOMAS)	177		300	1122	
202463	116	3311	504A	1104	
202609-0713	123A	3444	20	2051	
202617	199	3122	61	2010	
202862-518	519		514	1122	
202862-947	123A	3444	20	2051	
202907-047P1	123A	3444	20	2051	
202909-577	159	3466	82	2032	
202909-587	159	3466	82	2032	
202909-827	130		14	2041	
202911-737	159	3466	82	2032	
202913-057	128		243	2030	
202914-417	123A	3444	20	2051	
202915-627	123A	3444	20	2051	
202917-137	129		244	2027	
202922-237	123A	3444	20	2051	
202922-280	716		IC-208		
202925-047	219		74	2043	
203364	159	3466	82	2032	
203718	129		244	2027	
204117	199	3111	212	2010	
204201-001	128		243	2030	
204210-002	123A	3444	20	2051	
204211-001	130		14	2041	
204969	123A	3444	20	2051	
205032	159	3466	82	2032	
205048	159	3466	82	2032	
205049	159	3466	82	2032	
205367	159	3466	82	2032	
205782-103			20	2051	
205782-97			20	2051	
206180	5802	9005		1143	
206185	5802	9005		1143	
206190	5802	9005		1143	
206617	177	3100/519	300	1122	
0207046	1108		IC-87		
207119	124		12		
0207120	1087	3477	IC-103		
0207205	1155	3231	IC-179	705	
207417	312	3448	FET-2	2035	
209185-962	121	3717	239	2006	
209417-0714	108	3452	86	2038	
210074	123A	3444	20	2051	
210076	159	3466	82	2032	
211040-1	181		75	2041	
211083	519		514	1122	
212717	229		60	2015	
213217	199	3124/289	62	2010	
214105	116	3016	504A	1104	
214396	121	3009	239	2006	
215002			50	2004	
215008			50	2004	
215031			50	2004	
215038			50	2004	
215053			53	2007	
215071	131MP	3840	44(2)	2006(2)	
215072	123A		20	2051	
215074	123	3124/289	20	2051	
215075	128	3124/289	243	2030	
215081	123A	3444	20	2051	
215089	131		44	2006	
215669	116	3016	504A	1104	
216001	109	3088		1123	
216003	110MP	3088	1N60	1123(2)	
216014	116		504A	1104	
216020	116		504A	1104	
216024	5072A		ZD-8.2	562	
216445-2	123A	3444	20	2051	
216449	116	3016	504A	1104	
216817	116	3016	504A	1104	
216986	121	3009	239	2006	
217119	105	3012	4		
217230	105	3012	4		
217892	121	3009	239	2006	
218012DS	104	3009	16	2006	
218502	102A	3003	53	2007	
218503	102A	3004	53	2007	
218511	128		243	2030	
218537	129		244	2027	
218612	116	3016	504A	1104	
219016	101	3861	8	2002	
219245	117	3017B	300	1104	
219301	121	3009	239	2006	
219361	121	3009	239	2006	
219440	121	3009	239	2006	
219935	116	3016	504A	1104	
219940	121	3009	239	2006	
221128	116	3016	504A	1104	
221158	128	3024	243	2030	

Industry Standard No.	ECG	SK	GE	RS 276-	MOTOR.
221600	123A	3444	20	2051	
221601	101	3861	8	2002	
221602	121	3009	239	2006	
221605	121	3009	239	2006	
221856	126	3005	52	2024	
221857	123A	3122	20	2051	
221897	123A	3444	20	2051	
221918	123A	3444	20	2051	
221924	101	3861	8	2002	
221940	121	3009	239	2006	
221941	121	3009	239	2006	
222131	123A	3444	20	2051	
222509	102A	3004	53	2007	
222611	116	3016	504A	1104	
222867	125	3032A	510,531	1114	
222915	121	3009	239	2006	
223124	102A	3004	53	2007	
223215	116	3016	504A	1104	
223216	116	3016	504A	1104	
223323	116	3016	504A	1104	
223357	125	3032A	510,531	1114	
223358	116	3016	504A	1104	
223365	121	3009	239	2006	
223366	102A	3004	53	2007	
223367	101	3861	8	2002	
223368	101	3861	8	2002	
223369	160	3007	245	2004	
223370	101	3861	8	2002	
223371	102A	3003	53	2007	
223372	100	3005	1	2007	
223462	116	3016	504A	1104	
223467	116	3016	504A	1104	
223473	100	3005	1	2007	
223474	160	3007	245	2004	
223475	100	3005	1	2007	
223482	103A	3010	59	2002	
223483	102A	3004	53	2007	
223484	102A	3003	53	2007	
223485	102A	3003	53	2007	
223486	102A	3004	53	2007	
223487	126	3008	52	2024	
223489	117	3017B	504A	1104	
223490	121	3009	239	2006	
223576	121	3009	239	2006	
223684	101	3861	8	2002	
223720	117	3017B	504A	1104	
223724	116	3016	504A	1104	
223753	116	3512	504A	1104	
223810	102A	3009	53	2007	
224159	116	3016	504A	1104	
224503	121	3009	239	2006	
224506	123A	3124/289	20	2051	
224584	100	3005	1	2007	
224586	160	3006	245	2004	
224587	160	3006	245	2004	
224597	116	3016	504A	1104	
224696	102A	3004	53	2007	
224774	116	3016	504A	1104	
224780	142A	3062	ZD-12	563	
224820	103A	3011	59	2002	
224857	102A	3004	53	2007	
224873	121	3014	239	2006	
225200	116	3016		1104	
225265	5994	3501	5104		
225267	117	3017B	504A	1104	
225300	103A	3010	59	2002	
225301		3057	ZD-5.6	561	
225311	160	3007	245	2004	
225316	136A	3777/5011A	ZD-5.6	561	
225410	109	3311		1123	
225592	116	3016	504A	1104	
225593	102A	3003	53	2007	
225594	160	3007	245	2004	
225594A	126	3008	52	2024	
225595	121	3009	239	2006	
225596	121	3009	239	2006	
225600	160	3006	245	2004	
225925	121	3009	239	2006	
225927	121	3009	239	2006	
226058	116	3016	504A	1104	
226181	100	3005	1	2007	
226182	116	3016	504A	1104	
226237	116	3016	504A	1104	
226334	109			1123	
226338	100	3005	1	2007	
226344	109	3088		1123	
226441	101	3861	8	2002	
226517	312	3112	FET-1	2028	
226546	116	3016	504A	1104	
226634	121	3009	239	2006	
226788	116	3016	504A	1104	
226789	105	3012	4		
226791	103A	3010	59	2002	
226922	125	3032A	510,531	1114	
226924	102A	3004	53	2007	
226999	121	3009	239	2006	
227000	108	3452	86	2038	
227015	117	3017B	504A	1104	
227348	116	3016	504A	1104	
227517	123AP		211	2016	
227565	116	3016	504A	1104	
227566	121	3009	239	2006	
227675	5882	3500	5040		
227676	5994	3501	5104		
227720	116	3016	504A	1104	
227724	5882	3500	5040		
227744			504A	1104	
227752	100	3005	1	2007	
227801	5882	3500	5040		
227804	121	3009	239	2006	
228007	140A	3061	ZD-10	562	
228229	121	3009	239	2006	
228230	121	3009	239	2006	
228287	102A	3003	53	2007	
228417	229	3122	20	2051	
228558	121	3009	239	2006	
228559	121	3009	239	2006	
228560	125	3311	510,531	1114	
229017	123A	3444	211	2051	
229042	5882	3500	5040		
229045	105	3012	4		
229088	5994	3501	5104		
229133	160	3006	245	2004	
229392	108	3452	86	2038	
229522	116	3016	504A	1104	
229805	156	3051	512		
230084	175	3026	246	2020	
230199-1	HIDIV-12	3871/DIV-4	FR-8		
230199-2	HIDIV-1	3871/DIV-4			
230199-3	HIDIV-4	3871/DIV-4			
230199-4	HIDIV-4	3871/DIV-4			
230199-5	HIDIV-4	3871/DIV-4			
230208	121	3717	239	2006	
230209	101	3861	8	2002	
230214	128	3024	243	2030	
230218	116	3016	504A	1104	
230233	128		243	2030	
230253	102A	3004	53	2007	
230256	101	3861	8	2002	
230259	102A	3004	53	2007	
230523	121	3014	239	2006	
230524	102A	3004	53	2007	
230525	102A	3004	53	2007	
230756	5994	3501	5104		
230768	5529	6629			
230773	5882	3500	5040		
231017	297	3122	210	2051	
231140-01	108	3452	86	2038	
231140-04	127	3034	25		
231140-07	108	3452	86	2038	
231140-09	127	3034	25		
231140-11	121	3009	239	2006	
231140-15	123A	3444	20	2051	
231140-21	102A	3004	53	2007	
231140-23	108	3452	86	2038	
231140-26	127	3034	25		
231140-28	154	3040	40	2012	
231140-31	108	3452	86	2038	
231140-33	121	3009	239	2006	
231140-34	108	3452	86	2038	
231140-36	161	3019	39	2015	
231140-37	161	3019	39	2015	
231140-43	161	3019	39	2015	
231140-44	108	3452	86	2038	
231140-45	103A	3010	59	2002	
231150			504A	1104	
231339	116	3016	504A	1104	
231374	123A	3444	20	2051	
231375	128	3024	243	2030	
231378	130	3027	14	2041	
231517	294	3841			
231588	100	3004/102A	53	2007	
231665	116	3016	504A	1104	
231669	116	3016	504A	1104	
231672	121	3009	239	2006	
231706	102A	3008	53	2007	
231797	121	3009	239	2006	
231923	116	3016	504A	1104	
232017	123A		61	2038	
232194	121	3009	239	2006	
232203	116	3016	504A	1104	
232268	175		246	2020	
232359	130	3027	14	2041	
232359A	130	3027	14	2041	
232519	5882	3500	5040		
232520	5994	3501	5104		
232631	159	3466	82	2032	
232674	121	3009	239	2006	
232675	121	3009	239	2006	
232676	160	3006	245	2004	
232678	123A	3444	20	2051	
232680	160	3006	245	2004	
232681	126	3008	52	2024	
232840	108	3452	86	2038	
232841	128	3024	243	2030	
232949	101	3861	8	2002	
233011	116	3311	504A	1104	
233025			511	1114	
233062	506	3843	511	1114	
233117	229		61	2038	
233148	5022A	3788	ZD-13	563	
233150	145A	3063	ZD-15	564	
233305	105	3012	4		
233307	105	3012	4		
233507	100	3005	1	2007	
233508	105	3012	4		
233509	121	3014	239	2006	
233561	116	3016	504A	1104	
233597	116	3016	504A	1104	
233735	128	3024	243	2030	
233944	128	3024	243	2030	
233945	100	3005	1	2007	
233969	129	3025	244	2027	
234015	160	3006	245	2004	
234024	128	3024	243	2030	
234076	102A	3004	53	2007	
234077	121	3009	239	2006	
234078	105	3012	4		
234178	121	3009	239	2006	
234552	116	3016	504A	1104	
234553	125	3032A	510,531	1114	
234565	116	3016	504A	1104	
234566	121	3009	239	2006	
234611	116	3016	504A	1104	
234612	123A	3444	20	2051	
234630	102A	3004	53	2007	
234631	160	3006	245	2004	
234758	123A	3444	20	2051	
234761	116	3016	504A	1104	
234763	123A	3444	20	2051	
235157	5882	3500	5040		
235192	123A	3444	20	2051	
235194	102A	3004	53	2007	
235200	160	3006	245	2004	
235205	123A	3444	20	2051	
235206	123A	3444	20	2051	
235299	5994	3501	5104		
235312	121	3009	239	2006	
235313	116	3016	504A	1104	
235382	125	3311	510,531	1114	
235541	156	3051	512		
235543	116	3016	504A	1104	
235546	116	3016	504A	1104	
235997	162	3079	35		
236039	161	3039/316	39	2015	
236251	108	3452	86	2038	
236265	101	3861	8	2002	
236266	139A	3060	ZD-9.1	562	
236282	154	3040	40	2012	
236285	123A	3124/289	20	2051	
236286	123A	3124/289	20	2051	
236287	128	3024	243	2030	
236288	127	3035	245	2005	
236433	129	3025	244	2027	
236706	108	3452	86	2038	
236709	102A	3004	53	2007	
236854	181	3036	75	2041	
236907	108	3452	86	2038	
236935	121	3009	239	2006	
237020	108	3452	86	2038	
237021	108	3452	86	2038	
237024	108	3452	86	2038	
237025	123A	3444	20	2051	
237026	108	3452	86	2038	
237028	195A	3047	46		
237070	156	3051	512		
237075	191	3044/154	249		
237223	123A	3444	20	2051	

Industry Standard No.	ECG	SK	GE	RS 276-	MOTOR.
237227	116	3016	504A	1104	
237421	156	3051	512		
237450	124	3021	12		
237452	121	3009	239	2006	
237453	116	3016	504A	1104	
237509	5994	3501	5104		
237785	108	3452	86	2038	
237840	108	3452	86	2038	
237929	116	3016	504A	1104	
238208-002	175		246	2020	
238368	123A	3444	20	2051	
238417	102A	3004	53	2007	
238418	102A	3004	53	2007	
239097	156	3051	512		
239103	225	3045	256		
239219	138A	3059	ZD-7.5		
239221	116	3016	504A	1104	
239429	125	3033	510,531	1114	
239517	196	3054	241	2020	
239612	128	3024	243	2030	
239706	5074A	3786/5020A			
239713	181	3036	75	2041	
239917	235	3197	215		
239970	123A	3444	20	2051	
240003	102A		53	2007	
240006	102A		53	2007	
240055	156	3051	512		
240076	156	3051	512		
240077	5994	3501	5104		
240388	124	3021	12		
240401	123A	3444	20	2051	
240402	129	3025	244	2027	
240403	127	3034	25		
240404	162	3438	35		
240456	116	3016	504A	1104	
240564	177	3100/519	300	1122	
240594	116	3016	504A	1104	
240603	116	3016	504A	1104	
241052	129	3025	244	2027	
241184		3126	90		
241249	108	3452	86	2038	
241295	125	3031A	510,531	1114	
241302	125	3512	510,531	1114	
241420	5994	3501	5104		
241517	159		221	2032	
241657	130	3027	14	2041	
241778	108	3452	86	2038	
241960	108	3452	86	2038	
242029	116	3016	504A	1104	
242102	152	3893	66	2048	
242141	125	3033	510,531	1114	
242183	121	3009	239	2006	
242221	102A	3004	53	2007	
242226	116	3016	504A	1104	
242323	282	3024/128			
242422	129	3025	244	2027	
242460	129	3025	244	2027	
242590	108	3452	86	2038	
242758	123A	3444	20	2051	
242759	123A	3444	20	2051	
242838	121	3009	239	2006	
242958	129	3025	244	2027	
242960	108	3452	86	2038	
243028	724	3525	IC-86		
243115	156	3051	512		
243168	127	3035	25		
243215	105	3012	4		
243318	108	3452	86	2038	
243364	116	3016	504A	1104	
243645	108	3452	86	2038	
243815	105	3012	4		
243837	103A	3010	59	2002	
243843	5642	3506			
243939	102A	3004	53	2007	
244007	100	3005	1	2007	
244357	124	3021	12		
244817	123A	3444	211	2016	
245078-3	108	3452	86	2038	
245117	5072A		ZD-8.2	562	
245192	177		300	1122	
245217	139A		ZD-9.1	562	
245517	109	3087	1N34AS	1123	
245568-2	176	3123	80		
245917	306		276		
248017	297	3024/128	271	2030	
248717	295		270		
248817	177	3100/519	300	1122	
248917	116	3311	504A	1104	
249217	116	3100/519	504A	1104	
249508-3	177	3100/519	300	1122	
249588	126	3006/160	52	2024	
250400	103A	3010	59	2002	
252744	124		12		
252817	116	3311	504A	1104	
253704	105	3012	4		
255728	102A	3004	53	2007	
255821H8	193	3025/129	67	2023	
255903	5994	3501	5104		
256068	121	3009	239	2006	
256071	121	3009	239	2006	
256122	5994	3501	5104		
256126	100	3005	1	2007	
256127	101	3861	8	2002	
256217	123A	3122	20	2051	
256317	290A	3114/290	269		
256319			269	2032	
256417	229		212	2010	
256480	105	3012	4		
256517	123A		20	2051	
256617	229		60	2015	
256717	612	3325	90		
256728	5994	3501	5104		
256729	5882	3500	5040		
256730	5994	3501	5104		
256817	135A	3122	61	2038	
256917	123A	3122	61	2051	
257017	199		212	2010	
257242	105	3012	4		
257243	105	3012	4		
257317	5075A	3751			
257340	102A	3004	53	2007	
257341	121	3009	239	2006	
257385	101	3861	8	2002	
257403	121	3009	239	2006	
257470	100	3005	1	2007	
257473	102A	3004	53	2007	
257534	105	3012	4		
257536	121	3009	239	2006	
257540	108	3452	86	2038	
258017	222	3065	FET-4	2036	
258882	116	3016	504A	1104	
258884	5994	3501	5104		
258990	121	3009	239	2006	
258993	101	3011	8	2002	

Industry Standard No.	ECG	SK	GE	RS 276-	MOTOR.
259315	116	3311	504A	1104	
259368	116	3016	504A	1104	
259878	116	3016	504A	1104	
260429	116	3016	504A	1104	
260468	101	3861	8	2002	
260565	108	3452	86	2038	
261401	105	3012	4		
261463	116	3016	504A	1104	
261488	105	3012	4		
261586	160	3008	245	2004	
261596	116	3016	504A	1104	
261898	116	3016	504A	1104	
261970	121	3009	239	2006	
261975	116	3016	504A	1104	
262066	123A	3444	20	2051	
262111	140A	3061	ZD-10	562	
262112	116	3016	504A	1104	
262113	102A	3004	53	2007	
262114	121	3009	239	2006	
262116	175	3026	246	2020	
262309	105	3012	4		
262310	5994	3501	5104		
262370	121	3009	239	2006	
262417	128		243	2030	
262417-1	128		243	2030	
262546	140A	3061	ZD-10	562	
262638	129	3025	244	2027	
262648	5882	3500	5040		
262872	116	3016	504A	1104	
263424	140A	3061	ZD-10	562	
263561	124	3021	12		
263807	5882	3500	5040		
263856	121	3009	239	2006	
263857	124	3021	12		
265074	108	3452	86	2038	
265115	5994	3501	5104		
265164	116	3016	504A	1104	
265235	116	3016	504A	1104	
265236	117	3017B	504A	1104	
265240	123A	3444	20	2051	
265241	108	3452	86	2038	
265634	116	3016	504A	1104	
265771	160	3006	245	2004	
266583	117	3017B	504A	1104	
266685	123A	3444	20	2051	
266686	102A	3004	53	2007	
266702	100	3005	1	2007	
267272	146A	3064	ZD-27		
267611	177		300	1122	
267704	128		243	2030	
267791	130	3027	14	2041	
267797	108	3452	86	2038	
267838	159	3466	82	2032	
267878	181	3036	75	2041	
267898	123A	3444	20	2051	
267899	123A	3444	20	2051	
268003	199	3245	62	2010	
268044L	123A	3444	20	2051	
268717	129	3114/290	244	2027	
269367	103A	3010	59	2002	
269374	102A	3004	53	2007	
269922-001	177		300	1122	
270642	125	3031A	510,531	1114	
270744	121	3009	239	2006	
270745	121	3009	239	2006	
270746	121	3009	239	2006	
270779	5882	3500	5040		
270780	121	3009	239	2006	
270781	101	3861	8	2002	
270785	121	3009	239	2006	
270786	5529	6629			
270819	123A	3444	20	2051	
275131	123A	3444	20	2051	
275612	121	3009	239	2006	
275831	125	3033	510,531	1114	
275845	127	3034	25		
276097	116	3016	504A	1104	
276160	128	3024	243	2030	
276331	123A	3444	20	2051	
276413	128	3024	243	2030	
276415	128	3024	243	2030	
279317	103A		59	2002	
279417	229		20	2051	
279517	152		66	2048	
279617	295		270		
279717	235	3197	337		
279817	253	3180/263			
279917	123A		62	2051	
280017	519	3100	514	1122	
280117	601	3463			
280217	116	3032A	509	1114	
280317	139A	3060	ZD-9.1	562	
280417	612	3325			
281001-53	199	3245	62	2010	
281001-83	199	3245	62	2010	
281101-97	177		300	1122	
281917	138A	3059	ZD-7.5		
282217	390	3027/130		2041	
282317	123A	3122	210	2051	
282601	116(3)	3110/120	CR-3		
290458LGD	123A	3444	20	2051	
291509	181	3036	75	2041	
297065C03			51	2004	
297074C11	129	3025	244	2027	
297240-1	102A	3004	53	2007	
299371-1	123A	3444	20	2051	
299379	724		IC-86		
300008	102		2	2007	
300061-04	716		IC-208		
300113	123A	3444	20	2051	
300233	116	3016	504A	1104	
300312	5994	3501	5104		
300315	116	3016	504A	1104	
300486	101	3861	8	2002	
300524	116	3016	504A	1104	
300532	5882	3500	5040		
300536	101	3861	8	2002	
300538	102A	3004	53	2007	
300540	102A	3004	53	2007	
300541	102A	3004	53	2007	
300542	101	3861	8	2002	
300550	116	3016	504A	1104	
300732	140A	3061	ZD-10	562	
300733	5994	3501	5104		
300735	5882	3500	5040		
300774	101	3861	8	2002	
301586	116	3016	504A	1104	
301591	123A	3444	20	2051	
301606	128	3024	243	2030	
301915-1	941		IC-263	010	
302342	123A	3444	20	2051	
302540	137A	3058	ZD-6.2	561	
302865	129		244	2027	
304281-P1	177		300	1122	

Industry Standard No.	ECG	SK	GE	RS 276-	MOTOR.
304581B	123A	3444	20	2051	
304900	108	3452	86	2038	
308449	123A	3444	20	2051	
309412	121	3009	239	2006	
309418	128	3024			
309419	129	3114/290			
309421	102A	3004	53	2007	
309441	152	3893	66	2048	
309442	123A	3444	20	2051	
309449	130		14	2041	
309459	152	3054/196	66	2048	
309481	519		514	1122	
309683	234		65	2050	
309684	159	3466	82	2032	
309685	234		65	2050	
309689	196	3054	241	2020	
309690	197	3083	250	2027	
310017	102A	3004	53	2007	
310030	160	3007	245	2004	
310035	102A	3052	53	2007	
310110	128		243	2030	
310132	160	3006	245	2004	
310157	126	3008	52	2024	
310158	126	3008	52	2024	
310159	102A	3004	53	2007	
310160	102A		53	2007	
310162	160	3006	245	2004	
310201	102A	3004	53	2007	
310204	160	3006	245	2004	
310221	126	3006/160	52	2024	
310223	100	3005	1	2007	
310224	126	3006/160	52	2024	
310225	102A		53	2007	
310254	7408	7408		1822	
313309-1	123A		20	2051	
315930	128		243	2030	
315932	128		243	2030	
317208	142A	3062	ZD-12	563	
318835	130	3027	14	2041	
319304	129		244	2027	
320007	109	3091		1123	
0320031	123A	3122		2051	
0320051	154	3044	40	2012	
0320064	289A	3124/289			
320280	159	3466	82	2032	
320529	123A	3444	20	2051	
321006P1	703A		IC-12		
321145	128		243	2030	
321165	159	3466	82	2032	
321166	128		243	2030	
321264-2	195A	3765	46		
321517	123A	3444	20	2051	
321573	123A	3444	20	2051	
322968-140	121	3717	239	2006	
322968-141	127	3764	25		
322968-167	102A		53	2007	
322968-17	102A		53	2007	
323934	199	3245	62	2010	
324144	159	3466	82	2032	
325077	129		244	2027	
325079	123A	3444	20	2051	
325099	124		12		
325101	128		243	2030	
326309-10A	519		514	1122	
326809	909		IC-249		
326823	923		IC-259		
328785	123		20	2051	
330003	116	3016	504A	1104	
330018	116	3016	504A	1104	
330019	116	3016	504A	1104	
0330302	142A	3062	ZD-12	563	
330803	123A	3444	20	2051	
331383	128		243	2030	
332762	219		74	2043	
333060-1029	159	3466	82	2032	
333241	123A	3444	20	2051	
334724-1	123A	3444	20	2051	
335288-4	123A	3444	20	2051	
335613	129		244	2027	
335774	123A	3444	20	2051	
337342	159	3466	82	2032	
338307	519		514	1122	
339002	74H00	74H00			
339300	7400	7400		1801	
339300-2	7400	7400		1801	
339486	7486	7486		1827	
340866-2	128		243	2030	
346015-15	108	3452	86	2038	
346015-16	108	3452	86	2038	
346015-17	108	3452	86	2038	
346015-18	108	3452	86	2038	
346015-19	108	3452	86	2038	
346015-20	108	3452	86	2038	
346015-21	108	3452	86	2038	
346015-22	108	3452	86	2038	
346015-23	128	3024	243	2030	
346015-24	123A	3444	20	2051	
346015-25	108	3452	86	2038	
346015-30	128	3024	243	2030	
346015-37	108	3452	86	2038	
346016-1	102A	3004	53	2007	
346016-11	102A	3004	53	2007	
346016-14	123A	3444	20	2051	
346016-16	128	3124/289	243	2030	
346016-17	128	3024	243	2030	
346016-18	123A	3444	20	2051	
346016-19	123A	3444	20	2051	
346016-25	123A	3444	20	2051	
346016-26	123A	3444	20	2051	
346016-27	128	3024	243	2030	
346016-63	191	3044/154	249		
346607-4	160		245	2004	
348048-2	116	3016	504A	1104	
348053-3	116	3016	504A	1104	
348054-1	116	3016	504A	1104	
348054-10	116	3016	504A	1104	
348054-11	117	3017B	504A	1104	
348054-14	117	3017B	504A	1104	
348054-15	125	3311	510,531	1114	
348054-2	116	3016	504A	1104	
348054-5	116	3016	504A	1104	
348054-6	117	3017B	504A	1104	
348054-7	125	3032A	510,531	1114	
348054-9	116	3016	504A	1104	
348055-2	116	3016	504A	1104	
348055-3	116	3016	504A	1104	
348057-17	136A	3057	ZD-5.6	561	
348057-8	125	3031A	510,531	1114	
348057-9	125	3032A	510,531	1114	
348058-2	125	3311	510,531	1114	
348846-1	128		243	2030	
350109	311		277		
00351980	128		243	2030	
00352080	123A	3444	20	2051	
359810	788	3829			
373003	102A	3004	53	2007	
373117	102A	3004	53	2007	
373119	102A	3004	53	2007	
373401-1	7400	7400		1801	
373404-1	7404	7404		1802	
373405-1	7410	7410		1807	
373406-1	7420	7420		1809	
373407-1	7430	7430			
373408-1	7440	7440			
373409-1	7474	7474		1818	
373410-1	7486	7486		1827	
373414-1	7476	7476		1813	
373427-1	7490	7490		1808	
373429-1	7406	7406		1821	
373712-1	7492	7492		1819	
373713-1	7475	7475		1806	
373714-2	7454	7454			
373715-1	7451	7451		1825	
373718-1	7493A	7493			
374109-1	7408	7408		1822	
379005N	519		514	1122	
379101K	123A	3444	20	2051	
379102	123		20	2051	
386726-1	129		244	2027	
388060	123A	3444	20	2051	
395253-1	130		14	2041	
400108	128		243	2030	
400127	175		246	2020	
400909	128		243	2030	
401003-001 0	159			2032	
401003-0010		3466	82	2032	
401646	912		IC-172		
405101-1	194		220		
405192	106	3984	21	2034	
405457	123A	3444	20	2051	
405919-35AD	5882		5040		
405919-45AD	5883		5040		
405965-30A	179	3642	76		
405965-35A	129		244	2027	
405965-8A	128		243	2030	
417214	162	3438	35		
425411-01	128		243	2030	
425793	135A	3776/5010A			
433836	123A	3444	20	2051	
436119-002	159	3466	82	2032	
0440002-003	128		243	2030	
00444028-010	123A	3122	20	2051	
00444028-014	123A	3444	20	2051	
445023-P1	130		14	2041	
445111	910		IC-251		
450826-1	128		243	2030	
452077	312	3112		2028	
454549	100	3721	1	2007	
454760	160		245	2004	
462580-1	519		514	1122	
463984-1	7440	7440			
464010	116	3016	504A	1104	
464070	116	3016	504A	1104	
465002-30	799		IC-34		
475018	5802	9005		1143	
476171-18			511	1114	
476690-2	177		300	1122	
481335	128		243	2030	
485752-090	112		1N82A		
489751-001			1N60	1123	
489751-020		3089	1N82A		
489751-025	123A	3444	20	2051	
489751-026	123A	3444	20	2051	
489751-027	108	3452	86	2038	
489751-028	159	3466	82	2032	
489751-029	123A	3444	20	2051	
489751-030	123A	3444	20	2051	
489751-031	159	3466	82	2032	
489751-032	153	3025/129	69	2049	
489751-033	152	3893	66	2048	
489751-037	128	3018	243	2030	
489751-038	128	3018	243	2030	
489751-039	161	3018	39	2015	
489751-040	123A	3444	20	2051	
489751-041	123A	3444	20	2051	
489751-042	159	3466	82	2032	
489751-043	124	3021	12		
489751-044	152	3893	66	2048	
489751-045	102A	3004	53	2007	
489751-047	161	3018	39	2015	
489751-048			11	2015	
489751-049		3019	11	2015	
489751-052	107	3018	11	2015	
489751-058		3019	11	2015	
489751-097	159	3466	82	2032	
489751-107	123A	3444	20	2051	
489751-108	102A	3008	53	2007	
489751-109	102A	3004	53	2007	
489751-113	102A	3005	53	2007	
489751-114	102A	3004	53	2007	
489751-115	127	3034	25		
489751-119	130	3111	14	2041	
489751-120			11	2015	
489751-121	161	3039/316	39	2015	
489751-122	123A	3444	20	2051	
489751-123			20	2051	
489751-124	159	3466	82	2032	
489751-125	123A	3444	20	2051	
489751-127	161	3018	39	2015	
489751-128	161	3018	39	2015	
489751-129	128	3024	243	2030	
489751-130	159	3466	82	2032	
489751-131	108	3452	86	2038	
489751-137	108	3452	86	2038	
489751-143	108	3452	86	2038	
489751-144	128	3024	243	2030	
489751-145	108	3452	86	2038	
489751-146	159	3466	82	2032	
489751-147	108	3452	86	2038	
489751-148	108	3452	86	2038	
489751-162	108	3452	86	2038	
489751-163	131	3052	44	2006	
489751-164		3122	20	2051	
489751-165	108	3452	86	2038	
489751-166	123A	3444	20	2051	
489751-167	108	3452	86	2038	
489751-168	108	3452	86	2038	
489751-169	108	3452	86	2038	
489751-171	108	3452	86	2038	
489751-172	123A	3444	20	2051	
489751-173	107	3018	11	2015	
489751-174	175	3026	246	2020	
489751-175	1046	3026	IC-118		
489751-206	108	3452	86	2038	
489751-208	312	3448	FET-1	2035	
489752-001	109	3088		1123	
489752-003	109	3087		1123	
489752-005	116	3017B/117	504A	1104	

Industry Standard No.	ECG	SK	GE	RS 276-	MOTOR.
489752-013	116	3032A	504A	1104	
489752-014			504A	1104	
489752-015	116	3032A	504A	1104	
489752-016	506	3109/119	511	1114	
489752-017	114	3120	6GD1	1104	
489752-020	112	3089	1N82A		
489752-022	116	3031A	504A	1104	
489752-025	116	3031A	504A	1104	
489752-026	506	3043	511	1114	
489752-027	116		504A	1104	
489752-028	116	3032A	504A	1104	
489752-029	156	3051	512		
489752-031	109	3088		1123	
489752-035	116	3016	504A	1104	
489752-036	110MP	3709		1123(2)	
489752-037			504A	1104	
489752-038	166	9075	BR-600	1152	
489752-040	140A	3061	ZD-10	562	
489752-041	145A	3062/142A	ZD-15	564	
489752-042	109	3087		1123	
489752-043	116	3031A	504A	1104	
489752-044	113A	3119/113	6GC1		
489752-045	142A		ZD-12	563	
489752-049	109	3088		1123	
489752-050	116	3017B/117	504A	1104	
489752-051	116	3032A	504A	1104	
489752-052	112	3089	1N82A		
489752-054	503	3068	CR-5		
489752-066	116(3)	3032A	504A(3)	1104	
489752-072	120	3110	CR-3		
489752-073	116	3109/119	504A	1104	
489752-074	113A		6GC1		
489752-075	116	3031A	504	1104	
489752-076	109	3087		1123	
489752-077	501B		520		
489752-078	119		CR-2		
489752-08			1N34AS	1123	
489752-088		3126	90		
489752-089	112	3089	1N82A		
489752-090	112	3089	1N82A		
489752-091	5074A		ZD-11.0	563	
489752-092	116	3032A	504A	1104	
489752-094	116	3032A	504A	1104	
489752-095	161	3716	39	2015	
489752-096	506	3031A	511	1114	
489752-097	116	3031A	504A	1104	
489752-108	116	3311	504A	1104	
489752-109	140A	3061	ZD-10	562	
489752-123			504A	1104	
489752-125	110MP	3088	1N60	1123(2)	
489752-169		3126	90		
489765-005	113A	3119/113	6GC1		
489850-004	109	3087		1123	
497442	519		514	1122	
497616	519		514	1122	
497616-1	519		514	1122	
497616-2	519		514	1122	
500001	110MP	3091		1123	
500003	125		510,531	1114	
500009G	109	3088		1123	
500859	116	3017B/117	504A	1104	
500879	128		243	2030	
501010	116	3311	504A	1104	
501152	116	3017B/117	504A	1104	
501343	177		300	1122	
502349	130		14	2041	
503146-1	178MP		300(2)	1122(2)	
504720	177		300	1122	
504720-1	519		514	1122	
504833	5834		512		
505198	130		14	2041	
505256	175		246	2020	
505257	175		246	2020	
505287	128		243	2030	
505342	909	3590	IC-249		
505434	175		246	2020	
505469	175		246	2020	
505568	181		75	2041	
506902	199	3124/289	62	2010	
506911	5802	9005	510	1143	
508511	941		IC-263	010	
508590	7404	7404		1802	
508762	289A	3444/123A	20	2051	
0510006	116	3311	504A	1104	
510007	102A	3004	53	2007	
0510079	160	3006	245	2004	
0510079H	160	3006	245	2004	
510584	199	3245	62	2010	
5115348	110MP	3709		1123(2)	
511806	128		243	2030	
514023	123A	3444	20	2051	
514045	107	3018	11	2015	
5140678	129	3114/290	244	2027	
5140688	193	3114/290	67	2023	
5140728	129	3114/290	244	2027	
5150398	128	3122	243	2030	
5150418	107	3018	11	2015	
5150438	123A	3444	20	2051	
5150458	123A	3444	20	2051	
5160098	312	3448	FET-1	2035	
0517022	109	3088		1123	
0517132	177	3311	300	1122	
0517133	177	3100/519	300	1122	
0517261B		3126	90		
0517550-3	116		504A	1104	
0517750	116	3311	504A	1104	
0517750H	116	3311	504A	1104	
0517826	109	3087		1123	
0517828	109	3100/519		1123	
0517829	109	3087		1123	
517999	909	3590	IC-249		
5180228	720	9014	IC-7		
0518926	177	3170/731	300	1122	
524966	176	3845	80		
0525001		3126	90		
0525002	109	3087		1123	
0525002H	109	3091		1123	
0526224	109	3087		1123	
0526232	177	3100/519	300	1122	
529657	177		300	1122	
529658	519		514	1122	
530015-2	109			1123	
530043-1	113A		6GC1		
530045-1	114		6GD1		
530045-2	114	3120	6GD1		
530045-3	114	3120	6GD1	1104	
530045-4	114	3120	6GD1		
530051-2	116	3031A	504A	1104	
530057-1	116	3016	504A	1104	
530063-1	110MP			1123	
530063-10	109	3091		1123	
530063-11	116	3016	504A	1104	
530063-12	116	3016	504A	1104	
530063-13	177	3100/519	300	1122	
530063-14	110MP	3709		1123(2)	
530063-2			1N60	1123	
530063-4	109	3091		1123	
530063-6	116	3016	504A	1104	
530063-7	116	3016	504A	1104	
530065-1	109	3087		1123	
530065-1002	109	3087		1123	
530065-1002A	109	3088		1123	
530065-1003	110MP	3088		1123	
530065-2	109	3087		1123	
530065-3	109	3087		1123	
530071-1	5802	3016	504A	1104	
530071-2	116	3051/156	504A	1104	
530071-3	116	3017B/117	504A	1104	
530072-1	116		504A	1104	
530072-10			504A	1104	
530072-1001	110MP	3709	1N60	1123(2)	
530072-1002	177	3100/519	300	1122	
530072-1006	519	3100	514	1122	
530072-1008	177	3100/519	300	1122	
530072-1009	177	3031A	300	1122	
530072-1010	177	3100/519	300	1122	
530072-1011	177	3100/519	300	1122	
530072-1014	156	3032A	512		
530072-1015	177	3087	300	1122	
530072-1017	116		504A	1104	
530072-1018	519		514	1122	
530072-1019	116	3311	504A	1104	
530072-11	116	3016	504A	1104	
530072-14	116	3017B/117	504A	1104	
530072-15	116	3091	504A	1104	
530072-18	177	3100/519	300	1122	
530072-2	116		504A	1104	
530072-4	116		504A	1104	
530072-5	116	3311	504A	1104	
530072-6	116	3311	514	1122	
530072-7	109	3087		1123	
530072-8	109	3091	300	1122	
530072-9	116	3017B/117	504A	1104	
530073-0030	5133A	3399			
530073-1013	142A	3062	ZD-12	563	
530073-1015		3142	ZD-16	564	
530073-1016	5105A		ZD-200		
530073-1017	145A	3063	ZD-15	564	
530073-1020	151A	3099	ZD-110		
530073-1021	5135A	3401	ZD-20		
530073-1023	144A	3094	ZD-14	564	
530073-1028	5079A		ZD-20		
530073-1029	5137A	3403	ZD-24		
530073-1030	5133A	3399	ZD-18		
530073-1031	137A	3057/136A	ZD-6.2	561	
530073-1034	5071A	3058/137A	ZD-6.8	561	
530073-12	145A	3063	ZD-15	564	
530073-13	142A	3062	ZD-12	563	
530073-14	142A	3062	ZD-12	563	
530073-15	5075A	3751	ZD-16	564	
530073-16	5105A		ZD-200		
530073-17	145A	3063	ZD-15	564	
530073-18	146A	3064	ZD-27		
530073-2	5080A	3796/5030A	ZD-22		
530073-21	5079A		ZD-20		
530073-22	5077A	3752	ZD-18		
530073-23	144A	3094	ZD-14	564	
530073-24	5081A	3151	ZD-24		
530073-26	5075A	3751	ZD-16	564	
530073-28	5079A		ZD-20		
530073-3	140A	3061	ZD-10	562	
530073-30	5077A	3752	ZD-18		
530073-31	136A	3057	ZD-5.6	561	
530073-32	5127A	3393	5ZD-12		
530073-40		3056	ZD-5.0		
530073-5	137A	3058	ZD-6.2	561	
530073-6	142A	3062	ZD-12	563	
530073-8	125		510,531	1114	
530073-9	145A	3063	ZD-15	564	
530082-1	116	3017B/117	504A	1104	
530082-1002	125	3032A	510,531	1114	
530082-1003	116	3017B/117	504A	1104	
530082-1004	116	3032A	504A	1104	
530082-2	125	3016	510,531	1114	
530082-3	177	3032A	300	1122	
530082-4	156	3032A	512		
530084-4	116	3031A	504A	1104	
530085-2	109			1123	
530086-1	116	3016	504A	1104	
530087-2	120	3110	CR-3		
530088-1	125	3016	510,531	1114	
530088-1002	166	3032A	504A	1152	
530088-1003	116	3032A	504A	1104	
530088-1004	116	3031A	504A	1104	
530088-2	125	3017B/117	510,531	1114	
530088-3	156	3032A	512		
530088-4	116	3031A	504A	1104	
530089-2	358		42		
530091-1	5852	3500/5882	512		
530092-1	109	3091		1123	
530092-1001	178MP	3087	1N34AS	1123	
530092-1002	177	3100/519	300	1122	
530092-2	109	3087		1123	
530093-1	178MP	3119/113	6GC1	1122(2)	
530093-1001			504A	1104	
530093-2	114		6GD1		
530093-3	114		6GD1	1104	
530094-1	116	3016	504A	1104	
530095-1	116	3016	504A	1104	
530096	118	3066	CR-1		
530096-1	119	3066/118	CR-1		
530096-2	120		CR-3		
530096-3	119		CR-2		
530097-2	120	3110	CR-3		
530097-3	506	3109/119	CR-2	1114	
530098-1	116	3016	504A	1104	
530098-1001	116	3016	504A	1104	
530099-1	116	3016	504A	1104	
530099-3	116	3016	504A	1104	
530099-4	116	3016	504A	1104	
530099-5	113A		6GC1		
530099-6			504A	1104	
530104-1		3126	90		
530104-1001		3126	90		
530104-2		3323	90		
530105-1		3091		1123	
530105-1001	110MP	3091	1N34AS	1123	
530106-1	507	3016	511		
530106-1001	507	3033	509	1114	
530109-1	116	3016	504A	1104	
530111-1	116	3017B/117	504A	1104	
530111-1001	116	3031A	504A	1104	
530111-1002	125	3081	531	1114	
530113-1	172A	3156	64		
530113-1001	506	3130	511	1114	
530113-2	125	3033	510,531	1114	
530115-1001			1N60	1123	
530116-1	503	3087	CR-5		

Industry Standard No.	ECG	SK	GE	RS 276-	MOTOR.
530116-1001	177	3100/519	300	1122	
530116-1003	116	3031A	504A	1104	
530116-3	177		300	1122	
530118-2	142A	3062	ZD-12	563	
530119-1	503		CR-5		
530119-5	503	3068	CR-5		
530119-7	505	3108	CR-7		
530119-8	504	3108/505	CR-6		
530119-9	505	3108	CR-7		
530120-1	166	9075	504A	1152	
530122-1	116(3)	3017B/117	504A(3)	1104	
530122-2	506	3130	511	1114	
530123-1				1123	
530123-3	506	3125	511	1114	
530123-4	506	3016	511	1114	
530123-5	506	3125	511	1114	
530123-7	506	3125	511	1114	
530124-1	116	3017B/117	504A	1104	
530124-2	505		CR-7		
530124-3	116(4)	3108/505	504A(4)	1104	
530126-1	116	3017B/117	504A	1104	
530127-1	109			1123	
530127-4	116	3017B/117	504A	1104	
530127-5	113A	3119/113	6GC1		
530127-6	116	3017B/117	504A	1104	
530128-1	358		42		
530129-1		3057	ZD-3.6		
530130-1	199	3245	62	2010	
530132	118	3066	CR-1		
530132-1	118	3066	CR-1		
530135-1	125	3311	510,531	1114	
530135-1003	116	3311	504A	1104	
530135-2	519	3100	514	1122	
530135-3	116	3311	504A	1104	
530136-T	506	3032A	511	1114	
530140-1		3126	90		
530142-8		3126	90		
530144	506	3043	511	1114	
530144-1	177	3130	300	1122	
530144-1001	519	3043	514	1122	
530144-1002	177	3100/519	300	1122	
530144-1003	177	3100/519	300	1122	
530144-1004	519	3100	514	1122	
530144-3			300	1122	
530145-100	140A	3061	ZD-10	562	
530145-1100			ZD-10	562	
530145-120	142A	3062	ZD-12	563	
530145-130	5022A	3788	ZD-13	563	
530145-1470	5088A		ZD-47		
530145-1569	136A	3057	ZD-5.6	561	
530145-339	134A		ZD-3.6		
530145-569	136A	3057	ZD-5.6	561	
530145-689	5071A	3058/137A	ZD-6.8	561	
530146-1	178MP	3100/519	300(2)	1122(2)	
530146-2	178MP	3100/519	300(2)	1122(2)	
530148-1003	506	3843	511	1114	
530148-1004	506	3130	511	1114	
530148-3	506	3130	511	1114	
530148-4	506	3130	511	1114	
530149-9	505	3108	CR-7		
530150-1	177	3100/519	300	1122	
530151-1	177	3017B/117	504A	1104	
530151-1001	552	3313/116	504A	1104	
530152-1	166	9075	BR-600	1152	
530153-1	522	3303	523		
530154-1			300#(2)	1122	
530157-1100	140A	3061	ZD-10	562	
530157-130		3093	ZD-13	563	
530157-1569		3057	ZD-5.6		
530157-569	136A	3057	ZD-5.6	561	
530157-689	5071A	3058/137A	ZD-6.2	561	
530157-870	5094A	9024	ZD-87		
530162-1	116	3032A	504A	1104	
530162-1001		3032A	504A	1104	
530163-120	142A	3062	ZD-12	563	
530165-1	522	3303	523		
530165-10	534	3309/539	534		
530165-11	529	3307	529		
530165-12	539	3309	538		
530165-13	529	3307	529		
530165-14	535	3307/529	535		
530165-15	534	3305	534		
530165-16	534	3305	534		
530165-17	529	3307	529		
530165-18	535	3307/529	535		
530165-19	557	3307/529			
530165-2	522	3303	523		
530165-3	522	3303	523		
530165-4	522	3303	523		
530165-5	534	3305	517		
530165-6	529	3307	529		
530165-7	539	3309	527		
530165-8	534	3309/539	534		
530166-1004	137A	3058	ZD-6.2	561	
530166-1005	5097A		ZD-120		
530166-1006	5136A	3402	ZD-22		
530166-1007	5097A		ZD-120		
530166-1013	5097A		ZD-120		
530170-1	177	3175	300	1122	
530171-1	116	3311	504A	1104	
530171-1001	116	3311	504A	1104	
530171-1002	116	3017B/117	504A	1104	
530171-1003	116	3313	504A	1104	
530171-3	116	3311	504A	1104	
530179-1	177	3100/519	300	1122	
530179-1001	177	3175	300	1122	
530179-1002	506	3843	300	1122	
530179-2	177		300	1122	
530179-3	177		300	1122	
530180-1001	116	3311	504A	1104	
530181-1	177	3100/519	300	1122	
530181-1-1			300	1122	
530181-1001	177	3175	300	1122	
530181-1002	177	3100/519			
530181-1003	177	3100/519	300	1122	
530184-1001	506	3843			
530184-1002	506	3843	511	1114	
530191-3	549	3901/554			
530191-4	549	3901/554			
530191-5	549	3901/554			
530192-120			ZD-12	563	
530972-14	506	3031A	511	1114	
531298-001	106	3984	21	2034	
531841-002	199	3245	62	2010	
531972	128		243	2030	
532003	175	3026	246	2020	
532775	199	3245	62	2010	
533013			300	1122	
533034			504A	1104	
533038	116		504A	1104	
533802	123A	3444	20	2051	
534001H	177	3100/519	300	1122	
0535001	109	3087			
0535005	109	3088			

Industry Standard No.	ECG	SK	GE	RS 276-	MOTOR.
535151-1001	116		504A	1104	
537200	102A	3004	53	2007	
537428			50	2004	
0537640	116	3311	504A	1104	
0537640(DIO)			504A	1104	
0537640(XSTR)			1	2007	
537790	102	3722	2	2007	
0537820	109	3088		1123	
0539860		3126	90		
540204	161	3716	39	2015	
540205	154		40	2012	
543995	159	3466	82	2032	
547684	106	3984	21	2034	
551015	102A		53	2007	
551026	116	3016	504A	1104	
0551029	166	9075	504A	1152	
0551029H	166	9075		1152	
551051	102A	3003	53	2007	
552005	125	3017B/117	510,531	1114	
0552006	116	3016	504A	1104	
0552006H	116	3311	504A	1104	
552007	506	3016	511	1114	
0552007H	117	3843	504A	1104	
0552010	116	3016	504A	1104	
0552010H	116	3311	504A	1104	
552308	123A	3444	20	2051	
552503	159	3466	82	2032	
555606			20	2051	
558875	7400	7400		1801	
558876	7402	7402		1811	
558877	7410	7410		1807	
558878	7420	7420		1809	
558879	7430	7430			
558880	7440	7440			
558881	7473	7473		1803	
558882	7474	7474		1818	
558883	7490	7490		1808	
558885	7493A	7493			
559507	7401	7401			
559557	180	3513	74	2043	
560004	121MP	3015	239(2)	2006(2)	
560020S	156	3051	512		
562654	519		514	1122	
0563012H	102A	3004	53	2007	
564671	129	3025	244	2027	
567312	123	3020	20	2051	
568101			1N34AS	1123	
570000-5452	123A	3444	20	2051	
570000-5503	123A	3444	20	2051	
570004-503	123A	3444	20	2051	
570005-452	123A	3444	20	2051	
570005-503	123A	3444	20	2051	
570009-01-504	123A	3444	20	2051	
570029	130		14	2041	
570030	185	3083/197	58	2025	
570031	184	3190	57	2017	
0570519	109			1123	
572001	185	3512	58	2025	
572683	123A	3444	20	2051	
573001	102A	3004	53	2007	
0573001-14	102A	3004	53	2007	
0573001H	102A	3004	53	2007	
0573002		3004	52	2024	
0573003	102A	3004	53	2007	
0573003H	102A	3004	53	2007	
0573004	102A	3004	53	2007	
0573005	102A	3004	53	2007	
0573005-14	102A	3004	53	2007	
0573005H	102A	3004	53	2007	
0573011	102A	3004	53	2007	
0573012	102A	3004	53	2007	
0573012H	102A	3004	53	2007	
0573018	102A	3004	53	2007	
0573018H	102A	3004	53	2007	
0573022	102A	3004	53	2007	
0573022H	102A	3004	53	2007	
0573023	102A	3004	53	2007	
0573023A	102A		53	2007	
0573023H	102A	3004	53	2007	
0573024	102A	3087	53	1123	
0573024-14	102A	3004	53	2007	
0573025	102A	3004	53	2007	
573029	102A	3004	53	2007	
0573030	131	3052	44	2006	
0573030-14	131	3052	44	2006	
0573031	131MP	3840	44(2)	2006(2)	
0573034	102A	3003	53	2007	
0573036	102A	3004	53	2007	
0573036H	102A	3004	53	2007	
0573037	103A	3010	59	2002	
0573037H	103A	3010	59	2002	
0573040	121	3009	239	2006(2)	
0573055	158	3004/102A	53	2007	
0573056	102A	3004	53	2007	
0573066	123A	3444	20	2051	
573101	108	3452	86	2038	
573103	102A	3004	2	2007	
573110	102A		53	2007	
573114	102A	3004	53	2007	
0573114H	102A	3004	53	2007	
0573117	102A	3004	53	2007	
0573117-14	102A	3004	53	2007	
573118		3004	2	2007	
573119	102A	3004	53	2007	
573125	102A	3004	53	2007	
0573131	102A	3004	53	2007	
0573139	103A	3010	59	2002	
0573142	102A	3004	53	2007	
0573142H	102A	3003	53	2007	
0573152	102A	3004	53	2007	
573153	102A	3004	53	2007	
0573153H	102A	3004	53	2007	
0573166	104	3009	239	2006	
573184	102A	3004	53	2007	
0573185	176	3845	80		
0573187	102A	3004	53	2007	
0573199	127	3034	25		
0573199H	127	3034	25		
0573200	102A		53	2007	
0573202	123A	3444	20	2051	
0573204	158	3004/102A	53	2007	
0573205	121	3009	239	2006	
0573212	131	3035	25	2006	
0573212H	127	3035	25		
573303	126	3008	52	2024	
573328	102A	3004	53	2007	
573329	126	3008	52	2024	
573330	126	3006/160	52	2024	
0573335	160	3006	245	2004	
573336	160	3006	245	2004	
573356	100	3721	1	2007	
573366	126	3006/160	245	2004	
573371	160	3006	245	2004	
573398	160	3006	52	2024	

Industry Standard No.	ECG	SK	GE	RS 276-	MOTOR.
573402	126	3008	52	2024	
573405	160	3006	245	2004	
573406	160	3006	245	2004	
0573415	124	3021	12		
0573418	123A	3444	20	2051	
0573422	102A	3004	53	2007	
0573422H	102A	3004	53	2007	
0573427	126	3006/160	52	2024	
0573428	160	3006	245	2024	
0573429	102A	3004	53	2007	
0573430	123A	3444	20	2051	
573432	102A		53	2007	
0573460	123A	3444	20	2051	
573467	123A	3444	20	2051	
0573468	123A	3444	20	2051	
0573468(HITACHI)	107		11	2015	
0573469	123A	3444	20	2051	
0573469H	123A	3444	20	2051	
0573471	126	3006/160	52	2024	
573472	108	3452	86	2038	
0573474	161	3018	39	2015	
0573474H	161	3039/316	39	2015	
0573475	161	3018	39	2015	
573479	123A	3138/193A	268	2038	
0573479H	123A	3444	20	2051	
0573480	289A	3444/123A	268	2051	
0573480H	128	3124/289	243	2030	
0573481	289A	3444/123A	20	2051	
0573481H	123A	3444	20	2051	
0573485	107	3018	11	2015	
0573486	233	3018	210	2009	
0573486H	233	3452/108	86	20388	
0573487	233	3018	61	2009	
0573487H	233	3018	11	2015	
0573490	123A	3444	20	2051	
0573491	289A	3124/289	268	2038	
0573491H	289A	3444/123A	20	2051	
0573492	289A	3124/289	210	2038	
573494	108	3452	86	2038	
0573495	108	3452	86	2038	
0573501	154	3045/225	40	2012	
0573506	108	3452	86	2038	
0573506H	108	3452	86	2038	
0573507	107	3452/108	11	2015	
0573507H	108	3452	86	2038	
0573508	107	3018	11	2015	
0573509	229	3018	61	2038	
0573509H	107	3018	11	2015	
0573510	229	3018	11	2015	
0573510H	107	3018	11	2015	
0573511	108	3452	86	2038	
0573511H	107	3018		2015	
573515	124	3021	12		
0573515H	124	3021	12		
0573517	128	3047	243	2030	
0573518	126	3006/160	52	2024	
0573519	154	3047	40	2012	
0573523	123A	3444	20	2051	
0573525	175		246	2020	
0573526	163A	3439	36		
0573527	128	3024	243	2030	
0573529	123A	3444	53	2007	
573532	128	3024	243	2051	
0573541	237	3299	46		
0573542	129	3025	244	2027	
0573556	123A	3444	20	2051	
0573557	128		243	2030	
0573559	129	3025	244	2027	
0573560	129		244	2027	
0573562	130	3027	14	2041	
0573570	108	3452	86	2038	
0573607	107	3018	11	2015	
0573742	102A	3004	53	2007	
0573981	123A	3444	20	2051	
574003	102A	3512	53	2007	
0575001	109	3087		1123	
0575001H	109	3087		1123	
0575002	109	3087		1123	
0575002H	109	3087		1123	
0575004	109	3088		1123	
0575005	109	3088		1123	
0575005H	109	3087		1123	
0575007	109	3088		1123	
575009	109	3087		1123	
0575010H		3126	90		
0575019	110MP	3087		1123(2)	
0575019H	110MP	3709	1N60	1123(2)	
0575024		3126	90		
0575027		3126	90		
575028	116	3016	504A	1104	
575037	112	3089	1N82A		
575042	116	3016	504A	1104	
575047	506	3843	511	1114	
0575047H	506	3843	511	1114	
0575048		3110	511		
0575049	506	3843	511	1114	
0575049H	506	3843	511	1114	
0575050	125	3016	504A	1104	
575051	116	3017B/117	504A	1104	
0575054	506	3843	511	1114	
0575066	506	3125	511	1114	
0575067	109	3087		1123	
575091	109			1123	
0575099	109	3087		1123	
575995	120	3110	CR-3		
576001	102A	3004	53	2007	
576005	102A	3004	53	2007	
0576054	116	3100/519	300	1104	
0576054(BIAS)			504A	1104	
0576054(SW)			300	1122	
576063	109	3087		1123	
0577001	109	3087		1123	
580029	116	3016	504A	1104	
581005	102	3722	2	2007	
581024	101	3861	8	2002	
581034A	123A	3444	20	2051	
581042	102	3722	2	2007	
581054	123A	3444	20	2051	
581055	123A	3444	20	2051	
581070	130		14	2041	
595819-1	704		IC-205		
599995	358		42		
600080-413-001	123A	3444	20	2051	
600080-413-002	123A	3444	20	2051	
600096-413	123		20	2051	
600098-413-001	123A	3444	20	2051	
600115-413-001	130		14	2041	
601030	109	3087		1123	
601032	160		245	2004	
601040	160	3006	245	2004	
601052	160		245	2004	
601054	126	3008	52	2024	
601054(SHARP)	105		4		
601065	103A	3010	59	2002	
601113	108	3452	86	2038	
601122	123A	3444	20	2051	
602032	121	3009	239	2006	
602040	158	3004/102A	53	2007	
602051	158		53	2007	
602075	160	3006	245	2004	
602081	139A	3060	ZD-9.1	562	
602113	108	3452	86	2038	
602113(SHARP)	123A	3444	20	2051	
602122	128	3024	243	2030	
602909-2A	123A	3444	20	2051	
602909-3A	129		244	2027	
602909-7A	128		243	2030	
603020	160	3006	245	2004	
603030	160	3006	245	2004	
603031	121	3009	239	2006	
603040	160	3006	245	2004	
603112	160		245	2004	
603113	108	3452	86	2038	
603114	116		504A	1104	
603122	123A	3444	20	2051	
603312	160		245	2004	
604030	160	3006	245	2004	
604040	160		245	2004	
604080	160	3006	245	2004	
604112	126	3006/160	52	2024	
604113	108	3452	86	2038	
604122	123A	3444	20	2051	
604407	177		300	1122	
605030	102A	3004	53	2007	
605112	160		245	2004	
605113	128	3024	243	2030	
605122	131		44	2006	
605131	118	3066	CR-1		
606020	158	3004/102A	53	2007	
606112	102A		53	2007	
606113	116	3016	504A	1104	
606131			504A	1104	
607030		3124	5	2002	
607101	113A	3119/113	6GC1		
607113	119	3016	CR-2		
607122	124		12		
608030	116	3016	504A	1104	
608101	116	3016	504A	1104	
608112	127	3034	25		
608112(SHARP)	179		76		
608113	116	3016	504A	1104	
608122	128	3024	243	2030	
609020	160	3006	245	2004	
609030	116	3016	504A	1104	
609112	123A	3124/289	20	2051	
609113	116	3016	504A	1104	
610020	116	3017B/117	504A	1104	
610024-1			21	2034	
610030	110MP	3087		1123	
610030-5	100	3004/102A			
610031-1	289A	3019			
610031-2	289A	3019			
610031-3	289A	3019			
610031-4	289A	3019			
610035	102A	3004	53	2007	
610035-1	100	3004/102A	53	2007	
610035-2	102A	3004	53	2007	
610036	102A	3004	53	2007	
610036-1	100	3004/102A	53	2007	
610036-2	100	3004/102A	53	2007	
610036-3	100	3004/102A	53	2007	
610036-4	102	3004	53	2007	
610036-5	102A	3004	53	2007	
610036-6	102	3004	53	2007	
610036-7	102	3004	53	2007	
610036-8	102A	3004	53	2007	
610039	121	3009	239	2006	
610039-1	121	3009	239	2006	
610040	102A	3004	53	2007	
610040-1	102A	3004	53	2007	
610040-2	102A	3004	53	2007	
610041	108	3452	86	2038	
610041-1	108	3452	86	2038	
610041-2	161	3018	86	2015	
610041-3	108	3452	86	2038	
610042	161	3039/316	39	2015	
610042-1	108	3452	86	2038	
610043	102A	3004	53	2007	
610043-1	102A	3004	53	2007	
610043-2	102A	3004	53	2007	
610043-3	102A	3004	53	2007	
610043-4	102A	3004	53	2007	
610043-6	102A	3004	53	2007	
610043-7	102A	3004	53	2007	
610045	108	3452	86	2038	
610045-1	108	3452	86	2038	
610045-2	108	3452	86	2038	
610045-3	123A	3444	20	2051	
610045-4	123A	3444	20	2051	
610045-5	123A	3444	20	2051	
610046-7	108	3452	86	2038	
610049-1	121	3009	239	2006	
610050	160	3006	245	2004	
610050-1	160	3006	245	2004	
610050-2	160	3006	245	2004	
610050-3	160	3006	245	2004	
610051	160	3006	245	2004	
610051-1	160	3006	245	2004	
610051-2	160	3006	245	2004	
610051-4	160	3006	245	2004	
610052	126	3008	52	2024	
610052-1	102A	3008	53	2007	
610053	160	3006	245	2004	
610053-1	160	3006	245	2004	
610053-2	160	3006	245	2004	
610055	160	3006	245	2004	
610055-1	160	3006	245	2004	
610055-2	160	3006	245	2004	
610055-3	160	3006	245	2004	
610056	126	3008	52	2024	
610056-1	160	3008	245	2004	
610056-2	160	3008	245	2004	
610056-3	160	3008	245	2004	
610056-4	160	3008	245	2004	
610059-1	100	3004/102A			
610059-2	100	3004/102A	53	2007	
610061-1	160		245	2004	
610063-1	163A	3439	36		
610064-1	162	3438	35		
610067	121	3009	239	2006	
610067-1	121	3009	239	2006	
610067-2	121	3009	239	2006	
610067-3	121MP	3009	239(2)	2006(2)	
610067-D	121	3009	239	2006	
610068	121	3009	239	2006	
610068-1	121MP	3009	239(2)	2006(2)	
610069	108	3452	86	2038	
610069-1	108	3452	86	2038	

Industry Standard No.	ECG	SK	GE	RS 276-	MOTOR.
610070	123A	3444	20	2051	
610070-1	123A	3444	20	2051	
610070-2	123A	3444	20	2051	
610070-3	123A	3444	20	2051	
610070-4	123A	3444	20	2051	
610071	124	3021	12		
610071-1	124	3021	12		
610071-2	124	3021	12		
610072	108	3452	86	2038	
610072-1	108	3452	86	2038	
610072-2	108	3452	86	2038	
610073	108	3452	86	2038	
610073-1	161	3018	39	2015	
610073-13			ZD-12	563	
610074	126	3008	52	2024	
610074-1	159	3466	82	2032	
610074-2	100	3005	1	2007	
610075-1	154	3044	40	2012	
610076	123A	3444	20	2051	
610076-1	123A	3444	20	2051	
610076-2	123A	3444	20	2051	
610077	123A	3444	20	2051	
610077-1	123A	3444	20	2051	
610077-2	123A	3444	20	2051	
610077-3	123A	3444	20	2051	
610077-4	123A	3444	20	2051	
610077-5	123A	3444	20	2051	
610077-6	123A	3444	20	2051	
610078	123A	3444	20	2051	
610078-1	123A	3444	20	2051	
610078-2	123A	3444	20	2051	
610079	102A	3004	53	2007	
610079-1	102A	3004	53	2007	
610079-2	199	3245	62	2010	
610080	102A	3004	53	2007	
610080-1	102A	3004	53	2007	
610083	159	3466	82	2032	
610083-1	234	3466/159	65	2050	
610083-2	234	3247	65	2050	
610083-3	159	3466	82	2032	
610083-4	234	3114/290	65	2050	
610088	102A	3004	53	2007	
610088-1	102A	3004	53	2007	
610088-2	102A	3004	53	2007	
610091	108	3452	86	2038	
610091-1	108	3452	86	2038	
610091-2	108	3452	86	2038	
610092	108	3452	86	2038	
610092-1	108	3452	86	2038	
610092-2	108	3452	86	2038	
610093-1	159	3466	82	2032	
610094	123A	3124/289	20	2051	
610094-1	123A	3444	20	2051	
610094-2	123A	3444	20	2051	
610094-3		3122	85	2010	
610096	108	3452	86	2038	
610096-1	108	3452	86	2038	
610099	129	3025	244	2027	
610099-1	129	3025	244	2027	
610099-2	129	3025	244	2027	
610099-3	102A	3004	53	2007	
610099-5	129	3114/290	244	2027	
610099-6	159	3466	82	2032	
610100	108	3452	86	2038	
610100-1	161	3018	39	2015	
610100-2	107		11	2015	
610100-3	108	3452	86	2038	
610102-1	106	3984	21	2034	
610106	121	3009	239	2006	
610106-1	121	3009	239	2006	
610107-1	108	3452	86	2038	
610107-2	128	3122	243	2030	
610110	129	3025	244	2027	
610110-1	159	3466	82	2032	
610110-2	159	3466	82	2032	
610111	113A		6GC1		
610111-1		3009	6GC1		
610111-2	130	3027	14	2041	
610111-3	219	3009			
610111-4	130	3027	14	2041	
610111-5	104	3009	16	2006	
610111-6	152	3054/196	66	2048	
610111-7	104	3009	16	2006	
610111-8	196	3054	241	2020	
610112	116	3016	504A	1104	
610112-1	153		69	2049	
610113-1	172A	3156	64		
610113-2	172A	3156	64		
610120-1	159	3466	82	2032	
610121-1	6402	3628			
610121-2	6402	3628			
610121-3	6402	3628			
610122	140A	3061	ZD-10	562	
610122-2	283	3467	38		
610123-1	164	3133	37		
610124-1	123AP	3861/101	8	2002	
610125-1	159	3466	82	2032	
610126-1	102A		53	2007	
610126-2	103A	3835	59	2002	
610128-1	199	3245	62	2010	
610128-2	199	3018	62	2010	
610128-4	108	3452	86	2038	
610128-5	199	3245	62	2010	
610128-6	199	3245	62	2010	
610129-1	129	3025	244	2027	
610129-D			61	2038	
610131-2	191	3232	249		
610132	123A	3444	20	2051	
610132-1	123A	3444	20	2051	
610134-1	234	3025/129	65	2050	
610134-2	234	3247	65	2050	
610134-4	234	3118	65	2050	
610134-5	234	3247	65	2050	
610134-6	234	3114/290	65	2050	
610135-1	154	3044	40	2012	
610136-1	106	3984	21	2034	
610139-1	229	3018	61	2038	
610139-2	108		11	2015	
610140-1	130	3027	14	2041	
610142-1	289A	3018	268	2038	
610142-2	123A	3444	20	2051	
610142-3	123A	3444	20	2051	
610142-4	123A	3444	20	2051	
610142-5	123A	3444	20	2051	
610142-6	108	3452	86	2038	
610142-7	123A	3444	20	2051	
610142-8			20	2051	
610143-3	123A	3444	20	2051	
610144-1	171	3201	27		
610144-101	171/403	3104A	27		
610144-2	171	3104A	27		
610144-3	154	3104A	27	2012	
610144-4				2012	
610145-1	233	3018	210	2009	

Industry Standard No.	ECG	SK	GE	RS 276-	MOTOR.
610146-3	123A	3444	20	2051	
610146-5	123A	3444	20	2051	
610147-1	123A	3444	20	2051	
610147-2	159	3466	82	2032	
610148-1	128	3124/289	243	2030	
610148-2	123A	3444	20	2051	
610148-2A	123A	3444	20	2051	
610148-3	128	3045/225	243	2030	
610149-1	153		69	2049	
610149-2	152	3893	66	2048	
610149-3	153		69	2049	
610150	108	3452	86	2038	
610150-1	161	3018	39	2015	
610150-2	123A	3444	20	2051	
610150-3	161	3018	39	2015	
610151-1	199	3018	62	2010	
610151-2	123A	3444	20	2051	
610151-3	199	3018	62	2010	
610151-4	123A	3444	20	2051	
610151-5	199	3010	62	2010	
610152-1	104	3009	16	2006	
610153-1	152	3054/196	66	2048	
610153-2	152	3054/196	66	2048	
610153-3	152	3054/196	66	2048	
610153-4	152	3054/196	66	2048	
610153-5	152	3054/196	66	2048	
610153-6	152	3893	66	2048	
610155-1	196	3041	241	2020	
610157-3	188	3054/196	226	2018	
610157-4	189	3083/197	218	2026	
610158-1	129		244	2027	
610158-2	159	3466	82	2032	
610161-2	130	3027	14	2041	
610161-4	130	3036	14	2041	
610162-2	184		57	2017	
610162-4	152	3893	66	2048	
610162-7	153	3083/197	69	2049	
610162-8	152	3054/196	66	2048	
610162N2	184		57	2017	
610164-1	312	3834/132	FET-1	2035	
610165-1	123A	3444	20	2051	
610165-2	123A	3444	20	2051	
610166-1	222	3050/221	FET-4	2036	
610167-1	123A	3444	20	2051	
610167-2	123A	3444	20	2051	
610168-1	123A	3444	20	2051	
610168-2	123A	3444	20	2051	
610174-1	108	3452	86	2038	
610180-1		3132	61	2038	
610180-1(LAST IF)	233			2009	
610180-1(LAST-IF)			210	2051	
610181-1	319P	3117	283	2016	
610181-2	161	3117	39	2015	
610186-1	161	3710/238	39	2015	
610189-1	238	3710	259		
610189-2	238	3710	259		
610190-1	190	3202/210	217		
610190-3		3103A	57	2017	
610190-4	190	3201/171	217		
610194-1	165	3710/238	38		
610194-3	165	3710/238	38		
610195-1	196	3054	241	2020	
610195-2	197	3083	250	2027	
610195-3	152	3054/196	66	2048	
610195-4	153	3083/197	69	2049	
610202-1	189	3083/197	218	2026	
610202-2	189	3083/197	218	2026	
610203-1	222	3050/221	FET-4	2036	
610203-2	222	3050/221	FET-4	2036	
610203-3	222	3050/221	FET-4	2036	
610203-4	222	3050/221	FET-4	2036	
610203-5	222	3050/221	FET-4	2036	
610203-6	222	3050/221	FET-4	2036	
610209-1	159	3466	82	2032	
610213-1	128	3024	243	2030	
610216-1	165		38		
610216-2	165	3710/238	259		
610216-3	165	3710/238	38		
610217-1	280	3439/163A	36		
610217-2	280	3439/163A	36		
610217-5	163A	3115/165	73		
610223-1	159	3466	82	2050	
610224-1	199	3444/123A	62	2010	
610226-1	199	3124/289	62	2010	
610226-2		3444	62		
610226-5		3444	62		
610227-1	189	3083/197	29	2025	
610228-1	188	3199	226	2018	
610228-3	302	3178A	275		
610232-1	123A	3444	20	2051	
610232-2	123AP	3452/108	86	2038	
610233-1	165	3710/238	38		
610241-1	165		38		
610242-1	165	3710/238	38		
610245-1	280	3115/165	73		
610245-2	94		73		
610246-1	159	3118	21	2034	
610249-1	229	3246	61	2038	
610250-1		3201	27		
610358-1	222	3065	FET-4	2036	
610358-2	222	3065	FET-4	2036	
610358-3	222	3065	FET-4	2036	
610361-5		3203	253	2026	
610386-1		3452	86		
610392-2		3710	256		
610395-1		3440	233		
610395-2		3441	234		
611001-3	5650	3506/5642			
611001-4	5651	3506/5642			
611001-5	5642	3506			
611001-6	5642	3506			
611003-3	5404	3627			
611003-4	5404	3627			
611003-5	5404	3627			
611020	102A	3008	53	2007	
611064	7405	7405			
611065	7475	7475		1806	
611066	7486	7486		1827	
611071	7442	7442			
611111	116	3017B/117	504A	1104	
611132	113A	3119/113	6GC1		
611233	234		65	2050	
611428	123A	3444	20	2051	
611563	7400	7400		1801	
611564	7402	7402		1811	
611565	7404	7404		1802	
611566	7410	7410		1807	
611567	7420	7420		1809	
611568	7440	7440			
611571	7474	7474		1818	
611572	7490	7490		1808	
611573	7492	7492		1819	
611730	74193	74193		1820	
611731	74192			1831	

Industry Standard No.	ECG	SK	GE	RS 276-	MOTOR.
611844	7486	7486		1827	
611845	7405	7405			
611870	7476	7476		1813	
611872	74107	74107			
611900	74107	74107			
612002-2	717		IC-209		
612005-1	712	3072	IC-2		
612005-2	712	3072	IC-2		
612006-1	722	3161	IC-9		
612006-1M	722		IC-9		
612006-1Z	722		IC-9		
612006-2	722	3161	IC-9		
612007		3135	IC-10		
612007-1	708	3135/709	IC-10		
612007-2	708	3135/709	IC-10		
612007-3	708	3135/709	IC-10		
612008-1	725	3144/723	IC-19		
612008-2	725	3144/723	IC-15		
612020	179	3642	76		
612020-1	703A	3157	IC-12		
612020-2	703A	3157	IC-12		
612020-3	703A	3157	IC-12		
612021-1	718	3159	IC-8		
612021-2	718	3159	IC-8		
612024-1	731	3170	IC-13		
612025-1	730	3143			
612025-2	730	3143			
612025-3	730	3143			
612029-1	790	3454	IC-18		
612029-2	790	3454	IC-230		
612029-3	790	3077	IC-18		
612030-1	715	3076	IC-6		
612031-1	714	3075	IC-4		
612044-1	738	3167	IC-29		
612045-1	804	3455	IC-27		
612061-1	783	3215	IC-225		
612067		3076	IC-6		
612067-1	715		IC-6		
612069-1	715	3076	IC-6		
612070-1	714	3075	IC-4		
612072-1	739	3235	IC-30		
612075-1	801	3160	IC-35		
612076-1	742		IC-213		
612076-2	742	3453	IC-213		
612077-1	788	3829	IC-229		
612082-1	815	3255			
612082-2	815	3255	IC-244		
612103-3	960	3591			
612112	116	3016	504A	1104	
612130	116		504A	1104	
612132	116	3017B/117	504A	1104	
612194-1	74LS32	74LS32			
612200-1	74LS74A	74LS74			
613020	116	3016	504A	1104	
613112	108	3452	86	2038	
613130	116	3016	504A	1104	
613132	119		CR-2		
614010	116	3016	504A	1104	
614020	110MP			1123	
614112	112		1N82A		
615004-8	177		300	1122	
615010	109	3087		1123	
615093-2	123A	3444	20	2051	
615130	358		42		
615154-1	519		514	1122	
615179-1	123A	3444	20	2051	
615179-2	123A	3444	20	2051	
615180-1	159	3466	82	2032	
615180-2	159	3466	82	2032	
615180-3	159	3466	82	2032	
615180-4	159	3466	82	2032	
615246-1	941M		IC-265	007	
615268-101	941M		IC-265	007	
616010	113A	3119/113	6GC1		
617871-1	121	3009	239	2006	
618072	123A	3444	20	2051	
618126-1	123A	3444	20	2051	
618136-1	128		243	2030	
618139-1	121	3009	239	2006	
618150-3	177		300	1122	
618165-1	199	3245	62	2010	
618181-1	199	3245	62	2010	
618197	128		243	2030	
618217-2	123A	3444	20	2051	
618241-2	910		IC-51		
618483-1	941		IC-263	010	
618580	312	3112		2028	
618639-1	519		514	1122	
618810-2	123A	3444	20	2051	
618955-2	130		14	2041	
618960-1	198	3220	251		
618984-1	941		IC-263	010	
618986-2	218		234		
618986-4	218		234		
619006	123A	3444	20	2051	
619006-1	128		243	2030	
619006-7	123A	3444	20	2051	
619009-1	175		246	2020	
619010-1	198	3220	251		
619011	519		514	1122	
619050-1	910D		IC-252		
619087-1	519		514	1122	
619094	5626	3633			
619130	116	3017B/117	504A	1104	
619256	5697	3522			
619256-1	5695	3509			
619361-1	175	3131A/369	246	2020	
619526-1	519		514	1122	
619692-1	723	3144	IC-15		
619693-1	787	3146	IC-228		
619694-1	724	3525	IC-86		
620782	176	3845	80		
630063	5802	9005		1143	
633977	113A	3119/113	6GD1		
649002	130		14	2041	
650060	159	3466	82	2032	
650175	129		244	2027	
650196	102	3004	53	2007	
650845	177		300	1122	
650854	177		300	1122	
650859-1	102	3004	53	2007	
650859-2	102	3004	53	2007	
650859-3	102	3004	53	2007	
650860	101	3011	59	2002	
650970	121	3717	239	2006	
651012	102	3004	53	2007	
651030	116		300	1122	
651038	116	3031A	504A	1104	
651202	121	3717	239	2006	
651236	102	3004	53	2007	
651891	289A	3444/123A	20	2051	
651955	123A	3444	20	2051	
651955-1	289A	3124/289	17	2051	
651955-2	289A	3124/289	17	2051	
651955-3	289A	3019	17	2051	
651956	128	3192/186	28	2017	
651995-1	123A	3444	20	2051	
651995-2	123A	3444	20	2051	
651995-3	123A	3444	20	2051	
652072	289A	3018	20	2051	
652085	121	3009	239	2006	
652086	121	3717	239	2006	
652091	289A	3444/123A	20	2051	
652092	116	3048/329	504A	1104	
652230	289A	3444/123A	20	2051	
652231	128	3045/225	243	2030	
652321	154	3045/225	243	2030	
652615	116		504A	1104	
653406	128		243	2030	
654000	289A	3444/123A	20	2051	
654001	6401			2029	
654003	121	3009			
654007	100	3004/102A			
654008	289A	3018			
654010	121	3019			
654011	289A	3018			
654012	289A	3018			
654013	159	3025/129			
654032	114		300	1122	
654036	116	3016			
654041	100	3004/102A			
654420	5802	9005	510	1143	
655319	121	3717	239	2006	
656064	6410			2029	
656204	123A	3444	20	2051	
656524	289A	3444/123A	20	2051	
656719	289A	3444/123A	20	2051	
656746	289A		243	2030	
657179	185	3191	58	2025	
657180	184	3054/196	57	2017	
657181	185	3083/197	58	2025	
658577	123AP	3444/123A	20	2051	
658578	123AP	3444/123A	20	2051	
658583	909D	3590	IC-250		
658657	289A	3444/123A	20	2051	
659140	184	3054/196			
659141	183		74	2043	
659143	185	3083/197			
659174	130		14	2041	
660030	102A		53	2007	
660031	121MP	3013	239(2)	2006(2)	
660059	102A	3004	53	2007	
660060	102A	3004	53	2007	
660064	160		245	2004	
660070	128	3024	243	2030	
660072	102A	3004	53	2007	
660074	128	3024	243	2030	
660077	121	3009	239	2006	
660082	102A	3004	53	2007	
660084	160	3006	245	2004	
660085	160	3006	245	2004	
660094	121	3717	239	2006	
660095	121	3014	239	2006	
660097	121	3009	239	2006	
660100	128	3512	243	2030	
660103	121	3009	239	2006	
660138	218		234		
660144	105	3012	4		
660144A	105	3012	4		
660388-02	941M		IC-265	007	
661010	113A	3119/113	6GC1		
670850	121MP	3013	239(2)	2006(2)	
670850-1	121MP	3013	239(2)	2006(2)	
671077-6			20	2051	
681266	129		244	2027	
681266-1	129		244	2027	
696575-198	128		243	2030	
698941-1	159	3466	82	2032	
699291	199		62	2010	
699410-140	128		243	2030	
699414-164	124	3045/225	12		
699739	116(2)		504A(2)	1104	
700021-00	118	3066	CR-1		
700043-00	116	3017B/117	504A	1104	
700047-47	123A	3444	20	2051	
700047-49	123A	3444	20	2051	
700055-00	113A	3119/113	6GC1		
700063-00	116	3017B/117	504A	1104	
700080	130	3027	14	2041	
700080A	130	3027	14	2041	
700083	130	3027	14	2041	
700083A	130	3027	14	2041	
700180-00				1123	
700181	123A	3444	20	2051	
700191	175	3026	246	2020	
700195	175	3026	246	2020	
700230-00	123A	3444	20	2051	
700231-00	123A	3444	20	2051	
700647	116	3016	504A	1104	
700663	116	3016	504A	1104	
700664	116	3016	504A	1104	
701584-00	159	3466	82	2032	
701589-00	159	3466	82	2032	
701662-00	177		300	1122	
701678-00	161	3018	39	2015	
702407-00	128	3024	243	2030	
702415-00	210	3122	252	2018	
702884	123A	3444	20	2051	
702885	121	3717	239	2006	
702885-00	121	3717	239	2006	
702886	152	3893	66	2048	
705784-1	175		246	2020	
710206	910		IC-251		
710206-43	923		IC-259		
710398-28	909	3590	IC-249		
710838-1	177		300	1122	
717101	152	3054/196	66	2048	
717126-505	74H04	74H04			
717136-1	74S00	74S00			
717399-3	923		IC-259		
717399-4	941		IC-263	010	
717399-49	941		IC-263	010	
720236	123A	3444	20	2051	
720240	123A	3444	20	2051	
720453	5800	9003		1142	
720454	5801	9004		1142	
720455	5802	9005		1143	
720456	5804	9007		1144	
720457	5806	3848			
720458	5804	9007		1144	
720463	5806	3848			
720608-13	519		514	1122	
720609-1	519		514	1122	
721272	159	3466	82	2032	
723000-18	101	3861	8	2002	
723001-19	101	3861	8	2002	
723020-41	128		243	2030	
723043-1	129		244	2027	

Industry Standard No.	ECG	SK	GE	RS 276-	MOTOR.
723060-29	175		246	2020	
723423-16	175		246	2020	
723423-20	175		246	2020	
723423-7	175		246	2020	
723423-9	175		246	2020	
726654	519		514	1122	
730547	704	3023	IC-205		
731009	102A	3004	53	2007	
740183	116	3311	504A	1104	
740247	131MP		44(2)	2006(2)	
740289	116	3311	504A	1104	
740306	130	3124/289	62	2010	
740402	109	3087		1123	
740417	102A	3004	53	2007	
740437	123A	3444	20	2051	
740438	199	3245	62	2010	
740439	199	3245	62	2010	
740440	199	3245	62	2010	
740441	199	3444/123A	20	2051	
740442	199	3245	62	2010	
740443	131MP	3840	44(2)	2006(2)	
740461	289A	3444/123A	20	2051	
740462	289A	3444/123A	20	2051	
740463	289A	3444/123A	20	2051	
740466	289A	3444/123A	20	2051	
740470	289A	3444/123A	20	2051	
740471	131MP	3052	44	2006	
740502	1107	3526/947	IC-276		
740543	1107		IC-276		
740570	116	3311	504A	1104	
740583	1103	3281	IC-94		
740622	720	9014	IC-7		
740622(RCA)	720		IC-7		
740628	5071A		ZD-6.8	561	
740629	178MP		300(2)	1122(2)	
740630	116		504A	1104	
740728	161	3716			
740781	1037	3371	IC-170		
740782	1087	3364			
740828	177		300	1122	
740855	6400A		2N2160	2029	
740856	152	3893	66	2048	
740857	289A	3047	20	2051	
740885	176	3123			
740886	199	3122	62	2010	
740887	199	3124/289	62	2010	
740940	1005	3723	IC-42		
740946	126	3006/160	52	2024	
740947	126	3006/160	52	2024	
740948	126	3006/160	52	2024	
740949	107	3018	11	2015	
740950	107	3018	61	2038	
740951	107	3016	61	2038	
740952	109	3087		1123	
740953	5071A	3058/137A	ZD-6.8	561	
740954	109	3088		1123	
740955	614	3126			
741050	106	3008	245	2004	
741051	519	3008			
741052	612	3126	90		
741098	1006	3358	IC-38		
741100	177	3100/519	300	1122	
741101	116		504A	1104	
741114	199	3124/289	62	2010	
741115	186	3192	28	2017	
741116	136A	3058/137A	ZD-5.6	561	
741168	4011B	4011			
741340	1100	3223			
741341	1160	3243			
741342	703A	3157			
741343	7490	7490			
741348	7493A	7493			
741473	1170	3745			
741518	1082	3461			
741519	801	3160			
741673	1160	3243			
741686	1167	3732			
741687	1155	3231	IC-179	705	
741689	614	3325/612	90		
741726	107	3018	211	2051	
741727	123A	3018			
741728	297	3122			
741729	290A	3114/290	269	2032	
741730	199	3018			
741731	229		20	2051	
741732	235	3197	215		
741733	199	3018			
741735	199	3122			
741736	199	3122			
741737	199	3122	61	2010	
741738	135A	3056	ZD-5.0		
741739	139A	3060	ZD-9.1	562	
741740	116		504A	1104	
741741	177		300	1122	
741852	1082	3461	IC-140		
741853	1056	3458	IC-48		
741854	1058	3459	IC-49		
741855	107	3122	20	2051	
741856	297	3124/289	210	2009	
741857	199	3124/289	62	2010	
741858	315	3250	335		
741859	235	3197	338		
741860	222	3050/221	FET-1	2036	
741861	199	3122	210	2051	
741862	107		60	2015	
741863	234		65	2050	
741864	177	3175	300	1122	
741865	116	3311	504A	1104	
741866	109	3087	1N34AS	1123	
741867	140A	3061	ZD-10	562	
741868	614	3126	90		
741869	5072A	3136	ZD-8.2	562	
741870	5072A		ZD-8.2	562	
742004	109	3087		1123	
742008	116	3016	504A	1104	
742009	116	3016	504A	1104	
742362	1226	3763			
742363	1087	3477	IC-103		
742364	1155	3231	IC-179	705	
742510	1192		IC-181		
742512	199		212	2010	
742513	199		62	2010	
742537	107		11	2015	
742546	290A		269	2032	
742547	107		60	2015	
742548	199		212	2010	
742549	107		20	2038	
742723	1248	3497			
742726	977	3462			
742728	107	3122	210	2051	
742729	293	3024/128	47	2030	
742730	109	3088	1N60	1123	
742732	614	3126	90		
742920	165	3710/238			
742922	5072A		ZD-8.2	562	
742970	235	3197	337		
744002	113A	3119/113	6GC1		
744006	113A	3119/113	6GC1		
746003			504A	1104	
746004	116	3017B/117	504A	1104	
749002	107		11	2015	
749014	107		11	2015	
750746A	358		42		
752309	116	3016	504A	1104	
755722	109	3090		1123	
757008-02	123A	3444	20	2051	
759500			509	1114	
760005	222	3065	FET-4	2036	
760007			300	1122	
760011	7400	7400		1801	
760012	172A	3156	64		
760013	7490	7490		1808	
760015	7476	7476		1813	
760021	219	3173	74	2043	
760037	519	3100	514	1122	
760051	312	3834/132			
760101-0005	109	3087	1N34AS	1123	
760101-0006	109	3087	1N34AS	1123	
760142	123A	3122	20	2038	
760202-0003	116	3311	504A	1104	
760204-0001	610	3126	90		
760213-0002			39	2015	
760213-0005			39	2015	
760236	123A	3444	20	2051	
760239	123A	3444	20	2051	
760249	107	3122	11	2015	
760251	123A	3444	20	2051	
760253	107	3018	11	2015	
760268	312	3448	FET-1	2035	
760269	159	3466	82	2032	
760275	186	3192	28	2017	
760276	187	3193	29	2025	
760284	311	3048/329	277		
760298	135A	3056	ZD-5.1		
760304	142A	3062	ZD-12	563	
760309	5116A		5ZD-5.0		
760522-0002	722	3161	IC-9		
761113	116	3087	504A	1104	
765713	125	3081	510,531	1114	
765722	109	3087		1123	
770339(CRYSTAL)	358		42		
770523	102A	3004	53	2007	
770524	102A	3004	53	2007	
770525	158	3004/102A	53	2007	
770730	102A		53	2007	
770768-3170756	124	3021	12		
771907	116	3311	504A	1104	
771908	166	9075	BR-600	1152	
771909	109	3087		1123	
771910	109	3088		1123	
771911	109	3087		1123	
772712	109			1123	
772713	116	3311	504A	1104	
772714	166	9075	BR-600	1152	
772716	160			2004	
772717	160			2004	
772718	102A		53	2007	
772719	160		245	2004	
772720	102A		53	2007	
772721	102A	3004	53	2007	
772722	102A		53	2007	
772723	102A		53	2007	
772724	102A		53	2007	
772725	102A		53	2007	
772727	102A		53	2007	
772728	102A		53	2007	
772729	102A		53	2007	
772732	158	3004/102A	53	2007	
772733	158		53	2007	
772736	102A		53	2007	
772737	102A		53	2007	
772738	107	3039/316	11	2015	
772739	107	3039/316	11	2015	
772740	110MP	3087	1N34AS	1123(2)	
772768	126	3008	52	2024	
779821	197		250	2027	
785278-01	109	3087		1123	
785278-101	123A	3444	20	2051	
785897-01	160	3006	245	2004	
793356-1	130	3027	14	2041	
800019-001	128		243	2030	
800020-001	7420	7420		1809	
800021-001	7430	7430			
800022-001	7440	7440			
800023-001	7410	7410		1807	
800024-001	7400	7400		1801	
800033-1	5529	6629			
800073-6	128	3024	243	2030	
800073-7	128	3024	243	2030	
800080-001	7402	7402		1811	
800132-001	123A	3444	20	2051	
800382-001	7475	7475		1806	
800385-001	7442	7442			
800386-001	74193	74193		1820	
800387-001	7404	7404		1802	
800400-001	7474	7474		1818	
800491-001	74121	74121			
800651-001	7406	7406		1821	
800743-001	519		514	1122	
800747	102A		53	2007	
800946-001	128		243	2030	
801500	102		53	2007	
801501	102		53	2007	
801507	100		53	2007	
801509	102		53	2007	
801510	102		53	2007	
801511	102		53	2007	
801512	289A	3444/123A	20	2051	
801513	289A	3444/123A	20	2051	
801514	289A	3444/123A	20	2051	
801515	289A	3444/123A	20	2051	
801516	289A	3444/123A	20	2051	
801517	289A	3444/123A	20	2051	
801518	121	3009	239	2006	
801519	104	3009	239	2006	
801520	102A	3003	53	2007	
801522	121	3717	239	2006	
801523	179	3717/121	239	2006	
801524	289A	3444/123A	20	2051	
801525	6401			2029	
801527	192A	3124/289	62	2010	
801529	289A	3444/123A	20	2051	
801530	289A	3444/123A	20	2051	
801531	6401			2029	
801532	289A	3444/123A	20	2051	
801533	172A	3156	64		
801534	289A	3444/123A	20	2051	
801535	6402	3156/172A	64		

Industry Standard No.	ECG	SK	GE	RS 276-	MOTOR.
801536	289A	3444/123A	20	2051	
801537	130	3027	14	2041	
801538	179	3717/121	239	2006	
801539	121	3009			
801540	159	3466	82	2032	
801541	171	3201	27		
801543	289A	3444/123A	20	2051	
801545	5414	3954			
801546	289A	3122			
801547	459	3112			
801707	116		504A	1104	
801711	116	3016	504A	1104	
801712	177		300	1122	
801714	116	3016	504A	1104	
801715	125	3016	504A	1104	
801716	116	3016	504A	1104	
801718	116	3016			
801722	109	3091		1123	
801723	125	3031A	504A	1104	
801724	519		514	1122	
801725	289A	3018			
801726	139A		ZD-9.1	562	
801728	177		300	1122	
801729	289A	3444/123A	20	2051	
801730	5802	9005		1143	
801731	142A	3062	ZD-12	563	
801805	7401	7401			
801806	7404	7404		1802	
801807	7442	7442			
801808	7490	7490		1808	
802008	519		514	1122	
802032-2	102A	3004	53	2007	
802032-4	102A	3004	53	2007	
802033-3	102A	3004	53	2007	
802037-001	153		69	2049	
802054-0	102A	3004	53	2007	
802056-0	102A	3004	53	2007	
802189-7	102A	3004	53	2007	
802189-8	102A	3004	53	2007	
802263-0	102A	3004	53	2007	
802263-1	102A	3004	53	2007	
802389-2	102A	3003	53	2007	
802415-2	102A	3004	53	2007	
802425-0	127	3034	25		
802439-0	102A	3004	53	2007	
802560	102	3004	2	2007	
803182-5	123A	3444	20	2051	
803369-6	123A	3444	20	2051	
803372-0	123A	3444	20	2051	
803373-0	123A	3444	20	2051	
803696	199	3245	62	2010	
803733-0	123A	3444	20	2051	
803733-3	123A	3444	20	2051	
803735-3	123A	3444	20	2051	
810000-373	123A	3444	20	2051	
810002-733	129		244	2027	
810002-736	128		243	2030	
813362	727	3071	IC-210		
814044A	102A	3004	53	2007	
815003	102A	3004	53	2007	
815015	102A	3004	53	2007	
815020	100	3005	1	2007	
815020A	100	3005	1	2007	
815020B	100	3005	1	2007	
815021	100	3005	1	2007	
815021A	100	3005	1	2007	
815021B	100	3005	1	2007	
815022	102A	3004	53	2007	
815022A	102A	3004	53	2007	
815022B	102A	3004	53	2007	
815023	102A	3004	53	2007	
815023A	102A	3004	53	2007	
815023B	102A	3004	53	2007	
815024	102A	3004	53	2007	
815024A	102A	3004	53	2007	
815024B	102A	3004	53	2007	
815025	100	3005	1	2007	
815025A	100	3005	1	2007	
815025B	100	3005	1	2007	
815026	101	3861	8	2002	
815026A	101	3861	8	2002	
815026B	101	3861	8	2002	
815026C	101	3861	8	2002	
815026D	101	3861	8	2002	
815027	100	3005	1	2007	
815027A	100	3005	1	2007	
815027B	100	3005	1	2007	
815027C	100	3005	1	2007	
815028	100	3005	1	2007	
815028A	100	3005	1	2007	
815028B	100	3005	1	2007	
815028C	100	3005	1	2007	
815029	102A	3004	53	2007	
815029A	102A	3004	53	2007	
815029B	102A	3004	53	2007	
815029C	102A	3004	53	2007	
815030	102A	3004	53	2007	
815030A	102A	3004	53	2007	
815030B	102A	3004	53	2007	
815031	102A	3004	53	2007	
815031A	102A	3004	53	2007	
815031B	102A	3004	53	2007	
815033	102	3722	2	2007	
815034		3004	2	2007	
815034A	102A	3004	53	2007	
815034B	102A	3004	53	2007	
815034C	102A	3004	53	2007	
815036	100	3005	1	2007	
815036A	100	3005	1	2007	
815036B	100	3005	1	2007	
815036C	100	3005	1	2007	
815037	100	3005	1	2007	
815037A	100	3005	1	2007	
815037B	100	3005	1	2007	
815037C	100	3005	1	2007	
815038	102A	3005	53	2007	
815038A	102A		53	2007	
815038B	102A	3004	53	2007	
815038C	102A	3004	53	2007	
815041	100	3005	1	2007	
815041A	100	3005	1	2007	
815041B	100	3005	1	2007	
815041C	100	3005	1	2007	
815043	100	3005	1	2007	
815043A	100	3005	1	2007	
815043B	100	3005	1	2007	
815043C	100	3005	1	2007	
815055	100	3005	1	2007	
815056	100	3005	1	2007	
815057	100	3005	1	2007	
815058	102A	3004	53	2007	
815058A	102A	3004	53	2007	
815058B	102A	3004	53	2007	
815058C	102A	3004	53	2007	
815058X	102A	3004	53	2007	
815064	126	3008	52	2024	
815064A	126	3008	52	2024	
815064B	126	3008	52	2024	
815064C	126	3008	52	2024	
815065	100	3005	1	2007	
815065A	100	3005	1	2007	
815065B	100	3005	1	2007	
815065C	100	3005	1	2007	
815066	100	3005	1	2007	
815066A	100	3005	1	2007	
815066B	100	3005	1	2007	
815066C	100	3005	1	2007	
815067	160	3009	245	2004	
815067A	160	3008	245	2004	
815067B	160	3008	245	2004	
815067C	160	3008	245	2004	
815068	126	3008	52	2024	
815068A	126	3008	52	2024	
815068B	126	3008	52	2024	
815068C	126	3008	52	2024	
815069	102A	3004	53	2007	
815069A	102A	3004	53	2007	
815069B	102A	3004	53	2007	
815069C	102A	3004	53	2007	
815070	102A	3004	53	2007	
815070A	102A	3004	53	2007	
815070B	102A	3004	53	2007	
815070C	102A	3004	53	2007	
815070D	102A	3004	53	2007	
815074	102A	3004	53	2007	
815075	103A	3010	59	2002	
815076	103A	3010	59	2002	
815082	102A		53	2007	
815083	102A		53	2007	
815101	100	3005	1	2007	
815103	100	3005	1	2007	
815104	102A	3004	53	2007	
815105	100	3005	1	2007	
815107	100	3005	1	2007	
815108	100	3005	1	2007	
815109	100	3005	1	2007	
815114	102A	3004	53	2007	
815115	100	3005	1	2007	
815116	100	3005	1	2007	
815117	100	3005	1	2007	
815118	102A	3004	53	2007	
815120	102A	3004	53	2007	
815120A	102A	3004	53	2007	
815120B	102A	3004	53	2007	
815120C	102A	3004	53	2007	
815120D	102A	3004	53	2007	
815120E	102A	3004	53	2007	
815122	102A		53	2007	
815133	123A	3444	20	2051	
815134	123A	3444	20	2051	
815136	102A	3004	53	2007	
815137	121MP	3009	239(2)	2006(2)	
815137B			3	2006	
815137Y			3	2006	
815138	116	3009	504A	1104	
815139	102A	3003	53	2007	
815142	116	3016	504A	1104	
815158	102A	3004	53	2007	
815160	102A	3004	53	2007	
815160-C			2	2007	
815160-I	102A		53	2007	
815160-J	102A		53	2007	
815160-K	102A		53	2007	
815160-L	102A		53	2007	
815160-O	102A		53	2007	
815160-P	102A		53	2007	
815160-Q	102A		53	2007	
815160A	102A	3004	53	2007	
815160B	102A	3004	53	2007	
815160C	102A	3004	53	2007	
815160D	102A	3004	53	2007	
815160E	102A	3004	53	2007	
815160F	102A	3004	53	2007	
815160H	102A	3004	53	2007	
815164	108	3452	86	2038	
815165	108	3452	86	2038	
815166	124	3021	12		
815166-4	124	3021	12		
815167-3	124	3021	12		
815170	108	3452	86	2038	
815171	123A	3444	20	2051	
815171D	123A	3444	20	2051	
815172	108	3452	86	2038	
815172A	108	3452	86	2038	
815173	199	3018	62	2010	
815173A	108	3452	86	2038	
815173C	108	3452	86	2038	
815173F	108	3452	86	2038	
815174	123A	3444	20	2051	
815174L	123A	3444	20	2051	
815175	124	3021	12		
815175H	124	3021	12		
815177	102A	3004	53	2007	
815178	102A	3004	53	2007	
815179	102A	3004	53	2007	
815180-3	124	3021	12		
815180-4	124	3021	12		
815180-7	124	3021	12		
815181	102A	3004	53	2007	
815181-B		3004	2	2007	
815181A	102A	3004	53	2007	
815181B	102A	3004	53	2007	
815181C	102A	3004	53	2007	
815181D	102A	3004	53	2007	
815182	123A	3444	20	2051	
815183	123A	3444	20	2051	
815184	123A	3444	20	2051	
815184E	123A	3444	20	2051	
815185	129	3025	244	2027	
815185E	129	3025	244	2027	
815186	123A	3444	20	2051	
815186C	123A	3444	20	2051	
815186L	123A	3444	20	2051	
815189	102A	3444/123A	53	2007	
815190	123A	3444	20	2051	
815191	123A	3444	20	2051	
815193	160	3006	245	2004	
815195	102A	3004	53	2007	
815196	102A	3004	53	2007	
815197	160	3006	245	2004	
815198	123A	3444	20	2051	
815199	159	3466	82	2032	
815199-6	159	3466	82	2032	
815201	123A	3444	20	2051	
815202	123A	3444	20	2051	
815203-3	121MP	3015	239(2)	2006(2)	
815203-5	121MP	3013	239(2)	2006(2)	
815206	161	3018	39	2015	

Industry Standard No.	ECG	SK	GE	RS 276-	MOTOR.
815209	108	3452	86	2038	
815210	123A	3444	20	2051	
815211	159	3466	82	2032	
815212		3444	20	2051	
815213	159	3466	82	2032	
815215	711		IC-207		
815218-3	102A		53	2007	
815218-4	103A	3835	59	2002	
815227	123A	3444	20	2051	
815228		3004/102A	226,218		
815228A	102A	3004	53	2007	
815228A01	102	3722	2	2007	
815228A1	102A		53	2007	
815228B	102A		53	2007	
815228B1	102A		53	2007	
815229	159	3466	82	2032	
815232		3010	54	2002	
815233	123A	3444	20	2051	
815234	160	3006	245	2004	
815236	159	3466	82	2032	
815237	123A	3444	20	2051	
815243		3122	20,82		
815246-1	130	3027	14	2041	
815246-2	121	3717	239	2006	
815247	159	3466	82	2032	
815259	711		IC-207		
815260	724		IC-86		
815308A	100	3005	1	2007	
816135	124	3021	12		
817032	109	3087		1123	
817036	358		42		
817042	116	3017B/117	504A	1104	
817043	116	3017B/117	504A	1104	
817044	116	3017B/117	504A	1104	
817053	116	3017B/117	504A	1104	
817062	358	3120/114	6GD1		
817062(CRYSTAL)	114		6GD1		
817064	116	3017B/117	504A	1104	
817066	116	3016	504A	1104	
817067	116	3017B/117	504A	1104	
817068	116	3016	504A	1104	
817068P	116	3031A	504A	1104	
817074	113A	3119/113	6GC1		
817077	109	3087		1123	
817079	116	3017B/117	504A	1104	
817082			1N60	1123	
817088	116	3311	504A	1104	
817104	116	3016	504A	1104	
817109	116	3016	504A	1104	
817111	116	3016	504A	1104	
817112	116	3016	504A	1104	
817114	116	3016	504A	1104	
817117	116	3016	504A	1104	
817120	507		504A	1104	
817121	116	3016	504A	1104	
817122	116	3016	504A	1104	
817123	118	3066	CR-1		
817124	119	3109	CR-2		
817125	109	3087		1123	
817126	113A	3119/113	6GC1		
817127	113A	3110/120	6GC1		
817128	116	3016	504A	1104	
817129	358		42		
817130	116	3016	504A	1104	
817133	116	3016	504A	1104	
817134	116	3017B/117	504A	1104	
817135	116	3016	504A	1104	
817138	116	3016	504A	1104	
817140	116	3016	504A	1104	
817141	116	3016	504A	1104	
817143	116	3016	504A	1104	
817147	358		42		
817148	116	3016	504A	1104	
817149	120	3110	CR-3		
817155	142A	3062	ZD-12	563	
817156	116	3017B/117	504A	1104	
817157	116		42	1104	
817158	109	3087		1123	
817159	109	3087		1123	
817160	110MP	3709	1N34AS	1123(2)	
817161	116	3017B/117	504A	1104	
817164	116	3031A	504A	1104	
817166	116	3016	504A	1104	
817167	116	3017B/117	504A	1104	
817172	177	3016	300	1122	
817173	135A		ZD-5.1		
817175	358		42		
817177	110MP	3709	1N34AS	1123(2)	
817179	156	3311	512		
817179A	156	3032A	512		
817180	116	3311	504A	1104	
817190	177	3100/519	300	1122	
817193	125		510,531	1114	
817194	109			1123	
817195	116	3017B/117	504A	1104	
817197	139A	3060	ZD-9.1	562	
817199	109			1123	
817208	142A	3062	ZD-12	563	
817209	156		512		
817962	114		6GD1	1104	
0820220	233		210	2051	
824960-0	108	3452	86	2038	
825065	102A	3005	53	2007	
838105	159	3466	82	2032	
845050	123A	3444	20	2051	
848082	123A	3444	20	2051	
851759-3	160		245	2004	
851881	123		20	2051	
852158-7-1			504A	1104	
852158-7.1	116	3016		1104	
853640			51	2004	
853864-0	160	3007	245	2004	
860001-153	519		514	1122	
860001-8	128		243	2030	
860003-45	911		IC-253		
860003-46	910		IC-251		
860011	177	3100/519	300	1122	
870006	192		63	2030	
871125	109			1123	
880092	131		44	2006	
881916	7474	7474		1818	
882028	130		14	2041	
883802	123A	3444	20	2051	
889132	OBS-NLA		IC-277		
889302	744		IC-215	2022	
889303	OBS-NLA		IC-277		
889779P	736		IC-17		
891008	159	3466	82	2032	
891032	130		14	2041	
894876			17	2051	
900201-104	128		243	2030	
900201-105	128		243	2030	
900201-167	128		243	2030	
900201-81	128		243	2030	

Industry Standard No.	ECG	SK	GE	RS 276-	MOTOR.
900545-2	519		514	1122	
900546-2	519		514	1122	
900546-20	177		300	1122	
900552-17	159	3466	82	2032	
900552-20	123A		20	2051	
900552-30	123A		20	2051	
900552-6	123A		20	2051	
900552-8	123A		20	2051	
902521	160		245	2004	
908098	909		IC-249		
908703-1	177		300	1122	
908705-1	519		514	1122	
908705-2	519		514	1122	
908721-1	177		300	1122	
908742-1	519		514	1122	
908844-1	199		62	2010	
908864-2	159	3466	82	2032	
910050-2	102A	3004	53	2007	
910062-1	102A	3004	53	2007	
910070-6	102A	3004	53	2007	
910088	128		243	2030	
910094-4	102A	3004	53	2007	
910634	128		243	2030	
910799	108	3452	86	2038	
910807-11	175		246	2020	
910952	185	3041	58	2025	
911743-1	123A	3444	20	2051	
916009	123A	3444	20	2051	
916028	123A	3444	20	2051	
916029	108	3452	86	2038	
916030	123A	3444	20	2051	
916031	123A	3444	20	2051	
916031(CARTAPE)			62	2010	
916031(PENNYS)			10	2051	
916033	199	3024/128	62	2010	
916034	128	3024	243	2030	
916046	186	3192	28	2017	
916049	107	3122	11	2015	
916050	123A	3444	20	2051	
916051	159	3466	82	2032	
916052	199	3122	62	2010	
916055	199	3124/289	62	2010	
916059	123A	3444	20	2051	
916060	108	3452	86	2038	
916061	1029		IC-162		
916062	159	3466	82	2032	
916063	1092	3472	IC-130		
916064	1045		IC-2		
916067	1103	3281	IC-94		
916068	107		11	2015	
916069	108	3452	86	2038	
916070	1142	3485	IC-128		
916071	1003		IC-43		
916072	1006	3358	IC-38		
916082	312	3448	FET-1	2035	
916084	1052	3249	IC-135		
916085	1082	3461	IC-40		
916091	123A	3444	20	2051	
916092	1092	3472	IC-130		
916100	312	3116	62	2010	
916105	1104	3225	IC-91		
916106	1170	3745	IC-65		
916109	1153	3282			
916110	1155	3231	IC-179	705	
916111	746	3234	IC-217		
916112	709	3135	IC-11		
916113	801	3160	IC-35		
916114	184	3190	57	2017	
916118	235	3197	215		
916119	295	3047	270		
916121	1234		IC-181		
916122	1170	3745			
916125	1194	3484			
921150-021	519		514	1122	
921608	116	3016	504A	1104	
922021	109	3087		1123	
922092	116	3031A	504A	1104	
922094	116	3031A	504A	1104	
922114	128		243	2030	
922125	128		243	2030	
922183	116	3031A	504A	1104	
922214	611	3324	90		
922311	116	3311	504A	1104	
922358	139A	3060	ZD-9.1	562	
922359		3126	90		
922360			504A	1104	
922433	177	3100/519	300	1122	
922524	139A	3060	ZD-9.1	562	
922567	116	3031A	504A	1104	
922603	139A	3060	ZD-9.1	562	
922604	109	3087		1123	
922693	142A	3062	ZD-12	563	
922799	116	3017B/117	504A	1104	
922860	116	3311	504A	1104	
922873	177	3100/519	300	1122	
922896	100	3005	1	2007	
922943-1	519		514	1122	
922950	5074A	3139	ZD-11.0	563	
922969	116	3017B/117	504A	1104	
923147		3100	300	1122	
923233		3139	ZD-11	563	
923271	614	3327			
924605-3	116	3016	504A	1104	
924801-5	116	3016	504A	1104	
924805-5	116	3016	504A	1104	
924805-8	116	3016	504A	1104	
925075-501B	519		514	1122	
925252-1	519		514	1122	
925252-102	519		514	1122	
925252-2	519		514	1122	
925252-4	519		514	1122	
925253-3	519		514	1122	
925297	519		514	1122	
925297-1	519		514	1122	
925521-1B	177		300	1122	
925521-1C	177		300	1122	
925915-101	177		300	1122	
925939-101	177		300	1122	
925940-1B	177		300	1122	
925940-501B	177		300	1122	
928103-1	123A	3444	20	2051	
928291-101	128		243	2030	
928291-102	128		243	2030	
928408-101	159	3466	82	2032	
928506-101	910		IC-251		
928608-101	703A		IC-12		
930236	123A	3444	20	2051	
930293	519		514	1122	
930347-1	7420	7420		1809	
930347-10	7420	7420		1809	
930347-11	7402	7402		1811	
930347-12	7451	7451		1825	
930347-13	7404	7404		1802	
930347-15	7405	7405			

Industry Standard No.	ECG	SK	GE	RS 276-	MOTOR.
930347-2	7430	7430			
930347-3	7400	7400		1801	
930347-5	7440	7440			
930347-6	7454	7454			
930347-7	7473	7473		1803	
930419	923D		IC-260	1740	
932017-0001	128		243	2030	
932022-0001	519		514	1122	
932030-1	909	3590	IC-249		
932033-1	519		514	1122	
932040	159	3466	82	2032	
932050-1	519		514	1122	
932055-1	128		243	2030	
932081-1	175		246	2020	
932107-1	159	3466	82	2032	
936001	358		42		
941295-2	123A	3444	20	2051	
941295-3	123A	3444	20	2051	
942677-1			1N34AS	1123	
942677-2			1N60	1123	
942677-3			1N60	1123	
942677-4		3091	1N60	1123	
942677-6			1N34AS	1123	
942677-7			1N34AS	1123	
942677-9	177		300	1122	
943502	519		514	1122	
943720-001	123A	3444	20	2051	
944148	5695	3509			
945820-4	116	3017B/117	504A	1104	
954330-2	123A	3444	20	2051	
959492-2	123A	3444	20	2051	
960106-3	159	3466	82	2032	
960201	107		11	2015	
960202	107		11	2015	
960494-1	128		243	2030	
960494-2	128		243	2030	
961544-1	123A	3444	20	2051	
964158	112		1N82A		
964298			1N34AS	1123	
964547-2	199		62	2010	
964634	108	3452	86	2038	
964713	108	3452	86	2038	
964999	112		1N82A		
965000	123A	3444	20	2051	
965073	112		1N82A		
965074	108	3452	86	2038	
965632	161	3018	39	2015	
965633	108	3452	86	2038	
965634	108	3452	86	2038	
969099-1	519		514	1122	
970046	108	3452	86	2038	
970046-1	108	3452	86	2038	
970046-2	108	3452	86	2038	
970046A	108	3452	86	2038	
970047	112	3089	1N82A		
970107	129		244	2027	
970108	128		243	2030	
970244	108	3452	86	2038	
970245	108	3452	86	2038	
970246	159	3466	82	2032	
970247	123A	3444	20	2051	
970248	159	3466	82	2032	
970249	108	3452	86	2038	
970250	123A	3444	20	2051	
970251	159	3466	82	2032	
970252	123A	3444	20	2051	
970253	312	3834/132	FET-1	2035	
970254	159	3466	82	2032	
970255	124	3021	12		
970309	108	3452	86	2038	
970309-1	108	3452	86	2038	
970309-12	108	3452	86	2038	
970309-2	108	3452	86	2038	
970309-3	108	3452	86	2038	
970309-4	108	3452	86	2038	
970309-5	108	3452	86	2038	
970310	108	3452	86	2038	
970310-1	108	3452	86	2038	
970310-12	108	3452	86	2038	
970310-2	108	3452	86	2038	
970310-3	108	3452	86	2038	
970310-4	108	3452	86	2038	
970310-5	108	3452	86	2038	
970311			11	2015	
970332	108	3452	86	2038	
970332-12	108	3452	86	2038	
970659	123A	3444	20	2051	
970660	123A	3444	20	2051	
970661	123A	3444	20	2051	
970662	123A	3444	20	2051	
970663	159	3466	82	2032	
970759	112	3089	1N82A		
970762	159	3466	82	2032	
970762-6	129	3025	244	2027	
970911	108	3452	86	2038	
970916	123A	3444	20	2051	
970916-6	123A	3444	20	2051	
970939			11	2015	
970940			20	2051	
970962			11	2015	
971035			11	2015	
971059	159	3466	82	2032	
971457			504A	1104	
971458			504A	1104	
971459			39	2015	
971460	199	3018	62	2010	
971477			2	2007	
971526			11	2015	
971904			60	2015	
971905			22	2032	
972155	123A	3444	20	2051	
972156	123A	3444	20	2051	
972214	123A	3444	20	2051	
972215	123A	3444	20	2051	
972216	116	3311	504A	1104	
972217	5079A		ZD-20		
972258-1			1N295	1123	
972258-2			1N295	1123	
972258-3			1N295	1123	
972258-4			1N295	1123	
972258-5			1N295	1123	
972258-6	109	3087		1123	
972258-8	109	3090		1123	
972258-9	109	3090		1123	
972259-8	109	3088		1123	
972305	161	3018	39	2015	
972306	107	3018	11	2015	
972307	107	3018	11	2015	
972417	107	3018	11	2015	
972418	107	3018	11	2015	
972419	107	3018	11	2015	
972420	107	3018	11	2015	
972571-2	119	3109	CR-2		
972571-3	119	3109	CR-2		
972571-4	116	3016	504A	1104	
972571-5		3032A	509	1114	
972571-6			504A	1104	
972571-7		3016	504A	1104	
973935-20	116	3031A	504A	1104	
973936-1	116	3017B/117	504A	1104	
973936-10	116	3017B/117	504A	1104	
973936-11	116	3017B/117	504A	1104	
973936-12	116	3043	504A	1104	
973936-13	116	3017B/117	504A	1104	
973936-14	116	3017B/117	504A	1104	
973936-15	116	3017B/117	504A	1104	
973936-16	116	3017B/117	504A	1104	
973936-17	116	3017B/117	504A	1104	
973936-18	116	3031A	504A	1104	
973936-19			504A	1104	
973936-2	116	3017B/117	504A	1104	
973936-20	125	3031A	510,531	1114	
973936-21		3031A	504A	1104	
973936-3	116	3017B/117	504A	1104	
973936-4	116	3017B/117	504A	1104	
973936-5	116	3017B/117	504A	1104	
973936-6	116	3017B/117	504A	1104	
973936-7	116	3043	504A	1104	
973936-8	116	3017B/117	504A	1104	
973936-9			504A	1104	
973962-1	116	3016	504A	1104	
980052	105	3012	4		
980052A	105	3012	4		
980132	121	3009	239	2006	
980134	121	3009	239	2006	
980135	121	3009	239	2006	
980136	100	3005	1	2007	
980138	108	3452	86	2038	
980139	108	3452	86	2038	
980140	160	3006	245	2004	
980142	160	3006	245	2004	
980143	116	3016	504A	1104	
980144	102A	3004	53	2007	
980146	160	3006	245	2004	
980147	123A	3444	20	2051	
980148	102A	3004	53	2007	
980149	102A	3004	53	2007	
980150	127	3034	25		
980153	102A	3004	53	2007	
980155	121	3009	239	2006	
980164	116		504A	1104	
980316	100	3005	1	2007	
980372	126	3008	52	2024	
980373	126	3008	52	2024	
980374	126	3008	52	2024	
980375	102A	3003	53	2007	
980376	102A	3003	53	2007	
980426	100	3005	1	2007	
980432	100	3005	1	2007	
980434	100	3005	1	2007	
980435	160	3006	245	2004	
980437	121	3009	239	2006	
980438	100	3005	1	2007	
980439	100	3005	1	2007	
980440	123A	3444	20	2051	
980441	160	3006	245	2004	
980462	105	3012	4		
980462A	105	3012	4		
980463	105	3012	4		
980463A	105	3012	4		
980505	160	3006	245	2004	
980506	160	3006	245	2004	
980507	160	3006	245	2004	
980508	102A	3004	53	2007	
980509	160	3006	245	2004	
980510	102A	3004	53	2007	
980511	102A	3004	53	2007	
980514	109	3087		1123	
980514A	160	3006	245	2004	
980540	116		504A	1104	
980545A	160	3006	245	2004	
980626	126	3006/160	52	2024	
980636A	160	3006	245	2004	
980833	126	3008	52	2024	
980834	126	3008	52	2024	
980835	126	3008	52	2024	
980836	102A	3004	53	2007	
980837	102A	3004	53	2007	
980958	126	3008	52	2024	
980959	126	3008	52	2024	
980960	102A	3004	53	2007	
980961	102A	3004	53	2007	
980964	116	3016	504A	1104	
981143	160	3006	245	2004	
981144	160	3006	245	2004	
981145	160	3006	245	2004	
981146	160	3006	245	2004	
981147	102A		53	2007	
981148	102A	3004	53	2007	
981149	102A		53	2007	
981150	109	3087		1123	
981153	109	3087		1123	
981203	160		245	2004	
981204	160		245	2004	
981206	102A		53	2007	
981207	109	3087		1123	
981248	118		CR-1		
981249	119		CR-2		
981371	120		CR-3		
981445	116	3017B/117	504A	1104	
981522	109	3087		1123	
981672	102A	3004	53	2007	
981673	102A	3004	53	2007	
981674	102A	3004	53	2007	
981675	102A	3004	53	2007	
981676	109	3087		1123	
981739	116	3016	504A	1104	
981952	116	3017B/117	504A	1104	
981953	116	3031A	504A	1104	
981954	116	3017B/117	504A	1104	
981955	116	3032A	504A	1104	
981956	116	3017B/117	504A	1104	
981959	160	3032A	245	2004	
981969	105	3012	4		
981969A	105	3012	4		
982065	109	3087		1123	
982150	126	3008	52	2024	
982151	102A	3004	53	2007	
982152	102A	3004	53	2007	
982214	116	3017B/117	504A	1104	
982231	123A	3444	20	2051	
982244	102A	3004	53	2007	
982253	116	3031A	504A	1104	
982254	139A	3066/118	ZD-9.1	562	
982267	126	3008	52	2024	
982268	108	3452	86	2038	
982269	108	3452	86	2038	
982270	110MP	3091		1123	

Industry Standard No.	ECG	SK	GE	RS 276-	MOTOR.
982271	110MP	3091		1123	
982275	109	3091		1123	
982279	722		IC-9		
982283	102A	3004	53	2007	
982284	102A	3004	53	2007	
982285	102A	3004	53	2007	
982289	126	3008	52	2024	
982290	109	3091		1123	
982300	128	3004/102A	243	2030	
982307			30	2006	
982321	108	3452	86	2038	
982322	160	3006	245	2004	
982361	113A	3119/113	6GC1		
982374	160	3006	245	2004	
982375	102A	3004	53	2007	
982376	124	3021	12		
982377	116	3031A	504A	1104	
982497	160	3006	245	2004	
982510	123A	3444	20	2051	
982511	123A	3444	20	2051	
982512	123A	3444	20	2051	
982523	175	3021/124	246	2020	
982526	116	3311	504A	1104	
982528	124		12		
982531	102A	3004	53	2007	
982532	102A	3004	53	2007	
982815	108	3452	86	2038	
982816	108	3452	86	2038	
982817	108	3452	86	2038	
982818	108	3452	86	2038	
982819	108	3452	86	2038	
982820	102	3722	2	2007	
982822	109	3087		1123	
982823	116		504A	1104	
983011	138A	3059	ZD-7.5		
983012	176	3123	80		
983036	121	3009	239	2006	
983055	181	3510	75	2041	
983095	108	3452	86	2038	
983096	108	3452	86	2038	
983097	123A	3444	20	2051	
983099	109	3091		1123	
983101	116		504A	1104	
983233	107	3018	11	2015	
983234	107	3018	11	2015	
983235	107	3018	11	2015	
983236	126	3008	52	2024	
983237	102A		53	2007	
983238	102A	3123	53	2007	
983239	109	3087		1123	
983271	126	3008	52	2024	
983272	126	3008	52	2024	
983405	102A	3004	53	2007	
983406	102A	3004	53	2007	
983407	102A	3004	53	2007	
983408	102A	3004	53	2007	
983409	102A	3004	53	2007	
983411	102A		53	2007	
983413	116	3311	504A	1104	
983689			300	1122	
983742	107	3018	11	2015	
983743	123A	3444	20	2051	
983744	178MP		300(2)	1122(2)	
983795	121	3717	239	2006	
983874	121	3009	239	2006	
983945	105	3012	4		
983975	121	3009	239	2006	
983995	118	3066	CR-1		
984156	108	3452	86	2038	
984158	108	3452	86	2038	
984159	108	3452	86	2038	
984160	102A		53	2007	
984161	158		53	2007	
984162	177		300	1122	
984163	109	3090		1123	
984182	116	3017B/117	504A	1104	
984183	116	3017B/117	504A	1104	
984184	116	3017B/117	504A	1104	
984189	116		504A	1104	
984191	161	3018	39	2015	
984192	161	3018	39	2015	
984193	159	3466	82	2032	
984194	108	3452	86	2038	
984195	108	3452	86	2038	
984196	128	3024	243	2030	
984197	123A	3444	20	2051	
984198	123A	3444	20	2051	
984200	109	3087		1123	
984221	102A	3004	53	2007	
984222	123A	3444	20	2051	
984224	123A	3444	20	2051	
984226	109	3087		1123	
984227	154	3040	40	2012	
984228	102A	3004	53	2007	
984229	128	3024	243	2030	
984252	138A	3059	ZD-7.5		
984254	116	3311	504A	1104	
984259	130	3027	14	2041	
984259A	130	3027	14	2041	
984260	506	3125	511	1114	
984261	121	3009	239	2006	
984286	123A	3444	20	2051	
984431	121	3009	239	2006	
984521	131	3052	44	2006	
984522	116	3017B/117	504A	1104	
984577	108	3452	86	2038	
984590	123A	3444	20	2051	
984591	123A	3444	20	2051	
984593	123A	3444	20	2051	
984594	116	3017B/117	504A	1104	
984608	124	3021	12		
984666	109	3090		1123	
984667			1N60	1123	
984685	126		52	2024	
984686	123A	3444	20	2051	
984687	123A	3444	20	2051	
984690	166	9075	BR-600	1152	
984713	116		504A	1104	
984743	108	3452	86	2038	
984744	108	3452	86	2038	
984745	123A	3444	20	2051	
984746	102	3722	2	2007	
984794	156	3032A	512		
984795	156	3033	512		
984851	108	3452	86	2038	
984852	108	3452	86	2038	
984853	108	3452	86	2038	
984854	123A	3444	20	2051	
984875	108	3452	86	2038	
984876	107	3018	11	2015	
984877	107	3018	11	2015	
904878	107	3018	11	2015	
984879	123A	3444	20	2051	
984880	177		300	1122	
984881	110MP	3709	1N60	1123(2)	
984882	116	3032A	504A	1104	
984932	124	3021	12		
985036	121	3009	239	2006	
985087		3122	20	2051	
985096	108	3452	86	2038	
985097	108	3452	86	2038	
985098	123A	3444	20	2051	
985099	123A	3444	20	2051	
985100	123A	3444	20	2051	
985101	123A	3444	20	2051	
985102	123A	3444	20	2051	
985103	131		44	2006	
985104			504A	1104	
985105	116		504A	1104	
985106	177		300	1122	
985175	221		FET-4	2036	
985215	108	3452	86	2038	
985216	102A		53	2007	
985217	102	3722	2	2007	
985218	116		504A	1104	
985431	121	3009	239	2006	
985432	105	3012	4		
985442	107		11	2015	
985442A	108	3452	86	2038	
985443	121	3009	239	2006	
985443A	108	3452	86	2038	
985444	107		11	2015	
985444A	108	3452	86	2038	
985445	126	3007	52	2024	
985445A	160	3008	245	2004	
985446	126	3007	52	2024	
985446A	160	3006	245	2004	
985447	121	3009	239	2006	
985448	138A		ZD-7.5		
985449	121	3009	239	2006	
985453	121	3009	239	2006	
985455	121	3009	239	2006	
985468	102A		53	2007	
985468A	102A	3004	53	2007	
985469	102A		53	2007	
985469A	102A	3004	53	2007	
985470	158		53	2007	
985470A	102A	3004	53	2007	
985471	158		53	2007	
985472	116		504A	1104	
985543	123A	3444	20	2051	
985609	102A		53	2007	
985610	102A		53	2007	
985611	126	3006/160	52	2024	
985619	107		11	2015	
985621	110MP		1N60	1123	
985686	121	3009	239	2006	
985715	312	3834/132	FET-1	2035	
985735	103A	3835	59	2002	
985735A	103A	3835	59	2002	
985957		3126	90		
985961	5022A	3788	ZD-13	563	
986015	106	3984	21	2034	
986030	129		244	2027	
986302	102A	3004	53	2007	
986305	158	3004/102A	53	2007	
986414	116	3031A	504A	1104	
986542	123A	3444	20	2051	
986543	158	3004/102A	53	2007	
986576	107	3018	11	2015	
986577			300	1122	
986578	135A	3056	ZD-5.1		
986634	108	3452	86	2038	
986635	108	3452	86	2038	
986636	123A	3444	20	2051	
986637			504A	1104	
986693	107	3018	11	2015	
986694	107	3018	11	2015	
986766	102A	3004	53	2007	
986779	158	3004/102A	53	2007	
986930	222	3050/221	FET-4	2036	
986931	129	3114/290	244	2027	
986932	186	3192	28	2017	
986933	187	3193	29	2025	
986934	177	3100/519	300	1122	
986935	116	3311	504A	1104	
987010	123A	3444	20	2051	
987030	128		243	2030	
988000	108	3452	86	2038	
988001	108	3452	86	2038	
988002	108	3452	86	2038	
988003	123A	3444	20	2051	
988005	158	3004/102A	53	2007	
988048		3126	90		
988049	117	3031A	504A	1104	
988050	140A	3061	ZD-10	562	
988051	116	3311	504A	1104	
988080	121	3009	239	2006	
988336	121	3009	239	2006	
988413	121	3009	239	2006	
988414	105	3012	4		
988468	121	3009	239	2006	
988977	105	3012	4		
988985	108	3452	86	2038	
988986	108	3452	86	2038	
988987	108	3452	86	2038	
988988	108	3452	86	2038	
988989	108	3452	86	2038	
988990	159	3466	82	2032	
988991	123A	3444	20	2051	
988992	193	3025/129	67	2023	
988993	128	3024	243	2030	
988994	109	3087		1123	
988995	177	3100/519	300	1122	
988996	177		300	1122	
988997	109	3088		1123	
988998	109	3088		1123	
989171	105	3012	4		
989387	121	3009	239	2006	
989615	105	3012	4		
989692	105	3012	4		
989693	105	3012	4		
989709	909	3590	IC-249		
991064	125		510,531	1114	
991129	125		510,531	1114	
991421	125		510,531	1114	
991422	125		510,531	1114	
991429	125		510,531	1114	
992052	123A	3122	61	2038	
992066	199	3122	212	2010	
992108	229		60	2015	
992129	123A	3124/289	210	2051	
992143	109	3088	1N60	1123	
992150	177	3100/519	300	1122	
992157-1	519		514	1122	
992171	116	3016	504A	1104	
992248	1100	3223			
992269	74145	74145			

Industry Standard No.	ECG	SK	GE	RS 276-	MOTOR.
992289	102	3722	2	2007	
993570-4	129		244	2027	
993624-2	128		243	2030	
994634	108	3452	86	2038	
995001	121	3009	239	2006	
995002	102A	3003	53	2007	
995003	102A	3004	53	2007	
995014	121	3009	239	2006	
995015	121	3009	239	2006	
995016	123A	3444	20	2051	
995017	123A	3444	20	2051	
995030	124	3122	12		
995053-1	722	3161	IC-9		
995081-1	713		IC-5		
995870-1	123A	3444	20	2051	
995870-3	123A	3444	20	2051	
995928 1	128			2030	
995928-1			243	2030	
996746	123A	3444	20	2051	
996817	152	3893	66	2048	
999106-2	911		IC-253		
999106-3	911D		IC-254		
1000100	7492	7492		1819	
1000100-000	7492			1819	
1000101	7493A	7493			
1000101-000	7493A	7493			
1018715	124		12		
1018734-001	124	3021	12		
1022612	123A	3444	20	2051	
1043176-1	116	3017B/117	504A	1104	
1043176-2	116	3017B/117	504A	1104	
1043176-3	116	3017B/117	504A	1104	
1043176-4	116	3017B/117	504A	1104	
1043176-5	116	3017B/117	504A	1104	
1045013	116	3017B/117	504A	1104	
1045154-1	116	3017B/117	504A	1104	
1045154-2	116	3017B/117	504A	1104	
1045154-3	116	3017B/117	504A	1104	
1045154-4			509	1114	
1045154-5	156		512		
1045154-6	156		512		
1045494-1	114	3120	6GD1	1104	
1061854-2	126	3007	52	2024	
1107832-10	113A	3119/113	6GC1		
1107832-11	113A	3119/113	6GC1		
1107832-6	115	3121	6GX1		
1107832-7	113A	3119/113	6GC1		
1107832-8	113A	3119/113	6GC1		
1107832-9	113A	3119/113	6GC1		
1107863-1	358		42		
1107863-2	358		42		
1107863-3	358		42		
1121008/7611	109			1123	
1127859	123A	3444	20	2051	
1211016/7603/7608	102A		53	2007	
1211017/7603/7608	102A		53	2007	
1211018/7603/7608	102A		53	2007	
1221028	105	3012	4		
1221615	121	3717	239	2006	
1221625	104	3719	16	2006	
1221648	100	3123	1	2007	
1221649	100	3005	1	2007	
1221900	5072A		ZD-8.2	562	
1221962	123A	3444	20	2051	
1222123	123A	3444	20	2051	
1222133	123A	3444	20	2051	
1222136	160		245	2004	
1222314	160		245	2004	
1222371	160		245	2004	
1222424	123A	3444	20	2051	
1222463	108	3452	86	2038	
1223770	109	3087	1N34AS	1123	
1223771	519	3100	514	1122	
1223772	116	3311	504A	1104	
1223773	140A	3061	ZD-10	562	
1223781	108	3018	11	2015	
1223782	123A	3038	20	2051	
1223783	241	3188A/182	66	2048	
1223784	195A	3048/329	219		
1223785	195A	3047	219		
1223786	195A	3048/329	219		
1223909	1167	3732			
1223910	1155	3231		704	
1223911	107	3356	211	2016	
1223912	123A	3122	20	2051	
1223913	297	3122	210	2051	
1223914	290A	3114/290	269	2032	
1223915	229		212	2010	
1223916	123A		20	2051	
1223917	306		276		
1223918	235	3239/236	322		
1223919	229		60	2015	
1223920	123A	3122	61	2038	
1223921	199		1N60	1123	
1223925	614	3126	90		
1223926	135A		ZD-5.1		
1223927	139A	3060	ZD-9.1	562	
1223928	116	3311	504A	1104	
1223929	125	3311	504A	1104	
1223930	177	3175	300	1122	
1223931	109	3088	1N60		
1224031	294	3114/290	48		
1224076	1192	3445			
1224077	107	3356			
1224078	295	3253			
1224079	297	3124/289			
1224081	290A	3114/290			
1224085	199	3122			
1224086	199	3124/289			
1224278	233	3122			
1224279	137A	3058			
1261915-191	129		244	2027	
1261915-383	123A	3444	20	2051	
1288055	128		243	2030	
1289050	130		14	2041	
1303256	156	3032A	512		
1320135	123A	3444	20	2051	
1320135A	123A	3444	20	2051	
1320135BC	123A	3444	20	2051	
1320135C	123A	3444	20	2051	
1390655	506	3843	511	1114	
1407205-1	121	3717	239	2006	
1407206-1	121	3717	239	2006	
1408615-1	108	3452	86	2038	
1408640-1	108	3452	86	2038	
1408649-1	109	3019		1123	
1408694-1	222	3050/221	FET-4	2036	
1415721-1	506	3130	511	1114	
1415721-35	506	3843	511	1114	
1415742-5	797	3158	233		
1415762-1	230	3042	700		
1415762-2	231	3857	700		
1415780-1	5511	3683			
1417302-1	123A	3444	20	2051	
1417303	298	3450	272		
1417303-1	298	3466/159	244	2027	
1417303-2	298		244	2027	
1417306-1	128	3444/123A	243	2051	
1417306-2	123A	3444	20	2051	
1417306-3	123A	3444			
1417306-4	123A	3444	20	2051	
1417306-5	123A	3444	20	2051	
1417306-7	123A	3444			
1417306-8	123A	3444			
1417308-1	199	3245	62	2010	
1417308-2	199	3245	62	2010	
1417312-1	123A	3444	20	2051	
1417312-2	123A	3444	20	2051	
1417316-1	188	3199	226	2018	
1417317-1	189	3200	218	2026	
1417318	194		220		
1417318-1	123A	3275/194	20	2051	
1417318-2	297	3449	271	2030	
1417320	130		14	2041	
1417320-1	130	3027	14	2041	
1417321-1	194	3022/1188	220		
1417322-1	175	3538	246	2020	
1417324-1			20	2051	
1417325-1	128	3854/123AP	63	2030	
1417330-3	159	3466	82	2032	
1417330-4	159	3466	82	2032	
1417331-1	165	3041	38		
1417335-1	165	3115	38		
1417338-5	128		243	2030	
1417339-1	159		82	2032	
1417340-2	123A	3444	20	2051	
1417342-1	128	3024	243	2030	
1417344-1	128		243	2030	
1417344-2	128	3137/192A			
1417345-2	128	3024	243	2030	
1417346-1	123A	3444			
1417347-1	159	3466	82	2032	
1417349-2	297	3444/123A	243	2030	
1417352-5	171	3201	27		
1417356-2	130	3027	19	2041	
1417358-1	152	3893	66	2048	
1417359-1	185	3191	58	2025	
1417362-1	198	3201/171	251		
1417362-2	198	3201/171			
1417362-4	198	3220	27		
1417362-5	171	3201			
1417363-1	159	3466	82	2032	
1417364-1	196	3054	241	2020	
1417366-1	165	3115	38		
1417370-1	165	3115	38		
1417372-1	222	3065	FET-4	2036	
1417380-1	238	3710			
1417381-2	128	3024	243	2030	
1417398-1	190	3232/191			
1417399-4	152	3054/196			
1417401-1	159	3466			
1417402-1	159	3466			
1417872-11	519		514	1122	
1420427-1	102A		53	2007	
1420427-2	102A		53	2007	
1420427-3	102A		53	2007	
1421207-1	178MP	3100/519	300	1122	
1438917-1	1175	3212			
1440977	804	3066/118	CR-1		
1440977-1	118	3066	CR-1		
1440990	120		CR-3		
1442415-2	112	3089	1N82A		
1442415-3	112	3089	1N82A		
1443024-1	199	3245	62	2010	
1443200-3	102A		53	2007	
1443223-1		3126	90		
1444875-1	142A		ZD-12	563	
1445470-2	503		CR-5		
1445470-50	502		CR-4		
1445470-501	503	3068	CR-5		
1445470-502	506	3068/503	511	1114	
1445470-503	503	3068	CR-5		
1445470-504	504	3108/505	CR-6		
1445740-502	502	3067	CR-4		
1445829-501	189	3083/197	218	2026	
1445829-502	189	3083/197	218	2026	
1445829-503	188	3054/196	226	2018	
1445829-504	188	3054/196	226	2018	
1446149-1	506	3043	511	1114	
1446149-2	506	3043	511	1114	
1449098-1	154	3044	40	2012	
1462432-1	710	3102	IC-89		
1462434-1	710	3102	IC-89		
1462445-1	711	3070	IC-207		
1462506-1		3074	IC-23		
1462516	712	3072	IC-2		
1462516-001	712	3072	IC-148		
1462516-1	712	3072	IC-2		
1462554-1	730	3143			
1462554-2	730	3143			
1462554-3	730	3143			
1462554-4	730	3143			
1462559-1	728	3073	IC-22		
1462560	729	3074	IC-23		
1462560-001	729	3074	IC-23		
1462560-1	729	3074	IC-23		
1462577	780	3141	IC-222		
1463633-1	710		IC-89		
1463641-1	536A	3900			
1463641-2	536A	3900			
1463677-1	780	3141	IC-222		
1463677-2	780	3141	IC-222		
1463677-3	780	3141	IC-222		
1463681-1	904	3542	IC-289		
1463686-1	712		IC-2		
1464437-1	787	3146	IC-228		
1464437-2	787	3146	IC-228		
1464437-3	787	3146	IC-228		
1464438-1	788	3829	IC-229		
1464438-2	788	3829	IC-229		
1464438-3	788	3829	IC-229		
1464439-2	789	3078			
1464460-2	789	3078			
1464460-3	789	3078			
1464607-1	532	3301/531	525		
1464607-10	533	3302	526		
1464607-2	532	3301/531	525		
1464607-3	532	3301/531	525		
1464607-4	532	3301/531	525		
1464607-5	532	3301/531	525		
1464607-6	532	3301/531	525		
1464607-7	531	3300/517	524		
1464607-8	531	3300/517	524		
1464607-9	533	3302	526		
1464686-1	712	3072	IC-2		
1464778-9	507		504A	1104	
1464846-1	949		IC-25		
1464846-2	949		IC-25		

Industry Standard No.	ECG	SK	GE	RS 276-	MOTOR.
1464984-1	532	3301/531	525		
1464984-2	532	3301/531	525		
1465158-1	797	3158	IC-233		
1465158-2	797	3158	IC-233		
1465316-1	984	3185			
1465316-2	984	3185			
1465316-2A	984	3185			
1465345-1	1174	3186			
1465615-1	1176	3210			
1465615-2	1176	3210			
1465617-2	1175	3212			
1465617-3A	1175	3212			
1465629-1	985	3214			
1465629-2	985	3214			
1466701-1	1177	3214/985			
1466860-1	532	3302/533	525		
1466860-2	537	3302/533	536		
1466860-3	537	3302/533			
1466862-2	531	3300/517			
1466865-1	532		525		
1470872-6	177	3100/519	300	1122	
1470990	120		CR-3		
1470990-1	120	3110	CR-3		
1471036-14	121	3717	239	2006	
1471036-20	121	3009	239	2006	
1471072-4	177	3100/519	300	1122	
1471100-1	102A		53	2007	
1471100-8	102A		53	2007	
1471100-9	102A		53	2007	
1471101-15	102A	3004	53	2007	
1471101-2	102A		53	2007	
1471101-3	102A		53	2007	
1471101-4	102A		53	2007	
1471102-41	121	3717	239	2006	
1471104-5	126	3008	52	2024	
1471104-6	126		52	2024	
1471104-7	126		52	2024	
1471104-8	126		52	2024	
1471112-10	234	3114/290	65	2050	
1471112-12	129	3025	244	2027	
1471112-3	129		244	2027	
1471112-7	159	3466	82	2032	
1471112-8	159	3466	82	2032	
1471112-8-9	129	3025	244	2027	
1471112-9		3025	21	2034	
1471113-2	123A	3444	20	2051	
1471113-3	123A	3444	20	2051	
1471114-1	159	3466	82	2032	
1471115-1	123A	3444	20	2051	
1471115-10	161	3716	39	2015	
1471115-11	199	3245	62	2010	
1471115-12	123A	3444	20	2051	
1471115-13	108	3452	86	2038	
1471115-14	108	3452	86	2038	
1471115-2	199	3124/289	62	2010	
1471115-3	199	3245	62	2010	
1471115-4	199	3245	62	2010	
1471117-1	124	3021	12		
1471120-11	199	3245	62	2010	
1471120-14	128	3024	243	2030	
1471120-15	123A	3444	20	2051	
1471120-7	123A	3444	20	2051	
1471120-8	123A	3444	20	2051	
1471120-8-9	123A	3444	20	2051	
1471122	199	3245	62	2010	
1471122-6	199	3018	62	2010	
1471122-7	199	3018	62	2010	
1471123-2	128		243	2030	
1471123-3	128	3024	243	2030	
1471123-4	128	3024	243	2030	
1471123-5	128	3024	243	2030	
1471124-5	179	3642	76		
1471125-3	127	3034	25		
1471132-002	196	3041	241	2020	
1471132-2	184	3054/196	57	2017	
1471132-3	196	3054	241	2020	
1471132-4	152	3054/196	66	2048	
1471132-5	196	3054	241	2020	
1471132-6	181	3054/196	75	2041	
1471133-1	188	3054/196	226	2018	
1471134	153		69	2049	
1471134-1	189	3200	218	2026	
1471135-001	130		14	2041	
1471135-1	130	3027	14	2041	
1471135-2	181		75	2041	
1471136-3	172A	3156	64		
1471139-1	180		74	2043	
1471140-1	184	3190	57	2017	
1471141-1	185	3191	58	2025	
1471393-4	167	3647	300	1172	
1471405-1	156	3051	512		
1471822-11	110MP	3709		1123(2)	
1471858-1			504A	1104	
1471872-001		3311	504A	1104	
1471872-002	177	3100/519			
1471872-004		3311	504A	1104	
1471872-006		3175	504A	1104	
1471872-008		3311	504A	1104	
1471872-06	177	3175			
1471872-1	177	3311	300	1122	
1471872-10	178MP	3100/519	300(2)	1122(2)	
1471872-11	519	3088	514	1122	
1471872-12	177	3100/519	300	1122	
1471872-13	177	3311	300	1122	
1471872-14	177	3017B/117	511	1114	
1471872-15	109	3100/519		1123	
1471872-16	177	3100/519	300	1122	
1471872-17	109	3175		1123	
1471872-18	177	3175	300	1122	
1471872-2	177	9091	300	1122	
1471872-3	506	3043	511	1114	
1471872-4	177	3175	300	1122	
1471872-5	519	3031A	514	1122	
1471872-6	116	3313	504A	1104	
1471872-7	177	3175	300	1122	
1471872-8	177	3175	300	1122	
1471876-6	178MP	3100/519	300(2)	1122(2)	
1471878-3	5071A		ZD-6.8	561	
1471893-4	149A		ZD-62		
1471898-3	137A	3780/5014A	ZD-6.2	561	
1471898-4	149A		ZD-62		
1471898-5	136A		ZD-5.6	561	
1471898-8	140A	3785/5019A			
1471908-1	173BP	3999	305		
1471908-4		3999	305		
1471922-1	112	3089	1N82A		
1471922-2		3089	1N82A		
1472171-32	116	3311	504A	1104	
1472434-1	706	3101	IC-43		
1472446	121	3717	239	2006	
1472446-1	121	3009	239	2006	
1472450-1	108	3452	86	2038	
1472460-1	117		504A	1104	
1472460-13	506	3031A	511	1114	
1472460-14	156	3031A	512		
1472460-16	116	3031A	504A	1104	
1472460-2	177		300	1122	
1472460-4	117		504A	1104	
1472460-5	117		504A	1104	
1472460-6			300	1122	
1472460-7			300	1122	
1472460-8			300	1122	
1472474-2	154		40	2012	
1472475-1	199	3245	62	2010	
1472482-1	102A		53	2007	
1472494-1	127	3764	25		
1472495-1	123A	3444	20	2051	
1472500-1	127	3764	25		
1472501-1	159	3466	82	2032	
1472633	161	3716	39	2015	
1472634-1	108	3018	86	2038	
1472636-1	107		11	2015	
1472872-4			300	1122	
1473500-1	123A	3444	20	2051	
1473501-1	159	3466	82	2032	
1473502-1	704	3022/1188	IC-205		
1473502-2	704	3023	IC-205		
1473503	124		12		
1473503-1	124		12		
1473505-1	123A	3444	20	2051	
1473506-1	199		62	2010	
1473508-1	128		243	2030	
1473512-1	121	3717	239	2006	
1473514-1	127	3173/219	25		
1473515-1	121	3014	239	2006	
1473516-1	129	3025	244	2027	
1473519-1	123A	3122	20	2051	
1473520-1	124	3021	12		
1473521-1			11	2015	
1473523-1	159	3466	82	2032	
1473524-1		3018	11	2015	
1473524-2	108	3018	86	2038	
1473524-3			11	2015	
1473527-1	123A	3018	20	2051	
1473528-1	785	3254	IC-226		
1473529-1	107	3019	11	2015	
1473530-1	108	3452	86	2038	
1473530-2	108	3452	86	2038	
1473532-1	123A	3018	20	2051	
1473533-1	108	3452	86	2038	
1473535-1	319P	3246/229	39	2015	
1473535-1(RCA)			39	2015	
1473536-001	123A	3444	20	2051	
1473536-1	123A	3444	20	2051	
1473536-2	161	3124/289	39	2015	
1473537-1	108	3452	86	2038	
1473538-1	123A	3122	20	2051	
1473539-1	123A	3444	20	2051	
1473540-1	159	3114/290	82	2032	
1473541-1	154		40	2012	
1473543-1	161	3716	39	2015	
1473544-1	107	3122	11	2015	
1473545-1	128	3104A	243	2030	
1473546-1	123A	3018	20	2051	
1473546-2	123A	3018	20	2051	
1473546-3	123A	3444	20	2051	
1473547-1	128	3045/225	243	2030	
1473548-1	123A	3444	20	2051	
1473549-1	159	3118	82	2032	
1473549-2	159	3118	82	2032	
1473550-1	123A	3444	20	2051	
1473551-1	123A	3122	20	2051	
1473552-1	154	3044	40	2012	
1473553-1	128	3044/154	243	2030	
1473554-1	123A	3444	20	2051	
1473555-1	123A	3444	243	2051	
1473555-2	123A	3104A	20	2051	
1473555-3	128	3024	243	2030	
1473555-I	128			2030	
1473556-1	123A	3444	20	2051	
1473557-1	123A	3444	20	2051	
1473558-1	107	3050/221	11	2015	
1473559-001	159	3466	21	2034	
1473559-1	159	3466	243	2032	
1473560			18	2030	
1473560-002	123A	3122	20	2051	
1473560-1	128	3124/289	243	2030	
1473560-2	128	3122	243	2030	
1473560-3	123A	3444			
1473561-001	123A	3444			
1473561-1	128	3122	243	2030	
1473561-Y			20	2051	
1473562-1	159	3114/290	82	2032	
1473563-1	159	3114/290	82	2032	
1473564-1	162	3438	35		
1473565-001	128	3024	18	2030	
1473565-1	128	3122	243	2030	
1473566-1	128	3024	243	2030	
1473567-1	124	3045/225	12		
1473567-2	184	3045/225	57	2017	
1473567-4	152	3054/196	66	2048	
1473568-1	108	3452	86	2038	
1473569-1	123A	3122	20	2051	
1473570-1	159	3025/129	82	2032	
1473570-2	159	3466	82	2032	
1473571-1	161	3018	39	2015	
1473572-1	123A	3444	20	2051	
1473572-3	128	3124/289	243	2030	
1473573-1	103A	3835	59	2002	
1473574-1	159	3114/290	82	2032	
1473576	161	3018	39	2015	
1473576-1	233	3854/123AP	210	2009	
1473577-1	161	3018	39	2015	
1473578-1	102A	3004	53	2007	
1473579-1	161	3018	39	2015	
1473580-1	128		243	2030	
1473581-1	159	3466	82	2032	
1473582-1	123A	3444	20	2051	
1473583-7	230	3042			
1473583-8	230	3042			
1473584-1	124	3045/225	12		
1473585-5	230	3042			
1473585-7	230	3042			
1473585-8	231	3857			
1473586	161	3117	39	2015	
1473586-2	123A	3245/199	20	2051	
1473588-2	221	3050	FET-4	2036	
1473588-8		3042	FET-4	2036	
1473589-1	123A	3444	20	2051	
1473590-1	151A	3025/129	ZD-110		
1473591-1	159	3114/290	82	2032	
1473592-1	129	3118	244	2027	
1473593-1	128		243	2030	
1473595-1	123A	3018	20	2051	
1473597			21	2034	
1473597-1	159	3118	82	2032	
1473597-2	159	3118	82	2032	
1473598-1		3004	52	2024	

Industry Standard No.	ECG	SK	GE	RS 276-	MOTOR.
1473598-2	102A	3004	53	2007	
1473599-1	159	3466	82	2032	
1473601	163A	3439	36		
1473601-001	123A	3124/289	20	2051	
1473601-1	123A	3444	20	2051	
1473601-2	123A	3444	20	2051	
1473603-1	108	3452	86	2038	
1473604-1	172A	3156	64		
1473604-3	108	3452	86	2038	
1473606-1	108	3452	86	2038	
1473608-002	128	3020/123	18	2030	
1473608-1	128	3024	243	2030	
1473608-2	128	3024	243	2030	
1473608-3	128	3024	243	2030	
1473610-1		3018	11	2015	
1473611	152	3893	66	2048	
1473611-1	152	3054/196	66	2048	
1473612	152	3893	66	2048	
1473612-1	196	3054	241	2020	
1473612-11	152	3054/196	66	2048	
1473613-2	191	3232	249		
1473613-3	191	3232	249		
1473613-4	171	3104A	27		
1473613-5	128		243	2030	
1473614-1	123A	3122	20	2051	
1473614-2	199	3245	62	2010	
1473614-3	123A	3122	20	2051	
1473615-1	128	3024	243	2030	
1473616-1	129	3118	244	2027	
1473617-1	108	3019	86	2038	
1473617-2			11	2015	
1473618-1	222	3050/221	FET-4	2036	
1473619-1	5652	3506/5642			
1473619-2	5642	3506			
1473619-3	5641	3519/5643	1N60	1123	
1473620-001	159	3138/193A	21	2034	
1473620-1	159	3114/290	82	2032	
1473621-1	196	3054	241	2020	
1473622	128		243	2030	
1473622-002	128	3044/154	18	2030	
1473622-1	123A	3444	20	2051	
1473622-2	128	3024	243	2030	
1473623-1	196	3054	241	2020	
1473624-1	197	3083	250	2027	
1473625-001	128		63	2030	
1473625-1	128	3024	243	2030	
1473626-1	123A	3444	20	2051	
1473627-1	159	3114/290	82	2032	
1473628-1	188	3201/171	226	2018	
1473628-2	171	3104A	27		
1473628-3	188	3083/197	226	2018	
1473629-1	189	3200	218	2026	
1473629-3	189	3054/196	218	2026	
1473631-1	123A	3444	20	2051	
1473632-1	171	3054/196	27		
1473632-2	186	3192	28	2017	
1473632-3	196	3054	241	2020	
1473633-1	171	3104A	27		
1473634-1	165		38		
1473635	222			2036	
1473635-1	222	3050/221	FET-4	2036	
1473635-2	221	3050	FET-4	2036	
1473636-1	124	3054/196	12		
1473637-1	162	3438	35		
1473638-1	218	3085	234		
1473640-1	196	3054	241	2020	
1473647-1	165	3710/238	38		
1473648-1	127	3035	25		
1473649-1	165	3115	38		
1473651-1	106	3050/221	21	2034	
1473652-1	108	3452	86	2038	
1473656-1	182	3044/154	217		
1473656-2	182	3044/154	55		
1473656-4	154	3044	40	2012	
1473657-1	229	3018	61	2038	
1473657-2	229	3018	61	2038	
1473665-2	130	3027	14	2041	
1473666-1	129	3114/290	244	2027	
1473669-1	165	3111	38		
1473669-2	165	3535/181	38		
1473676-1	233	3246/229	39	2015	
1473676-1(3RD LF)				2009	
1473676-1(3RD-IF)			210	2051	
1473679-1	128	3232/191	243	2030	
1473679-2			40	2012	
1473680-1	319P	3018	39	2015	
1473681-1	152	3054/196	66	2048	
1473681-I	152		66	2048	
1473682-1	185	3191	58	2025	
1473683-1	130	3114/290	14	2041	
1473686-1		3018	52	2024	
1473687-1	191	3044/154	249		
1473688-1	5404	3627			
1473777-2	145A	3063	ZD-15	564	
1474178-10			504A	1104	
1474717-2	145A	3063	ZD-15	564	
1474777-002	145A	3063	ZD-15	564	
1474777-1	142A	3062	ZD-12	563	
1474777-11	144A	3063/145A	ZD-14	564	
1474777-2	145A	3063	ZD-15	564	
1474777-3	5103A		ZD-180		
1474777-4	147A		ZD-33		
1474777-7	149A	3064/146A	ZD-62		
1474777-8			504A	1104	
1474778			504A	1104	
1474778-013	156		504A	1104	
1474778-10	116	3031A	504A	1104	
1474778-11	116	3017B/117	504A	1104	
1474778-13	156	3051	512		
1474778-14	506	3051/156	511	1114	
1474778-16	177	3100/519	300	1122	
1474778-2	116	3311	504A	1104	
1474778-21	506	3052	511	1114	
1474778-3	116	3311	504A	1104	
1474778-4	156	3081/125	512		
1474778-5	506	3311	511	1114	
1474778-6	506	3843	511	1114	
1474778-7	116	3031A	300	1104	
1474778-8	156	3311	512		
1474778-9	507	3017B/117	504A	1104	
1474788-10	116	3031A	504A	1104	
1474872-1	177	3032A	300	1122	
1476049-2	506	3843	511	1114	
1476161-12	506	3843	511	1114	
1476161-13			504A	1104	
1476171-1			504A	1104	
1476171-11		3130	504A	1104	
1476171-11(TRANS)			16	2006	
1476171-12	515	9098	511		
1476171-13		3051	511	1114	
1476171-13(DIODE)	506		511	1114	
1476171-13(TRANS)			16	2006	
1476171-17	506	3843	511	1114	
1476171-18	506	3130	511	1114	
1476171-19	515	3175	504A	1104	
1476171-20	515	9098	511		
1476171-21	156	3081/125	512		
1476171-22	506	3130	511	1114	
1476171-24	515	9098	300	1122	
1476171-25	506	3843	511	1114	
1476171-26	506	3843	511	1114	
1476171-27	515	9098	504A	1104	
1476171-28	116	3175	504A	1104	
1476171-29	116	3016	504A	1104	
1476171-31	116	3175	504A	1104	
1476171-32	506	3175	511	1114	
1476171-33	177	3998/506	300	1122	
1476171-34	506	3998	511	1114	
1476171-35	515	3130	511		
1476171-8			504A	1104	
1476172-12			511	1114	
1476179-001	109	3087	1N34AS	1123	
1476179-01	109	3090			
1476179-1	109	3090		1123	
1476183-2	506	3843	511	1114	
1476183-3	503		CR-5		
1476183-5	506	3843	511	1114	
1476183-6	506	3843	511	1114	
1476183-8	119		511	1114	
1476188-1	123A	3444	20	2051	
1476689-1		3126	90		
1476690-1		3100	300	1122	
1476690-2	177	3031A	300	1122	
1476930	505	3100/519	CR-7		
1476933-501	505		CR-7		
1476933-502	505		CR-7		
1476933-503	505		CR-7		
1477022-1	553	3322	300	1122	
1477046-1	145A	3063	ZD-15	564	
1477046-10	5077A	3752	ZD-18		
1477046-4	142A	3094/144A	ZD-12	563	
1477046-5	5085A	3311	ZD-36		
1477080-501			ZD-15	564	
1477081-501	145A	3063	ZD-15	564	
1477949-1	616	3319			
1478164-1	147A	3095	ZD-33		
1478564-1	310	3856			
1478564-2	308	3855			
1478564-5	310	3856			
1478564-6	308	3855			
1478564-7	310	3856			
1478564-8	308	3855			
1501883	123A	3444	20	2051	
1502039	128		243	2030	
1503097-0	159	3466	82	2032	
1522237-20	123A	3444	20	2051	
1560048	912		IC-172		
1563295-101	123A	3444	20	2051	
1582501	925		IC-261		
1588035-42	177		300	1122	
1596408	123A	3444	20	2051	
1604609-2	909	3590	IC-249		
1611708-2	123A		20	2051	
1612738-1	199	3245	62	2010	
1616226-1	159	3466	82	2032	
1617032	159	3466	82	2032	
1617510-1	123A	3444	20	2051	
1638776-103JS1659	519		514	1122	
1641141-101	177		300	1122	
1641141-102	177		300	1122	
1642606B1	135A		ZD-5.1		
1642606B4	137A		ZD-6.2	561	
1642606B6	136A		ZD-5.6	561	
1662258-6	5987	3698	5097		
1680008-01	519		514	1122	
1690019-01	123A	3444	20	2051	
1700001	159	3466	82	2032	
1700008	123A	3444	20	2051	
1700019	123A	3444	223	2051	
1700020	229	3018	61	2038	
1700032	229	3018	61	2038	
1700033	229	3018	61	2038	
1700034	159	3466	82	2032	
1700035	311	3122	20		
1700036	152	3054/196	66	2048	
1700037	195A	3765	46		
1700038	195A	3765	46		
1701790-1	130		14	2041	
1702601-1	130		14	2041	
1780142	159	3466	82	2032	
1780145-1	123A	3444	20	2051	
1780145-2	123A	3444	20	2051	
1780145-2-001	123A	3444	20	2051	
1780169-1	177		300	1122	
1780174-1	519		514	1122	
1780522-1	159	3466	82	2032	
1780522-2	159	3466	82	2032	
1780522-2-001	159	3466	82	2032	
1780724-1	123A	3444	20	2051	
1780738-1	123A	3444	20	2051	
1800002	109	3087		1123	
1800006	116	3017B/117	504A	1104	
1800009	519	3100	514	1122	
1800012	136A	3057	ZD-5.6	561	
1800013	140A	3061	ZD-10	562	
1800017	156	3051	512		
1800018	116	3017B/117	504A	1104	
1800019	5092A		ZD-68		
1800020	5085A		ZD-36		
1802520-001	941		IC-263	010	
1802677-1	923D		IC-260	1740	
1802765	940		IC-262		
1810037	108	3452	86	2038	
1810038	108	3452	86	2038	
1810039	108	3452	86	2038	
1811119			504A	1104	
1813913-1	177		300	1122	
1815036	108	3452	86	2038	
1815037	108	3452	86	2038	
1815039	108	3452	86	2038	
1815041	123A	3444	20	2051	
1815042	123A	3444	20	2051	
1815043	123A	3444	20	2051	
1815045	108	3452	86	2038	
1815047	108	3452	86	2038	
1815054	123A	3444	20	2051	
1815067	108	3452	86	2038	
1815068	108	3452	86	2038	
1815139			8	2002	
1815153	128		243	2030	
1815154	123A	3444	20	2051	
1815154-9	128	3024	243	2030	
1815154/7			63	2030	
1815156	128	3024	243	2030	
1815157	128	3024	243	2030	
1815157-9			18	2030	
1815159	128	3024	243	2030	
1817004	108	3452	86	2038	

Industry Standard No.	ECG	SK	GE	RS 276-	MOTOR.
1817005	123A	3444	20	2051	
1817005-3	108	3452	86	2038	
1817006	128		243	2030	
1817006-3	108	3452	86	2038	
1817007	123A	3444	20	2051	
1817008	108	3452	86	2038	
1817017	103A	3010	59	2002	
1817045	108	3452	86	2038	
1817108	123A	3444	20	2051	
1819045	108	3452	86	2038	
1820829	128		243	2030	
1826065-1	532	3301/531	525		
1826065-2		3301	525		
1826065-3	531	3300/517	524		
1826860-1	532	3301/531			
1827322	129	3025	244	2027	
1833404	390		255	2041	
1835667	197		250	2027	
1840399-1	123A	3444	20		
1846282-1	123A	3444	20		
1846794-1	177		300	1122	
1851490	5995	3518	5105		
1851515	123A	3444	20	2051	
1851516	175	3026	246	2020	
1851517	181	3036	75	2041	
1851518	175	3026	246	2020	
1861223-1	159	3466	82	2032	
1872425-1	909	3590	IC-249		
1887048	120	3110	CR-3		
1905490-1	519		514	1122	
1908519	519		514	1122	
1944313A1	128		243	2030	
1944748	179	3642	76		
1945294	159	3466	82	2032	
1945295A1	145A		ZD-15	564	
1950003	519		514	1122	
1950039	123A	3444	20	2051	
1950052	159	3466	82	2032	
1950056-1	159	3466	82	2032	
1950060	177		300	1122	
1950160	152	3893	66	2048	
1956016	102A	3004	53	2007	
1956197	5072A	3066/118	ZD-8.2	562	
1956486	116	3016	504A	1104	
1960023	123A	3444	20	2051	
1960083-1	128	3024	243	2030	
1960085-2	154		40	2012	
1960177-2	123A	3444	20	2051	
1960584	121	3014	239	2006	
1960632	179	3642	76		
1960642	151A	3099	ZD-110		
1960643	102A	3004	53	2007	
1961479	121	3014	239	2006	
1961480	121	3009	239	2006	
1961835	121	3009	239	2006	
1961837	102A	3004	53	2007	
1961843	116	3016	504A	1104	
1962323	127	3035	25		
1962326	179	3642	76		
1962594	116	3016	504A	1104	
1965016	129		244	2027	
1965017	121	3009	239	2006	
1965019	139A	3066/118	ZD-9.1	562	
1966079	121	3014	239	2006	
1966808	116	3311	504A	1104	
1967784	181	3511	75	2041	
1967799	128	3024	243	2030	
1967799-1	128	3024	243	2030	
1967801	128	3024	243	2030	
1967813	116	3031A	504A	1104	
1968958	123A	3444	20	2051	
1968959	128	3044/154	243	2030	
1968977	130	3510	14	2041	
1969113	124	3021	12		
1969281	159	3466	82	2032	
1969497	116	3017B/117	504A	1104	
1971296	175	3026	246	2020	
1971487	390	3027/130	255	2041	
1971489	128	3512	243	2030	
1971503	390	3027/130	255	2041	
2000287-28	102A	3004	53	2007	
2000433-150	109			1123	
2000625-31	102A	3004	53	2007	
2000625-32		3004	53	2007	
2000625-33	102A	3004	53	2007	
2000625-34	102A	3004	53	2007	
2000625-36	116	3016	504A	1104	
2000626-32	116	3016	504A	1104	
2000646-103	123A	3444	20	2051	
2000646-104	160	3006	245	2004	
2000646-105	108	3452	86	2038	
2000646-106	126	3008	52	2024	
2000646-107	123A	3444	20	2051	
2000646-108	102A	3004	53	2007	
2000646-109	126	3034	52	2024	
2000646-110	127	3034	25		
2000646-111	158	3004/102A	53	2007	
2000646-113	121	3009	239	2006	
2000646-115	116	3016	504A	1104	
2000646-119	116	3016	504A	1104	
2000646-120	116	3016	504A	1104	
2000648-120	116	3017B/117	504A	1104	
2000648-21	160	3007	245	2004	
2000648-22	160	3007	52	2004	
2000648-23	126	3005	52	2024	
2000648-26	109	3091		1123	
2000752-80	128		243	2030	
2000757	113A		6GC1		
2000757-18	109	3087		1123	
2000757-79	113A		6GC1		
2000757-80	108	3452	86	2038	
2000804-7	108	3452	86	2038	
2000804-8	108	3452	86	2038	
2000804-9	102A	3004	53	2007	
2001653	102A		53	2007	
2001653-20	126	3008	52	2024	
2001653-21	126	3008	52	2024	
2001653-22	100	3721	1	2007	
2001653-23	102A	3004	53	2007	
2001653-24	102A	3004	53	2007	
2001653-58	160		245	2004	
2001653-59	160		245	2004	
2001786-134	109	3087		1123	
2001786-139	115	3121	6GX1		
2001786-141	116	3017B/117	504A	1104	
2001786-142	125	3032A	510,531	1114	
2001786-169	358		42		
2001786-207	116(3)	3017B/117	504A(3)	1104	
2001809-47	102A	3123	53	2007	
2001809-48	102A	3004	53	2007	
2001809-48A	102A	3004	53	2007	
2001809-48B	102A	3004	53	2007	
2001812-65	100	3005	1	2007	
2002151-020	109	3087		1123	
2002151-18	160	3007	245	2004	
2002151-18A	160	3008	245	2004	
2002151-19	102A	3004	53	2007	
2002151-20	109	3087		1123	
2002152-14	102A		53	2007	
2002153-58	160	3006	245	2004	
2002153-59	160	3006	245	2004	
2002153-60	160	3006	245	2004	
2002153-71	102A	3004	53	2007	
2002153-76	160	3006	245	2004	
2002153-77	123A	3124/289	20	2051	
2002153-78	102A	3004	53	2007	
2002153-83	103A	3835	59	2002	
2002207		3126	90		
2002207-2	614	3327	90		
2002209-1	139A	3060	ZD-9.1	562	
2002209-10	142A	3062	ZD-12	563	
2002209-11	5072A		ZD-8.2	562	
2002209-4	137A	3058	ZD-6.2	561	
2002209-5	140A	3061	ZD-10	562	
2002209-7	5077A	3752	ZD-18		
2002209-9	140A	3061	ZD-10	562	
2002210-110	102A	3004	53	2007	
2002211-24	102A	3004	53	2007	
2002211-25	102A	3004	53	2007	
2002331-46	143A	3750	ZD-13	563	
2002332-53	108	3452	86	2038	
2002332-54	108	3452	86	2038	
2002332-55	108	3452	86	2038	
2002332-56	108	3452	86	2038	
2002332-57	116	3016	504A	1104	
2002332-58	116	3017B/117	504A	1104	
2002336-115	116	3031A	504A	1104	
2002336-19	126	3004/102A	52	2024	
2002336-20	109	3088		1123	
2002402	177	3170/731	300	1122	
2002402-29	116	3017B/117	504A	1104	
2002403-19	126	3004/102A	52	2024	
2002620-18	108	3452	86	2038	
2002620-19	108	3452	86	2038	
2002621-2	123A	3124/289	20	2051	
2003069	116		504A	1104	
2003069-1	125	3031A	510,531	1114	
2003069-2	177	3100/519	300	1122	
2003069-4		3087	1N34AS	1123	
2003069-5	601	3100/519	504A	1104	
2003069-6	177	3100/519	300	1122	
2003073-0701	123A	3004/102A	20	2051	
2003073-0702	159	3025/129	82	2032	
2003073-10	123A	3124/289	20	2051	
2003073-11	160	3006	245	2004	
2003073-12	160	3006	245	2004	
2003073-13	160	3006	245	2004	
2003073-14	102A	3004	53	2007	
2003073-15	102A	3004	53	2007	
2003073-16	103A	3010	59	2007	
2003073-67	116	3016	504A	1104	
2003073-68	116	3311	504A	1104	
2003073-8	102A	3004	53	2007	
2003073-9	123A	3124/289	20	2051	
2003073-91	130	3027	14	2041	
2003073-91A	130	3027	14	2041	
2003168-135	123A	3124/289	20	2051	
2003168-136	123A	3124/289	20	2051	
2003229-25	123A	3122	20	2051	
2003239-65		3124	20	2051	
2003342-109	108	3452	86	2038	
2003342-244	107	3039/316	11	2015	
2003779-22	107	3018	11	2015	
2003779-23	107	3018	11	2015	
2003779-24	107	3018	11	2015	
2003779-25	107	3018	11	2015	
2003779-26	139A	3060	ZD-9.1	562	
2004107-40	113A	3119/113	6GC1		
2004357-106	109	3087		1123	
2004358-123	102A	3004	53	2007	
2004358-142	116	3311	504A	1104	
2004358-168	102A	3004	53	2007	
2004746-114	108	3452	86	2038	
2004746-115	108	3452	86	2038	
2004746-116	129	3025	244	2027	
2004746-117	129	3025	244	2027	
2004746-87	124	3021	12		
2005779-26	138A	3059	ZD-7.5		
2006226-14	102A	3004	53	2007	
2006227-51	123A	3122	20	2051	
2006334-115	123A		20	2051	
2006334-155	128	3024	243	2030	
2006334-31	158	3123	53	2007	
2006422-132	109	3088		1123	
2006422-133	116	3031A	504A	1104	
2006431-44	108	3452	86	2038	
2006431-45	123A	3004/102A	20	2051	
2006431-46	123A	3004/102A	20	2051	
2006431-49		3004	20	2051	
2006431-50	109	3087		1123	
2006436-35	186	3192	28	2017	
2006436-36	186	3192	28	2017	
2006436-37	129	3025	244	2027	
2006436-38	116	3017B/117	504A	1104	
2006436-40	128		243	2030	
2006436-89			504A	1104	
2006441-113	102A	3004	53	2007	
2006441-122	109	3088		1123	
2006441-123		3126	90		
2006441-132		3100	300	1122	
2006441-91	116	3031A	504A	1104	
2006463-89	116	3017B/117	504A	1104	
2006512-40	116	3017B/117	504A	1104	
2006512-79	113A	3119/113	6GC1		
2006512-80	113A	3119/113	6GC1		
2006513-133	158		53	2007	
2006513-19	108	3452	86	2038	
2006513-39	107	3018	11	2015	
2006514-59	131	3014	44	2006	
2006514-60	123A	3444	20	2051	
2006514-61	195A	3048/329	46		
2006582	116	3017B/117	504A	1104	
2006582-101	161	3018	39	2015	
2006582-20	109			1123	
2006582-21	125	3311	510,531	1114	
2006582-22		3032A	504A	1104	
2006582-23	116(4)	3031A	504A(4)	1104	
2006582-24	116(4)		504A(4)	1104	
2006582-25	123A	3444	20	2051	
2006607-59	121	3009	239	2006	
2006607-60	123A	3444	20	2051	
2006607-61	195A	3048/329	46		
2006607-62	237	3299	46		
2006607-63	138A	3059	ZD-7.5		
2006613-77	123A	3444	20	2051	
2006623-128	195A	3048/329			
2006623-145	123A	3444	20	2051	
2006623-148	128	3024	243	2030	

Industry Standard No.	ECG	SK	GE	RS 276-	MOTOR.
2006623-47	312	3834/132	FET-1	2035	
2006623-48	177		300	1122	
2006623-49	116	3017B/117	504A	1104	
2006623-88	184	3041	57	2017	
2006627-54	177		300	1122	
2006681-120	158	3004/102A	53	2007	
2006681-93	107		11	2015	
2006681-94	107	3018	11	2015	
2006681-95	107	3039/316	11	2015	
2006681-96	123A	3444	20	2051	
2008292-56	107		11	2015	
2008292-87	126		52	2024	
2008292-88	177		300	1122	
2008292-89	116(2)	3311	504A(2)	1104	
2008293-109		3004	2	2007	
2008293-111	158	3004/102A	53	2007	
2008299-1	107		11	2015	
2008299-2		3091	1N60	1123	
2008299-3		3311	504A	1104	
2008300-105	1003		IC-43		
2008302-41	116	3311	504A	1104	
2010088-49	123A	3444	20	2051	
2010499-52	123A	3444	20	2051	
2010952-14	103A	3010	59	2002	
2010957-49	116	3031A	504A	1104	
2010967-83	177		300	1122	
2010967-84	136A		ZD-5.6	561	
2013019-117	116	3016	504A	1104	
2014400	519		514	1122	
2041614	123A	3444	20	2051	
2047102	102	3004	2	2007	
2056606-0701	159	3466	82	2032	
2057013-0004	159	3466	82	2032	
2057013-0007	159	3466	82	2032	
2057013-0008	159	3466	82	2032	
2057013-0012	159	3466	82	2032	
2057013-0701	159	3466	82	2032	
2057013-0702	159	3466	82	2032	
2057013-0703	159	3466	82	2032	
2057062-0702	116	3031A	504A	1104	
2057199-0700	130		14	2041	
2057199-0701	130	3027	14	2041	
2057199-070BA	130	3027	14	2041	
2057199-701	130	3027	14	2041	
2057323-0500	130		14	2041	
2057323-0501	130	3027	14	2041	
2060041	116	3016	504A	1104	
2068491-704	221	3065/222		2036	
2076393	126	3007	52	2024	
2076403	160	3007	245	2004	
2076403-0703	160	3007	245	2004	
2076945-0701	102A	3004	53	2007	
2090056-1	104	3009	16	2006	
2090056-27	104	3009	16	2006	
2090056-5	104	3009	16	2006	
2090924-0008	102A	3004	53	2007	
2090924-008	102A		53	2007	
2090924-6	102A		53	2007	
2090924-8	102A	3004	53	2007	
2090924-8A	102A	3004	53	2007	
2090924B	102A		53	2007	
2091211-0014	100	3005	1	2007	
2091217-0014	160		245	2004	
2091241-0005	160	3008	245	2004	
2091241-0013	100	3005	1	2007	
2091241-0014	100	3008	1	2007	
2091241-0015	100	3008	1	2007	
2091241-0018	102A	3004	53	2007	
2091241-005	126		52	2024	
2091241-0719	126	3008	52	2024	
2091241-1	100	3008	1	2007	
2091241-10	100	3721	1	2007	
2091241-11	100	3721	1	2007	
2091241-12	100	3721	1	2007	
2091241-13	100	3005	1	2007	
2091241-13A	100	3005	1	2007	
2091241-14	100	3005	1	2007	
2091241-15	100	3005	1	2007	
2091241-15A	100	3005	1	2007	
2091241-2	100	3008	1	2007	
2091241-3	100	3008	1	2007	
2091241-4	100	3721	1	2007	
2091241-5	126		52	2024	
2091241-5A	126	3006/160	52	2024	
2091241-6	100	3721	1	2007	
2091241-7	100	3005	1	2007	
2091241-8	100	3721	1	2007	
2091241-9	102A		53	2007	
2091247-005	160	3006	245	2004	
2091260			8	2002	
2091260-1		3010	8	2002	
2091260-1(NPN)	103A		59	2002	
2091260-1(PNP)	102A		53	2007	
2091260-2		3010	8	2002	
2091260-2(NPN)	103A		59	2002	
2091260-2(PNP)	102A		53	2007	
2091260-3		3010	8	2002	
2091260-3(NPN)	103A		59	2002	
2091260-3(PNP)			53	2007	
2091578-0702	102A	3004	53	2007	
2091578-1	102A	3004	53	2007	
2091858-0712	121	3009	239	2006	
2091858-11	121	3009	239	2006	
2091859-0008	104	3009	16	2006	
2091859-0011	121	3009	239	2006	
2091859-0025	104	3009	16	2006	
2091859-0711	108	3452	86	2038	
2091859-0712	121	3009	239	2006	
2091859-0713	121	3014	239	2006	
2091859-0714	121	3009	239	2006	
2091859-0715	121	3014	239	2006	
2091859-0716	121	3009	239	2006	
2091859-0717	121	3014	239	2006	
2091859-0718	121	3009	239	2006	
2091859-0720	104	3009	16	2006	
2091859-0723	121	3009	239	2006	
2091859-10	104	3009	16	2006	
2091859-11	104	3009	16	2006	
2091859-16	121	3717	239	2006	
2091859-2	104	3014	16	2006	
2091859-25	121	3717	239	2006	
2091859-4	104	3009	16	2006	
2091859-6	121	3009	239	2006	
2091859-8	121	3009	239	2006	
2091859-9	121	3009	239	2006	
2091959-16	121	3717	239	2006	
2092055-0001	109	3087		1123	
2092055-0002	5072A		ZD-8.2	562	
2092055-0007	110MP	3087	1N34AS	1123	
2092055-001		3087	1N34AS	1123	
2092055-0010	109			1123	
2092055-0016	136A		ZD-5.6	561	
2092055-0017	5072A		ZD-8.2	562	
2092055-0018	5071A		ZD-6.8	561	
2092055-0024	5082A	3753	ZD-25		
2092055-0027	5072A		ZD-8.2	562	
2092055-0708	112	3087	1N82A		
2092055-0710	5071A	3334	ZD-6.8	561	
2092055-0711			1N34AS	1123	
2092055-0712	5072A		ZD-8.2	562	
2092055-0713	109	3087		1123	
2092055-0714	123A	3100/519	20	2051	
2092055-1	109	3087		1123	
2092055-5	138A	3059	ZD-7.5		
2092055-7	109	3087		1123	
2092117-0018			11	2015	
2092405-7			1N295	1123	
2092417-0017	107	3018	11	2015	
2092417-0018	107	3018	11	2015	
2092417-0019	107	3018	11	2015	
2092417-005	160		245	2004	
2092417-0704	160		245	2004	
2092417-0707	160	3006	245	2004	
2092417-0708	160	3006	245	2004	
2092417-0709	160	3006	245	2004	
2092417-0710	160	3006	245	2004	
2092417-0711	161	3018	39	2015	
2092417-0712	161	3018	39	2015	
2092417-0713	161	3018	39	2015	
2092417-0714	161	3117	39	2015	
2092417-0715	161	3018	39	2015	
2092417-0716	161	3018	39	2015	
2092417-0717	160		245	2004	
2092417-0719	123A	3444	20	2051	
2092417-0720	123A	3444	20	2051	
2092417-0721	123A	3444	20	2051	
2092417-0724	123A	3444	20	2051	
2092417-0725	123A	3444	20	2051	
2092417-1	160	3006	245	2004	
2092417-17	123A	3444	20	2051	
2092417-18	123A	3444	20	2051	
2092417-19	123A	3444	20	2051	
2092417-2	160	3006	245	2004	
2092417-3	160	3006	245	2004	
2092417-4	160		245	2004	
2092417-5	160		245	2004	
2092417-6	160	3006	245	2004	
2092417-7	160		245	2004	
2092417-8	160		245	2004	
2092417-9	160		245	2004	
2092418-0022	161	3018	39	2015	
2092418-0023	161	3018	39	2015	
2092418-0024	161	3018	39	2015	
2092418-0708			50	2004	
2092418-071	160		245	2004	
2092418-0710	160	3006	245	2004	
2092418-0711	101	3861	8	2002	
2092418-0712	160		245	2004	
2092418-0715	108	3452	86	2038	
2092418-0716	161	3018	39	2015	
2092418-0717	161	3018	39	2015	
2092418-0718	161	3018	39	2015	
2092418-0719	161	3018	39	2015	
2092418-0720	161	3018	39	2015	
2092418-0721	161	3018	39	2015	
2092418-0724	108	3452	86	2038	
2092418-1	160	3006	245	2004	
2092418-10	160		245	2004	
2092418-11	160		245	2004	
2092418-2	160	3006	245	2004	
2092418-5	160	3006	245	2004	
2092418-6	160	3006	245	2004	
2092418-7	160	3006	245	2004	
2092418-8	160		245	2004	
2092605-0705	123A	3444	20	2051	
2092608-22	123A	3444	20	2051	
2092609	123A	3344/5093A	20	2051	
2092609-0001	123A	3344/5093A	20	2051	
2092609-0002	123A	3344/5093A	20	2051	
2092609-001	123A	3344/5093A	20	2051	
2092609-0022	123A	3444	20	2051	
2092609-0023	123A	3444	20	2051	
2092609-0024	123A	3444	20	2051	
2092609-0025	159	3122	82	2032	
2092609-0026	123A	3444	20	2051	
2092609-0027	123A	3444	20	2051	
2092609-0028	123A	3444	20	2051	
2092609-0705	123A	3444	20	2051	
2092609-0706	123A	3444	20	2051	
2092609-0707	123A	3444	20	2051	
2092609-0713	123A	3444	20	2051	
2092609-0715	123A	3444	20	2051	
2092609-0718	123A	3444	20	2051	
2092609-0720	123A	3444	20	2051	
2092609-0721	123A	3444	20	2051	
2092609-1	123A	3444	20	2051	
2092609-2	123A	3444	20	2051	
2092609-3	123A	3444	20	2051	
2092609-5	123A	3444	20	2051	
2092693-0724	107	3039/316	11	2015	
2092693-0725	107	3039/316	11	2015	
2092693-0734	159		82	2032	
2092693-1	102A	3004	53	2007	
2092693-2	160	3007	245	2004	
2092693-3	160	3007	245	2004	
2092693-4	160	3007	245	2004	
2092693-8	102A	3004	53	2007	
2092693-9	160	3006	245	2004	
2093308-070	108	3452	86	2038	
2093308-0700	108	3452	86	2038	
2093308-0701	123A	3444	20	2051	
2093308-0702	123A	3444	20	2051	
2093308-0703	123A	3444	20	2051	
2093308-0704	107	3018	11	2015	
2093308-0704A	108	3018	86	2038	
2093308-0705	107	3018	11	2015	
2093308-0705A	108	3452	86	2038	
2093308-0706	107	3018	11	2015	
2093308-0706A	108	3452	86	2038	
2093308-0708	123A	3444	20	2051	
2093308-0725	107	3018	11	2015	
2093308-1	108	3452	86	2038	
2093308-2	108	3452	86	2038	
2093308-3	108	3452	86	2038	
2093308-8708			20	2051	
2095083	109	3088		1123	
2096700	129		244	2027	
2096700-TM18	129		244	2027	
2097013-0702	128	3024	243	2030	
2125310	199		62	2010	
2132523-1	159		82	2032	
2132524-1	519		514	1122	
2132763-1	912		IC-172		
2160153	725		IC-19		
2182124-1	519		514	1122	
2185494-1	519		514	1122	
2185494-2	519		514	1122	
2206582-22	116(4)		504A(4)	1104	

Industry Standard No.	ECG	SK	GE	RS 276-	MOTOR.
2227367	176	3845	80		
2243255-1	102A	3123	53	2007	
2311598	506	3130	511	1114	
2314009	127	3764	25		
2316177			11	2015	
2316183	161	3716	39	2015	
2320011	102A	3006/160	53	2007	
2320022	123A	3444	20	2051	
2320031	107	3018	11	2015	
2320041	233	3018	210	2009	
2320041H	161	3018	39	2015	
2320042	107	3122	11	2015	
2320043	107	3018	11	2015	
2320051	154	3040	40	2012	
2320051H	154	3040	40	2012	
2320052			18	2030	
2320062	108	3452	86	2038	
2320063	108	3452	86	2038	
2320073	108	3452	86	2038	
2320083	152	3893	66	2048	
2320084	175	3026	246	2020	
2320092	131	3052	44	2006	
2320111	123A	3444	20	2051	
2320123	123A	3444	20	2051	
2320141	233	3018	17	2015	
2320141H	161	3039/316	39	2015	
2320154	100	3005	1	2007	
2320161	159	3007	82	2032	
2320161(RCA)	160		245	2004	
2320162	159	3114/290	82	2032	
2320191	154	3040	40	2012	
2320201	127	3034	25		
2320221	124	3021	12		
2320222	124	3026	12		
2320223	124	3021	12		
2320228	124	3021	12		
2320233	128	3024	243	2030	
2320242	129	3025	244	2027	
2320243	129	3025	244	2027	
2320261	102A	3004	53	2007	
2320271	162	3438	35		
2320273	162	3438	35		
2320281	164	3079	37		
2320291	165	3710/238	38		
2320299	165	3710/238	38		
2320302	102A	3004	53	2007	
2320302H	102A	3004	53	2007	
2320331	103A	3010	59	2002	
2320413	123A	3444	20	2051	
2320422	158	3004/102A	53	2007	
2320422-1	102A	3004	53	2007	
2320423	102A	3004	53	2007	
2320432	52	3893/152	66	2048	
2320441	123A	3444	20	2051	
2320471	319P	3018	283	2016	
2320471(LAST IF)	233			2009	
2320471(LAST-IF)			210	2051	
2320471-1	107	3018	11	2015	
2320471H	107	3124/289	11	2015	
2320482	152	3893	66	2048	
2320482H	152	3054/196	66	2048	
2320483	152	3054/196	66	2048	
2320485	152	3054/196	66	2048	
2320486	152	3054/196	66	2048	
2320492	102A		53	2007	
2320512	100	3005	1	2007	
2320513	102A	3005	53	2007	
2320514	160	3005	245	2004	
2320514-1	100	3005	1	2007	
2320515	102A		53	2007	
2320541	292	3009	25	2006	
2320591	289A	3124/289	210	2038	
2320591-1	123A	3444	20	2051	
2320595	289A	3124/289	20	2051	
2320596	289A	3124/289	268	2038	
2320598	199	3124/289	268	2038	
2320602	152	3054/196	66	2048	
2320631	290A	3124/289	269	2032	
2320632	193	3114/290	67	2023	
2320637	290A	3114/290	221		
2320643	289A	3122	268	2038	
2320644	289A	3122	268	2038	
2320646	289A	3122	268	2038	
2320646-1	128	3024	268	2038	
2320647	289A	3024/128	268	2038	
2320647-1	128	3024	268	2038	
2320651	152	3893	66	2048	
2320652	152	3054/196	66	2048	
2320664	192	3122	63	2030	
2320671	159		82	2032	
2320681	159	3114/290	82	2032	
2320696	123A	3444	20	2051	
2320696-1	123A	3444	20	2051	
2320833	127	3035	25		
2320843	373	3190/184	57	2017	
2320845	184	3054/196	57	2017	
2320846	184		247	2052	
2320855	374	3191/185	58	2025	
2320884	153	3083/197	69	2049	
2320892	154	3040	40	2012	
2320931	124	3021	12		
2320946	128	3044/154	243	2030	
2320961	163A	3710/238	38		
2320963	165		38		
2320981	107		11	2015	
2320994	124		32		
2321001	124	3021	12		
2321095	198	3103A/396	251		
2321101	154	3045/225	40	2012	
2321111	288	3045/225	223		
2321112	288		223		
2321121	283	3439/163A	36		
2321211	198		251		
2321221	376		251		
2321241	283	3467	36		
2321264	5455	3597		1067	
2321281	153	3083/197	69	2049	
2321291	124		12		
2321302	291	3054/196	66	2048	
2321321	294	3114/290			
2321322	129	3114/290			
2321351	234	3114/290	244	2027	
2321372	162		36		
2321381	292	3083/197	69		
2321403	124	3021	12		
2321412	198		251		
2321472	376	3433/287	222		
2321511	108	3246/229	61	2038	
2321521	234	3025/129	221		
2321541	123AP		62	2051	
2321561	165		259		
2321652	280		262	2041	
2321763	292	3441			
2321764	153	3083/197			

Industry Standard No.	ECG	SK	GE	RS 276-	MOTOR.
2321791	375	3929			
2321881	293	3849	47	2030	
2321891	294	3841	48		
2326953	123A	3444	20	2051	
2326991	124		12		
2327022	128	3024	243	2030	
2327023	123A	3444	20	2051	
2327031	116	3017B/117	504A	1104	
2327041	116	3017B/117	504A	1104	
2327052	130	3027	14	2041	
2327053	130	3027	14	2041	
2327061	124		12		
2327071	142A		ZD-12	563	
2327073	5081A		ZD-24		
2327074	5022A	3788	ZD-13	563	
2327075	5084A	3755	ZD-30		
2327076	147A	3095	ZD-33		
2327077	5075A	3751	ZD-16	564	
2327078	5074A		ZD-11.0	563	
2327111			FET-4	2036	
2327122	123A	3444	20	2051	
2327132	312	3448	FET-1	2035	
2327142	312	3834/132	FET-1	2035	
2327142(JFET)			FET-2	2035	
2327142(MOSFET)			FET-4	2036	
2327152	186	3192	28	2017	
2327153	186	3192	28	2017	
2327172		3027	14	2041	
2327182	124	3021	12		
2327203	186	3192	28	2017	
2327206	152	3054/196	66	2048	
2327212	703A		IC-12		
2327232		3050	FET-4	2036	
2327262	159	3114/290	82	2032	
2327282	129	3025	244	2027	
2327283	129	3025	244	2027	
2327292	128	3024	243	2030	
2327293	128	3024	243	2030	
2327302(HITACHI)	703A		IC-12		
2327312	1039		IC-159		
2327332	128	3024	243	2030	
2327363	123A	3444	20	2051	
2327387	159	3114/290	82	2032	
2327403	128	3024	243	2030	
2327411	1041		IC-160		
2327422	720	9014	IC-7		
2327431	222	3050/221	FET-4	2036	
2330011	166	9075	504A	1152	
2330011H	166	9075	BR-600	1152	
2330021	138A	3059	ZD-7.5		
2330032	505	3067/502	CR-7		
2330033	503		CR-5		
2330034	504	3108/505	CR-6		
2330101	506	3843	511	1114	
2330191	506	3033	511	1114	
2330192	506	3843			
2330201	116	3311	504A	1104	
2330211	506	3017B/117	511	1114	
2330241	142A	3062	ZD-12	563	
2330251	125	3032A	504A	1104	
2330251(POWER)	156		512		
2330251H	116	3032A	504A	1104	
2330252	125	3017B/117	504A	1104	
2330253	506	3032A	511	1114	
2330254	125	3311	504A	1104	
2330256	116	3312	504A	1104	
2330302	5071A	3334	ZD-6.8	561	
2330302H	138A	3059	ZD-7.5		
2330305	144A	3094	ZD-14	564	
2330307	5072A		ZD-8.2	562	
2330332	156	3051	512		
2330351	519	3100	514	1122	
2330352	519	3311	514	1122	
2330356	506	3843	511	1114	
2330361	166	9075	BR-600	1152	
2330362	116	3017B/117	504A	1104	
2330381	504	3108/505	CR-6		
2330551	552	3998/506	510	1114	
2330553	116	3311	504A	1104	
2330561	116	3016	504A	1104	
2330562	552	3017B/117	504A	1104	
2330564	525	3017B/117	511	1114	
2330611	605	3031A		1104	
2330612	116	3016	504A	1104	
2330631	5014A	3334/5071A	ZD-6.8	561	
2330632	5071A	3334	ZD-6.8	561	
2330634	5014A	3059/138A	ZD-7.5		
2330643	5071A	3334	ZD-6.8	561	
2330721	116	3311	504A	1104	
2330771	156	3051	512		
2330773	156	3017B/117	504A	1104	
2330791	5071A	3334	ZD-6.8	561	
2330921	502	3067	CR-4		
2331121	506	3130	511	1114	
2331141	506		504A	1104	
2331142	116	3311	504A	1104	
2331152	142A	3093	ZD-12	563	
2331154	5021A			563	
2331161	5020A		ZD-10	562	
2331174	5074A		ZD-11.0	563	
2331351	519	3100	514	1122	
2331381	551	3843	511	1114	
2331491	5069A		ZD-4.7		
2331502	156		512		
2331991	125	3081	510	1114	
2332141	116			1104	
2332152	525	3843	511	1114	
2337011	177	3100/519	300	1122	
2337063	5080A	3336	ZD-22		
2337065	5075A	3751	ZD-16	564	
2337071	116	3311	504A	1104	
2337101	142A	3062	ZD-12	563	
2340331	601	3864/605			
2340332	601	3463			
2347021	116		504A	1104	
2360021	1061	3228			
2360042	712	3072	IC-2		
2360042(HITACHI)	712		IC-2		
2360092	712	3072	IC-2		
2360151	1213	3704	IC-150		
2360201	712	3072	IC-2		
2360221	1094		IC-157		
2360361	1004	3141/780			
2360391	712		IC-2		
2360401	1004	3102/710	IC-36		
2360431	1196		IC-73		
2360501	712		IC-148		
2360511	1004		IC-149		
2360631	1213		IC-150		
2360741	7400		7400		
2360924-5601	123A	3444	20	2051	
2361461	966	3592	VR-111		
2390022	5404	3627			
2391342	910		IC-251		

Industry Standard No.	ECG	SK	GE	RS 276-	MOTOR.
2391773	909	3590	IC-249		
2392152	941M		IC-265	007	
2396475	923		IC-259		
2402277	804	3455	IC-27		
2412275	804	3455	IC-27		
2412949-0001	129		244	2027	
2469749	123A	3444	20	2051	
2469755	123A	3444	20	2051	
2469936-1	128		243	2030	
2479692	123A	3444	20	2051	
2479836	123A	3444	20	2051	
2485076-2	128	3124/289	243	2030	
2485076-3	128	3124/289	243	2030	
2485077-2	128	3024	243	2030	
2485077-3	128	3024	243	2030	
2485078-1	123A	3444	20	2051	
2485078-2	123A	3444	20	2051	
2485078-2(SEARS)			20	2051	
2485078-3	123A	3444	20	2051	
2485079-1	123A	3444	20	2051	
2485079-2	123A	3444	20	2051	
2485079-3	123A	3444	20	2051	
2485080	177		300	1122	
2485080(DIODE)	109			1123	
2486836	116	3311	504A	1104	
2487305	177		300	1104	
2487340	159	3114/290	82	2032	
2487341	159	3114/290	82	2032	
2487424		3122	63	2030	
2487424(NPN)	192		63	2030	
2487424(PNP)	159		82	2032	
2495012	126	3008	52	2024	
2495013	126	3008	52	2024	
2495014	102A	3004	53	2007	
2495078	160	3006	245	2004	
2495079	160	3006	245	2004	
2495080	102A	3004	53	2007	
2495082	160	3006	245	2004	
2495083	110MP	3088	1N60	1123(2)	
2495083-2	109			1123	
2495084	109	3087		1123	
2495166-1	108	3452	86	2038	
2495166-2	123A	3444	20	2051	
2495166-4	123A	3444	20	2051	
2495166-8	123A	3444	20	2051	
2495166-9	123A	3444	20	2051	
2495200	126	3008	52	2024	
2495376	160	3006	245	2004	
2495377	160	3006	245	2004	
2495378	126		52	2024	
2495379	126	3006/160	52	2024	
2495380	109	3087		1123	
2495383	109			1123	
2495388	102A	3004	53	2007	
2495388-1	102A	3004	53	2007	
2495388-2	102A	3123	53	2007	
2495488-1	126	3006/160	52	2024	
2495488-2	126	3006/160	52	2024	
2495520	161	3132	39	2015	
2495521	161	3132	39	2015	
2495521-1	123A	3444	20	2051	
2495522-1	161	3132	39	2015	
2495522-4	123A	3444	20	2051	
2495523-1	161	3132	39	2015	
2495529(ARVIN)	128		243	2030	
2495529(DIODE)				1123	
2495567-2	102A	3123	53	2007	
2495567-3	102A	3123	53	2007	
2495568	158		53	2007	
2495568-2	102A	3004	53	2007	
2496125-2	123A	3444	20	2051	
2496436	109			1123	
2497094-1	161	3018	39	2015	
2497094-2	161	3018	39	2015	
2497473	102A	3004	53	2007	
2497473-1	102A	3004	53	2007	
2497496	102A	3123	53	2007	
2497888	158	3004/102A	53	2007	
2498163	128		243	2030	
2498456-2	107	3018	11	2015	
2498457-2	123A	3444	20	2051	
2498482-2	107	3018	11	2015	
2498507-1	108	3452	86	2038	
2498507-2	108	3452	86	2038	
2498507-3	108	3452	86	2038	
2498508-2	107		11	2015	
2498508-3	107		11	2015	
2498512	159	3114/290	82	2032	
2498513	116	3311	504A	1104	
2498530	109	3087		1123	
2498665-1	199		62	2010	
2498665-2	199	3245	62	2010	
2498665-3	199	3245	62	2010	
2498837	160	3008	245	2004	
2498837-4	102A		53	2007	
2498902-1	107	3018	11	2015	
2498902-2	107	3018	11	2015	
2498903-1	107	3018	11	2015	
2498903-2		3018	61	2038	
2498903-3	107		11	2015	
2498904-3	123A	3444	20	2051	
2498904-4	123A	3444	20	2051	
2498904-6	123A	3444	20	2051	
2499950	199	3124/289	62	2010	
2501549-341	703A		IC-12		
2503134-105	159		82	2032	
2505207	199		62	2010	
2505209	123A	3444	20	2051	
2509319	124		12		
2520063	123A	3444	20	2051	
2521108-1	159		82	2032	
2530733	123A	3444	20	2051	
2545989-2	128		243	2030	
2546059	519		514	1122	
2552007	156		512		
2596071	108	3452	86	2038	
02600043	109	3088			
2604688-3	909		IC-249		
2605022	128		243	2030	
2608169-1	128		243	2030	
2610023-01	910		IC-251		
2610023-02	910D		IC-252		
2610032	177	3506/5642	300	1122	
2610043-03	941M		IC-265	007	
2610153	725		IC-19		
2610154-01	923		IC-259		
2610783	7402	7402		1811	
2610786	7400	7400		1801	
2610788	74121	74121			
2621499-1	177		300	1122	
2621567-1	123A	3444	20	2051	
2621570	159		82	2032	
2621764	123A	3444	20	2051	
2621786-1	519		514	1122	
2621811	106	3984	21	2034	
2622284	123		20	2051	
2635012	177		300	1122	
2640830-1	123		20	2051	
2640843-1	123A	3444	20	2051	
2666307-1	128		243	2030	
2709759	309K	3629			
2710002	941		IC-263	010	
2712080	123A	3444	20	2051	
2729789		3123	80		
2760438-1	177		300	1122	
2777301	128		243	2030	
2780312	358		42		
2797658-616030A	941		IC-263	010	
2855296-01	102A	3004	53	2007	
2865101	192		63	2030	
2865141	519		514	1122	
2868536-1	7474	7474		1818	
2875493	152	3893	66	2048	
2887425RB	311		277		
2899018	911		IC-253		
2899414	923D		IC-260	1740	
2902798-2	915			017	
2903993-1	159		82	2032	
2903993-I	159			2032	
2904014	121	3009	239	2006	
2928054-1	128		243	2030	
2970038H05	100	3005	1	2007	
3004856	102A	3004	53	2007	
3005300	5882		5040		
3005861	123A	3444	20	2051	
3006206-00	784	3524	IC-236		
3006892-00	910		IC-251		
3006892-01	910D		IC-252		
3007579	912		IC-172		
3007579-00	912		IC-172		
3007680-00	941		IC-263	010	
3007680-01	941D		IC-264		
3008321-00	309K	3629			
3008340	923D		IC-260	1740	
3020061			255	2041	
3068305-2	123A	3444	20	2051	
3100017			ZD-15	564	
3110883-P01	177		300	1122	
3130006	121	3717	239	2006	
3130011	102	3722	2	2007	
3130025	102	3722	2	2007	
3130053	128		243	2030	
3130057	175	3026	246	2020	
3130058	130	3510	14	2041	
3130060	102	3722	2	2007	
3130090	181		75	2041	
3130091	130	3510	14	2041	
3130092	128		243	2030	
3130093	175		246	2020	
3130104	181		75	2041	
3130109		3717	239	2006	
3146977	130	3027	14	2041	
3146977A	130	3027	14	2041	
3152159	175	3026	246	2020	
3152170	130	3027	14	2041	
3152170A	130	3027	14	2041	
3160000-00-08A	519		514	1122	
3170717	154	3045/225	40	2012	
3170757	154	3045/225	40	2012	
3180006	519		514	1122	
3181972	108	3452	86	2038	
3195638	177		300	1122	
3201104-10	123A	3444	20	2051	
3263029-10	519		514	1122	
3311133	519		514	1122	
3403787		3444	20	2051	
3403866-3	128		243	2030	
3404114-1	102	3722	2	2007	
3404114-2	102	3722	2	2007	
3404520-301	128		243	2030	
3404520-601	221			2036	
3404520-81	126		52	2024	
3412004-1912	312			2028	
3412907-1	128		243	2030	
3430063	116	3017B/117	504A	1104	
3430063-1	5804	9007		1144	
3430063-I	5804			1144	
3438095	152	3893	66	2048	
3438854	163A	3439	36		
3438867	152	3054/196	66	2048	
3450842-10	154		40	2012	
3450842-20	154		40	2012	
3450842-30	154		40	2012	
3457107-1	123A	3444	20	2051	
3457632-5	123A	3444	20	2051	
3457633-1	128	9038/346	243	2030	
3457633-2	128	9038/346	243	2030	
3457936-1	234		65	2050	
3458267-1	128		243	2030	
3458573-1	192		63	2030	
3459332-1	161	3716	39	2015	
3460550-1	160		245	2004	
3460550-3	160		245	2004	
3460550-4	160		245	2004	
3460553-2	121	3717	239	2006	
3460553-4	121	3009	239	2006	
3460679-1	102A		53	2007	
3460758-2	5801			1142	
3462221-1	121	3717	239	2006	
3462306-1	121	3717	239	2006	
3463098-1	5873		5032		
3463098-2	5871		5033		
3463099-1	128		243	2030	
3463100-1	175	3510	246	2020	
3463101-1	130	3027	14	2041	
3463604-1	175		246	2020	
3463604-2	175		246	2020	
3463609-2	161	3716	39	2015	
3464482-1	102A		53	2007	
3464611	519		514	1122	
3464648-1	390		255	2041	
3464648-2	390		255	2041	
3468068-1	161	3716	39	2015	
3468068-2	161	3716	39	2015	
3468068-3	161	3716	39	2015	
3468068-4	161	3716	39	2015	
3468071-1	175		246	2020	
3468182-1	123A		20	2051	
3468182-2	123A		20	2051	
3468183-1	159		82	2032	
3468242-1	159		82	2032	
3468242-2	123A		20	2051	
3468841-1	175	3026	246	2020	
3520041-001	7400	7400		1801	
3520042-001	7410	7410		1807	
3520043-001	7473	7473		1803	
3520044-001	7440	7440			
3520045-001	7451	7451		1825	

Industry Standard No.	ECG	SK	GE	RS 276-	MOTOR.
3520046-001	7474	7474		1818	
3520047-001	7430	7430			
3520048-001	7404	7404		1802	
3539307-001	161	3018	39	2015	
3539307-002	161	3018	39	2015	
3596061	177	3100/519	300	1122	
3596062	109	3088		1123	
3596063	159	3114/290	82	2032	
3596067	108	3452	86	2038	
3596068	108	3452	86	2038	
3596069	108	3452	86	2038	
3596070	108	3452	86	2038	
3596071	108	3452	86	2038	
3596072	108	3452	86	2038	
3596091	196	3041	241	2020	
3596092	196	3041	241	2020	
3596100	197	3084	250	2027	
3596101	197	3084	250	2027	
3596116	128	3122	243	2030	
3596117	123A	3122	20	2051	
3596118	159	3114/290	82	2032	
3596242		3126	90		
3596260	108	3018	86	2038	
3596261	108	3018	86	2038	
3596338	123A	3122	20	2051	
3596339	123A	3122	20	2051	
3596340	129	3114/290	244	2027	
3596341	129	3114/290	244	2027	
3596353	736		IC-17		
3596354	723	3144	IC-15		
3596398	5071A	3334	ZD-6.8	561	
3596401	222	3050/221	FET-4	2036	
3596402	222	3050/221	FET-4	2036	
3596435	614		90		
3596440	107	3018	17	2015	
3596446	152	3054/196	66	2048	
3596447	152	3893	66	2048	
3596448	152	3054/196	66	2048	
3596449		3041	66	2048	
3596451	153	3083/197	69	2049	
3596452	153	3083/197	69	2049	
3596453	153	3083/197	69	2049	
3596453(ZENER)	151A		ZD-110		
3596454	153	3083/197	69	2049	
3596454(ZENER)	151A		ZD-110		
3596559	177		300	1122	
3596570	123A		20	2051	
3596809	789	3078			
3596810	788	3829	IC-229		
3597049	1115	3184	IC-278		
3597103	108	3018	86	2038	
3597104	108	3018	86	2038	
3597114	107	3018	17	2051	
3597260	108	3018	11		
3597261	108	3018	11		
3597280	1115	3184			
3598070			17	2051	
3598173	1115	3184			
3650238A	159	3114/290	221	2032	
3670724H	123A	3122	61	2051	
3671857B		3271	246		
3673351K	289A	3122	210	2038	
3673354G	289A	3122	210	2038	
3680322-3	5992	3501/5994	5104		
3700072	123A		20	2051	
3700085	121	3717	239	2006	
3700109	123A		20	2051	
3700135	130	3027	14	2041	
3700144	159		82	2032	
3700162	128		243	2030	
3700163	129		244	2027	
3700164	130		14	2041	
3700171	123A		20	2051	
3700219	130	3027	14	2041	
3700228	130	3027	14	2041	
3700249	159		82	2032	
3700258	129		244	2027	
3700279	123A		20	2051	
3720968-1	923		IC-259		
3731132-1	123A		20	2051	
3731133-1	159		82	2032	
3731313-1	121	3009	239	2006	
3731418-1	129		244	2027	
3731418-2	128		243	2030	
3731418-3	129		244	2027	
3731418-4	129		244	2027	
3755171	152	3893	66	2048	
3755862	199	3245	62	2010	
3755864	177		300	1122	
3939307-002	161	3132	39	2015	
04000655-1			514	1122	
4000921	199	3245	62	2010	
4002862-0001	123A		20	2051	
4010577-0701	519		514	1122	
4017621-0701	123A		20	2051	
4028839	161	3716	39	2015	
4031986-0701	159		82	2032	
4032122	716		IC-208		
4036392-P2	140A	3061	ZD-10	562	
4036598-P1	121	3009	239	2006	
4036598-P2	102A	3004	53	2007	
4036612-P1	100	3005	1	2007	
4036612-P2	100	3005	1	2007	
4036707-P1	100	3005	1	2007	
4036707-P2	100	3005	1	2007	
4036715-P1	160	3006	245	2004	
4036715-P2	160	3006	245	2004	
4036733-P1	121	3009	239	2006	
4036749-P1	101	3861	6	2002	
4036749-P2	101	3861	8	2002	
4036754-P1	103	3862	8	2002	
4036754-P2	103	3862	8	2002	
4036831-P1	105	3012	4		
4036832-P1	105	3012	4		
4036887-P8	123A	3122	20	2051	
4036923-P1	160	3006	245	2004	
4036923-P2	160	3006	245	2004	
4036924-P1	123A	3122	20	2051	
4036924-P2	123A	3122	20	2051	
4036937-P1	100	3005	1	2007	
4036937-P2	100	3005	1	2007	
4036962-P1	160	3006	245	2004	
4036962-P2	160	3006	245	2004	
4036963-P1	160	3006	245	2004	
4036963-P2	160	3006	245	2004	
4036965-P1	160	3006	245	2004	
4036965-P2	160	3006	245	2004	
4037145-P1	102A	3004	53	2007	
4037145-P2	102A	3004	53	2007	
4037289-P1	101	3861	8	2002	
4037289-P2	101	3861	8	2002	
4037325-P1	116	3017B/117	504A	1104	
4037410-P1	160	3006	245	2004	
4037410-P2	160	3006	245	2004	
4037413-P1	137A	3058	ZD-6.2	561	
4037586-P1	123A	3122	20	2051	
4037586-P2	123A	3122	20	2051	
4037594-P1	121	3717	239	2006	
4037607-P1	121	3009	239	2006	
4037647-P1	186	3192	28	2017	
4037647-P2	186	3192	28	2017	
4037764-P1	160	3006	245	2004	
4037764-P2	160	3006	245	2004	
4037800-P1	123A	3122	20	2051	
4037800-P2	123A	3122	20	2051	
4037804-P1	102A	3004	53	2007	
4037804-P2	102A		53	2007	
4037839-P1	101	3861	8	2002	
4037839-P2	101	3861	8	2002	
4037993-P1	100	3005	1	2007	
4037993-P2	100	3005	1	2007	
4038256-P1	103A	3010	59	2002	
4038256-P2	103A	3010	59	2002	
4038260-P1	100	3005	1	2007	
4038260-P2	100	3005	1	2007	
4038264-P1	101	3861	8	2002	
4038264-P2	101	3861	8	2002	
4038359-P1	160	3006	245	2004	
4038359-P2	160	3006	245	2004	
4038406-P1	160	3006	245	2004	
4038406-P2	160	3006	245	2004	
04040501-1	909D	3590	IC-250		
04040503-1	910		IC-251		
04040505	941D		IC-264		
04040505-001	941D		IC-264		
04040751-001	923D		IC-260	1740	
4041200-30110			300	1122	
4041200-40100			1N34AS	1123	
4080187-0502	152	3054/196	66	2048	
4080187-0504	130		14	2041	
4080187-0506	196	3054	241	2020	
4080187-0507	196	3054	241	2020	
4080320-0501	130	3027	14	2041	
4080320-0504	130		14	2041	
4080320-050B	130	3027	14	2041	
4080627-0501	175		246	2020	
4080835-0002	196	3054	241	2020	
4080838-0001	175	3026	246	2020	
4080838-0002	175	3026	246	2020	
4080838-002	196		241	2020	
4080838-2	175	3026	246	2020	
4080838-3	175	3026	246	2020	
4080866-0006		3054	28	2017	
4080866-0007	175		246	2020	
4080866-000A	130	3027	14	2041	
4080866-0012	196	3054	241	2020	
4080866-0013	152	3893	66	2048	
4080866-003		3893	66	2048	
4080866-006	152	3893	66	2048	
4080866-009	152	3893	66	2048	
4080866-1	175	3026	246	2020	
4080866-2	175	3026	246	2020	
4080866-4	152	3893	66	2048	
4080866-8006	130	3027	14	2041	
4080873-0001	184		57	2017	
4080879-0001	152	3054/196	66	2048	
4080879-0006	196	3041	63	2020	
4080879-0011	196	3054	63	2020	
4080879-0015	175	3054/196	246	2020	
4082501-0001	104MP	3720	16(2)	2006(2)	
4082501-0001A	121	3009	239	2006	
4082501-001	104	3009	16	2006	
4082626-0001	912	3543	IC-172		
4082665-0001	723	3144	IC-15		
4082665-0002	723		IC-15		
4082665-0003	723	3144	IC-15		
4082665-1	723	3144	IC-15		
4082665-2	723	3144	IC-15		
4082665-3	723	3144	IC-15		
4082671-0002	172A	3156	64		
4082748-0002	136A		ZD-5.6	561	
4082799-0001	789	3078			
4082799-0002	789	3078			
4082799-1	789	3078			
4082799-2	789	3078			
4082802-0001	781	3169	IC-223		
4082802-0002	781	3169	IC-223		
4082802-1	781	3169	IC-223		
4082802-2	781	3169	IC-223		
4082873-0001	197	3083	250	2027	
4082886-0002	184	3027/130	57	2017	
4082886-001	130	3027	14	2041	
4082886-002	130		14	2041	
4082886-3	130		14	2041	
4084114-0001	222	3050/221	FET-4	2036	
4084114-0002	222	3050/221	FET-4	2036	
4084117-0001	949		IC-25		
4084117-0002	949		IC-25		
4089549-0001	949		IC-25		
4089549-001	949		IC-25		
4090187-0502	130MP		15	2041(2)	
4101685	116	3016	504A	1104	
4350052-1	703A		IC-12		
4361620	130		14	2041	
04410025-001	519		514	1122	
04410025-005	519		514	1122	
04410042-001	519		514	1122	
4420022-P1	5995	3518	5105		
4420022-P2	5995	3518	5105		
04440028-001	123A	3444	20	2051	
004440028-002	123A	3122	20	2051	
04440028-003	123A	3444	20	2051	
04440028-006	123A	3122	20	2051	
004440028-007	123A	3444	20	2051	
004440028-008	123A	3444	20	2051	
004440028-010	123A	3444	20	2051	
04440028-013	123A	3444	20	2051	
004440028-014	123A	3444	20	2051	
04440032-002	159	3466	82	2032	
04440032-003	159	3466	82	2032	
04440032-004	159	3466	82	2032	
04440032-005	159	3466	82	2032	
04440032-006	159	3466	82	2032	
04440032-007	159	3466	82	2032	
04440032-008	159	3466	82	2032	
04440032-009	159	3466	82	2032	
04440052-001	123A	3444	20	2051	
04440052-002	123A	3444	20	2051	
04450002-001	123A	3444	20	2051	
04450002-004	123A	3444	20	2051	
04450002-005	123A	3444	20	2051	
04450016-001	159	3466	82	2032	
04450016-002	159	3466	82	2032	
04450016-003	129		244	2027	
4450023	130		14	2041	
04450023-001	130		14	2041	
04450023-005	181		75	2041	
04450023-006	130		14	2041	

Industry Standard No.	ECG	SK	GE	RS 276-	MOTOR.
4450023-007	130	3511	14	2041	
4450023-P3	130		14	2041	
4450023-P4	130		14	2041	
4450023-P6	130		14	2041	
4450023P5	130		14	2041	
4450026-P5	128		243	2030	
04450037-001	181		75	2041	
04450037-002	182	3188A	55		
4450040P1	390		255	2041	
4450300-1	5675	3508			
4450300P2	5626	3633			
4450300P3	5626	3633			
4450303-001	5642	3507			
4450303-002	5642	3507			
4511424	7432			1824	
4550106-001	152	3893	66	2048	
4550106-003	152	3893	66	2048	
4663001A905	7474	7474		1818	
4663001A909	7404			1802	
4663001A911	7440	7440			
4663001A912	7410	7410		1807	
4663001A915	7430	7430			
4663001D907	7400	7400		1801	
4813466	123A	3444	20	2051	
4822354	130		14	2041	
4822354J	130		14	2041	
4832800	181	3036	75	2041	
4906071	123A	3444	20	2051	
4906072	123A	3444	20	2051	
4906073	123A	3444	20	2051	
4906093	199	3245	62	2010	
4907976	102A	3004	53	2007	
4915702	7416	7416			
4999774	121	3717	239	2006	
4999885	213		254		
4999887	102A		53	2007	
5001266-3	177		300	1122	
5049911	128		243	2030	
5073004	102A	3004	53	2007	
5076054			300	1122	
5076204	74800	74800			
5113642	909	3590	IC-249		
5175460	7408	7408		1822	
5294477-1	123A	3444	20	2051	
5294477-2	123A	3444	20	2051	
5300073-15	145A	3063	ZD-15	564	
5300113-1	116	3017B/117	504A	1104	
5301710-001	116		504A	1104	
5320003	175	3026	246	2020	
5320004	289A	3138/193A	268	2038	
5320011	102A		53	2007	
5320022	289A	3138/193A	268	2038	
5320023	289A	3444/123A	20	2051	
5320023H	123A	3444	20	2051	
5320024	123A	3444	20	2051	
5320026	123A	3444	20	2051	
5320031	220	3990	FET-3	2036	
5320032	312	3448	FET-1	2035	
5320042H	159	3114/290	82	2032	
5320043H	159	3114/290	82	2032	
5320051	161	3018	17	2015	
05320064	289A	3124/289	210	2038	
5320064H	123A	3444	20	2051	
5320067	289A	3444/123A	268	2051	
5320074	123A	3038	20	2051	
05320074H	123A	3444	20	2051	
5320101	102A	3004	53	2007	
5320111	129	3025	244	2027	
5320141	102A		53	2007	
5320151	237	3299	46		
5320241	123A	3444	20	2051	
5320295	102A	3004	53	2007	
5320295H	103A	3835	59	2002	
5320296	102A	3004	53	2007	
5320305	103A	3835	59	2002	
5320305H	102A		53	2007	
5320306	103A	3010	59	2002	
5320326	123A	3124/289	20	2051	
5320326H	107	3018	11	2015	
5320328	108	3452	86	2038	
5320361	103A	3010	59	2002	
5320372	123A	3444	20	2051	
5320372H	123A	3444	20	2051	
5320373	123A	3444	20	2051	
5320386			8	2002	
5320422H	152	3054/196	66	2048	
5320432	152	3054/196	66	2048	
5320433	152	3054/196	66	2048	
5320475	103A	3010	59	2002	
5320485	102A	3004	53	2007	
5320492H	152	3054/196	66	2048	
5320501	195A	3048/329	46		
5320511	237	3299	46		
5320583	312	3112	FET-1	2028	
5320592	290A	3114/290	269	2032	
5320593	290A	3114/290	269	2032	
5320612	128	3024	243	2030	
5320613	289A	3122	268	2038	
5320622		3122	63	2030	
5320632	187	3193	29	2025	
5320642	186	3054/196	28	2017	
5320643	184	3188A/182	57	2017	
5320651	289A	3444/123A	20	2051	
5320671	152	3054/196	66	2048	
5320702	312	3834/132	FET-1	2035	
5320723	185		58	2025	
5320731			1N60	1123	
5320772	229		61	2038	
5320791	235		215		
5320813	199	3124/289	212		
5320851	123A	3018	20	2051	
5320861	107	3018	11	2015	
5320872	102A		53	2007	
5320921	184		57	2017	
5320942	312	3116	FET-2	2035	
5320943	312		FET-2	2035	
5321171	237	3299			
5321184	159	3114/290	82	2032	
5321204		3138	48		
5321214	287	3433	222		
5321235	186A	3248	28	2052	
5321252	294	3114/290	244		
5321253	129	3114/290	244	2027	
5321261	123A	3020/123	20	2051	
5321291	123AP	3020/123	20	2051	
5321301	152	3054/196	66	2048	
5321311	295	3253	270		
5321321	235	3197	215		
5321422	312	3116	FET-2	2035	
5321431	107	3293	86		
5321501	312	3112	FET-1	2028	
5321502	459	3448	FET-1	2028	
5321901	229	3246	61		
5330001	125	3017B/117	504A	1104	

Industry Standard No.	ECG	SK	GE	RS 276-	MOTOR.
5330011	139A	3060	ZD-9.1	562	
5330012	139A		ZD-9.1	562	
5330031	116		504A	1104	
5330041	116	3017B/117	504A	1104	
5330041(HR-5A)	125	3311	510,531	1114	
5330041H	116	3311	504A	1104	
5330042	116	3311	504A	1104	
5330051	5071A		ZD-6.8	561	
5330054	5022A	3788	ZD-13	563	
5330054H	5022A	3788	ZD-13	563	
5330059	5080A	3336	ZD-22		
5330101	116	3311	504A	1104	
5330101H	116	3031A	504A	1104	
5330102	116	3017B/117	504A	1104	
5330102H	116	3032A	504A	1104	
5330104	125	3017B/117	510,531	1114	
5330104H			504A	1104	
5330131	519	3100	300	1122	
5330133	177	3100/519	514	1122	
5330201		3126	90		
5330212H	177	3100/519	300	1122	
5330261	177	3100/519	300	1122	
5330261H	177		300	1122	
5330312	5071A	3058/137A	ZD-6.8	561	
5330322	5073A	3749	ZD-9.1	562	
5330331	109	3088		1123	
5330332	109	3088		1123	
5330334	177	3100/519	300	1122	
5330335	109	3087		1123	
5330336	116	3311	504A	1104	
5330341	116	3311	504A	1104	
5330371	116	3311	504A	1104	
5330372	116(2)		504A(2)	1104	
5330381	116	3311	504A	1104	
5330392	5012A	3777/5011A		561	
5330431	116	3311	504A	1104	
5330541	5024A	3790		564	
5330571	519		514	1122	
5330572	519	3100	514	1122	
5330661	614	3327	90		
5330721	109	3087	1N34AS	1123	
5330731	109	3088		1123	
5330732	109	3088		1123	
5330761	519		514	1122	
5330842	136A	3056/135A	ZD-5.1		
5330851	614	3327			
5330852	614	3327	90		
5331102	5312	3311	BR-206		
5340001	177	3124/289	300	1122	
5340001H	177		300	1122	
5340021	116	3311	504A	1104	
5340022	116	3100/519	504A	1104	
5340022H	177	3031A	300	1122	
5340051	177	3100/519	300	1122	
5340082		3126	90		
5340111	177		300	1122	
5350121	1029		IC-162		
5350132	1030		IC-163		
5350135	1030		IC-163		
5350136	1030		IC-163		
5350141	1031		IC-273		
5350151	1032		IC-164		
5350152	1032		IC-164		
5350161	1033		IC-165		
5350182	1075A	3877			
5350211	1043		IC-169		
5350231	1058	3459	IC-49		
5350251	1087	3477	IC-103		
5350491	1169		IC-64		
5350611	615	3095/147A	ZD-33		
5351041	722		IC-9		
5351042	722	3161	IC-9		
5351051	1041		IC-160		
5351061	1072		IC-59		
5351062	1072		IC-59		
5351351	712	3072	IC-148		
5351361	1004	3365	IC-149		
5359031	7400	7400	7400		
5359271	7493A	7493			
5406665-P2	102A		53	2007	
5406665-P5	102A		53	2007	
5490307-P2	140A		ZD-10	562	
5490307-P3	142A		ZD-12	563	
5490307-P4	145A		ZD-15	564	
5490415-P4	116		504A	1104	
5490459-P1	116		504A	1104	
5490459-P2	116		504A	1104	
5490459-P3	116		504A	1104	
5490459-P4	116		504A	1104	
5490804-P1	116		504A	1104	
5490804-P2	116		504A	1104	
5490804-P3	116		504A	1104	
5490804-P4	116		504A	1104	
5490804-P5	116		504A	1104	
5490804-P6	116		504A	1104	
5490860P1	105		4		
5491236-P1	109			1123	
5491236-P2	109			1123	
5491263-P1			1N34AS	1123	
5491338-P1	135A		ZD-5.1		
5492153-1	105		4		
5492639-P1	103	3862	8	2002	
5492639-P2	103	3862	8	2002	
5492653-P1	101	3861	8	2002	
5492653-P2	101	3861	8	2002	
5492655-P1	101	3861	8	2002	
5492655-P2	101	3861	8	2002	
5492655-P3	101	3861	8	2002	
5492655-P4	101	3861	8	2002	
5492655-P5	101	3861	8	2002	
5492655-P6	101	3861	8	2002	
5492659-P1		3861	8	2002	
5492659-P2	101	3861	8	2002	
5493158-1	121	3717	239	2006	
5493158-P1			3	2006	
5493159-P1	5874		5032		
5493957-P1	160		245	2004	
5493957-P2	160		245	2004	
5493957-P3	160		245	2004	
5493957-P4			51	2004	
5493957-P5	160		245	2004	
5493957-P6	160		245	2004	
5494922-P1	116		504A	1104	
5494922-P2	116		504A	1104	
5494922-P3	116		504A	1104	
5494922-P4	116		504A	1104	
5494922-P5	116		504A	1104	
5494922-P6	116		504A	1104	
5494922-P7	116		504A	1104	
5494922-P8	116		504A	1104	
5495178-P1	5874		5032		
5495922-P1	5874		5032		
5495922-P2	5874		5032		
5495923-P1	5545		MR-3		

Industry Standard No.	ECG	SK	GE	RS 276-	MOTOR.
5496438-P1	146A		ZD-27		
5496663-P1	121	3717	239	2006	
5496665-P1	102		2	2007	
5496665-P2	102	3722	2	2007	
5496665-P3	102	3722	2	2007	
5496665-P4	102	3722	2	2007	
5496665-P5	102	3722	2	2007	
5496665-P6	102	3722	2	2007	
5496666-P1	102	3722	2	2007	
5496666-P2	102	3722	2	2007	
5496666-P3	102	3722	2	2007	
5496666-P4	102	3722	2	2007	
5496666-P5	102	3722	2	2007	
5496666-P6	102	3722	2	2007	
5496666-P7	102	3722	2	2007	
5496666-P8	102	3722	2	2007	
5496667-P1	102	3722	2	2007	
5496667-P2	102	3722	2	2007	
5496668-P1	105		4		
5496774-P1	102	3722	2	2007	
5496774-P2	102	3722	2	2007	
5496774-P3	102	3722	2	2007	
5496774-P4	102	3722	2	2007	
5496774-P5			2	2007	
5496774-P6	102	3722	2	2007	
5496839-P1	102	3722	2	2007	
5496939-P1	121	3717	239	2006	
5496939-P2	121	3717	239	2006	
5496947-P3	146A		ZD-27		
5518017	220	3990		2036	
5573505-1	519		514	1122	
5573505-2	519		514	1122	
5729031	105		4		
5955748	102A	3004	53	2007	
5955749	116	3016	504A	1104	
5958539	176	3123	80		
5983945A	105	3012	4		
5985432	105	3012	4		
5988414	105	3012	4		
5988977	105	3012	4		
5989171	105	3012	4		
5989615	105	3012	4		
5989692	105	3012	4		
5989693	105	3012	4		
6088501-3	912		IC-172		
06100007	289A	3274/153	69		
06100008	292	3441	69		
06100016	290A	3114/290			
06100030		3025	244	2032	
06100033		3083	69	2049	
06100034		3114	221	2032	
06100035	290A	3114/290	221	2032	
06100047	290A	3114/290			
06100053	234	3114/290	244	2050	
06100054	234	3114/290			
06100055	290A	3114/290			
06100058	323	3932	221		
06100061	234	3247			
06100062	290A	3114/290			
6100724-2	100	3005	1	2007	
6101161-1	130		14	2041	
06110032	292	3441			
06120001	154	3040	235	2012	
06120005	123AP	3124/289	210	2051	
06120006	199		210		
06120008	233	3122	61		
06120009	107	3293	86		
06120012	283	3439/163A	36		
06120013	229	3246			
06120015	107	3132	17		
06120018	152	3893	66	2048	
06120021	152	3893	66	2048	
06120025		3018	61	2009	
06120026		3018	61	2015	
06120028		3024	81	2038	
06120030	152	3054/196	215	2053	
06120053	157	3747	232		
06120063	289A	3122	268		
06120064	284	3122			
06120073	229	3246	61	2038	
06120077	124	3021			
06120078	165	3710/238			
06120083	376	3219	251		
06120084	198	3220			
06120085	229	3132			
06120087	199	3124/289			
06120088	198	3219			
06120091	198	3219			
06120096	199	3122			
06120097	198	3220			
06120098	198	3220			
06130014	375	3929			
06130018	375	3440/291			
06130019	238	3710			
6142190	5883	3500/5882	5041		
06200001	109	3087	1N34AS	1123	
06200002	109	3088	1N60	1123	
06200003	110MP	3088	1N60	1123	
06200005	125	3081	510	1114	
06200009	125	3081	510	1114	
06200010	506	3843	511	1114	
06200012	5014A	3780	ZD-6.8	561	
06200013	519	3100	514	1122	
06200014	552	9000	511	1114	
06200017	109	3087	1N34AS	1123	
06200030	506	3998	511	1114	
06200031	5020A	3785/5019A			
06200034	5022A	3062/142A			
06200036	506	9000/552			
06200037	515	3314	512		
06200038	551	3925/525			
6200039	116	3312	504A	1104	
06200040	519	3100			
06200044	177	3175			
06200045	109	3087			
06200046	605	3864			
06200047	601	3463			
06200050	506	3843			
06200051	552	3081/125			
06200052	116	3311			
06200055	137A	3058			
06200056	144A	3094			
06200057	5071A	3334			
06200058	142A	3062			
06200059	145A	3063			
06200065	5801	3051/156	510	1114	
06200076	503		CR-5		
6207159	1243	3731	IC-113		
6208839	123A	3018	61	2038	
6212096	152	3054/196	66	2048	
6212839	123A	3444	20	2051	
6212922	108	3452	86	2038	
6218945	123A	3444	20	2051	
06300001	1173	3729			

Industry Standard No.	ECG	SK	GE	RS 276-	MOTOR.
06300002	1161	3968			
06300003	164	3727/1164			
06300004	1196	3725			
06300005	1046	3471			
06400001	5404	3627			
06400002	230	3042			
6460006	102	3722	2	2007	
6460037	102		2	2007	
6480000	130		14	2041	
6480001	121	3717	239	2006	
6480004	121	3717	239	2006	
6480006	128		243	2030	
6490001	175		246	2020	
6494030			255	2041	
06500004	605	3864			
6846503	116	3110/120	CR-3	1104	
6902021H25	152	3893	66	2048	
6902021H26	153		69	2049	
6933243-001	175		246	2020	
6984590	123A	3444	20	2051	
6984600	123A	3444	20	2051	
6993400	159		82	2032	
6993630	159		82	2032	
6993650	123A	3444	20	2051	
7002453	123A	3444	20	2051	
7011203-02	7475	7475		1806	
7011203-03	7475	7475		1806	
7011507	123A	3444	20	2051	
7011507-00	123A	3444	20	2051	
7011515	129		244	2027	
7012109-01	923D		IC-260	1740	
7012126	177		300	1122	
7012142-03	74193	74193		1820	
7012157	519		514	1122	
7012166	7402	7402		1811	
7012167-02	7490	7490		1808	
7012411	129		244	2027	
7020202	159		82	2032	
7020203	153		69	2049	
7023416-00	116	3017B/117	504A	1104	
7026011	108	3452	86	2038	
7026012	108	3452	86	2038	
7026013	108	3452	86	2038	
7026014	123A	3444	20	2051	
7026015	123A	3444	20	2051	
7026016	123A	3444	20	2051	
7026018	130	3027	14	2041	
7026019	129	3025	244	2027	
7026020	123A	3444	20	2051	
7026024	152	3893	66	2048	
7026036		3126	90		
7027002	116		504A	1104	
7027005	116	3311	504A	1104	
7027023	116	3016	504A	1104	
7027031	116	3311	504A	1104	
7027036	113A		6GC1		
7027038	116	3031A	504A	1104	
7027039	116	3017B/117	504A	1104	
7030105	102A		53	2007	
7043216	116	3016	504A	1104	
7043261	116	3017B/117	504A	1104	
7070692	116	3311	504A	1104	
7071021	123A	3444	20	2051	
7071031	123A	3444	20	2051	
7090612	519		514	1122	
7100460	109	3088		1123	
7100628	116	3016	504A	1104	
7100630	116	3016	504A	1104	
7116060	110MP	3709	1N60	1123(2)	
7121105-01	123A	3444	20	2051	
7129386-P2	109			1123	
7141123-1	177		300	1122	
7200351		3126	90		
7200953	116	3311	504A	1104	
7201090	5072A		ZD-8.2	562	
7201212	117	3100/519	504A	1104	
7202782	5072A		ZD-8.2	562	
7210027-003	102	3004	2	2007	
7210036-001	102	3722	2	2007	
7269847	103A	3835	59	2002	
7274653	121	3009	239	2006	
7276211	100	3005	1	2007	
7276605	105	3012	4		
7277066	103A	3835	59	2002	
7278421	100	3005	1	2007	
7278422	102A	3004	53	2007	
7278423	102A	3004	53	2007	
7279003	105		4		
7279005	105	3012	4		
7279007	105	3012	4		
7279009		3012	4		
7279011	105	3012	4		
7279013	105		4		
7279017	105	3012	4		
7279025	105	3012	4		
7279027	105	3012	4		
7279031		3012	4		
7279033	105	3012	4		
7279039	104	3009	16	2006	
7279049	104	3009	16	2006	
7279069	179	3642	76		
7279073	105	3012	4		
7279076	105	3012	4		
7279281	176	3123	80		
7279293	105	3012	4		
7279298	105	3012	4		
7279379	100	3005	1	2007	
7279497	116		504A	1104	
7279566	105	3012	4		
7279779	126	3006/160	52	2024	
7279779(G.M.)	123A		20	2051	
7279780	160	3008	245	2004	
7279781	160	3007	245	2004	
7279782	100	3005	1	2007	
7279788	100	3005	1	2007	
7279789	100	3005	1	2007	
7279793	105	3012	4		
7279893	109	3087		1123	
7279940	100	3005	1	2007	
7279941	102A	3004	53	2007	
7280281	105	3012	4		
7281307	100	3005	1	2007	
7281308	100	3005	1	2007	
7281309	100	3005	1	2007	
7281310	102A	3004	53	2007	
7281806	128		243	2030	
7281891	100	3005	1	2007	
7282315	105	3012	4		
7282358	109	3087		1123	
7284137	123A	3444	20	2051	
7284513	160	3006	245	2004	
7284751	102A	3004	53	2007	
7285354	5982	3609/5986	5096		
7285663	121	3009	239	2006	

Industry Standard No.	ECG	SK	GE	RS 276-	MOTOR.
7285774	104	3009	16	2006	
7285776	121	3717	239	2006	
7285778	104	3009	16	2006	
7286858	123A	3444	20	2051	
7287079	5072A		ZD-8.2	562	
7287107	179	3642	76		
7287110	121	3009	239	2006	
7287112	179	3642	76		
7287117	179	3642	76		
7287452	123A	3444	20	2051	
7287940	126	3008	52	2024	
7288072	105	3012	4		
7288073	105	3012	4		
7288076	105	3012	4		
7288079	105	3012	4		
7289041	121	3009	239	2006	
7289047	104	3009	16	2006	
7289079			3	2006	
7289097	179	3642	76		
7290593	121	3717	IC-7	2006	
7290594	104	3009	16	2006	
7291252	104	3009	16	2006	
7292308	160	3006	245	2004	
7292683	179		76		
7292684	127	3764	25		
7292689	179	3642	76		
7292690	121	3717	239	2006	
7292955	121	3717	239	2006	
7292956	330	3012/105			
7293818	162	3438	35		
7293819	162	3438	35		
7294133	102A	3004	53	2007	
7294796	121	3009	239	2006	
7294910	161	3018	39	2015	
7295195	108	3452	86	2038	
7295196	108	3452	86	2038	
7295197	123A	3444	20	2051	
7296314	123A	3444	20	2051	
7296476	5072A		ZD-8.2	562	
7296811	123A	3444	20	2051	
7297043	104	3009	16	2006	
7297053	123A	3444	20	2051	
7297054	123A	3444	20	2051	
7297092	104	3009	16	2006	
7297093	104	3009	16	2006	
7297258	5072A		ZD-8.2	562	
7297347	179	3642	76		
7297348	127	3764	25		
7297358			ZD-9.1	562	
7297980	161	3018	39	2015	
7298079	121	3009	239	2006	
7299325	116	3017B/117	504A	1104	
7299720	105	3012	4		
7299771	121MP	3718	239(2)	2006(2)	
7299780	179	3642	76		
7299803	121MP	3718	239(2)	2006(2)	
7301660	179	3642	76		
7301661	179	3642	76		
7301664	162	3438	35		
7301665	162	3438	35		
7301666	162	3438	35		
7302024	159		82	2032	
7302340	117		504A	1104	
7302699	162	3438	35		
7303105	102A		53	2007	
7303120	123A	3444	20	2051	
7303304	162	3438	35		
7304149	163A	3439	36		
7304380	123A	3444	20	2051	
7304971	5414	3954			
7305468	107		11	2015	
7305469	5072A		ZD-8.2	562	
7305476	726	3022/1188			
7305783			3	2006	
7306982	130	3027	14	2041	
7309160	130	3027	14	2041	
7311074	128	3044/154	243	2030	
7311325	709	3135	IC-11		
7311350	128	3512	243	2030	
7312294	718	3159	IC-8		
7313063	162	3438	35		
7313568	721		IC-14		
7314584	123A	3444	20	2051	
7397027	703A		IC-12		
7492377-5	142A		ZD-12	563	
7492377-P1	135A		ZD-5.1		
7492377-P10	135A		ZD-5.1		
7492377-P5			ZD-12	563	
7528156P1	910		IC-251		
7528157P1	909	3590	IC-249		
7570003	121		239	2006	
7570003-01	121	3009	239	2006	
7570004	123A	3444	20	2051	
7570004-01	123A	3444	20	2051	
7570005	123A	3444	20	2051	
7570005-01	123A	3444	20	2051	
7570005-02	123A	3444	20	2051	
7570005-03	123A	3444	20	2051	
7570008	123A	3444	20	2051	
7570008-01	123A	3444	20	2051	
7570008-02	123A	3444	20	2051	
7570009	128		243	2030	
7570009-01	123A	3444	20	2051	
7570009-21	123A	3444	20	2051	
7570011-01	125	3311	510,531	1114	
7570011-02	125	3031A	510,531	1114	
7570013	159		82	2032	
7570013-01	159	3025/129	82	2032	
7570014	5802	9005		1143	
7570014-02	109	3016		1123	
7570014-05	116	3031A	504A	1104	
7570014-34				1104	
7570015-34	116		504A	1104	
7570016	109			1123	
7570016-02	109	3087		1123	
7570016-03	109	3087		1123	
7570021-12	146A	3064	ZD-27		
7570021-21	146A		ZD-27		
7570022-01	175	3026	246	2020	
7570024-02	116	3311	504A	1104	
7570030-01	185	3083/197	58	2025	
7570031-01	152	3893	66	2048	
7570032-01	129	3025	244	2027	
7570215-21	177	3016	300	1122	
7570215-24	116	3017B/117	504A	1104	
7570215-34	116	3031A	504A	1104	
7571329-01	166			1152	
7576004-01	159	3114/290	82	2032	
7576015-01	123A	3122	20	2051	
7576015-02	199	3245	62	2010	
7576015-03	199	3245	62	2010	
7576015-04	199	3245	62	2010	
7576015-05	199	3245	62	2010	
7576016-01	128		243	2030	
7582438-01			1N60	1123	
7582439-01			1N34AS	1123	
7601010	128		243	2030	
7777146-P2	109			1123	
7777146-P23	109		1N60	1123	
7777146-P3	109			1123	
7777146-P4	109			1123	
7777146-PIO				1123	
7840540-1	123A	3444	20	2051	
7851316-01	102A		53	2007	
7851317-01	102A		53	2007	
7851318-01	103A		59	2002	
7851319-01	103A	3835	59	2002	
7851320-01	158		53	2007	
7851321-01	103A	3835	59	2002	
7851322-01	104	3719	16	2006	
7851323	160		245	2004	
7851324-01	123A	3444	20	2051	
7851325	123A	3444	20	2051	
7851326	123A	3444	20	2051	
7851327	123A	3444	20	2051	
7851328-01	169	3678	BR-600		
7851329-01	166	9075		1152	
7851379-01	123A	3444	20	2051	
7851380-01	123A	3444	20	2051	
7851441-01	116		504A	1104	
7851442-01	5069A		ZD-4.7		
7851467-01	103A	3835	59	2002	
7851650-01	161	3716	39	2015	
7851651-01	161	3716	39	2015	
7851652-01	159		82	2032	
7851654-01	109			1123	
7851655-01	109			1123	
7851657-01	116	3017B/117	504A	1104	
7851947-01	109	3087		1123	
7851949-01	123A	3444	20	2051	
7851950-01	123A	3444	20	2051	
7851952-01	123A	3444	20	2051	
7851953-01	123A	3444	20	2051	
7851954-01	159		82	2032	
7851955-01		3052	44,43		
7851956-01	128		243	2030	
7852223-01	109	3087		1123	
7852223-01A	109	3087		1123	
7852264-01	116	3031A	504A	1104	
7852438-01	109	3088		1123	
7852439-01	116(2)	3311	504A(2)	1104	
7852452-01	199	3245	62	2010	
7852454	123A	3444	20	2051	
7852454-01	123A	3444	20	2051	
7852455-01	123A	3444	20	2051	
7852457-01	289A		268	2038	
7852459-01	123A	3444	20	2051	
7852460-01	175		246	2020	
7852577-01	166	9075	504A	1152	
7852775-01	102A	3004	53	2007	
7852776-01	102A	3004	53	2007	
7852778-01	102A	3004	53	2007	
7852779-01	102A	3004	53	2007	
7852781-01	123A	3444	20	2051	
7852782-01	109	3087	1N34AS	1123	
7852897-01	160		245	2004	
7852899-01	126	3006/160	52	2024	
7852900-01	160	3006	245	2004	
7852902-01	109			1123	
7852903-01		3126	90		
7852904-01	135A	3056	ZD-5.1		
7852907-01	116		504A	1104	
7853090-01	312	3448	FET-1	2035	
7853091-01	229		61	2038	
7853092-01	123A	3444	20	2051	
7853093-01	229		61	2038	
7853094-01	123A	3444	20	2051	
7853098-01	109			1123	
7853099-01	166	9075	504A	1152	
7853351-01	102A	3004	53	2007	
7853352-01	102A	3004	53	2007	
7853354-01	102A	3004	53	2007	
7853356-01	102A	3004	53	2007	
7853357-01	109	3087		1123	
7853358-01	116	3031A	504A	1104	
7853463-01	123A	3444	20	2051	
7853464-01	123A	3444	20	2051	
7853465-01	123A	3444	20	2051	
7855282	109	3088		1123	
7855282-01	109	3087		1123	
7855283-01	177	3100/519	300	1122	
7855283-01 (AMPEX)	102A		53	2007	
7855284-01	116	3031A	504A	1104	
7855291-01	123A	3444	20	2051	
7855292-01	123A	3444	20	2051	
7855293-01	123A	3444	20	2051	
7855294-01	123A	3444	20	2051	
7855295-01	131	3052	44	2006	
7855296-01	102A	3004	53	2007	
7855297-01	102A	3004	53	2007	
7855298-01	152	3893	66	2048	
7855352-01	109			1123	
7855357-01	109			1123	
7902310	128	3024	243	2030	
7910070-01	102A		53	2007	
7910071-01	102A	3004	53	2007	
7910072-01	121	3717	239	2006	
7910073-01	184	3190	57	2017	
7910076-01	116	3017B/117	504A	1104	
7910108-01	161	3716	39	2015	
7910111-01	5072A		ZD-8.2	562	
7910112	1003	3288	IC-43		
7910112-01	1003	3288	IC-43		
7910122-01	1003	3288	IC-43		
7910134-01	312	3834/132	FET-1	2035	
7910267-01			23	2020	
7910268-01			FET-1	2028	
7910269-01			20	2051	
7910270-01			20	2051	
7910271-01			18	2030	
7910272-01			18	2030	
7910273-01			18	2030	
7910274-01			18	2030	
7910275-01			504A	1104	
7910278-01			ZD-9.1	562	
7910584-01	123A	3444	20	2051	
7910585-01	123A	3444	20	2051	
7910586-01	123A	3444	20	2051	
7910587-01	123A	3444	20	2051	
7910588-01	102A	3004	53	2007	
7910589-01	102A	3004	53	2007	
7910590-01			504A	1104	
7910780-01	108	3452	86	2038	
7910781-01	107		11	2015	
7910783-01	126	3006/160	52	2024	
7910801-01	102A	3004	53	2007	
7910803-01	158		53	2007	
7910804-01	123A	3444	20	2051	

Industry Standard No.	ECG	SK	GE	RS 276-	MOTOR.
7910805-01	116	3031A	504A	1104	
7910872-01	116	3031A	504A	1104	
7910873-01	116	3031A	504A	1104	
7910875-01			67	2023	
7913605	162	3438	35		
7914009-01	107	3018	11	2015	
7914010-01	107	3018	11	2015	
7914011-01	116	3031A	504A	1104	
7914014-01	177	3100/519	300	1122	
7914015-01		3126	90		
7932367	719		IC-28		
7932515	159		82	2032	
7932638	177	3100/519	300	1122	
7932980	709	3135	IC-11		
7935181	744	3146/787	IC-215	2022	
7936256	123A	3444	20	2051	
7936331	123A	3444	20	2051	
7937586	130	3027	14	2041	
7937762	721		IC-14		
7938318	312			2028	
7939165	161	3716	39	2015	
7939186	159		82	2032	
8000736	128	3024	243	2030	
8000737	128	3024	243	2030	
8002866	175		246	2020	
8004265	128		243	2030	
8010490	126	3008	52	2024	
8010520	100	3008	1	2007	
8010530	126	3008	52	2024	
8014711	160	3008	245	2004	
8014712	100	3008	1	2007	
8015613	129	3025	244	2027	
8020322	102A	3004	53	2007	
8020324	102A	3004	53	2007	
8020333	102A	3004	53	2007	
8020334	102A	3004	53	2007	
8020540	102A	3004	53	2007	
8020560	102A	3004	53	2007	
8021897	102A	3004	53	2007	
8021898	102A	3004	53	2007	
8022630	102A	3004	53	2007	
8022631	102A	3004	53	2007	
8023643	102A		53	2007	
8023892	102A	3003	53	2007	
8024152	158	3004/102A	53	2007	
8024250	127	3034	25		
8024390	160	3004/102A	245	2004	
8024390(AIWA)	102A		53	2007	
8024400	102A		53	2007	
8031825	108	3452	86	2038	
8031836	108	3452	86	2038	
8031837	107	3018	11	2015	
8031839	123A	3444	20	2051	
8033690	123A	3444	20	2051	
8033696	123A	3444	20	2051	
8033720	123A	3444	20	2051	
8033730	123A	3444	20	2051	
8033803	108	3452	86	2038	
8033804	108	3452	86	2038	
8033943	108	3452	86	2038	
8033944	123A	3444	20	2051	
8034903	152	3054/196	66	2048	
8036683	107	3018	11	2015	
8037330	123A	3444	20	2051	
8037332	123A	3444	20	2051	
8037333	123A	3444	20	2051	
8037343	128		243	2030	
8037353	123A	3444	20	2051	
8037722	107	3018	11	2015	
8037723	108	3452	86	2038	
8051060	109	3087		1123	
8052192	5071A	3058/137A	ZD-6.8	561	
8052446	109	3087		1123	
8110014	312			2028	
8111227	159		82	2032	
8111230	159	3114/290	82	2032	
8112023	102A	3004	53	2007	
8112027	102A	3004	53	2007	
8112028	102A		53	2007	
8112071	102A	3004	53	2007	
8112090	102A	3004	53	2007	
8112143	131	3052	44	2006	
8112146	102A	3004	53	2007	
8112161	102A	3004	53	2007	
8112162	102A	3004	53	2007	
8113024	192		63	2030	
8113034	123A	3444	20	2051	
8113051	123A	3444	20	2051	
8113052	123A	3444	20	2051	
8113060	123A	3444	20	2051	
8113102	152	3893	66	2048	
8113134	123A	3444	20	2051	
8113323	199	3122	62	2038	
8113327	199	3124/289	62	2010	
8114024	192		63	2030	
8114031	123A	3444	20	2051	
8114057	297	3024/128	88	2030	
8115009	159	3114/290	82	2032	
8120037	116	3311	504A	1104	
8120067	116		504A	1104	
8120069	139A		ZD-9.1	562	
8120126	519	3100	300	1122	
8120133	116	3311	504A	1104	
8121002	109	3087		1123	
8121014	109	3088		1123	
8130014	177		300	1122	
8131010	177		300	1122	
8249600	161	3018	39	2015	
8378759	128		243	2030	
8398315	128		243	2030	
8421133	128		243	2030	
8484771-1	5870		5032		
8484771-10	5877		5037		
8484771-3	5872		5032		
8484771-7	5874		5032		
8484771-8	5875		5033		
8505214-1	177	3764/127	300	1122	
8508309	129		244	2027	
8508331	784	3524	IC-236		
8508403-1	784	3524	IC-236		
8510671-1	160		245	2004	
8510671-2	160		245	2004	
8510671-4	160		245	2004	
8510694-1	130		14	2041	
8510744-1	101	3861	8	2002	
8510747-4	126		52	2024	
8511632-5	519		514	1122	
8511724-3	176	3845	80		
8511724-4	176	3845	80		
8511759-1	160	3004/102A	245	2004	
8512001-2	101	3010	8	2002	
8513070-1	519		514	1122	
8516861-1	102	3722	2	2007	
8516986	102	3005	2	2007	
8516986-1	102	3722	2	2007	
8516986-2	102		2	2007	
8518237-1	5870		5032		
8518237-10	5875		5033		
8518237-2	5872		5032		
8518237-3	5874		5032		
8518237-9	5873		5033		
8518382-1	5980	3610	5096		
8518382-10	5987	3698	5097		
8518382-2	5982	3609/5986	5096		
8518382-3	5986	3501/5994	5096		
8518382-9	5983	3698/5987	5097		
8518383-10	5862	3517/5883	5024		
8518383P1	5940		5048		
8518383P2	5944		5048		
8518383P3	5944		5048		
8519510-2	506		511	1114	
8521502-1	101	3861	8	2002	
8521502-2	101	3861	8	2002	
8521502-4	101	3861	8	2002	
8521587-1	116	3016	504A	1104	
8521587-101	5802	9005		1143	
8521587-102	5804	9007		1144	
8521692-1	5483	3942			
8522468-1	123A	3444	20	2051	
8523209-1	519		514	1122	
8524402-1	101	3861	8	2002	
8524402-4	101	3861	8	2002	
8524440-2	102	3722	2	2007	
8524457	128	3024	243	2030	
8526849-1	100	3123	1	2007	
8532787-2	519		514	1122	
8532787-3	519		514	1122	
8532787-4	519		514	1122	
8533519-1	519		514	1122	
8533588-1	908		IC-248		
8538640	160	3007	245	2004	
8549020-1	519		514	1122	
8556188	108	3452	86	2038	
8722248-2	199	3245	62	2010	
8820359	1003	3288			
8821008	1217	3700			
8821081	1052	3249			
8898302	744		IC-215	2022	
8945077-12	102	3004	2	2007	
8975103-2	101	3861	8	2002	
8975158-1	102A		53	2007	
8981399-1	116	3017B/117	504A	1104	
8981399-2	116	3017B/117	504A	1104	
8989441-2	160	3005	245	2004	
8989457-1	126		52	2024	
9000630	181	3036	75	2041	
9000940	159		82	2032	
9001225-02	910D		IC-252		
9001324	152	3893	66	2048	
9001345-02	7493A	7493			
9001349-02	7486	7486		1827	
9001349-03	7486	7486		1827	
9001549-02	7440	7440			
9001551-02	7404	7404		1802	
9001551-03	7404	7404		1802	
9001630	123A	3444	20	2051	
9001638	159		82	2032	
9001756	152	3893	66	2048	
9001757	153		69	2049	
9002097-03	74107	74107			
9002159	123A	3444	20	2051	
9003091-02	7410	7410		1807	
9003091-03	7410	7410		1807	
9003097-02	74107	74107			
9003151	7400	7400		1801	
9003151-03	7400			1801	
9003152	7474	7474		1818	
9003152-01	7474	7474		1818	
9003234-04	7438	7438			
9003398-03	7408	7408		1822	
9003398-04	7408	7408		1822	
9003445-03	7492	7492		1819	
9003642-03	7430	7430			
9004017-01	219		74	2043	
9004075-03	7401	7401			
9004076	7420	7420		1809	
9004076-03	7420	7420		1809	
9004076-04	7420	7420		1809	
9004093-03	7473			1803	
9004300-03	7476	7476		1813	
9004360-03	7454	7454			
9004508-01	159		82	2032	
9004915-04	912		IC-172		
9006527-01	219		74	2043	
9007038	159		82	2032	
9008964-01	123A	3444	20	2051	
9100502	102A	3004	53	2007	
9100621	102A	3004	53	2007	
9100706	102A	3004	53	2007	
9100944	102A	3004	53	2007	
9101600	915			017	
9176494	123A	3444	20	2051	
9246053	116	3016	504A	1104	
9246055	116	3017B/117	504A	1104	
9248015	116	3016	504A	1104	
9248055	116	3016	504A	1104	
9248058	116	3016	504A	1104	
09301020	102A	3004	53	2007	
09308004	703A	3157	IC-12		
09308038	1142		IC-128		
9330006			504A	1104	
9340311	130	3027	14	2041	
9340311-D5514	130		14	2041	
9340388	159		82	2032	
9340924	743		IC-214		
9341238	737	3375	IC-16		
9341510	152	3893	66	2048	
9341767	159		82	2032	
9341834	129		244	2027	
9341899	744	3734/806	IC-239		
9342291	159		82	2032	
9347173	743		IC-214		
9739636-20	116	3311	504A	1104	
10000020	123A	3444	20	2051	
10015595	123A	3444	20	2051	
10017663	102	3722	2	2007	
10022104-101	123A	3444	20	2051	
10027973-101	107		11	2015	
10041264-101	519		514	1122	
10041265-101	177		300	1122	
10106058	123A	3444	20	2051	
10112562	159		82	2032	
10180316	177		300	1122	
10180722	123A	3444	20	2051	
10182330	159		82	2032	
10183001	199	3245	62	2010	
10545502	123A	3444	20	2051	
10545506	199	3245	62	2010	

Industry Standard No.	ECG	SK	GE	RS 276-	MOTOR.
10641140	159		82	2032	
10644417	175	3026	246	2020	
10644433	123A	3444	20	2051	
10646032	130		14	2041	
10653086	175		246	2020	
10669652	519		214	1122	
10669666	159		82	2032	
10670399	159		82	2032	
10674068	5642	3507			
10676144	519		514	1122	
10713642	130	3027	14	2041	
10805364	130		14	2041	
10814788	128		243	2030	
10849792	123A	3444	20	2051	
10896074	123A	3444	20	2051	
11018181	124		12		
11019072	5526	6627/5527			
11041003-1	181		75	2041	
11056225	290A		269	2032	
11069934A	130		14	2041	
11076437	159		82	2032	
11083979	519		514	1122	
11103439	519		514	1122	
11116191	5547	3505			
11132875	102	3722	2	2007	
11159860	519		514	1122	
11166527	130		14	2041	
11166535	128	3024	243	2030	
11176377	177		300	1122	
11194628	181	3036	75	2041	
11198132	123A	3444	20	2051	
11210016/7825	109			1123	
11210024/7825	116		504A	1104	
11220009/7825	102A		53	2007	
11220018/7611	102A	3004	53	2007	
11220022/7611	102A		53	2007	
11220046/7825	123A	3444	20	2051	
11220061/7825	158		53	2007	
11220076/7825	102A		53	2007	
11220106/7611	158		53	2007	
11233335	175		246	2020	
11242096	909	3590	IC-249		
11253441	128		243	2030	
11268821	519		514	1122	
11290111	177		300	1122	
11292312	74H00	74Z00			
11322253	128		243	2030	
11667821	102	3722	2	2007	
11706998-2	199	3245	62	2010	
11707240	311		277		
11707240B	311		277		
11707240RA	311		277		
11718319	128		243	2030	
11728100	941		IC-263	010	
11729469	716		IC-208		
11783453	102	3722	2	2007	
11787173	519		514	1122	
11790334	5529	6629			
11800182	102	3722	2	2007	
11802400	123A	3444	20	2051	
11802500	123A	3444	20	2051	
11803000	177		300	1122	
11858909	102	3722	2	2007	
12023074	310	3856			
12090924-4	102A	3004	53	2007	
12363800	175	3026	246	2020	
12901503	159	3114/290	82	2032	
12947206	519		514	1122	
12965471	123A	3444	20	2051	
12994851	123A	3444	20	2051	
12997748	519		514	1122	
13020000	102A	3004	53	2007	
13020002	102A	3004	53	2007	
13023002-2			61	2038	
13030189	177	3031A	300	1122	
13030192	116	3311	504A	1104	
13030256	125	3031A	510,531	1114	
13030266	177		300	1122	
13030274	177	3100/519	300	1122	
13030285	5016A	3782	ZD-8.2	562	
13030301	109	3087		1123	
13030312	109			1123	
13030313	116	3311	504A	1104	
13030401	143A	3750	ZD-13	563	
13033801	116	3016	504A	1104	
13035807-1	123A	3444	20	2051	
13037215	123A	3444	20	2051	
13038115	107	3039/316	61	2038	
13038900	116	3016	504A	1104	
13040089	103A	3011	59	2002	
13040095	102A	3004	53	2007	
13040096	103A	3010	59	2002	
13040216	123A	3444	20	2051	
13040229	109	3088		1123	
13040236	102A	3004	53	2007	
13040304	108	3452	86	2038	
13040313	123A	3444	20	2051	
13040317	123A	3444	20	2051	
13040318	123A	3444	20	2051	
13040347		3010	53,59		
13040349	131		44	2006	
13040352	102A	3004	53	2007	
13040357	123A	3444	20	2051	
13040362	108	3452	86	2038	
13040421	108	3452	86	2038	
13040429	129	3025	244	2027	
13040456	158		53	2007	
13040459	108	3452	86	2038	
13050221	116(2)	3311	504A(2)	1104	
13073525	289A	3122	210	2038	
13083908	123A	3444	20	2051	
13094517	199	3124/289	62	2010	
13104367	130		14	2041	
13104560	128		243	2030	
13104561	129		244	2027	
13217724	102	3722	2	2007	
13217732	102	3722	2	2007	
13443767	102	3722	2	2007	
13447321	102	3722	2	2007	
14001376-01	5697	3522			
14001376-02	5697	3522			
14001376-03	5697	3522			
14317184	181		75	2041	
14500016-001	941D		IC-264		
14500016-002	941M		IC-265	007	
14500022-001	123		20	2051	
14628601	124		12		
14714661	519		514	1122	
14714703	519		514	1122	
14714729	5995	3518			
14714760	123A	3444	20	2051	
14714786	123A	3444	20	2051	
14734248	909		IC-249		
14735521	154		40	2012	

Industry Standard No.	ECG	SK	GE	RS 276-	MOTOR.
14736221	159		82	2032	
14769302	505	3108	CR-7		
14798029	159		82	2032	
15038433	123A	3444	20	2051	
15039456	123A	3444	20	2051	
15039464	123A	3444	20	2051	
15039472	519		514	1122	
15039480	519		514	1122	
15059850	124		12		
15077183	175	3026	246	2020	
15102718	102	3722	2	2007	
15105300	309K	3629			
15108046				1114	
15109200	74S04	74S04			
15348931	175		246	2020	
16233010	1006	3358	IC-38		
16270092	123A	3444	20	2051	
16270175	519		514	1122	
16353343	736		IC-17		
16520001-1	123A	3444	20	2051	
16658600	519		514	1122	
16763333-001	909	3590	IC-249		
16797300	123A	3444	20	2051	
16797301	123A	3444	20	2051	
17101881	109	3087		1123	
17101882	110MP	3087	1N34AS	1123(2)	
17184600	7405	7405			
17200000	177	3100/519	300	1122	
17200240	177	3100/519	300	1122	
17210010	116	3311	504A	1104	
17210430	614	3126	90		
17240010	156	3311	504A	1104	
17350060			82	2032	
17350061			82	2032	
17350066			82	2032	
17350068			244	2027	
17771400	128		243	2030	
17809000	130		14	2041	
17854700	519		514	1122	
17942800-01	159		82	2032	
17942800-801	159		82	2032	
18151417	128		243	2030	
18179900	123A	3444	20	2051	
18341300	519		514	1122	
18458117	941		IC-263	010	
18525200	126		52	2024	
19020065	152	3893	66	2048	
19020095	130		14	2041	
19080002	116	3016	504A	1104	
19901403	128	3024	243	2030	
19901503	159		21	2034	
19901606	229		211	2016	
19901607	229		211	2016	
19901806	108		86	2038	
19901807	108		86	2038	
20001786-134	109	3087		1123	
20001786-139	114		6GD1	1104	
20001786-141	116	3017B/117	504A	1104	
20001786-142	116	3017B/117	504A	1104	
20025153-77	123A	3444	20	2051	
20030703-0701	123A	3444	20	2051	
20030703-0702	159	3114/290	82	2032	
20052600	123A	3444	20	2051	
20080127			300	1122	
20088308	199	3245	62	2010	
20115070	109	3087		1123	
20130109	125	3081	510	1114	
20278600		3027	14	2041	
20912578-1	176	3123	80		
20918596-2	121	3009	239	2006	
22114210	192	3024/128	63	2030	
22114225	186	3041	28	2017	
22114253	159	3138/193A	82	2032	
22114254	187	3193	29	2025	
22115175		3126	90		
22115181	109	3087		1123	
22115182	109	3088		1123	
22115183	116	3031A	504A	1104	
22115192	106	3088		1123	
22115405	116	3311	504A	1104	
22130009	128	3024	243	2030	
22692012	177	3100/519	300	1122	
22901513	108	3452	86	2038	
22902046	108	3452	86	2038	
23109992	138A		ZD-7.5		
23111006	121	3717	239	2006	
23113056			20	2051	
23114001	108	3452	86	2038	
23114004	127	3034	25		
23114007	154	3018	40	2012	
23114009	127	3034	25		
23114011	121	3009	239	2006	
23114015	289A	3122	210	2038	
23114015(HOR OSC)	289A			2038	
23114015(HOR-OSC)			268	2038	
23114017	289A	3018	268	2038	
23114021	102A	3004	53	2007	
23114023	108	3452	86	2038	
23114026	127	3034	25		
23114028	154	3040	40	2012	
23114031	108	3452	86	2038	
23114033	121	3009	239	2006	
23114034	108	3452	86	2038	
23114036	108	3452	86	2038	
23114037	161	3019	39	2015	
23114038	286	3538	246	2020	
23114043	108	3452	86	2038	
23114044	233	3018	210	2009	
23114045	103A	3010	59	2002	
23114046	123A	3444	20	2051	
23114050	159	3025/129	82	2032	
23114051	159	3025/129	82	2032	
23114052	154	3040	40	2012	
23114053	154	3020/123	40	2012	
23114054	154	3020/123	40	2012	
23114056	108	3452	86	2038	
23114057	108	3452	86	2038	
23114058		3024	243	2030	
23114060	108	3452	86	2038	
23114061	102A	3004	53	2007	
23114070	130	3027	14	2041	
23114070A	130	3027	14	2041	
23114071			19	2041	
23114074	506	3125	511	1114	
23114078	233	3018	61	2009	
23114081	290A	3025/129	269	2032	
23114082	108	3452	86	2038	
23114083			246	2020	
23114084	175	3021/124	246	2020	
23114086	175		246	2020	
23114087			20	2051	
23114088	175		246	2020	
23114095	154	3020/123	40	2012	
23114097	127	3034	25		

Industry Standard No.	ECG	SK	GE	RS 276-	MOTOR.
23114100	130	3027	14	2041	
23114104	233	3004/102A	210	2009	
23114108	130	3027	14	2041	
23114109	108	3452	86	1123	
23114112			11	2015	
23114118	123A	3444	20	2051	
23114119	123A	3444	20	2051	
23114124	159	3025/129	82	2032	
23114125	154	3040	40	2012	
23114126	107	3245/199	11	2015	
23114127	108	3452	86	2038	
23114131	185	3191	58	2025	
23114132	184	3190	57	2017	
23114133	185	3191	58	2025	
23114134	184		57	2017	
23114136	159	3025/129	82	2032	
23114137	290A	3025/129	269	2032	
23114138	159	3114/290	82	2032	
23114148			11	2015	
23114155	123A	3444	20	2051	
23114157	107	3018	11	2015	
23114163	319P	3039/316	86	2016	
23114164	161	3132	39	2015	
23114165	107	3039/316	86	2016	
23114171	108	3452	86	2038	
23114172	108	3452	86	2038	
23114180	108	3452	86	2038	
23114181	108	3452	86	2038	
23114195	128	3024	243	2030	
23114200	124	3021	12		
23114208	162	3438	35		
23114211	102A	3004	53	2007	
23114212	123A	3444	20	2051	
23114213			20	2051	
23114214	123A	3444	20	2051	
23114215			20	2051	
23114216	199	3020/123	62	2010	
23114217	199	3245	62	2010	
23114218			20	2051	
23114220	175	3026	246	2020	
23114221	175	3026	246	2020	
23114232	128	3024	243	2030	
23114238	108	3452	86	2038	
23114248	198		251		
23114249	171	3044/154	27		
23114250	171	3044/154	27		
23114251	171	3044/154	27		
23114253	152	3054/196	66	2048	
23114254	152	3054/196	66	2048	
23114255	123A	3444	20	2051	
23114258	199	3245	62	2010	
23114259	289A	3137/192A	268	2038	
23114260	289A	3137/192A	268	2038	
23114261	289A	3138/193A	269	2032	
23114262	290A		269	2032	
23114266	198	3021/124	251		
23114267			20	2051	
23114268	124	3021	12		
23114275	123A	3444	61	2051	
23114276	123A	3444	61	2051	
23114277	124	3021	12		
23114280	229	3018	61	2038	
23114282	229	3018	61	2038	
23114293	290A	3025/129	82	2032	
23114296	123A	3444	20	2051	
23114297	289A	3122	210	2038	
23114299			10	2051	
23114300	159	3466	21	2034	
23114301	159	3466	221	2032	
23114302	159	3466	221	2032	
23114313	193	3025/129	67	2023	
23114314	192	3024/128	63	2030	
23114315	124	3021	12		
23114317	162	3438	35		
23114321	165	3114/290	38		
23114323	238	3710	259		
23114325	159		82	2032	
23114336	198	3104A	251		
23114343	165	3115	38		
23114344	198	3104A	251		
23114347	289A	3138/193A	269	2032	
23114349	199	3124/289	212	2010	
23114550	159	3114/290	82	2032	
23114864	328	3895/325	75		
23114900	198		255		
23114915	297	3137/192A	63	2030	
23114923	196	3054	66	2048	
23114939	197	3083	250	2027	
23114941	198	3245/199	251		
23114944	162		35		
23114945	238	3710	259		
23114954	198	3104A			
23114961	175	3538	233	2020	
23114962	238	3710	38		
23114966	229	3018	60	2038	
23114969	196	3054	32		
23114974	198	3104A	251		
23114975	198	3245/199	251		
23114982	198	3220	251		
23114983	190		217		
23114993	198	3104A	251		
23114994	192	3024/128	63	2030	
23114995	193	3025/129	67	2023	
23114999	198	3245/199	251		
23115012	125	3032A	510,531	1114	
23115019	116(2)		504A(2)	1104	
23115022	118		CR-1		
23115023			504A	1104	
23115042	116	3311	504A	1104	
23115044			504A	1104	
23115046	116	3016	504A	1104	
23115049	177	3100/519	300	1122	
23115050	177	3311	300	1122	
23115051	507	3311	300	1122	
23115052	116(3)		504A(3)	1104	
23115057	123A	3444	20	2051	
23115058	123A	3444	20	2051	
23115064	177	3016	300	1122	
23115065	178MP	3016	300(2)	1122(2)	
23115068	177	3016	300	1122	
23115070	109	3087		1123	
23115071	177	3016	300	1122	
23115072	177	3100/519	300	1122	
23115074	178MP		300(2)	1122(2)	
23115076	139A	3060	ZD-9.1	562	
23115078	115	3121	6GX1		
23115079	113A	3119/113	6GC1		
23115080	177	3311	300	1122	
23115085	116	3032A	504A	1104	
23115088	109	3088		1123	
23115094	506	3130	511	1114	
23115098	125		510,531	1114	
23115102			1N34AS	1123	
23115108	177	3100/519	300	1122	
23115109	177	3100/519	300	1122	
23115113	112	3089	1N82A		
23115115	109			1123	
23115117	125	3017B/117	510,531	1114	
23115118	125	3032A	510,531	1114	
23115120		3515	504A	1104	
23115120(DIODE)	177		300	1122	
23115121	112	3089	1N82A		
23115129	109	3087		1123	
23115130	125	3311	510,531	1114	
23115131	506	3017B/117	511	1114	
23115140	116	3017B/117	504A	1104	
23115142	116	3031A	504A	1104	
23115145	506	3017B/117	511	1114	
23115146	506	3130	511	1114	
23115157	502	3067	CR-4		
23115168	137A	3058	ZD-6.2	561	
23115185	116	3311	504A	1104	
23115192	506	3130	511	1114	
23115194	177	3100/519	300	1122	
23115199	116	3311	504A	1104	
23115200			504A	1104	
23115201			504A	1104	
23115215	504	3108/505	CR-6		
23115227		3089	1N82A		
23115249	177	3100/519	300	1122	
23115250	116		504A	1104	
23115263	116	3311	504A	1104	
23115273	116	3100/519	504A	1104	
23115277	5072A	3136	ZD-8.2	562	
23115285	177	3100/519	300	1122	
23115292	5081A		ZD-24		
23115294	116		504A	1104	
23115296	125	3017B/117	504A	1104	
23115298	506	3130	511	1114	
23115300	116	3016	504A	1104	
23115321	513	3443	513		
23115328	5081A	3151	ZD-24		
23115331	177	3100/519	30	1122	
23115334	513	3443	513		
23115337	116	3016	504A	1104	
23115338	506	3016	511	1114	
23115368	5074A	3092/141A	ZD-11.0	563	
23115374	5072A	3136	ZD-8.2	562	
23115377	506	3130	511	1114	
23115455	5116A		ZD-5.1		
23115884	5082A		ZD-25		
23115897	177	3100/519	300	1122	
23115903	6402	3638/5402			
23115908	5071A	3334	ZD-6.8	561	
23115912	506	3125	511	1114	
23115913	506	3051/156	511	1114	
23115914	506	3311	504A	1104	
23115919	142A	3062	ZD-12	563	
23115921	5116A	3056/135A	ZD-5.1		
23115938	504	3108/505	CR-6		
23115947	125	3016	510,531	1114	
23115955	137A		ZD-6.2	561	
23115960	506	3130	511	1114	
23115965	525	3925	530		
23115966	125	3051/156	510	1114	
23115978	177		300	1122	
23115981	506	3130	511	1114	
23115984	138A	3059	ZD-7.5		
23115987			ZD-8.2	562	
23115990	5084A	3755	ZD-30		
23115994	125		531	1114	
23115999	506	3017B/117	511	1114	
23119003	711	3070	IC-207		
23119004	710	3102	IC-89		
23119005	711	3070	IC-207		
23119007	710	3102	IC-89		
23119011	1101	3283	IC-95		
23119013	1109	3711	IC-99		
23119014	747	3279	IC-218		
23119016	1105	3285	IC-101		
23119017	1062		IC-52		
23119019	749	3168	IC-97		
23119022	1101	3283	IC-95		
23119023	1109	3711	IC-99		
23119025	748	3236			
23119028	1105	3285	IC-101		
23119030	1109	3711	IC-99		
23119031	1109	3711	IC-99		
23119032	1109	3711	IC-99		
23119033	1109	3711	IC-99		
23119960	7493A	7493			
23119971	1200	3714			
23119978	1236	3072/712	IC-148		
23119981	1128	3488	IC-105		
23119988	1133	3490	IC-107		
23119989	1134	3489	IC-106		
23119990	1128	3488	IC-105		
23119993	1130	3478	IC-111		
23119994	1132	3287	IC-110		
23119995	1131	3286	IC-109		
23119999	748	3236			
23124037	108	3452	86	2038	
23126177			11	2015	
23126183	161	3018	39	2015	
23126184			20	2051	
23126289	107		11	2015	
23126290	107		11	2015	
23126291	107		11	2015	
23126577	112		1N82A		
23126620	108	3452	86	2038	
23200596-1	128	3024	243	2030	
23311006	104	3009	16	2006	
23311066	121	3717	239	2006	
23576100	1083		IC-275		
24501000	123A	3444	20	2051	
24539800	159		82	2032	
24551302	128		243	2030	
24553500	519		514	1122	
24553501	519		514	1122	
24553600	123A	3444	20	2051	
24561601-E	5995	3518	5105		
24561602-E	5995	3518	5105		
24561603	5995	3518	5105		
24561604-E	5995		5105		
24562000	123A	3444	20	2051	
24562001	123A	3444	20	2051	
24562100	123		20	2051	
24562101	123A	3444	20	2051	
24562200	123A	3444	20	2051	
24562300	159		82	2032	
24733362	129		244	2027	
25114101	102A	3004	53	2007	
25114102	102A	3004	53	2007	
25114103	102A	3004	53	2007	
25114104	102A	3004	53	2007	
25114116	123A	3444	20	2051	
25114121	123A	3444	20	2051	
25114130	154	3045/225	40	2012	

Industry Standard No.	ECG	SK	GE	RS 276-	MOTOR.
25114143	124	3021	12		
25114161	108	3452	86	2038	
25115102	109	3087	86	1123	
25115108	116	3017B/117	504A	1104	
25115115	116	3311	504A	1104	
25175800	519		514	1122	
26004001	123A	3444	20	2051	
26004301	159		82	2032	
26004961	159		82	2032	
26005121	159		82	2032	
26006820	177		300	1122	
26006910	177		300	1122	
26008020	177		300	1122	
26010006	312	3834/132	FET-1	2035	
26010010	195A	3048/329	46		
26010011	507		300	1122	
26010016	159	3114/290	82	2032	
26010020	123A	3444	20	2051	
26010021	289A	3122	268	2038	
26010022	237	3299	46		
26010023	199	3124/289	62	2010	
26010024	152	3054/196	66	2048	
26010026	229	3018	61	2038	
26010027	290A	3114/290	269	2032	
26010028	177		300	1122	
26010029	137A	3058	ZD-6.2	561	
26010030	177	3100/519	300	1122	
26010031		3126	1N60	1123	
26010032	177	3100/519	300	1122	
26010033	116	3017B/117	504A	1104	
26010035	1102	3224	IC-93		
26010041	297	3137/192A	271	2030	
26010044	5072A	3136	ZD-8.2	562	
26010046	137A	3058	ZD-6.2	561	
26010047	177	3100/519	300	1122	
26010051	229	3018	61	2038	
26010052	297	3137/192A	271	2030	
26010053	236	3197/235	216	2053	
26010056	108	3452	86	2038	
26010057	177	3100/519	300	1122	
26010058	195A	3048/329	46		
26010059	152	3054/196	66	2048	
26011310	159		82	2032	
26316032	159		82	2032	
26501505	123		20	2051	
27113100	116	3311	504A	1104	
27119004	710		IC-89		
27123050	116	3016	504A	1104	
27123070	116	3016	504A	1104	
27123100	116	3016	504A	1104	
27123100A	116	3017B/117	504A	1104	
27123120	116	3016	504A	1104	
27123150	109	3087		1123	
27123220	116	3017B/117	504A	1104	
27123240	109	3087		1123	
27123270	109	3087		1123	
27123290	116(2)			1104	
27125080	123A	3444	20	2051	
27125090	123A	3444	20	2051	
27125110	102A		53	2007	
27125120	102A		53	2007	
27125140	123A	3444	20	2051	
27125150	123A	3004/102A	20		
27125160	123A	3444	20	2051	
27125170	158	3004/102A	53	2007	
27125210	108	3452	86	2038	
27125220			11	2015	
27125230	126		52	2024	
27125240	100	3005	1	2007	
27125250	123A	3444	20	2051	
27125260	102A		53	2007	
27125270	123A	3444	20	2051	
27125300	123A	3444	20	2051	
27125310	103A	3010	59	2002	
27125330	102A	3004	53	2007	
27125340	102A	3004	53	2007	
27125350	102A	3004	53	2007	
27125360	102A	3005	53	2007	
27125370	123A	3444	20	2051	
27125380	123A	3444	20	2051	
27125460		3444	20	2051	
27125470	123A	3444	20	2051	
27125480	158	3004/102A	53	2007	
27125490	103A	3010	59	2002	
27125500	123A	3444	20	2051	
27125540	176	3845	80		
27125550	102A	3004	53	2007	
27126090	121	3717	239	2006	
27126100	130	3027	14	2041	
27126130	121	3717	239	2006	
27126130-12	121	3014	239	2006	
27126220	108	3452	86	2038	
28100812-001	177		300	1122	
28101149	234		65	2050	
28102128-002	199	3245	62	2010	
28102128-004	199	3245	62	2010	
28102163-001	177		300	1122	
28105203-001	152	3893	66	2048	
28105203-002	152	3893	66	2048	
28105207-001	234		65	2050	
28105249-001	153		69	2049	
28105789-001	199	3245	62	2010	
29004423	9947		IC-268		
30000010	126		52	2024	
30000021	102A	3003	53	2007	
30200033	152	3054/196	66	2048	
30200053	123A	3444	20	2051	
30200061		3018	255	2041	
30200062	233	3018	210	2009	
30200091	108	3452	86	2038	
30200101	130	3027	14	2041	
30400021	312	3448	FET-1	2035	
30600010	109	3088		1123	
30600020	109			1123	
30600040	116	3311	504A	1104	
30600140	142A	3062	ZD-12	563	
30677435-001	909	3590	IC-249		
30682592-001	234		65	2050	
31210049-00	160	3006	245	2004	
31210077-44	160	3006	245	2004	
31210240-33	160	3006	245	2004	
31210240-44	160	3006	245	2004	
31210471-11	160	3006	245	2004	
31210506-11	160	3006	245	2004	
31210506-22	160	3006	245	2004	
31220054-00	102A	3004	53	2007	
32510001	116	3311	504A	1104	
32600025-01-08A	123A	3444	20	2051	
34621786-1	519		514	1122	
35002711			21	2034	
35002712			21	2034	
35002713			21	2034	
35002912			67	2023	
35024712	197		250	2027	
35044100			39	2015	
35044396			60	2015	
35044397			60	2015	
35044398			60	2015	
35044399			60	2015	
35045416			20	2051	
35045502			20	2051	
35045708			20	2051	
35045712			20	2051	
35046911			14	2041	
35046912			14	2041	
35046913			14	2041	
35047218			20	2051	
35047219			20	2051	
35049311			66	2048	
35050411	198	3245/199	251		
35050412	198	3245/199	251		
35062712	198	3245/199	251		
35393060-01	123A	3444	20	2051	
35393060-02	123A	3444	20	2051	
35393060-03	123A	3444	20	2051	
36001004			300	1122	
36001008			ZD-10	562	
36001009			1N60	1123	
36001200	128		243	2030	
36002003			1N60	1123	
36003034			ZD-8.2	562	
36107051			504A	1104	
36107062			504A	1104	
36107152			511	1114	
36107153			300	1122	
36107154			511	1114	
36107253			511	1114	
36107255			511	1114	
36171000		3984	21	2034	
36171100	123A	3444	20	2051	
36188000	7475	7475		1806	
36188700	74H00	74H00			
36771100	124	3271	32		
37000918	109	3088	1N60	1123	
37001003	1047		300	1122	
37001009	712	3072	IC-2		
37001011	712	3072	IC-2		
37002001	1046	3471	IC-118	1122	
37003001	1089*		300	1122	
37004001	1048		IC-122	1122	
37006001	1077		IC-124	1122	
37007001	1076		IC-125	1122	
37008002	1091*		IC-126	1122	
37009006	1086		IC-142	1122	
37246602	177	3100/519	300	1122	
37246711	177	3100/519	300	1122	
37275350	177	3100/519	300	1122	
37568200	506	3130	511	1114	
37568250	506	3051/156	510	1114	
37568800	116	3017B/117	504A	1104	
37582000	506	3130	511	1114	
38970300	123A		20	2051	
39126500	941		IC-263	010	
40306312	748		IC-183		
40306400	749	3168	IC-97		
40306604	1109	3711	IC-99		
41013566	116	3311	504A	1104	
41020618	116	3031A	504A	1104	
41020687	156	3051	512		
41021007	116	3311	504A	1104	
41023224	116	3031A	504A	1104	
41023225	116	3031A	504A	1104	
41025773	125	3016	510,531	1114	
41025849	116	3016	504A	1104	
41027063	116	3016	504A	1104	
41027612	109	3088		1123	
41027613	109	3088		1123	
41027991	125	3081	510,531	1114	
41027992	109	3088		1123	
41029009	109	3087		1123	
41029290	109	3088		1123	
41029498	5084A	3755	ZD-30		
41029499	116	3311	504A	1104	
41030618	156		512		
41520419	116	3311	504A	1104	
41522875	110MP	3087	1N34AS	1123	
41527380	109	3088		1123	
41622859	177	3100/519	300	1122	
41624836	109	3087		1123	
42020414	140A	3061	ZD-10	562	
42020737	140A	3061	ZD-10	562	
42023425	143A	3750	ZD-13	563	
42024925	5071A		ZD-6.8	561	
42027092	5075A	3751	ZD-16	564	
42029566	143A		ZD-13	563	
43020418	121MP	3718	239(2)	2006(2)	
43020731	128	3024	243	2030	
43021017	123A	3444	20	2051	
43021067	107	3018	11	2015	
43021083	123A	3444	20	2051	
43021168	129	3114/290	244	2027	
43021198	128	3024	243	2030	
43021415	128	3020/123	243	2030	
43022055	160	3006	245	2004	
43022134	312	3834/132	FET-1	2035	
43022458	159		82	2032	
43022577	121	3717	239	2006	
43022860	107	3039/316	11	2015	
43022861	123A	3444	20	2051	
43023190	181	3036	75	2041	
43023212	123A	3444	20	2051	
43023221	123A	3444	20	2051	
43023222	129	3025	244	2027	
43023223	128	3024	243	2030	
43023843	130	3027	14	2041	
43023844	123A	3444	20	2051	
43023845	159	3114/290	82	2032	
43024216	130	3027	14	2041	
43024218	184	3190	57	2017	
43024219	185	3191	58	2025	
43024225	312	3112	FET-1	2028	
43024306	190		217		
43024833	126	3005	52	2024	
43024834	126	3008	52	2024	
43024859	199	3124/289	62	2010	
43024873	199	3124/289	62	2010	
43024878		3124	61	2038	
43024879	128	3020/123	243	2030	
43024880	129	3025	244	2027	
43024972	123A	3444	20	2051	
43025055	123A	3444	20	2051	
43025056	123A	3444	20	2051	
43025059	123A	3124/289	20	2051	
43025538	199	3024/128	62	2010	
43025539	192	3024/128	63	2030	
43025620	160	3006	245	2004	
43025972	123A	3444	20	2051	
43026284	128	3020/123	243	2030	

Industry Standard No.	ECG	SK	GE	RS 276-	MOTOR.
43026285	129	3025	244	2027	
43027213	152	3054/196	66	2048	
43027214	152	3054/196	66	2048	
43027379	123A	3444	20	2051	
43027571	186	3192	28	2017	
43027614	108	3452	86	2038	
43027615	108	3452	86	2038	
43027616	108	3452	86	2038	
43027617	108	3452	86	2038	
43027618	107	3018	11	2015	
43027619	106	3118	21	2034	
43027620	123A	3444	20	2051	
43027987	152	3054/196	66	2048	
43029471	233	3018	210	2009	
43029472	233	3018	210	2009	
43029483	199	3122	62	2010	
43029484	297	3024/128	271	2030	
43029485	298	3450	272		
43029486	300	3464	273		
43040313	128	3122	243	2030	
43100103				1123	
43122874	726	3022/1188			
43126551	722	3161	IC-9		
43126551-A	722	3161	IC-9		
43200103	019	3088		1123	
43300203		3087		1123	
43540012	116	3016	504A	1104	
43600101	116	3311	504A	1104	
44007301	108	3452	86	2038	
44008401	108	3452	86	2038	
44011001	123A	3444	20	2051	
44079004	199	3122	62	2010	
44089001	107	3018	11	2015	
44090004	123A	3444	20	2051	
44211001	158	3004/102A	53	2007	
46156741P1	941		IC-263	010	
46156741P2	941D		IC-264		
48134666	123A		20	2051	
48134842	123A	3020/123	20	2051	
48137195A	159	3114/290	82	2032	
48869715	181		75	2041	
50210104	123A	3444	20	2051	
50210300-00	123A	3444	20	2051	
50210300-00 VF	123A	3444			
50210300-00VF			20	2051	
50210300-01	123A	3444	20	2051	
50210300-01 VF	123A	3444			
50210300-01VF			20	2051	
50210300-11	123A	3444	20	2051	
50210310-10	123A	3444	20	2051	
50210310-10 VF	123A	3444			
50210310-10VF			20	2051	
50210310-11	123A	3444	20	2051	
50210400-00	159		82	2032	
50210400-00VF			82	2032	
50210400-01	159		82	2032	
50210400-01VF			82	2032	
50210510	123A	3444	20	2051	
50210600-00 VF	159			2032	
50210600-00VF			82	2032	
50210600-01	159		82	2032	
50210600-01VF			82	2032	
50210600-03	159		82	2032	
50210610-10	159		82	2032	
50210610-13	159		82	2032	
50210700-00	199	3245	62	2010	
50210700-00 VF	199	3245			
50210700-00VF			62	2010	
50210700-01 VF	199			2010	
50210700-01VF			62	2010	
50210700-10	199	3245	62	2010	
50210710-10 VF	199			2010	
50210710-10VF			62	2010	
50210800-00	123A	3444	20	2051	
50210800-01	123A	3444	20	2051	
50210800-01 VF	123A	3444			
50210800-01VF			20	2051	
50210800-02	123A	3444	20	2051	
50210800-02 VF	123A	3444			
50210800-02VF			20	2051	
50210800-10	128		243	2030	
50210800-11	123A	3444	20	2051	
50211210	159		82	2032	
50211300-00	106	3984	21	2034	
50211300-00VF			21	2034	
50211300-01	106	3984	21	2034	
50211300-10	106		21	2034	
50211400-00	234	3247	65	2050	
50211400-01	234	3247	65	2050	
50211410-10	234	3247	65	2050	
50211410-11	234	3247	65	2050	
50211500	159		82	2032	
50211500-01	159		82	2032	
50211500-01VF			82	2032	
50211510-10	159		82	2032	
50211510-10VF			82	2032	
50211510-11	159		82	2032	
50211600-00	106	3984	21	2034	
50211600-00VF			21	2034	
50211600-01	106	3984	21	2034	
50211600-02	106	3984	21	2034	
50211600-02VF			21	2034	
50211600-10	106	3984	21	2034	
50211600-12	106	3984	21	2034	
50211610-10	106	3984	21	2034	
50211610-12	106	3984	21	2034	
50211900	128		243	2030	
50212100	159		82	2032	
50220601	175		246	2020	
50220602	175		246	2020	
50221300-01	130		14	2041	
50221400-01	218		234		
50221800	152	3893	66	2048	
50221800-01	152	3893	66	2048	
50251300	941		IC-263	010	
50254200	7406	7406		1821	
50254400	915			017	
50254600	74S00	74S00			
50613800	923		IC-259		
51003059	123A	3444	20	2051	
51003092	123A	3444	20	2051	
51003108	159		82	2032	
51010422A02		3159	IC-8		
51122245	123A	3444	20	2051	
51122870	128		243	2030	
51122880	128		243	2030	
51126540	175		246	2020	
51126850	130	3027	14	2041	
51161325	159		82	2032	
51320000	7400	7400		1801	
51320001	7402	7402		1811	
51320002	7404	7404		1802	
51320003	7410	7410		1807	
51320004	7420	7420		1809	
51320005	7440	7440			
51320011	7430	7430			
51320012	7400	7400		1801	
51320016	7451	7451		1825	
51320017	7405	7405			
51320018	7486	7486		1827	
51330005	7408	7408		1822	
51565600 VF	123A			2051	
51565600VF			20	2051	
51581300	123A	3444	20	2051	
51631000	519		514	1122	
51715100	910D		IC-252		
51753300	910		IC-251		
52335600	74H04	74H04			
52335600HL	74H04	74H04			
53015201	156	3051	512		
53020612			18	2030	
54967774-P5	102	3722	2	2007	
55310007-115	139A		ZD-9.1	562	
55335005-001	5870		5032		
55430001-001	159		ZD-82	2032	
55430028-001	218		234		
55440007-001	123		20	2051	
55440011-001	108	3452	86	2038	
55440023-001	108	3452	86	2038	
55440043-001	123		20	2051	
55440048-001	123A	3444	20	2051	
55440063-001	128		243	2030	
55440087-001	130		14	2041	
55440091-001	130		14	2041	
55440111-001	107		11	2015	
55460014-001	312	3448	FET-1	2035	
55517007	159		82	2032	
56301500	123A	3444	20	2051	
56301600	159		86	2032	
57000901-504	123A	3444	20	2051	
57276605	105	3012	4		
57279005	105	3012	4		
57279007	105	3012	4		
57279009	105	3012	4		
57279011	105	3012	4		
57279017	105	3012	4		
57279025	105	3012	4		
57279027	105	3012	4		
57279033	105	3012	4		
57279073	105	3012	4		
57279076	105	3012	4		
57279293	105	3012	4		
57279298	105	3012	4		
57279566	105	3012	4		
57279793	105	3012	4		
57280281	105	3012	4		
57282315	105	3012	4		
57288072	105	3012	4		
57288073	105	3012	4		
57288076	105	3012	4		
57288079	105	3012	4		
57299720	105	3012	4		
59700278	123A	3444	20	2051	
61260039	126	3008	52	2024	
61260039A	126	3008	52	2024	
62009616		3088		1123	
62013559	102A		53	2007	
62018526	102A		53	2007	
62020911	5804	9007		1144	
62034076	160	3006	245	2004	
62034297	112	3089	1N82A		
62034327	102A		53	2007	
62041528	102A	3004	53	2007	
62041536	102A	3004	53	2007	
62042575	126		52	2024	
62042583	123		20	2051	
62042591	102A		53	2007	
62042648	160		245	2004	
62042656	160		245	2004	
62042664	102A		53	2007	
62043679	160		245	2004	
62043695	160		245	2004	
62044888	126	3007	52	2024	
62045035	310	3856			
62045280	5012A		ZD-6.0	561	
62045531	102A		53	2007	
62045612	5018A	3784	ZD-9.1	562	
62045930	102A	3004	53	2007	
62046287	131		44	2006	
62046406	160	3006	245	2004	
62046414	160		245	2004	
62046422	160		245	2004	
62047119	116		504A	1104	
62047259	120	3110	CR-3		
62047364	126		52	2024	
62047895	125	3032A	510,531	1114	
62048158	108	3452	86	2038	
62049200	102A		53	2007	
62049626	126	3005	52	2024	
62049774	126	3006/160	52	2024	
62050217	101	3861	8	2002	
62051302	506	3125	511	1114	
62051310	506	3843	511	1114	
62051450	127	3034	25		
62051493	154	3040	40	2012	
62051509	104	3009	16	2006	
62052570	102	3004	2	2007	
62052791	116		504A	1104	
62052821	126		52	2024	
62052848	126		52	2024	
62054212	121MP	3013	239(2)	2006(2)	
62054298	128	3024	243	2030	
62054336	102	3004	2	2007	
62078502	102	3722	2	2007	
62081579	121	3717	239	2006	
62084103	121	3717	239	2006	
62084200	121MP	3009	239(2)	2006(2)	
62084936	121	3009	239	2006	
62084960	123A	3444	20	2051	
62085053	116		504A	1104	
62087025	126	3008	52	2024	
62087447	116		504A	1104	
62087498	102A		53	2007	
62087595	105		4		
62087609	103A	3010	59	2002	
62087617	101	3861	8	2002	
62087633	121	3717	239	2006	
62087641	160	3006	245	2004	
62087668	160(2)	3007	245(2)		
62087684	100	3005	1	2007	
62088502	102	3722	2	2007	
62094613	109			1123	
62103264	116		504A	1104	
62103272	116	3124/289	504A	1104	
62104678	108	3452	86	2038	
62104759	128		243	2030	
62105038	107		11	2015	
62105143	113A		6GC1		

Industry Standard No.	ECG	SK	GE	RS 276-	MOTOR.
62105275	123A	3018	20	2051	
62105291	107	3018	11	2015	
62111607	126		52	2024	
62114258	158		53	2007	
62116331	154		40	2012	
62118946	102A	3005	53	2007	
62118970	102	3004			
62119500	126		52	2024	
62119551	109			1123	
62119578	103A	3835	59	2002	
62123648	126	3006/160	52	2024	
62123672	5014A	3780	ZD-6.8	561	
62123788	126		52	2024	
62134143	102A		53		
62134496	109			1123	
62134941	109			1123	
62135182	160		245	2004	
62135301	102A		53	2007	
62139455	112	3089	1N82A		
62139633	160		245	2004	
62139668	160		245	2004	
62140844	116		504A	1104	
62140887	126		52	2024	
62140895	126		52	2024	
62140933	126		52	2024	
62141042	127	3764	25		
62154543	110MP	3709		1123(2)	
62166916	145A	3063	ZD-15	564	
62208945	128	3024	243	2030	
62208953	129	3025	244	2027	
62236302	123A	3444	20	2051	
62256877	116		504A	1104	
62256885	172A	3156	64		
62256893	159	3114/290	82	2032	
62258942	5019A	3785	ZD-10	562	
62260408	123		20	2051	
62278566	129	3025	244	2027	
62279664	159	3118	82	2032	
62360755	116	3017B/117	504A	1104	
62371566	121	3717	239	2006	
62372540	124	3021	12		
62379995	123A	3444	20	2051	
62380004	160		245	2004	
62380020	102	3722	2	2007	
62381620	110MP	3709		1123(2)	
62383976	703A	3157	IC-12		
62389478	185	3025/129	58	2025	
62389486	184	3024/128	57	2017	
62389494	128	3122	243	2030	
62389699	123A	3444	20	2051	
62389702	159	3114/290	82	2032	
62390174	142A		ZD-12	563	
62393661	153	3083/197	69	2049	
62393688	152	3893	66	2048	
62393696	129	3025	244	2027	
62438096	159	3114/290	82	2032	
62468211	177		300	1122	
62488131	5809	9010			
62506288	159		82	2032	
62506296	108	3452	86	2038	
62506318	108	3452	86	2038	
62506334	123A	3444	20	2051	
62506350	233	3132	210	2009	
62506377	123A	3444	20	2051	
62522666	136A	3057	ZD-5.6	561	
62522674	116	3031A	504A	1104	
62522682	113A	3119/113	6GC1		
62522704	120	3110	CR-3		
62522712	506	3130	511	1114	
62522720	177	3100/519	300	1122	
62522739	116	3031A	504A	1104	
62530898	126		52	2024	
62530915	128	3024	243	2030	
62537124	108	3452	86	2038	
62537140	123A	3444	20	2051	
62539283	123A	3444	20	2051	
62539291	159	3118	82	2032	
62539305	108	3452	86	2038	
62539313	106	3118	21	2034	
62539321	108	3452	86	2038	
62539348	161	3117	39	2015	
62540311	5081A	3151	ZD-24		
62541555	108	3452	86	2038	
62543299	109			1123	
62543892	108	3452	86	2038	
62543906	108	3452	86	2038	
62543914	154	3040	40	2012	
62562226	116		504A	1104	
62562625	506	3032A	511	1114	
62563265	159	3114/290	82	2032	
62563273	199	3018	62	2010	
62563281	199	3018	62	2010	
62563303	108	3452	86	2038	
62563309	154		40	2012	
62563311	159	3114/290	82	2032	
62563346	108	3452	86	2038	
62563354	108	3452	86	2038	
62563362	123A	3444	20	2051	
62563370	124	3021	12		
62564490	5071A		ZD-6.8	561	
62564504	142A	3092/141A	ZD-12	563	
62564520	506	3843	511	1114	
62564539	177	3087	300	1122	
62565122	108	3452	86	2038	
62565160	161	3716	39	2015	
62565963	159		82	2032	
62566369	159		82	2032	
62573741	130MP	3027/130	15	2041(2)	
62590661	159	3114/290	82	2032	
62593660	129	3025	244	2027	
62593687	123A	3444	20	2051	
62593707	123A	3444	20	2051	
62593717	161	3716	39	2015	
62593725	161	3124/289	39	2015	
62593733	160	3006	245	2004	
62596023	160		245	2004	
62596279	123A	3124/289	20	2051	
62604727	123A	3124/289	20	2051	
62605465	108	3452	86	2038	
62605618	118	3066	CR-1		
62608498	159	3114/290	82	2032	
62608528	312	3834/132	FET-1	2035	
62614765	160	3018	245	2004	
62618663	123A	3444	20	2051	
62636696	161	3018	39	2015	
62638214	123A	3444	20	2051	
62638222	191	3024/128	249		
62638230	123A	3444	20	2051	
62638249	199	3245	62	2010	
62638257	129	3025	244	2027	
62638265	128	3024	243	2030	
62638281	123A	3444	20	2051	
62638303	123A	3444	20	2051	
62638311	123A	3444	20	2051	
62638338	5080A	3336	ZD-22		
62638346	113A		6GC1		
62652381	123A	3444	20	2051	
62669462	5079A	3335	ZD-20		
62674954	712	3072	IC-2		
62674962	714	3075	IC-4		
62674970	715	3076	IC-6		
62674989	790	3077	IC-230		
62674997	147A	3045/225	ZD-33		
62675004	123A	3444	20	2051	
62675012	123A	3444	20	2051	
62691263	123A	3444	20	2051	
62691271	108	3452	86	2038	
62691298	128	3024	243	2030	
62695595	108	3452	86	2038	
62695919	107	3018	11	2015	
62695927	107	3018	11	2015	
62695943	108	3452	86	2038	
62697571	177	3100/519	300	1122	
62700122	102(2)		2(2)		
62707666	160		245	2004	
62711205	193	3025/129	67	2023	
62711728	102A		53	2007	
62711914	123A	3444	20	2051	
62713747	108	3452	86	2038	
62713755	102	3722	2	2007	
62714832	160		245	2004	
62723645	160		245	2004	
62727128	192	3137	63	2030	
62727772	116		504A	1104	
62728574	103	3862	8	2002	
62728590	160(2)		245(2)		
62728604	124		12		
62733942	123A	3444	20	2051	
62734450	153	3083/197	69	2049	
62734469	712	3072	IC-2		
62734596	196	3054	241	2020	
62734604	163A	3439	36		
62734612	124	3026	12		
62734639	154	3024/128	40	2012	
62734647	192	3044/154	63	2030	
62734655	233		210	2009	
62734663	193	3114/290	67	2023	
62734671	152	3054/196	66	2048	
62734728	506	3130	511	1114	
62736283	749	3171/744	IC-97		
62736291	713	3077/790	IC-5		
62736305	783	3215	IC-225		
62736976	121	3717	239	2006	
62737395	171	3040	27		
62737409	123A	3444	20	2051	
62737417	124	3131A/369	12		
62737425	130	3027	14	2041	
62737433	159	3114/290	82	2032	
62737441	165	3111	38		
62737468	159	3114/290	58	2032	
62737476	123A	3444	20	2051	
62737484	123A	3444	20	2051	
62737492	123A	3444	20	2051	
62737522	154	3040	40	2012	
62737646	124	3131A/369	12		
62737654	748	3236			
62741406	177	3100/519	300	1122	
62741422	177		300	1122	
62741449	165	3115	38		
62741457	159	3138/193A	82	2032	
62741465	199	3137/192A	62	2010	
62742585	104	3719	16	2006	
62743182	152	3054/196	66	2048	
62743794	153	3083/197	69	2049	
62744081	199	3137/192A	62	2010	
62744162	116	3311	504A	1104	
62746483	145A	3063	ZD-15	564	
62746491	149A	3097	ZD-62		
62747501	504	3108/505	CR-6		
62748273	161	3117	39	2015	
62748281	199	3132	62	2010	
62748451	171	3103A/396	27		
62749067	116	3032A	504A	1104	
62749558	116	3031A	504A	1104	
62752300	131		44	2006	
62752319	100	3721	1	2007	
62752327	100	3721	1	2007	
62752742	159	3114/290	82	2032	
62759720	159		82	2032	
62760087	116	3017B/117	504A	1104	
62761261	109	3087		1123	
62761288	5098A		ZD-130		
62761296	5079A	3335	ZD-20		
62761318	506	3130	511	1114	
62761326	506	3130	511	1114	
62761334	116	3016	504A	1104	
62761350	177	3100/519	300	1122	
62761377	177		300	1122	
62766212	184	3054/196	57	2017	
62766220	123A	3444	20	2051	
62766239	108	3452	86	2038	
62766247	159	3114/290	82	2032	
62766255	199	3024/128	62	2010	
62766263	199		62	2010	
62766271	159		82	2032	
62766603	160		245	2004	
62771208	171		27		
62771210	289A		268	2038	
62771224	290A		269	2032	
62771232	5404	3627			
62771240	161	3716	39	2015	
62771364	100	3721	1	2007	
62784822	109			1123	
62785225	123A	3444	20	2051	
62786272	188		226	2018	
62789204	165		38		
62789247	161	3018	39	2015	
62789255	161	3716	39	2015	
62789263	108	3452	86	2038	
62793163	123A	3444	20	2051	
62798564	107		11	2015	
62801328	506	3843	511	1114	
62806060	159		82	2032	
62887676	102A		53	2007	
62921967	142A		ZD-12	563	
63000061	358		42		
0063050118	116	3031A	504A	1104	
65372000	128		243	2030	
67283700	128		243	2030	
70260450	107	3018	11	2015	
70260490		3126	90		
70260540	139A	3060	ZD-9.1	562	
70270050	116	3031A	504A	1104	
70270250	177	3100/519	300	1122	
70270390	116	3031A	504A	1104	
70270490	5401	3638/5402			
70270500	5444	3634			
70270720	116	3031A	504A	1104	

Industry Standard No.	ECG	SK	GE	RS 276-	MOTOR.
70270730	1054	3457	IC-45		
70370050	116		504A	1104	
71773660	110MP	3709		1123(2)	
72035900	159		82	2032	
72729031	105		4		
73010780	5072A		ZD-8.2	562	
74140226-001	130	3027	14	2041	
74140226-002	130		14	2041	
74140226-003	130	3027	14	2041	
75857322	123A	3444	20	2051	
76276722	116		504A	1104	
80000900	358		42		
80001300	116	3031A	504A	1104	
80001400	116	3031A	504A	1104	
80002400	116	3017B/117	504A	1104	
80050100	123A	3444	20	2051	
80050300	103A	3010	59	2002	
80050400	160	3006	245	2004	
80050500	126	3008	52	2024	
80050600	129	3010	244	2027	
80050700	121	3009	239	2006	
80051001	130		14	2041	
80051500	123A	3444	20	2051	
80051600	129	3025	244	2027	
80052101	199	3245	62	2010	
80052102	123A	3444	20	2051	
80052201	128		243	2030	
80052202	123A	3444	20	2051	
80052301	128		243	2030	
80052302	123A	3444	20	2051	
80052402	130	3027	14	2041	
80052600	123A	3444	20	2051	
80052700	129	3025	244	2027	
80052800	103A	3010	59	2002	
80053001	123A	3444	20	2051	
80053300	152	3893	66	2048	
80053400	123A	3444	20	2051	
80053501	312	3448	FET-1	2035	
80053600	108	3452	86	2038	
80053702	130MP	3029	15	2041(2)	
80104900	126	3008	52	2024	
80105200	126	3008	52	2024	
80105300	126	3008	52	2024	
80146620	160	3006	245	2004	
80146630	160	3006	245	2004	
80203350	102A	3004	53	2007	
80205400	102A	3004	53	2007	
80205600	102A	3004	53	2007	
80222631U	102A	3004	53	2007	
80226310	102A		53	2007	
80236400	102A	3004	53	2007	
80236430	158	3004/102A	53	2007	
80243900	102A	3004	53	2007	
80285400	176	3123	80		
80318250	123A	3444	20	2051	
80337390	128		243	2030	
80338030	108	3452	86	2038	
80338040	108	3452	86	2038	
80339430	108	3452	86	2038	
80339440	108	3452	86	2038	
80366830	107	3018	11	2015	
80366840	107	3018	11	2015	
80383840	123A		20	2051	
80383930	123A		20	2051	
80383940	123A		20	2051	
80389910	123A		20	2051	
80414120	130		14	2041	
80414130	130		14	2041	
80421980	128	3024	243	2030	
80510600	109	3087		1123	
80521881	109	3087		1123	
81063405			29	2025	
81073304	290A	3118	269	2032	
82073206	199	3245	62	2010	
82409501	123A	3444	20	2051	
82410300	129	3025	244	2027	
83037204	123A	3444	20	2051	
83038004	229	3124/289	61	2038	
83039404			17	2051	
83073205	123A	3444	20	2051	
83073206	123A	3444	20	2051	
83073304	123A	3444	20	2051	
83073305	123A	3444	20	2051	
83073306	199	3245	62	2010	
83073504	123A	3444	20	2051	
83078402	229	3018	61	2038	
83093005	107	3018	11	2015	
83094502	123A	3444	20	2051	
83167403	229	3018	61	2038	
83167405	229	3018	61	2038	
83167503	229	3018	61	2038	
83167505	229	3018	61	2038	
84026103	192		63	2030	
84667800	909	3590	IC-249		
86140070			20	2051	
86401000	123A	3444	20	2051	
87096010	116		504A	1104	
87100600	109	3088		1123	
87100605	109	3088		1123	
87201881	109			1123	
87204260	109	3088		1123	
87205530		3126	90		
87219410	116	3031A	504A	1104	
87221394		3126	90		
87227880	177	3100/519	300	1122	
87227881	177		300	1122	
87302650	177		300	1122	
87400020	116	3311	504A	1104	
87500620	110MP	3088	1N60	1123(2)	
87500720	116	3031A	504A	1104	
87500920	139A	3060	ZD-9.1	562	
87600920	139A	3060	ZD-9.1	562	
87996010	614		90		
88125010-9	123A	3020/123	20	2051	
89941008	128		243	2030	
89942601	123A	3444	20	2051	
89942702	229		61	2038	
89946008	129		244	2027	
89960404	229	3018	61	2038	
89962306	123A	3444	20	2051	
89962307	123A	3444	20	2051	
89962308	123A	3444	20	2051	
89962404	123A	3444	20	2051	
89963008	123A	3444	20	2051	
89963009	123A	3444	20	2051	
90200090	1006	3358	IC-38		
90200100	1140	3473	IC-138		
90270020	5483	3942	MR-5	1067	
90270060	116	3311	504A	1104	
90270080				1067	
91011500	909	3590	IC-249		
91011900	923		IC-259		
91013300	941D		IC-264		
91013600	940		IC-262		

Industry Standard No.	ECG	SK	GE	RS 276-	MOTOR.
91014400	949		IC-25		
91015300	947		IC-267		
91056140	159	3025/129	82	2032	
91190032	116	3311	504A	1104	
92005600	102A	3004	53	2007	
92115000-23	177		300	1122	
93037230	123A	3444	20	2051	
93037240	123A	3444	20	2051	
93038040		3018	20	2051	
93039440		3018	17	2051	
93063240	123A	3444	20	2051	
93063270	123A	3444	20	2051	
93063280	123A	3444	20	2051	
93063470	123A	3124/289	20	2051	
93064440	199	3124/289	62	2010	
93064450	123A	3444	20	2051	
93073260	199	3020/123	62	2010	
93073340			20	2051	
93073350	128	3020/123	243	2030	
93073440	154	3020/123	40	2012	
93073540	123A	3444	20	2051	
93078420	107	3124/289	11	2015	
93082830	123A	3444	20	2051	
93082840	123A	3444	20	2051	
93082920	107	3018	11	2015	
93094502	123A	3444	20	2051	
93097120	128	3020/123	243	2030	
93938040	123A	3444	20	2051	
93939440	107		11	2015	
94029220	124	3026	12		
94650700-00	159		82	2032	
94650700-01	159		82	2032	
94742308	128		243	2030	
94810300	5547	3505			
94813000	181		75	2041	
94818400	390		255	2041	
94824100-00-PS203			20	2051	
94824100-01 PS203	123A			2051	
94824100-01-PS203			20	2051	
94824101	123A	3444	20	2051	
94846400-00	5695	3509			
94846400-01	5695	3509			
94846400-02	5695	3509			
94913000-01	941D		IC-264		
95114101-00	102A	3004	53	2007	
95114102-00	102A	3004	53	2007	
95114109-00	102A	3004	53	2007	
95114135-00	102A	3004	53	2007	
95114140-00	121	3009	239	2006	
95115115-00	125	3032A	510,531	1114	
95349700	197		250	2027	
95522800	128		243	2030	
95524600	124		12		
96421866	125	3031A	510,531	1114	
110654492-001	177		300	1122	
120001190	160	3006	245	2004	
120001192	100	3005	1	2007	
120001195	102A	3004	53	2007	
120001300	116	3016	504A	1104	
120002013	102A	3004	53	2007	
120002014	102A	3004	53	2007	
120002213	160	3007	245	2004	
120002214	160	3006	245	2004	
120002216	160	3007	245	2004	
120002513	160	3006	245	2004	
120002515	160	3006	245	2004	
120002518	160	3006	245	2004	
120002520	160	3006	245	2004	
120002521	102A	3004	53	2007	
120002656	160	3006	245	2004	
120002748	102A	3004	53	2007	
120004492	160	3006	245	2004	
120004493	102A	3004	53	2007	
120004494	102A	3004	53	2007	
120004495	102A	3004	53	2007	
120004496	108	3452	86	2038	
120004497	108	3452	86	2038	
120004880	108	3452	86	2038	
120004881	108	3452	86	2038	
120004882	108	3452	86	2038	
120004883	123A	3444	20	2051	
120004884	195A	3047	46		
120004885	224	3049	46		
120004886	224	3049	46		
120004887	121	3009	239	2006	
131000561	128		243	2030	
131000562	121	3717	239	2006	
131001007	130		14	2041	
131005353-1	197		250	2027	
131005807	18		243	2030	
131005808	129		244	2027	
134804204-101	116		504A	1104	
134804290-101	121	3717	239	2006	
134804596-301	116		504A	1104	
147187606	177	3843			
152221011	102A	3004	53	2007	
152221483	160		245	2004	
163547338	723		IC-15		
181515417		3024	63	2030	
226021014	108	3452	86	2038	
229005013	123A	3020/123	20	2051	
229005014	123A	3020/123	20	2051	
229005015	123A	3020/123	20	2051	
229018032	108	3452	86	2038	
229018033	108	3452	86	2038	
229018034	108	3452	86	2038	
229020423	108	3452	86	2038	
229021014	108	3452	86	2038	
229025010	108	3452	86	2038	
229510015V	108	3452	86	2038	
229510031V	108	3452	86	2038	
229510032V	108	3452	86	2038	
229510033V	108	3452	86	2038	
231150050	109	3087		1123	
238208001	175		246	2020	
260100602	611	3324			
261002302	910D		IC-252		
261004303	941M		IC-265	007	
261015401	923		IC-259		
312004900			245	2004	
312104732	160		245	2004	
312104733	160	3006	245	2004	
314743006	175		246	2020	
346190001	128		243	2030	
402003173	117	3017B	504A	1104	
415200050	1027	3153	721		
420208700	125		510,531	1114	
430203845	129	3025	244	2027	
430233843	152	3893	66	2048	
436004601	199	3245	62	2010	
436005001	123A	3444	20	2051	
436005901	234		65	2050	
436006101	128		243	2030	
436006201	129		244	2027	

Industry Standard No.	ECG	SK	GE	RS 276-	MOTOR.
436006401	199	3245	62	2010	
440008401	116		504A	1104	
440009905	519		300	1122	
440011001			20	2051	
450010201	108	3452	86	2038	
450010701	161	3716	39	2015	
485134922	123A	3444	20	2051	
485134923	123A	3444	20	2051	
485134924	123A	3444	20	2051	
485134925	123A	3444	20	2051	
485134926	123A	3444	20	2051	
485134956	160		245	2004	
485134997	199		62	2010	
485137002	128		243	2030	
485137370	197	3083	250	2027	
570004503	123A	3444	20	2051	
570005452	123A	3444	20	2051	
570005503	123A	3444	20	2051	
600000060	109	3088	1N60	1123	
600000150	177	3175	300	1122	
601100001	139A	3784/5018A		562	
601100002		3779		561	
602000002	116	3311	504A	1104	
603000002	613	3327/614	90		
630000003		3197	215	2053	
632000001	186A	3253/295	270		
632000002	235	3239/236	215		
632037218	123A	3245/199	61	2051	
632039418	229	3124/289	61	2038	
632049514		3247	65	2050	
632049518	159	3114/290	221	2032	
632073518		3122	210	2038	
632122617	186A	3197/235	215	2052	
632197406	235	3197	337		
640000002	1192	3445			
640000003	1232	3852		703	
661000001	601	3463			
661103200				1122	
661107300				1104	
662100200			246	2020	
802026310			2	2007	
881250108	123A	3444	20	2051	
881250109	123A	3444	20	2051	
885540026-3	7420	7420		1809	
910678870	159	3118	82	2032	
1348042041	116		504A	1104	
1348042901	121	3717	239	2006	
1348045963	116		504A	1104	
1500149001	109	3088			
1500169001	109	3087			
1510289001	519	3100			
1510409003	116	3311			
1510679001	177	3100/519			
1520759001	138A	3059			
1520909001	138A	3059			
1520939001	139A	3060			
1522210111	102A	3004	53	2007	
1522210131	126	3006/160	52	2024	
1522210300	102A	3006/160	53	2007	
1522210921	126	3007	52	2024	
1522211021	126	3006/160	52	2024	
1522211200	102A	3004	53	2007	
1522211221	102A	3004	53	2007	
1522211328	102A	3004	53	2007	
1522211921	160	3006	245	2004	
1522214400	160	3006	245	2004	
1522214411	126	3006/160	52	2024	
1522214435	126	3006/160	52	2024	
1522214821	126	3006/160	52	2024	
1522214831	160	3006	245	2004	
1522216500	102A	3004	53	2007	
1522216600	126	3007	52	2024	
1522217400	160	3006	245	2004	
1522223720	123A	3444	20	2051	
1522270100	109	3087		1123	
1522270101	109	3087		1123	
1522270208			1N60	1123	
1523270105	116	3311	504A	1104	
1540049001	613	3126			
1611819064	123A	3444	20	2051	
1679003608	1003		IC-43		
1720389001	295	3253			
1720389002	235	3197			
1760089001	123A	3444	20	2051	
1760099001		3245	62		
1760579001	123AP	3245/199	62		
1760589001	123AP	3245/199	62		
1760609001	107	3356	211		
1760609002	107	3356	211		
1760609003	123AP	3245/199	62		
1760609004	123AP	3124/289			
1760629001	199	3124/289			
1760829001	107	3356	211		
1760979001		3356	211		
1770209001	294	3138/193A			
2002010072	7473			1803	
2002100022	1004	3365	IC-149		
2002110206	712	3072	IC-2		
2002110269	712	3072	IC-148		
2002120012	749	3168	IC-97		
2002120022	1004	3365			
2002120033	1080	3284	IC-98		
2002500711	715	3076	IC-6		
2002500812	714	3075	IC-4		
2002500914	790		IC-5		
2002501006	714	3075	IC-4		
2002501300	791	3149	IC-231		
2002501708	1196	3725			
2002800034	941M		IC-265	007	
2003037227	123A	3038	61	2038	
2003037309	199	3122	62	2010	
2003038092	107	3245/199	61	2038	
2003038208	319P	3018	283	2016	
2003038812	233	3018	210	2009	
2003049614	295		57		
2003051513	124		12		
2003055813	163A	3439	36		
2003064314	238	3115/165	259		
2003073517	289A	3122	268	2038	
2003082823	199	3122	61		
2003082837	199	3122	61	2038	
2003098316		3044	27		
2003102529	175	3131A/369	246	2020	
2003110607	162		73		
2003117004	238	3710			
2003117044	238	3710	38		
2003121216	184	3190	55	2017	
2003144746	198	3104A			
2003144795	198		32		
2003145404	94	3560	35		
2003150702	198		32		
2003150710	198		32		
2003162538	291	3119/113	251		
2003162700	128		243	2030	
2003172208	376		251		
2003172216	198	3219	251		
2003172305	198	3219	325		
2003174006	289A	3122	62	2051	
2003174014	123AP	3122	62	2051	
2003185506	108	3246/229	61	2038	
2003189025	123A	3124/289	62	2010	
2003189109	165	3115	259		
2003189304	165	3710/238	38		
2003190604	107	3293	86	2038	
2003206800	198	3201/171	251		
2004047310	153	3083/197	69	2049	
2004049081	153	3274	69	2049	
2004049514	159	3118	82	2032	
2004056215	290A	3114/290	269	2032	
2004060858	159	3114/290	82	2032	
2004067350	290A	3114/290	269	2032	
2004074304	185	3191	58	2025	
2004082606	159	3138/193A	89	2032	
2004082614	159	3114/290	89	2032	
2004085415	288	3114/290	89	2050	
2004101500		3114	82		
2004101518		3114	82		
2004302210	158		53	2007	
2004305600	102A	3004	53	2007	
2004305618	102A	3004	53	2007	
2004341519	158	3004/102A	53	2007	
2004356122	159	3114/290	82	2032	
2004356220	288	3138/193A	48		
2004356801	292	3083/197	69	2049	
2004502439	124	3021			
2004503018	103A	3835	59	2002	
2004546826	287	3137/192A	47	2030	
2004547806	291	3054/196	66	2048	
2004700116	233	3114/290	210	2009	
2008000001	109	3088	1N60	1123	
2008000019	110MP	3088			
2008000026	109	3088	1N60	1123	
2008000064	109	3087	1N34AS	1123	
2008010028	177	3100/519	300	1122	
2008010094	177	3100/519	300	1122	
2008010102	605	3864			
2008010110	177	3100/519	300	1122	
2008010127	177	3100/519	300		
2008010131	177	3100/519	300	1122	
2008010165		3175	300		
2008100027	116	3100/519	504A	1104	
2008100076	506	3843	511	1114	
2008100093	125	3051/156	510	1114	
2008100130	116	3313	504A	1104	
2008130040	506	3843	511	1114	
2008130084	116	3016	504A	1104	
2008130171	506	3843	511	1114	
2008220125		3061	ZD-10	562	
2008220851	5019A	3785			
2008230042	5070A	9021	ZD-6.2	561	
2008230057	5071A		ZD-6.8	561	
2008300017	612	3325			
2008310019	612	3325			
2008310058	612	3325			
2009020019	116(3)	3017B/117			
2009120059	125	3016			
2009120101	116	3311	504A	1104	
2009120187	116	3311	504A	1104	
2009130017	125		510,531	1114	
2009130075	125	3017B/117	504A	1104	
2009700063	503		CR-5		
2012000092	109	3088	1N60	1123	
2012000100	109	3088	1N60	1123	
2012000118	109	3088	1N60	1123(2)	
2012010144	519	3100	514	1122	
2012010159	177	3100/519	514	1122	
2012010165	177	3100/519	300		
2012100119	525	3925	530		
2012100126	116	3081/125	504A	1104	
2012120009		3843	511	1114	
2012130234	552	3843	511	1114	
3121050622	160	3006	245	2004	
3122005400	102A	3004	53	2007	
3520743010	159		82	2032	
3539306001	123A	3444	20	2051	
3539306002	123A	3444	20	2051	
3539306003	123A	3444	20	2051	
4100006285	290A	3114/290	82		
4100104753		3004	53	2007	
4100204613	229	3122			
4100206443		3124	62	2010	
4100208271			60	2015	
4100208283	229	3124/289	61	2038	
4100208291		3018	60	2015	
4100208292	229	3018	61	2038	
4100208293	229	3018	61	2038	
4100209005	199	3899			
4100210472		3018	61	2038	
4100213591		3018	60	2015	
4100217172	289A	3122	268	2038	
4100900102	123AP	3356	211		
4100900103	123AP	3356	211		
4100900104	123AP	3356	211		
4100900116	199	3899	62		
4100903554	297	3849/293	271		
4100905254	298	3841/294	272		
4100907633	107	3122	61		
4102100220	102A	3004	53	2007	
4104007193	290A		269	2032	
4104007380	185	3084	58	2025	
4104103242	158	3004/102A	53	2007	
4104204602	107		11	2015	
4104204603	107	3018	11	2015	
4104204612	107	3018	11	2015	
4104204613	229	3122	61	2038	
4104205352		3018	20	2051	
4104206440	123A	3444	20	2051	
4104206442	199	3124/289	62	2010	
4104208282	199	3124/289	62	2010	
4104208283	199	3124/289	62	2010	
4104213173	289A		268	2038	
4104213421		3124	20	2051	
4108296047	108	3018	20	2051	
4108296237	123A	3444			
4108296238	123A	3018	20	2051	
4108296255	229	3018	20	2051	
4108296257	229	3018	20	2051	
4109208280		3124	18	2030	
4109208284	123A	3444	20	2051	
4109213174	289A	3024/128	268	2038	
4109213354	199	3020/123	62	2010	
4120100090		3087	1N34AS	1123	
4120100600	109	3088	1N60	1123	
4120129000	109	3088		1123	
4120200090		3087	1N34AS	1123	
4120200602	110MP	3088	1N60	1123	
4120500622	5013A	3779			
4120501200	142A	3062	ZD-12	563	

Industry Standard No.	ECG	SK	GE	RS 276-	MOTOR.
4120900010	177	9091	300	1122	
4120910010	116	3311	504A	1104	
4121200051	5069A		ZD-4.7		
4129100000	177	3100/519	300	1122	
4129100600	109	3088		1123	
4129100602	110MP	3088		1123(2)	
4129104002	116		504A	1104	
4129200602	110MP	3088	1N60	1123(2)	
4129227907		3126	90		
4129301329		3126	90		
4129321292		3126	90		
4129400003		3100	300	1122	
4129501200	142A	3062	ZD-12	563	
4139000005		3311	504A	1104	
4139104002		3016	504A	1104	
4150002031	1054	3457	IC-45		
4152000150	1027	3153			
4152032010	1179	3737	IC-75		
4154011370	788	3147	IC-229		
4154011510	1237	3707	IC-154		
4154011560	801	3160	IC-35		
4157013102	801	3160	IC-35		
4159761157	801	3160	IC-35		
4202003173	116	3032A	504A	1104	
4202003200	156	3311	512		
4202003500	109	3087	1N34AS	1123	
4202003700	506	3843	511	1114	
4202003900	506	3843	511	1114	
4202005000	113A	3119/113	6GC1		
4202005600	110MP	3709	1N34AS	1123(2)	
4202006100	177	3100/519	300	1122	
4202006200	177	3100/519	300	1122	
4202006500			504A	1104	
4202006700	506	3016	511	1114	
4202006800	116	3032A	504A	1104	
4202007600	116	3016	504A	1104	
4202007601	506	3130	511	1114	
4202007700	506	3130	511	1114	
4202007800	125		531	1114	
4202007801	506	3130	511	1114	
4202007802	116	3016	504A	1104	
4202007806			510	1114	
4202007900	113A	3100/519	6GC1		
4202008000	116		504A	1104	
4202008500	116	3016	504A	1104	
4202008600	109			1123	
4202008700	125	3081	510,531	1114	
4202009200	5077A	3752	ZD-18		
4202009400	119	3109	CR-2		
4202010100	177	3100/519	300	1122	
4202011200	5074A		ZD-11.0	563	
4202011300	5071A	3058/137A	ZD-6.8	561	
4202011400	5071A		ZD-6.8	561	
4202011500	5074A	3092/141A	ZD-11.0	563	
4202011900	5077A	3752			
4202012000	5077A	3752	ZD-18		
4202012501	5074A	3139	ZD-11	563	
4202012700	145A	3063	ZD-15	564	
4202012800	5077A	3752	ZD-18		
4202013300	5085A	3337	ZD-36		
4202013600	519	3100	300	1122	
4202014400	506	3017B/117	504A	1104	
4202014500	116	3017B/117	504A	1104	
4202014600	506	3017B/117	511	1114	
4202015100	506	3130	511	1114	
4202015600	177	3100/519	300	1122	
4202016000			300	1122	
4202016200	177		300	1122	
4202017200	125	3925/525	530		
4202017600	116	3125	511	1114	
4202018500	116	3016	504A	1104	
4202018700	177	3100/519	514	1122	
4202019400	5143A	3409			
4202019500	5085A	3409/5143A			
4202020100	5070A	9021	ZD-6.0	561	
4202020500	506	3125	511	1114	
4202020900	506	3130	511	1114	
4202021000	116	3313	504A	1104	
4202021100	156	3051	512		
4202021700	137A	9021/5070A	ZD-6.0	561	
4202021900	177	3100/519	514	1122	
4202022300	506	3311	504A	1104	
4202022900	525	3925	530		
4202023000	506	3125	511	1114	
4202023100	116	3313	504A	1104	
4202023200		3125	511	1114	
4202023300		3925	530		
4202023400	116	3311	504A	1104	
4202024200	138A	3334/5071A	ZD-6.8	561	
4202024300		3334	ZD-6.8	561	
4202093599	110MP	3709			
4202104170	167	3647		1172	
4202104470	125		510,531	1114	
4202104570	506	3032A	511	1114	
4202104770	102A	3088	53	2007	
4202104870	506	3043	511	1114	
4202105470	138A	3059	ZD-7.5		
4202106270	505	3108	CR-7		
4202106970	506	3043	511		
4202107370	503	3068	CR-5		
4202107470	177	3100/519	300	1122	
4202107570	504	3108/505	CR-6		
4202107670	519	3100	514	1122	
4202107970	506	3130	511	1114	
4202108070	505	3108	CR-7		
4202108170	505	3108	CR-7		
4202108270	116	3031A	504A	1104	
4202108570	5080A	3336	ZD-22		
4202109070	5074A	3092/141A	ZD-11.0	563	
4202109170	177	3100/519	300	1122	
4202109370	116	3017B/117	504A	1104	
4202110270	506	3311	504A	1104	
4202110370			504A	1104	
4202110470	552	3016	504A	1104	
4202110870	5074A	3139	ZD-11	563	
4202116670	5074A	3139	ZD-11	563	
4202970330	145A	3063	ZD-15	564	
4203008000	506	3130	511	1114	
4203970101	199	3122	210	2010	
4206002300	1004	3102/710			
4206002400	1004	3102/710	IC-149		
4206002600	1080	3284	IC-98		
4206002900	1080	3284			
4206003900	1080	3284	IC-98		
4206004000	712	3072	IC-148		
4206004400	1004	3102/710	IC-149		
4206004600	712	3072	IC-2		
4206004700	1122		IC-40		
4206005200	1159	3290	IC-72		
4206007500	1158	3289	IC-71		
4206009100	1183	3475/1050			
4206009200	738	3167	IC-29		
4206009700	738	3167	IC-29		
4206009800	1178	3480			

Industry Standard No.	ECG	SK	GE	RS 276-	MOTOR.
4206104970	712	3072	IC-2		
4206105170	749	3168	IC-97		
4206105370	1109	3711	IC-99		
4206105470	712	3072	IC-148		
4206105770	712	3072	IC-148		
4206105870	712	3072	IC-148		
4360021001	123A	3444	20	2051	
4803101109-02	121	3717	239	2006	
5147013102	801	3160	IC-35		
5700045452	123	3020	20	2051	
6611000900	506	3125	511	1114	
6611001000	506	3125	511	1114	
6611001100	506	3313/116	511	1114	
6611001200	506	3125	511	1114	
6611001500	506		511	1114	
6611001700	116	3311	504A	1104	
6611001800	116	3311	504A	1104	
6611002300	506	3125	511	1114	
6611003000	502	3067	CR-4		
6611003200	177	3100/519	300		
6611007300	116	3311	504A		
6612002200			511	1114	
6612004000	109	3088	1N60	1123	
6612006000	177	3100/519	300	1122	
6612009000	109	3087	1N34AS	1123	
6613002200	506	3125	511		
6615000100			ZD-6.0	561	
6615000200		3139	ZD-11	563	
6615005000	5018A	3060/139A	ZD-9.1	562	
6621001000	289A	3122	210	2038	
6621001100	124		210	2038	
6621001800	198	3104A	251		
6621002300	162	3438	35		
6621002400	238	3115/165	38		
6621002800	291		32		
6621003100	123A	3444	61	2051	
6621003200	123A	3444	61	2051	
6621003400	123A	3444	61	2051	
6621003500	289A	3122	210	2038	
6621004000	319P	3039/316	86	2016	
6621005000	233	3018	61	2009	
6621006200	287		27		
6621007200	152	3054/196	66	2048	
6621008100	289A	3122	210	2038	
6621009100	233	3122	61	2015	
6622000100	128		243	2030	
6623001100	159	3466	221	2032	
6623001900	290A	3114/290	269	2032	
6623002000	159	3466	221	2032	
6623002100	159	3466	221	2032	
6623002200	159	3466	221	2032	
6623002300	290A	3114/290	269	2032	
6624001100	129		244	2027	
6624002000	159	3466	221	2032	
6624003100	153	3083/197	69	2049	
6624003200	289A	3124/289	210	2038	
6644000100	712	3072	IC-148		
6644001100	1128	3488	IC-105		
6644001400	1134	3489	IC-106		
6644001500	1133	3490	IC-107		
6644001700	1131	3286	IC-61		
6644001800	1132	3287	IC-110		
6644001900		3478	IC-111		
8001100001	519		514	1122	
8001100003	519		514	1122	
8001100005	177		300	1122	
8001200001	123A	3444	20	2051	
8001200004	128		52	2030	
9511410100	102A	3004	53	2007	
9511410200	102A	3004	53	2007	
9511410900	102A	3004	53	2007	
9511413500	102A	3004	53	2007	
9511414000	131	3009	44	2006	
9511510200	109	3087		1123	
9511511500	125	3032A	510,531	1114	
13480084523	117		504A	1104	
13480084555	117		504A	1104	
13480204101			504A	1104	
13488240306	116		504A	1104	
13488240307	519		514	1122	
13488240308	116		504A	1104	
13488240309	177		300	1122	
13488240310	506	3843	511	1114	
13488240311	519		514	1122	
13488240312	506	3843	511	1114	
13488240313	177		300	1122	
13488240314	519		514	1122	
13488240315	177		300	1122	
13488240316	519		514	1122	
13488240317	519		514	1122	
13488240318	519		514	1122	
16000190201	126	3006/160	52	2024	
16009090545	193	3025/129	67	2023	
16100190668	108	3452	86	2038	
16102190693	123A	3444	20	2051	
16102190929	108	3452	86	2038	
16102190930	123A	3444	20	2051	
16102190931	186	3192	28	2017	
16103190668	108	3452	86	2038	
16103190930	107	3018	11	2015	
16104190668		3018	11	2015	
16104191168	123A	3444	20	2051	
16104191225	108	3452	86	2038	
16104191226	108	3452	86	2038	
16105190536	123A	3444	20	2051	
16105191229	128	3024	243	2030	
16106190537	123A	3444	20	2051	
16106190772		3018	17	2051	
16108190536	199	3124/289	62	2010	
16108290536	128		243	2030	
16109190536	199	3124/289	62	2010	
16109209536	123A	3444	20	2051	
16112190710	107		11	2015	
16112190772		3018	17	2051	
16114190772		3018	17	2051	
16116190634	123A	3444	20	2051	
16118190634			20	2051	
16147191229	123A	3444	20	2051	
16156197229	128	3024	243	2030	
16171190693	123A	3444	20	2051	
16171190858	123A	3444	20	2051	
16171191368	123A	3444	20	2051	
16172190693	123A	3444	20	2051	
16172190858	123A	3444	20	2051	
16200190186		3004	2	2007	
16201190022	102A	3004	53	2007	
16201190186		3004	2	2007	
16201190187	158	3004/102A	53	2007	
16204190457	158	3004/102A	53	2007	
16207190405	158	3004/102A	53	2007	
16208190187		3004	53	2007	
16211190022	158	3004/102A	53	2007	
16212190022	158	3004/102A	53	2007	
16256197228	129		244	2027	

Industry Standard No.	ECG	SK	GE	RS 276-	MOTOR.
16304190031	302	3252	275		
16304198000	195A	3048/329	219		
16304199400	182	3188A			
16307190632	123A	3444	20	2051	
16343190142	186	3192	28	2017	
16356179229	128		243	2030	
16377190632	123A	3444	20	2051	
16400690060	109	3088		1123	
16401190188	109	3088		1123	
16405990022	177	3100/519	300	1122	
16411190188	109	3087		1123	
16411992473	519	3100	300	1122	
16412190410		3087	1N34AS	1123	
16415490000	109	3088		1123	
16419990032	109			1123	
16501090016	116	3311	504A	1104	
16501190005		3311	504A	1104	
16501190016	116	3311	504A	1104	
16505090005	116	3311	504A	1104	
16516390010	117	3311	504A	1104	
16602995856	140A	3061	ZD-10	562	
16611190553		3126	90		
16613190555		3126	90		
16629291210	116	3311	504A	1104	
16740135809	1024	3152	720		
44202007800	506	3130	511	1114	
57000901504	123A	3444	20	2051	
62170835330	177		300	1122	
134800847274	109			1123	
134800849022	177		300	1122	
134800854093	177		300	1122	
134800855216	109			1123	
134800863030	109			1123	
134800867716	109			1123	
134800869525	123A	3444	20	2051	
134804204101	116			1104	
134804290101	121	3717	239	2006	
134804596301	116		504A	1104	
134882095301	116		504A	1104	
134882095302	116		504A	1104	
134882095303	116		504A	1104	
134882095304	116		504A	1104	
134882139701	109			1123	
134882139702	109			1123	
134882178101	109			1123	
134882178102	109			1123	
134882178103	109			1123	
134882178104	109			1123	
134882178105	109			1123	
134882178106	109			1123	
134882178107	109			1123	
134882178108	109			1123	
134882178110	109			1123	
134882178111	109			1123	
134882178112	109			1123	
134882178113				1123	
134882256301	5013A	3779	ZD-6.2	561	
134882256302	5014A	3780	ZD-6.8	561	
134882256303	5009A	3775	ZD-4.7		
134882256305	5045A	3811	ZD-68		
134882256306	141A		ZD-11.5		
134882256307	5009A	3775	ZD-4.7		
134882256308	5016A	3782	ZD-8.2	562	
134882256309	5016A	3782	ZD-8.2	562	
134882256310	144A		ZD-14	564	
134882256311	5019A	3785	ZD-10	562	
134882256312	5011A	3777	ZD-5.6	561	
134882256313	5023A	3789	ZD-14	564	
134882256314	5024A	3790	ZD-15	564	
134882256315	5010A	3776	ZD-5.1		
134882256316	5016A	3782	ZD-8.2	562	
134882256317	5021A	3787	ZD-12	563	
134882256318	5018A	3784	ZD-9.1	562	
134882256319	5014A	3780	ZD-6.8	561	
134882256320	5033A		ZD-27		
134882256321	146A		ZD-27		
134882256322	139A		ZD-9.1	562	
134882256323	5014A	3780	ZD-6.8	561	
134882256324	5027A	3793	ZD-18		
134882256325	5021A	3787	ZD-12	563	
134882256328	140A		ZD-10	562	
134882256329	5090A	3342	ZD-56		
134882256330	5096A		ZD-100		
134882256331	5031A	3797	ZD-24		
134882256332	5019A	3785	ZD-10	562	
134882256333	5002A	3768			
134882256334	5020A	3786	ZD-11	563	
134882256335	135A		ZD-5.1		
134882256336	5083A	3754	ZD-28		
134882256337	5071A	3334	ZD-6.8	561	
134882256338	5018A	3784	ZD-9.1	562	
134882256339	5079A	3335	ZD-20		
134882256340	140A		ZD-10	562	
134882256342	5082A	3753	ZD-25		
134882256343	139A		ZD-9.1	562	
134882256344	5015A	3781	ZD-7.5		
134882256345	5017A	3783	ZD-9.1	562	
134882256346	5067A	3331	ZD-3.9		
134882256347	5014A	3780	ZD-6.8	561	
134882256348	5022A	3788	ZD-13	563	
134882256349	5031A	3797	ZD-24		
134882256350	5022A	3788	ZD-13	563	
134882256351	135A		ZD-5.1		
134882256353	5077A	3752	ZD-18		
134882256354	142A		ZD-12	563	
134882256356	139A		ZD-9.1	562	
134882256357	5079A	3335	ZD-20		
134882256358	5086A	3338	ZD-39		
134882256359	145A		ZD-15	564	
134882392202	177		300	1122	
134882392203	177		300	1122	
134882392204	177		300	1122	
134882392206	177		300	1122	
134882392207	177		300	1122	
134882392208	177		300	1122	
134882392209	177		300	1122	
134882392210	177		300	1122	
134882392211	177		300	1122	
134882392212	177		300	1122	
134882392213	177		300	1122	
134882392214	177		300	1122	
134882392215	177		300	1122	
134882393301	177		300	1122	
134882393305	177		300	1122	
134882420301	177		300	1122	
134882420302	177		300	1122	
134882420303	519		514	1122	
134882420304	519		514	1122	
134882420305	177		300	1122	
134882466801	116		504A	1104	
134882466802	116		504A	1104	
134882466803	116		504A	1104	
134882466804	116		504A	1104	
134882466806	116		504A	1104	
134882466807	116		504A	1104	
134882466808	125		510,531	1114	
134882466812	506	3843	511	1114	
134882466813	506	3843	511	1114	
134882466814	506	3843	511	1114	
134882466815	506	3843	511	1114	
134882466816	506	3843	511	1114	
134882466817	506	3843	511	1114	
134882466818	506	3843	511	1114	
134882466821	506	3843	511	1114	
134882466825	506	3843	511	1114	
134882466827	506	3843	511	1114	
134882466828	506	3843	511	1114	
134882732304	5944		5048		
134882921701	109			1123	
134882921702	109			1123	
134882921704	109			1123	
134882921705	109			1123	
134883461501	5075A	3751	ZD-16	564	
134883461502	5022A	3788	ZD-13	563	
134883461503	5066A		ZD-3.3		
134883461504	5093A	3344	ZD-75		
134883461505	5070A	9021	ZD-6.2	561	
134883461506	5096A		ZD-100		
134883461507	5075A	3751	ZD-16	564	
134883461508	147A		ZD-33		
134883461509	5090A	3342	ZD-56		
134883461510	5010A	3776	ZD-5.1		
134883461511	5082A	3753	ZD-25		
134883461512	146A		ZD-27		
134883461513	5068A	3332	ZD-4.3		
134883461514	5096A		ZD-100		
134883461515	5018A	3784	ZD-9.1	562	
134883461516	5095A	3345	ZD-91		
134883461517	5001A	3767			
134883461518	5027A	3793	ZD-18		
134883461519	5067A	3331	ZD-3.9		
134883461520	146A		ZD-27		
134883461521	5129A	3395	5ZD-14		
134883461522	5029A	3795	ZD-20		
134883461523	5010A	3776	ZD-5.1		
134883461524	5028A	3794	ZD-19		
134883461525	5014A	3780	ZD-6.8	561	
134883461526	5031A	3797	ZD-24		
134883461527	5012A	3778	ZD-6.0	561	
134883461528	5141A	3407			
134883461530	148A		ZD-55		
134883461531	5022A	3788	ZD-13	563	
134883461532	5016A	3782	ZD-8.2	562	
134883461533	5027A	3793	ZD-18		
134883461534	5117A	3383			
134883461535	5137A	3403	5ZD-24		
134883461536	137A		ZD-6.2	561	
134883461537	5077A	3752	ZD-18		
134883461538	136A		ZD-5.6	561	
134883461539	5135A	3401			
134883461540	5010A	3776	ZD-5.1		
134883461541	135A		ZD-5.1		
134883461542	5071A	3334	ZD-6.8	561	
134883461543	5006A	3772	ZD-3.6		
134883875401	5454	6754		1067	
134883875402	5455	3597		1067	
134883875404	5431			1067	
134883875405	5422			1067	
134883875406	5455	3597		1067	
134884461544	5115A	3381			
202530001710	116(2)		504A(2)	1104	
301720490201	109			1123	
310720440170	121MP	3013	239#(2)	2006(2)	
310720490000	123A	3444	20	2051	
310720490010	123A	3444	20	2051	
310720490020	123A	3444	20	2051	
310720490060	185	3191	58	2025	
310720490070	121	3009	239	2006	
310720490080	108	3452	86	2038	
310720490100	123A	3444	20	2051	
310720490150	123A	3444	20	2051	
310720490190	121MP	3013	239#(2)	2006(2)	
310720490201			1N60	1123	
353511050009	804	3455	IC-27		
400100010160	102A		53	2007	
400100020120	102A		53	2007	
404100030180	158	3004/102A	53	2007	
404100900160	123A	3444	20	2051	
404120010180	109			1123	
404120030100	109			1123	
404120030110	177		300	1122	
404120040100	109			1123	
404130010150	116		504A	1104	
505130010150	116		504A	1104	
933000611112	116	3311	504A	1104	